Comprehensive Organic Functional Group Transformations

Comprehensive Organic Functional Group Transformations

Editors-in-Chief

Alan R. Katritzky, FRS
University of Florida, Gainesville, FL, USA

Otto Meth-Cohn
University of Sunderland, UK

Charles W. Rees, FRS
Imperial College of Science, Technology and Medicine, London, UK

Volume 2

SYNTHESIS: CARBON WITH ONE HETEROATOM ATTACHED BY A SINGLE BOND

Volume Editor

Steven V. Ley, FRS
University of Cambridge, UK

PERGAMON

UK	Elsevier Science Ltd., The Boulevard, Langford Lane, Kidlington, Oxford OX5 1GB, UK
USA	Elsevier Science Inc., 660 White Plains Road, Tarrytown, New York 10591-5153, USA
JAPAN	Elsevier Science Japan, Tsunashima Building Annex, 3-20-12 Yushima, Bunkyo-ku, Tokyo 113, Japan

First edition 1995

Library of Congress Cataloging in Publication Data
Comprehensive organic functional group transformations / editors-in-chief, Alan R. Katritzky, Otto Meth-Cohn, Charles W. Rees. — 1st ed.
p. cm.
Includes indexes.
Contents: v. 1. Synthesis: carbon with no attached heteroatoms / edited by Stanley M. Roberts — v. 2. Synthesis: carbon with one heteroatom attached by a single bond / edited by Steven V. Ley — v. 3. Synthesis: carbon with one heteroatom attached by a multiple bond / edited by Gerald Pattenden — v. 4. Synthesis: carbon with two heteroatoms, each attached by a single bond / edited by Gordon W. Kirby — v. 5. Synthesis: carbon with two attached heteroatoms with at least one carbon-to-heteroatom multiple link / edited by Christopher J. Moody — v. 6. Synthesis: carbon with three or four attached heteroatoms / edited by Thomas L. Gilchrist — v. 7. Indexes.
1. Organic compounds—Synthesis. I. Katritzky, Alan R. II. Meth-Cohn, Otto. III. Rees, Charles W.
QD262.C534 1995
547'.2—dc20 95–31088

British Library Cataloguing in Publication Data
A catalogue record for this book is available from the British Library.

ISBN 0–08–040604–1 (set : alk. paper)
ISBN 0–08–042323–X (Volume 2)

The paper used in this publication meets the minimum requirements of the American National Standard for Information Sciences—Permanence of Paper for Printed Library Materials, ANSI Z39.48–1984.

Printed and bound in Great Britain by Cambridge University Press, Cambridge, UK.

Contents

Preface

Some years ago the three of us met in a London club reviewing an ongoing publishing venture in Organic Synthesis. The conversation drifted to a consideration of volumes on the synthesis of key functional groups. No doubt the good wine helped since we actually broached the idea of a work on the synthesis of *all* functional groups. Would it be useful? Definitely. Would it be feasible? How would it be organized? Where do you start? We recognized that functionality was based on the coordination and heteroatom attachment of a carbon atom. But putting together a complete framework seemed particularly daunting. Two of us became very interested in the fascinating bouquet of the Muscat de Beaumes de Venise.

At our next dinner together Alan announced that he had solved the problems posed last time—problems that Charles and I hoped he had forgotten! He brought out a remarkable matrix analysis of *all* functional groups, analysed rigorously and logically. Even unknown functions were covered. Although we were all very impressed, the practicalities of the idea still seemed daunting. Those who know Alan's terrier instincts will appreciate that he would not give up such a challenge so easily. Our twice yearly club get-togethers, occasionally with friends from Pergamon, refined our thinking. Alan's cosmic vision was tempered by Charles's intuitive realism and fully supported by the publishers.

Another major problem remained: how to reduce our thinking into a practical handbook for authors—a dismaying task for three busy chemists. We settled on a seven-volume work and the indomitable ARK produced a rough breakdown to fit such a format. Putting flesh on these bones became feasible during a fortuitous three-month break between jobs by myself, and the largest handbook ever assembled by Pergamon (120 pages) was written and page allocations agreed—even for little or unknown functional groups. Sample chapters were commissioned and finally proved very encouraging, despite our first chosen topic uncovering virtually no known examples!

Contracts were defined and agreed, volume editors approached, and potential authors considered during a pleasant preconference stay in Grasmere. Following the sale of Pergamon to Elsevier Science Ltd there was a lull in the project but soon *Comprehensive Organic Functional Group Transformations* was back on track, and everyone adhered to a very businesslike timetable.

OTTO METH-COHN
Sunderland

CHARLES W. REES
London

ALAN R. KATRITZKY
Florida

Contributors to Volume 2

Dr K. I. Booker-Milburn
School of Chemical Sciences, University of East Anglia, Norwich NR4 7TJ, UK

Dr W. R. Bowman
Department of Chemistry, Loughborough University of Technology, Loughborough, Leicestershire LE11 3TU, UK

Dr C. K.-F. Chiu
Pfizer Inc. Central Research, Eastern Point Road, Groton, CT 06340, USA

Dr D. Choueiry
Department of Chemistry, Purdue University, West Lafayette, IN 47907-1393, USA

Dr E. W. Colvin
Chemistry Department, University of Glasgow, Glasgow G12 8QQ, UK

Dr I. G. C. Coutts
Department of Chemistry, Faculty of Sciences and Mathematics, The Nottingham Trent University, Clifton Lane, Nottingham NG11 8NS, UK

Dr D. G. Genov
School of Sciences, Staffordshire University, College Road, Stoke-on-Trent ST4 2DE, UK

Professor L. Hevesi
Laboratoire de Chimie Organique, Département de Chimie, Facultés Universitaires N.-D. de la Paix, Rue de Bruxelles 61, B-5000 Namur, Belgium

Dr A. D. Hobson
Department of Chemistry, University of Sheffield, Sheffield S3 7HF, UK

Dr P. R. Huddleston
Department of Chemistry, Faculty of Sciences and Mathematics, The Nottingham Trent University, Clifton Lane, Nottingham NG11 8NS, UK

Professor T. Kataoka
Department of Organic and Medicinal Chemistry, Gifu Pharmaceutical University, 5-6-1 Mitahora-higashi, Gifu 502, Japan

Professor W. Kitching
Department of Chemistry, The University of Queensland, Brisbane, Qld 4072, Australia

Dr C. Kouklovsky
Department of Chemistry, University of Cambridge, Lensfield Road, Cambridge CB2 1EW, UK

Dr K. Kwetkat
HÜLS AG, Paul-Baumann Strasse 1, D-45764 Marl, Germany

Professor S. V. Ley
Department of Chemistry, University of Cambridge, Lensfield Road, Cambridge CB2 1EW, UK

Dr S. D. Lindell
AgrEvo UK Ltd, Chesterfield Park Research Station, Saffron Walden, Essex CB10 1XL, UK

Dr R. T. Marmon
Department of Chemistry, Loughborough University of Technology, Loughborough, Leicestershire LE11 3TU, UK

Dr C. M. Marson
Department of Chemistry, University of Sheffield, Sheffield S3 7HF, UK

Professor T. Minami
Department of Applied Chemistry, Faculty of Engineering, Kyushu Institute of Technology, Sensui-cho, Tobata-ku, Kitakyushu 804, Japan

Professor E.-I. Negishi
Department of Chemistry, Purdue University, West Lafayette, IN 47907-1393, USA

Dr K. Okuma
Department of Chemistry, Faculty of Science, Fukuoka University, Nanakuma, Jonan-ku, Fukuoka 814-01, Japan

Dr P. C. B. Page
Department of Chemistry, University of Liverpool, PO Box 147, Liverpool L69 3BX, UK

Dr V. S. Rao†
Bristol Myers Squibb Pharmaceutical Research Institute, 100 Boulevard de l'Industrie, Candiac, PQ J5R 1J1, Canada

Dr R. W. Read
Department of Organic Chemistry, Heffron Building, Room 400D2, The University of New South Wales, Barker Street, Gate 14, Kensington, Sydney, NSW 2052, Australia

Dr D. Reynolds
Department of Chemistry, University of Liverpool, PO Box 147, Liverpool L69 3BX, UK

Professor G. Sauvé
Immunology Research Center, Institut Armand-Frappier, 531 Boulevard des Prairies, Case Postalc 100 Laval, PQ H7N 4Z3, Canada

Dr P. L. Spargo
Pfizer Central Research, Sandwich, Kent CT13 9NJ, UK

Profesor P. J. Stang
Department of Chemistry, University of Utah, Henry Eyring Building, Salt Lake City, UT 84112, USA

Dr J. B. Sweeney
School of Chemistry, University of Bristol, Cantock's Close, Bristol BS8 1TS, UK

Dr P. C. Taylor
Department of Chemistry, University of Warwick, Coventry CV4 7AL, UK

Professor J. C. Tebby
School of Sciences, Staffordshire University, College Road, Stoke-on-Trent ST4 2DE, UK

Dr C. J. Urch
ZENECA Agrochemicals, Jealott's Hill Research Station, Bracknell, Berkshire RG12 6EY, UK

Dr M.-X. Wang
Institute of Chemistry, Academia Sinica, Beijing 100080, China

Dr J. W. Wheeler
School of Sciences, Staffordshire University, College Road, Stoke-on-Trent ST4 2DE, UK

Dr R. D. Wilkes
Department of Chemistry, University of Liverpool, PO Box 147, Liverpool L69 3BX, UK

Dr M. Wills
School of Chemistry, University of Bath, Claverton Down, Bath BA2 7AY, UK

Dr M. Yoshimatsu
Department of Organic and Medicinal Chemistry, Gifu Pharmaceutical University, 5-6-1 Mitahora-higashi, Gifu 502, Japan

Dr V. V. Zhdankin
Department of Chemistry, University of Minnesota-Duluth, Duluth, MN 55812, USA

Contents of All Volumes

Introduction

OBJECTIVES, SCOPE, AND COVERAGE

Comprehensive Organic Functional Group Transformations (COFGT) aims to present the vast subject of organic synthesis in terms of the introduction and interconversion of functional groups. All organic structures can be considered as skeletal frameworks of carbon atoms to which functional groups are attached[a]; it is the latter which are mainly responsible for chemical reactivity and which are highlighted in COFGT. All known functional groups fit a logical and comprehensive pattern and this forms the basis for the detailed list of contents. The format of the present work was designed with the intention to cover systematically all the possible arrangements of atoms around a carbon, including those which are quite unfamiliar. The work also considers the possibility of as yet unknown functional groups which may be constructed in the future and prove to be important; thus COFGT also indicates what is not known and so points the way to new research areas.

The philosophy of the present work has been to rationalize this enormous subject within as logical and formal a framework as possible, in a scholarly and critical fashion. COFGT is designed to provide the first point of entry to the literature for synthetic organic chemists, together with an unrivalled source for anyone interested in less common, obscure, or unknown functional groups.

All functional groups are viewed as being carbon based (even if the group contains no carbon). Thus, a nitro compound is considered from the standpoint of the immediately attached carbon atom, whether di- (*sp*), tri- (sp^2), or tetracoordinated(sp^3). The work is organized on the basis of formation or rupture of bonds to a carbon atom and it is the nature of the carbon atom left after the transformation that determines the classification of the overall sequence. Several key criteria have been used to organize the work and to minimize overlap. These are, in order of priority:

1. the number of attached heteroatoms;
2. the coordination of the carbon atom involved in the functional group;
3. the nature of the immediately attached heteroatom(s); and
4. the Latest Placement Principle.

These four key principles have been used to determine the content of each volume, and to develop the detailed chapter breakdown within each volume.

Thus, according to the number of attached heteroatoms:

Volume 1 deals with synthetic reactions which result in the alteration of bonding at carbon atoms which are left with *no* attached heteroatoms.

Volume 2 deals with syntheses which result in carbon atoms attached to *one* heteroatom by a single bond.

Volume 3 deals with syntheses which result in carbon atoms attached to *one* heteroatom by a double or by a triple bond.

Volume 4 deals with syntheses which result in carbon atoms attached to *two* heteroatoms, each by a single bond.

Volume 5 deals with syntheses which result in carbon atoms attached to *two* heteroatoms by one single and one double bond, or by two double bonds, or by one single and one triple bond.

Volume 6 deals with syntheses which result in carbon atoms attached to *three* or *four* heteroatoms.

Volume 7 comprises the author and subject indexes.

Certain key principles apply to all the volumes because all functional groups are viewed as carbon based (e.g. a nitro group is either alkyl-, vinyl-, aryl-, or alkynyl-); these are:

(a) Volumes are subdivided according to the *coordination* of the carbon atom which is the product of the reaction, i.e., tetra- coming before tri- before di- before monocoordinated carbon functions.

[a] The major exception to this lies in heterocyclic compounds, where the cyclic heteroatoms are more logically considered as part of the framework. The subject of heterocycles has been treated elsewhere in the companion work *Comprehensive Heterocyclic Chemistry* published in 1984 with a second edition to be published in 1996.

In Volumes 1 and 6, reactions producing four-coordinated carbon are considered first, followed by three- and then two-coordinated carbon. The other volumes contain a more limited range of coordination types (Volumes 2 and 4 only four-coordinated, Volumes 3 and 5 only two- or three-coordinated). Each type of coordination is allocated a separate section in each volume.

(b) Attached *heteroatoms* are discussed in the following order of priority:

Halogens—F, Cl, Br, I
Chalcogens—O, S, Se, Te
Nitrogen—N
Other group 15 elements—P, As, Sb, Bi
Metalloids—B, Si, Ge
Main group metals—Sn, Pb, Al, Ga, In, Tl, Be, Mg, Ca, Sr, Ba, Li, Na, K, Rb, Cs
Transition metals—Cu, Ag, Au, Zn, Cd, Hg, Ti, Zr, Hf, Cr, Mo, W, Mn, Fe, Co, Ni, Pd, Pt, and others.

Higher coordination of heteroatoms is treated after lower. Thus, in sections dealing with iodo compounds, monocoordinate (e.g. iodides) are discussed before dicoordinate (e.g. iodoxyls) and tricoordinate functions.

(c) The *Latest Placement Principle* (or Last Position Principle) is used to avoid undue overlap in the work. Thus, the carbon attached to the heteroatom is discussed at the last possible position in the above prioritizing of heteroatoms. Examples of its application are noted later. On this basis, for example, when both C—C and C—H bonds are formed the reaction will appear in the latest chapter (i.e. Chapters 1.04–1.10 rather than in the earlier Chapters 1.01–1.03), and when both C—C and C═C bonds are formed, this will be found in the later Chapter 1.17. The Latest Placement Principle is particularly important in determining where to find electrocyclic reactions in Volume 1. If C—H, ═C—C, and C═C bonds are all formed in a reaction, then the latest appropriate chapter will deal with the reaction. Only if a change in heterofunction occurs is the reaction left to a later volume.

Exceptions to the above principles are rare. However the reactions of heteroarenes are mentioned along with those of arenes. If on reduction no change in the heterofunction occurs (e.g. in going from thiophene to tetrahydrothiophene or from pyridine to 2,3,4,5-tetrahydropyridine) the reaction is found in Volume 1. However, when the function changes (e.g. pyridine to piperidine), the conversion is considered in Volume 2. Conversion of methyl phenyl sulfone into methyl cyclohexyl sulfone appears in Volume 2, whereas the formation of cyclohexyl methyl ketone is treated in Volume 1, since the coordination of the carbon atom to the heteroatoms is changed in the first but not in the second hydrogenation.

Some further exceptions to the rigorous ordering of the work have been made for the purpose of easy reference. Thus, in Volume 1, a special chapter on ions, radicals, and carbenes is added: this chapter is limited to the treatment of species capable of more than a transitory existence. Throughout the work aspects of the Latest Placement Principle are occasionally ignored for reasons of clarity. Thus metal ligands that are incidental to the chemistry under discussion are not considered when prioritizing. Also, references to aromatic substituents, some of which involve a heteroatom (e.g. pyridyl, thienyl, etc.) but are incidental to the chemistry being described, are not viewed as changing the priority.

Within each section, we have endeavored to explain the influence of important secondary effects such as inclusion in a ring, degree of strain, degree of substitution, various types of activation, influence of stereochemistry, and so on, on the transformation under consideration. General synthetic methods are treated before specific methods.

Transient intermediates, as such, do not fall within the scope of this work. Although there is clearly no sharp division, we have attempted to restrict coverage of radicals, etc., to more stable, longer lived species. It is the aim of this work to consider all organic functional groups provided that the molecules which incorporate them, though they may be unstable, can have a finite lifetime and chemistry. The whole work deals with the generation and transformation of functional groups, *not* of molecules such as CO_2, COS, CS_2, ClCN, etc. Such simple carbon derivatives are not treated unless a further carbon is attached (e.g. RN═C═O).

VOLUME 1 SYNTHESIS: CARBON WITH NO ATTACHED HETEROATOMS

Volume 1 deals solely with the formation of nonheteroatom functional groups and as such is different in style to the remaining volumes.

In addition to the general principles, Volume 1 is further organized as follows:

1. By the type of bond formed (i.e. C—H before C—C).
2. By the type of reaction involved (i.e. substitution, then addition, then rearrangement). With C=C bond formation the order is addition, elimination, condensation, then electrocyclic and other methods. One rearrangement chapter only is devoted to each of the Parts I and II.
3. In Parts II and III the treatment of formation of ions, radicals, and carbenes is added at the end of the section dealing solely with those species with a significant rather than a transient lifetime.

In Volume 1, the heteroatom sequence is a secondary feature since only remote heteroatom functions are involved in the products: but the standard order pertains in reactants that contain heteroatoms (see, e.g. Chapters 1.01 and 1.02).

All the major structural influences that are treated throughout this work apply equally (or perhaps more importantly) in Volume 1. Thus the effects of conjugation, remote substituents, rings, stereochemistry, strain, kinetic or thermodynamic factors, solvation, primary, secondary and tertiary nature, etc., are mentioned whenever relevant.

VOLUME 2 SYNTHESIS: CARBON WITH ONE HETEROATOM ATTACHED BY A SINGLE BOND

Volume 2 is arranged in three parts: I, II and III, dealing respectively with sp^3, sp^2, and *sp* carbon linked to the heteroatom. In each chapter we have endeavored to explain important effects due to such features as the primary, secondary, tertiary nature, ring effects, strain activation, effect of beta, gamma, and more remote functionality, stereochemical effects, and so on. Methods that are common to a larger group are dealt with at their first appearance and suitably cross-referenced.

Volumes 2–6 all deal with the synthesis of functions involving at least one heteroatom. To avoid major overlap we have applied the Latest Placement Principle; that is, the chemistry is discussed at the last possible position based on the prioritization of the carbon attached to the heteroatom. Thus the compound CH_3ONH_2 is treated under "Alkyl Chalcogenides" in the subsection "Functions Based on the RON-Unit" (i.e. 2.02.6). However, CH_3ONHCH_3 appears under "Alkyl Nitrogen Compounds" (2.06.2.3) since the Latest Placement Principle prevails. Also, dialkyl ethers appear in Part I of Volume 2 (Functions Linked by a Single Bond to an sp^3 Carbon Atom), while alkyl aryl ethers appear in Part II of Volume 2 (Functions Linked by a Single Bond to an sp^2 Carbon Atom). Exceptions to the rule are:

(a) When a fully unsaturated heterocyclic substituent (e.g. thienyl, pyridyl, etc.) is used as an example of an aryl group, the ring heteroatom(s) is (are) not taken into account (e.g. 2-methoxypyridine should strictly appear in Volume 6, but is covered in Volume 2 along with 3- and 4-methoxypyridine).

(b) Carbon-based metal ligands that are incidental to the synthesis under discussion (e.g. carbonyls, cyclopentadienyls, etc.) are not taken into consideration.

VOLUME 3 SYNTHESIS: CARBON WITH ONE HETEROATOM ATTACHED BY A MULTIPLE BOND

Volume 3 follows the logical development indicated in Volume 2. Thus, according to the Last Placement Principle, the imines, RCH=N—R, appear in Volume 3 rather than in Volume 2 (where functions singly bonded to carbon are treated). Furthermore, acetophenone, $PhCOCH_3$, is treated under α,β-unsaturated ketones (3.05) rather than saturated ketones (3.04). Chloronitroacrylonitriles would appear under the section "α,β-Vinylic Nitriles with Nitrogen-based Substituents" (3.19.2.7), not under the related earlier section dealing with halo-substituents (3.19.2.3).

VOLUME 4 SYNTHESIS: CARBON WITH TWO HETEROATOMS, EACH ATTACHED BY A SINGLE BOND

Volume 4 is in three parts. Part I deals with tetracoordinated carbon bearing two heteroatoms, Part II with tricoordinated carbon bearing two heteroatoms, and Part III (a brief chapter) with stabilized radicals, ions, and the like bearing two heteroatoms. The material is arranged according to the Latest Placement Principle: thus, the synthesis of $CHBr_2CHI_2$ would appear in the section dealing with diiodo, not dibromo functions (i.e. in 4.01.5, not 4.01.4), and the synthesis of CF_3CH-BrCl is discussed in Volume 6 (carbons bearing three heteroatoms), rather than in Volume 4.

VOLUME 5 SYNTHESIS: CARBON WITH TWO ATTACHED HETEROATOMS WITH AT LEAST ONE CARBON-TO-HETEROATOM MULTIPLE BOND

Volume 5 is in three parts. Part I deals with functions with one doubly bonded and one singly bonded heteroatom, Part II with functions containing two doubly bonded heteroatoms and Part III with one triply bonded and one singly bonded heteroatom. Part I constitutes the bulk of Volume 5.

The arrangement of the chemistry in each part follows the same logical sequence. The multiply bonded heteroatom is focused on first and then the other heteroatom in a secondary classification, both following the priority rules already described. Each section excludes the coverage of the previous sections. Thus, all carbonyl derivatives will appear in Chapters 5.01–5.10 but not in Chapters 5.11, *et seq*.

According to the Latest Placement Principle structure RC(O)OC(S)R is discussed in the chapter dealing with carbons bearing a doubly bonded sulfur and singly bonded oxygen (5.12.3), *not* in that dealing with doubly and singly bonded oxygen (5.04.1). Another effect of the Latest Placement Principle is that the amides RCONMePh are discussed under *N*-arylalkanoamides (5.06.2.4), rather than *N*-alkylalkanoamides (5.06.2.2). Again, exceptions are made to the latest placement rules for: (a) hetaryl rings used as examples of aryl substituents which are not viewed as functional groups. Thus, 2-methylimidazole is not considered as an example of an amidine function and 2-methoxypyridine is not an example of a doubly bonded nitrogen, singly bonded oxygen function; (b) metal ligands that are incidental to the organic chemistry under discussion are not viewed as functions in priority considerations.

VOLUME 6 SYNTHESIS: CARBON WITH THREE OR FOUR ATTACHED HETEROATOMS

Volume 6 is in four parts. Part I deals with tetracoordinate carbons bearing three heteroatoms. Part II covers tetracoordinate compounds bearing four heteroatoms, i.e. substituted methanes, and Part III deals with tricoordinate systems bearing three heteroatoms, i.e. where one heteroatom is attached by a double bond. Part IV is brief and deals with stabilized radicals and ions. Not surprisingly, the coverage of Volume 6 is very large—and also shows that many gaps in the development of organic chemistry still exist.

The organization within the three sections not only follows the same broad logic developed in the previous volumes, but also has a structure unique to the multiheteroatom volume. According to the Latest Placement Principle $CF_3C(NR_2)_3$ appears in the section dealing with carbons bearing three nitrogens (6.05.1.1), not that dealing with carbons bearing three halogens (6.01.2), while $(CF_3CH_2O)_2CO$ appears in Part III, not in Part I.

In the chapter dealing with iminocarbonyl functions in Part III, the substituents on nitrogen are discussed in each appropriate subsection in the order outlined above. Thus, the RN= group would be first considered with R = H, then alkyl, alkenyl, aryl and hetaryl, alkynyl and then heteroatom substituents in the usual order.

In each relevant section, we have endeavored to explain the influence of important secondary effects on the synthesis such as structure (primary, secondary, etc.), ring effects, strain, activation, stereochemistry, remote substituent effects, etc.

The arrangement of the chemistry in each of Parts I–III follows a similar pattern. Thus, each section commences with functions containing at least one halogen. This section deals with all combinations of halogen with other heteroatoms in the described order. The next section deals with functions containing at least one chalcogen in combination with any other heteroatoms except halogens. Subsequent sections each exclude the previous title heteroatom functions.

VOLUME 7 INDEXES

Subject Indexes are included in each of Volumes 1–6 and Cumulative Subject and Author Indexes appear in Volume 7. Most entries in the Subject Index consist of two or three lines: the first line is the entry itself (e.g. Lactones) and the second line is descriptive of that entry (e.g. reduction); in many cases more detail is given (e.g. with 9-BBN).

REFERENCES

The references are handled by the system previously used successfully in *Comprehensive Heterocyclic Chemistry*. In this system reference numbers appear neither in the text, nor as footnotes, nor at the end of chapters. Instead, each time a reference is cited in the text there appears (in parentheses) a two-letter code assigned to the journal being cited, which is preceded by the year (tens and units only for twentieth-century references) and followed by the page number. For example: "It was shown ⟨80TL1327⟩ that . . .". In this phrase, "80" refers to 1980, "TL" to *Tetrahedron Letters*, and "1327" to the page number. For those journals which are published in parts, or which have more than one volume number per year, the appropriate part of the volume is indicated, e.g. as in ⟨73JCS(P2)1594⟩ or ⟨78JOM(162)611⟩, where the first example refers to *J. Chem. Soc., Perkin Trans 2*, 1973, page 1594, and the second to *J. Organomet. Chem.*, 1978, volume 162, page 611.

This reference system is adopted because it is far more useful to the reader than the conventional "superscript number" system. It enables readers to go directly to the literature reference cited, without first having to consult the bibliography at the end of each chapter.

References to the last century quote the year in full. Books have a prefix "B-" and if they are commonly quoted (e.g. *Organic Reactions*) they will have a code. Otherwise, as with uncommon journals, they are given a miscellaneous code (MI) and numbered arbitrarily *abb1*, *abb2*, etc., where *abb* refers to the volume and chapter number and *1*, *2*, etc., are assigned sequentially. Patents are assigned appropriate three-letter codes.

The references are given in full at the end of each volume. They include *Chemical Abstract* references when these are likely to help; in particular, they are given for all patents, and for less accessible sources such as journals in languages other than English, French, or German, company reports, obscure books, and theses.

2.01
Alkyl Halides

PETER L. SPARGO
Pfizer Central Research, Sandwich, UK

2.01.1 GENERAL METHODS FOR ALKYL HALIDES

The chemistry and preparation of halogen-containing compounds have been reviewed in *Houben-Weyl* ⟨60HOU(5/4)1, 62HOU(5/3)1⟩, in *Comprehensive Organic Chemistry* ⟨79COC(1)493⟩, and in an excellent review by Hudlicky and Hudlicky in *The Chemistry of Functional Groups* series ⟨B-83MI 201-01⟩. The latter review includes some useful 'Halogenation Tables' (reproduced from an earlier publication ⟨78OPP181⟩ which correlate starting materials, halogenating agents and products in such a way that the reader can quickly identify generally useful methods, as well as the compatibility of functional groups with halogenating agents. A review in *Comprehensive Organic Synthesis* ⟨91COS(6)203⟩ provides an account of nucleophilic halogenation methods, while the synthesis and reactivity of α-halogenated ketones, aldehydes and imines is the subject of an update volume of the Patai series ⟨B-88MI 201-01⟩. Many classical methods for the synthesis of alkyl halides are still widely used, and the *Houben-Weyl* volumes ⟨60HOU(5/4)1, 62HOU(5/3)1⟩, despite their age, provide detailed procedures and numerous tables from which much useful information may be gleaned. Literature procedures up to and including 1987 have been clearly tabulated in easily accessible form in Larock's *Comprehensive Organic Transformations* ⟨B-89MI 201-01⟩. In addition, an annual review of the synthesis of organic halides can be found in the new journal *Contemporary Organic Synthesis* ⟨94MI 201-01⟩.

It would be impossible here to provide a truly comprehensive review of alkyl halide synthesis, so coverage has been restricted primarily to those methods which would appear to have the greatest general synthetic utility. Mechanistic details have necessarily been kept to a minimum and are only discussed where they have a direct bearing on regio-, stereo- or chemoselectivity. Brief mention of some less well used methods is also made.

Because of the large differences in reactivity of fluorides, chlorides, bromides and iodides, there are very few methods which are generally applicable to all four halogens. In particular, the unique properties of fluorine mean that special methods have had to be developed for this halogen (Section 2.01.2). Alkyl chlorides and bromides are synthetically the most widely used alkyl halides and their chemistry is often closely related (Sections 2.01.3 and 2.01.4). Although alkyl iodides are often prepared using methods similar to those used to prepare alkyl bromides, they are much less common synthetic targets or intermediates (Section 2.01.5).

In this section a range of general synthetic approaches to alkyl halides is described. Certain transformations are discussed in detail in this section, while others are expanded in the later sections specific to each halogen. The reader is therefore encouraged to consult the relevant subsection within each of the five sections in this chapter for a balanced coverage.

2.01.1.1 Alkyl Halides from Alkanes

Direct halogenation of unactivated alkanes with elemental halogen, often in the presence of visible or ultraviolet light ⟨see reviews B-69MI 201-01, B-69MI 201-02, B-73MI 201-02, B-73MI 201-03⟩, is generally indiscriminate and therefore not preparatively useful, except in cases where symmetry dictates that all of the replaceable hydrogens are equivalent (e.g., cyclohexane, ethane). There are scattered reports of halogenations of unactivated hydrocarbons with a variety of different reagents ⟨B-89MI 201-01⟩, but yields are often low, and none of the methods appears general. The most recent work in this area has been by Barton *et al.* in the early 1990s, and their chemistry, which can be used to prepare chlorides, bromides and iodides (but not fluorides), is exemplified by Equation (1) ⟨92T9195, 92TL3413, 93TL1871, 93TL5689, 94T31⟩. For a short review on this and related chemistry see ⟨92ACR504⟩.

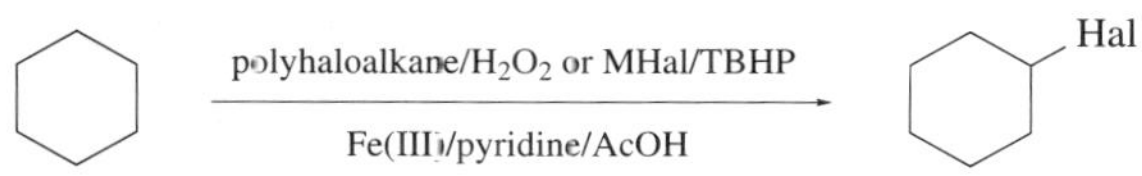

(1)

polyhaloalkane = CCl_4, CBr_4, $CBrCl_3$, CBr_2Cl_2, etc.
M = Li, Na
Hal = Cl, Br, I

While the existence of radical intermediates in the processes above has been the source of some dispute ⟨94TL1427, 94TL1431⟩, the radical nature of halogenation at allylic and benzylic sites is universally accepted ⟨B-72MI 201-01⟩ The latter reaction is most commonly applied in the synthesis of allylic and benzylic bromides using *N*-bromosuccinimide (the Wohl–Ziegler reaction) (Section 2.01.4.1).

Alkane activation by an electron-withdrawing group greatly widens the scope of reagents and reaction conditions for halogenation, since ionic mechanisms may then operate. Aldehydes and ketones (often in an enol form such as silyl enol ether or an enol acetate) can be halogenated in the α-position with a variety of reagents, including elemental fluorine, chlorine, bromine and iodine (Scheme 1). The most difficult of these is fluorination, but a range of useful procedures have been devised to overcome this problem (Section 2.01.2.1).

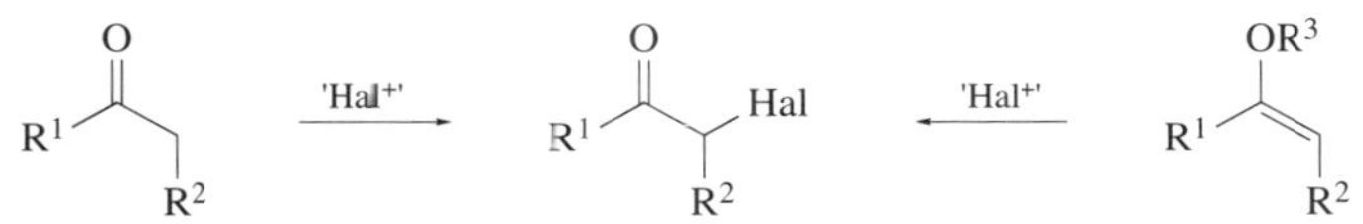

R^3 = alkyl, acyl, silyl

Scheme 1

As a general rule, clean monohalogenation (with minimal dihalogenated by-product formation) is more easily achieved under acidic rather than basic conditions, although there are nevertheless many examples of the latter. For unsymmetrical ketones, halogenation under acidic conditions generally occurs at the more substituted α-carbon, because the reaction proceeds under thermodynamic control through the more stable enol tautomer.

Halide ions can also be used to α-halogenate carbonyl compounds and their enol derivatives in the presence of a suitable oxidant such as lead tetraacetate ⟨82S1021⟩, benzoyl peroxide, hydrogen peroxide or mcpba ⟨76CPB820⟩. α-Chloro-, bromo- and iodocarbonyl compounds have all been prepared using these methods. For a detailed review of the preparation of α-halo aldehydes, ketones and imines, see ⟨B-88MI 201-01⟩. Ketals have been brominated and occasionally chlorinated (but apparently not fluorinated or iodinated) at the β-carbon, probably via transient enolic intermediates (Section 2.01.4.1). Carboxylic esters, amides and acids are also straightforwardly α-halogenated, as are nitriles ⟨48JA165⟩. Thionyl chloride converts acid chlorides to α-chloro-, α-bromo- or α-iodoacid chlorides when combined with NCS, NBS or iodine respectively ⟨75JOC3420⟩. A surprisingly little-used alternative approach to α-haloketones exploits the reactivity of the active methylene group in β-ketoesters or malonates by halogenation with NBS, NCS, SO_2Cl_2 or Br_2, followed by hydrolysis and decarboxylation (Scheme 2) ⟨83TL163, 87S188⟩ or deacetylation ⟨49JA3107, 72TL4067, 87TL5505⟩. A related method for preparing α-fluoroketones has also been described ⟨89CL577⟩.

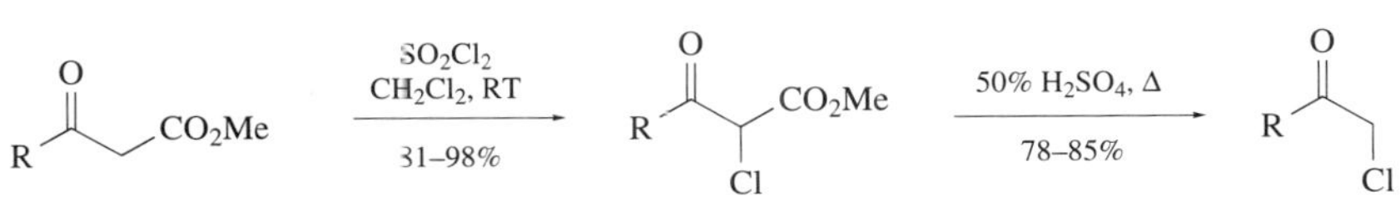

Scheme 2

2.01.1.2 Alkyl Halides from Alkenes

A wide variety of 2-functionalised alkyl halides can be prepared by addition of Hal—Y (Y = O, N, S, Se, etc.) to alkenes ⟨93S1177⟩. In accordance with the 'rule of latest placement' applied to the organisation of this publication, most of these are covered in later chapters. In this chapter the discussion focuses on the addition of halogen–hydrogen and halogen–halogen only.

2.01.1.2.1 *Alkyl halides by hydrohalogenation of alkenes*

The direct addition of HHal (Hal = F, Cl, Br, I) to alkenes is not a particularly widely used synthetic approach to alkyl halides, and there are a number of reasons for this. Among these is the fact that mixtures of regioisomers and rearranged products are often obtained (see reviews ⟨40CRV351, 62CRV599⟩ and hydrobromination ⟨63OR(13)150, B-77MI 201-01, 91COS(4)269⟩). Commonly, the reaction proceeds through an ionic mechanism via the more stable of the two possible carbocation intermediates to give the Markovnikov product as indicated in Scheme 3 for a terminal alkene.

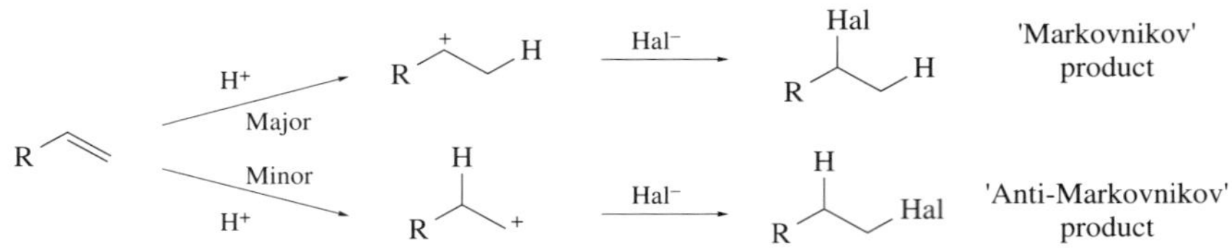

Scheme 3

A general method for Markovnikov addition of HHal (Hal = Cl, Br, I) to alkenes using phase-transfer catalysis has been reported ⟨80JOC3527⟩ and a polymer-supported phase transfer catalyst can be conveniently used for this purpose ⟨88IJC(B)1129⟩. It has also been shown that Markovnikov hydrohalogenation can be facilitated by performing the reaction in the presence of an inorganic support such as silica or alumina. Furthermore, under these latter conditions there is no need to use HHal itself, since it can be generated *in situ* from species such as $SOCl_2$, $(COCl)_2$, TMS-Cl, TMS-Br, TMS-I or PI_3 ⟨90JA7433, 93JA3071⟩. Hydrohalogenation of alkenes bearing an electron-withdrawing group gives the β-halogenated product exclusively, as expected on electronic grounds. Anti-Markovnikov addition to alkenes is often observed in hydrobromination with HBr, and suggests a free-radical or four-centre addition mechanism. Indeed, if Markovnikov addition of HBr is required, it is often necessary to take precautions to exclude peroxides or to add free radical inhibitors ⟨40CRV351⟩.

Anti-Markovnikov addition of HCl, HBr or HI is generally achieved via hydrometallation, usually hydroboration ⟨81JCR(S)376, 81JOC2582, 81JOC3113, 83HCA1018⟩ or hydroalumination ⟨76JOM(122)C25, 78CL833⟩, followed by treatment with an electrophilic halogen source (Scheme 4).

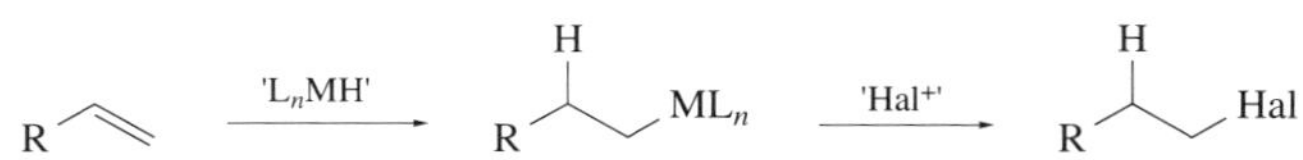

M = B, Al, Zr, Si
L = Ligand (including carbon-bonded ligands)
Hal = Cl, Br, I

Scheme 4

The halogenolysis of organoboranes has been briefly reviewed ⟨85OR(33)1, 91COS(7)593⟩, as has its applications to the incorporation of radioactive halogen isotopes ⟨84ACR215⟩. BCl_3 and BBr_3 are recent additions to the list of reagents suitable for this purpose ⟨93S973⟩. Hydrosilylation followed by treatment with Cl_2, Br_2, I_2, NBS or copper(II) chloride or bromide also gives access to the anti-Markovnikov products ⟨78JA290, 78TL1809, 82OM355, 82OM369⟩. In addition, it has been shown that hydrozirconation of a substituted alkene leads to migration such that, on quenching with NCS, NBS, iodine, bromine or iodobenzene dichloride, the terminal primary alkyl halide is obtained (Scheme 5, ⟨74JA8115⟩; see also ⟨76AG(E)333, 81JOC1821⟩).

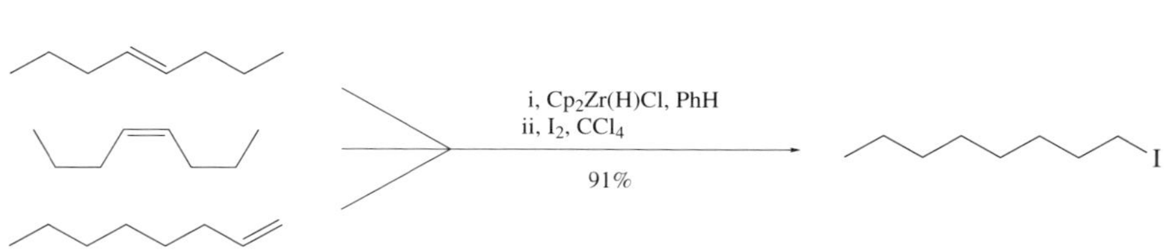

Scheme 5

2.01.1.2.2 Alkyl halides by halogen–halogen addition to alkenes

The addition of two halogens (X—X or X—Y) across a double bond is a commonly used strategy in synthetic organic chemistry, and can be achieved in a number of ways. In most cases, the addition proceeds by the ionic mechanism depicted in Scheme 6, giving the product of Markovnikov addition. Although the reaction is believed to proceed via the cyclic halonium ion (**1**), the fact that there is a preference for Markovnikov addition suggests the transient intermediacy of a species such as (**2**). The regioselectivity is frequently not as high as is usually observed in hydrohalogenation ⟨72RCR740, B-76MI 201-01, B-77MI 201-01, 81RCR151, B-89MI 201-02, 91COS(4)329⟩. For the purposes of this presentation, this chemistry has been divided up according to the net addition products obtained, i.e., Section 2.01.2.2.2, F—F, F—Cl, F—Br, F—I; Section 2.01.3.2.2, Cl—Cl, Cl—Br, Cl—I; Section 2.01.4.2.2, Br—Br, Br—I; Section 2.01.5.2.2, I—I.

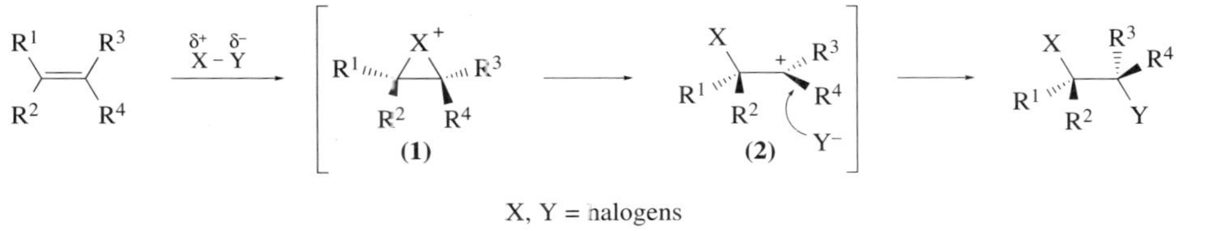

Scheme 6

2.01.1.3 Alkyl Halides from Alkyl Halides

Nucleophilic interconversion of halides (Equation (2)) is an equilibrium process, and despite the wide range of C—Hal bond energies (C—F > C—Cl > C—Br > C—I), methods exist for the preparation of almost any alkyl halide from almost any other. It is therefore quite surprising that there are almost no general methods which are genuinely applicable to *all* halides! For the purposes of clarity, therefore, each product halide is considered separately in Sections 2.01.2.3 (fluorides), 2.01.3.3 (chlorides), 2.01.4.3 (bromides) and 2.01.5.3 (iodides). The reader should be aware that this necessarily results in a fair degree of overlap between these sections.

$$\mathrm{RHal^1} \xrightarrow{\mathrm{MHal^2}} \mathrm{RHal^2} \qquad (2)$$

$\mathrm{Hal^1, Hal^2 = F, Cl, Br, I}$
$\mathrm{M = metal, R_4N}$

2.01.1.4 Alkyl Halides from Alcohols and their Derivatives

Alcohols and their derivatives are by far the most widely used precursors to alkyl halides, and a vast array of procedures for this transformation can be found in the literature. Hydroxide ion is a poor leaving group and cannot be directly displaced with halide ion. However, activation (either *in situ* or in discrete steps) by protonation, sulfonation, or by formation of an oxyphosphonium intermediate, allows easy access to all four halides. If a stable carbocation can be formed on loss of the oxygen-containing functionality, the reaction may proceed in an S_N1 fashion with no stereocontrol (Scheme 7).

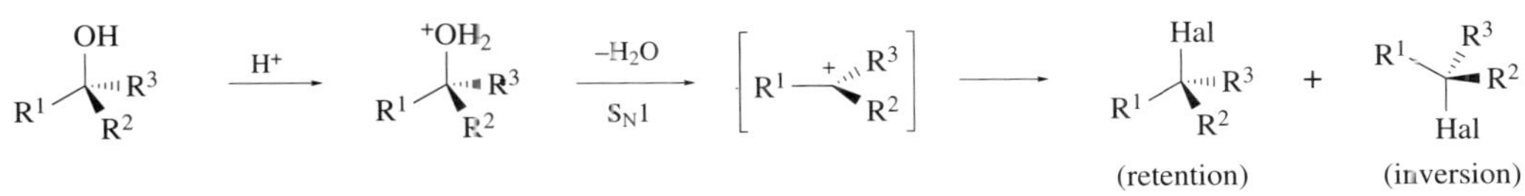

Scheme 7

More commonly, the reaction proceeds via an S_N2 mechanism with inversion of configuration at carbon (*vide infra*). *In situ* product epimerisation by S_N2 halide exchange can sometimes be a problem with the more nucleophilic halides, iodide and bromide, but is rarely seen with chloride.

The conversion of alcohols to alkyl halides has been discussed in sections of wider reviews ⟨60HOU(5/4)1, 62HOU(5/3)1, B-71MI 201-01, 80T1901, B-83MI 201-01, 91COS(6)203⟩, and extensive tabulations of reagents covering the literature up to 1987 can be found in Larock's compilation of references ⟨B-89MI 201-01⟩. The use of oxyphosphonium intermediates for this transformation has been thoroughly reviewed ⟨83OR(29)1⟩, as has the synthesis of optically active alkyl halides from optically active alcohols ⟨69S112⟩, although the latter was at a time when meaningful and reliable methods for measuring enantiomeric purity were not available.

Space restrictions mean that priority has been given to the more widely applied procedures, together with those which, at the time of writing, have yet to be covered in review publications.

2.01.1.4.1 Alkyl halides directly from alcohols

Alcohols can be converted to alkyl halides by treatment with hydrogen halide HX under a variety of conditions. The reaction is rapid for alcohols which form a stable carbocation on protonation and loss of water (e.g., tertiary or benzylic), but side reactions such as elimination to alkenes or rearrangement of the carbon skeleton to a more stable carbocation are quite common. Not surprisingly, the stereochemical integrity of the carbon centre is often lost under these conditions (Scheme 7).

The high reactivity and corrosiveness of hydrogen fluoride means that it is rarely used in this context. Much more useful (at least for secondary and tertiary alkyl fluorides) is a combination of HF with organic bases ⟨73S786, 91T5329⟩, particularly Olah's pyridine poly(hydrogen fluoride) reagent (PPHF), either in solution or in a poly(vinylpyridine)-polymer-supported form (PVPHF) ⟨90SL267, 93S693⟩. Indeed, by incorporating NaCl, NH_4Br or KI in the reaction mixture, it is possible to use Olah's reagent to prepare the corresponding alkyl chlorides, bromides and iodides from an extremely wide range of alcohols, including neopentyl systems ⟨74S653, 79JOC3872⟩. This is one of the few procedures useful for the preparation of all four halogens from alcohols.

Inorganic acid halides such as $SOCl_2$, $POCl_3$, PCl_5 and PBr_3 can often be used to overcome some of the side reactions associated with the use of hydrogen halides. Discussion of chloride synthesis using phosphorus chlorides and $SOCl_2$ can be found in Section 2.01.3.4, while bromide and iodide synthesis using phosphorus halides is covered in Sections 2.01.4.4 and 2.01.5.4. While these reagents are of fairly wide applicability, they are still quite aggressive, and a wide range of alternative phosphorus-based reagent systems has been developed ⟨83OR(29)1⟩. Many of these are applicable to more than one halogen, and are discussed in this section.

Triphenylphosphine and diethyl azodicarboxylate (dead) can be used to activate alcohols towards nucleophilic attack of halide ion, as outlined in Equation (3) and Scheme 8. This is a Mitsunobu-like procedure ⟨81S1, 92OR(42)335⟩. Using zinc(II) halide as halide source, chlorides, bromides and iodides can be prepared with clean inversion of configuration ⟨84JOC3027, 90SC3039⟩, while the use of LiF can also give access to fluorides ⟨85SC663⟩. A mild variation of this procedure, suitable for use with sensitive substrates, uses amine hydrohalide salts such as pyridine hydrochloride and hydrobromide instead of metal halides ⟨85G393⟩. Alternatives to dead include the more stable diisopropyl azodicarboxylate (diad) ⟨90SC3039⟩ and the cyclic diazodicarboxylate (**3**), which is used in combination with PPh_3 and MeI or MeBr ⟨84BCJ2675⟩. A driving force in all of these reactions, and related reactions described below, is the formation of the strong phosphorus–oxygen double bond.

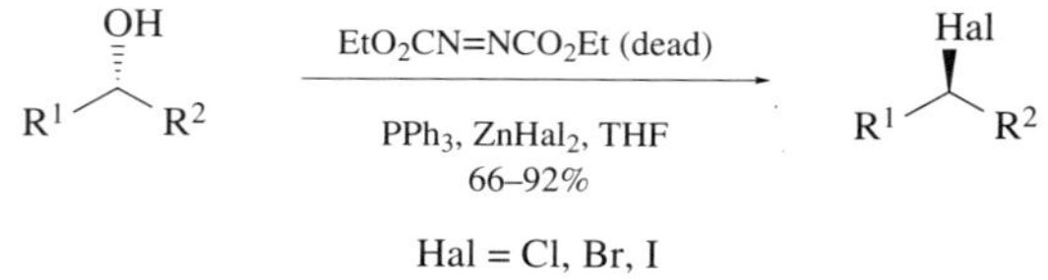

(3)

Hal = Cl, Br, I

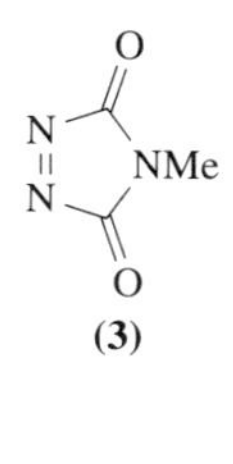

(**3**)

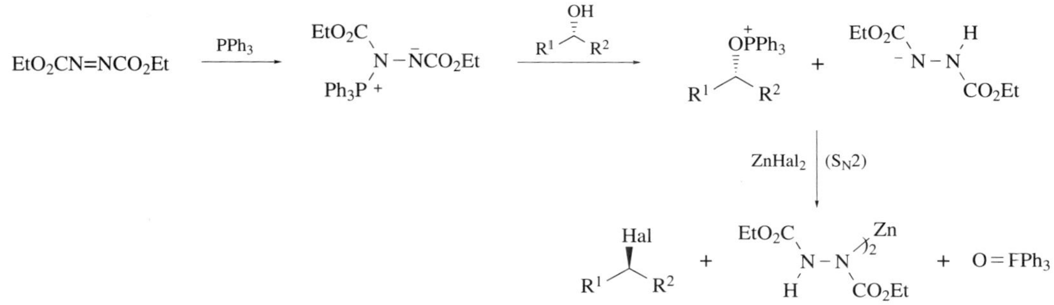

Scheme 8

Adducts of triphenylphosphine with elemental halogen, Ph_3PHal_2, either commercially available or prepared *in situ*, can be used to convert alcohols cleanly with inversion to fluorides ⟨68CFB1C09⟩, chlorides, bromides and iodides (Equation (4)) ⟨64JA964⟩.

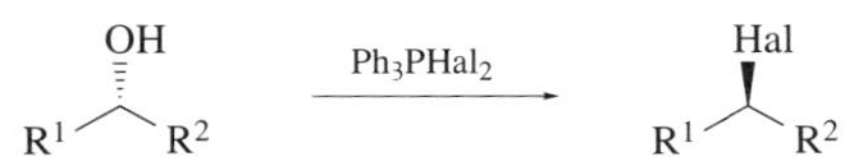

(4)

Hal = F, Cl, Br, I

While Ph_3PBr_2 usually gives the best results ⟨65JOC2635, 73OSC(5)249, 84JOC431, 92SC2945⟩, it has been found that the addition of imidazole to Ph_3PCl_2 ⟨84S168⟩ or Ph_3PI_2 ⟨79CC978, 90SC1473⟩ or, alternatively, the use of triiodoimidazole instead of iodine ⟨79CC978, 80JCS(P1)2866, 82JCS(P1)681⟩, all lead to significantly improved yields. Other phosphines such as Bu_3P ⟨82AJC517⟩, $Ph_2PCH_2CH_2PPh_2$ ⟨87TL767⟩ or a triaryl phosphite $(ArO)_3P$ ⟨92CL1505⟩ have proved advantageous in certain circumstances. The use of a polymer-supported triphenylphosphine dibromide ⟨84JCS(P1)195⟩ greatly facilitates the reaction workup, since the phosphine oxide by-product is simply removed by filtration. Alternatively, placing a dimethylamino group on one of the phenyl rings of triphenylphosphine enables removal of the phosphine oxide by-product by an aqueous acid wash ⟨88JOC6126⟩. Another practical modification is the replacement of PPh_3 with Ph_2PCl, which enables the phosphorus-containing by-product to be removed by an aqueous base wash ⟨88JOC6126⟩. In a logical extrapolation of this approach, $PhPCl_4$ has been introduced as a new reagent for alkyl chloride synthesis from alcohols ⟨90JOC3415⟩.

The combination of triphenylphosphine and carbon tetrahalide has proved to be a very powerful but mild method, particularly for the preparation of alkyl chlorides, once again with inversion of configuration (Equation (5)) ⟨73CC110, 76CB3446, 84S168, 88OSC(6)634, 91TL3977⟩. For reviews on this chemistry see ⟨75AG(E)801, B-79MI 201-01, 83OR(29)1⟩. This reagent system is particularly useful for the conversion of allylic alcohols to halides without allylic rearrangement ⟨72JOC1466⟩. As with the Ph_3PHal_2 method described above, the use of a polymer-supported triphenylphosphine gives easier workups ⟨75CC622, 75JOC1669, 84JCS(P1)195, 85CC337⟩, as does the use of a phosphine carrying a water-soluble pyridyl side chain ⟨87JOC4999⟩. Another alternative phosphine is tris(dimethylamino)phosphine, $(Me_2N)_3P$, which generates the water-soluble (but carcinogenic) by-product hexamethylphosphoramide (HMPA) ⟨68CC1350, 75BSF607⟩. CBr_4 is commonly used in combination with PPh_3 ⟨77LA806, 87CS477, 93JOC4382⟩ or other phosphines ⟨86JOC789, 86TL1607⟩, and is particularly useful for allylic, benzylic and saturated primary alkyl systems. CI_4–PPh_3 is less commonly used ⟨78CAR(61)511, 91TL3977⟩.

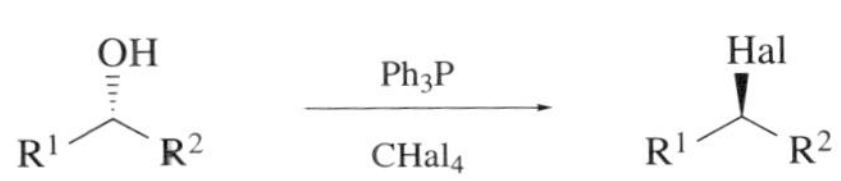

(5)

Hal = Cl, Br, I

Other related halide sources for triphenylphosphine-mediated conversions of alcohols to alkyl halides include a range of polyhalogenoalkanes, such as C_2Cl_6, C_2Br_6, $BrCCl_2CCl_2Br$ and ICH_2CH_2I ⟨83S139⟩, ethyl trichloroacetate and ethyl tribromoacetate ⟨89JOU642⟩, trichloroacetonitrile (which even works on neopentyl alcohols) ⟨89JOU587⟩, and hexachloroacetone, which is most effective in converting allylic alcohols to allylic chlorides without rearrangement ⟨77TL2999, 79JOC359, 81JOC824, 84JOC431⟩. Some of these reagents give higher-boiling by-products, enabling low-boiling product halides to be distilled from the reaction mixture without contamination with the haloform $CHHal_3$ by-product obtained when carbon tetrahalide is used as halogen source. The introduction in 1993 of triphosgene as an easily handled solid form of phosgene has rekindled interest in phosgene chemistry, and when combined with triphenylphosphine it has been shown to give excellent yields of alkyl chlorides from a wide range of alcohols ⟨93SC711⟩.

Related to the above methods is the triphenylphosphine/*N*-halosuccinimide combination (Equation (6)), which is particularly good for selecting primary over secondary alcohols (although secondary chlorides, bromides and iodides can nevertheless all be prepared this way ⟨73TL3937⟩). It is also particularly useful for the preparation of bromides in sensitive systems (e.g., ⟨86JOC2637, 93OPP249⟩.

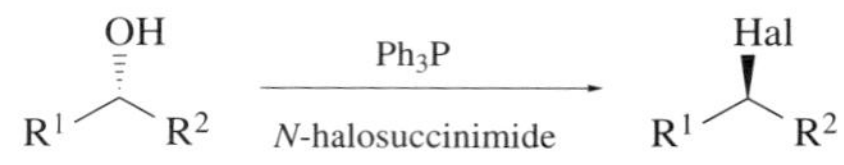

Hal = Cl, Br, I (6)

Triphenyl phosphite $P(OPh)_3$ has been used in combination with Cl_2, Br_2 and I_2 ⟨54JCS2281⟩, with NCS, NBS and NIS ⟨73TL3937⟩, and with benzyl chloride and bromide ⟨53JCS2224⟩, but is most commonly used in combination with methyl iodide for the preparation of alkyl iodides. The latter reagent system, using the so-called triphenyl phosphite–methiodide reagent, $(PhO)_3PMeI$, converts a wide range of primary (including neopentyl), secondary, tertiary, allylic and benzylic alcohols to iodides (Equation (7)) ⟨53JCS2224, 88OSC(6)830, 93TL2757⟩. It is particularly widely used in carbohydrate and nucleoside chemistry ⟨70JOC2319⟩.

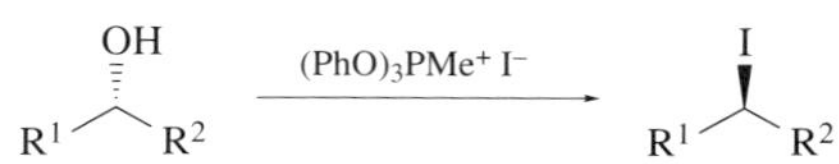

(7)

An interesting and perhaps surprisingly little-used method of phosphorus activation of alcohols entails the preparation of the nicely crystalline azaphospholane intermediates (**4**). Treatment of (**4**) with SO_2Cl_2, Br_2 or MeI yields chlorides, bromides and iodides, respectively ⟨82TL4411⟩.

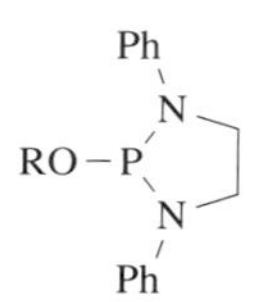

(**4**)

Moving away from phosphorus-based activation, a method applicable to all four halogens uses the haloenamines (**5**) derived from dialkylisobutyrylamides (Equation (8)). Using these reagents, primary, secondary, allylic and propargylic alcohols are converted to alkyl halides in high yields with inversion of configuration under neutral conditions at room temperature ⟨89TL3077⟩. (See also ⟨60JA909⟩ for related work.) Vilsmeier salts (generalised as formula (**6**)) are widely used to activate alcohols towards halide displacement, and this approach has been applied to the synthesis of chlorides, bromides and iodides. A range of different haloiminium salts has been used for the same transformation ⟨80S746⟩, as well as the benzothiazolium salts (**7**) ⟨76CL619⟩ and benzoxazolium salts (**8**) ⟨77CL383⟩. The latter works on alicyclic alcohols for which the former fails. Other procedures useful for the preparation of alkyl chlorides, bromides and iodides (including optically active ones), involve activation of the alcohol with trifluoroacetic anhydride ⟨87S511⟩, or with carbodiimides ⟨72AG(E)229, 87TL4445⟩. Meanwhile, chlorides, bromides and iodides can all be readily prepared from xanthates (**9**) ⟨76JCS(P1)2112⟩.

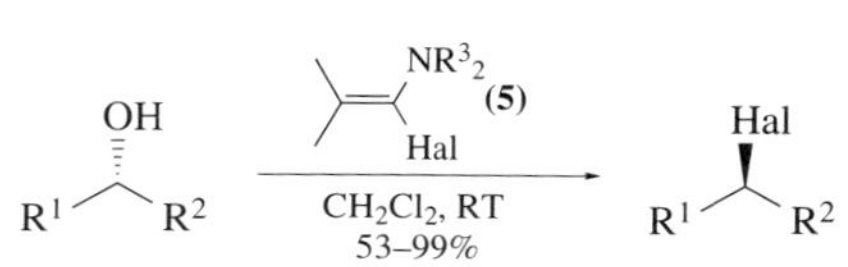

Hal = F, Cl, Br, I
R^3 = Me, Pr^i (8)

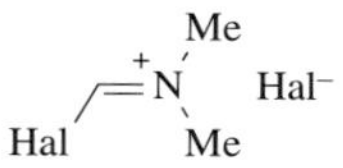

(6)

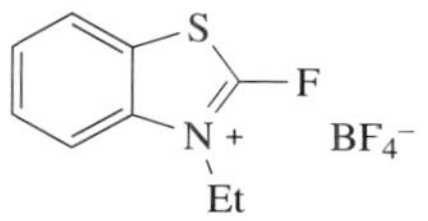

(7)

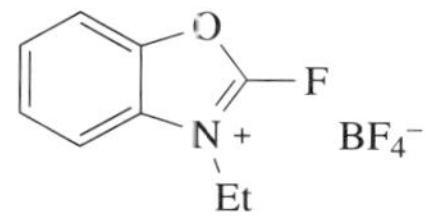

(8)

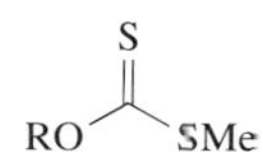

(9)

2.01.1.4.2 Alkyl halides from alcohols via sulfonates

Activation of an alcohol towards nucleophilic displacement with halide ion can be achieved by converting it to a sulfonate ester, commonly tosylate (*p*-toluenesulfonate), mesylate (methanesulfonate) or triflate (trifluoromethanesulfonate) (Scheme 9). These are readily prepared and are often stable enough to be isolated and purified. In the case of allylic or benzylic alcohols, the intermediate sulfonate may react further under the reaction conditions to generate the alkyl chloride directly ⟨84TL2295, 87TL723⟩.

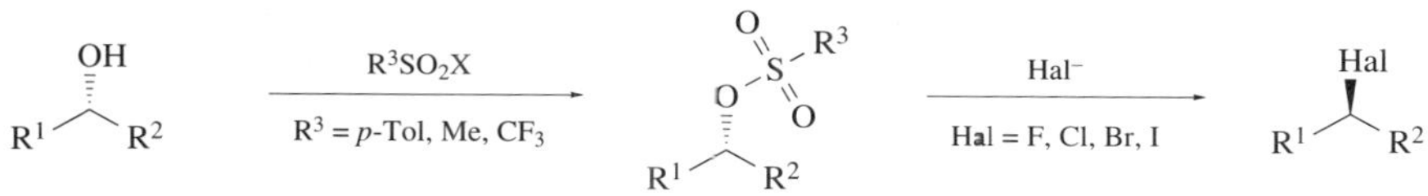

Scheme 9

Treatment of sulfonate esters of primary and secondary alcohols with alkali metal halides or tetrabutylammonium halides gives high yields of the corresponding alkyl halides, generally with inversion of configuration ⟨75S430⟩. Indeed, a comparison of the stereochemical course of substitution of homochiral 2-octyltosylate with all four halides has been made. Good inversion was observed in all except the iodide case, where it is believed that the product racemises by reversible S_N2 iodides substitution ⟨75JOC1514⟩. In the early literature, sulfonate displacement was almost exclusively applied to the formation of iodides by displacement of tosylates with sodium iodide, but many other halide sources have been successfully used since then, including KF ⟨74CC879, 87S920, 93TL293⟩ tetra-*n*-butylammonium fluoride (tbaf) ⟨87JOC658, 87LA617⟩, $Et_3N \cdot 2HF$ ⟨90TL6527⟩, CsF ⟨94TL1851⟩, tris(diethylamino)sulfonium (trimethylsilyl)difluoride (tas-f) ⟨87CJC412⟩, LiCl ⟨86JOC5291⟩, NaBr ⟨63OSC(4)753⟩, KBr ⟨74CC879⟩ and pyridinium halides ⟨52M1398⟩. Resin-bound quaternary ammonium fluoride and chloride sources ⟨79JCS(P1)2248, 93CCC902⟩ have also been applied to sulfonate displacement reactions. Even neopentyl tosylates can be converted to chlorides, bromides and iodides without rearrangement, provided the correct solvent is used (often HMPA) ⟨72JA4184⟩. Other generally applicable procedures include the use of calcium(II) halides in DMSO, which have been shown to be an effective way of converting allylic and benzylic mesylates to chlorides, bromides and iodides ⟨93JOC272⟩. Magnesium(II) bromide and iodide can also be used ⟨73CC821, 76BSF169⟩, and are particularly good for bridgehead sulfonates ⟨75AG(E)824, 91S353⟩ as are halides of aluminum(III) ⟨89TL4555⟩, iron(III) and titanium(IV) ⟨75AG(E)824⟩. In these cases, an S_N2 mechanism is of course impossible, and a bridgehead carbocation is the presumed intermediate.

Fluorosulfonates $ROSO_2F$ can be converted to chlorides, bromides and iodides by treatment with the corresponding tetrabutylammonium halide ⟨86JOC3217⟩, while fluorides, chlorides, bromides and iodides can all be obtained from secondary mesylates by treatment with potassium halide under phase transfer conditions ⟨75CC950, 75S430⟩. Imidazolyl sulfonates $ROSO_2Im$ have been shown to be less prone to hydrolysis than triflates, and give less elimination and lower reaction temperatures than do conventional sulfonates on substitution with fluoride, chloride or iodide ⟨81TL3579⟩.

Less commonly used sulfonates are the aminoalkanesulfonates (known as 'betylates') (**10**) which give chlorides, bromides or iodides with inversion in a so-called substrate–reagent ion pair (SRIP) reaction (Equation (9)) ⟨79TL3615⟩.

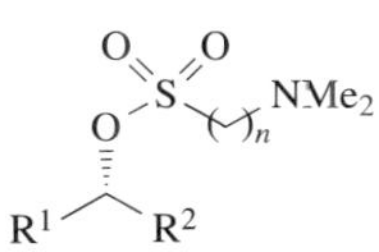

(10)

MeHal or HHal

71–90%

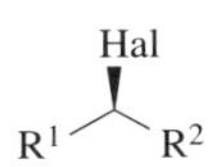

(9)

Hal = Cl, Br, I
$n = 2, 3$

2.01.1.4.3 Alkyl halides from ethers

The conditions required to cleave alkyl ethers to alkyl halides are generally quite vigorous and this approach is therefore of limited synthetic use. Treatment of an ether with halide ion under Brønsted or Lewis acidic conditions can result in cleavage to 1 mole of alcohol and 1 mole of alkyl halide (Equation (10)). Except for the cases R^2 = aryl (when cleavage of the alkyl–oxygen bond always occurs) and R^1 = Me (in which cleavage usually occurs to give the methyl halide), ether cleavage is usually nonselective, giving mixtures of both possible alcohols and alkyl halides. By using an excess of hydrogen halide, the alcohol produced is also converted to alkyl halide, and this procedure therefore often finds use when applied to symmetrical and/or cyclic ethers, particularly tetrahydrofurans, tetrahydropyrans and epoxides. The reaction is most commonly used to prepare bromides and iodides. Chlorides can also be thus prepared, but there are few reports of alkyl fluoride preparation by ether cleavage ⟨68BCJ1724⟩ other than fluorohydrin formation by epoxide opening. Diethylammonium sulfur trifluoride, Et_2NSF_3 (DAST), has been used to make 1,2-difluorides from epoxides but in low yield ⟨87JFC(36)373⟩.

$$R^1OR^2 \xrightarrow{\text{HHal}} R^1Hal + R^2OH \qquad (10)$$

Hal = F, Cl, Br, I

Ether cleavage (particularly for the formation of iodides) has been reviewed ⟨83S249⟩, as has cleavage of alkylaryl ethers ⟨88S749⟩. Examples of alternative reagents for the conversion of ethers to alkyl halides can be found in Sections 2.01.3.4 (chlorides), 2.01.4.4 (bromides) and 2.01.5.4 (iodides). Note that the conversion of cyclic ethers (including epoxides ⟨94S225⟩) to α,ω-halohydrins is covered in later chapters, as are ring openings of lactones to ω-haloacids.

Tetrahydropyranyl (and sometimes tetrahydrofuranyl) ethers are easily converted directly to alkyl bromides on treatment with bromine and PPh_3 (Equation (11)) ⟨75JOC2410, 76SC21, 79RTC371⟩ or a related bisphosphine, the latter also allowing access to iodides ⟨87TL767⟩. The use of PPh_3/CBr_4 gives inversion with low levels of elimination by-products, even in tertiary systems ⟨89TL557⟩.

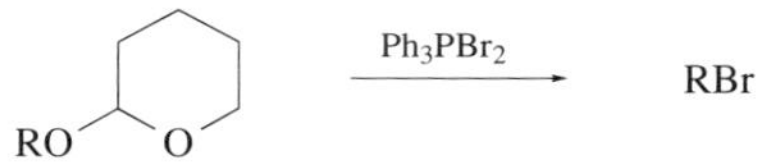

(11)

In the early 1990s, alkali metal halides were shown to convert tetrahydropyranyl ethers into alkyl bromides and iodides under Lewis acid catalysis, predominantly with retention of configuration ⟨91TL1081⟩. An S_Ni mechanism involving a tight ion pair (**11**) has been proposed to explain the stereoselectivity.

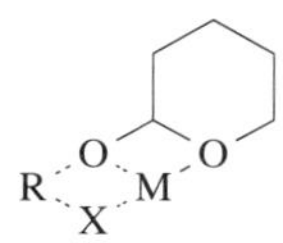

(11)
M = TMS or BF_3
X = Br, I

Silyl ethers are converted directly to alkyl fluorides using either phenylphosphorus tetrafluoride, $PhPF_4$ (Equation (12)) ⟨72TL847, 74T3793, 78TL4507, 79JOC3406, 81JFC(17)127⟩ or sulfonyl fluorides (e.g., TsF) in the presence of tetraalkylammonium fluorides (e.g., TBAF) ⟨85TL4207⟩.

$$R^1OSiR^2{}_3 \xrightarrow{\text{PhPF}_4} R^1F \qquad (12)$$

Conversion of silyl ethers to bromides can be achieved via alkoxyphosphonium intermediates using Bu_3PFBr ⟨82S326⟩, or more generally Ph_3PBr_2 ⟨86JOC4941⟩ or PPh_3/CBr_4 in the presence of acetone ⟨87TL1697⟩, the latter two proceeding with inversion and being applicable to primary, secondary, benzylic, allylic, cyclic and acyclic *t*-butyldimethylsilyl (TBDMS) ethers. Boron tribromide has also been shown to convert benzylic, allylic, primary, secondary and tertiary TBDMS ethers to bromides, but in this case, up to 71% retention of configuration was observed at secondary centres ⟨88JOC3111⟩. Alkyl iodides are formed on treatment of silyl ethers with TMS-Cl/NaI ⟨79S379⟩.

2.01.1.4.4 *Rearrangement of cyclopropyl carbinols*

Homoallylic halides can be prepared by treatment of cyclopropyl carbinols with halides under a variety of conditions (Equation (13)). Suitable reagents include HCl ⟨51JA2509⟩, HBr or HI ⟨79S37⟩, $MgCl_2$, $MgBr_2$ or MgI_2 ⟨75CC303, 80JOC2566⟩ with or without the corresponding zinc(II) halide, and trimethylsilyl chloride, which, on its own, gives chlorides but can give other halides when combined with the appropriate lithium ⟨86TL1907⟩ or zinc ⟨90OPP215⟩ halide. The only reported procedure for fluoride preparation from cyclopropyl carbinols uses Olah's pyridine poly(hydrogen fluoride) (PPHF) reagent in combination with KHF_2 ⟨87TL663, 89BCJ2024⟩.

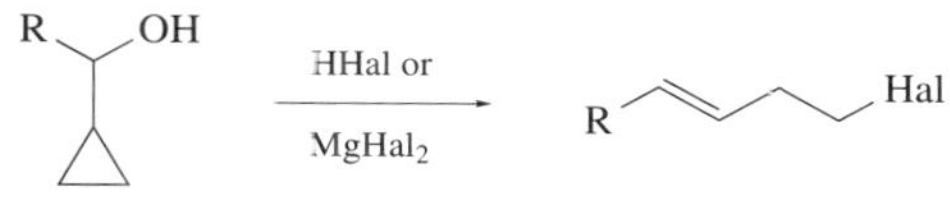

(13)

2.01.1.5 Alkyl Halides from Amines and their Derivatives

There are very few general methods for this transformation, since direct nucleophilic displacement of an amine with halide ion is not possible. However, the desired transformation can be achieved by converting the amine to a better leaving group, for example the bis(triflate) $RNTf_2$, before reacting with halide ion (Scheme 10) ⟨78JOC2291⟩.

Tf = triflyl; HMPT = hexamethylphosphoric triamide

Scheme 10

An interesting and powerful way of turning an amine nitrogen into a leaving group is by conversion into a pyridinium salt, as depicted in a generalised manner in Scheme 11. This chemistry has been used to prepare fluorides ⟨79CC238, 80JCS(P1)2901⟩, chlorides ⟨79JCS(P1)436, 79S437⟩, bromides ⟨79JCS(P1)436, 80JCS(P1)1890⟩ and iodides ⟨77S634, 79JCS(P1)433⟩, and even works in neopentyl systems without rearrangement of the carbon skeleton ⟨83IJC(B)421⟩. For reviews of this type of chemistry see ⟨80T679, 84AG(E)420⟩ and for a related procedure see ⟨80S853⟩.

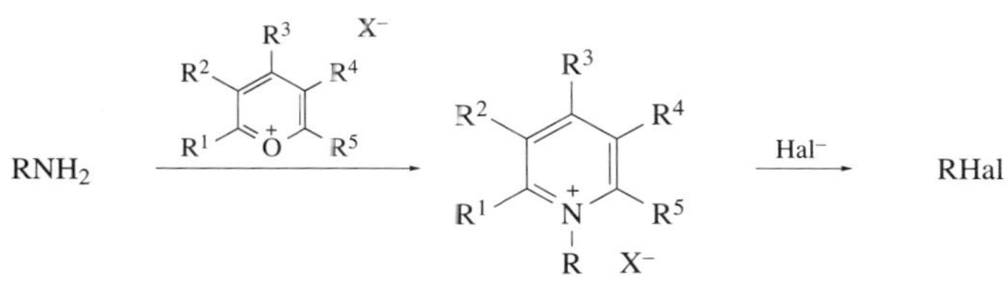

Scheme 11

Another well-known way of converting an amine to a leaving group is by diazotisation. While unactivated primary amines can indeed be converted to alkyl chlorides by treatment with NOCl ⟨78JOC4120⟩, there are few reports of applications of this chemistry, presumably because simple alkyl chlorides are more easily accessed by other routes. In contrast, diazotisation of α-amino acids has proved to be a good method for preparing a range of α-halocarboxylic acids. Treatment of α-amino acids with $NaNO_2$ in the presence of Olah's reagent, pyridine poly(hydrogen fluoride) (PPHF), results in conversion into α-fluorocarboxylic acids ⟨74S652, 79JOC3872, 81HCA2528, 82TL1549⟩. When carried out in the presence of excess KCl or KBr, the corresonding α-chloro- and α-bromocarboxylic acids are formed ⟨83HCA1023⟩. It has been shown that essentially complete retention of configuration can be obtained in a related preparation of α-chlorocarboxylic acids ⟨87OS(66)151⟩. A double-inversion mechanism via an α-lactone (**12**) can account for this observation (Scheme 12).

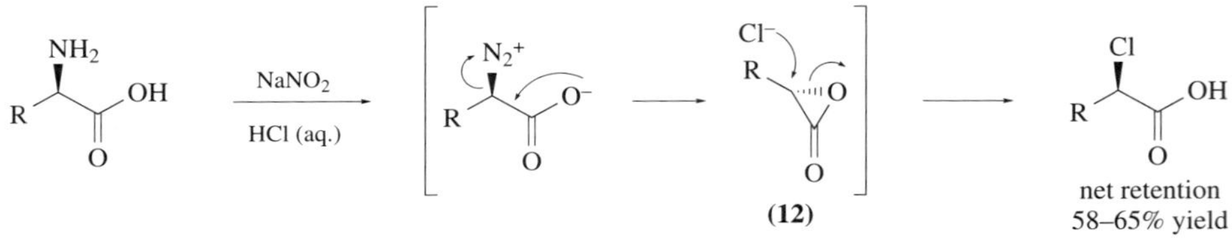

Scheme 12

Diazoketones are good substrates for conversion to α-haloketones by treatment with aqueous HCl ⟨55OSC(3)119⟩, or HBr ⟨53JOC868, 75LA1252⟩, and a simple and general procedure for the preparation of chlorides, bromides and iodides has been reported ⟨91CE751⟩ (Equation (14)). α-Fluoroketones can also be prepared using this chemistry ⟨74S896, 76HCA1027⟩.

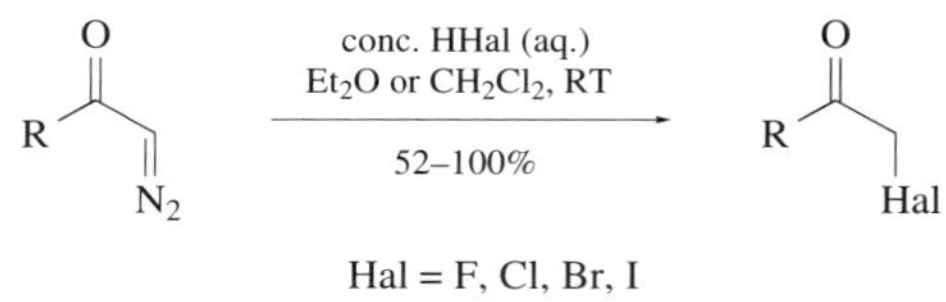

(14)

Hydrazines have been used as precursors to alkyl halides by treatment with NBS or NIS or iodine (Equation (15)) ⟨67JCS(C)252⟩. More recently, a method for converting tosylhydrazines to alkyl halides under basic conditions has been described (Equation (16)) ⟨92JOC3772⟩. Further methods apparently unique to particular halogens are described in Sections 2.01.3.5 (chlorides), 2.01.4.5 (bromides) and 2.01.5.5 (iodides).

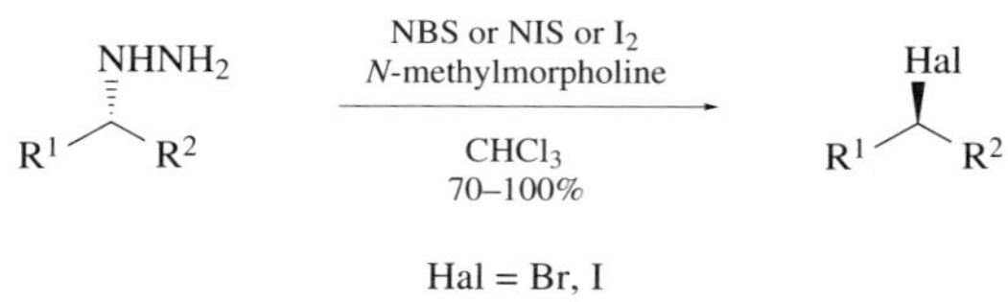

(15)

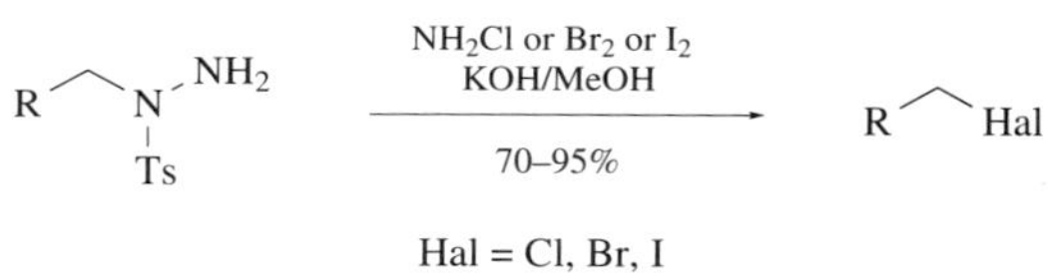

(16)

2.01.1.6 Alkyl Halides by Halodecarboxylation of Carboxylic Acids and their Derivatives

Halodecarboxylation has been briefly reviewed ⟨91COS(7)723⟩ and is one of the best methods for preparing bridgehead tertiary alkanes. The first reported halodecarboxylation procedure was the classical Hunsdiecker reaction, which has been the subject of several early reviews ⟨47CRV381, 56CRV219, 57OR(9)332⟩. In this process, silver carboxylates react with chlorine, bromine or iodine to give the corresponding acyl hypohalite (**13**) which undergoes thermal cleavage to give the alkyl halide (Scheme 13). This method suffers from the practical drawbacks of expense and the need to isolate the silver salt dry. Various modifications of the Hunsdiecker reaction have been reported, including the widely used Cristol–Firth modification using mercuric oxide ⟨61JOC280, 65JOC415, 72JOC664, 73OSC(5)126, 79JOC3405, 87JA7230⟩. More recent modifications include the use of thallium salts ⟨81JCS(P1)2608⟩, or lead tetraacetate in combination with LiCl or NCS, LiBr or iodine. The latter method has been reviewed ⟨72OR(19)279⟩ and is most frequently applied in the preparation of secondary and tertiary alkyl chlorides ⟨73S493⟩, sometimes on quite large laboratory scale ⟨90SC1011⟩.

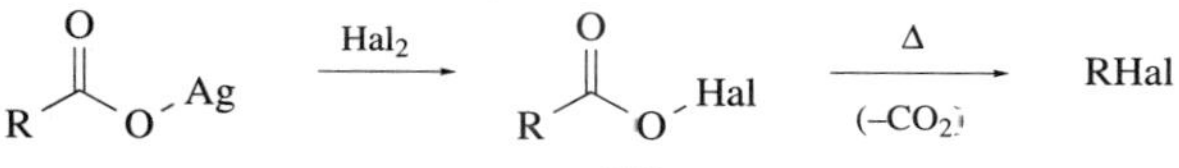

(13)

Hal = Cl, Br, I

Scheme 13

An alternative which avoids the use of heavy metals is the radical halodecarboxylation of thiohydroxamate ('Barton') esters (**14**) in the presence of the halogen donors CCl_4, $BrCCl_3$ or CHI_3 for alkyl chlorides, bromides and iodides, respectively (Equation (17)). This transformation can be achieved thermally ⟨83TL4979⟩, photochemically ⟨89TL2461, 89TL6967⟩, and ultrasonically ⟨89JOC6101⟩ and has been applied to a wide range of primary, secondary and tertiary carboxylic acids.

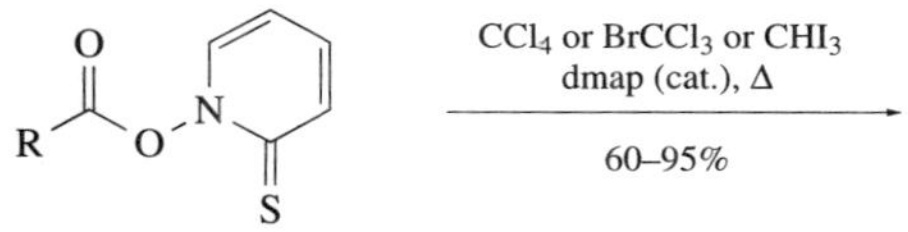

(17)

(14)

Hal = Cl, Br, I
dmap = 4-dimethylaminopyridine

Alternative radical precursors for related photochemical chlorodecarboxylation include hydroxyphthalimide esters (**15**) ⟨89CC1636⟩ and benzophenone oxime esters (**16**) ⟨88TL6287⟩. Radical iododecarboxylation can be achieved by direct photolysis of primary, secondary and tertiary carboxylic acids in the presence of *t*-butyl hypoiodite, Bu^tOI ⟨65JCS2438, 80JOC4226⟩ or $PhI(OAc)_2$ and iodine ⟨86JOC402, 87CC675⟩.

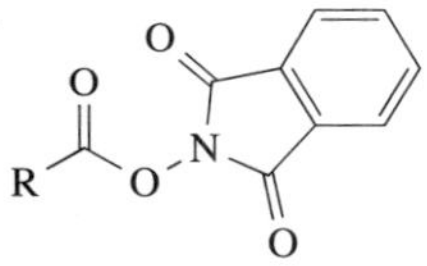

(15)

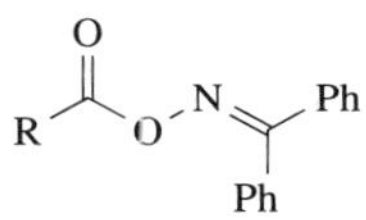

(16)

There are very few reported methods for fluorodecarboxylation. The earliest involved the treatment of sodium or potassium carboxylates with elemental fluorine, but generally gave low yields ⟨69JOC2446⟩. More synthetically useful is the application of xenon difluoride, which can be used to prepare primary, tertiary and benzylic fluorides from carboxylic acids in generally good yields ⟨83JOC4158, 86CJC138⟩.

2.01.1.7 Alkyl Halides by Haloalkylation of Arenes

Treatment of an aromatic hydrocarbon with an aldehyde and hydrohalic acid HHal effects the transformation outlined in Equation (18), and has been reviewed ⟨42OR(1)63, B-64MI 201-01, 77RCR891⟩. Although this reaction has been mostly the preserve of hydrochloric acid and formaldehyde (in its various forms, e.g., paraformaldehyde or trioxane), bromomethylation and iodomethylation ⟨40JA3098⟩ are nevertheless possible, as are chloroethylation, chloropropylation and chlorobutylation ⟨42OR(1)63⟩. Although further catalysis is often not needed, rate enhancements can be achieved using zinc chloride, acetic acid, phosphoric acid ⟨55OSC(3)195⟩ or sulfuric acid. In accordance with the ionic mechanism, electron-rich aromatics react better than electron-poor ones. For less-activated aromatic rings, chloromethyl methyl ether, $ClCH_2OMe$, has been shown to be useful ⟨70BCJ3299⟩, while methoxyacetyl chloride, $MeOCH_2COCl$, ⟨83TL1933⟩ and 1-chloro-4-(chloromethoxy)butane ⟨74S560, 76JOC1627⟩ provide useful and mechanistically intriguing alternative reagents for chloromethylation. High-yielding phase-transfer conditions for both chloromethylation ⟨91S1003⟩ and bromomethylation ⟨89SL55⟩ have been reported.

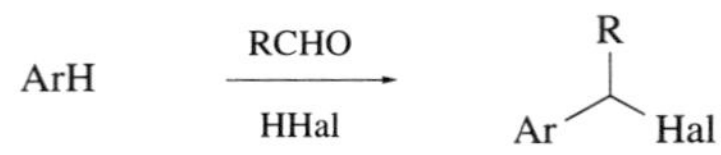
(18)

2.01.1.8 Alkyl Halides by Miscellaneous Methods

The reduction of vinyl halides to alkyl halides is rarely used synthetically because of problems with over-reduction. Modest yields can nevertheless be achieved using diimide generated *in situ* (Equation (19)) ⟨65JOC3985⟩. More spectacular is the enantioselective yeast reduction shown in Equation (20) ⟨89JOC4989⟩.

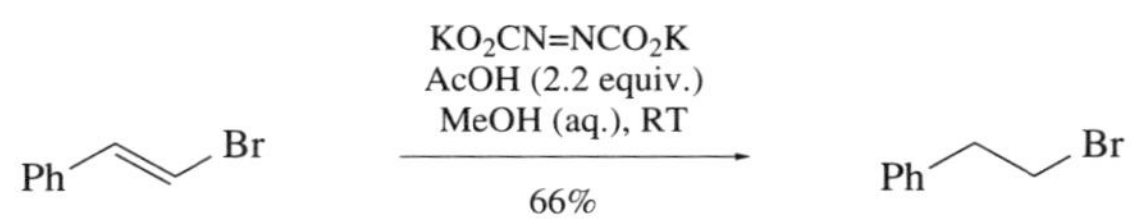
(19)

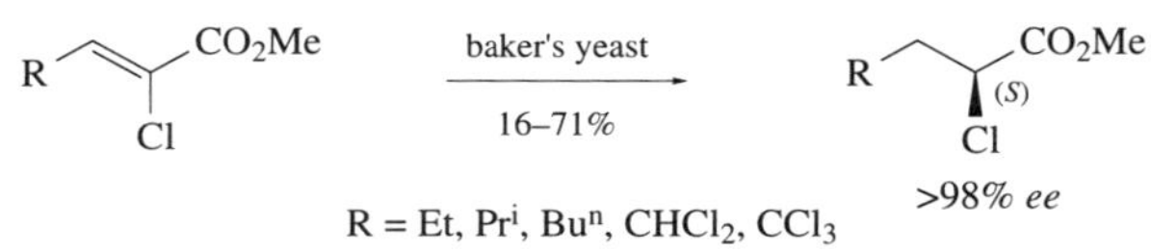
(20)

Another reductive approach to alkyl halides is the partial reduction of geminal dihalo compounds, and is applied almost exclusively to cyclopropyl systems as illustrated in Equation (21). A number of different reagents and conditions for this transformation have been reported ⟨74JOC2300, 83BCJ1881, 83CL1877, 84BCJ303, 93JOC6529⟩.

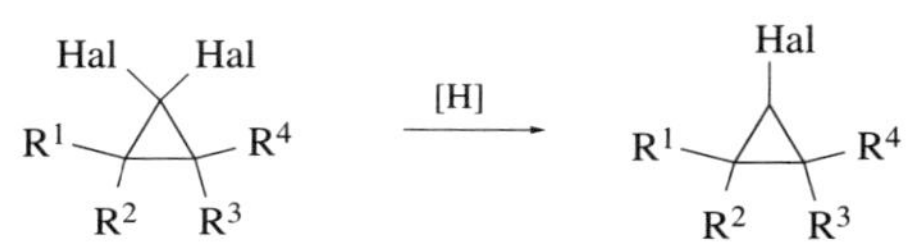
(21)

Reductive halogenation of carbonyl compounds (including acids, aldehydes, ketones and even ketals) can be achieved under a variety of conditions, using boranes and silanes as *in situ* reducing agents (Scheme 14) ⟨84TL1103, 85T4657, 86CJC2342, 89CC313, 89TL5663, 90JOC2927, 92T8329⟩.

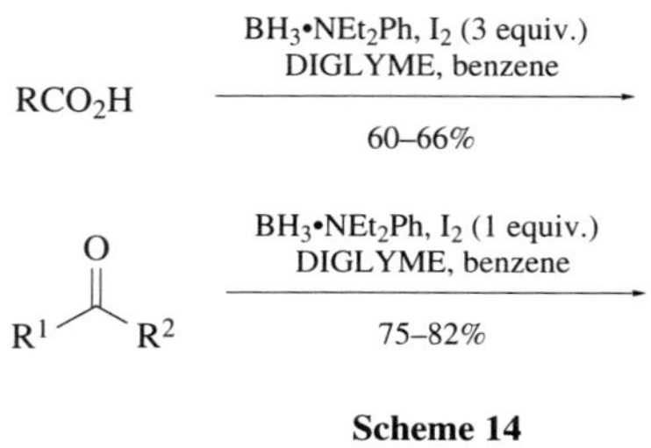

Scheme 14

Alkyl halides can be derived from halogen atom transfer reactions such as that shown in Equation (22) ⟨94TL2763⟩. For related chemistry, including cyclisations, see ⟨80BCJ770, 85PAC1827, 87TL2477, 91COS(7)593, 91COS(4)779, 91SL659, 92TL213⟩.

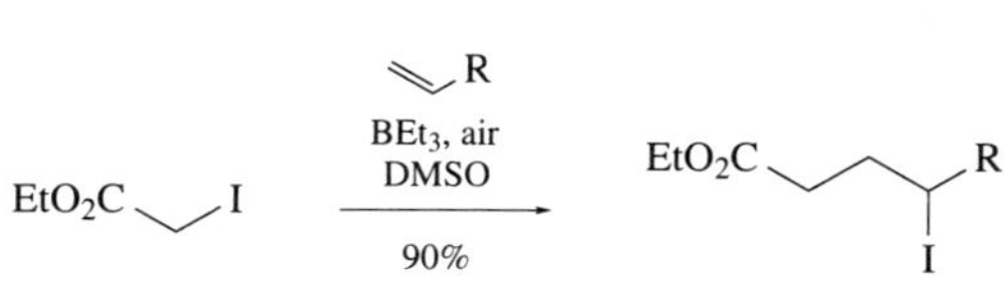
(22)

Finally, a range of homologation procedures for alkyl halides has been reported and can be found in Larock's survey ⟨B-89MI 201-01⟩. A recent example is shown in Equation (23) ⟨90TL937⟩.

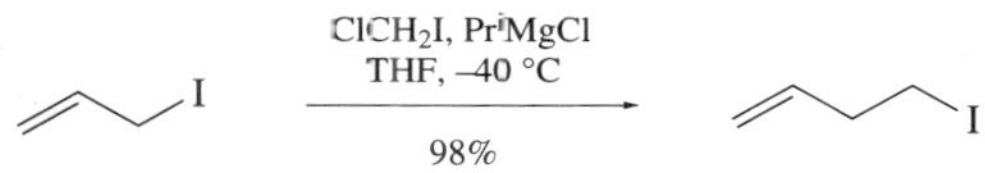

(23)

2.01.2 ALKYL FLUORIDES: RF

The high dissociation energy of the C—F bond (>450 kJ mol^{-1}) makes the use of alkyl fluorides as synthetic intermediates very rare. It also means that methods for the formation of C—F bonds are generally quite different from those used for the preparation of other carbon–halogen bonds. Considerable efforts have been made to facilitate the incorporation of fluorine into organic molecules because organofluorine compounds often exhibit interesting biological, chemical and physical properties. The number of reviews and monographs relating specifically to fluorine-containing compounds far outweighs those relating to the other halogens, and the reader is directed to some reviews in this area for more detailed treatments than can be afforded here ⟨74OR(21)125, 78IJ71, 78T3, 81AG(E)647, 85C134, 86JFC(33)227, 86JFC(33)377, 87CSR381, 87T3123, B-89MI 201-03, 90TA661, 91RCR850, 91RCR1050, 92CRV505⟩. The use of elemental fluorine in organic synthesis has been reviewed ⟨85C305, 86CRV(86)997, 88ACR307⟩, as has the use of combinations of hydrogen fluoride with organic bases ⟨91T5329⟩.

2.01.2.1 Alkyl Fluorides from Alkanes

Direct replacement of unactivated aliphatic hydrogen with fluorine using elemental fluorine is usually nonselective, often resulting in polyfluorination and fragmentation owing to the high heat of reaction. Notable exceptions include the use of elemental fluorine in the regioselective fluorinations of adamantane (Equation (24)) ⟨76JA3034, 88JOC2803⟩ and branched alkanes ⟨87JOC4928⟩ at the more substituted positions. These reactions are believed to be electrophilic in nature ⟨87JOC2769⟩.

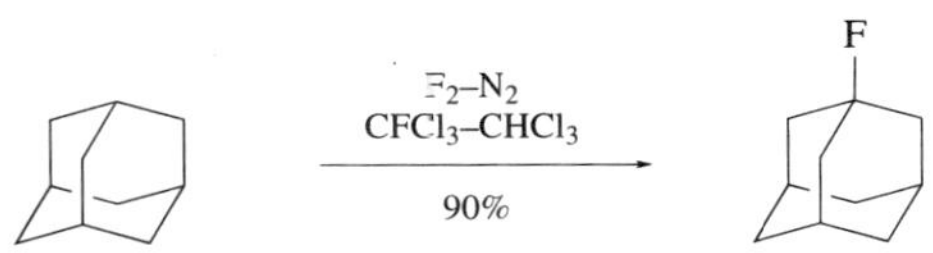

(24)

Notwithstanding these successes, the high electronegativity of fluorine makes its use as an electrophilic reagent somewhat challenging. One way to overcome this is to incorporate an excellent leaving group adjacent to fluorine, such as in CF_3OF, which has been successfully applied to the fluorination of rigid systems such as adamantanes and steroids ⟨70JA7494, 70PAC285, 76JA3034, 76JA3036, 80NJC239⟩. Olah's reagent, pyridine poly(hydrogen fluoride) (PPHF), can 1-fluorinate adamantane in the presence of $NOBF_4$ ⟨83JOC3356⟩, while other reagents used for fluorination at tertiary carbon centres include ClF_3 ⟨87JOC798⟩, BrF_3 ⟨89JOU1835⟩ and XeF_2 ⟨86BCJ1659⟩.

A class of reagents developed in the early 1990s is that of the 'Selectfluor' reagents ⟨92CC595, 93JOC2791, 94CC343⟩, exemplified by F-TEDA-BF_4 (**17**), a stable solid which has been shown to convert aryl-alkyl substituted alcohols directly to vicinal fluorohydrins (Equation (25)) ⟨94CC149⟩. Benzylic fluorination has been reported using $CsSO_4F$ ⟨91JOC7347⟩ and HF/PbO_2 ⟨77JFC(10)375, 79JOC1252⟩.

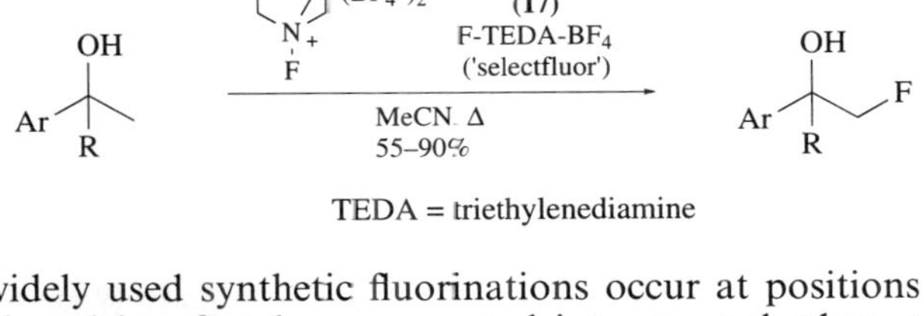

(25)

TEDA = triethylenediamine

By far the most widely used synthetic fluorinations occur at positions α to a carbonyl group, usually after the carbonyl has first been converted into an enol ether, enol silane, enol ester or enamine (Equation (26)) (reviews ⟨85T1111, B-88MI 201-01⟩. For example, enol silanes (including those derived from carboxylic acids, esters, amides and *β*-dicarbonyl compounds) can be converted to *α*-fluorocarbonyl compounds by treatment with fluorine ⟨86TL2715, 87JOC4307, 90JOC3423⟩, trifluoromethyl hypofluorite, CF_3OF ⟨80JA4845⟩, XeF_2 or $PhIF_2$ ⟨82TL1165⟩, or *N*-fluoropyridinium triflate ⟨86TL4465, 90OS(69)129⟩. Although $FClO_3$ can be used, it is not recommended because of its explosive nature ⟨75JOM(91)C20⟩. Enol esters are fluorinated by a range of fluoroxy compounds ⟨72CC122⟩, as well as by XeF_2 ⟨82JOC573, 88JOC5153, 89T6003⟩, $CsSO_4F$ ⟨89T6003⟩, perfluoroacyl hypofluorites ⟨79CC479, 85JOC3698⟩ or electrochemically with HF/Et_3N ⟨83TL903, 86BSF955, 87TL2359⟩.

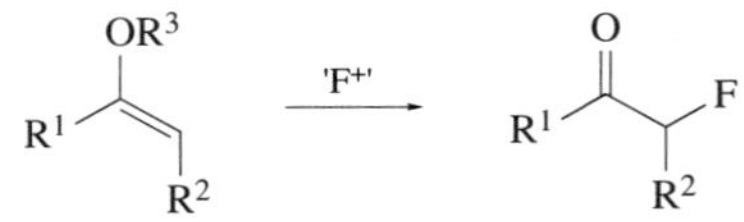

(26)

R^3 = alkyl, acyl, silyl

Direct *α*-fluorination of ketones or *β*-dicarbonyl compounds with AcOF has been reported ⟨83JOC724, 85S665⟩, and can also be achieved electrochemically ⟨89T4431⟩. A range of *N*-fluorosulfonamide derivatives **(18)**–**(22)** have been shown to fluorinate enolate anions and derivatives ⟨84JA452, 89HCA1248, 91CC179, 91JOC4925, 91SL187, 91TL1631, 91TL1779⟩, while *N*-fluoropyridinium salts **(23)** have also been exploited in some depth ⟨86TL4465, 91JOC5962⟩. Indeed, by changing the substituents on the pyridinium ring of **(23)**, the reactivity of the *N*-fluoro group can be 'tuned' to fit a range of substrates, including enolates and other enol derivatives ⟨90JA8563⟩.

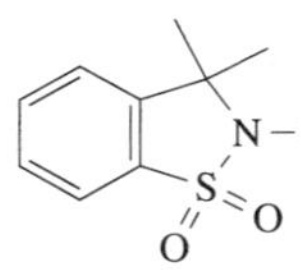

(18)
⟨89HCA1248, 91TL1779⟩

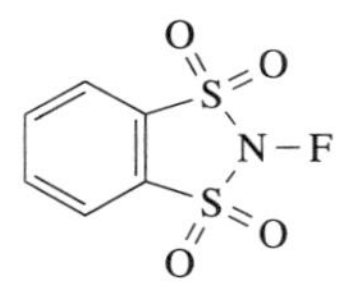

(19)
⟨91TL1631⟩

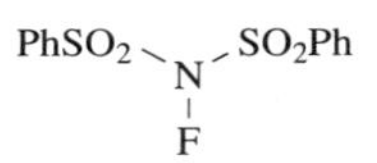

(20)
⟨91SL187⟩

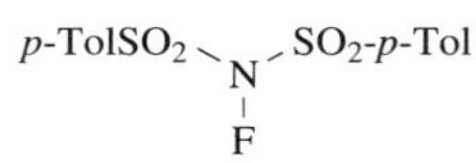

(21)
⟨84JA452⟩

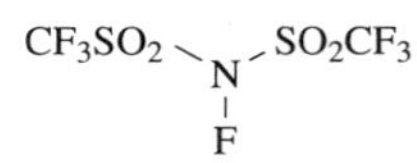

(22)
⟨91CC179, 91JOC4925⟩

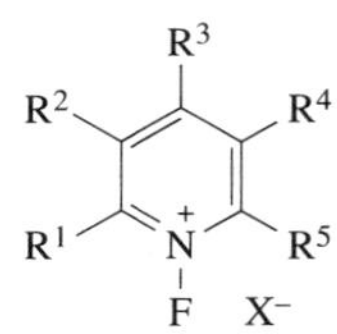

(23)

The previously mentioned 'Selectfluor' reagents such as **(17)** have been successfully reacted with enol acetates, enol silanes and enamines. They are an extension of previously reported *N*-fluoroquinuclidinium reagents **(24)** ⟨88JCS(P1)2805, 88JFC(41)297⟩.

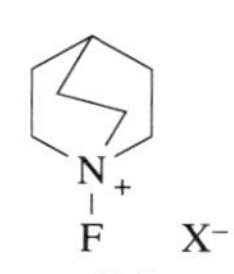

(24)

Enantioselective *α*-fluorination of carbonyl compounds is possible using a homochiral *N*-fluorosultam (Equation (27) in which HMDS = hexamethyldisilazide) ⟨88TL6087⟩, but the 75% enantiomeric excess thus obtained has yet to be improved upon ⟨93TL3971⟩. Higher levels of asymmetric induction have been achieved using the achiral fluorinating agent **(19)** on a substrate with a covalently bound chiral auxiliary (Equation (28)) ⟨92H(33)105, 92JCS(P1)221, 92TL1153⟩.

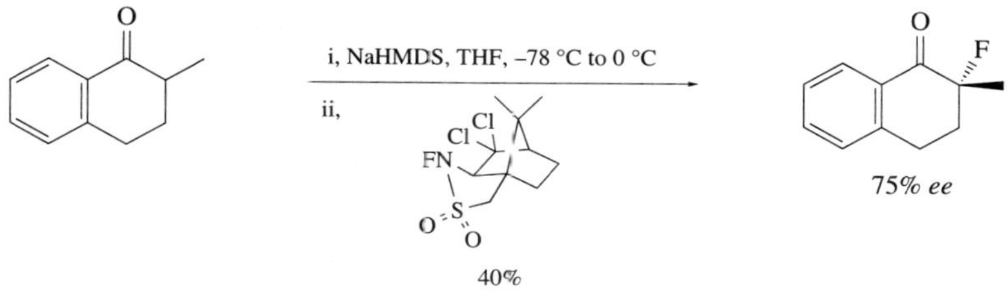

(27)

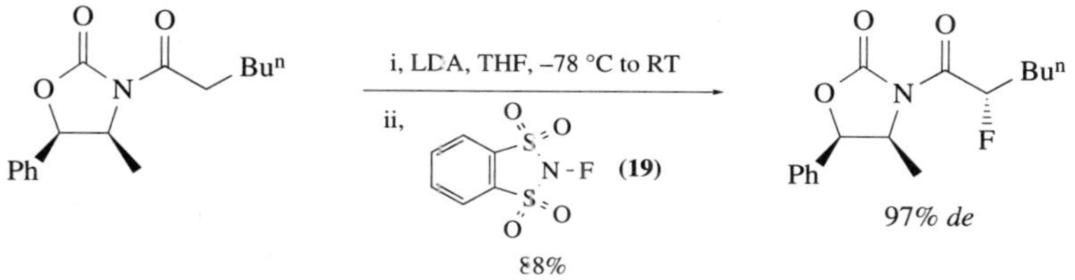

(28)

LDA = lithium diisopropylamide

NAHMDS = sodium hexamethyldisilazide

2.01.2.2 Alkyl Fluorides from Alkenes

2.01.2.2.1 Alkyl fluorides by hydrofluorination of alkenes

All recorded additions of hydrogen fluoride to alkenes occur in a Markovnikov sense, i.e., the proton attacks the more electronegative (generally less-substituted) end of the double bond (Equation (29)). The practical difficulties of handling highly corrosive anhydrous hydrogen fluoride, and the fact that it tends to cause alkene polymerisation, means that this reaction is better carried out with the commercially available pyridine poly(hydrogen fluoride) (PPHF) (Olah's reagent) ⟨73S779, 79JOC3872, 86JFC(33)377⟩. This has been further developed into an easily handled solid polymer-bound form, poly(4-vinylpyridinium) poly(hydrogen fluoride) (PVPHF) ⟨90SL267, 93S693⟩. One other hydrofluorination procedure of note is the combination of HF with melamine ⟨83CL1135⟩, which, with the use of co-solvents such as pentane or CCl_4, gives simple workups and is reusable ⟨84CL1241⟩.

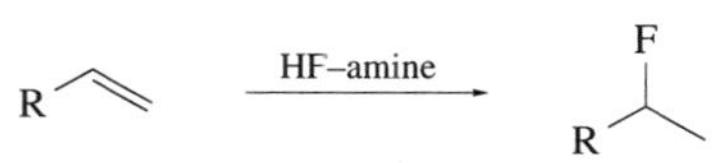

(29)

2.01.2.2.2 Alkyl fluorides by fluorine–halogen addition to alkenes (F—F, F—Cl, F—Br, F—I)

The high reactivity of elemental fluorine makes its use in synthesis somewhat problematic ⟨85C305, 86CRV(86)997, 88ACR307⟩. In contrast to the other halogens, which generally add in a *trans* (or *anti*) fashion (Scheme 6), fluorine adds in a stereospecifically *syn* fashion, suggesting either a four-centre transition state (**25**) or the rapid collapse of a tight ion pair (**26**) ⟨86JOC3607, 94CPB459⟩.

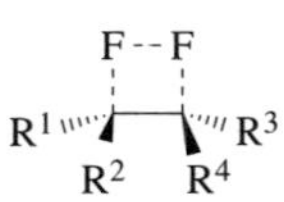

(25)

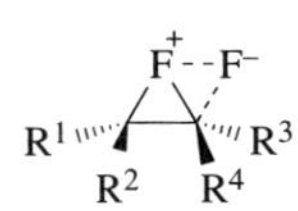

(26)

A number of papers discuss the use of xenon difluoride for the difluorination of phenyl-substituted alkenes ⟨76JOC4002, 77JOC1559, 78IJ71, 93JCS(P1)2851⟩ and simplified benchtop procedures for handling XeF_2 have been described ⟨77TL363⟩. One of the many electrophilic fluorinating agents developed since the 1980s is the 'Selectfluor' reagent (**17**) which, when combined with pyridine poly(hydrogen fluoride) (PPHF) allows phenyl-substituted alkenes to be vicinally difluorinated ⟨93JOC2791⟩. Other methods for alkene fluorination are known ⟨87JOC919, B-89MI 201-01⟩ but space restrictions preclude further discussion of these.

Fluorine–halogen addition (halogen = Cl, Br, I) is achievable in a very large number of ways and has been reviewed ⟨72RCR740, 84RCR1178⟩. In essentially all cases, the addition proceeds in a *trans* (*anti*) fashion as indicated in Scheme 15, and Markovnikov regioselectivity is usually observed.

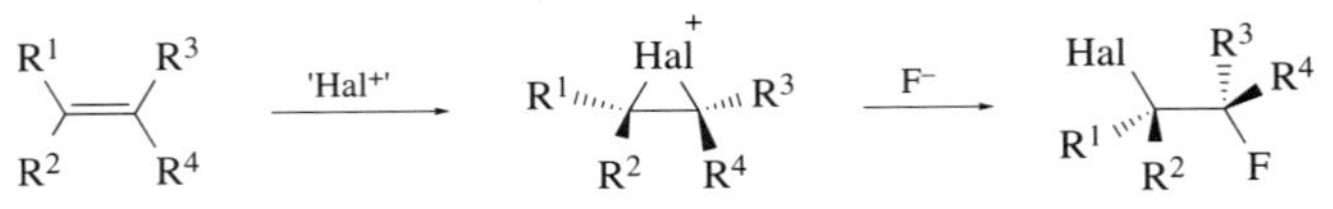

Scheme 15

The most widely used active halogen sources are *N*-chloro-, *N*-bromo- and *N*-iodosuccinimide, *N*-bromoacetamide ⟨73OSC(5)136, 78S217⟩, and dibromodimethylhydantoin (dbh), although mixtures of elemental fluorine with iodine or bromine have also been used ⟨80TL4543, 85JOC3342⟩. The fluoride source is commonly hydrogen fluoride, usually in combination with organic bases such as indicated in Table 1 ⟨91T5329⟩. One of these, pyridine poly(hydrogen fluoride) (PPHF) ⟨73S780⟩ has been prepared in a more convenient poly(vinylpyridine) polymer-supported form (PVPHF) which would appear to offer practical advantages ⟨93S693⟩. Alternatively, a mild, and as yet little exploited way of generating *in situ* HF for bromo- or iodofluorination of alkenes involves reaction with Ishikawa's reagent (**27**) (hexafluoropropene–diethylaniline complex, sometimes referred to as FPA) in the presence of water (Equation (30)) ⟨91BCJ2596⟩.

Table 1 Halofluorination of alkenes using amine–HF adducts.

	NCS	*NBS*	*NIS*	*DBH*
HF (no amine)			60JA4001, 60JA4007	86JFC(31)99
$Et_3N \cdot 3HF$		87S562, 86TL4449, 89TL3167	87S562	
PPHF		86TL4449		86JFC(31)99
PVPHF		93S693	87BCJ3083	93S693
Bu_4NF		87JFC(34)337	91SL185	
Bu_4NHF_2	89JOC4294	89JOC4294	91SL185, 89JOC4294	
$Bu_4NH_2F_3$			91TL1215	91TL1215
NH_4HF_2/AlF_3		90TL3167	90TL3167	

(30)

(27)
(Ishikawa's reagent)

Alternative reagent combinations not referred to in Table 1 or discussed above include bis(pyridine)iodine tetrafluoroborate, IPy_2BF_4 ⟨85AG(E)319, 91JOC2234⟩, and its collidine analogue ⟨87S551⟩, XeF_2/Br_2 ⟨78IJ71⟩, BrF_3/Br_2 ⟨87JOU230⟩, AgF/I_2 ⟨93JCS(P1)1795⟩, $BF_3/ROBr$ (or $BF_3/ROCl$) ⟨83JOC3195⟩, $AgF–CaF_2/I_2$ (or $AgF–CaF_2/DBH$) ⟨88CL1877⟩, $Pb(OAc)_4/CuCl_2/HF$ ⟨90JOU1784⟩, SiF_4/DBH ⟨89CC1881⟩ and $Bu_4PH_2F_3$/(NIS or NBS or DBH) ⟨93CL673⟩.

2.01.2.3 Alkyl Fluorides from Alkyl Halides

Nucleophilic halogen exchange is one of the most widely used approaches to alkyl fluorides, and a very large number of procedures for the conversion of alkyl chlorides, bromides and iodides to alkyl fluorides (Equation (31)) have been reported. The reader is therefore referred to the relevant sections in *Houben-Weyl* ⟨62HOU(5/3)1⟩, Larock ⟨B-89MI 201-01⟩ and *Organic Reactions* ⟨44OR(2)49, 74OR(21)125⟩ for more comprehensive listings of references than can be afforded here.

$$\mathrm{RHal} \xrightarrow{F^-} \mathrm{RF} \qquad (31)$$

Hal = Cl, Br, I

Calcium fluoride is a useful support for halide displacement with potassium ⟨86CC791, 86CC793⟩, caesium ⟨86CC791⟩ or silver ⟨88CL1877⟩ fluorides. Although copper fluoride ⟨87CL1675⟩ and lead fluoride ⟨92CL1161⟩ also have some utility, displacement of alkyl chlorides and bromides with fluoride is most commonly achieved with potassium fluoride in anhydrous polar solvents such as glycols ⟨57JA2311, 63OSC(4)525, 74CCC2616, 86JCR(S)300⟩, acetonitrile ⟨81CL761, 86CC793⟩, formamide ⟨86TL1499, 89BSF256⟩ or sulfolane ⟨86CC791⟩. Anhydrous conditions are usually essential for the success of the reaction, since water hydrogen bonds to the fluoride ion, decreasing its nucleophilicity. Formation of solubilised 'naked fluoride' ion from potassium fluoride can be enhanced by 'spray-drying' the KF ⟨81CL761⟩, by using crown ethers ⟨74JA2250, 82JCS(P1)85⟩, by way of a CuF_2–bipyridine complex ⟨85CL233⟩, or by phase-transfer catalysis ⟨74S428, 86TL1499, 88SC1661, 89BSF256⟩. Even bridge-head halides can be substituted with fluoride using XeF_2 ⟨92JOC2850⟩ or amine–HF adducts ⟨83S713⟩, although an S_N2 mechanism is clearly impossible in these cases.

Polymer-bound tetraalkylammonium fluorides can also displace chlorides, bromides and iodides ⟨76S472, 79JCS(P1)2248, 89JOC5380⟩. Indeed, the stoichiometric use of organic-soluble quaternary ammonium or quaternary phosphonium fluorides or hydrogen fluorides has become increasingly popular, examples of reagents including Bu_3MePF ⟨79TL353⟩, Bu_4NF ⟨84JOC3216⟩, Bu_4NHF_2 ⟨87TL4733⟩, Ph_4PHF_2 ⟨85CC672⟩, $Bu_4PF \cdot nHF$ ($n = 0,1,2$) ⟨91CL1185⟩ and polymer-supported HF_2^- or $H_2F_3^-$ ion ⟨89JOC5380⟩.

One of the drawbacks of fluoride ion is its basicity, which can lead to elimination rather than substitution with some secondary (particularly cyclic) and tertiary alkyl halides. This has been overcome by the use of amine–HF adducts ⟨80JFC(15)423, 90TL6527⟩ or Cu_2O/HF mixtures ⟨85CL1693⟩, which work particularly well on cyclic or tertiary systems ⟨87CL1675⟩.

2.01.2.4 Alkyl Fluorides from Alcohols and their Derivatives

Procedures for the conversion of alcohols to alkyl fluorides are quite specialised and, with the exceptions described in Section 2.01.1.4, cannot be applied to the other halides. Sulfur tetrafluoride can be used for this transformation ⟨74OR(21)1, 85OR(34)319⟩, but its gaseous and corrosive nature makes it unattractive. Although selenium tetrafluoride offers some advantages ⟨74JA925⟩, both SF_4 and SeF_4 have been superseded by a more convenient liquid reagent, diethylaminosulfur trifluoride, Et_2NSF_3 (DAST) (review ⟨88OR(35)513⟩), which readily converts primary, secondary, tertiary, allylic and benzylic alcohols to alkyl fluorides (Scheme 16) ⟨75JOC574, 88OSC(6)835⟩. Rearrangements and eliminations sometimes occur with DAST, suggesting an S_N1 mechanism, but there are many examples which proceed with complete inversion of configuration, suggesting an S_N2 mechanism ⟨88T2875⟩. Retention has sometimes been observed as a result of neighbouring group participation ⟨79TL1823, 91CPB1088⟩. DAST has been extensively applied in sugar chemistry ⟨88JCS(P1)549⟩, and a modified form of DAST, 4-morpholinosulfur trifluoride, has been reported to give high yields in reactions with cyclohexanols provided the solvent is chosen carefully ⟨89JFC(43)405⟩.

$$\underset{}{ROH} + \underset{\text{(DAST)}}{F_3SNEt_2} \longrightarrow ROSF_2NEt_2 \xrightarrow{F^-} RF$$

Scheme 16

Prior to the discovery of DAST, the most popular reagent for converting alcohols to fluorides was the less reactive fluoroalkylamine reagent (sometimes referred to as FAR), 2-chloro-1,1,2-trifluorotriethylamine ('Yarovenko's reagent') (**28**) ⟨59JGU2125⟩, giving fluorides generally with inversion of configuration (Equation (32)). A polymer-supported version of Yarovenko's reagent has been reported ⟨81JFC(17)93⟩, and Ishikawa's reagent (**27**) has also been used for this purpose ⟨79BCJ3377, 90JOC5291⟩. Alcohols can also be converted to alkyl fluorides with a related perfluoro-alkene/Et_3N reagent combination ⟨81BCJ1151⟩.

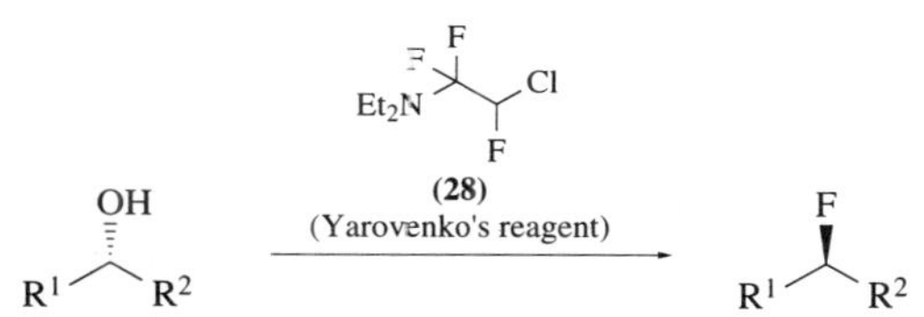

(32)

Less well-used reagents for the fluorodehydroxylation of alcohols include the phenylfluorophosphoranes $Ph_mPF_{(5-m)}$ ($m = 1,2,3$) ⟨68CPB1784⟩, which, with the exception of $PhPF_4$ ⟨73CPB867⟩, require elevated temperatures. Finally, primary alcohols have been shown to be selectively converted to fluorides by treatment with tosyl fluoride and tbaf in the presence of molecular sieves ⟨85TL4207⟩.

Ethers are rarely used as direct precursors to alkyl fluorides (although see Section 2.01.1.4.3).

2.01.2.5 Alkyl Fluorides from Amines and their Derivatives

See Section 2.01.1.5.

2.01.2.6 Alkyl Fluorides by Fluorodecarboxylation of Carboxylic Acids and their Derivatives

See Section 2.01.1.6.

2.01.2.7 Alkyl Fluorides by Fluoroalkylation of Aromatic Rings

Fluoroalkylation of aromatic rings has not been reported.

2.01.3 ALKYL CHLORIDES: RCl

Chlorine is the cheapest and most readily available of all the halogens. Alkyl chlorides are not as reactive as bromides (or iodides) and are therefore not quite so widely used as synthetic intermediates. Nevertheless, methods for their preparation abound.

2.01.3.1 Alkyl Chlorides from Alkanes

The direct chlorination of unactivated alkanes using Barton's 'Gif' GoAgg-type systems has been discussed briefly in Section 2.01.1.1, and is complemented by a high-yielding 1-chlorination of adamantane using $Bu^tCl/AlCl_3$ ⟨89SC1697⟩. *N*-Chloroamines chlorinate alkanes, often quite regioselectively, under strongly acidic conditions in the presence of a radical initiator such as iron(II) or light, in reactions that are closely related to the classical Hoffman–Löffler–Freytag reaction ⟨70CRV639, 71JA438, 76AG(E)306, 79JOC3728⟩ (Scheme 17).

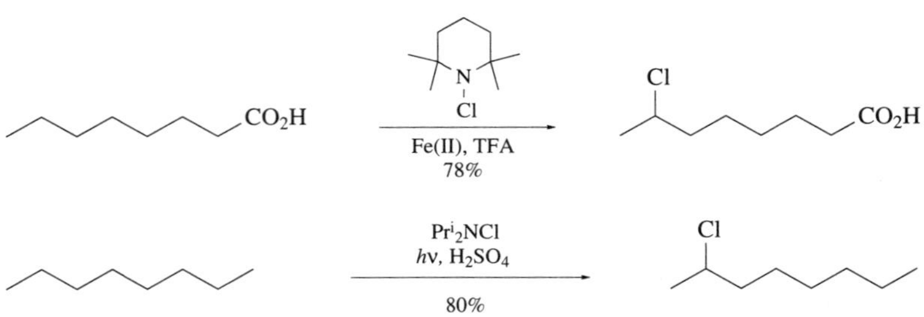

Scheme 17

Somewhat less vigorous conditions for alkane chlorination involve the use of iodobenzene dichloride, $PhICl_2$ ⟨79CL961⟩, which has been elegantly applied to the regioselective photochlorination of steroids, under the direction of a covalently bound template (Equation (33)) ⟨77JA905, 91COS(7)43, 93TL1107⟩.

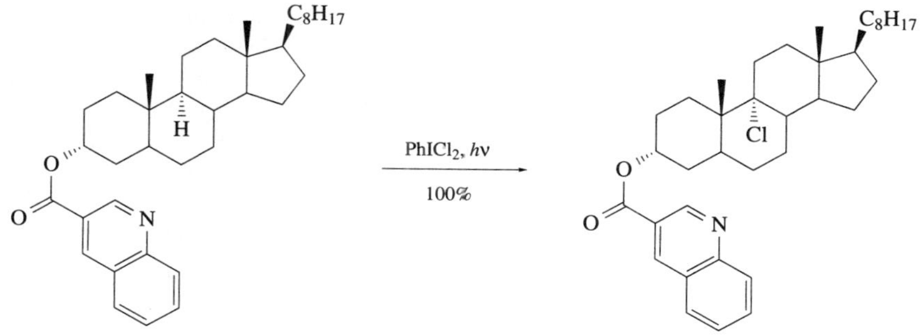

(33)

Photochlorination of alkanes with Cl_2 ⟨70S7, 90ACR219, 92LA559⟩, and chlorination with sulfuryl chloride (SO_2Cl_2) in the presence of a radical initiator such as dibenzoyl peroxide are well documented ⟨B-74MI 201-01⟩.

SO_2Cl_2 can also be used for benzylic chlorination using benzoyl peroxide as radical initiator. Palladium catalysis has been shown to reduce the levels of dichlorinated by-product ⟨78CL223⟩, and milder radical conditions using a tetraalkylammonium tetrachloroiodate, $BnMe_3NICl_4$, have been described ⟨88TL5783⟩. While Cl_2 can give allylic or benzylic chlorination under certain conditions ⟨88T6525, 92TL495⟩, allylic chlorination quite often requires reagents such as Bu^tOCl ⟨80TL781⟩ or HOCl ⟨80TL441, 81TL5019⟩, which is conveniently generated *in situ* from $Ca(OCl)_2$ and acetic acid ⟨90JOC3880⟩. Unlike the corresponding bromination reaction with *N*-bromosuccinimide, allylic halogenation with *N*-chlorosuccinimide usually requires extra catalysis. Using PhSeCl or PhSeSePh or TsNSO as catalysts, allylic chlorination can be controlled to give the rearranged or unrearranged product (Equation (34)) ⟨79JOC4204, 79JOC4208⟩.

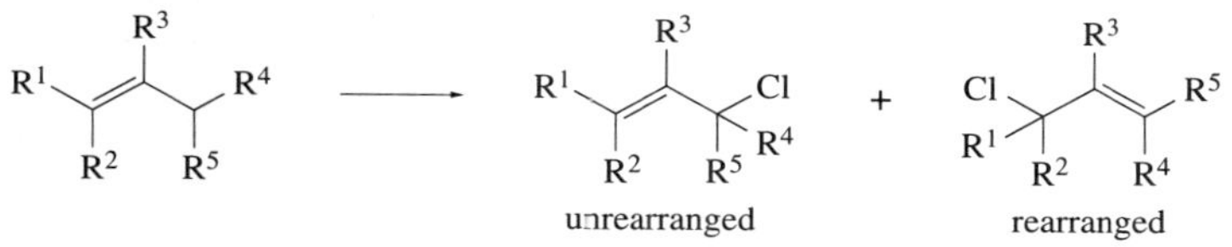

(34)

Other procedures which have found limited use in the chlorination of allylic and benzylic systems include the use of chlorine oxide, Cl_2O ⟨84CL877⟩, a Vilsmeier-type reagent in the presence of peroxide ⟨91SL477⟩, and electrochemical methods using sodium chloride as halogen source ⟨81TL2291, 81TL3193⟩. In many cases, the reaction is believed to proceed by way of an ene-type mechanism depicted in a generalised manner in Equation (35).

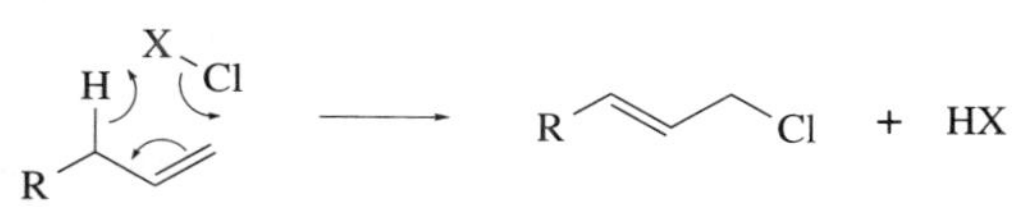

(35)

α-Chlorination of enolisable carbonyl compounds is quite readily achieved, often without the need to convert the carbonyl to an enol silane or similar derivative ⟨B-88MI 201-01⟩. The classical methods which have been applied to ketones, aldehydes, esters and particularly carboxylic acids involve the treatment of the carbonyl compound with Cl_2 under various oxidative conditions (Equation (36)) ⟨55OSC(3)188, 75JOC2960, 77CI(L)538, 79BCJ255, 83JOC1364, 85SC977, 88OSC(6)90⟩ or SO_2Cl_2 ⟨63OSC(4)162, 81JOC4486, B-88MI 201-01, 90S595⟩.

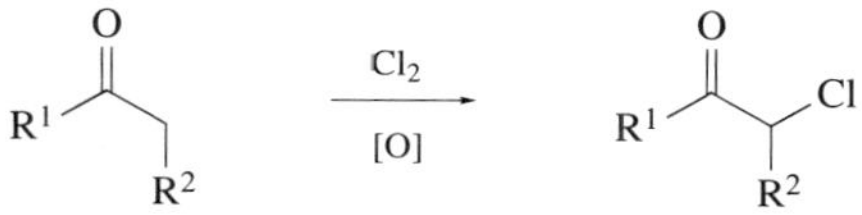

(36)

A more convenient reagent for α-chlorination of carbonyl compounds is benzyltrimethylammonium dichloroiodate, $BnMe_3NICl_2$, either in solution ⟨88S545, 90S212⟩ or in a polymer-bound form ⟨80CC1278⟩, while a related polymer-bound tetrachloroiodate has been reported for a similar application ⟨84T2865⟩.

Direct chlorination of carbonyl compounds can be readily achieved with copper(II) chloride ⟨B-73MI 201-02⟩, and the use of this method in malonate chemistry has been discussed in some depth

⟨93JOC4596⟩. Chloride ion can also be used to α-chlorinate carbonyl compounds when in the presence of manganese(III) ⟨87BCJ809, 93JCS(P1)3171⟩ or manganese(IV) ⟨90JCR(S)188⟩ species. In the latter case, TMS-Cl and MnO_2 combine to generate reactive $MnCl_4$ *in situ*. A related procedure for α-chlorination of acetals has been described, which is believed to proceed through an enol ether intermediate ⟨92T4579⟩. Also related is a method which combines TMS-Cl with $KBrO_3$ ⟨92BAU356⟩, but a method which may find wider application makes use of DMSO as the oxidant (Scheme 18) ⟨86JCR(S)426⟩. A chlorosulfonium intermediate (**29**) is postulated as the reactive halogenating species. Interestingly, this reaction can be enhanced by bromide ion catalysis ⟨88SC1071⟩.

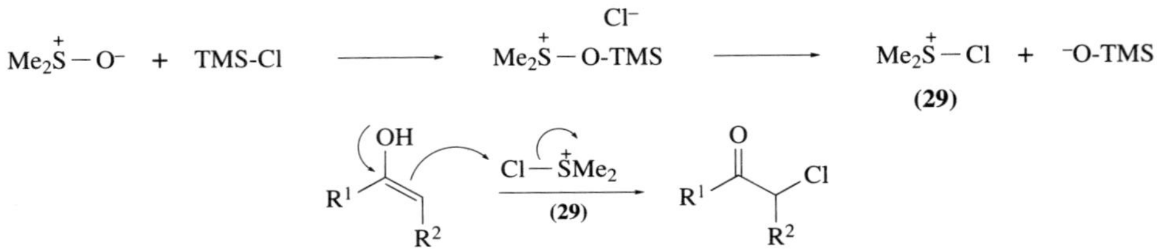

Scheme 18

Although direct chlorination of ketones with *N*-chlorosuccinimide has not been described, the related trichloroisocyanuric acid (**30**) has been shown to chlorinate ketones in the presence of $BF_3 \cdot OEt_2$ ⟨85SC385⟩. *N*-Chlorosuccinimide can, nevertheless, be used to chlorinate enol silanes without Lewis acid catalysis ⟨86TL2563, 89JCS(P1)1544⟩, and this chemistry has been applied in asymmetric synthesis using covalently-bound chiral auxiliaries ⟨85TL5037, 91SL807⟩. *N*-Chlorosuccinimide is one of a few reagents which can be used to α-chlorinate enamines and imines ⟨77JA6672, 78ACS(B)(32)646, 79OPP115⟩, and is used in an improved method for the α-chlorination of imines (and therefore, indirectly, aldehydes) which avoids unwanted dichlorination (Scheme 19) ⟨92BSB237⟩.

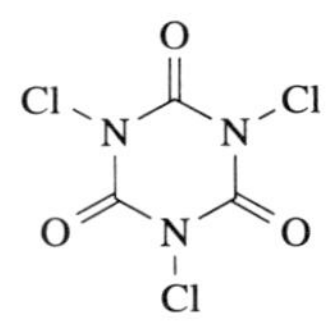

(30)

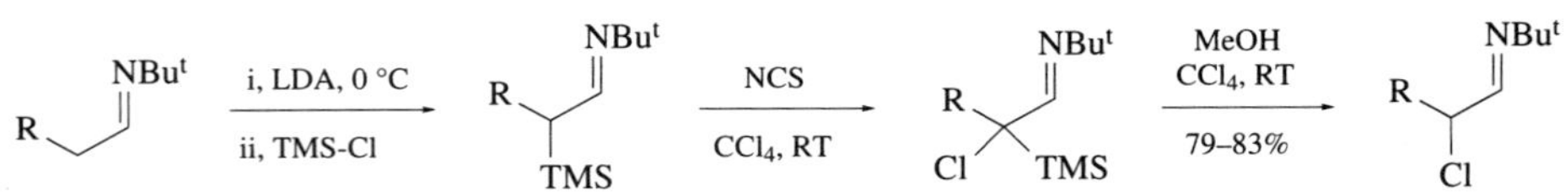

LDA = lithium diisopropylamide

Scheme 19

Other reagents useful for chlorination of enol silanes include SO_2Cl_2 ⟨84JOC2032⟩, $CuCl_2$ ⟨80JOC2022, 91SC11⟩, Cl_2 ⟨74JOC1785⟩, and chlorine–triphenylphosphine adduct, Ph_3PCl_2, in the presence of bis(trimethylsilyl) hydroperoxide, although the yield in the latter case is modest ⟨91BCJ3749⟩. Enolate anions can be quenched with CCl_4 ⟨78JOC3687, 87JOC307, 92CC1200⟩, C_2Cl_6 ⟨87OS(66)194⟩, CF_3SO_2Cl ⟨85JA7771⟩, or *N*-chlorosuccinimide ⟨84JOC1286⟩ (Equation (37)).

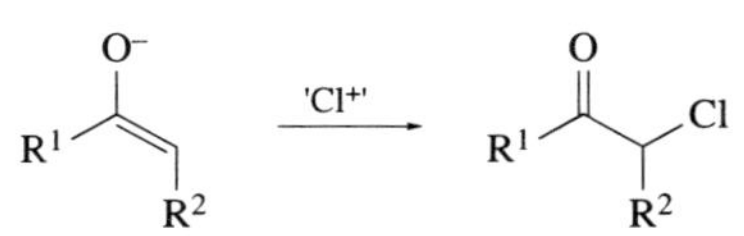 (37)

'Cl+' = CCl_4, C_2Cl_6, CF_3SO_2Cl, NCS

Although amides are not easily α-chlorinated, $ClSO_2Ph$ has been used to quench enolate anions of lactams ⟨93T3193⟩. An interesting alternative approach to α-chloroamides is shown in Scheme 20 ⟨92JA6262, 92JOC5700⟩.

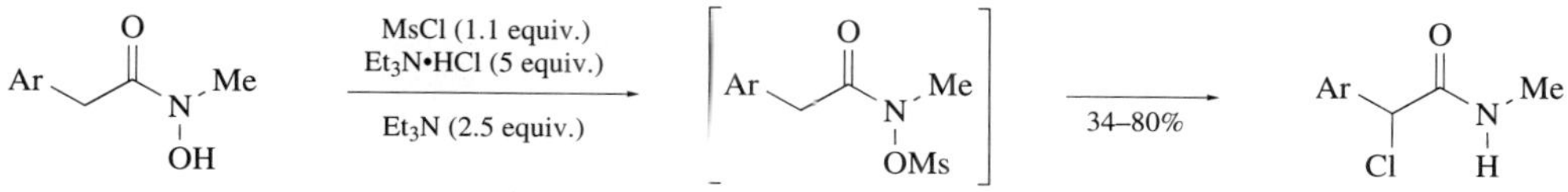

Scheme 20

2.01.3.2 Alkyl Chlorides from Alkenes

2.01.3.2.1 Alkyl chlorides by hydrochlorination of alkenes

Many of the general methods for alkene hydrohalogenation have already been discussed in Section 2.01.1.2.1, and a review of alkene hydrochlorination has been published ⟨83RCR259⟩. The addition of HCl to alkenes generally occurs in a Markovnikov sense, but is very slow, except in strained, highly substituted or styryl systems ⟨e.g., 43OSC(2)336⟩. Synthetically useful reaction rates can only be achieved by using either a phase-transfer catalyst ⟨80JOC3527⟩, or a solid inorganic support such as silica or alumina ⟨90JA7433, 93JA3071⟩. The latter approach not only tends to improve the regioselectivity ⟨91TL3705⟩, but has the added advantage of working with more convenient halide sources, such as $SOCl_2$, $(COCl)_2$ or TMS–Cl, which are presumed to generate HCl *in situ*. Hydrochlorination can give *syn* or *anti* addition to alkenes, depending on the reaction conditions ⟨73S789⟩, but the surface-mediated procedure described above gives *syn* addition stereoselectively ⟨93JA3071⟩.

If anti-Markovnikov addition of HCl is required, the most reliable approach is via hydroboration (Scheme 4). Although a number of reagents can be used for the conversion of trialkylboranes to alkyl chlorides see ⟨88JOC5664⟩ for a comparison, one of the most effective would appear to be NCl_3 ⟨88T2785⟩. (Note, however, that nitrogen trichloride is explosive and shock sensitive.) Hydroalumination, using $LiAlH_4$ with titanium or zirconium catalysis, has also been used to good effect in preparing alkyl chlorides from alkenes ⟨76JOM(122)C25, 77JOM(142)71, 78CL833⟩.

Finally, compounds derived from boron trichloride have been found to be effective in promoting the conjugate addition of chloride ion to α,β-unsaturated carboxylic acid derivatives, and this has been realised diastereoselectively (Equation (38)) ⟨94CC735⟩.

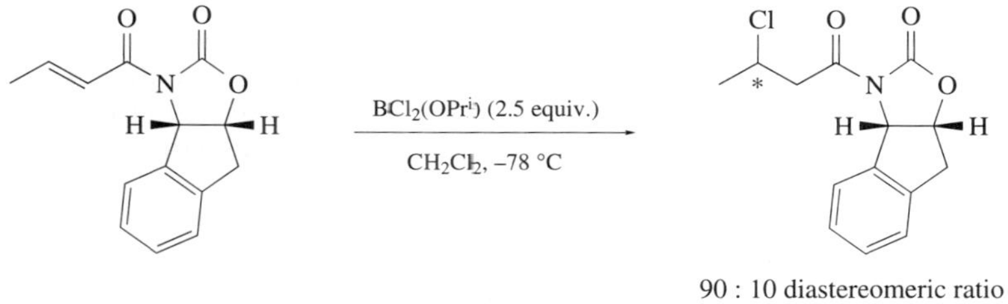

(38)

2.01.3.2.2 Alkyl chlorides by chlorine–halogen addition to alkenes (Cl—Cl, Cl—Br, Cl—I)

The reaction of elemental chlorine with alkenes is conceptually the most obvious method for chlorine–chlorine addition, but it is often a messy reaction giving many by-products. Furthermore, the gaseous and noxious nature of Cl_2 makes it experimentally unappealing. Although alternatives such as $SOCl_2$, PCl_5 ⟨39JA940⟩, SO_2Cl_2 ⟨39JA3432, 68JCS(C)416⟩, and NCl_3 ⟨69S135⟩ can be used, a variety of more attractive alternatives have been reported. Among these are the tetraalkylammonium chloroiodate salts R_4NICl_2 and R_4NICl_4, which are less reactive than elemental chlorine, thus giving cleaner reactions. They can be used in solution ⟨60JOC20, 91CE185⟩, or in a polymer-bound form ⟨80CC1278⟩, and give the product of *anti* addition to the alkene. Iodobenzene dichloride, $PhICl_2$ ⟨66CJC2339, 68JOC38⟩, and, perhaps more conveniently, copper(II) chloride ⟨71JOC3324⟩ also give *trans* dichlorination of alkenes. In 1994, hexachloroethane was shown to chlorinate alkenes in the presence of catalytic amounts of $RuCl_2(PPh_3)_2$, but the overall stereochemistry of the products was not discussed in detail ⟨94TL737⟩.

In the 1980s and 1990s, the use of chloride ion in the presence of manganese(III) species was shown to be very useful in effecting *trans* addition of chlorine to alkenes ⟨84TL607, 89JCR(S)108, 89JCR(S)360, 91SC489, 91SL733, 91TL1831⟩. The mechanism by which these processes occur is not fully understood, but this approach is synthetically quite flexible, in that different manganese sources (including permanganate, MnO_2, $Mn(OAc)_3$), and different chloride sources (including AcCl, $(COCl)_2$, TMS-Cl) can be used. This methodology allows electron-rich alkenes to be selectively chlorinated in the presence of less-activated double bonds ⟨91TL1831⟩.

Using DMSO as oxidant, TMS-Cl can be used to vicinally chlorinate styrenes, the active electrophilic chlorine source being the chlorosulfonium species (**29**) ⟨91G559⟩. The latter procedure is related to one using DMSO and *t*-butyl bromide to generate a dimethylsulfide–bromine adduct, which, in the presence of $CaCl_2$, generates vicinal dichlorides from alkenes ⟨85G29⟩. Hydrogen peroxide can also be used as oxidant, and a convenient phase transfer method using HCl/H_2O_2 has been described ⟨77S676⟩.

Cis-addition of chlorine to alkenes is readily achieved using molybdenum pentachloride and antimony pentachloride ⟨74BCJ692, 74BCJ3121, 75JA1599⟩. An extension of this approach makes use of an octamolybdate species, $(Bu_4N)_4Mo_8P_{26}$, which enables even tetrasubstituted alkenes to be vicinally dichlorinated in acceptable yields using acetyl chloride as halide source ⟨78TL3427⟩. Meanwhile, a one-pot method for achieving overall *cis*-chlorination via reaction with phenylselenenyl chloride (PhSeCl) has been reported (Scheme 21) ⟨84TL1197⟩.

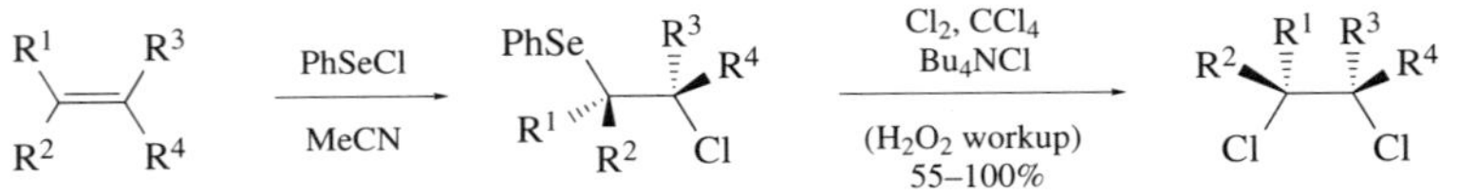

Scheme 21

Asymmetric chlorination of methacrylic acid has been achieved with 100% optical purity by prior inclusion of the starting material into α-cyclodextrin (Equation (39)). Using β-cyclodextrin, the opposite enantiomer was obtained in 88% optical purity ⟨83CC947⟩.

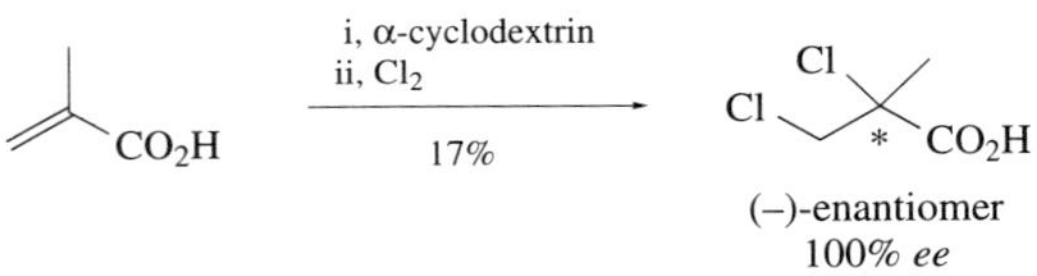

(39)

Chlorobromination of alkenes can be achieved in a variety of ways, and all of the reported methods give the expected Markovnikov product (i.e., bromine as the positive halogen) resulting from *trans* (*anti*) stereochemical addition to the double bond. The addition can be effected with BrCl ⟨73MI 201-01⟩ or with a mixture of Cl_2 and Br_2 ⟨60JOC24⟩. Other suitable reagent combinations include *N*-bromoacetamide/HCl ⟨51JA998, 52JA4891, 59JA2191⟩ and $MoCl_5/Br_2$ ⟨74BCJ3121⟩ or $SbCl_5/Br_2$ ⟨74BCJ143⟩. From the late 1950s, a report of chlorobromination of a steroidal alkene using phenyltrimethylammonium dichlorobromate ($PhMe_3NBrCl_2$) ⟨59TL24⟩ has been followed more recently with methods using tetrabutylammonium dichlorobromate ⟨84BCJ2111, 86BCJ3519⟩. The regiochemistry of addition with the latter reagent has been shown to be influenced more by steric than electronic factors ⟨86BCJ2547⟩. Addition of benzyltriethylammonium chloride to a mixture containing dimethylsulfide–bromine adduct generated *in situ* also results in net chlorobromination of alkenes ⟨85G29⟩.

Chlorine–iodine addition is similarly regio- and stereoselective under a range of conditions. It is possible to use iodine monochloride ⟨B-83MI 201-02⟩ either directly ⟨69T4283, 77JCS(P1)226, 87CC1577⟩ or generated *in situ* from iodine and a wide range of metal halides ⟨71JOC3324⟩, including $CuCl_2$ ⟨71JOC2088, 71JOC3324⟩, $HgCl_2$ ⟨86JCR(S)274⟩, $MoCl_5$ ⟨74BCJ3121⟩ and $SbCl_5$ ⟨80BCJ1390⟩. The latter is particularly good for deactivated alkenes. Other reports of chlorine–iodine addition to alkenes describe the combination of chloride ion with either $I_2/CuO \cdot HBF_4$ or the easily-handled solid reagent bis(pyridine)iodine tetrafluoroborate, IPy_2BF_4 ⟨85AG(E)319⟩. Regio- and stereoselective chloroiodination can also be achieved with iododichloride ion (ICl_2^-) either as its benzyltrimethylammonium salt ⟨90BCJ3033⟩ or in a polymer-bound pyridinium salt form ⟨90T2503⟩.

2.01.3.3 Alkyl Chlorides from Alkyl Halides

Substitution of alkyl halides with chloride ion is not a particularly widely used method for alkyl chloride preparation. Alkyl fluoride substitution can be achieved using somewhat vigorous conditions (12 M HCl (aq.)) ⟨90TL4973⟩ and is facilitated by phase-transfer catalysis ⟨92JCS(P1)2309⟩. Alkyl bromide substitution is more readily achieved, either under the same conditions, or more classically using lithium chloride in acetone ⟨55JCS3173⟩. Other phase-transfer catalytic procedures for this transformation are known ⟨77JOC875, 79JOC1961, 81TL4409, 84TL5949, 86CC1250⟩. The conversion from alkyl bromides can also be achieved using a polymer-supported chloride source (Amberlyst A26) ⟨76S472⟩, or in more extreme cases (e.g., adamantyl systems) using tin tetrachloride ⟨89S614⟩ or silver chlorodifluoroacetate ⟨70TL3447⟩. Both bromides and iodides are converted to chlorides on treatment with antimony pentachloride on graphite ⟨74TL763⟩, while primary iodides can be converted to chlorides with PCl_5 in $POCl_3$ ⟨85AJC1879⟩. Iodobenzene dichloride $PhICl_2$ is useful for bridgehead iodide-to-chloride conversions ⟨86TL6055, 89TL791⟩, while a little-used iodide-to-chloride conversion uses HCl in the presence of HNO_3, which oxidises the liberated iodide ion to I_2 ⟨69JOU971⟩.

2.01.3.4 Alkyl Chlorides from Alcohols and their Derivatives

A number of methods for the conversion of alcohols to halides (including chlorides) have already been discussed in Section 2.01.1.4.1. Classical methods using hydrochloric acid often work well on tertiary systems ⟨41OSC(1)144⟩ where an S_N1 mechanism is favoured, but primary and secondary alcohols react more slowly, and often need catalysis with $ZnCl_2$ ⟨41OSC(1)142, 43JCS636⟩ or phase-transfer catalysts ⟨74S37, 87JCS(P1)2157, 88S868⟩. Use of HCl in hexamethylphosphoramide (HMPA) gives good yields of alkyl chlorides from primary, secondary and tertiary alcohols without rearrangement ⟨75CJC3620⟩. In many cases, hydrogen chloride causes in elimination, rearrangement and loss of stereochemical integrity at carbon, so alternative reagents are frequently used. Probably the most commonly used inorganic acid chloride for this purpose is thionyl chloride ($SOCl_2$) ⟨63OSC(4)169, B-74MI 201-01, 75JOC134⟩, which can give chlorides with inversion, retention or racemisation, depending on the substrate and conditions ⟨52JA308, 53JCS1709, B-74MI 201-01, 76JCS(P1)104⟩. A rarely used alternative to $SOCl_2$ is $SeOCl_2$ (generated *in situ* from SeO_2 and TMS-Cl), which is effective on primary, secondary, tertiary, allylic and benzylic alcohols ⟨88JOC3634⟩. PCl_5 usually gives inversion ⟨46JCS1138⟩, but tertiary chlorides can be obtained with retention of configuration provided $CaCO_3$ is present to buffer the HCl produced ⟨76AJC133⟩. PCl_5 adsorbed onto a polymer-supported tertiary amine has been applied to alkyl chloride synthesis from alcohols ⟨83S306⟩. Although $POCl_3$ is rarely used for this purpose, $Me_2NP(O)Cl_2$ is effective in primary systems ⟨78CL923⟩. DMF catalyses the decomposition of alkyl chloroformates derived from alcohols as shown in Scheme 22 ⟨83JOC2625⟩.

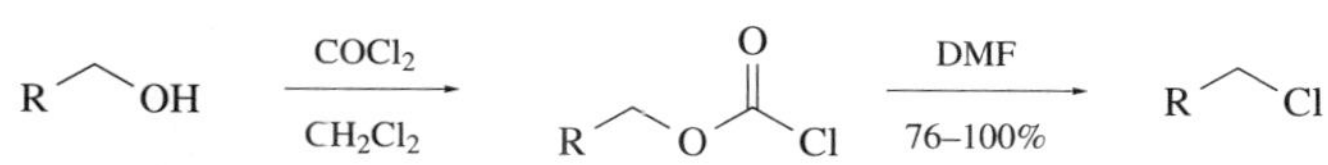

Scheme 22

Further reductions in the levels of rearrangement products can be achieved using reactive Vilsmeier species such as (**6**) and (**31**), which can be prepared *in situ* by reaction of amides (usually DMF) with PCl_3 ⟨76S398⟩, $SOCl_2$ ⟨76JCS(P1)104, 76JCS(P1)754⟩, PCl_5 ⟨74CI(L)664, 76JCS(P1)754⟩, $POCl_3$ ⟨93SC2199⟩ or $(COCl)_2$ ⟨69JOC2163, 85JA3285⟩. Vilsmeier's salt (**31**) and the related Viehe's salt (**32**) have been shown to be selective for primary over secondary alcohols in their conversion to chlorides ⟨92TL3129, 92TL4901⟩. A range of less common iminium salts have also been reported for alkyl chloride preparation from alcohols ⟨80S746, 84CL1173⟩ (see also Section 2.01.1.4.1).

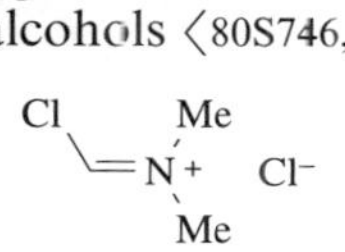

(31)

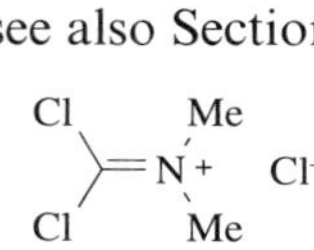

(32)

Allylic and benzylic alcohols are readily chlorinated with either $COCl_2$/pyridine ⟨91JOC7186⟩, or a combination of dimethylsulfide and *N*-chlorosuccinimide (Scheme 23) ⟨72TL4339⟩, but note that under the latter conditions saturated alcohols give carbonyl compounds if Et_3N is present (Swern-type oxidation). Allylic rearrangement can be minimised by converting the alcohol to a phosphate ester before treatment with halide ion (Equation (40)) ⟨84S841⟩.

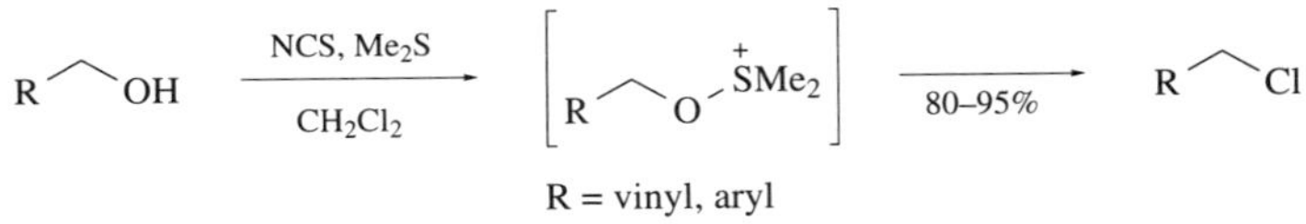

Scheme 23

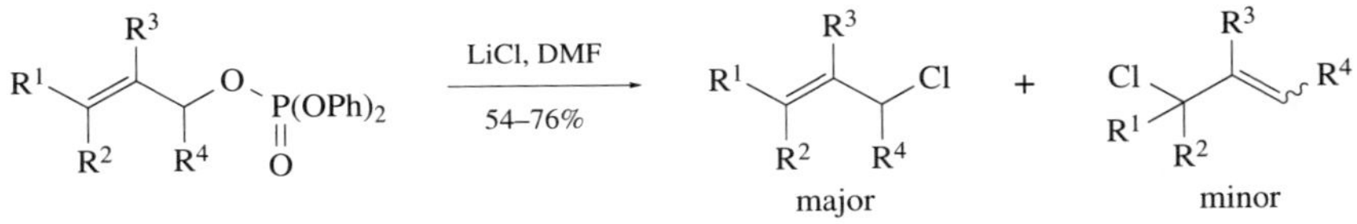
(40)

Other reagent systems useful for the preparation of allylic and benzylic chlorides from alcohols include TMS-Cl/K_2CO_3 ⟨83S314⟩ and the somewhat more vigorous tin tetrachloride ⟨89S614⟩. The latter reagent also converts tertiary (including adamantyl) and benzylic alcohols to chlorides. Milder conditions for tertiary and adamantyl systems use radical methods as illustrated in Scheme 24 ⟨75JA2281, 82JOC132, 87S35⟩.

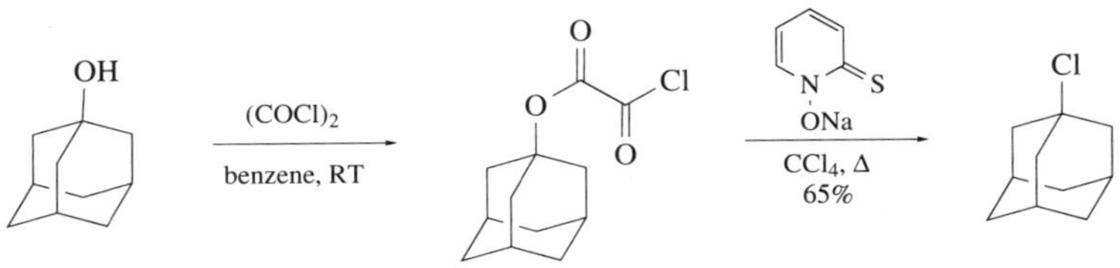

Scheme 24

Cyanuric chloride ⟨70JOC3967⟩ and even stoichiometric $Pd(PhCN)_2Cl_2$ ⟨78TL4575⟩ have been reported for alkyl chloride preparations from alcohols, but these methods are little used.

Cleavage of dialkyl ethers with HCl is of little preparative use, but the use of surfactants has facilitated the HCl-mediated conversion of arylalkyl ethers to alkyl chlorides, and of cyclic ethers to α,ω-dichlorides (Equation (41)) ⟨89JCR(S)284⟩. Cyclic ethers can also be converted to α,ω-dichlorides by treatment with the Vilsmeier reagent derived from thionyl chloride and DMF ⟨62CB2976⟩. Acetyl chloride cleaves ethers to alkyl chlorides in the presence of $ZnCl_2$ ⟨54JCS2819⟩ or $SnCl_4$ ⟨67JOC31, 78T1251⟩, the former being restricted to benzylic systems and the latter being particularly effective in bridgehead systems. BCl_3 has also been used to cleave secondary ethers to secondary alkyl chlorides ⟨63CI(L)609⟩.

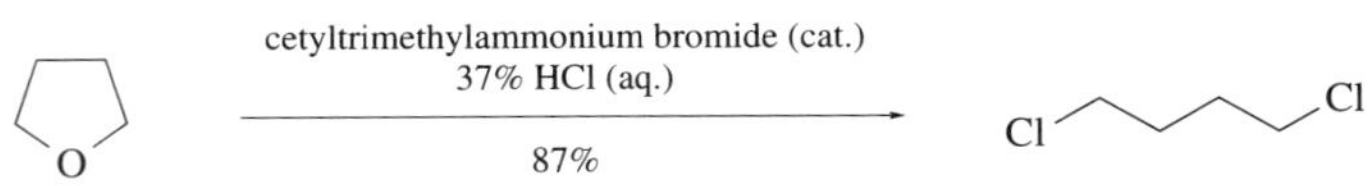
(41)

Finally, epoxides can be converted directly into 1,2-dichlorides by treatment with SO_2Cl_2/pyridine ⟨66CJC2339⟩, PCl_5 ⟨68CC976⟩, PPh_3/Cl_2 ⟨71T2617, 76JOC3279, 88OSC(6)424⟩ or PPh_3/CCl_4 ⟨72TL3869, 83JOC3353⟩. Inversion occurs at both carbons, so the stereochemistry of the product is dictated by the stereochemistry of the epoxide (Equation (42)). An alternative procedure for 1,2-dichloride synthesis from epoxides is depicted in Equation (43) ⟨77CL1013⟩.

(42)

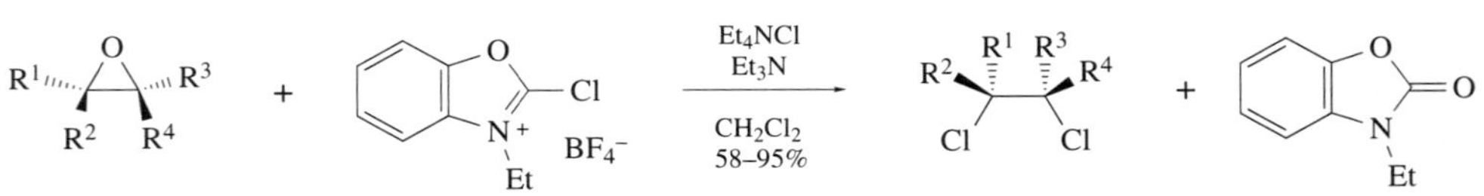
(43)

2.01.3.5 Alkyl Chlorides from Amines and their Derivatives

In addition to those methods described in Section 2.01.1.5, it has been shown that treatment of allylic or benzylic tertiary amines with ethyl chloroformate gives rise to the corresponding allylic or benzylic chlorides (Equation (44)) ⟨71JMC982, 77CL1025, 77S786⟩. By using α-chloroethyl chloroformate, substituted piperidines can be converted to saturated primary alkyl chlorides in high yields (Equation (45)) ⟨84JOC2795⟩. Extension of this procedure to secondary alkyl chloride preparation is problematic, not so much because of competing piperidine ring cleavage, but because the secondary alkyl halide produced is prone to HCl elimination under the reaction conditions. In a procedure sometimes called the von Braun reaction (but see also Section 2.01.4.5), benzylic amides are converted to benzylic chlorides by treatment with thionyl chloride (Equation (46)) ⟨62JA769⟩.

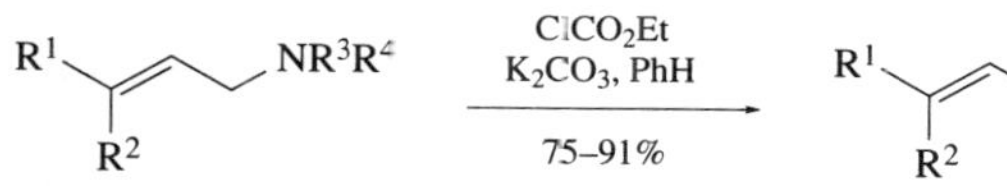

(44)

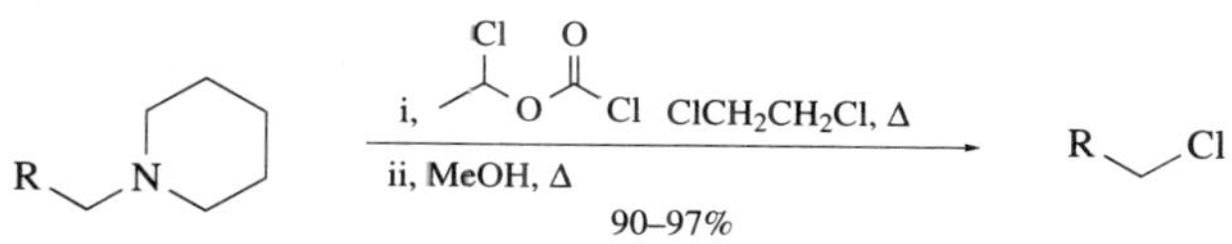

(45)

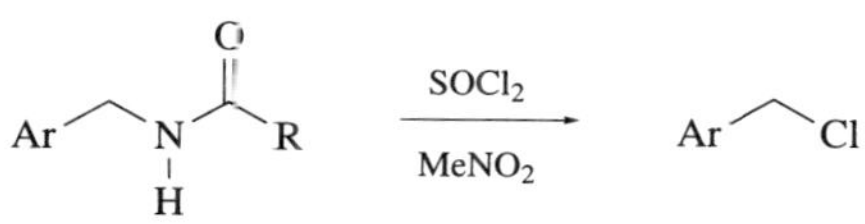

(46)

2.01.3.6 Alkyl Chlorides by Chlorodecarboxylation of Carboxylic Acids and their Derivatives

See Section 2.01.1.6.

2.01.3.7 Alkyl Chlorides by Chloroalkylation of Arenes

See Section 2.01.1.7.

2.01.4 ALKYL BROMIDES: RBr

The chemistry of alkyl bromides closely resembles that of alkyl chlorides, but bromides are far more widely used as synthetic intermediates since they undergo a large range of transformations under relatively mild reaction conditions.

2.01.4.1 Alkyl Bromides from Alkanes

The bromination of unactivated alkanes by free-radical means has been discussed in a review ⟨70S7⟩. In addition to the methods of Barton *et al.* described in Section 2.01.1.1, the bromination of symmetrical alkanes can be achieved with elemental bromine in the presence of HgO (forming Br_2O *in situ*) ⟨72CJC3109⟩. While elemental bromine itself can halogenate benzylic positions ⟨63OSC(4)984, 82BSF(2)327, 93TL5725⟩, *N*-bromosuccinimide (NBS) is far and away the reagent of choice for allylic and benzylic bromination (without bromination of aromatic rings) and this reaction (known as the Wohl–Ziegler bromination) has been the subject of a number of reviews ⟨48CRV271, 59AG349, B-74MI 201-02⟩. The reaction usually (but apparently not always, ⟨90TL7509⟩) requires radical initiators such as irradiation ⟨92BCJ345⟩ or dibenzoyl peroxide ⟨63OSC(4)108, 63OSC(4)921, 73OSC(5)328, 73OSC(5)825, 93JOC4382⟩, and is typically carried out in a nonpolar solvent such as CCl_4 from which the succinimide by-product is readily removed by filtration. It has been suggested that the reaction involves molecular bromine which is formed at very low concentrations ⟨74JA5616⟩. Little used alternatives to NBS include 1,2-dibromotetrachloroethane ⟨63CI(L)1954⟩ and *N*-bromosaccharin

⟨76S736, 82JOC1588⟩. A polymer-supported poly(vinylpyridine)–Br_2 complex has been used for benzylic bromination ⟨86JOC929⟩, as has $CuBr_2$ in the presence of *t*-butyl hydroperoxide ⟨81SC669⟩. A comparison of photochemical methods of benzylic bromination has been made ⟨88T6525⟩.

A whole host of different brominating agents have been applied to the α-bromination of various ketones and aldehydes, as indicated in Larock's tabulations ⟨B-89MI 201-01⟩. In addition to this, a detailed discussion of the synthesis of α-bromoketones has been published ⟨B-88MI 201-01⟩. Notwithstanding the plethora of alternative reagents, molecular bromine is probably still the most commonly used reagent for this transformation. Direct bromination of ketones can be achieved with Br_2 in acetic acid (e.g., Equation (47)) ⟨86JOC3380⟩, or conc. HBr(aq.) ⟨88OSC(6)520, 88OSC(6)711⟩. Nonacidic solvents can also be used ⟨88OSC(6)193, 88OSC(6)991⟩, and the addition of trimethylborate has been shown to boost yields in certain circumstances ⟨87IZV2395⟩. Even aldehydes can be controllably α-brominated ⟨73BSF1465⟩. HBr is of course generated during the reaction, and this can be scavenged by incorporating an epoxide into the reaction mixture ⟨77JCS(P1)501⟩. Other milder reagent systems include pyridinium hydrobromide perbromide ($PyHBr_3$) ⟨88TL6835⟩, its 4-dimethylamino analogue ⟨84SC939⟩, 2-pyrrolidone hydrotribromide ⟨85BCJ1599, 87JOC5624⟩, $PhMe_3NBr_3$ ⟨88OSC(6)175⟩, Bu_4NBr_3 ⟨87BCJ1159⟩ and some polymer-supported tetraalkylammonium perbromides ⟨79S64, 80S143, 92SC1923⟩. In the latter case, even acid-sensitive furyl ketones were successfully brominated (Equation (48)) ⟨92SC1923⟩. A polymer-supported pyridine–Br_2 complex has also been reported for ketone bromination ⟨89SC2481⟩.

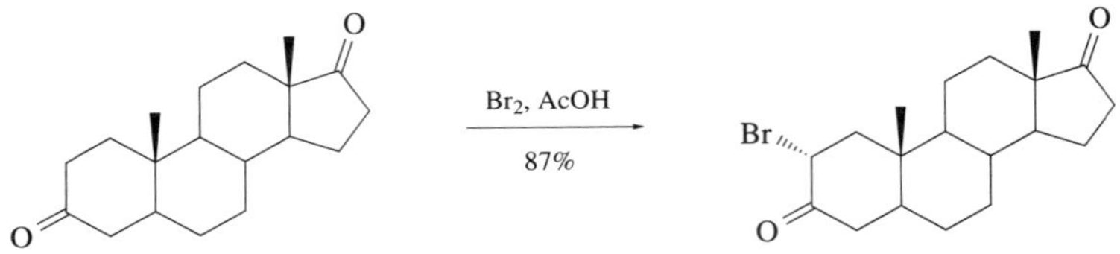

(47)

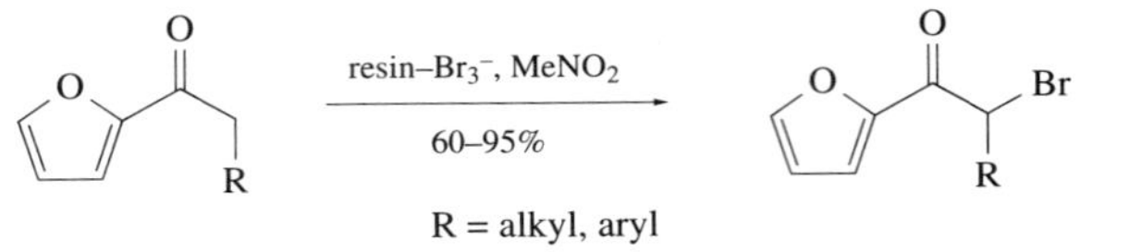

(48)

For substrates which contain a double bond or aromatic ring which might be susceptible to electrophilic bromination, the use of $CuBr_2$ is recommended ⟨62JOC4937, 64JOC2025, 64JOC3459, B-73MI 201-04, 75JOC1990⟩, and this is a particularly useful reagent for the bromination of malonates ⟨93JOC4596⟩. Using DMSO as oxidant, TMS-Br ⟨86JCR(S)428⟩ or Bu^{t}Br ⟨84T2035⟩ can be used to brominate aldehydes and ketones. In both cases, a bromosulfonium ion (analogous to the chlorosulfonium ion (**29**)) is believed to be the active halogenating species. Other reagents which have found occasional use in the α-bromination of ketones include NBS ⟨87JA7230⟩, *N*-bromosaccharin ⟨82JOC1588⟩ and 4-(tribromomethyl)quinazoline ⟨87G83⟩.

While direct α-bromination of carbonyl compounds is often quite readily achieved, side reactions such as self-condensation or dibromination can occur and, in the case of unsymmetrical dialkyl ketones, mixtures of regioisomers may be obtained. These problems can usually be overcome by prior conversion of the carbonyl compound to an enol ester ⟨53JA3513, 55OSC(3)127, 80JOC4702⟩, enol silane ⟨74JOC1785, 76S194, 87JA2435, 87JA7230, 89JCS(P1)1544, 91SC11⟩ or enamine ⟨78ACS(B)(32)646⟩. Scheme 25 gives an indication of the selectivities achievable in the bromination of 1-methylcyclohexanone by judicious choice of reaction conditions ⟨73JOC2576⟩.

Electrochemistry has found applications in the α-bromination of enol esters, enol ethers and enol silanes ⟨80JOC2731⟩, and also in the α′-bromination of α,β-unsaturated ketones (Equation (49)) ⟨91CC1418⟩.

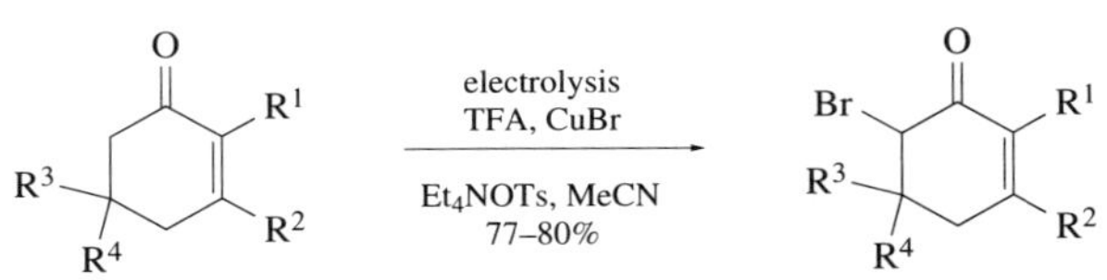

(49)

The α-bromination of carboxylic acids by the Hell–Volhard–Zelinski procedure is one of the oldest synthetic methods ⟨1881CB891⟩ and is still used today. In this process, phosphorus and

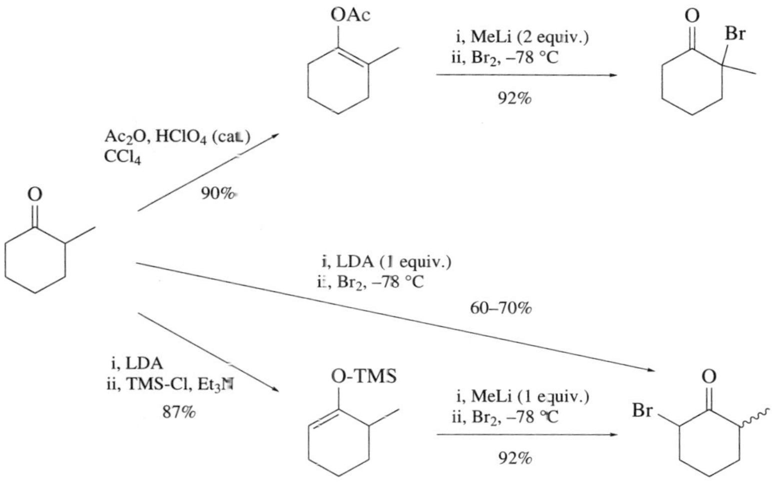

LDA = lithium diisopropylamide

Scheme 25

bromine are used to generate PBr_5, which converts the acid to its acid bromide, which in turn is α-brominated, then hydrolysed back to the α-bromoacid during workup ⟨43OSC(2)74⟩. The phosphorus has been used catalytically ⟨22JCS1161⟩ and can be replaced with phosphorus trichloride ⟨41OSC(1)115, 55OSC(3)523, 55OSC(3)348, 63OSC(4)398, 88OSC(6)403⟩ or PBr_3 ⟨63OSC(4)608, 88OSC(6)512⟩. While nitriles are readily α-brominated under similar conditions ⟨53JOC1112⟩, application of the P/Br_2 conditions to esters causes cleavage to the acid bromide. Nevertheless, this can be turned to advantage by incorporation of an alcohol during the workup to enable an α-bromoester to be isolated ⟨69S39⟩. Alternative approaches to α-bromoesters involve treatment of the ester with LDA at low temperature and quenching of the resultant anion with Br_2 ⟨71TL3995, 73JOC2576⟩ or carbon tetrabromide ⟨78JOC3687⟩. Asymmetric bromination of enolates is readily achieved using oxazolidinone chiral auxiliaries (Equation (50)) ⟨87TL1123, 93TA2315, 93TL2561, 94T2391⟩.

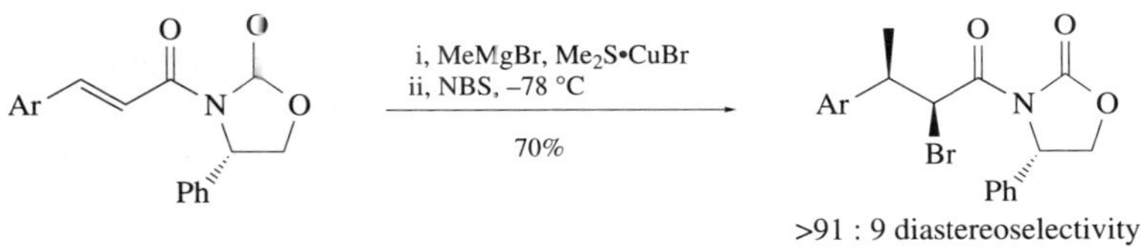

(50)

Finally, unlike most of the halogens, bromine has the ability to halogenate the α-position of acetals and ketals, either thermally ⟨69BSF4169, 93JOC2501, 93SC993⟩, photochemically ⟨62CB803⟩ or electrolytically ⟨80SC821⟩, and under acidic ⟨86AG(E)259⟩ or basic ⟨27JA2517, 42JA1963, 42JA1966⟩ conditions. Indeed, using a ketal derived from a chiral diol, such as dimethyl tartrate, asymmetric bromination can be effected ⟨86AG(E)259, 87JOC3018, 87S1039, 90JOC5400⟩, and this chemistry forms the basis of an industrial asymmetric synthesis of the anti-inflammatory drug naproxen (Scheme 26) ⟨89T4243⟩. The transient intermediacy of a ring-opened enol ether (**33**) enables the reaction to proceed. Bu_4NBr_3 is also capable of brominating ketals ⟨84BSB157⟩.

2.01.4.2 Alkyl Bromides from Alkenes

2.01.4.2.1 Alkyl bromides by hydrobromination of alkenes

As stated in Section 2.01.1.2.1, addition of HBr to alkenes may proceed via ionic or radical mechanisms; as a result of this, much of the early literature is somewhat confusing. Anti-Markovnikov addition can be effected using radical initiators such as peroxides or light, whereas careful exclusion of such radical initiators allows addition to proceed in an ionic fashion to yield products with Markovnikov regiochemistry ⟨62CRV599, 63OR(13)150⟩. More reliable methods for

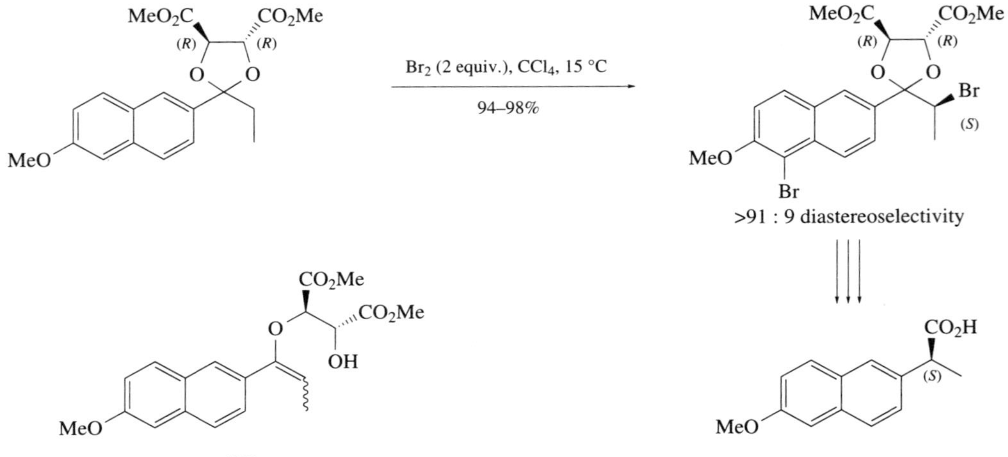

(33) (*S*)-naproxen

Scheme 26

effecting Markovnikov addition utilise either phase-transfer catalysis ⟨80JOC3527⟩, or a silica- (or alumina-)mediated reaction with HBr or TMS-Br ⟨90JA7433, 93JA3071⟩. For reliable anti-Markovnikov addition, hydroboration followed by reaction with bromine or BrCl is recommended (Scheme 4) ⟨70JA6660, 81JOC3113, 88T2763, 88T2773⟩. The same net result can be achieved via hydro-alumination ⟨77CL1117, 86JA6036, 87TL5793⟩.

2.01.4.2.2 Alkyl bromides by bromine–halogen addition to alkenes (Br—Br, Br—I)

Addition of elemental bromine to a wide variety of electron-rich and electron-poor alkenes is usually fast and quantitative ⟨41OSC(1)521, 43OSC(2)171, 43OSC(2)177, 43OSC(2)270, 55OSC(3)105, 55OSC(3)209, 55OSC(3)350, 55OSC(3)785, 63OSC(4)195, 63OSC(4)851, 63OSC(4)969⟩, so much so that it is sometimes exploited as a means of double-bond protection during synthesis ⟨79S964, 79S966, 81TL623⟩. Bromine addition generally proceeds in a *trans–anti* fashion via a cyclic bromonium ion as depicted in Scheme 6 (X = Br). For discussions of possible mechanisms of electrophilic bromination of alkenes, see ⟨90ACR87⟩ and ⟨93H(35)701⟩. Bromine itself is a fuming liquid and, because of this, more convenient solid 'perbromide' reagents, such as the commercially available crystalline pyridinium hydrobromide perbromide ($PyHBr_3$), have been developed ⟨52JCS2669, 56JA1218, 79S964, 79S966, 81TL623⟩. Related reagents include 4-dimethylaminopyridinium hydrobromide perbromide ⟨84SC939⟩ and Bu_4NBr_3 ⟨84BSB157⟩, which gives higher yields and shorter reaction times under ultrasonication ⟨91TL4135⟩, Me_3PhNBr_3 ⟨91TL3217, 94T6433⟩, and a polymer-bound tetraalkylammonium tribromide ⟨80S143⟩. Some of these give improved chemo- and stereoselectivities when compared with bromine itself ⟨78JOC2793, 84BSB157, 86JOC4224⟩. By pre-adsorbing bromine onto a molecular sieve, selective bromination of a terminal alkene can be achieved in the presence of an internal one, because of the restricted pore size offered by the zeolite (Equation (51)) ⟨78JA4919⟩.

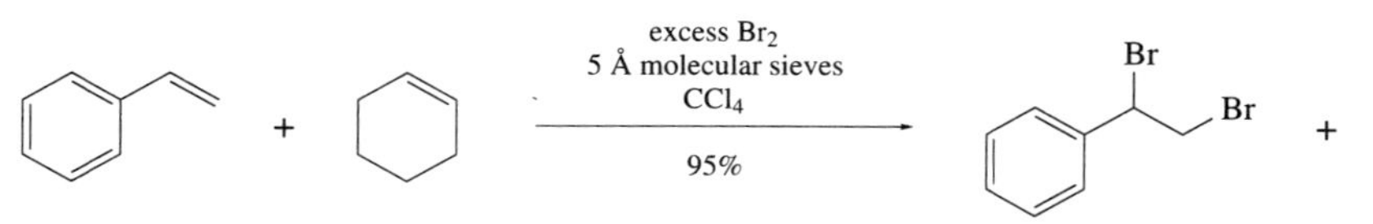

(51)

Bromide ion can be used to dibrominate alkenes, for example electrochemically ⟨81JOC3312⟩, or in combination with acidic hydrogen peroxide under phase transfer catalysis ⟨77S676⟩. Indeed, copper(II) bromide can be used to dibrominate alkenes directly, provided catalytic amounts of a complexing ligand (e.g., phosphine, nitrile) are present ⟨71JOC3324⟩. Other less widely used procedures for 1,2-dibromination of alkenes make use of 2,4,4,6-tetrabromocyclohexadienone (**34**) ⟨76CL283, 80JCS(P1)1051⟩ and *in situ* generated dimethylsulfide–bromine adduct ⟨85G29⟩.

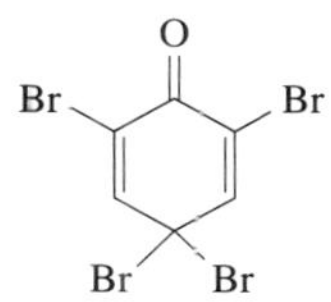

(34)

Stereoselective bromine–iodine addition to alkenes can be achieved using bromide ion in combination with either I_2/CuO·HBF_4 ⟨87CC1491, 90JCS(P1)2807⟩ or the stable solid reagent bis(pyridine)iodine tetrafluoroborate, IPy_2BF_4 ⟨85AG(E)319⟩. A less attractive method (from a toxicity standpoint) requires stoichiometric quantities of $HgBr_2$ and iodine ⟨85CC1422, 86JCR(S)274⟩.

2.01.4.3 Alkyl Bromides from Alkyl Halides

Conversion of alkyl fluorides to bromides is rare, but can be achieved with excess concentrated aqueous HBr solution ⟨90TL4973⟩, a process which is enhanced by phase-transfer catalysis ⟨92JCS(P1)2309⟩. These conditions also convert alkyl chlorides to bromides, as does anhydrous hydrogen bromide under iron(III) bromide catalysis ⟨87CC1013, 89JOC3028⟩, the latter conditions being particularly good for secondary, tertiary and even bridgehead bromides. BBr_3 can also convert alkyl chlorides to bromides ⟨89S614⟩. HBr will convert iodides to bromides, provided that the product iodide ion is oxidised *in situ* to I_2 using HNO_3 ⟨69JOU971⟩. The counter-thermodynamic conversion of primary alkyl chlorides to bromides of high (>99%) purity can be achieved by repeated treatment (the second after aqueous workup) with 10 equiv. (each) of lithium bromide in refluxing 3-pentanone ⟨94SC733⟩. This transformation can also be effected under phase transfer catalytic conditions ⟨89BSF247⟩ using sodium ⟨77JOC875, 89JCR(S)104⟩, potassium ⟨82TL4651, 89JCR(S)104⟩, lithium ⟨86CC1250, 88SC1275⟩, calcium ⟨84S34⟩, copper ⟨89JCR(S)104⟩ or resin-bound tetraalkylammonium ⟨76S472⟩ bromides. Microemulsions are believed to play a role in the success of this reaction with surfactant phase-transfer catalysts ⟨82TL4651, 89JCR(S)104, 92JCS(P1)2309⟩. Alternative bromide sources for metal halide-catalysed halide substitution include ethyl bromide ⟨76BCJ1989⟩, propyl bromide ⟨84TL5949⟩ and dibromomethane ⟨84SC1313⟩, but the vast excesses of alkyl halide needed, and their inherent toxicity, somewhat reduces their appeal. A short overview of alkyl bromide synthesis via halogen exchange processes has been published ⟨90MI 201-01⟩.

2.01.4.4 Alkyl Bromides from Alcohols and their Derivatives

A number of methods for the conversion of alcohols to halides (including bromides) have already been discussed in Section 2.01.1.4.1. As is the case with the other halides, conversion of alcohols to alkyl bromides using HBr works well (particularly in systems where the S_N1 pathway is facilitated by stabilisation of the intermediate carbocation ⟨41OSC(1)25, 87JOC5560⟩, but rearrangements of the carbonium ion sometimes occur. While the use of phase transfer catalysis can reduce rearrangements considerably ⟨87TL1223⟩, and has even facilitated the preparation of primary bromides using HBr ⟨87JCS(P1)2157⟩, a far more popular reagent, which also leads to less rearrangement, is phosphorus tribromide, PBr_3. Although PBr_3 can be generated *in situ* from phosphorus and bromine ⟨41OSC(1)36, 43OSC(2)308⟩, it is more common to use commercially available liquid PBr_3 ⟨43OSC(2)358⟩. The presence of pyridine can help ⟨55OSC(3)793⟩, or the PBr_3 can be conveniently adsorbed onto a resin-bound tertiary amine ⟨83S306⟩. The PBr_3 reactions generally proceed with inversion via an S_N2 mechanism ⟨76JOC1071⟩, although epimerisation of product by reversible bromide ion attack can sometimes cloud the stereochemical issue. Other reagents which find occasional use include solid phosphorus pentabromide (PBr_5), $POBr_3$ ⟨50JA1753⟩ and $SOBr_2$ ⟨50JCS406, 71TL87, 75JOC1?4⟩. Vilsmeier reagents (**6**) are also effective for this transformation ⟨74CI(L)664, 76JCS(P1)754⟩. A cheap and attractive process for the preparation of alkyl bromides from alcohols with clean inversion is outlined in Equation (52) ⟨92SC2187⟩.

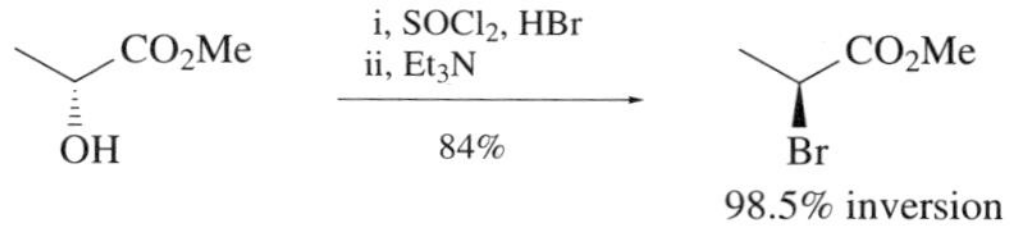

(52)

Work in the mid-1960s highlighted the synthetic utility of bromoboranes derived from Br_2 and BH_3 or $NaBH_4$ ⟨65CI(L)223⟩. In the 1980s and 1990s, boron tribromide has been found to convert secondary, tertiary and benzylic alcohols (but not simple primary aliphatic alcohols) to bromides ⟨89S614, 94TL1051⟩. This procedure can be viewed as a logical extension of an earlier method using boron trifluoride etherate and bromide ion ⟨85TL3863⟩.

While the above methods and those discussed in Section 2.01.1.4.1 satisfy most chemists' needs, there are many more, including a high-yielding general procedure using trimethylsilyl bromide ⟨78TL4483⟩. Improved reaction rates are observed using a TMS-Cl/LiBr combination, suggesting that bromide ion catalysis might be important ⟨80JOC1638⟩. In another modification of this procedure, using pyridinium bromide perbromide/$(TMS)_2$, tertiary alcohols can be selectively converted to bromides in the presence of primary and secondary ones ⟨80JOC1638⟩. In these reactions, the alcohol is initially converted to the trimethylsilyl ether, which is then cleaved (predominantly with inversion) by the HBr generated during the reaction. Crystalline dimethylbromosulfonium bromide (Me_2SBr_2) ⟨73CC212⟩ and related species derived from dimethyl sulfide and *N*-bromosuccinimide ⟨72TL4339⟩ have also found use in converting alcohols to bromides, often with clean inversion of configuration.

While conversion to a sulfonate ester is a highly effective way of accessing alkyl bromides from alcohols (see Section 2.01.1.4.2 for details), there are numerous other ways of activating alcohols towards displacement with bromide ion. These include activation as the trichloroacetamidate (**35**) ⟨70TL2517⟩, reaction with carbodiimides ⟨87TL4445⟩, and reaction with carbonyl diimidazole in the presence of allyl bromide, which acts both as activating agent and bromide source as shown in Scheme 27 ⟨83CPB4189⟩. Note that in the latter case, racemic product is obtained from optically active starting materials. Finally, an indirect alcohol-to-bromide conversion via a selenide has been reported which proceeds with net retention of configuration (Scheme 28) ⟨80CC656⟩.

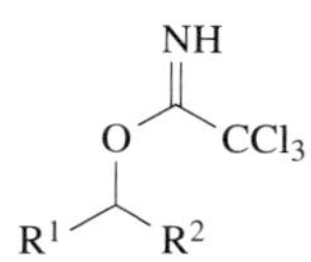

(35)

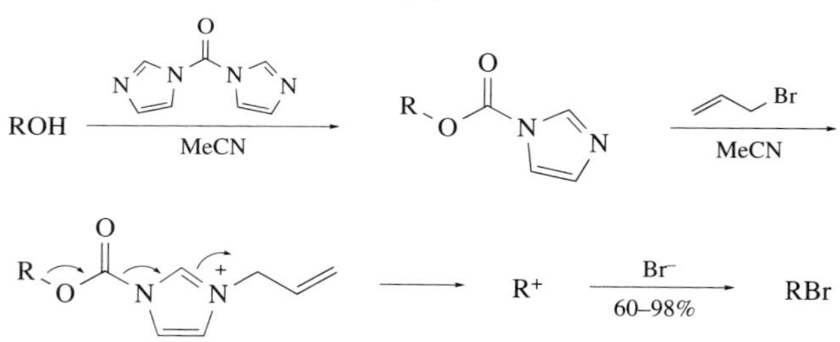

Scheme 27

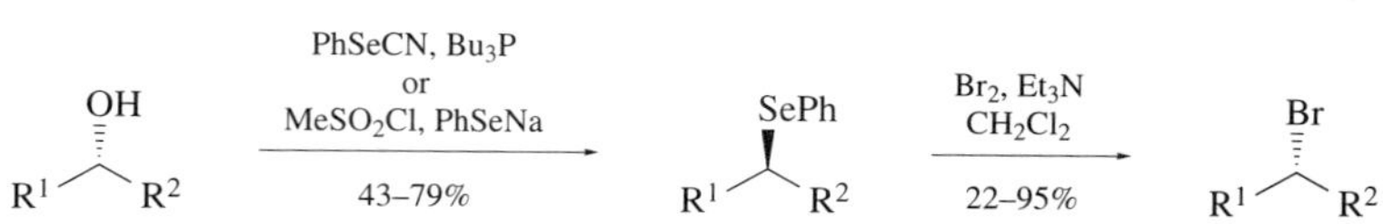

Scheme 28

Ethers are less widely used precursors to alkyl bromides, but this approach is nevertheless worthy of mention. Concentrated HBr can be used to cleave alkyl ethers to the corresponding bromides, and this is most commonly applied to cyclic ⟨55OSC(3)692⟩ or benzylic ⟨56JA3135⟩ ethers. In the latter case, the benzylic C—O bond is cleaved. A phase transfer method for cleavage of ethers with HBr has been reported ⟨78S771⟩, while a milder reagent combination, $Ph_3P–Br_2$, has also proved useful in this regard ⟨64JA5037, 71T2617, 72JOC626, 76JCS(P1)1446⟩, even in sterically demanding situations (Equation (53)) ⟨69CI(L)200⟩.

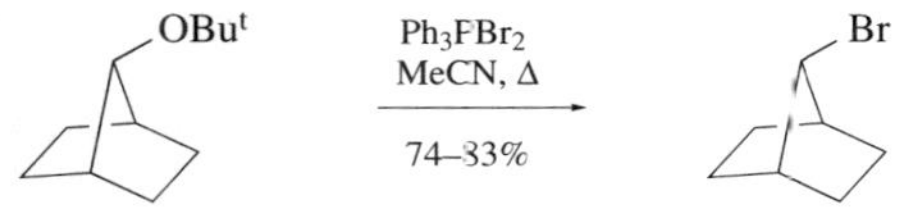

(53)

Bromoboranes such as BBr_3 ⟨42JA1128, 63CI(L)609⟩ and *B*-bromo-9-BBN ⟨78JOM(156)221⟩ have found occasional use, the former for the replacement of the methoxy group, the latter giving best results on tertiary and secondary alkyl ethers. In addition, iodine monobromide (IBr) has been shown to be an effective catalyst for the cleavage of ethers with TMS-Br ⟨83JOC1678⟩.

Epoxides give 1,2-dibromides with inversion at each carbon when treated with PPh_3/Br_2 ⟨76JOC3279⟩ or a molybdenum bromide complex ⟨83JOM(243)175⟩.

2.01.4.5 Alkyl Bromides from Amines and their Derivatives

In addition to the methods described in Section 2.01.1.5, alkyl bromides are generated by the treatment of tertiary amines with cyanogen bromide (Equation (54)) (the von Braun cyanogen bromide reaction, for a review see ⟨53OR(7)198⟩), but this transformation is rarely used for the synthesis of alkyl halides (although see ⟨81JOC3349⟩). This should not be confused with a related procedure, also referred to as the von Braun reaction, in which cyclic benzamides are converted to α,ω-bromides on treatment with PBr_5 or PBr_3/Br_2 (Equation (55)) ⟨41OSC(1)428, 46JA2402, 86JOC2206, 87JOC5466⟩.

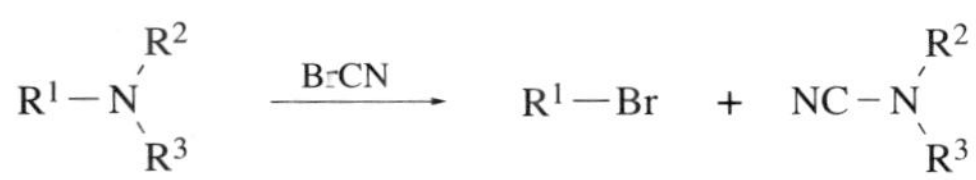

(54)

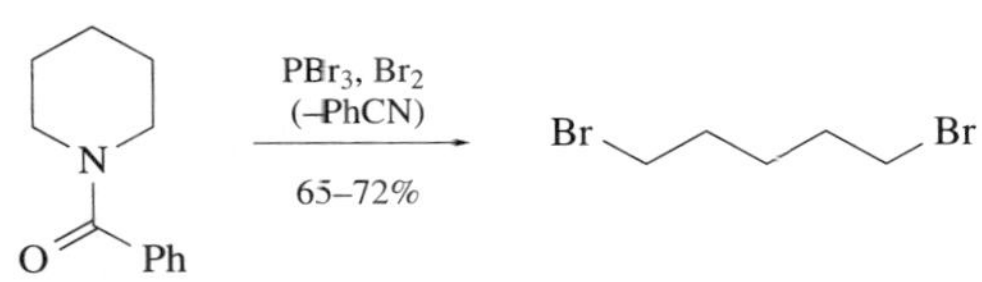

(55)

2.01.4.6 Alkyl Bromides by Bromodecarboxylation of Carboxylic Acids and their Derivatives

See Section 2.01.1.6.

2.01.4.7 Alkyl Bromides by Bromoalkylation of Arenes

See Section 2.01.1.7.

2.01.5 ALKYL IODIDES: RI

Alkyl iodides are less stable than fluorides, chlorides and bromides, and iodine is the most expensive of the common halogens. As synthetic intermediates, they offer few advantages over alkyl bromides and are therefore less frequently used. Much of their chemistry is similar to that used to make bromides.

2.01.5.1 Alkyl Iodides from Alkanes

Photochemical iodination of alkanes with elemental iodine ⟨60JA3816⟩ or Bu[t]OI ⟨68JA808⟩ is known but is of little synthetic use. In contrast, α-iodination of carbonyl compounds and their enol derivatives (enol silanes, enol esters) is much more readily achieved, for example using elemental iodine in combination with $Cu(OAc)_2$ ⟨81S312, 84CL1509⟩, $Cu(NO_3)_2$ ⟨91JOC6708⟩, AgOAc

⟨79JOC1731⟩, TlOAc ⟨78JCS(P1)126⟩, pyridinium chlorochromate ⟨82SC1127⟩, ceric ammonium nitrate ⟨88CL31, 94BCJ271⟩, or trimethylsilyl chlorochromate ⟨85T2903⟩. In particular, a direct α-iodination of carboxylic acids has been described using simple copper salts and elemental iodine ⟨84CL1509⟩. For activated methylene groups such as those in malonates, iodination is straightforwardly achieved under phase transfer catalysis using K_2CO_3 as base and I_2 as halogen source ⟨93T5133⟩. Indeed, direct treatment of enolate ions with elemental iodine can be a very effective method for the α-iodination of a range of carbonyl compounds including aldehydes and esters ⟨71TL3995, 78JOC3687, 79TL2817, 87TL2477⟩. Amides are readily α-iodinated with I_2 using 2,4,6-collidine as base, both in achiral systems ⟨92TL1299⟩ and in an asymmetric sense using a covalently bound C_2-symmetric chiral auxiliary (Equation (56)) ⟨93TL2165⟩. Note that iodolactonisation does not occur under these conditions.

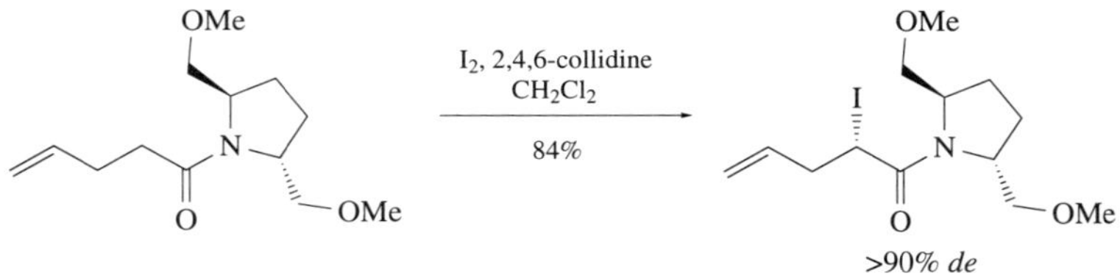

(56)

N-Iodosuccinimide can be used to iodinate enol esters ⟨53JA3493, 55JA3826⟩ and enol silanes ⟨84TL233⟩, although in the latter case a combination of *N*-chlorosuccinimide and sodium iodide is used. Finally, iodide ion in combination with either *m*-chloroperbenzoic acid or KIO_3 yields α-iodocarbonyl compounds from enol silanes or aryl ketones, respectively ⟨87JOC3919, 92BCJ1731⟩.

2.01.5.2 Alkyl Iodides from Alkenes

2.01.5.2.1 Alkyl iodides by hydroiodination of alkenes

The reaction of alkenes with hydrogen iodide affords the Markovnikov addition products exclusively. Classically, the HI is generated *in situ* by treatment of potassium iodide with phosphoric acid ⟨63OSC(4)543⟩, but more modern methods of *in situ* HI generation include combinations of iodine with alumina ⟨87TL4497, 88JOC4477⟩ and BH_3–diethylaniline complex with iodine in acetic acid ⟨90TL1919⟩. Controlled amounts of water can also generate HI when in the presence of either AlI_3 ⟨91CI(L)175⟩, or TMS-Cl/NaI ⟨88S366⟩. Indeed, water is not needed at all if the substrate can donate an acidic proton, as in the example shown in Equation (57) ⟨83S461⟩. Markovnikov addition of HI to enones can also be achieved by reaction with tetraethylammonium iodide and trifluoroacetic acid ⟨83T1529⟩.

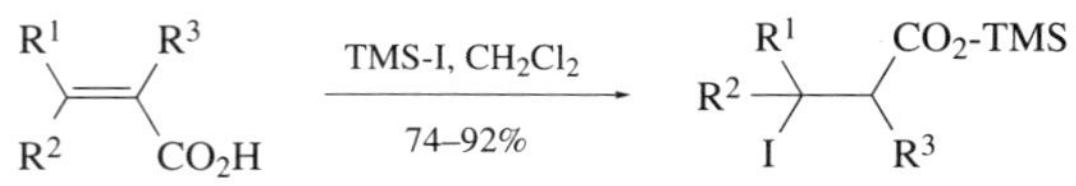

(57)

Straightforward access to the anti-Markovnikov addition products is via hydroboration (Scheme 4), quenching with iodine ⟨68JA5038, 76S114, 85JA3915, 88T2751⟩ or iodine monochloride ⟨80JOC3578, 81SC521⟩ under basic conditions. For base-sensitive substrates, alternative iodine-quench protocols have been described ⟨71JA1508, 82SC101⟩. Enantiomerically enriched secondary alkyl iodides can be obtained if a chiral hydroborating reagent is used (Equation (58)) ⟨76JA1290⟩. Interestingly, under the right conditions it is possible to hydroborate–iodinate more highly substituted double bonds in the presence of less substituted ones (Equation (59)) ⟨88JOC187⟩. Other related procedures for anti-Markovnikov addition include hydroalumination ⟨86JA6036⟩ and hydrozirconation ⟨81JOC1821⟩. Finally, a novel umpolung approach has been described using IPy_2BF_4 as positive iodine source and Et_3SiH as hydride source ⟨85AG(E)319⟩.

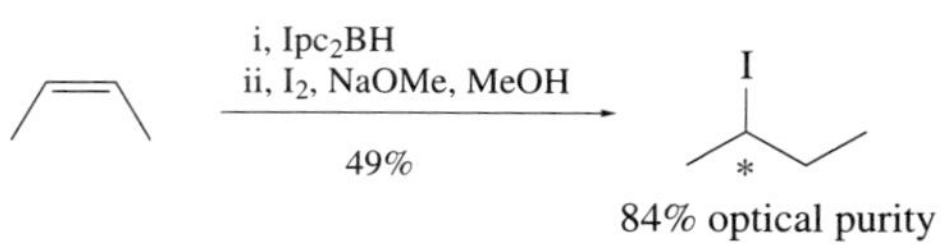

(58)

Ipc_2BH = diisopinocampheylborane

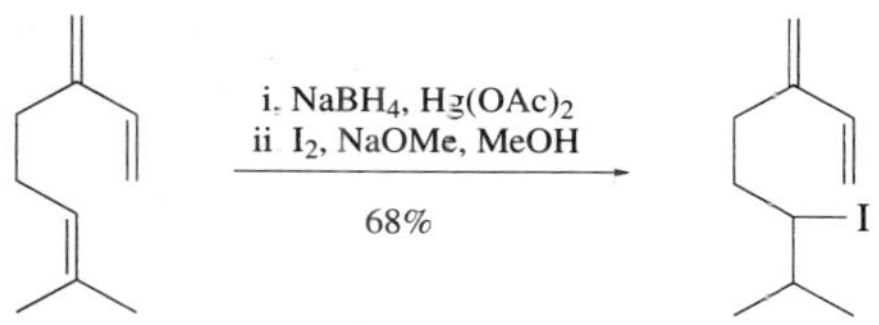

(59)

2.01.5.2.2 *Alkyl iodides by iodine–iodine addition to alkenes*

It is possible to add elemental iodine across double bonds to give vicinal diiodo compounds, but this is of little preparative use since the reverse reaction is thermodynamically favoured ⟨64CJC2710, 64JA2956, 71JA1389, 75JOC248⟩. Isolated reports of alternative procedures include reaction of alkenes with an iron–halide–dialkylthiourea complex ⟨85CC1404⟩, or with iodine, sodium iodide and CuO–HBF_4 ⟨87CC1491, 90JCS(P1)2807⟩. A combination of Bu^tBr, DMSO and KI also effects this transformation ⟨85G29⟩.

2.01.5.3 Alkyl Iodides from Alkyl Halides

Alkyl iodides are the least stable of the alkyl halides but are readily prepared by S_N2 halide exchange under the classical conditions of Finkelstein. Although halide exchange is reversible, treatment of an alkyl chloride (or bromide) with a solution of sodium iodide in acetone at reflux effects conversion to the alkyl iodide, owing to a shift of the equilibrium position caused by the precipitation of the by-product sodium chloride (or bromide), which is less soluble in acetone than is sodium iodide (Equation (60)).

$$\text{RHal} + \text{NaI} \xrightarrow{\text{acetone}} \text{RI} + \text{NaHal}\downarrow \qquad (60)$$

$$\text{Hal} = \text{Br, Cl}$$

Because of the S_N2 nature of halide substitution, secondary and tertiary halides are slow to react with iodide ion, and usually require different conditions such as zinc or iron halide catalysis ⟨74TL2691, 76JCS(P1)416, 89JOC3028⟩. In fact, alkyl fluorides, chlorides and bromides can be converted to iodides simply by heating with excess aqueous HI with ⟨92JCS(P1)2309⟩ or without ⟨90TL4973⟩ phase-transfer catalysis. Exploitation of the strength of the silicon–fluorine bond enables alkyl iodides to be prepared from a wide range of alkyl fluorides (including tertiary ones) by treatment with trimethylsilyl iodide (or its *in situ* equivalents TMS-Cl/NaI and $(TMS)_2/I_2$) ⟨81JOC3727⟩. By adding catalytic tetrabutylammonium iodide, TMS-I can be used to convert primary and secondary alkyl chlorides and bromides to iodides, whereas secondary and tertiary alkyl chlorides and bromides need iodine to catalyse this transformation ⟨82JOM(226)143⟩. For the conversion of alkyl bromides (and to a lesser extent chlorides) to iodides, the generally poor solubility of sodium or potassium iodides has been overcome in a variety of ways, including the use of dipolar aprotic solvents such as HMPA ⟨68JA6225⟩ or DMF ⟨69JOC3519⟩, addition of crown ethers to solubilise the metal counterion ⟨74CC879, 76S516, 79TL5055⟩, and application of phase transfer catalysis ⟨79BCJ1653, 79JOC3436, 79S952, 81TL1017, 85JOC5828, 88S403, 89JCR(S)104, 92JCS(P1)2309⟩. A particularly interesting development is the use of a polymeric sulfone as a soluble but recoverable phase-transfer catalyst ⟨85CC400⟩. An 'inverse phase transfer' catalytic procedure for converting alkyl bromides to iodides has been reported using a β-cyclodextrin–epichlorohydrin copolymer ⟨87CL715⟩. Amberlyst resin-bound tetraalkylammonium iodide is also a useful reagent for the iodination of alkyl bromides ⟨76S472⟩.

2.01.5.4 Alkyl Iodides from Alcohols and their Derivatives

A number of methods for the conversion of alcohols to halides (including iodides) have already been discussed in Section 2.01.1.4.1. One of the difficulties associated with iodide preparation using inorganic acid iodides such as HI (often generated *in situ* from KI and phosphoric acid, ⟨63OSC(4)323⟩, or occasionally from TMS-I in the presence of water ⟨89S283⟩), is rearrangement of the carbon skeleton. While this can be partially overcome by the use of PI_3 (prepared *in situ* from white or red phosphorus and iodine ⟨43OSC(2)322⟩), the other side reaction, reduction of the iodide to the alkane ⟨69JCS(C)1046⟩, can still be a problem. A better reagent is diphosphorus tetraiodide P_2I_4 ⟨79TL1801, 83CC229, 83SC387, 90SC41⟩, which works particularly well on tertiary and allylic alcohols, giving the more stable allylic iodide in the latter case. Indeed, there is a plethora of reagents for the conversion of allylic, benzylic or tertiary alkyl alcohols to iodides, including MgI_2 ⟨84S608, 87TL6441, 88S58⟩, AlI_3 ⟨89T3569⟩ and $NaI/BF_3 \cdot OEt_2$ ⟨85TL2717, 85TL3863⟩. Reports from the mid 1960s described the combination of borane ⟨64CI(L)1582⟩ or $NaBH_4$ ⟨65CI(L)223⟩ with iodine for the conversion of ethers or alcohols to alkyl iodides, but this method has lain dormant until the late 1980s ⟨89TL5663, 92T8329⟩. Boron triiodide can also be used to generate alkyl iodides by ether cleavage ⟨67TL4131, 90TL6977⟩.

Trimethylsilyl iodide ⟨82T2225, B-91MI 201-01⟩ is used to prepare alkyl iodides from primary and secondary alcohols ⟨77TL2659⟩ and ethers ⟨77JOC3761⟩, generally with inversion of configuration. It can be generated cheaply *in situ* from trimethylsilyl chloride and sodium iodide ⟨78CC874, 79JOC1247, 79S379⟩ or hexamethyldisilane and iodine ⟨79AG(E)612⟩. Slightly improved selectivity is noted with Cl_3SiMe/NaI ⟨83JOC3667⟩, while diiodosilane H_2SiI_2 has also been applied successfully ⟨87JOC4846⟩ to the iodination of both alcohols and ethers. Further modification of these conditions enables iodides to be prepared by reductive cleavage of epoxides (Equation (61)) ⟨84TL3123⟩.

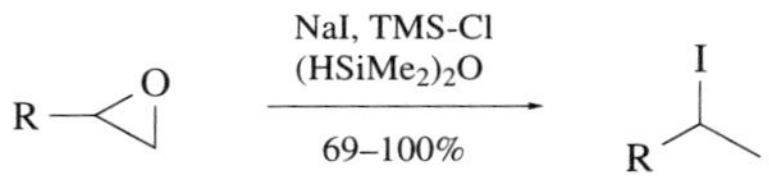

(61)

A less well-used reagent, trimethylsilyl polyphosphate (PPSE), can be used to prepare primary, secondary and benzylic iodides from alcohols in the presence of sodium iodide ⟨83S460⟩. Nonacidic methods for the activation of alcohols towards iodide displacement include reaction with Vilsmeier reagent systems ⟨93SL489⟩, fluorobenzothiazolium salts **(7)** ⟨76CL619⟩, carbodiimides ⟨72AG(E)229, 87TL4445⟩, chloroethyl chloroformate ⟨85TL5445⟩ and carbonyl diimidazole/MeI (cf. Scheme 27) ⟨83CPB4189⟩.

In addition to the ether cleavage methods described above, alkyl iodides can also be obtained from ethers simply by treatment with hydrogen iodide. This cleaves ethers faster than hydrogen bromide or chloride, although the reducing power of this reagent can sometimes lead to undesirable side reactions. Typically the HI is generated *in situ* from KI and phosphoric acid, and is most commonly applied to cyclic ethers to generate α,ω-diiodides ⟨63OSC(4)321⟩. Milder conditions for the same transformation utilise phenyl dichlorophosphate in the presence of sodium or lithium iodide ⟨86H(24)3043, 88SC119⟩.

2.01.5.5 Alkyl Iodides from Amines and their Derivatives

In addition to the methods described in Section 2.01.1.5, it has been shown that tertiary alkyl nitro compounds are converted to the corresponding iodides by reaction with trimethylsilyl iodide (Equation (62)) ⟨83JOC2766⟩. This is restricted to tertiary systems, since primary and secondary nitroalkanes give nitriles and oximes, respectively, under these conditions.

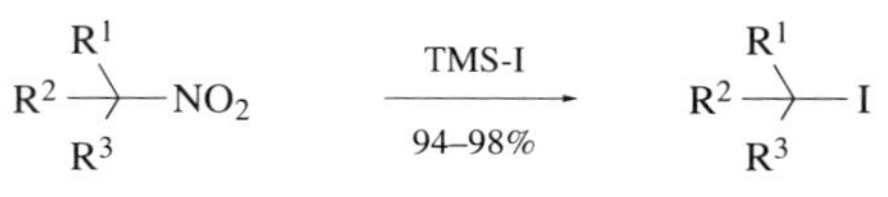

(62)

2.01.5.6 Alkyl Iodides by Iododecarboxylation of Carboxylic Acids and their Derivatives

See Section 2.01.1.6.

2.01.5.7 Alkyl Iodides by Iodoalkylation of Arenes

See Section 2.01.1.7.

2.02
Alkyl Chalcogenides: Oxygen-based Functional Groups

JOSEPH B. SWEENEY
University of Bristol, UK

2.02.1 ALCOHOLS

The alcohol functional group is widespread in natural and unnatural products and has, therefore, long been the target of synthetic organic chemists ⟨B-65MI 202-01, B-68MI 202-01, B-71MI 202-01, B-77MI 202-01, 79COC(1)594, B-89MI 202-01, 91COS(6)1⟩.

2.02.1.1 From Biotransformations

There is a wealth of literature dating from early in the twentieth century concerning microbial oxidation of hydrocarbons, in a controlled manner such as synthetic chemists could never hope to be able to parallel. The utility and efficiency of such reactions is demonstrated by their widespread employment in industry and the subject area has been well documented by several texts to which the reader is directed ⟨91COS(8)183, B-94MI 202-01, 94MI 202-02⟩. Two representative examples are shown in Equations (1) ⟨83AG(E)406⟩ and (2) ⟨21MI 202-01⟩.

(1)

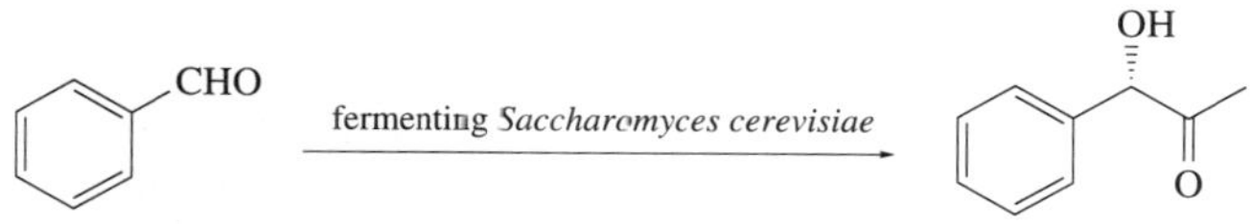

(2)

The hydroxylation reactions of arenes in the presence of oxidases such as those found in *Pseudomonas putida* strains has attracted much attention. Benzene is oxidized to 1,2-*cis*-dihydroxycyclohex-3,5-diene ⟨70B1626, 74MI 202-01⟩, while substituted benzenes are converted in variable enantiomeric excesses to similar 1,2-*cis*-dihydroxycyclohexadienes (Equation (3)) ⟨90JA9439, 92CC234⟩. *Pseudomonas* strains will also hydroxylate nonaromatic substrates (Equation (4)) ⟨76JOC3505, 81HCA1158⟩.

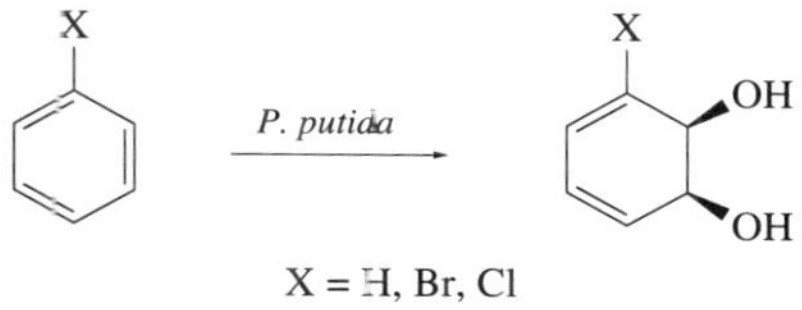

(3)

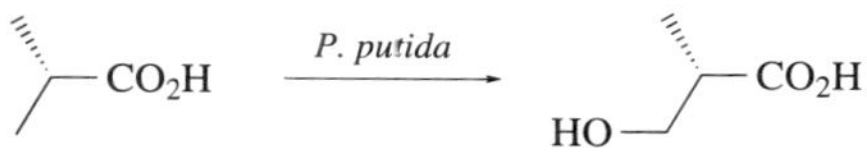

(4)

The use of baker's yeast as a carbonyl reducing agent is ubiquitous ⟨94MI 202-03, 90S1⟩. 2-Ketoesters are the most-studied substrates ⟨94TL4985⟩, but the reduction of many prochiral carbonyl groups has been reported. Most reductions employ the extremely inexpensive whole-cell system, but some improvements may be gained by the use of cell-free extracts ⟨94TL4569⟩. 3- and 4-Ketosulfones may be reduced with excellent enantiomeric excess ⟨90TL5575⟩, for example, in 98% *ee* when methyl ketones are substrates, although ethyl ketones are reduced with inferior selectivity (<67% *ee*) (Equation (5)). Baker's yeast also reduces acetoacetylated Meldrum's acid derivatives with high selectivity (Scheme 1). Decarboxylation of the resulting compounds leads to 3-oxo-δ-lactones of greater than 99% enantiopurity ⟨90TL7463⟩. An efficient baker's yeast-mediated reduction of 3-aryl-2,3-epoxy ketones has been reported (Equation (6)). The epoxide ring is cleaved stereospecifically by fermenting yeast after the carbonyl has been stereoselectively reduced; only one of the eight diastereoisomeric triols possible is produced in the reaction. The authors suggested that the carbonyl reduction was followed by a regiospecific epoxide ring-opening by a hydrolytic moiety present in the yeast. Water then displaced the endogenous nucleophilic species to give the product ⟨93CC119⟩.

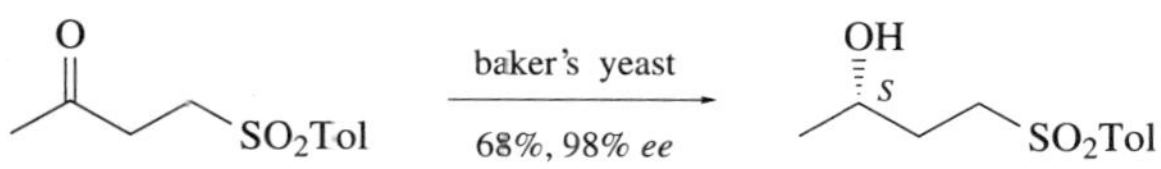

(5)

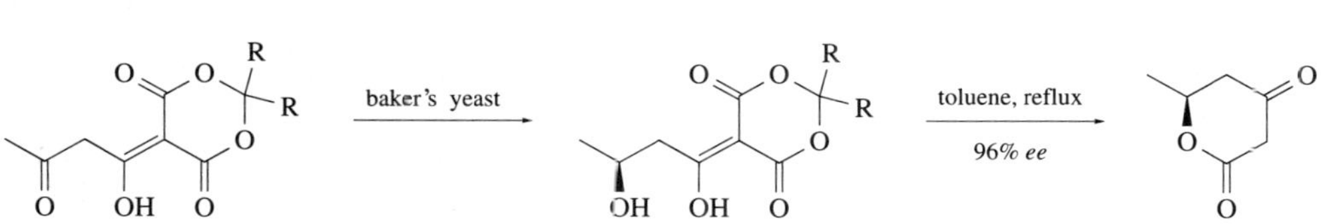

Scheme 1

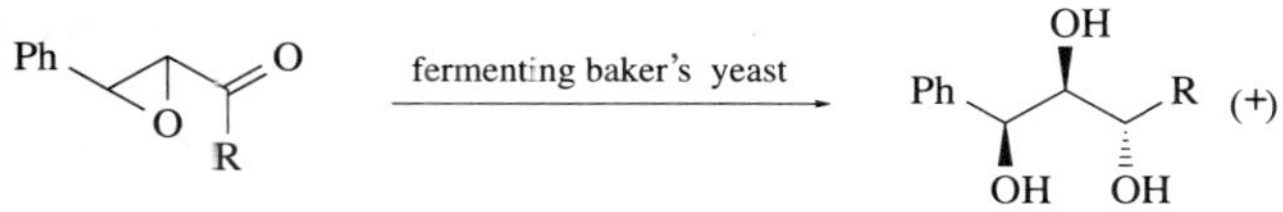

(6)

There are methods available for controlling the observed enantioselectivity of baker's yeast reductions ⟨91JOC4778⟩; for instance, certain 2-ketoesters, are reduced with higher levels of enantiocontrol when the reaction is carried out in benzene containing one drop of water ⟨91TL7075⟩.

2.02.1.2 By Addition to Alkenes

The addition reactions of alkenes provide many synthetically efficient routes to alcohols. Simple acid-catalysed hydration of alkenes is one of the earliest reported and most widely studied synthetic processes known ⟨18JA822, 71JA4907⟩. Such reactions famously proceed via the so-called Markovnikoff pathway, leading to an alcohol with the greatest possible substitution. The reaction, although of past commercial importance ⟨35ZAAC282⟩ is limited by the natural propensity for many carbenium ions to rearrange, which often leads to extremely complex product mixtures ⟨79COC(1)594⟩.

To a large extent, the acid-catalysed hydration of alkenes has been superseded by mercuration reactions ⟨71S527, B-86MI 202-01⟩. The general reaction ⟨67JA1522⟩ involves addition of a suitable mercuric cation to an alkene, followed by hydration of the subsequent carbocation. The resulting 2-hydroxyalkylmercury species may be smoothly and mildly protio-demercurated by a reductive workup to give, overall, a hydrated alkene. The reaction conditions are extremely mild; both mecuration and demercuration reactions may be carried out in the same vessel, and the pronounced sensitivity of the reaction to steric influences often leads to high stereoselectivity (Equation (7)).

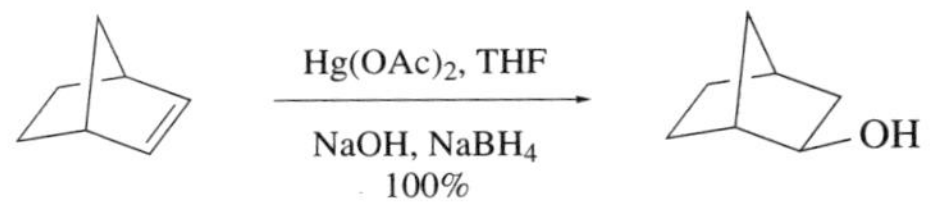

(7)

When the mercury carboxylate employed contains asymmetry, enantioselective oxymercuration is feasible, but reported enantiomeric excesses are poor ⟨71TL3361⟩. Certain cyclopropanes may be ring-opened in an oxymercuration process ⟨83JA6882, 86JA2094⟩. Thallium sulfate adds to cyclic alkenes to give (after hydrolysis) *trans*-diols ⟨71CJC2586⟩.

The acid- and metal-catalysed hydration reactions of alkenes are beautifully complemented by the addition of borane and its derivatives, the so-called 'hydroboration reaction' ⟨63OR(13)1⟩. Thus, addition of diborane to an alkene proceeds to give a *syn*-addition of boron and hydrogen, thereby producing a monoalkylborane. The remaining hydrides may be added in similar fashion to two further moles of alkene, to give a trialkylborane which undergoes an oxidative rearrangement reaction when reacted with either basic hydrogen peroxide (see below), or amine *N*-oxides ⟨75JOC1776⟩, to give the corresponding trialkyl borate. These borates may, in turn, be hydrolysed to liberate the parent alcohols; overall, the sequence is an *anti*-Markovnikoff hydration. In practice, the hydroboration reaction and subsequent oxidation reaction are carried out in the same vessel; yields are good and the reaction is one of the most widely-used synthetic techniques. Although the product of the reaction is, formally, that of *anti*-Markovnikoff addition, the reaction obeys that convention because the initial reaction involves addition of *hydride* rather than proton. In any case, the preparation of *anti*-Markovnikoff alcohols is uniformly straightforward.

Since the original report of hydroboration ⟨56JA5694, 57JOC1136⟩ which contradicted the previous reports of the sluggish reaction with alkenes ⟨48JA2053⟩ and the subsequent report describing the oxidation of organoboranes to give alcohols ⟨59JA4108⟩, hydroboration reactions have become ubiquitous in organic synthesis ⟨B-62MI 202-01⟩. The use of diborane has been superseded by the commercial availability of the ether and thioether complexes of monomeric borane, and there are a wide variety of alkylboranes available for use in synthesis. As there is an increase in the steric demands of the alkyl groups present, there is a corresponding increase in the regioselectivity of the hydroboration reaction of the organoborane. For example the organoborane prepared by hydroboration of 2-methylbut-2-ene (disiamylborane, (**1**)) will react preferentially with terminal alkenes (Equation (8)); in hydration of hex-1-ene, disiamylborane gives hexan-1-ol with at least 99% selectivity, whereas the corresponding reaction with diborane gives at least 6% hexan-2-ol ⟨60JA3222⟩. Δ^1-Cholestene reacts with similar selectivity (Equation (9)) ⟨61JOC630⟩.

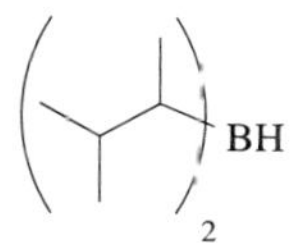

disiamylborane

(1)

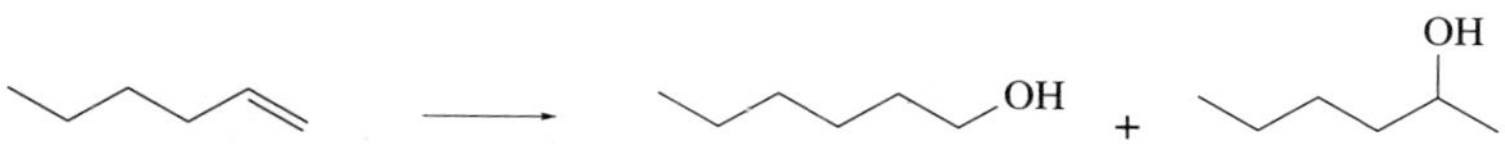

(8)

with disiamylborane 99% 1%
with diborane 94% 6%

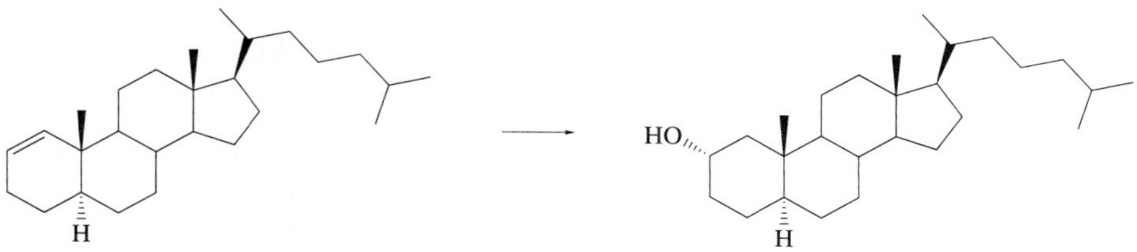

(9)

with disiamylborane 100%
with diborane 50%

The even more hindered organoborane prepared by hydroboration of 2,3-dimethylbut-2-ene (thexylborane (**2**)) is more selective in its reactions. Haloboranes derived from thexylborane are also employed in alcohol synthesis ⟨94TL2⟩. Vinyl allyl ethers react with thexylborane to give 1,3-diols in good to excellent yield (30–90%) and with, in some cases, superb diastereoselectivity (up to 200 : 1) in favour of the *syn*-isomer ⟨89CC1429⟩ as predicted by theoretical studies ⟨84T2257⟩. Mesitylborane has been claimed to offer similar advantages to thexylborane, with an enhanced stability ⟨94AG(E)851⟩. The borane derived from hydroboration of cycloocta-1,5-diene is 9-borabicyclononane, 9-BBN (**3**), which is air-stable, commercially available and widely used in synthesis ⟨73JOC4092⟩. Monochloroborane is less sensitive to steric influence and more susceptible to electronic influence; as its ether complex the reagent is as selective as disiamylborane in its reaction with terminal alkenes ⟨73JOC182⟩.

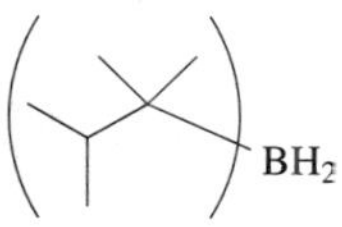

thexylborane

(2)

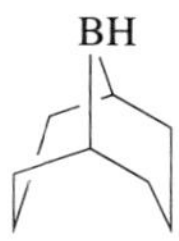

9-borabicyclononane

(3)

Alkynes are converted to 1,1-diboranes upon reaction with diborane. These diboranes are converted to primary alcohols upon usual peroxidative workup (Scheme 2) ⟨67JA291⟩.

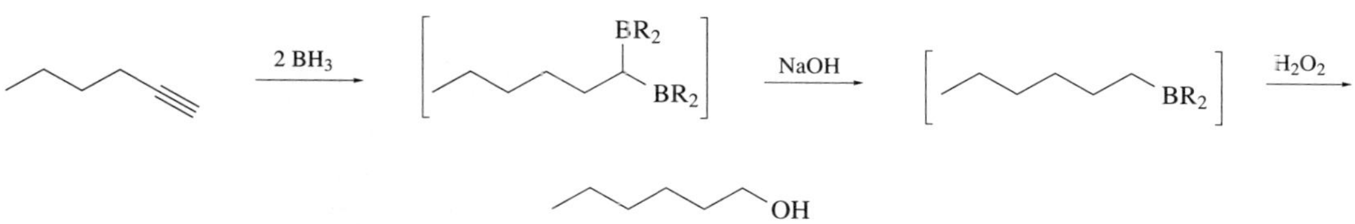

Scheme 2

2.02.1.2.1 *Stereo- and enantiocontrol in hydroboration*

Hydroboration is a *syn*-addition process, with addition occurring from the alkenic face of least hindrance. The reaction of borane with 2-pinene is a good example of the regio- and stereoselectivity seen in such reactions; the initial product of addition is diisopinocampheylborane (Ipc_2BH), arising from regio- and diastereoselective addition (Scheme 3). When obtained using the enantiopure alkene, the resulting borane is capable of carrying out asymmetric hydroboration reactions ⟨B-83MI 202-01⟩; indeed, the reaction of *cis*-but-2-ene with Ipc_2BH was probably the first example of *reagent-controlled* asymmetric synthesis ⟨64JA393, 64JA397⟩. Reasonable levels of enantioselectivity are usually only possible when *cis*-alkenes are employed; the area of asymmetric hydroboration has been thoroughly reviewed. Brown has reported 2-isopropylapopinocampheylborane as an improved reagent for asymmetric hydroboration ⟨94TL4715⟩. Hydroboration of allylic ketones by Ipc_2H is followed by an intramolecular reduction, leading to 1,4-diols in excellent enantiomeric excess ⟨94JOC2676⟩. Monoisopinocampheylborane ($IpcBH_2$) reacts with 1-heterocycloalkenes at −25°C to give *trans*-2-heteroarylcycloalkenyl isopinocampheylboranes, which are oxidised to the corresponding alcohols of moderate optical purity ⟨88BCJ93⟩. Enantiopure boranes are available from other terpenes; for instance, (1*S*)-di-2-isocaranylborane (2-$dIcr_2BH$) and the corresponding regioisomer (4-$dIcr_2BH$) crystallise in high enantiomeric purity from the reaction of (+)-2- and (+)-3-carene (respectively) with $BH_3 \cdot Me_2S$ in THF at 0°C. In hydroboration of representative alkenes, caranylboranes are less reactive than Ipc_2BH, with 4-$dIcr_2BH$ being much less reactive than 2-$dIcr_2BH$; the latter was found to be only slightly inferior to Ipc_2BH in terms of the enantiomeric purity of the alcohols produced after oxidation, but gave products of *opposite* configuration to that obtained with the campheyl borane ⟨88JOC2911⟩. Dithiane-substituted monoalkylboranes have been reported as useful reagents in asymmetric hydroboration and carbonyl reduction reactions (Equation (10)). Thus, reaction of borane with the dithianylmyrtanyl derivative (**4**) leads to the preparation of crystalline stable (+)-(1*R*, 2*S*, 3*R*, 5*R*)-[2-(1,3-dithianyl)]myrtanylboranes (**5**) in good yield, which hydroborate trisubstituted alkenes in acceptable enantiomeric excess (83%). Prochiral ketones are reduced with good diastereoselectivity (up to 99%) ⟨90JOC2855⟩.

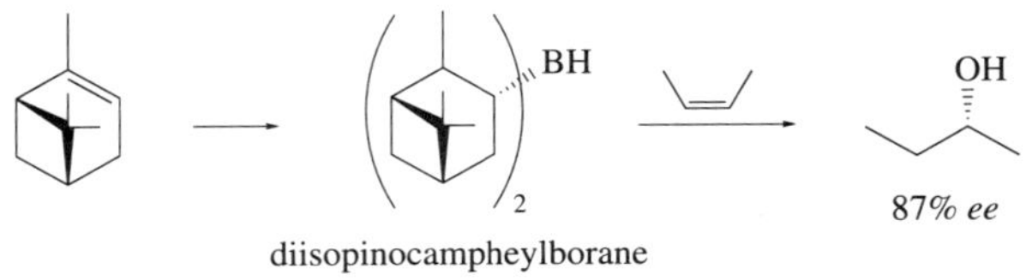

Scheme 3

(10)

A reversal in the usual regiochemistry of hydroboration may be catalysed by transition-metal complexes. The observation that Wilkinson's catalyst catalysed the addition of catecholborane to alkenes at ambient temperature was first made in 1985 ⟨85AG(E)878⟩; this observation led workers to examine the reaction of allylic alcohols with the same borane using rhodium catalysis in the expectation that the presence of catalyst would allow a directing effect. Thus the reaction of O-TBDMS-cyclohex-2-en-1-ol with 9-BBN gave mainly the *trans*-1,2-diol as product, whereas the catalysed reaction afforded mainly the *trans*-1,3-diol. A reversal of stereoselectivity was observed in the hydroboration of acyclic allylic alcohols (Equation (11)) ⟨88JA6917⟩. When the process was modified to allow deuterium incorporation into the product alcohols, it was found that deuterium was incorporated into *both* positions of the double bond of the substrate. The authors suggest that the hydride/deuteride transfer step of the proposed mechanism is reversible, and that the degree of reversibility is closely related to the structure of the alkenic substrate ⟨90JOC2280⟩. The reactions of alkenes with catecholborane in the presence of enantiopure rhodium catalysts have been studied. Yields of alcohols were generally excellent, but enantiomeric excesses were only moderate ⟨88JOC5178⟩.

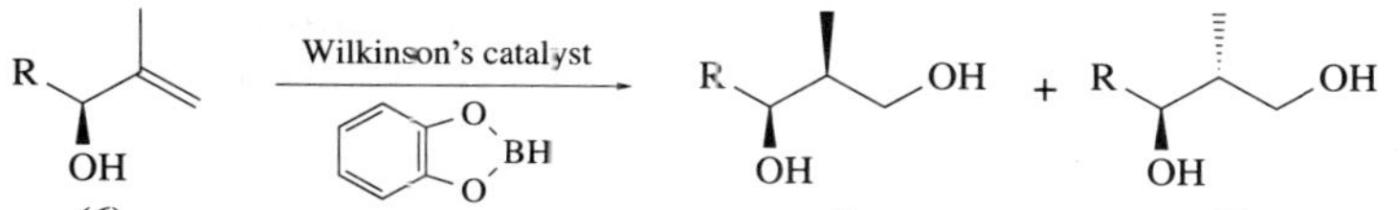

(6) 8 : 92 (11)

The influence of the R group in **(6)** on the face selectivity of the transformation has been investigated ⟨89TL395⟩. Interestingly, oxidation of benzylic alcohols (R = Ph) confers an unusual diastereoselectivity, with a dominance of *anti*-isomer rather than the more often found *syn*-isomer. The authors offer a description of the transition state of the reaction and conclude that the *shape* of the R group is rather more important than the *size* of R when assessing diastereoselectivity. In any case, the good to excellent diastereoselectivities seen in this reaction make the process powerful and of great utility to the synthetic chemist.

Allylic amines react with catecholborane under catalysis to provide (after oxidative workup) 1,3-amino alcohols with a *syn*-selectivity in good to excellent yield ⟨89TL5857⟩. Enantiopure allylic amines may be obtained from (*S*)-*Z*-amino acids, allowing easy access to optically pure aminols. A model to allow accurate prediction of the sense of diastereoselectivity is available ⟨89TL5861⟩. Cationic iridium catalysts also facilitate hydroboration, and under these reaction conditions the presence in the alkenic substrate of groups capable of ligating leads to highly directed addition to cycloalkenes ⟨92JA4042⟩; for structural analysis of such systems, see ⟨93CJC930⟩. A variety of lanthanide complexes mediate hydroboration; a common feature of all the catalysts is the use of polymethylated cyclopentadienyl ligands. The fulcrum of the process is the formation of a dicyclopentadienyllanthanide hydride which adds regiospecifically across the alkene, producing an alkyllanthanide which is converted by a second equivalent of catecholborane into an alkyl borate ⟨92JA9220⟩. Samarium(III) ⟨93JOC5307⟩ and nickel species ⟨94SC1019⟩ also catalyse hydroborations using catecholborane. Hydroboration of 1,3-dienes using catecholborane and enantiopure cationic rhodium catalysts has been reported, but the enantiomeric excesses of the resulting 1,3-diols were low (~ 50%) ⟨91TL3387⟩. The catalytic asymmetric hydroboration of styrenes using rhodium catalysis produces predominantly (*R*)-2-methylbenzyl alcohols in good yield (54–98%) and with moderate to good enantioselectivity ⟨89JA3426⟩. Asymmetric rhodium-catalysed hydroborations using axially asymmetric isoquinolinylnaphthyl ligands show good enantioselectivity ⟨93CC1673⟩.

Lithium borohydride is also an efficient catalyst for hydroborations using catecholborane in that reaction of trisubstituted alkenes is rapid, in contrast to the reaction mediated by rhodium catalysts ⟨91CC205⟩. Zinc borohydride supported on silica gel also mediates the hydroborative hydration of alkenes ⟨94T6579⟩.

Alkenes may be hydroborated using a mixture of sodium perborate and acetic anhydride to give good yields of vicinal acetoxy alcohols. Performing the reactions in dichloromethane as solvent gives the corresponding epoxide ⟨88TL2967⟩.

The reaction between sodium borohydride and mercury acetate at 0 °C leads to acetoxyborane, $AcOBH_3$, which performs as a selective hydroborating reagent, preferentially reacting at more substituted alkenes. Intermediate boranes may be converted to iodides ⟨88JOC187⟩.

Enamines are hydroborated to give 2-aminoalcohols; pyrolytic elimination of the amine species gives allylic alcohols in good overall yield (Equation (12)) ⟨71BSF1649⟩.

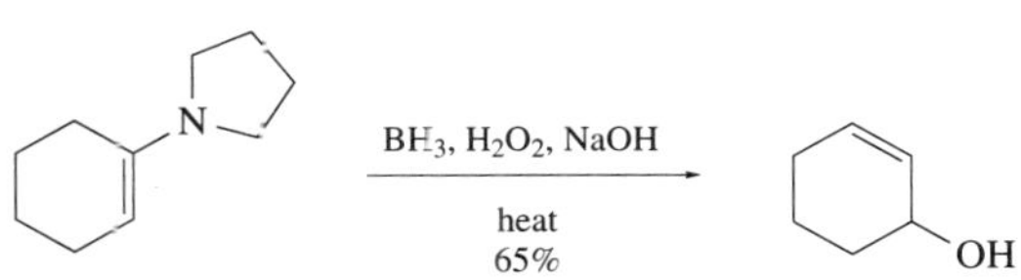

(12)

Vinyl boranes undergo Diels–Alder cycloadditions, leading to cycloalk-3-enylboranes difficult to prepare by direct hydroboration ⟨92TL1017⟩. Fullerenes may be hydroborated to fullerols ⟨94CC463⟩.

Alane (AlH_3) additions to alkenes are also feasible, but the process is not widely used for preparation of alcohols. Ziegler telomerisation of ethylene with triethylaluminum is the major industrial method for preparation of *n*-alkan-1-ols ⟨60LA(629)241⟩. The reaction proceeds via a hydroalumination reaction of ethene, with the resulting alkyl aluminum species being readily oxidised by air to give aluminum alkoxides, which are hydrolysed to alcohols.

Oxidation of alkenes by molecular oxygen is an efficient process in the presence of bis(trifluoroacetylacetonato)cobalt(II) ($Co(tfa)_2$) ⟨89CL757⟩. The products of the reaction are the more substituted alcohols in good to excellent yield.

The reaction of alkenes with selenenyl chlorides may be controlled ⟨89TL3175⟩; use of excess phenylselenenyl chloride can directly furnish *trans*-1,2-hydroxychloride while one equivalent of selenenyl chloride gives hydroxyselenides.

The Sharpless kinetic resolution of secondary allylic alcohols has been examined with a view to facilitating preparations of enantiopure *cis*-allylic alcohols. Kinetic resolution of *cis*-substituted allylic alcohols is often troublesome (as in the reaction of dicyclohexyl alcohol, Scheme 4) but an isomerisation reaction of the *trans*-isomer allows efficient preparations of *cis*-isomers via a tellurium-mediated transposition reaction. The course of the reaction is predictable, with *erythro*-epoxy alcohols always giving *cis*-allylic alcohols ⟨90TL5603⟩.

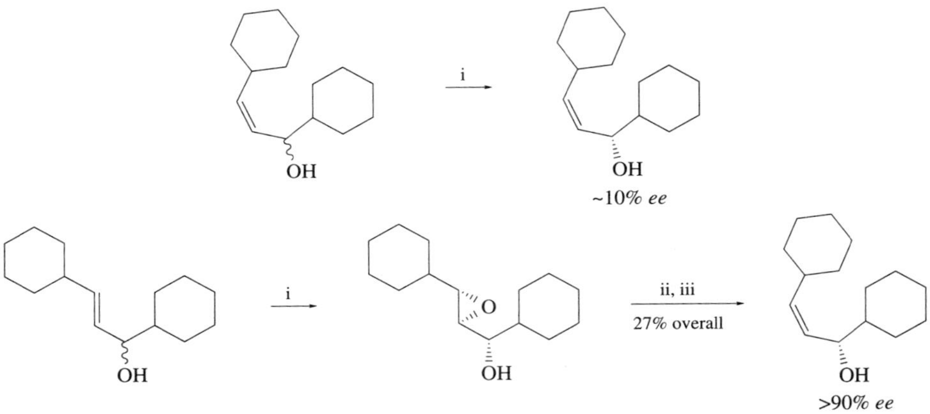

i, Bu^tO_2H, (+)-dipt, $Ti(OPr^i)_4$; ii, MsCl; iii, Te, $NaBH_4$, DMF

Scheme 4

The Reppe alcohol synthesis involves the addition to alkenes of carbon monoxide in the presence of an iron carbonyl catalyst ⟨53LA(582)133⟩.

Enantiopure oxaziridines have been demonstrated as excellent reagents for asymmetric oxidations of enolates to 2-hydroxyketones ⟨92CRV807, 93PAC633⟩. The molybdenum pentoxide–pyridine–HMPA (MoOOPH) complex has been widely used in the hydroxylation of enolate anions. For instance, alkylations of dianions derived from aspartic acid diesters have been employed in the stereodivergent preparation of (3*R*)- and (3*S*)-3-hydroxyaspartates ⟨92TL4637⟩. An alternative procedure which avoids the use of HMPA has been devised ⟨90SL107⟩.

(*S*)-2-Hydroxycycloalkanones may be prepared with virtually complete enantioselectivity via the rhenium-mediated oxidation of cycloalkanone ketals derived from enantiopure 1,2- and 1,3-diols ⟨92TL7823⟩. The reaction proceeds through the (2-hydroxyalkyl)enol ether. Ketones may be oxidised to the corresponding 2-hydroxy compounds by reaction with hypervalent iodine reagents ⟨86ACR244⟩. Dioxiranes will also oxidise enolates to hydroxycarbonyl compounds ⟨91TL715, 94AG(E)69⟩.

(i) Dihydroxylations

There many reagents which will convert alkenes to 1,2-diols; stereocontrol is usually possible, enabling specific preparation of either *cis* or *trans* isomers. Reviews of dihydroxylation have been compiled ⟨60MI 202-01, 80CRV187⟩.

$KMnO_4$ is a well-known dihydroxylating agent, but this reagent does not generally give good yields of *cis*-hydroxylation with a wide range of substrates, although certain alkenes (particularly lipophobic alkenes, such as alkenoic acids) react to give diols in good yield (Equation (13)). Phase-transfer reagents may facilitate reaction where inhomogeneity is a hindrance ⟨72TL4907, 80TL177⟩. The reaction of alkenes with iodine in the presence of silver salts of carboxylic acids is known as the Prévost reaction. The product of the reaction is a *trans*-dicarboxylate, which may be hydrolysed to give a *trans*-diol. In the *Woodward modification*, water is added to the reaction mixture, thereby causing an alteration in mechanism and production of the *cis*-isomer.

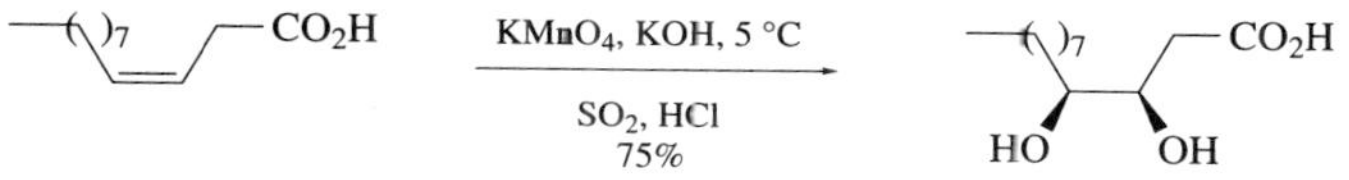

(13)

Both Prévost and Woodward reactions proceed via an iodonium ion to formation of a dioxolanylium cation, formed by silver-enhanced internal nucleophilic displacement of iodide by carboxylate (Scheme 5): the fate of this cation decides the stereochemical nature of the diol formed. Under anhydrous conditions, in nonpolar solvent, the dioxolanylium cation is displaced by a second equivalent of carboxylate to give a *trans*-dicarboxylate (**7**). If water is added to the reaction mixture, hydration of the cation occurs to give a hemi*ortho*-ester, which collapses to give *cis*-diol monoester (**8**). To avoid the expense of silver salts, thallium carboxylates may be employed to good effect ⟨74JCS(P1)1858, 80OS(59)169⟩. Other variations upon the reaction involve the use of *N*-iodosuccinimide in the presence of a carboxylic acid ⟨76TL3661⟩, KIO_3 ⟨75G377⟩ and iodine(III) trifluoroacetate (Equation (14)) ⟨73AG(E)163⟩.

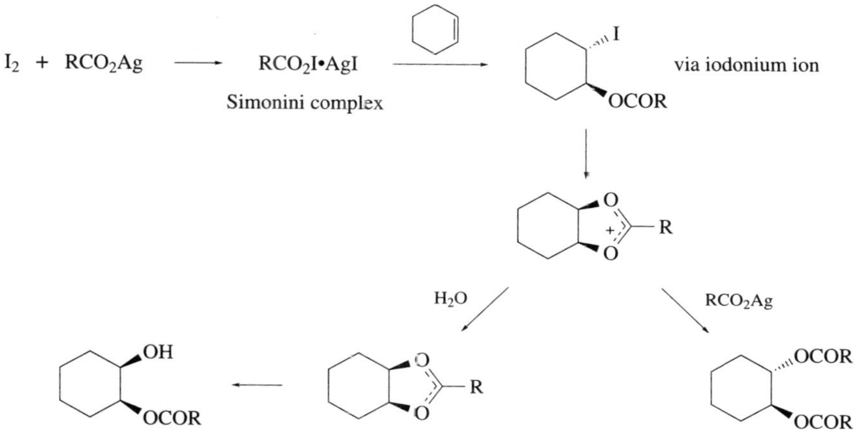

(8) **(7)**

Scheme 5

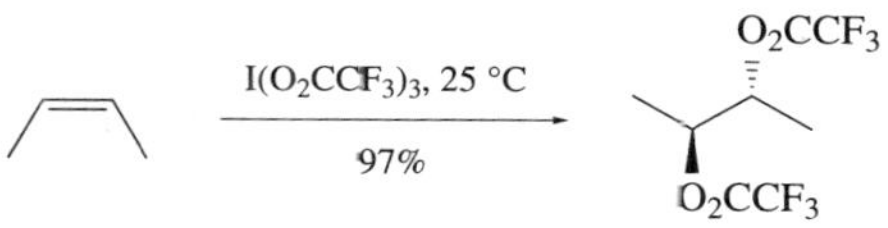

(14)

Reviews covering dihydroxylation involving iodonium ions have appeared ⟨57OR(9)350⟩. Conceptually similar to dihydroxylation via iodonium ions is the epoxidation/hydrolytic ring-opening reaction of alkenes ⟨55OSC(3)217⟩. Reagents capable of carrying out this reaction to give diols directly are *o*-sulfoperbenzoic acid ⟨71TL691⟩ and disuccinoyl peroxide, the latter reagent possessing the not inconsiderable advantages of crystallinity and stability ⟨73S156⟩.

OsO_4 is the reagent of choice for *cis*-hydroxylation of alkenes, but there are problems of toxicity and cost associated with using this substance ⟨80CRV187⟩. The reagent is most often used catalytically in the stoichiometric presence of a cooxidant, such as *t*-butyl hydroperoxide or *N*-methyl morpholine *N*-oxide. Owing to the extremely high vapour pressure, sublimation of OsO_4 is always a problem and this ready sublimation is a common hindrance to even the simplest techniques, such as weighing the reagent. Nevertheless, OsO_4 remains ubiquitous in synthetic methodology and developments in *asymmetric* osmylations have served to emphasise the reagent's utility (see below).

Cyanoferrates may be employed as stoichiometric oxidants in the osmium-catalysed vicinal hydroxylation of higher alkenes. The use of potassium hexacyanoferrate(III) allows preparation of diols in good to excellent yields (Equation (15)) ⟨90JOC766⟩; the reaction is accelerated by addition of equimolar amounts of tertiary amines (such as quinuclidine or dabco).

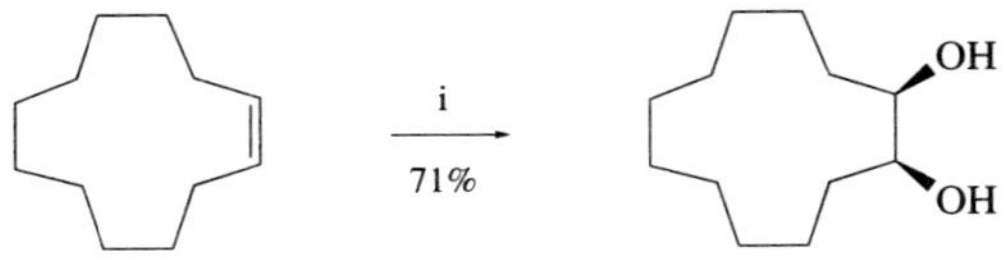

(15)

i, OsO_4 (1.25 mol%), $K_3Fe(CN)_6$ (3 equiv.), Bu^{t}OH, H_2O, K_2CO_3, RT

2.02.1.2.2 Substrate control in osmylation

Where there is considerable stereochemical influence present in a substrate, extensive stereocontrol in osmylation reactions is possible ⟨83TL3951⟩. For instance, cyclohex-2-en-1-ol reacts to give only 1,2-*anti*-2,3-*anti*-cyclohexane-1,2,3-triol ⟨83TL3947⟩. Other allylic alcohols also show selectivity in dihydroxylation ⟨94TL2495⟩. Reaction of secondary allylic alcohols with osmium tetroxide proceeds in high yield to give products whose stereochemistry is greatly influenced by the nature of the allylic substituent R (Scheme 6). Thus, 1,1-disubstituted alkenes such as (**8a**) undergo highly diastereoselective osmylations when the allylic substituent R contains a hydroxyl group, and the hydroxyl of the allylic alcohol is unprotected. Protection of the hydroxyl groups leads to severe diminution of diastereomeric excess ⟨90JOC1698⟩.

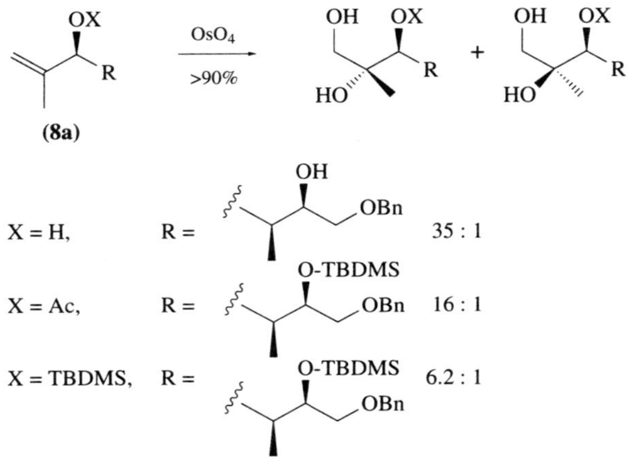

Scheme 6

A similar vicinal hydroxylation of 2-hydroxycrotylsilanes gives good yields of 1,2-*anti*-2,3-*syn*-triols (**9**) of high diastereomeric purity (Equation (16)). The selectivity is reduced if the 2-hydroxy moiety of the allylsilane is protected ⟨90JA4873⟩. Osmylations of 4-hydroxyenoates proceed stereoselectively in the case only of (*E*)-isomers (Equation (17)) ⟨88T491⟩. The predominance of the *arabino*- or *xylo*-triol in the product mixture depends on the substituent R^2 present upon the pre-existing hydroxyl group. (*Z*)-Alkenes give 1 : 1 mixtures of stereoisomers. The model first suggested by Stork ⟨83TL3951⟩ rationalises this observation. A hypervalent Sn–O interaction has been invoked to explain the reactivity of certain stannylalkenes in the presence of osmium tetraoxide (Equation (18)) ⟨88JA4606⟩.

(16)

(**9**)

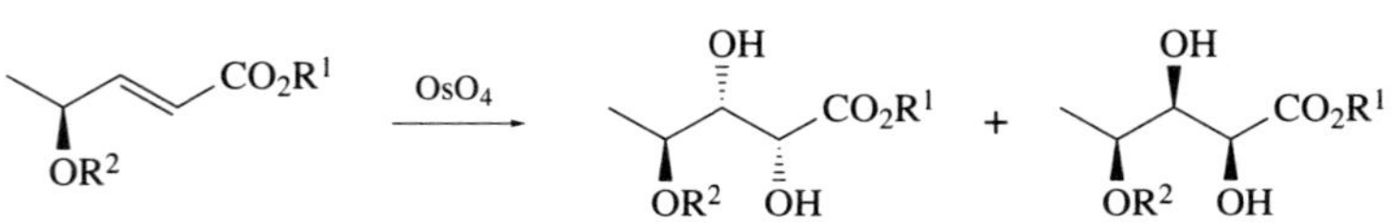

(17)

arabino *xylo*

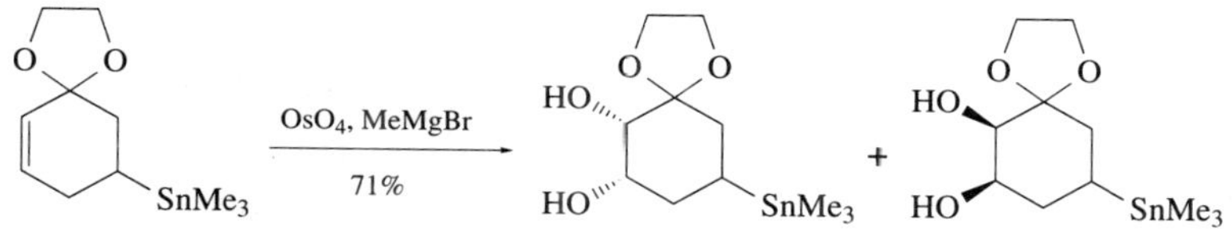

(10)

(18)

The presence of a dimethylchlorostannyl group causes very high stereoselectivity in the *cis*-hydroxylation reaction, while a trimethylstannyl group exerts virtually no stereocontrol. The authors suggest that a structure such as (**11**) must be involved; this hypervalent interaction would shield the lower face of the alkene, it is claimed, thereby leading to high discrimination.

Dienes, but not trienes and tetraenes, undergo highly stereoselective osmylation reactions. Thus (*E*,*E*)-1,4-diphenylbutadiene reacts with osmium tetroxide to give almost exclusively the 1,2-*syn*-2,3-*anti*-3,4-*syn* tetrol (Equation (19)) ⟨91TL1003⟩.

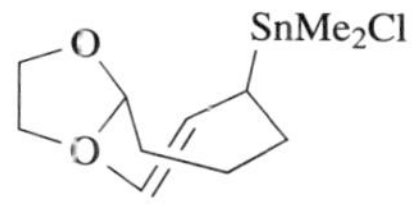

(**11**)

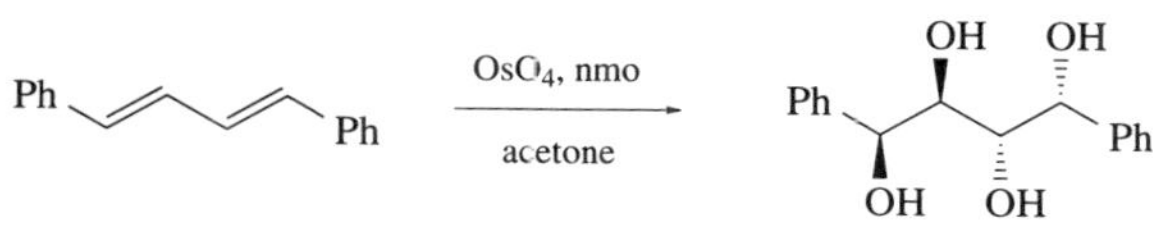

(19)

Highly diastereoselective hydroxylations are reported to occur in the reaction of triene organometallic complexes such as (**12**) (Equation (20)). The uncomplexed alkenic bond of the triene is regioselectively oxidised with high diastereoface selectivity to the *cis*-1,2-diol (**13**). The stereochemistry of the alkene is crucial, since the (*Z*)-isomer is converted in excellent yield to a single diastereomer (**13**) while the (*E*)-isomer shows only 90% diastereoface selectivity. The reason for this phenomenon is the lack of rotation possible in (*Z*)-(**12**); rotation about the adjacent allylic single bond leads to conformers in which there are severe steric interactions which would destabilise this conformation. The (*E*)-isomer has not as severe a steric interaction, and this allows a small amount of flexibility of rotation. Since both stereoisomers are available enantiopure, this method allows for enantiospecific functionalisation of 1,3,5-hexatriene-containing molecules ⟨89AG(E)755⟩.

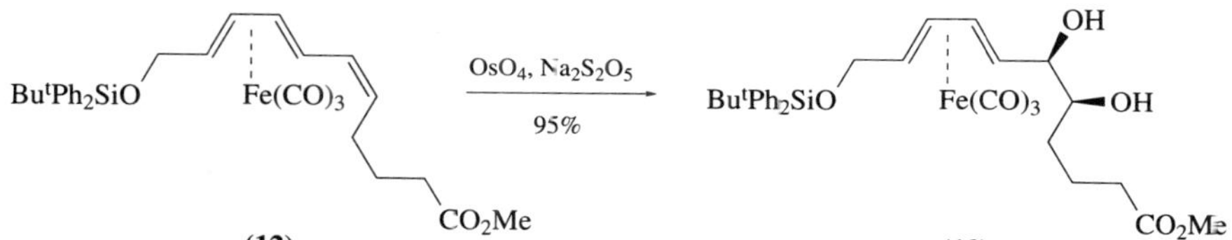

(**12**) (**13**)

(20)

Torsional effects have been proposed to exert stereochemical influence on the course of *cis*-hydroxylation of flexible 3-substituted cyclopentenes. In particular, introduction of a methyl group in place of a hydrogen atom leads to a startling inversion of diastereoselectivity, as shown below (Equation (21)) ⟨89TL7385⟩.

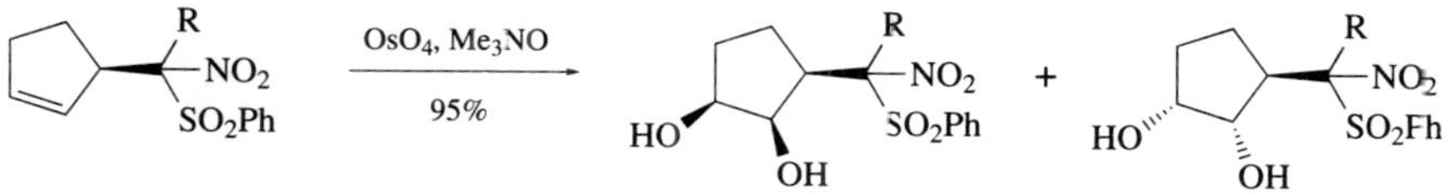

(21)

R = H	92 : 8
R = Me	≥5 : ≥95

2.02.1.2.3 *Reagent control in osmylation*

Sharpless has developed an asymmetric *cis*-dihydroxylation (ADH) of alkenes by means of *ligand accelerated catalysis* ⟨88JA1968, 92TA1317, 94CRV2483, 94T3639⟩. The osmylation of prochiral alkenes proceeds in good yields (>80%) and enantiomeric excesses of 20–88% when carried out in the presence of chinchona alkaloid derivatives (Equation (22)). Very small amounts of osmium are required because of the accelerating effect of the chiral ligands. The sense of asymmetric induction was found to be highly predictive (Scheme 7): quinidine derivatives lead to (*R*)-diols, while quinine ligands produced mainly (*S*)-diols ⟨94TL425⟩. Ferrate oxidants have been found to be optimal cooxidants ⟨90TL2990⟩. In particular, the use of potassium ferricyanide as oxidant for osmium-catalysed dihydroxylation leads to an increase in levels of asymmetric induction for all alkenes (Scheme 8). It is suggested that the effect is caused by suppression of the so-called second cycle (which leads to decreased selectivity); the need for slow addition of alkene to the reaction mixture (a vital requirement in an nmo-oxidised reaction (nmo = *N*-methylmorphine *N*-oxide)) is thereby eliminated. The structures of the OsO_4–alkaloid complexes responsible for the transformation have been examined in detail ⟨89JOC2263⟩. The use of aryl *ethers* of dihydroquinidine, rather than chlorobenzoates, leads to improvement in the asymmetric dihydroxylation of dialkyl-substituted alkenes, substrates whose oxidation proceeds with enantioselectivity inferior to their diaryl counterparts ⟨90TL3817⟩. Polymer-bound alkaloids are available to carry out the ADH reaction ⟨90TL3003⟩. Dramatic improvements in enantioselectivity of ADH reactions were observed when phthalazine derivatives of cinchona alkaloids were employed. Together with the discovery thatmethanesulfonamide greatly accelerates the attainment of reaction completion, this improvement means that highly enantioselective asymmetric *cis*-dihydroxylation of a huge range of alkenes is possible ⟨92JOC2768⟩. Two so-calledAD-mixes (reagent mixtures containing the requisite asymmetric catalyst, a catalytic quantity of potassium osmate and stoichiometric ferrate cooxidant) are now commercially available. A comparison of the various ligands available for asymmetric dihydroxylation has been made ⟨94TL543⟩.

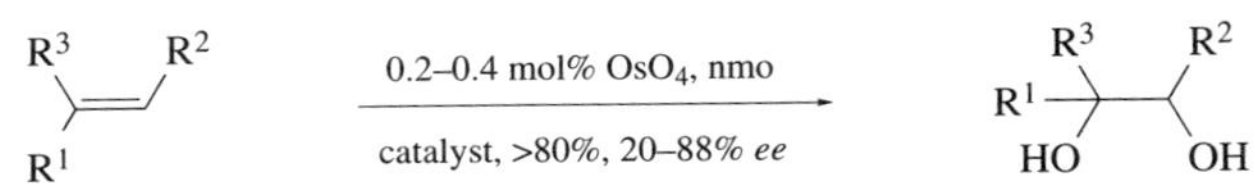

(22)

catalyst = *p*-chlorobenzoyl derivatives of dihydroquinine and dihydroquinidine

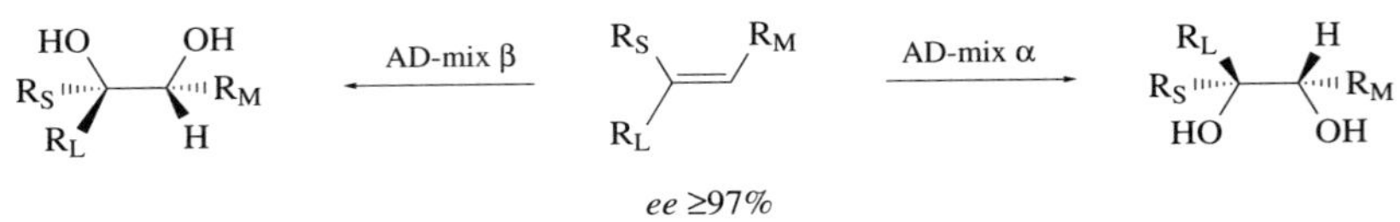

Scheme 7

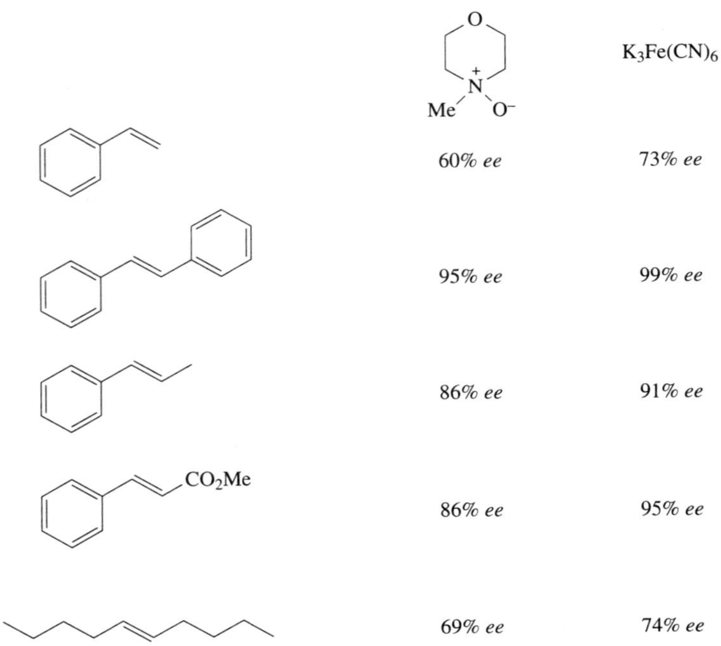

Scheme 8

Full details of the preparation and crystal structures of the cinchona alkaloids used in AD-mix reactions are available ⟨93JOC844⟩.

ADH may be applied to synthesis of chiral 1,2-dihydroxyesters and (ultimately) enantiopure β-lactams (Scheme 9) ⟨90TL4317⟩.

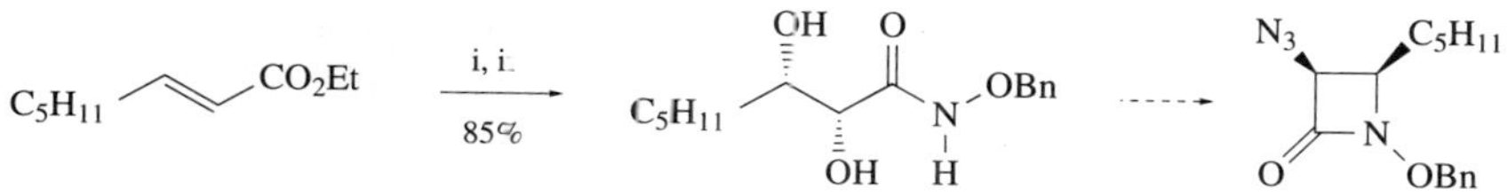

i, ADH; ii, H_2NOBn, water-soluble diimide, LiOH, THF/H_2O

Scheme 9

The AD-mix reaction of β,γ- and γ,δ-unsaturated esters has been shown to be highly enantioselective ⟨92TL6407⟩ and has been used to synthesise all four isomers of disparlure ⟨92TL6411⟩.

To accomplish highly selective hydroxylation of α,β- and β,γ-unsaturated *amides*, a modified AD-mix employing a fivefold increase in the ligand and potassium osmate is required ⟨93TL2079⟩. When this new, improved mix was used, both α,β- and β,γ-unsaturated amides reacted in good chemical yield and with high levels of enantiomeric excess. The products of these reactions were easily dehydrated to give enantiopure γ-hydroxy-α,β-unsaturated amides ⟨93TL2083⟩. An interesting effect is in operation when the AD protocol is applied to enol ethers. 2-Hydroxyketones are produced in high enantiomeric purity, but the reaction of enol ethers derived from phenyl benzyl ketone gives the same absolute stereochemistry at the new chiral centre, *regardless* of enol ether geometry (Scheme 10). Thus, the (*Z*)-enol ether (**14**) gives the (*R*)-hydroxyketone, as does the (*E*)-enol ether (**15**). These substrates are expected to give enantiomeric products ⟨92JOC5067⟩.

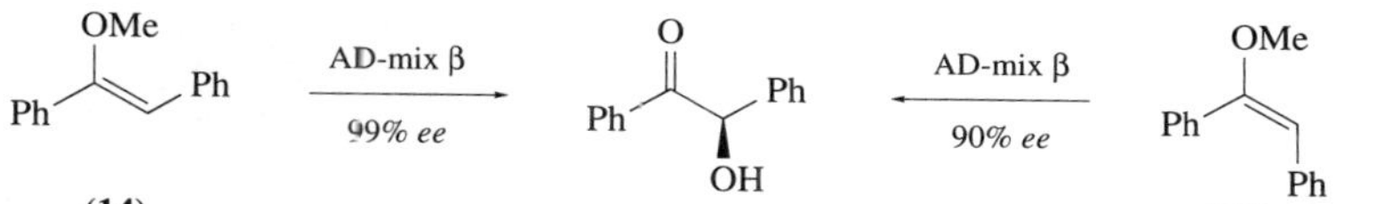

(**14**) (**15**)

Scheme 10

The selective ADH of dienes is a highly viable and valuable synthetic process ⟨92JA7570⟩. Both conjugated and nonconjugated dienes and trienes react selectively, with the alkene of greater substitution reacting preferentially. When an electron-withdrawing substituent lies in conjugation with the diene, the alkene most distant from the substituent is hydroxylated.

The mechanism of AD is not proven. Frontier molecular orbital treatments have been used to consider the factors involved ⟨90TL6417⟩, while Corey has experimented with the origins of the high enantioselectivity in the AD reaction ⟨90TL2665, 93JA3828, 93JA12579⟩. To this end, monomeric derivatives of cinchona alkaloids (e.g., (**16**)) were prepared and used in dihydroxylations. AD reactions using (**16**) were 100-fold slower than the reactions in which dimeric alkaloids were employed. The Harvard workers used this evidence to support their suggestion of bridged osmium complex (**17**) as the active ingredient of AD-mixes.

Other ligands have been revealed as highly effective in executing enantioselective dihydroxylation

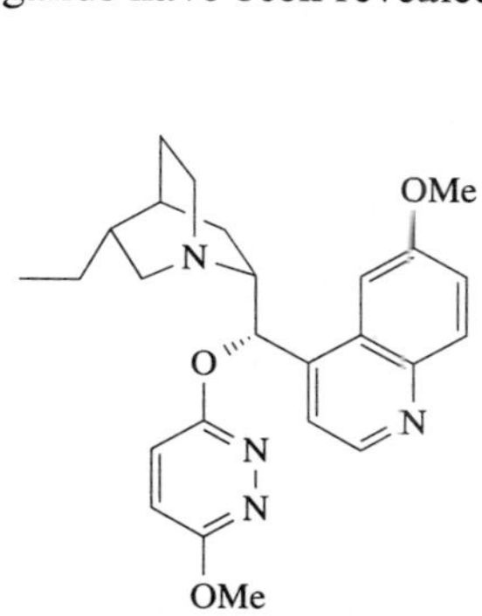

(**16**)

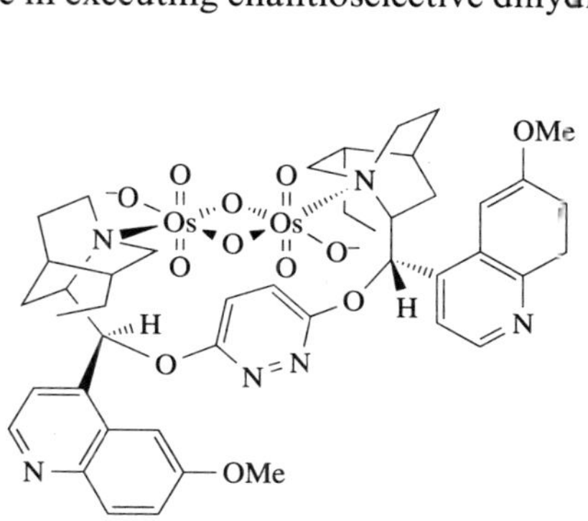

(**17**)

reactions. Thus, the use of C_2-symmetric chiral diamines (derived from *trans*-2,3-diarylpyrrolidines) has been reported ⟨89CC1921, 90TL1741⟩. The enantiomeric excesses of this reaction are moderate.

Corey has employed a chiral controller molecule to good effect in ADH of various prochiral alkenes ⟨89JA9243⟩ with osmium tetroxide at very low temperature in the presence of the chiral diamine (**18**) furnishing *cis*-1,2-diols with good to excellent enantiomeric excess (82–98%) and in high chemical yield (75–95%) (Scheme 11). The amine is present in stoichiometric amounts. The author suggests a rational model to explain the observed selectivity involving an octahedrally coordinated OsO_4–diamine complex as the reactive species.

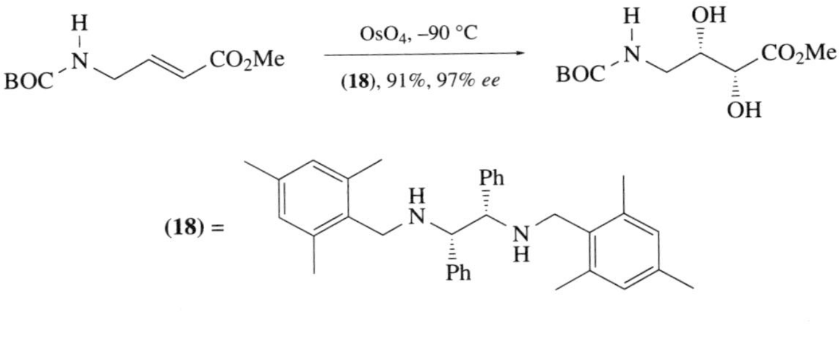

Scheme 11

Bis-piperazine (**19**) mediates dihydroxylation of representative alkenes in good chemical yield and with good to excellent enantiomeric excesses. The best substrates for the reaction were *trans*-1,1-disubstituted alkenes in which at least one of the substituents is aromatic. The limiting feature of this work was the need for stoichiometric quantities of ligand ⟨92TL4021⟩, which also applies to the otherwise impressive enantioselective osmylation of *trans*-stilbene in the presence of enantiopure *trans*-[*N*,*N*′-di-(3,3-dimethylbutyl)]cyclohexane-1,2-diamine at a very low temperature (Equation (23)). Although in this reaction *cis* substrates reacted poorly (as is often the case), there were several alkenes which reacted with greater efficiency than with previously reported systems. Thus, fumarates react with high enantiomeric excess (96%) as do other previously unreactive AD substrates. The authors isolated a derivative (**20**) of the first product of the reaction and solved its x-ray structure, thereby demonstrating the origins of enantioselectivity ⟨93JOC1991⟩. Bis-aziridines have been utilised as ligands for asymmetric dihydroxylations ⟨94TL4631⟩. Aryl borates may be used to assess the optical purities of diols formed from such reactions ⟨94AG(E)1182⟩.

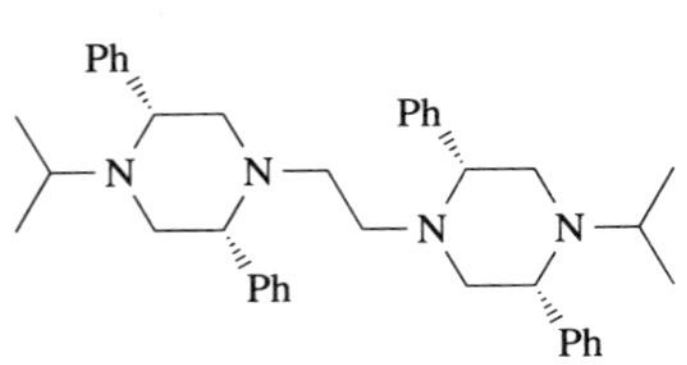

(19)

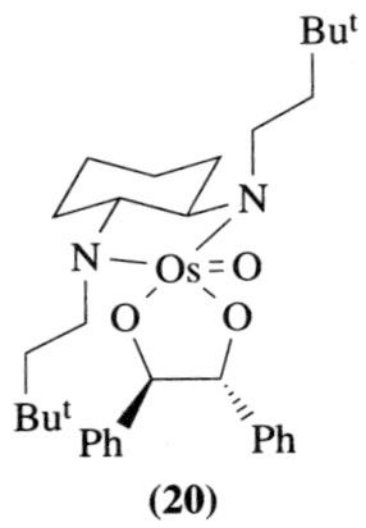

(20)

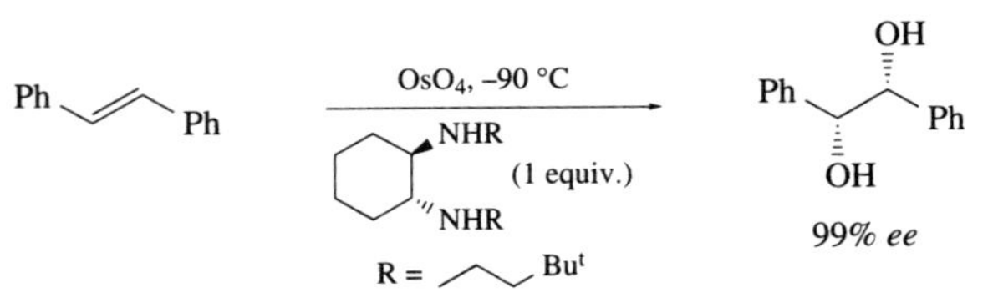

(23)

2.02.1.2.4 *Other additions to alkenes producing diols*

The directed hydrosilylation reaction ⟨88MI 202-01⟩ is a highly selective method for preparation of 1,3-diols ⟨88JA3712⟩. Excellent yields of 2,3-*syn*-1,2,3-triols are obtained from 2-hydroxy enol ethers by oxidation of the initially produced cyclic oxysilanes (**21**) (Scheme 12). Diastereoselectivity is good. Binaphthyl-based catalysts have been used to effect catalytic enantioselective 1,4-disilylation of α,β-unsaturated ketones ⟨88JA5579⟩. The initial products of the reactions were 3-silylenol ethers (**22**) which could be alkylated and oxidised to give 1-substituted 2-hydroxyketones in reasonable enantiomeric excess (74–85%) (Scheme 13).

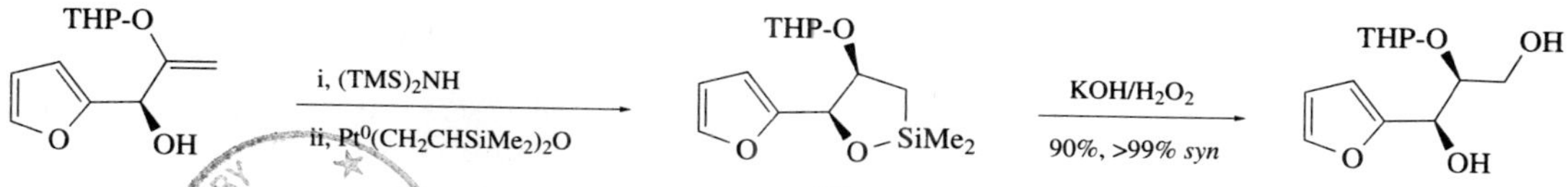

(21)

Scheme 12

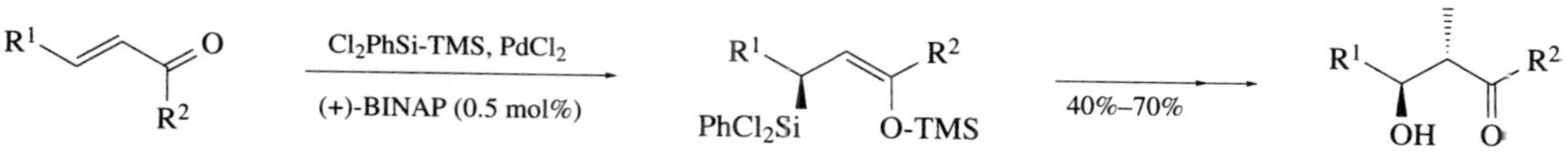

(22)

BINAP = binaphthylphosphine (**27**)

Scheme 13

The intramolecular desymmetrising hydrosilylation reaction of prochiral diallyl alcohols (**23**) has been reported ⟨90TL7333⟩. Cyclic oxysilane (**24**) is produced with complete diastereoselectivity (*syn* : *anti* = >99 : 1) and with high enantioselectivity; such compounds may routinely be converted into (2*S*,3*R*)-2,4-dimethyl-4-pentene-1,3-diols (Scheme 14).

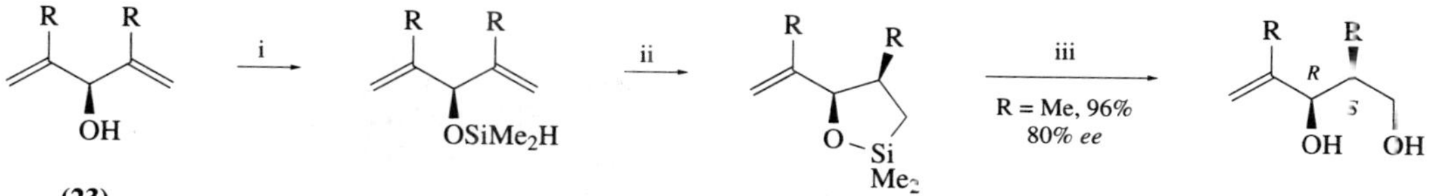

(23) **(24)** *syn* : *anti* = 95 : 5

i, $(HMe_2Si)_2NH$ (**9**); ii, $[RhCl(CH_2{=}CH_2)_2]_2$/(*R*)-BINAP (2 mol%)
iii, 30% H_2O_2, KF, $KHCO_3$

BINAP = binaphthylphosphine (**27**)

Scheme 14

Tungsten reagents (present in catalytic quantity) effect 1,2-dihydroxylation of alkenes, as shown in Equation (24) ⟨89S295⟩, providing 1,2-diols in good yield (71–88%). The extremely high molecular weight of the catalyst may be a drawback. A highly enantioselective and regiospecific dihydroxylation reaction occurs when urethanes derived from geraniol are incubated with *Aspergillus niger* (Equation (25)) ⟨89JOC4686⟩. The authors have no evidence to confirm which of the possible mechanisms (direct dihydroxylation, epoxide formation followed by ring-opening or oxetane formation followed by ring-opening) lead to the observed product, but they comment upon the total absence of any epoxide products which could lead to the diol.

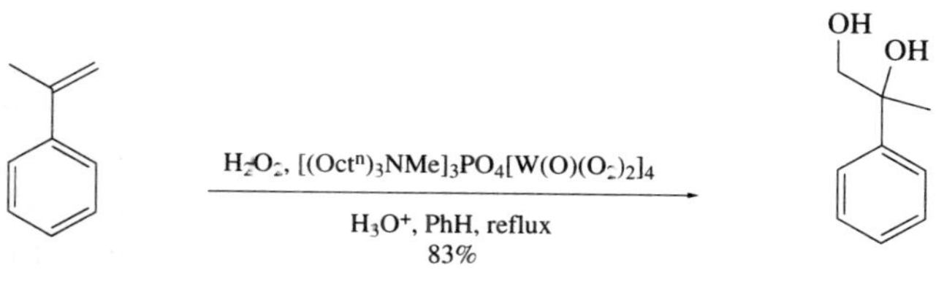

(24)

(25)

(25)

The reaction of oxygen with dienes in the presence of photosensitisers is a cycloaddition allowed under Woodward–Hoffmann rules. The products of such cycloadditions (so-called singlet oxygen reactions) are 1,2-diox-4-enes, which may be reduced to give *syn*-1,4-cyclohex-2-enes in good yield. For instance, *cis-p*-menth-3-ene-1,2-diol has been synthesised from a commercially available diene, via a 1,2-diox-4-ene intermediate (Scheme 15) ⟨89CC1916⟩.

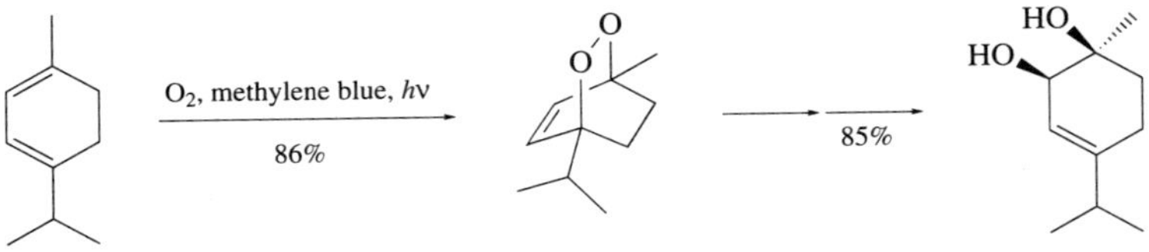

Scheme 15

The reported synthesis of the highly potent nonopioid analgesic epibatidine also employed such a cycloaddition-reduction as its key step ⟨94CC1775⟩.

2.02.1.3 By Addition to Carbonyl Compounds

2.02.1.3.1 Reductive addition

(i) Hydrogenation

This subject has been comprehensively reviewed ⟨91COS(8)139⟩. Aldehydes are relatively easily reduced, for example, by hydrogen in the presence of metal catalysts, such as PtO_2 ⟨23JA1071, 24JA1675⟩, nickel ⟨48JA695⟩ and copper chromite ⟨B-65MI 202-01⟩. Reduction of ketones generally requires more forcing conditions. Esters may also be reduced, to primary alcohols, using copper chromite ⟨54OR(8)1⟩. Conditions required are harsher than for the corresponding reduction of aldehydes.

(a) Asymmetric hydrogenation under heterogeneous catalysis. The asymmetric hydrogenation of ketones using heterogeneous catalysts has been reviewed ⟨91COS(8)139⟩; the best results have been obtained using tartaric acid-modified catalyst. Most reports concerning such reductions are concerned with ketoesters as substrates, since these compounds give best results under the reaction conditions. The absence from a substrate of another group capable of acting as a Lewis base (in addition to that undergoing reduction) is generally found to cause often drastic diminution in enantioselectivity (but see below). The enantioselective reduction of other substrates has also been reported but, for simple alkyl ketones, enantiomeric excesses are moderate (49–74%) ⟨82CL315⟩ and, generally, metal hydride reagents offer better prospects for preparation of enantiopure alcohols.

Enantioselective hydrogenation of ketoesters may also be effected by Raney nickel which has been treated with tartaric acid in a hot aqueous solution of sodium bromide under acidic conditions. Enantiomeric excess is not as impressive as with binaphthyl catalysis, but the reaction may be carried out on kilogram scale (Equation (26)) ⟨91CC795⟩.

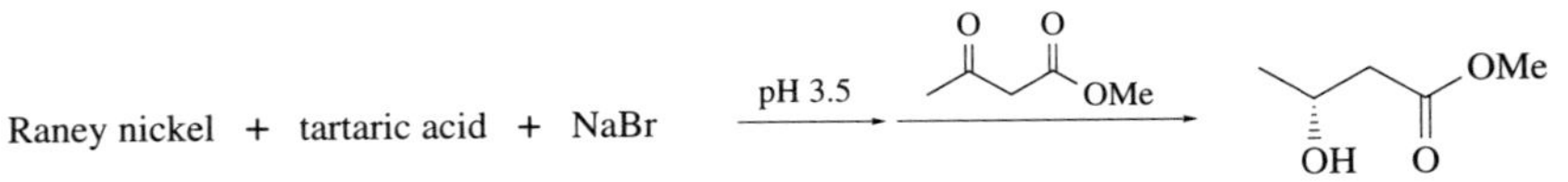

(26)

(b) Asymmetric hydrogenation under homogeneous catalysis. This area of research is rapidly developing and attracts the interest of many workers. The use of soluble rhodium and ruthenium

complexes allows highly enantioselective reduction of many prochiral ketones to be a facile process ⟨91COS(8)139⟩. Almost without exception, the ligands present in such catalysts are enantiopure phosphines.

Binaphthylphosphines (**26**) possess asymmetry due to the restricted C—C bond rotation and may be obtained enantiomerically pure. Such phosphines exert great steric demands when present as ligands in transition-metal complexes. The use of ruthenium complexes containing such phosphines was pioneered by Noyori ⟨87JA5856⟩. 2-Ketoesters, 1,2-diketones and many other substrates are reduced to the corresponding alcohols in excellent yield and (in many cases) with virtually complete enantiocontrol. The availability of both antipodes of the ligand allows great flexibility in use of this synthetic methodology. The stereoselective hydrogenation of ketoesters proceeds via dynamic kinetic resolution (Scheme 16) ⟨89JA9134⟩. Thus, hydrogenation in the presence of binaphthyl catalysts gives 2-hydroxyesters with excellent diastereoselectivity and good enantiomeric excess (~92%). The dynamic kinetic resolution involves equilibration of the starting materials *in situ*, which allows for complete conversion of racemate to product, of either totally *syn*- or totally *anti*-configuration. The application of binaphthyl catalysts to reduction of ketones containing no other ligating group is problematic and enantiomeric excesses of such reactions with ordinary BINAPs are poor. To solve the difficulty associated with hydrogenation of ketones which do not contain another coordinating group, a combination of the cationic BINAP-derived iridium complex (**27**) and a mixed P,N-donor ligand (**28**) has been prepared ⟨93JA3318⟩. Using this combination, the hydrogenation at high pressure of bicyclic ketones proceeded with good enantiomeric excess (often ≥92%); the reduction of monocyclic ketones was less selective. When ligand (**28**) was omitted from reaction mixtures, lower enantioselectives were obtained. *p*-2-Cymene derivatives (**29**) of bisphenylphosphine similarly catalysed the hydrogenation of phenyl glyoxalate and glyoxamides to mandelates and mandelamides ⟨93TL2351⟩ with high enantiomeric excess (≥93%). Glyceryl ethers may be efficiently hydrogenated with high selectivity in the presence of H_2[RuBINAP] to give optically active 2-hydroxy-1,3-diethers (Equation (27)) ⟨91TL4381⟩.

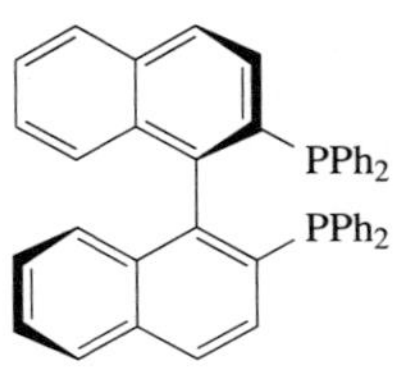

(*R*)-BINAP
(26)

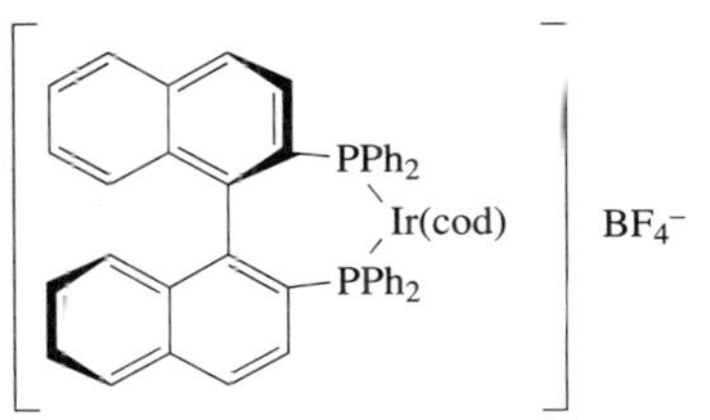

(27)

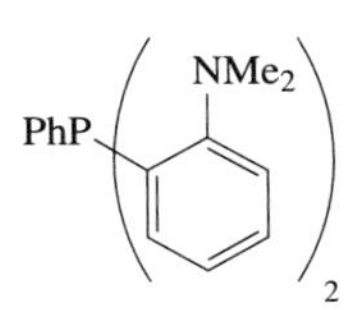

(28)

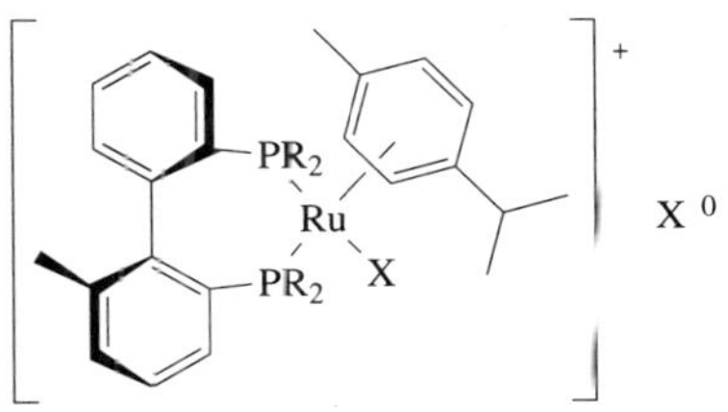

R = cyclohexyl, X = Cl or I

(29)

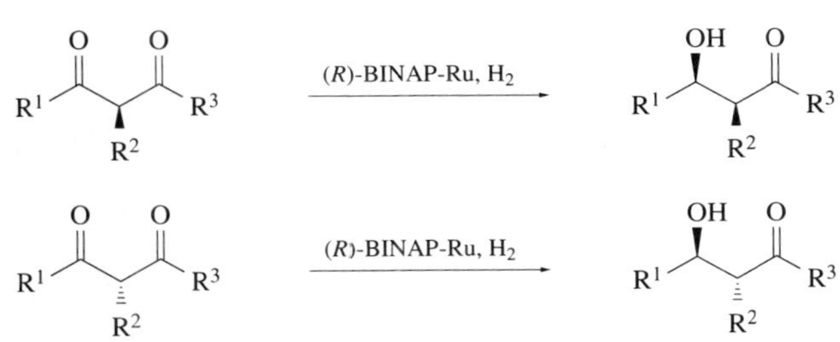

Scheme 16

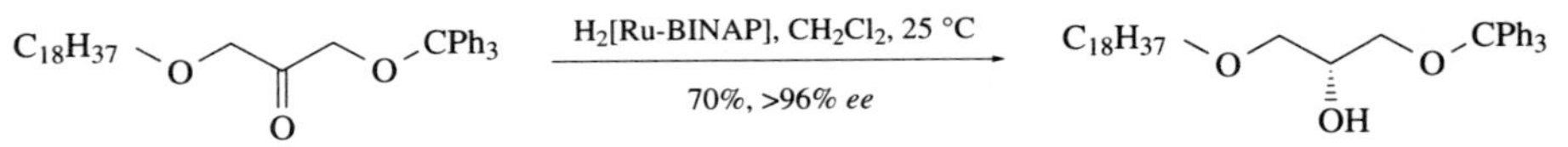

(27)

Modifications to the original protocol using the Noyori catalyst are known: the Ru-binaphthyl species may be prepared *in situ*, and lower pressures may be employed without detrimental effect upon yield or enantioselectivity (Equation (28)) ⟨91TL4227⟩.

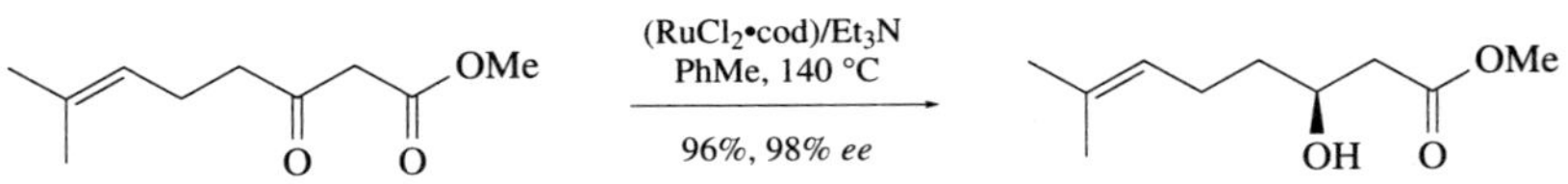

(28)

Rhodium catalysis is also vital in the reduction of 2-amino ketones to the corresponding amino alcohols, using the enantiopure hydroxyproline-derived diphosphine ligand (**30**). Enantioselective reductions of 2-aminoketones proceed to give predominance of (*S*)-configuration at the newly created stereogenic centre in all cases. Enantiomeric excesses are good (81–91%) and the reactions are quantitative ⟨89TL363⟩.

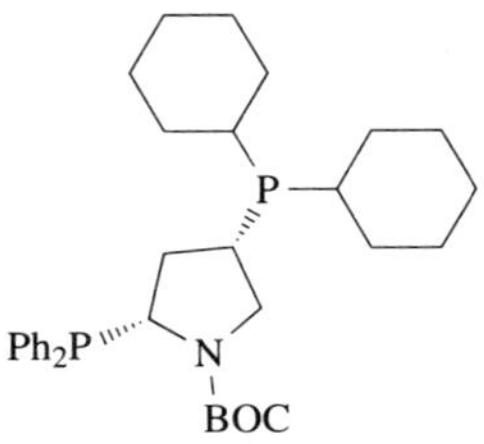

(30)

The preparation of (*S*)-propranolol and a variety of related compounds has been reported in the reaction sequence involving a rhodium-catalysed hydrogenation of *N*-isopropyl-1-phenoxy-3-aminopropaneone (Equation (29)) ⟨90JA5876⟩.

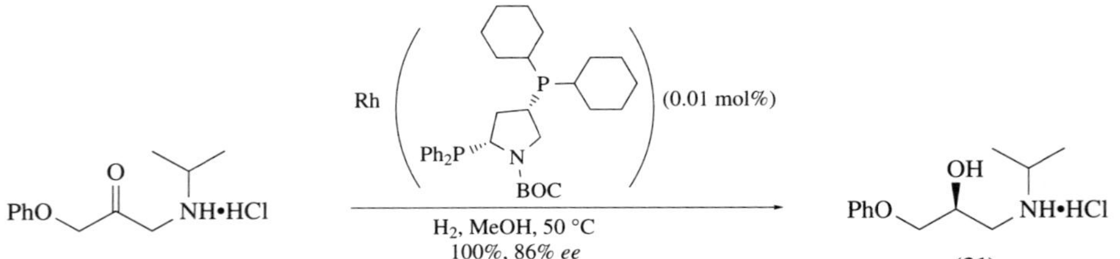

(31) (29)

(*ii*) *Dissolving-metal reduction*

Esters are reduced by sodium to alcohols in the Bouveault–Blanc reaction ⟨03CR(136)1676, 03CR(137)60, 04BSF748, 91COS(8)107⟩. If the reaction is carried out in the presence of an acid, lower temperatures may be used ⟨23RTC1050⟩. Darzens introduced sodium hydride as an effective alternative to metallic sodium ⟨47CR(224)570⟩. Amides also are reduced, to alcohols and amines ⟨04CR(138)148⟩. Acid anhydrides and chlorides are converted to primary alcohols in low yield by reduction with sodium amalgam ⟨1872LA(161)178⟩. Electron-transfer initiated reductions of acyclic ketones have been shown to favour formation of '*anti*-Cram' isomers (Equation (30)) ⟨88JA4475⟩. The mechanistic rationale for this phenomenon suggests that the preferred conformers of the intermediate anions (**32**) are those which lead to such '*anti*-Cram' products. Essentially, this argument means that the products are actually formed by a Cram-like process and that the difference in mechanism of reduction accounts for the different product distribution from that seen where nucleophilic hydride is the reductant.

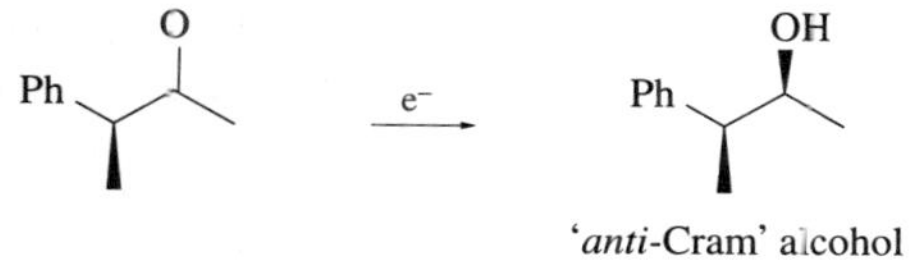

(30)

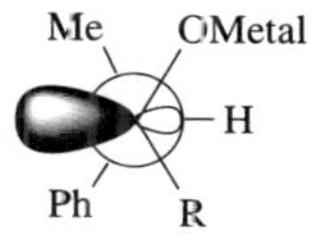

(32)

A combination of antimony trichloride and metallic aluminum or zinc serves as an efficient reducing agent for the conversion of aromatic and aliphatic aldehydes to alcohols (Equation (31)) ⟨90TL1185⟩. Reduction of ketones, aldehydes and acid halides may be mediated by the combination of cadmium chloride and magnesium in water. Simple carbonyl compounds were reduced to alcohols, while acid chlorides were reduced to aldehydes in good yield. Certain epoxides may be regioselectivity reductively cleaved ⟨93TL1681⟩.

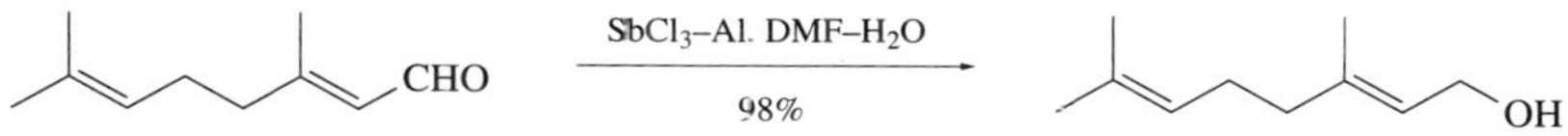

(31)

(iii) Reduction using carbon hydrides

The reduction of carbonyl compounds by delivery of hydride from carbon is a well-known process ⟨91COS(8)79⟩. The reduction of alcohols using aluminum isopropoxide is better known as the Meerwein–Ponndorf–Verley reaction. The reaction is carried out by refluxing the components in isopropanol. The reaction is in equilibrium and the reaction is driven to completion by removal of acetone by distillation. The reduction proceeds through a six-membered cyclic transition state (Equation (32)). Aluminum alkoxides are ideally suited to the reaction owing to their high solubility in organic solvents and their low basicity.

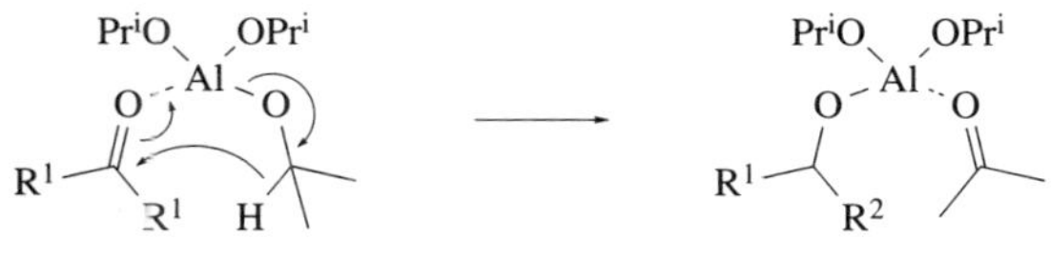

(32)

The replacement of aluminum triisopropoxide with hydrous zirconium oxide has been studied in the gas phase and in solution (Equation (33)) ⟨88BCJ3283, 88CL1633⟩.

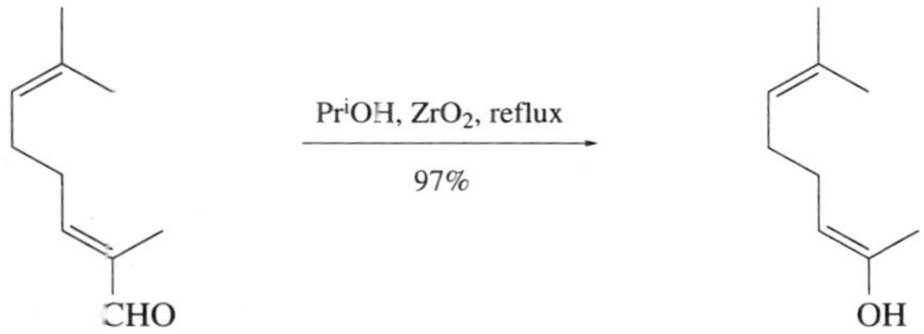

(33)

Reduction of α,β-unsaturated systems proceeded to give products of 1,2-reduction. Cyclopentadienyl hydride complexes of zirconium and hafnium efficiently effect Meerwein–Ponndorf–Verley reductions. The reaction is chemoselective with aldehydroketones being reduced exclusively to hydroxyketones, while α,β-unsaturated carbonyl groups give allylic alcohols ⟨86JOC240, 88JOC3752⟩. Organometallic complexes of titanium have been used as catalysts in the reduction of aliphatic, aromatic and α,β-unsaturated ketones to secondary alcohols ⟨88TL4113⟩. Yields of alcohols are up to 67%. The reagent is claimed to offer improvement over other methods ⟨80TL2171⟩.

Magnesium diisopropylamide (MDA) may be prepared from commercially available *n*-butyl-*s*-butylmagnesium and used to effect the reductions of aliphatic and aromatic aldehydes and ketones (Scheme 17) ⟨88TL139⟩. MDA is essentially unsolvated; the mechanism suggested by the authors is analogous to the known reductive decomposition of alkyl Grignard reagents. The fragmentation of THF in the presence of butyllithium is well known as a useful method for preparing the lithium enolate of acetaldehyde. This enolate has been reported to effect the reduction of nonenolisable aldehydes in good yield. Yields are poor for the reduction of ketones (< 30%) and at low reaction temperature (< 35 °C) aldol reactions predominate. The mechanism proposed invokes formation of ketene as a by-product; on workup this would produce acetic acid and the isolation of such a product was taken as confirmation of the postulated reaction ⟨87T2181⟩. The preparation and utility in carbonyl reduction of several alkyl metal compounds **(33)**–**(35)** derived from terpenoid natural products has been reported ⟨89JOC2383⟩.

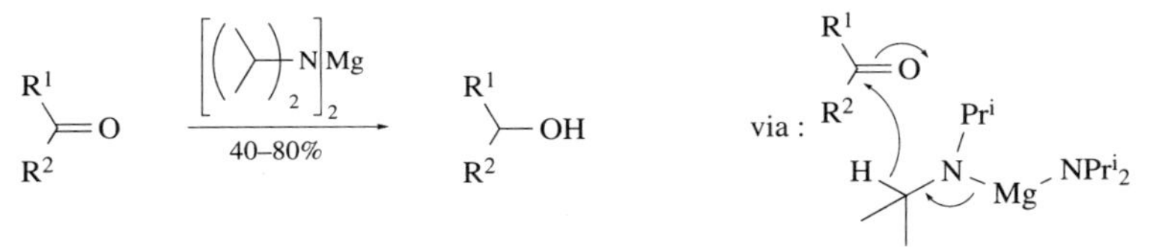

Scheme 17

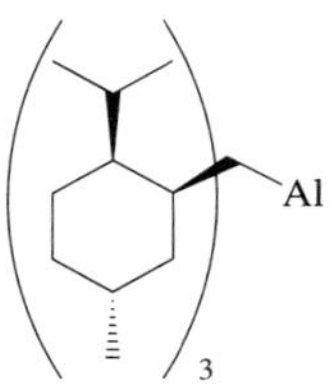

(33)

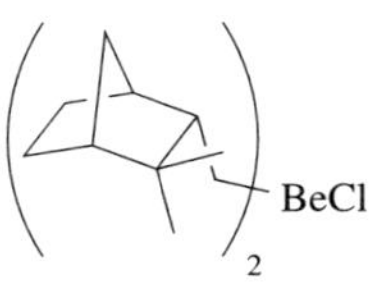

(34)

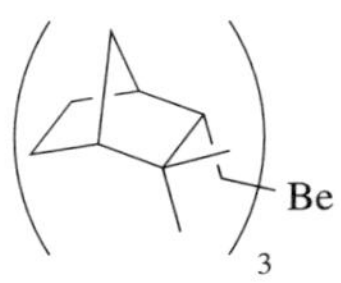

(35)

Enantioselective reductions are feasible using a modification of a samarium-mediated system ⟨84JOC2047⟩. The use of enantiopure oxazolinyl pyridine and amino alcohol ligands (Equation (34)) allows enantioselective lanthanide-mediated Meerwein–Ponndorf–Verley reduction of aryl alkyl ketones in up to 97% *ee* ⟨93JA9800⟩.

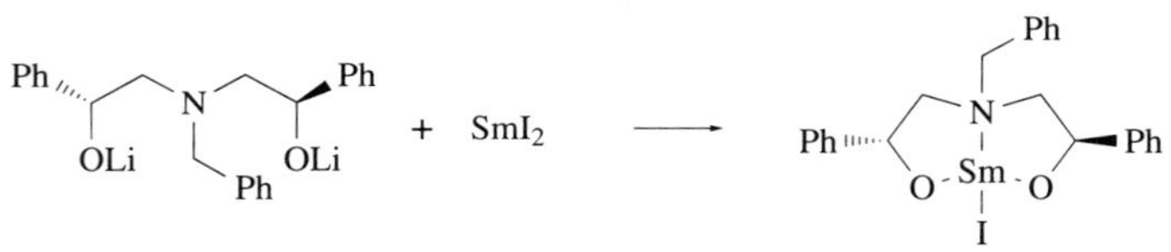

(34)

(iv) Reduction using metal hydrides: boron-based reagents

The majority of reductions of carbonyl groups employ hydrides of group 13 elements, in particular, hydrides of boron and aluminum ⟨91COS(8)1, 91COS(8)259⟩.

There are two main types of boron reducing agents: neutral and anionic. The parent compounds in both cases are boranes and borohydrides, respectively; in theory, multiple hydrides may be transferred, but in many cases only one equivalent of hydrogen is delivered.

(a) Boranes. Borane itself exists as a gaseous dimer and is a powerful reducing agent which possesses a complementary activity to other hydride reducing reagents. The commercially available THF and DMS complexes are commonly used to reduce a variety of carbon–oxygen multiple bonds to alcohols. The mechanism of the reductions makes use of the inherent electronic deficiency of boron due to its vacant *p*-orbital. The first step in reduction using a borane is a Lewis acid–base interaction between oxygen and boron. Hydride is then transferred to the electrophilic carbon atom in an intramolecular process. Carboxylic acids, aldehydes, ketones, anhydrides, epoxides and esters react with borane and its complexes to give alkoxyboranes which are hydrolysed upon treatment with mild proton sources. Borane reacts in a fundamentally different manner from other metal

hydrides and this is entirely due to its strong Lewis acidity. This means that electron-rich multiple bonds react fastest with the reagent, while electron-deficient bonds are inert. In particular, carboxylic acids (often slow to react with metal hydrides) are rapidly reduced to primary alcohols (double bonds often react at the same rate as acids, however), while acid halides (often very reactive with metal hydrides) are inert under usual reaction conditions. The reduction of aldehydes and ketones by borane proceeds by (reversible) formation of an oxygen–boron complex to which is then (irreversibly) transferred a hydride from boron to carbon. The complete lack of reactivity of acid halides may be attributed to the much-diminished Lewis base character of the carbonyl lone pair in such compounds owing to the electron-withdrawing influence of the halogen. Analogously and predictably, esters of carboxylic acids are reduced very slowly by borane. The parent acids, however, are reduced extremely readily and often with complete chemoselectivity. The reaction of amine–borane complexes with α-, β- and γ-cyclodextrins gives crystalline complexes in good yield. These complexes when suspended in water are capable of reducing carbonyl compounds in good yield and with moderate enantioselectivity (generally ~40% *ee*) ⟨89JOC3482⟩. Although not strictly speaking a reduction of carbonyl compounds, the unusual reactivity shown by borane towards tetrahydropyranyl ethers is a useful reductive method for the direct preparation of 5-hydroxyethers ⟨92TL5045⟩.

(b) Alkylboranes. Less reactive reductants are obtained when either or both of the excess hydrides of borane are replaced by sterically-demanding alkyl groups. Often, such alkylboranes (e.g., disiamylborane and thexylborane) are prepared by the action of borane upon alkenes. The great steric demands of these reagents means that their reactivity with, in particular, ketones is strongly influenced by the structure of the substrate. These reagents do not reduce acid chlorides, acid anhydrides, esters and carboxylic acids. Epoxides are reduced very slowly. 9-BBN effects highly chemo- and regioselective reduction of α,β-unsaturated carbonyl compounds to allylic alcohols in the presence of many other multiple bonds. Cyclic ketones are reduced with a reversed selectivity when a hindered alkylborane is utilised in place of borane itself.

Much attention has been concentrated on the preparation of alkylboranes derived from enantiopure alkenes and on the subsequent use of such boranes in enantioselective reduction of carbonyl compounds. Of these enantiopure reductants, greatest selectivities have been observed in the reaction of 'Alpine Borane' (see below) with α,β-alkenic ketones and in the reactions of Ipc_2BH. Indeed, the use of Ipc_2BH was probably the first reported asymmetric reaction ⟨64JA393, 64JA397⟩. The product from the reaction between 9-BBN and diacetone glucose (also known as diisopropylidene-glucofuranoside) is an alkoxy borane which upon treatment with potassium hydride gives the chiral borohydride **(36)** ⟨88JOC1231⟩. This asymmetric reagent reduced ketones enantioselectivity under reagent control and best results were obtained with 2-keto esters (enantiomeric excess of products 79–113%) and alkylaromatic ketones (78–100% *ee*). There has been a great interest in the development of terpene-derived boranes as asymmetric reductants ⟨89CRV1553⟩. The greatest interest has been shown for preparation of pinene-derived boranes, such as *B*-3-pinanyl-9-borabicyclo-[3.3.1]nonane (commercially available under the trade name Alpine-Borane) and eapine borane **(37)**. Alpine-Borane reduces α,β-alkynyl ketones with high enantiomeric excess, but high enantioselectivity is not generally observed in reduction of aliphatic ketones.

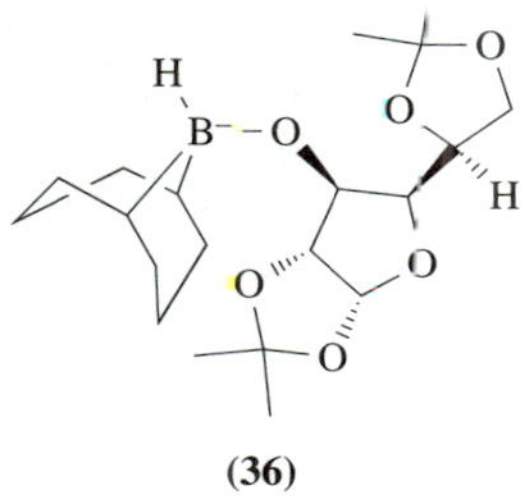

(36)

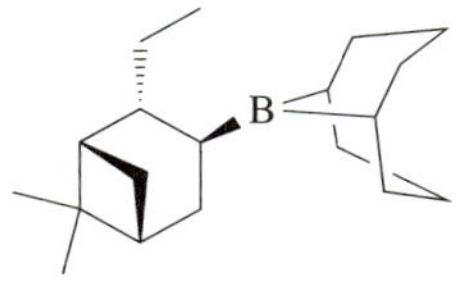

Eapine borane

(37)

Similar results are obtained using apopinene-derived boranes (Scheme 18) ⟨90JOC6328⟩. There is an effect on enantiomeric excesses of carbonyl reductions using asymmetric boranes at pressures greater than atmospheric, but it is small. Thus, Alpine-Borane reacts with prochiral ketones under 2000–6000 atmospheres with only modest increases in enantiomeric excess of the alcohols obtained compared to the ambient pressure reaction (Equation (35)) ⟨89JOC159⟩.

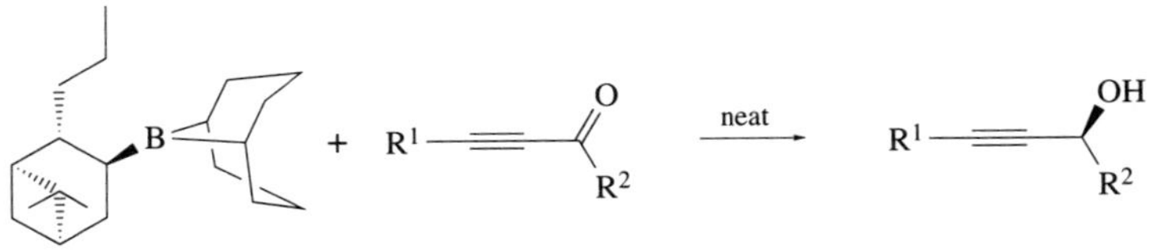

R¹	*R²*	*ee* (%)
H	Me	82
H	n-C_5H_{11}	99
Et	n-C_5H_{11}	88
n-C_5H_{11}	Me	91

Scheme 18

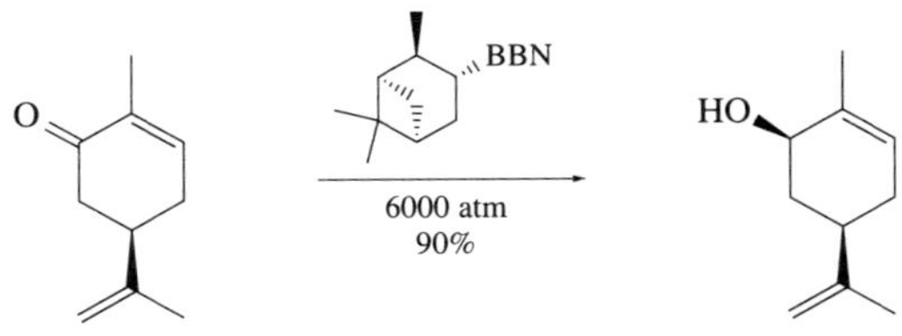

(35)

64% *ee*

The same group also investigated the use of aminoboranes (**38**) derived from pinene: enantioselectivity in reductions of prochiral ketones were moderate at best ⟨92JOC2953⟩.

Alkylboranes generally exhibit low reactivity towards ketones. This considerable defect may be remedied by use of chloroboranes, in which boron is more Lewis acidic, thereby aiding the reaction mechanism (which is general for pinanylborane reductions) shown in (**39**). Thus, Brown was led to examine the use of chlorodiisopinocampheylborane (**40**) in asymmetric carbonyl reduction ⟨88JA1549⟩. Ketones bearing aryl or tertiary substituents were reduced with high levels of selectivity using this reagent, but aliphatic ketones gave much lower enantioselectivity ⟨94TA1075⟩. Modified variants on this theme have appeared; for example, the use of substituted apopinocamphenyl-*t*-butylchloroboranes (**41**) in enantioselective carbonyl reductions ⟨89JOC4504⟩. Reaction of (**41**) produced alcohols in moderate yields (55–70%) and enantiomeric excesses (generally <90%). An improvement is obtained when *β*-chlorodiiso-2-ethylapopinocampheylborane (**42**) ⟨89JOC1577⟩ is utilised in reduction of ketones possessing 2-branching (Equation (36)).

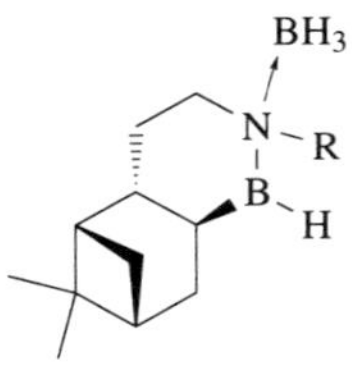

(38)

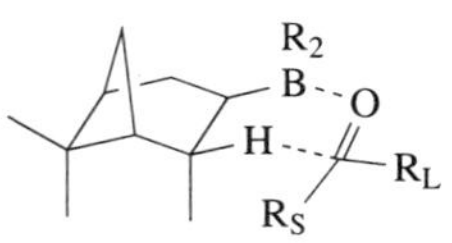

(39)

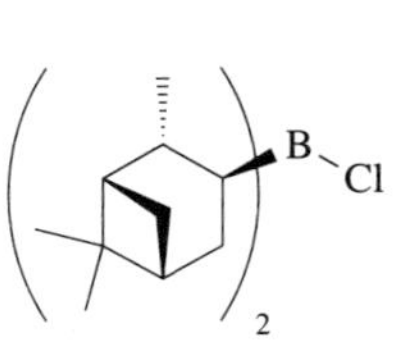

(40)

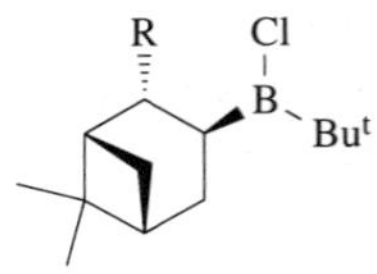

(41)

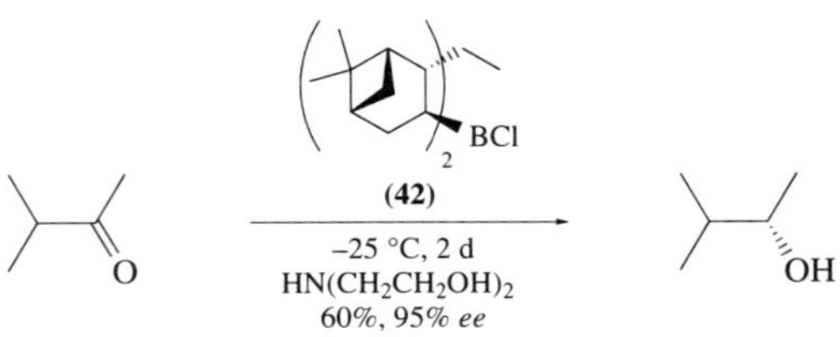

(36)

Asymmetric chloroboranes mediate a highly enantioselective preparation of 2-silylalcohols ⟨88T4403, 90TL4677⟩. The previously documented hydroboration of vinylsilanes with isopinocampheylborane has been usurped by an asymmetric reduction of acylsilanes using chlorodiisopinocampheylborane (Scheme 19) to give alcohols in excellent enantiomeric excess (>96%), and in good yields. C_2-symmetric borolane (**43**) executes highly enantioselective reductions of ketones ⟨86JA7404⟩.

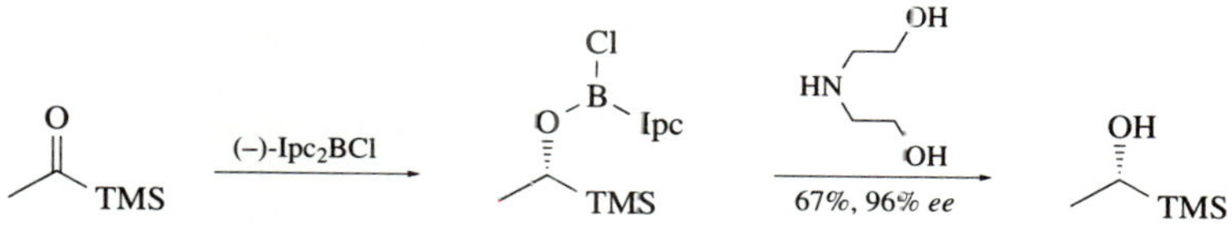

Scheme 19

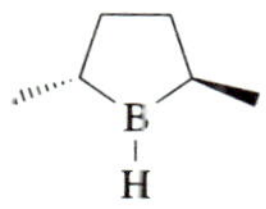

(43)

(c) Borohydrides. These reagents are among the most widely used hydride sources with Li and Na as the main counterions, though each possesses subtly different reactivities. With both reagents all four hydrides *may* be transferred in a simple stepwise manner, although there is some evidence for a more complex process ⟨70JA6894⟩. Each intermediate produced in the successive reduction steps is *more reactive* when transferring a hydride ion than its predecessor, because of the strong Lewis acid–base interaction which occurs between oxygen and boron. This electron donation to boron increases the nucleophilicity of the remaining hydrides, and the greater the number of oxygens attached to boron the greater is the augmentation of nucleophilicity. Since the majority of borohydride reductions are carried out in aqueous alcoholic solvent, the intermediate alkoxyboranes are usually rapidly hydrolysed.

Sodium borohydride is a mild reducing agent and will reduce only ketones and aldehydes rapidly. In the presence of aluminum trichloride, however, the reagent converts esters to primary alcohols in good yields, and in the presence of other metal salts (particularly lanthanides) the reducing properties are substantially modified. In particular, the use of cerium(III) chloride allows atypical chemoselectivity in the reduction of compounds containing carbonyl groups of different reactivity. The selectivity lies in the transient formation of a cerium-stabilised hydrate at the carbonylic function. Cerium chloride also suppresses the frequently troublesome competitive 1,4-reduction of α,β-unsaturated ketones ⟨79JA5848⟩.

The reaction of sodium borohydride with acids produces the corresponding sodium carboxyborane, which is normally inert with respect to further reduction. If, however, the reaction mixture is then treated with elemental iodine, a smooth reduction occurs to give primary alcohols in excellent yield ⟨91JOC5964⟩. This method represents a highly selective and high-yielding reduction of otherwise difficult to reduce carboxylic acids; in particular, alkenes are unreactive under the conditions (Scheme 20).

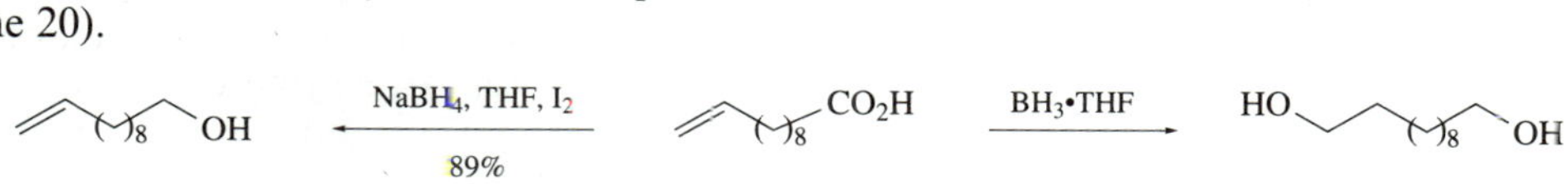

Scheme 20

$NaBH_4$ will reduce 2-keto-*N*-methoxy-*N*-methylamides to the corresponding diols in good yield. The presence of the *N*-methoxy group is vital to the success of the reaction ⟨92SL497⟩.

Sodium trimethoxyborohydride reduces acid chlorides in the presence of esters in the same molecule ⟨53JA6263⟩. The reagent, prepared *in situ* from sodium borohydride and zirconium(IV) chloride, reduces many other types of carbonyls (aldehyde, ketone, acid, ester) to the corresponding alcohols in excellent yield (~85%) ⟨88S995⟩. Sodium tris(pentafluorophenoxy)borohydride also exhibits high selectivity in reduction of substituted cyclohexanones ⟨94JFC(66)123⟩.

Lithium borohydride is a more reactive reducing agent than its sodium counterpart owing to the greater Lewis acid complexing power of Li^+. Its reactivity lies in between that of $NaBH_4$ and $LiAlH_4$. $LiBH_4$ reduces aldehydes or ketones in the presence of acids and esters at 0°C, but under more forcing conditions carboxy derivatives are also reduced. The reagent has the significant advantage of solubility in ethereal solvents. Acid chlorides may be selectively reduced in the presence of esters ⟨54NAT125, 55NAT346⟩.

Lithium aminoborohydrides were recently reported as a new class of powerful reducing agents. They exhibit similar levels of reactivity to lithium aluminum hydride but they are nonpyrophoric and are air-stable, and may prepared by reaction of borane–amine complexes with butyllithium ⟨92TL4533⟩. Lithium pyrrolidinoborohydride is a highly useful reagent for the chemoselective 1,2-reduction of enones ⟨93TL257⟩. Aminoborohydrides have been claimed as reagents of choice for stereoselective reductions of alkylcyclohexanones ⟨94TL5201⟩.

Calcium borohydride possesses similar reactivity to $LiBH_4$, again due to the enhanced Lewis acidity of the counterion, and is also more selective than LAH ⟨54NAT125, 55NAT346⟩.

(d) Zinc borohydride. $ZnBH_4$ has been employed particularly to effect chelation-controlled reductions (see below). The use of zinc borohydride in the presence of TMEDA has been reported to allow chemoselective reductions of acyl chlorides to alcohols in the presence of other unsaturated functional groups ⟨88BCJ2684⟩. Alkenes react under the same conditions to give products arising from hydroboration, owing to the presence of a borane–TMEDA complex which exists in small amounts under the reaction conditions.

(e) Trialkylborohydrides. Metal trialkylborohydrides are more powerful and sterically discriminating reducing reagents than lithium borohydride; esters are readily reduced. In particular, lithium, potassium and sodium tri-*s*-butylborohydrides (commercially available as Selectrides) react with ketones with high selectivity, due to the increased steric bulk compared to the parent compounds. Cyclohexanones are reduced almost exclusively from the *equatorial* direction (thereby minimising 1,3-diaxial-type interactions), leading to *axial* alcohols in good yields. Several alkylated borohydride-based reductants are effective for the asymmetric reduction of ketones. NB-Enantride, an anionic borane derived from pinene, may be used to reduce prochiral ketones in moderate enantiomeric excess.

(v) Reduction using metal hydrides: aluminum-based reagents

Aluminum hydrides are generally more reactive than the corresponding boron hydrides. The parent compound is alane, which reduces by a mechanism similar to borane, viz. by intramolecular hydride transfer within a Lewis acid–base complex ⟨66JA1464⟩. AlH_3 is more reactive than borane, reducing (for instance) acyl chlorides and esters to primary alcohols.

(a) Aluminum hydrides. Diisobutylaluminum hydride (dibal-H) ⟨55AG424, 55AG425⟩ is a widely used reducing agent capable of selective reductions owing to its single hydride equivalence ⟨75S617⟩. Sodium bis(2-methoxyethoxy)aluminum hydride (Red-Al, also known as SMEAH ⟨91COS(8)1⟩ is a highly soluble hydride reagent and is stable enough to be handled (briefly!) in ambient atmosphere. Both reagents will reduce lactones to lactols, for which transformation they are usually the reagent of choice. Sodium diethylpiperidinohydroaluminate (prepared from commercially available sodium diethylaluminum dihydride) has also been prepared as a new selective reducing agent ⟨93JOC1941⟩.

Lithium aluminum hydride (LAH) is an extremely powerful reductant and has been widely used for preparation of alcohols ⟨B-95MI 201-01⟩. The reagent must be used under anhydrous conditions under inert atmosphere, although the *N*-methylpyrrolidine complex is air-stable ⟨94TL1515⟩, and is most often used in ethereal solvent. Aldehydes, ketones, carboxylic acids, esters and lactones are all reactive at low temperature. Unsymmetrical epoxides are regioselectively ring-opened at the least-substituted carbon atom by LAH alone (see below). When the same reactions are carried out in the presence of Lewis acids, the regiochemistry may be reversed. Since the reaction is an S_N2 displacement, inversion of configuration is observed at the hydrogenated carbon atom.

(b) Lithium hydridoalkoxyaluminates. Although LAH is an extremely effective and versatile reducing agent, it is *highly* reactive and as a result several modified reducing agents have been prepared by addition of stoichiometric amounts of alcohols to LAH itself, leading (see above) to deactivated reductants. One, two or all three excess hydrides present in LAH may be replaced by alkoxy groups, thereby leading to a range of hydride donors of precise reactivity and, therefore, high selectivity. Particularly useful is the reagent derived from reaction of three equivalents of *t*-butanol with LAH: lithium hydrido(tri-*t*-butoxy)aluminate. This species will reduce epoxide, carboxylic esters and acids only slowly ⟨91COS(8)1, 91COS(8)259⟩.

2.02.1.3.2 Stereocontrol in carbonyl reduction

(i) Substrate control

Reduction of prochiral ketones containing adjacent asymmetry proceeds to give products according to Cram's rule. In the reduction of highly substituted and rigid cyclic ketones, the main product of reduction is again that arising from attack from the least-hindered side of the carbonyl group. Thus, norbornane-type ketones undergo reduction from the *exo*-face, leading to *endo*-alcohols as major products. Simpler and less rigid cyclic ketones generally give a mixture of products upon reduction; with unhindered ketones it is found that the more stable equatorial isomer predominates, but in the reduction of hindered ketones the axial product is dominant owing to the predominant attack of hydride from the sterically less-demanding direction. The selectivity of the reduction is due not only to the steric demands of the substrate but also of the reagent. Thus, reduction of conformationally locked (but relatively unhindered) *trans*-4-*t*-butylcyclohexanone by $LiAlH_4$ proceeds via almost exclusive axial attack to give the *trans*-alcohol; reduction using L-Selectride gives only the *cis*-alcohol, resulting from equatorial attack.

In reduction of the more encumbered 3,3,5-trimethylcyclohexanone by LAH, the ketone reacts to give a mixture of products in which the *axial* alcohol dominates. When the more selective (and less bulky) reagent lithium hydrido(tri-*t*-butoxy)aluminate is used in the reaction, the axial alcohol is the only product. As might be expected, it is generally more difficult to obtain the equatorial alcohol by reduction of a hindered ketone. Cyclic ketones are reduced with some selectivity in the presence of crown ethers: when poly-dibenzo-18-crown-6 is employed as a solid support for sodium borohydride, the observed selectivities are greater than when the monomeric crown ether is employed ⟨91TL2157⟩. 2-Cyclopropyl ketones undergo highly stereoselective reductions with lithium triethylborohydride when a trimethylsilyl group is present in the ring (Scheme 21). Thus, there is an almost exclusive preference for *anti* product, while the reduction of the corresponding protiodesilylated ketone furnishes a product of much lower stereochemical purity ⟨89TL5447⟩.

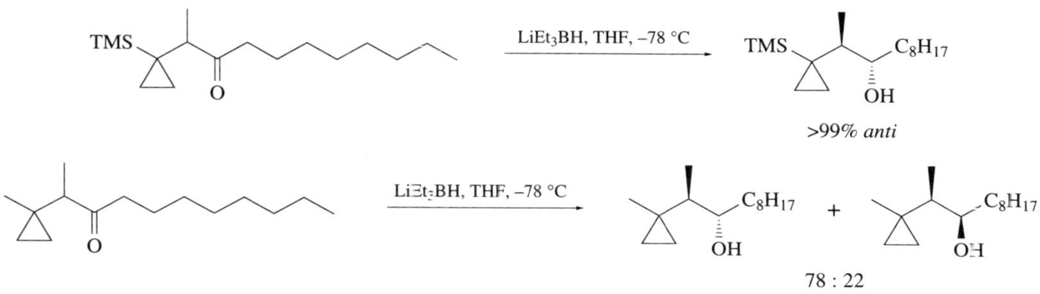

Scheme 21

(ii) Chelation control

When the carbonyl group undergoing reduction possesses an adjacent heterogroup, Cram's rule is often *not* obeyed owing to the dipole–dipole repulsion forces present in such molecules. Furthermore, when the heteroatom group is capable of chelation, a relatively rigid cyclic chelate may be formed prior to reduction and the reduction usually occurs so as to direct the nucleophilic hydride to approach from the face of the chelate which exhibits lesser hindrance and a so-called *chelation-controlled* reduction is feasible. Such reactions are capable of proceeding with great stereocontrol. The reaction of 2-hydroxy- and alkoxyketones with zinc borohydride proceeds under such control, to give *anti*-1,2-diols (and derivatives thereof) with high selectivity ⟨83TL2653⟩. When reducing agents incapable of inducing chelation are employed in the reactions, *syn*-diols are obtained via a Felkin–Anh transition state. A method for the diastereoselective reduction of *β*-alkoxyketones utilises a mixture of LiI and $LiAlH_4$ as the reducing agent. *syn*-1,3-Alkoxyalcohols are the favoured product of the reaction ⟨88TL5419⟩. Highly diastereoselective reductions of hydroxyketones are facilitated by a 1 : 1 complex of lithium (tris-*t*-butoxy)aluminum hydride and lithium iodide, involving a *syn*-reduction, thereby leading to *anti*-diol ⟨89TL4383, 89TL4387⟩. Directing effects due to relatively distant ligating species are also feasible, for instance in the selective reduction of 5-hydroxyketones ⟨94TL4891⟩. Functional groups other than carbonyls may also undergo chelation-controlled reduction. 1,3-Diols may be obtained by the sequential stereoselective reduction of enantiopure 2,4-diketosulfoxides using dibal-H (Scheme 22) ⟨92TL1605⟩.

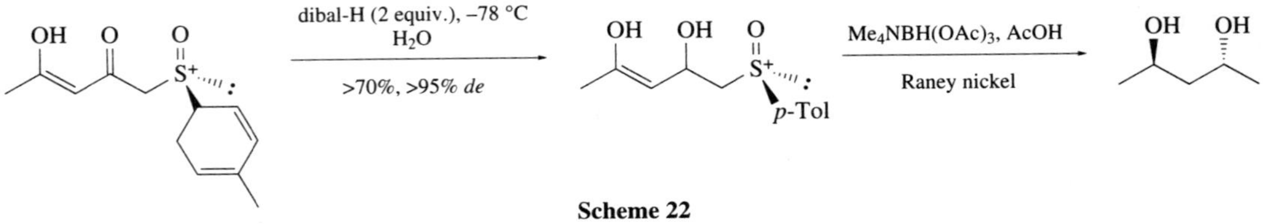

Scheme 22

(iii) Directed reduction of carbonyl groups

The use of *directed* hydride reduction is an extremely powerful synthetic tool ⟨91COS(8)9⟩. This description applies to the directing influence of a pre-existing coordinating group (such as a carbonyl, hydroxyl or thiol moiety) upon the stereochemical outcome of reduction of a carbonyl contained in the same molecule. There are two general possibilities for such reductions: formation of a rigid chelate followed by *internal* hydride delivery, and formation of a rigid chelate followed by *external* hydride delivery. These generalisations are represented in Figure 1.

Figure 1

As a general rule, external hydride is the reductant when M = B, L = alkyl or when M = Li and L = solvent. When M = B and L = OAc and when M = Si and L = alkyl (i.e., a hydrosilylation reaction), internal delivery of hydride is expected and is observed. Intramolecular hydrosilylation reductions of *β*-hydroxyketones are controllable, to allow either *anti*- or *syn*-selectivity ⟨86CC831, 88T3761⟩. The process is *anti*-selective when the reaction is carried out in the presence of Lewis acids other than $ZnBr_2$, which favours dominance of *syn*-diols (Scheme 23).

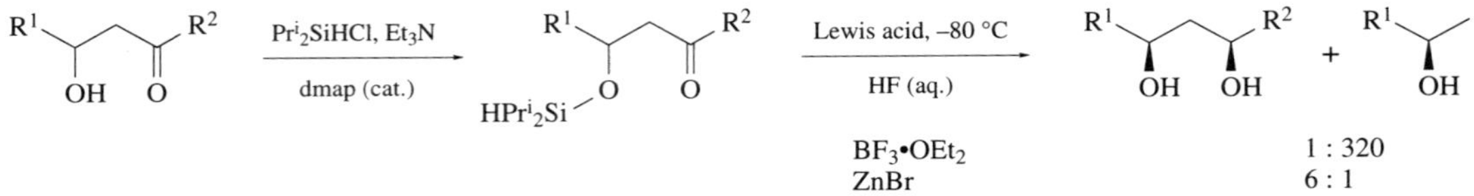

Scheme 23

When the reaction is carried out in the presence of transition-metal complexes bearing enantiopure phosphine ligands, good enantioselectivity may be observed ⟨91JA8518, 92TL2099⟩. The iterative sequence of reduction of products derived from enantiospecific aldol reaction followed by further specific aldol reaction allows flexible synthesis of polyoxygenated arrays. Using tetramethylammonium triacetoxyborohydride, *multiple* stereoselective sequential reduction of 3-hydroxyketones (Equation (37)) and 4-(3-hydroxyketo)-3-ketoesters is possible (Equation (38)) ⟨88JA3560⟩.

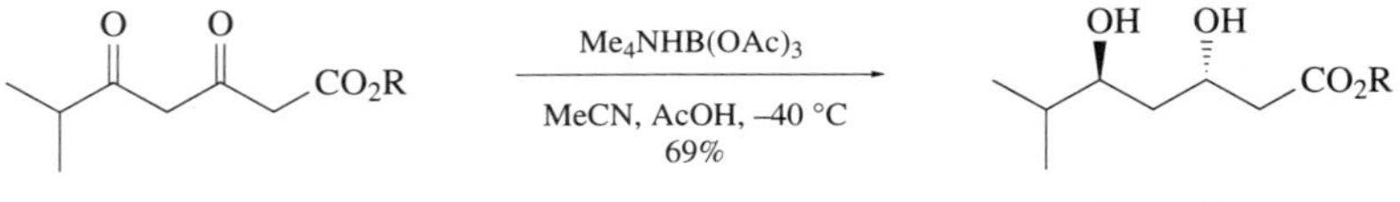

(37)

92% *anti*-isomer

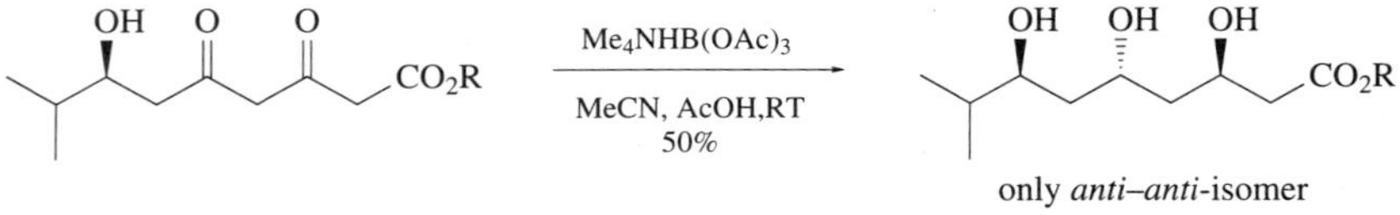

(38)

only *anti–anti*-isomer

Catecholborane allows stereoselective synthesis of 1,3-diols from 3-hydroxyketones ⟨90JOC5190⟩. Thus, reaction of such ketones with excess catecholborane leads to good yields of 1,3-diols of predominantly *syn*-stereochemistry. The highest level of diastereoselectivity was observed when the substrate contained other oxygenation; the use of rhodium catalysts improved the selectivity of the process only in certain cases (Equation (39)).

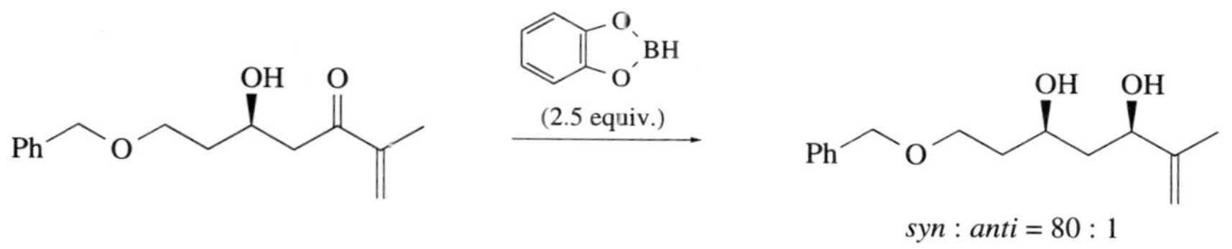

(39)

(*iv*) *Reagent control*

(*a*) *Asymmetric $LiAlH_4$ modifications*. The most successful aluminum hydride-based reagent ⟨91COS(8)159⟩ for asymmetric reduction of carbonyl compounds is BINAL-H, a modified lithium aluminum hydride formed by reaction of axially asymmetric BINOL (1,1-bi-2,2-naphthol), available in both enantiomeric forms by enzymatic hydrolysis of the dibutyrate) with LAH and one equivalent of ethanol (essential for high levels of selectivity). Given this precise blend of reagents, very high levels of enantiocontrol are possible. Some examples (and the proposed transition state (**44**) responsible for selectivity) are shown in Scheme 24 and Equation (40).

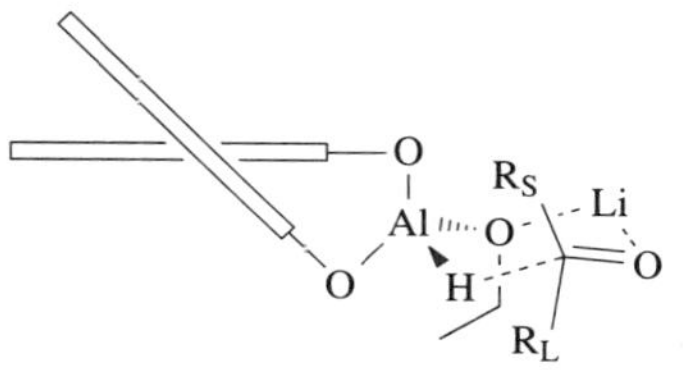

bulkier group
pseudoequatorial
(**44**)

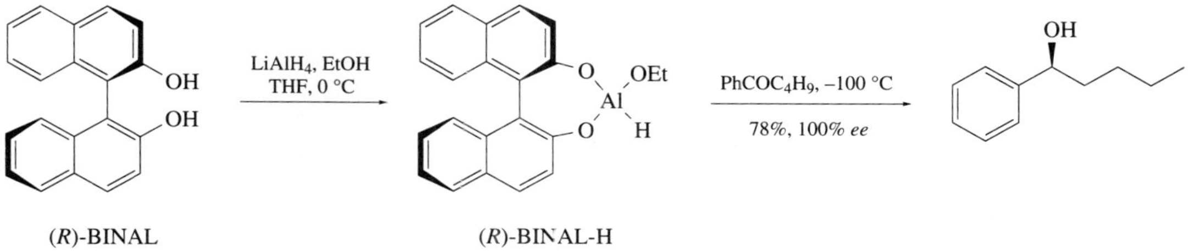

Scheme 24

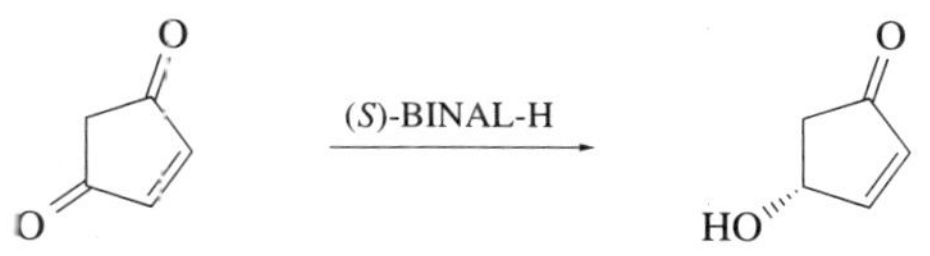

(40)

As is frequently the case in asymmetric carbonyl reductions, the best substrates for the BINAL-H reduction are those containing aryl substituents, although an impressive contradiction of this generalisation was seen in preparation of a key prostaglandin intermediate (**45**), obtained as a single enantiomer ⟨84JA6709⟩.

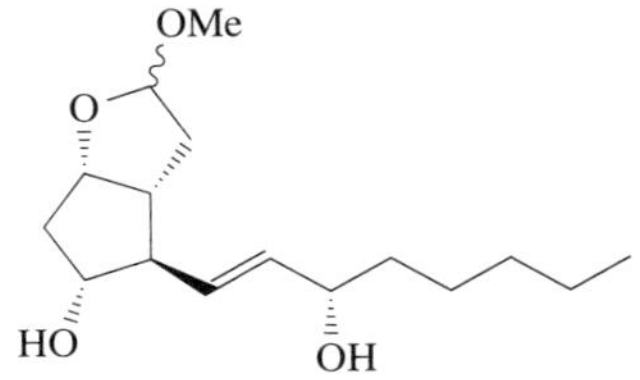

(**45**)

Asymmetric reduction has also been mediated by the use of enantiopure amino alcohols. For instance, the reduction of 3-aminoketones has been accomplished with asymmetric induction using an acyclic chiral ligand as the source of asymmetry. Thus, the Darvon alcohol-modified lithium aluminum hydride reagent ⟨73JOC1870, 73TL1505⟩ effects reduction of such ketones in excellent yield and with moderate to good enantioselectivity (Equation (41)).

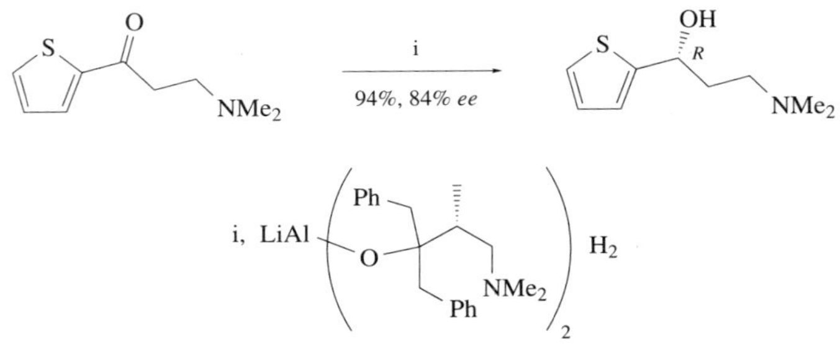

(41)

The modified lithium aluminum hydride prepared by combination of (*R*)-(−)-2-(2-iso-indolinyl)butan-1-ol with $LiAlH_4$ has been used as an asymmetric reducing agent for benzophenones (Equation (42)) in up to 95% *ee* ⟨92TA841⟩. Enantioselective transition-metal-catalysed hydro-silylation of carbonyl compounds is possible when such reactions are carried out in the presence of ligands containing asymmetry. The use of rhodium catalysts in the presence of enantiopure bidentate oxazolines (**46**) led to an asymmetric reduction, with predominance of (*R*)-enantiomer in the mixture of product alcohols. The stereochemical purity of the products varied from poor to moderate (11–80% *ee*). A variety of oxazolines (readily available on 100 g scale) were examined ⟨89TL5141⟩. Other catalytic enantioselective hydrosilylations of aryl and alkyl ketones are known ⟨88JOM(346)413, 88TL3235⟩.

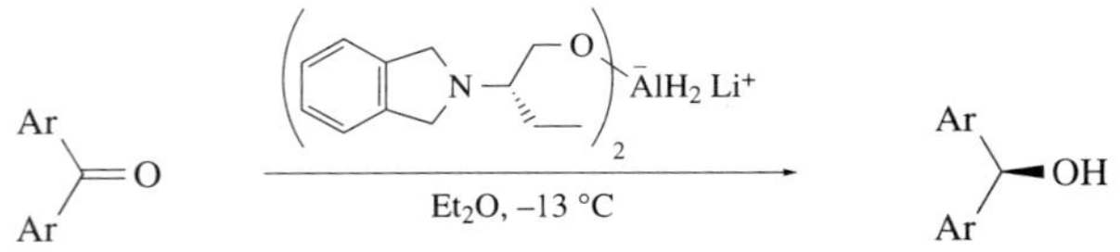

(42)

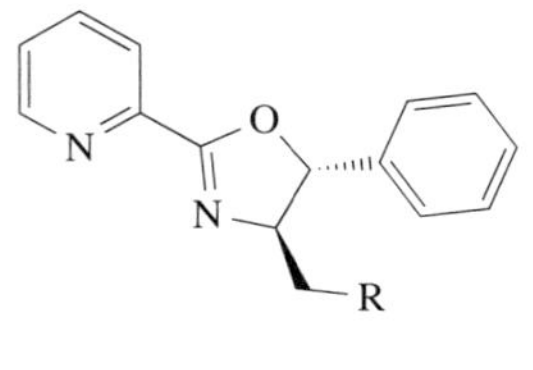

(**46**)

(b) Oxazaborolidines. Advances in asymmetric reduction of ketones has been achieved through the use of enantiopure oxazaborolidines prepared by reaction of enantiopure 1,2-aminoalcohols with boranes ⟨92TA1475⟩. The first reports of such reactions using enantiopure naturally occurring amino acids in the mediation of enantioselective borane reductions of ketones came in the mid-1980s from Itsuno, who utilised (*S*)-2-amino-3-methyl-1,1-diphenylbutan-1-ol. Aryl alkyl ketones could be reduced with high enantioselectivity using this stoichiometric asymmetric additive

⟨87BCJ395⟩. A more effective (and catalytic) asymmetric mediator in the borane reduction of prochiral ketones has been developed by Corey and co-workers; the enantiopure oxazaborolidine derived from diphenylprolinol is an extremely effective catalyst in such reactions: in particular, boroxazolidines **(47)** and **(48)** have been used in the enantioselective preparation of secondary alcohols (and deuterated primary alcohols). The general catalytic cycle explaining the selectivity shown by these reagents is shown in Scheme 25. The enantiopure additive activates both substrate and reagent while simultaneously constraining the attainable conformations of both.

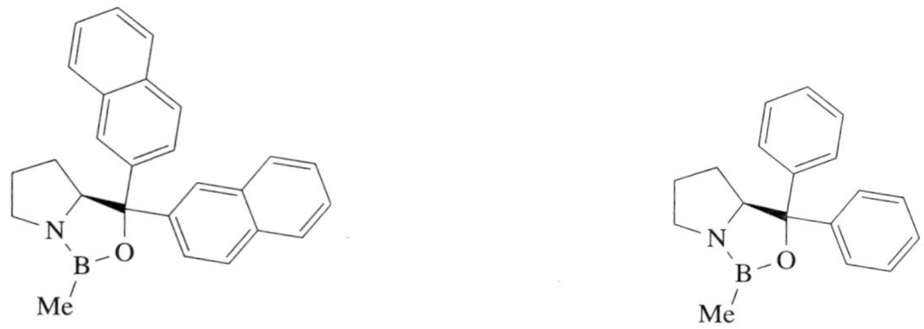

(47) **(48)**

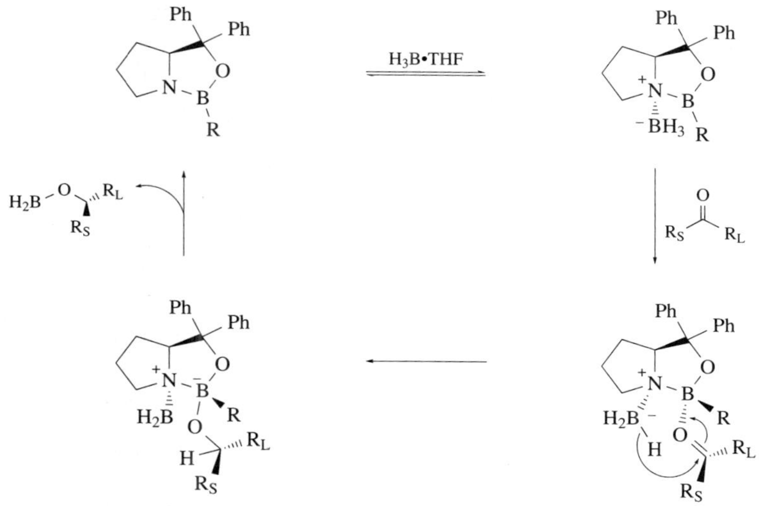

Scheme 25

An example of the utility of these reagents was shown in a synthesis of the antidepressant pharmaceutical Prozac (Scheme 26). Thus, **(47)** (derived from (*S*)-(−)-2-(di-*β*-naphthylhydroxymethyl)pyrrolidine and boronic acids) was demonstrated to catalyse the reduction of ketones to give secondary alcohols with enantiomeric excesses of 82–98% in >90% yield. Similarly, the reductive incorporation of deuterium into primary alcohols was accomplished highly selectively at very low temperature (−126°C) using (^{2}H)-catecholborane as deuterium source ⟨89TL6275⟩. The reaction allowed the first enantioselective preparation of fluoxetines; the crucial step was the enantioselective reduction of (2-chloroethyl)phenyl ketone in 99% yield and with 94% *ee*. This alcohol was transformed into (*R*)-(+)-fluoxetine by a series of very high-yielding reactions ⟨89TL5207⟩.

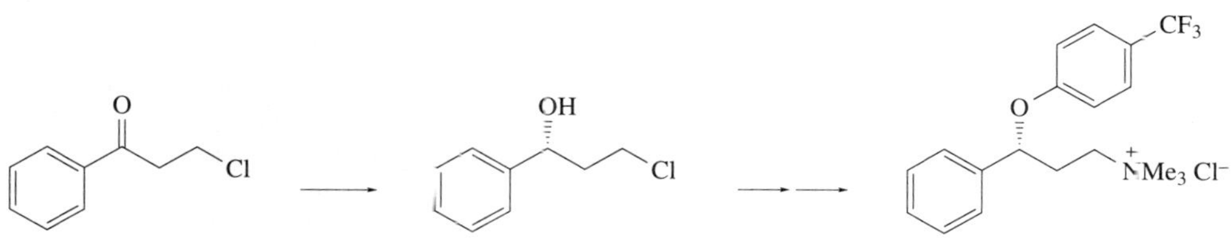

Scheme 26

The preparation of such 1,3,2-oxazaborolidines has been optimised ⟨92TL4141⟩. Previous methods for preparing these heterocyclic catalysts used the reaction of aminoalcohols with the borane–THF complex or substituted boronic acids, and such processes require long reaction times. Bis(2-trifluoromethylethyl) alkylboronates are more reactive alkylboronic acid equivalents, and the use of these species thereby speeds catalyst formation (Scheme 27).

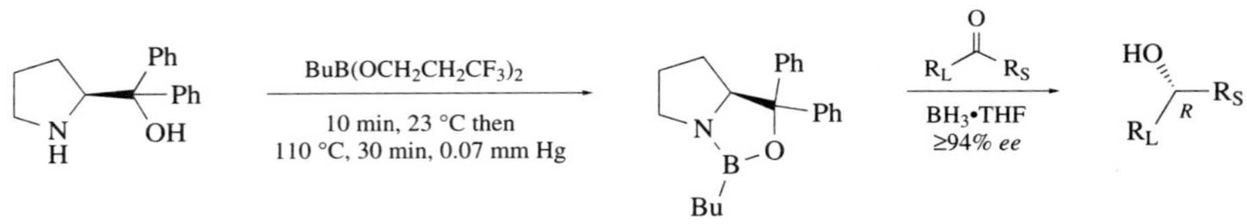

Scheme 27

The addition of amines to such oxazaborolidine reductions has been found to have an influence upon the levels of enantiomeric excess (Scheme 28). The reason for the improved selectivity was in the availability of more than one equivalent of hydride from the enantiopure intermediate. It was proposed that the product (**49**) arising from the first hydride transfer then itself acted as a possible less-selective reducing agent. When triethylamine was added to the product, intermediate (**49**) was converted to (**50**), which could be isolated and fully characterised, thus preventing a second, less-selective hydride transfer occurring ⟨93TL3243⟩. Reduction of ketodithianes with borane in the presence of oxazaborolidine (**47**) gives 2-(2-dithianyl) alcohols in excellent enantiopurity. Reductive removal of the dithiane then yields enantiopure alcohols (Scheme 29) ⟨90TL7415⟩.

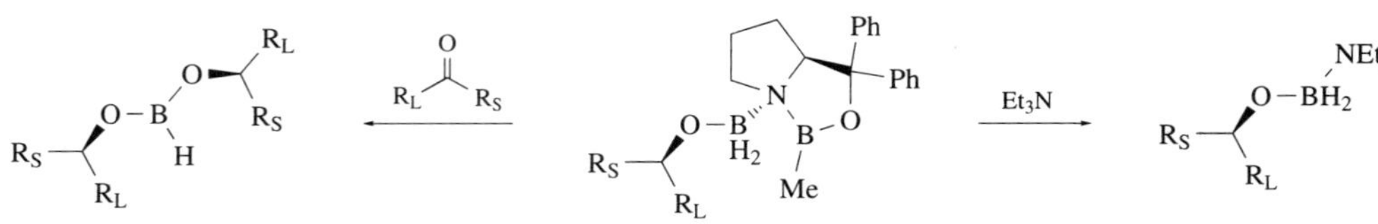

(49) **(50)**

Scheme 28

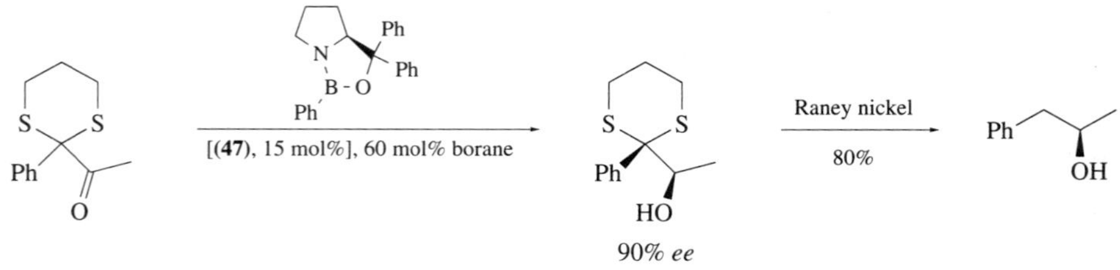

Scheme 29

MNDO Hamiltonian calculations have been performed in an attempt to ascertain the origins of the enantioselectivity observed in oxazaborolidine ketone reduction. These calculations showed that the most favoured situation was that in which hydride transfer occurred via a chair transition state, with the oxazaborolidine and ketone substituent effects reinforcing each other ⟨93JOC799⟩. Quantum-chemical modelling of chiral catalysis by oxazaborolidines has been carried out ⟨92TA921, 92TA933, 92TA1441, 92TA1563⟩. Oxazaborolidine catalyst (**51**), available in both antipodes, reduces aryl alkyl ketones with high enantioselectivity. The sense of chiral induction is predictive: the selectivity was proposed to arise from the highly ordered transition state arising from π-stacking ⟨93TL4145⟩.

Enantioselective reductions of prochiral ketones have been achieved using the oxazaborolidine derived from (*S*)-2-indoline carboxylic acid ⟨88TL4453⟩.

1,3-Diols may be enantioselectively prepared by a sequential one-pot aldol reaction/asymmetric reduction. Thus, silyl enol ethers react with aldehydes to give aldol products which undergo reduction

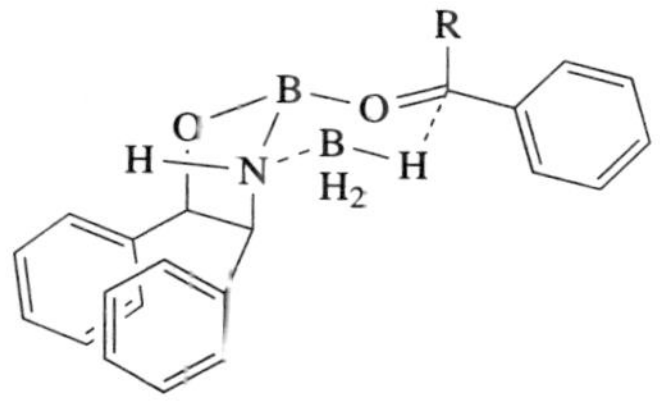

(51)

to give diols; both reactions are controlled by an enantiopure oxazaborolidine (Equation (43)) ⟨94TL4107⟩.

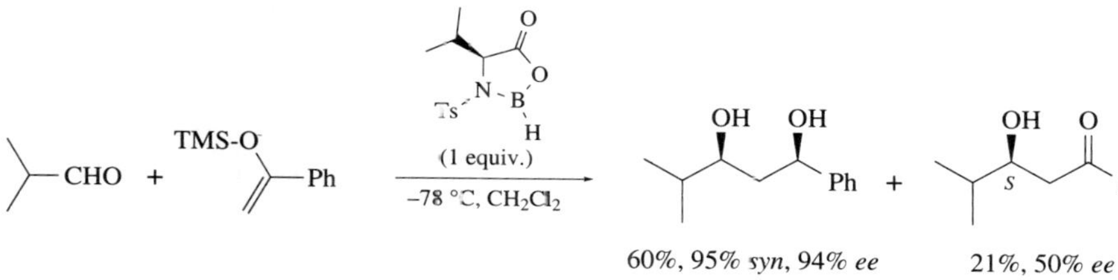

(43)

2.02.1.3.3 *Alkylative addition*

The addition of carbon nucleophiles is one of the most valuable methods available for synthesis of alcohols ⟨B-66MI 202-01, B-87MI 202-01, 91COS(1,2)⟩. The specific properties of particular types of reagents have been reviewed: Grignards ⟨B-54MI 202-01⟩, organolithiums ⟨B-88MI 202-02⟩, organozincs ⟨74AOC(67)83, 93CRV2117⟩, organocadmiums ⟨78CRV491⟩, organotitaniums ⟨82TCC1, 83AG(E)31, B-86MI 202-02, 92CRV807⟩ and organozirconiums ⟨83AG(E)31⟩.

The diastereoselective addition of organometallic reagents to carbonyl groups has been reviewed ⟨82H(18)357⟩. Chelation-controlled alkylations have also been reviewed ⟨84AG(E)556, 91COS(1)49⟩. Reviews have appeared concerning the enantioselective nucleophilic alkylation of carbonyl compounds ⟨B-83MI 202-02, B-86MI 202-03, 91AG(E)49, 92CRV833⟩.

(i) *Organomagnesium nucleophiles*

The most widely used organomagnesium reagents are those first prepared by Grignard, which bear his name. Grignard reagents react readily with the carbonyl moiety through a cyclic transition state (Equation (44)) ⟨80PAC545⟩.

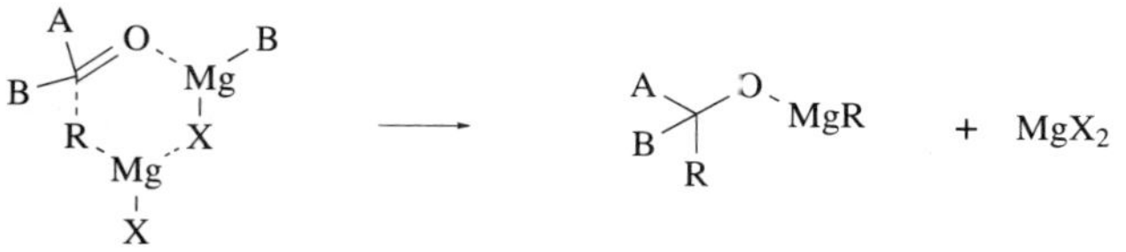

(44)

Ketones and aldehydes react to give secondary and primary alcohols, respectively, and esters (and acid derivatives in general) react with two equivalents to give tertiary alcohols. When the reaction of Grignard reagents is subject to significant steric influence, reduction of the carbonyl group by a process analogous to the Meerwein–Ponndorf–Verley reduction is often observed (see, e.g., ⟨62JOC1⟩).

Grignard reagents react with cyclic ketones to give predominantly the axial alcohol, arising from attack from the equatorial direction ⟨75CRV521⟩. Enantiomerically pure 2,3-cyclopropylcycloalkanones undergo highly selective alkylation with a variety of organometallic reagents ⟨90TL4565⟩. In particular, these reactions allow for reaction of nucleophiles with medium ring cyclopropyl substrates in excellent diastereoselectivity (Equation (45)).

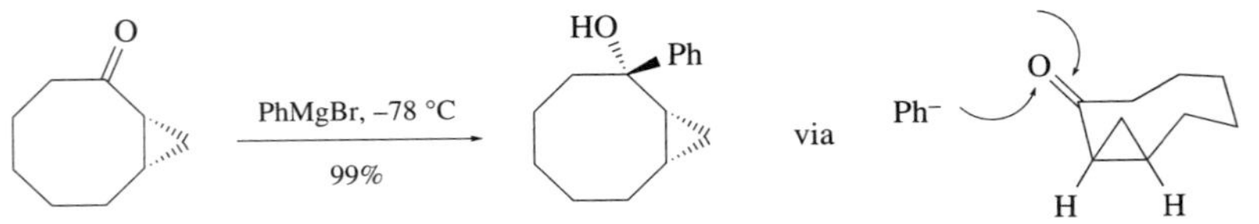

(45)

In the reaction of acyclic substrates, factors similar to those described earlier for reduction of carbonyls are in operation. Thus, chelation control may allow high selectivity, as is the case for the reaction of 2-alkoxyketones ⟨80TL1031⟩, but the effect diminishes rapidly as the ligating species becomes more removed from the ketone: 3-alkoxyketones exhibit low stereoselectivity in reaction with Grignards ⟨80TL1035⟩.

Grignard reagents may be prepared and stored as powders; for example, ethyl magnesium bromide forms a stable complex with the polyether amine TDA-1 (tri-(2-(2-methoxy)ethoxy)ethylamine). This complex reacts in a highly chemoselective fashion, reacting only with an aldehyde in the presence of both aldehyde and ketone ⟨89T171⟩.

(ii) Organolithiums

These species ⟨B-88MI 202-02⟩ undergo similar reactions to Grignard reagents, with the exception that reductive side reactions are much less frequently encountered ⟨72JA4628⟩. The addition of organolithiums to cyclic ketones is subject to great influence by addition of metal salts. Thus, 4-*t*-butylcyclohexanone undergoes almost exclusive equatorial attack when reacted with methyllithium in the presence of stoichiometric lithium perchlorate (Equation (46)) ⟨79JOC4371⟩.

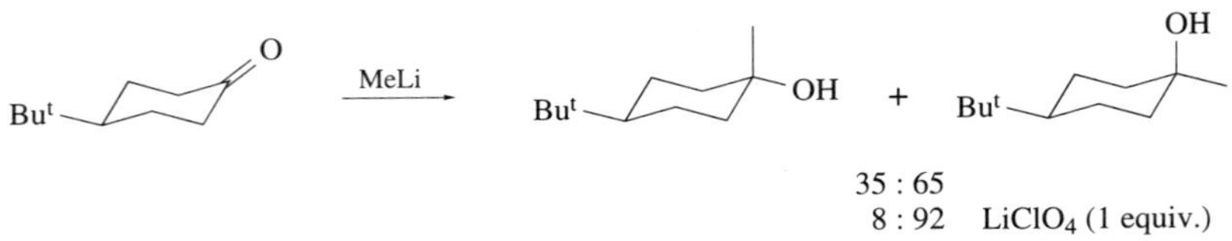

(46)

The transmetallation of enantiopure 1-alkoxystannanes occurs to give enantiopure alkyllithiums with retention of configuration and allows carboxylation reactions to occur in good yields ⟨90TL1985⟩. This method, therefore, furnishes an excellent route to protected 2-hydroxy acids of high optical purity.

Certain unsaturated alkyllithiums undergo intramolecular cyclisation reactions prior to carbonyl alkylation. For instance, 5-alkyn-1-yllithium reagents, obtained from reaction of the iodoalkynes with *t*-BuLi leads to vinyl anions at temperatures higher than −25 °C; these anions may be reacted efficiently with aldehydes to give trisubstituted allylic alcohols (Scheme 30) ⟨89TL3901⟩.

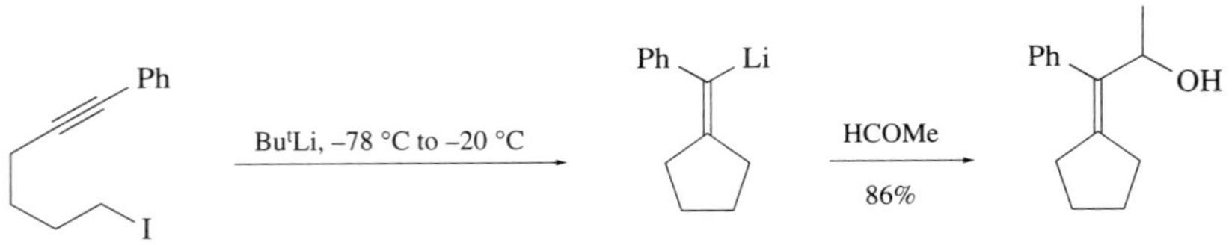

Scheme 30

The reactions of vinyl sulfides with lithium 4,4′-(di-*t*-butyl)biphenyl (LDBB) are known to give vinyl anions which may then be reacted with aldehydes and ketones. In particular, the reactions allow efficient preparations of cycloalkenyllithium reagents previously difficult to obtain (Equation (47)) ⟨90JOC4784⟩.

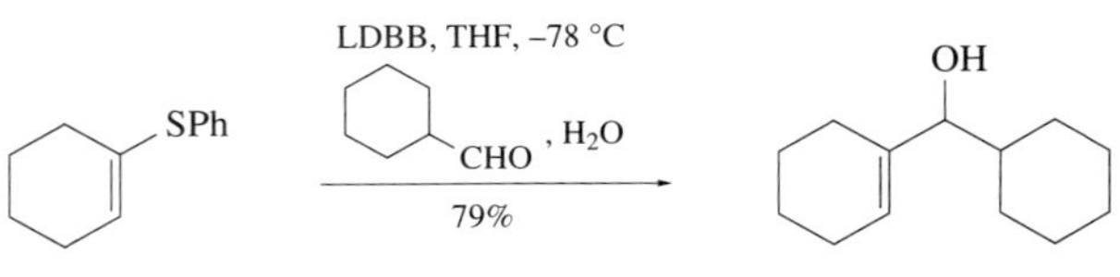

(47)

(iii) Organozinc reagents

Diethylzinc was the first organometallic compound prepared ⟨1849LA(79)171⟩, and alkylzinc reagents are of great importance in carbonyl alkylation reactions. Ketones tend to react with dialkylzinc reagents alone to give products of reduction, and so some type of activation is essential to allow efficient reaction. This inherently poor reactivity has been used to advantage to allow highly selective reactions to take place. Thus, the use of dialkylzincs in enantioselective alkylations of ketones has received considerable attention (see below). Frequently, however, the *in situ* preparation of organozincs by transmetallation of Grignard reagents with $ZnCl_2$ allows sufficient activation for the alkylation of ketones to proceed in good yield ⟨71JOC3311⟩.

The Reformatsky reaction of 2-bromoesters with carbonyl compounds to give 3-hydroxyesters is mediated by zinc metal and involves the intermediacy of an organozinc species, stabilised by typical enolate resonance ⟨75OR(22)423⟩. TMS-Cl has been used to activate zinc in the Reformatsky reaction ⟨87JOC4796⟩; the use of copper acetate (to form a zinc–copper couple) also allows activation ⟨85TL981⟩. Usually unreactive alkylzinc reagents have been found to undergo successful reactions with aldehydes in the presence of trimethylsilyl chloride ⟨88CC610⟩. An enantioselective Reformatsky reaction mediated by the naturally occurring diamine (−)-sparteine was first reported in the early 1970s, although the stereoselectivity of this pioneering work was poor (7–39% *ee*) ⟨73T3659⟩. The Reformatsky reaction of aryl methyl ketones with *t*-butyl bromoacetate has been found to be more selective than the previously documented work when the reaction is carried out in the presence of *N*,*N*-diallylephedrine. The use of (1*S*, 2*R*)-ephedrine gave (*S*)-2-hydroxy esters with moderate enantioselectivity ⟨93CC811⟩. A feature of organozinc chemistry not shared by other organometallics is the tolerance of other functional groups within the structure of the organic group. Thus, organozincs derived from 3-, 4- and 5-iodoesters are stable reagents and undergo bimolecular rather than intramolecular reactions ⟨87JOC4418⟩. Similar reagents may be prepared from strained acetals (Equation (48)) ⟨87JA8056⟩. Other polyfunctional organozincs may be prepared by reaction of diethylzinc in the presence of catalytic amounts of copper(I) cyanide ⟨92JOC1956; 93TL3115⟩

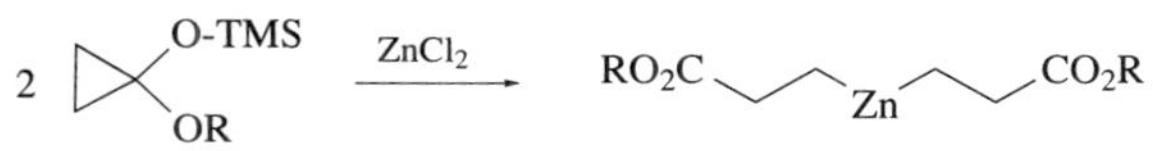

(48)

Organocadmium reagents share many of the features of organozinc chemistry ⟨78CRV491⟩.

(*iv*) *Organolanthanide reagents*

The chemistry of lanthanides has received considerable attention; one of the major features of this interest has been in the conversion of organolithium reagents to organocerium reagents, with particular emphasis placed upon those reactions in which the proposed substrate is prone to enolisation ⟨84JOC3904⟩. Cerium(II) chloride mediates reaction of Grignard organolithium reagents with carbonyl compounds ⟨85TL4763, 89JA4392⟩. For example, the dichlorocerium anion derived from acetonitrile is prepared by transmetallation of the lithium equivalent at low temperature. This organocerium is highly nucleophilic and relatively abasic, allowing alkylation of carbonyl substrates to proceed smoothly ⟨91TL5473⟩. Ytterbium triflate enhances the stereoselectivity of nucleophilic attack of alkyllithiums on cycloalkanones ⟨90JOC4990⟩ to give *trans*-1,2-dialkylcycloalkanols in good yield.

(*v*) *Organocopper reagents*

Organocuprates are famously unreactive with the majority of carbonyl compounds, this being an essential feature of much of their chemistry, though aldehydes will react. Zinc alkylcyanocuprates (prepared from organozinc reagents) react with aldehydes in the presence of Lewis acids to furnish secondary alcohols in good yield (Scheme 31); the process is chemoselective and the organozinc reagent may, therefore, contain functional groups of a wide variety ⟨88TL3887⟩. 2-Chlorosulfides may be transformed into the corresponding 2-thiophenylalkylcyanocuprates, which may then be reacted with aldehydes (Scheme 32) ⟨90TL7575⟩.

Scheme 31

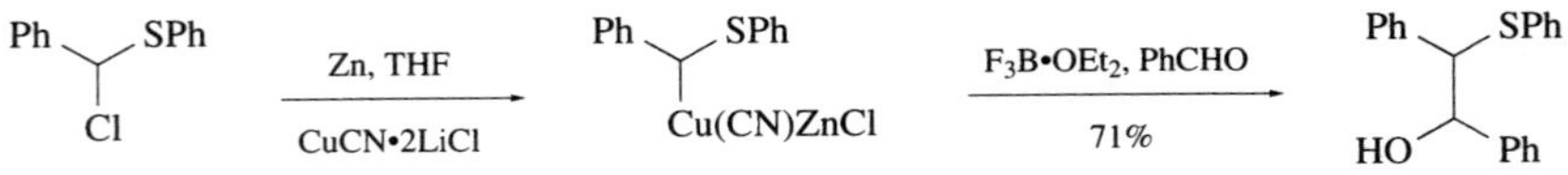

Scheme 32

(vi) Organosilanes

The carbon–silicon bond is relatively inert, but may be activated by the generation of positive charge β to the C—Si bond, owing to the great stabilisation provided for such cationic character by hyperconjugation ⟨87JA7838⟩. Alternatively, the great silaphilicity of fluoride ion may be employed, in what has been termed a 'push' mechanism (where the generation of a β-cation would be a 'pull' mechanism). When considering nucleophilic attack at carbonyl groups to generate alcohols, the most frequently employed silanes are allyltrialkylsilanes (see below).

2-Halocarbanions may be generated by reaction of 2-silylalkyl halides with tris(diethylamino)sulfonium difluorotrimethyl silicate (TASF). Such anions undergo nucleophilic addition to aldehydes leading to 3-haloalcohols. The same reagent facilitates the reaction of polyhaloaryl- and polyhaloalkenylsilanes with aldehydes ⟨88T4134⟩. The reaction of 2-trimethylsilylbut-2-ene with glycolates has been reported to yield 2-hydroxy-3-methyl-4-trimethylsilylalk-4-enoic esters in good yield and in high selectivity ⟨90JA6737⟩.

2-Trifluoromethyl alcohols may be prepared by reaction with (trimethylsilyl)trifluoromethane (TMSTFM) in the presence of tbaf ⟨89JA393⟩. TMSTFM is prepared from bromofluoromethane ⟨84TL2195⟩.

(vii) Organotitaniums

Vinyltitaniums react chemoselectively with aldehydes to give allylic alcohols in acceptable yields (Equation (49)) ⟨89TL3271⟩.

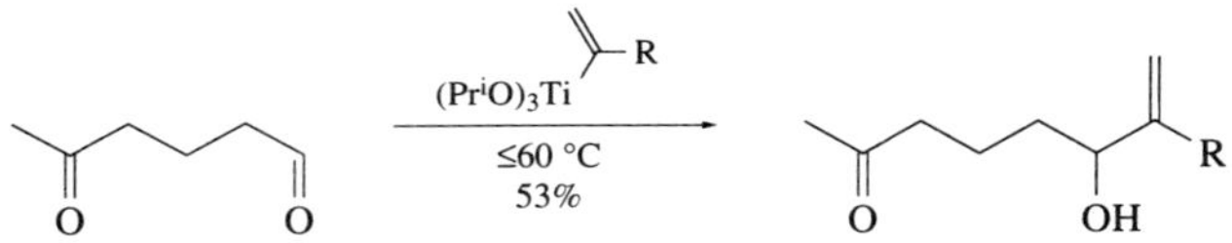

(49)

(viii) Miscellaneous organometallics

2-Alkoxyorganoplumbanes react with aldehydes in the presence of titanium tetrachloride at low temperature to give alcohols ⟨90JA6118⟩. Plumbanes may be prepared by reaction of 2-methoxystannanes sequentially with alkyllithium and tributylplumbyl bromide. Plumbanes containing asymmetry react with retention of configuration via an S_E2 reaction pathway.

Organoiron(II) reagents exhibit high levels of selectivity in addition reactions with cyclohexanones. The species involved are octahedral iron reagents which deliver alkyl groups with high facial selectivity to substituted cyclohexanones via an equatorial line of attack. Essentially complete stereocontrol was observed. Organomagnesium reagents exhibited lower selectivity ⟨93CC328⟩. The addition of alkenyl- and alkylzirconocene chlorides to aldehydes is subject to a remarkable acceleration in rate in the presence of silver perchlorate. As an example, under normal conditions the zirconium reagents react with carbonyl compounds to give low yields of addition product. When silver perchlorate (0.1 mol%) is present, yields of alcohols are excellent ⟨92TL5965⟩. Dibutylzirconocene reacts with imines in tetrahydrofuran or toluene to give a zirconoaziridine (**52**). Such metalloheterocycles react with aldehydes to yield exclusively *threo*-1,2-aminoalcohols in good yield ⟨92TL4469⟩. Unusually, the authors found that the facial selectivity could be completely reversed by *increasing* the temperature of the reaction.

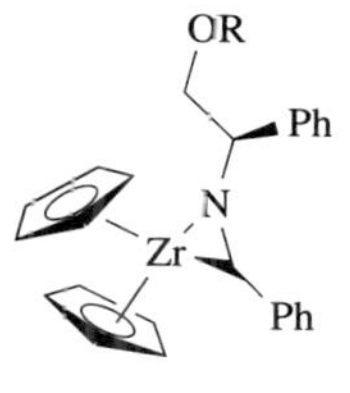

(52)

2.02.1.3.4 *Enantioselective alkylation*

The development of methods to allow for enantioselective nucleophilic attack at carbonyl groups has occupied researchers for many years, since Prelog's alkylation of a steroidal ester of phenylglyoxalate during the synthesis of atrolactic acid ⟨53HCA325⟩. The use of substrate control in such alkylations has been usurped by the quest for reagent control, and in particular for *catalytic reagent control.* Bisproline ligands (**53**) allow stoichiometric reagent control in the enantioselective delivery of organometallic reagents ⟨79BCJ3371, 79JA1455⟩. An equivalent of enantiopure ligand is necessary, because the nonligated organometallic reagent is still nucleophilic enough to undergo noncatalysed addition.

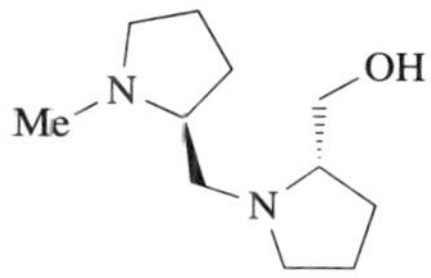

(53)

Alkylzinc reagents have inherently low reactivity and their reactions with carbonyl compounds are vastly accelerated by the presence of ligating species. This has allowed the development of catalysts containing ligating groups of differential facial steric requirements, thereby allowing for enantioselective alkylation of ketones ⟨92CRV833⟩.

Enantiopure catalysts derived from camphor sulfonic acid facilitate the enantioselective alkylation of aldehydes by alkylzinc reagents ⟨88TL5645⟩. The multichelate chiral additive need be present in less than stoichiometric amounts (typically <20 mol% is required). Enantiomeric excesses of the product alcohols are 82–96%. Catalysis of alkylzinc alkylations of aldehydes by polymer-bound catalyst is known ⟨88JOC927, 89JCS(P1)109, 90JOC304, 90JOC4832⟩. The presence in reaction mixtures of 10 mol% of a polystyrene-linked ephedrine catalyst results in enantioselectivities up to 89%, slightly better than with monomeric ephedrine catalysis.

Other camphor-derived catalysts are highly effective as mediators of asymmetric alkylations ⟨88PAC1597⟩. Enantioselectivities obtained using the enantiopure additive DAIB were good generally and in some cases excellent (Equation (50)). Indeed, even when the asymmetric additive was of low optical purity, excellent enantiomeric excesses were realised owing to a double asymmetric induction ⟨89JA4028⟩.

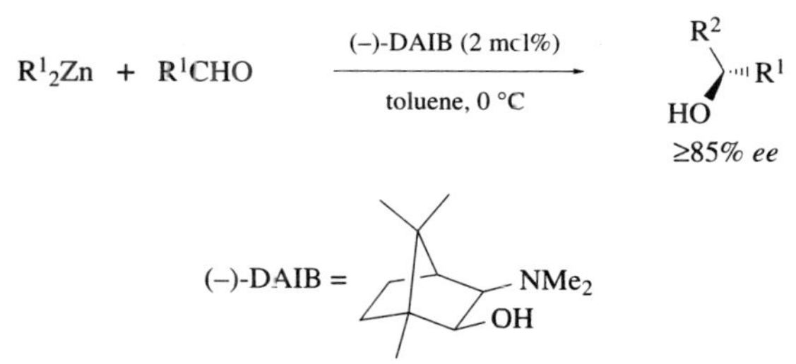

(50)

The combination of norbornane and proline structural motifs into the same ligand has been studied. Thus, the reaction of diethylzinc and benzaldehyde gives (*R*)-alcohol in 92% *ee* and high chemical yield when a catalytic amount of (**54**) is present ⟨89CC1700⟩. Corey has developed chiral catalysts to enhance the selectivity of delivery of dialkylzincs to aromatic aldehydes ⟨90JOC784⟩. Thus, bisamino alcohol (**55**) acts as enantiopure mediator in diethylzinc alkylations of ketones.

Similar reactions between ketoaldehydes and diethylzinc in the presence of proline- and norephedrine-derived catalysts demonstrate the chemoselectivity of organozinc reagents, as well as the enantioefficiency, since only the aldehyde moiety suffers alkylation. Acceptable enantiomeric excesses are obtained when only 8 mol% catalyst is present (Equation (51)) ⟨87TL5237⟩.

Brown's ephedrine-derived chiral companion (**56**) allows alkylation by diethylzinc of aromatic aldehydes with good selectivity (91% *ee*) ⟨89TL5551⟩.

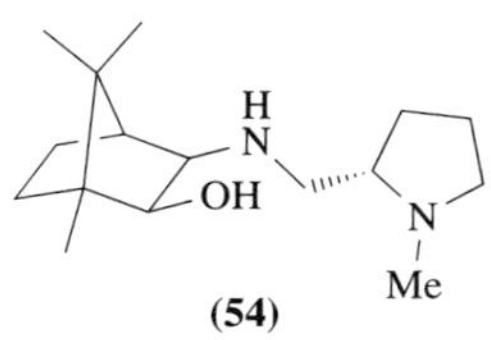

(54)

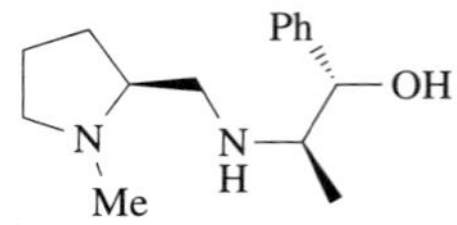

(55)

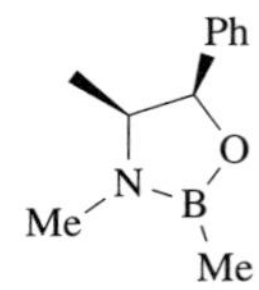

(56)

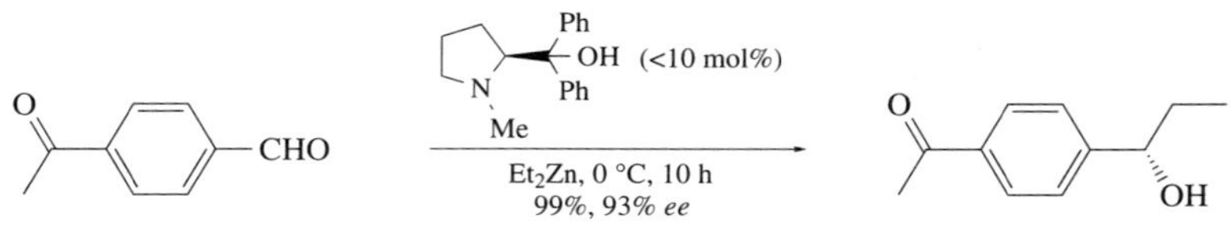

(51)

Studies using bistriflamido-*trans*-1,2-diaminocyclohexane have shown this compound to be effective in combination with titanium tetraisopropoxide in furnishing alcohols of good enantiomeric excess from the addition of alkylzincs to aldehydes. The active species is postulated to be (**57**) (Equation (52)) ⟨89TL1657, 89TL7095⟩. The same reagent combination also allows polyfunctionalised dialkylzinc reagents to take part in enantioselective alkylation reactions. When a functional alkyl iodide is reacted with diethylzinc and a catalytic amount of a copper(I) salt (either CuI or CuCN), a functionalised dialkylzinc is formed which reacts with α,β-unsaturated aldehydes to give functionalised secondary allylic alcohols with moderate to excellent chemical yields and enantiomeric excesses (Equation (53)) ⟨92JOC1956, 93TL3115⟩. The enantiopure bipyridyl catalyst (**58**) has been found to deliver dialkylzinc to aldehydes in good yield and with high enantiomeric excess ⟨90AG(E)205⟩.

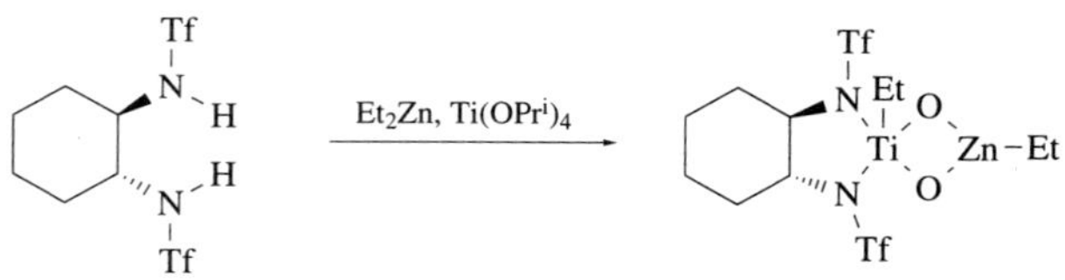

(52)

(57)

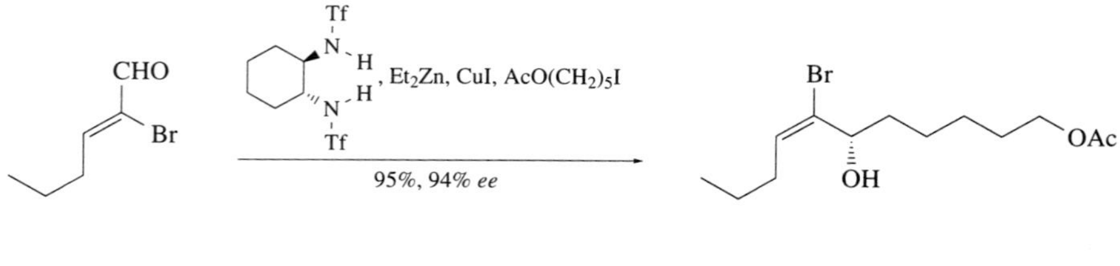

(53)

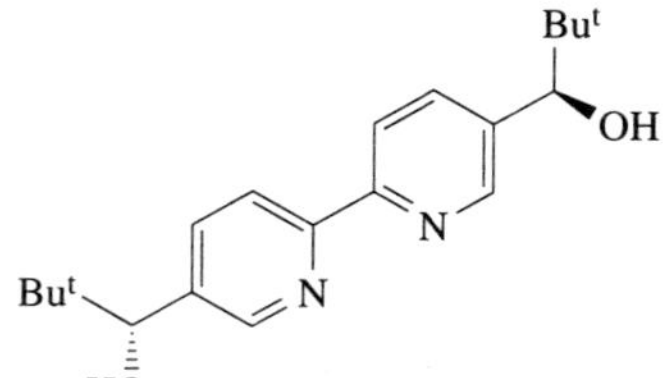

(58)

2.02.1.3.5 *Nucleophilic allylation*

The allylation of carbonyl compounds is a well-investigated area of synthetic organic chemistry, owing to the highly stereoselective reactions allylic anions (and their equivalents) exhibit. Many allylic Grignard reagents are unstable and often it is more satisfactory to prepare such reagents *in situ*. The reaction of allyl halides with carbonyl compounds in the presence of magnesium is known as the Barbier reaction ⟨77S18⟩. Many other allyl anion equivalents undergo similar reactions ⟨B-89MI 202-02⟩: e.g., allyl metal reagents ⟨93CRV2207⟩; allylsilanes ⟨89OR(37)57⟩.

Tin dichloride mediates allylation of aldehydes by allylic iodides (Equation (54)) ⟨81T3873⟩. Palladium catalyses the reaction of allylic alcohols with 1,2-diketones in the presence of tin dichloride ⟨89CL1647⟩. Double allylation is effected by the reaction: the process is suggested to involve conversion of the alcohol into an allylic stannane, which then allylates one of the carbonyl groups of the diketone, and then the second in another step. The reaction is tunable, in that complete diastereoselectivity in favour of *either syn* or *anti* isomer is possible.

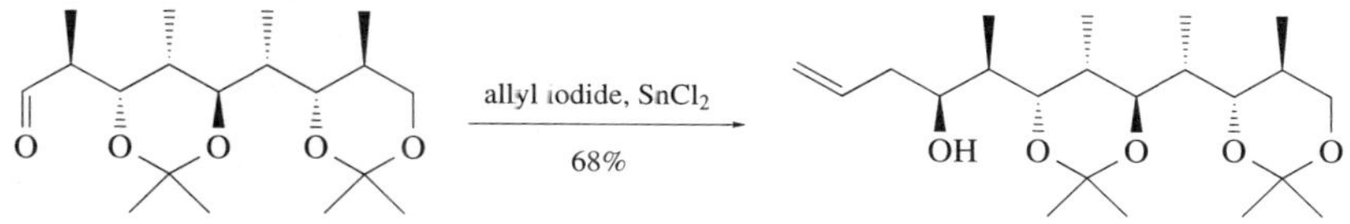

(54)

Indium metal is an effective mediator of Barbier reactions, and requires no initiation for complete reaction to take place ⟨91TL7017⟩. Indium(I) iodide has also been found to mediate Barbier and Reformatsky reactions. Thus, the reaction of allyl iodide with aldehydes at room temperature gives good yields of homoallylic alcohols, while the reaction of 2-iodoesters gives 3-hydroxyesters ⟨89JOM(369)291⟩.

Barium metal is also effective in such reactions. 2-Alkylation is always preferred over γ-alkylation and the alkenic geometry is always retained, in contrast to reactions using allylmagnesium and allyllithium reagents, which often give mixtures of products (Scheme 34) ⟨91JA8955⟩.

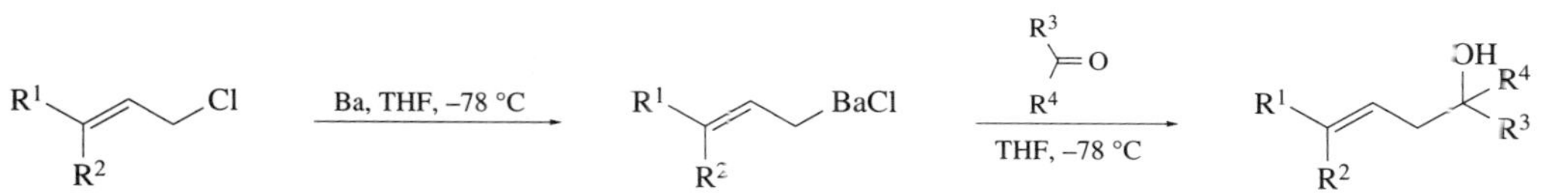

Scheme 34

Allyl(cyclopentadienyl)iron(II) is reported as effective in allylation of aldehydes and ketones ⟨91TL3001, 91TL4639⟩.

Allylchromium species are postulated to be the active species in the chromium-mediated reduction of acrolein dialkyl acetals, leading to a chemoselective reaction with aldehydes at low temperature to produce 1,2-*erythro*-3,4-butene-1,2-diol derivatives (Equation (55)) ⟨88TL5263⟩.

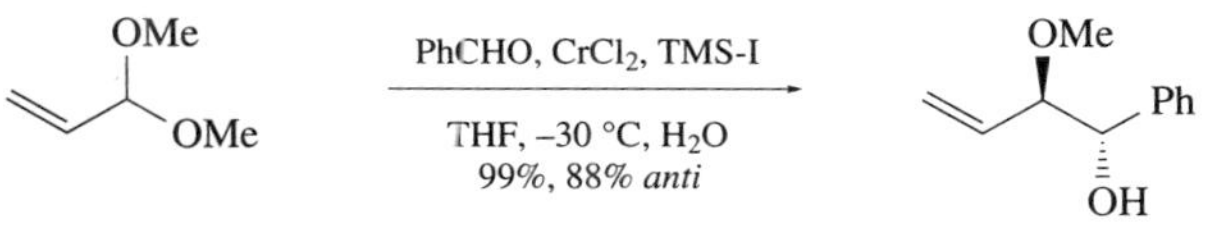

(55)

2-Monosubstituted allylic chromium reagents undergo *anti*-selective addition reactions with aldehydes ⟨82BCJ561⟩; this propensity may be inverted by the use of allylic phosphates to prepare the chromium reagents ⟨92JOC6384⟩.

Reaction of allyl bromide with trialkylstibines and an aldehyde gives the allylated product in good yield ⟨88TL1395⟩. Oxidative addition to antimony leads to a pentavalent stibine, which is the actual allylating reagent. Similar pentavalent organoantimony reagents execute propargylations of aldehydes ⟨91TL6579⟩. Trialkylstibines ⟨88TL1395⟩ mediate the alkylation of aldehydes by 2-bromoketones, to give 3-hydroxyketones in good to excellent yield ⟨88JCS(P1)2855⟩. Iridium metal-mediated reactions of allyl halides and phosphonates with carbonyl compounds at room temperature also proceed to give good yields of allylation ⟨88JOC1831⟩. A similar reaction using cadmium metal is also known ⟨88JOM(347)5⟩. Carbonyl compounds undergo efficient Barbier-type reaction with allylic halides with zinc in aqueous solution ⟨87JOM(322)177⟩. The use of a solid organic support (such as reversed-phase silica or styrene/divinyl benzene copolymers) instead of THF as cosolvent has been reported to improve the process in that there are no side products due to self-coupling, and hydroxyl groups present in the substrate do not need to be protected ⟨89JOC3087⟩. Manganese metal is efficient in mediating the reactions of allylic halides and 2-haloesters with carbonyl compounds. The intermediate organomanganese reagents must be prepared in the presence of the carbonyl compound to avoid Wurtz-type coupling and maximise yields. In the Reformatsky-type reaction, one equivalent of acetic anhydride must be used to prevent the reversibility of the transformation. With this proviso satisfied, the reaction proceeds in good yield to furnish 3-acetoxy esters ⟨89TL7373⟩.

Iodoketones react with aldehydes in the presence of chromium(II) chloride and a cobalt–phthalocyanine catalyst to give hydroxyketones ⟨89JOC4732⟩.

A regioselective propargylation reaction between acylsilanes and propargylzinc reagents has been reported ⟨89JOC5198⟩. Allylic acetates and carbonates act as allylating reagents when reacted with aldehydes in the presence of palladium catalysts ⟨88TL3563⟩. Allylcuprates may be prepared by reaction of vinylcopper reagents with iodomethylzinc iodide at low temperature. Reaction of such reagents with aldehydes furnishes homoallylic alcohols in good to excellent yield ⟨90JA6146⟩.

A highly active form of copper(0), generated *in situ* from copper(I) cyanide–lithium chloride complex, reacts at very low temperature with 1,2-dichloropropene to generate 1,2-dimetallated propene. This underwent sequential reaction with active electrophiles to give homoallylic alcohols in moderate to good yields ⟨92JA5110, 92TL6575⟩.

2-Lithioallyl chloride reacts nonregioselectively with electrophiles ⟨82JOM(236)139⟩, unless freshly fused zinc chloride is present, in which case, exclusive 2-reactivity is observed in reaction with carbonyl compounds ⟨93TL3145⟩. The product chlorohydrins were exclusively *syn* isomers when the carbonyl substrates were aromatic aldehydes.

Allylations of 2-aminoaldehydes using a variety of allylmetal reagents have been reported ⟨90TL1803⟩. The highest diastereomeric ratios were obtained when trimethylallylsilane was employed in the presence of tin(IV) chloride. Enantioselective allylation of such aldehydes is also possible when asymmetric allyl titanates are employed (see below) ⟨92JA2321⟩.

2.02.1.3.6 Allylations using boranes, silanes and stannanes

Allylboranes always undergo allylic transposition reaction, indicating a cyclic transition state, which arises from the initial coordination of a carbonyl lone pair to the Lewis acidic boron atom ⟨90JOC1868⟩. A generalised diagram representing the usual stereochemical features of the reaction is shown in Scheme 35. The reaction occurs when L = alkyl or alkoxy. Diastereoselectivity in such allylation reactions using cyclic boronates is high (>95%) (Equation (56)) ⟨81JOC1309⟩.

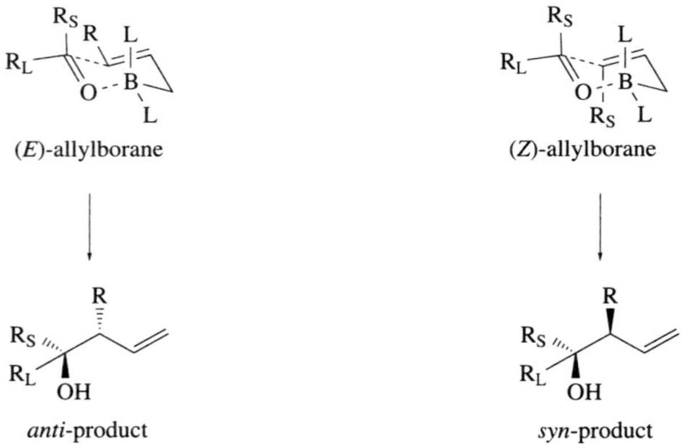

Scheme 35

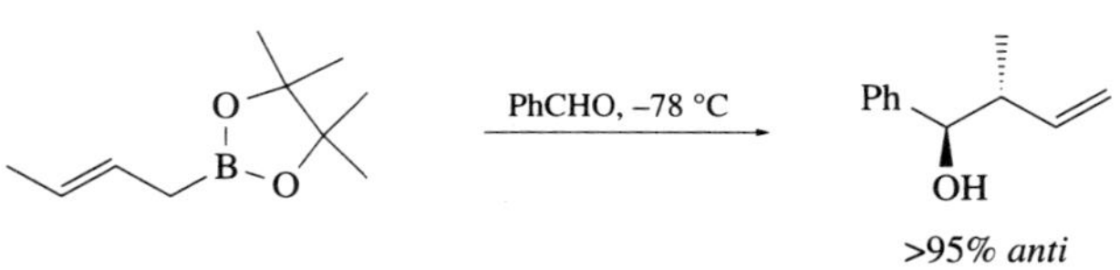

(56)

When the boronate contains asymmetry, diastereo- and enantioselectivity are feasible. In particular, boronates derived from tartaric acid have been shown to exert considerable asymmetric induction ⟨88JA3979, 90JA6339, 90JA6348⟩. Asymmetry present in either the allyl group (Equation (57)) *or* the boronate may also cause effective asymmetric induction ⟨81CB375⟩. The combination of chirality in both boronate and allyl units is highly effective (Equation (58)) ⟨88CB1501, 88CB2225, 88PAC123⟩.

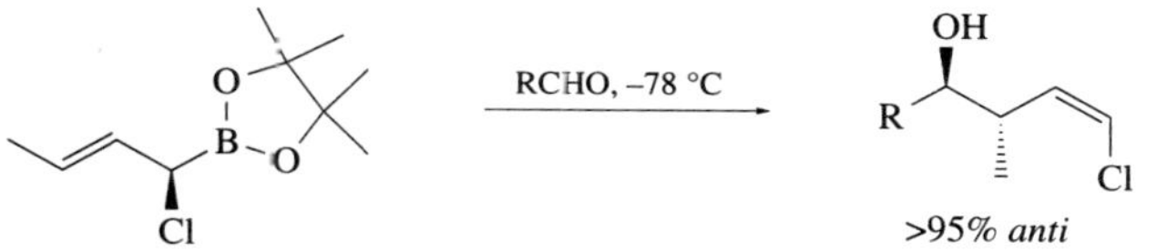

(57)

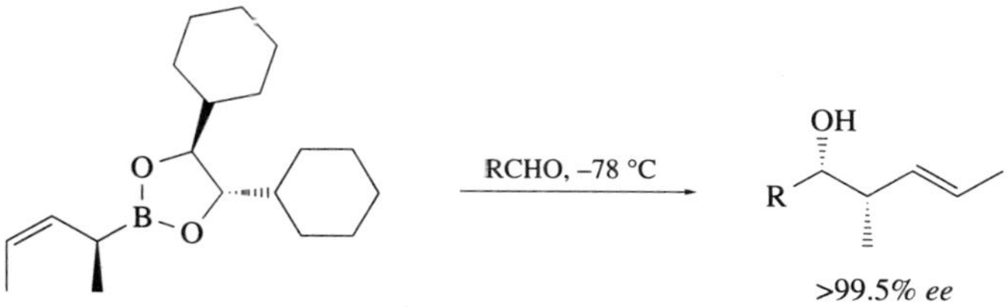

(58)

Enantiopure 3-silylpropenylboranes may be used for the efficient and enantioselective preparation of *anti*-alk-1-ene-3,4-diols ⟨91JOC5243⟩. The structurally related B-[(*E*)-3-(diphenylamino)prop-2-enyl]diisopinocampheylboranes (**59**) are also highly selective allylating reagents, allowing the synthesis of *anti*-3-(diphenylamino)-2-hydroxy-1-alkenes ⟨93CC339⟩. (*Z*)-3-Alkoxyallylboranes prepared from pinene have also been widely used by Brown ⟨88JA1535⟩. The reactions are highly *syn*-selective (Equation (59)).

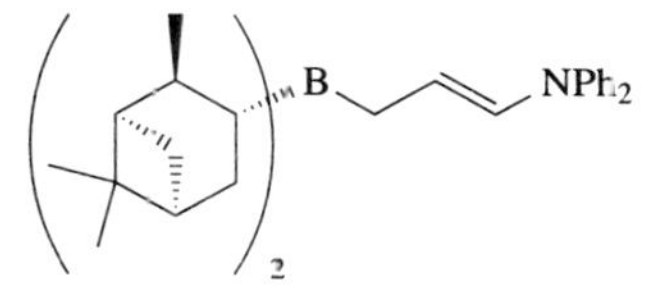

(**59**)

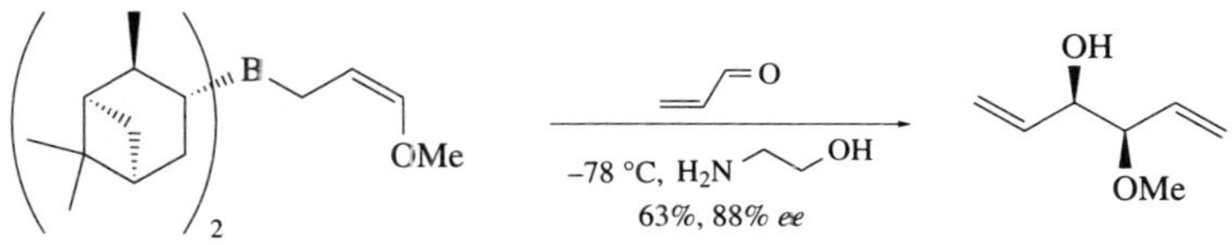

(59)

Reactions of pinene-derived allylboranes with enantiopure aldehydes are highly face-selective ⟨89JOC1570⟩. From a different corner of the chiral pool come caranyl allylboranes. Enantioselectivities with these reagents are uniformly high (>99% *de* and >94% *ee*) when they are used to allylate aldehydes (Scheme 36) ⟨90JA2389, 90T4457⟩.

The use of dienylboranes derived from diisopinocampheylborane has also been reported in the preparation of 2-(2-but-2-1,4-dienyl)alkanols in good yield and with high enantiomeric excess (Scheme 37) ⟨90T4463⟩.

3-Borolenes react as equivalents of 3-anions of homoallylic alcohols, giving (*E*)-alkene-1,3-diols upon sequential reaction with aldehydes and peroxidation (Equation (60)) ⟨88JA5578⟩

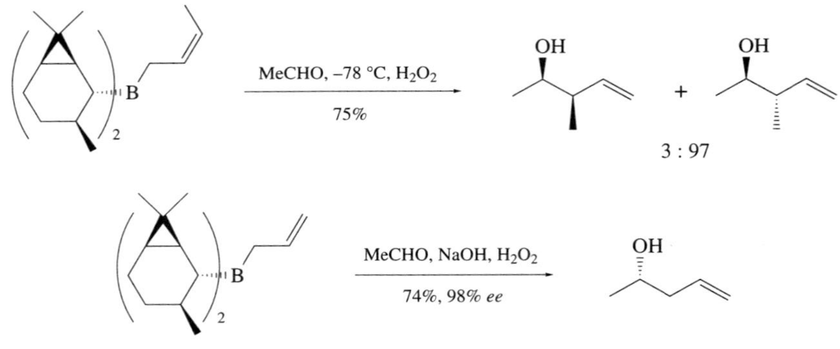

Scheme 36

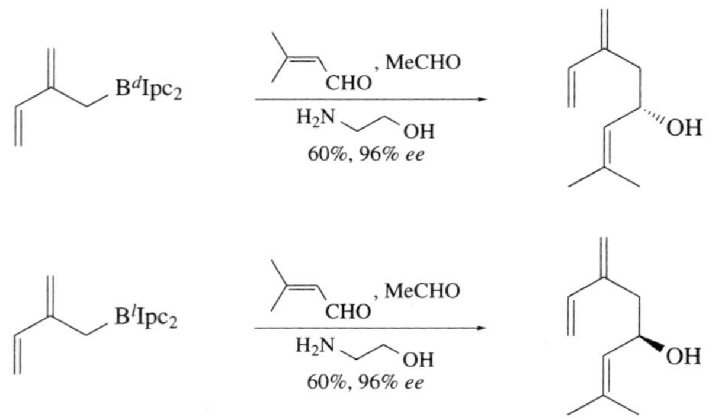

Ipc = isopinocamphenyl

Scheme 37

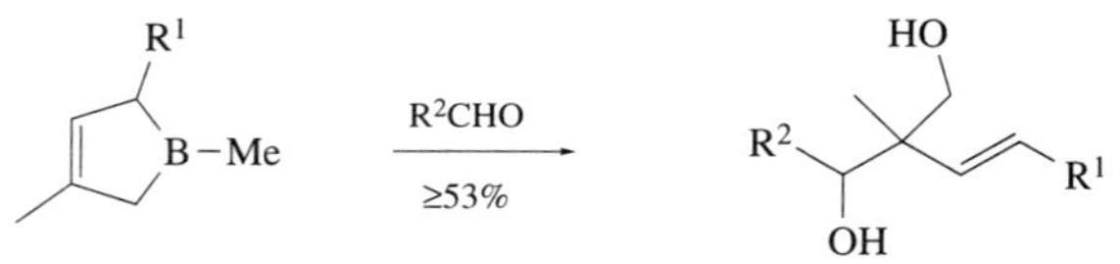

(60)

Diphenoxyallylboranes may be prepared *in situ* from allene, tribromoborane and anisole and are useful reagents for allylation of aldehydes ⟨91TL6749⟩.

Optically active η^3-allylmolybdenum complexes (**60**) allylate aldehydes with high enantiomeric excess ⟨89JA1937⟩.

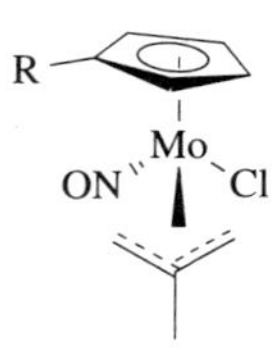

R = neomenthyl

(60)

Highly diastereoselective allylation reactions are possible using crotyltitanocenes (Equation (61)) ⟨88OM2289⟩. An increase in size of the cyclopentadienyl substituent R (**61**) increases the observed diastereoselectivity; a similar effect is observed when the aldehydic substituent is made more sterically demanding. The reaction may also be extended to allow incorporation of asymmetric auxiliaries ⟨89JOC4154⟩. Enantiopure *B*-allyldiazaborolane (**62**) (which may be prepared *in situ*) gives homoallylic alcohols in good yield and with good to excellent stereochemical integrity (79–99% *ee*) upon reaction with aldehydes ⟨89JA5495⟩. The *B*-allylborolane (**63**), developed by Masamune and co-workers ⟨89JA1892⟩ allylates representative aliphatic aldehydes in good yield and with high enantioselectivity (>96% *ee*). Generally, alcohols with (*S*)-configuration are obtained; the arrangement shown (**64**) is postulated to rationalise the observations.

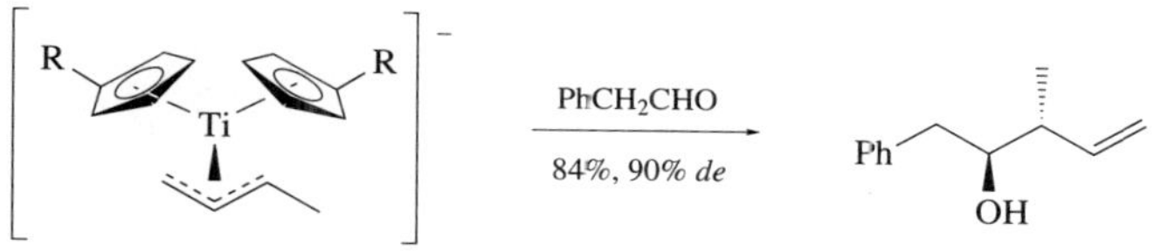

(61)

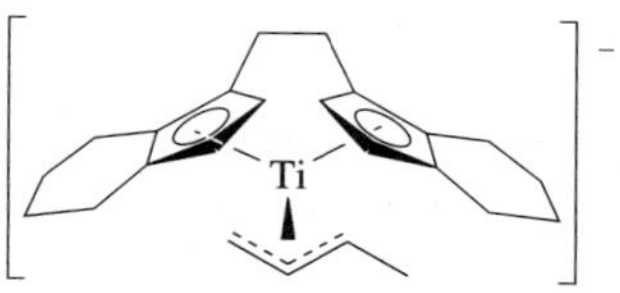

(61)

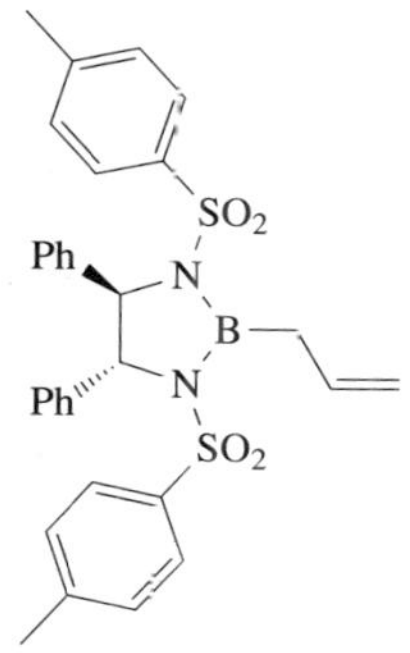

(62)

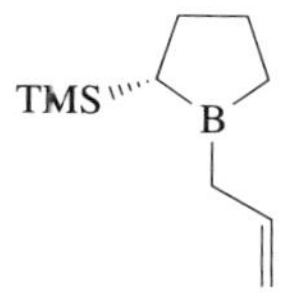

(63)

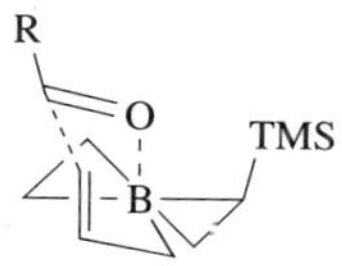

(64)

The mechanism of allylation using allylsilanes differs from that of boranes: the reaction does not occur through a cyclic transition state, because silicon is not Lewis acidic, and the stereochemical control of such reactions, therefore, relies on more subtle effects. Lewis acid activation is essential to successful reaction.

Both (*E*)- and (*Z*)-allylsilanes react to give *syn*-products preferentially, owing to the open transition state which allows both isomers to place bulky substituents *anti* to each other (Scheme 38). 2-(Trialkylsilyloxy)allylsilanes react as a synthetic equivalent of a 2,2′-dianion of acetone with two equivalents of aldehyde in the presence of Lewis acid to give dihydroxyketones in acceptable yield ⟨89JOC3254⟩. 2-(Chlorodifluoromethyl)allylsilanes react with aldehydes in the presence of copper and zinc to give 2,2-difluorohomoallylic alcohols with excellent 2-selectivity. The reaction is highly chemoselective: under usual reaction conditions, ketones are recovered unchanged (Equation (62)) ⟨91TL1069⟩.

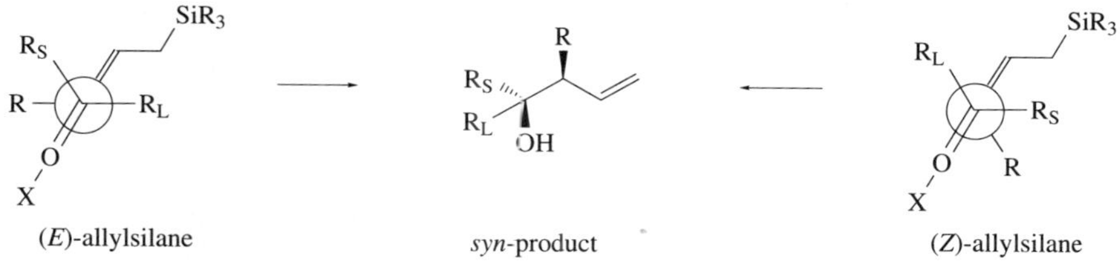

X = Lewis acid

Scheme 38

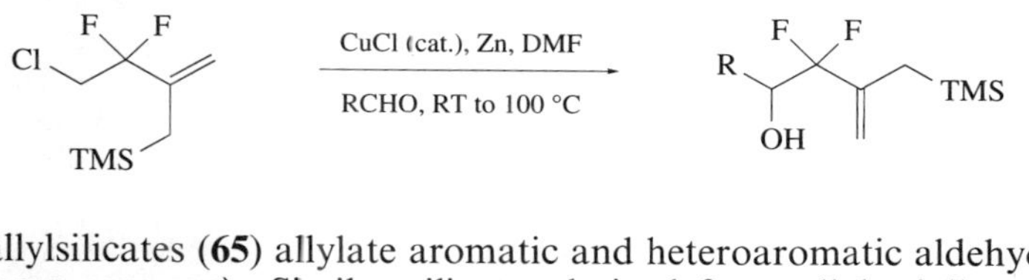

(62)

Pentacoordinate allylsilicates (**65**) allylate aromatic and heteroaromatic aldehydes in good yield ⟨87CC1517, 87JOM(328)C17, 88JA4599⟩. Similar silicates derived from allyl trialkoxysilanes will also allylate aldehydes ⟨90JOC2415⟩. Enantioselective allylation of aldehydes is feasible using optically active allyl siliconates ⟨88TL5667⟩.

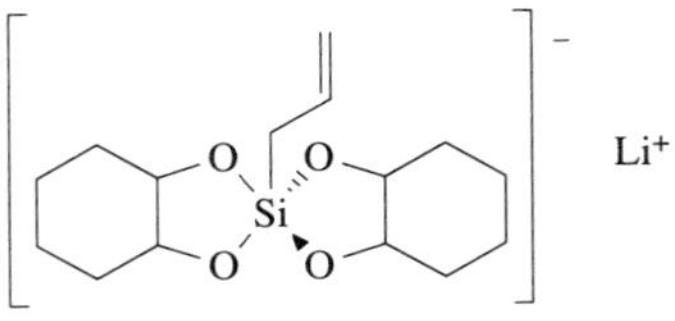

(65)

Sakurai has reported preparation of (*E*)- and (*Z*)-crotyl trifluorosilanes and has studied their reaction with aldehydes ⟨89TL1099⟩. Reaction of the (*E*)-isomer with aldehydes using caesium fluoride activation (so-called 'push' mechanism) gave good yields of almost exclusively *anti*-allylated products. (*Z*)-Crotyltrifluorosilane gave mainly *syn* products although the diastereoselectivity was not as pronounced as in the reaction of the (*E*)-crotyl compound. The reaction of these reagents and allyltrialkoxysilanes with 2-hydroxyketones has also been reported ⟨89JA6429⟩. (*E*)-Crotyltrifluorosilanes led to *syn* products, while the (*Z*)-isomer led to *anti* products, in contrast to the results reported above. A chelated cyclic transition state (**66**) is suggested to rationalise these observations.

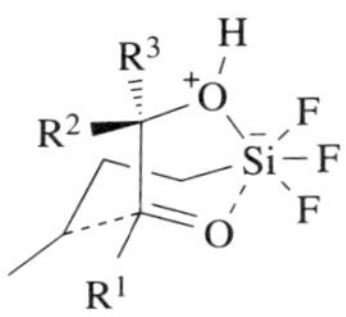

(66)

The stereoselective allylation of aldehydes occurs without the need for a catalyst upon reaction with allyltrichlorosilanes in DMF ⟨93TL3453⟩.

In an analogous fashion to allylboron reagents, asymmetry may be introduced into allylsilanes by means of a chiral ligand on silicon and so used for enantioselective allylations ⟨89JOC5768, 89TL3041⟩. Thus, reaction of proline- and pinene-derived trialkylallylsilanes (Scheme 39) with aldehydes in the presence of excess titanium(IV) chloride leads to homolallylic alcohols in mediocre enantiomeric excess (~50% *ee*).

Scheme 39

Allylstannanes are also widely used in allylation reactions of carbonyl compounds. Either thermal or Lewis acid initiation is required for high-yielding reaction of allyltrialkylstannanes with aldehydes; under thermal conditions, there is a cyclic transition state ⟨72JOM(35)C20⟩ and under Lewis acidic conditions an open-chain mechanism is involved ⟨84T2239⟩. The stereochemistry of reaction of allylstannanes with aldehydes depends on the Lewis acid employed. When BF_3 is used, both (*E*)- and (*Z*)-isomers react to give *syn*-allylated products, in direct analogy to the reaction of silanes ⟨84T2239, 87ACR243⟩. When $TiCl_4$ is used, exclusively *syn* or *anti* isomers may be obtained, depending on the order of addition of reagents; premixing of Lewis acid and stannane leads to *anti* isomer, via a titanium–tin exchange reaction ⟨84TL3927⟩. In the presence of $SnCl_4$, there is a rapid exchange of ligands, leading to highly reactive trichloroallylstannanes as the actual reacting species ⟨88JA984⟩. Reaction of such stannanes is often highly stereoselective (see below), because a cyclic transition state is feasible due to the Lewis acidity of alkyltrichlorotin. Chelation-controlled allylations using stannanes are also reported ⟨84TL1879⟩.

The use of cobalt(II) chloride in the reaction of allylstannanes with aldehydes leads to a reversal in the observed regiochemistry compared to the same reaction using conventional Lewis acids

⟨89TL2421⟩. Thus, the products of the reaction are those of *retention* of configuration rather than those of allylic transposition. Reaction of tributylallylstannanes with aldehydes in the presence of dibutyltin dichloride in water has been reported to give products of allylation ⟨88JOM(341)345⟩. Remote directing effects have been observed in allylations by allyltributyltin (Equation (63)) ⟨93SL81⟩.

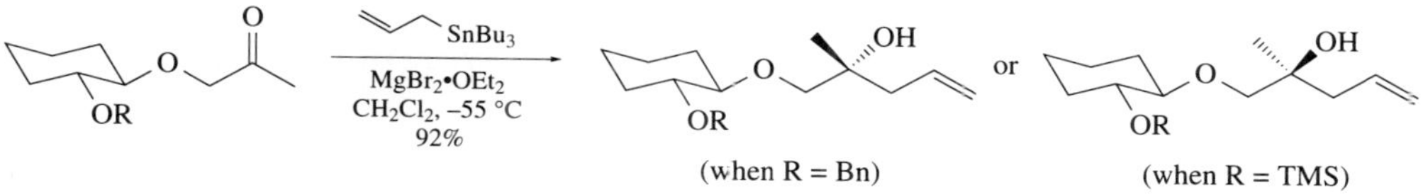

(63)

Enantiopure 5-alkoxyallyl stannanes react with aldehydes under thermal ⟨88TL2479⟩ and Lewis acid initiation ⟨90TL6239⟩. Owing to the high temperatures (130°C), poor ratios of diastereomers were obtained in the thermal reactions; in the presence of tin(IV) chloride, however, these reagents react with a wide range of aldehydes to give 5-alkoxypent-3-enols as single diastereomers (Equation (64)) ⟨94S322⟩. 2-Oxygenated allylstannanes are effective allylating reagents under thermal initiation and are *anti* selective ⟨89JCS(P1)1521, 89JCS(P1)1529⟩. Enantiopure 4-alkoxyallylstannanes execute enantioselective allylations of aldehydes in high enantiomeric excess (>95%) with good *syn* selectivity (85% *syn*) ⟨89TL2183⟩.

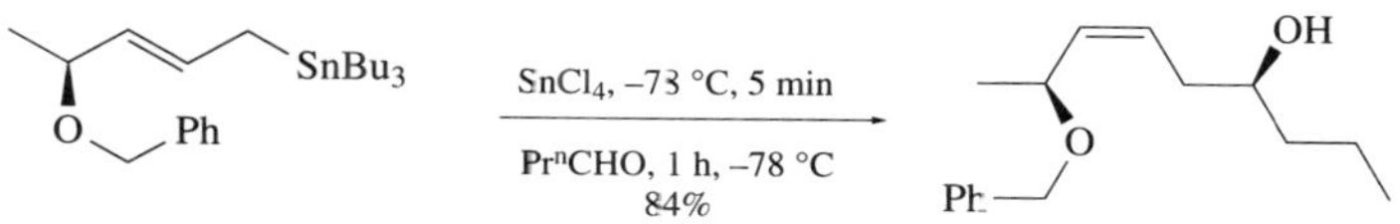

(64)

Reactions of 2,3-*syn*-alkoxy acylsilanes with allyltributyltin in the presence of zinc chloride lead to 4,5,6-*syn*-5,6-dialkoxy-4-hydroxy-4-trimethylsilylhexenes. Cleavage of the carbon–silicon bond with CsF furnishes the corresponding dialkoxyalkenols, with retention of configuration ⟨90JOC6071⟩. In contrast to the analogous boron reagents, tartrate-modified allylstannanes exhibit only moderate enantioselectivities in reactions with aldehydes ⟨88JOM(353)177⟩. Corey has described the uses of his chiral controller reagents in the propargylation and allenylation of aldehyde by stannanes ⟨90JA878⟩. In the alkylation of carbonyl compounds using (**67**) and (**68**), enantioselectivities are uniformly high (>91% *ee*) with enantioselection in the allenylation reaction virtually complete.

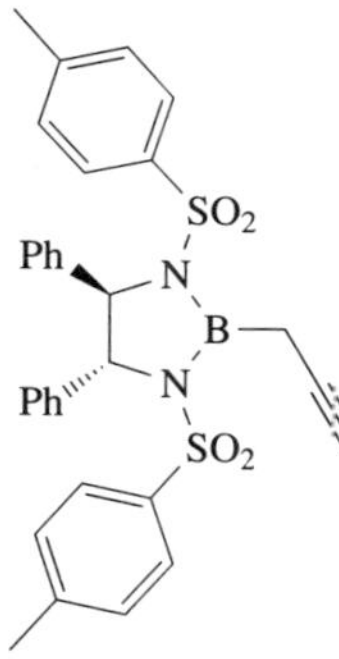

(67)

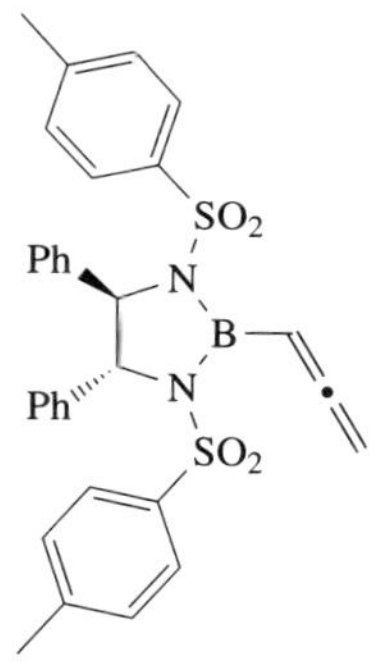

(68)

A selective allylation of aldehydes in moderate to good enantiomeric excess was effected by use of enantiopure 3-(tetrahydropyranyloxy)propenylstannanes (Equation (65)) ⟨92JOC7003⟩.

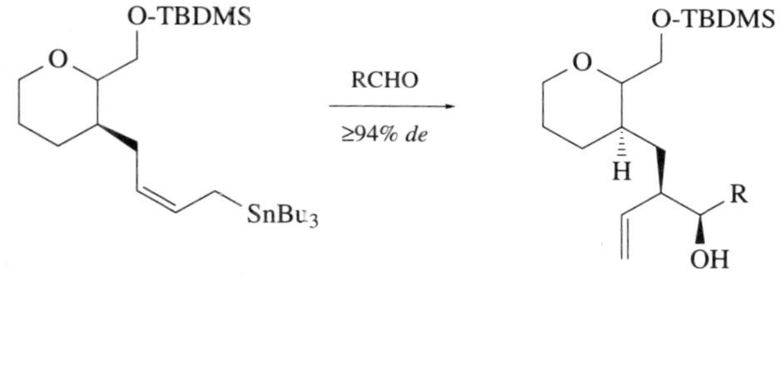

(65)

Allylic acetates and carbonates act as nucleophilic allylating reagents when reacted with aldehydes in the presence of palladium catalysts ⟨88CC44, 88JA4473, 88TL3563⟩. Polar solvents are essential for good yields of products; *anti*:*syn* ratios were variable.

2-Chloroallylicchromiums are reported to add to aldehydes with great stereoselectivity. In the product homoallylic alcohols, there is always a *Z*-configured alkene (>93% *Z*) and a predominance of *anti* isomer (>88% *anti*) ⟨89TL4389⟩.

An enantiopure allylating reagent derived from a combination of $CpTiCl_3$, diacetone glucose and allylmagnesium chloride reacts with aldehydes to give homoallylic alcohols of good enantiomeric excess (Equation (66)) ⟨89AG(E)494⟩.

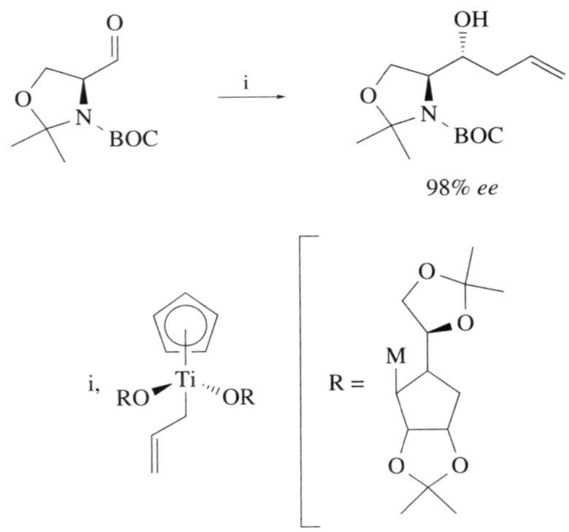

(66)

Simple alkenes will allylate aldehydes via an ene reaction; asymmetric variants using, in particular, glyoxalate esters have been reported. The reactions utilise titanium-chelated (*R*)-BINOL as asymmetric educt and simple alkenes produce predominantly *R*-configured homoallylic alcohols ⟨89JA1940, 90JA3949⟩.

Vinyl selenides and sulfides ⟨93CC327⟩ and symmetrical bis-alkenes ⟨92JA6566⟩ also undergo efficient reaction. The latter reaction offers a protocol for desymmetrisation of doubly allylic alcohols (Equation (67)).

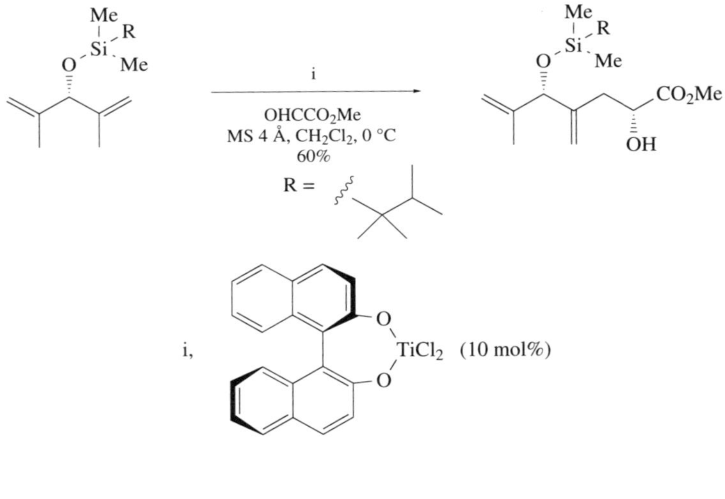

(67)

2.02.1.3.7 *Electrolytic and reductive coupling reactions*

The area of cathodic reduction of carbonyl compounds has been reviewed ⟨91COS(8)129⟩. The preparation of pinacols by such processes is a favoured reaction under electrolytic conditions because the mechanism for formation of such compounds requires only one electron to be transferred, whereas simple reduction requires a double electron transfer (Scheme 40).

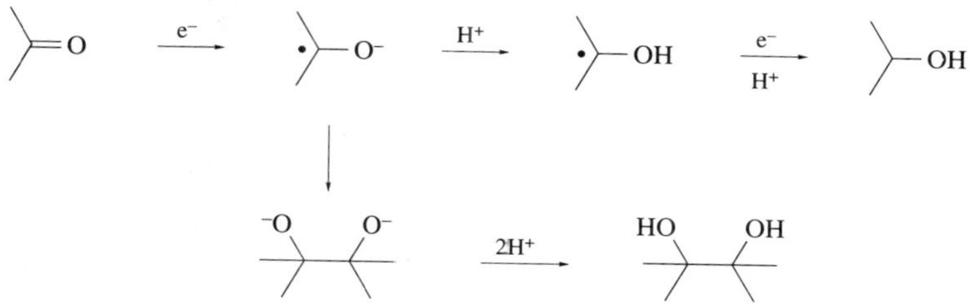

Scheme 40

The use of tetraalkylammonium salts as mediators allows simple reduction to be the exclusive reaction pathway ⟨89JA1211⟩. Similar types of single-electron reductions may lead to cyclisation if the structure contains unsaturation in addition to the carbonyl unit. Reductive tandem cyclisation of allyl pentenyl ketones is executed by electrolysis using cathodic reduction ⟨89JOC4022⟩. Thus, a series of bicyclo[3.3.0]octanols could be prepared from the corresponding allyl pentenyl ketones in acceptable yields. Electroreductive intermolecular coupling of ketones with alkenes provides access to tertiary alcohols in good yield (~80%) ⟨89JOC6001⟩. The reductive addition of polyhalomethanes to aldehydes using a lead–aluminum bimetal redox system has been reported to give good yields (>75%) of 2-polyhalomethyl alcohols ⟨89JOC444⟩.

Radical additions to carbonyl carbon centres may be promoted by a variety of low-valent metal species; such reaction of 2-dicarbonyl compounds with aldehydes is promoted by aqueous titanium trichloride ⟨89JOC3872⟩. Titanium trichloride in the presence of zinc–copper couple effects intramolecular reductive cyclisation of dialdehydes ⟨89TL1169⟩. Titanium tetrachloride may, in the presence of zinc–copper couple, also efficiently promote such reactions; thus reductive cross-coupling of glyoxalates with carbonyl compounds has been achieved ⟨89CL1401⟩. Diastereomeric excesses of the product diols range from poor (equimolar *syn*/*anti*) to moderate (82/18) with the *syn* isomer always dominant. Electronically similar aldehydes may be coupled to give diols in the presence of vanadium(II) complexes ⟨89JA8014⟩. Under similar conditions, reaction of 2-((alkoxycarbonyl)amino)aldehydes with aliphatic aldehydes provides a stereoselective method for the syntheses of 3-amino-1,2-diols, while reaction of 1,2-disubstituted 2-(diphenylphosphinoyl)-acetaldehydes and saturated aldehydes leads stereospecifically to 3,3-disubstituted allylic alcohols ⟨90JOC4506⟩. Diastereoselectivities of up to 200 : 1 (*syn* isomer always favoured) are possible.

Intramolecular reductive cyclisation reactions of pentenyl aldehydes and ketones bearing an electron-deficient δ,ε-alkene cyclopentanol are mediated by samarium diiodide ⟨89TL1063⟩. Reaction of hex-5-enal under the same conditions led only to products resulting from dimerisation. Other radical initiators will also promote such reactions ⟨89TL4939⟩. Thus, reaction with tributyltin hydride in refluxing benzene also led to products of reductive cyclisation in yields of up to 88% via addition of tributyltin radicals to the carbonyl bond followed by 5-exo cyclisation and hydride capture.

2-(But-4-enyl) keto esters undergo highly diastereoselective reductive cyclisation reactions in the presence of SmI_2 ⟨89JA8236⟩.

An intramolecular SmI_2-induced cyclisation reaction of alkenic carbonyl compounds is reported to lead to the facile preparation of 2-methylenecycloalkanols ⟨90TL4765⟩. Where substitution is present at the alkynic terminus, (*E*)-alkenes are usually the only products (Equation (68)).

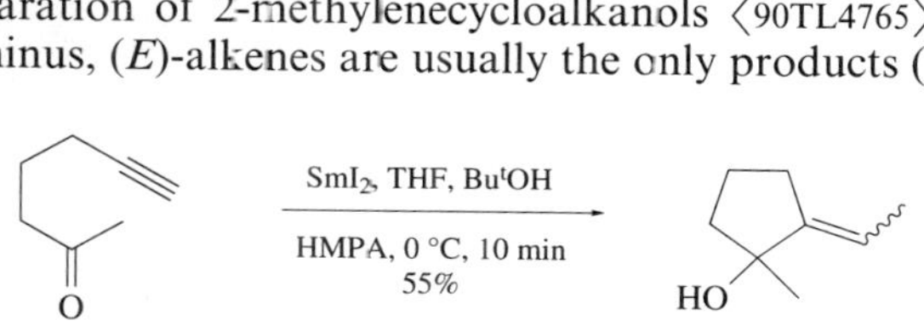

(68)

Electrolytic reduction of cyanoketones at a tin cathode allows cyclisation to occur in moderate to good yield (Equation (69)) ⟨90TL1303⟩.

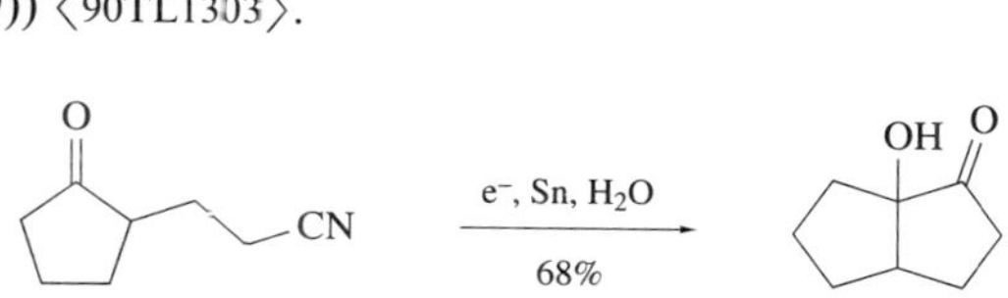

(69)

2.02.1.3.8 *Silicon and tin hydrides*

Esters may be reduced to primary alcohols via hydrosilylation using titanocene as catalyst. The reaction of esters with triethoxysilane under such conditions leads to triethoxysilylethers of primary alcohols, easily converted to alcohols upon treatment with mild acid or base. A modified 'second generation catalyst system' for ester hydrosilylation employing tetraisopropyloxytitanium as the transition-metal component has also been developed. The method possesses several advantages over the reagents available for such transformations; the catalyst is self-activating, the process requires no solvent, and the catalyst may be used and generated without the need for an inert atmosphere ⟨91JA5093, 92JOC3751⟩. Some selectivity is feasible, as demonstrated by the reduction of a methyl ester in the presence of a tertiary ester. Alkyl bromides are unaffected by the process. As an alternative to transition-metal catalysis, heterogeneous acid or base will enable the reduction of ketones and aldehydes to alcohols by triethoxysilane. Hydroxyapatite is a typical catalyst for the reaction ⟨91TL4741⟩. Tin hydrides are effective in reduction of ketones ⟨91COS(8)9⟩.

2.02.1.4 By Ether Cleavage

2.02.1.4.1 *Ring-opening of epoxides*

The ring-opening of epoxides is a much-used and valued synthetic procedure ⟨83PAC589, 84S629⟩. In asymmetrically substituted epoxides, there are two possible sites for ring-cleavage reactions and such reactions are known to occur under acidic, basic or neutral conditions ⟨71COC(5)123⟩.

(a) Under acidic conditions. The presence of Lewis or Brönsted acids favour that transition state which contains a relatively high proportion of carbocationic character and so these reactions tend to give products arising from ring-opening at the carbon of greater substitution. This is accounted for by a transition state in which the epoxide C—O bond is nearer cleavage than under nonacidic conditions.

(b) Under basic or neutral conditions. In the absence of acids, the ring-opening of epoxides proceeds via nucleophilic attack at the carbon of lesser hindrance.

(c) Epoxides derived from cyclic epoxides. Cyclohexene oxides undergo highly predictable ring-opening reactions owing to the precise orbital overlap requirements of the reaction. Thus, *trans*-diaxial ring-opening is the norm, with the reacting conformation often being that in which the number of equatorial or pseudoequatorial substituents is maximised (Scheme 41).

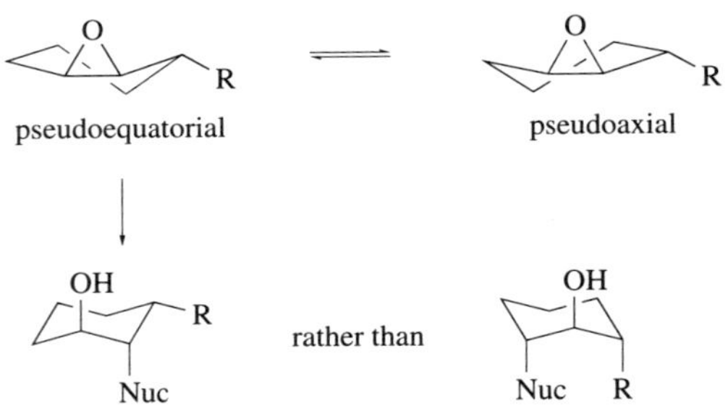

Scheme 41

(d) Ring-opening of 2,3-epoxy alcohols. The ready availability of enantiopure 2,3-epoxy alcohols by asymmetric epoxidation has led to the development of many methods for regioselective nucleophilic ring-openings of such epoxides ⟨91COS(6)1⟩. Both carbon nucleophiles and heteroatom nucleophiles may be constrained to react via either C-2 or C-3 attack (Scheme 42).

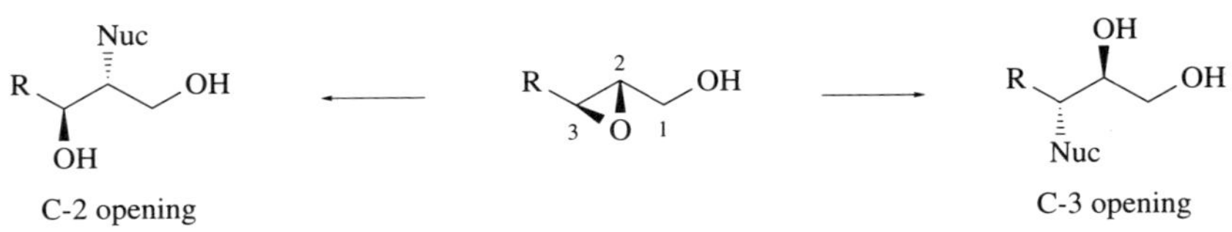

Scheme 42

(i) Reductive

Asymmetric hydrogenolysis of chiral *meso* epoxy acids is reported to allow preparation of malic acids in moderate enantiomeric excess ⟨91CC535⟩.

Reductive lithiation of epoxides with LDBB (lithium-4,4′-di-tert-butylbiphenyl) furnishes 2-lithio lithium alkoxides which may be reacted with a variety of carbonylic electrophiles to give alcohols ⟨90JOC1528⟩.

Photochemical ring-cleavage reactions of epoxides are known. Such reactions are regioselective but stereocatholic ⟨92CC1133⟩.

Epoxides may be reduced selectively in the presence of carbonyl groups using a mixture of methyllithium, copper(I) bromide and tributylphosphine to give alcohols resulting from attack of hydride at the least-substituted carbon atom ⟨90JA1286⟩.

An electroreductive ring-opening of 1,2-epoxy carbonyl compounds is mediated by a mixture of diphenyldiselenide and diphenylditelluride ⟨90JOC1548⟩.

(a) Reduction by metal hydrides. LAH reduces epoxides at the carbon atom of lesser substitution (Equation (70)); the presence of a Lewis acid inverts this inherent preference (Equation (71)).

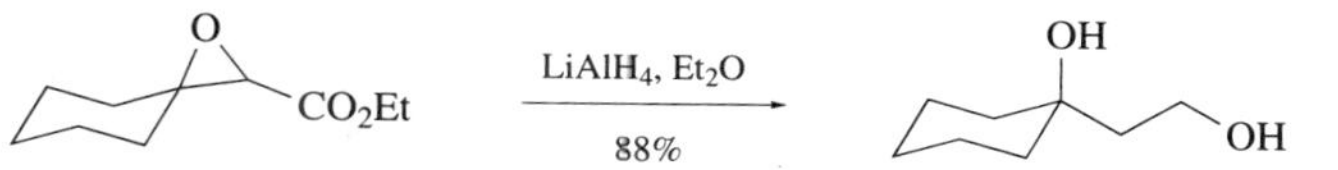

(70)

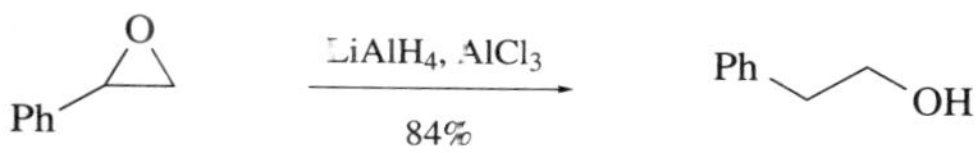

(71)

The reduction of epoxides by BH_3 is contrary to that often seen with other hydride reductants, since the product of the reaction is often the alcohol of *lesser* substitution and must, therefore, be formed by a nucleophilic attack of hydride at the *more* hindered carbon centre of the epoxide (Equation (72)) ⟨66JA2201, 68JA2686⟩. The reaction is high-yielding and is much accelerated by the presence of catalytic amounts of either lithium or sodium borohydride, or boron trifluoride. The fact that the reduction of epoxides derived from 1-alkylcycloalkenes therefore gives *cis*-2-alkylcyclo-alkanols which has been used to complement the preparation of *trans* isomers by hydroboration of the same alkene substrate (see above).

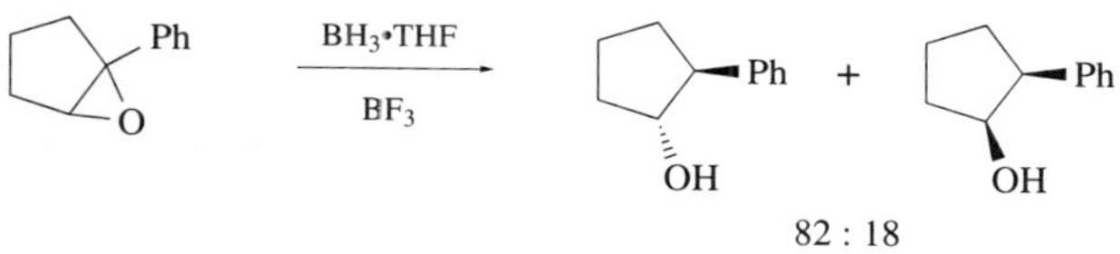

(72)

Trialkylborohydrides are more powerful *and* more selective reductants for such reactions (Equation (73)) ⟨73JA8486, 76JA3383⟩.

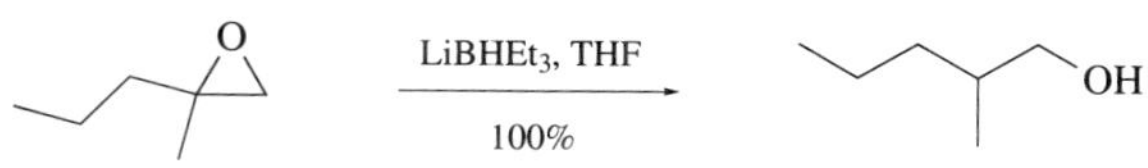

(73)

1-Substituted 1-alkynyl epoxides react with dibal-H regiospecifically to give products arising from nucleophilic hydride attack at the more hindered carbon. Even extremely hindered substrates react regioselectively and this is the method of choice for the preparation of such alcohols ⟨92TL7507⟩.

The ring-opening reactions of enantiopure 2,3-epoxy alcohols by reducing agents is a widely employed method for preparation of enantiopure 1,2- or 1,3-diols ⟨91COS(6)1⟩. Red-Al is a highly selective reagent which enables this reaction; 1,3-diols are generally obtained ⟨82TL2719⟩. The reaction of epoxy alcohols derived from allylstannanes is anomalous, with regiospecific C-3 reduction being observed (Equation (74)) ⟨92JOC46⟩.

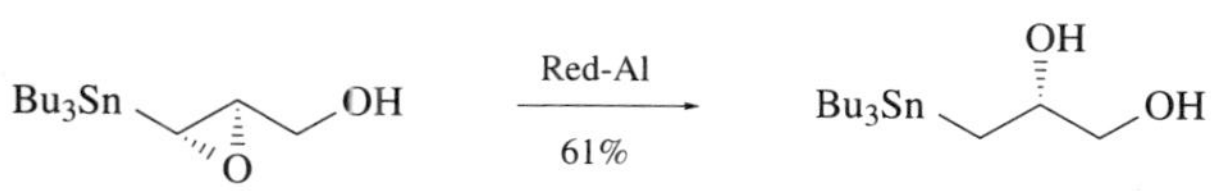

(74)

Lithium borohydride has been used in the regioselective ring-opening of 2,3-epoxy alcohols. The solvent employed in this reaction exerts a great effect on the selectivity observed: THF gives poor selectivity, while nonpolar solvents such as hexane and benzene allow the most selective transformations ⟨90TL7467⟩.

The reductive ring-opening of epoxy tetrahydrofurans is indiscriminate when LAH is employed, but proceeds with regiospecificity when heterogeneous catalysis is used (Equation (75)) ⟨91TL5273⟩.

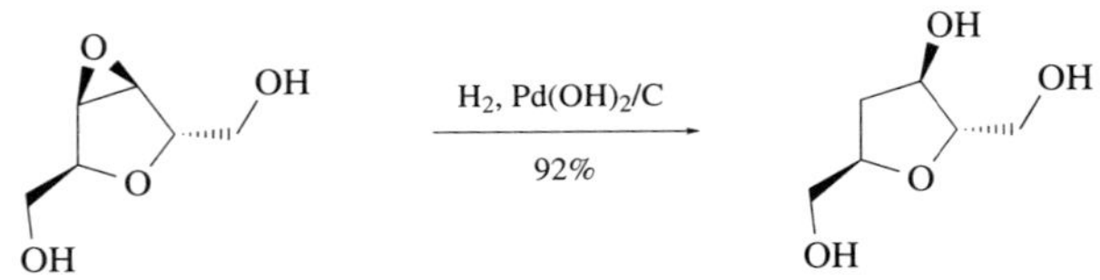

(75)

Pattenden and co-workers have devised an interesting reductive ring-cleavage of epoxy alcohols which allows subsequent cyclisation reaction. Treatment of linalool epoxide with Co-DMG (cobalt dimethylglyoxime) leads to a reductive formation of an alkylcobalt which undergoes photolytic Baldwinesque cyclisation to give 1,2-dihydroxy-1-methyl-4-(2-propenyl)cyclohexane in excellent yield (Scheme 43) ⟨91TL243⟩.

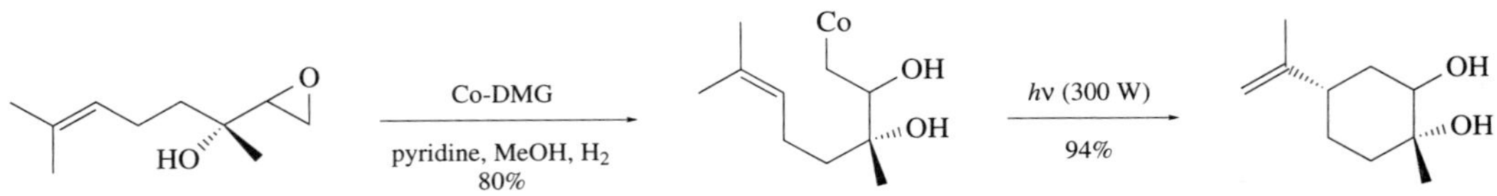

Scheme 43

(ii) Alkylative

(a) By heteroatomic nucleophiles. High pressure may be used to catalyse the hydrolysis of epoxides ⟨93TL4031⟩. Optically active ether-linked lipids have been prepared by regioselective C-2-opening of 1-tosyloxy-2,3-epoxides (tosylglycidols) with long-chain alcohols ⟨89JA3077⟩. Ring-opening of 2,3-epoxy alcohols by primary amines in the presence of alkoxytitanates is highly regioselective (in favour of C-3 attack) ⟨91TL6931⟩. ddq Catalyses the ring-opening of epoxides by alcohols (Equation (76)) ⟨90TL735⟩. $Ti(OPr^i)_2(N_3)_2$ has been used in the regioselective ring opening of epoxy alcohols derived from the asymmetric epoxidation reaction. The reagent is generally preferable to sodium azide (yields are higher, ratios of C-3 : C-2 opening are better) ⟨88JOC5185⟩. Tributylstannyl azide has been used to prepare 1,2-azido alcohols via least-hindered ring-opening of epoxides ⟨90TL221⟩. *trans*-2,3-Epoxy esters may be cleaved by hydrazoic acid to give 2-amino-3-hydroxy acids with high regioselectivity (Equation (77)) ⟨91TL667⟩.

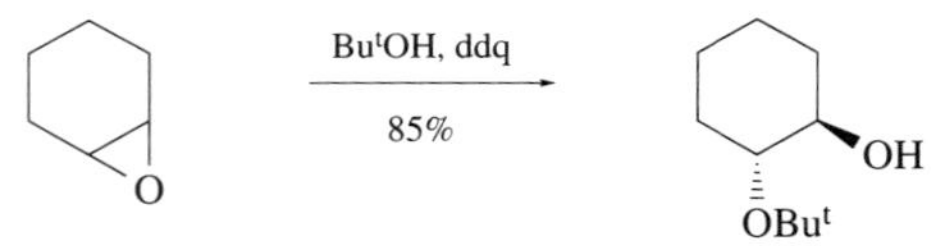

(76)

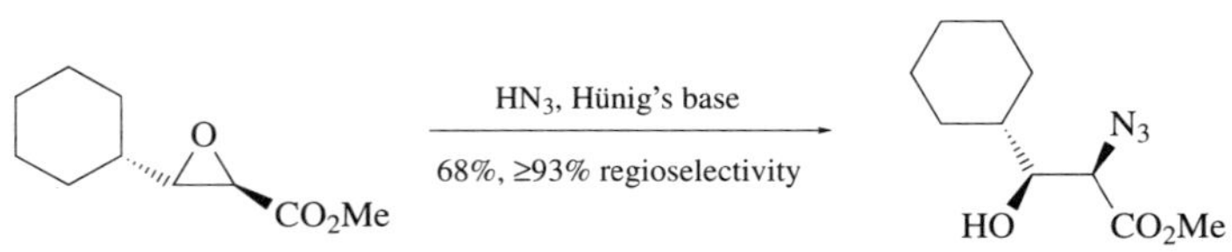

(77)

Regioselective ring openings of epoxyalcohols may be accomplished using $Ti(OPr^i)_4$ and benzoic acid to produce the diol benzoates, which can be converted without racemisation to 2-benzoyloxy acids ⟨88TL2701⟩. Dilithium tetrabromocuprate effects regiospecific C-3 ring-opening of glycidyl tosylates ⟨91SL248⟩.

Epoxides and cyclic ethers have been shown to undergo ring-opening when treated with triphenylphosphine in the presence of strong acids. The resulting hydroxyalkylphosphonium salts may take part in Wittig alkenation reactions to give unsaturated alcohols ⟨88BCJ4476⟩.

Asymmetric ring openings of *meso*-2,3-disubstituted epoxides with thiols, aniline and trimethylsilylazide may be carried out using metal salts of tartrates ⟨88BCJ1213, 88JOM(346)C7⟩.

Palladium(0) species catalyse the reaction of vinyl epoxides with sodium azide to give good yields of 1,4-ring opened products. Without catalyst, 1,2-ring opening predominates ⟨88TL4851⟩.

The ring opening of epoxides with amine–lead species (R_3PbNHR) is reported to give selective attack at the least-hindered carbon, even over attack at allylic positions ⟨89TL4255⟩.

(b) By carbon nucleophiles. Enantiopure *syn*-1,3-diols may be prepared by a sequence which has as its fulcrum the regiospecific ring-opening of disubstituted epoxides by proline-derived silyl anions. Stereochemical control in the reaction was complete and subsequent oxidation delivers diols ⟨92JOC6107⟩.

The presence of a tributylstannyl substituent can invert the regiochemistry of epoxide ring opening by directing the nucleophile to attack at the carbon atom adjacent to tin ⟨92JOC46⟩. Thus, 2,3-epoxy-3-trialkylstannyl propanoates are ring-opened by organocuprates at the 2-position, (Equation (78)), in contrast to the reaction of the simple epoxy ester, which gives products arising from nucleophilic attack at the 3-position ⟨84JA2949⟩. A convergent synthesis of polyol chains via alkylation of epoxides involves treatment of thioacetals with LDBB, thereby leading to 2-lithioalkoxy species which undergo alkylative ring-opening reactions with epoxides. The anions may be equilibrated to provide high diastereoselectivity. If the epoxide also bears a polyhydroxy functionality, the preparation of polyols is feasible ⟨89JOC4982⟩. Epoxides will add to activated alkenes in the presence of titanocene chlorides giving 5-hydroxyesters via a radical ring cleavage of the epoxide, following chelation of oxirane oxygen with low-valent titanium ⟨89JA4525⟩.

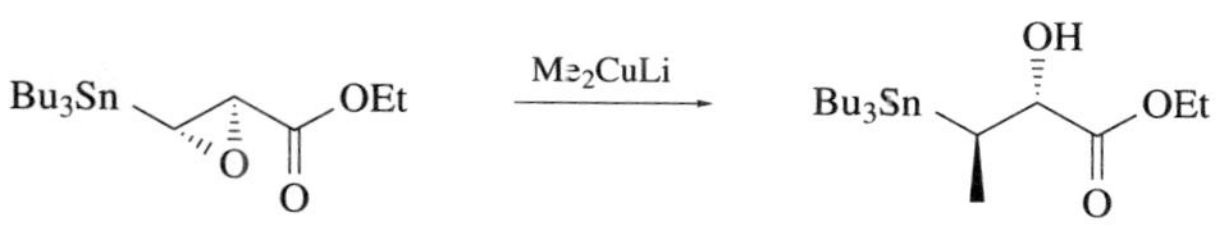

(78)

Ring-opening of enantiopure trifluoromethyl epoxides with organometallic reagents in the presence of copper(I) iodide gives good yields of 1-trifluoromethylalkanols via attack of the nucleophile at the least-hindered carbon. Friedel–Crafts type reactions of the same epoxides with arenes in the presence of aluminum trichloride give good yields of 2-aryl-1-trifluoromethyl ethanols, which is the opposite preference to the regiochemistry of ring-opening in the identical reaction of methyl epoxides ⟨90TL7031⟩.

(iii) Eliminative

The eliminative ring-opening of epoxides by amide bases has been reviewed ⟨83OR(29)345⟩. Epoxides derived from asymmetric 3-hydroxy sulfoxides react with strong base to give homoallylic alcohols in high optical purity ⟨88TL2851⟩. Ring-opening of epoxides by phenylselenide anions gives hydroxyselenides, which undergo oxidative elimination to give allylic alcohols. This process is, therefore, analogous to the base-mediated elimination but avoids strongly basic conditions ⟨73JA2697⟩. The reaction of 2,3-epoxy sulfones with alkyllithium reagents leads stereospecifically to 1-sulfonyl-1-alkenyllithium species, resulting from eliminative ring-opening (Equation (79)) ⟨89JOC1491⟩. Such anions react with aldehydes in the presence of TMEDA to give 1,4-dihydroxyalk-2-enes.

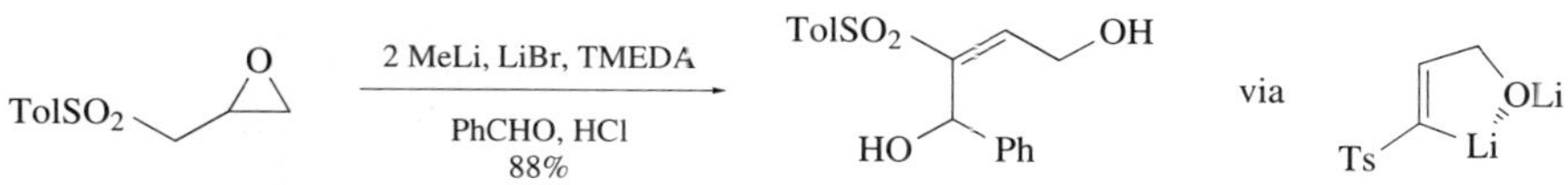

(79)

Ring cleavage of Sharpless-derived chloroepoxides gives good yields of optically active 3-hydroxyalkynes ⟨89CC1344⟩. Enantiopure 2,3-epoxyalkyl chlorides undergo eliminative cleavage to give either chlorovinyl alcohols or acetylenic alcohols simply by controlling the relative amount of the base used (Scheme 44) ⟨90TL4495⟩. Reaction of racemic epoxides with enantiopure lithium amides derived from Mukaiyama-type diamines allows a kinetic resolution to occur; the enantiomeric purity of the allylic alcohols thus obtained is poor ⟨89TL2125⟩.

2.02.1.4.2 *Vinyl epoxides*

Unsaturated epoxides may react with nucleophiles in a '1,2-' (S_N2) or '1,4-' (S_N2') fashion, leading to allylic or homoallylic alcohols, respectively. The palladium-catalysed reactions of

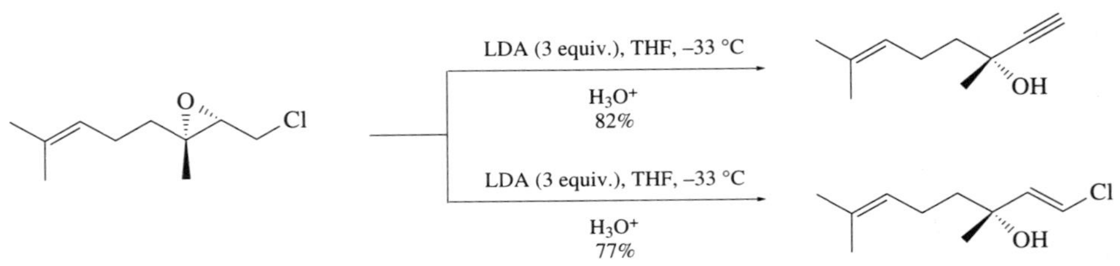

Scheme 44

organostannanes with vinyl epoxides proceed by a 1,4-mechanism ⟨89T979⟩ and are found to undergo improvement when water (10 equiv.) is present in the reaction medium ⟨88JA4039⟩. Organocuprates are similarly selective in reaction with vinylic epoxides ⟨88JOC4274, 90JOC1540⟩. Interesting reports of ring-opening of 1,2-epoxyalk-3-ynes have been documented ⟨89TL2387, 89TL2391⟩. Addition of Grignard reagents to 2-ethynyl-1,2-epoxycyclohexane may proceed to *syn* or *anti* allenic alcohols; the *anti* isomer predominates when a phosphine ligand is added to the reaction mixtures. The similar reaction of silylaluminum reagents with acyclic alkynic alcohols produces silylallenyl alcohols in good yield ⟨89JOC484⟩. Metallated ylides will ring-open unsaturated epoxides in an S_N2' fashion. Butadiene monoepoxide reacts with the anion derived from methylenetriphenylphosphorane to give 3-hydroxyhexa-1,5-dienes via alkenation of the initially formed 3-alkoxyphosphorane ⟨90JOC324⟩. Quaternary centres may be prepared by a metallative ring-opening reaction of vinyl epoxides. Thus, allylic chromium reagents may be prepared from vinyl epoxides and $CrCl_2$ at 0 °C in the presence of lithium iodide. These allylchromium reagents react with aldehydes to give 2-dialkyl-1,3-diols in good yield and with high stereopurity ⟨90JOC1705⟩. Vinylic epoxides may be coupled with aryl or vinyl halides to give allylic alcohols ⟨89TL3487, 90JOC6244, 93JOC804⟩. Vinylic oxetanes also undergo ring-opening by organometallic species in the presence of palladium ⟨93SL145⟩.

2.02.1.5 Miscellaneous Methods of Alcohol Preparation

2.02.1.5.1 Oxidative methods

Oxidation of simple hydrocarbons is not a generally useful process. Such nonenzymatic oxidations tend to be highly uncontrolled unless the substrate possesses symmetry or activated C—H bonds. The intramolecular activation of C—H bonds is a more controlled reaction and has been used to great effect. Lead tetra(trifluoroacetate) is capable of oxidising unactivated alkanes to trifluoroacetates which may be hydrolysed to alcohols ⟨67JA3662⟩. When alkanes are reacted with boric acid in air at elevated temperatures, a similar oxidation reaction occurs, giving alkyl borates which may be converted to alcohols in good yield. Tertiary alcohols may be synthesized by oxidation of the corresponding hydrocarbons ⟨B-89MI 202-03⟩. In particular, ozone absorbed on silica gel is an effective reagent ⟨75JOC2141, 88OSC(6)43⟩.

The oxidation of alkenes to give allylic alcohols is more fruitful and the reaction is of much greater utility. In such reactions, the major ambiguity of the transformations is the multiple sites capable of undergoing oxidation and the possibility of rearrangement reactions taking place; all these factors lead to the possibility of mixtures of products being obtained, but many reagent systems have been developed which overcome these hindrances ⟨B-79MI 202-01⟩. In particular, SeO_2 is a highly useful reagent for oxidation of allylic C—H bonds ⟨76OR(24)261⟩; SeO_2 may be present in only catalytic amounts when *t*-butyl hydroperoxide is used as stoichiometric co-oxidant, although cyclic alkenes give mixtures of products using this reagent combination ⟨77JA5526⟩.

2.02.1.5.2 From other alcohols

Treatment of alcohols with acid, particularly hydriodic acid, gives alkenes, which under the reaction conditions are converted to the most-substituted alkyl iodide. Treatment of these iodides with wet silver oxide gives the corresponding alcohol ⟨B-65MI 202-01⟩.

Secondary alkyl tosylates may be converted to the epimeric isomer by a two-step inversion procedure involving acetolysis using tetraalkylammonium acetate salts, followed by reductive or hydrolytic removal of acetate (Equation (80)) ⟨69JCS(C)1605⟩.

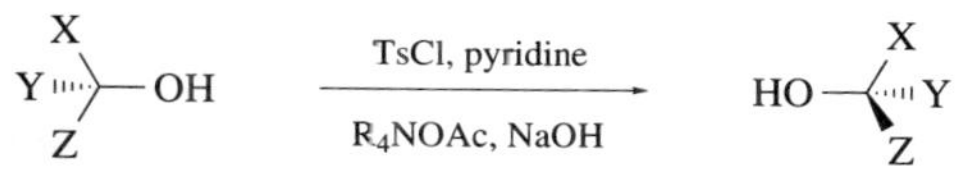

(80)

This procedure has been virtually completely superseded by the use of azodicarboxylates and triphenylphosphine as a reagent combination for reliable inversion of alcohols ⟨71BCJ3427, 72JA679⟩. This reaction (the Mitsunobu reaction ⟨81S1⟩) proceeds through formation of an oxyphosphonium species, which may be efficiently displaced by nucleophilic attack of a suitable acid. Benzoic acid is often employed in the reaction, leading to the benzoate derivative of the alcohol of opposite absolute configuration to that of the starting material. There are exceptions, but the reaction is widely recognised as the process of choice for inversion; application of the technique to nonasymmetric substrates is, of course, also a highly feasible process.

Superoxide radical anion is highly nucleophilic and its potassium salt may be used to effect nucleophilic displacement of leaving groups. The intermediate hydroperoxide which is the first product of the reaction is reduced *in situ* by the solvent to give alcohols in good yield. The reaction is applicable to primary, secondary and allylic halides and sulfonates (Equation (81)) ⟨75JOC1678, 75TL3183⟩. When certain metal alkoxides (particularly those of sodium or barium) are heated with the corresponding alcohol, alcohols containing two or three times as many carbon atoms are formed in variable yield (Equation (82)) ⟨1899CR(128)511, 01CB2346, 01LA(318)157⟩.

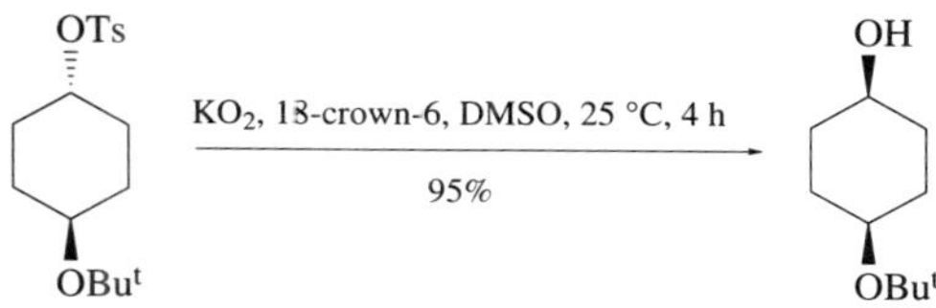

(81)

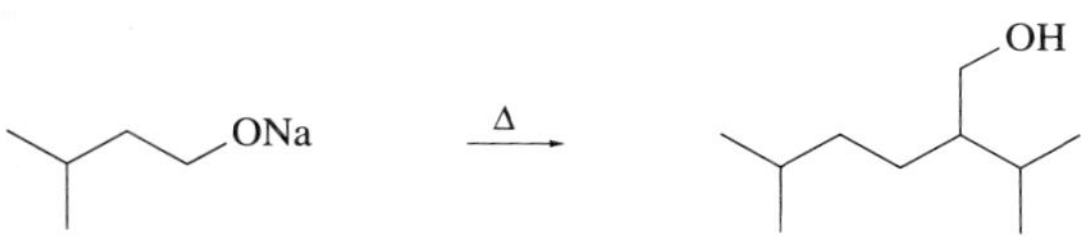

(82)

2.02.1.5.3 *Hydrolytic reactions*

There are a number of reliable methods for conversion of halides, sulfonates and other such groups into alcohols. Such reactions are favoured when there is activation, but the use of unmodified hydroxide is fraught with problems of eliminative side reactions. Where such reactions are problematic, milder alternatives (e.g., the use of carboxylates to effect a two-step conversion) or suitable activation (e.g., the use of silver salts to enhance nucleofugacity of halide) is essential. In particular, the use of lead oxide or wet silver oxide (Equations (83) and (84)) is effective in suppressing elimination, as is the use of silver or potassium acetate to displace halide ⟨23LA(431)231⟩. The general area has been reviewed ⟨B-65MI 202-02, B-68MI 202-01, B-77MI 202-04, B-79MI 202-03⟩ and a few methods are worth consideration here. Carbonate salts are reported as efficient reagents for hydrolysis of primary alkyl and benzyl halides to the corresponding alcohols ⟨86JOC3762⟩ and the process may be simplified by the use of polymer-supported reagents ⟨81S793⟩. Bis(tributyltin) oxide is an effective reagent for hydrolysis of primary alkyl iodides and bromides ⟨89TL279⟩. Allylic halides may be homolytically oxygenated to give the corresponding alcohols; hydroperoxides are intermediates (Equation (85)) ⟨91JA8981⟩.

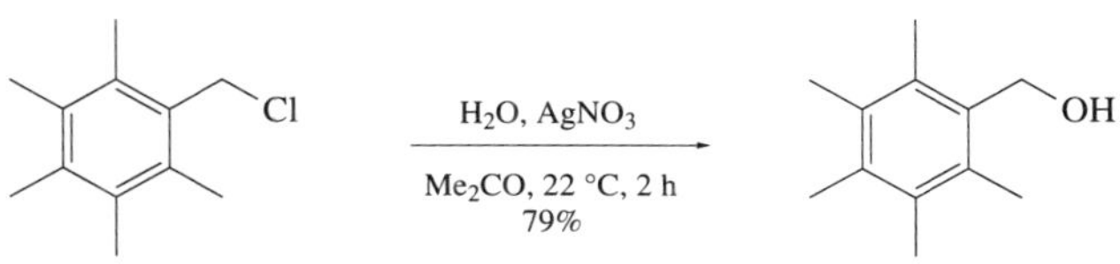

(83)

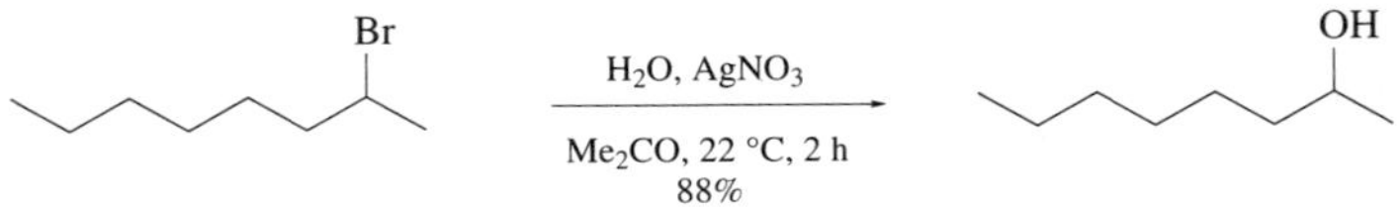

(84)

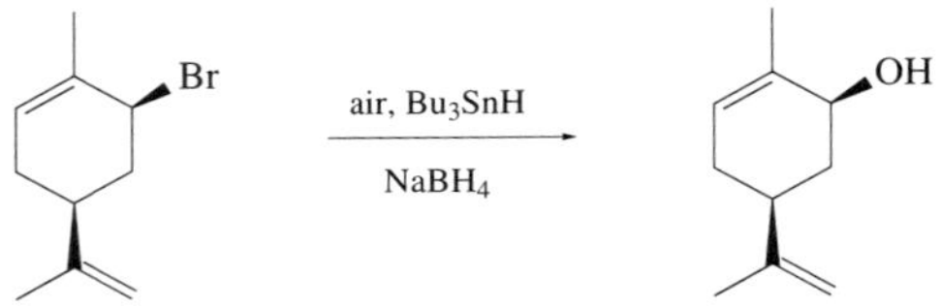

(85)

The preparation of *aliphatic* alcohols by displacement of amines is a reaction of limited utility, unlike the reaction of aryl amines. The primary means of activation of amines is diazotisation; aliphatic diazonium ions are notoriously capricious and highly reactive species and their reactions are consequently unselective; ethanol is prepared in 60% yield from ethylamine, but methylamine and propylamine give complex mixtures of alcohols ⟨41JA1118, 60NAT(188)1086⟩. Only in rigid systems, whose restricted conformational mobility diminishes the possibility of side reactions, is the process uniformly high yielding. Primary and secondary amines may be converted to the corresponding alcohols by sequential reaction with dinitrogen tetroxide and a reducing agent ⟨77JCS(P1)1114⟩. 2-Amino acids undergo high-yielding diazotisation reactions owing to the intramolecular trapping of the intermediate diazonium ion by the neighbouring carboxyl group, followed by displacement of carboxylate by incoming aqueous nucleophile. Overall retention of configuration is observed at the 2-carbon owing to the double inversion process ⟨64JA5326⟩.

Hydrolytic displacement of quaternised amines is sometimes possible when no elimination is allowed. Iron(III) oxidation of 1- and 3-amino esters leads to hydrolytic displacement of an amine moiety ⟨86JOC3913⟩.

2.02.1.5.4 *Alcohols from rearrangements*

Allyl iodides react with peracids to give iodonium oxides which undergo a sigmatropic rearrangement to give allylically transposed alcohols in good yield ⟨83JA2908⟩.

The sigmatropic rearrangement of anions derived from ethers leads to alcohols: the process is known as the Wittig rearrangement ⟨71BSF1921, 91COS(3)975⟩. Where one of the groups is allylic (the [2,3]-Wittig rearrangement ⟨86CRV885, 91COS(3)913⟩), high selectivity may be observed because of the concerted mechanism. The rearrangement of allylic ethers of glycolates proceeds with high specificity in the presence of dibutylboron triflate (Equation (86)) ⟨91TL4647⟩.

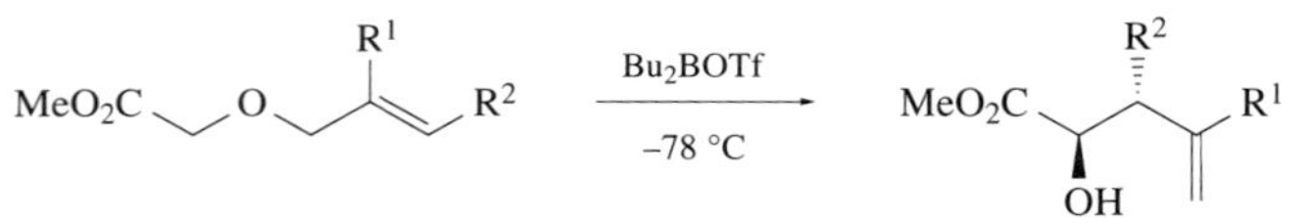

(86)

Lithiated cyclic enol ethers undergo alkylative cleavage reaction upon treatment with organolithiums. The products of cleavage are alkoxyvinyllithiums, which may be treated with suitable electrophiles to give 2- and 3-hydroxy di- and trisubstituted alkenes (Equation (87)) ⟨91TL5903⟩.

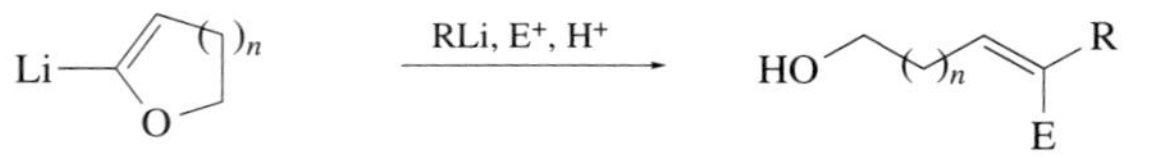

(87)

2.02.2 ETHERS

2.02.2.1 Acyclic Ethers

2.02.2.1.1 *General methods*

The best and most general method for preparation of ethers is that originally devised by Williamson, involving alkylation of alkoxides or their equivalents ⟨1852JCS106, 1852JCS229⟩. The Purdie modification introduced silver oxide as base to enhance the reaction by promoting etherification at the expense of eliminative side reactions seen when hydroxide is employed. The Lewis acidity of silver ion also assists the nucleofuge in its departure ⟨03JCS1021⟩. An example of the sensitivity of the Purdie modification is the preparation of ethers from asymmetric benzyl alcohols, where partial racemisation is observed when standard Williamson conditions are used. Thus, the use of silver oxide allows preparation of enantiomerically pure compounds ⟨51JA4043⟩. Similarly effective in suppressing racemisation during etherification of base-labile alcohols is thallium ethoxide ⟨75AG(E)762⟩. Acid catalyses promote the formation of symmetrical ethers from alcohols ⟨26CR(182)612⟩; asymmetric ethers may be prepared by the same method only when one of the alkyl substituents is primary and the other tertiary ⟨32JA2088⟩. Calixarenes have been used to catalyse Williamson-type syntheses of ethers. In some cases the yields of ether are superior to those produced using quaternary ammonium salts as catalysts ⟨88CL1773⟩. An improvement in yields of products of Williamson reactions is observed when cationic micellar catalysis is employed ⟨88T6681⟩. Suitable electrophiles for the reaction are alkyl halides/sulfonates and epoxides ⟨63HOU(6/3)1), B-65MI 202-01, B-91COS(1)819⟩.

Of the simple dialkyl ethers, only primary ethers may routinely be prepared in high yield. Even primary–secondary ethers are somewhat difficult to prepare.

Ethers may be obtained by addition of alcohols to alkenes under a variety of conditions. Under acidic conditions, ethers are obtained by a Markovnikoff addition ⟨51IEC1596⟩. A similar reaction is observed when alkenes are reacted with alcohols in the presence of mercuric salts, and in this latter reaction there are relatively few examples of the types of carbonium ion rearrangement frequently seen under acid catalysis. The products of the reaction are 2-alkoxyalkylmercury species, which may be protiodemercurated in high yield upon reaction with sodium borohydride ⟨69JA5646⟩. Ethers may be prepared by a two-step process involving reaction of alkenes with, firstly, NBS (to give the 2-alkoxyalkylbromide) and, secondly, with tributyltin hydride ⟨72S483⟩. Several methods are available for the preparation of ethers by reduction of acetals using a variety of hydride sources ⟨91COS(1,2), 91COS(8)211⟩. Ethers may be prepared by alkylation of acetals and their sulfur and nitrogen hemianalogues. A wide range of nucleophiles undergo the reaction ⟨91COS(8)235⟩. Allylic acetals react in the presence of a rhodium catalyst with TMS cyanide to give 2-cyanoethers. In the presence of silyl enol ethers, the products are 2-alkoxy ketones ⟨89CL1273⟩. Allylation of dimethylacetals occurs upon reaction of such compounds with TMS triflate and lithium triallylbutyl borate ⟨89TL2013⟩.

The reactions of substituted vinyl alkoxymethyl ethers with aluminum trichloride have been used to prepare 2-ketoethers ⟨88JOC2737⟩. The reactions of alkyl bromides with dibutyltin oxide in the presence of fluoride ion give good yields of symmetrical ethers. When the reactions are carried out using an *O*-tributylstannyl alkyl ether in place of bis(tributyltin) oxide, an unsymmetrical ether can be prepared, also in good yield ⟨88JA7737⟩.

Esters of carboxylic acids have been converted to the corresponding ethers by conversion to thionoesters (with Lawesson's reagent) and subsequent reduction with tributyltin hydride ⟨88PS(37)25⟩.

Esters can be converted into 1,2-difluoroethers via thionation with Lawesson's reagent followed by reaction with diethylaminosulfur trifluoride. Yields are good (68–88%) ⟨90JOC768⟩.

2.02.2.1.2 *Specific methods*

(i) Methyl ethers

Diazomethane will etherify alcohols in the presence of Lewis acids such as boron trifluoride etherate; yields are satisfactory with primary and unhindered secondary alcohols ⟨58JA2584, 59T36, 73OSC(2)245⟩. The same reaction using higher diazoalkanes is less reliable, except in the special case of diphenyldiazomethane, which reacts with alcohols in the absence of catalyst to give 1,1-diphenylmethyl (benzhydryl) ethers in good yield ⟨69JCS(3)749⟩.

Triethylsilyldiazomethane is a convenient reagent for the *O*-methylation of alcohols ⟨90TL5507⟩. This reaction represents an improvement over the previous method of choice for preparation of methyl ethers. Difluorocarbene alone from the family of dihalocarbenes reacts with alcohols to give 1,1-difluoromethyl ethers ⟨65JHC152⟩. Certain dialkyl peroxides or diacyl peroxides react with Grignard reagents to give ethers, but yields are variable ⟨50JA2735, 59JA4230⟩.

Trifluoromethyl ethers may be prepared by desulfurative fluorination of xanthates ⟨92TL4173⟩. Certain ethers possess selective lability which has allowed their use as protecting groups and there are correspondingly more reports of the synthesis of such ethers than their more inert counterparts.

(ii) Benzyl ethers

These ethers may be prepared in good yield from the reaction of alcohols with metal hydrides and benzylic halides. Frequently, benzyl chlorides are sufficiently electrophilic to give good yields in the reaction (in direct contrast to the Williamson reaction of nonactivated chlorides). Alternatively, benzylic bromides are widely applicable and benzyl triflates (generated *in situ* from the corresponding benzyl alcohols and triflic anhydride) may be used to effect benzylation under extremely mild conditions. Alcohols which are exceptionally difficult to benzylate may be etherified upon reaction with benzyl trichloroacetimidate ⟨81CC1240⟩. Similarly useful is *p*-methoxybenzyl trichloroacetimidate, which is sufficiently reactive to allow the benzylation reaction to occur in the presence of other acid- or base-labile groups ⟨88TL4139⟩. Triphenylmethyl (trityl) ethers of primary alcohols may be selectively formed in the presence of more substituted analogues.

t-Butyl ethers may be prepared in good yield by the reactions of alcohols with *O*-*t*-butyltrichloroacetimidate in the presence of boron trifluoride ⟨88TL2483⟩.

2.02.2.2 Cyclic Ethers

2.02.2.2.1 Oxiranes (epoxides)

Application of methods for the preparation of this specially reactive class of cyclic ether have received particular attention.

(i) Direct synthesis

The reaction of alkenes with peracids, the Prilezhaev reaction (Equation (88)) ⟨09CB4811, 53OR(7)378, 73TS93⟩ is a widely employed and well-investigated method for preparation of these three-membered ring ethers.

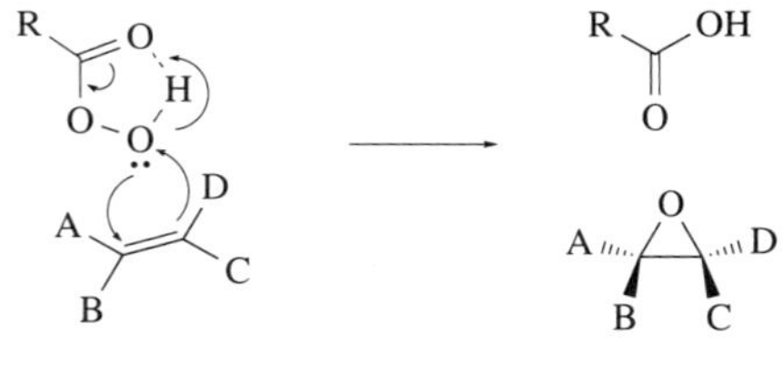

(88)

m-2-Chloroperoxybenzoic acid has enjoyed widespread popularity as an epoxidizing agent owing to its relative stability and ready commercial availability. A modern alternative is the magnesium salt of monoperoxyphthalic acid (MMPP) ⟨B-89MI 202-06, 93MI 202-02⟩. Frequently, peracids are prepared *in situ* by reaction of the corresponding carboxylic acid with hydrogen peroxide. H_2O_2 alone will not epoxidise electron-rich or electroneutral alkenic bonds, preferring to react with α,β-unsaturated alkenes bearing electron-withdrawing groups, but epoxidation of alkenes is catalysed by certain tungsten salts ⟨91COS(7)38⟩. Dioxiranes, especially dimethyldioxirane and its fluorinated analogues, have also been used as highly effective and selective epoxidising reagents ⟨89CRV1187⟩.

Metal-catalysed epoxidations using a combination of transition metals and alkyl hydroperoxides have been spectacularly effective in mediating stereoselective epoxidations ⟨91COS(7)357⟩. The general procedure involves reaction of (usually) *t*-butyl hydroperoxide or cumyl hydroperoxide with an alkene (electron-rich or electroneutral) in the presence of a transition-metal catalyst; often the reaction is regioselective, with a bias in favour of non-electron-deficient alkenes of higher substitution (Equation (89)) ⟨70JOC1839, 80JMOC107, B-81MI 202-01⟩.

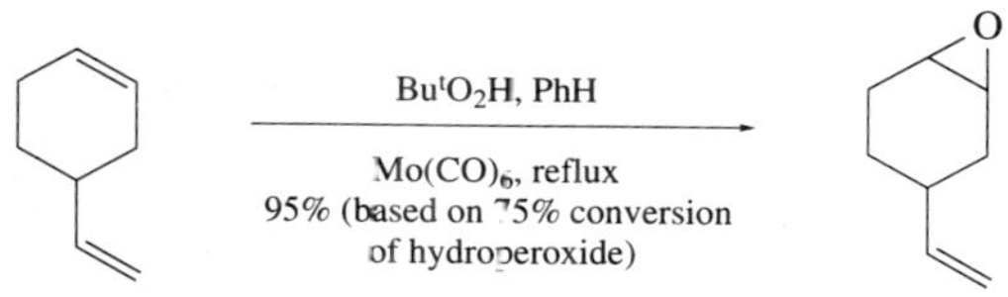

(89)

The reaction has been most widely studied using molybdenum, titanium, tungsten and vanadium catalysis, and stereoselectivities of these processes are impressive. When the alkenic substrate possesses no chelating groups, molybdenum catalysts are most effective (Equation (90)).

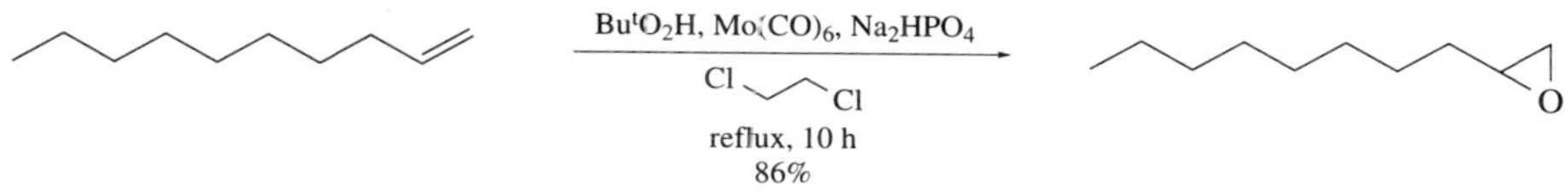

(90)

When the alkene contains ligating groups, directing effects are usually observed in transition-metal-catalysed epoxidations; most frequently, hydroxyl groups are used to direct face-selective delivery, but other oxygenated subunits may exert stereocontrol (Equation (91)) ⟨86JOC2505⟩.

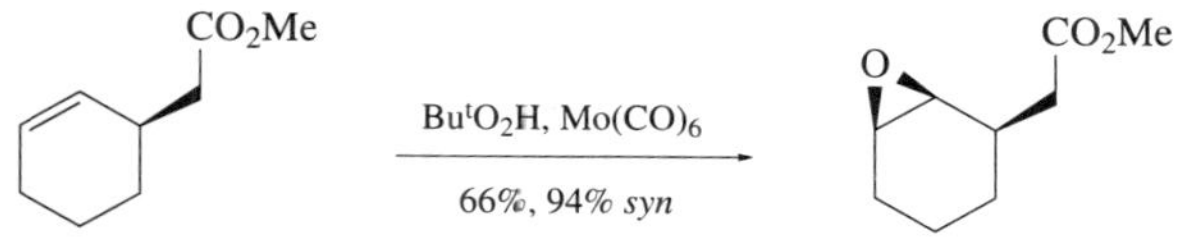

(91)

In the reaction of cyclic alkenes, the *syn*-directing effect of metal-catalysed epoxidations is much more predictable and powerful than in corresponding peracid epoxidations (Equation (92)) ⟨79JA159⟩. Epoxidation of highly substituted cyclopentenol (A) was a key step in studies directed towards vitamin B_{12} synthesis (Equation (93)) ⟨83JA7719⟩.

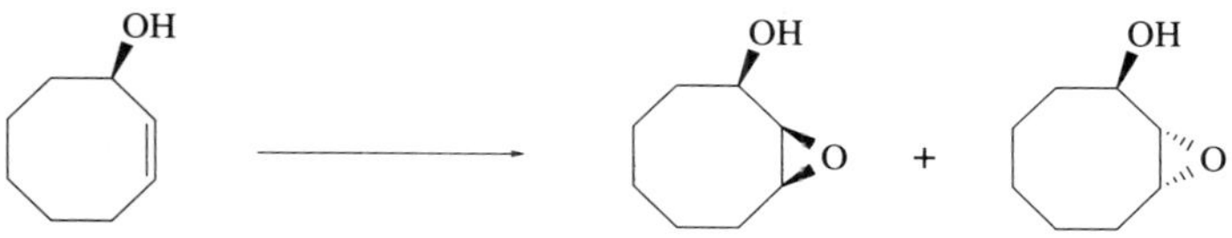

(92)

mcpba	0.2%	99.8%
$VO(acac)_2$, Bu^tO_2H	97%	3%

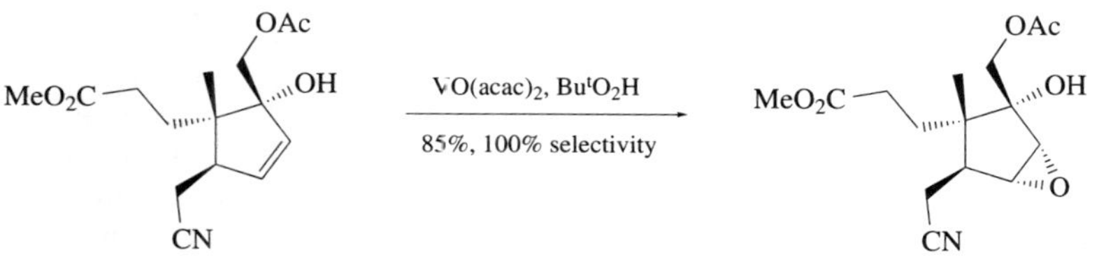

(93)

(A)

The epoxidation of acyclic allylic alcohols using vanadium catalysis is also highly stereoselective (Equation (94)) ⟨82TL4199, 83T2323⟩. There is a conformational rationale which allows accurate prediction of the stereochemistry of the product of vanadium-mediated epoxidations ⟨79MI 202-02⟩.

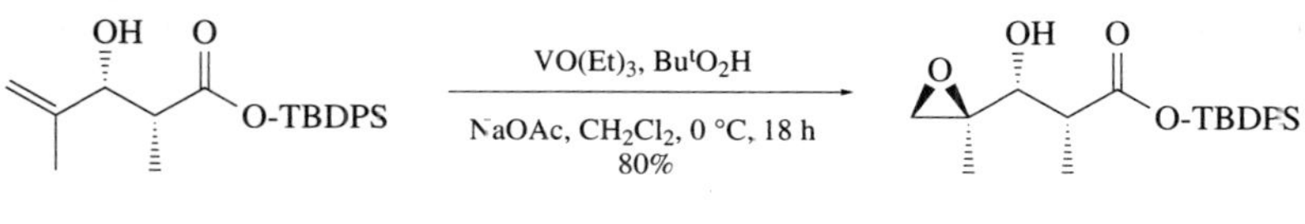

(94)

The *asymmetric* epoxidation (AE) of allylic alcohols is probably the most widely used and most reliable asymmetric transformation available to the practising synthetic chemist. First reported in 1980 ⟨80JA5974⟩, the reaction is basically a metal-catalysed, hydroxyl-directed *t*-butyl hydroperoxide epoxidation in which asymmetry is generated by the use of tartrate esters as ligands. Thus, when an allylic alcohol is reacted with Bu^tOOH in the presence of $Ti(OPr^i)_4$ and either (+)- or (−)-dialkyl tartrate, a 2,3-epoxy alcohol is produced with enantiomeric excess of usually more than 90%. The reaction exhibits superb chemoselectivity (Equation (95)) ⟨88JA4818⟩. Furthermore, the enantiocontrol of the process is entirely predicted by the mnemonic shown in Scheme 45 ⟨91COC(7)389⟩. Thus, use of D-tartrates leads to epoxidation from above the alkene (which must be represented as shown in the scheme, with the C—OH bond *syn* to and coplanar with the alkene), and L-tartrate shows an inverse preference.

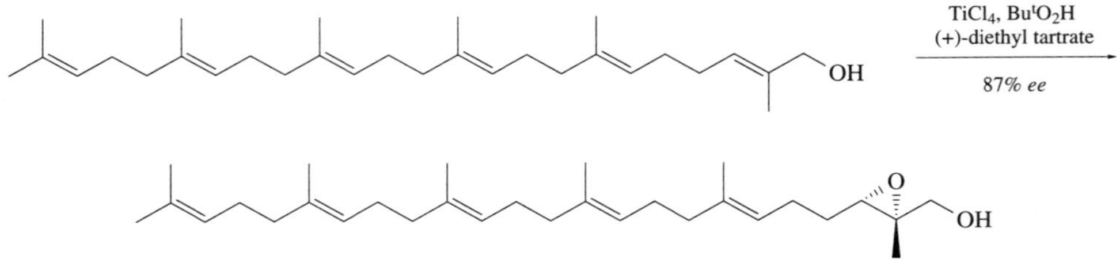

(95)

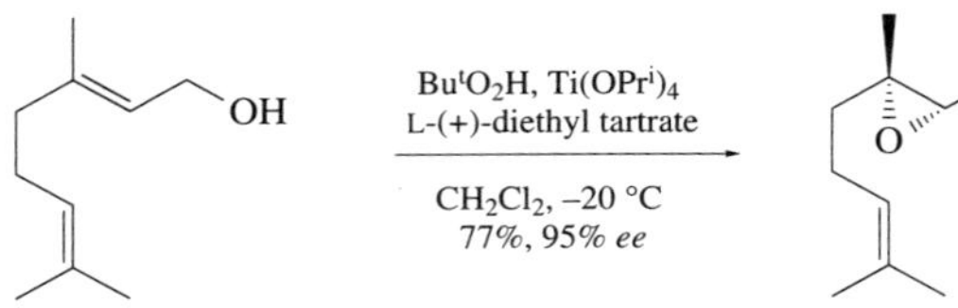

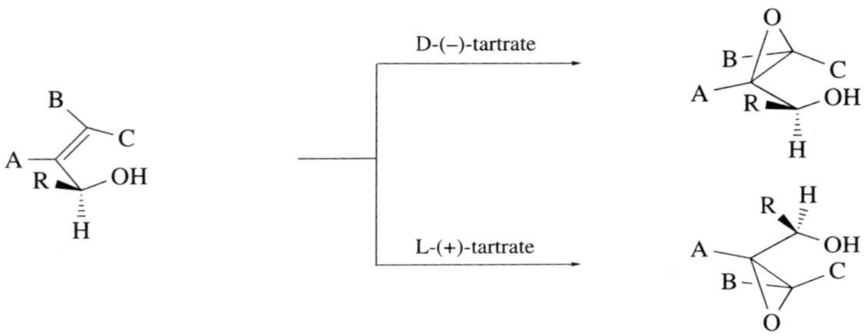

Scheme 45

When the allylic alcohol contains an α-substituent R, the rate of the reaction depends upon the absolute configuration of the α-centre. When the configuration is as shown in Scheme 45, the reaction with D-tartrate is slower than that with L-tartrate. Extrapolating this notion, when *racemic* α-substituted allylic alcohols undergo AE, one of the enantiomers present usually reacts faster than the other: in the case of (*E*)-1-cyclohexyl-1-hydroxybut-2-ene, the (*S*)-isomer reacts with L-(+)-diisopropyltartrate 104 times faster than the (*R*)-enantiomer. In many cases this difference in rate due to double asymmetric induction allows for a *kinetic resolution* of enantiomeric allylic alcohols (Scheme 46). Thus, if a mixture of enantiomeric α-substituted allylic alcohols is subjected to an AE reaction *run to 50–60% conversion*, it is possible to isolate both epoxy alcohol and unreacted enantiomeric alcohol, both of high enantiomeric excess. When operating under kinetic resolution, the bulk of the ester substituent exerts a pronounced effect, with larger alkyl groups giving better selectivity ⟨81JA6237⟩.

*Homo*allylic alcohols also undergo AE, but usually with much lower enantiomeric excess and with inverted enantioselectivity compared to that seen for the lower homologue ⟨91COC(7)411⟩.

Homochiral (salen)manganese(II) complexes are highly selective catalysts for asymmetric epoxidation of conjugated (*Z*)- di- and trisubstituted alkenes ⟨94JOC4378⟩.

2.02.2.2.2 Oxetanes

(i) Direct synthesis

The Paterno–Büchi photocyclisation (Equation (96)) ⟨68MI 202-03, 91COS(5)151⟩ of aldehydes and ketones with alkenes is a major preparative route to oxetanes.

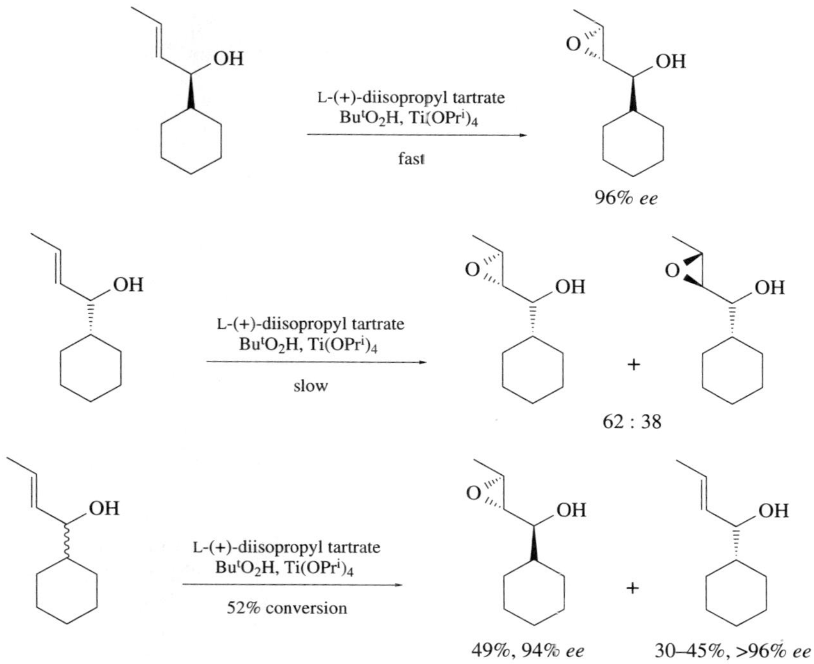

Scheme 46

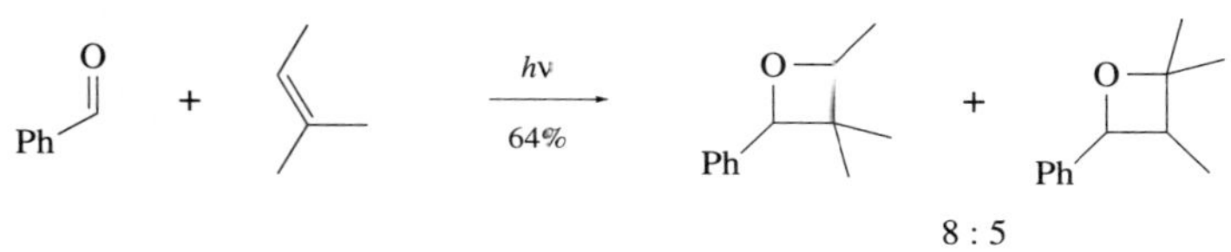

(96)

(ii) Indirect synthesis

The synthesis of oxetanes by two-step reactions is usually considerably less straightforward than that of epoxides. This is due to the absence of kinetic acceleration of rate which is present in epoxide formation. The rate of formation of oxetanes has been estimated to be 1/100 that of the corresponding oxirane reaction, even when the reaction is carried out at 80 °C ⟨54ACS135⟩.

2.02.2.2.3 *Other cyclic ethers*

The synthesis of cyclic ethers by cyclisation of hydroxyalkenes is a highly useful route to cyclic ethers, especially tetrahydrofurans and tetrahydropyrans (Equation (97)). The selectivity in such processes is frequently high ⟨87TL523, B-89MI 202-01, 91COS(4)363⟩.

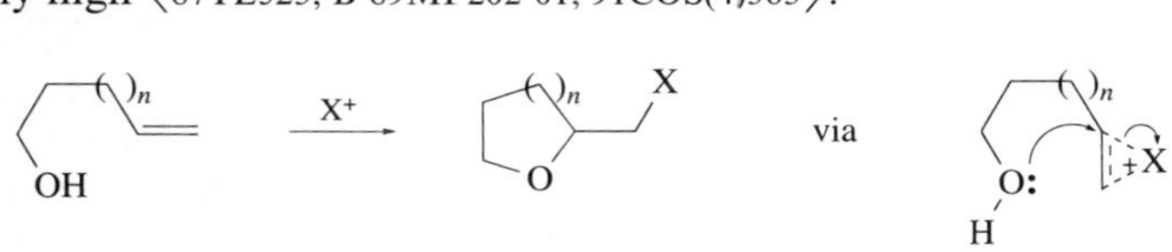

(97)

The insertion into C—H bonds of oxy radicals is a versatile method for preparation of cyclic ethers; lead tetraacetate is particularly useful (Equation (98)) ⟨70S209⟩.

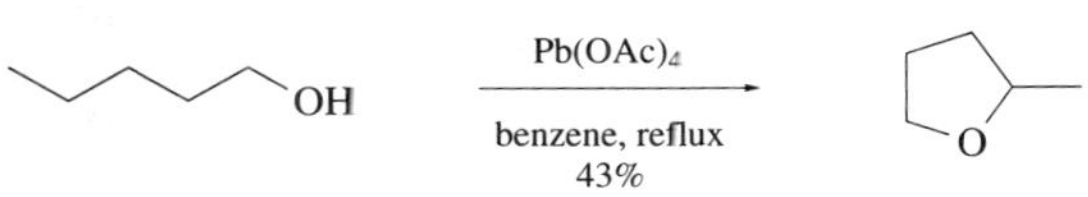

(98)

The insertion reaction of oxy radicals derived from alkyl hypohalites is mediated by soluble silver salts (Equation (99)). Both hypobromites ⟨71CC474⟩ and hypoiodites ⟨71S501⟩ may be employed in the reaction.

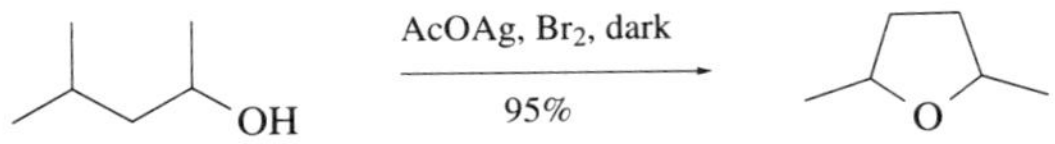

(99)

2.02.3 ALKYL HYPOHALITES

2.02.3.1 Hypofluorites

Hypofluorites have been known for many years; however, it is only in the 1960s that *simple* alkyl hypofluorites have been prepared. Perfluorinated alkyl hypofluorites are vastly more stable than the corresponding alkyl analogues; such compounds may routinely be prepared by reaction of perfluoroketones with elemental fluorine at low temperature ⟨66JA4531⟩. The compounds are stable at room temperature (certain derivatives are reported to possess extreme stability ⟨65JA230⟩), but are extremely powerful oxidants and fluorinating species. Trifluoromethylhypofluorite has been prepared by fluorination of methanol and the mono- and dioxides of carbon ⟨90JOC5155, 91JA2648⟩. The simplest member of the family, methyl hypofluorite, was first prepared in 1990; characterisation was later reported. The instability of the compounds arises from the propensity to evolve HF and has precluded synthesis of other hypofluorites bearing (potentially acidic) β-hydrogens. The synthesis of *t*-butyl hypofluorite (in which, of course, no elimination is directly possible) reported in 1993: the compound may be prepared by reaction of the alcohol with elemental fluorine at low temperature in nitrile solvents ⟨93JA1379⟩.

2.02.3.2 Hypochlorites and Higher-coordinate Derivatives

There are several general methods for synthesis of alkyl hypochlorites: the first involves the reaction of an alkaline solution of the corresponding alcohol with elemental chlorine ⟨23JCS2999⟩, the second requires reaction of the same alcohol with hypochlorous acid in inert solvents ⟨25JA395, 54BSF1109⟩, while the third utilises household bleach (sodium hypochlorite) as chlorine source in the presence of alkanol and acetic acid ⟨63OSC(4)125⟩. As generally observed for this class of compound, instability increases with lessening substitution: primary and secondary hypochlorites are unstable in the presence of bases or in light (methyl hypochlorite is an explosive gas), while tertiary hypohalites require the application of heat for their decomposition. *t*-Butyl hypochlorite is indefinitely stable at temperatures lower than ambient, when stored under darkness in inert solvents ⟨63OSC(4)125⟩. Perfluoroalkylhypochlorites may be obtained from the reaction of chlorine monofluoride with perfluoroketones ⟨69JA1310⟩.

2.02.3.2.1 Alkyl perchlorates

The reaction of alkyl iodides with silver perchlorate has long been known to yield alkyl esters of perchloric acid ⟨36ZAAC(228)341⟩. The method first described in 1961 by Radell is the most frequently used procedure: this involves the reaction at room temperature of a large excess of benzene-azeotroped silver perchlorate with the particular alkyl iodide required ⟨61JA3985⟩. The primary, secondary and tertiary alkyl perchlorates produced by this process are freed from solid by-products and then distilled *in vacuo* or (more safely) used in dilute solution. Details of thermal and drop-weight stabilities of representative alkyl perchlorates are reported. *It should be taken as read that these esters explode extremely readily* and this violent instability has precluded their characterisation. Alkylation of aromatics in the presence of methyl perchlorate has been taken to provide some evidence of ionic character in alkyl perchlorates ⟨56JCS3933⟩. The complexes formed between triethyl thiophosphate and metal perchlorates are reported to decompose in an ethanol–triethylorthoformate mixture at temperatures ranging from room to 90 °C to give ethyl perchlorate. The authors do not indicate whether the reaction is of preparative utility ⟨86ICA179⟩.

2.02.3.3 Hypobromites

There are few reports of synthesis of alkyl hypobromites. Walling and Padwa reported the preparation of *t*-butyl hypobromite as a distillable liquid from the reaction of hypobromous acid and the tertiary alcohol ⟨62JOC2976⟩, while French workers prepared the same species from the reaction of sodium hypobromite, sodium bromide and *t*-butanol at room temperature ⟨63BSF470⟩.

2.02.3.4 Hypoiodites

Alkyl hypoiodites have not been isolated in purity, but are fairly easily prepared *in situ* by the reaction of alcohols with either acetyl hypoiodite or *N*-iodosuccinimide ⟨61E475, 83JOC3126⟩. Also fruitful in this respect is the reaction of *t*-butoxide with elemental iodine ⟨84JA5261⟩. Ethyl hypoiodite has been prepared by the reaction of ethyl hypochlorite with iodine or methyl iodide ⟨75JOC2536⟩, while *t*-butyl hypoiodite is given by the reaction of the corresponding hypochlorite with mercuric iodide ⟨68JA808⟩. Although all of these procedures deliver substances which react as hypohalites, there is some evidence that these may not be the *actual* reagents involved; in the light of the lack of characterisation, the issue remains unconcluded. A report has described the preparation of methyl, ethyl, 2-propyl, *t*-butyl, 1-pentyl, cyclopentyl and cyclohexyl hypoiodites from the reaction of the corresponding alkanols with a solution of acetyl hypoiodite prepared *in situ* by disproportionation of iodosobenzene diacetate. The data obtained from *t*-butyl hypoiodite generated in this fashion is in accord with Tanner's findings ⟨94TL1003⟩.

2.02.4 PEROXIDIC FUNCTIONS

A three-volume review by Swern covers the vast majority of the chemistry of preparation of organic peroxides and to this the reader is directed for the details of preparative methods available ⟨B-71MI 202-02, 79COC(4)909⟩.

2.02.4.1 Alkyl Hydroperoxides

2.02.4.1.1 From hydrogen peroxide, by displacement of alkyl sulfates or sulfonates in basic solution

Hydrogen peroxide, and the anion derived therefrom, are highly nucleophilic moieties and may be used to directly introduce the hydroperoxy function in a single step, in contrast to other reagent systems. A wide range of the usual types of electrophile serve as substrates for the reaction. The most reliable method for preparation of primary and secondary hydroperoxides (the least stable of the family) is the displacement of mesylates by alkaline hydrogen peroxide ⟨54JA2984, 54JA2987, 60JOC621⟩. Allyl hydroperoxide (notoriously unstable) may also be prepared by this method ⟨57JA3474⟩. The reaction is S_N2 and, therefore, leads to inversion of configuration where appropriate. Tertiary alcohols may be converted to hydroperoxides by reaction with hydrogen peroxide and concentrated (70%) sulfuric acid ⟨79COC(4)909⟩. The reaction proceeds through *in situ* conversion to alkyl hydrogensulfates and subsequent displacement.

2.02.4.1.2 By displacement of alcohols and ethers

Concentrated hydrogen peroxide (a most hazardous substance) reacts with certain alcohols in the presence of catalytic amounts of mineral acid to give hydroperoxides in high yield ⟨79COC(4)909⟩. As might be expected, the most suitable substrates for the reaction are those alcohols whose structure allows for stabilisation of carbenium ion character ⟨53JCS1541, 54JCS2200⟩. Certain esters and ethers ⟨56JCS4669⟩ also undergo the reaction. Optically active alcohols normally undergo complete racemisation ⟨62JCS4288⟩.

2.02.4.1.3 By displacement of halides

Reactive halides may be converted to hydroperoxides upon reaction with hydrogen peroxides, but the reaction is somewhat underemployed because it is usual to prepare halides from alcohols and therefore it is simpler to start from alcohols as described above. The use of silver salts in this reaction greatly enhances its scope. Primary, secondary and tertiary alkyl hydroperoxides may be obtained by reaction of alkyl bromides or iodides with hydrogen peroxide in the presence of silver trifluoroacetate ⟨76CC1022⟩.

2.02.4.1.4 By addition to multiple bonds

Hydrogen peroxide is known to add to alkenes, carbonyls and imines to give hydroperoxides in variable yield, in some cases excellent. Electron-rich alkenes undergo S_N1 addition reactions with hydrogen peroxide ⟨53JCS1541, 54JCS2200⟩.

Imines react with hydrogen peroxide at low temperature to give 2-hydroperoxy amines, but at higher temperatures another mole of imine reacts with the initial product to give a symmetrical bis-peroxide ⟨58CB1212⟩. Some imines react at low temperature to give only 2-hydroperoxy amines ⟨65AG(E)524⟩.

Bis-peroxides are also obtained when aldehydes and ketones react with hydrogen peroxide: the first product of the reaction is the simple addition compound, which reacts with another molecule of carbonyl compound to give symmetrical 1,1-dihydroxydialkyl peroxides ⟨58AG251⟩. The reaction is in equilibrium, the attainment of which is much accelerated by the presence of an acid catalyst. It is usual to obtain 2-hydroxyhydroperoxides only from aldehydes; only ketones containing certain structural features (such as 2-halo substituents ⟨58JOC1322⟩) give hydroperoxides, and simple ketones give 2,2-dihydroxydialkyl peroxides (*inter alia*). Furthermore, most ketones give a wide variety of peroxides, all of which are extremely shock-sensitive and thermally unstable.

2.02.4.1.5 By oxidation of alkanes, alkenes, ethers, alcohols and ketones

The reaction of molecular oxygen with organic compounds (so-called autoxidation reactions) is a widely used process of general utility for preparation of alkyl hydroperoxides. As with most uncontrolled free-radical reactions, initiation is required and major by-products (often problematic to remove) may result, and this is the major limitation to the method. Alkanes may be converted to alkyl hydroperoxides, but the reactions must be carried to low completion levels, since it is often easy to separate products from unreacted starting material. The industrial preparation of cumene hydroperoxide is the classic example of the utility of this process. Because radical stabilisation aids the reaction, allylic, benzylic and tertiary hydroperoxides are most easily prepared by autoxidation: the reaction is generally unsuitable for preparation of primary and secondary hydroperoxides, although there are some exceptions.

The regiochemical course of the oxidation of alkenes depends on the presence or absence of photosensitizers: addition to an alkenic carbon takes place when sensitization is employed and insertion of oxygen into an allylic C—H occurs when no sensitizer is used.

The formation of peroxides from ethers is well known and the reaction can present a problem in the use of ethers as solvent. The majority of 2-peroxy ethers are highly explosive. Peroxides derived from autoxidation α to ketones are highly unstable ⟨79COC(4)909⟩.

Ozone has been used to prepare a 2-methoxy hydroperoxide ⟨65JA3286⟩.

2.02.4.1.6 By oxidation of organometallic compounds

Organometallic reagents react with oxygen to give metal salts of alkyl hydroperoxides, from which the corresponding protiated species may routinely be obtained. The product from autoxidation of a Grignard reagent could be converted to a hydroperoxide (by hydrolysis), a dialkyl peroxide (by alkylation) and esterified to a peroxyester ⟨53JA4372, 65MI 202-02⟩—thus the intermediate was an alkylperoxymagnesium species. Although the reaction is a general one, organolithiums and Grignards react less well than organozinc and organocadmium reagents, from which even long-chain alkyl hydroperoxides may be obtained in excellent yields ⟨62LA(652)7⟩.

2.02.4.1.7 *From organoboranes*

Oxidation of organoboranes yields alkyldi(alkylperoxy)boranes (see below), which may be further oxidised to alkyl hydroperoxides and in superior yield to the oxidation of other organometallics. In the three-step procedure, either one or two moles of alkyl hydroperoxide may be obtained from trialkylboranes according to how many oxidations are employed. Thus, reaction of trialkylborane with oxygen gives an alkylbis(alkylperoxy)borane, which may be treated in two ways; hydrolysis yields 1 mol of alkyl hydroperoxide and 2 mol of alcohol, while peroxide oxidation gives an alkoxybis(alkylperoxy)borane (in direct analogy to the usual hydroboration/oxidation sequence to produce alcohols). Hydrolysis of this moiety liberates 2 mol of alkyl hydroperoxide and 1 mol of alcohol. This methodology has been used to prepare alkyl hydroperoxides which are not readily accessible using other reactions ⟨62LA(652)7⟩.

Alkyl dichloroboranes may be converted to alkyl hydroperoxides upon reaction with oxygen and subsequent hydrolysis. This reaction allows complete conversion of alkyl substituent to alkyl peroxide and is, therefore, more efficient (and operationally simpler in terms of workup and purification) than the oxidation of trialkylboranes ⟨87T4059⟩.

Yields are good, when considering that the entire conversion of alkene to alkyl hydroperoxide may be carried out without isolation of intermediates.

2.02.4.2 Dialkyl Peroxides

These are generally obtained by extension of the methods used for alkyl hydroperoxides:

(i) From alkyl sulfates

Dialkyl peroxides are obtained in good yield by the reaction of alcohols or alkenes with hydrogen peroxide in concentrated sulfuric acid. As in the preparation of alkyl hydroperoxides, alkyl hydrogen sulfates are intermediates which may, in some cases, be isolated, although this is unnecessary in the majority of situations. There is frequently produced a mixture of di- and monoalkyl peroxides, which may be fractionated either by extraction or by (*extremely cautious*) distillation or by chromatographic means ⟨B-71MI 202-02⟩.

Under basic conditions, dialkyl sulfates are converted in variable yield to symmetrical dialkyl peroxides when reacted with H_2O_2 ⟨B-71MI 202-02⟩.

(ii) From alkyl hydroperoxides

Unsymmetrical dialkyl peroxides are produced when alkyl hydroperoxides are reacted with alkyl sulfates. Primary and secondary alkyl mesylates are also converted to the corresponding dialkyl peroxides when treated with basic hydrogen peroxide—this is the method of choice for preparation of these compounds ⟨B-71MI 202-02⟩.

Decomposition of alkyl hydroperoxides by metal ions (cuprous, rhodium, cobalt, manganese, lead) generates alkoxy radicals which abstract hydrogen radicals from positions of a substrate capable of stabilising the resulting carbon-centred radical. Oxidation of these radicals by the metal ion present leads to a carbenium ion which is trapped by alkyl hydroperoxide to give an unsymmetrical dialkyl peroxide. Alkenes, ketones and amines undergo the reaction; yields can be excellent (90%) ⟨62JA2785, B-71MI 202-02⟩.

(iii) By other nucleophilic displacements

Alkali salts of alkyl hydroperoxides are alkylated efficiently by alkyl halides; intramolecular variants of this procedure allow preparation of cyclic peroxides ⟨B-71MI 202-02⟩. Phase-transfer catalysis has been employed ⟨90SC349, 93SC1025⟩. Diazoalkanes will alkylate metal hydroperoxides ⟨55CB1544, 70T3279⟩.

(iv) By addition to carbenium ions

Substances capable of generating stabilised carbenium ions undergo reaction with hydrogen peroxide or alkyl hydroperoxides. This method allows preparation of either symmetrical or unsymmetrical dialkyl peroxides. Of course, there may be concomitant problems of carbenium ion rearrangement in those systems bearing appropriate constraints.

Bisperfluorodi-*t*-butyl peroxide may be prepared from perfluoro-*t*-butanol by reaction with chlorine trifluoride ⟨70CC216⟩.

(v) From t*-butyl hydroperoxide*

Both primary and secondary triflates react with *t*-butyl hydroperoxide in weakly basic solutions to give unsymmetrical dialkyl peroxides in good yield ⟨76JOC3983⟩. *t*-Butylacetimidates react with *t*-butyl hydroperoxide to give unsymmetrical alkyl *t*-butyl peroxides ⟨93T2477⟩. Yields are fair to good.

(vi) Miscellaneous methods

Dialkyl peroxides are prepared at room temperature by a peroxide transfer reaction between alkyl triflates and trialkyltin peroxides. Germanium peroxides also accomplish the transformation ⟨79JA4290⟩.

When primary or secondary alkyl bromides are reacted with potassium superoxide in the presence of a suitable metal ion complexing agent, symmetrical dialkylperoxides are obtained in good yield ⟨78JA7960⟩.

Primary, secondary and tertiary halogenoalkanes react with alkyl hydroperoxides in the presence of silver salts to give dialkyl peroxides in good yield ⟨76CC1022⟩.

Peroxymercuration of phenylcyclopropane occurs when the cycloalkane is treated with mercury trifluoroacetate and *t*-butyl hydroperoxide; reductive workup with sodium borohydride gives *t*-butyl-(2-phenyl)ethylperoxide ⟨82JCS(P1)1807⟩.

The reaction of singlet oxygen with dienes in the presence of a photosensitizer leads to cyclic peroxides (1,2-dioxenes), which are well-characterised substances ⟨B-71MI 202-02⟩. A similar reaction is known for highly electron-rich mono-alkenes, but the resulting 1,2-dioxetanes are highly unstable, except at low temperatures ⟨70JA3223⟩.

Alcoholysis of poly(alkylperoxy)silanes leads to dialkylperoxides; both symmetrical and unsymmetrical peroxides may be obtained. The reaction has been included in a review of the chemistry of metalloid organoperoxides ⟨66CRV529⟩.

The reaction at high temperature of carbonyl difluoride and trifluoromethyl hypofluorite under high pressure offers a high-yielding entry to bis(trifluoromethyl) peroxide ⟨57JA5628⟩. The photolytic decarboxylation of fluoroformyl peroxide gives the same product in somewhat lower yield ⟨68JOC2095⟩.

The reaction of hydroperoxyorganomercurials obtained by addition of hydrogen peroxide to alkenes in the presence of mercuric acetate may be turned to the synthesis of *t*-butylalkyl peroxides by substituting *t*-butyl peroxide as the peroxide component. The resulting alkylperoxyalkylmercurials may be converted under mild conditions into demercurated compounds ⟨71JCS(C)945⟩.

2.02.4.3 R^1OOOR^2 Functions

The preparation of organic trioxides and tetroxides has proved to be extremely demanding. The synthesis and chemistry of trioxides has been reviewed ⟨84MI 202-01⟩.

Strictly speaking, the products obtained from addition of ozone to alkenes are cyclic trioxides; these compounds are, however, more commonly known as molozonides. Such species exhibit greater stability than linear trioxides, and many ozonides are stable enough to withstand complete characterisation.

Di-*t*-butyl trioxide is obtained in low yield when di-*t*-butyl peroxycarbonate is photolysed in the solid state at low temperature. As might be expected, the product is unstable at even subzero temperatures ⟨72IJ101⟩.

Carbonyl difluoride reacts with oxygen difluoride to give bis(trifluoromethyl) trioxide in good yield ⟨67JA4313⟩; the same compound may also be prepared by fluorinative dimerisation of sodium trifluoroacetate ⟨67JA4316⟩.

2.02.4.4 ROOB Functions

The preparation of organoperoxyboranes has been reviewed ⟨65MI 202-02⟩. There are three main classes: peroxyborates, diperoxyboronates and peroxyborinates. The following section is concerned with preparation of peroxyborates.

2.02.4.4.1 Peroxyborates

The reaction of alkyl hydroperoxides with boron trihalides gives tri(alkylperoxy)boranes, which are isolable and characterisable compounds. Alternatively, alkyl borates may be converted into peroxyboranes by removal, by distillation, of alkanol from a mixture of borate and alkyl hydroperoxide ⟨57CI(L)1622, 58JCS2372⟩. Although the reaction proceeds by exchange of alkoxy and peroxy groups, the intermediate mixed alkoxyperoxyboranes are usually unstable. Only mixed alkoxyperoxyboranes arising from reaction involving at least one equivalent of *o*-nitrophenol are stable enough to be isolated. Thus, di-*t*-butylperoxy-*o*-nitrophenoxyborane and *t*-butylperoxydi-(*o*-nitrophenoxy)borane are isolable from the reaction of *t*-butyl hydroperoxide with the corresponding chloro-*o*-nitrophenoxyborane ⟨57CI(L)1622, 58JCS2372⟩. The particular stability of these compounds has been ascribed to the intramolecular association between the nitro group and the boron atom ⟨56JCS3006⟩.

Di(*t*-butylperoxy)hydroxyborane, a low melting solid, may be prepared by reaction of *t*-butylhydroperoxide with oxybis(diacetoxy)borane ⟨57CI(L)1622, 58JCS2372⟩.

2.02.5 FUNCTIONS BASED ON ROS, ROSe AND ROTe UNITS

2.02.5.1 *O*-Alkylsulfenic Acids and their Se Analogues

The parent compound of this class of chalcogenides is hydrogen thioperoxide, HOSH. The alkylated derivatives of this compound are alkylsulfenic esters. There are no general methods for synthesis of this class of compound; this may be due to the instability of the parent, which has been only transiently prepared in an argon matrix from the reaction of ozone and H_2S. There are no reports of preparation of selenium or tellurium analogues.

2.02.5.2 Alkyl Sulfenyl, Sulfinyl and Sulfonyl Halides and their Se Analogues

Sulfoxylic acid, HO—S—OH, forms dialkyl esters. These esters are also known as alkoxysulfanes. All methods for their preparation allow only symmetrical diesters to be made. These compounds were first prepared in the 1930s by the reaction of the corresponding dialkoxy disulfides with sodium alkoxides ⟨35CB1011, 36CB937, 38CB1873, 47CB219⟩. The method remained the only one until 1965, when the reaction of disulfur dichloride with alcohols at low temperature was shown to allow preparation of simple dialkylesters of sulfoxylic acid in 56–70% yield (Equation (100)) ⟨65JOC2703⟩.

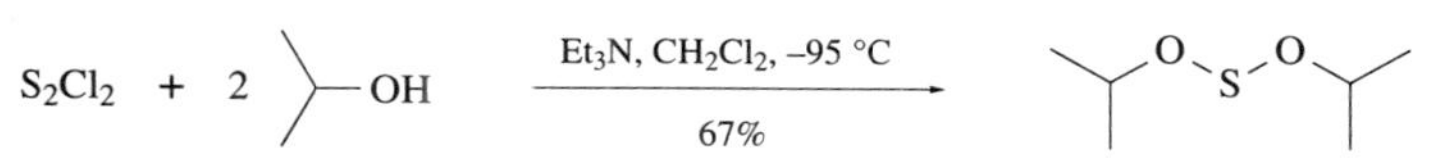

(100)

The low temperature is essential for a clean reaction because even above −75 °C significant amounts of by-products are observed. Benzylic and allylic alcohols react anomalously to give only symmetrical dialkyl esters of sulfurous acid. The reaction of diisopropyl sulfides with trimethylsilyl chloride gives the corresponding symmetrical sulfoxylate ⟨86SUL161⟩. *N*-Chloro-*N*-methylsulfonyl methylsulfonamide reacts with elemental sulfur to give the symmetrical diaminosulfane, which reacts with *n*-butanol to give di-*n*-butyl sulfoxylate in 63% yield (Scheme 47) ⟨91ZN(B)5⟩.

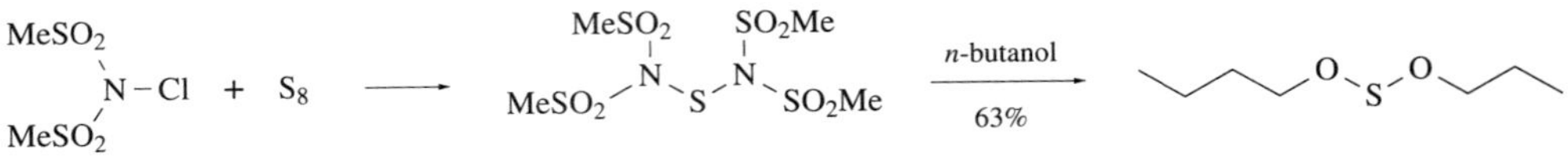

Scheme 47

The reaction of dialkoxy disulfides with sulfur dichloride gives an alkylsulfenylchloride and an alkoxychlorosulfane; this exotic species decomposes at room temperature to dialkyl sulfoxylates (*inter alia*). Reaction of the chlorosulfane with titanocene pentasulfide gives dialkoxy polysulfides in which many species are present, up to at least 18. The reaction is reported for methyl, isopropyl and stearyl esters. No yields were given (Scheme 48) ⟨90ZN(B)557⟩. ROSCl is prepared by the method referred to above, but this is not claimed as a general method for synthesis of these compounds.

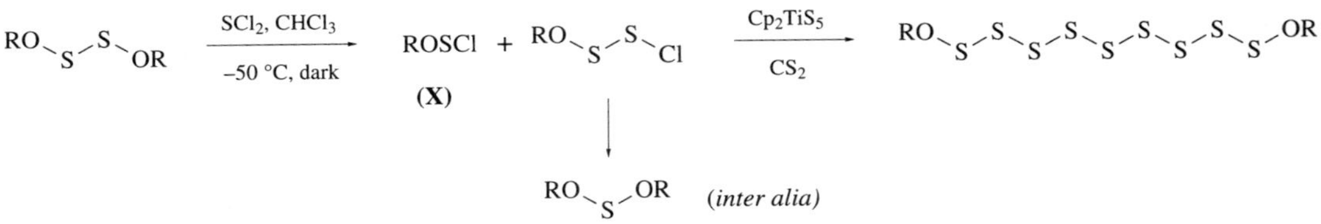

Scheme 48

2.02.5.3 Esters of HO—S—S—OH

Such compounds (diesters of the so-far hypothetical dihydroxydisulfide) are dialkoxy disulfides; these compounds were first prepared in 1895, by Lengfeld, from the reaction of sodium ethoxide and methoxide with disulfur dichloride. The yield of dimethoxy disulfide was quantitative; the yield of the ethyl analogue was not so good (Equation (101)) ⟨1895CB449⟩.

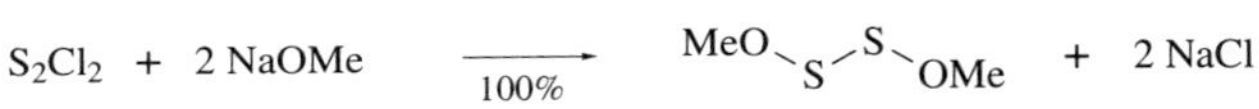

(101)

As previously mentioned, disulfur dichloride reacts at low temperature with alcohols to give dialkyl sulfoxylates. A wide range of dialkoxy disulfides may be prepared when the same reaction is carried out at room temperature ⟨65JOC2692⟩. The reaction of di(disulfonamido) disulfides with alcohols is a modern preparative route to dialkoxy disulfides (see below) ⟨91ZN(B)5⟩.

ROS(O)OR diesters of sulfurous acid are also known as dialkyl sulfites. The methods available for synthesis of these compounds have been reviewed ⟨63CRV557⟩. The first generally applicable synthetic procedure reported involves reaction of alcohols with disulfur dichloride ⟨24CR(178)1827⟩; this method was improved by the realisation that the intermediate in this reaction was the alkyl chlorosulfite, which class of compound had already been prepared by use of thionyl chloride ⟨1858LA(106)334, 1859LA(111)93, 31LA(485)258, 34JA1204, 39JCS99, 43OSC(2)112, 47JA2158⟩. The intermediate chlorosulfinates are stable in the cold and may be reacted with a second equivalent of another alcohol to give unsymmetrical dialkyl sulfites ⟨33BSF1070⟩. Diazoalkanes esterify alkylsulfinic acids ⟨60CB1249⟩. *N*-Sulfinylsulfonamides are reported to react violently with alcohols to give dialkyl sulfites in quantitative yield, although no detailed experimental procedures were given ⟨62AG(E)97⟩.

Cyclic sulfites are well known and are prepared along the same lines as their acyclic counterparts. Their preparation and reactivity have been reviewed ⟨92S1035⟩. Cyclic sulfites may also be prepared by the reaction of diols with sulfur tetrafluoride ⟨88JOU1473⟩.

2.02.5.4 ROSe(O)OH, Alkyl Hydrogenselenites

The first preparation of these compounds was reported by Hinsberg in 1890; his experimental observation that an alcoholic solution of selenium dioxide, when dried over calcium chloride, gave rise to $C_2H_6O_3Se$ ⟨1890LA(260)40⟩. Subsequently, it was confirmed that the product of this reaction was the hydrogenselenite when selenium dioxide was dissolved in methanol and excess solvent was removed *in vacuo* to give methyl hydrogenselenite ⟨35JCS901⟩. Both ethyl and methyl hydrogenselenites were later prepared by essentially the same methods ⟨60ZAAC(303)53⟩.

2.02.5.5 $(RO)_2SeO$, Dialkyl Selenites

These compounds are diesters of selenious acid and were first prepared by reaction of selenium oxychloride with excess ethanol ⟨1887LA(241)150⟩. Strecker and Daniel later showed that alkoxides also undergo the reaction and they extended the methodology by demonstrating that silver nitrate stoichiometrically mediates the alkylation of selenious acid by alkyl iodides. The alkoxide reactions proceed in good yield, while the alkylation reactions proceed in poor yield (30%) ⟨28LA(462)186⟩. The literature contains very few examples of the preparation of dialkyl selenites; bis(2-chloroethyl)esters have been prepared by essentially the method of Michaelis ⟨1887LA(241)150⟩. No yield was quoted and the product must be distilled under an atmosphere of carbon dioxide to prevent decomposition ⟨50JCS3125⟩. The reaction of selenious acid with alcohols gives symmetrical dialkyl selenites. The reaction of alkyl hydrogenselenites with diazoalkanes or alcohols gives unsymmetrical dialkyl selenites. There is no need to isolate the intermediate (unstable) hydrogen selenite: if diazoalkane is added to an alcoholic solution of selenium dioxide, dialkyl selenites are obtained in greater than 80% yield ⟨60ZAAC(303)53⟩.

2.02.5.6 ROSe(O)Cl, Alkyl Chloroselenites

These were first prepared by Michaelis in 1887 by reaction of selenium dioxide with chloroethane, or by reaction of selenium oxychloride with ethanol ⟨1887LA(241)150⟩. The procedure was subsequently refined and the yield of ethyl chloroselenite from this reaction is 83% ⟨50JCS3125⟩.

2.02.5.7 Alkyl Chlorosulfites

These species are generally prepared by the reaction of alcohols with sulfur chloride or thionyl chloride ⟨63CRV557⟩. The reactions are applicable to a wide range of alkanols: from primary alcohols ⟨36JCS688⟩, chloroalkanols ⟨51USP2529493⟩ and 2-hydroxyesters ⟨55MI 202-01⟩. The chlorosulfite derived from neopentyl alcohol is remarkably stable ⟨50JCS2088⟩, though many alkyl chlorosulfites are unstable: this is of no great surprise, because the chlorination of alcohols using thionyl chloride is the best-known example of this inherent reactivity. To enable good yields of alkyl chlorosulfite to be obtained, several features must be borne in mind: (1) The order of addition of reactants is crucial to the success of the preparation. Alkanol *must* be added to excess thionyl chloride, otherwise dialkyl sulfites are produced. (2) Pyridine and tertiary nitrogenous bases must be avoided. The presence of such bases leads to chlorination of alkanol, with release of sulfur dioxide. (3) Benzylic chlorosulfinates are too unstable to be isolated and decompose, even in the absence of base, to give benzylic chlorides.

2.02.5.8 Monoalkyl Sulfates (Alkyl Hydrogensulfates)

These partial esters of sulfuric acid may be prepared by the reaction of alkanols with sulfuric acid or chlorosulfonic acid or sulfur trioxide; the latter substance may safely be employed in the shape of its dioxan complex ⟨38JA538⟩. Extreme hygroscopicity limits the preparation of anhydrous monoesters and the compounds are further complicated by the existence of equilibria between the mono- and diesters (Equation (102)). Ethyl hydrogensulfate cannot be obtained in a pure state because of this equilibrium.

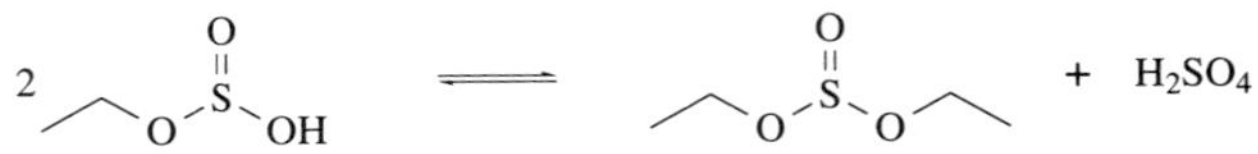

(102)

The sodium salts of long-chain monoalkyl sulfates have been employed commercially for use as detergents ⟨44IEC610⟩.

2.02.5.9 Alkyl Chlorosulfonates

These are, of course, the acid chlorides of alkyl hydrogensulfates; many methods allowing their preparation, therefore, parallel those usually seen for preparation of acid chlorides. Sulfuryl chloride reacts with alkanols to give the corresponding alkyl chlorosulfonates in good yield ⟨38JA2810⟩. Sandmeyer prepared methyl and ethyl chlorosulfonates by reaction of the corresponding alkyl hypochlorite with sulfur dioxide ⟨1888CB857⟩.

2.02.5.10 Dialkyl Sulfates

In the research laboratory, these compounds are not cleanly prepared from sulfuric acid and alcohols, with elimination and dimerisative etherification reactions often a problem, although in industry this has been the most important method for preparative reaction ⟨21MI 202-02⟩. Thus, it is frequently simpler to prepare dialkyl sulfates by reaction of the corresponding dialkyl sulfites with sulfuryl chloride ⟨43OSC(2)111⟩. The reaction proceeds through the intermediacy of the alkyl chlorosulfite, which reacts with further dialkyl sulfite to give the product. Similar methods for preparation of dialkyl sulfites employ the reaction of alkyl chlorosulfites with alkoxides ⟨36AC(P)560⟩ and the decarboxylative alkylation of alkyl hydrogen sulfates by chloroformate esters ⟨31ZOB190⟩.

Alkyl iodides react with silver sulfate to give dialkyl sulfates. Dimethyl sulfate ⟨02MI 202-01⟩ may be prepared on a large scale from dimethyl ether and sulfur trioxide or methanol and sulfuric acid ⟨19CR(164)795⟩. Esterification mediated by dcc allows preparation of dialkyl sulfates from sulfuric acid and alkanols in good yield ⟨66M 202-02, 69JA4273⟩.

2.02.5.11 $(RO)_2SeO_2$, Dialkyl Selenates

These compounds have been reported as arising in poor yield (<20%) from the reaction of silver selenate with alkyl iodides ⟨28LA(462)186⟩.

2.02.6 FUNCTIONS BASED ON THE RON UNIT

2.02.6.1 *O*-Alkylhydroxylamines, $RONH_2$

Owing to the inherently greater nucleophilicity of nitrogen when compared with oxygen, the direct *O*-alkylation of hydroxylamine is not a practical method for preparation of these compounds. It is necessary firstly to mask the nitrogen lone pair, alkylate at oxygen, and then release the amine. There exist four methods which have been shown to offer an efficient entry to *O*-alkylhydroxylamines (also known as *amino xy* compounds):

(i) Via N-*hydroxyphthalimides and* N-*hydroxyurethanes*

N-Hydroxyphthalimide may be alkylated by usual means and then converted to *O*-alkylhydroxylamine by treatment with hydrazine hydrate, in a direct analogy to the Gabriel amine synthesis. The reaction was first reported in 1960 ⟨60CJC343⟩ and is of general utility ⟨76BSF833, 88JOC2997⟩.

N-Hydroxyurethane (*N*-ethoxycarbonylhydroxylamine) may be converted to *O*-alkylhydroxylamines by a similar reaction pathway ⟨14JA2202, 47JCS963, 62HCA358⟩. The reaction has been employed in a total synthesis of the amino acid-derived *O*-alkylhydroxylamines canaline and canavanine ⟨57JA1222⟩.

(ii) Via oximes

Acetone oxime is benzylated by reaction with benzyl chloride in the presence of sodium ethoxide. The resulting *O*-benzyl acetoxime is hydrolysed in acceptable yield to *O*-benzylhydroxylamine upon treatment with concentrated acid at reflux ⟨90OM301⟩. This is a modification of the original procedures of Janny ⟨1883CB175⟩ and Behrend ⟨1890CB203⟩.

(iii) Via hydroxamic acids

Hydroxamic acids (*N*-acylated derivatives of hydroxylamine) may be alkylated by a variety of electrophiles to give alkyl hydroxamates which give hydrochloride salts of *O*-alkylhydroxylamines when hydrolysed in hydrochloric acid. Benzhydroxamic acid is generally employed in the reaction ⟨1893CB1567, 1894CB3350⟩. A wide range of *O*-alkylhydroxylamines may be prepared by these reactions ⟨36MI 202-01, 60JCS225, 60MI 202-03⟩.

(iv) Via dipotassium hydroxylamine disulfonate

Dipotassium hydroxylamine disulfonate undergoes *O*-alkylation when reacted with alkyl halides in refluxing KOH solutions. Acidification and desulfonylation using dilute sulfuric acid allow preparation of *O*-alkylhydroxylamines in good overall yield; the process is highly suitable for multigram preparations (Equation (103)) ⟨32JA977⟩.

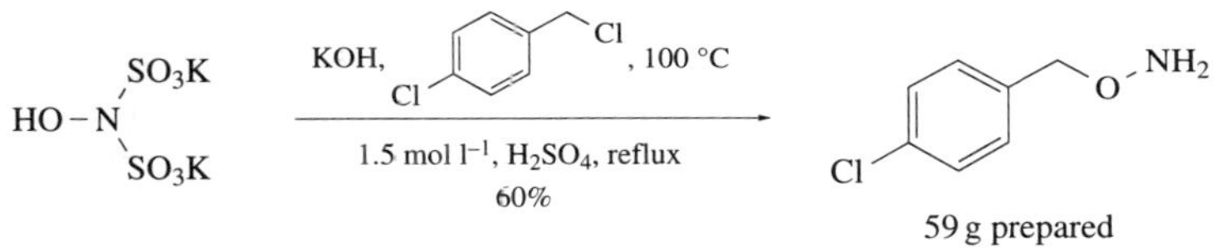

(103)

2.02.6.2 Further *N*-Functionalised RON Functions of Type RONY

2.02.6.2.1 Nitrite and nitrate esters, RONO and RONO$_2$

The preparative methods for nitrate esters have been reviewed ⟨55CRV485, B-89MI 202-05⟩. Nitrate esters are commonly prepared from the corresponding alcohols by reaction with nitrating agents; for instance, reaction of alcohols with concentrated nitric acid in the presence of urea (to prevent the oxidative side reactions which may lead to explosion) ⟨33JA766⟩. A drawback to this reaction is the preparation of nitrites as by-products. Reaction of finely powdered silver nitrate with alkyl halides in inert solvent ⟨52JCS1193⟩ is applicable to the synthesis of primary, secondary and tertiary nitrates. Mercury(II) nitrate reacts in similar fashion with bromoalkanes ⟨74T2467⟩. Ammonium and ion exchange resin nitrates convert alkyl sulfonates to the corresponding nitrates with inversion of configuration ⟨85TL3369⟩. Alcohols may be converted through the corresponding iodides to nitrate esters in good yield. The alcohols are reacted with Ph$_3$P, iodine and imidazole and to this mixture is added silver nitrate ⟨92SC677⟩. The combination of acetic anhydride and sodium nitrate (presumably leading to an *in situ* formation of acetyl nitrate) will nitrate alcohols in unreported yield (Equation (104)) ⟨69T761⟩.

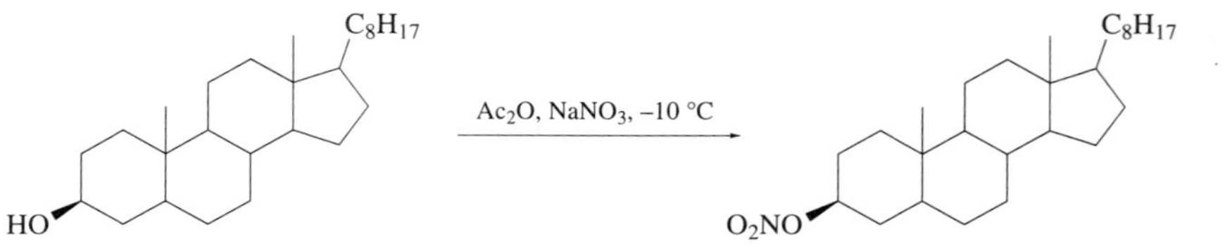

(104)

Alcohols react with potassium nitrate in the presence of boron trifluoride to give good yields of alkylnitrates ⟨93S207⟩. *O*-Nitration of alcohols may also be effected by reaction with other *O*-nitrated species. Thus, benzoyl and acetyl nitrates are useful reagents for preparation of nitrate esters ⟨06JCS1, 67ACS1257⟩, while *N*-nitrocollidinium tetrafluoroborate will effect *O*-nitration in preference to *C*-nitration (Equation (105)) ⟨78S452⟩. Similar selectivity for preference of *O*-nitration is exhibited by thionyl chloride nitrate ($ClS(O)ONO_2$), and thionyl nitrate ($O_2NOS(O)ONO_2$) which are prepared *in situ* from thionyl chloride and silver nitrate ⟨84HCA906⟩. These reagents will selectively nitrate primary alcohols in the presence of secondary and tertiary ones.

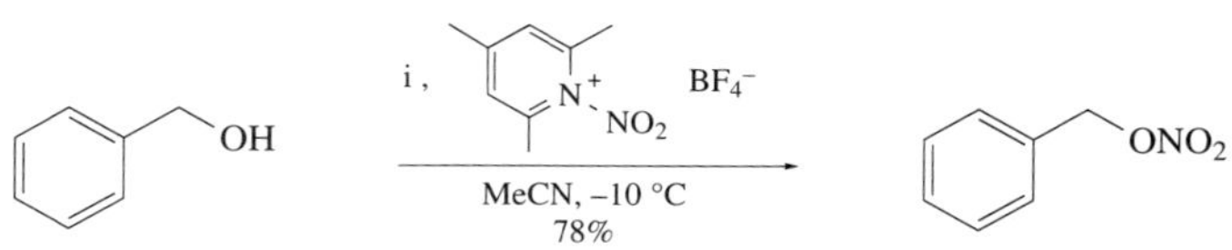

(105)

Triphenylpyrylium nitrate reacts with primary amines to give alkyl nitrates ⟨80JOC2515⟩. Dinitrogen tetroxide reacts with primary amines in the presence of strong amine bases to give primary alkyl nitrates ⟨77JCS(P1)1114⟩, while the pentoxide is reported to ring-open epoxides and aziridines to give hydroxynitrates. Owing to the highly explosive nature of the products of the reactions (many of which are explosives), the authors suggest that the reactions are carried out in armoured fume cupboards ⟨93T7037⟩.

2.02.6.2.2 *Alkyl nitrites*

The most frequently employed method for preparation of alkyl nitrites is that of Noyes, in which an alcohol is treated with nitrous acid (formed *in situ* from $NaNO_2$ and concentrated H_2SO_4) at 0°C ⟨1888G434, 33JA3888, 43OSC(2)108, 63HOU(6/2)325⟩. Thus, for instance, was isopulegol converted to its nitrite in 50% yield ⟨91BSF222⟩. Nitrosyl chloride (gaseous) is also effective in preparation of nitrite esters ⟨82JCS(P1)111, 82JCS(P1)1963⟩. *t*-Butyl nitrite has been used to effect synthesis of steroidal nitrite esters in quantitative yield ⟨83JOC3379⟩.

2.02.7 FUNCTIONS BASED ON THE ROP UNIT

2.02.7.1 Alkyl Phosphites, $(RO)_{3-n}PH_n$

2.02.7.1.1 *Mono- and dialkylhydrogen phosphites*

These compounds may be prepared from reaction of alcohols with phosphorus trihalides, though preparation of monoalkyl esters by this process is not usually high-yielding. The dialkyl phosphites may be obtained in good yield if more than one equivalent of alcohol is employed ⟨45JCS380⟩. Indeed, unless an excess of alcohol is employed, dialkyl hydrogenphosphites are the principal products, formed by reaction of three equivalents of an alkanol with a phosphorus trihalide; the dialkyl phosphite results from hydrogen halide-induced dealkylation of the initially formed trialkyl phosphite ⟨B-73MI 202-01⟩.

Monoalkyl phosphites may be prepared in certain cases by reaction of phosphorous acid with the alkanol ⟨64HCA1647⟩. The reaction of certain alkenes with white phosphorus under an oxygen atmosphere leads to a polymeric phosphorate which reacts with alkanols to give dialkyl hydrogenphosphites ⟨58JA4546⟩.

2.02.7.1.2 *Trialkyl phosphites*

The esters may be prepared in good yield by the reaction of phosphorus trihalides with *excess* alcohol, or with sodium alkoxides, in a process analogous to that described above. This is the most widely used method for preparation of trialkyl phosphites, and few other methods are used. Alcohols react with white phosphorus in the presence of triethylamine to give trialkyl phosphites in good yields ⟨85EGP231074, 90ZAAC(580)217⟩.

Epoxides and other cyclic ethers may be cleaved by phosphorus trichloride to give ω-chlorotrialkyl phosphites ⟨1879BSF(2)551⟩.

Triethyl and triphenyl phosphites may be *trans*-esterified by reflux in a large excess of an alcohol to give trialkyl phosphites ⟨B-73MI 202-01⟩. Tris(2-chloroethyl) phosphite may also be employed ⟨59CJC1498⟩. The reaction is catalysed by alkoxides ⟨17MI 202-01⟩.

2.02.7.2 Alkyl Phosphates, $(RO)_nP(OH)_{3-n}(O)$

The general methods for preparation of alkyl phosphates have been reviewed, in particular the Specialist Periodical Report of the Royal Society of Chemistry offers exhaustive coverage of the wide interest in synthesis of esters of pentavalent phosphorus. The following section deals with general preparative methods towards such compounds ⟨B-63MI 202-01, B-72MI 202-01, B-73MI 202-02, B-76MI 202-01, 77S737⟩.

2.02.7.2.1 *Monoalkyl dihydrogenphosphates*

There are many convenient procedures for preparation of di- and trialkyl phosphates, but relatively few for synthesis of monoalkyl dihydrogenphosphates. Reaction of one equivalent of an alkanol with phosphoric trichloride (phosphorus oxychloride, $POCl_3$) gives the corresponding monoalkyl phosphorodichloridate. This species may be hydrolysed to give a monoalkyl phosphate. The reaction is easily modified to allow preparation of di- and trialkyl phosphates. Reaction of pyrophosphoryl chloride ($Cl_2P(O)OP(O)PCl_2$) with alcohols at low temperatures gives alkyl phosphorodichloridates, which may be hydrolysed to monoalkyl dihydrogenphosphates or reacted with further equivalents of alcohol to give trialkyl phosphates ⟨59CB850⟩. Alcohols react with amine salts of phosphoric acid in the presence of nitriles to give monoalkyl esters of phosphoric acid. For instance, 2-cyanoethanol reacts with tris(triethylamino) phosphate in the presence of trichloroacetonitrile to give monoester in good yield ⟨64BBR(17)146⟩. This ester may itself be used for preparation of other monoalkyl phosphates because of the ready base-lability ⟨59CI(L)542⟩ of the cyanoethyl group. Thus, reaction of an alcohol with mono-(2-cyanoethyl) phosphate in the presence of (for instance), dcc gives the corresponding mixed dialkyl phosphate, which may be converted to monophosphate by cleavage of the cyanoethyl group with alkali under mild conditions ⟨61JA159⟩. Cyclopropylmethyl groups may also be easily removed from phosphate triesters upon reflux for 30 min in 80% formic acid, thus providing a two-step route to monoalkyl phosphates by sequential reaction of dicyclopropyl phosphorochloridate with alcohol and subsequent de-esterification ⟨66T399⟩.

Monoalkyl phosphates may be exclusively prepared via reaction of dibenzyl phosphochloridate with an alcohol; the resultant mixed triester may be hydrogenolysed to the monoalkyl ester ⟨45JCS382⟩. The reaction allows some selectivity, because it is possible to remove only one of the benzylic groups to give an alkylbenzyl phosphoric acid. The limitation of this process is the relative instability of dibenzyl phosphochloridate; at low temperatures, where the compound is stable, reactivity is low. Bis-(4-nitrobenzyl) phosphochloridate is more stable at higher temperature and has the advantage of photolytically labile blocking groups ⟨74JOC3767⟩. Bis-(2,2,2-trichloroethyl) phosphochloridate may be converted to monoalkyl dihydrogenphosphates by sequential reaction with alcohol and reductive cleavage by metallic zinc ⟨67AG(E)362⟩. The dibenzyl phosphate may also react as a nucleophile in formation of the intermediate dibenzyl triester, as in the preparation of phosphorylated carbohydrates by a two-step reaction, involving displacement of primary tosylates by lithium dibenzylphosphate and subsequent debenzylation by dissolving metal ⟨68CAR(6)253⟩.

Where the alcohol to be incorporated into the ester is readily available, oxidation of an alcoholic solution of phosphorous acid with iodine gives good yields of monoesters of phosphoric acid ⟨63CI(L)1877⟩.

Cyanovinyl phosphate ($NCCH{=}CHOPO_3H_2$) may be converted upon reaction with alcohols (particularly nucleoside alcohols) to monoalkyl phosphates ⟨68SCI3836⟩.

Monoalkyl dihydrogenphosphates may be prepared in good yield by the reaction of alcohols with dry phosphorous acid in the presence of mercuric chloride and other mercury salts (Equation (106)) ⟨67JOC1063⟩. The same reaction is mediated by catalytic amounts of cupric chloride ⟨91PS(55)195⟩. Reaction of phosphorodichloridates with silver nitrate in aqueous MeCN gives monoalkyl dihydrogenphosphates in good yield ⟨92OPP57⟩. Diethyl phosphochloridodithionite ($ClP(SEt)_2$) reacts with alcohols to give dithiotrialkyl phosphates which, upon treatment with aqueous iodine, are converted to monoalkyl phosphates ⟨72TL411⟩. The reaction of diphosphorus pentoxide with hexamethylsiloxane gives trimethylsilyl polyphosphate, which is highly effective in preparation of monoalkyl phosphates ⟨85BCJ3393⟩.

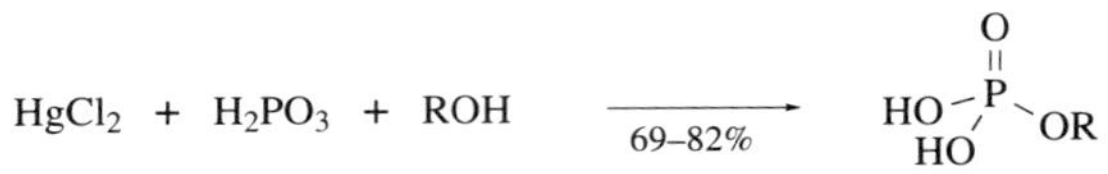

(106)

2.02.7.2.2 *Dialkyl phosphates*

The preparation of diesters of phosphoric acid is of huge importance owing to the presence of such functionality in nucleic acids. The most widely used method for preparation of diesters derived from nucleosides involves preparation of phosphoramidites ($(R^1O)_2PNR^2{}_2$), such as those shown in Scheme 49 ⟨91T8717, 92MI 202-01⟩.

Scheme 49

Phosphoramidites undergo reaction with alcohols to give trialkyl phosphites, which are easily oxidised to the corresponding phosphate (in the case of nucleoside phosphoramidites, reaction with another nucleoside gives a dinucleoside phosphate diester). 2-Cyanoethyl groups are particularly used because their base-lability allows deprotection to give dialkyl hydrogenphosphates under mild conditions ⟨59CI(L)542, 61JA159⟩. The oxidation step in this sequence may be carried out at low temperature by peracids, alkyl hydroperoxides ⟨91TL6021⟩, a bis(trimethylsilyl) peroxide in the presence of TMS-OTf and base ⟨86TL4191⟩. Use of benzylic alcohols allows subsequent C—O bond cleavage to give dialkyl hydrogenphosphates. When phosphoramidites derived from benzylic diols are employed, facile preparation of both mixed dialkyl and monoalkyl phosphates is possible (Scheme 50) ⟨92TL1313⟩.

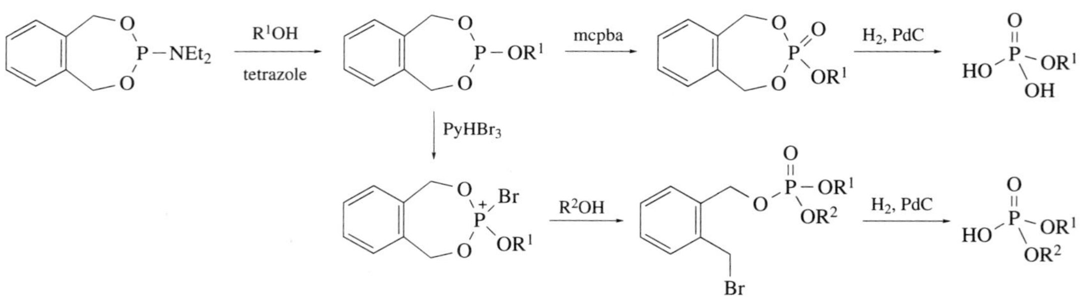

Scheme 50

Dialkyl phosphates may be prepared by the dcc coupling reaction of alcohols and monoalkyl phosphates: the reaction is believed to proceed through cyclic trialkyl trimetaphosphates ⟨58JA6212⟩. The combination of 2-bromo-2-cyanoacetamide activates monoalkyl phosphates and allows esterification by a dissimilar alcohol to give mixed dialkyl phosphates in good yields ⟨58CB1181, 62CB1664⟩.

2.02.7.2.3 *Trialkyl phosphates*

Symmetrical trialkyl phosphates may be generally prepared by reaction of three equivalents of alkanol with phosphorus oxychloride ⟨43OSC(2)109⟩.

Triethyl phosphate may be prepared by reaction of tetraethyl hypophosphate with aluminum ethoxide and ethanol ⟨63JOC481⟩.

Silver trimetaphosphate is alkylated by alkyl iodides to give trialkyl phosphates ⟨62CB1664⟩.

Alkylation of dialkylphosphates by alkyl halides is effected by fluoride ion in acetonitrile (Equation (107) and Table 1) ⟨83CPB2157⟩.

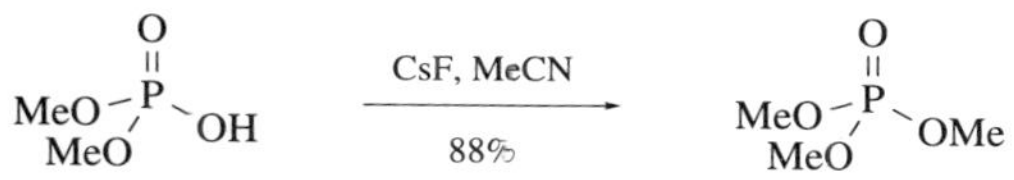

(107)

Table 1 Preparation of trialkyl phosphates by fluoride ion-mediated alkylation of dialkyl phosphates.

X = I, Br

R^1	R^2	*Yield* (%)
Me	Et	81
Me	Pr^i	88
Me	Bu^n	84
Me	n-C_6H_{13}	84
Me	$PhCH_2$	75
Bu	Me	88
Bu	Et	89
Bu	Pr^i	96
Bu	Bu^n	86
Bu	n-C_7H_{15}	79
Bu	n-C_9H_{19}	75
Bu	$CH_2{=}CHCH_2$	76

Tetrabutylammonium salts of dialkyl phosphates may be alkylated in similar fashion ⟨71T3163⟩. The method of Ogilvie, whereby *mixed* trialkyl phosphates may be prepared by a *trans*-esterification reaction, is widely used (Equation (108)) ⟨77JA1277⟩.

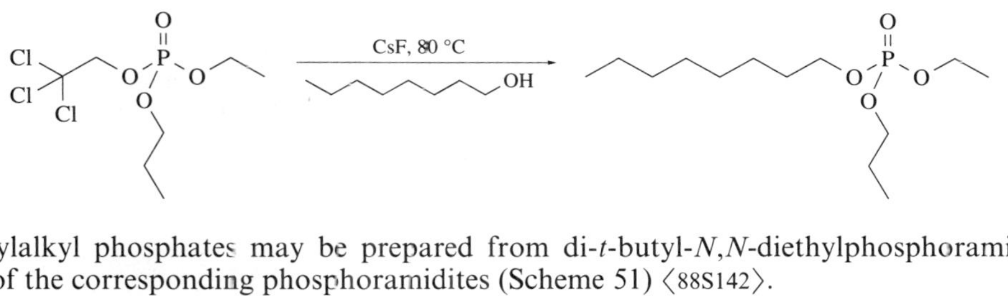

(108)

Di-*t*-butylalkyl phosphates may be prepared from di-*t*-butyl-*N*,*N*-diethylphosphoramidite via oxidation of the corresponding phosphoramidites (Scheme 51) ⟨88S142⟩.

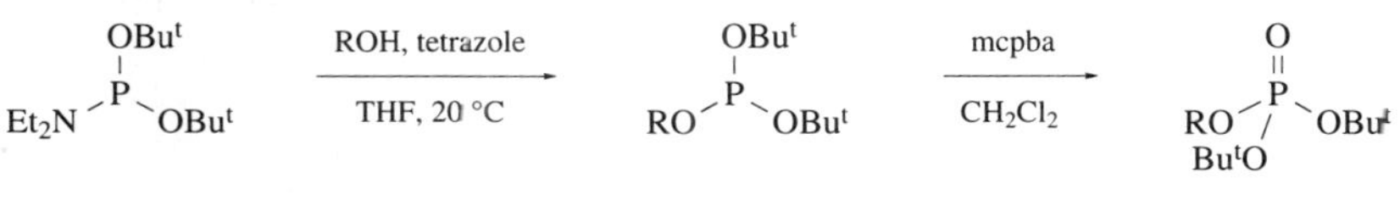

Scheme 51

Mixed trialkyl phosphates may be prepared directly from symmetrical dialkyl phosphites utilising a copper-mediated oxidation reaction. The alcohol used as solvent is incorporated into the product ⟨88BCJ3359⟩. Dialkyl hydrogenphosphites are converted to mixed trialkyl phosphites upon electrolysis in an alcohol at constant current ⟨79CPB1700⟩. Symmetrical trialkyl phosphites may be oxidised to the corresponding phosphates by reaction with bis(trimethylsilyl) peroxide in the presence of a Lewis acid ⟨85TL4965⟩.

Trialkyl phosphates may be prepared from *alkanes* in a process catalysed by iron(II) species. Thus, alkanes react with trimethyl phosphite in the presence of $FeCl_2 \cdot 4H_2O/Zn^0/O_2$ to give alkyl dimethylphosphates via iron-catalysed oxidative insertion of oxygen into a C—H bond ⟨91TL4671⟩. Reaction of red phosphorus with alkanols may be controlled by the presence of catalytic cupric chloride, allowing preparation of trialkyl phosphates in very high yield ⟨91JGU999⟩.

Enantiopure trialkyl phosphates may be prepared by reaction of phosphoramidates **(69)** with alkoxides ⟨90JA6936⟩.

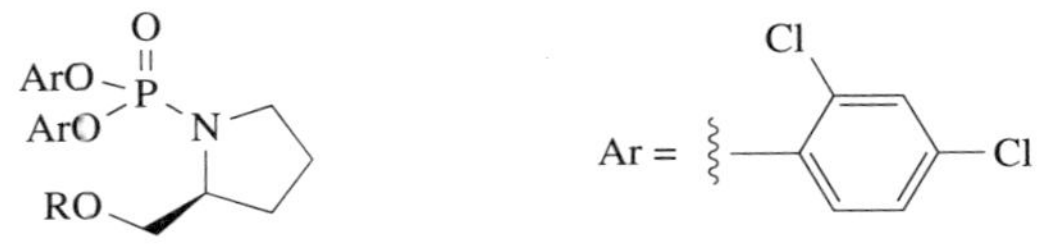

(69)

Phosphofluoridates are reported to selectively alkylate alcohols in the presence of amines; in particular, amino alcohols react efficiently to give mixed trialkyl phosphates ⟨67CJC495⟩. The synthesis of nucleoside phosphorothioates and -dithioates has been reviewed ⟨91SR670⟩.

2.02.7.2.4 Alkyl pyrophosphates

These compounds are anhydrides of phosphoric acid. dcc Mediates the synthesis of symmetrical di- and tetraalkyl pyrophosphates from mono- and dialkyl hydrogenphosphates, respectively ⟨53JCS2257⟩. Unsymmetrical tetraalkyl pyrophosphates may be prepared by reaction of dialkyl phosphates with dialkyl phosphochloridates ⟨48JA3882⟩.

2.02.8 RO METALLOID FUNCTIONS WITHOUT FURTHER ATOMS ATTACHED TO THE METALLOID

2.02.8.1 Silicon Derivatives, ROSi

2.02.8.1.1 Trialkoxysilanes, $(RO)_3SiH$

The first reported preparation of a trialkoxysilane was that of triethoxysilane, prepared by the reaction of trichlorosilane with anhydrous ethanol ⟨1867LA(143)118⟩. Subsequent workers found that trialkoxysilanes decompose with concomitant evolution of hydrogen on standing at room temperature; this tendency to decomposition is slightly diminished by use of benzene as reaction solvent, but is still extremely problematic. The products of decomposition are tetraalkoxysilanes and hexaalkoxydisiloxanes ⟨48JOC230⟩.

2.02.8.1.2 Tetraalkoxysilanes (alkyl silicates), $(RO)_4Si$

Silicon tetrachloride reacts with ethanol to give ethyl silicate, $Si(OC_2H_5)_4$. The reaction was first performed by Mendeleef, and is used on an industrial scale. While this reaction is preparatively useful for lower silicates, the hydrogen chloride liberated by the reaction causes the decomposition of higher-molecular-weight esters of silicic acid. Such esters are more conveniently obtained by *trans*-esterification reactions ⟨B-73MI 202-03⟩. The reaction of trichlorosilane with ethanol also gives tetraethoxysilane ⟨05CB53⟩. This reaction was extended to include examples of higher alcohols ⟨05CB1661, 24CB795⟩.

Silicon disulfide reacts with alcohols to give alkyl silicates, while ethylene oxide reacts with $SiCl_4$ to give tetra-(2-chloroethoxy)silane. Silicon dichloride diethoxide has been prepared ⟨63JCS2101⟩. Tetraalkoxysilanes may be prepared by reaction of silica with dialkyl carbonates ⟨93MI 202-01⟩.

2.02.8.2 Boron Derivatives, ROB

2.02.8.2.1 *Alkoxyboranes, $(RO)_{3-n}BH_n$*

(i) From borane and alcohols

These compounds are derived from borane by sequential alcoholysis of hydride bonds. Diborane reacts with alcohols to give dialkoxyboranes in good yield ⟨60JPC682⟩. In these reactions there is no measurable presence of monoalkoxyborane, according to infrared measurements. Even when employing a large excess of diborane in its reaction with ethanol, no monoethoxyborane was observed ⟨59JCP(30)1222, 59JPC1319⟩. Only diethoxyborane was produced until all diborane was consumed; after this stage of the reaction, triethoxyborane was produced rapidly. It is probable that the monoalkoxyborane is disproportionated to give dialkoxyborane and borane.

Higher hydrides of boron will also react with alkanols to give di- and trialkoxyboranes ⟨53JA228⟩.

(ii) From borane and other oxygenated compounds

The initial products of the diborane reduction of carbonyl compounds and epoxides are di-alkoxyboranes ⟨39JA673, 60JA3866⟩. Once again, attempts to isolate monoalkoxyboranes from such reactions are not routinely successful; such products are obtained only as polymeric solids ⟨39JA673⟩. The reduction of benzoyl chloride in the presence of *t*-butylamine allows preparation of the corresponding monoalkoxyborane–amine complex ⟨60CB1078⟩.

(iii) From borane and trialkoxyboranes

Diborane and trialkoxyboranes equilibrate slowly at room temperature to give dialkoxyboranes ⟨33JA4020, 53JA199⟩.

(iv) From dialkoxyhaloboranes

Trialkoxyboranes react with sodium hydride and dialkoxychloroboranes to give dialkoxyboranes (*inter alia*) ⟨65MI 202-02⟩. It is supposed that the actual reducing agent is a sodium trialkoxy-hydroborate, formed *in situ* from reaction of sodium hydride with trialkoxy borane (see below).

(v) From sodium borohydride and tertiary alcohols

The reaction of tertiary alcohols with sodium borohydride in the presence of 1 equiv. of acetic acid allows facile preparation of dialkoxyboranes. The reaction proceeds via alcoholysis of borane liberated by proteolysis of borohydride ion. This procedure is only applicable to the preparation of tertiary dialkoxyboranes, for, under the same conditions, primary and secondary alcohols react to give good yields of trialkoxyboranes ⟨56JA3613⟩.

(vi) From sodium trimethoxyborohydride

Unlike simple borohydrides, trimethoxyborohydride reacts with hydrogen chloride at low temperature (−80 °C) to give dimethoxyborane. The reaction is believed to proceed via protonation of a methoxy group, leading to loss of methanol and production of dimethoxyborane ⟨53JA192⟩. Dimethoxyborane may be prepared in much inferior yield, as a by-product of the thermal decomposition of sodium trimethoxyborohydride, in which reaction the major products are sodium borohydride and sodium tetramethoxyborate. Similarly, a trace of diethoxyborane may be obtained from the pyrolysis of ethoxyborohydrides ⟨53JA205⟩.

(vii) From reduction of trialkoxyboranes

Trimethoxyborane reacts with sodium hydride, sodium borohydride or alkoxyborohydrides to give dimethoxyborane ⟨B-64MI 202-01⟩.

(viii) From reduction of boroxines

Reducing agents react with trialkylboroxines in the absence of solvent to give dialkoxyboranes ⟨61USP2992072, 61USP3014060⟩. The stoichiometry of the reaction is complex and depends upon the relative proportions of boroxine and reductant. In particular, sodium borohydride reduction of trialkylboroxines in high-boiling solvents is reported to be the method of choice for preparation of dialkoxyboranes ⟨61USP2992266⟩.

2.02.8.2.2 *Alkoxyhaloboranes, $(RO)_{3-n}BHal_n$*

(i) From boron trihalides

The first reports of preparation of haloboranes concerned the reactions of boron trifluoride with methanol and ethanol. The products were monoalkoxyfluoroboranes, which may be converted into dialkoxyfluoroboranes upon reaction with alkoxides. The reaction of boron trifluoride with trialkoxyboranes allows preparation of mono- and dialkoxyfluoroboranes in what amounts to a redistribution reaction ⟨1894MI 202-01⟩. In addition to the analogous reaction of alcohols with boron trichloride ⟨31ZAAC(202)1⟩, the dealkylation reaction of ethers using boron trichloride leads to either monoalkoxydichloroboranes or dialkoxymonochloroboranes, depending upon the stoichiometry of the reaction ⟨30CB1136, 31ZAAC(202)22⟩.

While alkoxydichloroboranes derived from primary alcohols are quite stable at ambient temperatures, those bearing secondary radicals are highly unstable, even at low temperatures (less than $-70\,^{\circ}C$). Dibromomethoxyborane is unstable and undergoes a disproportionation reaction ⟨55ZAAC(282)86⟩; the stabilities of other bromoalkoxyboranes are analogous.

Diols react with boron trichloride to give cyclic chlorodialkoxyboranes, also known as 2-chloro-1,3,2-dioxaborolanes. Treatment of diols with two equivalents of boron trichloride leads transiently to a linear bis(dichloroalkoxyborane), which disproportionates upon distillation to give the chlorodioxaborolane ⟨61JOC3250⟩.

(ii) From boron trihalides and cyclic ethers

Epoxides undergo cleavage reactions with boron trichloride to give dichloro ω-chloroalkylalkoxyboranes, though in the case of ethylene oxide, the product of the reaction is the monochloro*di*alkoxyborane ⟨55JCS1470⟩. Depending upon the structure of the epoxide, a mixture of regioisomers (where possible) is often obtained.

When the boron trichloride complex of tetrahydrofuran is heated under vacuum, 4-chlorobutoxydichloroborane is obtained ⟨55JCS1470⟩. The same product may be obtained by treating the complex with pyridine at ambient temperature ⟨57JCS348⟩. When a similar reaction is carried out with the tetrahydropyranyl complex, however, an elimination reaction occurs to give dichloropent-4-enoxyborane ⟨54JCS1212⟩. A similar reaction occurs when the dioxanyl complex is reacted under the same conditions; the product of the reaction is dichloro(2-vinyloxy)ethoxyborane ⟨52JCS11⟩. 1,3-Dioxolanes react to give 3-chloromethoxyethoxychloroboranes of variable stoichiometry ⟨61JCS5545⟩.

(iii) From boron trihalides and aldehydes and ketones

Boron trichloride adds to the carbonyl group of aldehydes to give dichloro-2-chloroalkoxyboranes, which immediately undergo further reaction to give trialkoxyboranes as final products ⟨57JCS739⟩. Similar reactions occur with ketones.

(iv) From boron trihalides and silyl ethers

Silyl ethers of alcohols are cleaved by reaction with boron trihalides to give alkoxydihaloboranes and the corresponding silyl halide ⟨53ZN608⟩. Monoalkoxysilanes react similarly ⟨61JA3384⟩.

2.02.8.2.3 *Alkoxyhydroxyboranes, $(RO)_{3-n}BOH_n$*

(i) Dialkoxyhydroxyboranes (n = 1)

These partial esters of boric acid have not been prepared. However, cyclic hydroxydioxaborolanes result from the reaction of diols and boric acid. The reaction is a general one. When diols are treated with alkali in the presence of boric acid, metals salts of hydroxydioxaborolanes are produced ⟨25ZAAC(142)83, 61JPS193⟩. Dimerisation occurs when hydroxydioxaborolanes are treated with dehydrating agents, to give 2,2-oxybis(1,3,2-dioxaborolanes) ⟨59JCS1529, 61JCS922⟩.

(ii) Alkoxydihydroxyboranes (n = 2)

Trimenthoxyborane (menthyl metaborate) reacts on standing in air to give menthoxydihydroxyborane; attempts to prepare such a compound by partial esterification of boric acid with menthol gave only the metaborate ⟨55JA1578⟩. There are no general methods for direct preparation of this class of compound from monoalcohols.

(iii) From polyhydroxy compounds

Aureomycin forms a dihydroxyborane derivative, upon reaction with boric acid ⟨55MI 202-02⟩. Dihydroxyboranes are also formed by the reaction of mannitol with boric acid ⟨40USP2223948⟩.

(iv) Borates, esters of boric acid

Trialkyborates (also known as *ortho*-borates) are generally formed by reaction of alcohols with boron trihalides, boric acid, metaboric acid, boric oxide or borax ($Na_2B_4O_7$). The subject area of boric acid derivatives has been included in a review of boron chemistry ⟨56CRV959⟩.

(v) Symmetrical orthoborates

The most efficient and usual method for preparation of trialkylborates directly parallels the esterification reactions of carboxylic acids, viz. dehydrative reaction of boric acid in the presence of three equivalents of alkanol. The reaction equilibrium lies far to the left and therefore requires removal of water to give synthetically useful yields. The esterification of boric acid is covered in great detail in the two-volume work by Steinberg ⟨B-64MI 202-01⟩. Borates derived from carbohydrates have particularly attracted the attention of researchers in the 1990s ⟨91CAR(225)113⟩.

(vi) From alkoxyhaloboranes

Unsymmetrical trivalent boron compounds are thermodynamically unstable with regard to their symmetrical counterparts; as an example of this property, disproportionation of alkoxychloroboranes occurs, upon heating, to give symmetrical borates. Thus, chlorodimethoxyborane reacts with trimethylamine to give trimethylborate ⟨35ZAAC(222)92⟩. Similar reactions occur with chlorodimethoxyborane and other haloboranes. Dialkoxyboranes also disproportionate to give diborane and the corresponding trialkylborates ⟨33JA4020, 39JA673⟩. In a few cases, esters of boronic acid undergo thermal disproportionation to give borates ⟨57AG677⟩. The use of boron trifluoride greatly facilitates the reaction ⟨56JCS824⟩. The disproportionation of borinates is rare, but the ethyl ester of diphenylborinic acid gives rise to triethylborate (*inter alia*) ⟨58JCS1451⟩.

THF reacts with diborane in a sealed tube to give tri-*n*-butoxyborane in 61% yield; other tetrahydrofurans react in a similar fashion ⟨61JA1515⟩.

Reduction of carbonyl compounds with trihydridic boranes necessarily leads to trialkylborates; these species are most usually decomposed by treatment with some protic source to give the corresponding alcohols, but the interception and isolation of borates is possible (and sometimes a troublesome problem, if the alcohol is desired!).

Mono- and dialkoxyboranes react at ambient temperature to give trialkoxyboranes ⟨60MI 202-02⟩.

Both boranes and borinates may be converted to borates ⟨55JOC9, 58JCS1451⟩. Oxidation of trialkylboranes give trialkylborates ⟨36JPR251, 38JA121⟩.

Trialkoxyboroxines give trialkylborates upon heating; however, since these compounds are most often prepared from the borates, the reaction is of little preparative value.

Trialkylboranes react with tertiary amine oxides to give trialkyl esters of boric acid in quantitative yield ⟨66AG(E)580⟩. Metaborates result from the oxidation of alkanes in the presence of boric acid at high temperature ⟨68MI 202-02⟩.

(vii) Unsymmetrical orthoborates

The methods available for preparation of these compounds parallel those outlined briefly above for the preparation of symmetrical compounds. The reader is directed to the detailed review previously alluded to ⟨B-64MI 202-01⟩.

Haloboranes may also be autoxidised to give peroxyboranes ⟨59CB2716⟩.

2.02.9 RO METAL DERIVATIVES

Alkoxide synthesis has been excellently and comprehensively reviewed ⟨B-78MI 202-01⟩. The general principles are as follows.

(1) Highly electropositive metals react with alcohols to give alkoxides directly. All group 1 metals will undergo the reaction, and calcium, strontium and barium are also electropositive enough to react in this fashion.

(2) Less-electropositive metals form alkoxides upon reaction with alcohols in the presence of a catalyst (such as iodine or mercury salts). The preparative routes to other metal alkoxides mimic those already mentioned. Thus, magnesium alkoxides may also be prepared by reaction of alcohols with ethylmagnesium halides in THF (Equation (109)) ⟨94T6117⟩.

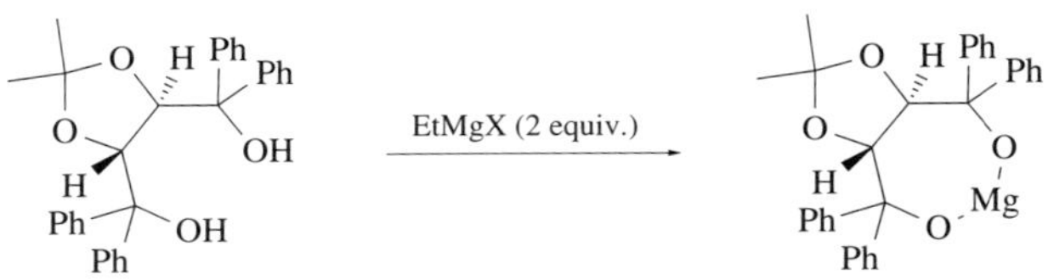

(109)

(3) Reaction of alcohols with metal chlorides (particularly those of boron, silicon, titanium and tin) and hydroxides allows preparation of alkoxides.

(4) Deprotonation of alcohols by metal dialkylamides and alkyl metals allows preparation of alkoxides. Thus, for instance, zirconium tetra-*t*-butoxide may be prepared by reaction of $Zr(NEt_2)_4$ with *t*-butanol ⟨61CJC1386⟩. Of course, in some circumstances it may be tedious to prepare the metal dialkylamide, and this is the primary limitation of this method.

When the alkyl substituent has particularly high steric demand, deprotonation using the corresponding metal hydride is often successful. Boron and aluminum alkoxides may be prepared by this method ⟨68ZAAC(358)44⟩.

The reaction of alkyl metals with alcohols is a reliable method for preparation of metal alkoxides ⟨94JA6577⟩. The ready availability of alkyllithiums allows for particularly facile preparation of lithium alkoxides ⟨94T529⟩. Dialkylzincs are oxidised in air to give dialkoxyzincs ⟨60JCS4130⟩. Alkylboranes and alanes also undergo such reactions ⟨59JCS429, 71USP3600417⟩.

Transmetallation of alkoxides is also feasible; treatment of sodium alkoxides with Grignard reagents, for instance, allows preparation of (crystallisable) alkyl- and arylmagnesium alkoxides ⟨93JOM(452)1⟩.

Aluminum alkoxides may be prepared via a hydroalumination reaction of aluminum hydride (alane) and ethene, with the resulting alkylaluminum species being readily oxidised by air to give alkoxides ⟨60LA(629)241⟩.

2.03
Alkyl Chalcogenides: Sulfur-based Functional Groups

PHILIP C. BULMAN PAGE, ROBIN D. WILKES and DOMINIC REYNOLDS
University of Liverpool, UK

2.03.1 ALKANETHIOLS

A variety of well-known and long-established procedures exist for the preparation of alkanethiols, and indeed, this topic has been the subject of several detailed reviews, with even the most up-to-date procedures being covered ⟨B-68MI 203-01, B-74MI 203-01, B-77MI 203-01, 94MI 203-01⟩.

Preparations of alkanethiols are discussed here in a number of subsections. The most extensively used routes to alkanethiols originate from alkene (Section 2.03.1.1), alcohol (Section 2.03.1.2), and alkyl halide (Section 2.03.1.3) starting materials, although many other procedures are available. It should be noted that some of the less well-known routes can offer the best methods of preparation of a particular alkanethiol.

2.03.1.1 Formation from Alkenes

Alkanethiols can be readily prepared from alkene starting materials either in a direct or an indirect manner. The direct preparation of thiols from alkenes involves the addition of hydrogen sulfide to the starting material, and ranks alongside the substitution of alkyl halide or alcohol substrates by hydrogen sulfide as conceptually the simplest route to alkanethiols (Scheme 1) ⟨38JA2452, 40CRV351, 42JOC472, 51JOC524⟩.

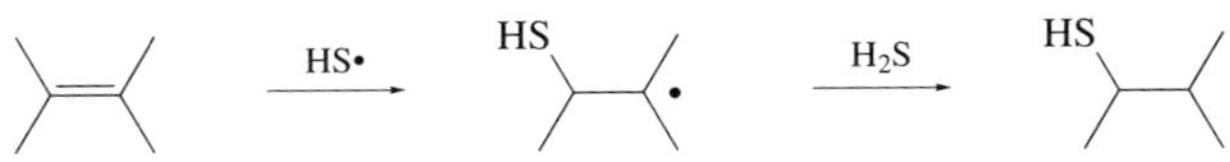

Scheme 1

The shortcoming of this particular direct preparation is a high probability that the thiol product will undergo further radical reaction to furnish a dialkyl sulfide under the reaction conditions. This side reaction is not easily controlled, and indeed becomes a serious problem when the alkanethiol formed in the reaction is more reactive towards the alkene than is the hydrogen sulfide itself. Formation of sulfide can be minimized by the use of a high ratio of hydrogen sulfide to alkene substrate ⟨56JA348⟩.

Hydrogen sulfide addition to substituted alkenes can occur with selectivity for either regiochemistry, depending upon the reaction conditions. Under certain conditions, including high temperatures, the two regioisomers from reaction of an alkene with hydrogen sulfide can exist in equilibrium. However, if the reaction is performed at lower temperatures in the absence of a radical initiator, a wide variety of catalysts promote the preferential formation of the Markovnikov addition product ⟨66JOC776⟩. Typical catalysts include metal sulfides ⟨47JCS1532⟩, metal oxides, a variety of acids, and sulfur ⟨38JA2452, 47JCS1532⟩. In contrast, the anti-Markovnikov product of hydrogen sulfide addition is readily and predominantly formed under free-radical reaction conditions, such as ultra violet or other irradiation, often in the presence of a photosensitizer, or in the presence of radical initiators such as azonitriles ⟨40CRV351, B-69MI 203-01⟩. As a result, reactions can often be tailored to produce either regiochemistry of the thiol product by appropriate choice of reaction conditions. A development in 1994 uses triphenylsilanethiol as a hydrogen sulfide equivalent; addition takes place to alkenes in the presence of radical initiators to give thiols after mild hydrolysis of the silyl thioether products in good to excellent yields (Scheme 2) ⟨94TL1837⟩.

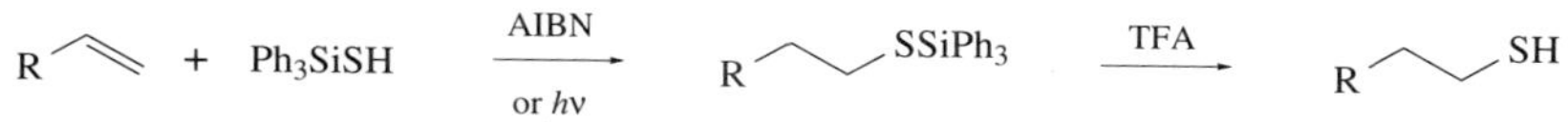

R = alkyl
AIBN = 2,2'-azobisisobutyronitrile

Scheme 2

The addition of hydrogen sulfide to alkenes containing electron-withdrawing substituents is usually carried out under basic catalysis, and then usually takes place by conjugate addition of thiolate anion ⟨66CJC2315⟩. Examples of additions of hydrogen sulfide to unsaturated compounds containing nitro ⟨47JCS1477⟩, cyano ⟨51ACS690, 69JOC2955, 47JA2328, 48JOC97⟩, carboxyl ⟨51ACS690, 48JCS1683⟩, carbonyl ⟨35JA1098⟩, and ester ⟨48JCS1683, 69JOC2955⟩ functionalities have all been reported (Scheme 3).

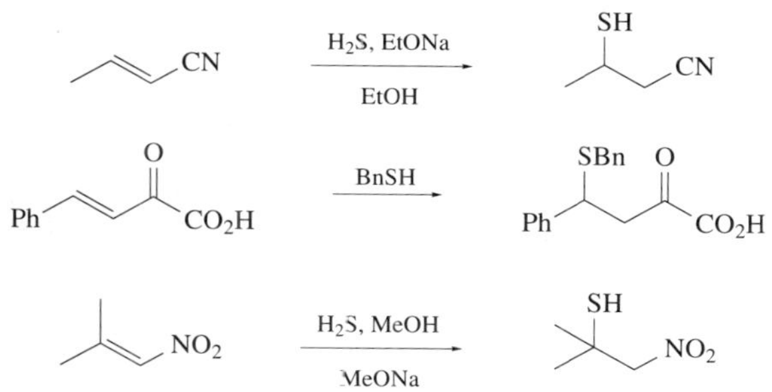

Scheme 3

The propensity for the thiol products to undergo further reaction, such as oxidation to disulfide, coupled with the toxicity and difficulty of handling of hydrogen sulfide has inevitably resulted in relatively few applications of this very direct method of thiol preparation. Less direct methods, such as the addition of thiolcarboxylic acids to alkenes followed by hydrolysis or reduction are more often employed, despite the extra step involved.

Thiolcarboxylic acids of various structures can be employed in addition reactions with alkenes, particularly those with strongly electron-withdrawing substituents, by far the most commonly used reagent being thiolacetic acid. The addition of thiolacetic acid to alkenes, a radical process, is exothermic and usually yields the anti-Markovnikov intermediate, which is readily converted into the corresponding alkanethiol by any of a wide variety of moderate reaction conditions, typically treatment with alcoholic alkali (Scheme 4) ⟨47JCS134, 51JCS2123, 57JA3493, 65ACS1490⟩.

One interesting example of this type of reaction is the action of thiolacetic acid upon an allyl chloride. As shown, the only product isolated following alkaline hydrolysis of the expected

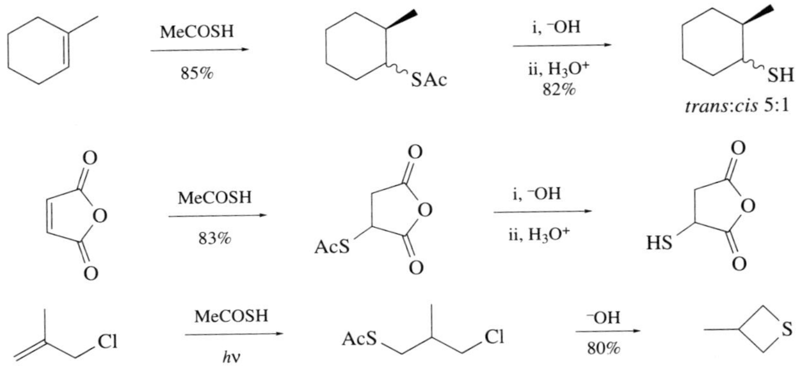

Scheme 4

intermediate is the four-membered ring cyclic sulfide, arising from an internal nucleophilic substitution ⟨58JOC636, 60JA5354⟩. The reaction is also successful for vinyl and homoallyl substrates, yielding the three- and five-membered ring products respectively.

2.03.1.2 Formation from Alcohols

Alcohols, like alkenes, are among the most readily available starting materials for the synthesis of alkanethiols. Alcohols can also be converted directly into the desired thiols by reaction with hydrogen sulfide, normally in the presence of a suitable catalyst, although, once again, sulfide formation is a competing reaction. One such catalytic system uses basic alumina as a support and a heated thoria catalyst (Equation (1)) ⟨21JA880⟩. Basic conditions are essential as acidic conditions result in dialkyl sulfide formation. The drawbacks of such catalytic processes are the high temperature and pressures required, and the variable yields obtained.

$$\text{ROH} \xrightarrow[\text{basic promoter}]{H_2S,\ \text{basic alumina}} \text{RSH} \qquad (1)$$

$$R = \text{alkyl}$$

Tertiary thiols, which are generally best prepared from alcohols by the action of hydrogen bromide and thiourea (see below), have however also been successfully prepared by the action of hydrogen sulfide gas upon a solution of the alcohol in a mixture of acetic and sulfuric acids (Equation (2)) ⟨50JCS3309⟩. Ethanethiol has also been prepared using a similar method, albeit in low yield and with significant by-product formation.

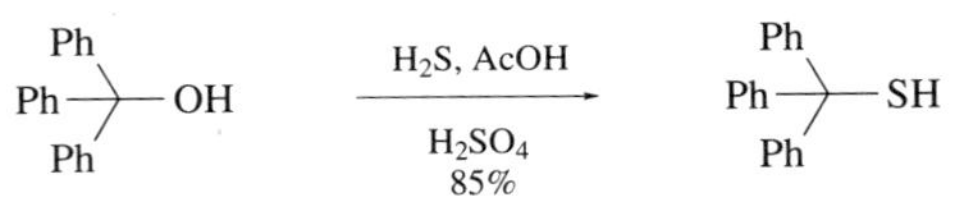

(2)

Although there are a limited number of procedures for the direct conversion of alcohols to thiols, it must be remembered that alcohols are potential precursors of a large range of other compound types, which may themselves serve as precursors to alkanethiols. For example, alcohols may be readily converted into alkyl xanthates using hydroxide ion and carbon disulfide (see Section 2.03.1.7), or into thiolesters via sulfonates. Both thiolesters and alkyl xanthates are easily converted into thiols by reduction.

Thiolesters, as indicated above, can also be produced by conversion of an alcohol into a leaving group which undergoes subsequent nucleophilic substitution, for example using thiourea or thioacetate, as shown for the preparation of the postulated intermediate thiol in biotin biosynthesis (Scheme 5) ⟨90BSF758⟩, or by using a Mitsunobu reaction (Scheme 6) ⟨67BCJ2380, 68TL5907, 71BCJ3427, 72JA679, 81S1, 93JMC460, 93SL209⟩. Several modifications of this Mitsunobu inversion procedure, such as preformation of the triphenylphosphine–azodicarboxylate adduct, have been developed, and generally exhibit the expected high level of stereochemical integrity and high yields ⟨81TL3119⟩. The

Mitsunobu reaction has also been used in the preparation of enantiomerically pure benzylic thiols ⟨92TL4099⟩.

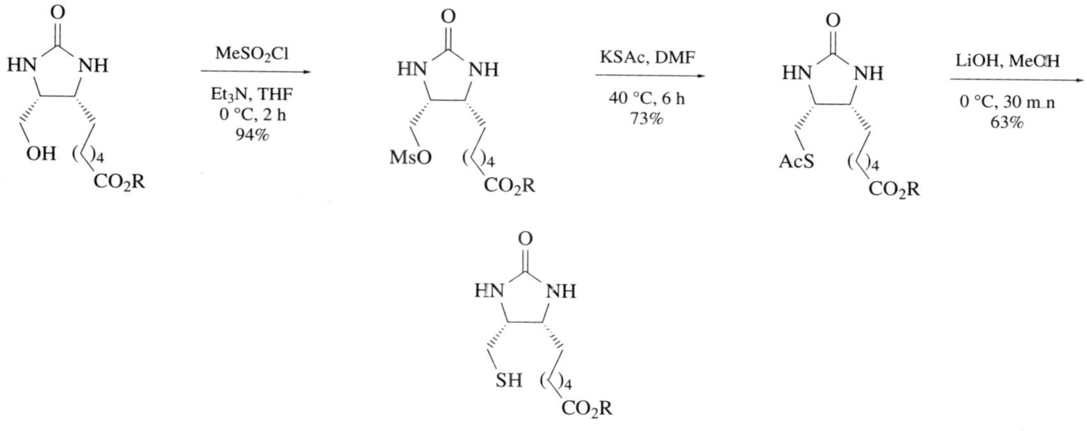

Scheme 5

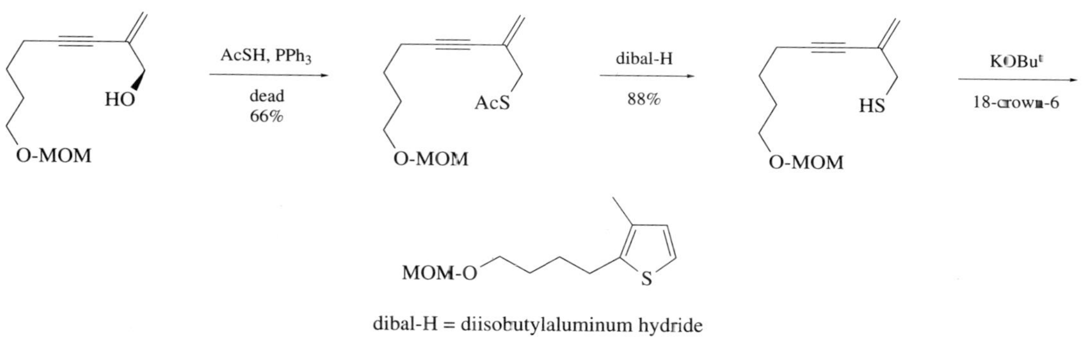

Scheme 6

Alkanethiols may also be prepared from alcohols by conversion of the alcohol into a sulfate ⟨1834LA(11)1⟩ or a sulfonate ⟨53JOC1591⟩, and subsequent reaction with hydrogen sulfide. Both of these reactions have been long established. A less direct, although very efficient route to alkanethiols from alcohol starting materials involves initial reaction with phosphorus pentasulfide. Alcohols are converted directly into dialkyldithiophosphonates, which upon acidic hydrolysis furnish the thiols in greater than 70% yields ⟨63BRP917921⟩.

As discussed below (Section 2.03.1.5), thiourea in combination with halogen acids has been used to form isothiouronium salts directly from alcohols ⟨46JA2103⟩.

2.03.1.3 Formation from Alkyl Halides and Hydrogen Sulfide

Alkyl halides are probably the single most important readily available starting materials for alkanethiol preparation. Metal salts of hydrogen sulfide, in particular the sodium salt, react efficiently with alkyl chlorides, bromides, and iodides in alcoholic solution to produce thiols (Equation (3)). Primary and secondary alkanethiols (up to C_9) are readily prepared in this way from alkyl chlorides and sodium hydrogen sulfide ⟨32JA1674⟩. For the higher alkanethiols, alkyliodides, sodium hydrogen sulfide, and autoclave conditions are preferable. Dithiols can be prepared from the corresponding dibromides using hydrogen sulfide under high temperature sealed tube conditions ⟨47JCS592⟩. Nonetheless, all the procedures described in this section suffer from two drawbacks. The most common is dialkyl sulfide formation, favored by higher temperatures. Lower temperatures are therefore to be preferred for thiol formation. Unwanted alkene formation by elimination can also be a problem, particularly in the synthesis of tertiary alkanethiols.

$$RX \xrightarrow{NaSH} RSH \qquad (3)$$

$$X = Cl, Br, I; R = alkyl$$

An indirect method of thiol formation from alkyl halides involves reaction with disodium disulfide in liquid ammonia to furnish the disulfides, which can then readily be converted into the thiols in high yields (Section 2.03.1.10) (Equation (4)). In a more complicated example of hydrosulfide alkylation, fluoride-mediated desilylation/hydrolysis of α-trimethylsilyloxy sulfides, a special case of oxathioacetals prepared either by direct alkylation of α-silyloxy thiols with alkyl halides ⟨86TL3975⟩ or by other methods ⟨76TL319⟩, leads to aliphatic thiols, as illustrated in Scheme 7.

$$RX \xrightarrow{Na_2S_2} RSSR \longrightarrow RSH \qquad (4)$$

$$X = Cl, Br, I; R = alkyl$$

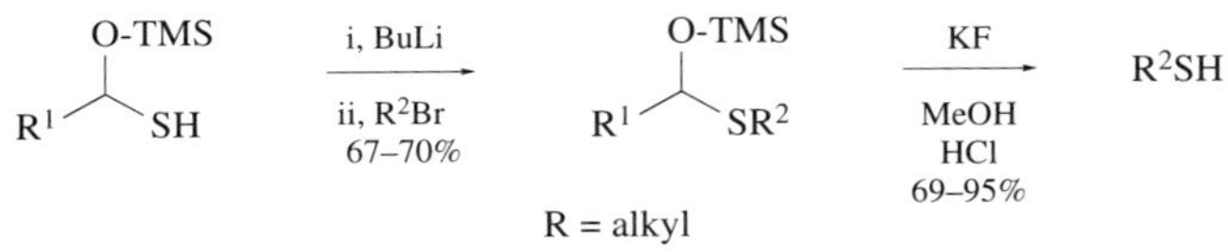

Scheme 7

The utility of alkyl halides as starting materials for the preparation of alkanethiols is further illustrated below.

2.03.1.4 Formation using Phosphorothiolate Ion

Acidic hydrolysis of alkyl-*S*-phosphorothiolates, formed by the treatment of an alkyl halide with trisodium or trilithium phosphorothiolate, is an efficient and mild method of alkanethiol preparation, as shown in Scheme 8.

The best method of preparation of the alkali metal phosphorothiolates is by alkaline hydrolysis of thiophosphoryl chloride, and an improved procedure to give trisodium phosphorothiolate in 90% yield has been reported by Åkerfeldt ⟨60ACS1980⟩.

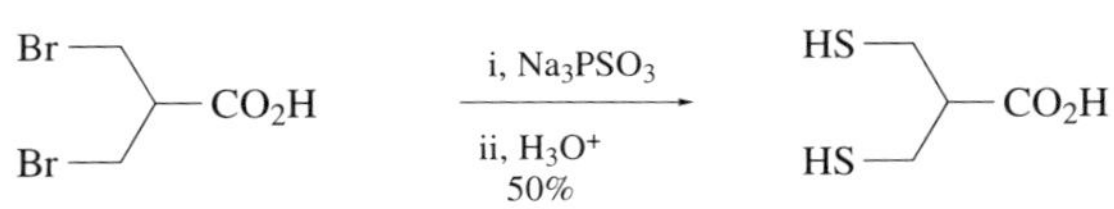

Scheme 8

2.03.1.5 Formation via Isothiouronium Salts, and Related Procedures

One of the most common laboratory methods for the preparation of thiols involves the reaction of an alkyl halide with thiourea to form an isothiouronium salt which is subsequently hydrolyzed to give the desired thiol product (Scheme 9) ⟨55OSC(4)363, 63OSC(4)401, 63OSC(4)491, 63OSC(4)638⟩. Yields are highest for primary alkyl halides, and poorest for tertiary alkyl halides, where there is competing elimination ⟨58JOC1525⟩, and with haloalkylcarboxylic acids and unsaturated halides. Thiourea in combination with halogen acids has also been used to form isothiouronium salts directly from alcohols (Scheme 9) ⟨46JA2103, 87OS(66)150⟩. This transformation was initially described by Stevens ⟨02JCS79⟩ and subsequently developed by Johnson and Sprague ⟨36JA1348, 37JA1837⟩. Yields of thiols obtained by this procedure are comparable to those obtained from alkyl halides.

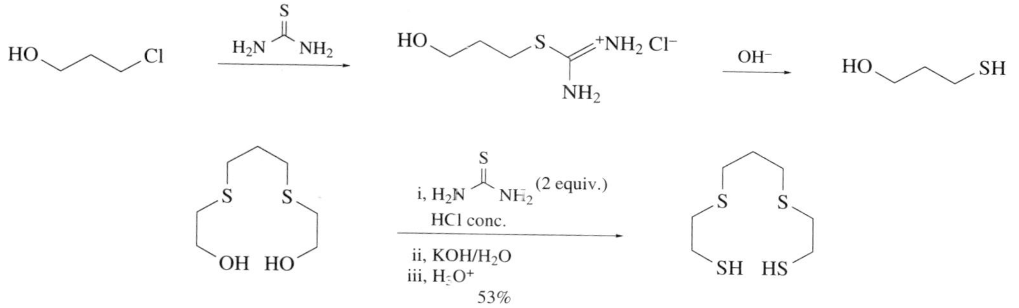

Scheme 9

While these procedures are generally efficient, (preparations of batches of up to 44 grammes of dithiol have been described ⟨50OS(30)35⟩), the one severe limitation of this route is the harsh conditions often required for the hydrolysis of the intermediate salt. These include alkaline hydrolysis followed by steam distillation or extraction; the use of a high-boiling solvent, such as triethylene glycol, to furnish the isothiouronium salt, which is decomposed by a high-boiling amine to give the desired thiol *in situ*; or even thermal decomposition in the presence of bicarbonate ⟨62JOC93⟩.

One interesting variation is the preparation of a thiol by ring opening of an oxetane by thiourea in the presence of a strong acid ⟨61JOC3024⟩.

While the thiourea method is of general utility in the preparative scale production of simple thiols, the method cannot be applied indiscriminately. For example, α-thiolcarbonyl compounds cannot be obtained by the process due to favored thiazole formation ⟨43OSC(2)31⟩. As a result of such limitations, a range of alternative sulfur nucleophiles has been employed, including 1-methylpyridine-2-thione ⟨77JOC2180⟩, 1,2-benzisothiazole-3(2*H*)-thione, 1,1-dioxide sodium salt ⟨81CL1457⟩, and 1-(2-hydroxyethyl)-4,6-diphenylpyridine-2-thione ⟨85TL469⟩. For example, reaction of 1-methylpyridine-2-thione with alkyl halides gives the corresponding 2-methylthiopyridinium salts, which are readily hydrolyzed to give thiols under alkaline conditions. Primary and secondary alkyl halides, α-halo ketones, α- and β-halocarboxylic esters and halosugars were all successful substrates, furnishing the intermediate salts in yields of 81–84%. Scheme 10 shows an interesting example using a monosaccharide alkyl halide substrate. In one case, however, using phenethyl β-bromopropionate, no thiol product was formed due to preferential elimination from the intermediate salt.

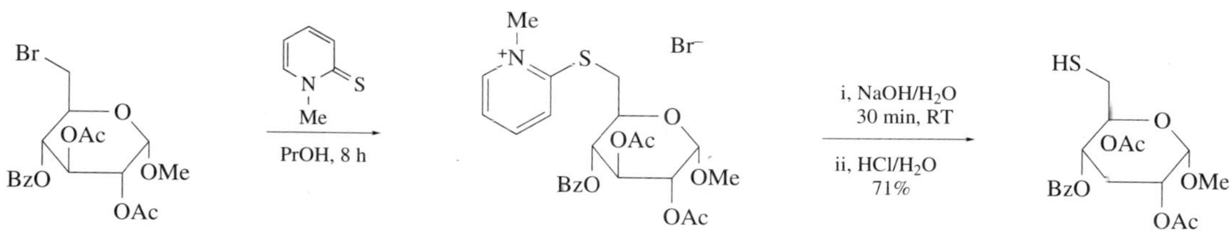

Scheme 10

The sodium salt of thiosaccharin (1,2-benzisothiazol-3(2*H*)-thione 1,1-dioxide) readily reacts with alkyl halides at ambient temperature in DMF solution to furnish the thioalkylated products, which can easily be precipitated using water, in better than 80% yields (Scheme 11). Further reaction with a secondary amine, in this case piperidine, furnished the desired thiol in high or quantitative yield ⟨81CL1457⟩.

In a clever alternative procedure, 1-(2-hydroxyethyl)-4,6-diphenylpyridine-2-thione ⟨70JCS(C)1578⟩ has been found to react with alkyl chlorides, bromides, and iodides to furnish the corresponding thiols directly, as shown in Scheme 12, after a facile intramolecular nucleophilic displacement. Reaction occurs with alkyl iodides in dry benzene at 20 °C to give thiols in yields of up to 89%, although for alkyl chloride and bromide substrates it was necessary to employ acetonitrile as solvent together with tetraethylammonium iodide as a halide exchange agent. This route to alkanethiols is particularly attractive as high yields are obtained, the reaction conditions are mild and neutral, and the workup is simple.

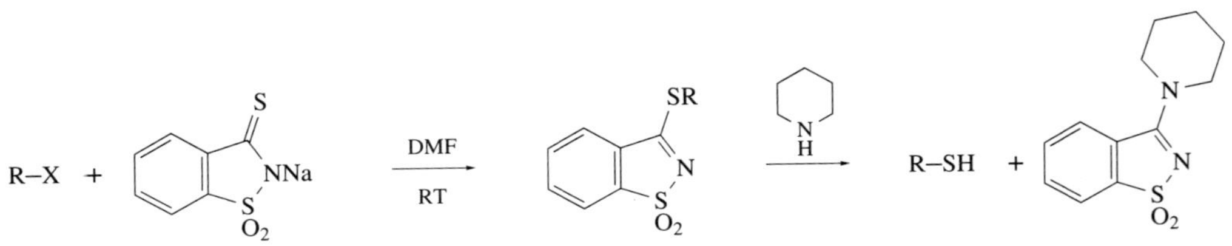

X = Cl, Br, I; R = alkyl

Scheme 11

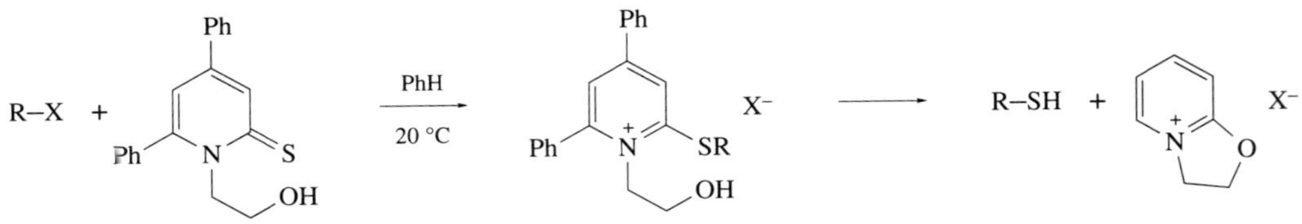

RX	*Yield (%)*
$C_6H_{13}I$	69
$PhCH_2Br$	80
$C_5H_{11}Cl$	61

Scheme 12

An analogous procedure, and perhaps the mildest method of alkanethiol preparation of this type, involves the decomposition of alkylhalide–dimethylthioformamide adducts in neutral non-aqueous solvents (Scheme 13) ⟨83YGK776⟩.

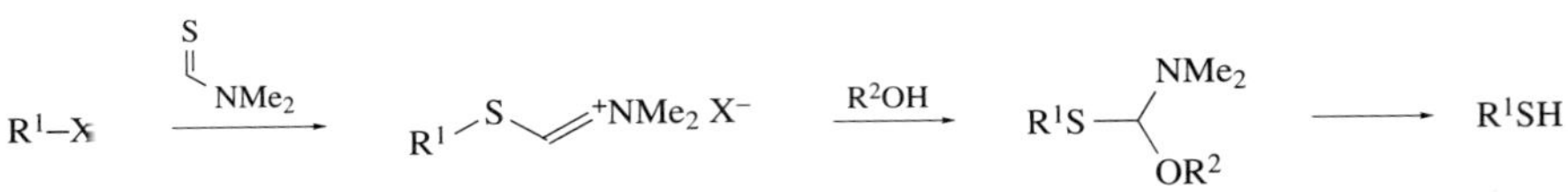

X = Cl, Br

R¹	*Yield (%)*
Bu	95
CH_2CO_2Me	94
$PhCH_2$	91

Scheme 13

2.03.1.6 Formation via Bunte Salts

Organic thiosulfites, commonly known as Bunte salts ⟨1874CB646, 67AG(E)544, 88OSC(6)235⟩, are traditional and convenient precursors to thiols. The free thiol is liberated by acid-mediated hydrolysis of the Bunte salt ⟨09JCS1725, 63JOC957⟩, which is formed in turn by treatment of an alkyl halide with thiosulfate in aqueous conditions (Scheme 14). An alternative fate for the Bunte salt is conversion into a disulfide or sulfide ⟨63JOC3608⟩. The Bunte salt can usually be generated and hydrolyzed without isolation, to produce alkanethiols in yields of greater than ca. 75% in most cases ⟨89SC3143⟩. Reducing agents such as sodium hydrogen telluride can also react with Bunte salts to form thiols.

Sodium thiosulfate is the most commonly used reagent for the preparation of Bunte salts, although thallous thiosulfate has also been employed, as the thallous halide by-products are insoluble in the aqueous medium, allowing ready isolation of the Bunte salt ⟨55JOC475⟩. Bunte salts are significantly easier to generate from primary and secondary alkyl halides than from tertiary ones. The preparation of alkanethiols via Bunte salts has several advantages over other routes, including improved yields ⟨50JA1687⟩ and particularly ready hydrolysis of the intermediates to give the thiol products ⟨66JA5245, 66JOC957⟩.

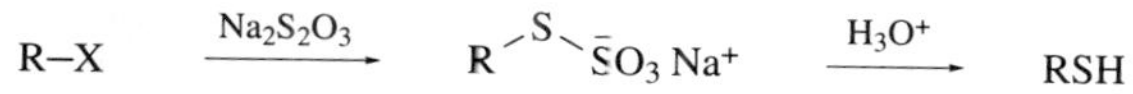

Bunte salt

RX	*Yield* (%)
$C_9H_{19}Br$	77
$PhCH_2Br$	89
$Br(CH_2)_7Br$	85 (dithiol)

Scheme 14

2.03.1.7 Formation via Xanthates and Related Esters

Alkyl xanthates, which are readily prepared from alkyl halides ⟨45JCS666, 55JA568⟩, serve as useful precursors to alkanethiols, the transformation from xanthate to thiol being effected by basic hydrolysis or reduction. The alkyl xanthates are prepared as shown in Equation (5) from alkyl halides using an alkali metal xanthate reagent, such as potassium ethylxanthate, derived in turn from the action of carbon disulfide on alkoxides ⟨45JCS666⟩, or from the reaction of carbon disulfide with alcohols in the presence of base ⟨72ACS279, B-89MI 203-01⟩. Cleavage of the alkyl xanthate intermediate to give alkanethiol is generally a high yielding (ca. 80%) process ⟨69JOC4170, 69TL3631⟩. Basic hydrolysis of alkyl xanthates can be carried out using hydroxide ion, ammonia, 2-aminoethanol ⟨69TL3631⟩, ethylenediamine ⟨69JOC4170, 81JOC3141⟩, or phenylhydrazine ⟨45JCS666⟩. Reduction, which is more appropriate than hydrolysis for alkali-sensitive ⟨57JOC561⟩ or easily oxidized thiols, is usually carried out with lithium aluminum hydride or similar reagents.

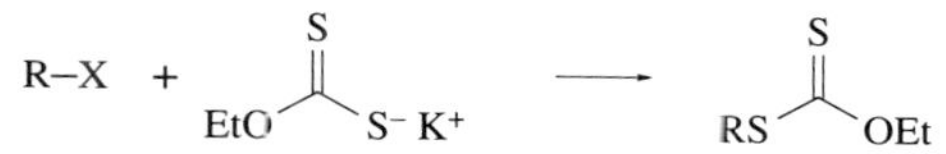

(5)

X = Cl, Br, I; R = alkyl

An interesting example of alkyl xanthate hydrolysis using nitrogen bases is the synthesis of optically pure benzhydryl thiols ⟨76S326⟩, where other standard routes resulted in racemization. A related preparation of optically active thiols has been reported using trifluoroacetic acid catalyzed rearrangement of dialkyl xanthates to dithiocarbonates with inversion of configuration. Hydrolysis furnished thiols with ca. 89% *ee* and up to 99% yield (Scheme 15) ⟨81JOC3141⟩. An alternative procedure uses pyridine *N*-oxide, or a solid phase copolymer of 4-vinyl pyridine *N*-oxide, as efficient catalysts for the conversion of xanthates to thiols via the corresponding *S*,*S*-dialkyldithiocarbonates ⟨87H(26)2583⟩.

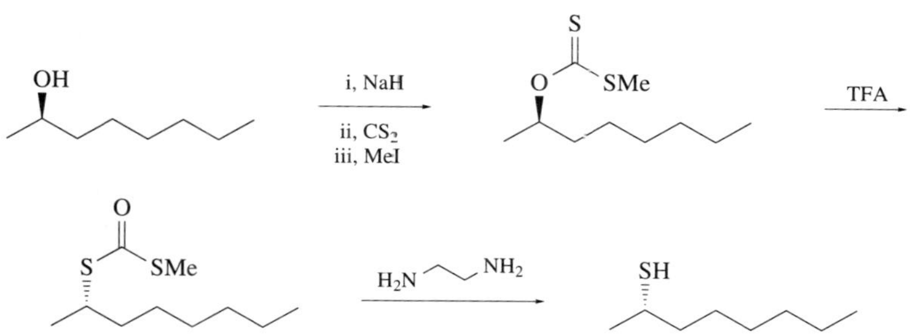

Scheme 15

Monoalkyl and dialkyl trithiocarbonates can, like xanthates, be hydrolyzed readily either under mildly acidic ⟨68JOC1275⟩ or basic ⟨69TL3631⟩ conditions, or may alternatively be reduced, to give alkanethiols. Both cyclic and acyclic trithiocarbonates can be prepared by the action of sodium trithiocarbonate on the alkyl dihalide or alkyl halide substrates respectively, as illustrated in Equation (6) and Scheme 16. Cyclic trithiocarbonates, useful as dithiol precursors, may be prepared by treatment of dihalides, episulfides ⟨60JCS1024, 60JCS1030, 64JOC360⟩, and epoxides with alkyl xanthates. In some cases (Scheme 16), use of an epoxide can result in a single displacement, leading to β-hydroxythiol formation ⟨60JCS1030⟩. Acyclic trithiocarbonates are typically cleaved to give thiols using mild acidic hydrolysis. For cyclic trithiocarbonates, lithium aluminum hydride reduction is the most efficient method for dithiol production ⟨60JCS1030⟩. Yields of thiol from all of these procedures are generally around 70–80% overall, although for optimum results solvent choice is crucial.

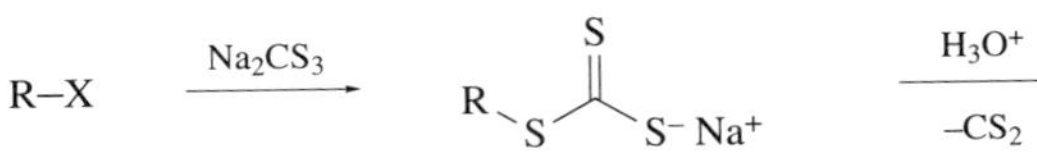

(6)

RX	*Yield* (%)
EtBr	85
$PhCH_2Cl$	85

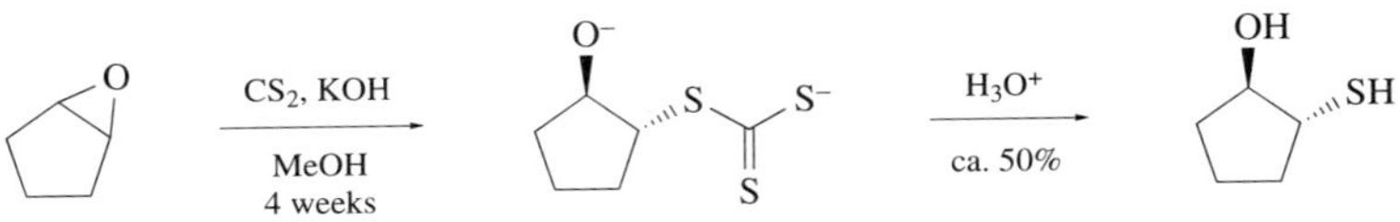

Scheme 16

Thiolesters are extremely versatile precursors to thiols, due in part to the wide variety of preparative methods available, and in part to their easy conversion into thiols. Three principal methods exist for the synthesis of thiolesters: by thiolcarboxylate substitution of halide or sulfonate groups, as illustrated in Scheme 17 (often carried out in the presence of a weak base such as triethylamine) ⟨93JOC633⟩; by the thermal rearrangement of thionocarboxylates, most commonly used for aromatic substrates (Scheme 18); and through the addition of thiolcarboxylic acids to alkenes, as described above (Section 2.03.1.1).

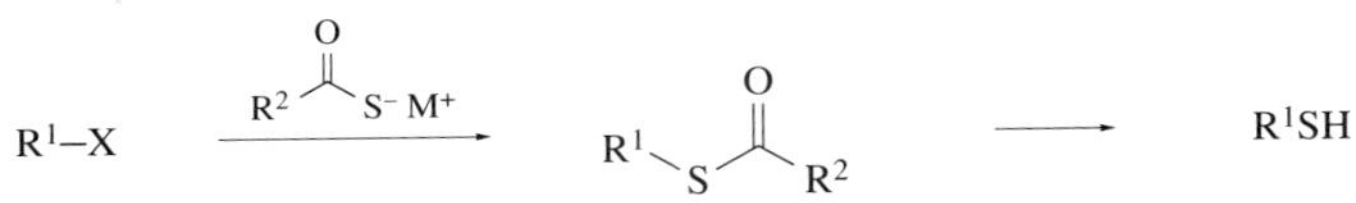

X = halide, sulfonate
M = Na, K
R^1 = alkyl

Scheme 17

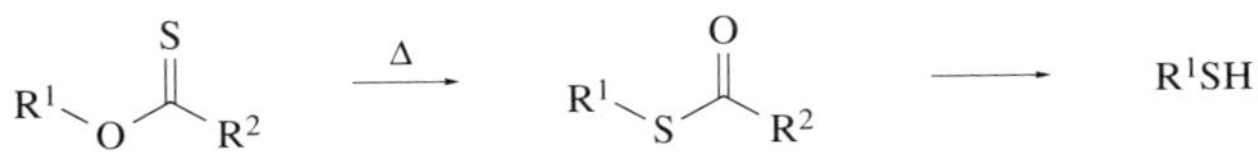

Scheme 18

Simple nucleophilic substitution of halides or sulfonates has the advantage that sulfide by-product formation is not observed. Usually a polar aprotic solvent such as DMF ⟨72HCA481⟩ or DMSO ⟨87S1009⟩ is employed with either the sodium or potassium salt of thiolacetic acid. Caesium thiolcarboxylates have been used in the preparation of optically active thiols to help prevent product racemization ⟨86JOC3664⟩. A variation of this method is the Mitsunobu reaction, as discussed above in Section 2.03.1.2 (Scheme 6) ⟨67BCJ2380, 68TL5907, 71BCJ3427, 72JA679, 81S1, 92TL4099, 93JMC460, 93SL209⟩. Conversion of thiolesters to thiols is normally carried out by simple hydrolysis with acid ⟨93JOC633⟩ or aqueous hydroxide ⟨90BSF758⟩. Other reagents may be employed, for example ammonia and *p*-chloroaniline ⟨65ACS1490⟩, used as shown in Scheme 4 to produce 2-thiolaevulinic acid, without the elimination of thiolacetic acid which was observed upon treatment with sodium hydroxide ⟨65ACS1490⟩. Reductive methods are also common. Lithium aluminum hydride gives particularly good yields (usually above 90%), and milder reagents such as dibal-H have also given good results ⟨93SL209⟩. Irradiation with UV light effects the same transformation by initial sulfur-carbonyl group bond cleavage followed by hydrogen abstraction.

Alkyl dithiocarbamates, generally prepared by the nucleophilic substitution of halides ⟨56CJC1093⟩ or other leaving groups ⟨77CL437⟩ with sodium *N*,*N*-dimethyldithiocarbamate are also useful precursors to thiols, as shown in Scheme 19. They are very stable and easily purified. Conversion to thiol is again readily achieved using basic hydrolysis, in yields of around 70% ⟨56CJC1093⟩, or reduction, generally in good yields ⟨77CL437⟩. An illustrative example is the preparation of cholestenethiol, where use of either 3*β*-cholesteryl chloride and potassium thiocyanate ⟨53JA2118⟩, or 3*β*-cholesteryl tosylate and thiourea ⟨48JA1176, 64T1265⟩, furnished only the *β* isomer, with retention of configuration. Use of sodium *N*,*N*-dimethyldithiocarbamate, however, gave a 5 : 1 mixture of isomeric dithiocarbamates, with the major component being the *α* isomer, formed with inversion of configuration ⟨77CL437⟩.

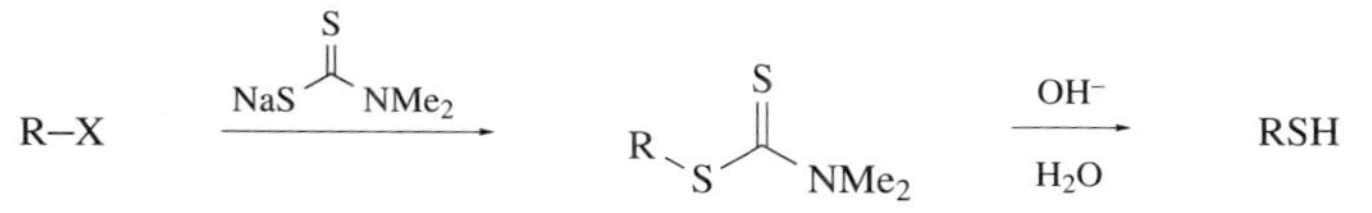

RX	*Yield* (%)
$PhCH_2Cl$	86
$Me(CH_2)_5Br$	61

Scheme 19

2.03.1.8 Formation using Sulfur Insertion Reactions of Organometallics

The reaction of inorganic sulfur with organometallic compounds provides metal thiolates (Scheme 20). Organomagnesium ⟨04CR(138)982⟩ and organolithium ⟨09BSF405, 49CB594, 49JA1478⟩ compounds are the usual reagents. This method is particularly useful for preparation of *t*-alkanethiols ⟨33JA2932⟩ and cycloalkanethiols, which are not easily synthesized by other routes, but is poor for primary and secondary substrates. The choice of organometallic reagent and its method of preparation are obviously important considerations. In addition, the magnesium-mediated reactions suffer from considerable sulfide and disulfide formation, so that a large excess of sulfur is generally employed. An interesting modification employs the use of thiiranes as controled sulfur releasing agents ⟨54JA1082⟩. This approach is highly successful for organolithium reagents, but is poor for organomagnesium species.

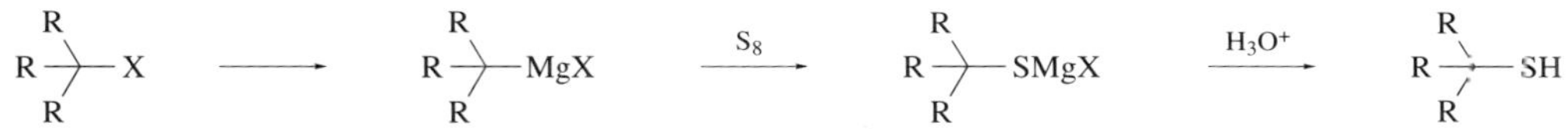

R_3C–X	*Yield* (%)
Bu^tCl	75
Ph_3CBr	70

Scheme 20

2.03.1.9 Formation from Sulfonyl Halides and Other Sulfonic Acid Derivatives

Several reagents may be used for the reduction of sulfonyl halides to the corresponding thiols, the most common being lithium aluminum hydride (Scheme 21). An excess of the reducing agent is generally employed, to prevent the formation of sulfinic acids, disulfides and so on. Yields can be modest; for example reduction of butanesulfonyl chloride gave only a 45% yield of butanethiol ⟨50JA1033⟩. Another method of thiol formation from the corresponding sulfonyl halide is deliberate conversion into the disulfide and subsequent reductive cleavage (see Section 2.03.1.10). One particularly colorful method involves the use of hexacarbonyl molybdenum in anhydrous tetramethylurea (TMU) ⟨69AG(E)677⟩ to furnish disulfide in greater than 70% yields.

These methods are most commonly used for aromatic thiol preparation, perhaps because alkyl sulfonyl halides are relatively difficult to prepare. One of the most efficient routes uses thiourea followed by aqueous chlorine ⟨36JA1348, 37JA1837⟩, yielding sulfonyl chlorides in up to 96% yields.

Scheme 21

2.03.1.10 Formation by Reductive Cleavage of Disulfides

The conversion of disulfides into thiols is readily achieved by any of a host of reagents, to the extent that easily oxidized thiols are commonly stored as the disulfides. Particularly reactive alkanethiols are best stored as unsymmetrical disulfides formed from the corresponding sulfenyl chloride and an α-thiolcarboxylic acid, as shown in the example in Equation (7) ⟨59JOC1029⟩. The thiol is liberated from this "storable intermediate" by reductive or nucleophilic cleavage, as described below.

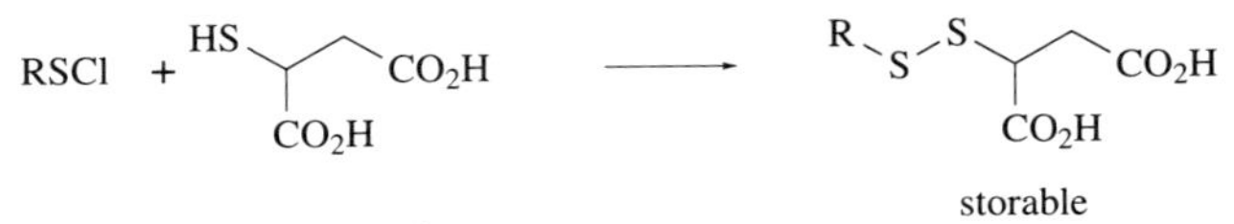

(7)

The reductive cleavage of disulfides to liberate thiols is achievable by most common reducing agents or reducing processes (Scheme 22). Lithium aluminum hydride furnishes thiols in high yields after workup, although yields can fall off dramatically with increasing steric hindrance around the S—S bond ⟨50JA731, 50RTC638, 57JA2553, 66JA1458⟩. Sodium borohydride furnishes thiols in particularly high yields, especially when employed in conjunction with a Lewis acid ⟨56JA2582, 57ANC154⟩. Potassium triisopropoxyborohydride (KIPBH) has been shown to be selective for the reduction of disulfides in the presence of other functional groups, and is also selective for reduction of aromatic disulfide in the presence of aliphatic disulfide ⟨84S498⟩. For example, up to 99% yield can be achieved for the reduction of diphenyldisulfide in the presence of dibutyldisulfide. Reaction of a disulfide with one molar equivalent of KIPBH produces one equivalent each of thiol and of the potassium thiolate. The free thiol is liberated from its potassium salt by acid hydrolysis to give high overall yields, usually greater than 80%.

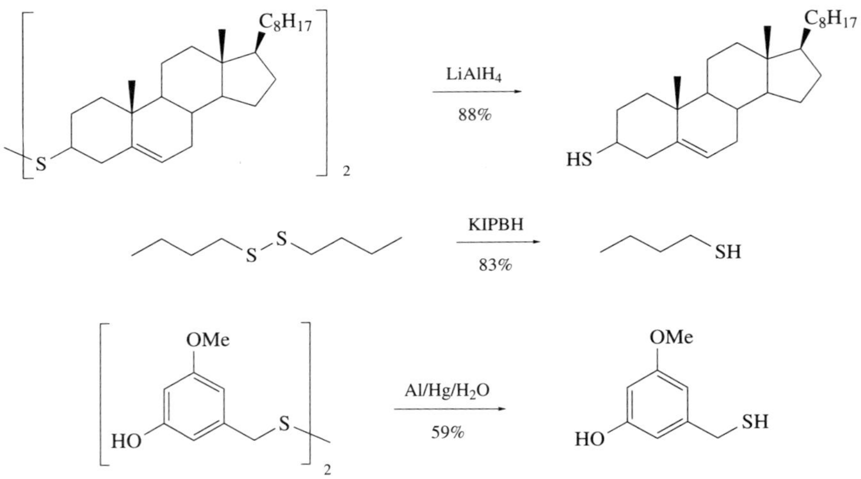

Scheme 22

Reducing metal conditions also induce disulfide cleavage, and generally employ either zinc or tin in the presence of an acid, although metal amalgams have also been used ⟨46MI 203-01, 49JA2270⟩, even in the presence of water. Very fine zinc dust is the standard choice. Reductions of this type are most frequently encountered in aromatic thiol synthesis. Dissolving metal conditions, particularly sodium in liquid ammonia, have been employed for disulfide reduction, for example in the well-known transformation of cystine to cysteine ⟨30JA4500⟩. Alternative solvents include ether ⟨26JA776⟩ and xylene ⟨33JA1242⟩. Yields are good, and can be enhanced by the use of alternative metals such as sodium/potassium alloy (more reactive) and sodium amalgam (less reactive).

Electrolytic reduction is a fairly uncommon process for disulfide reduction, but has been shown to reduce cystine-^{35}*S*-hydrochloride to cysteine-^{35}*S*-hydrochloride ⟨59BSF1544⟩ in very high yield, the process being cathodic in nature. Another interesting example is the reduction of dithiodiglycollic acid ⟨28CB1439⟩.

Reductive cleavage of disulfides by organophosphorus compounds is generally performed in aqueous methanolic solution, usually under acidic conditions. Once again, however, this method of reduction has few examples in alkanethiol chemistry. Yields are very dependent upon the choice of organophosphorus reagent and temperature, the more nucleophilic tributylphosphine usually being employed ⟨65ANC164, 65MI 203-01⟩.

Other less common methods for the reductive cleavage of disulfides include irradiation of solutions of disulfides in proton donor solvents ⟨60T76⟩, the conversion occuring via free radicals. One process employs a low pressure mercury arc as the irradiation source and cumene as solvent ⟨59JA1137⟩, illustrated for isobutylthiol in Scheme 23. The reaction is however far from efficient for thiol production.

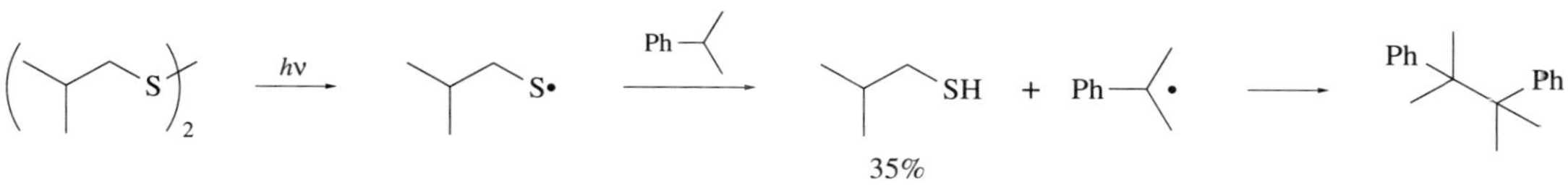

Scheme 23

A number of disulfides, mostly aromatic, have been reduced using glucose in basic ethanolic solution ⟨22CB2423, 53JA6019⟩. Other reducing agents occasionally encountered for disulfide to thiol conversion include sodium sulfide, hypophosphorous acid in the presence of catalytic diselenide ⟨66JOC1202⟩, and sodium hydrogen telluride ⟨89SC3143⟩. *meso*-2,5-Dithiol-*N*,*N*,*N'*,*N'*-tetramethyladipamide (*meso*-DTA), which has been described as a biochemical reducing agent, reduces disulfides in aqueous solution at pH 7 ⟨91JOC7328⟩.

The ubiquitous presence of disulfide and thiol moieties in biologically active compounds ⟨B-72MI 203-01⟩ such as biotin and its biosynthetic intermediates ⟨92JCS(P1)255, 93JA2139⟩, peptides ⟨93JOC2216⟩, and ribosides ⟨90JCS(P1)969⟩ has led to many new and varied methods for their interconversion. Many of these, however, are biotransformations and are beyond the scope of this review.

2.03.1.11 Formation from Thiocyanates

Thiocyanates, prepared by nucleophilic substitution of the corresponding halides, sulfates, and sulfonates ⟨37JA2473, 53JA2118, 54JA585⟩ can be efficiently reduced by most common reducing agents to the corresponding thiols. For example, lithium aluminum hydride has been used to reduce cholesteryl thiocyanate to the corresponding cholesteryl thiol as shown in Equation (8) ⟨53JA2118⟩, and sodium borohydride has been employed in the study of biotin biosynthesis ⟨93JA2139⟩. Sodium in liquid ammonia has also frequently been used to reduce alkyl thiocyanates. Acid hydrolysis is also successful ⟨1868CB177, 50RTC638⟩.

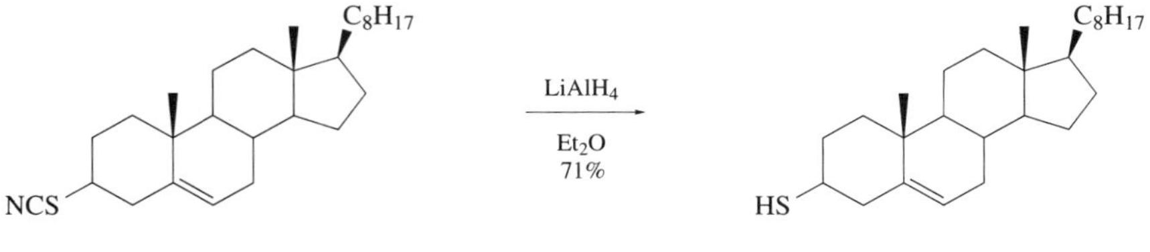

(8)

2.03.1.12 Formation by Dealkylation (Carbon–Sulfur Bond Cleavage) of Sulfides

Thiols are highly susceptible to oxidation, and may be stored as derived sulfides as an alternative to the corresponding disulfides. Indeed, alkanethiol formation has often been achieved from reaction of an alkyl halide/sulfonate with a thiol to furnish the sulfide, which is subsequently dealkylated to give the desired thiol. This necessitates ready and selective cleavage of one of the carbon–sulfur bonds present in the sulfide to liberate the desired thiol. The benzyl group is particularly useful as an alkyl group for this purpose ⟨52JCS800, 55JA416⟩. Liberation of the thiol can be surprisingly efficient, as illustrated in Scheme 24. Sulfide cleavage can, in most cases, be performed by treatment with a dissolving metal–amine system, and is commonly selective for cleavage of one carbon–sulfur bond. The ease of carbon–sulfur bond cleavage increases in the order 3° > 2° > 1° ⟨60JA2872, 62JA1621⟩. Benzyl sulfide carbon–sulfur bonds are among those most easily cleaved ⟨60JA2872⟩.

Alternative reagent systems such as sodamide in piperidine, calcium hexamine in ether, and sodium in ammonia have also found use.

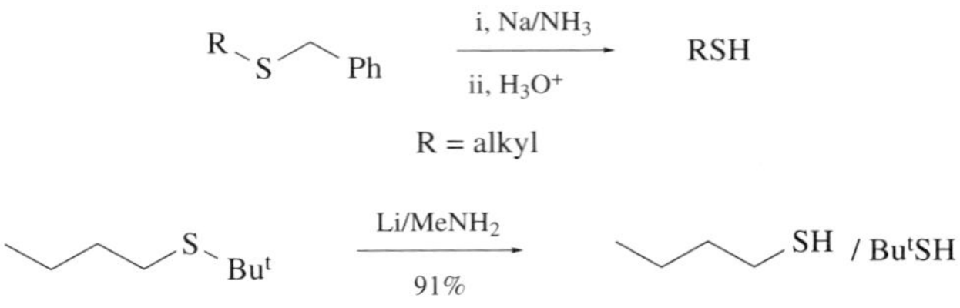

Scheme 24

One particularly interesting application is thioacetal/thioketal cleavage to furnish the corresponding alkoxy or thioalkoxythiol, as shown in Scheme 25 ⟨50JCS2938, 63CJC2671, 66JA1828, 66JCS(C)415, 70JOC2716, 70JOC3641⟩. Excellent yields of β-alkylthioethane thiols and γ-alkylthiopropane thiols can be achieved by carbon–sulfur bond cleavage in 1,3-dithiolanes and 1,3-dithianes respectively, again using dissolving metal systems ⟨50JCS2938, 66JA1828, 66JCS(C)415, 70JOC3641⟩. The reaction tolerates the presence of hydroxy groups. Curiously, for oxathioacetals, carbon–oxygen bond cleavage is preferred over carbon–sulfur bond cleavage when lithium aluminum hydride/aluminum(III) chloride is employed. In general, where the desired product is the thiol, dissolving metal conditions are more suitable for oxathianes and for ketals than for oxathiolanes and for acetals. Both mechanistic and stereochemical investigations on this metal–amine reductive process have been carried out; the lower yields obtained from acetals were attributed to the formation of alkene by-products.

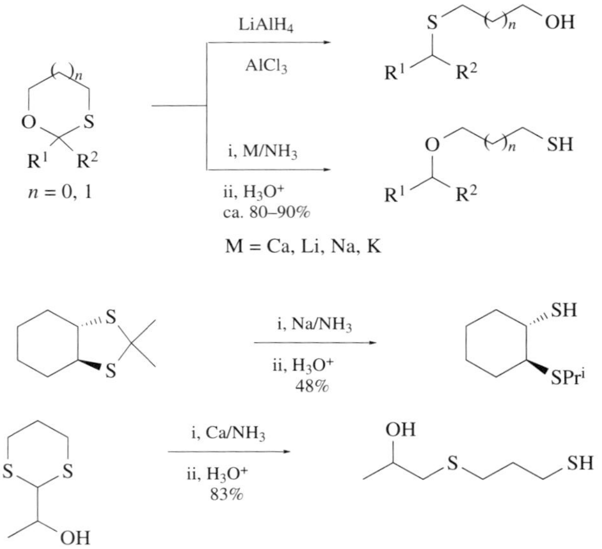

Scheme 25

Alkoxythiols, although sometimes difficult to prepare, can be generated by a number of alternative routes including many which are successful for simple thiols, such as reaction of thiourea with alkoxyalkyl halides ⟨45JA594⟩, reaction of hydrogen sulfide with alkoxyalkylhalides ⟨30JA651, 48JCS2177⟩, reaction of sulfonate esters with thioacetic acid ⟨50JCS579⟩, reaction of vinyl ethers with hydrogen sulfide, or the reaction of thiiranes with alcohols ⟨47JA2675⟩ (Scheme 26). Other methods of preparing β- and γ-alkylthioalkanethiols include the addition of hydrogen sulfide to vinyl sulfides ⟨56JCS437⟩, the addition of thiols to thiiranes ⟨49JCS282, 47JA2675⟩, and the nucleophilic substitution of alkyl thioalkyl halides ⟨56JCS437⟩.

The cleavage of carbon–sulfur bonds to furnish thiols is an important area of organosulfur chemistry, in part due to the presence of sulfur functionality in biologically important compounds. Two examples involving carbon–sulfur bond cleavage in penicillin-related compounds (Scheme

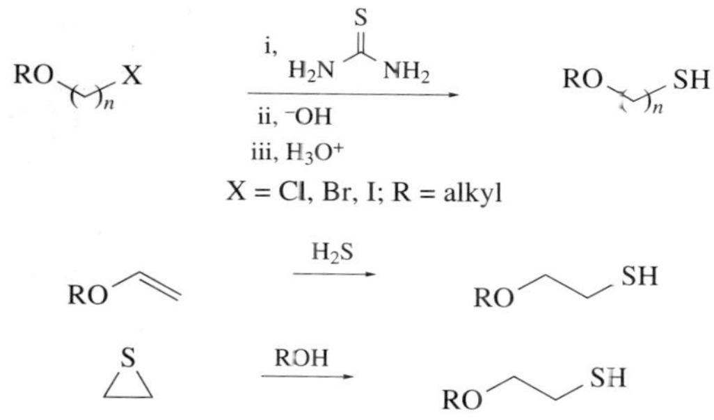

Scheme 26

27) ⟨67JOC677⟩, and peptide-related compounds (Equation (9)) ⟨91CC934⟩ serve to illustrate the importance of the disulfide–thiol interchange.

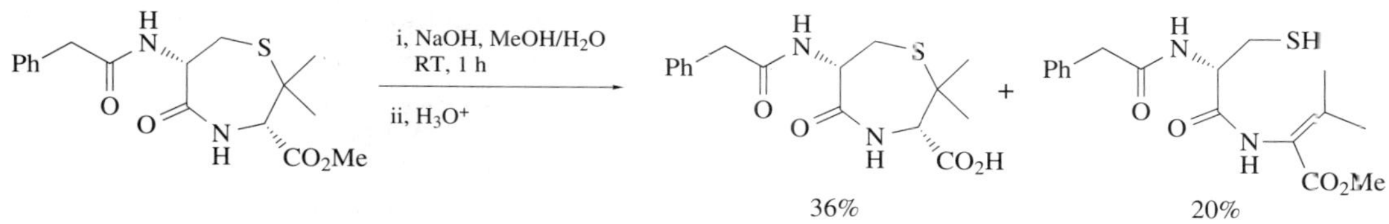

Scheme 27

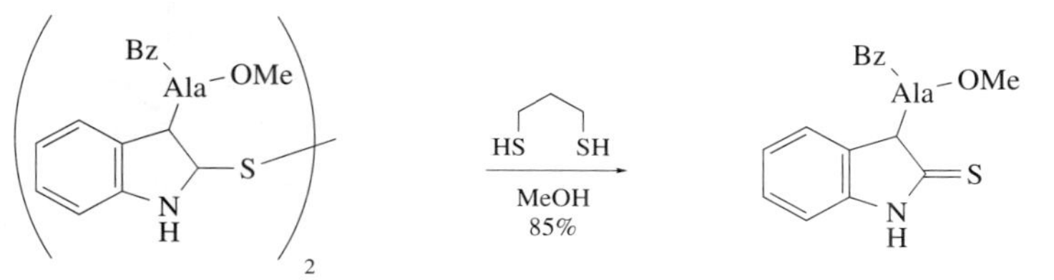

(9)

2.03.1.13 Formation by Heterocyclic Ring Cleavage

The ring opening of oxiranes, thiiranes, and aziridines by nucleophilic attack with hydrogen sulfide or an equivalent directly furnishes β-substituted thiols ⟨45JA1844, 47JA2672, 50LA(566)210⟩. The sodium and potassium salts of hydrogen sulfide are often used as alternative reagents (Scheme 28) ⟨40JA1173, 48JCS1892, 49JCS278, 58JA1680, 72JMC968⟩. In the case of aziridines, further reaction with a hydrosulfide ion can occur, leading to the formation of dithiols by nitrogen displacement ⟨65JOC3689⟩. Sulfide and disulfide formation are both also observed with thiirane, oxirane, and aziridine substrates. Cleavage of thiiranes to give thiols can be accomplished with any of a range of nucleophiles (Scheme 29) ⟨47JA2672, 69JOC97, 70JOC4209⟩.

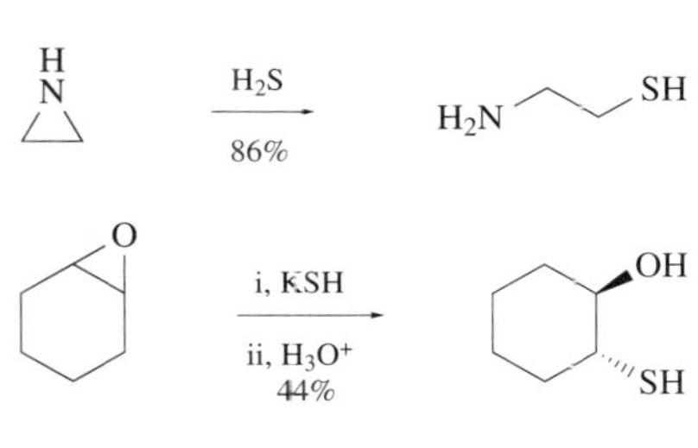

Scheme 28

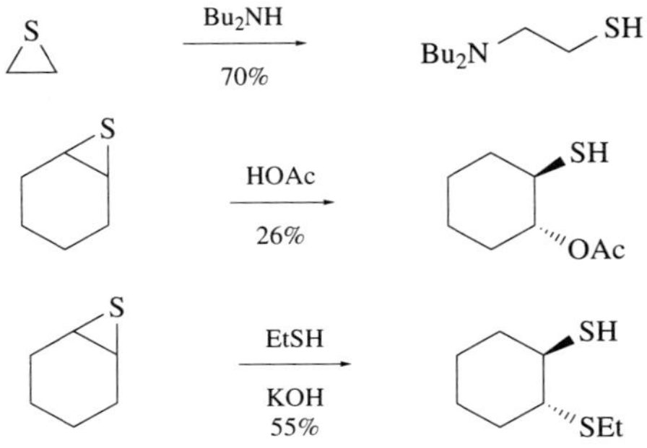

Scheme 29

2.03.1.14 Formation from Aldehydes, Ketones and Carboxylic Acid Derivatives

Aldehydes and ketones may be reduced using hydrogen and hydrogen sulfide, or hydrogen and sulfur, in the presence of a suitable catalyst, to give the corresponding thiols directly ⟨67JOC282⟩. Typical catalysts are cobalt and molybdenum sulfides. In the presence of the catalyst and hydrogen, sulfur is converted into hydrogen sulfide (Equation (10)). Unfortunately, many by-products can be formed and are often observed, including polysulfides, hydroxysulfides, trithianes, and even geminal dithiols ⟨67BSF1213⟩. Low temperature reactions are necessary for isolation of the geminal dithiol, as trithiane formation occurs at higher temperature. Basic catalysts have also been employed ⟨62CB1764⟩, including morpholine and butylamine ⟨62ACS1536, 63AG(E)370⟩. It is also possible to induce dithiol formation by high pressures and temperatures in the absence of a catalyst ⟨52JA3982⟩. Other routes for geminal dithiol preparation include the reaction of ketimines and enamines, derived from the corresponding carbonyl compounds, with hydrogen sulfide at low temperature ⟨60BCJ1743, 62BSF2126, 62JOC1041, 63ACS273⟩, and the reaction of tetrathianes with sodium in liquid ammonia ⟨62ACS772, 65CB998⟩ (Scheme 30).

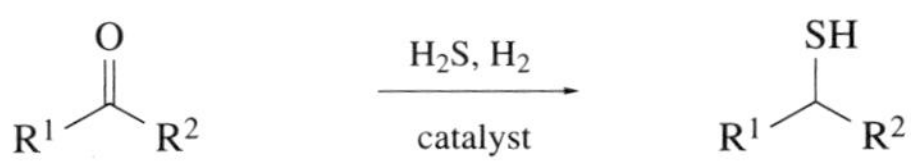

(10)

R = alkyl

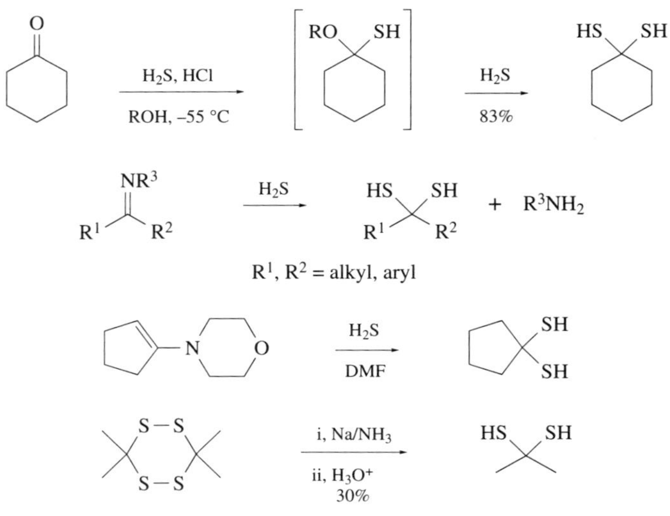

Scheme 30

Acyl halides and thioacids have also been used as starting materials for thiol synthesis. Photolysis of acyl xanthates, derived from acyl halide starting materials, furnishes alkyl xanthates, which can be hydrolyzed to give the corresponding thiols (Scheme 31) 〈62JCS1967〉. Reduction of thioacids with lithium aluminum hydride or sodium borohydride–Lewis acid also furnishes thiols, but frequently contaminated with the corresponding alcohol.

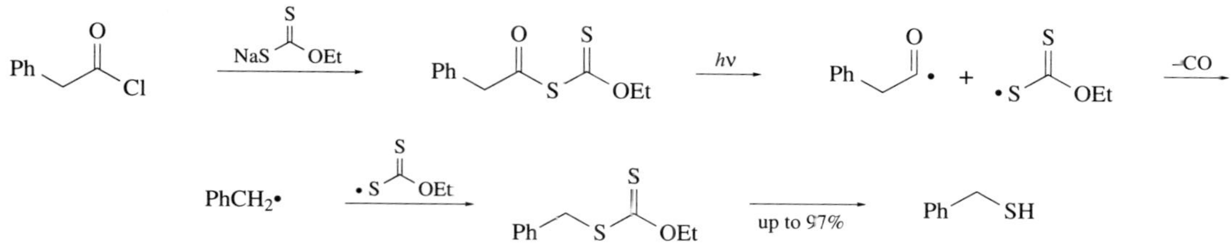

Scheme 31

2.03.1.15 Miscellaneous

Several unusual methods of thiol preparation have been reported in addition to those described above. Thiols can be formed by reaction of an alkene with inorganic sulfur (Scheme 32) 〈69JA5415〉 and from primary organic halides by treatment with bis(triorganotin)sulfides 〈90TL1397〉 in the presence of fluoride ion in aqueous acetonitrile at 20 °C (Scheme 33). The advantages of this one-step fluorodestannylation methodology are the mild and neutral conditions involved.

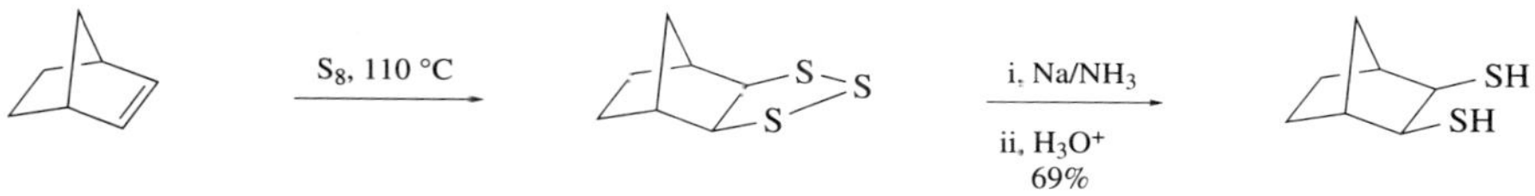

Scheme 32

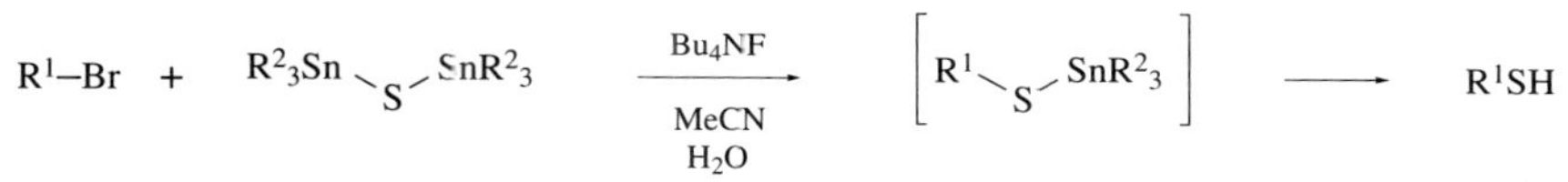

R^1	R^2	*Yield* (%)
C_6H_{13}	Ph	36
$PhCH_2CH_2$	Bu	82

Scheme 33

The reaction of amines with 2-thiolbenzothiazole in the presence of *t*-butylnitrite affords the 2-(alkylthio)benzothiazole condensation products, which were converted into the corresponding alkanethiols, as shown in Scheme 34 〈84TL2675〉. Mechanistic investigation suggested that the thiolbenzothiazole attacks *t*-butylnitrite to form a thionitrite. Efficient nitroso transfer to amines prevents any decomposition of thionitrite leading to disulfide.

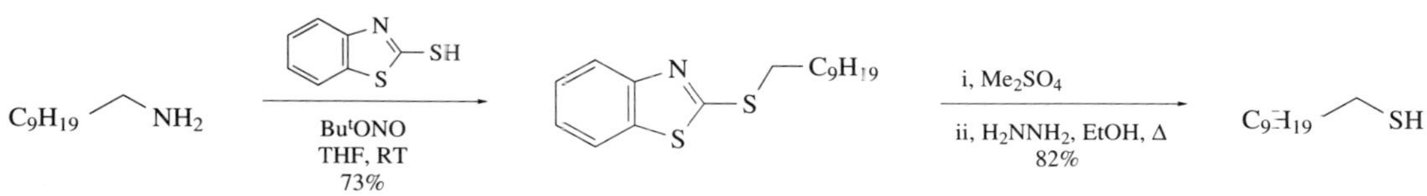

Scheme 34

Thiols may also be prepared by using a Wittig type 1,2-anionic rearrangement of silyl thioethers such as benzylthiotrimethylsilane 〈72JA4784〉. Deprotonation with organolithium reagents, usually *t*-butyllithium, induces migration of the silyl group from sulfur to the negatively charged carbon atom (Scheme 35). A similar remote rearrangement has also been reported in an aromatic system ⟨70JOM(25)389⟩. Excess organometallic reagent is normally required to effect efficient rearrangement.

In an interesting synthesis, ω-hydroxythiols have been prepared by the sulfoboration of cyclic ethers using a diborylsulfide followed by hydrolysis ⟨92LA835⟩ (Scheme 36).

Scheme 35

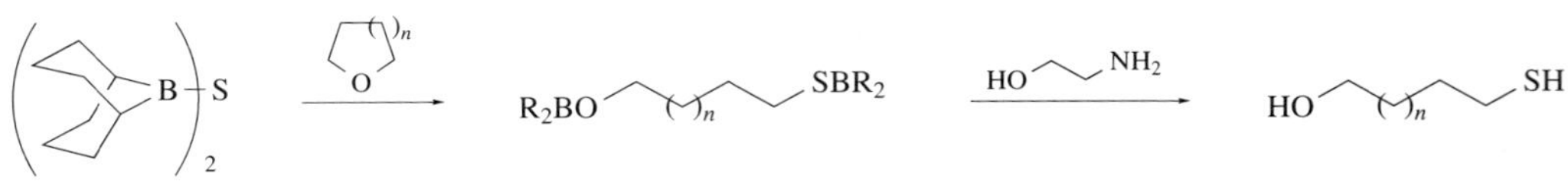

Scheme 36

2.03.2 DIALKYL SULFIDES AND THEIR HIGHER COORDINATED DERIVATIVES

Dialkyl sulfides and their oxidized analogues, sulfoxides, and sulfones, are abundant in the natural world and are widely utilized in organic synthesis. The greater nucleophilic character of sulfur over oxygen helps to explain the relative ease of sulfide formation relative to ether formation.

Symmetrical sulfides have commonly arisen historically as the major by-products of thiol preparation. Many methods for the production of sulfides, including unsymmetrical and highly functionalized sulfides, have also been developed.

The ready oxidation of sulfides to sulfoxides and sulfones remains the principal synthetic route for these compounds, although other methods are available, including rearrangements and the alkylation of sulfur stabilized carbanions.

2.03.2.1 Dialkyl Sulfides

2.03.2.1.1 Formation from alkyl halides and metal sulfides

Sulfides are often a significant by-product of thiol synthesis, and indeed by modification of the synthetic procedure the sulfide can often become the major product ⟨B-60MI 203-01, B-68MI 203-01, B-92MI 203-01⟩. The reaction of alkyl halides with metal sulfides is obviously applicable only for symmetrical sulfides, but a wide range of substituents are tolerated ⟨30JA651, 43OSC(2)547, 43OSC(2)576, 39JA2971, 91S557⟩. The method has also been applied to the synthesis of a cyclic sulfide, using 1,5-dibromopentane, in 34% yield (Equation (11)) ⟨47JCS1106, 78CB2445, 81SC409⟩. The synthesis of cyclic sulfides by this method is, in general, not a high yielding process ⟨B-90MI 203-01⟩.

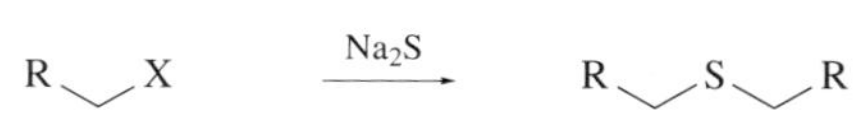

(11)

R = alkyl

2.03.2.1.2 Formation from alkyl halides and alkyl thiolates

This is probably the most general method for formation of both symmetrical and unsymmetrical sulfides, having the advantage that each alkyl group can in principle form either the halide or thiolate component. The thiolate is normally prepared first by treatment of the thiol with base, the halide being added subsequently, although oxidation to the disulfide can occur using this sequence ⟨B-58MI 203-01⟩. Aqueous or alcoholic solutions of sodium or potassium hydroxide or of the corresponding alkoxides are the typical bases used; alternatively, sodium hydride can be used in DMF solution ⟨91T5185⟩. The halide component is usually the bromide or chloride (Scheme 37) ⟨29JA3131, 49JA84, 91T5185, 92HCA2265⟩. As with ether formation by the Williamson ether synthesis, phase-transfer catalysis can often improve yields ⟨B-77MI 203-02⟩. The wide range of reaction conditions which may be used to form the thioethers is particularly well demonstrated by comparing syntheses of famesyl cysteinyl thioethers by several different groups (Equation (12)) ⟨91JA3176, 91JA3177, 92TL1435⟩. A sulfide intermediate in the preparation of chiral pentenoic acids was prepared from an optically active thiol and allylic bromide in 93% yield using 1,5-diazabicyclo[5.4.0]undec-5-ene (dbu) as base ⟨90JOC2286⟩.

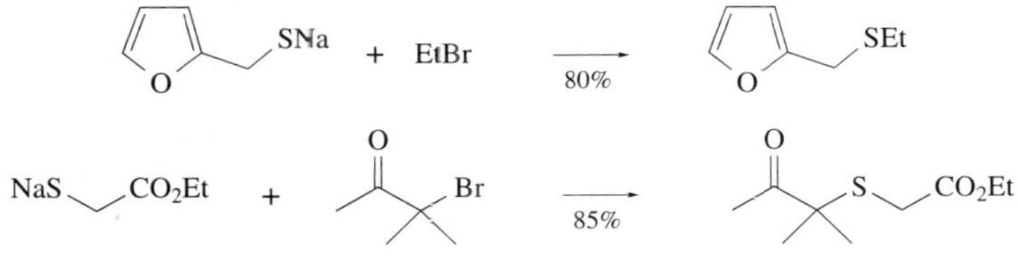

Scheme 37

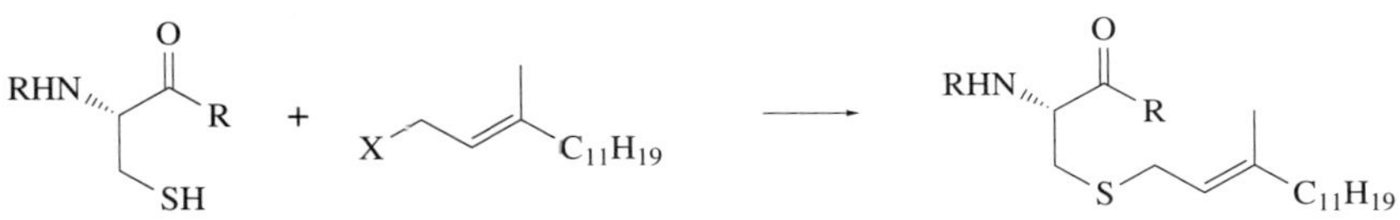

(12)

X = Cl, Br

Conditions	*Yield* (%)
NH_3 (l)	40–60
NH_3/MeOH	80
$Pr^i{}_2NEt$/DMF	90
$Zn(OAc)_2$/MeCN/H_2O	74–88

Basic conditions are not always required; for example trityl chloride will react directly with a thiol in ether solution ⟨15CB524⟩, and Lewis acids such as aluminum trichloride, boron trifluoride, and tin tetrachloride have been used to promote sulfide formation ⟨49USP2472471⟩.

The analogous cyclization reaction using a dithiol is known, although choice of metal counterion and solvent appear to be important. Caesium carbonate in DMF is often used, providing yields of 70–80% of various macrocycles in the reaction of caesium dithiolates with 1,3-dichloroacetone (Equation (13)) ⟨91CC90⟩.

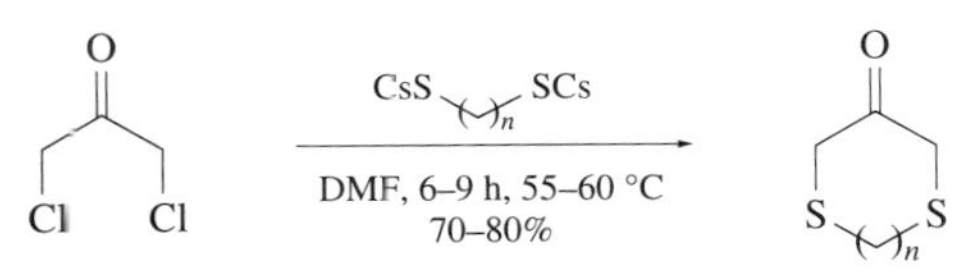

(13)

Reasonable yields were obtained in the synthesis of thia-crown compounds using 1,3-dithiopropan-2-ol in an ethanolic solution of sodium hydroxide (Equation (14)) ⟨76JOC134⟩. A related example makes use of a molybdenum template to form a macrocyclic thioether from a substituted 1,2-dibromooctane and the bis(tetramethylammonium) salt of a dithiol, in yields of 6–25% (Equation (15)) ⟨91CC475⟩.

An interesting alternative to the deprotonation of thiol with base is electrochemical reduction with a platinum or glassy carbon electrode ⟨90S841⟩. Yields are in the range 70–100%.

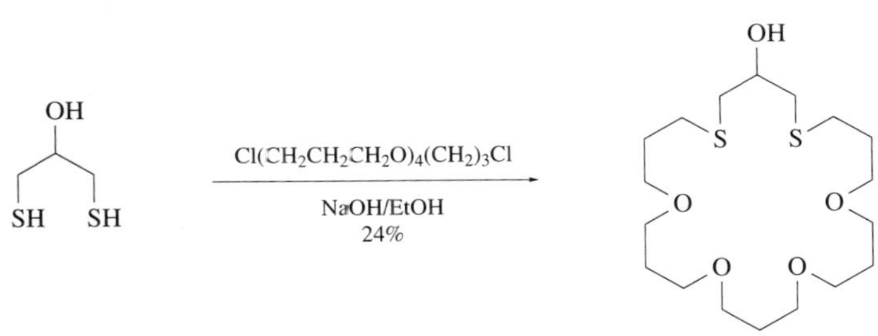

(14)

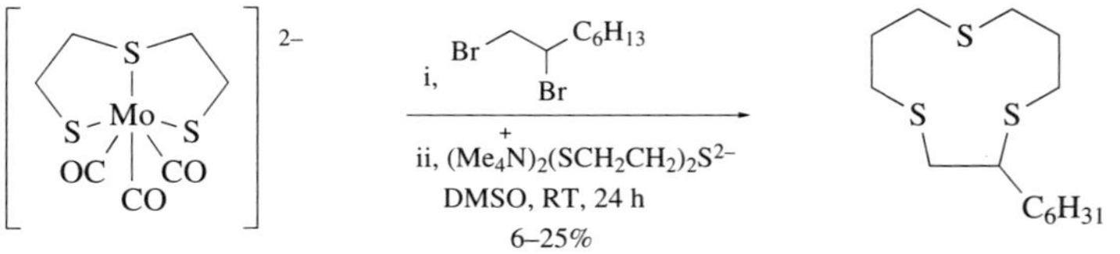 (15)

The need to preform the thiolate anion can be circumvented by a simple one-pot reaction in which an alkanethiol and an alkyl iodide are heated under reflux in acetone solution in the presence of a platinum catalyst and sodium carbonate (Scheme 38) ⟨88TL4477, 92T5933⟩. The reaction was shown to be applicable to primary and secondary thiols and dithiols. Yields of 69–80% were realised, no product being observed in the absence of catalyst. The application to thioacetal formation, using a dithiol and a 1,1-dihalide, is particularly interesting as it is complementary to the more usual dithiol/ketone route, and is successful for formaldehyde thioacetal formation. *S,S*-Dialkyl dithiocarbonates have been used as sources of thiolate ion in reactions with alkyl halides ⟨83S630⟩.

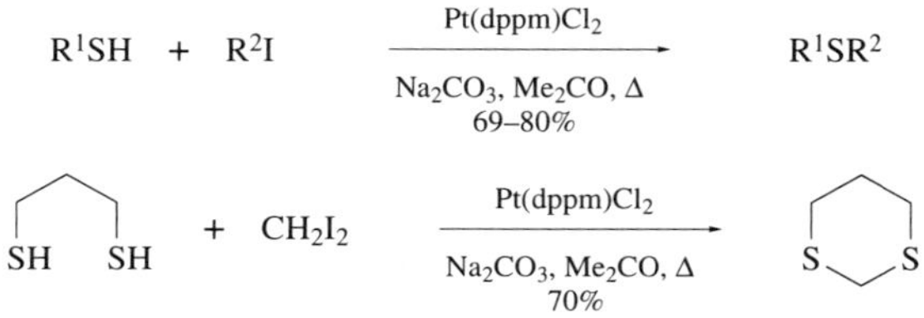

Scheme 38

Sulfides can also be formed from alkyl halides via an organometallic intermediate, usually a Grignard reagent, by treatment with an electrophilic source of sulfur. Typical sulfenylating agents include elemental sulfur, which produces the symmetrical sulfide, or a dialkyl disulfide, which produces an unsymmetrical sulfide. The use of organometallics in sulfide formation has been widely studied (Scheme 39) ⟨B-74MI 203-02, B-80MI 203-01⟩.

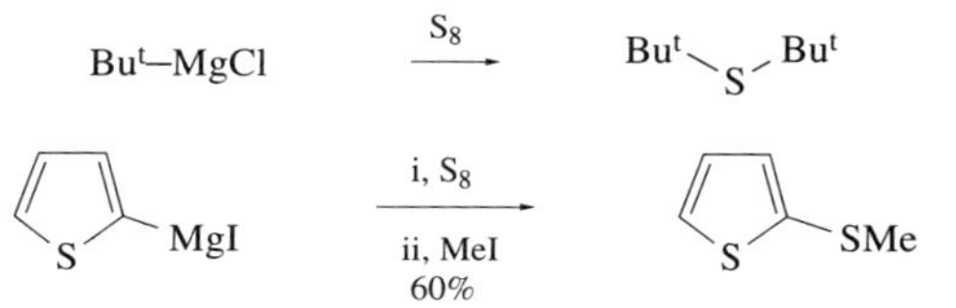

Scheme 39

2.03.2.1.3 *Formation from alkyl sulfates*

Although monoalkyl sulfate salts are normally sluggish and low yielding alkylating agents for alkyl thiols, reaction with highly reactive metal sulfides, such as sodium sulfide, can produce good yields of symmetrical sulfides. Metal thiolates, including sodium ethanethiolate and sodium *t*-butanethiolate, can be similarly alkylated to give unsymmetrical sulfides (Scheme 40) ⟨20JA856⟩. The much more reactive dialkyl sulfates may be similarly treated to give symmetrical or unsymmetrical sulfide products ⟨28JPR198⟩. For example, *t*-butanethiol may be treated with dimethyl or diethyl sulfate under alkaline conditions to give the unsymmetrical sulfides ⟨32JPR257⟩. The trisulfide has been similarly prepared from 1,2,3-propanetrithiol and dimethyl sulfate ⟨37CB675⟩.

$$2\ BuOSO_3Na \xrightarrow{Na_2S} Bu_2S + Na_2SO_4$$

$$BuOSO_3Na + BuSH \xrightarrow[\Delta\ 53\%]{NaOH} Bu_2S + Na_2SO_4$$

Scheme 40

2.03.2.1.4 *Formation from sulfonic esters*

Esters of sulfonic acids, particularly those of *p*-toluenesulfonic acid (tosylates), easily prepared from alcohols, have frequently been successfully employed in the synthesis of alkyl sulfides. The significance of the tosylate group as a good leaving group for S_N2 reactions is well known, and is demonstrated by its use in the preparation of a great variety of complex sulfides, illustrated in (Equation (16)) ⟨25JA1449⟩.

$$\text{BuSH} + \text{EtOTs} \xrightarrow[\text{H}_2\text{O},\ 78\%]{\text{NaOH}} \text{BuSEt} \qquad (16)$$

2.03.2.1.5 *Formation from alcohols*

Hydroxy groups can be displaced by thiol nucleophiles under acidic conditions to form the sulfide. Many functional groups, however, cannot survive such harsh treatment, and this reaction has only been of real use in the synthesis of *t*-butyl sulfides ⟨49JA84⟩. For most alcohols, prior conversion into a sulfonic ester provides a milder and more reliable route (Section 2.03.2.4). Allylic or benzylic alcohols can be converted into unsymmetrical sulfides by reaction with thiol in the presence of zinc iodide ⟨83JOC1357⟩.

Aliphatic, particularly primary, alcohols can be smoothly converted into alkyl aryl sulfides by treatment with *N*-phenylsulfenylimides and tributylphosphine ⟨77TL4475⟩. The reaction appears to be general for primary and secondary alcohols, but few examples involve nonaromatic thiols, although *N*-(benzylthio)succinimide, benzoin, and tributyl phosphine form the dialkyl sulfide in 64% yield in benzene at room temperature (Equation (17)). An example of a related reaction from nucleoside chemistry shows the selective synthesis of a sulfide from a primary alcohol in good yield under mild, neutral conditions using tributylphosphine and dimethyldisulfide in DMF solution (Equation (18)) ⟨75TL1409⟩. The reaction is reported to proceed by reaction of a phosphonium salt with the nucleoside to form a 5′-*O*-phosphonium salt which is attacked by thiolate to give the sulfide.

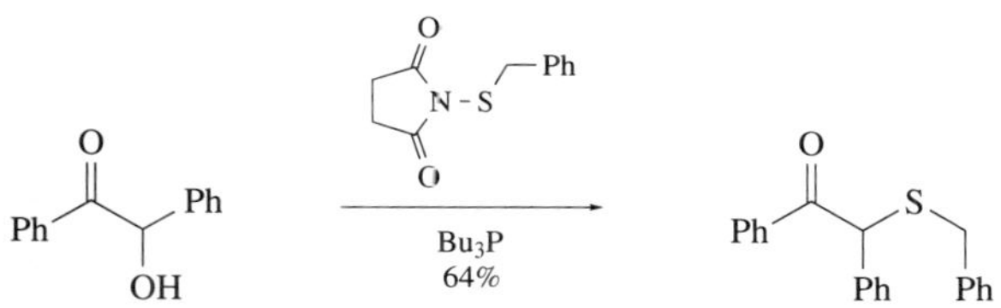

(17)

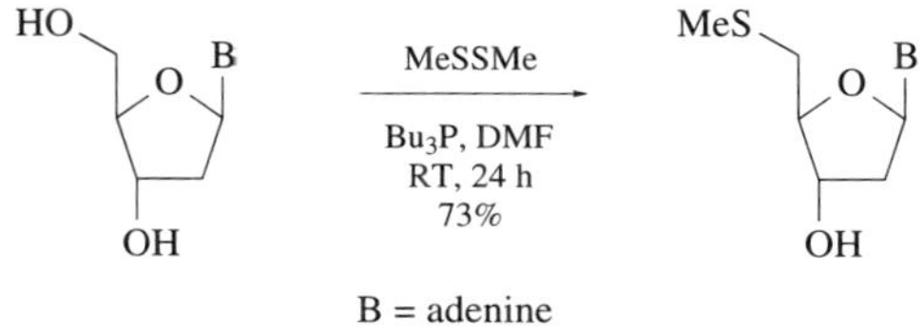

(18)

B = adenine

In the preparation of butanethiol by passage of butanol over a thoria catalyst at 380 °C in the presence of hydrogen sulfide, a low yield of dibutylsulfide is produced ⟨21JA880⟩. When ethanol vapour is passed over an aluminum sulfide catalyst at >300 °C, diethylsulfide is formed in ca. 10% yield together with ethanethiol and dehydration products ⟨25G973⟩. The reaction of methanol with ethyl dithiophosphate produces ethylmethylsulfide in the absence of any catalyst ⟨1861LA(119)313⟩.

2.03.2.1.6 *Formation from epoxides*

The most commonly used procedure for the ring opening of epoxides by sulfur nucleophiles to form sulfides involves the use of thiols in protic solvents in the presence of base (Equation (19)) ⟨49JA1478, 80JA1436, 80JA3663, 80TL3143, 91T9861, 92SL303⟩. Variations include the use of metal catalysis, allowing the possibility of enantioselective control induced by optically pure ligands ⟨88BCJ1213⟩.

Yields of up to 88% and enantiomeric excesses of up to 85% have been obtained using tartrate ligands and zinc as a heterogeneous catalyst, which is recoverable by filtration (Equation (20)).

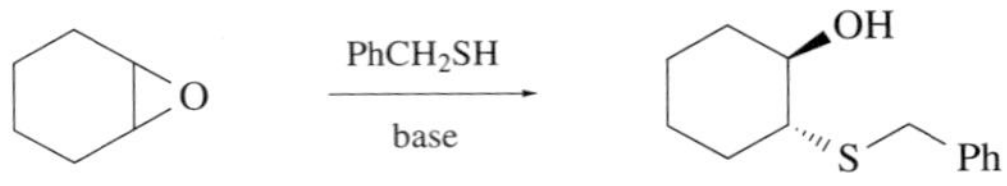

(19)

Base	*Yield* (%)
$LiClO_4$	95
Et_3N	70

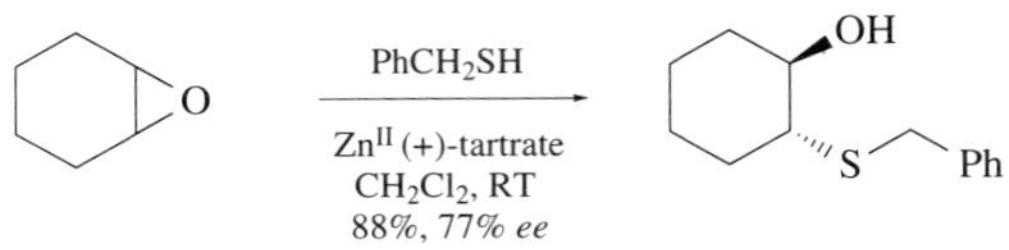

(20)

The ring opening of an epoxide by a thiol, in this case glutathione, under basic conditions is a key step in one synthesis of leukotrienes (Equation (21)) ⟨80TL3143⟩. In a similar manner, aziridines and thiiranes can react with sulfur nucleophiles to give ring opened sulfide products ⟨41JA2361, 47JA2675⟩.

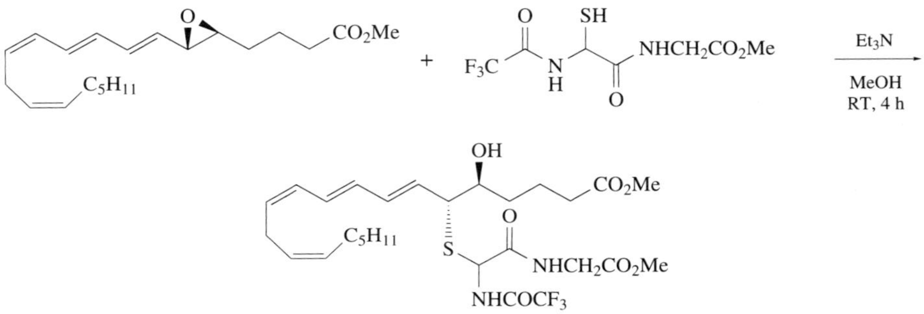

(21)

It is also possible to replace the oxygen atom of an epoxide with sulfur, so producing an thiirane. A number of different reagents have been used for this transformation including thiourea, thiocyanates, and thiocarbanilide (Equation (22)) ⟨46JCS1050, 47JA2672⟩. Cyclohexene sulfide has been produced from cyclohexene oxide on a large scale by the use of potassium thiocyanate in water/ethanol in 60% yield. Isobutylene sulfide has been prepared similarly in 73% yield ⟨47JA2672⟩. A detailed comparison of reagents for this transformation using cyclohexene oxide and epichlorohydrin as substrates has been published by Culvenor ⟨46JCS1050⟩. It notes that while simple starting materials give good yields of episulfide product, more complex materials can result in a mixture of products, especially when thiourea is used as reagent. In an interesting variant, the potassium thiocyanate is supported on silica gel, allowing isolation of the thiirane product in good yield by simple filtration and solvent evaporation. The authors indicate a necessity for the presence of a small amount of water in the preparation of the reagent.

Larger ring cyclic sulfides have been prepared from halo-substituted epoxides by tandem epoxide ring opening and halide displacement by sulfide reagents ⟨90KGS1477, 92CL1655⟩.

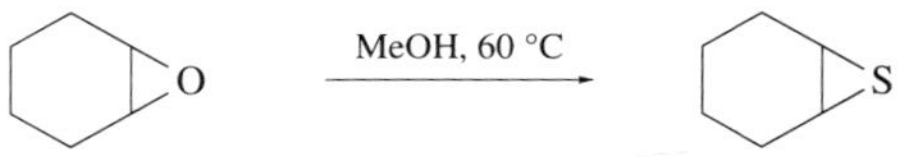

(22)

Reagent	*Yield* (%)
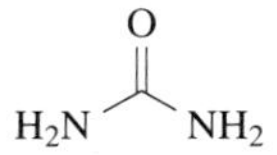	58
KSCN	61
NH_4CNS	60

2.03.2.1.7 Formation from alkenes

Additions of inorganic sulfur, of hydrogen sulfide, and of thiols to alkenes have all been thoroughly studied. Sulfide products can often ultimately be formed from each of these three reaction types. For example, sulfur can react with ethene producing hydrogen sulfide which, in turn, can react with more ethene to form the thiol. This product can react further with ethene to give diethylsulfide ⟨38JA2452⟩.

The direct addition of a thiol to an unactivated alkene often requires vigorous reaction conditions, even with catalysis. For example, the reaction of ethene with ethanethiol was only successful (93%) after heating to 180 °C for ten hours in a bomb with sulfur as a catalyst. In the absence of sulfur the product was not isolated. Concentrated sulfuric acid and aluminum trichloride ⟨89JOC1230⟩ were also used as catalysts, giving the Markovnikov products. In contrast, the free radical mediated addition of thiols to alkenes proceeds much more readily and yields the anti-Markovnikov product as expected. Thiols are much more sensitive to the presence of peroxides than the analogous free radical addition of hydrogen bromide to alkenes, so that while the product regiochemistry can be selected by choice of ionic or radical chemistry—a useful synthetic feature—the starting materials must be carefully purified if the ionic addition is required. The reaction has been widely studied, perhaps due to its facility (Scheme 41) ⟨59JA714, B-60MI 203-01, 70AG(E)273, B-74MI 203-03⟩. The stereochemistry of addition of thiol radicals to substituted cyclohexenes has been examined for each of hydrogen sulfide, methanethiol, and benzyl thiol ⟨69JOC3112⟩. Additions to nonconjugated and conjugated dienes have also been investigated, as has addition to allene systems ⟨63JOC1952⟩.

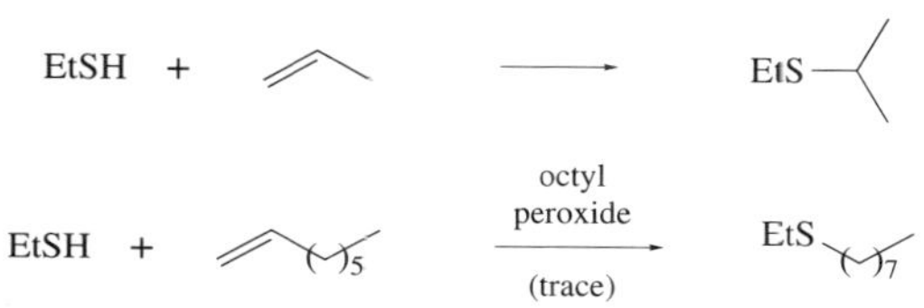

Scheme 41

As indicated above, it can be difficult to stop the addition of hydrogen sulfide to alkenes at the thiol stage, although bulky substituents tend to give only low yields of the sulfide ⟨38JA2452⟩. The outcome of heating hydrogen sulfide with various alkenes at 180 °C for ten hours is summarized below (Scheme 42). As with the addition of thiols to unactivated alkenes, it is necessary to add to the reaction a catalyst such as sulfur if any product is to be isolated. Under UV illumination, the addition of hydrogen sulfide is greatly accelerated, as is the case with thiols, although the product ratio is often in favour of the thiol and not the desired sulfide. For example, butene and hydrogen sulfide produce butanethiol in approximately 64% yield and dibutyl sulfide in 16% yield.

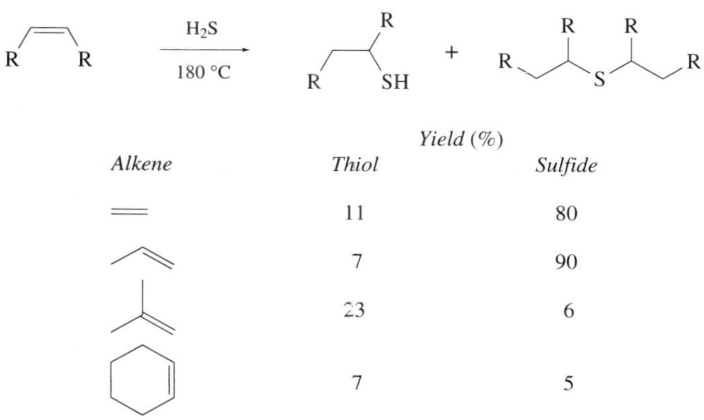

Scheme 42

Direct addition of inorganic sulfur to alkenes is of limited synthetic utility. Different products can be isolated depending upon whether the alkene is treated with sulfur or with a compound which produces sulfur when heated, such as diethyl tetrasulfide. The use of high temperature (> 300 °C) limits the applicability of this method when using inorganic sulfur, although when diethyl tetrasulfide is used as sulfur donor, the temperature can be reduced to about 150 °C ⟨38JA2452⟩. Yields are typically low (less than 20%), with some thiol product also obtained.

The addition of the elements of sulfenyl halide or disulfide across a carbon–carbon double bond is a useful method for producing *β*-substituted sulfides ⟨71S563, 82JA3225, 82JA3231, 85JOC4390, 87TL523, 92TL947⟩. One notorious example is the formation of *β*-chlorosulfides, the mustards derived from the addition of *β*-chlorosulfenyl chlorides to alkenes (Scheme 43) ⟨46JOC475⟩.

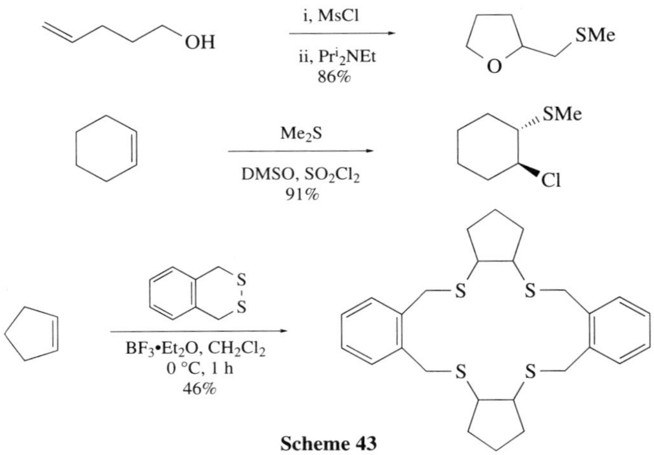

Scheme 43

The unfavorable nature of the reaction of a sulfur-containing species with an isolated double bond can be circumvented by hydroboration followed by substitution of boron by sulfur. The reaction occurs under free radical conditions through attack of a dialkyl disulfide by the alkyl radical produced from the borane. The reaction of norbornene with dimethyl disulfide under these conditions gave methyl norbornyl sulfide in 75% yield on a 10 g scale (Equation (23)) ⟨71JA3291⟩.

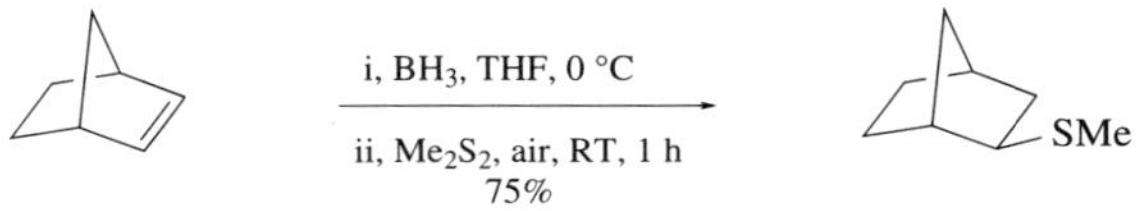

(23)

2.03.2.1.8 *Formation from electron-deficient alkenes*

In contrast to the harsh conditions usually required for ionic addition of sulfur species to simple alkenes, addition to electron-deficient alkenes such as *α*,*β*-unsaturated carbonyl compounds can be a reasonably facile and high yielding process.

Thiols add smoothly to *α*,*β*-unsaturated aldehydes under basic conditions or in the presence of a suitable catalyst ⟨48JA1450, 49JCS2723⟩. Optimization of the addition of thiols to *α*,*β*-unsaturated carbonyl compounds demonstrated that the choice of solvent and base used can be important ⟨48JA3667⟩. For the addition of benzenethiol to ethyl acrylate, Triton B (benzyl trimethylammonium hydroxide) was found to be the preferred base; use of benzene as solvent allowed a very simple workup to provide product in 90% yield, although use of dioxane gave a slightly improved yield (94%). Scheme 44 shows some unsaturated substrates that can act as electrophiles in the reaction with benzylthiol. Methanethiol and ethanethiol were found to give similar results, the reactions with ethyl acrylate producing yields of 95% and 93% respectively. Addition of various thiols to acrylonitrile also takes place in good yields (85–96%) ⟨47JA2328⟩. Interestingly, reaction of cinnamaldehyde with benzenethiol did not result in the product of conjugate addition, but gave the dibenzyl thioacetal.

α,*β*-Unsaturated cyclic ketones can undergo regiospecific reductive sulfenylation to give the *α*-substituted product ⟨77JOC3233⟩. Treatment of an *α*,*β*-unsaturated ketone with a solution of lithium in liquid ammonia/*t*-butanol followed by sulfenylation with dimethyl disulfide gives a moderate yield of the product of conjugate reduction and enolate sulfenylation, the major isomer arising from a slight preference for pseudoaxial attack of the enolate (Scheme 45).

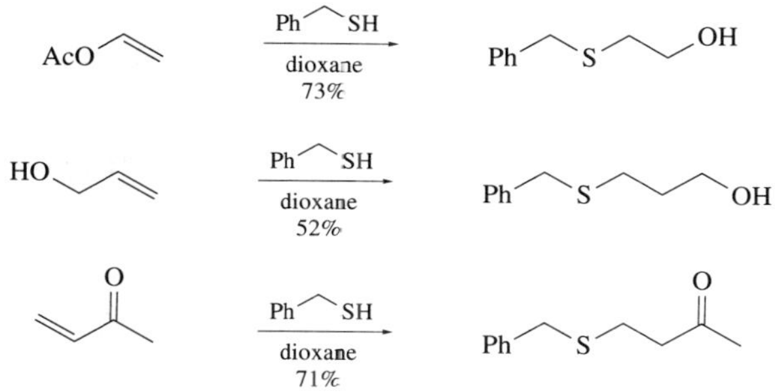

Scheme 44

Danishefsky and co-workers have used the addition of a sulfur nucleophile to an α,β-unsaturated carbonyl system in a synthesis of enediyne antibiotics. The process also triggers a cascade cyclo-aromatization ⟨90JOC2576⟩.

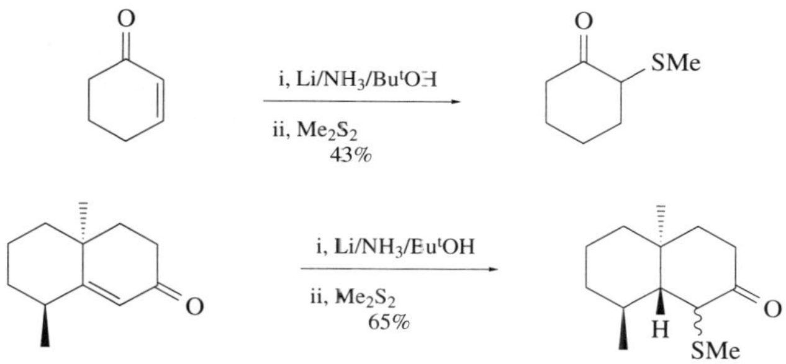

Scheme 45

2.03.2.1.9 *Formation from carbonyl compounds*

An α-ketosulfide can easily be formed by reaction of the enolate derived from a carbonyl compound with a sulfenylating agent such as a disulfide (Equation (24)) ⟨73JA6840, 74TL1097, 75PAC563, 76JA4887, 76JOC3587, 77JOC3236, 83S944, 91COS(7)119, 91JOC6697⟩.

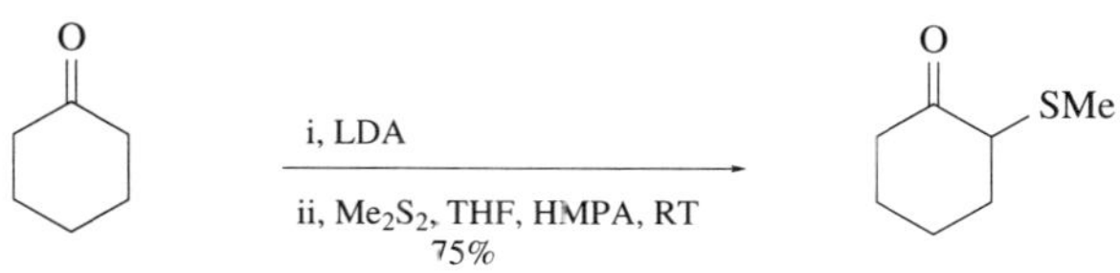

(24)

Direct replacement of the oxygen atom of a carbonyl group by reductive alkylation with a thiol can be effected in the presence of borane (Scheme 46) ⟨81CL1157⟩. Sulfides can be produced from aldehydes and ketones in a high yielding one-pot process by treatment with trimethylsilylsulfides, triethylsilane, and chlorotrimethylsilane in the presence of indium chloride ⟨91BCJ2524⟩.

β-Hydroxy-substituted sulfides can be produced by the addition of bis(trimethylsilylmethyl)sulfide to a carbonyl compound under fluoride ion catalysis (Equation (25)) ⟨91SL557⟩.

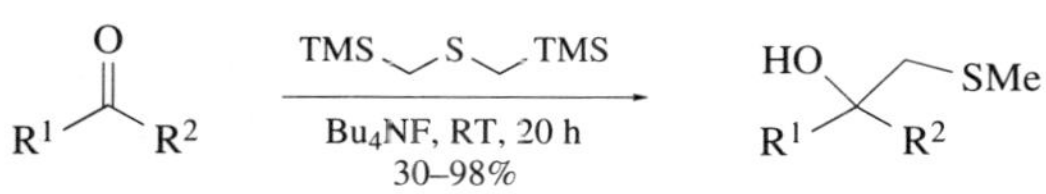

(25)

R¹	*R²*	*Yield (%)*
Ph	H	98
Ph	Ph	30
Bu^t	H	70
–(CH$_2$)$_5$–		65

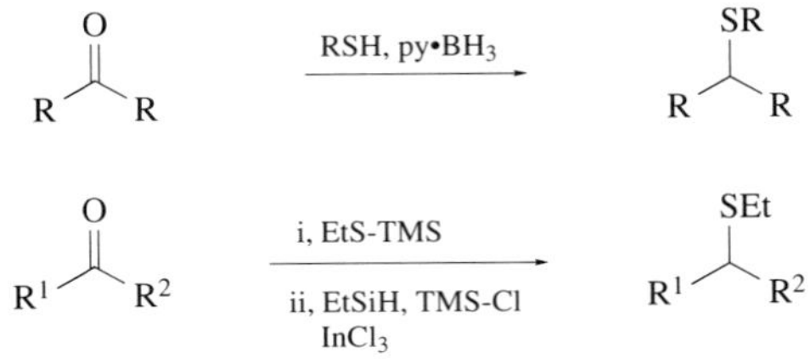

R[1]	*R[2]*	*Yield (%)*
Ph	H	87
Ph	Me	98
$Ph(CH_2)_2$	H	81
$-c-C_5H_{10}-$		93

Scheme 46

Fluoride ions also promote the transformation of aldehydes into thiiranes using *S*-methyl-*S*′-trimethylsilylmethyl *N*-tosyl dithioiminecarbonate or 2-(trimethylsilylmethylthio) thiazoline ⟨92TL85⟩. These compounds are synthetic equivalents of iminothiocarbonyl ylides and take part in 1,3-dipolar additions to aldehydes. The intermediate oxathiolanes so formed undergo ring contraction to give thiiranes (Equation (26)).

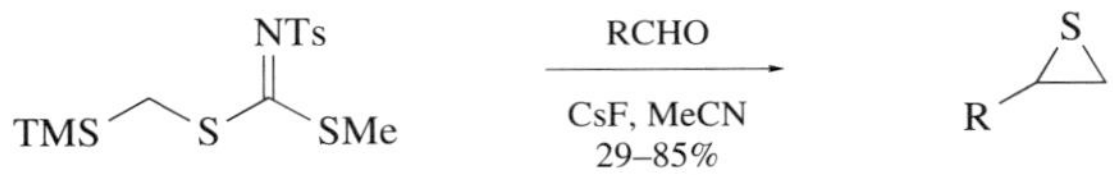

(26)

R	*Yield (%)*
Ph	66
$4\text{-}MeOC_6H_4$	85

2.03.2.1.10 *Formation from carboxyl compounds*

As with simple carbonyl compounds, deprotonation of a range of carboxylic derivatives, including esters, lactones, amides, and lactams, followed by treatment with a disulfide gives the corresponding α-substituted sulfide product ⟨73JA6840, 75PAC563, 77JOC3236⟩.

The acetoxy group can be displaced from protected α-acetoxyglycine by thiolate ion to give the corresponding sulfide, in an approach to the synthesis of α-substituted amino acids (Equation (27)) ⟨91TL3163⟩.

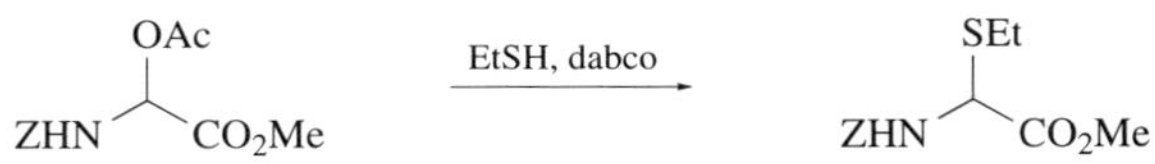

(27)

2.03.2.1.11 *Formation from thiocarbonyl compounds*

Advances in the preparation of the highly reactive thioketone and thioaldehyde species have led to their use as intermediates in sulfide synthesis, particularly through cycloaddition reactions ⟨93CSR199⟩. Illustrative examples with products containing a range of functionalities are shown in Scheme 47 ⟨91JOC7323⟩.

Many thiocarbonyl compounds have been characterized through their ready cycloaddition reactions with dienes, producing cyclic sulfides. These reactions have been widely reviewed ⟨79COC(3)373, 89CRV1413, 91PS(58)223, 91PS(59)295⟩. The difference in reactivity between carbonyl and thiocarbonyl groups is nicely demonstrated by the reaction of acetaldehyde and thionoethanal with various dienes (Equation (28)) ⟨90JOC2596⟩. A more complex diene, the alkaloid thebaine, reacted with a thioaldehyde generated *in situ* from the Bunte salt to furnish the Diels–Alder adduct in 73% yield ⟨91JCS(P1)2329⟩.

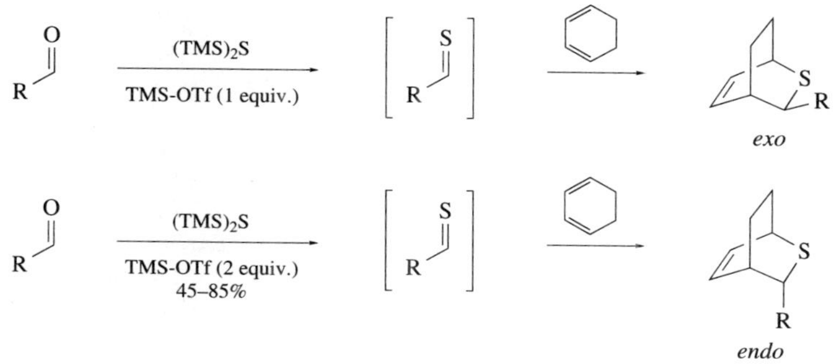

Scheme 47

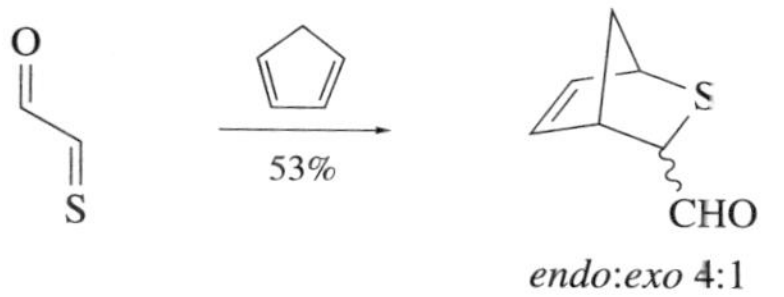

(28)

Thioketones undergo generally similar reactions to form sulfide products ⟨91S785⟩.

2.03.2.1.12 *Formation from thioacetals*

The dithioacetal moiety is frequently used as both a protecting group and, by deprotonation, as a masked acyl anion. However, the reaction of both cyclic and acyclic dithioacetals of acetophenone with alkyllithium reagents takes place at −78 °C with single carbon–sulfur bond cleavage to give α-thioalkyllithium species which react readily with electrophiles (Equation (29)) ⟨91TL2509⟩. One carbon–sulfur bond of dithioacetals can also sometimes be selectively cleaved under free radical conditions, for example with tris(trimethylsilyl)silane ⟨91TL2853⟩ or tributyltin hydride ⟨88TL897⟩ in the presence of AIBN. The stabilization of the carbon centered free radical by sulfur allows further reaction to take place, for example an intramolecular reaction with alkenes (Scheme 48) ⟨90JOC6248⟩.

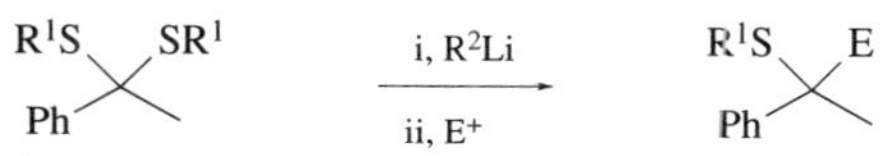

(29)

R^1	R^2	*Yield (%), $E^+ = H^+$*	*Yield (%), E = PhCHO*
Me	Bu	80	85
$-(CH_2)_3-$	Bu^s	82	80

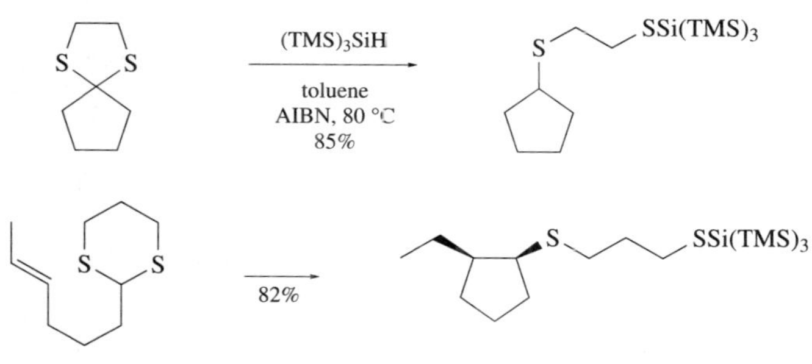

Scheme 48

The allylation of dithioacetals at the carbon atom with displacement of one thioalkyl group has been performed with allyl stannanes using gallium chloride as a Lewis acid catalyst (Equation (30)) ⟨90CL1097⟩. Optimization of conditions gave yields of 60–87%. The reaction of dithioacetals derived from ketones was shown to be much faster than that of those derived from aldehydes.

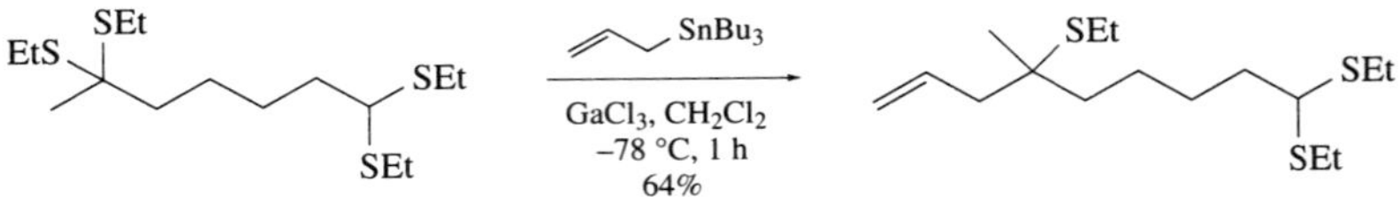

(30)

Silyloxy monothioacetals (acyclic oxathioacetals) may be converted into sulfides with cleavage of the carbon–oxygen bond using triethylsilane catalyzed by chlorotrimethylsilane and indium(III) chloride ⟨90CL2239⟩. The reaction has been extended to carbon–carbon bond formation with silyl enol ethers or allyl trimethylsilane as the nucleophile in place of triethylsilane (Scheme 49).

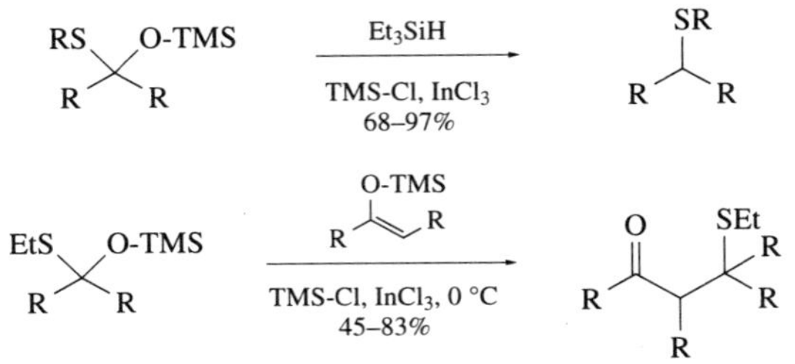

Scheme 49

2.03.2.1.13 *Formation from sulfoxides*

Several methods are available for the reduction of sulfoxides to sulfides. Traditional reducing agents such as lithium aluminum hydride are well known, while other reagents exploit the high oxygen affinity of phosphorus, boron, and silicon reagents, for example tris(dimethylamino)phosphine/iodine, phosphorus triiodide, boron tribromide, and iodotrimethylsilane (Equation (31)) ⟨78JOC4503, 78S137, 80CC544, 80CL143, 82SC903, 83JOC3667, 84JOC4538, B-88MI 203-01⟩.

(31)

R^1	R^2	*Reagent*	*Yield* (%)
Bu	Bu	PI_3	90
Bu	Bu	9-Br-BBN	95
Bu	Bu	$NaBH_4/CoCl_2$	80
Bu	Bu	BBr_3	90
c-C_6H_{11}	Et	TMS-Cl/NaI	85

Sulfoxides can also furnish sulfides through the well known Pummerer rearrangement, a useful process which has been thoroughly reviewed (Scheme 50) ⟨79COC(3)121, 90CPB3331, 91COS(7)193, 91OR(40)157⟩.

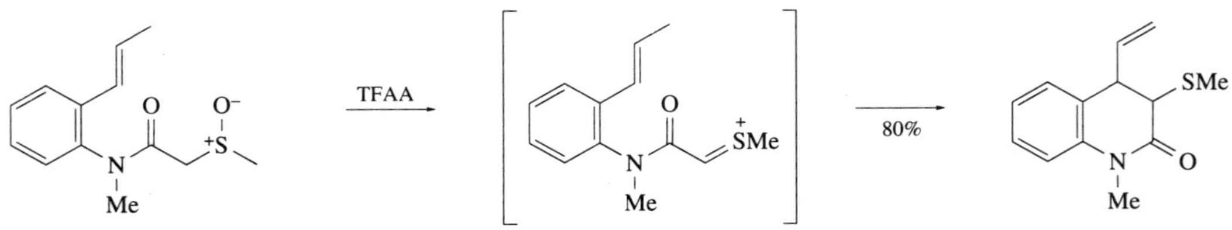

TFAA = trifluoroacetic acid

Scheme 50

2.03.2.1.14 *Formation from sulfones*

The reduction of sulfones to sulfides is a difficult process, as carbon–sulfur bond cleavage is often a more favourable reaction. Dialkyl sulfones, especially with small alkyl groups, are particularly resistant, even towards solutions of lithium aluminum hydride under reflux. Cyclic sulfones are however more readily converted into sulfides ⟨51JA2251⟩.

2.03.2.1.15 *Formation from disulfides*

Reduction of a disulfide to a sulfide with expulsion of one sulfur atom can be performed by reaction with hexamethyl phosphorous triamide ⟨78OS(58)138⟩. The mechanism of the reaction has been extensively studied ⟨71JA2437⟩.

2.03.2.1.16 *Formation from thiol esters*

Thiol esters can be reduced to sulfides in the same way as carboxylic esters and lactones are reduced to ethers, using reagents such as lithium aluminum hydride–boron trifluoride ⟨64JA1630⟩. The transformation can also be performed in good yields using lithium aluminum hydride and aluminum trichloride (Equation (32)). *S,S*-Dialkyl dithiocarbonates have been used as sources of thiolate ion in reactions with alkyl halides ⟨83S630⟩.

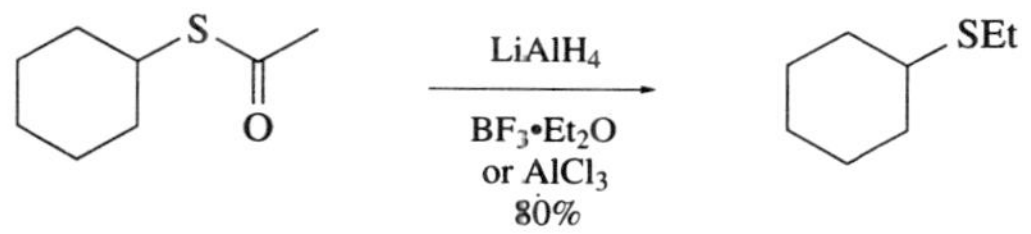

(32)

2.03.2.1.17 *Formation from sulfonium salts*

Medium ring (eight- to ten-membered ring) cyclic sulfides have been prepared by treatment of sulfonium salts with magnesium metal or even sodium borohydride in yields of 27–87% (Equation (33)) ⟨90TL3027⟩. Unsaturated sulfonium ylides can also undergo a sigmatropic rearrangement to give cyclic sulfides (Equation (34)) ⟨81JOC3315⟩.

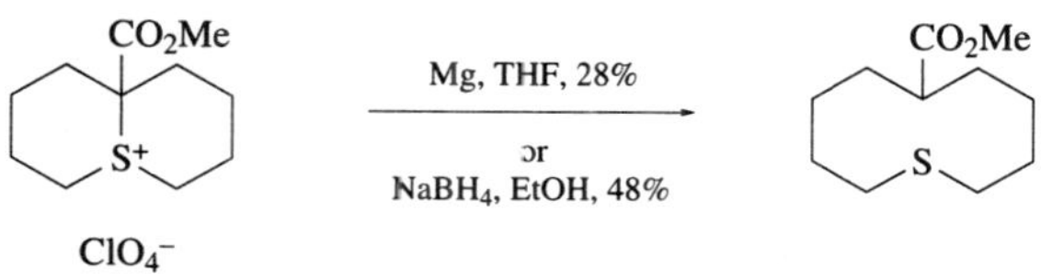

(33)

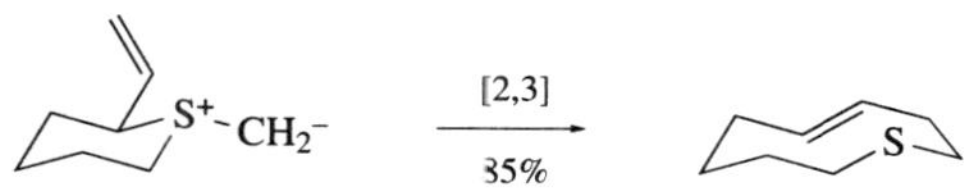

(34)

2.03.2.1.18 *Formation from α-sulfenyl anions*

The ability of a sulfur atom to stabilize an adjacent negative charge obviously allows further functionalization of sulfides by either carbon or heteroatom electrophiles ⟨B-90MI 203-02⟩; a strong base is required to deprotonate α to a single sulfur atom. Higher analogues of dimethyl sulfide tend to undergo elimination resulting in alkene formation, although thioacetal deprotonation and alkylation is of course well known ⟨89T7643⟩. Substitution can however be achieved in high yields with allylic or benzylic sulfides ⟨68TL5629, 69TL3707, 71T5861, 72CL259⟩.

2.03.2.1.19 *Miscellaneous reactions*

Saponification of trialkyl phosphorodithioates $(R^1O)_2PSSR^2$ can lead to the formation of sulfides in varying yields, together with the corresponding thiols. These reactions are not however widely applicable. Attack of cyclic phosphate esters by methanethiolate anion causes cleavage of a C—O bond of the ester and gives a moderate yield of the sulfide (Equation (35)) ⟨93CC70⟩.

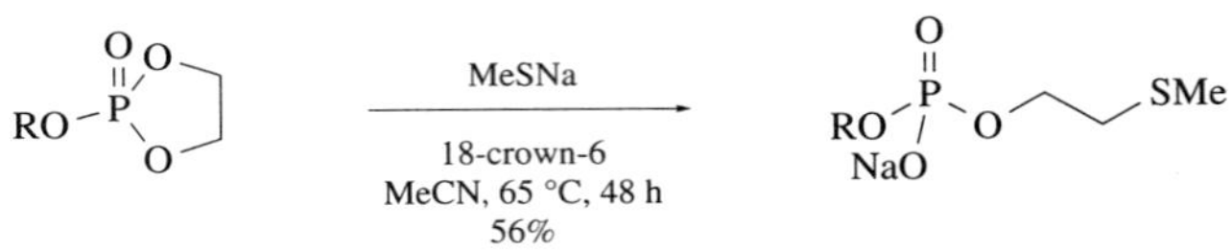

(35)

Alkanes can be sulfurated by the direct action of inorganic sulfur in very strong acid such as trifluoromethanesulfonic acid (Equation (36)) ⟨90JA3697⟩. Although yields are not exceptional and the conditions are particularly harsh, this can be a useful method for unfunctionalized sulfides where a longer or more involved synthesis may result in a poor overall yield.

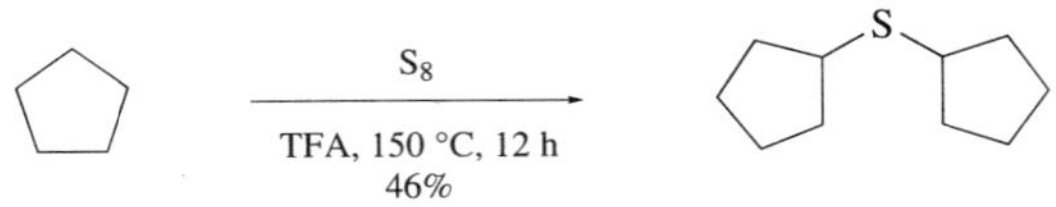

(36)

2.03.2.2 Dialkyl Sulfoxides

2.03.2.2.1 *Oxidation of sulfides*

Oxidation of sulfides is the oldest, most obvious, and most simply applied method of sulfoxide synthesis. The ease of sulfur oxidation by numerous reagents allows this transformation to be performed under very mild conditions, the only drawback being the ready overoxidation to sulfone, although with careful choice of reagent, quantity of oxidant and control of reaction conditions this limitation can readily be avoided ⟨79COC(3)121, 86T5459, B-88MI 203-02, 91COS(6)133, 91COS(7)757⟩.

A selection of the range of oxidants available is shown below for the oxidation of a cyclic sulfide (Equation (37)) ⟨65JA1109⟩. Widely varying stereoselectivities are observed.

(37)

Conditions	*cis:trans*
O_3, –40 °C	10:90
CrO_3•py, 25 °C	27:73
Bu^tO_2H, MeOH, 50 °C	27:73
H_2O_2, Me_2CO, 25 °C	37:63
mcpba, CH_2Cl_2, 0 °C	36:64
$PhIO_2$, C_6H_6, 80 °C	46:54
$NaIO_4$, H_2O, 0 °C	75:25
Bu^tOCl, MeOH, –70 °C	100:0

(i) Hydrogen peroxide

The oxidation of sulfides to sulfoxides and sulfones using hydrogen peroxide has been performed since 1908. Acetone was originally used as the solvent, but use of methanol can dramatically increase the rate of reaction ⟨08JCS1833, 53LA(579)23, 81SC1025⟩. The mild reaction conditions are aptly demonstrated by the successful oxidation of substrates such as allyl sulfides, where the product is acid sensitive (Scheme 51) ⟨61JCS5339⟩.

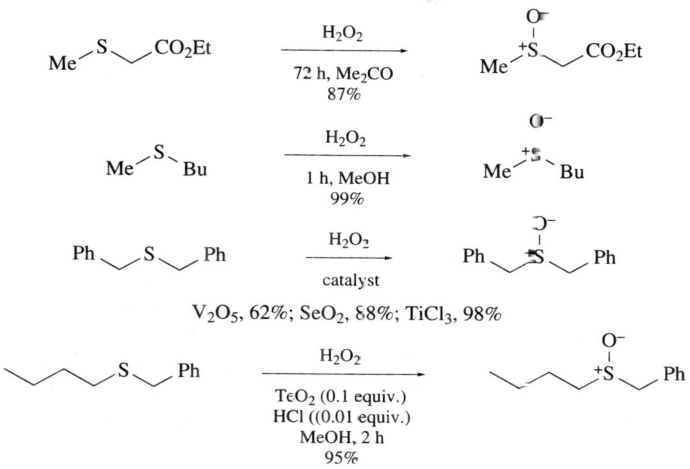

Scheme 51

(ii) Hydrogen peroxide and a catalyst

The oxidation of sulfides to sulfoxides by hydrogen peroxide is susceptible to catalysis, particularly by acids and metal salts (Scheme 51). Acetic acid, either as solvent or in stoichiometric quantities, is known to catalyze sulfide oxidation by hydrogen peroxide, as are sulfuric and perchloric acids ⟨60G383⟩. The oxidizing species is a peracid derived from the acid catalyst.

In the case of metal salts also, it is thought that the oxidizing species is generally a peracid derivative of the catalyst; for example selenium dioxide and hydrogen peroxide rapidly oxidize sulfides, the presumed intermediate being perselenic acid ⟨78S758⟩. The catalytic oxidation of methyl butyl sulfide to its sulfoxide in the presence of tellurium oxide was greatly accelerated by the addition of 0.01 equivalents of hydrogen chloride. Under these experimental conditions a 95% yield was achieved in one hour, with hydroxyl, carbonyl, and alkene functionalities being unaffected ⟨90TL2893⟩. Other metal salts that have been used include vanadium pentoxide and titanium trichloride ⟨69JCS(C)2334, 81S204⟩.

Titanium silicate in the form of molecular sieves has also been found to catalyze the oxidation of dialkyl sulfides by hydrogen peroxide in better than 85% yields ⟨92CC84⟩.

A development in 1993 has involved the use of hydrogen peroxide in the presence of a nitrile forming the peroxyimidic acid *in situ* as the oxidizing species. A range of substrates has been selectively oxidized to the sulfoxides and sulfones, selective oxidation to sulfoxide occuring at 0°C. This method is particularly applicable on a large scale as the oxidizing species is present only in low concentration (Scheme 52) ⟨93JCS(P2)2161, 93SC1507⟩. This work has been extended to a catalytic system using imine derivatives, including a catalytic asymmetric oxidation driven by hydrogen peroxide, giving up to 98% *ee* and quantitative yields ⟨94TL9629, 95SL773⟩.

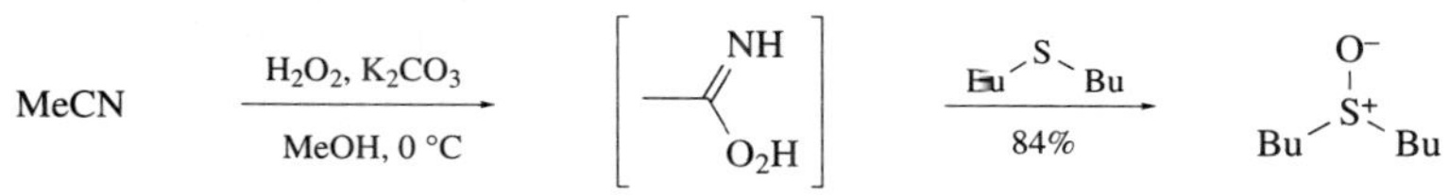

Scheme 52

A biomimetic catalyst based on a flavin has been developed for use in conjunction with hydrogen peroxide; dibutyl sulfide and dibenzylsulfide were oxidized to the corresponding sulfoxides in yields of 99% and 97% respectively (Equation (38)) ⟨89JA5002⟩.

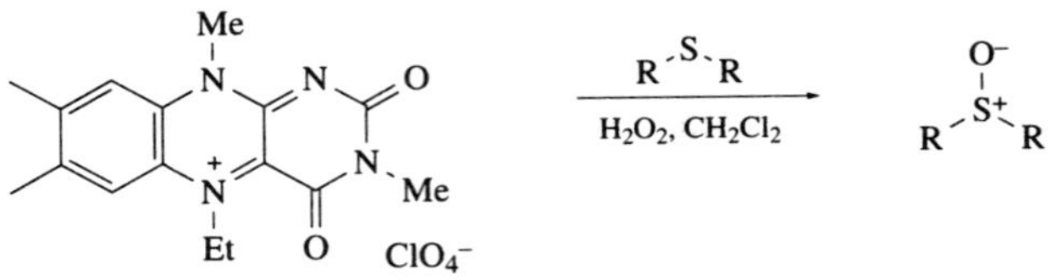

(38)

R = Bu, 99%; $PhCH_2$, 97%

(iii) Organic peroxides and hydroperoxides

Replacement of one or both of the hydrogen atoms of hydrogen peroxide by alkyl groups allows a wider range of organic solvents to be used (Scheme 53). The oxidizing strength of the peroxide is also increased. Organic hydroperoxides have been used since 1954 for sulfide oxidations; saturated sulfides were oxidized to the sulfoxides in quantitative yield when cyclohexyl or *t*-butyl hydroperoxide was used ⟨54PRS(A)389⟩. *t*-Butyl hydroperoxide has also been used in aqueous solution to good effect ⟨93SC3157⟩.

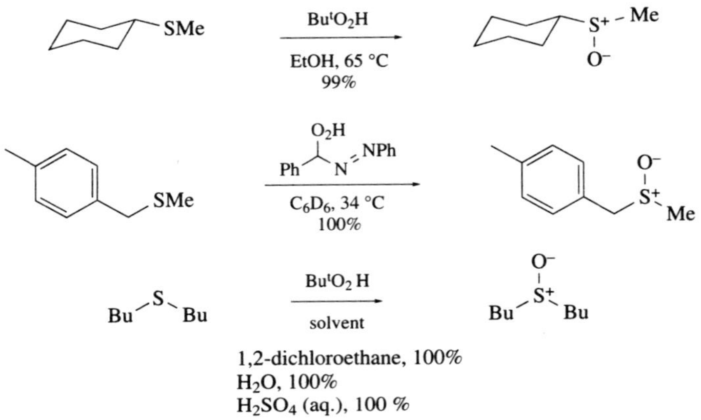

Scheme 53

It is possible to oxidize a sulfide by generation of an organic hydroperoxide *in situ* from a ketone and hydrogen peroxide ⟨80TL689⟩.

A peroxide produced by the photochemical-induced oxidation of a hydroxy indolinone has been used for the oxidation of a number of dialkyl sulfides in dichloromethane solution under reflux (Scheme 54) ⟨91JCS(P1)1717⟩.

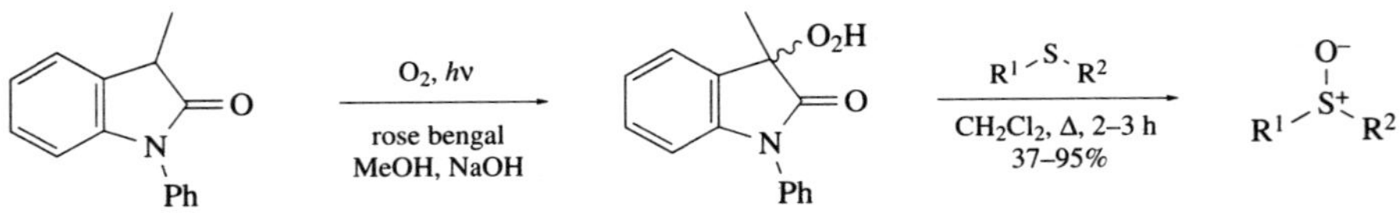

R^1, R^2 = alkyl

Scheme 54

By using a chiral hydroperoxide it is possible to induce optical activity in the sulfoxide product, albeit with generally poor and variable enantiomeric excesses, typically of the order of 30% *ee* at best (Equation (39)) ⟨86TL1591, 86BCJ1275⟩. An alternative is to use a peroxide in optically active solvent (Equation (40)) ⟨76TL4637⟩.

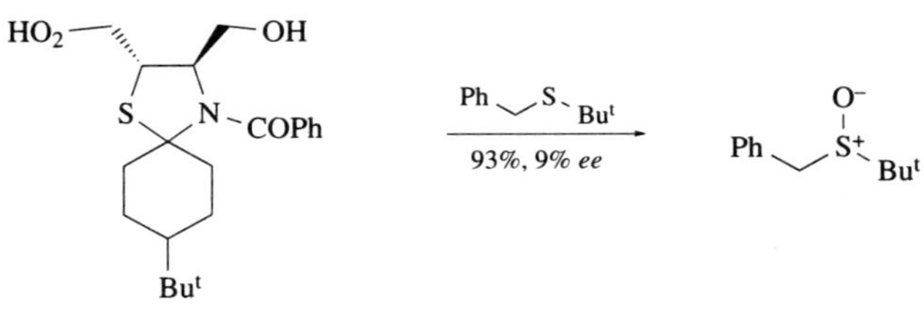

(39)

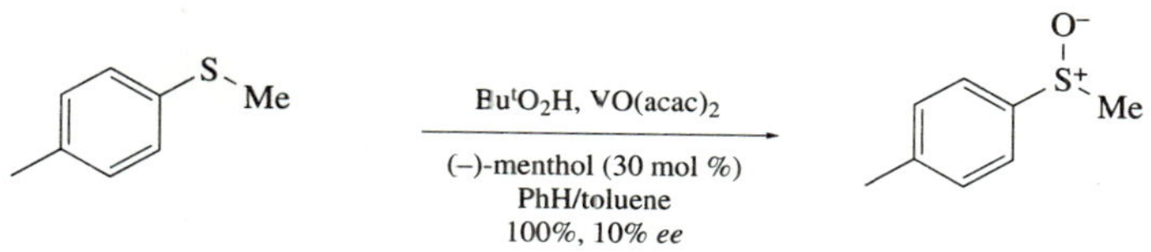

(40)

Higher enantiomeric excesses can be achieved by use of a metal bearing a chiral ligand in conjugation with organic hydroperoxides. Kagan and Modena have independently modified the Sharpless asymmetric epoxidation system to provide two most useful methods for asymmetric sulfide oxidation which gives reasonable enantiomeric excesses for dialkyl sulfoxide formation. In the Kagan modification one equivalent of water is normally added (Equation (41)) ⟨84JA8188, 85PAC1911, 90SL943⟩. Modena has also studied the effect of ligand, metal, and solvent in modified Sharpless oxidations. The Modena system, which is also effective for asymmetric dialkyl sulfide oxidation, does not involve the addition of water, but uses a larger quantity of tartrate cofactor ⟨84S325, 86TL6257, 89TL2575, 89TL4859⟩. Applications of the Kagan/Modena system include the production of epoxysulfoxides in good diastereomeric and enantiomeric purity (Equation (42)) ⟨92TL7237⟩, and the oxidation of acyl dithiane systems with up to 97% enantiomeric excess ⟨90SL457, 91SL80, 93TA2139⟩. Both systems also allow over oxidation to sulfone; over oxidation of dithiane systems leads to *trans*-disulfoxides.

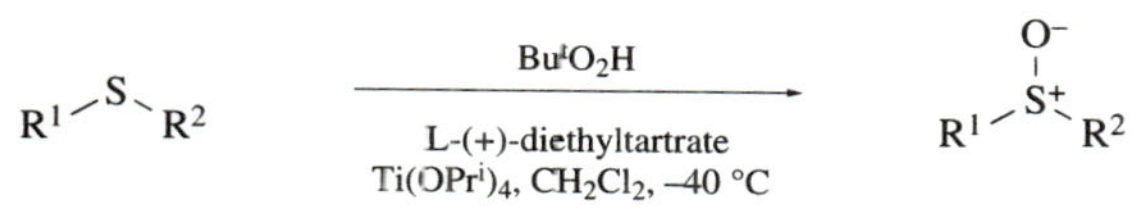

(41)

R^1	R^2	*Yield (%)*	*ee (%)*
Me	n-C_8H_{17}	77	71
Me	Bu^t	72	53
Me	c-C_6H_{11}	67	54

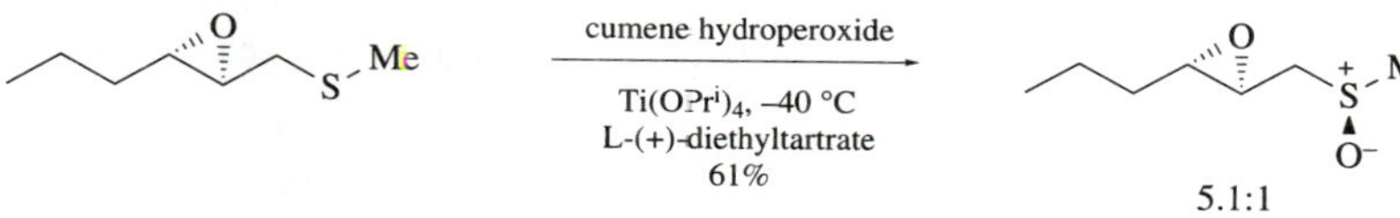

(42)

Jørgensen has investigated the mechanism of these metal-catalyzed peroxide oxidations using a theoretical approach ⟨94JCS(P2)117⟩. He postulates that the sulfur atom first coordinates to the peroxide bound metal atom, and that this is followed by transfer of the sulfide from the metal to oxygen. The geometries adopted in the initial coordination of the sulfide to the metal and in the subsequent transfer determine the enantioselectivity of the oxidation.

(iv) Peracids

The greater oxidizing ability of peracids over hydrogen peroxide allows them to be used at lower temperatures, permitting the oxidation of sensitive compounds such as di(α-bromobenzyl) sulfide by mcpba at 0 °C ⟨79JA390⟩.

A variety of peracids have been used to effect the oxidation of sulfides to sulfoxides, the most commonly used including mcpba, perbenzoic acid, peracetic acid, and monoperoxyphthalic acid (Scheme 55) ⟨27JCS1798, 63JCS4864, 82JOC3773, 87S1015⟩. The highly reactive nature of peracids can easily lead to overoxidation to sulfones, and therefore only one equivalent of reagent is generally used in conjuction with reaction monitoring and temperature control where necessary.

The use of chiral peracids has allowed enantioselective oxidation to be achieved, although the level of asymmetric induction is normally low, typically only of the order of 5–10% enantiomeric excess ⟨60TL20, 61CI(L)469, 61TL607, 68JCS(C)1317⟩.

(v) Halogens and halogen compounds

Oxidation of sulfides to sulfoxides can be achieved by reaction with molecular halogens followed by hydrolysis. The various competing reactions such as halogenation and carbon–sulfur bond

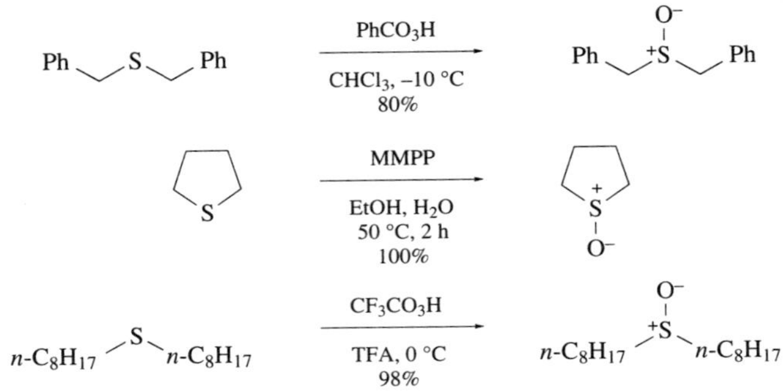

Scheme 55

cleavage can be suppressed by introducing an acid scavenger into the reaction, for example by the addition of amine or by using a two-phase system such as dichloromethane/aqueous potassium bicarbonate ⟨66BCJ364, 79S39⟩. The slow oxidation to sulfoxide by iodine can be accelerated by the addition of nucleophiles, such as phthalate ion or β-cyclodextrin phosphate ion (Equation (43)) ⟨66JA3874, 80CL1063⟩.

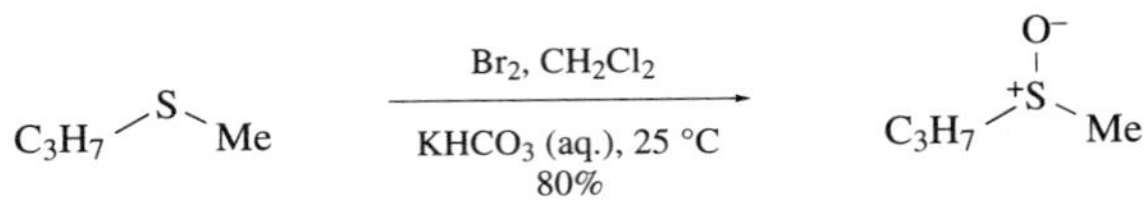

(43)

Hypervalent halogen compounds, including iodobenzene dichloride ($PhICl_2$), iodosobenzene (PhIO), iodobenzene diacetate ($PhI(OAc)_2$), and 2-iodosyl benzoic acid, have also been used for sulfide oxidation, with yields of dialkyl sulfoxides of up to 98% (Scheme 56). In an interesting modification, the use of ^{18}O labelled water allows the formation of ^{18}O isotopically enriched sulfoxides ⟨49JCS2126, 85SUL127, 92SC1799⟩. A chiral menthyloxyiodinane has been used to oxidize *t*-butyl methyl sulfide to the sulfoxide in 71% yield and up to 81% enantiomeric excess ⟨90JA5672⟩.

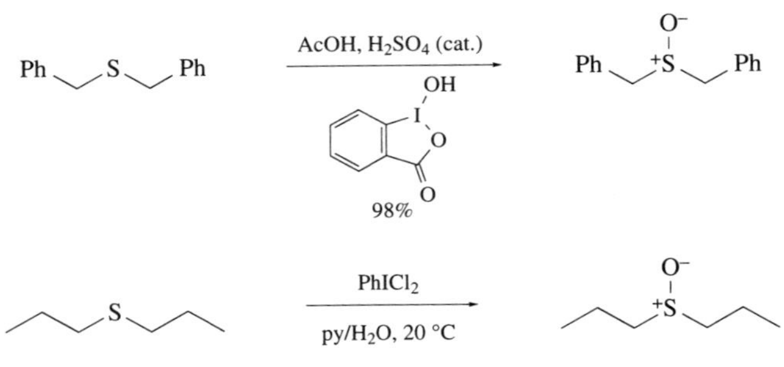

Scheme 56

The oxidation of sulfide to sulfoxide can be readily achieved using electrophilic halogen sources such as *N*-bromosuccinimide, chloramine-T and 1-chlorobenzotriazole, although in some cases, particularly di-*t*-butyl and dibenzylsulfides, carbon–sulfur bond cleavage can compete, and reactions are therefore often performed at low temperature ⟨68JOC3976, 69CC365, 77S791, 82JCS(P2)1075⟩. In a similar reaction, efficient oxidation of sulfides can be achieved using 2,4,4,6-tetrabromocyclohexa-2,5-dienone in aqueous dioxane or tetrahydrofuran (Equation (44)) ⟨71IJS(A)139⟩.

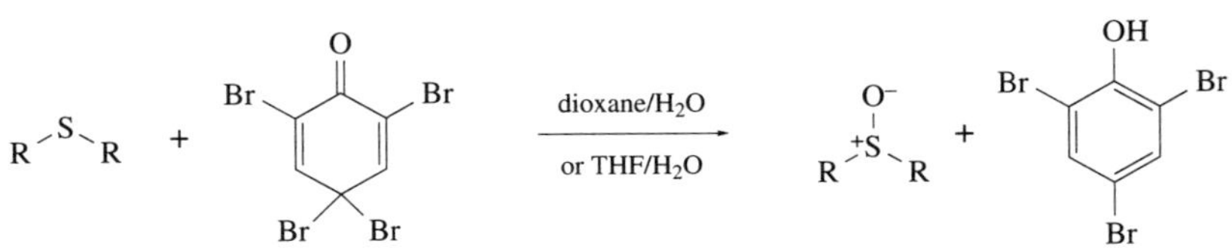

(44)

R	*Yield* (%)
Bu	96
$PhCH_2$	98

Oxidation to dialkyl sulfoxide can be performed in a two-phase system of dichloromethane/aqueous sodium hydroxide using benzyl trimethylammonium tribromide, with yields typically greater than 90% in less than 150 minutes at room temperature. A useful feature of this reagent is that it is decolorized when reduced, and hence a titration of the oxidant can be used to prevent overoxidation to the sulfone (Equation (45)) ⟨89BCJ3376⟩.

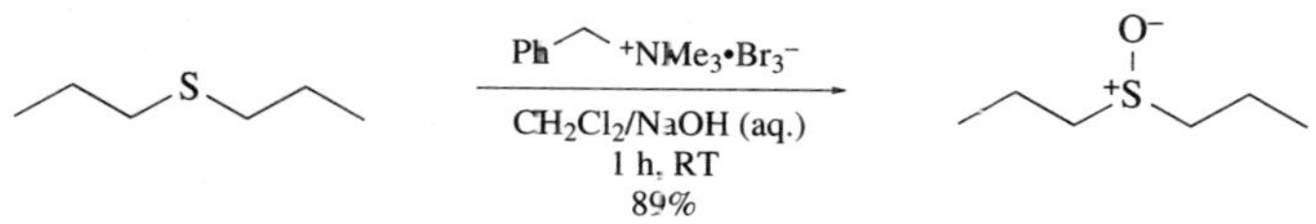

(45)

Sodium bromite, in the presence of the wet clay minerals montmorillonite or Kaolin, oxidizes sulfides in good yields under mild and neutral conditions (Equation (46)) ⟨92BCJ1744⟩. Hypohalites have also found widespread use in the oxidation of sulfides to sulfoxides. Sodium hypochlorite, hypochlorous acid, and *t*-butylhypochlorite have all been used chemoselectively, and sometimes show a degree of stereoselectivity. Of these, *t*-butylhypochlorite has proved the most useful. In alcoholic solution, cyclic, acyclic and β-hydroxysulfides have been selectively oxidized to the corresponding sulfoxides (Scheme 57) ⟨65JA1109, 67JOC3111, 76CJC3012⟩.

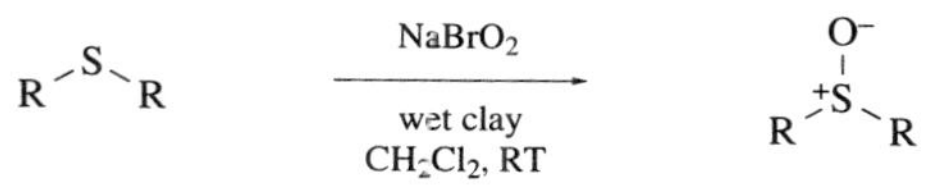

(46)

R	*Yield* (%)
$PhCH_2$	74
Pr^i	80

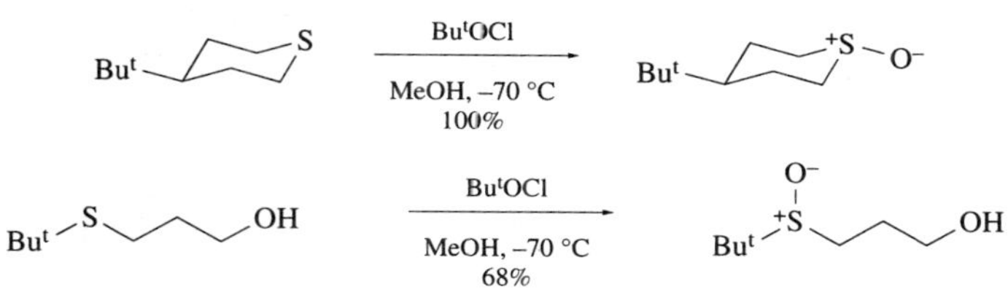

Scheme 57

One of the most common reagents used for the oxidation of sulfides to sulfoxides and sulfones is sodium metaperiodate, usually used in methanol/water solution at 0°C. This reagent is widely applicable, with yields of 90% or more in 3–24 h, and may also be used supported on silica or alumina (Scheme 58) ⟨62JOC282, 73OSC(5)791, 78JOC2717⟩. Periodate oxidations can be performed in other solvents, especially when tetrabutylammonium periodate, prepared from sodium metaperiodate and tetrabutylammonium hydrogen sulfate, is used ⟨80S563⟩.

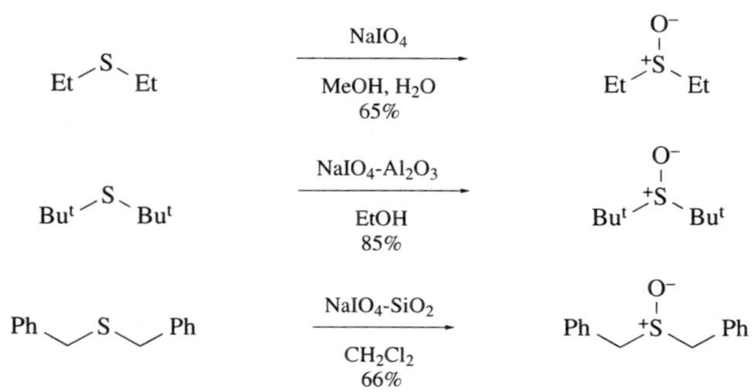

Scheme 58

(vi) Nitrogenous compounds

Nitric acid was first used as an oxidant in sulfoxide formation in 1865, and since that time many other nitrogen containing reagents have been developed ⟨1865LA(136)891, 57JA717⟩.

Organic nitrates have been reported to oxidize dialkylsulfides at $-78\,°C$ in high yields (Equation (47)) ⟨76CC496⟩. Inorganic nitrates have also been used, although overoxidation can be problematical. One solution is to employ a catalytic quantity of nitrate and a stoichiometric cooxidant (Equation (48)) ⟨79SC237⟩.

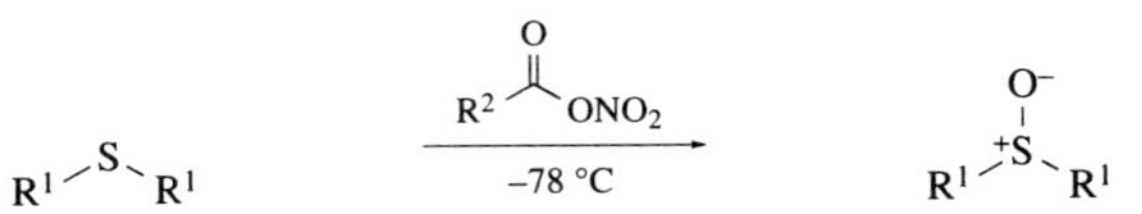

(47)

R¹	*R²*	*Yield (%)*
Me	Me	100
Me	Ph	95

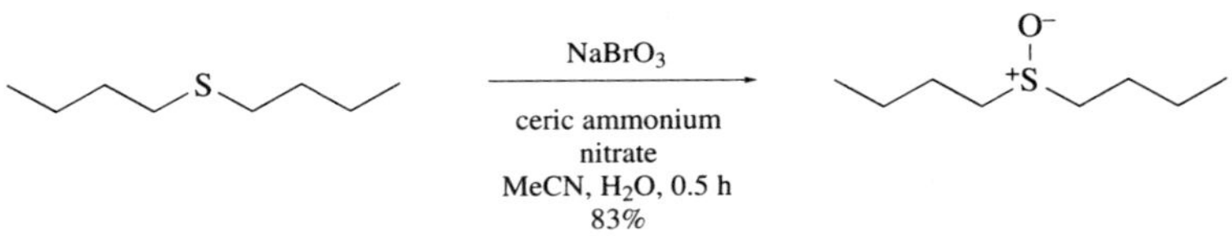

(48)

Tetranitromethane, when used as a photochemically initiated oxidant, selectively oxidizes sulfides to the sulfoxides with no overoxidation ⟨93S1057⟩. Nitronium salts perform the oxidation in moderate to high yield at $-78\,°C$ in dichloromethane solution (Equation (49)), as can dinitrogen tetroxide ⟨56JCS2705, 77TL1345, 79JA5317⟩.

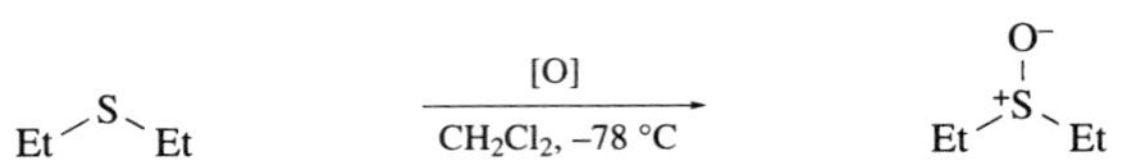

(49)

[O]	*Yield (%)*
NO_2PF_6	90
$Te(NO_3)_3$	86
N_2O_4	95

(vii) Other chemical oxidants

A number of other oxidants can provide a useful level of chemoselectivity and stereoselectivity for the oxidation of sulfides to sulfoxides.

One such reagent is oxone, potassium hydrogen persulfate ($KHSO_5$) ⟨86SC1207⟩. For example, a cyclic sulfide has been oxidized to the sulfoxide using oxone in 95% yield to give a *cis*/*trans* ratio of 1:15 (Equation (50)). A series of solvents was examined, and most water miscible solvents gave similar yields and selectivity, including methanol, butanol, acetonitrile, THF, and acetic acid.

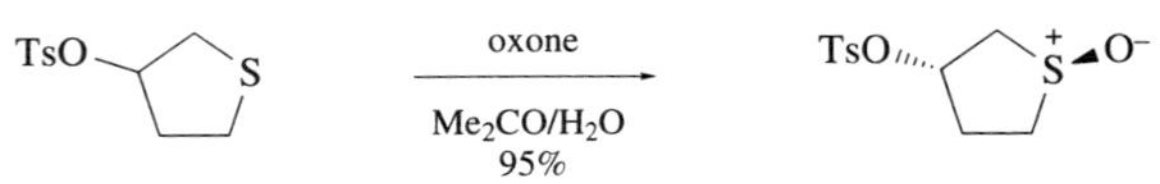

(50)

Oxaziridines have been developed as chiral oxidants by Davis, and applied to the enantioselective synthesis of chiral sulfoxides with good results. These compounds are normally used in stoichiometric quantities, but have been used catalytically with a reoxidant (Scheme 59) ⟨82JA5412, 84JOC1465, 88JOC5004, 92JA1428, 92JOC7274⟩.

Potassium superoxide in conjunction with 2-nitrobenzene sulfinyl chloride has been used for sulfide oxidation. It is thought that a reactive sulfinyl peroxy intermediate is formed which is the active oxidant (Scheme 60). Dibenzyl sulfoxide was formed in 97% yield at $-25\,°C$ in acetonitrile ⟨88TL6453⟩. Other metal oxides have also found applications in sulfur oxidation. Two publications describe the use of zinc bismuthate ($Zn(BiO_3)_2$) and manganese(IV) oxide/trimethylchlorosilane ⟨92BCJ1131, 93SC1759⟩.

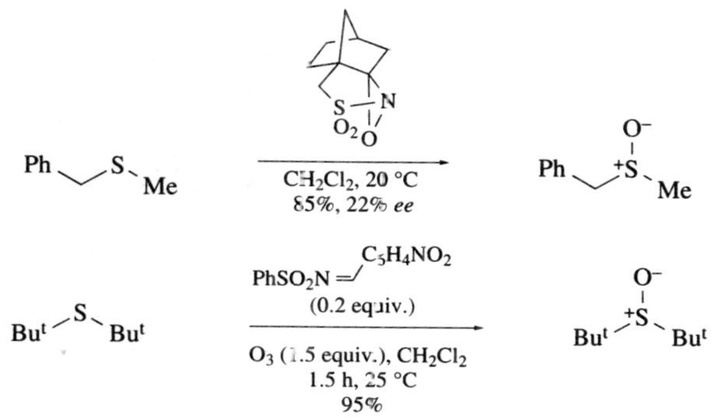

Scheme 59

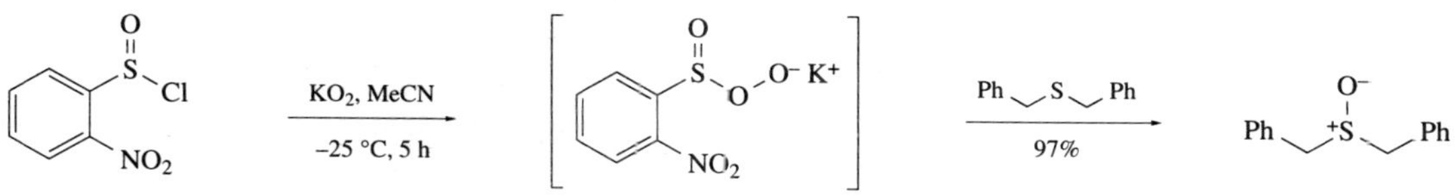

Scheme 60

(viii) Enzymatic oxidation

The use of enzymes is an attractive proposition for sulfide oxidation, particularly in view of the stereoselectivities and enantioselectivities that enzymes can confer during reaction ⟨88CRV473⟩. Unfortunately the selectivity of enzyme-mediated reactions is at the expense of substrate generality. Some excellent yields and enantiomeric excesses have however been observed (Scheme 61) ⟨68JCS(C)2374, 88CC1451, 92TA95, 92TA1063⟩.

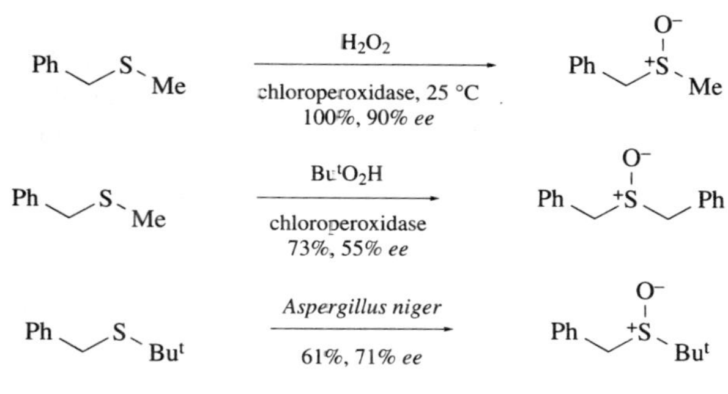

Scheme 61

(ix) Sulfinyl group transfer

Before the advent of enantioselective sulfur oxidation one of the very few realistic methods of preparing nonracemic chiral sulfoxides was by sulfinyl transfer from an optically active sulfinate ester, usually prepared from menthol and a sulfinyl halide followed by resolution, by treatment with an organometallic reagent, known as the Andersen procedure ⟨62TL93, 76JOC3987⟩. The stereochemistry of this displacement reaction has been examined in detail ⟨92CC1500⟩.

Most examples cited in the literature are toluene sulfinates which are significantly more easily handled, particularly in the resolution step, than alkyl sulfinate esters. It was originally found that this method was only suitable for the formation of aryl alkyl sulfoxides as the displacement does not occur in the corresponding alkylsulfinate esters. Later developments have shown however that the aryl group in the aryl alkyl sulfoxide product may be displaced using certain alkyllithium reagents, again with an inversion of configuration ⟨74CJC761, 73S485⟩. Limitations to the process can arise through competing deprotonation; the substituents in the sulfoxide have a powerful effect upon the acidity of the α-protons and also upon the leaving ability of the sulfinate group. The nature of the nucleophile can also have a profound effect upon the reaction; for example the small and highly reactive methyllithium provides only limited success in displacement, while butyllithium is more effective. The overall transformation of methyl-*p*-tolylsulfinate into an optically active dialkyl sulfoxide is nicely exemplified by a stereospecific synthesis of (+)-disparlure (Scheme 62) ⟨77TL4009⟩.

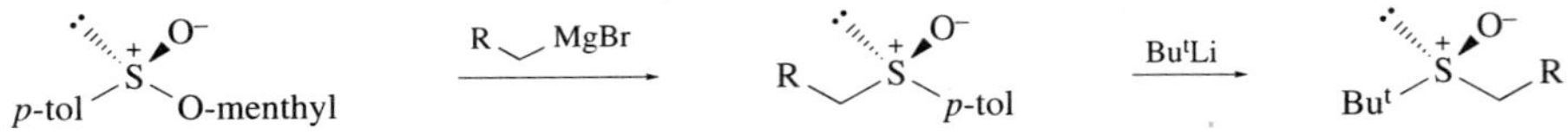

Scheme 62

A variation involves the use of sulfinate esters prepared from *trans*-2-phenylcyclohexanol in place of menthol, which improves this method considerably in practical terms as the diastereoisomers are readily separable by chromatography and the major products are all crystalline ⟨91JOC4552⟩. This provides a route to either enantiomer of the sulfoxide, as both enantiomers of the alcohol are available.

The use of a cheap, readily available chiral auxiliary is a common goal in synthetic chemistry, often fulfilled by suitably protected sugar derivatives. This theme has been explored in the preparation of optically active sulfoxides from diacetone D-glucose, with the formation of either (+) or (−) sulfinate governed by the choice of the base used in the reaction (Equation (51)) ⟨92JOC6789⟩.

(51)

Base	*Solvent*	*Ratio*
imidazole	THF	82:18
Me_2NPh	THF	42:58
Et_3N	toluene	2:98

DAG = diacetone D-glucose

Evans has reported the use of chiral sulfinyl transfer reagents based on the oxazolidinone auxiliary. After separation of diastereoisomers by chromatography, the displacement by Grignard reagents proceeded in better than 78% yield and with enantiomeric excesses of up to 79% (Equation (52)) ⟨92JA5977⟩.

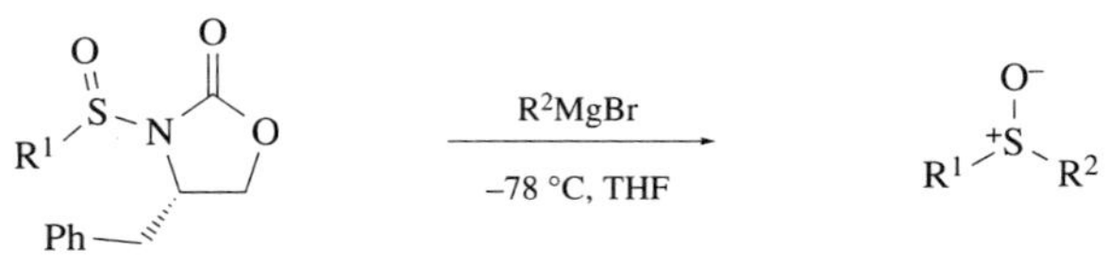

(52)

R^1	*R^2*	*Yield (%)*	*ee (%)*
Me	Bu^t	78	93
Me	$PhCH_2$	82	91
Bu^t	Me	92	100
Bu^t	Bu^n	91	100

An extension of the sulfinyl group transfer methodology has been investigated by Kagan. A chiral sulfite derived from a diol and thionyl chloride has two groups which can undergo displacement by carbon nucleophiles, leading to enantiomerically pure sulfoxides (Equation (53)) ⟨89TL3659, 91JOC5991⟩.

A similar approach utilising the oxathiazolidine sulfoxides prepared from ephedrine has also been described, with observed enantiomeric excesses of greater than 99% ⟨91TL5885⟩.

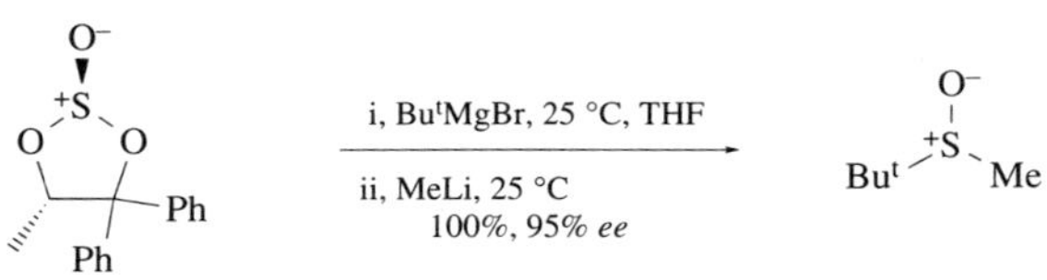

(53)

2.03.2.2.2 *Addition to unsaturated compounds*

Although the addition of aryl sulfinyl chlorides or of thionyl chloride to alkenes or enol ethers, forming the α-chloro sulfoxides, is not known, the corresponding addition of alkyl sulfinyl chlorides is well documented. The addition of an alkyl sulfenic acid to an activated alkene is also a known process. For example, the reaction of *t*-butanesulfenic acid, generated by thermolysis of di-*t*-butylsulfoxide, with ethyl acrylate at room temperature gives ethyl-β-*t*-butane sulfinyl propionate (Scheme 63) ⟨67JA718⟩. Addition to unactivated alkenes has been reported for both cyclic and acyclic systems ⟨77JCS(P1)1574⟩.

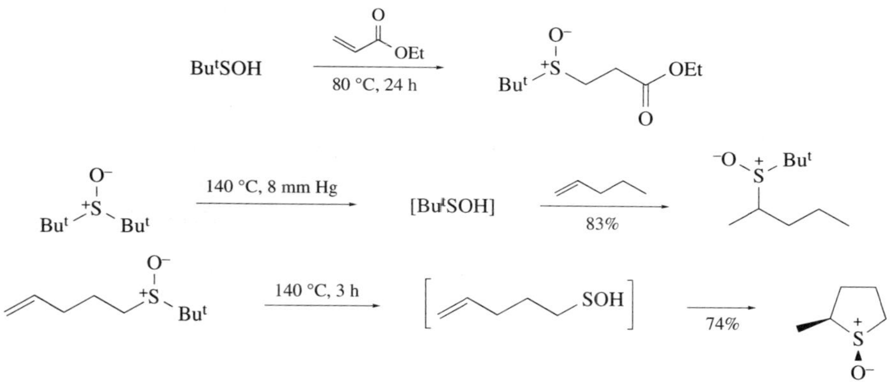

Scheme 63

Sulfur monoxide can also be trapped by dienes and trienes to form the cyclic sulfoxides. The yields of sulfoxide are not spectacular, but nonetheless it is still a useful method (Scheme 64). The thermal decomposition of ethylene sulfoxide is a convenient source of sulfur monoxide ⟨67CC1189, 69CC1159, 71CC979, 73JA920, 73JA922⟩.

Conjugate addition and cycloaddition to alkyl vinyl sulfoxides, to give sulfoxide products, are poorly represented in the literature compared with aryl vinyl sulfoxide substrates.

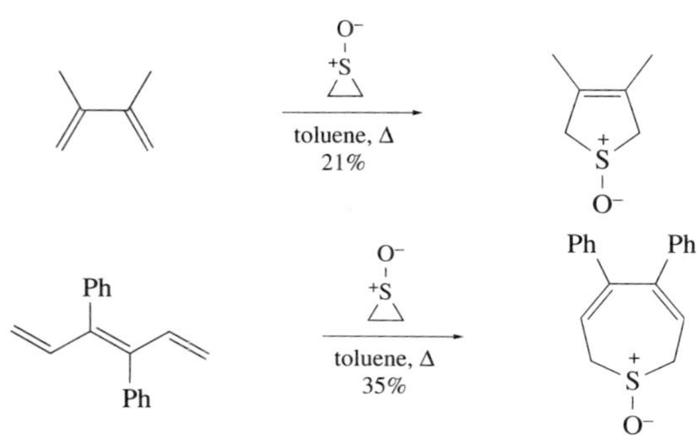

Scheme 64

2.03.2.2.3 *From sulfur-stabilized carbanions*

The ability of a sulfinyl group to stabilize an α carbanion has lead to its widespread use in organic synthesis and provides an obvious method for sulfoxide synthesis ⟨65JA1345, 79COC(3)121, 91COS(1)505, 92TA961⟩. A great variety of electrophiles is successful, including alkyl halides, carbonyl compounds, α,β-unsaturated carbonyl compounds, imines, and alkenes (Scheme 65).

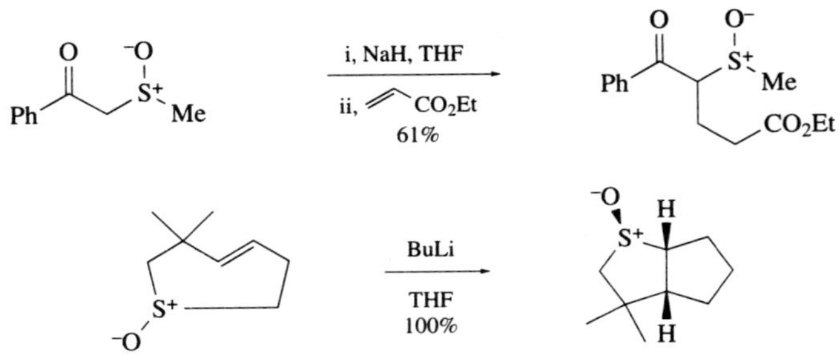

Scheme 65

2.03.2.2.4 By rearrangements

Allyl alkyl sulfenate esters, formed by reaction of an allylic alcohol and alkyl sulfenyl halide, undergo a well-known rearrangement to allyl alkyl sulfoxides. This is however an equilibrium process, and the sulfoxide is normally the starting material. Particularly reactive compounds such as allyl trichloromethyl sulfenate rearrange spontaneously at low temperature (−70 °C) ⟨67CC270, 74ACR147, 79AG(E)563, 91COS(6)873⟩.

2.03.2.2.5 Reduction of sulfones

The partial reduction of sulfone to sulfoxide is difficult to achieve. One successful approach is to activate one of the sulfone oxygen atoms before reduction. Reaction of a sulfone with *p*-chlorophenyl diazonium tetrafluoroborate generates an intermediate oxysulfonium ion which can subsequently be reduced by sodium borohydride and alumina in 91–95% yield (Scheme 66) ⟨79SC923⟩.

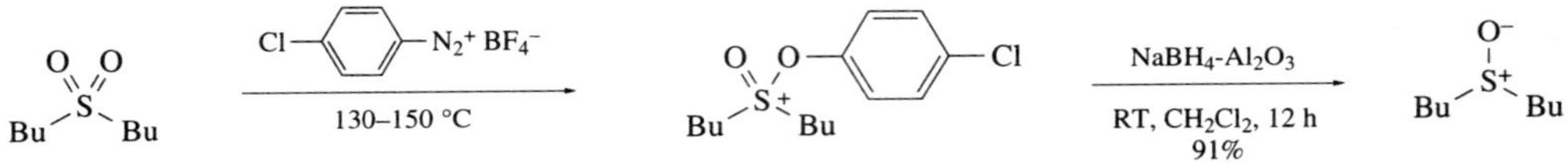

Scheme 66

2.03.2.3 Dialkyl Sulfones

The increasing use of sulfones in synthesis has provided the driving force for new selective methods for their synthesis ⟨79COC(3)71, B-88MI 203-03, 91COS(6)133, B-93MI 203-01⟩.

The ease of sulfide synthesis and their ready oxidation allows a simple route to sulfones from the corresponding sulfide either directly or via the sulfoxide. As is noted above, however, many more examples exist for diaryl or alkyl aryl sulfones than for dialkyl sulfones.

2.03.2.3.1 By oxidation of sulfides and sulfoxides

The most common method of dialkyl sulfone formation is oxidation of the corresponding dialkyl sulfide. Although many oxidants will perform this transformation on simple substrates, as discussed above for sulfoxide formation, the number of reagents that are both mild and tolerant of other oxidizable functionality is more limited.

(i) Oxidation of sulfides

There have been several reports describing the selective oxidation of sulfides to sulfones mediated by transition metals (Scheme 67). Of particular interest are those involving a catalytic quantity of the metal salt with stoicheiometric oxidant. Ruthenium tetroxide and periodic acid in a two-phase solvent system is one procedure which has been used to oxidize α-carboxyl alkyl sulfides to the corresponding sulfones in yields of 84–90% ⟨92T3571⟩.

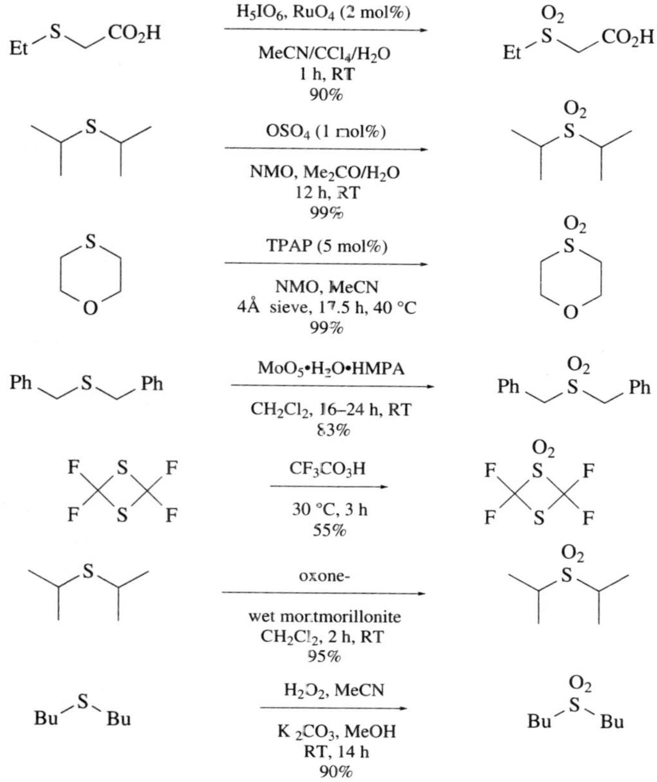

Scheme 67

Another ruthenium salt, tetrapropylammonium perruthenate (TPAP), may be used catalytically (5 mol%), with *N*-methyl morpholine-*N*-oxide (NMO) as the cooxidant. This system has the advantage that isolated double bonds are unaffected, sometimes a drawback with other oxidants during the sulfoxide to sulfone oxidation stage. High yields of sulfone product are isolated (>80%) ⟨93TL5369⟩.

A range of dialkyl sulfides has been oxidized in excellent yields using 1 mol% osmium tetroxide in the presence of NMO. Curiously, although osmium tetroxide does oxidize sulfoxides, oxidation of sulfides does not occur in the absence of NMO, even with up to two equivalents of osmium tetroxide. In the absence of osmium tetroxide no oxidation was observed ⟨91TL5043⟩.

Peroxymolybdenum complexes, used for various oxidation reactions, have been used for sulfone formation, and the effects of different ligands investigated. The anhydrous conditions used allow the oxidation of water sensitive substrates ⟨92ACS867⟩.

Oxidation of sulfides to sulfones by manganate salts under both heterogeneous phase-transfer and acid-catalyzed conditions has been reviewed ⟨87S85⟩.

Peroxyacids are ubiquitous and useful oxidants which are widely used either alone or in the presence of a catalyst, for example the manganese compound illustrated (Scheme 67) ⟨80BSF411, 80JOC2887, 82JOC3773, 85CB116⟩.

Other commonly used oxidants include sodium perborate and oxone (potassium hydrogen persulfate, $KHSO_5$) ⟨81TL1287, 83TL1505⟩. The use of oxone typically necessitates an aqueous solvent system, although this can be avoided by the addition of wet montmorillonite clay, allowing conventional organic solvent to be used, so facilitating workup. Typically, yields of 92–97% in dichloromethane are obtained in 2–6 h at ambient temperature ⟨91CL523⟩.

As indicated above, hydrogen peroxide has long been used for sulfide oxidation. It is traditionally used in an acetic acid solution, although in one report, which highlights its utility in sulfone formation, hydrogen peroxide is used under basic conditions in conjunction with acetonitrile to produce a reactive intermediate, formulated as a peroxyimidic acid, to oxidize dialkyl sulfides to sulfones in excellent yields ⟨93SC1507⟩.

Sulfides are also oxidized to sulfones in good yield by electrophilic halogen compounds, although side reactions involving halogenation can be troublesome ⟨73TL289, 80BCJ3615⟩.

(ii) Oxidation of sulfoxides

Sulfoxides are intermediates in the oxidation of sulfides to sulfones, and can obviously themselves be used as oxidation substrates (Scheme 68) ⟨91COS(7)757⟩.

The reduced nucleophilicity of the sulfoxide sulfur atom compared with that of sulfides results in slower reaction with electrophilic oxidants. Peroxyacids and dialkyl peroxides, as well as hydrogen peroxide, still find use in this oxidation, particularly under transition metal catalysis by iron, tungsten, vanadium, or manganese compounds ⟨63JA3992, 63JOC1140, 81CJC718, 84JA4283, 85JOC1784⟩. Nitronium salts have also been used ⟨83JOC3585⟩. For many years the standard reagent was mcpba, which has found many applications in sulfone formation, but it is now commonly replaced by mangesium monoperoxyphthalate (MMPP), which is not shock sensitive when purified ⟨87S1015, 88JOC5689⟩. Organic hydroperoxides are also useful in sulfoxide oxidation under catalysis by vanadium, molybdenum, and titanium ⟨66AG(E)893, 81JCS(D)447, 83JCS(D)415⟩. Hydrogen peroxide itself may be used in the presence of a nitrile, forming peroxyimidic acid *in situ* as the oxidizing species ⟨93JCS(P2)2161, 93SC1507⟩.

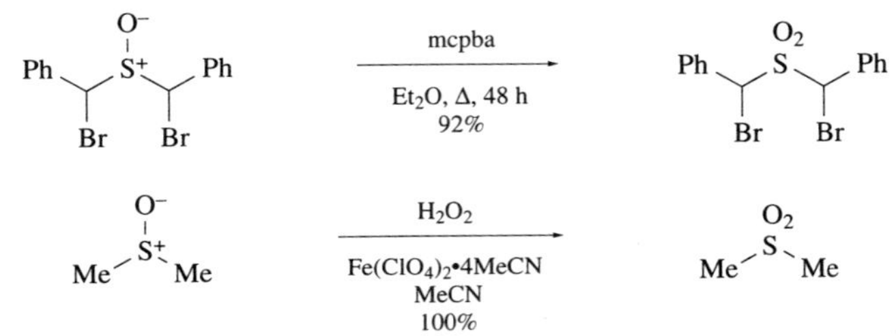

Scheme 68

2.03.2.3.2 *From sulfinic acids and their derivatives*

Alkylation of sulfinic acids or their salts provides, in theory at least, a facile synthesis of dialkyl sulfones. Two factors reduce the practicality of this route: first, the instability of alkane sulfinic acids, and secondly, the ambident nature of sulfinates, which necessitates proper choice of reaction conditions and alkylating agent to prevent *O*-alkylation to form the sulfinate ester (Scheme 69) ⟨85HOU(E11)1131⟩. An alternative is to recycle the *O*-alkylated sulfinic ester to the sulfinate salt by *in situ* hydrolysis, leading to the sulfone as the major product ⟨63OSC(4)674⟩. Despite these difficulties it is possible to synthesize sulfones by this method, often in good yields ⟨68JOC3422, 88T6687⟩. The most favourable conditions appear to involve an alkyl halide as the alkylating agent in a polar solvent, often ethanol or polyethyleneglycols, and the sodium, silver, or potassium salt of the sulfinic acid ⟨84BCJ613⟩. Tetraalkylammonium salts of sulfinic acids have also been used either stoicheiometrically or in catalytic amounts under phase transfer conditions ⟨79S733, 88JCR308⟩.

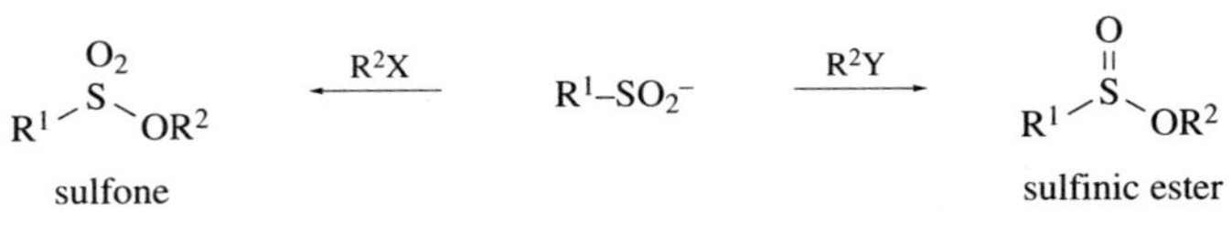

Scheme 69

The salts of the weaker sulfinic acids generally react to provide highest yields of the sulfones, presumably due to their greater nucleophilicity ⟨84S320⟩.

Sulfinic acids can also be alkylated by addition to an alkene, but in practice little tendency is seen for this reaction to occur with simple substrates, except under forcing conditions unsuitable for most alkanesulfinic acids. Perfluorobutanesulfinic acid has been found to undergo addition to polyenes under strong acid or palladium catalysis (Equation (54)) ⟨78CC367, 81GEP2926671⟩. If the double bond is activated by conjugation to a carbonyl group, addition of sulfinic acids is more facile, although only examples involving aryl sulfinic acids seem to have been reported.

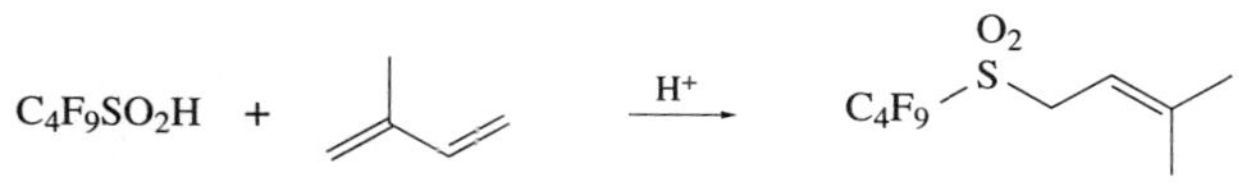

(54)

Alkylation of sodium dithionate or sodium hydroxymethyl sulfinate with two equivalents of a benzyl halide, in the presence of the phase-transfer catalyst aliquat 336 (methyltrioctylammonium chloride) or of potassium carbonate leads to reasonable yields of dibenzyl sulfone (Scheme 70) ⟨89SC2939⟩.

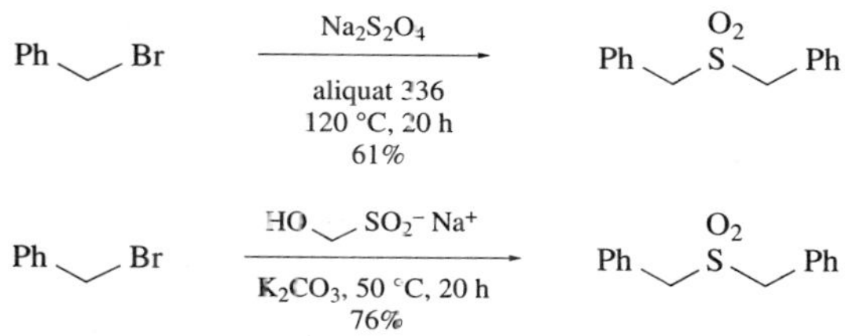

Scheme 70

2.03.2.3.3 *From sulfonic acid derivatives*

Sulfonyl halides (except sulfonyl fluorides) readily add across carbon–carbon double bonds to give the α-halosulfones. The reaction is often carried out under copper(I) or (II) salt catalysis which reduces the competing polymerization ⟨64JCS4962, 70JOC4217, 72JCS(P1)1543⟩.

Sulfonyl halides can be alkylated by reaction with carbon nucleophiles to furnish dialkyl sulfones in good yields (Equation (55)). For example, methanesulfonyl chloride undergoes substitution of chloride by the sodium salt of dimethyl malonate in 58% yield. Alkanesulfonyl fluorides have also been successfully substituted ⟨71LA(744)15, 86JA2358⟩. Activated sulfonic acid esters such as trifluoromethyl methanesulfonate are also alkylated to form the sulfone, as are some sulfonic acid anhydrides ⟨63JOC1945, 64ACS954, 81S229⟩.

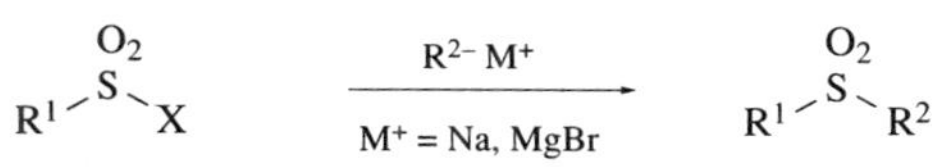

(55)

R^1 = alkyl; X = Cl, F, OCF_3

Enolates have been sulfonated by reaction with sulfur dioxide followed by treatment of the intermediate sulfinyl anion with an alkyl halide (Scheme 71). Unfortunately this process is not particularly high yielding ⟨88SC1061⟩. Enamines have been directly sulfonated by reaction with trifluoromethyl trifluoromethanesulfonate (Equation (56)).

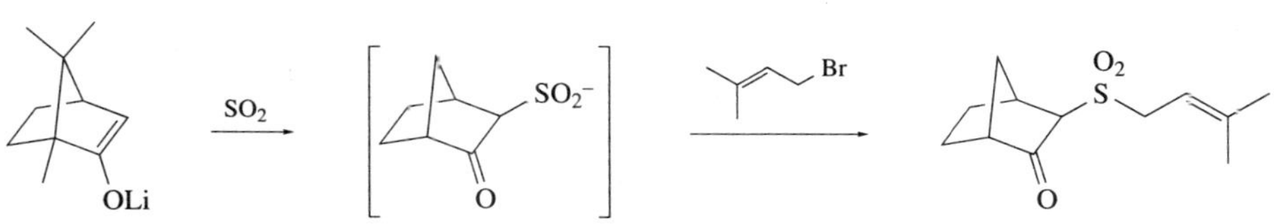

Scheme 71

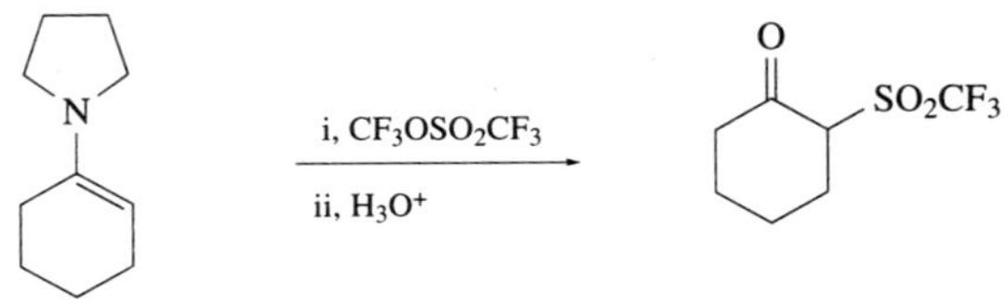

(56)

2.03.2.3.4 *Via α-sulfonyl carbanions*

One of the best known features of sulfone chemistry is their ability to stabilize an adjacent carbanion in a diverse range of substrates, while remaining unaffected by nucleophilic attack at sulfur. All normal carbanion reactions are typically successful, including alkylation, conjugate addition, and epoxide ring opening. Such reactions, producing more highly functionalized sulfones, have been reviewed in detail ⟨B-93MI 203-01⟩.

2.03.2.3.5 *Addition to unsaturated sulfones*

(i) Cycloadditions

Although many examples of cycloaddition reactions with aryl vinyl sulfones are recorded in the literature, those involving alkyl vinyl sulfones are less well represented. Nevertheless alkyl vinyl sulfones do partake in [2 + 2], [2 + 3] and [2 + 4] cycloadditions with ease (Scheme 72) ⟨83T689, 91COS(6)133, B-93MI 203-01⟩.

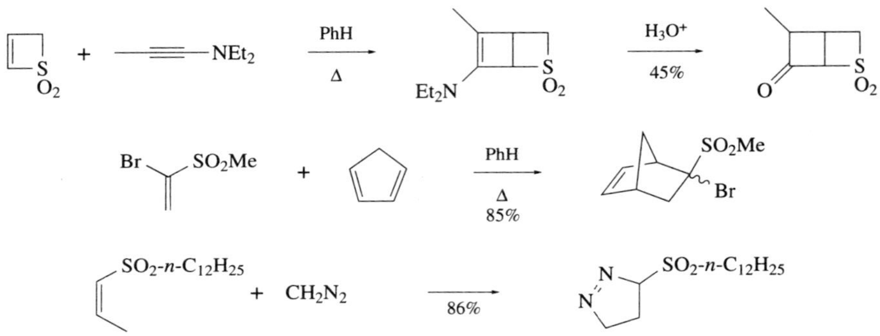

Scheme 72

A [2 + 2] intramolecular cycloaddition of a vinyl sulfone leads to a highly functionalized dialkyl sulfone product (Equation (57)) ⟨82JOC3121⟩. Vinyl sulfones may also add by [2 + 2] mechanism to enamines ⟨70JOC905⟩.

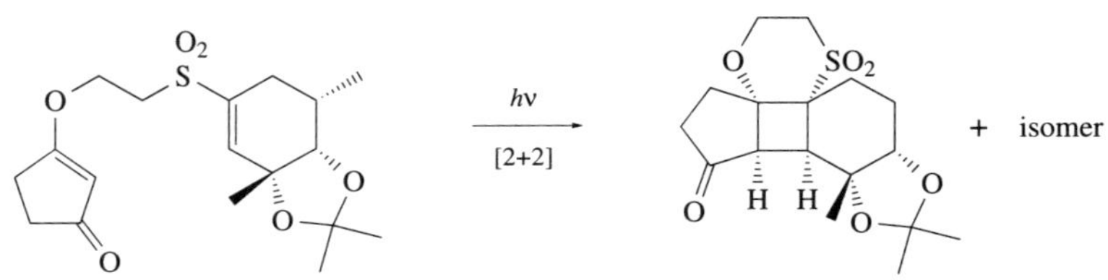

(57)

Divinyl sulfones react with oximes or imines by an initial conjugate addition to form an intermediate 1,3 dipolar species which subsequently attacks the second vinyl group, formally to give an alkyl sulfone cycloaddition product (Equation (58)) ⟨89JCS(P1)1550, 91T8297⟩.

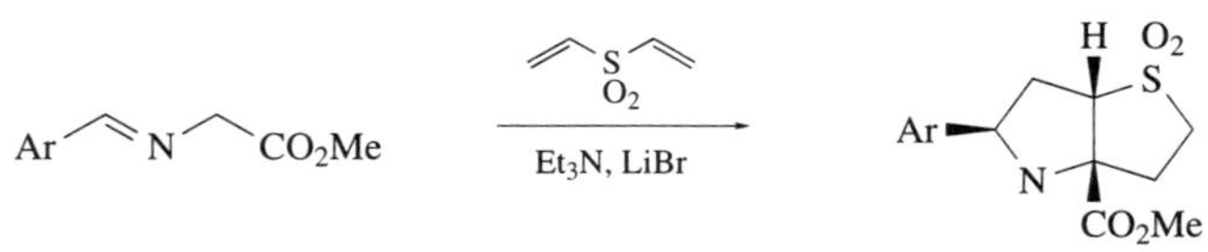

(58)

(ii) Conjugate addition

Additions of oxygen, nitrogen, sulfur, and carbon nucleophiles to vinyl sulfones to form alkyl sulfones are abundant throughout the literature, emphasizing how versatile are unsaturated sulfones as Michael acceptors. Again, however, there are relatively few examples of addition to alkyl vinyl sulfones. An intramolecular addition of an oxygen nucleophile has been used in the formation of a spirocyclic ether in 81% yield (Equation (59)) ⟨85TL4455⟩.

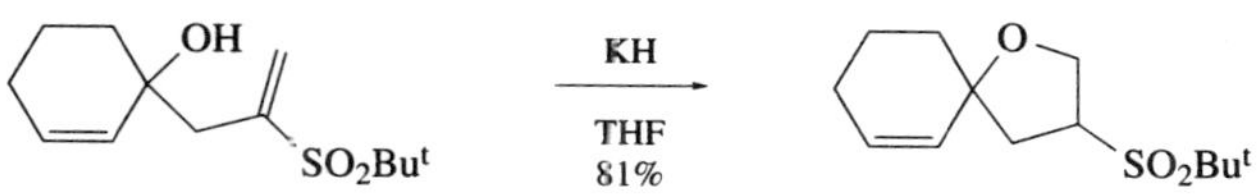

(59)

2.03.2.3.6 *Rearrangement*

Allylic sulfinate esters undergo thermal rearrangement to give allylic sulfones, by both 1,2- and 2,3-sigmatropic rearrangement mechanisms (Scheme 73) ⟨67CC131, 76JA3332, 83CC1470, B-88MI 203-04⟩. This reaction was used in the synthesis of the natural product agelasidine from a sulfinate derived from famesol (Scheme 74) ⟨92JCS(P1)1497⟩.

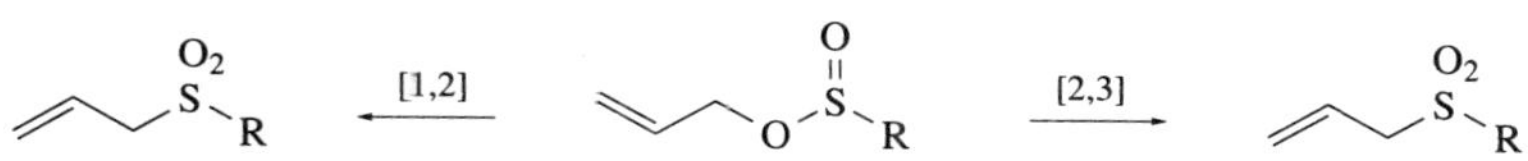

Scheme 73

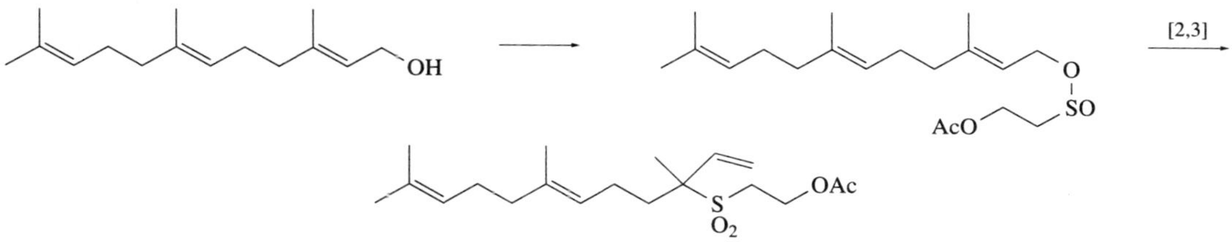

Scheme 74

2.03.2.4 Sulfimines

These compounds, also known as sulfimides, sulfilimines, or iminosulfuranes, are the nitrogen analogues of sulfoxides. Although widely known, they have not been as extensively studied as sulfoxides, and a much more restricted range of methods is available for their synthesis ⟨77CRV409⟩. They are often found as the *N*-(4-toluenesulfonyl) derivatives, the electron-deficient sulfonyl group imparting additional stability.

2.03.2.4.1 *The reaction of sulfides with* N*-halo compounds*

Treatment of dialkyl sulfides with *N*-bromoacetamide in carbon tetrachloride leads to good yields of the sulfonium bromides, in a reliable but highly solvent dependent reaction. The addition of triethylamine in dichloromethane at 0°C furnishes the *N*-acylsulfimines in high yields (Scheme 75)

⟨72JOC1121, 77PS(3)9⟩. *N*-Acyl and *N*-sulfonylsulfimines are particularly stable members of this class of compounds, the *N*-sulfonyl derivatives generally being prepared in an analogous manner using *N*-halosulfonamides such as chloramine-T ⟨22JCS1052⟩. The kinetics of the reaction have been studied ⟨74S277, 76T2763⟩.

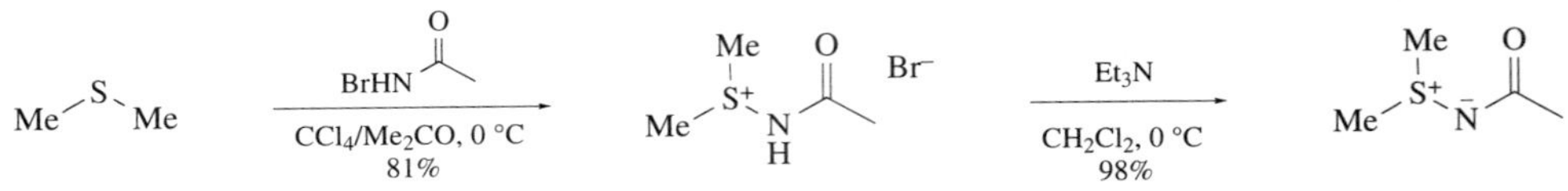

Scheme 75

In a similar reaction, *N*-ethoxycarbonylsulfimines have been produced in high yields by the action of ethyl *N*-chlorocarbamate upon dialkyl sulfides followed by triethylamine (Scheme 76) ⟨74JOC2148⟩. Dimethyl sulfide reacts in the same way with *N*-chloro-*O*-methylisourea to form the sulfonium salt, but attempts to generate the iminosulfimine by treatment of this intermediate with bases were unsuccessful. A sulfimine could however be produced through *N*-halogenation with sodium hypochlorite (Scheme 77) ⟨81JOC5215⟩.

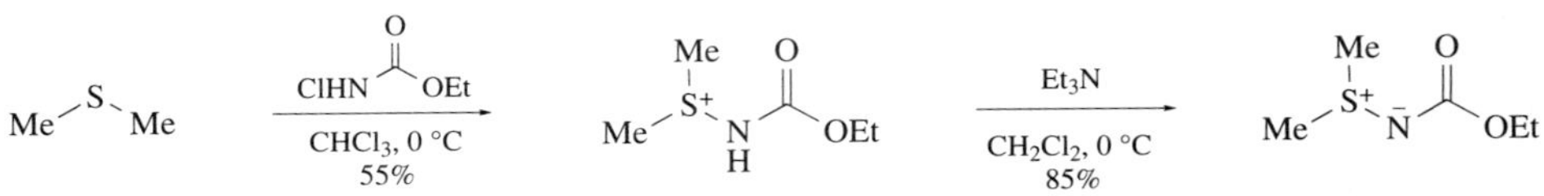

Scheme 76

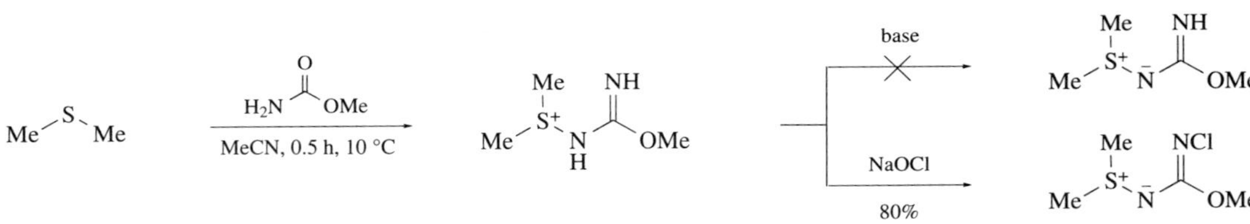

Scheme 77

2.03.2.4.2 *The reaction of sulfides with amines*

A proposed intermediate in the reaction of *N*-haloamines with sulfides is the *S*-halosulfonium salt, the halide group of which is rapidly replaced by the nitrogen nucleophile to give the *S*-aminosulfonium salt (Scheme 78). Separate sources of electrophilic halogen and amine may therefore be used to give the sulfimine product from a sulfide substrate. *N*-Chlorosuccinimide has found wide application in this reaction (Equation (60)) ⟨74TL3319, 76CC1003, 82JOC552, 84JMC1634, 84JOC2500⟩. Alternative electrophilic halogen reagents successful for this reaction include *N*-chlorobenzotriazole, *t*-butyl hypochlorite, and sulfuryl chloride (Scheme 79) ⟨75T505, 76JCS(P1)2166, 77JOC593⟩. An intramolecular variation used iodine to form a cyclic sulfimine (Scheme 79) ⟨79JOC2632⟩.

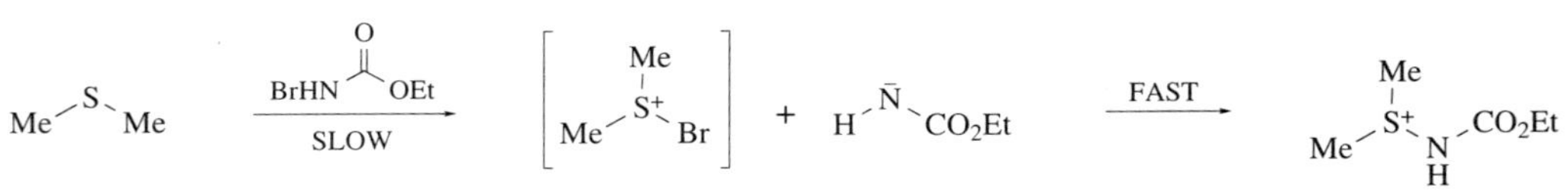

Scheme 78

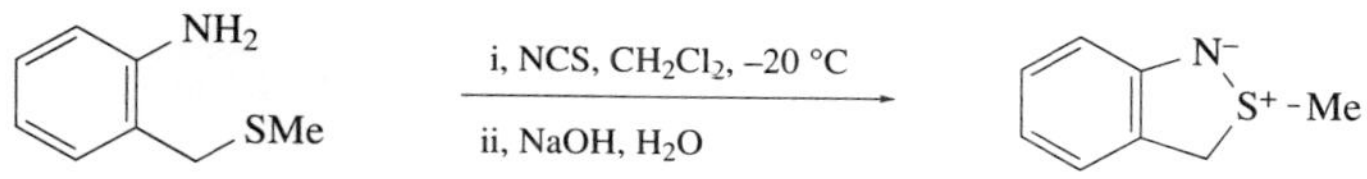

(60)

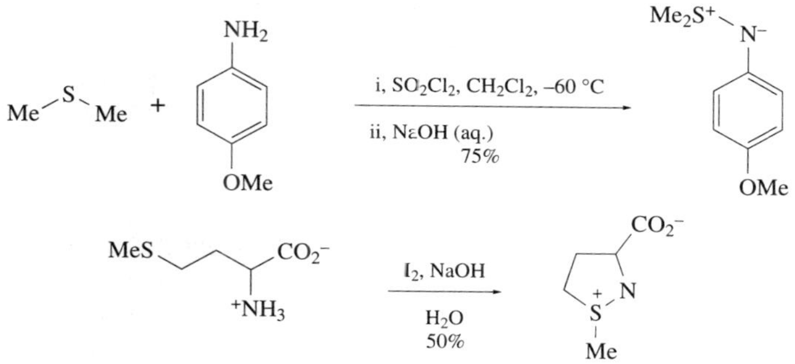

Scheme 79

2.03.2.4.3 The reaction of sulfides with hydroxylamine derivatives

Hydroxylamine-*O*-sulfonic acid reacts with thiane to give an intermediate sulfonium salt which is decomposed to the sulfimine by the action of ammonia (Equation (61)) ⟨72JA208⟩. *O*-Mesitylenesulfonyl hydroxylamine has found use in this reaction with both cyclic and acyclic sulfides, furnishing sulfimines upon treatment of the sulfonium salts with a basic ion exchange resin (Scheme 80) ⟨72TL4137, 74S277⟩.

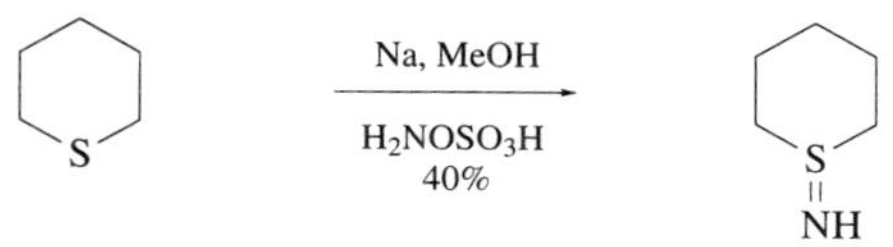

(61)

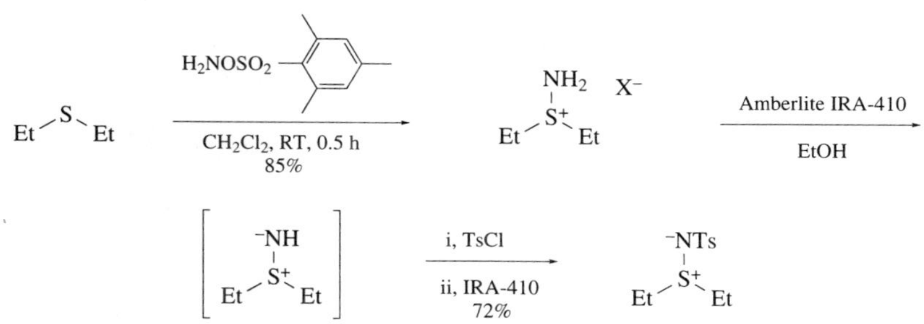

Scheme 80

2.03.2.4.4 The reaction of sulfides with azides

Photolysis of perfluorosulfonyl azides in the presence of dimethyl sulfide leads to generation and trapping of the nitrene to form *N*-sulfonyl sulfimines (Equation (62)) ⟨92TL6503⟩. The photochemical decomposition of ethyl azidoformate and 4-toluenesulfonyl azide have been studied and the nitrenes similarly trapped with sulfides to give sulfimines in low yields (Equation (63)) ⟨80JCS(P2)385⟩.

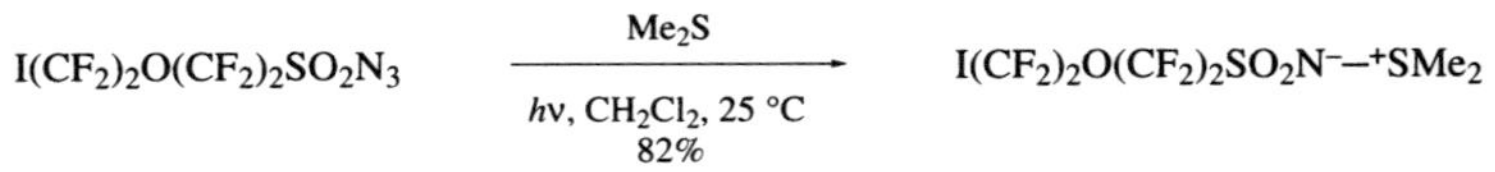

(62)

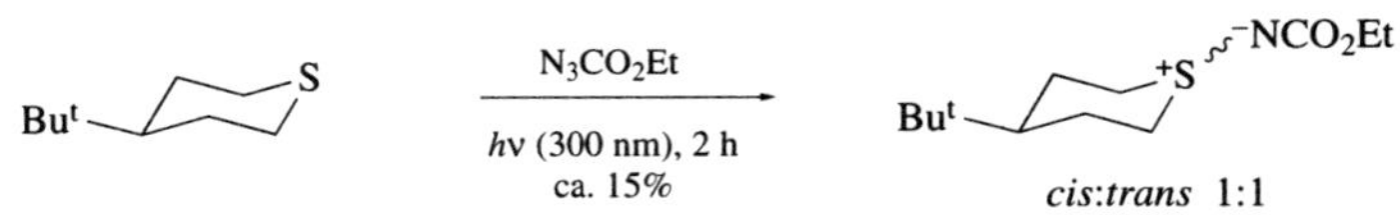

(63)

2.03.2.4.5 *The reaction of sulfides with sulfamides*

The oxidation of *N,N*-disubstituted sulfamides with lead tetraacetate in dimethyl sulfide leads to reasonable yields of the dimethyl sulfimine derivative, as shown in Equation (64). The reaction may be monitored by the color change due to the reduction of lead(IV) to lead(II) ⟨72S203⟩.

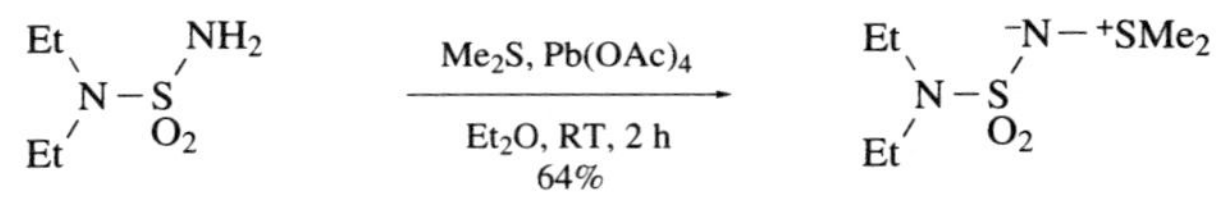

(64)

2.03.2.4.6 *From sulfoxides*

Activation of the oxygen atom of a sulfoxide allows displacement by a nitrogen nucleophile, so forming a sulfimine (Scheme 81). As in the analogous DMSO-mediated oxidation of alcohols, many activating reagents have proved successful, including sulfur trioxide, phosphorus pentoxide, dicyclohexyl carbodiimide, acetic anhydride, trifluoroacetic anhydride, oxalyl chloride, and trifluoromethanesulfonic anhydride ⟨41JA2938, 68TL3607, 71JOC3687, 71JOC3861, 74JOC3365, 75JOC2759, 78JOC4537, 83TL1011, 84JMC1634⟩.

Sulfoxides may be converted directly into sulfimines by reaction with 4-toluenesulfonyl isocyanate ⟨60JOC352⟩. A cyclic mechanism is proposed, as shown in Scheme 82.

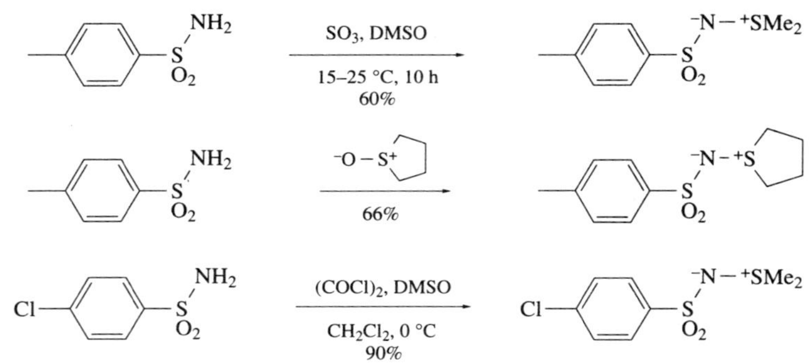

Scheme 81

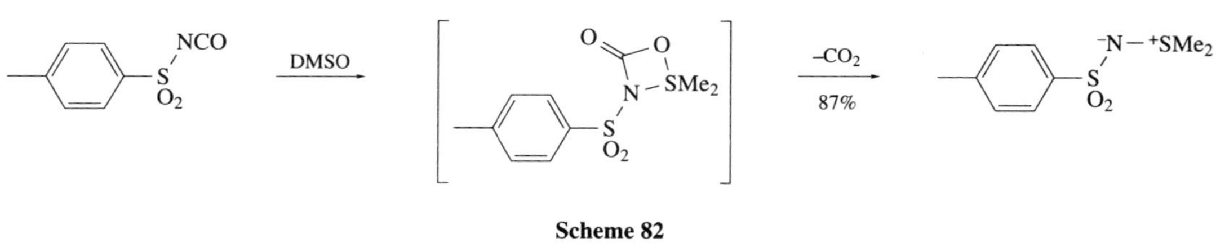

Scheme 82

Sulfoxides have also been shown to react readily with *N*-sulfinyl perfluoroalkanesulfonamides, prepared from sulfonamides by treatment with thionyl chloride, to produce sulfimines (Equation (65)) ⟨91CC732⟩. Dimethyl sulfoxide reacts with an oligomeric sulfur-nitrogen compound,

1,7-bis(tosyl)trisulfur tetranitride, furnishing *N*-toluenesulfonyl dimethylsulfimine in 81% yield (Equation (66)) ⟨89LA605⟩.

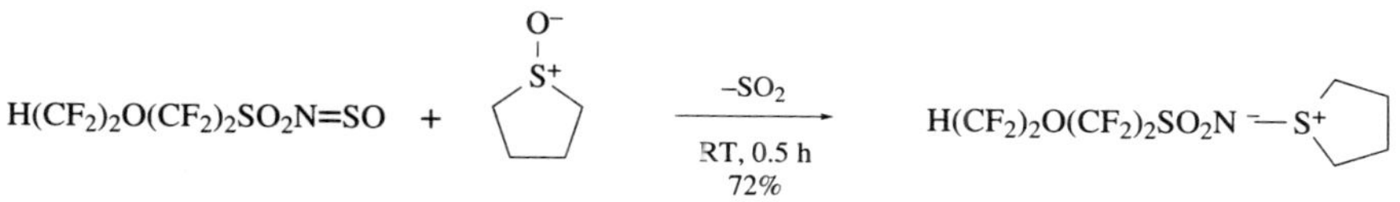

(65)

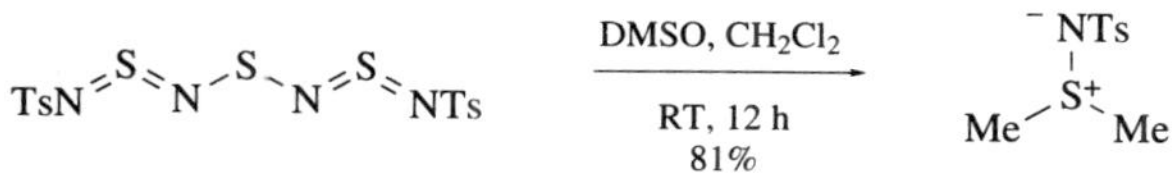

(66)

2.03.2.5 Sulfoximines

2.03.2.5.1 From sulfoxides

Chloramine-T reacts with dimethyl sulfoxide under copper(II) catalysis to form the sulfoximine (also known as sulfoximide) in good yield at room temperature (Equation (67)) ⟨69TL477, 76S731, 80CSR477⟩. Sulfoximines are also generated by the low temperature reaction of dimethyl sulfoxide and arylamines in the presence of *t*-butyl hypochlorite (Equation (68)) ⟨71JOC2207⟩.

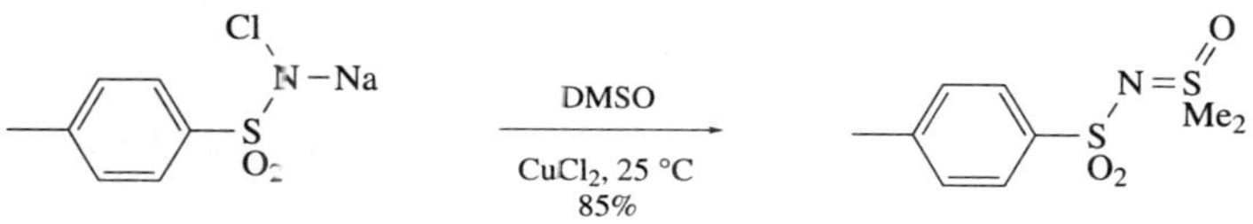

(67)

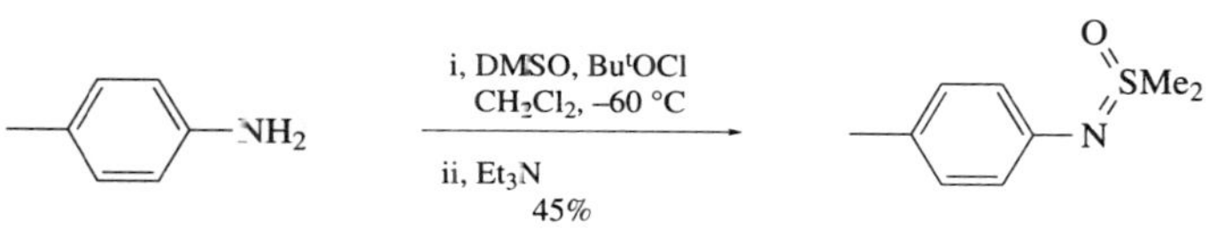

(68)

Treatment of a sulfoxide with concentrated sulfuric acid and hydrazoic acid (generated *in situ* from sodium azide) produces moderate yields of the corresponding sulfoximine (Equation (69)) ⟨51JA459, 71CB3234, 84JMC228⟩.

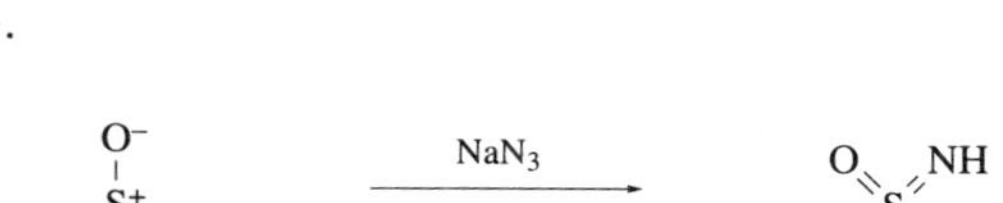

(69)

A number of other reactions of sulfoxides have been shown to generate sulfoximines. For example, dimethyl sulfoxide reacts with imides derived from an *o*-quinone to form the sulfoximines in very high yields (Equation (70)) ⟨86JOC829⟩. Isoxazolones and benzisoxazolequinones undergo attack by dimethyl sulfoxide to furnish sulfoximines in high yields upon prolonged heating (Equation (71)) ⟨83JHC899, 85JHC701⟩. Upon thermolysis, dioxazolinones give rise to acyl nitrenes, which can react with dimethyl sulfoxide to form sulfoximines (Scheme 83) ⟨68TL319⟩. In direct analogy to the generation of sulfimines from sulfides with *O*-mesitylenesulfonylhydroxylamine, the same reagent converts sulfoxides into sulfoximines at 0 °C in dichloromethane solution (Equation (72)) ⟨93TL133⟩.

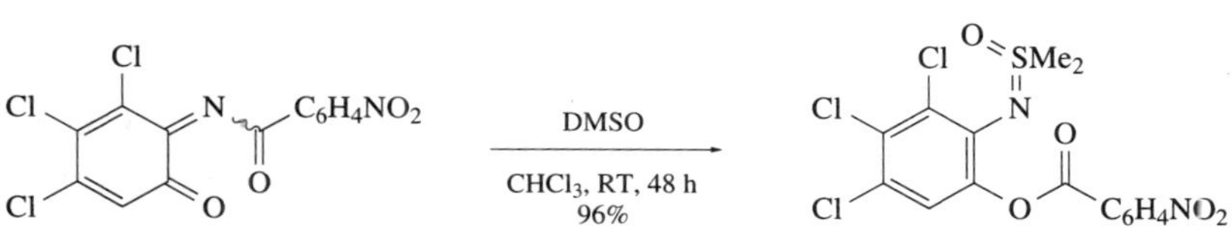

(70)

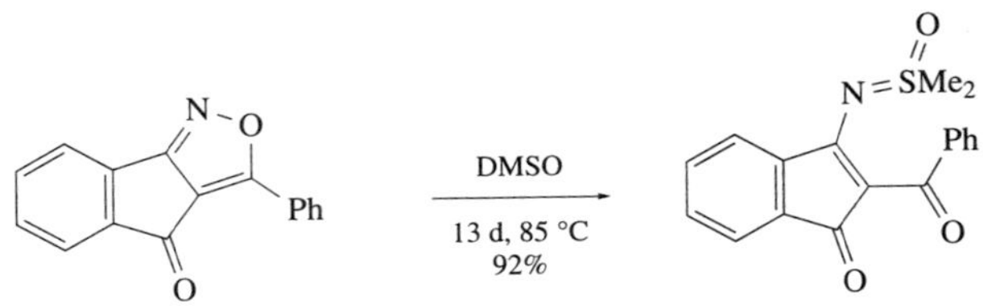

(71)

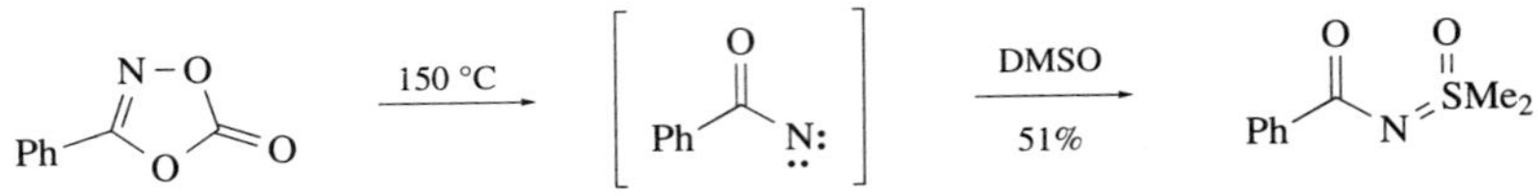

Scheme 83

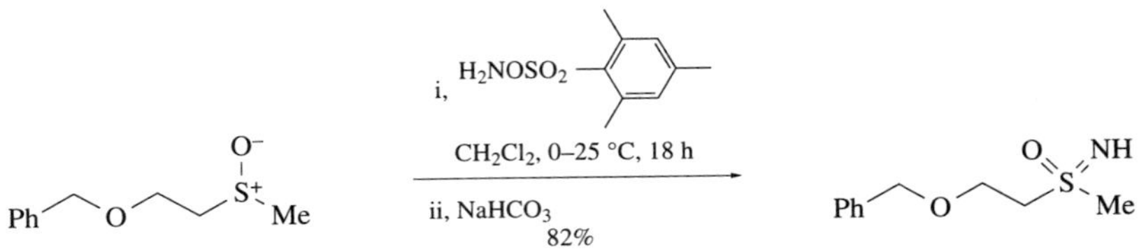

(72)

2.03.2.5.2 *From sulfimines and sulfonimidoyl chlorides*

Sulfimines may be oxidized to sulfoximines using typical sulfur oxidants including mcpba, hydrogen peroxide, potassium permanganate, and sodium periodate or peracetic acid catalyzed by ruthenium dioxide (Scheme 84) ⟨74TL3319, 78TL503, 79JOC2280, 79JOC2510, 84SC915⟩.

Sulfonimidoyl chlorides eliminate HCl upon treatment with base to give iminosulfene intermediates which can be trapped with electron-rich alkenes to give cycloaddition products including sulfoximines (Scheme 85) ⟨79JOC2055⟩.

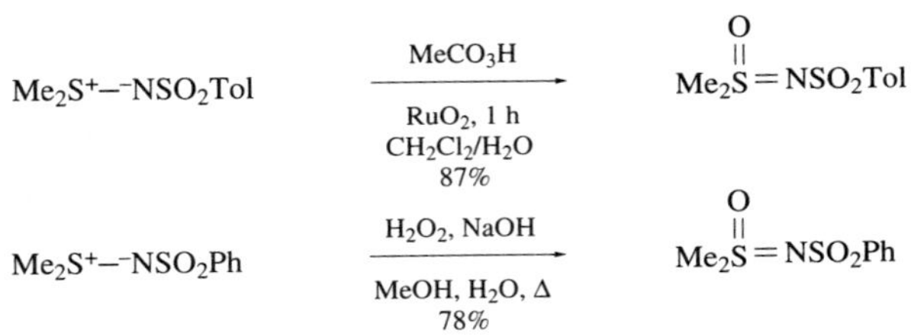

Scheme 84

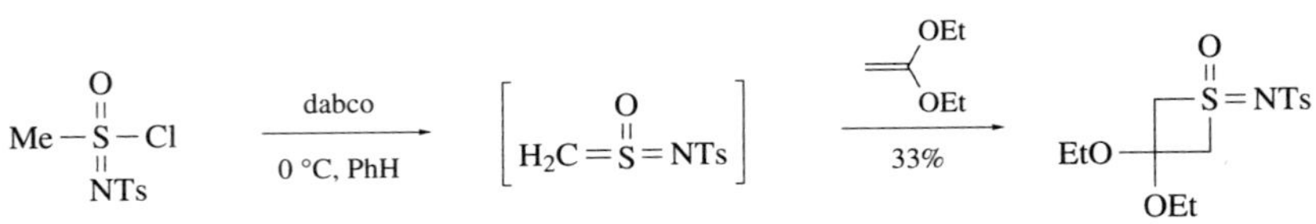

dabco = 1,4-diazabicyclo[2.2.2]octane

Scheme 85

2.03.2.6 Sulfur Diimines

Dialkyl sulfur diimines, or sulfone diimines, are formed in the reaction of dialkyl sulfides with ammonia and chloramine, generated *in situ* from ammonia and chlorine or *t*-butyl hypochlorite (Equation (73)) ⟨67JA2435, 70TL4449⟩.

Dimethyl sulfimine reacts with primary amines in the presence of bromine to give sulfur diimines in moderate yields, as illustrated in Equation (74) ⟨71CB2023⟩.

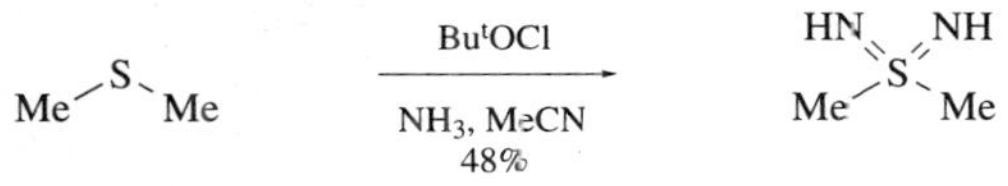

(73)

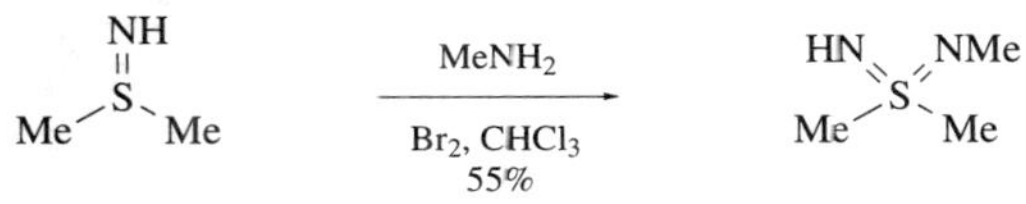

(74)

2.03.3 ALKANESULFENYL HALIDES AND THEIR HIGHER COORDINATED DERIVATIVES

2.03.3.1 Alkanesulfenyl Halides

Sulfenyl halides, other than the iodides, are moderately stable molecules, especially in comparison to the parent acids. Sulfenyl chlorides are the most stable and most common of the sulfenyl halides, while the stability of sulfenyl iodides is so poor that they are in general only postulated as intermediates, few having been characterized ⟨70S561, 71S563, 71S617, B-90MI 203-03, B-90MI 203-04⟩.

2.03.3.1.1 Alkanesulfenyl fluorides

Very few stable sulfenyl fluorides have ever been isolated, and many unsuccessful synthetic approaches have been reported ⟨21JA591, 46JCS1126, 55JA1041, 58JOC381⟩. Evidence for their existence mainly derives from low temperature spectroscopic examination. The usual methods applied for the preparation of sulfenyl chlorides and bromides, such as the halogenation of various sulfur functional groups, do not usually bring about sulfenyl fluoride formation, and instead produce only highly fluorinated sulfur derivatives. However, examples of direct fluorination of sulfur functionality to furnish sulfenyl fluorides are known. Scheme 86 illustrates direct fluorination of sulfide substrates with trifluoromethane hypofluorite ⟨73JA4064, 73JA8191⟩. Hydrolysis of the products gave diastereoisomeric mixtures of sulfoxides. Again, polyfluorinated sulfur species usually resulted.

A further interesting reagent, SF_5Cl, has been reported to introduce fluorinated sulfur functionality into alkene substrates (Scheme 87) ⟨61JCS2066, 61JCS2070, 89IZV197⟩.

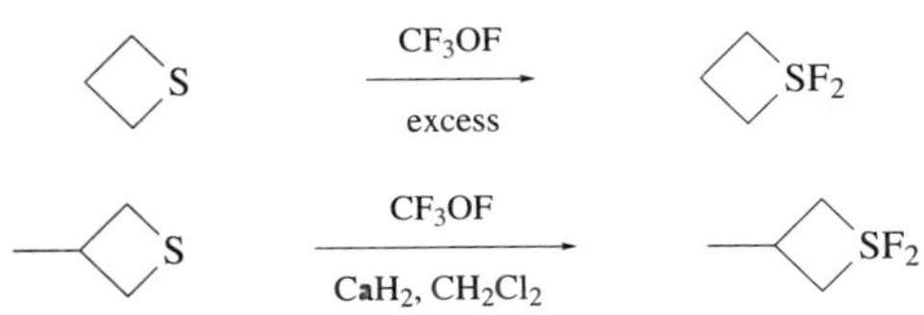

Scheme 86

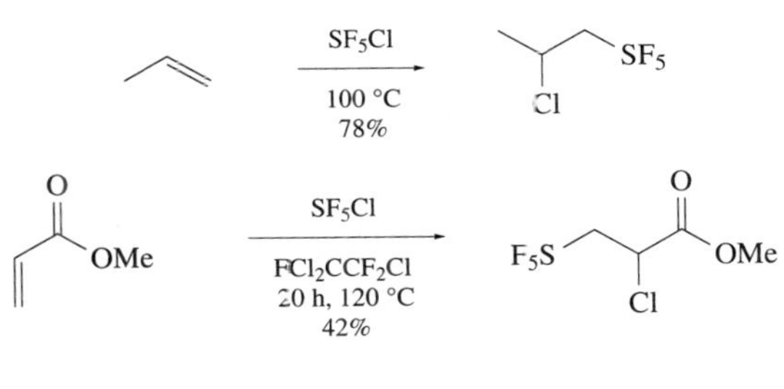

Scheme 87

Halogen exchange reactions of the relatively stable sulfenyl chlorides provide the most important preparative method for sulfenyl fluoride production, and have been used for the preparation of a number of very unstable sulfenyl fluorides (Scheme 88). Trichloromethanesulfenyl fluoride spontaneously undergoes isomerization to form dichlorofluoromethanesulfenyl chloride, whereas trifluoromethanesulfenyl fluoride dimerizes at temperatures greater than 0 °C ⟨67AG(E)706, 68JA5403, 69AG(E)773⟩. The halogen exchange is normally effected by activated potassium fluoride, although other fluoride salts, such as mercury(II) fluoride and silver(I) fluoride, can also be employed to equal effect. One example is the preparation of methanesulfenyl fluoride, a compound which immediately undergoes decomposition regardless of the fluoride salt employed for its preparation (Scheme 89) ⟨76JFC(7)115, 76ZN(B)727⟩. It should however be noted that one report in the literature describes a preparation of perfluoropropanesulfenyl fluoride, using silver(I) fluoride, giving a yield of only 5% ⟨59JA4810⟩.

$$Cl_3CSCl \xrightarrow{KF} Cl_3CSF \longrightarrow Cl_2FCSCl$$

$$F_3CSCl \xrightarrow{KF} F_3CSF \xrightarrow{>0\ ^\circ C} F_3CS{-}SCF_3$$

Scheme 88

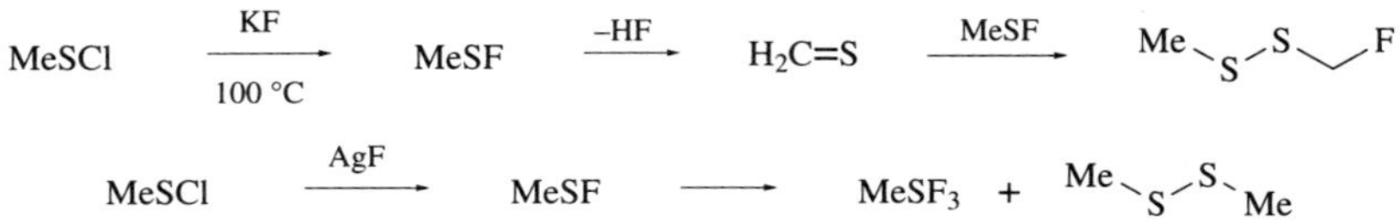

Scheme 89

Other examples of sulfenyl fluoride formation by halogen exchange are illustrated by preparations of dichloro(difluoroamino)methanesulfenyl fluoride and chlorodifluoromethanesulfenyl chloride, where two alternative fluoride sources are used (Scheme 90) ⟨71ICA311, 78ZAAC1937⟩.

A number of other methods of generation of sulfenyl fluorides are known, such as the pyrolysis of difluorosulfuranes ⟨62IC756⟩, and the photochemical addition of tetrafluorohydrazine to dihalothiocarbonyl compounds ⟨65ZOB1412, 71ICA311⟩.

$$F_2NCCl_2SCl \xrightarrow[25\ ^\circ C]{AgF_2} F_2NCCl_2SF$$

$$F_2ClCSCl \xrightarrow[HgF_2,\ 90\ ^\circ C]{BrF_3} F_2ClCSF$$

Scheme 90

2.03.3.1.2 Alkanesulfenyl chlorides

The first synthesis of a sulfenyl chloride was reported in 1870 by Rathke, and involved the chlorination of carbon disulfide in the presence of an iodine catalyst ⟨1870CB858⟩.

Since 1870, a wide range of general routes to sulfenyl chlorides have been developed, most involving the chlorination of compounds already containing sulfur functionalization. The chlorination of alkanethiols has been performed with a range of chlorinating agents, the formation of a disulfide normally being the first step. *In situ* chlorination of the disulfide intermediate then produces the sulfenyl chloride ⟨38USP2123082, 51CB911, 53JCS3219, 63JA749, 65JOC1384, 69JHC627, 80JCS(P1)2228⟩. A representative example of the chlorination of thiol substrates, in this case using sulfuryl chloride at low temperature, is shown in Equation (75) ⟨72JOC887⟩.

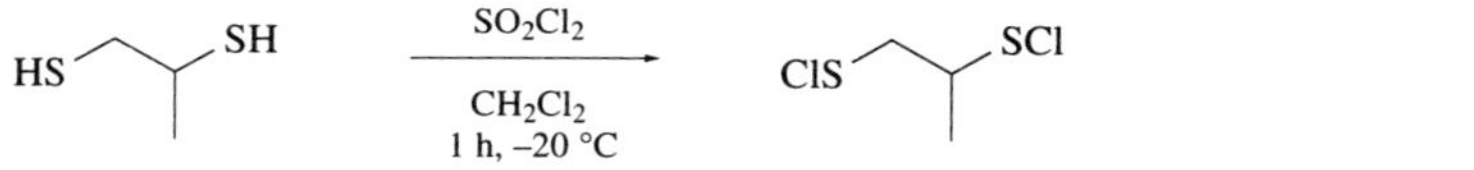

(75)

Disulfides are alternative precursors of sulfenyl chlorides. Generally, chlorine gas ⟨46JOC469, 46JOC499, 57JA6248, 59JOC973, 60CB782, 60OS(40)62, 61JOC1996, 64JCS5482, 65JOC175, 87GEP3543567⟩ or, less commonly, sulfuryl chloride ⟨55JA181⟩ is used as the chlorinating agent, usually at low temperature to increase chemoselectivity. A number of examples are shown in Scheme 91 ⟨50CB87, 50USP2530695, 53CB557, 53JCS3219, 54CB325, 60CB782, 60LA(637)146, 85ZOK1835, 86ACS609⟩.

$$RSH \xrightarrow{Cl_2} RSCl$$

R	*Yield* (%)
CF_3	72
Bu^t	100
$(CF_3)_2CClSH$	57

$$RSSR \xrightarrow{Cl_2} RSCl$$

R	*Yield* (%)
CF_3	100
$CF_3OCFClCF_2$	83
$EtO_2CCH_2CH_2$	80

Scheme 91

The chlorination of sulfides with cleavage of one carbon–sulfur bond can also provide a route, albeit very restricted, for sulfenyl chloride preparation. As with disulfides, overchlorination of sulfides can occur with some reagents, although the chlorination of cycloalkyl sulfides has been successfully employed for the preparation of ω-chloroalkanesulfenyl chlorides (Scheme 92) ⟨53JA243, 64ZOB1948⟩. Chlorinating agents have also been employed to effect ring opening of cyclic-substituted sulfides to furnish alkanesulfenyl chloride products. Scheme 93 illustrates such reactions of thiiranes using *t*-butyl hypochlorite ⟨74JA3146⟩, thietanes ⟨66DOK(170)96, 68IZV1889⟩, and thiazoles ⟨71JA6267, 76JOC2561⟩.

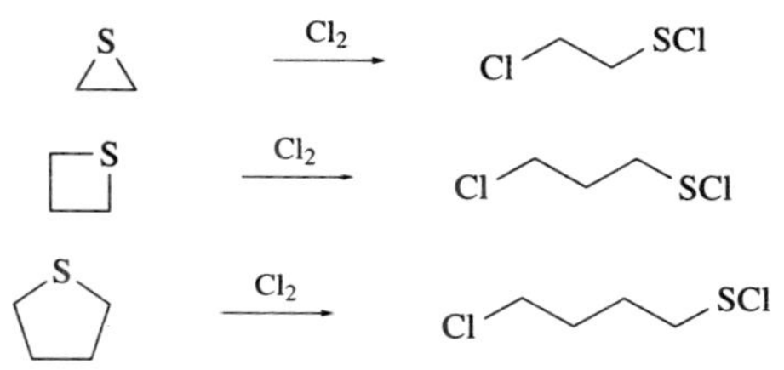

Scheme 92

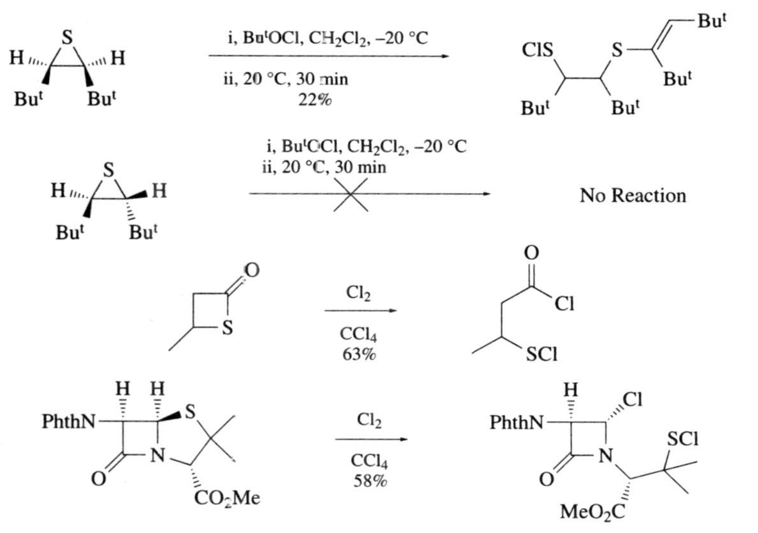

Scheme 93

Sulfenyl chlorides can result from the chlorination of dithioacetals with cleavage of one carbon–sulfur bond using chlorine or sulfuryl chloride, although this method, illustrated in Scheme 94, is unreliable. For example, upon chlorination, formaldehyde dimethyl dithioacetal gives 90% of the desired sulfenyl chloride plus a small amount of sulfide by-product. 1,3,5-Trithiane can be converted into a dichlorosulfenyl chloride in 85% yield if a large excess of the chlorinating agent is used ⟨49JOC272, 50JOC795, 61CB1682⟩.

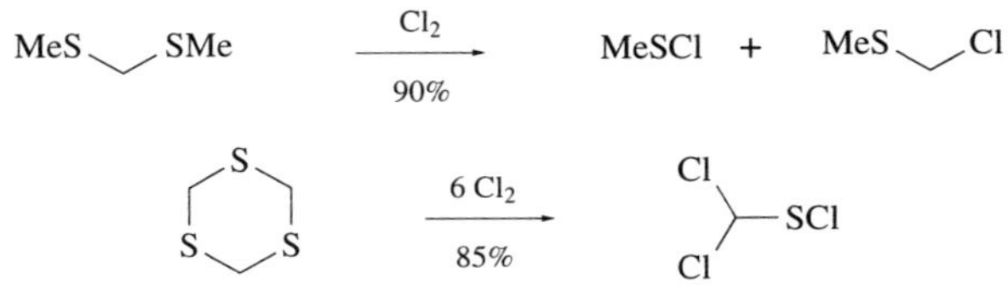

Scheme 94

Benzyl sulfides such as 1-benzylthioperfluorocyclobutene and benzyl trityl sulfide have been shown to undergo chlorination with standard chlorinating agents to furnish sulfenyl chlorides. Chlorination of 1-benzylthioperfluorocyclobutene results in the expulsion of benzyl chloride, whereas upon chlorination of benzyl trityl sulfide, benzylsulfenyl chloride and trityl chloride are produced (Scheme 95) ⟨61JOC2478, 82IZV1123⟩.

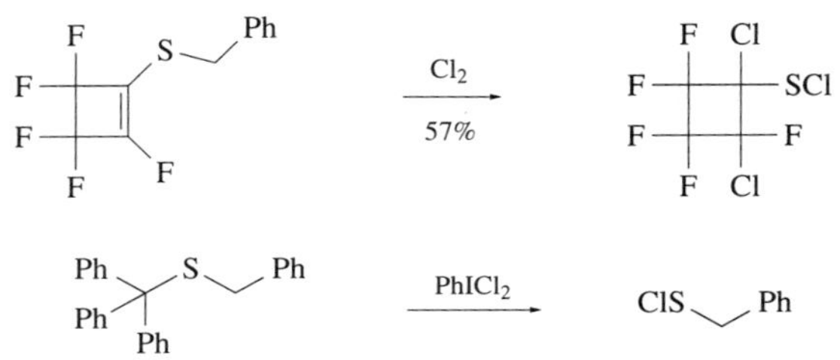

Scheme 95

Other examples of benzyl sulfide chlorination include that of the β-thioalkoxyamide shown in Equation (76) with sulfuryl chloride ⟨71JOC3145⟩.

Sulfides located α to carboxylic acid derivatives also undergo facile chlorination at sulfur ⟨74JA5508, 84JMC1379⟩. In the example shown (Equation (77)), *t*-butyl hypochlorite was used as the chlorinating agent.

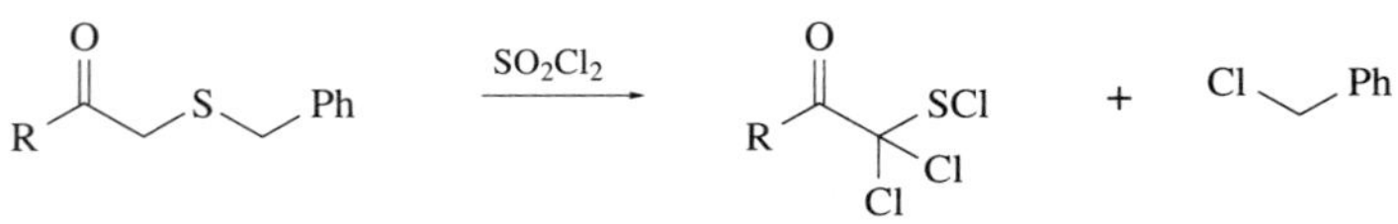

(76)

R	*Yield* (%)
Me_2N	37
$Pr^i{}_2N$	65

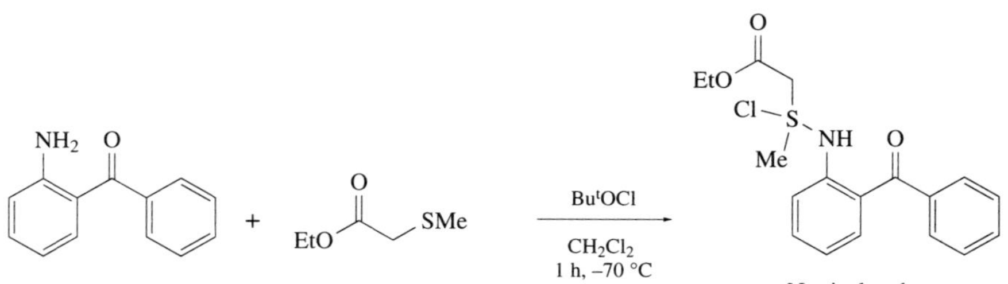

(77)

Not isolated

Thiolesters also undergo chlorinolysis with chlorine to give sulfenyl chlorides ⟨53JA4582⟩. Thea and Cavasco have found that sulfuryl chloride, when used in equimolar amounts, reacts with thiolesters smoothly and cleanly to form sulfenyl halides ⟨88TL2865⟩. Other suitable substrates include thiolactones and bisthiolesters, which furnish the bis-sulfenyl chlorides (Scheme 96) ⟨63JOC1901, 65CB1455, 69JHC627, 70ZOB1692⟩.

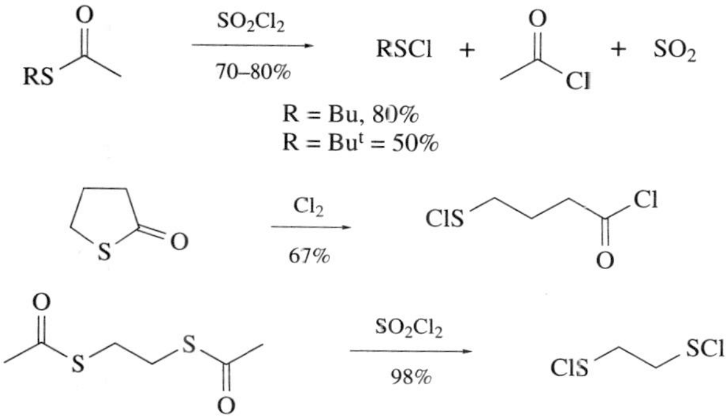

Scheme 96

Thiocarbonyl compounds are perhaps obvious substrates for the preparation of sulfenyl chlorides by reductive chlorination. Equation (78) illustrates the first known example of such a conversion; the trichloromethanesulfenyl chloride product is the precursor of a range of other sulfenyl chlorides, such as trifluoromethane sulfenyl chloride, prepared by further reactions as shown in Equation (78a) ⟨60JOC2016⟩.

$$CS_2 \xrightarrow[I_2 \text{ (cat.)}]{Cl_2} Cl_3CSCl + S_2Cl_2 \quad (78)$$

$$Cl_3SCl \xrightarrow[47\%]{NaF} F_3CSCl \quad (78a)$$

Thiophosgene ⟨1887CB2376⟩, thiocarbonyl difluoride ⟨62JCS4361, 64CB2614⟩, and thiocarbonyl fluoride chloride all readily undergo chlorine addition to furnish sulfenyl chlorides. Thiocarbonyl isothiocyanates have also been shown to undergo chlorination in a selective manner, with the chlorination occurring only at the thiocarbonyl functionality (Scheme 97). No chlorination occurs at the isothiocyanate functional group in this case.

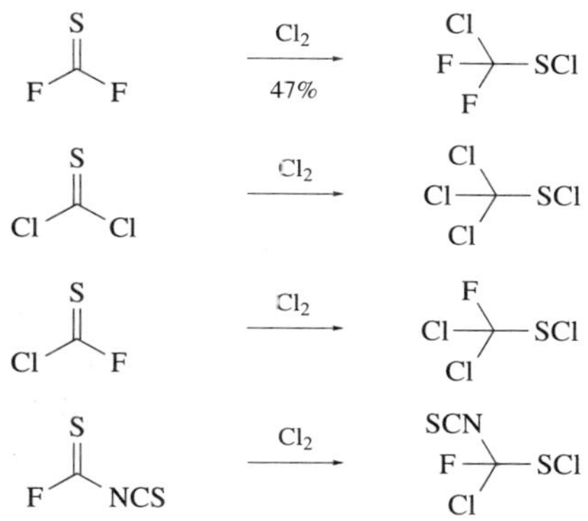

Scheme 97

Other sulfenic acid derivatives, such as sulfenic esters and sulfenamides, can also be converted into sulfenyl halides by thermal rearrangements or similar processes ⟨68JCS(C)2713⟩. Such transformations are not however of general utility.

Other thiocarbonyl compounds which undergo chlorination to give sulfenyl halides include dithiocarboxylates, dithiocarbonates, and dithiothiocarbonyl chlorides ⟨86ACS609⟩. Selected examples are shown in Scheme 98. It should be noted that of all thiocarbonyl compounds, thioketones cannot usually be converted into sulfenyl halides due to the occurrence of a more favorable desulfurization reaction. One exception to this rule is hexafluorothioacetone, which can be converted into sulfenyl chlorides and even bromides ⟨65JOC1384⟩.

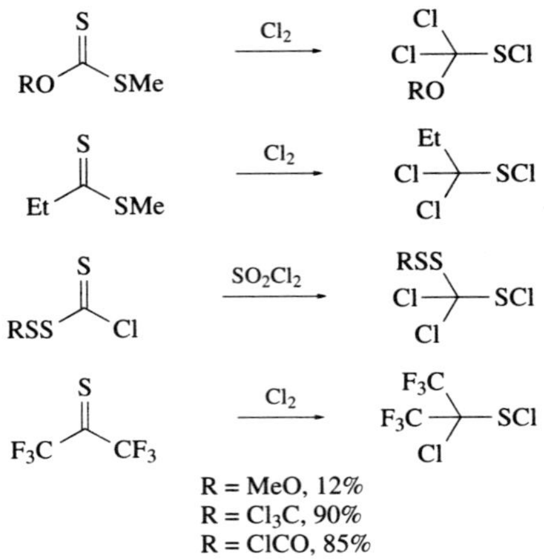

Scheme 98

Sulfur dichloride has proved to be an extremely useful reagent for sulfenyl chloride preparation. It adds to alkenes to furnish sulfenyl chlorides directly, and can even add to alkanes, including cyclohexane and simple acyclic hydrocarbons, under UV irradiation. Sulfur dichloride is compatible with a variety of functional groups including fluorides, and carbonyl and phosphoryl groups (Scheme 99) ⟨22JCS594, 63CB3050, 64CB2614, 65USP3454631, 66HCA2344, 67CB1482, 80EGP139258, 82JPR631, 88IZV2367⟩.

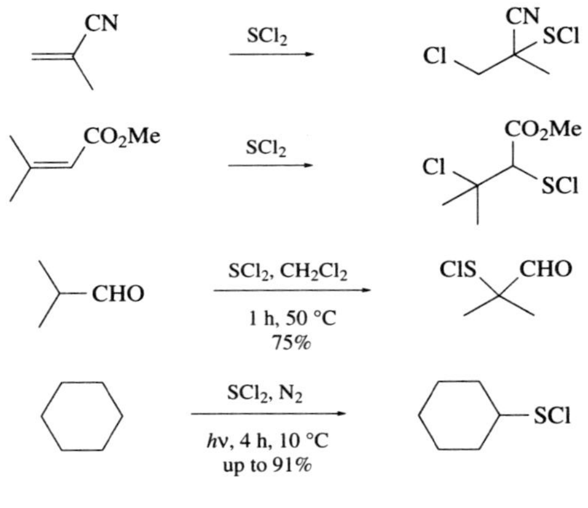

Scheme 99

A number of other methods of sulfenyl chloride preparation may be found in the literature. One example involves the photolysis of thiophosgene and tetrafluorohydrazine to furnish dichloro(difluoroamino)methanesulfenyl chloride in ca. 30% yield, along with many by-products (Equation (79)) ⟨65ZOB1412, 71ICA311⟩.

Finally, the reaction of a ketone, benzylic acid, or even sulfinyl chloride with thionyl chloride in the presence of pyridine can furnish sulfenyl chlorides ⟨67JA5838, 68TL5149, 80SC245⟩.

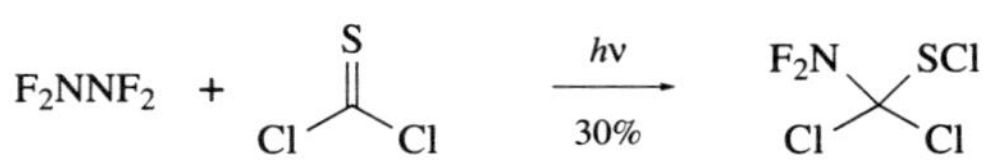

(79)

2.03.3.1.3 *Alkanesulfenyl bromides*

Alkanesulfenyl bromides are considerably less stable than are alkanesulfenyl chlorides, and, although they share a number of common methods of preparation, examples of isolated and fully characterized sulfenyl bromides are relatively few. Most methods of preparation of alkanesulfenyl bromides are analogous to those described above for alkanesulfenyl chlorides, and several are illustrated in Scheme 100 ⟨29JCS1048, 51CB911, 59ZOB3792, 63JOC2992⟩, together with the transformation of sulfenyl chlorides into sulfenyl bromides by halogen exchange ⟨55USP2821554, 70JOC2006⟩.

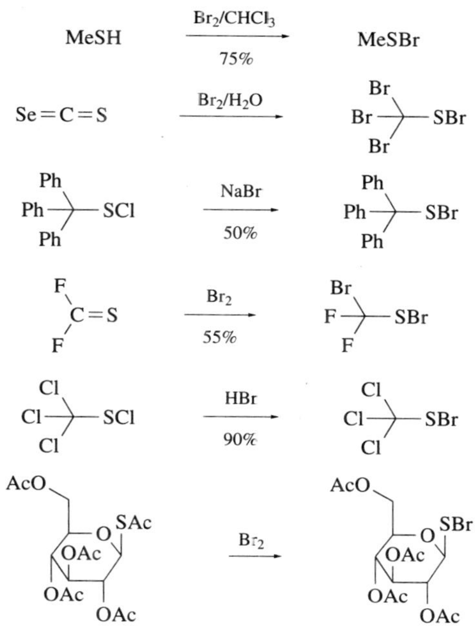

Scheme 100

The bromination of thiocarbonyl compounds such as dithioesters and thioketones can lead directly to α-bromosulfenyl bromides or α,α-dibromosulfenyl bromides respectively (Scheme 101) ⟨29JCS1048, 59ZOB3792, 69CAR(9)201, 81JOC289⟩. Bromination of dithioesters has also been used as a route to alkanesulfenyl bromides.

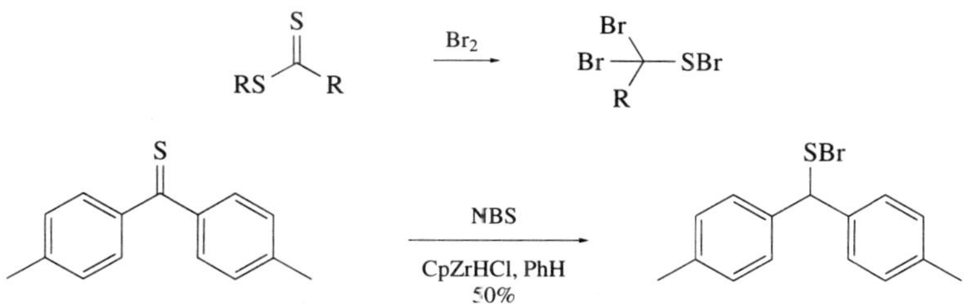

Scheme 101

Direct bromination of sulfides with bromine also generates sulfur–bromine bonds (Equation (80)) ⟨79LA784⟩. Treatment of the sulfoxide-substituted amino acid shown in Equation (81) with hydrobromic acid led to the bromosulfonium bromide in 88% yield ⟨61HCA61⟩.

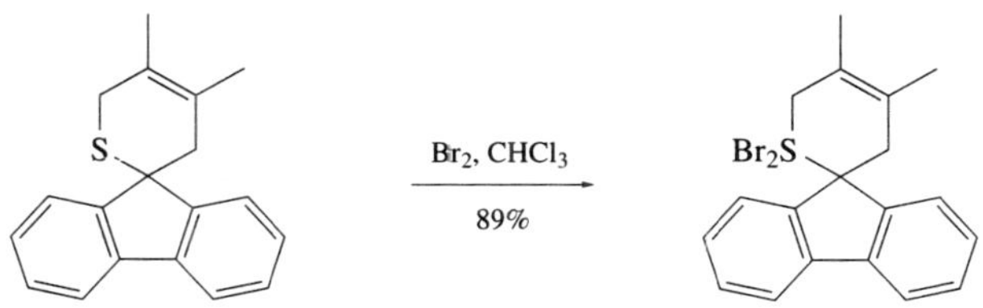

(80)

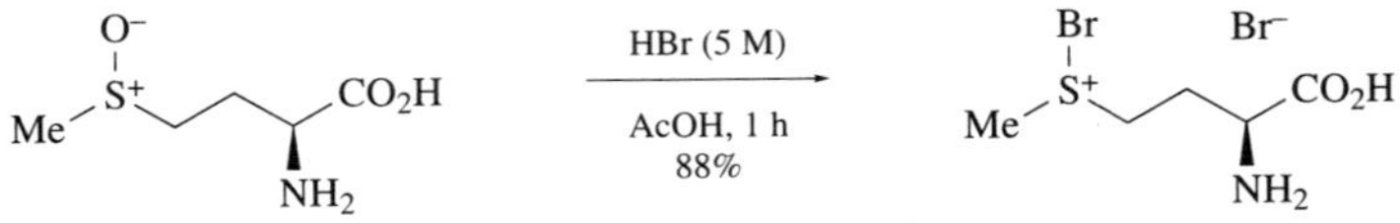

(81)

2.03.3.1.4 *Alkanesulfenyl iodides*

Alkanesulfenyl iodides are the most unstable of the alkanesulfenyl halides, and are often only proposed as reactive intermediates. Isolation of sulfenyl iodides is a particularly rare occurrence ⟨49MI 203-01, 53JCS3219, B-61MI 203-01, 62JCS2172, 63JCS2636, 65JOC933⟩. Sulfenyl iodides first appeared in the literature in 1939, with the preparation of *t*-butanesulfenyl iodide, by treatment of the mercury salt of *t*-butyl thiol with iodine ⟨39CB657⟩. Since 1939, however, mercuric compounds have been replaced by silver(I) salts. Triphenylmethanesulfenyl iodide has been prepared by treatment of the corresponding sulfenyl chloride with sodium iodide in acetonitrile. It could be kept for several hours at dry ice temperature under a nitrogen atmosphere in sealed vials and was used in a study of the effect of leaving group on the reaction between triphenylmethanesulfenyl derivatives and butylamine ⟨70JOC2006⟩. The existence of sulfenyl iodides as reactive intermediates has been proved by trapping experiments, for example with *p*-chlorobenzene thiol (Scheme 102) ⟨71JOC2530⟩. The mixed disulfide was obtained in 36% yield. Sulfenyl iodides are important in protein iodination ⟨70JOC3267⟩. Perhaps partly as a result, the synthesis of sulfenyl iodides has become an area of considerable interest, and in 1973 Field and White achieved a landmark with the preparation of the first stable crystalline sulfenyl iodide (Equation (82)) ⟨73PNA(70)328⟩. In direct contrast to triphenylmethanesulfenyl iodide, this sulfenyl iodide is more stable in the solid phase than in solution.

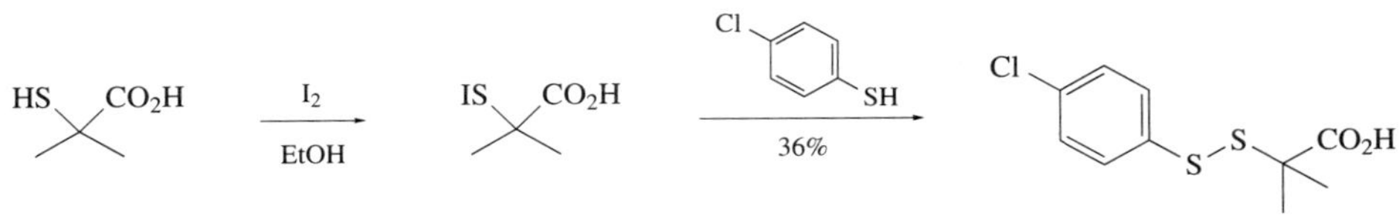

Scheme 102

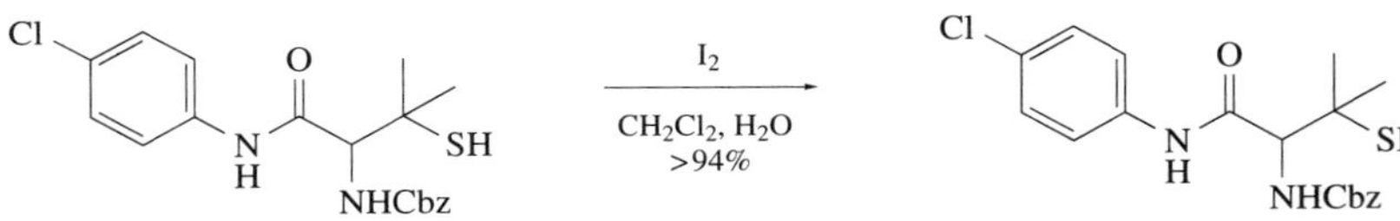

(82)

2.03.3.2 Alkanesulfinyl Halides

Of the possible alkanesulfinyl halides, the literature describes only the preparation and isolation of alkanesulfinyl chlorides and fluorides ⟨55HOU(9)343, 55HOU(9)547, 71IJS(B)277, B-90MI 203-05⟩. Examples of alkanesulfinyl fluoride preparation are relatively few in number, although many sulfinyl chlorides are known. Care must be exercised when handling sulfinyl chlorides due to their explosive nature ⟨68JOC2104⟩. Care must also be taken in their storage: alkanesulfinyl chlorides decompose readily with disproportionation at room temperature upon standing ⟨64JOC951⟩ with the formation of several by-products, including hydrogen chloride, and the development of exceedingly high pressure ⟨51CB911⟩. Alkanesulfinyl chlorides are also extremely moisture sensitive.

The first reported synthesis of a sulfinyl chloride appeared in 1908, by the reaction of thionyl chloride with a sulfinic acid (Equation (83)) ⟨08CB4115⟩. Reactions of sulfinic acids or their salts with an excess of thionyl chloride generally furnish sulfinyl chlorides in moderate to good yields ⟨10JCS2585, 23CB553, 30CB2856, 36CB937, 53JCS549⟩. For example, perfluoromethanesulfinyl chloride and perfluorobutanesulfinyl chloride can be obtained from the corresponding perfluoroalkanesulfinic acids by this method. Silylated derivatives of alkanesulfinic acids react with thionyl chloride at room temperature to furnish sulfinyl chlorides in high yields (Equation (84)) ⟨80ZOB1059⟩. Interestingly, in the reaction shown, the sensitive silyl carboxylate functionality was unaffected. Phosphorus trichloride or phosphorus pentachloride can be used as alternatives to thionyl chloride ⟨74CB508⟩. This method of synthesis is, however, of limited synthetic value, with the disadvantages of poor reproducibility of results, poor availability of the desired sulfinic acid starting material, and difficult product isolation ⟨35JA2172⟩.

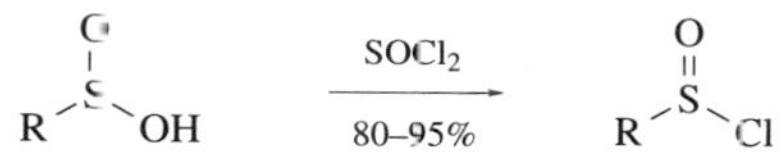

(83)

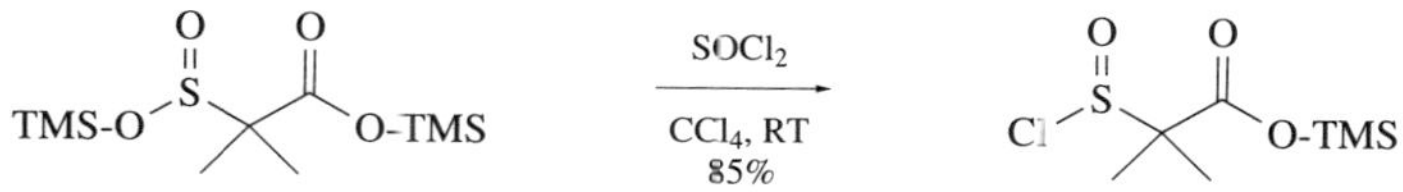

(84)

Perhaps the most useful method of sulfinyl chloride preparation is the oxidative chlorination of thiols, disulfides, and thiolesters. The most conveniently handled chlorinating agent used for such reactions is sulfuryl chloride. Herrmann has demonstrated that, under carefully controlled low temperature reaction conditions, clean and essentially quantitative thiol to sulfinyl chloride transformation can be achieved provided that acid-sensitive functional groups are suitably protected (Equation (85)) ⟨87S72⟩. One interesting example is the synthesis of 2,2,2-trifluoro-1,1-diphenylethanesulfinyl chloride by the treatment of the corresponding thiol with sulfuryl chloride followed by exactly one equivalent of trifluoroperacetic acid (Scheme 103) ⟨87S683⟩. Disulfides also react with sulfuryl chloride in the presence of acetic acid to form the corresponding sulfinyl chlorides in nearly quantitative yields ⟨86TL1493⟩. This reaction is particularly useful in the preparation of chiral sulfoxides by the Andersen procedure. The use of acetic anhydride as the solvent for chlorination prevents the formation of gaseous products such as hydrogen chloride ⟨68JOC2104⟩. Chlorine gas is also widely used for the oxidative chlorination of disulfides, usually in the presence of glacial acetic acid (Equation (86)) ⟨59JOC973, 60OS(40)62, 61JOC1996, 65JOC175⟩.

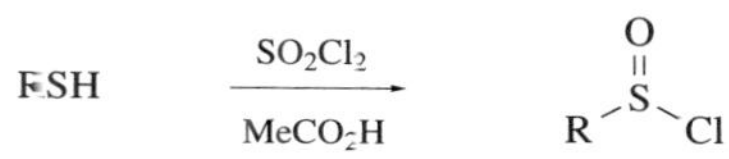

(85)

R	*Yield (%)*
Et	95
Pr^i	95
Bu^s	95

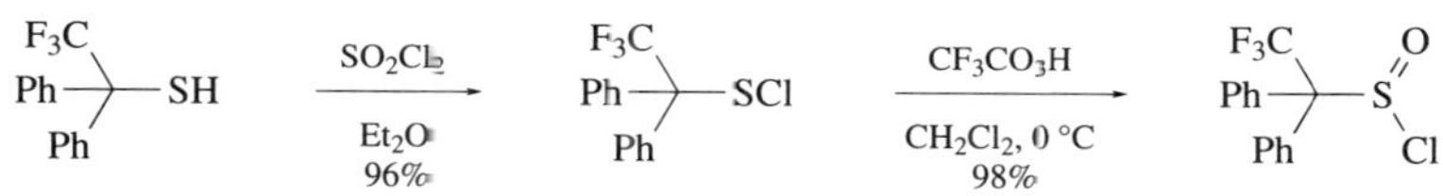

Scheme 103

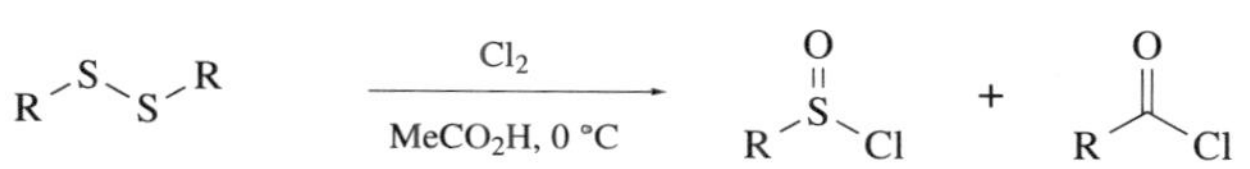

(86)

R	*Yield (%)*
Pr^i	81
C_5H_{11}	90

Thiolesters also react well with sulfuryl chloride to furnish the corresponding sulfinyl chlorides ⟨87TL5193⟩. The reaction is performed in the presence of acetic anhydride (Equation (87)). It is noteworthy however that *t*-butylsulfinyl chloride cannot be prepared by the treatment of *t*-butyl thiolacetate with sulfuryl chloride. *t*-Butylsulfinyl chloride has proved particularly difficult to prepare, although it is furnished in excellent yield by reaction of the disulfide with sulfuryl chloride and acetic acid ⟨86TL1493⟩, and alternatively can be synthesized by the reaction of thionyl chloride with *t*-butanesulfinic acid ⟨61JCS5339⟩ or by the chlorination of the oxidation product of di(*t*-butyl)disulfide ⟨74JA3921⟩ (Scheme 104). Thiolacetates also undergo chlorination efficiently with chlorine in dichloromethane solution (Equation (88)) ⟨81IZV159⟩.

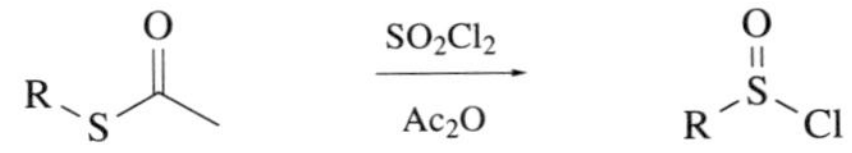

(87)

R	*Yield* (%)
Bu	86
Bu^t	0
$Me(CH_2)_{11}$	97

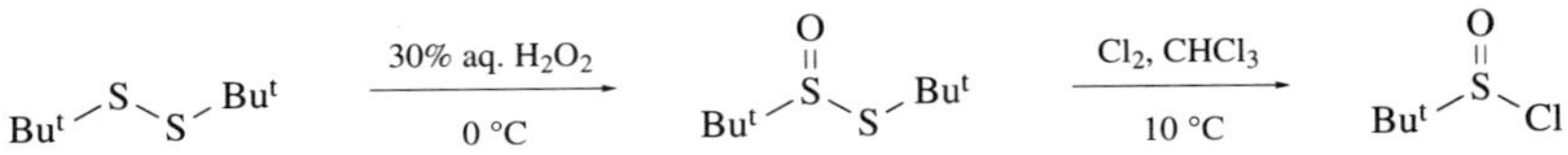

Scheme 104

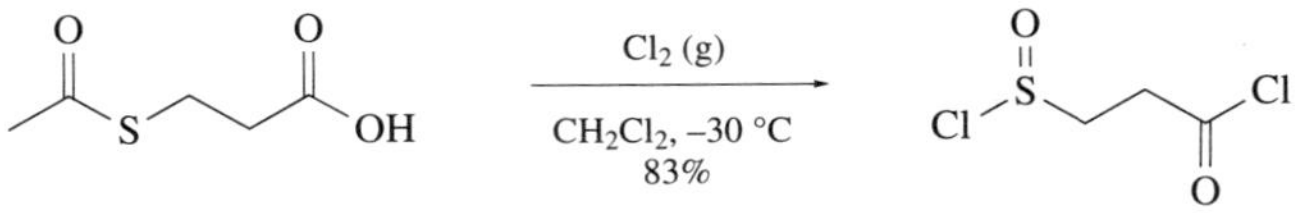

(88)

The reaction of thiolesters, easily prepared from alcohols, with sulfuryl chloride is valuable as a synthetic route for the preparation of optically active sulfinyl chlorides, as many enantiomerically pure chiral alcohols are commercially available, in contrast to thiols and disulfides.

Sulfuryl chloride also reacts with 2-trimethylsilylethylsulfoxides to furnish sulfinyl chlorides ⟨92TL3973⟩. The trimethylsilyl group precludes oxidation α to the sulfur atom ⟨B-68MI 203-02⟩, and instead, a fragmentation occurs to furnish sulfinyl chlorides, as shown in Scheme 105. An analogous product to that shown in (Equation (88)) is furnished by sulfuryl chloride mediated ring opening of a thietane (Equation (89)).

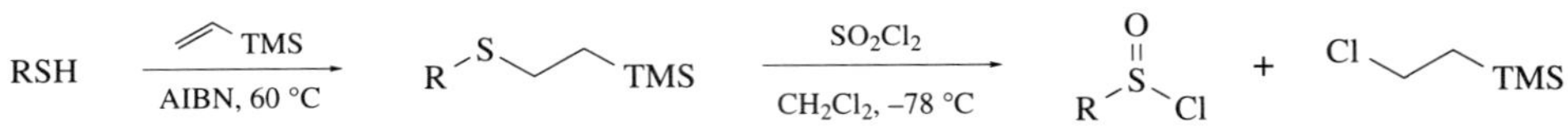

R	*Yield* (%)
Pr^i	46
$PhCH_2$	69

Scheme 105

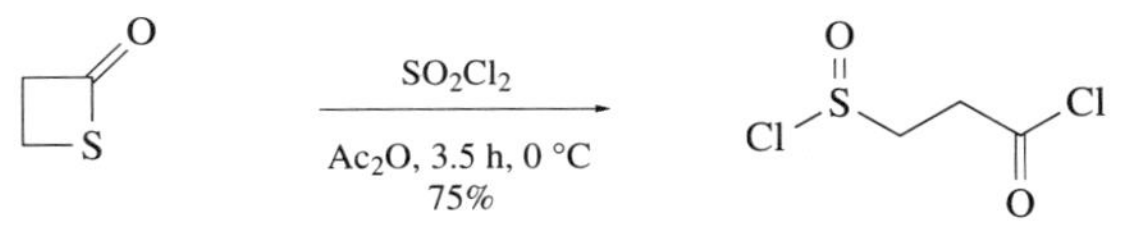

(89)

Prior to the development of sulfuryl chloride as a reagent capable of effecting the oxidative chlorination of thiols, disulfides, and the like, chlorine, which is difficult to handle and can lead to overoxidation to sulfonyl chlorides, was the reagent of choice ⟨61JOC1996⟩. Simpler preparative routes for sulfinyl chlorides were first investigated by Douglass, and involved reaction of organosulfurtrichlorides with equimolar amounts of water, alcohols, or organic acids to furnish the desired sulfinyl chlorides in good yields (Equation (90)) ⟨57JOC536⟩. The very reactive nature of the organosulfur trichloride is however a clear drawback to the process, but can be circumvented by *in situ* generation of the organosulfur trichloride via the sulfenyl chloride, by treatment of the disulfide or thiol with chlorine in the presence of two moles of carboxylic acid without additional solvent (Scheme 106) ⟨58JOC330, 61JOC1996, 68JOC2104⟩. Yields are of the order of 90%. A further improvement was made by use of thiolesters rather than disulfides or thiols as starting materials ⟨70OPP235⟩.

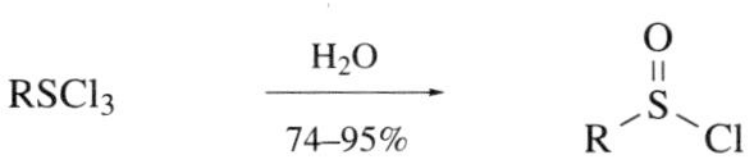

(90)

Thionyl chloride has been shown to react with adamantane to furnish a sulfinyl chloride in the presence of aluminum trichloride (Equation (91)) ⟨68AG970, 69CB3357⟩. In the absence of thionyl chloride, only alkyl chlorides were produced. Hydrocarbons have also been shown to combine with sulfur dioxide and chlorine under irradiation to give sulfinyl chlorides.

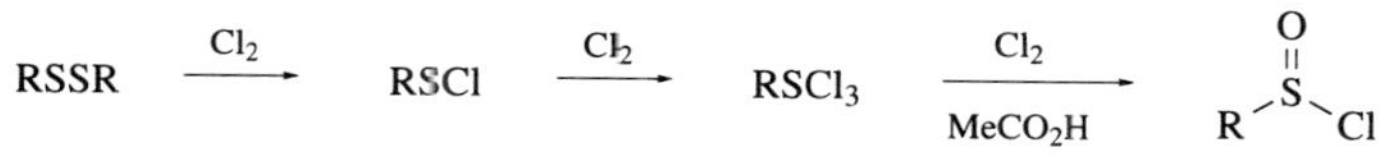

R	*Yield* (%)
Me	92
Pr^i	81
Ph	90

Scheme 106

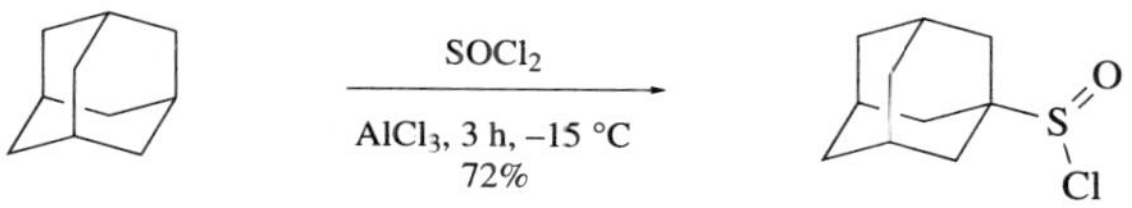

(91)

As indicated above, the preparation and reactions of some sulfinyl fluorides have been reported ⟨74IC2710, 75CL307⟩, including trichloromethylsulfinyl fluoride, given by the hydrolysis of trifluoromethyl sulfur trifluoride, itself resulting from a reaction involving bis(trifluoromethyl)sulfide and silver(II) fluoride ⟨68IC391⟩. Perfluoropropene and sulfur dioxide react under UV irradiation to give good yields of 1 : 1 addition products, the α-fluorosulfinylacyl fluorides ⟨70TL1313⟩. Scheme 107 shows the proposed reaction pathway, proceeding via a cyclic adduct. The sulfinyl fluorides were produced in 80–90% yields.

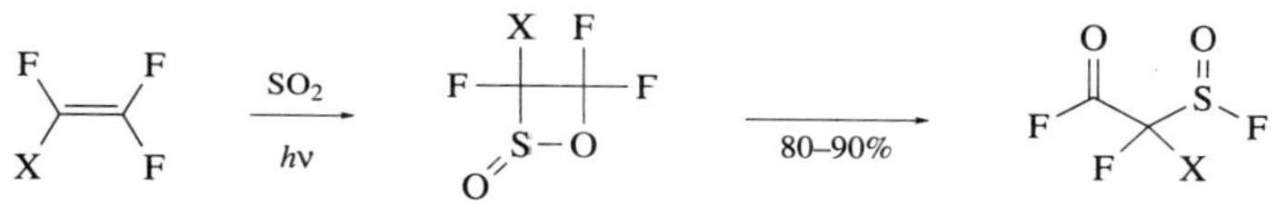

X = F, CF_3

Scheme 107

2.03.3.3 Alkanesulfonyl Halides

Of the alkanesulfonyl halides, the chlorides are by far the most common, although sulfonyl fluorides, bromides, and iodides are well known and characterized in the literature. In common with alkanesulfenyl and -sulfinyl halides, of the sulfonyl halides known, the chlorides and fluorides are the most stable, the bromides and iodides being much less so. The majority of methods of sulfonyl halide synthesis have been known for many years, and have been thoroughly reviewed ⟨55HOU(9)343, 55HOU(9)407, 55HOU(9)547, B-65MI 203-02, 69S3, B-75MI 203-01, B-91MI 203-01⟩.

The principal methods of sulfonyl halide preparation include formation of a carbon–sulfur(VI) bond by the use of a suitable source of sulfur, the oxidation of lower order sulfur containing moieties, and preparation from other sulfur(VI) functionalities.

2.03.3.3.1 Carbon–sulfur(VI) bond formation

Sulfonyl halides can be formed by carbon–sulfur(VI) bond formation employing sulfur dioxide, sulfuryl halides, or halosulfonic acids as the source of sulfur. Sulfur dioxide is normally used in conjunction with chlorine: the electrophilic chlorine atom abstracts a hydrogen atom from the site of highest electron density, thereby initiating a sulfochlorination chain mechanism, shown in Equation (92) ⟨42CB34, 55JOC1165, 58CB2130, 58JOC363, 63CB3068, 76BSF483⟩. This method of introduction of the chlorosulfonyl group by sulfur dioxide and chlorine is used as an industrial process, but is however of limited laboratory use as a mixture of isomeric products is usually formed, and reaction conditions must be strictly monitored in order to obtain the preferred product. Tertiary carbon atoms are not usually sulfochlorinated under these conditions. Steric factors can also be important ⟨58CB2130⟩.

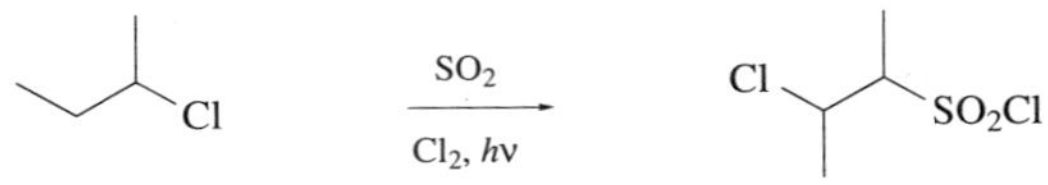

(92)

Sulfuryl halides are used to generate sulfonyl halides from alkenes or alkyl halides. Once again, the reaction proceeds through a radical mechanism, and usually a base is present to scavenge the HCl produced in the reaction. Sulfuryl fluoride ⟨68JOC344⟩ and sulfuryl chloride ⟨48JA3526, 68JCS(C)992⟩ have both been used to furnish sulfonyl halides from alkene starting materials. Thiolacetic acid together with chlorine has also been used to prepare sulfonyl chlorides from alkenes ⟨57HCA2148, 57JOC980, 63JOC3537⟩. In the example shown in Equation (93), the sulfonyl chloride was isolated from the aqueous reaction mixture in 60% yield.

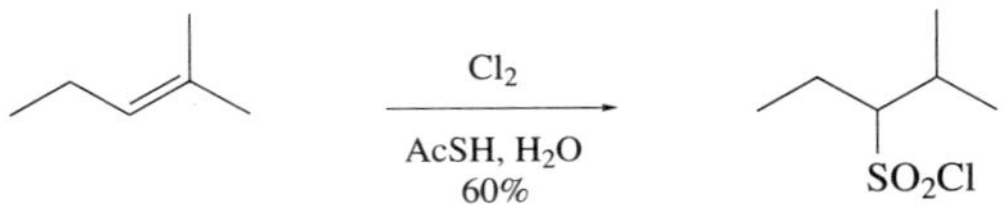

(93)

Scheme 108 illustrates the reactions of sulfur dioxide with various organometallic reagents to give sulfinic acid salts, which can be directly treated with chlorine or bromine, or with sulfuryl chloride, to give the corresponding sulfonyl chlorides ⟨69S3⟩. Grignard reagents react particularly well with sulfuryl chloride to give alkanesulfonyl chlorides in good yields, as exemplified by the benzyl and *t*-butyl derivatives shown in Scheme 108 ⟨23HCA249, 68JCS(C)1265⟩. Alkyllithium reagents also react well with sulfuryl chloride at low temperature to form sulfonyl halides ⟨74S489⟩.

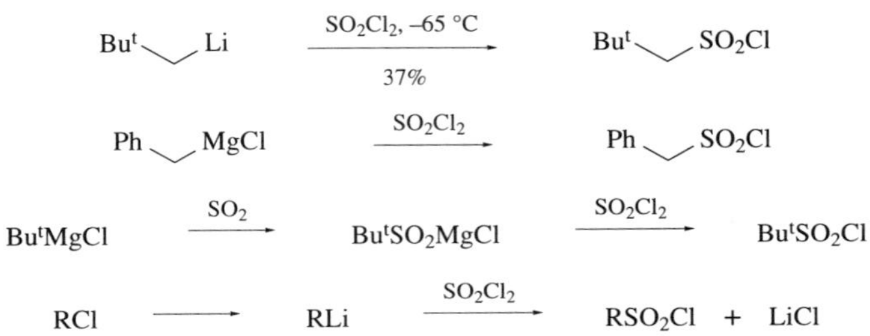

R	*Yield* (%)
Bu	61
Bu^s	52
Bu^i	48

Scheme 108

Alkyl halides are also readily chlorosulfonylated by sodium disulfite (Na_2SO_3) in the presence of phosphorus pentachloride ⟨55CB201, 57RTC129, 74S876, 87JOC2162⟩ or chlorine ⟨51JCS3292, 51JOC621, 56JA5580, 59JCS2969⟩. Thiourea in conjunction with chlorine in aqueous solution is an effective system for alkanesulfonyl chloride formation from alkyl halides (Scheme 109) ⟨62AP(295)615⟩.

Halosulfonic acids have been used to furnish alkanesulfonyl chlorides by reaction with a suitable aliphatic compound. The yields tend to be poor and the reaction is infrequently used ⟨75JA1492⟩. One exception is the use of chlorosulfonic acid in the presence of phosphorus oxytrichloride, which has been found to furnish geminal bis-sulfonyl chlorides in excellent yields upon reaction with the corresponding aliphatic carboxylic acids (Equation (94)) ⟨76CZ391⟩.

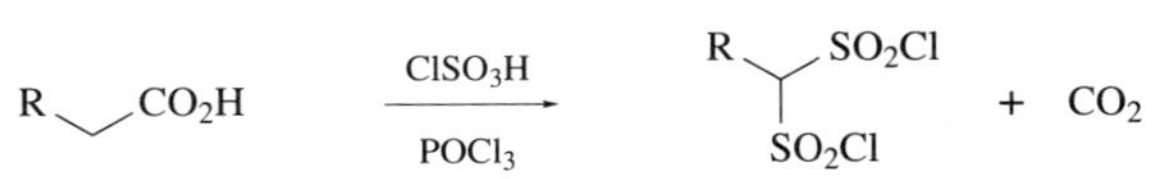

(94)

Sulfonyl halide formation by carbon–sulfur(VI) bond formation is not limited to those methods described above. Other methods include the reaction of alkanes with sulfur trioxide and chlorine ⟨76JPR1019⟩ and of carboxylic acids with sulfur trioxide and phosphorus oxychloride ⟨73BSF210⟩, and the photochemical reaction of alkylcobaloximes with trichloromethanesulfonyl chloride to give high yields of alkanesulfonyl chlorides in a reaction which tolerates a wide range of functionality (Scheme 110) ⟨82JCS(P1)849⟩.

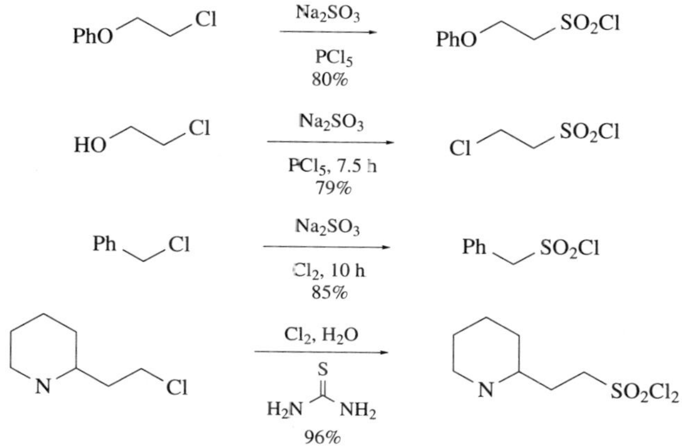

Scheme 109

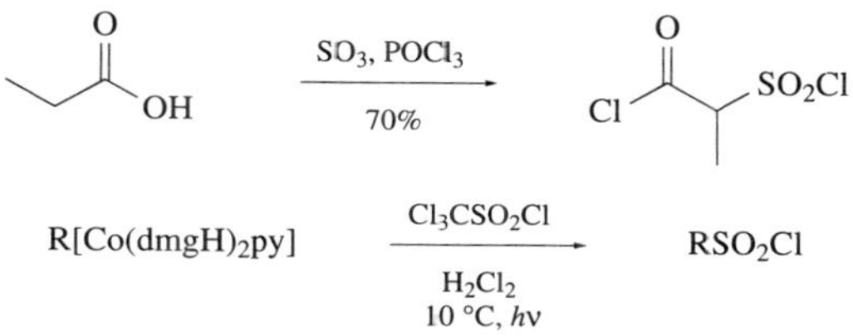

R	*Yield (%)*
Bu	95
$PhCH_2CH_2$	75
$I(CH_2)_3$	24
$NC(CH_2)_3$	26

Scheme 110

2.03.3.3.2 Oxidation of sulfur-containing moieties

Sulfonyl halides can be prepared by the oxidation of diverse sulfur-containing compounds. Suitable substrates for such oxidation include disulfides, thiols, sulfinyl and sulfenyl derivatives, sulfides, sulfoxides and sulfones, xanthates, thiosulfates, isothiocyanates, and isothiouronium salts.

Thiols, disulfides, and other thiol derivatives may be oxidized to the corresponding sulfonyl halides by the use of molecular halogen in the presence of either water, ⟨38JA1486, 42JOC81, 65BSB21⟩, acetic acid ⟨44ZPC(281)156, 50JA4890, 51JOC621⟩, or nitric acid (Scheme 111). Of the thiol derivatives, sodium alkylthiosulfates ⟨51JCS3292, 51JOC621, 56JA5580, 59JCS2969⟩ and alkylisothiouronium salts are the most suitable for oxidation by direct chlorination in aqueous media. This procedure has however been reported to be dangerous ⟨41JA3530⟩. Even so, treatment of isothiouronium salts with aqueous chlorine is a method of sulfonyl chloride preparation favoured by many chemists (Scheme 112) ⟨37JA2439, 62AP(295)615, 80S285⟩. Other suitable chlorinating agents include chlorine in the presence of hydrochloric acid and potassium acetate ⟨52JCS3334⟩, and silica gel treated with hydrochloric acid together with iodosobenzene ⟨92CL891⟩. Aqueous chlorination has been used in the preparation of trimethylsilylmethanesulfonyl chloride, which is used to prepare sulfene, as shown in Scheme 113 ⟨82TL4203⟩. One further interesting reaction allows a one-pot preparation of dichloromethanesulfonyl chloride in 45% yield, as shown in Scheme 114 ⟨73ACS1451⟩. Similarly, alkanesulfonyl fluorides (Equation (95)) ⟨74S887⟩ and alkanesulfonyl bromides ⟨48EC1129⟩ can be synthesized from alkanethiols.

$$RSH \xrightarrow[NO_2,\ 30\text{–}85\%]{HF} RSO_2F + 3NO + H_2O \qquad (95)$$

The action of chlorine on thiocyanates has also been shown to furnish alkanesulfonyl halides in good yields ⟨39JA2548⟩. These reactions are generally performed in an aqueous suspension at low

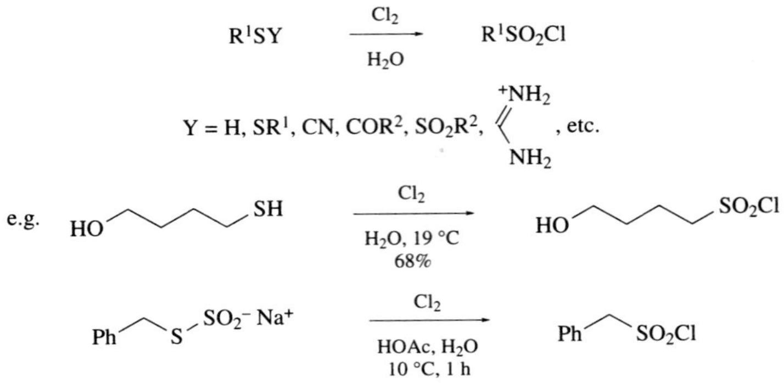

Scheme 111

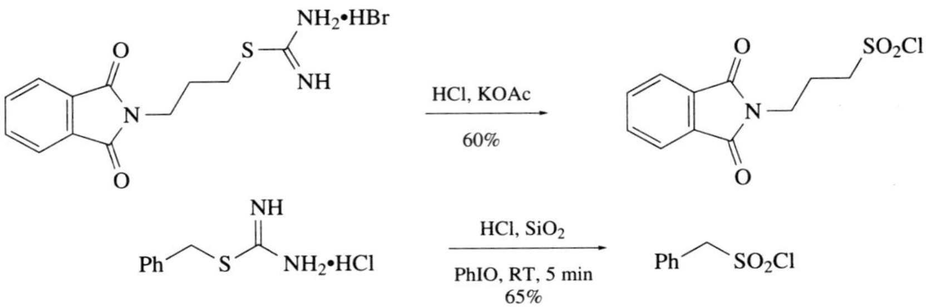

Scheme 112

Scheme 113

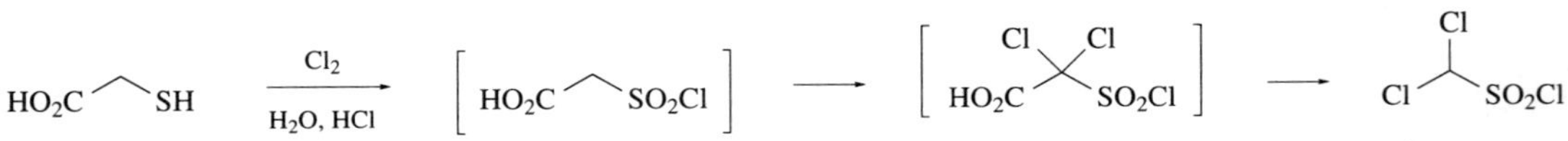

Scheme 114

temperature (0–5 °C), one exception being benzyl sulfonyl chloride formation, which is achieved at higher temperature, lower temperature leading to benzyl sulfinic acid (Equation (96)) ⟨39JA2548, 42CB344, 48JGU729⟩.

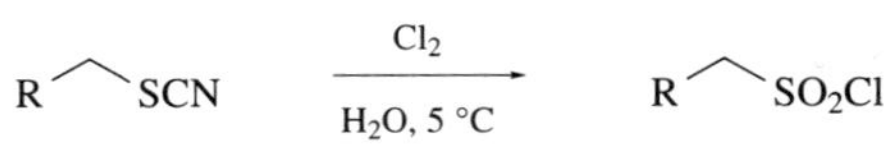

(96)

R	*Yield* (%)
Me	79
Pr^i	91

Oxidation of sulfenyl halides can be readily accomplished in good yields by a range of oxidants, including concentrated nitric acid ⟨41JA1764⟩, aqueous chlorine ⟨42JOC81⟩, and nitrogen dioxide/oxygen. Oxidation of alkanesulfinyl derivatives occurs just as readily and is better documented in the literature. Once again, water or acetic acid is the preferred reaction medium. Copper halides have also been used for the conversion ⟨77CJC407⟩. Methyl methanesulfinate has been similarly oxidized with bromine to furnish methanesulfonyl bromide in 75% yield ⟨65JOC633⟩, and alkanesulfonyl iodides have been prepared by oxidation of thiol sulfinic acids and salts of sulfinic acids (Equation (97)) ⟨64JOC918⟩.

$$RSO_2Na \xrightarrow[\text{or } I_2/KI]{I_2} RSO_2I \qquad (97)$$

The oxidation of sulfinyl derivatives can also offer a route to sulfonyl halides otherwise difficult to synthesize. For example, tertiary alkanesulfonyl chlorides have proved to be synthetically somewhat inaccessible, but can be prepared by *in situ* formation of a sulfinate salt, for example from an alkyllithium or alkylmagnesium halide, followed by oxidative chlorination using chlorine ⟨55JOC1165, 66JOC2357⟩. Tertiary alkanesulfonyl bromides may be prepared in a similar manner using oxidative bromination.

Sulfides, sulfoxides, and sulfones have all been used as forerunners of sulfonyl halides. Both benzylated and halogenated sulfides have been shown to undergo a sulfohaloform reaction, also known for sulfoxides, brought about by reaction with chlorine gas ⟨42JOC81, 76CJC498, 77CJC407, 79JOC1708⟩. The course of the generalized sulfohaloform reaction is outlined in Schemes 115 and 116. Scheme 115 shows the formation of sulfonyl chlorides by the reaction of sodium benzyl sulfide with chlorofluoromethane and chlorodifluoromethane. Scheme 116 illustrates the high efficiency of the sulfohaloform reaction (yields 74–97%).

$$CH_xF_yCl \xrightarrow[\text{DMF}]{PhCH_2SNa} CH_xF_ySCH_2Ph \xrightarrow[H_2O]{Cl_2} CH_xF_ySO_2Cl$$

$x = 2, y = 1$
$x = 1, y = 2$

Scheme 115

$$RSR \xrightarrow[H_2O,\ HOAc]{Cl_2} RSO_2Cl$$

R	*Yield* (%)
Et	97
Bu	86
$PhCH_2$	74

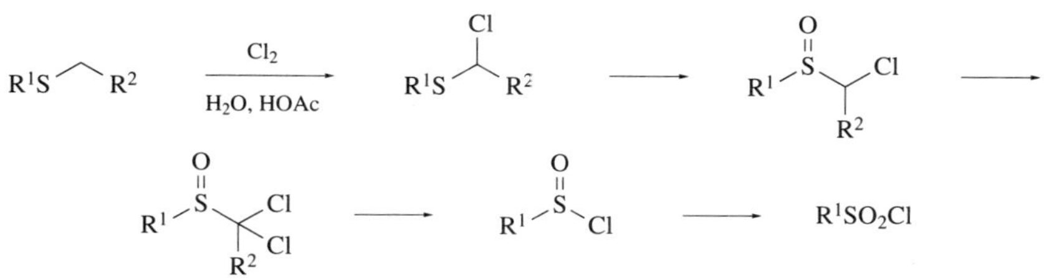

Scheme 116

One drawback to the sulfohaloform reaction is the use of chlorine gas as the chlorinating agent, which inevitably introduces difficulties in both reagent equivalence control and reaction versatility. Several modifications are known which overcome this problem, including the use of the convenient, solid chlorinating agent formed by pulverization of iodosobenzene and silica gel treated with hydrogen chloride ⟨92CL891⟩. The reagent has been found to cause both oxidation and chlorination ⟨91CL1391⟩ and appears ideal for the conversion of sulfides into sulfonyl chlorides. Further advantages are that the reaction can be performed in the absence of a solvent, and that the reaction is over within five minutes (Scheme 117). Sulfides bearing benzylic carbon–sulfur bonds were converted into the desired sulfonyl chlorides in 89–98% yield, and sulfoxides and disulfides could also be transformed in reasonable yields, as shown. Less satisfactory yields were obtained with benzylamidinium, dialkyl, and alkylaryl sulfides. Indeed diarylsulfides gave no sulfonylchloride product at all.

Sulfinyl chlorides can be conveniently synthesized from thioacetals by the action of NCS in aqueous acetic acid (Equation (98)) ⟨92S1203⟩. Chlorine in aqueous systems has also been employed for the oxidative chlorination of sulfide substrates ⟨52JA5880, 71CJC943⟩, including thiiranes, thietanes, and thiolanes (Scheme 118).

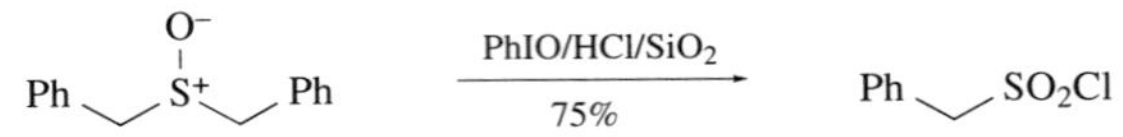

$R^1SR^2 \xrightarrow{PhIO/HCl/SiO_2} R^1SO_2Cl + R^2SO_2Cl$

R¹	*R²*	*Yield* (%)
$PhCH_2$	$PhCH_2$	98
$PhCH_2$	Ph	96
$PhCH_2$	Me	91
Bu	Bu	56
Ph	Me	4

$PhSSPh \xrightarrow[56\%]{PhIO/HCl/SiO_2} PhSO_2Cl$

Scheme 117

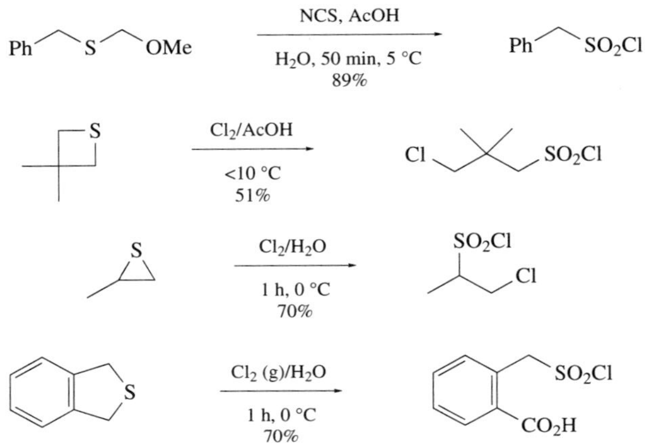

(98)

Scheme 118

Sulfoxides can also undergo the sulfohaloform reaction, for example using reagents such as halogen in aqueous acetic acid ⟨63JOC2485, 77CJC407⟩, but sulfones require more forcing conditions ranging from use of the neat halogen gas to use of the anhydrous hydrogen halide in conjunction with electrolysis. The use of hydrogen fluoride can induce overfluorination; generally, polyhalogenated sulfones undergo oxidation to the sulfonyl halides much more easily than non-halogenated sulfones.

Thiolesters are suitable substrates for oxidative chlorination, the preferred method consisting of bubbling chlorine gas through an aqueous solution of thiolester (Scheme 119) ⟨57HCA2148, 57JOC980, 58JCS2834, 58JOC1525, 81S149, 89S957⟩.

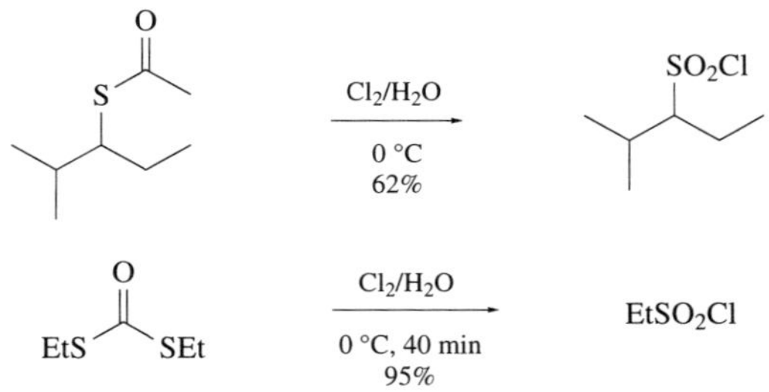

Scheme 119

2.03.3.3.3 Preparation of sulfonyl halides from other sulfur(VI) functionality

Two principal methods exist for the conversion of sulfur(VI) functional groups into alkanesulfonyl halides. These methods are the conversion of sulfonic acid derivatives to sulfonyl halides, and the halide exchange reaction ⟨1898BSF120, 57RTC129, 73OSC(5)196, 74S876⟩.

The conversion of sulfonic acid derivatives to sulfonyl halides can be effected by a broad range of reagents, the wide choice allowing many types of functionality to be tolerated within the substrate. Perhaps the most commonly used reagent is phosphorus pentachloride, which can be used alone or in conjunction with other reagents ⟨55CB201, 84TL4553⟩, but will not tolerate hydroxy, alkoxy, or amino functions (Scheme 120) ⟨74S877⟩.

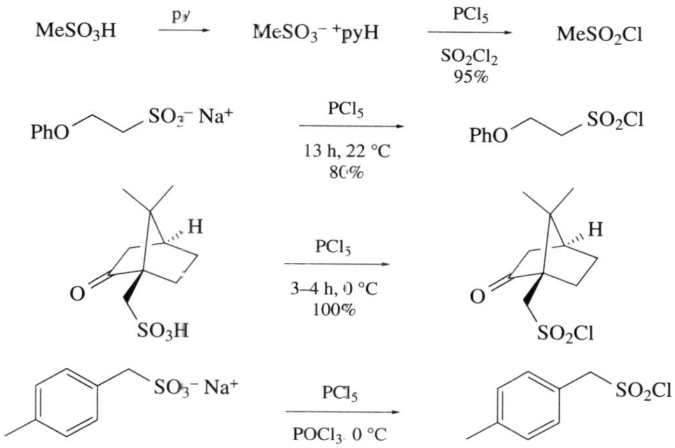

Scheme 120

Use of phosphorus pentachloride alone can result in anhydride formation ⟨56JCS173, 57JCS2640, 63JOC1244⟩. In a study using perfluorinated alkanesulfonic acid substrates, use of phosphorus pentachloride resulted in reasonable yields by perfluoralkanesulfonyl chlorides; perfluorocycloalkanesulfonyl chlorides however could only be formed in poor yields, the major component of the reaction mixture being the corresponding sulfonic anhydride ⟨63JOC1244⟩. Unwanted anhydride formation is thought to proceed through initial salt formation, the salt decomposing or dissociating under the conditions required for the reaction (Scheme 121). Cyclic fluorocarbon sulfonates are not sufficiently reactive towards phosphorus pentachloride up to its dissociation temperature (162 °C), and, as a result, sulfonic acid salts are formed leading ultimately to anhydrides, rather than sulfonyl chlorides. A search for a more efficient chlorodeoxygenation agent resulted in development of a complex of phosphorus pentachloride with molten zinc chloride which was capable of converting the metal salts of fluorocarbon sulfonic acids into sulfonyl chlorides (Equation (99)).

$$RSO_3H \xrightarrow{PCl_5} RSO_3^-\,PCl_4^+ \xrightarrow{\Delta} (RSO_2)_2O$$

Scheme 121

$$RSO_3^-\,M^+ \xrightarrow{PCl_5 \cdot 2ZnCl_2} RSO_2Cl \qquad (99)$$

R	*Yield (%)*
CF_3	94
n-C_8H_{17}	85

Phosphorus pentabromide has similarly been used for the conversion of sodium salts of sulfonic acids into sulfonyl bromides, although disulfide formation is in this case a serious handicap ⟨32JA2441, 32JA3364⟩. In addition, the alkanesulfonyl bromide products can in many cases readily be reduced to disulfides by the reagent. This problem is not encountered with sulfonyl chloride preparation when using phosphorus pentachloride.

Phosphorus oxytrichloride and phosphorus trichloride are much less aggressive reagents than phosphorus pentachloride, and both have found use in the synthesis of sulfonyl chlorides. For example, methanesulfonyl chloride has been prepared in 57% yield by the use of phosphorus trichloride, the reaction comparing favorably with the preparation of methanesulfonyl chloride using phosphorus pentachloride ⟨1884JPR280, 38CB712, 48MI 203-01⟩. Phosphorus oxytrichloride has also been used in conjunction with phosphorus pentachloride ⟨24JCS1463⟩.

Of other reagents used for conversion of sulfonic acid to sulfonyl halide, thionyl chloride is favoured for the preparation of methanesulfonyl chloride, in a reaction, illustrated in Equation (100), which is successful on a large scale ⟨48JA3955, 50OS(30)58⟩. Sulfonic acids or sodium sulfonates, which either do not react with thionyl chloride under normal conditions or which react sluggishly, can be induced to react by the addition of DMF ⟨47BSF351, 59HCA1653, 62HCA717⟩.

$$MeSO_3H \xrightarrow[\substack{95\ ^\circ C,\ 3\ h \\ 83\%}]{SOCl_2} MeSO_2Cl \qquad (100)$$

One interesting example of the use of sulfuryl chloride to furnish a sulfonyl chloride involves the carbohydrate sulfonate substrate shown in Equation (101). Relatively few methods of sulfonyl chloride preparation could hope to cope with the complex functionality found in carbohydrates and nucleosides. However, in direct analogy with the use of triphenylphosphine and a halogen donor to form alkyl halides from alcohols, the action of triphenylphosphine and sulfuryl chloride in methylene chloride in this case gave the desired sulfonyl chloride in 87% yield ⟨92TL2657⟩. A number of sulfonate salts containing a variety of functionalities have been converted into the corresponding sulfonyl chlorides using this reagent system. Yields are generally high ⟨92TL2657⟩.

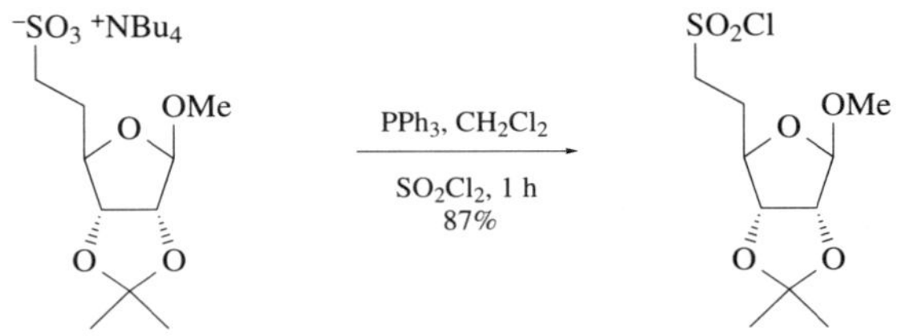

(101)

Dichloromethyl methyl ether has been used for the preparation of alkyl sulfonyl chlorides ⟨59CB83⟩. The reagent is used in conjunction with zinc chloride to effect the desired transformation. The sulfonyl chlorides are, however, only formed in moderate yields. Trimethylsilyl derivatives have also been converted to sulfonyl chlorides using this reagent.

Alkanesulfonyl bromides can be prepared by the reaction of a bromosulfonylating agent with a suitable substrate such as an alkene. Equation (102) illustrates the reaction using an unsaturated boronic ester, part of a synthesis of electron-deficient alkenes ⟨92SL581⟩.

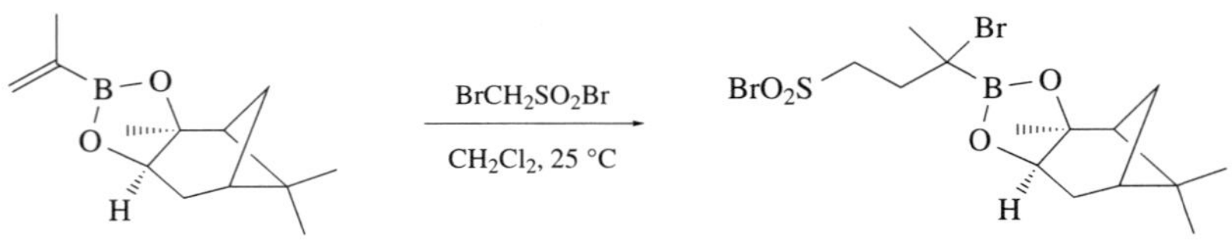

(102)

Many examples of sulfonyl halide preparation from various sulfonate derivatives are scattered throughout the literature. These include the reaction of sulfonate esters with phosphorus trichloride ⟨58JCS692⟩, the use of methylmagnesium iodide in conjunction with sulfuryl chloride to form methanesulfonyl chloride ⟨23HCA256⟩, and the preparation of sulfonyl iodides from iodine and the sodium alkylsulfinate ⟨74JOC238, 74JOC245⟩. Sulfonyl iodides are very unstable and in many cases are not isolated, but instead may be used *in situ*. Commonly their preparation involves either addition of a concentrated ethanolic solution of iodine to a very dilute solution of sodium alkylsulfinate in water, or addition of a benzene solution of iodine to an aqueous solution of sodium sulfinate; a visible color change from purple (iodine) to a bright yellow/orange color (sulfonyliodide) indicates their formation; the solution is then immediately used, as incidence of light or a delay in use leads to decomposition. Sulfonyl fluorides, usually available only by electrochemical fluorination, ⟨56JCS173, 57JCS2574, 57JCS2640⟩ can also be furnished from sulfonic anhydrides by reaction with a metal fluoride salt ⟨63JOC1244⟩. While appropriate choice of metal salt can be difficult, the method has proved particularly effective for the formation of fluorocarbon sulfonyl fluorides, as shown in Equation (103). A further method of sulfonyl fluoride preparation involves reaction of the corresponding sulfonyl anhydride with anhydrous hydrogen fluoride ⟨62JOC2667⟩.

$$(RSO_2)_2O \xrightarrow[X\,=\,Cl,\,F]{M^+X^-} RSO_2X \qquad (103)$$

R	*X*	*Yield* (%)
CF_3	Cl	94
C_8H_{17}	Cl	85

The second broad method of sulfonyl halide formation from existing sulfur(VI) functionality involves the halide exchange reaction. This reaction is of particular use for the preparation of sulfonyl halides which are difficult or tedious to synthesize by other routes, such as sulfonyl fluorides. In general, sulfonyl chlorides are easier to form than the other sulfonyl halides, and so the chloride to fluoride, bromide or iodide conversions are extremely useful. Techniques employed to effect the exchange include the use of alkali metal salts, as shown in Scheme 122 ⟨32JCS483, 56JCS173, 66JCS(C)1171⟩, where yields are good, but are dependent upon the presence of traces of water in the reaction. For example, the use of caesium fluoride/water gives an 80% yield of sulfonyl fluoride, while the yield drops to 7% under rigorously anhydrous conditions ⟨66JCS(C)1171⟩. Yields can be enhanced by the use of DMF as a solvent and by the employment of phase transfer systems ⟨77JOC2031⟩. Use of a crown ether phase-transfer catalyst in association with potassium fluoride has allowed complete reaction within four hours, with product purification possible by aqueous extraction (Equation (104)).

$$MeSO_3H \longrightarrow MeSO_2Cl \xrightarrow[H_2O,\ 84\%]{KHF_2} MeSO_2F \xrightarrow[\text{electrolysis}]{HF} CF_3SO_2F$$

Scheme 122

$$RSO_2Cl \xrightarrow[\text{18-Crown-6, RT, 4 h}]{KF,\ MeCN} RSO_2F \qquad (104)$$

R	*Yield* (%)
Me	84
Ph	93

Dialkylaminosulfur trifluorides are extremely efficient chloride–fluoride exchange agents when used for the preparation of sulfonyl fluorides (Equation (105)) ⟨75S801⟩. Sulfonyl chlorides can also be converted into sulfonyl bromides by the use of sodium or potassium salts, followed by bromine, as shown in Scheme 123 ⟨65JOC38⟩.

$$R^1SO_2Cl \xrightarrow{R^2{}_2NSF_3} R^1SO_2F \;+\; R^2{}_2NSClF_2 \qquad (105)$$

$$Cl_3CSO_2Cl \xrightarrow[SO_2\ (l)]{KCN} Cl_3CSO_2K \xrightarrow[\text{PhH, 15 min, 49\%}]{Br_2,\ H_2O} Cl_3CSO_2Br$$

Scheme 123

Other preparative methods for sulfonyl fluorides include heating of the corresponding sulfonyl chloride with antimony trifluoride, xenon difluoride, or potassium fluorosulfinate ⟨55AG32⟩, and the use of an ion exchange resin ⟨72JOC3549⟩. The use of a resin furnishes essentially pure products in high yield. For example, methanesulfonyl fluoride can be generated from the corresponding sulfonyl chloride in 71% yield. The resin used was prepared by diluting the strongly basic quaternary amine anion exchange resin AG1-X10 with aqueous potassium fluoride, and its use is based upon the principle that the resin has a lower affinity for fluoride ion than chloride. One further interesting and high yielding example of alkanesulfonyl fluoride preparation involves the ring opening of a difluorooxathietane with concomitant migration of a fluorine atom (Equation (106)) ⟨75IZV2530⟩.

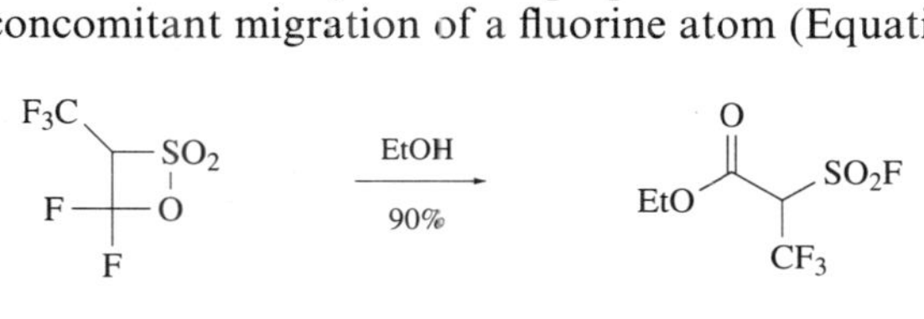

(106)

2.03.4 ALKANESULFENIC ACIDS AND THEIR HIGHER COORDINATED DERIVATIVES

2.03.4.1 Alkanesulfenic Acids

Sulfenic acids are derived from sulfur monoxide, and demonstrate similar properties of high reactivity and low stability, often only being postulated as transient intermediates. The rearrangement of penicillin sulfoxides to cephalosporins is one well-known example (Scheme 124) ⟨76ACR293, 76CRV113, 79COC(3)261, B-90MI 203-06⟩.

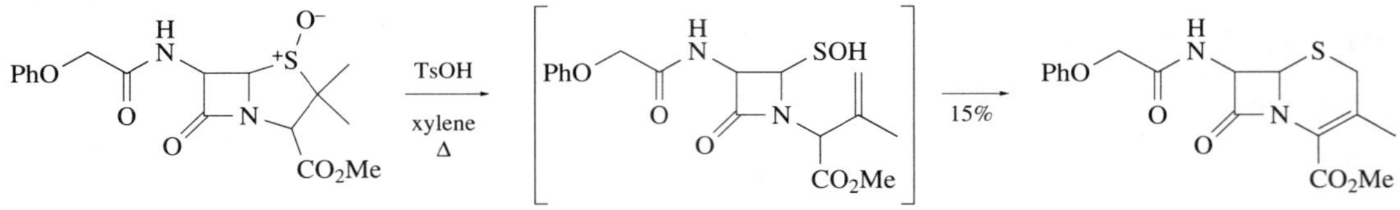

Scheme 124

The few stable alkanesulfenic acids which have been isolated include aromatic and heterocyclic examples and those derived from penicillins. The occurrence of stable alkanesulfenic acids is rare. Two routes to a very sterically hindered alkyl sulfenic acid have however been described in the literature (Scheme 125) ⟨83JA7172, 89AG(E)97⟩.

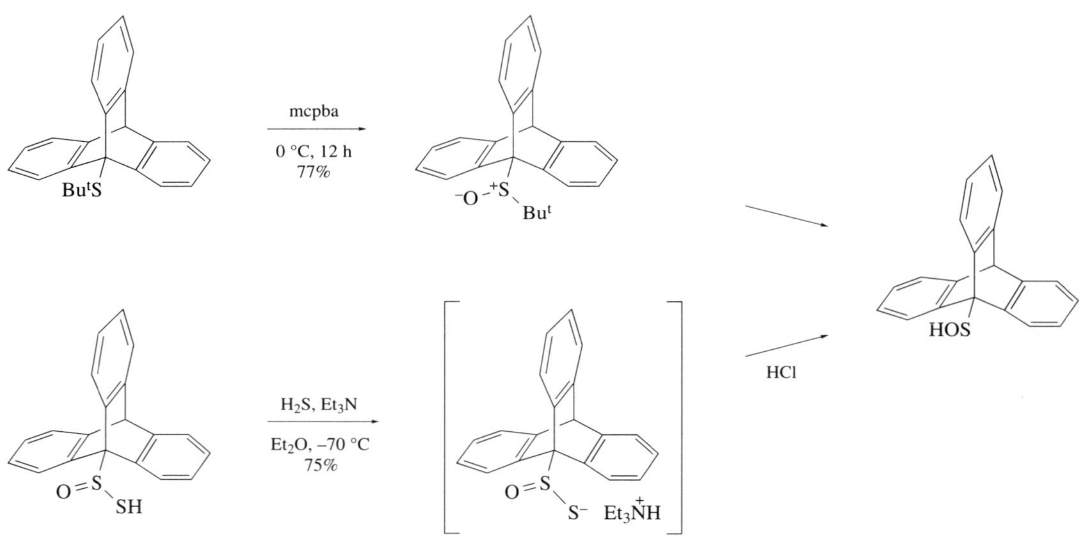

Scheme 125

Unstable sulfenic acids can be trapped by reaction with alkynes or activated alkenes, and this method has often been used in proving their existence, although direct observation by IR or microwave spectroscopy at low temperature has been used in structural studies ⟨67JA718, 78JA3622, 86JOC1033⟩.

2.03.4.1.1 Thermolysis of sulfoxides

The elimination of isobutene from an alkyl *t*-butyl sulfoxide to form the alkanesulfenic acid is one of the most generally applicable methods for their synthesis, although other alkyl groups containing a β-hydrogen atom may also undergo elimination (Scheme 126) ⟨67JA718, 81JCS(P1)2106, 91COS(6)1011⟩. The use of flash vacuum pyrolysis (FVP) allows the unstable sulfenic acid intermediates to be trapped in reasonable concentrations, and is widely used for this reaction ⟨78JA3622, 78TL97, 80JA7967⟩. By introduction of an electron-withdrawing group such as nitrile or carbonyl in a

β position relative to the sulfoxide sulfur atom, the elimination of the proton is made more facile, and hence still lower temperatures are required ⟨79TL4977, 88CC40, 88SC2083⟩.

Baldwin has reported formation of an allylic sulfenic acid by rearrangement of a strained episulfoxide at 35 °C (Scheme 127) ⟨71JA2810⟩.

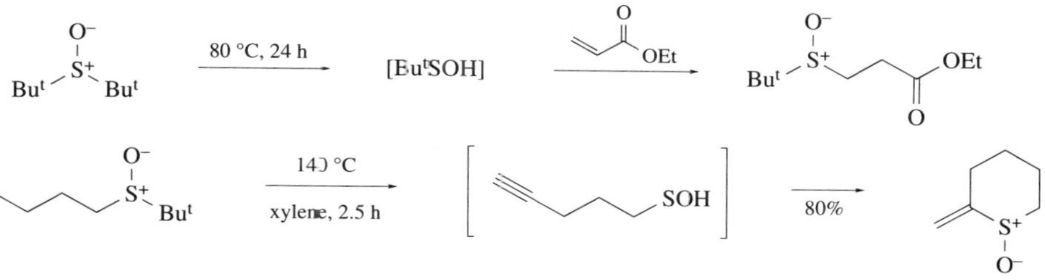

Scheme 126

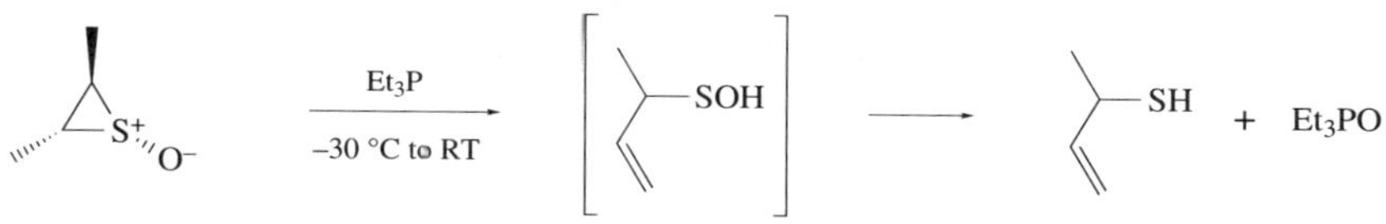

Scheme 127

2.03.4.1.2 *Hydrolysis of sulfenyl derivatives*

In contrast to alkanesulfenic acids, their derivatives—halides, esters and amides—are relatively stable, and formation of the parent acid is often possible through hydrolysis of these derivatives. For example, methanesulfenic acid is generated as an intermediate during the hydrolysis of methyl sulfenyl chloride with water, ultimately giving dimethyl disulfide and *S*-methyl methanethiosulfonate through a disproportionation process (Scheme 128) ⟨59JOC2004⟩.

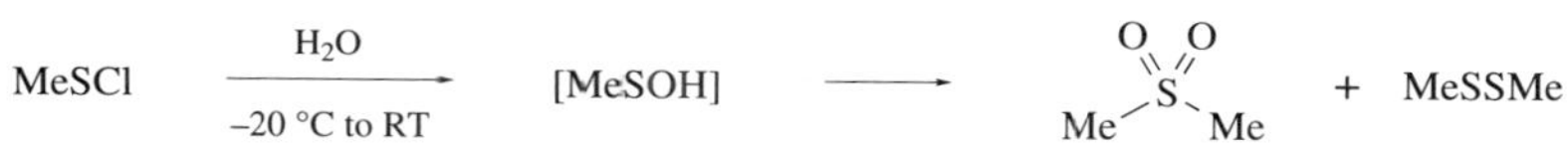

Scheme 128

2.03.4.1.3 *Oxidation of thiols*

Oxidation of *t*-butanethiol with an oxaziridine generated the corresponding *t*-butanesulfenic acid, which was trapped *in situ* by methylpropiolate (Scheme 129), giving credence to the idea that sulfenic acids are involved in the oxidation of thiols ⟨81JA7016⟩.

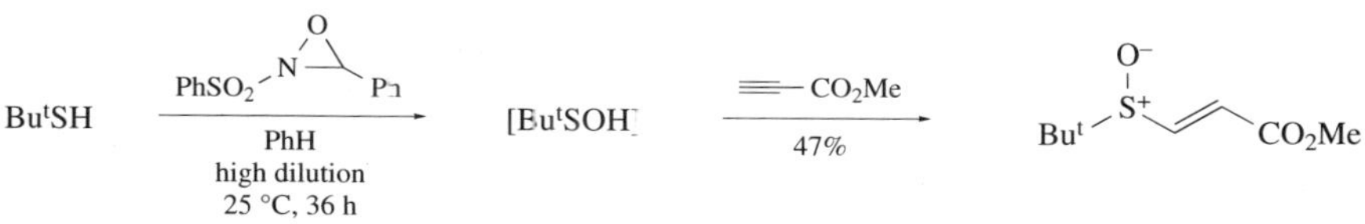

Scheme 129

2.03.4.1.4 Thermolysis of dialkyl thiosulfinates

The formation of alkanesulfenic acids by the decomposition of dialkyl thiosulfinates has been demonstrated by trapping with alkynes and alkenes (Scheme 130). These reactions often occur under milder conditions than are required for the thermolysis of sulfoxides ⟨72JA642, 74JA3929⟩.

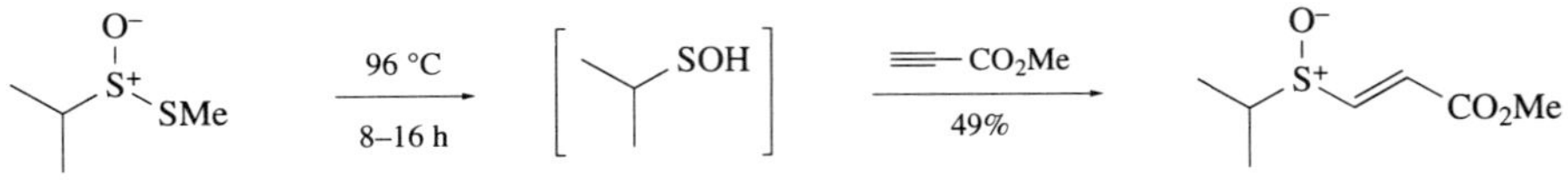

Scheme 130

2.03.4.2 Alkanesulfinic Acids

The position of the sulfinyl group between divalent thiols, sulfides and sulfenyl compounds, and hexavalent sulfones and sulfonyl compounds theoretically allows the formation of sulfinic acids by either oxidation or reduction procedures. However, in practice the moderate to poor stability of alkanesulfinic acids prejudices some synthetic routes in favor of others. Alkanesulfinic acids are often isolated as their more stable metal salts ⟨79COC(3)317, B-90MI 203-07⟩.

2.03.4.2.1 Oxidation of thiols

The direct oxidation of widely available thiols is the most attractive method of synthesis of alkanesulfinic acids. The "interval" oxidation of alkanethiols by mcpba, involving periodic addition of the reagent to keep the concentration of oxidant low, gave good yields of sulfinic acids, via the sulfenic acids, in high purity, with no disulfide or sulfonic acid being observed ⟨73JOC4070⟩. A 1993 paper describes a similar oxidation using dimethyl dioxirane (DMD) as the initial oxidant at $-40\,^{\circ}C$ followed by an aerial oxidation (Scheme 131) ⟨93TL67⟩. The use of two equivalents of DMD led to ready overoxidation, but the use of atmospheric oxygen as secondary oxidant gave yields of 73–96% of the sulfinic acids.

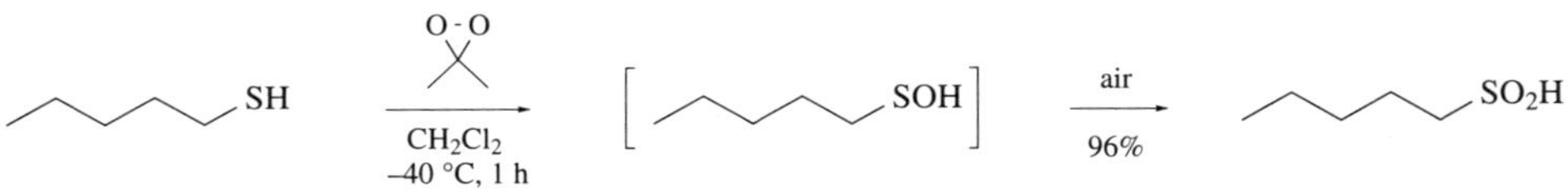

Scheme 131

2.03.4.2.2 Reaction of organometallics with sulfur dioxide

The reaction of a Grignard reagent with sulfur dioxide is one of the best-known procedures for the formation of a carbon–sulfur bond, and provides a direct entry to the sulfinic acid upon acidic workup (Equation (107)). For example, the Grignard reagent prepared from 1-bromododecane and magnesium produced the sulfinic acid in 65% yield upon treatment with sulfur dioxide at $-40\,^{\circ}C$ in diethyl ether ⟨38JA544, 42JOC23, 69JOM(121)123⟩.

$$\mathrm{BuMgBr} + \mathrm{SO_2} \xrightarrow[\substack{-50\,^{\circ}C \\ 69\%}]{\mathrm{Et_2O}} (\mathrm{BuSO_2})_2\mathrm{Mg} \qquad (107)$$

Inverse addition of the organometallic reagent to excess sulfur dioxide is preferred in order to suppress by-products such as sulfoxides, formed by reaction of the sulfinate with excess organometallic reagent ⟨79JOC160⟩. Organolithium reagents have found use in sulfinic acid synthesis: butyllithium reacts with sulfur dioxide at $-78\,^{\circ}C$ to form butanesulfinic acid in quantitative yield ⟨79JOC160⟩. Other organometallic species have also been used, including trialkylaluminum compounds ⟨60BRP916751⟩.

2.03.4.2.3 Reduction of sulfonyl halides

Sulfonyl halides can be reduced to give the sulfinic acid salts by the action of zinc dust or iron in aqueous alkaline solution (Scheme 132) ⟨48OSC(1)492⟩. The rather harsh reaction conditions do not favor alkanesulfinic acid production, although 1-dodecanesulfinic acid has been isolated as crystalline product ⟨48JOC822⟩. Lithium aluminum hydride has been used to reduce alkanesulfonyl chlorides to the corresponding sulfinic acids ⟨51JOC964⟩. Triethylaluminum and ethylaluminum sesquichloride have also each been reported to reduce alkanesulfonyl chlorides to the sulfinic acids in nearly quantitative yields ⟨66CB1718⟩.

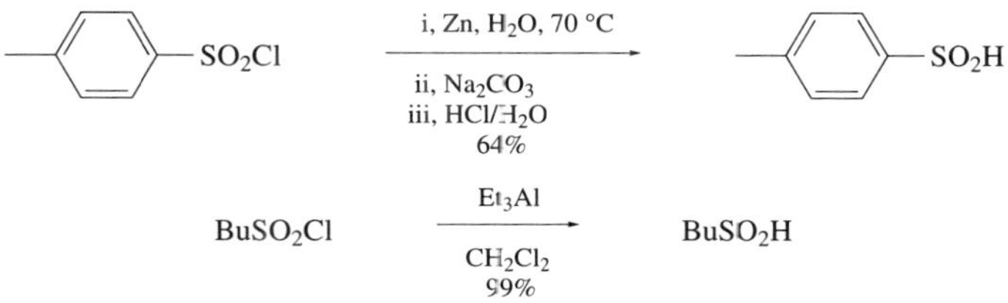

Scheme 132

2.03.4.2.4 Hydrolysis of sulfinic acid derivatives

The hydrolysis of alkanesulfinic esters to the parent acids is a facile if usually somewhat indirect process for the synthesis of sulfinic acids. The reaction is typically complete in several minutes in dilute aqueous sodium hydroxide at 0 °C ⟨65JOC633⟩. The hydrolysis of alkanesulfinyl chlorides is a more useful route, as the starting materials can be obtained by direct chlorination of dialkyl disulfides or alkanethiols with chlorine in aqueous solution (Scheme 133) ⟨58JOC330, 65BCJ1309, 67JA4099, 68BCJ3015, B-77MI 203-03⟩.

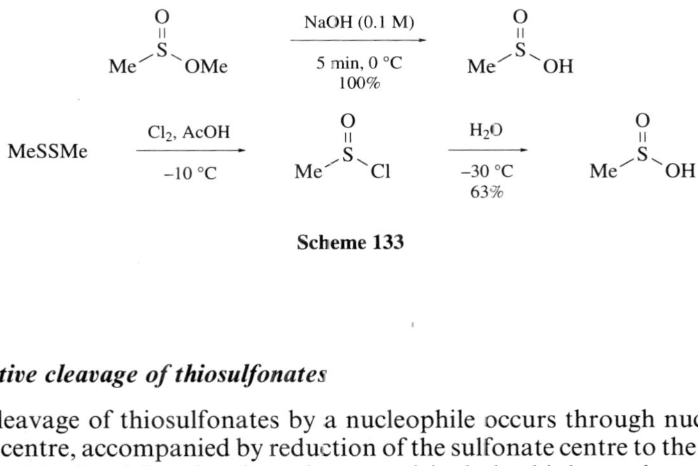

Scheme 133

2.03.4.2.5 Reductive cleavage of thiosulfonates

The reductive cleavage of thiosulfonates by a nucleophile occurs through nucleophilic attack at the divalent sulfur centre, accompanied by reduction of the sulfonate centre to the sulfinate oxidation level (Scheme 134). Nucleophiles that have been used include thiolate salts, sodium polysulfide, sodium amide, and sodium dialkyl phosphites ⟨60JCS1665, 63JOC870, 81JOC2025⟩.

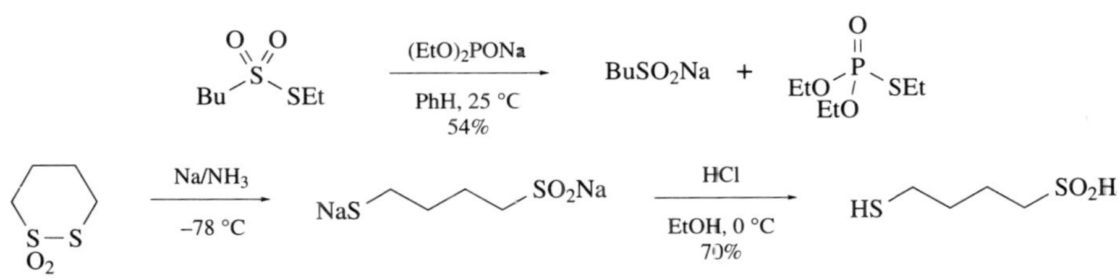

Scheme 134

2.03.4.2.6 *Addition to double bonds*

Thionyl chloride reacts with alkenes in the presence of aluminum chloride to give the 2-chloroalkanesulfinic acid after hydrolysis (Equation (108)) ⟨64DOK(157)139⟩.

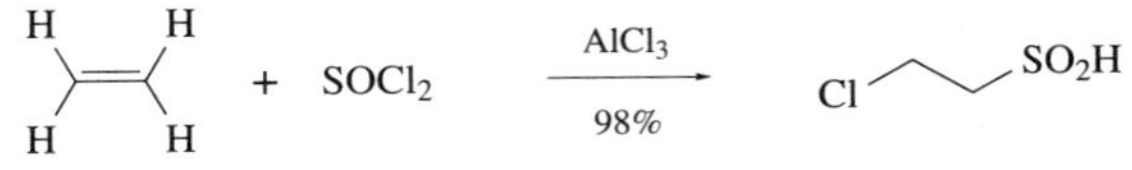

(108)

2.03.4.2.7 *From sulfones*

(i) *Reductive cleavage*

As the sulfinate group can behave as a leaving group, the carbon–sulfur bond of a sulfone can be cleaved to give the sulfinic acid. Electrochemical reduction of propyl benzyl sulfone with a mercury cathode in the presence of tetramethylammonium chloride gave propanesulfinic acid in good yield. In aryl alkyl or vinyl alkyl sulfones, the alkyl–sulfur bond is cleaved ⟨65CB1715⟩. Similar alkyl–sulfur bond cleavage is seen upon treatment with sodium amalgam ⟨52JCS4887⟩.

The use of metallic lithium in methylamine or metallic sodium in ammonia allows the formation of alkanesulfinic acids from aryl alkyl or dialkyl sulfones and alkyl benzyl sulfones respectively (Equation (109)) ⟨60JA2872, 70JCS(C)270⟩. Reductive cleavage of alkyl benzothiazolyl sulfones with sodium borohydride in alcoholic solution gives good yields of the alkanesulfinic acids as the sodium salts (Equation (110)) ⟨84CL2125⟩.

$$PhSO_2C_{10}H_{21} \xrightarrow[\text{MeNH}_2\ 95\%]{\text{Li}} C_{10}H_{21}SO_2Li + C_6H_6 \qquad (109)$$

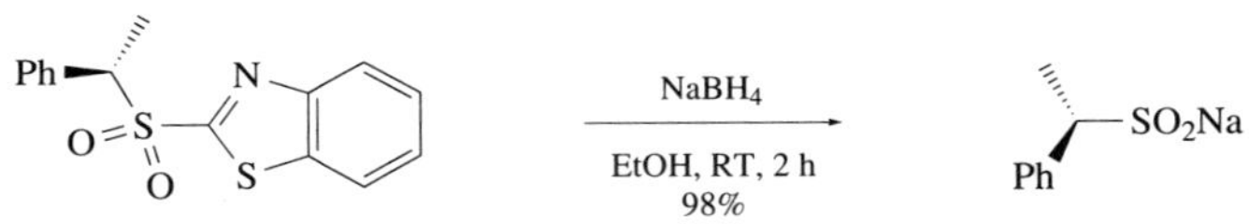

(110)

(ii) *Base-induced cleavage*

Alkyl sulfones can undergo displacement of alkanesulfinic acid upon treatment with base (Scheme 135); the presence of an electron-withdrawing group facilitates this process ⟨63JOC593⟩. Cyanide ion and alkanethiolates have been used as the base ⟨77LA1457⟩. Symmetrical β-substituted sulfones can be used to give a β-substituted sulfinic acids. Simple dialkyl sulfones require rather more drastic basic conditions to effect this cleavage; consequently, poor yields and low purity are often observed. The reaction is improved by displacement of the phthalimidomethyl group from phthalimidomethyl sulfones by treatment with sodium ethoxide ⟨77S794⟩. If the ethoxide nucleophile is replaced with an appropriate thiolate, then the sulfide by-product can be recycled by oxidation. The greater nucleophilicity of sulfur provides a faster reaction than is observed with ethoxide (Scheme 136). Similar cleavage of 2-sulfonyl pyridines and their *N*-oxides can be accomplished with ethoxide ion, and again, in a direct analogy to the phthalimdomethane sulfones, the use of a thiolate nucleophile generates a useful reaction cycle (Scheme 137) ⟨86H3019⟩.

(iii) *Cleavage of small ring cyclic sulfones*

Treatment of a thietane dioxide with base forms a cyclopropane sulfinic acid, with carbon–sulfur bond cleavage, in good yield (Equation (111)) ⟨71JOC2698⟩.

Cleavage of the three-membered rings of thiirane dioxides can be performed with organometallic reagents, or by sulfur bases such as thiolate, metal sulfide, or thiourea, to give the β-substituted alkanesulfinic acids in good yields ⟨74TL3987, 75S55⟩. In the reaction of the organometallic species, the thiirane dioxide acts as a source of sulfur dioxide, and it is the sulfinic acid of the alkyl Grignard reagent (or alkyllithium) that is the product (Scheme 138).

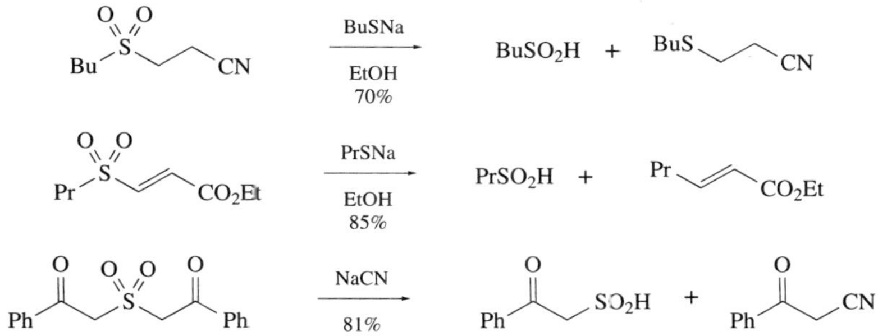

Scheme 135

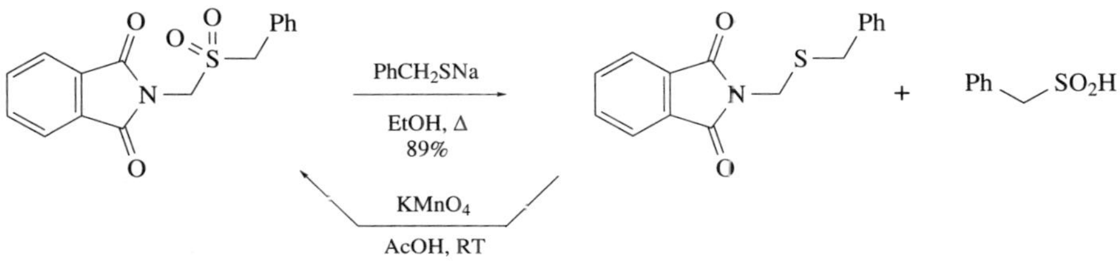

Scheme 136

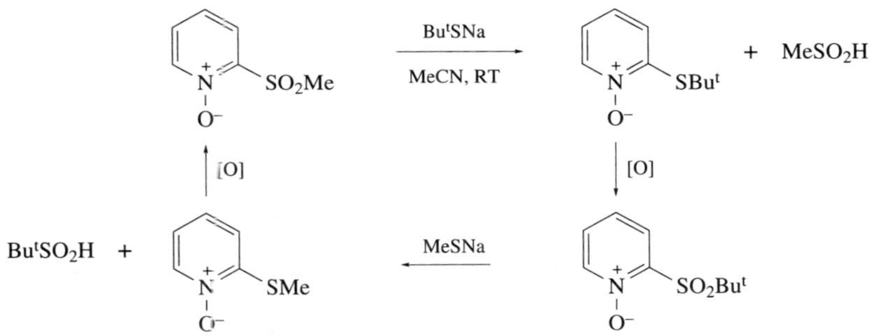

Scheme 137

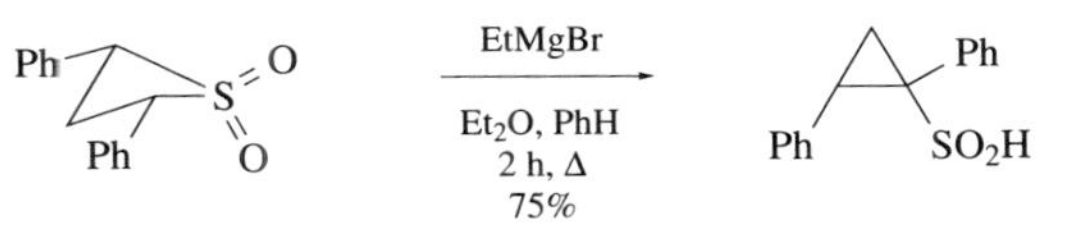

(111)

2.03.4.3 Alkanesulfonic Acids

Sulfonic acids are generally stable, polar, and water soluble compounds which exhibit a strongly acidic nature. The synthesis of sulfonic acids has been reviewed, although not specifically that of alkanesulfonic acids ⟨79COC(3)317, B-90MI 203-08⟩. There are three principal synthetic routes for alkanesulfonic acids: addition of a sulfonyl group to a carbon fragment, the oxidation of an alkyl sulfur compound, and manipulation of an existing sulfonic acid or its derivatives. Each of these approaches is discussed in turn below.

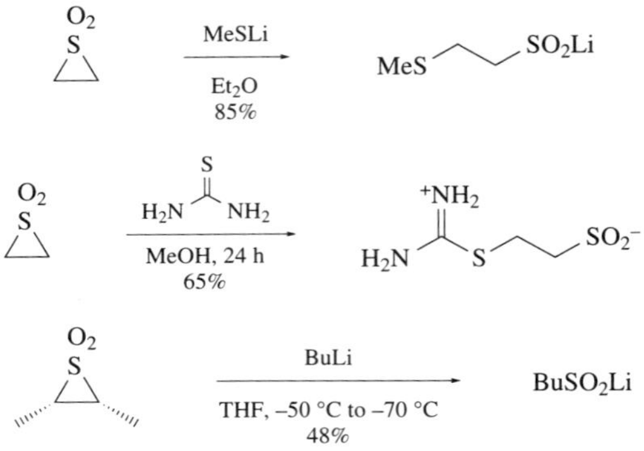

Scheme 138

2.03.4.3.1 *Addition of a sulfonyl group*

(i) With sulfuric acid derivatives

Direct sulfonation of an aliphatic compound with sulfuric acid is not a widely used method, although, under forcing conditions, sulfonic acids are produced, along with a host of other by-products. Derivatives of sulfuric acids can however produce good results, particularly in the substitution of a trimethylsilyl group. For example, the reaction of trimethylsilylcyclopropane with trimethylsilyl chlorosulfonate gave cyclopropanesulfonic acid in 70% yield (Equation (112)) ⟨76TL1197⟩.

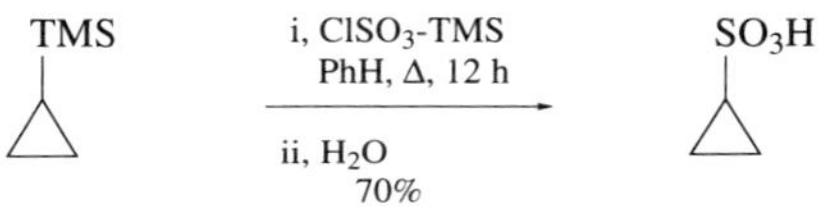

(112)

(ii) With sulfur trioxide

Sulfur trioxide is more reactive than sulfuric acid or its derivatives and is often used for aromatic sulfonation, particularly when modified as its adducts with dioxane or pyridine. The addition of sulfur trioxide across a carbon–carbon double bond to give an aliphatic sulfonic acid is possible, but commonly gives a mixture of products ⟨54JA3945, 65BSF787⟩. Addition to an alkyne leads to α-ketosulfonic acids with two equivalents of sulfur trioxide, and to α-ketodisulfonic acids with four equivalents of sulfur trioxide (Scheme 139) ⟨55ZOB1943⟩.

Sulfonation with sulfur trioxide occurs readily at the α position of carbonyl, carboxyl, dialkylamino, and sulfonyl compounds (Scheme 140) ⟨52JA2526, 63OSC(4)8623, 68TL4479, 74JCS(P1)771, 79T1315⟩.

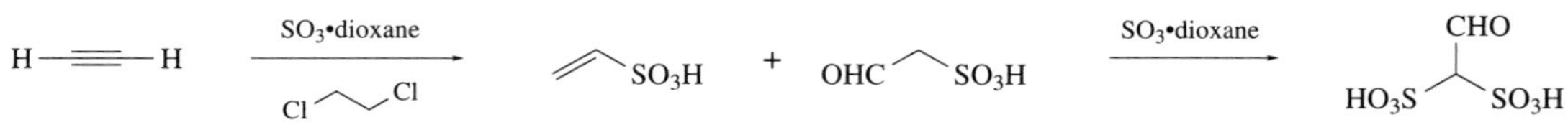

Scheme 139

(iii) With hydrogen sulfite ion and sulfites

In the presence of an oxidant, usually oxygen or hydrogen peroxide, hydrogen sulfite ion (HSO_3^-) adds to alkenes and alkynes with formation of the alkanesulfonic acid salt or disulfonic acid salt respectively. The anti-Markovnikov product normally results from this free radical addition reaction (Equation (113)) ⟨34JA1628, 38JOC175, 39JA3093⟩.

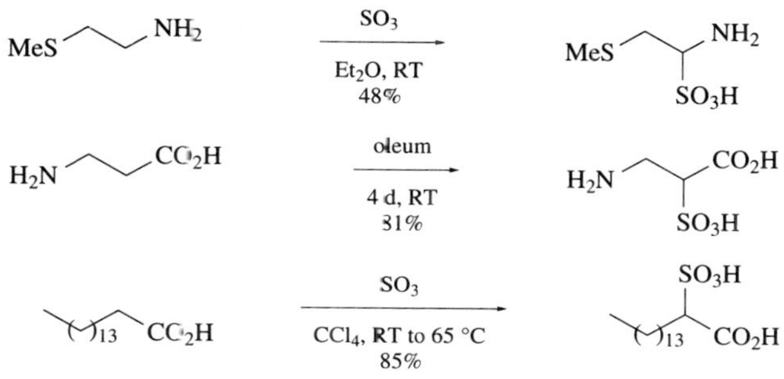

Scheme 140

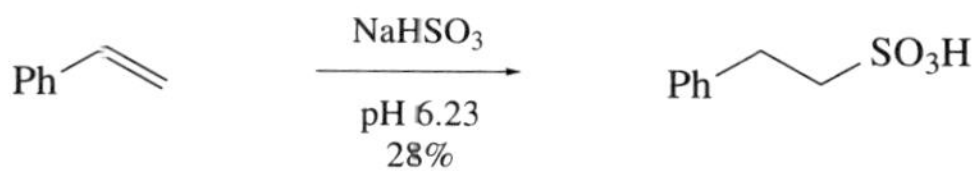 (113)

The reaction of an alkyl halide and inorganic sulfite in aqueous media to form the sulfonic acid is a well-known process, the Strecker synthesis, first reported in 1868 (Scheme 141). This method has in the past been widely applied to the synthesis of primary, secondary, and tertiary alkanesulfonic acids ⟨31JA3407, 40JA1044, 40JOC24, 43OSC(2)558, 45JA1578, 55JA6231⟩. Disulfonic acids have also been produced ⟨45JA1578⟩.

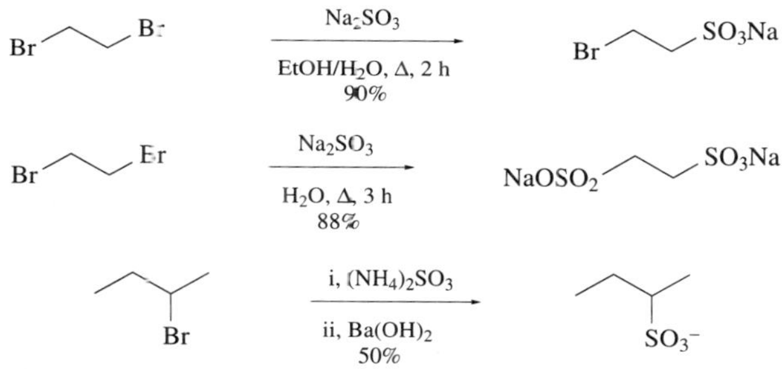

Scheme 141

It is possible to use a leaving group other than halide in these reactions; for example, sulfate has been displaced in a synthesis of the *β*-aminosulfonic acid taurine (Equation (114)) ⟨43JCS4⟩. The mild reaction conditions tolerate other functionalities such as esters and ketones ⟨64CB2903, 75S321⟩.

Inorganic sulfite salts have also been used to cleave an epoxide ring, forming a hydroxysulfonic acid salt (Equation (115)) ⟨54JA3957, 64CB2903⟩.

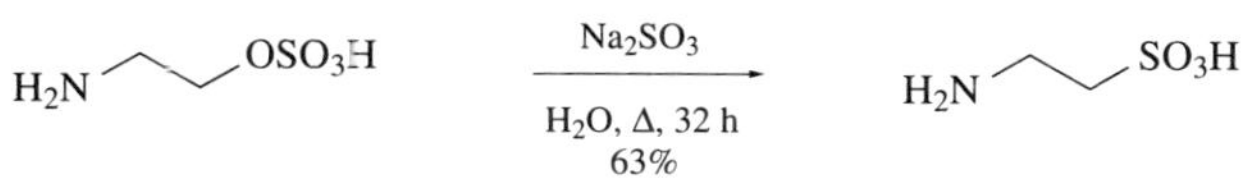 (114)

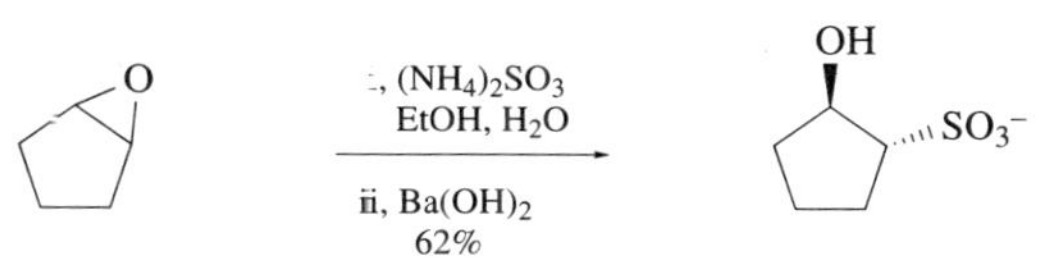 (115)

(iv) With sulfur dioxide

The sulfonation of alkanes with sulfur dioxide in the presence of oxygen is a well-known process for the synthesis of alkanesulfonic acids (Equation (116)). Discussions of the mechanism and of the effect of the sulfur dioxide/oxygen ratio have appeared ⟨B-65MI 203-02, 73ACS3180⟩.

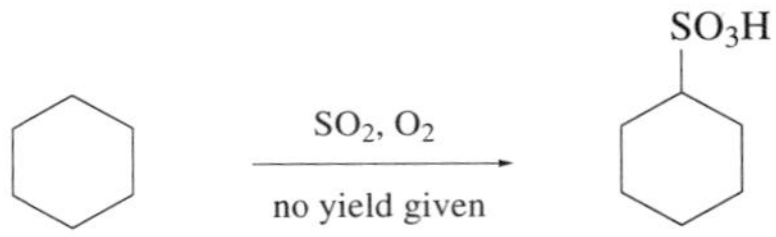

(116)

A later publication describes the synthesis of sulfonic acids from alkenes and sulfur dioxide in the presence of hydrogen and a palladium catalyst, in a similar process to the hydroformylation of alkenes (Equation (117)) ⟨93CC1592⟩.

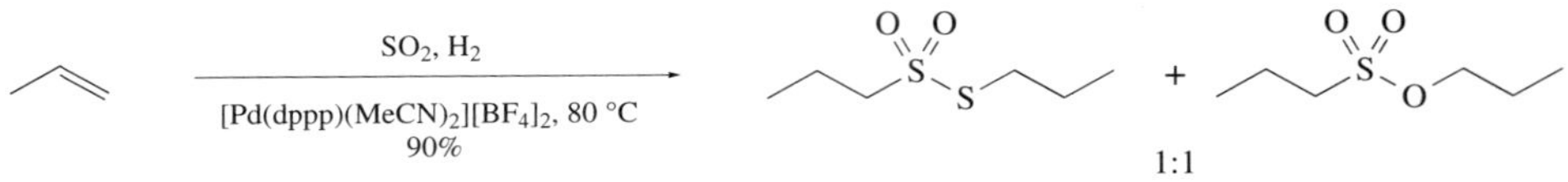

(117)

Sulfonic acids are also formed in the reaction of sulfur dioxide with alkenes when treated with mercury and then photosensitized, producing sulfinic acids which undergo oxidative workup with performic acid to give alkanesulfonic acids in up to 99% yield ⟨91JOC5503⟩.

2.03.4.3.2 *By oxidation*

In principle both divalent and tetravalent sulfur compounds can be oxidized to the corresponding sulfonic acids, although in practice it is normally only divalent species not requiring carbon–sulfur bond cleavage which are used. Thiols, disulfides, and related compounds such as thiocyanates and xanthates have been found to be generally useful (Scheme 142). The forcing conditions needed for oxidative carbon–sulfur bond cleavage of sulfides, sulfoxides, and sulfones usually rule out these reactions as synthetic methods.

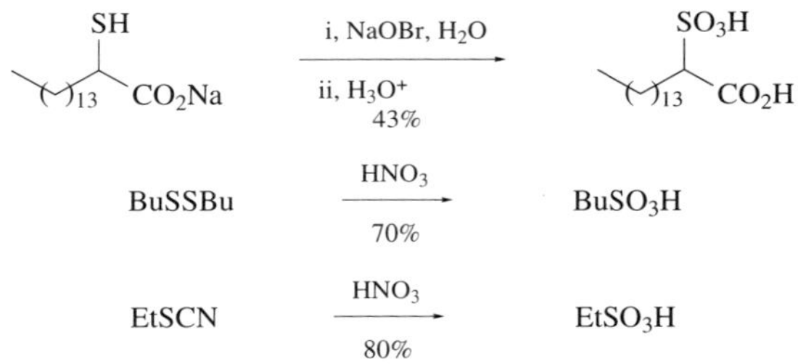

Scheme 142

(i) From thiols, disulfides, and related compounds

The oxidation of thiols and disulfides is a standard method for alkanesulfonic acid synthesis, and many oxidants have been used successfully (Scheme 143) ⟨B-74MI 203-04⟩. Nitric acid has been used as an inexpensive oxidant in sulfonic acid formation to good effect ⟨33JCS739, B-65MI 203-03⟩. Other useful reagents include hydrogen peroxide, peroxyacids, and oxone ⟨60JOC1901, 67JOC3032, 74S295⟩. Oxygen in DMF or hexamethylphosphoramide (HMPA) solution oxidizes thiols and disulfides to the sulfonic acid salts under basic conditions ⟨63TL1131, 65TL2271⟩. Oxygen has also been used under irradiation to oxidize thiols in good yields ⟨63RTC773⟩, and ozone has been shown to oxidize thiols quantitatively to the sulfonic acid ⟨57JCS4547⟩.

Thioacetates are oxidized to sulfonic acids by hydrogen peroxide, peroxy acids or oxone ⟨41OSC(1)84, 62JOC2853, 87SC1129⟩.

(ii) From sulfides and sulfones

As indicated above, the oxidation of these compounds is rarely of synthetic value: oxidants such as chromate, permanganate, and nitric acid produce only low yields of sulfonic acids from sulfides

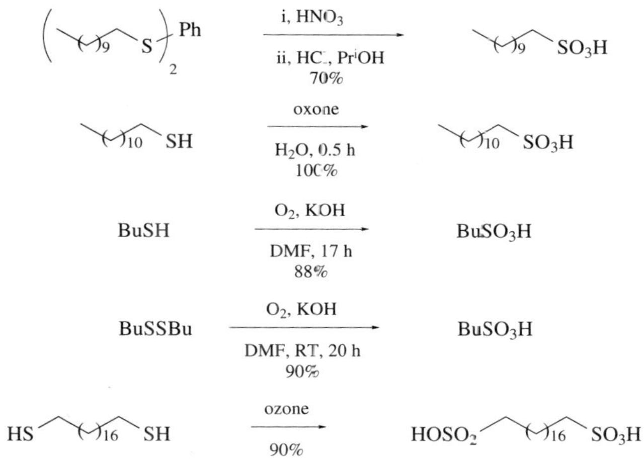

Scheme 143

⟨05CB646, 53JOC36⟩. Methyl heptafluoropropyl sulfone is however converted into the methanesulfonic acid by cleavage with aqueous sodium hydroxide at 100 °C. Methyl perfluoroalkyl sulfones can be similarly cleaved to give the perfluoroalkyl sulfonic acids using aqueous potassium permanganate at reflux (Scheme 144) ⟨73JCS(P1)676⟩.

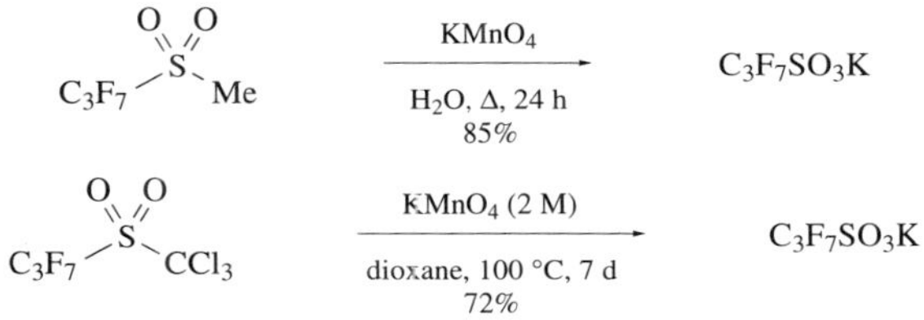

Scheme 144

(iii) From sulfinic acids

The instability of most alkanesulfinic acids dictates that this is a much less important route for alkanesulfonic acid perparation than for the aromatic analogues. Atmospheric oxygen oxidizes alkanesulfinic acids to the corresponding sulfonic acids by an autocatalytic radical mechanism (Scheme 145) ⟨58AC(P)108⟩.

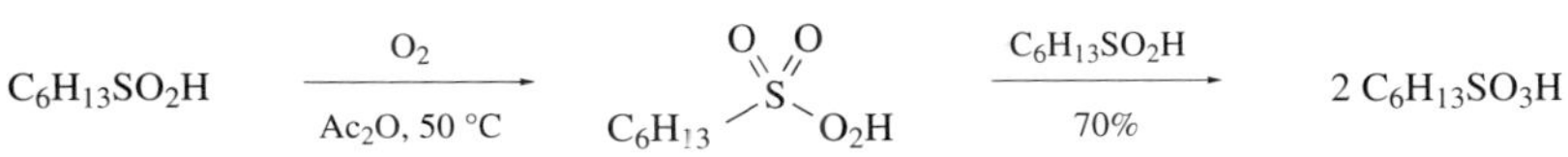

Scheme 145

Sulfinic acids may be oxidized under mild conditions by potassium superoxide (KO_2) in the presence of a crown ether (Equation (118)) ⟨79TL821, 81T37⟩. Aqueous potassium permanganate has also been used to perform the oxidation, but not without side product formation, mainly of the α-disulfone. Careful control of the reaction conditions can however limit unwanted products; for example, performing the reaction in cold glacial acetic acid favors sulfonic acid formation ⟨42JOC23, 49JA2221, 51JOC767⟩.

$$\mathrm{PhSO_2Na} \xrightarrow[\text{18-Crown-6, 25 °C, 2.5 h, 80\%}]{\mathrm{KO_2,\ py}} \mathrm{PhSO_3Na} \qquad (118)$$

2.03.4.3.3 From sulfonic acid derivatives

Sulfonate esters and amides are readily hydrolyzed to give the parent acid, but, as these must usually be prepared from the acid to start with, this is not normally a useful method ⟨52JA2006, 56JCS173, 85JOC5592⟩. In contrast, alkanesulfonyl halides may be synthesized by various methods and their hydrolysis is a useful preparative method for alkanesulfonic acids. The hydrolysis takes place in acidic or basic solution or in boiling water (Equation (119)) ⟨55JCS2901, 63JCS3692, 66JCS(C)1171⟩.

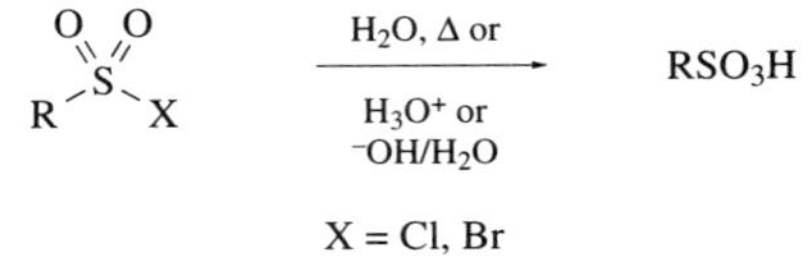

(119)

2.03.5 ALKANESULFENIC ESTERS AND THEIR HIGHER COORDINATED DERIVATIVES

2.03.5.1 Alkanesulfenic Esters

Less elusive than the parent acids, alkanesulfenic esters are isolable though still highly reactive. The instability of the acids generally precludes their use in the synthesis of alkanesulfenic esters, and sulfenyl chlorides are often the most useful starting materials ⟨B-90MI 203-06⟩.

2.03.5.1.1 From alkanesulfenyl halides

The reaction of an alkanesulfenyl chloride with an appropriate alcohol is probably the method of choice for the synthesis of alkanesulfenic esters, and indeed it has been used for many years. The reaction is normally carried out in the presence of a base, commonly pyridine or hydroxide ion ⟨19CB413⟩. The synthesis of esters of trihalomethanesulfenates, especially the trichloro derivatives, has been extensively studied, commonly using the reaction of trichloromethanesulfenyl chloride and an alcohol in pyridine (Equation (120)) ⟨56JCS3139, 58JOC1694, 73TL3563⟩. Trifluoromethane sulfenates have been prepared similarly ⟨63USP3081350, 67FRP1493679⟩.

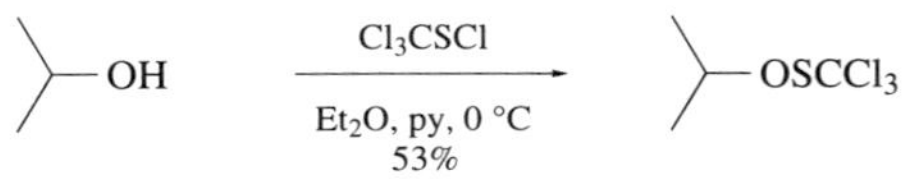

(120)

Alternatively, preformation of the lithium alkoxide from various alcohols before reaction with methanesulfenyl chloride at −40 °C to −60 °C in dimethoxyethane leads to reasonable yields of the *O*-alkylmethane sulfenates (40–70%) ⟨62JOC1398⟩.

Sulfenyl iodides have been used to prepare alkanesulfenic esters. For example, *t*-butanesulfenyl iodide reacts with sodium ethoxide to form *O*-ethyl-*t*-butane sulfenate (Equation (121)) ⟨39CB657⟩.

$$\mathrm{Bu^tSI} \xrightarrow[\text{Et}_2\text{O, 0 °C; no yield given}]{\text{EtONa}} \mathrm{Bu^tSOEt} \qquad (121)$$

Sulfenyl halides have also been reported to react with epoxides, with ring opening, to form the *β*-chloroalkyl sulfenates. Again, trichloromethanesulfenyl chloride has been widely used, in reactions with a range of epoxides (Equation (122)) ⟨58JOC1694⟩.

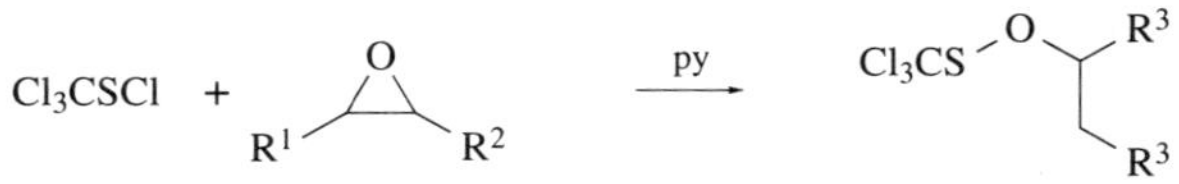

(122)

R¹	*R²*	*Yield (%)*
H	H	85
H	Me	83

2.03.5.1.2 *From alkanesulfenic esters*

O-Benzyl trichloromethane sulfenates undergo selective cleavage of the oxygen–sulfur bond when treated with an alcohol under acid conditions to form a new sulfenate ester (Equation (123)) ⟨73TL3563⟩. Other moieties can be similarly replaced, particularly aryl groups ⟨70JA3035⟩.

Cyclic sulfenates can form the ring-opened sulfenates as reactive intermediates upon similar treatment with an alcohol (Scheme 146) ⟨85JA1783⟩.

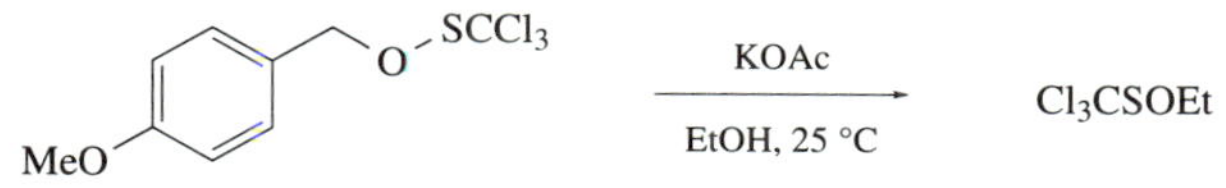

(123)

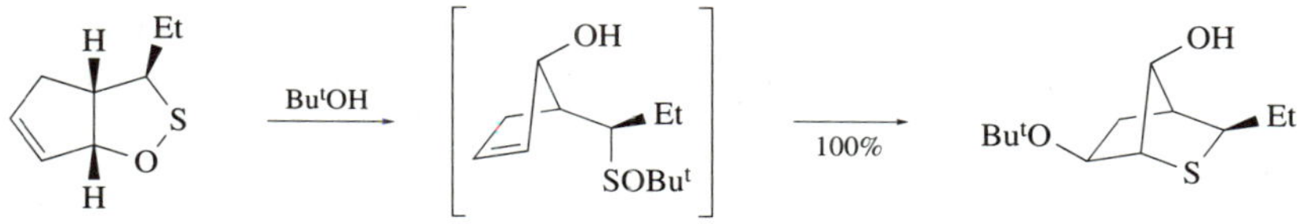

Scheme 146

2.03.5.1.3 *From thiirane-S-oxide*

The reaction of thiirane-*S*-oxide with alkyl chloromethyl ethers results in formation of *β*-chloroethylsulfenic esters in good yields (Equation (124)) ⟨71S590⟩.

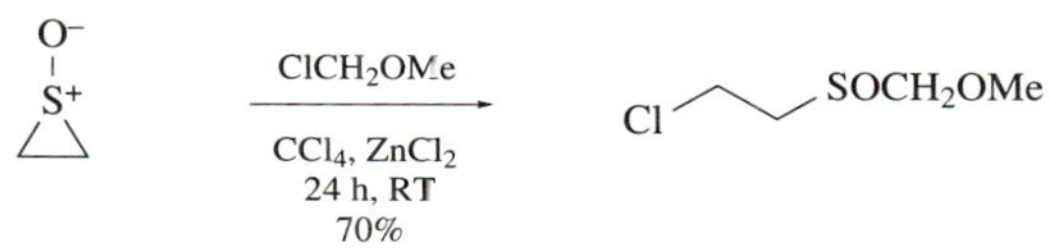

(124)

2.03.5.1.4 *From N-(alkanethio)phthalimides*

The reaction of a sodium alkoxide with *N*-*t*-butylthiophthalimide produces the *O*-alkyl *t*-butane sulfenate in moderate yield (Equation (125)) ⟨70CC1466⟩.

An intramolecular variation of this reaction has been used to form a cyclic sulfenate (sultene), which was characterized by NMR and mass spectroscopy. The product was formed by thermal decomposition of 3-(phthalimidothio)-propan-1-ol in vacuo (Equation (126)) ⟨81CC741, 81CC742⟩.

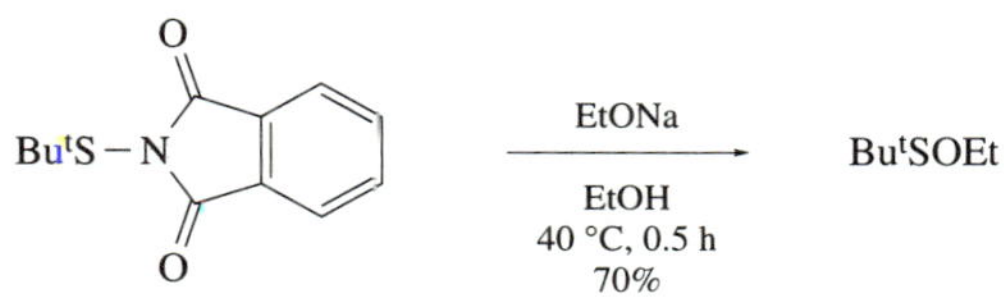

(125)

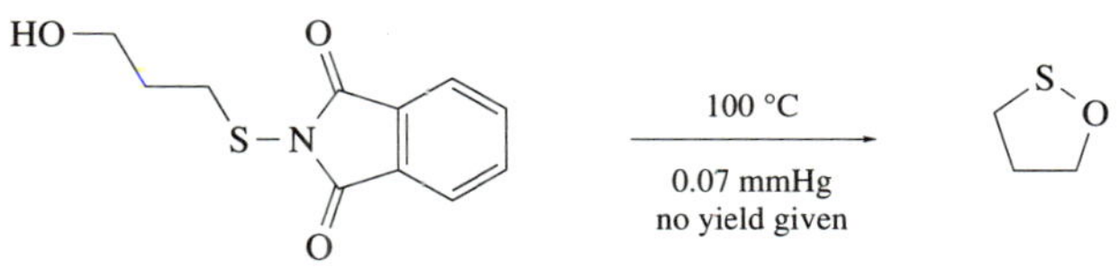

(126)

2.03.5.1.5 *By rearrangements*

Allylic sulfoxides undergo a well known [2,3] rearrangement to give the allylic sulfenates. The rearrangement is an equilibrium process with a low energy barrier, and is one mechanism of allylic sulfoxide racemization. The equilibrium generally favors sulfoxide, although the presence of a thiophile such as trimethylphosphite or hydrolytic conditions can drive the reaction via the sulfinate ultimately to give allylic alcohol (Scheme 147) ⟨74ACR147, B-78MI 203-01, B-88MI 203-04, 91COS(6)873⟩.

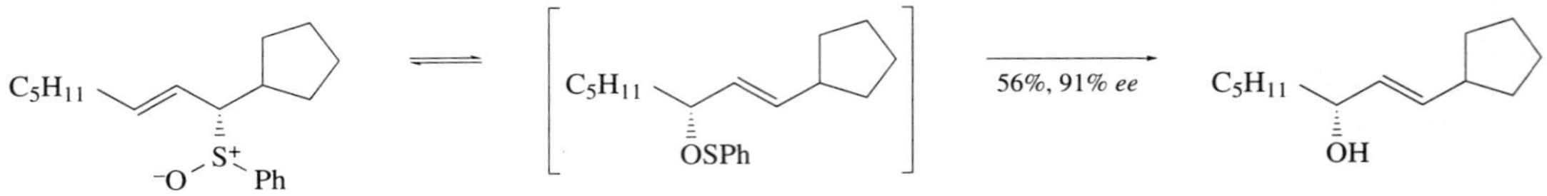

Scheme 147

Many examples are known of this reversible rearrangement, in one or both directions, although the vast majority involve aryl substrates. In one interesting example, the rearrangement occurs in a cyclic system to form the sultene (Equation (127)) ⟨85JA1783⟩. The rearrangement has been used in several particularly elegant natural product syntheses by Grieco, which take advantage of the reaction equilibrium (Scheme 148) ⟨82JA5781, 86JA5908⟩.

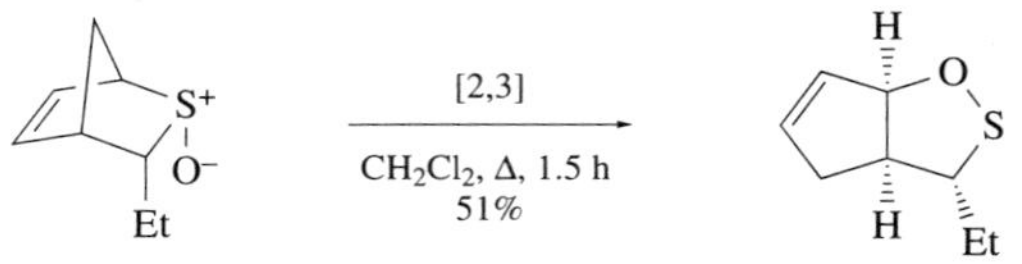

(127)

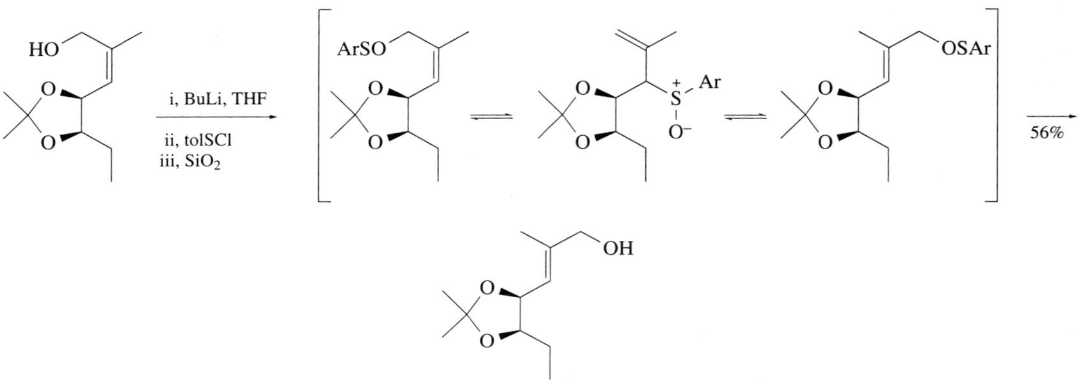

Scheme 148

2.03.5.2 Alkanesulfinic Esters

The extra stability of sulfinic esters over the corresponding acids mirrors that seen for sulfenate esters, and indeed the configurational stability of the tetrahedral structure has led to their use in asymmetric synthesis, particularly for the formation of optically active sulfoxides (the Andersen synthesis, Equation (128)).

(128)

2.03.5.2.1 From sulfinyl halides

The reaction of sulfinyl chlorides with an alcohol constitutes the most widely used and convenient preparation of sulfinate esters ⟨65JOC633, 79COC(3)317, B-90MI 203-07, B-90MI 203-09⟩. The method has been applied to give optically active sulfinates for use in the Andersen synthesis by using an enantiomerically pure chiral alcohol such as menthol (Equation 128). The diastereomeric sulfinates are usually readily separated.

Asymmetric induction has also been observed in the reaction of a sulfinyl chloride with an achiral alcohol in the presence of a chiral tertiary amine at low temperature (Equation (129)) (8.9–43.6% *ee*) ⟨74CC547⟩.

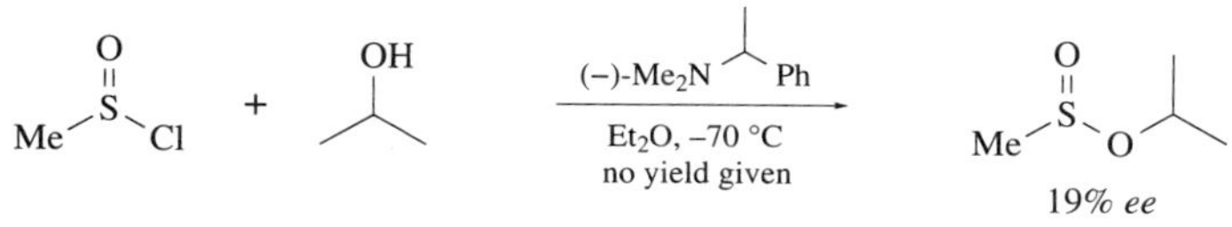

(129)

2.03.5.2.2 *From sulfinic acids*

The direct esterification of sulfinic acids with an alcohol requires activation of the sulfur–oxygen bond. The instability of alkyl sulfinic acids is one drawback to this method, although it has proved successful in many cases. Dicyclohexylcarbodiimide has been used as an activator to good effect ⟨89PS(45)35⟩, as have diethyl azodicarboxylate–triphenylphosphine, 2-chloro-1-methyl pyridinium iodide–triethylamine, and diphenyl phosphochloridate–pyridine (Scheme 149) ⟨80CPB134⟩. The acid-catalyzed reaction often used for esterification of carboxylic acids results in disproportionation to sulfonates and thiosulfonates.

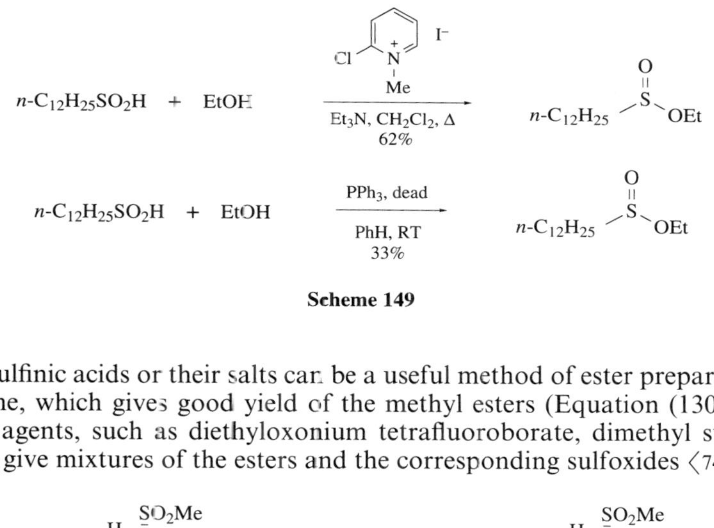

Scheme 149

Alkylation of sulfinic acids or their salts can be a useful method of ester preparation, particularly with diazomethane, which gives good yield of the methyl esters (Equation (130)) ⟨92JCS(P1)2371⟩. Other alkylating agents, such as diethyloxonium tetrafluoroborate, dimethyl sulfate and methyl toluenesulfonate, give mixtures of the esters and the corresponding sulfoxides ⟨74CB3026, 75S519⟩.

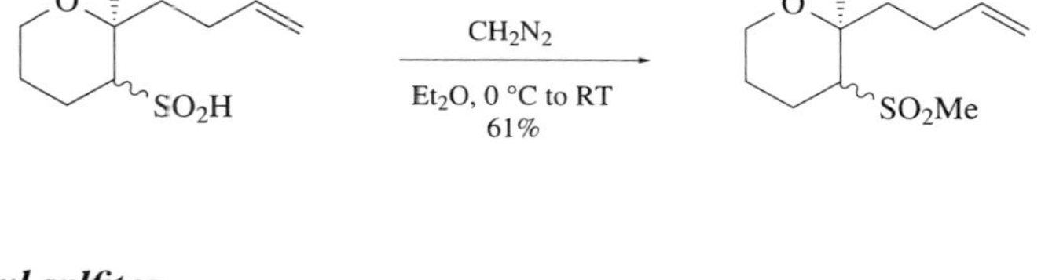

(130)

2.03.5.2.3 *From alkyl sulfites*

The preparation of sulfinic esters from sulfites has become important since the 1970s, particularly for stereocontrolled sulfinate synthesis and subsequent transformation into optically active sulfoxides.

It has been reported that reaction of a tertiary alkyl Grignard reagent with a dialkyl sulfite only proceeds as far as the sulfinate ester, even using an excess of the reagent in refluxing tetrahydrofuran. Yields of 50–90% were realised for a range of tertiary alkyl magnesium chlorides and various symmetrical dialkyl sulfites (Equation (131)) ⟨74S124⟩. The same authors have investigated an asymmetric version of this reaction by the introduction of a series of chiral amino alcohols. Highest enantiomeric excesses ($\approx 70\%$) were obtained when quinine was used as the chiral mediator, and the absolute configuration at the sulfur center could be predicted by choice of amino alcohol:

(+)-(*S*)-isopropyl-*t*-butanesulfinate was formed in the presence of quinine, while the (−)-(*R*)-enantiomer predominated when quinidine was used (Scheme 150) ⟨85CC1670⟩.

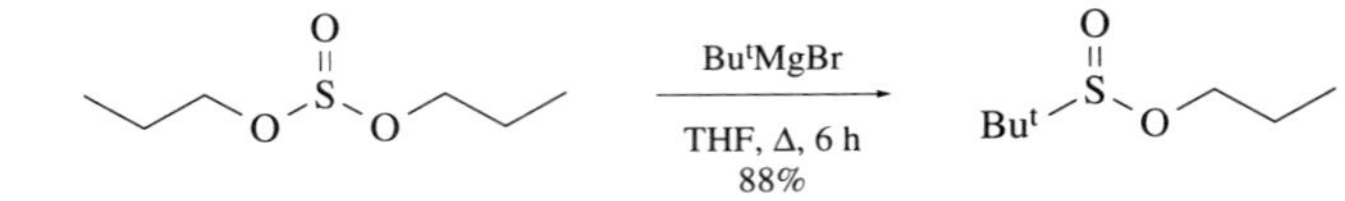

(131)

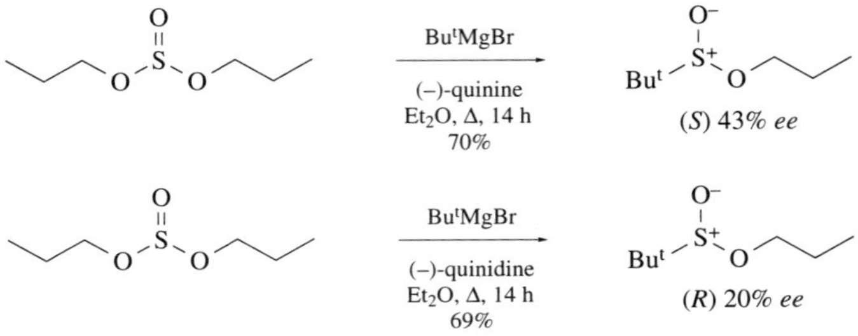

Scheme 150

The use of cyclic sulfites derived from a chiral diol and thionyl chloride has been investigated. Curiously, the regiochemistry of the displacement is dependent upon the organometallic reagent used (Scheme 151). The high crystallinity of the products allows ready crystallization to optical purity ⟨89TL3659, 91JOC5991⟩. The starting materials in this case, thionyl chloride and (*S*)-ethyl lactate, are readily available and inexpensive, and the methodology provides a versatile avenue for the synthesis to chiral molecules with predictable stereochemistry and high optical purity.

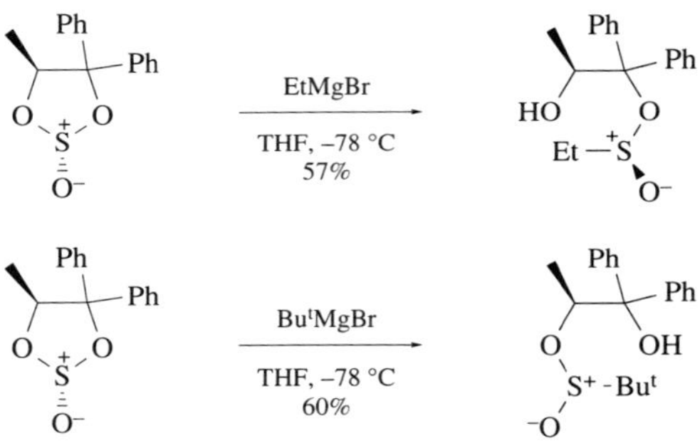

Scheme 151

2.03.5.2.4 *From disulfides*

The oxidation of a disulfide in the presence of an alcohol can lead to the formation of a sulfinic ester. Diaryl sulfides can undergo this transformation upon treatment with lead tetraacetate, but this method has not been applied to dialkyldisulfide synthesis, and in any case can be low yielding ⟨61JA1256⟩.

The use of *N*-bromosuccinimide (NBS) as an oxidant in alcoholic solvents leads to oxidation and alcoholysis at both sulfur centers, so that one mole of disulfide gives two moles of sulfinic ester. Yields are in the range of 32–92% for a variety of disulfides and alcohols. Reactions with solid or nonvolatile alcohols are performed in dichloromethane instead of alcohol solvent; the addition of potassium carbonate increases the yield for secondary alcohols (Scheme 152).

An alternative strategy is required for tertiary or poorly nucleophilic alcohols, such as menthol. In such cases the disulfide is initially oxidized to the intermediate thiosulfinate with mcpba, and subsequent addition of NBS and potassium carbonate furnishes the sulfinate ester in 17–90% yield (Equation (132)) ⟨88S252⟩.

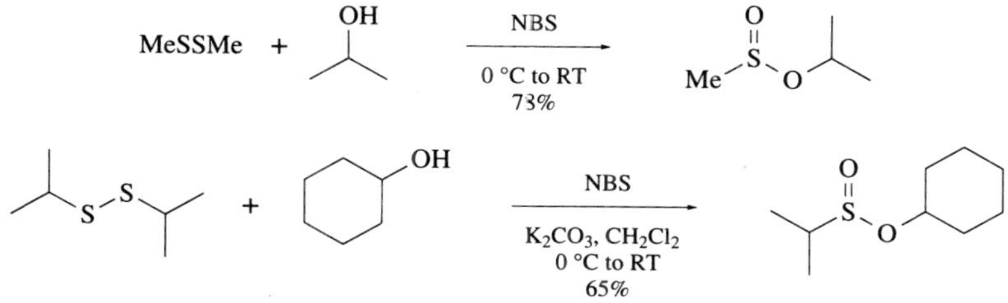

Scheme 152

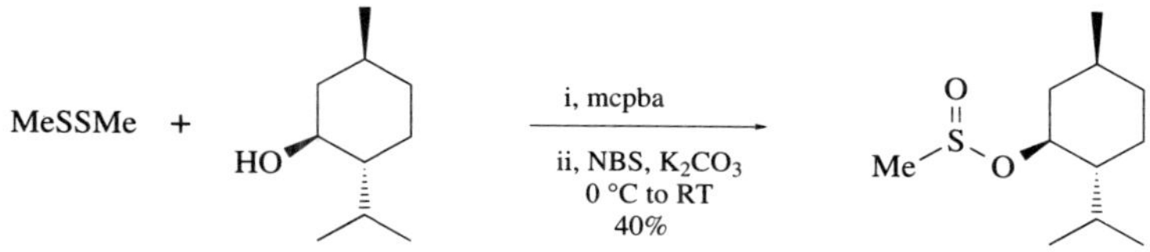

(132)

N-Chlorosuccinimide (NCS) has also been used to carry out this transformation with a hydroxy disulfide to form the cyclic sulfinate ester (sultine) in 90% yield (Scheme 153). The use of chlorine results in attack of the initially formed ester to give the ring opened sulfonyl chloride ⟨81JOC5408⟩.

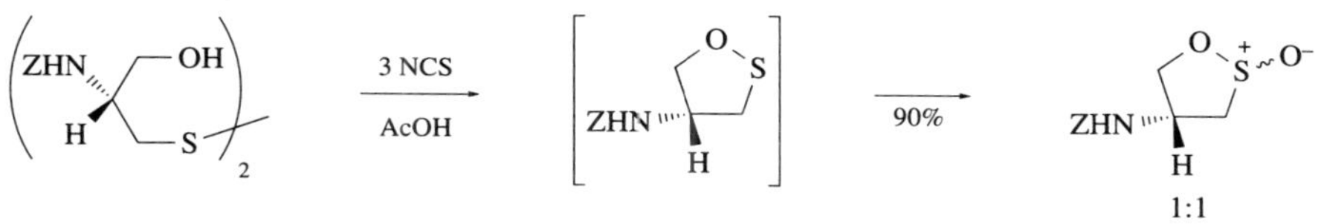

Scheme 153

2.03.5.2.5 From thiosulfonates

It is possible to synthesize sulfinic esters from thiol sulfonic esters, their more stable higher coordinated analogues, by a reductive desulfurization process. Reaction of thiol sulfonates with trialkylphosphites results in nucleophilic attack by the phosphite at the sulfonyl center with displacement of alkanesulfinate. Alkylation of these ambident anions can occur at either sulfur or oxygen, but in this case only the products of *O*-alkylation, alkanesulfinate esters, are observed (Equation (133)) ⟨60JCS1665⟩.

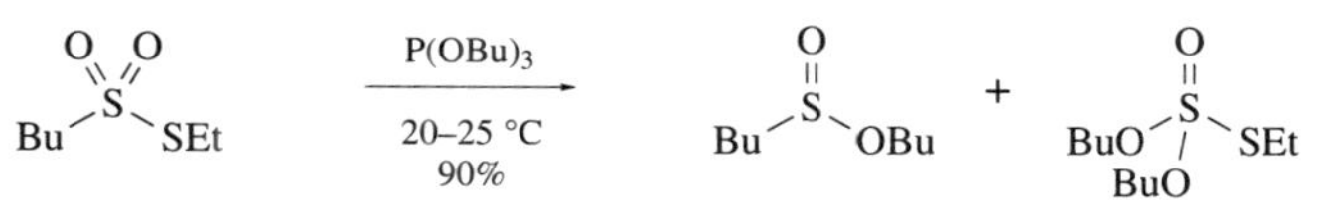

(133)

When a different phosphorus reagent, tris(diethylamino)phosphine, is used, the yield of sulfinate ester drops to 39% at most, with sulfone formation predominating. If the thiolsulfonate is cyclic, however, excellent yields of the cyclic sulfinate ester are realised (Equation (134)) ⟨69TL1447, 71JOC322⟩.

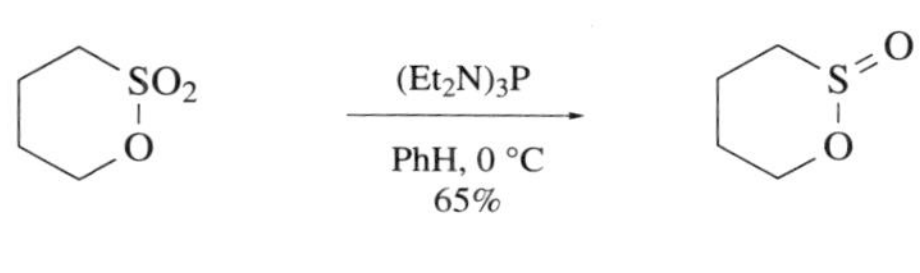

(134)

2.03.5.2.6 From sulfoxides

When an alkyl *t*-butyl sulfoxide is treated with *N*-bromo- or *N*-chlorosuccinimide (NCS) in chloroform/ethanol solution, the *t*-butyl carbon–sulfur bond is cleaved and ethyl alkanesulfinate produced (Scheme 154). Similar reaction occurs when sulfuryl chloride is used in place of NCS. The reactions presumably proceed through initial electrophilic chlorination of sulfur followed by displacement of chloride by ethanol. The ethoxysulfonium ion decomposes to form the sulfinate and *t*-butyl chloride ⟨73CC4⟩. If the alkyl *t*-butyl sulfoxide contains a hydroxyl group, then an intramolecular reaction may take place to form the sultine ⟨73TL2863, 74JA935, 76CJC3012⟩.

The ring-opening reaction of episulfoxides by copper(II) halides in absolute alcohol at 0 °C leads to the formation of alkyl-β-haloethane sulfinates (Equation (135)) ⟨69TL2743⟩.

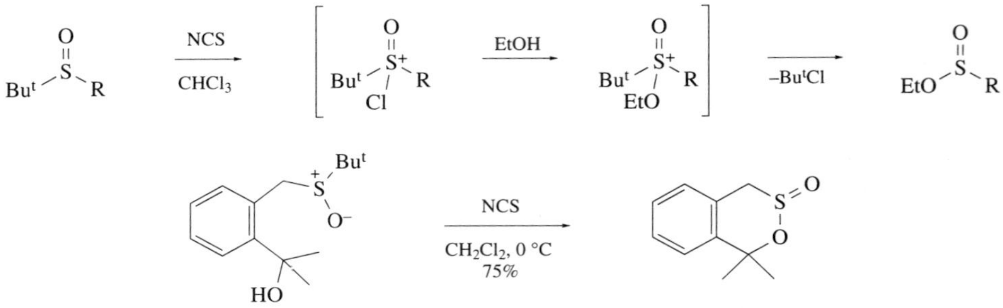

Scheme 154

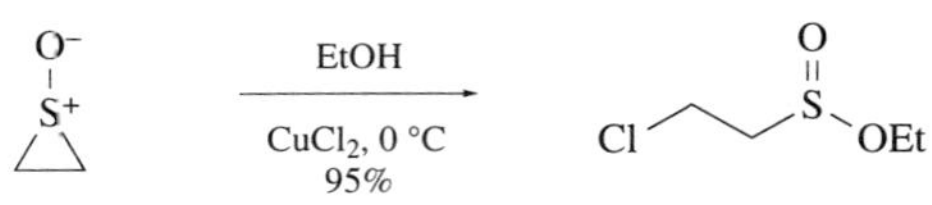

(135)

2.03.5.2.7 *From sulfones*

Treatment of a hydroxysulfone with a base induces carbon–sulfur bond cleavage to give a carbonyl sulfinate anion. Reduction of the carbonyl group with sodium borohydride, to give the hydroxysulfinate, followed by acidification results in formation of a cyclic ester (Scheme 155) ⟨84TL5287⟩.

Thietane dioxides have been found to rearrange to give the five-membered ring sultines on treatment with *t*-butoxymagnesium bromide (Equation (136)) ⟨68CC9, 71JOC2693, 71JOC2703⟩.

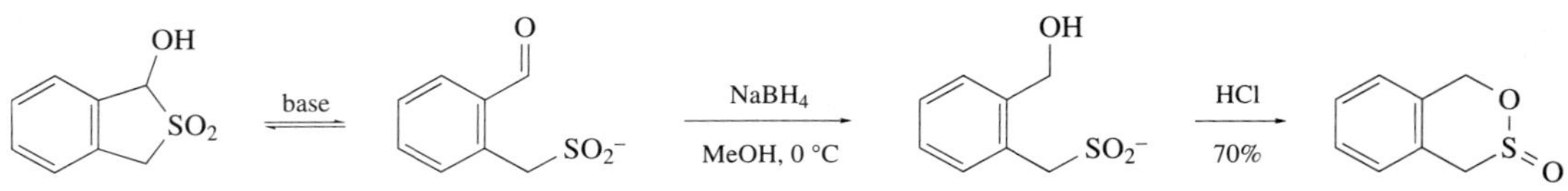

Scheme 155

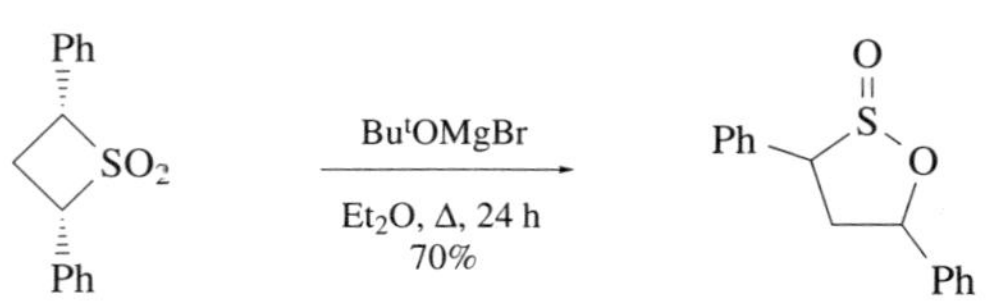

(136)

2.03.5.2.8 *From sulfinamides*

The known acid-promoted reaction between an aryl sulfinamide and an alcohol has been applied to the synthesis of alkyl alkanesulfinates. Using an allylic sulfinylmorpholine and a range of alcohols, the sulfinate esters were prepared in 83–91% yield using 1.5 equivalents of boron trifluoride etherate as catalyst in toluene solution (Equation (137)) ⟨88TL3255⟩.

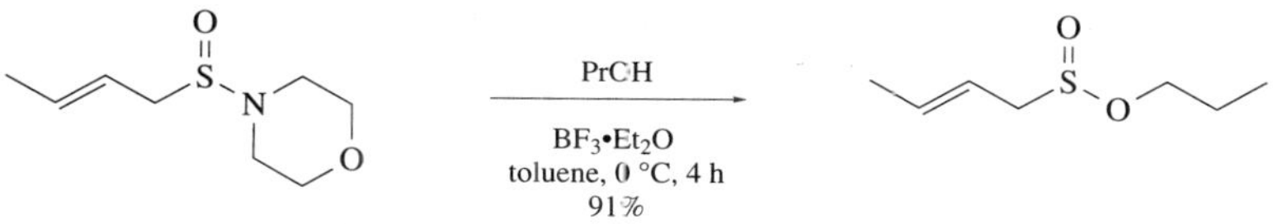

(137)

2.03.5.2.9 *Miscellaneous*

(i) From sulfoxylic dialkyl esters

Sulfoxylic dialkyl esters are alkylated in reasonable yields by methyl or ethyl iodide, giving the alkyl methanesulfinate or alkyl ethanesulfinate respectively (Equation (138)) ⟨76S829⟩.

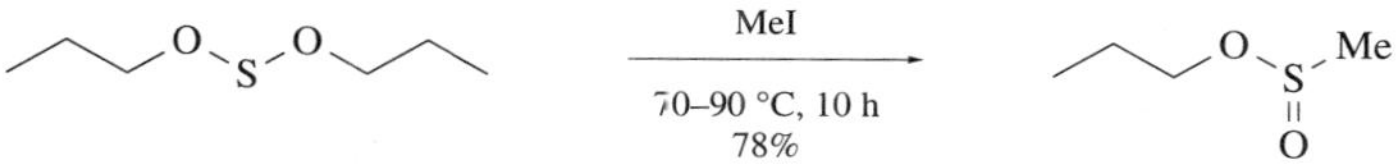

(138)

(ii) From a hydroxy thiol

Chlorination of a hydroxy thiol at −78 °C followed by addition of water leads to formation of a sulfinate ester. The structure of the product is determined by the number of carbon atoms between thiol and hydroxy groups: ring formation occurs to give five- and six-membered sultines (Equation (139)), whereas longer chain lengths can results in polymeric sulfinate esters. Yields of 70–100% are observed for the cyclic products ⟨89TL2763⟩.

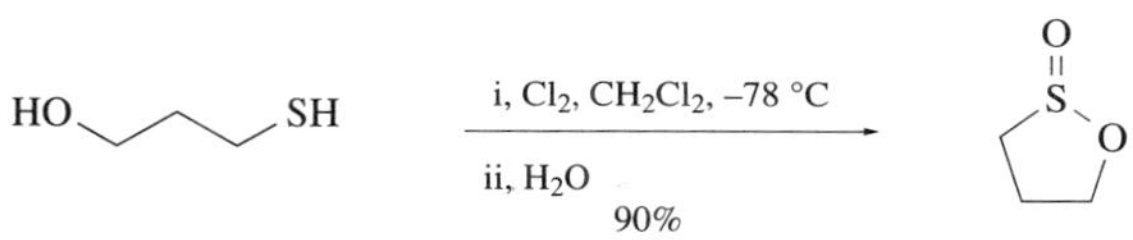

(139)

(iii) From dienes and sulfur dioxide

Sulfur dioxide can be trapped by a diene in a $[2n + 4\pi]$ cycloaddition, usually to form the 2,5-dihydrothiophene, however it is also possible to produce the 1,2-oxathian-2-oxide (i.e., cyclic sulfinate) product by an alternative cycloaddition (Scheme 156). Only a few examples are recorded of this transformation, perhaps as a result of the instability of the products, which undergo facile cycloreversion to the 1,3-diene and sulfur dioxide ⟨78TL2353⟩.

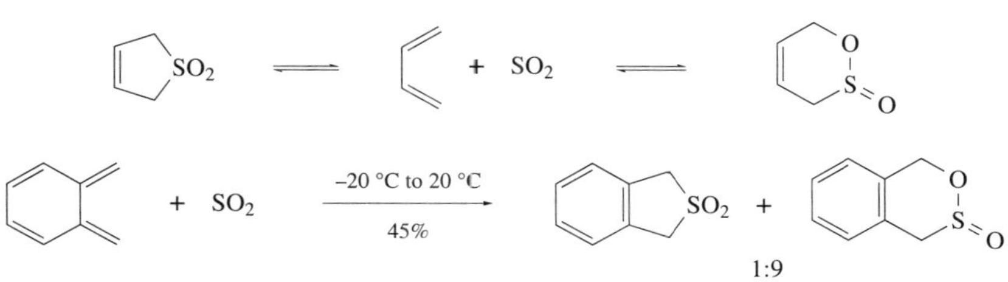

Scheme 156

(iv) From unsaturated alcohols

Cyclic sulfinate esters have been produced by the reaction of an unsaturated alcohol with *N*-sulfinyl-*p*-toluenesulfonamide. This interesting reaction is promoted by the addition of 0.5 equivalents of boron trifluoride etherate (Equation (140)) ⟨93CC1195⟩.

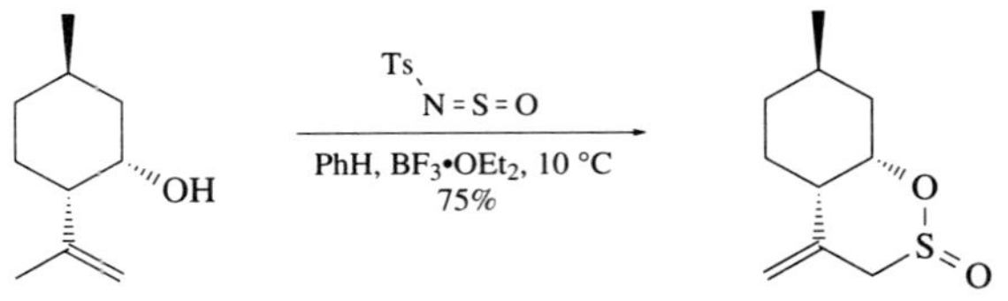

(140)

(v) From thioketenes

The oxidation of thioketenes by singlet oxygen when performed in the presence of an alcohol can lead to the formation of sulfinate esters. By cooling the reaction to −78 °C yields are considerably improved; for example at room temperature the product derived from the reaction between 1,1,3,3-tetramethyl-2-(thiocarbonyl)cyclohexane and methanol is produced in only 25% yield with several by-products. At −78 °C the reaction proceeds in 65% yield with little contamination from other materials (Equation (141)) ⟨84JOC615⟩.

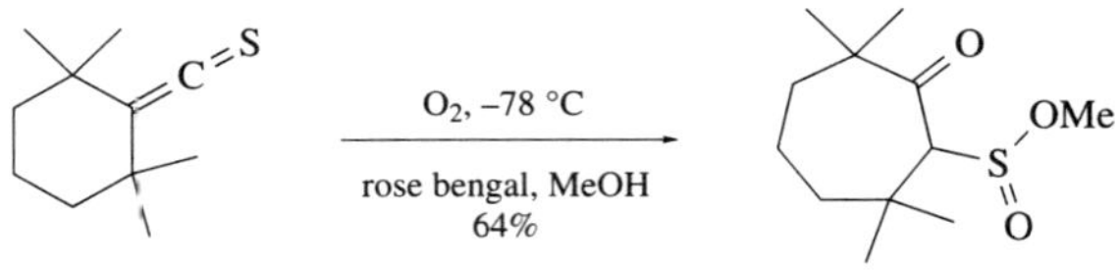

(141)

2.03.5.3 Alkanesulfonic Esters

The utility of the sulfonate ester as a leaving group is widely recognised, particularly in mesylates, tosylates, triflates, etc. Their use as protecting groups for alcohols capable of chemoselective removal has also been developed ⟨79COC(3)331, B-90MI 203-08, B-90MI 203-01⟩.

2.03.5.3.1 From sulfonyl chlorides

The most widely used method for the synthesis of sulfonyl esters is the reaction between a sulfonyl chloride and an alcohol in the presence of a base, no doubt due to the ready accessibility of the starting materials. The reaction is usually performed at low temperature under anhydrous conditions, the reactive intermediate, at least in some cases, being the sulfene generated by the action of the base on the sulfonyl chloride (Equation (142)) ⟨70CJC845, 75ACR10⟩.

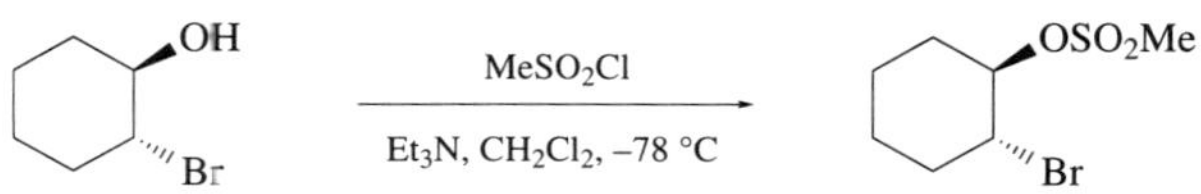

(142)

Chiral sulfonyl chlorides, such as that derived from camphor, have found use in the resolution of racemic mixtures of alcohols. The diastereoisomeric mixture of sulfonic esters can often easily be separated by simple crystallization ⟨54JA5011⟩. The chirality in the sulfonic ester can be exploited in further transformations of the molecule, including rearrangements and cycloadditions (Equation (143)) ⟨82TL5427, 84TL5889⟩.

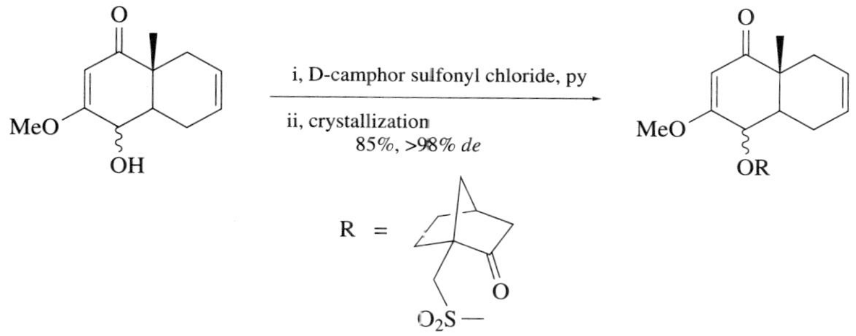

(143)

2-Arylethanesulfonic esters have found use in the nucleoside field as 2′-hydroxyl protecting groups ⟨89MI 203-02⟩. The allylsulfonyl group has also been used to protect alcohols as the sulfonic ester, by reaction of the alcohol with allylsulfonyl chloride in the presence of pyridine at −78 °C in dichloromethane solution. Yields of 55–97% were realised for a series of carbohydrate alcohols. This sulfonyl group can be cleaved using a palladium(0)-catalyzed transfer of the allyl group to morpholine, and is stable to conditions used for desilylation, deacetylation, and deacetalization ⟨91SL163⟩.

The reaction of allylsulfonyl chloride with ethanol under strongly basic conditions (potassium hydroxide in ether at room temperature) causes not only formation of the ethyl sulfonate ester but also addition of ethanol across the double bond (Equation (144)) ⟨55JCS1307⟩.

Alkanesulfonyl chlorides not only react with alcohols, but also with aryl methyl ethers, with loss of the methyl group, in the presence of zinc chloride (Equation (145)) ⟨45JCS14⟩.

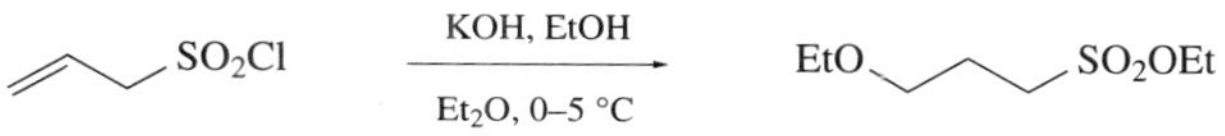

(144)

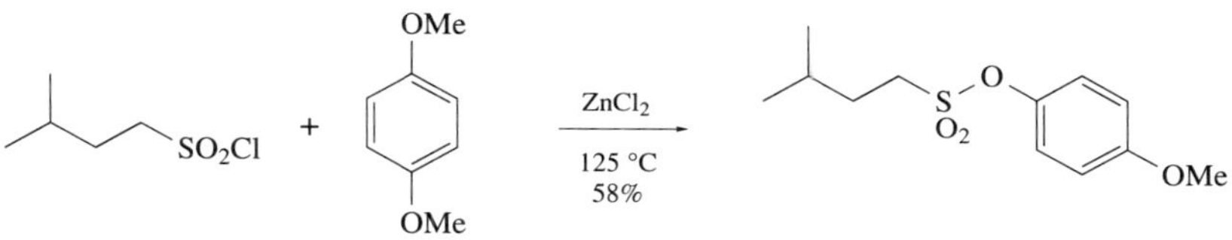

(145)

2.03.5.3.2 From sulfonic acids

Esterification of alkanesulfonic acids with alcohols requires azeotropic removal of water, but can be a useful method of preparation ⟨77AJC2479⟩. If a hydroxysulfonic acid is used, then intramolecular cyclization can take place, particularly for five- and six-membered ring formation (Equation (146)) ⟨54JA5357⟩.

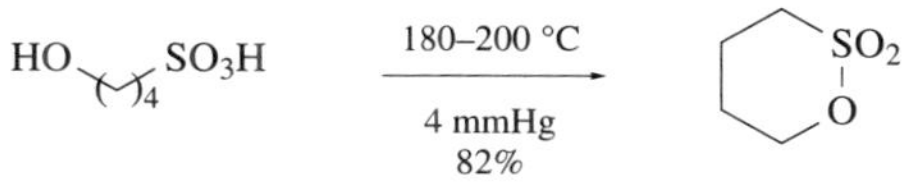

(146)

Direct alkylation of alkanesulfonic acids can give useful yields of the sulfonic esters. Diazoalkanes react cleanly in high yield. 2,2,2-Trifluorodiazoethane selectively alkylates sulfonic acids in the presence of carboxylic acids to form stable crystalline products (Equation (147)). As the trifluoroethyl group can be selectively removed, for example with sodium methoxide in methanol, this reagent has potential as a protecting group for sulfonic acids ⟨48JCS849, 84S1041⟩. Dimethyl sulfate has also been used to prepare the methyl esters of alkanesulfonic acids in good yields ⟨61ZOB1869⟩.

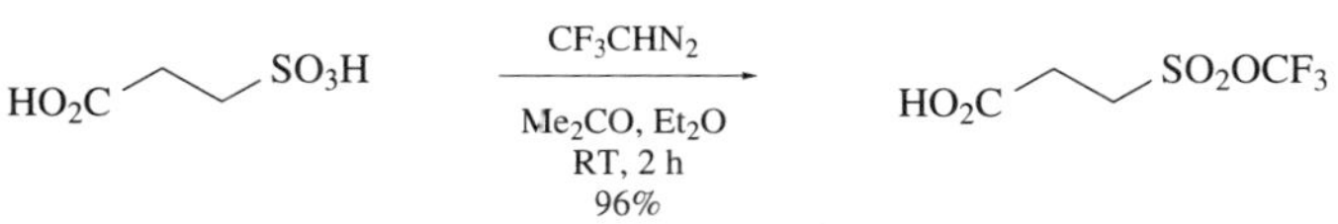

(147)

Hydroxyalkyl sulfonic esters are formed by reaction of an alkanesulfonic acid with an epoxide to give the product of *trans* ring opening (Equation (148)) ⟨36CB2753, 49JCS315⟩.

The addition of sulfonic acids to alkenes constitutes a further method of sulfonate ester formation, often in near quantitative yields (Equation (149)) ⟨65CRV385, 70JA5457, 71CC1614⟩.

A cyclic sulfonate ester (sultone) has been prepared from an ether disulfonate by heating under reduced pressure (Equation (150)) ⟨63OSC(4)529⟩.

Alkanesulfonic acids have also been *O*-alkylated by heating with 2-ammoniumethanesulfonates in toluene (Scheme 157) ⟨81TL3573⟩.

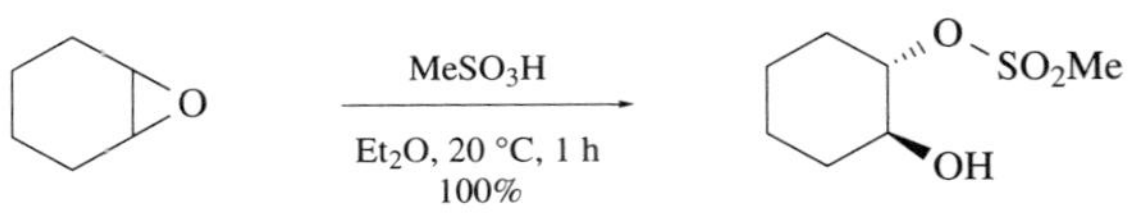

(148)

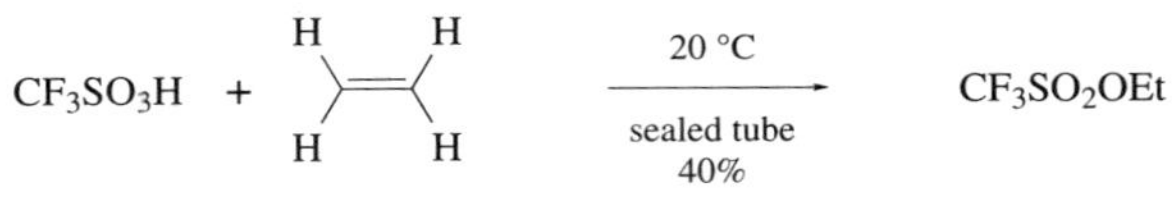

(149)

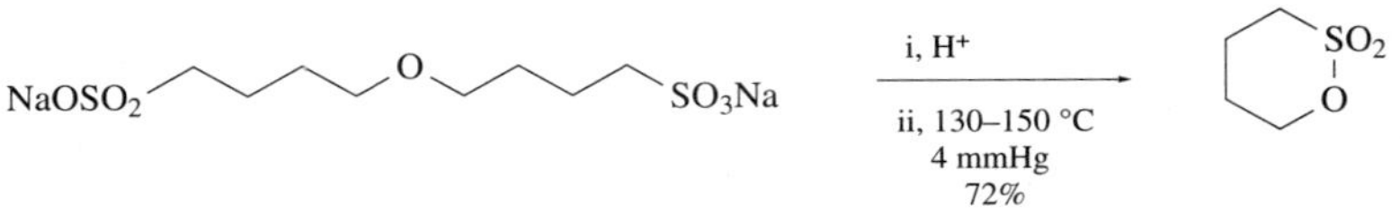

(150)

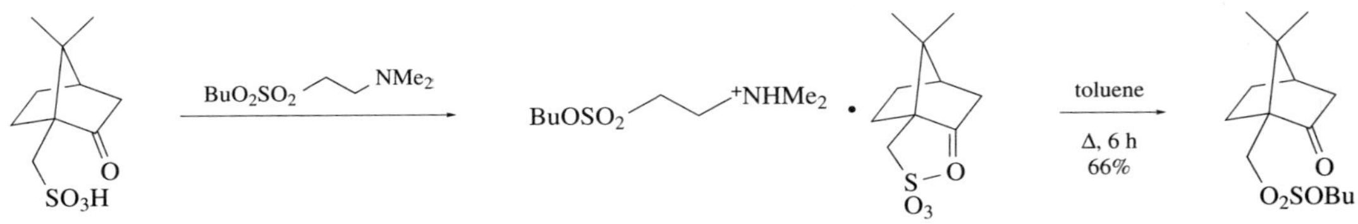

Scheme 157

2.03.5.3.3 *From sulfenes*

As indicated above, sulfenes can be formed by the action of a base upon sulfonyl chlorides. These compounds react *in situ* to give various cycloadducts (Scheme 158) ⟨67AG(E)107⟩. For example, using the reaction of a sulfene with an unsaturated aminoketone, a series of sultones have been prepared in yields of 73–91% ⟨57M1518, 83JHC839⟩. The 1,3 cycloaddition of diaryl nitrones to sulfenes however does not proceed as expected, but is followed by a further rearrangement to form an aryl sulfonate (Scheme 159) ⟨69JOC3097⟩.

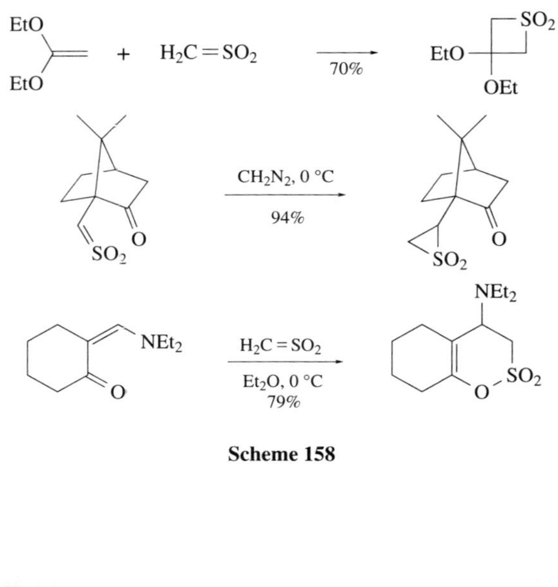

Scheme 158

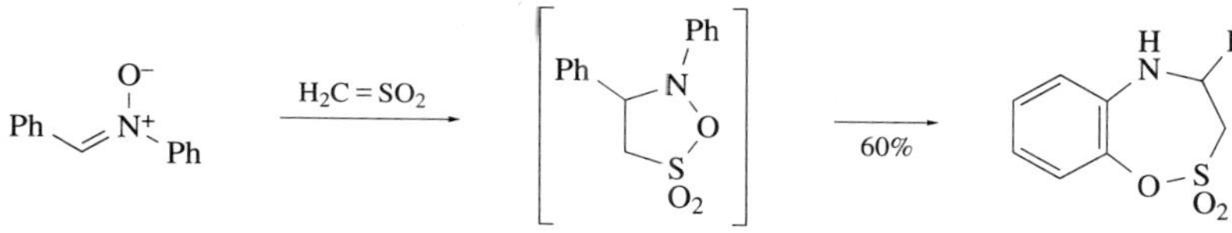

Scheme 159

2.03.5.3.4 *From sulfonamides*

Imidazolylsulfonamides have been used in the synthesis of alkanesulfonate esters through displacement of the imidazole group by an alcohol (Equation (151)) ⟨60CB2902⟩. In a related example, attack by an alcohol on an *N*-methyl ammonium sulfonamide cleaved the sulfur–nitrogen bond, forming an alkanesulfonate ester (Equation (152)) ⟨78JA1637⟩.

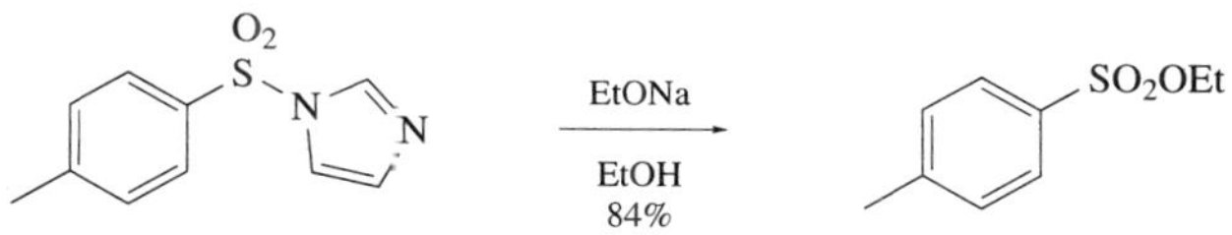

(151)

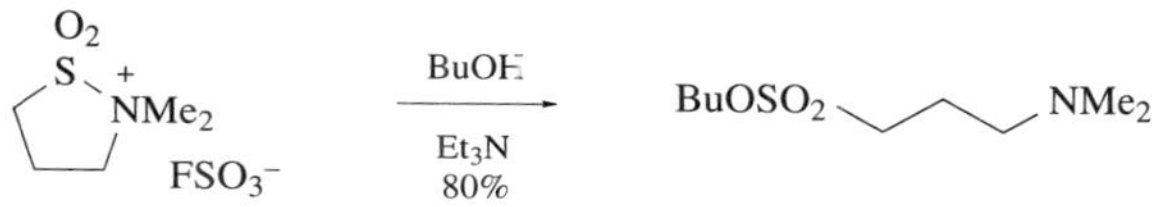

(152)

Addition of an α-lithiated sultam to an enone does not lead to the expected simple addition product, but an amino sultone, due to nucleophilic attack of the enolate oxygen atom at the sulfur centre (Scheme 160) ⟨75JOC1343⟩. Attempts to hydrolyze a phenolic sulfonamide with dilute hydrochloric acid caused unexpected cyclization to give the sultone (Equation (153)) ⟨82CPB332⟩.

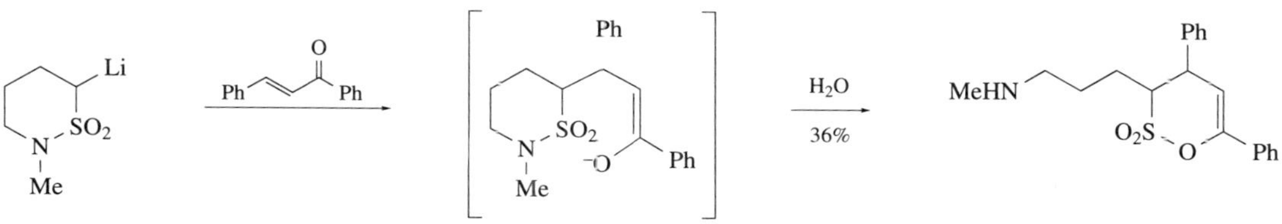

Scheme 160

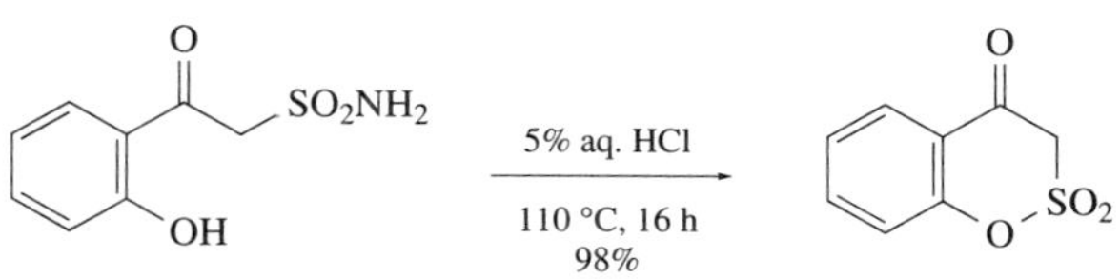

(153)

2.03.5.3.5 *From sulfones*

Photolysis of cyclic sulfones in methanol causes ring opening, possibly via a sulfene intermediate, and reaction with the solvent to give the methyl alkanesulfonate (Equation (154)) ⟨74TL4781, 75CJC3656⟩.

Thermal rearrangement of ketosulfones can lead to almost quantitative conversion to the sulfonate esters (Equation (155)) ⟨80CPB1131⟩.

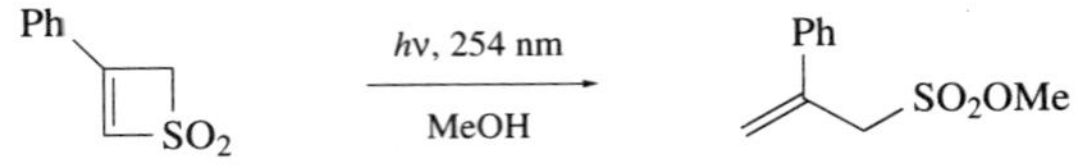

(154)

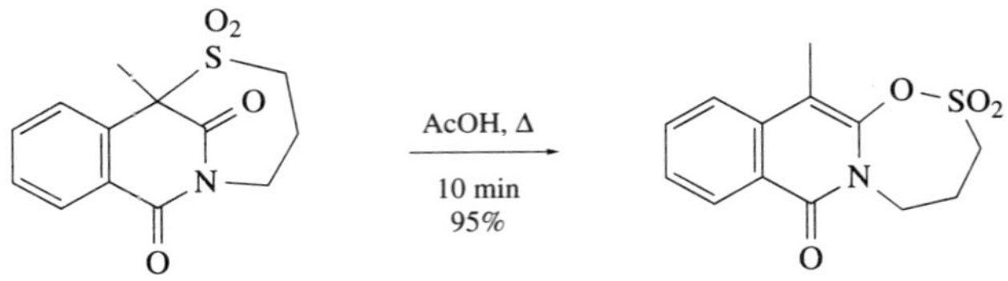

(155)

2.03.5.3.6 *From sulfur trioxide and alkenes*

The addition of sulfur trioxide to an alkene can lead to sultone or alkanesulfonic acid products. Depending upon the structure of the alkene and its ability to stabilize the intermediate carbocation, the intermediate can be trapped by the sulfonyl oxygen atom to form the sultone, or can lose a proton to give a vinyl sulfonic acid as the major product (Scheme 161) ⟨59JA2002, 74JOC2459⟩.

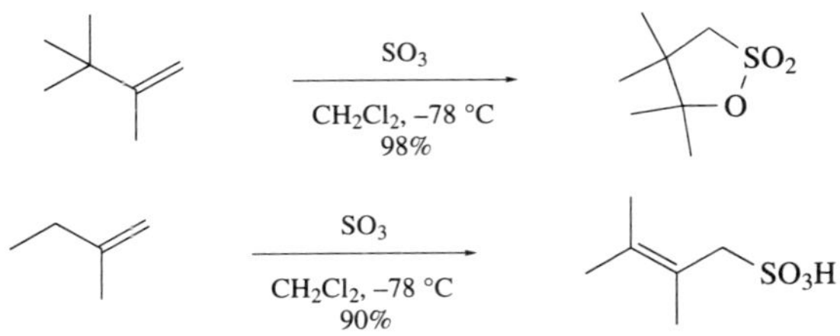

Scheme 161

2.03.5.3.7 *From allylic sulfites*

Allylic sulfites have been found to undergo a facile [2,3] sigmatropic rearrangement upon catalysis by a novel palladium complex to furnish the sulfonates in reasonable yields (Equation (156)) ⟨90JOC1823⟩. Both cyclic and acyclic sulfites react under these conditions. Unsymmetrical allylic groups provide the product with greater substitution at the allylic position. A 1993 publication describes a related rearrangement in cyclic systems, catalyzed by Lewis acid (Equation (157)) ⟨93TL3667⟩.

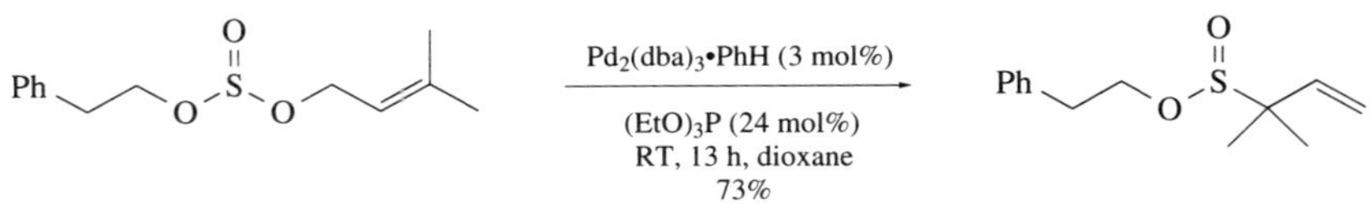

(156)

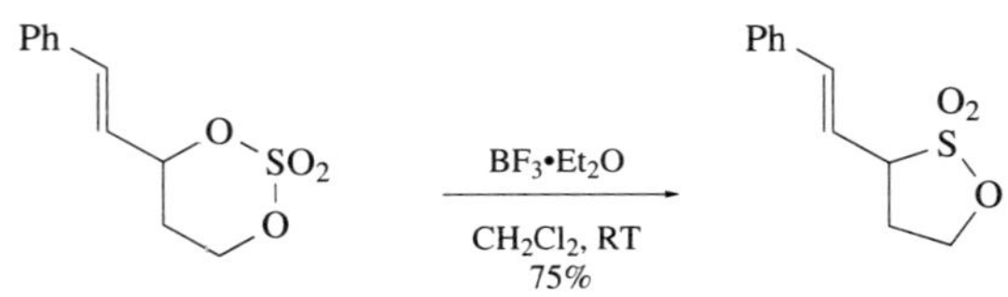

(157)

2.03.5.3.8 *From unsaturated sulfonic esters*

Vinyl sulfonate esters are extremely good dienophiles, cycloadditions occurring at much lower temperatures than their carboxylate counterparts. Metz has investigated the use of these dienophiles in the intramolecular Diels–Alder reaction with cyclic dienes and furans to furnish the cyclic sulfonate esters in excellent yields (Scheme 162). The intermediate trienes often cannot be isolated as the cycloaddition is so facile. Further functionalization is possible by conversion of the sulfonyl group into a ketone by oxidative desulfurization via the α-borinate, the vinyl sulfonate ester then acting as a masked ketone ⟨89AG(E)202, 92SL985, 93SL399⟩.

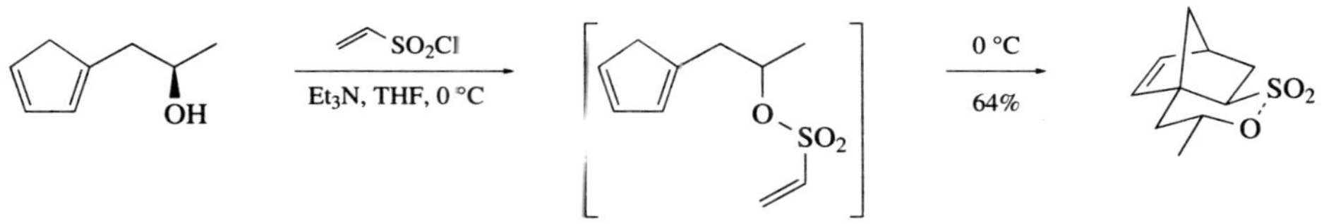

Scheme 162

2.03.5.3.9 *From diazo compounds*

The reaction of a diazoalkane and sulfur dioxide in ethanol can result in formation of ethyl alkanesulfonates. When a diaryl-substituted starting material is used, yields of 87–89% are observed (Scheme 163) ⟨78BCJ1459⟩.

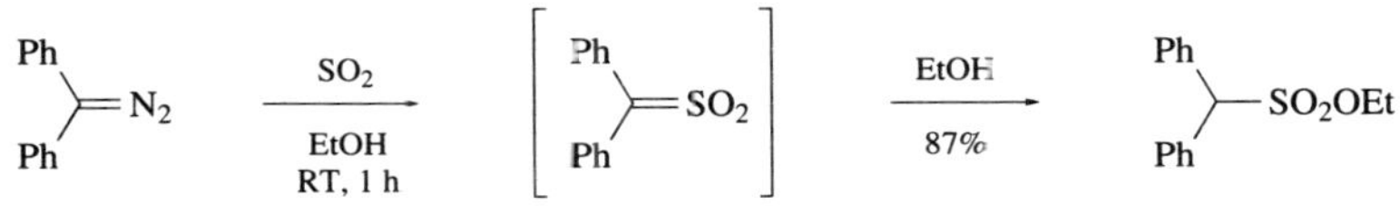

Scheme 163

2.03.5.3.10 *From sulfinate esters*

Oxidation of sulfinate esters to the sulfonyl esters is not a particularly useful method of synthesis, as the sulfinyl starting materials are in general more difficult to prepare than are sulfonates, for which more direct synthetic methods exist. Nonetheless it is still a viable route to alkanesulfonates. Aqueous potassium permanganate is an excellent oxidant ⟨71JOC1314⟩.

2.03.5.3.11 *From sulfonyl esters*

Facile deprotonation adjacent to the sulfonyl group allows easy functionalization of simple sulfonates by alkylation ⟨70JOC1226⟩. In an unusual example, deprotonation of a disulfonate ester with butyllithium at −78 °C followed by warming to 0 °C caused an intramolecular displacement of one of the sulfonate groups and cyclization to form the sultone in 68–88% yield (Scheme 164) ⟨70CJC845⟩.

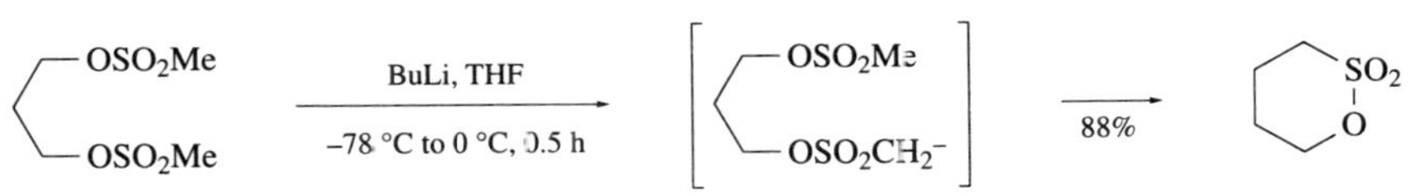

Scheme 164

2.03.6 RSOX FUNCTIONS

2.03.6.1 X = Sulfur

The first successful synthesis of a sulfinic anhydride was reported in 1968, formed in the reaction between the silver salt of *t*-butane sulfinic acid and *t*-butane sulfinyl chloride ⟨68JA7378⟩. Performing the reaction in diethyl ether at $-10\,^{\circ}C$ led to a 50% yield of the anhydride (Equation (158)). Shortly afterwards a cyclic sulfinic anhydride was prepared from 1,2-ethanedisufinyl chloride by a controlled hydrolysis in THF at room temperature to furnish the product in reasonable yield (Scheme 165) ⟨69CC1205⟩.

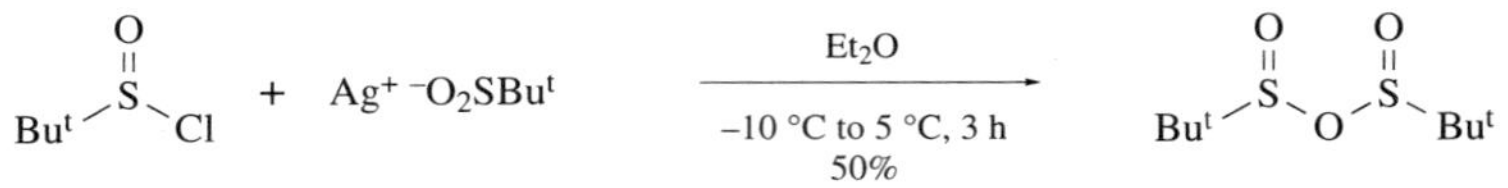

(158)

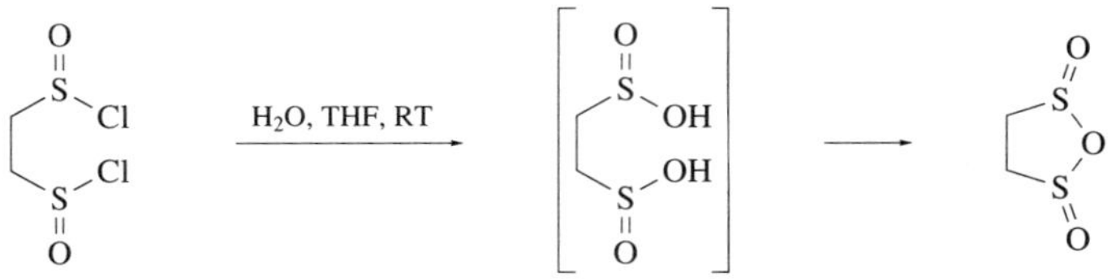

Scheme 165

The direct sulfonation of alkenes with an excess of sulfur trioxide generally leads to formation of cyclic sulfonate–sulfate anhydrides, also known as carbyl sulfates, which have been isolated for simple alkenes such as ethene and propene ⟨57JA5000, 65BSF787⟩. The mechanism of the reaction of sulfur trioxide with octenes has been studied by NMR to deduce the origin of the observed stereochemistry. Initially a β sultone is formed, a rapid reaction even at $-20\,^{\circ}C$, followed by a slow insertion of another molecule of sulfur trioxide at room temperature (Scheme 166) ⟨87TL1703⟩.

The reaction of perfluorinated alkenes with sulfur trioxide has also been investigated. Heating 2-trifluoromethyl perfluoropropene with sulfur trioxide to 158–180 °C in a sealed tube gave a 58% yield of the anhydride ⟨72IZV2510⟩.

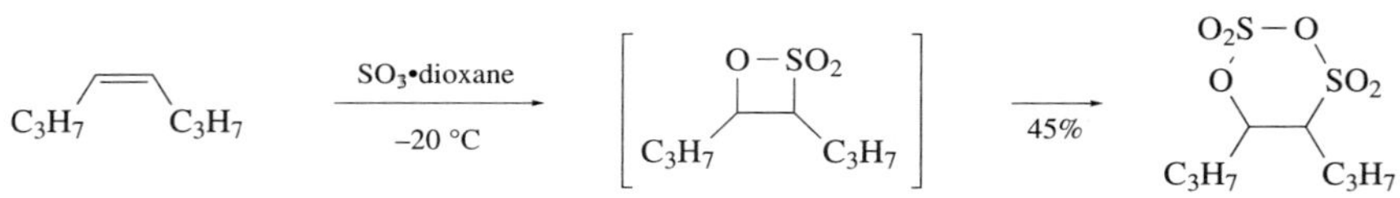

Scheme 166

2.03.6.2 X = Acyl

Mixed anhydrides of carboxylic and sulfonic acids are known, and are powerful acylating agents ⟨71JOC532, 71JOC540⟩. These compounds are very susceptible to hydrolysis and decompose on heating, but can be easily obtained by oxidation of thiolesters with ozone or ruthenium tetroxide. The reaction is proposed to occur by initial sulfur oxidation followed by a rearrangement involving migration of an acyl group from oxygen to sulfur to give an acyl sulfenate anhydride, which is readily oxidized to the acyl sulfonic anhydride (Scheme 167) ⟨76TL2171⟩. The intermediate sufenic anhydride could not be isolated using either ozone or ruthenium tetroxide.

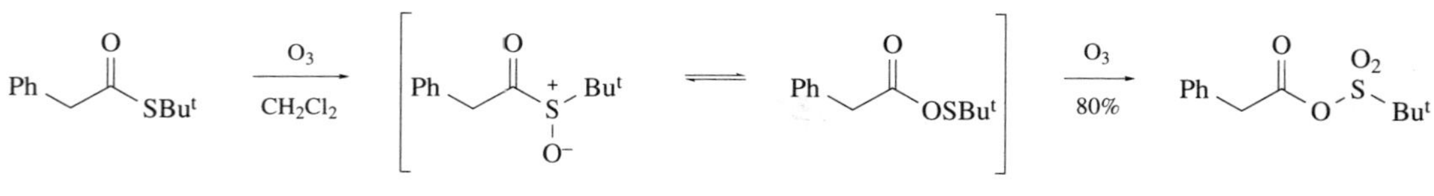

Scheme 167

2.03.6.3 X = Nitrogen

The reaction of sulfines with 1,3-dipoles leads to cycloadducts in good yields. Stable dipolarophiles have been used to study the regio- and stereochemistry of the cycloaddition process. The addition usually observed with benzonitrile oxide forms a carbon–sulfur bond and oxygen–carbon bond (Equation (159)), however, with thiofluorenone-*S*-oxide, the reverse regioselection occurs (Equation (160)) to form the sulfinate product ⟨77JCS(P1)1469⟩. A similar result is obtained when bis(trifluoromethyl) sulfine is used (Equation (161)) ⟨88CB75⟩.

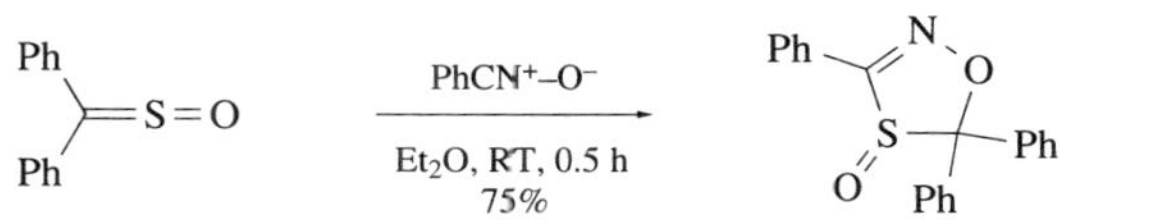

(159)

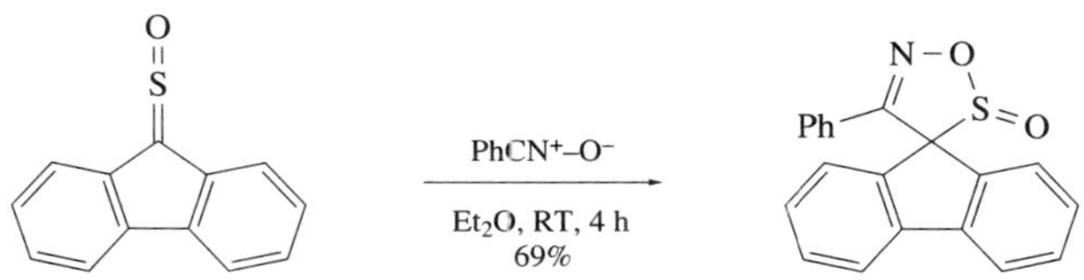

(160)

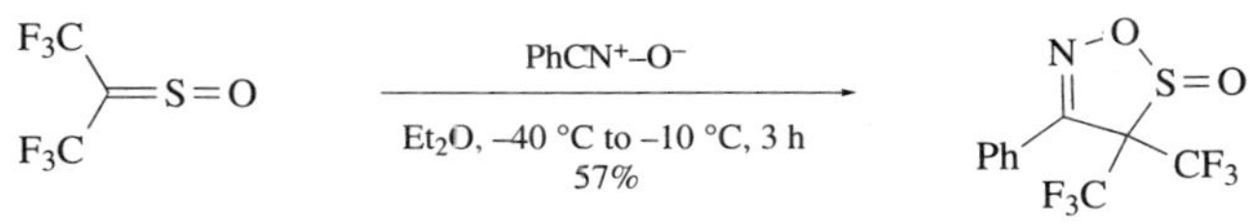

(161)

2.03.6.4 X = Silicon

The reaction of sulfur dioxide with a trimethylsilyl ketene acetal at 0 °C has led to the formation of a trimethylsilyl sulfinic ester in 96% yield (Equation (162)). A similar reaction of ketene acetals with trimethylsilychlorosulfonate produces the trimethylsilyl sulfonate esters ⟨84LA39⟩.

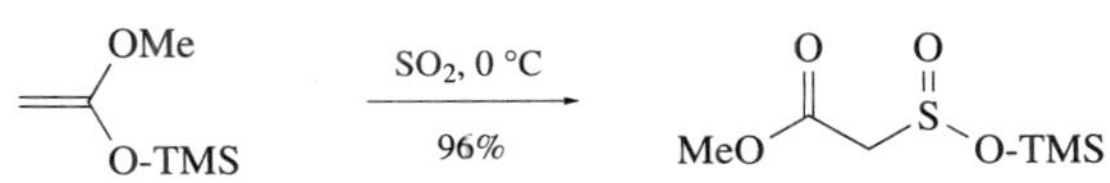

(162)

Aldehydes and ketones react with trimethylsilyl chlorosulfonate either directly or as the derived trimethylsilyl enol ethers to give trimethylsilyl 2-oxosulfonates, while trimethylsilyl esters undergo decarboxylation to give the trimethylsilyl sulfonate esters (Scheme 168). Three- or four-membered cyclic silanes undergo attack by sulfur dioxide and sulfur trioxide to give the cyclic insertion products ⟨69BSF3609, 75JOM(100)237⟩. At −78 °C the addition to dimethyl silacyclobutane furnishes the product in 80% yield in one hour. A more complex silacyclopropane derivative undergoes insertion of sulfur dioxide at room temperature (Scheme 169).

Silyl sulfonates can also be prepared through the reaction of trimethylsilyl halides with sultones, to afford ring opened bromoalkyl sulfonic esters (Equation (163)) in good yield ⟨84ZOB 842⟩.

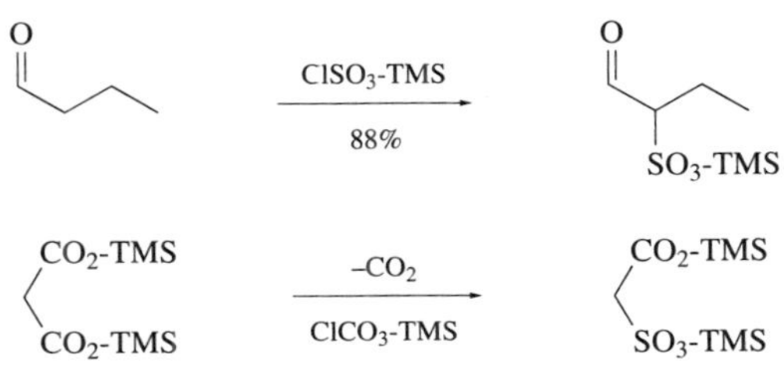

Scheme 168

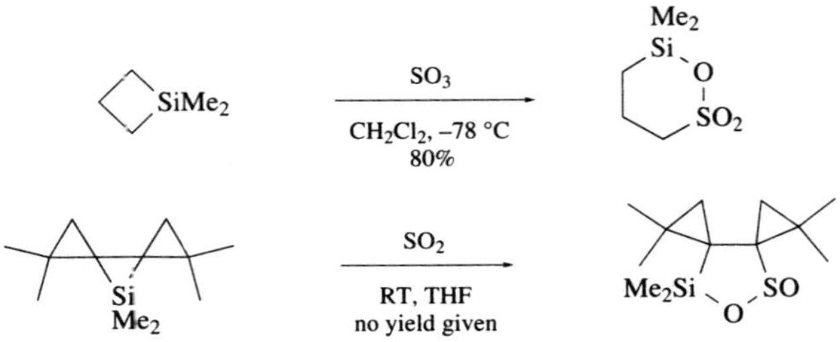

Scheme 169

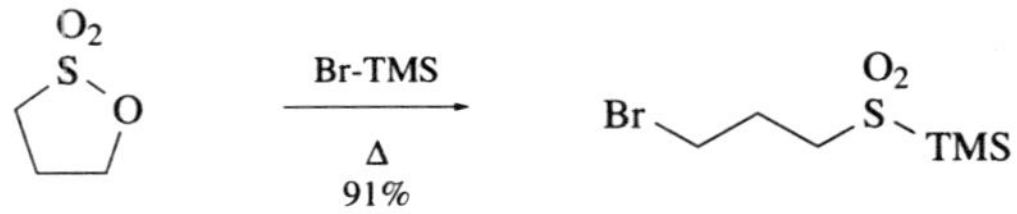

(163)

2.03.6.5 X = Tin

Tin compounds react particularly well with sulfinyl oxygen to give tin sulfinates. The displacement of a sulfinyl group from benzothiazole with two equivalents of tributyltin hydride under radical conditions gives excellent yields of the tin sulfinate (Equation (164)) ⟨86TL215⟩.

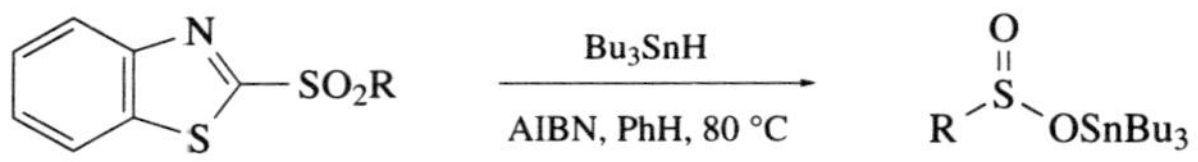

(164)

R	*Yield* (%)
$Me(CH_2)_{15}$	89
$PhCH_2CH_2$	90

AIBN = 2,2'-azobisisobutyronitrile

Cyclopropylcarbinyltrialkyl stannanes react with sulfur dioxide in chloroform or methanol solution, forming allylic tin sulfinates (Equation (165)). In chloroform a concerted mechanism is proposed, although the reaction in methanol occurs without stereoselectivity, as does the corresponding reaction with allyl stannanes ⟨88OM1196, 91TL807⟩.

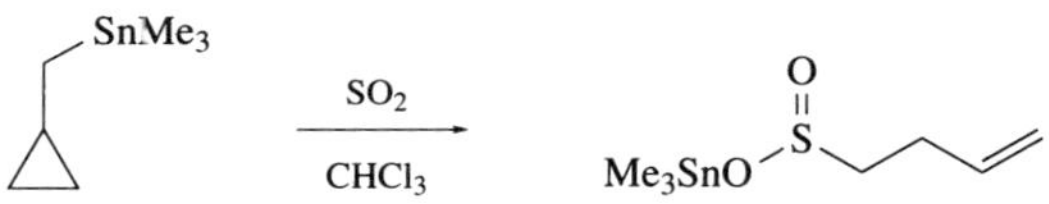

(165)

2.03.6.6 X = Phosphorus

Oxidation of a phosphorothiolate with excess mcpba in chloroform solution at room temperature has been found to produce the mixed phosphoric–sulfonic anhydride in 86% yield (Equation (166)) ⟨B-81MI 203-01⟩.

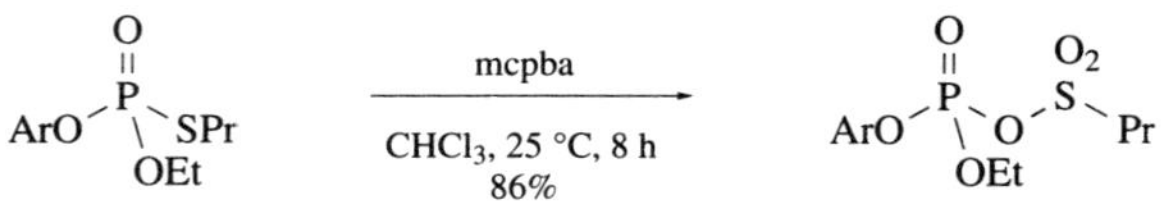

(166)

2.03.7 RSSH, R¹SSR², AND RSSX FUNCTIONS AND THEIR HIGHER COORDINATED DERIVATIVES

This section describes the preparation of aliphatic compounds containing an disulfide linkage, of which dialkyl disulfides are the most common. A comprehensive survey of this vast area is not possible within the constraints of this chapter, and this section therefore combines coverage of examples of classical methods of disulfide formation with an introduction to the more important of those methods of disulfide synthesis which have been developed. Many reviews on the subject of disulfide production have been reported in the past, and these provide a thorough overview of the area ⟨B-77MI 203-04⟩.

2.03.7.1 Hydrodisulfides and Hydropolysulfides

Hydrodisulfides and hydropolysulfides have received little synthetic attention. These compounds contain both disulfide and thiol linkages and they display reactivity akin to that of each of these units. Initial work in this area was reported by Bohme ⟨54LA(585)142, 55HOU(9)49⟩ whose favored entry to hydrodisulfide or hydrotrisulfide compounds was through chlorinolysis of an acetyl–sulfur bond, followed by acid-catalyzed ester exchange, as shown in Scheme 170.

$$Ac_2O + H_2S \xrightarrow{py} Ac_2S \xrightarrow[-20\ ^\circ C]{Cl_2} AcCl + AcSCl \xrightarrow{PhCH_2SH}$$

$$AcSSCH_2Ph \xrightarrow[HCl,\ 88\%]{EtOH} PhCH_2SSH + AcOEt$$

Scheme 170

Alkyl hydrodisulfides have also been prepared from dialkyl thiosulfones, higher coordinated derivatives of disulfides ⟨B-68MI 203-03, 70BCJ1890⟩, and from haloalkylated acetyldisulfides (Scheme 171) ⟨57LA(606)75⟩. Bohme has also described the synthesis of alkyl hydrodisulfides from sulfenyl chloride analogues ⟨58LA(617)62⟩.

$$RSO_2SR + AcSH \longrightarrow RSSAc \xrightarrow[H^+]{ROH} RSSH$$

$$AcSCl + CH_2N_2 \longrightarrow AcSS(CH_2)_2Cl \xrightarrow[H^+]{ROH} HSS(CH_2)_2Cl$$

Scheme 171

In a similar manner to that shown in Schemes 170 and 171, Tsurugi and Nakaboyashi were able to show that treatment of an alkyl acetyldisulfide with an ethanolic solution of hydrochloric acid resulted in the formation of an alkyl hydrodisulfide ⟨59JOC807, 71JOC3677, 72IJS199⟩. Hydrodisulfides have been treated with iodine or with sulfur monochloride or dichloride to furnish pentasulfides and hexasulfides ⟨54LA(585)142⟩.

A range of other procedures for the preparation of alkylhydrodisulfides is scattered throughout the literature. These include the ring opening of 2,3-dithietanones by diethylamine, which furnishes the desired alkylhydrodisulfides in quantitative yields (Equation (167)) ⟨81IZV1850⟩, and the reaction of a sulfinic acid derivative with flowers of sulfur in sodium hydroxide ethanol ⟨63MI 203-01⟩. The alkyl hydrodisulfide compound formed by this procedure contains a sulfonate–sulfide bond (Equation (168)). Bohme has shown the sulfur linkage in these alkyl hydrodisulfides and polysulfides to be linear.

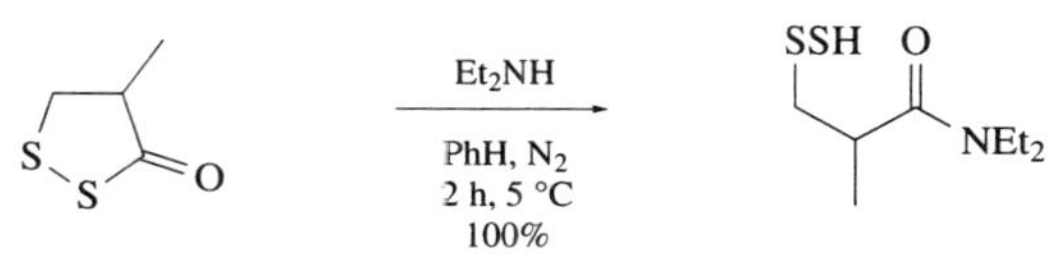

(167)

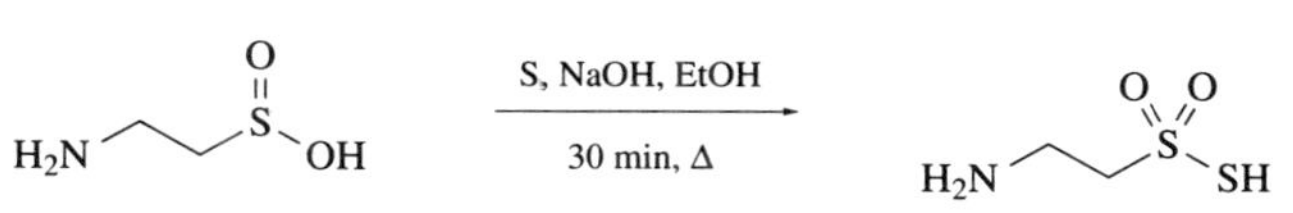

(168)

2.03.7.2 Alkyl Disulfides

Preparative methods efficient for the preparation of symmetrical disulfides are often ineffective for the preparation of unsymmetrical disulfides. By far the most important methods are the oxidation of thiols and thiol derivatives, and the thioalkylation or thiolysis of sulfur-containing compounds. Other methods of preparation include reduction of sulfonyl and sulfenyl halides, and rearrangements of sulfur containing species.

2.03.7.2.1 Oxidation of thiols

Perhaps the single most utilized method for the preparation of disulfides is the oxidation of the appropriate corresponding thiols. Oxidation occurs using any of a wide variety of oxidizing agents, a small cross-section of the more useful and convenient of which are discussed here ⟨B-74MI 203-04⟩. The ease of oxidation of thiols usually decreases from aromatic thiols, through primary and secondary alkanethiols to tertiary. Reaction conditions often need to be tailored in order to prevent overoxidation, ultimately to sulfonic acids. Mixed disulfides may be prepared using this method, although a statistical ratio of disulfide products is to be expected. However, when sufficient structural differentiation between the two thiols is present, for example the oxidation of a primary thiol in the presence of a tertiary alkanethiol, then the mixed disulfide can become the favored product. Separation of mixtures is often possible by distillation. The most common oxidizing agents are discussed below.

(i) Halogens

The products of the oxidation of thiols by halogens vary with both the halogen employed and the reaction medium. As a general rule, however, the use of bromine or iodine in an inert solvent such as ethanol results in the conversion of thiol into disulfide. Iodine is the most frequently used, and has been employed under a variety of conditions, commonly in alcoholic solvent with no other reagent ⟨47JA1710, 61JA4357, 74JOC253, 80JOC1839, 92T1883⟩, an unusual example of which has been described by Trost (Equation (169)), but also in the presence of an additive such as ethylene oxide ⟨55JA704, 61JOC3401, 78S50⟩ or sodium hydroxide ⟨51JA2647, 51JA3627, 51JA4464, 54CB1661⟩. In each of these cases a wide range of reactive functionalities is tolerated, as illustrated in Equation (170). Iodine has also been used in ethereal solution in conjunction with carbon disulfide ⟨67AG(E)703⟩, in the presence of triethylamine ⟨70JOC3259, 72JOC2710, 78S138, 82TL323, 84JA5271, 85TL4093⟩, an example of which is shown in Equation (171), and in conjunction with both triethylamine and tributyltin chloride ⟨86TL441⟩. Iodine has also been used with triethylamine under high dilution techniques for the preparation of molecules containing two disulfide linkages (Equation (172)) ⟨91CL141⟩.

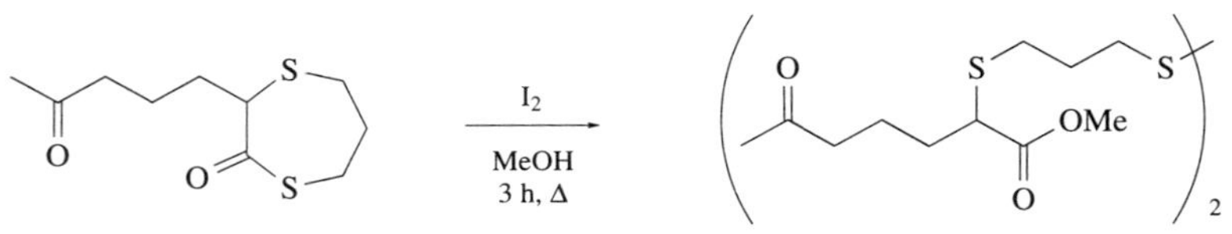

(169)

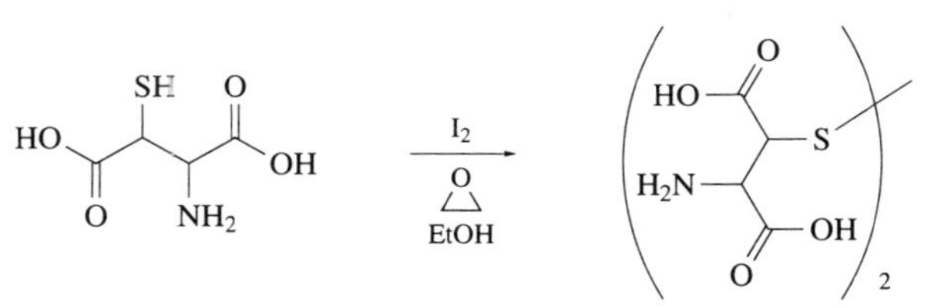

(170)

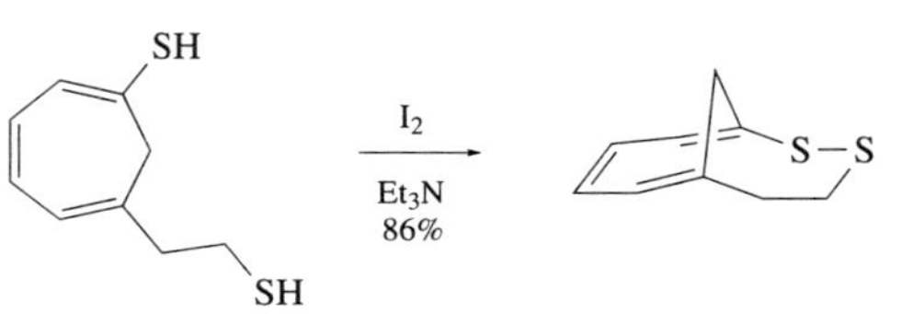

(171)

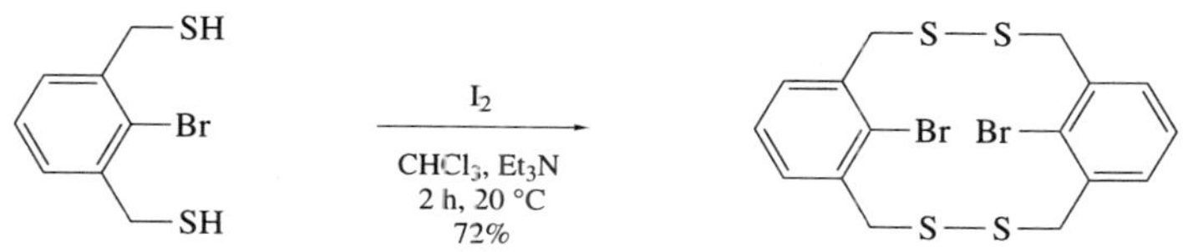

(172)

Bromine is also very commonly used for the oxidation of thiols to disulfides. Once again a wide variety of conditions may be used to effect the transformation, although bromine alone is effective in appropriate solvents ⟨50G180⟩. Bromine is also frequently used in conjunction with potassium hydrogen carbonate ⟨80S32⟩.

Polymeric amine/halogen complexes have been developed for thiol oxidation. An example of such a complex is the 1 : 1 2-polyvinyl-pyridine/bromine complex developed by Christensen ⟨79S8⟩. This polymeric complex has proved to be particularly useful for the oxidation of L-cysteine to cystine. Oxidation of cysteine by molecular bromine occurs with substantial conversion to cysteric acid ⟨61MI 203-02⟩.

N-bromosuccinimide can also be used to effect thiol oxidation. In a study of the synthesis of 3-substituted-3*H*-1,2-benzodithioles, Chan used NBS to effect a disulfide formation involving carbon–sulfur bond cleavage, perhaps through the mechanism shown in Scheme 172.

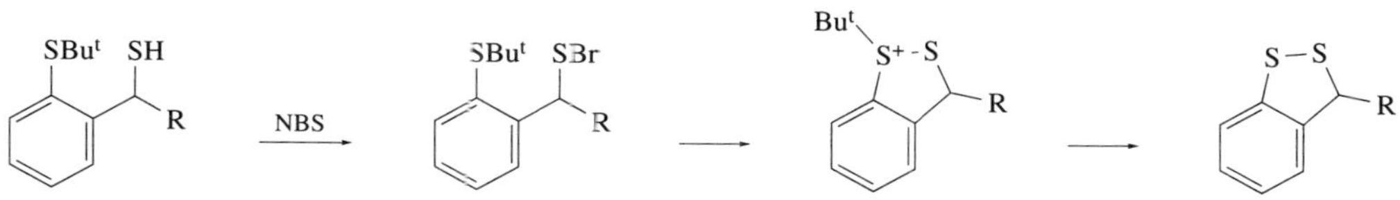

Scheme 172

As indicated above, the conditions under which the halogen are employed can dramatically affect the product obtained from the thiol substrate. For example, in aqueous solvents, chlorine and bromine each react with thiols to produce the corresponding sulfonyl halides or sulfonic acids ⟨37JA811, B-61MI 203-03⟩. Sulfur trihalides, halogenated sulfenyl halides, sulfinic acids, sulfinyl halides, and many more species can be formed if the thiol is not oxidized under suitable conditions.

(ii) Peroxides

Peroxides are also frequently used for the oxidation of thiols. Once again however, further oxidation of the initially formed disulfide product occurs rapidly in the presence of excess oxidant, and the reaction requires careful monitoring. Hydrogen peroxide itself is common, and can be used as an aqueous solution, ⟨48CB210⟩ in conjunction with acetic acid ⟨69JOC36, 71JOC1314⟩, or in alkaline ethanolic solution ⟨65CI(L)680, 66JOC615, 66JOC1217, 76CL857⟩. Potassium peroxide has also been used for thiol oxidation ⟨84AJC2331⟩.

(iii) Diethyl azodicarboxylate

Diethyl azodicarboxylate (dead) is a commonly used reagent for the oxidation of thiols to both symmetrical and unsymmetrical disulfides. The oxidation is best performed in the dark at room temperature or under reflux in dry solvent ⟨66JA2328, 67JOC727, 68TL5907⟩. The reaction is catalyzed by triphenylphosphine ⟨70JOC4227⟩. Equation (173) illustrates the dead-mediated oxidation of thiols to yield unsymmetrical disulfides ⟨68TL5907, 70JOC4227, 70TL1203, 75CL813, 79JOC3230⟩.

(173)

(iv) Aerial oxidation

Aerial oxidation of alkanethiols to disulfides is in some cases extremely facile, although tertiary alkanethiols have been found to be fairly resistant ⟨B-58MI 203-01, B-66MI 203-01⟩. Selected examples of aerial oxidation, and of the oxidation of thiols by oxygen, are shown in Scheme 173 ⟨64IZV644, 81BEP886595, 88USP4721813⟩. The reaction of thiols with elemental oxygen or air is particularly sensitive to catalysis, by metal ions, UV light or radical initiators. These processes are further described below.

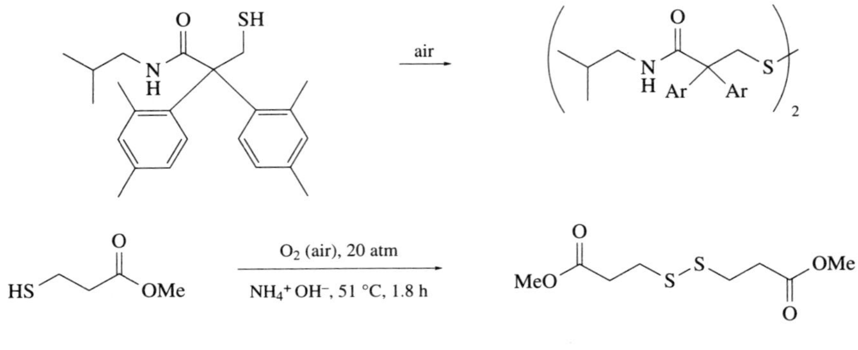

Scheme 173

(v) Miscellaneous organic reagents (chemical methods)

In addition to the methods of thiol oxidation described above, and those methods based upon inorganic metal salts discussed below, there are a number of chemical-based methods for the preparation of disulfides from thiols. Reagents which have been used successfully include trichloromethylsulfenyl chloride ⟨61JOC3780⟩, cyanogen bromide ⟨67JPS292, 73SC317, 74JOC253, 77SC363⟩, dimethyl sulfoxide and other sulfoxides ⟨63JOC3246⟩, dinitrogen tetroxide ⟨77CC407, 77CL893, 79CL1077⟩, 2-(4-methylphenyl)-1,2-benzselenazolidin-1,4-dione ⟨86SUL1⟩, and many others ⟨84H(22)1017⟩.

(vi) Oxidation by metal ions/metal oxides

Another frequently employed method for the oxidation of thiols to disulfides involves transition metals, used either in catalytic amount in autocatalytic processes or in conjunction with an oxidizing agent, or, as is more often the case, quantitatively to effect disulfide formation in the absence of oxygen. Examples of these processes are described below.

Of the transition metal complexes employed for this purpose, ferric compounds are the most extensively used. Early work in this area featured the ferric hexacyanide complex anion, $Fe(CN)_6^{3-}$, and ferric octanoate for thiol oxidation ⟨58CRV439, 66JOC3071⟩, and indeed these were the ferric salts used for the production of synthetic rubber from thiols ⟨58CRV439⟩. Procedures described later have involved ferric chloride or ferric sulfate. Transition metal salts can be employed quantitatively in the absence of other oxidants for the oxidation of thiols to disulfides. An example is shown in Equation (174), where ferric chloride was used to oxidize a range of dithiols to their cyclic disulfide products ⟨57AG713, 58LA(614)66⟩. The nine-, ten-, twelve- and eight- (shown) membered ring systems were all prepared under high dilution conditions, although in only moderate yields.

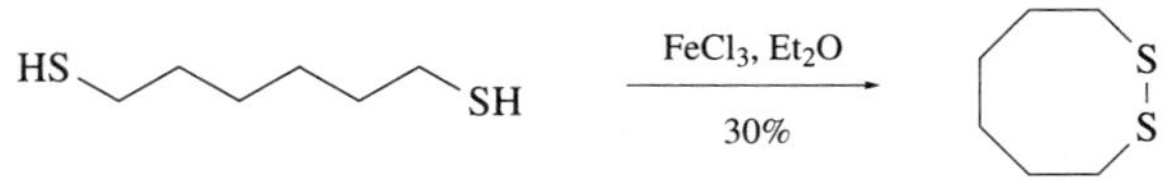

(174)

In the presence of oxygen, ferric chloride has been used in catalytic quantity, usually ≤10%, to effect thiol oxidation ⟨54JA1828, 55JA416⟩. Both amine and carboxylic acid functionality can be tolerated by the reaction conditions ⟨57MI 203-01⟩. Potassium carbonate is sometimes added to the reaction mixture ⟨59CB1177, 78CB3159⟩ (Equation (175)). Remarkably, potassium carbonate alone is able to mediate the oxidation ⟨54JA1828, 55JA416⟩. Ferric chloride in aqueous hydrochloric acid may also be used ⟨55JA5446, 57JA3503, 73JOC2965⟩. Ferric chloride is also effective for thiol oxidation in other acidic media, such as acetic acid and methanol ⟨55AG304, 73E389, 73TL655⟩, and in aqueous ammonia ⟨90JHC1361⟩. Iron sulfate salts have also been used (Equation (176)) ⟨56JCS4066, 84JOC1434⟩.

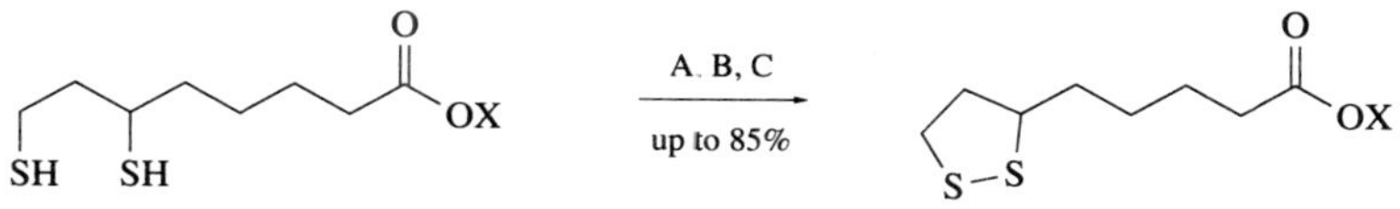

(175)

A: X = H; $FeCl_3$ (1%), O_2, K_2CO_3, H_2O, HCl/EtOH, pH 7
B: X = H; K_2CO_3
C: X = K; $FeCl_3$ (cat.), O_2, EtOH/H_2O
D: X = H; $FeCl_3$, HCl (aq.)

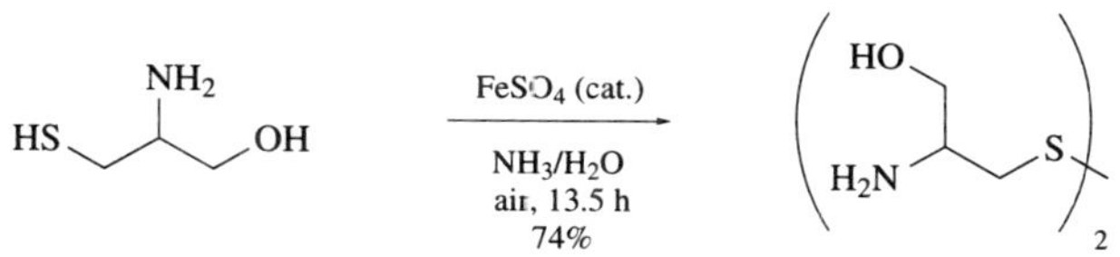

(176)

The use of an alkoxystannane and ferric chloride reagent for disulfide production from thiols under mild conditions has been developed by Otera ⟨90TL3591⟩. Thiostanne is formed upon addition of tributyltin methoxide to an acetonitrile solution of the thiol substrate; the ferric chloride present then cleaves the tin–sulfur bond of the thiostannane. For alkanethiols, conversion into the corresponding disulfides were observed in yields of 82–97%, and the method proved applicable for primary, secondary and tertiary alkane thiols, as well as aryl thiols. The utility of this reagent combination was illustrated by the cyclization of an α,ω-alkanedithiol in good yield. With most common thiol oxidizing systems the yields of the cyclic disulfide product were only moderate ⟨58LA(614)66, 69JOC36, 70JOC749, 81S457, 86TL441⟩. For example, the eight-membered cyclic disulfide shown in Equation (177) was produced in 50% yield with the tin alkoxide/ferric chloride system. In this procedure stoicheiometric ferric chloride was necessary for the reaction to reach completion, supporting reduction from iron(III) to iron(II) as a key step in the sequence.

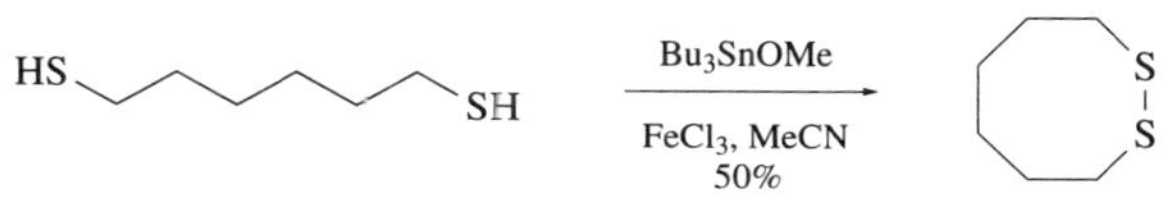

(177)

Lead tetraacetate has been used to great effect for the oxidation of alkanethiols. Early usage of this reagent employed benzene as solvent ⟨58JA838, 73TL655, 80JOC710⟩. Interestingly, while lead tetraacetate oxidizes thiols effectively to the disulfides when the reagent is treated with two equivalents of thiol ⟨58JA838, 62JA847⟩, when four equivalents of thiol are used, the lead(II) thiolate species is formed along with the disulfide ⟨67BCJ2388⟩.

Other metallic species have been shown to react with thiols to give disulfides, including cerium(IV), cobalt(III), and vanadium(V), which react, usually by single electron transfer, to form the thiyl radicals. Subsequent dimerization furnishes the disulfides. Barium permanganate has been shown to be a mild, stable oxidizing agent for use under nonaqueous, aprotic conditions, capable of affording the desired disulfide products in extremely high yields (80–99%) ⟨89S378⟩. Barium ferrate monohydrate, formed by the addition of a saturated solution of barium nitrate to a solution of sodium ferrate, has also been shown to effect the transformation of thiols into disulfides ⟨88BCJ2185⟩. Alkanethiol oxidation with this reagent proceeded rapidly under reflux to give the dialkyldisulfides in 85–90% yield.

The oxidation of thiols by copper complexes is also known, but is more susceptible to the formation of by-products such as the sulfides, presumably formed via a cupric thiolate intermediate ⟨66JOC3071⟩. However, successful reactions are known, and disulfides were produced in up to quantitative yields by the use of a bipyridine copper permanganate species (Equation (178)). The complex was used in 50 mol% stoichiometry ⟨83SC611⟩.

$$\mathrm{RSH} \xrightarrow[\mathrm{CH_2Cl_2,\ 10\ min,\ 22\ ^\circ C}]{\mathrm{(bipy)_2Cu(MnO_4)_2}} \mathrm{RSSR} \qquad (178)$$

R	*Yield* (%)
Bu	100
Bu^s	100
Bu^t	60
$PhCH_2$	90

Sodium tellurite (Na_2TeO_3) displays the interesting chemoselective ability to differentiate between different types of thiol under mild phase transfer conditions ⟨92BCJ626⟩. Aromatic and benzylic thiols are oxidized rapidly, while primary and secondary thiols reacted sluggishly, and tertiary thiols not at all. Metallic tellurium is produced during the oxidations. Many functionalities sensitive to other oxidants are inert to sodium tellurite. Both symmetrical and, unusually, unsymmetrical disulfides may be prepared in excellent yields using this reagent (Scheme 174). Sodium borate hydrate has also been reported to induce efficient, high-yielding oxidation of thiols and selenols to disulfides and diselenides respectively ⟨90TL5007⟩.

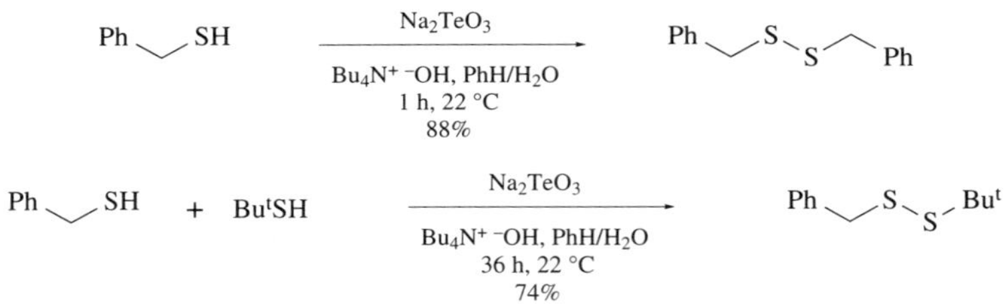

Scheme 174

(vii) Miscellaneous methods for the oxidation of thiols to disulfides

While the procedures described above are the best known for the preparation of disulfides from thiols, there exist many alternative methods. These include photochemical and electrochemical oxidation; oxidation by molecular oxygen or air, catalyzed by metal ions, UV light, strong bases and aliphatic amines ⟨B-74MI 203-04⟩; and biological oxidation methods. Flavins oxidize thiols to the corresponding disulfides in an anaerobic manner, the flavin itself being reduced to dihydroflavin. Introduction of oxygen to this system generates a catalytic cycle. Equation (179) illustrates a system wherein a disulfone species oxidizes a thiolate anion ⟨69JOC1792, 69TL3901⟩. Bakers' yeast has been shown to perform thiol oxidation to yield disulfides in excellent yields (Equation (180)) ⟨91BMC507⟩. Other biological oxidation agents are known, including diazenecarboxylic acid bis(N,N'-dimethylamide), which oxidizes glutathione in red blood cells ⟨69BBR(37)593⟩.

(179)

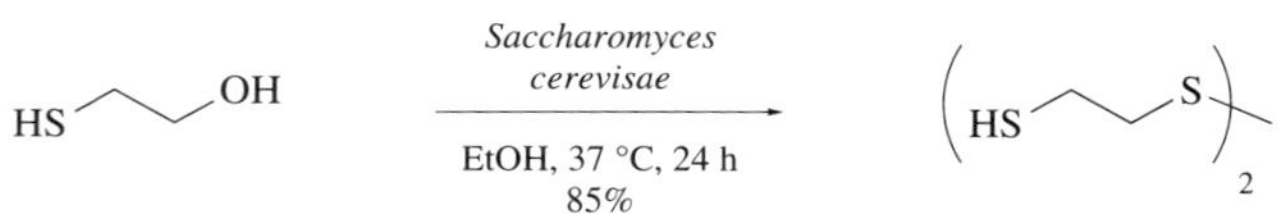

(180)

2.03.7.2.2 *Thioalkylation of thiols (thiolysis)*

With the oxidation of thiols excepted, perhaps the most widely used method of disulfide formation involves the thiolalkylation of a thiol by a sulfenyl derivative, that is nucleophilic displacement of a suitable leaving group from the sulfenyl compound by either a thiol or a thiolate anion, a thiolysis reaction. Thiolysis is favored for the synthesis of unsymmetrical disulfides, although yields may be compromised by rapid thiol–disulfide exchange reactions. The usefulness of thiolysis for disulfide preparation is enhanced by the wide range of sulfenyl derivatives available ⟨47ACS307, B-61MI 203-01, 66QRS285⟩. Of the thioalkylating agents described, sulfenyl chlorides, Bunte salts and thiosulfenates are the traditional reagents ⟨55HOU(9)55⟩. Other thiolysis reagents such as disulfides, thionitronates, thiocyanates, sulfenyl thiocyanates, and thiosulfonates and thiosulfinates have been used for the disulfide preparation. Thioalkylation methodology to yield disulfides is discussed below according to the sulfenyl derivative type used as substrate.

(i) With sulfenyl halides

The formation of disulfides by thiolysis of sulfenyl halides, in particular sulfenyl chlorides, has long been known ⟨B-60MI 203-02, B-61MI 203-01⟩. The thiolysis reaction with sulfenyl halides is usually performed in the presence of a base such as triethylamine (Scheme 175) ⟨80MI 203-02, 80MI 203-03, 81CL737⟩. Use of the optically pure (*S*)-(−)-*N*,*N*-dimethylamphetamine at −70 °C furnishes the thiosulfinate product with a degree of asymmetric induction (11.3% *ee*) at the sulfoxide sulfur atom ⟨76CC220⟩.

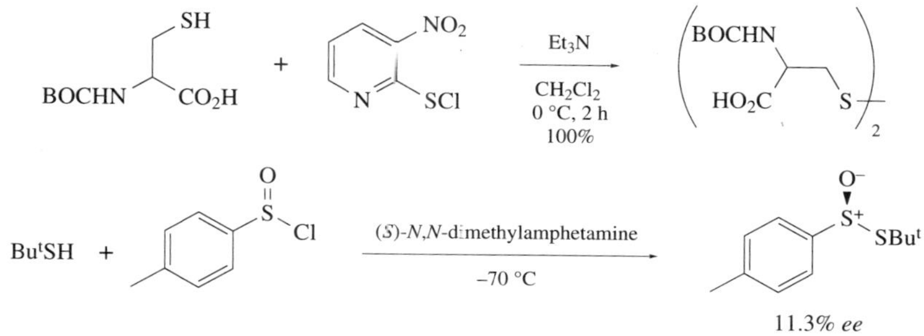

Scheme 175

Nucleophilic displacement of bromide ion from alkanesulfenyl bromide or arenesulfenyl bromide by an alkanethiol has been reported as an efficient route to alkyl disulfides. Similar use of higher coordinated sulfur substrates, such as the sulfinyl or sulfonyl bromides, have also been described (Equation (181)) ⟨88SC227⟩. Interestingly, the use of sulfonyl chlorides in this reaction under the same conditions resulted in the formation of disulfides.

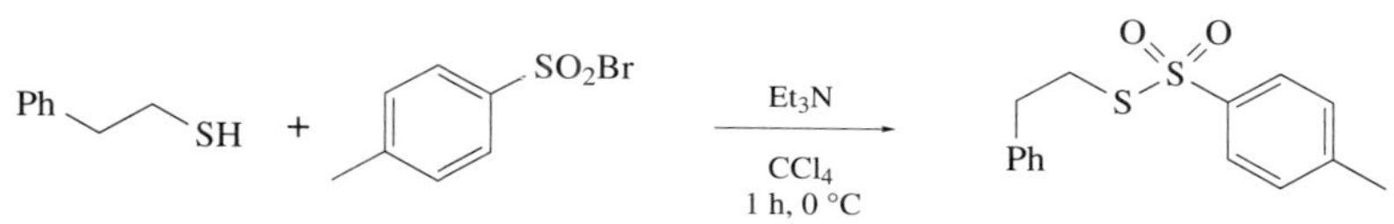

(181)

(ii) With sulfenamides

Thiolalkylation reactions with sulfenamides represent a popular method of unsymmetrical disulfide preparation. One of the earliest examples of the thiolysis process was reported by Mukaiyama, who generated a sulfenamide, capable of further reaction with a thiol to generate a disulfide, by reaction of a thiol with dead (Scheme 176) ⟨68TL5907⟩.

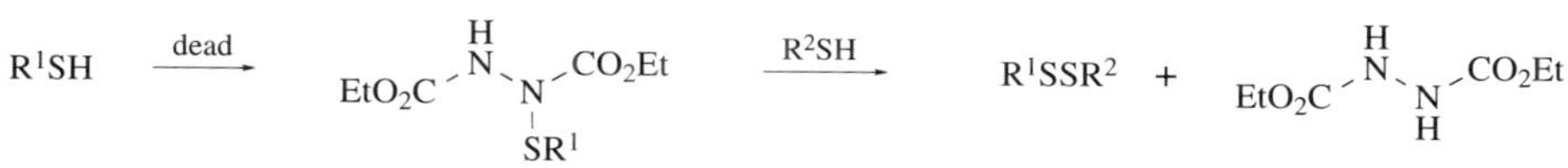

Scheme 176

Perhaps the sulfenamide derivatives most widely used for this purpose are *N*-sulfenylphthalimides, which have the advantage that they are relatively shelf-stable ⟨70TL3547, 70TL3551⟩ (Equation (182)) ⟨71JOC3828, 74S59⟩. Among the advantages offered by this method of disulfide preparation are, as mentioned above, the stability of the precursor, and also, the minimal disulfide interchange, and the excellent yields achieved. Unfortunately, diaryl disulfides cannot be prepared by this method. One interesting example is the reaction between *N*-thiophthalimide and the α-amino acid cysteine, present in the reaction mixture as its hydrochloride monohydrate. The reaction gave an 89% yield of the disulfide. A similar derivative of cysteine, the 2-pyridyldisulfide derivative, has been prepared by reaction of cysteine with the bis 2-pyridyldisulfide ⟨81JBC(256)5064⟩. Sulfenamides derived from other imides, such as succinimide and maleimide, have also proved effective for this transformation ⟨70TL3547⟩.

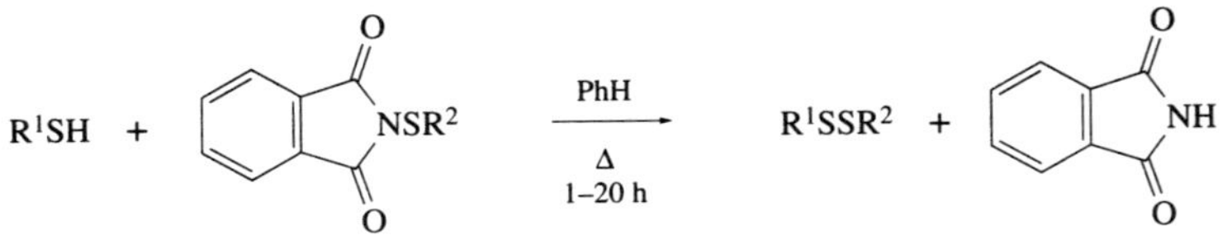

(182)

R^1	R^2	Yield (%)
Pr	Ph	71
$PhCH_2$	$PhCH_2$	98
p-MeC_6H_4	Pr^i	88

Cysteine is a widely chosen thiol with which to perform thiolysis reactions ⟨82ZPC(363)1461⟩. Mikolaczjk has applied this chemistry to the synthesis of optically enriched alkyl disulfide derivatives, in this case alkanethiosulfinates, by the reaction of an optically pure sulfenamide with an alkanethiol in the presence of trifluoroacetic acid, as shown in Equation (183) ⟨85TL5703⟩. The alkanethiosulfinate product shown was produced with an *ee* of 60% in favor of the (*S*)-configuration at sulfur, and in 91% yield.

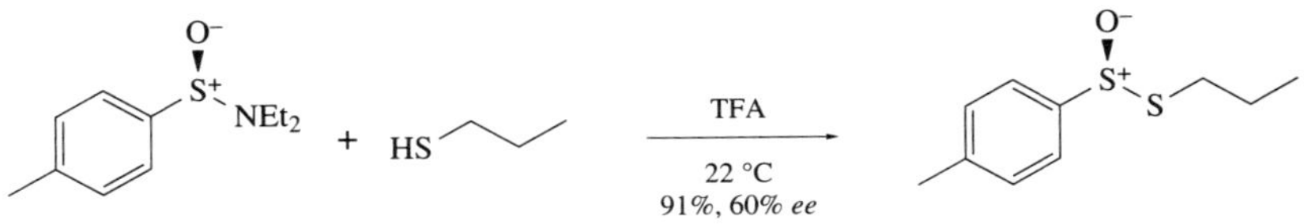

(183)

(iii) *With thiocarbamates*

The use of thiocarbamate functionality for nucleophilic displacement has also been described. In 1964, Watson reported the reaction of the thiocarbamate shown in Equation (184) with the sodium salt of a dithiocarbamate ⟨64JCS2100⟩. A similar transformation has been subsequently reported by Graber, who subjected a sulfinyl derivative to reaction with a range of thiols ⟨87JOC4620⟩. Interestingly, only the disulfides were obtained, in 60–90% yield, and not the corresponding thiosulfinate which might have been expected.

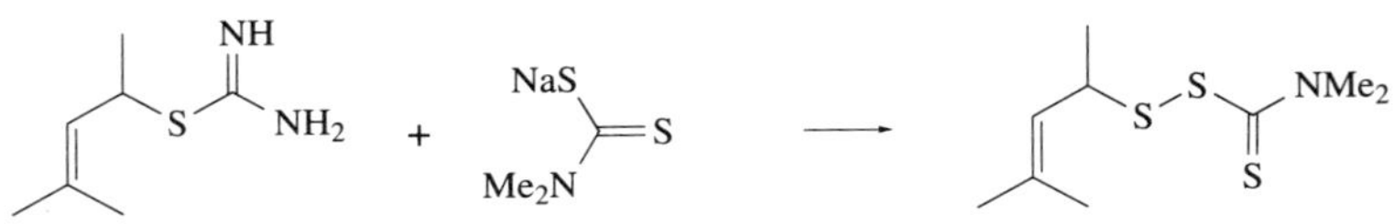

(184)

(iv) *With thiosulfonates and thiosulfinates*

In common with sulfenylphthalimides, thiosulfonates are stable precursors to disulfides which can be stored for long periods. Unsymmetrically substituted thiosulfonates have achieved wide usage in organic synthesis. Equation (185) shows an example where the substrate thiosulfonate is cyclic, so that the sulfonate moiety of the starting material is retained in the product as a sulfonic acid moiety ⟨69JOC1792⟩. This reaction has also been reported with an α-thioketone ⟨71JMC995, 72JOC2710⟩.

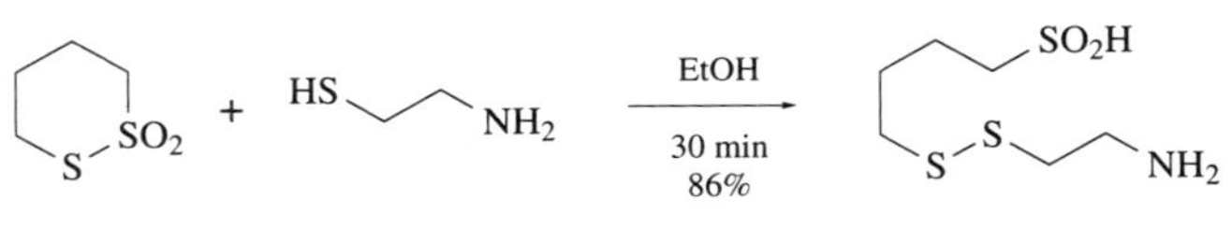

(185)

In spite of the loss of half of the often carefully prepared thiosulfonate substrate in the nucleophilic displacement reaction, symmetrical thiolsulfonates have been utilized in thiolysis reactions (Equation (186)) ⟨71JOC309⟩. The authors comment that the disulfide products precipitate from solution readily, and as a result this procedure has some merit as a method of thiol characterization and purification. Thiols could easily be regenerated from the disulfide products by reduction. Methods for the preparation of alkanethiosulfonates are described in detail below.

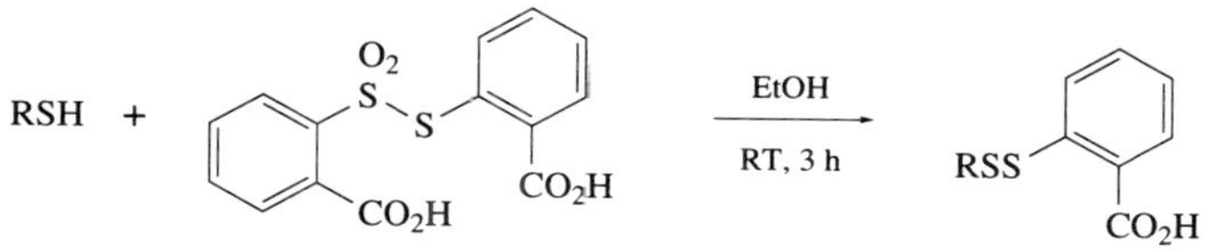

(186)

R	*Yield (%)*
Bu^t	84
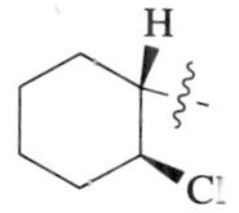	97

Thiosulfinates have also been reported as substrates for thiolysis in the formation of unsymmetrical disulfides (Equation (187)) ⟨58LA(617)71, 67BCJ2913⟩. Two displacements can take place in the reaction of cyclic thiosulfinates with thiols, to give bis-disulfides as the major products ⟨58LA(617)71⟩.

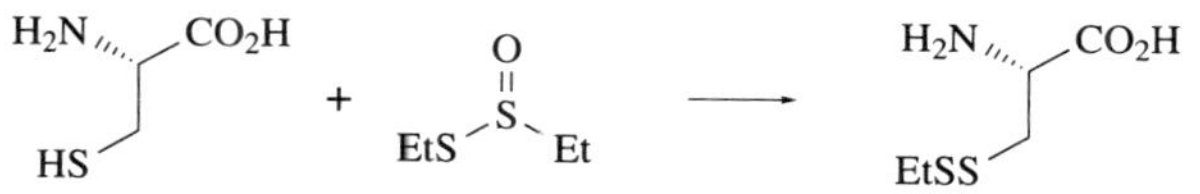

(187)

(v) With disulfides

Thiolysis of disulfides with thiols, in effect a thiol–disulfide exchange reaction, may also be used for disulfide preparation, and indeed is a popular method for the preparation of unsymmetrical disulfides. In many cases the unsymmetrical product can be purified and isolated in one distillation step by removal of the less volatile thiol by-product ⟨51JA3627⟩. The reaction is promoted by a range of different organic catalysts, including 4-dimethylaminopyridine (dmap) ⟨92TL451⟩, perchloric acid and triphenylphosphine ⟨84CC843⟩, and acetic acid ⟨72BJ723, 89JOC244⟩ (Scheme 177).

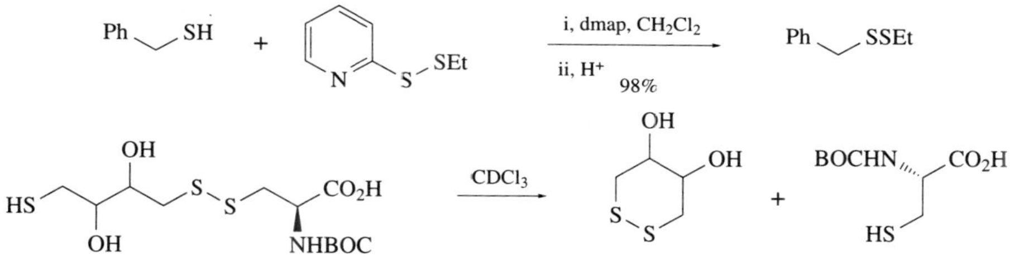

Scheme 177

Thiols can react with dimethyl disulfide in the presence of sulfuryl chloride and pyridine to generate the unsymmetrical disulfides by nucleophilic displacement of a chloride ion from the methanesulfenyl chloride generated *in situ* by reaction of dimethyl disulfide and sulfuryl chloride. An example from the Japanese patent literature is shown in Equation (188).

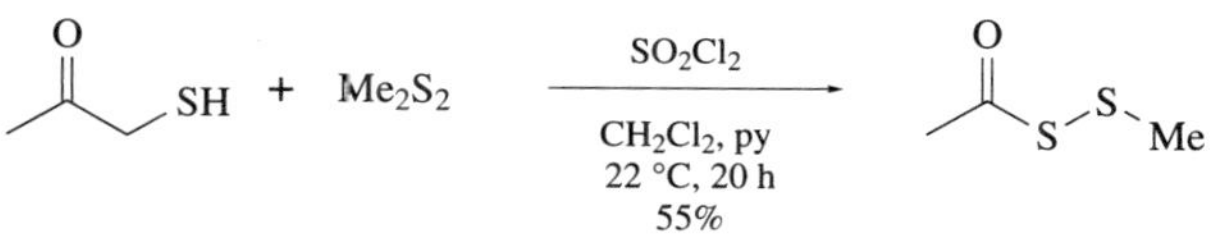

(188)

(vi) With sulfenyl thiocyanates

The use of sulfenyl thiocyanates for disulfide formation is again a widely used synthetic process. The instability of the sulfenyl thiocyanates under some conditions does however hinder their use, and they are generally not isolated, but instead used *in situ* for reaction with the thiol (Equation (189)) ⟨22CB1424, 61JOC1152, 67JOC97, 69JA7525, 70JOC4148, 71JOC488⟩. Formation of a thiosulfenylcyanide has also been observed in dichloromethane in the presence of sodium acetate (Equation (190)) ⟨65JA3965, 68JA2677, 69JA7525, 70JOC1118⟩.

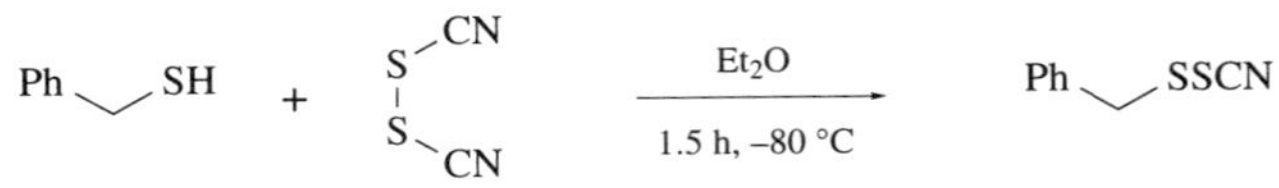

(189)

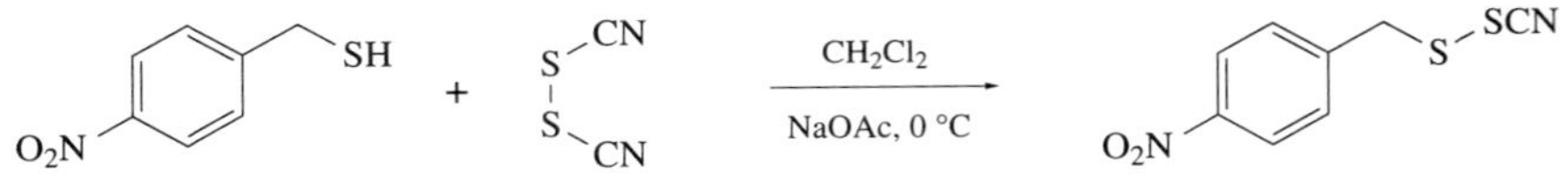

(190)

This thiolysis reaction is particularly effective for the formation of the disulfide linkages in polypeptides, perhaps because the reaction proceeds even with the thiol group protected as a trityl, a benzhydryl, an α-pyranyl, or an isobutoxymethyl thioether ⟨70JOC1118⟩.

(vii) With thiocyanates

The reaction of a thiol with a thiocyanate, although a rarely used synthetic operation, has been used to generate disulfides. The transformation is usually performed at elevated temperature in the presence of a base ⟨69RTC513⟩.

(viii) With thionitronates

Thionitronates have proved to be effective substrates for thiolysis reactions, as illustrated in Equation (191). The reactions of alkanethiols with phenyl and other thionitrates provide a high yielding method of disulfide synthesis ⟨77CC407, 77CL893⟩.

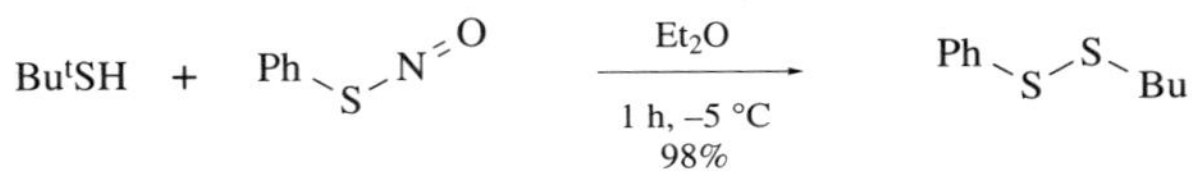

(191)

(ix) With thiosulfonic acid salts (Bunte salts)

Bunte salts are one of the traditional substrates for thioalkylation reactions. The salts of thiosulfonic acids have been shown to be particularly effective for the practical, large scale synthesis of disulfides (Equation (192)) ⟨68QRS189, 88OSC(4)235⟩. The sodium salt of the thiol was employed in this reaction, allowing the transformation to be performed in aqueous medium. A Japanese group has described an analogous synthesis of a structurally more complex disulfide.

A similar reaction takes place with equal facility between sulfenyl chlorides and thiosulfonic acid derivatives (Equation (193)) ⟨62AJC570, 62JCS2172, 78OS(58)147⟩.

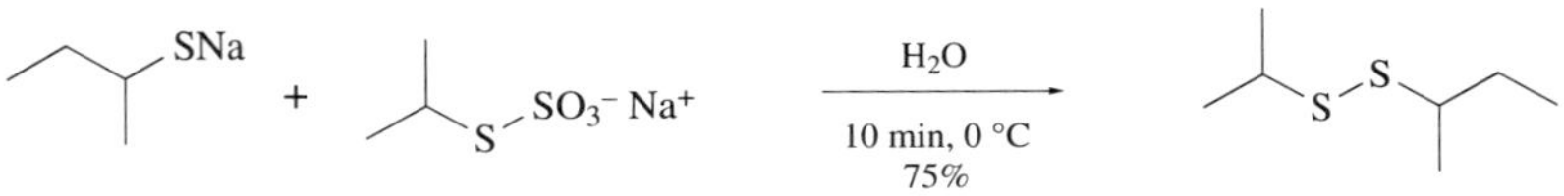

(192)

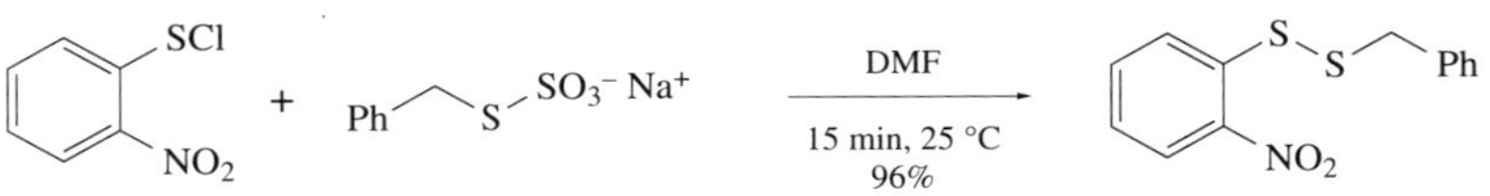

(193)

(x) With sulfenyl thiocarbonates

Sulfenyl thiocarbonates, prepared by the reaction of a thiol with a suitable acylsulfenyl chloride, such as methoxycarbonyl sulfenyl chloride, have been shown to undergo further reaction with thiols at room temperature to yield unsymmetrical disulfides ⟨70JA7629, 75JOC250, 79JOC2654⟩ (Scheme 178). The route is both simple and clean, and appears to be widely applicable.

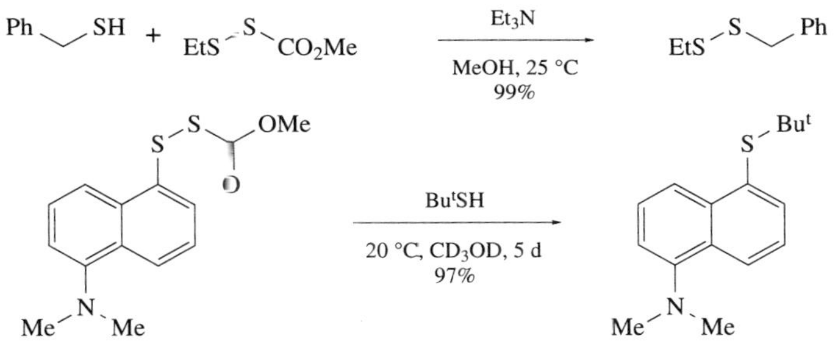

Scheme 178

2.03.7.2.3 *By reductive coupling of sulfonyl and sulfinyl derivatives*

Reductive coupling of sulfonyl chlorides is another important method for the synthesis of disulfides. Reducing agents commonly used include trichlorosilane and tripropylamine ⟨70JA7224⟩, and an acetic acid/hydriodic acid mixture. Molybdenum hexacarbonyl has also been found to induce the same transformation ⟨69AG(E)677⟩. As shown in Equation (194), a borane–iodine–alkylamine reagent is also effective in reductive coupling of sulfonyl chlorides to give the corresponding disulfides ⟨91SL125⟩. Other sulfonyl derivatives are reductively coupled by this reagent to give the disulfides, including sulfonic acids, sulfonate esters, and sulfinyl chlorides. Sulfoxides were reduced to the corresponding sulfides by the same reagent.

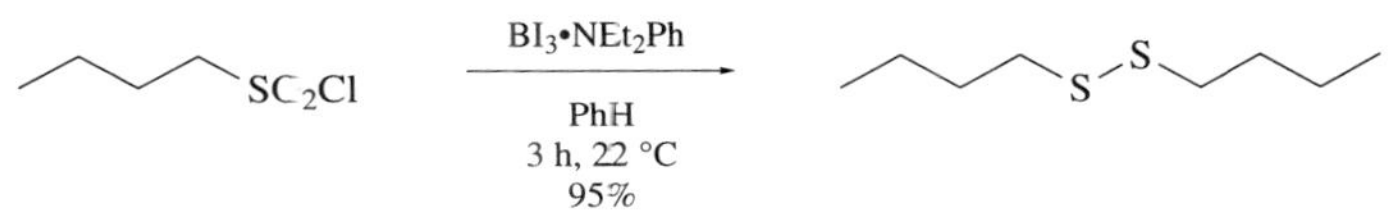

(194)

An interesting variation on this reaction has been reported by Freeman and Keindl, who, upon reaction of a sulfinyl chloride with activated powdered zinc in carbon tetrachloride, observed a 46–78% yield of the corresponding coupled thiosulfonyl products, where formation of the disulfide linkage was accompanied by reduction of one of the sulfoxide moieties present, and, in addition, further oxidation of the other ⟨83S913⟩. A similar result was achieved upon use of the sulfinyl chloride in DMF, although the major product isolated contained a chlorine substituent α to the sulfenyl sulfur atom (Equations (195) and (196)) ⟨85S500⟩.

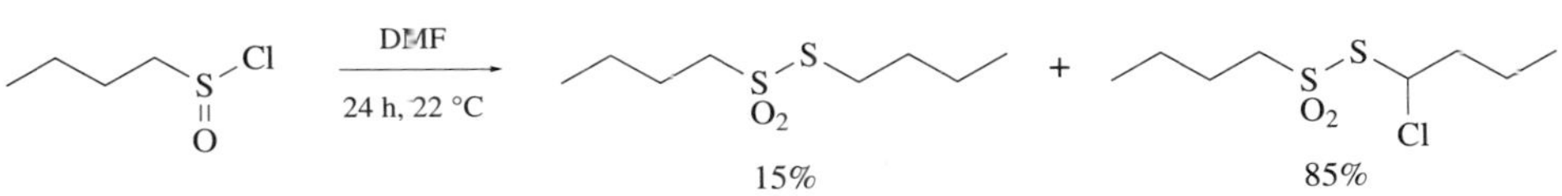

(195)

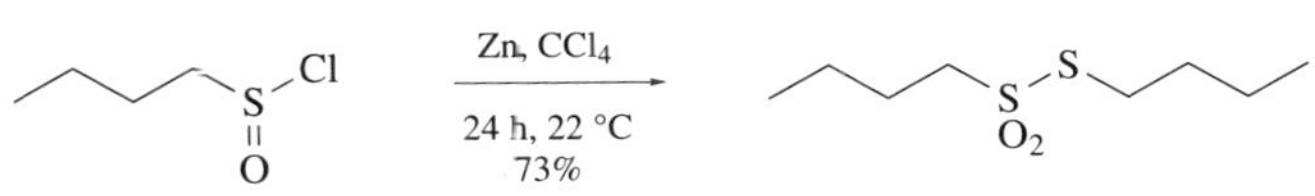

(196)

(i) *By reaction of alkenes with sulfenylating agents*

There are a wide range of reagents available for the direct introduction of a sulfur atom into an organic molecule. This section and those following draw attention to several of these reagents with illustrative examples of the substrates with which these sulfenylating agents react. Perhaps one of the most widely used substrate types is alkenes, which react with sulfur monochloride to furnish symmetrical disulfides ⟨69JOC3991⟩, and with trityl sulfenyl chloride to give the unsymmetrical disulfides (Equation (197)) ⟨81TL3659⟩.

Dienes undergo formal [4 + 2] cycloaddition to disulfur monoxide (S_2O) ⟨72JOC2367⟩, and cycloaddition to bis(triphenylsilyl)trisulfide ⟨84JA759⟩ (Scheme 179).

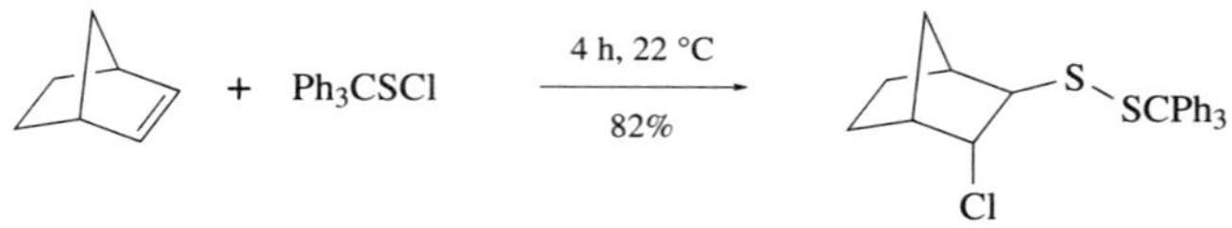

(197)

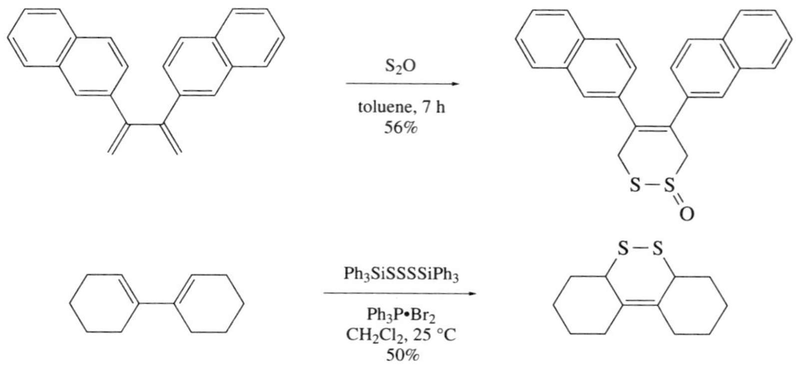

Scheme 179

(ii) Via ring opening and coupling of thiiranes

Disulfide and thiosulfinate formation by ring opening of a thiirane or thiirane sulfoxide has appeared throughout the literature. Of the simpler methods of effecting the ring-opening reaction, bromine ⟨52JA5880⟩ and ceric ammonium nitrate (can) ⟨91T149⟩ have both been used for disulfide formation; one further advantage of the ring opening process is the generation of additional functionality from the nucleophile at the β-position to both sulfur atoms (Scheme 180).

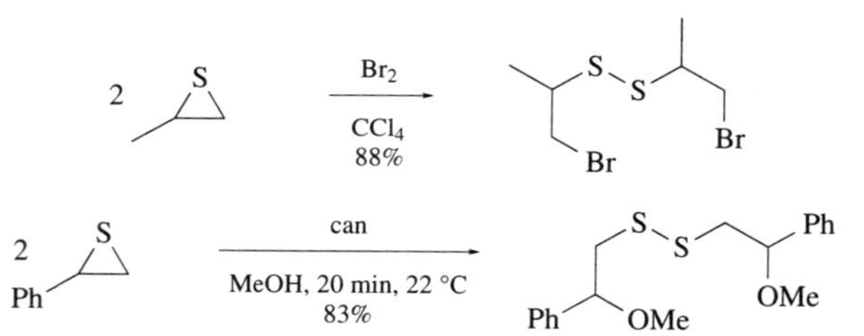

Scheme 180

McCabe and Stewart have also reported the use of bromine for thiirane ring opening and subsequent disulfide formation (Equation (198)). The use of one equivalent of bromine resulted in a rearrangement, while the use of 50 mol% gave the desired disulfide in 95% yield ⟨80CC100⟩.

Copper(II) chloride has been used for the ring opening of thiirane sulfoxides (episulfoxides), generating the thiosulfinates, as shown in Equation (199) ⟨69TL2743⟩.

The ring opening of thiirane sulfoxides has also been performed in acidic media, for example methanol/sulfuric acid, and acetic acid/sulfuric acid (Scheme 181) ⟨72JA5786⟩.

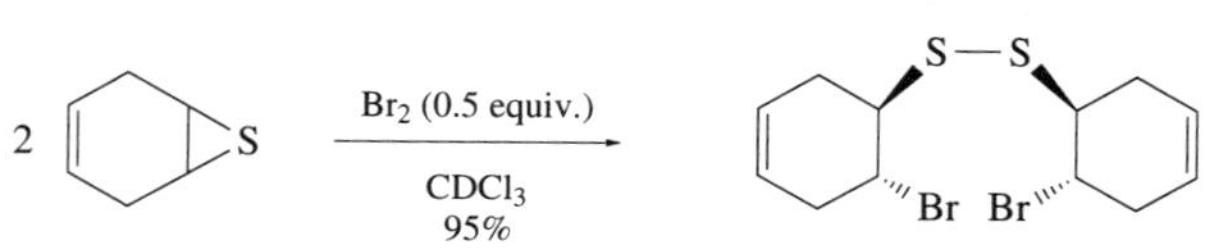

(198)

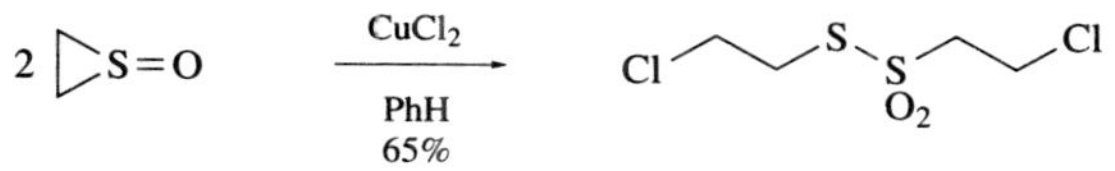
(199)

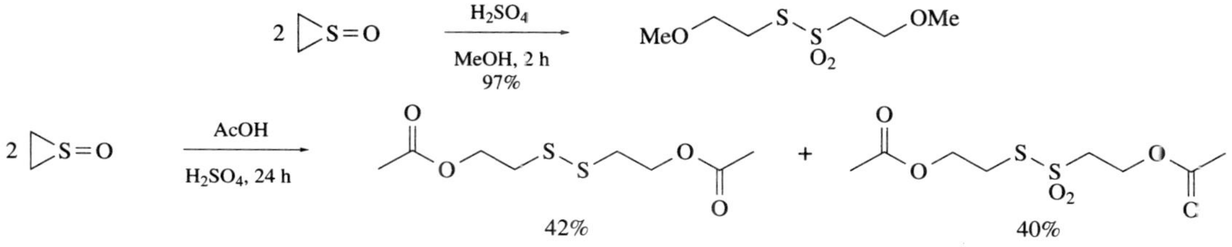

Scheme 181

Thiirane ring cleavage can occur merely by stirring the reaction mixture in a heated solvent. Baldwin has described the thermal decomposition of episulfoxides to give thiosulfinate and thiosulfoxylate products 〈71JA2870〉. In this case, the thiosulfinate product is formed initially, but rearranges to the thiosulfoxylate (Equation (200)). The episulfoxide starting material was found to be stable in the *cis* form under the reaction conditions. Oae has also reported a single example of ring opening of thiiranes induced by heating 〈75BCJ1665〉. In this reaction, 1,2-epithiocyclohexane was heated to 210°C for one hour in a sealed tube, and gave a 12% yield of the desired disulfide.

Thiirane ring opening to give the disulfide products (30–90%) has been observed upon treatment with 2,3-dichloro-5,6-dicyano-1,4-benzoquinone (ddq) in methanol under reflux (Equation (201)) 〈90SC1047〉.

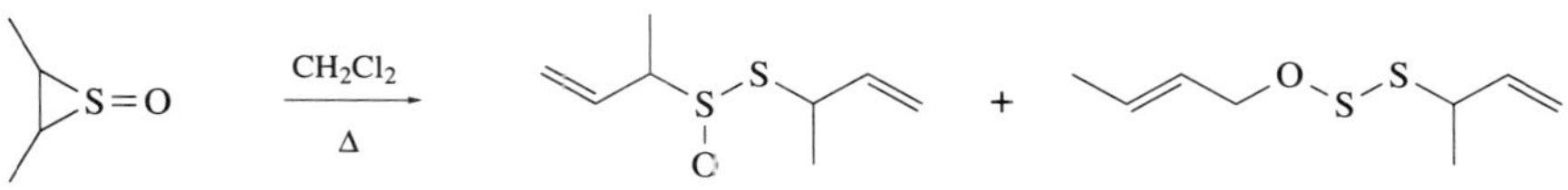
(200)

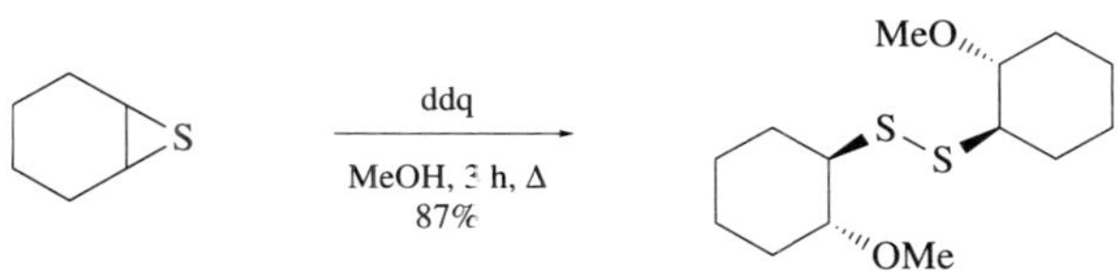
(201)

2.03.7.2.4 *By reaction with hydrogen sulfide*

Hydrogen sulfide has been found to react with a plethora of functional groups to furnish disulfide products. Both thiols 〈64JCS3832, 68TL3057, 74JPR719, 78JOC1270〉 and sulfonyl chlorides 〈88OSC(6)1016〉 have been employed for disulfide preparation. Equations (202) and (203) depict the synthesis of an amino acid related disulfide from a thiol starting material using hydrogen sulfide in the presence of ethanethiol (Equation (202)), and the synthesis of an alkanethiosulfonate, formed from reaction of the corresponding sulfonyl chloride with hydrogen sulfide in the presence of potassium hydroxide (Equation (203)). Ketones 〈56JA2469, 76HCA840〉 and aldehydes 〈49JA2270, 76HCA840〉 may also be transformed into the corresponding dialkyldisulfides by reaction with hydrogen sulfide, introduced either as its ammonium salt, or as the gas (Equation (204)).

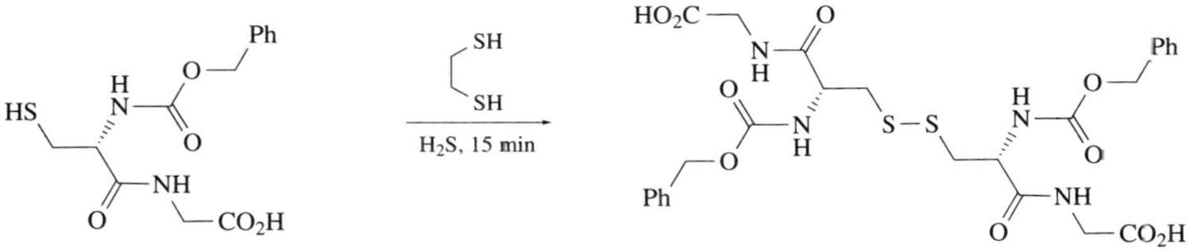
(202)

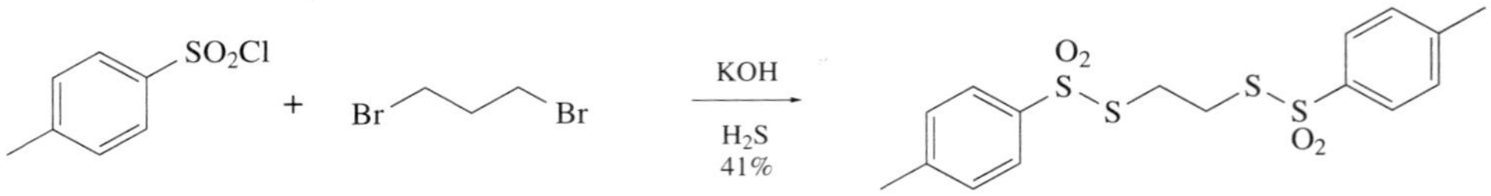

(203)

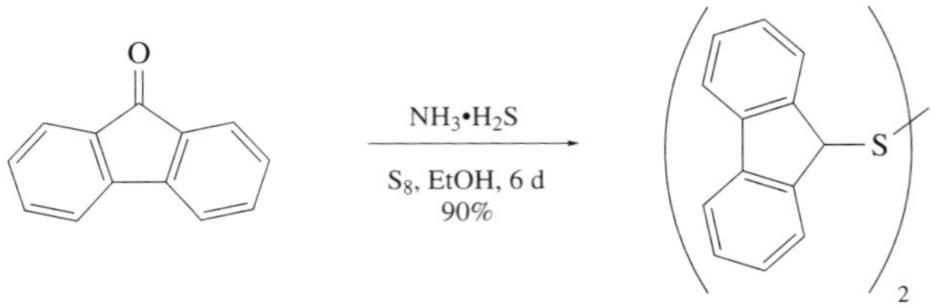

(204)

Reaction of imines with hydrogen sulfide in the presence of triethylamine and pyridine gave rise to the product shown in Equation (205) ⟨76HCA840⟩, in a very general reaction. Under the same conditions, disulfides were obtained from aldehydes, and cyclic and acyclic ketones.

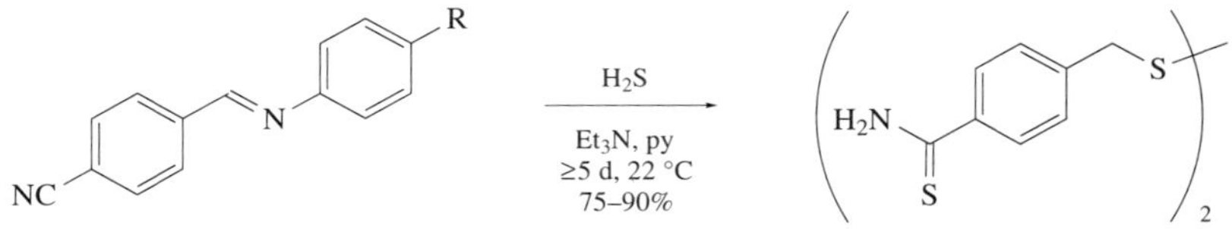

(205)

The conversion of thietanes into dithiolanes, with the formation of a disulfide linkage within the thietane, can be accomplished with DMF saturated with hydrogen sulfide (Scheme 182) ⟨70TL441, 71CPB1022⟩.

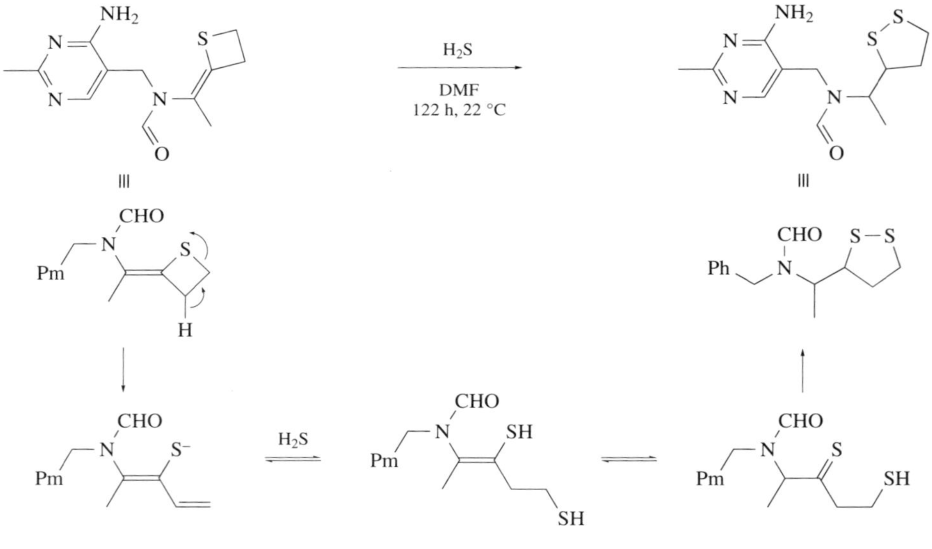

Scheme 182

2.03.7.2.5 *By reaction with sodium sulfide*

Sodium sulfide, often used as its hydrate, can react with dihalides to yield a cyclic disulfide such as a dithiolane, or with monohalides to give a symmetrical disulfide. Examples of the cyclization reaction are well illustrated by synthetic approaches to both racemic and optically pure lipoic acid ⟨57JA6483⟩. As shown in Equations (206), (207), and (208) good leaving groups such as mesylates ⟨90JCS(P1)1615⟩ and tosylates ⟨75JOC524⟩ may also be employed for the reaction. The reactions involve displacement of a suitable leaving group by reaction with disulfide anion generated *in situ* from sodium sulfide and sulfur. This methodology has also been used for polysulfide synthesis.

Syntheses of lipoic acid by mesylate displacement (Equation (207)), have been reported ⟨75JOC524, 83CC1051, 88JCS(P1)9⟩.

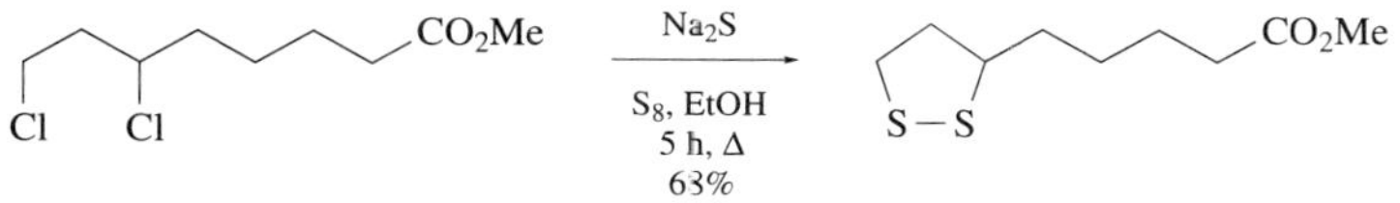

(206)

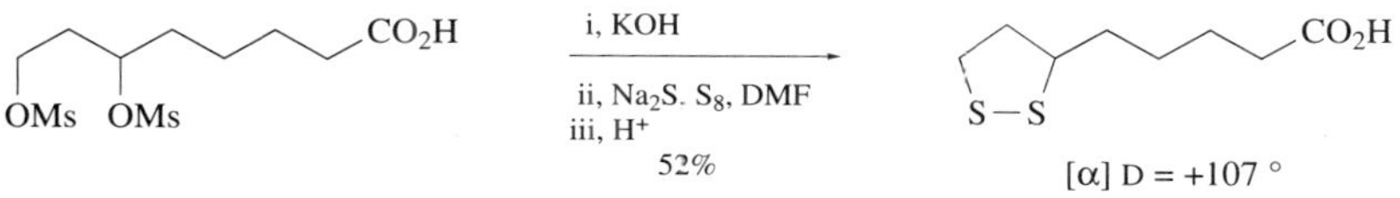

(207)

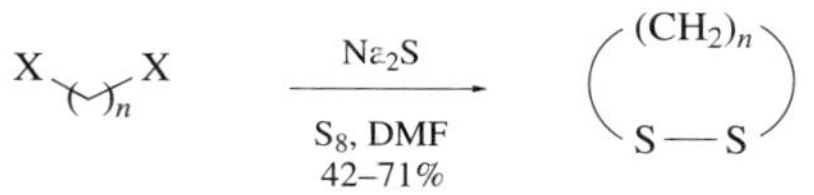

(208)

X = OMs, OTs, Cl

Another example of this type of double displacement reaction is the formation of an unsaturated 1,2-dithiane from a dibromide, although in rather low yield ⟨72JOC2367⟩.

Equations (209), (210), and (211) illustrate examples of simple tosylate ⟨51M970, 53JOC1591, 62JA2938⟩ and bromide ⟨92ZOR927⟩ displacements. An extremely interesting example of the use of sodium sulfide as a source of sulfur for disulfide formation has been described by Barton ⟨92T1883⟩, in the iron-catalyzed insertion of sulfur into a nonactivated carbon–hydrogen bond (Equation (212)). Sodium disulfide is also used for the synthesis of disulfides, and can be readily alkylated with alkyl halides, tosylates, mesylates, or other sulfonates (Equation (213)) ⟨55HOU(9)55, 90JCR160⟩. Potassium disulfide has also been employed to prepare symmetrical disulfides from dihalides, in methanolic solution under reflux ⟨74G455⟩.

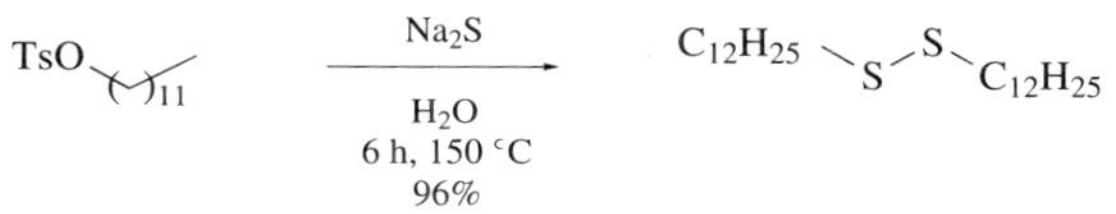

(209)

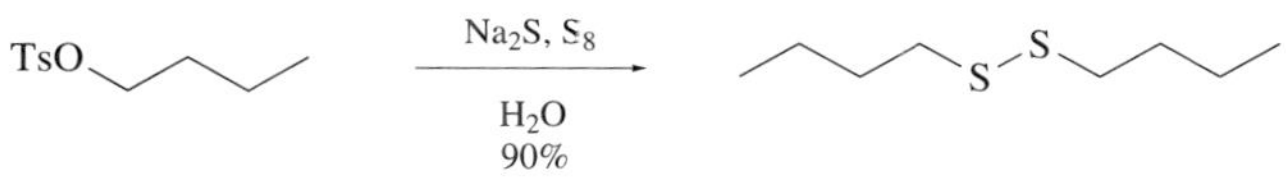

(210)

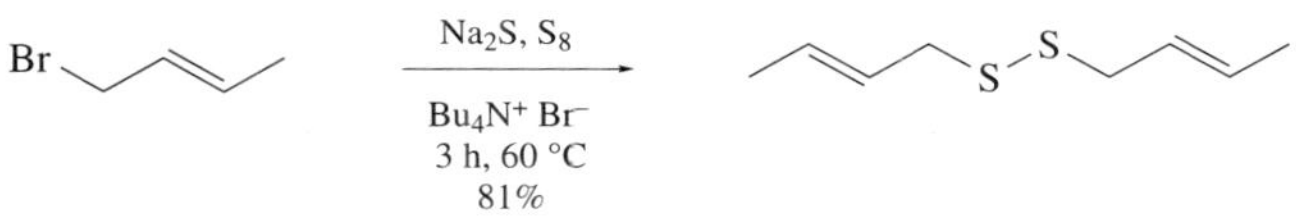

(211)

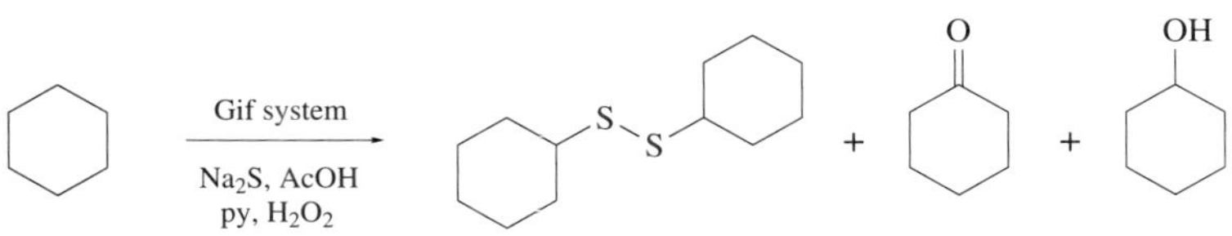

(212)

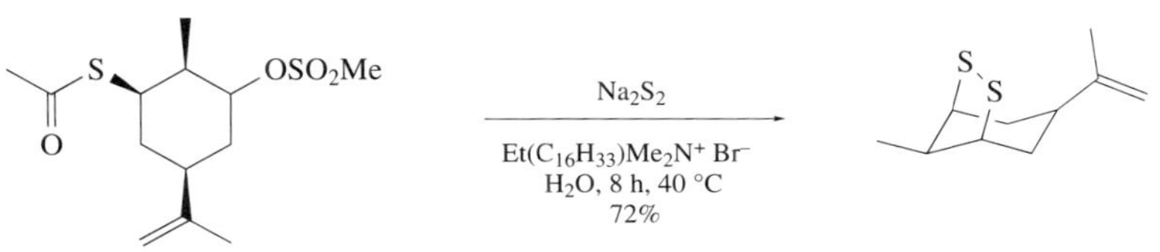

(213)

2.03.7.2.6 *By the use of miscellaneous sources of sulfur*

A survey of the literature reveals a wide range of reagents capable of providing a source of sulfur to a chemical reaction. Some of those more widely applicable for alkyldisulfide preparation are described below.

(i) Elemental sulfur

Elemental sulfur has been used to prepare dialkyl disulfides by reaction with thiocarbonyl compounds (Equation (214)) ⟨92CC600⟩. Sulfur has also been shown to react with alkyl bromides in the presence of sodium hydroxide to give dialkyldisulfides in good yields ⟨77JPR1036⟩.

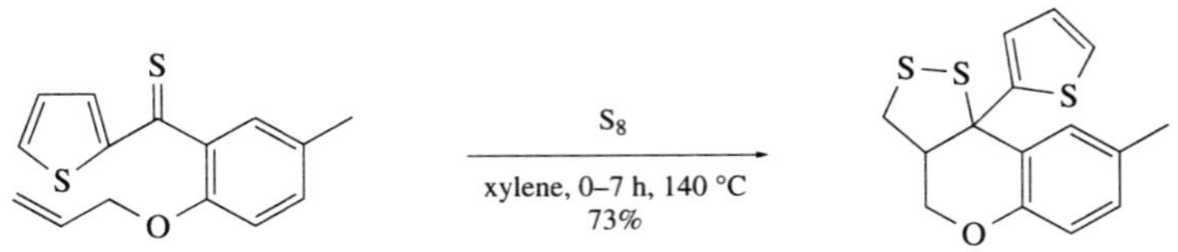

(214)

(ii) Sulfur dioxide

Sulfur dioxide has been used to prepare disulfides from a range of substrate types. Equations (215) and (216) illustrate its use for thiosulfonate preparation by reaction of a thiol with a diazo species (Equation (215)) ⟨53RTC612⟩ and with an *N*-carboxy-2-pyridyl thione under irradiation ⟨88T1153⟩, the reaction in the latter case proceeding by a radical mechanism from the carboxylic acid derivative (Equation (216)).

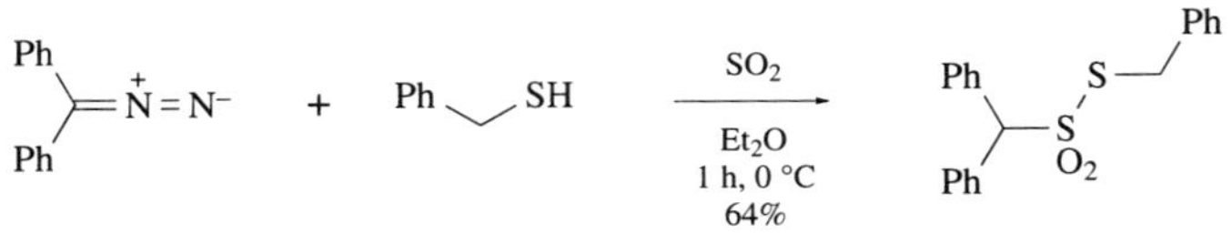

(215)

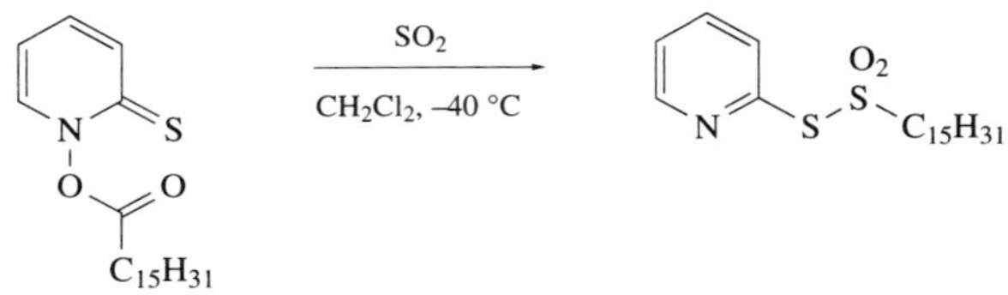

(216)

(iii) Sulfur monochloride and dichloride

Sulfur monochloride has been shown to undergo reaction with alkenes, alkynes, and activated aromatic compounds to furnish symmetrical disulfides, generally in excellent yields (Equation (217)) ⟨69JOC3991, 72CL935, 72TL3567⟩. Sulfur dichloride can also be used to effect disulfide formation. Equation (218) shows an example in which a lead thiolate, found to be more successful than the free thiol, was converted by sulfur dichloride into the corresponding disulfide, in which one of the sulfur atoms is present in a higher coordinated form, reportedly containing a sulfur–sulfur double bond.

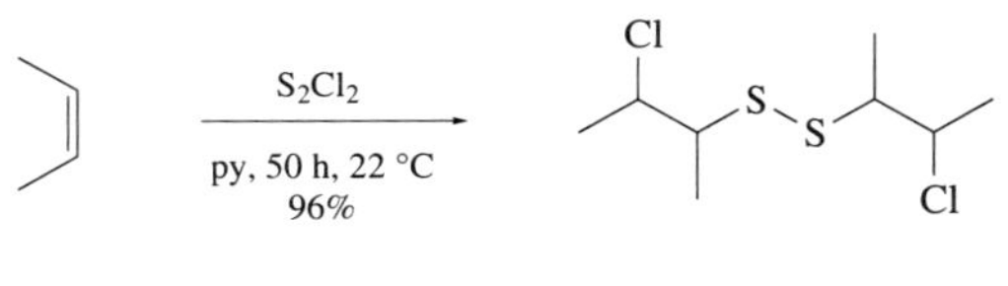

(217)

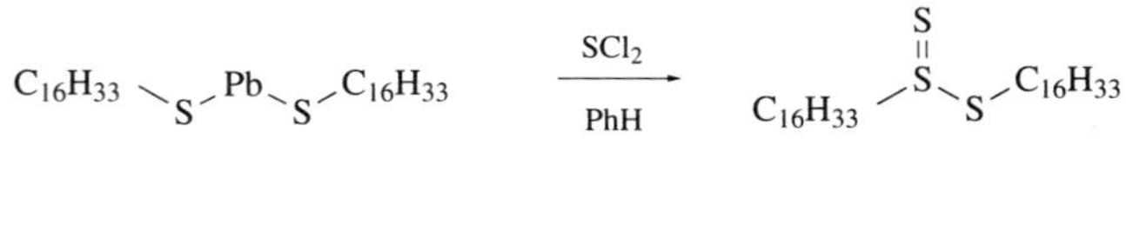

(218)

(iv) Sodium borohydride trisulfide (sulfurated sodium borohydride)

Aldehydes and ketones are converted by sulfurated sodium borohydride into symmetrical disulfides. Analogously, epoxides are transformed into bis(2-hydroxyl-alkyl) disulfides ⟨71CJC4047,

73TL1401⟩. Sulfurated sodium borohydride is used *in situ*, and is prepared from sodium borohydride by stirring with sulfur in an appropriate solvent such as THF. The reagent can act as either a reducing or sulfenating agent, depending upon reaction temperature and stoichiometry.

(v) Sodium thiosulfate

Sodium thiosulfate is widely used in conjunction with an oxidizing agent such as iodine for the synthesis of symmetrical disulfides from alkyl halides or of cyclic disulfides from the corresponding dihalides ⟨41JA658, 42JA149, 50JOC865, 54HCA375, 57JOC145⟩.

(vi) Thiolacetic acid

Thiolacetic acid can be employed for the preparation of disulfides from alkyl halides. Generally this reaction is used for the preparation of symmetrical disulfides (Equation (219)). Thiolacetic acid has also been employed with thionyl chloride for the preparation of cyclic disulfides from mono-sulfide starting materials (Equation (220)) ⟨80IZV2111⟩.

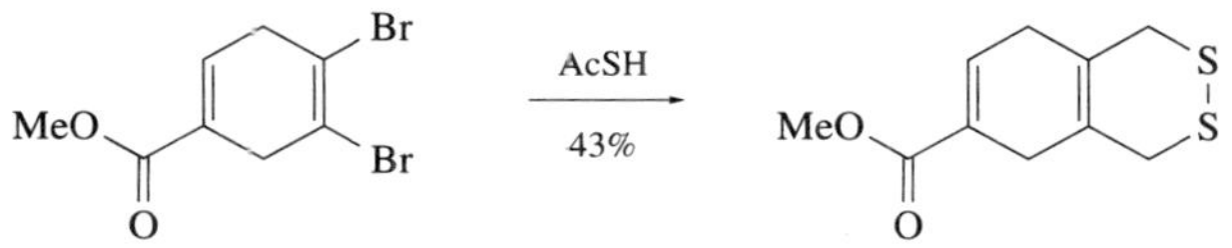

(219)

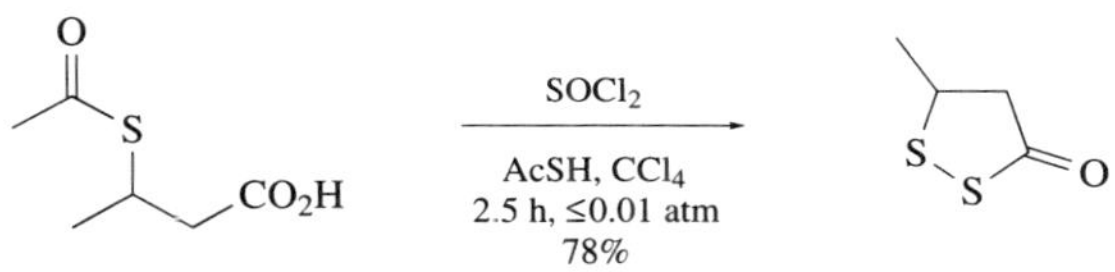

(220)

(vii) Trithioalkyl phosphine reagents

Trithioalkylphosphine reagents have been reported as species capable of acting as sources of sulfur, the sulfur usually being transferred in the form of an alkylthio ligand ⟨88CPB791, 88ZOR1088⟩. The phosphorus atom in these reagents can be in either the phosphorus(III) or phosphorus(V) oxidation states (Scheme 183). Trialkylphosphine sulfide reagents have also been reported for the conversion of thiols to disulfides ⟨84CC843⟩.

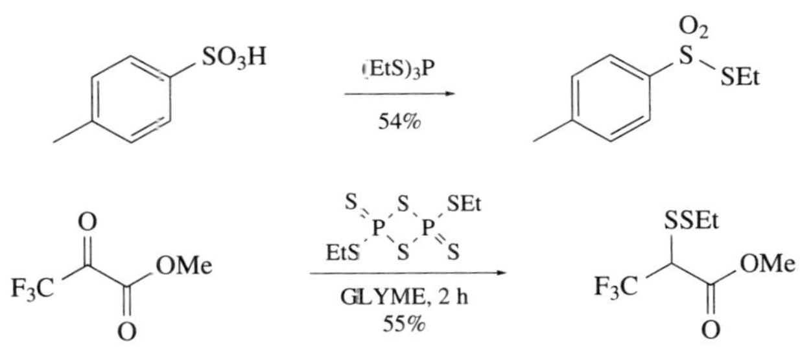

Scheme 183

(viii) Miscellaneous sulfur sources for disulfide formation

Reagent systems involving sulfoxides and silyl derivatives have been used for disulfide synthesis on many occasions. Various combinations of sulfoxide oxidants (e.g., DMSO and Ph_2SO) and silyalting agents (e.g., TMS-Cl, $SiCl_4$, TMS-OTf, $TMS\text{-}Cl_3$) have been employed ⟨91CC167, 91SL345⟩, and some of the more successful are illustrated in Equation (221). The reactions are commonly performed in trifluoroacetic acid as solvent.

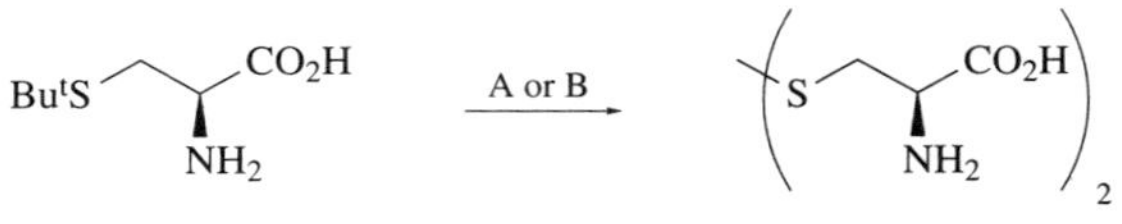

(221)

A: TMS-Cl, Ph_2SO, TFA, 4 °C, 10 min, 99%
B: TMS-OTf, TFA, 4 °C, 4 h, 87%

Dimethylsulfoxide has also been used in conjunction with hydrogen bromide and acetic acid for the oxidative cleavage of alkyl *t*-butyl sulfides and aryl sulfides, forming either symmetrical alkyl or aryl disulfides respectively ⟨93S573⟩.

Harpp has reported the reaction of silyl sulfides with symmetrical disulfides to produce unsymmetrical disulfides, and with sulfenyl chlorides to produce unsymmetrical disulfides. In both cases the reaction was mediated by sulfuryl chloride (Scheme 184) ⟨78JOC3481⟩.

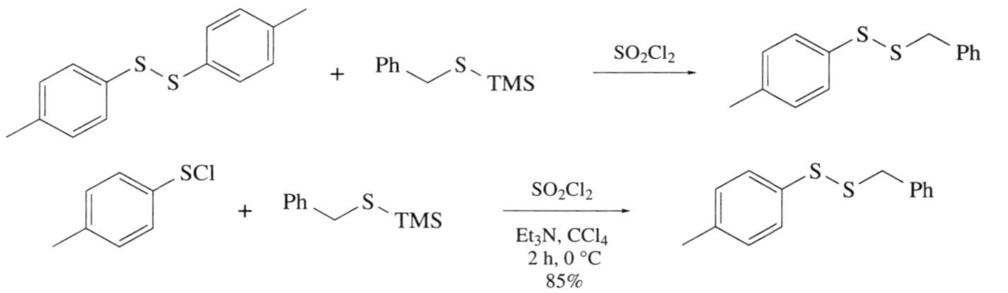

Scheme 184

Piperidinium tetrathiotungstate has been used by Chandrasekaran for the preparation of cyclic disulfides from dihalo compounds ⟨92JOC1699⟩. The same group have reported the use of tetrathiometallates of both molybdenum and tungsten for the formation of acidic disulfides from alkyl halides ⟨89JOC2998⟩. The reaction has been performed on a range of alkyl halide substrates, and has proved to be an effective alternative to the more traditional methods of disulfide synthesis, such as thiol oxidation and thiolysis. Chandrasekaran has reported the use of the benzyltriethylammonium tetrathiomolybdate reagent as an efficient sulfur transfer reagent for the synthesis of disulfides ⟨92SC3277⟩. Molybdenumhexacarbonyl has also been reported for the synthesis of disulfides from sulfonyl chlorides ⟨69AG(E)677⟩.

Other useful sources of disulfide include 5,5-dimethyl-1,2-dithia-3,7-diselenacycloheptane, a storable reagent effective for disulfide preparation ⟨87AG(E)887⟩, and lithium sulfide, effective when used in conjunction with sulfur ⟨82SC947⟩. Thiourea has been used to convert a bis-Bunte salt into a cyclic disulfide ⟨65JCS2901, 70CPB235⟩. Many other sources of sulfur are known, and can be used to furnish disulfides from a range of substrates, including thiols, sulfides, alkyl halides, and carbonyl compounds. Some reagents have proven to be general, while others are of very restricted application, and a thorough search of the relevant literature is recommended before selecting a reagent system.

2.03.7.2.7 Formation of alkyl disulfides by photochemical and thermal reactions

Photochemical reactions provide an alternative, often simple method of disulfide preparation. Nishio has reported that the irradiation of trithiane thiocarbonyl trimers in cyclohexane gives rise to mixtures of thiophenes, dithietanes, and the disulfides (Equation (222)) ⟨73BCJ2253⟩. Irradiation of thioketones has also been reported to furnish complex disulfide products, although again the yields of disulfide were generally poor (Equation (223)) ⟨73TL1561⟩.

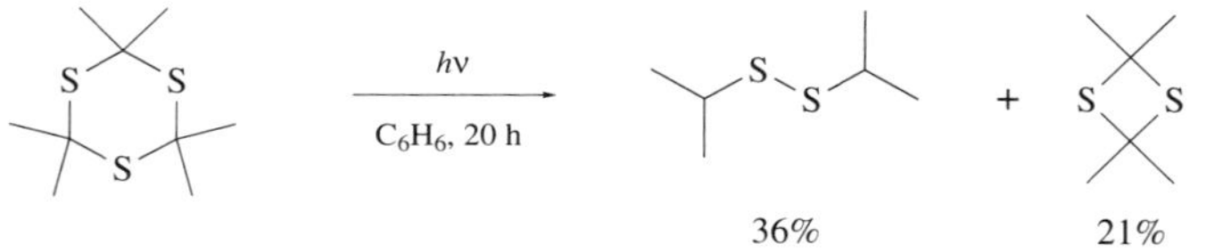

(222)

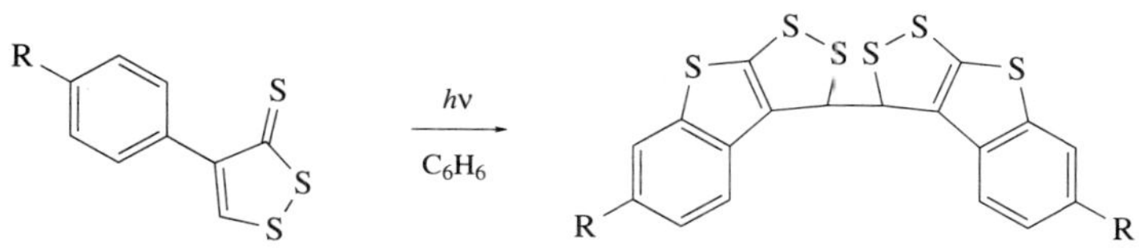

(223)

R	*Yield* (%)
Me	44
H	40

Other photochemical methods of disulfide synthesis, such as that reported by Still ⟨75CJC455⟩, shown in Equation (224), are often low yielding, although generally clean reactions. One exception was reported in 1978 by Yates and Toong, who were able, under photochemical conditions, to couple two molecules of a saturated thiophene with ring cleavage in 81% yield (Equation (225)) ⟨78CC205⟩. Interestingly, this reaction did not proceed when *t*-butanol was used as the solvent.

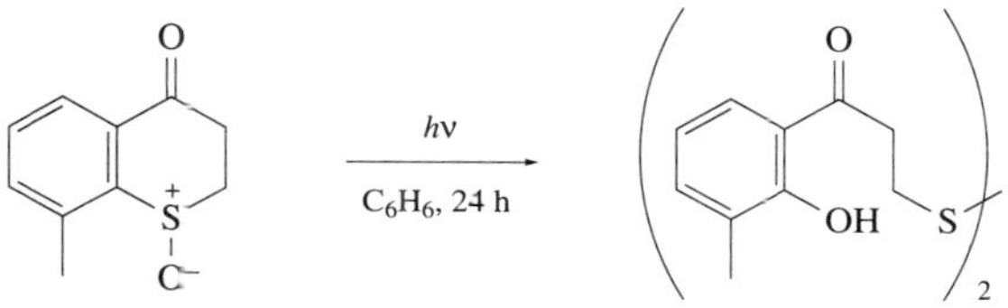

(224)

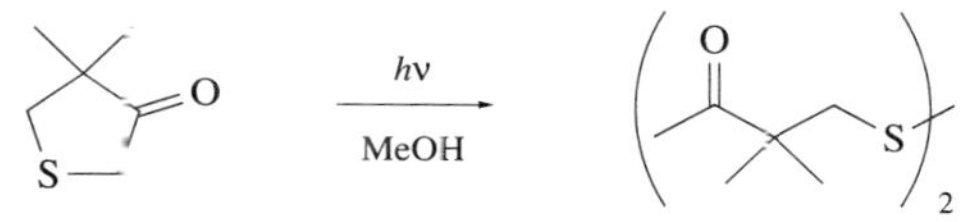

(225)

Similar α-keto tetrahydrothiophenes have also been converted into the corresponding disulfides under photochemical conditions ⟨79BCJ3601⟩.

Thermally mediated methods of disulfide synthesis have proved to be considerably more efficient than the photochemical processes. Eisner and Krishnamurthy have reported the conversion of the 2,6-dihydro-1,3-dithiin shown in Equation (226) into the 5,6-dihydro-1,2-dithiin in quantitative yield by brief heating under reflux in the absence of solvent ⟨71T5753⟩. The mechanism of this unusual rearrangement was proposed by the authors to involve fission of a carbon–sulfur bond with formation of an allylic diradical. Other unusual thermal rearrangement processes have been reported to furnish disulfide products ⟨83T3839⟩.

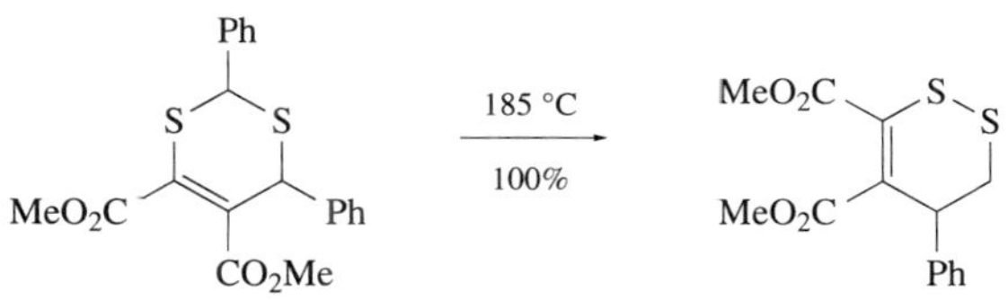

(226)

2.03.7.2.8 From miscellaneous sulfur functionalities

(i) Sulfenyl halides

In addition to thioalkylation reactions, sulfenyl halides can undergo a range of other transformations which result in unsymmetrical disulfide formation. Perhaps the best known is displacement of chloride by a sulfide nucleophile, with cleavage of a carbon–sulfur bond. The reaction is usually performed in acetic acid ⟨75CC976, 77MI 203-05, 79HCA2507, 81JOC1868, 89JOC3640⟩, although other acidic media such as trifluoroacetic acid ⟨87JOC3154⟩ have been used as solvent, as have common organic solvents ⟨82CL921, 89JOC244, 91T5603⟩. The reaction is high yielding, and is most often seen in amino acid synthesis, used for the preparation of cysteine and cystine residues (Equation (227)).

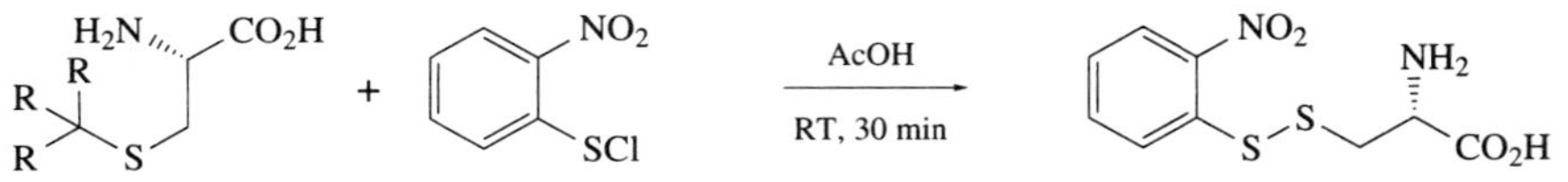

(227)

R	*Yield* (%)
Me	82
Ph	100

(ii) *Thiocarbonyl compounds*

Thiocarbonyl compounds are efficient precursors to symmetrical disulfides. The reaction can be brought about most successfully under reducing conditions (Equation (228)) ⟨75JCS(P1)2513⟩.

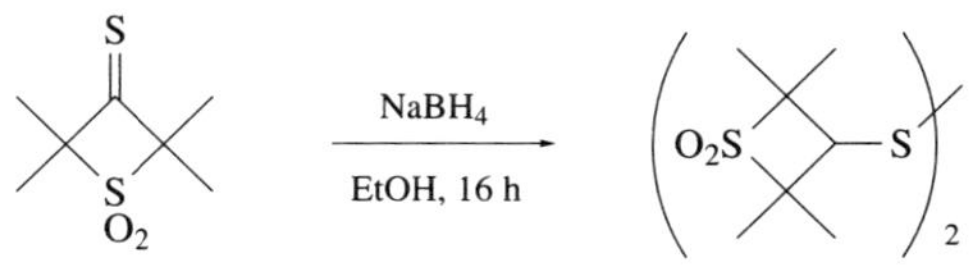

(228)

(iii) *Oxidative coupling of sulfides*

The conversion of two sulfide molecules into a symmetrical disulfide occurs readily under oxidative conditions. For example, cysteines are readily converted into cystines using phenyliodoso diacetate (Equation (229)) ⟨84TL4153⟩. Iodine in acetic acid transforms thiols protected as their trimethylacetamidomethyl thioethers into the coupled disulfides (Equation (230)). In a similar process, 4-nitrobenzyl sulfides are converted into the symmetrical disulfides in high yields by nitro group reduction with hydrogen and treatment with mercuric sulfate (Equation (231)) ⟨72JOC3550⟩. Thiolacetates may also be cleaved to produce disulfides in high yields and under mild conditions, using clayfen (ferric nitrate on K-10 bentonite clay) in the absence of solvent ⟨93TL2521⟩.

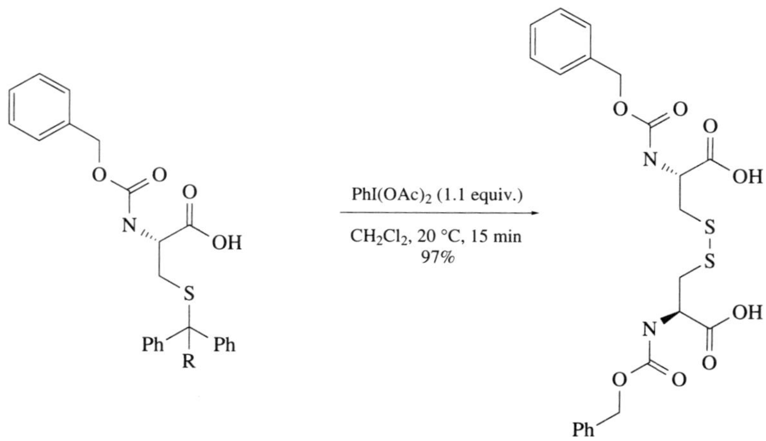

(229)

R	*Yield* (%)
CPh_3	97
$CHPh_2$	92
$CH_2NC(O)Me$	85

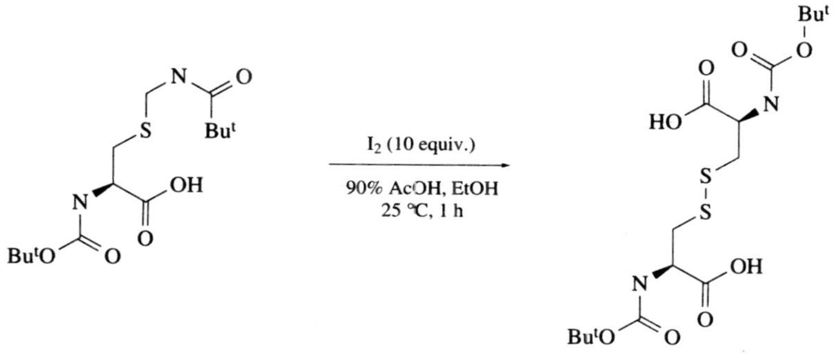

(230)

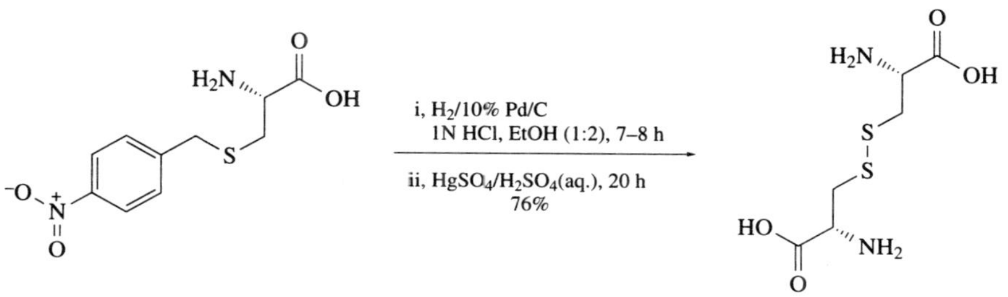

(231)

(iv) Alkylthiosilanes and alkylthiostannanes

Harpp has described the preparation of unsymmetrical disulfides from alkylthiostannanes. The process may be carried out as an intermolecular process (Equation (232)), in which case another suitable sulfur-containing molecule, typically a sulfinyl halide, is present in the reaction mixture ⟨83TL5173⟩, or as an intramolecular process to furnish a cyclic disulfide (Equation (233)) ⟨86TL441⟩. Alkylthiosilanes have also been used as precursors of symmetrical disulfides, typically by reaction with halogens ⟨64JCS5975⟩.

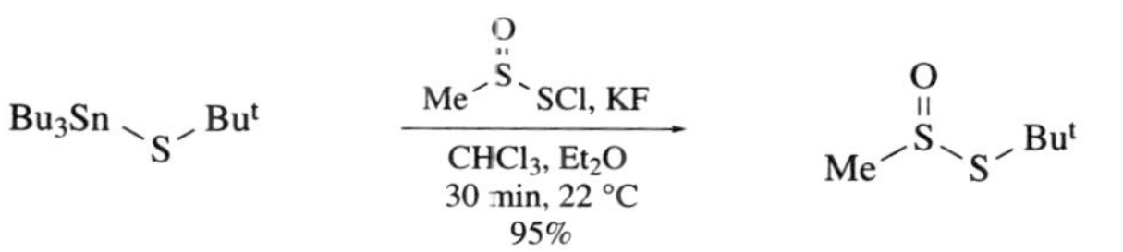

(232)

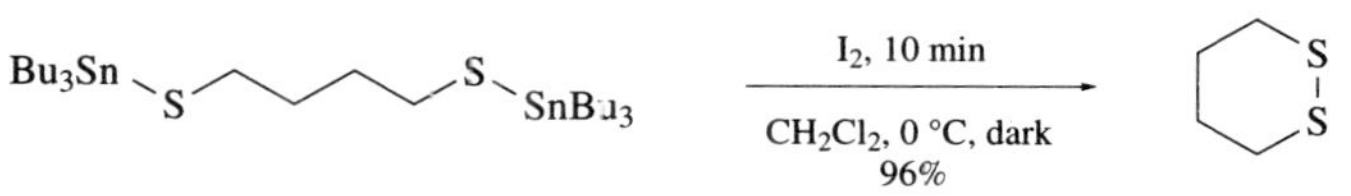

(233)

(v) Thiocyanates

Thiocyanates are valuable precursors of symmetrical disulfides. The powerful one-electron reducing agent samarium diiodide has been shown to couple two molecules of thiocyanate to afford the corresponding disulfide via the radical anion in excellent yield (Scheme 185) ⟨94TL8835⟩.

$$\text{RSCN} \xrightarrow{\text{SmI}_2} [\text{RSCN}]^{\bullet -} \longrightarrow \text{RS}^{\bullet} + \text{CN}^- \longrightarrow \text{RSSR}$$

R	*Yield (%)*
C_5H_{11}	80
$C_{10}H_{21}$	85
$C_{16}H_{33}$	76

Scheme 185

(vi) Vinyl sulfides

Electron-deficient vinyl sulfides can react with alkyllithium reagents in the presence of copper(I) salts to produce disulfides ⟨80S910⟩.

2.03.7.3 Alkanethiosulfinates and Thiosulfonates

In many cases, preparative methods successful for disulfides are also suitable for thiosulfinate and thiosulfonate synthesis. The principal additional route is through oxidation of disulfides, described here for the more common oxidizing agents.

Two interesting systems which have been used for this oxidation are oxygen in the presence of methylene blue ⟨72JOC3516, 73IJS307⟩, and tetrabutylammonium iodate in the presence of an iron porphyrin catalyst ⟨83TL3631⟩. Both have been used for the oxidation of di-*t*-butyl disulfide to the thiosulfinate (Equation (234)).

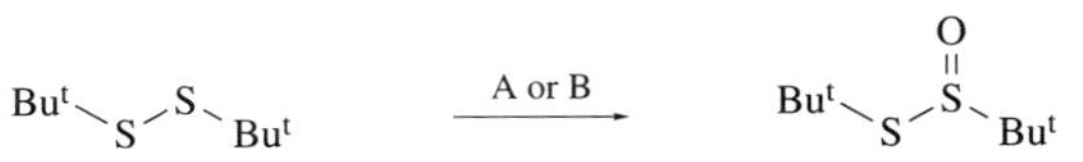

(234)

A: O_2, methylene blue, MeOH, 3 h, 75%
B: Bu_4NIO_4, TPPFe^{III}Cl, RT, 20 h, 90%

Perbenzoic acid is capable of selective monooxidation of disulfides ⟨47JA1710⟩, as is mcpba ⟨91JOC904⟩.

Thiosulfonates may also be produced from disulfides using mcpba, although oxidations are not always very regioselective. Scheme 186 shows an example in which a cyclic disulfide was oxidized in 55% overall yield, providing a 5:2:3 mixture of the thiosulfonate and the two thiosulfinates ⟨78JOC3374⟩. Use of sodium metaperiodate gave a 1:1 mixture of monosulfoxides in high yield. Disulfoxides are not uncommon as oxidation products of disulfides, as is illustrated by oxidation of the cysteine disulfide shown in Equation (235) ⟨59BSF1887⟩.

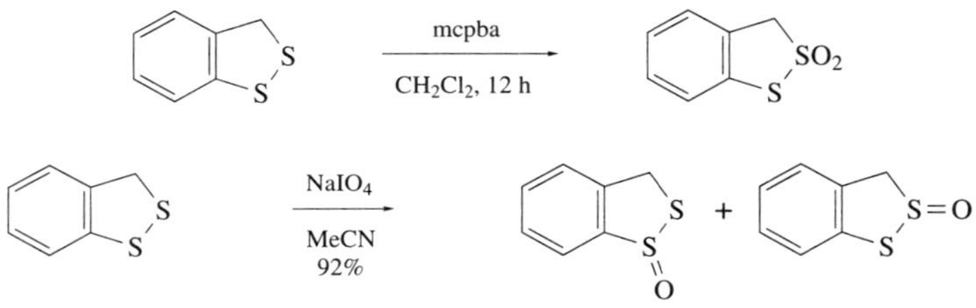

Scheme 186

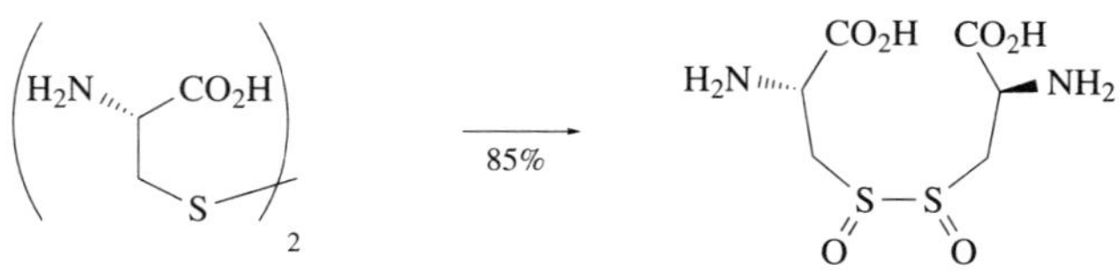

(235)

Alkanethiosulfonates have been produced by treatment of disulfides with a sulfenyl halide, chlorine and aqueous acid (Equation (236)) ⟨59JOC973, 65JOC175⟩, by treatment of a sulfonyl chloride with sodium sulfide (Equation (237)) ⟨69BSB191⟩, by the reaction of an alkene with a halothiosulfonate (Equation (238)) ⟨92MI 203-02⟩, and by the reaction of a disulfide with a sulfinic acid salt in the presence of silver nitrate (Equation (239)) ⟨72JOC333⟩.

Alkanesulfonesulfonates, the fully oxidized derivatives of disulfides, may be prepared by the coupling of two molecules of a sulfonyl chloride ⟨93S490⟩.

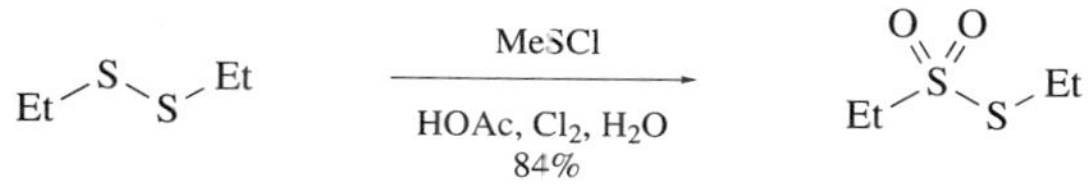

(236)

(237)

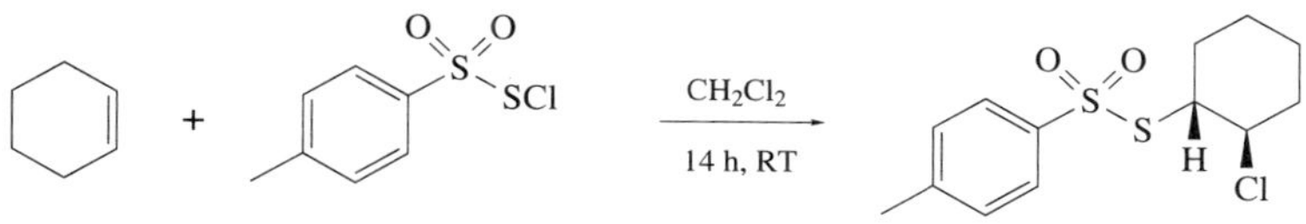

(238)

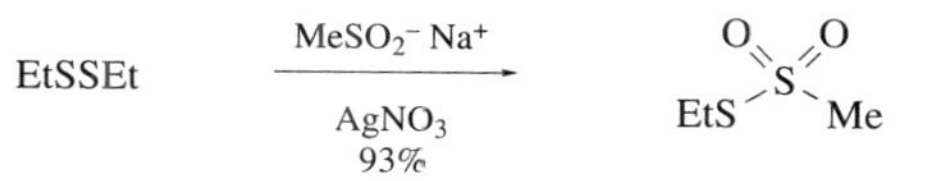

(239)

2.03.7.4 Trisulfides and Polysulfides

The preparation of trisulfides and polysulfides is usually not trivial. These compounds are relatively uncommon in the literature, with the notable exception of the trisulfide functionality found in the tumor inhibitors calichaemycin and esperamycin, members of the enediyne group of antibiotics.

Trisulfide preparation may be carried out by treatment of a suitable dithiol with sulfur (Equation (240)) ⟨86CL349⟩. Similar treatment of alkanethiols can also give rise to symmetrical trisulfides.

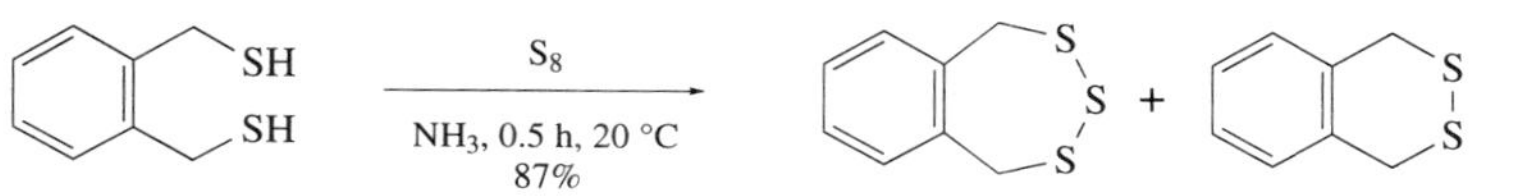

(240)

Thioalkylation of various thiosulfenate species can produce trisulfides. Suitable substrates include Bunte salts, metal sulfides, and thiosulfenyl chlorides (Scheme 187) ⟨84S657⟩. The methoxycarbonyltrisulfides produced could be efficiently converted into unsymmetrical dialkyl trisulfides by further treatment with alkanethiols.

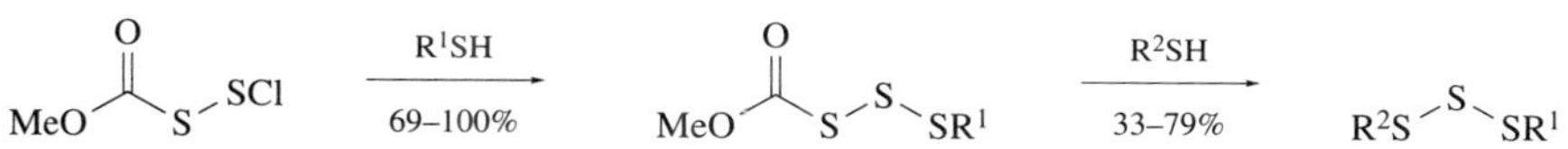

Scheme 187

Disulfides may undergo sulfurization to give trisulfides. An interesting example is given in Equation (241), in which a 1,3,4-trithiolane furnished ring-cleaved disulfides and trisulfides upon treatment with secondary amine and hydrogen sulfide at elevated temperature ⟨59LA(627)195⟩.

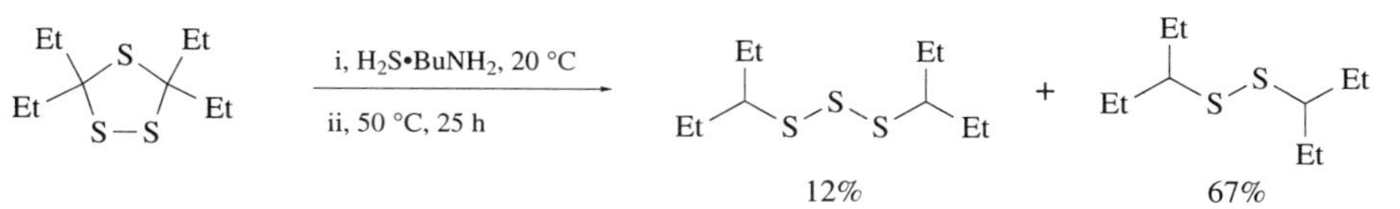

(241)

Interest in the conversion of trisulfides into disulfides has been stimulated with the development of trisulfide-containing antitumor agents such as the calichaemycins. Carbon–sulfur bond cleavage in these materials appears to be important in the mode of action of these compounds. In addition to traditional phosphorus reagents, newer methods for desulfurization of trisulfides include treatment with bis(triphenylstannyl)tellurides and -selenides, the telluride reagent being the most effective. The authors propose an intermolecular mechanism ⟨93TL903⟩. Hexamethylphosphoric triamide (HMPT) has also been used for this transformation ⟨80JOC5155⟩.

The rare higher polysulfides may also be converted into disulfides, the polysulfide acting as a sulfur transfer agent ⟨89CL2111, 92TL2517⟩. Benzopentathiepin is such a sulfur transfer agent capable of transferring sulfur to an alkene in the presence of Lewis acid (Equation (242)).

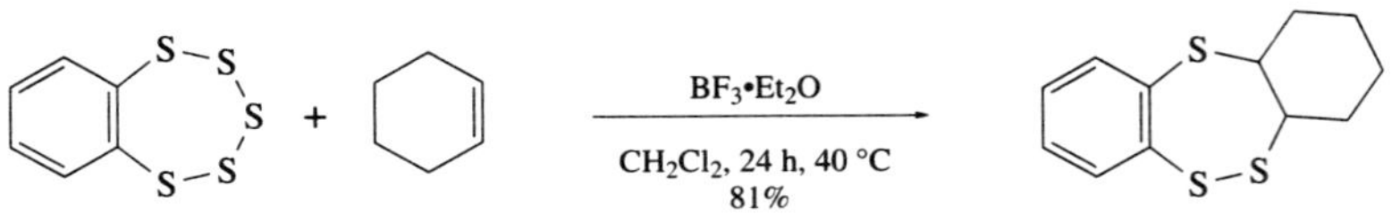

(242)

2.03.8 RSSe AND RSTe FUNCTIONS AND THEIR HIGHER COORDINATED DERIVATIVES

Methods of preparation of alkane sulfenyl selenide and alkane sulfenyl telluride functions and their higher coordinated derivatives are relatively few and far between. Existing methods usually involve the use of an alkanethiol as one of the starting materials. The most common compounds of the type are alkane sulfenyl selenides, and this section therefore deals mainly with their preparation. Higher coordinated derivatives of alkane sulfenyl selenides, alkane sulfinyl selenides, and sulfonyl selenides, are very uncommon in the literature, as are all alkane thiotellurides.

2.03.8.1 Sulfenyl Selenides

The principal method of alkane sulfenyl selenide preparation involves the reaction of a selenium-containing reagent with an alkanethiol. Selenium compounds which have been used for this purpose are discussed below.

2.03.8.1.1 Selenocarbonyl compounds

Electron-deficient selenoaldehydes have been shown to react with alkanethiols in the presence of triethylamine to furnish selenylsulfides in yields of up to 55% ⟨93TL4925⟩. As shown in Scheme 188, the desired selenocarbonyl compounds are prepared from their 9,10-dimethylanthracene adducts. In the presence of two equivalents of thiol, in direct analogy to the reaction of thioaldehydes with thiols to form disulfides ⟨93CC806⟩, these electron-deficient selenoaldehydes react further to form sulfenyl selenides. Selenocarbonyl reactivity is known to mirror that of the thiocarbonyl group ⟨B-87MI 203-01⟩. In the reaction, two equivalents of thiol were employed to minimize selenoaldehyde oligomerization. Even so, yields were only moderate for tertiary thiols (35–55%) and worse for primary thiols (0–18%) Carbon–selenium bond cleavage was troublesome, as was the expected carbon–sulfur bond cleavage, enhanced by the use of excess thiol.

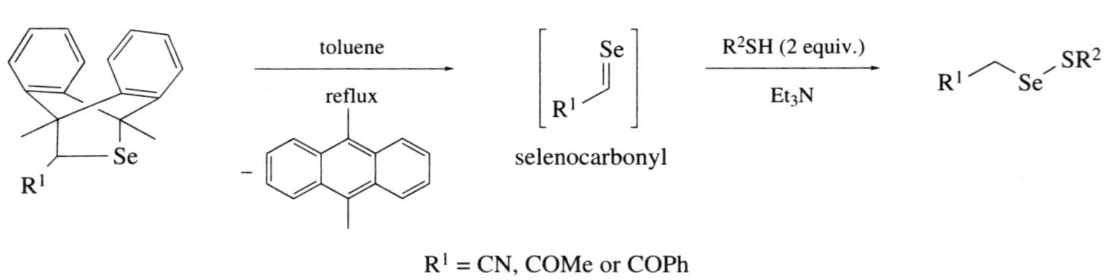

R^1 = CN, COMe or COPh

Scheme 188

Scheme 189 shows the reaction of selenocarbonyl *N*-benzenesulfonylimide with a thiol to furnish an addition compound containing a tetravalent selenium atom. The reaction proceeds via a number of proposed intermediates containing R—S—Se functionality, ultimately to furnish the desired five-membered ring, sulfur- and selenium-containing sulfonylimine product ⟨74TL1059⟩.

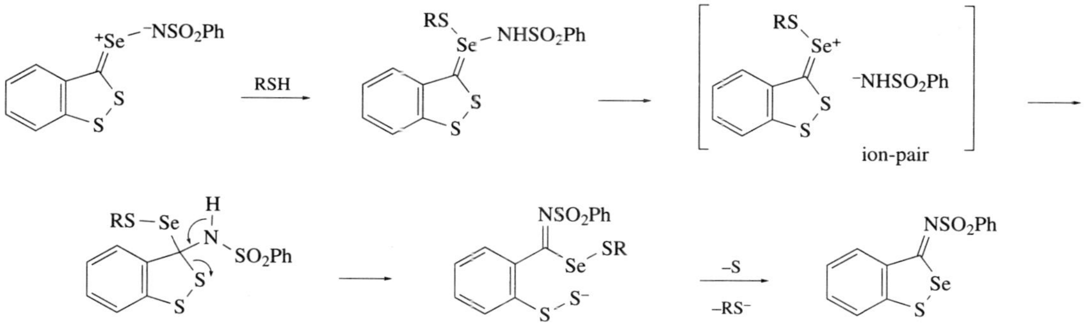

Scheme 189

2.03.8.1.2 *Phenylselenenyl chloride*

In a reaction sequence geared towards the synthesis of a thiomuscarone analogue, the extremely stable thioselenide shown in Equation (243) was formed by the reaction of a *β*-ketothiol with phenylselenenylchloride at low temperature. The unusual stability of the thioselenide product is presumably related to the presence of the carbonyl group ⟨90H(32)2099⟩.

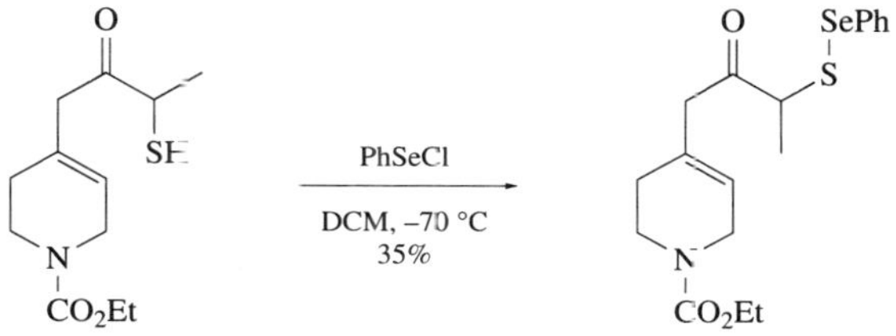

(243)

2.03.8.1.3 *Selenium–nitrogen bond containing heterocycles*

The reaction of heterocycles containing a selenium–nitrogen bond, such as that shown in Equation (244), with a thiol results in ring opening to furnish selenosulfides. The nitrogen selenium heterocycle ebselen PZ 51 ⟨86BCJ2179⟩, a good catalyst for the oxidation of glutathione to the disulfide by hydroperoxide, and a compound of possible relevance to the mechanism of action of the selenium-containing enzyme glutathione peroxidase, undergoes such a ring-opening reaction at room temperature in chloroform, as shown in Equation (244) ⟨89JOC1092⟩. Similarly, 2-aryl-1,2-benzisoselenazol-3(2*H*)-one can be ring opened by alkanethiols to give selenylsulfides in excellent yields (ca. 95%) (Scheme 190) ⟨86H(24)303⟩. This reaction proceeds through nucleophilic attack of the carbonyl oxygen atom at the hydrogen atom of the thiol, giving a 3-hydroxy-2-aryl-1,2-benziso-selenazolium ion and a thiolate anion. The nucleophilic nature of the carbonyl oxygen atom results from electron donation from the nitrogen lone pair. Similar ring-opening reactions have previously been reported ⟨69AJC765, 70T1463⟩. The thiolate anion formed attacks the selenium atom of the intermediate, resulting in ring opening to furnish the sulfenyl selenide product which can readily tautomerize. Other examples are known ⟨86EUP165534, 86GEP3443467, 88GEP3626554⟩.

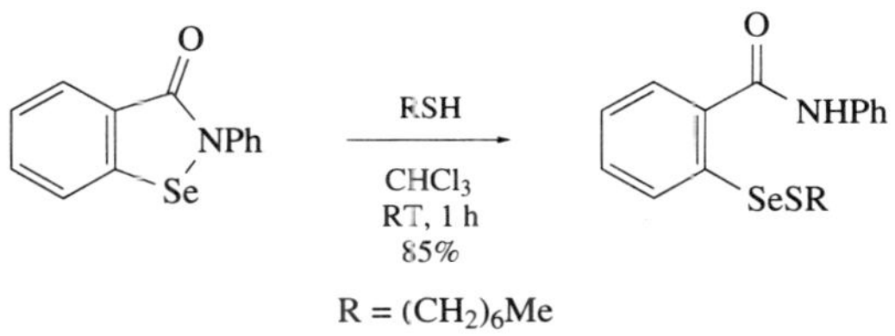

(244)

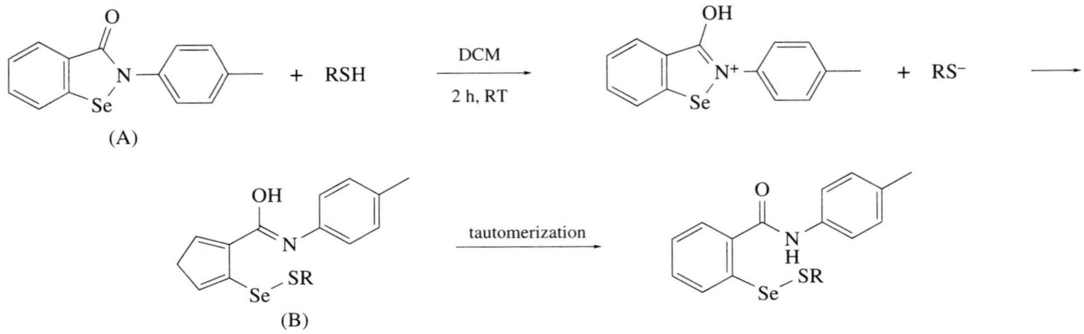

R = $MeCH_2CH_2$, 96%
R = $Me(CH_2)_3$, 94%

Scheme 190

2.03.8.1.4 *Selenites*

Selenites react with thiols to form bis(alkylthiol)selenides (Equation (245)) ⟨91JOC3520⟩. Such reactions are known between selenites and the thiol groups in cysteine residues, glutathione, and coenzyme A ⟨68B2898, 69B2257, 71B4089⟩. The ideal ratio of thiol to selenite is 4 : 1, giving the reaction stoichiometry shown in the equation. This reaction is thought to be an important pathway for the incorporation of inorganic selenium into living systems.

$$4\,RSH + H_2SeO_3 \longrightarrow RSSeSR + RSSR + 3\,H_2O \qquad (245)$$

R = alkyl

2.03.8.1.5 *Selenium transfer reagents*

A selenium transfer agent, formed as shown in Scheme 191 by the reaction of *N*-chlorobenzotriazole with a bisallylselenide, reacts with a thiol to furnish the desired sulfenyl selenide in ca. 90% yield. Equation (246) illustrates the reaction for trityl thiol which was converted to ditrityl-2-selentrisulfide ⟨92TL2129⟩. These selenium transfer reagents are relatively new reagents for organic synthesis, their development initiated by the unstable nature of selenyl halides, and indeed by the lack of selenyl halides available for synthetic use. A similar, earlier research effort led to the preparation of nitrogen-blocked analogues of sulfur chlorides, which were easy to use, stable alternatives to the sulfenyl chlorides ⟨78JA1222, 90JOC2576⟩.

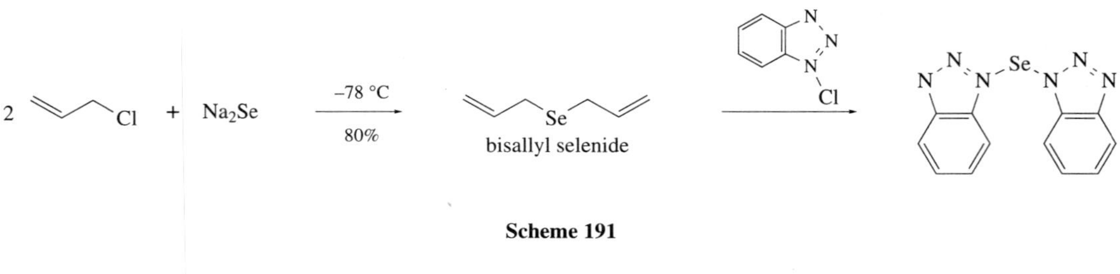

Scheme 191

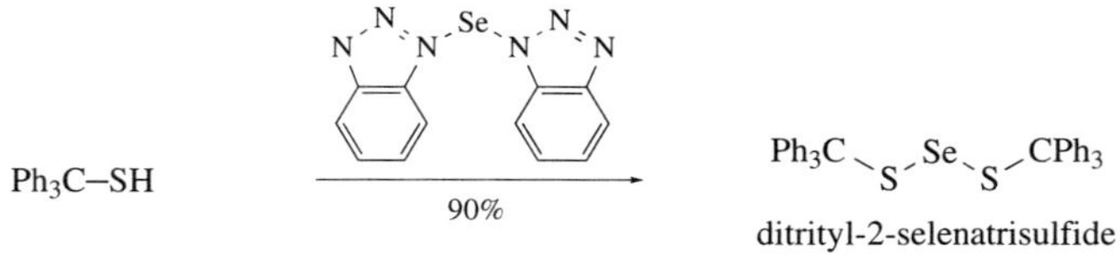

ditrityl-2-selenatrisulfide (246)

2.03.8.1.6 *Selenosulfates*

2-Aminoethane thiol derivatives of cysteamine and their selenium analogues have found use as antiradiation agents. The preparation of such compounds generally begins from 2-aminoethane selenosulfuric acid ⟨65JOC2454, 84JMC228⟩, which is itself the precursor of many selenium compounds. Reaction of the sodium salt of this acid with a thiol, often also present in the form of its sodium salt, furnishes a selenylsulfide (Equation (247)) ⟨65JOC2454⟩.

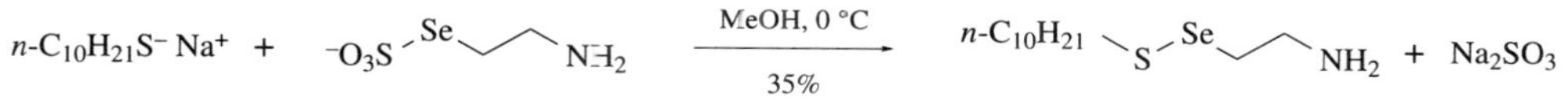

(247)

2-Aminoethane selenosulfuric acid can be prepared by reaction of potassium selenosulfate with ethylenimine in water, or by reaction of potassium selenosulfate with a 2-aminoethyl halide. Potassium selenosulfate is formed by the reaction of selenium powder with aqueous potassium sulfite ⟨48JA1115⟩ or by the reaction of an alkyl selenyl halide with potassium sulfite ⟨14JCS1672, 47JA2236⟩. Numerous aliphatic selenosulfates have been prepared in this manner.

2.03.8.1.7 *By oxidation of a thiol*

An interesting example of the preparation of an alkanesulfenyl selenide involves the oxidation of a thiolselenol starting compound by the use of oxygen in the presence of ferric chloride to give the desired 1-thia-2-selena-cyclopentan-4-carboxylic acid (Equation (248)) ⟨61AK(18)1143⟩.

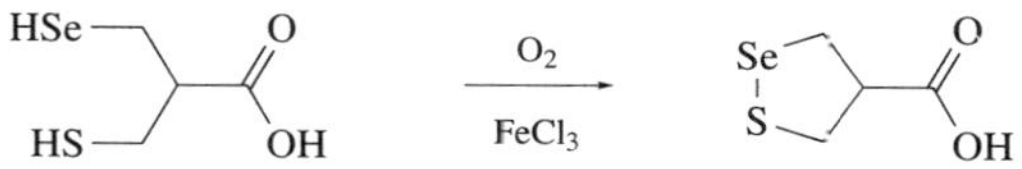

(248)

2.03.8.1.8 *Diaryl selenides*

Alkanesulfenyl selenides can be formed by the reaction of diaryl selenides with alkanethiols in the presence of hydrogen peroxide (Equation (249)) ⟨92JA9737⟩. Yields obtained from this reaction are not exceptional, with only 43% of *t*-butylthiophenyl selenide isolated. For 1-octylthio phenyl selenide, the desired product was contaminated with 1-octyl disulfide.

$$\text{PhSeSePh} + 2\text{RSH} + H_2O_2 \xrightarrow[CH_2Cl_2]{\text{MeOH}} 2\text{PhSeSR} + 2H_2O \quad (249)$$

$$R = Bu^t \text{ or } C_8H_{17}$$

2.03.8.1.9 *Seleneous acid*

1,2-Dithiols are reported to undergo reaction with seleneous acid to furnish selenotetrasulfides (Equation (250)) ⟨71USP3544593⟩.

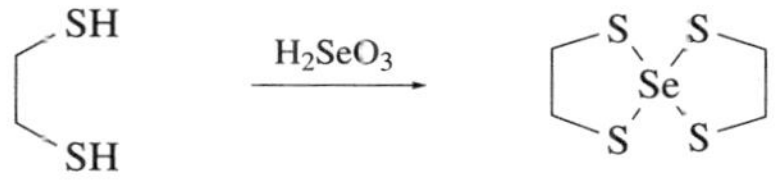

(250)

2.03.8.1.10 *Sulfides*

Alkanesulfenyl selenides can be prepared from sulfide starting materials as well as from thiols. For example, selenium can be inserted into strained carbon–sulfur bonds, such as those in small saturated ring systems. For example, reaction of 2-thia-spiro-[3.5]-nonane, prepared from sodium sulfide and either 1,1-bisbromomethyl-cyclohexane or cyclohexane-1,1-dimethylbenzenesulfonate, with elemental selenium in diethylene gylcol containing a trace amount of potassium cyanide gave a 75% yield of 2-thia-3-selenaspiro-[4.5]-decane (Equation (251)) ⟨64ACS815⟩. The product was isolated as a red crystalline solid; this color is characteristic of thioselenacyclopentanes.

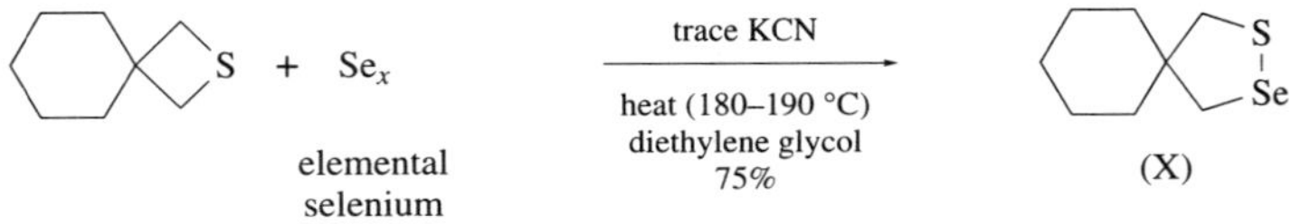

(251)

A cyclic alkanesulfenyl selenide is also formed, as shown in Equation (252), by a Birch reduction–oxidation sequence, to give a 1-thia-2-selenenacyclopentane-4-carboxylic acid in 63% yield. As in Equation (248), oxidation takes place in the presence of ferric chloride ⟨61AK1143⟩.

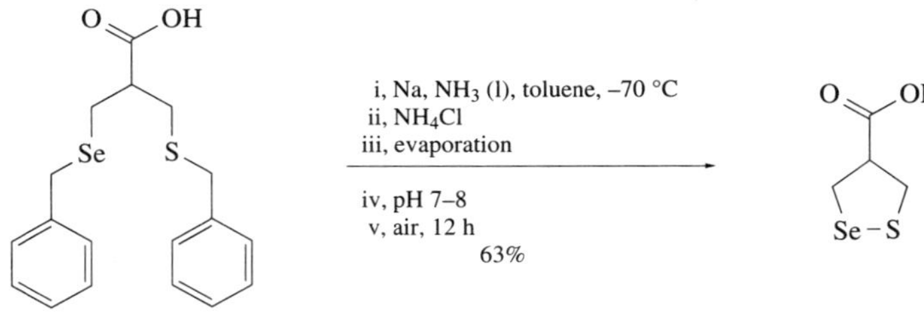

(252)

Alkanesulfenyl selenides have been reported to be minor by-products of the reaction of Grignard reagents with selenium disulfide. In contrast to the reaction with selenium dioxide, which cleanly gives alcohol products, thiols were not isolated, dialkyl selenide and dialkyl sulfide being the major products ⟨91PS281⟩.

Reaction of dimethylsulfide with phenylselenenyl bromide in the presence of silver tetrafluoroborate results in the formation of phenylselenenodimethylsulfonium tetrafluoroborate, a compound found to be a highly reactive selenenating agent for electron-rich aromatic compounds including amines, phenols, and ethers (Equation (253)) ⟨82JOC951⟩. The tetrafluoroborate salt is stable at 0 °C, but decomposes slowly at room temperature. The preparation of a sulfonium salt containing a sulfenylselenide moiety, dimethyl(selenocyanato)sulfonium hexafluoroarsenate, by halide–pseudohalide exchange in dimethylbromosulfonium hexafluoroarsenate has also been reported ⟨90ZN(B)1625⟩.

Alkylsulfenyl selenides have found use as spin trapping agents in ESR spectroscopy ⟨92JOM(434)151⟩, and are known to disproportionate readily to give diselenides and disulfides.

$$\mathrm{PhSeBr} + \mathrm{MeSMe} + \mathrm{AgBF_4} \xrightarrow[92\%]{0\ ^\circ\mathrm{C,\ MeCN}} \mathrm{PhSe{-}{}^{+}S(Me)_2}\ \ \mathrm{BF_4^-} + \mathrm{AgBr} \qquad (253)$$

2.03.8.2 Higher Coordinated RSSe Functionality

Higher coordinated analogues of alkylsulfenyl selenides are exceedingly rare in the literature. Examples of higher coordinated aromatic sulfenyl selenides are also relatively limited in number, although a few can be found ⟨75T2099, 80TL1697, 80TL2213, 88OS(67)157⟩.

2.03.8.3 Alkanesulfenyl Tellurides

As noted above, the chemistry of sulfenyl tellurides and the higher coordinate derivatives of sulfenyl tellurides is sparse in the literature for alkyl derivatives. Some examples do exist for non-sp^3 hybridized compounds ⟨51ACS115, 82IJC(A)305⟩ and for arylsulfenyl tellurides ⟨53ACS227, 77JA7743, 83ZOR343⟩. It is however reported that alkylsulfenyl tellurides can be prepared by reaction of a dialkyl telluride with a thiol in the presence of hydrogen peroxide as shown in Equation (254) ⟨92JA9737⟩. This reaction, part of a study of thiol peroxidase activity, reported the formation of *t*-butylthiophenyltelluride in 44% yield. A yield for the preparation of 1-octylthiophenyl telluride is not reported, as product isolation presented a problem due to contamination from dioctylsulfide. Other possible methods for the formation of akanesulfenyltellurides include reaction of a dithiocarbonate with an organic tellurium(II) chloride ⟨93CJC42⟩, and reaction of an alkanetellurenyl halide with the silver salt of a pseudochalcogenide ⟨91CB1131⟩. Such reactions are not however reported for simple alkylsulfenyl telluride synthesis.

$$\text{PhTeTePh} + 2\text{RSH} \xrightarrow[\text{MeOH, CH}_2\text{Cl}_2]{\text{H}_2\text{O}_2} 2\text{PhTeSR} + 2\text{H}_2\text{O} \quad (254)$$

$$R = Bu^t, C_8H_{17}$$

2.03.9 RSN FUNCTIONS AND THEIR HIGHER COORDINATED DERIVATIVES

2.03.9.1 Alkanesulfenamides

Since their initial appearance in the literature in 1873 ⟨1873LA(167)212⟩, sulfenamides have become important intermediates in organic chemistry. They are usually prepared by modification of existing sulfur-containing functionality ⟨B-90MI 203-06⟩. A number of such methods are described below.

2.03.9.1.1 From sulfenyl halides

Sulfenyl halides are frequently used as precursors of alkanesulfenamides of all types, the conversion usually being effected simply by treatment with ammonia (primary sulfenamides) or amines (secondary and tertiary sulfenamides). The reaction is successful even for tritylsulfenamides, with both ammonia (Equation (255)) ⟨19CB413, 60JCS1103, 83JOC3531⟩ and amines (Scheme 192) ⟨83JOC3538, 88CL857, 89JOC2321, 91T8145⟩.

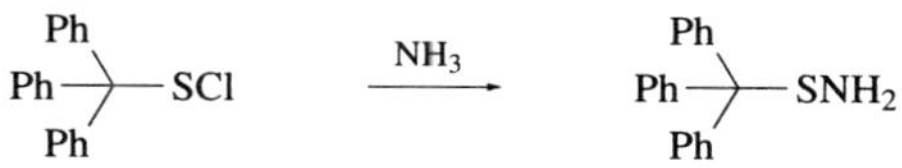

(255)

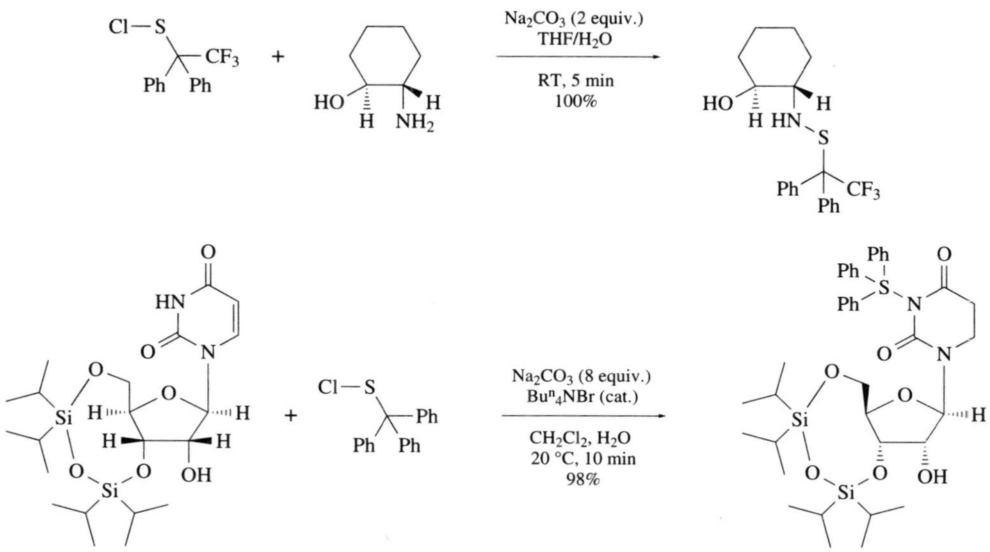

Scheme 192

An alternative approach is seen in the reaction of halosulfenamides, especially succinimide and phthalimide derivatives, with nucleophiles, including alkenes (Equation (256)) ⟨79JCS(P1)3013⟩ and ketones (Scheme 193) ⟨86ZOB898, 90ZOR2083⟩.

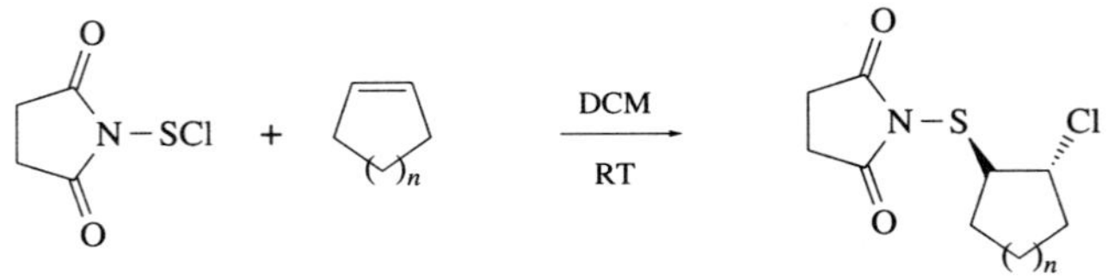

(256)

n	*Yield* (%)
1	99
2	99
3	85

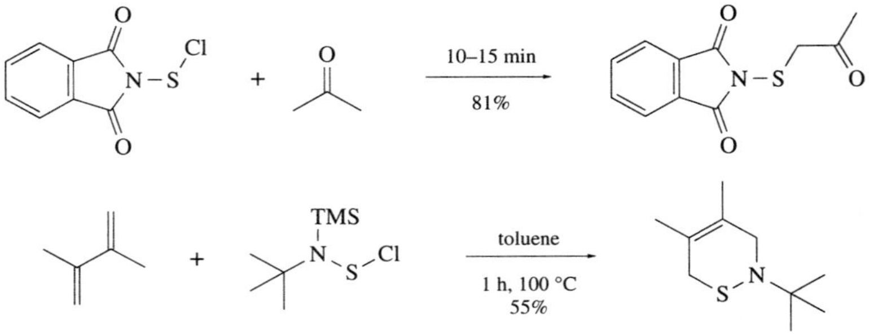

Scheme 193

In a particularly interesting related reaction, *N*-benzyliminosulfur difluoride was used as a precursor of benzonitrile *N*-sulfide, which underwent dipolar addition to maleic anhydride to give a cyclic sulfenimine (Scheme 194) ⟨75TL1739⟩.

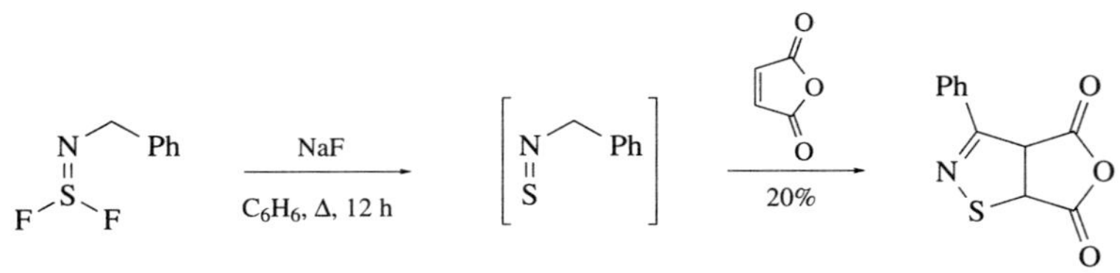

Scheme 194

2.03.9.1.2 From thiols

Thiols can undergo reaction with amines and other nitrogen-containing materials to afford sulfenamides directly. The reaction with amines is normally effected under chlorinating conditions, and no doubt proceeds via *in situ* formation of sulfenyl chloride (Equation (257)) ⟨91JOC6702⟩. Yields can be excellent. In another interesting reaction, treatment of alkanethiols with azides under copper catalysis leads to the isolation of sulfenamides (Scheme 195) ⟨70JOC2979⟩.

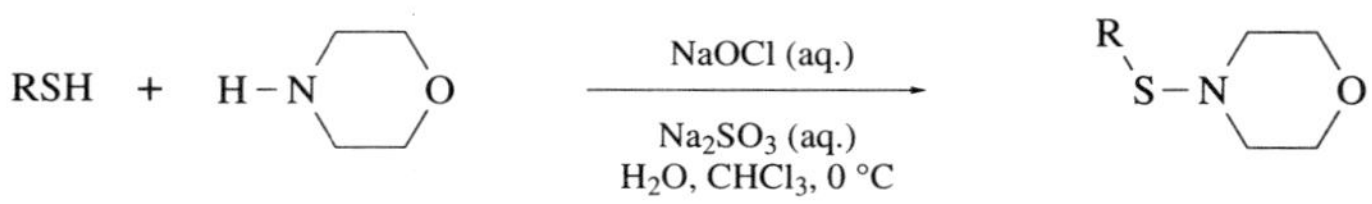

(257)

R	*Yield* (%)
Bu^t	97
Bn	96

Further examples of sulfur(II)–nitrogen single bond formation are seen in the nitrosation of thiols by nitrous acid or alkyl nitrites (Scheme 196) ⟨59CB1048, 91EUP412699⟩.

Thiolates can add to the nitrogen atom of nitriles to give sulfenimines ⟨73LA1644⟩. Sodium *t*-butylthiolate has been shown to react with azoxybenzene to give sulfenamides ⟨94TL301⟩ (Scheme 197).

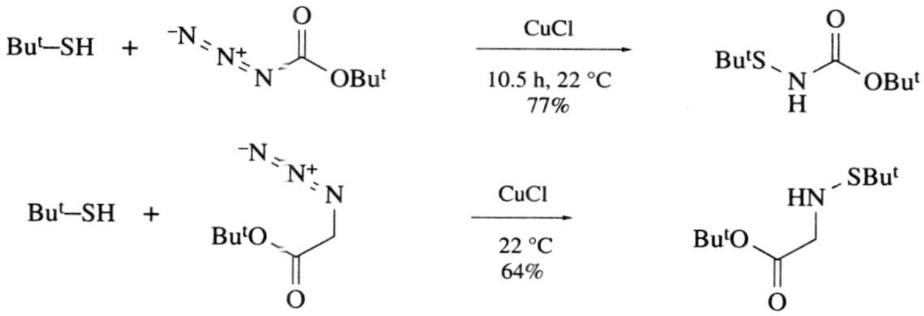

Scheme 195

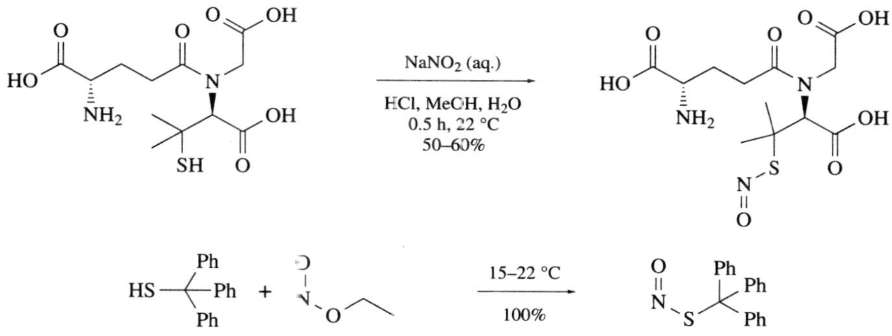

Scheme 196

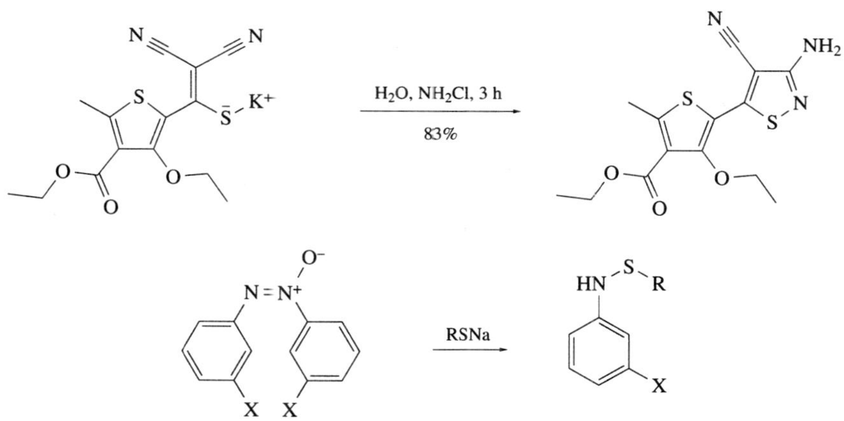

X	*R*	*Yield* (%)
Cl	Bu^t	65
H	Bu^t	44

Scheme 197

2.03.9.1.3 From N*-thiophthalimides*

Nucleophilic displacement of the phthalimide moiety from an *N*-thiophthalimide by a nitrogen nucleophile is a simple and efficient method of sulfenamide synthesis (Scheme 198) ⟨71TL4953⟩, although the alternative pathway of imide ring cleavage has also been reported (Scheme 199) ⟨70TL4983⟩. The reaction is successful even with *β*-lactam nucleophiles ⟨85TL3891⟩.

Three other reactions in which *N*-thiophthalimides are used for sulfenamide synthesis are shown in Equations (258)–(260); involving formal cycloaddition to an amine sulfide (Equation (258)) ⟨88CC950⟩; addition to an alkene (Equation (259)) ⟨85GEP3332269⟩; and desulfurization of a phthalimidodisulfide, accomplished using triphenylphosphine (Equation (260)) ⟨82JOC2785⟩.

R^1	Amine	Yield (%)	Product
Bu	morpholine	90	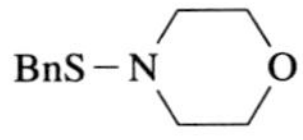
Et	*N*-methyl-*n*-butylamine	91	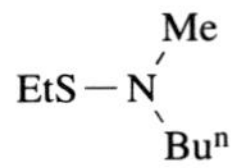
Bu^n	piperidine	100	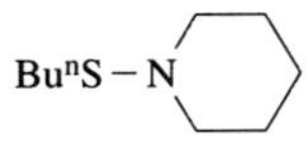
Bn	cyclohexylamine	87	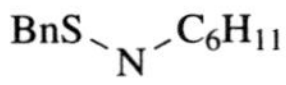
Et	piperazine	90	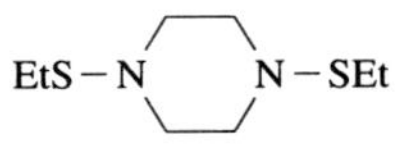

Scheme 198

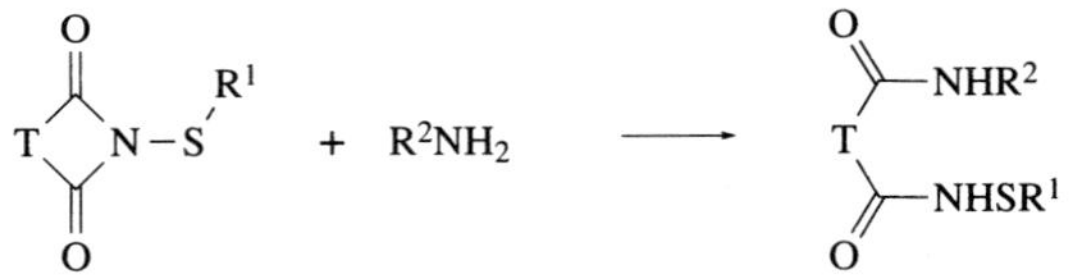

T = σ-C_6H_4 or CH_2CH_2

Scheme 199

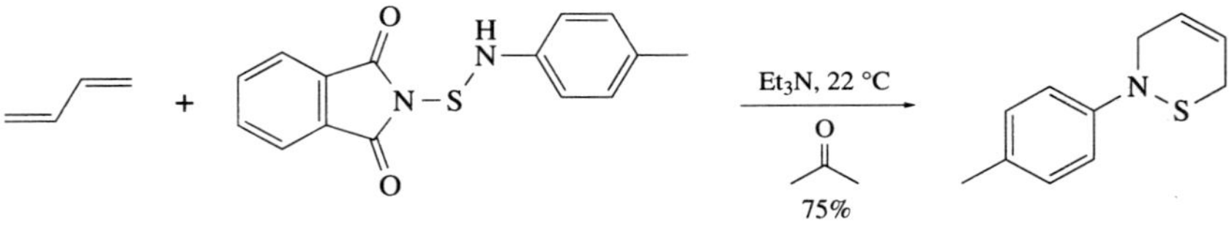

(258)

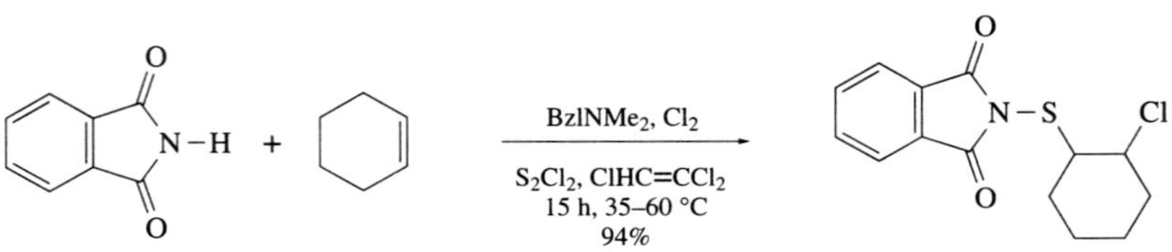

(259)

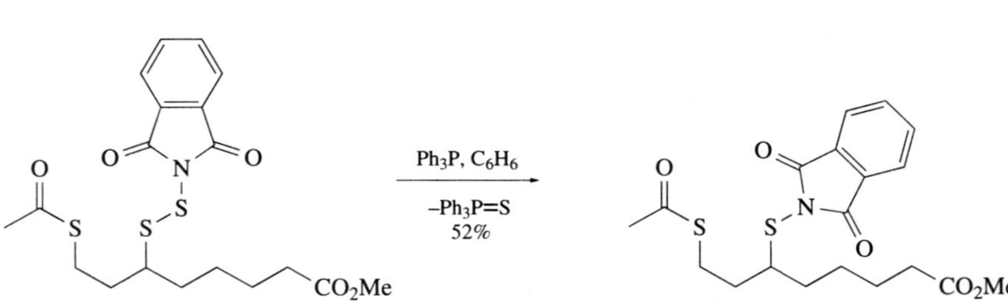

(260)

2.03.9.1.4 *From sulfenylthiocyanates*

Amines react with sulfenylthiocyanates to generate sulfenamides. The reaction proceeds most efficiently with secondary amines (Equation (261)) ⟨56JA6181⟩, and has found use in the decomplexation of a lysine derivative from its copper complex (Equation (262)) ⟨73S795⟩.

$$R^1S\text{–}SCN \; + \; 2\,R^2{}_2NH \longrightarrow R^1SNR^2{}_2 \qquad (261)$$

R¹	*R²*	*Yield (%)*
Et	Et	31
Pr^i	Pr^n	30
Pr^i	Bu^n	70
c-C_6H_{11}	Bu^n	77
Bu^t	Bu^n	68

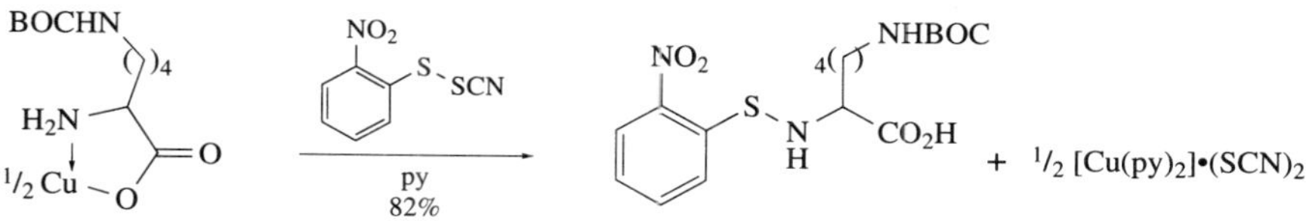
(262)

2.03.9.1.5 *From disulfides*

Symmetrical disulfides react readily and cleanly with lithium dialkylamides to furnish alkanesulfenamides in high yields (Equation (263)) ⟨83S716⟩. An electrochemical procedure has also been used to induce a reaction between disulfides and amines ⟨78JOC3223, 79JOC1554⟩.

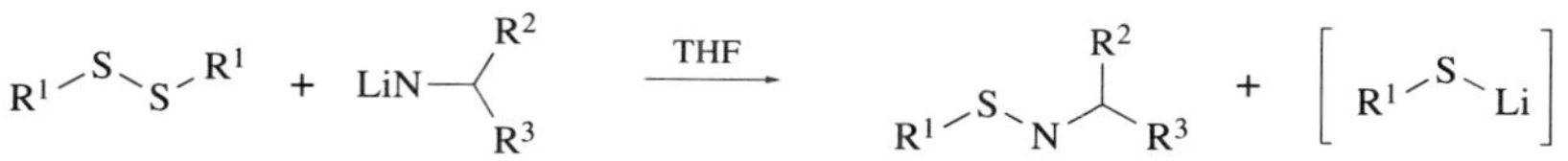
(263)

R¹	*R²*	*R³*	*Yield (%)*
Pr^n	Et	Et	69
Pr^n	Pr^n	Pr^n	75
Pr^n	Pr^i	Pr^i	88
Pr^n	Pr^i	Pr^i	96
Pr^n	Bu^n	H	92

2.03.9.1.6 *From thionitrites*

Amines react with thionitrites to give sulfenamides (Equation (264)) ⟨80BCJ2023⟩.

$$n\text{-}C_8H_{17}NH_2 \; + \; Bu^tSNO \xrightarrow[\Delta,\ 81\%]{C_6H_6} n\text{-}C_8H_{17}NHSBu^t \qquad (264)$$

2.03.9.1.7 *From thiocarbonyl compounds*

Thioketones have been shown to undergo 1,3-dipolar cycloaddition to diazomethane to furnish thiadiazolines (Equation (265)) ⟨93JOC82⟩.

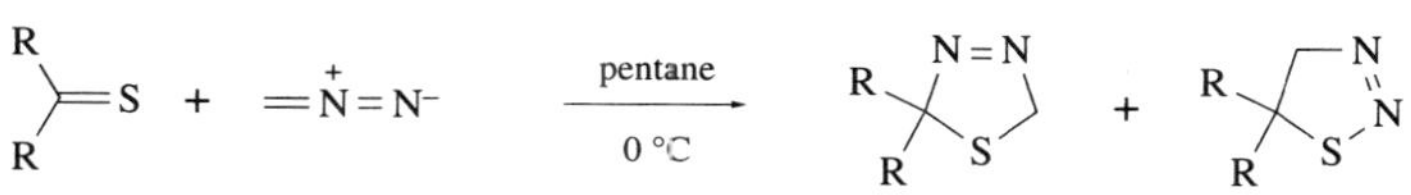
(265)

2.03.9.1.8 *From sulfides and thiiranes*

Three interesting reactions illustrate how sulfides and thiiranes have been used as precursors of sulfenamides. Allyl sulfides have been shown to react with chloramine-T with rearrangement to form *N*-tosylsulfenamides (Equation (266)) ⟨84H(22)1303⟩. Cyclohexene sulfide can undergo an addition reaction with a chloroimine to provide a sulfenimine (Equation (267)) ⟨80JOC1691⟩, and can act as a source of sulfur in a reaction with butadiene and an oxaziridine to form a dihydrothiazine (Equation (268)) ⟨74TL837⟩.

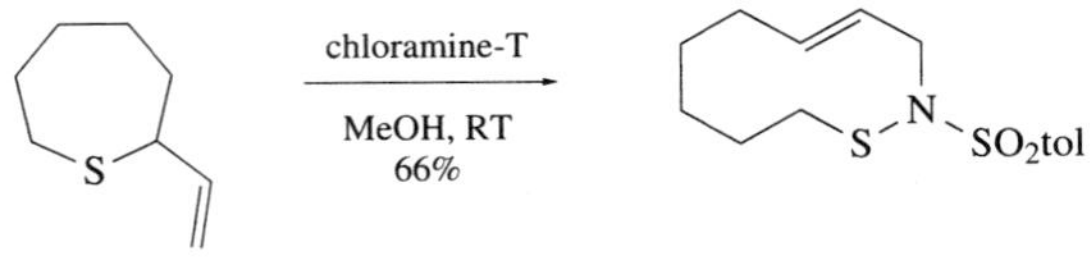

(266)

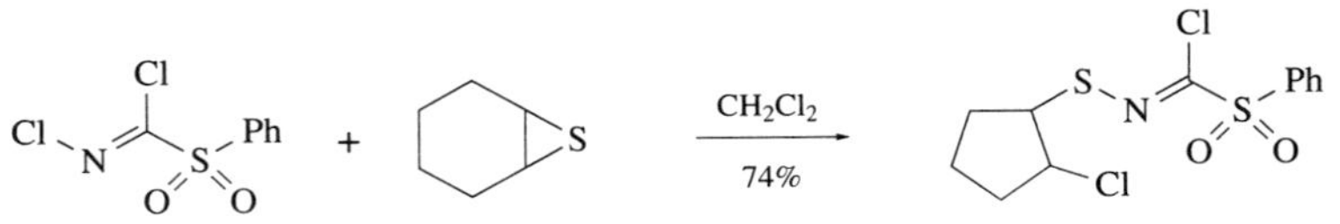

(267)

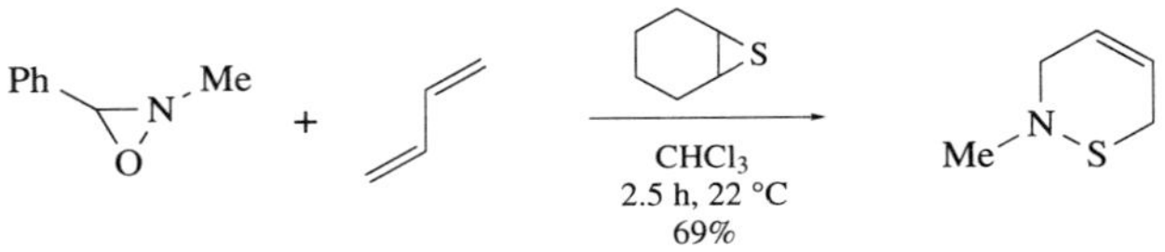

(268)

2.03.9.1.9 *From thiostannanes*

Harpp has described the preparation of *N*-thiosuccinimides by the reaction of thiostannanes with NBS at 0 °C. The reaction takes place in high yield and over a short period of time (Equation (269)) ⟨84S1037⟩.

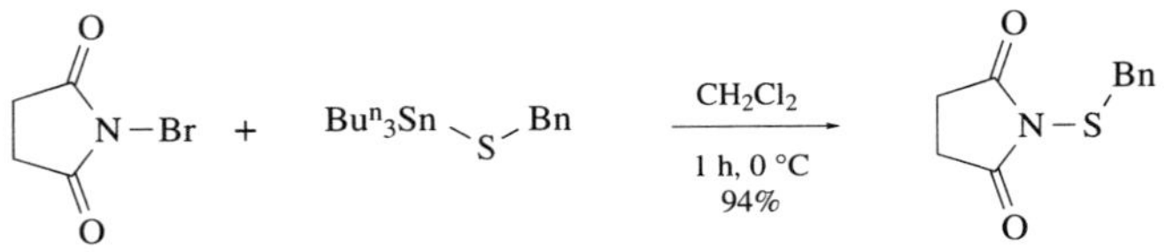

(269)

2.03.9.1.10 *Miscellaneous*

In addition to the methods described above involving the transformation of other sulfur functionality into sulfenamides, a wide range of reactions generating sulfenamides are known.

Sulfenimines have proved to be useful precursors to sulfenamides. One example, described by Rees, involves irradiation of a thiazole azide in the presence of cyclopentadiene, giving rise to a bicyclic sulfenamide (Equation (270)) ⟨84CC374⟩. Meth-Cohn has shown that thionitroso compounds can also react with alkenes and dienes to give sulfenamide containing cycloadducts (Scheme 200) ⟨84CC1144⟩. Efficient 1,3-dipolar cycloadditions to related species are also known (Equation (271)) ⟨78CL591⟩, and sulfur diimides undergo cycloaddition reactions with diphenylketene ⟨69TL447, 70TL1427, 72JOC3810⟩ and dienes ⟨84JA7867⟩ (Equation (272)).

Allylic sulfoxides containing a nucleophilic nitrogen atom can undergo cyclization to give cyclic sulfenamides (Equation (273)), perhaps through initial [2,3] sigmatropic rearrangement to sulfenic esters ⟨85H(23)1897⟩.

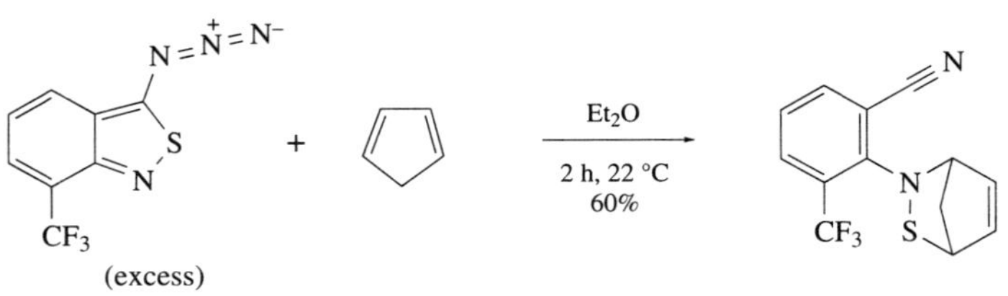

(270)

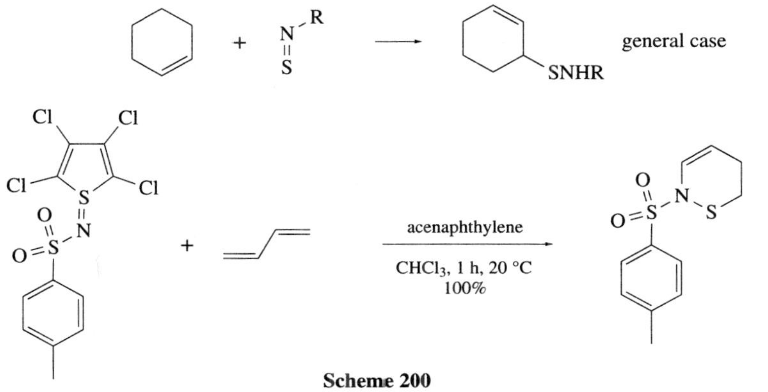

Scheme 200

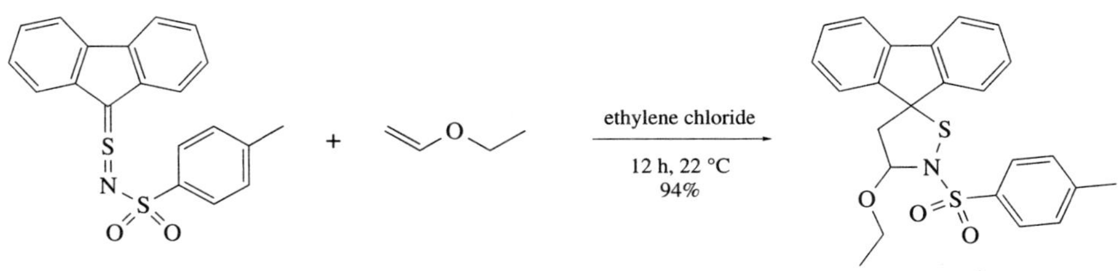

(271)

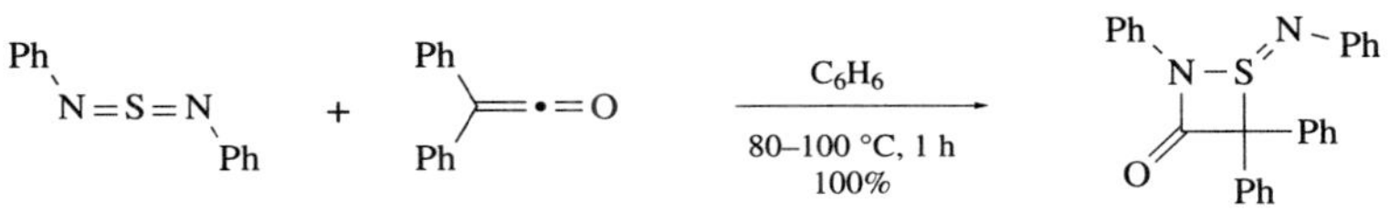

(272)

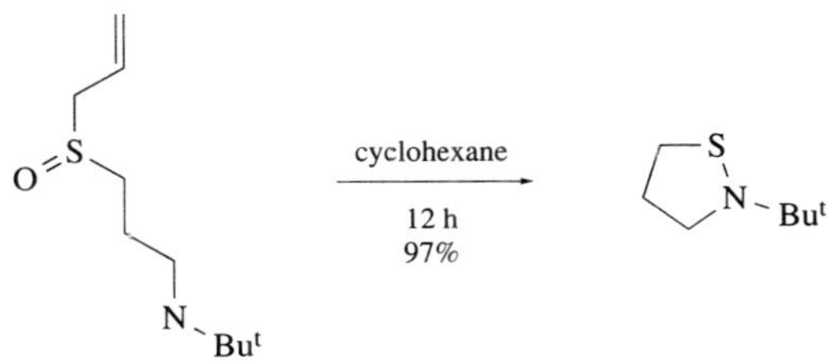

(273)

2.03.9.2 Alkanesulfinamides

Alkanesulfinamides are much less well known than either sulfenamides or sulfonamides, but a few general methods are known and are discussed here together with several examples of other alkanesulfinamide preparations.

2.03.9.2.1 Cycloaddition reactions of sulfoximines

Cyclic sulfinamides are among the best known of alkanesulfinamides, and many have been prepared by various cycloaddition reactions of sulfoximines, for example with dienes ([4 + 2]), and alkenes (ene reaction, [2 + 2]) and ketenes ([2 + 2]). A number of examples of the [4 + 2] process are illustrated in Scheme 201 ⟨75JA6521, 82T3087, 84JA7867, 84JOC3243, 85CC1449, 86TL1853, 89TL3061, 88TL4233⟩; in one case the sulfoximine is generated *in situ* (see the first equation) ⟨83JA4499, 84JA7861⟩. A related reaction of a thiocarbonyl *S*-oxide is also shown ⟨81JCS(P1)2322⟩.

A range of other cycloaddition and related processes successful for alkanesulfinamide preparation from sulfoximines is known (Scheme 202) ⟨64JOC1688, 67AG109, 67AG(E)951, 67CB2159, 68CB3567, 80LA629, 80LA843, 88T4941, 90JA7653, 91JA3526⟩, including one reaction in which a reasonable level of asymmetric induction was observed upon addition on phenylglycine to the reaction mixture (see the first equation). A selection of [2 + 2] reactions is shown in Scheme 203 ⟨62T1527, 81TL583⟩.

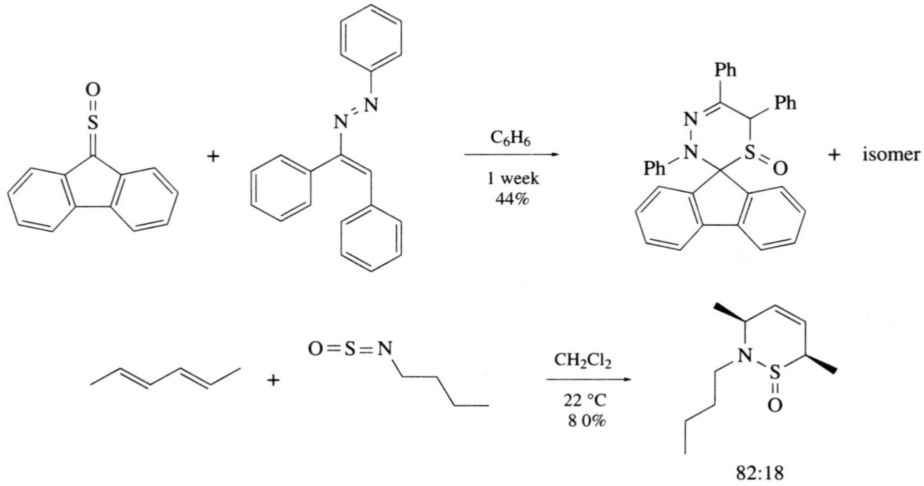

Scheme 201

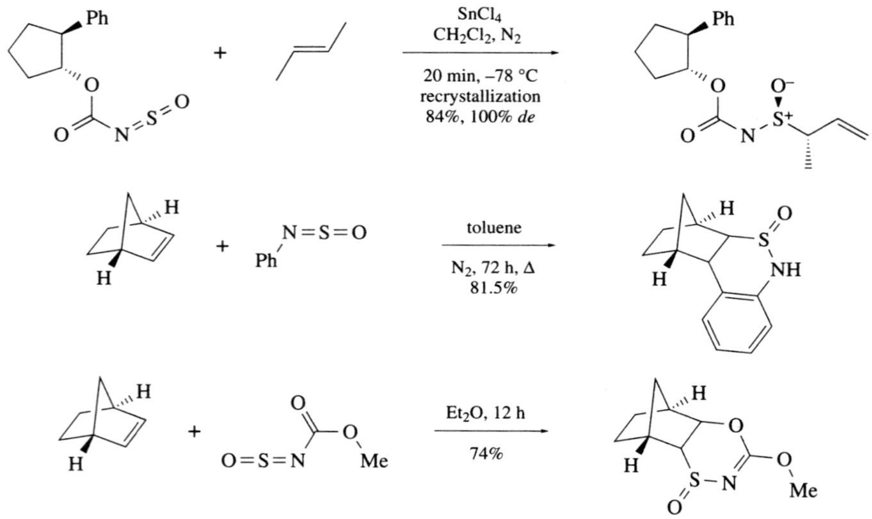

Scheme 202

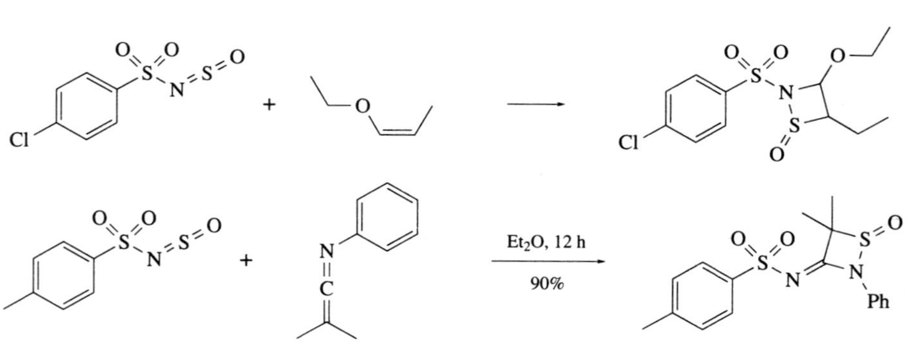

Scheme 203

2.03.9.2.2 Miscellaneous methods of alkanesulfinamide preparation

Sulfoximines undergo nucleophilic attack by enols and enolates to produce alkanesulfinamides in high yields, both with Reformatsky-type organozinc reagents (Scheme 204) ⟨82S281⟩ and with ketones (Scheme 205) ⟨80AG(E)737⟩.

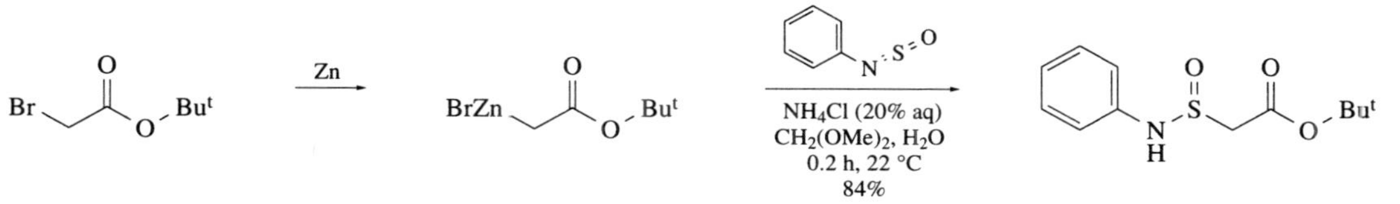

Scheme 204

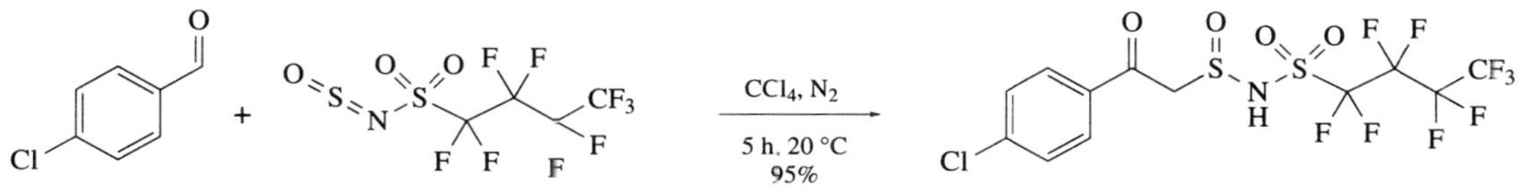

Scheme 205

Allylic sulfinamides have been prepared from allylic alcohols through a variation of the well-known [2,3] sigmatropic rearrangement of sulfenic esters (Scheme 206) ⟨88TL3251⟩. As expected, the rearrangement is subject to a high degree of stereocontrol.

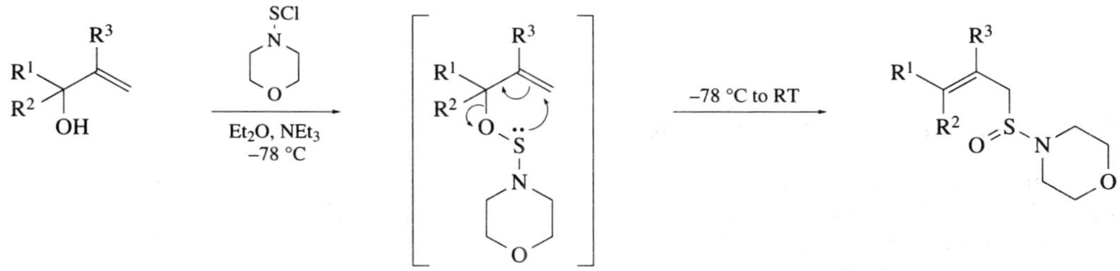

R¹	*R²*	*R³*	*Yield (%)*	*(E):(Z)*
H	H	H	83	
Me	H	H	87	excl. (E)
Me	Me	H	81	
n-C_8H_{17}	H	H	82	excl. (E)
$BnOCH_2$	H	Me	86	9:1

Scheme 206

In perhaps the most direct route possible, alkanesulfinamides may be produced in excellent yields by treatment of amines with sulfinyl chlorides (Equation (274)) ⟨79PS(7)325⟩.

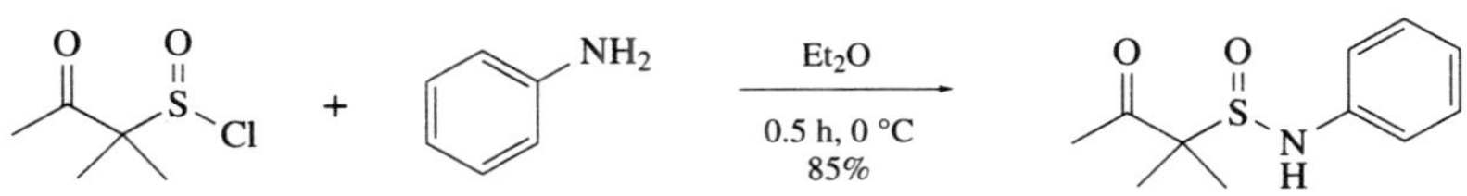

(274)

In an interesting reaction, the treatment of 1,3-dithiolane 1,3-dioxides with lithium diisopropylamide (LDA) ultimately results in sulfinamide formation ⟨86BCJ3897⟩, presumably through deprotonation and fragmentation to produce the thiocarbonyl *S*-oxide, which subsequently suffers nucleophilic attack by imide (Scheme 207).

Dondoni has observed [2 + 2] cycloadditions of sulfur dioxide to ketenimines to give unstable cyclic sulfinamide products in excellent yields (Equation (275)) ⟨81CC350⟩.

In another interesting transformation, treatment of a disulfide derived from a γ-amino thiol with iodine and sodium sulfite in the presence of potassium hydroxide gave a cyclic alkanesulfinimide in high yield (Equation (276)) ⟨90JOC4156⟩.

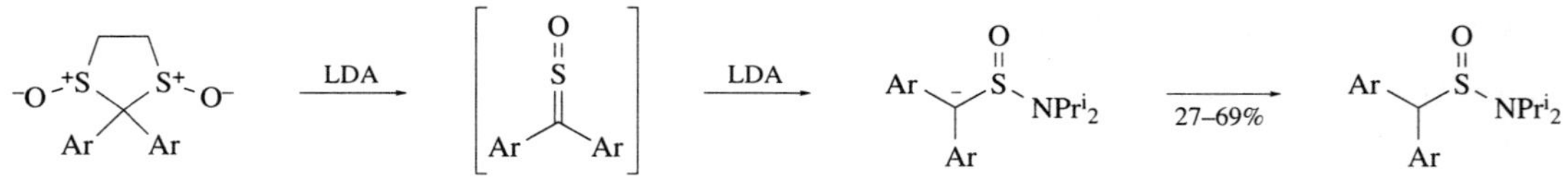

Scheme 207

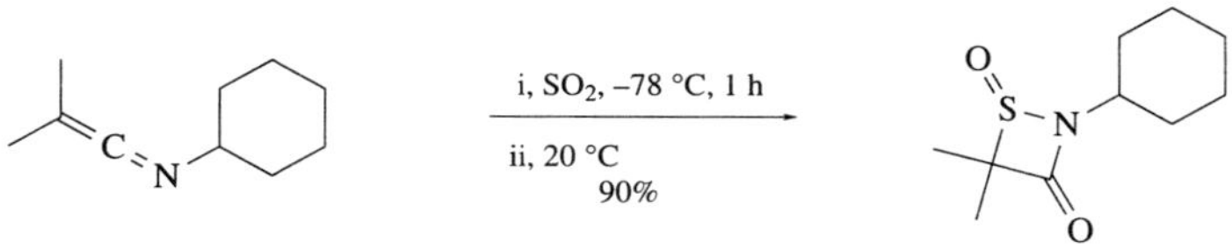

(275)

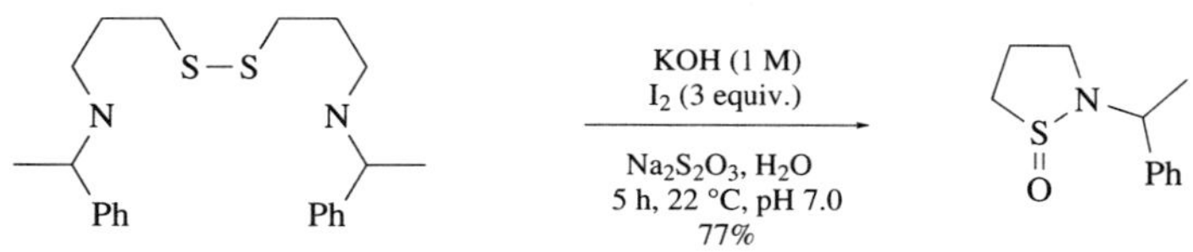

(276)

2.03.9.3 Alkanesulfonamides

Alkanesulfonamides are important compounds, and their synthesis may be achieved by a number of different procedures. These may be broadly divided into manipulation of other sulfur(VI) species, the oxidation of sulfenyl and sulfinyl species, and other methods ⟨79COC(3)331, B-91MI 203-01⟩.

2.03.9.3.1 From other sulfur(VI) functionality

One of the most widely reported methods of alkanesulfonamide preparation is the reaction of sulfonyl halides with ammonia or amines. Equations (277) and (278) show two examples of the reaction with ammonia in which cyclization takes place by additional intramolecular displacement of chloride ⟨62HCA717, 76BSF277⟩. A related example involving an internal nitrogen nucleophile is shown in Equation (279) ⟨82TL5389⟩. Alkanesulfonic esters can also react with ammonia to produce sulfonamides (Equation (280)) ⟨47HCA1650⟩.

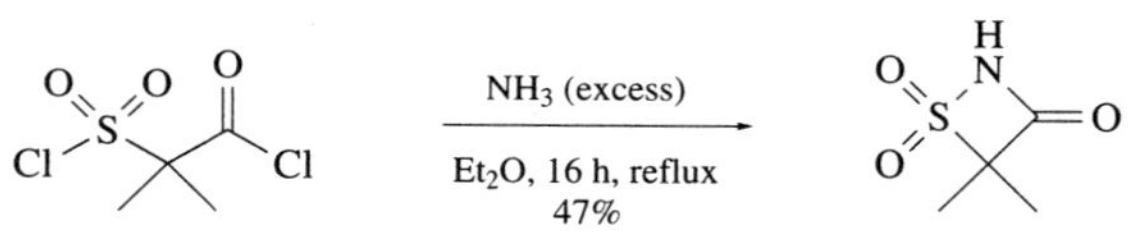

(277)

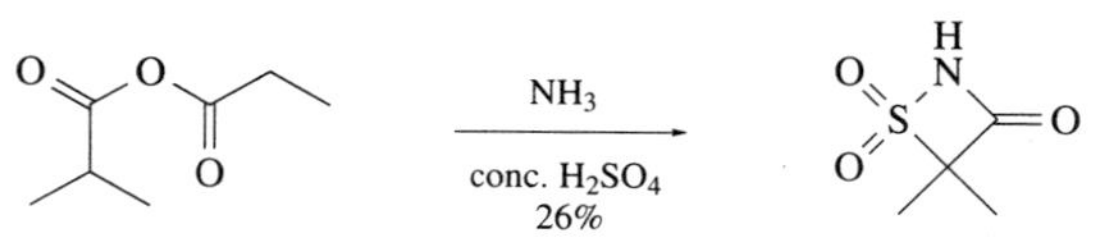

(278)

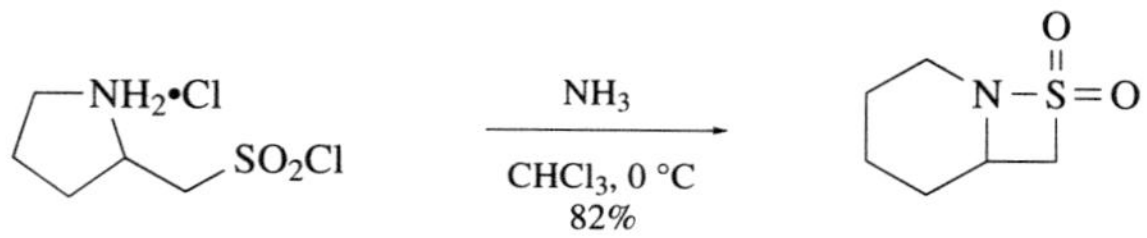

(279)

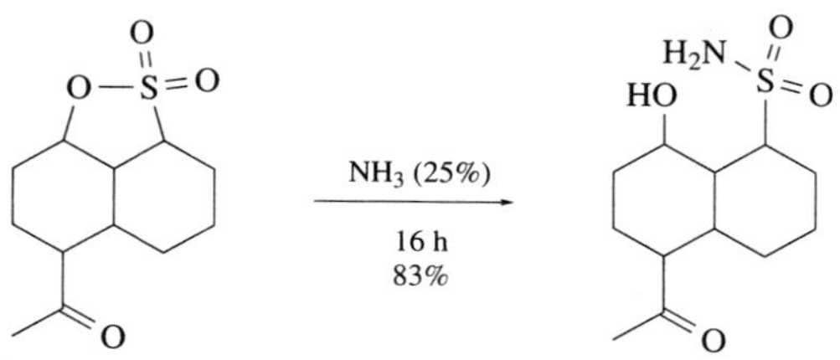

(280)

Similarly, alkanesulfonyl chlorides, fluorides, and esters react with primary and secondary amines to give the corresponding sulfonamides. A range of examples is shown in Scheme 208 ⟨62LA(657)86, 75IZV2530, 81AP(314)799, 85BSF463, 86TL2099⟩. Sulfonyl chlorides react with imines to provide cyclic sulfonamides (Scheme 209) ⟨70BCJ3543, 71BCJ2750, 75BCJ480, 89TL2869⟩. Sulfenes may also react with amine and imine derivatives to form alkane sulfonamides ⟨70CC305, 70S545, 78CJC1183, 87TL2641⟩.

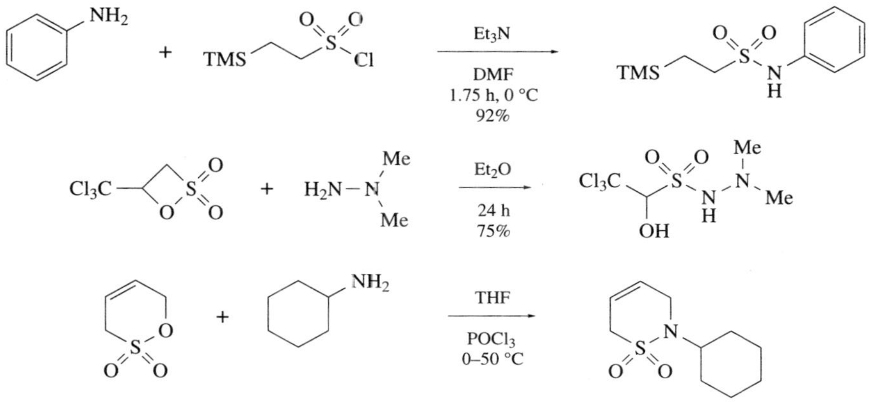

Scheme 208

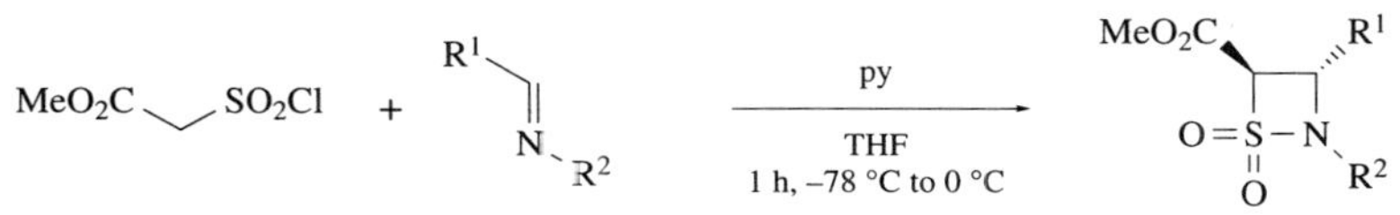

R^1	R^2	*Yield* (%)
Ph	Bn	93
Et	Bn	55

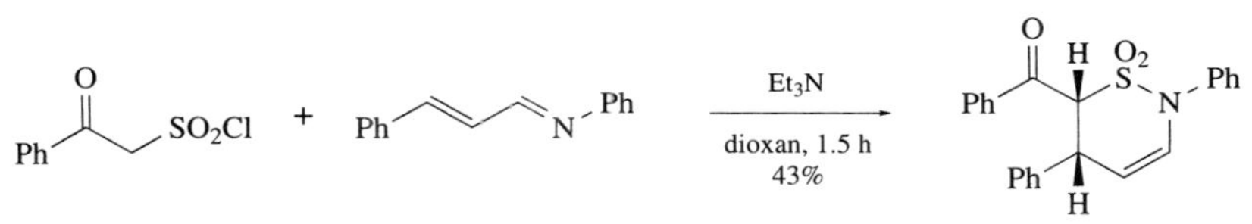

Scheme 209

In an unusual variation of this approach, β-ketoalkanesulfonamides have been prepared by the reaction of silyl enol ethers with *N*-methylsulfamoyl chloride (Equation (281)) ⟨92TL3677⟩, in only the second synthesis of these compounds ⟨85S66⟩. Enamines also react with this reagent to give cyclic alkanesulfonamides (Equation (282)) ⟨79BCJ1102⟩.

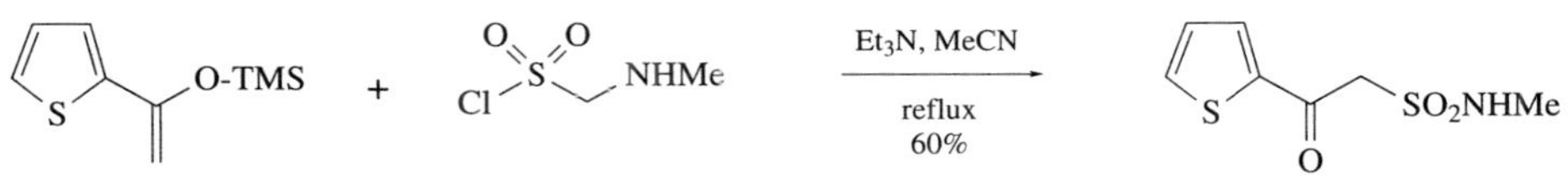

(281)

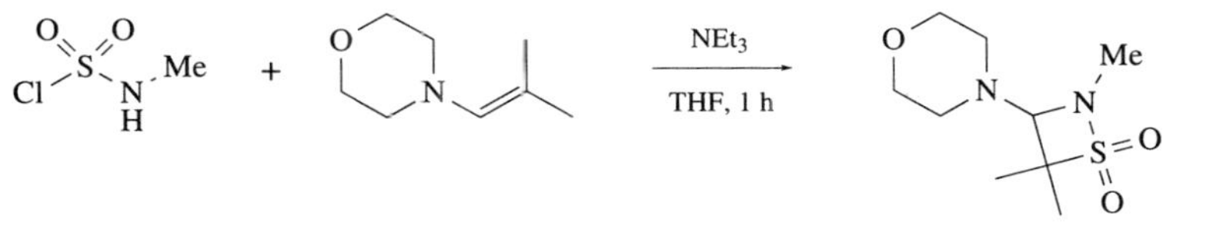

(282)

Sulfonyl isocyanates are common precursors of alkanesulfonamides, by hydrolysis or other reactions (Scheme 210) ⟨04CB690, 93SC121⟩. In a different approach, sulfonyl azides may be converted into sulfonamides under any of a range of reducing conditions (Equation (283)) ⟨55JA80, 65CJC3407, 65ZOR1256, 68JA4096, 74TL381⟩. Interestingly, lithium aluminum hydride is not efficient for this transformation.

Nucleophilic addition to the imine unit of an alkanesulfonyl imine has been used for sulfonamide generation in good yields (Equation (284)) ⟨93SL465⟩.

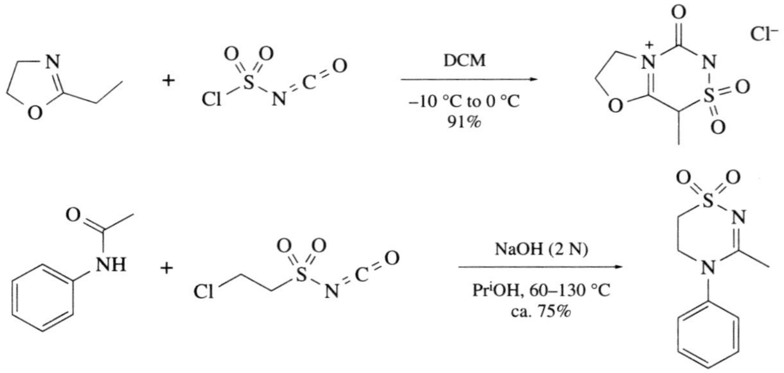

Scheme 210

$$R^1SO_2N_3 \xrightarrow[\text{or ii, Zn, HOAc; iii, } \Delta,\ R^2OH]{\text{i, } h\nu,\ R^2OH} R^1SO_2NH_2 \qquad (283)$$

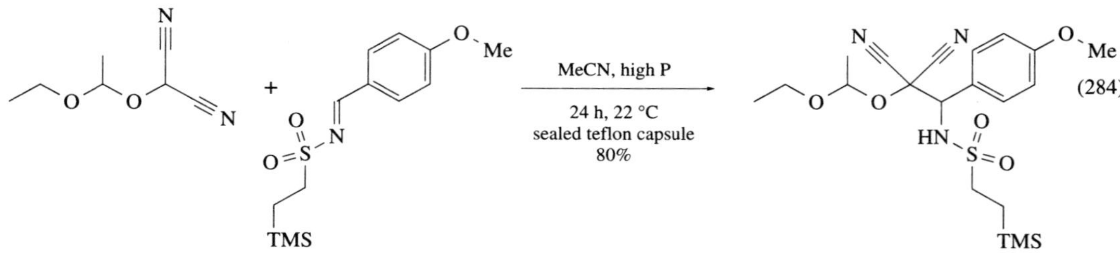

(284)

2.03.9.3.2 *By oxidation of sulfenyl and sulfinyl derivatives*

Although this is a relatively rare process, alkanesulfenamides have been shown to undergo oxidation to sulfonamides ⟨75JA6521, 77S798⟩. Sulfinamides are rather more common substrates for oxidation, which has been accomplished with peracid reagents ⟨82S281, 81TL583⟩, and hydrogen peroxide ⟨68CB3567⟩ (Scheme 211). Alkanesulfinyl chlorides may be directly converted into sulfonamides by the action of chlorine and ammonia ⟨48JCS147⟩ or even by treatment with a hydroxylamine, in a reaction effective even for tertiary sulfonamides (Equation (285)) ⟨72TL181⟩. The mechanism of the reaction is unclear. Sulfinic acids have been converted into sulfonamides in the presence of ammonia ⟨78JOC4659⟩.

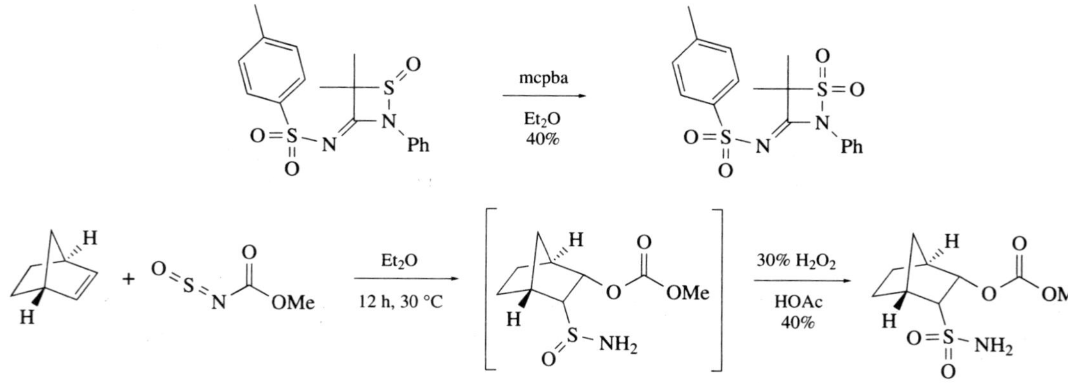

Scheme 211

$$R^1SOCl + 2R^2R^3NOH \xrightarrow[\text{DCM, } N_2 \text{, 6.25 h, dark}]{Et_2O} R^1SO_2NR^2R^3 + R^2R^3NOH{\bullet}HCl \quad (285)$$

R¹	*R²*	*R¹*	*Yield* (%)	*R¹*	*R²*	*R¹*	*Yield* (%)
Bu^t	H	H	59	Bu^t	Et	Et	13
Bu^t	H	Bu^n	39	1-adamantyl	H	H	56
Bu^t	H	Bu^t	45	1-adamantyl	H	c-C_6H_{11}	68

2.03.9.3.3 Miscellaneous methods

Sulfur dioxide has been reported to react efficiently with diazo compounds in the presence of excess amine to generate alkanesulfonamides, including tertiary examples (Equation (286)) ⟨52RTC1228⟩.

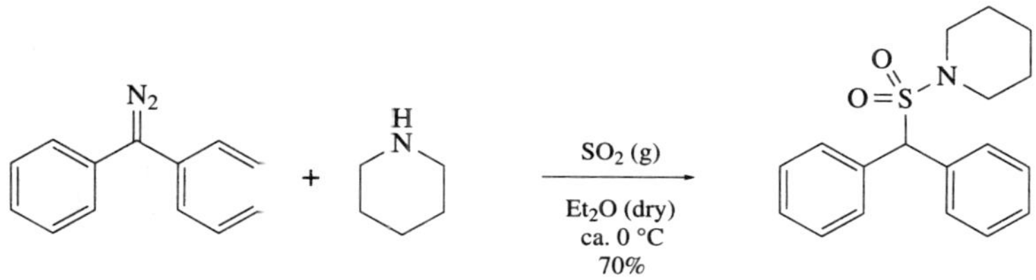

(286)

Several *N*-functionalized alkanesulfonamides, including selenylimines and *N*-silylsulfonamides, are known (Scheme 212) ⟨83TL3191, 92S582⟩.

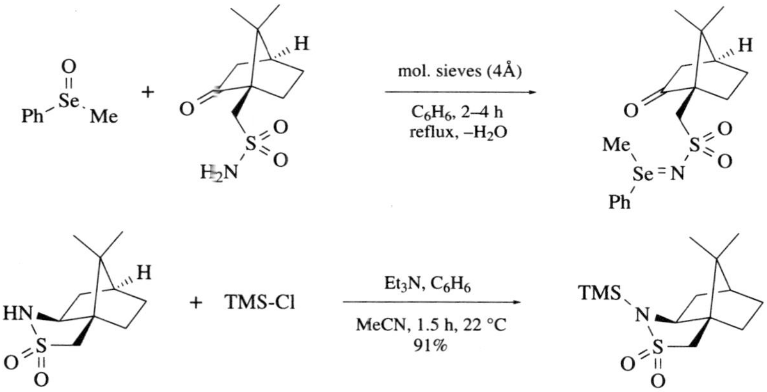

Scheme 212

Finally, alkanesulfonamides may also of course be generated from other alkanesulfonamides by, for example, deprotonation at the α position and alkylation. One such case involving a cyclization is given in Equation (287) ⟨73S667⟩. Hydrolysis of alkanesulfonyl imines has also been carried out (Equation (288)) ⟨90TL1653⟩.

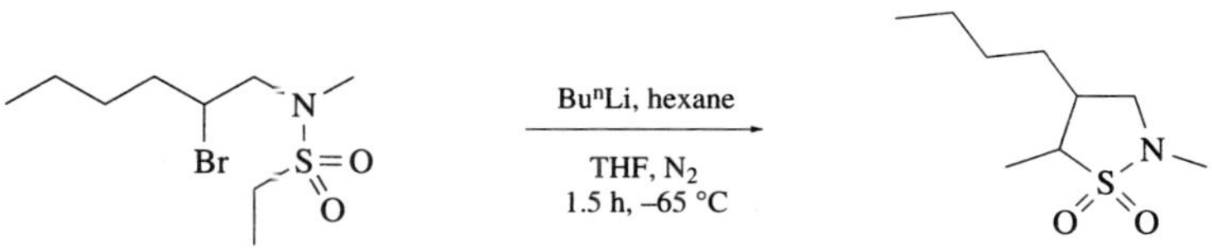

(287)

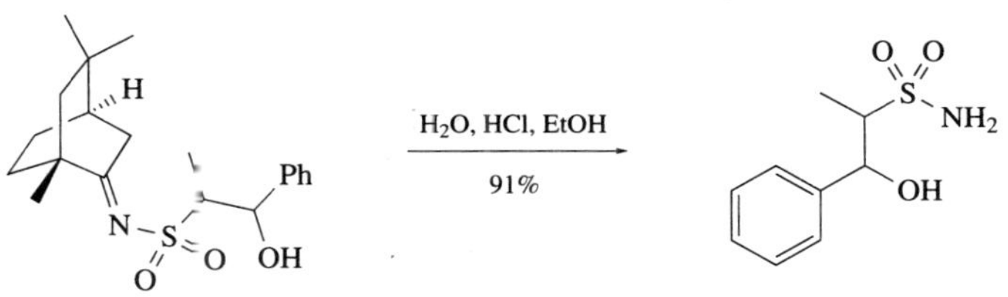

(288)

2.03.10 RSP, RSAs, RSSb, AND RSBi FUNCTIONS AND THEIR HIGHER COORDINATED ANALOGUES

2.03.10.1 Alkanethiophosphines and their Higher Coordinated Analogues

Phosphorus–sulfur compounds, often the thiol esters of phosphoric acid derivatives, exhibit biological activity and are widely used in nucleoside chemistry and as pesticides ⟨79JOC2938, 86JMC2114⟩. The two principal routes to their synthesis are reaction of a thiol with a phosphorus reagent, and alkylation of a thiophosphate. Phosphorus–sulfur compounds in which sulfur exists in a higher oxidation state are extremely rare and are not discussed here.

2.03.10.1.1 From thiols

The simplest synthetic route possible is the reaction between a thiol and elemental phosphorus, as described in a German patent ⟨86GEP231075⟩. White phosphorus reacts with three equivalents of ethanethiol in DMF in the presence of triethylamine and tetrachloromethane at 50 °C to give a 90% yield of the product (Equation (289)). More commonly, a thiol or thiolate is used to displace a leaving group from phosphorus to form a sulfur–phosphorus bond by substitution. The group that is usually displaced is halide, normally chloride or bromide. Simple reaction of an alkane thiol and thiophosphoryl tribromide ($P(S)Br_3$) gives a monosubstituted product in 82–93% yield (Equation (290)) ⟨84ZOB2545⟩.

$$\mathrm{EtSH} \xrightarrow[\mathrm{CCl_4,\ DMF,\ Et_3N,\ 50\ °C}\ \ 90\%]{\text{white phosphorus}} \mathrm{P(SEt)_3} \qquad (289)$$

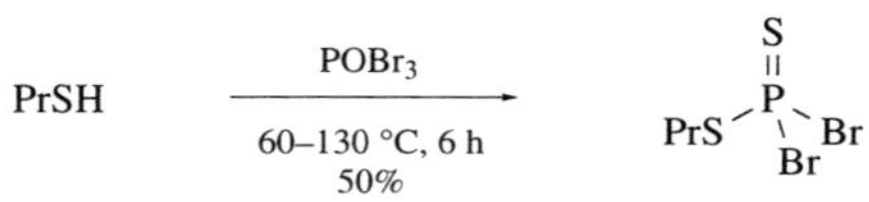

(290)

A protecting group for thiols has been developed from a dialkyl phosphinthioyl chloride, which reacts with thiol groups, for example in cysteine derivatives, to form a phosphorus–sulfur bond (Scheme 213). The protecting group has been applied in the synthesis of peptides, and its stability under different conditions examined ⟨79BCJ1431, 83BCJ1187⟩. The thiol group is selectively protected in the presence of the amino group, although with two equivalents both groups are functionalized. Another example of an amino acid thiol group reacting with a phosphorus compound has been seen in the formation of metal complexes with cysteine methyl ester ⟨83ZN(B)365⟩.

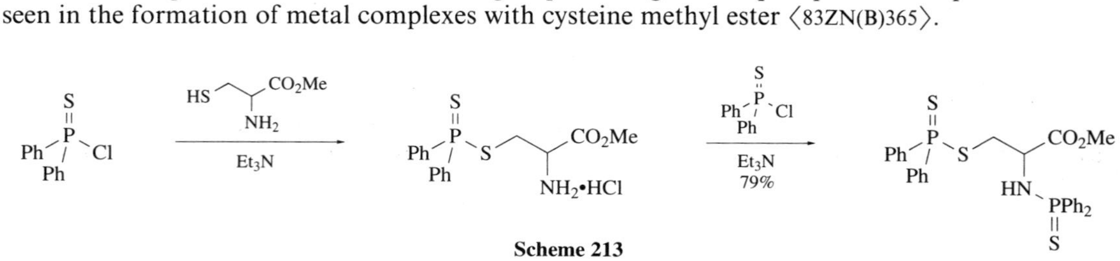

Scheme 213

In the synthesis of cyclic phosphorylating agents, a dichlorophosphine reacts with *N*-methyl thioacetamide at low temperature to give the phospholidine product in 63% yield ⟨90S361⟩. Further reaction with sulfuryl chloride followed by two equivalents of alcohol (or two different alcohols) leads to the ring opened thiophosphate (Scheme 214). The industrial importance of these compounds is demonstrated by the many patents in the area, two examples of which describe the synthesis of insecticidal thiophosphate derivatives from tertiary thiols (Scheme 215) ⟨83USP4383991, 84USP4450158⟩.

With more than one leaving group attached to phosphorus, it is possible for reaction to take place with more than one equivalent of the thiol, or with a dithiol, thus forming cyclic compounds. For example, tetrabutylphosphonium chloride has been used to catalyze the reaction between a dichlorophosphorothiolate and two equivalents of butanethiol to furnish the disubstituted product in 90% yield ⟨92PS(68)147⟩. 1,3-Dithiols react with dichlorophosphines or dichlorophosphine oxides

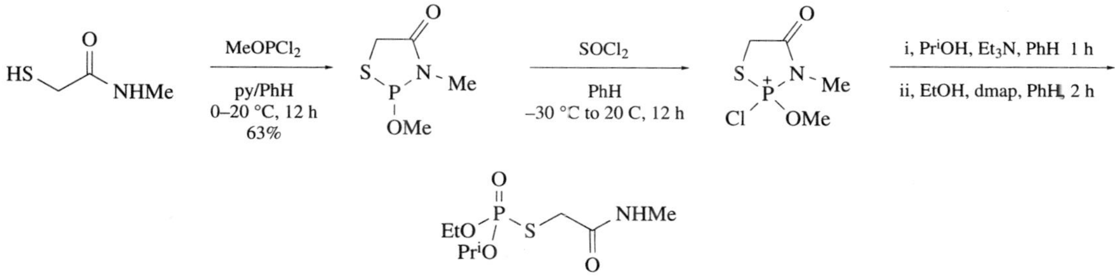

Scheme 214

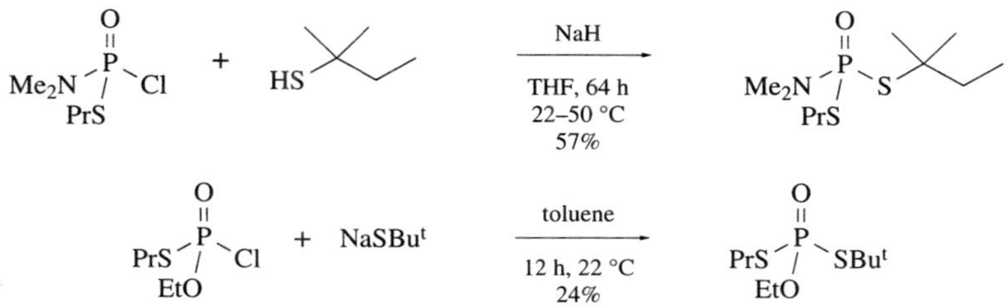

Scheme 215

to form the six-membered cyclic compounds (Scheme 216) ⟨81JA4432⟩. The structure and conformation of these compounds was extensively studied by NMR and crystallography. Reaction of a cyclic trichlorophospholine with two equivalents of an alkanethiol results in the formation of dithiophosphoric ester with concomitant shift of the double bond (Equation (291)) ⟨63T1553⟩.

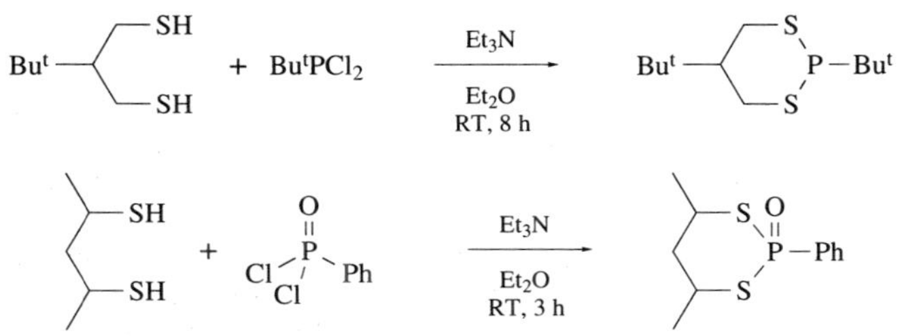

Scheme 216

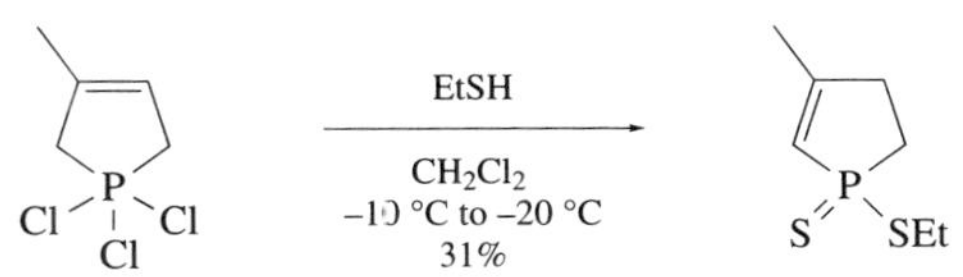

(291)

Leaving groups other than halide have been used in the formation of a sulfur–phosphorus bond. In nucleoside chemistry much use is made of phosphoramidites, in which the leaving group is a dialkylamino moiety. The simplest reaction is illustrated by the substitution of one dipropylamino group by benzylthiol from hexapropyl phosphorus triamide (Equation (292)) ⟨56JA976⟩. Dithiols can react in a similar manner to give the cyclic products on heating in toluene under reflux ⟨82TL1345⟩. A cyclic phosphonium salt with two dimethylamino groups can be consecutively substituted by addition of two equivalents of thiol (Scheme 217). As indicated above, this reaction has found applications in nucleoside research: activation of a dialkylamino group by a weak acid, usually tetrazole, allows displacement to occur using a thiol as nucleophile, often, as in the example shown, in quantitative yield (Equation (293)) ⟨89JA2321, 90TL1953⟩.

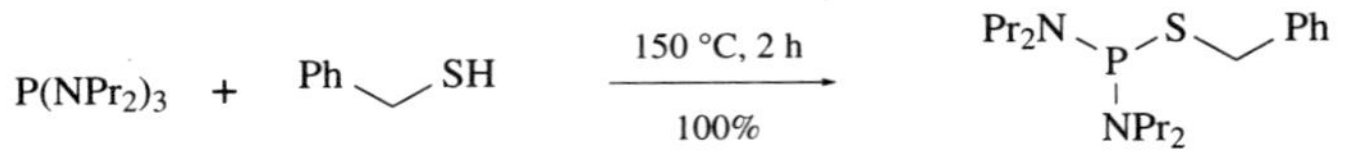

(292)

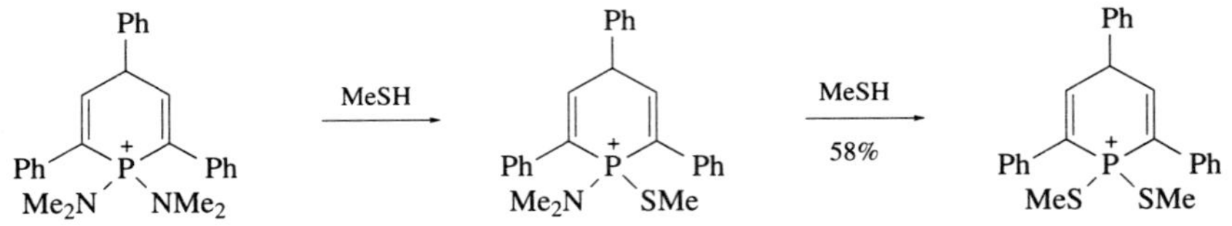

Scheme 217

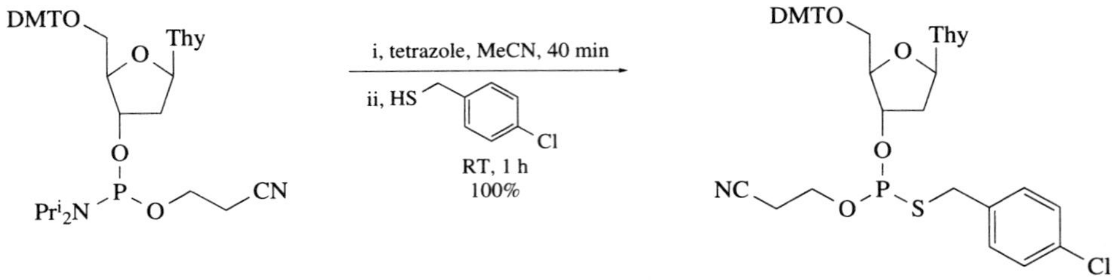

(293)

Thiaphosphonium salts prepared from tris(dimethylamino)phosphine and a thiol using *t*-butyl hypochlorite have been used to effect desulfurization in high yield by nucleophilic displacement (Scheme 218) ⟨85TL4867⟩.

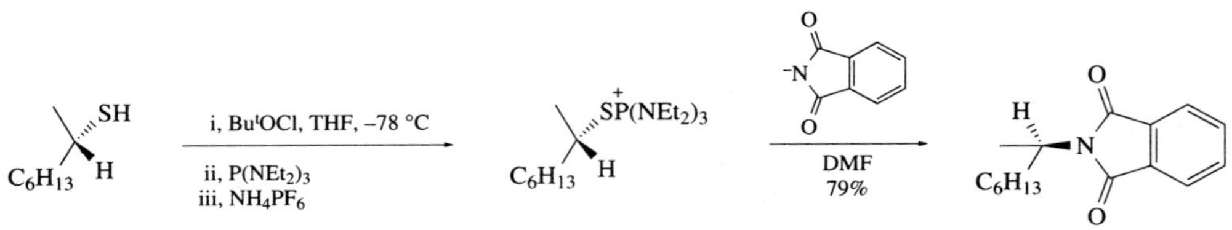

Scheme 218

Other leaving groups used to form sulfur–phosphorus bonds include nitrile, and the enolate of a 1,3-diketone ⟨81PS(11)349, 87ZOB1429⟩. The substitution of butanethiolate for ethoxide, seen in triethylphosphite in the presence of bromotrichloromethane, is presumed to proceed through a radical mechanism ⟨62JCS2953⟩.

Other phosphorus-containing starting materials that have been used in reactions with thiols to form phosphorus–sulfur bonds are phosphorus pentasulfide and diethyldiphosphorus tetrasulfide. Good yields of products are isolated from reaction with thiols or trimethylsilyl thioethers (Scheme 219) ⟨84EUP118109, 93PS(79)179⟩.

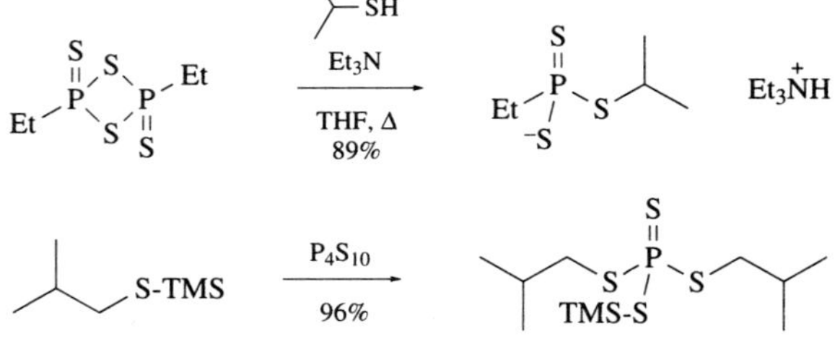

Scheme 219

2.03.10.1.2 From alkyl halides

Alkylation of the sulfur atom of a phosphorothioate is a viable synthesis of *S*-alkyl phosphorothioates. In reactions of alkyl halides, including primary, secondary, tertiary, allylic, and benzylic, with triethylammonium di-*t*-butyl phosphorothioate in refluxing dimethoxyethane, yields of up to 95% were realized (Equation (294)) ⟨71ZN(B)386⟩.

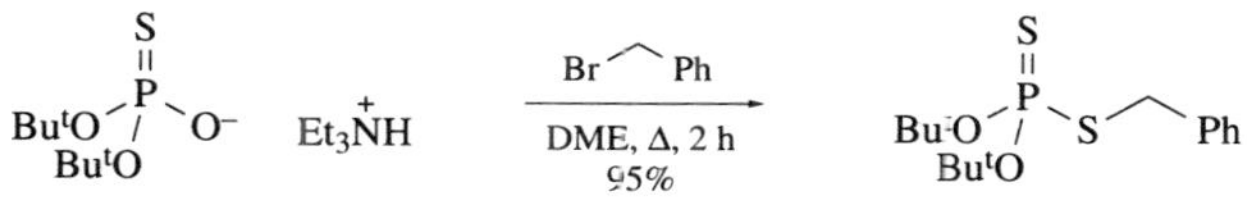

(294)

A similar procedure has been reported for the synthesis of dialkylthiophosphinic esters, in which the free acid is treated with an alkyl bromide under phase transfer catalysis in the presence of potassium carbonate as a base. Exclusive *S*-alkylation is observed, with yields approaching quantitative in 15 minutes. The use of tetrabutylammonium bromide as a phase-transfer catalyst increases both rate and yield over the uncatalyzed reaction. Bromochloromethane reacts with the triethylammonium salt of a phosphorothioate derivative in THF under reflux to give monoalkylation at sulfur, with only the bromide being displaced, forming the chloromethyl derivative (Equation (295)) ⟨84EUP118109⟩.

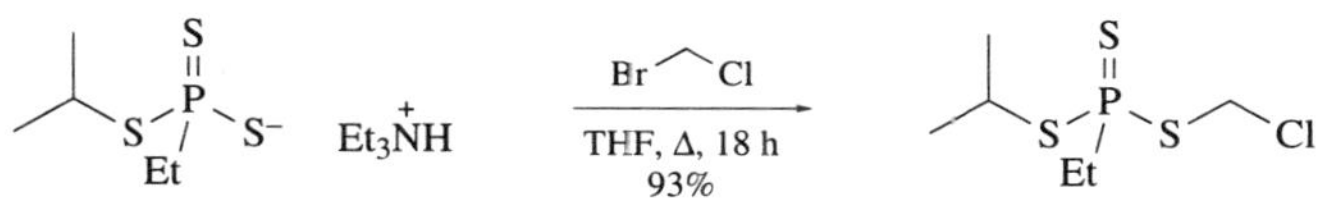

(295)

Alkyl halides produce phosphorothioates in high yield on treatment with an aqueous solution of sodium thiophosphate (Na_3SPO_3). Substituted alkyl halides containing amino or carboxyl groups have been used ⟨67JOC1261, 79JMC631⟩. A related cyclic product has been produced from intramolecular alkylation; the reaction took place over 120 hours *in vacuo* (Equation (296)) ⟨70CR(C)2086⟩.

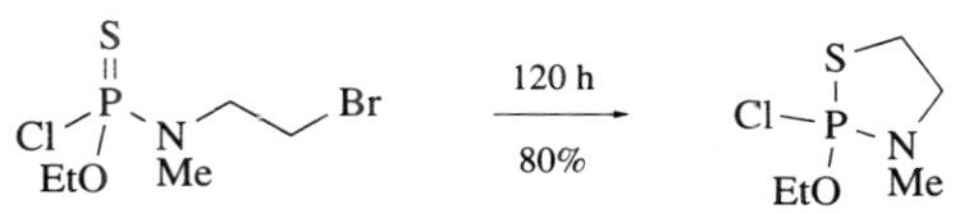

(296)

Alkylation of bis(*O*-trimethylsilyl) derivatives of alkyl phosphinothioic acids with a range of alkyl bromides gives moderate yields of *S*-alkylated products (Equation (297)) ⟨87ZOB786⟩. Dithiophosphonic anhydrides can also be alkylated to good effect (Equation (298)) ⟨80PS(9)175⟩. These reactions have again been applied to nucleoside chemistry to produce the phosphorodithioate derivatives ⟨70JA190, 90TL1319⟩.

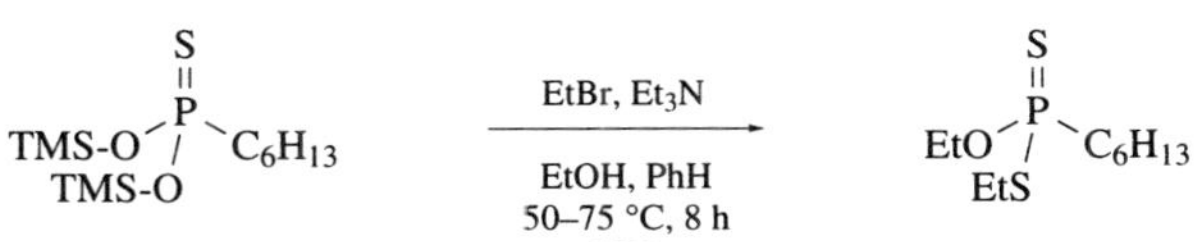

(297)

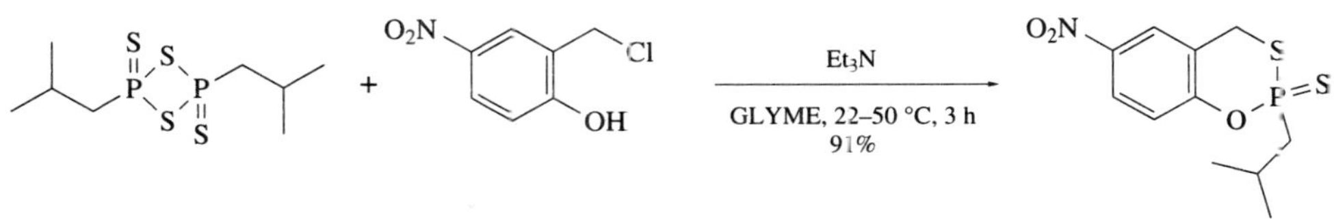

(298)

2.03.10.1.3 *Addition to carbon–carbon double bonds*

Alkylation of phosphinodithioic acids can also be performed by addition to alkenes or other unsaturated systems including enol ethers and enol acetates. Addition to higher boiling alkenes takes place readily under atmospheric pressure, while volatile, low molecular weight analogues require pressure of 0.69–6.9 MPa (100–1000 psi). Addition takes place to give the Markovnikov product (Equation (299)) ⟨52JA161, 65JA4447, 92PS(73)49⟩.

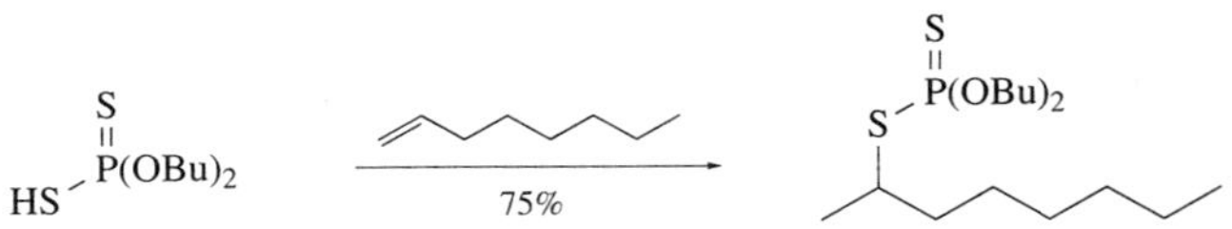

(299)

The corresponding reaction with 1,3-dienes can take several courses. Cycloaddition, perhaps through a diradical mechanism, occurs when dithiophosphinic anhydrides react with dienes (Equation (300)) ⟨73M503⟩. In another example, addition of the potassium salt of a dithiophosphate takes place to the terminus of butadiene with tandem coupling to a benzene diazonium salt (Equation (301)) ⟨80ZOB2459⟩. Free radical mediated addition, initiated by dibenzoyl peroxide, leads to a 97% yield of the anti-Markovnikov product by addition to the terminal position of the diene ⟨87ZOB1078⟩.

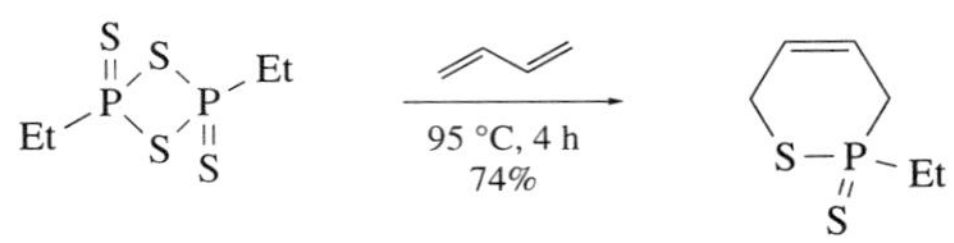

(300)

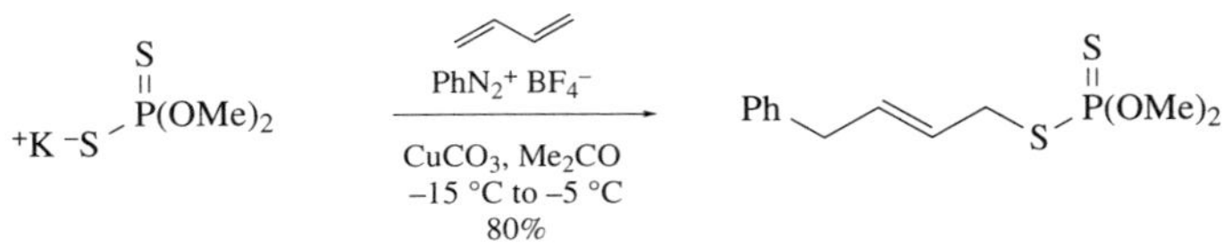

(301)

Phosphoranesulfenyl bromides undergo electrophilic addition to carbon–carbon double bonds. The ambident electrophilicity of these reagents, that is their ability to act as sources of positive bromine or positive sulfur, allows two pathways to occur, by initial formation of either an epibromonium or episulfonium intermediate. The course of the reaction is largely dependent upon the alkene. Electron-donating substituents favour formation of the episulfonium ion, opened by bromide to give stereospecific *trans* addition as expected ⟨82CC1362⟩.

Conjugate addition to α,β-unsaturated esters, amides and ketones is a further viable route to the formation of alkyl sulfur-phosphorus compounds, usually in good yields (Equation (302)) ⟨52JA161, 81S547⟩.

The addition of Lawesson's reagent to an unsaturated amide results in the formation of a cyclic product (Equation (303)) ⟨81ACS(B)239⟩.

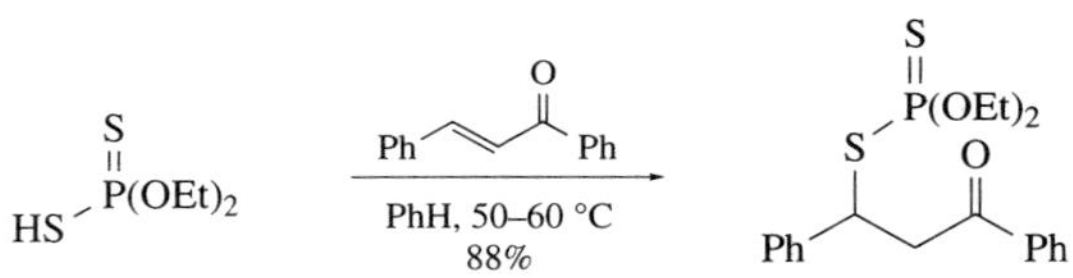

(302)

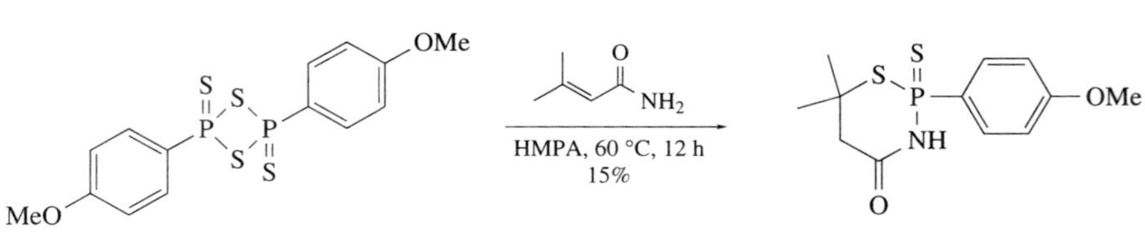

(303)

2.03.10.1.4 *From disulfides*

Cleavage of the sulfur–sulfur bond of a disulfide with a phosphorus reagent can result in formation of a sulfur–phosphorus bond. Alkylthiophosphates have been so prepared from disulfides by reaction with triethyl phosphite in an Arbuzov-type reaction (Equation (304)) ⟨55JA6064, 93JCS(P1)1091⟩. A related example proceeds by a radical mechanism on irradiation with UV light ⟨87JA1227⟩ (Equation (305)).

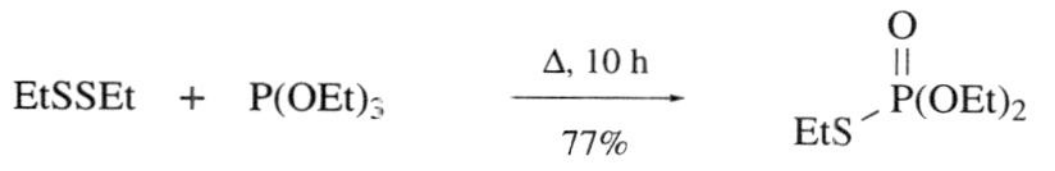

(304)

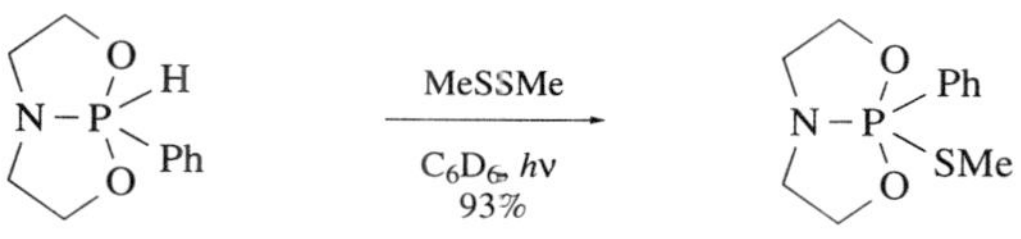

(305)

2.03.10.1.5 *From sulfenyl chlorides or sulfenates*

Sulfenyl chlorides readily react with alkyl phosphites at −78 °C to give good yields of thiophosphates. The sulfenyl chlorides are usually generated *in situ* from the corresponding disulfides by treatment with sulfuryl chloride ⟨55JA181, 91ZOB1115⟩. A 1991 example demonstrates the utility of this method in the synthesis of thiophospholipids (Equation (306)) ⟨91LA1361⟩.

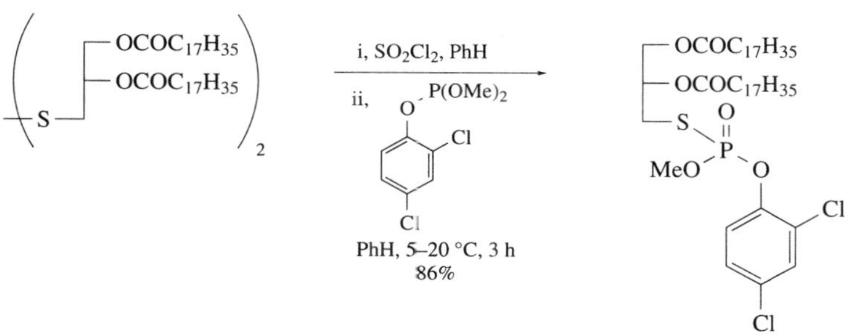

(306)

Trimethylphosphite has been used in the cleavage of a sulfenate derived from the rearrangement of an allylic cysteine sulfoxide. The reaction gave the thiophosphate in 67% yield on heating to 80 °C in benzene for 24 hours ⟨92JA1521⟩.

2.03.10.1.6 *By rearrangement*

O-Aryl-*O*,*O*-dialkylthiophosphates rearrange on heating with phosphorus oxychloride to give the *S*-alkyl chlorophosphates in reasonable yields (Scheme 220) ⟨91S454⟩. This chlorination–isomerization reaction also occurs for *O*,*O*,*O*-trialkylthiophosphates ⟨83JAP58113195⟩.

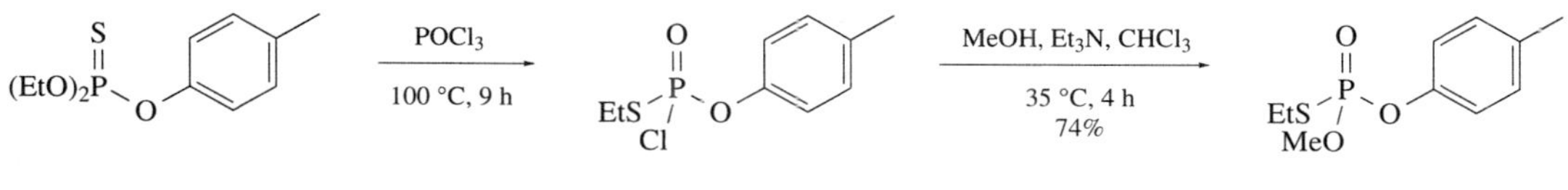

Scheme 220

Allylic phosphorothionates and phosphonothionates rearrange thermally or under palladium(0) catalysis to give the *S*-allylphosphoro or phosphonothiolates respectively (Equation (307)) ⟨79TL5015, 83JOC1293⟩. Thiopyrophosphates are known to disproportionate in the presence of salts such as triethylammonium hydrochloride to give the thiophosphate (Equation (308)), although yields are poor ⟨56CJC1819⟩.

Treatment of the oxazaphospholidine-2-thione with *t*-butylmagnesium bromide or magnesium bromide, causes a rearrangement to give the *S*-alkylated product (Equation (309)) ⟨80TL4959⟩.

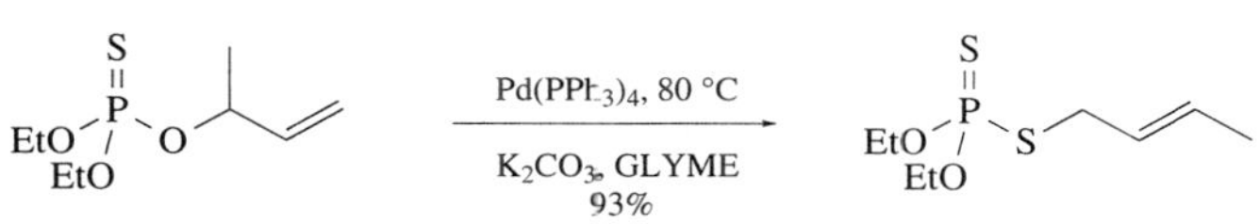

(307)

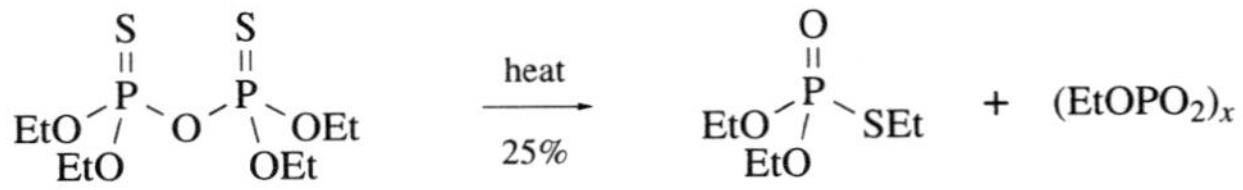

(308)

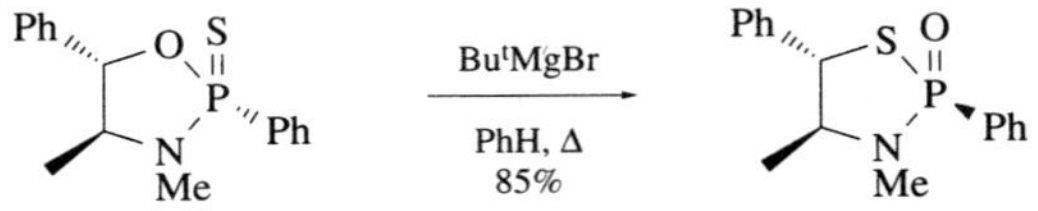

(309)

2.03.10.1.7 *Miscellaneous*

(i) From alcohols

Direct esterification of phosphinodithioic acids or derivatives leads directly to *S*-alkyl phosphinodithioates with elimination of water rather than hydrogen sulfide (Equation (310)). The reaction of alcohols with phosphorus pentasulfide at 180–190 °C also results in formation of the *S*-alkylated product (Equation (311)) ⟨56JA4447⟩.

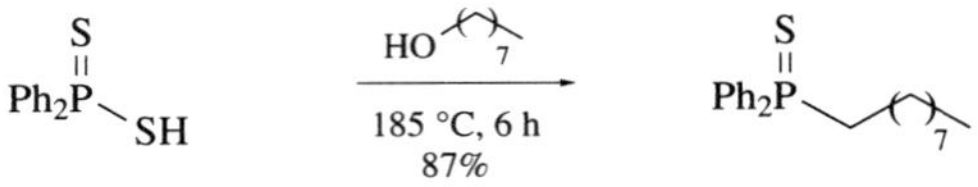

(310)

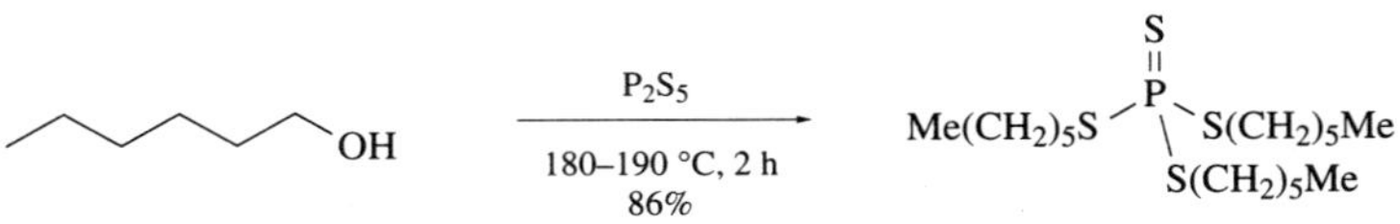

(311)

(ii) From carbonyl compounds

1,3-Dicarbonyl compounds and silyl enol ethers are alkylated by thiophosphates in good yields (Equation (312)). Silyl enol ethers react with bromo- or chlorodithiophosphonates at −78 °C in up to quantitative yields (Equation (313)) ⟨90S609, 93PS(79)131⟩.

The oxygen atom of benzophenone is lost in the reaction with metallic sodium and *O*,*O*-diethyl thiophosphite, to give the *S*-alkyl thiophosphate in 85% yield (Equation (314)) ⟨71CB3740⟩.

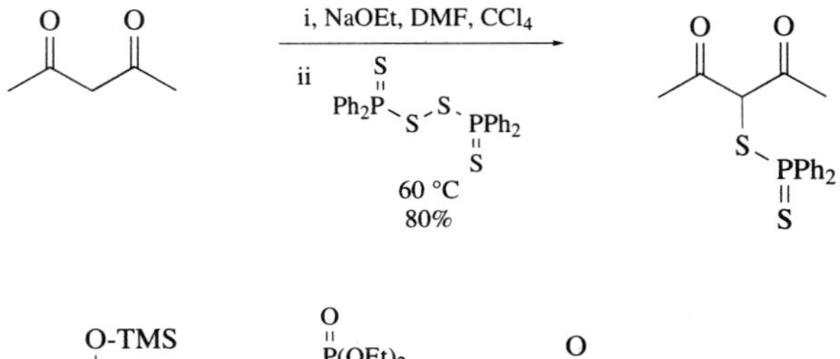

(312)

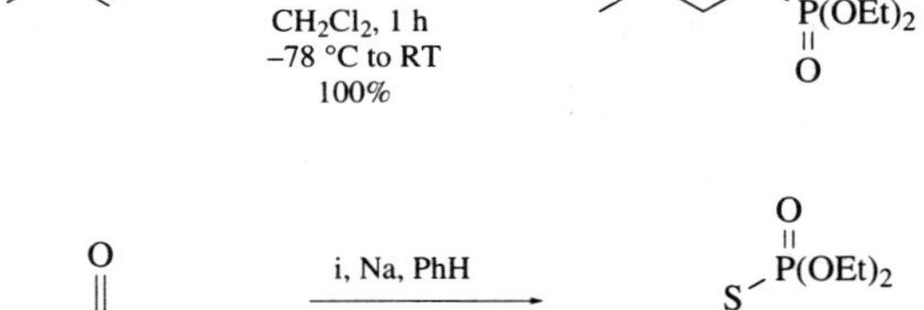

(313)

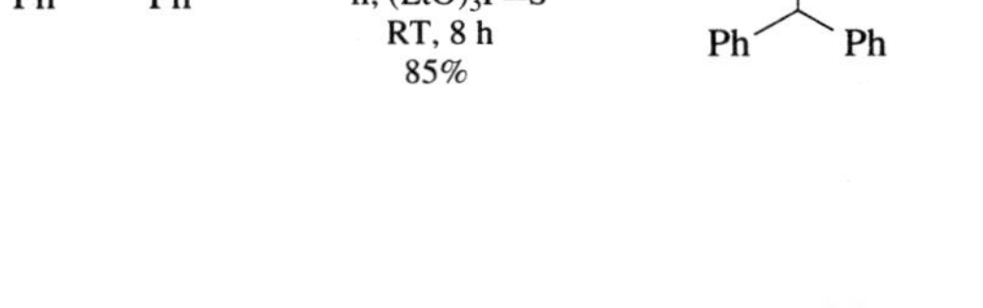

(314)

(iii) From thiocyanates

Alkyl thiocyanates act on phosphites or thiophosphites to form *S*-alkylthiophosphates, or *S*-alkyldithiophosphates, in reasonable yields (Equation (315)) ⟨66JHC14, 90ZOB476⟩.

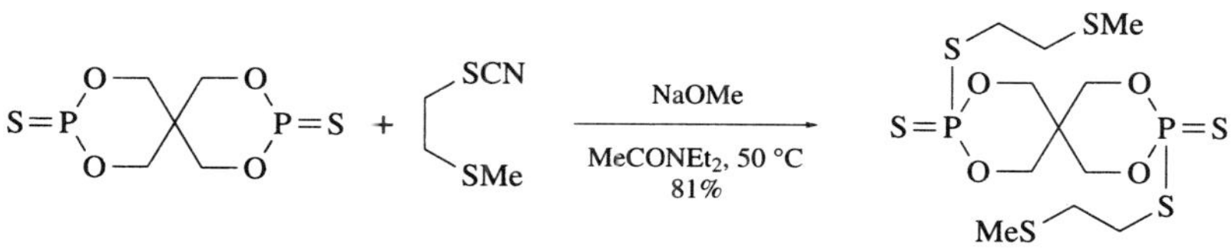

(315)

(iv) From thiomides

S-Alkylthiosuccinimides and *S*-alkylthiophthalimides react spontaneously with trialkyl phosphites, by an Arbuzov-type mechanism, to furnish the thiophosphates in good yield, avoiding use of the labile sulfenyl halide which can give halogenated side products (Equation (316)) ⟨90TL501⟩.

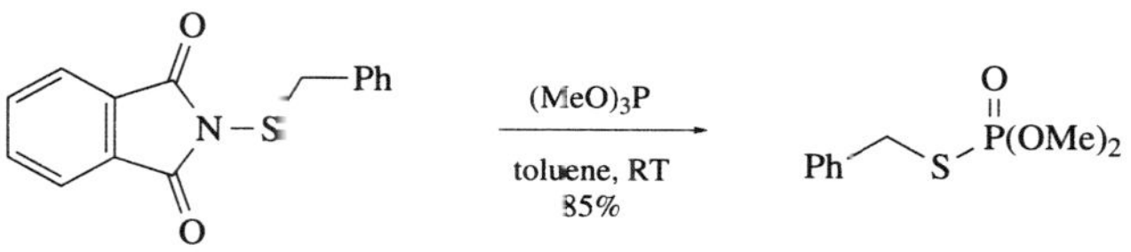

(316)

(v) From ortho-*esters*

Phosphorus pentasulfide reacts with ethyl *ortho*-formate to produce a triethyl dithiophosphate. Similarly, the reaction between ethyl *ortho*-thioformate and phosphorus pentasulfide is reported to form a triethyl tetrathiophosphate. No evidence is however offered for the structure of these compounds ⟨51JA4953⟩.

(vi) From epoxides

The trimethylsilyl derivative of a dithiophosphate has been reported to react with ethylene oxide with ring cleavage to give the *S*-alkylated product in 96% yield ⟨82URP891678⟩.

(vii) From thiiranes

Triethyldichlorophosphorane cleaves thiiranes, with formation of a phosphorus–sulfur bond, to produce a phosphonium salt (Equation (317)) ⟨81ZN(B)447⟩.

(317)

2.03.10.2 Alkanethioarsenates and their Higher Coordinated Analogues

Interest was initially shown in alkanethioarsenates shortly after the Second World War, when large quantities of unwanted arsenical chemical weapons became available, and alternative uses or disposal was necessary ⟨46JA1860⟩. The toxicity of these arsenic compounds was found to be due to formation of stable, cyclic sulfur–arsenic compounds through reaction with thiol groups in proteins. Antidotes to these chemicals were therefore based on dithiols ⟨47JCS592⟩.

2.03.10.2.1 From arsenic halides

Replacement of the chloride ligands in arsenic trichloride with a thiol is the simplest procedure for production of these compounds. For example, reaction of arsenic trichloride and ethanedithiol in carbon tetrachloride at room temperature gives the product in greater than 90% yield on a molar scale (Equation (318)) ⟨46JA1860⟩. A range of substituted thiols has been treated in a similar way, but with pyridine present in the reaction mixture, to give yields of 30–78% ⟨47JCS592⟩.

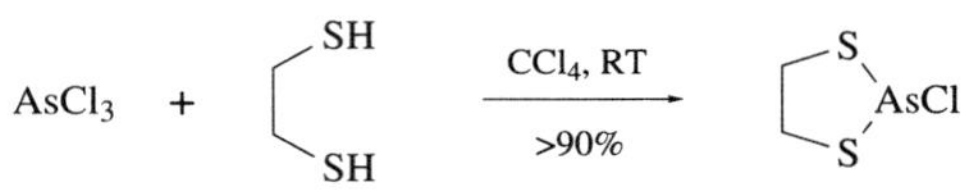

(318)

Other arsenic halides have been used in forming arsenic–sulfur bonds. A German patent describes the reaction between dimethyl arsenic iodide and a sugar thioacetate (Equation (319)) ⟨84GEP3229565⟩. The ethyl derivative of this arsenic reagent has been used to good effect in the reaction of a sugar thiolate to give the anomeric thioarsenate in 90% yield ⟨80PS(9)107⟩.

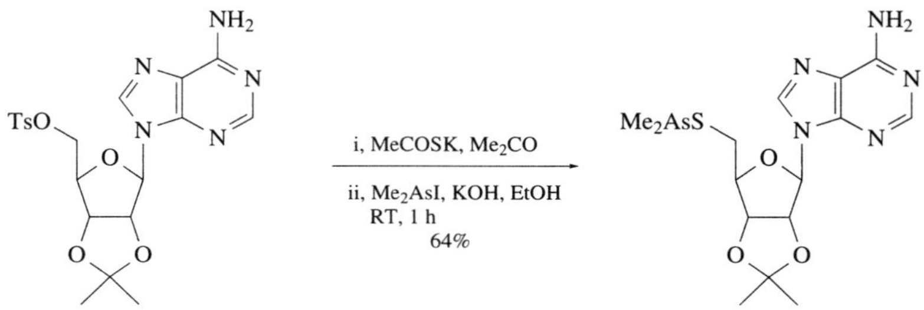

(319)

2.03.10.2.2 From arsenoxides

There are several examples of reactions between arsenoxides and thiols resulting in alkyl thioarsenate formation. Aryl arsenoxides generally give good yields by reaction with a thiol in either aqueous or ethanolic solution under reflux ⟨46JA1864, 66JOC221⟩. Alternatively, the reaction can be performed at room temperature at pH7 by the addition of sodium hydroxide (Equation (320)) ⟨48JA1762⟩.

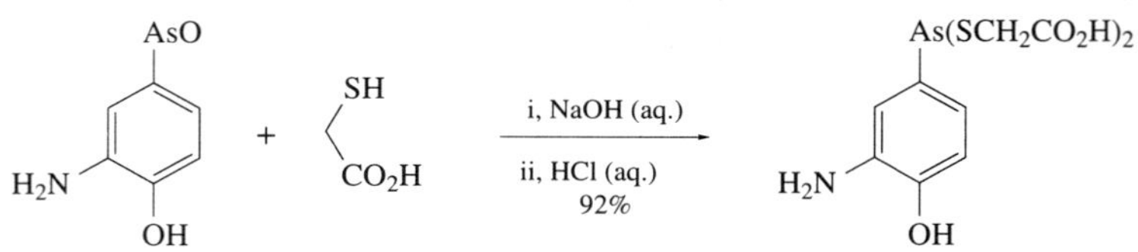

(320)

2.03.10.2.3 From aminoarsines

Aminoarsines have only limited coverage in the literature compared with their phosphorus analogues, although they appear to be versatile intermediates. They have been successfully used in the synthesis of thioarsenates by reaction with dithiols or thioalcohols in petroleum ether under reflux to give the cyclic products in good yields (Equation (321)) ⟨70CC425, 82TL1345⟩.

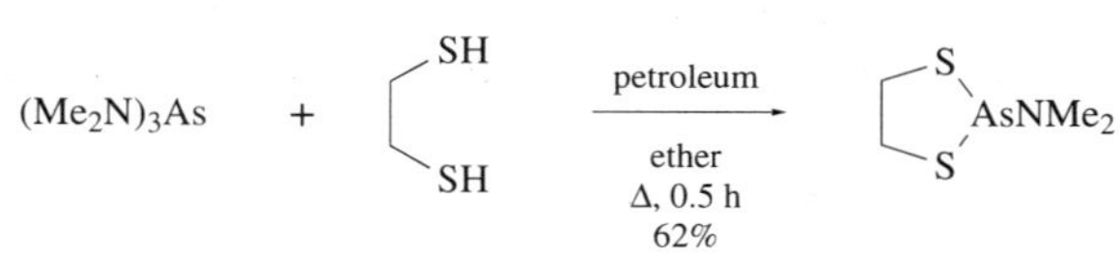

(321)

2.03.10.2.4 From tetraalkyldiarsines

Sulfur–arsenic bonds have been formed from tetramethyldiarsine and disulfides in the synthesis of functionalized glucopyranoses to produce compounds with potential anticancer activity (Equation (322)) ⟨80JCS(P1)2287⟩. The reaction shown was performed in a dry box at room temperature, in dichloromethane solution, to give the product in quantitative yield.

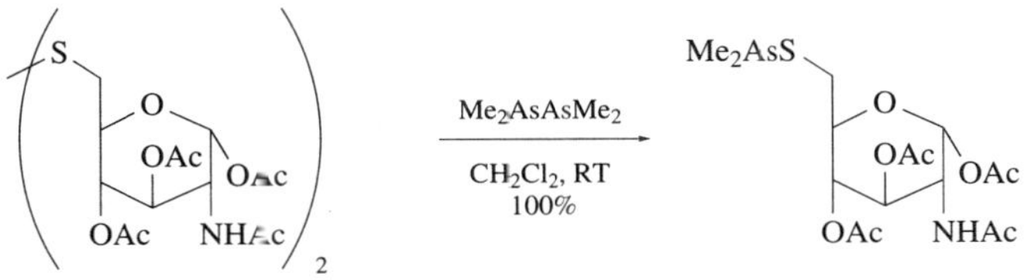

(322)

2.03.10.3 Alkanethioantimonates and their Higher Coordinated Analogues

Organic antimony compounds have found some use in the treatment of protozoan diseases, but suffer from high toxicity. Reports in the literature only give examples for which sulfur is in the divalent oxidation state, that is thiol derivatives. Synthesis of these compounds involves simple addition of antimony trichloride to a solution of the thiol, or a sodium salt of the thiol, to give up to quantitative yields of products (Equation (323)) ⟨38JA2370, 48JA2434⟩. A range of thiols of up to C-18 chain length were investigated. Other thioantimonates which possess greater functionality in the alkyl group, for example an α-amido group, have been prepared in a similar manner ⟨27JA1017⟩.

Little work had followed these early examples, but in 1991 Buchwald described the synthesis of cyclic thioantimonates by transmetalation of a zirconocene heterocycle (Equation (324)) ⟨91JA165⟩.

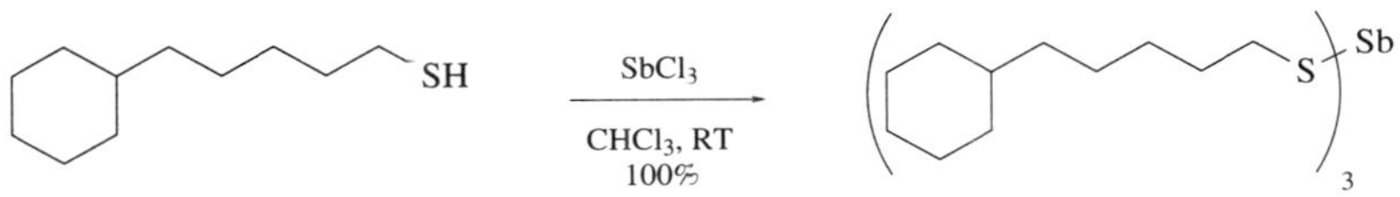

(323)

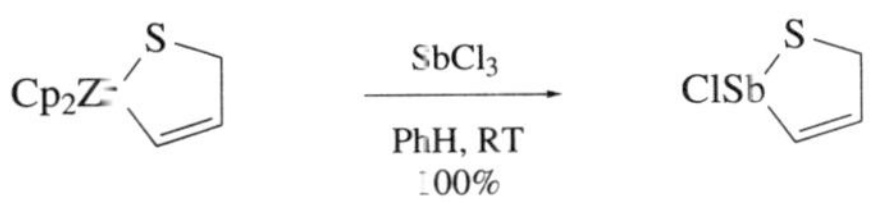

(324)

2.03.10.4 Alkanethiobismuthates and their Higher Coordinated Analogues

Few examples can be found in the literature of this class of compounds. An α-carboxyl thiobismuthate has been prepared from ethyl thioglycolate and bismuth oxide in alcoholic solution ⟨27JA1017⟩. Tris(ethylthio)bismuthate has been prepared by reaction of ethanethiol with bismuth nitrate (Scheme 221) ⟨38JA2370⟩.

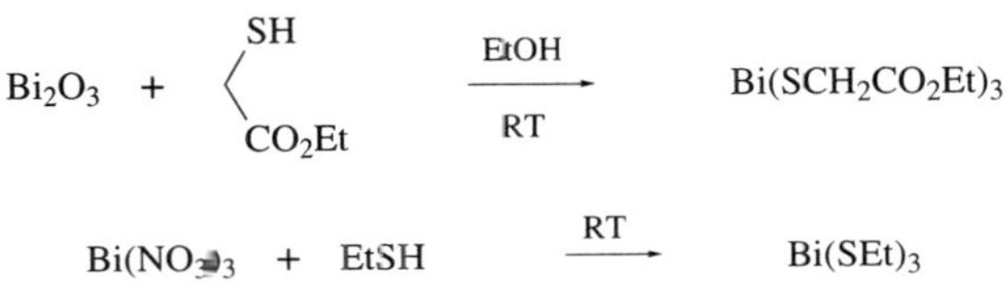

Scheme 221

2.03.11 RSSi, RSB, AND RELATED METALLOID FUNCTIONS AND THEIR HIGHER COORDINATED DERIVATIVES

2.03.11.1 Alkanethiosilanes

Given the great synthetic utility of compounds containing either a sulfur or silicon functionality, it is perhaps not surprising that the chemistry of mixed organosulfur–silicon compounds has received much attention ⟨88T281⟩. The vast majority of these bifunctional compounds however contain the sulfur and silicon functionality attached to the same or neighbouring carbon atoms; examples of alkanethiosilanes, containing sulfur–silicon bonds, are still relatively rare. Some of the preparative methods for these compounds are described below. By far the most popular methods of synthesis involve alkanethiol starting materials, although syntheses from sulfides, thiocarbonyl compounds, disulfides, and other procedures are known.

2.03.11.1.1 From alkanethiols

Perhaps the most extensively used route for the synthesis of alkanethiosilanes involves the reaction of an alkanethiol with a silicon compound containing a suitable leaving group. A wide range of silylating agents is of course known from protecting group chemistry, the most common of which are chlorotrialkylsilanes. Indeed, a number of reports of the preparation of alkanethiosilanes by reaction of an alkanethiol with a chlorosilane have appeared. In 1951, Wolinski reported the preparation of a range of alkylthiosilanes by the reaction of appropriate alkanethiols with trichlorosilane in benzene in the presence of pyridine ⟨51JOC395⟩. Yields for the reaction were however generally poor, although in the case shown in Equation (325), tripropylthiosilane was produced in a satisfactory 63% yield.

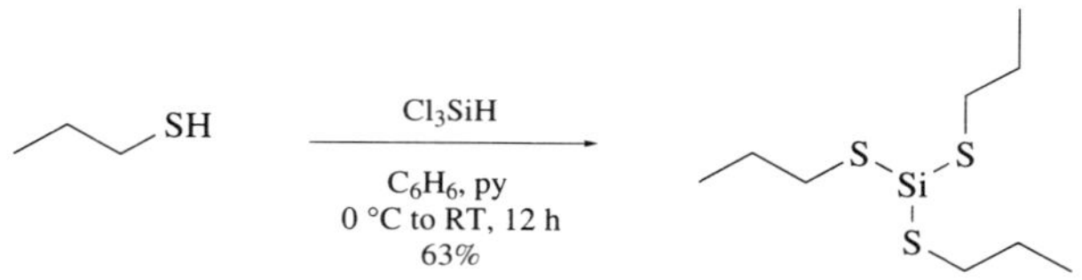

(325)

Chlorotrialkylsilanes react equally efficiently with alkanethiols. For example, *t*-butyldimethylsilyl chloride reacts readily with alkanethiols in the presence of the base dbu ⟨85TL475⟩. Arylthiols are silylated significantly more slowly and in lower yields. A particular interesting example of thiol silylation in the field of peptide synthesis appeared in 1961 ⟨61CB1263⟩. In this report, silylation of cysteine occurred at the thiol moiety, with concomitant silylation of the amino and carboxyl groups, as shown in Equation (326). The process was repeated for a number of amino acids. Thexyldimethylsilyl chloride, formed by the reaction of chlorodimethylsilane with 2,3-dimethyl-2-butene, was shown to react with the sodium salt of butanethiol at room temperature to give the alkanethiosilane in 63% yield ⟨85TL5515⟩.

Cycloalkanethiosilanes can be formed by the use of dithiols with chlorotrialkylsilanes in which one of the alkyl groups contains an additional leaving group (Equation (327)) ⟨63CB1019, 63CB1561, 63CB2822⟩.

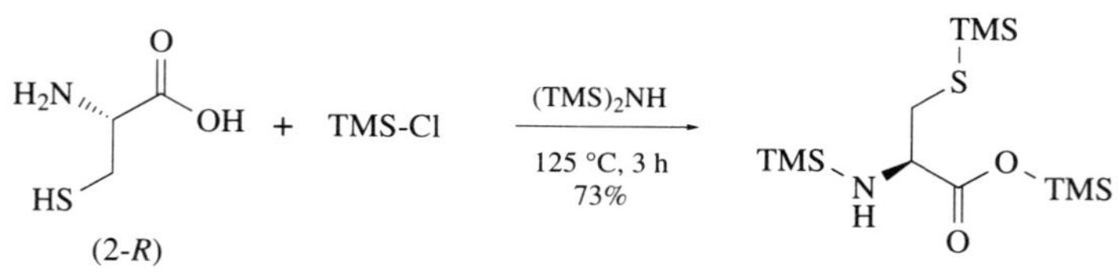

(2-*R*)

(326)

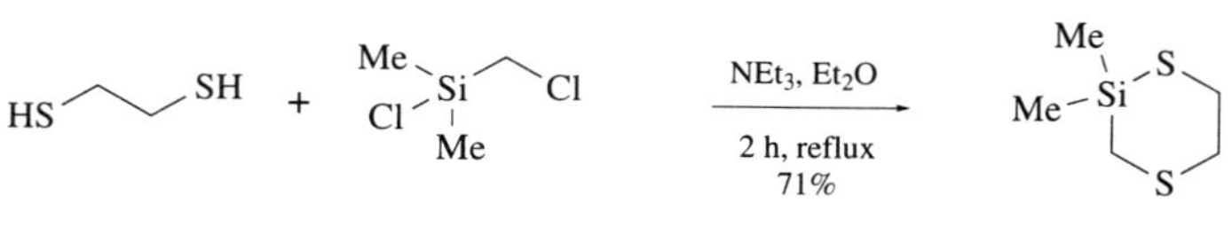

(327)

Silyl amides have also been used for the silylation of alkanethiols. For example, cyclic alkanethiolsilanes containing two RSSi moieties can be formed by the reaction of bis(diethylamino)dimethylsilane with ethanethiol (Equation (328)) ⟨65JCS7098⟩. Similarly, in a reaction catalyzed by imidazole, trimethylsilyl thioethers have been prepared in 50–93% yields by the reaction of hexamethyldisilazane with alkanethiols (Equation (329)) ⟨73JOM(61)83⟩. The presence of imidazole in catalytic amounts dramatically accelerated the reaction. The mechanism of alkanethiosilane formation by the reaction of a thiol with 1-(trimethylsilyl)-imidazole has been established, and *N*-methyl-*N*-(*t*-butyldimethylsilyl)trifluoroacetamide (MTBSTFA) and the related *N*-*t*-butyldimethylsilylamides have been reported as silylating agents for thiols ⟨82JOC3336⟩. These reagents are perhaps more useful for thiol substrates than the usual TBDMS-protecting group, with which yields are often only moderate (Equations (330) and (331)). In both cases shown, reaction of the thiol substrates with TBDMS-Cl did not yield alkanethiosilane product. This type of silylation reaction has been carried out on the amino acid cysteine ⟨61CB1876⟩.

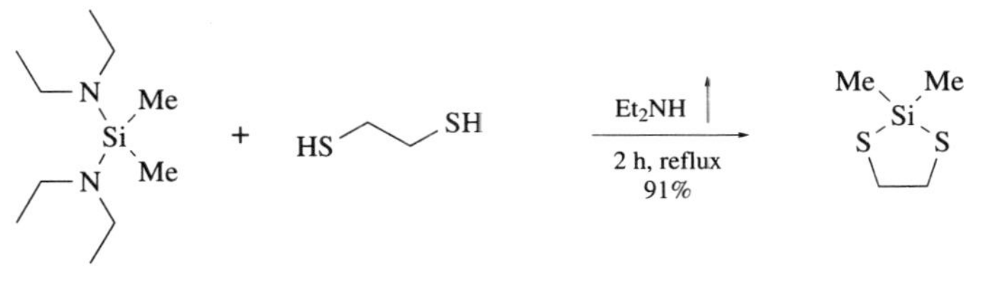

(328)

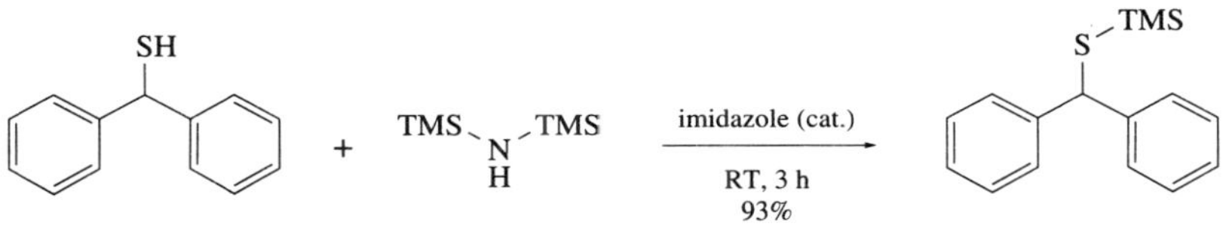

(329)

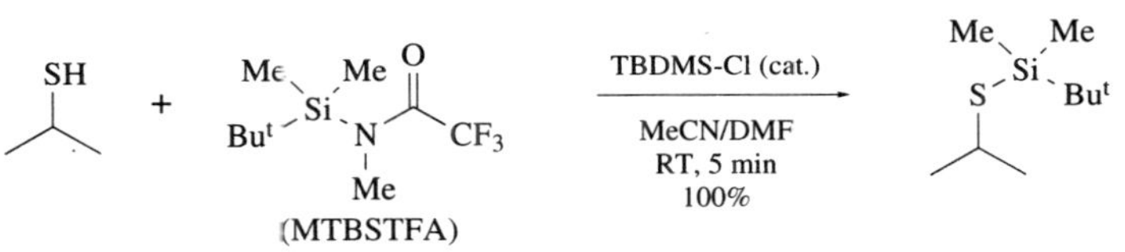

(330)

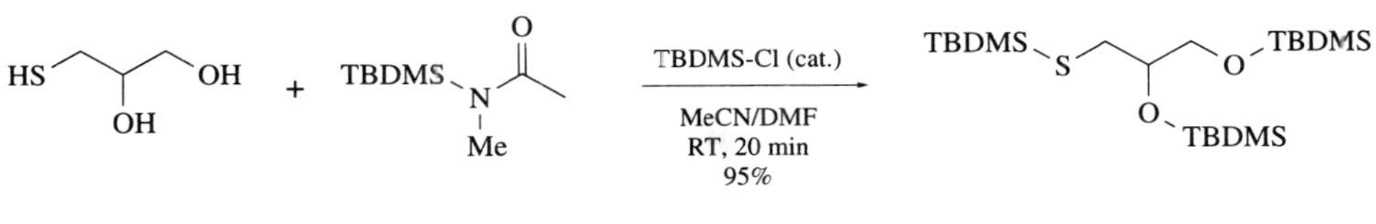

(331)

Other reagents effective for the silylation of thiols include ketene methyltrialkylsilyl acetals (Equation (332)) ⟨79TL4311⟩; bis(1-methoxy-2-methyl-1-propenyloxy) silanes (Equation (333)) ⟨83TL1273⟩; and alkyltrimethylacetates (Equation (334)) ⟨81BCJ805⟩. An unusual reagent is hexamethyldisilathiane, prepared from sodium sulfide and chlorotrimethylsilane (Equation (335)) ⟨61JCS4933⟩. Reaction of this substance with butanethiol in the absence of solvent under reflux gave butanethiosilane in 76% yield. Polymer-supported silylating agents are known. Naflon-TMS, prepared by the immobilization of trimethylsilyl trifluoromethanesulfonate on a resin, has been used for the silylation of a thiol ⟨80TL767⟩.

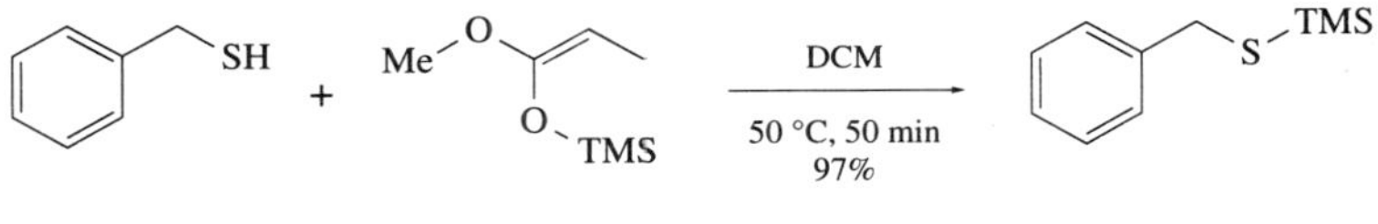

(332)

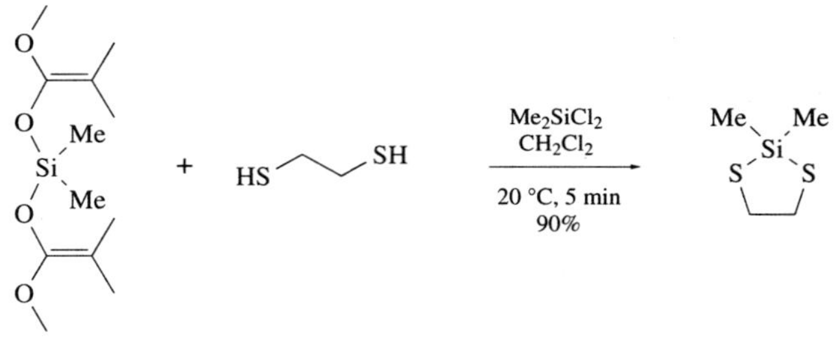

(333)

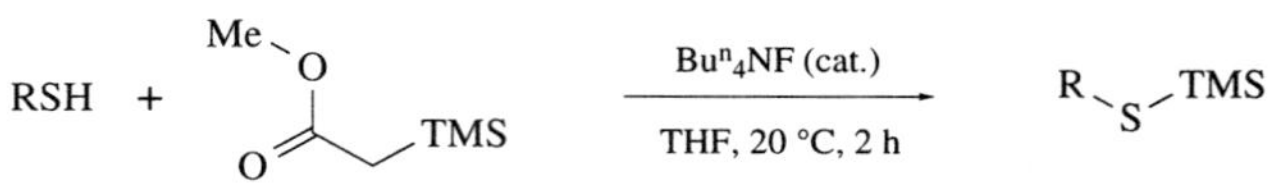

(334)

R	*Yield* (%)
Hex^n	91
Bn	87
Bu^t	82

(335)

The tris(triphenylphosphine)chlororhodium-catalyzed dehydrogenative condensation of various hydrosilanes with alkanethiols also provides alkanethiosilanes (Equation (336)) ⟨73JOM(50)26, 80BSB783⟩.

Alkanethiosilanes have been prepared by silylation of lead thiolates with chlorotrimethylsilane, the by-product of the reaction being lead(II) chloride. The reaction is faster than that of thiols with trialkylchlorosilanes, generally requiring heating under reflux for less than two hours (Equation (337)) ⟨60JCS4406⟩.

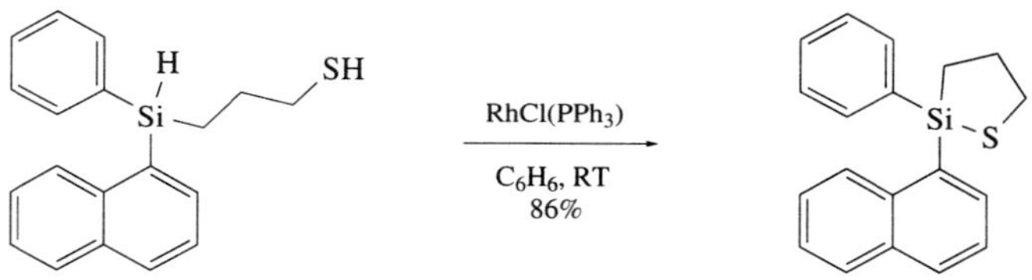

(336)

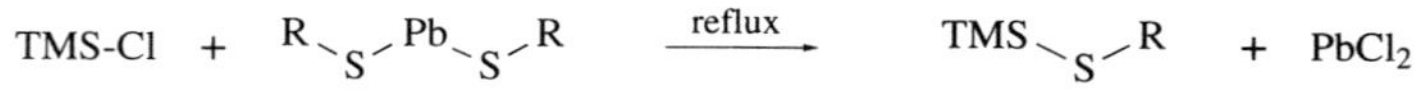

(337)

R	*Yield* (%)
Et	76
Pr^n	92
Pr^i	73
Bu^n	81
Bu^t	70

2.03.11.1.2 From silyl thiolate anions

Kraus has studied the preparation of silyl thiolates for access to alkanethiosilanes ⟨91TL2189⟩. The trimer shown in Equations (338) and (339) underwent reaction with organolithiums to give silyl thiolates as shown ⟨54MI 203-01, 63AG(E)328⟩. Reaction with primary halides resulted in the formation of silyl thioethers in good to excellent yields. The silyl sulfides formed could be directly converted into the corresponding thiols by treatment with HF in acetonitrile. This route therefore has applications for the direct introduction of a protected thiol.

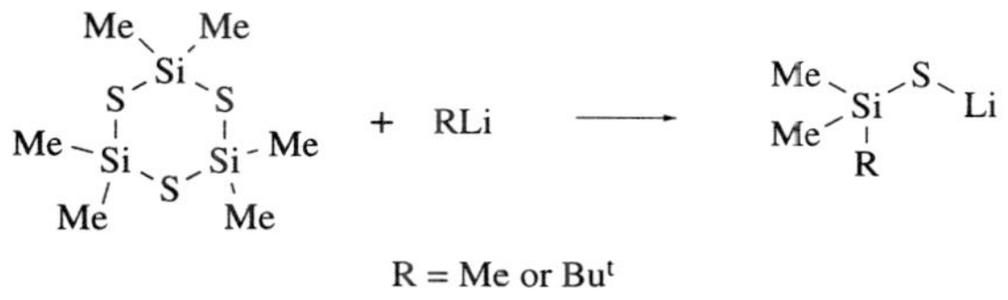

(338)

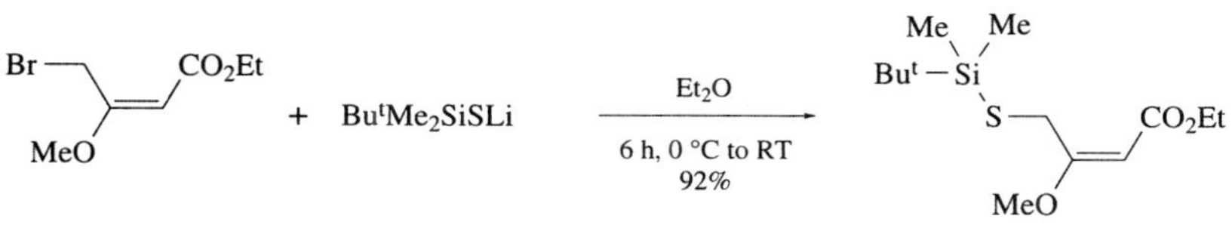

(339)

2.03.11.1.3 From alkyl sulfides

Alkyl sulfides or compounds containing an alkyl sulfide linkage can also be used to prepare alkanethiolsilanes. Arya has reported the use of tris(trimethylsilyl)silane (TTMSS) as an effective reagent for the conversion of carbon–sulfur bonds into sulfur–silicon bonds, in a free radical process ⟨90JOC6248, 91TL2853⟩. The effectiveness of the TTMSS reagent is due to its relatively weak silicon–hydrogen bond. TTMSS has been used to carry out the reductive cleavage of a range of sulfide substrates including 1,3-dithiolanes, 1,3-dithianes, oxathiolanes and thiazolidines (Equation (340)). In all cases, AIBN was employed as the radical initiator.

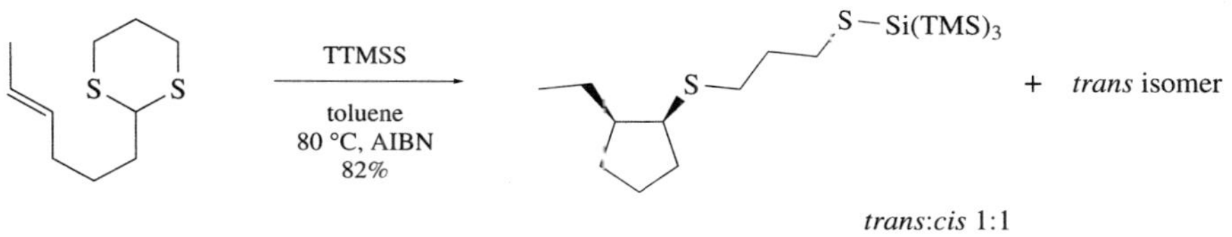

(340)

A further interesting procedure for the preparation of alkanethiosilanes is the reaction of a sulfide with a trialkylsilane in the presence of a catalytic nickel, freshly prepared nickel(II) chloride and triethylsilane (Equation (341)) ⟨87ZOR1981⟩.

Thiiranes can undergo regioselective ring cleavage to produce alkanethiosilanes by treatment with trimethylsilyl cyanide, in the presence of aluminum trichloride as catalyst (Equation (342)) ⟨83TL2311⟩.

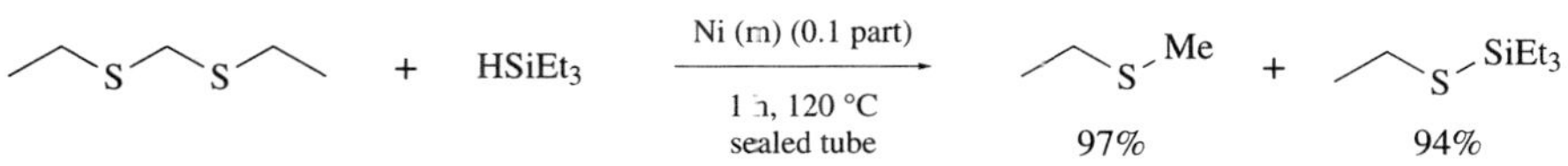

(341)

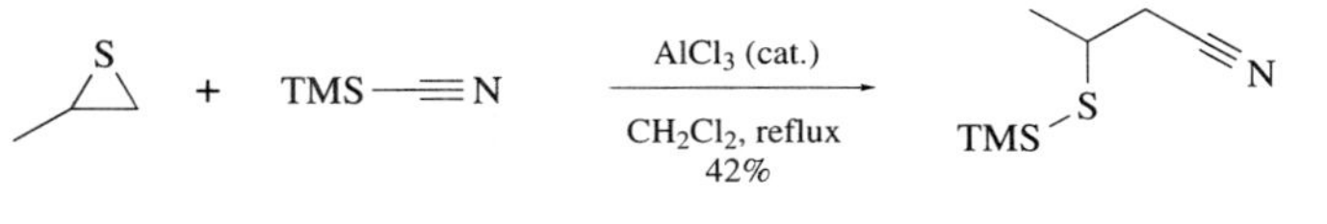

(342)

2.03.11.1.4 From disulfides

Alkanethiotrimethylsilanes can be conveniently prepared from disulfides by reaction with a solution of chlorotrimethylsilane induced by suspended sodium metal (Equation (343)) ⟨78BCJ2183⟩. While this reaction has only been performed with symmetrical disulfides, yields were good to excellent, and simple isolation of the desired product by filtration and distillation render the process particularly useful. Alkanethiosilane derivatives were best prepared using nonpolar solvents such as pentane or hexane. In more polar solvents such as ether, the corresponding dialkylsulfide was also formed in substantial quantity. In the case of dimethyldisulfide, bis(trimethylsilyl)sulfide was obtained as the major product. Weber has also reported the preparation of alkanethiosilanes from disulfides by irradiation-mediated processes ⟨81JOM(210)163⟩.

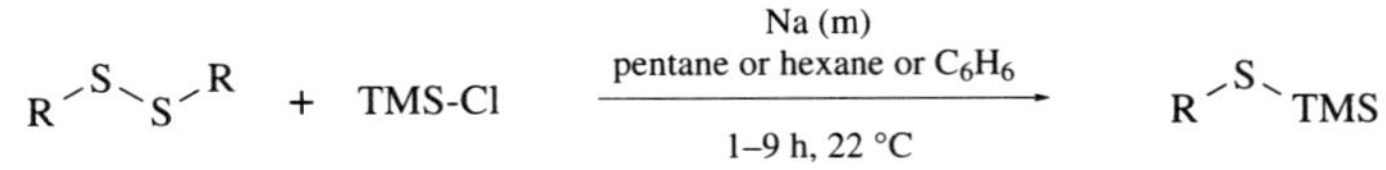

(343)

R	*Yield* (%)
Et	83
Pr^i	80
Bu^t	80
n-C_6H_{13}	78

2.03.11.1.5 From thiocarbonyl compounds

An unusual silathiacyclopropane has been prepared by the irradiation of a thiocarbonyl compound in the presence of a silylene ⟨83TL4033⟩. The structure was confirmed by x-ray analysis (Equation (344)). A similar silathiocyclopropane was isolated in 29% yield by irradiation of a trisilane with adamantanethione.

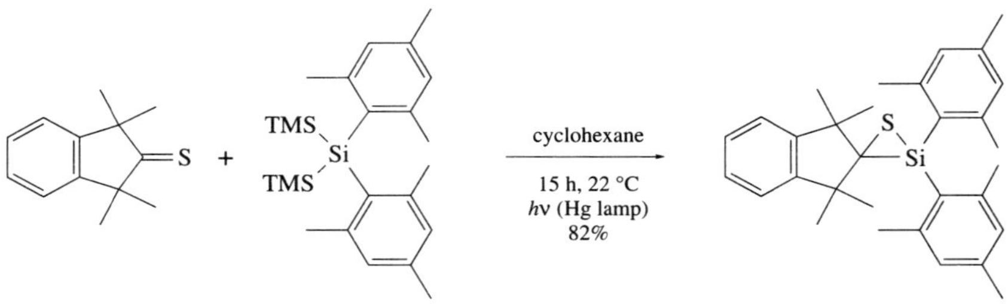

(344)

Alkanethiosilane preparation has been reported by the reaction of a thiocarbonyl compound with a silyloxyphosphinate, as shown in Equation (345), to provide an interesting, highly functionalized phosphonate product in 97% yield ⟨83URP956484⟩. A further example of the combination of thiocarbonyl compounds and phosphinates is shown in Equation (346). In this case the reaction, which took place in excellent yield, was carried out in the absence of solvent ⟨81ZOB2140⟩.

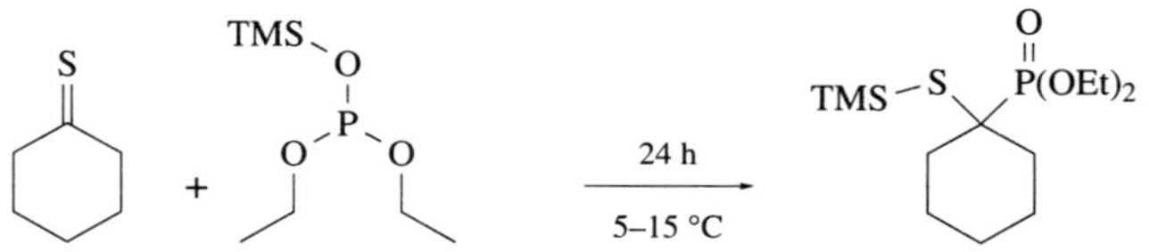

(345)

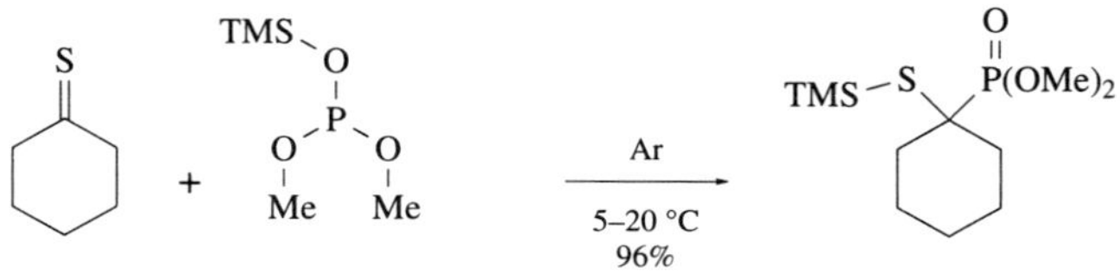

(346)

2.03.11.1.6 From thermally mediated processes

Thermally mediated processes such as isomerization can effect formation of an alkanethiosilane. Pyrolysis of 3,3-diethyl-2,4-dimethyl-3-silathietane at ca. 300–350 °C has been shown to cause isomerization to the unsaturated thiosilane-containing product (Equation (347)), although this was accompanied by some decomposition ⟨81JOM(214)415⟩.

(347)

Sulfur-insertion reactions have also resulted in alkanethiosilane formation ⟨79JOM(164)305⟩. Upon heating of the extremely reactive silacyclobutane compound with elemental sulfur in a sealed tube, a rare insertion reaction took place to give the cyclic alkanethiosilane product in high yield (Equation (348)). Analogous insertion reactions were also carried out with the more reactive silacyclopropanes (siliranes). This however resulted in the formation of dithiosilacyclopentanes in only moderate yields.

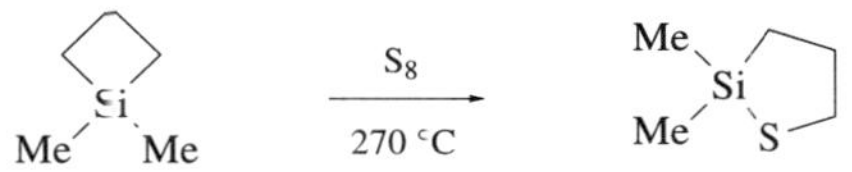

(348)

2.03.11.1.7 Higher coordinated RSSi functions

Higher coordinated alkanethiosilane derivatives are rare in the literature. One notable example has been reported by Chou involving a sulfenic acid trimethylsilylester ⟨74TL725⟩. Upon reaction of the penicillin sulfoxide ester shown in Equation (349) with the silylating agent derived from trimethylsilyl chloride and hexamethyldisilazane in benzene under reflux over a period of two hours, the sulfinyl silane was obtained in quantitative yield.

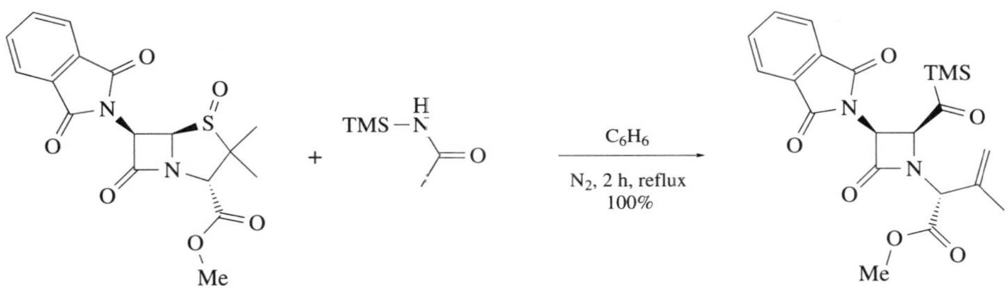

(349)

2.03.11.2 Alkanethioboranes

In comparison to their silyl counterparts, alkanethioboranes are rare in the literature. Those which are known possess a high reactivity towards unsaturation and acidic species. Examples of the reaction of bis(ethylthio)phenylborane with isocyanates ⟨68JCS(A)2962⟩, double bonds, and alcohols ⟨61JA1345⟩ have been described. Alkanethioboranes react readily and in high yield with difunctional organic compounds such as diols, amino alcohols, and thiolalcohols at ambient temperature to afford the corresponding organoboron heterocycles (Equation (350)) ⟨69CC832⟩.

Relatively few procedures are known for the synthesis of alkanethioboranes. Those which are described in the literature normally commence from an alkanethiol, although alternative materials such as boranesulfides have also been employed. These methods are discussed below.

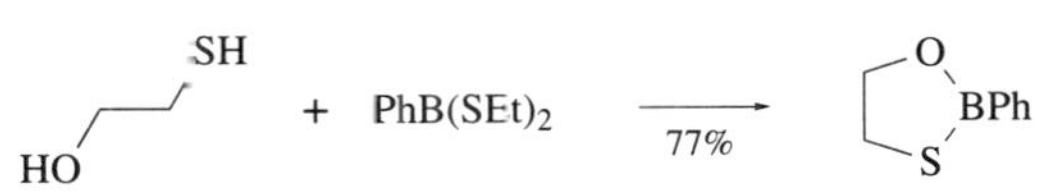

(350)

2.03.11.2.1 Boron mono-, di- and trichlorides

Reaction of a boron monochloride with a thiol generates an alkylthioborane (Equation (351)) ⟨59JCS1529⟩.

Alkylthioboranes can also be indirectly synthesized from dialkylchloroboranes by their reaction with masked acyl carbanions ⟨74JOC600⟩. These lithium aldimine reagents react with dialkylchloroboranes to give tertiary alcohols upon treatment with thioglycollic acid, followed by aqueous alkaline hydrogen peroxide ⟨74TL793⟩. The reaction proceeds through an alkylthioborane intermediate (Scheme 222) ⟨75BCJ3682, 75JOC3644, 75TL2689⟩.

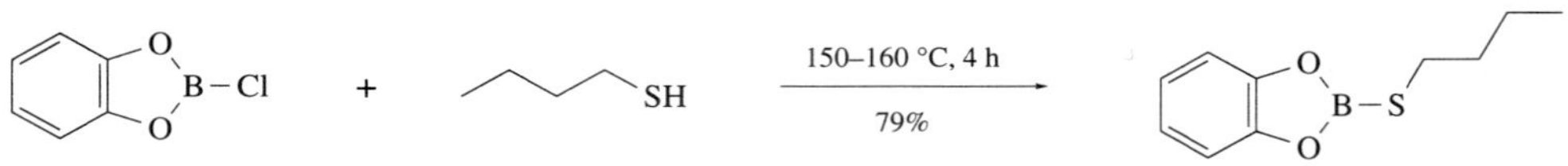

(351)

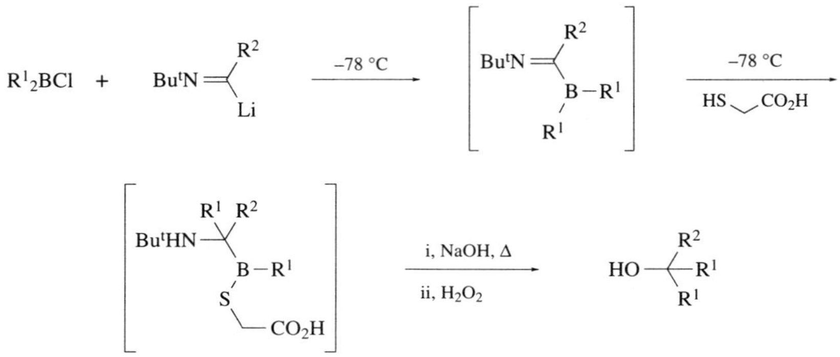

Scheme 222

Boron dichlorides and trichlorides react with alkanethiols to yield alkylthioboranes. The addition of an appropriate dithiol to boron trichloride or phenyl boron dichloride in dichloromethane, as shown in Equation (352), results in the formation of the cyclic alkylthioboranes in 80–92% yields ⟨64T173⟩. Interestingly, reaction does not take place at low temperature, unlike the analogous reaction with alcohols, which occurs at −80 °C. A related species, formally containing a nitrogen–boron ylide, has been prepared in 47% yield by the related reaction of a nitrogen-containing dithiol with phenylboron dichloride in the presence of triethylamine (Equation (353)) ⟨64M33⟩.

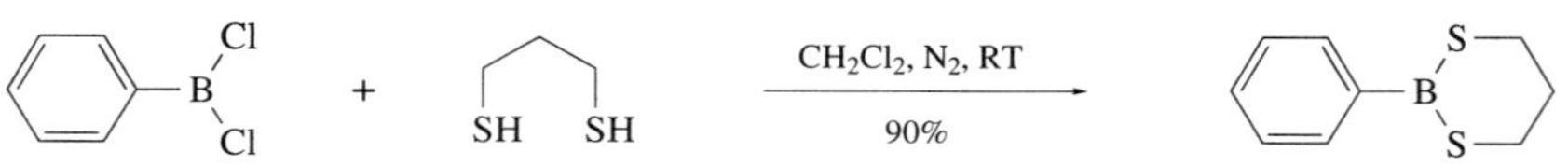

(352)

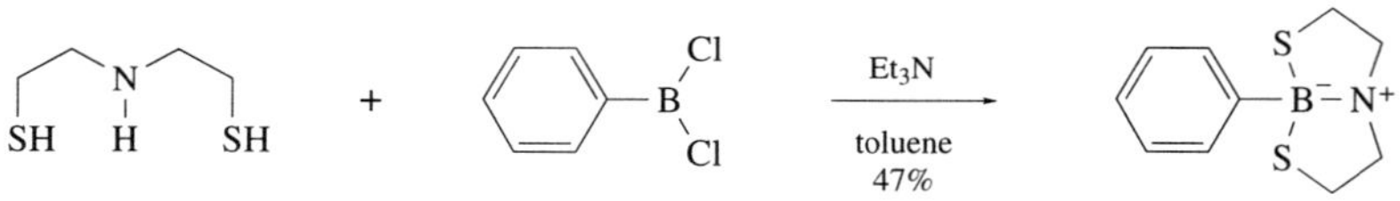

(353)

Boron–sulfur ylides have also been reported in an improved method for the preparation of $BH_2Cl \cdot SMe_2$. The procedure involved the reaction of commercially available borane-DMS complex with lithium borohydride in an inert atmosphere. After one hour, ^{11}B NMR showed the product distribution to be ca. 72% of the alkylthioborate, $BH_2Cl \cdot SMe_2$, 14% $BHCl_2 \cdot SMe_2$ and 14% ylide, as shown in Equation (354) ⟨85JA7167⟩. Related compounds have been reported by both Brown and Zweifel in reactions of the thexylchloroborane-DMS complex ⟨80JA5919, 84JA8001⟩.

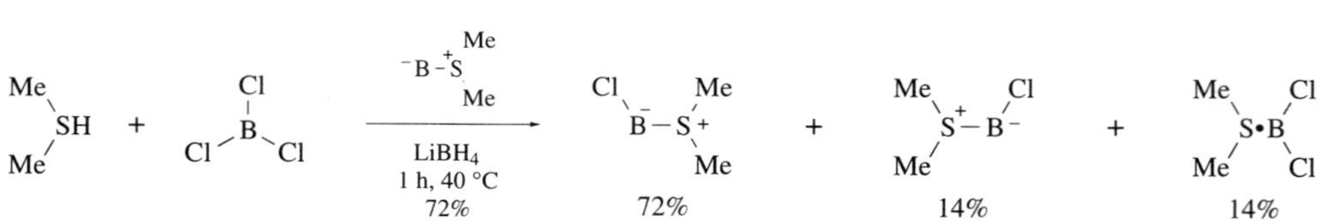

(354)

2.03.11.2.2 Alkylboranes

Simple trialkylboranes have been reported to react with thiols. Equation (355) illustrates one example, in which the reaction of triisopropylborane with butanethiol resulted in the formation of the alkylthioborane in 77% yield. The reaction was performed with heating in the absence of solvent ⟨60IZV386⟩.

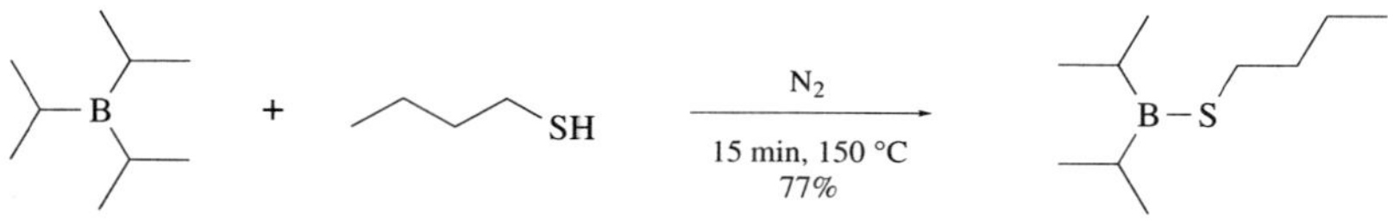

(355)

2.03.11.2.3 *Trimethylamineboranes*

Trimethylamine complexes of alkylboranes, prepared by reduction of the corresponding alkylboroxines in the presence of trimethylamine ⟨59JA5836⟩, have been found to undergo reaction with alkanethiols at 60–100 °C to yield the corresponding dialkyl alkylthioboronates in yields of up to 72% ⟨60JA748⟩. The use of trimethylamine borane for reaction with high boiling thiols has also been shown to afford the corresponding trialkylthioboronates. The thioboronates were easily hydrolyzed to furnish the corresponding thiols (Equations (356) and (357)).

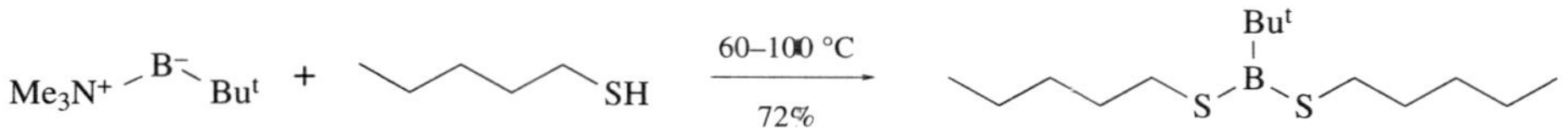

(356)

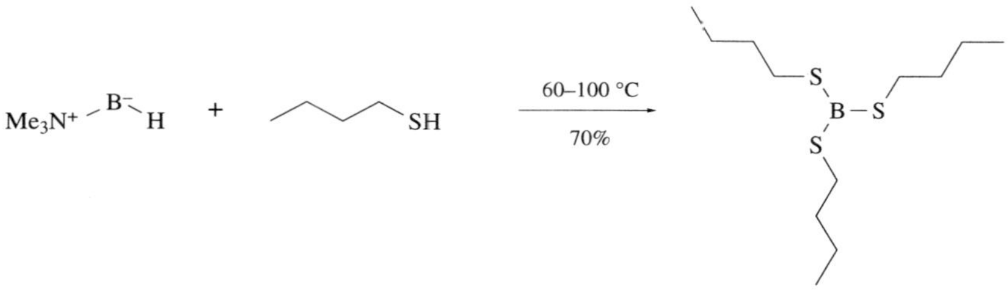

(357)

2.03.11.2.4 *Miscellaneous boron-containing compounds*

An interesting example of an alkanethioborane species appeared in 1984 in a study on the asymmetric reduction of alkyl aryl ketones ⟨84CL251⟩. (*S*)-*N*-Benzoylcysteine was used as the chiral ligand for the lithium borohydride-mediated enantioselective reduction of ketones. The reagent is believed to be a borohydride complex having the structure shown in Scheme 223, containing a sulfur–boron bond.

Diborane reacts with certain heteroaromatic compounds to give ring cleaved thiol products (Scheme 224). In this reaction, the stable boron intermediates were the boroles shown; the reaction proceeds via a bridged alkanethioborate.

In an example published in 1992, the preparation of an alkylthioborane involves the reaction of

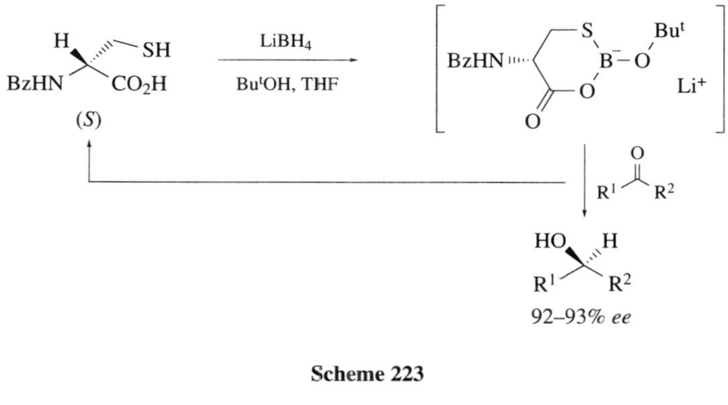

Scheme 223

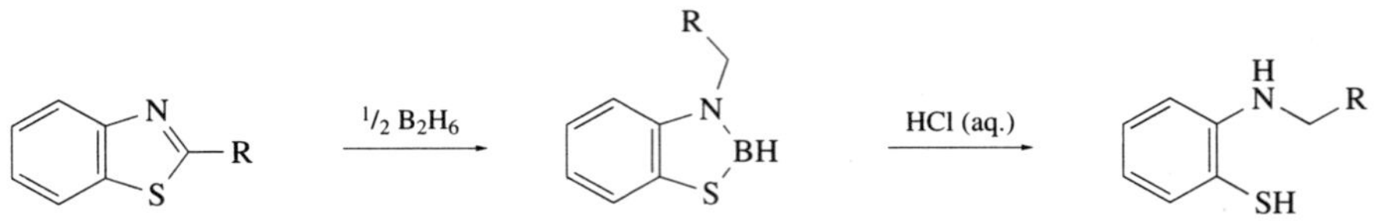

Scheme 224

bis(1,5-cyclooctanediylbornyl) sulfide with cyclic ethers under reflux ⟨92LA835⟩. *O,S*-bis(9-BBN) thioalkanols were produced in very high yields (Equation (358)). The preparation of a number of similar adducts was also discussed by the authors. Higher coordinated RSB functions appear to be unknown in the chemical literature.

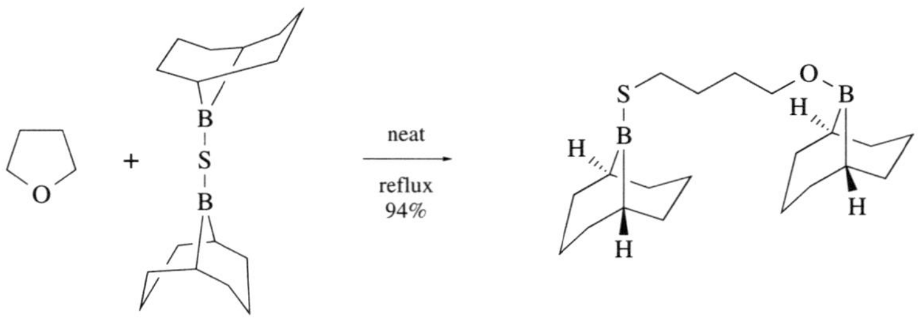

(358)

2.03.12 RS–METAL FUNCTIONS AND THEIR HIGHER COORDINATED DERIVATIVES

Metal thiolates are generally used as nucleophiles in reactions described variously above, and are often not isolated. Even so, a number of methods exist for their formation. Sulfinate salts on the other hand are widely used as stable derivatives of the parent acid.

2.03.12.1 M = Sodium

Sodium thiolates are readily prepared by addition of the thiol to an aqueous solution of sodium hydroxide at room temperature. An inert atmosphere is recommended to prevent oxidation of the thiol to disulfide, which can be a problem under alkaline conditions. In one example an unsymmetrical disulfide was prepared from the sodium thiolate so generated (Scheme 225) ⟨88OSC(4)235⟩. More complex thiolates derived from cysteine and glutathione have been prepared in a similar manner from the thiol and sodium hydroxide in methanol, and subsequently used to form sulfides of polyaromatic compounds ⟨62ZN(B)807, 76JMC1423⟩.

The thiolactone of cysteine has been treated with sodium methoxide to give the ring opened sodium thiolate methyl ester which was alkylated *in situ* to give the sulfide in yields of 75–96% (Scheme 226) ⟨69CJC3271⟩.

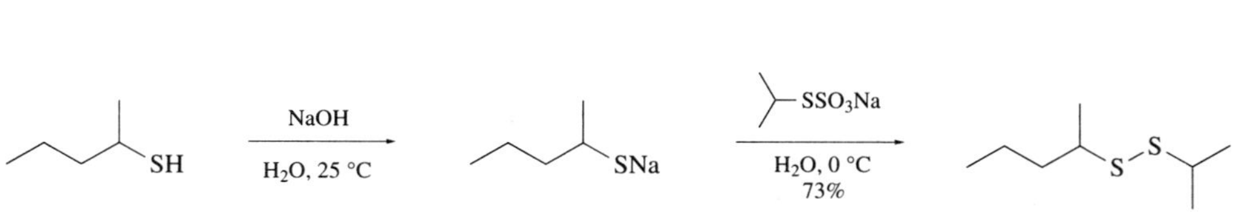

Scheme 225

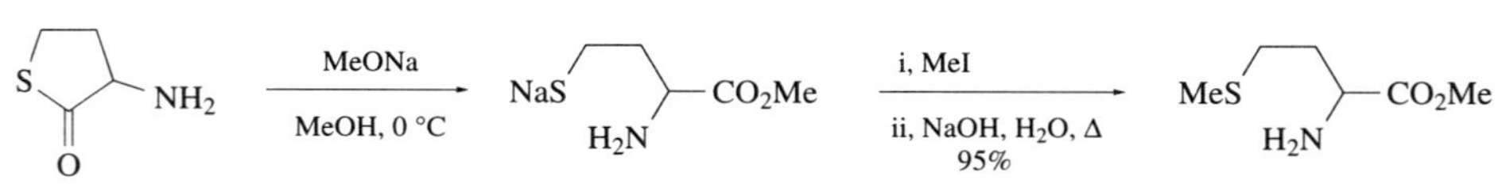

Scheme 226

Sodium thiolates have also been formed by the action of sodium metal on a thiol in ethanol. The high yields of the subsequent reactions demonstrate the near quantitative yield of the intermediate sodium thiolates ⟨63JOC593, 70JCS(C)2409, 77S795⟩. Sodium metal in liquid ammonia has been used for this reaction, also forming the sodium thiolates in near quantitative yields. Dimethyldisulfide may be used instead of methanethiol to give an identical yield of sodium methane thiolate ⟨78S359⟩.

Sodium hydride in THF effectively reacts with thiols to form the thiolates at −5 °C ⟨88OS(67)108⟩. Alternatively DMF can be used as solvent, in this case providing good yields of the products from subsequent reaction of the thiolate with bromobenzaldehydes (Scheme 227) ⟨86SC565⟩.

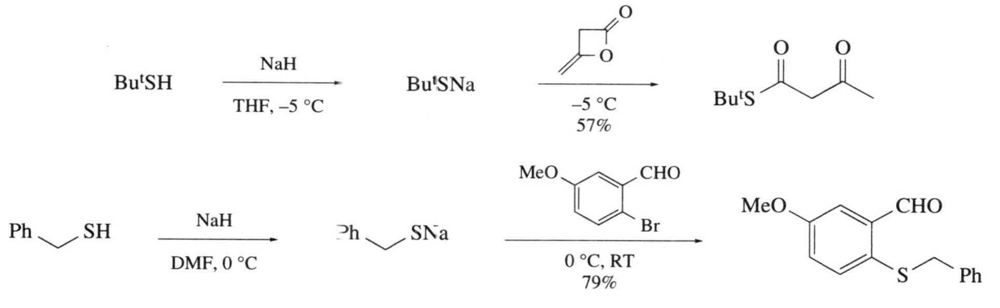

Scheme 227

Displacement of a sulfinyl group by a metal thiolate or alkoxide has proved a successful method for synthesis of sodium sulfinate. For example, 2-sulfonyl pyridine-*N*-oxides react with sodium ethoxide in refluxing ethanol to produce sodium sulfinates in quantitative yields. The displacement of the sulfinyl group could be equally well performed using a thiolate nucleophile. Use of acetonitrile as solvent simplifies isolation of the sulfinate to a straightforward filtration ⟨86H(24)3019⟩. Similar yields of sodium sulfinates of up to 16 carbon chain length are observed in the displacement of phthalimidomethyl sulfones by sodium ethoxide (Scheme 228). Displacement of the sulfinate is as high yielding with sodium thiolates, regeneration of starting material after an oxidation step then being possible ⟨77S795, 78CPB1837⟩. The displacement of a sulfinate from the *β*-sulfonyl propionitrile, again using the corresponding sodium thiolate, leads to good yields (70–75%) of the sodium sulfinate (Equation (359)) ⟨63JOC593⟩.

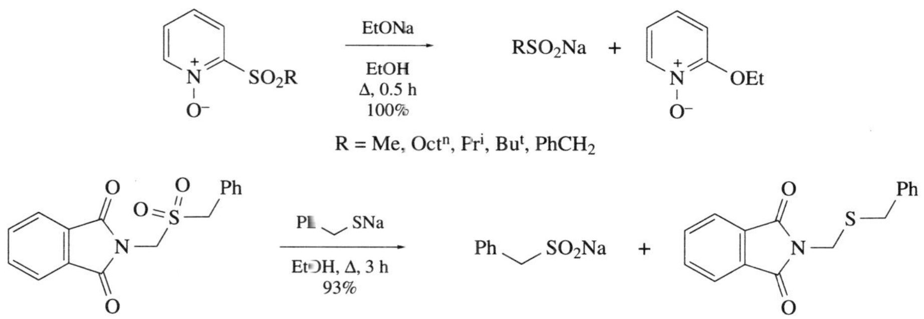

Scheme 228

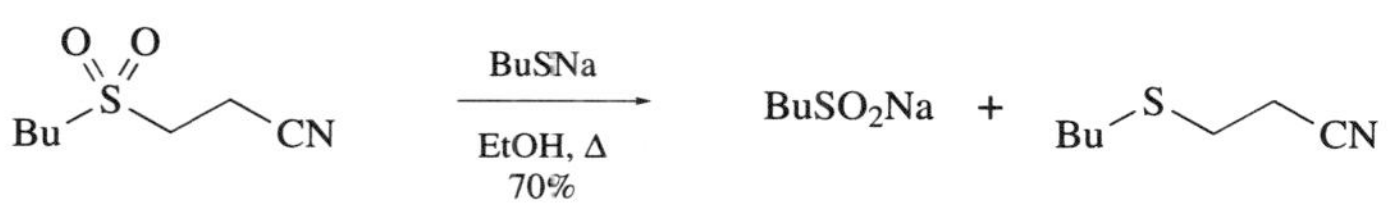

(359)

Sodium sulfinates are also formed in the reduction of sulfonyl chlorides with sodium sulfite in aqueous sodium hydroxide solution at 30–40 °C ⟨57RTC129⟩. The disodium salt of 1,4-butanedisulfinic acid was formed in the reduction of the corresponding 1,4-disulfonyl chloride with sodium sulfite and sodium bicarbonate solution at 45–50 °C to give the free diacid after acidification (Scheme 229) ⟨59JA5430⟩.

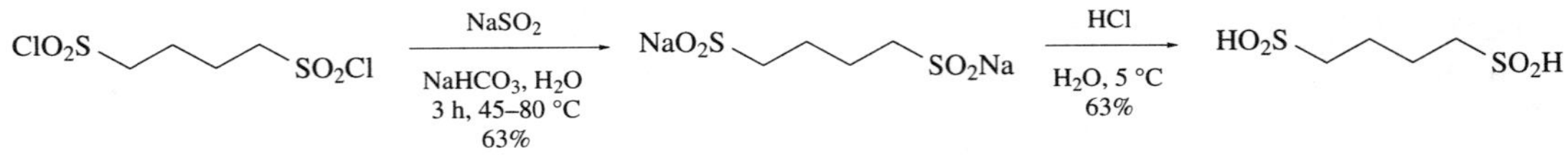

Scheme 229

2.03.12.2 M = Lithium

Lithium thiolates have been prepared by treatment of thiols with methyllithium in methanol and subsequently used in the ring opening reaction of thiirane-1,1-dioxide. Cleavage of the sulfur–carbon bond occurs to yield a lithium sulfinate product (Equation (360)). Primary, secondary, and tertiary alkane thiols gave yields of 69–85% of the metal sulfinates. Methyllithium has also given a lithium sulfinate by carbon–sulfur bond cleavage in an α-hydroxy sulfone. Acidification produced the sulfine in 80% yield (Scheme 230) ⟨84TL5287⟩.

The reaction of excess butyllithium with benzyl thiol at −5 °C produces a lithium dianion, which can be trapped with various electrophiles in high yields (Scheme 231). The use of TMEDA decreases reaction time from 4 hours to 2.5 hours ⟨74AG(E)202⟩.

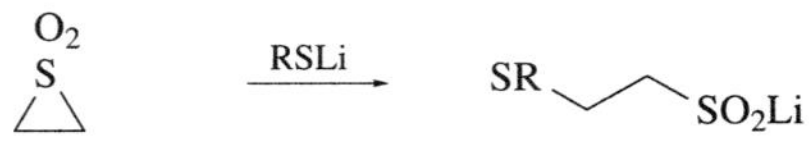

(360)

R	*Yield* (%)
Me	85
Et	75
Pr^i	69
Bu^t	69

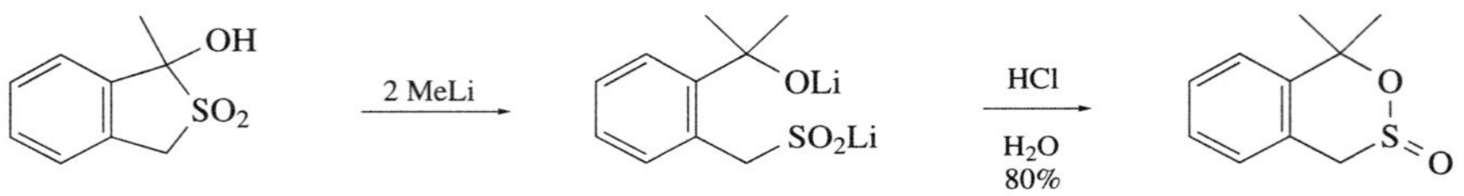

Scheme 230

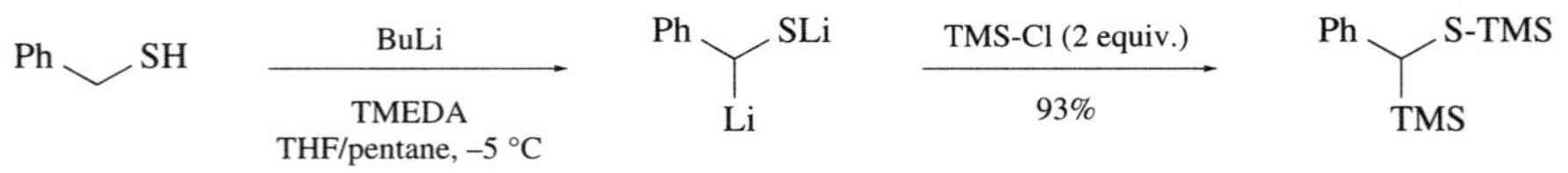

Scheme 231

2.03.12.3 M = Potassium

Treatment of trichloromethanesulfonyl chloride with potassium cyanide in sulfur dioxide furnished the potassium sulfinate in 91% yield. The compound was used in the preparation of trichloromethane sulfonyl bromide (Equation (361)) ⟨65JOC38⟩.

Alkyl sulfonates are widely used as alkylating agents due to their excellent leaving group ability. Although it is the alkylated product which is usually required, a side product is of course the metal sulfonate (Equation (362)) ⟨64JOC3055⟩.

$$Cl_3CSO_2Cl \xrightarrow[SO_2,\ 91\%]{KCN} Cl_3CSO_2^-\ K^+ \qquad (361)$$

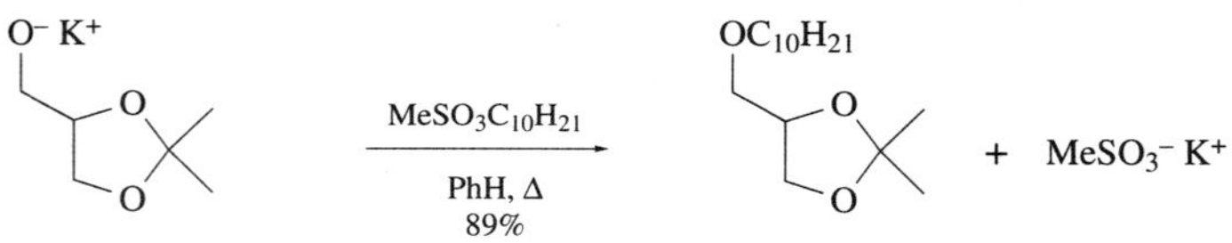

(362)

2.03.12.4 X = Magnesium

Magnesium thiolates have been prepared by the slow addition of an alkanethiol to ethylmagnesium bromide in THF. Again, these reactive intermediates are not normally isolated. The reactions in which they have been used proceed in up to quantitative yields ⟨85CC715⟩.

One of the most general methods for formation of magnesium sulfinates, and therefrom the sulfinic acids, is by addition of an alkyl Grignard reagent to sulfur dioxide. Indeed, this was the method used for synthesis of the first stable sulfinic acid, 1-dodecane sulfinic acid (Scheme 232) ⟨48JOC822⟩. Magnesium 2,3-dimethylbutane sulfinate, prepared via the Grignard reagent, was used to form the sulfonyl chloride by treatment with chlorine ⟨54JOC1165⟩. A 1979 publication recommends the use of an excess (≈10 equiv.) of sulfur dioxide to prevent formation of sulfoxide side products which arise from the reaction of the Grignard reagent with the newly formed sulfinate ⟨79JOC160⟩.

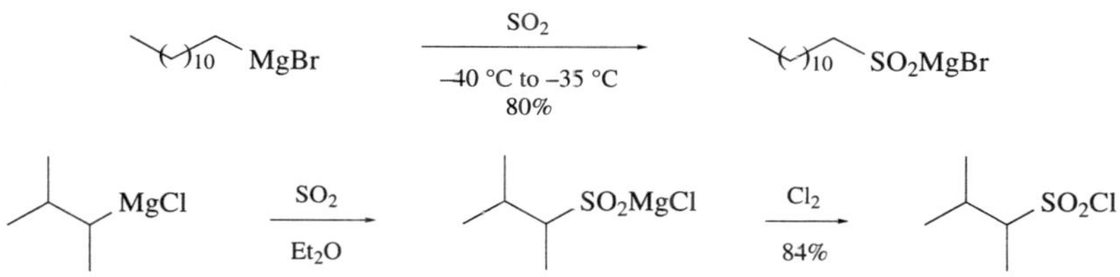

Scheme 232

2.03.12.5 M = Copper

A cuprous dithiolate has been used in the preparation of an aryl alkyl sulfide from the aryl bromide in greater than 73% yield. The copper thiolate was formed by reaction of butanethiol with cuprous oxide (Cu_2O) in refluxing ethanol, in essentially quantitative yield (Equation (363)) ⟨73OSC(5)107⟩.

$$\mathrm{BuSH} \xrightarrow[\text{EtOH, }\Delta\text{, 12 h; 97\%}]{\mathrm{Cu_2O}} \mathrm{(BuS)_2Cu} \qquad (363)$$

2.03.12.6 M = Zinc

Zinc thiolates can be prepared by reductive cleavage of disulfides using zinc dust in aqueous hydrochloric acid (Equation (364)). The zinc thiolates were treated with an aqueous solution of *p*-toluene sulfonic acid to give the product as its amine salt. Yields depend upon the quality of the starting disulfide, but are in the range 75–90% for cysteine derivatives ⟨62JA3887⟩.

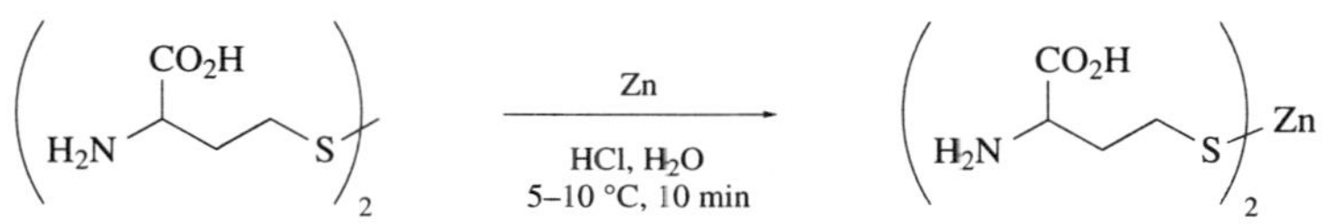

(364)

The reaction of a dithiocarbonate with zinc chloride in a neutral, aqueous solvent system forms the corresponding isothiocyanate in reasonable yield (65%). A by-product of this reaction is a zinc thiolate (Scheme 233) ⟨73OSC(5)223⟩.

The zinc salts of disulfinic acids have been formed by the action of zinc chloride on sodium sulfinates. The insoluble zinc salts precipate from the aqueous solution ⟨59JA5430⟩.

2.03.12.7 M = Mercury

Mercury salts of sulfur compounds are well known due to the thiophilic nature of mercury. For example, the deprotection of *S*-trityl derivatives proceeds in high yields to give the mercury thiolates

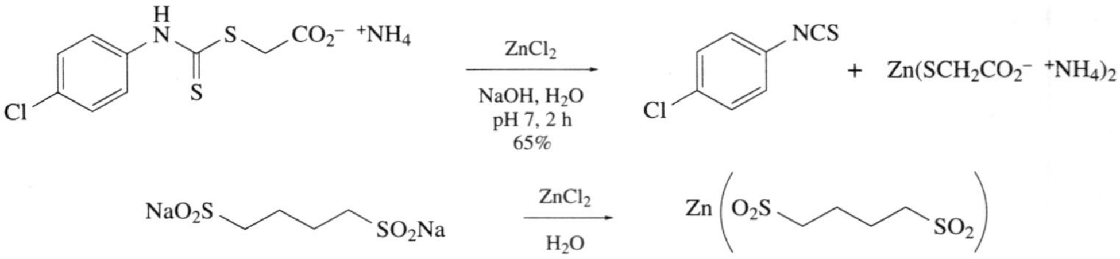

Scheme 233

on treatment with mercuric chloride ⟨64JOC3561, 65JOC33⟩. The selective removal of *S*-trityl groups in cysteine-containing peptides to form the mercury thiolates leaves *S*-diphenylmethane *N*-Z and *N*-BOC protecting groups unaffected (Equation (365)). The free thiols were obtained by bubbling hydrogen sulfide through a solution of the mercuric thiolate in DMF ⟨65JA4922, 78CPB1576⟩.

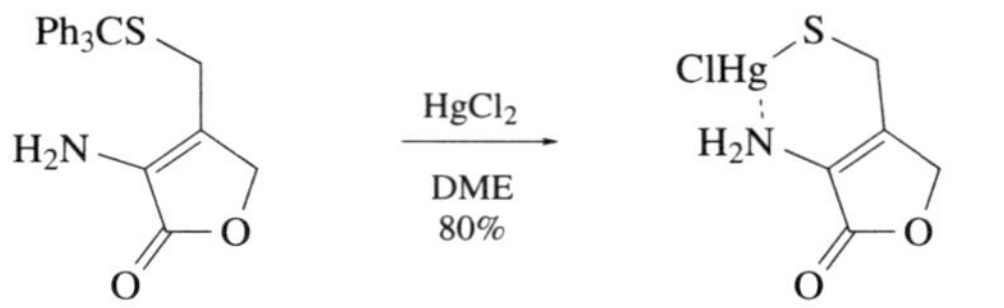

(365)

S-Methoxybenzyl and *S*-*t*-butyl groups have also been deprotected to form the mercuric thiolates, in this case using mercury trifluoroacetate. The synthesis of a diuretic drug, Mercaptomerin, involves the formation of a sulfur–mercury bond, by reaction of an alkyl mercury compound with a sodium thiolate ⟨50JA3595, 58JOC1449⟩.

The addition of mercuric chloride to a sulfur-containing nitrone in aqueous acetonitrile furnishes a mercuric thiolate by carbon–sulfur bond cleavage and hydrolysis (Scheme 234) ⟨75JCS(P1)1925⟩.

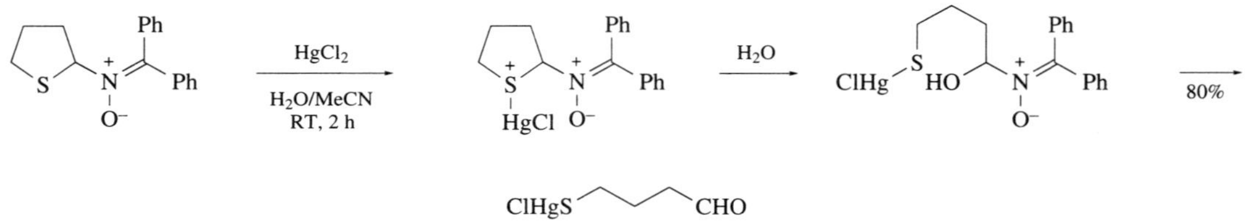

Scheme 234

2.03.12.8 M = Gold

It has long been known that gold thiolates are readily prepared from thiols in high yields. Trimethyl gold, prepared *in situ*, reacts with thioglycolic acid to produce the gold thiolate in 94% yield at −65 °C after recrystallization (Scheme 235) ⟨48JA550⟩.

Diethyl gold bromide reacts directly with 2-aminoethanethiol to give the gold thiolate, again in good yield (Equation (366)).

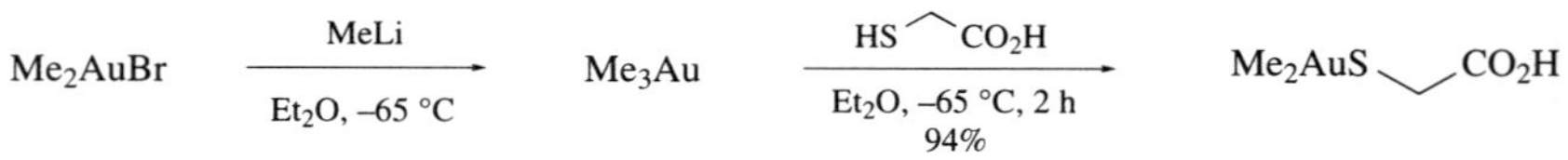

Scheme 235

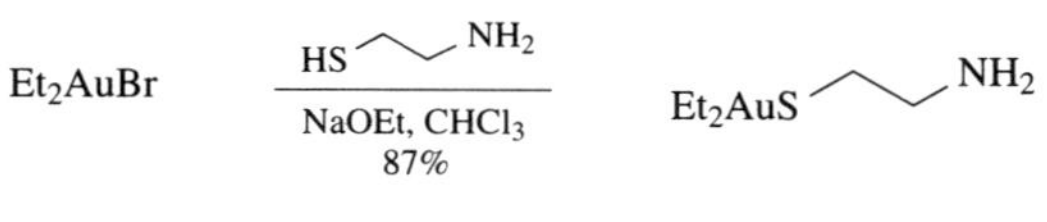

(366)

2.03.12.9 M = Silver

Silver salts have been used in a similar manner to mercury salts in the deprotection of cysteine derivatives, selectively removing *S*-trityl groups in ca. 80% yield and leaving *S*-diphenylmethane, *O*-Bu^t and *N*-Z groups unaffected. The free thiol is again released upon treatment of the silver thiolate with hydrogen sulfide ⟨62JA3887, 66JOC2179⟩.

2.04 Alkyl Chalcogenides: Selenium- and Tellurium-based Functional Groups

TADASHI KATAOKA and MITSUHIRO YOSHIMATSU
Gifu Pharmaceutical University, Japan

2.04.1 ALKANESELENOLS (RSeH) AND ALKANETELLUROLS (RTeH)

Reactions leading to the synthesis of alkaneselenols can be classified roughly as (a) those in which selenium is directly introduced into the organic molecule with the formation of selenols, and (b) those in which selenium-containing organic compounds are reduced to selenols by one of a number of reducing agents.

Methods of the first type (Equation (1)) involve the substitution of alkyl halides ⟨55JOC1019, 57JCS799⟩ by alkali selenides. The yields are satisfactory in all cases. The reactions of elemental selenium with Grignard reagents ⟨69JCS(A)890, 71JOC1700⟩, organolithiums ⟨73JCS(D)1716, 83JOC4289⟩ or trialkylaluminum compounds ⟨78JOC2735⟩ give alkaneselenols in good yields (Scheme 1). These reactions are carried out under an atmosphere of dry nitrogen or dry argon and are always accompanied by the formation of dialkyl diselenides, because the alkaneselenols are easily oxidized. In most cases, the alkaneselenols are not isolated but oxidized to the diselenides or converted to other more stable compounds.

$$R^1X + SeH^- \longrightarrow R^1SeH \qquad (1)$$
$$X = \text{halogen, } R^2SO_4, R^2SO_3$$

$$RMgX + Se \longrightarrow RSeMgX \xrightarrow{H^+} RSeH$$
$$RLi + Se \longrightarrow RSeLi \xrightarrow{H^+} RSeH$$
$$R_3Al + Se \longrightarrow R_2AlSeR \xrightarrow{H^+} RSeH$$

Scheme 1

In the second classification, selenium-containing organic compounds such as diselenides (Equation (2)) and selenocyanates (Equation (3)) are easily reduced to selenols using hypophosphorus acid ⟨64ANC1768, 65JMC845, 66JOC1202, 85T4793⟩, lithium aluminum hydride ⟨63MI 204-01, 94JOC1011⟩, sodium-liquid ammonia ⟨63RTC68, 66CI(L)379, 67JCS(A)2014⟩, sodium borohydride ⟨65JOC2454, 68JMC446, 70JOC3086⟩, and Rongalite (formaldehyde sodium bisulfite addition compound) ⟨61ACS938⟩.

$$RSeSeR \underset{[O]}{\overset{[H]}{\rightleftharpoons}} RSeH \qquad (2)$$

$$RSeCN \xrightarrow{[H]} RSeH \qquad (3)$$

The selenolate ions (RSe^-) obtained by the above methods are generally employed without isolation of the selenols and are used for further syntheses as the sodium or other salts. This procedure has some advantages, because alkaneselenols are toxic and notorious for their foul odor, and must be handled with considerable caution in a well-ventilated hood. Alkanetellurols are synthesized by the reaction of aluminum telluride with alcohols (Equation (4)) ⟨69JCS(D)1023, 72JCP(57)4434⟩. Tellurols are much more sensitive to oxygen than the corresponding selenols, and are readily oxidized to ditellurides. Therefore, metal tellurolates are generated from the reaction of elemental tellurium with organometallic reagents and employed directly for subsequent reactions (Equation (5)).

$$Al_2Te_3 + ROH \xrightarrow{300\text{–}350\,^\circ C} RTeH \qquad (4)$$

$$R^1M + Te \longrightarrow R^1TeM \longrightarrow R^1TeR^2 \qquad (5)$$
$$M = MgX, Li$$

2.04.2 DIALKYL SELENIDES (R^1SeR^2), DIALKYL TELLURIDES (R^1TeR^2) AND THEIR HIGHER COORDINATED DERIVATIVES

2.04.2.1 Alkyl Selenides and Alkyl Tellurides

Selenides are the largest class of selenium compounds. The aliphatic selenides are subdivided into symmetrical and unsymmetrical dialkyl selenides and aryl alkyl selenides. Alkylation of sodium selenide (Scheme 2) is the most useful synthetic method for symmetrical dialkyl selenides ⟨27JA2011, 56JOC1217, 67NKK991, 73JA197⟩. The yields are very good in all cases.

$$\mathrm{Se} \xrightarrow[\text{i, ii or iii}]{\text{reduction under basic conditions}} \mathrm{Na_2Se} \xrightarrow{\mathrm{RX\ or\ R_2SO_4}} \mathrm{R_2Se}$$

i, Rongalite, NaOH (aq.) — $R = Me$, 88%
ii, $NaBH_4$ — $R = Pr^n$, 54%
iii, Na-liq. NH_3 — $R = PhCH_2$, 86%

Scheme 2

Unsymmetrical dialkyl selenides are synthesized by alkylation of alkaneselenolates with alkyl halides (Equation (6)) ⟨65JMC845, 69JCS(A)890, 92S933⟩. The selenolates can be prepared by reduction of diselenides (see Sections 2.04.1 and 2.04.7) with Rongalite in basic media, hydrazine ⟨84S439⟩, sodium borohydride ⟨64JMC229⟩ or lithium aluminum hydride ⟨94JOC1011⟩, by alkaline hydrolysis or reduction of selenocyanates, or by treatment of alkaneselenols with sodium ethoxide or alkylmagnesium iodide (Scheme 3) ⟨09CB49⟩. The latter method can be advantageously used for two or more substitutions at the same time (Equations (7) and (8)) ⟨82JCR(S)212, 89CC1789, 89T1611⟩.

$$\mathrm{R^1SeH} \xrightarrow[\text{ii, } \mathrm{R^2X}]{\text{i, base}} \mathrm{R^1SeR^2} \qquad (6)$$

$$\mathrm{R^1SeSeR^1} \xrightarrow{\text{reducing agent}} \mathrm{R^1Se^-} \xrightarrow{\mathrm{R^2X}} \mathrm{R^1SeR^2}$$

$$\mathrm{R^1SeCN} \xrightarrow{\text{NaOH (aq.)}} \mathrm{R^1Se^-}$$

$R^1 = R^2 = Me, Et, Pr, Pr^i, Bu^n, Bu^t$, 5–90%

$$\mathrm{RX} + \mathrm{ArSe^-} \xrightarrow[58\text{–}72\%]{} \mathrm{ArSeR}$$

Scheme 3

(7)

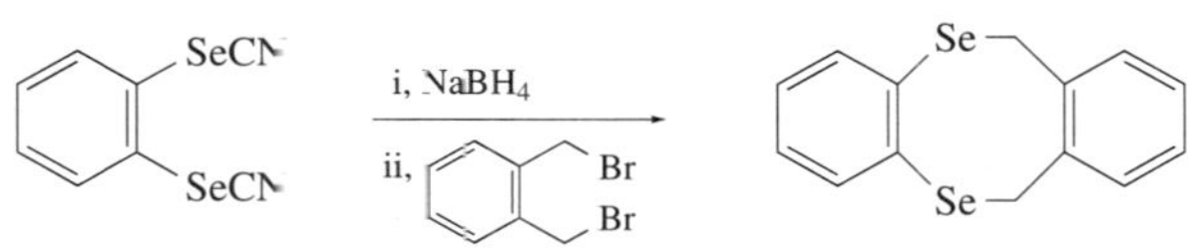

(8)

Aryl alkyl selenides are similarly prepared by the reaction of areneselenolate ion or diaryl diselenides with a variety of organic compounds such as alkyl halides, methylates or acetates ⟨77JOC1773, 79CL873, 79JA6638, 79JA6648, 90CL927⟩. Other useful methods for the preparation of unsymmetrical selenides include the use of an alcohol and a selenol in the presence of an acid (Equation (9)) ⟨84TL3625⟩, or an alcohol, *N*-phenylselenophthalimide (NPSP) and tributylphosphine (Scheme 4) ⟨76JOC1485, 81JOC1215⟩.

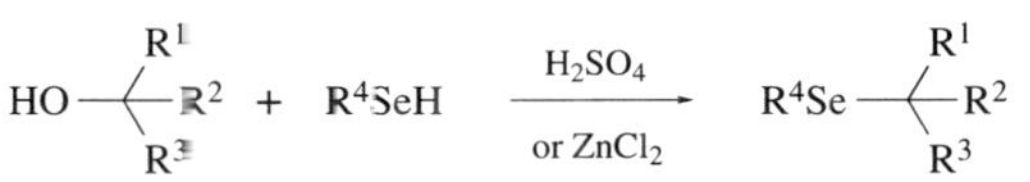

(9)

$R^1 = Bu, R^2 = R^3 = Bu$, 88%
$R^1 = Ph, R^2 = R^3 =$ alkyl, 65–87%
$R^1 = CH{=}CH_2, R^2 = R^3 =$ alkyl or Ph, 45–80%

Symmetrical and unsymmetrical tellurides are prepared in similar ways to the corresponding selenides ⟨75CB314, 75JHC423, 75JOM(88)175, 83JOM(253)31, 83JOM(255)61, 83OM1106⟩. The reduction of elemental tellurium can be achieved by Rongalite in aqueous sodium hydroxide solution ⟨14CB1269⟩ or by sodium in liquid ammonia ⟨6[illegible]RTC68⟩ and then the sodium telluride is allowed to react with

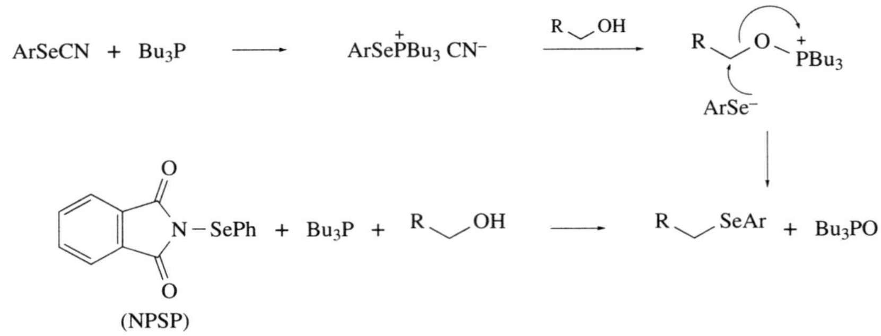

R = Me(CH$_2$)$_{10}$, 94%; R = *n*-hexyl, 85%; R = Ph, 97%; R = propargyl, 93%

Scheme 4

alkyl halides or alkyl sulfate to give the symmetrical tellurides (Scheme 5). Elemental tellurium inserts into the carbon–metal bonds of RLi ⟨70BSB353, 70BSB367, 70BSB383⟩. Since the RTeM compounds (M = MgBr, Li) thus formed are unstable, they are not isolated but treated with alkyl halides to give the unsymmetrical tellurides (Equation (10)). The RTe^- functional group is also generated by reduction of ditellurides with sodium borohydride or lithium aluminum hydride and then treated with alkyl halides (Scheme 6). The latter method is attractive for alkyl aryl tellurides ⟨83CB1001, 83JCS(D)2143, 86JCS(D)2003⟩.

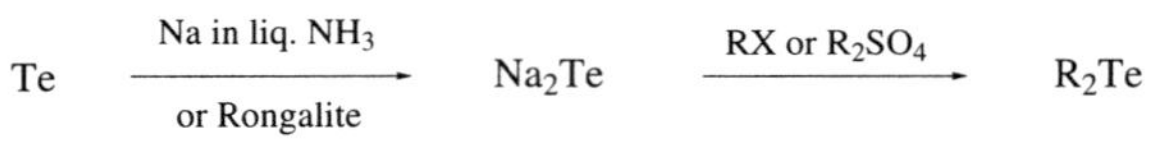

Scheme 5

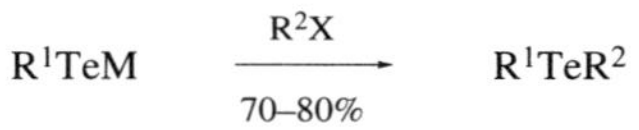 (10)

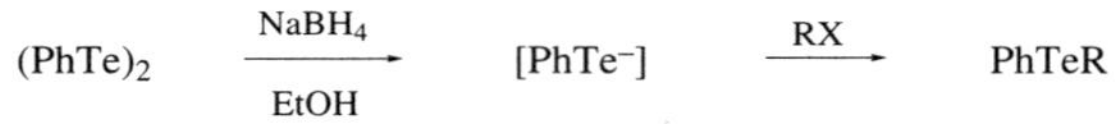

Scheme 6

Unsymmetrical selenides bearing a functional group possess the potential for a number of interesting synthetic transformations ⟨B-78MI 204-01, 78T1049, 79ACR22, B-84MI 204-01, 88TL6787⟩. Addition of RSeX (or RSe^+) to alkenes is very useful for the synthesis of the *β*-functionalized selenides (Scheme 7) ⟨B-86MI 204-01, B-87MI 204-01, 91COS(7)515⟩. *β*-Chloro- and bromoselenides are prepared using areneselenenyl chlorides or bromides ⟨77JOC1776, 77TL3909, 78JOC4885⟩. Addition of benzeneselenenyl bromide to terminal alkenes initially produces *anti*-Markovnikov adducts which, on standing, isomerize to the thermodynamically more stable Markovnikov adducts (Scheme 8) ⟨77JOC2950⟩. The reaction with internal alkenes usually gives a mixture of Markovnikov and *anti*-Markovnikov adducts, and their ratio is influenced by both the electronic and steric effects of the substituents ⟨78T2869, 80CJC1030⟩. Although the adducts are often unstable at room temperature, the formation of regioselective adducts can be achieved if the addition is carried out below $-50\,^{\circ}C$ (Equation (11)) ⟨78TL4977⟩. Fluoroselenenylation is achieved using benzeneselenenyl fluoride generated *in situ* from benzeneselenenyl bromide and silver fluoride (Equation (12)) ⟨89CC1881, 89CL1235⟩, diphenyl diselenide and XeF_2 ⟨93TL703⟩, NPSP and triethylamine tris(hydrofluoride) ⟨90TL663, 90TL2127⟩, or phenylseleniumtrifluoride ⟨93TL703⟩.

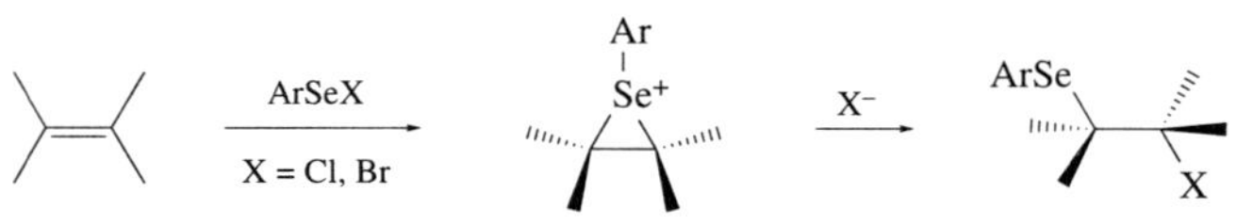

Scheme 7

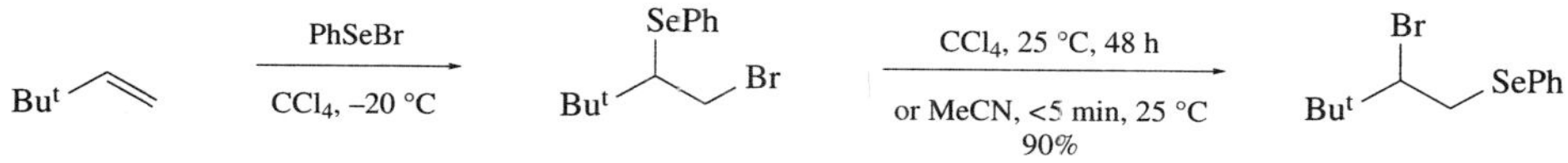

Scheme 8

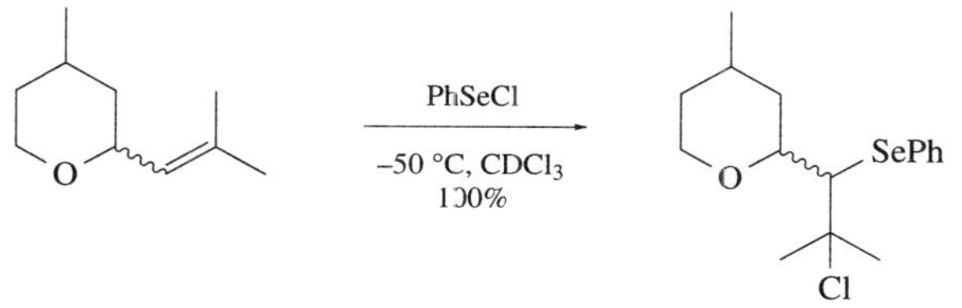

(11)

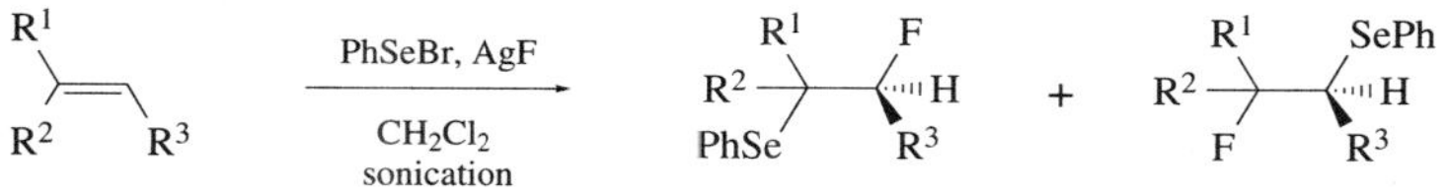

(12)

R^1–R^3 = $(CH_2)_4$, 57%
R^1 = Bu^n, R^2 = R^3 = H, 59%
R^1 = R^3 = Me, R^2 = H, 49%

Hydroxyselenides are important intermediates for the synthesis of allyl alcohols, α-seleno carbonyl compounds, alkenes, epoxides, and halohydrins, and can be prepared from selenocyanates (Scheme 9) ⟨88S133⟩, epoxides (Equation (13)) ⟨81TL2491, 88CC1283, 92SL317⟩ or cyclic ethers (Equation (14)) ⟨88S175, 89S434, 90H(31)889, 92SL320⟩.

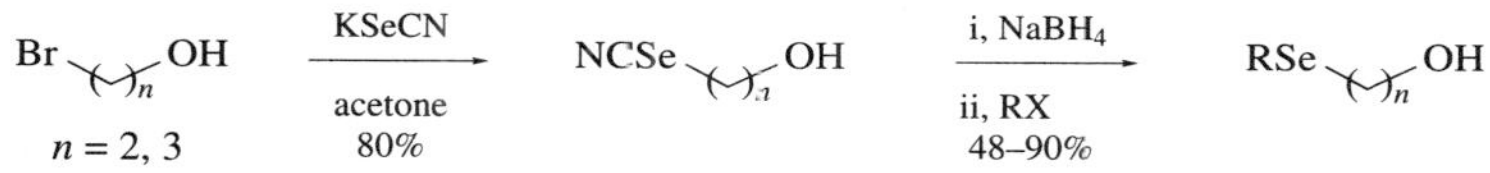

R = Et, 2,4-$(NO_2)_2C_6H_3$, $PhCH_2$, EtO_2CCH_2, 3-pyridylCH_2

Scheme 9

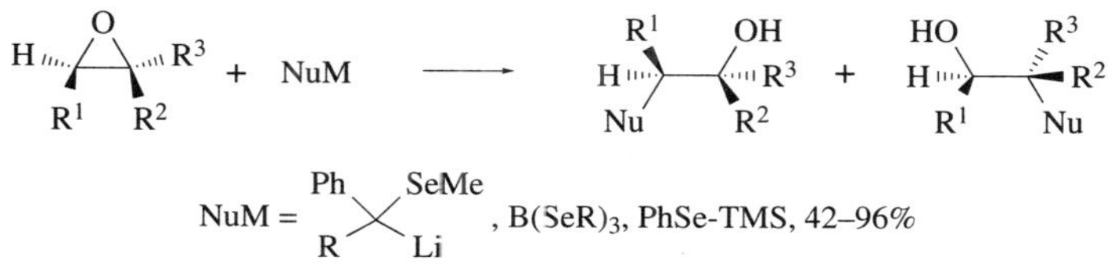

(13)

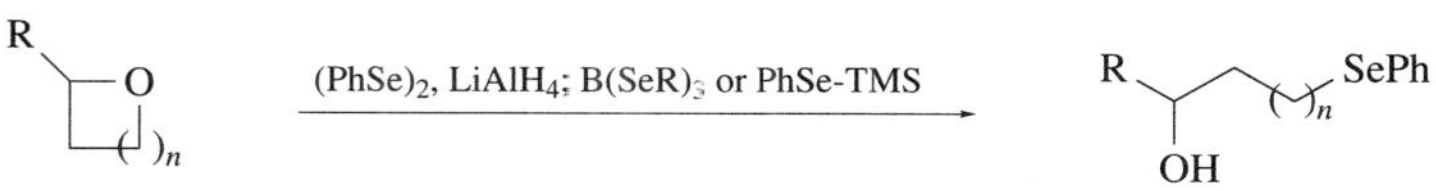

(14)

n = 1–3, 65–86%

β-Hydroxyselenides are available through the reaction of selenium-stabilized carbanions with carbonyl compounds (Equation (15)) ⟨69AG(E)450, 74AG(E)804, 74AG(E)806, 75AG(E)350, 75AG(E)700, 75CC790, 77LA846, 78JOC4252, 89T2005, 89T2023⟩ or through hydroxyselenation of alkenes. The hydroxyselenenylation of alkenes has been achieved by a variety of procedures: (1) using benzeneselenenyl chloride in a water-acetonitrile mixture (1 : 5) (Equation (16)) ⟨80CC412, 85T5301⟩, (2) using benzeneselenenyl bromide followed by TFA ⟨79TL2697⟩, (3) using NPSP or *N*-Phenylsuccinimide (NPSS) and 2–3 equivalents of water in the presence of an acid (Equation (17)) ⟨79JA3704⟩, (4) using benzeneselenenic acid generated *in situ* by the reaction of diphenyl diselenide, benzeneseleninic acid and water ⟨78JOC1689, 78JOC1697⟩ or by the reduction of seleninic acid with hypophosphorus acid (Equation (18)) ⟨78TL3967⟩.

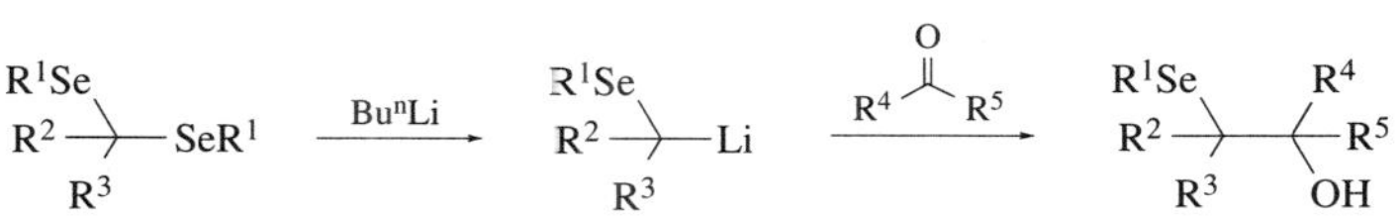

(15)

R^1 = Ph, R^2 = R^3 = Me, R^4 = Ph, R^5 = H, 58%

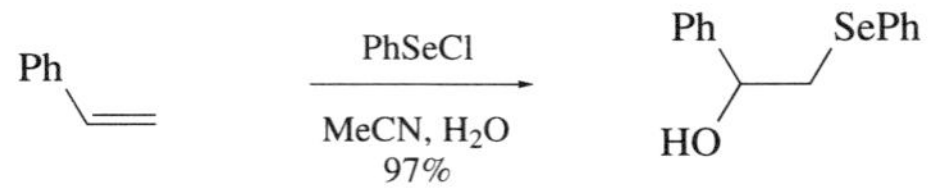

(16)

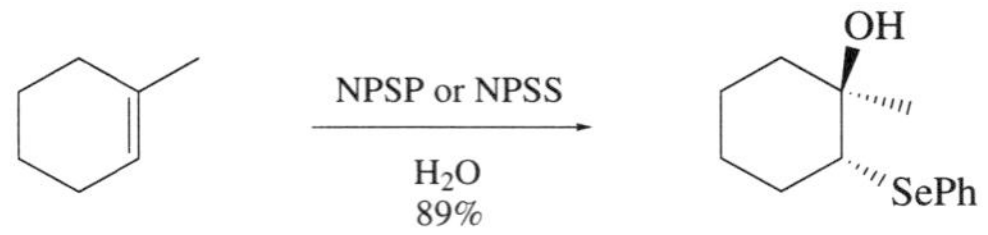

(17)

(18)

$PhSeO_2H$, PhSeSePh, $MgSO_4$, H_2O, $RSeO_2H$, H_3PO_2, 63–91%

Alkoxyselenenylation of alkenes has been achieved by the reaction of an alcohol and either potassium selenocyanate and copper(II) chloride (Equation (19)) ⟨78JCS(P1)1273, 79JCS(P1)1206⟩, or phenylselenocyanate in the presence of metal salts (Equation (20)) ⟨80JOC1953, 81JOC3021⟩. Alternatively, pyridineselenenyl bromide in an alcohol will effect this transformation (Equation (21)) ⟨85JCS(P1)373⟩.

(19)

R = Me, Et, Pr^n, Pr^i, Bu^t, 15–99%

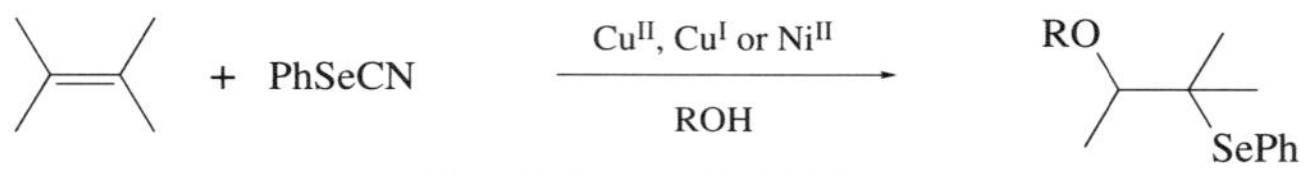

(20)

R = alkyl, Ac, H 80–100%

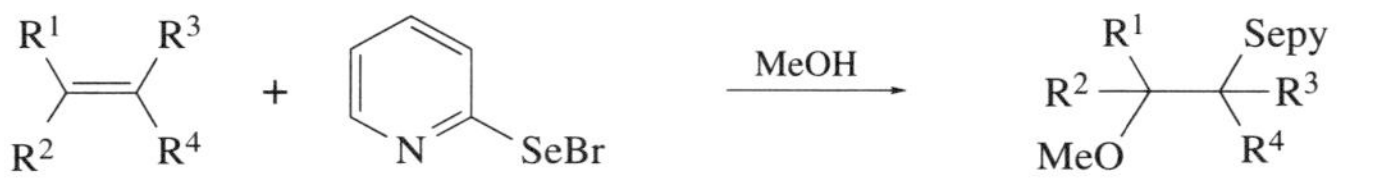

(21)

(pySeBr)

R^1 = Me, Et, Pr^n, OAc; R^2 = H, Me; R^3 = H, Me, Pr^n; R^4 = H, Me, Pr^n, 36–100%
cycloalkenes, 70–93%

α-Selenoaldehydes or -ketones can be prepared directly from enolizable aldehydes or ketones and areneselenenyl chloride ⟨73JA6137, 82TL2105, 84JOC3796⟩ or *N*,*N*-dialkylbenzeneselenenamides (Scheme 10) ⟨81TL3561, 82TL1557, 84TL1983⟩. More efficient methods for the synthesis of these compounds are selenylation of enol acetates ⟨58HCA593, 73CC695, 74JOC428, 75JA5434⟩, enol ethers (Equation (22)) ⟨79HCA1406, 79S982, 80JA1404, 80JA7950⟩, silyl enol ethers (Scheme 11) ⟨77S874, 82TL2105⟩ or enamines (Scheme 12) ⟨80TL4417, 82TL1361, 82TL1557, 84TL1983⟩. Metal enolates can be selenenylated with areneselenenyl halides ⟨88OS(67)23⟩ or diselenides (Equation (22)) ⟨84JOC3796⟩ directly, or allowed to react stepwise with selenium and then alkyl halides ⟨80TL3643⟩ to give α-selenoketones (Scheme 13). The reaction using selenium metal has an advantage for large-scale laboratory synthesis, in that selenium metal is much less expensive than the selenenyl halides. α-Seleno carbonyl compounds can also be obtained from the inversely charged fragments, that is by acylation of α-lithioselenides with DMF, acid chlorides or anhydrides ⟨76TL453⟩. α-Selenocarboxylic acids or esters are similarly obtained by carbonation or alkoxycarbonylation of α-lithioselenides (Equation (23)) ⟨76TL453⟩. ω-Selenocarboxylic acids are prepared by the reactions of lactones with selenolates on the basis of the property that selenolate ions are soft nucleophiles and cleave the C—O bond of esters by a preferential nucleophilic attack on the alcoholic carbon (Equation (24)) ⟨78TL5087, 79CL873, 81JOC2605⟩. The nucleophilicity of a selenolate ion is altered by a counter ion or a solvent. Liotta has proposed the following nucleophilicity gradient for benzeneselenolate ion: NaSePh/18-crown-6/THF > NaSePh/HMPA/THF > LiSePh/HMPA/THF > LiSePh/THF or ether > PhSeSePh/$NaBH_4$/THF/EtOH ⟨84ACR28⟩.

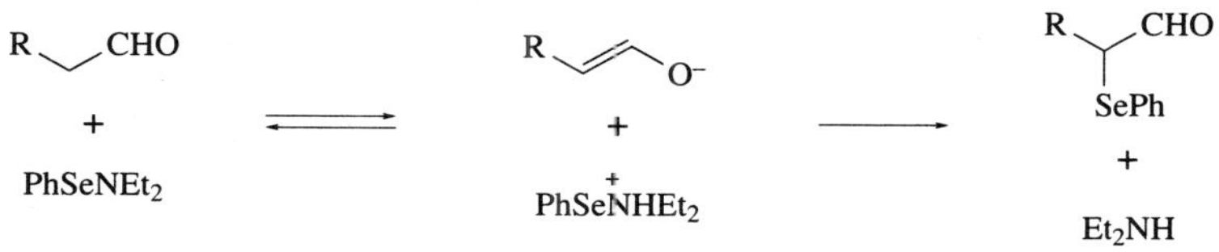

R = $PhCH_2CH_2$, 78%
R = $MeCO(CH_2)_3$, 72%

Scheme 10

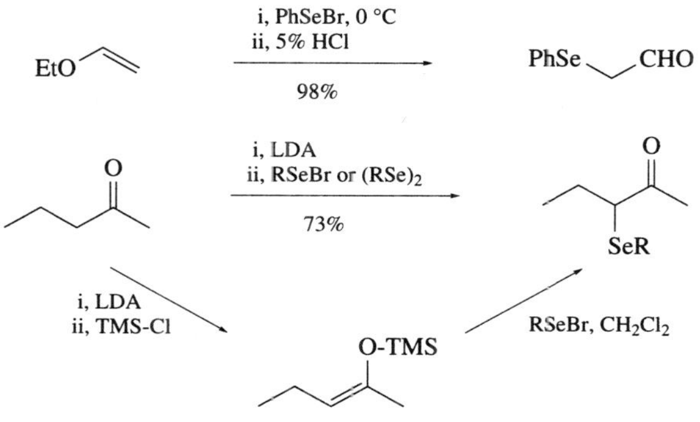

(22)

LDA = Lithium diisopropylamide

Scheme 11

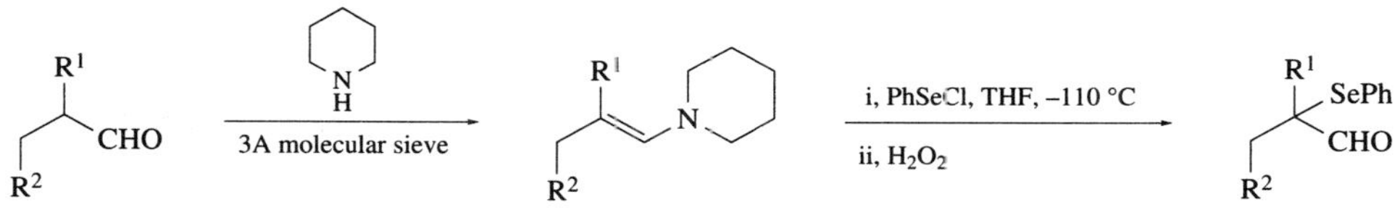

R^1 = H, R^2 = Ph, 99%
R^1–R^2 = $(CH_2)_4$, 99%

Scheme 12

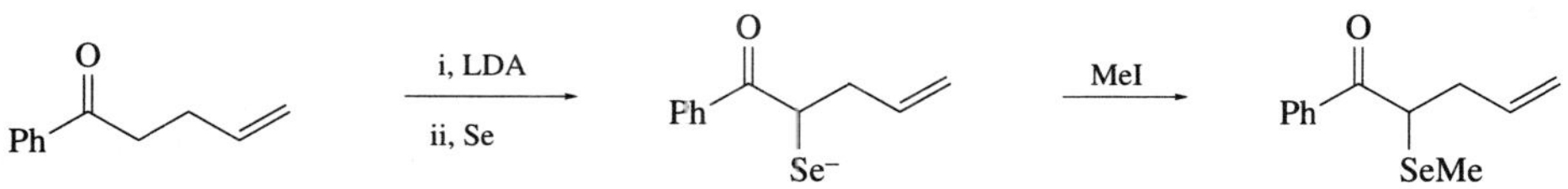

Scheme 13

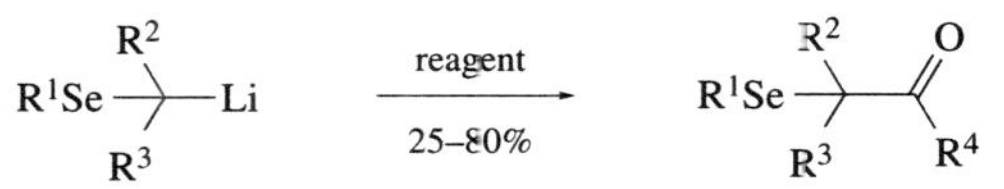

(23)

R^1 = Me, Ph; R^2, R^3 = H, Me, C_6H_{13}, Bu^t– 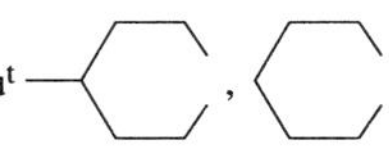

R^4 = H, Ph, C_5H_{11}, Me, OMe, OH
Reagent: DMF, acid chloride, acid anhydride, MeOCOCl, CO_2

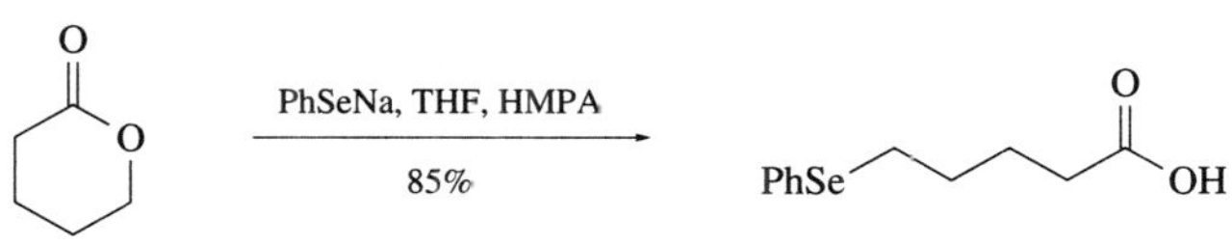

(24)

Selenocyanation is an important method for introducing a carbon substituent to alkenes. Reactions of enamines with phenyl selenocyanates in ethanol proceed very quickly via *trans*-addition to give α-amino-β-(phenylseleno)nitriles in high yields (Equation (25)) ⟨82TL1361⟩. Ketene acetals undergo selenocyanation to give α-cyano-β-phenylselenoacetals (Equation (26)) ⟨82CL1733⟩. Silyl enol ethers are less reactive than enamines and ketene acetals, but do react with phenyl selenocyanate in the presence of tin(II) chloride ⟨81CL1715⟩. Simple alkenes do not react with phenyl selenocyanate, except in the presence of tin(IV) chloride, and give predominantly the Markovnikov product (Equation (27)) ⟨82CC871⟩.

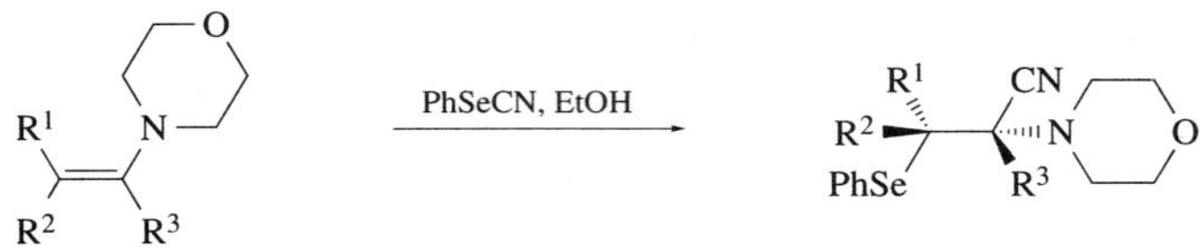

(25)

$R^1 = R^3 = H, R^2 = Me$, 95%
$R^1 = R^3 = H, R^2 = Et$, 98%
$R^1 = R^2 = Me, R^3 = H$, 84%
$R^1 = H, R^2 = Me, R^3 = Ph$, 81%

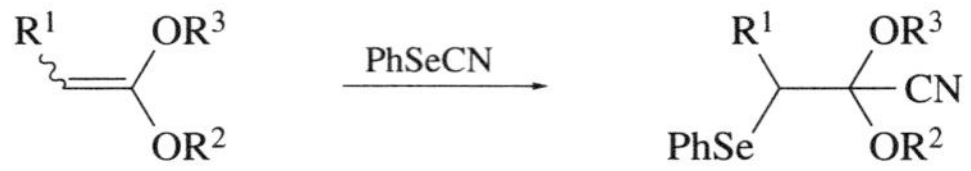

(26)

$R^1 = H, R^2 = R^3 = Et$, 73%
$R^1 = H, R^2 = Me, R^3 = TBDMS$, 64%

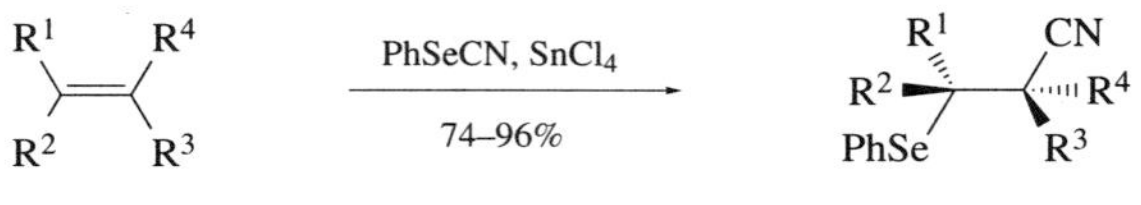

(27)

R^1–$R^4 = (CH_2)_4$, R^1–$R^4 = (CH_2)_3$, R^1–$R^4 = (CH_2)_5$, $R^1 = H, Pr^n$, $R^3 = H, Bu^n$, $R^4 = H, Et, Pr^n$

Electrophilic selenenylation can be applied to the conversion of alkenes to selenides bearing various β-nitrogen substituents such as amido, amino, azido, and nitro groups. Toshimitsu *et al.* have reported β-amidoselenenylation of alkenes with benzeneselenenyl chloride in aqueous acetonitrile in the presence of trifluoromethanesulfonic acid (Scheme 14) ⟨81JOC4727⟩. The amides are formed via incorporation of acetonitrile and hydrolysis of the intermediate imine. Both mono- and disubstituted alkenes may be used, and nitriles such as propionitrile, butyronitrile, and ethyl cyanoacetate are also effective. Similar products are provided by anodic oxidation of a mixture of an alkene and diphenyl diselenide in acetonitrile ⟨80TL3827⟩. Introduction of a sulfonamido group has been achieved by treatment of alkenes with a mixture of diphenyl diselenide and chloramine-T, followed by reduction with sodium borohydride (Scheme 15) ⟨78JCS(P1)1682⟩. The attacking electrophile is believed to be the intermediate (**1**) but has not been isolated. β-Azido selenides can be prepared from the reaction of alkenes with selenenyl bromides followed by treatment with sodium azide in either trifluoroethanol or DMF (Scheme 16) ⟨79TL2697⟩. β-Nitroselenenylation has been carried out by treatment of alkenes with benzeneselenenyl bromide and silver nitrate. Yields are enhanced in the presence of mercury(II) chloride, which prevents the formation of β-hydroxy selenides (Equation (28)) ⟨82TL4733⟩. β-Nitro-α-silylselenides can be prepared by the nitroselenenylation of vinylsilanes ⟨84JOC3235⟩.

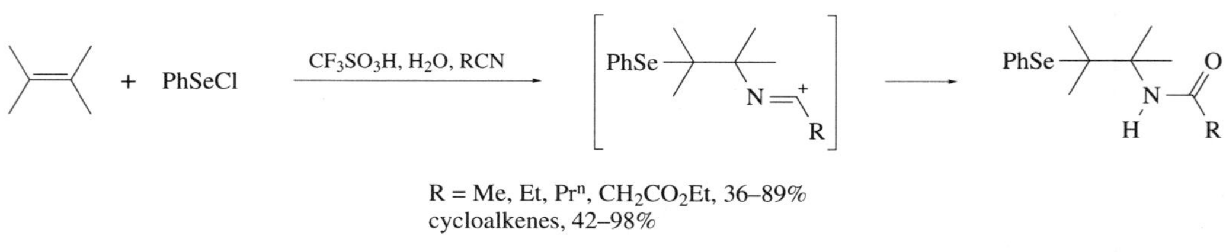

$R = Me, Et, Pr^n, CH_2CO_2Et$, 36–89%
cycloalkenes, 42–98%

Scheme 14

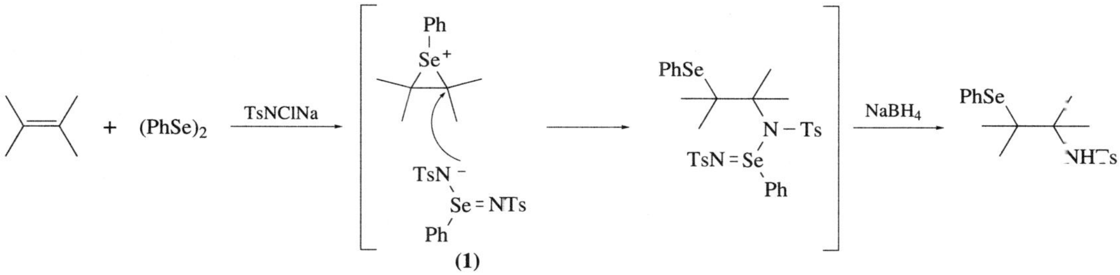

Scheme 15

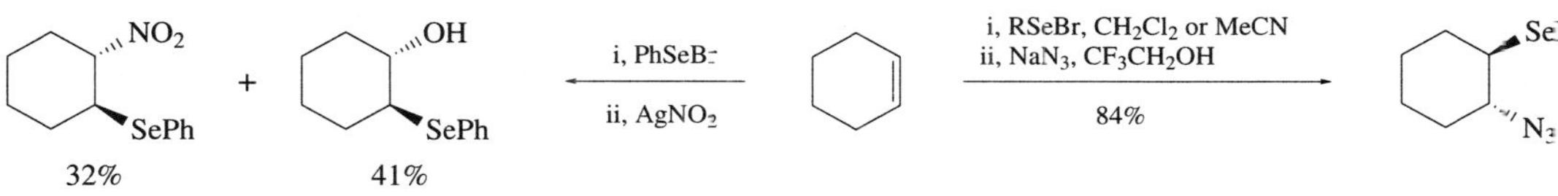

Scheme 16

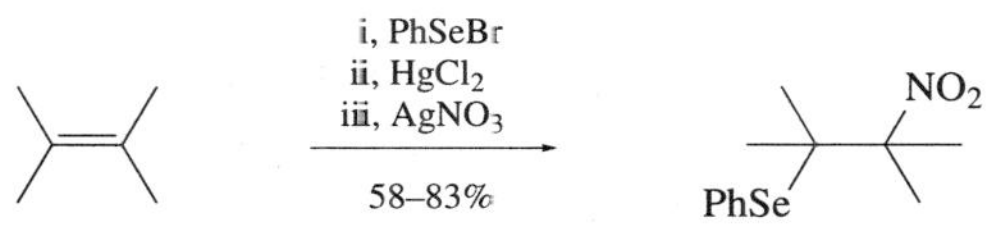 (28)

alkenes: cyclohexene, cyclopentene, (*E*)- and (*Z*)-4-octene, (*Z*)-2-pentene, 1-hexene, indene, styrene

Selenosulfonates alone are not sufficiently reactive towards alkenes, but they do react in the presence of boron trifluoride etherate (Scheme 17). The reactions proceed stereospesifically *anti* and the products are sulfones rather than sulfinate esters ⟨80TL2215, 81JOC3249⟩. Gancarz and Kice found that phenylselenenyl sulfonates decompose homolytically upon exposure to UV or visible light, or when heated ⟨80TL4155, 81JOC4899⟩. Gancarz and Kice, and Back *et al.* independently studied the selenosulfonation of alkenes ⟨80TL2215, 81JOC3249, 81JOC4899⟩, alkynes ⟨81TL5111, 82CC438, 83JOC3077, 83JOC4776⟩, and allenes (Equation (29)) ⟨82TL5373, 85T4739⟩. The reactions occur with *anti*-Markovnikov orientation, are accelerated by AIBN (2,2′-azobisisobutyronitrile) and suppressed by inhibitors such as 2,6-di-*t*-butyl-4-cresol. Similar radical addition was investigated in the reaction of alkenes with (phenylselenothioperoxy)benzoate (PhCOSSePh) ⟨88JCS(P1)575, 89JCS(P1)1927⟩.

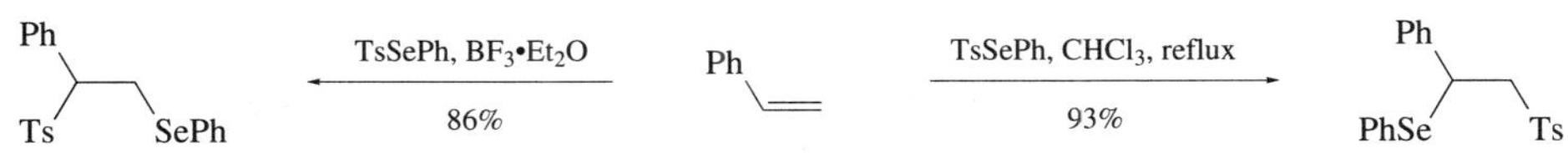

Scheme 17

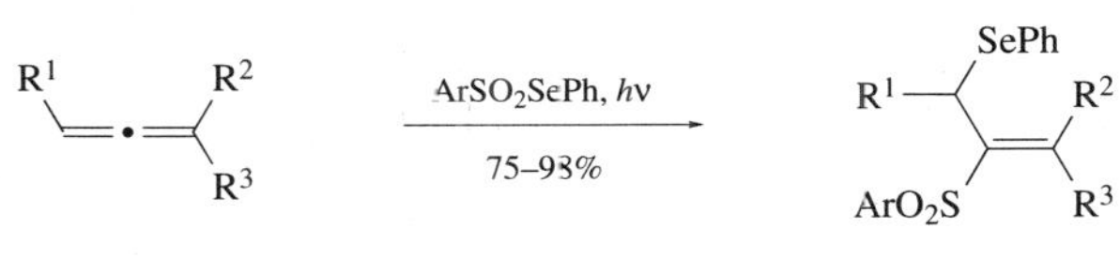 (29)

R^1 = H, Me; R^2, R^3 = H, Me, Et, *n*-Hex, Ph

The Se—Se bond is very weak (bond energy 184.2 kJ mol^{-1}) and is easily homolytically cleaved on irradiation with a tungsten lamp. Although seleno radicals do not readily add to alkenes, they can be trapped by other radicals. This forms the basis of an elegant method of thioselenenylation of alkenes ⟨92JOC111⟩, developed by Sonoda *et al.*, in which irradiation of a mixture of an alkene, diphenyl disulfide, and diphenyl diselenide through pyrex with a tungsten lamp without solvent affords *β*-thioseleno-alkanes (Scheme 18). Terminal alkenes afford regioselectively 1-(phenylthio)-2-(phenylseleno)alkanes as the sole product, and cycloalkenes give *trans*-isomers stereoselectively. A conjugated diene undergoes 1,4-thioselenation. The regioselectivity is attributable to the high reactivity of PhS· toward the C=C bond and the ease with which diphenyl diselenide can be captured by carbon radicals. 1,2-Bis(selenides) have been prepared using dimethyl diselenide and tin(IV) chloride (Equation (30)) ⟨92TL4629⟩. Diphenyl diselenide is inefficient for this electrophilic addition.

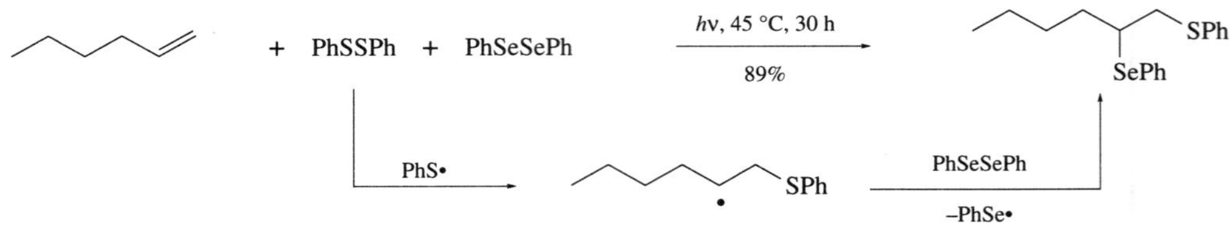

Scheme 18

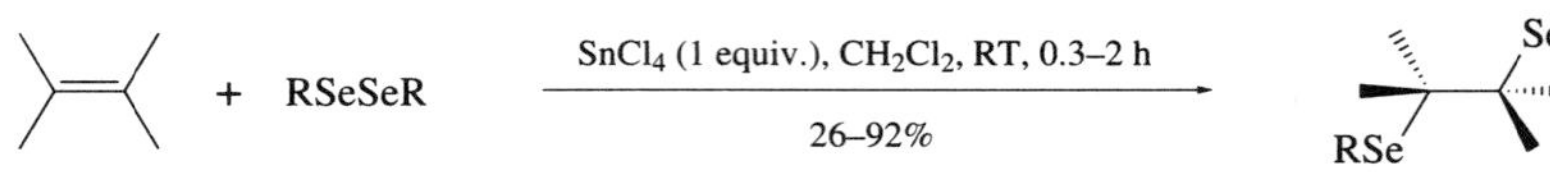

(30)

When organoselenium electrophiles are used in addition reactions to alkenes bearing an internal nucleophile such as OH, CO_2H, SH, SCOR, NHR or a double bond, in an appropriate position, the intramolecular cyclization of the type shown in Scheme 19 proceeds smoothly ⟨B-86MI 204-02, 87TL4343⟩. This methodology makes it possible to synthesize not only *N*-, *O*-, and *S*-heterocycles but also carbocyclic compounds ⟨78CC441, 81T4097, B-87MI 204-02⟩; this selenium-based ring closure has been termed "cyclofunctionalization" by Clive ⟨77CC725⟩.

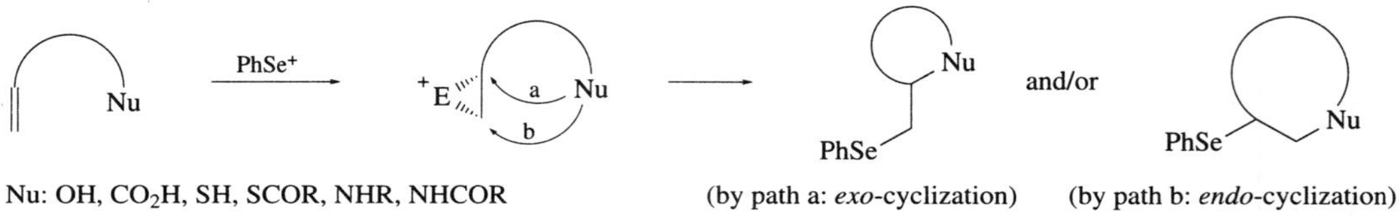

Scheme 19

2.04.2.2 Alkyl Selenoxides and Alkyl Telluroxides

Alkyl selenoxides are easily formed by oxidation of the corresponding selenides. Alkyl selenoxides bearing a hydrogen on the β-carbon are generally unstable and undergo *syn*-β-elimination under very mild conditions to give alkenes (Scheme 20) ⟨78T1049⟩. The isolable selenoxides are those without a β-hydrogen, for instance dimethyl selenoxide, aryl methyl selenoxides, and aryl benzyl selenoxides ⟨28JCS2293, 42JCS570⟩, or those without an electron-withdrawing group on the α-carbon. Selenoxides with an α-electron-withdrawing group cannot be isolated because they undergo the Pummerer reaction to give selenoacetals ⟨82TL4371, 94CPB811⟩. Selenoxides with hydrogen bonding between the seleninyl oxygen and an intramolecular hydroxy or amino group are stable and can be isolated (Equation (31)) ⟨80JOC274, 81CC546⟩. Allyl or propargyl selenoxides are difficult to be isolated because they easily undergo [2,3]-sigmatropic rearrangement rather than β-elimination to afford allylic alcohols or enones ⟨73TL1979, B-86MI 204-03⟩.

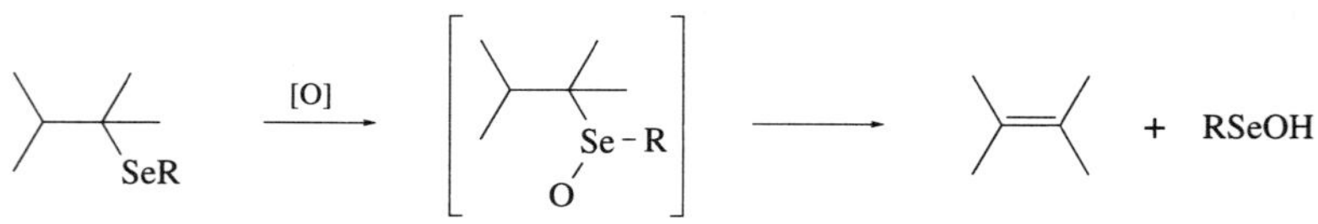

Scheme 20

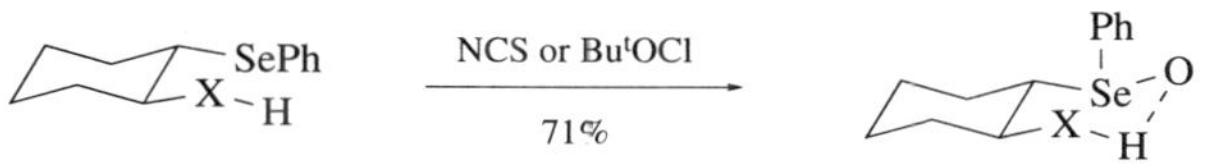

X = O, NOR (31)

The oxidation of selenides can be accomplished with a variety of reagents, including hydrogen peroxide, peroxycarboxylic acids, sodium periodate, ozone, 3-aryl-2-benzenesulfonyloxaziridines ⟨78T1049⟩. Hydrogen peroxide is cheap and commonly used for the selective oxidation of selenides containing a sulfide, amino or sulfinyl group. A solution of the selenide in THF or dichloromethane can be oxidized at lower than 0 °C ⟨75JOC2570⟩. Peracetic acid and mcpba are also used for oxidation at −78 °C in THF or dichloromethane ⟨51JA3196, 73JA6137⟩. Sodium periodate can be used to oxidise selenides in aqueous methanol, and is suitable for selenides bearing a formyl group ⟨71JOC2561, 73JA6137, 75JA5434⟩. Ozone can be used for the chemoselective oxidation of the selenide group in molecules containing double bonds or sulfide groups, at low temperature ⟨57JCS4547, 62JCS2089, 74TL2091⟩. Oxidation with oxaziridines is achieved in aprotic solvents at room temperature in good yields ⟨83TL1213⟩. An interesting indirect method is the treatment of selenides with NBS, NCS or *t*-butyl hypochlorite followed by hydrolysis of the intermediate adduct ⟨77CL835, 80JOC274⟩.

n-Alkyl phenyl telluroxides and cyclohexyl phenyl telluroxides are stable at room temperature, but s-alkyl phenyl telluroxides and other cycloalkyl phenyl telluroxides are unstable. Alkyl aryl telluroxides are prepared by addition of bromine in carbon tetrachloride to the alkyltellurides and subsequent alkaline hydrolysis of the tellurium dibromide. All the telluroxides are isolated as their hydrates (Equation (32)) ⟨80JOC274, 81CC277, 83JA2748⟩.

(32)

2.04.2.3 Alkyl Selenones and Alkyl Tellurones

The oxidation of selenium(II) compounds to selenium(VI) compounds is more difficult than the oxidation to selenium(IV), in contrast to the corresponding sulfur species. Therefore, while conversion of selenides to selenoxides can be achieved with a variety of oxidizing agents (see Section 2.04.2), selenones are difficult to prepare, requiring strong oxidizing agents and careful control of reaction conditions.

The isolable selenones are obtained by oxidation of stable selenoxides with ozone ⟨68ZAAC(360)293⟩ or by the direct oxidation of selenides (Scheme 21) ⟨46JA2671, 73JA4634, B-88MI 204-01, 88TL3269⟩. Alkyl aryl selenones are formed by oxidation of the selenides with mcpba or trifluoroperacetic acid ⟨68ZOB2509, 85CC569, 88TL3265⟩. When alkyl aryl selenides are oxidized with an excess of mcpba in an alcohol at room temperature, the selenones are not obtained; instead, an interesting displacement product is obtained (Scheme 22) ⟨83CC1501, 84CC426, 85JCS(P1)471, 86JCS(P1)1983⟩. The intermediates of the reactions are thought to be carbocations, because *β*-phenyl group migration and the ring-contraction have been observed. Similar reactions are observed in oxidation of alkyl phenyl tellurides with mcpba.

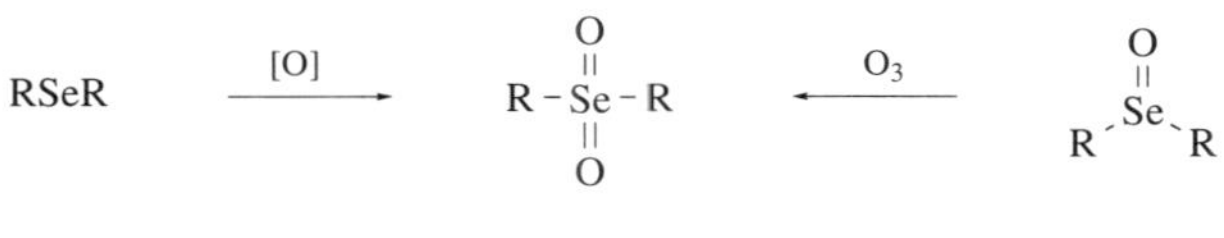

Scheme 21

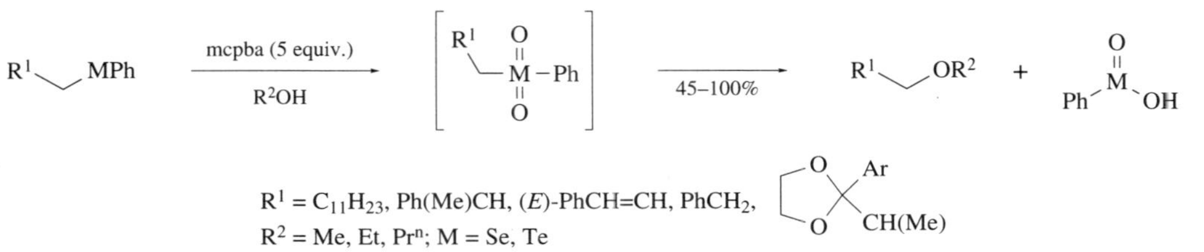

Scheme 22

The oxidation product of dimethyl telluride with hydrogen peroxide has been reported as dimethyl tellurone (**2**) ⟨20JCS889⟩, but was found to be a "hydroxy perhydrate" (**3**) ⟨38JCS341⟩. Diethyl telluride ⟨28JCS3179⟩, tellurocyclohexane ⟨28JCS321⟩, and telluroisochroman ⟨45JCS37⟩ provide similar "hydroxy perhydrate" products. Cava prepared the first tellurones to be definitely characterized by oxidation of alkyl aryl tellurides with sodium periodate (Equation (33)) ⟨81CC277⟩. Oxidation of tellurides with mcpba causes C—Te bond cleavage via tellurones as mentioned above ⟨85JCS(P1)471, 86JCS(P1)1983⟩.

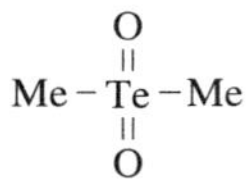

(2)

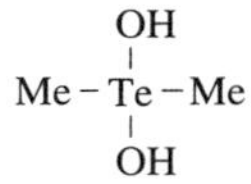

(3)

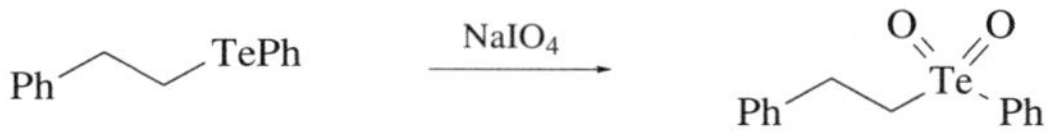

(33)

2.04.2.4 Alkyl Selenonium Salts and Alkyl Telluronium Salts

Alkylation of selenides gives trivalent selenium compounds, selenonium salts, which are more easily prepared than the corresponding sulfonium salts (Equation (34)). Kinetic studies show that selenides are 2.6 to 5.8 times more reactive than sulfides and the polarity of the solvent plays an important role in the alkylation ⟨83JCS(P2)1605⟩. The lone pair electrons of dialkyl selenides are more reactive than those of alkyl aryl selenides, and therefore the experimental conditions depend on the two substituents of the selenide. Trialkylselenonium halides are formed by treatment of dialkyl selenides with alkyl halides ⟨67NKK991, 69JA1649⟩. Selenonium salts with a functional group can also be obtained by alkylation with α-bromocarboxylic acids or α-bromoesters ⟨36BSF2310, 42JCS570, 68NKK83⟩. Alkylation of alkyl aryl selenides ⟨58CB2345, 67JOC2580, 68BSB153⟩ and cyclic selenides ⟨45JCS37, 87JOC1397, 92JCR(S)393⟩ also proceeds smoothly to give selenonium salts in good yields. Trialkyloxonium tetrafluoroborates (Meerwein reagents), dimethoxycarbenium tetrafluoroborate and methyl triflate have been used as more active alkylating agents ⟨74AG(E)274, 79TL4233, 94CPB811⟩, and alkyl halides can be activated by addition of silver tetrafluoroborate or silver perchlorate ⟨87JOC1397, 92JCR(S)393⟩. The exchange of counteranions in selenonium salts is possible by utilizing solubility differences between the selenonium salts or by taking advantage of the insolubility of silver salts ⟨67NKK991, 68NKK83⟩.

(34)

R^1, R^2 = alkyl or aryl group; R^3 = alkyl group, Me_3O^+, Et_3O^+, $(MeO)_2CH^+$
X = Cl, Br, $MeSO_4$, BF_4, ClO_4, FSO_4, CF_3SO_4

2.04.3 ALKANESELENENYL HALIDES AND THEIR HIGHER COORDINATED DERIVATIVES (RSeHal), SELENINYL HALIDES (RSe(O)Hal), SELENONYL HALIDES (RSe(O)$_2$Hal) AND CORRESPONDING TELLURIUM COMPOUNDS

Alkaneselenenyl halides are prepared by the treatment of selenocyanates or diselenides with a stoichiometric amount of halogen or a halogenating agent (Scheme 23) ⟨58JCS2939, 59ZOB942, 60ZOB4069, 67ZAAC(352)167⟩. When an excess of halogen is used, trihalides are formed as stable compounds. The halogenation of diselenides is very fast and must be carried out just before use, without isolation. The trihalides can be transformed to monohalides by heating under reduced pressure. Alkaneselenenyl chlorides are prepared from diselenides and chlorine ⟨50JCS1371⟩ or sulfuryl chloride (Equation (35)) ⟨88JCS(P1)1913⟩.

$$\text{RSeCN} \ \text{ or } \ \text{RSeSeR} \xrightarrow{X_2} \text{RSeX} \underset{-X_2}{\overset{X_2}{\rightleftharpoons}} \text{RSeX}_3$$

Scheme 23

$$\text{RSeSeR} + SO_2Cl_2 \longrightarrow \text{RSeCl} + SO_2 \qquad (35)$$

Alkanetellurenyl halides are generally unstable, but can be isolated as complexes with thiourea or selenourea (Equation (36)) ⟨59ACS2155, 63ACS1806, 66ACS123, 66ACS132⟩. Alkanetellurenyl trichlorides can be prepared by the treatment of ditellurides with halogens (Equation (37)) or by the addition of alkenes to tellurium tetrachloride (Equation (38)) ⟨81JA2715, 81TL1919⟩.

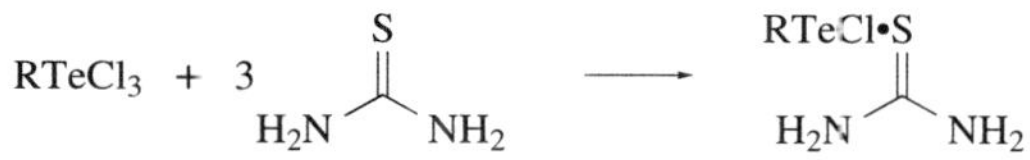

(36)

$$R_2Te_2 + X_2 \longrightarrow RTeX_3 \qquad (37)$$

(38)

R^1 = Me, Et, *n*-Oct; R^2 = H, Me; R^3 = H, D, Me

Alkaneseleninyl chlorides are formed by oxidation of the selenenyl chloride with ozone or nitrogen dioxide, or by treatment of the diselenide with chlorine followed by oxidation with ozone (Scheme 24) ⟨67ZAAC(352)167⟩. Alkanetellurinyl halides are formed by hydrolysis of tellurium trihalides with water but have not been isolated (Equation (39)) ⟨29JCS2076⟩.

Alkaneselenoyl halides and alkanetelluronyl halides have not been prepared.

$$\text{RSeCl} \xrightarrow{O_3 \text{ or } NO_2} \text{RSe(}\rightarrow\text{O)Cl} \xleftarrow[\text{ii, } O_3]{\text{i, } Cl_2} (\text{RSe})_2$$

Scheme 24

$$RTeX_3 \xrightarrow{H_2O} [\text{RTe(}\rightarrow\text{O)X}] \qquad (39)$$

2.04.4 ALKYL SELENOCYANATES (RSeCN) AND ALKYL TELLUROCYANATES (RTeCN)

Alkyl selenocyanates are prepared by the nucleophilic displacement of alkyl halides or sulfonates with selenocyanate anion in either organic solvents or water, in good yields (Equation (40)). The selenocyanate anion can also displace a dialkylborane group from trialkylboranes in the presence of iron(III) ⟨76CL785, 76CL1115⟩. A secondary or a tertiary alkyl group is transferred more easily than a primary group. Conversion of alcohols to alkyl selenocyanates has been achieved by reaction with triphenylphosphine and selenocyanogen ⟨79TL2251⟩. Numerous papers have reported the synthesis of alkyl selenocyanates ⟨79JMC572, 80SC595, 86JA1314⟩, including those with functional groups on the carbon skeleton, such as electron-withdrawing groups ⟨88JA8679, 88JCS(P1)1913⟩ and silyl groups (Scheme 25) ⟨88JA8671⟩.

$$R^1X + KSeCN \longrightarrow R^1SeCN + KX \qquad (40)$$

X = halogen, OSO_2R^2

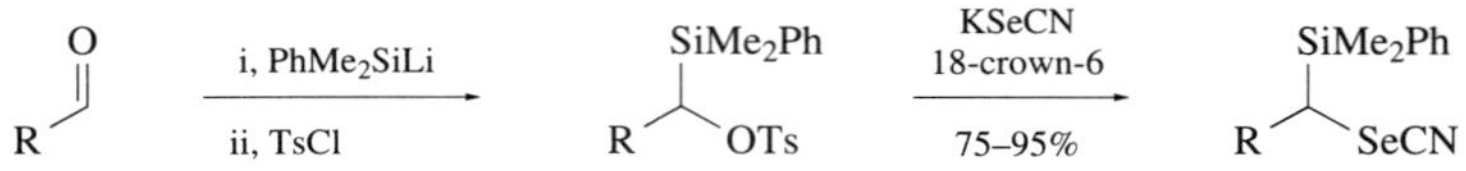

Scheme 25

Alkyl tellurocyanates are little known compared with selenocyanates. Alkali metal tellurocyanates are very unstable to water and decompose with liberation of elemental tellurium ⟨68CC1290⟩. However, solutions of alkali metal tellurocyanates can be prepared from elemental tellurium and alkali metal cyanides, in dry DMSO at 100 °C, under an inert atmosphere (Scheme 26) ⟨77JA1470⟩. Some benzyl tellurocyanates can be prepared using potassium tellurocyanate in DMSO ⟨81ACS(A)151, 81NKK472⟩. Benzyl tellurocyanates are also obtained by cyanation of tellurenyl halides with potassium cyanide (Equation (41)) ⟨81JOC4194⟩.

$$\text{Te} + \text{KCN or NaCN} \xrightarrow[\text{DMSO}]{100\,°\text{C}} [\text{KTeCN or NaTeCN}] \xrightarrow[60\%]{\text{ArCH}_2\text{Cl}} \text{ArCH}_2\text{TeCN}$$

Scheme 26

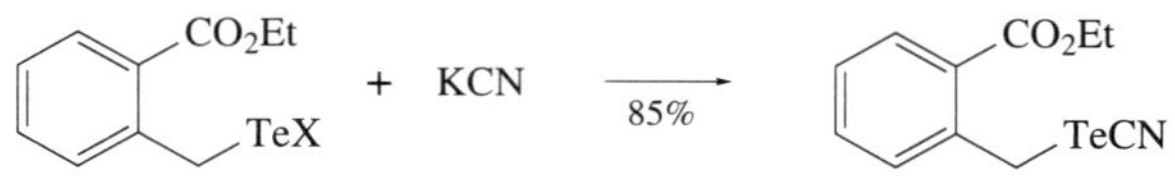 (41)

X = Cl, Br, I

2.04.5 FUNCTIONS BASED ON THE RSeOH UNIT AND ITS HIGHER COORDINATED DERIVATIVES—SELENENIC, SELENINIC AND SELENONIC ACIDS (RSeOH, $RSeO_2H$, $RSeO_3H$), AND CORRESPONDING TELLURIUM ACIDS

2.04.5.1 Alkaneselenenic Acids and Alkanetellurenic Acids

Alkaneselenenic acids are unstable and are presumed intermediates in reactions such as selenoxide *syn*-elimination and oxidation of selenols and diselenides. Selenenic acids disproportionate into the corresponding diselenides and seleninic acids (Equation (42)). Methaneselenenic acid has been generated *in situ* by reduction of methaneseleninic acid with hypophosphorus acid (Equation (43)) ⟨78TL3967⟩. 2-Aminoethaneselenenic acid has been formed by oxidation of the corresponding diselenide with hydrogen peroxide or mcpba (Equation (44)) ⟨66JOC3396, 82TL3319⟩.

$$3\,\text{RSeOH} \longrightarrow \text{RSeSeR} + \text{RSeO}_2\text{H} + \text{H}_2\text{O} \qquad (42)$$

$$\text{MeSeO}_2\text{H} + \text{H}_3\text{PO}_2 \longrightarrow [\text{MeSeOH}] \qquad (43)$$

$$(\text{H}_2\text{NCH}_2\text{CH}_2\text{Se})_2 \xrightarrow[\text{or mcpba, 97–99\%}]{\text{H}_2\text{O}_2} \text{H}_2\text{NCH}_2\text{CH}_2\text{SeOH} \qquad (44)$$

Alkanetellurenic acids are also unstable and undergo thermal disproportionation to tellurinic acids and tellurols, which are oxidized to ditellurides (Scheme 27).

$$2\,\text{RTeOH} \longrightarrow \text{RTeO}_2\text{H} + \text{RTeH} \left[\xrightarrow{[\text{O}]} (\text{RTe})_2\right]$$

Scheme 27

2.04.5.2 Alkaneseleninic Acids and Alkanetellurinic Acids

Alkaneseleninic acids are generally prepared by oxidation of the corresponding selenols, diselenides or selenocyanates. The simplest and most reliable preparation involves oxidation of diselenides with concentrated nitric acid (Scheme (28)) ⟨60ZAAC(305)88, 65ZAAC(335)289, 65ZAAC(338)195, 67ACS652⟩. The product of this reaction is a crystalline hydronitrate salt and is easily converted to the free acid by treatment with water or neutralization by base. Oxidation using hydrogen peroxide or *t*-butyl hydroperoxide is also very common ⟨62JCS2089, 65JOC2454, 83JCS(P1)471⟩, and can also be achieved using bromine water ⟨61T1⟩. In addition, seleninic acids are formed by hydrolysis of selenium trihalides ⟨58JCS2939, 51T1⟩.

$$(RSe)_2 \xrightarrow{HNO_3} [RSe(OH)_2]^+ NO_3^- \xrightarrow{base} RSeO_2H$$

Scheme 28

Alkaneseleninic anhydrides can be synthesized by dehydration of seleninic acids (Equation (45)) ⟨62JCS2089, 65ZAAC(335)289⟩, by oxidation of diselenides with ozone at low temperature ⟨62JCS2089⟩ or by oxidation of selenols with *t*-butyl hydroperoxide (Equation (46)) ⟨83JCS(P1)471⟩.

$$2RSeO_2H \xrightarrow{\Delta} (RSeO)_2O + H_2O \qquad (45)$$

$$RSeSeR,\ RSeH \xrightarrow{[O]} (RSeO)_2O \qquad (46)$$

$R = Bu^n$, *n*-dodecane (quantitatively)

Alkane tellurinic acids are prepared by oxidation of tellurides with hydrogen peroxide–urea adduct or air (Equation (47)) ⟨38JCS341, 41JCS70⟩.

$$RTeR \xrightarrow[\text{or air}]{H_2O_2\text{-urea}} RTeO_2H \qquad (47)$$

2.04.5.3 Alkaneselenonic Acids

Alkaneselenonic acids have been prepared by oxidation of the seleninic acids with potassium permanganate in aqueous potassium hydroxide solution (Equation (48)) ⟨42JCS570⟩.

$$RSeO_2H \xrightarrow{KMnO_4} RSeO_3H \qquad (48)$$

No telluronic acids have been prepared. The product of such a preparation appears to be polymeric.

2.04.6 FUNCTIONS BASED ON THE R¹SeOR² UNIT AND ITS HIGHER COORDINATED DERIVATIVES—SELENENATES, SELENINATES, SELENONATES AND RELATED FUNCTIONS (R^1SeOR^2, $R^1Se(O)OR^2$, $R^1Se(O)_2OR^2$), AND CORRESPONDING TELLURIUM ESTERS

Alkaneselenenic acid esters are generated *in situ* in the [2,3]sigmatropic rearrangement of allyl selenoxides and are usually too unstable to be isolated (Scheme 29) ⟨73CC695, 76CC61⟩.

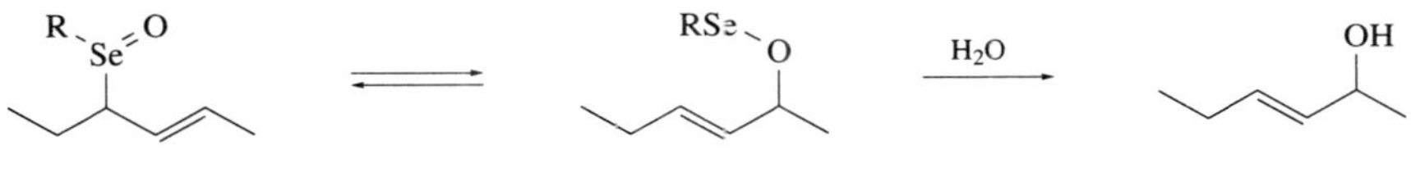

Scheme 29

Alkaneseleninic acid esters are formed by treatment of seleninic anhydrides with methanol, treatment of seleninic acid silver salts with alkyl halides, or treatment of seleninyl chlorides with sodium methoxide in methanol ⟨65ZAAC(338)195, 67ZAAC(352)167⟩. Cyclic seleninic acid esters, seleniolactones, can be isolated from the selenium dioxide oxidation of suitably substituted alkenes (Equation 78) ⟨73JA7917⟩.

Esters of alkaneselenonic acids, tellurenic acid, tellurinic acid, and telluronic acid have not been reported.

2.04.7 ALKYL DISELENIDES (R^1SeSeR^2) AND ALKYL DITELLURIDES (R^1TeTeR^2)

There are three common methods of preparing alkyl diselenides; one is alkylation of the diselenide anion and the others are oxidation or reduction of selenium-containing compounds (Scheme 30). The alkylation of sodium or potassium diselenides can be achieved with alkyl halides or dialkyl sulfates (Scheme 30) ⟨27JA2011, 73JA197, 94CPB811, 94JOC1011⟩. The reactions can be carried out in aqueous medium, in ethanol or in liquid ammonia. Potassium selenosulfate, prepared from a potassium sulfite solution and elemental selenium, reacts with alkyl halides to afford dialkyl diselenides. Potassium organic selenosulfates can be transformed to diselenides by oxidation with iodine or hydrogen peroxide, by heating in aqueous solution, or by refluxing the acidified reaction mixture ⟨64JMC229, 65JMC845, 65JOC2454, 66JOC1202⟩.

$$Na_2Se_2 + 2RX \xrightarrow[77\text{–}92\%]{} RSeSeR + 2NaX$$

$$2RSeCN \xrightarrow{H_2O} RSeSeR$$

$$2RSeH \xrightarrow{[O]} RSeSeR$$

$$RSeOH,\ RSeO_2H \xrightarrow{[H]} RSeSeR$$

Scheme 30

The best method for preparing a symmetrical diselenide is oxidation of selenols or selenolate salts. Any of the methods described in Section 2.04.1 for the production of selenols will give diselenides when the reaction mixture is treated with a mild oxidizing agent, for example by bubbling air through the selenol solution or by the addition of dilute hydrogen peroxide ⟨57JCS799⟩.

Alkyl ditellurides are synthesized in similar ways to alkyl diselenides. The commonest ways are alkylation of the ditelluride anion, Te^{2-}, generated by reduction of elemental tellurium with Rongalite ⟨14CB1269, 57ACS571⟩ or reduction with sodium in liquid ammonia ⟨68JOM(12)401, 69ZOB1804, 70IC106⟩, or oxidation of metal alkanetellurolates (see Section 2.04.1). Another route using elemental tellurium involves insertion of tellurium into C—metal bond, followed by aerobic oxidation ⟨80IC1879⟩. However, alkyl magnesium halides do not react with tellurium ⟨72JOM(32)97⟩.

2.04.8 RSeN FUNCTIONS AND THEIR HIGHER COORDINATED DERIVATIVES

Alkane selenenamides (selenenyl amides) can be prepared from alkylselenenyl halides and amines ⟨63JCS1268, 65JCS7511, 79ZAAC(448)35⟩ or phthalimide (Equation (50)) ⟨88JCS(P1)1913⟩.

$$R^1SeX + R^2R^3NH \xrightarrow[62\text{–}92\%]{} R^1SeNR^2R^3 \qquad (50)$$

$R^1 = CF_3, C_2F_5, C_3F_7, CH_2CO_2Et, CH_2CO_2Me, CH_2COPh, CH_2CONHPh, CH_2CN$

R^2, R^3 = Me, Ph, 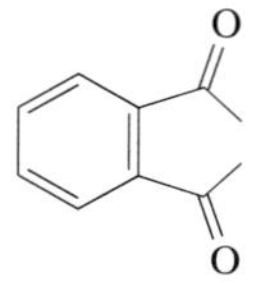

The preparation of alkane seleninamides and selenonamides has not been reported.

N-Substituted selenimides (selenilimines) have been prepared from selenides or selenoxides as shown in Equation (51), Scheme 31 and Equation (52) ⟨75TL649, 81PS153⟩.

(51)

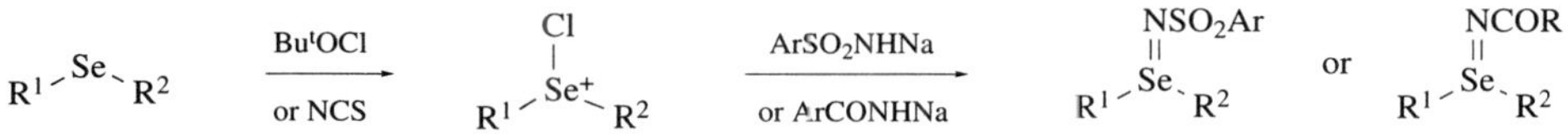

Scheme 31

(52)

Alkanetelluramides (tellurenyl amides) have not been reported. Alkanetellurimides can be generated by the reactions of dialkyl tellurides with *t*-butyl hypochlorite and chloramine-T (Scheme 32) ⟨81CL447⟩ or with tosyl azide (Equation (53)) ⟨85CL679⟩.

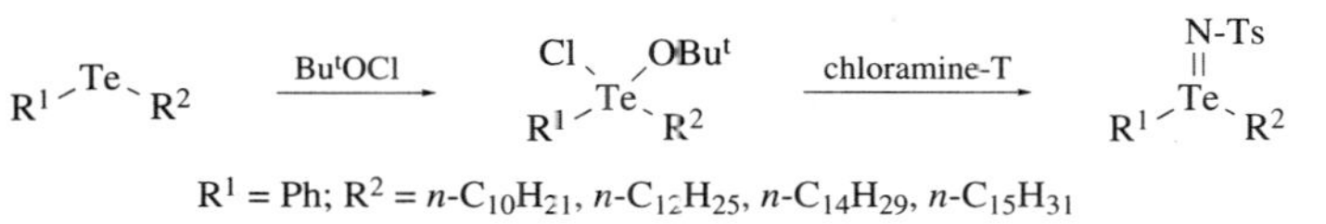

Scheme 32

$$(Bu^i)_2Te \xrightarrow{TsN_3} (Bu^i)_2TeNTs \quad (53)$$

2.04.9 RSeP, RSeAs, RSeSb, AND RSeBi FUNCTIONS AND THEIR HIGHER COORDINATED ANALOGUES

Bond formation between selenium or tellurium and phosphorus or arsenic can be effectively achieved by reactions of diorganyl dichalcogenides with tetraorganyl diphosphanes or diarsanes ⟨79ZN(B)1646⟩. Alkylseleno and alkyltelluro phosphinous acid esters and arsinous acid esters are prepared by this method (Equation (54)).

$$R^1{}_2XXR^1{}_2 \; + \; R^2YYR^2 \longrightarrow R^1{}_2XYR^2 \quad (54)$$

$$R^1, R^2 = Me, CF_3; X = P, As, Bi; Y = Se, Te$$

Alkylseleno esters of phosphorus acid and arsenous acid are synthesized by the reactions of PCl_3 (Equation (55)) or $AsCl_3$ with (methylseleno)trimethylsilane (Equation (56)) ⟨75JIN233⟩.

$$PCl_3 \; + \; 3\,MeSe\text{-}TMS \longrightarrow P(SeMe)_3 \; + \; 3\,TMS\text{-}Cl \quad (55)$$

$$AsCl_3 \; + \; 3\,MeSe\text{-}TMS \longrightarrow As(SeMe)_3 \; + \; 3\,TMS\text{-}Cl \quad (56)$$

Alkylseleno and alkyltelluro phosphonium salts can be synthesized by alkylation of seleno- or tellurophosphoranes (Equation (57)) or by *O*-alkylation of alkylselenophosphinates, -phosphonates or -phosphates (Equation (58)) ⟨69ZN(B)179, 77JCS(D)527⟩.

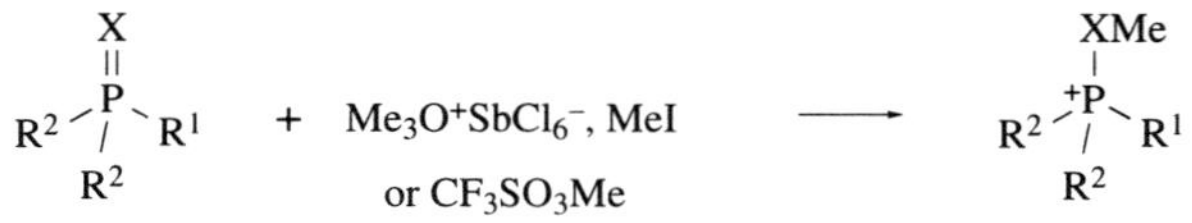

(57)

R^1, R^2 = Ph, Bu, Pr^i, OMe, NMe_2; X = Se, Te

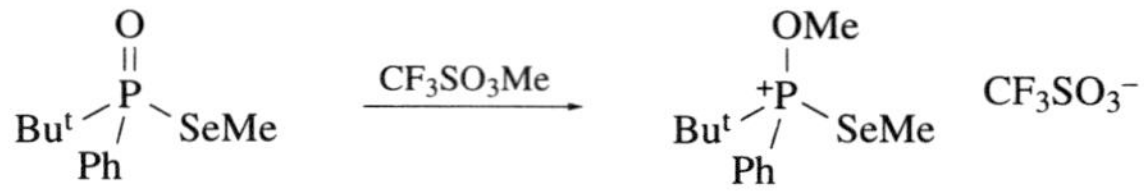

(58)

For a review on the organoselenium and -tellurium chemistry of RSeP, RSeAs, RSeSb, and RSeBi functions, see ⟨B-87MI 204-03⟩.

2.04.10 RSeSi, RTeSi, RSeB, AND RELATED METALLOID FUNCTIONS AND THEIR HIGHER COORDINATED DERIVATIVES

2.04.10.1 RSeSi Functions

Methylselenosilanes are synthesized by the reaction of methaneselenol with chloromethylsilanes and trimethylamine (Equation (59)) ⟨73JCS(D)1716⟩. Lithium tetra(methylseleno)aluminate reacts with trimethylsilyl halides to give methylseleno(trimethylsilane) (Scheme 33) ⟨73JCS(D)1716, 85T4793⟩. This method can be applied to the synthesis of methylseleno derivatives of the other elements of Group 14 germanium and tin. Synthesis of the germanium and tin derivatives can also be achieved via methylseleno(trimethylsilane) (Equation (60)).

$$n\text{-MeSeH} + Me_{4-n}SiCl + n\text{-}Me_3N \xrightarrow[70\%]{} (MeSe)_nSiMe_{4-n} + n\text{-}Me_3NHCl \quad (59)$$

n = 1–4

$$4MeSeH + LiAlH_4 \longrightarrow LiAl(SeMe)_4 + 4H_2$$

$$LiAl(SeMe)_4 + 4Me_3MX \longrightarrow 4Me_3MSeMe + LiX + AlCl_3$$

M = Si, Ge, Sn; X = Cl, Br ca. 70%

Scheme 33

$$n\text{-MeSe-TMS} + Me_{4-n}MX_n \longrightarrow Me_{4-n}M(SeMe)_n + n\text{-TMS-X} \quad (60)$$

M = Ge, Sn; n = 1–4; X = Cl, Br 5–100%

Methyltelluro(trimethylsilane) is prepared from lithium methanetellurolate and chlorotrimethylsilane. Methyltellurogermane and -stannane derivatives are obtained by the exchange reaction of fluorogermane and chlorostannane, respectively, with methyltellurosilane (Scheme 34) ⟨80IC1879⟩.

$$MeTeLi + TMS\text{-}Cl \longrightarrow MeTe\text{-}TMS + LiCl$$

$$MeTe\text{-}TMS + H_3GeF \longrightarrow H_3GeTeMe + TMS\text{-}F$$

$$MeTe\text{-}TMS + Me_3SnCl \longrightarrow MeTeSnMe_3 + TMS\text{-}Cl$$

Scheme 34

2.04.10.2 RSeB Functions

Tris(alkylseleno)boranes are prepared by the reactions of alkaneselenols or alkaneselenolates with boron trihalides (Equation (61)) ⟨70JOM(25)17, 75ZN(B)642, 79JOC4279, 85T4793⟩. Alkylseleno-(diiodo)boranes and (alkylseleno)alkyliodoboranes are similarly synthesized from selenols, diselenides or alkylselenoboranes (Scheme 35) ⟨72CB454⟩.

$$BX_3 + 3RSeNa \xrightarrow[62-93\%]{} B(SeR)_3 + 3NaX \qquad (61)$$

$R = Me, Et, Pr^i, Bu^n$; $X = Cl, Br, I$

$$\left.\begin{array}{r} 2RSeH \\ RSeSeR \\ B(SeR)_3 \end{array}\right] \xrightarrow[34-45\%]{2BI_3} 2RSeBI_2$$

$$\left.\begin{array}{r} R^1SeSeR^1 \\ (R^1Se)_2BR^2 \end{array}\right] \xrightarrow[44-89\%]{R^2BI_2} R^1SeBIR^2$$

$$\left.\begin{array}{r} RSeSeR \\ (RSe)_3B \end{array}\right] \xrightarrow{BI_3} (RSe)_2BI$$

Scheme 35

2.05
Alkylnitrogen Compounds: Amines and their Salts

C. M. MARSON and A. D. HOBSON
University of Sheffield, UK

2.05.1 INTRODUCTION

The aim of this chapter has been to provide an overview of a number of key major areas in the synthesis of alkylamines, rather than to attempt an exhaustive coverage of the field. The need to be selective means that illustrative examples have been chosen, chiefly from the years 1978–1993, although some important publications from this period will inevitably have been omitted. Several excellent reviews of syntheses form a useful adjunct to the chapter ⟨57HOU(11/1)1, B-65MI 205-01, B-65MI 205-02, B-68MI 205-01, B-82MI 205-01⟩.

2.05.2 ILLUSTRATIVE METHODS FOR THE INTRODUCTION OF NITROGEN

2.05.2.1 Replacement of Hydrogen by Nitrogen using Aminating Reagents

The amination of a variety of organometallic reagents is described in Section 2.05.4.1. In contrast, direct replacement of hydrogen in aliphatic hydrocarbons by nitrogen is generally confined to tertiary positions. One of the most effective aminating systems is a mixture of trichloramine and $AlCl_3$ (1 : 2 equiv.) which is reacted with 1.5 equivalents of the hydrocarbon; decalin is converted into *cis*-9-aminodecalin (50%), and adamantane into 1-aminoadamantane (95%) ⟨70CR639⟩. However, many cycloalkanes lacking a tertiary centre suffer rearrangements and give mixtures of products ⟨70CR639⟩. The toxicity of trichloramine and its potentially explosive nature especially in an aqueous medium represent substantial disadvantages in its use.

Carbethoxynitrene, EtOOCN ⟨B-84MI 205-01⟩, inserts into C—H bonds to form *N*-substituted carbamates, EtOOCNHR, hydrolysis of which affords a route to RNH_2 ⟨63JA1200⟩. Although insertion of oxycarbonylnitrenes ⟨87ACR18⟩ into C—H bonds has limited synthetic application, carbethoxynitrene generated by α-elimination of ethyl *p*-nitrobenzenesulfonoxycarbamate has been shown to convert α-methylstyrene into 3-carbethoxyamino-2-phenyl-1-propene (11%), and cyclohexane into ethyl cyclohexylcarbamate (13%) ⟨78JOC3345⟩. A potentially useful observation is the insertion of carbethoxynitrene into the tertiary C—H bond of optically active 3-methylhexane with essentially complete retention of configuration ⟨69JA5107⟩ irrespective of the means by which the nitrene was generated, or the reaction conditions.

Alkenes can be aminated at the allylic position by treatment with imido selenium compounds RN═Se═NR (R = Bu^t or Ts). For example, TsN═Se═NTs is prepared by reacting selenium with chloramine-T; reaction with cyclohexene in dichloromethane affords the allylic sulfonamide (45%) ⟨76JA269⟩. Such aminations usually proceed in modest yields and are unsatisfactory where attack at an alternative site can provide regioisomers.

2.05.2.2 Use of Ammonia and Amines

The conversion of R^1X into $R^1NR^2R^3$ depends greatly on the nature of R^2 and R^3 for its success. The most straightforward route, at least in principle, involves ammonia ($R^2 = R^3 = H$) or another amine. Where R^1X is an alkyl halide (see Section 2.05.4.3), this direct approach is often of limited value (owing to further alkylation) for the preparation of primary and secondary amines, although it is possible if an appropriate excess of ammonia or primary amine is used. However, tertiary amines can be prepared efficiently by reacting one molar equivalent of the alkylating agent (R^1X) with the secondary amine. A generally satisfactory method of preparing primary amines from primary alkyl halides uses sodamide. The alkylation of tertiary amines, the Menschutkin reaction, is usually efficient ⟨78AHC(22)71⟩.

2.05.2.3 Gabriel Synthesis

Primary amines can often be usefully prepared by means of the Gabriel synthesis ⟨1887CB224, 68AG(E)919⟩. In the classical reaction, the phthalimide anion, particularly its potassium salt, is treated with an alkylating agent followed by a second step involving removal of the phthaloyl group (Scheme 1).

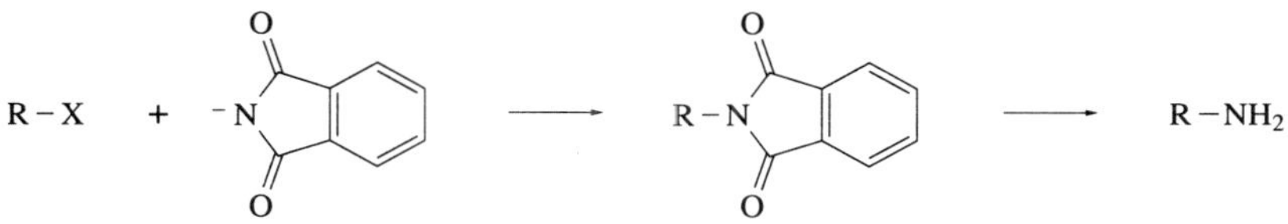

Scheme 1

Alkyl halides and alkyl sulfonates are commonly employed as alkylating agents. Phosphonium salts ⟨76S389⟩ or ammonium salts ⟨78CC742, 80SC611, 81JCS(P1)574⟩ have been used to effect alkylation of the phthalimide anion under conditions of solid–liquid phase transfer. Using hexadecyltributylphosphonium bromide, alkyl chlorides were found to be less reactive than alkyl bromides or alkyl methanesulfonates; however, octyl bromide and chloride are more reactive than octyl iodide ⟨76S389⟩. (*S*)-2-Octylamine was obtained (92.5% inversion) from (*R*)-2-octyl methanesulfonate ⟨76S389⟩. [2.2.2]Cryptands have been used in the preparation of *N*-alkylphthalimides from alkyl halides at room temperature ⟨80T1223⟩.

The *N*-alkylation step has been widened by the use of Mitsunobu reactions ⟨81S1⟩. For example, the (*S*)-alcohol (**1**) can be converted into the (*R*)-imide (**2**) by reaction at 0 °C (30 min), then keeping overnight at room temperature (Equation (1)) ⟨85JCS(P1)2201⟩. Such reactions usually proceed with excellent inversion of configuration, although loss of stereocontrol occurs at benzylic positions ⟨87JOC1956⟩, presumably through a carbocationic process. Thiols can be converted into their corresponding alkylthiophosphonium salts which undergo clean inversion of configuration with phthalimide anion ⟨85TL4867⟩.

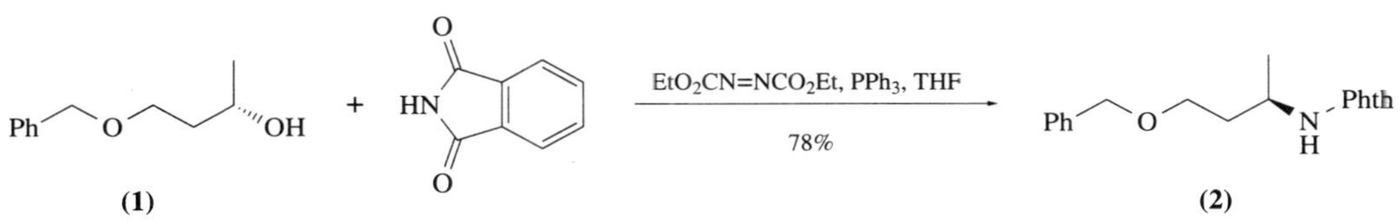

(**1**) (**2**) (1)

The mild alkylation conditions described above have led to a search for milder conditions for the deprotection step, which traditionally requires the use of strong acid or base, or alternatively treatment with hydrazine at elevated temperature (owing to the stability of the phthaloyl group) ⟨68AG986⟩.

A review ⟨91ACR285⟩ describes notable developments, including reagents that allow dialkylation to occur specifically. For the preparation of primary amines, alternatives to phthalimide derivatives include imidodicarbonates and carbamates which bear either an *N*-carboxy, *N*-phosphoryl or *N*-sulfonyl group. Dicarboximide and sulfenimide groups have also been examined.

Modified Gabriel syntheses have been developed using methyl *t*-butyl iminodicarboxylate and di-*t*-butyl iminodicarboxylate ⟨77CC758, 78JCS(P1)1088, 83JCS(P1)2983, 87S275⟩. Alkylation can be conducted in DMF or DMSO (20–60 °C). Specific monodeprotection can be accomplished under either basic conditions or acidic conditions(trifluoroacetic acid). The use of the sodium salt of diethyl *N*-(*t*-butoxycarbonyl)phosphoramidate (**3**) allows the corresponding *N*-alkyl derivatives to be deprotected by treatment overnight with benzene saturated with dry HCl (Scheme 2). Secondary α-halo esters also undergo efficient aminations using that procedure ⟨82S922⟩. In a related procedure, both primary and secondary alcohols can be converted into the corresponding amines by reaction with diethyl azodicarboxylate, triphenylphosphine and *N*-(*t*-butoxycarbonyl)phosphoramidate at room temperature, followed by cleavage with HCl to give the amine hydrochlorides ⟨86LA402⟩.

A sequence ⟨84S941⟩ involving reaction at around room temperature (Scheme 3) involves alkylation of trifluoroacetamide by alkyl halides or methanesulfonates. The resulting *N*-alkylamides are converted into the primary amines either by alkaline hydrolysis or by reduction. Competitive elimination limits the use of relatively unreactive primary alkyl chlorides and secondary halides or methanesulfonates.

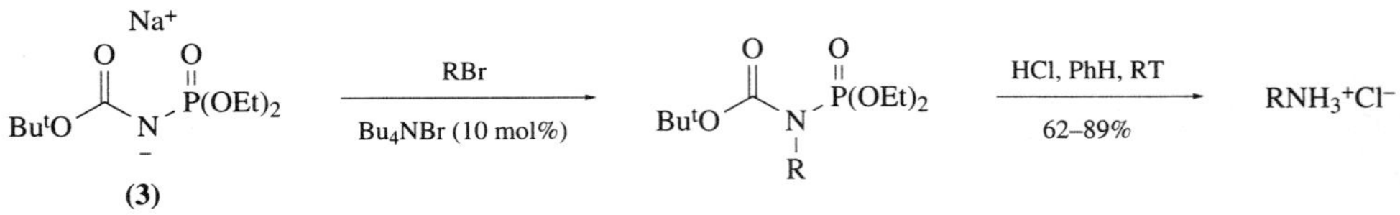

(3)

Scheme 2

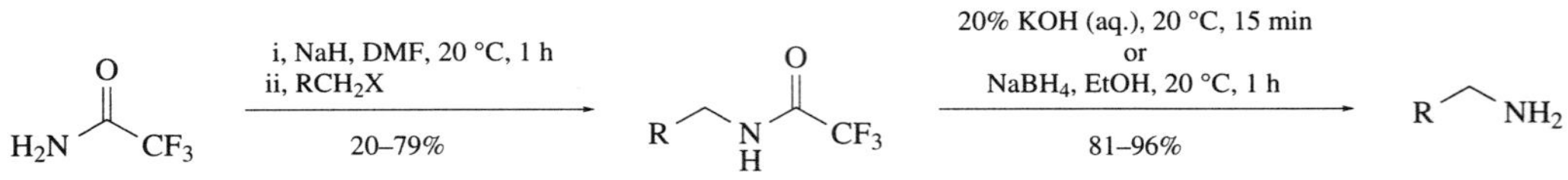

Scheme 3

2.05.3 AMINES FROM OTHER AMINES

2.05.3.1 Alkylation at Nitrogen

In many cases, direct *N*-alkylation is either unsatisfactory or not possible, usually because subsequent alkylation occurs and cannot be readily prevented. Exceptions to this are described in Sections 2.05.2.2 and 2.05.4.3; additionally, intramolecular alkylation of amines provides a usually efficient general method of preparing cyclic amines (Section 2.05.5). However, primary and secondary amines are often prepared by a sequence of blocking (protection which may also involve activation of the nitrogen atom), monoalkylation and lastly deprotection. Suitable blocking groups include phthalimido (Gabriel reaction, Section 2.05.2.2) resulting in primary amines; aryl sulfonyl ⟨B-68MI 205-01, B-75MI 205-01⟩ (Hinsberg synthesis), trifluoromethylsulfonyl ⟨73TL3839⟩, and phosphoramidate ⟨77AG(E)107⟩ groups have all been used to convert primary amines into secondary amines by controlled monoalkylation.

Useful alternatives to the reductive methylation of primary and secondary amines with formaldehyde and formic acid, the Eschweiler–Clarke procedure, have been developed. Borohydride reagents have been used in place of formic acid ⟨80JOC357, 90TL5595⟩. In one procedure, primary and secondary amines can be *N*-methylated to give di- and monomethyl tertiary amines, respectively, using a combination of paraformaldehyde, sodium borohydride and trifluoroacetic acid at room temperature (Scheme 4) ⟨87S709⟩.

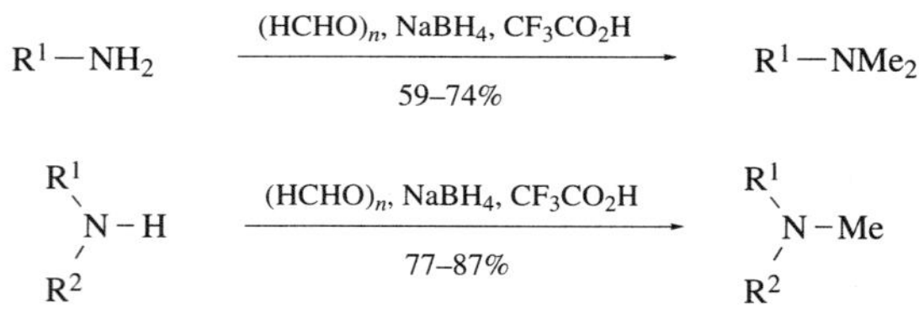

Scheme 4

By using phosphinic acid and a dialkylsulfoxide, some primary and some secondary amines can be reductively alkylated; for example, ephedrine is converted by $(PhCH_2)_2SO$ and phenylphosphinic acid into *N*-benzylephedrine ⟨91SC1425⟩.

Primary amines are conveniently prepared from potassium 2,2,6,6-tetramethyl-2,6-disilapiperidide (prepared from 2,2,6,6-tetramethyl-2,6-disilapiperidine and potassium hydride) by *N*-alkylation followed by acidic hydrolysis. The disilapiperidide anion acts as a synthetic equivalent of the amide anion (Scheme 5) ⟨88CPB2342⟩. Reaction of primary amines with trimethylsilyl chloride gives the monosilylated amine which, after treatment with butyllithium followed by chloromethyl ether, furnishes unsymmetrical secondary amines (Scheme 6) ⟨87TL1659⟩.

Secondary amines can be *N*-phenylated using bromobenzene and lithium in dry THF at room temperature ⟨93JOC3156⟩. The reaction was also successful for some primary amines, including those with a chiral centre adjacent to nitrogen.

Dimethylborane can be used to hydroborate alkenes regiospecifically to form dimethyl-

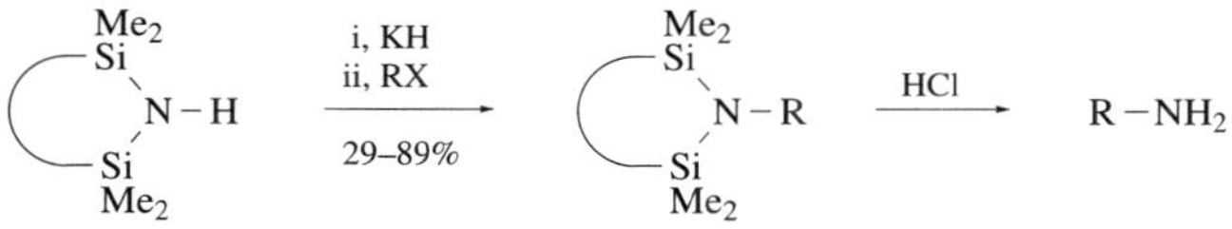

Scheme 5

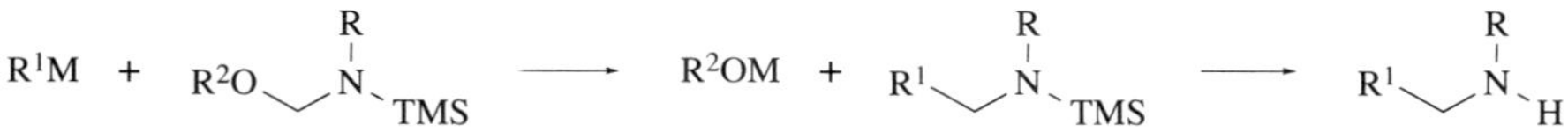

Scheme 6

alkylboranes which were then reacted with primary alkylamines and sodium hypochlorite to yield isomerically pure dialkylamines ⟨90SC231⟩.

Palladium-catalysed amination of allylic substrates has been shown to be regio- and stereoselective ⟨83TL2745⟩. The use of a palladium complex, bearing a chiral (hydroxyalkyl)ferrocenylphosphine ligand, as the catalyst for the asymmetric allylic amination of 2-butenyl acetates leads to high regio- and stereoselectivities; enantiomeric excesses of up to 84% were reported ⟨90TL1743⟩.

Trialkylamines react with electron-deficient aryl and heteroaryl bromides and chlorides at very high pressures to give dialkylarylamines; thus, 2-chloroquinoline reacts with triethylamine at 0.8 GPa and 100 °C to give 2-(diethylamino)quinoline (51%) ⟨91CC306⟩.

Alkylation of *N*-*s*-butylaniline (prepared by reductive alkylation of aniline with butanone) with *s*-butyl iodide in the presence of methyllithium in diethyl ether affords *N*,*N*-bis(*s*-butyl)aniline in moderate yield ⟨91OPP399⟩.

Propargylic amines are formed from 1,1-dibromo-1-alkenes using Triton B and secondary amines (Equation (2)) ⟨90S931⟩.

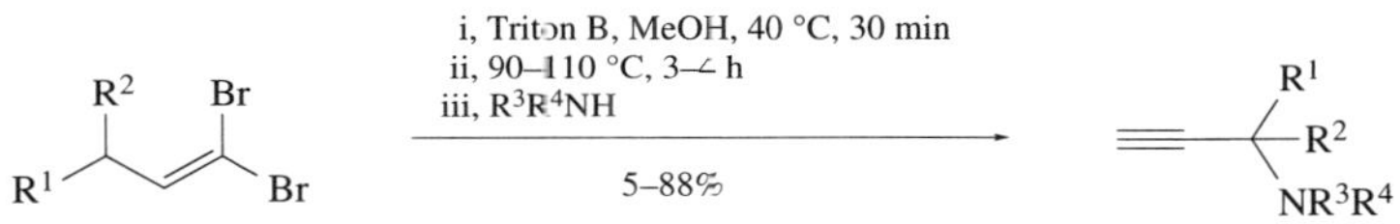

(2)

Secondary amines can be formed by the alkylation of primary amines using trialkylboranes and sodium hypochlorite in THF ⟨89OM1093⟩.

β-Nitroethylamines can be prepared in near quantitative yield by Michael addition of secondary amines to nitroethene or *α*-nitrostyrene, generated *in situ* ⟨91S479⟩.

Primary and secondary amines have been converted into bis- and monopropargylated tertiary amines using the Nicholas reaction (Equation (3)) ⟨93TL2919⟩.

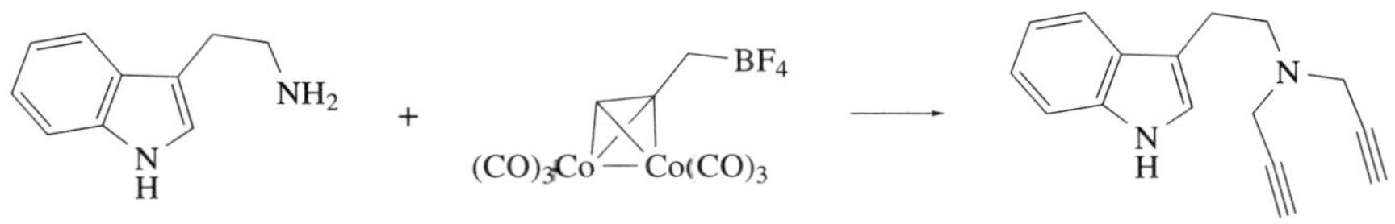

(3)

A 2-aminoethyl group can be attached to the 1-position of a pyrrole ring by alkylation with 2-chloroethylamine in the presence of $Bu_4N^+HSO_4^-$ under conditions of phase-transfer catalysis ⟨91SC535⟩.

Various secondary amines were shown to add at the *α*-position of *exo*-7-halobicyclo[3.2.0]hept-2-en-6-ones such as (**4**) by a cine substitution involving displacement of chloride (Equation (4)) ⟨91JCS(P1)483⟩.

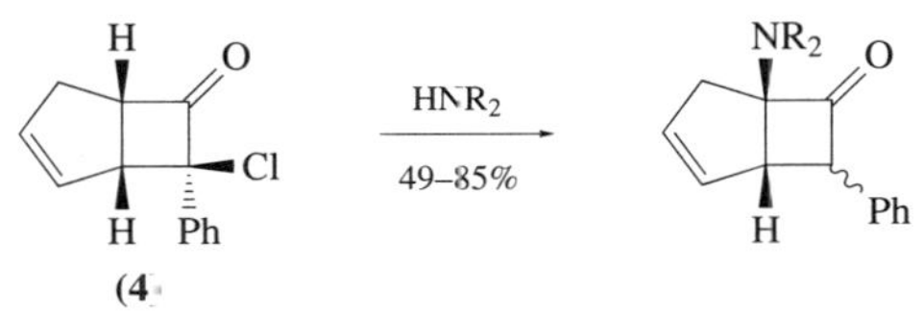

(**4**)

(4)

R = Me, Et, Pr^i

2.05.3.2 Alkylation at Carbon

Since direct carbanion formation adjacent to nitrogen is not favoured, modification of an amine C_α carbon is often conducted by nucleophilic attack on an electrophilic α-carbon, as in the Mannich reaction ⟨53OR(7)99⟩. However, masked secondary α-aminocarbanions have been prepared via nitrosation of secondary amines and subsequent metallation, and can be readily alkylated (Scheme 7). Metallation of an amide at a carbon atom adjacent to nitrogen allows the introduction of an α-substituent ⟨84CR471⟩.

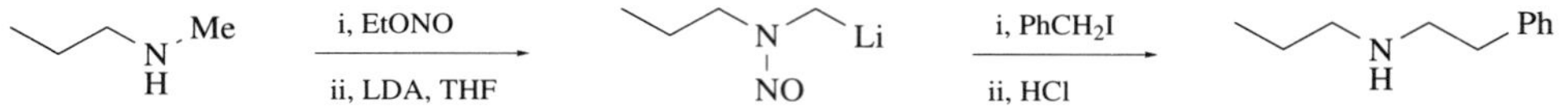

Scheme 7

Primary amines have been substituted at the α-position by the alkylation of 2-azaallyl anions (Scheme 8) ⟨89TL5853⟩. Reaction of a primary amine with mesityl 2-pyridyl ketone gives the corresponding imine which is deprotonated using lithium diisopropylamide (LDA) and treated with an alkyl halide. The α-alkylated amine is liberated upon hydrolysis.

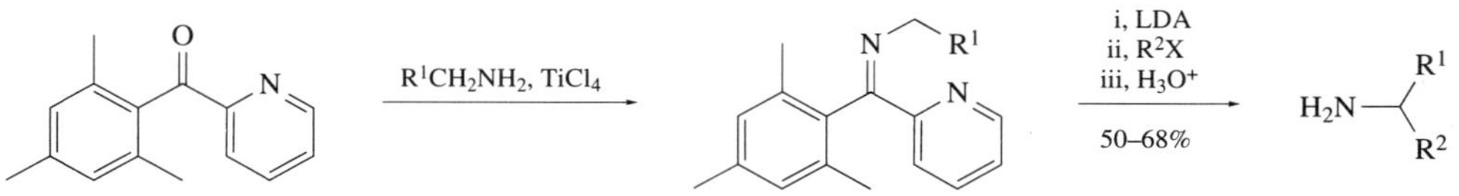

Scheme 8

Secondary methylamines can be functionalised at the methyl carbon atom by *in situ* conversion into the lithium *N*-lithiomethyl-dithiocarbamate and subsequent reaction with one equivalent of an electrophile, followed by *N*-deprotection at ambient temperature ⟨88S775⟩.

The reaction of alkyl and allyl halides with *N*-(alkylamino)benzotriazoles at 20 °C in THF–water, in the presence of bismuth(III) chloride and metallic aluminum, furnishes the corresponding homoalkylated secondary and tertiary amines in high yield ⟨91TL4247⟩.

Alkylation of secondary amines with aldehydes in the presence of benzotriazole gives tertiary amines in good to essentially quantitative yield ⟨89OPP135⟩.

Mannich aminoalkylations can be efficiently conducted in anhydrous DMF ⟨92SC2305⟩. Whereas saturated cycloalkanones are alkylated alpha to the carbonyl group, the corresponding α,β-unsaturated ketones can be aminoalkylated at the γ-position via their *O*-silylated dienolates. Thus reaction of the *O*-silylated dienolate (**5**) with formaldehyde *N*,*N*-dimethyliminium chloride in DMF at 50 °C affords the tertiary amine (**6**) in 75% yield (Equation (5)) ⟨92SC2305⟩.

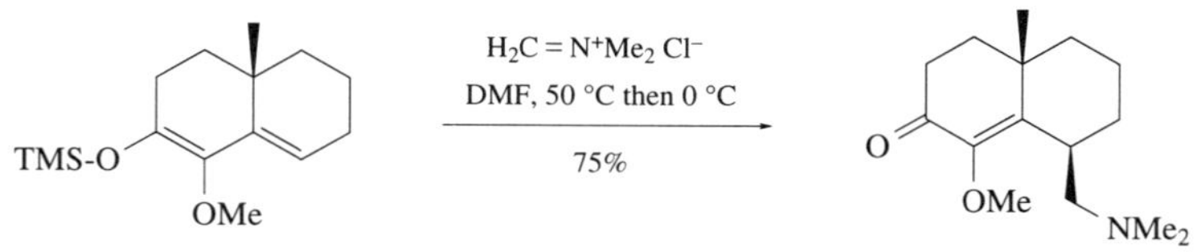

(5) **(6)** (5)

The *N*,*N*-dimethylaminomethylation of various carbonyl compounds has been achieved in moderate to good yields by a Mannich reaction of boron enolates and *N*,*N*,*N*′,*N*′-tetramethyldiaminomethane in dichloromethane ⟨90CC1532⟩.

2.05.3.3 Dealkylation

Dealkylation of amines, particularly demethylation, has been of value in the synthesis and elucidation of structures, particularly alkaloids ⟨57HOU(11/1)961⟩. Controlled dealkylation is a complementary approach to the preparation of amines by alkylation. Primary amines can be prepared by the Délépine reaction, in which a primary halide (RX) is reacted with hexamethylenetetraamine to give the (mono) quaternary salt which is then hydrolysed using HCl—EtOH to liberate RNH_2 (after neutralisation) ⟨B-68MI 205-01, 73TL3839, 75MI 205-01⟩.

Secondary amines have long been prepared by the dealkylation of tertiary amines using cyanogen bromide, the von Braun reaction (Scheme 9) ⟨57HOU(11/1)1, B-68MI 205-01, 73TL3839, B-75MI 205-01, 89S1⟩. A hydrolysis step of some kind is required, unless the intermediate is sufficiently unstable. Secondary amines can also be prepared by the demethylation of tertiary amines in alkaline methanol and dichloronaphthalene using ultraviolet light (Equation (6)) ⟨90TL1199⟩. The *N*-benzyl group of secondary and tertiary amines can be selectively removed by hydrogenation over 20% palladium hydroxide on carbon in high yields without cleaving benzyl ethers ⟨90SC1209⟩.

$$R_3N + BrCN \longrightarrow RBr + R_2NCN \xrightarrow{H_3O^+} R_2NH$$

Scheme 9

$$\text{Ph–N(Me)Et} \xrightarrow[\text{dichloronaphthalene}]{h\nu\text{, MeOH, KOH, } N_2\text{, 12 h}} \text{Ph–N(H)Et} \qquad (6)$$

One of the best methods of dealkylating tertiary amines uses α-chloroethyl chloroformate (Scheme 10) ⟨84JOC2081⟩. An order of cleavage of various groups by such chloroformates has been established as: *N*-debenzylation > *N*-dealkylation > deamination of cyclohexylamines > *N*-demethylation > *N*-deethylation ⟨83TL3233⟩; this is exemplified by Equation (7). Selective demethylations have been used in the synthesis of opiate analogues such as nalbuphine ⟨84JOC2081⟩. The *N*-benzyl group of tertiary amines is cleaved by treatment with α-chloroethyl chloroformate followed by reaction with hydroxide, affording secondary amines in good yields ⟨89TL7383⟩. Phenyl chloroformate has been shown to be more effective than alkyl esters, and compares well with the von Braun reaction ⟨67JCS(C)2015⟩. Tropinol has been demethylated by allyl chloroformate in 77% yield, a yield far superior to the use of either ethyl chloroformate or the von Braun reaction.

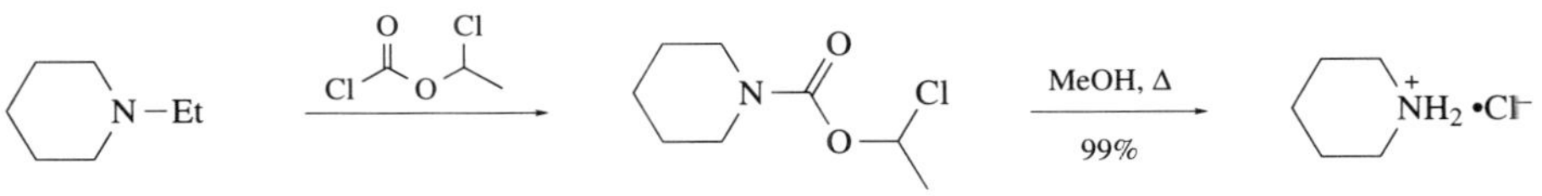

Scheme 10

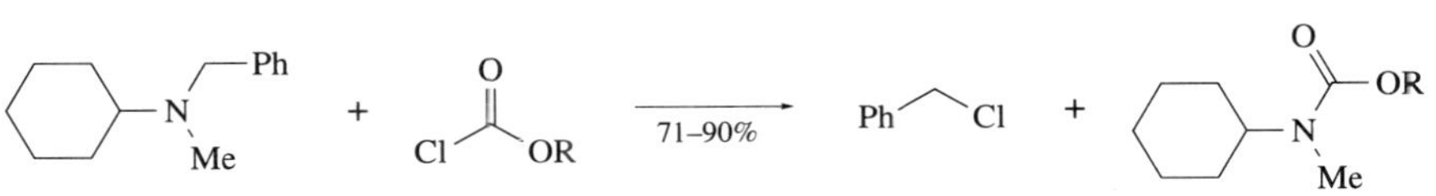

(7)

Selective dequaternisation of quaternary ammonium salts can provide a useful route to tertiary amines, as in the Emde reduction (Scheme 11) or in other particular examples, including Equation (8) ⟨71S89⟩. Hofmann elimination of some quaternary ammonium hydroxides of 1,2,5,6-tetrahydropyridines provides stereoselective syntheses of (*Z*,*E*)-penta-2,4-dienylamines ⟨92JCR(S)309⟩. The Stevens rearrangement ⟨70OR(18)403⟩ is a specific intramolecular dealkylation of a quaternary ammonium salt, typically $ZCH_2N^+R^1R^2R^3X^-$ to give $ZCH(R^1)NR^2R^3$. The electron-withdrawing group Z facilitates the required deprotonation by strong base (NaOR or $NaNH_2$).

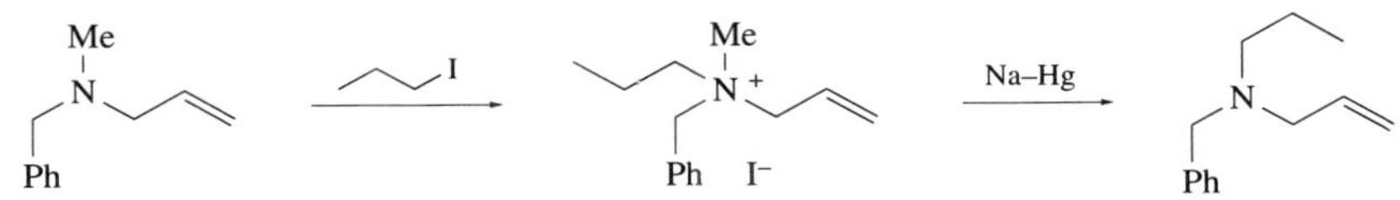

Scheme 11

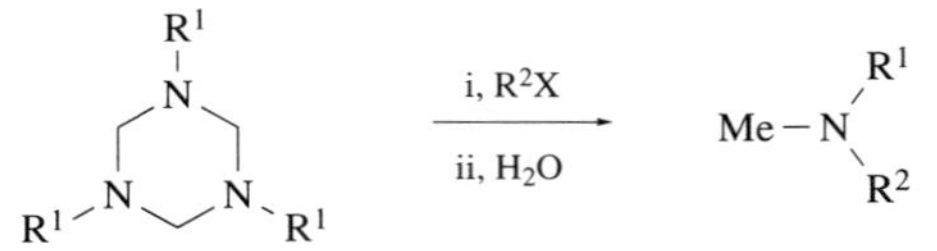

(8)

Tertiary methylamines are efficiently demethylated to the corresponding secondary amines by 9,10-dicyanoanthracene-sensitised photooxidation in the presence of $LiClO_4$ ⟨89TL2927, 89TL3977⟩.

N-Benzyl-protected arylamines can be deprotected under physiological conditions by biological transformation using resting cultures of the aerobic fungi *Mortierella isabellina* NRRL 1757 and *Helminthosporium* species NRRL 4761 ⟨88TL6393⟩.

2.05.3.4 Deprotection

In this section, deprotections, other than dealkylation reactions, are considered. Primary and secondary amines are obtained in excellent yield by the mild and selective palladium(0)-catalysed dealkylation of allylic amines ⟨93JOC6109⟩. The removal of allyloxycarbamate protecting groups using a palladium catalyst furnishes primary and secondary amines in essentially quantitative yield ⟨93SL680⟩. Phthalimides can be deprotected by treatment with anion exchange resins, e.g., DIANION WA20, at 80–90 °C. Advantages over conventional methods (N_2H_4) include good chemoselectivities and the noncontamination by metal ions ⟨90S735⟩.

Selective removal of *t*-butoxycarbonyl groups from secondary amines can be accomplished using zinc dibromide; similarly protected primary amines are unaffected under these conditions ⟨89SC3139⟩.

Methyl carbamates are reduced chemoselectively to the corresponding amines, in the presence of ketones, nitriles or halogens, by treatment with sodium hydrogen telluride in DMF ⟨89SC1347⟩. An efficient method for the cleavage of benzyl carbamates involves a hard acid (boron trifluoride etherate)–soft nucleophile (ethane thiol) system. The method avoids the reduction of alkenes, alkynes, imines, halides and nitro groups, as well as carboxylic ester hydrolysis ⟨90TL6903⟩.

1,2-Diamino groups can be unmasked by the reductive opening of imidazoles using palladium on carbon (Equation (9)) ⟨89S825⟩.

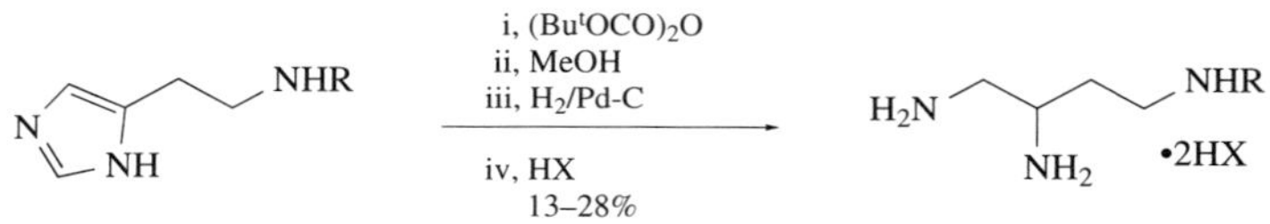

(9)

An iminium salt, generated by alkylation of an imine, can be 'deprotected' by hydrolysis to reveal a secondary amine (Scheme 12). A similar strategy can be adopted using phosphinimines (Scheme 13) ⟨70JOC2826⟩.

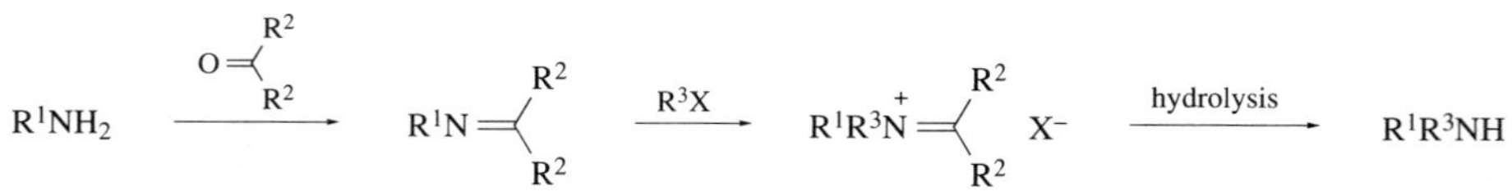

Scheme 12

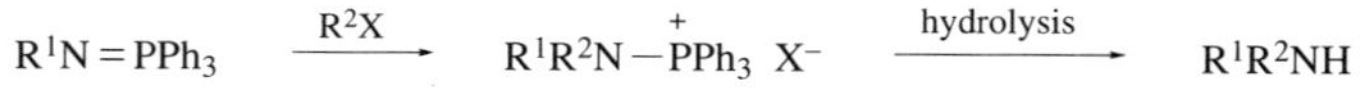

Scheme 13

2.05.3.5 Miscellaneous

Grignard reagents, RMgX, can be aminoalkylated by $R^1R^2NCH_2OBu$ to give the tertiary amines $R^1R^2NCH_2R$ ⟨89TL7383⟩. *N*,*N*-Dialkylarylamines are formed in good yields by the reaction of Grignard reagents with *N*,*N*-bis(benzotriazolylmethyl)arylamines ⟨90CJC456⟩. Arylamines can be

N,N-dialkylated with two different alkyl groups: the first group is introduced by *N*-benzotriazolylmethylation of an amine followed by a Grignard reaction; repetition of that procedure is used to introduce the second alkyl group.

Other procedures involving the addition of Grignard reagents to iminium salts or their equivalent have been reported; the cyano group of certain α-cyano amines can be displaced. Thus, several *N,N*-dialkyl[1-(trimethylsilyl)alkyl]amines have been prepared by the reaction of 2-dialkylamino-2-(trimethylsilyl)acetonitriles with Grignard reagents ⟨90S36⟩.

The reductive decyanation of α-aminonitriles has been achieved using borane–THF complex in THF at room temperature in excellent yields ⟨91JOC2920⟩. An efficient photoinduced α-cyanation of various amines can be effected by cyanotrimethylsilane in acetonitrile, sensitised by *N,N'*-dimethyl-2,7-diazapyrenium bis(tetrafluoroborate) ⟨90TL4735⟩.

Efficient asymmetric hydrogenation of α-aminoketones to the corresponding amino alcohols has been accomplished using neutral chiral rhodium complexes ⟨89TL363, 89TL367⟩. Mixed anhydrides, obtained from chiral *N*-protected α-amino acids and isobutyl chloroformate in 1,2-dimethoxyethane, react cleanly with 1.5 molar equivalents of sodium borohydride in an organic/aqueous medium to afford chiral *N*-protected β-amino alcohols (Scheme 14) ⟨91TL923⟩.

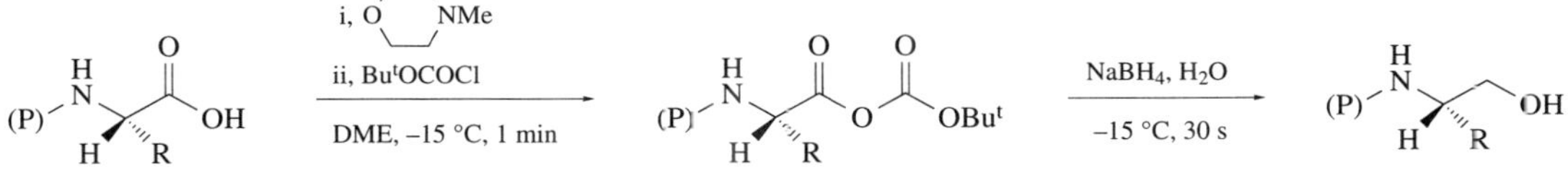

Scheme 14

The free-radical decomposition of arenediazonium salts catalysed by aqueous titanium trichloride and conducted in the presence of aldehydes and arylamines affords secondary amines (Equation (10)) ⟨90TL2069⟩.

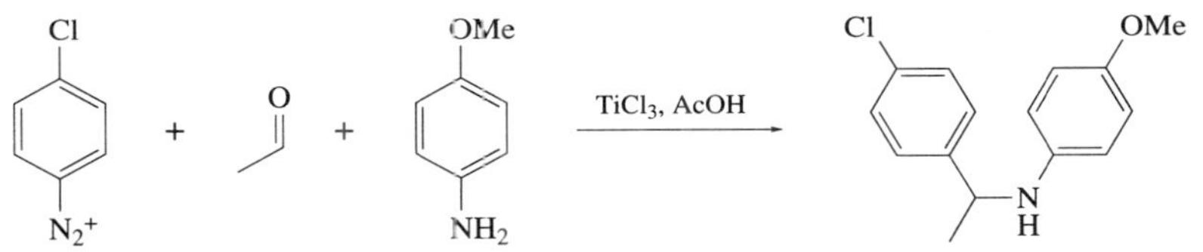

(10)

Oxidation of 4-substituted *N,N*-dimethylanilines with molecular oxygen in the presence of benzoyl cyanide and a catalytic quantity of $FeCl_3$, afforded the corresponding *N*-cyanomethyl-*N*-methylanilines together with *N*-methylformanilides ⟨93BCJ1297⟩.

2.05.4 FUNCTIONAL GROUP MANIPULATION

2.05.4.1 By Amination of Organometallic Reagents

A wide variety of carbanions derived from organometallic species can be aminated ⟨89CR1947⟩. Organolithiums and dialkylzinc reagents can be aminated, but yields are not usually good ⟨70CR639⟩. Alkyllithiums react with methoxyamine and methyllithium in ether at $-78\,°C$ to give primary amines ⟨82JOC2822, 83TL3795⟩. Primary amines can also be prepared by reacting organolithiums with two equivalents each of methoxyamine and methyllithium in hexane-ether ⟨82JOC2822⟩. Yields of the isolated amides (obtained by quenching the reaction with PhCOCl) are typically 70–80%. This avoids the requirement of two equivalents of organometallic reagent, as employed in early work ⟨38JGU1825⟩. Grignard reagents may also be used and they react with a variety of *N*-haloamines including NH_2Cl and NCl_3 to give the corresponding primary amines, although yields are not high ⟨70CR639⟩. In contrast, organoboranes, which are conveniently generated by hydroboration of an alkene, react with either hydroxylamine-*O*-sulfonic acid or chloramine to give the amine usually in about 60% yield ⟨64JA3565⟩. Organoboranes react with a mixture of aqueous NH_3 and NaOCl to produce primary amines ⟨81JOC4296⟩. Reaction of $RBCl_2$ with R^1N_3 affords RR^1NBCl_2, which on alkaline hydrolysis gives $RNHR^1$. Such secondary amines can also be prepared in good yield by reaction of R_2BCl with R^1N_3 ⟨87T4079⟩. A valuable feature of the use of organoboron reagents is the retention of configuration; use of optically active $RBCl_2$ has been shown to give secondary amines of extremely high enantiomeric purity ⟨91JOC1170⟩.

A route to tertiary amines, RNR_2^1, involves reaction of dialkylcopperlithium reagents R_2CuLi with RNH_2 ⟨80JOC2739⟩; however, yields are usually modest.

N-Triphenylphosphorylidene-1-(benzotriazol-1-yl)methylamine (**7**), a synthetic equivalent of $^+CH_2NH_2$, reacts with Grignard reagents and organolithiums to give after workup the hydrochloride salts of the corresponding homologous primary amines (Equation (11)) ⟨89TL3303⟩. This is a useful extension of the use of Grignard reagents to prepare aliphatic amines.

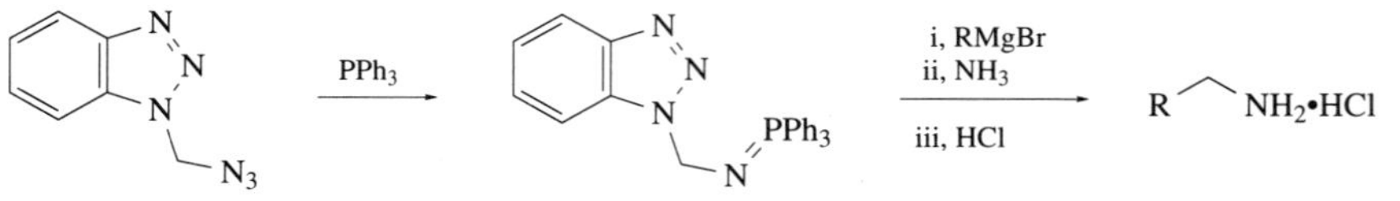

(**7**) (11)

A chiral centre adjacent to nitrogen is created by the diastereoselective ring opening by Grignard reagents of 2-substituted-*N*-benzyl-4,4,7α-trimethyl-*trans*-octahydro-1,3-benzoxamines, prepared from [1R(1α,2β,5α]-5-methyl-2-(1-benzylamino-1-methylethyl)cyclohexanol and aldehydes ⟨90SL763⟩. This route provides a highly enantioselective synthesis of primary amines (Equation (12)).

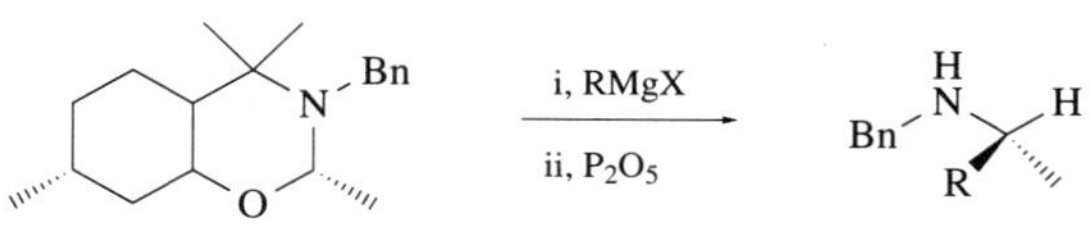

(12)

Unsaturated amines are valuable intermediates in synthesis. Reaction of lithium trimethylsilylalkylamide and alkynes in the presence of the zirconocene complex $Cp_2ZrMeCl$ affords β,γ-unsaturated primary amines in up to 80% yield (Equation (13)) ⟨89JA4486⟩. Tertiary propargylamines can be prepared from 1-alkynyllithiums by using *N*-substituted benzotriazoles, such as (**8**), as synthetic equivalents of substituted iminium cations (Scheme 15) ⟨89S31⟩.

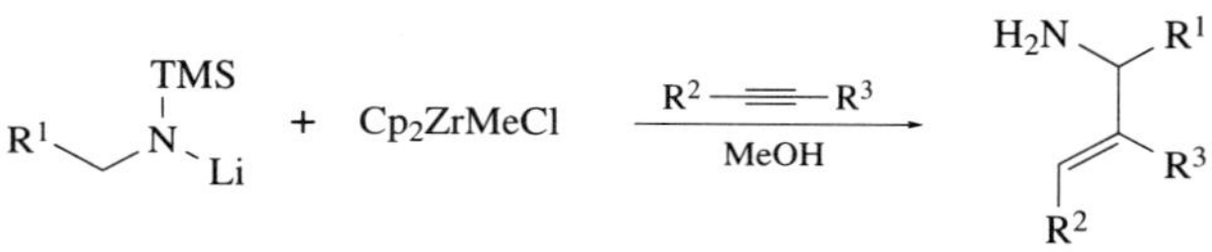

(13)

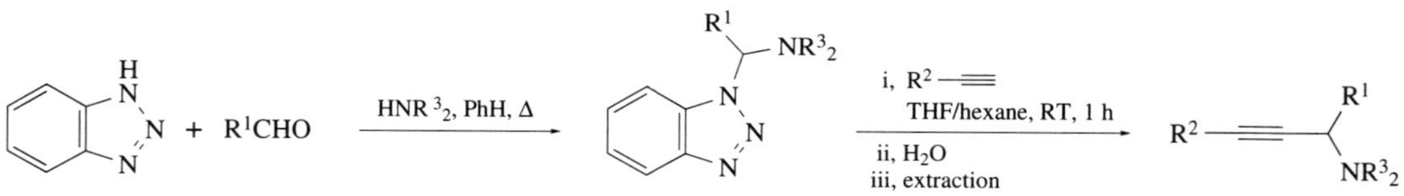

(**8**)

$R^1 = H, Pr^i, Ph; R^2 = Ph, C_6H_{13}; NR^3_2 = N(Bn)_2, NMe_2, N(CH_2CH_2)_2O$

Scheme 15

2.05.4.2 Amines from Alkanes and Alkenes

The tertiary position of an alkane, arylalkane or cycloalkane can be aminated using trichloramine and aluminum chloride at 0–10 °C ⟨76JA5678⟩.

The amination of alkenes has been reviewed ⟨83T703⟩. Intramolecular amination of alkenes is a versatile procedure for the synthesis of cyclic amines (see Section 2.05.5). Primary amines can be prepared by the regioselective *anti*-Markovnikov amination of terminal alkenes. Thus, alkylboration affords the corresponding dimethylalkylboranes which are then reacted with chloramine to give isomerically pure alkylamines ⟨89SC2409⟩.

Aliphatic and aromatic aldehydes react with *N*-sulfinyl-*para*-toluenesulfonamide and an allylsilane in the presence of a Lewis acid to furnish homoallylic sulfonamides, from which the corresponding α-substituted primary amines can be obtained (Equation (14)) ⟨90SC573⟩.

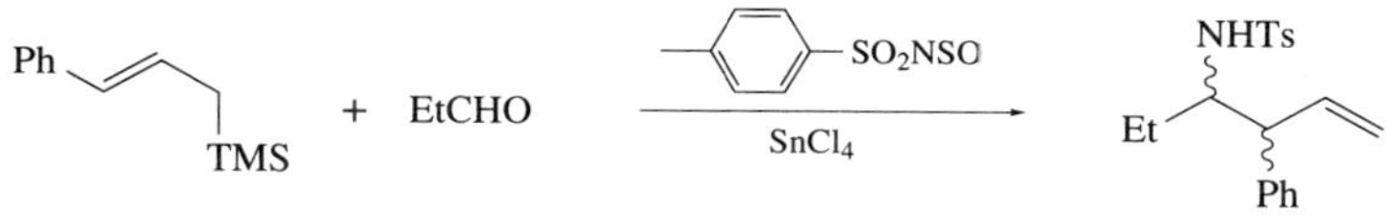

(14)

Alkenes react with isothiocyanic acid to give isothiocyanates. Such primary- and secondary-alkyl as well as aryl isothiocyanates react with 4-methyl-1,2-benzenedithiol in methanol at room temperature to give the corresponding primary amines ⟨92TL3599⟩.

Trialkylboranes, prepared from alkenes and boron trifluoride etherate, react with trimethylsilyl azide in a protic solvent under neutral conditions to give primary amines in moderate yields ⟨88SC1363⟩.

Terminal alkenes react with *N*,*N*-dibromophosphoramidate and boron trifluoride etherate to give diethyl *N*-(β-bromoalkyl)phosphoramidates. Subsequent treatment with sodium azide in DMSO furnishes various vicinal diamines (Scheme 16) ⟨90S505⟩. 2,5-Dihydrofuran is transformed selectively into *trans*-tetrahydro-3,4-furandiamine by an efficient four-step sequence ⟨88JHC1035⟩. Nitromercuration of 2,5-dihydrofuran with sodium nitrite and mercury(II) chloride followed by elimination with aqueous sodium hydroxide afforded 3-nitro-2,5-dihydrofuran which was treated with aqueous ammonia, and the resulting *trans*-tetrahydro-4-nitro-3-furanamine was catalytically reduced over palladium on carbon. Heating a mixture of an allylamine with a nitrone in an inert solvent affords 1,2,5-oxadiazinanes ⟨93CC169⟩. Reduction of the oxadiazinanes furnishes the vicinal diamines and hence achieves overall amination of the alkene moiety of the allylamine.

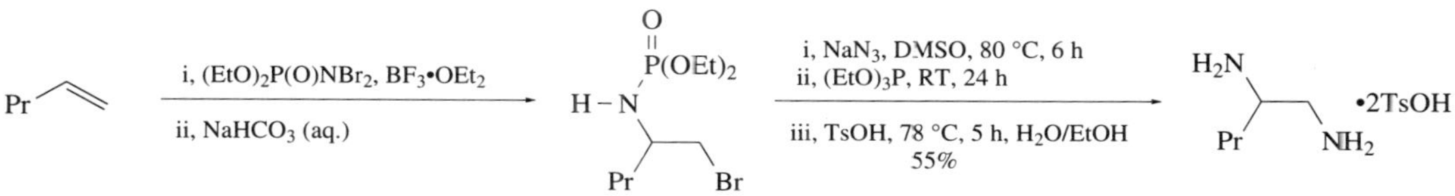

Scheme 16

The enantioselective synthesis of allylic amines (> 95% ee) from alkynes has been achieved using trifluoromethanesulfonic acid, an appropriately substituted lithium anilide and a catalyst derived from [1,2-ethylenebis(η^5-4,5,6,7-tetrahydro-1-indenyl)]zirconium dichloride (Equation (15)) ⟨91JA2321⟩. The pathway has been shown to involve ring expansion of an intermediate zirconaziridine by the alkyne, giving a metallapyrroline.

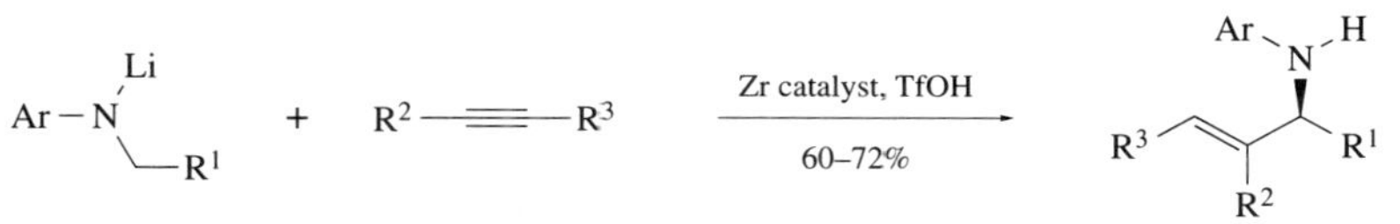

(15)

R^1 = Me, Pr^i, Bu; R^2 = Me, Et; R^3 = Me, Et, TMS >95% *ee*

The palladium catalysed cross-coupling of aryl acetylenes with 2-bromo-1,1-dichloroethane affords the 4-aryl-1,1-dichloro-1-buten-3-ynes (**9**); treatment of the latter with 2 equivalents of lithium dialkylamide furnishes the (4-aryl-1,3-butadiynyl)amines (Equation (16)) ⟨90S125⟩.

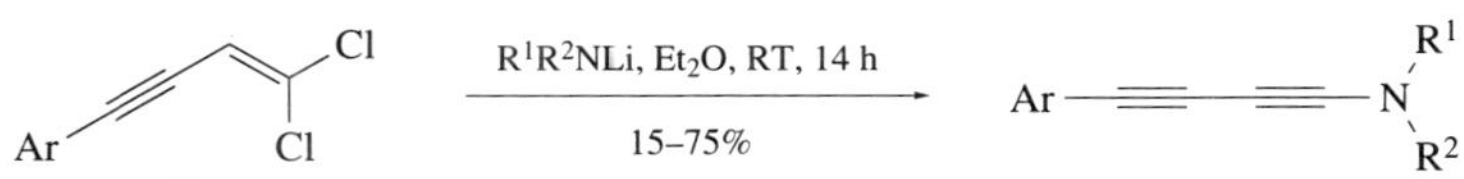

(16)

(9)

Dioxymolybdenum(VI) complexes catalyse the allylic amination of alkenes by phenyl hydroxylamine; the corresponding β,γ-unsaturated amines are obtained in moderate yields ⟨92CC853⟩.

2.05.4.3 Amines from Halides

2.05.4.3.1 From alkyl halides

The order of reactivity of alkyl halides with amines is generally I > Br ≫ Cl ≫ F. Iodide ion can be added to promote halogen exchange and hence increase the reactivity of other halides as

N-alkylating agents. Increasing substitution in the amine or halide decreases reactivity; the greater the alkyl branching in the halide, the more likely is *β*-elimination to occur (assuming there are one or more suitably placed hydrogen atoms). Tertiary amines can usually be prepared by alkylation of the lithium salt of a secondary amine ⟨69BCJ3606⟩.

A convenient preparation of R_2NH free from primary or tertiary amines involves reaction of an alkyl halide with the sodium or calcium salt of cyanamide; the disubstituted cyanamide is then hydrolysed and decarboxylated to give the secondary amine ⟨78S882⟩.

Optimised procedures for the one-pot conversion of primary and secondary alkyl bromides into the corresponding primary amines via the Staudinger reaction have been described ⟨92S1063⟩. The reduction using sodium borohydride in methanol of crude imines, prepared from *N*-alkyltriethoxyiminophosphoranes and aldehydes via an aza-Wittig reaction, furnishes secondary amines as their hydrochloride salts in moderate to good yields ⟨92SC1929⟩.

The Gabriel synthesis has been discussed in Section 2.05.2. Sodium diformylamide has been employed as a convenient substitute for phthalimide in the Gabriel synthesis; primary alkyl iodides have been converted into primary amines in up to 96% yield (Scheme 17) ⟨90S122⟩.

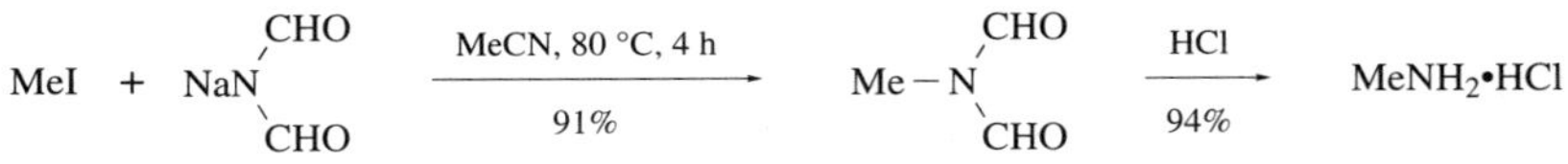

Scheme 17

Several *N,N*-diformylaminoketones, isolated as their hydrochloride salts, have been obtained by treating the corresponding bromomethylketone with sodium diformylamide in acetonitrile at room temperature (Equation (17)) ⟨90S615⟩.

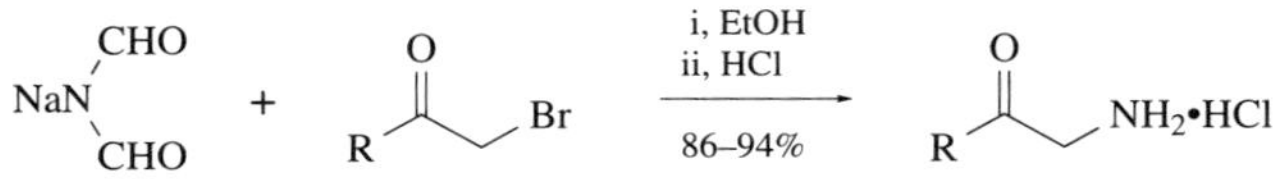

(17)

The treatment of enamines with an excess of diethylamine or piperidine in dichloromethane affords the corresponding *N,N*-dialkylaminomethyl derivatives ⟨93SC817⟩. The dichloromethane acts as a one-carbon source in the reaction.

2.05.4.3.2 *From allylic halides and acetates*

Active halides such as allylic and benzylic halides, and primary iodides are converted into primary amines by reaction with hexamethylenetetraamine (the Délépine reaction) ⟨79S161⟩.

Alkylation of primary and secondary amines with allyl halides using a mixture of copper(II) perchlorate and copper metal affords the corresponding secondary and tertiary allylic amines in good yields ⟨91T9449⟩.

N-Boc-protected allylamines can be prepared directly from allylic acetates using a palladium catalyst and $HN(Boc)_2$ ⟨88JOC3845⟩. *N*-Allylhydroxylamines are produced by the palladium-catalysed reaction of hydroxylamines with allyl esters. Conversion into the corresponding secondary allylamines is accomplished using zinc powder in dilute aqueous hydrochloric acid (Scheme 18) ⟨88TL2973⟩.

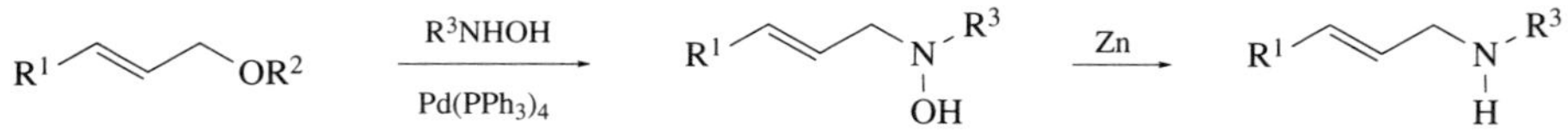

Scheme 18

2.05.4.4 Amines from Hydroxyl Compounds and their Derivatives

The Ritter reaction (Scheme 19) provides a longstanding means of converting an alcohol, via a nitrilium salt, into a primary amine ⟨48JA4045, 60RCR331, 66AHC(6)95, 69OR(17)213⟩. The intermediate amide need not be isolated. Alcohols that do not give stable carbocations can be treated with

$Ph_2CCl^+SbCl_6^-$ in the nitrile as solvent ⟨74JCS(P1)2101, 81JOC78⟩. The Ritter reaction applies particularly to tertiary alkylamines, *t*-RNH_2, which is useful since there are few alternative direct routes to such amines.

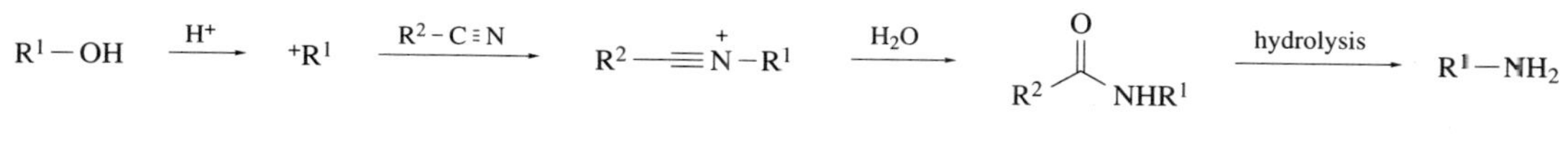

Scheme 19

Using a modified Staudinger procedure, primary alcohols undergo a one-pot conversion into the corresponding primary amines via the intermediate azide in moderate to good yield ⟨89JCR(S)296⟩. A one-pot procedure involving the treatment of alcohols with hydrazoic acid, diisopropyl azodicarboxylate and an excess of triphenylphosphine in THF, followed by hydrolysis, furnishes primary amines in moderate or better yields (Scheme 20) ⟨87S190⟩. A number of amino protected 3-substituted-1,2-diaminopropanes have been synthesized from the corresponding *N,N′*-bis-protected 2-(aminomethyl)aziridine derivatives, prepared from 2-hydroxy-1,3-diaminopropane, by nucleophilic ring opening of the aziridine ring ⟨89JOC1940⟩. Subsequent nucleophilic attack allows the incorporation of various functional groups. Regioselective palladium(0)-catalysed azidation, followed by reduction of 1-alkenylcyclopropyl tosylates, affords a useful route to α-substituted allylamines (Scheme 21) ⟨93TL4193⟩. Additionally, these tosylates undergo S_N2' attack to give (2-cyclopropylidineethyl)amines.

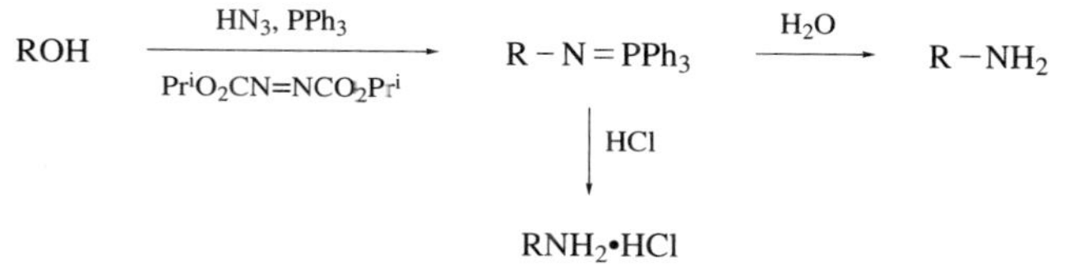

Scheme 20

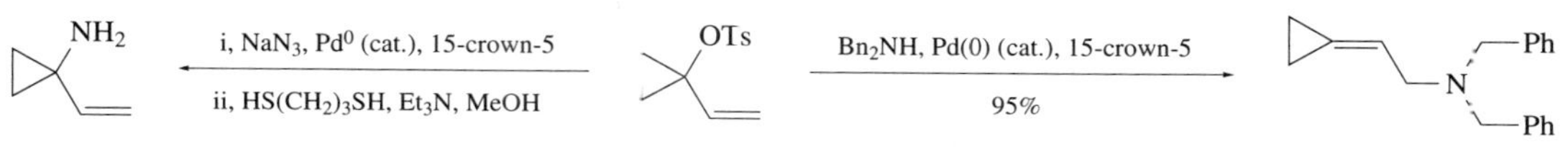

Scheme 21

Similarly, reaction of *N*-methyltrifluoromethanesulfonamide, triphenylphosphine and diethyl azodicarboxylate allow the conversion of alcohols to secondary amines ⟨90TL3417⟩. Mitsunobu conditions have been employed in the coupling of *N*-methyl *para*-toluenesulfonamide and *N*-Boc *para*-toluenesulfonamide with primary and secondary alcohols to furnish various sulfonamide-protected amines (Equation (18)) ⟨89TL5709⟩. Mesylation of the hydroxyl moiety of a variety of substituted oxazolidinones and pyrrolidinones and subsequent reaction with aqueous methylamine or dimethylamine furnishes the secondary or tertiary amines as their hydrochloride salts (Scheme 22) ⟨93JMC3606⟩.

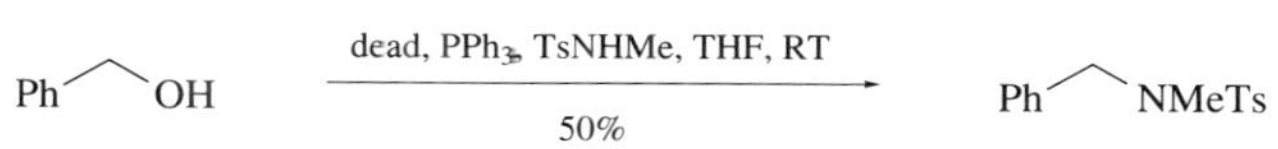

(18)

Reaction of primary alcohols with dimethylamine hydrochloride in refluxing acetic acid proceeds smoothly giving the corresponding secondary amine; with 2-(1-hydroxyethyl)pyridine, for example, the amine is obtained in 64% yield ⟨89AP(322)181⟩. The reaction of 5-hydroxy-1-naphthalenesulfonic acid with tris(2-aminoethyl)amine affords 5-tris(2-aminoethyl)amino-1-naphthalene sulfonic acid in 57% yield ⟨93BMC2883⟩.

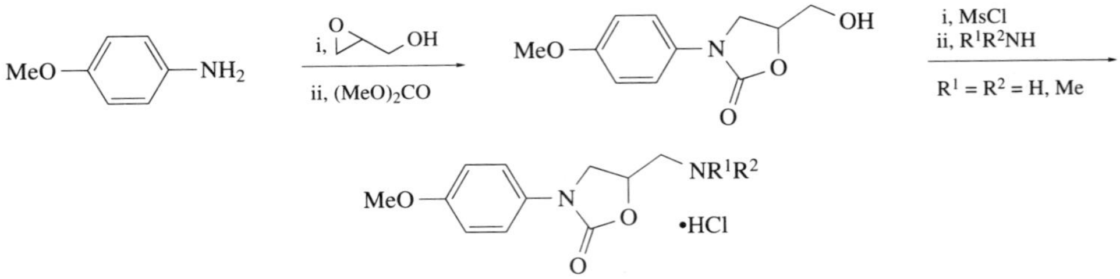

Scheme 22

2.05.4.5 Amines from Acetals and Ketals

Some β-amino ethers can be prepared by the action of sodium cyanoborohydride on hemiaminals, prepared from hemiacetals and benzylamine ⟨93S1089⟩.

The reaction of isothiosemicarbazide hydrohalides with lactam acetals and lactim ethers affords 3-(ω-aminoalkyl)1,2,4-triazoles as their hydrohalide salts ⟨93JHC1061⟩.

2.05.4.6 Amines from Aldehydes and Ketones

Reductive amination of an aldehyde or ketone with ammonia or an amine in the presence of a reducing agent, frequently hydrogen and a catalyst, is a longstanding route to amines ⟨80RCR14⟩. Other reducing agents that can be used include sodium borohydride ⟨87S709⟩, sodium cyanoborohydride ⟨75S135⟩, and iron pentacarbonyl and alcoholic KOH ⟨74TL1879⟩. The Wallach procedure employs formic acid as the reducing agent ⟨49OR(5)301⟩. The Eschweiler–Clarke procedure uses a mixture of formaldehyde and formic acid, thereby affording $RNMe_2$ and R_2NMe from RNH_2 and R_2NH, respectively ⟨71JOC829⟩. The same conversions can be accomplished by treating the amine with aqueous formaldehyde and $NaBH_4$ or $NaBH_3CN$ ⟨77TL3469⟩.

The reductive amination of aldehydes and saturated aliphatic ketones with primary and secondary amines using sodium triacetoxyborohydride has been reported ⟨90SL537, 90TL5595⟩. Several α-amino nitriles have been prepared by reacting a primary amine with an aldehyde and trimethylsilyl cyanide ⟨92TL6295⟩.

Reductive amination of cyclic ketones with piperidine using sodium cyanoborohydride in the presence of titanium(IV) isopropoxide affords tertiary amines ⟨90JOC2552, 90LA795⟩. The overall reductive amination of an aldehyde with a secondary amine can be achieved by reductive cleavage of the benzotriazol-1-yl moiety in (**10**) using sodium borohydride (Scheme 23) ⟨89JCS(P1)225, 89JCS(P1)639⟩. By using a Grignard reagent in place of borohydride in the above sequence, a very useful route to α-substituted tertiary amines is provided. The reductive amination of aliphatic and heterocyclic aldehydes with various substituted piperidines has been reported using a borane–pyridine complex in ethanol ⟨93SC789⟩.

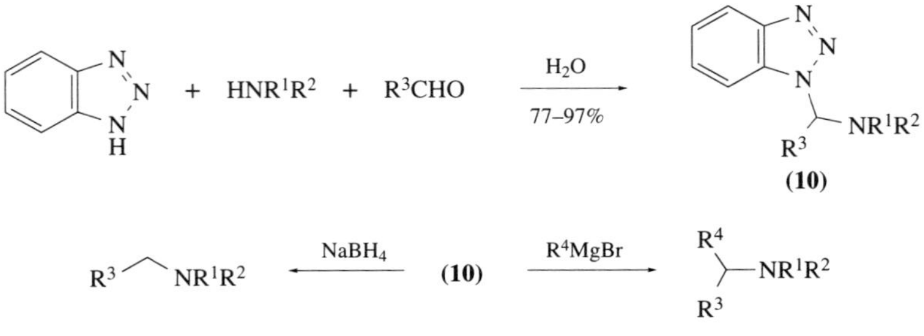

Scheme 23

The reaction of α-tetralones with dibenzyl azodicarboxylate affords protected α-hydrazino ketones which, depending on the reduction conditions, furnish predominantly either *trans*- or *cis*-hydrazino alcohols. Subsequent hydrogenolysis affords the α-amino alcohols ⟨92LA273⟩.

Treatment of aldehydes with a mixture of *N*-sulfinyl-*para*-toluenesulfonamide, triethylsilane and boron trifluoride etherate as the catalyst affords sulfonamides (Equation (19)) ⟨90JOC2563⟩.

α-Cyanobenzylamines can be made by treating arylaldehydes with a primary amine in the presence of KCN (Equation (20)) ⟨89S616⟩. Secondary amines are obtained with complete *erythro* selectivity from ketones and primary amines using $Mg(ClO_4)_2$ and sodium borohydride ⟨90TA163⟩. Tosylmethyl isocyanide and aldehydes in the presence of chiral silver(I) catalysts undergo an asymmetric aldol reaction giving optically active 5-alkyl-4-tosyl-2-oxazolines in up to 86% ee (Scheme 24) ⟨90JOC5935⟩. Subsequent reduction with lithium aluminum hydride affords optically active α-alkyl-β-(*N*-methylamino)ethanols.

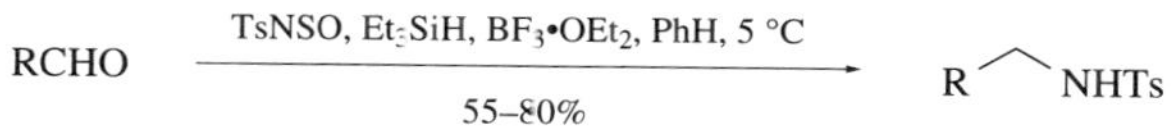

(19)

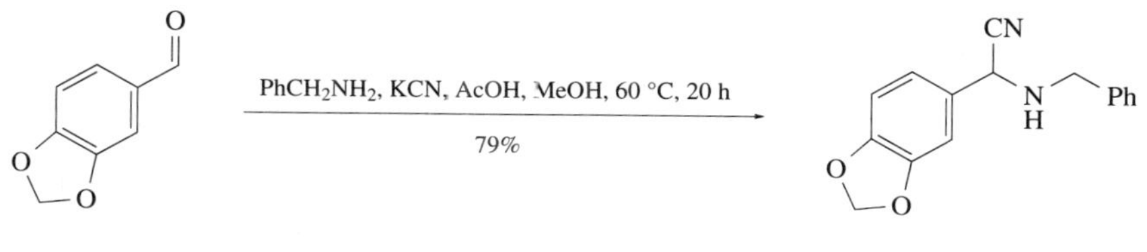

(20)

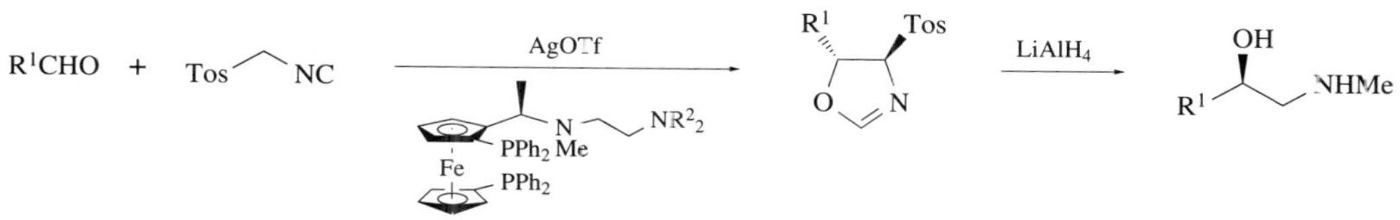

Scheme 24

Hydrogen telluride reduces imines and enamines to the corresponding amines under mild conditions ⟨87CL1275⟩. This procedure provides a new method for the reductive alkylation of primary and secondary amines with aldehydes and ketones. Perfluoroalkylorganotitanium reagents ($R^1Ti(NEt_2)_3$) have been employed in the fluoroalkylative amination of aldehydes (R^2CHO); the resulting tertiary amines $R^1CH(NEt_2)R^2$ are obtained in moderate to good yields ⟨90JCR(S)268⟩.

3-Aminocyclopent-2-en-1-one had previously been prepared by the use of very high pressure. It can, however, be prepared by conversion of cyclopentane-1,3-dione into 3-ethoxycyclopent-2-en-1-one (ethanol–benzene, cat. TsOH at reflux) which is then treated with ammonia, in a Parr pressure vessel, first at −78 °C, then by gradual heating to 100 °C; the overall yield is 69% ⟨91S176⟩.

2.05.4.7 Amines from Imines

The reaction of [*N*-(*para*-toluenesulfonyl)imino]phenyliodinane, PhI=NTs, with trialkylboranes in THF at room temperature furnishes *N*-toluenesulfonyl protected primary amines in good to excellent yield ⟨93S481⟩.

Reduction of azomethines using a catalytic amount of di-(*ortho*-aminophenyl)disulfide or *ortho*-aminothiophenol in alcohol in the presence of hydrogen bromide furnishes the corresponding arylamines ⟨91SL668⟩.

(Perfluoroalkyl)lithiums, generated *in situ* from primary perfluoroalkyl iodides and MeLi—LiBr, have been reacted with imines in the presence of boron trifluoride etherate to give perfluoroalkylated amines in good yields (Equation (21)) ⟨92JOC1504⟩.

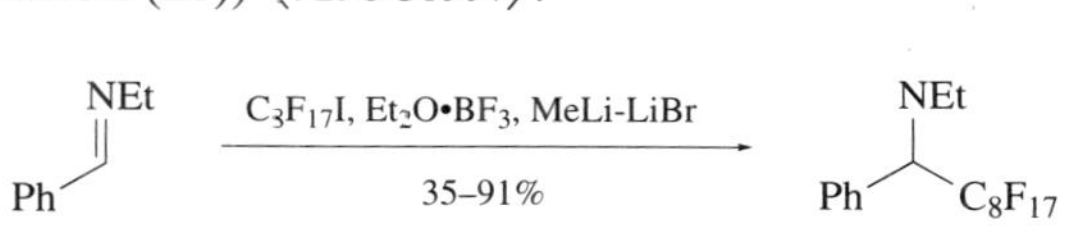

(21)

Reductive dimerisation of aldimines using an excess of samarium(II) iodide in refluxing THF gives vicinal diamines ⟨90SC981⟩.

The naphthalene-catalysed lithiation of phenone imines in the presence of carbonyl compounds gives 1,2-amino alcohols ⟨93T7761⟩.

2.05.4.7.1 By reduction

Imines are reduced to amines by catalytic transfer hydrogenation using isopropanol and $RuCl_2(PPh_3)_3$ ⟨92CC980⟩. Imines can be reduced to the corresponding amines, with high chemo- and stereoselectivities, using Hantzsch's ester in the presence of silica gel ⟨89BCJ3845⟩. Under similar conditions allylic amines can be prepared from α,β-unsaturated imines by selective 1,2-reduction with Hantzsch's ester ⟨89BCJ3845⟩. 4-Amino-1-aza-1,3-dienes are reduced to 1,3-diamines in near quantitative yield by $Na\text{-}Pr^iOH$ ⟨83JOC2255⟩.

Imines are reduced using sodium borohydride and cobalt chloride in methanol/THF at −10°C to the corresponding secondary amines ⟨89SC565⟩. Borane-methyl sulfide in dichloromethane has been applied to the reduction of α-chloro-, α,α-dichloro- and α,α,α-trichloroimines ⟨91T3407⟩. The corresponding β-chloramines were obtained in 80–96% yield. A convenient preparation of allyl 2-benzylaminoalkyl ethers involves the *O*-allylation of 2-(benzylidenamino)alcohols followed by reduction with sodium borohydride ⟨89S963⟩. The reduction of imines with zinc borohydride affords secondary amines in excellent yields, in some cases the amine–borane complexes require treatment with 6M hydrochloric acid to liberate the amine ⟨90S401⟩. This procedure has also been applied to the alkylation–reduction of nitriles; thus reaction of R^1CN with R^2MgBr or R^2Li followed by reduction with $Zn(BH_4)_2$ afforded $R^1R^2CHNH_2$ ⟨90S401⟩. Aldimines, synthesized from aldehydes and primary amines, can be α-halogenated and subsequently treated with base to afford the corresponding α,β-unsaturated aldimines ⟨88S587⟩. The selective reduction of the imino function with sodium borohydride furnishes secondary allylic amines with a high preponderance of the (*E*)-isomer (Scheme 25). Ketimines are reduced to the corresponding primary and secondary amines by treatment with ytterbium metal followed by water ⟨90CL203⟩. The catalytic enantioselective hydrogenation of imines is achieved using iridium(III) hydride complexes. The reaction is essentially quantitative in chemical yield and shows high enantiomeric purity ⟨90JA9400⟩. The asymmetric reduction of *N*-substituted ketimines with Itsuno's reagent (**11**) has been achieved; secondary amines are obtained in excellent yield with good optical purity (Equation (22)) ⟨90JCS(P1)3200⟩. A lead/aluminum bimetallic redox system effects pinacol-type coupling of imines, affording symmetrical *vic*-diamines ⟨88TL3811⟩.

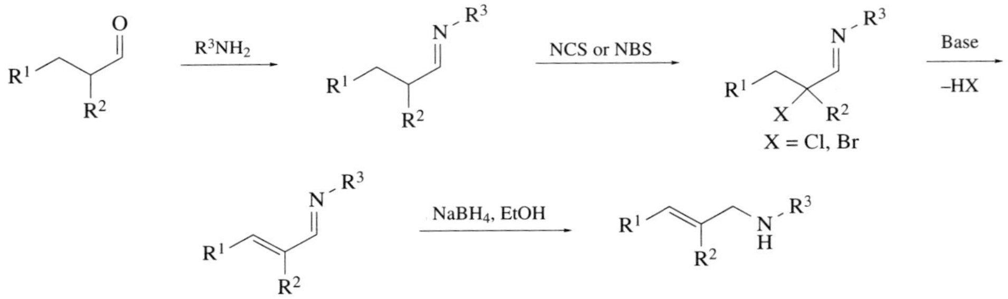

Scheme 25

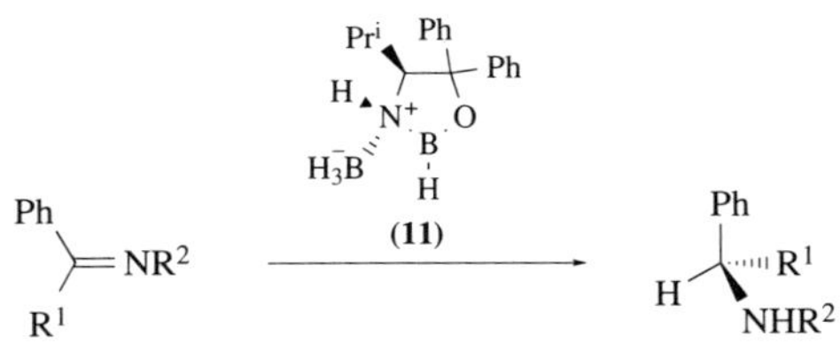

(22)

R^1 = Me, Et, Pr, Pr^i; R^2 = Ph, Bn, C_7H_{13}

2.05.4.7.2 By addition of an organometallic reagent

Organolithiums add to aldimines and ketimines to give the corresponding α-branched amines ⟨64BSF952⟩. Grignard reagents also add to aldimines to afford secondary substituted amines, but the reactions with ketimines usually lead to reduction rather than addition ⟨B-70MI 205-01⟩. Tertiary carbinamines are obtained in excellent yields by the two-fold addition of organocerium reagents to nitriles. The organocerium reagents are prepared by the reaction of aromatic and primary and secondary alkyllithium reagents with anhydrous cerium chloride (Equation (23)) ⟨92JOC4521⟩.

(23)

Secondary substituted amines result from Barbier-type allylation of imines using allyl bromide and aluminum and titanium(IV) chloride in THF (Equation (24)) ⟨90TL3023⟩.

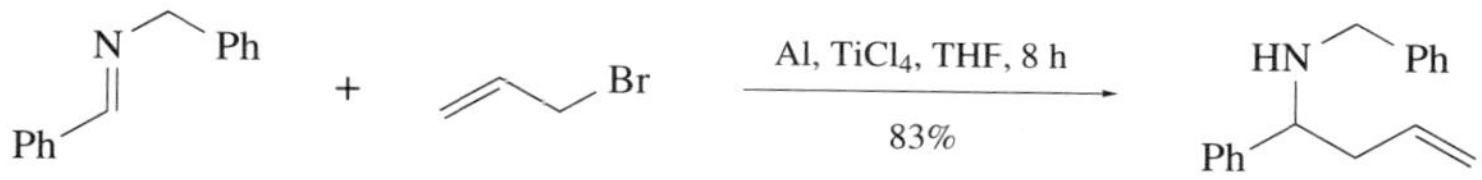

(24)

Primary amines react with (*Z*)-*β*-iodo acrylaldehyde to afford the intermediate imines which, on treatment with various organometallic reagents, furnish the corresponding (*Z*)-*γ*-iodo allylic amines (Scheme 26) ⟨93SL386⟩.

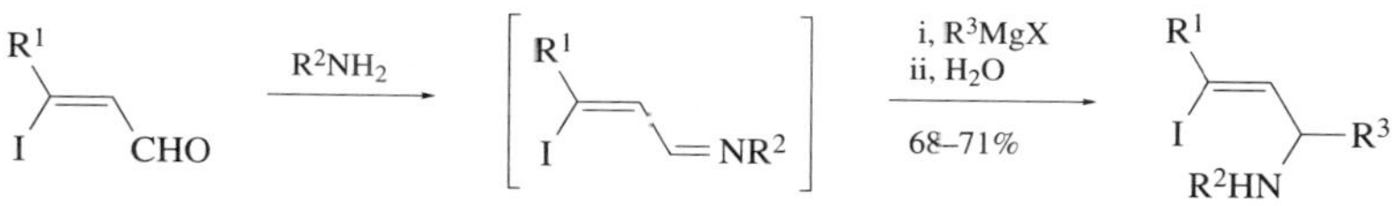

R^1 = H, Me; R^2 = $PhCH_2$, *p*-MeOPh; R^3 = allyl, butyl

Scheme 26

A reductive allylation of imines with allyl bromide can be effected in THF in the presence of $PbBr_2$ and tetrabutylammonium bromide using a platinum cathode and an aluminum anode ⟨89TL4161⟩.

Amphiphilic reaction systems, e.g., $BF_3 \cdot OEt_2$/RLi, RMgX, R_2CuLi, or trimethylsilyl trifluoromethanesulfonate/RLi, RMgX have been used in a procedure for *α*-addition to the C=N double bond of 3,4-dihydro-*β*-carboline (Scheme 27) ⟨93CPB287⟩. The corresponding 1-substituted-1,2,3,4-tetrahydro-*β*-carbolines are obtained in moderate to good yield.

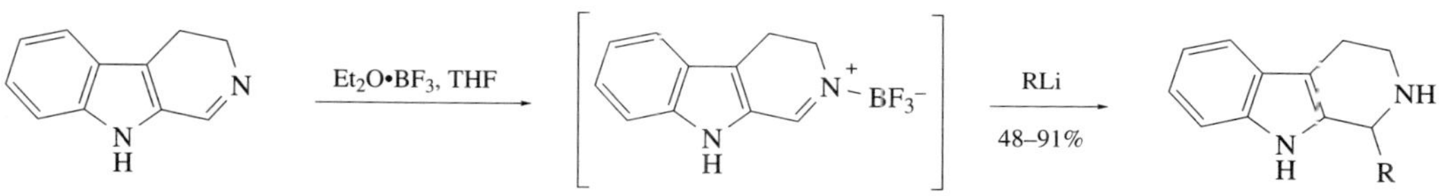

R = Me, Bu^n, Bu^s, Bu^t, Ph, CH_2CN. $CH=CH_2$

Scheme 27

The reaction of trimethylsilyl trifluoromethanesulfonate with imines affords *N*-trimethylsilyliminium salts which react with a variety of organolithium and Grignard reagents to give secondary amines in moderate yields ⟨88SC893⟩.

The achievement of high levels of asymmetric induction during the addition of organometallic reagents to imines is a very significant goal. Amines can be prepared enantioselectively by the addition of alkyllithium reagents to *N*-(trimethylsilyl)benzaldimine in the presence of a variety of chiral modifiers ⟨91JCS(P1)1341⟩. Reaction of aromatic aldehydes with (*R*)-phenylglycinol affords imines that react with organolithiums or Grignard reagents to give the corresponding (*R*,*R*)-amines with good diastereoselectivity. In contrast, the same aldehydes were reacted with (*R*)-*N*-methylphenylglycinol to give the corresponding 1,3-oxazolidines, which then underwent ring opening with the same organometallic reagents, but to give the corresponding (*R*,*S*)-amines ⟨92TL235⟩.

2.05.4.7.3 By addition of other carbon nucleophiles

In the presence of Lewis acid catalysts, the reaction of mono-*S*-trimethylsilyl ketene dithioacetals with imines furnishes the corresponding secondary *β*-amino dithioesters with *anti*-selectivity (Equation (25)) ⟨92TL5339⟩. 2-Hydroxy-2-allyl-1-imines and 2-hydroxy-2-propargyl-1-imines rearrange in refluxing DIGLYME to α-aminoketones ⟨90TL2277⟩; those processes are examples of the normally unfavourable 1-aza-Cope rearrangement being assisted by an hydroxy substituent, as is the case in oxy-Cope rearrangements.

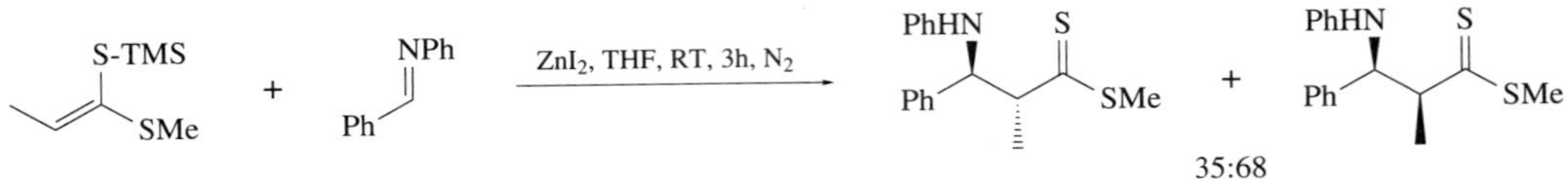

(25)

2.05.4.8 Amines from Oximes and their Derivatives

Primary amines can be obtained from both aldoximes and ketoximes by reduction with $LiAlH_4$ ⟨50CR(230)304⟩. Many other reagents have been used, of which diborane ⟨69JOC1817⟩ and $NaBH_3CN$—$TiCl_3$ ⟨88SC777⟩ are notable. Catalytic hydrogenation is also effective ⟨B-67MI 205-01⟩.

A mild and convenient method for the reduction of oximes to primary amines in a single step has been developed using titanium trichloride and sodium cyanoborohydride ⟨88SC777⟩. The procedure is useful for substrates in which the intermediate imine is readily hydrolysed.

The electroreduction of *O*-methyl oximes with ketones furnishes the intermolecularly coupled products ⟨91TL525⟩. The initially formed 2-methoxyamino alcohols are easily reduced to 2-amino alcohols. Reduction of O-mesyloximes, prepared from α,β-unsaturated ketones, furnishes allylic amines in moderate yield ⟨90SL677⟩. Amines have been prepared by the selective deoxygenation of various N—O bonds, the process being catalysed by rhodium carbonyl clusters in the presence of water and carbon monoxide ⟨88CL1759, 91BCJ602⟩.

α-Bromo oxime ethers react with sodium azide to yield α-azido oxime ethers which on treatment with lithium aluminum hydride furnish 1,2-ethanediamine derivatives. Hydrogenolysis of the intermediate α-azido oxime ethers using 5% palladium on calcium carbonate as the catalyst gives α-amino oxime ethers ⟨92LA1005⟩.

Enantioselective reduction of oximes have been performed with baker's yeast ⟨90TL5555⟩, and with Ph_2SiH_2 in the presence of an optically active rhodium catalyst ⟨86OM739⟩.

Reduction of oxime ethers using sodium borohydride, zirconium tetrachloride and chiral amino alcohols affords primary amines in excellent yields but with only moderate enantioselectivity (Equation (26)) ⟨89JCS(P1)1548, 90JCS(P1)1859⟩.

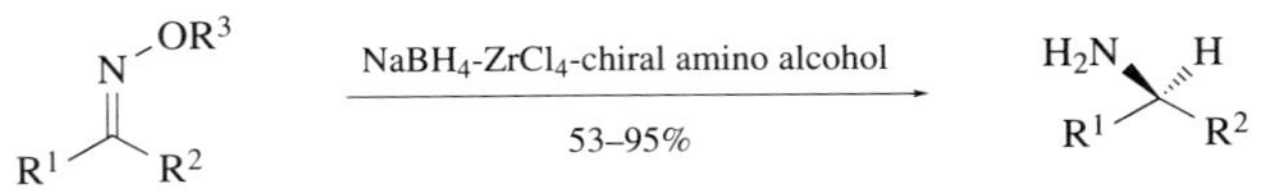

(26)

Prochiral ketoxime ethers are reduced with a chiral reducing agent prepared from (−)-norephedrine and two equivalents of BH_3 with modest to good enantioselectivity (Equation (27)) ⟨88TL223⟩. The absolute configuration of the amine produced is dependent on the geometry of the oxime ether; *anti*-ketoximes give (*S*)-amines, whereas *syn*-ketoximes give the corresponding (*R*)-isomers.

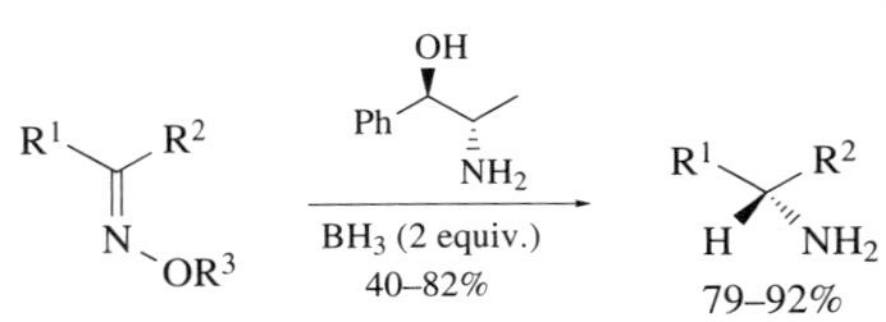

(27)

R^1, R^2 = Me, CH_2Ph, Ph; R^3 = Me, CH_2Ph, Ph, TMS

2.05.4.9 Amines by Addition to α,β-unsaturated Carbonyl Compounds

Sufficiently basic nitrogen nucleophiles such as ammonia and methylamine have long been known to add to electrophilic alkenes such as mesityl oxide in the absence of catalysts ⟨03M773⟩. Since the 1980s, asymmetric addition of nitrogen nucleophiles to electrophilic alkenes has received attention, chiefly owing to the number of natural products that contain β-amino carbonyl moieties ⟨93JCS(P1)1375⟩. The addition of chiral amines to α,β-unsaturated carbonyl compounds is appealing, although initial results were not promising; for example, conjugate addition of α-methylbenzylamine to methyl crotonate in alcoholic solvents proceeds with poor diastereoselectivity ⟨87TL3103⟩. A major advance in levels of the asymmetric induction has been made by the use of chiral lithium amides (Equation (28)). By using C_2-symmetric chiral lithium amides de of greater than 99% have been obtained ⟨91TA183⟩.

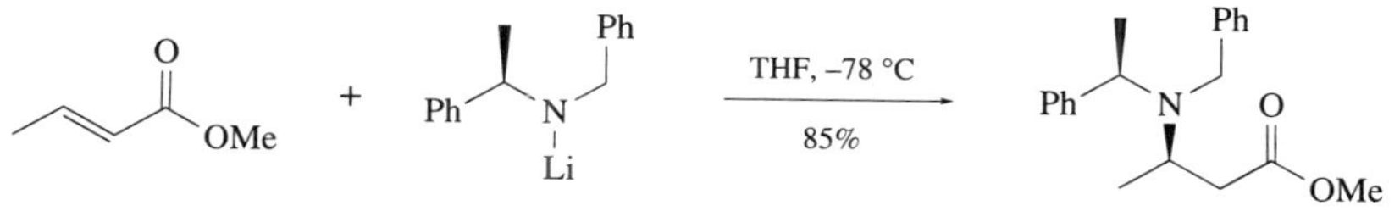

95% *de*

(28)

The above conjugate additions can be extended to functionalisation *alpha* to the carbonyl compound. Thus, conjugate addition of lithium (*R*)- and (*S*)-(α-methylbenzyl)benzylamide (**13**) to *t*-butyl cinnamate affords, after *in situ* hydroxylation with (+)-(camphorsulfonyl)oxaziridine (**14**), the tertiary amine (**12**) with high diastereoselection (Equation (29)) ⟨93SL731⟩. The primary amines are furnished by debenzylation. Michael addition of a nitroalkane to an α,β-unsaturated ester followed by catalytic hydrogenation using 10% palladium on carbon in acetic acid and acidic workup affords a variety of 3-alkyl-4-aminobutanoic acids (Equation (29)) ⟨89S953⟩.

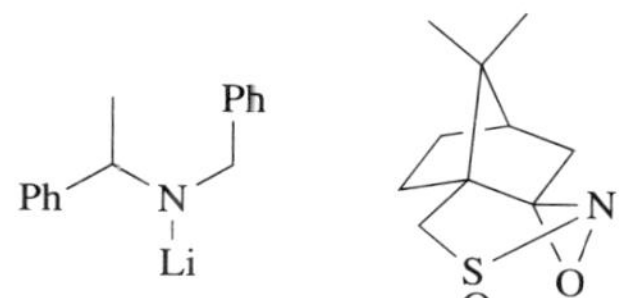

(**13**) (**a**) (*R*)
(**b**) (*S*)

(**14**)

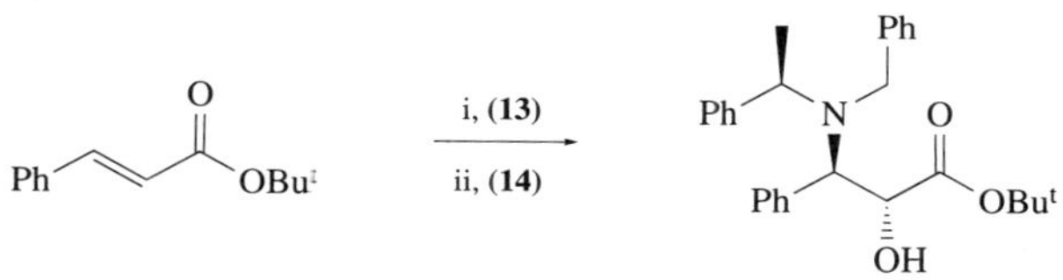

(**12**)

(29)

Conjugate addition reactions of α-aminoalkylcyanocuprates, prepared from organostannyl-*t*-butylcarbamates, with both simple and sterically hindered enones affords the Boc-protected γ-amino ketones ⟨93SL407⟩.

2.05.4.10 Amines from Hydrazines and Related Compounds

Primary amines are obtained by the reductive N—N cleavage of (sterically hindered) *N*-substituted *N'*,*N'*-dimethylhydrazines using hydrogen and Raney nickel (Scheme 28) ⟨91SL625⟩. The reaction is performed under an atmospheric pressure of hydrogen with ultrasonication of the reaction vessel.

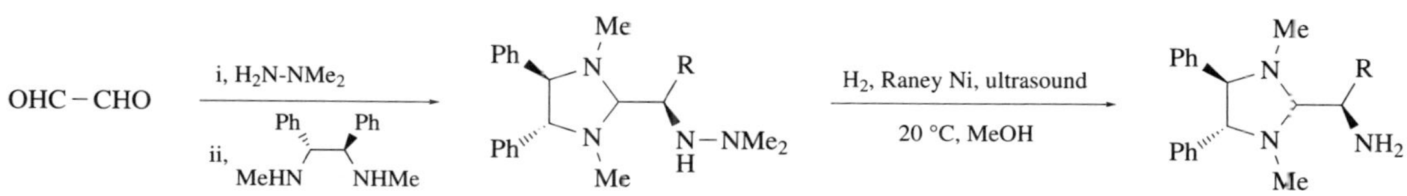

R = Me, Bu^i, Bu^t, Ph

Scheme 28

Ketones, R^1COR^2, have been reacted with 1-aminopyrazoles to give the corresponding alkenylaminopyrazoles which undergo hydrogenolysis with H_2/Pd—C in acid to liberate salts of the amines $R^1R^2CHNH_2$ ⟨92JHC321⟩.

2.05.4.11 Amines from Iminium Salts

In the Mannich reaction ⟨73S703⟩ performed under acid-catalysed conditions, iminium species are derived from the secondary amine and the aldehyde, usually formaldehyde. Attack of an enol intermediate (of the carbonyl compound employed) results in the *β*-dialkylamino carbonyl compound. The reaction of 2-methylfuran with iminium salts in acetonitrile leads to secondary amines ⟨90TL4229⟩. The Mannich reaction of furans with preformed iminium chlorides gives the 2-aminomethylated analogue under mild conditions ⟨88TL2377⟩. A new method for the introduction of amino groups at the α-position of carboxylic acids via an anionic hetero[3,3]rearrangement of *N*-acylhydroxylamine-*O*-carbamates has been reported (Equation (30)) ⟨91SL649⟩.

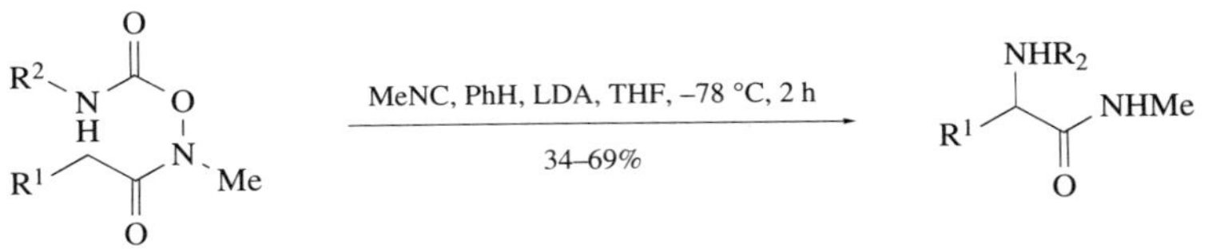

(30)

2.05.4.12 Amines from Nitriles

2.05.4.12.1 By reduction of nitriles

Many efficient methods are available for the reduction of nitriles ⟨B-89MI 205-02⟩. Secondary amines, $(RCH_2)_2NH$, have been prepared in good yields by the rhodium-catalysed hydrogenation of RCN at 1 atm of hydrogen at 20 °C ⟨91JOC452⟩; the reaction is of good scope for both aliphatic and aromatic R substituents. The reduction of nitriles has been accomplished using a mild and neutral reducing system consisting of Al—$NiCl_2 \cdot 6H_2O$—THF, e.g., 5-bromovaleronitrile affords the corresponding primary amine in 80% yield ⟨91T8587⟩. Reduction of nitriles with sodium borohydride–iodine in THF affords the corresponding amines (70–75%) ⟨92T4623⟩.

2.05.4.12.2 By reduction, then addition

N-diisobutylaluminum imines, obtained from nitriles and diisobutylaluminum hydride in pentane, have been reacted with a variety of alkyl and aryl Grignard and lithium reagents, thereby affording primary amines ⟨90JOC4199⟩. *N*-Boryl imines, prepared from benzonitrile and various alkylboranes, have been converted into secondary carbinamines by treatment with organolithium and Grignard reagents ⟨92TL627⟩. α-Substituted-α-phenyl primary amines are secured by the partial reduction of benzonitrile with borane–THF and subsequent addition of an organolithium or Grignard reagent ⟨91JCS(P1)1767⟩.

2.05.4.12.3 By addition, then reduction

Primary amines can be prepared by reaction of silyl-protected α-hydroxy nitriles with Grignard reagents and subsequent reduction using sodium borohydride. High yields and good diastereoselection are reported ⟨91LA269⟩. Tertiary carbinamines can be prepared by the twofold addition of organocerium reagents (prepared from alkyllithium reagents and anhydrous cerium chloride) to nitriles in excellent yields (Equation (31)) ⟨92JOC4521⟩. Treatment of (*R*)-(+)-α-[*t*-butyldimethylsilyl)oxy]-phenylacetonitrile with an excess of methylmagnesium iodide affords the imine intermediate which on reaction with 2 equivalents of alkylamine and subsequent reduction furnishes secondary amines. The procedure is notable for being a one-pot reaction and for the high level of *erythro* selectivity that is observed (Equation (32)) ⟨91RTC25⟩.

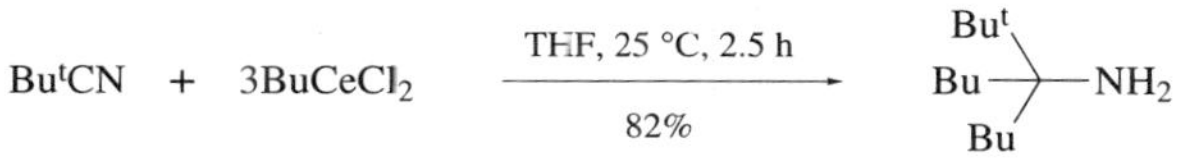

(31)

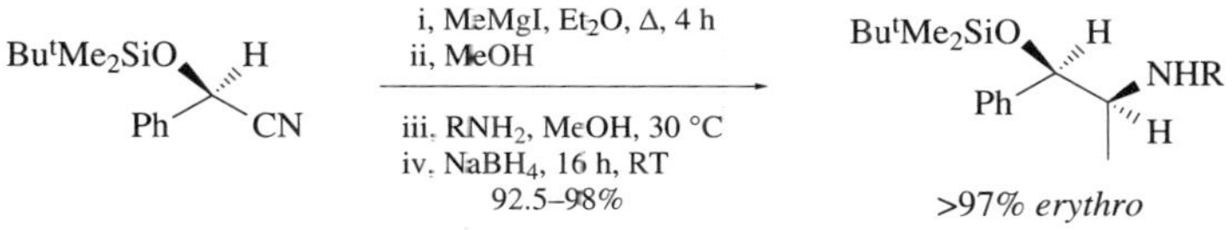

(32)

1,2-Aminoalcohols are formed with moderate selectivity in favour of the *erythro* isomer by reacting 2-silyloxy nitriles with Grignard reagents, followed by *in situ* reduction of the imine complex with zinc borohydride (Scheme 29) ⟨90TL1447⟩.

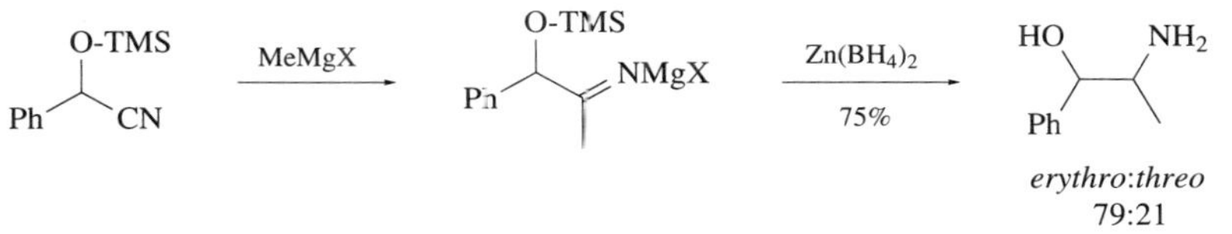

Scheme 29

2.05.4.13 Amines from Azides

Reduction of azides using hydrogen and 10% palladium on carbon or sodium borohydride affords the primary amine ⟨92JCR(S)391⟩. Numerous alkyl azides have been reduced to the corresponding primary amines in high yields by reaction with hydrazine in refluxing methanol in the presence of a catalyst, chiefly 20% palladium hydroxide on carbon ⟨89S450⟩. Aliphatic azides can also be reduced to the corresponding primary amines in excellent yields using a borohydride exchange resin–nickel acetate system in methanol ⟨93SC3047⟩.

The hydrochloride salt of a primary amine can be obtained by reaction of the corresponding azide with trimethyl phosphite and subsequent treatment with hydrogen chloride gas ⟨92BSF71⟩.

The catalytic reduction of alkyl azides using a heterogenised homogeneous catalyst, interlamellar montmorillonite diphenyl phosphine palladium(II) complex, affords primary amines in moderate to good yields ⟨89SC3289⟩. The tin(II) complex derived from the condensation of bis(thiophenyl)tin with thiophenol in the presence of triethylamine is highly efficient in reducing azides to primary amines ⟨87TL5941⟩.

Reaction of *t*-alkyl chlorides with trimethylsilyl azide in the presence of tin tetrachloride affords the *t*-alkyl azides which on treatment with triethyl phosphite give the corresponding primary amines as their hydrochloride salts (Equation (33)) ⟨87S487⟩. However, severely hindered chlorides cannot be transformed into the corresponding amine hydrochlorides under these conditions.

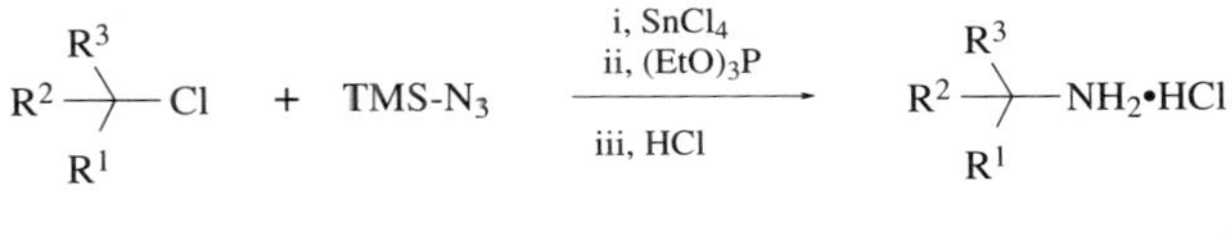

(33)

Amination at a benzylic carbon atom can be achieved by oxidative nucleophilic substitution using trimethylsilyl azide or hydrazoic acid in chloroform in the presence of 2,3-dichloro-5,6-dicyano-1,4-benzoquinone (ddq) to give 1-arylalkyl azides which are then reduced to the corresponding benzylic amines (Scheme 30) ⟨88S900⟩.

Reaction of α-chiral organyldichloroboranes with organic azides provides a convenient route to

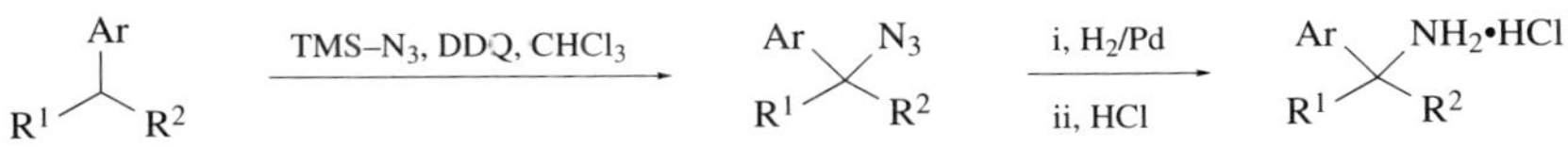

Scheme 30

secondary carbinamines with exceptionally high enantiomeric purities ⟨91JOC1170⟩. The reaction of various functionalised azides with cyclohexyldichloroborane can provide an efficient and chemoselective route to secondary amines ⟨87T1799⟩.

2.05.4.14 Amines from Nitro Compounds

An extremely wide variety of agents has been used to reduce nitro groups to amines ⟨B-89MI 205-01⟩. Primary amines are selectively obtained in fair yields from 1-nitroalkenes using a mercury or graphite cathode at a cathodic reduction potential of -1.1 V ⟨91CB2303⟩. β-Arylethylamines are obtained by catalytic hydrogenation of the corresponding nitrostyrenes using palladium on carbon ⟨90BCJ1252⟩.

Deoxygenation of nitro compounds is mediated by carbon monoxide and catalysed by polymeric rhodium carbonyl complexes ⟨88CL1759⟩. Reduction of aliphatic nitro compounds to amines by sodium borohydride in THF is facilitated by using 10% palladium on carbon as catalyst; ester groups are unaffected ⟨87S713⟩.

A general route to *N*-allylarylamines involves 1,2-addition of allyl Grignard reagents to nitroarenes, followed by reduction with lithium aluminum hydride in the presence of palladium on carbon (Equation (34)) ⟨88TL2251⟩.

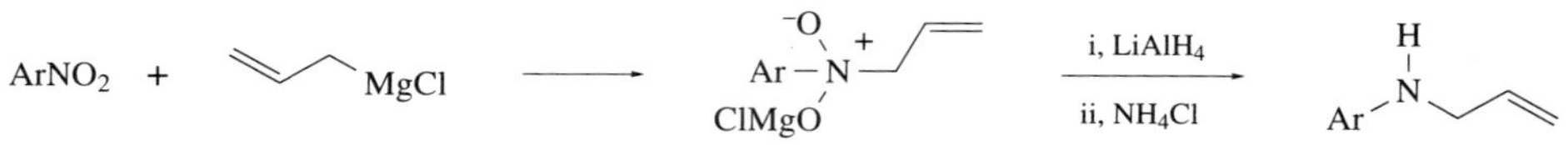

(34)

2.05.4.15 Amines from Epoxides

erythro-1-Amino-pent-4-ene-2,3-diols are formed in excellent yields by reaction of 1,2-epoxypenten-4-ol with amines including ammonia ⟨89TL1517⟩.

1,2-Epoxy-3-ols, prepared by asymmetric Sharpless epoxidation of the corresponding allylic alcohols, are attacked by amines at the 1-position giving 1-amino-2,3-diols in excellent yields (Equation (35)) ⟨91S769⟩.

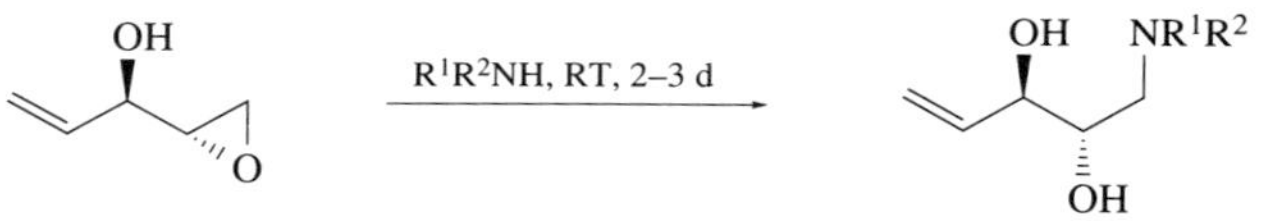

(35)

The regioselective ring opening of chiral epoxyalcohols by primary amines using titanium isopropoxide provides a practical and enantioselective route to 3-amino-1,2-diols ⟨91TL6931, 91TL6935⟩. Lithium aluminum amides, prepared from lithium aluminum hydride and primary amines, open aryl epoxides to afford β-amino-α-arylethanols with high regioselectivity and in excellent yields ⟨92JOC5831⟩. Reaction of aniline with epoxides using cobalt(II) chloride in acetonitrile at room temperature gives secondary amines in moderate yield ⟨90TL575⟩. Reaction of primary amines with epoxides in the presence of the enzyme porcine pancreatic lipase affords secondary amines. Enantiomeric excesses of greater than 99% are reported ⟨92TA1361⟩. The reaction of primary and secondary amines with 2-(1-bromoalkyl)oxiranes furnishes 2-aminomethyloxiranes in good yields ⟨93CL1787⟩.

Secondary amines (and some primary amines) react with epoxides in acetonitrile in the presence of a wide variety of Lewis acids, to give excellent yields of β-amino alcohols ⟨90TL4661⟩.

The ring opening of epoxides with primary or secondary amines is catalysed by tetraphenylstibonium triflate and proceeds under mild conditions furnishing the corresponding β-amino alcohols with high regio- and chemoselectivity (Equation (36)) ⟨89TL739⟩. *cis*-Amino alcohols have been reported from the regio- and stereoselective ring opening of 4,5-epoxy-2,3,4,5-tetrahydro-1-benzoxepines with secondary amines ⟨93TL4403⟩.

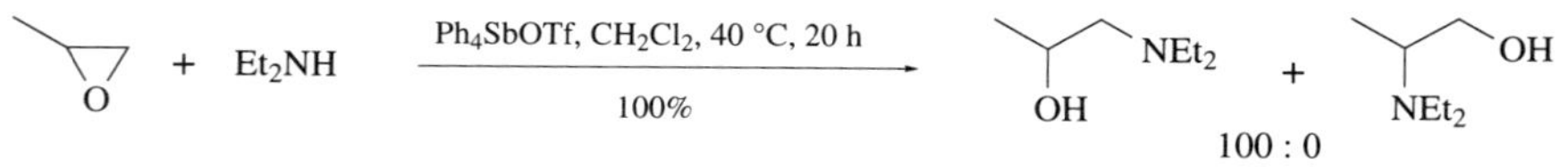

(36)

2.05.4.16 Amines from Aziridines

Ring opening of aziridines can give useful routes to amines. Both primary and secondary amines can be formed in good yields by addition of diorganocopperlithium reagents to *N*-substituted aziridines; the reaction is promoted by boron trifluoride etherate (3 equiv.), and an excess of the organocuprate (3 equiv.) per equivalent of the aziridine was employed ⟨85TL1153⟩. Ring opening of 2,2-dimethylaziridines by Grignard reagents has been studied, but other products in addition to those of normal ring cleavage are common, thereby limiting the synthetic utility ⟨87T565⟩.

N-Tosyl *β*-amino nitriles are formed by a lanthanide tricyanide-catalysed ring opening of *N*-tosylaziridines with cyanotrimethylsilane. The reaction proceeds by selective attack of cyanide at the less substituted carbon atom of the ring; an optically pure 2-substituted aziridine furnishes the corresponding amino nitrile without loss of stereochemical purity at an adjacent stereocentre to that which is attacked (Equation (37)) ⟨90TL6379⟩.

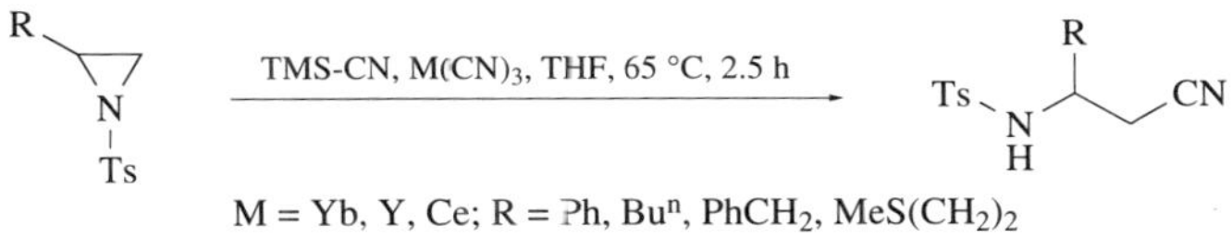

(37)

A one-pot synthesis of unsymmetrical 1,2-diamines from aziridines and amine oxides, using lithium iodide and triiron dodecacarbonyl in THF, has been reported (Equation (38)) ⟨90JOC2531⟩.

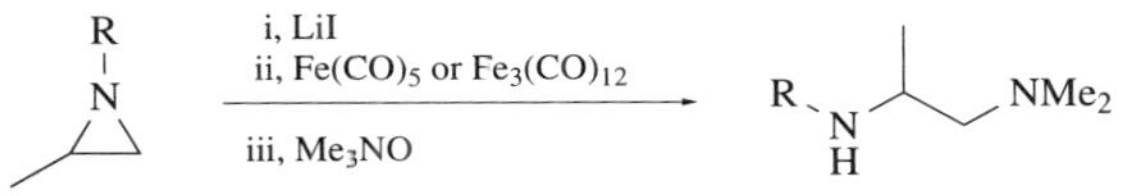

(38)

The regioselective ring opening of *N*-tosyl aziridines by sulfur-stabilised anions, derived from 1,3-dithiane, affords secondary amines in excellent yields ⟨93SL675⟩. Attack of the nucleophile occurs at the carbon of lesser substitution. In a similar manner, dianions derived from *β*-ketoesters have been used in the ring opening of *N*-tosyl aziridines ⟨93SL764⟩.

2.05.4.17 Amines from Carboxylic Acids and their Derivatives

Among the conventional methods for the conversion of RCOOH into RNH_2, the Schmidt reaction is prominent, involving the addition of hydrazoic acid to give the isocyanate RNCO which is hydrolysed *in situ* ⟨78RCR1084⟩. The same isocyanate intermediate is involved in the Curtius reaction, arising by decomposition of $RCON_3$, prepared from RCOOH ⟨47OR(4)337⟩.

The reduction of acylsulfonamides with borane–dimethyl sulfide affords sulfonamides which, on treatment with hydrogen bromide in phenol, are cleaved to the corresponding primary amines in high yields (Scheme 31) ⟨88SC29⟩.

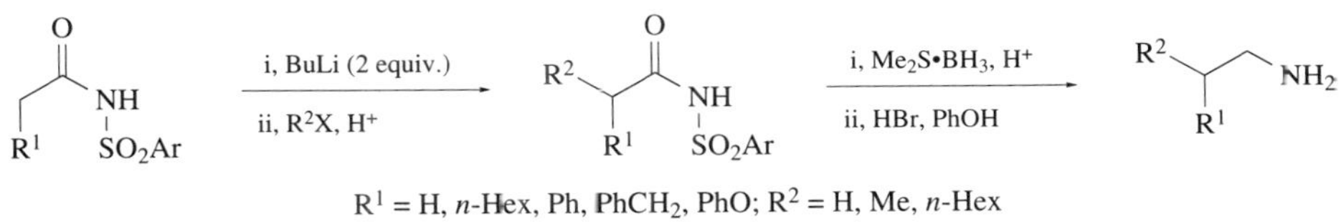

Scheme 31

The reaction of a lithium ester enolate, prepared from a ketene silyl acetal, with *N*-(alkylamino)benzotriazoles derived from benzotriazole, an aldehyde and a primary or secondary amine, affords a wide variety of secondary and tertiary *β*-amino esters ⟨90TL3999⟩.

2.05.4.18 Amines from Amides

Amides have been used to prepare amines by a number of routes, of which direct reduction is one of the most important. Amides are reduced to amines by $LiAlH_4$, but although the reactions are usually efficient, they are relatively slow; moreover, other functional groups are sometimes reduced without reduction of the amide group ⟨B-70MI 205-02⟩. Primary, secondary and tertiary amides can all be satisfactorily reduced to the corresponding amines with $LiAlH_4$, although notice must be taken of the stoichiometry and the reaction conditions ⟨B-70MI 205-02⟩. Borane–dimethyl sulfide ⟨81S441, 91SC1579⟩ and sodium in propan-2-ol ⟨90CI547⟩ are also good reducing agents for all three types of amides. The reaction of a formamide, R_2NCHO, with a Grignard reagent, R^1MgX, affords $R_2NC(R^1)_2H$ and R^1CHO (Bouveault reaction) ⟨B-65MI 205-01⟩.

Benzotriazole, an aldehyde (R^1CHO) and an amide (R^2CONH_2) condense with elimination of water to give an adduct which can be cleaved reductively by $LiAlH_4$ to give a secondary amine (Scheme 32) ⟨88JCS(P1)2339⟩. By starting with a secondary amide, R^2CONHR^3, and an aldehyde R^1CHO, application of the above strategy affords the tertiary amine $R^2CH_2NR^3(CH_2R^1)$ ⟨93JOC2086⟩.

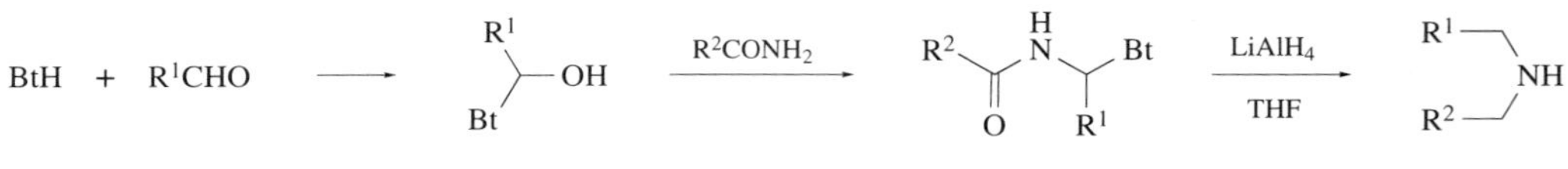

Scheme 32

The Hofmann rearrangement ⟨46OR(3)267⟩ of an unsubstituted amide, $RCONH_2$ to RNH_2 is usually achieved using sodium hypobromite (or NaOH and Br_2); for alkyl chains of six or more carbon atoms, improved yields are obtained using NaOMe and Br_2 ⟨74S290⟩. Hofmann rearrangement of acid amides to the corresponding amines under mild conditions and at ambient temperature is effected by quaternary ammonium salts ⟨89CL463⟩.

2-Arylethylamines have been prepared via the Heck reaction of iodobenzenes and acrylamide using a polymeric palladium catalyst. Hydrogenation of the initially formed α,β-unsaturated amide with palladium on carbon gives the corresponding saturated amide, treatment of which with sodium hypobromite affords the arylethylamine ⟨90SC3563⟩.

Conversion of acid amides, $RCONH_2$, into their *N*-phenyliodonio carboxamide tosylates and degradative hydrolysis of the latter in acetonitrile, via the isocyanate, RNCO, affords the corresponding alkyl ammonium tosylates, $RNH_3{}^+TsO^-$, in high yields ⟨87JOC476⟩.

2.05.4.19 Amines from Heterocycles by Cleavage of a Heterocyclic Ring

Azetidines are cleaved using palladium on carbon or Raney nickel to give the corresponding acyclic amino compounds ⟨91JOC5263⟩. The reaction of tetrahydro-2H-1,3-oxazin-2-one or its *N*-methyl analogue with aniline salts or thiophenols at 180 °C furnishes the corresponding *N*-aryl-1,3- or 3-(arylthio)propanamines in up to 78% yield ⟨93SC1329⟩. Reaction of *N*-benzyl tetrahydro-1,3-oxazines with various Grignard reagents affords 3-*t*-aminopropanol derivatives in excellent yields (Equation (39)) ⟨90SC1149⟩.

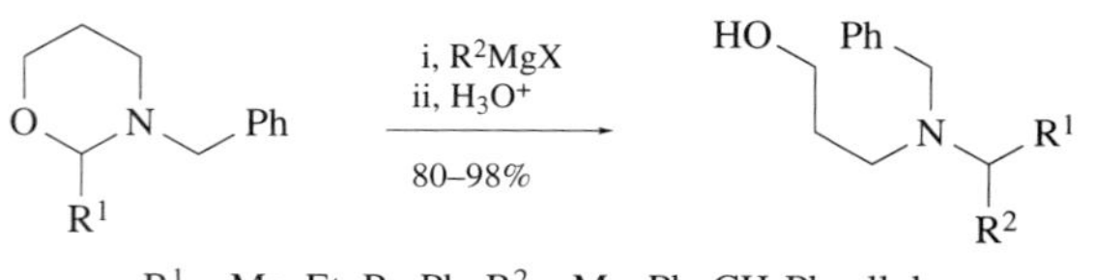

(39)

R^1 = Me, Et, Pr, Ph; R^2 = Me, Ph, CH_2Ph, allyl

Cyclic ethers and primary amines can be reacted to give *N*-alkylated or *N*-arylated pyrroles, pyrrolidines or piperidines. Thus, aniline reacts with THF in the presence of titanium dioxide at 250–300 °C to give *N*-phenylpyrrolidine in 97% yield ⟨90TL2991⟩. Aniline also reacts with tetrahydropyran in the vapour phase when passed over alumina at 330–340 °C; 1-phenylpiperidine is obtained in 83–90% yield ⟨63OSC(4)795⟩.

Treatment of isoxazolidines with molybdenum hexacarbonyl in water/acetonitrile at reflux gives secondary β-hydroxy amines ⟨90TL3351⟩. Reductive ring opening of tetrahydro-1,3-oxazines affords *N*,*N*-disubstituted 3-aminoalcohols via C—O bond cleavage. However, if the tetrahydro-1,3-oxazines are quaternised (MeI) and then treated with lithium aluminum hydride, N—C bond cleavage occurs giving disubstituted 3-amino ethers (Scheme 33) ⟨90S153⟩. The reaction of 3,4-dihydro-4,4-dimethyl-2H-pyran-2-one with primary amines in the presence of a catalytic amount of *para*-

toluenesulfonic acid, followed by treatment with lithium aluminum hydride furnishes 1-alkyl-4,4-dimethyl-1,4,5,6-tetrahydropyridines ⟨93SC2355⟩.

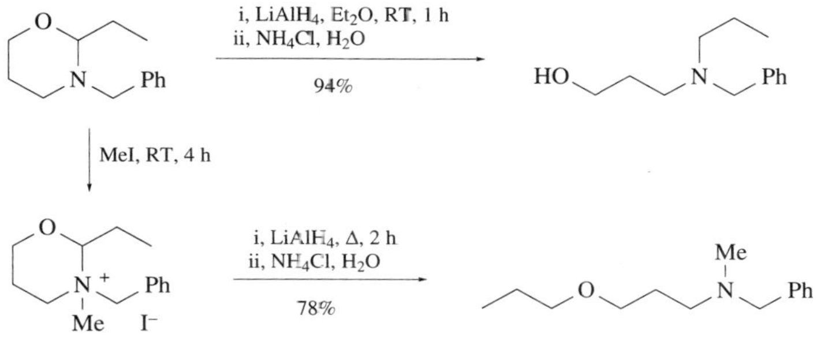

Scheme 33

α-Disubstituted primary amines can be prepared by alkylation of oxazolidinones with various alkyl halides followed by acidic hydrolysis (Scheme 34) ⟨89JOC3002⟩.

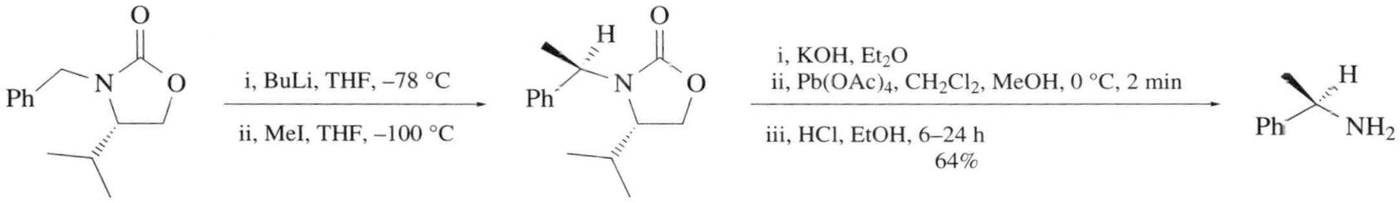

Scheme 34

The reaction of triamines with benzil and subsequent acylation affords acylated fused piperazines which can be hydrolysed to the terminal *N*-monoprotected triamines (Scheme 35) ⟨90CC1385⟩.

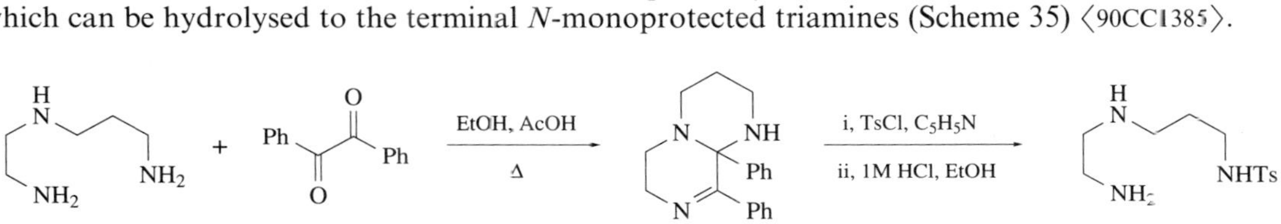

Scheme 35

Aldehydes undergo enantioselective conversion into the (*R*)-cyanohydrins when treated with hydrogen cyanide and a catalytic quantity of (*R*)-oxynitrilase. The resulting optically active cyanohydrins are reduced to *β*-amino alcohols with lithium aluminum hydride (Scheme 36) ⟨90S575⟩.

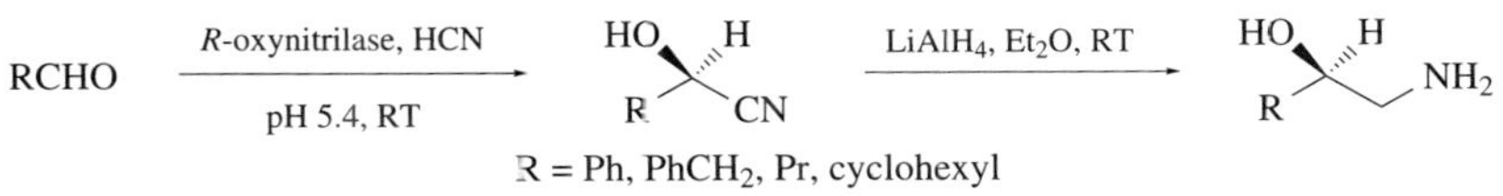

Scheme 36

Phenylethylamines have been synthesized via Heck arylation of the vinylamine equivalent 3-ethenyl-4,5-diphenyl-2(3H)oxazolone, followed by hydrogenolysis at 3 atm and 50 °C over palladium on carbon in ethanol containing acetic acid ⟨93JOC3299⟩.

2.05.4.20 Amines from Other Functional Groups

Secondary amines are obtained as their hydrochloride salts by the treatment of organyl-bis(di-isopropylamino)boranes with dry hydrogen chloride in the presence of alkyl azides (Equation (40)) ⟨93SL519⟩.

$$R^1BN(Pr^i{}_2)_2 \;+\; R^2N_3 \xrightarrow[\text{iii, HCl}]{\text{i, HCl gas; ii, KOH (aq.)}} R^1R^2NH{\bullet}HCl \quad (40)$$

R^1 = Me, Bu, Ph; R^2 = $PhCH_2$, C_7H_{15}, $Cl(CH_2)_3$, $EtO_2C(CH_2)_4$, $TosO(CH_2)_2$

Reaction of 4-amino-1-azadienes with an excess of a mixture of 3 : 1 lithium aluminum hydride and aluminum chloride or $Bu^i{}_2AlH$ affords the primary allylamine ⟨89CC1132⟩; the process is highly stereoselective and is considered to involve 1,4-addition of an alane species, followed by elimination of the 4-amino group.

N-Cyano-O-methylpseudoureas are obtained by the reaction of *N*-cyanoimidocarbonate with concentrated aqueous ammonia, methylamine or ethylamine at 0–25 °C ⟨91OPP721⟩.

The aminomethylstannanes (**16**), prepared by alkylation of allylic amine (**15**) using iodomethyltrialkylstannane in acetonitrile with potassium carbonate, undergo an aza-Wittig rearrangement on treatment with organolithium reagents. The ratio of the allylic amine (**18**) to the homoallylic amine (**17**) depends upon the nature of the alkyllithium reagent and the solvent (Scheme 37) ⟨93JCS(P1)1275⟩.

Primary amines react with cyclic sulfates to furnish *β*-aminosulfates which, on treatment with base or aqueous acid, afford aziridines or aminoalcohols ⟨89TL2623⟩.

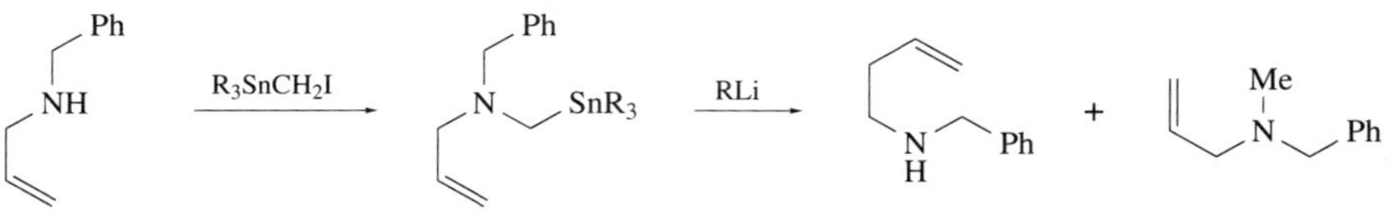

(15) **(16)** **(17)** **(18)**

Scheme 37

2.05.5 CYCLIC AMINES

2.05.5.1 Aziridines

α-Tosyl nitriles can be cyclised using lithium aluminum hydride to give aziridines ⟨91AG(E)873⟩.

A stereocontrolled synthesis of *cis*-*N*-arylaziridines proceeds by addition of the anion of an *α*-chlorosulfoxide to an imine, subsequent ring closure effected with *t*-butoxide, followed by desulfinylation (Equation (41)) ⟨88TL4093⟩.

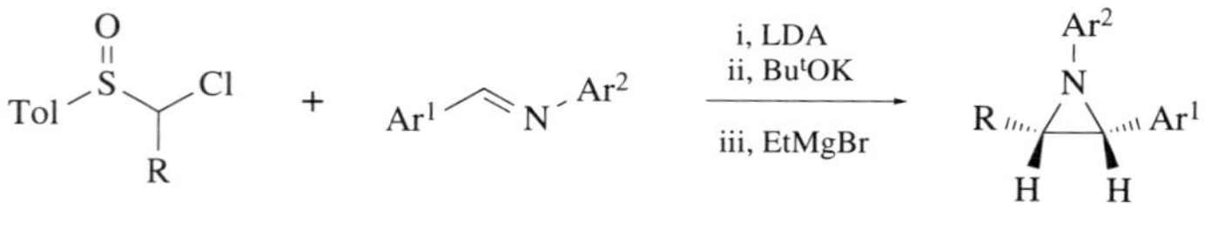

(41)

6*H*-1,2-Oxazines are converted into aziridines by reduction with lithium aluminum hydride in ether at room temperature ⟨93LA1155⟩.

The reaction of cyclic sulfates with azide followed by lithium aluminum hydride gives *N*-unsubstituted aziridines ⟨89TL2623⟩.

2.05.5.2 Azetidines

Kilogram quantities of azetidine (62% yield) can be prepared via the direct alkylation of benzhydrylamine with 1-bromo-3-chloropropane and subsequent hydrogenolysis ⟨88SC205⟩. 2-Methyleneazetidines can be prepared in high yield via closure of *N*-aryl *β*-chloro ketimines induced by potassium *t*-butoxide (Equation (42)) ⟨88JOC4462⟩.

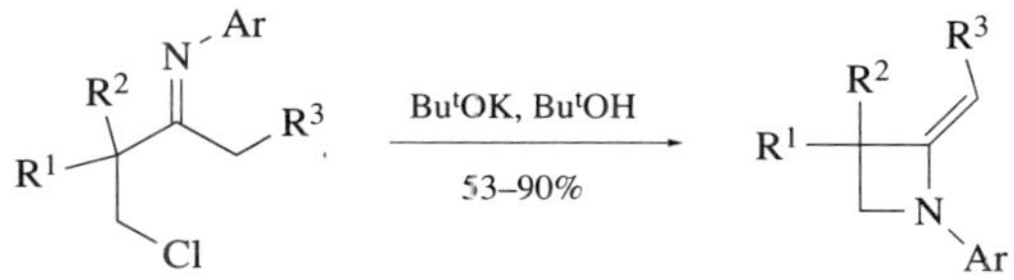

(42)

Azetidinium salts can be formed by cyclisation of 3-aminopropanols with trifluoromethanesulfonic anhydride ⟨87SC469⟩.

A route to 2-substituted azetidines involves anodic acetoxylation of *N*-tosylazetidine, giving (**19**) (Equation (43)). Subsequent nucleophilic displacement of the acetate can be effected with nucleophiles such as trimethylsilyl cyanide, allyltrimethylsilane, 2-methoxyfuran and trimethyl phosphite ⟨88BCJ3029⟩.

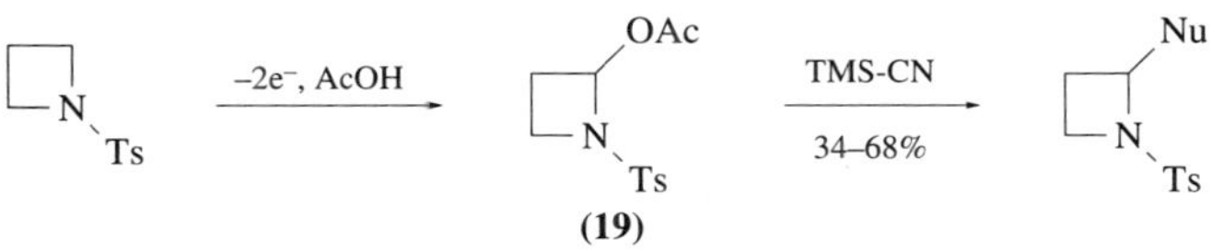

(43)

2.05.5.3 Pyrrolidines

Cyclisations that afford pyrrolidine rings are numerous and are particularly relevant to the synthesis of pyrrolizidine alkaloids ⟨93MI 205-01⟩. The Hofmann–Loeffler–Freytag reaction provides a longstanding route to pyrrolidines from δ-halo amines, by thermal cyclisation in the presence of acid ⟨83AG(E)337⟩. The stereoselectivity of the well-known mercury(II)-induced cyclisation of 4-alkenylamines has been investigated in regard to the effect of the counterion of the mercury salt and of the solvent (Equation (44)) ⟨88CL1289⟩. For 2,5-dialkylpyrrolidines, $Hg(OAc)_2$ in either THF or $CHCl_3$ leads predominantly to the *trans*-isomer, whereas with $HgCl_2$ the *cis*-isomer was selectively formed.

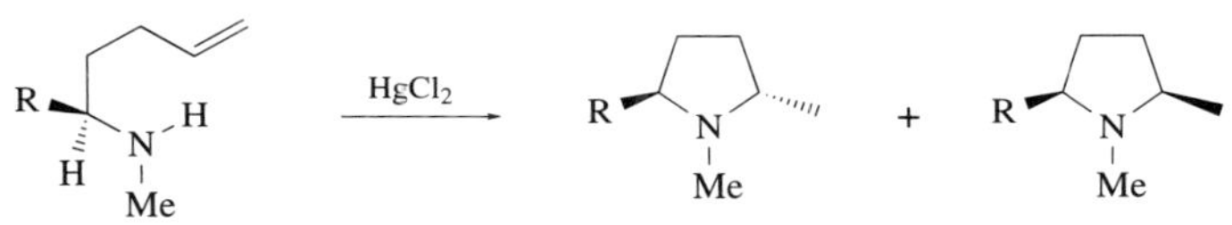

(44)

The anodic oxidation of *N*-alkyltosylamides in methanol containing potassium bromide gives pyrrolidine derivatives or α-(tosylamino) aldehyde acetals, each of which could be formed selectively by modifying the reaction conditions ⟨90JA2868⟩.

Irradiation of *N*-hydroxypyridine-2-thione carbamates provides aminyl or aminium radicals that can attack intramolecularly a δ,ε C=C double bond, resulting in cyclisation to form a 2-substituted pyrrolidine ring ⟨87JA3163⟩.

A one-pot tandem S_N2 displacement–Michael addition sequence has been developed for the synthesis of pyrrolidines and pyrrolines. Reaction of an ω-halo-2-alkenoate ester with a primary amine in the presence of triethylamine furnishes the nitrogen heterocycle in moderate yield ⟨92JOC1727⟩.

A stereoselective route to 2-substituted pyrrolidines involves the cyclisation of amido alkenes by treatment with iodine in aqueous acetonitrile (Equation (45)) ⟨89H(29)1861⟩. The 4-*O*-benzoate centre is obtained as 2:1 mixture of epimers.

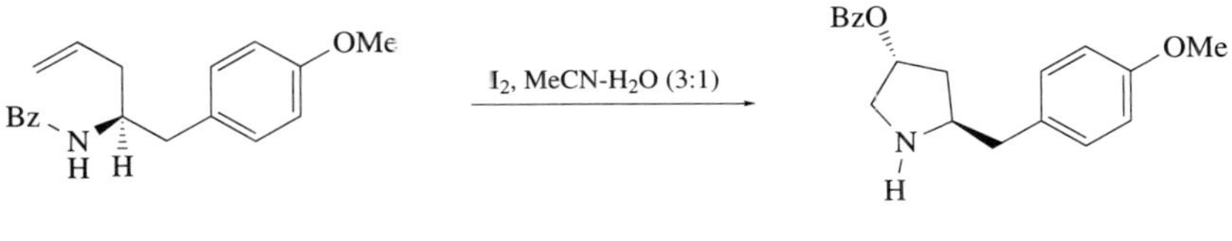

(45)

Derivatives of proline have been prepared via acyliminium cyclisations generated from the adducts of formaldehyde and *n*-butyl glyoxylate or methyl glyoxylate with amines containing an allyl or propargyl silane terminator. The cyclisations are initiated under thermal or Lewis acid conditions

⟨87TL3285⟩. For propargyl silane terminated cyclisations, the Lewis acid method is more efficient, but for allyl silane cyclisations thermal initiation is usually the preferred method.

Intramolecular Michael addition of primary amines, generated *in situ* by the action of triphenylphosphine upon an azide, to the unsaturation of an electrophilic alkenic site provides a route to pyrrolidines and piperidines ⟨87TL1757⟩. The cyclisations are independent of the alkene geometry, occur under very mild conditions, and with good diastereoselectivity.

A stereoselective intramolecular Michael-type addition of the enolates of α-amino esters provides an approach to pyrrolidines required for conversion into the kainoid class of excitatory amino acids ⟨88TL2195⟩.

An efficient multigram synthesis of 3-pyrrolidone from ethyl glycinate by sequential Michael addition and Dieckmann condensation has been reported (Scheme 38) ⟨92SC1249⟩. Ring expansion of *N*-*t*-butoxycarbonyl-4-piperidone allows access to 4-perhydroazepinone.

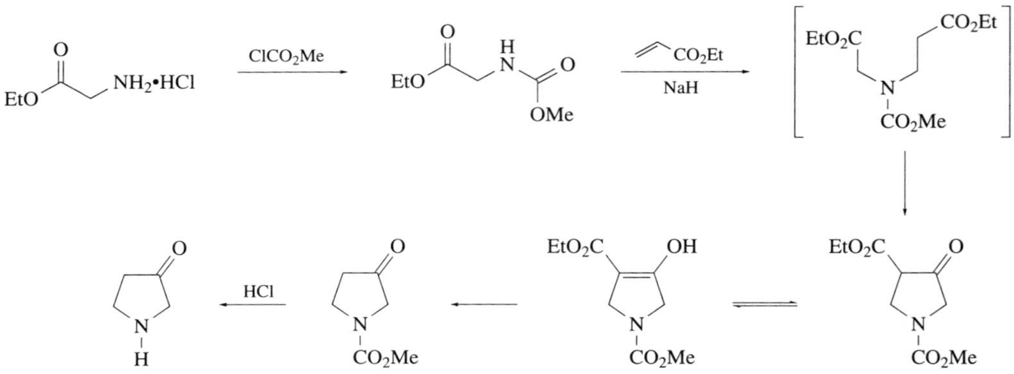

Scheme 38

3-Methylenepyrrolidines are formed by the addition of 2-aryloxymethylallylzinc bromides to imines to give aminozinc intermediates (**20**) which undergo palladium(0)-catalysed cyclisation via anionic displacement of the aryloxy group (Scheme 39) ⟨88TL3579⟩.

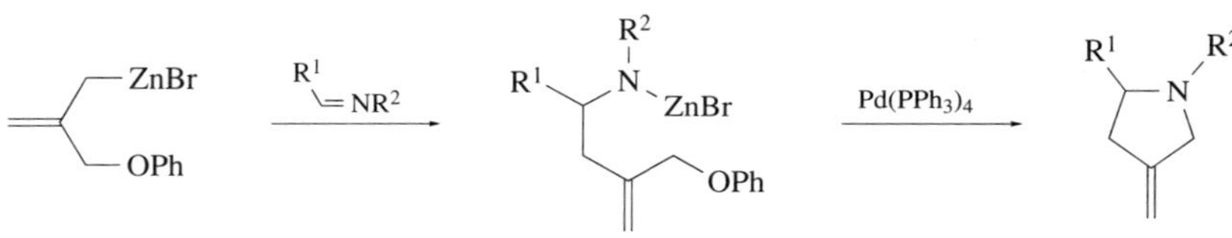

(20)

Scheme 39

Treatment of δ-azido epoxides with triphenylphosphine affords 1-azabicyclo[3.1.0]hexanes in enantiomerically and diastereoisomerically pure form, via a Staudinger-type aminocyclisation reaction. Derivatives of prolinol result by subsequent opening of the aziridine ring with benzoic anhydride (Scheme 40) ⟨89JA7500⟩.

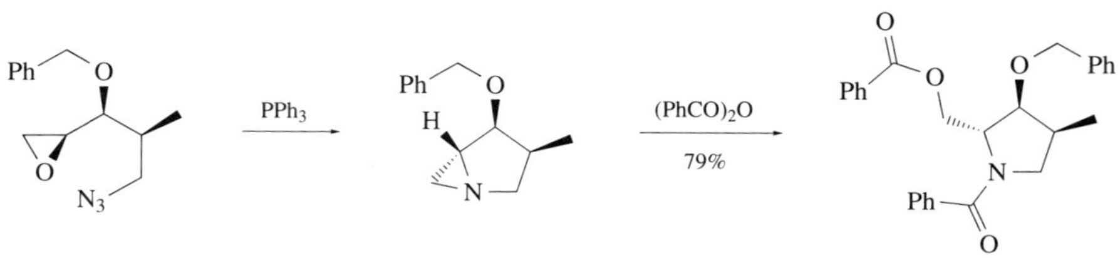

Scheme 40

The preparation of pyrrolidines by [3 + 2] cycloaddition of azomethine ylides has been reviewed ⟨86CRV941⟩. The stereoselectivity and yields of intramolecular pyrrolidine annulation of diene–azide cycloadditions have been markedly improved by protection of an allylic alcohol moiety as its thexyldimethylsilyl derivative. A subsequent thermolysis of the 1-azabicyclo[3.1.0]hexanes delivers bicyclic precursors which can be transformed into pyrrolizidine alkaloids (Equation (46)) ⟨88JOC2094⟩.

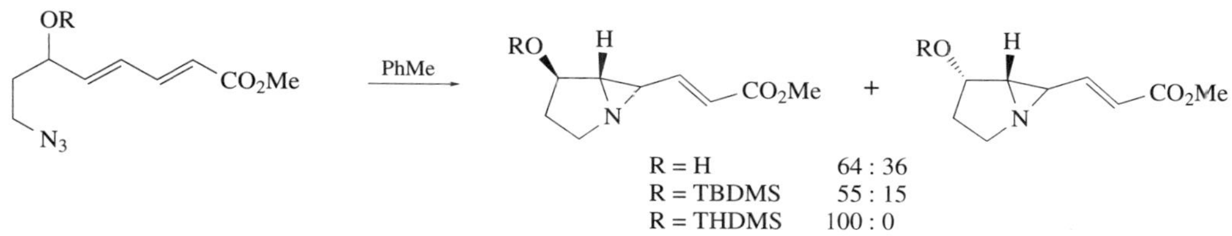

(46)

The synthesis of the chiral auxiliary (2S,5S)-2,5-dimethylpyrrolidine has been reported ⟨91SL573⟩. The six-step procedure starts from L-alanine, each step proceeding in at least 83% yield. A phenyl glycinol moiety acts as a highly stereoselective chiral auxiliary in the synthesis of *trans*-2,5-disubstituted indolizidine alkaloids ⟨88T2457⟩.

New pyrrolidines can be conveniently prepared by modification of an existing pyrrolidine ring; appropriate imides are reduced directly to pyrrolidines by a mixture of sodium borohydride and iodine in THF (Equation (47)) ⟨92T4623⟩. An aminoselenation procedure provides a route to α-methylene substituted pyrrolidine derivatives ⟨87H(26)929⟩.

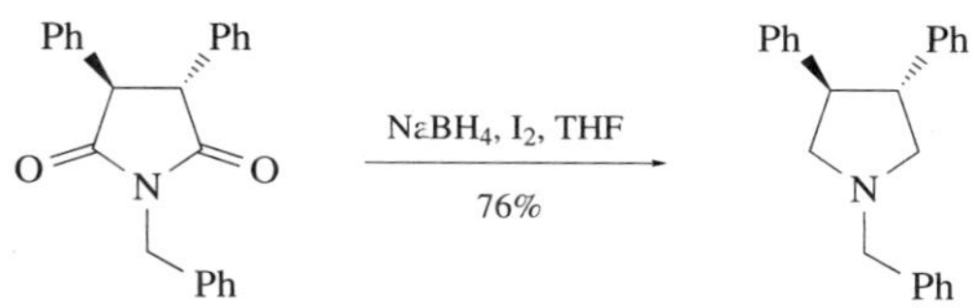

(47)

2.05.5.4 Pyrrolines

A convenient synthesis of 3-pyrroline, via the Délépine reaction ⟨79S161⟩, proceeds in four steps in an overall yield of 74% ⟨88S347⟩. Thus (*Z*)-1,4-dichloro-2-butene was reacted with hexamethylenetetraamine to give the mono-quaternary salt which was hydrolysed by ethanolic HCl to give (*Z*)-4-chloro-2-butenylammonium chloride. The latter was then reacted with K_2CO_3 (to effect cyclisation) followed by pentaethylenehexamine (to effect deprotonation).

Upon heating at 70 °C, various azidodienes undergo diastereoselective cycloaddition to give 3-pyrrolines (Equation (48)) ⟨89TL6661⟩. Thus, the azide (**21**) reacts to give chiefly the amine (**22**) which possesses the tricyclic skeleton and correct relative stereochemistry of gephyrotoxin.

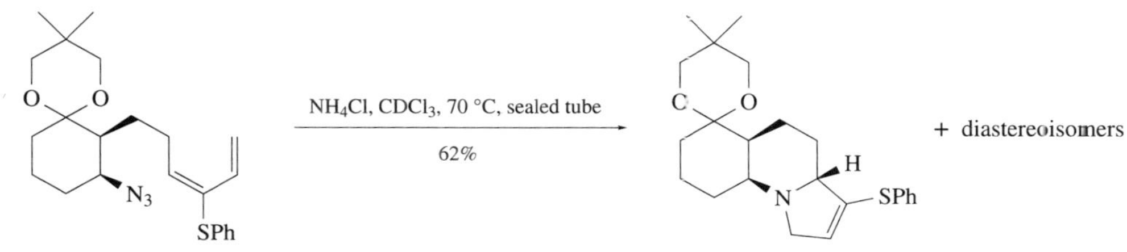

(21) **(22)**

(48)

2.05.5.5 Pyrrolizidines

Numerous pyrrolizidines have been prepared via the cyclisation of α-acylamino radicals generated by homolytic cleavage of a phenylthio- or phenylseleno-group onto acetylenic, allenic and vinyl silane ethers ⟨88JOC6022, 88T7037⟩. A stereoselective route to (−)-trachelanthamidine and related alkaloids has been reported; the key step is a tin-mediated ring closure at the α-position of an iodoamide onto a vinyl group ⟨88JA7536⟩. A route to pyrrolizidine alkaloids is illustrated in Scheme

41 ⟨87SC1155⟩. The ester (**24**) is obtained by the alkylative ring opening of a vinyl aziridine (**23**) with trimethylsilyl iodide followed by cyclisation. Ester (**24**) can be reduced to pyrrolizidine alkaloids including supinidine and isoretronecanol.

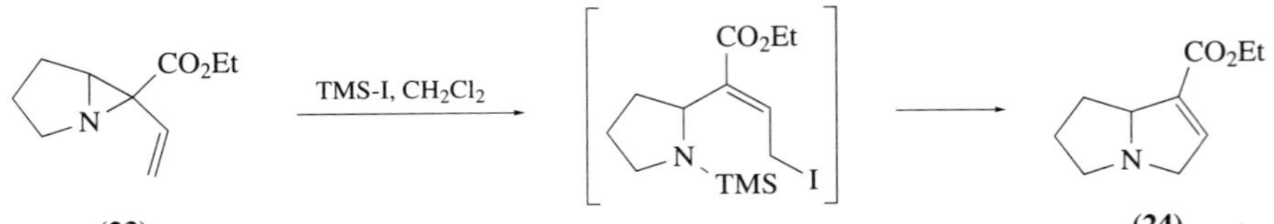

Scheme 41

2.05.5.6 Piperidines

Substituted piperidine $\Delta^{1,2}$imines can be reduced stereoselectively to *cis*-2,6-disubstituted piperidines by catalytic hydrogenation over palladium on carbon ⟨88JOC1380⟩. Such reductions are crucial in the synthesis of *trans*-decahydroquinoline alkaloids. A reductive condensation of anilines with glutaraldehyde in the presence of sodium borohydride in dilute acid/methanol/THF affords *N*-arylpiperidines ⟨91S447⟩. 2,6-Disubstituted piperidines can be synthesized stereoselectively by rearrangement of ω-epoxy ketimines followed by reduction with hydride ⟨88TL4977⟩. Reduction of ω-alkoxycarbonyl alkylamides with borane–dimethyl sulfide affords *N*-substituted piperidines and pyrrolidines in high yields, but the reaction is sensitive to steric bulk in the substituent attacked (Equation (49)) ⟨88S985⟩.

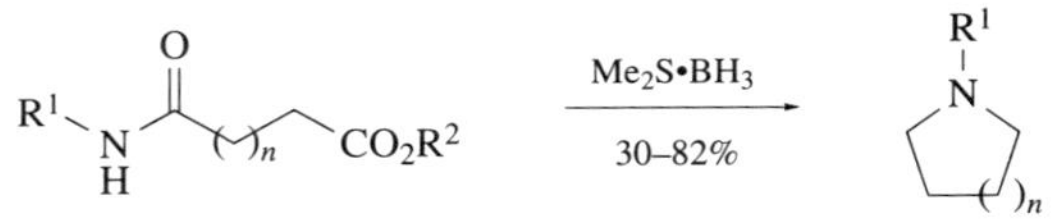

(49)

Various piperidines, including 2-alkyl derivatives, can be prepared by the Hofmann–Loeffler–Freytag reaction involving the cyclisation of ε-haloamines ⟨83AG(E)337⟩. 2,6-Dialkylated tetrahydropyridines can be prepared by cyclisation of allylsilane nitrones, induced thermally, or by TMS-OTf ⟨88JOC5989⟩. A stereoselective route to functionalised piperidines involves the thermolytic cyclisation of an azido alkene followed by reduction with $LiAlH_4/Me_3Al$ ⟨88JOC2968⟩. The electrophilic cyclisation of an alkyne moiety onto an iminium ion promoted by the nucleophilic attack of azide provides a one-pot conversion of amines into piperidines which possess at the 3-position an β-exocyclic double bond of defined geometry ⟨88JA612⟩. Various (*R*)-1,2,3,4-tetrahydrobenzothieno[2,3-c]pyridine derivatives have been synthesized from 3-substituted benzothiophenes, as depicted in Scheme 42 ⟨93JMC3526⟩. Kinetic resolution was achieved by the deacetylation of amide (**25**) with acylase. An intramolecular Mannich reaction of amine (**26**) provides a succinct route to the spirocyclic ring system of the nitramine alkaloids (Equation (50)) ⟨87CC509⟩. Derivatives of glutamic acid are converted enantioselectively into 5-hydroxypipecolic acids via the cyclisation of a protected chlorohydrin (Equation 51) ⟨88TL2231⟩. Derivatives of pipecolic acid ((**27**; $n = 2$), Equation (52)) have been prepared via acyliminium cyclisations. The cyclisations are initiated under thermal or Lewis acid conditions ⟨87TL3285⟩. For propargyl silane terminated cyclisations the Lewis acid method is more efficient, but for allylsilane cyclisations thermal initiation is simpler. The first enantioselective synthesis of (+)-conhydrine has been reported (Equation (53)) ⟨89TL6395⟩.

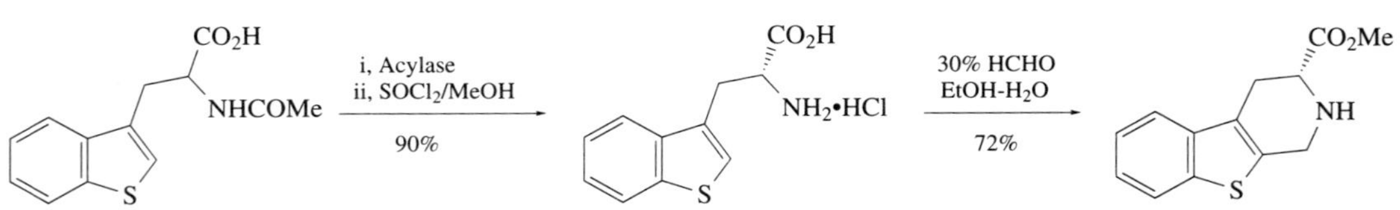

Scheme 42

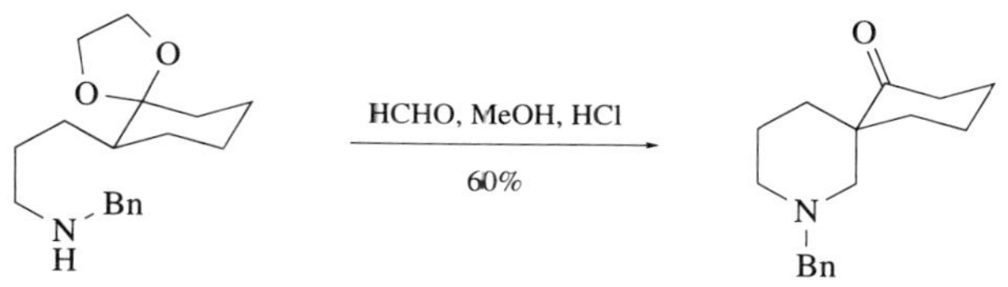

(26)

(50)

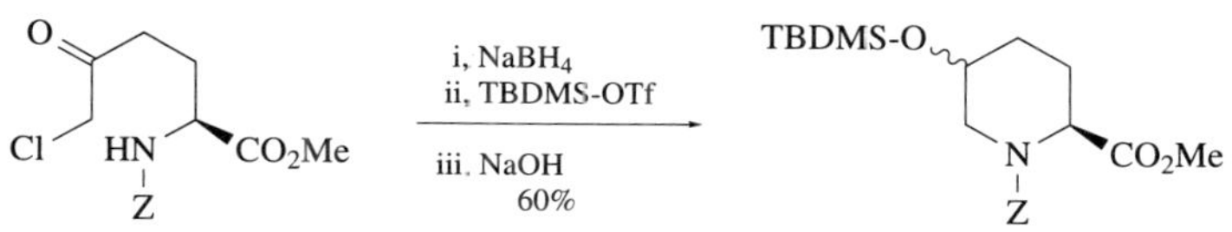

(51)

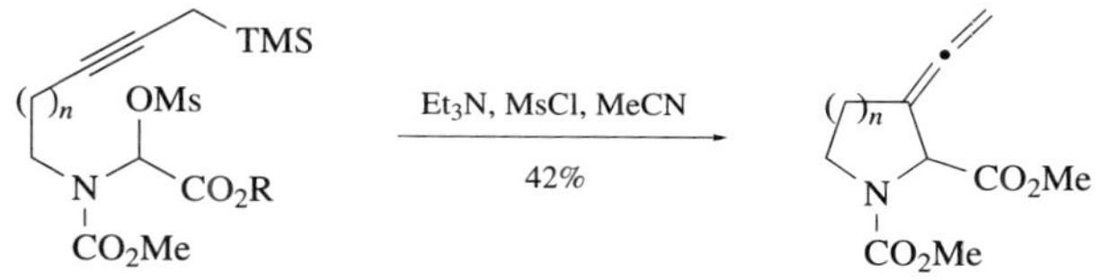

(52)

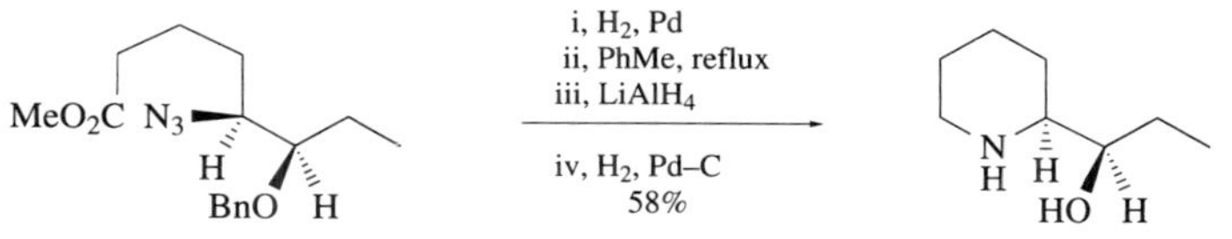

(+)-Conhydrine

(53)

Imino-ester (**28**), formed by the thermal reaction of methyl glyoxylate with *para*-toluenesulfonyl isocyanate, undergoes *in situ* hetero-Diels–Alder reactions with a variety of 1,3-dienes (Scheme 43), to afford substituted pipecolic esters ⟨91SL29⟩.

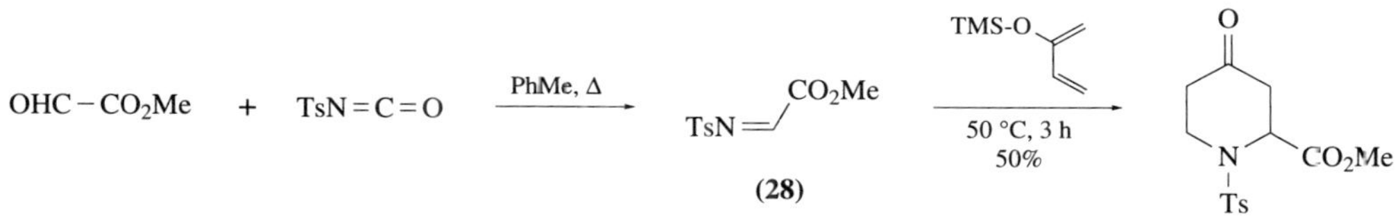

Scheme 43

2.05.5.7 Miscellaneous

A nickel–aluminum alloy in aqueous potassium hydroxide effects chemoselective reduction of nitrogen-containing heteroaromatic rings in the presence of benzene rings ⟨87JOC1043⟩. Reduction of isoquinolines, quinolines and pyridines with lithium triethylborohydride in THF at room temperature affords 1,2,3,4-tetrahydroisoquinolines, 1,2,3,4-tetrahydroquinolines and piperidines, respectively (40–93% yields) ⟨93TL7239⟩. The heteroaromatic ring of quinolines and isoquinolines can be hydrogenated selectively, giving the corresponding benzopiperidines, by reaction with carbon monoxide and water in the presence of a rhodium carbonyl catalyst ⟨87TL77⟩; quinolines are directly reduced, but isoquinolines require further reduction of the initially formed *N*-formyl compounds. Detailed studies on the rhodium-mediated reduction of the heteroaromatic ring of quinolines have shown that the nitrogen atom is involved in binding to the rhodium metal centre in these regioselective reductions ⟨88OM2250⟩. A variety of both aliphatic and aromatic amines undergo selective bis-reductive amination with adipaldehyde to form perhydroazepines (Equation (54)) ⟨88JHC1383⟩. The cyclisation is effected in the presence of ethanolic tetracarbonylhydridoferrate and carbon monoxide, the ferrate anion acting as an efficient and selective reducing agent.

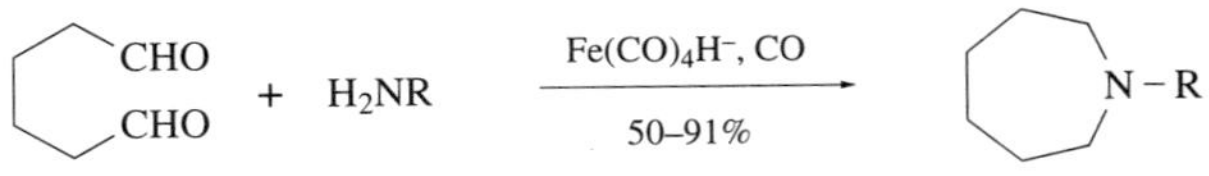

(54)

Acyliminium vinyl silane cyclisations (Equation (55)) have been used in a regiocontrolled synthesis of indolizidinones and quinolizidinones, including the elaeocarpus alkaloids, elaeokanines A and B ⟨87JA6097⟩. Oxazalo[2,3-a] tetrahydroisoquinolines have been prepared by reaction of D-phenylglycinol with a 2-formylphenethyl bromide. The D-phenylglycinol provides the nitrogen atom and also acts as a chiral auxiliary ⟨88TL6949⟩. A trimethylsilyl group placed on an aromatic ring provides sufficient activation for ipso-attack in a Pictet–Spengler cyclisation synthesis of 8-methoxytetrahydroisoquinolines ⟨88TL6715⟩. The silicon group overrides the effect of the methoxy group, which by itself directs cyclisation to the position *para* to the methoxy group. A biomimetic synthesis of the skeleton of the alkaloid goniomitine features a reductive cyclisation of an indole nitrogen atom onto a piperidinium ion (Equation (56); $R^1 = Me$, $R^2 = H$ 23%) ⟨88TL4563⟩. Lithiation that results in the displacement of an alkoxy group of a alkoxyamine is an established procedure for the amination of alkyl groups (Equation (57)). Further applications have been developed including aminations of aromatic lithium reagents ⟨89JOC5574⟩. *N*-Boc arylamines can be converted by a one-pot procedure into 1,2,3,4-tetrahydroquinolines by directed *ortho*-lithiation and subsequent alkylation with 1-chloro-3-iodopropane ⟨88TL5725⟩. Cyclic hemiaminals can be ring cleaved by diethyl cyanophosphate with concomitant introduction of a nitrile group alpha to the nitrogen atom. Subsequent reaction with potassium and 18-crown-6 induces a stereocontrolled ring contraction to indolizidines ⟨88H(27)1575⟩. In the absence of the silyl group, only the 6-methoxy isomer is formed. A total and asymmetric synthesis of (+)-reticuline proceeded by means of a chiral formamidine which was metallated and then alkylated with a benzyl bromide. Three subsequent conventional steps afforded the isoquinoline alkaloid (Equation (58)) ⟨89H(28)295⟩. In an enantioselective synthesis of (−)-tylophorine, the enamide group of (**29**) undergoes an intramolecular double Michael addition involving the chiral α,β-unsaturated ester moiety, to give the bicyclic amide (**30**) (Equation (59)) ⟨89H(28)63⟩. Iminomethano-dibenzo[a,e]cyclooctenes, intermediates in the synthesis of analogues of the *N*-methyl-D-aspartate receptor antagonist, MK-801, have been prepared by a transannular cycloaddition of a labile nitrone intermediate ⟨88TL6985⟩.

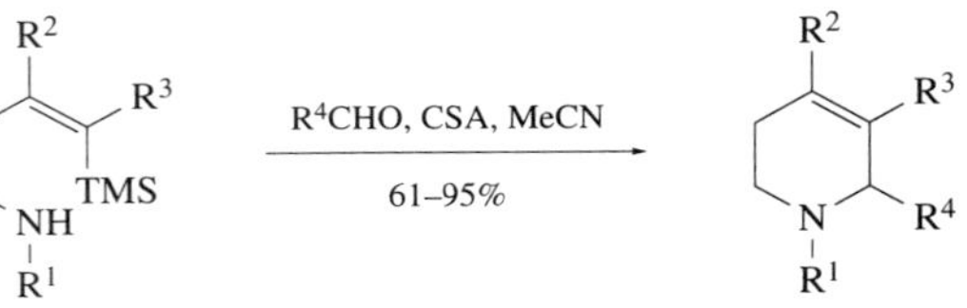

(55)

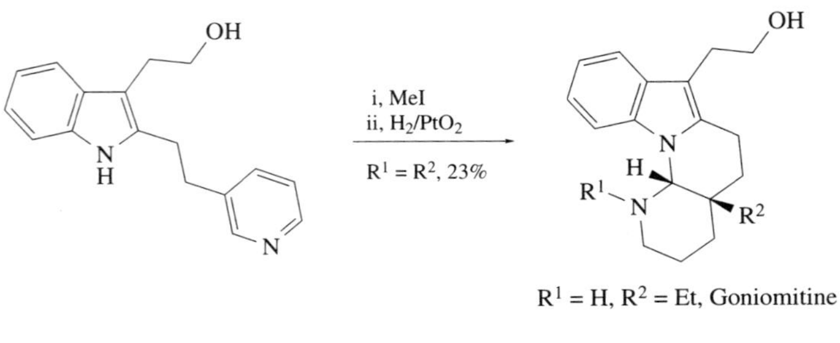

(56)

$R^1 = H$, $R^2 = Et$, Goniomitine

(57)

$n = 1$, 21%; $n = 3$, 64%; $n = 4$, 24%

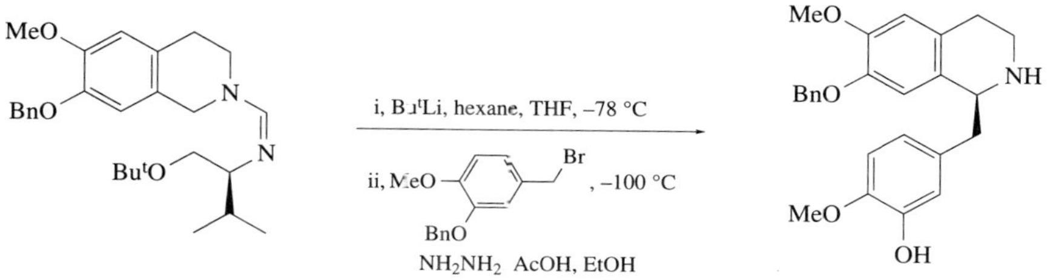

(58)

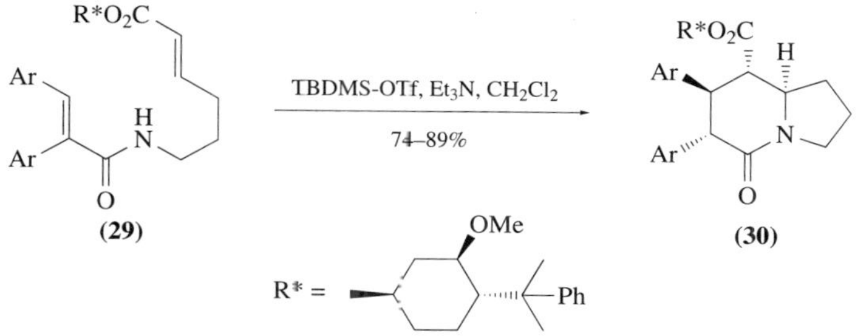

(59)

(±)-Gephyrotoxin 223AB has been synthesized by using an intermolecular 1,3-dipolar cycloaddition involving the cyclic nitrone (**31**), which results in control of the stereochemistry of the alkyl substituents (Scheme 44) ⟨89JOC1748⟩. A range of 1,2-dihydroisoindolines has been prepared from allylfurfurylamine hydrochlorides via [4 + 2] cycloadditions (Scheme 45) ⟨88SC2055⟩.

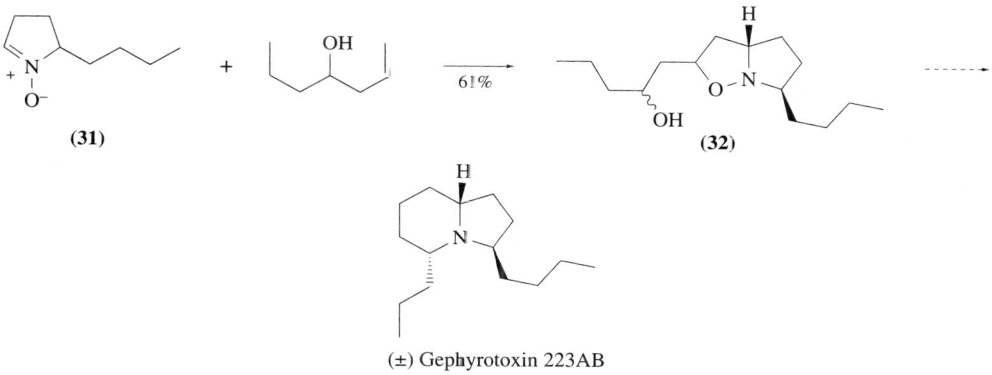

Scheme 44

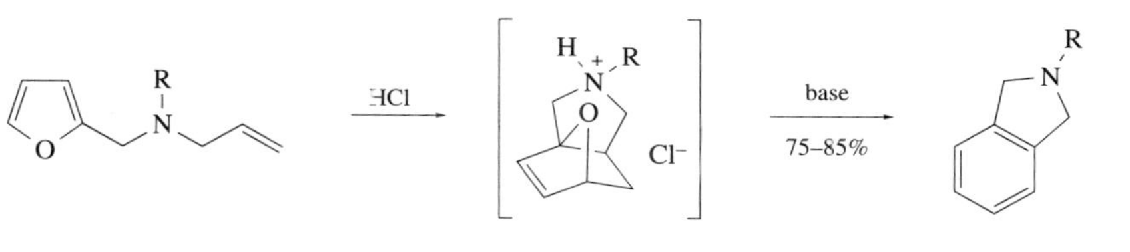

R = H, alkyl, phenyl

Scheme 45

Iminium ions, generated from silylated 3-butenylamines and aldehydes in the presence of a sulfonic acid or silver salt catalyst, undergo cyclisation to give 1,2,5,6-tetrahydropyridines containing substituents at the 1-, 2-, 3- or 4-positions (Equation (60)) ⟨87JA6097⟩. α,β-Unsaturated *N*-benzenesulfonylimines can be used as the 4π-component in inverse electron demand Diels–Alder reactions with a range of electron-rich dienophiles ⟨89JA1517⟩. Reaction at 12 kbar pressure affords substituted *N*-benzenesulfonyl-1,2,3,4-tetrahydropyridines in moderate yields.

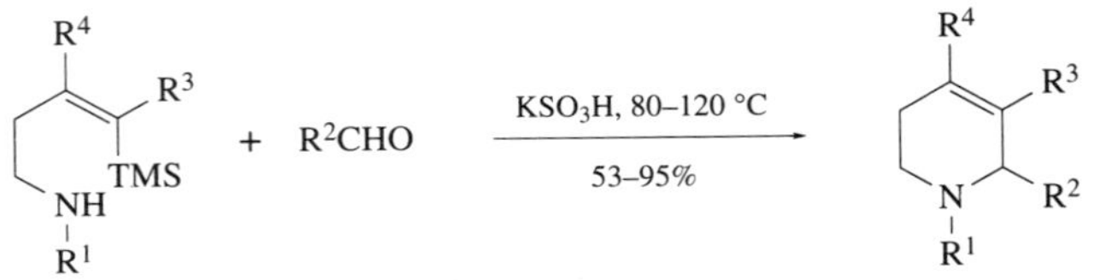

(60)

A total synthesis of (±)-methyl homosecodaphniphyllate features a remarkable tetracyclisation of an imine ⟨88JA8734⟩. The cyclisation involves condensation of ammonia with a 1,5-dialdehyde to give a 3,4-dihydropyridine which in its protonated form undergoes an intramolecular Diels–Alder reaction with an alkenic site. The cyclic iminium species so produced then undergoes a cationic cyclisation with another alkenic site to give the tetracyclic system. Dienamides have been used in both Diels–Alder cycloadditions and tandem intramolecular Michael reactions for the synthesis of polycyclic amines. For example, dienamide (**33**) underwent Diels–Alder cyclisation to afford the tricyclic enone (**34**) which was subsequently converted into (±)-dendrobine (Scheme 46) ⟨89JOC265⟩.

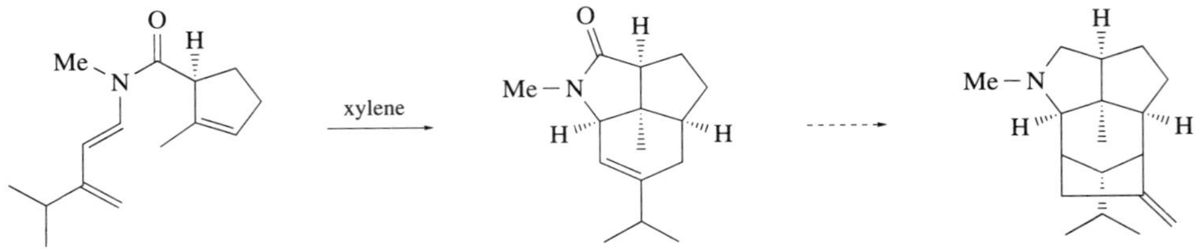

(33) **(34)** (±) Dendrobine

Scheme 46

2.05.6 ASYMMETRIC SYNTHESIS OF AMINES

The asymmetric reduction of imines using chiral dialkoxyboranes in the presence of $MgBr_2 \cdot OEt_2$ furnishes optically active tertiary amines in up to 73% enantiomeric excess ⟨92TA227⟩. The kinetic resolution of racemic α-methylbenzylamine has been achieved by conversion into an imine which is then hydrogenated over a rhodium catalyst under conditions of kinetic resolution. The enantiomeric imine that remains is hydrolysed to liberate the α-methylbenzylamine (98% ee using (2*S*,4*S*)-2,4-bis(diphenylphosphino)pentane) ⟨93TA215⟩. Work using sulfonic acid derivatives of that catalyst showed that such catalytic asymmetric hydrogenations of imines could be conducted in two-phase solvent systems ⟨91CC1684⟩. The reaction of optically pure *t*-butyldimethylsilyl protected cyanohydrins with Grignard reagents followed by reduction of the intermediate imine with sodium borohydride and subsequent deprotection affords 2-amino-1-arylethanols of high optical purity (Scheme 47) ⟨90T1653⟩. The chiral induction gives a high preponderance of the *erythro* isomers. Enantioselective hydrogenations of ketimines to amines (up to 98% ee) have been achieved using a titanocene-derived catalyst ⟨92JA7562⟩. Oximes have been reduced by hydrogen in the presence of a ruthenium–BINAP complex and 3 Å molecular sieves at 90°C and about 100 atm to the corresponding amine, but the greatest enantiomeric excess reported was only 29% ⟨92TA1283⟩.

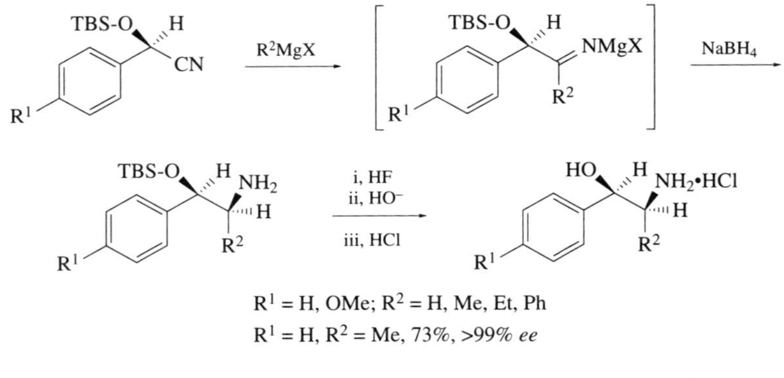

R^1 = H, OMe; R^2 = H, Me, Et, Ph

R^1 = H, R^2 = Me, 73%, >99% *ee*

Scheme 47

Enantiomerically pure, alanine-derived, α-amino aldehyde acetals have been prepared in 91–94% *ee* by hydrogenation of chiral imines over a Raney nickel catalyst ⟨89S608⟩.

The highly enantioselective addition to the carbon–nitrogen double bond of *N*-diphenylphosphinoylimines using dialkylzinc reagents in the presence of a chiral amino alcohol furnishes optically active phosphoramides with good enantiomeric excess. The use of (1*S*,2*R*)-2-(di-*n*-butyl)amino-1-phenylpropan-1-ol gave a phosphoramide in 75% *ee*. Acidic hydrolysis of the phosphoramides affords the optically active primary amines without racemisation ⟨92CC1097⟩. The nucleophilic addition of several primary and secondary amines to 1-methoxycarbonyl-3-methyl-(η^3-allyl)-tetracarbonyliron tetrafluoroborate (**35**) and subsequent removal of the tetracarbonyliron moiety furnishes 4-aminoenoates of high enantiomeric purity (Scheme 48) ⟨93SL401⟩. Asymmetric syntheses of both (*R*)- and (*S*)-α-substituted benzylamines of high optical purity have been achieved by the alkylation of a chiral ketimine prepared from benzylamine and pinanone (Scheme 49) ⟨89SC1423⟩. The observed diastereoselectivity during the alkylation was not found to depend on the alkyl halide, in contrast to the use of (+)-camphor ketimine as the chiral template.

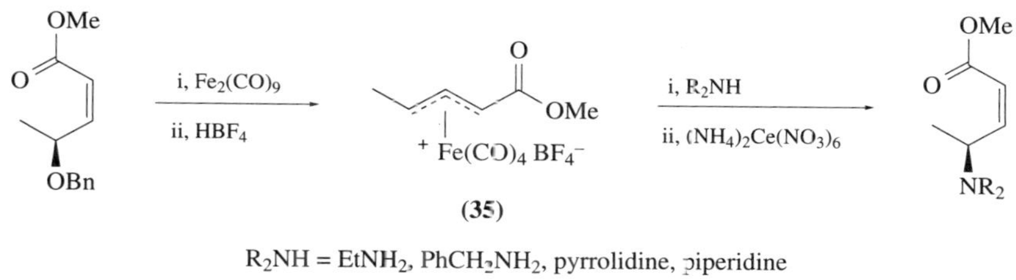

Scheme 48

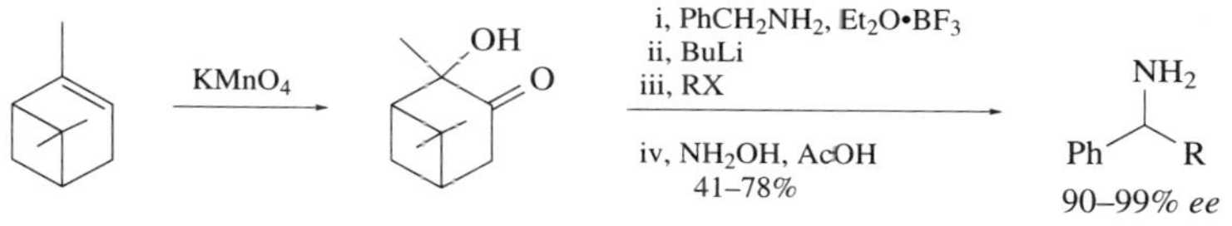

Scheme 49

An asymmetric synthesis of amines employs the reductive amination of ketones using (+)- or (−)-norephedrine followed by periodate oxidation ⟨93TA2095⟩. Chiral ferrocenylphosphine–transition metal complexes have been successfully used in the palladium-catalysed asymmetric allylic amination of 1,3-disubstituted allylic substrates ⟨89JA6301⟩. The use of chiral α-chloro-α-nitroso reagents derived from camphor-10-sulfonamides for the asymmetric α-amination of the enolates of ethyl ketones leads to *erythro*-β-amino alcohols with excellent enantioselectivity ⟨92JA5900⟩. Overall allylic amination of *cis*-alkenes with an enantiomerically pure form of the *N*-sulfinylcarbamate of *trans*-2-phenylcyclohexanol in the presence of $SnCl_4$ afforded allylic sulfinamides (from which chiral primary amines can be prepared) with high stereocontrol ⟨91JA3526⟩.

The kinetic resolution of primary amines *via* enantioselective *N*-acetylation in aqueous solution has been accomplished using *cylindrocarpon radicicola* ⟨93BMC681⟩.

An enantioselective synthesis of the natural antibiotic (+)-streptazolin involves the stereoselective vinylsilane-terminated cyclisation of a tartrate-derived *N*-acyliminium ion ⟨87JA6115⟩. The final stage of a short stereoselective synthesis, the spirocyclic alkaloid (±)-nitramine, involves the sulfone-directed spirocyclisation of epoxy sulfone (**36**) (Scheme 50) ⟨88TL6493⟩. Both enantiomers of isonitramine can be synthesized from a common chiral intermediate (**37**) which is used to form differing spirocyclic systems that are eliminated stereoselectively (Scheme 51). The intermediate is prepared by the condensation of glutaraldehyde (2 equiv.) with (−)-phenylglycinol ⟨88TL3311⟩. Each enantiomer of *trans*-2,5-dimethylpyrrolidine has been prepared starting from either D- or L-alanine as outlined in Scheme 52 ⟨87TL2083⟩. Enantiomers of 3-substituted pyrrolidine carboxylic acids including cucurbitine can be prepared by 1,3-dipolar cycloaddition of azomethine ylides to enones, followed by pig liver esterase-catalysed hydrolysis ⟨87CPB3845⟩.

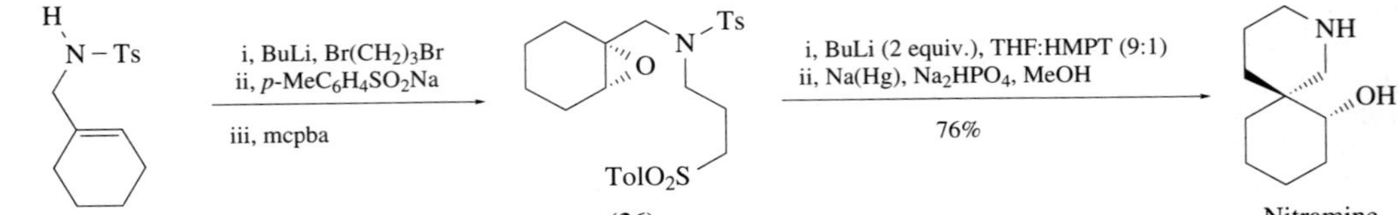

(36)

Scheme 50

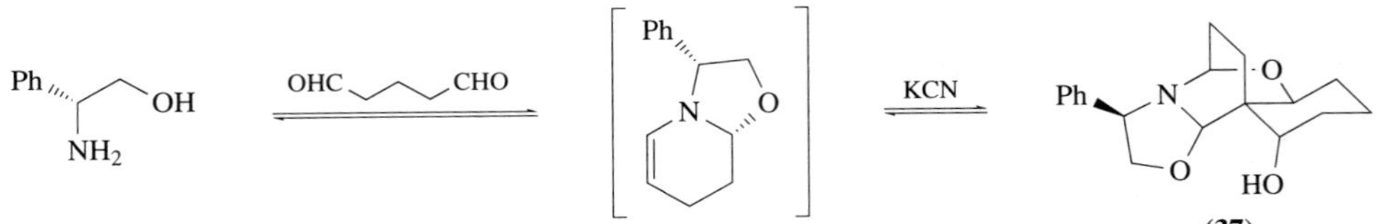

(37)

Scheme 51

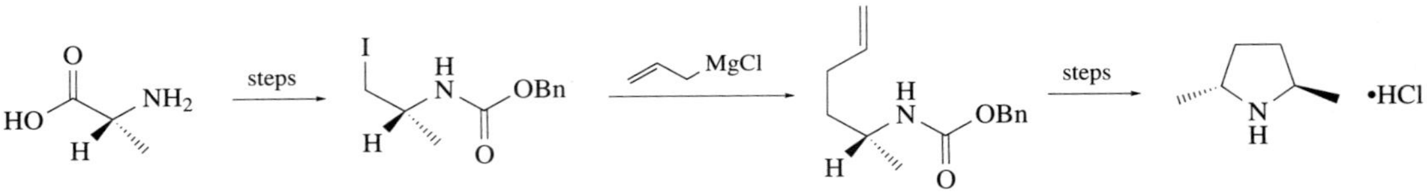

Scheme 52

2.06
Alkylnitrogen Compounds: Compounds with N—Halogen, N—O, N—S, N—Se and N—Te Functional Groups

W. RUSSELL BOWMAN and ROBERT J. MARMON
Loughborough University of Technology, UK

2.06.1 *N*-HALOGENOALKYLAMINES; RNHHal, R^1R^2NHal, $RNHal_2$, AND SALTS THEREOF

Halogenation of amines can be regarded as an oxidation reaction. Most syntheses of *N*-halogenoalkylamines involve the reaction between the amine and a positive source of the halogen. The most obvious source of positive halogen is from the respective dihalogens X_2, but a range of other common halogen sources are used and are referred to in the following sections. Reaction between primary and secondary amines and dihalogens X_2 in suitably buffered solutions yield the respective *N*-halogenoalkylamines, whereas reaction between tertiary amines and halogens yield *N*-halogeno quaternary halides $R_3N^+X\ X^-$. The latter normally give elimination of HX to yield iminium salts, which readily hydrolyze to aldehydes and secondary amines.

No reports of mixed *N,N′*-dihalogenoalkylamines (e.g., RNXX′ (X = F, X′ = Cl, Br, I; X = Cl, X′ = Br, I)) were located in *Chemical Abstracts*. Syntheses should be feasible from primary *N*-fluoro- or *N*-chloroalkylamines, RNHF or RNHCl.

2.06.1.1 *N*-Fluoroalkylamines; RNHF, R_2NF, RNF_2, R_3N^+F

Methods of synthesis of *N*-fluoroalkylamines have been reviewed as parts of larger reviews. The synthesis of *N*-fluoroamines has been covered in a review on nitrogen fluorides ⟨67CRV665⟩, and in reviews on *N*-halogenoalkylamines ⟨62HOU(5)796, 91COS(7)735⟩. Fluorinating agents have been reviewed ⟨86CRV997, B-89MI 206-01⟩, and most should react with amines, although many have not been reported in the literature. However, certain reagents (e.g., caesium fluoroxysulfate $Ce^+\ {}^-OSO_2F$) have been shown not to yield *N*-fluoroamines, even though they work well with amides ⟨91T7447⟩.

2.06.1.1.1 Primary and secondary N*-fluoroalkylamines; RNHF, R_2NF*

Perfluorination is a common problem with fluorinations of amines using fluorine, and perfluoroamines are readily *N*-fluorinated ⟨67CRV665⟩. For instance, perfluoro-*N*-fluoropiperidine is used as a fluorinating agent ⟨91JFC(52)389⟩. However, trifluoroamine oxide NF_3O has been reported to fluorinate secondary amines successfully, to yield the corresponding *N*-fluoroamines R_2NF (Equation (1)) ⟨90JA2383⟩. The reactions also yield the corresponding *N*-nitrosoamines which is a disadvantage because of their toxicity. The method allows selective fluorination of the amino group, and perfluorinated dialkylamines are not required. A range of dialkylamines (R_2NH, R = Me, Et,

Pr^n, Pr^i, cyclohexyl and other cycloalkylamines) give good yields of *N*-fluoroamines. The addition of fluorine to imines to yield secondary *N*-fluoroamines has been reported ⟨67JOC416⟩.

$$4\,R_2NH + NF_3O \xrightarrow{\leq 0\,^\circ C} R_2NF + R_2N{-}NO + 2\,R_2NH{\bullet}HF \qquad (1)$$

No useful syntheses of *N*-fluoro primary amines RNHF are located in the literature, and fluorination of primary amines yields *N*,*N*-difluoroalkylamines.

2.06.1.1.2 N,N-*Difluoroalkylamines; RNF_2*

In contrast to monofluoroamines, a number of methods for synthesizing *N*,*N*-difluoroalkylamines have been reported, and the earlier methods have been reviewed ⟨67CRV665, B-68MI 206-01⟩. Primary alkylamines have been fluorinated successfully in bicarbonate buffered aqueous solutions (HCO_3^-, H_2O, CCl_3F) using fluorine to give *N*,*N*-difluoroalkylamines (e.g., R = cyclohexyl, Bu^t, Bu^n) ⟨68JOC1008⟩. Fluorine gas in nitrogen gas was bubbled through the solutions at 0 °C.

The fluorination of amides is also known, and provides several methods for synthesizing *N*,*N*-difluoroamines ⟨69JA1679, 70JOC1545⟩. Fluorination of secondary amides with fluorine yields intermediate *N*-fluoroamides which react further with displacement of acylium ions to yield *N*,*N*-difluoroamines. Sulfamides, urethanes, and ureas give similar results. The use of fluoroxytrifluoromethane CF_3OF, developed as a fluorinating reagent by Barton *et al.* ⟨74JCS(P1)732⟩, gives high yields of *N*,*N*-difluoroamines for a wide range of substrates. *N*-Fluoroamides are initially formed and undergo further fluorination with cleavage of the amide bond (Scheme 1). The use of CF_3OF is compared to bis(fluoroxy)difluoromethane $CF_2(OF)_2$, and fluoroxysulfur pentafluoride SF_5OF, for example the yields of 1-(*N*,*N*-difluoroamino)adamantane from 1-acetamidoadamantane are 75% (CF_3OF), 77% (SF_5OF) and 96% ($CF_2(OF)_2$). CF_3OF reacts faster with electron-rich arenes, such as phenols, than with amides which limits its use. A second procedure using CF_3OF also gives high yields of *N*,*N*-difluoroamines from primary amines, (e.g., *N*,*N*-difluoroaminoadamantane (70%) and 4-(*N*,*N*-difluoroamino)butyric acid (64%)) ⟨75CC97⟩. In this protocol, reaction between the primary amine and benzaldehyde yields the respective imine, which is treated with CF_3OF in MeOH (Scheme 2). The method was also used for the difluorination of aminomonosaccharides ⟨75CC97⟩.

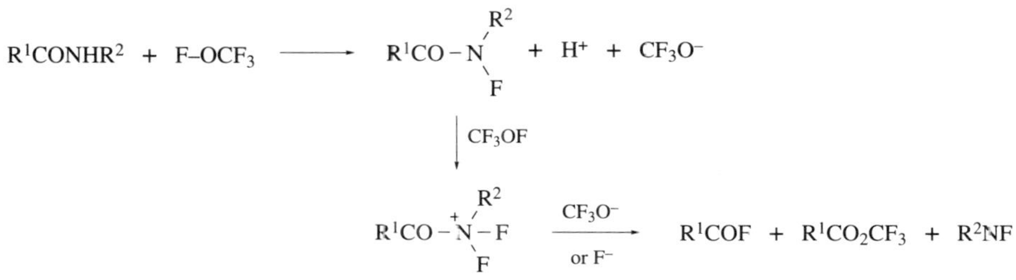

Scheme 1

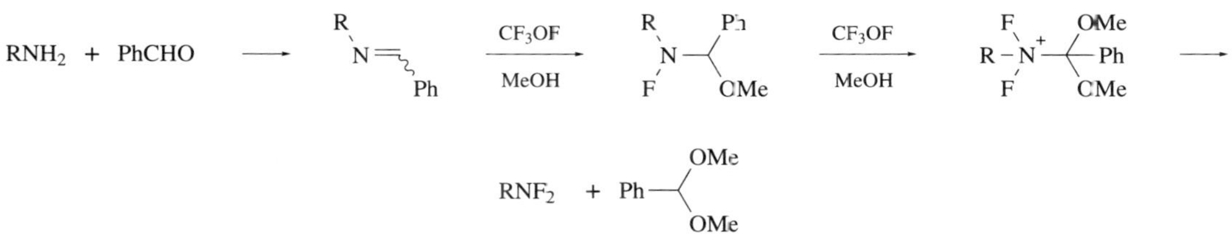

Scheme 2

Nitrogen fluorides ⟨89MI 206-02⟩ have been used for several synthetic procedures. The acidic properties of the hydrogen in difluoroamine have been exploited to react with alkenes, bromides, and alcohols, but the conditions are not very practical ⟨67JA716⟩. Tetrafluorohydrazine has been extensively used for addition to alkenes ⟨67CRV665, 67JOC4034, B-69MI 206-01⟩. A wide range of alkenes (e.g., Equation (2)) have been converted to vicinal bis(*N*,*N*-difluoroamino) compounds, including cycloalkenes, arylalkenes and olefinic steroids. Evidence indicates a radical mechanism as shown in Scheme 3 ⟨67JOC4034⟩.

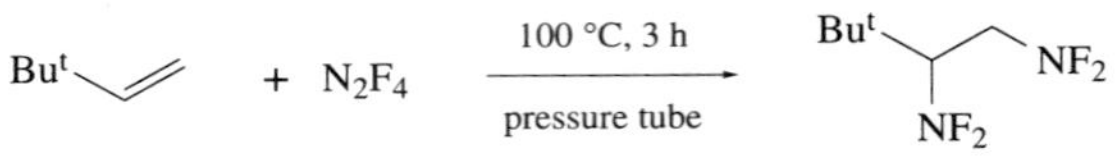

(2)

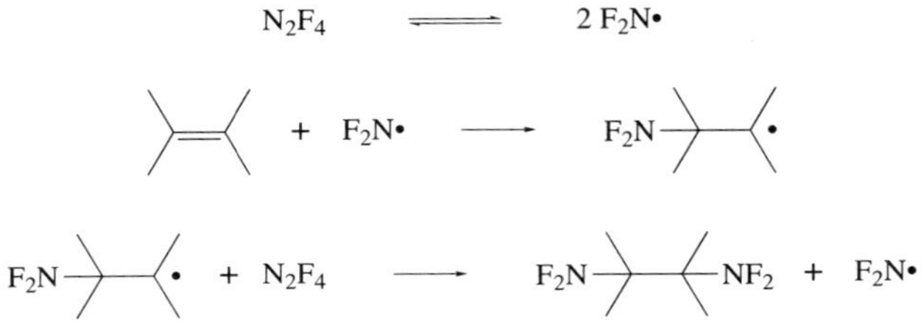

Scheme 3

2.06.1.1.3 N-*Fluoro quaternary salts;* R_3N^+F

The reaction between tertiary amines and fluorine yields quaternary fluorides R_3N^+F. The latter normally give elimination of HF to yield iminium salts, which readily hydrolyze to aldehydes and secondary amines. Some important exceptions are reported, for example the preparation of the fluorinating reagent *N*-fluoroquinuclidinium fluoride (NFQNF) (**1**) (Equation (3)), an extremely hygroscopic white solid ⟨88JCS(P1)2805⟩. Treatment of quinuclidine in trichlorofluoromethane with fluorine at low pressure, in a specially designed glass vacuum system, gives good yields of the fluorinating agent (Equation (3)). A range of 1-alkyl-4-fluoro-1,4-diazabicyclo[2.2.2]octane salts (**2**) have been prepared by fluorination of the 1-alkyl-1,4-diazabicyclo[2.2.2]octane salt with fluorine. These compounds are used as reactive fluorinating reagents; 1-chloromethyl-4-fluoro-1,4-diazabicyclo[2.2.2]octane bis(tetrafluoroborate) is available commercially (Selectfluor) ⟨92USP5083178⟩.

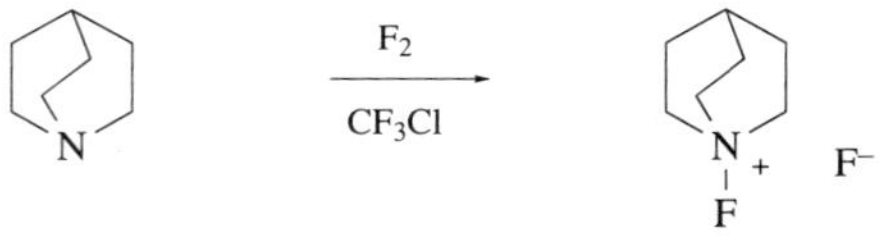

(3)

(1)

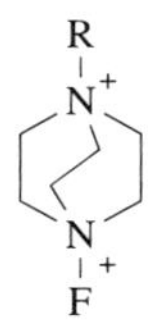

(2) selectfluor R = CH_2Cl

N-Fluoroammonium salts, for example (**3**), have been synthesized using perchloryl fluoride, $FClO_3$, from the corresponding tertiary amine ⟨83JOU1203⟩. *N*-Fluoro secondary ammonium salts, $R_2NH^+F\ ClO_3^-$ (**4**), have been reported to be formed in the reaction between the 2,2,6,6-tetramethylpiperidin-4-one and perchloryl fluoride, and further reaction with excess amine gives the corresponding secondary *N*-fluoro-amine ⟨67JOC1115⟩. The gas $FClO_3$ is explosive and must be used in dilute solutions.

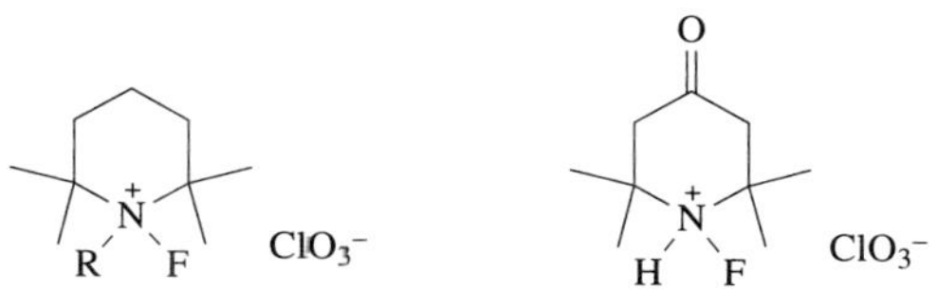

(3) R = Me, Et, Bu **(4)**

2.06.1.2 *N*-Chloroalkylamines; RNHCl, R_2NCl, $RNCl_2$, R_3N^+Cl

The syntheses and reactions of *N*-chloroamines have been reviewed ⟨62HOU(5)796, B-68MI 206-01, 70CRV639, 91COS(7)735⟩. Most sources of positive chlorine have been reported for the chlorination of primary, secondary and tertiary amines, for example chlorine, hypochlorous acid, hypochlorite ion, *N*-chlorophthalimide, *N*-chlorosuccinimide (NCS), *N*-chloroacetamide, *N*-chloro- and *N*,*N*-dichlorourea, *N*,*N*,*N*-trichloroisocyanuric acid, *N*-chlorosaccharin, *N*-chloro- and *N*,*N*-dichloro-*p*-toluenesulfonamide and chloroamine.

2.06.1.2.1 *Primary and secondary* N-*chloroalkylamines; RNHCl, R_2NCl*

Monochlorination of primary amines has not been commonly reported but the use of one equivalent of chlorinating agent (e.g., NCS or *t*-butylhypochlorite) allows the synthesis of primary *N*-chloroamines ⟨91COS(7)735⟩. Volatile primary *N*-chloroamines have been synthesized in high yields (>95%) using solid NCS under high vacuum (0.1 tor (mm Hg)) at room temperature ⟨85S1131⟩. Chloroamine NH_2Cl acts as a source of positive chlorine and has been used to chlorinate a range of amines including α-amino acids, to yield primary *N*-chloroamines, but the reaction is in equilibrium and not satisfactory for synthesis ⟨82IC2545⟩.

Secondary *N*-chloroamines are common synthetic targets (e.g., as precursors for forming aminyl radicals ⟨83AG(E)337⟩) and are readily prepared using various sources of positive chlorine. This latter review provides references to the syntheses of a wide range of secondary *N*-chloroamines. Chlorination of amines using sodium hypochlorite has been known for over 100 years, but has been supplanted largely by more convenient modern reagents such as NCS. Practical details of the use of sodium hypochlorite are given in *Houben-Weyl* ⟨62HOU(5)796⟩. Alkenyl secondary amines can be chlorinated selectively to yield chloroamines without reaction of the alkene ⟨81TL61⟩. The synthesis of chloroamines by addition of amine salts to strongly alkaline sodium hypochlorite (10–14%) is high yielding (Equation (4)) ⟨72CJC1167⟩. The use of hypochlorous acid in Et_2O allows the chlorine equivalents to be measured accurately.

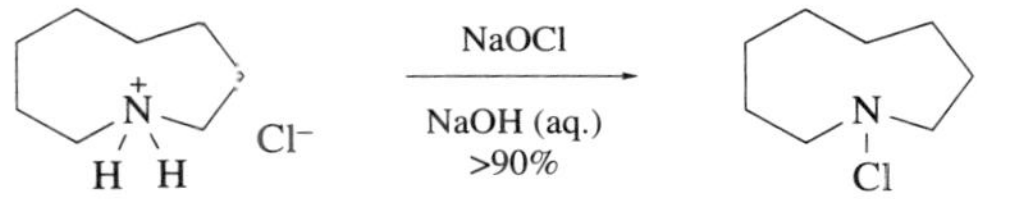

(4)

In the 1990s, the most common method for synthesizing secondary *N*-chloroamines is the use of NCS. Many *N*-chloroamines do not possess long-term stability, and are often made *in situ* and then reacted immediately ⟨94JCS(P1)777⟩. *N*-Chloroamines formed by selective chlorination of alkenyl secondary amines have been used in synthesis, for example the indolizidine gephyrotoxin-223AB (Equation (5)) ⟨86JOC5045, 88JOC2144⟩.

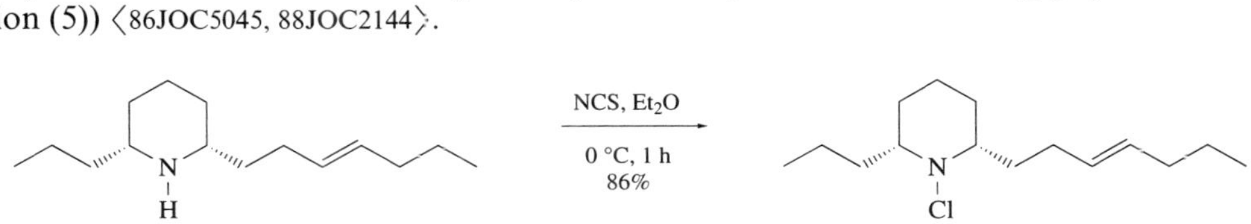

(5)

N-Chlorination of homochiral 2-substituted aziridines using NCS or *t*-butyl hypochlorite (Bu^tOCl) in the presence of various chiral trifluoromethylcarbinols (e.g., (*S*)-(+)-$PhCH(OH)CF_3$) gave mixtures of (*E*) and (*Z*) diastereoisomers with little stereoselectivity ⟨83JOC2640⟩. The nitrogen inversion barrier in *N*-chloroaziridines is *ca.* 115 kJ mol^{-1} and therefore, the invertomers (1*R*,2*S* and 1*S*,2*S*) formed by NaOCl chlorination of 2(*S*)-methylaziridine can be separated easily by GLC and characterized ⟨88JA1719⟩.

Ethyl hypochlorite, and more commonly Bu^tOCl, have been used as chlorinating reagents. *t*-Butyl hypochlorite is not available commercially although the preparation is reported ⟨73OSC(5)184⟩. *N*-Chloroamines of *N*-BOC-protected α-amino acid esters are synthesized using Bu^tOCl and sodium methoxide in methanol, but prolonged exposure to sodium methoxide leads to elimination of HCl, to yield α,β-unsaturated amino acids ⟨75CB2917, 76AG(E)294⟩. Chlorination of 2,6-diphenylpiperidine (with labile benzylic hydrogens) using Bu^tOCl, yields the *N*-chloro derivative without elimination of HCl. However, several other amines with α-benzylic hydrogens readily undergo loss of HCl during chlorination ⟨83CB3931⟩.

The sodium salts of dichloroisocyanuric acid ⟨88SC2135⟩ and trichloroisocyanuric acid ⟨91CJC1482⟩ have been reported to be better chlorinating reagents than NCS or NaOCl. Both reagents are stable and commercially available. Hindered chloroamines have also been synthesized by direct reaction between the sodium salt of dichloroisocyanuric acid and the amine in water or benzene. Secondary trimethylsilylamines R_2N-TMS react with chlorine to yield secondary *N*-chloroalkylamines ⟨70CRV639⟩.

2.06.1.2.2 N,N-*Dichloroalkylamines; $RNCl_2$*

Most of the methods used for monochlorination of secondary amines have also been used for dichlorination of primary amines to yield *N,N*-dichloroalkylamines, for example sodium hypochlorite, NCS, Bu^tOCl. Two equivalents or an excess of chlorinating reagent with suitable buffering are required ⟨91COS(7)735⟩.

Trichloroamine NCl_3 can be used to add across alkenes to yield 1-(*N,N*-dichloro-amino)-2-chloroalkanes ⟨28JA2739⟩. NCl_3 must be prepared in solution because of its very explosive properties. The use of $NCl_3/AlCl_3$ as an aminating system gives synthetically useful yields of *N,N*-dichloroamines from tertiary alkanes (e.g., adamantane and methylcyclohexane) via the tertiary cation of the alkane and the anion NCl_2^- ⟨68OS(48)4, 70CRV639⟩. Skeletal rearrangements expected of intermediate cations take place prior to dichloroamination. This aminating system has also been used with tertiary chloroalkanes to yield *N,N*-dichloroamines formed via the intermediate cation, but the results are variable with several by-products ⟨69JOC911⟩.

2.06.1.2.3 N-*Chloro quaternary salts; R_3N^+Cl*

Stable *N,N*-dichloroalkylamines (e.g., with electron-withdrawing α-substitutents) can be isolated as their hydrochloride salts $R_2C(CN)NCl_2 \cdot HCl$ ⟨89RCR250⟩.

N-Chloroammonium salts are normally unstable, and their formation from the tertiary amine and hypochlorite is reversible. In general, elimination of HCl takes place to yield intermediate iminium salts, which rapidly undergo hydrolysis to the corresponding aldehydes and secondary amines (Scheme 4). *N*-Chloro-1-azobicyclo[2.2.2]octane does not undergo oxidative elimination and is stable even in aqueous acid at 25 °C ⟨69JCS(C)626⟩. *N*-Chlorotrimethylammonium chloride $Me_3N^+Cl\ Cl^-$ is stable for several days, but $Et_3N^+Cl\ Cl^-$ is unstable and cannot be isolated ⟨54JCS1750⟩. *N*-Chlorotrialkylammonium chlorides (*N*-chloro derivatives of 1-azobicyclo[2.2.2] octane, *N*-methylmorpholine, *N*-methylpiperidine, and *N*-methyl-2,2,6,6-tetramethylpiperidine) have been stabilized by conversion to the corresponding perchlorate (Scheme 5) ⟨81JCS(P2)545⟩.

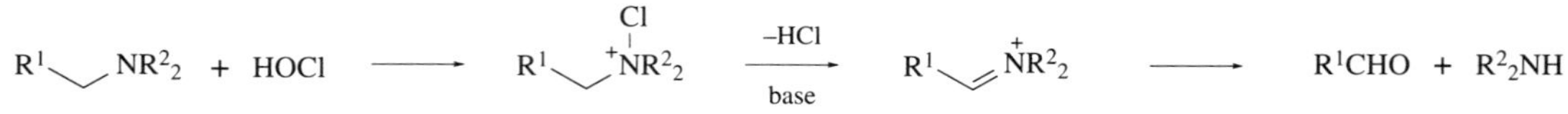

Scheme 4

Scheme 5

2.06.1.2.4 N-*Perchlorylamines*; R_2NClO_3

N-Perchlorylamines are explosive and need to be handled with care. Primary and secondary amines have been reacted with dichlorine heptoxide Cl_2O_7 in CCl_4 solution to yield *N*-perchlorylamines (Scheme 6) ⟨74JA3237, 87ZAAC(547)233⟩. Examples include *N*-perchlorylamines from aziridine, adamantane, azetidine, piperidine (73%), 2-ethylaziridine (83%) and isopropylamine (60%). Salts can be formed from primary *N*-perchlorylamines. Treatment of 1-(*N*-perchlorylamino)adamantane with $HgCl_2$ gives the mercury(II) salt (Scheme 6), which is explosive when heated, and treatment with hydroxides ZOH also yields the respective salt $RNClO_3^-Z^+$ (Z = Ag, Na, Ba, K, Et_4N). Primary *N*-perchlorylamines have been chlorinated with NaOCl to give *N*-chloro-*N*-perchlorylamines (Scheme 7) ⟨74JA3237⟩. *N,N′*-Disubstituted diamines react with Cl_2O_7 to give the respective *N,N′*-diperchlorylamines ⟨85ZAAC(526)186⟩. Tertiary amines give unstable perchloryl derivatives, for example the reaction between triethylamine and perchloryl fluoride in dilute solution gave *N*-perchloryltriethylamine fluoride which, with excess amine, eliminated $HClO_3$, to yield the corresponding iminium salt ⟨67JOC1115⟩.

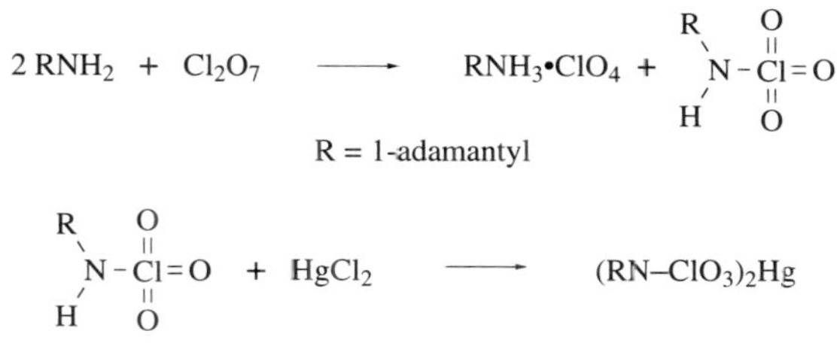

Scheme 6

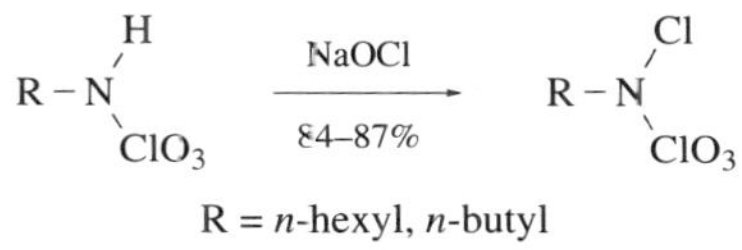

Scheme 7

2.06.1.3 *N*-Bromoalkylamines; RNHBr, R_2NBr, $RNBr_2$, R_3N^+Br

Alkylamines are brominated by a range of brominating agents such as bromine, aqueous sodium hypobromite, NBS, *N*-bromophthalimide, etc. ⟨B-68MI 206-01, 70CRV639, 91COS(7)735⟩. There has been some discussion on the mechanism of NBS bromination, about whether the brominating agent is a low dilution of Br_2, as for allylic bromination with NBS. Reactions between NBS and alkenyl secondary amines show that amines are brominated much faster than alkenes, but that the bromine is slowly transferred from the amine to the alkene, probably via a six-membered ring transition state ⟨94UP⟩. Therefore, the selective synthesis of alkenyl *N*-bromoamines is difficult. Trimethylsilylamines are also readily brominated as with chlorination.

Tertiary amines are brominated with Br_2 in CCl_4 to yield tertiary ammonium bromides, which readily revert to the amine and Br_2 ⟨51CB170, 81JCS(P2)545⟩. Even the *N*-bromoammonium salt of 1-azabicyclo[2.2.2]octane, unlike the stable *N*-chloro analogue, reverted to amine and Br_2.

2.06.1.4 *N*-Iodoalkylamines; RNHI, R_2NI, RNI_2

There is little mention in the literature of *N*-iodoamines. *N*-Iododimethylamine Me_2NI and *N,N*-diiodoethylamine and -methylamine are known ⟨91COS(7)735⟩. *N*-Iodobistrifluoromethylamine $(CF_3)_2NI$ has been prepared by treatment of di(bistrifluoromethylamino)mercury with iodine in CCl_4 ⟨68JCS(C)796⟩. Iodination of tertiary amines, as for chlorination and bromination, yields *N*-iodoammonium salts, which give elimination of HI to yield iminium salts, which undergo hydrolysis.

2.06.2 HYDROXYLAMINES AND RELATED FUNCTIONS

The synthesis and chemistry of hydroxylamines and related functions have been extensively reviewed and readers are referred to these reviews for further detailed information and older references ⟨B-68MI 206-01, 71HOU(10/1)1091, B-74MI 206-01, 79COC(2)185, B-89MI 206-03, 90HOU(E16)1⟩. Attempts have been made here to survey methodology in general and not to repeat previous reviews. More recent examples of synthetic methods have been cited where possible.

2.06.2.1 *N*-Alkylhydroxylamines; RNHOH, R^1R^2NOH, and Salts Thereof

Hydroxylamines have been reviewed ⟨91COS(6)103⟩, and while not part of this chapter, reviews on *N*-hydroxy-α-amino acids ⟨86CRV697⟩ and α-hydroxylamino oximes ⟨86S704⟩ detail further synthetic methods.

2.06.2.1.1 N-Alkylhydroxylamines; RNHOH

(i) Oxidative methods

The oxidation of primary amines with aqueous hydrogen peroxide yields *N*-substituted hydroxylamines RNHOH but further oxidation to nitroso and nitro compounds is a problem. 2,2-Dimethyldioxirane smoothly oxidizes a variety of primary amines to hydroxylamines in high yield ⟨90JOC1981⟩. The methodology has been applied to amino mono- and disaccharides in particular (Equation (6)). Some amines did not react or were oxidized further to yield oximes.

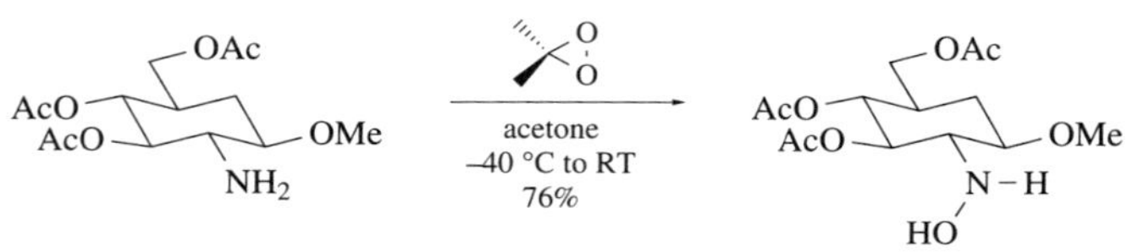

(6)

Reaction between primary and secondary amines and dibenzyl peroxide involves nucleophilic displacement on the peroxide bond to yield *O*-benzyloxyamines, which can be subjected to hydrogenolysis to yield *N*-alkylhydroxylamines ⟨83S537, 84S217⟩. *N*-Benzoyloxyamines RNHOCOPh prepared by treatment of primary amines with dibenzoyl peroxide, are readily hydrolyzed to *N*-alkylhydroxylamines (Scheme 8) ⟨83S537⟩. Because of the dangers of explosion, the procedure should only be carried out on small scale. Oxidation of primary amines with diphenylphosphinylperoxide $(Ph_2PO_2)_2$ yields diphenylphosphinyloxyamines (Scheme 9) ⟨86T2703⟩ which, after two rearrangements facilitated by further heating of the reaction and treatment with methoxide, yield *N*-substituted hydroxylamines RNHOH.

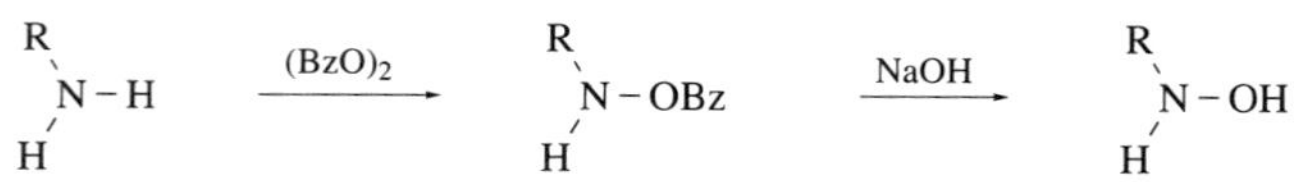

Scheme 8

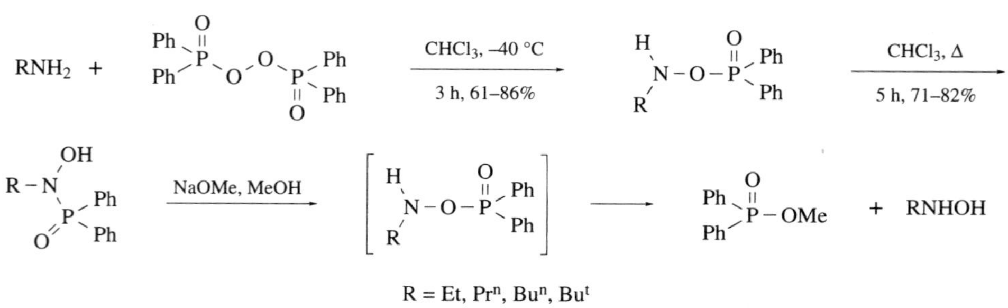

Scheme 9

(ii) Reductive methods

The most common and satisfactory protocol for the synthesis of *N*-substituted hydroxylamines RNHOH is by reduction of a nitrogen precursor which is at a higher oxidation state, for example nitro and nitroso compounds and oximes. Various nitro compounds are reduced by diborane to *N*-substituted hydroxylamines ⟨91COS(8)373⟩. The reduction of nitro alkanes with four equivalents of samarium diiodide provides a facile and mild method with good yields (Equation (7)) ⟨91TL1699⟩. The reduction is compatible with acetals, arenes, esters and silyl protected alcohols. The use of di-(benzenesulfenyl)tin $(PhS)_2Sn$ for the reduction of tertiary nitro compounds, which have no α-hydrogens, gives excellent yields of *N*-alkylhydroxylamines (Scheme 10) ⟨90T587⟩. The reagent is best made *in situ* from benzenethiol, $SnCl_2$, and triethylamine. Secondary nitro compounds can also be reduced to *N*-alkylhydroxylamines if the intermediate nitroso compound does not tautomerize rapidly to the corresponding oxime, and a polar solvent is used.

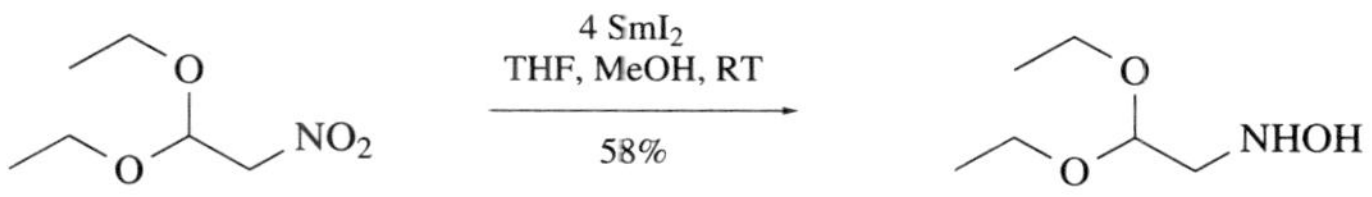

(7)

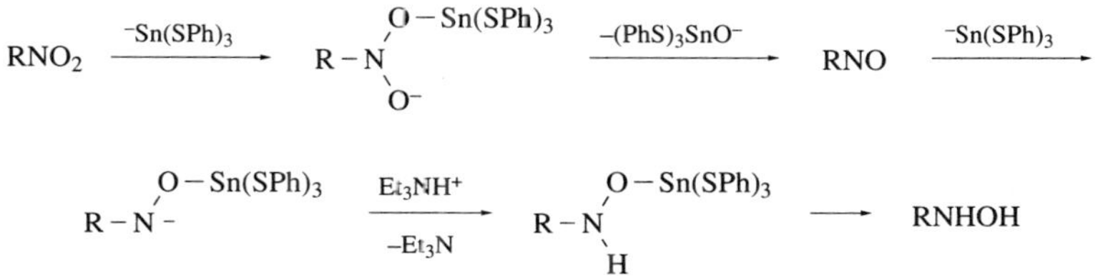

$Sn(SPh)_2$ (2.5 equiv.), PhSH, Et_3N in PhH, RT, 10 min

Scheme 10

Primary and secondary nitronate anions $R^1R^2C{=}NO_2^-$ (R^1 = alkyl, R^2 = H or alkyl) are reduced by diborane to *N*-substituted hydroxylamines $R^1R^2CHNHOH$ ⟨65JOC2880⟩. Sodium borohydride, which is commonly used to reduce nitroalkenes to nitronate anions, which are subsequently protonated to yield nitroalkanes, can be used in catalytic amounts with borane complexes to produce *N*-substituted hydroxylamines (Scheme 11) ⟨85JOC133⟩. A recent example uses $BH_3 \cdot THF$ with catalytic $NaBH_4$ to reduce 3-nitrochromenes to 3-hydroxylamino-2*H*-1-benzopyrans in high yield ⟨87JHC767⟩.

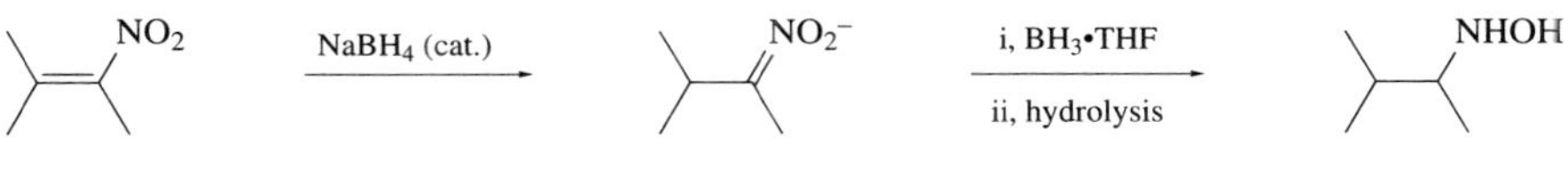

Scheme 11

The reduction of oximes to *N*-substituted hydroxylamines RNHOH is more reliable and is fully reviewed, and tables of examples are given ⟨91COS(8)60⟩. Aldoximes and ketoximes are reduced to *N*-substituted hydroxylamines by a wide range of reagents including $LiBH_4$, $NaBH_4$ in basic or acidic media, $NaBH_3CN$ in acidic medium ⟨75T2595, 86TL4961⟩, BH_3/THF and $BH_3 \cdot$pyridine/HCl, for example in the synthesis of hydroxylamino nucleotides (Equation (8)) ⟨90TL531⟩.

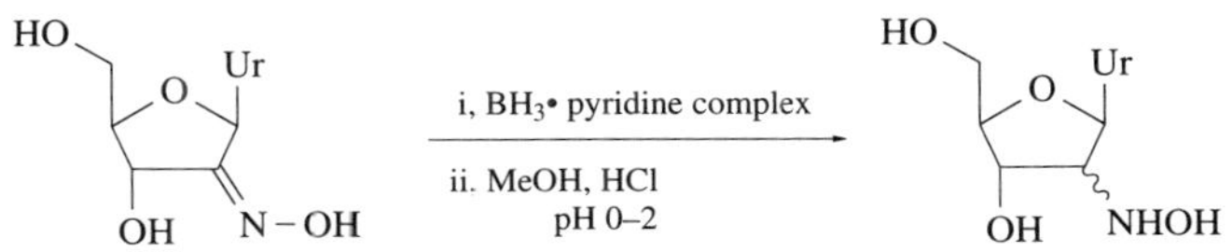

(8)

Ur = uracil

(iii) From alkylation, C—N bond formation

Monoalkylation of hydroxylamine is not normally a good procedure, owing to problems of further alkylation, and because the products are difficult to predict ⟨B-89MI 206-03, 91COS(6)103⟩.

S_N1 Substitution has been reported for the reaction between NH_2OH and (*p*-chlorophenyl) diphenylmethyl chloride to yield *N*-[(*p*-chlorophenyl)diphenylmethyl]-hydroxylamine ⟨B-89MI 206-03⟩. Acid-catalyzed S_N1 substitution of benzylic hydroxyl groups by hydroxylamine has been reported ⟨88JOC1768⟩.

Development of specific alkylating agents allows selective formation of the *N*-alkylhydroxylamines. Reaction between *O*-benzyloxylamine $BnONH_2$ and alkyl halides yield *N*-alkylated products. The benzyl group is removed by hydrogenolysis to yield the *N*-alkyl-hydroxylamine (Scheme 12). The *N*-alkylation of the isoxazole (**5**) with alkyl halides provides an excellent method for the synthesis of *N*-alkylhydroxylamines RNHOH (Scheme 13) ⟨87TL2993⟩.

Scheme 12

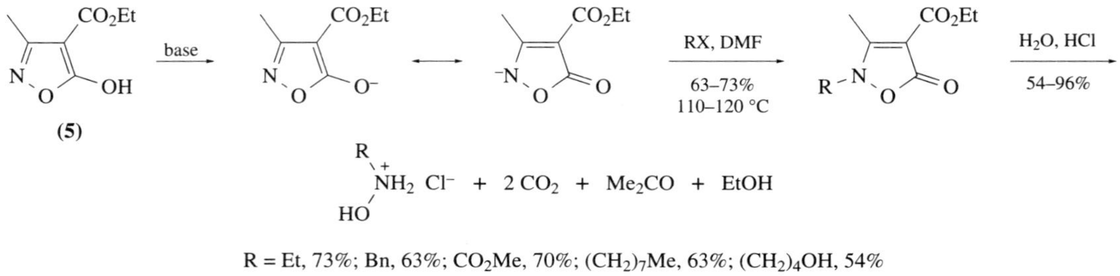

Scheme 13

Other less common alkylating reagents have been reported but not extensively investigated, for example the reaction between oxiranes and NH_2OH or *N*-alkylhydroxylamines ⟨B-89MI 206-03⟩. Hydroxylamine isodisulfonate HO_3S—NH—OSO_3H can be alkylated in strongly basic media to yield an intermediate, $RN(SO_3^-)OSO_3^-$, which on acid hydrolysis gives *N*-alkylhydroxylamines ⟨B-89MI 206-03, 79COC(2)504⟩.

(iv) From hydrolysis of nitrones

The hydrolysis of nitrones to yield *N*-alkylhydroxylamines and aldehydes or ketones has been used for some 100 years (Scheme 14) ⟨B-82MI 206-01, B-89MI 206-03, 90HOU(E16)1⟩. Alkylation of the respective oximes provides the most facile route to the required nitrones. Nitrones may also be cleaved by hydroxylamine or hydrazine to yield *N*-alkylhydroxylamines and oximes or hydrazones, respectively.

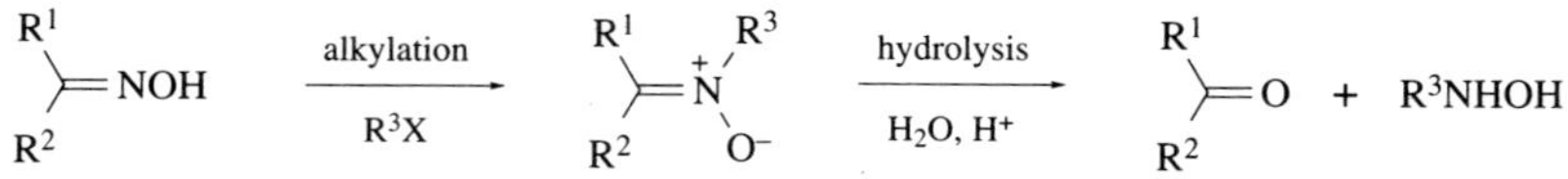

Scheme 14

The use of oxaziridines, including Davis reagents, provides a protocol for the conversion of amines to the corresponding *N*-substituted hydroxylamine ⟨74TL2453, 89T5703⟩. The methodology is particularly important because it has been used as the main method for converting homochiral amines to the corresponding homochiral *N*-substituted hydroxylamines ⟨75T2595, 85T3455, 91TL4431⟩. An example using (*S*)(−)-2-phenylethylamine is shown in Scheme 15 ⟨85T3455⟩. The amine is converted to an imine using an aromatic aldehyde. The imine is peroxidized to the corresponding oxaziridine which is ring opened with acid to yield the nitrone. The nitrone may be hydrolyzed by various reagents, but the use of hydroxylamine hydrochloride works well, giving the *N*-substituted hydroxylamine and the oxime of the aldehyde.

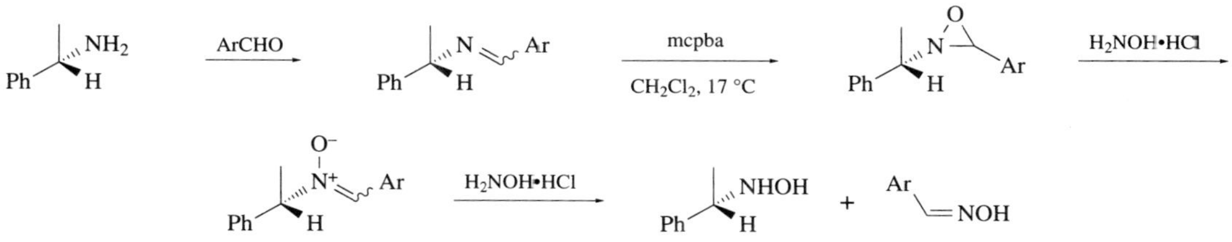

Scheme 15

(v) *From addition of alkyllithium reagents to oximes and nitrones*

Alkyllithium reagents add to oximes to yield sterically hindered *N*-alkylhydroxylamines (Scheme 16) ⟨76TL233⟩. Higher yields can be obtained using *O*-benzyl oxime ethers followed by removal of the benzyl group ⟨88TL3455⟩. Heteroarene lithium reagents are added to *N*-THP protected nitrones in THF to yield *N*-THP protected hydroxylamines which are hydrolyzed with acid to yield *N*-alkylhydroxylamines in moderate yield ⟨91TL3783⟩.

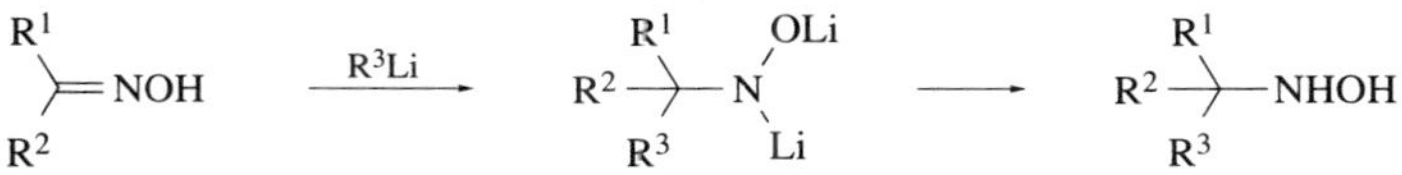

Scheme 16

(vi) *From addition of hydroxylamine to alkenes*

Hydroxylamine undergoes Michael reactions and adds to electron-deficient alkenes (α,β-unsaturated ketones, esters, amides and sulfones, and nitroalkenes). The addition to α,β-unsaturated ketones is used in the synthesis of 1,2-oxazoles. When two equivalents of the electron-deficient alkenes are used, secondary *N,N*-dialkylhydroxylamines result ⟨71HOU(10/1)1091, B-89MI 206-03⟩.

(vii) *From nitroxides*

The disproportionation of nitroxides, $R_2NO\cdot$, yields *N*-alkylhydroxylamines RNHOH and nitroso compounds RNO, but useful synthetic methods have not been located ⟨B-82MI 206-02⟩.

2.06.2.1.2 N,N-*Dialkylhydroxylamines; R_2NOH*

(i) *Oxidative methods*

The oxidation of secondary amines with peroxide to yield *N,N*-disubstituted hydroxylamines R_2NOH is more useful than the equivalent reaction for synthesizing *N*-substituted hydroxylamines RNHOH from primary amines, but can also lead to further oxidation products. Oxidation of secondary amines with hydrogen peroxide normally gives nitrones but in the presence of selenium dioxide, the *solanum* steroid alkaloids, spirosolane and tomatidine, it gives low yields of the corresponding *N,N*-disubstituted hydroxylamines ⟨92LA1091⟩. 2,2-Dimethyldioxirane oxidation of secondary amines provides the first general method for oxidizing secondary amines to the corresponding hydroxylamines in high yield (Scheme 17) ⟨89SC3509⟩. The method allows oxidation of amines selectively over alcohols and of sterically hindered amines. The method is facile, and at the end of the reaction the solvent is removed leaving crystalline hydroxylamines. Secondary amines with an α-benzylic hydrogen are oxidized further to the corresponding nitrone if two equivalents of 2,2-dimethyldioxirane are used ⟨90JOC2954⟩.

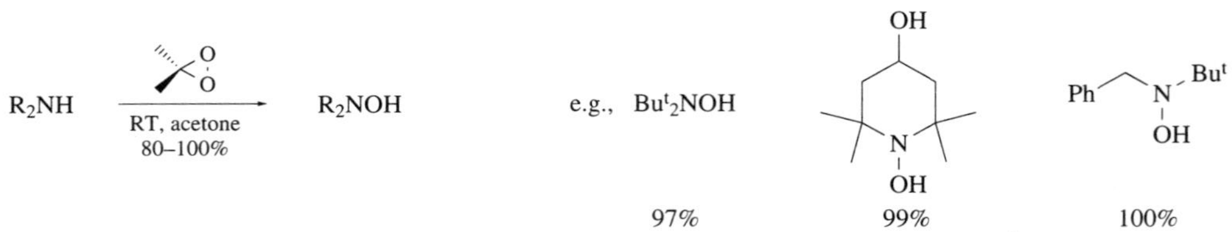

Scheme 17

N-Acyloxyamines $R_2NOCOPh$, prepared by treatment of secondary amines with dibenzoyl peroxide, are readily hydrolyzed to *N*,*N*-dialkylhydroxylamines as shown for primary amines in Scheme 8 ⟨83S537, 90JOC3464, 90JOC3475⟩. This was the standard method of preparing hydroxylamines from secondary amines until the dioxirane route was developed, although the yields were low (25–35%). The procedure requires the use of two equivalents of amine, one for neutralizing the benzoic acid that is formed. The use of added base (Na_2HPO_4 or polyvinylpyridine) to prevent reversibility increases the yield two- to three-fold. The use of potassium methoxide in anhydrous methanol instead of sodium hydroxide in water also improves the yields. Because of the dangers of explosion, the procedure should only be carried out on a small scale. The use of 2-chlorobenzoyl peroxide in place of benzoyl peroxide gave improved yields ⟨90JOC3464, 90JOC3475⟩.

The reaction between secondary amines and diphenylphosphinylperoxide $(Ph_2PO_2)_2$ yields diphenylphosphinyloxyamines, which are hydrolyzed to yield *N*,*N*-disubstituted hydroxylamines ⟨85S807⟩.

The use of Davis reagents, 2-(phenylsulfonyl)-3-aryloxaziridines, to oxidize secondary amines to hydroxylamines gives variable yields, and often mixtures which contain nitrone impurities ⟨89T5703⟩. In the synthesis of novel polyamines from the venom of *Agelnopsis aperta* spiders, the use of Davis reagent gave better yields of the required *N*,*N*-disubstituted hydroxylamines than the use of mcpba or 2,2-dimethyldioxirane ⟨90JA6696⟩.

(ii) Reductive methods

Nitrones are reduced to *N*,*N*-disubstituted hydroxylamines with LAH, $NaBH_4$, or $NaBH_3CN$ in acid ⟨71HOU(10/1)1091, 79COC(2)185, 91COS(8)60⟩. A 1990 example illustrated the use of LAH (Equation 9) ⟨90JOC3007⟩. More unusual reducing agents, such as trichlorosilane, also give good yields of hydroxylamines from nitrones ⟨78JOC2289⟩.

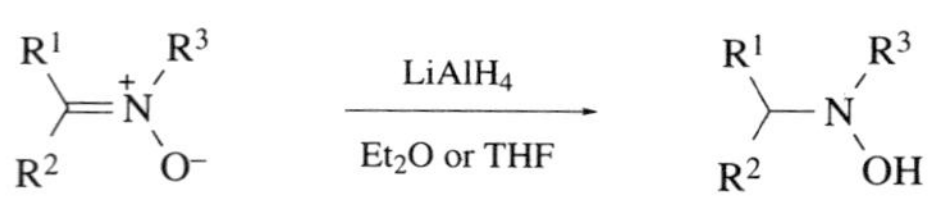

(9)

N-Substituted hydroxylamines RNHOH can be reacted with aldehydes to give intermediate nitrones which can be further reduced with $NaBH_4$ to yield *N*,*N*-disubstituted hydroxylamines ⟨91COS(8)60, 90TL531⟩.

(iii) From alkylation; C—N bond formation

Monoalkylation of hydroxylamine is troublesome, but dialkylation has been commonly used. Stepwise alkylation takes place, and the alkylation of the oxygen takes place last. A large number of examples and procedures for the monoalkylation of *N*-alkylhydroxylamines RNHOH into *N*,*N*-dialkylhydroxylamines R_2NOH are reviewed ⟨B-68MI 206-01, 71HOU(10/1)1091, 78CL1057, B-89MI 206-03⟩. Palladium(0)-catalyzed reactions between hydroxylamines and allylic acetates yield the corresponding *N*-alkyl-*N*-allylic hydroxylamines, (Equation (10)) ⟨88TL2973⟩. A new and more rigorous procedure uses benzyloxy primary amines to form benzyloxy secondary amines ⟨91TL2711⟩. The benzyl protecting group is selectively removed by standard methods to yield *N,N*-dialkylhydroxylamines in high yield; an example is shown in Scheme 18.

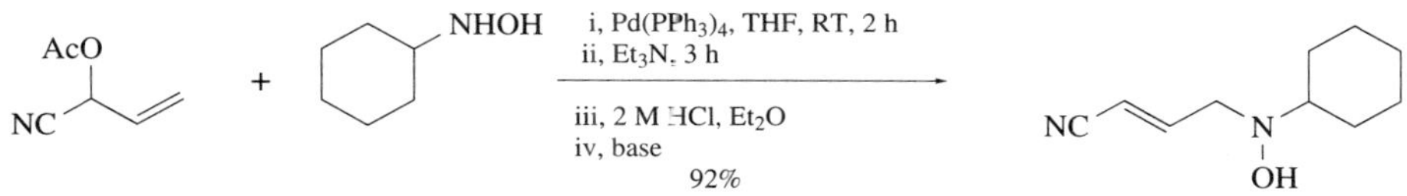

(10)

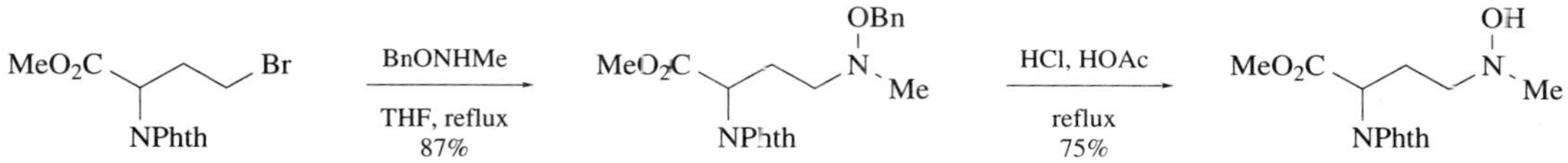

Scheme 18

(iv) From nucleophilic addition to nitrones

The addition of organometallic reagents to aldonitrones has been reviewed ⟨79COC(2)504, B-88MI 206-01, 91COS(1)355⟩. Acyclic ketonitrones are resistant to addition by organometallic reagents. The addition of Grignard reagents yields *N,N*-dialkylhydroxylamines. The addition of Grignard reagents to cyclic aldonitrones has been used in several natural product syntheses, for example the Grignard (**6**) was added to cyclic nitrone (**7**) in the synthesis of (±)-porantheridine (Equation (11)) ⟨80TL2229⟩.

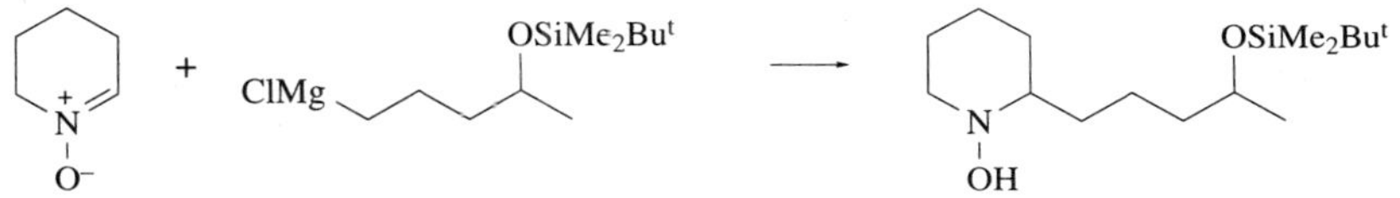

(11)

(**7**) (**6**)

Diastereoselectivity is observed in the addition of Grignard reagents to nitrones with a chiral α-centre to yield *N,N*-dialkylhydroxylamines in high yield (Scheme 19) ⟨88CC1503⟩. Attack by the Grignard reagent takes place preferentially on the least hindered side of the nitrone to yield (**8**) over (**9**) (ca. 2:1) as predicted by the Felkin–Anh modification of Cram's rule (Scheme 19). The diastereoselectivity in the formation of hydroxylamines can be improved dramatically by using nitrones bearing stereogenic *N*-substituents (e.g., β-methoxyalkyl, β-(silyloxy)alkyl, α-arylethyl ⟨90JOC3464, 90JOC3475⟩). The high facial diastereoselectivity observed with the β-methoxyalkyl substituents has been explained by a simple chelation transition state model (**10**) (Scheme 20). The diastereoselectivity has been fully investigated with a variety of examples and is discussed fully. This methodology has been applied to the enantioselective synthesis of amines ⟨90JOC3475, 92TL1689⟩.

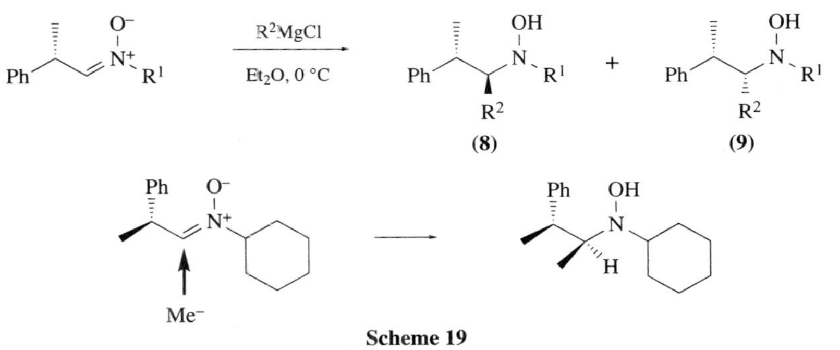

(**8**) (**9**)

Scheme 19

The addition of allylamines to nitrones yields *N,N*-disubstituted hydroxylamines ⟨93CC169⟩. The hydroxylamines rapidly undergo retro-Cope rearrangements to tertiary amine *N*-oxides, and finally yield 1,2,5-oxadiazinanes.

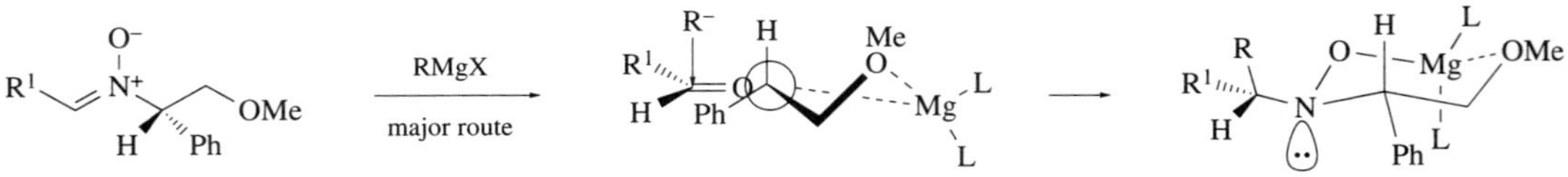

(10)

Scheme 20

(v) From organometallic addition to nitro and nitroso compounds

Methyl and propyl Grignard reagents have been added to alkylnitro compounds in the presence of anhydrous $CeCl_3$ in THF at $-70\,^{\circ}C$ to give *N*,*N*-dialkylhydroxylamines in good yield ⟨90JCS(P1)2133, 93CC1373⟩. Grignard reagents also add to N═O double bond of alkylnitroso compounds to yield *N*,*N*-dialkylhydroxylamines ⟨71HOU(10/1)1091, 79COC(2)185⟩.

(vi) From addition of N-*alkylhydroxylamines to alkenes*

N-Substituted hydroxylamines are readily added to α,β-unsaturated esters ⟨84T4363⟩ and to the double bond of 5,6-dihydroquinolines in acid-catalyzed reactions (Equation (12)) ⟨91TL6789⟩.

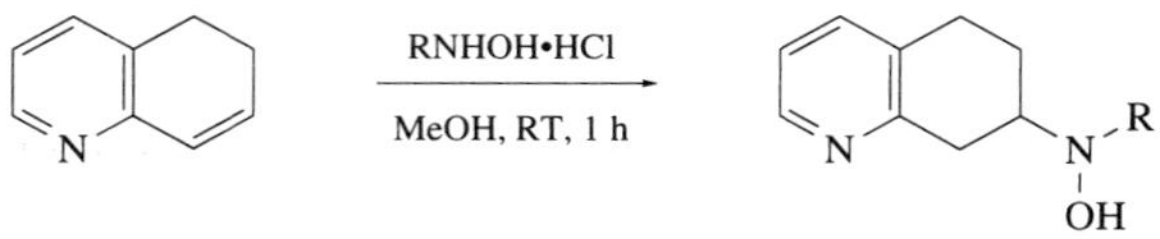

R = PhCHMe, 56%
R = Me, 65%
R = Pr, 49% (12)

Retro-Cope reactions involving cyclization of *N*-alkylhydroxylamines onto alkenes in *N*-4-alkenylhydroxylamines give cyclic *N*,*N*-disubstituted hydroxylamines (Scheme 21) ⟨78AJC2317, 79TL4391, 88JOC1768, 93CC169, 94JA3139⟩. Evidence indicates that the cyclizations are thermal, and proceed stereospecifically in a suprafacial manner as exemplified in Scheme 21 ⟨94JA3139⟩. The (*E*)-diastereoisomer (**11**) cyclizes suprafacially to yield (**12**), and the (*Z*)-diastereoisomer cyclizes to yield the opposite diastereoisomer. The methodology has been applied to the synthesis of (±)-α-lycorane ⟨94JA3139⟩.

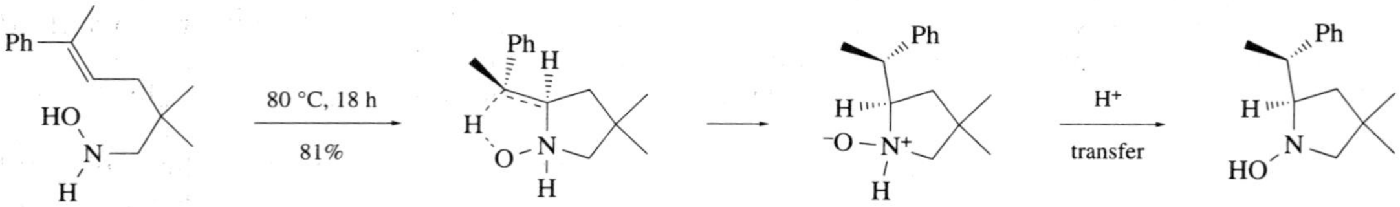

(11) **(12)**

Scheme 21

(vii) From nitroxides

Nitroxides have been used as oxidizing agents and are reduced to hydroxylamines R_2NOH in the process ⟨B-82MI 206-02⟩.

(viii) From rearrangements

The Cope rearrangement of tertiary amine *N*-oxides provides a useful synthetic method with yields up to 90% with few side products (Equation (13)). The methodology and mechanism are reviewed and procedures reported ⟨60CRV448, 60OR(11)317, B-89MI 206-03, 93S263⟩. The rearrangements are carried out by pyrolysis under vacuum or by heating in solution in solvents such as DMSO, THF, or DIGLYME. The tertiary amine *N*-oxide is commonly generated from the amine and aqueous hydrogen peroxide, and reacted without full purification. The nature of the β-hydrogen is crucial to the rate and yield, for $R^1{}_2R^2NO$; $R^2 = PhCH_2CH_2 \gg Bu^t > 2$-methylpropyl $>$ $Pr^i \sim Bu^n > Et > Pr^n$. The β-hydrogen must be lined up such that a planar five-membered transition state for *syn* elimination is permitted. The regiochemistry of elimination is explained by steric reasons.

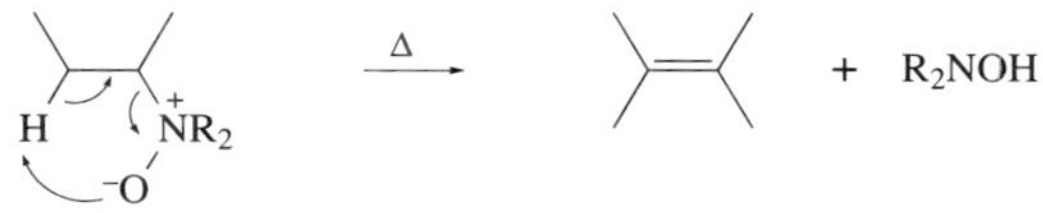

(13)

Hydroxylamines have been synthesized from the rearrangement of hydrazone *N*-oxides ⟨93TL141⟩.

(ix) From hydrolysis of O-*substituted* N,N-*disubstituted hydroxylamines*

Hydrolysis of *N*-acyloxyamines ⟨83S537, 90JOC3464, 90JOC3475⟩, $R_2NOCOPh$ (Section 2.06.2.1.2.(i) and Scheme 8), and diphenylphosphinyloxyamines ⟨85S807⟩ $R_2NOP(O)Ph_2$ yield *N,N*-dialkylhydroxylamines (Section 2.06.2.1.2(i)). Hydrolysis of the *N*-aroyl derivatives of 1,2,5-oxadiazinanes yields β-aroylamino *N,N*-disubstituted hydroxylamines ⟨93CC169⟩.

2.06.2.1.3 *Alkylhydroxylamine salts; RN^+H_2OH, R_2N^+HOH*

Hydroxylamines are protonated in acid and have acid/base properties similar to amines and hydrazines. However, there are reports of difficulty in forming stable salts ⟨88CC1503⟩. The salts of hydroxylamines are stable at ambient temperatures, and are normally neutralized without any decomposition to yield the corresponding hydroxylamine (Equation (14)). Reports suggest that the salts should be treated with care, as explosions are possible at elevated temperatures ⟨B-89MI 206-03⟩. The salts of hydroxylamines are initially formed in a number of synthetic reactions, for instance salts of *N*-alkylhydroxylamines result from the synthetic method using isoxazoles (Scheme 13). The salts are also used for purification and storage if the hydroxylamines are unstable, for example homochiral *N*-substituted hydroxylamines are converted to oxalate salts ⟨85T3455⟩.

(14)

2.06.2.2 *N*-Oxides; R_3NO and Salts Thereof

The synthesis and synthetic applications of tertiary amine *N*-oxides have been reviewed ⟨58HOU(11)190, 91COS(7)735, 90HOU(E16/1)404, 93S263⟩. Only a few common methods of synthesis are used, most involve an oxidizing agent, and are facile with good yields.

2.06.2.2.1 N-*Oxides; R_3NO*

(i) From oxidation of tertiary amines with hydrogen peroxide

Oxidation of tertiary alkylamines to tertiary amine *N*-oxides using aqueous hydrogen peroxide is the oldest method and works well in most cases, commonly in >90% yields (Scheme 22) ⟨58HOU(11)190, 71JCS(B)1330⟩. Methanol and ethanol are used as co-solvents with the water to improve the solubility of the tertiary alkylamines. A standard procedure is reported in *Organic Synthesis* ⟨63OSC(4)612⟩. A number of more modern hydrogen peroxide reagents, which are either safer or more efficient, are now available and have been thoroughly researched. Urea-hydrogen peroxide (UHP) is the safest and most applicable of these reagents and has been reviewed ⟨93MI 206-01⟩. This review contains an up-to-date survey and comparison of modern peroxidizing reagents.

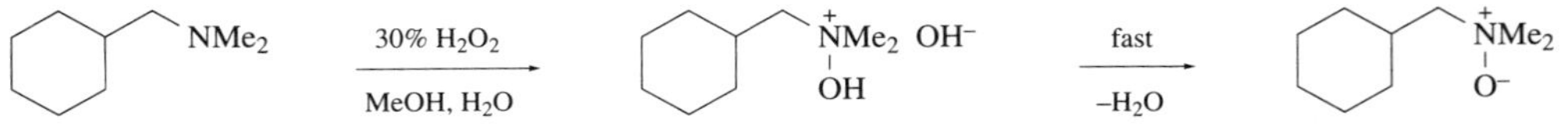

Scheme 22

(ii) From oxidation of tertiary amines with alkyl peroxides

The use of alkyl peroxides provides one of the common and successful methods of oxidizing tertiary amines to amine *N*-oxides, and good yields are obtained. However, yields can be improved by using metal catalysis, such as vanadium and molybdenum salts ⟨66CB3384, 70OS(50)56, B-89MI 206-04⟩. The Sharpless peroxide methodology using *t*-butyl peroxide allows stereoselective oxidation and kinetic resolution of one isomer of a racemic mixture of β-hydroxy tertiary amines (Equation (15)) ⟨83JOC3608, 90CL547⟩. The (+) or (−) isomers of diisopropyl tartrate can be used to give the respective enantiomers of the *N*-oxide. Trituration is used to separate the amine *N*-oxide and chromatography is not required.

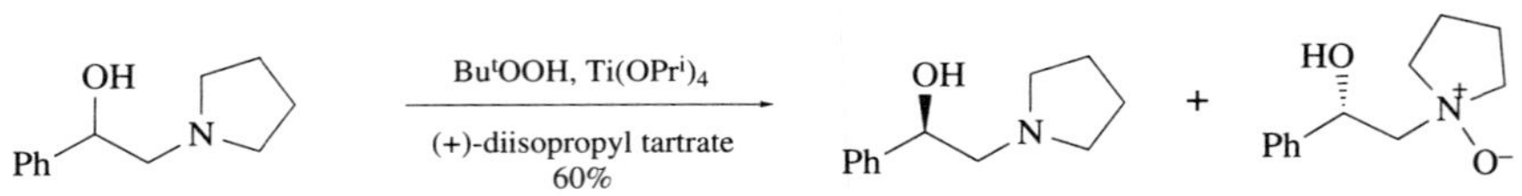

(15)

(iii) From oxidation of tertiary amines with peroxy acids

A variety of organic and inorganic peroxy acids readily convert tertiary amines into amine *N*-oxides even at low dilution ⟨49CRV34, B-68MI 206-01⟩. The mechanism is considered to be nucleophilic attack by the amine on the electrophilic oxygen in a reaction which is first order in each reagent. The resulting salt is neutralized by the carboxylate anion in a fast reaction (Scheme 23). *m*-Chloroperoxybenzoic acid is most commonly used ⟨91TL4481⟩.

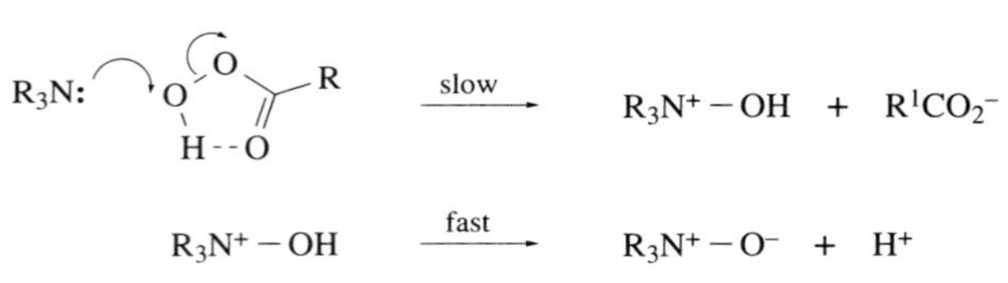

Scheme 23

(iv) From oxidation of tertiary amines with ozone and oxygen

The oxidation of tertiary amines with ozone has been reviewed ⟨58CRV925, 58HOU(11)190⟩, but reports suggest that mixtures of products are formed, and the reactions are not synthetically useful ⟨78JOC2662⟩. Yields are reasonable if protic solvents and tertiary amines with no β-hydrogens are used.

Autoxidation of tertiary amines occurs under high oxygen pressures (up to 150 atm) and temperature (100 °C) in polar solvents such as MeOH or water via single electron transfer to yield tertiary amine *N*-oxides as the major product ⟨83AJC719, 85JOC1563, 88JOC1695, 90JOC3233⟩. The mechanism is uncertain but the initial single electron transfer between the amine and oxygen to yield the radical cation $R_3N^{\cdot +}$ and superoxide $O_2^{\cdot -}$ is rate determining. The latest study suggests that hydrogen peroxide, which is eventually formed in the reaction, further oxidizes the tertiary amine to the

N-oxide ⟨90JOC3233⟩. Only one atom of the O_2 molecule is incorporated into the *N*-oxide. This surprising reaction has not yet been applied synthetically but has important implications for redox reactions carried out with O_2 in the presence of tertiary amines.

(v) Use of Davis reagents

Oxidation of a range of tertiary amines with Davis reagents, 2-(phenylsulfonyl)-3-aryloxaziridines, gives >95% yields of pure amine *N*-oxides ⟨88JOC5856⟩. The reaction is facile, and the *N*-oxide precipitates out of the chloroform solution of the reaction and is filtered. The use of Davis reagents in synthesis have been reviewed ⟨89T5703⟩.

(vi) From alkylation of hydroxylamines

The alkylation of hydroxylamine with three equivalents of alkyl halide to yield amine *N*-oxides has been known for nearly 100 years. However, treatment of *N*,*N*-dialkylhydroxylamines with alkyl halides to yield the corresponding tertiary amine *N*-oxides is a more satisfactory reaction (Scheme 24) ⟨B-89MI 206-03⟩.

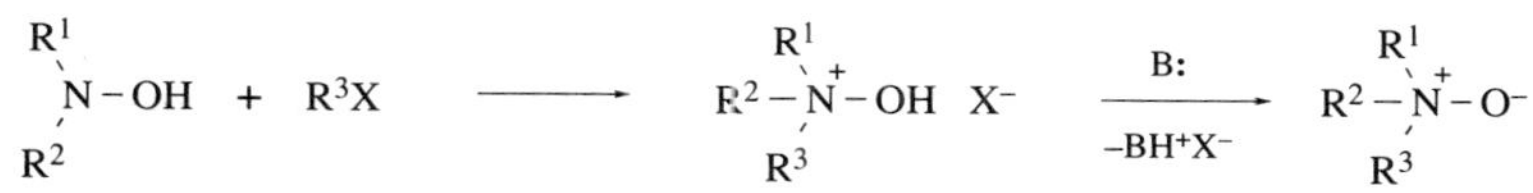

Scheme 24

(vii) From retro-Cope rearrangements

Retro-Cope rearrangements of ω-alkenyl *N*,*N*-disubstituted hydroxylamines, for example (**13**), proceed rapidly at ambient temperatures to yield cyclic tertiary amine *N*-oxides (Scheme 25) ⟨79TL4391, 90JOC3007⟩. The retro-Cope rearrangements proceed thermally and stereospecifically in a suprafacial manner, for example (**14**) to (**15**) ⟨90JOC3007⟩. The Cope rearrangement of the cyclic tertiary amine *N*-oxides back to the alkenyl *N*,*N*-disubstituted hydroxylamines is often possible at high temperature. Retro-Cope rearrangements of intermediate *N*,*N*-disubstituted hydroxylamines, formed from the addition of allylamines to nitrones, yield tertiary amine *N*-oxides ⟨93CC169⟩.

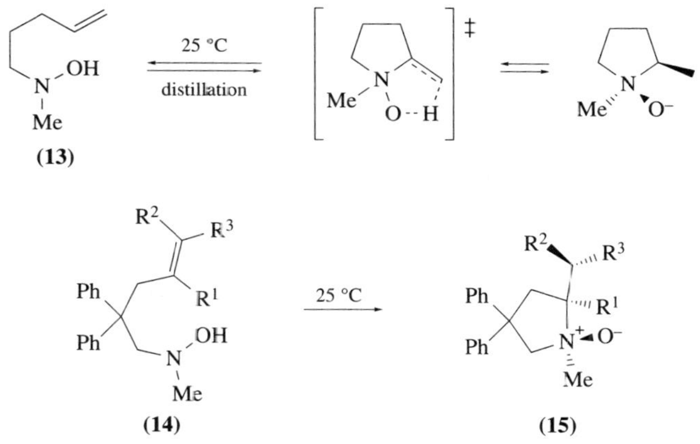

Scheme 25

2.06.2.2.2 *Salts of* N-*oxides;* R_3N^+OH, $R^1{}_3N^+OR^2$

Amine *N*-oxides are weak bases and when treated with acid form stable salts, $R_3N^+OH\ X^-$, which can be isolated and in most cases are crystalline ⟨91COS(7)735, 93S263⟩. The $pK_{BH}{}^+$ of the salts range between pH 4 and 5. The salts $R_3N^+OH\ X^-$ are readily neutralized by base to yield the corresponding tertiary amine *N*-oxides without problem. The salts are also formed initially in many synthetic procedures. The salts can also be formed by direct oxidation of tertiary amines by the anion of peroxysulfuric acid $HO—OSO_3{}^-$ to give high yields of the *N*-oxide sulfate $(R_3N^+OH)_2\ SO_4{}^{2-}$ ⟨76MI 206-01⟩.

N-Alkoxy-*N,N,N*-trialkylamine salts $R^1{}_3N^+OR^2$ can be formed by alkylation of either *N*-alkoxy secondary amines $R^1{}_2NOR^2$ ⟨59JA5800, 79JA7114, 83JHC1207⟩, or *N,N,N*-trialkylamine *N*-oxides $R_3N^+O^-$ ⟨89MI 206-03⟩. Intramolecular alkylation of an intermediate *N*-alkoxy secondary amine has been used in the synthesis of (±)-epilupinine ⟨76TL4037⟩.

2.06.2.3 *O*-Substituted *N*-Alkylhydroxylamines: R^1NHOR^2, $R^1N(OR^2)_2$

The *Chemical Abstracts* rules of nomenclature are used, R^1NHOR^2 and $R^1{}_2NOR^2$ are named *N*-alkoxyamines and not *N,O*-dialkylhydroxylamines and $R^1N(OR^2)_2$ are named *N,N*-dialkoxyamines. The related hydroxylamine derivatives, *N*-silyloxyamines, $R^1{}_2NOSiR^2{}_3$ and $R^1N(OSiR^2{}_3)_2$, and *N*-acyloxyamines, $R^1{}_2NOCOR^2$ and $R^1N(OCOR^2)_2$, are reported in later chapters. The synthesis and chemistry of *O*-substituted *N*-alkylhydroxylamines have been reviewed ⟨71HOU(10/1)1091, B-89MI 206-03, 90HOU(E16)1⟩.

2.06.2.3.1 N-*Alkoxy primary amines:* R^1NHOR^2

(i) Reductive methods

Similar reagents which are used to reduce oximes to *N*-substituted hydroxylamines RNHOH can be successfully applied to the reduction of *O*-alkyl-, and *O*-aryloximes to *N*-alkoxy- and *N*-aryloxyamines, for example $LiBH_4$, $NaBH_4$ in basic or acidic media, $NaBH_3CN$ in acid ⟨75T2595⟩, BH_3/THF and $BH_3\cdot$pyridine/HCl ⟨92JOC2726⟩. The methods are fully reviewed ⟨91COS(8)60⟩. Most of the reagents are used under mild conditions, and selectively reduce the *O*-substituted oximes in the presence of groups such as nitriles, acetals, amides, and esters. Diastereoselective reduction of α-hydroxyoximo benzyl ethers using tetramethylammonium triacetoxyborohydride (TABH) to acyclic 1,2-*syn*-*N*-benzyloxyamino alcohols (e.g., **(16)**) has been rationalized via a Felkin model for external hydride delivery (e.g., **(17)**) (Scheme 26) ⟨93TL3271⟩.

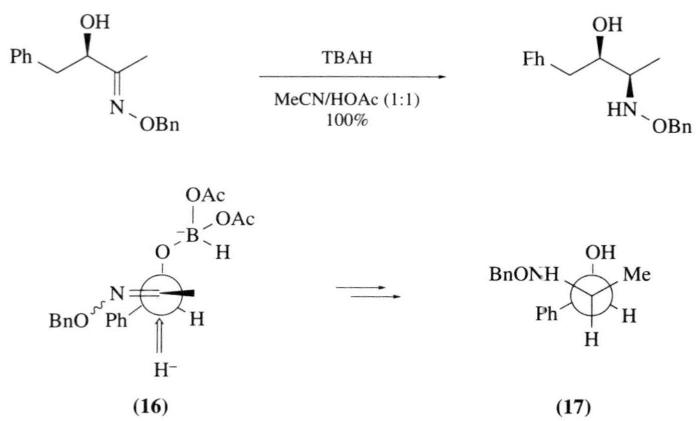

(16) **(17)**

Scheme 26

(ii) By C—N bond formation

Dialkylation of *N*-hydroxyurethanes and -ureas followed by hydrolysis has been a common procedure for the synthesis of *N,O*-dialkylamines since the turn of the century ⟨B-89MI 206-03⟩. A more modern procedure uses *N*-benzyloxyamine $BnONH_2$ which is converted to the corresponding *N*-benzyloxyamine RNHOBn via alkylation of an intermediate urethane (Scheme 27) ⟨89TL31⟩. The anion of the urethane formed from the reaction between *O*-benzylhydroxylamine and di-*t*-butyl dicarbonate was alkylated by primary and secondary mesylates and iodides. The resulting urethanes are hydrolyzed to yield the corresponding *N*-benzyloxyamine. *O*-Benzylhydroxylamine has also been alkylated using allyl esters with palladium(0) catalysis (Equation (16)) ⟨88TL2973⟩. The reaction is particularly useful because of the high regioselectivity. S_N1 Substitution of 1-aryl-1-alkylethylalcohols with *N*-benzyloxyamine $BnONH_2$ in TFA gives good yields of benzyloxy-(1-aryl-1-alkylethyl)amine ⟨89JOC1221⟩.

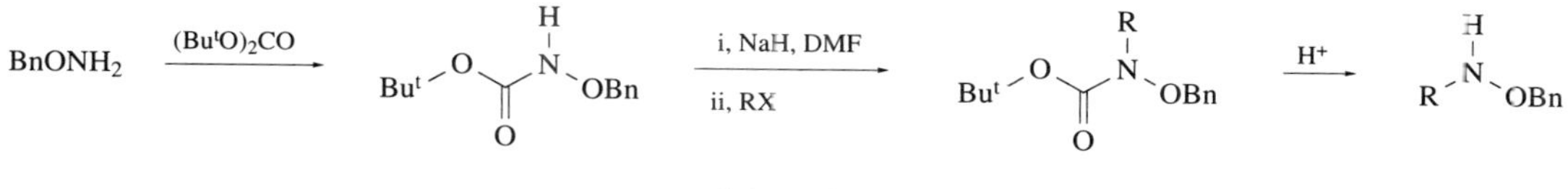

Scheme 27

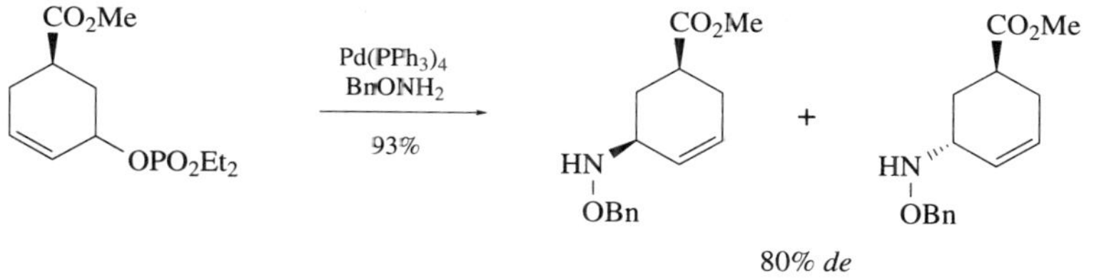

(16)

(iii) From N—O bond formation

Reaction between primary and secondary amines and dibenzyl peroxide involves nucleophilic displacement on the peroxide bond to yield *O*-benzyloxylamine (see Section 2.06.2.1.1.(i)) (Scheme 8).

(iv) From addition of organometallic reagents to oxime ethers

Reasonable yields are obtained when a range of unstabilized aryllithium reagents are added to methyl and benzyl oxime ethers in the presence of $BF_3 \cdot Et_2O$; an example is shown in Equation (17) ⟨88TL3455⟩.

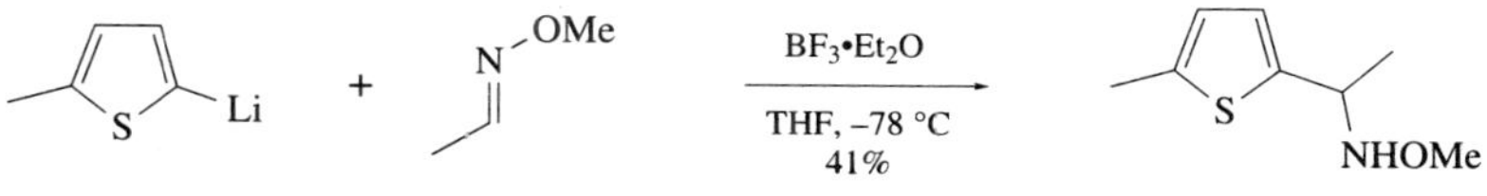

(17)

(v) From radical addition to oxime ethers

Radical cyclization on to oxime ethers is more favored than cyclization on to aldehydes, and yields cyclic *exo*-*N*-alkoxyamines. The radicals have been generated from a range of precursors using a variety of reagents used in radical chemistry. Early studies used alkyl radicals, generated from ketones using zinc and TMS—Cl ⟨83TL2821⟩, and from alcohols via their phenyl thionocarbonate derivatives PhOC(=S)OR, using tributyltin hydride ⟨88JA1633⟩. Radicals generated from vinyl bromides using tributyltin hydride undergo 6-*exo*-cyclization (Scheme 28) ⟨91CC1248⟩. β-Allenic *O*-methyl oximes undergo radical hydrostannylation to yield cyclopentene *N*-methoxyamines (Equation (18)) ⟨92TL1057⟩. Aryl radicals resulting from enediyne cyclization have been trapped using oxime ethers in tandem radical cyclizations ⟨93JOC6559⟩. Intermediate ketyl radicals, generated using samarium diiodide ⟨91TL2555⟩ and electroreduction ⟨91TL525⟩, have been added to oxime ethers to yield acyclic *β*-hydroxy-*N*-alkoxyamines. Oxime ethers connected by a tether to aldehydes or ketones undergo Bu_3SnH-mediated radical cyclization to yield cyclic amino alcohols (Scheme 29) ⟨94TL2205⟩.

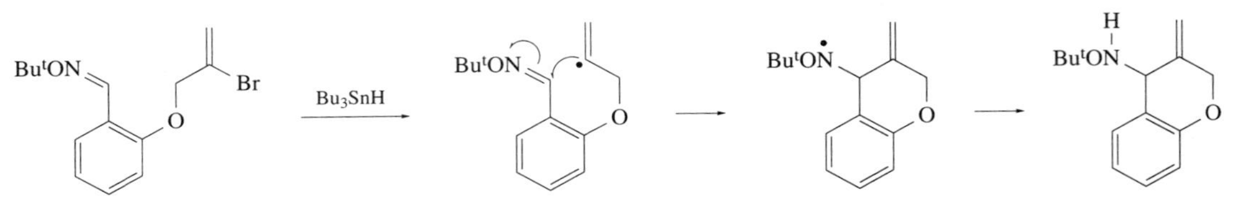

Scheme 28

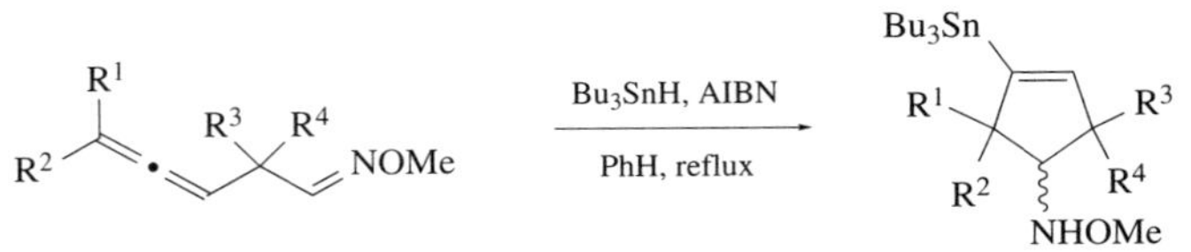

(18)

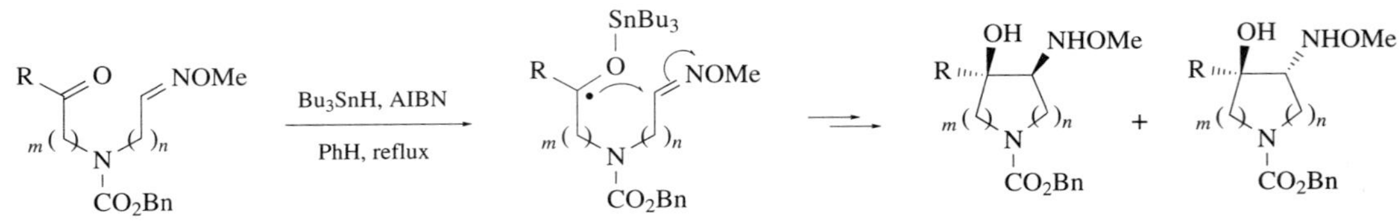

R = H, Me
m = 1, 2, 3
n = 1, 2

Scheme 29

(vi) From 1,3-dipolar addition of nitrones to alkenes

1,2-Prototopy in aliphatic oximes generates a small equilibrium concentration of the corresponding NH-nitrones which act as 1,3-dipoles and undergo cyclization to yield cyclic *N*-alkoxyamines, isoxazolidines (Scheme 30) ⟨B-88MI 206-02, 91T4007, 92T6929⟩. The factors which favor and disfavor the prototropic generation of NH-nitrones are discussed in the publication.

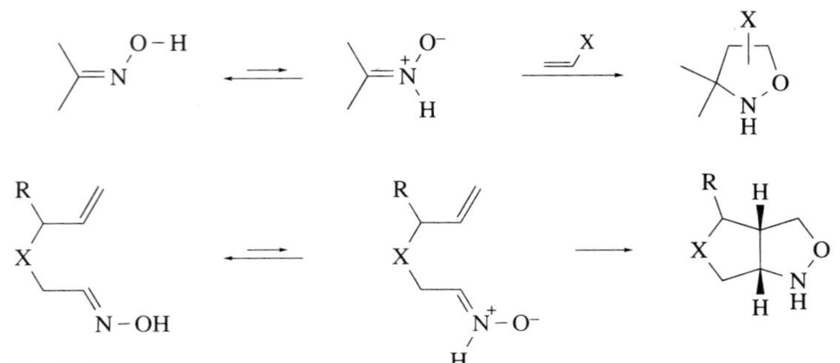

R = Ph, H
X = O, S, CH_2

Scheme 30

(vii) From cycloaddition of alkylnitroso compounds and alkenes

Chiral acylnitroso compounds have been developed as chiral auxiliaries in cycloaddition reactions with alkenes (Scheme 31) ⟨92TL3583⟩. The acylnitroso compound stereoselectively undergoes Diels–

Alder addition with dienes to yield a cyclic acyl alkoxyamine. The chiral auxiliary (**18**) is crucial for the selectivity of the cycloaddition reaction. The acyl group is removed by hydrolysis or reduction to yield cyclic *N*,*O*-dialkylamines with good enantiomeric excesses. α-Chloronitro compounds undergo cycloaddition followed by elimination to yield cyclic *N*,*O*-dialkylamines (Equation (19)) ⟨76AG(E)760⟩.

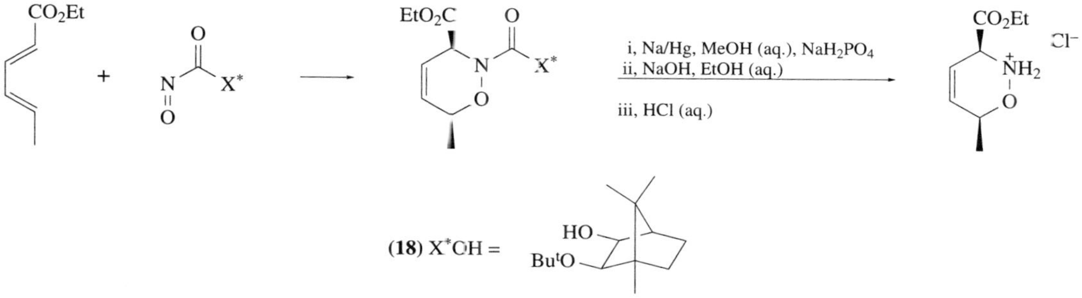

Scheme 31

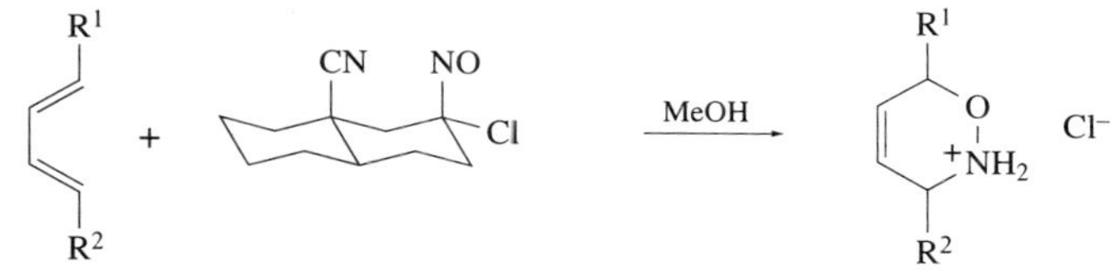

(19)

2.06.2.3.2 N-*Alkoxy secondary amines*; $R^1{}_2NOR^2$

(i) From O-*alkylation of* N,N-*dialkylhydroxylamines*

Selective methylation of the hydroxylamine function over β-alcohols in *N*,*N*-dialkylhydroxylamines (e.g., (**19**)) with BuLi followed by alkylation with methyl iodide yields the corresponding *N*-methoxyamine without racemization of chiral α-centres (Equation (20)) ⟨90JOC3464⟩. The oxygen anions of dialkylhydroxylamines can be added to electron-deficient alkenes; a 1988 example is shown in Equation (21) ⟨88JOC5776⟩.

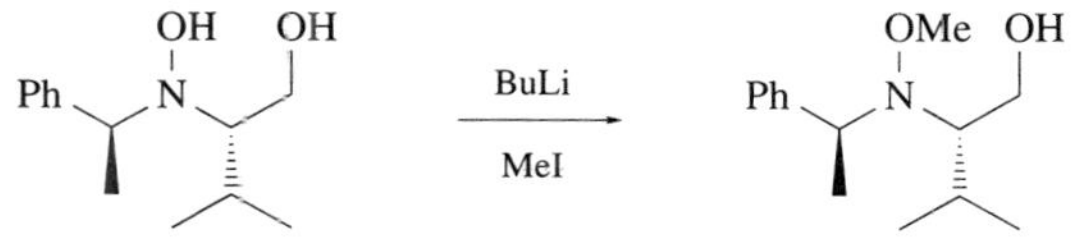

(20)

(**19**)

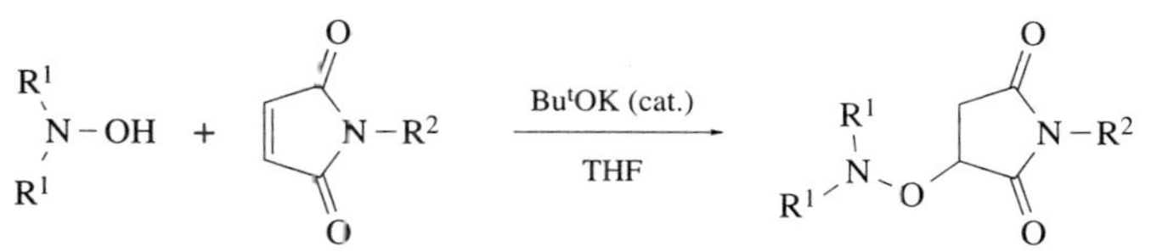

(21)

R^1 = Bn, Et
R^2 = H, Ph, alkyl

(ii) From C—N bond formation

N-Alkoxyamines $R^1{}_2NOR^2$ can be synthesized by *N*-alkylation of *N*-alkoxyamines, $RONH_2$ and R^1NHOR^2, by a variety of methods, for example with dimethylsulfate, oxiranes and Mannich reactions ⟨71HOU(10/1)1091, 90HOU(E16)1⟩. Electrophilic nitrogen atom cyclization of alkenyl

N-alkoxyamines using electrophiles such as $HgCl_2$ or iodine provides a useful protocol for the synthesis of *trans*-2,3 and 3,5-disubstituted pyrrolidino iodides which is analogous to the better known cyclization using amines, commonly called the Giese reaction ⟨91COS(4)363⟩. An example of the protocol is shown in Equation (22) ⟨89TL1327⟩.

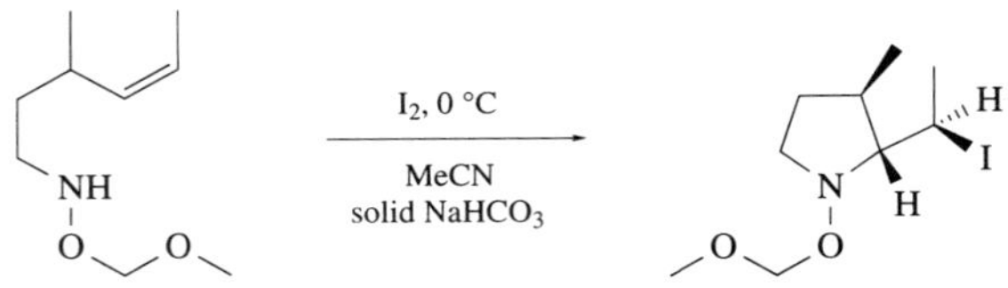

(22)

(*iii*) *From* 1,3-*dipolar addition of nitrones to alkenes*

The 1,3-dipolar cycloaddition of nitrones on to alkenes has been widely investigated and provides a major synthetic method for carbon–carbon bond formation. The general protocol is shown in Equation (23). The isoxazolidines which are formed in the cycloaddition reactions are regarded as synthetic relays, and are reductively cleaved to yield a range of hydroxylated and aminated intermediates in synthesis. The methodology has been used in a large number of natural products syntheses, and has been extensively reviewed ⟨68OSC(6)670, 75S205, 76AG(E)123, 77AG(E)10, B-84MI 206-01, B-87MI 206-01, B-88MI 206-03⟩. For example, δ'-pyrroline-*N*-oxides have been used for the synthesis of pyrrolizidines and indolizidines ⟨93TL3939⟩. The nitrones may be preformed or made *in situ*, which is particularly useful for intramolecular cycloadditions (Scheme 32). The cycloadditions are concerted *cis*-additions (suprafacial). The factors influencing the stereoisomeric ratios in 1,3-dipolar cycloadditions vary widely and are yet to be clearly elucidated ⟨91TL4431⟩. Evidence indicates that secondary orbital interactions are not important ⟨90JOC3427⟩. Further stereoselectivity can be induced by the use of chiral nitrones, for example the use of chiral nitrones derived from 5-alkenylaldehydes and chiral hydroxylamines in the cycloadditions shows diastereoselectivity ranging between 1.7 : 1 and 16 : 1 (Scheme 32) ⟨91TL4431⟩.

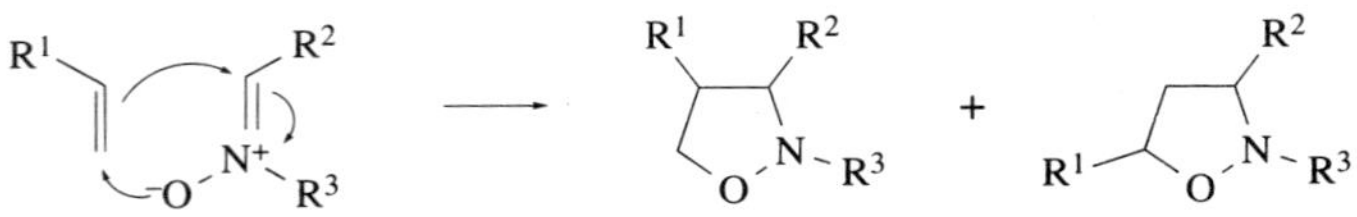

(23)

R^1 = Ar, OR, CN, CO_2R, CH_2OH, NO_2, OCOR, SO_2Ar, alkyl
R^2 = Ar, H, electron withdrawing group (EWG)
R^3 = Me, Bu^t

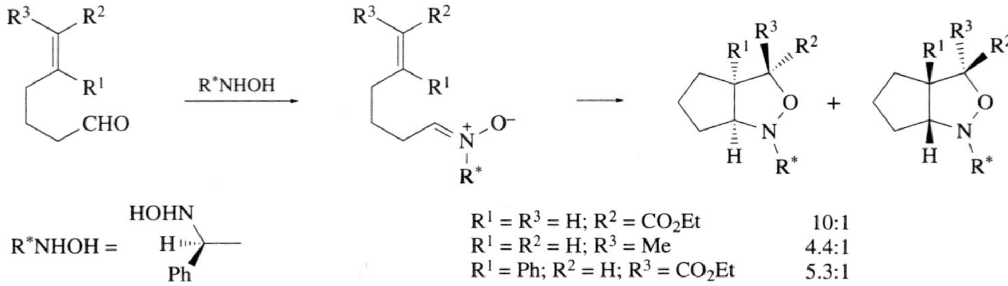

$R^1 = R^3$ = H; $R^2 = CO_2Et$ 10:1
$R^1 = R^2$ = H; R^3 = Me 4.4:1
R^1 = Ph; R^2 = H; $R^3 = CO_2Et$ 5.3:1

Scheme 32

(*iv*) *Cycloaddition of alkylnitroso compounds to alkenes*

The cycloaddition of alkylnitroso compounds to dienes to form six-membered ring *N*-alkoxyamines, tetrahydro-1,2-oxazines, has been reviewed ⟨71HOU(10/1)1091, 77CSR1, 87MI 206-01⟩.

(v) From rearrangements of amine N-*oxides*

Thermal rearrangement of amine *N*-oxides in which one alkyl group is allylic or benzylic yields *N*-alkoxyamines $R^1_2NOR^2$ ⟨71HOU(10/1)1091, B-89MI 206-03, 90HOU(E16)1, 93S263⟩. The rearrangements have been reviewed ⟨93S263⟩, and only representative examples are shown. The benzylic rearrangement, called the Meisenheimer rearrangement, is considered to proceed by a homolysis to benzylic radicals and nitroxides and recombination to yield the *N*-benzyloxyamine (Scheme 33) ⟨73JOC1813, 93S263⟩. Extensive racemization is observed in this protocol as would be expected for the radical mechanism, and the more stable the intermediate benzylic radical the faster the rearrangement. Synthetically, the most useful application is for ring expansion reactions in which aryl and allyl substituted cyclic amine *N*-oxides undergo expansion to 1,2-oxazaheterocycles. However, most allylic compounds proceed by a concerted [2,3] sigmatropic shift with 1,3-transfer of chirality ⟨76JOC300⟩. An example is shown in Scheme 34 ⟨91TL4481⟩. An interesting example demonstrates the competition between the two mechanisms; the allyl substituent directs a sigmatropic rearrangement, whereas in the absence of an allyl group the radical 1,2-rearrangement takes place (Scheme 35).

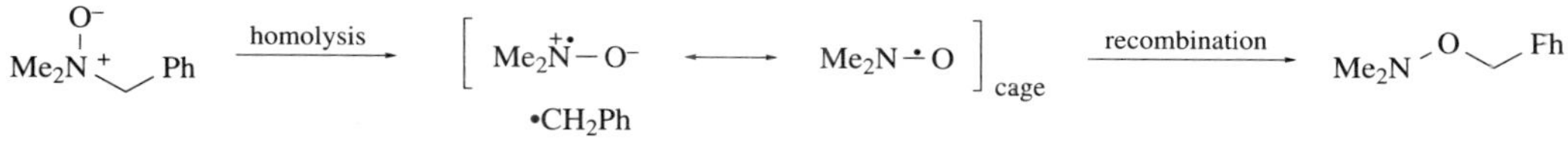

Scheme 33

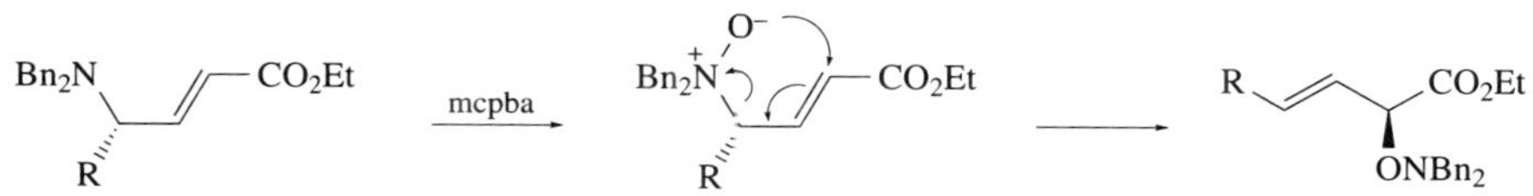

Scheme 34

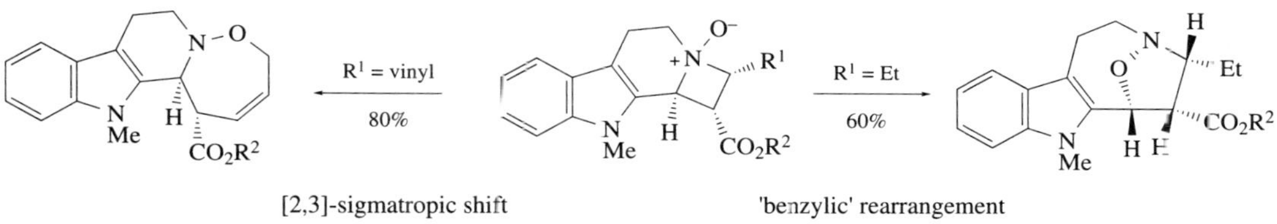

Scheme 35

(vi) From nitroxides

The trapping of alkyl radicals $R^2\cdot$ by nitroxides $R^1_2NO\cdot$ yields *N*-alkoxyamines $R^1_2NOR^2$, but no useful synthetic strategies have been located ⟨B-82MI 206-02⟩. Dimerization of nitroxides to hydroxylamine peroxides R^1_2NO—ONR^1_2 is energetically unfavorable ⟨B-82MI 206-02⟩.

2.06.2.3.3 N,N-*Dialkoxyamines: RN(OR¹)₂*

The synthesis and reactions of *N,N*-dialkoxyamines $R^1N(OR^2)_2$ have been reviewed ⟨93CRV725⟩ and therefore the synthetic methods are only briefly discussed. *N,N*-Dialkoxyamines are also called nitrosoacetals.

(i) From cycloaddition of nitronic esters and alkenes

The 1,3-dipolar [3 + 2]-cycloaddition of nitronic esters $R^2R^3C{=}NO_2R^4$ on to alkenes provides the main protocol for synthesis of cyclic *N,N*-dialkoxyamines **(20)** (i.e., *N*-alkoxyisoxazolidines

(Scheme 36)) ⟨B-88MI 206-04, 93CRV725⟩. Five- and six-membered ring cyclic alkyl nitronates (isoxazoline *N*-oxides) can also be used to add to alkenes to form bicyclic *N,N*-dialkoxyamines (**21**) and (**22**) (Scheme 36). The cycloadditions proceed fastest when the alkene or alkyl nitronate substituents (R^2, R^3, and R^4 in Scheme 36) are electron-withdrawing groups such as nitro or ester groups. Silyl esters of nitronic acids $R^1{}_2C{=}NO_2SiR^2{}_3$ also make suitable substrates for [3 + 2]-cycloadditions and yield the more stable *N*-silyloxyisoxazolidines, but the reactions are slower ⟨B-88MI 206-04, 93CRV725⟩. A number of variations allow intermediate formation of isoxazoline *N*-oxides which react *in situ* with further alkene. An example which has been exploited synthetically is the inter- and intra-[4 + 2]-heterodiene cycloaddition of nitroalkenes to alkenes to yield unstable isoxazoline *N*-oxides, which rapidly react by [3 + 2] cycloadditions to yield bicyclic *N,N*-dialkoxyamines ⟨85HCA319, 86JA1306, 90JA311, 90T4857, 91TL2517⟩. Another example is shown for the stereoselective intermolecular [4 + 2]-cycloaddition followed by a [3 + 2] 1,3-dipolar cycloaddition to yield the tricyclic *N,N*-dialkoxyamine (**23**) (Scheme 37) ⟨91JOC6738⟩. In this example, chiral auxiliaries have been used with the best *de* (>98%) resulting from the use of *trans*-2-phenylcyclohexanol.

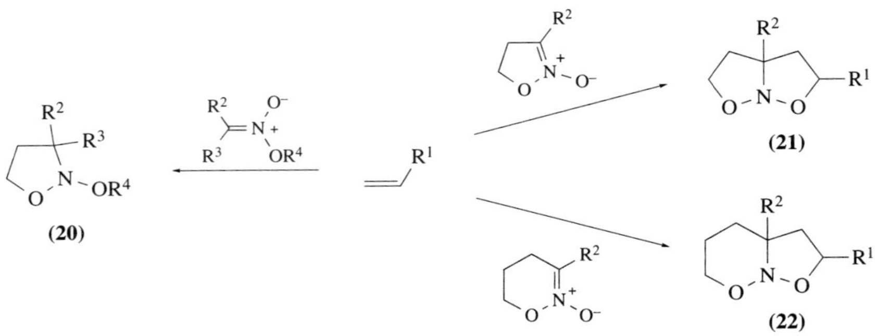

Scheme 36

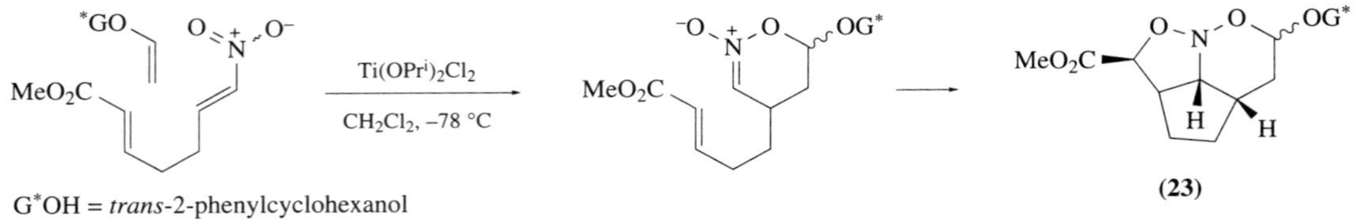

Scheme 37

(ii) From substitution in N-*substituted*-N-*alkoxylamines; $R^1N(X)OR^2$*

N-Chloro-*N*-alkoxyamines $R^1N(Cl)OMe$ readily undergo substitution of chlorine by alkoxides, or by alcohols in Et_3N, under mild conditions to yield *N,N*-dialkoxyamines $R^1N(OR^2)_2$ in 21–88% yield ⟨79ZV924, 81T4245, 91IZV1069⟩. *N*-Chloro-*N*-alkoxyamines with tertiary alkyl substituents are suitable, but those with primary or secondary alkylamines lose hydrogen chloride by elimination. Acid-catalysed *trans*-esterification of the alkoxy groups in *N,N*-dialkoxyamines by alcohols has been reported ⟨93CRV725⟩.

2.06.2.3.4 *Salts of* O-*substituted* N-*alkylhydroxylamines; $R^1N^+H_2OR^2$, $R^1N^+H(OR^2)_2$*

O-Substituted *N*-alkylhydroxylamines are protonated in acid, and have properties similar to amines and hydrazines ⟨71HOU(10/1)1091, B-89MI 206-03, 90HOU(E16)1⟩. The salts of *O*-substituted *N*-alkylhydroxylamines are formed in a number of synthetic reactions, and are neutralized by standard procedures or used for purification. *N*-Alkyl salts $R^1{}_3N^+OR^2$ are regarded as salts of *N,N,N*-trialkylamine *N*-oxides $R_3N^+O^-$ (Section 2.06.2.2.2).

2.06.2.3.5 N-*Chloro*-N-*alkoxylamines*; $R^1N(Cl)OR^2$

N-Chloro-*N*-alkoxyamines $R^1N(Cl)OR^2$ have been formed in moderate yields by chlorination of the corresponding *N*-alkoxyamines R^1NHOR^2 with sodium hypochlorite or *t*-butylhypochlorite at temperatures between $-20\,°C$ and $0\,°C$ ⟨79ZV924, 81T4245, 91IZV1069⟩. Good yields are obtained only when no α-hydrogens are present, otherwise hydrogen chloride is eliminated to yield oxime ethers. The methoxy group in acylic or monocyclic *N*-alkoxy-*N*-methoxyamines $R^1N(OR^2)OMe$ can be substituted by chlorine using MeCOCl, TMS-Cl, $SOCl_2$ or HCl to yield *N*-chloro-*N*-alkoxyamines ⟨93CRV725⟩.

2.06.2.4 *N*-Sulfonyloxyamines $R^1{}_2NOSO_2R^2$ and Related Compounds

Primary and secondary amines react with diarylsulfonyl peroxides $(ArSO_3)_2$ at low temperature to yield *N*-(arenesulfonyloxy)amines (Equation (24)) ⟨73JOC1239, 76JA6702, 77S1, 83S34, 84S217, 85JA4731, 91T1109⟩. These compounds are of synthetic interest as sources of 'positive amines'. *N*-Sulfinyloxyamines $R^1{}_2NO{-}SOR^2$ have been prepared from the corresponding hydroxylamines $R^1{}_2NOH$ with $R^1 = Me$, Et, and Bn, and sulfinyl chlorides R^2SOCl with $R^2 = Me$ and Ph, in dichloromethane at $-78\,°C$ with Et_3N as the base ⟨86JCS(P2)151⟩. Sulfitoamines $R_3N^+OSO_2^-$ are prepared by addition of SO_2 to tertiary amine *N*-oxides. In particular, bridgehead amine *N*-oxides (e.g., strychnine and quinuclidine) give the most stable derivatives in anhydrous solvents ⟨79JOC3867, 90OR85⟩. Sulfitoamines rearrange on standing to aminesulfamates $R_3N^+SO_3^-$. Most sulfitoamines are unstable and rapidly undergo Polonovski reactions in acid solution ⟨90OR(39)85⟩. Certain sulfitoamines are able to transfer SO_3 to further amine *N*-oxide to yield sulfatoamines, for example $Et_3N^+OSO_2^-$ changing to $Et_3N^+OSO_3^-$.

$$R_2NH + (ArSO_3)_2 \xrightarrow[-78\,°C,\ 2-4\ h]{Et_2O} R_2N{-}O{-}SO_2Ar + R\overset{+}{N}H\ ArSO_3^- \qquad (24)$$

2.06.2.5 *N*-Phosphinyloxyamines; RNH—O—P(=O)R_2, R_2N—O—P(=O)R_2

The reaction between primary and secondary amines and diphenylphosphinylperoxide $(Ph_2PO_2)_2$ yields *N*-(diphenylphosphinyloxy)amines (Scheme 9) ⟨86T2703, 85S807⟩. Treatment of *N*,*N*-dialkylhydroxylamines with BuLi/$(EtO)_2$POCl gave reasonable yields of the corresponding phosphate derivatives (Equation (25)) ⟨90JOC3464⟩.

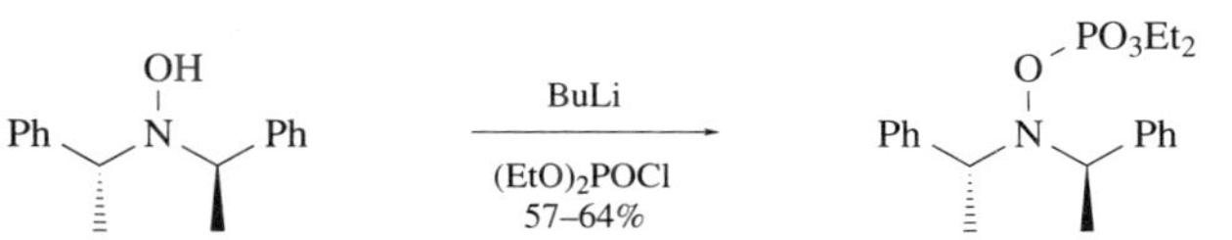

(25)

2.06.3 THIOHYDROXYLAMINES AND RELATED FUNCTIONS

Certain reviews have dealt comprehensively with the chemistry and preparation of *S*-alkyl and *S*-aryl derivatives of thiohydroxylamines (sulfenamides), and therefore a brief survey is detailed ⟨46CRV269, 73IJS71, B-73MI 206-01, 79OPP33, 85HOU(E11)68, 89CRV689, 90RCR832, B-90MI 206-01⟩.

2.06.3.1 Thiohydroxylamines; R_2NSH

The reduction of disulfides $R^1{}_2NSSR^2$ gives thiohydroxylamines $R^1{}_2NSH$ ⟨75CC855⟩. These thiohydroxylamines are unstable above $-40\,°C$ and decompose to give the secondary amine and sulfur. Thiohydroxylamines have been postulated as reactive intermediates but have not been previously isolated.

2.06.3.2 Sulfenamides; R^1NHSR^2, $R^1{}_2NSR^2$

The best known method of preparing sulfenamides is by the reaction between a sulfenyl halide and an amine (Equation (26)) ⟨B-90MI 206-01⟩. One study ⟨70JOC3012⟩ of this reaction found that amines of lower basicity and higher substitution pattern gave better results in both synthesis and purification. Excess amine or another base (e.g., triethylamine) must be present to react with acid generated during the reaction as sulfenamides are unstable to even traces of acid. Sulfenamides can be purified by distillation, crystallization, or alumina chromatography. Alkylsulfenyl chlorides often give complex mixtures of products ⟨61CB634, 71JCS(C)2840⟩. Arenesulfenyl chlorides normally give good yields of sulfenamides; however, it should be noted that sulfenyl halides can react with hydroxyls, active methylene compounds and alkenes ⟨89CRV689⟩. Sulfenamides containing reactive double bonds have, however, been prepared by this method (Equation (26)) ⟨94T1275, 94T1295⟩. The reaction between amino acids and *o*-nitrobenzenesulfenyl chloride to yield the corresponding sulfenamides is used as a method of protecting the amino group ⟨64JA3660⟩.

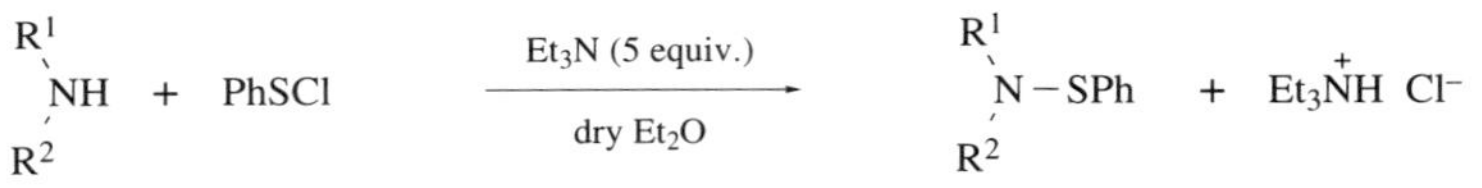

(26)

R^1 = *p*-methylbenzyl, R^2 = cyclopropyl, 96%
R^1 = butyl, R^2 = 4-pentenyl, 95%

Reaction between amines and *N*-sulfenyl substituted phthalimides gives sulfenamides (Equation (27)) ⟨70TL4983, 71TL4953, 94T1275⟩. The conditions are mild, and the highly insoluble phthalimide byproduct is precipitated from solution. Work-up is simply by filtration of the reaction mixture and evaporation of the solvent. Yields are normally above 80% after purification. This method is effective for both alkane- and arenesulfenylphthalimides, but does not work with bulky alkyl groups such as isopropyl, *t*-butyl or cyclohexyl. Reaction with sterically hindered amines is very slow even after refluxing in toluene for long periods.

(27)

R^1 = Me, R^2 = cyclohexyl, 93%, FCH_2Cl_2, RT, 30 min
R^1 = allyl, R^2 = 4-pentenyl, 86%, benzene, reflux, 3 h

A number of sulfenamide syntheses involve the oxidation of a thiol or its metal salt in the presence of an amine. The oxidation of thiazolyl thiolate salts by iodine in potassium iodide solution, chlorine or sodium hypochlorite and reaction with an amine has been described ⟨49JOC921⟩. Thiols have been condensed oxidatively with amines in the presence of $K_2S_2O_8$ ⟨79ZOB841⟩. 2-Mercaptobenzothiazole has been cross-coupled electrolytically in the presence of tetraethylammonium perchlorate with primary and secondary amines. The yields of cross-coupled products are generally above 90%, except for sterically hindered amines such as diisopropylamine. The thiol must also contain electron-withdrawing groups for the reaction to work.

Disulfides have been utilized in a number of syntheses of sulfenamides. The reaction of alkyl or aryl disulfides with primary or secondary amines in the presence of a metal activator such as silver nitrate, silver acetate or mercuric chloride has been reported ⟨77JOC967, 94T1275⟩. Reasonable yields are obtained under mild conditions, and the method is tolerant of many functional groups but the method requires at least two equivalents of amine. In a variant of this method, lithium salts of primary or secondary amines are reacted with alkyldisulfides to give the corresponding sulfenamides in excellent yield (Equation (28)) ⟨83S716⟩. Several research groups ⟨71JCS(C)3867, 72JOM(44)37⟩ have described the use of arenesulfinic esters in the preparation of sulfenamides. The reaction between arenesulfenates ArSOMe and either primary or secondary amines gives high yields of sulfenamides, as did reaction with *N*,*N*-dialkylamino silanes or aminoboranes $R^1R^2N—BR^3R^4$. Thiolsulfonates $ArSO_2—SR$ have also been reacted with primary and secondary amines (Equation (29)) ⟨66JOC842, 69ZOR1088, 70JOC279⟩. It was found that, as the electronegative character of the thiol component increased, formation of the sulfenamide became easier.

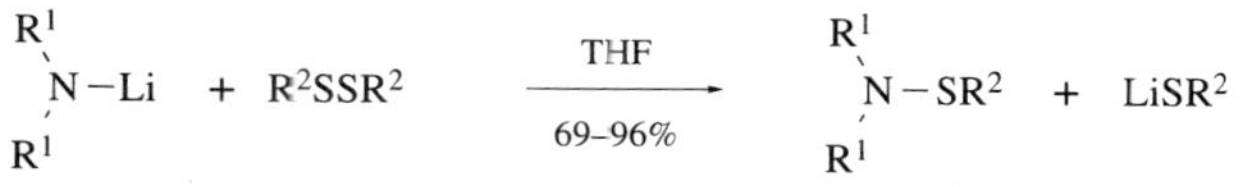

(28)

R^1 = Pr, R^2 = Et, Pr, Bu^i

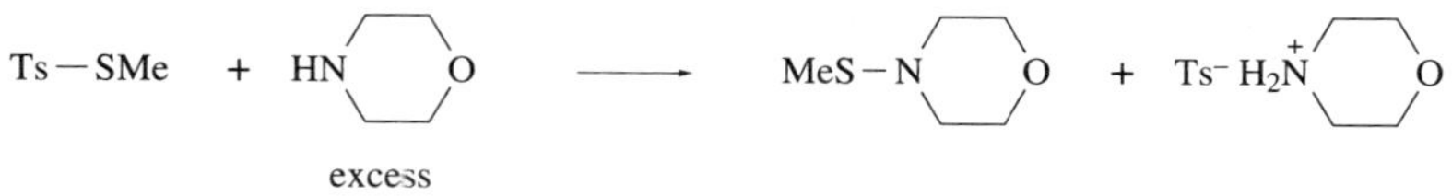

(29)

Alkylthionitroso compounds RN=S undergo Diels–Alder addition to dienes and addition to alkenes to yield sulfenamides (Scheme 38) ⟨92AHC(55)1⟩.

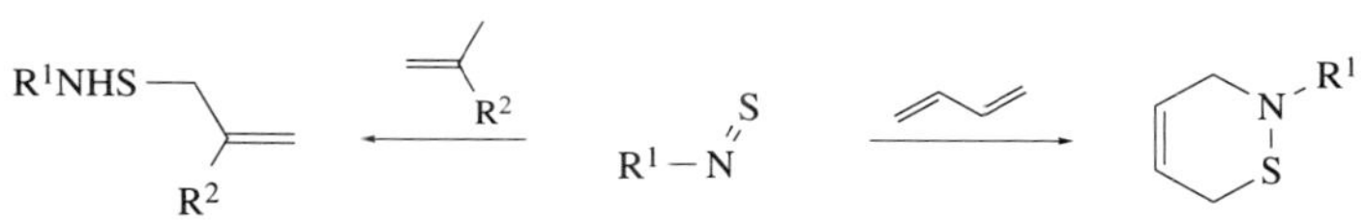

Scheme 38

2.06.3.2.1 *Primary sulfenamides; R^1NHSR^2*

Tritylsulfenimines can be reduced with sodium cyanoborohydride at pH 3–6 giving the corresponding triphenylmethanesulfenamides (Equation (30)) ⟨83JOC3531⟩. The addition of Grignard or organolithium reagents to sulfenylimines has been reported to give primary sulfenamides in good yields ⟨81TL5073, 94JOC914⟩. An example which involves complete diastereoselectivity is shown in Equation (31) ⟨94JOC914⟩. When an azide is treated with a thiol in the presence of copper(I) salts, primary sulfenamides are isolated in good yield ⟨70JOC2979⟩. The reaction is limited to nonaromatic thiols.

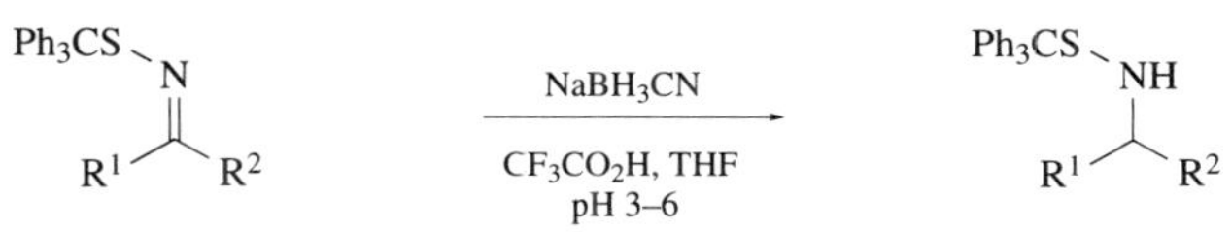

(30)

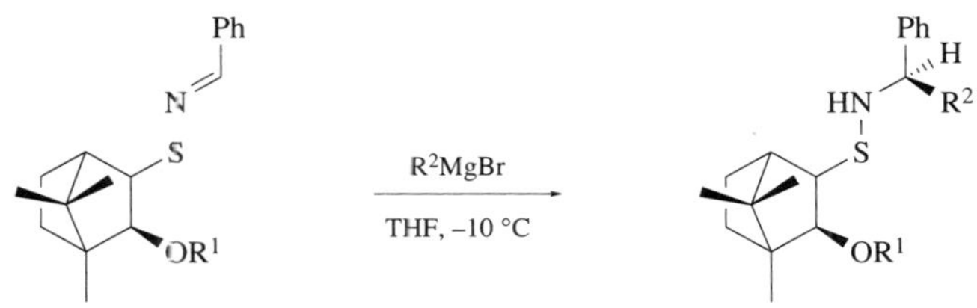

(31)

R^1 = H, Bn. neopentyl; R^2 = Me, Bu^n, Bu^t, Pr^i, allyl

2.06.3.2.2 *Secondary sulfenamides; $R^1{}_2NSR^2$*

The reaction of aminosulfenyl chlorides with organometallic reagents has been reported to give sulfenamides in good yields (Scheme 39) ⟨87BSF181, 87T881⟩. Dimethylaminosulfenyl chloride will react with an alkene at low temperatures to give a β-chlorosulfenamide, which reverts to starting materials when distillation is attempted at 80 °C ⟨68JOC2111⟩. *N*-Sulfenyl substituted lactams and hydantoins can be used in *trans*-amination reactions to give the corresponding sulfenamides in good yield ⟨79S228, 82LA121⟩. The *trans*-amination reactions of 2-benzothiazole sulfenamide have also been described; the reaction is slowed as the amine becomes more basic, or if water or other basic amines are added. Acids increase the rate of this reaction ⟨81ZOB609⟩. Thiols or disulfides react with amines in the presence of an oxidizing agent to give the corresponding sulfenamides ⟨78JOC3223, 79JOC1554⟩. Chloroamines react with thiols to provide sulfenamides in reasonable yields (Equation

(32)) ⟨91JOC6702⟩. A disadvantage is that a further equivalent of the amine precursor is required as a base, but the advantage is that the method can be used for synthesizing sulfenamides when the corresponding sulfenyl chloride cannot be used. Certain disulfides have also been reported to react with chloroamines in the presence of excess amine, which acts as a base ⟨80EUP10477⟩. Secondary sulfenamides have been prepared by adding sulfenyl thiocyanates RS—SCN to cold ethereal solutions containing two equivalents of the respective secondary amines ⟨56JA6181⟩. α-Halomethyl sulfoxides react with four equivalents of various secondary amines to give the corresponding sulfenamides in nearly quantitative yields ⟨71IJS215⟩.

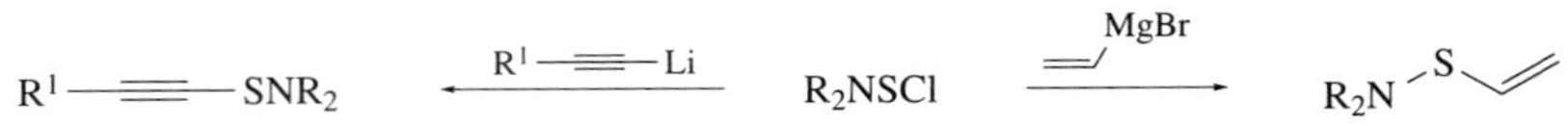

Scheme 39

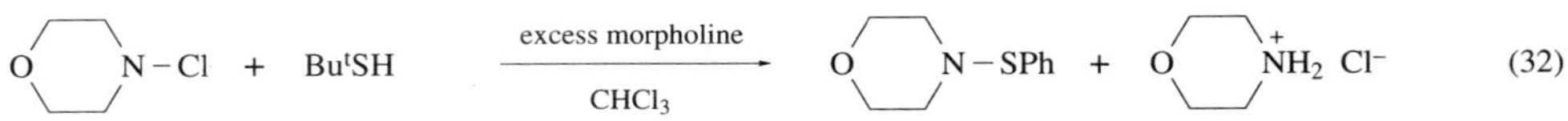

2.06.3.2.3 *Sulfenamide salts; $R^1_3N^+SR^2$, $R^1_2N—S^+R^2_2$*

Unstable salts $R_3N^+SAr\ X^-$ are formed from reaction between tertiary amines and arenesulfenyl chlorides or thiosulfonates, and react rapidly with nucleophiles ⟨75CC413⟩.

Azasulfonium salts (**24**) can be prepared by the reaction between secondary amines, dimethyl sulfide and Bu^{t}OCl ⟨84JCS(P1)1401⟩. *S*-Amination of sulfenamides by reaction with *N*-chlorosuccinimide in nonhydrolytic conditions gives the corresponding azasulfonium salts (**25**), which are stable enough to be isolated as their chlorides or tetrafluoroborates ⟨76S308, 78JOC652⟩. The reaction of *N*-alkyl-*N*-sulfinylalkaminium salts with sulfoxides and subsequent extrusion of SO_2 furnishes the analogous azasulfonium salts (Scheme 40) ⟨82TL1243⟩. The reaction of sulfinamides with strong methylating agents such as $CF_3SO_2CH_3$ converts them to the corresponding methoxyazasulfonium salts (**26**) ⟨75CL991⟩.

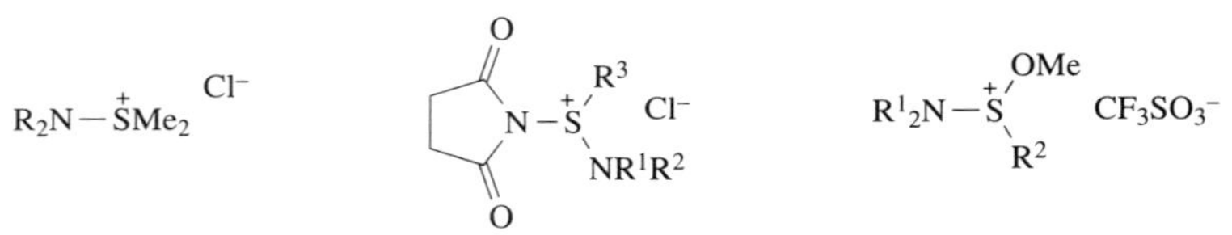

(24) **(25)** **(26)**

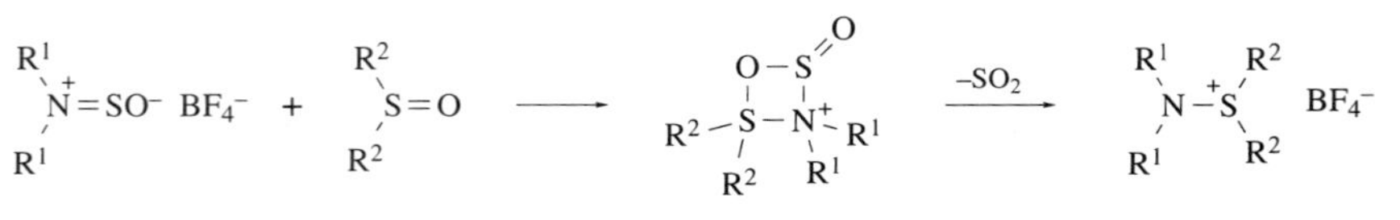

Scheme 40

2.06.3.3 Derivatives of Thiohydroxylamines; RN(X)SY

2.06.3.3.1 *RNH—SY and R_2N—SY, with Y = halogen*

Amino sulfenyl chlorides R_2NSCl have been prepared by treatment of thiobisamines R_2NSNR_2 with chlorine, hydrogen chloride or sulfur dichloride, or by reaction between dithiobisamines R_2NSSNR_2 and sulfuryl chloride ⟨71CC1413⟩, SO_2Cl_2, or chlorine ⟨58CB2528, 77ZN(B)721⟩. Secondary alkyl amines react with sulfur dichloride in the presence of a hydrogen chloride acceptor to produce aminosulfenyl chlorides ⟨64JCS5892, 74ZN(B)277⟩.

2.06.3.3.2 RNH—SY, R_2N—SY

Aminosulfenyl chlorides react with nucleophiles such as ethoxide at sulfur, displacing chloride ⟨69CB1489, 88TL3251⟩. The formation of amino disulfides $R^1{}_2NSSR^2$ occurs when aminosulfenyl chlorides react with silylated sulfides ⟨71CC1413⟩. Amino disulfides R_2NSSCF_3 are formed when dialkylaminosulfenyl chlorides react with mercury(II) trifluoromethanethiolate ⟨75CB1365⟩.

The reaction between secondary amines and sulfur dichloride and S_2Cl_2 in chlorinated solvents at low temperature gives *N,N*-thiobisamines R_2NSNR_2 and *N,N*-dithiobisamines R_2NSSNR_2, respectively ⟨63CB1131⟩. *t*-Butylamine reacts with sulfur dichloride to give the corresponding thiobisamine $(Bu^tNH)_2S$ ⟨76JOC1333, 77ZAAC(432)173⟩. Unsymmetrical *N,N′*-thiobisamines have been made by the reaction of aminosulfenyl chlorides with secondary amines in the presence of added base ⟨56GEP948330⟩. Without added base this reaction gives exclusively the symmetrical thiobisamine because hydrogen chloride is generated in the reaction and causes disproportionation ⟨60JCS5079⟩. Aminosulfenyl chlorides react with various diamines to give the corresponding cyclic thiobisamines (Equation (33)) ⟨84JCS(P1)2591⟩. Thiobisamines have also been prepared from the reaction of amines with *N*-sulfenyl substituted hydantoins ⟨82LA121⟩. Aminobistrimethylsilanes react with aminosulfenyl chlorides to give aminobisthioamines (Equation (34)) ⟨71CC1413, 78TL4003⟩.

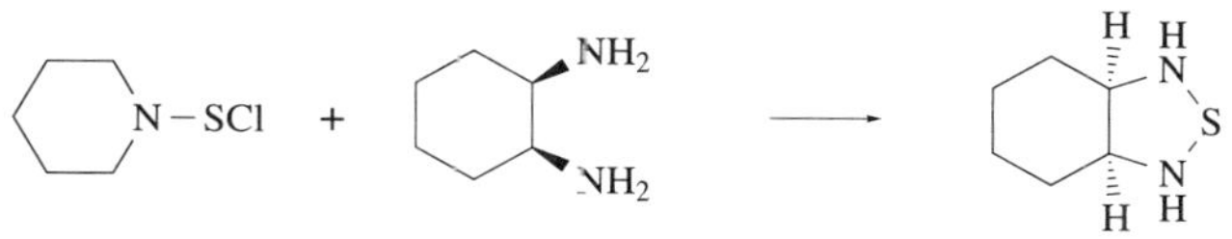
(33)

$$R^1-N(TMS)_2 \; + \; 2\,R_2N-SCl \longrightarrow R^1-N(SNR_2)_2 \; + \; 2\,TMS\text{-}Cl \qquad (34)$$

The entire series of compounds S_7NR to $S_4(NR)_4$ are known for R = Me. All of these compounds have puckered eight-membered rings like S_8 ⟨73CJC2504⟩. The chemistry of *N,N*-polythiobisamines $R_2NS_nNR_2$ has been reviewed ⟨70QRS245⟩. Primary amines react with S_2Cl_2 to give cyclic tetrathiobisamines in reasonable yield ⟨59CB1149, 61CR(252)4007⟩. Sulfenamides phosphorylated on sulfur can be prepared from the reaction between bis(dialkoxyphosphoryl) disulfides and primary or secondary amines in good yields (Equation (35)) ⟨75ZOB1005⟩. The synthesis of a benzenesulfonyl substituted sulfenamide R_2NSSO_2Ph has been reported ⟨82SUL33⟩.

$$\left((R^1O)_2P(=O)-S\right)_2 \; + \; R^2NH_2 \longrightarrow (R^1O)_2P(=O)-S-NHR^2 \; + \; (R^1O)_2P(=O)-SH \qquad (35)$$

R^1 = Et, Pr^i; R^2 = Me, Et, Pr^i

2.06.3.3.3 Disulfenamides: $R^1N(SR^2)_2$

Disulfenamides $R^1N(SR^2)_2$ are synthesized by several routes. The most common method is the reaction between a primary amine and two equivalents of a sulfenyl chloride in diethyl ether at low temperature with triethylamine as a base (Equation (36)) ⟨84JOC2724, 89CRV689, 94UP⟩. Disulfenamides are also formed from the reaction between primary amines and arenesulfinic esters, unless the amine is in excess ⟨71JCS(C)3867⟩. The reaction of acetic anhydride with sulfenamides is reported to give disulfenamides and other products ⟨60ZOB1431⟩. Another method of making disulfenamides utilizes the disulfenamide of ammonia $(PhS)_2NH$ (Scheme 41) ⟨70TL3411, 94UP⟩. The anion $(PhS)_2NLi$ is made using butyllithium and is then reacted with bromides or tosylates to give the analogous disulfenamides in moderate yields.

$$RNH_2 \; + \; 2\,PhSCl \xrightarrow[\text{dry } Et_2O]{Et_3N} RN(SPh)_2 \; + \; 2\,Et_3\overset{+}{N}H\;Cl^- \qquad (36)$$

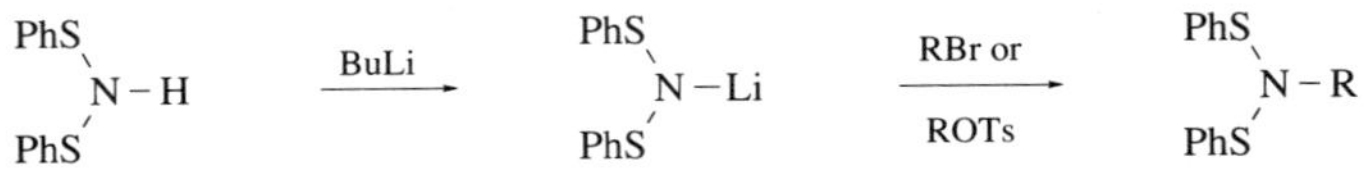

Scheme 41

2.06.3.4 Dialkylaminosulfur Trifluorides; R_2NSF_3

The syntheses and synthetic applications of a range of NSF_x fluorinating agents have been reviewed ⟨88OR513⟩. Dialkylaminosulfur trifluorides R_2NSF_3 of which diethylaminosulfur trifluoride (DAST) is the most common, bis(dialkylamino)sulfur difluorides $(R_2N)_2SF_2$ and tris(dialkylamino)sulfonium difluorotrimethylsilicates $(R_2N)_3S^+Me_2{}^-SiF_2$ are easily handled fluorinating agents and can be used under very mild conditions ⟨88OR513⟩. Reaction between dialkylaminotrimethylsilanes and sulfur tetrafluoride is used to synthesize these compounds (Scheme 42) ⟨88OR513, 88OSC(6)440⟩.

$$Et_2N\text{-}TMS + SF_4 \xrightarrow[-65\ °C\text{ to RT},\ 84\%]{CCl_3F} Et_2NSF_3 + TMS\text{-}F$$

$$Et_2NSF_3 + Et_2N\text{-}TMS \xrightarrow[92\%]{} (Et_2N)_2SF_2 + TMS\text{-}F$$

Scheme 42

2.06.3.5 Sulfinamides; $R^1NHS(O)R^2$, $R^1{}_2NS(O)R^2$, and Derivatives Thereof

Methods for the preparation of sulfinamides have been reviewed ⟨B-90MI 206-02, B-90MI 206-03, B-90MI 206-04⟩.

2.06.3.5.1 *Redox methods*

Sulfenamides can be oxidized to sulfinamides using a number of reagents including monopercamphonic acid ⟨67CC985⟩ which gives low stereoselectivity, $Ti(O^iPr)_4/DET/H_2O/Bu^tOOH$ which gives low *ee* ⟨86NJC761⟩, $(Bu_3Sn)_2O/Br_2$ ⟨77TL2413⟩, and mcpba ⟨77S798, 91JCS(P1)2169⟩. Optically active primary sulfinamides have been made by the reduction with retention of configuration of optically active sulfoximines $R_1R_2S(O){=}NMe$, with aluminum amalgam ⟨71JA5305, 86JOC81⟩. Sulfinamides of primary amines have been prepared by reduction of *N*-sulfinylimines $R^1R^2C{=}N{-}S(O)R^3$ with a range of reductants including LAH ⟨77CC723⟩ and dibal ⟨91JOC4, 93TA2159⟩. Reduction of these imines substituted on nitrogen by a chiral sulfoxide group gives substantial chiral induction at the resulting amino carbon of the primary sulfinamide (Equation (37)) ⟨93TA2159⟩.

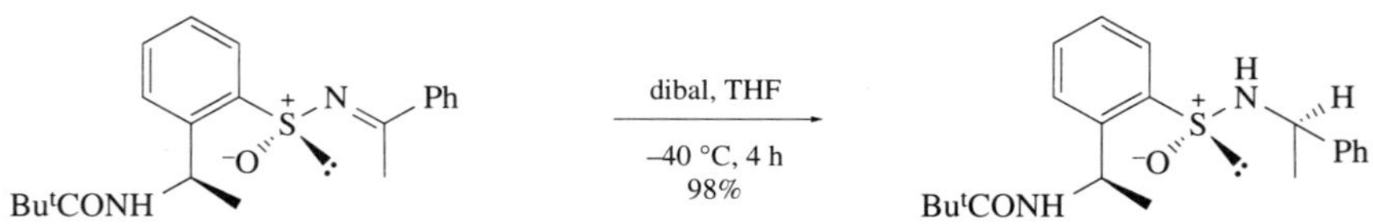

86% *ee* (37)

2.06.3.5.2 *From substitution at sulfur(IV)*

The reaction between alkylsulfinyl chlorides RS(O)Cl and secondary amines has been reported to give low yields of the corresponding sulfinamides ⟨69JOC2397⟩. The anion of an optically active amine has been added to a sulfoxyl chloride RS(O)Cl to give a sulfinamide ⟨68JA3869⟩. An excellent general method for preparing sulfinamides uses mcpba oxidation of *N*-(alkane- and arenesulfenyl)phthalimides to yield the corresponding sulfinylphthalimides, which on reaction with primary or secondary amines give the corresponding sulfinamides and phthalimide as a precipitate (Scheme 43) ⟨72TL5313, 73JOC4328, 91JCS(P1)2169⟩. Succinimides have also been used under these conditions ⟨77TL963⟩. The yields of sulfinamides prepared by this method are generally above 80%. Reaction between sulfinic acids and amines in the presence of activating agents have been reported but with moderate yields ⟨69JOC2397, 76S339, 82CPB1646⟩. The use of 2-chloropyridinium iodide as a coupling agent gives reasonable yields of primary sulfinamides (Scheme 44) ⟨76S339⟩.

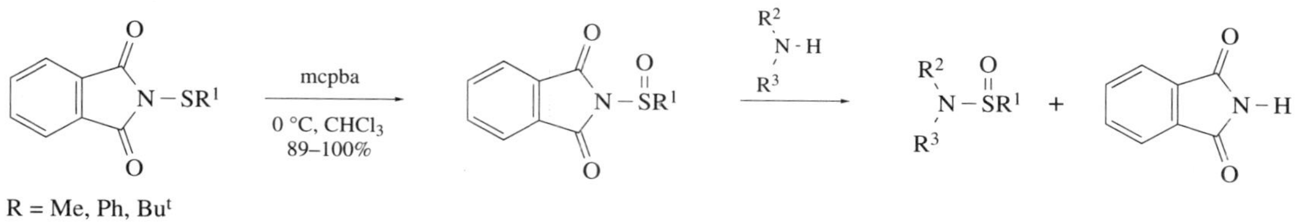

Scheme 43

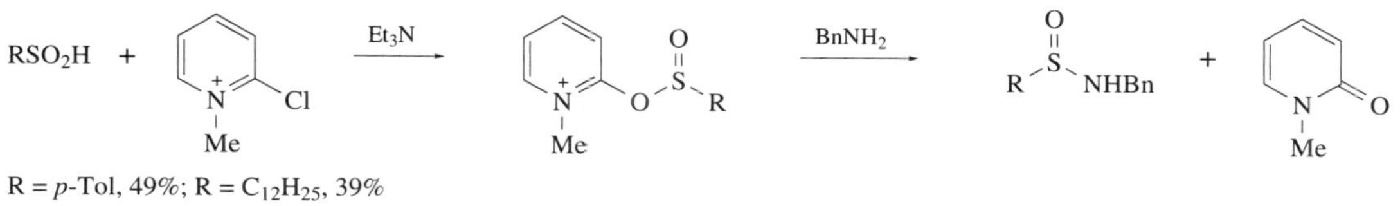

Scheme 44

Sulfinamides can be synthesized by reaction between organometallic reagents (e.g., R_2NMgBr) and suitable sulfinyl compounds RS(O)X, for example sulfinates, $R^1S(O)OR^2$ and *N*-acylsulfinamides $R^1S(O)N(R^2)COR^3$. The reaction between chiral nonracemic *N*-sulfinyloxazolidinones (**27**) and metallated amides R_2NMgBr has been applied to the asymmetric synthesis of sulfinamides (Equation (38)) ⟨92JA5977⟩. The reaction proceeds with inversion at sulfur in high yield and enantioselectivities.

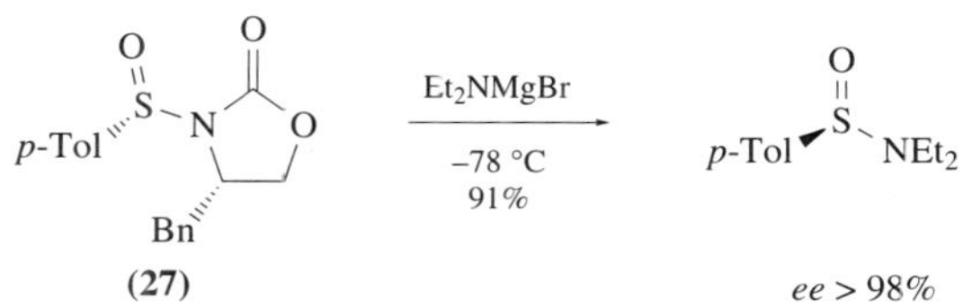

(27) (38)

2.06.3.5.3 *Other general methods*

S-Alkoxysulfenamides, prepared from reaction between allylic alcohols and amino sulfenyl chlorides, undergo a [2,3]-sigmatropic rearrangement to give the corresponding allylic sulfinamide (Scheme 45) ⟨88TL3251, 91T6655⟩. Some of these rearrangements proceed with complete diastereoselectivity ⟨91T6655⟩. *N*-Alkylation of lithium salts of primary sulfinamides gives the analogous secondary sulfinamides ⟨70JPR129⟩. The addition of Grignard reagents to sulfinylimines $R^1CH{=}NS(O)R^2$ has been reported to give primary sulfinamides with complete diastereoselectivity and good yields, using the chiral auxiliary shown in Equation (31) ⟨94JOC914⟩. Grignard reagents add to the nitrogen of *N*-sulfinylamines, RN=S=O, to yield sulfinamides ⟨59CB1910⟩.

2.06.3.5.4 *Derivatives of sulfinamides*

Amino oxosulfoxonium tetrafluoroborate salts have been synthesized by alkylation of sulfoximines (Scheme 46) ⟨73ACR341, 94JA2390⟩. Sulfoxonium methylides derived from these salts are used synthetically for cyclopropanation.

N-Alkoxysulfinamides $R^1N(OR^2)S(O)R^3$ are prepared from reaction between alkanesulfinyl

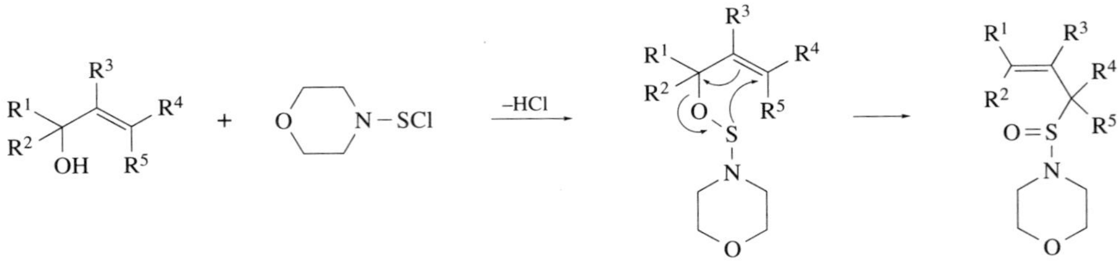

Scheme 45

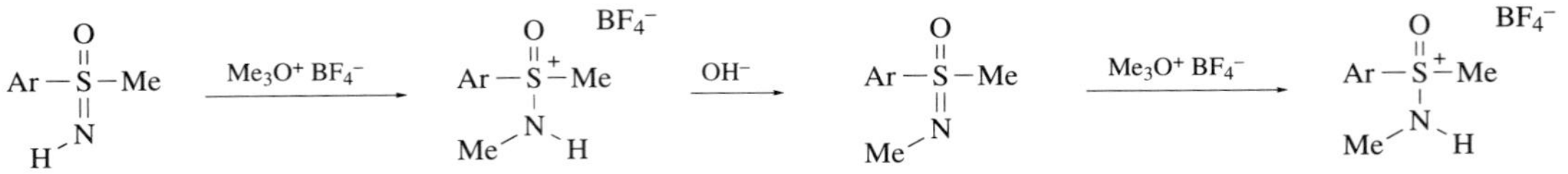

Scheme 46

chlorides $R^3S(O)Cl$ and *N*-alkoxyamines R^1NHOR^2 ⟨72TL181, 79PS(7)325, B-90MI 206-03⟩. The equivalent arene *N*-alkoxysulfinamides can be prepared using arenesulfinyl chlorides ⟨73JA5831⟩.

Synthesis of aminosulfites R^1_2N—$S(O)OR^2$ has been reported, for example, α-hydroxyaziridines react with thionyl chloride in the presence of sodium hydride to yield cyclic aminosulfites ⟨76TL1093⟩.

2.06.3.6 Sulfonamides; $R^1NHSO_2R^2$, $R^1_2NSO_2R^2$

No references were located for aminosulfinic acids R_2NSO_2H but certain amines form crystalline 1 : 1 charge transfer complexes with sulfur dioxide; one of the most stable complexes is formed by trimethylamine ⟨76JA7617⟩.

The synthesis of sulfonamides is well known, and only a brief survey is detailed. Further detail of synthesis can be obtained from the review ⟨B-90MI 206-05⟩.

2.06.3.6.1 From reaction at sulfur(VI)

In general, activated sulfonyl compounds, RSO_2X with X = leaving group, can be reacted with amines to yield sulfonamides. The most common and facile route to sulfonamides is the reaction between sulfonyl chlorides and amines (Equation (39)) ⟨87PS(31)245⟩. The reaction stoichiometry should be carefully controlled when primary amines are used, to prevent the formation of disulfonamides. Other activated sulfonyl compounds RSO_2X include sulfonyl fluorides ⟨88JHC1857⟩, sulfonic anhydrides ⟨57JCS2574⟩, sulfonate esters ⟨28RTC942⟩, and α-disulfones RSO_2SO_2R at elevated temperatures ⟨51RTC254⟩. Sulfonamides can exchange their nitrogen substituents with amine hydrochlorides in a transamination reaction ⟨53LA(581)182⟩. A very mild method for sensitive substrates uses the reaction between 2-pyridylthiolsulfonates and amines in which the sulfonyl halide is generated *in situ* (Equation (40)) ⟨86TL1153⟩.

$$R^1R^2N{-}H + R^3SO_2Cl \xrightarrow{\text{base}} R^1R^2N{-}SO_2R^3 \quad (39)$$

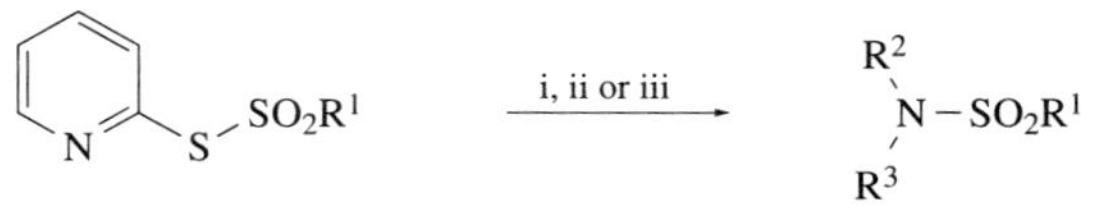

(40)

i, SO_2Cl_2; ii, R^2R^3NH; iii, R^2R^3NH, DMF, 1,2-dibromo-1,1,2,2-tetrachloroethane

Several methods have been developed for the reaction of primary amines with sulfonic acids to give the corresponding sulfonamides (e.g., the use of $POCl_3$) in which case the reaction probably proceeds via a sulfonyl chloride ⟨62LA(657)86⟩. Reaction of alkyl substituted arenes with sulfamoyl chlorides R_2NSO_2Cl in the presence of a Lewis acid leads to the formation of arenesulfonamides ⟨77S39⟩. Sulfonimidoyl chlorides $R^1N{=}S(O)(Cl)R^2$ react rapidly with alcohols in the presence of base to give sulfonamides ⟨70JA3825⟩.

2.06.3.6.2 Other general methods

Sulfenamides $R^1{}_2NSR^2$ can be oxidized by mcpba to sulfonamides $R^1{}_2NSO_2R^2$ ⟨74JA5000⟩. Unsubstituted and monosubstituted sulfonamides are acidic and readily deprotonated to yield anions, which can be alkylated by a range of electrophiles, such as alkyl halides (Scheme 47) ⟨75JCS(P1)2445, B-90MI 206-05⟩. Intramolecular alkylation to yield five- and six-membered ring sultams is facile. Standard preparative methods are unsuccessful for tertiary alkyl sulfonamides, although they may be made by reaction of sulfinyl chlorides with hydroxylamines ⟨72TL181⟩.

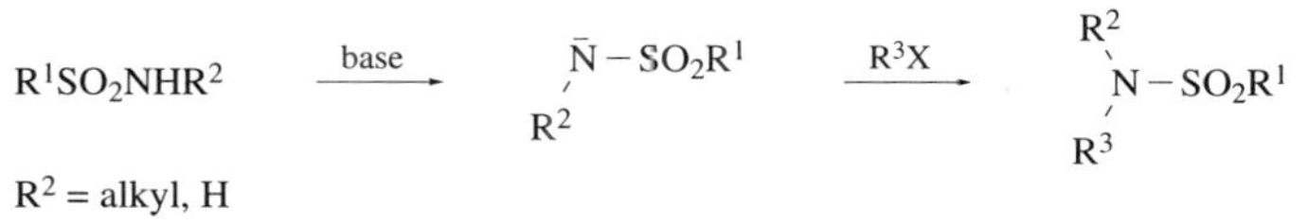

Scheme 47

Decomposition of sulfonyl azides to intermediate nitrenes can be used for the synthesis of sulfonamides by insertion into C—H bonds ⟨75JA676⟩ and addition to alkenes to form *N*-sulfonylaziridines ⟨B-90MI 206-05⟩. *N*-Alkyl sulfonamides cyclize via radical intermediates to give *N*-sulfonyl pyrrolidines, when they are treated with sodium thiosulfate in aqueous copper chloride (Equation (41)) ⟨85T4279⟩.

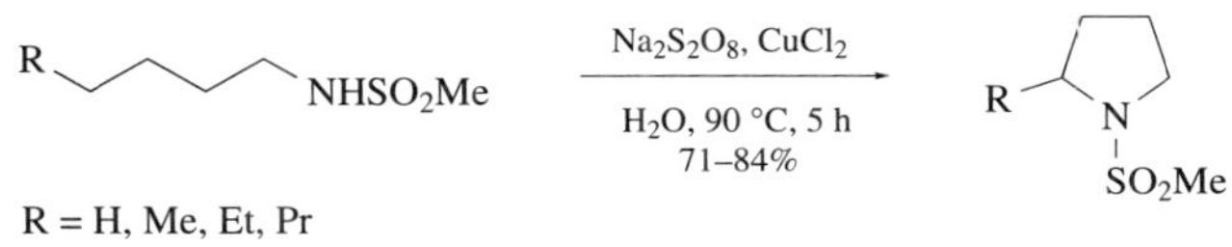

(41)

2.06.3.7 Sulfonamide Salts; $R^1SO_2N^+R^2{}_3$

Forced methylation of *N*,*N*-disubstituted methanesulfonamides gives salts (e.g., $CH_3SO_2N^+(Et)_2Me \cdot FSO_3^-$) which are effective methanesulfonating agents towards amines and alcohols under mild conditions ⟨75JA2566⟩. *N*-Methylation of *N*,*N*-disubstituted arenesulfonamides with dimethoxycarbonium hexachloroantimonate gives crystalline sulfonamidium salts ⟨70CC777⟩. The synthesis of zwitterions $R_3N^+SO_3^-$ has been reported ⟨79JOC3867⟩.

2.06.3.8 *N*-Substituted Sulfonamides; $R^1N(X)SO_2R^2$

The synthesis and reactions of *N*-halogenosulfonamides have been reviewed ⟨89RCR250⟩. The anions of sulfonamides are readily halogenated to yield *N*-chloro- and *N*-bromosulfonamides. The preparation of *N*-fluorosulfonamides has also been described ⟨84JA452, 86JA2445⟩. The method used is to bubble nitrogen-diluted fluorine through a solution of an alkyl sulfonamide in trichlorofluoromethane at low temperature; yields are generally around 50%. This method has also been used to prepare *N*-fluorosultams ⟨88TL6087, 89HCA1248⟩. *N*-Fluorosulfonamides can also be prepared by fluorination of sulfonamides with CF_3OF ⟨78IJ60⟩ and cesium fluoroxysulfate in acetonitrile ⟨91T7447⟩.

The preparation of sulfonamide *N*-sulfenyl chlorides $R^1N(SCl)SO_2R^2$ has been described in several patents. Sulfonamides react with sulfur dichloride in the presence of triethylamine to give the analogous sulfonamide *N*-sulfenyl chlorides ⟨61GEP1156403⟩. Dithiobissulfonamides are also reported to give sulfonamide *N*-sulfenyl chlorides when they react with gaseous chlorine ⟨57GEP1101407⟩.

The synthesis of *N*-sulfonylsulfenamides $R^1N(SR^2)SO_2R^3$ has been reported ⟨71JA2692⟩. Compounds of this type have also been prepared by [2,3]-sigmatropic rearrangement of allylic sulfilimines;

this still occurs if the allyl group is conjugated to phenyl and rearrangement will lead to deconjugation (Equation (42)) ⟨56JCS1755, B-90MI 206-02⟩.

(42)

N-Tosyl cyclic sulfinamides, 3,6-dihydro-2-(*p*-toluenesulfonyl)-2*H*-1,2-thiazine-1-oxides, are prepared from hetero Diels–Alder reactions between dienes and *N*-sulfinyl-*p*-toluenesulfonamide TsN=S=O (Equation (43)) ⟨72LA(762)93, 88ACR313, B-90MI 206-04⟩.

Disulfonamides can be prepared by treating the anion of a monosubstituted sulfonamide with sulfonyl chlorides or a primary amine with an excess of sulfonyl chloride ⟨74JOC3525⟩. Disulfonamides can be synthesized by oxidation of *N*-sulfonyl sulfenamides $R^1N(SO_2R^2)SR^2$ with mcpba ⟨77S798⟩, and *N*-sulfinyl-*N*-sulfonamides $R^1N(SO_2R^2)S(O)R^3$ with performic acid ⟨75JA6521⟩. Stable thallium(I) derivatives of sulfonamides undergo *N*-substitution to give disulfonamides ⟨75S39⟩.

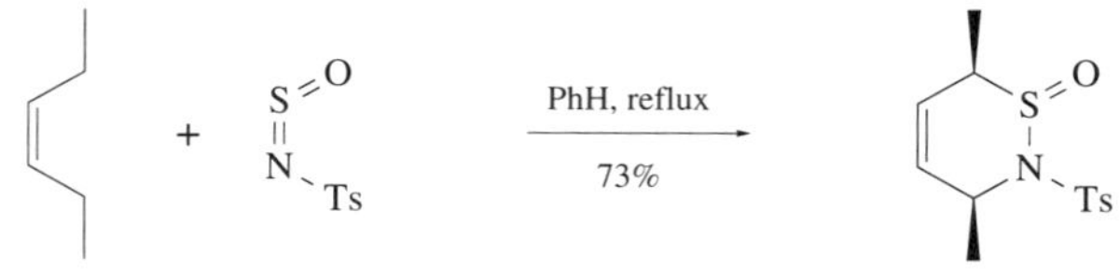

(43)

2.06.3.9 Sulfamic Acids and Derivatives Thereof; $R^1NHSO_3R^2$, $R^1{}_2NSO_3R^2$

The synthesis and chemistry of sulfamic acid derivatives has been reviewed ⟨73IJS469, 79COC(3)363, 80CRV151, B-91MI 206-01⟩.

2.06.3.9.1 *Sulfamic Acids; R_2NSO_3H*

The main methods for synthesizing sulfamic acids are by the sulfamation of alkyl amines with sulfur trioxide complexes $R_3N \cdot SO_3$ ⟨58JOC1133⟩, and by treatment of amines with chlorosulfonic acid ⟨44JOC89, B-91MI 206-01⟩. For example, the sulfamic acid of 5′-amino-5′-deoxyadenosine was prepared by treatment of 5′-amino-5′-deoxyadenosine with $Me_3N \cdot SO_3$ in anhydrous methanol (Equation (44)) ⟨78JMC704⟩. A number of other preparative methods for sulfamic acids have been reported, for example the treatment of isocyanates or urethanes with oleum ⟨76JOC4028⟩, and the reaction of hydroxylamines or amine *N*-oxides with sulfur dioxide ⟨72CJC3502⟩. Catechol sulfate reacts with amines to give 2-(hydroxyphenyl) sulfamate esters, which can be hydrolyzed to yield potassium salts of sulfamic acids ⟨82USP4290975⟩. A large number of heterocyclic sodium sulfamates, het-$NHSO_3{}^-Na^+$, have been prepared for testing as artificial sweeteners ⟨B-91MI 206-01⟩. Of interest, although not part of this chapter, the *N*-acylsulfamate monobactam antibiotics are *β*-lactams with a sulfamate moiety on the *β*-lactam nitrogen, and are normally prepared using sulfur trioxide complexes.

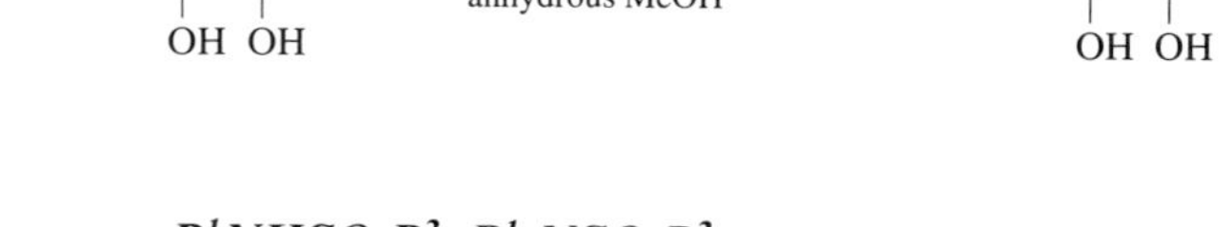

(44)

2.06.3.9.2 *Sulfamate esters; $R^1NHSO_3R^2$, $R^1{}_2NSO_3R^2$*

The principal methods for synthesizing sulfamate esters are by treatment of sulfamoyl chlorides with alkoxide or phenoxide anions, but the yields are poor (Equation (45)) ⟨39JA3250, B-91MI 206-01⟩. Better yields of sulfamate esters are achieved from reactions between alkyl- and aryl-fluorosulfates,

$ROSO_2F$, and amines ⟨84JA7496⟩. Much improved yields are also obtained by the use of phase transfer agents for the reactions between sulfamoyl chlorides and alcohols or phenols (Equation (45)) ⟨82JCS(P1)677, 84PS(20)371⟩.

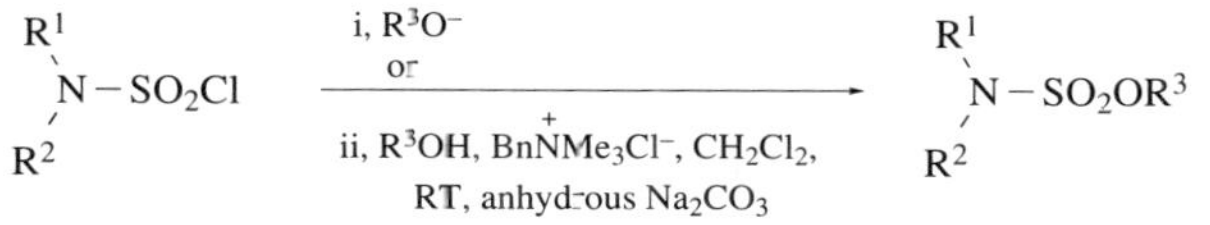

(45)

Sulfamate esters $R^1NHSO_3R^2$ can be alkylated under phase transfer conditions to give *N,N'*-disubstituted sulfamate esters $R^1R^3NSO_3R^2$ (Equation (46)) ⟨85S935⟩. Primary sulfamate esters $ROSO_2NH_2$ can also be readily alkylated by this general protocol ⟨86PS(27)293⟩. Sulfamate esters are of interest because of their pharmaceutical activity and properties as artificial sweeteners, and a large number of syntheses are reported in patents ⟨B-91MI 206-01⟩.

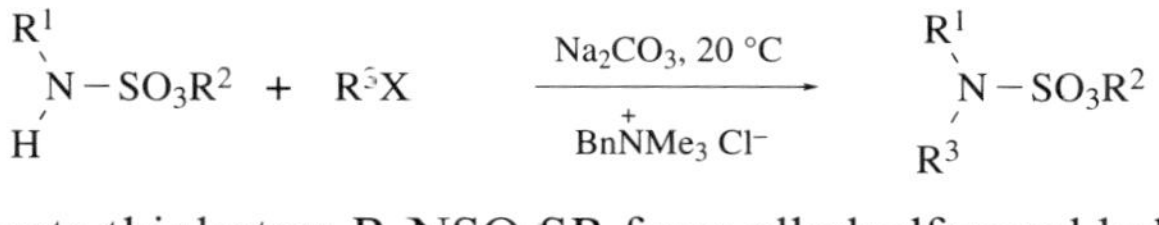

(46)

The synthesis of sulfamate thiolesters R_2NSO_2SR from alkylsulfamoyl halides has been reported ⟨69LA(729)40⟩.

2.06.3.9.3 N,N'-*Bisalkyl sulfamides; $R^1NHSO_2NHR^2$, $R^1{}_2NSO_2NR^2{}_2$*

N,N'-Bisalkyl sulfamides are prepared by several main protocols. Symmetrical sulfamides are most easily synthesized by treatment of amines with sulfuryl chloride. Homochiral *N,N*-bisalkyl sulfamides, prepared by treatment of chiral amines with sulfuryl chloride, have been used to make chiral reducing agents (Equation (47)) ⟨84JOC3861⟩. Unsymmetrical *N,N'*-bisalkyl sulfamides are commonly prepared by treatment of amines with sulfamoyl chlorides ⟨67JA2502⟩. Reaction between silylated amines and sulfuryl chloride yields bisalkyl sulfamides ⟨B-91MI 206-01⟩. Reaction between 2-hydroxyphenyl sulfamate esters and amines gives bisalkyl sulfamides in high yields ⟨80JOC5373⟩. Transamination reactions of suitable *N,N'*-bisalkyl sulfamides gives a range of new sulfamides ⟨B-91MI 206-01⟩. Sulfamide itself can be used but better leaving groups give higher yields. For instance, the use of the cyclic sulfamide (**28**) gives high yields of symmetrical sulfamides when treated with alkylamines (Scheme 48) ⟨82JCR(S)84⟩. Exchange of mono-substituted sulfamides (**29**) with alkylamines gave unsymmetrical sulfamides at lower temperatures and with equal amounts of amine and sulfamide; at higher temperatures, and with two equivalents of amine, symmetrical sulfamides were obtained (Scheme 49) ⟨83S192⟩. The synthesis of bis-*N,N'*-dichloro-*N,N'*-bisalkyl sulfamides has also been reported ⟨64JPR128⟩.

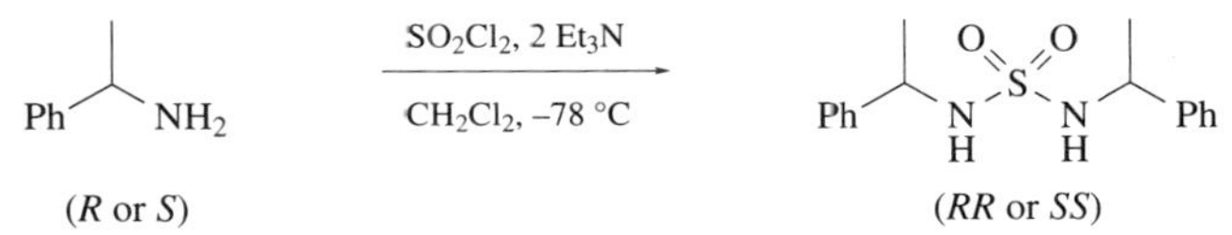

(47)

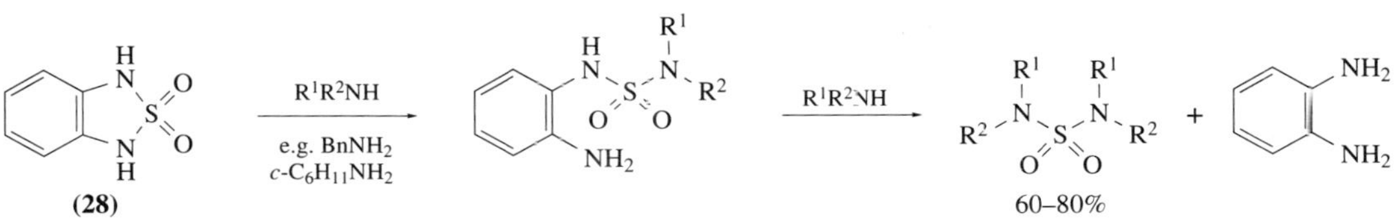

(**28**) 60–80%

Scheme 48

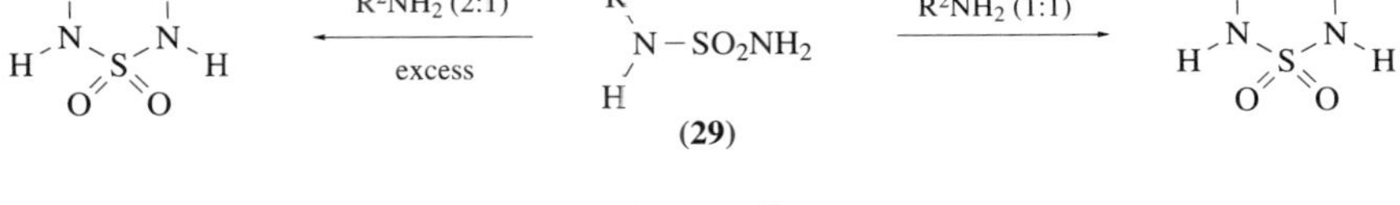

(**29**)

Scheme 49

2.06.3.9.4 Sulfamoyl halides; $R^1R^2NSO_2X$

The synthesis and uses of sulfamoyl halides has been reviewed ⟨81AG(E)151⟩. Amines ⟨B-91MI 206-01⟩, amine hydrochlorides ⟨69LA(729)40⟩ and *N,N′*-bisalkyl sulfamides ⟨B-91MI 206-01⟩ react with sulfuryl chloride to yield sulfamoyl halides (Equation (48)). These methods are limited to amines not having another reactive functional group. Sulfuryl chlorofluoride SO_2ClF has been used to prepare sulfamoyl chlorides ⟨65CB2429⟩. Sulfamoyl chlorides have also been prepared by treatment of sulfamic acids and their salts with phosphorous pentachloride (Equation (48)) ⟨76JOC4028⟩. Treatment of *N*-chloroamines with sulfur dioxide yields sulfamoyl chlorides ⟨63ACS2141⟩.

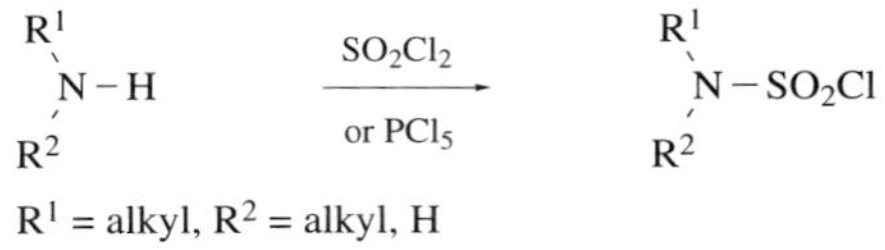

(48)

2.06.3.9.5 Other derivatives of sulfamic acids

Sulfamoyl azides $R_2NSO_2N_3$ have been prepared by reaction between sulfamoyl halides and alkali azides in aqueous alcoholic solutions and by treatment of amines with the explosive chlorosulfonyl azide $ClSO_2N_3$ ⟨B-91MI 206-01⟩.

N-Sulfonylsulfamate esters $R^1N(SO_2R^2)SO_2OR^3$ can be prepared by reaction between sulfonyl halides and sulfamate esters ⟨86S1021⟩.

A large number of tertiary amine–sulfur trioxide complexes have been prepared, and one example uses the reaction between triethylamine and chlorosulfonic acid to form the complex Et_3N—SO_3 ⟨87OPP466⟩.

2.06.4 SELENIUM AND TELLURIUM ANALOGS OF THIOHYDROXYLAMINES AND THEIR DERIVATIVES

The chemistry and synthesis of compounds with Se—N and Te—N bonds have been comprehensively reviewed ⟨B-87MI 206-02⟩. The naming of the selenium compounds is as follows: $R^1R^2NSeR^3$ selenenamide, $R^1R^2NSe(O)R^3$ seleninamide, and $R^1R^2NSeO_2R^3$ selenonamide. Few references to the corresponding tellurium compounds could be located, and it appears only tellurenamides $R^1R^2NTeR^3$ are known. A monograph on the use of selenium reagents and intermediates in organic synthesis, which covers the preparation of some selenenamides, has been published ⟨B-86MI 206-01⟩.

2.06.4.1 Selenenamides; $R^1_2NSeR^2$

Selenenamides can be prepared by the reaction between alkyl amines and benzeneselenenyl chloride ⟨75JOC3313⟩ and arylselenenyl bromides ⟨93JOC917⟩. Chiral selenenamides have been prepared using areneselenenyl bromides and chiral amines ⟨85S635⟩. Benzeneselenenyl chloride reacts rapidly with trimethylsilyl amines to give the corresponding selenenamides in good yield ⟨85S212⟩. Selenothiocyanates also react with amines to give selenenamides ⟨B-87MI 206-02⟩. A sterically hindered selenenate ester and the corresponding acid have been prepared; these both react with benzylamine to give the corresponding selenenamide in about 95% yield (Equation (49)) ⟨88JOC2389⟩. Benzeneselenenic acid has been generated via a selenoxide elimination reaction and reacted *in situ* with an alkyl amine to give a selenenamide ⟨B-87MI 206-02⟩.

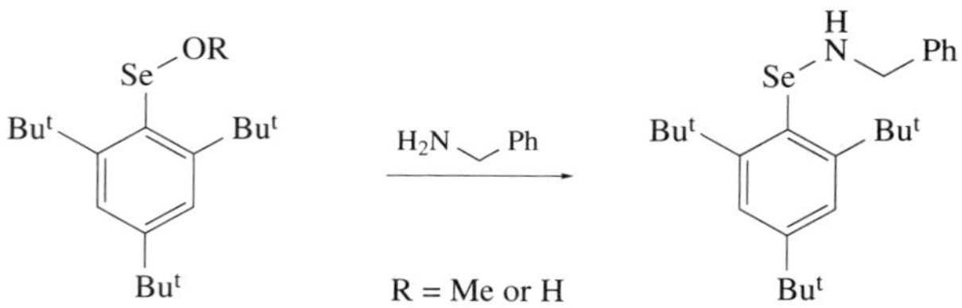

(49)

Partial decomposition of diaminoseleninamides $(R_2N)_2SeO$ yields diaminoselenenamides $(R_2N)_2Se$ ⟨65ZAAC(338)22⟩. The preparation of aminotetraselanes and aminotriselanes has been described ⟨80JCS(D)620⟩. For example, piperidine reacts with selenium and Pb_3O_4 to give an aminotetraselane $(R_2N)_2Se_4$. Selanes are compounds with three or more consecutive selenium atoms. Primary amines can be selenenylated twice when they are reacted with arylselenenyl halides to give diselenenamides $RN(SeAr)_2$ ⟨B-86MI 206-01⟩.

2.06.4.2 Amino Selenium(IV) Derivatives: R_2NSe^{IV}

When *N*-methyl-5*H*,7*H*-dibenzo[b,g][1,5]selenazocine (**30**) is treated with *t*-butyl hypochlorite followed by KPF_6, a chloroselenurane salt (**31**) is formed (Equation (50)) ⟨91JA6337⟩. The same salt is obtained when the selenoxide of (**30**) is treated with thionyl chloride followed by NH_4PF_6. The chloroselenurane salt (**31**) reacts with either lithium dimethyl cuprate or lithium diphenyl cuprate to give the corresponding *Se*-methylammonioselenurane and *Se*-phenyl-ammonioselenurane. Reaction of the selenoxide of (**30**) with one equivalent of trimethylsilyl triflate gives the corresponding hydroxylamino selenurane ⟨92JCS(P1)2247⟩. A related cyclic diazaselenurane has also been prepared ⟨93JA9826⟩.

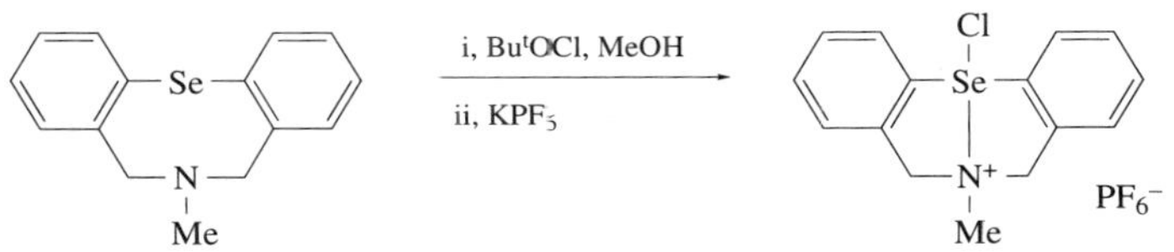

(**30**) (**31**)

(50)

Selenium tetrafluoride reacts readily with silylated amines to form aminoselenium trifluorides, which are reasonably stable (Equation (51)) ⟨74ZOR2619⟩. Diaminoselenium difluorides can be formed either by reaction of aminoselenium trifluorides with a further equivalent of silylated amine or by reacting selenium tetrafluoride with two equivalents of silylated amine (Scheme 50). Diaminoselenium difluorides are very unstable and decompose vigorously within an hour of their preparation. When selenenamides are treated with sulfuryl chloride, dichloroselenenamides $R^1{}_2NSeCl_2R^2$ result ⟨B-86MI 206-01⟩. Dichloroselenenamides can also be prepared by the reaction between the corresponding selenenyl halides and chloroamines ⟨B-86MI 206-01⟩.

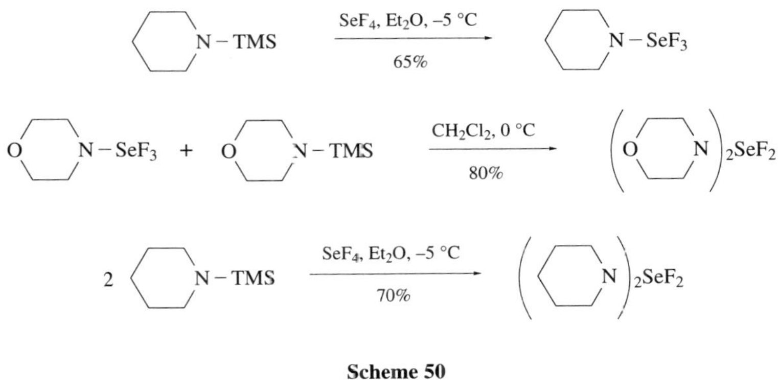

(51)

Scheme 50

2.06.4.3 Seleninamides and Derivatives; $R^1{}_2NSe(O)R^2$

No reports of simple seleninamides were located, though it seems likely that the methods used to make seleninamide derivatives of ammonia could be applied to the synthesis of these compounds ⟨B-87MI 206-01⟩. $(Me_2N)_2SeO$ is formed in the reaction between Cl_2SeO and Me_2NH; further reaction with Cl_2SeO gives $Me_2NSeOCl$, and reaction with HCl then gives $Me_2NH_2{}^+SeOCl_3{}^-$ ⟨65ZAAC(338)22⟩. Diaminoseleninamides are prepared by reaction between the esters $(MeO)_2SeO$ and a primary amine or via a transamination reaction with another diaminoseleninamide ⟨73ZAAC(401)179⟩. These compounds are reported to be stable only at low temperature. Trimethylsilyl esters of seleninamides R_2NSeO_2TMS have been prepared by the reaction of selenium oxide with *N*-trimethylsilyl amines ⟨78ZOB1110⟩.

2.06.4.4 Selenonamides and Derivatives; $R^1{}_2NSeO_2R^2$

Selenonamides have been prepared by the reaction of methyl selenonic esters $PhSeO_3Me$ with amines ⟨77CCC1466⟩.

2.06.4.5 Tellurenamides and Derivatives; $R^1{}_2NTeR^2$

Arenetellurenamides R_2NTeAr (Ar = Ph, *p*-anisyl, 2-naphthyl; R = Me_2CH, Me_2CHCH_2) have been prepared by the action of lithium amides on arenetellurenyl iodides ⟨91OM1095⟩. These *N,N*-dialkyl derivatives generally decompose above 0 °C. Tellurium hexafluoride reacts with silyl amines R_2N-TMS (R = Me, Et, —$(CH_2)_4$—) to produce dialkylaminotellurium pentafluorides R_2NTeF_5 ⟨71JCS(A)1125⟩. When the silyl amine is used in excess, diaminotellurenamide tetrafluorides are formed, for example $(Me_2N)_2TeF_4$ is produced in 45% yield.

2.07
Alkylnitrogen Compounds: Compounds with N—N, N—P, N—As, N—Sb, N—Bi, N—Si, N—Ge, N—B and N—Metal Functional Groups

PATRICK R. HUDDLESTON and IAN G. C. COUTTS
The Nottingham Trent University, UK

2.07.1 ALKYLHYDRAZINES AND RELATED FUNCTIONS

The preparation of alkylhydrazines has been reviewed ⟨59AG321, 67HOU(10/2)169, 68ZC41, B-75MI 207-01, 79COC(2)219, B-83MI 207-01, B-84MI 207-01, 90HOU(E16a/1)425, 90HOU(E16a/1)529⟩.

2.07.1.1 Monoalkylhydrazines

2.07.1.1.1 By alkylation

The direct alkylation of hydrazine is often difficult because of polyalkylation which occurs as with amines; in a detailed investigation Westphal found that only higher alkyl halides led to satisfactory yields of monoalkylhydrazines ⟨41CB759⟩. Later it was found that a large excess of hydrazine was beneficial ⟨65CB1588⟩; with primary benzyl halides yields were 60–75% but fell to 35–45% with secondary halides ⟨59JA2805⟩. Phenoxyalkyl halides give good yields ⟨84JOC336⟩ as do oxiranes ⟨85MI 207-01⟩; alkyl chlorides bearing substituted amino groups give poor to satisfactory yields of aminohydrazines on treatment with hydrazine and K_2CO_3 ⟨64JMC493⟩.

Protecting groups largely solve this difficulty and have made this probably the most popular method for making alkylhydrazines. Azines were one of the earliest protecting groups to be used ⟨43OSC(2)395, 72JOC2259⟩ and more recently two phosphorus-containing blocking groups have been introduced ⟨78TL4731, 87S485⟩. An interesting group derived from DMF has also been used (Scheme 1) ⟨80JOU244⟩.

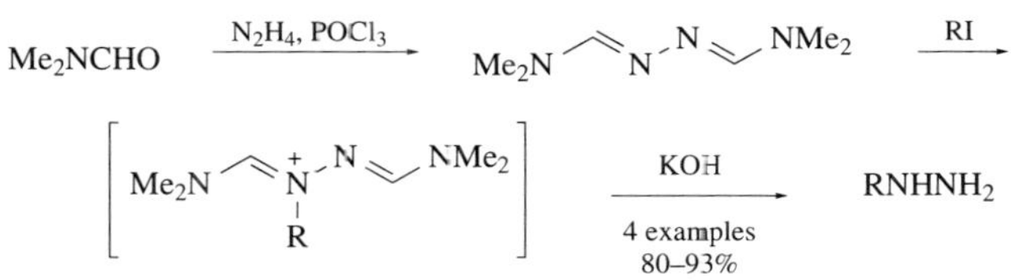

Scheme 1

2.07.1.1.2 *By* N-*amination*

N-Amination is another widely used method for making alkylhydrazines. Hydroxylamine-*O*-sulfonic acid (HOSA) was one of the first amino-transfer reagents to be used for converting amines into hydrazines ⟨49JOC813, 73OSC(5)43⟩, as was chloramine ⟨54JA4869, 55JA3131, 56JA3874⟩; the use of chloramine has been reviewed ⟨B-64MI 207-01⟩. The action of HOSA and a ketone on an amine yields a hydrazine via a diaziridine (Scheme 2) ⟨68JPR257, 71CB1088⟩.

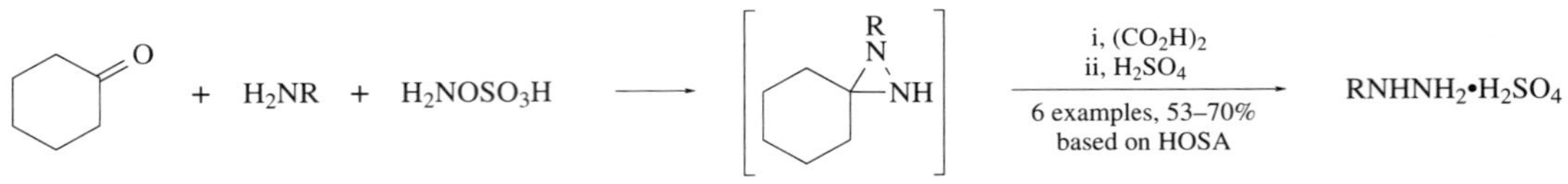

Scheme 2

A closely related method uses a diaziridine formed from an azomethine and chloramine (Scheme 3) ⟨61CB2166, 62CB680⟩, an oxaziridine can also act as an *N*-aminating agent ⟨88S529⟩; these three-membered rings containing two heteroatoms have been reviewed ⟨64AG(E)333⟩.

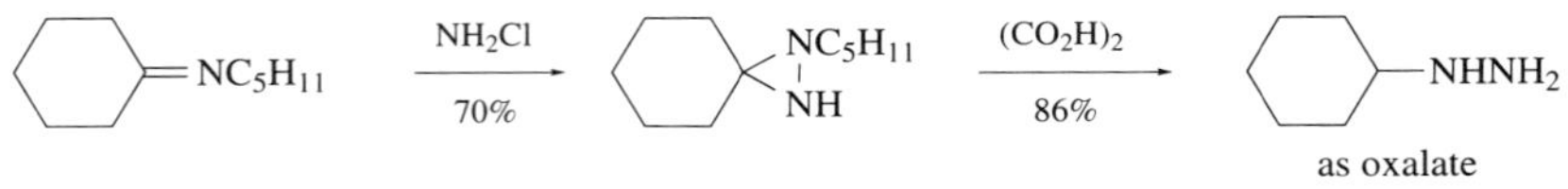

Scheme 3

Diaziridinones may be intermediates in the process shown in Scheme 4 ⟨70JPR349⟩; substituted sulfamides give hydrazines with NaOCl in a closely related procedure ⟨68LA(713)74⟩.

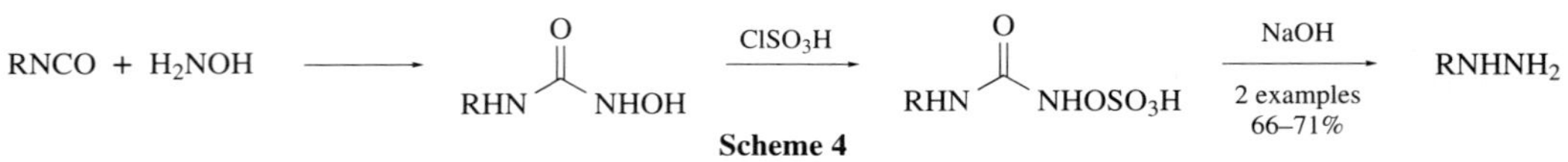

Scheme 4

2.07.1.1.3 *By reduction of hydrazones*

Hydrazones may be easily reduced ⟨59JA2805⟩ using either hydrogenation or diborane; hydrazones of ethyl ⟨85JOC5391⟩ or *t*-butyl ⟨76JOC3805, 81JOC5413⟩ carbazate have proved popular. After reduction the former are converted into hydrazines with alkali and in the latter case with acid (Scheme 5). The preparation of ethyl carbazate (H_2NNHCO_2Et) has been described ⟨88OSC(6)936⟩.

2.07.1.1.4 *By reduction of hydrazides*

An example of the reduction of a hydrazide with $LiAlH_4/AlCl_3$ has been given ⟨71CB1088⟩.

2.07.1.1.5 *By the hydrolysis of sydnones*

Sydnones are mesoionic compounds readily prepared from primary amines; on hydrolysis they yield hydrazines (Scheme 6) ⟨55JA1843, 57HCA918⟩.

2.07.1.1.6 *By the action of hypochlorite on ureas*

The reaction of hypochlorite with ureas is a variation of the Hofmann reaction of amides; an example is shown in Scheme 7 ⟨74JPR347⟩. α-Aminoacids have been converted into the corresponding α-hydrazinoacids by this method; as in the Hofmann reaction there is no racemisation ⟨87T891⟩.

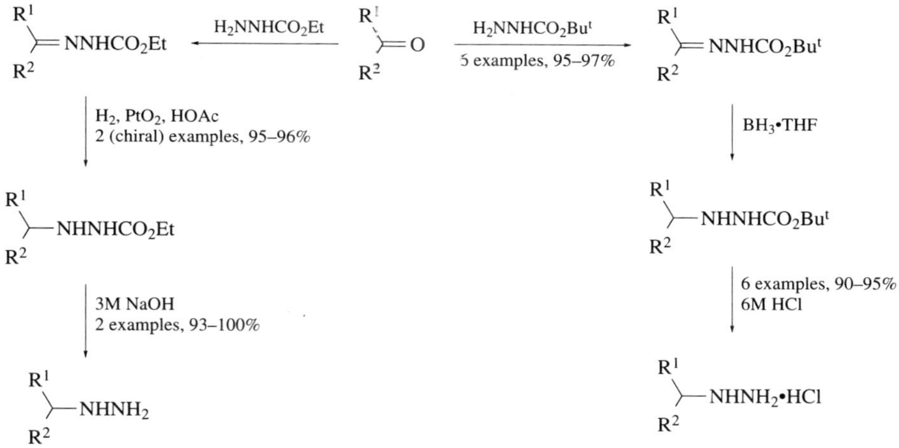

Scheme 5

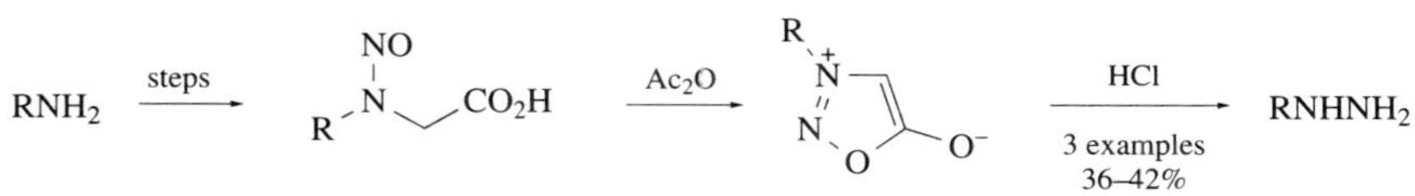

Scheme 6

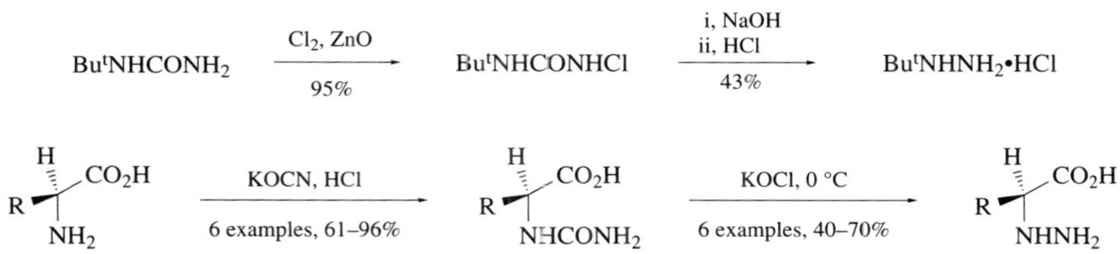

Scheme 7

2.07.1.1.7 By addition to azoesters

A number of α-hydrazinoacids have been made by the addition of suitable chiral precursors to $Bu^tO_2CN{=}NCO_2Bu^t$ followed by hydrolysis ⟨86JA6394, 86JA6395a, 86JA6395b⟩; the reaction has also been carried out with achiral starting materials (Scheme 8) ⟨92TL935⟩. This method can be regarded as a variation of an earlier procedure in which hydrazines were obtained by the hydrolysis of the adducts of diethyl azodicarboxylate (dead) with suitably activated hydrocarbons ⟨54LA(590)1⟩.

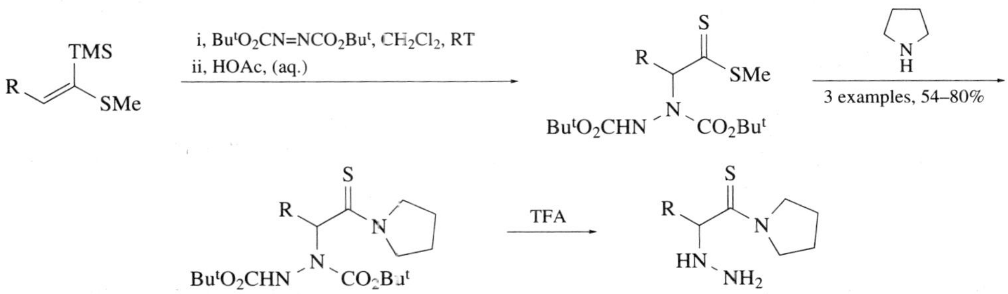

Scheme 8

2.07.1.1.8 By the action of Grignard reagents on diazocompounds

The action of Bu^{t}MgCl on Ph_2CN_2 gave a hydrazone in 88% yield which on hydrolysis afforded Bu^{t}NHNH$_2$ hydrochloride in 60% yield ⟨58JOC1595⟩.

2.07.1.1.9 By miscellaneous methods

Hydrazines of predictable stereochemistry can be prepared from alkenes by the procedure outlined in Scheme 9 ⟨85HCA220⟩.

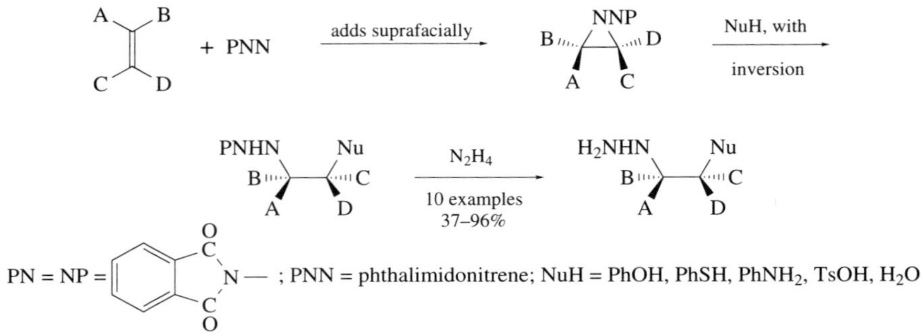

Scheme 9

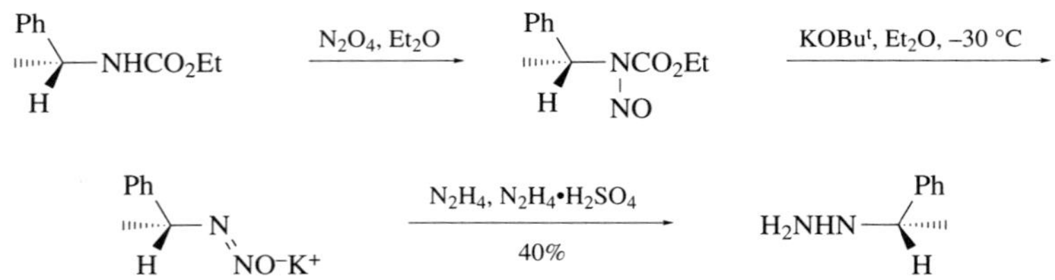

Scheme 10

Homochiral hydrazines can be prepared by the novel process shown in Scheme 10 ⟨75JOC1213⟩.

The addition of α-substituted acrylonitriles to hydrazine yields β-cyanoethylhydrazines ⟨84JOC336⟩. The classical hydrazine synthesis involving nitrosation of a urea followed by reduction and hydrolysis has been applied to the preparation of ethylhydrazine in 67% yield ⟨73OPP219⟩.

2.07.1.2 *N,N′*-Disubstituted Hydrazines

2.07.1.2.1 By alkylation

Direct alkylation is usually not a successful process because of problems of regioselectivity (alkylation occurs mainly on the already alkylated nitrogen atom) and/or polyalkylation. Normally, protection must be used to obtain the required orientation and degree of alkylation. The commonest form of protection is with an acyl group, for example, dibenzoylation of ethylhydrazine followed by methylation and hydrolysis gives 1-ethyl-2-methylhydrazine ⟨72JOC2259⟩; 1,2-dimethylhydrazine is prepared similarly (Scheme 11) ⟨43OSC(2)208⟩.

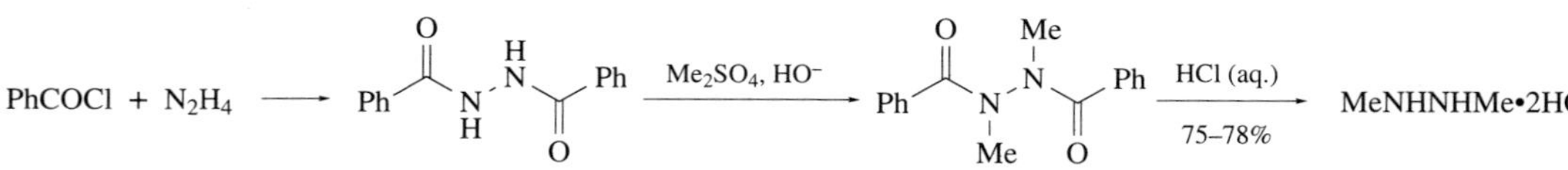

Scheme 11

The formyl group will also serve as a protecting group ⟨69CCC4010⟩ and esters have also been used: both diethyl hydrazodicarboxylate ⟨67JA1417, 70CJC2745⟩ (see an example of its use in the preparation of a cyclic hydrazine ⟨81JOC442⟩) and the corresponding di-*t*-butyl ester have been utilised ⟨63JA2144⟩. By using a phosphorylated hydrazine and an acyl group two different alkyl groups can be introduced, see Scheme 12 ⟨81S537⟩.

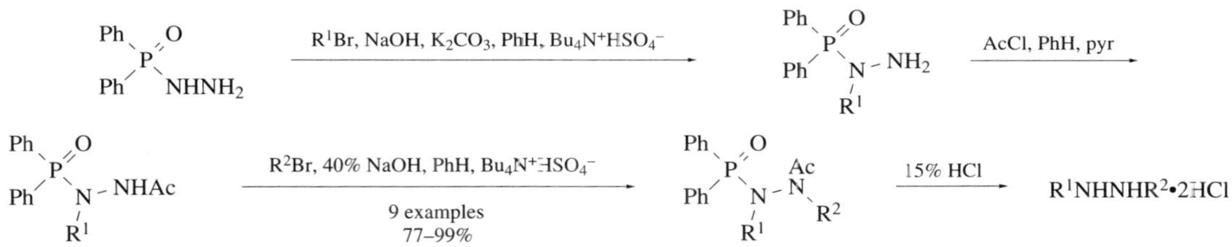

Scheme 12

2.07.1.2.2 *By reduction of hydrazones*

Hydrazones can be hydrogenated with a palladium ⟨62CB2012⟩ or platinum ⟨72ZC129⟩ catalyst. The addition of a Grignard reagent to one double bond of an azine followed by reduction of the resulting hydrazone with PtO_2/H_2 gave hydrazines in 30–50% yields, although the products had to be purified via their picrates ⟨58JA6562⟩. The hydrazide $MeAcNNH_2$ has been condensed with a variety of carbonyl compounds and the products reduced with $NaBH_4$; acid hydrolysis removed the acetyl group to give a hydrazine salt (Scheme 13) ⟨72JOC3615⟩. $NaBH_3CN$ has been used to reduce hydrazones ⟨83JOC4116⟩ and reduction has also been achieved with diborane ⟨70JCS(C)1714⟩ and with $LiAlH_4$ ⟨60JA2728⟩.

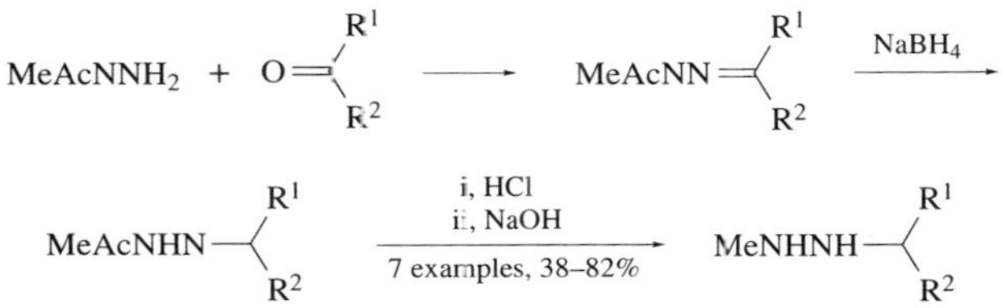

Scheme 13

2.07.1.2.3 *By reduction of hydrazides*

$LiAlH_4$ can be used for reduction of hydrazides ⟨56JA1645⟩ as can diborane ⟨70JOC1468⟩; $LiAlH_4$ will reduce simultaneously an acyl and a hydrazone group (Scheme 14) ⟨65JOC3278⟩. Diethyl hydrazodicarboxylate can be reduced to 1,2-dimethylhydrazine in 84% yield with $LiAlH_4$ ⟨68JHC397⟩.

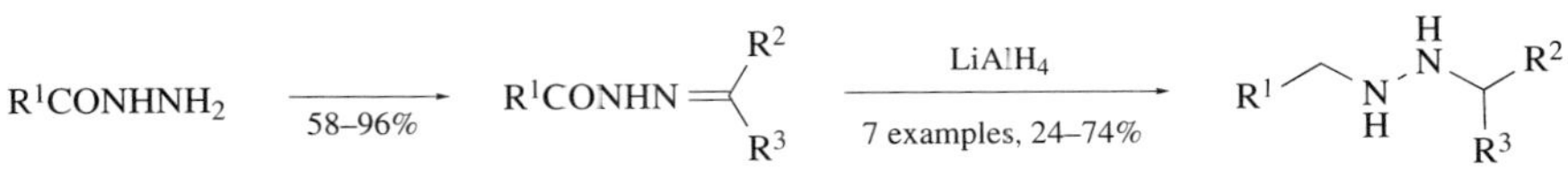

Scheme 14

2.07.1.2.4 *By reduction of azo compounds*

Hydrogenation with palladium has been the most widely utilised method of reducing azo-compounds ⟨67JOC2360, 73TL2859, 81LA943⟩, but hydrazine and Raney nickel can also be used ⟨69ZOR1735⟩.

2.07.1.2.5 *By reduction of azines*

The reduction of azines has been accomplished in good yield with $LiAlH_4$ ⟨54CJC545⟩.

2.07.1.2.6 *By scission of diaziridine derivatives*

These three-membered ring compounds have served to give disubstituted compounds in a similar manner to their use in the preparation of monoalkylhydrazines (see Section 2.07.1.1.2). Reaction of an *N*-chloro primary amine with an azomethine gives a diaziridine which can be hydrolysed with HCl to afford an *N,N'*-disubstituted hydrazine (Scheme 15) ⟨64CB49⟩.

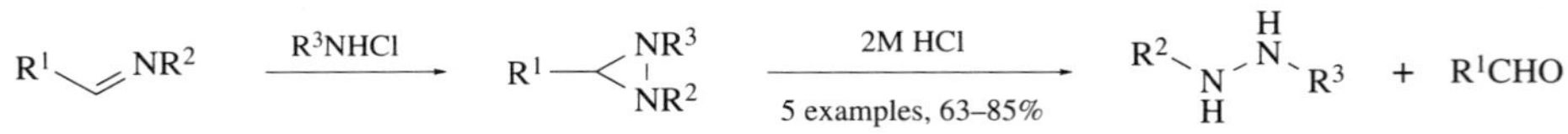

Scheme 15

N,N'-Di-*t*-alkylsubstituted ureas can be converted into mono-*N*-chloro derivatives with Bu^{t}OCl and these give diaziridinones with Bu^{t}OK. Treatment of the latter with HCl gives *N,N'*-disubstituted hydrazines ⟨69JOC2254⟩. A reaction which probably involves a diaziridine-type intermediate is that in which an *N,N'*-disubstituted sulfamide is treated with hypochlorite and alkali to give an *N,N'*-disubstituted hydrazine (Scheme 16) ⟨68LA(713)74⟩.

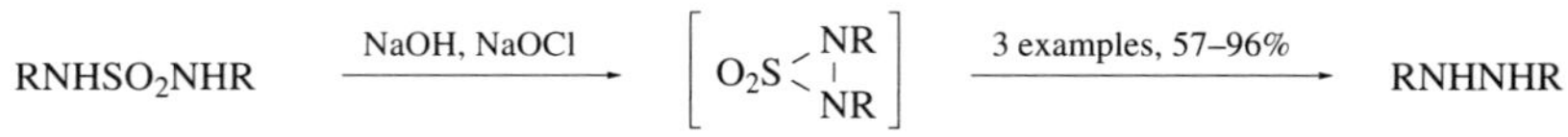

Scheme 16

2.07.1.2.7 *By cycloaddition reactions*

Diethyl azodicarboxylate and its relatives have been used in cycloaddition reactions many times; removal of the ester groups from the adduct yields a hydrazine. An example is shown in Scheme 17 ⟨73TL2859⟩.

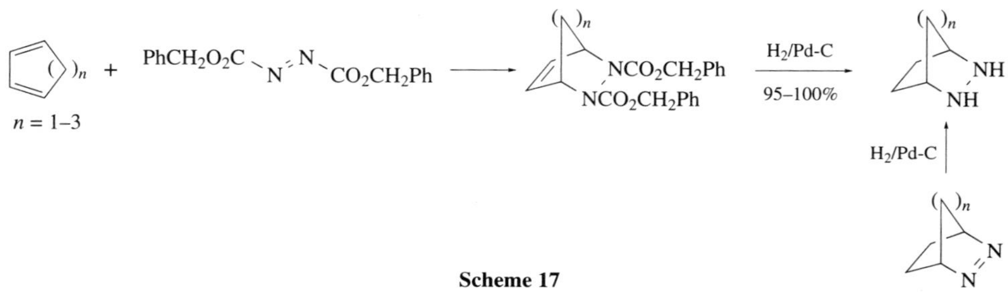

Scheme 17

2.07.1.2.8 *By miscellaneous methods*

α-Cyanohydrazines can be obtained from NaCN, ketones and hydrazine (Equation (1)) ⟨63OSC(4)274⟩.

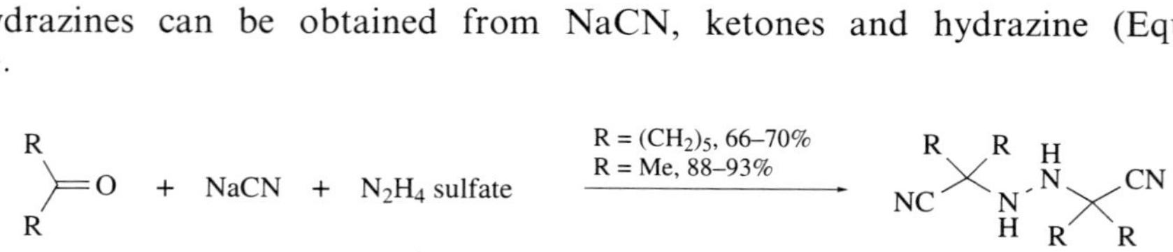

(1)

2.07.1.3 N,N-*Disubstituted Hydrazines*

2.07.1.3.1 *By alkylation*

Direct alkylation of hydrazine itself or its monoalkyl derivatives usually leads to *N,N*-disubstituted hydrazines but care must be taken to prevent polyalkylation ⟨41CB759, 41CB1365⟩. Hydrazine has been alkylated with allyl chlorides ⟨68JMC504⟩ and with benzyl chloride ⟨73JA1562⟩. *o*-Substituted benzyl halides will alkylate methylhydrazine ⟨71JMC1052⟩ as will 2-diethylaminoethyl chloride ⟨64JMC493⟩. Oxirane will react with both hydrazine itself and its monosubstituted derivatives ⟨68JMC504⟩. Benzylhydrazine has been alkylated in reasonable yields with benzyl halides in DMSO containing K_2CO_3 ⟨85ZOR124⟩. Where electrophilic groups are suitably arranged, dialkylation of hydrazine leading to ring closure can be achieved (Equation (2)) ⟨79JA2076⟩.

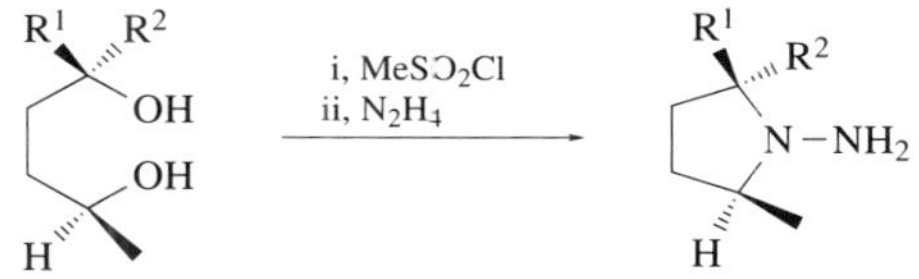

(2)

R^1 = Me, R^2 = H; R^1 = H, R^2 = Me

2.07.1.3.2 *By reduction of* N*-nitrosoamines*

Reduction of *N*-nitrosoamines has been a very popular method for the preparation of the target compounds but it must be borne in mind that many nitrosoamines are suspected or confirmed carcinogens. Whilst the same applies to many hydrazines, the greater volatility of the nitrosoamines makes them considerably more hazardous to health.

The classical procedure for the reduction uses Zn and HOAc ⟨43OSC(2)211⟩ and a modified version utilises Zn and ammonium carbonate in aqueous ammonia ⟨70JCS(C)1088⟩; zinc amalgam and HCl has also been employed ⟨66JCS(C)813⟩.

A sequence for the preparation of ethylmethylhydrazine in 51% overall yield from ethylamine and benzaldehyde without isolating intermediates and culminating with a Zn and HOAc reduction has been described ⟨73OPP225⟩. $LiAlH_4$ has also been widely used; an induction period for the reaction was noted in an early paper ⟨52JA3693⟩. Conditions for the avoidance of this potentially dangerous effect have been described ⟨71JMC1052⟩; no induction period was noted when a nitroamine was reduced instead of a nitrosoamine ⟨60JA2728⟩. There are further examples of the use of $LiAlH_4$ for nitrosoamine reduction ⟨55JA790, 58JOC529, 59JA2805, 64JA2395, 74JMC948, 83JHC121⟩ including an instance of a difficult reduction because of steric hindrance ⟨82JA766⟩. Other ways in which this reduction has been achieved include the use of Al/Hg ⟨60JA2728, 63JA2144⟩, Na in ethanol or liquid ammonia ⟨55JA790⟩, dithionite and alkali ⟨83JHC121⟩, electrolysis ⟨71ACS2337⟩ and reduced Ti species ⟨82T419, 84JOC3470⟩.

2.07.1.3.3 *By* N*-amination*

N-Amination can be achieved by chloramine ⟨55JA3131, 81JINC3389⟩, hydroxylamine-*O*-sulfonic acid ⟨73OSC(5)43⟩ or its mesityl derivative ⟨77S1⟩, *O*-mesitoylhydroxylamine ⟨60JA3133, 63JA2144⟩ and by a suitable oxaziridine ⟨77JPR195, 85JPR445⟩.

2.07.1.3.4 *By reduction of acyl compounds*

An *N*-aminoimide has been reduced to a hydrazine with $NaBH_4$/LiBr ⟨60JA2728⟩ and *N*-aroyl-*N*-methylhydrazines yield benzylmethylhydrazines with $LiAlH_4$ (Scheme 18) ⟨71JMC1052⟩.

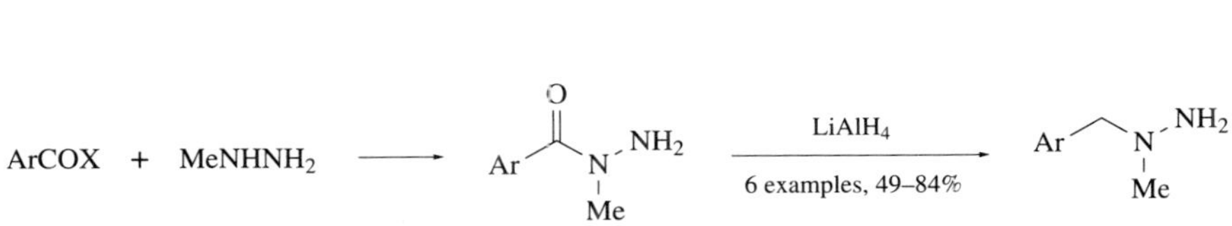

Scheme 18

2.07.1.3.5 *By the action of hypochlorite on ureas*

The action of hypochlorite on ureas has been used several times ⟨70JPR349, 83CPB423, 87OS(65)173⟩; in the last case the chiral auxiliary (*S*)-(−)-1-amino-2-methoxymethylpyrrolidine (SAMP) was prepared (Scheme 19). A related method uses $R_2NSO_2NH_2$ as starting material ⟨68LA(713)74⟩.

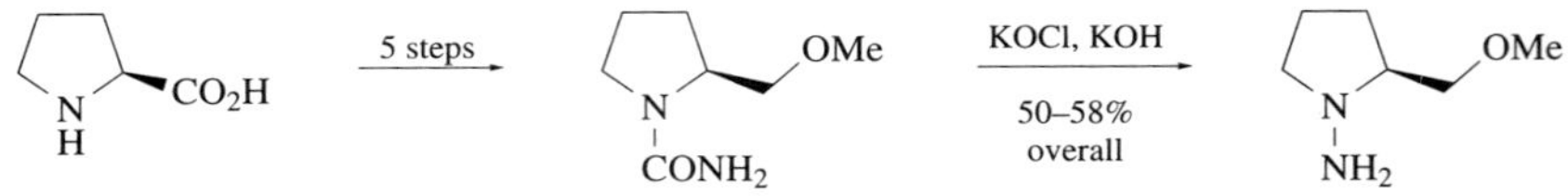

Scheme 19

2.07.1.3.6 *By Michael addition*

Hydrazine hydrate is dialkylated on one nitrogen in 82% yield when treated with 2 mol of acrylonitrile ⟨69ZC105⟩ whilst alkyl hydrazines react at the alkylated nitrogen ⟨65CB3377, 67ZC151⟩.

2.07.1.3.7 *By miscellaneous methods*

N-aminoaziridines can be made by the procedures shown in Scheme 20 ⟨88OSC(6)56⟩, and quaternary hydrazinium salts give hydrazines with strong alkali ⟨59CB2521⟩.

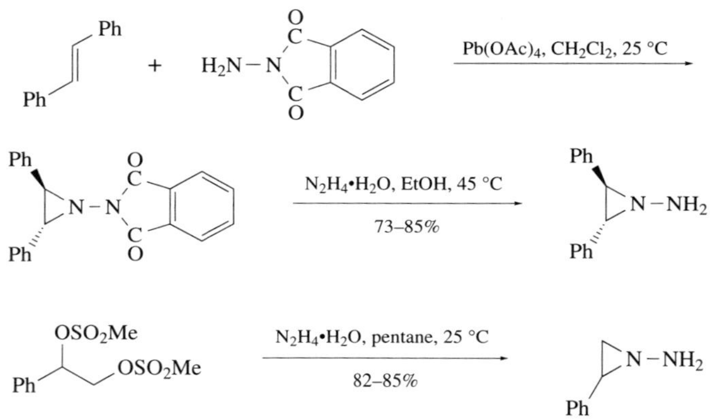

Scheme 20

2.07.1.4 Trisubstituted Hydrazines

2.07.1.4.1 *By alkylation*

Again, by alkylation, problems of regioselectivity and polyalkylation arise but satisfactory yields are sometimes possible ⟨41CB759, 41CB1365⟩.

2.07.1.4.2 *By addition of organometallic reagents to hydrazones*

Addition of organometallic reagents to hydrazones has proved to be a popular method ⟨41LA(547)1, 59ZOB3154, 73OPP233⟩ and it has been used to prepare hydrazines of known stereochemistry

⟨81CPB3387, 90TL1429⟩. An intriguing case has been reported in which the change from a Grignard reagent to an organolithium compound reverses the diastereoselectivity (Scheme 21) ⟨91TL1171, 92JOC4563⟩.

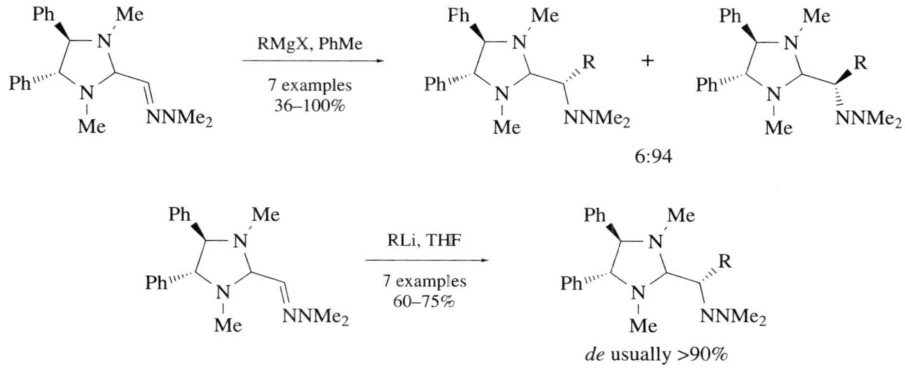

Scheme 21

2.07.1.4.3 *By reduction of hydrazones*

A variety of reducing agents have been used for the reduction of hydrazones: sodium amalgam ⟨20CB1329⟩, $LiAlH_4$ ⟨64JMC493, 73OPP233, 73ZN(B)829⟩, Red-Al (for *o*-aminophenyl derivatives ⟨84JHC1709⟩, $NaBH_4$ ⟨61JOC2740⟩, $NaBH_3CN$ ⟨73JMC736⟩, and hydrogen with a Pt catalyst ⟨59JA2805⟩; a comparison of H_2/Pt with $NaBH_4$ showed that the latter is generally better ⟨66JPR96⟩.

2.07.1.4.4 *By reduction of hydrazides*

$LiAlH_4$ has been used to effect reduction of hydrazides ⟨56JA1645, 56JA2467, 73JA1562⟩; Red-Al was utilised when an *o*-aminophenyl substituent was present as mentioned in Section 2.07.1.4.3 ⟨84JHC1709⟩.

2.07.1.4.5 *By Michael addition*

Acrylonitrile has been added to *N,N*-disubstituted hydrazines giving *β*-cyanoethylhydrazines which were then reduced with $LiAlH_4$ to afford *γ*-aminohydrazines ⟨64JMC493⟩. In experiments using both acrylonitrile and an acrylate it was later found that the nature of the product depended both on the solvent and the acceptor, hydrazinium salts being formed in certain cases ⟨78BSF(2)602⟩; a similar result was obtained with acrylamide ⟨75BSF1627⟩. Addition of ethyl acrylate or methacrylate to *N,N′*-disubstituted hydrazines furnished *β*-hydrazinoesters in variable yields ⟨68JHC397⟩.

2.07.1.4.6 *By the action of organometallic reagents on* N-*nitrosoamines*

Grignard reagents and MeLi give mixtures with *N*-nitrosoamines from which acceptable yields of trisubstituted hydrazines can be isolated ⟨73JOC2412, 75JOC1070⟩.

2.07.1.4.7 *By diaziridinone ring opening*

$LiAlH_4$ reduction of 1,2-di-*t*-butyldiaziridinone gave 1,2-di-*t*-butyl-1-methylhydrazine in 85% yield ⟨69JOC2254⟩.

2.07.1.4.8 By miscellaneous methods

t-Butyl lithium has been added to an N=N bond in >90% yield (Equation (3)) ⟨73JA2719⟩.

In a variant of the usual *N*-amination procedure $MeNHOSO_3H$ has been allowed to react with piperidine and its 2-methyl derivative when an MeNH group was introduced in moderate yield ⟨73ZN(B)829⟩. Certain hydrazinium salts containing an allyl or benzyl group have been rearranged to trisubstituted hydrazines with NaOH (Equation (4)) ⟨70CB2052⟩.

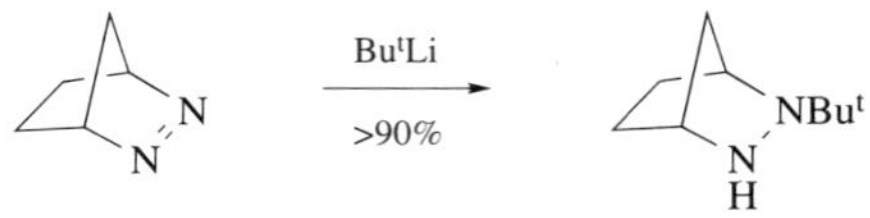

(3)

$$Me_2\overset{+}{N}(R)-NH_2\ Cl^- \xrightarrow[\text{8 examples, 50–80\%}]{NaOH} Me_2NNHR \quad (4)$$

R = allyl or benzyl

2.07.1.5 Tetrasubstituted Hydrazines

2.07.1.5.1 By alkylation

The direct alkylation of hydrazine to a tetrasubstituted product is possible ⟨73JA1562⟩; mono- and dialkyl hydrazines have also been used. *N,N′*-Disubstituted hydrazines can be cyclised to pyrazolidines with oxiranes (Equation (5)) ⟨79JHC1485⟩; pyridazolidine is dialkylated to a bicyclic compound with 1,3-dibromopropane ⟨76JA5269⟩.

(5)

Trisubstituted hydrazines can be alkylated in variable yield ⟨20CB1329, 41CB759⟩ and *N*-*β*-hydroxyethylhydrazines have been cyclised with $SOCl_2$ (Equation (6)) ⟨69JOC2720⟩.

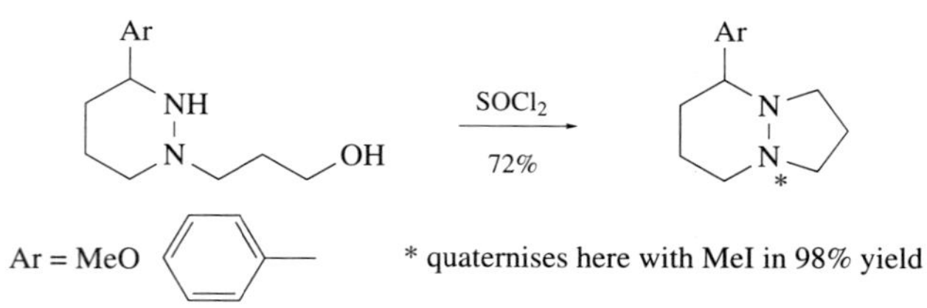

(6)

2.07.1.5.2 By reduction of hydrazides

$LiAlH_4$ has been used most frequently for the reduction of hydrazides in which both acyl ⟨56JA1645, 56JA2467, 69JOC2720, 73OPP233⟩ and ester groups ⟨73CB1052, 76JA5269⟩ have been reduced. Diborane has also found use in the reduction of acyl compound ⟨70JOC1468⟩.

2.07.1.5.3 By reductive alkylation

$NaBH_3CN$ has been used in MeCN to reductively alkylate amines, hydrazines and hydrazones (Scheme 22) ⟨73TL2321⟩.

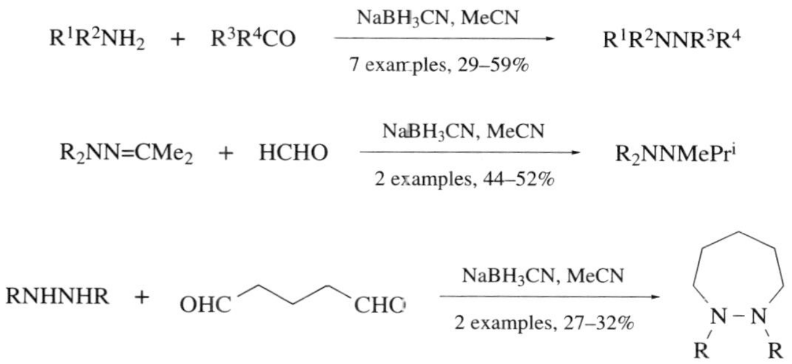

Scheme 22

2.07.1.5.4 By Michael addition

Acrylic esters have been added to both *N,N'*-disubstituted ⟨68JHC397⟩ and *N,N*-disubstituted hydrazines ⟨71ZOR258⟩; acrylamide has been made to react with trisubstituted hydrazines ⟨75BSF1627⟩.

2.07.1.5.5 By cycloaddition reactions

Bicyclic azo compounds will add to suitable dienes after being protonated with HBF_4 (Scheme 23) ⟨84JA3366, 86T1769⟩.

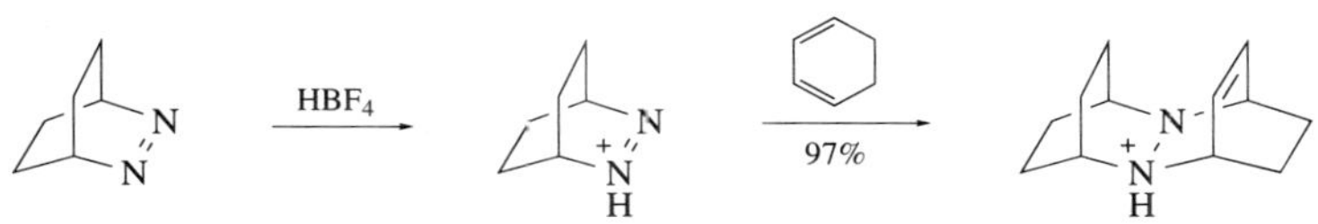

Scheme 23

2.07.1.5.6 By miscellaneous methods

After the addition of Bu^{t}Li to the azo compound (Equation (3)) treatment with MeI gives an *N*-methyl derivative in 86% yield ⟨76JA5269⟩. A large number of bicyclic tetrasubstituted hydrazines have been made by criss-cross addition ⟨90HOU(E16a/1)561⟩; an example is shown in Equation (7) ⟨71JCS(C)2404⟩. Pyrolysis of the tetrazene $Me_2NN{=}NNMe_2$ gives Me_2NNMe_2 ⟨56JCS3677⟩.

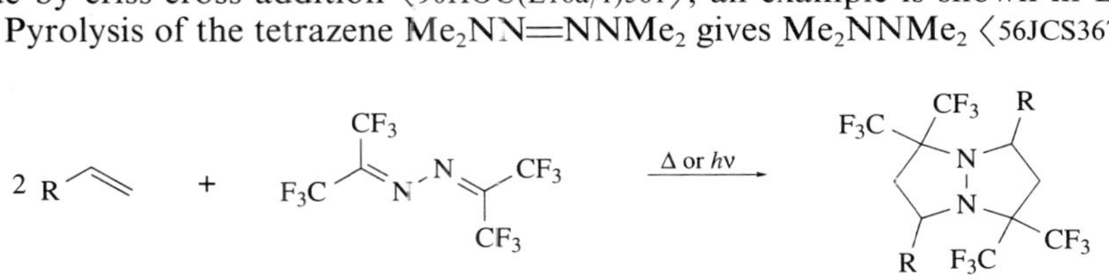

(7)

2.07.1.6 Quaternary Hydrazinium Salts

Quaternary hydrazinium salts have been reviewed ⟨57CRV1021, 90HOU(E16a/2)1035⟩.

2.07.1.6.1 By alkylation

The alkylation of hydrazine itself ⟨41CB1365, 59CB1842⟩, its *N,N*-di- ⟨59CB2521, 70CB2052, 82SC801⟩, *N,N'*-di- ⟨20CB1329⟩ and tetra- ⟨56JA2467, 59CB1842, 82JCS(P1)603, 84JCS(P2)411⟩ substituted derivatives yields hydrazinium salts. MeI quaternises the nitrogen atom furthest from the aryl group of the bicyclic hydrazine shown in Equation (6) in excellent yield ⟨69JOC2720⟩. *N,N*-Dialkylated hydrazines can be quaternised on the alkylated nitrogen atom by 3-bromopropanoic acid and its ester ⟨80AJC1365⟩.

2.07.1.6.2 *By* N*-amination*

Chloramine is a common reagent for *N*-amination ⟨56JA1211, 61JOC1819⟩, and a method using its *in situ* generation has been devised ⟨70CB3733⟩. Hydroxylamine-*O*-sulfonic acid (HOSA) has also been widely employed ⟨59CB2521, 73OSC(5)43, 74JOC1588, 92JOC1585⟩. The use of *O*-mesitylenesulfonylhydroxylamine has been reviewed ⟨77S1⟩.

2.07.1.6.3 *By Michael addition*

The addition of acrylamide to *N,N*-disubstituted hydrazines gives quaternary hydrazinium salts ⟨75BSF1627⟩.

2.07.1.6.4 *By miscellaneous methods*

Quaternisation of acetophenone dimethylhydrazone with MeI followed by hydrolysis with HI gave trimethylhydrazinium iodide in 85% yield ⟨57JOC358⟩. Dications can be prepared from quaternary hydrazinium salts if a sufficiently strong acid is used, for example, $Et_3N^+NH_2\ SbCl_6^-$ is converted into $Et_3N^+NH_3^+\ 2SbCl_6^-$ in 77% yield with $SbCl_5$ and HCl ⟨59CB1842⟩; tricyclic dications have also been prepared ⟨82JCS(P1)603⟩. Reaction of propiolic and acrylic acids with *N,N*-disubstituted hydrazines yields betaines (Equation (8)) ⟨80AJC1365⟩.

(8)

2.07.1.7 Other Alkylated Two-nitrogen Functions

2.07.1.7.1 *Amine-imines*

Treatment of trimethylhydrazinium chloride with potassamide gave a very unstable amine-imine which, however, could be easily handled as its *t*-butanol adduct; the latter reacted with several Lewis acids to give the products shown (Scheme 24) ⟨66CB3118⟩.

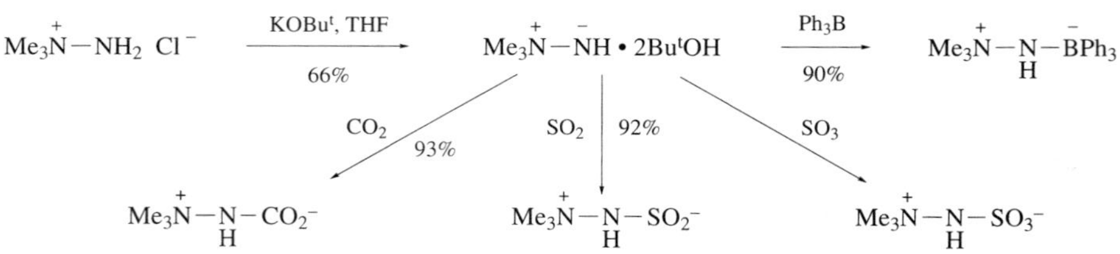

Scheme 24

It was later found that oxiranes will react with *N,N*-disubstituted hydrazines to afford β-hydroxyamine-imines ⟨68JOC1374⟩. Dehydration of hydrazinium nitrates $R_3N^+-NH_2NO_3^-$ with TFAA/TFA gives $R_3N^+-NO_2^-$ which can be regarded as oxidised amine-imines ⟨92JOC1585⟩.

2.07.1.7.2 N-*Nitrosoamines and* N-*thionitrosoamines*

When working with nitrosoamines it must be remembered that they are extremely carcinogenic.

The preparation of nitrosoamines has been reviewed ⟨58HOU(11/2)178, 79COC(2)363, B-82MI 207-01, 90HOU(E16a/2)1137⟩. The classical method for making *N*-nitrosoamines uses $NaNO_2$ in acid, which may be HCl ⟨43OSC(2)211⟩ or HOAc ⟨55OSC(3)244⟩ or both ⟨71CB1088⟩; there is an alternative procedure used in a difficult case ⟨82JA765⟩. The reaction is faster in neutral or alkaline aqueous solution with N_2O_3 or N_2O_4 ⟨76CC877⟩; the latter compound is probably the reagent responsible when nitrosation is effected with NCS and $NaNO_2$ (2 equiv.) under phase transfer conditions ⟨84TL2619⟩. Electrolysis with $NaNO_2$ is also efficient ⟨88CPB459⟩. Anhydrous conditions may be used: the sodio-derivative is made from R^1R^2NH and NaH and treated with NOCl in THF, alternatively, the amine is allowed to react with NOCl in pyridine (Scheme 25) ⟨76S462⟩. Nitrosoamines may be made from secondary amines by using a variety of other reagents: $BrCH_2NO_2$ ⟨90CC1598⟩; Cl_3CNO_2 ⟨92SC2607⟩ and $C(NO_2)_4$ with $h\nu$ ⟨89JCR(S)12⟩. *N*-Aminopiperidine is oxidised to the corresponding nitroso compound with $C(NO_2)_4$ and a nitroxide but the reaction appears not to be general ⟨91CL569⟩.

$$R^1R^2NH \xrightarrow[\text{6 examples, 67–100\%}]{\text{i, NaH, THF; ii, NOCl}} R^1R^2NNO \qquad R^1R^2NH \xrightarrow[\text{5 examples, 94–100\%}]{\text{NOCl, pyridine, THF}} R^1R^2NNO$$

Scheme 25

Tertiary amines containing a methylene group can be converted into nitrosoamines with loss of one alkyl group under the action of Ac_2O/HNO_3 (Equation (9)) ⟨85S677⟩.

$$R^1R^2NCH_2R^3 \xrightarrow[\text{7 examples, 13–73\%}]{Ac_2O,\ HNO_3} R^1R^2N{-}NO \;+\; R^3CHO \qquad (9)$$

N-Chloroamines react with two equivalents of $NaNO_2$ to give nitrosoamines, apparently by a nucleophilic substitution on nitrogen; however, the reaction actually involves the intermediacy of NO_2Cl ⟨84CC451⟩.

Nitrosoamines can be lithiated at the α-position and the resulting anions will react with electrophiles such as alkyl halides and carbonyl compounds ⟨74JMC1225, 75CB1293, 88OSC(6)542⟩. Reaction of azomethines with NOCl gives α-chloronitrosoamines which undergo nucleophilic displacement of chlorine with AgOAc ⟨84LA1468⟩ or with thiols; the sulphides resulting from the latter reaction have been oxidised to sulphones with $KMnO_4$ ⟨87LA583⟩. Cyclic *N*-nitrosoenamines will undergo addition reactions to the double bond providing further functionalised nitrosoamines (Equation (10)) ⟨80JOC2919⟩.

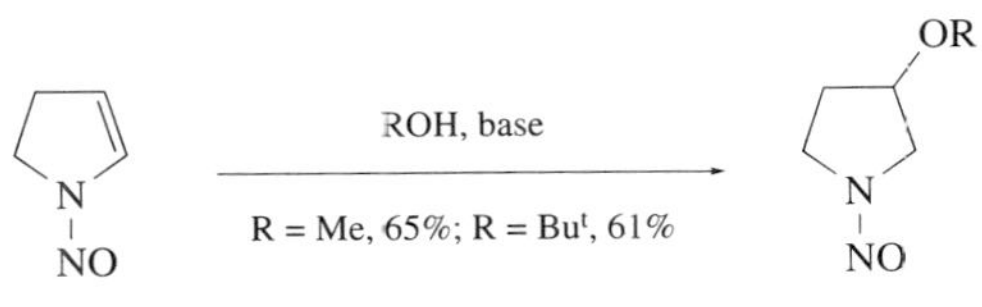

(10)

N-Thionitrosoamines, such as $Me_2NN{=}S$, are made either by prolonged stirring of an *N,N*-disubstituted hydrazine with sulfur or by the action of $LiAlH_4$ on a sulfinylhydrazine, $R_2NN{=}S{=}O$ ⟨66JA3842⟩. *N*-Thionitrosoamines are compounds having limited lifetimes but many stable complexes have been prepared.

2.07.1.7.3 N-*Nitroamines*

Compounds of the types $RNHNO_2$ and $R^1R^2NNO_2$ are known; their preparation has been reviewed ⟨51QR75, 58HOU(11/2)99, 69RCR640, 79COC(2)370, 90HOU(E16a/2)1147⟩. A great many nitroamines have been made over the years because of their importance as explosives, hence only a summary can be given here.

The most common method of preparation is by direct nitration of an amine or amine derivative; free amines can be nitrated with acetone cyanohydrin nitrate ⟨55JA4387⟩, with $CF_3CMe_2ONO_2$

⟨87JOC2292⟩ or with N_2O_5 ⟨58JOC311⟩. The last reagent will also convert cyclic tertiary amines into open-chain derivatives (Equation (11)) ⟨88TL2735⟩.

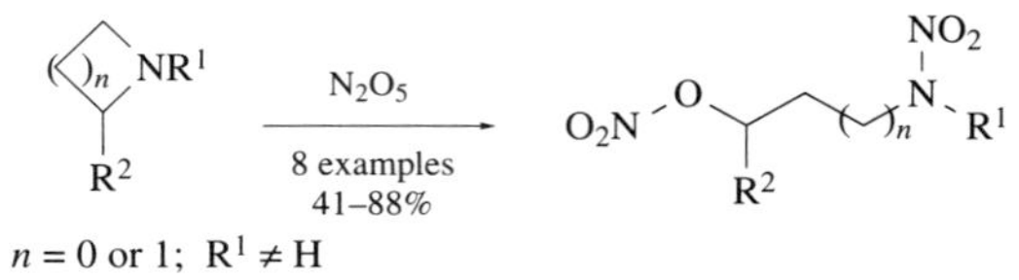

(11)

Primary amines with straight or branched alkyl groups can be lithiated with Bu^nLi and then nitrated with ethyl nitrate ⟨65JOC2471⟩. Amine derivatives which have been nitrated to give *N*-nitroamines include amides in TFAA with HNO_3 ⟨55JA2453⟩ or NH_4NO_3 ⟨88S743⟩; amides with $NO_2^+BF_4^-$ in MeCN ⟨78ZOR240⟩ and sulfonamides with HNO_3 in TFAA ⟨55JA2453⟩. The *N*-nitration of *p*-toluenesulfonamide and alkylation of its silver salt followed by hydrolysis has been used to make nitroamines from alkyl halides ⟨49JCS1883⟩.

β-Methylaminopropionitrile has been converted into its *N*-nitroderivative by treating its HNO_3 salt with $Ac_2O/HCl/HNO_3$; the product was converted into the corresponding ester, acid and acid chloride ⟨56JA5428⟩. An interesting reaction which gives excellent yields is the nitro exchange process shown in Equation (12) ⟨55JA6061⟩.

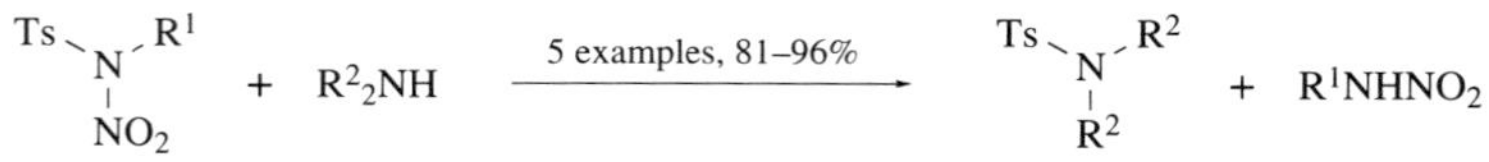

(12)

N-Nitrosoamines can be oxidised to nitroamines with CF_3CO_3H ⟨54JA3468⟩ or with H_2O_2 in HOAc ⟨85S677⟩. The action of $K_3Fe(CN)_6$ on $PhCH_2N{=}NONa$ gives benzylnitroamine ⟨10LA(376)239⟩.

Reduction of nitrimines, ($R_2C{=}NNO_2$, from oximes with NOX) with $LiAlH_4$ in THF gives primary nitroamines in fair to good yield ⟨61JOC4190⟩. Nitrocarbamates can be alkylated with diazoalkanes and hydrolysis with ammonia gives the ammonium salts of the nitroamines in reasonable yields (Scheme 26) ⟨49JCS1883⟩.

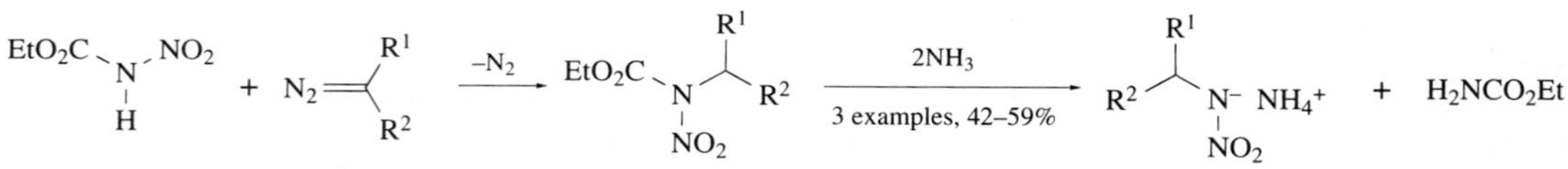

Scheme 26

Primary nitroamines give salts with amines ⟨51JA5041⟩; *S*-benzylisothiouronium salts have also been reported ⟨49JCS1883⟩. Nitroamines undergo Michael addition with acrylonitrile and the products have been converted into acid derivatives ⟨75AP(308)954⟩; *N*-alkylation is also possible ⟨55JCS3997⟩.

2.07.1.7.4 *Sulfinylhydrazines (thionylhydrazines)*

Sulfinylhydrazines, $R^1R^2NN{=}S{=}O$, have been reviewed ⟨85HOU(E11)558⟩. They are usually made by the action of $SOCl_2$ on hydrazines either alone ⟨1893CB310⟩ or in the presence of pyridine ⟨68JMC504, 79JOC1881⟩ or triethylamine ⟨70JCED356, 81JCS(P2)1449⟩. An alternative method uses 'trans-thionylation' in which an aliphatic *N,N*-disubstituted hydrazine is treated with an aromatic sulfinylamine; some examples are shown in Scheme 27 ⟨66JA3842⟩.

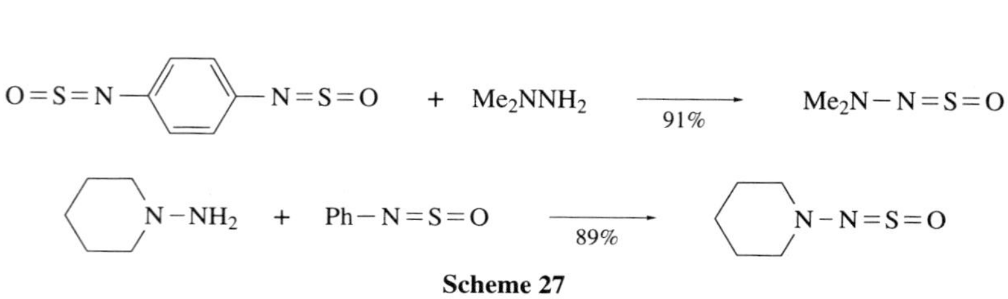

Scheme 27

2.07.1.7.5 *Miscellaneous hydrazine derivatives*

It seems that hydroxyhydrazines are too unstable to be isolated; however, two stable compounds which can be considered to be alkoxyhydrazines have been prepared (Equation (13)) ⟨23CB679⟩. A borylhydrazine has been made ⟨64CR(259)3737⟩ and dimethylhydrazine has been combined with Et_3Al to give $Et_2AlNHNMe_2$ in 100% yield ⟨62JCS3818⟩.

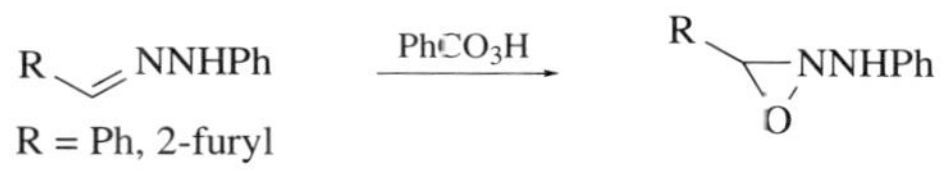

(13)

2.07.1.8 Alkylated Three-nitrogen Functions

2.07.1.8.1 *Triazanes*

Triazanes, which contain the system R_2NNRNR_2, have been reviewed ⟨79COC(2)219, B-83MI 207-01⟩. It appears that saturated chains of nitrogen atoms are not stable in the absence of electron-withdrawing groups, possibly because of lone pair/lone pair repulsions. Treatment of an *N*,*N*-disubstituted hydrazine with a salt of hydroxylamine-*O*-sulfonic acid (Equation (14)) ⟨62AG(E)405, 71AG(E)813⟩ or with chloramine ⟨66IC1835, 68IC698⟩ yields a triazanium salt quaternised on N2; these compounds are reasonably stable. They can also be prepared by the action of an oxaziridine on a hydrazine; triazanium betaines are available this way as well (Scheme 28) ⟨69LA(725)1⟩.

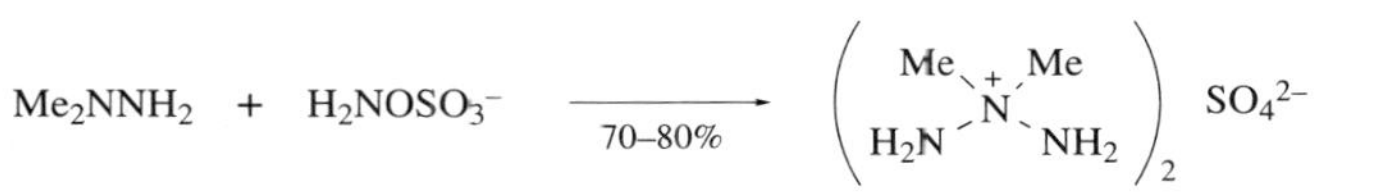

(14)

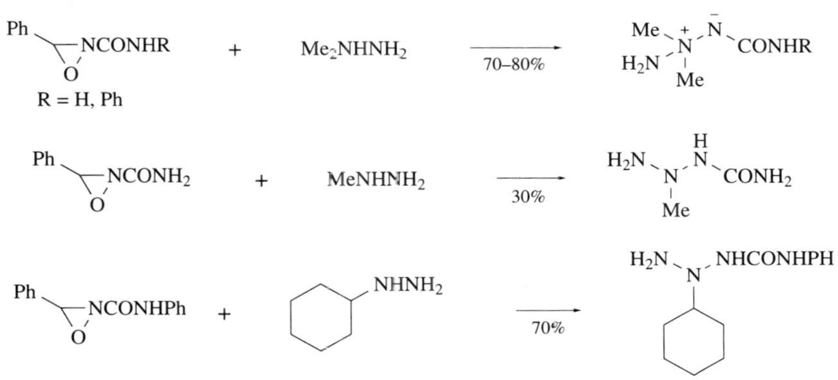

Scheme 28

2.07.1.8.2 *Triazenes*

Triazenes have been reviewed ⟨67HOU(10/2)827, 79COC(2)219, B-83MI 207-01, 90HOU(E16a/2)1182⟩. The presence of the double bond makes these compounds reasonably stable but there remains the hazard of an explosion. Triazenes are usually made by the action of a Grignard reagent ⟨05CB670, 06CB3905, 74CB745, 80JA3883⟩ or an alkyl lithium (Equation (15)) ⟨80JA3883, 81JA442, 83S476, 88JOC1467⟩ on an alkyl azide.

$$RN_3 + RLi \xrightarrow[\text{R = Me, 60\%; R = Et, 45\%}]{} RN{=}NHR \qquad (15)$$

Alkylation of triazenes can be carried out with an alkyl halide and Bu^tOK ⟨80JA3883⟩ or KH/

dicyclohexano-18-crown-6 ⟨83S476⟩. Triazenes of the type $R^1{}_2N{-}N{=}NZ$ where $Z = SO_3R^2$ and CN have been made from the action of Z^- on $R^1{}_2N{-}N{\equiv}N^+$ which in turn derives from a suitable tetrazene (Scheme 29) ⟨78AG(E)203⟩.

$$Me_2NN{=}NNMe_2 \xrightarrow{R^1X} Me_2NN{=}N\overset{+}{N}Me_2R^1\ X^- \longrightarrow \left[Me_2N\overset{+}{N}{\equiv}N\right] \xrightarrow{M^+Z^-} Me_2NN{=}NZ$$

$Z^- = R^2SO_2^-$ 3 examples, 78–92%

$Z^- = CN^-$ 1 example, 73%

Scheme 29

2.07.1.8.3 *Nitrosohydrazines*

Compounds of the type $RN(NO)NH_2$ and $R^1N(NO)NHR^2$ were first made by Thiele from the action of HOAc and $NaNO_2$ on a hydrazine ⟨10LA(376)239⟩. Further examples have since been prepared by the same method ⟨22JA2556⟩ or by a variant using a hydrazine hydrochloride ⟨22MI 207-01, 58JOC1230⟩. The action of aqueous Pr^iONO on methylhydrazine gave $MeN(NO)NH_2$ in 55% yield ⟨64ACS2199⟩. Both types of nitrosohydrazine can be alkylated on the nonnitrosated nitrogen atom ⟨10LA(376)239⟩.

2.07.1.9 Alkylated Four-nitrogen Functions

2.07.1.9.1 *Tetrazenes*

Reviews of tetrazenes, which contain the $R^1R^2N{-}N{=}N{-}NR^3R^4$ system, are available ⟨67HOU(10/2)827, B-83MI 207-01, 90HOU(E16a/2)1227⟩. The classical method for the preparation of tetrazenes, published by Emil Fischer in 1879, is the oxidation of a hydrazine. A great variety of reagents has been used, the most popular being HgO ⟨41CB774, 55JA4100, 56JCS3677, 58JA2516, 59JA3294, 66JA1078, 66JCS(C)813, 69JOC2997⟩. Other oxidants which have been utilised are: halogens ⟨57JA572, 66JA1078, 73JA3228, 81JOC2402⟩; $KMnO_4$ ⟨55JA4100⟩; quinone ⟨12LA(392)133, 59JA3294⟩; $KBrO_3$ ⟨57JA572, 66JA1078⟩ (this reagent has also been used to make unsymmetrical tetrazenes ⟨59JA5546⟩); KIO_3 ⟨66JA1078⟩; $MeCO_3H$ ⟨61CB712⟩; dead ⟨23CB1933⟩; lead tetraacetate (LTA) ⟨72JCS(P1)1692⟩; MnO_2 ⟨68JOC2407⟩; $K_3Fe(CN)_6$ ⟨73CB3097⟩; TMS-Cl/KH ⟨89JOC1070⟩ and oxygen ⟨82JA766⟩. It would seem from this list that the ideal general reagent has not been found; the most promising candidate appears to be benzeneseleninic acid, $PhSe({=}O)OH$, which oxidised six hydrazines in yields of 28–96% ⟨82CJC2711⟩.

Tetrazenes have also been prepared by the base-induced degradation of sulfonylhydrazides, for example, (**1**) gives (**2**) (Equation (16)) ⟨64JA2395⟩. Some cyclic tetrazene-*N*-oxides have been reduced to tetrazenes with $P(OMe)_3$ and with $LiAlH_4$ ⟨78HCA1622⟩.

$$\underset{(\mathbf{1})}{Me_2NNHSO_2Ph} \xrightarrow[76\%]{\text{NaH, tetraglyme, 110–120 °C}} \underset{(\mathbf{2})}{Me_2NN{=}NNMe_2} \qquad (16)$$

2.07.1.9.2 *Tetrazenium salts*

Tetrazenes have been quaternised in the usual way with alkyl halides, see Equation (17) ⟨66JCS(C)813, 78AG(E)203⟩.

$$Me_2NN{=}NNMe_2 + MeI \xrightarrow{60\%} Me_2NN{=}N\overset{+}{N}Me_3\ I^- \qquad (17)$$

2.07.1.9.3 N-*Azidoamines*

The dangerously explosive compound Me_2NN_3 has been made in 25% yield from Me_2NCl and NaN_3 ⟨64ZAAC(332)238⟩.

2.07.1.9.4 Dinitrosohydrazines

Thiele prepared (**3**) by the action of $NaNO_2/H_2SO_4$ on the corresponding hydrazine hydrochloride ⟨10LA(376)239⟩.

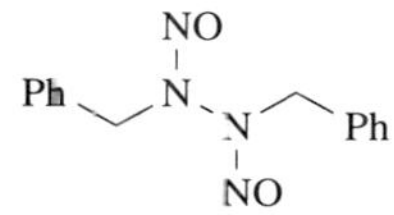

(**3**)

2.07.1.10 Alkylated Five-nitrogen Functions

Compound (**4**) has been made by the action of N_2O_4 or N_2O_3 on the corresponding tetrazene at $-20\,^{\circ}C$ ⟨69JOC2997⟩.

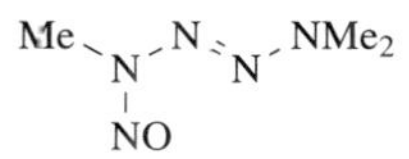

(**4**)

2.07.2 *N*-ALKYL COMPOUNDS OF THE TYPE RN=Y

2.07.2.1 Nitroso and Nitro Compounds

2.07.2.1.1 Nitroso compounds

(i) Introduction

The synthesis of nitroso compounds is complicated by two factors. First, there is the possibility of primary and secondary nitroso compounds undergoing a rearrangement to oximes (isonitroso compounds). This reaction is affected by a number of factors, for example, heat, light and the action of acids or bases ⟨70RTC164⟩. Second, nitroso compounds frequently dimerise; this can be detected by the change in colour from the blue monomer to the colourless dimer. These dimers are *N,N′*-azodioxides (which can exist in *cis* and *trans* forms) and therefore should strictly be dealt with in Section 2.07.2.3.3; however, since many synthetic methods give mixtures of these possible products, and since the dimerisation may usually readily be reversed by heating ⟨69JCS(C)1073, 70RTC164⟩ some are considered here. These problems are exemplified in Equation (18) ⟨92JOC6759⟩. The preparation of nitroso compounds has been reviewed ⟨B-69MI 207-01, 71HOU(10/1)891, 90HOU(E16a/2)950⟩.

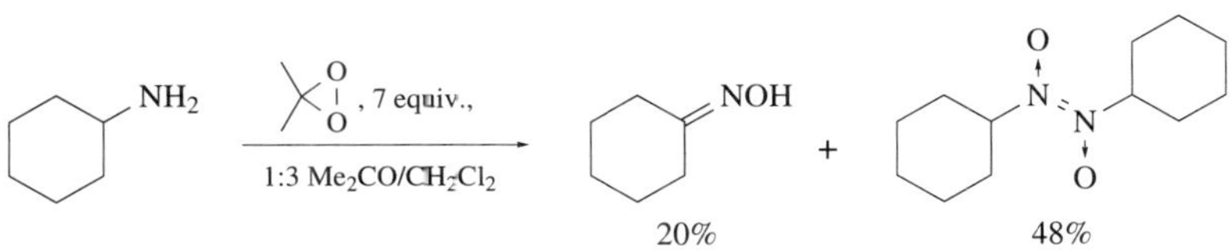

(18)

(ii) By the oxidation of primary amines

The first use of mcpba to make nitroso compounds was in 1966 ⟨66JOC524⟩. Later, eight amines were oxidised in yields from 30% to 85% with some formation of nitro derivatives being noted ⟨69JCS(C)1073⟩. This overoxidation can be avoided by keeping reactions at room temperature, leading to yields of 85–100% ⟨79JOC659⟩. 2-Methyl-2-nitrosopropane, like some other monomeric nitroso compounds, is prone to codistil with the solvent in which it is made. This difficulty can be overcome by the use of a convenient vapour phase method ⟨70JCS(C)2195⟩. Oxidation of chiral amines with mcpba occurs without racemisation under normal conditions ⟨76T1267⟩. Hydrogen peroxide with a tungstate catalyst has also been used as an oxidant ⟨71JOC3055⟩; detailed directions are available ⟨87OS(65)166⟩.

Two closely related procedures, introduced by Zajac *et al.*, may prove to be the methods of choice for this oxidation. An oxidant, percarbonate ⟨88S808⟩ or perborate ⟨89TL6495⟩, is used to generate peracetic acid from *N,N,N,N*-tetraacetylethylenediamine in a two-phase system; the results from the two methods are compared in Table 1. Oxidation with dimethyldioxirane has been used to prepare milligram amounts of a number of nitroso compounds ⟨92JOC6759⟩.

Table 1 Nitroso compounds from primary amines: comparison of percentage yields from two methods.

Amine	*Na percarbonate*	*Na perborate*
n-dodecyl	78	65 + oxime
benzyl	100 oxime only	100 oxime only
2-phenethyl	44 + azoxy + oxime	44 + azoxy + oxime
2-*endo*-norbornylmethyl	56 + azoxy	
cyclohexyl	100	75
2-*exo*-norbornyl	100	
2-*endo*-norbornyl	85	78
1-adamantyl	97	78
2-adamantyl	63 + oxime	80
3-noradamantyl	54 + azoxy	36

(iii) By the oxidation of amine derivatives

Azomethines may be oxidised to nitroso compounds by way of oxaziridines (Scheme 30) ⟨57JA6522⟩. Unfortunately, the process requires the use of 90% hydrogen peroxide which is reported to be no longer commercially available.

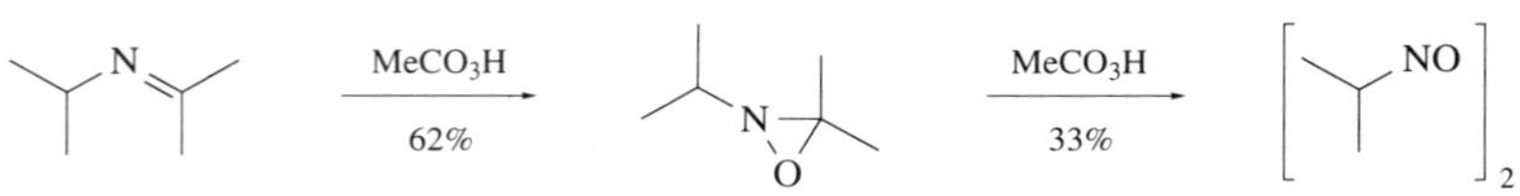

Scheme 30

A nitrone has also been oxidised with 80% peracetic acid ⟨58CB1057⟩ and nine bicyclic azo compounds have been converted into nitrosodimers in yields of 33–87% with hydrogen peroxide or mcpba (Scheme 31). Heating to 250 °C failed to dissociate these bicyclic dimers ⟨75JOC1395⟩.

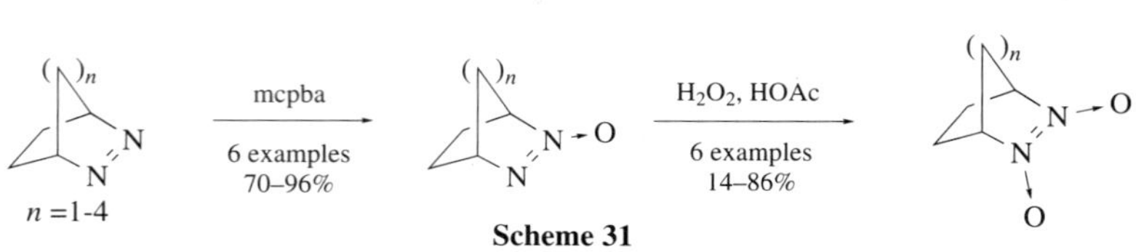

Scheme 31

(iv) By the oxidation of hydroxylamines

Oxidation of six hydroxylamines with silver carbonate on Celite ⟨71RTC373⟩ gave nitroso compounds in 57–95% yield and appears to be the method of choice. Other oxidants which have been used are: NaOBr ⟨88OSC(6)803⟩; benzeneseleninic anhydride ⟨80JCS(P1)1212⟩; periodic acid ⟨60JA4903⟩; phenyliodine trifluoroacetate ⟨75S445⟩; $Na_2Cr_2O_7$/HCl ⟨71CB917⟩ and controlled electrolysis in a buffer of pH 8.8 ⟨84CPB2609⟩. The last method (four examples; yields 26–73%) was used to make two 2-nitrosoalcohols.

(v) By addition reactions to double bonds

The addition of NOCl to alkenes has long been known; the use of NOCl as such, and of NOBr generated *in situ*, has been described ⟨64JA4074⟩. N_2O_4 will add to give nitroso-nitrates ⟨65JCC1488⟩ whilst N_2O_3 yields nitrosonitro compounds ⟨70JOC2785⟩. Thionitrites yield vicinal nitrosothioethers (Equation (19)) ⟨79CC392⟩. The elements of HNO can be added by an interesting new reaction which can be applied to alkenes (Equation (20)) ⟨92CL1137⟩ and α,β-unsaturated amides ⟨90CL1395⟩. Unsaturated esters, however, yield only oximes.

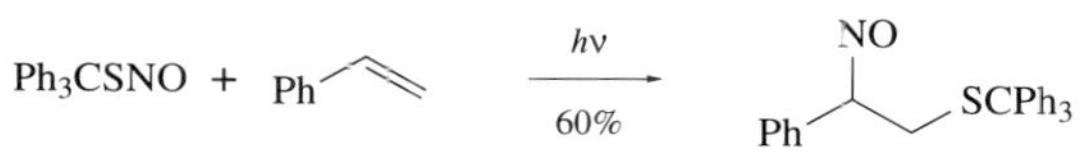

(19)

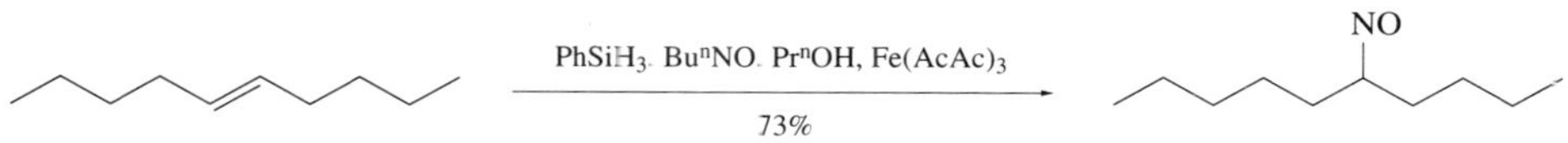

(20)

Nitrite esters react with enol ethers to afford α-nitrosoacetals ⟨79JOC275⟩ and, in the presence of $TiCl_4$, with the closely related silylated ketenehemiacetals to yield α-nitrosoesters (Equation (21)) ⟨88S805⟩.

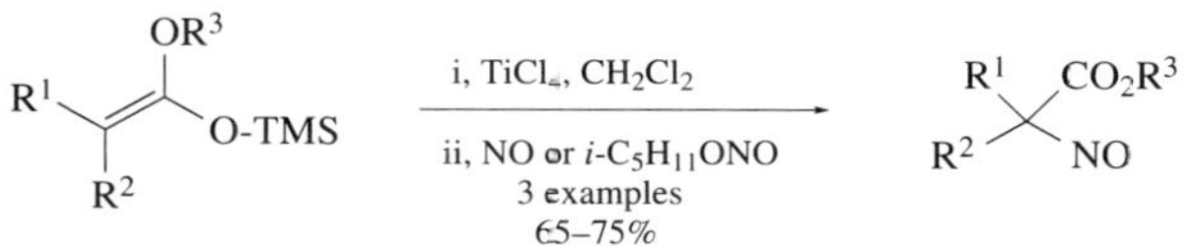

(21)

(vi) By the photolysis of nitrites

Hydrocarbons containing primary or secondary hydrogens can be converted to nitroso compounds in 34–77% yield by irradiation at ca. 400 nm in the presence of *t*-butyl nitrite ⟨70RTC151⟩. Tertiary hydrogens are attacked using a special Soxhlet-type reactor ⟨70RTC159⟩. It should also be noted that the Barton reaction ⟨60JA2640⟩ gives nitroso compounds in certain cases; refluxing with 2-propanol was necessary to cause isomerisation to the oxime in the example shown in Scheme 32.

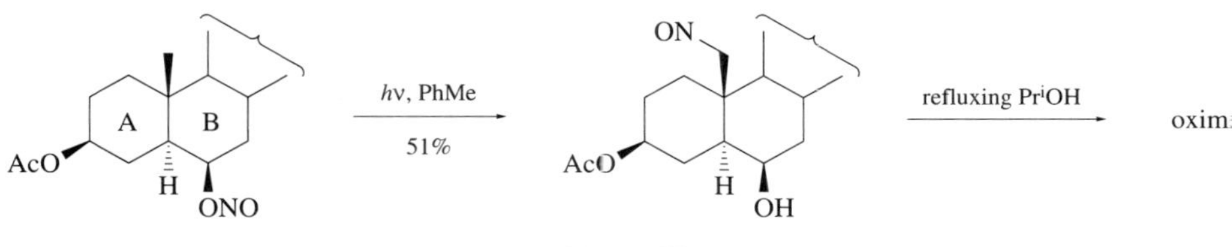

Scheme 32

2.07.2.1.2 *Nitro compounds*

(i) Introduction

The chemistry of aliphatic nitro compounds has greatly increased since the 1980s because of their use in the synthesis of many varied types of compound ⟨88CSR283, B-90MI 207-01⟩. The synthesis of nitro compounds in general has been reviewed ⟨B-65MI 207-01, B-69MI 207-02, 71HOU(10/1)12, 88S333⟩ as has that of α-nitroesters ⟨79S666⟩, α-nitronitriles ⟨89RCR470⟩ and nitroketones ⟨80S261, 90OPP707⟩.

(ii) By direct nitration

The nitration of alkanes is not generally preparatively important since the reaction, whether carried out in the vapour or liquid phase, is not selective and leads to complex mixtures of regioisomers and products from chain scission. If, however, there is an activating group in the molecule, selective nitration is possible ⟨62OR(12)101, B-69MI 207-02, 71HOU(10/1)12, 80S261, 90OPP707⟩; an example is given (Equation (22)) ⟨49JA3079⟩ in which the high yield is no doubt partly due to the lack of possibility of regioisomerism.

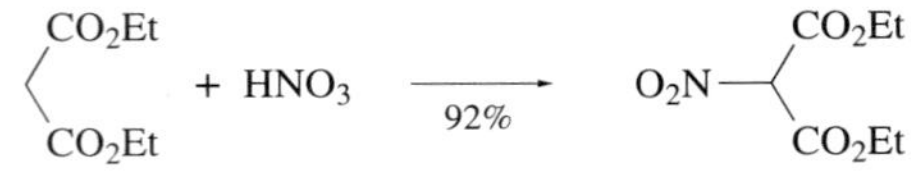
(22)

Silyl enol ethers and enol acetates can be nitrated under suitable conditions; an example of the latter is given in Equation (23) ⟨83S545, 72JPR891⟩.

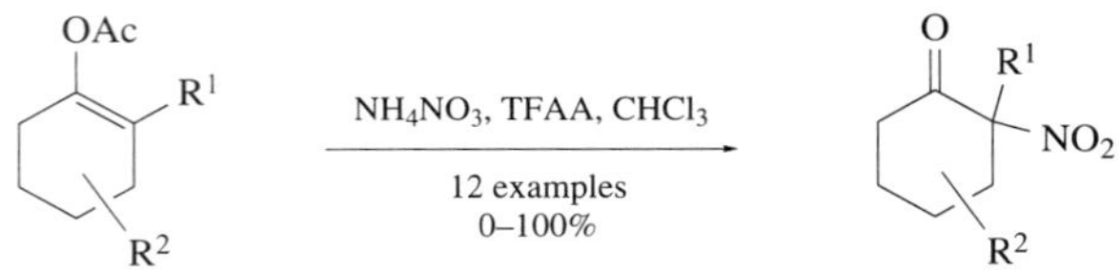
(23)

TFAA = trifluoroacetic anhydride

Some silanes can be nitrated ⟨87JOC701⟩ whilst side chain nitration of aromatics is sometimes possible, an example being shown in Equation (24) ⟨77JOC563⟩.

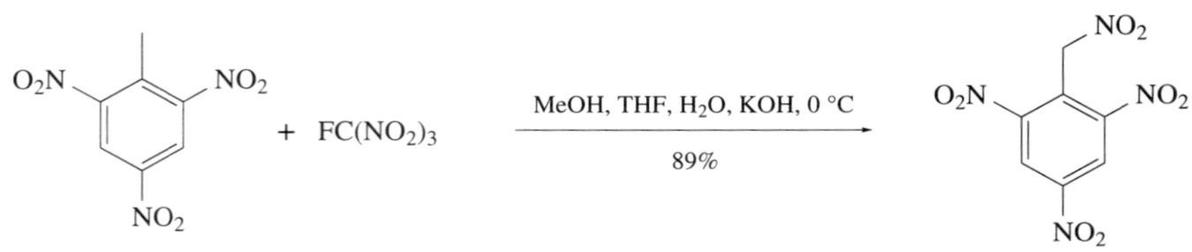
(24)

Alkaline nitration of the α-position of acids can be achieved when they have been protected as oxazolines (Equation (25)); reaction of the products with R^1OH/TFA gives nitroesters ⟨86JHC825⟩.

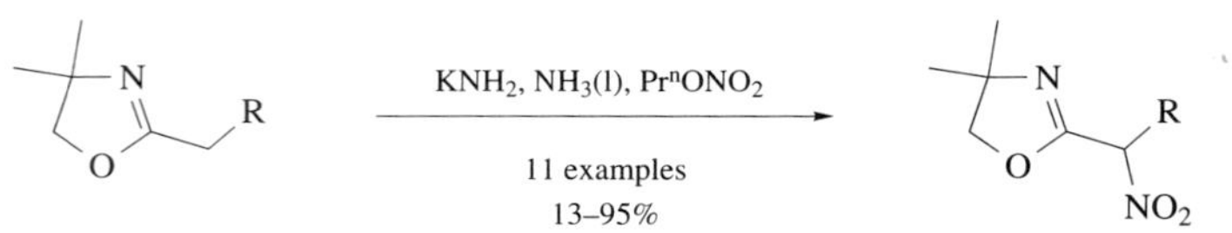
(25)

(iii) By addition to double bonds

The addition to double bonds of species of the type NO_2X yields nitro compounds. The addition of NO_2F has been achieved relatively recently (Equation (26)) ⟨79JOC3872⟩ whilst NO_2Cl has been used for many years ⟨63OR(13)150, B-69MI 207-02, 71HOU(10/1)12⟩; note that an excess of NOCl also leads to the elements of NO_2Cl being added in some cases ⟨64JOC1350⟩. A mixture of iodine and N_2O_4 ⟨88OSC(6)799⟩ or $AgNO_2$ ⟨85TL1193⟩ is equivalent to NO_2I.

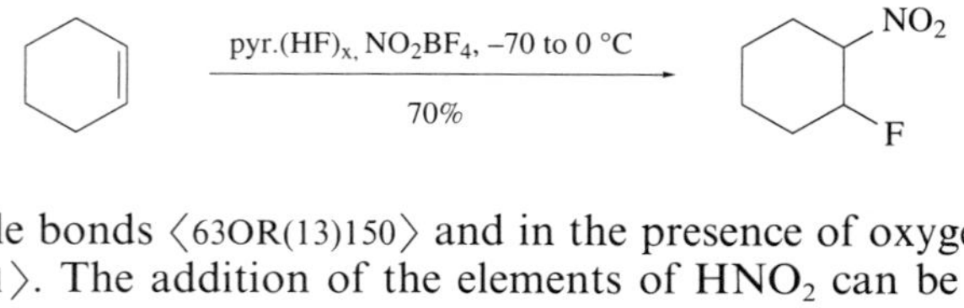
(26)

N_2O_4 will add to double bonds ⟨63OR(13)150⟩ and in the presence of oxygen gives α-nitroketones with cycloalkenes ⟨80S261⟩. The addition of the elements of HNO_2 can be achieved with $NaNO_3$ and polyphosphoric acid (ppa) ⟨90JOU583⟩. Nitroselenylation involves the (stereospecific) addition of the equivalent of $RSeNO_2$; this reaction has been achieved with both alkenes ⟨82TL4733⟩ and conjugated dienes ⟨90TL4199⟩. In the latter case an unusual nitroepoxide was prepared from the initial adduct (Scheme 33). Nitrite ion has been added to propenal to afford β-nitropropanal ⟨86S535⟩.

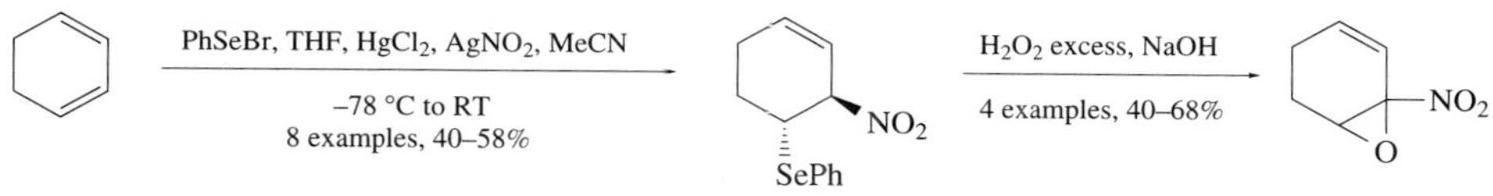

Scheme 33

(iv) By the action of nitrites on alkyl halides

Primary mono- ⟨63OSC(4)724⟩ and dihalides ⟨63OSC(4)368⟩ have been converted into nitro compounds with $AgNO_2$. $NaNO_2$ can be used either with or without the addition of phloroglucinol as a nitrite ester scavenger. An example of the latter, using DMF as solvent, is given ⟨86S849⟩ whilst both mono- ⟨63OSC(4)454⟩ and diesters ⟨78ZC214⟩ have been made by the former process. $NaNO_2$ in DMSO has been used without a scavenger to make a nitroester ⟨73JOC1418⟩ and a single example has been recorded where a tertiary benzylic halide was converted into a nitro compound using this mixture in a single electron transfer (SET) reaction ⟨67JA725⟩.

α-Nitroesters have been prepared using $NaNO_2$ in ethanol ⟨67M1944⟩; acetonitrile was used as solvent in a solid/liquid phase transfer method, using 18-crown-6, for making simple nitro compounds ⟨75TL71⟩. Both bromoesters and -alkanes yield nitro compounds with nitrite ion on Amberlite IRA 900 resin; S_N2 conversion was demonstrated in one case ⟨77S113⟩.

(v) By decarboxylation of nitro acids

Full details of the classic nitromethane synthesis of Kolbe have been given ⟨41OSC(1)401⟩ and a phenylnitromethane preparation is also available ⟨43OSC(2)512⟩. Treatment of an alkanoic acid (not α-branched) with an excess of LDA leads to a dianion which may be nitrated with an organic nitrate (Scheme 34) ⟨70TL699⟩; this procedure has been extended to phenylacetic acids ⟨88JOC2872⟩.

Scheme 34

(vi) By conjugate addition to nitroalkenes

Reviews of the conjugate addition to nitroalkenes are available ⟨B-73MI 207-01, 86CRV751, 87OPP283⟩. The addition of a large number of nucleophiles to nitroalkenes was investigated in a series of papers in 1947 ⟨47QR358⟩. Since that time many other examples have been recorded including instances of the use of the following nucleophiles: alkoxides ⟨88SC21, 93TL1259⟩, phenoxides ⟨88SC21⟩, enolates ⟨75CB1924⟩, thiols ⟨88SC21, 90JOC2437⟩, ammonia ⟨78JA6294⟩ as well as amines and hydrazines ⟨88SC21⟩. The addition of the elements of water to β-nitrostyrenes leads to α-nitroketones (Equation (27)) ⟨88S229⟩ whereas the action of alkoxides on β-phenylthionitroalkenes gives α-nitroacetals (Equation (28)) ⟨90S694⟩.

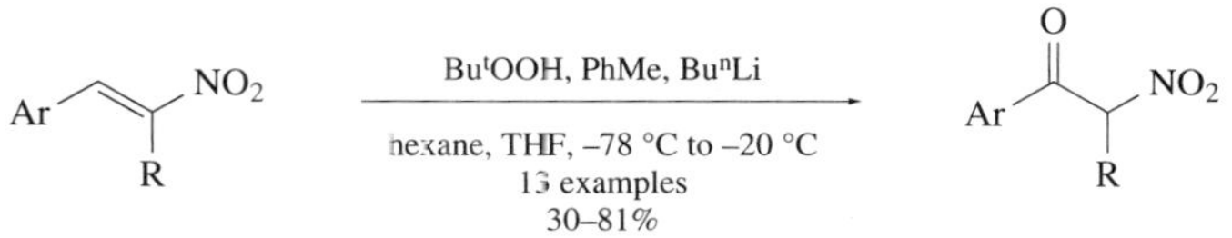

(27)

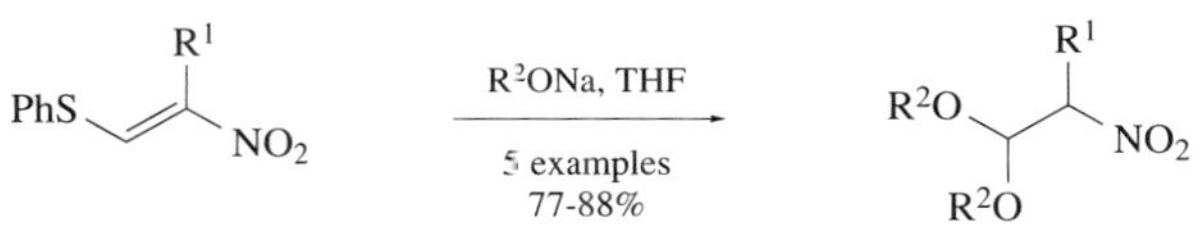

(28)

Conjugate addition of the di- and trianions of β-diketones has been achieved ⟨75CB1961⟩; alanes also add to give, in effect, a carbanion attack (Equation (29)) ⟨87T5411, 88JOC45, 90T7497⟩ Complex zinc cuprates also behave as carbanion equivalents and can contain a wide variety of functionality ⟨92JOC5431⟩. An example of addition to a nitroalkene with allylic shift is given in the next section.

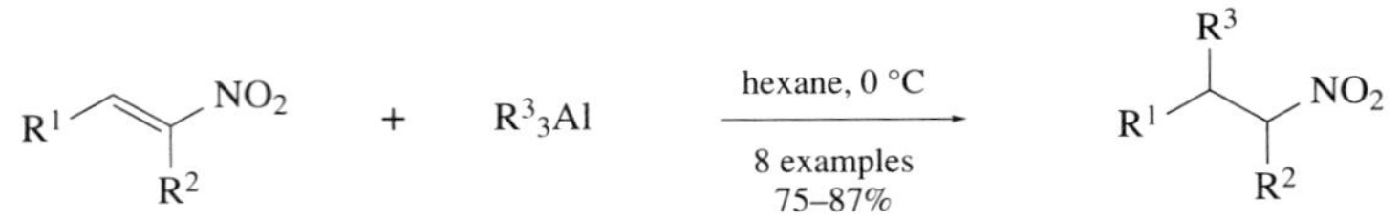

(29)

(vii) By carbon–carbon bond formation

Alkylation of α-nitroesters at the α-position has been described ⟨87SC1421, 88OSC(6)503⟩ as has the bimolecular homolytic reaction with allylic shift (S_H2' reaction) shown in Equation (30) ⟨85BCJ1069⟩.

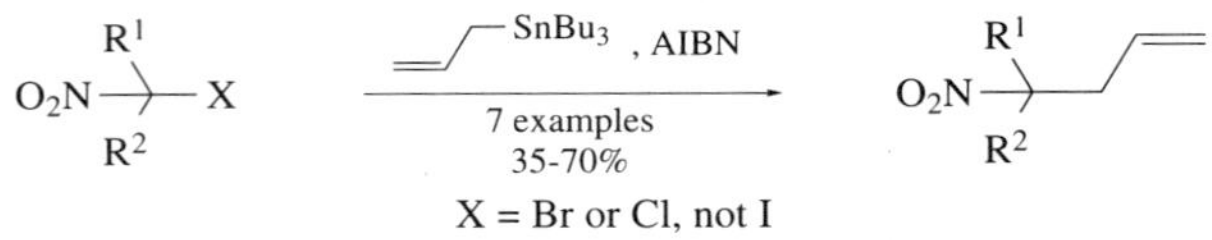

(30)

The Mannich reaction has been used many times to form nitroamines by aminoalkylation of nitroalkanes ⟨71HOU(10/1)12, B-73MI 207-01⟩. The alkylation of nitroalkenes with aldehydes or with Michael acceptors proceeds with allylic shift (Scheme 35) ⟨87S258⟩.

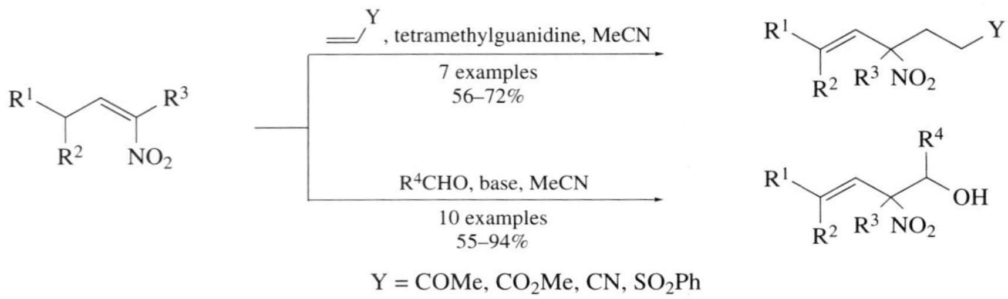

Scheme 35

The Henry (nitroaldol) reaction has been reviewed ⟨67OR(15)204, B-73MI 207-01, 91COS(2)321⟩. Examples of this reaction are the condensation of aldehydes with nitro compounds employing as catalysts KOH ⟨73OSC(5)833⟩, Amberlyst A-21 ⟨86S849⟩, $Bu_4NF/NEt_3/Bu^tMe_2SiCl$ (where some *threo*/*erythro* stereoselectivity was seen ⟨91TL3225⟩, Al_2O_3/no solvent ⟨83S1014⟩, KF/Al_2O_3 ⟨86TL493⟩ and KF/Pr^iOH ⟨91OS(70)68⟩. Ketones have been condensed with a large excess of nitromethane using $Me_2N(CH_2)_2NH_2$ to give the otherwise inaccessible allyl nitro compounds ⟨86JOC4368⟩.

The Henry reaction has been modified by preparing a lithium nitronate before the condensation (Scheme 36) ⟨82JOC365, 82JOM(225)43, 82TL2965, 88AJC1063, 89JOC1233⟩.

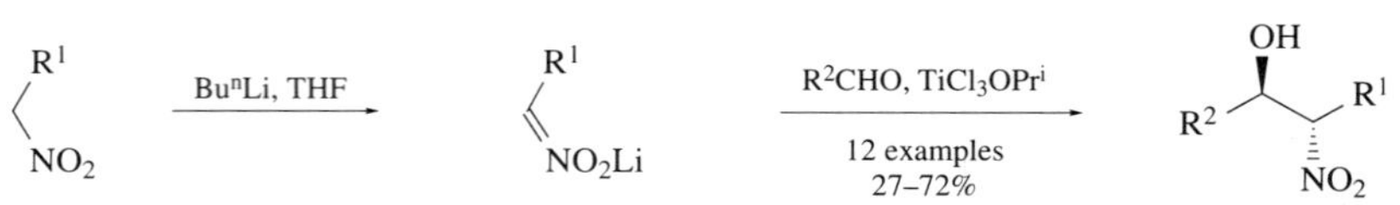

Scheme 36

The dilithium nitronate can also be used both with nitroalkanes (where the method can be adapted to give mainly *erythro* or *threo* products ⟨82HCA1101⟩) and with protected nitroalcohols ⟨85JA3601⟩; in the latter case a variety of electrophiles can be used and the reaction proceeds with a high degree of stereospecificity.

Silyl nitronates have been condensed with aldehydes in the presence of fluoride ion to give silylated β-nitroalcohols (Scheme 37) ⟨81HCA2264, 89SC3055⟩ and their reaction with the silyl ethers of ketones yields β-nitroketones ⟨91CL423⟩; with silylated carbinolamines they give β-nitroamines ⟨86BAU2442⟩.

A double Henry reaction has been described (Scheme 38) ⟨90OS(69)19⟩.

Examples of the reaction of nitroalkanes with α,β-unsaturated carbonyl compounds are collected in Table 2 ⟨71HOU(10/1)12, B-73MI 207-01⟩. Suitably activated alkynes have been made to yield tertiary allylic nitro compounds (Scheme 39) ⟨90JOC511⟩.

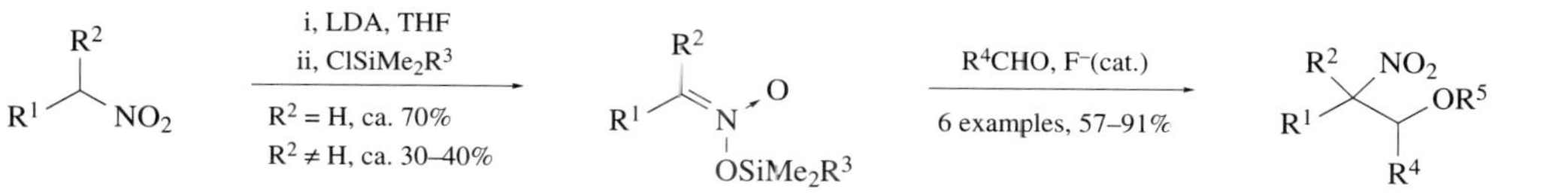

$R^5 = SiR^3Me_2$ or H

Scheme 37

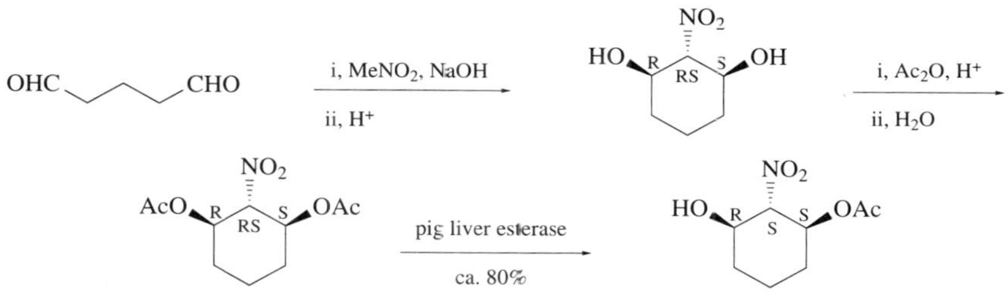

Scheme 38

Table 2 Some examples of nitroalkanes made by the Michael reaction.

Acceptor	*Donor*	*Reagent(s)*	*Ref.*
$R^1CH{=}C(R^2)COR^3$	$R^4R^5CHNO_2$	Al_2O_3, RT	86S237
$CH_2{=}CHCOMe$	$EtNO_2$	Al_2O_3, no solvent	92T327
$R^1CH{=}CR^2COR^3$	$R^4_2CHNO_2$	KF, basic Al_2O_3, THF	87JOC1601
$CH_2{=}CHCOMe$	Bu^nNO_2	Pr^i_2NH	73JOC1418
$CH_2{=}CHCOMe$	Pr^nNO_2	Pr^i_2NH	88OSC(6)648
$CH_2{=}CHCHO$	RCH_2NO_2	Al_2O_3	86S1024
$CH_2{=}CHCO_2Me$	RCH_2NO_2	Amberlyst A21	87S711
$CH_2{=}CHCO_2Me$	Me_2CHNO_2	Triton B	63OSC(4)652
$R^1R^2C{=}CHCO_2Me$	$MeNO_2$	dbu, MeCN, 60 °C	87OPP471
$R^1CH{=}CHX^a$	$R^2R^3CHNO_2$	dbu, MeCN	84S226
$R^1R^2C{=}CR^3CO_2R^4$	$MeNO_2$	Tetramethylguanidine	72S44
$R^1CH{=}CR^2CO_2R^3$	$R^4R^5CHNO_2$	K_2CO_3, Aliquat366, sonication	91SL907
$H_2C{=}CHCO_2Et$	$R^1R^2CHNO_2$	Bu_3P	73TL3597
$R^1R^2C{=}CHC(SO_2Ph){=}CH_2$	$R^3CH_2NO_2$	dbu	92TL131

[a] X = CO_2Me, COPh, CN, SO_2Ph, SOPh.

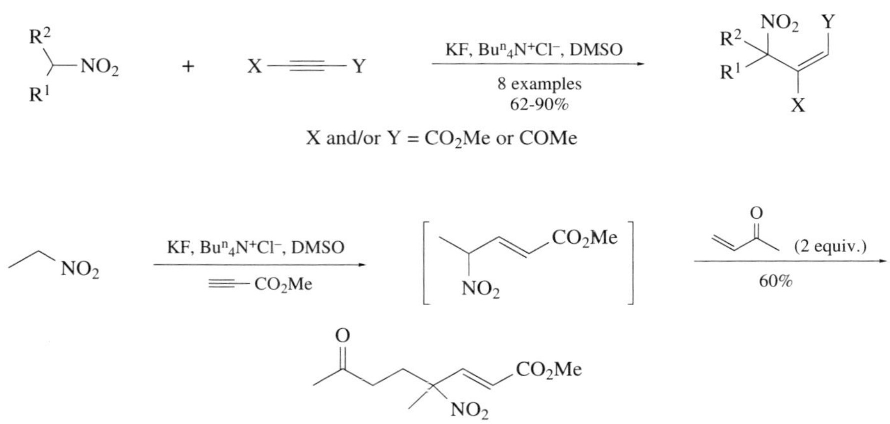

X and/or Y = CO_2Me or COMe

Scheme 39

Enamines have been condensed with nitroacetic acid whereby the CH_2NO_2 group is introduced at the α-position ⟨69JCS(C)986, 71JCS(C)1812⟩. Allyl ethers have been reacted with nitroalkanes in the presence of a palladium catalyst (Equation (31)) ⟨82T1321⟩.

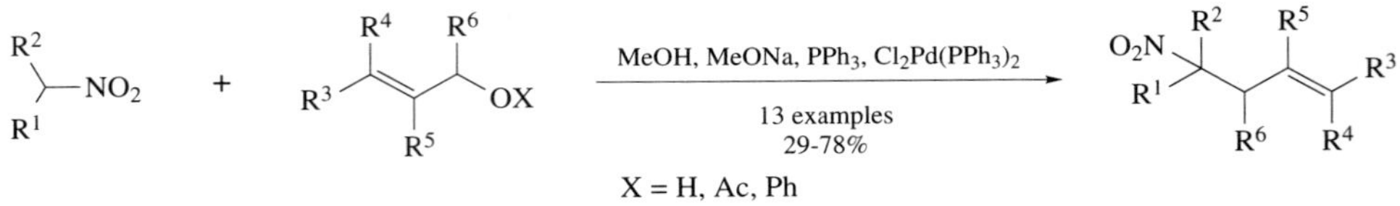

(31)

Addition with ring opening occurs with enol γ-lactones yielding γ-diketones (Equation (32)) ⟨91S629⟩.

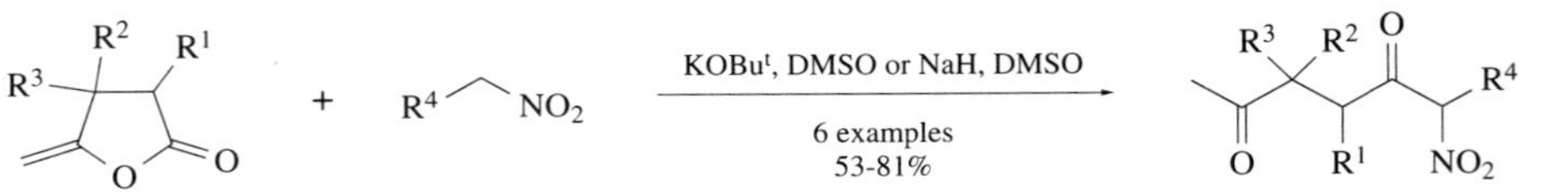

(32)

Reaction of a lithium nitronate with an acylimidazole furnishes an α-nitroketone; α-nitroesters can also be made this way ⟨82JOC4040⟩. Coupling of nitroacids to give α,ω-di-nitroalkanes has been achieved by Kolbe electrolysis ⟨73OSC(5)445⟩.

(viii) By carbon–carbon bond scission

Cyclic nitroketones can undergo ring cleavage with KF to give ω-nitroacids or esters ⟨86SC1781⟩ and the latter can also be obtained with Amberlyst-A21 and methanol ⟨92S355⟩; the action of $NaBH_4$ in aqueous MeCN produces ω-nitroalcohols ⟨90T7531⟩.

(ix) By cycloaddition reactions

The cycloadditions of nitroalkenes have been reviewed ⟨B-73MI 207-01, 86CRV751⟩. They undergo the Diels–Alder reaction both inter- and intra-molecularly (Equations (33) and (34)) ⟨78JA6294, 86JOC2139, 92JOC5431⟩.

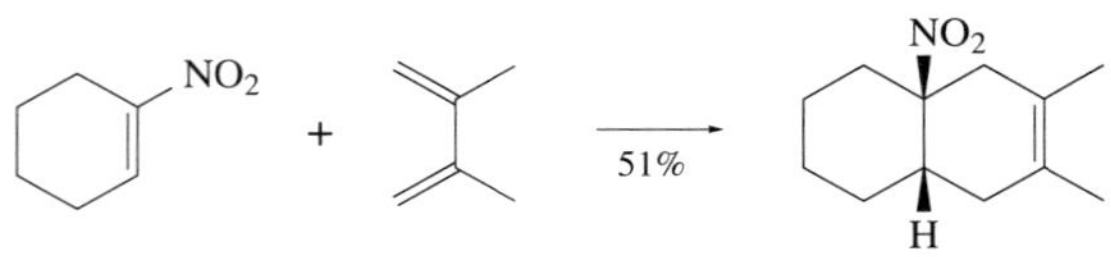

(33)

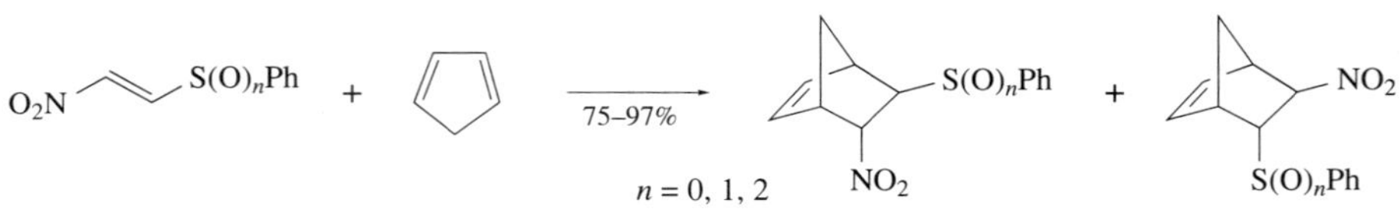

(34)

The nitroalkene can be generated and used *in situ* (Scheme 40) ⟨61JOC1357⟩.

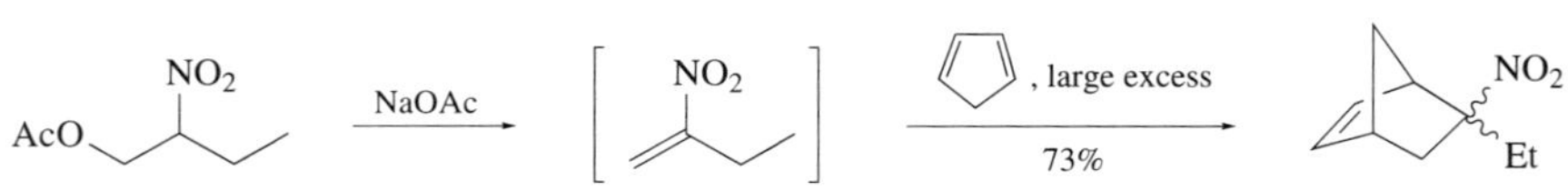

Scheme 40

Enamines yield γ-nitroketones by a cycloaddition followed by ring opening (Scheme 41) ⟨89T2099⟩ and tandem A + B + C couplings have been achieved (Equation (35)) ⟨90T7509⟩.

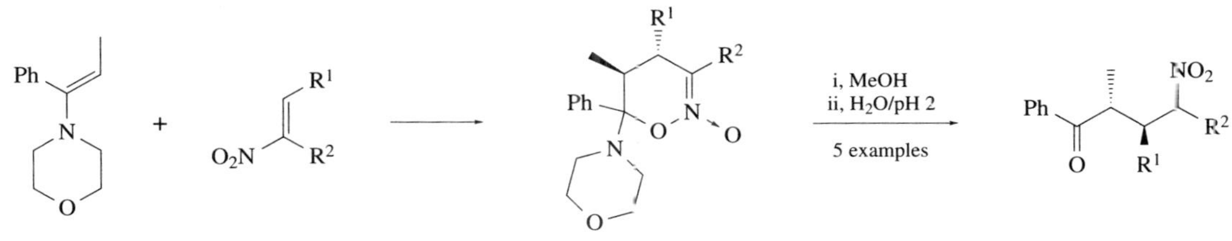

Scheme 41

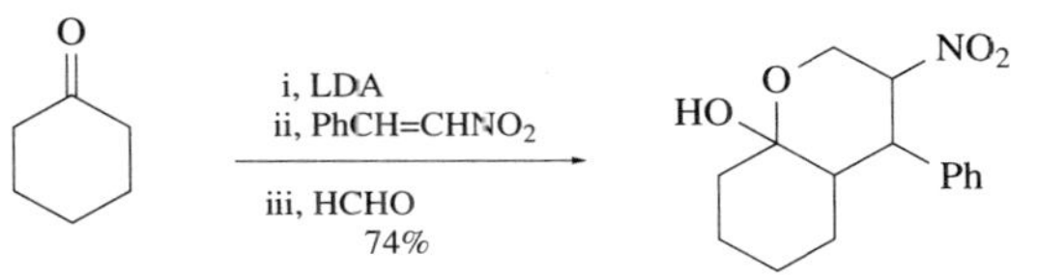

(35)

(x) By oxidation of amines and their derivatives

The oxidation of amines and their derivatives is a good method for making tertiary nitroalkanes ⟨56JOC4003, 71CB917, 73OSC(5)845, 88OSC(6)803⟩. *M*-Chloroperbenzoic acid is useful in refluxing 1,2-dichloroethane ⟨79JOC659⟩ and HOF (from F_2 in aqueous MeCN) gives 70–95% yields ⟨92JOC7342⟩. Dimethyldioxirane has been used with both amines ⟨89JOC5783⟩ and isocyanates ⟨88JOC5353⟩. Ozone alone gives poor yields ⟨68JOC313⟩ but these are much improved in the presence of silica gel ⟨77JOC844⟩. An unusual method to make nitroalkanes involves ozonolysis of phosphazenes which are derived from azides and phosphines by the Staudinger reaction (Scheme 42) ⟨84JA3682⟩.

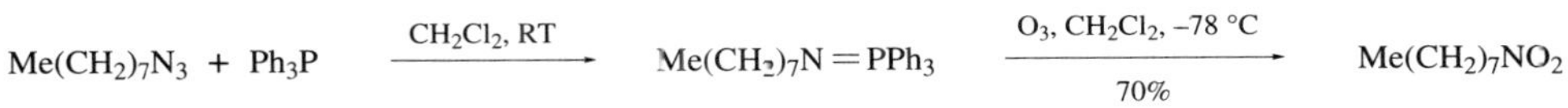

Scheme 42

(xi) By oxidation of oximes and related compounds

Most oxidations of oximes have used H_2O_2, either with alkali ⟨68BAP567⟩ or with TFA ⟨55JA4557, 71TL2053, 75BCJ3413⟩. The latter process requires 90% H_2O_2 and a recent improvement generates anhydrous solutions of CF_3CO_3H from trifluoroacetic anhydride (TFAA) and urea peroxide; this modification works with aldoximes only ⟨92TL4835⟩. Ketoximes can be converted into nitroalkanes in 77–95% yield by the process shown in Scheme 43 ⟨80TL1117⟩. The most convenient method of oxidation seems to be that using $NaBO_3 \cdot 4H_2O$ ⟨92SL337⟩. Nitrones give nitro compounds with ozone at −78 °C ⟨83HCA789⟩.

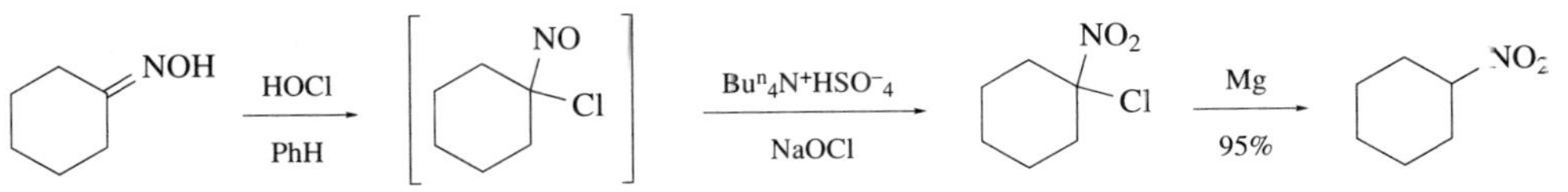

Scheme 43

(xii) By the pyrolysis of nitrates

Both alkyl ⟨69JOC4121⟩ and acyl nitrates ⟨70JOC4229⟩ can be converted into nitroalkanes by pyrolysis. The latter procedure is carried out in the vapour phase and several methods are given for making the acyl nitrates.

(xiii) By the reduction of nitroalkenes

Nitroalkenes are usually readily prepared by the Henry reaction and their reduction thus provides a good route for making nitroalkanes; the process has been reviewed ⟨86CRV751, 87OPP283⟩. $NaBH_4$ has long been used for the reduction but in some cases, such as β-nitrostyrenes, dimerisation is a problem. Several solutions to this have been proposed and some of the methods used are shown in Table 3.

Table 3 Reduction of nitroalkenes.

Reagents	*Yield* %	*Ref.*
H_2/Pd-C	87	65CJC840
H_2/Pd-C	64	65CJC3074
$NaBH_4$/NaOH/pH 3–6	48–85	67JOC4134
$NaBH_4$/MeOH	56–76	84S1054
$NaBH_4$/THF/MeOH	62–82	85SC151
$NaBH_4$/DMSO	75–80	72JOC2810
$NaBH_4$/SiO_2 gel/$CHCl_3$/Pr^iOH	90–98	83TL227
$NaBH_4$/dioxan/EtOH	80–95	85S886
IRA 400—BH_4^-/MeOH	78–83	89SC805
$Zn(BH_4)_2$/DME	86–93	92T5317
$NaBH_3CN$/EtOH/HCl	67	76JOC3328
Hantzsch ester/HOAc	70–97	88BCJ3020
Hantzsch ester/SiO_2 gel	45–100	88BCJ4029
Resin-bound dihydronicotinamide	60–100	87BCJ4492
$LiBHR_3$ (R = Bu^s or Et)	65–81	84SC1093
$LiBHBu^i_3$; $NaBH_4$/MeOH/THF; $NaBH_4$/resin[a]	62–83	90T7443
Bu^n_3SnH/CH_2Cl_2	80–95	90JOC2070
o-Phenylenediamine/PhCHO/Bu^nOH	70–93	85SC527
Baker's yeast[b]	48–57	87CL191
Rhodococcus rhodocrous[c]	~100	85ABC2331

[a] Comparison of the three methods in five cases. [b] Gives an *ee* of 89–98%.
[c] Gives an *ee* of 30–40%.

(xiv) By miscellaneous methods

Compounds of the type $R^1R^2CXNO_2$ have been reductively converted to $R^1R^2CHNO_2$ for X = Br ⟨86S826⟩ and X = SO_2R ⟨88JOC3897, 92JCS(P1)601⟩ whilst the nitrite group can be removed from 1,2-nitronitrates with $NaNO_2$ ⟨71JOC2574⟩. On a few occasions tertiary nitroalkanes have been made by the oxidation of the corresponding nitrosoalkanes ⟨87CC1405⟩. Optically active nitroalcohols can be prepared by the reduction of the corresponding ketones with baker's yeast ⟨88TL4769, 90T7471⟩. Some nitrohydrazines have been made by reduction of the corresponding hydrazones with $NaBH_4$ ⟨88JOC1251⟩. PhSCl reacts with the sodium salt of nitromethane to give $PhSCH_2NO_2$ in 50% yield ⟨89OS(68)8⟩ whilst $NaNO_2$ combines with mucobromic acid ($HO_2CCBr{=}CBrCHO$) to afford the sodium salt of nitromalondialdehyde ⟨63OSC(4)844⟩. The useful synthetic reagent methyl nitroacetate can be made from nitromethane in 66–70% yield ⟨88OSC(6)797⟩.

2.07.2.2 Compounds Containing a N=S Bond

2.07.2.2.1 Thionitroso compounds and related substances

In spite of intensive efforts, no compounds of the types RN=S, R(N=S)→O, or R(N=S)→S appear to have been prepared, although in the first case adducts were made with 2,3-dimethylbutadiene ⟨78ZC323, 85PS(23)277⟩. Modified neglect of diatomic overlap (MNDO) calculations indicate that the compounds will be highly reactive towards nucleophiles and that substances of the type R(N=S)→S should have comparable stability to those having the RN=S=S structure ⟨88CCC2116⟩.

2.07.2.2.2 Substances having two atoms coordinated to sulfur

(i) Sulfinylamines (N-thionylamines) RN=S=O

The preparation of sulfinylamines has been reviewed ⟨58HOU(11/2)738, 79COC(3)241, 82ZC237, 85HOU(E11)558⟩. They can be made by the action of $SOCl_2$ on an amine, either alone ⟨59CB1910, 66BCJ1778, 80CB1950⟩ or in the presence of pyridine which substantially improves the yield in some cases ⟨59CB1910⟩. An interesting variant of this reaction uses $RNCl_2$ instead of the amine; Cl_2 is released during the process ⟨72ZOR2057⟩. The action of $SOCl_2$ ⟨64JCS3122⟩ or $Me_2N(S{=}O)Cl$ ⟨74JINC993⟩ on *N*-substituted hexamethyldisilazanes also leads to sulfinylamines. *N*-Lithiated amines combine with SO_2 to give RNSO derivatives in poor to fair yields ⟨75JOC3291⟩ but the results can be improved by using a silylated lithium compound ⟨81S368⟩. Sulfinylamines can also be made by the reaction between an *N*-sulfinylsulfonamide and a carbodiimide ⟨73BCJ2156⟩, by ozonolysis of a sulfurdiimine ⟨70CB2152⟩ and by 'transthionylation' (Scheme 44) ⟨66BCJ1778⟩.

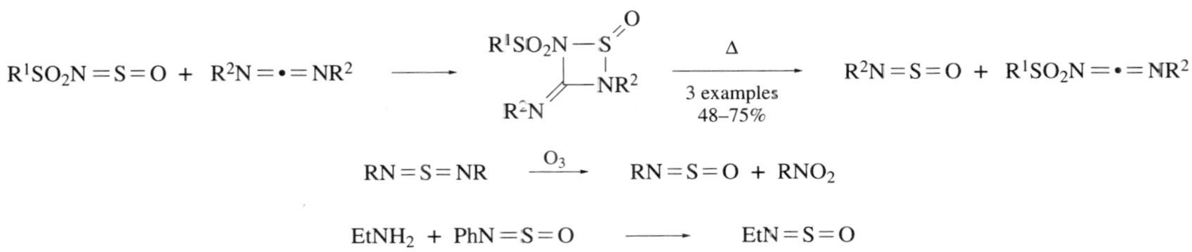

Scheme 44

(ii) Thiosulfinylamines RN=S=S

Thiosulfinylamines have been reviewed ⟨85HOU(E11)579⟩. Aliphatic representatives of this type seem to be very labile; they are made by the action of $RN{=}SCl_2$ on S-TMS_2 ⟨77ZOR727⟩.

(iii) Sulfurdiimines RN=S=NR

Reviews dealing with the preparation of sulfurdiimines are available ⟨79COC(3)233, 85HOU(E11)584⟩. These compounds were first prepared by the action of SCl_4 on primary amines ⟨56AG678⟩ and subsequently SF_4 has been employed ⟨66JCS(A)1780, 74JOM(74)289⟩. Sulfur dichloride can also be used either on amines ⟨65TL1487, 77ZAAC(432)173⟩ or on *N*-substituted hexamethyldisilazanes (Equation (36)) ⟨76ZC437⟩.

$$RN(TMS)_2 + SCl_2 \xrightarrow{} RN{=}S{=}NR \qquad (36)$$

R = Bu^n, 50%
R = Pr^n, 53%

Sulfuriminodifluorides, $RN{=}SF_2$, will react with salts of the type $[Ph_3PNHR]^+ Cl^-$ to give sulfurdiimines ⟨76CB2442⟩, whilst sulfuriminodichlorides combine with amines in a process which gives access to diimines with different alkyl groups on the two nitrogen atoms ⟨67ZOR1282⟩. Primary amines also yield sulfurdiimines with $Cl_2/(Me_2N)_2S$ ⟨70CB2152⟩ or Br_2/H_2S ⟨71CB2648⟩. Cyclic sulfurdiimines result from the reaction of diamines with TsN=S=NTs by an exchange reaction (Equation (37)) ⟨69TL4117⟩.

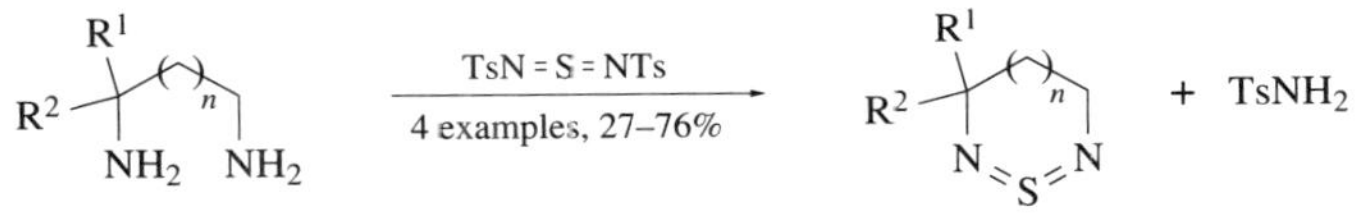

(37)

Sulfinylamines can be converted into *N*-alkyl-*N'*-trimethylsilylsulfurdiimines by reaction with the lithium derivative of hexamethyldisilazane ⟨75CB2329⟩; an alternative route to substances of this type involves the combination of $(TMS)_2NLi$ with $Bu^tN{=}SCl_2$ to give $Bu^tN{=}S{=}N$-TMS in 82% yield ⟨75AG(E)485⟩. The salt $Bu^tN{=}S{=}N^-\ K^+$ has been made in 100% yield by the action of KNH_2/NH_3(l) on *N,N′*-di-*t*-butylsulfurdiimine ⟨81ZAAC(473)80⟩.

(iv) Thione-S-*imines* $RN{=}S{=}CR_2$

Thione-*S*-imines have been reviewed ⟨85HOU(E11)943⟩ and are made by the methods shown in Equation (38) ⟨76CB1069⟩ and Scheme 45 ⟨83JOC4582⟩.

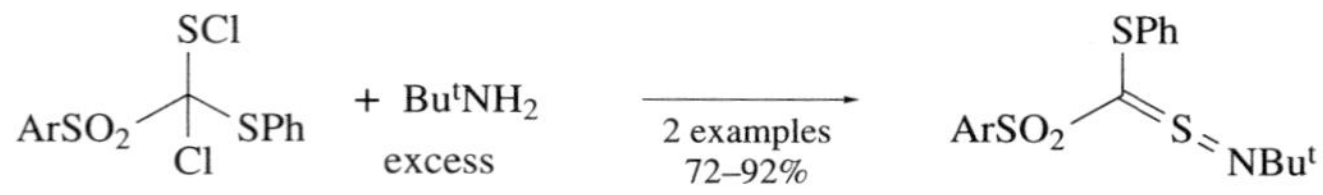

(38)

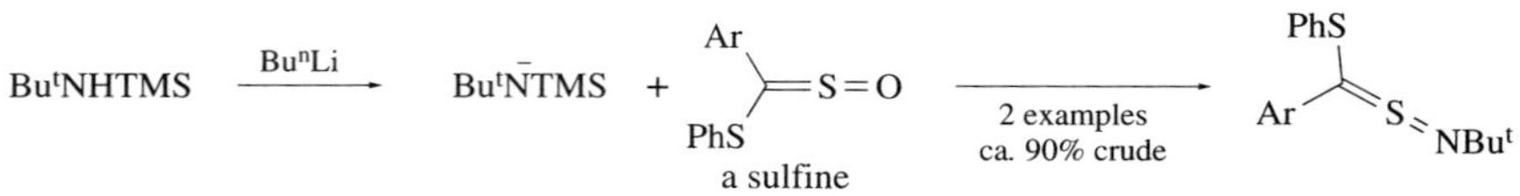

Scheme 45

2.07.2.2.3 *Substances having three atoms coordinated to sulfur*

A host of compounds of the type $RN{=}SXY$ (where X, Y = Hal, OR, NR_2) are known ⟨85HOU(E11)532, 85HOU(E11)510⟩. However, many of them seem to need the stabilising influence of an electron withdrawing group, for example, R = perfluoroalkyl, aryl, acyl or sulfonyl, to exist.

(i) Sulfuriminodifluorides, $RN{=}SF_2$, *and related compounds*

A review of sulfuriminodifluorides can be found ⟨85HOU(E11)532⟩. They have been prepared by the action of SF_4 on amines ⟨66JCS(A)1780⟩ (the use of SF_5Cl is also described) or on phosphonium salts ⟨71ZN(B)73, 76CB2442⟩. The unstable compound $MeN{=}SF(NEt_2)$ has been made by the reaction of Et_2NSF_3 with $MeN(TMS)_2$ at −40 °C ⟨71CB1247⟩.

(ii) Sulfuriminodichlorides, $RN{=}SCl_2$, *and related compounds*

The action of SCl_2 on primary amines ⟨67ZOR1282, 77ZAAC(432)173⟩ or $RNCl_2$ ⟨67ZOR1282⟩ yields sulfuriminodichlorides which are also available by the reaction of PCl_5 with sulfinylamines ⟨72ZOR2057⟩. The reaction of a sulfurdiimine with an acid chloride followed by an amine has furnished a substance of the type $RN{=}S(NHR)NR(COR)$ (Scheme 46) ⟨69M959⟩ whilst disulfides were shown to react with *N,N*-dichloroamines to give $RN{=}SCl(R)$ ⟨78ZOR1846⟩.

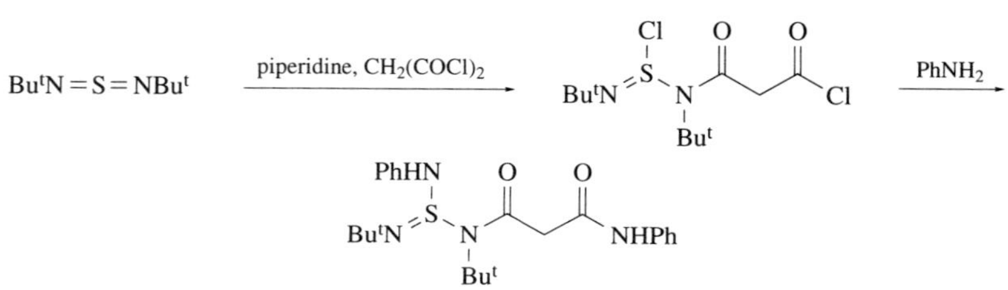

Scheme 46

(iii) Sulfonylamines RN═SO$_2$

The preparation of sulfonylamines has been reviewed ⟨79COC(3)241, 85HOU(E11)1031⟩; they are rather unstable and are used without isolation. They are usually made by the dehydrochlorination of sulfamyl chlorides $RNHSO_2Cl$ with a base ⟨72JA6135, 79BCJ1102, 79JOC305, 81CB774⟩; in one case a sulfonylurea was used instead of a sulfamyl chloride ⟨77JCR(S)237⟩.

(iv) Sulfurtriimines (RN═)$_3$S

Sulfurtriimines have been reviewed ⟨85HOU(E11)1031⟩. Aliphatic derivatives of this type have been made in poor yield by the reaction between NSF_3 and a lithiated silylamine (Scheme 47) ⟨79ZAAC(452)103⟩.

Scheme 47

(v) Sulfilimines RN═SR$_2$

Sulfilimines have been reviewed ⟨77CRV409, 79COC(3)215, 85HOU(E11)887, 90RCR819⟩. Primary amines react with $Ph_2S(OC(CF_3)_2Ph)_2$ to give sulfilimines ⟨75JA583, 75JA6137⟩. The *N*-unsubstituted sulfilimine $Ph_2S{=}NH$ has been treated with alkyl halides ⟨76T431⟩ and with Michael acceptors ⟨76S30⟩ to give a number of alkylated products.

2.07.2.2.4 Substances having four atoms coordinated to sulfur

(i) Sulfurylfluoridealkylimides RN═S(═O)F$_2$

Aliphatic representatives of this class have been made by the action of SOF_4 on primary amines ⟨61JOC4010⟩.

(ii) Sulfonimidoyl halides RN═S(═O)(R)Hal

Sulfonimidoyl halides have been reviewed ⟨85HOU(E11)1123⟩. The reaction of MeSOCl with Bu^nNCl_2 gives $Bu^nN{=}S({=}O)(Me)Cl$ in 91% yield but the product is unstable. Compounds having an aryl group attached to the sulfur atom are more stable; they are made by treating a sulfinamide, ArS(═O)NHR, with chlorine or *N*-chlorobenzotriazole ⟨79JOC2055⟩ (Bu^tOCl was the reagent of choice in another case but the halide was not isolated ⟨79JOC2278⟩). Sulfonimidoyl fluorides can be prepared by treating the corresponding chloride with fluoride ion ⟨83JOC1⟩.

(iii) Alkyl or aryl N-*alkylbenzenesulfonimidates RN═S(═O)(OR)R*

Alkyl or aryl *N*-alkylbenzenesulfonimidates are made by reacting the corresponding sulfonimidoyl halides with alkyl- or aryloxides ⟨79JOC2055, 79JOC2278⟩.

(iv) Sulfonimidamides RN═S(═O)(NHR)R

Benzenesulfinylcyclohexylamine was treated with a secondary amine and bromine or nbs to yield a sulfonimidamide (Equation (39)) ⟨65BCJ1989⟩.

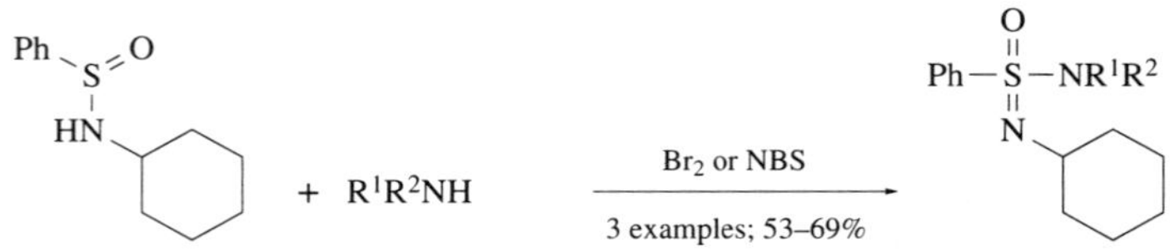

(39)

They can also be made from sulfonamidebisimides, $(RN{=})_2S(NHR)R$, either by acid hydrolysis ⟨68JA2651⟩ or by the action of bromine and K_2CO_3 ⟨71CB3875⟩. Reaction of sulfonimidoyl halides with amines has been used to make a variety of sulfonimidamides ⟨79JOC2055, 79JOC2278⟩. The NHR group has been alkylated in one case ⟨68JA2651⟩.

(v) Sulfoximines $RN{=}S({=}O)R_2$

Sulfoximines have been reviewed ⟨55HOU(9)260, 79COC(3)223, 85HOU(E11)1299⟩. The action of a primary amine on an aryloxysulfonium salt, $R_2S^+({=}O)OAr$, gives sulfoximines in variable yield ⟨78JCS(P1)1580, 78TL3435⟩. The latter may also be made from phenyl *N*-alkylbenzenesulfonimidates (Equation (40)) ⟨79JOC2061⟩ or sulfonimidoyl fluorides ⟨83JOC1⟩ with an organolithium.

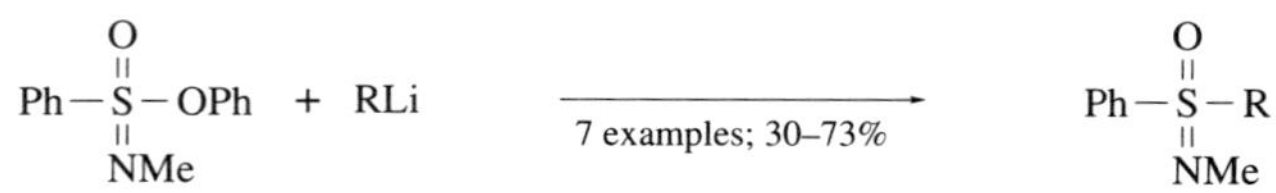

(40)

(vi) Sulfonamidebisimides $(RN{=})_2S(NHR)R$

Sulfonamidebisimides are prepared by treating R^1SH with R^2NH_2 and R^2NHCl ⟨68JA2651⟩ or R^2NHBr ⟨71CB3875⟩ to give $(R^2N{=})_2S(NHR^2)R^1$. The first reference also describes the alkylation of the alkylamino group in the compound and the second details the preparation of a quaternary ammonium salt.

(vii) Sulfodiimides $(RN{=})_2SR_2$

Sulfodiimides have been reviewed ⟨79COC(3)237, 85HOU(E11)1321⟩. An aliphatic representative of the class was made by the reaction of RNHBr with the sulfilimine $Me_2S{=}NH$; the products were isolated as the hemihydrates of their sodium salts ⟨71CB2023⟩.

2.07.2.2.5 Substances having five atoms coordinated to sulfur

Treatment of $MeN{=}SF_3{}^+AsF_6{}^-$ with NaF at 160 °C led to the formation of $MeN{=}SF_4$ in 67% yield ⟨78AG(E)530⟩.

2.07.2.3 Azo and Azoxy Compounds and Their Derivatives

2.07.2.3.1 Azo compounds

Aliphatic azo compounds have been reviewed ⟨67HOU(10/2)757, B-75MI 207-01, 79COC(2)219, B-83MI 207-01⟩ and cyclic derivatives have also received attention ⟨B-75MI 207-02⟩.

The simplest azo compounds are the diazenes, $RN{=}NH$, which have been reviewed ⟨71ACR193⟩. $MeN{=}NH$ was made by the reaction of hydroxylamine-*O*-sulfonic acid with MeNHOH ⟨68JA7173⟩ and a few other examples were prepared from $TsNRNH_2$ with NaOEt ⟨71JA1992⟩. The last reference records excellent yields of these highly unstable compounds from the action of imidazole hydrochloride on the salts $RN{=}NCO_2{}^-M^+$.

Substances of the type $RN{=}NO^-M^+$ were made many years ago by Thiele; reaction of ethyl nitrite and NaOMe with $MeN(NO)NH_2$ gave the methyl derivative whilst the benzyl compound resulted from the action of Al/NaOH on $PhCH_2NHNO_2$ ⟨10LA(376)239⟩.

The most popular method for making azo compounds is the oxidation of *N,N′*-disubstituted hydrazines but the procedure often needs to be tailored to the specific compound. Mercury(II) oxide is the oxidant most widely used ⟨52JCS4191, 54CJC545, 58JA6562, 62JA2922, 65JOC3278, 81JOC442⟩. In some cases the red form is specified ⟨70CJC2745, 72JA7406, 70JOC3401⟩ whilst in others the yellow form was used ⟨69JOC2263⟩; with one set of compounds only freshly precipitated HgO was satisfactory ⟨71S204⟩. Oxygen has also been utilised for the oxidation, either alone ⟨69CJC2371, 81JOC442⟩ or with a Pd/C ⟨73TL2859⟩ or Cu_2Cl_2 catalyst ⟨74TL3847⟩. Other oxidants used for this purpose include H_2O_2 ⟨10LA(376)239, 55JA2457, 58JA6562⟩, copper oxide ⟨22JA2556⟩, $AgNO_3$ ⟨58JA6562⟩, Br_2 ⟨63OSC(4)66⟩, $K_3Fe(CN)_6$ ⟨77S339, 78JOC2921⟩ and CCl_4/Na_2CO_3 under phase transfer conditions ⟨88BSB461⟩.

Azo compounds can also be made by oxidation of NH_2 groups attached to tertiary carbon atoms; IF_5 ⟨61JOC2531, 66JA137, 68CJC1149⟩ and H_2O_2 with a tungstate catalyst have been used ⟨75JOC1409⟩. An azo compound was prepared in one case by the action of H_2O_2 on an isocyanate but the method seems of limited applicability ⟨56CB685⟩; electrolytic oxidation of *N*-lithiated amines has been used but the yields were poor and the procedure appears to be demanding ⟨78AG(E)202⟩.

An elegant method for making azo compounds, usually in good yield, involves the oxidation of *N,N′*-disubstituted sulfamides with NaOCl ⟨67JOC2360, 70JOC2482, 88OSC(6)78⟩ or by electrolysis ⟨78AG(E)370⟩. Sometimes difficulties are encountered in dissolving the sulfamides in the former process and two methods of solving this problem have been recommended ⟨72S632⟩. The first uses a homogeneous medium containing $KOBu^t$ and Bu^tOCl in Bu^tOH, the second is heterogeneous employing NaH, Bu^tOCl, Et_2O and pentane. Five examples are given of the second method; yields are 36–85%.

Azo compounds have been synthesised by the reduction of azoxy derivatives with $(EtO)_3P$ ⟨65LA(687)236⟩ or $LiAlH_4$ ⟨70JOC2482, 75JOC1409⟩ and of ketazines with H_2/Adkins catalyst ⟨64LA(679)83⟩. A cyclic nitroso dimer has been reduced to an azo compound with Si_2Cl_6 ⟨75JOC1409⟩. Some bicyclic substances have been alkylated to give diazenium salts ⟨74JA1788⟩; for an example see Equation (41).

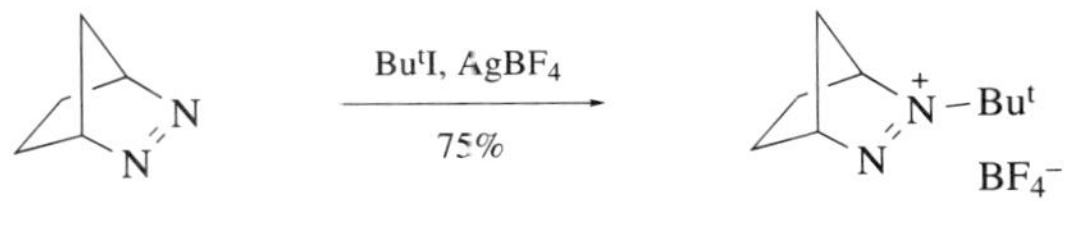

(41)

Reaction of a diaziridinone with an *N,N′*-disubstituted hydrazine has afforded azo compounds ⟨69JOC2263⟩; for an example see Equation (42).

(42)

The equilibrium between a hydrazone (**5**) and an azo compound (**6**), established on treatment with base, has been investigated ⟨69ZOR268⟩. Silver oxide reacted with Bu^tNHBr to give the corresponding azo derivative in 13% yield ⟨53RTC993⟩; however, the *N*-bromoadamantylamines gave good yields of azo compounds with $KOBu^t$ ⟨71CB917⟩.

The dinitrosohydrazine (**3**) furnished (**7**) in unspecified yield on heating to 40°C in a stream of oxygen-free hydrogen ⟨10LA(376)239⟩.

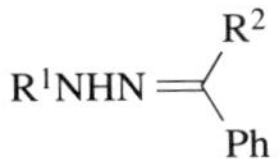

(**5**)

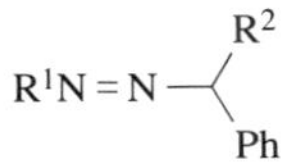

(**6**)

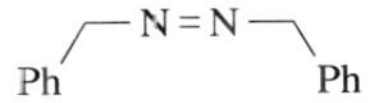

(**7**)

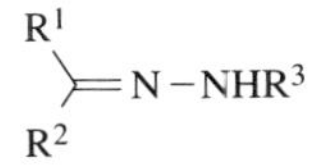

(**8**)

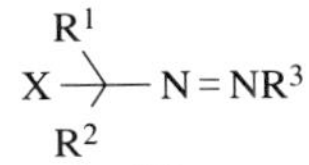

(**9**)

Functionalised azo compounds have been made on many occasions. The reaction between LTA and a hydrazone (**8**) leads to (**9**) where X = OAc ⟨61JA747⟩ whilst the same starting material gives (**9**; X = OOH) with oxygen ⟨84JOC336⟩. Hydroxylamine-*O*-sulfonic acid yields an unstable diol of the type >C(OH)N=NC(OH)< with cyclohexanone by way of a spirooxaziridine ⟨63AG(E)157⟩. Reaction of suitable nitrosoamines with Me_3OBF_4 gives the alkoxydiazenium salts (**10**) and these on mild hydrolysis afford α-hydroxydialkyldiazenes (**11**); further transformations of these compounds are shown in Scheme 48 ⟨71CB1088, 71CB1104⟩.

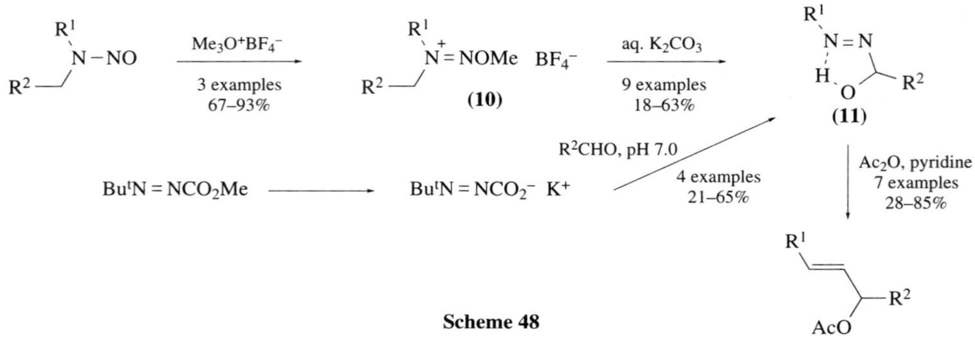

Scheme 48

Ketazines are chlorinated with Cl_2 gas to the dichloroazo compounds (**12**) (Scheme 49) ⟨58LA(618)173, 60LA(631)1, 70JA4586, 74TL265, 87JOC2911⟩. The chlorine atoms in the product can be replaced by alkyl (using R_3Al) ⟨74TL265⟩, OAc, SR, N_3, ⟨60LA(631)1⟩, CN, ⟨60LA(631)1, 87JOC2911⟩ or O_2CR groups ⟨58LA(618)173⟩.

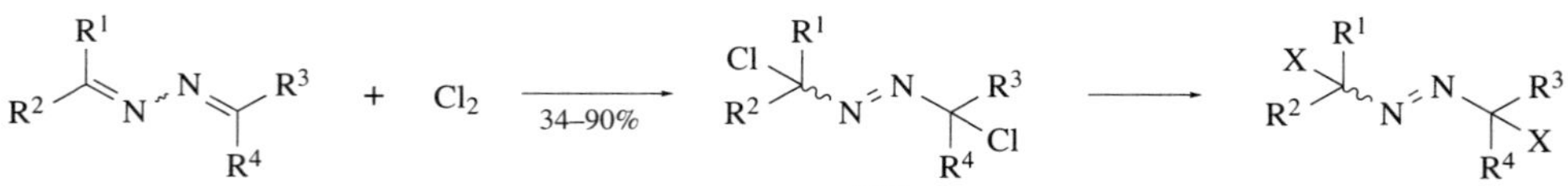

(12)

NaOAc, HOAc→ X = OAc, 6 examples, 35–96%
RS^-→ X = SR, 8 examples, 27–72%
CN^-→ X = CN, 1 example, 31%
R^5CO_2Ag→ X = O_2CR^5, 4 examples, ~70%
$R^5{}_3Al$→ X = R^5, 13 examples 50–100%

Scheme 49

Several azo derivatives of dimedone have been made as shown in Equation (43) ⟨65LA(687)214⟩.

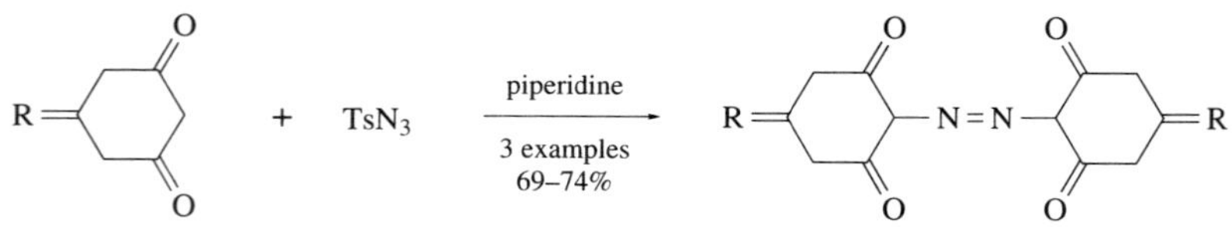

(43)

2.07.2.3.2 *Azoxy compounds*

The preparation of azoxy compounds has been reviewed ⟨67HOU(10/2)757, B-75MI 207-01, 79COC(2)219, B-83MI 207-01⟩. They have been made most commonly by the oxidation of azo compounds with a peroxyacid ⟨52JCS4191, 63JOC2508, 64LA(679)83, 70JOC2482, 71JA3816, 71JA7337⟩; in one case a cyclic azo compound was oxidised with H_2O_2/tungstate ⟨75JOC1409⟩. Direct oxidation of *N,N′*-disubstituted hydrazines with H_2O_2 has been employed ⟨69TL4643⟩ and diamines have been converted into cyclic azoxy compounds with the same reagent in the presence of tungstate, for example, Equation (44) ⟨75JOC1409⟩.

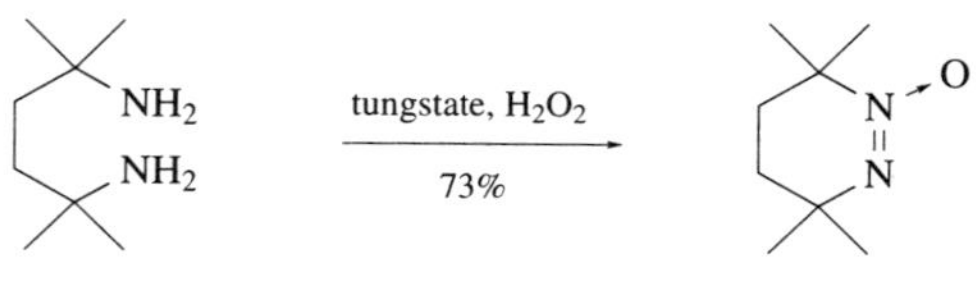

(44)

Reduction of azo-*N,N′*-dioxides (nitroso dimers) to azoxy compounds has been accomplished with H_2/Pd—C ⟨65LA(687)236⟩ and with Si_2Cl_6 ⟨75JOC1409⟩.

Condensation of a nitroso compound with a hydroxylamine using KOH ⟨38JA1930, 63JOC2508, 71CB917⟩ or refluxing HOAc or, better, cyclohexane ⟨64LA(679)83⟩ affords azoxy compounds. An amine in the presence of an oxidant ($PhI(OAc)_2$) was used instead of a hydroxylamine in one case ⟨90SC2353⟩; mixed azoxy compounds result from the use of R^1NCl_2 and R^2NO ⟨74CC217⟩. The conversion of *trans*- to *cis*-azoxy compounds has been achieved photochemically by way of an oxadiaziridine ⟨71JA7337⟩.

Alkylation of the salts $RN{=}NO^-M^+$ has been achieved as shown in Scheme 50 ⟨72JA4392, 73JA3070⟩.

Formation of the silver salt of an *N*-nitroamine ⟨55JCS3997⟩ or of a nitrosohydroxylamine ⟨69JCS(C)397⟩ followed by alkylation yields isomeric azoxyethers (Scheme 51).

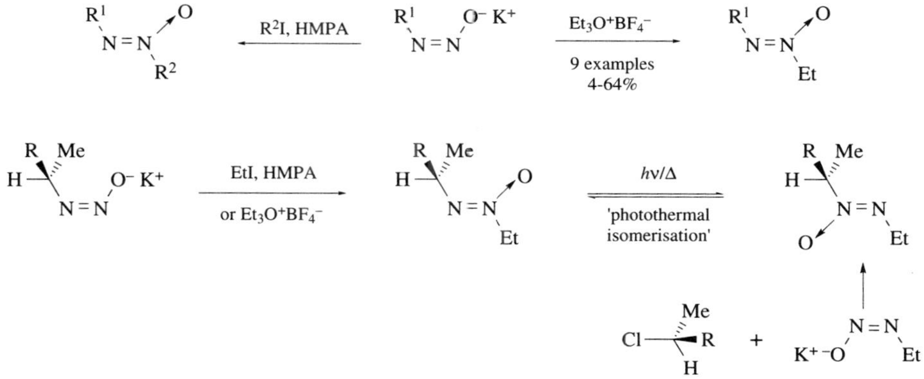

Scheme 50

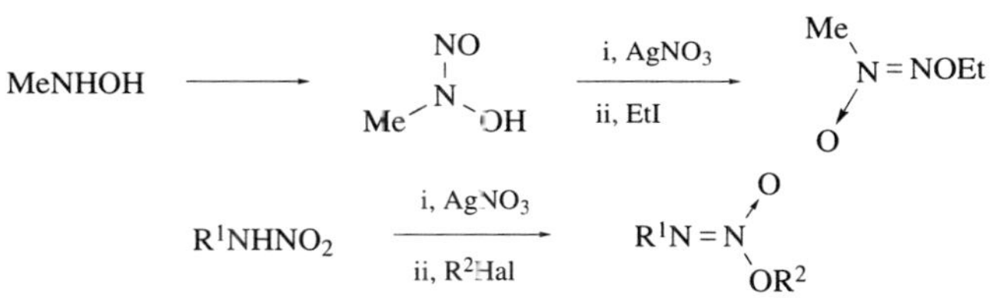

Scheme 51

2.07.2.3.3 N,N'-*Azodioxides (nitroso dimers)*

Some *N,N'*-azodioxides have already been mentioned (see Section 2.07.2.1.1). The equilibrium between monomeric and dimeric nitroso compounds has been investigated in detail in some cases (Scheme 52) ⟨75JOC1409⟩.

These materials are usually made by oxidation. Hydroxylamines have been oxidised with oxygen using a cobalt naphthenate catalyst ⟨64LA(679)83⟩ or with bromine ⟨72JA5077, 75JOC1409⟩. Azoxy compounds have been oxidised using mcpba ⟨73JA1677, 75JOC1409⟩ whilst H_2O_2/tungstate has been employed for the conversion of a diamine into a cyclic nitroso dimer ⟨75JOC1409⟩.

2.07.2.3.4 *Azamines*

Azamines are only stable at −78 °C and are made by the reaction of Bu^tOCl on highly hindered *N,N*-dialkylhydrazines (Scheme 53) ⟨82JA766⟩.

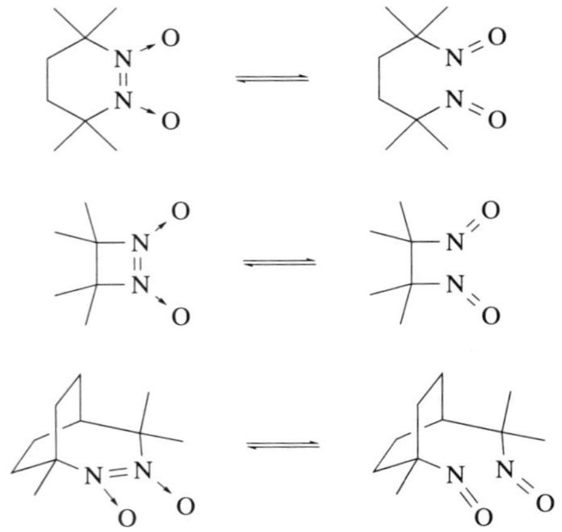

Scheme 52

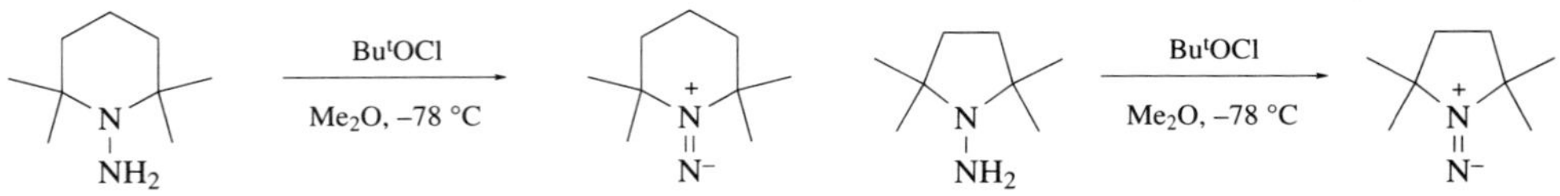

Scheme 53

2.07.2.4 Compounds Containing a N=P, N=As, N=Sb or N=Bi Bond

2.07.2.4.1 Compounds containing a N=P bond

There are many reviews of organophosphorus compounds but this section is concerned with substances lacking a carbon atom attached to phosphorus; this field has been reviewed ⟨82HOU(E2)798, 91AG(E)217⟩.

(i) Substances of the type RN=PHal

Compounds of the type RN=PHal seem not to exist as monomers; an example is Bu^{t}N=PCl which exists as a dimer (Scheme 54) ⟨89IS8⟩.

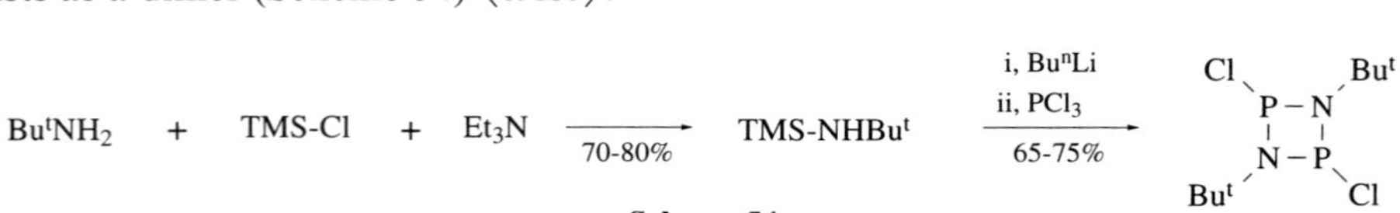

Scheme 54

(ii) Substances of the type RN=PNR$_2$

Substances of the type RN=PNR$_2$ may be made by the reaction between PBr$_3$ or PCl$_3$ and an amine (Scheme 55) ⟨79ZOB1908⟩ or a *N*-lithiated amine (Equation (45)) ⟨74AG(E)811⟩.

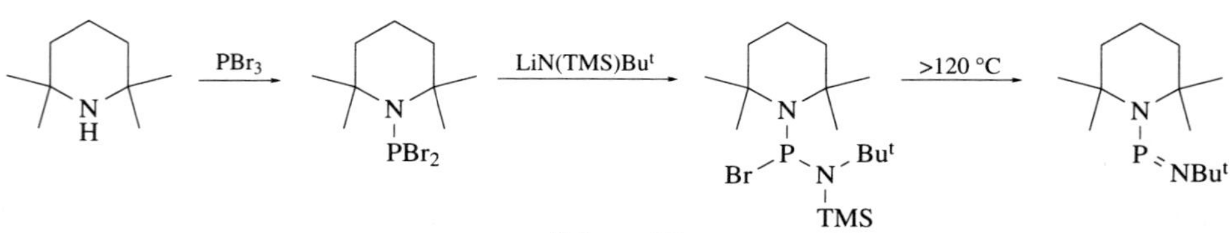

Scheme 55

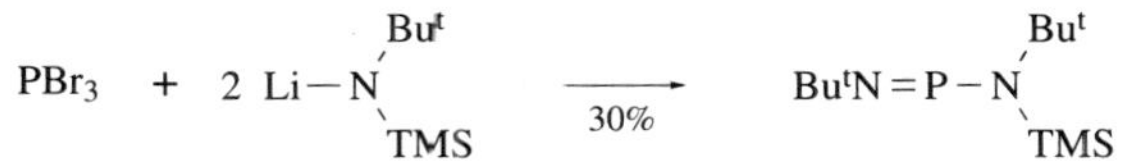 (45)

It is also possible to prepare them by way of an intermediate of the type $R^1R^2NPCl_2$ (Scheme 56) ⟨77CB3874; see also 74JOM(82)C3, 80ZOB2621, 81ZN(B)515⟩. Similar methods can be used to make $R^1N{=}PNR^2NR^3R^4$ ⟨75AG(E)629, 77CB3874⟩.

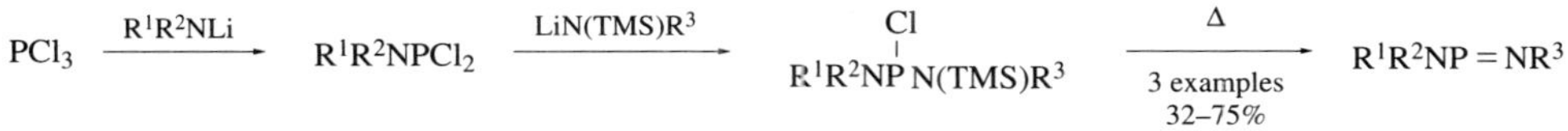

Scheme 56

(iii) Substances of the type RN═PHal₃

Substances of the type $RN{=}PHal_3$ usually rapidly dimerise when R = alkyl (e.g., for R = Me, ⟨61JCS1825⟩) but monomers can be isolated when the group is sufficiently branched ⟨64ZOB1441⟩. They are made by the action of PCl_5 on amine hydrochlorides ⟨61JCS1825, 64ZOB1441, 64ZOB3055, 65ZOB344, 67ZOB852, 67ZOB856, 67ZOB896, 67ZOB2693⟩. It is possible to convert the trichloro compounds into the tribromo derivatives by treatment with TMS-Br ⟨90ZOB465⟩.

(iv) Substances of the type RN═P(OR)₃

Reactions of phosphites, $P(OR)_3$, with R_2NCl ⟨56ZOB3378⟩ or with $RNCl_2$ ⟨75ZOB2394⟩ yield compounds of the type $RN{=}P(OR)_3$; the second reference describes the synthesis of materials having the formulae $RN{=}P(OR)_{3-n}Cl_n$ where n = 0, 1 or 2. A variant of the Staudinger reaction has also been employed in which an azide reacts with a phosphite with loss of nitrogen (Equation (46)) ⟨75JCS(D)12, 92S1063⟩.

$$(R^1O)_3P + R^2N_3 \xrightarrow[\text{3 examples; 80–90\%}]{20\ °C} (R^1O)_3P{=}NR^2 \qquad (46)$$

(v) Substances of the type RN═P(NR₂)₃

A number of methods has been employed to make compounds of the type $RN{=}P(NR_2)_3$; in the first, another variant of the Staudinger reaction, an azide is allowed to react with $(Me_2N)_3P$ ⟨75JCS(D)12⟩. The second procedure allows an amine to add to one of the P═NR groups of compounds of the type **(13)** ⟨80PS(8)221⟩. Two alternative methods are shown in Scheme 57 ⟨89ZOB2135⟩ and Scheme 58 ⟨89ZOB2136⟩.

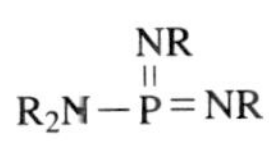

(13)

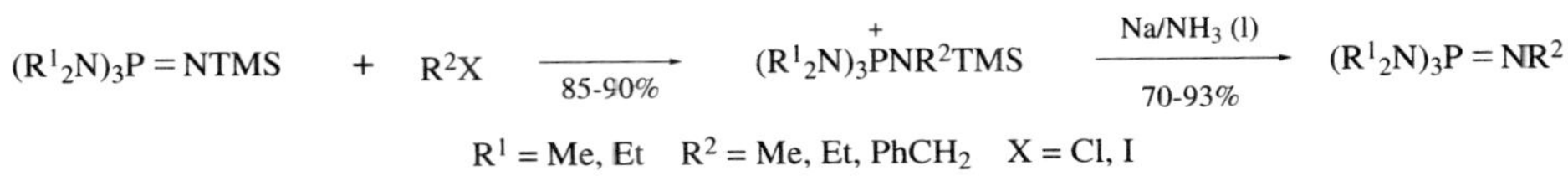

R^1 = Me, Et R^2 = Me, Et, $PhCH_2$ X = Cl, I

Scheme 57

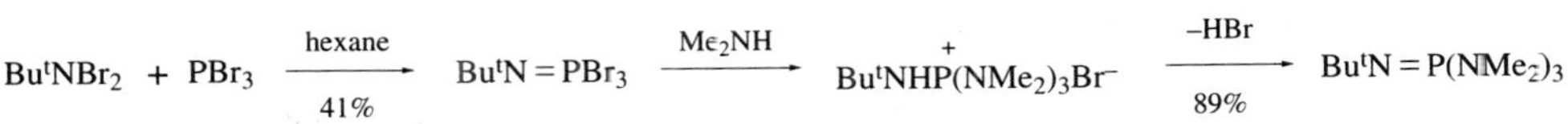

Scheme 58

(vi) Miscellaneous substances containing N=P bonds

Compounds of the type (**14**) are available from the reaction of R^1NCl_2 with $R^2{}_2NPCl_2$ ⟨78ZOB2686⟩; the chlorine atoms can be exchanged for bromine ⟨90ZOB465⟩. Heating in vacuum is sufficient to eliminate a molecule of alkyl halide from (**15**) to afford (**16**) (Equation (47)) ⟨80ZOB951⟩ whilst compounds of the type (**14**) where one Cl atom has been replaced by OR can be made by the processes of Scheme 59 ⟨81ZOB331⟩.

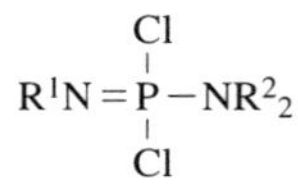

(14)

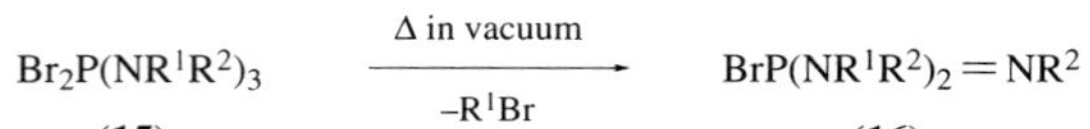

(47)

(15) **(16)**

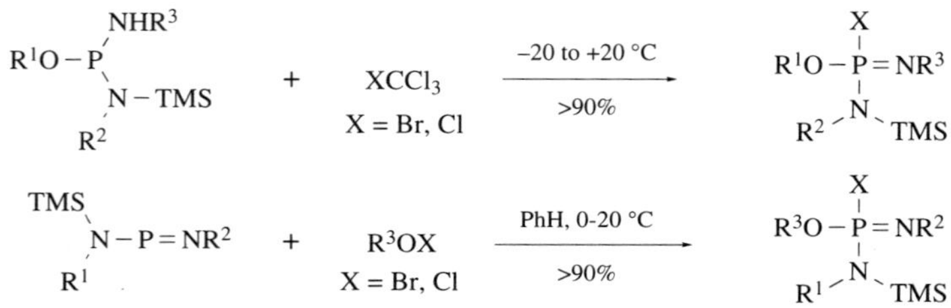

Scheme 59

Treatment with sulfur or selenium suffices to convert (**17**) to (**18**) (Equation (48)) ⟨74AG(E)811, 74JOM(82)C3⟩ and substances of type (**13**) can be made as shown in Scheme 60 ⟨82CB185⟩.

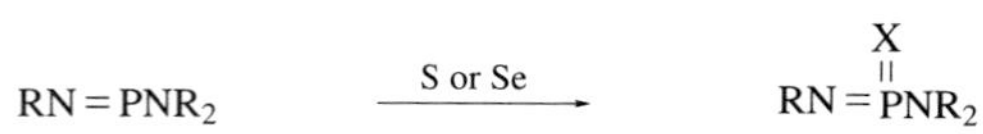

(48)

(17) **(18)**

X = S or Se

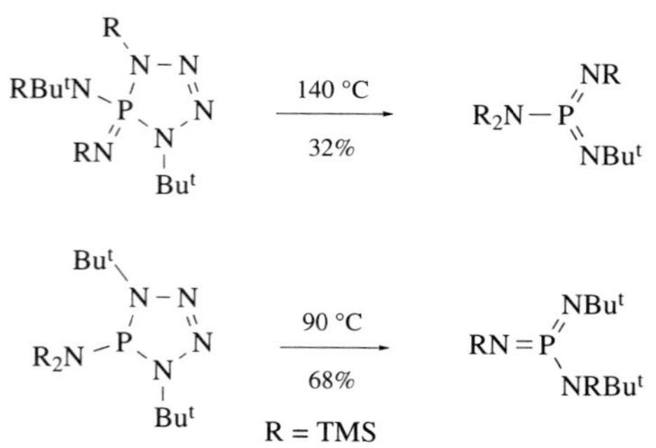

R = TMS

Scheme 60

2.07.2.4.2 *Compounds containing a N=As bond*

Substances containing a N=As bond have been reviewed ⟨78HOU(13/8)220⟩. Most simple compounds containing a N=As bond exist as oligomers, for example, MeAs=NH is a hexamer ⟨29CB598⟩ and both RN=AsCl ⟨60CJC1428, 77ZOB56⟩ and RN=$AsCl_3$ ⟨77ZOB56⟩ exist as dimers as does Bu^tNHAs=NBu^t ⟨63AG(E)218⟩; substances of the type RN=AsF are oligomers ⟨60CJC1431⟩.

2.07.2.4.3 Compounds containing a N=Sb bond

As in the case of arsenic, antimony compounds which should contain a nitrogen–element double bond exist as dimers. Examples are RN=SbX, where X = Cl or Br ⟨79ZN(B)888⟩ or X = R ⟨90CB975⟩.

2.07.2.5 Compounds Containing a N=Metal Bond

Very few substances containing a N=metal bond seem to be known; a group of tantalum and niobium compounds apparently containing a N=metal link has been described ⟨62CJC449, 62CJC1355, 69JCS(A)980⟩.

2.07.3 COMPOUNDS WITH A N—P BOND

Excluded from this section are compounds containing phosphorus–carbon bonds.

The synthesis of compounds containing nitrogen–phosphorus bonds has been comprehensively reviewed ⟨63HOU(12/1)1⟩ with a useful update ⟨(82HOU(13/3b)1⟩. The most widely used routes involve aminolysis of phosphorus halides and of alkoxy- and aryloxyphosphorus compounds.

2.07.3.1 Compounds Related to Phosphorous Acid, $(HO)_3P$

2.07.3.1.1 N,N-Dialkylaminophosphinic dihalides, R_2NPX_2

Secondary amines in ether or petroleum solution react at 5°C or lower with PCl_3 to give dialkylaminodichlorophosphines ⟨03LA(326)149, 58JA1107, 86TL199⟩; with primary amines mixtures of products are obtained.

By treatment with antimony trifluoride R_2NPCl_2 (R = Me, Et) afford the corresponding difluorides ⟨64HOU(12/2)97⟩. The mixed halides Me_2NPFX (X = Cl, Br) can be obtained in a pure form by reaction of dimethylamine with PFX_2 in the gas phase ⟨73IC2482⟩.

2.07.3.1.2 N,N-Dialkylhalogenophosphoramidites, $R^1{}_2NP(Hal)(OR^2)$

The introduction by Caruthers ⟨81TL1859⟩ of chloro-*N,N*-dimethylaminomethoxyphosphine as a monofunctional phosphitylating agent in nucleotide synthesis has led to the preparation of a variety of *N,N*-dialkylchlorophosphoramidites of differing stabilities. They may be prepared by the reaction of an alkoxydichlorophosphine with a secondary amine ⟨49JCS2921, 81TL1859, 91JOC4608⟩. For the synthesis of chloramidites derived from cyclic amines such as morpholine or pyrrolidine the reaction of $ROPCl_2$ with the *N*-trimethylsilyl derivative of the amine has been shown to be effective ⟨83TL245, 84TL1307, 84TL5513, 89S497⟩, while 2,2,6,6-tetramethylpiperidine was *N*-lithiated before treatment with $MeOPCl_2$ ⟨83TL245⟩.

Reaction of $(RO)PCl_2$ with primary amines leads to ill-defined products ⟨63HOU(12/1)1⟩.

In an alternative approach the thymidine derivative (**19**) was obtained from $Pr^i{}_2NPCl_2$ as shown in Equation (49) ⟨86TL199⟩.

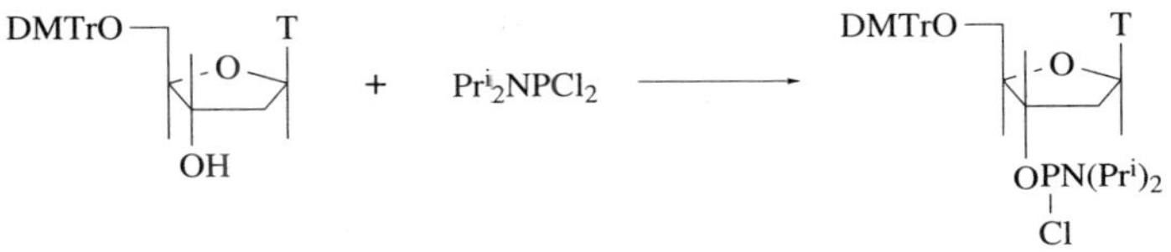

(**19**)

(49)

DMTr = dimethoxytrityl
T = thyminyl

The chlorine atom in $(EtO)P(Cl)NMe_2$ may be exchanged for a fluorine atom by heating under reflux with KF in toluene ⟨49JCS2921⟩.

2.07.3.1.3 N,N-*Dialkylphosphoramidites*, *$(R^1O)_2PNR^2_2$*

(i) Reaction of $(RO)_2PCl$ with 2 mol of a secondary amine in an inert solvent such as ether gives reasonable yields of the corresponding amidites ⟨03LA(326)149, 49JCS2921, 50JA5491⟩.

(ii) A major route to phosphoramidites, intermediates in the synthesis of oligonucleotides, is the treatment of a dialkylaminochlorophosphoramide with an alcohol ⟨92JOC5535⟩ or a suitably protected nucleoside in the presence of a tertiary amine ⟨81TL1859, 83TL245, 84TL5513, 89S497, 91JOC4608, 93JOC4606⟩.

(iii) Although very high yields of phosphoramidites are obtained by method (ii), acid hydrolysis of the product may be catalysed by the amine hydrochloride also formed in the reaction. To avoid this, an alternative approach is to treat a bis(dialkylamino)alkoxyphosphine with the 3′-hydroxyl of a deoxyribonucleoside. The reaction is catalysed by imidazole ⟨93JOC4606⟩ or 4,5-dichloroimidazole ⟨85JOC2019⟩. In an interesting variation of this method (Scheme 61) a bistriazolylalkoxyphosphine (**20**) is treated sequentially with a nucleoside and then with a secondary amine to give (**21**) in good yields ⟨84TL4511⟩.

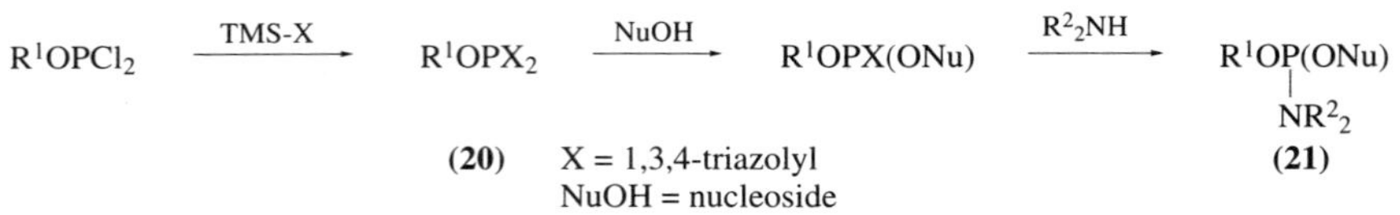

Scheme 61

2.07.3.1.4 *Thiophosphoramidites*, *$R^1_2NP(SR^2)(OR^3)$*

Thiophosphoramidites are useful intermediates in the synthesis of nucleoside phosphorodithioates and may be obtained as shown in Scheme 62. The relative inactivity of $Pr^i_2NPCl_2$ towards thiols is overcome by the use of $AlCl_3$ and KI as cocatalysts ⟨88TL6843⟩.

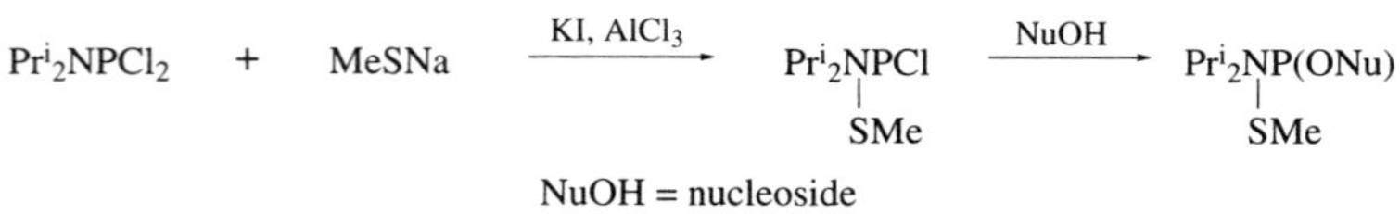

Scheme 62

2.07.3.1.5 *Bis(dialkylamino)chlorophosphines*, *$(R_2N)_2PCl$*

(i) By the reaction of 4 mol of the appropriate amine in ether and PCl_3 at 0–10°C there have been prepared $(Me_2N)_2PCl$ ⟨64HOU(12/2)104⟩ and $(Pr^i_2N)_2PCl$ ⟨86CL1401⟩. The latter compound is stable over long periods at −20°C, making it a useful phosphitylating agent.

(ii) Hexamethylphosphorous triamide and PCl_3 react exothermically to give $(Me_2N)_2PCl$ in 91% yield ⟨61CB1505⟩. Note that the product reacts violently with water ⟨64HOU(12/2)104⟩.

2.07.3.1.6 *Alkoxy- (or aryloxy-) bis(dialkylamino)phosphines*, *$(R^1_2N)_2P(OR^2)$*

(i) Alkoxydichlorophosphines in ether or benzene react with 4 mol of secondary amines to afford $(R^1_2N)_2P(OR^2)$ ⟨03LA(326)160⟩. A similar reaction occurs with phenyl esters ⟨67IZV1606⟩ but 2,6-disubstituted phenyl esters are inactive ⟨80ZOB1702⟩.

(ii) Cyclic products (**22**) are obtained from the reaction of 2- and 3-alkylaminoalcohols with dialkylaminodichlorophosphines as shown in Equation (50) ⟨71BSF136, 76ZOB477⟩.

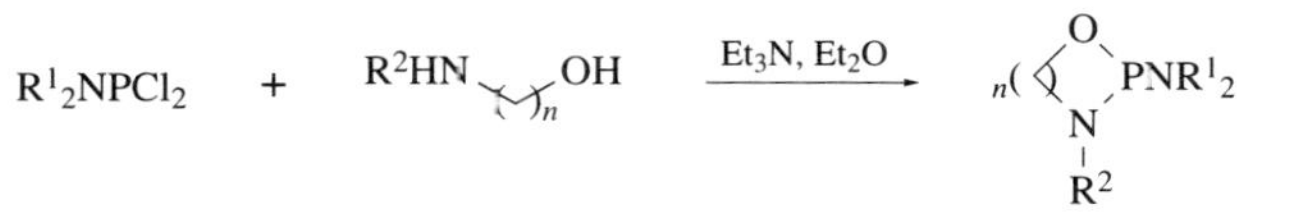

(22) $n = 2, 3$ (50)

(iii) Treatment of a wide range of bis(dialkylamino)chlorophosphines with alcohols or phenols provides the corresponding esters ⟨71BSF136, 86CL1401⟩. The purity of non-cyclic products is affected by a tendency to disproportionate, attempted purification by distillation must be carried out at very low pressures. The silyl ester (**23**), however, readily obtained from $(Pr^i_2N)_2PCl$ (Scheme 63) distils without decomposition, and can be stored at room temperature ⟨90AG(E)522⟩.

$$(Pr^i_2N)_2PCl \xrightarrow{H_2O,\ Et_3N} (Pr^i_2N)_2POH \xrightarrow[90\%]{TMS\text{-}Cl,\ Et_3N} (Pr^i_2N)_2POTMS$$

(**23**)

Scheme 63

(iv) When tris(dialkylamino)phosphines are mixed with a molar equivalent of an alcohol, one amino group is replaced by alkoxide (Equation (51)) ⟨65BSF2368⟩. Again, the preparation of pure open-chain compounds by this route is complicated by easy disproportionation of the products but this decreases with increasing substituent bulk. Heating $(Et_2N)_3P$ with thiophenol and diethylamine gives a moderate yield of $(Et_2N)_2PSPh$ ⟨62ZOB3070⟩.

$$(Et_2N)_3P \quad + \quad Pr^iOH \longrightarrow (Et_2N)_2P(OPr^i) \qquad (51)$$

2.07.3.1.7 Tris(dialkylamino)phosphines, $(R_2N)_3P$

(i) The preparation of hexamethylphosphorous triamide from the reaction of PCl_3 and dimethylamine has been described in detail ⟨73OSC(5)602⟩; the method is also suitable for higher homologues.

(ii) Treatment of the *N,N′*-bis(trimethylsilyl) derivatives of piperazines and homopiperazines with PCl_3 effected the formation of polymeric phosphorus triamides; *N*-trimethylsilyltetrahydroisoquinoline yields the aminophosphine (**24**) ⟨78TL3989⟩.

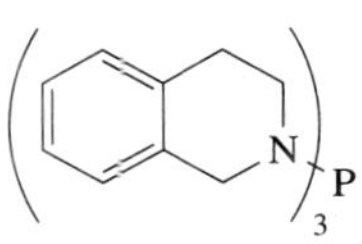

(**24**)

2.07.3.2 Compounds Related to Phosphoric Acid, $(HO)_3P(O)$

As with the compounds of trivalent phosphorus discussed in Section 2.07.3.1, the majority of substances in this series are obtained by successive replacement of halide from $P(O)X_3$ with amines or with alcohols and phenols. The substitution of the lone pair on phosphorus by oxygen (or sulfur) leads to a lower reactivity of the halide, and more vigorous conditions are common in the formation of this class of compound.

2.07.3.2.1 N,N-Dialkylphosphoramidic dihalides, $R_2NP(O)X_2$

The preparation of $Me_2NP(O)Cl_2$ by the treatment of $POCl_3$ with 2 mol of dimethylamine has been described in detail ⟨59IS69⟩. The mixed halides $Et_2NP(O)FX$ ($X = Cl$, Br) can be prepared similarly from $P(O)FX_2$ ⟨60CB1323⟩, while a number of dialkylphosphoramidic difluorides were obtained in 60–84% yields by the low temperature reaction of the secondary amine with a solution of gaseous POF_3 in toluene ⟨59LA(625)88⟩.

A number of *N*-monoalkyl compounds, $RNHP(O)F_2$, have been prepared by reaction of the appropriate dichlorides with KF ⟨64HOU(12/2)393⟩.

A quantitative yield of the cyclophosphamide precursor $Cl_2P(O)N(CH_2CH_2Cl)_2$ is obtained by heating $POCl_3$ and $(ClCH_2CH_2)_2NH_2 \cdot HCl$ together under reflux without solvent ⟨93S311⟩.

2.07.3.2.2 N,N-*Dialkylhalogenophosphoramidates*, $R^1{}_2NP(O)Hal(OR^2)$

(i) $R^1{}_2NP(O)Cl_2$ reacts slowly with alcohols and phenols to give the corresponding amidophosphochloridates; faster reactions may sometimes be obtained by heating under reflux in benzene solution or by use of the alkali metal salts of the hydroxy compound ⟨57JA1970, 57RZC879, 90AG(E)782⟩. Fluorophosphoramides $(R^1O)P(O)FNR^2{}_2$ may be obtained from $P(O)Cl_2F$ by a similar reaction ⟨60CB1323⟩ or by heating a chlorophosphoramide with KF ⟨49JCS2921⟩. Reaction of the phosphorinane (**25**) with poly(HF-pyridine) causes displacement of the ethyleneimine ligand by fluoride, with stereochemical inversion at the phosphorus atom (Equation (52)) ⟨85JOC1815⟩.

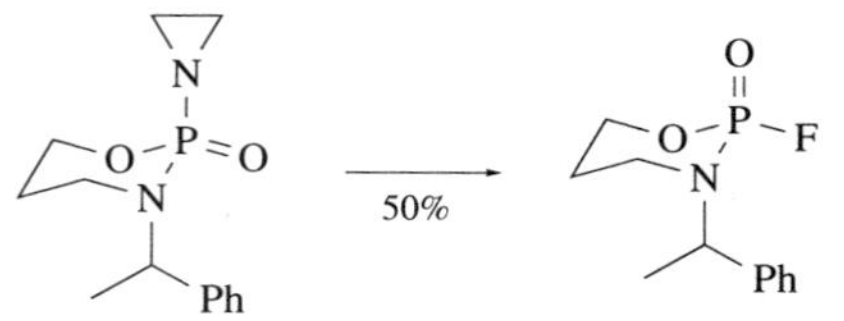

(**25**)

(52)

(ii) Diethyl *N,N*-dichlorophosphoramidate in boiling benzene adds to styrene and α,β-unsaturated esters (Equation (53)). The acceleration of the reaction by light and the anti-Markovnikov nature of the products suggest a free-radical mechanism for the addition ⟨70T3527, 74TL3383, 76T1649⟩.

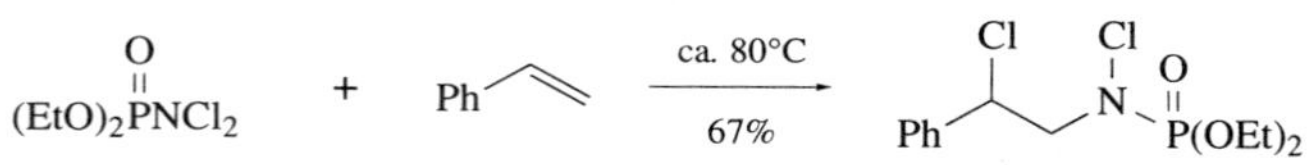

(53)

2.07.3.2.3 *Phosphoramidates*, $(R^1O)_2P(O)NR^2{}_2$

(i) The most useful and common method for obtaining phosphoramidates is by the reaction of $(R^1O)_2P(O)Cl$ with primary or secondary amines ⟨42JA1337, 48JCS699, 51JA5202, 61JOC1533⟩. It has been used to prepare a variety of phosphoramides from amino acids ⟨55JA1866, 55JA5351, 61CB2644⟩.

(ii) In the Atherton–Todd procedure ⟨45JCS660, 47JCS674⟩ the dialkyl chlorophosphonate, created *in situ* by the reaction of $(RO)_2P(O)H$ with CCl_4 or $CBrCl_3$, gives high yields of the phosphoramidate on addition of primary or secondary amines (Scheme 64). Chloral cleaves the dioxaphosphaline (**26**) to form the amidate (**27**) (Equation (54)) ⟨55JA2871⟩.

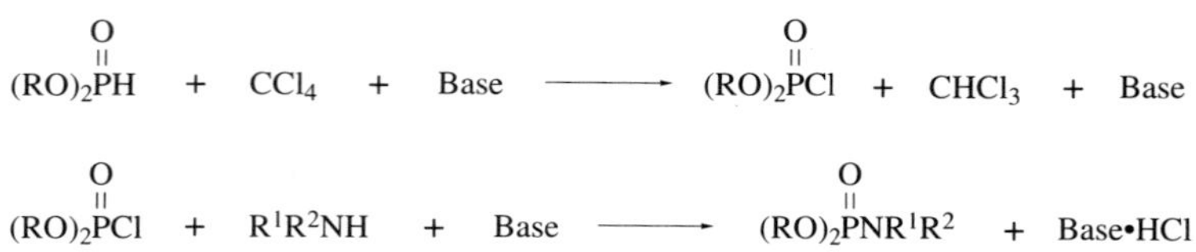

Scheme 64

(**26**)

(**27**)

(54)

(iii) Using a modified version of the Appel reaction, a variety of nucleotide salts (**28**) were reacted with Ph_3P and CCl_4 and then primary or secondary amines to give good yields of amidates (**29**) (Equation (55)) ⟨85JOC1271⟩.

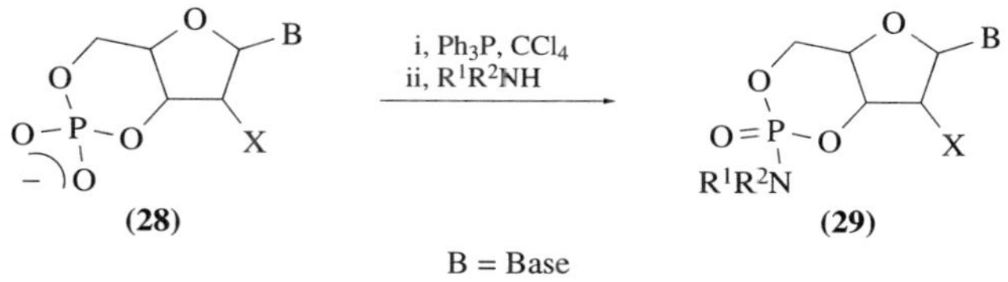

(55)

2.07.3.2.4 N,N,N′,N′-*Tetraalkylphosphorodiamidic halides or bis(dialkylamino)phosphoryl halides, $(R_2N)_2P(O)Hal$*

A solution of $POCl_3$ in ether reacts with 4 mol of a dialkylamine ⟨49JCS2921, 58JCS1963⟩ or with 2 mol of amine in the presence of base ⟨87JOC5320⟩ to afford $(R_2N)_2P(O)Cl$. By heating in benzene with zinc fluoride, $(Me_2N)_2P(O)Cl$ is converted to the highly toxic fluoride ⟨49JCS2921⟩. Higher homologues have been prepared by the action of dialkylamines on $POCl_2F$ ⟨48JCS1313⟩. The unsymmetric diamide $(C_6H_{11}N)P(O)FNEt_2$ is obtained by reacting cyclohexylamine with $(Et_2N)P(O)BrF$ ⟨60CB1323⟩.

2.07.3.2.5 N,N,N′,N′-*Tetraalkylphosphorodiamidates, $(R^1{}_2N)_2P(O)(OR^2)$*

(i) When dialkylaminophosphoryl chlorides are heated with sodium alkoxides, moderate yields of the diamidate esters are obtained ⟨50JCS1769⟩.

(ii) As before, a better approach is to treat the more reactive $(R^1O)P(O)Cl_2$ with 2 mol of a secondary amine ⟨42JA1337, 50LA(566)210, 57JOC1186, 59JOC1419⟩. For the reaction of ethyl phosphorodichloridate with hindered aziridines, prior formation of the *N*-lithio derivative proved advantageous (Equation (56)) ⟨88JOC1761⟩.

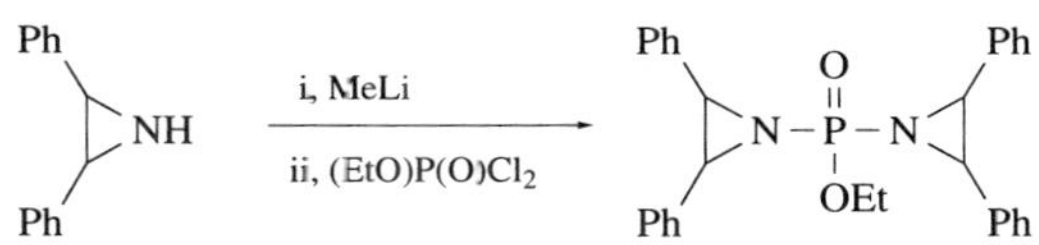

(56)

2.07.3.2.6 *Hexaalkylphosphoric triamides, $(R_2N)_3P(O)$*

Symmetrical hexaalkyl triamides are conveniently prepared by reacting $POCl_3$ with at least 6 mol of primary or secondary amines ⟨42JA1553, 59JOC1419, 88JOC1761⟩. Thiophosphoric triamides can be similarly obtained from $P(S)Cl_3$ ⟨59JOC1419⟩.

Unsymmetrical products can be obtained by treating $R_2NP(O)Cl_2$ with 2 mol of a secondary amine in the presence of base ⟨50LA(566)210⟩. This method, and the reaction of phosphorodiamidic chlorides with amines were both used in the synthesis of chiral triamides (Scheme 65) ⟨87JOC5320⟩.

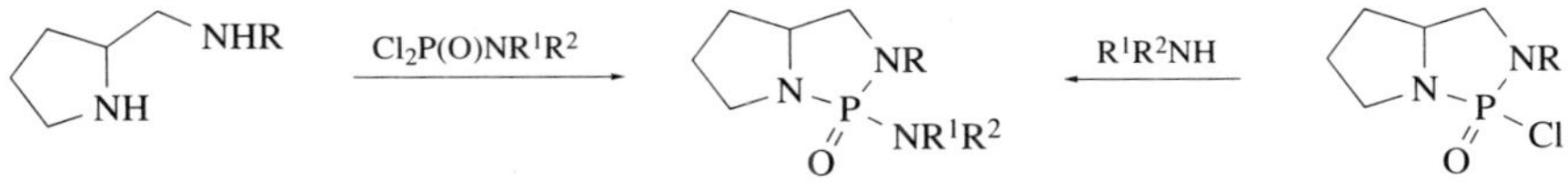

Scheme 65

2.07.3.2.7 *Miscellaneous compounds containing a N—P bond*

(i) Octamethylpyrophosphoramide is prepared by the reaction shown in Equation (57) ⟨59IS73⟩.

$$(Me_2N)_2P(O)Cl \quad + \quad H_2O \quad + \quad 2\,Et_3N \xrightarrow[95\%]{45\text{–}50\,^\circ C} (Me_2N)_2P(O){-}O{-}P(O)(NMe_2)_2 \qquad (57)$$

(ii) Bis(dialkylamino)phosphines are conveniently prepared by the reduction of $(R_2N)_2PCl$ with LAH ⟨84JOC1784⟩. Diamides $(R_2N)_2P(O)H$ may be obtained by the hydrolysis of hexaalkylphosphoric triamides ⟨66JCS(C)2176⟩ or of $(R_2N)_2PCl$ ⟨80ZOB1236⟩. The diamides are cleaved to $R_2NP(O)HOH$ by reaction with acetic acid ⟨72ZOB1936⟩ or enolisable ketones ⟨70BSF192⟩. Amide esters $(R^1{}_2N)PH(O)(OR^2)$ result from the hydrolysis of $(R^1{}_2N)P(Cl)O(OR^2)$ or the aminolysis of dialkyl pyrophosphates ⟨67T2243⟩.

(iii) Hydrogenation of $PhOP(O)(NR_2)_2$ over platinum results in selective cleavage of the ester bond to yield $HOP(O)(NR_2)_2$ ⟨54JA655, 57CB2075⟩.

(iv) Monoester monoamides of phosphoric acid, $(R^1R^2N)P(O)OH(OR^3)$, may be obtained by hydrolysis of intermediate phosphorochloramidates (Equation (58)) ⟨58HCA244⟩ or by partial hydrogenolysis of dibenzyl esters ⟨58JCS3039⟩. They have also been prepared by the reaction of nucleoside-5′-phosphates with secondary amines in the presence of dicyclohexylcarbodiimide (dcc) ⟨61JA649⟩.

$$PhOP(O)Cl_2 \quad + \quad H_2NCH_2CO_2H \xrightarrow{Ba(OH)_2} [PhOP(O^-)(O){-}NHCH_2CO_2^-]\ Ba^{2+} \qquad (58)$$

(v) *N*-Phosphoglycine has been isolated as its barium salt after hydrogenolysis of a bisbenzyl ester ⟨55JA5354⟩.

2.07.4 COMPOUNDS WITH A N—As, —Sb OR —Bi BOND

2.07.4.1 Compounds Containing a N—As Bond

Compounds containing an N—As bond have been reviewed ⟨78HOU(13/8)33, B-80MI 207-01⟩; the latter reference is particularly useful for the compounds, not considered here, containing a C—As bond.

2.07.4.1.1 (Dialkylamino)dihaloarsines, $(R_2N)AsHal_2$

(Dialkylamino)dihaloarsines are usually prepared by the replacement of one halogen atom in $AsHal_3$; the first example seems to have been made by Michaelis and Luxembourg in a pioneering paper published in 1896, when they treated $AsCl_3$ with isobutylamine ⟨1896CB710⟩. Further developments followed six decades later when Kamai and Khisamova showed that the three chlorine atoms in $AsCl_3$ could be replaced successively by dimethylamino ⟨55DOK(105)489⟩ or diethylamino ⟨56ZOB126⟩ groups. It was demonstrated in 1960 that one of the fluorine atoms in AsF_3 could also be replaced by groups derived from both primary and secondary amines ⟨60CJC1431; see also 73IC2119⟩. $AsCl_3$ could also be employed (Scheme 66) ⟨60CJC1428, 64ZAAC(326)280⟩ and a new method of preparation utilised the reaction between a tris(dialkylamino)arsine, $As(NMe_2)_3$, and $AsCl_3$ or $AsBr_3$ ⟨64ZAAC(330)233⟩.

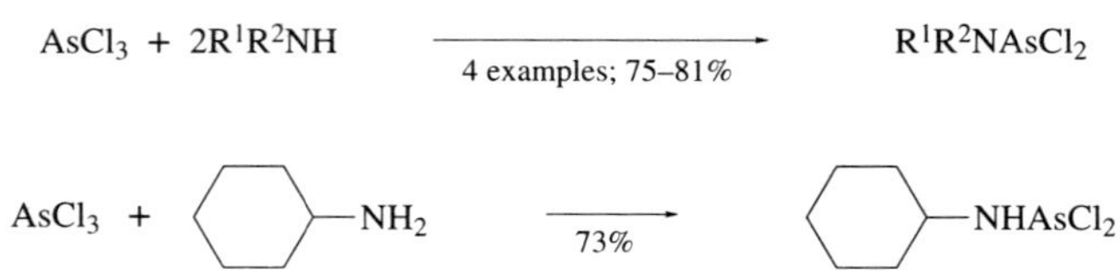

Scheme 66

2.07.4.1.2 Bis(dialkylamino)haloarsines, $(R_2N)_2AsHal$

Bis(dialkylamino)haloarsines, where Hal = Cl, can be made by the reaction between $AsCl_3$ and an amine using the appropriate molar ratio ⟨68JOM(12)377⟩; the use of a diamine yields a cyclic product ⟨64AG(E)702⟩. Alternatively, the (dialkylamino)dihaloarsines described above may be treated with further amine ⟨55DOK(105)489, 56ZOB126⟩. A tris(dialkylamino)arsine can be made to react with hydrogen, benzoyl or benzyl chloride to give the chloro compound or with $AsBr_3$ to give the bromo derivative (Scheme 67) ⟨64ZAAC(330)233⟩.

$$(Me_2N)_2AsBr \xleftarrow[82\%]{AsBr_3} As(NMe_2)_3 \xrightarrow{XCl} (Me_2N)_2AsCl$$

X = H, 76%
X = PhCO, 82%
X = $PhCH_2$, 100%

Scheme 67

2.07.4.1.3 *Tris(dialkylamino)arsines, $(R_2N)_3As$*

Compounds of the type $(R_2N)_3As$ have been made by the reaction of a dialkylamine with $AsCl_3$ ⟨59CB2637⟩ or with a bis(dialkylamino)chloroarsine ⟨55DOK(105)489⟩. *N*-lithiated amines have also been used with arsenic halides; (**30**) reacted with $AsBr_3$ to yield (**31**) (Equation (59)) ⟨66JOM(6)259⟩ whilst a similar reaction is depicted in Equation (60) ⟨68ZAAC(361)296⟩.

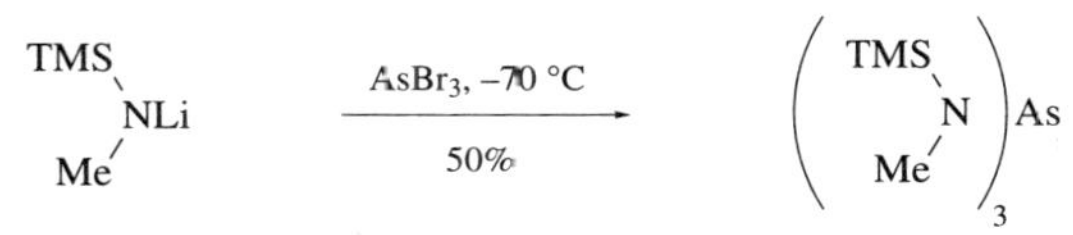

(**30**) (**31**) (59)

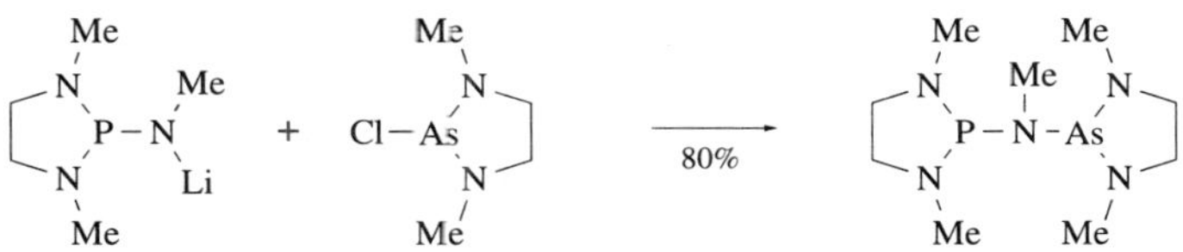

(60)

Treatment of a secondary amine R_2NH with tris(dimethylamino)arsine results in an amine exchange process, the rate of which decreases as the bulk of R increases; primary amines yield cyclic aminoarsines ⟨63AG(E)218, 64ZAAC(331)35⟩.

In an alternative approach, which seems to be very versatile, a stannane is used to deliver the dialkylamino group; thus compound (**32**) reacts with AsF_3 to afford (**33**) (Equation (61)) ⟨69JCS(A)992⟩.

$$\underset{(\mathbf{32})}{Me_2NSnMe_3} + AsF_3 \xrightarrow{42\%} \underset{(\mathbf{33})}{(Me_2N)_3As} \qquad (61)$$

2.07.4.1.4 *Miscellaneous compounds containing a N—As bond*

A number of interesting compounds can be made by the reactions shown in Scheme 68 ⟨64ZAAC(330)233⟩.

$$(Me_2N)_3\overset{+}{As}Br\ Br^- \xleftarrow[93\%]{Br_2,\ -70\ ^\circ C} \underset{(\mathbf{33})}{(Me_2N)_3As} \xrightarrow{Cl_2,\ -70\ ^\circ C} \text{quaternary salt, isolated as } (Me_2N)_3\overset{+}{As}Cl\ SbCl_6^-$$

$(\mathbf{33}) \rightarrow (Me_2N)_3AsNi(CO)_3$

$(\mathbf{33}) \xrightarrow[72\%]{Me_2NCl} (Me_2N)_4\overset{+}{As}Cl\ Cl^-$

Scheme 68

2.07.4.2 Compounds Containing a N—Sb Bond

Reaction of (**30**) with $SbBr_3$ gives the antimony analogue of (**31**) in 59% yield ⟨66JOM(6)259⟩ whilst the use of $SbCl_3$ with Me_2NLi at $-78\,^{\circ}C$ affords a 62% yield of tris(dimethylamino)stibane ⟨64IC609⟩. A homologue of compound (**30**) in which the methyl group has been replaced by a *t*-butyl residue reacts with $SbCl_3$ to give (**34**; X = Cl) and with $SbBr_3$ to afford (**34**; X = Br) in yields of 49% and 44% respectively ⟨79ZN(B)888⟩. Reaction of (**32**) with SbF_3 gives the antimony analogue of (**33**) in 60% yield ⟨69JCS(A)992⟩. Replacement of the ethoxy groups in the fluorinated compounds (**35**) and (**36**) by substituted amino groups leads to (**37**) and (**38**) respectively ⟨70ZC33⟩.

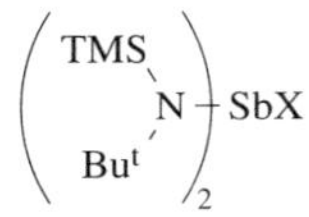

(34)

$F_2Sb(OEt)$ **(35)** $FSb(OEt)_2$ **(36)** $F_2Sb(NHR)$ **(37)** $FSb(NHR)_2$ **(38)**

2.07.4.3 Compounds Containing a N—Bi Bond

Reaction of (**30**) with $BiBr_3$ gives the bismuth analogue of (**31**) in 54% yield ⟨66JOM(6)259⟩.

2.07.5 COMPOUNDS WITH A N—METALLOID BOND

2.07.5.1 Compounds Containing a N—Si Bond

Compounds containing a N—Si bond have been extensively reviewed ⟨79COC(3)541⟩ but most reviews deal with substances having an Si—C bond which are not included here; although there is an exception ⟨B-80MI 207-01⟩. A large number of cyclic silazanes is known but these are considered to be heterocycles rather than functional groups; these have been reviewed ⟨82COMC-I(2)1⟩.

In the headings below the expression 'dialkylamino' should be taken to include both mono- and dialkylamino residues.

2.07.5.1.1 *Dialkylaminosilanes, $(R_2N)SiX_3$*

Two dialkylaminotrichlorosilanes were made in 1896 by the reaction of dialkylamines with $SiCl_4$ ⟨1896CB710⟩. Breederveld and Waterman carried out extensive studies in the early 1950s on the synthesis of substances having N—Si bonds. One chlorine atom in $SiCl_4$ was replaced by a diethylamino or piperidino group by reaction with the appropriate amine ⟨52MI 207-01⟩; later, similar reactions were achieved with *t*-butylamine ⟨53MI 207-01⟩ and diisopropylamine ⟨54MI 207-01⟩. By heating the complex between a primary or secondary amine and SiF_4 to 100–200 °C one fluorine atom could be replaced to give substances of type (**39**) in variable yield ⟨70CI(L)499⟩. The combination of trichlorosilane with dimethylamine in ether at $-70\,^{\circ}C$ gave a 74% yield of (**40**) ⟨69MI 207-03⟩ whilst iodosilane gave dialkylaminosilanes with dimethylamine, diethylamine ⟨67JCS(A)652⟩, pyrrolidine and piperidine (Equation (62)) ⟨67JCS(A)1918⟩.

R_2NSiF_3 **(39)** Me_2NSiCl_2H **(40)**

$$R_2NH + ISiH_3 \xrightarrow{\text{4 examples; 68–99\%}} R_2NSiH_3 \qquad (62)$$

2.07.5.1.2 *Bis(dialkylamino)silanes, $(R_2N)_2SiX_2$*

By using the correct molar ratio of dialkylamine it is possible to replace two of the chlorine atoms in $SiCl_4$ with diethylamino, piperidino, ⟨52MI 207-01⟩ or *t*-butylamino residues ⟨53MI 207-01⟩. Heating the trichlorosilane (**41**) with diisopropylamine in benzene at 200 °C in a sealed tube led to replacement of one of the chlorine atoms by a diisopropylamino group in 25% yield but the other two halogen atoms could not be replaced because of steric hindrance ⟨54MI 207-01⟩. Compound (**43**) is available by the disproportionation reaction shown in Equation (63) ⟨52MI 207-01⟩. Trichlorosilane reacts with dimethylamine in the correct molar ratio to give (**44**) in 76% yield ⟨69MI 207-03⟩; diethylamine and *t*-butylamine gave similar results ⟨59ACS35⟩.

$(Pr^i)_2NSiCl_3$

(41)

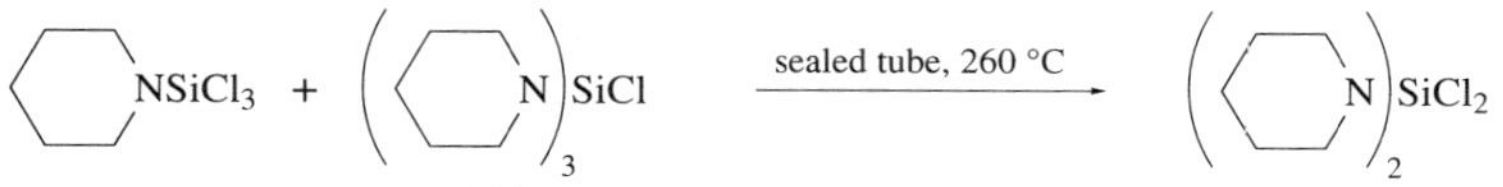

(42) **(43)** (63)

$(Me_2N)_2SiClH$

(44)

2.07.5.1.3 *Tris(dialkylamino)silanes, $(R_2N)_3SiX$*

Refluxing the dichlorosilanes (**43**) and (**45**) with the corresponding amines in benzene gave the expected tris(dialkylamino)silanes; three chlorine atoms in $SiCl_4$ could be replaced by treatment with dimethylamine in ether ⟨52MI 207-01⟩. Replacement of one of the chlorine atoms in (**46**) by reaction with *t*-butylamine in benzene at 150 °C was possible but the remaining chlorine atom failed to react because of steric hindrance ⟨53MI 207-01⟩. Under the correct conditions the reaction of methylamine with trichlorosilane led to tris(dimethylamino)silane in 82% yield ⟨69MI 207-03⟩.

$(Et_2N)_2SiCl_2$ **(45)** $(Bu^tNH)_2SiCl_2$ **(46)**

2.07.5.1.4 *Tetrakis(dialkylamino)silanes, $(R_2N)_4Si$*

The dimethylamino representative of this group, $(Me_2N)_4Si$, has been made by the action of dimethylamine on $SiCl_4$ ⟨52JA1421, 69IC279⟩; it has also been prepared from this amine and chlorotris(dimethylamino)silane in a sealed tube at 100 °C ⟨52MI 207-01⟩. It was found that the reaction corresponding to the latter process could not be carried out in the diethylamine series because of steric hindrance ⟨52MI 207-01⟩. Tetrakis(diethylamino)silane could be prepared in 53% yield from Et_2NMgBr and bromotris(diethylamino)silane at 150–175 °C ⟨53MI 207-02⟩. Tetrakis(*t*-butylamino)silane was made similarly from *t*-butylamine and $SiBr_4$ ⟨53MI 207-03⟩; the corresponding piperidine derivative was available by the reaction between the amine and (**42**) at 250–270 °C ⟨52MI 207-01⟩.

2.07.5.1.5 *Miscellaneous compounds containing a N—Si bond*

The siloxane (**47**) reacts with diethylamine to give (**48**) ⟨52JA1421⟩ whilst (**49**) gives (**50**) ⟨53MI 207-04⟩. Chlorosilane reacts with isopropylamine and with *t*-butylamine to give disilylamines, $RN(SiH_3)_2$ ⟨89JCS(D)779⟩.

Si_2OCl_6 **(47)** $Si_2O(Et_2N)_6$ **(48)**

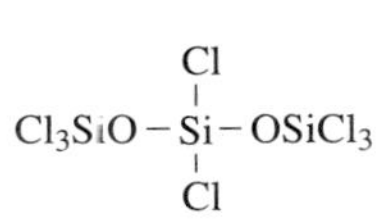

(49)

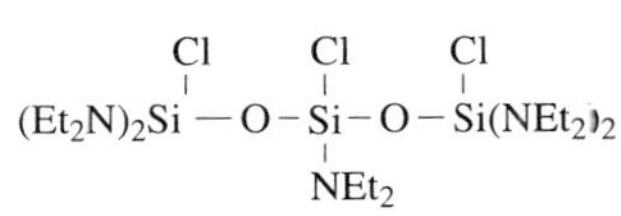

(50)

2.07.5.2 Compounds Containing a N—Ge Bond

The germane derivative (**51**) was prepared as shown in Equation (64) ⟨81CB527⟩.

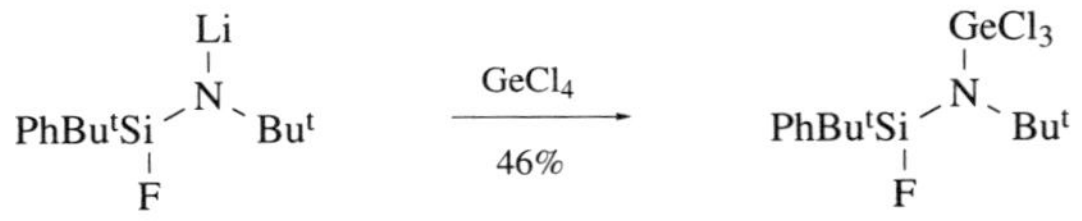

(64)

(**51**)

Tetrakis(dialkylamino)germanes have been made; the dimethylamino derivative is obtained in about 80% yield by the action of $GeBr_4$ on dimethylamine in cyclohexane at $-60\,^{\circ}C$ ⟨52JA1421⟩; $GeCl_4$ can also be used ⟨69IC279⟩. Pyridine forms a complex with $GeCl_4$ which reacts with diethylamine in cyclohexane to give tetrakis(diethylamino)germane in good yield ⟨58JCS3746⟩.

2.07.5.3 Compounds Containing a N—B Bond

A great many compounds containing a N—B bond exist but only those not containing a B—C bond will be considered here; cyclic borazines are also not dealt with (although they have been reviewed ⟨79COC(3)925, 82COMC-I(1)311⟩). Many amine–borane donor–acceptor complexes have been made by the reaction between diborane and the appropriate amine; there is an exhaustive review of N—B compounds ⟨83HOU(13/36)1⟩.

In the headings below the expression 'dialkylamino' should be taken to include both mono- and dialkylamino residues.

2.07.5.3.1 Dialkylaminoboranes, R_2NBX_2

Dialkylaminoboranes were first prepared almost a century ago from dialkylamines and BCl_3 ⟨1896CB710⟩. The reaction between HCl and tris(diethylamino)borane gives the hydrochloride (**52**) in 93% yield ⟨57JCS381⟩ and alane (**53**) reacts with BCl_3 to afford dichloro(dimethylamino)borane ⟨61JA2835⟩. The versatile tin compound (**32**) can also be used instead of (**53**), a yield of 81% is then obtained ⟨69JCS(A)992⟩.

$Et_2NBCl_2{\cdot}HCl$ (**52**) $(Me_2N)_3Al$ (**53**)

Dimethylamineborane can be persuaded to lose hydrogen to give an almost quantitative yield of (dimethylamino)borane ⟨49JA3451⟩; a similar process uses triethylamineborane with a dialkylamine to form a (dialkylamino)borane with loss of triethylamine and hydrogen (Equation (65)) ⟨68LA(720)1⟩.

$$R_2NH + H_3B \leftarrow NEt_3 \xrightarrow{\text{10 examples; 56-ca. 100\%}} R_2N{-}BH_2 + NEt_3 + H_2 \qquad (65)$$

2.07.5.3.2 Bis(dialkylamino)boranes, $(R_2N)_2BX$

Tris(diethylamino)borane reacts with BCl_3 at $-80\,^{\circ}C$ to give a 36% yield of chlorobis(diethylamino)borane ⟨57JCS381⟩ (the yield has been improved to 77% ⟨60JA6242⟩) and BBr_3 has been used in a similar process ⟨60JA6242⟩; dialkylamines can be used with BCl_3 instead of boranes ⟨62JCS1786⟩. Triethylamineborane with dialkylamines gives bis(dialkylamino)boranes with loss of hydrogen and triethylamine ⟨68LA(720)1⟩ whilst hydrogenation is used in an interesting reaction in which the gas, at 3000 psig, reacts with a dialkylamine, aluminum and triphenyl borate to give good yields of the title compounds ⟨71IC900⟩. The latter reaction proceeds through an alane intermediate as shown by the 96% yield of bis(diethylamino)borane which results on heating the corresponding alane with triphenyl borate at $55\,^{\circ}C$ (Scheme 69).

$$(PhO)_3B + R_2NH + Al \xrightarrow[\text{3 examples, 67–85\%}]{H_2\text{(3000 psig), benzene, 180 °C}} (R_2N)_2BH$$

$$(PhO)_3B + (Et_2N)_2AlH \xrightarrow[96\%]{55\ °C} (Et_2N)_2BH$$

Scheme 69

Bis(dialkylamino)chloroboranes will react with AgCNO to afford the corresponding isocyanates; substances with an alkoxy group on boron (borinates) can also be made this way ⟨65JCS4256⟩.

2.07.5.3.3 *Tris(dialkylamino)boranes, $(R_2N)_3B$*

Dialkylamines and BCl_3 give tris(dialkylamino)boranes ⟨57JCS381, 62JCS1786⟩ and primary amines may also be used ⟨59JCS2927⟩. Unsymmetrically substituted compounds may be made by the reaction of an amine with a bis(dialkylamino)chloroborane; however, both of these reactions are subject to steric hindrance ⟨62JCS1786⟩. Tris(dimethylamino)borane may be prepared by the reaction of the stannane (**32**) with trimethyl borate (98%), $BF_3 \cdot Et_2O$ (92%) or triethylamineborane (80%) ⟨69JCS(A)992⟩; the same compound, and related species, can be made from $BF_3 \cdot Me_2O$ and an alkylamine which yield a complex which is then treated with EtMgBr ⟨58ZAAC(294)81⟩.

In different approaches the dimethylamino derivative is prepared by the reaction of LAH with trimethylamine hydrochloride followed by treatment with tributyl borate ⟨62JOC1020⟩ and the diethylamino compound by the hydrogenation of a mixture of triphenyl borate, aluminum and diethylamine ⟨71IC900⟩. Exchange reactions between alkylamino groups can be carried out: *s*-butylamine reacts with tris(isopropylamino)borane on heating to 205 °C to give tris(*s*-butylamino)borane in 75% yield and this product in turn can be converted into tris(methylamino)borane in 40% yield ⟨61JA2122⟩.

2.07.5.3.4 *Miscellaneous compounds containing a N—B bond*

Replacement of the chlorine atoms of *n*-butoxydichloroborane by treatment with an amine gives a borinate (**54**) ⟨57JCS381, 59JCS2927⟩; reaction of *n*-butanol with tris(diethylamino)borane or chlorobis(diethylamino)borane in the presence of triethylamine leads to (**54**;$R^1R^2 = Et$) ⟨57JCS381⟩. When (**55**) reacts with ethylamine in pentane at −78 °C the boronate (**56**) is produced in 28% yield (Equation (66)) ⟨59JCS2927⟩.

$Bu^nOB(NR^1R^2)_2$
(54)

$$\underset{\textbf{(55)}}{(Bu^nO)_2BCl} \xrightarrow[28\%]{EtNH_2\text{, pentane, −78 °C}} \underset{\textbf{(56)}}{(Bu^nO)_2BNHEt} \qquad (66)$$

Treatment of dimethylamine with diborane at 135 °C leads to the substituted diborane (**57**) ⟨49JA3451⟩ and reaction of (**58**) with molten sodium in refluxing toluene gives the related substance (**59**) in 81% yield ⟨60JA6242⟩.

$Me_2NBH_2 \cdot BH_3$	$(Me_2N)_2BBr$	$(Me_2N)_2BB(NMe_2)_2$
(57)	**(58)**	**(59)**

2.07.6 COMPOUNDS WITH A N—METAL BOND

2.07.6.1 Compounds Containing a N—Alkali Metal Bond

2.07.6.1.1 Compounds containing a N—Li bond

Substances of this type, for example, LDA, have found extensive use in organic synthesis. Homochiral lithium amides are finding increasing synthetic use; they have recently been reviewed ⟨91TA1⟩.

Lithium amides have most often been made by the action of an alkyl or aryl lithium on the amine; some examples of this reaction are collected in Table 4. Slight alterations in the procedure for making LDA have been made for special purposes; thus, solutions free from LiCl ⟨62CJC449⟩ or hexane ⟨76JA2868⟩ have been prepared as has a solution with good storage stability ⟨78JOC700⟩.

Table 4 The preparation of R^1R^2NLi by the action of R^3Li on R^1R^2NH.

R^1R^2NH	*R^3 in R^3Li*	*Ref.*
Me_2NH	Bu^n	74JOM(78)107, 75TL1859
Et_2NH	Me	48JA3945
	Bu^n	75TL1859, 88OSC(6)948
	Ph	50JOC162, 53CB1483
Pr^i_2NH	Bu^n	75CB1293, 75TL1859, 92OS133, 93OS38, 93OS62
	Ph	50JOC162
Pr^iNHBu^t	Bu^n	81ZN(B)515
Oct^tNHBu^t	Me[a]	84TL491
$Pr^iNHSiMe_3$	Bu^n	81ZN(B)515
$Bu^tNHSiMe_3$	Bu^n	88SC253
Aziridine	Me	71LA(746)149
Pyrrolidine	Me	71LA(746)149
	Bu^n	75TL1859
Piperidine	Bu^n	75JOM(97)325, 75TL1859
Tetramethylpiperidine	Me	73JA581
	Bu^n	73JA581, 75TL1859, 92OS146
Cyclohexylamine	Bu^n	78AG(E)202
$CyclohexylNHPr^i$	Bu^n	71JA2318, 75CB1293
$(Cyclohexyl)_2NH$	Bu^n	67JOC510
Dihydroindole	Me	71LA(746)149
Tetrahydroisoquinoline	Me	71LA(746)149
$(PhCH_2)_2NNH_2$	Bu^n	70JOC960
$MeNH(CH_2)_2NHMe$	Bu^n	75TL1859
$Me_2N(CH_2)_2NHMe$	Bu^n	84JOC1078
$EtNH(CH_2)_3NHEt$	Bu^n	75TL1859
N-Methylpiperazine	Bu^n	84JOC1078

[a] There is hardly any reaction with Bu^nLi; the product is even more hindered than LITMP.

The disadvantage of the above method is that half of the lithium metal initially taken is lost as LiCl etc. during the preparation of the alkyl- or aryllithium. This is not important for small scale reactions but it becomes disadvantageous with molar amounts or higher and several alternative procedures have been suggested. Refluxing a mixture of ethylenediamine and lithium in benzene gives a 97% yield of the metal derivative which can be isolated as a crystalline solid ⟨63JOC2775, 73JOC489⟩ whilst diethylamine reacts at room temperature with lithium in benzene containing hexamethylphosphoric triamide ⟨75LA719⟩. The aziridine (**60**) is converted by lithium in the presence of a catalytic quantity of naphthalene via the naphthalenide radical-anion to give (**61**) (Equation (67)) ⟨93TL1649⟩.

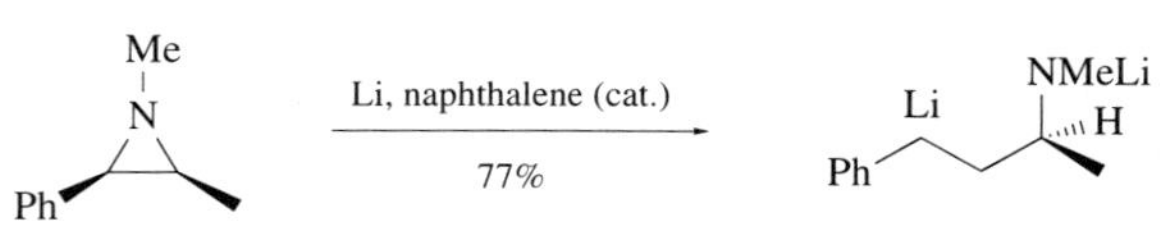

(**60**) (**61**) (67)

In reactions such as this, removal of the naphthalene at the end of the process may prove difficult and this problem may be avoided by the use of the radical anion (lithium 1-dimethylamino-

Table 5 The preparation of some metal amides.

Product		*Reagents*	*Ref.*
Me_2NBeH	P	Pr^iBeNMe_2, 220°C	54JCS22
$(Me_2N)_2Be$	T	$Pr^i{}_2Be + Me_2NH$, heat	54JCS22
$(Me_2N)_2Be$	T	$Et_2Be + Me_2NH$, RT	65CJC1884
$(Me_2N)_2Be$		$BeH_2 + Me_2NH$	54JCS2526
R_2NMgH		$(R_2N)_2Mg + MgH_2$	78IC1862
R_2NMgH		$R_2NH + MgH_2$	78IC1862
$R^1{}_2NMgH$		$R^1{}_2NMgR^2 + H_2$	71IC906
$Pr^i{}_2NMgH$		$Pr^i{}_2NMgEt + LiAlH_4$	71IC906
R_2NMgH		$R_2NMgBr + KH$	71IC906
R_2NMgBr		$R_2NH + EtMgBr$	49JOC310
Et_2NMgBr		$Et_2NH + EtMgBr$	53MI 207-02
Et_2NMgI		$Et_2NH + EtMgI$	55JA1796
$(R_2N)_2Mg$		$R_2NH + Me_2Mg$	78IC1862
$(Me_2N)_2Mg$	P	$Me_2NH + Et_2Mg$	67JCS(A)56
$(Pr^i{}_2N)_2Mg$		$Bu^nMgBu^s + Pr^i{}_2NH$	88TL139
$(Bu^n{}_2N)_2Mg$		$Bu^n{}_2NLi + MgBr_2$	69CR(C)834
R_2NCaCl		$Ph_3CCaCl \cdot 2THF + R_2NH$	69ZAAC(364)315
$(Bu^n{}_2N)_2Ca$		$Bu^n{}_2NLi + CaI_2$	69CR(C)834
$(R_2N)_4V$		$R_2NLi + VCl_4$	69JCS(A)980
$(R_2N)_4V$		$R_2NLi + VCl_3$	69CC1064
$Et_2NVO(OPr^i)_2$		$ClVO(OPr^i)_2 + Et_2NLi$	78ICA163
$(Et_2N)_2VO(OPr^i)$		$Cl_2VO(OPr^i) + Et_2NLi$	78ICA163
$(Et_2N)_3VO$		$Cl_3VO + Et_2NLi$	78ICA163
Et_2NNbF_4	P	Et_2N-TMS $+ NbF_5$	72JCS(D)1766
$(Et_2N)_2NbF_3$	P	Et_2N-TMS large excess $+ NbF_5$	72JCS(D)1766
$(Me_2N)_2NbCl_3$		$Me_2NH + NbCl_5$	57JCS2078
$(MeNH)_3NbCl_2$		$MeNH_2 + NbCl_5$	57JCS2078
$(R_2N)_4Nb$		$R_2NLi + NbCl_5$	62CJC449
$((CH_2)_5{=}N)_4Nb$		$(Et_2N)_4Nb$ + piperidine, prolonged boiling	62CJC449
$(Bu^nMeN)_4Nb$		$(Bu^nMeN)_5Nb$, heat in vacuum	62CJC449
$(Et_2N)_2((CH_2)_5{=}N)_2Nb$		$(Et_2N)_4Nb$ + piperidine	62CJC449
$(Et_2N)_3((CH_2)_5{=}N)Nb$		$(Et_2N)_2((CH_2)_5{=}N)_2Nb$, distil	62CJC449
$(Me_2N)(Et_2N)_3Nb$		$(Me_2N)_5Nb + Et_2NH$, prolonged boiling	62CJC449
$(R_2N)_5Nb$		$R_2NLi + NbCl_5$	62CJC449
$((CH_2)_5{=}N)_5Nb$		$(Me_2N)_5Nb$ + piperidine	62CJC449
$(Me_2N)_3(Et_2N)_2Nb$		$Et_2NH + (Me_2N)_5Nb$	62CJC449
Et_2NTaF_4	P	Et_2N-TMS $+ TaF_5$	72JCS(D)1766
$(Et_2N)_2TaF_3$	P	Et_2N-TMS large excess $+ TaF_5$	72JCS(D)1766
$(Bu^nMeN)_4Ta$		$(Bu^nMeN)_5Ta$, 160–180°C	62CJC1355
$(R_2N)_5Ta$		$R_2NLi + TaCl_5$	62CJC1355
$((CH_2)_5{=}N)_5Ta$		$(Me_2N)_5Ta$ + piperidine	62CJC1355
$(Et_2N)_3Cr$		$Et_2NLi + CrCl_3$	71JCS(A)1433
$(Pr^i{}_2N)_3Cr$		LDA $+ CrCl_3$	68CC495
$(Et_2N)_4Cr$		$(Et_2N)_3Cr$, 40–60°C, 10^{-3} mm	71JCS(A)1433
$(R_2N)_4Mo$		$R_2NLi + MoCl_3$	71JCS(C)2741
$(Et_2N)_2WF_4$		Et_2N-TMS $+ WF_6$	73JCS(D)1876
$(Et_2N)_2WF_4$		$(Et_2N)_4WF_2 + WF_6$	73JCS(D)1876
$(Et_2N)_4WF_2$		Et_2N-TMS $+ WF_6$	73JCS(D)1876
$(RNH)_4WCl_2$		$RNH_2 + WCl_6$	62JCS1330
$(Me_2N)_6W$		$Me_2NLi + WBr_5$	76JA4477
$(Me_2N)_6W$		$Me_2NLi + WOCl_4$	76JA4477
$(R_2N)_6W_2$		$R_2NLi + WCl_4$	76JA4477
$(Et_2N)_3Fe$ and/or $(Et_2N)_2Fe$		$Et_2NLi + FeCl_3$	65JOM(3)295
R^1R^2NCu		$R^1R^2NLi + (CuCl)_2$	67AG(E)633
$(Me_2N)_2Zn$	P	$Me_2NH + Me_2Zn$	65JCS1870
$PhCH_2NHZnI$		$PhCH_2NH_2 + MeZnI$	68JCS(C)1704
$(Et_2N)_2Zn$		$Et_2NLi + ZnCl_2$	69CR(C)834
$(Bu^n{}_2N)_2Cd$		$Bu^n{}_2NLi + CdCl_2$	69CR(C)834
Me_2NAlH_2	D	$Me_2NN{=}NNMe_2 + Me_3N \rightarrow AlH_3$	62CJC342
Me_2NAlH_2	D	$Me_2NH \cdot HCl + LiAlH_4$	62CJC342
Me_2NAlH_2		$Me_3N \rightarrow AlH_3 + (Me_2N)_3Al$	61JA2835
R_2NAlH_2		$R_2NH + Me_3N \rightarrow AlH_3$	71IC893
Bu^tNHAlH_2	D	$Bu^tNH_2 + AlH_3$	76ZN(B)697
$(Me_2N)_2AlH$	T	$Me_2NH + Me_3N \rightarrow AlH_3$	61JA2835
$(Me_2N)(Pr^iN)AlH$		$Pr^i{}_2NH + Me_2NAlH_2$	61JA2835
$(Me_2N)_2AlH$		$Me_2AlH_2 + NaH$	61JA535
$(Et_2N)_2AlH$		$Et_2NH + Al + H_2$	71IC893
$(R_2N)_2AlH$		$R_2NH + Me_3N \rightarrow AlH_3$	71IC893

Table 5 Continued.

Product		*Reagents*	*Ref.*
$(Me_2N)_2AlCl$		$Me_3N \rightarrow AlH_3 + AlCl_3$	61JA2835
$Bu^tNHAlHCl$	D	$Bu^tNHAlH_2 + HgCl_2$	76ZN(B)697
Me_2NAlCl_2		$Me_2NAlH_2 + HgCl_2$	70IC146
Me_2NAlCl_2		$(Me_2N)_3Al + AlCl_3$	70IC146
Me_2NAlCl_2		$Me_3N \rightarrow AlH_3 + AlCl_3$	61JA2835
$Bu^tNHAlCl_2$	D	$Bu^tNHAlH_2 + HgCl_2$	76ZN(B)697
$(Me_2N)_3Al$		$Me_2NH + Me_2NAlH_2$	70IC146
$(Me_2N)_3Al$	D	$Me_2NH + Me_3N \rightarrow AlH_3$	61JA2835
$(Me_2N)_3Al$		$Me_2NH + Me_3N \rightarrow AlH_3$	71IC893
$(Et_2N)_3Al$		$Et_2NH + Al + H_2$	71IC893
$(Et_2N)_3Al$		$Et_2NH + (Et_2N)_2AlH$	71IC893
$(Pr^i_2N)_3Al$		$Pr^i_2NH + Me_3N \rightarrow AlH_3$	61JA2835
$((CH_2)_5{=}N)_3Al$		piperidine + $Me_3N \rightarrow AlH_3$	71IC893
$(Me_2N)_4AlLi$		$Me_2NH + LiAlH_4$	61JA2835
$(Et_2N)_4AlLi$		$Et_2NH + LiAlH_4$	71IC1888
$((CH_2)_5{=}N)_4AlLi$		piperidine + $LiAlH_4$	71IC1888
$(Et_2N)_3HAlLi$		$Et_2NH + LiAlH_4$	71IC1888
$Me_2NAl_2H_5$	M	Me_2NAlH_2 or $Me_2NAlHCl + NaAlH_4$	64JA5359
$Et_2NAl_2H_6Li$		$Et_2NAlH_2 + LiAlH_4$	71IC1888
Me_2NGaH_2	D	$Me_2NH + GaH_3$	66JCS(A)706
$MeNHGaH_2$	T	$MeNH_2 + Me_3N \rightarrow GaH_3$	68JCS(A)2605
$RNHGaH_2$		$RNH_2 + Me_3N \rightarrow GaH_3$	71JCS(A)1237
$(CH_2)_n{=}NGaH_2$		$(CH_2)_n = NH + Me_3N \rightarrow GaH_3$	72JCS(D)326
$(Me_2N)_2GaCl$		$Me_2NLi + GaCl_3$	75ZN(B)681
$(Me_2N)_2GaCl$		$(Me_2N)_3Ga + GaCl_3$	75ZN(B)681
$(Me_2N)_3Ga$	D	$Me_2NLi + GaCl_3$	75ZN(B)681
$(Me_2N)_3Ga$		$LiGa(Me_2N)_4 + GaCl_3$	75ZN(B)681
$(Me_2N)_4GaLi$		$Me_2NLi + GaCl_3$	75ZN(B)681
$(Me_2N)_4Sn$		$Me_2NLi + SnCl_4$	65JCS1944
$(Et_2N)_4Sn$		$Et_2NLi + SnCl_4$	61CJC1386
$(Et_2N)_4Sn$		$Et_2NLi + SnCl_4$	65JCS1944
$(Et_2N)_4U$	M	$Et_2NLi + UCl_4$	56JA4285

D = dimer, M = monomer, P = polymer, T = trimer.

naphthalenide (LDMAN)) derived from the readily available 1-dimethylaminonaphthalene since the latter may be easily removed by means of an acid wash ⟨80SC311⟩.

Lithium, isopropylamine and styrene react in refluxing ether to give LDA ⟨80LA1471⟩ and this Ziegler-type procedure, which is suitable for large scale use, has been modified to use α-methylstyrene ⟨84T2211⟩. An earlier method which also gave excellent results used α-methylstyrene or phenanthrene with the amine and lithium in ether/THF ⟨79S463⟩.

Excellent yields of LDA and lithium tetramethylpiperidide (LITMP) can be made by a Barbier-type process in which a mixture of the amine, lithium and *n*-butyl chloride is sonicated ⟨87JOC4124⟩; the procedure has been extended to the lithiation of a diamine ⟨88CC1350⟩.

2.07.6.1.2 Compounds containing a N—Na bond

These have usually been made by the reaction of sodamide or sodium hydride on the amine. The sodium equivalent of LDA was prepared, however, from phenyl sodium and isopropylamine ⟨60JA472⟩. Heating dialkylamines with NaH in THF produced the required sodio derivatives ⟨76S462⟩ and the same strong base was used to make the interesting substituted sodium borohydride (**62**) ⟨84JOC2438⟩. The reaction between sodium or sodamide and an amine or diamine has been used to prepare many sodio-derivatives ⟨73JOC489⟩. The sodium analogue (NAPA) of the very powerful base potassium 3-aminopropylamide (KAPA) can be made by treating 1,3-diaminopropane, in liquid ammonia, with sodamide (from Na and $NH_3(1)$ in the presence of Fe^{+++}) ⟨77RTC160⟩ or by sonicating a mixture of the amine and the metal at 90 °C with ferric ion present ⟨84JOC2494⟩.

$NaBH_3(NMe_2)$

(62)

Alternatively, NAPA can be prepared from the amine and NaH at 70°C without the foaming problems encountered in the preparation of KAPA using KH; it appeared that NAPA was equally as effective as KAPA in the reactions examined ⟨80JOC734⟩.

2.07.6.1.3 Compounds containing a N—K bond

The potassium derivative of pyrrolidine can be made by the action of KH on the amine ⟨74JOC3913, 73JA982⟩. The reaction between potassium *t*-butoxide, *n*-butyllithium and diisopropylamine allows the potassium analogue of LDA to be prepared ⟨78JOC3794⟩. KAPA can be made from KH and 1,3-diaminopropane ⟨76TL2565, 87OS(65)224⟩ or by the liquid ammonia or sonication methods mentioned in connection with the preparation of NAPA. A potassium-containing strong base which behaves like KAPA can be made from the lithiated diamine and $KOBu^t$ ⟨87OS(66)127⟩.

2.07.6.2 Compounds Containing a N—Metal Bond (not N—Alkali Metal)

The metal amides have been comprehensively reviewed ⟨B-80MI 207-01⟩. The methods used to make these compounds are limited in number; they are illustrated by the examples in Table 5 which are mainly derived from the above reference.

2.08
Alkylphosphorus Compounds

JOHN C. TEBBY, DANIEL G. GENOV and JOHN W. WHEELER
Staffordshire University, Stoke-on-Trent, UK

2.08.1 INTRODUCTION

Phosphorus is introduced into organic molecules using relatively few basic inorganic reagents. Probably the most important of these reagents are the phosphorus halides, usually phosphorus trichloride, and also phosphorus pentachloride and phosphoryl trichloride. Phosphorous acid (H_3PO_3), its esters and phosphoric anhydride are also important. White phosphorus and phosphine (PH_3) are both readily available, but have found less utility. There are a number of commercially available organic phosphorus compounds, for example the trialkyl phosphites, trialkylphosphines and various, mixed functionality compounds, such as mono- and dialkyl chlorophosphites, which are used extensively as starting materials.

2.08.2 ALKYLPHOSPHINES—RPH_2, R_2PH, R_3P—AND SALTS THEREOF—$R_4P^+X^-$, ETC.

Methods for the synthesis of phosphines were reviewed comprehensively by Maier in *Organic Phosphorus Compounds* ⟨B-72MI 208-01⟩ and also by Elsner ⟨82HOU106⟩. Also, the preparation and chemistry of primary, secondary and tertiary phosphines have been reviewed in *The Chemistry of Organophosphorus Compounds* ⟨B-90MI 208-01⟩. The synthesis and chemistry of quaternary phosphonium salts has been covered by Beck, also in *Organic Phosphorus Compounds* ⟨B-72MI 208-03⟩. In addition, material since 1969 has been reviewed on an annual basis in the Specialist Periodical Reports of the Royal Society of Chemistry entitled *Organophosphorus Chemistry* ⟨70MI 208-01⟩.

2.08.2.1 Primary Phosphines—RPH_2

2.08.2.1.1 Primary phosphines from PH_4I and alkylating agents

Among the oldest known methods for the preparation of phosphines are the reactions of phosphonium iodide with alkyl halides in the presence of zinc oxide or powdered zinc ⟨1871CB430, 1871CB605, 1873CB292⟩. A complex mixture of primary, secondary and tertiary phosphines, as well as phosphonium salts, is formed, but these compounds can easily be separated, due to their different basicities.

2.08.2.1.2 Primary phosphines from metallated phosphines

Monometallated phosphine will react with an organohalide to give a primary phosphine (Equation (1)).

$$MPH_2 + RX \longrightarrow RPH_2 + MX \qquad (1)$$

The metal used is usually sodium, potassium, calcium or lithium ⟨48JA2295, 53JA3869, 82HOU106⟩, and the reaction is often carried out in the liquid ammonia used as the solvent for phosphide generation. The phosphide may also be prepared by using an aryl- or alkyllithium base ⟨54CB919, 59CB1118, 59JA6273⟩. The metal phosphide may react with alkyl halides to yield simple primary phosphines ⟨48JA2295, 76ZAAC(425)209⟩. For example, monopotassium phosphide reacts with allyl(2-chloroethyl)amine, to give the primary phosphine (**1**), which, on treatment with formaldehyde, is converted into the bicyclic tertiary phosphine (**2**) (Scheme 1) ⟨85ZAAC(523)7⟩. Metal phosphides can also react with sodium salts of halocarboxylic acids ⟨67CB3331⟩, β-chloroethylamines ⟨68CB3612⟩, epoxides ⟨60ZOB784⟩ and episulfides ⟨65FRP1401930⟩ to form primary phosphines.

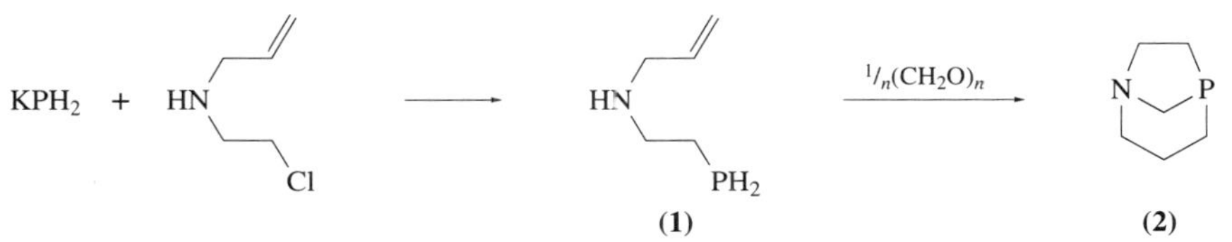

Scheme 1

The most successful modification of this method is the selective alkylation of phosphine with alkyl halides, in the presence of concentrated potassium hydroxide in DMSO ⟨90ZN(B)203⟩. Hence, reaction of PH_3 with methyl iodide under these conditions produces a high yield of methylphosphine (75%) ⟨68IS124⟩.

The synthesis of primary phosphines by the selective monoalkylation of the phosphine–aluminum chloride complex is an alternative strategy ⟨62M230, 66JOC3817⟩. For example, heating 1,3-disubstituted haloadamantanes with the phosphine–aluminum chloride complex, provides a high-yielding route to the functionally substituted adamantylphosphines (**3**) ⟨91ZOB1021⟩.

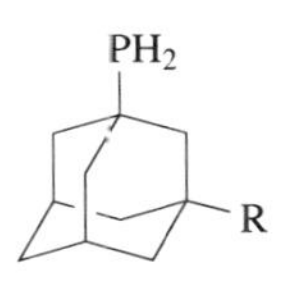

(3)
R = Cl, Br, Me

2.08.2.1.3 *Primary phosphines by addition of P—H to unsaturated compounds*

The addition of P—H-containing compounds to multiple bonds may be achieved using acid catalysis ⟨52USP2584112, 59JOC356⟩, but, generally, either base-catalysed or free radical methods are used. Addition reactions of this type are often employed as a route for the preparation of tertiary phosphines (see Section 2.08.2.3.4), although several examples have been reported for the synthesis of primary (and secondary) phosphines.

The base-catalysed process has been known for a long time ⟨B-72MI 208-01⟩, an example of which is the reaction of PH_3 with acrylonitrile. Although mixtures of primary, secondary and tertiary phosphines are always obtained, the degree of substitution may be controlled to a high degree—tris(2-cyanoethyl)phosphine (80% yield) was obtained by adding PH_3 to an excess of acrylonitrile, bis(2-cyanoethyl)phosphine (63% yield) was prepared by the slow addition of acrylonitrile to a mixture kept saturated with phosphine, and the reaction carried out under a phosphine pressure of 28–32 atm gave 2-cyanoethylphosphine (52% yield) ⟨59JA1103⟩.

The free radical reaction is often the most useful strategy that may be adopted and, again, the distribution of products may be, to a certain extent, controlled. For example, the reaction of a high concentration of PH_3 with oct-1-ene gives a good yield of the primary phosphine product ⟨61JOC5138⟩. Also, bulky alkenes shift the product ratio in favour of the less substituted phosphine ⟨61JOC5138⟩. The addition of trimethylsilylphosphine to various alkenes (including vinylphosphine) and dienes leads to a wide range of acyclic and cyclic phosphines, from which the trimethylsilyl group may easily be removed by hydrolysis, thus allowing the selective synthesis of primary and secondary phosphines ⟨84IC4130⟩.

It has been shown that a series of zirconium phosphido complexes, for example $[(ZrCp^*_2(PH_2)(H))_2]$, promote the formation of primary alkylphosphines under extremely mild conditions ⟨92CC699⟩.

2.08.2.1.4 *Primary phosphines by reduction of phosphonous and phosphonic compounds*

A highly exploited route for the synthesis of primary phosphines is the reduction of phosphonous (Equation (2)) and phosphonic (Equation (3)) compounds by a variety of reducing agents.

$$RPX_2 \longrightarrow RPH_2 \qquad (2)$$

(3)

The reduction of phosphonous dihalides with LAH is one of the most widely used methods for the preparation of primary phosphines ⟨59AG574, 59M148⟩. If the reduction is undertaken at low temperature with a slow addition of LAH, the yields may be greatly enhanced ⟨78ZAAC(443)42⟩. The reducing agent should be used in excess to avoid the formation of by-products ⟨63MI 208-01⟩.

LAH has also been used to reduce alkylphosphonic dichlorides ⟨58CB1583, 59M148, 78ZAAC(443)42, 88ZOB33⟩, alkylphosphonates ⟨58CB1583, 59M148, 69JINC3684, 78ZAAC(443)42, 87ZN(B)579⟩ and acylphosphonates ⟨90ZOB697⟩ to the corresponding primary phosphines. Primary phosphines have also been prepared by the reduction of alkylphosphonates with diphenylsilane ⟨84ZC261, 65CB1681⟩.

The synthesis of bis(primary phosphines)alkanes (**4**) has been achieved by the LAH reduction of bis(dichlorophosphino)alkanes ⟨84CB3400, 84ZC261⟩, alkylene bis(phosphonites) ⟨62CB473⟩ and alkylene bis(phosphonates) ⟨66HCA842, 80CB1356, 90ZAAC(589)187⟩. LAH reduction of alkylene bis(phosphonic dichlorides) yielded the *gem*-diphosphine alkanes (**5**) ⟨66JOC3391⟩.

$H_2P(CH_2)_nPH_2$

(4)

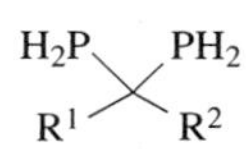

(5)

R^1 = H, *n*-alkyl
R^2 = H, Me, Bu^n

2.08.2.2 Secondary Phosphines—R_2PH

Methods for the synthesis of secondary alkylphosphines are generally similar to those adopted for the synthesis of primary alkylphosphines, since care must be exercised in order to avoid oversubstitution.

2.08.2.2.1 Secondary phosphines from phosphines and alkylating agents

Secondary alkylphosphines can be obtained in reasonable yields by the reaction of methylphosphine with alkyl halides ⟨66JOC3817, 69JINC3684⟩, the products being obtained in the form of HI salts, from which they are liberated by alkali. Primary phosphines react with chloroformate esters, for example (**6**), to yield the secondary phosphines (**7**), which may subsequently undergo free radical intramolecular cyclization to yield the tertiary phosphine heterocycles (**8**) (Scheme 2) ⟨82ZN(B)965⟩.

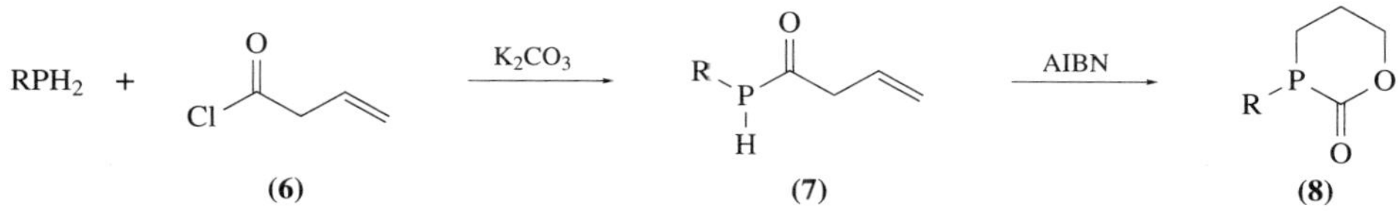

(6) **(7)** **(8)**

$R = c\text{-}C_6H_{11}$, Ph, CH_2CH_2CN
AIBN = 2,2'-azobisisobutyronitrile

Scheme 2

2.08.2.2.2 Secondary phosphines from metallated phosphines

Two approaches are available for the synthesis of secondary alkylphosphines using the metallated phosphine method. The reaction of dimetallated phosphine with a suitable alkylating agent will yield a symmetrical secondary phosphine (Equation (4)), or alkylation of a monometallated primary phosphine may lead to a symmetrical or asymmetrical secondary phosphine, depending upon the reactants chosen (Equation (5)).

$$M_2PH + 2RX \longrightarrow R_2PH + 2MX \quad (4)$$

$$MR^1PH + R^2X \longrightarrow R^1R^2PH + MX \quad (5)$$

The second of these two strategies is by far the most useful and commonly exploited. The metal phosphide required for the reaction is made by the action of an alkali metal, usually sodium or potassium, on a primary phosphine ⟨48USP2437797, 53JA3869, 59M792, 61CB2664⟩. The subsequent reaction with an organohalide may be carried out in the liquid ammonia in which the phosphide is generated. A more convenient procedure is to reflux finely divided metal with the phosphine in an inert solvent, keeping the temperature between 50 °C and 80 °C, since at temperatures greater than 90 °C a second substitution may occur. However, if the dimetallated phosphine is the only product obtained, then interaction with another equivalent of primary phosphine will yield the monosubstituted phosphide (Equations (6) and (7)) ⟨59M792⟩.

$$RPH_2 + 2Na \longrightarrow RPNa_2 + H_2 \quad (6)$$

$$RPNa_2 + RPH_2 \longrightarrow 2RPHNa \quad (7)$$

Monometallated phosphides have also been generated by the use of sodium dihydrobis(2-methoxyethanolato)aluminate ⟨83BCJ219⟩.

Treatment of monoalkali phosphides with various alkylating agents allows a diverse range of secondary phosphines to be accessed. Hence, the reaction with alkyl halides yields simple dialkylphosphines ⟨48USP2437795, 53JA3869, 59M792⟩, and bis(secondary phosphines) are produced in the reaction of alkali phosphides with α,ω-dihaloalkanes ⟨61CB2656, 61CB2664, 63BRP921463, 87CB1707⟩.

Metal phosphides also react with sodium salts of halocarboxylic acids ⟨65AG172⟩, β-chloroethylamines ⟨67CB2685, 68CB3612⟩, epoxides ⟨66ZOB484⟩ and episulfides ⟨65FRP1401930, 67ZN(B)1238, 82ZAAC(491)163⟩ to yield functionalized secondary phosphines. Also, metallated primary phosphines have been used in the phosphinylation of haloalkylpyridines to give the phosphines (**9**) ⟨87ZN(B)579⟩. Phosphiranes (**10**) have been prepared by the reaction of sodium derivatives of phosphine or primary phosphines with 1,2-dichloroalkanes in liquid ammonia ⟨67JA1102, 69T1097⟩. The synthesis and chemistry of phosphiranes have been reviewed ⟨90CRV997⟩.

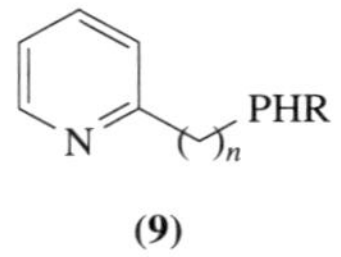

(9)

R = Pr^i, Bu^t, Ph

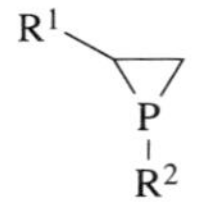

(10)

R^1 = H, Me

R^2 = H, D, Me, Ph

Secondary phosphines may alternatively be synthesized by the alkylation of primary phosphines with alkyl halides, in the presence of concentrated potassium hydroxide in DMSO ⟨86S198, 90ZN(B)203⟩. Dimethylphosphine has been prepared in 65% yield by the reaction of PH_3 with methyl iodide under these conditions ⟨68IS126⟩.

2.08.2.2.3 Secondary phosphines by addition of P—H to unsaturated compounds

The addition of P—H-containing compounds to unsaturated bonds generally results in the formation of a range of products that represent all possible degrees of substitution, although the reaction conditions often may be optimized to favour the preparation of a particular phosphine. See Sections 2.08.2.1.3 and 2.08.2.3.4 for further details.

The synthesis of a series of bis(secondary phosphines) (**11**) was achieved by the AIBN-initiated free radical reaction of phenylphosphine with several dialkenyl compounds ⟨69JOM(17)201⟩. The addition of phosphine to cod gave a mixture of isomeric secondary phosphines, corresponding to the 1,4- and 1,5-addition products. Subsequent alkylation, followed by treatment of the intermediate phosphonium salts with sodium hydride, yielded bicyclic tertiary phosphines, for example (**12**) ⟨85IC4437⟩.

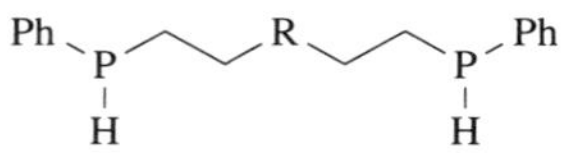

(11)

R = $-(CH_2)_6-$, $-C_6H_4-$, $-CH_2OCH_2-$

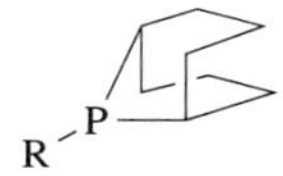

(12)

R = n-C_6H_{13}

2.08.2.2.4 Secondary phosphines by reduction

(i) Reduction of phosphinous and phosphinic compounds

The reduction of phosphinous (Equation (8)) and phosphinic (Equation (9)) compounds using a variety of reducing agents is a widely adopted route for the synthesis of secondary phosphines. The same reducing agents and reaction conditions as those for the reduction of phosphonous and phosphonic compounds are employed, LAH being the most commonly used reagent.

$$R_2PX \longrightarrow R_2PH \qquad (8)$$

$$R_2P(=O)X \longrightarrow R_2PH \qquad (9)$$

Secondary phosphines have been obtained by the reduction of alkylphosphinous halides ⟨70ZAAC(379)56, 83IC547, 84CB3400, 85TL1717⟩, alkylphosphinic halides ⟨58CB1583, 59JOC2013, 90ZOB1795⟩, alkylphosphinates ⟨63CB3184, 68CB2197, 77PS(3)203, 80CB1356⟩ and alkylphosphinous acids ⟨58CB1583⟩, using LAH.

Secondary phosphines, such as the cyclic phosphine (**13**), have been prepared by the reduction of aminophosphines using $NaBH_4$ ⟨60JA2145, 60JA2148⟩, the reaction proceeding via the formation of a borine adduct, which decomposes upon heating to give the required product.

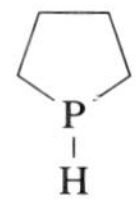

(13)

Silanes have been used in the preparation of secondary phosphines from phosphinous acids ⟨71T4245⟩ and phosphinates ⟨65CB1681⟩. The final step in the synthesis of the cyclic bisphosphine (**14**) was achieved by the reduction of the corresponding diphosphinate ester using diphenylsilane ⟨90CZ211⟩.

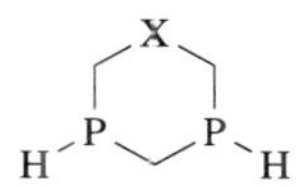

(14)
X = CH_2, S

Combination reagents are also effective in the reduction of phosphinic compounds. For example, LAH has been used in combination with TMS-Cl ⟨83OM1877⟩ in the synthesis of the macrocyclic secondary phosphine (**15**) ⟨85IC1613⟩. Sodium dihydrobis(2-methoxyethoxy)aluminate has been employed for the reduction of phosphinates to the corresponding secondary phosphines ⟨91ECJ2398⟩.

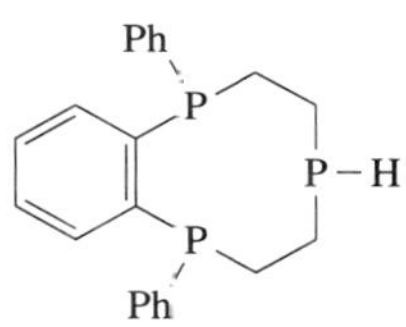

(15)

Prior to the development of silanes as reducing agents, alkali metals were extensively used for the production of dialkylphosphines. The reduction of phosphinous halides produces, initially, the tetraalkylbiphosphine which reacts with excess metal to give the phosphide. Subsequent hydrolysis yields the secondary phosphine ⟨59BRP823483, 60USP2959621, 62CB64⟩.

(ii) Reduction of organosubstituted biphosphines and biphosphine sulfides

A convenient method for the preparation of secondary phosphines, due to the accessibility of the starting material ⟨B-72MI 208-02⟩, is the reduction of organo-substituted biphosphine disulfides with LAH ⟨59CB704, 61CB3056, 62USP3031509, 68IS157⟩. Homolytic cleavage of P—P bonds in biphosphines ⟨60AG210, 60CB1852, 62CB64⟩ and biphosphine disulfides ⟨60CB1852⟩ is also a source of alkali dialkylphosphides, which, upon hydrolysis, yield secondary phosphines. Good yields of secondary phosphines are afforded by the desulfurization of alkylbiphosphine disulfides using tributylphosphine in a protic medium ⟨79ZN(B)642, 84IC4170⟩. Catalytic reduction of biphosphine disulfides with hydrogen and Raney copper has been used to produce secondary phosphines in high yields ⟨60AG210⟩.

(iii) Reductive cleavage of P—C bonds

The treatment of a tertiary phosphine with an alkali metal may result in the reductive cleavage of a P—C bond, provided that the phosphine contains at least one aromatic substituent ⟨61CB392⟩. The reaction produces initially the diorganophosphide (Equation (10)), which may be hydrolysed to the secondary phosphine ⟨74MI 208-01⟩ or used directly in a subsequent reaction to synthesize a new tertiary phosphine ⟨87JOM(323)C1⟩.

(10)

Lithium, sodium and potassium metals have been used for the reaction, either in liquid ammonia ⟨67AG(E)355, 70ZAAC(376)37, 92RTC170⟩ or in an inert solvent such as THF, dioxan or naphthalene ⟨61CB392, 84JOC2906, 86JOM(312)53, 92RTC170⟩. Ultrasound has been found to accelerate the production of lithium and potassium diorganophosphides by this method ⟨85JCR(S)18, 85JOC4329⟩. By employing this route, a phenyl substituent has been used as a protecting group, its cleavage resulting in the liberation of the desired secondary phosphine, for example in the synthesis of the cyclic phosphine **(16)** from **(17)** (Equation (11)) ⟨84JOC2906⟩.

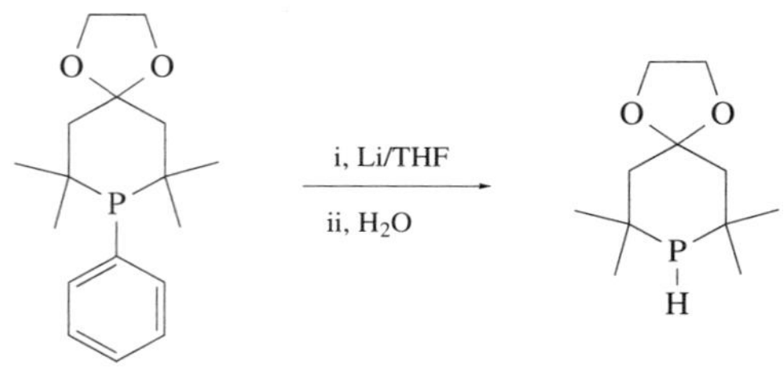

(11)

(17) **(16)**

The reductive cleavage of substituents from diarylalkyl- or aryldialkylphosphines may produce a mixture of products. It has been proposed that lithium/THF favours the cleavage of an aromatic substituent, whereas an alkyl substituent may be cleaved using potassium/dioxan, although this depends largely upon the nature of all three substituents of the tertiary phosphine ⟨74MI 208-01⟩. Lithium/THF and sodium/NH_3 often give complementary results in the formation of preparatively interesting secondary phosphines, by the selective cleavage of benzyl, 2-methoxyphenyl and 2,6-dimethoxyphenyl substituents ⟨92RTC170⟩. It has been proposed that the substituent undergoing cleavage is the one able to form the most stable radical, and, hence, based upon experimental results, a cleavage series was deduced (Figure 1).

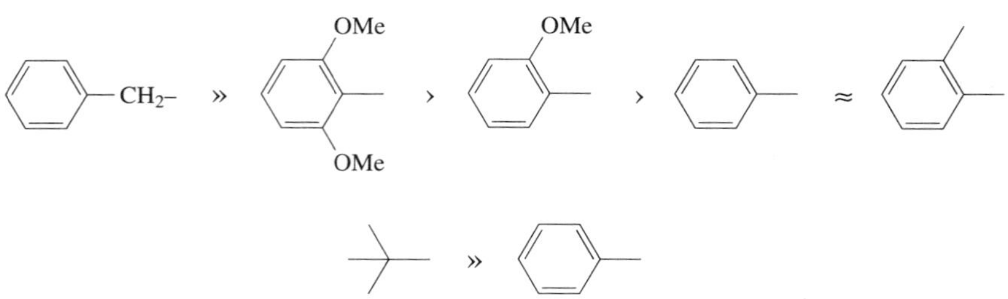

Figure 1 Cleavage series for substituents indicating the most stable radicals.

2.08.2.3 Tertiary Phosphines—R_3P

2.08.2.3.1 Tertiary phosphines from phosphines and alkylating agents

While alkylation of PH_3 or PH_4I with alkylating agents usually gives a mixture of products (see Section 2.08.2.1.1), high yields of tertiary phosphines have been obtained when primary (R^1 = alkyl, R^2 = H) ⟨66JOC3817⟩ or secondary phosphines (R^1, R^2 = alkyl) ⟨69JINC3684, 83OM979⟩ were used as starting materials. The reaction yields initially the phosphonium salt (Equation (12)), from which the phosphine is liberated upon treatment with alkali (Equation (13)).

$$R^1R^2PH + R^3X \longrightarrow [R^1R^2R^3PH]^+X^- \quad (12)$$

$$[R^1R^2R^3PH]^+X^- + NaOH \longrightarrow R^1R^2R^3P + NaX + H_2O \quad (13)$$

Alkyl iodides are generally used as alkylating agents, since the stability of the trialkylphosphonium salt formed disfavours dissociation of a proton and further reaction to form the quaternary phosphonium species ⟨60JA5794, 66CB1134⟩. Dialkylphosphino-substituted carboxylic esters have been prepared using this alkylation procedure ⟨61CB2244, 87URP1294811⟩. Reaction of α,ω-dichloroalkanes with primary phosphines may yield cyclic tertiary phosphines, such as the preparation of (**18**), which proceeds via the formation of the phosphonium salt (**19**) (Scheme 3) ⟨63CB2186⟩.

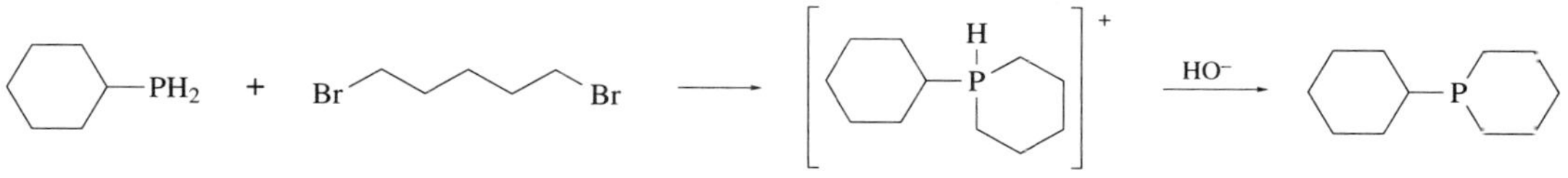

(19) **(18)**

Scheme 3

Cyanoalkyl- ⟨91ZN(B)1578, 92ZN(B)155⟩ and aminoalkylphosphines ⟨91ZOB1016⟩ may be synthesized by the reaction of (trimethylsilyl)phosphines with suitable alkylating agents. For example, the reaction of tris(trimethylsilyl)phosphine with chloromethylamines yields the tris(aminomethyl) phosphines (**20**) (Equation (14)) ⟨91ZOB1016⟩.

$$(TMS)_3P + ClCH_2NR_2 \longrightarrow P(CH_2NR_2)_3 \quad (14)$$

(20)

R = Et, Pr, Bu^i

2.08.2.3.2 *Tertiary phosphines from electrophilic phosphorus and organometallic reagents*

This strategy is most useful for the synthesis of tertiary phosphines (Equations (15)–(18)), and has been employed extensively for this purpose.

$$PX_3 + 3R^- \longrightarrow R_3P + 3X^- \quad (15)$$

$$R^1PX_2 + 2(R^2)^- \longrightarrow R^1R^2{}_2P + 2X^- \quad (16)$$

$$R^1{}_2PX + (R^2)^- \longrightarrow R^1{}_2R^2P + X^- \quad (17)$$

$$R^1R^2PX + (R^3)^- \longrightarrow R^1R^2R^3P + X^- \quad (18)$$

Both Grignard reagents and alkali organic compounds have been used widely in reactions following this route, and examples of the use of various other organometallic reagents have also been reported. This method is far less useful for the preparation of primary and secondary phosphines, as inaccessible halophosphines would be required as starting materials (Equations (16)–(18); R^1 = H).

(i) Phosphines of the form R_3P

The synthesis of symmetrical tertiary phosphines occurs according to Equation (15). Following standard Grignard procedures, phosphorus trichloride is added to an appropriate Grignard reagent (often used in excess) in diethyl ether. Refluxing and subsequent hydrolytic workup yields the required trialkylphosphine (Equation (19)) ⟨29JCS1262, 54ZAAC(277)258, 59USP2912465, 67NEP6614945, 79JCS(P1)3029⟩.

$$PCl_3 + 3RMgX \longrightarrow R_3P + 3MgXCl \quad (19)$$

Primary Grignard reagents usually give the highest yields, whereas those reagents derived from secondary ⟨29JCS1262⟩ and tertiary ⟨66CB1134, 67CB692⟩ alkyl halides may give little or no tertiary phosphine, due to the difficulties associated with the formation of the sterically hindered product. Often, products are isolated in which full substitution has not occurred, for example as seen in the reaction of *t*-butylmagnesium chloride with phosphorus trichloride (Equation (20)) ⟨63RTC302⟩. However, further reaction of di-*t*-butylphosphinous chloride with *t*-butylmagnesium chloride affords the required tri-*t*-butylphosphine ⟨67CB692⟩.

$$PCl_3 + Bu^tMgX\ (excess) \longrightarrow Bu^t_2PCl \quad (20)$$

Tertiary phosphines bearing ether (**21**) ⟨86ZOB2794, 87ZOB884⟩ and polyether (**22**) ⟨82CL977⟩ substituents have been prepared by the reaction of phosphorus chlorides with Grignard reagents.

$R^1_nP(CH_2OR^2)_{3-n}$

(21)

R^1 = Me, Et; R^2 = Ph, Pr^i; n = 0–2

$P(CH_2CH_2CH_2OR)_3$

(22)

R = Me, CH_2CH_2OMe

A modification of the original method is to use an alternative source of electrophilic phosphorus. Hence, the reaction of triphenylphosphite with methylmagnesium bromide ⟨90IS305⟩ or iodide ⟨87GEP3612629⟩ affords high yields of trimethylphosphine in large quantities and high purity, respectively. Also, high yields of tertiary alkylphosphines have been obtained from the reactions of Grignard reagents with the phosphorochloridite (**23**) ⟨90IC3502⟩, as in the preparation of triethylphosphine (**24**) (Equation (21)).

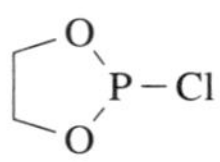

$$+ 3EtMgBr \longrightarrow PEt_3 \quad (21)$$

(23) **(24)**

The reaction of organolithium compounds with phosphorus trichloride has been used extensively for the preparation of tertiary phosphines (Equation (22)) ⟨62JOC2573, 73JINC2659⟩.

$$3RLi + PCl_3 \longrightarrow R_3P + 3LiCl \quad (22)$$

The procedures adopted are similar to those involving Grignard reagents, and give comparable yields. Due to the greater reactivity of organolithium compounds, this method may be used when a Grignard route has failed. Thus, for the synthesis of sterically hindered phosphines (e.g., tri-*t*-butylphosphine) the use of organolithium compounds is essential to yield the required product ⟨67CB692, 85JCR(S)330⟩.

Alternative sources of electrophilic phosphorus have been employed in reactions with organolithium compounds. The reaction of pentachlorophosphorane with isopropyllithium affords triisopropylphosphine in good yield ⟨78URP635102⟩. Tricyclopropylphosphine has been synthesized by the reaction of cyclopropyllithium with triphenyl phosphite ⟨67JOC2445, 69JA2915⟩. In a similar reaction, tris(methylthiomethyl)phosphine (**25**) is prepared by the reaction of methylthiomethyllithium with triphenyl phosphite (Equation (23)) ⟨67JOC1717⟩.

$$P(OPh)_3 + 3MeSCH_2Li \longrightarrow P(CH_2SMe)_3 + 3LiOPh \quad (23)$$

(25)

Organometallic reagents other than Grignard and organolithium have occasionally been used for the preparation of symmetrical tertiary phosphines. Trialkylaluminum compounds react with PCl_3 to give a range of trialkylphosphines ⟨58MI 208-01, 68USP3392180⟩, and bis(trifluoromethyl) cadmium converts phosphorus triiodide to the fully substituted tris(trifluoromethyl)phosphine at ambient temperature ⟨80CC671⟩.

(ii) Phosphines of the form $R^1{}_2R^2P$

Unsymmetrical tertiary phosphines of the form $R^1{}_2R^2P$ may be synthesized using organometallic reagents according to Equations (16) and (17), depending upon whether the phosphonous (Equation (16)) or phosphinous (Equation (17)) starting materials are utilized. The reaction of alkyl- or phenylphosphonous dihalides with two equivalents of alkyl Grignard reagents yields a diverse range of trialkyl- and dialkylphenylphosphines, respectively (Equation (24)) ⟨29JCS1262, 67JOC781, 78IS169⟩.

$$R^1PCl_2 + 2R^2MgX \longrightarrow R^1R^2{}_2P + 2MgXCl \qquad (24)$$

$$R^1 = \text{alkyl, Ph}$$

The treatment of a phosphonous dichloride with methylmagnesium halides has been employed as the ultimate step in the introduction of a tertiary phosphine into a molecule ⟨85JOM(296)411⟩, as in the synthesis of dimethyl(7-norbornyl)phosphine (**26**) ⟨78JOC3508⟩. Also, alternative sources of electrophilic phosphorus compounds, such as diethyl and diphenyl phosphinites, undergo alkylation by Grignard reagents ⟨69JOC2684, 83ZN(B)1027⟩.

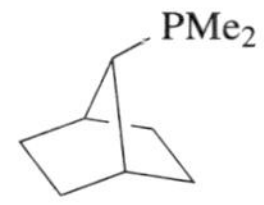

(26)

Cyclic tertiary phosphines have been prepared by the reaction of alkylphosphonous dichlorides with bis-Grignard reagents. For example, phosphoranes ⟨58USP2853518⟩ and larger phosphorus heterocyclic systems, for example (**27**) ⟨91PS(61)261⟩, have been accessed by this route. The reaction of *t*-butylphosphonous dichloride with ethynylmagnesium bromide yields *t*-butyldiethynylphosphine, which, upon double deprotonation of the alkynyl substituents and further treatment with *t*-butylphosphonous dichloride, leads to the formation of macroheterocyclic phosphine systems, such as (**28**) ⟨90JA7823⟩.

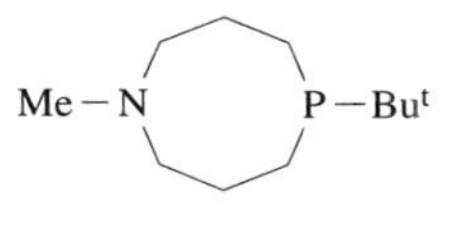

(27)

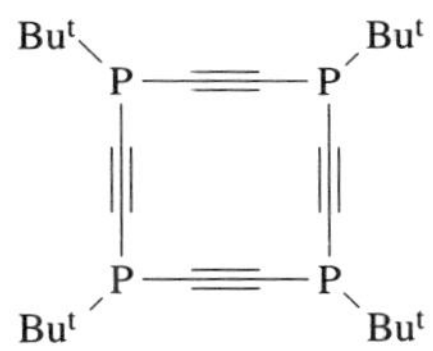

(28)

Phosphinous chlorides undergo reaction with alkyl Grignard reagents according to Equation (25) ⟨67JOC781, 74IS129, 79ZN(B)1171⟩.

$$R^1{}_2PCl + R^2MgX \longrightarrow R^1{}_2R^2P + MgXCl \qquad (25)$$

The synthesis of tertiary phosphines bearing a chiral substituent (e.g., (**29**)), may be achieved by reaction of a phosphinous chloride with the appropriate Grignard reagent ⟨87JOM(334)369⟩.

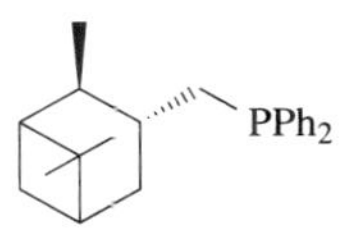

(29)

Organolithium reagents may react with phosphonous (Equation (26)) and phosphinous (Equation (27)) halides to yield tertiary alkylphosphines ⟨68JCS(A)1909, 83JOM(250)617⟩.

$$R^1PCl_2 + 2R^2Li \longrightarrow R^1R^2{}_2P + 2LiCl \qquad (26)$$

$$R^1{}_2PCl + R^2Li \longrightarrow R^1{}_2R^2P + LiCl \qquad (27)$$

For example, the reaction of pentadienyllithium with phenylphosphonous dichloride yields the phosphine (**30**) ⟨85TL4067⟩. Also, tertiary phosphines bearing cyanomethyl ⟨87S938⟩, TMS-methyl ⟨83JCS(D)905⟩ and Cp ⟨77CB1269, 90POL2901⟩ substituents have all been prepared by treatment of phosphorus chlorides with organolithium reagents.

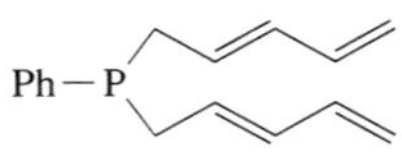

(30)

Organometallic reagents other than Grignards and organolithiums have been used in the synthesis of unsymmetrical phosphines. Tertiary phosphines bearing a range of protected substituents have been synthesized using organo(trialkyl)tin reagents ⟨67ZOB2080, 68LA(714)1, 68MI 208-01⟩. Trialkylboranes react with diphenylphosphinous chloride to yield alkyldiphenylphosphines ⟨70TL1687⟩. The transfer of heterosubstituted-1,3-dienes from a Cp–zirconium complex to phosphorus affords the corresponding tertiary phosphines ⟨88JOC4425⟩. Phosphonous and phosphinous chlorides react with magnesium butadiene ⟨82AG(E)292, 83CB3293⟩ and magnesium cot ⟨85CB97⟩ to yield acyclic and cyclic tertiary phosphines (e.g., (**31**) and (**32**)) ⟨83CB3293⟩.

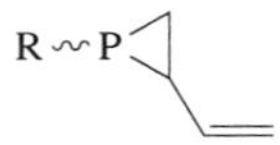

(31)
R = Me, Bu^t, c-C_6H_{11}, menthyl, Ph

(32)
R = Bu^t, c-C_6H_{11}, menthyl

(iii) Phosphines of the form $R^1R^2R^3P$

The most common method for the synthesis of the desirable asymmetric phosphines, when employing the electrophilic phosphorus route, is the reaction of phosphinous halides (e.g., R^1R^2PCl) with an organometallic compound (RMgX, RLi) (see Equation (18)) ⟨62ZOB2848, 76JOM(118)C8⟩. The reaction of phosphonous dihalides (e.g., $RPCl_2$) and a sequence of organometallics may also be appropriate ⟨87USP4668823⟩, or, alternatively, the same result may be achieved by using a mixture of organometallics, as long as the substituents are sufficiently different to allow separation of the asymmetric phosphine from the unsymmetrical by-products (Equation (28)) ⟨64JA2378⟩.

$$R^1PCl_2 + R^2M + R^3M \longrightarrow R^1R^2R^3P + R^1R^2{}_2P + R^1R^3{}_2P \qquad (28)$$

The reaction of both organolithium and Grignard reagents with a phosphonous starting material, in which one of the leaving groups is other than halide, is illustrated by the reaction of an *o*-lithioaromatic with $Ph(Cl)PNEt_2$ and subsequent reaction with an alkyl Grignard to yield the phosphines (**33**) ⟨84PS(19)65⟩. Also, another case in which the leaving group is not a halide is the use of phosphinites to generate asymmetric phosphines (e.g., (**34**)) ⟨82IZV2390⟩. A combination of an organocadmium reagent with a Grignard reagent or an organolithium can be used to replace sequentially (in a one-pot reaction) the two chlorines of $PhPCl_2$ with different groups ⟨84JOM(269)107, 85IC533⟩. These methods generally lead to the production of a racemic mixture of asymmetric phosphines, the separation of which may be possible by various methods.

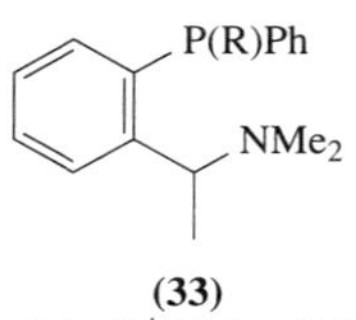

(33)
R = Me, Pr^i, Bu^t, c-C_6H_{11}

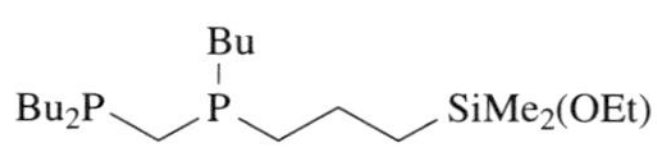

(34)

The directed synthesis of a single enantiomeric phosphine is a more challenging task, with varying degrees of success having been reported (see also Section 2.08.2.3.6). Phenylphosphonous acid dimenthyl and dibornyl esters react with bulky nucleophiles to give diastereoisomeric phosphinous acid esters with a high degree of asymmetric induction (up to 95% *de*) (Scheme 4) ⟨86JOM(301)289⟩. Further substitution yields the optically active phosphine, although this second step, in which an inversion of the stereochemistry at phosphorus occurs ⟨79T1531⟩, is accompanied by partial racemization.

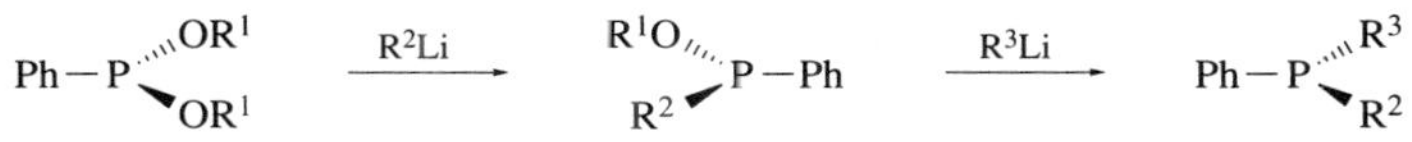

R^1 = bornyl, menthyl: $R^2 = Pr^i, Bu^t, c\text{-}C_6H_{11}$; R^3 = Me

Scheme 4

Other phosphinites have been utilized in the synthesis of chiral phosphines ⟨79TL1069⟩, as have chiral phosphinite–borane complexes, in which the BH_3 ligand is intended to decrease the possibility of racemization at phosphorus ⟨90TL6357⟩.

2.08.2.3.3 Tertiary phosphines from metallated phosphines

Several possibilities are available for the synthesis of tertiary phosphines following the metal phosphide route. Symmetrical, $R^1{}_3P$ (Equation (29)) ⟨59CB1118, 65FRP1401930⟩, unsymmetrical, $R^1{}_2R^2P$ (Equations (30) and (31)), and asymmetrical, $R^1R^2R^3P$ (Equation (32)), phosphines are all accessible, depending upon the starting materials used.

$$PM_3 + 3R^1X \longrightarrow R^1{}_3P + 3MX \quad (29)$$

$$R^1PM_2 + 2R^2X \longrightarrow R^1R^2{}_2P + 2MX \quad (30)$$

$$R^1{}_2PM + R^2X \longrightarrow R^1{}_2R^2P + MX \quad (31)$$

$$R^1R^2PM + R^3X \longrightarrow R^1R^2R^3P + MX \quad (32)$$

The metal phosphides required for the reaction may be generated in a number of ways. As discussed previously (see Sections 2.08.2.1.2 and 2.08.2.2.2), PH_3, primary and secondary phosphines may be metallated by alkali metals in liquid ammonia or an inert solvent ⟨59CI(L)541, 61JCS896, 82ZAAC(491)163⟩, by organometallic reagents ⟨86CB2921, 90ZN(B)1548⟩ or *in situ* using KOH in DMSO ⟨86S198⟩. Also, other available methods include the metallation of phosphinous halides using alkali metal or organometallic reagents ⟨88JOM(342)263⟩, cleavage of tetraalkylbiphosphines by an alkali metal ⟨85BCJ2793⟩ and the treatment of a tertiary phosphine with alkali metal to cleave a P—C bond and form a diorganophosphide (see Section 2.08.2.2.4(iii)). Lithium, sodium and potassium phosphides, and also phosphorus Grignard reagents ⟨52JCS3039⟩ have all been used for the synthesis of tertiary phosphines. The generation and use of diorganophosphide reagents have been reviewed ⟨91CZ7⟩.

Although reactions with all the usual types of carbon electrophiles are possible, organic halides are those most commonly employed. Alkali diorganophosphides react with alkyl halides to yield tertiary alkylphosphines ⟨59CB3175, 76IS155, 90CB1827⟩. For example, tetrakis(diphenylphosphinomethyl)methane (**35**) is produced by the reaction of sodium diphenylphosphide with tetrakis(bromomethyl)methane ⟨92ZN(B)640⟩. The introduction of unsaturated substituents may be achieved by the reaction of the corresponding organohalide with an alkali phosphide ⟨62CB268, 82JCS(D)1915, 90ZOB313⟩. Tertiary phosphines containing functionalized substituents, such as ether ⟨87JOM(335)59, 92ZN(B)693⟩, thioether ⟨83CJC2214, 91BCJ2942⟩, amino ⟨83JOC2195, 90POL2413⟩ and amide ⟨87PS(31)33⟩, may also be synthesized by the alkylation of an alkali metal phosphide. Also, the preparation of chiral phosphines by the metal phosphide route may be achieved by the alkylation of a diorganophosphide of the form R^1R^2PM (see Equation (26)) ⟨84ZN(B)115, 88ZN(B)295, 90BCJ2051⟩. For example, the chiral bis(phosphino)pyrrolidine ligands (**36**) were synthesized as a diastereoisomeric mixture at phosphorus by alkylation of the potassiophosphide precursors ⟨83CB1123, 92CB1061⟩.

Reactions by this route generally proceed by an S_N2 mechanism involving displacement of the halide leaving group by the phosphide nucleophile, accompanied by the inversion of stereochemistry at the carbon centre. However, tertiary phosphines have also been prepared by a photostimulated radical reaction, $S_{RN}1$, in which the metal phosphide is reactive towards bridgehead radicals

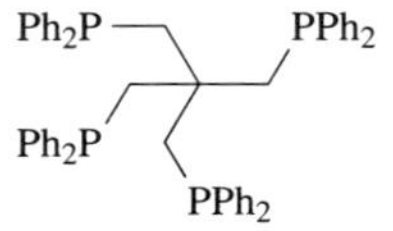
(35)

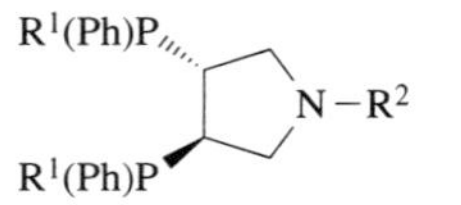
(36)
R^1 = Me, CH_2Ph; R^2 = H, CH_2Ph

generated *in situ* ⟨84JOC4609, 88JOC3016⟩, an example of which is the synthesis of (7-norcaranyl)diphenylphosphine (**37**) ⟨84JOC3387⟩.

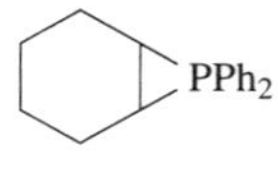
(37)

Dimetallo-organophosphide reagents have proven very useful for the synthesis of phosphorus heterocycles, such as phosphiranes ⟨69T1097, 85S291, 90CRV997, 94CC475⟩ and larger systems ⟨61CB113, 84JCS(D)575⟩, by dialkylation at the phosphorus centre. The reaction of dilithiophenylphosphide with enantiomerically pure dimesylate synthons afforded both enantiomers of the chiral phospholanes (e.g., (**38**)) ⟨90OM2653⟩. Also, the cyclization of phosphorus heterocycles has been demonstrated using phosphorus–transition metal complexes to catalyse and direct the stereochemical outcome of the reaction ⟨90PAC1139, 92JCS(D)1751, 92S157⟩.

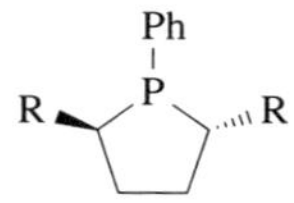
(38)
R = Me, Et, Pr^i

Alternative electrophiles that may react with metallophosphides include epoxides ⟨65CB2091, 84URP1035028, 90ZOB1243⟩ and oxetanes ⟨73JA4194⟩, episulfides ⟨82ZAAC(491)163, 86IC3396⟩ and ethylenimine ⟨65ZN(B)916⟩, which, upon nucleophilic ring opening, result in the formation of tertiary alkylphosphines incorporating hydroxyl, thio and amino functionalities, respectively. Also, the use of electrophiles with leaving groups has been reported. For example, the chiral 1,4-diphosphine (**39**) was prepared from the corresponding ditosylate derivative ⟨83JOM(253)249⟩, and mesylate esters ⟨91JOC5919⟩ and acyl chlorides ⟨66ZAAC(346)241, 68CB3438⟩ have also been utilized.

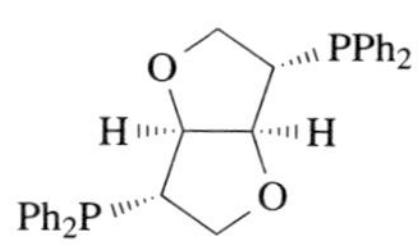
(39)

2.08.2.3.4 Tertiary phosphines by addition of P—H to unsaturated compounds

Reactions that involve the addition of P—H to multiple bonds are more useful for the synthesis of tertiary phosphines than for the preparation of either primary (see Section 2.08.2.1.3) or secondary (see Section 2.08.2.2.3) phosphines, since in the latter cases oversubstitution may be difficult to avoid.

(i) Addition to alkenes and alkynes

As discussed previously (see Section 2.08.2.2.3), the addition reaction may be catalysed in three different ways, that is, under acidic, basic or free radical conditions. Of these, procedures involving basic or free radical catalysts are those most commonly employed.

Two approaches to the base-catalysed reaction are possible, either by reaction of PH_3, primary or secondary phosphines with the unsaturated compound in a basic medium ⟨71JA4158, 76JOC972, 79JOC3095, 85IC2344⟩, or reaction of a formal metallophosphide with the chosen unsaturated starting material ⟨65ZN(B)181, 67CB412⟩. For example, the reaction of diphenylphosphine with cyclic α,β-unsaturated nitriles under basic conditions results in the formation of a mixture of *cis*- and *trans*-2-diphenylphosphino-1-cyanocycloalkanes, for example **(40)** and **(41)** ⟨90JOM(393)143⟩. Base-catalysed addition of diphenylphosphine to alkyne in the presence of a phase-transfer catalyst gives high yields of dppe ⟨86AKZ258⟩. The route to polydentate ligands may be improved by initially coordinating the phosphines to a transition metal and proceeding *via* a metal template-assisted addition ⟨84IC778⟩.

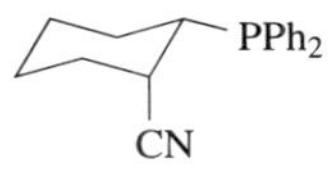

(40)

(41)

The free radical addition of phosphines to alkenes and alkynes has been used more widely and often gives higher yields and purer products than either acid- or base-catalysed procedures. The reaction follows a chain mechanism ⟨40CRV351⟩ which can be initiated by radical initiators, such as ultraviolet light ⟨83JCS(P1)861, 87JOM(320)339, 91ZAAC(592)121⟩, peroxides ⟨52BRP673451, 52JA3282⟩ and AIBN ⟨80GRP2936210, 85CB2365, 90CZ353⟩. The tetraphosphine ligand **(42)** was synthesized by the radical-initiated addition of dimethylphosphine to tris(but-3-enyl)phosphine ⟨86PS(26)143⟩. In a similar manner, diethylvinylphosphine and diphosphinomethane yielded the hexa-tertiary phosphine **(43)** ⟨85JA7423⟩.

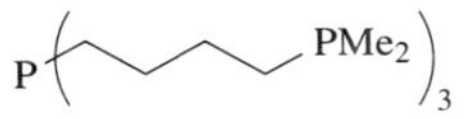

(42)

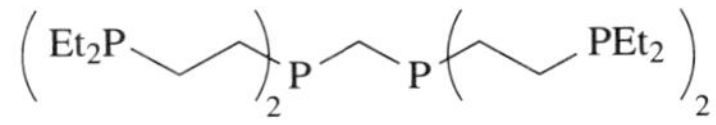

(43)

Intramolecular cyclization reactions of alkenyl-substituted phosphines have been utilized for the synthesis of several mono- and bicyclic tertiary phosphines ⟨66JCS(C)245, 83PS(17)73, 87ZAAC(553)136, 91ZAAC(600)195⟩. For example, the diphosphinomethane **(44)** undergoes AIBN-catalysed intramolecular cyclization to form 1,5-diphosphabicyclo[3.3.1]nonane **(45)** (Equation (33)) ⟨91ZAAC(594)66⟩.

(33)

(44) **(45)**

Generally, the addition reaction carried out under acid-catalysed conditions yields the Markovnikov product, whereas a radical-initiated reaction results in a phospine of the anti-Markovnikov type, as in Scheme 5 ⟨78JOC3408⟩. Therefore, the reaction conditions may be chosen to produce the desired tertiary phosphine.

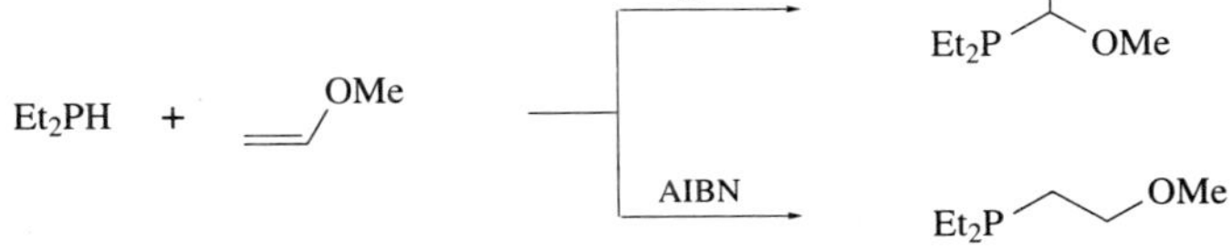

Scheme 5

In some cases, if the multiple bond is activated by electron-withdrawing substituents, the use of a catalyst may not be necessary ⟨59DOK(127)1217, 68IZV190, 86EUP184872⟩. For example, no catalyst is required for the reaction of diphenylphosphine with vinyl sulfonyl fluoride to yield $Ph_2PCH_2CH_2SO_2F$ ⟨77ZAAC(436)20⟩. The addition reaction between phosphines and alkenes has also been reported at high temperatures and pressures without the use of a catalyst ⟨86EUP194716⟩.

(ii) Addition to other unsaturated compounds

The reaction of phosphines with carbonyl-containing compounds can afford a variety of tertiary phosphines, the precise nature of the product being dependent upon the starting materials and the reaction conditions employed ⟨85CZ317⟩. Reactions of phosphines with aliphatic ⟨61JA168, 61DOK(139)1359, 62T1231, 63ZOB1919⟩ and aromatic ⟨60CCC2191, 62T1231⟩ aldehydes, ketones ⟨61ZOB3411, 61ZOB3417⟩, isocyanic acid ⟨66JOC588, 68JOC3088⟩, isocyanates ⟨61USP2969390, 62T1221⟩ and isothiocyanates ⟨64CB3430, 67ZAAC(351)18, 87PS(29)373⟩ all result in the formation of α-functionalized tertiary phosphines. However, in many cases, the appearance of more complex products, either in addition to, or in place of those required, is observed ⟨B-72MI 208-01⟩. Also, care must be taken to avoid oversubstitution and the formation of quaternary phosphonium products, although if this does occur, reduction back to the tertiary phosphine is possible ⟨62LA(659)49⟩ (see also Section 2.08.2.3.5(ii)).

The reactions may be undertaken in the presence of an acid or using no catalyst, depending upon the reactivity of the substrate. The reaction of phosphine with formaldehyde has been carried out in aqueous solution using activated nickel and cobalt catalysts ⟨85URP1145022⟩. The tetrakis(hydroxymethyl)diphosphine ligand (**46**) was prepared by the uncatalysed reaction of formaldehyde with diphosphinoethane at 100–110°C ⟨84ZC224⟩. Macrocyclic ligands have been synthesized by the metal template-assisted cyclization of bis(secondary phosphines) with dialdehydes and diketones ⟨83IC3624, 84IC3304⟩.

(46)

Aminomethylphosphines may be prepared by the Mannich reaction of primary and secondary phosphines with formaldehyde and either ammonia or primary or secondary amines ⟨66HCA842, 83JA6355, 84ZC365⟩. Phosphines may also react with carbon–nitrogen multiple bonds ⟨65ZAAC(336)234⟩. For example, primary, secondary and tertiary aminoalkylphosphines are synthesized by the addition of methylphosphines to the C=N bond of hexafluoroisopropylideneimine ⟨84ZN(B)356⟩.

2.08.2.3.5 Tertiary phosphines by reduction

(i) Reduction of tertiary phosphine oxides and sulfides

As with the preparation of primary and secondary phosphines by reduction (see Sections 2.08.2.1.4 and 2.08.2.2.4(i)), a number of reducing agents may be used for the synthesis of tertiary phosphines from the corresponding phosphine oxides and sulfides (Equations (34) and (35)), the choice being dependent upon the sensitivity of the substrate to reduction and the stereochemical outcome required.

$$R_3P{=}O \longrightarrow R_3P \qquad (34)$$

$$R_3P{=}S \longrightarrow R_3P \qquad (35)$$

Silanes are the most commonly used reagents for the reduction of tertiary phosphine oxides since they are easy to use and give clean stereochemistry and high yields. The most popular reagents are trichlorosilane ⟨65CB171, 83CJC21, 88ICA(146)111, 91TA667⟩, phenylsilane ⟨82JOC905, 83JA4727, 92JOC511⟩ and hexachlorodisilane ⟨69JA2788, 69JA7027, 80JOC3691⟩. The procedure typically involves simply mixing the phosphine oxide and reductant in an inert solvent, under nitrogen. A secondary or tertiary amine may be added to scavenge hydrogen chloride if chlorosilanes are used. Following refluxing, the product is obtained by distillation or recrystallization.

In cases where enantiomeric phosphines are desired, the stereospecificity of the reduction is dependent upon the silane used, the type and strength of base added and whether the phosphine is cyclic or acyclic. Reduction of acyclic substrates with trichlorosilane in the presence of a weakly basic tertiary amine, such as pyridine or diethylaniline, proceeds with retention of configuration. However, the use of more basic amines yields phosphines with inversion of configuration in high enantiomeric excess ⟨65TL1157, 69JA7012, 71FCF61⟩. Generally, phenylsilane reduces acyclic phosphine oxides with retention of configuration ⟨74JOC265, 83JA4727⟩, whereas reduction with hexachlorodisilane proceeds rapidly with inversion of configuration ⟨69JA7012, 80JOC3691⟩. The reduction of cyclic substrates follows a different pattern, with retention usually being observed with all silane reagents, especially for smaller ring sizes ⟨67JOC4066, 69JA4724, 69JA7027, 83JOC4466⟩, although there are a few exceptions to this ⟨84JA7021, 92JOC511⟩. An example of the use of trichlorosilane is shown in Scheme 6. Reduction of the phosphine oxide (**47**) gives a phosphine which undergoes spontaneous intramolecular cyclization to the phosphonium salt (**48**), reduction of which yields 1-phosphaadamantane (**49**) ⟨83T4225⟩.

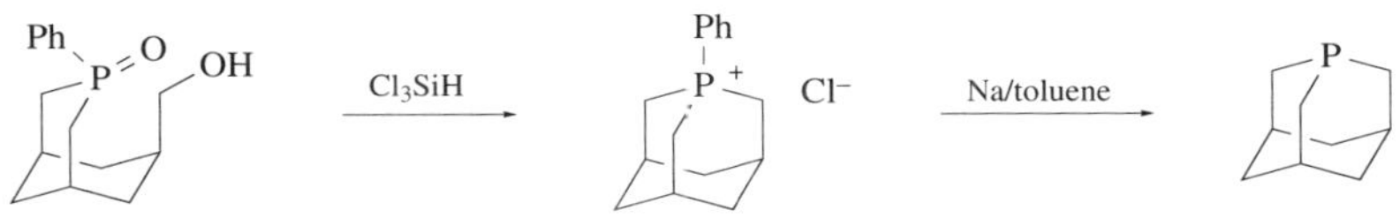

(**47**) (**48**) (**49**)

Scheme 6

Phosphine oxides have also been reduced using LAH ⟨56ZAAC(287)208, 91PS(56)27⟩, although the stereochemical outcome of the reduction is not as consistent as that of silane reductions. Both retention of configuration ⟨73MI 208-01⟩ and racemization ⟨69JA5645, 70CB318⟩ have been reported. Also, the asymmetric reduction of racemic phosphine oxides using LAH in conjunction with an optically active diol or diamine has been attempted with a moderate degree of success ⟨84JCR(S)32⟩.

Other reagents used to reduce phosphine oxides are sodium aluminum hydride ⟨85USP4507501, 85USP4507502⟩, calcium aluminum hydride ⟨56ZAAC(287)208⟩, lithium hydride in combination with titanium complexes ⟨80IZV734⟩, alanes in conjunction with a boron trihalide ⟨85USP4507504⟩ or ester ⟨85USP4507503⟩, and various reducing metals or metal complexes ⟨64HCA2137, 79JOM(177)255, 80TL2525⟩.

Phosphine sulfides and selenides may be reduced by a similar range of reagents to those used in phosphine oxide reductions, silanes ⟨69JA7023⟩ and LAH ⟨58CB1583, 71AG(E)494, 71TL2177⟩ being most commonly utilized. The stereospecificity of the reduction contrasts with that observed for phosphine oxides, phosphine sulfides being reduced with retention of configuration by both hexachlorodisilane ⟨69JA7023⟩ and LAH ⟨71TL2177⟩. Alternative reducing agents include copper ⟨82JOM(231)237⟩ and nickel ⟨79TL3081⟩ complexes. Also, alkylation of the phosphine sulfide to form a quasi-phosphonium salt, and subsequent treatment with potassium cyanide, yields the tertiary phosphine with retention of configuration ⟨80PS(8)215⟩.

(ii) Reduction of phosphonium salts

There are a variety of methods by which the reduction of a quaternary phosphonium salt may be achieved, the general aim being to cleave a (possibly specific) P—C bond.

(a) Electrolytic reduction. Electrolytic reduction may be undertaken with various electrodes, the most popular being lead and mercury cathodes ⟨61GEP1114190, 61LA(646)65, 67JA4784, 82HOU106⟩. The product of the reduction depends upon the precise nature of the phosphonium salt. Some groups, such as benzyl, are more easily cleaved than others, since they form more stable radicals upon reduction ⟨61LA(646)65⟩. Electrolytic reduction is observed to proceed with retention of configuration at phosphorus ⟨64PAC225, 67JA4784⟩.

(b) Base-induced cleavage. Generally, the treatment of a phosphonium salt with an alkali metal hydroxide or alkoxide yields a tertiary phosphine oxide and a hydrocarbon. However, if a β-elimination pathway is available, the base-induced cleavage process can result in the direct formation of the tertiary phosphine (Equation (36)).

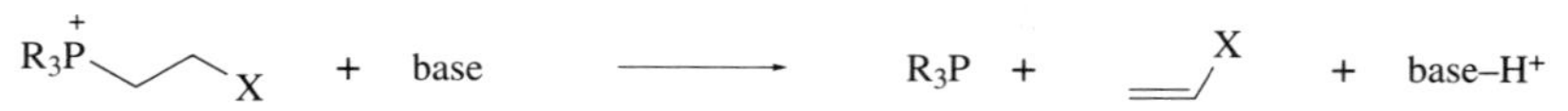 (36)

Commonly, phosphonium salts containing 2-cyanoethyl ⟨59JA4803, 63ZOB2631⟩ and 1-hydroxyalkyl ⟨62ZOB553, 63JA79, 66USP3243450, 67ZOB2522⟩ substituents may be reduced in this manner. By employing this methodology, it is also possible to synthesize asymmetric tertiary phosphines by the sequential base-induced cleavage and alkylation of a phosphorus centre (Scheme 7) ⟨62LA(659)49⟩.

$$[R^1P(CH_2OH)_3]^+ X^- \xrightarrow{HO^-} R^1P(CHOH)_2 \xrightarrow{R^2X} [R^1R^2P(CH_2OH)_2]^+ X^- \xrightarrow{HO^-}$$

$$R^1R^2P(CH_2OH) \xrightarrow{R^3X} [R^1R^2R^3P(CH_2OH)]^+ X^- \xrightarrow{HO^-} R^1R^2R^3P$$

Scheme 7

(c) Thermal decomposition. Quaternary phosphonium salts decompose at high temperatures with the loss of one substituent, to give tertiary phosphines ⟨32JGU524, 33JCS989, 53DOK(92)69, 56ZOB3426⟩. Although a range of products is often seen, good yields of a single product have been obtained if ethyl is one group on the phosphonium centre, due to the facile elimination of ethene.

(d) Hydride reduction. LAH is used to reduce quaternary phosphonium salts to tertiary phosphines with the cleavage of one substituent ⟨61JCS4263, 83URP1046248, 84ZOB2550⟩. There is a preference for the loss of a benzyl group and, since only one group is removed at a time, this method may also form the basis of a route to asymmetric phosphines by a sequence of cleavage and alkylation reactions starting from benzylphosphonium salts ⟨57JA3567, 60JOC1996⟩. In many cases, racemized phosphines are produced by the LAH reduction of chiral phosphonium salts ⟨64JA2378, 70CB318⟩. However, those possessing a sterically hindered or dialkylamino group are converted to phosphines with a predominant retention of configuration ⟨73MI 208-01, 80PS(8)225⟩. The cleavage of a furfuryl group from a series of furfurylphosphonium salts has also been achieved ⟨86ZN(B)10⟩.

2.08.2.3.6 Chiral tertiary phosphines

Chiral tertiary phosphines may derive their chirality by virtue of having either a chiral centre at the phosphorus atom, or at another atom within one or more of the substituent groups. Phosphines that are not chiral at phosphorus will not be considered here, since the phosphinylation of a chiral organic fragment should be relatively straightforward by one of the previously discussed methods (see Sections 2.08.2.3.1–2.08.2.3.4) ⟨B-85MI 208-01, B-85MI 208-02⟩ as long as racemization does not occur under the chosen reaction conditions. The majority of methods that have been employed for the synthesis of chiral phosphines involve either the resolution of chiral precursors into an enantiomerically pure state prior to conversion into the phosphine, or resolution of the phosphines themselves. Syntheses in which an achiral starting material, or one in which the phosphorus atom has no chirality, is used to prepare an optically active phosphine are far less common ⟨B-84MI 208-01, B-90MI 208-02⟩.

(i) Complexation method

One route to optically active phosphines is their resolution via diastereoisomeric transition metal complexes. For example, coordination of a racemic phosphine to a palladium complex containing an optically active amine ligand affords the diastereoisomeric complexes (**50**), which may be separated by fractional crystallization and the enantiomeric phosphine liberated subsequently by treatment of the complex with dppe ⟨71JA4301⟩. Similar palladium complexes incorporating related optically active ligands have been used to resolve both mono- ⟨77JA7876⟩ and bidentate ⟨79JA6254, 82IC1007⟩ chiral phosphines.

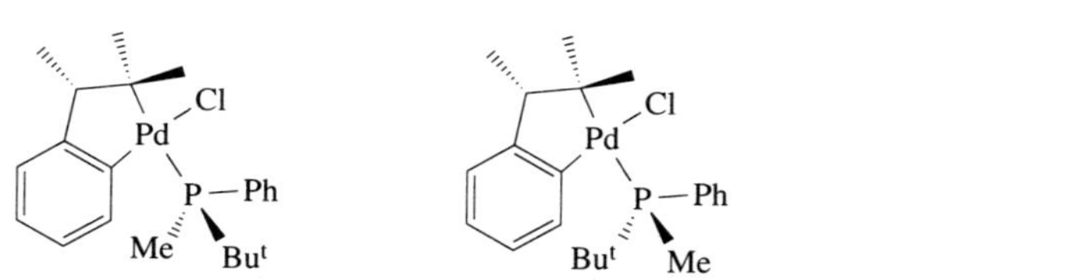

(*R,S*)-(**50**) (*S,S*)-(**50**)

(ii) Phosphonium salt method

The resolution of chiral quaternary phosphonium salts by fractional crystallization is possible with the use of an optically active counterion ⟨80JOC3691⟩. Dibenzoyltartrate salts are those most often used ⟨59JA248, 64JA2378⟩. Once the salts have been separated, they may be converted to the optically active phosphines indirectly, by hydrolysis to the phosphine oxide (inversion), followed by reduction (inversion or retention) ⟨B-84MI 208-01⟩ (see also Section 2.08.2.3.5(i)), or directly by electrolytic cleavage, or β-elimination as appropriate (see Section 2.08.2.3.5(ii)).

(iii) Phosphinate ester method

This widely used method for the generation of chiral phosphines ⟨B-84MI 208-01⟩ is dependent upon the availability of chiral phosphinate esters (**51**). For example, diastereoisomeric menthylphosphinate esters (**51**; R^3 = menthyl) may be separated by fractional crystallization and the diastereoisomers subjected independently to a sequence leading to both enantiomers of the chiral phosphine (**52**) (Scheme 8) ⟨71FCF61⟩. The reduction of the phosphine oxide (**53**) can also be undertaken by a method giving retention of configuration, yielding the alternative enantiomer of the phosphine.

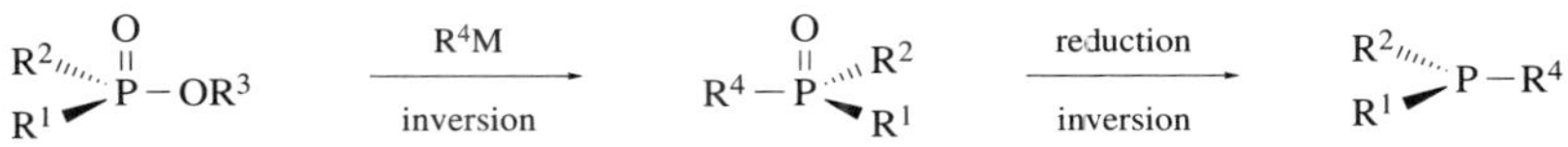

(51) (53) (52)

Scheme 8

One elegant route that has been developed in the quest to generate chiral phosphinate esters, involves the synthesis of diastereoisomeric oxazaphospholidines ⟨83EUP82057, 84CB2328, 85GEP3512781⟩. These have been prepared by the reaction of phosphonous compounds, such as $RPCl_2$ or $R^1P(NR^2{}_2)_2$, with chiral amino alcohols, for example (−)-ephidrine. The chiral phosphinate esters are produced in high optical purity upon ring opening, since only one diastereoisomer of the oxazaphospholidine was synthesized initially (e.g., Scheme 9).

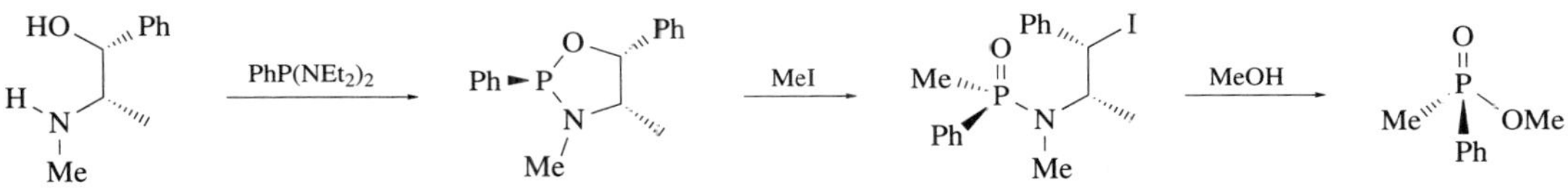

Scheme 9

(iv) Phosphinite method

The phosphinite method for the synthesis of chiral phosphines proceeds by the electrophilic phosphorus route (see Section 2.08.2.3.2). Diastereoisomeric phosphinites are prepared by the action of an organometallic reagent on phenylphosphonous acid dimenthyl and dibornyl esters. Further nucleophilic substitution affords the chiral phosphines with an inversion of stereochemistry (see Scheme 4) ⟨79T1531, 86JOM(301)289⟩. Similarly, phosphonous chloroesters of cinchonine and cinchonidine have been used for the synthesis of chiral phosphinites and, hence, chiral phosphines ⟨79TL1069, 84JOM(273)C55⟩. Optically active phosphinites, prepared from quasiphosphonium salts, have yielded chiral phosphines on treatment with organolithium compounds (Scheme 10) ⟨79JA7292, 80PAC959⟩.

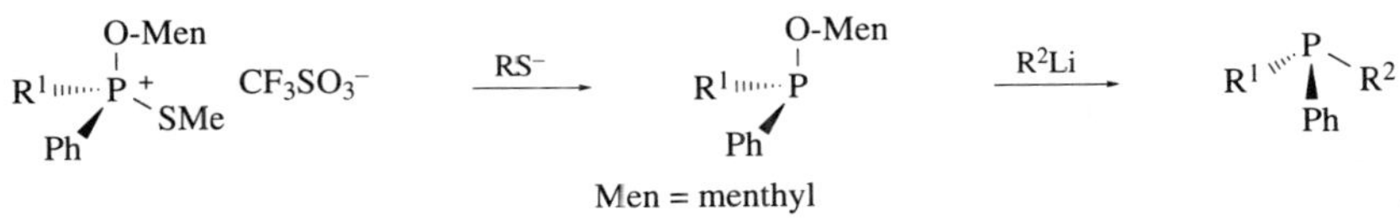

Scheme 10

Optically active phosphine–boranes, prepared from the corresponding phosphinite derivatives, react with amines to give phosphines in a completely stereospecific manner ⟨90JA5244, 90TL6357, 91TL3371⟩.

(v) *Reduction of phosphine oxides*

Often it is more convenient to separate diastereoisomeric phosphines by conversion to the more optically stable phosphine oxides. Following resolution by chromatographic methods, the phosphine oxides may be reduced with either retention or inversion of configuration as discussed in Section 2.08.2.3.5(i).

2.08.2.4 Quaternary Alkylphosphonium Salts—$[R_4P]^+X^-$

Quaternary phosphonium salts have many uses, both as a final product and as intermediates in synthetic procedures, an important example being as precursors to Wittig reagents (ylides). A number of routes are available for their preparation from a variety of starting materials. Reactions that have been developed specifically for the synthesis of arylphosphonium salts will not be considered in this section, only those in which the formation of an aliphatic carbon–phosphorus bond will be discussed. Also, due to the great number of reports available in the published literature ⟨B-70MI 208-01, B-72MI 208-03⟩, the references provided for each of the described methods will be limited to a few representative examples.

2.08.2.4.1 *Quaternary alkylphosphonium salts from phosphines and alkylating agents*

The conventional method for the synthesis of quaternary phosphonium salts is the alkylation of tertiary phosphines using organohalides (Equation (37)) ⟨57JA5884, 60JA3919, 60JOC1996, 87PS(29)111, 88JAP63119491⟩. The reaction is also possible with PH_3 and primary and secondary phosphines, although, following initial alkylation, further substitution is dependent upon the ease of dissociation of a proton (in the form of HX) from the phosphorus centre ⟨66JOC3817⟩. Also, metal phosphides may react with organohalides to give quaternary phosphonium salts ⟨64ZAAC(328)21, 88S911⟩, as will both the red and white forms of elemental phosphorus ⟨65JCS3354, 66HCA2488, 91ZOB1075⟩. The phosphine–carbon disulfide complex may be alkylated by primary alkyl bromides, and thus the isolation of the phosphine from a previous synthetic step may be avoided ⟨89S204⟩.

$$R^1{}_3P \;+\; R^2X \longrightarrow [R^1{}_3R^2P]^+\,X^- \qquad (37)$$

In its most basic form the procedure involves simply mixing the reactants, often using the organohalide as the solvent. The exclusion of oxygen is imperative, due to the facile oxidation of phosphines to phosphine oxides. With lower alkyl halides the reaction may be vigorous, but with the higher alkyl halides the reaction becomes sluggish, and heating is necessary. The use of a polar solvent, such as acetonitrile, nitromethane, DMF or DMSO, considerably enhances the rate of reaction. Also, reactions undertaken at high pressures show greater rates of reaction and may be carried out at lower temperatures ⟨84JOC4293⟩. As expected, there is a marked difference in reactivity of the halides. Iodides are more reactive than bromides, and these in turn are more reactive than chlorides ⟨60JA5794⟩. Fluorides are used for quaternization in special cases only, with fluorohalomethanes often yielding rearranged products ⟨90PS(48)279⟩.

In many cases, alkyl halides with additional functional groups, such as alcohols ⟨61ACS438⟩, aldehydes and ketones ⟨75ZOB1228, 85GEP3332716⟩, ethers ⟨78S588⟩, thioethers ⟨82MI 208-01, 85JOC4252⟩, carboxylic acids ⟨84MIP30022, 91ZN(B)775⟩, amines ⟨82PS(13)97⟩ and heterocycles ⟨87PS(33)41, 91USP4997952⟩, react smoothly to the corresponding quaternary phosphonium salts. Fluorinated phosphonium salts, bisphosphonium salts and phosphoranium salts may be prepared by the reaction of phosphines with fluorinated halomethanes ⟨83JFC(23)339, 85JFC(27)85, 88JOC366⟩.

Phosphonium salts with phosphorus in the ring are accessible by treating either a mono- or bisphosphine with an organodihalide ⟨83CB3567, 89AG(E)1700, 91T403⟩. For example, treatment of dppp with 1,3-diiodopropane yielded the diphosphonium salt (**54**) ⟨91PS(56)27⟩. Also, reaction of 1-methylphosphorane with an equimolar amount of 1,4-dibromobutane gave the salt (**55**), which, on treatment with potassium hydride followed by an acid, afforded the spirobicyclophosphonium salt **(56)** ⟨80ZN(B)990⟩.

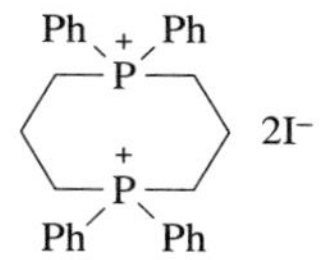
(54)

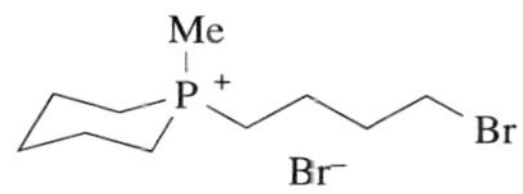
(55)

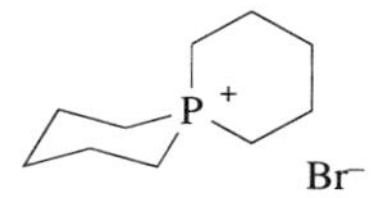
(56)

Reagents with leaving groups other than halide may also be used to quaternize phosphines. Aryl ⟨89JOC2580, 91JAP70374395⟩ and trifluoromethyl ⟨79JA6452, 90JOC5033⟩ sulfonates have been used successfully in the preparation of quaternary phosphonium salts. Also, reports have demonstrated that iodonium salts may be used for the same purpose ⟨91S1073, 92JOC1861⟩.

2.08.2.4.2 Quaternary alkylphosphonium salts from phosphines, alcohol and hydrohalides

The preparation of quaternary phosphonium salts by the reaction of a phosphine with an alcohol and hydrohalide occurs in a similar manner as that discussed in Section 2.08.2.4.1, the alkylation being facilitated by organohalide formation *in situ* (Equation (38)) ⟨75JOC2801, 81USP4296042, 84ZOB1192⟩. It is also possible to conduct the reaction with phosphonium hydrohalides and an appropriate alcohol (Equation (39)) ⟨75BRP1391806, 78HCA1708⟩. In several cases, the reaction may proceed more quickly and smoothly and the yields may be better than with the corresponding alkyl halides. The water arising during the reaction is removed by azeotropic distillation. In certain cases, ethers may be used instead of alcohols ⟨54JCS2832, 81S53⟩.

$$R^1_3P + R^2OH + HX \longrightarrow [R^1_3R^2P]^+ X^- + H_2O \quad (38)$$

$$[R^1_3PH]^+ X^- + R^2OH \longrightarrow [R^1_3R^2P]^+ X^- + H_2O \quad (39)$$

2.08.2.4.3 Quaternary alkylphosphonium salts from phosphines and cyclic compounds

Quaternary phosphonium salts may be synthesized by the ring-opening nucleophilic addition of tertiary phosphines to cyclic compounds. Initially, a betaine **(57)** is formed, which undergoes protonation in the presence of acid to yield the phosphonium salt **(58)**. The reaction of phosphines with epoxides yields β-hydroxyalkylphosphonium salts (Scheme 11) ⟨65JCS1539, 68JOC2590, 88BCJ4476⟩. Quaternary phosphonium salts are also formed in the reactions of tertiary phosphines with oxetane and THF ⟨88BCJ4476⟩, substituted cyclopropanes ⟨87S648, 87ZOR2120⟩ and oxadiazines ⟨76GEP1768461⟩.

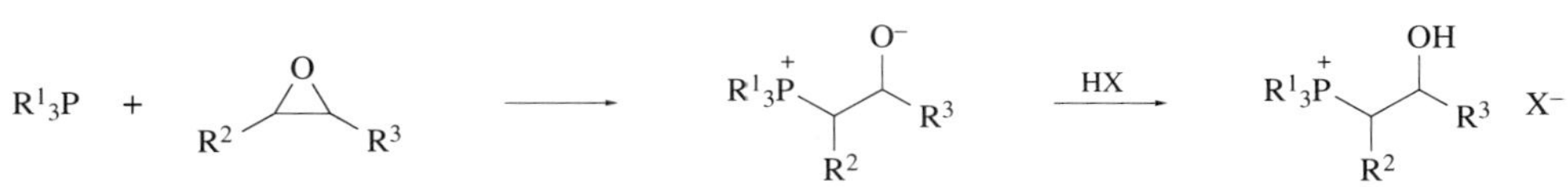
(57) **(58)**

Scheme 11

2.08.2.4.4 Quaternary alkylphosphonium salts from phosphines and unsaturated compounds

The addition of PH_3 and primary, secondary and tertiary phosphines to compounds containing multiple bonds may, depending upon the reaction conditions employed, result in the formation of quaternary phosphonium salts (see also Sections 2.08.2.1.3, 2.08.2.2.3 and 2.08.2.3.4). The addition of a phosphine to a strongly polarized alkene double bond results in the initial formation of betaine-like structures, which subsequently yield quaternary phosphonium salts upon treatment with acid

(Scheme 12) ⟨61CB1331, 72MI 208-05⟩. The addition to more weakly polarized double bonds is possible, if the reaction is carried out under acidic conditions.

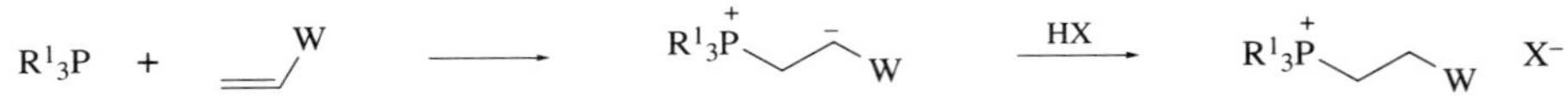

W = electron-withdrawing group

Scheme 12

Quaternary phosphonium salts may be synthesized by the reaction of tertiary phosphines with α,β-unsaturated ketones ⟨79S538, 88S395, 89MI 208-01⟩, α,β-unsaturated ethers ⟨84TL5903, 88PS(35)349⟩, thiodienes ⟨77S628⟩ and dienyl cations coordinated to iron–carbonyl acceptors ⟨83HCA1774⟩. An interesting example is the double nucleophilic substitution reaction of tributylphosphine with 2,3-dichlorobuta-1,3-diene to yield the bis(phosphonio)alkyne (**59**) (Equation (40)) ⟨90AKZ610⟩. Phosphenium ions may also undergo reaction with compounds containing alkenyl double bonds to form phosphonium salts ⟨87TL603⟩. In a similar fashion to alkenes, activated alkynes yield quaternary phosphonium salts on treatment with tertiary phosphines ⟨65CB363, 80AKZ408, 86AKZ463, 90T6381⟩.

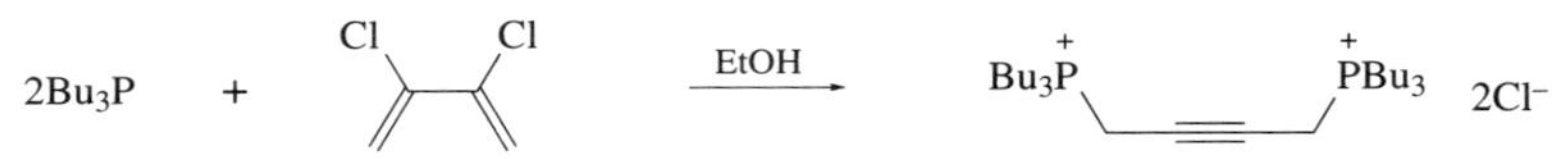 (40)

(59)

Carbonyl compounds, such as aldehydes ⟨61JA168, 62LA(659)49, 87ZOB296⟩ and ketones ⟨61CB186⟩, may react with phosphines to afford quaternary phosphonium salts. Again, betaine-like compounds are the initial products, which may be transformed into the salts by protonation. A recent example is the reaction of triphenylphosphine with an alkyl aldehyde and a trimethylsilylhalide to give the silyloxyalkylphosphonium salts (**60**), the trimethylsilylhalide being used in place of a mineral acid ⟨92S577⟩. It has also been reported that tertiary phosphines may react with the C=N bond of imines to give quaternary phosphonium salts ⟨84IVZ719⟩.

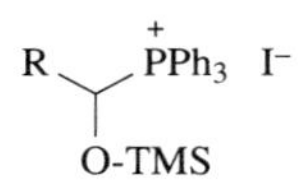

(60)

R = H, Me, Et, Pr^i

2.08.2.4.5 Quaternary alkylphosphonium salts from ylides and polar reagents

Phosphorus ylides, usually prepared from phosphonium salts by the abstraction of the α-hydrogen atom, readily add polar reagents to yield quaternary phosphonium salts (Equation (41)).

 (41)

Alkyl halides add to ylides to give α-substituted phosphonium salts ⟨75IZV2845, 82JOC2768, 85CB3105⟩. Further to this, quaternary phosphonium salts, incorporating sterically demanding substituents, that are difficult to prepare by direct alkylation may be synthesized by construction of the substituent at the phosphorus centre. For example, tetrakis(isopropyl)phosphonium iodide (**61**) was prepared as shown in Scheme 13. Quaternization of triisopropylphosphine with ethyl iodide gave the phosphonium salt (**62**), which, on treatment with base, was converted to the ylide (**63**), to which subsequent addition of methyl iodide yielded the desired quaternary phosphonium salt **(61)** ⟨86JA976⟩. Tetrakis-*t*-butylphosphonium iodide has been prepared in a similar fashion ⟨80CB1612⟩.

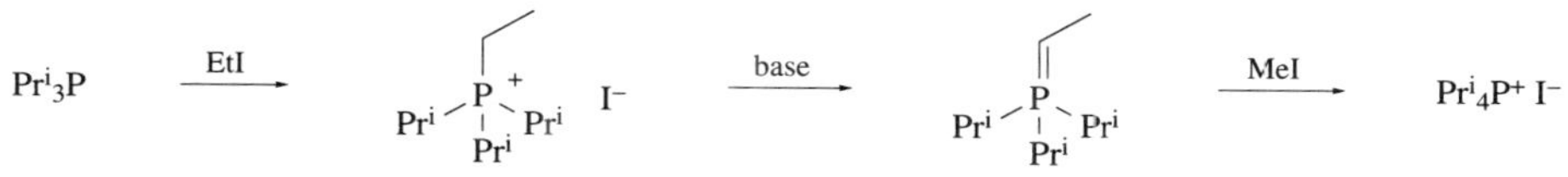

(62) **(63)** **(61)**

Scheme 13

Other reagents that have been used to prepare quaternary phosphonium salts on addition to ylides include epoxides ⟨83C10⟩, acid chlorides ⟨77AG(E)45⟩, quaternary nitrile salts ⟨85IZV1200⟩, methyl sulfonyl fluoride ⟨81JCS(P1)3059⟩, silicon ⟨59CI(L)849⟩ and selenium ⟨85JAP60152493, 88JOC2937⟩ reagents. The addition of 3-borahomoadamantane to methylenetriphenylphosphorane afforded the betaine **(64)** ⟨91MI 208-01⟩.

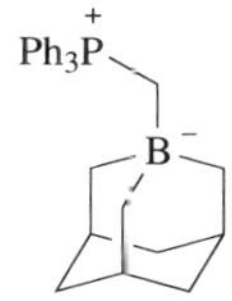

(64)

Simple ylides **(65)**, derived from methylphosphonium salts, undergo oxidative dimerization on treatment with copper(II) chloride to yield bis(phosphonium salts) **(66)** (Equation (42)) ⟨84ZN(B)518⟩.

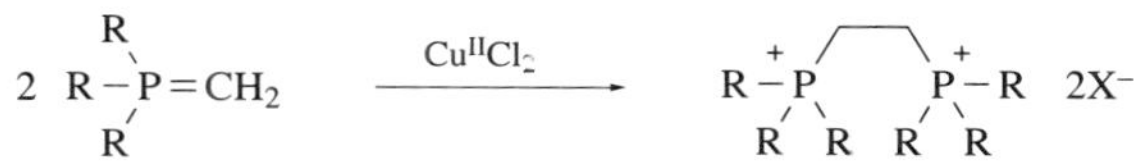

(42)

(65) **(66)**

R = Me, Et, Ph

2.08.2.4.6 *Quaternary alkylphosphonium salts by transquaternization*

As tertiary phosphines are more readily quaternized than comparable amines, when a solution of a quaternary ammonium salt and a tertiary phosphine is heated the ammonium salt acts to alkylate the phosphine. The result is an amine and a quaternary phosphonium salt. This is possible due to the equilibrium in solution between the quaternary ammonium salt and the corresponding amine and alkyl halide, it being assumed that the latter is the actual alkylating agent. Since no such equilibrium exists for quaternary phosphonium salts, the reaction can go to completion ⟨61LA(640)79, 65AJC903⟩. An example is the reaction between tributylphosphine and the quaternary ammonium salt **(67)**, giving the quaternary phosphonium salt **(68)** and the corresponding tertiary amine **(69)** (Equation (43)) ⟨86AKZ457⟩.

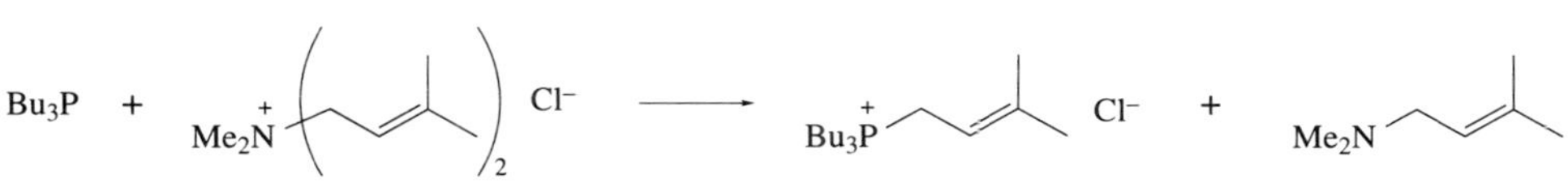

(43)

(67) **(68)** **(69)**

2.08.3 ALKYLPHOSPHORUS HALIDES

2.08.3.1 Alkylphosphorus Compounds with a P—Cl Bond—$RPCl_2$ and R_2PCl

The chlorophosphines are key intermediates in the synthesis of a wide range of organophosphorus compounds. Whilst trichlorophosphine is commonly used as the starting material, the methods used to synthesize aryl analogues are generally not suitable for the aliphatic series, its reactions with alkylmetals being the most common synthetic approach.

The reaction of trichlorophosphine with dialkylcadmium in ether at −20 °C gives yields ranging from 26% to 50% of the alkyldichlorophosphines ($RPCl_2$), depending on the size of R ⟨50JA4147⟩. The synthesis of $MePCl_2$ and Me_2PCl has been reviewed ⟨73AG(E)877⟩, and improved syntheses of each based on the aluminum chloride complex and tetramethyldiphosphine disulfide, respectively, together with purification procedures, reported ⟨74JINC503⟩.

For large-scale syntheses, it has been reported that methane in a CO_2 atmosphere reacts with PCl_3 at 500–650 °C for 0.1–0.9 s to give 95% $MePCl_2$ ⟨83GEP(O)3149937⟩. It has also been shown that methane or ethane, when passed over PCl_5 at 200 °C and passed through silica chips in a silica tube at 200 °C, gives the corresponding dichlorophosphines in about 10% conversion ⟨68ZOB1551⟩. $MePCl_2$ can also be prepared in 70–80% yield by consecutive treatment of PCl_3 with $AlCl_3$, MeI and KCl/Fe ⟨76SR115⟩. By using appropriate reactant ratios, either $MePCl_2$ or Me_2PCl may be synthesized by passing chloromethane through red phosphorus, copper powder and glass wool at 350 °C ⟨76SRI115⟩. Dichloromethane is converted (67%) to the tetrachlorobisphosphine Cl_2P-CH_2PCl_2 via an organoaluminum intermediate which is treated with PCl_3 followed by $POCl_3$ and KCl workup ⟨77JGU707⟩. Commercially available phenyldichlorophosphine is methylated by dimethylcadmium to give PhMePCl in 72% yield, provided pyridine is used for the workup ⟨78JOM-(149)C7⟩. Chloromethyldichlorophosphine ($ClCH_2PCl_2$) was prepared in 83% yield from the corresponding phosphoryl compound ($ClCH_2P(O)Cl_2$), via triphenyl phosphite reduction of an intermediate sulfide ⟨61JA4381⟩.

Whilst tetraethyllead and phosphorus trichloride give excellent yields of ethyldichlorophosphine ⟨49JOC429⟩, the poor availability of other tetraalkyllead compounds limits the scope of this reaction. The various routes available for the synthesis of diethylchlorophosphine utilizing tetraethyllead and ethylmagnesium halide have been reviewed ⟨83CZ77⟩. Carbanions, $RO_2CCH_2^-$, react with PCl_3 to give esters such as $Cl_2PCH_2CO_2R$ ⟨76JGU572⟩.

The reaction of dialkylaminochlorophosphines with organometallic reagents followed by HCl conversion of the amino groups to the chloro derivative is a very reliable route to the less common chlorophosphines ⟨86PS(26)307⟩ such as dicyclopropylchlorophosphine ⟨83S372⟩ (Scheme 14). Bu^tPCl_2 can be prepared from Bu^tCl in three steps; thus, following an initial reaction with PCl_3 using $AlCl_3$ as catalyst, the resultant Bu^tPCl_4 salt may be converted to the sulfide $Bu^tP(S)Cl_2$, which is desulfurized by tributylphosphine ⟨70JCS(A)2359⟩. Note that it has been reported that $ArOPCl_2$ will convert salts $RPCl_3^+PCl_6^-$ to the dichlorophosphines $RPCl_2$ ⟨91ZOB2161⟩. Mono- and di-*t*-butylchlorophosphines ⟨70JCS(A)2359⟩ as well as the mono- and di-*t*-pentylchlorophosphines ⟨70JCS(C)2529⟩ can be prepared directly from PCl_3 and the appropriate Grignard reagent. High yields (69–90%) of mono- and dichlorophosphines are obtained from phosphine and an alkene, which initially gives the primary or secondary alkylphosphines, under radical or base-catalysed conditions, followed by chlorination with phosgene at −50 °C in chloroform solution ⟨61JOC4770⟩ (Scheme 15). Photolytic addition of phosphorus trichloride to cyclohexene gives dichlorocyclohexylphosphine ⟨72ZOB1285⟩. The dihalo-2-methylbut-2-enylphosphines, their 3-methyl isomers and 2,3-dimethyl analogues may be synthesized by photochemical addition of PCl_3 and PBr_3 to butadienes ⟨90ZOB2394⟩. Radical conditions can also be achieved by the use of oxygen for the addition of phosphorus trihalides to 1-alkenes ⟨76JGU1213⟩. White phosphorus, benzyl chloride, phosphorus trichloride and an iodine catalyst at 200–235 °C give benzyldichlorophosphine (80% yield) ⟨66CA-P2297⟩. This phosphine may be used to desulfurize $MeP(S)Cl_2$ at 150–200 °C, to give $MePCl_2$ ⟨68CA-P106877⟩. The unsymmetrical chlorophosphine (**70**) was prepared by heating dichloromethylphosphine with norbornadiene ⟨65JCS541⟩ and adamantylmethylhalophosphines (adamantyl-MePCl) may be prepared via the addition of $MePCl_2$ to 1,3-dihydroadamantane ⟨91ZOB1906⟩.

$$R^1{}_2N{-}PCl_2 \xrightarrow{R^2MgX} R^1{}_2N{-}PR^2{}_2 \xrightarrow{HCl} Cl{-}PR^2{}_2$$

Scheme 14

2.08.3.2 Alkylphosphorus Compounds with a P—F, P—Br or P—I Bond

Mono- and di-*t*-butylchlorophosphine react with AsF_3, SbF_3 or NaF in sulfolan to produce the corresponding fluorophosphines ⟨70JCS(A)2359⟩. Potassium fluoride in acetonitrile, with a crown

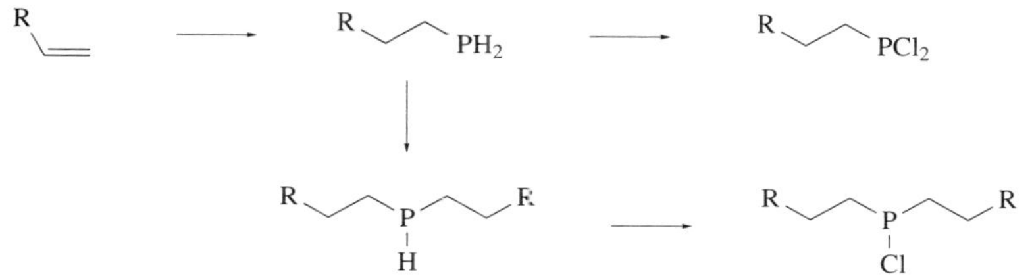

Scheme 15

(70)

ether, has been used to convert dichloromethylphosphine to the chlorofluoro- and difluoromethylphosphines ⟨91ZN(B)887⟩. Chlorophosphines have also been converted to the corresponding fluorides by treatment with the triethylamine–hydrogen fluoride adduct in the presence of free triethylamine ⟨91ZAAC(603)145⟩. This method allows thermally sensitive fluorides to be prepared. Chloride-to-fluoride conversion via reduction is also possible ⟨91ZN(B)887⟩. The reverse conversion using HBr and allyldifluorophosphine may be used to give allyldibromophosphine ⟨75IC571⟩. A mixture of mono- and dioctylbromophosphines is formed in the high-temperature equilibration between trioctylphosphine and PBr_3 ⟨84ZOB2703⟩. Unsymmetrical bromodialkylphosphines such as EtMePBr may be prepared in 65% yield by the exothermic reaction of bromine with the corresponding diphosphine ⟨62JINC275⟩. Alkyldichlorophosphines are converted to the diiodophosphines by the action of magnesium diiodide in ether ⟨76JGU428⟩ or with lithium iodide in benzene ⟨78JGU329⟩. 1,2-Bis(iodocyclohexyl)diphosphine (**71**) was formed by the reaction of cyclohexyldichlorophosphine with iodide ⟨69ZOB2139⟩.

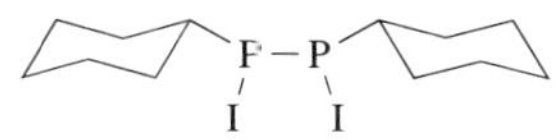

(71)

2.08.3.3 Cyclic Alkylphosphorus Halides

There are a number of specific methods available for the synthesis of cyclic halophosphines. Chlorophosphirenes such as (**72**) were obtained from phosphaalkynes and chlorovinylcarbenes ⟨91SL433⟩. *P*-Chloro-3-phospholenes (**73**) may be synthesized via the reaction of butadiene with PCl_3, followed by elimination of chlorine from the adduct (**74**) by triphenylphosphine (Scheme 16) ⟨71JOC1285⟩,. Alternatively the corresponding oxychloride (**75**) can be reduced by $(SiCl_3)_2$.

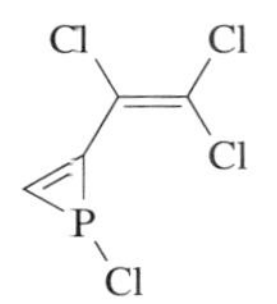

(72)

Conjugated imines (e.g., (**76**)) and PCl_3 give the azaphospholene (**77**) ⟨44JA82⟩. Benzochlorophospholene (**78**) is produced in 38% yield by the reaction of white phosphorus, PCl_3 and 1,2-dichloromethylbenzene at 340 °C ⟨68DOK(182)337⟩.

Dialkylaminochlorophosphines have found considerable utility as intermediates for the synthesis of compounds such as the *P*-chlorophospholane (**79**) ⟨74S119⟩ and *P*-chlorophosphorinanes ⟨83ZC249⟩ as well as $Br_2P(CH_2)_nPBr_2$ ⟨82CB1947⟩. Phosphaalkynes, $Bu^tC{\equiv}P$, have been converted to 1,3-dihalo-1,3-diphosphetanes ⟨91AG(E)207⟩.

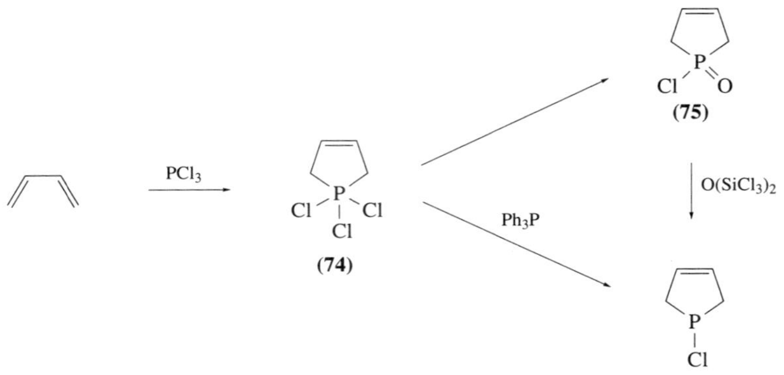

Scheme 16

(76)

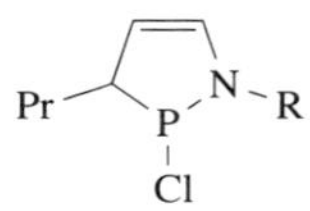

(77)

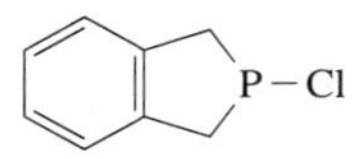

(78)

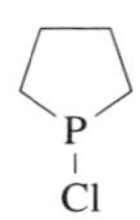

(79)

2.08.4 ALKYLPHOSPHORUS COMPOUNDS WITH A P—O BOND

2.08.4.1 Alkylphosphinous Acid Derivatives—RPHOH and R_2POH, etc.

The reaction of phosphorous acid dialkyl esters ⟨68JOC3690⟩ or phosphonous acid monoalkyl esters ⟨68JA2459⟩ with Grignard reagents provides a major route for the synthesis of phosphinous acids (Scheme 17). Alkoxydialkylphosphines ⟨60DOK(135)323⟩ and sterically hindered phosphinous acids ⟨68JCS839⟩ can be prepared from the corresponding alkoxydichlorophosphines or alkyldichlorophosphines by treatment with a Grignard reagent. Phosphinous acids have been obtained by the reduction of disubstituted phosphinic acids, or phosphinic halides ⟨52JA5418⟩, with LAH (or $NaBH_4$), and oxidation of secondary phosphines with dry air ⟨61JOC4626⟩. Controlled hydrolysis of dialkylalkoxyphosphines ⟨63IZV1227⟩, or chlorodialkylphosphines ⟨86JCS(P2)1029⟩ in the presence of a catalytic amount of mineral acid, produces phosphinous acids (Equation (44)). Phosphinous acids can be obtained by acid-catalysed addition of primary phosphines ⟨63USP3116334⟩ to carbonyl compounds (Equation (45)), or by base-catalysed addition of primary phosphine oxides ⟨62T1221⟩ to alkenes. The most satisfactory laboratory synthesis of alkoxydialkylphosphines involves the addition of an alcohol to a disubstituted chlorophosphine in the presence of a tertiary base ⟨48ZOB2008⟩ (Equation (46)). The reaction of chlorodialkylphosphines with methanol or propanol in the presence of optically active amines gives alkoxydialkylphosphines with an optical purity of at least 10% ⟨75CC382⟩. Alkoxydialkylphosphines have been prepared from alkyldialkoxyphosphines in up to 95% *de* when R^2 is a bulky group (Equation (47)) ⟨86JOM(301)289⟩.

$$(R^1O)_2P(O)H \;+\; 3R^2MgX \longrightarrow R^2{}_2P(O)MgX \xrightarrow{H^+} R^2{}_2P(O)H$$

Scheme 17

$$R^1{}_2POR^2 \;+\; H_2O \xrightarrow{H^+} R^1{}_2P(O)H \;+\; R^2OH \qquad (44)$$

$$R^1PH_2 + R^2{}_2CO \longrightarrow R^1(H)P(O)CHR^2{}_2 \quad (45)$$

$$R^1{}_2PX + R^2OH \xrightarrow{B} R^1{}_2POR^2 + B{\cdot}HX \quad (46)$$

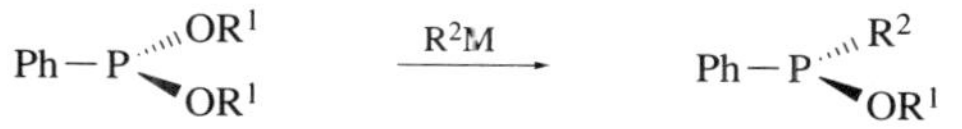
(47)

2.08.4.2 Alkylphosphonous Acid Derivatives—RP(OH)$_2$, etc. Also Halides—RP(Hal)(OH), etc.

Phosphonous acids have been prepared in high yields by the addition of hypophosphorus acid and its salts to alkenes ⟨60USP2957931⟩ (Equation (48)). The presence of a free radical catalyst is required. The hydrolysis of alkyldichlorophosphines ⟨1874CB1688⟩ (Equation (49)), alkyldialkoxyphosphines ⟨60CB1220⟩ or alkyldiaminophosphines ⟨55RZC960⟩ provides an alternative route to phosphonous acids. Ethoxytrimethylsilyloxy(diethoxymethyl)phosphine has been used to prepare a series of α-, β- and γ-aminoalkylphosphonous acids ⟨89T3787⟩. α-Aminoalkylphosphonous acids are also obtainable by the addition of hypophosphorus acid to diphenylmethylimines ⟨84JCS(P1)2845⟩. The addition of hypophosphite esters to alkenes (Equation (50)) ⟨67ZOB1692⟩ or carbonyl compounds ⟨67ZOB758⟩ yields phosphonous acid monoesters. Phosphonous acid monoesters are also the products of the hydrolysis of alkoxyalkylaminophosphines (Equation (51)) ⟨62ZOB1974⟩ or alkyldialkoxyphosphines ⟨63HCA2464⟩, or the alcoholysis of alkyldichlorophosphines ⟨52MI 208-01⟩. Phosphonous acids have been esterified by alcohols (Equation (52)) ⟨53USP2659714⟩, epoxides ⟨54USP2686803⟩, diazomethane ⟨54USP2693482⟩ and by dialkyl sulfates ⟨33JCS776⟩. Grignard reagents react with chlorodialkoxyphosphines (Equation (53)) ⟨57DOK(117)817⟩ or trialkoxyphosphines (Equation (54)) ⟨55JCS2039⟩, producing alkyldialkoxyphosphines. These compounds are also obtainable by alcoholysis of alkoxyalkylaminophosphines ⟨63ZOB1246⟩, alkyldiaminophosphines ⟨63HCA2667⟩ and alkyldichlorophosphines, in the presence of a tertiary amine ⟨45MI 208-01⟩, or from alkyldichlorophosphines and epoxides ⟨61ZOB949⟩.

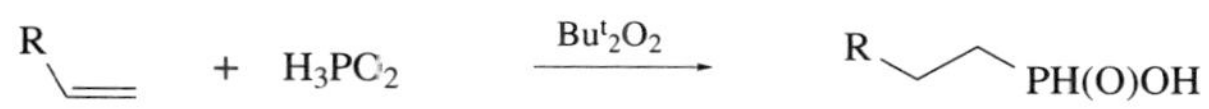
(48)

$$RPX_2 + 2H_2O \longrightarrow RPH(O)OH + 2HX \quad (49)$$

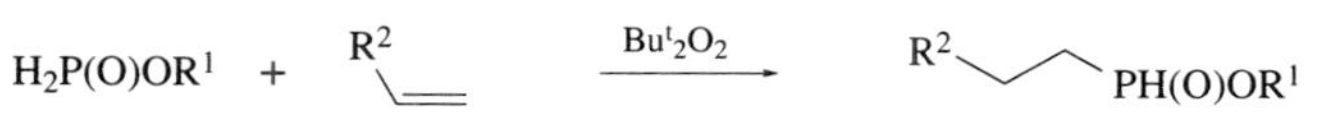
(50)

$$R^1P(NR^1{}_2)OR^2 + H_2O \longrightarrow R^1PH(O)OR^2 + R^1{}_2NH \quad (51)$$

$$R^1PH(O)OH + R^2OH \overset{H^+}{\rightleftharpoons} R^1PH(O)OR^2 + H_2O \quad (52)$$

$$(R^1O)_2PCl + R^2MgCl \xrightarrow[60\text{–}70\%]{} R^2P(OR^1)_2 + MgCl_2 \quad (53)$$

$$(R^1O)_3P + R^2MgCl \longrightarrow R^2P(OR^1)_2 + Mg(OR^1)X \quad (54)$$

Dibutoxyphosphine reacts with orthoformates when catalysed by boron trifluoride, to give dibutoxy(dialkoxymethyl)phosphines ⟨84JGU2237⟩, and with $BuOCH_2NR_2$, giving dibutoxy-(dialkylaminomethyl)phosphines ⟨89JGU2130⟩. The synthesis of alkoxyalkylchlorophosphines has been achieved by the reaction of alkyldichlorophosphines with alkyldialkoxyphosphines (Equation (55)) ⟨62CB2993⟩. Alkoxydialkylaminoalkylphosphines are readily prepared by the reaction of lithium alkoxide with the corresponding chlorodialkylaminoalkylphosphines ⟨85PS(24)217⟩.

$$R^1PCl_2 + R^2P(OR^3)_2 \xrightarrow[70\text{–}80\%]{} 2R^1P(OR^3)Cl \quad (55)$$

2.08.4.3 Alkylphosphine Oxides—R_3PO, etc.

Several methods have been applied to the synthesis of tertiary phosphine oxides. The oxidation of tertiary phosphines is one of the most important routes (Equation (56)). Various oxidizing agents have been used, including hydrogen peroxide ⟨78ZN(B)1556⟩, ozone ⟨61JA845⟩, sulfur dioxide ⟨65JCS5516⟩, sulfur trioxide and sulfuryl chloride fluoride ⟨83JOC1760⟩, and nitric acid or dinitrogen tetraoxide ⟨70CC1495⟩. Bis(trimethylsilyl)peroxide can be used to produce phosphine oxides stereospecifically with retention of the configuration at phosphorus ⟨85TL4965⟩.

$$R_3P \xrightarrow{[O]} R_3P(O) \qquad (56)$$

Dihalophosphoranes, prepared by halogenation of tertiary phosphines (Scheme 18), can be converted to tertiary phosphine oxides; the hydrolysing agent may be water ⟨1891JCS1931⟩, aqueous sodium hydroxide ⟨31JPR357⟩ or alcohol ⟨56ZAAC(288)201⟩.

$$R_3P + Br_2 \longrightarrow R_3PBr_2 \xrightarrow{HO^-} R_3P(O)$$

Scheme 18

Tertiary phosphine oxides have been obtained in very good yields by decomposition of phosphonium hydroxides (Equation (57)), derived from the phosphonium halide and moist silver oxide ⟨1888JCS636⟩, or by heating the phosphonium salt with 20–40% aqueous NaOH ⟨68JOC579⟩. Tetraalkylphosphonium salts, which are extremely difficult to hydrolyse with aqueous NaOH, react smoothly upon heating with anhydrous NaOH ⟨67JOC1060⟩.

$$R_4P^+ \ ^-OH \longrightarrow R_3P(O) + RH \qquad (57)$$

Alkoxydialkylphosphines, in general, undergo the Michael–Arbuzov rearrangement readily, producing tertiary phosphine oxides in high yields (Equation (58)). Different alkylating agents have been used, such as primary and secondary alkyl chlorides, bromides and iodides, 1-haloketones, 1-haloesters, carbon tetrachloride, acetyl chloride and heteroatom-substituted alkyl halides ⟨62JA3442⟩. The syntheses of compounds containing two, three or four phosphorus atoms have also been achieved by this reaction ⟨68HCA405⟩. 3-Ketophosphol-1-oxides have been obtained from alkyldialkoxyphosphines and 2-bromoethylpropanoate ⟨74ZC477⟩.

$$R^1{}_2POR^2 + R^3X \longrightarrow R^1{}_2R^3P(O) + R^2X \qquad (58)$$

An important route for preparing tertiary phosphine oxides is the reaction of phosphorus halides or esters with organometallic compounds. Symmetrical trialkylphosphine oxides (Equation (59)) are obtainable from phosphorus oxychloride and an excess of alkylmagnesium halide ⟨51JA4590⟩. Unsymmetrical phosphine oxides of the type $R^1R^2{}_2P(O)$ and $R^1R^2R^3P(O)$ can be prepared in excellent yields from phosphonic dichlorides (Equation (60)), phosphinic chlorides (Equation (61)) ⟨52JA4526⟩ or phosphonates (Equation (62)) ⟨65JCS6638⟩. The reaction of the diethyl ester of phosphorous acid, with a Grignard reagent is the first step for preparing 2-hydroxyalkyl ⟨68JOC3690⟩, alkoxymethyl and alkylthiomethyl tertiary phosphine oxides.

$$P(O)Cl_3 + 3RMgX \longrightarrow R_3P(O) + 3MgXCl \qquad (59)$$

$$R^1P(O)Cl_2 + 2R^2MgX \longrightarrow R^1R^2{}_2P(O) \qquad (60)$$

$$R^1R^2P(O)Cl + R^3MgX \longrightarrow R^1R^2R^3P(O) \qquad (61)$$

$$R^1P(O)(OR^2)_2 + 2R^3MgX \longrightarrow R^1R^3{}_2P(O) \qquad (62)$$

Red phosphorus reacts with alkyl iodides (Scheme 19) in the presence of catalytic quantities of iodine, producing hexaalkyliodophosphoranephosphonium pentaiodide, which, upon decomposition with alcohol, gives the trialkylphosphine oxide ⟨78BRP1517876⟩. Alcohols have also been employed as alkylating agents ⟨66ZOB564⟩.

$$P_4 + RI \longrightarrow [R_3P(I){-}PR_3]^+ I_5^- \longrightarrow R_3P(O)$$

Scheme 19

(2-Chloroalkyl)phosphine oxides are formed in a Wittig reaction (Equation (63)) ⟨81TL1231⟩.

(63)

Cyclic phosphine oxides have been obtained by the hydrolysis of the cycloaddition products of chlorophosphines with dienes (Scheme 20) ⟨82JA1893⟩.

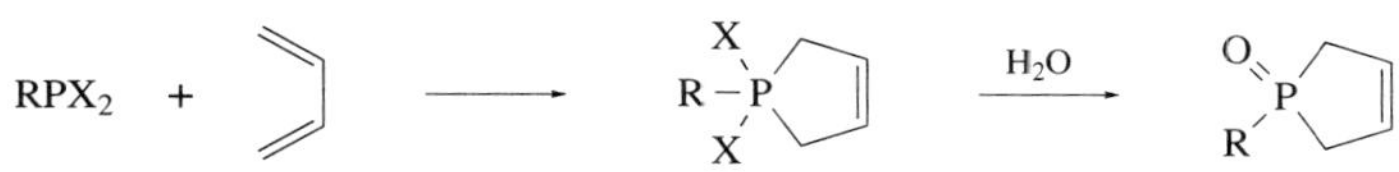

Scheme 20

Routes to optically active phosphine oxides have been developed using a menthyl unit to control the stereochemistry (Equation (64)) ⟨85TL783⟩. An Arbuzov rearrangement of the optically active 1,3,2-dioxaphosphacycloheptane ⟨87CL1915⟩ or 1,3,2-oxazaphospholane ⟨89TL2783⟩ and subsequent conversion of the intermediate acyclic phosphinate by reaction with the appropriate Grignard reagent provides an important method for a stereoselective synthesis of tertiary phosphine oxides. Optically active phosphine oxides have been obtained from *O*-isopropylmethylphosphorothioate (Scheme 21), which is available as either enantiomer. Alkylation at sulfur, followed by successive treatment with Grignard reagents, provides the corresponding phosphine oxide in high optical purity ⟨83JA4727⟩. Chiral phosphine oxides, having a functional group, have been prepared by the conjugate addition of alcohols or amines to optically active vinylphosphine oxides ⟨87JOC2170⟩.

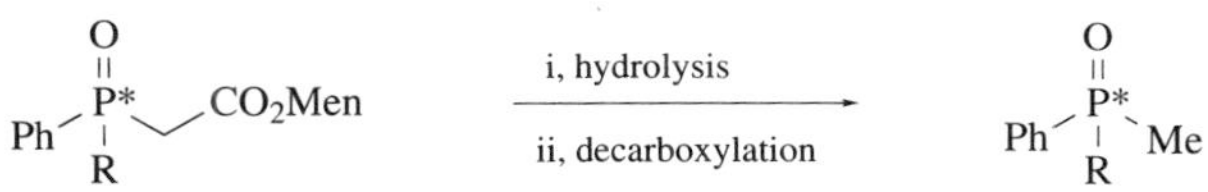

(64)

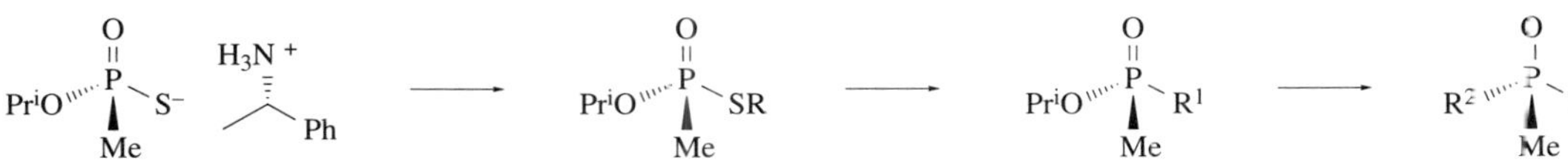

Scheme 21

2.08.4.4 Alkylphosphinic Acid Derivatives—$R_2P(O)OH$, etc. Also Halides—$R_2P(O)Hal$, etc.

The Arbuzov reaction of phosphorus(III) trimethylsilyl esters ⟨89PS(45)151⟩, alkoxydialkylphosphines ⟨76CB805⟩ and alkyldialkoxyphosphines (Equation (65)) ⟨56ZOB2463⟩ is a common route for the formation of phosphinic acids and phosphinates. Diphosphinates may be prepared by the Arbuzov reactions of alkyl dihalides and alkyldialkoxyphosphines ⟨63CB3184⟩, or alkyldialkoxydiphosphines and alkyl halides ⟨67JOC2172⟩. Esters of cyclic phosphinic acids are obtained by intramolecular reactions of ω-haloalkylphosphonites ⟨66ACS2508⟩.

$$R^1P(OR^2)_2 + R^3X \longrightarrow R^1R^3P(O)OR^2 \qquad (65)$$

The reaction of organometallic compounds with dialkyl esters of phosphorous acid, followed by oxidation (Scheme 22), is one of the best general methods for preparing symmetrical unsubstituted dialkyl phosphinic acids ⟨59ZOB1450⟩. These compounds are also obtainable from ester chlorides ⟨60ZOB2995⟩ or amide chlorides ⟨49JA369⟩ of phosphoric or phosphonic acids and organometallic compounds, and subsequent hydrolysis of the intermediate phosphinic esters and amides. Phosphonic dichlorides react with Grignard reagents and give phosphinic acids, which could contain bulky groups ⟨53JA3379⟩.

$$(R^1O)_2P(O)H \; + \; 3R^2MgX \longrightarrow [R^2{}_2P(O)MgX] \longrightarrow [R^2{}_2P(O)H] \longrightarrow R^2{}_2PO_2H$$

Scheme 22

An effective method for the conversion of long-chain dialkylphosphinous acids to phosphinic acids is the halogenation of the former with phosphorus pentachloride and hydrolysis of the phosphinic chlorides *in situ* (Scheme 23) ⟨55JA3411⟩. Esters have been prepared from crude bromides and alcohols ⟨61JOC4628⟩.

$$R_2PH \xrightarrow{PCl_5} [R_2P(O)X] \xrightarrow{[O]} R_2PO_2H$$

Scheme 23

Dimethylphosphinic acid has been prepared from dimethylphosphine by oxidation with H_2O_2 ⟨67CB2685⟩ or Na_2O_2 ⟨68CB2197⟩ (Equation (66)).

$$R_2PH \xrightarrow{[O]} R_2PO_2H \qquad (66)$$

Tertiary phosphine oxides react with hydroxides ⟨58CB64⟩ or sodium hydride ⟨61CB1317⟩, giving phosphinic acids (Equation (67) and Scheme 24).

$$R^1R^2R^3PO \; + \; HO^- \longrightarrow R^2R^3PO_2H \; + \; R^{1-} \qquad (67)$$

$$R_3PO \; + \; NaH \longrightarrow [R_2P(O)H] \longrightarrow R_2PO_2H$$

Scheme 24

The reaction of phosphinic halides with alcohols (Equation (68)) is a major route to phosphinates ⟨51JA5392⟩.

$$R^1{}_2P(O)Cl \; + \; R^2OH \longrightarrow R^1{}_2P(O)OR^2 \; + \; HCl \qquad (68)$$

Monoesters of phosphonous acids add to alkenes, producing phosphinates (Equation (69)) ⟨60ZOB2348⟩. Products containing a hydroxy group on the α carbon have been obtained by reactions of phosphonous acids and esters with carbonyl compounds ⟨69ZOB2418⟩.

(69)

α-Aminophosphinic acids and their ester derivatives have been prepared by the reactions between phosphorus(III) chlorides ⟨88SC425⟩ or phosphorus(III) esters ⟨86S821⟩, including trimethylsilyl esters, and a mixture of aldehydes or ketones and a nitrogen source (e.g., amides, amines or imines).

N-Sulfonylaziridines react with sodium ethylmethylphosphinate (or sodium diethylphosphonate) to give 2-aminophosphinic (or 2-aminophosphonic) acids (Equation (70)) ⟨83CB2936⟩. Starting from optically active 2-aminoalcohols, pure enantiomeric products may be obtained ⟨83TL2935⟩.

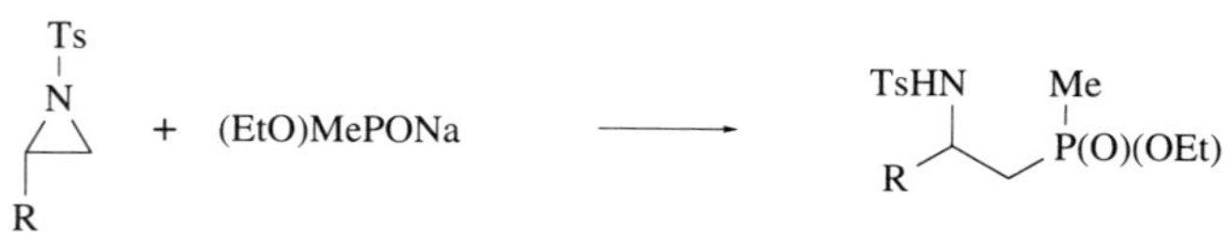

(70)

Cyclic diesters of phosphorus acid, in which the third bond from phosphorus is to chlorine ⟨62MI 208-01⟩, or an alkoxy ⟨68DOK(181)877⟩ or acyloxy ⟨69ZOB2367⟩, carbon, for example, react with conjugated dienes, yielding esters of cyclic phosphinic acids (Scheme 25). Cyclic phosphinates have also been obtained by reaction of 1,5-diketones and 1,4-dienes with $ROP(O)H_2$ ⟨85JGU2198⟩ (Equation (71)). Oxidation of the initial products from phosphine and 1,4-diketones produces 1-hydroxyphospholane 1-oxides ⟨88JGU693⟩ (Scheme 26). Esters of phospholanic acids have been obtained by the reaction between 1,4-butanediyldimagnesium dibromide and an alkyl phosphorodichloridate. Ring-substituted derivatives were prepared by the alkylation of ring-lithiated compounds ⟨91JOC3137⟩.

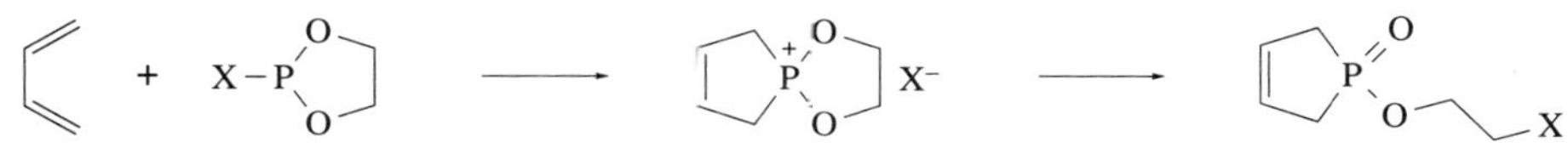

Scheme 25

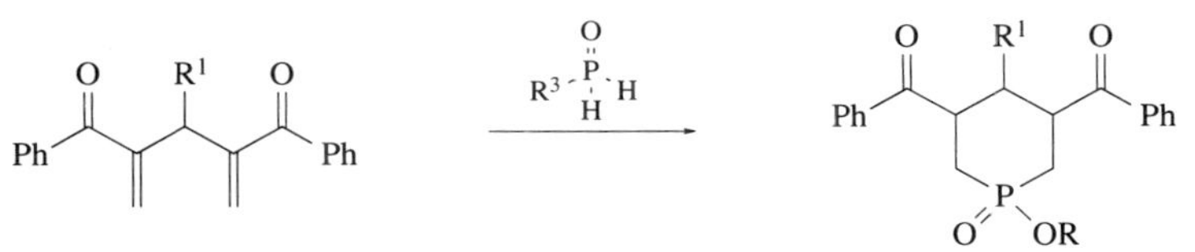

(71)

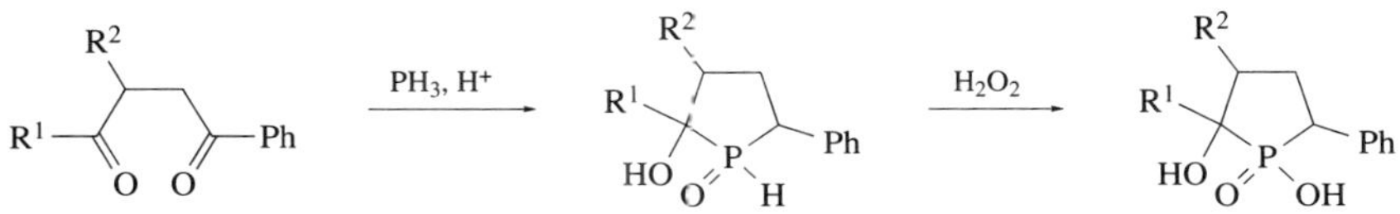

Scheme 26

Dialkylphosphinic chlorides are obtainable from dialkylphosphinous acids ⟨87ZAAC(548)55⟩, phosphinic acids or their esters ⟨51JA5466⟩ and phosphorus pentachloride (Equation (72)). Dialkyltrihalophosphoranes react with P_2O_5 or H_2O, producing phosphinic chlorides ⟨01LA(315)43, 55JA1864⟩.

$$R_2P(O)H + PCl_5 \longrightarrow R_2P(O)Cl \quad (72)$$

The reactions of phosphinic chlorides with phosphinates provide a satisfactory route to phosphinic anhydrides (Equation (73)) ⟨51JA4101⟩.

$$R_2P(O)Cl + R_2P(O)(OEt) \longrightarrow R_2P(O)\text{–}O\text{–}P(O)R_2 + EtCl \quad (73)$$

The 1,3,2-oxazaphospholan in Scheme 27, obtained from (−)-ephedrine, reacts with alkyl halides to give phosphinamides with predominant retention of the configuration at phosphorus. Acid methanolysis gave methyl phosphinates with a high degree of inversion ⟨89TL2783⟩.

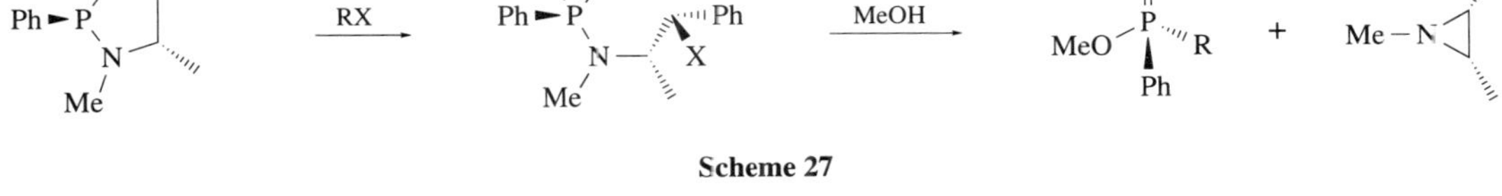

Scheme 27

Chiral alkyl methylphenylphosphinates have been obtained by treatment of the phosphinic chloride with L-proline ethyl ester, followed by the acid-catalysed alcoholysis of the separated diastereomeric phosphinic amides ⟨79S110⟩.

A convenient preparation of simple optically active phosphinates employs the trifluoroacetolysis of the quaternary salts ⟨75JOC1523⟩ (Scheme 28).

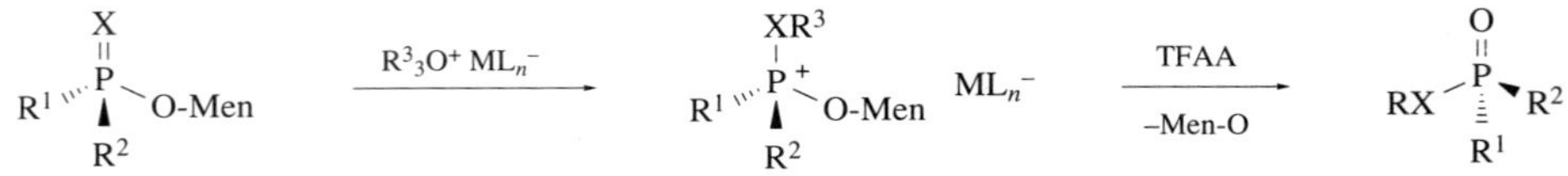

R^3 = Me, Et; ML_n^- = PF_6^-, BF_4^-, $SbCl_6^-$; X = O, S; Men = menthyl; TFAA = trifluoroacetic anhydride

Scheme 28

2.08.4.5 Alkylphosphonic Acid Derivatives—$RP(O)(OH)_2$, Also Halides—RP(O)(Hal)OH, $RP(O)(Hal)_2$, etc.

The hydrolysis of phosphonic acid dichlorides (Equation (74)) ⟨56GEP1023033⟩, and alkyltetrachlorophosphoranes (Equation (75)) ⟨32JCS2880⟩, is a convenient route to alkylphosphonic acids. Phosphonic acid esters and amides can be hydrolysed with mineral acid, producing the free phosphonic acids ⟨45JA1180⟩. Phosphorus trichloride reacts with aldehydes and α,*β*-unsaturated ketones, giving products which upon hydrolysis yield the corresponding 1-hydroxyalkylphosphonic acids ⟨53CJC976⟩ and 3-oxoalkylphosphonic acids ⟨20JA830⟩.

$$RPOCl_2 + 2H_2O \longrightarrow RPO(OH)_2 + 2HCl \qquad (74)$$

$$RPCl_4 + 3H_2O \longrightarrow RPO(OH)_2 + 4HCl \qquad (75)$$

Phosphonic acid monoesters can be produced by the treatment of phosphonous acid with an alcohol in the presence of dicyclohexylcarbodiimide and 4-dimethylaminopyridine, followed by the oxidation of the phosphonous acid monoester with $NaIO_4$ ⟨86TL1751⟩. Phosphonic acid anhydrides (Equation (76)) are easily cleaved by alcohols to give phosphonic acid monoesters ⟨61HCA1812⟩, which are also obtainable by the base-catalysed hydrolysis of phosphonic acid diesters ⟨57JOC972⟩.

$$(R^1PO_2)_n + nR^2OH \longrightarrow nR^1PO(OH)(OR^2) \qquad (76)$$

Phosphonic acid monoesters ⟨70MIP1945146⟩ and diesters ⟨45USP2471472⟩ can be prepared by the alcoholysis of phosphonic dihalides (Equation (77)).

$$R^1POCl_2 + 2R^2OH \longrightarrow R^1PO(OR^2)_2 + 2HCl \qquad (77)$$

Mixed phosphonate diesters can be synthesized in high yields by 1*H*-tetrazole-catalyzed monoaddition of alcohols to phosphonic dichlorides ⟨93T363⟩. With the exception of trimethyl phosphate, which suffers dimethylation, trialkyl phosphates can be converted into dialkyl alkylphosphonates by the action of alkyllithiums ⟨87TL405⟩.

The alkylation of phosphonous acid dialkyl esters using alkyl halides in the presence of crown ethers (Equation (78)) provides a route to dialkyl alkylphosphonates ⟨87MI 208-01⟩.

$$(R^1O)_2P(O)H \xrightarrow{R^2X,\ \text{crown ethers}} (R^1O)_2P(O)R^2 \qquad (78)$$

The Michaelis–Arbuzov reaction (Scheme 29) is one of the most important methods for the preparation of phosphonates. Alkyl halides and acyl halides ⟨1898CB1048⟩, *gem*-dibromocyclopropanes ⟨85BCJ3104⟩, α-chlorooxiranes ⟨80TL2687⟩, among other compounds, have been used as alkylating agents. An alternative to the Arbuzov reaction is to use a catalytic amount of iodine (Equation (79)) instead of RX ⟨90SC239⟩.

$$P(OR^1)_3 + R^2X \longrightarrow [R^2P(OR^1)_3]^+ X^- \longrightarrow R^2PO(OR^1)_2 + R^1X$$

Scheme 29

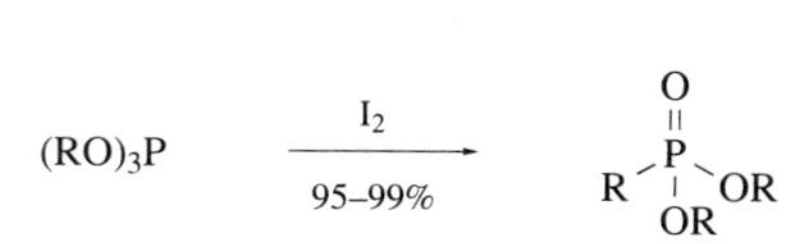

(79)

R = Me, Et, Bu

A procedure for the *C*-phosphorylation of 1-bromoketones (Scheme 30) has been applied in cases when the Arbuzov reaction is unsuccessful ⟨86JOC4342⟩. The method allows the use of secondary haloketones and phosphoryl halides with highly electronegative substituents.

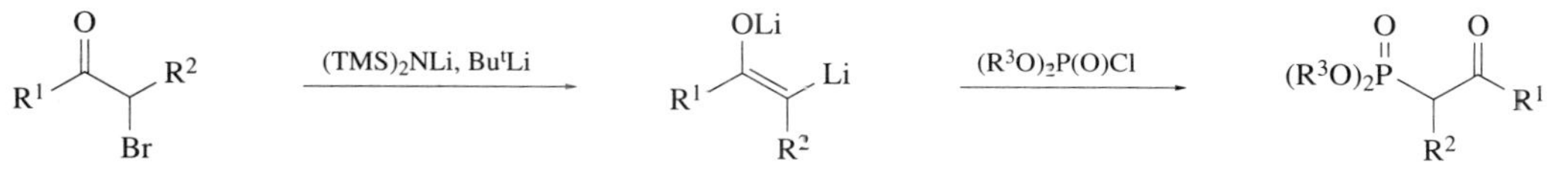

Scheme 30

2-Oxoalkylphosphonates have been obtained starting from the hydrazones of 1-chloroalkylket-2-ones ⟨85S1048⟩.

Chlorodiethylphosphine reacts with enolates ⟨91JOC5556⟩ and α-hydroxyketones ⟨89JOC627⟩, giving 2-oxoalkylphosphonates (Equation (80)).

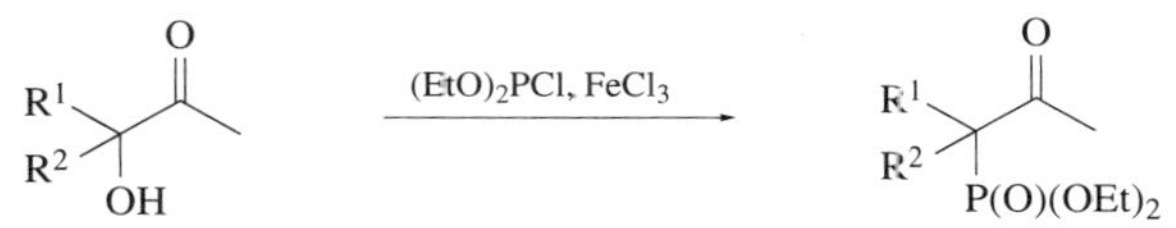

(80)

The addition of phosphorous acid dimethyl ester to polarized C═C bonds (Scheme 31) leads to dimethyl alkylphosphonates ⟨79BCJ466⟩. The reaction of phosphorous acid dialkyl esters ⟨86TL3515⟩ with aldehydes and ketones produces dialkyl 1-hydroxyalkylphosphonates (Equation (81)). The reaction of hydrazine derivatives of carbonyl compounds with $(MeO)_2P(O)H$ provides a route to secondary alkylphosphonates ⟨77S179⟩.

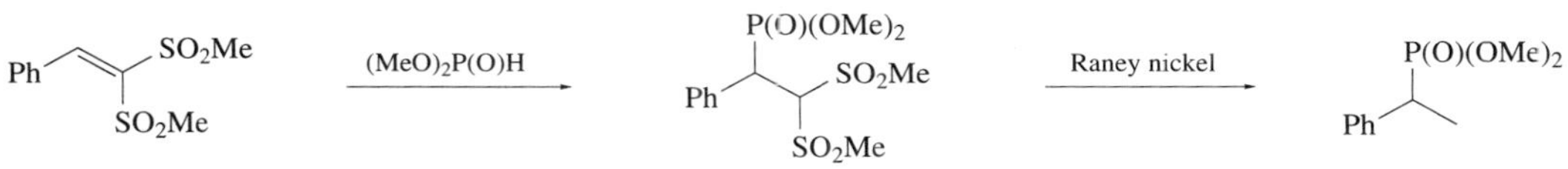

Scheme 31

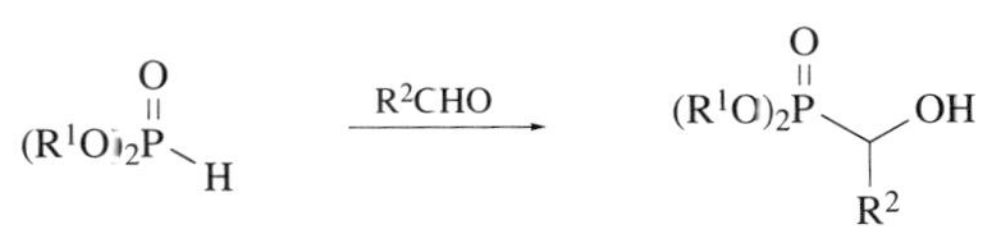

(81)

Dialkyl 2-hydroxyalkylphosphonates (Equation (82)) can be synthesized from oxiranes and diethyloxytrimethylsilyloxyphosphine ⟨83S916⟩ or phosphorous acid dialkyl esters ⟨79PS(7)101⟩.

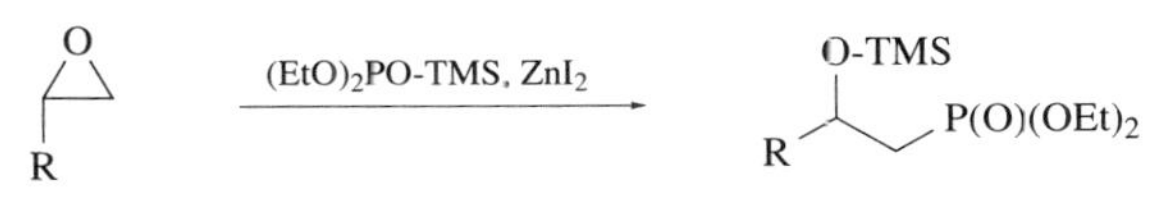

(82)

Mannich-type reactions (Equation (83)) have been used in the synthesis of dialkyl α-aminoalkylphosphonates ⟨92T2911⟩. A general route to α-aminophosphonic acids is the reaction of esters of appropriate bromo-acids with triethoxyphosphine, followed by a Curtius reaction ⟨72JOC4396⟩. In the presence of Lewis acids, α-methoxyurethanes react with trialkoxyphosphines, yielding α-aminoalkylphosphonates ⟨81TL3249⟩.

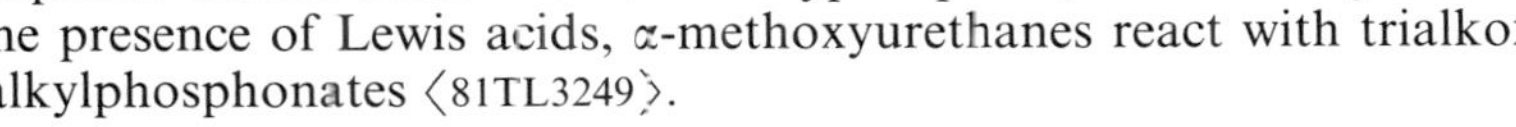

(83)

Aziridines react with dialkyl esters of the phosphorous acid ⟨83CB2936⟩ to give 2-aminoalkylphosphonates (Equation (84)).

(84)

ω-Aminoalkylphosphonic acids (Scheme 32) have been prepared from ω-aminoalkyl bromides by their reaction with chlorodiethoxyphosphine, followed by treatment with triethoxyphosphine and hydrolysis ⟨79T1345⟩.

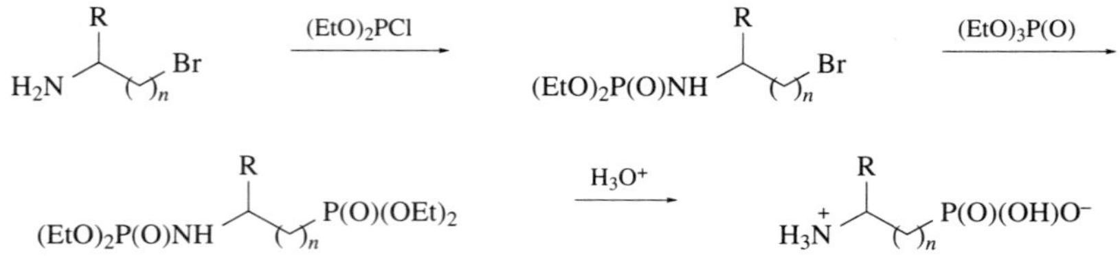

Scheme 32

Polyphosphonic acid anhydrides react with epoxides (Equation (85)) to form cyclic esters ⟨65ZOB732⟩.

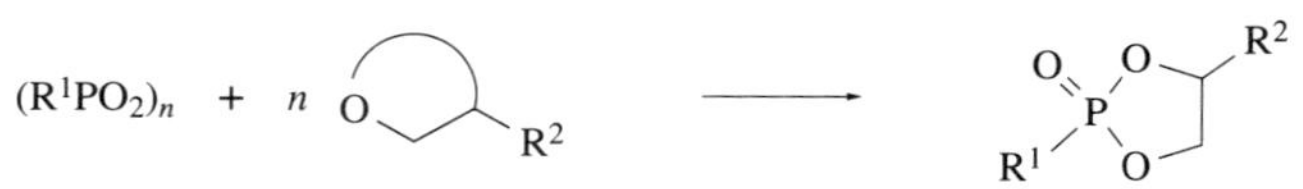
(85)

Cyclic phosphonic acids have been obtained from 1,5-diketones and phosphorus trichloride ⟨79JGU1714⟩, hypophosphites ⟨79JGU1721⟩ or H_3PO_2 ⟨79JGU285⟩.

Dialkyl alkylphosphonates (Equation (86)) react with halogenating agents, such as PCl_5, $POCl_3$ or $(COCl)_2$, to give phosphonic dichlorides ⟨61JCS238⟩ and alkylchlorophosphonates ⟨90PS(47)465⟩; the latter are also obtainable by the action of dichlorotriphenylphosphorane on dialkyl alkylphosphonates ⟨88SC285⟩. Aluminum halide complexes of alkyltetrachlorophosphoranes (Scheme 33) react with H_2O ⟨63HOU(12)396⟩ or alcohols ⟨57JA3570⟩, producing phosphonic acid dichlorides and monochlorides (as their monoesters). Phosphonic acid monochlorides ⟨54BRP810930⟩ and dichlorides ⟨63HOU(12)404⟩ can be synthesized from dialkoxyhalophosphines (Scheme 34), which react with alkyl halides in the fashion of an Arbuzov reaction. Chlorophosphonium complexes, obtained from alkenes and PCl_5, can be decomposed to give phosphonic dichlorides (Scheme 35) by the action of boric acid, sodium acetate ⟨83JGU1059⟩, SO_2 ⟨63HOU(12)394⟩ or P_2O_5 ⟨51JA855⟩.

$$R^1PO(OR^2)_2 + PCl_5 \longrightarrow R^1PO(OR^2)_2 + R^2Cl_2 + POCl_3 \quad (86)$$

$$PCl_3 + RCl + AlCl_3 \longrightarrow RPCl_4{\bullet}AlCl_3 \xrightarrow{(x+1)H_2O} RPOCl_2 + AlCl_3{\bullet}xH_2O + 2HCl$$

Scheme 33

$$(R^1O)_2PX + R^2Y \longrightarrow [R^2P(OR^1)_2X]^+\,Y^- \longrightarrow R^2PO(OR^1)X + R^1Y$$

Scheme 34

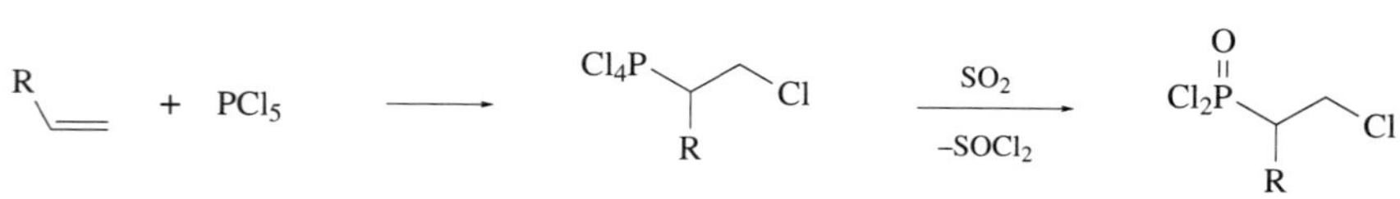

Scheme 35

Chloroalkylphosphonic dichlorides have been formed from butanol and 2-methylbutan-2-ol on treatment with PCl_5, followed by SO_2 ⟨74ZOB2621⟩. PCl_3 ⟨74ZOB1261⟩ and PBr_3 ⟨73CL599⟩ add to alkenes, yielding the corresponding phosphonic dihalides.

Optically active phosphonates have been obtained, utilizing the chirality of (−)-(*S*)-methylbenzylamine (Scheme 36) ⟨81S811⟩. L-Proline ethyl ester has been used as a chiral template in the synthesis of optically active phosphonates ⟨78TL3741⟩.

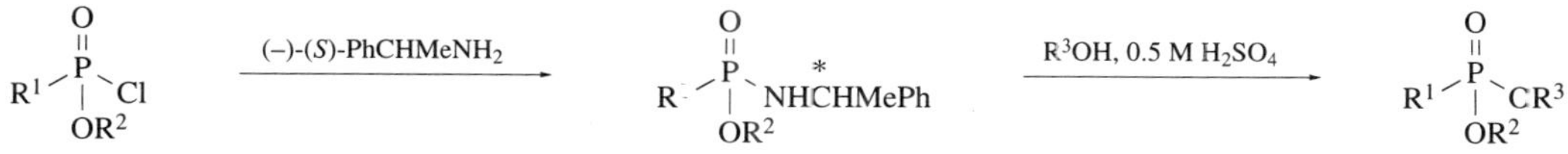

Scheme 36

Chiral 1-hydroxyalkylphosphonic acids ⟨92TA377⟩ and 1-chloroalkylphosphonic acids ⟨90TL6461⟩ have been prepared, employing chiral acetals (Scheme 37) and (*R*,*R*)-1,2-bis(methylamino)-cyclohexane (Scheme 38).

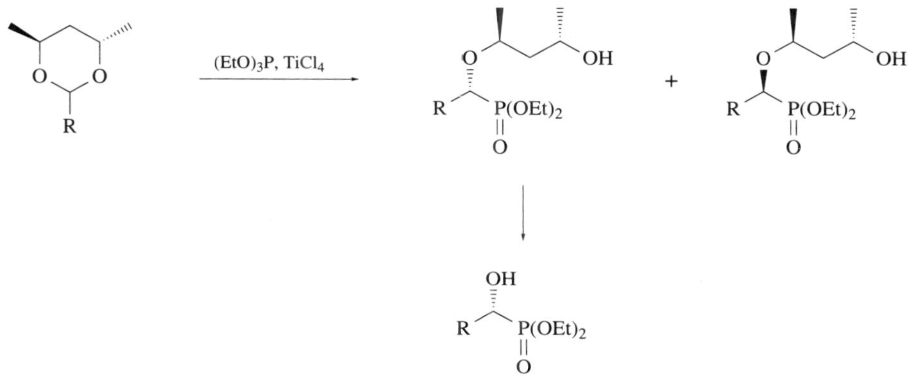

Scheme 37

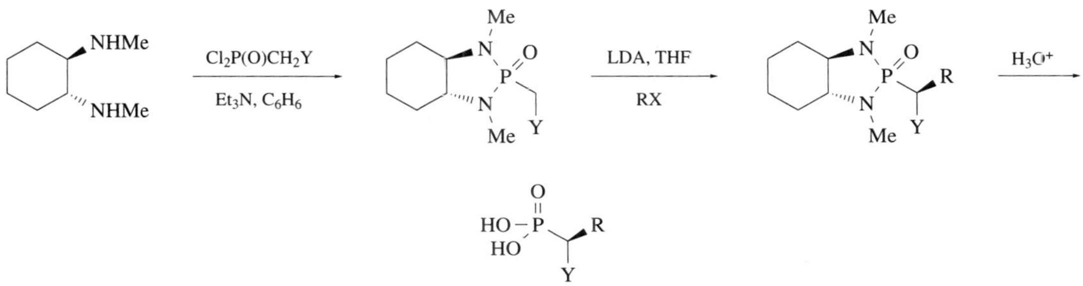

Scheme 38

Chiral α-aminoalkylphosphonic acids have been obtained using methodologies, based on the chemistry of 1,3,2-oxazaphospholidines ⟨90S132⟩ and sugar amines ⟨92S90⟩. Enantiomeric forms of 2-aminoalkylphosphonic acids have been prepared in a sequence of reactions (Scheme 39) ⟨90BCJ2728⟩.

2.08.5 ALKYLPHOSPHORUS COMPOUNDS WITH A P—S BOND

2.08.5.1 Alkylthiophosphinous Acid Derivatives—R_2PSH, etc.

Thiophosphinous acids can be obtained in yields higher than 90% by the reaction of dialkyl thiophosphonates with Grignard reagents (Equation (87)) ⟨66HCA1249⟩. Organolithium compounds can be used instead of Grignard reagents.

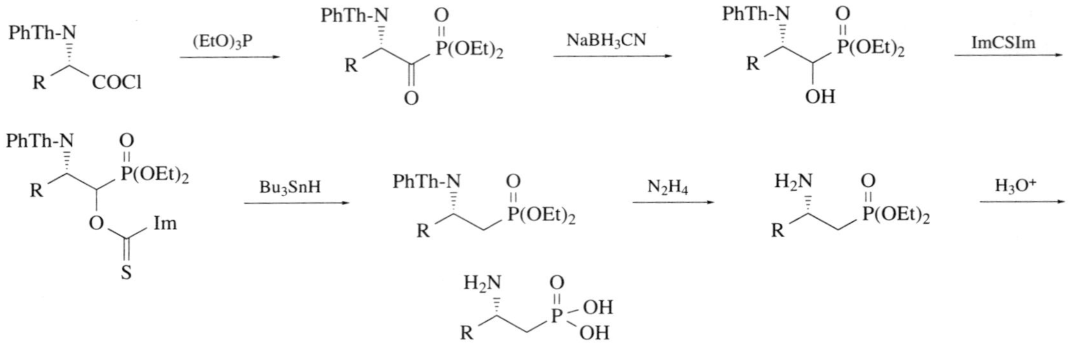

Phth = phthaloyl
Im = Imidazolyl

Scheme 39

$$(EtO)_2P(S)H + 3R^1MgX \longrightarrow R^1_2P(S)H + 2EtOMgX \qquad (87)$$

The addition of a stoichiometric amount of sulfur to secondary phosphines produced thiophosphinous acids in nearly quantitative yields ⟨62JOC2198⟩. Dithiophosphinous acids have also been prepared by this technique (Equation (88)) ⟨83ZAAC(504)147⟩. Aminodialkylphosphines react with anhydrous hydrogen sulfide, giving thiophosphinous acids in high yields (Equation (89)) ⟨66HCA1249⟩.

$$R(H)P(CH_2)_nP(H)R + 2S \longrightarrow R(H)P(S)(CH_2)_n(S)P(H)R \qquad (88)$$

$$R^1_2PNR^2_2 + H_2S \longrightarrow R^1_2P(S)H + R^2_2NH{\bullet}H_2S \qquad (89)$$

Alkylthiodialkylphosphines have been prepared in quantitative yields by the reaction of dialkylchlorophosphines with mercaptans (Equation (90)) ⟨66DOK(168)807⟩.

$$R^1_2PCl + R^2SH \xrightarrow{\text{base}} R^1_2PSR^2 + HCl \qquad (90)$$

Disulfides react with diphosphines on irradiation with ultraviolet light ⟨59JA6177⟩, or with secondary phosphines in the presence of free radical inhibitors ⟨67JCS(C)2131⟩, forming alkylthiodialkylphosphines (Equations (91) and (92)).

$$R^1_2PPR_2 + R^2SSR^2 \longrightarrow 2R^1_2PSR^2 \qquad (91)$$

$$R^1_2PH + R^2SSR^2 \xrightarrow{\text{inhibitor}} R^1_2PSR^2 + R^2SH \qquad (92)$$

Optically active thiophosphinous acids have been prepared by the reaction of optically active dithiophosphinates with butyllithium, followed by treatment with electrophiles (Equation (93)) ⟨89CL849⟩.

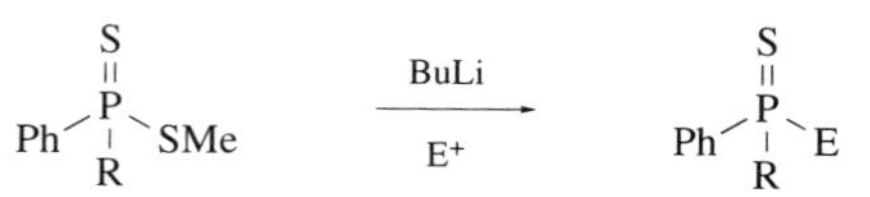

(93)

2.08.5.2 Alkylthiophosphonous Acid Derivatives—$RP(SH)_2$, RPHal(SH), etc.

Alkyldi(alkylthio)phosphines have been obtained in 80–90% yields by the reaction of primary phosphines with sulfenyl chlorides ⟨62ZOB4017⟩ (Equation (94)) or disulfides ⟨67JOC236⟩ (Equation (95)).

$$R^1PH_2 + 2R^2SCl \longrightarrow R^1P(SR^2)_2 + 2HCl \quad (94)$$

$$R^1PH_2 + 2R^2SSR^2 \longrightarrow R^1P(SR^2)_2 + 2R^2SH \quad (95)$$

The thermal decomposition of cyclotetraphosphines in the presence of dialkyl disulfides leads to alkyldi(alkylthio)phosphines (Equation (96)) in high yields ⟨69JA5609⟩.

$$[R^1P]_4 + 4R^2SSR^2 \longrightarrow 4R^1P(SR^2)_2 \quad (96)$$

Alkyldichlorophosphines react with mercaptides ⟨29MI 208-01⟩ or with thiols in the presence of a tertiary amine ⟨65JA2113⟩ to give alkyldi(alkylthio)phosphines.

Di(ethylthio)phosphine adds to aldehydes, producing (1-hydroxyalkyl)di(alkylthio)phosphines (Equation (97)) ⟨88JGU2161⟩.

$$(EtS)_2PH + RCHO \longrightarrow (EtS)_2PCH(OH)R \quad (97)$$

Alkoxyalkyl(alkylthio)phosphines can be prepared by the reaction of alkoxyalkylchlorophosphines with thiols ⟨62CB2993⟩ (Equation (98)) or alkyl(alkylthio)chlorophosphines with alcohols (Equation (99)) ⟨66USP3261743⟩. The presence of a base is often required.

$$R^1P(OR)Cl + R^2SH \xrightarrow{\text{base}} R^1P(OR)SR^2 + B{\bullet}HCl \quad (98)$$

$$R^1P(SR)Cl + R^2OH \xrightarrow{\text{base}} R^1P(OR^2)SR + B{\bullet}HCl \quad (99)$$

Alkyl(alkylthio)halogenophosphines can be synthesized by the reaction of the corresponding alkyldihalogenophosphines with alkyldialkoxyphosphines ⟨67ZOB1173⟩ (Equation (100)).

$$RPCl_2 + RP(SR)_2 \longrightarrow 2RP(SR)Cl \quad (100)$$

Alkoxyaminoalkylphosphines ⟨62ZOB3052⟩ and alkoxyalkylchlorophosphines (Equation (101)) ⟨62RZC1781⟩ react with hydrogen sulfide, producing *O*-esters of thiophosphonous acids, which are also obtainable by the reaction of monoesters of phosphonous acids with phosphorus pentasulfide ⟨62BEP609076⟩, carried out at boiling water-bath temperature.

$$RP(OR)Cl + H_2S \longrightarrow RPH(S)OR + HCl \quad (101)$$

2.08.5.3 Alkylphosphine Sulfides—R_3PS, etc.

Tertiary phosphine sulfides have been prepared by the addition of sulfur to tertiary phosphines (Equation (102)) ⟨1857(101)LA1⟩. The reaction proceeds with retention of configuration, when optically active phosphines are used ⟨64TL175⟩. Also, by this method diphosphine disulfides ⟨61CB2656⟩ and cyclic phosphine sulfides ⟨59JCS2835⟩ have been obtained.

$$R_3P + S \longrightarrow R_3PS \quad (102)$$

Tributylphosphine sulfide ⟨61JOC5205⟩, diphosphine disulfides ⟨66ZOB1436⟩ and several functionally substituted tertiary phosphine sulfides ⟨65URP173230⟩ have been prepared by the reaction of the corresponding dihalogenotrialkylphosphoranes with hydrogen sulfide (Equation (103)).

$$R_3PX_2 + H_2S \longrightarrow R_3PS + 2HCl \quad (103)$$

Symmetrical tertiary phosphine sulfides can be produced by the reaction of $PSCl_3$ with alkyllithium (Equation (104)) ⟨68JCS(B)1416⟩, triethylaluminums ⟨64HCA27⟩ and Grignard reagents ⟨62CB64⟩. With alkyl Grignard reagents, dialkylthiophosphinic halides gave products of the type $R^1{}_2R^2PS$ ⟨63JOC3430⟩.

$$PSCl_3 + 3RLi \longrightarrow R_3PS + 3LiCl \quad (104)$$

The Michaelis–Arbuzov reaction provides a route for the synthesis of triethylphosphine sulfide (Equation (105)) ⟨57MI 208-01⟩ and dimethylethylphosphine sulfide ⟨68ZOB609⟩.

$$EtP(SEt)_2 + EtBr \longrightarrow Et_3PS \qquad (105)$$

Tertiary phosphine sulfides have been obtained in high yields by the addition of dialkylthiophosphinous acids to alkenes (Equation (106)) ⟨66HCA1249⟩.

(106)

Dialkylthiophosphinous acids react with carbonyl compounds, giving α-hydroxyalkyl tertiary phosphine sulfides (Equation (107)) ⟨60CRV243⟩, and with Schiff bases, α-aminoalkyl tertiary phosphine sulfides ⟨71PS(1)71⟩.

$$R_2P(S)H + R^1R^2CO \longrightarrow R_2P(S)C(OH)R^1R^2 \qquad (107)$$

Tertiary phosphine oxides can be converted to tertiary phosphine sulfides by reaction with P_2S_5 ⟨59CB2088⟩.

Diphosphine disulfides have been obtained from the corresponding diphosphine dioxides by heating with elemental sulfur ⟨53USP2642461⟩.

The treatment of optically active phosphine oxides with boron trisulfide leads to the corresponding phosphine sulfides, with retention of the configuration at phosphorus ⟨73CC273⟩.

3-Phospholene sulfides have been prepared by [1 + 4] cycloaddition reactions (Equation (108)) ⟨76JOC1578⟩.

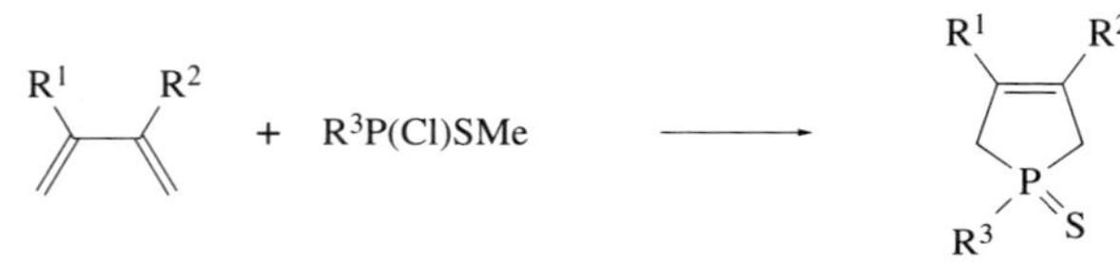

(108)

The reaction of optically active dithiophosphinates with butyllithium and subsequent treatment with electrophiles provides a method for the synthesis of tertiary phosphine sulfides (see Equation (93)) ⟨89CL849⟩.

2.08.5.4 Alkylthiophosphinic Acid Derivatives—$R_2P(S)OH$, $R_2P(O)SH$, $R_2P(S)SH$, $R_2P(S)Hal$, etc.

Thiophosphinic and dithiophosphinic acids can be prepared by the addition of sulfur to dialkylphosphinous acids (Equation (109)) ⟨66CB1134⟩ and secondary phosphines ⟨68CB3612⟩. Addition of sulfur to alkoxydialkylphosphines ⟨55JA3526⟩ and alkylthiodialkylphosphines (Equation (110)) ⟨66IZV493⟩ leads to thiophosphinic and dithiophosphinic esters.

$$R_2P(O)H + S \longrightarrow R_2P(O)SH \qquad (109)$$

$$R^1{}_2PSR^2 + S \longrightarrow R^1{}_2P(S)SR^2 \qquad (110)$$

Thiophosphinic (Equation (111)) and phosphinic halides react with sodium hydrosulfide, producing dithiophosphinic ⟨61ZOB507⟩ and thiophosphinic acids ⟨62ZOB3579⟩. The corresponding thiophosphinate *O*-esters ⟨66ZAAC(343)268⟩ and dithiophosphinates ⟨63T1563⟩ have been prepared by the reaction of thiophosphinic chlorides with alcohols and thiols in the presence of a tertiary amine, or with sodium alkoxides and thiolates.

$$R_2P(S)Cl + HS^- \longrightarrow R_2PS_2H \qquad (111)$$

The fusion of biphosphine disulfides with sodium sulfide and sulfur gives excellent yields of thiophosphinic acids as their sodium salts (Equation (112)) ⟨68CB3454⟩.

$$R_2P(S){-}P(S)R_2 + Na_2S + S \longrightarrow 2R_2PS_2^-Na^+ \qquad (112)$$

The reaction of secondary phosphines with disulfides, usually carried out in benzene, under reflux, gives dithiophosphinates (Equation (113)) in excellent yields ⟨67JOC236⟩.

$$R^1_2PH \; + \; R^2SSR^2 \longrightarrow R^1_2P(S)SR^2 \; + \; R^2H \qquad (113)$$

(α-Hydroxyalkyl)thiophosphinates have been obtained by the addition of monoesters of alkylthiophosphonous acids to aldehydes and ketones (Equation (114)) ⟨69ZOB1543⟩.

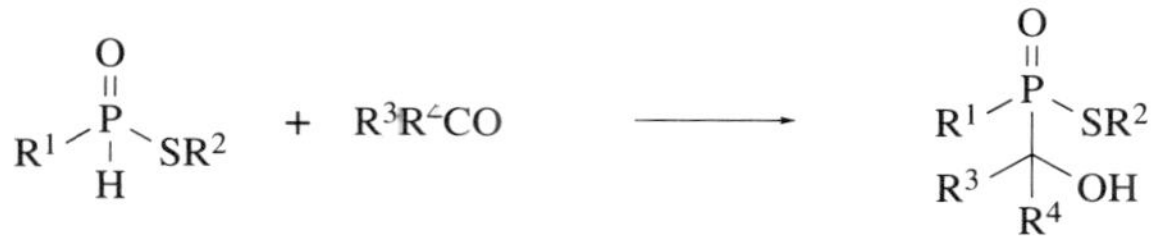

(114)

The reaction of alkyl(alkylthio)chlorophosphines with α,β-unsaturated carboxylic acids yields *S*-alkyl β-carboxythiophosphinates ⟨67IZV1976⟩. Mono- and dithioanhydrides of phosphinic acids have been formed in very good yields by the reaction of thiophosphinyl chlorides with thiophosphinates ⟨64HCA1448⟩. Phosphinic acids have been converted to thiophosphinic chlorides when heated with $PSCl_3$ ⟨82JGU195⟩.

Dialkylthiophosphinic iodides ⟨82JGU202⟩ and chlorides ⟨63HOU(12)274⟩ (Equation (115)) have been prepared from diphosphine disulfides and iodine, or chlorine.

(115)

A facile, one-step synthesis of thiophosphinic chlorides involves the decomposition of chlorophosphine–alkyl halide–$AlCl_3$ complexes with imidazolidine-2-thione ⟨81IJC(B)932⟩.

Thiophosphinic chlorides have also been formed by the addition of sulfur to chlorophosphines ⟨63HOU(12)275⟩.

The treatment of 2-methyl-2-thione-1,2,3-phosphilidine with alkyllithiums leads to stereospecific ring opening, giving after acidolysis methylphenylthiophosphinic acids ⟨79TL2729⟩ (Scheme 40).

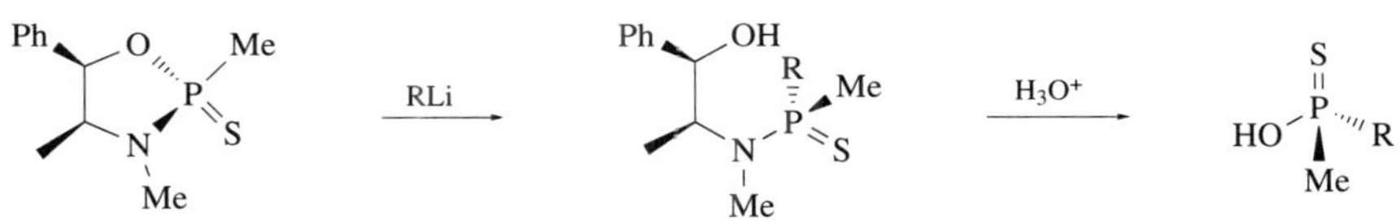

Scheme 40

2.08.5.5 Alkylthiophosphonic Acid Derivatives—RP(S)(OH)$_2$, RP(O)(OH)(SH), RP(O)(SH)$_2$, RP(S)(OH)(SH), RP(O)Hal(SH), RP(S)(Hal)$_2$, etc.

Thiophosphonic acids have been prepared by base-catalysed hydrolysis (Equation (116)) of thiophosphonic dichlorides ⟨61IZV604⟩. The hydrolysis of alkyl alkylchlorothiophosphonates ⟨59JA148⟩ and dialkyl alkylthiophosphonates ⟨59JA148⟩ leads to thiophosphonic acid monoesters.

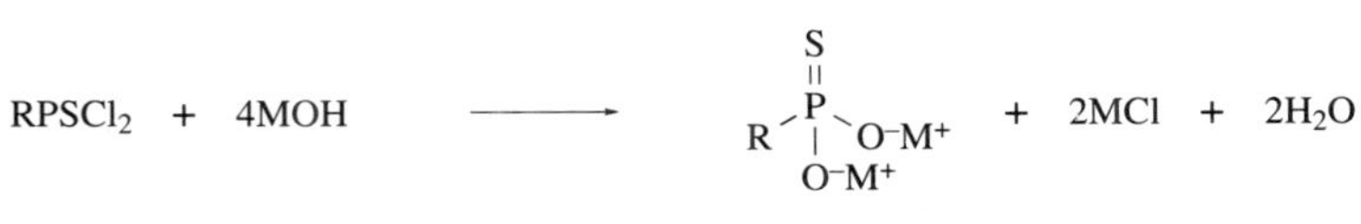

(116)

Good yields of dialkyl alkylthiophosphonates are obtained by the reaction of thiophosphonic acid dihalides with alcoholates ⟨58JA3945⟩. The synthesis of asymmetric diesters is achieved from $R^1PS(OR^2)Cl$ (Equation (117)) ⟨57MIP1057112⟩.

$$R^1PS(OR^2)Cl \; + \; R^3OH \longrightarrow R^1PS(OR^2)(OR^3) \; + \; HCl \qquad (117)$$

In the presence of a tertiary amine, *S*-alkyl alkylchlorothiophosphonates and *S*-alkyl alkylchlorodithiophosphonates (Equation (118)) react with alcohols, yielding the corresponding *O*,*S*-dialkyl alkylthiophosphonates ⟨54MIP1032247⟩.

$$R^1PS(SR^2)Cl + R^3OH \xrightarrow{R^4{}_3N} R^1PS(SR^2)(OR^3) + HCl \qquad (118)$$

S,*S*-Dialkyl alkyldithiophosphonates have been synthesized by the reaction of phosphonic dihalides or *S*-alkyl alkylhalogenothiophosphonates with alkali thiolides ⟨55IZV823⟩. Alkyl alkylchlorophosphonates with thiolides give *O*,*S*-dialkyl alkylthiophosphonates ⟨59ZOB1671⟩ (Equation (119)).

$$R^1PO(OR^2)Cl + NaSR^3 \longrightarrow R^1PO(OR^2)(SR^3) + NaCl \qquad (119)$$

The reaction between dithiocarboxylic acids and dialkoxychlorophosphines produces *O*,*S*-dialkyl alkylthiophosphonates (Equation (120)) ⟨86JGU1567⟩.

$$(R^1O)_2PCl + R^2CS_2H \xrightarrow{Et_3N} R^2PO(SR^1)(OR^1) \qquad (120)$$

Thiophosphonic acids can be obtained by the addition of sulfur to alkylbis(trimethylsiloxy)phosphines, followed by the hydrolysis of the products under acidic conditions (Scheme 41) ⟨80JGU2159⟩.

$$RP(O\text{-}TMS)_2 \xrightarrow{S} RPS(O\text{-}TMS)_2 \xrightarrow{H^+} RPS(OH)_2$$

Scheme 41

Dialkyl α-aminoalkylthiophosphonates can be formed by a Mannich-type reaction (Equation (121)) ⟨60IZV651⟩.

$$R^1R^2NH + R^3R^4C{=}O + HPS(OR^5)_2 \longrightarrow R^1R^2NCR^3R^4PS(OR^5)_2 \qquad (121)$$

Tri(alkylthio)phosphines react with ketones (Equation (122)), under acidic conditions, to yield *S*,*S*-dialkyl (α-alkylthio)alkyldithiophosphonates ⟨88JGU1131⟩.

$$(EtS)_3P + R^1R^2CO \xrightarrow{H^+} (EtS)_2P(O)C(SEt)R^1R^2 \qquad (122)$$

Alkyldichlorophosphines react with sulfenylchlorides, in liquid SO_2, to give *S*-alkyl alkylchlorothiophosphonates (Equation (123)) ⟨66ZOB500⟩.

$$R^1PCl_2 + R^2SCl + SO_2 \longrightarrow R^1PO(SR^2)Cl + SOCl_2 \qquad (123)$$

O-Alkyl alkylchlorothiophosphonates can be prepared by the oxidizing chlorination of thiophosphonous acid monoesters (Equation (124)) ⟨58MIP1067011⟩.

$$R^1PS(OR^2)H + SO_2Cl_2 \longrightarrow R^1PS(OR^2)Cl \qquad (124)$$

The preparation of thiophosphonic dichlorides can be achieved by the treatment of phosphorus trichloride, alkyl chloride or aluminum trichloride complexes with thiourea ⟨80JOC2270⟩.

Cyclic dithiophosphonates have been formed by the reaction of dithiophosphonic acid anhydrosulfides with epoxides (Equation (125)) ⟨61GEP1104520⟩.

$$(RPS_2)_n + n\overline{H_2CCHRO} \longrightarrow n\overline{R(S)POCHRCH_2S} \qquad (125)$$

2,9-Dithia-1-phosphabicyclo[4.3.0]nona-3,7-diene-1-sulfides are obtained when enones are treated with P_4S_{10} and triethylamine in CS_2 (Equation (126)) ⟨82TL1263⟩.

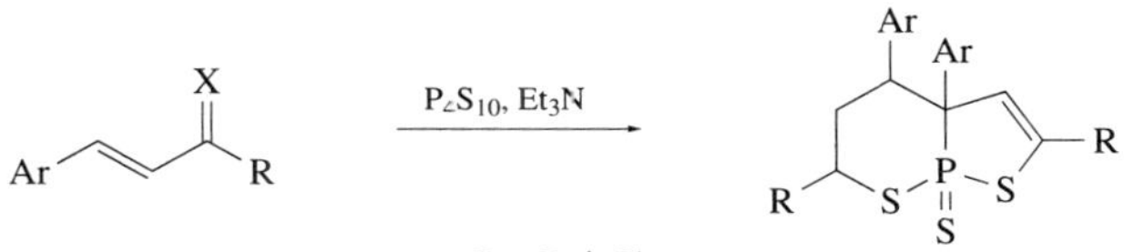 (126)

R = Bu^t, Ph

The reaction between tervalent phosphorus chloride and an alcohol in the presence of brucine, followed by the addition of sulfur to the optically active tervalent ester, provides a route to chiral *O*,*S*-dialkyl alkyldithiophosphonates (Scheme 42) ⟨84JGU1704⟩.

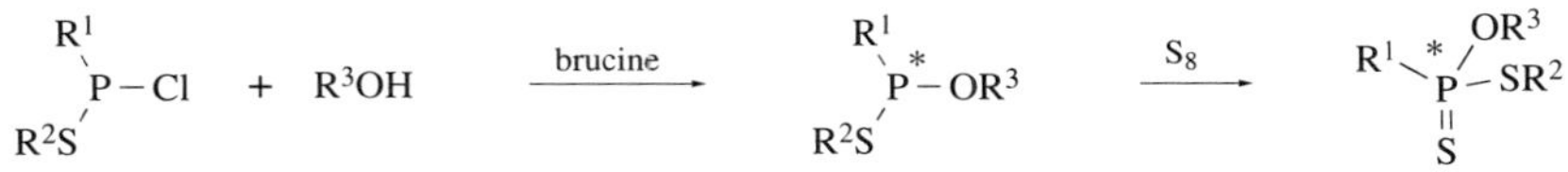

Scheme 42

2.08.6 ALKYLPHOSPHORUS COMPOUNDS WITH A P—Se AND/OR A P—Te BOND

The addition of selenium or tellurium to tervalent alkylphosphorus compounds is a general route to four-coordinated phosphorus compounds, containing P═Se or P═Te bonds. In this way, secondary phosphine selenides ⟨66HCA1000⟩ and tellurides ⟨83PS(18)73⟩, tertiary phosphine selenides ⟨64JCS1020⟩ and tellurides ⟨65JOM(4)320⟩, selenophosphinates ⟨1885CB2109⟩, selenophosphonates ⟨60DOK(135)1150⟩ (Equation (127)) and tellurophosphonates ⟨60BSF1794⟩ have been prepared. The synthesis of cyclic tertiary phosphine selenides has also been achieved by direct addition of selenium to the corresponding cyclic phosphine ⟨61JA4406⟩. The reactions are often carried out in an inert solvent, under reflux.

$$RP(OR)_2 + Se \longrightarrow RPSe(OR)_2 \qquad (127)$$

Dialkylalkyltellurophosphines have been synthesized by the reactions of dialkyl ditellurides with tetraalkyldiphosphines (Equation (128)) ⟨83PS(18)73⟩.

$$R^1{}_2PPR^1{}_2 \xrightarrow{R^2TeTeR^2} R^1{}_2PTeR^2 \qquad (128)$$

Tertiary phosphines react with KSeCN (Equation (129)) in acetonitrile, to give high yields of tertiary phosphine selenides ⟨66IC1297⟩.

$$R_3P + KSeCN \longrightarrow R_3PSe + KCN \qquad (129)$$

Tertiary phosphine selenides are also formed by the photolysis of dialkyldiselenides ⟨77JCS(D)159⟩.

The reaction of Grignard reagents with dialkyl esters of the phosphorous acid, followed by the addition of selenium to the crude phosphinous acid, leads to selenophosphinic acids (Scheme 43) ⟨60RZC1675⟩.

$$(R^1O)_2P(O)H + 3R^2MgX \longrightarrow [R^2{}_2P(O)MgX] \longrightarrow [R^2{}_2P(O)H] \xrightarrow{Se} R^2{}_2P(Se)OH$$

Scheme 43

Selenophosphinic chlorides react with sodium hydrosulfide ⟨66CB1663⟩ and sodium hydroselenide ⟨65AG(E)244⟩, to give the corresponding selenothiophosphinic and diselenophosphinic acids as their sodium salts.

Dialkyltellurochalcogenophosphinic acids can be prepared by the reaction of tellurium with chalcogenophosphinides (Equation (130)) ⟨90PS(47)341⟩.

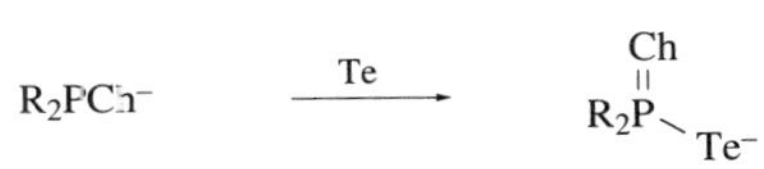 (130)

Ch = O, S, Se, Te

Selenophosphonates have been synthesized by the reaction of selenophosphonic dihalides with alcohols in the presence of tertiary amines ⟨52IZV894⟩.

The base-catalysed addition of $HPSe(OR)_2$ to activated double bonds (Equation (131)) provides a route to selenophosphonates ⟨57RZC715⟩.

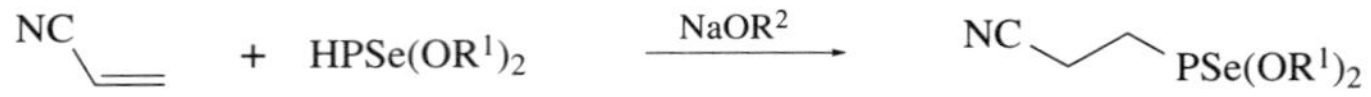

(131)

Optically active selenophosphonates have been obtained employing (−)-ephedrine ⟨76TL3645⟩ as a chiral template.

Alkyl ethylselenophosphonic acids have been resolved using quinine or brucine ⟨76MI 208-01⟩.

2.08.7 ALKYLPHOSPHORUS COMPOUNDS WITH A P—N BOND

2.08.7.1 Amides of Alkylphosphinous Acid

2.08.7.1.1 Amides of the type $R^1{}_2PNR^2$

The dialkylphosphinous acid amides are readily prepared by the reactions of dialkylhalophosphines with a primary or secondary amine ⟨77JCS(D)1674⟩. An alternative approach involves the reaction of a dialkylaminochlorophosphine, R_2NPCl_2, with a Grignard reagent ⟨74S119⟩; a reaction which has also been used to synthesize bis(phosphino)alkanes, (**80**) ⟨83PS(15)155⟩. The aminodichlorophosphine intermediates (**81**), derived from the more hindered amines, are available from the reaction of lithium amides with phosphorus trichloride ⟨77S580, 79IC347⟩, and if necessary the chlorine atoms can be replaced by iodine, using NaI in benzene ⟨74ZOB2357⟩. $(TMS)_2NPR_2$ is prepared in 75% yield using the Wilburn reaction, which involves the reaction of $(TMS)_2NLi$ with PCl_3 at −78 °C, followed by reaction with an appropriate Grignard reagent ⟨82IC3568⟩. Chiral analogues have also been synthesized ⟨84IC2062⟩.

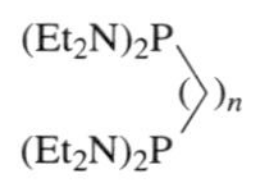

(80)

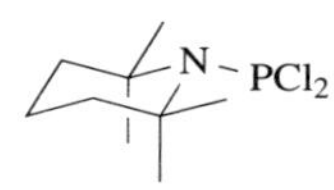

(81)

A series of cyclic amides (**82**; Y = TMS, H) were made by the radical-initiated reaction shown in Equation (132) ⟨86CC850⟩. The cyclic aminophosphine (**83**) ⟨90T6715⟩, and the diazadiphosphacyclopentane (**84**; Y = R, Cl) were synthesized via amine displacement of chloride at phosphorus ⟨86CB2767⟩.

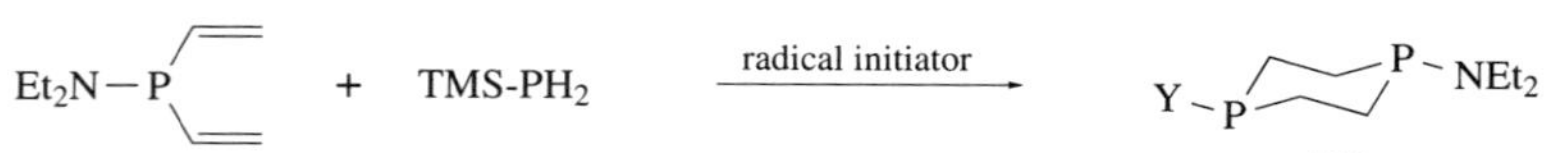

(132)

(82)

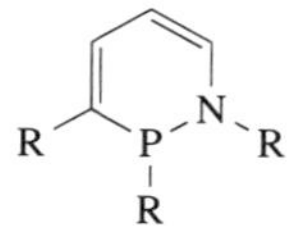

(83)

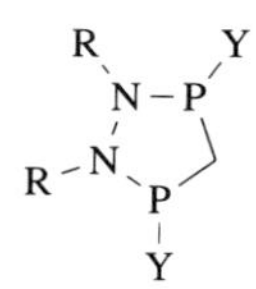

(84)

2.08.7.1.2 Amides of the type R^1 $PHNR^2$

The P—Cl bonds in aminochlorophosphines are selectively reduced by LAH ⟨85IC3136⟩, as in the conversion of $Bu^tPClN(TMS)_2$ to $Bu^tPHN(TMS)_2$ ⟨83IC814⟩. Excess isopropylmagnesium halide can also act as a reducing agent and convert an aminochloro compound into the secondary aminoalkylphosphine ⟨83IC814⟩.

2.08.7.2 Amides of Alkyl Phosphonous Acid

2.08.7.2.1 *Diamides—$R^1P(NR^2_2)_2$*

The syntheses of this class of compound usually involve the reaction of bis(dialkylamino)chlorophosphines with an organometallic reagent such as those shown in Equations (133) ⟨92TL2357⟩ and (134) ⟨80PS(15)155⟩. The precursors are readily prepared; thus, $(Pr^i_2N)_2PCl$ is obtained (67%) from trichlorophosphine and diethylamine ⟨86JOC3559, 86TL199⟩. The diaminochlorophosphine intermediates may also be conveniently prepared by the reaction of tris(dialkylamino)phosphines with chlorotrimethylsilane ⟨85T2471⟩. Another route involves the addition of secondary amines to phosphaalkenes, $R_2NP{=}CHR$ ⟨86IC361⟩. Predominant metallation of TMS-$CH_2P(NMe_2)_2$ at carbon by BuLi/ethyltetramethyldiamine ⟨86ZAAC(535)47⟩ produces a useful synthon for making substituted methylphosphonates. Various routes for the synthesis of diazaphosphaboretidines (**85**) have been investigated ⟨85CB2030⟩. The six-membered heterocycle (**86**) was prepared from halodiphosphine and Me-TMS-NCONMe-TMS ⟨91CB2705, 91ZN(B)258⟩.

$$ClP(NPr^i)_2 \xrightarrow{\text{RLi or RMgX}} RP(NPr^i)_2 \qquad (133)$$

R = Me, Bu, octyl

(134)

n = 4–10

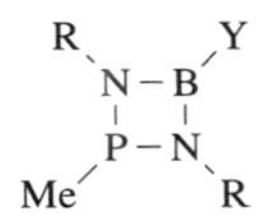

(85)

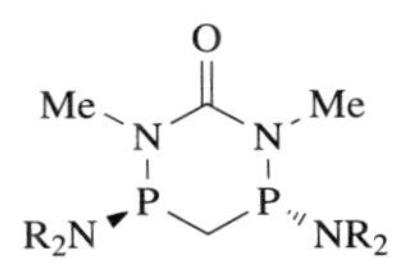

(86)

2.08.7.2.2 *Amidohalides—$R^1P(NR^2_2)Hal$*

It is possible to displace one of the two chlorine atoms of alkyldichlorophosphines by an amino group. Thus, in the presence of Et_3N, the reaction between $MePCl_2$ and Pr^i_2NH in THF at $-10\,°C$ for 1 h gave $Pr^i_2NP(Cl)Me$ ⟨91TL4981⟩. Similarly Bu^tPCl_2 and diethylamine gave $Bu^tP(Cl)NEt_2$ ⟨85PS(24)217⟩, and the reaction of the dichlorophosphine $Cl_2PCH_2PCl_2$ and Et_2NH led to bis(chlorodiethylamino)phosphinomethane ⟨79JGU616⟩. The unsymmetrical compounds $R^1R^2PN(TMS)_2$ were synthesized by the stepwise substitution of chlorine atoms in $(TMS)_2NPCl_2$ ⟨84IC2062⟩. The secondary aminoalkylphosphine $(TMS)_2PHR$ was also prepared from this starting material.

2.08.7.2.3 *Amidoesters—$R^1P(NR^2_2)ChR^3$*

This type of compound is available through the reaction of RChLi with the above chlorodialkylaminoalkylphosphines ⟨85PS(24)217⟩. On the other hand, the 1,3,2-oxazaphospholene (**87**) was made directly from $MePCl_2$ and a β-aminoketone ⟨79JGU34⟩.

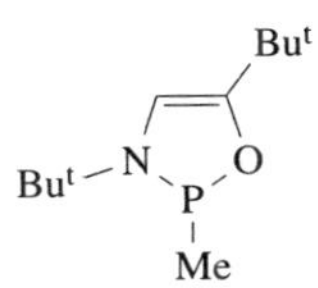

(87)

2.08.7.3 Alkylphosphine Imides—$R^1{}_3P{=}NR^2$

There are two widely used methods for the synthesis of this class of phosphazene; one is the Staudinger reaction, which involves the reaction of three-coordinate compounds such as phosphines or phosphites with organic azides (RN_3), the other is the Kirsanov reaction, which involves the reaction of a dichlorophosphorane with primary amines, ammonia and other nucleophilic nitrogen compounds.

The Staudinger reaction has been reviewed ⟨92OPP209, 92T1353⟩. It is reliable and applicable to a wide range of substituents ⟨89AG(E)1675, 89CZ349, 92TL251⟩. Some unusually *P*-substituted phosphazenes may be prepared via two-coordinate intermediates such as $(TMS)_2P{=}CH\text{-}TMS$ ⟨86IC361⟩. Thus, cyclic phosphazenes (Equation (135)) are available via cycloaddition reactions of alkenes with iminophosphines ⟨72ZN(B)769⟩. A great deal has been published on the synthesis of polyphosphazenes and, in particular, cyclopolyphosphazenes. In this series most of the methods which are used to make aryl compounds can also be applied to the synthesis of the alkyl compounds. The Kirsanov reaction has been used to produce a variety of ring systems. Thus, the reaction of methyltetrachlorophosphorane and ammonium chloride gives up to 58% of cyclic products, such as the cyclotrimer **(88)** ⟨71JGU475⟩, whereas the reaction of dimethyltrichlorophosphorane and ammonium chloride gives over 60% yields of the cyclotetramer **(89)** ⟨71IC2362⟩. The reaction (Equation (136)) between dichlorophosphoranes and primary amines is useful for the synthesis of unsymmetrical ring systems ⟨69JINC271⟩, and similar compounds can be prepared from alkylphosphonous acid diphenyl esters (Equation (137)) ⟨72ZAAC(394)171⟩. Methods such as that used to prepare a six-membered cyclodiphosphazene (Equation (138)) allow a wide variety of cyclic systems to be synthesized ⟨68CB3883⟩.

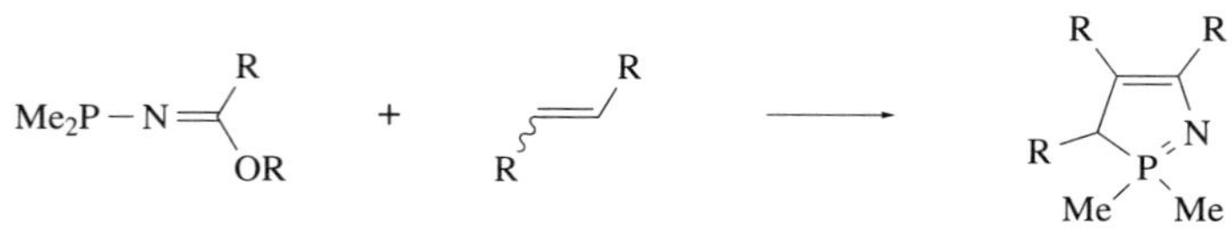
(135)

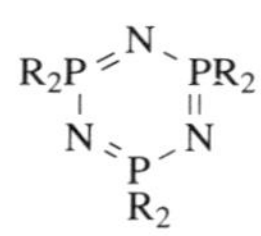
(88)

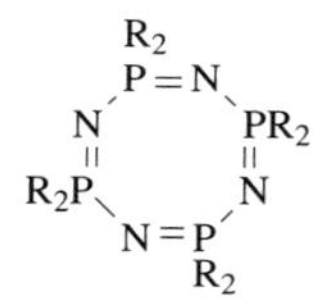
(89)

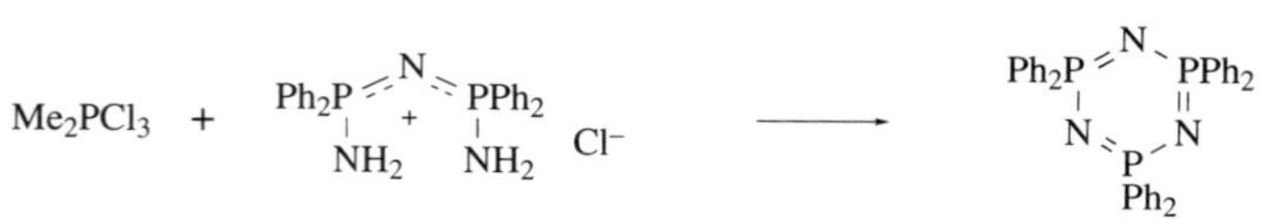
(136)

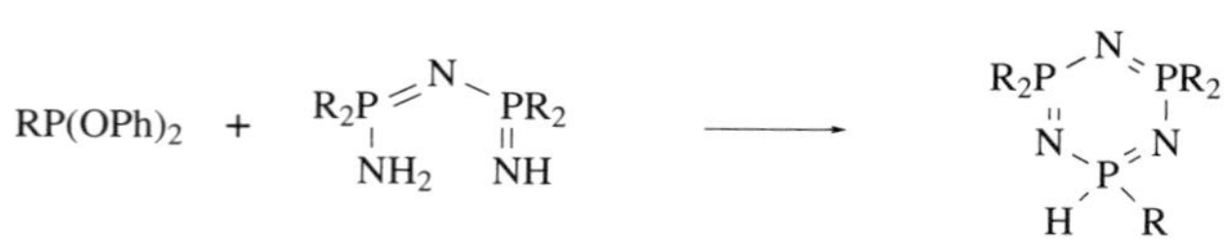
(137)

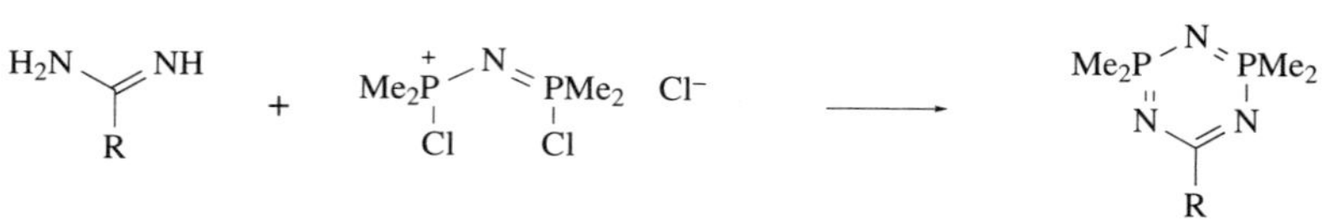
(138)

2.08.7.4 Amides of Dialkyl Phosphinic Acid—$R^1{}_2P(Ch)NR^2{}_2$

Phosphinic acid amides may be prepared directly from phosphinic acid chlorides ⟨71ZOB1644⟩. An *N*-disilylated thiophosphinic acid amide ⟨74ZN(B)611⟩ has been made by the reaction shown in Equation (139), and is a useful intermediate for conversion to the corresponding esters

⟨75ZAAC(413)266⟩. Organic azides (RN_3) and dialkylphosphinous acid methyl esters (R_2PChMe) give phosphazenes, which may be hydrolysed to phosphinic acid amides ($R^1{}_2P(O)NHR^2$) ⟨92BSF71⟩.

(139)

2.08.7.5 Amides of Alkyl Phosphonic Acid

2.08.7.5.1 Diamides—RP(O)(NR$_2$)$_2$

Phosphonic acid diamides may be prepared from phosphonic acid dihalides ⟨71ZOB1644⟩ and by ester exchange ⟨67ZOB2787⟩. The Arbuzov rearrangement ⟨81CRV415⟩ of alkoxybis(dialkylamino)phosphines may also be employed, except for bulky alkyl groups ⟨86JCR(S)26⟩. The reaction may also be used to synthesize azaphosphetidenes ⟨78ZOB2624⟩. The reaction of phosphonous acid methyl esters with *N*-chlorodialkylamines provides another route to phosphonic acid amides ⟨77JOC782⟩. 2-Dimethylamino-(2-hydroxymethyl)phosphonic acid amides are also the products of the reactions of tris(dimethylamino)phosphine with aldehydes ⟨71ZOB2372⟩.

2.08.7.5.2 Amidohalides—R^{1}P(Ch)Hal(NR$^2{}_2$)

One dialkylamino group can be introduced into alkyldichlorophosphines, and the resultant phosphine oxidized to the corresponding chalcogenide ⟨85PS(24)217⟩.

2.08.7.5.3 Amidoesters—RP(O)(NH$_2$)OR

This type of ester has been prepared by amine displacement of the halogen in MeP(O)Cl(OMe), which in turn was made by the reaction of methylphosphonic acid dimethyl ester with oxalyl chloride in ether ⟨83JA1613⟩. Another approach was used to make $EtP(O)(OEt)NR_2$, which involved the reactions of ethylphosphonous acid ethyl ester with dialkylamines in the presence of CCl_4 and Et_3N ⟨77IZV1437⟩. Cyclic derivatives can also be formed directly by the reaction shown in Equation (140) ⟨75ZOB2554⟩. The dialkoxydimethylaminophosphines may also be converted to the amidoesters (e.g., Equation (141)) ⟨70BCJ3232⟩.

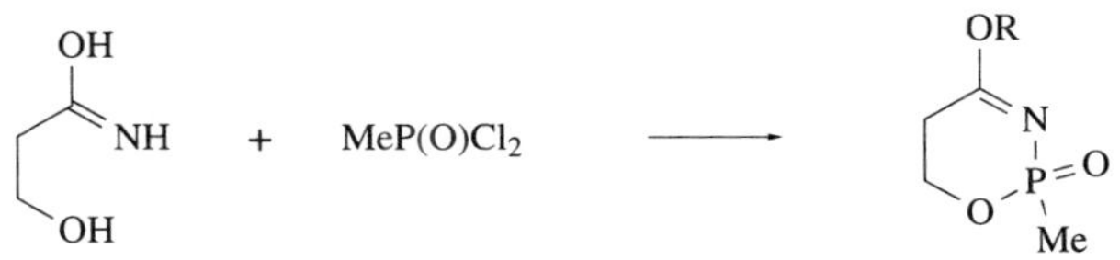

(140)

(141)

2.08.7.6 Alkylphosphorus Compounds Incorporating an NY Function—R$_3$PN=O, etc.

2.08.7.6.1 N-Chalcogenide derivatives

Alternative *N*-substituents can be introduced by using the appropriately substituted amine. Thus, *p*-toluenesulfonamide reacts with chlorodimethylphosphine to produce the phosphazene $Me_2PH{=}NSO_2Tol$ ⟨71ZN(B)81⟩. A similar approach can be utilized using a lithium amide and a dichlorophosphorane (Equation (142)) ⟨70CA88061⟩. It is also possible to replace a phosphoryl oxygen by a fluorosulfonyl group using FSO_2NCO ⟨69M311⟩.

(142)

2.08.7.6.2 N-*Azo and* N-*diazo derivatives*

The hydrazine group may be introduced at phosphorus by the same methods as those employed for the amides but using a hydrazine instead of an amine ⟨71ZOB1644, 69CJC4281⟩. Azido groups are introduced likewise ⟨86IC2695⟩. One of the products of hydrazoic acid, HN_3, with di-*t*-butylaminophosphine is the ylide (**90**) ⟨68CB4148⟩.

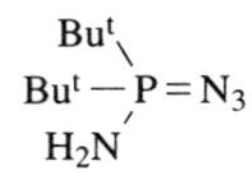
(90)

2.08.7.6.3 N-*Phospho and* N-*arseno derivatives*

N-Phosphorylated phosphazenes may be synthesized by the reaction of PCl_5 with a phosphinamide as shown in Equation (143) ⟨69JGU2161⟩. Unusually, *N*-substituted phosphine imines such as $R^1{}_3P{=}NP(OR^2)_2$ may be made prepared by the substitution of the silyl group in a silylated phosphine imine by trialkyl phosphite ⟨78ZAAC(447)97⟩ (Equation (144)), whereas chlorodialkylarsine was used to react with $Me_3P{=}N$-TMS to give $Me_3P{=}NAsR_2$ ⟨74MI 208-02⟩.

(143)

(144)

2.08.7.6.4 N-*Silyl derivatives*

The silylated phosphine imides, such $R_3P{=}N$-TMS, which are prepared from *t*-phosphines and TMS-N_3 ⟨67CB1000⟩, are useful intermediates to other *N*-substituted compounds. A silylated phosphazene was formed in the reaction of aminodi-*t*-butylphosphine with TMS-N_3 ⟨68CB4184⟩. An alternative approach involves the reaction of an azidophosphine with chlorotrimethylsilane—for example, $Bu^t{}_2P(Cl){=}N$-TMS was obtained in 80% yield ⟨86IC2695⟩. The azide $Bu^t{}_2PN_3$ may be prepared in 60% yield from sodium azide and $Bu^t{}_2PCl$ ⟨86IC2695⟩. The phosphiridine (**91**) was produced in 65% yield by the action of diazomethane on a diazophosphaalkene ($R^1{}_2NP{=}NR^2$) ⟨75AG(E)363⟩.

(91)

2.08.7.7 Other P—N Compounds

Dialkylchlorophosphines and dialkoxyalkylaminophosphines react to give the corresponding phosphine imides, which generally rearrange immediately to the aminophosphines (**92**) ⟨83JGU2244⟩. The cyclic azadiphosphiranes (**93**) were prepared from the reaction of an iminophosphine ($R^1P{=}NR^2$) and a silylated chlorophosphine (R^3PCl-TMS) ⟨91CB765⟩.

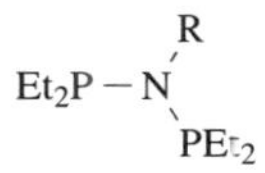

(92)

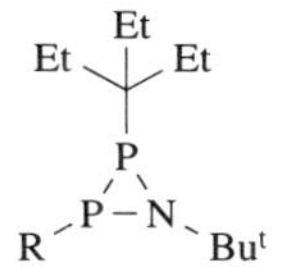

(93)

2.08.8 ALKYLPHOSPHORUS COMPOUNDS WITH A P—P, P—As, P—Sb OR P—Bi BOND

Whilst the trihalides of phosphorus, arsenic, antimony and bismuth are convenient starting materials for the synthesis of these compounds via their reactions with phosphides, there are many other methods also available.

2.08.8.1 Alkylphosphorus Compounds with a P—P Bond

2.08.8.1.1 Acyclic compounds

There have been extensive studies of acyclic alkylphosphorus compounds since the early 1980s. Tetraalkyldiphosphines may be prepared by the reaction of a dialkylphosphine with a chlorodialkylphosphine ⟨72CB406⟩. An interesting alternative involves the reaction of trialkylphosphine with chlorodialkylphosphine, followed by dealkylation of the resultant salt ⟨69IC1006⟩. The symmetrical diphosphines are synthesized (50–70% yields) by the action of disilylmercury on chlorodialkylphosphines ⟨78JGU207⟩. The dialkyldisilyldiphosphines are produced by the reaction of alkyldichlorophosphines, chlorotrimethylsilane, magnesium and bromine ⟨76ZN(B)1305⟩. Another silylated diphosphine, F_2P—P(TMS)Me, has been synthesized from $(TMS)_3P$ and $PBrF_2$ ⟨90ZAAC(589)23⟩.

A very convenient and well-studied method of forming the P—P bond involves the reactions of anions of dialkylphosphinites (R_2PHO), or the corresponding thiophosphinites, with chlorodialkylphosphine ⟨78CB890, 79JGU2143⟩. The corresponding dioxides and disulfides are readily produced, and the diphosphine monosulfides may be desulfurized. The reaction of alkyldialkoxyphosphines with chlorodiisopropylphosphine in the presence of $NiCl_2$, or $SnCl_2$ at 150 °C for 30 min gives an almost quantitative yield of Pr^i_2P—P(O)R^1(OR^2) ⟨91JGU460⟩. It has been shown that Bu_2PF and Et_3N will give Bu_2P—PF_2Bu_2 ⟨92ZAAC(607)161⟩.

2.08.8.1.2 Cyclic polyphosphines

Alkyldichlorophosphines can be dechlorinated by magnesium to produce three- and five-membered cyclic polyphosphines $(PR)_n$ ⟨83ZAAC(493)57⟩. Higher yields are obtained (60–80%) if lithium is used in place of magnesium ⟨63JA2462⟩. Sodium in dioxan has also been used ⟨77ZN(B)1244⟩. The four-membered cyclic polyphosphines, $(PR)_4$, are prepared (80–90% yields) through the reactions of alkyldichlorophosphines with the corresponding primary phosphines ⟨63JA2462⟩ and also through the reaction of white phosphorus, *t*-butyllithium and TMS-Cl ⟨83ZAAC(504)47⟩. Phosphorus trichloride converts the monocyclopolyphosphines to tricyclic pentaalkyl P_9 polyphosphines ⟨83MI 208-01⟩. These polyphosphines may also be made directly from alkyldichlorophosphines, phosphorus trichloride and magnesium ⟨83MI 208-01⟩. Polycycloalkylphosphines can be synthesized by the alkylation of inorganic polyphosphides such as Li_3P_7 ⟨91ZAAC(549)36⟩ and Li_2P_{16} ⟨85CB3268⟩. Similar polycyclic phosphines are produced in high yields via the reaction of white phosphorus (P_4), which is the most reactive form of the element, with alkyldichlorophosphines in the presence of magnesium ⟨91CB69, 92ZAAC(611)47⟩, for example $P_{11}Pr^i_5$ ⟨91CB1153⟩.

The three-membered tri-*t*-butylcyclotriphosphines $(Bu^tP)_2PR$ (e.g., (**94**; R = Bu^t)) ⟨85ZAAC(523)7⟩, when heated with potassium, undergo P—C cleavage to give the phosphides, which hydrolyse to the stable secondary phosphines (e.g., (**94**; R = H)) ⟨85ZN(B)1274⟩. Electrolysis of the diphosphine $Bu^t_2PP(NPr^i)_2$ causes disproportionation with the formation of the $Bu^t_2PPBu^t_2$ ⟨91CB1563⟩. P—P bonds may be formed in high yields by an Arbuzov-type reaction (Equation (145)) ⟨91JGU460⟩. Through the reaction of *t*-butyldichlorophosphine with PCl_3 and magnesium, the bicycloheptaphosphine (**95**) was prepared ⟨85AG(E)504⟩. The tetraphosphetane (**96**) may be prepared from Me_5CpPCl_2 and $Pr^i_2PP(TMS)_2$ ⟨91MI 208-02⟩. The diphosphirane (**97**) can be produced by cyclodehalogenation of bis(halagenophosphino)methanes by $Fe(CO)_9$ ⟨90JOM(399)C4⟩. The diphosphacyclopentane (**98**) may be formed by the dehydrogenation of the diphosphine $(H_2P(CH_2)_3PH_2)$ by the action of butyllithium ⟨78PS(4)137⟩. The tetraphosphabicyclopentane (**99**) is produced (55% yield) by the reaction of $MePCl_2$ and $H_2PCH_2CH_2PH_2$ ⟨78CB3838⟩. Dimerization of phospholes gives diphosphatricyclodecadienes ⟨83JA6871⟩.

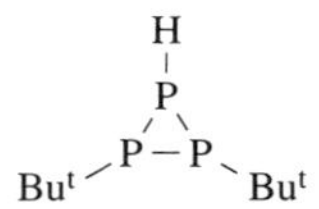

(94)

$$Pr^i_2PCl + (EtO)_3P \xrightarrow{NiCl_2,\ 150\ ^\circ C} Pr^i_2P{-}P(O)(OEt)_2 \qquad (145)$$

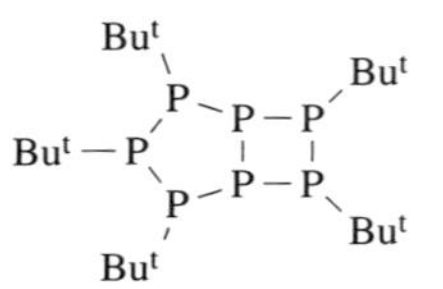

(95)

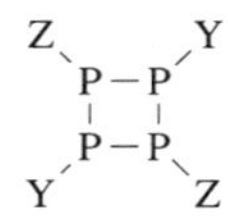

(96)

Y = $P(NPr^i_2)_2$, Z = $CpMe_5$

(97)

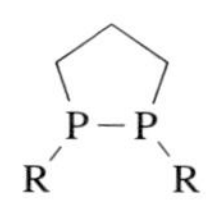

(98)

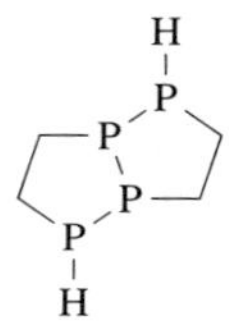

(99)

2.08.8.2 Alkylphosphorus Compounds with a P—As Bond

There have been very few reports of the synthesis of compounds with a P—As bond. The reaction between dimethylphosphine and $As_2(CF_3)_4$ can be controlled to give either a 90% yield of $Me_2PAs(Tf)_2$ (Tf = trifyl) or a 96% yield of Me_2AsPTf_2 ⟨68JCS(A)1406⟩. A P—As bond is also formed in the reaction of an *N*-metallated aminophosphine with chloroarsine, as shown in Scheme 44 ⟨69ZAAC(370)171⟩.

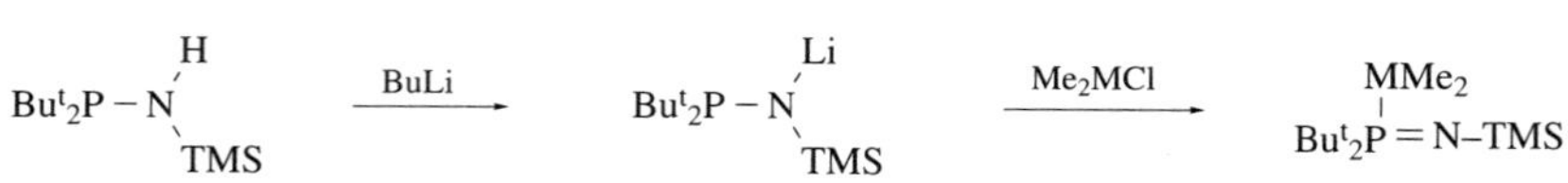

M = As, Sb

Scheme 44

2.08.8.3 Alkylphosphorus Compounds with a P—Sb Bond

The above reaction (Scheme 44), using chlorodialkylstibine instead of the chloroarsine, follows a similar pathway ⟨69ZAAC(370)171⟩. Bis(*t*-butylphosphine)potassium phosphide and Bu^tSbCl_2 gave the cyclic heterocycles (**100**) and (**101**) ⟨83MI 208-02⟩.

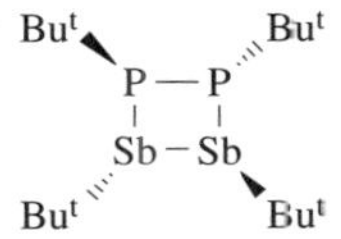

(100)

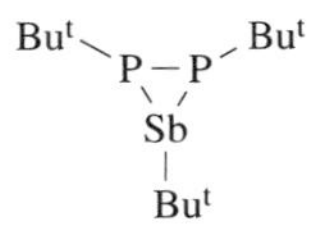

(101)

2.08.8.4 Alkylphosphorus Compounds with a P—Bi Bond

There is an increasing trend for complex formation rather than substitution as the atomic number of the acceptor increases. Thus, trimethylphosphine and tribromobismuth form the crystalline adduct Me_3P—$BiBr_4$ ⟨91CB2457⟩.

2.08.9 ALKYLPHOSPHORUS COMPOUNDS WITH A P—Si, P—Ge OR P—Sn BOND

The main synthetic routes to this class of compound have involved the reaction of chlorosilanes, -germanes or -stannanes with dialkylphosphides, silylphosphines or alkylchlorophosphines.

2.08.9.1 Alkylphosphorus Compounds with a P—Si Bond

Chlorotrimethylsilane and magnesium react in THF with PCl_3, Bu^tPCl_2 and Bu^t_2PCl to give $(TMS)_3P$, $(TMS)_2PBu^t$ and TMS-PBu^t_2, respectively ⟨74CB854⟩. TMS-Cl reacts with $LiAl(PH_2)_4$ (made from PH_3 and LAH) in TRIGLYME at $-30\,^\circ C$, to give TMS-PH_2 in high yields ⟨70IC870⟩. The di- and tribromosilanes have also been utilized ⟨68JA6556⟩. Allylphosphine reacts with TMS-Cl in the presence of triethylamine to give the secondary allyl phosphines ⟨85IC1107⟩. Dialkylphosphinolithium and chlorosilanes give a series of phosphinosilanes ⟨70ZAAC(372)171⟩. Likewise, conversion of phospholanes into phosphorus–lithium derivatives and their reaction with TMS-Cl produce the silyl compound (**102**; M = Si) ⟨84PS(20)81⟩. Also, the reaction of RPK—PKR and $SiCl_4$ gives tetraphosphaspirosilanes ⟨83AG(E)775⟩. The condensation of $Bu^tP(TMS)_2$ with the disilyl compounds (**103**; M = Si) gives the phospholane (**104**; M = Si) in 85% yield (Equation (146)); related oxaphospholans have also been synthesized ⟨82PS(12)265⟩. The reaction of chlorodiallylsilane with $NaAl(PH_2)_4$, followed by radical cyclization, gives the bicyclic phosphasilylbicyclooctane (**105**) ⟨85PS(21)367⟩. Cyclic silylphosphines (e.g., (**106**; M = Si), 76% yield) may be made by the reaction of lithium phosphides with dialkyldichlorosilanes ⟨86ZAAC(535)106⟩. Silylphosphines such as $(TMS)_2PCH_2NR_2$ have been synthesized via the reaction of $(TMS)_3P$ with chloromethylamines ⟨91ZOB1016⟩. Et_2PLi inserts into the Si—O bond of polydimethylsiloxane, to give $(Et_2PSiMe_2OLi)_6$ ⟨92CC726⟩, and various methods for the synthesis of $Me_2PSiMe_2(CH_2)_2PMe_2$ have been investigated ⟨92ZAAC(608)43⟩.

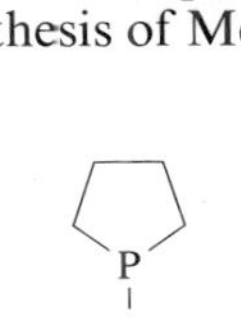

(102)

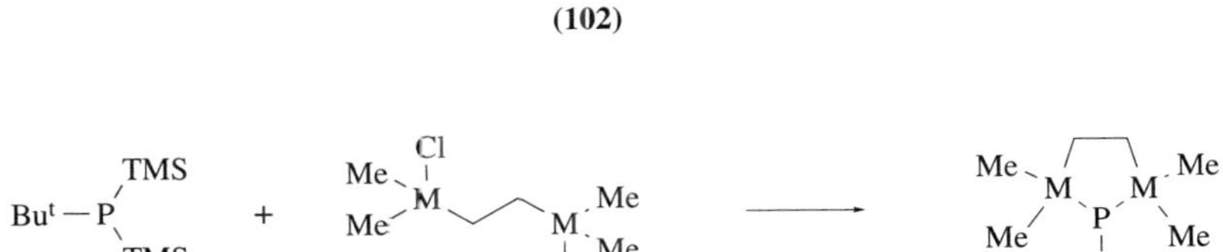

(103) **(104)** (146)

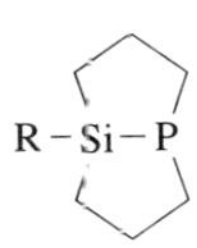

(105)

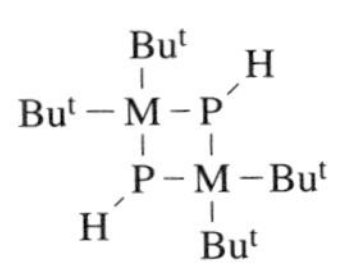

(106)

2.08.9.2 Alkylphosphorus Compounds with a P—Ge Bond

The methods used to make P—Si compounds are usually applicable to the synthesis of P—Ge compounds as well ⟨74CB854⟩. Mono- and dialkylchlorophosphines react with Me_3GeCl in the presence of magnesium to give corresponding germylphosphines ⟨74CB854⟩. Another approach involves the reaction of either a phosphine in the presence of a tertiary amine ⟨85IC1107⟩, or a lithium phosphide with a halogermane ⟨84PS(20)81, 82PS(12)279⟩. The primary germylphosphines (e.g., Me_3GePH_2) are readily prepared ⟨70IC870⟩, and can be used as intermediates to a wide variety of germylphosphorus compounds. Another convenient route to digermylphosphines is the reaction of silylphosphines with halogermanes ⟨69JCS(A)2704⟩—a method which has also been used to make cyclic digermylphosphines such as (**104**; M = Ge) ⟨82PS(12)265⟩.

2.08.9.3 Alkylphosphorus Compounds with a P—Sn Bond

The above methods are usually also applicable to the formation of a P—Sn bond ⟨74CB854⟩. The methods used include the reaction of a primary phosphine and chlorotrimethylstannane (Me_3SnCl) in the presence of a base ⟨70JOM(21)411⟩, the reaction of lithium diethylphosphide and an alkyltrichlorostannane ⟨70ZAAC(372)171⟩, as well as the reaction between di-*t*-butylphosphine and the aminostannane Me_3SnNMe_2 ⟨70JOM(21)351⟩. The stannophosphine Me_3SnPH_2, which may be prepared in 62% yield ⟨71JOM(28)81⟩, is also a useful intermediate for the synthesis of secondary and tertiary phosphines. Other approaches include the reaction of di-*t*-butyldichlorostannane with $NaPH_2$ in liquid ammonia to give (48% yield) the diphosphadistannylbutane (**106**; M = Sn) ⟨90CB1837⟩. The reaction of lithium phosphide with $SnCl_2$ to give a tetraphosphinostannane (**107**) ⟨85AG(E)402⟩, and the reaction of silylphosphines with chlorostannanes, for example the synthesis of (**104**; M = Sn) ⟨82PS(12)265⟩, are also used.

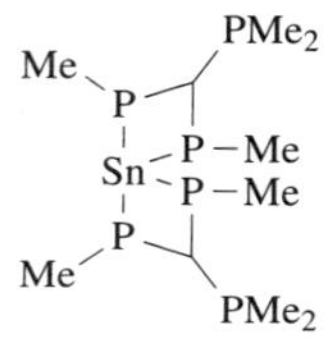

(107)

2.08.10 ALKYLPHOSPHORUS COMPOUNDS WITH A P—B, P—Al, P—Ga OR P—M BOND

2.08.10.1 Alkylphosphorus Compounds with a P—B Bond

The dimethylphosphineborane adduct and the corresponding dimethylborane adduct each extrude H_2 at 150 °C to give the six-membered trimers (**108**) ⟨53JA3872⟩. Dimethyl- or diethyllithium phosphides and chlorobis(dialkylamino)borane or dichloro(dimethylamino)borane give (in 40–60% yields) the boraphosphines $R^1{}_2PB(NR^2{}_2)_2$ and $(R_2P)_2BNMe_2$, respectively ⟨72CB1962⟩. Likewise, lithium phosphides from primary phosphines react with a variety of haloboranes to produce acyclic, four- or six-membered boraphosphines ⟨88CB871⟩. Three-membered boradiphosphines can also be synthesized in high yields ⟨86ZAAC(533)146⟩. Alternative approaches include the reaction of an iminoborane with Bu^{t}PCl-TMS to give a phosphaborazirine ⟨91CB765⟩, and the reaction of tertiary phosphines with borohydride and iodine in THF to give phosphine–borane adducts ($R_3P \cdot BH_3$) ⟨86SRI67⟩. The latter are also produced by the action of LAH/borohydride on phosphine oxides in the presence of a $CeCl_3$ catalyst ⟨85JA3501⟩.

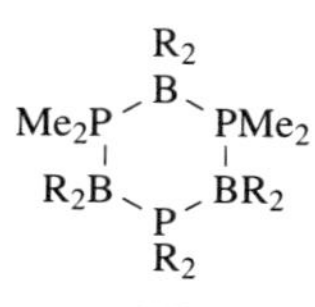

(108)

2.08.10.2 Alkylphosphorus Compounds with a P—Al or P—Ga Bond

Lithium phosphide reacts with $AlCl_3$ to give a hexaphosphino complex ⟨85AG(E)402⟩. On the other hand, gallylphosphines have been prepared from organogallium dichlorides and dilithium alkylphosphides ⟨90AG(E)1150, 91JA3597⟩.

2.08.10.3 Alkylphosphorus Compounds with a P—M Bond

A variety of methods are available for the preparation of the alkali metal phosphides ⟨B-72MI 208-04⟩. Primary and secondary phosphines react with concentrated aqueous alkali in dipolar solvents (e.g., DMSO) to give phosphide anions. The preferential cleavage of the phosphorus–aryl carbon bonds of aryldialkylphosphines by alkali metals can also be utilized to make dialkylphosphides. However, whilst lithium metal converts phenyldialkylphosphines in THF to lithium dibutylphosphide, potassium cleaves a phosphorus–alkyl P–C bond ⟨74MI 208-01⟩.

2.08.11 PENTACOORDINATE ALKYLPHOSPHORUS COMPOUNDS (PHOSPHORANES)

2.08.11.1 Pentaalkylphosphoranes—R_5P

The synthesis of pentaalkylphosphoranes not only require high-energy precursors, but the incorporation of the phosphorus atom into small rings is also needed for their isolation. Thus, both the known pentaalkylphosphoranes (**109**) ⟨71JA4065⟩ and (**110**) ⟨77AG(E)722⟩ are sterically strained. The final step of their syntheses involved the reaction of a cyclic phosphonium salt with methyl lithium. For the latter compound, a temperature of $-60\,°C$ was required throughout the 6 h addition of the salt, since at higher temperatures the phosphonium ylide was produced.

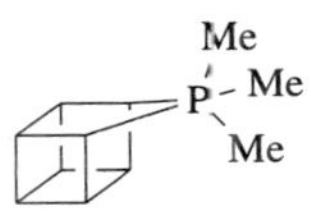

(109)

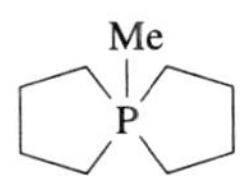

(110)

2.08.11.2 Tetraalkylphosphoranes—R_4PX

If the noncarbon substituent X is stable as an anion, there is a strong tendency for the phosphorane to dissociate to the phosphonium salt, especially in a polar solvent. However, the incorporation of the phosphorus atom into small rings and the use of highly electron-accepting groups such as fluorine or trifluoromethyl stabilizes the pentacoordinate state considerably. Thus, the dimethyl fluorophospholane (**111**) is quantitatively produced by the action of ammonium fluoride upon the phosphonium ylide $(CH_2)_4PMe{=}CH_2$ ⟨79ZAAC(458)249⟩.

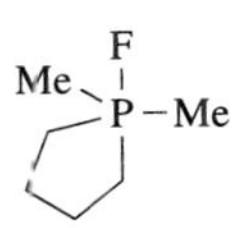

(111)

2.08.11.3 Trialkylphosphoranes—R_3PX_2

Trialkyldifluorophosphoranes may be prepared (60–85% yields) either by the reaction of bromotrialkylphosphonium bromides with NaF in MeCN ⟨82JFC(20)85⟩ or by the action of FSO_2Cl on trialkylphosphine oxides ⟨82JA290⟩. The corresponding dichlorides dissociate to the phosphonium chlorides ⟨84JCS(D)1911⟩. The use of fluorinated alkyl groups enabled tris(trifluoromethyl)-

dihydrogenphosphorane (Tf_3PH_2) to be prepared from Tf_3PF_2 and TMS-H ⟨73CC813⟩. Neat tributylphosphine and diethylperoxide give the unstable tributyldiethoxyphosphorane ⟨68TL85⟩, but trimethylphosphine reacts with TfCOCOTf to give the stable product 2,2,2-trimethyl-1,3,2-dioxyphosphole ⟨68JA6726⟩. Difluorooxyphosphoranes are produced by the addition of an alcohol to difluorophosphonium ylides ⟨91JGU464⟩, and through the reaction of CsF with tris(pentafluoroethyl)phosphine oxide ⟨89JGU469⟩. Bridged phosphoranes such as (**112**) have also been prepared ⟨89JGU87⟩.

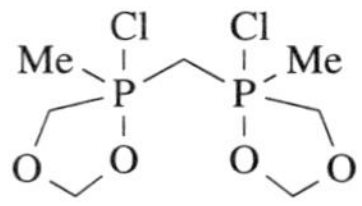

(112)

2.08.11.4 Dialklyphosphoranes—R_2PX_3

Dimethylfluoro-1,3,2-dioxyphospholanes may be prepared (60–90% yields) from Me_2PF_3 using TMS-$OCOCH_2O$-TMS ⟨92CB801⟩. Bicyclic dimethyloxydiazaphosphoranes are prepared from dimethyltrichlorophosphorane and an acylamidohydrazine ⟨76CB3581⟩.

2.08.11.5 Monoalkylphosphoranes—RPX_4

Alkyltetrafluorophosphoranes may be prepared from a number of starting materials, namely phosphonic acids or their derivatives by reaction with SF_4 or SbF_3 ⟨60JA6176⟩, and alkylfluorides and PF_3 ⟨76CC643⟩, and they can be converted to trifluorooxa-, trifluorothio- and trifluoroaminophosphoranes ⟨70JCS(A)1049⟩. Chloromethyltetrachlorophosphorane has been prepared by the addition of chlorine to chloromethyldichlorophosphine ⟨91CC234⟩. *t*-Butyltetraiodophosphorane has also been described ⟨78JGU1088⟩. Bicyclictetraoxyphosphoranes are prepared from methyltetraphenoxyphosphorane and diols ⟨89JGU634⟩, from dichloromethylphosphine and catechol ⟨68TL5333⟩, as well as from the addition of biacetyl to diethylaminofluoromethylphosphine at 0°C ⟨68ZOB1906⟩. Aldehydes have been added to trifluoromethylethyl phosphites to produce phosphoranes ⟨91MI 208-03⟩ and *ortho*-quinones have also been added to the chiral phosphine $MeP(NPr^i_2)O$-menthyl ⟨90JA8575⟩. Tetraoxyphosphorane (**113**) ($Z = CH(NMe)_2P(O)(OEt)_2$), was prepared by the reaction of (**113**) (Z = H) with $(EtO)_2P(O)CH(OMe)(NMe_2)$ ⟨90JGU398⟩, and the phosphorane (**114**) was generated from ethanediol and a *P*-aminooxazaphosphole ⟨90JGU397⟩. A P—C bond is also formed by the addition of aldehydes to the P—H bond of the bicyclic tetraoxyhydridophosphorane (**115**) ⟨68CRC270⟩.

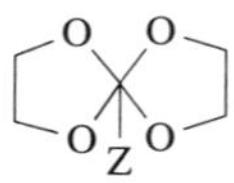

(113)

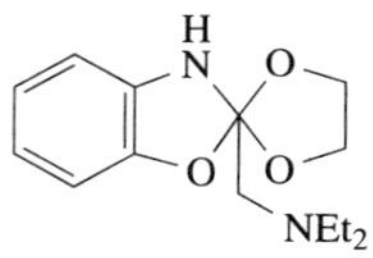

(114)

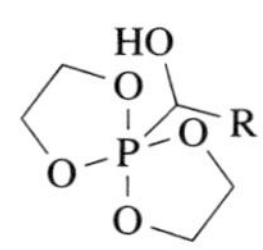

(115)

2.08.12 HEXAVALENT ALKYLPHOSPHORUS COMPOUNDS

The stabilities of hexavalent compounds are even more dependent on the influence of electronegative atoms or groups and the incorporation of phosphorus in small rings. Nevertheless, it is possible to prepare acyclic compounds with six P—C bonds (i.e., $EtP(CN)_5$) by the reaction of ethylpentachlorophosphate with silver cyanide ⟨88JCS(D)2077⟩. Carbamate-, thiocarbamate- and dithiocarbamatemethyltris(trifluoromethyl)phosphates may be prepared from aminophosphoranes with CO_2, COS or CS_2, under pressure, to give the betaine (**116**) in 98% yield ⟨77JA7379⟩. Similar compounds may be produced by the reaction of fluorophosphoranes with *O*-silylated carbamates ⟨83PS(18)89⟩. *N*,*N*′-Dialkyldiimides will react with Tf_3PCl_2 in tetrachloromethane to give a related betaine (**117**) ⟨90IC5081⟩. The addition of catechol to the benzodioxyphospholene (**118**) at 0 °C in the presence of a tertiary amine gives the alkylhydridophosphate (**119**) ⟨74CB639⟩; whereas addition of catechol to a dimethylamino bicyclic phosphorane displaces the amino group with the formation of the tricyclic phosphate (**120**) ⟨77ZOB1432⟩. Dimethylamine will displace a fluorine atom of methyltetrafluorophosphorane, the product of which disproportionates to give $MePF_5^-$ ⟨66IC164⟩. A quaternary ammonium chloride will convert methyltetrachlorophosphorane to the pentachlorophosphate ⟨84JCS(D)1911⟩. Unlike the dialkyltrifluorophosphoranes, the alkyltetrafluorophosphoranes dissociate to RPF_5^- ⟨65IC119, 67IS63⟩.

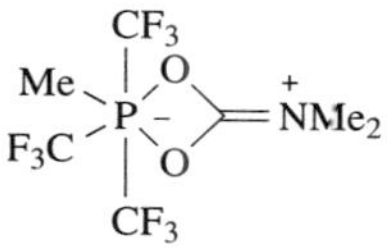

(116)

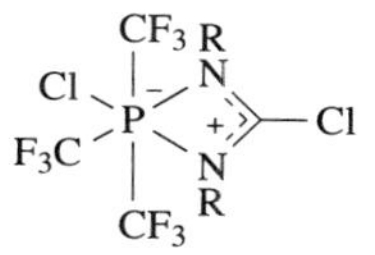

(117)

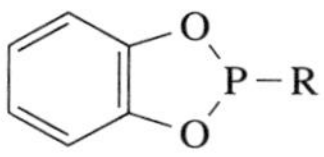

(118)

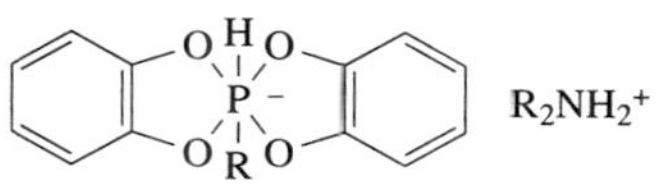

(119)

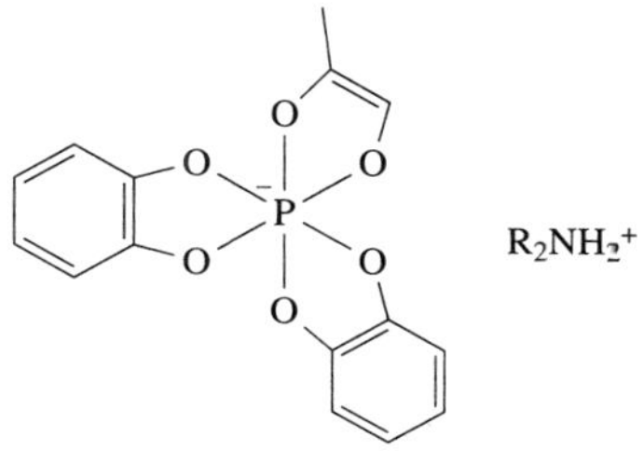

(120)

2.09
Alkylarsenic, -antimony, and -bismuth Compounds

MEI-XIANG WANG
Academia Sinica, Beijing, China

2.09.1 INTRODUCTION

The chemistry of, and synthetic methods for, organoarsenic, -antimony, and -bismuth compounds, a class of compounds referred to as pnicogens, have been reviewed previously. The most important of these reviews are by Doak and Freedman ⟨B-70MI 209-01⟩, Samaan ⟨78HOU(13/8)1⟩, Aylett ⟨B-79MI 209-01⟩, and Patai ⟨B-94MI 209-01⟩. In addition, a compilation of preparative methods for organoantimony compounds is given in the *Gmelin Handbooks*, 1981–1990 ⟨B-81MI 209-01, B-82MI 209-01, B-83MI 209-01, B-86MI 209-01, B-90MI 209-01⟩. The reviews by Wieber ⟨B-77MI 209-01⟩ and by Freedman and Doak ⟨82CRV15⟩ contain methods of preparation of organobismuth compounds.

Many organoarsenic, -antimony, and -bismuth compounds are toxic, and some are extremely harmful to health; it is strongly recommended that their preparation, storage, and manipulation should be done in an efficient hood, and that protective gloves should be worn in all manipulations. As most organoarsenic, -antimony, and -bismuth compounds are sensitive to moisture and oxygen, an inert gas atmosphere and carefully dried oxygen-free solvents are usually required.

2.09.2 ALKYLARSINES, -STIBINES, AND -BISMUTHINES AND SALTS THEREOF AND ARSORANES, STIBORANES, AND BISMUTHORANES

2.09.2.1 Primary and Secondary Alkylarsines, Alkylstibines, and Alkylbismuthines

2.09.2.1.1 Reduction of alkylarsenic, -antimony, and -bismuth halides

A general method for the preparation of these compounds is the reduction reaction of the alkylarsenic, -antimony, and -bismuth halides. Both primary and secondary alkylarsines have been obtained from alkylarsenic dihalides and dialkylarsenic halides, respectively. The yields are variable and dependent upon the reagents and reaction conditions employed. Lithium aluminum hydride is a less satisfactory reducing agent, especially when it is used to reduce perfluoroalkylarsenic halides ⟨53JCS1552, 54JA386⟩. Zinc amalgam, zinc–copper amalgam, and mercury combined with an acid such as hydrochloric acid or hydroiodic acid usually give better results. Examples of preparation of alkylarsines by reducing the corresponding alkylarsenic halides are summarized in Table 1.

Primary and secondary alkylstibines are unstable compounds, and only a few have been prepared at low temperature. Bromodimethylstibine reacts with $NaBH_4$ to give a mixture of $MeSbH_2$ and Me_2SbH, while secondary stibine Me_2SbH is afforded in fair yield (35%) when it is reduced by $LiBH(OMe)_3$ ⟨59JA1⟩. Other secondary stibine compounds R_2SbH have been prepared ($R = c\text{-}C_6H_{11}$, 93%; Et; Bu^t, 17%) from reduction of the corresponding dialkylantimony halides with $LiAlH_4$ ⟨64ZAAC(332)179, 65ZAAC(339)289, 65ZAAC(339)298⟩. Methylbismuthine and dimethylbismuthine, the only known primary and secondary bismuthines, are formed by reducing dichloro(methyl)bismuthine and chlorodimethylbismuthine, or the corresponding bromobismuthines, with $LiAlH_4$ at low temperature ⟨61CB1447⟩.

A variety of preparative methods has been reported for the primary and secondary arsine compounds, and they are listed below. Of general importance is the reduction of alkylarsonic and alkylarsinic acids.

Table 1 Primary and secondary alkylarsines from alkylarsenic halides by reduction method.

Starting material	*Reagents*	*Product*	*Yield* (%)	*Ref.*
Et_2AsCl	$LiAlH_4/Et_2O$	Et_2AsH	80	64ZAAC(326)280
$(Pr^n)_2AsCl$	$LiAlH_4/Et_2O$	$(Pr^n)_2AsH$	57	64ZAAC(326)280
$(Bu^n)_2AsCl$	$LiAlH_4/Et_2O$	$(Bu^n)_2AsH$	73	64ZAAC(326)280
$(Bu^t)_2AsCl$	$LiAlH_4/Et_2O$	$(Bu^t)_2AsH$	41	65ZAAC(336)36
(*c*-Hexyl)$_2$AsCl	$LiAlH_4/Et_2O$	(*c*-Hexyl)$_2$AsH	77	64ZAAC(326)280
$((CH_3)_3SiCH_2)_2AsCl$	Zn/Cu amalgam, HCl(aq.)/THF	$((CH_3)_3SiCH_2)_2AsH$	86	86JOM(308)281
$(CF_3)_2AsCl$	Hg, HCl (gas)	$(CF_3)_2AsH$	86	61CJC1855
$(CF_3)_2AsI$	Zn/Cu, 5N HCl	$(CF_3)_2AsH$	43	53JCS1552
$(CF_3)_2AsI$	Hg, HI	$(CF_3)_2AsH$	88	67JCS(A)1308
Bu^tAsCl_2	$LiAlH_4/Et_2O$	Bu^tAsH_2	65	65ZAAC(336)36
$(CH_3)_3SiCH_2AsCl_2$	$LiAlH_4/Et_2O$	$(CH_3)_3SiCH_2AsH_2$	47	90POL319
CF_3AsI_2	$LiAlH_4/(Bu^n)_2O$	CF_3AsH_2	49	53JCS1552
CF_3AsI_2	Zn/Cu, 5N HCl	CF_3AsH_2	98	53JCS1552
CF_3AsI_2	Hg, HI	CF_3AsH_2	92	67JCS(A)1308

2.09.2.1.2 *Reduction of alkylarsonic and alkylarsinic acids*

Alkylarsonic and alkylarsinic acids are the precursors of primary and secondary arsines. Reduction of them by zinc or zinc amalgam in hydrochloric acid leads to the formation of the desired products. For example, butylarsine is prepared (41%) by reducing butylarsonic acid with zinc amalgam and hydrochloric acid ⟨67LA(709)248⟩. The procedure for synthesizing dimethylarsine through reduction of dimethylarsinic acid with zinc dust and hydrochloric acid has been described in detail (Scheme 1) ⟨67IS160⟩.

$$Me_2AsO(OH) \xrightarrow[87\text{–}94\%]{Zn,\ HCl} Me_2AsH$$

$$Bu^nAsO(OH)_2 \xrightarrow[41\%]{Zn/Hg,\ HCl} Bu^nAsH_2$$

Scheme 1

2.09.2.1.3 *Alkylation of arsine, primary alkylarsines, and metal derivatives thereof*

Sodium, potassium, and calcium arsenides, which are formed by passing arsine into a solution of the corresponding metals in liquid ammonia, undergo alkylation readily with alkyl halides to afford alkylarsines in good yields ⟨37JA2068, 54JA386⟩. Methylarsine is transformed into dimethylarsine in 40% yield by further metallation and methylation reactions ⟨54JA386⟩. Methyl(2-hydroxyethyl)arsine has been prepared in high yield from the reaction of alkali metal methylarsenide with 2-chloroethyl alcohol (Equation (1)) ⟨83ZAAC(501)89⟩.

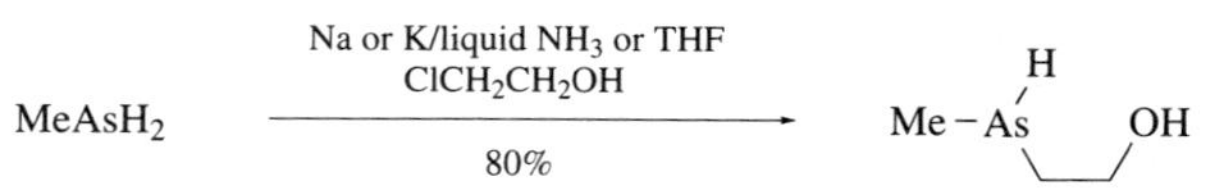

(1)

A specific example has been reported: methylarsine reacts readily with hexafluoroacetone to afford 2-(methylarsino)hexafluoro-2-propanol ⟨65JOM(4)151⟩.

2.09.2.1.4 *Other methods*

Cleavage of the As—As bond of tetrakistrifluoromethyldiarsine or polytrifluoromethylarsine $(CF_3As)_n (n = 4, 5)$ by anhydrous hydrogen chloride and hydrogen iodide in the presence of mercury leads to the formation of bistrifluoromethylarsine or trifluoromethylarsine, respectively ⟨61CJC1855, 67JCS(A)1308⟩.

2.09.2.2 Tertiary Alkylarsines, Alkylstibines, and Alkylbismuthines

2.09.2.2.1 Reaction of organometallics with arsenic, antimony, and bismuth halides

The classic method for preparing tertiary alkylarsines and their antimony and bismuth analogues is the reaction of organometallic compounds with arsenic, antimony, and bismuth trihalides. A variety of reactive organometallic compounds has been employed as alkylating reagents, and among them the Grignard reagent is probably utilized the most extensively.

Arsenic trihalides react with Grignard reagents in a 1 : 3 molar ratio to produce tertiary alkylarsines in moderate to high yields. Ether and petroleum ether are normally used as solvents, and products can easily be isolated from the reaction mixture and purified by distillation. To get pure trimethylarsine, a procedure involving the formation of its silver or copper nitrate complex and decomposition of it with thiourea has been used ⟨72SRI53⟩. Some tertiary alkylarsine compounds prepared from arsenic trihalides and the Grignard reagents are given in Table 2. Similar reaction of antimony trihalides with Grignard reagents lead to the formation of trialkylstibines in good yield, and examples are also shown in Table 2.

Table 2 Tertiary alkylarsines and alkylstibines prepared from Grignard reagents.

Product	*Yield* (%)	*Ref.*
Me_3As	63	64IC580
Et_3As	83	30JA376, 69CB3955
$Pr^n{}_3As$	86	69CB3955
$Bu^n{}_3As$	79	39JA530, 69CB3955
$(C_6H_{13}{}^n)_3As$	81	69CB3955
$(C_8H_{17}{}^n)_3As$	74	64IC580, 69CB3955
$(C_{12}H_{25}{}^n)_3As$	56	44JOC211
$(C_6H_{11})_3As$	87	69CB3955
$(Me_3SiCH_2)_3As$	80	58JA1336
Et_3Sb	80–85	30JCS463
$Pr^n{}_3Sb$	60–70	30JCS463
$(C_6H_{13}{}^n)_3Sb$		30JCS463
$(Me_3CCH_2)_3Sb$	80	92OM2163
$(TMS\text{-}CH_2)_3Sb$	80	92OM2163, 58JA1336

The alkylation of bismuth trihalides by Grignard reagents, usually in diethyl ether or THF, provides the most important synthetic method for tertiary alkylbismuthines. Compounds such as trimethyl, triethyl, tri-*t*-butyl, and tris(trimethylsilylmethyl)bismuthines are obtained in acceptable yields ⟨37JA2068, 58JA1336, 75JOM(87)83, 86OCS638⟩.

Organometallic compounds other than Grignard reagents have been used as alkylation agents. For example, cyclopentadienylsodium undergoes alkylation with arsenic, antimony, and bismuth trichlorides to yield $(C_5H_5)_3As$ (71%), $(C_5H_5)_3Sb$ (89%) and $(C_5H_5)_3Bi$ (80%), respectively ⟨70CB799⟩. Reaction of arsenic triiodide and bis(trifluoromethyl)mercury gives rise to the formation of tris(trifluoro)methylarsine (75%) ⟨88IC4535⟩. A better yield (88%) has been achieved by utilizing bis(trifluoromethyl)tellurium as an alkylation agent (Equation (2)) ⟨90IC33⟩. Triethylstibine (65%) and triethylbismuthine (85%) are obtained when antimony and bismuth trichlorides are treated with lithium tetraethylaluminate ⟨62AJC710⟩. In addition, organocadmium reagents $(R_f)_2Cd \cdot 2MeCN$ have been used to alkylate antimony and bismuth trihalides yielding perfluoroalkyl-substituted stibines and bismuthine ⟨87JOM(334)323⟩.

$$AsX_3 \xrightarrow{(CF_3)_2M} (CF_3)_3As \qquad (2)$$

$$M = Hg, 75\%; M = Te, 88\%$$

2.09.2.2.2 Reaction of organometallics with alkylarsenic, -antimony, and -bismuth halides

This is a modification of the previous procedure, which produces asymmetric tertiary alkylarsines, -stibines, and -bismuthines.

Reaction of alkylarsenic halides with organometallic compounds provides an important route to asymmetric tertiary alkylarsines. Compounds of the type $R^1R^2_2As$ (**1**) can be prepared either from a monohalide R^2_2AsX or from a dihalide R^1AsX_2 by treatment with organomagnesium reagents (Scheme 2). A range of asymmetric tertiary alkylarsines $R^1R^2_2As$ have been obtained with substituents R^1, R^2 being Me, Et, Pr^n, Bu^n, isobutyl, *n*-amyl, *n*-hexyl, cyclohexyl, and Bz, $PhCH_2CH_2$ ⟨28CB1911, 32JCS2284⟩. Cyclopentadienyl(dimethyl)arsine (**2**) has been synthesized from cyclopentadienyllithium and chlorodimethylarsine ⟨75JOM(93)191⟩. It has been prepared alternatively by using cyclopentadienyltrimethylsilane (Scheme 3) ⟨74CB1228⟩. When *t*-butyldichloroarsenic reacts with di-Grignard reagents such as $BrMg(CH_2)_nMgBr$ in ether at 0°C, heterocyclic compounds (**3**; $n = 4$, 50–60%) and (**4**; $n = 5$, 65–70%) are obtained as the major products (Equation (3)) ⟨90PS(48)157⟩. A similar example is the preparation of 3-arsolene compounds (**6**) from alkyldichloroarsines and zirconocene diene complexes (**5**) (Equation (4)) ⟨90JOM(396)C8⟩.

$$R^1AsX_2 \xrightarrow{2R^2MgX} R^1R^2_2AsX \xleftarrow{R^1MgX} R^2_2AsX$$

(1)

Scheme 2

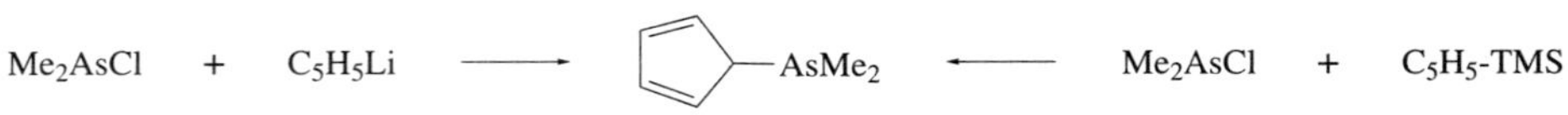

(2)

Scheme 3

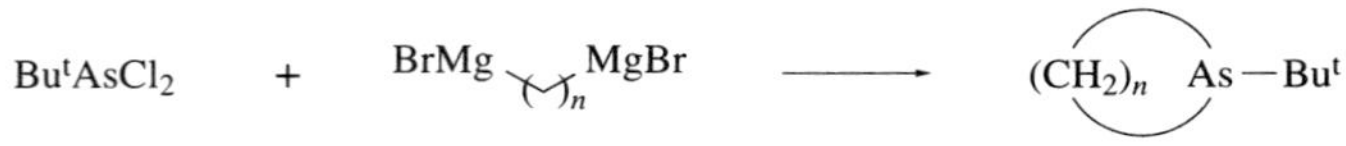

(3), $n = 4$

(4), $n = 5$

(3)

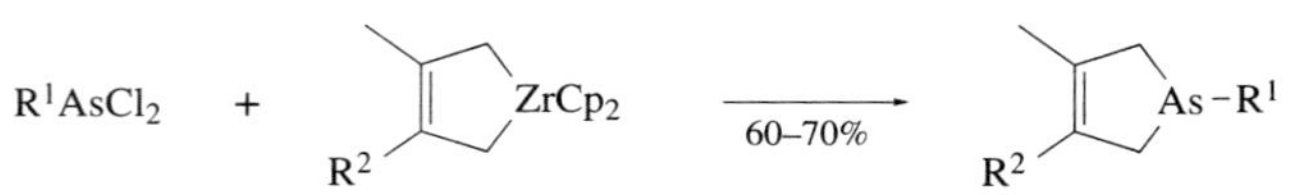

(5) **(6)**

(4)

Asymmetric tertiary alkylstibines are synthesized similarly ⟨32JCS1815⟩. *t*-Butyldimethylstibine (**7**) has been prepared in a one-pot reaction from antimony trichloride on successive treatment with one mole of *t*-butylmagnesium chloride at −50°C and two moles of methylmagnesium bromide at 0°C (Scheme 4) ⟨92OM967⟩. Methyldichlorostibine is converted into a heterocyclic stibine by use of a di-Grignard reagent ⟨32CB409⟩ in a similar manner to the preparation of compound (**4**).

$$SbCl_3 \xrightarrow[-50\ ^\circ C]{Bu^tMgCl} Bu^tSbCl_2 \xrightarrow[0\ ^\circ C,\ 64\%]{2MeMgCl} Bu^tSbMe_2$$

Scheme 4

2.09.2.2.3 *From aminoarsines, -stibines, and -bismuthines*

Aminoarsine serves as a good precursor of tertiary alkylarsines. The displacement of amino from aminoarsines by an alkyl group of organometallics, which has been developed, results in the formation of alkylarsines. Trimethylarsine, along with other alkylarsine analogs, has been obtained in high yield from tris(dimethylamino)arsine $As(NMe_2)_3$ and organoaluminum or organomagnesium reagents, while the similar reaction starting from dimethylaminodimethylarsine Me_2AsNMe_2 affords the asymmetric arsine product Me_2AsR in excellent yield ⟨90POL289, 92SRI461, 93POL89⟩. A relevant example is the preparation of cyclopentadienyldimethylarsine Me_2AsCp (50%) by treatment of dimethylaminodimethylarsine with cyclopentadiene in the presence of equimolar trimethylchlorostannane ⟨75JOM(92)181⟩.

Reactions of aminostibines and aminobismuthines are less exploited. Examples are the reactions of monomeric cyclopentadiene with tris(dimethylamino)stibine $Sb(NMe_2)_3$, with dimethyl(dimethylamino)stibine Me_2SbNMe_2, and with the silyl-substituted aminobismuthine $Me_2BiN(Me)TMS$ giving, respectively, Cp_3Sb (100%) ⟨81JOM(208)161⟩, Me_2SbCp (76%) ⟨75JOM(88)329⟩, and Me_2BiCp (50%) ⟨75JOM(88)329⟩.

2.09.2.2.4 *Reactions of organometallics with other arsenic, antimony, and bismuth derivatives*

Alkylation of arsenic oxide (As_2O_3) by aluminum alkyls (R_3Al) leads to the formation of the tertiary alkylarsine compound (Et_3As, 72%; $Bu^i{}_3As$, 32%). Preparation is limited to lower aluminum alkyls (C < 4), since higher aluminum alkyls give mainly the bis(dialkylarsine) oxide product. Good yields of tertiary alkylstibines (Et, 89%; Bu^n, 87%; Bu^i, 71%; $C_6H_{13}{}^i$, 62%) are obtained when antimony oxide is used, while the reaction of bismuth oxide gives very poor yields ⟨64AG(E)66⟩.

A relevant synthetic route to highly pure tertiary alkylarsines has been reported. Starting from 2-chloro-1, 3, 2,-tetrahydrodioxarsolane (**8**) instead of arsenic trihalides, a series of tertiary alkylarsines are conveniently prepared in the yield of 60–95%. Although organometallic reagents including lithium, zinc, and aluminum alkyls can be used in this reaction, Grignard reagents are strongly recommended owing to their wide availability and low cost (Equation (5)) ⟨90IC3502⟩.

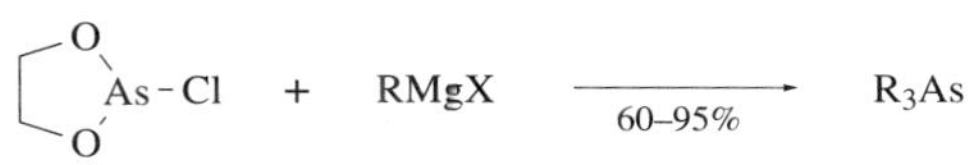

(5)

(**8**)

R = Et, Pr^n, Pr^i, Bu^n, Bu^i, Bu^t, Bz, allyl, $TMSCH_2$

2.09.2.2.5 *Alkylation of alkylarsenides and alkylstibides*

Alkylation of the metal derivatives of the secondary alkylarsines and alkylstibines constitutes a very important method for preparing tertiary alkylarsines and -stibines, especially for preparing asymmetric compounds. Alkali metallic arsenides and stibides, usually lithium and sodium arsenides and stibides, are strongly nucleophilic reagents, and they readily react with a wide variety of electrophiles to yield tertiary alkylarsines and alkylstibines. For example, lithium dicyclohexylarsenide derived from dicyclohexylarsine attacks epoxy compounds to give, after hydrolysis, the alkylated products. Both oxetane and tetrahydrofuran rings are opened by lithium dicyclohexylarsenide with the formation of 3-hydroxypropyl- and 4-hydroxybutyl-substituted arsine compounds, respectively (Scheme 5) ⟨65CB977⟩. Similar reactions of sodium dialkylstibides have been reported with the formation of tertiary stibines bearing an additional functional group such as the hydroxy, amino or carboxylic moiety (Scheme 5) ⟨73JOM(51)223⟩.

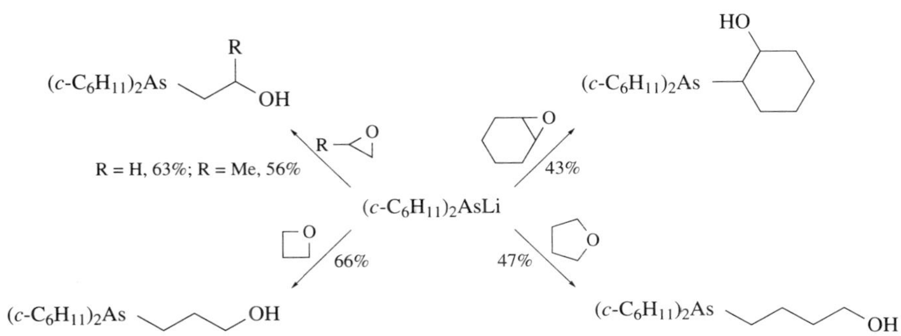

Scheme 5

A range of tertiary arsine and stibine compounds has been prepared simply from the reaction of alkylarsenic and alkylantimony anions with alkyl halides ⟨64JOC986, 67CJC675, 83ZAAC(501)89⟩. The synthesis of butylethylpropylstibine (**9**), the first tertiary stibine with three different alkyl groups, represents an excellent example utilizing the metallation and alkylation reactions in preparation (see

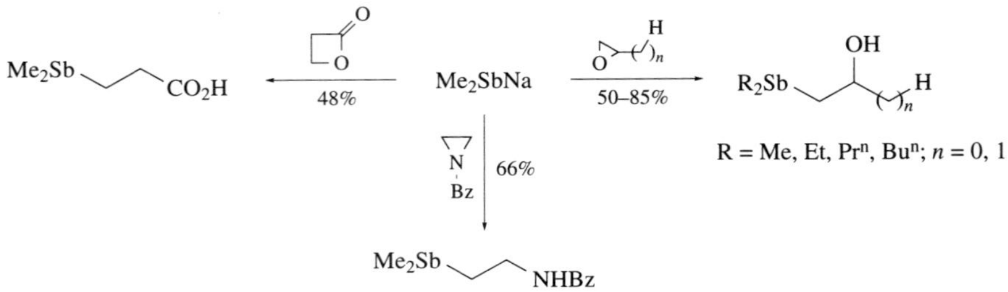

Scheme 6

Section 2.09.9.1.3) (Scheme 7) ⟨73JOM(51)223⟩. When reacted with an excess of α,ω-dichloroalkanes, alkylarsenides are converted into tertiary arsines containing an ω-chloroalkyl group ⟨70ZC118⟩, while α,ω-bis(dialkylarseno)alkanes are produced when dichloroalkanes react with two equivalents of lithium dialkylarsenides. The same results are obtained from the reaction of lithium dialkylstibides (Scheme 8) ⟨64ZAAC(332)179, 65ZAAC(339)289, 65ZAAC(339)298, 69MI 209-01⟩. 1,2-Bis(dialkylarsino) ethane, however, is obtained in very low yield, and 1,2-bis(dialkylstibo)ethane is not obtained at all, owing to the competitive reaction pathway that gives diarsine and distibine under the reaction conditions (see Section 2.09.7.1.2) ⟨61AG26, 70ZAAC(377)278, 73JOM(51)223⟩.

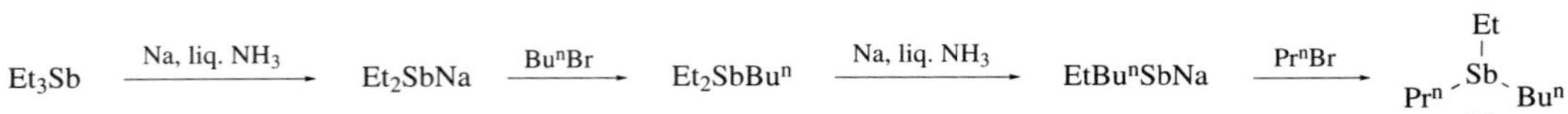

Scheme 7

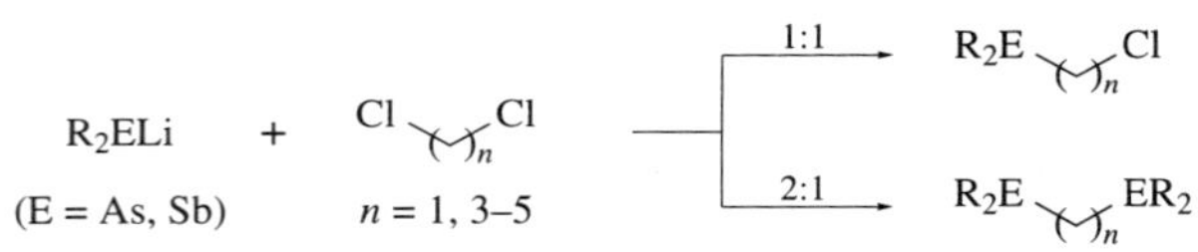

Scheme 8

The reaction between a dihaloalkane and a dimetallic alkylarsenide compound provides a useful method for preparing the heterocyclic arsine derivative; 1-methylarsolane (**10**) and 1-methylarsenane (**11**) are obtained in moderate yield by the action of disodium methylarsenide on 1,4-dichlorobutane and 1,5-dichloropentane (Equation (6)) ⟨70ZA(377)278⟩.

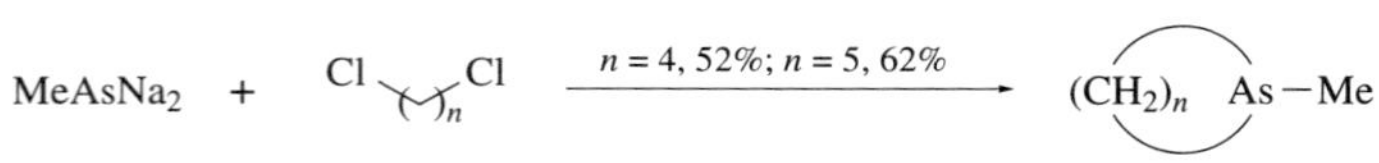 (6)

(**10**) $n = 4$
(**11**) $n = 5$

2.09.2.2.6 *Miscellaneous methods*

The reaction between arsenic or antimony and alkyl halides is an old and direct synthetic method. Tris(trifluoromethyl)arsine and tris(trifluoromethyl)stibine are formed as the major product, respectively, by heating arsenic and trifluoromethyl iodide at 220–240 °C ⟨52JCS2552, 53CB272, 53JCS1552⟩ and by heating antimony and trifluoromethyl iodide at 165–170 °C ⟨57JCS3708⟩. Alternatively, the former has been obtained in a lower yield (46%) by the use of bis(trifluoromethyl)tellurium instead of CF_3I ⟨90IC33⟩. Asymmetric tertiary arsines, stibines and bismuthines bearing a perfluoroalkyl substituent such as trifluoromethyl group have been formed through the exchange reaction of tertiary alkylarsines, alkylstibines, and bismuthines with trifluoromethyl iodide ⟨53JCS1552, 56JCS3631, 57JCS3708, 63AJC636⟩. Specific reactions are summarized as follows under the headings of alkylarsines and stibines.

(i) Alkylarsines

Preparations of tertiary alkylarsines from the alkylation of the secondary arsines are known. Since secondary arsines are weaker nucleophiles, strongly electron-deficient reagents are usually required. Alkylation of dimethylarsine by fluorinated reagents including CF_3COCF_3 ⟨65JOM(4)151⟩, $(CF_3)_2C{=}N_2$ ⟨70CJC1885⟩, $CF_2{=}CFCF_3$ ⟨66JOM(6)364⟩ and $CF_2{=}CClCF_2Cl$ ⟨68JOM(12)133⟩ has been described. As illustrated in Equation (7), methyl(2-hydroxyethyl)arsine reacts with unsaturated compounds such as acrylic acid, methyl acrylate, and acrylonitrile to give the alkylated products (**12**) ⟨83ZAAC(501)89⟩.

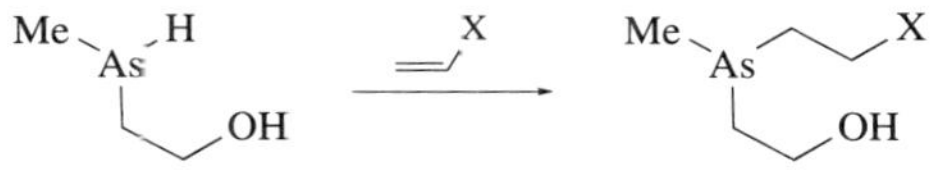

(**12**) (7)

X = CO_2H, 35%; CO_2Me, 70%; CN,70%

Biotransformation of organoarsenic compounds has been reported ⟨89CRV713⟩. Of special interest is the formation of chiral products. The mold *Scopulariopsis brevicaulis*, when growing on bread containing ethyl-*n*-propylarsinic acid, produces *R*-ethylmethyl-*n*-propylarsine of 60% optical purity, although the chemical yield is quite low (Equation (8)) ⟨90CC1169⟩.

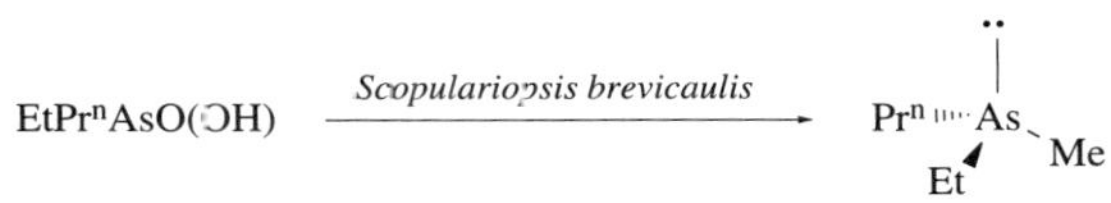

(8)

(ii) Tertiary alkylstibines

A wet method for preparation of trimethylstibine has been described from aqueous $(NH_4)_2(MeSiF_5)$ and SbF_3 ⟨66CB1609⟩. Purification of trimethylstibine is carried out by converting it into the dibromide derivative followed by reduction with zinc, $LiAlH_4$, or $LiBH_4$ ⟨56ZN(B)750⟩.

2.09.2.3 Alkylarsonium and Alkylstibonium Salts

2.09.2.3.1 Alkylation of tertiary arsines and tertiary stibines

The most straightforward method for preparing alkylarsonium and alkylstibonium salts is the alkylation of tertiary alkylarsines and -stibines by an alkyl halide. The reaction of arsines and alkyl halides proceeds readily when reactants are mixed in 1 : 1 ratio in ether solution ⟨31JCS185⟩. In most cases, when heat is required, the reaction is carried out in a sealed tube ⟨30JA376, 58JA1336, 65IC1458⟩. Compound (**13**) is converted into heterocyclic arsonium salt (**14**) (80%) at high temperature through an intramolecular alkylation reaction (Equation (9)) ⟨70ZC118⟩. Analogously, the reaction of tertiary alkylstibines and methyl iodide proceeds rapidly giving stibonium iodide in excellent yield ⟨53ACS1293, 65ZAAC(339)289⟩. Asymmetric tetraalkylstibonium iodide MeEtPr^{n}Bu^{n}SbI is prepared in a similar way ⟨73JOM(51)223⟩. Tertiary alkylstibines such as tributylstibine and triethylstibine react smoothly with active alkyl halides including benzyl bromide, allyl halides, α-haloacetic acid esters, amide, and α-haloacetonitrile to yield the corresponding alkylstibonium salts which have been demonstrated as useful intermediates in organic synthesis ⟨89JOM(366)87, 90CC85, 91JOC1381, 92ACR182, 92JOC774⟩.

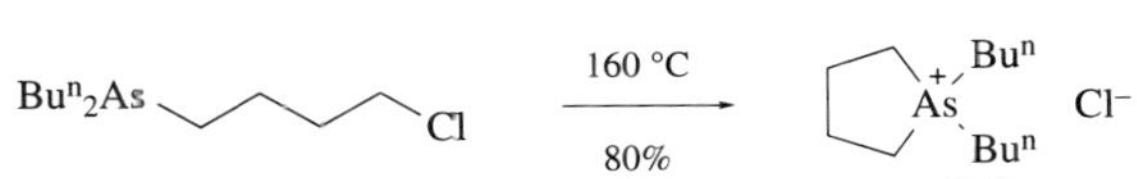

(**13**) (**14**) (9)

2.09.2.3.2 Alkylarsonium and alkylstibonium salts by other methods

The cleavage of carbon–arsenic and carbon–antimony bonds of the respective pentamethylarsorane and pentaalkylstiborane by hydrochloric acid, halogens, and other compounds leads to the formation of the corresponding tetramethylarsonium salt ⟨73CB3645⟩ and alkylstibonium salts ⟨53ACS1293, 67JOM(8)225, 69CB4136, 71ZAAC(386)139, 73CB1226⟩. The method for preparing stibonium salts by exchanging the halogen anion for other halogen or other anion species is known ⟨53ZAAC(273)161, 89JOM(366)87⟩. Scheme 9 shows the preparation of a spiro-arsonium salt through an intermediate salt at high temperature with loss of methyl bromide ⟨45JCS30, 45JCS45⟩.

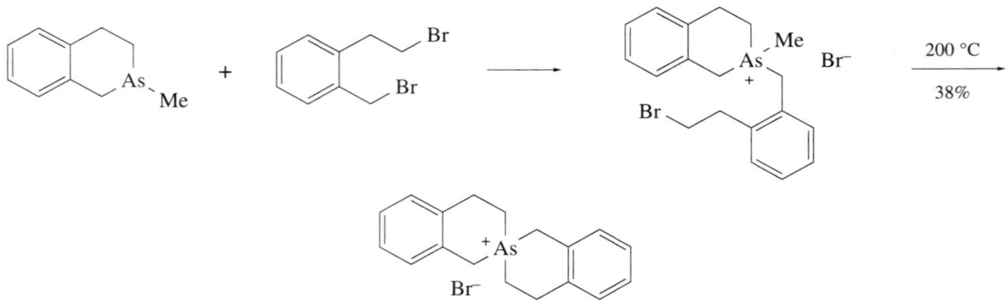

Scheme 9

2.09.2.4 Alkylarsonanes and Alkylstiboranes

2.09.2.4.1 Alkylarsonanes

Pentamethylarsonane Me_5As, the only known alkylarsonane until now, is prepared from trimethylarsenic dichloride and methyllithium in diethyl ether at low temperature ⟨73CB3645⟩.

2.09.2.4.2 Alkylstiborane

Alkylstiboranes are usually prepared from the reaction between trialkylantimony dihalides and organometallic reagents. Examples are the syntheses of pentamethyl- (63%), pentacyclopropyl- (10%), and pentakis(trimethylsilylmethyl)stiborane (89%) from alkyllithium and trialkylantimony dibromide ⟨53ACS1293, 71JA2150, 78CB2702⟩, and of pentaethylstiborane (80%) from triethylantimony dichloride and diethylmagnesium. The last compound has also been obtained by another route (85%) from the alkylation of tetraethylstibonium chloride by ethyllithium ⟨67JOM(8)225⟩. The alkylation of alkylstibonium salts with butyllithium has also been reported, and products of the form $Bu^n{}_4SbCH_2E$ (E = electron-withdrawing group) are useful intermediates in organic synthesis ⟨92ACR182⟩.

Attempts to prepare pentaalkylantimony compounds of the type Me_nSbEt_{5-n} ($n = 1$–4) have been made. It is difficult to get pure products because of the ready exchange of alkyl groups between the pentaalkylstiborane initially formed and the alkylation agent ⟨70JOM(22)653⟩. However, pentaalkylstiboranes with mixed methyl and trimethylsilylmethyl substituents have been prepared in good yield (Equation (10)) ⟨78CB2702⟩.

$$R^1{}_3SbX_2 + 2R^2Li \longrightarrow R^1{}_3SbR^2{}_2 \qquad (10)$$

$R^1 = Me$, $R^2 = CH_2TMS$, 83%; $R^1 = CH_2TMS$, $R^2 = Me$, 78%

2.09.3 ALKYLARSINE HALIDES AND THEIR ANTIMONY AND BISMUTH ANALOGUES

2.09.3.1 Alkylarsine Halides (R_2AsX, $RAsX_2$) and their Antimony and Bismuth Analogues

2.09.3.1.1 Reaction of organometallics with arsenic, antimony, and bismuth trihalides

It is difficult to get high yields of dialkylhalo- and alkyldihaloarsines, -stibines, and -bismuthines by selective alkylation of antimony trihalides with Grignard reagents. The reaction usually goes to completion with the formation of tertiary alkylarsines, stibines, and bismuthines as mentioned before. Examples include diiodoethylstibine ⟨86OCS633⟩, dibromo(methyl)bismuthine ⟨22BSF545⟩ and dibromoethylbismuthine ⟨15CB1473⟩ which are obtained in low yields. Nevertheless, reaction of organometallic compounds with arsenic, antimony, and bismuth trihalides is a useful pathway to alkylhaloarsines, -stibines and -bismuthines. Better yields are achieved when a sterically hindered alkyllithium or alkylmagnesium reagent is employed, exemplified by the preparation of bis(trimethylsilylmethyl)-substituted alkylarsenic chlorides ⟨65ZAAC(336)36, 80JCS(D)2428⟩ and chlorodi-*t*-butylstibine (Equation (11)) ⟨65ZAAC(339)298⟩. Less nucleophilic organometallic reagents also produce good yields of mono- and dihaloarsines and their antimony and bismuth analogues. Examples include $EtAsCl_2$ (95–97%) from Et_4Pb ⟨49JOC429⟩, $ClCH_2AsCl_2$ (60–70%) from $(ClCH_2)_2Hg$ ⟨70ZAAC(377)128, 89CB1247⟩, Me_2SbCl (70–80%) from Me_4Pb or Me_4Sn ⟨83ZAAC(505)134, 86OCS633⟩ and Et_2BiCl from Et_4Pb ⟨39JOC162⟩. NaC_5H_5 reacts with $BiCl_3$ at low temperature to give the thermolabile, light-sensitive product $CpBiCl_2$, which possesses a π-bond between the bismuth and cyclopentadienyl moieties identified by low temperature crystal structure analysis ⟨90JOM(386)177⟩.

$$Bu^tMgCl + SbCl_3 \xrightarrow[71\%]{-25\ ^\circ C} Bu^t_2SbCl \qquad (11)$$

A similar approach is the use of an organosilicon reagent which reacts with arsenic trichloride readily at ambient temperature to give functionalized alkylarsine dichlorides in quantitative yield (Scheme 10) ⟨89TL349, 90TL1147⟩.

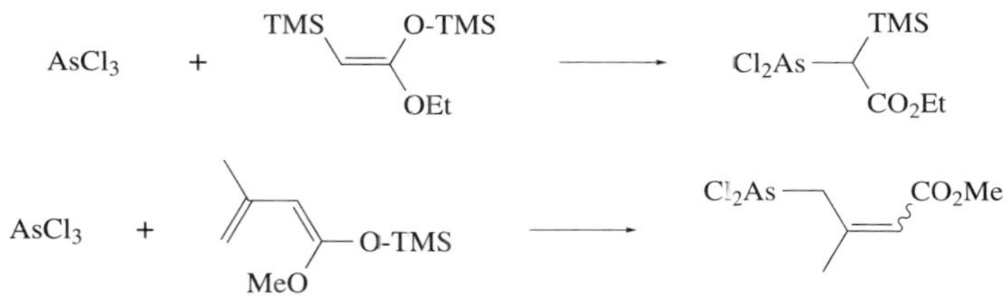

Scheme 10

2.09.3.1.2 *Pyrolysis of R_3EX_2 and R_2EX_3 (E = As, Sb)*

Pyrolysis of triorganoarsenic dihalides (R_3AsX_2) and diorganoarsenic trihalides (R_2AsX_3) under controlled conditions gives rise to the formation of dialkylhaloarsines and alkyldihaloarsines, respectively, in good yields with the elimination of an alkyl halide. This method is especially useful for the preparation of alkylarsine dihalide derivatives, since dialkylarsenic trihalides undergo thermolysis reaction more readily at a lower temperature. Some examples are shown in Scheme 11 ⟨28CB1911, 89OM482⟩.

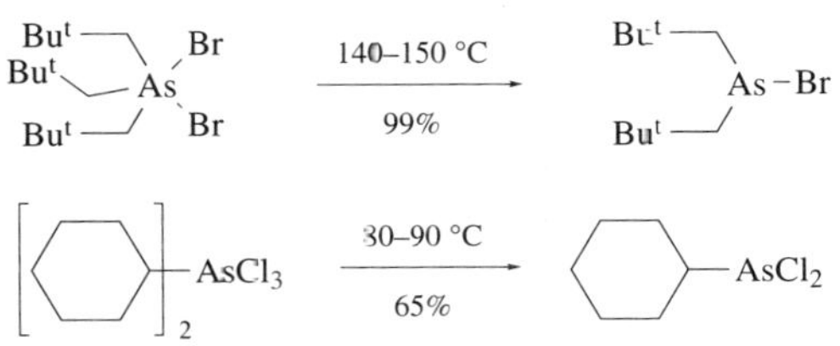

Scheme 11

Similar to the cases of arsenic halides, trialkylantimony dihalides undergo thermal decomposition at elevated temperatures, while the thermolysis of dialkylantimony trihalides is effected under mild conditions. A few pure dialkylantimony halides R_2SbX (R = Me, 71%; Et, 79%; Pr^i, 76%; cyclohexyl, 36%) ⟨61ZAAC(312)186, 82PS(12)149⟩ and alkyldihalostibines $RSbX_2$ (R = Me, 60–83%; Et, 90%; isopropyl, 54%) ⟨26PRS523, 66JOM(6)259, 82PS(12)149, 89JOM(364)67⟩ have been prepared by this method.

2.09.3.1.3 *Redistribution reaction*

Some dialkylhaloarsine and alkyldihaloarsine compounds have been prepared by the redistribution reactions of the corresponding trialkylarsine and arsenic trihalides in different molar ratios. For example, benzyldichloroarsine $BzAsCl_2$ was obtained more than 100 years ago by heating a 1:2 mixture of tribenzylarsine and arsenic trichloride at 160–180°C ⟨1886LA(233)60⟩. Phenyl-substituted arsine compounds (**15**) and (**17**) undergo redistribution with arsenic trichloride to give mono- and dichloroarsine products (**16**) and (**18**) (Scheme 12) ⟨70ZAAC(377)278, 70ZAAC(376)150⟩. The method is not generally applicable to other arsine halide compounds since only adduct formation and redox reactions occur between R_3As and AsX_3 ⟨70IC862, 91PS(57)1⟩. Similar redox reactions between tertiary alkylstibines and antimony trihalides, and between tertiary alkylstibines and alkyldihalostibines, make it less attractive to prepare compounds of the type R_2SbX by redistribution reactions. However, the redistribution reaction between dialkylhalostibine and antimony trihalide compounds in 1 : 1 ratio has been used successfully to synthesize alkyldihalostibines $RSbX_2$ (R = Me, Pr^n; X = Cl, Br) ⟨89JOM(364)67⟩.

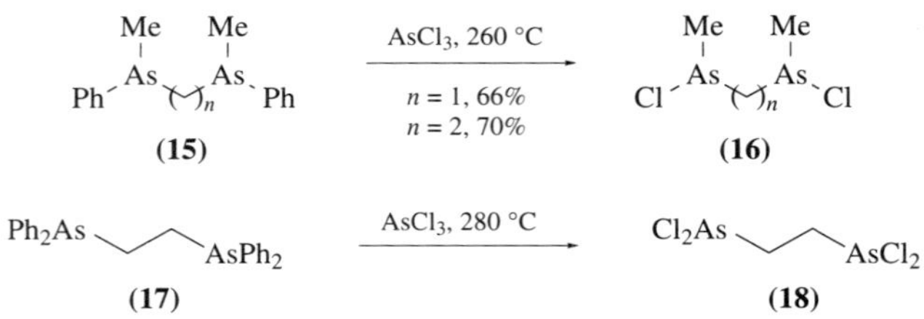

Scheme 12

In contrast to its arsenic and antimony analogs, alkylbismuth(III) halides are best prepared by the redistribution reactions between tertiary alkylbismuthines and bismuth trihalides. The chloro- and bromo(dialkyl)bismuthine compounds are formed in good yields when R_3Bi and BiX_3 (X = Cl and Br) react in 2:1 molar ratio. Reversing the ratio of the reactants leads to the formation of alkyldihalobismuthines. Compounds prepared by this method include chloro- and bromo(dimethyl)bismuthines ⟨61CB1447⟩, chloro(diethyl)bismuthine, dichloro(ethyl)bismuthine ⟨1887CB1516⟩, dibromo(isobutyl)bismuthine, and dibromo(isoamyl)bismuthine ⟨1888CB2035⟩.

2.09.3.1.4 *Reactions involving cleavage of As—C, Sb—C, and Bi—C bonds*

Cleavage of the C—As bond of a triorganoarsine by hydroiodic acid or bromine results in the formation of an alkylarsine halide. Bromo(benzyl)methylarsine is obtained in 74% yield on treatment of dibenzylmethylarsine with bromine ⟨75PS(5)265⟩, and reaction of 1,3-bis(diphenylarsino)propane with hydroiodic acid affords 1,3-bis(diiodoarsino)propane (76%) ⟨86JCS(D)997⟩. Alkyldiphenylstibine (**19**) has been reported to give dichlorostibine (**20**) in excellent yield when reacted with anhydrous hydrogen chloride ⟨82AG(E)375, 85JOM(290)307⟩ (Scheme 13). Halogens cleave trialkylbismuthines with the formation of dialkylhalobismuthines (e.g., chloro(dimethyl)bismuthine is formed from the reaction of trimethylbismuthine and chlorine at 0°C), while mineral acids cleave all three alkyl groups of tertiary alkylbismuthines ⟨1887CB1516, 1888CB2035⟩.

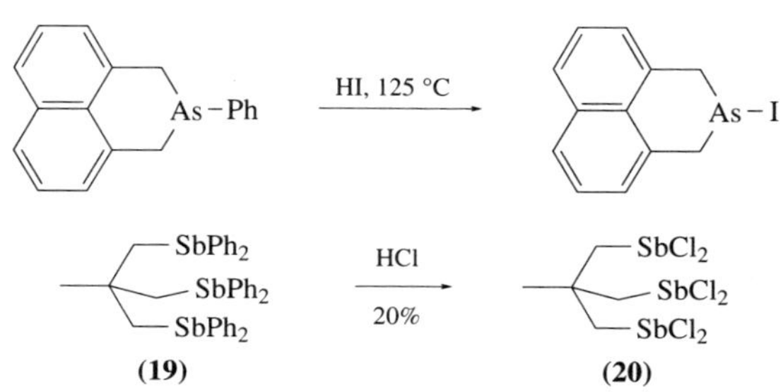

Scheme 13

2.09.3.1.5 *Cleavage of As—As, Sb—Sb, and Bi—Bi bonds*

Tetramethyldiarsine and tetramethyldistibine undergo As—As and Sb—Sb cleavage with iodine to give the corresponding dimethyl pnicogen iodide ⟨35JCS366, 52JCS4331⟩. Other halogens or halogen-based reagents such as PCl_5, SO_2Cl_2, HCl and MeI have also been used for the reaction ⟨59JA1, 77JOM(136)173, 83JOM(243)417, 86ZAAC(539)50⟩. An interesting example is the preparation of 1-chlorostibolane (**22**) and 1-chloroantimonane (**24**) from distibine compound (Equation (12)) ⟨77JOM(136)173⟩. Tetramethylbismuthine, however, gives a mixture of compounds when it reacts with iodine, anhydrous hydrogen chloride, or benzyl bromide ⟨83OM1859⟩.

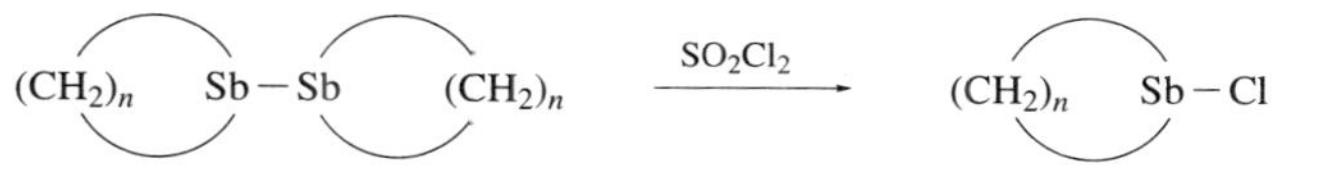

(12)

(**21**) $n = 4$ (**22**) $n = 4$, 97%
(**23**) $n = 5$ (**24**) $n = 5$, 76%

2.09.3.1.6 *Miscellaneous methods for alkylarsine halides*

(i) Reduction of alkylarsinic and alkylarsonic acid derivatives

Reduction of alkylarsinic and alkylarsonic acids and their salts, a frequently used method, provides an important entry to the alkylarsine halide compounds (Scheme 14). Although sodium hypophosphite and a phosphorus trihalide ⟨61JCS1398⟩ can be used as reducing agents, sulfur dioxide in concentrated hydrochloric acid is most frequently employed. A trace amount of potassium iodide has been found to facilitate the reaction ⟨47JA927⟩. Details for the preparation of dimethyliodoarsine and diiodomethylarsine have been reported ⟨60IS113, 60IS116⟩.

$$R^1R^2AsO(OH) \xrightarrow{SO_2/HCl,\ trace\ KI} R^1R^2AsCl$$

$$R^1 = Et;\ R^2 = Et,\ Pr^n,\ Bu^n$$

$$RAsO(OH)_2 \xrightarrow{SO_2/HCl\ or\ HBr} RAsX_2$$

$$R = Me,\ Et,\ Pr^n,\ Bu^n;\ X = Cl,\ Br$$

Scheme 14

(ii) From alkylarsonous acid derivatives

A method for the large-scale production of dichloromethylarsine has been developed. Polymeric methylarsaoxane $(MeAsO)_n$ formed from reduction of sodium methylarsonate is converted to dichloro(methyl)arsine by treatment with hydrogen chloride. Dichloro(ethyl)arsine has been prepared analogously from ethylarsaoxane $(EtAsO)_n$ ⟨19IEC105⟩.

(iii) Conversions of chloride to bromide derivatives, etc.

The halogen group of alkylarsine halides is interchangeable with other halide anions from metal halides or ammonium fluoride; this method has been used frequently for the preparation of alkylfluoroarsine compounds. For example, difluoro(methyl)arsine, dimethylfluoroarsine and its trifluoromethyl analog have been isolated from corresponding chlorides after interaction with ammonium fluoride, sodium, or silver fluoride ⟨53CB272, 53JCS1552, 66JOM(5)446⟩.

(iv) Cleavage of As—N and As—O bonds

Arsenic–heteroatom bonds such as the As—N bond of aminoarsines and the As—O bond of alkoxyarsines are easily cleaved by HX and other halogen-based electrophiles to give alkylarsine halides. For example, chloro(dimethyl)arsine is obtained from the reactions of dimethylaminodiethylarsine with HCl ⟨59CB2637⟩ and of $Me_2AsOTMS$ or $Me_2AsOAsMe_2$ with $SOCl_2$ (85–89%) ⟨64CB449⟩; bis(trifluoromethyl)fluoroarsine (85%) and -chloroarsine (100%) are obtained when bis(trifluoromethyl)methoxyarsine and bis[bis(trifluoromethyl)arsine] oxide are reacted, respectively, with BF_3 and HCl ⟨65JA1213⟩.

(v) *Alkylbismuth halides*

The cleavage of the Bi—Ge bond of $[(C_6F_5)_3Ge]_nBiEt_{3-n}$ by HCl gas in THF at room temperature leads to Et_2BiCl and $EtBiCl_2$ in good yield ⟨78JOM(162)289⟩.

2.09.3.2 Arsenic(V) and Antimony(V) Halides–R_3EX_2, R_2EX_3 (E = As, Sb)

2.09.3.2.1 R_3AsX_2 and R_3SbX_2 (X = Cl, Br, I) by reaction of tertiary arsines and stibines with halogens

The reaction of tertiary alkylarsines and -stibines with halogens, which has been used widely, provides a straightforward method for the preparation of trialkylarsenic and trialkylantimony dihalides. Their synthesis is easily performed by treating alkylarsines and alkylstibines with a dilute solution of chlorine and bromine in carbon tetrachloride, or iodine in petroleum ether. A range of trialkylarsenic dihalides R_3AsX_2, most being solid and purified by recrystallization ⟨30JCS2426, 31JCS185, 67ICA34⟩ and trialkylantimony dihalides R_3SbX_2 ⟨30JCS1920, 32JCS1815, 64JA209, 67IS92, 71JCS(C)1660⟩ have been synthesized (R = Me, Et, Pr^n, Bu^n, isobuty, *n*-amyl, etc.) Tris(trifluoromethyl)arsenic and -antimony dichloride are prepared similarly by the action of chlorine on the parent arsine and stibine in a sealed tube at room temperature, although other halogens give products with the loss of the trifluoromethyl group ⟨53JCS1552, 57JCS3708⟩.

2.09.3.2.2 R_3AsF_2 and R_3SbF_2

In practice, triorganoarsenic and -antimony difluorides are prepared conveniently by conversion of other dihalide compounds with silver fluoride or sodium fluoride ⟨53JCS1552, 64JA209, 67ICA34, 68IC834⟩. Direct oxidative fluorination of tertiary alkylarsines and -stibines has been developed by utilizing fluorinating reagents such as XF_2 and IF_5. For example, reaction of trimethylarsine and xenon difluoride in dilute solution of $CFCl_3$ ($Me_3As : XeF_2 : CFCl_3$ = ca. 2 : 1 : 30) proceeds smoothly at low temperature to afford trimethylarsenic difluoride in excellent yield with very high purity ⟨85POL1625, 88JCS(D)451⟩. Tributylarsenic difluoride is prepared (64%) when tributylarsine reacts with IF_5 at $-35\,^\circ C$ ⟨86JFC(34)129⟩.

2.09.3.2.3 R_2AsX_3 by reaction of R_2AsX with halogens

Similar to the preparation of triorganoarsenic dihalides, the straightforward reaction of halogens with dialkylhaloarsine serves as a main synthetic method for the dialkylarsenic trihalides ⟨58LA(107)257, 89OM482⟩. For example, chlorodicyclohexylarsine is converted into dicyclohexylarsenic trichloride when reacted with chlorine ⟨28CB1911⟩.

2.09.3.2.4 Miscellaneous methods

It has been reported that dimethylarsinic acid reacts with PCl_3 to produce dimethylarsenic trichloride ⟨58LA(617)257⟩. Interaction between trimethylstibine oxide and hydrogen chloride has been reported to form trimethylantimony dichloride ⟨30JCS1920⟩, which has also been obtained from the reaction of trimethylbis(trimethylsiloxy)antimony with HCl or SO_2Cl_2.

2.09.4 ALKYLARSENIC COMPOUNDS WITH AN As—O BOND AND THE ANALOGOUS ANTIMONY AND BISMUTH COMPOUNDS

Free alkylarsinous acids R_2AsOH readily form their anhydrides $R_2AsOAsR_2$, which are named as bisdialkylarsine oxides, while free alkylarsonous acids $RAs(OH)_2$ form anhydrides $(RAsO)_n$ which exist in oligomeric and/or polymeric states. The latter compounds are known as alkylarsoxanes or alkylarsaoxanes. Preference is given to the latter name, due to its great clarity when spoken. Similarly, free alkylstibinous and alkylstibonous acids and bismuth analogues are unknown.

2.09.4.1 Alkylarsinous Acid Derivatives ($R^1_2AsOAsR^1_2$, $R^1_2AsOR^2$) and their Analogous Antimony and Bismuth Compounds

2.09.4.1.1 *Bisdialkylarsine oxides and their analogous antimony and bismuth compounds–R_2EOER_2 (E = As, Sb, Bi)*

(i) Hydrolysis of R_2AsX and R_2SbX

A simple and general method for preparing bisdialkylarsine and bisdialkylstibine oxides is the hydrolysis of respective dialkylhaloarsines and -stibine compounds. Bisdimethylarsine oxide $Me_2As OAsMe_2$ has been obtained from the hydrolysis of chlorodimethylarsine with a solution of sodium carbonate ⟨70JA3969⟩. Heterocyclic diarsine oxides such as bisarsolane oxide (**26**) and bisarsenane oxide (**28**), have been prepared similarly from chlorides (**25**) and (**27**) (Equation (13)) ⟨71ZAAC(383)136⟩. Examples of distibine oxides are bisdimethylstibine oxide, a colorless oil, spontaneously inflammable in the air from hydrolysis of bromodimethylstibine ⟨26PRS523, 82ZN(B)1104⟩ and bis(di-*t*-butylstibine) oxide (85%) from reaction of di-*t*-butylchlorostibine with sodium hydroxide ⟨81ZN(B)1105⟩.

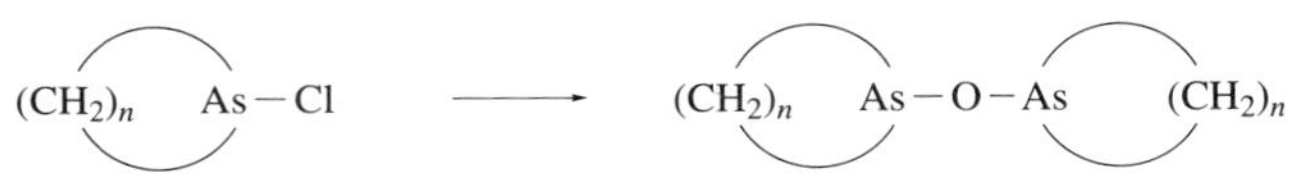

(13)

(**25**) $n = 4$
(**27**) $n = 5$

(**26**) $n = 4$
(**28**) $n = 5$

(ii) From tetraalkyldiarsines, -distibines and -dibismuthines

Bis(dialkylbismuthine) oxides were synthesized successfully in the 1980s by oxidation of dibismuthines (R_2BiBiR_2) with oxygen. At low temperature, tetramethyl- and tetrapropyldibismuthines are transformed into the corresponding oxides in petroleum ether ($Me_2BiOBiMe_2$, 95%; $Pr_2BiOBiPr_2$, 100%) ⟨84ZN(B)887, 86ZN(B)1129⟩. Interaction of tetraalkyldistibines and air has been reported to form distibine oxides ⟨82ZN(B)1104⟩.

(iii) Miscellaneous methods for tetraalkyldiarsine oxides

(a) Hydrolysis of aminoarsines. Aminoarsines are good starting materials, and they undergo hydrolysis readily to form diarsine oxide compounds ⟨82S173⟩. Heterocyclic diarsine oxides (**26**) and (**28**) have also been obtained by the hydrolysis of the corresponding aminoarsines ⟨71ZAAC(383)136⟩. Another example shown in Equation (14) is the preparation of the eight-membered heterocyclic compound (**30**) starting from methylenebis(methyldimethylaminoarsine) (**29**) ⟨70ZAAC(377)120⟩.

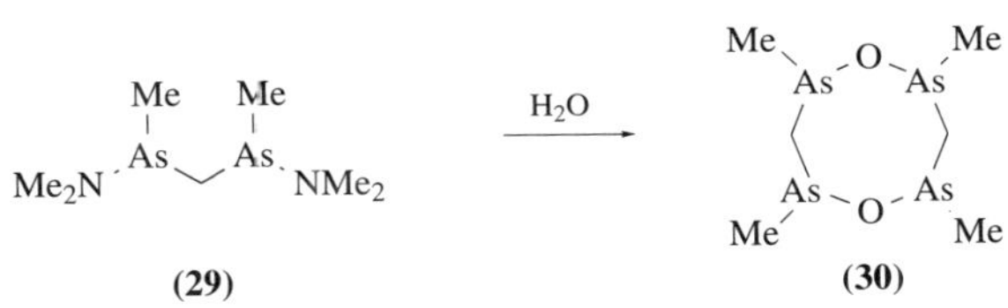

(14)

(**29**) (**30**)

(b) Cadet reaction. Bisdimethylarsine oxide, well known as cacodyl oxide, was first prepared in 1760 as one of the major components of "Cadet's fuming arsenical liquid", prepared by the pyrolysis of a mixture of arsenic(III) oxide and potassium acetate at about 325 °C ⟨1760MI 209-01⟩. Later investigations of Cadet's liquid show that it contained 40% of bisdimethylarsine oxide, 56% of tetramethyldiarsine and some other compounds ⟨27CR(185)779, 27CR(185)956⟩. Cadet's liquid has been produced in good yield by passing a mixture of arsenic trioxide and acetic acid over an alkali metal acetate catalyst at 300–400 °C. The process has also been applied to prepared tetraethyldiarsine and bisdiethylarsine oxide $Et_2AsOAsEt_2$ ⟨47JA559⟩.

2.09.4.1.2 Esters of arsinous acid -$R^1{}_2AsOR^2$ and their analogous antimony and bismuth compounds

(i) Reaction of R_2EX (E = As, Sb, Bi) with alcohols

The reaction of dialkylhaloarsines, -stibines and -bismuthines with alcohol derivatives provides a general approach to alkoxyarsine, -stibine and -bismuthines. Sodium alkoxides are usually employed to displace halogen from dialkylhaloarsines ⟨61JCS3208⟩, dialkylhalostibines ⟨74JOM(78)107⟩, and dialkylhalobismuthines ⟨42JA392⟩. For example, trimethylsiloxy-substituted dimethylarsine and dimethylstibine have been obtained in good yield from reactions of the sodium salt of trimethylsilanol with chloro(dimethyl)arsine and with bromodimethylstibine, respectively ⟨64CB449⟩. Dimethyltrimethylsiloxybismuthine has been prepared in a similar reaction, but the yield is low (Equation (15)) ⟨68ZAAC(363)84⟩.

$$Me_2EX \quad + \quad TMS\text{-}ONa \xrightarrow{-NaX} Me_2EO\text{-}TMS \qquad (15)$$

$$E = As,\ 84\%;\ Sb,\ 79\%;\ Bi,\ 10\%$$

(ii) Alkoxy(dialkyl)arsines by other methods

The most important and widely used preparative method for alkoxy(dialkyl)arsines is the reaction of aminoarsines $R^1{}_2AsNR^2{}_2$ and alcohols. In a three-fold molar excess of the alcohol, dialkyl(diethylamino)arsines are transformed into alkyoxy(dialkyl)arsines easily and completely under reflux with the release of diethylamine, and progress of the reaction can be followed by monitoring the evolution of the amine. The mechanism of the reaction is not clear, although the formation of an ammonium ion as an intermediate is most likely ⟨72JOM(39)301⟩. When reacted with diols and polyethylene glycols, dimethyl(dimethylamino)arsine gives monoester and diester products in excellent yield under controlled conditions (Scheme 15), and some of them are listed below ⟨80ZAAC(466)179, 81ZAAC(472)75⟩.

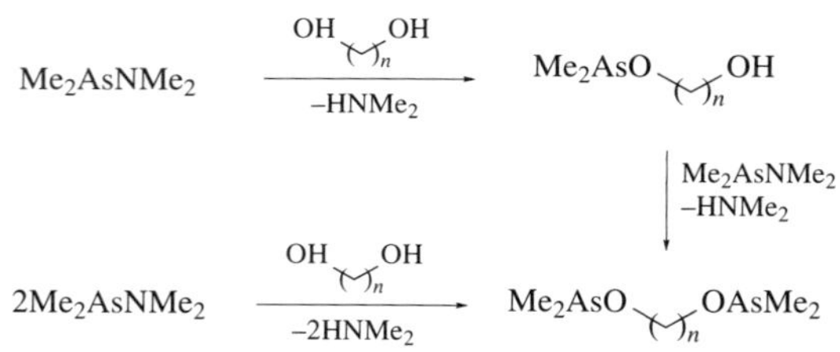

Scheme 15

A specific preparation for alkoxy(dialkyl)arsine has been reported: 2-chloro-1,2-oxarsolane (**35**) derived from dichloro(3-hydroxypropyl)arsine undergoes alkylation with methyllithium at low temperature to yield 1-methyl-1,2-oxarsolane (**36**) (Scheme 16) ⟨70ZAAC(375)55⟩.

(iii) Alkoxy(dialkyl)stibines by other methods

Reactions of alkoxydialkylstibines and amino(dialkyl)stibines with enol acetates or enol trifluoroacetates have been investigated. Normally, both reactions give tertiary alkylstibines with an aldehyde or ketone functionality. Exclusive alkoxy(dialkyl)stibines, however, have been obtained

(31) 75% **(32)** 90% **(33)** 85% **(34)** 95%

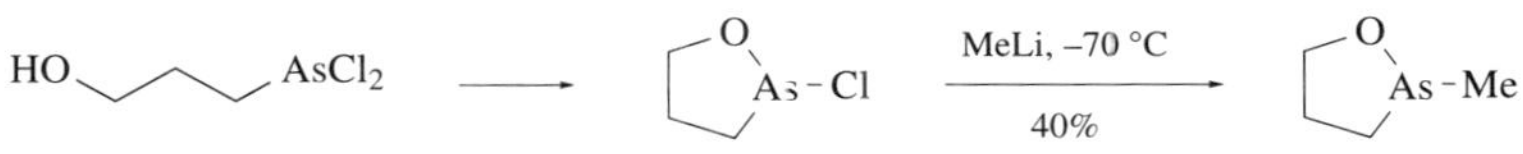

(35) **(36)**

Scheme 16

from sterically hindered reactants. For example, compound (**37**) is prepared as the sole product in 79% yield when di-*t*-butylmethoxystibine or di-*t*-butyl(dimethylamino)stibine reacts with the enol acetate of cyclohexanone (Equation (16)) ⟨74JOM(78)107⟩.

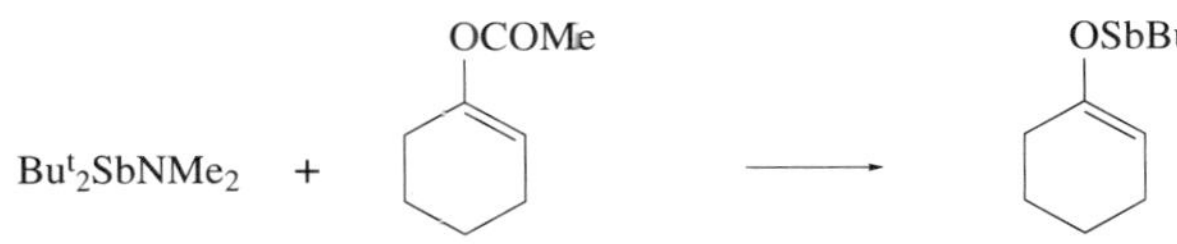

(37) (16)

2.09.4.2 Alkylarsonous Acid Derivatives $(RAsO)_n$, $R^1As(OR^2)_2$, $R^1As(OR^2)X$ and their Analogous Antimony and Bismuth Compounds

2.09.4.2.1 $(RAsO)_n$ and $(RSbO)_n$

Alkylarsaoxanes $(RAsO)_n$ have been prepared by a variety of procedures; the most important methods include hydrolysis of dihalo(alkyl)arsines, reduction of arsonic acids and oxidation of primary arsines and cyclopolyarsines. Different methods, which often lead to complex mixtures of cyclic and linear oligomers, are described below. Only a few alkylstibaoxanes are known from the hydrolysis of alkyldihalostibines.

(i) Hydrolysis of $RAsX_2$ and $RSbX_2$

Hydrolysis of alkyldihaloarsines, an older and simpler method, gives alkylarsaoxanes ⟨03CR(137)925, 20CB1013⟩. For example, treatment of dibromo(methyl)arsine in benzene with potassium carbonate and a trace of water leads to the formation of methylarsaoxane ⟨70JA3969⟩. Evidence available indicates that alkylarsaoxanes $(RAsO)_n$ exist predominantly as cyclic trimers ($R = Et, Pr^n, Bu^n$) and tetramers ($R = Me$) with bridging oxygen atoms ⟨74JOM(77)225⟩.

Cyclo-*t*-butylstibaoxane $(Bu^tSbO)_n$ ($n = 3,4$) is obtained in good yield by hydrolysing butyldichlorostibine in aqueous sodium hydroxide ⟨81ZN(B)1105⟩.

(ii) Reduction of alkylarsonic acids $(RAsO(OH)_2)$

Alkylarsaoxanes are obtained when alkylarsonic acids are reduced by sulfur dioxide ⟨03CR(137)925⟩. The procedure for large-scale production of ethylarsaoxane $(EtAsO)_n$ has been described ⟨19IEC105, 19IEC817⟩.

(iii) Oxidation of primary alkylarsines $(RAsH_2)$ and polyarsines $(RAs)_n$

Primary methylarsine is oxidized by oxygen in the presence of mercury to afford methylarsaoxane ⟨05MI 209-01⟩. Oxidation of polyarsines provides another useful route to alkylarsaoxane compounds ⟨70JA3969⟩: cyclic methylarsaoxane (**38**) has been synthesized in 42% yield from the aerobic oxidation of homocyclic pentamethyl-*cyclo*-pentaarsine via a metal–carbonyl template reaction. Another interesting example is the preparation of a 12-membered heterocyclic compound, *cyclo*-$(MeAsO)_6$ (**39**), which is obtained as a complex of dimolybdenum hexacarbonyl in high yield (Scheme 17) ⟨86OM393, 90JA4037, 91OM3764⟩.

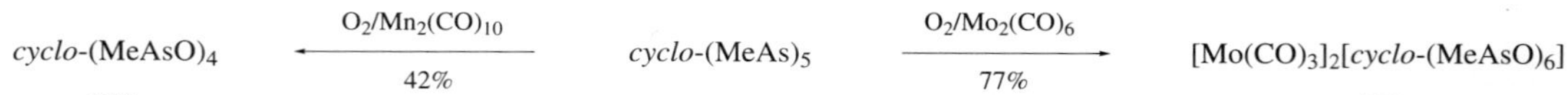

(**38**) (**39**)

Scheme 17

2.09.4.2.2 *Esters of alkylarsonous acids–$R^1As(OR^2)_2$ and their analogous antimony and bismuth compounds*

(i) Reaction of REX_2 (E = As, Sb, Bi) with alcohols

Preparation of alkyldialkoxyarsines, -stibines and -bismuthines has been effected by a general reaction between sodium alkoxides and the corresponding alkylarsenic, alkylantimony, and alkylbismuth dihalides ⟨57JOC452, 75ZAAC(418)167, 76ZAAC(423)40, 78ZAAC(439)139⟩. A relevant example is the preparation of diacetoxy(propyl)arsine $PrAs(OCOMe)_2$ (51%) from the reaction of dichloro(propyl)arsine and silver acetate ⟨47JA927⟩. Utilization of alcohols with the aid of an acid absorbent such as triethylamine, an improved procedure, has been reported to give a better yield than by use of an alkoxide. Dialkoxy(methyl)arsines $MeAs(OR^2)_2$ have been prepared ($R^1 = Pr^i$, 62%; $R^1 = C_6H_{11}$, 78%) when dichloro(methyl)arsine reacts with a mixture of an alcohol and an equimolar quantity of triethylamine ⟨57JOC452⟩. Attempts to prepare mixed esters have been made from reaction of alcohols with a alkyl(chloro)alkoxyarsine $R^1As(OR^2)Cl$, but no pure products are isolated, owing to disproportionation in the course of distillation ⟨63JGU2831⟩.

(ii) Reaction of $R^1E(NR^2{}_2)_2$ and $R^1E(OR^2)_2$ (E = As, Sb, Bi) with alcohols

In a similar manner to the preparation of esters of alkylarsinous acid from aminoarsines, diaminoarsine reacts with two equivalents of an alcohol to give the corresponding ester of an alkylarsonous acid in good yields ⟨91ZAAC(605)151⟩. Various heterocyclic alkyldialkoxyarsines have been synthesized starting from α,ω-diol substrates, and some examples are listed below ⟨79CZ119, 76ZAAC(420)74⟩. For example, transesterification of *t*-butyldiethoxystibine and diethoxy(methyl)bismuthine with diols takes place readily giving the heterocyclic products (**43**) in high yield (Scheme 18) ⟨75ZAAC(418)167, 76ZAAC(423)40⟩.

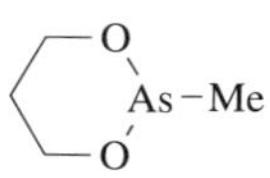

(**40**)

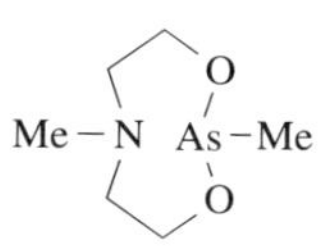

(**41**)

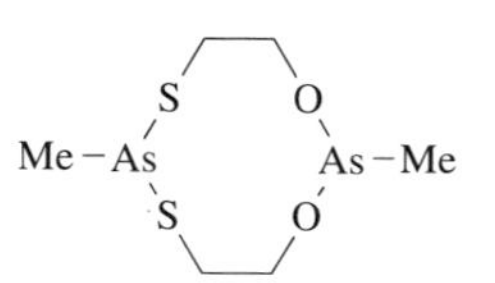

(**42**)

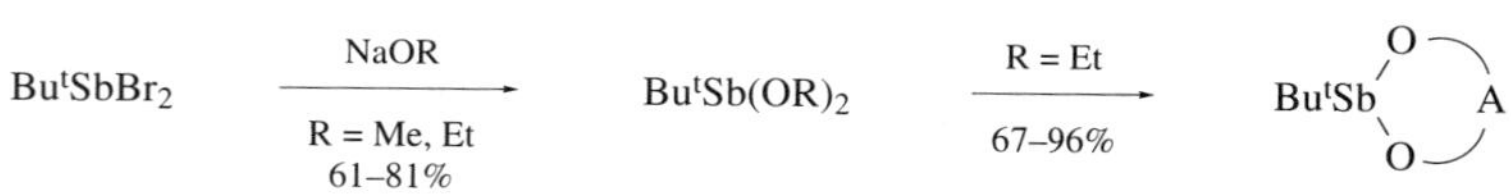

(**43**)

A = –$CHMe_2CHMe$–, CMe_2CMe_2–, *o*-phenylene

Scheme 18

2.09.4.3 $R^1As(OR^2)X$

2.09.4.3.1 Reaction of $RAsX_2$ with an alcohol

Although alkyl(alkoxy)haloarsines are formed through the reaction of alkyldihaloarsines and an alcohol, purification of this class of compounds has proved difficult, owing to disproportionation. For example, pure products (**44**) are obtained in low yield after repeated distillation from the reaction shown in Equation (17) ⟨63JGU2831⟩.

$$R^1AsCl_2 + R^2OH \xrightarrow[21\text{–}45\%]{\text{pyridine}} \underset{\mathbf{(44)}}{R^1As(OR^2)Cl} \qquad (17)$$

$$R^1 = Et,\ Pr^n,\ Bu^n,\ R^2 = n\text{-}C_6H_{13}$$

2.09.4.3.2 Reaction of $R^1As(OR^2)_2$ with halogenating agents

A high yield production of alkylarsonous acid halide has been reported; bis(1,2-oxarsolanyl) oxide (**45**) reacts with arsenic trichloride to afford 2-chloro-1,2-oxarsolane (**35**) in 95% yield (Equation (18)) ⟨70ZAAC(375)55⟩.

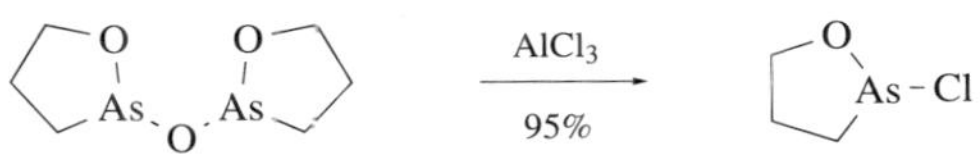

(18)

(45) **(35)**

2.09.4.4 Alkylarsine and Alkylstibine Oxides–R_3EO (E = As, Sb)

2.09.4.4.1 Oxidation of tertiary alkylarsines and tertiary alkylstibines

Oxidation reaction provides a straightforward method for preparing tertiary alkylarsine and -stibine oxides. Both trialkylarsines and -stibines are very easily oxidized, but atmospheric oxidation gives a mixture of products. *t*-Butylhydroperoxide has been used successfully to convert arsines and stibines to oxides under controlled conditions ⟨66IC187, 71JCS(C)1660⟩. A more convenient and practical method is by use of mercuric oxide. Preparation is carried out simply by refluxing trialkylarsines in acetone solution or by shaking trialkylstibines in dry ether with mercuric oxide; and a number of trialkylarsine oxides ⟨66IC187⟩ and trialkylstibine oxides ⟨70JOM(22)637⟩ have been obtained accordingly. Both trialkylarsine oxides and trialkylstibine oxides are extremely hygroscopic; an earlier preparation of trialkylstibine oxides in absolute ethanol ⟨30JCS1920, 32JCS1815⟩ has been shown to form diethoxytrialkylstiboranes ⟨70JOM(22)637⟩.

2.09.4.4.2 From R_3EX_2 (E = As, Sb)

Trialkylarsenic and trialkylantimony dihalides can be converted to the oxides by the treatment with suitable metallic oxides. For example, tri-*n*-butylarsine oxide has been prepared when the dibromide is treated with either mercuric or silver oxide in alcohol ⟨31JCS185⟩, while colorless crystals of trimethylstibine oxide and its perdeuterated analogue are obtained when their parent dibromides are reacted with silver oxide ⟨75JOM(91)295⟩.

2.09.4.5 Alkylarsonic Acids ($RAsO(OH)_2$) and Alkylarsinic Acids ($R_2AsO(OH)$)

2.09.4.5.1 The Meyer reaction

Alkyl halides react with alkali metal arsenites M_3AsO_3 to form alkylarsonic acids $RAsO(OH)_2$ and the alkali halide. This reaction was discovered by Meyer in 1883 ⟨1883CB1439⟩ and has been the most widely used method for preparing alkylarsonic acids. The reaction proceeds satisfactorily in the cold in an aqueous solution containing sufficient alcohol to dissolve the alkyl halide. Primary alkyl halides react rapidly and smoothly with alkali arsenites; in contrast, secondary halides react only slowly while tertiary halides do not react. Under optimal conditions with appropriate alkyl halides, products are formed in very high yield ⟨05MI 209-01, 61MI 209-01⟩. Dimethyl sulfate has also been used to prepare methylarsonic acid ⟨47JA927⟩.

When alkali alkylarsonites are used instead of alkali metal arsenites, the Meyer reaction gives alkylarsinic acids. This method is general for the synthesis of both dialkylarsinic acids bearing the same or different alkyl substituents ($R_2AsO(OH)$ and $R^1R^2AsO(OH)$). For example, butylpropylarsinic acid resulted from the reaction between propyl bromide and sodium butylarsenite. A range of alkylarsinic acids have been prepared by this procedure and examples are $R_2AsO(OH)$ ($R = Me$, Et, Pr^n, Bu^n, *n*-amyl) and $R^1R^2AsO(OH)$ ($R^1 = Bu^n$, $R^2 = Pr^n$, *n*-amyl, *n*-heptyl) ⟨22JA805, 61MI 209-01, 64MI 209-01⟩.

2.09.4.5.2 *Oxidation of various derivatives (R_2AsX, $RAsX_2$)*

Trivalent arsenic compounds containing two alkyl groups directly attached to arsenic are readily oxidized to arsinic acids or to the products which can be hydrolyzed to arsinic acids. Some alkylarsinic acids have been prepared by the oxidation of chloro(dialkyl)arsines ⟨30JPR223⟩. Oxidation reactions of dimethylarsine and bisdimethylarsine oxide lead to the same dimethylarsinic acid compound ⟨06MI 209-01, 63AC(R)1869⟩. Bis(trifluoromethyl)iodoarsine undergoes oxidation readily with hydrogen peroxide to give the corresponding bis(trifluoro)methylarsinic acid in high yield ⟨53JCS1552⟩. Alkoxyarsines are oxidized similarly with selenium dioxide (SeO_2) with the formation of alkylarsinic acids ⟨59DOK(128)299⟩. The preparation of C_5–C_{20} alkylarsinic acids in the yield of 50–80% from dichloro(diethylamino)arsine has been described, involving the reaction of Cl_2AsNEt_2 with RMgX, followed by the hydrolysis and oxidation reactions ⟨66JOM(6)17, 71JINC3177⟩.

Trivalent arsenic compounds bearing an alkyl substituent to arsenic undergo similar reactions to give either an alkylarsonic acid or a product which is readily hydrolyzed to an alkylarsonic acid. Examples shown in Scheme 19 indicate that primary alkylarsines, alkyldihaloarsines and alkyldialkoxyarsines are converted to the corresponding arsonic acids with an oxidant. Successive treatment of bis(diethylamino)chloroarsine with the Grignard reagent of 2-ethylhexylchloride, hydrochloric acid and hydrogen peroxide led to the formation of 2-ethylhexylarsonic acid in 78% yield ⟨05MI 209-01, 54JCS881, 57JOC452, 71JINC3177⟩.

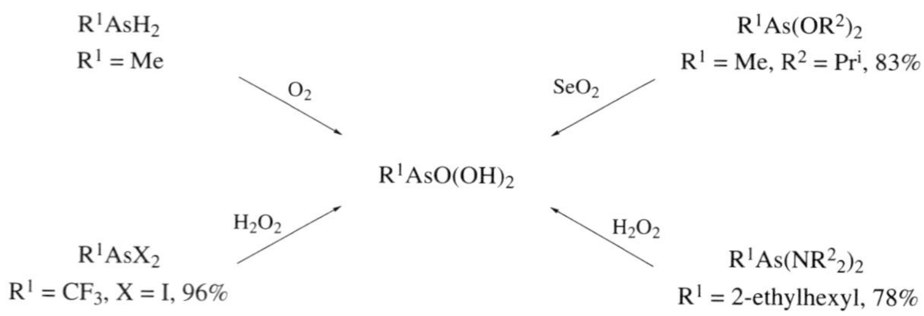

Scheme 19

2.09.4.6 Derivatives of Alkylantimony(V) Acids

2.09.4.6.1 *Oxidative hydrolysis of antimony(III) halides*

Although arylstibonic and arylstibinic acids are the well-known organoantimony compounds, fewer alkyl counterparts have been reported. Dimethylstibinic acid results from hydrolysis of diethylantimony trichloride ⟨26PRS523⟩. A useful method for dialkylstibinic acids is the oxidative hydrolysis of dialkylhalostibines: bis(1-chloroethyl)stibinic acid is obtained from the reaction of bis(1-chloroethyl)chlorostibine with aqueous hydrogen peroxide ⟨50DOK303, 52ZOB1528⟩.

2.09.5 ALKYLARSENIC COMPOUNDS WITH AN As—S, As—Se, OR As—Te BOND, AND THE ANTIMONY AND BISMUTH ANALOGUES

2.09.5.1 Bis(dialkyl)arsine Chalcogenides ($R_2AsE'AsR_2$; E′ = S, Se, Te) and their Antimony and Bismuth Analogues

2.09.5.1.1 Reaction of dialkylhaloarsines and -stibines with metal sulfides

Few bis(dialkyl)arsine sulfides are reported. Although treatment of secondary alkylarsines with sulfur results in the formation of diarsine sulfides ⟨06MI 209-01, 63CJC2424⟩, a convenient method is the reaction of dialkylhaloarsines with metal sulfides. For example, dimethyliodoarsine reacts with an aqueous solution of sodium sulfide to give bis(dimethyl)arsine sulfide (62%) as a water-insoluble liquid which has been purified by vacuum distillation under nitrogen ⟨71JA5677⟩, and $(CF_3)_2As$-$SAs(CF_3)_2$ has been obtained by interacting bis(trifluoromethyl)iodoarsine with silver sulfide at 20 °C ⟨63CJC2424⟩. Chlorodi-*t*-butylstibine undergoes an analogous reaction with sodium sulfide to give bis(di-*t*-butylstibine) sulfide ⟨81ZN(B)1105⟩.

2.09.5.1.2 Insertion reaction of tetraalkyldipnicogens by a chalcogen element

Tetraalkyldistibines and tetraalkyldibismuthines undergo ready reaction with chalcogens at room temperature with the formation of the products of the type $R_2EE'ER_2$ (E = Sb, Bi; E′ = S, Se, Te). A variety of compounds, as shown in Equation (19), have been prepared in excellent yield ⟨82ZN(B)1105, 84ZN(B)1668, 88PS(38)97⟩.

$$R_2EER_2 \xrightarrow{E'} R_2EE'ER_2 \qquad (19)$$

E = As, Sb; E' = S, Se, Te; R = Me, Pr^n

2.09.5.2 Dialkyl(organochalcogeno)arsines ($R^1_2AsE'R^2$; E′ = S, Se, Te) and their Antimony and Bismuth Analogues

2.09.5.2.1 $R^1_2EE'R^2$ (E = As, Sb, Bi; E′ = S, Se, Te) from exchange reaction between tetraalkyldipnicogens and diorganodichalcogenides

An exchange reaction between tetraalkyldipnicogens $R^1_2EER^1_2$ (E = As, Sb, Bi) and diorganodichalcogenides $R^2E'E'R^2$ (E′ = S, Se, Te) provides a general synthetic method for the compounds containing a bond between all pnicogen and chalcogen atoms (Equation (20)). The reaction proceeds rapidly and entirely at, or below, room temperature to yield clean products, quantitatively ⟨67CJC379, 79ZN(B)1646, 84ZN(B)1668, 85ZN(B)848, 86JOM(308)289, 86ZN(B)1387, 88OCS587, 88PS97⟩.

$$R^1_2EER^1_2 \quad + \quad R^2_2E'E'R^2_2 \longrightarrow R^1_2EE'R^2_2 \qquad (20)$$

R^1 = Me, Pr^n, CF$_3$; R^2 = Me, CF$_3$, Ph, 4-MePh; E = As, Sb, Bi; E' = S, Se, Te

2.09.5.2.2 $R^1_2AsSR^2$ and $R^1_2AsSeR^2$ from reactions of R^1_2AsX with R^2SH and R^2SeH derivatives

Displacement of a halogen group from dialkylhaloarsines by a sulfur-based nucleophile, usually a mercaptide, leads to the formation of As—S bonded compounds, a method popular for the preparation of alkylthioarsines. Examples are the high yield synthesis of dimethyl(ethylthio)arsine (75%) ⟨63CJC2424⟩ and bis(trifluoro)methyl(trifluoromethylthio)arsine (90%) ⟨62JCS2529⟩ through the reactions of alkylarsenic halides with bis(ethylthio)lead and bis(trifluoromethylthio)mercury. Sodium alkanethiolates RSNa have also been used in preparation. An improved procedure has been described: dimethyl(methanethio)arsine is produced in 80% yield by reacting chlorodimethylarsine and methane thiol in the presence of trimethylamine. Alkylselenoarsines $R^1_2AsSeR^2$ have been synthesized similarly by reacting dialkylhaloarsines with either a bis(alkylseleno)mercury or alkaneselenol in presence of an acid absorbent (Scheme 20) ⟨62JCS2529, 79ZN(B)1646⟩.

$$Me_2AsCl + MeEH \xrightarrow[E = S, 80\%; Se, 72\%]{} Me_2AsEMe$$

$$(CF_3)_2AsI + (CF_3Se)_2Hg \xrightarrow[100\%]{} (CF_3)_2AsSeCF_3$$

Scheme 20

2.09.5.2.3 *$R^1_2AsSR^2$ and $R^1_2AsSeR^2$ from reactions of aminoarsines with RSH and RSeH*

Aminoarsines are valuable precursors of organoarsenic compounds containing As—S and As—Se bonds. Dialkyl(diethylamino)arsines react readily with thiols (a three-fold molar excess of the thiols is used in this case) under reflux to produce dialkyl(alkanethio)arsines in high yields ⟨72JOM(39)301, 73ZAAC(401)243, 75ZAAC(412)202⟩. Organoselenoarsines have been prepared successfully when aminoarsines react with stoichiometric amounts of a selenol at 50 °C (Scheme 21) ⟨72JOM(39)301⟩.

$$R^1_2AsNEt_2 + R^2SH \longrightarrow R^1_2AsSR^2$$
R^1 = Et, Pr^n; R^2 = Me, Et, Pr^n

$$R^1_2AsNEt_2 + R^2SeH \xrightarrow[40\text{–}83\%]{} R^1_2AsSeR^2$$
R^1 = Me, Et, Pr^n; R^2 = Et, Bz, Ph, 1-naphthyl

Scheme 21

2.09.5.3 Alkylthioarsonous Acid Anhydrides $(RAsS)_n$ and their Antimony Analogues $(RSbS)_n$

Methylarsine sulfides $(MeAsS)_n$ are formed as oligomeric mixtures from the reactions of methylarsine with hydrogen sulfide ⟨05MI 209-01, 63JCS(C)798⟩ and of methylarsaoxane $(MeAs)_n$ with sulfur ⟨90IC798⟩. A method for preparation of cyclic methylarsine sulfide has been established; $(MeAsS)_n$ (n = 3,4) has been obtained in 53% yield through a metal-mediated reaction of *cyclo*-$(MeAs)_5$ (Equation (21)) ⟨90IC798⟩. *t*-Butylstibine sulfide $(Bu^tSbS)_n$ (n = 2, 3, 4, 5), the first example of thioanhydride of *t*-butylstibonous acid, has been obtained by the reaction of *t*-butyldichlorolstibine with sodium sulfide ⟨81ZN(B)1105⟩.

$$cyclo\text{-}(MeAs)_5 + S_8 \xrightarrow[53\%]{Mo(CO)_6/\text{toluene, } 125\,°C} cyclo\text{-}(MeAsS)_n \;\; (n = 3, 4) \qquad (21)$$

2.09.5.4 Alkyldiorganochalcogenoarsines ($R^1As(E'R^2)_2$; E′ = S, Se, Te) and their Antimony and Bismuth Analogues

2.09.5.4.1 *From alkyldihaloarsines, -stibines and -bismuthines*

Alkyldihaloarsines, -stibines, and -bismuthines react with thiols and selenols or their salts to produce the corresponding dithio- and diselenoarsine, -stibine, and -bismuthine compounds ⟨57DOK(115)305, 57IZV69, 73ZAAC(402)43, 79ZAAC(448)89, 87POL1593⟩. Examples are the synthesis of heterocyclic compounds (**46**) and (**47**) from the reactions of dithiols with dibromo(methyl)stibine ⟨73ZAAC(402)43⟩, and of dilithium *o*-phenylenediselenide with dibromo(methyl)bismuthine (Scheme 22) ⟨87POL1593⟩.

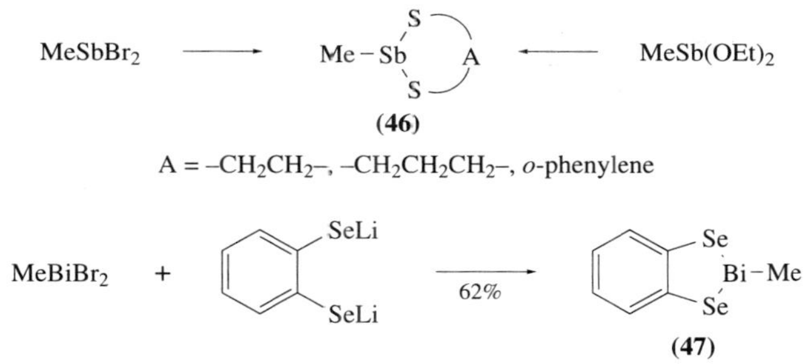

Scheme 22

2.09.5.4.2 *From diamino- and dialkoxy-arsines, -stibines, and -bismuthines*

A better method for alkyldi(alkylthio)arsines is the displacement reaction of diaminoarsines with thiols. For example, cyclic compound (**48**) is obtained from bis(dimethylamino)methylarsine and ethylenedithiol with the elimination of dimethylamine (Scheme 23) ⟨73ZAAC(397)97⟩. Alkanethio–alkoxy and alkanethio–alkanethio exchange reactions have also been applied synthetically ⟨73JGU825, 76ZAAC(423)40⟩. One example is the reaction of diethoxy(methyl)stibine with *o*-phenylenedithiol giving methyl(benzene-1,2-dithio)stibine in 95% yield ⟨74ZAAC(408)261⟩.

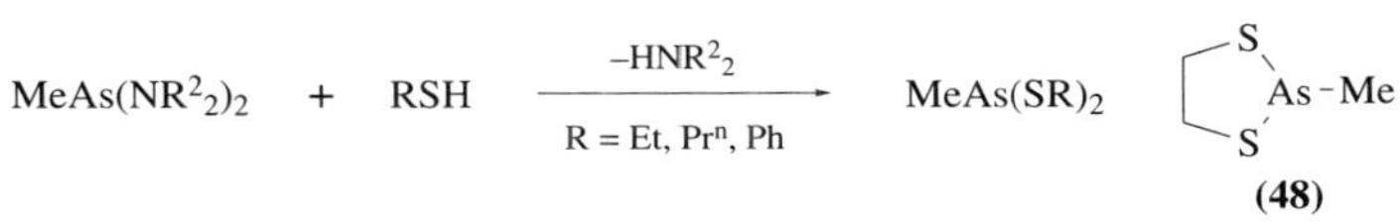

Scheme 23

2.09.5.4.3 *Other methods*

The redistribution reaction of organochalcogeno(dialkyl)stibines takes place at elevated temperatures forming dichalcogenostibine and trialkylstibine compounds. The equilibrium moves to the side of dichalcogenostibine by removing trialkylstibine from the reaction mixture through distillation (Scheme 24). Tellerium analogues have not been successfully prepared by this method, since a mixture of several products are formed when an organotellurodialkylstibine is heated ⟨88PS(38)97, 88ZN(B)998⟩.

$$2R^1{}_2SbER^2 \longrightarrow R^1Sb(ER^2)_2 + R^1{}_3Sb$$

R^1 = Me, R^2 = Ph, E = S, 84%; R^1 = Et, R^2 = Ph, 52%
R^1 = Me, R^2 = Ph, E = Se, 78%; R^1 = Et, R^2 = Ph, E = Se, 48%; R^1 = R^2 = Me, E = Se, 74%

Scheme 24

Dibenzenethiostibines $RBi(SPh)_2$ (R = Me, 92%; Et, 85%) have been obtained from the reaction of benzenethiol and tertiary alkylbismuthines in which two alkyl groups are displaced by mercaptan substituents ⟨70JCS(B)735, 81ZN(B)70⟩.

2.09.5.5 Tertiary Alkylarsine Sulfides and Selenides -(R_3AsS, R_3AsSe)- and their Antimony Analogues

2.09.5.5.1 Reaction of tertiary alkylarsines and -stibines with chalcogens

The reaction of trialkylarsines with sulfur and selenium, which has been used widely, is a convenient and efficient synthetic method to the tertiary alkylarsine sulfides and selenides. A range of trialkylarsines, both symmetric and asymmetric, has been transformed into the corresponding sulfides and selenides by refluxing arsines with an equivalent or slight excess of sulfur and selenium in alcohol ⟨31JCS185, 63JCS(C)798, 64IC580, 66IC187, 80ZAAC(466)183⟩. Trialkylstibine sulfides and selenides are prepared analogously ⟨64JOM(1)369, 65ZAAC(339)289⟩. An improved procedure, which gives pure products in good yield, is to shake reactants in the absence of any solvent, heat or light, and freeze out the liquid or solid products from ether or petroleum ether ⟨70JOM(22)637⟩.

2.09.5.5.2 Other methods

Several methods are available for the preparation of tertiary alkylstibine sulfides. For example, treatment of trimethylstibine oxide with H_2S in methanol gives trimethylstibine sulfide. Triethyl- and tricyclohexylantimony dibromides are converted into sulfides when they are reacted with hydrated sodium sulfide in methanol ⟨68JOM(11)299⟩.

2.09.5.6 Other Alkylantimony(V) Derivatives with Sb—S and Sb—Se Bonds

In the presence of triethylamine, trimethylantimony dichloride reacts with mercaptanes at low temperature to afford dialkanethiotrimethylantimony $Me_3Sb(SR^2)_2$ (Me, 74%; Et, 78%; Bz, 79%; Ph, 79%) ⟨71CB1842⟩. $Me_3Sb(SCOR)_2$ derivatives, which are thermally labile, are obtained (R = Me, 71%; R = Ph, 92%) from the reaction of diethoxytrimethylantimony with thioacetic acid and thiobenzoic acid ⟨67MI 209-02⟩.

When refluxed with excess selenium in benzene, trimethylstibine is transformed into a cyclic diselenide (**49**) (Equation (22)) ⟨64JOM(1)369⟩.

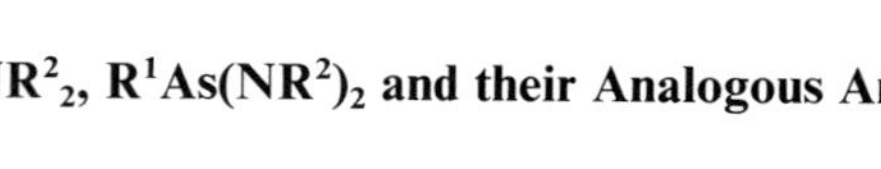

(49)

(22)

2.09.6 ALKYLARSENIC COMPOUNDS WITH AN As—N BOND AND THE CORRESPONDING Sb—N AND Bi—N ANALOGUES

2.09.6.1 Aminoarsines $R^1_2AsNR^2_2$, $R^1As(NR^2)_2$ and their Analogous Antimony and Bismuth Compounds

Aminoarsines, being important intermediates, have been studied extensively, and a review regarding their preparation and reactions has appeared ⟨82S173⟩. Aminostibines and aminobismuthines, however, have been investigated less. A general synthetic method for aminoarsines, -stibines and -bismuthines is the reaction of amines or metal amides with the corresponding alkylarsenic, -antimony and -bismuth halides.

2.09.6.1.1 Reaction of alkylarsenic, -antimony and -bismuth halides with ammonia, amines and metal amides

Aminoarsines of the type R_2AsNH_2 and $(R_2As)_2NH$ are known when their alkyl group R is sterically bulky or is electronegative. The reaction between chlorodi(*t*-butyl)arsine and ammonia gives aminodi(*t*-butyl)arsine in 75% yield. Further reaction of aminoarsine with further organoarsenic chloride, with the aid of *n*-butyllithium, produces bis(di-*t*-butylarsino)amine (**50**) (Scheme

25) ⟨69JOM(16)P69⟩. Ammonolysis of bis(trifluoro)methylchloroarsine in the gas phase leads to the formation of both aminobis(trifluoro)methylarsine and bis(bistrifluoromethylarsino)amine; the latter compound is obtained as the sole product when the reaction is carried out in liquid ammonia ⟨59JCS372⟩.

$$Bu^t_2AsCl \xrightarrow[75\%]{NH_3,\ -NH_4Cl} Bu^t_2AsNH_2 \xrightarrow{BuLi,\ -BuH} Bu^t_2AsNHLi \xrightarrow[65\%]{Bu^t_2AsCl,\ -LiCl} \underset{\textbf{(50)}}{Bu^t_2AsNHAsBu^t_2}$$

Scheme 25

Reaction between alkylarsenic halides and amines is the most convenient and efficient method for preparing aminoarsine compounds. A range of aminoarsines has been obtained according to the reactions depicted in Scheme 26. Tetra-*N*-alkyl-2,4,6,8-tetraaza-1,3,5,7-tetraarsaadamantanes (**51**) (R = Me, Et, Pr^n, Pr^i, Bu^n, Bu^i) have been synthesized by the interaction of primary amines with $H_2C(AsCl_2)_2$ ⟨75JOM(94)393⟩.

$$R^1{}_2AsX + R^2{}_2NH \longrightarrow R^1{}_2AsNR^2{}_2$$

$$R^1{}_2AsX + R^2NH_2 \longrightarrow (R^1{}_2As)_2NR^2$$

$$R^1AsX_2 + R^2{}_2NH \longrightarrow R^1As(NR^2{}_2)_2$$

R^1, R^2 = Me, Et, Pr^n, Pr^i, Bu^n, Bu^i, Bu^t

Scheme 26

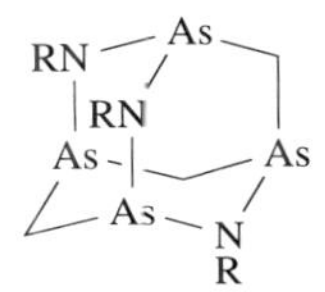

(51)

Alkali metal amides such as lithium methyltrimethylsilylamide have been used to react with dimethylhalo- and methyldihaloarsines, -stibines, and -bismuthines, giving the corresponding aminoarsines, -stibines, and -bismuthines in moderate yield (Scheme 27) ⟨66JOM(6)259⟩.

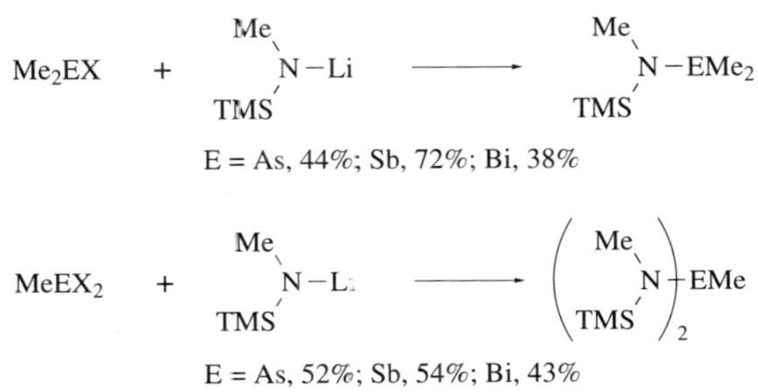

Scheme 27

2.09.6.1.2 *Reaction of R_2NEX_2, $(R_2N)_2EX$ (E = As, Sb) with organometallics*

Aminoarsenic halides, which have been readily prepared from amines and arsenic trihalides, undergo reaction with organometallic reagents to yield aminoarsine products. For example, alkylation of dichloro(diethylamino)arsine with Grignard reagents results in the formation of dialkyl(dimethylamino)arsines R_2AsNEt_2 (R = Et, 51%; Pr^n, 60%; Bu^n, 57%) ⟨64ZAAC(326)280, 69JOM(19)367⟩. The cyclic aminostibine (**53**; R = Me, Bu^t) has been prepared by reacting chloride (**52**) with organolithium reagents (Equation (23)) ⟨90CB975⟩.

(52) **(53)** (23)

It may be noted that aminoarsines are important precursors of other organoarsenic compounds, and in many cases these intermediates are formed *in situ* and used directly for the next synthesis. Examples include the preparation of organoarsinic and organoarsonic acids and organoarsenic halides, in which aminoarsines are formed from the reaction of aminoarsenic halides with Grignard reagents ⟨66JOM(6)17, 68JOM(12)377, 77SRI195, 90POL319⟩.

2.09.6.1.3 *Other methods*

An amino-exchange reaction of aminoarsines with other amine substrates has been used to prepare anilino and *cyclo*hexylamino substituted alkylarsines from dimethyl(dimethylamino)arsine (Scheme 28) ⟨73ZAAC(401)243⟩. It has been reported that bis(dimethylamino)trifluoromethylarsine is obtained (70%) by heating trifluoromethyl iodide and excess tris(dimethylamino)arsine in a sealed ampoule at 145 °C ⟨68JINC1715⟩.

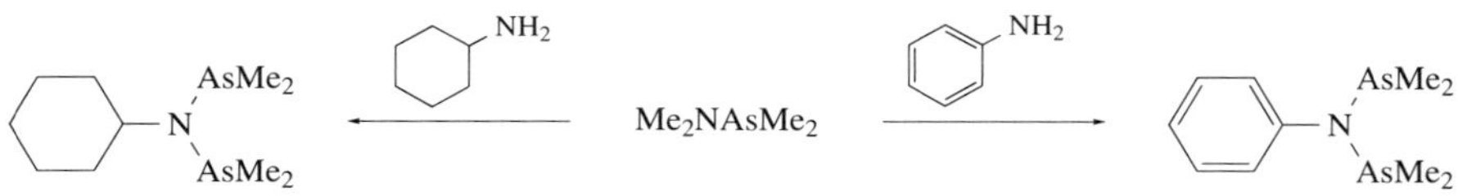

Scheme 28

2.09.6.2 Other Compounds with As—N, Sb—N and Bi—N Bonds

Azidoarsines and azidostibines have been synthesized by reacting alkylhaloarsines and alkylhalostibines with a metal azide. For example, treatment of dimethyliodoarsine and bromodiethylstibine with silver azide yields azidodimethylarsine Me_2AsN_3 (90%) ⟨71ZAAC(381)103, 72JCS(D)847⟩ and Et_2SbN_3 (65%) ⟨85ZN(B)1320⟩, the compound being easily converted into arsinophosphazene derivatives through a Staudinger reaction with triorganophosphine (**54**) (Scheme 29) ⟨68CB1271, 93JCS(D)663⟩. Azidodimethylbismuthine, however, is prepared from the reaction of trimethylbismuthine with hydrazoic acid ⟨71ZAAC(381)103⟩.

$$R^1{}_2AsX \xrightarrow{AgN_3} R^1{}_2AsN_3 \xrightarrow{PR^2{}_3} R^1{}_2AsN{=}PR^2$$

(54)

$R^1 = R^2 = Me$; $R^1 = CF_3$, $R^2 = Ph$

Scheme 29

2.09.6.3 Arsenic(V) and Antimony(V) Compounds with As—N, Sb—N Bonds

Preparation of iminotriorganoarsoranes has been reported: treatment of aminotrialkylarsonium halides, which are formed either from ammonolysis of trialkylarsenic dihalides or from reaction of trialkylarsines and chloroamine, with potassium amide in liquid ammonia gives iminotrialkylarsoranes $R_3As{=}NH$ (R = Me, 83%; Bu^t, 75%; Pr^i, 92%; C_6H_{11}, 85%) ⟨80CB2928⟩.

Cleavage of a carbon–antimony bond of pentamethylstiborane by hydrazoic acid leads to the formation of azidotetramethylstiborane in high yield ⟨71ZAAC(386)139⟩. Trimethylantimony pseudohalides of the type Me_3SbX_2 ($X = N_3$, NCO and NCS) have been prepared by the metathetical reaction between trimethylantimony dichloride and the corresponding nitrogen-containing reagents ⟨68CB3976, 71MI 209-01⟩.

2.09.7 ALKYLARSENIC, -ANTIMONY, OR -BISMUTH COMPOUNDS WITH THE HETEROATOM BONDED TO P, As, Sb OR Bi

2.09.7.1 Tetraalkyldiarsines, -distibines, and -dibismuthines–R_2EER_2 (E = As, Sb, Bi)

Tetraorganodiarsines, -distibines, and -dibismuthines are well-known species. Tetramethyldiarsine was obtained in 1760 ⟨1760MI 209-01, 27CR(185)779⟩, and the first report of tetramethyldibismuthine may date back to 1935 ⟨35JCS366⟩. Increasing interest in these compounds is largely due to their thermochromic characteristics. A review of thermochromic distibines and dibismuthines has appeared ⟨90AOC77⟩. The most important preparative methods are the reduction reaction of halide compounds and coupling reaction of metal derivatives of the arsenic, antimony and bismuth.

2.09.7.1.1 Reduction of R_2EX (E = As, Sb, Bi) derivatives

Reduction of dialkylhaloarsines and their antimony and bismuth analogs, one of the most frequently used methods, provides a straightforward route to the diarsine, distibine and dibismuthine compounds. Reactive metals employed in the reaction include lithium, sodium, magnesium, mercury and zinc. Examples are listed in Table 3. It should be noted that reaction can lead to different types of products, depending on the particular metal used and, in some cases, on the nature of the solvent. For example, when dimethyliodoarsine reacts with other alkali metals such as sodium and potassium in THF, only sodium and potassium arsenides are formed ⟨67CJC675⟩.

Table 3 Diarsines, distibines, and dibismuthines from reduction of the halides.

Reactant	*Reagents*	*Product*	*Yield* (%)	*Ref.*
Me_2AsI	Li/Et_2O or THF	$Me_2AsAsMe_2$	80	67JCS675
$(CF_3)_2AsCl$	Hg	$(CF_3)_2AsAs(CF_3)_2$	99	52JCS2552, 63CJC322
Me_2SbBr	Mg/THF	$Me_2SbSbMe_2$	85	86OCS625
Et_2SbBr	Mg/THF	$Et_2SbSbEt_2$	64	77CZ399
$(C_6H_{11})_2SbCl$	Na/Toluene	$(C_6H_{11})_2SbSb(C_6H_{11})_2$	96	64ZAAC(332)179
$(CF_3)_2SbI$	Hg or Zn	$(CF_3)_2SbSb(CF_3)_2$	quant.	57JCS3708
Me_2BiBr	Na/liquid NH_3	$Me_2BiBiMe_2$	78	84ZN(B)887

2.09.7.1.2 From arsenic, antimony, and bismuth metal derivatives

As mentioned in Section 2.09.2.2.5, the reaction of metal dialkylstibides with 1,2-dihaloethanes does not afford expected alkylation products, and distibines are formed instead. This reaction has been developed as another general synthetic route; a variety of tetraalkyldistibines (e.g., R = Me, 79% Et, 88% Pr^n, 69% cyclohexyl 78% along with 1,1′-bistibolane, 82% and 1,1′-biantimonane, 79%) ⟨65ZAAC(339)289, 65ZAAC(339)298, 73JOM(51)223, 77JOM(136)173⟩ and tetraalkyldibismuthines (e.g., R = Me, 70%; Et, 63% Pr^n, 75% Pr^i, 53% Bu^n, 81% and 1,1-bibismolane, 58%) ⟨82AG(E)439, 82OM1408, 83OM1859, 83ZN(B)125, 86OCS636⟩ have been prepared by this method. The reaction proceeds through pathways involving halogen–metal exchange and E—E (E = As, Sb, Bi) coupling reactions (Scheme 30). In most cases, preparation of distibines is accomplished by reacting an alkali metal dialkylstibide (formed *in situ* by metallation of dialkylstibine or dialkylhalostibine) with 1,2-dichloro- or 1,2-dibromoethane. Dibismuthine compounds are prepared starting from trialkylbismuthines in liquid ammonia on successive treatment with sodium and 1,2-dihaloethane. Ammonium chloride is added in some cases before addition of 1,2-dihaloethane in order to destroy alkyl sodium (Equation (24)).

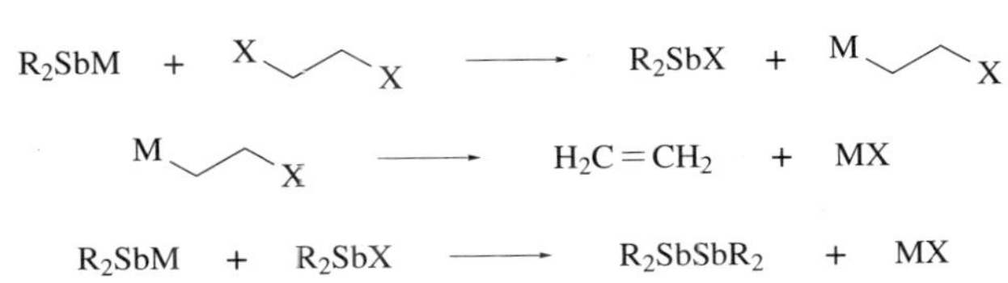

Scheme 30

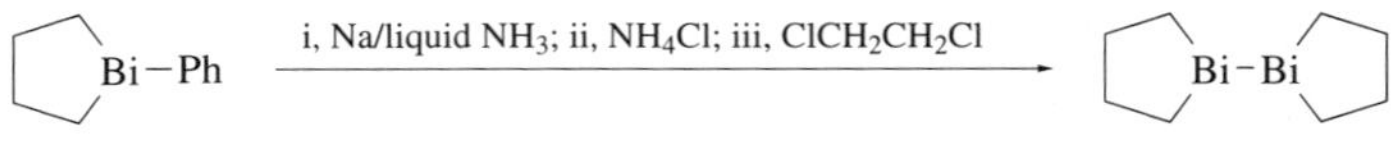

(24)

Reagents other than 1,2-dihaloethane have been used in preparation. For example, lithium dicyclohexylarsenide is converted into tetracyclohexyldiarsine when treated with iodine ⟨61AG26⟩ and tetra-*t*-butyldistibine is obtained (96%) when lithium di-*t*-butylstibide dioxane complex reacts with 1,4-dichlorobutane in dioxane at 65 °C ⟨65ZAAC(339)298⟩. In addition, ammonium bromide has also been reported to accomplish the reaction ⟨65ZAAC(339)298, 77CZ399⟩.

2.09.7.1.3 *Miscellaneous methods*

(i) Reaction of alkylarsinic acid derivatives

Dimethylarsinic acid undergoes reduction, an old preparative method, with an excess of hypophosphorous acid in hydrochloric acid solution to give tetramethyldiarsine ⟨06CR(142)1151⟩. The same compound has also been obtained by electrolytic reduction ⟨16CB239⟩.

(ii) Cadet reaction

The oldest method for preparation of this class compounds is the Cadet reaction (see Section 2.09.4.1.1). Tetramethyldiarsine is the major component (56%) of "Cadet's fuming arsenical liquid," formed by the pyrolysis of a mixture of arsenic oxide and potassium acetate ⟨1760MI 209-01, 27CR(185)779⟩. An improved procedure for preparing diarsine and diarsine oxide is to pass a mixture of arsenic oxide and acetic acid vapor over an alkali metal acetate catalyst at 300–400 °C. Tetraethyldiarsine has been prepared by using propionic acid in place of acetic acid ⟨47JA559⟩.

(iii) From secondary alkylarsines and polyarsines

Condensation of a dialkylarsine and a dialkylhaloarsine yields the diarsine, an example being the preparation of tetrakis(trifluoromethyl)diarsine by heating $(CF_3)_2AsCl$ and $(CF_3)_2AsH$ at 100 °C ⟨63CJC322⟩. It has been reported that tetramethyldiarsine is obtained (96%) directly from dimethylarsine when it is treated with small amount of chloroamine ⟨69IC1032⟩.

Cyclic polyarsine adds across the diene to give the product (**55**), an interesting heterocycle having an As—As bond (Equation (25)) ⟨68CB1381⟩.

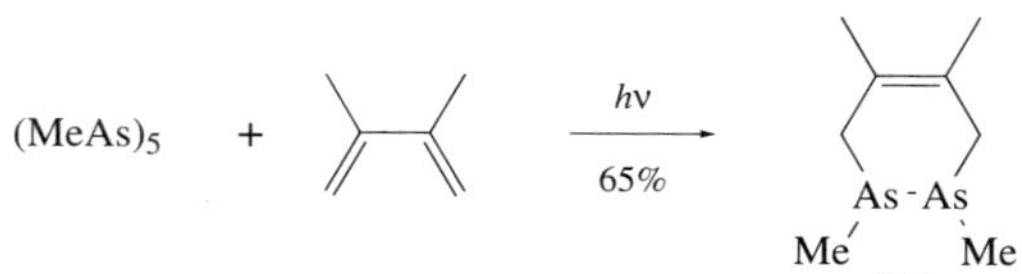

(25)

(55)

2.09.7.2 Polyalkylarsines ($(RAs)_n$) and Polyalkylstibines ($(RSb)_n$)

Polyarsines exist in both polymeric and oligomeric forms, and much attention has been given to the cyclic polyarsines mainly because of their novel structure. There is a variety of preparative methods for polyarsines. A literature survey on cyclopolyarsines up to mid-1974 is available ⟨75JOM(84)1⟩. In contrast to polyalkylarsines, only a few examples of polyalkylstibines are known, which are formed as by-products in certain reactions.

2.09.7.2.1 *Reduction of $RAsX_2$ and $RSbX_2$ derivatives*

Reduction of alkyldihaloarsines is one widely employed method for preparing polyalkylarsines. Active metals such as sodium, magnesium, and mercury have been used as the reducing agents. A few cyclic polyalkylarsine tetramers, and pentamers are obtained in moderate yield and some examples are $(Bu^nAs)_5$, (67%); $(Bu^iAs)_5$, (64%), and $(Bu^tAs)_4$, (54%) ⟨71JPR259, 82ZAAC(485)129, 66JA3178⟩.

Reduction of *t*-butyldichlorostibine by magnesium in THF, the most successful example of preparation of polystibine, gives tetra-*t*-butylcyclotetrastibine as the major product (20%), a folded four-membered ring compound with the *t*-butyl groups in the pseudoequatorial all-*trans* configuration ⟨78CZ263, 82ZAAC(486)70, 83ZAAC(502)175, 86OCS621⟩.

When $MeC(CH_2AsI_2)_3$ and $MeC(CH_2SbI_2)_3$ are reacted with sodium in THF, the polyarsine (**56**) ⟨78AG(E)135⟩ and the polystibine (**57**) ⟨82AG(E)375, 85JOM(290)307⟩ with all *cis*-configuration, are obtained in good yield (Equation (26)).

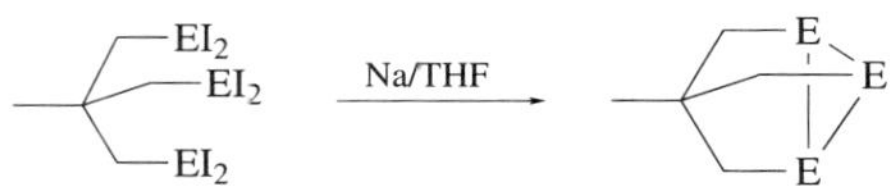

(26)

(**56**), E = As
(**57**), E = Sb

2.09.7.2.2 *Reduction of alkylarsonic acid derivatives*

Alkylarsonic acids or their salts undergo reduction with hypophosphorous acid providing another valuable route to polyalkylarsines. Both cyclic and linear polyalkylarsines are formed when methylarsonic acid or its sodium salt is reduced by this method ⟨04CR(138)1705, 28JA536, 45JA2014, 69JA4986, 86OCS618⟩. Other cyclic polyalkylarsines such as $(EtAs)_5$, $(Pr^nAs)_5$ and pentakis(chloromethyl)cyclopentaarsine $(ClCH_2As)_5$, a more stable polyarsine compound, have been synthesized through hypophosphorous acid reduction of the corresponding alkylarsonic acids or their salts ⟨28CB1906, 70AJC1559, 75JOM(102)445⟩.

2.09.7.2.3 *Oxidation of primary alkylarsines*

Reaction of primary alkylarsines with dibenzylmercury produces quantitatively a cyclic pentamer of polyalkylarsine of high purity together with metallic mercury. Examples are the preparation of methyl, ethyl and 2,2,2-trideuterioethyl polyarsines ⟨77JOM(128)155, 82JOM(231)119, 86OCS618⟩.

2.09.7.2.4 *Miscellaneous methods*

In addition to the oxidation reaction discussed above, primary alkylarsines have been used frequently as building blocks in the synthesis of polyalkylarsines. Examples listed in Scheme 31 demonstrate that alkylarsines react with alkydihaloarsine, alkylarsaoxane and pentamethyl-*cyclo*-pentaarsine to give the desired products ⟨08MI 209-01, 71JPR259, 87IC1638⟩. Synthesis of tetrameric poly-*t*-butylarsine (46%) has been accomplished by the coupling reaction between dilithium *t*-butylarsenide and *t*-butyldichloroarsine ⟨82ZAAC(486)70⟩.

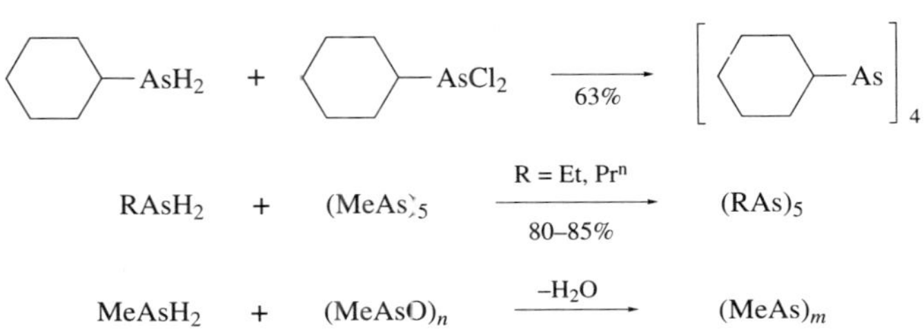

Scheme 31

Arsenide derivative (**58**) undergoes direct [2 + 1]-*cyclo*-condensation with *t*-butyldichloroarsine at low temperature to give a cyclic triarsine product (**59**) (Equation (27)) ⟨81AG(E)123⟩.

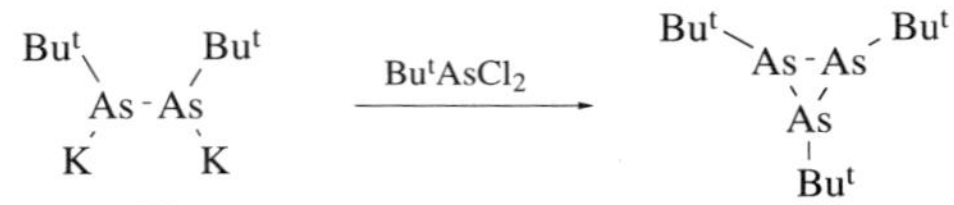

(58) **(59)** (27)

2.09.7.3 Mixed Tetraalkyl Dipnicogens – R_2AsSbR_2, R_2SbBiR_2, etc.

2.09.7.3.1 Exchange reaction

Mixed tetraalkyl dipnicogens are formed rapidly in equilibrium with reactant tetraalkyl dipnicogens, both in solution and as neat liquids at ambient temperature. The exchange reaction is limited to the compounds in which different pnicogens are near neighbors in the Periodic Table. However, isolation of mixed tetramethyldipnicogens from this reaction is difficult, owing to the mobility of the equilibrium between reactants and products (Equation (28)) ⟨86JOM(303)197⟩.

$$\text{MeE}-\text{EMe} \quad + \quad \text{MeE'}-\text{E'Me} \rightleftharpoons 2\text{MeE}-\text{E'Me} \qquad (28)$$

$$\text{E} = \text{P, E'} = \text{As; E} = \text{As, E'} = \text{Sb; E} = \text{Sb, E'} = \text{Bi}$$

Exchange reactions between dimethylarsine and tetrakis(trifluoro)diphosphine, and between tetrakis(trifluoromethyl)diarsine and dimethylphosphine, produce corresponding phosphinoarsine compounds $R^1{}_2AsPR^2{}_2$ ($R^1 = Me$, $R^2 = CF_3$; $R^1 = CF_3$, $R^2 = Me$) in excellent yields ⟨68JCS(A)1406⟩.

2.09.7.3.2 Miscellaneous methods

Reaction of tetramethyldiarsine with *cyclo*-tetraphosphine $(CF_3P)_4$ yields the product Me_2As $P(CF_3)AsMe_2$ almost quantitatively with a CF_3P moiety inserting into the As—As bond ⟨69JA6609⟩. A relevant compound containing As—P bonds, benzodiphosphaarsole (**60**), has been synthesized from the reaction of dichloro(methyl)arsine with *o*-phenylenebis(lithiophenyl)phosphine (Scheme 32) ⟨72JCS(P1)2548⟩.

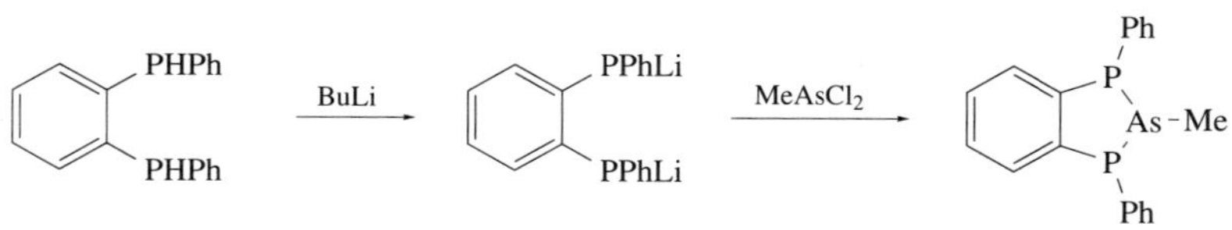

(60)

Scheme 32

2.09.8 ALKYLARSENIC, -ANTIMONY, OR -BISMUTH COMPOUNDS WITH THE HETEROATOM BONDED TO A METALLOID

2.09.8.1 Compounds with the Heteroatom Bonded to Silicon

Alkylarsinosilanes and alkylstibosilanes are mainly prepared by reactions of an arsenide and a stibide with a chlorosilane ⟨64JOC986, 66AG(E)418, 68JCS(A)2623, 80ZAAC(462)113, 86OM1266⟩. For example, interesting ligands $Me_2AsSi(Me_2CH_2CH_2EMe_2$ are obtained (E = N, 19%; P, 75%; As, 85%) from the reaction of $LiAsMe_2$, derived *in situ* from butyllithium and dimethylarsine, and ClSi-$Me_2CH_2CH_2EMe_2$ ⟨92ZAAC(608)43⟩. A simpler procedure involves reaction of organoantimony chlorides, magnesium and chlorotrimethylsilane in THF giving stibinosilane products in acceptable yield (Scheme 33). It should be noted that the use of *t*-butylantimony bromides under same conditions leads to the exclusive formation of the Sb—Sb bond ⟨82ZN(B)395⟩.

$$\mathrm{Bu^t{}_2SbCl} + \mathrm{TMS\text{-}Cl} + \mathrm{Mg} \xrightarrow[54\%]{} \mathrm{Bu^t{}_2Sb\text{-}TMS}$$

$$\mathrm{Bu^tSbCl_2} + \mathrm{TMS\text{-}Cl} + \mathrm{Mg} \xrightarrow[31\%]{} \mathrm{Bu^tSb(TMS)_2}$$

Scheme 33

Alkylation of metal derivatives of silyl-substituted arsines and stibines provides another useful approach to arsinosilanes and stibosilanes ⟨80ZAAC(462)113, 89ZAAC(569)62⟩. Lithium bis(trimethylsilyl)arsine complex $(TMS)_2AsLi \cdot 2THF$ prepared from tris(trimethylsilyl)arsine and methyllithium in THF undergoes alkylation with a range of alkyl halides to produce good yields of alkylbis(trimethylsilyl)arsines $(TMS)_2AsR$ (R = Me, 91%; Et, 88%; Pr^i, 81%; Bz, 82%; Ph_2CH, 58%). The reaction is limited to primary and secondary alkyl halides, as only a β-elimination product, bis(trimethylsilyl)arsine, results when *t*-butyl chloride is used ⟨80ZAAC(462)113⟩.

A metal silane derivative $Mg(TMS)_2 \cdot DMF$ has been applied in the preparation of arsinosilanes. Di-*t*-butylchloroarsine and *t*-butyldichloroarsine are converted into the corresponding arsinosilanes ⟨89JOM(373)307⟩.

2.09.8.2 Compounds with the Heteroatom Bonded to Germanium

Dimethylarsinotrimethylgermane has been produced (91%) from lithium dimethylarsenide and chlorotrimethylgermane ⟨68JCS(A)2623⟩. Metathetical reactions between chlorotrimethylgermane and arsinosilanes lead to the formation of arsinogermanes (Scheme 34) ⟨89JOM(373)307⟩.

$$\mathrm{Bu^t{}_2As\text{-}TMS} + \mathrm{Me_3GeCl} \xrightarrow[78\%]{} \mathrm{Bu^t{}_2AsGeMe_3}$$

$$\mathrm{Bu^tAs(TMS)_2} + \mathrm{Me_3GeCl} \xrightarrow[67\%]{} \mathrm{Bu^tAs(GeMe_3)_2}$$

Scheme 34

Preparation of organobismuthinogermanes have been reported from the reaction of triethylbismuthine and tris(pentafluorophenyl)germane, in which the carbon–bismuth bond is cleaved ⟨78JOM(162)289⟩.

2.09.8.3 Compounds with the Heteroatom Bonded to Boron

Reaction of secondary methylarsine and diborane forms an adduct which is converted into dialkylarsinoborane (80%) on heating ⟨54JA386⟩. Bis(trifluoromethyl)arsinoborane is obtained (70%) conveniently by heating the reactants below 50–60 °C ⟨67JA1040⟩. Both compounds exist in the form of a trimer $(R_2AsBH_2)_3$ (R = Me, 80%; CF_3, 70%).

It has been reported that lithium diethylarsenide reacts with bis(dimethylamino)chloroborane to give the dimethylarsinoborane $Et_2AsB(NMe_2)_2$ in 71% yield ⟨72CB1962⟩.

2.09.9 ALKYLARSENIC, -ANTIMONY, OR -BISMUTH COMPOUNDS WITH THE HETEROATOM BONDED TO A METAL

2.09.9.1 Alkali Metal Derivatives of Arsenic, Antimony, and Bismuth Compounds

Alkali metal arsenides, stibides and bismuthides are very important intermediates for the synthesis of organoarsenic, -antimony and -bismuth compounds. Much more effort has been focused on the alkylarsenide and alkylstibide compounds, and synthetic methods for them are well documented. The review by Doak and Freedman ⟨74S328⟩ provides a valuable survey of the literature up to the early 1970s which contains experimental details of preparations and reactions. No new general synthetic methods have been reported since then. Reactive alkali metal arsenides and antimony analogues can be isolated, and in most cases, however, they are generated *in situ*, and used for further reaction without isolation.

2.09.9.1.1 From primary and secondary arsines and stibines

Secondary alkylarsines and -stibines are readily converted into their lithium salts in almost quantitative yield when they are treated with organolithium such as methyl-, *n*-butyl-, and phenyllithium in petroleum ether or ether, for example, the reaction of dicyclohexylarsine with phenyllithium in ether to give lithium dicyclohexylarsenide (92%) ⟨61AG26, 65CB977⟩. Other arsenides R_2AsLi (R = Me, Et, Pr^n, Bu^n and TMS-CH_2) ⟨64ZAAC(332)179, 70ZC118, 87JOM(325)-C7⟩ and stibides R_2SbLi in which R is ethyl, *t*-butyl and cyclohexyl ⟨64ZAAC(332)179, 65ZAAC(339)289, 65ZAAC(339)298⟩ have been prepared similarly. Dilithium alkylarsenide, *n*-$BuAsLi_2$, has been formed (75%) in petroleum ether starting from primary alkylarsine *t*-$BuAsH_2$ ⟨67LA(709)248⟩.

Alternatively, alkylarsines have been reported to undergo direct metallation with alkali metals in liquid ammonia or THF providing alkali metal alkylarsenides such as sodium and potassium methylarsenides (MAsHMe; M = Na, K) ⟨83ZAAC(501)89⟩, and sodium and caesium bis(trifluoro) methylarsines ($MAs(CF_3)_2$; M = Na, Ce) ⟨89IC1627⟩.

2.09.9.1.2 From dialkylhaloarsines and -stibines

Reaction of dialkylhaloarsines and dialkylhalostibines with an alkali metal provides a useful route to their metal derivatives, but in no case has the intermediate alkylarsenide and alkylstibide been isolated. Formation of dialkylarsenides and -stibides is effected by reacting sodium or potassium and dialkylhaloarsines and -stibines in liquid ammonia or THF ⟨59JA1, 64JOC986, 67CJC675, 69MI 209-01⟩.

2.09.9.1.3 From tertiary alkylarsines, -stibines, and -bismuthines

Although the cleavage of triarylarsines and triarylstibines with alkali metals are well-known methods for preparation of diarylarsenides and -stibides, few examples have been reported concerning the preparation of corresponding dialkyl analogs by the same reaction. A study conducted by Meinema and co-workers has shown that trialkylstibines are readily converted into sodium dialkylstibides on treatment with sodium in liquid ammonia. The rate of antimony–alkyl cleavage decreases in the order Me > Et > Pr^n > Bu^n. This has been applied successfully in the preparation of trialkylstibine with three different alkyl groups through a consecutive metallation and alkylation reaction sequence illustrated in Scheme 7 ⟨73JOM(52)223⟩.

Sodium dialkylbismuthides are readily formed from the interaction of tertiary alkylbismuthines with sodium in liquid ammonia, which has been applied in the synthesis of tetraalkyldibismuthines (see Section 2.09.7.1.2).

2.09.9.1.4 Miscellaneous methods

There are some other preparative methods with less generality. Cleavage of the As—As bond tetramethyldiarsine by sodium and butyllithium leads to sodium and lithium dimethylarsenide, respectively ⟨67CJC675⟩. The former compound has also been obtained from bisdimethylarsine sulfide and sodium ⟨70ZAAC(377)278⟩. Tetraethyldistibine undergoes reaction with phenyllithium to produce lithium diethylstibide in high yield ⟨65ZAAC(339)289⟩.

2.09.9.2 Other Main Group Metal Derivatives of Arsenic, Antimony, and Bismuth Compounds

There is considerable interest in compounds containing the heavier group 13 and group 15 elements due to their potential importance as precursors to semiconductors such as gallium arsenide and indium phosphide ⟨89AG(E)1208⟩. This has prompted development of synthetic methods for forming bonds between arsenic and antimony with gallium or indium. Other compounds containing As—Sn and Sb—Sn bonds are known.

2.09.9.2.1 *From secondary alkylarsines*

The original investigation conducted by Coates and his co-workers ⟨65JCS3241⟩ established a preparative method for compound $(Me_2MAsMe_2)_3$ (M = Al; Ga; In) by heating a mixture of Me_3M and Me_2AsH (Equation (29)). Similarly prepared is $(Pr^i_2AsGaMe_2)_3$ from Pr^i_2AsH and Me_3Ga at 130 °C ⟨91OM1635⟩. Compounds obtained are usually in the form of the polymer and cyclic trimer, owing to the proclivity of gallium(III) toward tetracoordination ⟨65JCS3241, 88JOM(341)C1⟩. The yield of arsinogallanes prepared by the thermal reaction decreases to the point of impracticality as the steric bulk of the substituents is increased.

$$Me_2AsH + Me_3M \xrightarrow[-CH_4]{\substack{M = Al,\ 200\text{–}215\ °C \\ M = Ga,\ 30\text{–}50\ °C \\ M = In,\ 15\text{–}25\ °C}} (Me_2AsMMe_2)_n \qquad (29)$$

2.09.9.2.2 *From arsenide and stibide derivatives*

The reaction of arsenides and stibides with gallium and indium halides provides an efficient method to form As—Ga, Sb—Ga and Sb—In bonds, and a few compounds have been prepared accordingly ⟨86CC1543, 87JOM(325)C7, 93JCS(D)1737⟩. For example, lithium di-*t*-butylarsenide undergo a facile reaction with chlorodi-*t*-butylgallane at ambient temperature to give an excellent yield of arsinogallane $Bu^t_2AsGaBu^t_2$, a monomeric product because of the bulky substituents ⟨90OM275⟩. The preparation of Sb—Ga and Sb—In bonded compounds from lithium stibide is illustrated in Equation (30) ⟨90CM221⟩.

$$Bu^t_2SbLi + Me_2MCl \longrightarrow (Bu^t_2SbMMe_2)_3 \qquad (30)$$

M = Ga, 74%; M = In, 46%

2.09.9.2.3 *Metathetical reaction*

The metathetical reaction between an arsinosilane or a stibinosilane and gallium or indium trihalides has been used to prepare compounds containing bonds between arsenic or antimony and gallium or indium (Scheme 35) ⟨86OM1266, 88JOM(341)C1, 88JOM(354)287, 88POL77⟩. The method has also been used to synthesize benzylbis(trimethylstannyl)stibine ⟨89ZAAC(569)62⟩.

$$(TMS)_2As\text{-}TMS \xrightarrow{GaCl_3\ (1:1)} [(TMS)_2AsGaCl_2]_3$$

$$[(Bu^t_2Sb)_2InCl]_2 \xleftarrow{InCl_3} Bu^t_2Sb\text{-}TMS \xrightarrow{GaCl_3} (Bu^t_2SbGaCl_2)_3$$

Scheme 35

2.09.9.3 Transition Metal Derivatives of Arsenic, Antimony, and Bismuth Compounds

2.09.9.3.1 *Reaction of a transition metal complex anion with alkylarsenic, -antimony, and -bismuth halides*

Reaction of dialkylhaloarsines, -stibines, and -bismuthines with the anions of transition metal complexes provides the most important method for forming bonds between pnicogen elements and transition metals. Organometallic compounds containing transition metals such as Cr, Mo, W, Fe, Mn, Co, and Re have been prepared in good yield ⟨74AG(E)84, 74JOM(76)C7, 80CB3318, 80CB3517,

82JOM(228)C29, 84ICAL49⟩. For example, organotransition metal compounds (**62**; E = As, Sb, Bi) have been synthesized by nucleophilic metallation of dimethylarsenic, -antimony, and -bismuth halides with carbonylmetallate (**61**) (Equation (31)) ⟨76CB692, 80CB1180, 82JOM(228)C29⟩. $R_2EM(CO)_3Cp$ in which M = Cr, Mo, W; E = As, Sb has been obtained from $Na[M(CO)_3Cp]$ (**63**) ⟨74AG(E)670, 74JOM(76)C7, 80CB3318⟩. *t*-Butyldichloroarsine reacts with (**63**; M = W) to give product (**64**) which is transformed into (**65**) by methylation with methyllithium (Scheme 36) ⟨84ICAL49⟩. In most cases, dialkylarsino and dialkylstibino moieties of the compounds readily coordinate with another transition metal atom with the formation of complexes of the type $MAsR_2M'$. Four-membered heterocycles (**66**) are obtained via an intramolecular CO substitution reaction when compound (**65**) is heated in toluene at 100 °C. When alkyldihaloarsines and -stibines are used in the reaction, complexes of the formulae MERM (E = As, Sb; M = transition metals) are obtained ⟨75CB700, 84JOM(260)281, 85ZN(B)1430⟩.

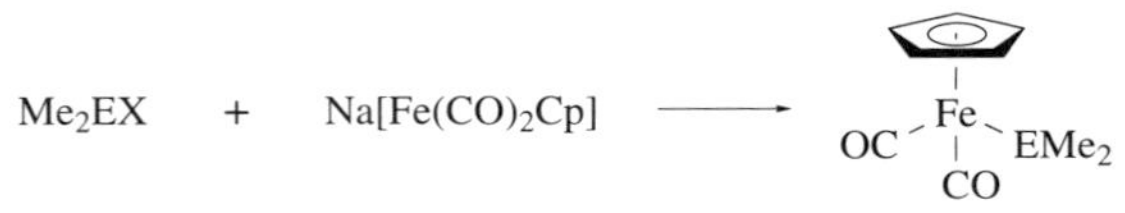

(31)

(61) **(62)**

E = As, Sb, Bi

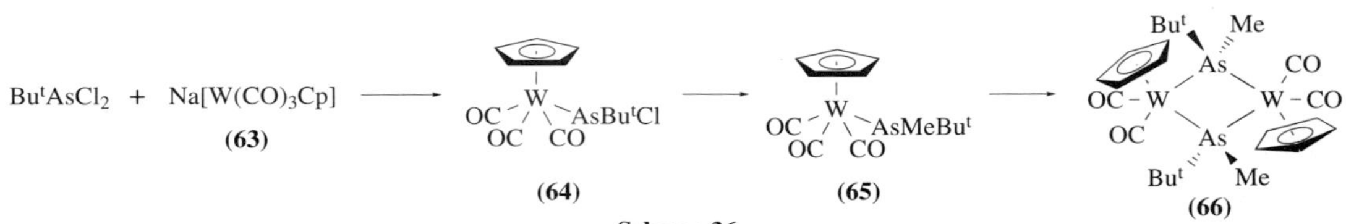

Scheme 36

2.09.9.3.2 *Miscellaneous methods*

Arsenic–transition metal bond formation has been achieved by several other reactions. Utilization of an alkali metal alkylarsenide is a useful method for preparation of transition metal complexes: diethylarsenide reacts with tris(dimethylamino)titanic bromide to produce compound $(Me_2N)_3TiAsMe_2$ (58%) ⟨70MI 209-02⟩. Secondary alkylarsines have been reported to react with cobalt and iridium complexes with the formation of As—Co ⟨67MI 209-01⟩ and As—Ir ⟨86JA1703, 87IC4065⟩ bonds. Diarsine $(CF_3)_2AsAs(CF_3)_2$ affords $Cp(CO)_3MoAs(CF_3)_2$ (64%) when it reacts with $[Cp(CO)_3Mo]_2$ ⟨64JA1030⟩. Another interesting approach to transition metal complexes is the alkylation of arsine (**67**) (Equation (32)) ⟨84AG(E)812, 85CB2472⟩. Binuclear complex (**70**) containing Bi—Fe bonds has been obtained by alkylation of bismuth bromide intermediate (**69**) with triethyl aluminum (Equation (33)) ⟨87JOM(325)159, 90JOM(399)C21⟩.

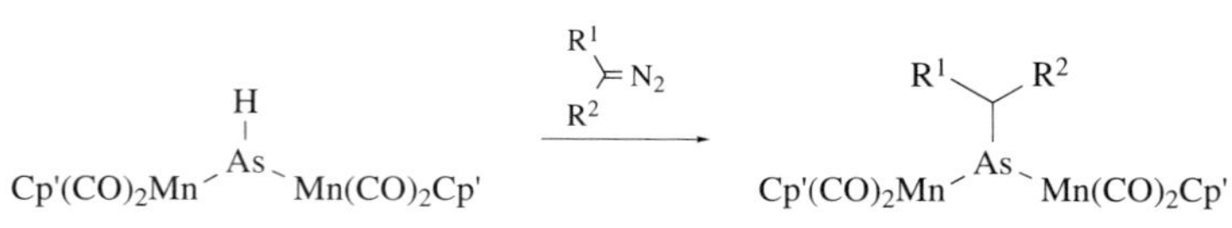

(32)

(67) **(68)**

Cp' = pentamethyl-η-cyclopentadienyl; R^1, R^2 = H, Me, CF_3

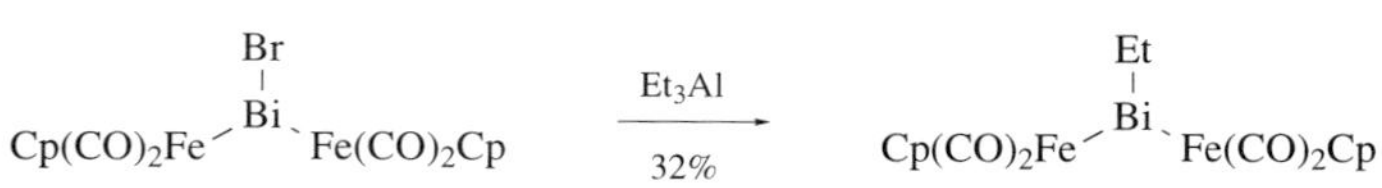

(33)

(69) **(70)**

ACKNOWLEDGMENT

The author is grateful to Professor Otto Meth-Cohn for his encouragement. Without his help, this chapter would not have been finished.

2.10
Alkylboron and -silicon Compounds

MARTIN WILLS
University of Bath, UK

and

ERNEST W. COLVIN
University of Glasgow, UK

2.10.1 ALKYLBORON DERIVATIVES

2.10.1.1 Alkylboranes

2.10.1.1.1 Hydroboration of alkenes

(i) General comments

While every effort has been made to be comprehensive, the wide scope of organoborane chemistry precludes inclusion of all related references in a chapter of this type. However, there are a number of excellent texts and reviews dealing with various aspects of organoborane chemistry which should be referred to when further information and practical procedures are required ⟨B-62MI 210-01, B-72MI 210-01, B-73MI 210-01, 74CSR443, B-75MI 210-01, B-79MI 210-01, 82COMC-I(7)111, B-83MI 210-01, B-83MI 210-02, B-84MI 210-01, B-85MI 210-01, 85OR(33)1, 85TCC23, B-86MI 210-01, 87MI 210-01, B-88MI 210-01, 91COS(8)703⟩.

The reaction of borane, supplied either in the form of diborane $(BH_3)_2$ or as a complex with a sulfide or an amine, with an alkene results in addition of B—H across the double bond. In principle one mole of borane may react with three moles of alkene although in practice this is only generally observed for primary, 1,1-disubstituted and the least hindered 1,2-disubstituted alkenes. More hindered 1,2-disubstituted and trisubstituted alkenes generally stop reacting at an appreciable rate after addition of the second mole of the alkene, whilst borane adds only one equivalent of a tetrasubstituted alkene. The borane–THF complex is rather more reactive than the borane–dimethylsulfide (10M when neat) and borane–triethylamine complexes, and they therefore have complementary applications ⟨B-75MI 210-01, B-88MI 210-01⟩. Hydroboration reactions can be accelerated by ultrasound ⟨85TL2187⟩.

In the absence of electron-donating or withdrawing effects the stereochemistry and regiochemistry of addition of borane across a double bond is largely driven by steric factors. Addition will always take place on the least hindered face and the boron atom will append itself to the least substituted carbon atom of an unsymmetrical alkene. An example which demonstrates all of the factors described above is the reaction of borane with α-pinene to give essentially a single product (**1**) from the addition of just two molecules of alkene (Equation (1)) ⟨60JA4708, B-75MI 210-01, 82JOC5065, B-88MI 210-01⟩; this reaction is discussed in more detail later.

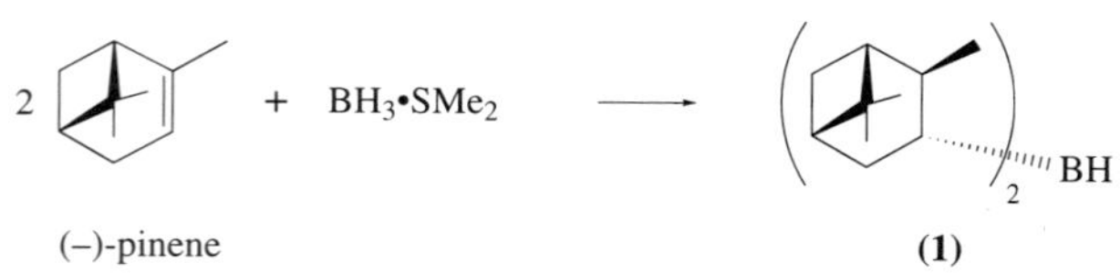

(1)

Polar functional groups may have a significant effect on the regiochemistry of hydroboration. These are readily predictable in a qualitative sense if it is assumed that, whilst steric effects dominate, the electron-poor boron atom will prefer to add to the most electron-rich terminus of the alkene. In the case of substituents with an inductive effect the regioselectivity may be significantly changed (compare Equations (2) and (3)). More significant changes can be achieved when substituents have a mesomeric effect (Equation (4)). Hydroboration of enamines gives 1,2-amino alcohols after oxidation ⟨87JOC4014⟩. Note that the values given in Equations (2)–(4) are for averaged positions of addition in a number of mixed products. Apart from the regiochemical influence, it is important that any functional groups in the substrates are compatible with borane—aldehydes, ketones and carboxylic acids are readily reduced for example, while chlorides can give elimination products ⟨B-88MI 210-01⟩.

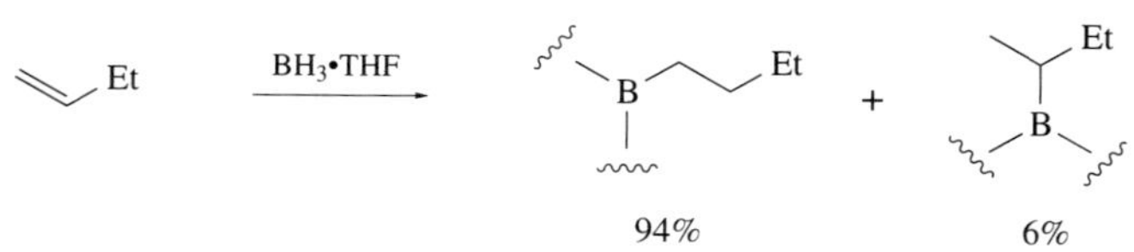

(2)

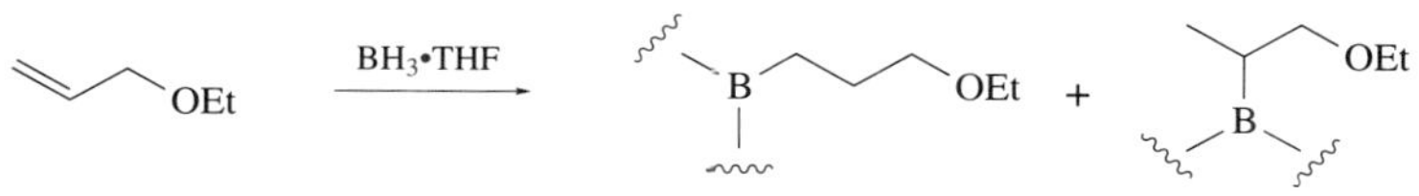

(3)

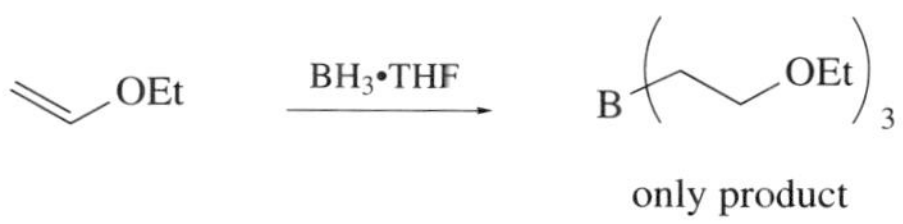

(4)

(ii) Monoalkylboranes by hydroboration

Only the most hindered monoalkylboranes such as 1,2,2-trimethylpropylborane (thexylborane) (**2**) are sufficiently stable to be of synthetic use (Equation (5)) ⟨74S77⟩. Even this compound slowly isomerises to the primary borane and is therefore best prepared *in situ* just before use. In general, monoalkylboranes will disproportionate rapidly to a mixture of dialkylborane and borane, however this can be prevented by isolating the amine complexes, which are considerably more stable ⟨78TL4715, 79JOC465⟩. The reaction of thexylborane with α-pinene gives a large percentage of mono-isopinocampheylborane (**3**) due to elimination of 2,3-dimethyl-2-butene which serves to illustrate an important side reaction in organoborane preparation (Equation (6)) ⟨74S77, 77JOM(135)C10⟩. This sequence can be made more synthetically useful if an amine (usually TMEDA or triethylamine) complex of thexylborane is employed. In this case the corresponding amine complex of mono-alkylborane, which is reasonably stable, is formed (Equation (7)) ⟨78S146, 79JOC465⟩. A rather more effective approach to monoisopinocampheylborane (**2**), which involves the reaction of diiso-pinocampheylborane with TMEDA, is described in Section 2.10.1.1.3.2. Other monoalkylboranes may be kept for reasonably long periods in the form of complexes with amines ⟨61JA831, 61JA2541, 75JA2791⟩. Boron trifluoride etherate can be conveniently employed to remove the amine when the free borane is required.

(**2**)
(5)

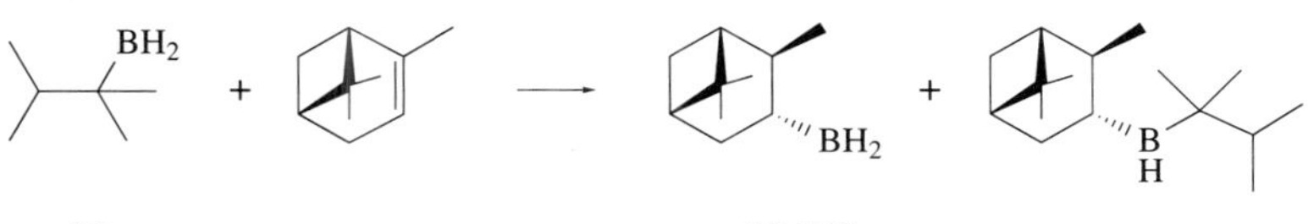

(**2**) (**3**) 75% 25%
(6)

(7)

(iii) Dialkylboranes by hydroboration

The reaction of borane with exactly two moles of an alkene can be achieved in a number of cases. Perhaps the most important example of such a direct approach is the preparation of the asymmetric hydroboration agent diisopinocampheylborane (**1**) (Equation (1)) ⟨82JOC5065, 84JOC945, 87MI210-01⟩. This is one of several ways of preparing (**1**) which achieves an enantiomeric enrichment via formation of a single diastereoisomer. Enantiomerically pure, crystalline (**1**) may be prepared from α-pinene of only 93% enantiomeric excess (*ee*) in this process, in which dimethyl sulfide must be removed prior to an equilibration. Several other important dialkylboranes may be prepared by direct hydroboration including dicyclohexylborane (**4**) ⟨63JA2072, B-75MI 210-01⟩ and disiamylborane (**5**) ⟨80S151, B-75MI 210-01, B-88MI 210-01⟩.

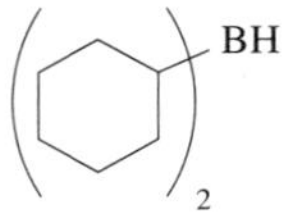

(4)

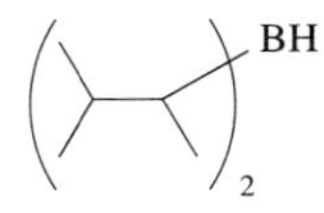

(5)

Direct reaction of certain dienes leads directly to the dialkylborane and this method has been employed for the synthesis of the important reagent 9-borabicyclo[3.3.1]nonane (9-BBN) (**6**) from 1,5-cyclooctadiene (Equation (8)) ⟨68JA5280, 81JOC4599⟩. Whilst in some cases this selectivity is due to the steric hindrance involved with adding a further group to the boron atom, in other cases a trialkylborane is initially formed, which equilibrates to the final product under the reaction conditions ⟨72JA3561⟩. 3,5-Dimethylborinane (**7**) and 3,6-dimethylborepane (**8**) have been prepared in this way ⟨71JOM(28)C1, 73JA6757⟩.

(6) (8)

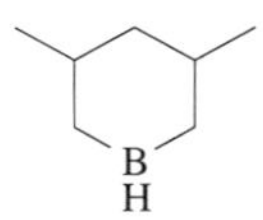

(7)

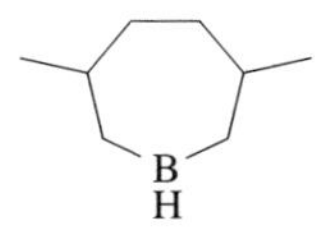

(8)

Reaction of a monoalkylborane with an alkene to give a dialkylborane is generally difficult to control since elimination competes (Equation (6)). In some cases thexylborane will react to give a mixed dialkylborane in this way (Equation (9)) ⟨74S77, 85TL6297⟩, however a strategy involving an intermediate chloroborane provides a more dependable strategy and will be discussed in Section 2.10.1.2.1. More remarkably, under certain conditions methylborane, which is a uniquely unhindered and reactive borane, can effect a monoboration of alkenes ⟨87TL3771⟩.

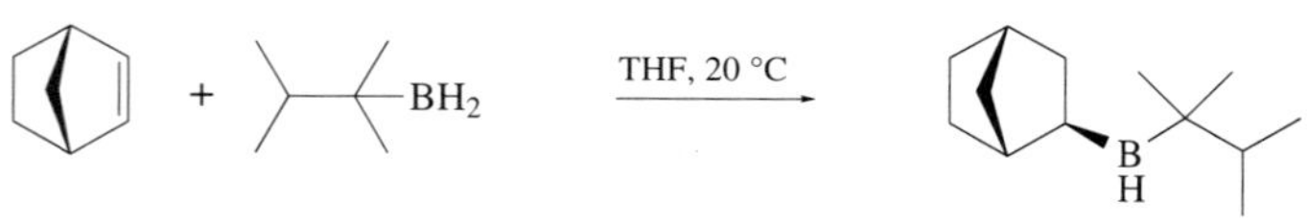

(9)

Whilst it is not as synthetically useful as the dialkyl analogue, monoisopinocampheylborane ($IpcBH_2$) (**3**) may be used as a reagent for asymmetric hydroboration of certain alkenes. Reaction of enantiomerically pure $IpcBH_2$ (**3**), formed by reaction of the TMEDA adduct with boron trifluoride as discussed in Section 2.10.1.1.3.2, with trisubstituted aliphatic alkenes gives adducts which yield the corresponding alcohols after oxidation with enantiomeric excesses (*ee*'s) up to 72% ⟨77JA5514, 78JOC4395, 79TL4777, 81JOC5047⟩. With trisubstituted alkenes bearing an aromatic group, even higher *ee*'s are possible (Scheme 1) ⟨80JOC3543, 87MI 210-01⟩.

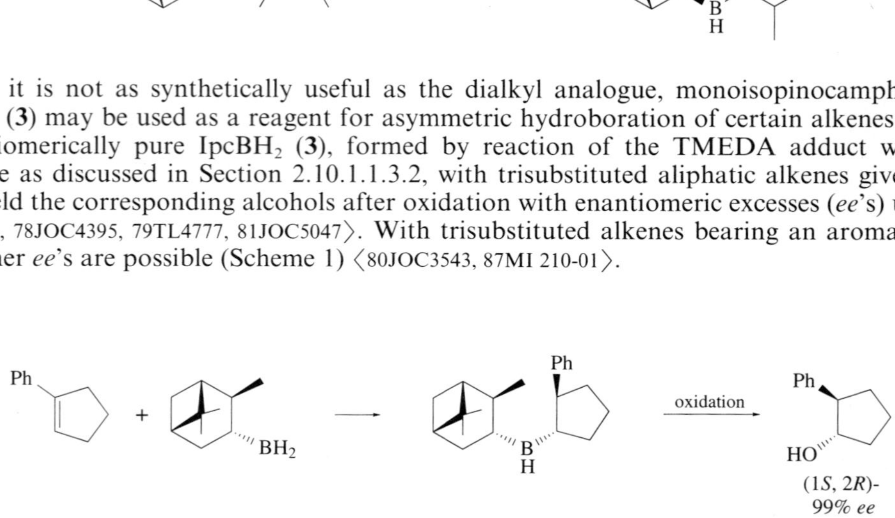

Scheme 1

(iv) Trialkylboranes by hydroboration

Several dialkylboranes have been used for the synthesis of trialkylboranes ⟨85JOC1582⟩. 9-Borabicyclo[3.3.1]nonane (9-BBN-H) (**6**), which is commercially available, is particularly popular due to its wide substrate scope and lack of susceptibility to side reactions such as disproportionation. Although it is a relatively unhindered dialkylborane, it retains the increased regioselective hydroboration properties relative to borane typical of this class of compound ⟨76JA5297, 78JOM(156)213, 81JOC3978, 81JOC4599, 85JOC5583, 88MI 210-01⟩. Hydroboration of pent-1-ene with 9-BBN gives a 99.9:0.1 ratio of boron attachment between primary and secondary carbon atoms; the same reaction of borane gives only a 94:6 selectivity. 9-BBN (**6**) also displays excellent stereoselectivity in hydroboration reactions (Equations (10) and (11)) ⟨77JA3427⟩.

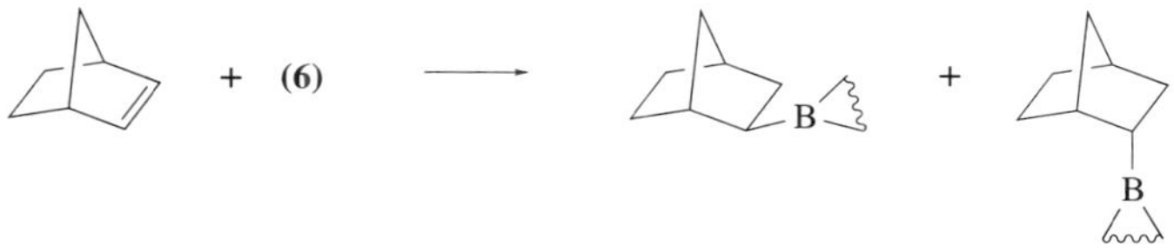

99.5 : 0.5 (10)

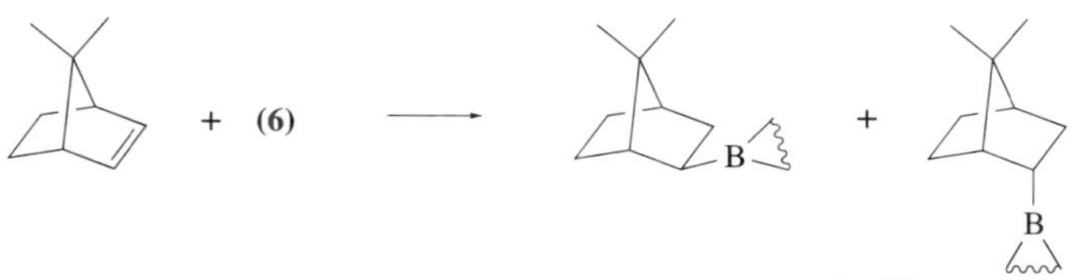

3 : 97 (11)

In contrast to borane and several other organoboranes, 9-BBN (**6**) shows a marked preference for hydroboration at the terminal carbon of allenes, providing a good method for the synthesis of allylborane derivatives (Equation (12)). Substituted allenes tend to work rather more effectively than allene itself, the adduct of which readily undergoes a double addition ⟨74TL3435, 79JA2966, 89JOC5814⟩. Dialkylallylboranes are highly air-sensitive and are prone to rearrangement ⟨80CL993, 81JA1969, 81JA3229⟩, although work on more stable hindered reagents has led to the development of many important synthetic applications, particularly in asymmetric synthesis (see Sections 2.10.1.1.1.5 and 2.10.1.1.2). Dialkoxyallylboranes are also important for synthesis and will be described in detail in a later section. Hydroboration of certain 1,3-dienes with 9-BBN can also be employed for the synthesis of allylboranes, provided the substrate steric control is in favour of this adduct (Equation (13)) ⟨78JOC1058⟩. The asymmetric carbonyl reduction reagent *B*-isopinocampheyl-9-BBN (**9**) (Alpineborane) may be prepared by the direct reaction of 9-BBN-H with α-pinene ⟨79JA2352, 85JOC1384, 87MI 210-01, 88T5597⟩.

(12)

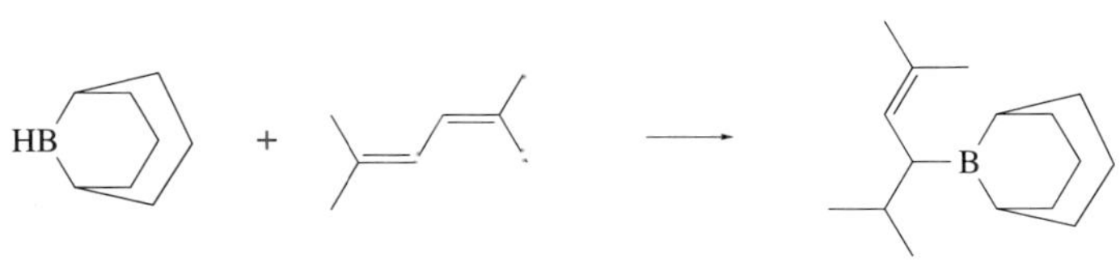

(13)

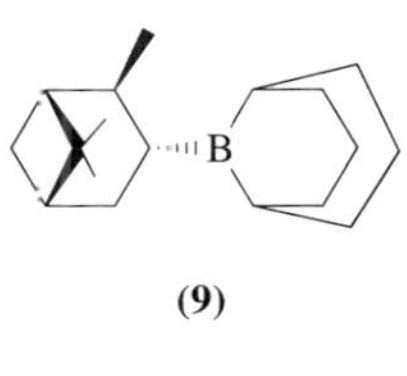

(9)

Most other common dialkylboranes, such as disiamylborane and dicyclohexylborane, are more hindered than 9-BBN, which dramatically influences their synthetic applicability ⟨63JA2072,

66JOC4178⟩. Although the regioselectivity observed with hydroboration of an unsymmetrical double bond with disiamylborane is often not as good as with 9-BBN, the former reagent shows a significantly higher level of differentiation between two double bonds of different substitution level in a substrate ⟨61JA1241, 61JA3417, 62JA190, 63JA2063⟩. In many cases a single product can be isolated (Equation (14)).

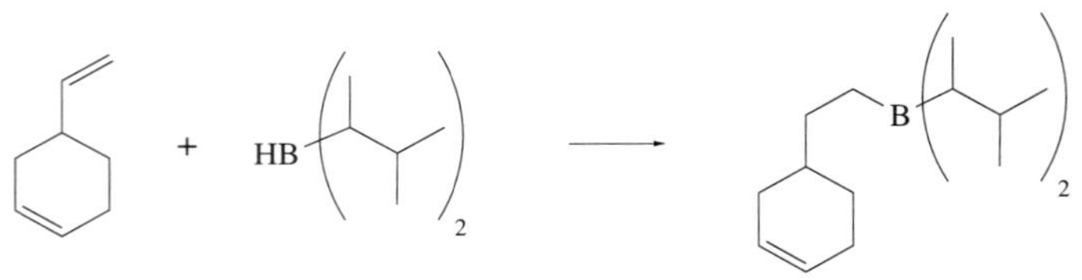

(14)

The reactions of thexylborane (**2**) with alkenes tend to show a rather higher regioselectivity than the corresponding reactions of borane itself, which is mainly due to the increased effect of the extra bulk. This is particularly acute when it is remembered that the overall regioselectivity of hydroboration is the result of two additions, and that in this case the second addition will involve a *very* hindered borane complex ⟨63JA2066⟩. In practical terms the hydroboration of dienes with thexylborane (**2**) is a rather more synthetically useful reaction which gives single products—five- or seven-membered rings usually predominate over six-membered rings (Equation (15) ⟨63JA2066, 74S77, B-79MI 210-01, B-88MI 210-01⟩. Some impressive examples of intramolecular cyclisation can be achieved using thexylborane (**2**) as illustrated by the example shown in Scheme 2 ⟨91TL1479⟩. In this example the borane adduct is converted to the ketone by a well established cyanation sequence, which involves two boron to carbon migrations. In other examples the large bulk of thexylborane can result in the formation of a different product to that obtained using borane alone ⟨89TL2187⟩. Thexylborane (**2**) is almost unique in its ability to add in a controlled stepwise fashion to two different alkenes (Scheme 3) ⟨74S77⟩.

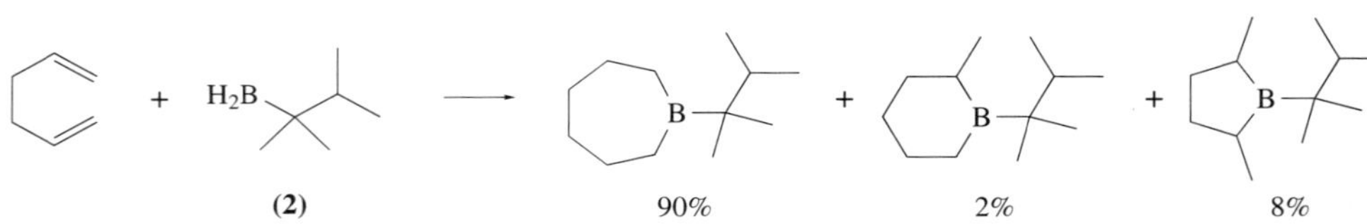

(15)

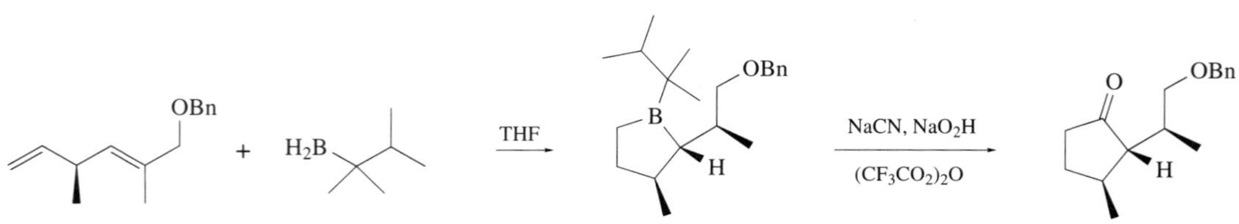

Scheme 2

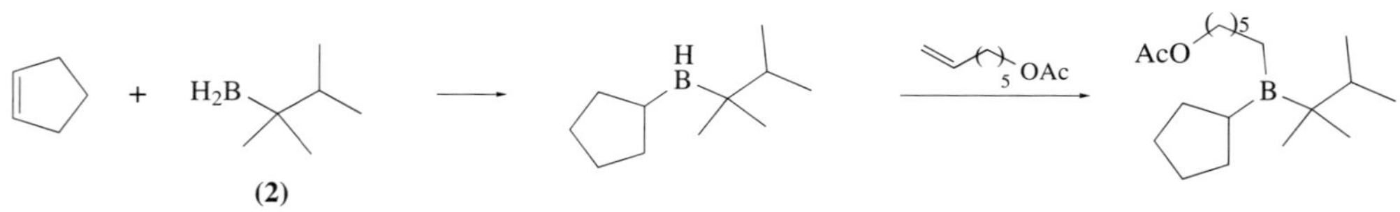

Scheme 3

Borane will react with up to three equivalents of an alkene to give a trialkylborane. When these alkenes are part of the same molecule some attractive intramolecular cyclisations become possible (Equation (16)) ⟨67JA5478, 69JA1224, 69JA1226, 77T2331⟩. Polymerisation can be a problem in these reactions, however heating the reaction mixture to an elevated temperature for several hours serves to isomerise polymeric products to monomeric ones.

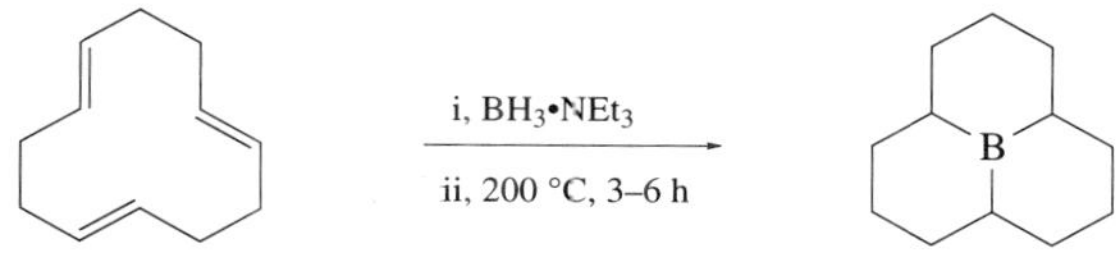

(16)

(v) Asymmetric hydroboration with chiral dialkylboranes

Methods for the preparation of enantiomerically pure Ipc_2BH (**1**) from pinene of rather lower enantiomeric purity have already been described. This reagent has proved to be excellent for the asymmetric hydroboration of *cis*-alkenes. In contrast, the related reagent $IpcBH_2$ (**3**) is rather more effective with *trans*-disubstituted and trisubstituted alkenes (Equation (17)) ⟨87JOC310, 89JA1754⟩. A great deal of work has now been carried out on this reagent ⟨64JA397, 81T3547, 82JOC5065, 87MI 210-01, B-88MI 210-01⟩ and structurally related dialkylboranes, and a full discussion is beyond the scope of this chapter. However a number of representative examples are given in Equations (18) to (20) ⟨64JA397, 73JA7171, 82JOC5065, 86JA2049, 86JOC4296, 87MI 210-01⟩. Note that in all cases the trialkylboranes are not isolated but are converted to the corresponding alcohols by oxidation; the *ee* value therefore refers to that of the derived alcohol. Asymmetric hydroboration of enamines using Ipc_2BH (**1**) gives *β*-amino alcohols after oxidation with alkaline hydrogen peroxide with good levels of enantiomeric induction (Equation (21)) ⟨93TL7693⟩.

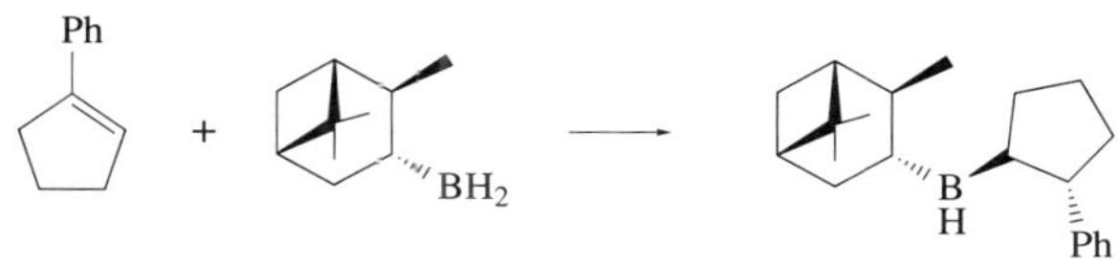

(17)

98% *ee*
(derived alcohol)

(18)

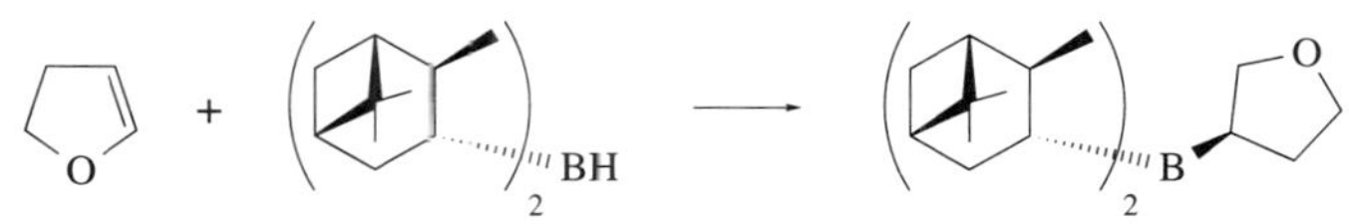

>99% *ee*
(derived alcohol)

(19)

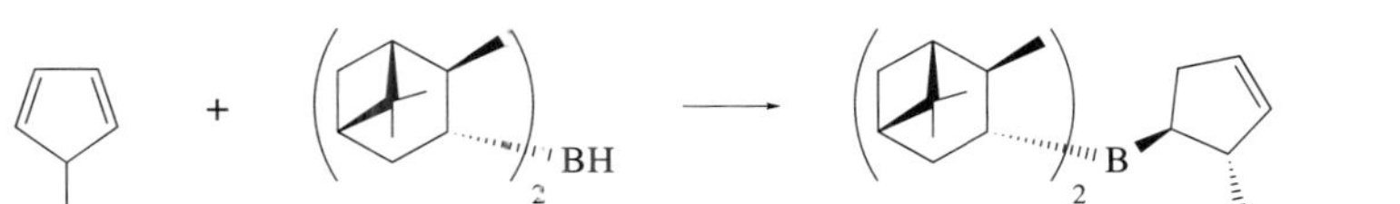

92% *ee*
(derived alcohol)

(20)

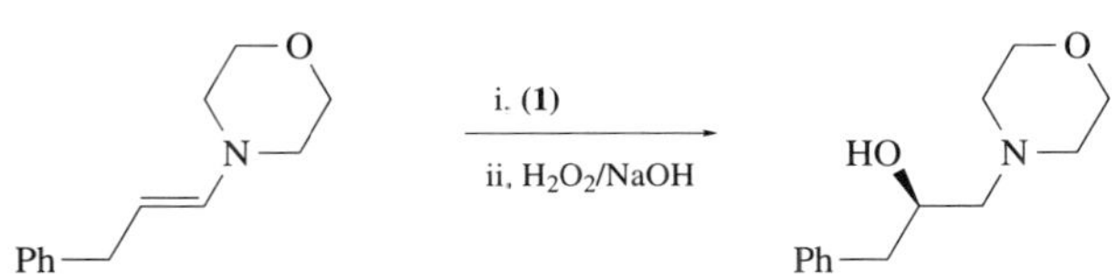

86% *ee*

(21)

Hydroboration of cyclic 1,3-dienes such as cyclohexadiene proceeds with similar reactivity to 9-BBN (Equation (13)) to give allylborane derivatives. In contrast to 9-BBN the products are formed in high diastereoisomeric excess. Unlike many allylboranes the adduct (**10**) does not epimerise on storage or in the reaction with electrophiles, such as aldehydes, which react via an S_N2' mechanism (Scheme 4) ⟨85JA2564, 86JOC445, 91JCS(P1)2633⟩.

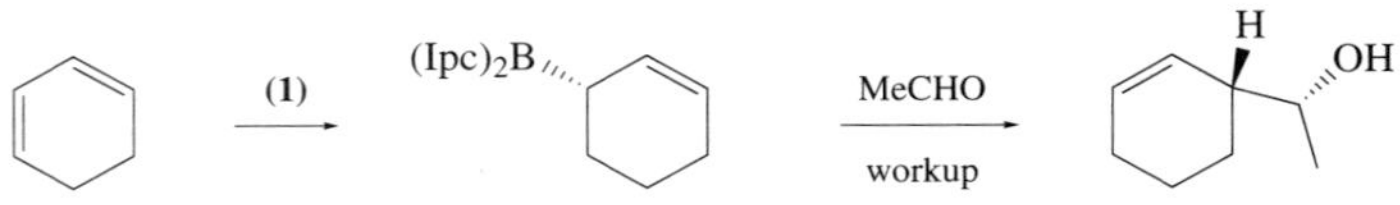

(10)

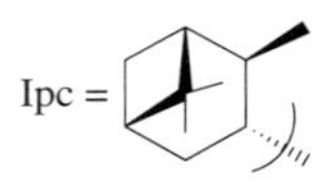

Scheme 4

The homochiral reagent (**11**) derived from the reaction of (+)-longifolene is less hindered than (**1**) and has been employed in the asymmetric hydroboration of more hindered alkenes ⟨81JOC2988⟩.

The asymmetric hydroboration of alkenes using (**1**) and other pinene derived reagents has been extensively reviewed ⟨87MI 210-01, 92ACR16⟩.

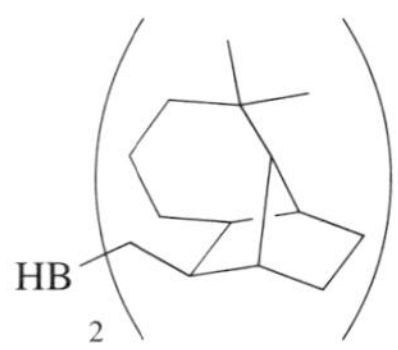

(11)

2.10.1.1.2 *Reaction of an organometallic compound with a haloborane, alkoxyborane or dialkylaminoborane*

The first reported synthesis of organoboranes was achieved by the reaction of diethylzinc with triethoxyborane in 1859 by Frankland ⟨1859PRS568⟩.

The reaction of alkylmetal reagents with boron halides is a highly versatile and effective method for the preparation of mixed organoboranes ⟨B-67MI 210-01, B-67MI 210-02, B-75MI 210-02, B-88MI 210-01⟩. The mechanism is likely to involve a two-step process (Scheme 5), therefore the ability of the halide or other functionality to act as a leaving group may have as great an affect on the rate of the reaction as the reactivity of the organometallic reagent. In general the leaving group ability follows the order Cl > RO > R_2N, hence reactions of chloroboranes with excess organolithium reagent or Grignard reagent results in ready replacement of all the halides. In contrast, the substitution of a dialkylamino group may not be complete until the reasonably stable anionic intermediate is treated with a Lewis acid or heated ⟨63JCS1582, 70JA2467, 71MI 210-01⟩. Hence in the latter case it may prove possible to replace only one of several leaving groups (Equations (22) and (23)) ⟨63JCS1582, 70JA2467, 74IS134, 76JOC1078⟩. In some cases thioalkylboranes have been used as electrophilic reagents. The synthesis of dimesitylborane requires the reaction of two equivalents of mesitylmagnesium bromide with boron trifluoride and is reasonably easy to control due to the steric congestion in the initially formed product (Equation (24)). Reduction with lithium aluminum hydride completes the synthesis (Equation (25)) ⟨B-75MI 210-01, 83TL1433, B-88MI 210-01⟩. The use of ultrasound has been reported to accelerate this class of substitution reaction ⟨86JOC427⟩.

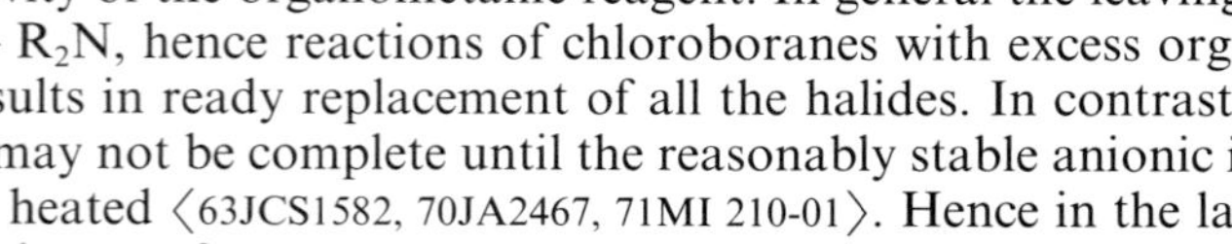

R = alkyl, aryl; X = Cl, OMe, NMe_2; M = Li, MgBr

Scheme 5

$$BCl_3 + 3\,RLi \longrightarrow R_3Li + 3\,LiCl \qquad (22)$$

(23)

$$8\,MesBr + 8\,Mg + 4BF_3{\bullet}OEt_2 \longrightarrow 4\,Mes_2BF \qquad (24)$$

$$4\,Mes_2BF + LiAlH_4 \longrightarrow 4\,Mes_2BH + LiF + AlF_3 \qquad (25)$$

Such reactions are not without problems and redistribution or isomerisation can be competitive side reactions ⟨57JA5190, 58JA617⟩. Alkyllithium reagents have most commonly been employed for preparation of alkylboranes by this method, although this may be a reflection of their general synthetic popularity, and alkoxyboranes are obviously more convenient to work with than most other electrophilic borane sources ⟨70CR(C)589, 74JOM(73)1⟩.

Enantiomerically pure, C2 symmetric, 2,5-dimethylborolanes (**12**) have proved to be valuable reagents for asymmetric synthesis. These compounds are prepared by a nucleophilic displacement reaction using a bis-Grignard reagent (**13**), which initially gives an adduct (**14**) as a mixture of racemic diastereoisomers (Scheme 6). A lengthy sequence of isomer separations, followed by reduction affords the borane ⟨85JA4549, 86JA7402, 86JA7404⟩. Diboranes (**12**), like (**1**), react with high diastereoisomeric excess with prochiral alkenes and show some interesting reactivity patterns. Of particular interest is the observation that a small amount of trifluorosulfonic acid (5 mol%), which presumably forms a boron-triflate species, is essential for rapid reaction ⟨86JA7402, 86JA7404⟩.

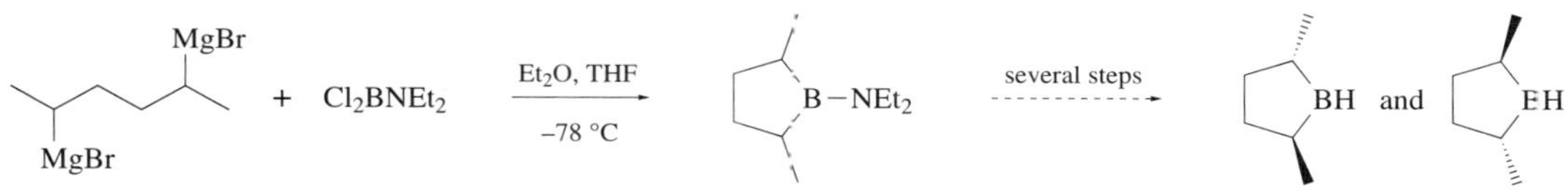

(13) **(14)** (*S*,*S*)-**(12)** (*R*,*R*)-**(12)**

Scheme 6

Allyldialkylboranes are accessible by a nucleophilic displacement strategy, which has been applied most notably to the synthesis of asymmetric allylating agents such as the diisopinocampheylboranes (**15**) ⟨86JOC432⟩ and the (2-isocarenyl)boranes (**16**) (Equation (26)) ⟨90JA2389, 90T4457, 91JOC401, 93TL4655⟩. The latter are reported to be superior in terms of enantiomeric induction in allyl addition reactions ⟨94JOC365⟩. Allyl Grignard reagents are most commonly used as the allylic anion source although allylpotassium reagents, which retain their geometric integrity well, are favoured in many cases (e.g., Equations (27) and (28)) ⟨87JOC4831, 89JA1892, 90T4457, 90TL455⟩. Allyllithiums are also used where an appropriate stabilising group is available (Equations (29) and (30)) ⟨88JA1535, 91JOC5243, 93T7857⟩ and in some cases allylaluminum reagents have been used ⟨85TL2797⟩. In almost all cases, an alkoxyborane provides the electrophilic component of this versatile reaction.

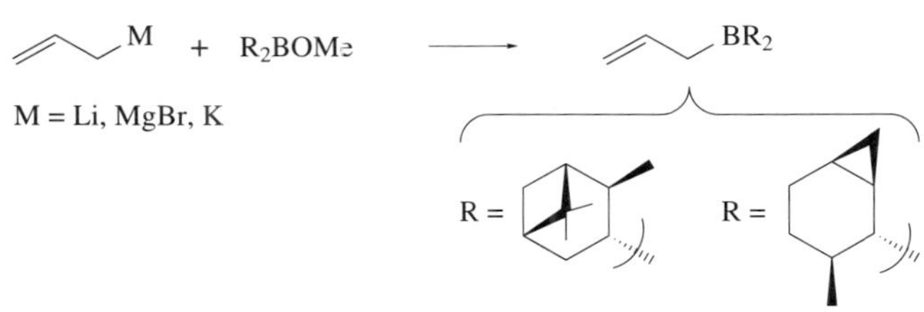

(26)

(15) **(16)**

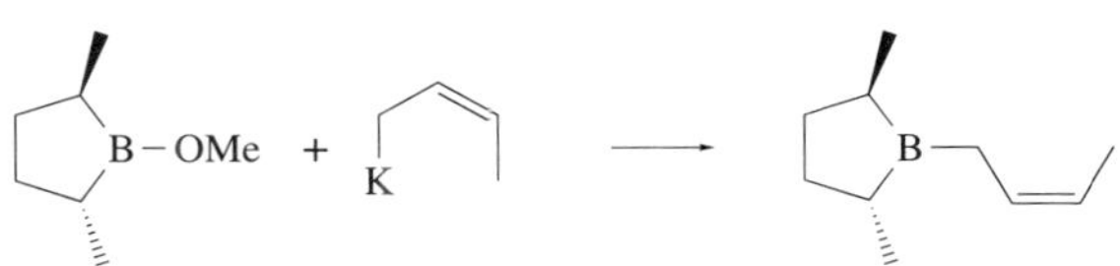

(27)

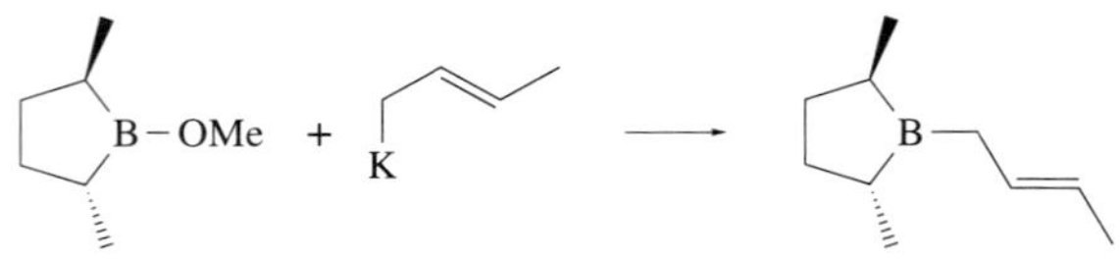

(28)

(29)

(30)

The reaction of an organocuprate with 9-BBN has been reported to give organoboranes via what is presumably a reductive (for the metal) process, although this method has not become generally adopted by the synthetic community ⟨85JOC1969⟩.

2.10.1.1.3 *Miscellaneous methods*

(i) Organoboranes by hydride reduction of chloroboranes or alkoxyboranes

Reduction of chloroboranes with a source of hydride forms an integral part of a method for the selective synthesis of mixed di- or trialkylboranes. Reaction of thexylchloroborane (**17**) (see Section 2.10.1.2.1) with an alkene gives the corresponding dialkylchloroborane, and subsequently the mixed dialkylborane after reduction with lithium aluminum hydride. Finally, hydroboration of a second alkene gives a mixed trialkylborane (Scheme 7). Although not explicitly shown in the scheme, the dimethylsulfide complex is generally employed in this sequence ⟨80JA5919, 80JOC4542, 82JOC863, 82JOC872⟩. Mono- and dialkylboranes may be prepared by the reduction protocol described above, which acts as a good method for the *in situ* preparation of boranes which tend to undergo further reactions, such as disproportionation (Equation (31)) ⟨59JA5836, 61JA2541, 71JA4062⟩. A reductive carbonylation protocol, involving the use of carbon monoxide and lithium aluminum hydride, permits the synthesis of alkylboranes via a one-carbon homologation procedure ⟨78S676, 80JOC4067⟩.

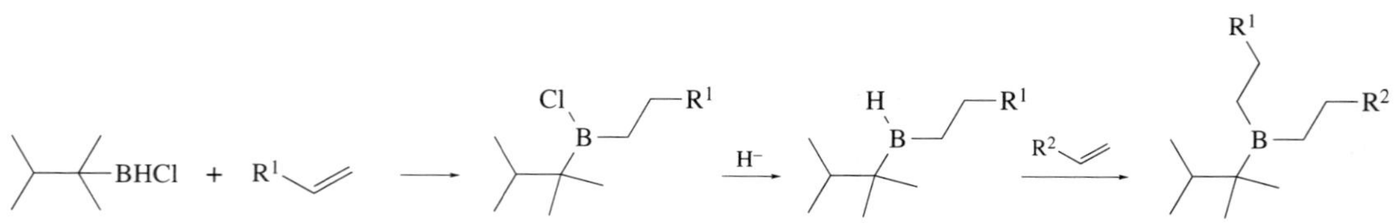

(**17**)

Scheme 7

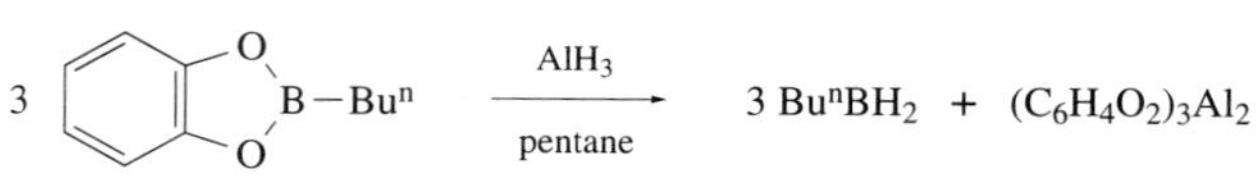

(31)

(ii) Organoboranes by equilibration (redistribution) and isomerisation

Treatment of (**18**) (formed by the reaction of 1,4-pentadiene with borane) with excess borane results in formation of the dialkylborane (**19**) via a redistribution mechanism ⟨72JA3561⟩. Other classes of organoborane may be prepared by such redistributions, and will be discussed in later sections. Isomerisation of organoboranes proceeds readily at elevated temperature which in many cases leads to a complex mixture of products. However, in situations where one possible isomer is of significantly higher thermodynamic stability it can be formed predominantly in the reaction

(Equation (32) ⟨67JA561, 81JOC4313⟩). Diethylborane is an example of a dialkylborane which may be prepared by a direct equilibration method with borane ⟨74IS141⟩. Heat-promoted cycloelimination processes can be synthetically useful in some cases, however a suitable 'energy sink' isomer is usually essential. Treatment of trioctylborane at high temperature leads to formation of **(20)** ⟨65LA(682)21⟩. Arylboranes also undergo cyclisation reactions if a sufficient energetic advantage is afforded (Equation (33) ⟨73JOM(47)33⟩).

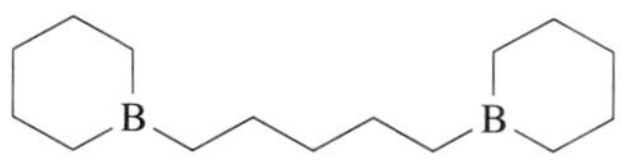

(18)

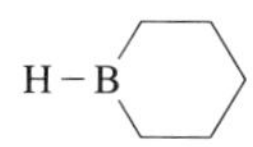

(19)

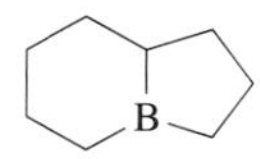

(20)

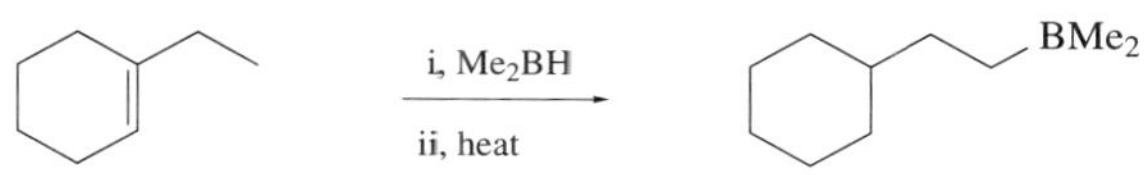

(32)

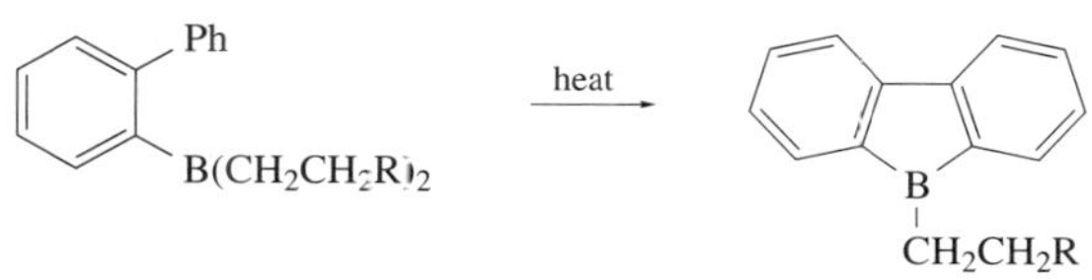

(33)

An important asymmetric reagent which may be made via a redistribution reaction is monoisopinocampheylborane (**3**), which is prepared by the reaction of diisopinocampheylborane (**1**) with TMEDA, with release of a molecule of pinene from each borane component (Equation (34)). In this case the amine adduct (**21**) has higher optical purity than the α-pinene precursor (see above) ⟨78JOC4395, 82JOC5069, 87MI 210-01⟩ and may be recrystallised to 100% enantiomeric purity. Reaction of the recrystallised complex with boron trifluoride etherate releases homochiral monoalkylborane (Equation (35)). This strategy has been extended to other monoalkylboranes ⟨94JOC2365⟩.

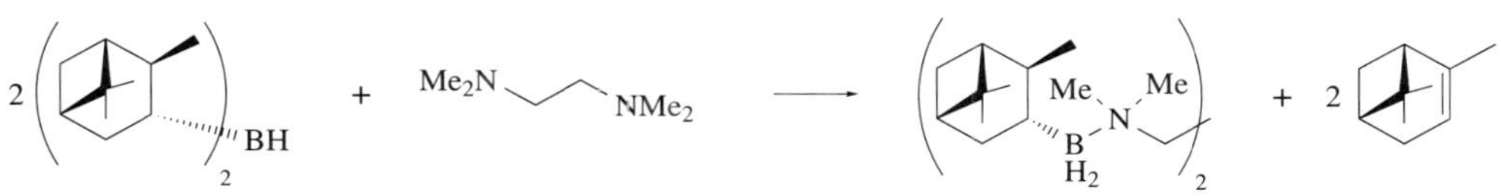

(34)

(1) **(21)**

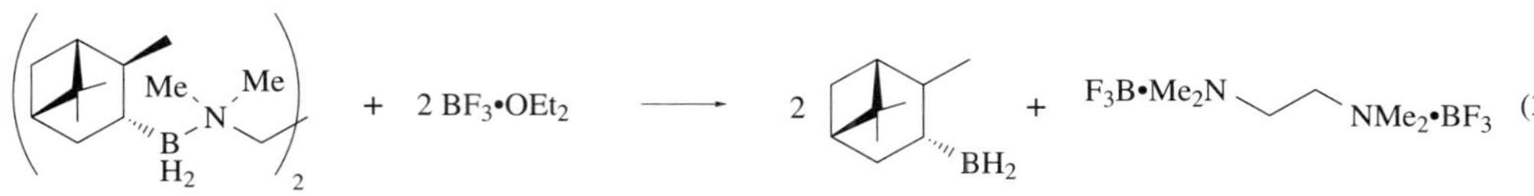

(35)

(3)

Organoboranes which contain a leaving group on a carbon atom adjacent to the boron react with nucleophiles to give initially anionic boron species which rearrange by migration of a group from boron to carbon (Scheme 8). This reaction, in various forms, is of central importance to the organic chemistry of boron, however it does not represent a general method for the preparation of these compounds because the rearrangement step is usually part of the process of decomposition of the organoborane to the organic product, for example in the oxidation of organoboranes with basic hydrogen peroxide (Scheme 9) ⟨B-88MI 210-01⟩. This section will feature only methods which give isolable boron-containing products in good yield and high selectivity where several products may be formed.

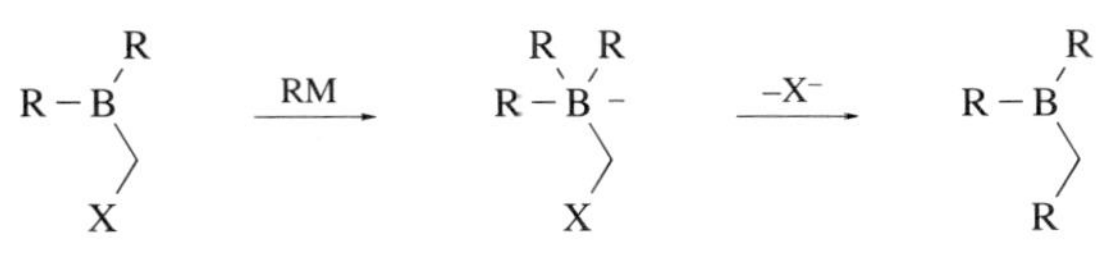

R = alkyl, aryl: X = Cl, OMe, NMe_2; M = Li, MgBr

Scheme 8

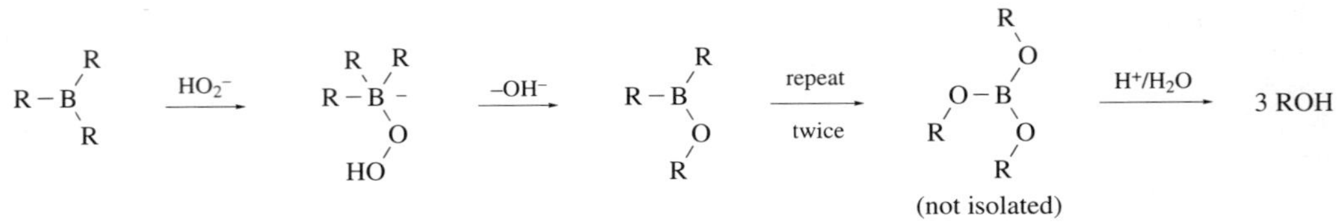

Scheme 9

An example of a rearrangement promoted by a nucleophile is provided by the conversion of α-bromoboranes (from bromination of trialkylboranes) to the hydroxy dialkyl derivatives (Scheme 10). However, this process is frequently complicated by a second migration, particularly if excess bromine is used. In fact even the use of only one equivalent of bromine can result in a mixture if the initial migration gives a secondary (and hence more reactive) centre adjacent to the boron ⟨72CC71, 72IC1150, B-72MI 210-01, 79JA248, 79MI 210-01⟩. The boron has an activating effect greatly facilitating the first radical bromination, which may be achieved using bromine or NBS ⟨72S699, 74JA311⟩. Some interesting rearrangements can be achieved (Scheme 11); in the example shown the rearranged product is converted to a diol by oxidation ⟨B-79MI 210-01⟩.

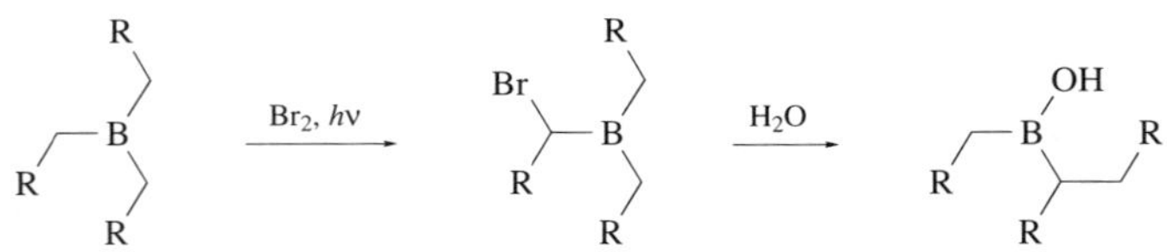

Scheme 10

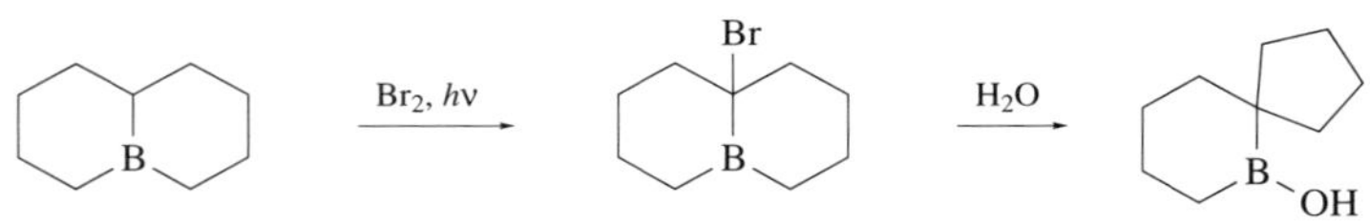

Scheme 11

The precursor for rearrangement may also be prepared by the reaction of nucleophiles which contain a leaving group at the α-position (Scheme 12). When α-chloro ketones or diazoketones are the nucleophiles, the initial product rearranges to a boron enolate, which is usually hydrolysed to a ketone ⟨70TL215⟩, therefore this cannot be considered as an effective way of preparing organoboranes. However, the reactions of haloform anions, carbenoids and ylides are rather different. Suitable leaving groups can be sulfoxides (from stabilised sulfur ylides ⟨66JA4757⟩), triphenylphosphine (from phosphorus ligands ⟨61JA367, 61JA1613, 67JA2782⟩), amines (from trimethylammonium methylide ⟨69IC255⟩) and halides. As for the preceding section, difficulties can be observed in controlling the reaction, since multiple addition/migration steps can occur. Several strategies exist for counteracting this problem, for example the addition of lithiated methylthiophenyl ether gives an organoborate salt which, upon treatment with methyl iodide, gives a sulfonium salt which rearranges ⟨72BCJ2244, 75JOC814⟩. A rather more successful application of this process, which involves the intermediacy of dialkoxyalkylboranes, is described in a later section. Ring contractions of alkylborates, promoted by hydride abstraction, can be used to make organoboranes ⟨75JOM(90)C1, 76JA1964⟩.

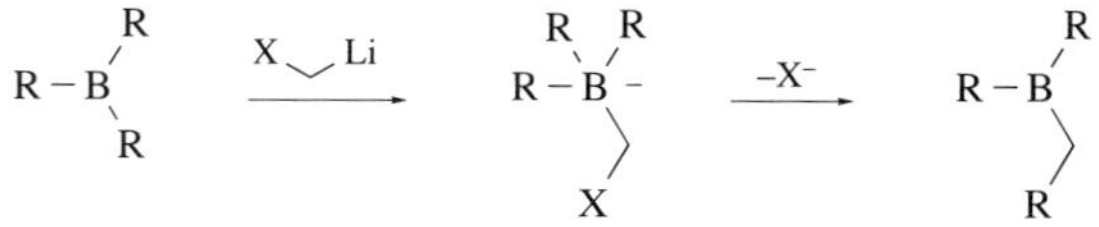

Scheme 12

(iii) Organoboranes by alkylation α to boron

With the use of a hindered lithium alkoxide, it is possible to selectively deprotonate at a carbon atom adjacent to borane in a trialkylborane. The anion thus formed may be alkylated to give another alkylborane or participate in a number of further reactions (Scheme 13) ⟨72JA6854, 80JOM(186)297, 81JOC1693, 83TL623⟩. The use of a bulky borane and base are essential to prevent addition of the amide to the boron atom. Hindered diarylboranes may be elaborated in this way; dimesityl derivatives are particularly valuable in this respect, since the high degree of steric hindrance favours deprotonation (Scheme 14) ⟨83TL627, 83TL631, 89TL5643, 89TL5647⟩.

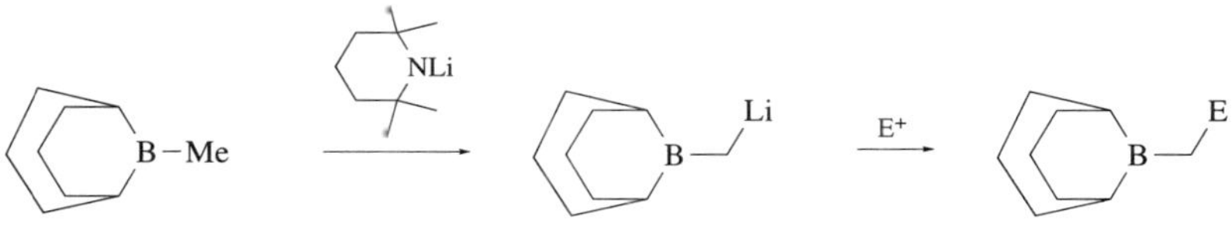

Scheme 13

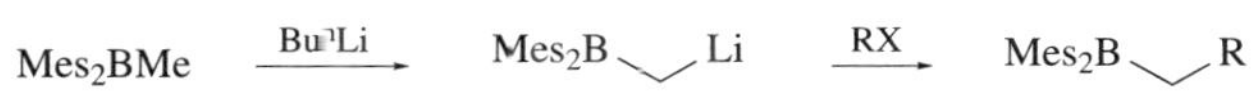

Scheme 14

In a related approach the dimesitylborane group serves to shield a boron atom from attack by an alkyllithium in an addition to a double bond. Initially this yields an anion adjacent to boron, which is then converted to the required borane (Scheme 15) ⟨87JA931⟩.

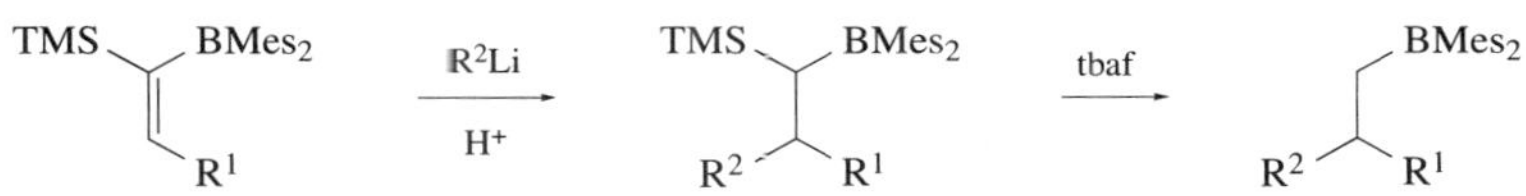

Scheme 15

2.10.1.2 Alkylhaloboranes

2.10.1.2.1 Hydroboration of alkenes

The most direct method for the preparation of alkylhaloboranes is by the direct hydroboration of an alkene using a haloborane. Monochloro(or bromo)borane and dichloro(or bromo or iodo)-borane are most conveniently prepared and used as dimethylsulfide complexes ⟨77IC2938, 77JOC2533, 80CC1169, 88MI 210-01, 88PAC1705⟩.

As would be expected of a reagent with added electron-withdrawing groups, haloboranes are more powerful Lewis acids than borane itself. However, whilst this would initially suggest that they should be more powerful hydroborating agents, in donor solvents such as THF they associate so closely with the solvent that their reactivity is reduced dramatically. For monochloroborane, ether is the favoured solvent ⟨76JA1785⟩, however, for dichloroborane only hydrocarbon solvents are compatible if a reasonable reaction rate is to be achieved ⟨76JA1798⟩. This solvent effect can be used to advantage; in ether, chloroborane will readily react with two molecules of a primary alkene (or both double bonds of a diene) ⟨76JA4917⟩, while the addition of 1–2 equivalents of THF can slow the reaction down sufficiently for the addition of a single equivalent of the alkene to be achieved ⟨76JA1785⟩. The regioselectivities observed in hydroborations with monohaloboranes are generally somewhat higher than the selectivities achieved with borane alone ⟨76JA1785, 77S695⟩.

It is essential that excess alkene is not used in the reactions with dihaloboranes since acid generated during workup can give rise to the hydrated product. Since the regioselectivity of hydroboration opposes that of acid-catalysed hydrolysis, this will result in apparent reversal of regioselectivity to give the unexpected product alcohol ⟨80JOC384, 80JOC389, 83JOC1389, 86JOC895⟩. Provided that there is not an excess of alkene, the regioselectivity of hydroborations with dihaloboranes can be similar to that obtained using 9-BBN (**6**) ⟨83JOC1389⟩. Hydroboration using chloroborane has been used in the synthesis of the excellent reducing agent *B*-chlorodiiso-2-ethylapo-pinocampheylborane (**22**) (Equation (36)) ⟨91TL6691⟩. Dichloroborane can be generated *in situ* by the reaction of boron trichloride with triethyltin hydride as demonstrated in the synthesis of the homochiral alkyldichloroborane (**23**) which has been used for the asymmetric catalysis of Diels–Alder reactions ⟨91JA7794⟩.

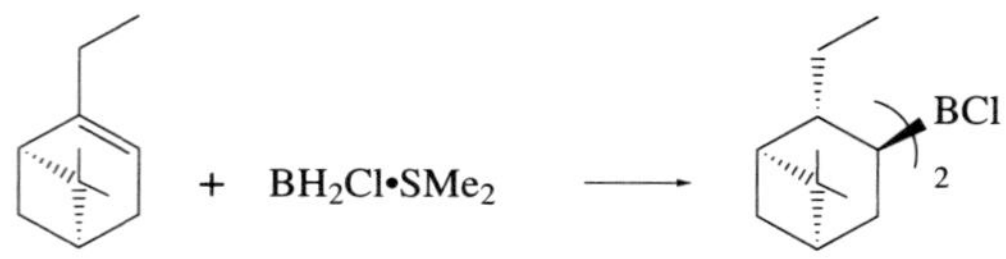

(36)

(22)

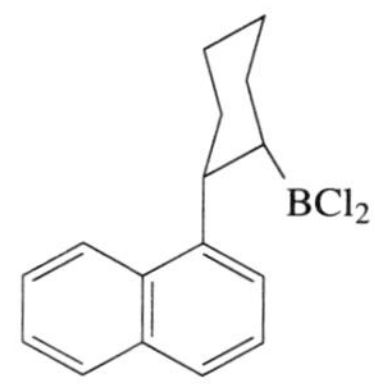

(23)

Thexylchloroborane (**17**) is an important product of this class of reaction and is readily prepared with high selectivity for the monoadduct due to the bulk of the thexyl side chain (Equation (37)) ⟨80JOC4540, 86JOC5264⟩. The bulky thexylchloroborane (**17**) (in the form of its dimethylsulfide complex) is a useful hydroborating agent—it has a single B—H bond which ensures that only one further alkene can react (Scheme 7), and due to its bulk it reacts much faster with nonhindered alkenes than with hindered ones and with *cis*-alkenes much more rapidly than with *trans*-alkenes ⟨82JOC863, 82JOC872⟩. Reduction of the product dialkylchloroborane to the dialkylborane may be achieved using one of a number of hydride reducing agents, and the product dialkylboranes may be employed in subsequent hydroboration steps to give mixed trialkylboranes (Scheme 7). The same class of products may be prepared by reaction of the chloroboranes with an alkyllithium (Scheme 16) ⟨80JA5919, 80JOC4542⟩) (Section 2.10.1.1.3).

(37)

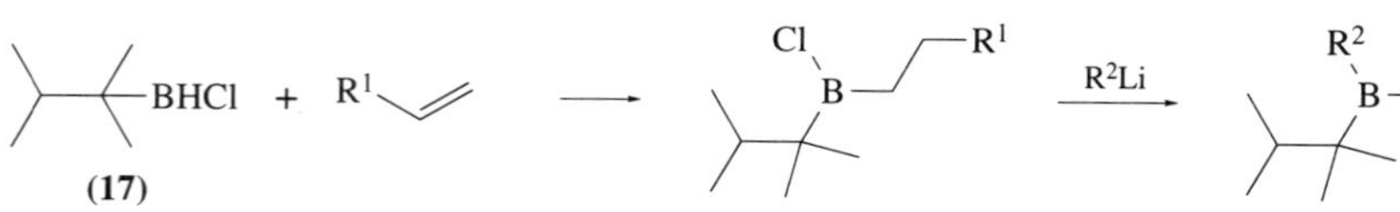

(17)

Scheme 16

2.10.1.2.2 *Miscellaneous methods*

The preparation of haloalkylboranes by the reaction of a haloborane with an organometallic compound is generally difficult to control. However in some cases, for example where the alkyl group is large, a useful product can be obtained (Equation (24)).

Redistribution reactions have been used in the synthesis of haloallylboranes, although there are relatively few examples which are truely synthetically valuable ⟨68LA(719)169, 72JOM(44)233⟩.

The important carbonyl reducing agent, chlorodiisopinocampheylborane (**24**), has been prepared by the action of hydrochloric acid on the dialkylborane precursor (**1**) (Equation (38)) ⟨85JOC5446, 86JOC3394, 87MI 210-01, 92JOC2379⟩. This reagent has been used extensively for asymmetric reductions, having a high reactivity relative to other boranes which permits its use with hindered substrates such as trialkylsilyl ⟨90TL4677⟩ and *t*-butyl substituted ketones ⟨88JA1539, 88JOC2916⟩ (see Section

2.10.1.3.4). The asymmetric ring-opening of *meso*-epoxides using this reagent has also been described ⟨88JA6246⟩.

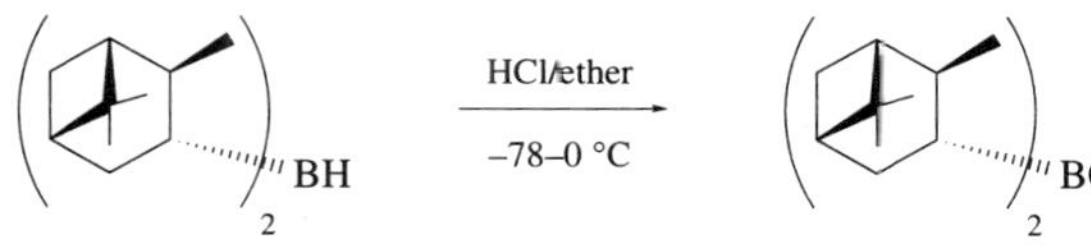

(1) (24) (38)

2.10.1.3 Alkyloxyboranes

2.10.1.3.1 Hydroboration of alkenes

Catecholborane (1,3,2-benzodioxaborole) (**25**) stands out as a valuable and perhaps the most versatile synthetic reagent for the synthesis of alkyloxyboranes. Rather less reactive than most dialkylboranes, it requires elevated temperatures to achieve alkene hydroboration ⟨71JA1816, 73JOC158, 75JA5249⟩ in generally high regioselectivity (Equation (39)). Hydroboration of alkynes to give vinyl boranes is a valuable application of catecholborane, which will be described in more detail elsewhere (see Chapter 2.18.1 ⟨72JA4370, 75JA5249, 76JOC3031, 76T981⟩). Catecholborane adducts may be converted to boronic acids, which have been used extensively in synthesis ⟨75JA5249⟩, by hydrolysis. Allyldialkoxyboranes may be prepared by 1,4-addition of catecholborane to 1,3-dienes ⟨89TL3789⟩.

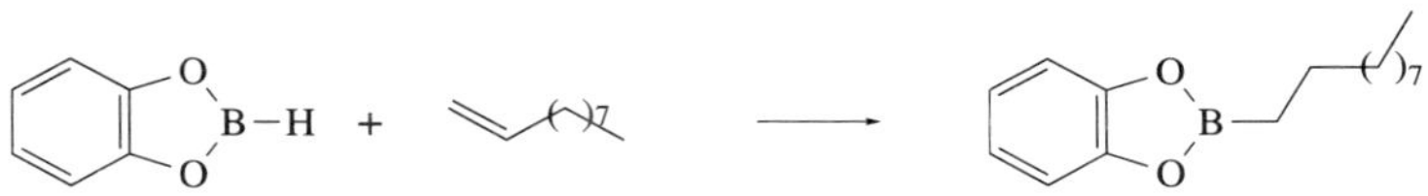

(**25**) 98% this regioisomer (39)

A very important development in this area is the use of transition metal catalysts for the mediation of hydroboration ⟨91CRV1179⟩, a reaction which was first achieved in 1975 using Wilkinson's catalyst ⟨75CL1095⟩. Many catalysts have been used for this reaction, however the rhodium(I) and iridium(I) based materials have proved to be the most effective. As well as generating an acceleration, transition metal catalysts can significantly alter the regio- and stereochemical outcome of hydroboration reactions relative to uncatalysed additions of dialkylboranes (Scheme 17) ⟨89TL5857, 89TL5861⟩. In most cases of alkene hydroboration, the expected products of boron addition at the least hindered end of the alkene are obtained, often in higher selectivity than for uncatalysed reactions. A number of reports have appeared concerning the mechanism of the reaction, which is complex ⟨91JA6139, 91JOC1020, 92JA9350⟩. The reaction is highly amenable to directing-group effects, which permits highly diastereoselective reactions to be achieved in a predictable fashion (Equation (40)) ⟨91JA4042, 92JA6671⟩.

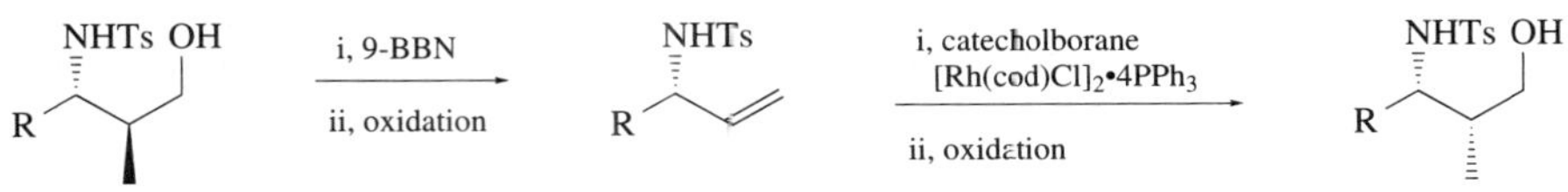

Scheme 17

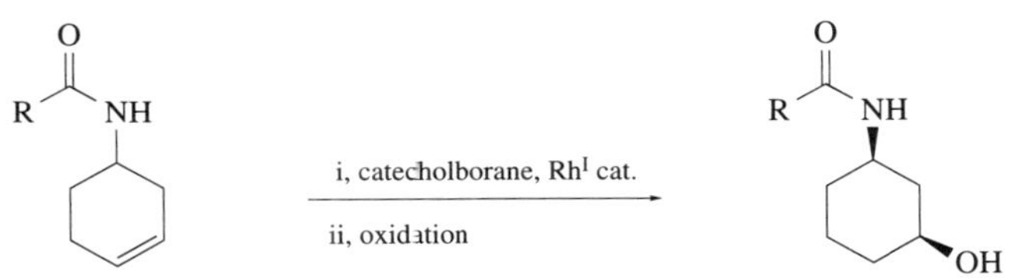

(40)

Another key observation regarding the rhodium(I) catalysed hydroboration was that styrene hydroboration gave almost exclusively the benzylic alcohol. This has led to the development of asymmetric modifications, through the use of homochiral ligands (Equation (41)) ⟨93CC1673⟩. The designed ligand shown in Equation (41) is one of the best for this application. Although other more established ligands have been employed, few have given comparable results ⟨89JA3426, 90TL231, 91TA601, 91TA613, 91TL3387, 94JA866⟩.

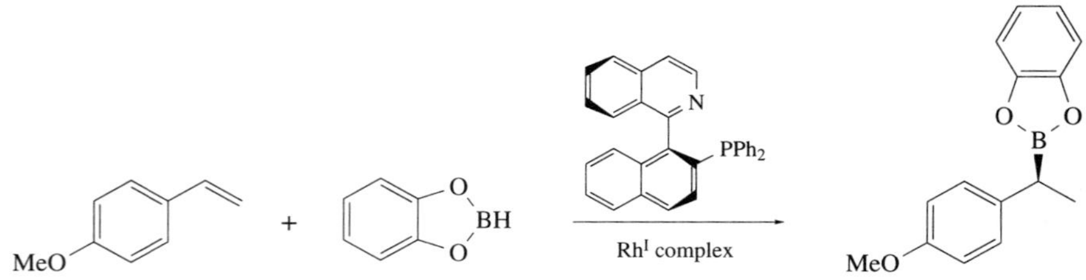

(25)

(41)

94% *ee* after oxidation

2.10.1.3.2 Reaction of alcohols with a borane derivative

1,3,2-Benzodioxaborole (**25**) (catecholborane) is a widely exploited alternative to borane which may be conveniently prepared by the reaction between borane and catechol (Equation (42)) ⟨71JA1816⟩. Nucleophilic substitution of a suitable leaving group by an alcohol (Equation (43)) provides a versatile method for the synthesis of alkoxy substituted boranes ⟨B-88MI 210-01⟩. The leaving group (X) may be NR_2 OR, OH, OAc in which case a simple displacement is involved, or hydrogen, in which case a molecule of hydrogen is produced.

(25)

(42)

(43)

X = Cl, Br, OMe, NMe_2, SMe

A very important class of boron reagents which have recently emerged are the oxaborolidines ⟨87BCJ395, 92TA1475, 93CR763⟩. In particular the proline-derived compound (**26**) has found extensive application as a catalyst for the asymmetric reduction of ketones by borane ⟨87JA5551⟩. While the compound (**26**; R = H) may be made by the direct reaction between an amino alcohol at borane and elevated temperature, the derivatives where (R = alkyl) are prepared by combining the aminoalcohol and the appropriate boronic acid with removal of water ⟨87JA7925⟩. It is essential that the compound is perfectly dry for optimum results—this can often present difficulties and use of the boron-complex (**27**) may be favoured ⟨91JOC751, 91JOC763, 93JOC2880⟩. A similar procedure has been employed for the preparation of related catalysts (**28**) for Diels–Alder and aldol reactions ⟨92JOC7372, 92TL6907, 93JA11490⟩.

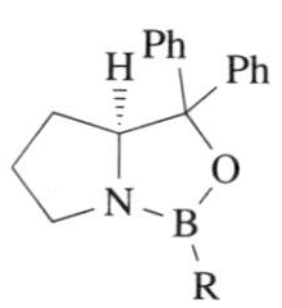

(26)

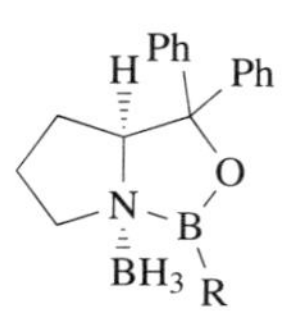

(27)

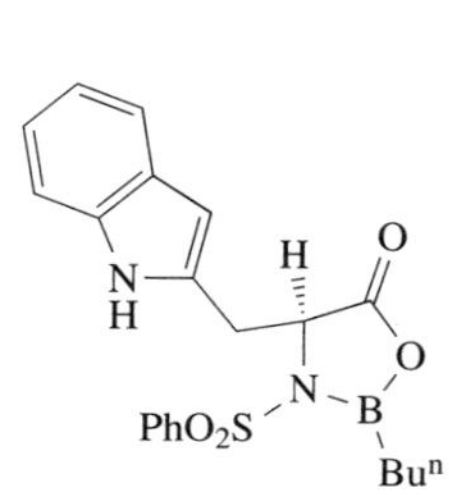

(28)

2.10.1.3.3 *Reaction of an organometallic reagent with a borane derivative*

Displacement reactions of alkoxyboranes using organometallic reagents may be readily carried out in a sequential fashion, since each subsequent displacement occurs less readily (Scheme 18) ⟨B-88MI 210-01⟩.

$$B(OR^1)_3 \;+\; R^2M \longrightarrow R^2B(OR^1)_2 \xrightarrow{R^2M} (R^2)_2B(OR^1)$$

Scheme 18

Allyldialkoxyboranes may be prepared by the reaction of allylmetals with trialkoxyboranes which either give the required product directly ⟨82JOC2498, 87LA881, 88PAC123⟩ or after hydrolysis to the boronic acid followed by esterification with a diol. The latter procedure has proved successful in the preparation of homochiral allyldialkoxyboranes (**29**) derived from dialkyl tartrates (Scheme 19) ⟨85JA8186, 86JA294, 90JA6339, 90TL7563, 93SL927⟩. A useful alternative strategy requires the treatment of chlorobis(dimethylamino)borane with crotyl-potassium followed by treatment with a 1,2-diol to give 2-allyl-1,3,2-dioxaborolanes (**30**) (Scheme 20) ⟨78JA3258, 80JOM(187)321, 81JOC1309⟩. This is a highly versatile reaction and numerous substituted derivatives have been prepared this way ⟨84T2219, 85LA2246, 87LA215, 92T1981⟩. Similar reaction conditions using a corresponding Grignard reagent may also be employed ⟨81JOC1309⟩. Allylstannanes may be used for the synthesis of allylboranes in an analogous method, although the geometric control is less predictable in this reaction ⟨80AG(E)218, 81TL2751, 82AG(E)555, 82CB2357⟩.

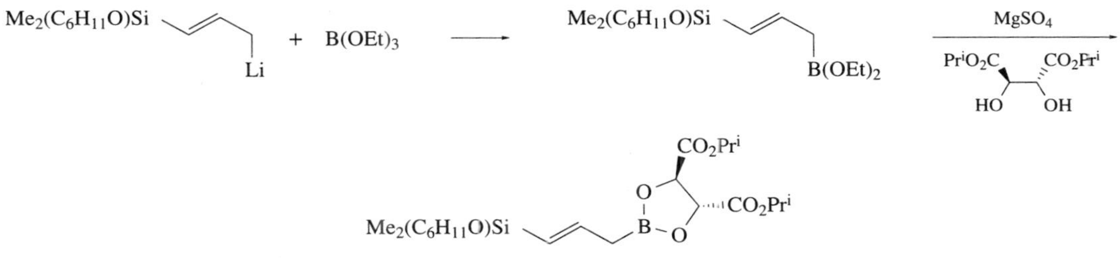

(**29**)

Scheme 19

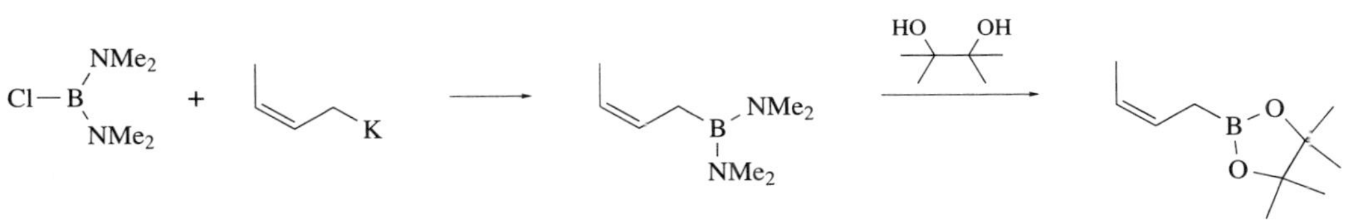

(**30**)

Scheme 20

A variation of the α-chloromethyl displacement process is applicable to allylborane synthesis ⟨89CRV1535, 89T1859⟩. The reaction of dichloromethyllithium with alkyldialkoxyboranes offers some intriguing synthetic possibilities ⟨86S973, 91JOC3286⟩. Following the first chloride displacement by migration from boron, a second displacement may be achieved using an organometallic reagent (again via addition to the boron atom followed by migration). Furthermore the use of a chiral diol permits the synthesis of enantiomerically pure products via diastereoselective migration reactions. An application which summarises all of the key points of the chemistry of these compounds is the synthesis of *exo*-brevicomin (**31**) (Scheme 21) ⟨83JA2077⟩. The use of zinc dichloride in the rearrangement step is essential for high selectivity. In an example from 1993 of this prolific chemistry the synthesis of stegobiol (**32**) has been achieved via a similar series of addition–migration/displacement reactions controlled in an absolute sense entirely by the cyclic dialkoxyborane (**33**) ⟨93JOC6545⟩. Dehalogenative reduction of the α-chloroalkylborane with potassium tri-(isopropoxyborohydride) provides a convenient method for one-carbon homologation of organoboranes ⟨85JOC5586, 86JOC3150⟩. A sequence similar to that shown in Scheme 21 has been used for the synthesis of α-amino alcohols. The key chiral centre is created by the use of an azide or amine as nucleophilic displacement of the diastereomerically pure α-chloroalkylborane ⟨87TL4499, 92TL4209⟩. Enolates may also be used as electrophiles in these reactions, giving β-boronate esters as products ⟨93T177⟩.

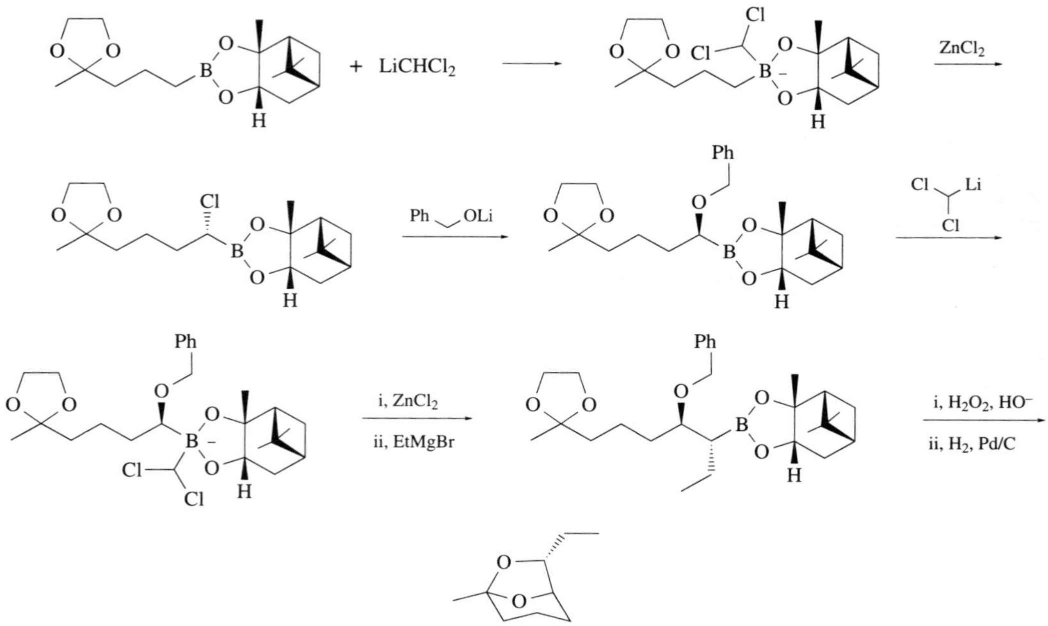

(31) *exo*-brevicomin

Scheme 21

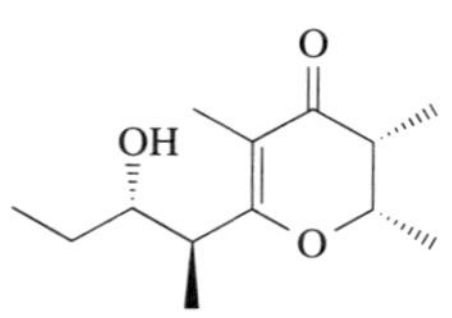

(32)

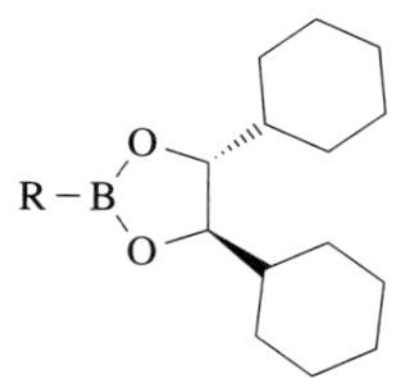

(33)

Provided the alkenyllithium is geometrically pure, either isomer of an allylborane may be prepared (Equation (44) ⟨81JOC1309, 81TL2751, 83JOC5398, 86JA3422, 93TL6899⟩. The order of addition of the reaction components can be reversed to make the chemistry more accessible; therefore the lithiated dichloromethane may be added to the vinylic dialkoxyborane ⟨93TL7845⟩. An attractive application of this methodology is in the synthesis of diastereoisomerically pure α-chloroallyldialkoxyboranes (**34**) and subsequently, further isomerically pure derivatives following the method illustrated in Scheme 22 ⟨83TL3209, 89CB1783, 89LA883, 90SL759⟩.

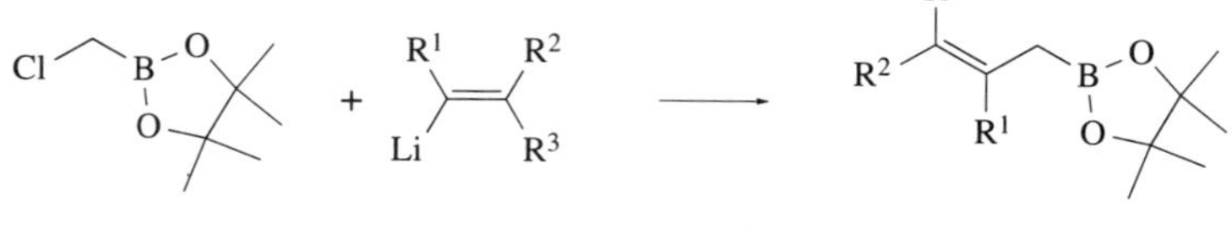

(44)

Three alkyl migrations from boron to carbon can be employed in a synthetically useful process to give functionalised alkylboranes as intermediates. As shown in Scheme 23, the reaction of lithiodichloromethoxymethane with a trisubstituted alkylborane gives the alkylchloromethoxyborane as intermediate ⟨73JOC2422, 73JOC3968, 73JOM(54)61⟩.

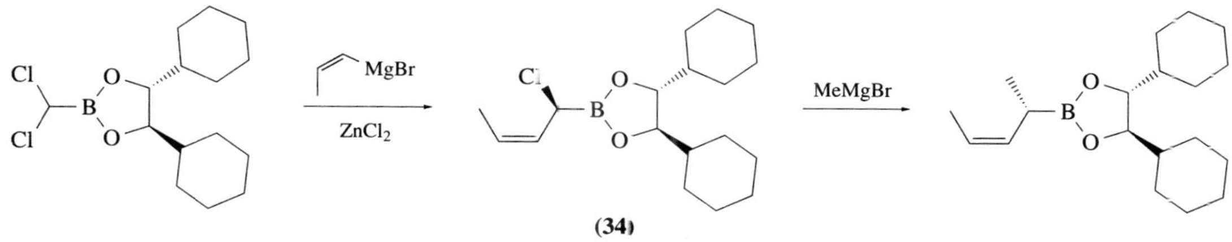

(34)

Scheme 22

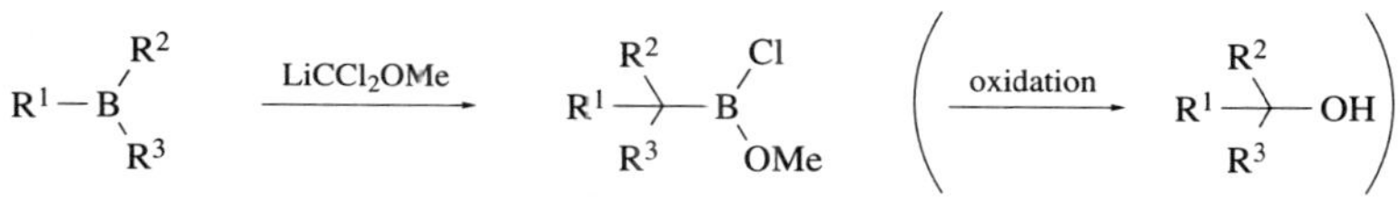

Scheme 23

2.10.1.3.4 *Reaction of an organoborane with an aldehyde or ketone*

Boranes readily reduce aldehydes and ketones to alcohols by the transfer of a hydride, either directly from the boron atom or via a β-hydride transfer from an alkyl group, in the case of Alpine-borane (**9**) ⟨85JOC1385⟩. This has been developed into a convenient method for the conversion of certain adducts from asymmetric hydroboration reactions. As illustrated in Scheme 24, reaction of adducts from monoisopinocampheylborane and prochiral alkenes react with ethanol at room temperature to give the corresponding boronic esters (**35**). Note that the bulky isopinocampheyl groups react much faster than the other alkyl residue in this process. The boronic esters are generally crystalline solids of high stability which permit purification to 100% *ee* by recrystallisation before they are converted into alcohols by oxidation or a variety of other compounds. A further bonus of this chemistry is that pinene is regenerated in the process, and may be recovered and reused ⟨85JA4980, 87JOC310, 89JA1754⟩. While the example in Scheme 24 is of an adduct of $IpcBH_2$ (**3**) similar chemistry is possible using adducts of the related Ipc_2BH (**1**) ⟨86JA2049, 86JOC4526⟩. Methods for the regeneration of borane reagents in this way have been extensively reported ⟨92JOC6608⟩.

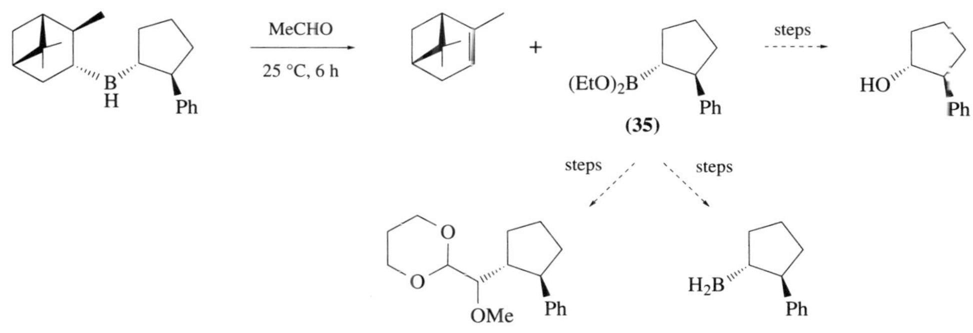

Scheme 24

The asymmetric reduction of carbonyl compounds is an important synthetic method initially yielding alkoxyboranes which are generally not isolated and has been described in great detail B-79MI 210-01, 87MI 210-01, B-88MI 210-01⟩. Perhaps the most versatile reagent reported as of the mid-1990s is chlorodiisopinocampheylborane (**24**) (see Equation (35) for preparation) which combines high reactivity (due to the electron-withdrawing chlorine atom) with high diastereoselectivity in the additions to carbonyl groups ⟨85JOC5446, 86JOC3394, 94TL2141⟩ (Scheme 25).

2.10.1.3.5 *Miscellaneous methods*

In some cases it is possible to prepare alkoxyboranes by a redistribution process (Equation (45)) ⟨B-64MI 210-01, 70JA6983, 71JA2802⟩. Oxidation of alkylboranes represents a very important process

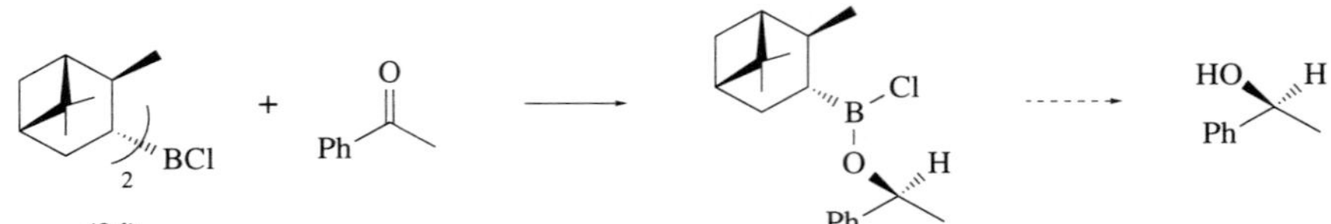

(24)

Scheme 25

from the viewpoint of the synthetic chemist, since it is often employed to convert them into organic products, such as alcohols. However, it is often difficult to prepare alkyloxyboranes in this way because excess oxidant is usually employed, and isolation of intermediates is impossible. The reaction of alcohols with sodium borohydride, in the presence of an equivalent of acetic acid, is a method for the synthesis of trialkoxyboranes ⟨56JA3613⟩.

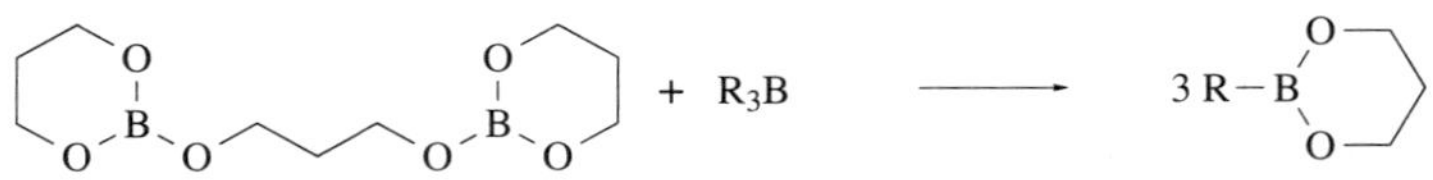

(45)

Boron enol ethers (vinyloxyboranes) are important materials for synthesis having been employed in numerous aldol reactions ⟨68AG(E)735, 71JOC1790⟩. The enol ethers thus generated are generally not isolated but are used directly in subsequent reactions, therefore the discussion in this section will be limited to cases where specific syntheses are possible.

The generation of vinyloxyboranes from carbonyl compounds represents the simplest approach to this class of compound ⟨B-82MI 210-01, 82OR(28)203, 85AG(E)1⟩. The most successful class of reagents for this purpose is the combination of a dialkylboron triflate with a tertiary amine ⟨76CL559, 77CL153, 80BCJ174⟩, the correct choice of which can allow complete regioselectivity in many cases (Scheme 26). The geometry of the enol ether can also be controlled by appropriate choice of reaction conditions, and a great deal of research has been carried out on the subsequent reactions with aldehydes ⟨79JA6120, 79TL1665, 79TL2229, 81JA1566, 81JA3099⟩. Provided that geometrically pure enol ethers are available, the reactions proceed with very high and predictable diastereoselectivity, however very minor changes to the reaction conditions can dramatically affect the enolisation selectivity (Scheme 27) ⟨89JA3441, 93JOC147, 93JOC7162⟩. The two latter references form part of an excellent series of papers on enolboration by Brown, who has found after investigation the optimum combinations of base and diborane source for a given target enolate. In particular, the use of di(cyclohexyl)boron chlorides and iodides with triethylamine is recommended as a versatile combination, a mixture which has been used by other researchers ⟨92JOC499, 92JOC2716, 92T2127, 92TL3421, 94JOC2336⟩.

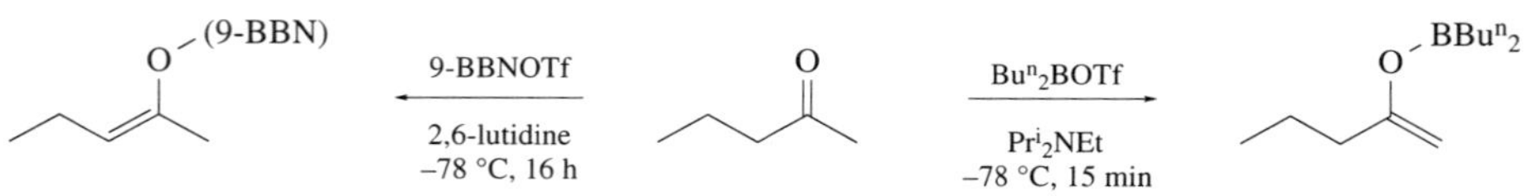

Scheme 26

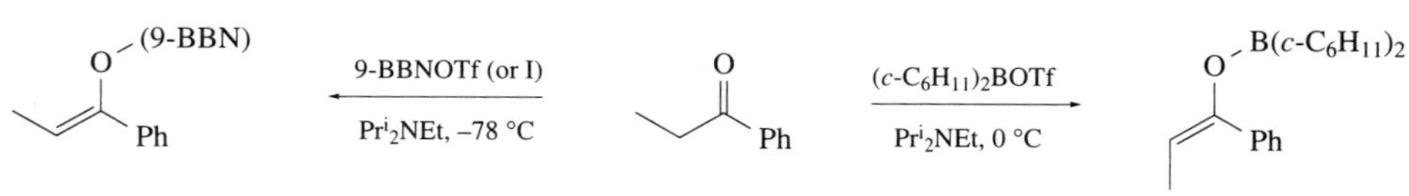

(*Z*) (*E*)

Scheme 27

Homochiral boron enol ethers have been used extensively as reagents for asymmetric synthesis. In some cases these are derivatives of amides attached to a chiral auxiliary ⟨90JA2767, 90JOC173⟩ or homochiral derivatives containing chiral alkyl groups directly attached to the boron atom. Of the latter class those derived from diisopinocampheylborane, for example (**36**), vastly predominate in the literature ⟨88TL585, 89TL1293, 89TL7121, 90T4663⟩ although some other classes, such as (**37**), are also important reagents ⟨89TL7357, 90TL3863⟩.

Boron enol ethers may also be prepared, in some cases, by reductive addition of dialkylboranes to enones ⟨91JOC5820⟩ or to α-diazo esters ⟨87JOC1347⟩.

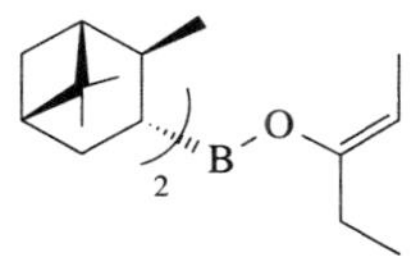

(36)

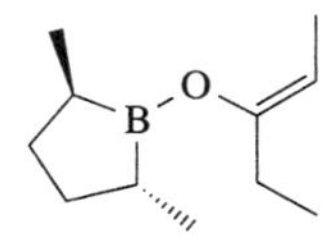

(37)

2.10.1.4 Alkylboranes with a B—S, B—Se or B—Te Bond

Members of this class of compound are generally most easily prepared from the reaction between a thiol and an alkylborane ⟨64CJC2903, B-88MI 210-01⟩. In some cases the presence of oxygen, to act as an inducer of radical formation, is valuable ⟨76JCS(D)2087⟩. Disproportionation reactions have been employed to make (dialkylthio)alkyboranes ⟨61DOK(139)385⟩.

As far as can be determined there are no examples of alkylboranes with B—Se bonds or alkylboranes with B—Te bonds which have significant synthetic applications.

2.10.1.5 Alkylaminoboranes and Related Compounds with a B—N Bond

Compounds of this class may be prepared most conveniently from the reaction of boron trichloride with an amine. If the tris(mono- or dialkylamino)borane is required, excess amine must be used to neutralise the hydrochloric acid which is generated ⟨59JCS2927, 62JCS4088⟩. An alternative method which avoids the use of excess amine is the reaction of an amine with tris(dimethylamino)borane.

The reaction of organometallic reagents with a suitable precursor serves as a convenient method for the synthesis of alkylaminoboranes. One example is the preparation of compound (**38**), an asymmetric allenylating agent, which is prepared using an organotin nucleophile as shown in Equation (46) ⟨90JA878⟩. The related allylborane may be prepared in a similar manner ⟨89JA5493⟩.

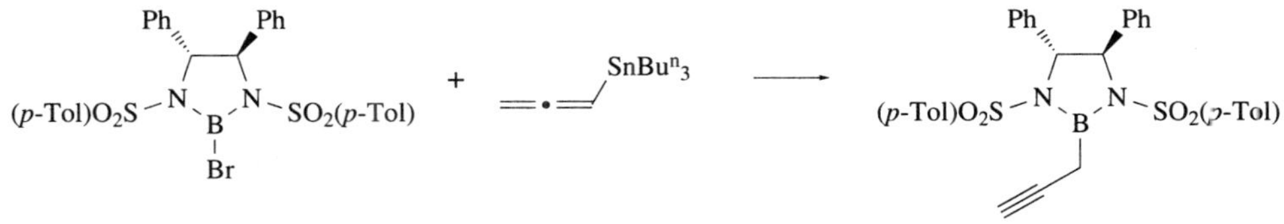

(38) (46)

2.10.1.6 Alkylboranes with a B—P, B—As, B—Sb or B—Bi Bond

As far as can be determined there are no examples of reagents in this class which have significant synthetic applications.

2.10.1.7 Alkylboranes with a B—Si Group

As far as can be determined there are no examples of reagents in this class which have significant synthetic applications.

2.10.1.8 Alkylboranes with a B—B Bond

As far as can be determined there are no examples of reagents in this class which have significant synthetic applications.

2.10.1.9 Alkylboranes with a B—Metalloid Bond (other than Si or B)

As far as can be determined there are no examples of reagents in this class which have significant synthetic applications.

2.10.1.10 Alkylboranes with a B—Metal Bond

As far as can be determined there are no examples of reagents in this class which have significant synthetic applications.

2.10.1.11 Alkylboranes with Boron having a Valency Higher than Three

Reaction of a hydride donor with an alkylborane results in the formation of the corresponding hydroborates (Equations (47)–(49)). The hydroborates thus formed are generally stable in solution for several months under an inert atmosphere but are unstable to air and moisture. Exposure to either can result in violent decomposition ⟨81JOC2712⟩.

(47)

(48)

(49)

Some valuable reducing agents have been prepared by the reaction of potassium hydride with organoboranes or alkoxyboranes ⟨85JOC549, 86JOC3278, 86JOC3396⟩. Such reactions generally proceed rapidly at ambient temperature to furnish potassium hydroborates of high quality. In some cases it is essential to add a small quantity of an alkoxyborane to the reaction mixture ⟨86JOC238⟩. Of the reducing agents formed the potassium 9-thexyloxy-9-borabicyclo[3.3.1]nonane **(39)** appears to be one of the most diastereoselective (Equation (50)) ⟨85JOC549⟩. Asymmetric reducing agents derived from chiral alcohols such as (−)-1,2,5,6-di-*O*-isopropylidene-α-D-glucofuranose **(40)** may also be prepared using the potassium hydride method ⟨86JOC3278, 86JOC3396⟩.

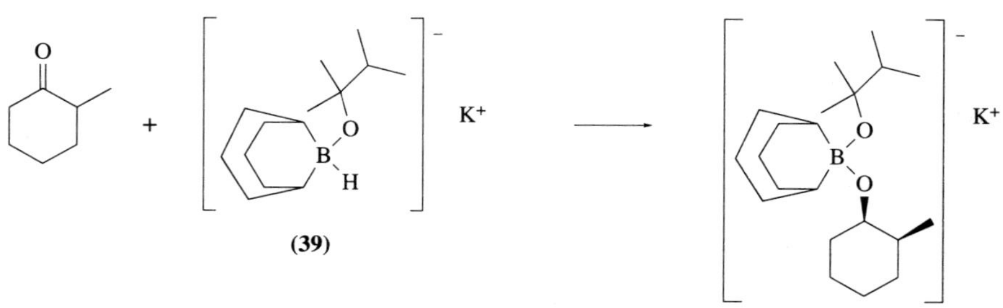

(39)

98.5% this isomer

(50)

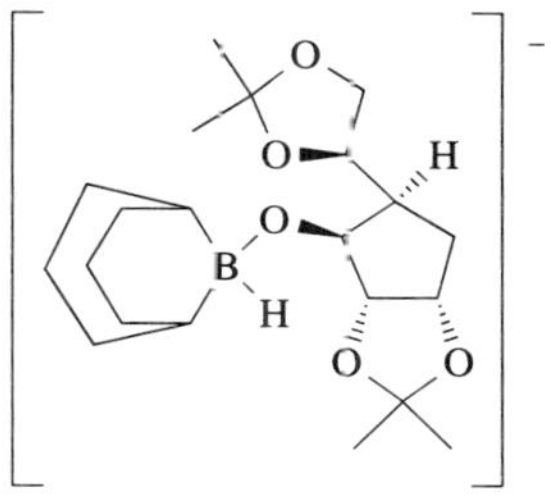

(40)

The reaction of alkoxyboranes with lithium aluminum hydride results in full reduction to the corresponding lithium hydride ⟨85JA460, 87JOC728, 87JOC4020⟩. This represents a convenient method for the conversion of boronic esters, perhaps derived from asymmetric hydroboration (see Section 2.10.1.3.4) to primary boranes (Scheme 28), which may then be used in further synthetic reactions.

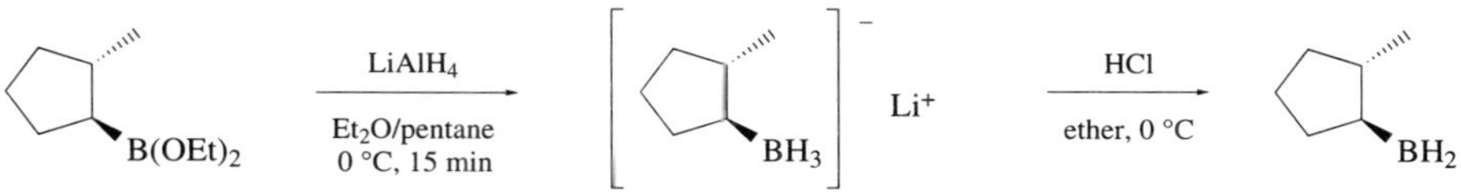

Scheme 28

2.10.2 ALKYLSILICON DERIVATIVES

This chapter deals with the synthesis of functionalised alkylsilicon derivatives, mainly of those with demonstrable synthetic utility in organic chemistry. Several books and reviews have been published which deal more generally with organosilicon chemistry ⟨79COC(3)541, 80EOC(13/5), B-81MI 210-01, 82COMC-I(2)1, 82COMC-I(7)515, B-83MI 210-03, B-87MI 210-02, B-88MI 210-02, B-89MI 210-01, B-91MI 210-01⟩. Very many, if not all, of the alkylsilicon derivatives discussed in this chapter are commercially available.

In addition, detailed listings of the preparation of individual organosilicon compounds are also available ⟨B-65MI 210-01, B-73MI 210-02, B-77MI 210-01, B-84MI 210-02⟩.

2.10.2.1 Alkylsilanes

2.10.2.1.1 Hydridosilanes

Hydridosilanes, compounds with silicon—hydrogen bonds, can be prepared by treatment of metal silicides or compounds with a silicon—metal bond with aqueous acid or another source of protons (Scheme 29). Silane itself and the higher silanes are of little importance in synthetic organic chemistry.

$$Ca_2Si \xrightarrow{H_3O^+} SiH_4$$

$$Mg_2Si \xrightarrow{H_3O^+} SiH_4 + Si_2H_6 + Si_3H_8 \quad \text{etc.}$$

$$Ph_3SiK + Ph_3CH \longrightarrow Ph_3SiH + Ph_3CK$$

Scheme 29

Trichlorosilane, first prepared by Wöhler in 1857 by passing HCl over Si—Fe, is also formed in the direct synthesis (see Section 2.10.2.2) of dimethyldichlorosilane, as is methyldichlorosilane. It is the usual starting material for the large scale preparation of hydridosilanes, since its chlorine atoms can readily be displaced by other groups (Equation (51)).

$$Cl_3SiH + 3\ EtMgBr \longrightarrow Et_3SiH \qquad (51)$$

The other main route to hydridosilanes, and the one well suited to the laboratory, is reduction of the corresponding chlorosilane, normally using lithium aluminum hydride. However, there are two limitations to this usually straightforward technique. It is not possible to achieve selective partial reduction of a polychloride, and the aluminum chloride formed as the reduction proceeds can polymerise reactive silanes such as allyltrichlorosilane; this can be reduced successfully using lithium hydride. Some hindered Grignard reagents with β-hydrogens can reduce chlorosilanes, while hindered halosilanes undergo reductive condensation with *t*-butyllithium (Scheme 30).

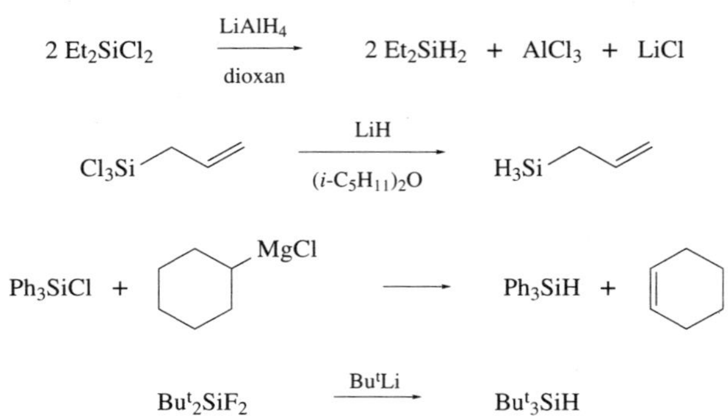

Scheme 30

The major uses of hydridosilanes are in hydrosilylation, ionic hydrogenation (especially using triethylsilane) and as radical based reducing agents. For the last purpose, phenylsilane, diphenylsilane ⟨91SL435, 92TL6629⟩, triphenylsilane ⟨90JOC5413⟩, triethylsilane in the presence of thiols ⟨89CC1387⟩, tris(alkylthio)silanes ⟨90SL257⟩ and tris(trimethylsilyl)silane ⟨92ACR188⟩ all show considerable utility (see also Sections 2.10.2.4 and 2.10.2.8.1).

2.10.2.1.2 Tetraorganosilanes

There are several methods for the preparation of tetraorganosilanes. The two most useful laboratory methods are reaction of organomagnesium or organolithium reagents with chloro- or alkoxysilanes, and the addition of hydridosilanes to alkenes (hydrosilylation); these are interlinked when hydrosilylation is performed using chlorosilanes. Both methods have been comprehensively surveyed ⟨B-68MI 210-01, B-72MI 210-02, B-92MI 210-01⟩. A third valuable route, the reaction of silyl anions or their equivalents with carbon electrophiles, is discussed in Section 2.10.2.9.

Hydrosilylation of alkenes can be achieved under a wide variety of reaction conditions. It can be initiated thermally, when relatively high temperatures and pressures are required. More conveniently, the process can be initiated by catalysts including radical initiators, Lewis acids and transition metals. Of the last, the most commonly used catalyst is hexachloroplatinic acid, $H_2PtCl_6 \cdot 6H_2O$, also known, particularly when used in isopropanol, as Speier's catalyst ⟨79AOC(17)407⟩. With simple alkenes, terminal addition is highly favoured, even when the alkene is internal (Scheme 31) ⟨57JA974⟩.

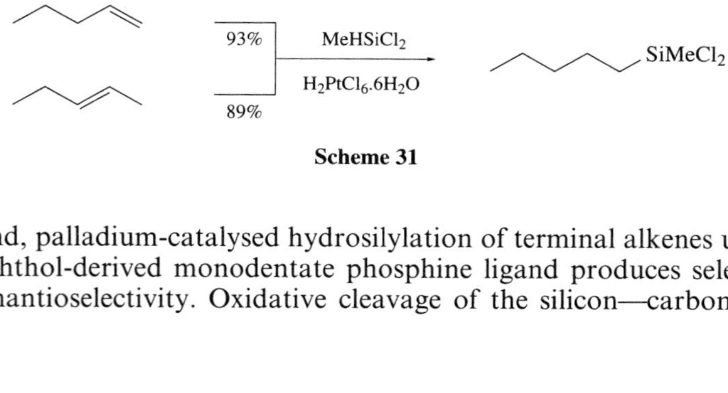

Scheme 31

On the other hand, palladium-catalysed hydrosilylation of terminal alkenes using trichlorosilane and a chiral binaphthol-derived monodentate phosphine ligand produces selectively the internal silane with high enantioselectivity. Oxidative cleavage of the silicon—carbon bond ⟨88MI 210-03,

91COS(7)641⟩ then provides chirally enriched secondary alcohols (Scheme 32) ⟨91JA9887, 92TL7185, 93TL2335⟩.

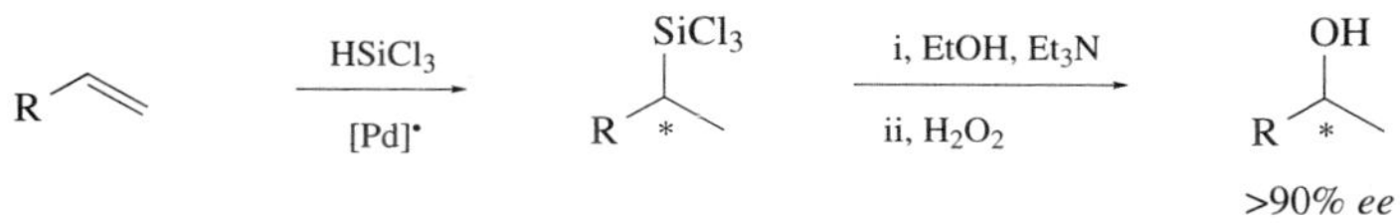

Scheme 32

Of the Lewis acid-catalysed hydrosilylations, one of the most valuable synthetically is the hydrosilylation of tetrasubstituted alkenes with chlorodimethylsilane, as exemplified by the synthesis of the useful alcohol protective reagent, thexyldimethylchlorosilane (**41**) ⟨85TL5511, 85TL5515⟩. Although immaterial in this example, this addition has been shown to proceed in a *trans* manner (Scheme 33) ⟨90SL259⟩.

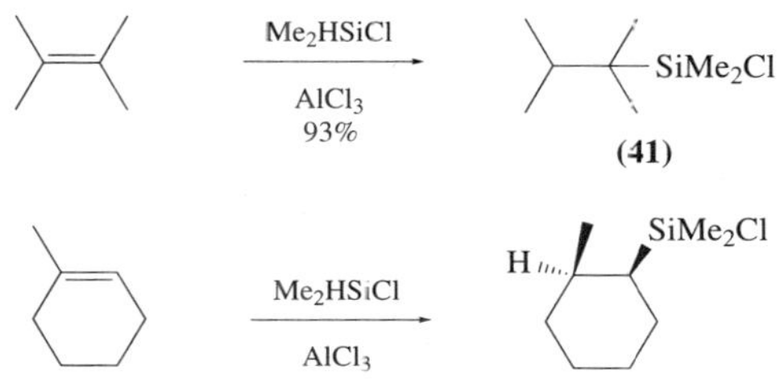

Scheme 33

The high reactivity of chlorosilanes towards Grignard reagents normally ensures that other reactive groups bonded to carbon are left unaffected, as exemplified by an efficient synthesis of the synthetically valuable (chloromethyl)trimethylsilane (**42**) (Equation (52)) ⟨47JA1976, 85S717⟩.

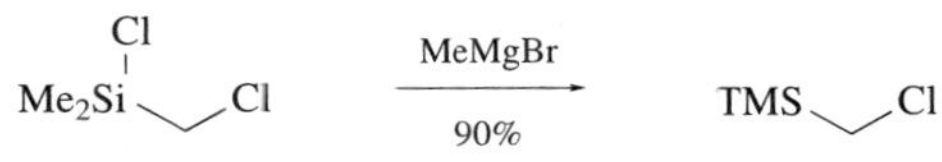

(42) (52)

Tetraalkylsilanes are extremely stable compounds, and only undergo silicon—carbon bond fission under quite vigorous conditions unless also bearing a suitable activating group, such as vinyl, allyl, alkynyl or aryl (see Chapters 2.18.2 and 2.23.2).

For example, alkyltrimethylsilanes undergo regioselective cleavage of the silicon—methyl bond when treated with powerful electrophiles such as ICl ⟨86BSF413⟩. Cleavage can also be achieved using acetyl chloride and $AlCl_3$, to give chlorosilanes ⟨66TL5493⟩. This acylium cation induced cleavage has been put to good effect in the preparation of acyclic ⟨77S677⟩ and cyclic ⟨84JOC1140⟩ ketones (Scheme 34).

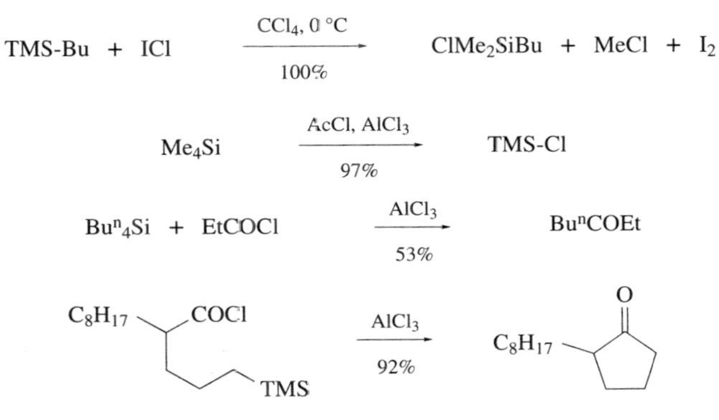

Scheme 34

2.10.2.2 Alkylsilyl Halides

Alkylsilyl halides, in particular the chlorides, are by far the most important intermediates in organosilicon chemistry. Prior to the 1940s, silicon tetrachloride was the main starting material. It can be prepared directly from silicon and chlorine, or from a high temperature reaction between coke, silica and chlorine; in the latter process, the cheapness of the starting materials offsets the energy costs. Reaction with Grignard reagents then provides alkylsilyl chlorides. However, with simple Grignard reagents, such as MeMgBr or EtMgBr, this is not a good method for the preparation of a particular chloride, with mixtures of products being obtained, even though the reactivity of the silyl chloride decreases with increasing substitution. However, stepwise substitution becomes more controlled when more hindered Grignard and organolithium reagents are employed. For example, *t*-butyldimethylchlorosilane, the source of *t*-butyldimethylsilyl protection of alcohols, is cleanly produced from *t*-butyllithium and dichlorodimethylsilane ⟨54JA1030⟩. Similarly, *t*-butyldiphenylchlorosilane is prepared from *t*-butyllithium and dichlorodiphenylsilane (Scheme 35) ⟨75CJC2975⟩.

$$Me_2SiCl_2 \xrightarrow[\text{pentane}]{Bu^tLi} Bu^tMe_2SiCl$$

$$Ph_2SiCl_2 \xrightarrow[\text{pentane, 95\%}]{Bu^tLi} Bu^tPh_2SiCl$$

Scheme 35

Large scale production of alkylsilyl chlorides became possible with the advent of the direct synthesis, discovered by Rochow in 1940 and, independently, by Müller in 1942 ⟨50IS50, 64MI 210-01, B-67MI 210-03, 68AG(E)112, B-68MI 210-01⟩. This process involves reaction of alkyl or aryl chlorides with silicon in the presence of a catalyst, usually copper, at 250–300 °C to give dialkyl- or diaryldichlorosilanes. Such vigorous conditions are necessary because silicon is only mildly metallic. With chloromethane, and under appropriate reaction conditions, the product is composed of dichlorodimethylsilane (65%, b.p. 70 °C), trichloromethylsilane (25%, b.p. 65.7 °C) and chlorotrimethylsilane (5%, b.p. 57.3 °C) (Equation (53)). By-products, of which there can be up to forty, include hydridosilanes, such as the synthetically useful chlorodimethylsilane, and disilanes. Careful fractional distillation separates these, which are then available from the large scale production of dichlorodimethylsilane.

$$MeCl \xrightarrow[250\text{–}300\ ^\circ C]{Si/Cu} Me_2SiCl_2 + MeSiCl_3 + \text{TMS-Cl} \qquad (53)$$

The third valuable method for the production of chlorosilanes involves hydrosilylation using chlorosilanes. For example, the alcohol protective reagent thexyldimethylchlorosilane is prepared by Lewis acid-catalysed reaction between chlorodimethylsilane and 2,3-dimethylbut-2-ene (Section 2.10.2.1) ⟨85TL5511⟩, and 1,2-bis(chlorodimethylsilyl)ethane (**43**), used in the protection of primary amines (Section 2.10.2.6), is prepared by hexachloroplatinic acid catalysed hydrosilylation of chlorodimethylvinylsilane with chlorodimethylsilane (Equation (54)) ⟨60JA1883⟩.

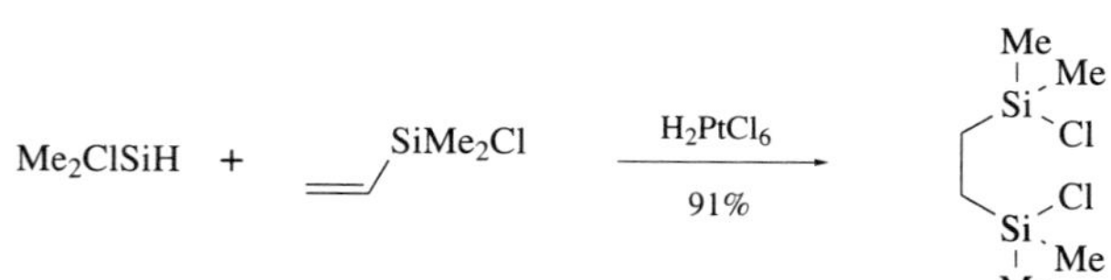

(43) (54)

Chlorotriisopropylsilane is best prepared by reaction between triisopropylsilane ⟨80JOC4797⟩ and copper(II) chloride (Equation (55)); chlorotriethylsilane can be prepared similarly ⟨74SRI23⟩. Iodotrimethylsilane ⟨80S861, 82T2225⟩ is readily prepared from hexamethyldisiloxane and iodine in the presence of aluminum powder ⟨79OS(59)35⟩, while bromotrimethylsilane results from the reaction of hexamethyldisiloxane with phosphorus(III) bromide in the presence of a catalytic amount of iron(III) chloride (Scheme 36) ⟨46JA1161⟩.

$$Pr^i_3SiH + CuCl_2 \xrightarrow[92\%]{MeCN} Pr^i_3SiCl \qquad (55)$$

$$\text{TMS-I} \xleftarrow{Al,\ I_2} \text{TMS-O-TMS} \xrightarrow[FeCl_3 \cdot 6H_2O\ (cat.)]{PBr_3} \text{TMS-Br}$$

Scheme 36

2.10.2.3 Alkylsiloxanes

The silicon—oxygen bond is stronger than that of any other element to silicon apart from fluorine. However, despite its strength, it is readily cleaved either hydrolytically or by nucleophilic attack by fluoride ion, providing an excellent range of protective groups for alcohols, phenols and carboxylic acids (see also Chapter 5.04.10). The preparation, cleavage and stability profiles of such ethers and esters have been tabulated ⟨85S817, B-88MI 210-04, B-91MI 210-02⟩.

The most convenient precursors to alkyl and aryl silyl ethers and esters are the chlorosilanes, alkylsilazanes and silyl trifluoromethylsulfonates ⟨81T3899, 82S1⟩. However, direct reaction of hydridosilanes and disilanes with alcohols, catalysed by transition metals ⟨90JOC6082, 92TL5041⟩ or fluoride ion ⟨94TL8413⟩ presents a viable alternative to the use of the frequently more expensive chlorosilanes.

The two most common silyl ethers used in alcohol protection are trimethylsilyl (**44**) and *t*-butyldimethylsilyl (**45**). For simple primary and secondary alcohols, direct reaction with hexamethyldisilazane with chlorotrimethylsilane catalysis is a convenient procedure for the preparation of trimethylsilyl ethers. For tertiary and other hindered alcohols, the combination of equimolar amounts of hexamethyldisilazane and chlorotrimethylsilane in a non-polar solvent such as hexane is very effective. For most purposes, one or other of these two procedures should be satisfactory ⟨58JOC50⟩.

RO-TMS	$ROSiMe_2Bu^t$
(44)	**(45)**

However, trimethylsilyl ethers show high acid- and base-catalysed solvolytic lability. For practical protection, use of more highly substituted silyl groups, such as *t*-butyldimethylsilyl ⟨68JA4462⟩, is preferred. The steric bulk which confers increased protection also resists formation of the silyl ether in the first place, and other methods for their formation have been devised. Both primary and secondary alcohols can be readily converted into their *t*-butyldimethylsilyl ethers using the chlorosilane and imidazole in DMF ⟨72JA6190⟩. Alternatively, use of the chlorosilane in the presence of stoichiometric amounts of triethylamine and catalytic amounts of dmap (4-dimethylaminopyridine) gives equally good yields and additionally shows a pronounced kinetic preference for primary alcohols ⟨79TL99⟩.

A third general method makes use of *t*-butyldimethylsilyl perchlorate. This reacts with hindered alcohols, including tertiary alcohols, in the presence of pyridine to give the corresponding silyl ethers in yields superior to either of the other methods ⟨82S1⟩.

Other silyl ethers which have been employed to confer differing levels of protection include triisopropylsilyl (**46**) ⟨80JOC4797, 81TL3455⟩, *t*-butyldiphenylsilyl (**47**) ⟨79ACR159⟩, dimethylthexylsilyl (**48**) ⟨85TL5511, 85TL5515⟩ and dimethyltritylsilyl (**49**) ⟨77JCR(S)6⟩. Methyldiphenylsilyl ethers (**50**) show promise as protective groups; they are readily prepared from primary, secondary and tertiary alcohols, with their stability lying between trimethylsilyl and *t*-butyldimethylsilyl ethers ⟨87JOC165⟩. The value of *t*-butylmethoxyphenylsilyl ethers (**51**) has also been explored ⟨84TL4717⟩.

$ROSiPr^i_3$	$ROSiPh_2Bu^t$	$ROSiMe_2C_6H_{13}$	$ROSiMe_2CPh_3$	$ROSiMePh_2$	$ROSi(OMe)PhBu^t$
(46)	**(47)**	**(48)**	**(49)**	**(50)**	**(51)**

Bromomethylchlorodimethylsilane in the presence of a base converts alcohols into the corresponding bromomethyldimethylsilyl ethers. Although such ethers have not been used as protective groups, when derived from allylic or homoallylic alcohols they undergo a most useful radical-induced cyclisation. Oxidative cleavage of the silicon—carbon bond ⟨88MI 210-03, 91COS(7)641⟩ then results in the overall *anti* addition of a hydrogen and a hydroxymethyl group to an alkene (Scheme 37) ⟨85JA500, 86JA6826⟩ and in the stereocontrolled syntheses of 1,3- and 1,4-diols (Equation (56)) ⟨84JOC2298, 90JA8175⟩.

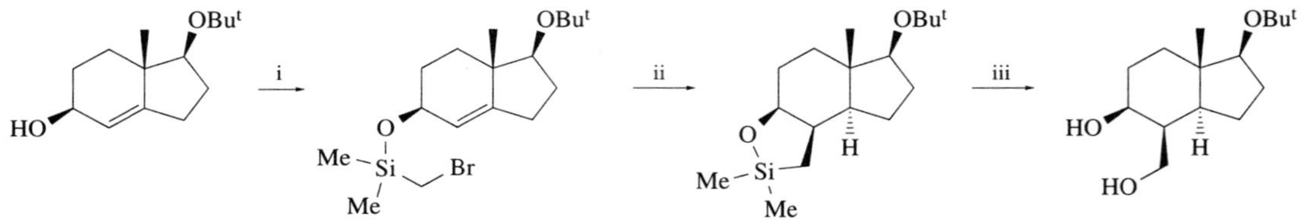

i, $Me_2Si(Cl)CH_2Br$, Et_3N, dmap; ii, Bu_3SnH, AIBN (cat.), PhH or $NaCNBH_4$, Bu_3SnCl (cat.), AIBN (cat.), Bu^tOH;
iii, H_2O_2, KF, DMF

Scheme 37

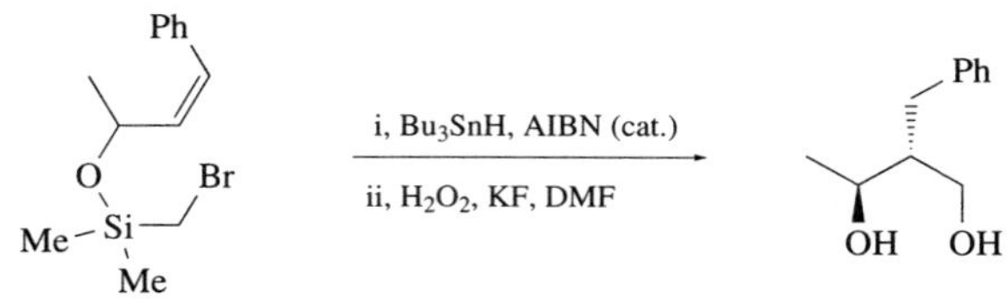

(56)

A variety of difunctionalised silanes and disiloxanes has been described for the protection of 1,2-, 1,3- and 1,4-diols. Di-*t*-butylsilylene protection can be achieved using di-*t*-butyldichlorosilane and triethylamine with 1-hydroxybenzotriazole as catalyst ⟨81TL4999, 83JOC3252⟩. Similar protection can be achieved using di-*t*-butylsilyl or diisopropylsilyl bistrifluoromethanesulfonate in the presence of 2,6-lutidine, when milder conditions can be used ⟨82TL4871⟩.

1,3-Dichloro-1,1,3,3-tetraisopropyldisiloxane (**52**) has found extensive use in carbohydrate and nucleotide chemistry, in particular for the simultaneous protection of the 3′- and 5′-hydroxyl groups of nucleosides ⟨80TL4523⟩, and also in serine-containing peptides ⟨87JCS(P1)1107⟩. Related protection of nucleosides as their dimethylsilylene, diphenylsilylene, methylphenylsilylene and di-*t*-butylsilylene derivatives has also been explored ⟨83MI 210-02, 85MI 210-02⟩.

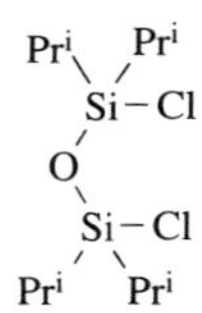

(52)

2.10.2.4 Alkylsilathianes

The preparation and properties of alkylsilathianes have been reviewed ⟨80S861, 88RCR447⟩.

In addition to the normal preparative routes to alkylsilathianes involving reaction of a chlorosilane with a metal thiolate or by substitution of a silylamine with thiols, a wide range of alkylsilathianes can be readily prepared by tris(triphenylphosphine)rhodium chloride-catalysed dehydrogenative condensation between thiols and hydridosilanes (Equation (57)). Silathianes prepared in this manner include TMS-SPh, Et_3SiSPh, $Pr^n{}_3SiSPh$, $PhMe_2SiSPh$, Et_3SiSEt, Et_3SiSBu^t, Et_3SiSCH_2Ph, $Ph_2HSiSPh$, PhMeHSiSPh, $Et_2HSiSPh$, $Ph_2HSiSEt$, PhMeHSiSEt and $Et_2HSiSEt$; all are obtained in essentially quantitative yield ⟨73JOM(50)C26⟩.

$$R^1{}_3SiH + R^2SH \xrightarrow{(Ph_3P)_3RhCl} R^1{}_3SiSR^2 \quad (57)$$

Bis(trimethylsilyl) sulfide can be readily prepared by an imidazole-catalysed reaction between hexamethyldisilazane and H_2S ⟨76S721⟩ or by reaction of H_2S with 1-(trimethylsilyl)imidazole ⟨68IC1253⟩. Indeed, this imidazole-catalysed reaction with hexamethyldisilazane provides a general method for the conversion of thiols, including ethanedithiol and propane-1,3-dithiol, into trimethylsilyl thioethers ⟨73JOM(61)83, 77JA5009⟩. Alternatively, and avoiding the use of H_2S, bis(trimethylsilyl) sulfide can be obtained in 83–90% yield by lithium borohydride reduction of elemental sulfur in THF, and quenching the dilithium sulfide with chlorotrimethylsilane (Scheme 38) ⟨82JOC1354⟩.

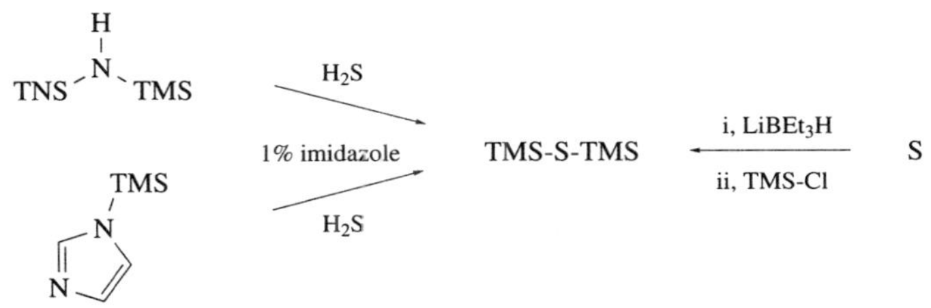

Scheme 38

In the presence of BCl_3, this reagent converts carbonyl compounds into the corresponding thiocarbonyl compounds via the intermediate formation of B_2S_3; aldehydes, ketones, lactones and lactams all proved suitable substrates ⟨82JA3104⟩. Use of $CoCl_2 \cdot 6H_2O$ or trimethylsilyl triflate with bis(trimethylsilyl) sulfide on aldehydes allows the generation and *in situ* trapping of thioaldehydes; only the latter catalyst was able to promote the thionation of ketones (Equation (58)), some of which were isolated as the corresponding trimers ⟨93TL873⟩.

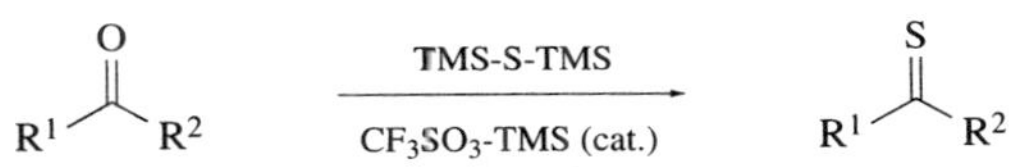

(58)

Alkylthio- and arylthiotrimethylsilanes react smoothly with aldehydes under the catalytic influence of cyanide or fluoride ions to form *O*-silyl hemithioacetals ⟨77JA5009⟩. Ketones are unreactive under such conditions, offering the potential of selective carbonyl group protection. However, under ZnI_2 catalysis and employing at least two equivalents of thiosilane, both aldehydes and ketones are transformed into dithioacetals and dithioketals, respectively; reaction can be stopped, if desired, at the intermediate hemithioacetal or hemithioketal stage (Scheme 39). With α,β-unsaturated aldehydes and ketones, exclusive 1,4-addition is observed.

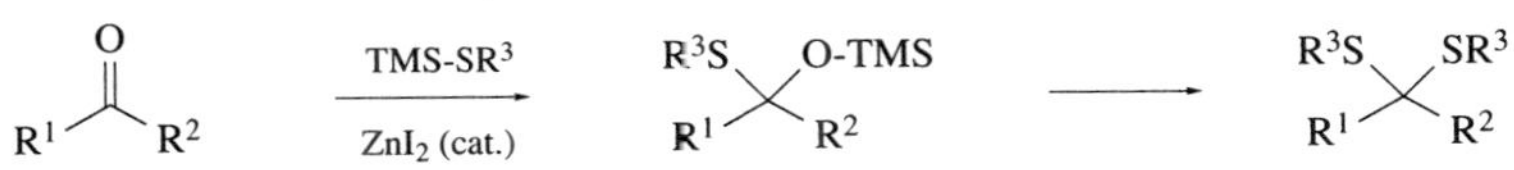

Scheme 39

Allyl tris(trimethylsilyl)sulphide (**53**) can be prepared in 60% yield in a one-pot reaction by addition of tris(trimethylsilyl)silane to carbon tetrachloride (which forms tris(trimethylsilyl)silyl chloride) followed by addition of allyl thiol and triethylamine; this compound acts as a radical allylating agent ⟨92TL6931⟩.

Tris(methylthio)silane (**54**) and tris(isopropylthio)silane (**55**) can be prepared in yields of 40% and 60% respectively using trichlorosilane, the appropriate alkanethiol and pyridine. These compounds act as radical-based reducing agents; however, they are very unpleasant, and should be handled carefully in a well-ventilated area ⟨90SL219⟩.

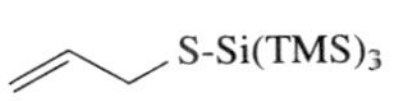

(53)

$(MeS)_3SiH$

(54)

$(Pr^iS)_3SiH$

(55)

Triphenylsilanethiol, readily prepared in 72% yield from triphenylsilane and elemental sulfur ⟨63CB3289⟩, has been recommended as a solid H_2S equivalent for the opening of oxiranes (Scheme 40) ⟨93TL3363⟩.

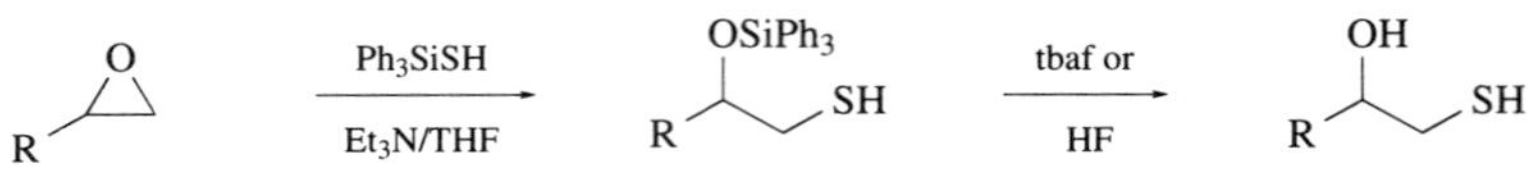

Scheme 40

2.10.2.5 Alkylsilicon Compounds with an Si—Se and/or Si—Te Bond

Bis(trimethylsilyl) selenide can be prepared conveniently in 63–95% yield ⟨82JOC1354⟩ by lithium triethylborohydride reduction of selenium shot in THF, and quenching the dilithium selenide with chlorotrimethylsilane. It is an air-sensitive, colourless oil which can be stored at −20 °C in argon-flushed vials for up to one month without major decomposition (Scheme 41). Bis(trimethylsilyl) telluride can be prepared analogously in 53–75% yield from tellurium shot, lithium triethylborohydride and chlorotrimethylsilane; it is extremely sensitive to both air and light. Bis(*t*-butyldimethylsilyl) telluride, prepared similarly in 79% yield as a low-melting, white solid, can also be stored in argon-flushed vials at −20 °C for several weeks.

$$\mathrm{Se} \xrightarrow{\mathrm{LiBEt_3H}} \mathrm{Li_2Se} \xrightarrow{\mathrm{TMS\text{-}Cl}} \mathrm{(TMS)_2Se}$$

$$\mathrm{Te} \xrightarrow{\mathrm{LiBEt_3H}} \mathrm{Li_2Te} \xrightarrow{\mathrm{TMS\text{-}Cl}} \mathrm{(TMS)_2Te}$$

Scheme 41

(Phenylseleno)trimethylsilane itself can be prepared by a variety of methods, including the reaction of the phenylselenide anion with chlorotrimethylsilane and dehydrogenative coupling of benzeneselenol with trimethylsilane. Application of one or other of these methodologies has also allowed the preparation of (phenylseleno)triethylsilane, (phenylseleno)-*t*-butyldimethylsilane and (phenylseleno)-*t*-butyldiphenylsilane (Scheme 42) ⟨81JOC1283⟩. They all undergo selenium—silicon bond cleavage with halogens to give the corresponding halosilanes.

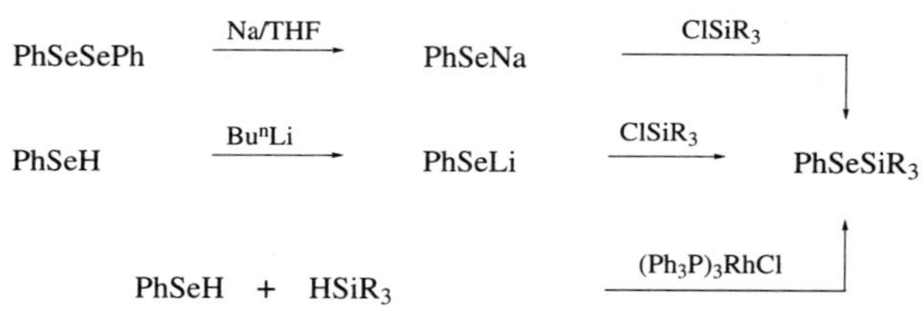

Scheme 42

(Phenylseleno)trimethylsilane reacts with saturated aldehydes to give *O*-silyl hemiselenoacetals, paralleling the reactivity of (phenylthio)trimethylsilane (Section 2.10.2.4). Once again, ketones are unreactive under such conditions. With α,β-unsaturated carbonyl compounds, the products of 1,4-addition are formed. Suitable catalysts include triphenylphosphine or iodotrimethylsilane ⟨78TL5087, 78TL5091, 79TL4189⟩.

Selenoaldehydes can be prepared by reacting aldehydes with bis(trimethylsilyl) selenide in the presence of catalytic amounts of Bu^nLi; the reaction is carried out in the presence of cyclopentadiene to trap the reactive product (Scheme 43) ⟨88JA1976⟩.

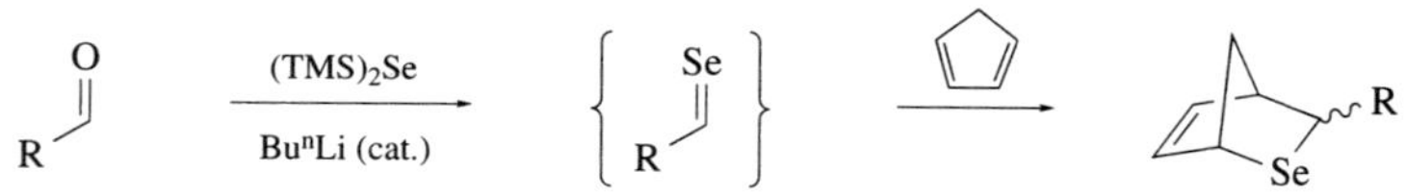

Scheme 43

Alternatively, reaction of bis(trimethylsilyl) selenide with aldehydes in the presence of $BF_3 \cdot OEt_2$ produces 1,3,5-triselenanes; as trimers of selenoaldehydes, they undergo thermal or Lewis acid induced fragmentation to the parent selenoaldehydes, which are again trapped by suitable dienes (Equation (59)) ⟨89TL6047⟩.

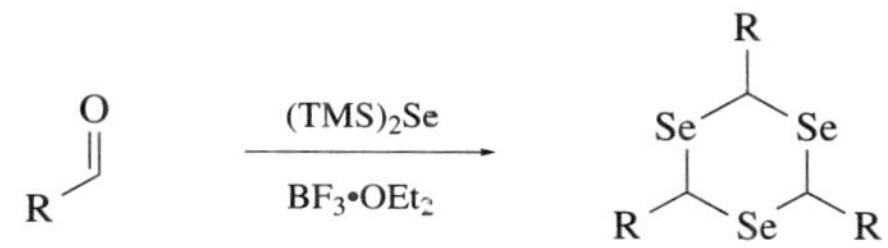

(59)

2.10.2.6 Alkylsilazanes

The preparation and properties of alkylsilazanes have been reviewed ⟨61CRV361⟩. Of the simple silazanes, hexamethyldisilazane (bis(trimethylsilyl)amine) is by far the most useful. Its applications are three-fold: as a silyl transfer reagent (Section 2.10.2.3), as a bulky ligand for transition metals, particularly when in low coordinative environments, and as a source of hindered strong bases, the alkali metal bis(trimethylsilyl)amides ⟨76MI 210-01, B-80MI 210-01⟩.

The commercially available dichlorosilane 1,2-bis(chlorodimethylsilyl)ethane (**43**) (Section 2.10.2.2) converts primary amines into their cyclic 'stabase' derivatives, thus providing highly useful protection of such amines. For primary amines in the pKa range of 10–11, triethylamine suffices as base; for less basic amines such as the anilines, more vigorous conditions, using Bu^nLi (2 equivalents) must be used ⟨81TL1787⟩. Alternatively, anilines can be protected in high yield using 1,1,4,4-tetramethyl-1,4-bis(*N*,*N*-dimethylamino)disilethylene with zinc iodide catalysis at 140 °C (Scheme 44) ⟨84TL1253⟩.

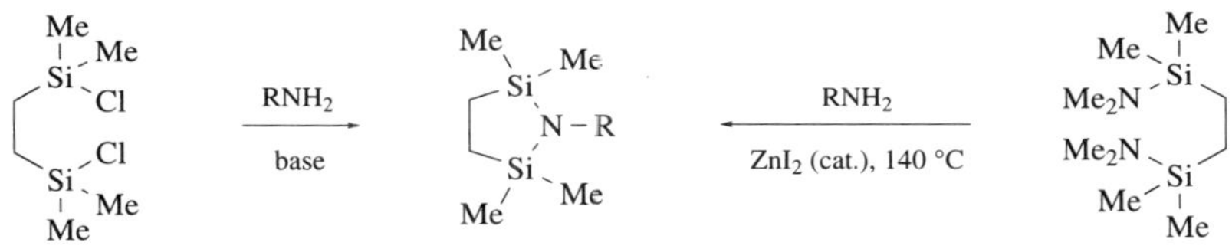

(**43**)

Scheme 44

Anilines can also be protected as 'benzostabase' derivatives by dehydrogenative silylation with 1,2-bis(dimethylsilyl)benzene (Equation (60)) ⟨90TL6721⟩. Such benzostabase derivatives are considerably more stable than the corresponding 'stabase' analogues towards protic hydrolysis and silica gel chromatography. Similar protection of primary aliphatic amines confers a lesser, but still useful, improvement in stability ⟨90TL6725⟩.

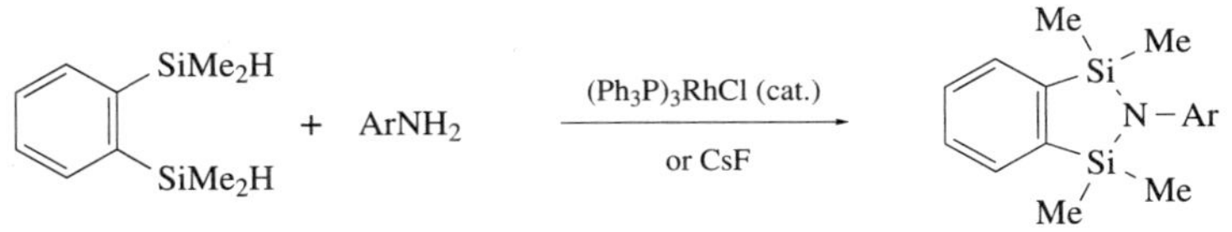

(60)

Primary aliphatic amines can also be protected as their *t*-butyldiphenylsilyl derivatives by reaction with *t*-butylchlorodiphenylsilane in the presence of triethylamine; secondary amines are unreactive under these conditions, thus allowing, for example, selective silylation of *N*-methyl-1,3-propanediamine (Scheme 45) ⟨86TL4391⟩.

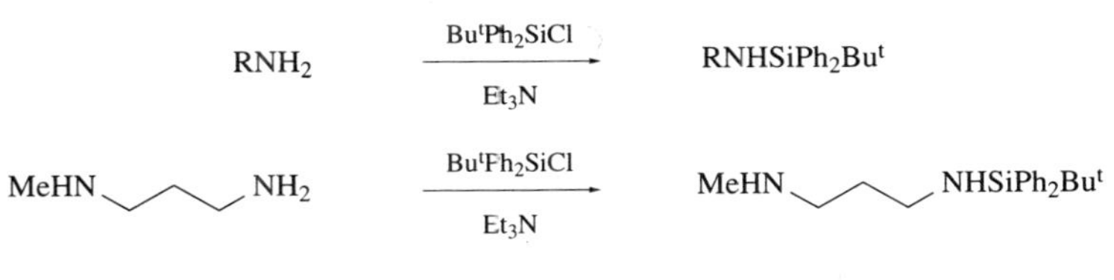

Scheme 45

Full details have been published on the preparation and reactions of *N*,*N*-bis(silyl)enamines ⟨90JOC2878⟩.

The preparation and some reactions of *N*-silyl aldimines have been reviewed ⟨90PAC605⟩, as have their condensation reactions with ester enolates and synthetic equivalents to produce *N*-silyl β-lactams ⟨89CRV1447⟩. A frequently used route to such aldimines involves reaction of an aldehyde, even if enolisable, with lithium bis(trimethylsilyl)amide (Scheme 46).

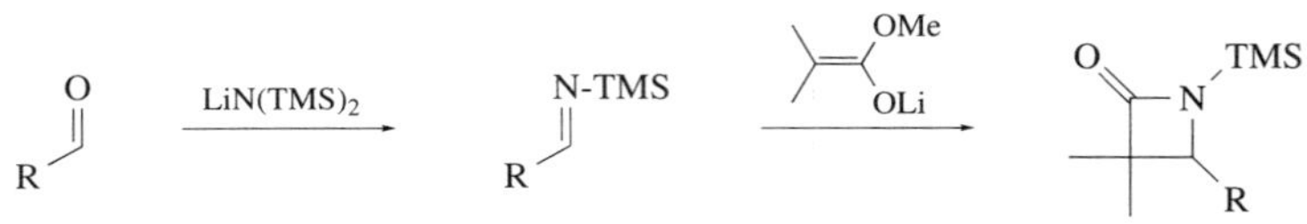

Scheme 46

N-Trimethylsilyl benzaldehyde imine undergoes alkylation with chirally modified Bu^nLi to give chiral primary amines in up to 62% *ee*. Remarkably, use of diethyl ether as solvent favours production of the (*S*) enantiomer, while use of hexane favours the (*R*) ⟨91JCS(P1)1341⟩.

A range of silyl isoselenocyanates, including isoselenocyanato-trimethylsilane, TMS-NCSe, have been prepared by *in situ* reaction between chlorosilanes and excess KSeCN ⟨89JOC2015⟩. TMS-NCSe shows high chemselectivity in its conversion of carbonyl compounds into *O*-trimethylsilyl cyanohydrins. Aliphatic aldehydes react smoothly, whereas aromatic and α,β-unsaturated aldehydes react only very slowly; ketones are unreactive. This is in marked contrast to isothiocyanatotrimethylsilane, which reacts with aldehydes under Lewis acid-catalysis to yield dimeric α,α'-isocyanato ethers (Scheme 47) ⟨87BCJ2289⟩.

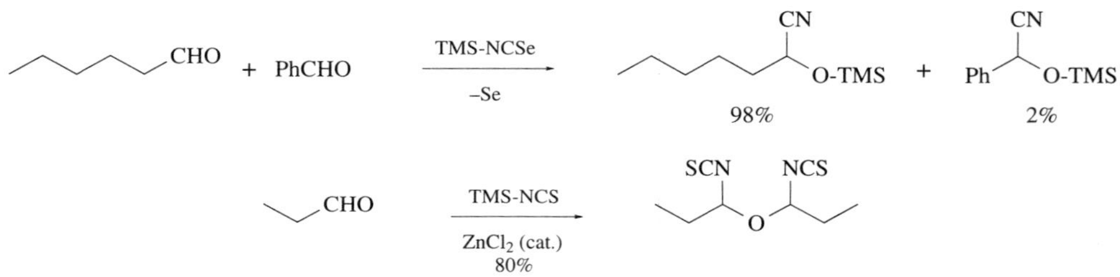

Scheme 47

2.10.2.7 Alkylsilicon Compounds with an Si—P, Si—As, Si—Sb or Si—Bi Bond

The preparation and properties of phosphines, arsines, stibines and bismuthines containing silicon have been comprehensively reviewed ⟨70MI 210-01⟩. As far as can be determined, there are no examples of reagents in these classes which have significant synthetic applications in organic chemistry.

2.10.2.8 Alkylsilicon Compounds with an Si—Metalloid Bond

2.10.2.8.1 Compounds with Si—Si bonds

The direct synthesis (Section 2.10.2.2) of dichlorodimethylsilane also produces a variety of compounds with Si—Si bonds, including TMS-TMS, $ClMe_2SiSiMe_2Cl$ and $Cl_2MeSiSiMeCl_2$, which can be processed further. However, the most frequently employed route to disilanes involves Wurtz coupling of organohalosilanes ⟨72JOM(36)C13⟩. For example, hexamethyldisilane can be prepared in 97% isolated yield by the simple process shown (Equation (61)) ⟨79SC451⟩. Hexamethyldisilane and hexaphenyldisilane can also be prepared electrochemically from the corresponding chlorosilanes ⟨76AG(E)370⟩.

$$\text{TMS-Cl} + \text{Li dispersion} \xrightarrow[\text{8 h}]{\text{THF, reflux}} \text{TMS-TMS} \tag{61}$$

The most convenient preparation of tris(trimethylsilyl)silane involves reaction of trichlorosilane with lithium metal and chlorotrimethylsilane (Equation (62)) ⟨69JOM(18)299⟩; this process also produces tetrakis(trimethylsilyl)silane in 50% yield. The use of tris(trimethylsilyl)silane as a radical-based reducing agent has been reviewed ⟨92ACR188⟩.

$$HSiCl_3 + TMS\text{-}Cl \xrightarrow{Li,\ THF} \underset{30\%}{(TMS)_3SiH} + \underset{40\%}{Me_4Si} \qquad (62)$$

The disilanes (**56**) and (**57**) and the trisilanes (**58**) and (**59**) have been prepared ⟨92JA3989⟩ by simple manipulation of the (aminosilyl)lithiums (**60**) and (**61**) (Section 2.10.2.9).

$ClPh_2Si\text{-}SiMe_2X$

(**56**) X = H
(**57**) X = Cl

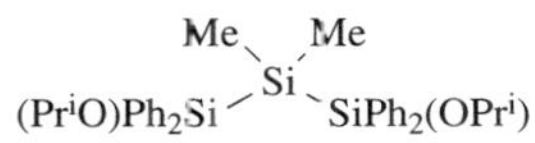

(**58**)

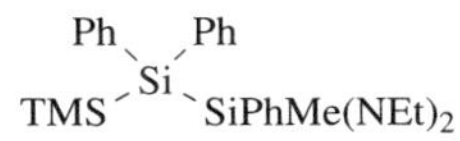

(**59**)

2.10.2.8.2 *Compounds with Si—Ge bonds*

As far as can be determined there are no examples of reagents of this class which have significant synthetic applications in organic chemistry.

2.10.2.8.3 *Compounds with Si—B bonds*

As far as can be determined there are no examples of reagents of this class which have significant synthetic applications in organic chemistry.

2.10.2.9 Alkylsilicon Compounds with an Si—Metal Bond

Triphenylsilyllithium can be prepared by the reaction of lithium with chlorotriphenylsilane in THF. Similarly, dimethylphenylsilyllithium, triphenylsilylpotassium, triphenylsilylrubidium, triphenylsilylcaesium, tri-*o*-tolylsilyllithium and tri-*o*-tolylsilylcaesium can be prepared by treatment of the respective chlorosilanes with the appropriate alkali metal. This process is believed, in most cases, to involve the intermediate generation of the symmetric disilane, which can indeed be isolated by stopping the reaction at an intermediate stage (Scheme 48) ⟨60JA403⟩.

Ph_3SiLi	$PhMe_2SiLi$	Ph_3SiK	Ph_3SiRb	Ph_3SiCs	(*o*-tolyl)$_3$SiLi	(*o*-tolyl)$_3$SiCs
89.5%	33%	75%	69%	38%	55%	21.5%

$$Ph_3SiCl \xrightarrow{Li} Ph_3SiLi \xrightarrow{Ph_3SiCl} Ph_3SiSiPh_3 \xrightarrow{Li} Ph_3SiLi$$

Scheme 48

t-Butyldiphenylsilyllithium has been prepared from the corresponding chlorosilane and metallic lithium; it has also been converted into the corresponding cuprate by reaction with CuCN ⟨89SC275⟩.

A second cleavage process, which involves the use of HMPA (hexamethylphosphoramide) as solvent or cosolvent, employs sodium or potassium hydride (Scheme 49); if the corresponding silyl hydride is readily available, cleavage of the silicon–hydrogen bond can be achieved under similar conditions ⟨80CC168⟩.

$$R_3SiSiR_3 + 2\,MH \xrightarrow[\text{HMPA}]{\text{DME or}} R_3SiM + [R_3SiH] \longrightarrow 2\,R_3SiM$$

R = Me, Et, Ph; M = Na, K

$$R_3SiH + KH \xrightarrow[\text{HMPA}]{\text{DME or}} R_3SiK$$

R = Et, Ph

Scheme 49

A third process involves disilane cleavage using methoxide ion, a good nucleophile for silicon. In unfunctionalized cases, HMPA is required, but methoxydisilanes can be cleaved in its absence (Scheme 50) ⟨71TL1511, 75JOM(92)C46, 77CC534⟩.

$$\text{TMS-TMS} + \text{MeOM} \xrightarrow{\text{HMPA}} \text{TMS-M} + \text{TMS-OMe}$$

M = Na, K

$$(MeO)_mSi_2Me_{6-m} + NaOMe \xrightarrow{\text{THF}} (MeO)_nMe_{3-n}SiNa$$

$m = 2, 4$ $\qquad$ $n = 1, 2$

Scheme 50

Trimethylsilyllithium can be obtained conveniently by cleavage of hexamethyldisilane (Section 2.10.2.8) with methyllithium in HMPA (Equation (63)) ⟨76JOC3063⟩. Interestingly, it has been observed that when this process is performed on a large scale, carbon—silicon, and not silicon—silicon, bond cleavage occurs, generating TMS-$SiMe_2Li$ ⟨94AG(E)99⟩.

$$\text{TMS-TMS} \xrightarrow[\text{THF/HMPA}]{\text{MeLi}} \text{TMS-Li} \;\text{ or }\; \underset{\text{large scale}}{\text{TMS-SiMe}_2\text{Li}} \qquad (63)$$

Trimethylsilyllithium, phenyldimethylsilyllithium and triphenylsilyllithium have been converted into the corresponding cuprates by reaction with CuCN ⟨84JCS(P1)1805, 92JCS(P1)3277⟩. Thorough studies on the diastereoselectivity of electrophilic attack on enolates having a β-silyl group, prepared by conjugate addition of such silyl-cuprates to enone systems, have been reported ⟨92JCS(P1)3277, 92JCS(P1)3295, 92JCS(P1)3363⟩.

The preparation of the (aminosilyl)lithium species **(60)**, **(61)** and **(62)** has been reported ⟨92JA3989⟩. These species are the first stable functional silyl anions to be prepared; an example of the use of **(60)** is given in Scheme 51.

$$(Et_2N)_nPh_{3-n}SiCl \xrightarrow[\text{THF, 0 °C, 4 h}]{\text{Li dispersion}} (Et_2N)_nPh_{3-n}SiLi$$

(60) $n = 1$, 98%
(61) $n = 2$, 97%

$$(Et_2N)PhMeSiCl \xrightarrow[\text{THF, 0 °C, 4 h; 80\%}]{\text{Li dispersion}} (Et_2N)PhMeSiLi$$

(62)

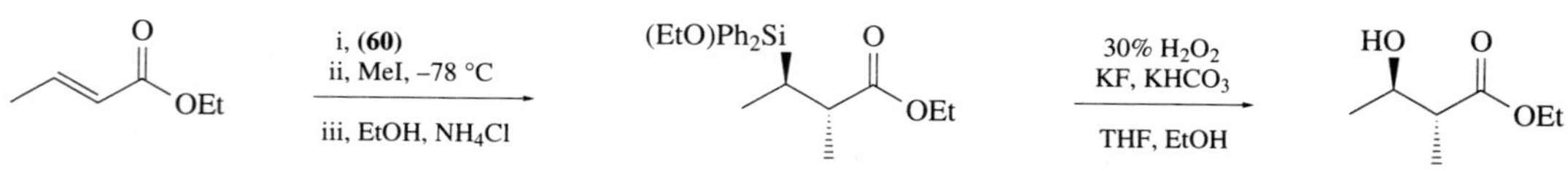

Scheme 51

2.10.2.10 Alkylsilicon Compounds with a Silicon Valency Higher than Four

This has been an intensively studied area ⟨B-89MI 210-01, 89SL1, B-91MI 210-03, 93CRV1371, B-94MI 210-01⟩.

Facile syntheses of the pentavalent compounds (**63**) have been described (Equation (64)) ⟨79JA1591⟩.

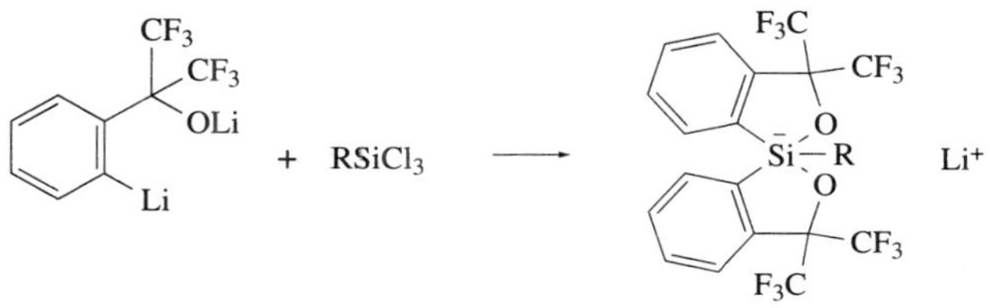

(**63**) R = Me, Ph

(64)

From analogy with the Wittig reaction and the *syn*-stereochemistry of base-induced elimination, a pentacoordinate 1,2-oxasiletanide has been proposed as a reaction intermediate or transition state in the Peterson reaction; however, until 1992, neither isolation nor spectroscopic observation of such a species had been reported. Treatment of the *β*-hydroxysilane (**64**), derived from the same fluorinated ligand as above, with Bu^nLi gave the oxasiletanide (**65**), as observed by ^{19}F and ^{29}Si NMR spectroscopy; gentle heating gave alkene (**67**) and lithium silanoxide, indicating that (**65**) is a real intermediate of the Peterson reaction (Scheme 52) ⟨92JA7598⟩. The use of KH in the presence of 18-crown-6 gave colourless needles of the potassium salt (**66**), whose structure was determined by x-ray crystallographic analysis.

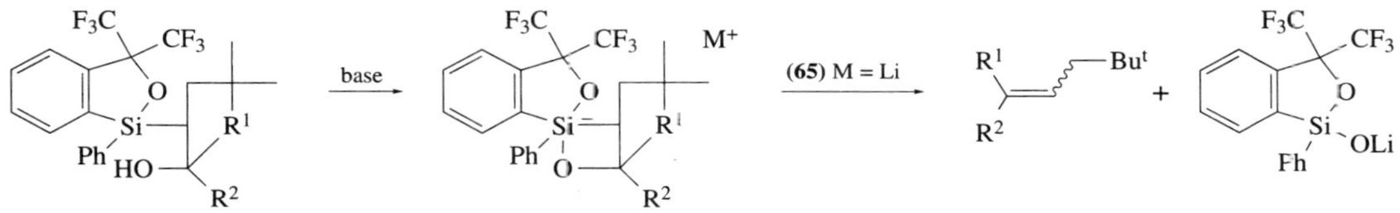

(**64**)

(**65**) M = Li
(**66**) M = K/18-crown-6

(**67**)

Scheme 52

There is detailed spectroscopic and other evidence for a high degree of transannular interaction between silicon and nitrogen in compounds such as the biologically active silatranes (**68**), which have Si—N separations of 0.202–0.234 nm, depending on the electronegativity of X ⟨61JA996, 66PAC35, 79TCC1, 79TCC77⟩.

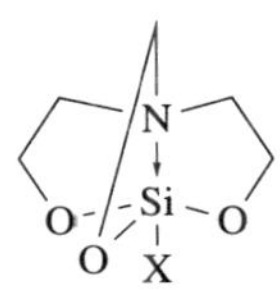

(**68**) X = R, Ar

2.11
Alkyl Metals

STEVEN V. LEY and CYRILLE KOUKLOVSKY
University of Cambridge, UK

2.11.1 INTRODUCTION

The chemistry of alkyl metal reagents (σ-bonded organometallic reagents) first started in 1849 when Frankland reported the preparation of dimethylzinc and diethylzinc. By the turn of the century, the major work from Barbier and Grignard emphasised their use in organic synthesis, particularly as a source of carbon nucleophiles. In fact, alkyl metal compounds are now routinely used as sources of carbon nucleophiles, and as strong bases or catalysts for various organic transformations. All these properties are directed towards carbon–carbon bond formation, and it would be fair to say that little synthetic work can be accomplished without their application. The systematic preparation of alkyl metals started as early as the late nineteenth century, and alkyl derivatives of nearly all metals have been synthesized using various methods. These compounds may be divided into two classes: those which are nonstabilized, highly reactive compounds which are used *in situ* with electrophilic substrates, and those which are stabilized complexes which may be isolated.

The purpose of this review is to survey preparative methods for the synthesis of alkyl metal reagents and as such it deals only with compounds in which the metal is single-bonded to an sp^3 carbon substituted only with carbon or hydrogen atoms. Within the limits of this chapter, it is impossible to cite all the structures and methods used; consequently the authors limited themselves to examples which are representative, with a particular emphasis on reports from the last 15 years. Comprehensive lists of structures, with corresponding references, may be found in the monographs *Comprehensive Organometallic Chemistry* or *Houben-Weyl*. The structure, stability or reactivity of these compounds will be discussed only if they have influence on the method for their preparation. Moreover, reactions in which the carbon–metal bond is established will be discussed in priority to ligand or counteranion exchange. The general methods for the synthesis of alkyl metal derivatives are shown in Table 1. Other aspects of carbanion and organometallic chemistry are covered throughout this work.

2.11.2 ALKYL GROUP 1 METAL DERIVATIVES

2.11.2.1 Alkyllithium Derivatives

The central importance of alkyllithium derivatives in organic and organometallic chemistry is illustrated by the large number of compounds which have been prepared, and by the extensive range of synthetic methods that have been used for their preparation. The chemistry of organolithium compounds has been extensively studied and reviewed ⟨70HOU(13/1)1, B-73MI 211-01, B-74MI 211-01, 82COMC-I(1)43, 82COMC-I(7)1, B-87MI 211-01, B-87MI 211-02, B-88MI 211-01, B-90MI 211-01, 91CCS(1)1, 93AG(E)1501, 93HOU(E19d)713⟩, and most of these reviews contain detailed experimental procedures.

2.11.2.1.1 Preparation by halogen–metal exchange

(i) From alkyl halides

(a) With Li metal. Treatment of an alkyl halide with lithium metal in a hydrocarbon solvent or in ether is the easiest method for the preparation of alkyllithiums (Equation (1)). This reaction is broad in scope, and suitable for primary, secondary and tertiary halides; however, tertiary systems react more slowly. The order of reactivity for halides is, as expected, I > Br > Cl; fluorides are virtually inert. The lithium metal can be used in various forms, although it is recommended that 1% of sodium should be incorporated for the best results, owing to its greater reactivity. The yields and reactivity may also be increased with the help of ultrasound techniques, which have proved to be very useful for reactions occurring at metal surfaces ⟨B-89MI 211-01, 89S787⟩. Indeed, sonication of a solution of alkyl halide in undried diethyl ether in the presence of lithium wire gives the alkyllithium solution after very short reaction times. This method is particularly useful in the Barbier-type reactions when the organolithium is prepared *in situ* in the presence of an electrophile ⟨80JA7926⟩.

$$\mathrm{RX} + 2\mathrm{Li} \longrightarrow \mathrm{RLi} + \mathrm{LiX} \qquad (1)$$

$$\mathrm{R} = \text{alkyl};\ \mathrm{X} = \mathrm{Cl, Br, I}$$

The mechanism of the reaction has been investigated, and is believed to involve successive single electron transfers from the metal to the carbon–halogen bond, giving a radical anion and subsequently a radical R•. Evidence for this process is given by racemisation of some chiral substrates (e.g., cyclopropyl derivatives) and the nature of the by-products, the main one being the Wurtz-type coupling product R—R. Wurtz coupling occurs readily when iodides are used as starting materials (except for MeI) and for some very reactive substrates such as benzylic or allylic halides.

Despite the generality of this method, it has some restrictions: for example, long reaction times and high temperatures are often required, which are not suitable when sensitive functionalities are present. THF is also slowly cleaved at room temperature by organolithiums, which therefore limits its use as a solvent for the reaction.

(b) With Li-arene complexes. Mixing lithium metal and some arenes produces a lithium-arene complex in which the aromatic species exists as a radical anion. These complexes are able to promote single electron transfer reactions and therefore they strongly activate the halogen–metal exchange process in the reaction of Li with halides. Li-naphthalenide (LiN, (**1**) ⟨78JOC1064⟩, Li-4,4′-di-*t*-butylbiphenylide (LiDBB, (**2**)) ⟨80JOC1924⟩ or Li-1-dimethylaminonaphthalenide (LiMAN, (**3**)) ⟨80SC311⟩ (Scheme 1) promote metallation in good to high yields. An important side reaction is radical attack onto the arene, which occurs particularly with LiN. LiDBB is more resistant towards radical attack and generally gives the best yields. In 1991 it was shown that catalytic amounts of naphthalene (1–10%) were sufficient to promote metallation with LiN ⟨91CC398⟩. This is a very practical enhancement of the procedure since the separation of the arene from the reaction products can sometimes be tedious. Halogen–lithium exchange occurs cleanly and rapidly at low temperatures, thus allowing the presence of a wide range of functionalities and the use of THF as solvent

Table 1 General methods for the synthesis of alkyl metals (M* denotes activated metal).

Preparative methods	*Reaction equation*	*Metals involved*	*Remarks*
Halogen–metal exchange	$RX + 2M \longrightarrow RM + MX$	Group 1 metals (especially Li), Cu*	Wurtz coupling is the main side reaction
Halogen–metal interconversion	$R^1X + R^2M \rightleftharpoons R^1M + R^2X$	Li, Zn	
Oxidative addition	$RX + M \longrightarrow RMX$	Group 2 metals, Eu, Yb, Pd^0, Pt^0, Os^0 Ru^0, Rh^I, Ir^I, Zn, Cd*, Al, Sn	Activation of the metal is often necessary
	$RX + MX_n \longrightarrow RMX_{n+1}$	Sm^{II}, Cr^{III}, Ga^I, In^I, Tl, Pb	
Transmetallation	$n\,RM^2 + M^1X_n \longrightarrow R_nM^1$	M^2 = Li, Na, Mg, Al, Sn, Zn M = nearly all metals	Li and Mg are the most used. Li is preferred for the synthesis of peralkylated metals and ate complexes. Competing reduction may occur with transition metals in higher oxidation states.
	$n\,RM^2 + M^1 \longrightarrow R_nM^1$	M^2 = Hg M = all metals (except Zn)	
Alkylation	$LM^- + RX \longrightarrow LMR$	Fe, Co, Re, Os, Ru, R_3Sn	S_N2 or radical mechanism
Carbometallation	$-CH{=}CH- + RM \longrightarrow$ 320CH(R)CH(M)–	Li, Mg, Sc, Ti, Zn, Al	Main applications in polymer synthesis
Hydrometallation	$-CH{=}CH- + HM \longrightarrow -CH_2CH(M)-$	Mg, Ti, Zr, Sc, Co^I, Ni^0, Rh^I, Pt^{II}, Al	Main applications in polymer synthesis and the *oxo* synthesis
Oxymetallation	$-CH{=}CH- + MX_n \longrightarrow -CH(X)CH(MX_{n-1})-$	Hg, Tl	
Deprotonation	$RH + M\ (\text{or } R^2M) \longrightarrow RM$	Group 1 metals	Benzylic or allylic positions only
Orthometallation	$ML_n \longrightarrow LMCH_2$	Pd^{II}, Pt^{II}, Rh^{III}, Th^{IV}	Phosphine or nitrogen ligands for Pd, Pt or Rh; neopentylic alkyl groups for Th

(Equations (2)–(4)) ⟨88TL2859, 91CC398, 93JOC5976⟩. β-Amino-, hydroxy- or amidolithium derivatives have also been prepared in this way, giving useful intermediates for synthesis ⟨90PAC595, 93JOC5976⟩.

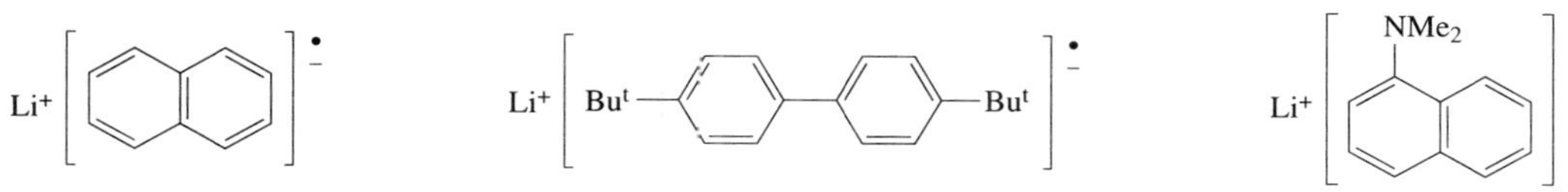

(1) LiN **(2)** LiDBB **(3)** LiMAN

Scheme 1

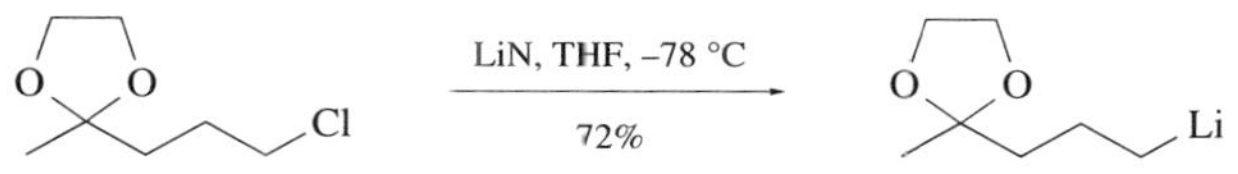

(4) **(5)** (2)

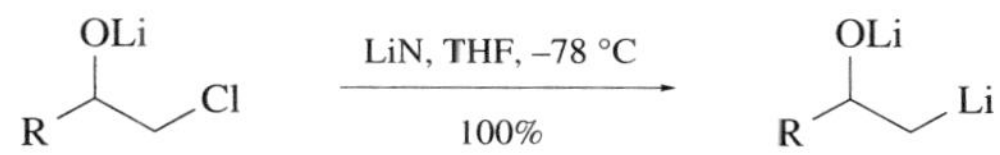

(6) **(7)** (3)

R = Me, Ph

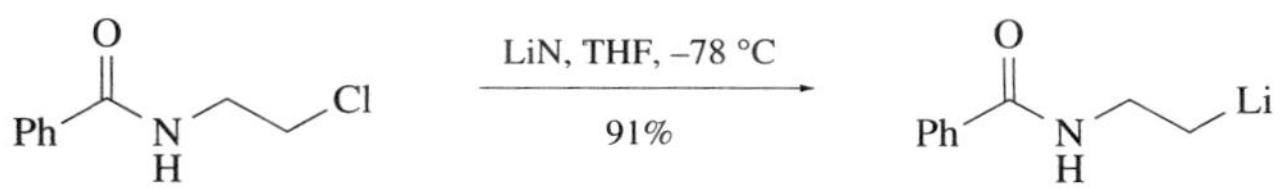

(8) **(9)** (4)

(ii) From other leaving groups

Alternative groups to halides have been used for the preparation of alkyllithiums using similar methods to those described above. For example, cleavage of ethers with Li metal in THF at low temperature affords an alkyllithium together with an alkoxide, and this method is very useful for the preparation of benzyllithium derivatives ⟨58JOC2044, 94SC591⟩. LiDBB in ether has also been used in these reactions. The cleavage of mesylates ⟨92T4593⟩ or sulfates ⟨92TL5597⟩ (Equation (5)) with Li and catalytic naphthalene has been reported to give organolithium compounds in good yields, although the reaction with mesylates seems to be limited to allylic or benzylic substrates.

(5)

R = Me, Et, Pr^i, Bu^n, $C_6H_{13}CH(Me)$–

Cyclic ethers, which are inert to Li at low temperatures, are conveniently opened with Li-arene complexes; for example, epoxides ⟨86AG(E)653, 90JOC1528⟩ and oxetanes ⟨89JOC5657, 90JA6389, 91JOC5760, 92CC1623⟩ are opened with LiDBB, whereas the cleavage of tetrahydrofurans to γ-alkoxy-organolithium compounds requires addition of a Lewis acid at low temperatures (Equation (6)) ⟨91JA1866, 92T3585⟩. Cyclic amines such as aziridines ⟨93TL1649⟩ and azetidines ⟨94T5775⟩ are cleaved with LiDBB.

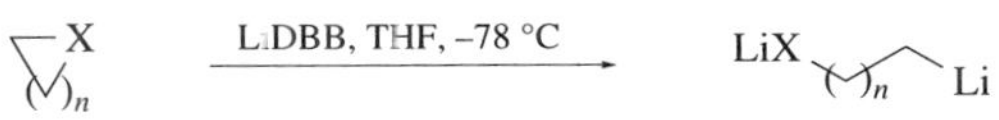

(6)

X = O, n = 1,2; X = NR, n = 1,2

The cleavage of dialkyl or alkyl aryl sulfides with Li metal is not always satisfactory, although Li-arene complexes give excellent yields for a wide range of substrates, including allylic and benzylic species (Equation (7)) ⟨89ACR152, 91CC398, 92JOC3680⟩. Other chalcogenides can be transformed into organolithium reagents, provided that stronger conditions are employed. Selenides ⟨85TL1093, 86TL1719, 86TL1723, 92TL3381⟩ and tellurides ⟨87AG(E)1187⟩ are cleaved with alkyllithium reagents.

$$R^1SR^2 \xrightarrow{\text{Li-arene, THF, }-78\ ^\circ\text{C}} R^1Li \quad (7)$$

R^1 = alkyl; R^2 = Me, Ph

2.11.2.1.2 *Preparation by halogen–metal interconversion*

Reaction between an organic halide and a strong lithium base such as *n*-butyl- or *t*-butyllithium to give an organolithium compound is called the Wittig–Gilman reaction ⟨51OR(6)339⟩. This reaction is particularly useful for the synthesis of aryl- or vinyllithiums (see Chapter 2.19), but has also been successful in the preparation of some alkyllithiums (Table 2). The reaction proceeds at low temperature, many functional groups are inert and Wurtz coupling is minimised. In 1990, the use of alkyl iodides as starting materials in pentane/diethyl ether as the solvent was recommended ⟨90JOC5404⟩. The by-product in the reaction with *t*-butyllithium, *t*-butyl bromide, can be destroyed by addition of a second equivalent of *t*-butyllithium to give unreactive species.

Table 2 Preparation of alkyllithium reagents by halogen–metal interconversion.

$$R^1X + R^2Li \longrightarrow R^1Li + R^2X$$

Entry	*Alkyl halide*	*Conditions*	*Product*	*Yield (%)*	*Ref.*
1	 **(10)**	$2Bu^tLi$, pentane, Et_2O, –78 °C	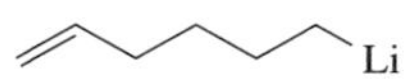 **(11)**	>99	90JOC5404
2	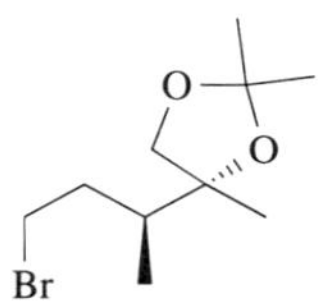 **(12)**	$2Bu^tLi$, Et_2O, –78 °C	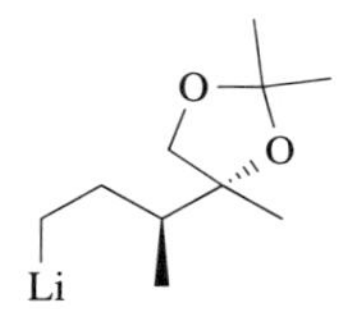 **(13)**	86	88JA2506
3	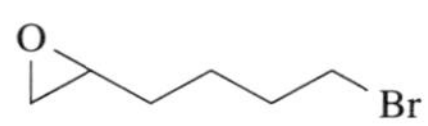 **(14)**	$2Bu^sLi$, Et_2O, –78 °C	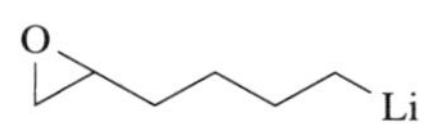 **(15)**	90	84TL4323
4	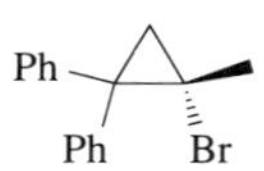 **(16)**	$2Bu^nLi$, Et_2O, 8 °C	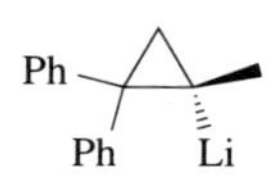 **(17)**	50	64JA3283

2.11.2.1.3 *Preparation by deprotonation*

Alkanes are generally not sufficiently acidic to be directly deprotonated by lithium bases, although exceptions can arise with some bridged polycyclic compounds. Deprotonation is usually successful only if the protons are activated by conjugation, by a heteroatom, or when additives are used. Tertiary amines are often used as additives, the most popular being TMEDA ⟨92ACR448⟩. Strongly coordinating solvents such as THF or DME have similar effects. Allylic or benzylic substrates are good candidates for deprotonation; however, toluene is converted into benzyllithium only if TMEDA is added to the mixture ⟨70JOC10⟩, and metallation of the aromatic ring is often a side reaction ⟨68JOM(11)615⟩. Bis-allylic or bis-benzylic substrates such as diphenylmethane are readily converted into their lithio derivatives by deprotonation.

The importance of intramolecular heteroatom-assistance in the lithiation reaction has been recognised for many years and has been the subject of several reviews ⟨79OR(26)1, 86RTC1, 90CRV879⟩.

The presence of oxygen or nitrogen in γ or δ positions activates the base by precoordination and stabilises the formed organolithium by intramolecular coordination. Allylic or polycyclic substrates are thus easily deprotonated where the heteroatom not only facilitates but also directs the reaction. Two examples are shown in Equations (8) ⟨84OS(62)58, 92S191⟩ and (9) ⟨85JA724⟩.

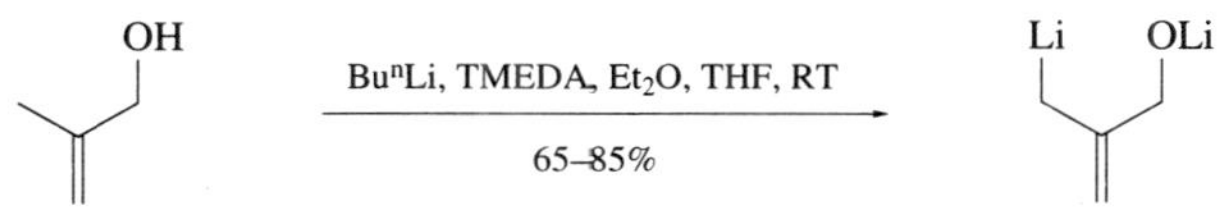

(8)

(18) **(19)**

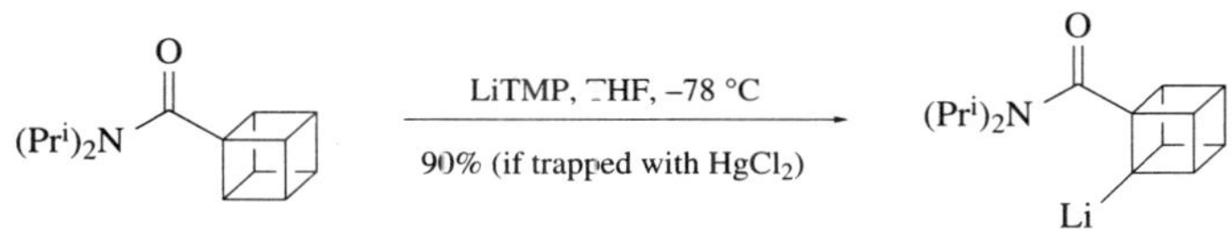

(9)

(20) **(21)**

2.11.2.1.4 *Preparation from other alkyl metals: transmetallation*

The preparation of alkyllithium reagents by transmetallation from other alkyl metals is not of great use, since most of these compounds are themselves prepared from alkyllithiums. However, transmetallation can be useful for preparing alkyllithiums which are difficult to access by direct methods and especially when salt-free solutions are desired. This reaction is an equilibrating process which must be driven in the desired direction by precipitation. The most popular transmetallations are mercury–lithium and tin–lithium exchange reactions.

Treatment of a solution of dialkyl mercury with Li wire affords clean and salt-free solutions of the alkyllithium. In fact, this was the first method for the production of an organolithium reagent ⟨17CB262⟩ and it is suitable for virtually any alkyl group substituent.

Tin–lithium exchange has been a popular method for the preparation of allyllithium compounds or β-lithioketone enolates, which are difficult to obtain using traditional methods, or for the preparation of cyclopropyllithium compounds with complete retention of configuration (Equations (10) ⟨92JOC3680⟩ and (11) ⟨91AG(E)177⟩). Cleavage of stannanes occurs readily at low temperature using lithium bases (generally n-butyllithium) to afford the organolithium in good yield together with an inert tetraalkyltin species.

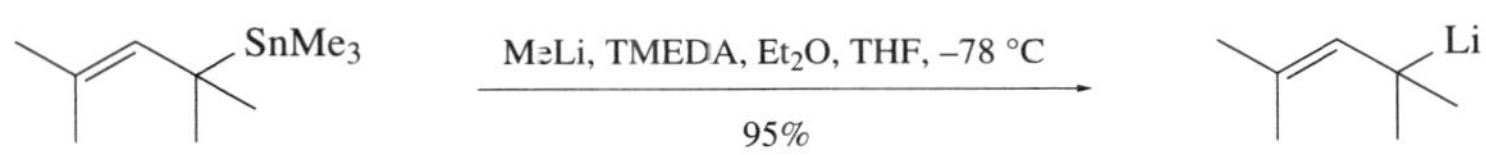

(10)

(22) **(23)**

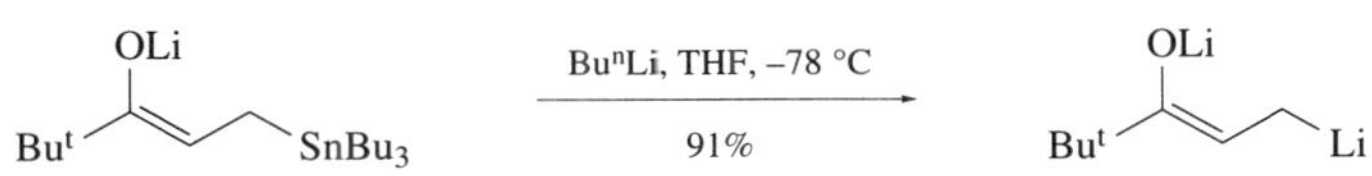

(11)

(24) **(25)**

2.11.2.1.5 *Addition to carbon–carbon double bonds: carbolithiation*

Addition of organolithium compounds to a carbon–carbon double bond generates new organolithium species. This reaction, often named carbolithiation, can be performed in an inter- or intramolecular fashion and has been widely applied to polymer synthesis ⟨B-85MI 211-01⟩. The reactivity of organolithiums towards double-bond insertion is rather low but may be enhanced by the use of coordinating reagents such as ether, THF and tertiary amines ⟨94T5933⟩. Carbolithiation of ethene itself has been achieved under pressure ⟨50LA(567)195⟩, although in pentane solution at $-25\,^{\circ}$C alkyllithiums add to ethene in the presence of diethyl ether ⟨94T5933⟩. The reactivity of alkyllithiums towards alkenes is greatly increased when the latter possess a heteroatom capable of promoting intramolecular coordination, and indeed carbolithiation of bis-homoallylic ethers, acetals or amines proceeds cleanly under mild conditions ⟨86RTC1, 94JOC2925⟩. The regioselectivity once again is directed by the heteroatom, and is found to be opposite to that observed with simple alkenes, generally affording secondary organolithium species (Equation (12)).

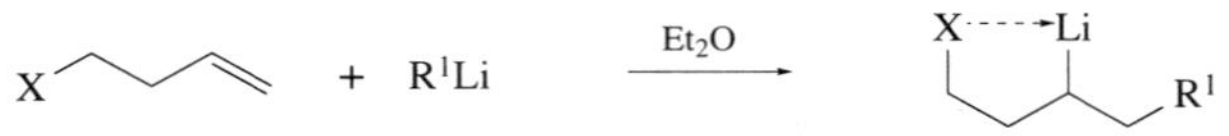

$X = OR^2, NR^2_2; R^1 = alkyl$ (12)

The intramolecular carbolithiation, or anionic cyclisation, of 5-hexenyllithium derivatives has also been studied extensively ⟨75JOM(91)123, 88JA4788, 91JA5720, 92JA8053, 94T5957, 94T7177⟩. These compounds are stable at $-78\,^{\circ}C$ but cyclise at room temperature to give cyclopentylmethyllithium derivatives in good to high yields (Equation (13)). Stepwise cyclisations leading to polycyclic products are known ⟨89JA765⟩ and even heterocycles can be synthesised using this method ⟨89JA2981⟩. Vinyllithiums can also be used, to afford methylene cyclopentane derivatives ⟨88JA4788⟩. This reaction is particularly interesting because it is more selective than its radical counterpart: 5-*exo-trig* cyclisation occurs exclusively, and stereoselectivities are generally high, although they can be greatly affected by the presence of substituents bearing heteroatoms. The reaction seems to be restricted to carbolithiation of terminal double bonds and disubstituted double bonds bearing electron-withdrawing groups ⟨92JOC1495⟩. However, in 1994 it was demonstrated that phenyl-substituted alkenes may be similarly cyclised in good yields, giving secondary benzylic organolithium reagents ⟨94T5957⟩.

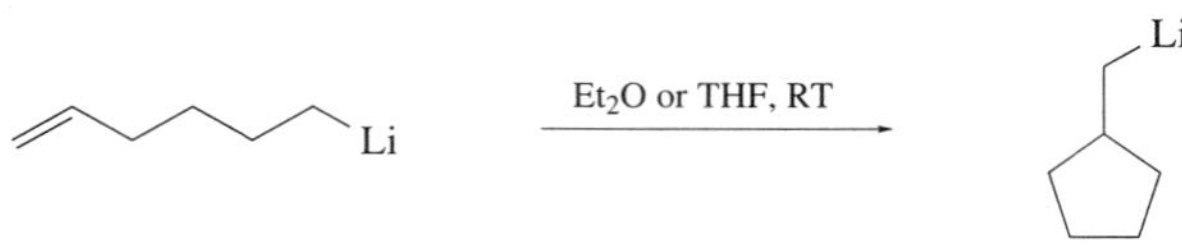

(13)

2.11.2.2 Alkylsodium and Alkylpotassium Derivatives

In theory, most of the methods for the synthesis of alkyllithiums reported above are suitable for the preparation of alkylsodium and alkylpotassium reagents ⟨70HOU(13/1)261, B-87MI 211-01, 93HOU(E/19d)713⟩. However, there are appreciable differences in some cases. Sodium and potassium tend to be more reactive towards alkyl halides than lithium, and also more prone towards Wurtz coupling, so their preparation from alkyl halides is limited to some specific examples. Sodium-arene complexes are more sensitive towards radical attack than their lithium counterparts and, in addition, organosodium and organopotassium species are less stable and therefore more difficult to handle. Consequently, despite having been prepared historically prior to alkyllithium reagents, their chemistry has not been extensively developed. Transmetallation from organomercury compounds is probably the most convenient way to prepare nonconjugated metallo derivatives. Some examples of the preparation of various alkylsodium and alkylpotassium reagents are illustrated in Table 3.

Table 3 Preparation of various alkylsodium and alkylpotassium reagents.

Entry	*Starting material*	*Conditions*	*Product*	*Yield (%)*	*Ref.*
1	n-$C_5H_{11}Cl$	Na sand, pentane, RT	n-$C_5H_{11}Na$	89	53JA134
2	Bu^n_2Hg	Na/K alloy, pentane, RT	2 Bu^nK		64AG(E)326
3	Et_2Hg	Na, Ligroin, RT	2 EtNa		17CB262
4	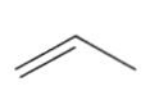	n-$C_5H_{11}Na$, pentane, RT	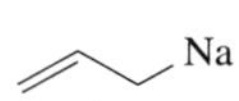	69	47JA1675
5	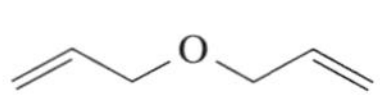	K dispersion, decane	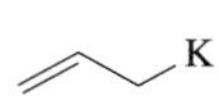	45	47JA950

Organosodium and organopotassium compounds are stronger bases than their lithium counterparts. In the presence of the K/Na_2O system, toluene is cleanly transformed into benzylpotassium (Equation (14)) ⟨55JOC981⟩. Deprotonation at allylic positions is more difficult, although *n*-pentylsodium has been shown to be effective in the deprotonation of various alkenes. The addition of TMEDA can be useful, but the best results have been obtained with a mixture of *n*-butyllithium and potassium *t*-butoxide ⟨88PAC1627⟩. This strongly basic system converts a great variety of substrates into allylpotassium derivatives in good to high yields. Side reactions such as carbometallation do not occur, and the presence of functional groups such as ethers and acetals is tolerated. Likewise, THF is not metallated below −60 °C and can thus be used as the solvent below this temperature. The most appealing feature of the reaction is the low temperature metallation of both (*Z*)- and (*E*)-2-butene to give the potassium derivatives with complete retention of stereochemistry (Scheme 2) ⟨90JA6339⟩. This high reactivity cannot be explained in terms of simple transmetallation since *n*-butylpotassium prepared by other methods possesses different properties ⟨84TL741, 84TL745⟩. Trimethylsilylmethylpotassium has also been reported to convert alkenes into allylpotassium compounds. These reagents are prepared *in situ* by transmetallation from bis-trimethylsilylmethylmercury in the presence of the alkene ⟨87S645, 91JOM(409)103⟩.

$$2PhMe + 2K + Na_2O \xrightarrow[87\%]{110\,^{\circ}C} 2PhCH_2K + NaOH + NaH \qquad (14)$$

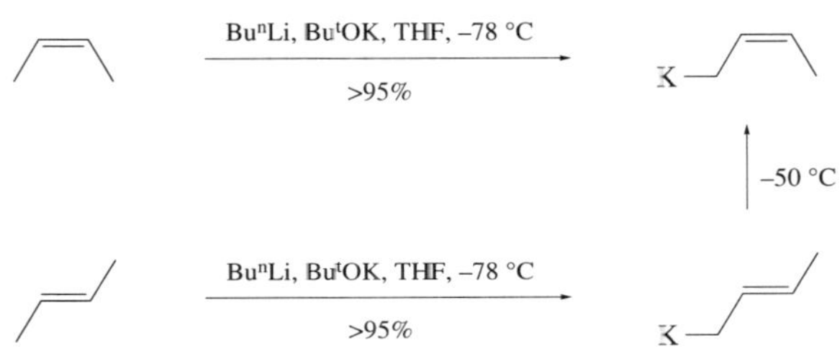

Scheme 2

2.11.2.3 Other Derivatives

Rubidium and caesium are the most reactive group 1 metals and consequently the problems encountered with potassium and sodium derivatives are even more pronounced for these elements. Caesium is able to deprotonate toluene with hydrogen evolution ⟨75JOM(96)139⟩, and trimethylsilylmethylcaesium has been reported to deprotonate various substrates ⟨B-73MI 211-01⟩. Saturated alkylrubidium and alkylcaesium compounds are generally prepared by transmetallation from dialkylmercury or by metal-mediated ether cleavage. Because of their high instability, these compounds have received little attention.

2.11.3 ALKYL GROUP 2 METAL DERIVATIVES

Some general reviews about the chemistry of alkyl group 2 metal derivatives are available ⟨B-67MI 211-01, B87MI 211-03⟩. Other reviews concerning one or several metal derivatives will appear in the corresponding subsections.

2.11.3.1 Alkylmagnesium Derivatives

Together with alkyllithium reagents, alkylmagnesium compounds are the most important alkyl metal derivatives in organic and organometallic chemistry. Since their discovery in 1900 ⟨1900CR1322⟩ they have been used extensively as a source of carbon nucleophiles and for the preparation of other organometallic species. Their preparation, reactions and properties have been comprehensively reviewed ⟨B-54MI 211-01, 73HOU(13/2a)47, 82COMC-I(2)155⟩. Alkylmagnesium compounds exist in two forms, either alkylmagnesium halides RMgX, the commonly known 'Grignard reagents', or dialkylmagnesium species R_2Mg. Specific preparations for both types and for their interconversion are known. Grignard compounds may also be transformed into alkylmagnesium alkoxides by displacement ⟨93JOM(452)1⟩ or into alkylmagnesium hydrides by reduction ⟨77CC169⟩. However, Grignard reagents are by far the most widely used magnesium derivatives in organic synthesis, and the most commonly used are commercially available.

Alkylmagnesium compounds are more stable than the corresponding alkyllithiums and tend to be less basic. As they do not cleave ethereal solvents at room temperature, they can be prepared and stored in diethyl or THF. Despite their enormous potential in synthesis, there are fewer methods for their preparation than for the corresponding alkyllithium species. The most convenient methods of preparation are the well-known Grignard reaction or transmetallation methods.

2.11.3.1.1 *Preparation by oxidative addition: the Grignard reaction*

(i) Preparation from alkyl halides

The Grignard reaction, namely treatment of an organic halide with magnesium turnings in an ethereal solvent, is the main method for the preparation of organomagnesium compounds and has been thoroughly studied since the first report in 1900 (Equation (15)) ⟨1900CR1322⟩. The mechanism, scope and reactivity have been investigated and are similar to the lithiation reaction, the main difference between the two being that diethyl ether or THF are used as solvents (Grignard compounds are not very soluble in hydrocarbons), in which Grignard compounds generally exist as solvates. Wurtz coupling is again the main side reaction, although it is less problematic than with alkyllithium reagents.

$$RX + Mg \xrightarrow{Et_2O \text{ or THF}} RMgX \qquad (15)$$

$$R = \text{alkyl; } X = Cl, Br, I$$

The success of the reaction depends critically on its initiation, and prior activation of magnesium may often be necessary to start the reaction ⟨81S585⟩. Activators such as iodine or 1,2-dibromoethane are often added to initiate these reactions. Sonication strongly activates magnesium towards alkyl halides; for example, reaction of 2-chlorobutane in dry diethyl ether is induced within seconds while it takes a few minutes in wet solvents ⟨83ICAL241⟩. Another simple and convenient method for activation consists of stirring magnesium turnings without solvent under an inert atmosphere to form highly reactive powdered magnesium particles ⟨91JOC698⟩. This method secures high and reproducible yields for a wide range of substrates.

These methods for magnesium activation are useful for reactive alkyl halides but poor for unreactive or unstable substrates, or when Wurtz coupling is predominant. Therefore, various methods for the preparation of highly reactive magnesium have been developed. These generally consist of the preparation of suspensions of finely divided magnesium powder in ethereal solvents, by both physical or chemical methods. Condensation of magnesium vapour into THF or diethyl ether gives a slurry of finely divided magnesium powder ⟨81HCA2606⟩. Using this system, it is possible to metallate cyclopropylmethyl bromide at $-75\,^{\circ}C$ without fragmentation ⟨81HCA2606⟩ or several allylic bromides without side reactions taking place ⟨84TL3305⟩. Another route uses the reduction of magnesium halides to generate highly reactive magnesium, the best known procedure being the preparation of Rieke's magnesium ⟨72JA7178, 77ACR301⟩. In this method, anhydrous magnesium chloride is reduced with potassium in boiling THF to give a very reactive black slurry of magnesium/potassium chloride. The reactivity may be further increased by the addition of alkali salts, usually potassium iodide. Rieke's magnesium metallates in quantitative yield, alkyl chlorides or bromides at room temperature, and even unreactive alkyl fluorides can be converted into Grignard compounds ⟨74JA1775⟩. This system is still reactive at temperatures as low as $-100\,^{\circ}C$, thus allowing the preparation of unstable Grignard reagents such as β-aminoethylmagnesium bromide derivatives which decompose above $-90\,^{\circ}C$ ⟨79JOM(182)313⟩. Sodium or lithium naphthalenide ⟨77SC223, 87JOC3674⟩ have also been used for reduction. Other methods involve reduction of magnesium chloride (prepared *in situ* from magnesium and 1,2-dichloroethane) with Li wire in THF under sonication conditions ⟨85MI 211-02⟩, or dehydrogenation *in vacuo* of magnesium hydride ⟨90CB1517⟩. This latter method probably gives the most reactive form of magnesium metal, able to promote

Grignard reactions with a wide range of organic halides and in hydrocarbon or aromatic solvents. However, this method suffers practical disadvantages owing to the requirement of special apparatus for the dehydrogenation.

A mild and efficient method for delivering highly reactive magnesium metal has been developed using the magnesium–anthracene complex as a source of magnesium metal ⟨83AG(E)728, 88ACR261⟩. Sonication of magnesium powder in the presence of anthracene in THF at 65°C gives an orange, sparingly soluble addition complex, Mg-anthracene-$(THF)_3$, which acts as a source of soluble magnesium (Equation (16)). This system cleanly metallates allylic, benzylic and propargylic halides at −78°C in high yield, without any trace of Wurtz coupling ⟨84CC1702, 84TL3305, 90CB1507⟩. A catalytic amount of anthracene may be used, which simplifies workup procedures. Simple alkyl halides can be metallated only if toluene is used as solvent, since in THF solution the intermediate alkyl radical tends to react with anthracene to give additional compounds; moderate yields are obtained in diethyl ether ⟨90CB1507⟩.

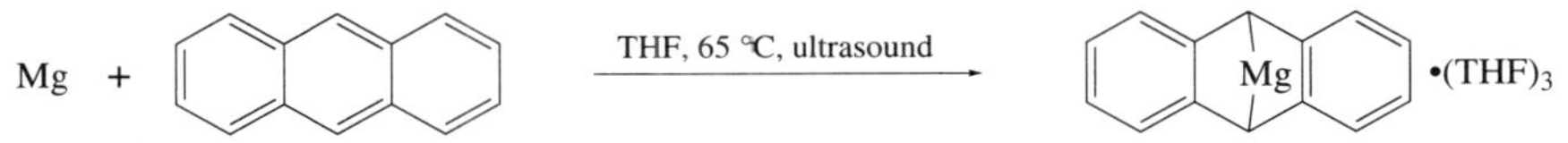

(16)

These various methods for magnesium activation by chemical or physical methods have greatly increased the generality of the Grignard reaction and nearly any kind of substrate can be metallated using appropriate procedures. For normally reactive alkyl halides, overnight dry stirring of magnesium turnings is recommended. This can be combined, if necessary, with the addition of an activator (1,2-dibromoethane or iodine). For less reactive substrates, highly reactive magnesium must be used. A comparative study between three different sources of activated magnesium (condensed magnesium, Rieke's magnesium and magnesium-anthracene) for the metallation of an allylic halide has been undertaken ⟨84TL3305⟩, the best yields were obtained using the magnesium-anthracene method. This is also the most practical procedure. A summary of representative compounds and procedures is listed in Table **4**.

(ii) The Schlenk equilibrium

Grignard reagents in solution are in equilibrium with their dialkylmagnesium counterparts; this is referred to as the Schlenk equilibrium. It is therefore possible to prepare either of the two forms by driving the equilibrium process. Dialkylmagnesium compounds may be obtained by adding strongly coordinating solvents which precipitate magnesium halide. Dioxane is the most common solvent for this purpose, although DIGLYME, pyridine and TMEDA have also been used. On the other hand, mixing dialkylmagnesium compounds in solution with magnesium halides (except MgF_2) affords the corresponding Grignard reagent.

(iii) Preparation from other leaving groups

Cleavage of ethers does not generally occur on treatment with magnesium. Highly activated metal, however, is able to promote such reactions. Rieke's magnesium cleaves dibenzyl ether only slowly at room temperature ⟨87JOC3674⟩, but magnesium-anthracene readily generates benzylmagnesium species from the corresponding benzyl methyl ethers ⟨88CC289⟩. At 65°C, THF is cleaved by Rieke's magnesium, but this reaction is not very useful in synthesis since it affords mainly decomposition products.

2.11.3.1.2 *Preparation by transmetallation*

Owing to the great versatility of the Grignard reaction, transmetallation is not commonly used for the preparation of organomagnesium derivatives. Transmetallation from alkyllithium compounds (or other group 1 derivatives) may be useful in some cases, especially in small-scale syntheses. As magnesium salts are more soluble in ethereal solvents than are lithium salts, the equilibrium is driven towards the formation of Grignard species. Transmetallation from mercury derivatives has

Table 4 Grignard reactions of various alkyl halides.

Entry	*Alkyl halide*	*Conditions*	*Product*	*Yield (%)*	*Ref.*
1	Bu^nBr	Mg turnings, I_2, Et_2O, reflux	Bu^nMgBr	94	29JA1576
2	Bu^sBr	Mg turnings, I_2, Et_2O, reflux	Bu^sMgBr	77	29JA1576
3	Bu^tCl	Mg turnings, I_2, Et_2O, reflux	Bu^tMgCl	85	32JA3714
4	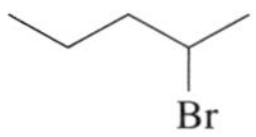	Mg turnings, Et_2O, 50 °C, ultrasound	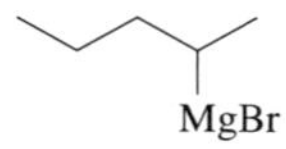	67	83ICAL241
5	Bu^tCl	Rieke's Mg, THF, RT	Bu^tMgCl	100	74JA1775
6	n-$C_8H_{17}Br$	Rieke's Mg, THF, RT	n-$C_8H_{17}MgBr$	100	74JA1775
7	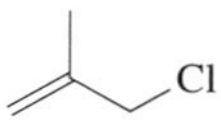	Rieke's Mg, THF, RT	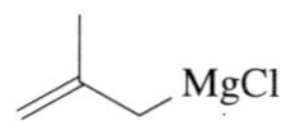	82	74JA1775
8	n-$C_6H_{13}F$	Rieke's Mg, KI, THF, RT	n-$C_6H_{13}MgX$	89	74JA1775
9	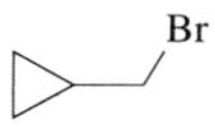	Condensed Mg, THF, –75 °C	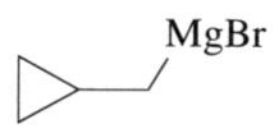	78	81HCA2606
10	BnBr	Dry-stirred Mg turnings, Et_2O, 0 °C	BnMgBr	95	91JOC698
11	AllylBr	Dry-stirred Mg turnings, Et_2O, 0 °C	AllylMgBr	95	91JOC698
12	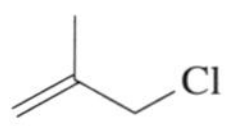	Condensed Mg, THF, –70 °C	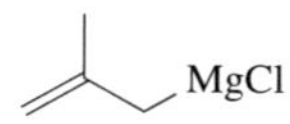	92	82TL3901
13	EtBr	Mg(anthracene)$(THF)_3$, toluene, RT	EtMgBr	80	90CB1507
14	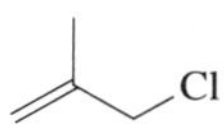	Mg(anthracene)$(THF)_3$, THF, –78 °C	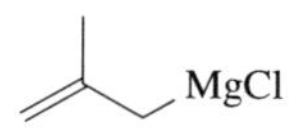	66	90CB1507
15	BnCl	Mg(anthracene)$(THF)_3$, THF, RT	BnMgCl	67	90CB1507
16	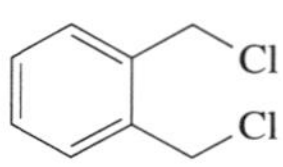	Mg(anthracene)$(THF)_3$, THF, RT	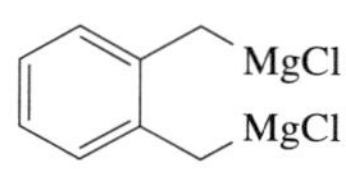	92	84CC1702
17	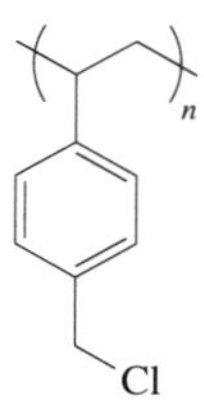	Mg(anthracene)$(THF)_3$, THF, RT	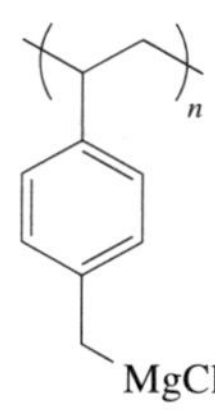	100	87JOC4644

been used for the preparation of substrates which are difficult to obtain by nonactivated magnesium methods and for the preparation of salt-free solutions of dialkylmagnesium reagents, which may be obtained by treatment of dialkylmercury derivatives with magnesium in THF. Alkylmagnesium halides are obtained if magnesium halide is added.

Other transmetallations from boron and zinc have also been performed, but are less common.

2.11.3.1.3 Addition to carbon–carbon double bonds

(i) Hydromagnesation

Addition of hydride and magnesium to a carbon–carbon double bond affords dialkylmagnesium compounds when magnesium hydride MgH_2 is added, or a Grignard reagent when magnesium halohydride MgHX (X = Br, Cl) is used. The reactivity is rather low even under forcing conditions but can be increased using transition-metal catalysts. Titanocene dichloride ⟨78CC30, 83JOM(250)1⟩ and zirconium tetraiodide ⟨82AG(E)199⟩ are efficient catalysts for this reaction. The reaction is regioselective for terminal alkenes, giving the primary organomagnesium derivative, with the exception of styrene, which leads to the secondary benzylic derivative. Internal double bonds generally do not react. Other magnesium sources may also be used, for example, reaction between 1-octene and a Grignard compound bearing a β-hydrogen in the presence of titanium tetrachloride ⟨62JOC1493, 71BSF3571⟩, or titanocene dichloride ⟨83JOM(250)1⟩, gives *n*-octylmagnesium bromide in moderate to good yields. The reaction is, however, not totally regioselective and isomerisation of the double bond can occur as a side reaction. Dialkylmagnesium compounds and alkylmagnesium hydrides may alternatively be used as magnesium sources. An interesting tandem hydromagnesation–telomerisation of butadiene has been described, using *n*-propylmagnesium bromide and a nickel catalyst ⟨75CC242⟩.

(ii) Diene magnesation

Magnesium metal adds to 1,3-dienes to give cyclic 2-butene-1,4-diylmagnesium derivatives (Equation (17)) ⟨85AOC(24)1, 85N789⟩. The reaction of magnesium with anthracene is also an illustration of this process (see Section 2.11.3.1.1(i)). The use of highly reactive magnesium is recommended since magnesium powder reacts only very slowly with dienes. Thus, magnesium prepared by reduction of a magnesium halide with lithium biphenylide reacts readily with various 1,3-dienes to give the corresponding metallacycles in good yields ⟨89JOC3247⟩. Since these compounds can react with various electrophiles, the process seems to have great potential in organic synthesis.

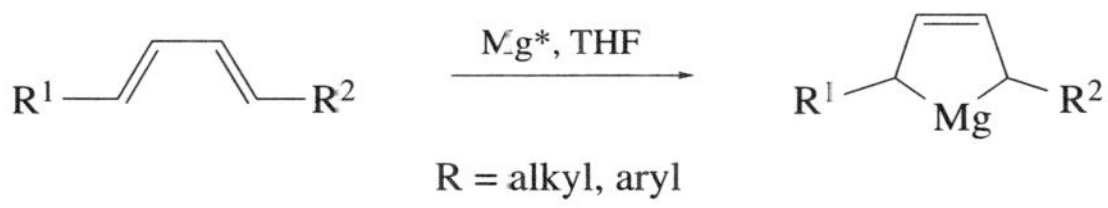

(17)

(iii) Carbomagnesation

Alkylmagnesium derivatives are less reactive towards carbon–carbon double bonds than are alkyllithium species and, unless activation is provided by a heteroatom or a transition-metal catalyst, high temperatures and pressures are required ⟨81BSF(2)87⟩. Nevertheless, this reaction has great potential in organic and polymer synthesis and has been thoroughly investigated. Allylmagnesation of allyl alcohols (as their magnesium alkoxides) occurs readily under mild conditions and has been studied extensively ⟨83JOM(259)1, 89TL6753⟩. The high reactivity is due to intramolecular activation of the double bond by the magnesium alkoxide, giving cyclic magnesium derivatives (Equation (18)). The regioselectivity is substrate-dependent but is generally in favour of attack at the carbon β to the oxygen, thus leading to a five-membered metallocycle; however, α-addition occurs predominantly when the alcohol is tertiary. Mechanisms for the formation of both isomers have been proposed ⟨65JA4194, 66TL875⟩. Allylmagnesation of allylic amines has also been described ⟨71T2183, 81JOC3773⟩.

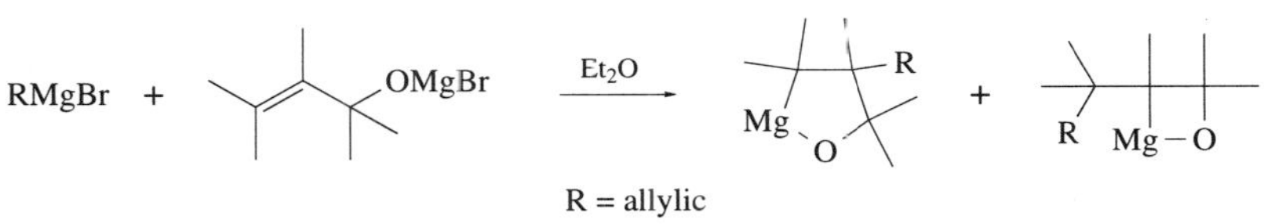

(18)

Transition-metal-catalysed carbomagnesation has become a popular and important tool for carbon–carbon bond formation and zirconium-mediated ethylmagnesation of alkenes has been studied extensively ⟨83IZV218, 83IZV2053, 85JOM(285)43, 91JA5079, 91JA8950⟩. Reaction of ethylmagnesium bromide (or diethylmagnesium) with various terminal alkenes in the presence of 1–5% mol of zirconocene dichloride affords primary Grignard reagents in good yields (Equation (19)). Regioselectivities are high, many functional groups are compatible with the reaction conditions and stereoselectivities are often excellent for allylic alcohols. The reaction is, however, sensitive to steric hindrance and disubstituted double bonds are generally unreactive, although the more strained norbornene derivatives are an exception. Mechanistic studies were reported in the early 1990s ⟨92JA6692, 93JA6614⟩ and a chiral catalyst for enantioselective carbomagnesation was proposed ⟨93JA6997⟩.

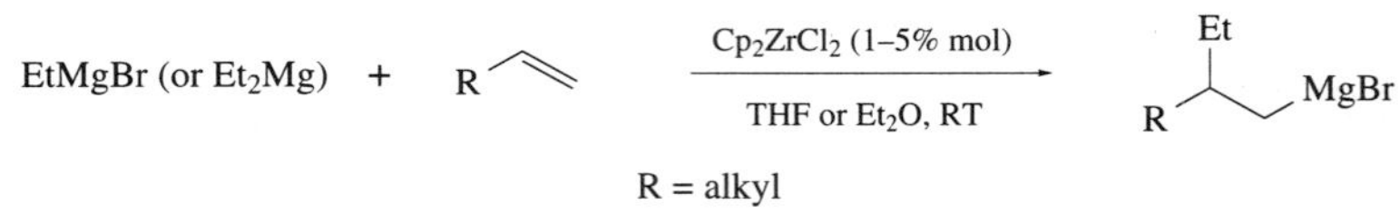

(19)

(iv) The magnesio-ene reaction

The magnesio-ene reaction was first discovered when it was found that carbomagnesation of alkenes using allylic Grignard reagents occurred with γ regioselectivity ⟨81BSF(2)87⟩. The intramolecular version of this reaction has been developed and shows great potential in synthesis because of its high regio- and stereoselectivity ⟨72TL2285, 89AG(E)38⟩. The reaction occurs cleanly on gentle heating of the preformed allylic Grignard compound to give the rearranged magnesium derivative in good yield. This process has been used as the key step in several total syntheses of natural products (Equations (20)–(22)). Other intramolecular carbomagnesation reactions are less useful as they occur with equilibration.

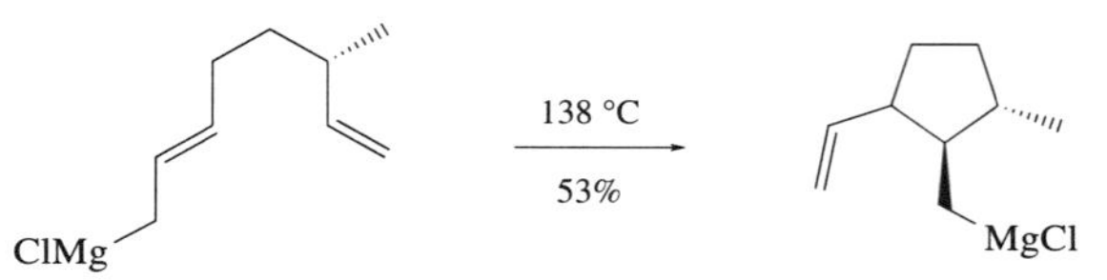

(20)

(26) **(27)**

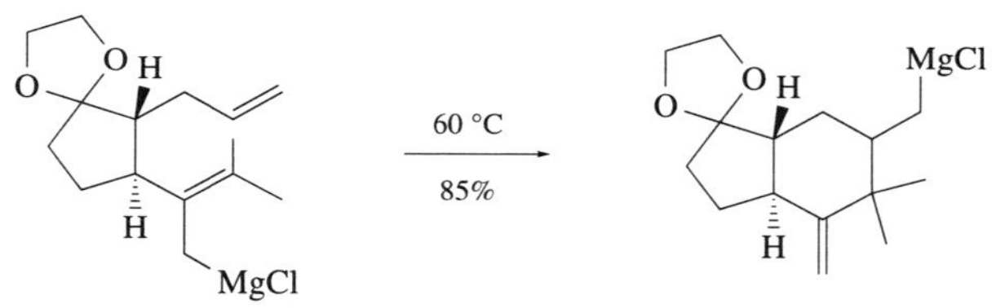

(21)

(28) **(29)**

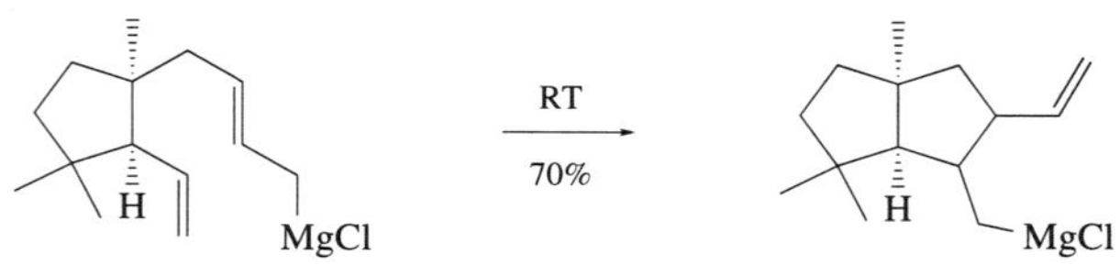

(22)

(30) **(31)**

3 : 2 mixture of stereoisomers

2.11.3.2 Other Derivatives

The chemistry of organoberyllium, -calcium, -strontium and -barium derivatives has been the subject of several reviews, which often treat them together with magnesium derivatives (see previous section). In general, these compounds have received little attention in the synthetic literature since they offer little advantage over the readily available Grignard reagents. Organocalcium compounds have been most widely studied, whereas beryllium and barium derivatives have been used as catalysts for polymer synthesis. Depending on the metal and the mode of preparation, both RMX and R_2M forms can be obtained.

2.11.3.2.1 *Alkylberyllium derivatives*

Alkylberyllium derivatives are generally of the type R_2Be; very few alkylberyllium halides have been prepared. The chemistry of alkylberyllium species has been reviewed ⟨73HOU(13/2a)4, 82COMC-I(2)121⟩. Dialkylberyllium derivatives are prepared by transmetallation from alkyllithium or -magnesium compounds with beryllium halides, from dialkylmercury compounds using beryllium powder, or from trialkylboron derivatives using diethylberyllium (Scheme 3). Interestingly, mixed species of the type R^1BeR^2 have been prepared by redistribution. Alkylberyllium halides may be obtained by reacting Be powder with an alkyl halide, or by mixing dialkylberyllium compounds with beryllium halides, in analogy to the Schlenk equilibrium (see Section 2.11.3.1.1(ii)). Some organoberyllium derivatives have been used as carbon nucleophiles but their utilisation in synthesis is hampered by the very high toxicity of beryllium.

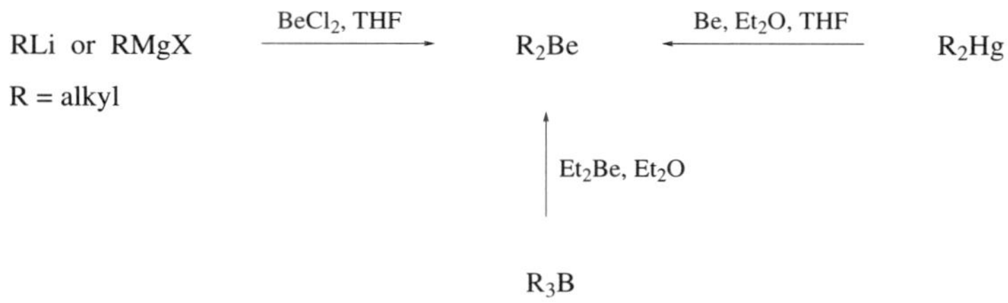

Scheme 3

2.11.3.2.2 *Alkylcalcium, -strontium and -barium derivatives*

In contrast to beryllium compounds, alkylcalcium, -strontium and -barium derivatives are generally of the RMX type and their chemistry has been reviewed ⟨73HOU(13/2a)533A⟩. They are generally prepared by direct reaction between the metal and an alkyl halide ⟨73T1069, 78JCS(D)657⟩. Iodides are often used as starting materials and the metallation is performed at low temperature to avoid Wurtz coupling. Highly reactive metal may also be prepared as described for magnesium. Reduction of anhydrous calcium bromide ⟨90JOC5045⟩ or barium iodide ⟨91JA8955⟩ with lithium biphenylide affords highly reactive calcium and barium metals, which react with various alkyl halides at low temperature to give the corresponding organometallic reagents in good yields. Calcium may also add to 1,3-dienes in an analogous fashion to magnesium. Highly reactive barium is particularly useful for the preparation of allylbarium derivatives (Scheme 4). These latter reagents have been the subject of great interest because of the high α-regioselectivity of their reactions with electrophiles ⟨91JA8955, 92JOC6386, 92TL6435⟩. Activated calcium has also been prepared by condensation of calcium vapours into a suitable solvent ⟨87JOM(332)247⟩.

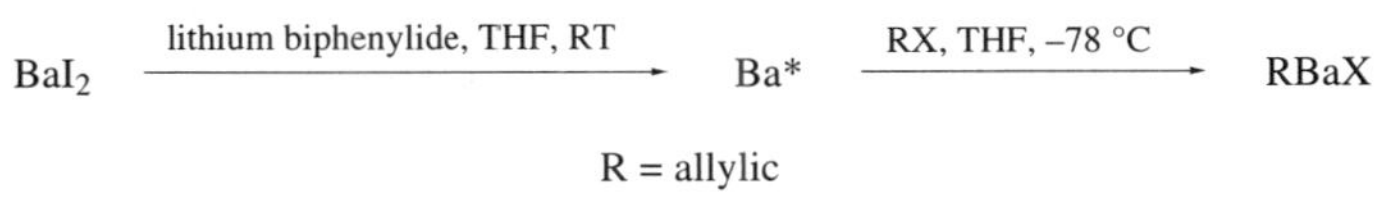

Scheme 4

2.11.4 ALKYL TRANSITION METAL DERIVATIVES

Alkyl derivatives are known for almost all transition metals and several have been incorporated into the basic repertoire of synthetic organic chemistry. Transition-metal derivatives are characterised by a great variety of ligands, and a number of possible oxidation states and coordination numbers. The organisation of this section will be further subdivided according to the oxidation state of the metal. For a general review of organotransition-metal chemistry (including alkyl metals), see ⟨B-80MI 211-01⟩.

2.11.4.1 Alkylscandium, -yttrium, -lutetium and -lanthanide Derivatives

The importance of lanthanides in organic synthesis has grown since the 1970s and has been covered by many reviews ⟨86T6573, B-89MI 211-02, 92CRV29, B-94MI 211-01⟩. Organocerium reagents, the most important of the organolanthanides, have been subject of a separate review ⟨91COS(1)231⟩, together with chemistry of samarium and ytterbium (including the corresponding alkyl metal derivatives) ⟨91COS(1)251⟩. Group 3a elements, Sc, Y and Lu are discussed generally in the lanthanide series since they have similar properties. For a general review of all these elements (plus actinides) see ⟨82COMC-I(3)173⟩. The most stable oxidation state is (III) although some lanthanides have relatively stable oxidation state of (II) (Sm, Eu and Yb) and (IV) (especially Ce).

2.11.4.1.1 M^{II} derivatives

Lanthanide metals, as powders, react with alkyl iodides in THF to give the corresponding alkyl metal iodides in a Grignard-type reaction ⟨71JCS(A)1931⟩. The success of this reaction depends on the stability of the (II) oxidation state of the metal. Yields are good with Eu and Yb, moderate with Sm, while Ce gives mixtures of trivalent species. The other lanthanides are unreactive. As for the Grignard reaction, activation (using 1,2-dibromoethane) may be necessary (especially for Yb).

Intramolecular coordination with an oxygen can greatly stabilise the alkyl metal reagent. Methyl 3-iodopropionate (**32**) reacts with various lanthanides in THF to give the corresponding organometallic reagents (**33**), which can be reacted with electrophiles (Equation (23)) ⟨86CC475⟩. These results are remarkable since even metals with inaccessible (II) oxidation states such as Ce react.

(**32**) (**33**)

M = La, Ce, Nd, Sm

(23)

2.11.4.1.2 M^{III} derivatives

(i) Preparation by transmetallation

Monoalkylated derivatives are the most important compounds for organic synthesis. Reaction between an organometallic reagent and anhydrous lanthanide salts in THF at low temperature affords the organolanthanide species cleanly and quantitatively (Equation (24)) ⟨84JOC3904⟩. Organolithium reagents are preferred for transmetallation, although Grignard reagents may also be used. Lanthanide chlorides are used as starting materials since they are readily available and can be conveniently dried. They must be finely ground and give a free-flowing suspension in THF. In 1990, the use of triflates as a starting material was recommended both for La ⟨90JOC3565⟩ and Yb ⟨90JOC4990⟩ with transmetallation achieved in 30 min in both cases (Equation (25)).

$$\text{RLi} + \text{MCl}_3 \xrightarrow[\text{quantitative}]{\text{THF, }-78\,^\circ\text{C}} \text{RMCl}_2 \qquad (24)$$

R = primary, secondary, tertiary alkyl, allyl, benzyl; M = Sc, Yb, Ln

$$\text{RLi} + \text{M(OTf)}_3 \xrightarrow[\text{quantitative}]{\text{THF, }-78\,^\circ\text{C}} \text{RM(OTf)}_2 \qquad (25)$$

R = Me, Bu^n; M = La, Yb

In order to increase the thermal stability, ligands other than halides or triflate may be used. The anionic π ligands such as Cp are known to stabilise transition-metal complexes by saturation of empty *d* or *f* orbitals. Such complexes may be prepared by two methods: reaction between Cp_2LnCl and organolithium reagents is restricted to alkyllithiums containing bulky alkyl groups ⟨81CC292⟩; using an excess of alkyllithium, anionic tetracoordinated compounds are obtained ⟨86OM1296⟩.

Alternatively, treatment of the chloride with lithium tetraalkylaluminate gives a bimetallic derivative which is decomposed with THF or pyridine to give the organolanthanide compound ⟨79JCS(D)54⟩.

Dialkylated derivatives are prepared by treatment of $CpLnCl_2$ with two equivalents of alkyllithium reagents. These compounds are reported to be stable, but have not been studied further.

Trialkylated lanthanides are the most unstable alkyl derivatives; however, some compounds of this kind have been prepared and characterised. Transmetallation with the corresponding alkyllithium reagents has been used for the preparation of trimethyl, tri-*t*-butyl or tris(neopentyl) derivatives starting from lanthanide trichlorides or *t*-butoxides. Anionic tetra- or hexaalkyl metal derivatives have been prepared using the appropriate number of equivalents of alkyllithium reagents in the presence of TMEDA ⟨84OM69⟩. These compounds are stable enough to be isolated and characterised. The hexamethyl derivatives are interesting reagents since they react with high 1,2 selectivity with α,β-unsaturated carbonyls.

(ii) Preparation by oxidative addition

The great reducing potential of Sm^{II} has been applied extensively in organic synthesis and some reactions involving Sm^{II} salts and alkyl halides are believed to proceed through organosamarium(III) species via oxidative addition ⟨92SL943⟩ (Equation (26)). Alkylsamarium(III) diiodides have been identified as the reactive intermediates in the samarium diiodide-mediated Barbier reaction ⟨80JA2693, 90SL773, 91JOC4112⟩. More stable Sm^{III} compounds are prepared starting from $SmCp_2$ ⟨87JOM(328)81⟩. Reaction with benzyl chloride gives stable benzylsamarium(III) which can be further reacted with electrophiles ⟨92T3877⟩.

$$RI + 2SmI_2 \xrightarrow[R = \text{alkyl}]{\text{THF}} RSmI_2 + SmI_3 \qquad (26)$$

2.11.4.1.3 M^{IV} derivatives

Cerium has the most stable (IV) oxidation state and some alkylcerium(IV) derivatives have been synthesised. Cp_3CeR and $(Indenyl)_2CeR_2$ have been obtained by treatment of the corresponding chlorides with the appropriate organolithium reagent in refluxing THF ⟨73JINC311⟩. These results are noteworthy since reduction to Ce^{III} would normally be expected under these conditions.

2.11.4.2 Alkyltitanium, -zirconium and -hafnium Derivatives

The chemistry of alkyl group 4 metal derivatives has been dominated by its applications to polymer synthesis (especially for Ti). However, their use in carbon–carbon bond formation has been developed extensively since the 1970s, and organotitanium and organozirconium reagents are now very important tools for organic synthesis. As a result, many reviews are available: ⟨75HOU(13/7)265⟩ (Ti, Zr, Hf); ⟨82COMC-I(3)281, 82COMC-I(3)331, 82COMC-I(3)433⟩ (Ti); ⟨82COMC-I(3)559, 82COMC-I(3)635⟩ (Zr, Hf); ⟨B-83MI 211-01⟩ (Ti, Zr); ⟨B-86MI 211-01⟩ (Ti); ⟨88S1⟩ (Zr); ⟨91COS(1)139⟩ (Ti, Zr). Alkyl derivatives of these metals in the (II), (III) and (IV) oxidation states are known, the latter being the most stable.

2.11.4.2.1 M^{II} and M^{III} derivatives

Very few σ-bonded divalent metal compounds are known. MeTiCl has been prepared, as a catalyst for polymer synthesis, by transmetallation from the corresponding Grignard reagent ⟨57BEP560503⟩. The (III) oxidation state is more accessible to Ti than to the other group 4 elements. Mono-, di- and trialkylated derivatives of these elements are known. They are very unstable species which decompose easily either by β-elimination or to give alkyl radicals. Not surprisingly, methyl derivatives or bulky alkyl group derivatives have been the most studied. They are prepared either by transmetallation or by reduction of metal(IV) derivatives.

(i) Preparation by transmetallation

Reaction of methyl magnesium bromide with titanium trichloride in the presence of pyridine gives a precipitate of $MeTiCl_2(Py)$, which is stable only at low temperature ⟨67JOM(10)P9⟩. Further stabilisation is provided using Cp ligands. Reaction of $(Cp_2MCl)_2$ with alkyllithiums gives $RMCp_2$ complexes (M = Ti, Zr, Hf) ⟨85JOM(295)159⟩. For reasons of stability, this reaction is restricted to very bulky alkyl groups devoid of β-hydrogen atoms. Stable monoalkyl derivatives are isolated, albeit in very low yield, when the alkyl chain possesses a heteroatom in the δ or ω position able to coordinate the low-valent metal. Some bis(alkyl) derivatives of this kind have also been prepared.

(ii) Preparation by reduction

Alkyllithium or alkylmagnesium compounds have been used for the reduction of various metal derivatives. Thus, treatment of Cp_2TiCl_2 with a twofold excess of benzyllithium gives Cp_2TiBn ⟨77JA6645⟩. Zr and Hf have less accessible (III) oxidation states and undergo alkylation without reduction. Alternatively, reduction may also occur by loss of a Cp ligand from Cp_2TiR_2 (R = Me, Bn), either by treatment with benzyllithium ⟨72SRI33⟩ or by electrochemical or sodium amalgam reduction ⟨80CC214⟩. This latter method has also been used for the reduction of $Cp_2Zr(Cl)(CH_2Bu^t)$ to $Cp_2Zr(CH_2Bu^t)$. Finally, reduction is the most convenient method for the preparation of the highly unstable trialkylderivatives. Treatment of MBn_4 (M = Ti, Zr, Hf) with ethyllithium gives the corresponding tribenzyl metal compounds ⟨77JOM(142)145⟩.

2.11.4.2.2 M^{IV} derivatives

The chemistry of M^{IV} derivatives has been reviewed ⟨91COS(1)139⟩. Alkyltitanium and alkylzirconium compounds as carbon nucleophiles have been developed primarily through the work of Seebach and Reetz ⟨B-83MI 211-01, B-86MI 211-01⟩; they are prepared by transmetallation. Also, preparation of zirconium derivatives by hydrozirconation has become an important tool in organic synthesis. Alternative approaches to the synthesis of alkyl metal(IV) derivatives include oxidation of metal(III) compounds.

(i) Preparation by transmetallation

Treatment of anhydrous metal salts MX_4 (M = Ti, Zr, Hf) with an alkyl metal reagent provides an easy entry to mono-, di, tri- and tetraalkyl derivatives, depending on the number of equivalents of reagent used. The ligand X can be a halide (generally Cl), alkoxide (generally isopropoxide) or dialkylamide. Ligand homogeneity is not mandatory, and mixed halide–alkoxide compounds are known. However, in the transmetallation reaction, a halide group is necessary to obtain neutral species, since the addition of an alkyl metal to titanium tetraisopropoxide or tetra(dialkylamide) gives ate complexes. The latter, however, have interesting chemical properties and are often used in synthesis, especially allyl derivatives. A wide variety of organometallic reagents may be used for transmetallation. Alkyllithium or Grignard reagents are often used, but transmetallation from zinc, aluminum, lead or tin compounds has been described. For the preparation of the parent compound $MeTiCl_3$, transmetallation from dimethylzinc is recommended as the by-products are easily removed and the product can be isolated in high purity. Another advantage in the use of zinc reagents is that the reaction may be performed in nonethereal solvents such as hexane or dichloromethane to give unsolvated species. However, when isolation is not a concern, transmetallation from alkyllithium reagents in THF at $-78\,^{\circ}C$ is an easy and convenient method. Preparations of some representative alkyl derivatives are listed in Table 5. Yields are given only for isolated compounds.

The main concern is the stability of the alkyl compounds, and this often dictates the choice of metal and ligand. Titanium compounds are preferred since they are far cheaper and more reactive than their zirconium counterparts. However, alkyltitanium reagents exhibit poor thermal stability and easily undergo β-elimination. In general, instability increases with the number of alkyl groups on the metal. Only methyl and allyl derivatives are relatively stable (it should, however, be pointed out that $MeTiCl_3$ and $MeTi(OPr^i)_3$ enjoy greater stability than most other methyl metal derivatives and may be isolated and purified), and α-branched alkyl groups such as *t*-butyl undergo elimination. Zirconium derivatives, which are less prone to β-elimination, offer a convenient alternative. The

Table 5 Preparation of various alkyltitanium, alkylzirconium and alkylhafnium(IV) derivatives by transmetallation.

Entry	*Organometallic reagent* (equivalents)	*Metal salt*	*Conditions*	*Product*	*Yield* (%)	*Ref.*
1	Me_2Zn (0.5)	$TiCl_4$	hexane, RT	$MeTiCl_3$	60–70	75IS120
2	MeMgBr (1)	$TiCl_4$	hexane, –20 °C to 0 °C	$MeTiCl_3$		68ZAAC(356)195
3	MeLi (1)	$TiCl_4$	Et_2O, –78 °C to –30 °C	$MeTiCl_3 \cdot OEt_2$	100	86T2931
4	Me_2Zn (1)	$CpTiCl_3$	benzene, RT	$MeCpTiCl_2$	60	84JOM(267)265
5	MeLi (1)	$ClTi(OPr^i)_3$	Et_2O, –35 °C to 0 °C	$MeTi(OPr^i)_3$	69	74JOM(74)85
6	MeLi (1)	$ClZr(OBu^n)_3$	Et_2O, –10 °C	$MeZr(OBu^n)_3$		81HCA1552
7	Bu^nLi (1)	$ClZr(OBu^n)_3$	Et_2O, –10 °C	$Bu^nZr(OBu^n)_3$		81HCA1552
8	Bu^tLi (1)	$ClZr(OBu^n)_3$	Et_2O, –10 °C	$Bu^tZr(OBu^n)_3$		81HCA1552
9	MeLi (1)	$BrTi(NMe_2)_3$	THF, –20 °C to RT	$MeTi(NMe_2)_3$	67	69JOM(20)129
10	EtMgBr (1)	$BrTi(NMe_2)_3$	THF, –20 °C to RT	$EtTi(NMe_2)_3$	23	69JOM(20)129
11	AllylMgBr (1)	$Ti(OPr^i)_4$	THF, –78 °C	$AllylTi(OPr^i)_4^- MgBr^+$		82TL5259
12	AllylMgCl (1)	$Ti(NMe_2)_4$	THF, –78 °C	$AllylTi(NMe_2)_4^- MgCl^+$		82TL5259
13	AllylMgCl (1)	$ClTi(OPr^i)_3$	THF, –78 °C	$AllylTi(OPr^i)_3$		80AG(E)1011
14	AllylMgCl (1)	$ClTi(NEt_2)_3$	Et_2O, 0 °C to RT	$AllylTi(NEt_2)_3$	75	71JOM(32)313
15	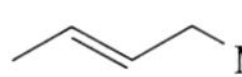 MgCl	Cp_2TiCl_2	Et_2O, –30 °C	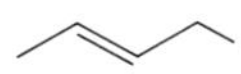 $TiClCp_2$		81CC1140
	or					
16	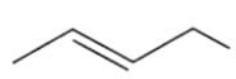 Li	Cp_2ZrCl_2	THF temperature not given	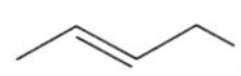 $ZrClCp_2$		81TL2895
17	Me_2Zn (3)	$TiCl_4$	CH_2Cl_2, –30 °C	Me_2TiCl_2		80AG(E)900
18	MeLi (2)	Cp_2ZrCl_2	Et_2O, –20 °C to 0 °C	Me_2ZrCp_2	79	73JA6263
19	MeLi (2)	Cp_2HfCl_2	Et_2O, –20 °C to 0 °C	Me_2HfCp_2	60	73JA6263
20	MeLi (4)	$TiCl_4$	Et_2O, hexane, –78 °C	Me_4Ti		63ZAAC(319)230
21	BnMgCl (4.1)	$TiCl_4$	Et_2O, hexane, –15 °C to RT	Bn_4Ti		71JOM(32)335
22	Bu^tCH_2Li (4)	$ZrCl_4$	Et_2O, hexane, 0 °C	$(Bu^tCH_2)_4Zr$	68	73JOM(57)269
23	Bu^tCH_2Li (4)	$HfCl_4$	Et_2O, hexane, 0 °C	$(Bu^tCH_2)_4Hf$	59	73JOM(57)269

preparation of dialkyl- (of the type Cp_2TiR_2) and tetraalkyltitanium derivatives is complicated by reduction (see previous sections). The nature of the ligand also has an influence on stability. Alkoxide and dialkylamide derivatives are generally more stable than the halide counterparts, and Cp ligands are known to confer further stabilisation (Table 5, entry 4). One advantage of the alkoxide ligands is the possibility of ligand exchange with chiral alcohols (or diols) for the preparation of chiral titanium species. Scheme 5 ⟨86TL5711, 92T5691, 92T5719⟩ shows some chiral alkyltitanium compounds used for enantioselective addition of alkyl groups to aldehydes ⟨92CRV807⟩, although the absolute configuration at titanium was not determined.

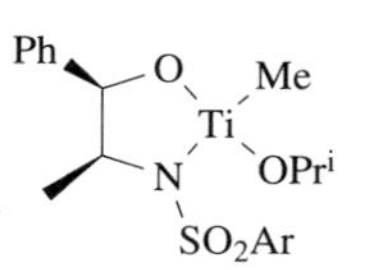

(34)

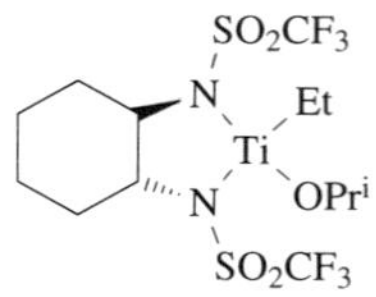

(35)

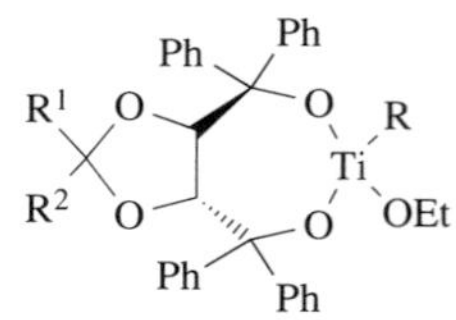

(36) $R^1 = R^2 = Me$
(37) $R^1 = H, R^2 = Ph$

Scheme 5

(ii) Preparation by hydrometallation

The most important application of zirconium in organic synthesis is the preparation of alkylzirconium species by hydrozirconation ⟨76AG(E)333⟩. Zirconocene chloride hydride ('Schwartz's reagent') cleanly inserts various alkenes and dienes to give, regioselectively, the primary alkylzirconium species in good yield (Equation (27)). In contrast to most metals, mild conditions are employed, generally in THF at room temperature, and polymerisation does not normally occur. Sonication greatly increases the reaction rate ⟨91JOC6494⟩. Schwartz's reagent is air-, moisture- and light-sensitive, and reactions should be carried out in the dark under an inert atmosphere. Since a metal hydride is involved, care should be taken to protect functional groups susceptible to reduction (carbonyl, nitriles) or deprotonation (hydroxyls). The reaction is suitable for mono- or disubstituted alkenes; however, internal double bonds are isomerised to terminal ones via successive hydrozirconation and elimination. Dienes are regioselectively hydrometallated on the least-substituted double bond. Owing to the weakness of the carbon–zirconium bond, alkylzirconocene species may be cross-coupled with various electrophiles (especially halides) or transmetallated to other alkyl metals. Zirconocene dichloride has also been used as a catalyst for hydrometallation with other metals, especially Mg and Al (see corresponding sections).

R = alkyl (27)

The Zr^{II} derivative zirconocene (prepared *in situ* by reduction of zirconocene dichloride with Grignard reagents) reacts with 1,6 dienes or enynes to give bicyclic metallacycles (Equation (28)) which can be hydrolysed or reacted with carbon monoxide to give bicyclic ketones ⟨86TL2829⟩. This reaction proceeds under mild conditions and with good stereoselectivity. The mechanism and synthetic applications have been detailed in a recent review ⟨91COS(5)1163⟩.

(28)

(iii) Preparation by oxidation

Oxidation of alkyl M^{III} to M^{IV} derivatives is hampered by the weakness of the carbon–metal bond. Oxidation of $MeZrCp_2$ with iodine has been reported to give Cp_2ZrI_2 as the main product along with $MeZr(I)Cp_2$ ⟨85JOM(295)159⟩.

2.11.4.3 Alkylvanadium, -niobium and -tantalum Derivatives

Alkyl derivatives of group 5 metals have received less attention than the other transition metals, probably because of their instability. Alkyl derivatives of the metal in the (II) to (V) oxidation states are known and some reviews covering their preparation and reactions are available ⟨75HOU(13/7)355, 82COMC-I(3)647, 82COMC-I(3)705⟩.

2.11.4.3.1 M^{II} and M^{III} derivatives

The preparation and characterisation of alkylvanadium(II) compounds was reported in 1987 ⟨87OM2354⟩. They are stabilised by coordination with a phosphine ligand, and are prepared by transmetallation from the corresponding alkyllithium or alkylmagnesium reagent. These compounds are reported to be stable below 0 °C.

The stability of M^{III} derivatives depends on the nature of the counteranion and the presence or absence of coordinating neutral ligands. The preparation of mono-, di- and triallylvanadium species from vanadium and allylmagnesium bromide has been described ⟨93CB2093⟩. These reagents are prepared *in situ* in THF solution and used as a source of carbon nucleophiles. The simplest stable compounds have the formula $RVCp_2$ and are prepared by transmetallation (Equation (29)) ⟨76JOM(110)327⟩. Other compounds of moderate stability such as $RV(acac)_2$ or $RV(OBu^t)_2$ have also been synthesised ⟨79DOK(245)198, 82JOM(239)335⟩ by similar routes.

$$Cp_2VCl \quad + \quad RMgX \xrightarrow[30\text{–}80\%]{Et_2O,\ pentane} RVCp_2 \qquad (29)$$

$$R = Me, Et, Pr^n, Bu^n, C_5H_{11}, CH_2Bu^t$$

Finally, transmetallation has been used for the synthesis of dialkyl derivatives in the presence of coordinating phosphine ligands ⟨87OM2354⟩.

2.11.4.3.2 M^{IV} derivatives

Various mono-, di- or tetraalkyl derivatives are known. They are generally prepared by transmetallation using chloride as a leaving group on the metal. Monoalkylvanadium species with the general formula RVX_3 have been prepared with $X = NEt_2$ from alkyllithium reagents (Equation (30)) ⟨79ZAAC(458)89⟩, or with $R = (OR^1)_3$ (R^1 = Me, Pr^i, Ph) from alkylzinc reagents ⟨77ZAAC(431)88⟩. The latter, however, are very unstable and decompose even at low temperature.

$$RLi \quad + \quad ClV(NEt_2)_3 \xrightarrow[\substack{R = Pr^n,\ 80\% \\ R = Bu^n,\ 64\%}]{heptane,\ -25\ ^\circ C\ to\ RT} RV(NEt_2)_3 \qquad (30)$$

Dialkyl compounds with the general formula R_2MX_2 have been prepared for X = Cl (M = V) using alkylzinc reagents ⟨72PAC575⟩ or with X = Cp (M = V, Nb, Ta) using various organometallic reagents (Equation (31)) ⟨81JCS(D)80, 88MI 211-02⟩. Dialkoxydialkylvanadium species $R_2V(OR^1)_2$ may be prepared using two different routes: reaction between VR_4 and Bu^tOH, or transmetallation from either $V(OBu^t)_4$ or $Cl_2V(OBu^t)_2$ using Grignard reagents ⟨77JOM(131)43⟩. Several alkoxide ligands have been used, but the *t*-butoxide ligand gives the most stable compounds.

$$2RM^1 \quad + \quad Cp_2M^2Cl_2 \longrightarrow R_2M^2Cp_2 \qquad (31)$$

$$\begin{gathered} M^2 = V,\ R = Me, Et, Bn \\ M^2 = Nb,\ R = Me, CD_3 \\ M^2 = Ta,\ R = Me, CD_3 \\ M^1 = Li, Mg, Zn, Cd \end{gathered}$$

Very few tetraalkylated derivatives of vanadium are known. Tetramethylvanadium may be prepared *in situ* at −70 °C, but is highly unstable and must be used immediately. However, tetrabenzyl- and norbornylvanadium, also prepared by transmetallation, are stable enough to be isolated. It is noteworthy that the choice of reagent is crucial for the preparation of $V(Bn)_4$: while reaction of dibenzylmagnesium with VCl_4 affords the expected product, the use of benzylmagnesium chloride results in reduction.

2.11.4.3.3 ***M*^{V} *derivatives***

M^{V} derivatives are the most widely known and studied alkyl derivatives in this series. The synthesis of alkylvanadium(V) compounds, however, may be complicated by competitive reduction, although this problem may be countered by the incorporation of oxo or imino ligands.

(i) Alkyloxo- and alkyliminovanadium(V) complexes

The first synthesis of an alkyloxovanadium compound was achieved by reaction between dimethylzinc and $VO(OR^1)_3$ (R^1 = Me, Pr^i, Bu^t) to give the moderately stable methyl complex ⟨77ZAAC(431)88⟩. This reaction has been extended to other alkyl groups ⟨82ZN(B)380⟩. In 1993, mono-, di- and triallyloxovanadium species were prepared *in situ* by the reaction at −70 °C of $VOCl_3$ with 2, 4 or 6 equivalents, respectively, of allylmagnesium bromide and condensed with various aldehydes ⟨93CB2093⟩.

Compounds bearing bulky imino ligands enjoy greater stability and their preparation is less complicated by side reactions. Reaction of $VOCl_3$ with bulky amines gives compounds with formula RN=VClX(NHR) (X = Cl, Cp) which may be mono- or dialkylated using alkyllithium or alkylmagnesium reagents (Equation (32)) ⟨90OM2207, 90ZN(B)191⟩.

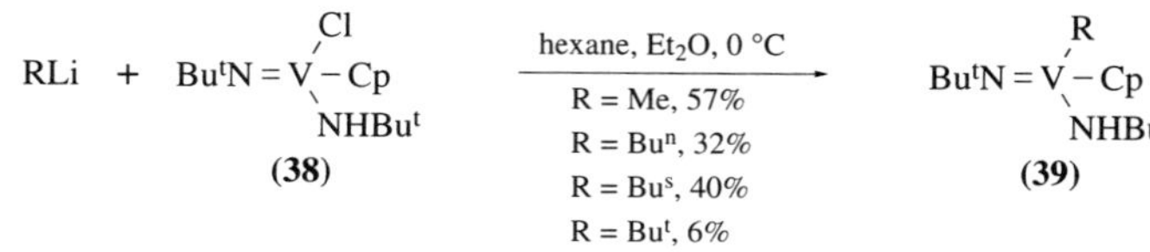

(32)

(ii) Other derivatives

Treatment of MCl_5 (M = Nb, Ta) with methyl metal reagents affords mono-, di- or trimethylated species, depending on the reagent used. Monomethylation (to $MeMCl_4$) is best achieved using $HgMe_2$ or $SnMe_4$ ⟨73JCS(D)2436⟩, whereas the use of dimethylzinc results in di- or trimethylation, depending on the number of equivalents used ⟨72JCS(D)2313, 73JCS(D)961⟩. All these compounds may be used as starting materials for ligand interchange. Methylation using dimethylzinc does not go beyond trialkylation; however, treatment of Me_3MCl_2 with two equivalents of methyllithium affords the highly unstable and explosive pentamethylated compounds, which may be isolated as their 1,2-(bis)dimethylphosphinoethane adducts ⟨74JA5288, 76JOM(122)209⟩. Treatment of the same starting material with a greater excess of methyllithium provides anionic hexa- or heptamethylated species, which are stable. Other alkyl derivatives such as benzyl or neopentyl are known. All mono- to pentabenzyltantalum compounds may be prepared, either by successive transmetallations or by substitution of an alkyl group with a chloride ion (Scheme 6) ⟨76JOM(122)209, 80JA6744⟩.

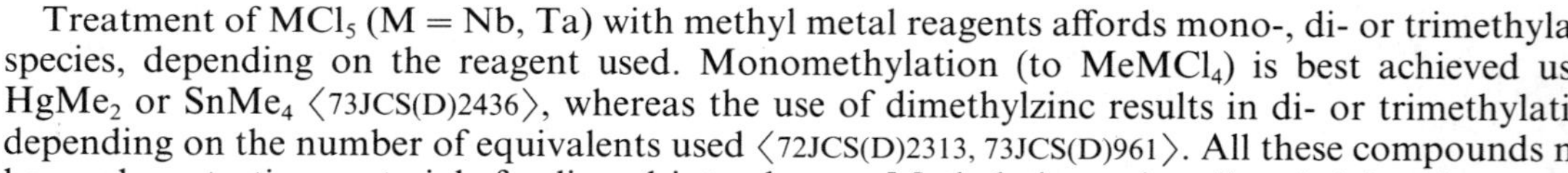

Scheme 6

2.11.4.4 Alkylchromium, -molybdenum and -tungsten Derivatives

These elements play an important role in organometallic chemistry. Various oxidation states, from (0) to (IV) are known for alkyl derivatives. The most important reagents for organic synthesis are alkylchromium(III) derivatives.

For general reviews, see: (i) organochromium chemistry ⟨75HOU(13/7)379, B-75MI 211-01, 82COMC-I(3)783⟩; (ii) organomolybdenum chemistry ⟨75HOU(13/7)433, 82COMC-I(3)1079⟩; (iii) organotungsten chemistry ⟨75HOU(13/7)489, 82COMC-I(3)1255⟩.

2.11.4.4.1 Cr^{III} derivatives

Alkylchromium derivatives, especially alkylchromium dihalides, have been widely used as efficient sources of carbon nucleophiles in organic synthesis. They are more stable than other alkyl transition-metal reagents and display interesting chemoselectivity in their reactions with electrophiles ⟨82AG(E)144⟩. The allylic compounds are amongst the best reagents for stereoselective allylation of aldehydes ⟨91COS(1)173⟩ and they may be prepared either by transmetallation or by oxidative addition reactions.

(i) Preparation by transmetallation

Treatment of a suspension of anhydrous chromium trichloride in THF with various organometallic reagents cleanly affords monoalkylated derivatives, generally in quantitative yield. Alkylaluminum, alkyllithium or alkylmagnesium reagents may be used, depending on the alkyl group to be transferred (Equation (33)). In contrast with most of the other transition metals, transmetallation is achieved at, or just below, room temperature without substantial decomposition, and some alkylchromium derivatives are stable enough to be isolated. Studies carried out in the early 1990s on allylchromium dichloride have demonstrated the existence of a Schlenk-type equilibrium which lies in favour of the diallylated species ⟨93CB2093⟩. No evidence has been found so far for this type of equilibrium in alkyl derivatives.

$$\text{RM} + \text{CrCl}_3(\text{THF})_3 \xrightarrow[\text{ii, RT, 5 h}]{\text{i, THF, }-20\,^\circ\text{C, 1 h}} \text{RCrCl}_2(\text{THF})_3 \qquad (33)$$

R = Me, Et, Pr^n, Pr^i, Bu^n, Bu^s, Bu^i, n-C_8H_{17}; M = Li, MgX, AlR_2, RAl(OEt)

Grignard or alkyllithium methods are suitable for the preparation of di-, tri- or peralkylated species. These compounds, however, are very unstable and as a result very few trialkylchromium compounds have been obtained Peralkylated species are more stable, owing to their greater coordination number, and the hexamethyl compound Li_3CrMe_6 has been prepared by addition of an excess of methyllithium to chromium trichloride.

In a search for stabilised alkylchromium species, the synthesis of alkylchromium compounds with Cp ligands has been investigated (Equation (34)) ⟨90ACR263⟩. $CpCrCl_2$ is obtained from chromium trichloride and may undergo successive alkylation with alkyllithium reagents to give stable mono- or dialkylated species. Anionic trialkylated species R_3CrCp may also be prepared.

$$n\,\text{RLi} + \text{CpCrCl}_2 \xrightarrow{\text{THF, RT}} \text{R}_n\text{CrCpCl}_{2-n} \qquad (34)$$

R = Me, Et, Bu^n, Bn, CH_2Bu^t; n = 1, 2

(ii) Preparation by oxidative addition

Oxidative addition, for example, addition of alkyl halides to chromium(II) salts to give alkylchromium(III) derivatives, is probably the most useful method for synthetic applications. Hiyama and co-workers prepared anhydrous chromium dichloride by reduction of the trichloride with $LiAlH_4$ in THF and the resulting chromium salt was reacted directly with allylic halides in the presence of an electrophile to give the coupled products in good yield ⟨77JA3179⟩ (anhydrous $CrCl_2$ is now commercially available). Further investigations showed that the reaction was suitable for allylic and propargylic halides or tosylates ⟨82BCJ561⟩. The reaction is performed at room temperature in THF and in some cases, especially with tosylates, the use of a dipolar aprotic solvent such as DMF is necessary. In order to avoid side reactions such as Wurtz coupling, the alkylchromium reagent is prepared *in situ* in the presence of the electrophile. Chromium is less reactive in the first stage of the reaction than other metals and only substrates giving rise to stable radicals undergo metallation with success. This limitation has been countered by the use of additives which facilitate and accelerate homolytic carbon–halogen bond cleavage. In 1989, Nakaï and co-workers prepared various alkylchromium(III) derivatives in good yields by oxidative addition in the presence of a catalytic amount of vitamin B_{12} or cobalt phthalocyanin in DMF (Equation (35)) ⟨89JOC4732⟩. The reaction was noted to be sensitive to steric hindrance, as yields were lower with branched or secondary halides. The chromium-mediated Barbier reaction has great potential in synthesis, owing to its mild conditions and its selectivity. Functional groups such as esters, nitriles and in some cases ketones, are not affected. Indeed, alkylchromium reagents are among the most functional-group-tolerant organometallic reagents in synthesis.

$$\mathrm{RI} + \mathrm{CrCl_2} \xrightarrow[\text{THF or DMF, RT}]{\text{vitamin } B_{12} \text{ (cat.) or Co phthalocyanin}} \mathrm{RCrCl_2} \qquad (35)$$

$$\mathrm{R} = \mathrm{C_{12}H_{25}, MeCO(CH_2)_5, EtO_2C(CH_2)_5, Cl(CH_2)_{12}}$$

2.11.4.4.2 *Other derivatives*

(i) Group 6 M^0 and M^{II} derivatives

The chemistry of group 6 metal(0) carbonyl complexes has found many applications in organic synthesis, but few of them involve the preparation of alkyl metal reagents. In separate studies, complexes with formula $RM(CO)_5^-$ (M = Cr, W) have been prepared either by decarbonylation of acyl metal complexes, alkylation of the dianionic complexes $M(CO)_5^{2-}$, or by transmetallation from $BrM(CO)_5^-$ using alkyllithium reagents ⟨77IC1357, 78JA7565⟩. These methods have been used for the synthesis of neutral M^{II} complexes with formula $RMCp(CO)_3$ (M = Cr, Mo, W), the most convenient being alkylation of $MCp(CO)_3^-$ with various alkyl halides. Other representative metal(II) complexes are the dimeric species with two quadruply bonded metals, $Li_2M_2Me_8$, which are generally prepared by reaction of an excess of methyllithium with metal(II) chlorides.

(ii) M^{III} (other than Cr^{III}) and M^{IV} derivatives

Alkyl derivatives of molybdenum(III) and tungsten(III) generally possess a dimeric structure M_2R_6 with a metal–metal triple bond. They are prepared by reductive transmetallation, starting from $MoCl_5$ or WCl_6, using an excess of alkyllithium or alkylmagnesium reagents.

The structure of alkyl metal(IV) complexes depends on the nature of the metal. Chromium compounds are tetraalkylated and may be prepared by two routes: treatment of chromium trichloride with alkyllithium or alkylmagnesium reagents followed by oxidation, or direct transmetallation starting from a chromium(IV) salt. These compounds are very resistant towards β-elimination; however, their reactivity is low and therefore they are not useful for organic synthesis. Molybdenum derivatives may be mono- or dialkylated. They are very unstable and only methyl derivatives have been isolated. $MeMoCl_3$ is prepared by transmetallation starting from $MoCl_4$ and an alkyl metal reagent ⟨72ZAAC(394)293⟩, whereas Cp_2MoCl_2 may be mono- or dialkylated using Grignard reagents ⟨74JCS(D)1324⟩. Very few tetraalkyltungsten(IV) compounds have been prepared. Reaction between WCl_4 and dibenzylmagnesium affords the stable tetrabenzyltungsten ⟨74ZAAC(412)11⟩.

(iii) M^{VI} derivatives

These derivatives may be of two kinds, either alkyl or alkyloxo metal complexes. Simple alkyl derivatives are known only for tungsten and may be mono-, hexa- or peralkylated, methyl derivatives being the most studied. Treatment of WCl_6 with alkylzinc, alkyltin, alkylmercury or alkylboron reagents affords monoalkylated products only, whereas reaction with trimethylaluminum gives Me_6W. Treatment of Me_6W with methyllithium gives the complex Li_2WMe_8.

Metal oxo complexes are prepared using the same transmetallation techniques, with complexes of formula $RWOCl_3$ being prepared from $WOCl_4$. Dialkyl derivatives are stabilised with Cp ligands and are prepared from Cp_2WOCl_2 using two equivalents of alkyllithium or alkylmagnesium reagents. Bis(oxo) derivatives may be synthesised using various routes, including oxygen oxidation of dialkyl metal(IV) complexes ⟨88OM1877⟩ and mono- or dialkylation of LMO_2Br_2 (L = bipyridine ligand, M = Mo, W) (Equation (36)) ⟨86OM2452, 91OM244⟩. Interestingly, monoalkylated derivatives, when treated with an aqueous base, give rise to anionic trioxo derivatives $(L)RMO_3^-$ ⟨82ZN(B)380⟩. Oxo derivatives show remarkable stability and may be handled in air. They are also more resistant towards β-elimination and alkyl derivatives other than methyl or benzyl may be synthesised.

$$n\,RMgX + LMO_2Br_2 \xrightarrow[\text{'good yields'}]{\text{THF, }-78\text{ °C to }0\text{ °C}} R_nLMO_2Br_{2-n} \qquad (36)$$

R = Me, Et, Pr^n, Pr^i, Bu^n, Bu^i, Bn, CH_2Bu^t, $(CH_2)_4CH{=}CH_2$
L = bipyridine, 4,4'-dimethylbypyridine
n = 1, 2

2.11.4.5 Alkylmanganese Derivatives

Despite only being developed in the 1980s, the chemistry of alkylmanganese derivatives has found many applications in synthesis and two reviews are available ⟨82COMC-I(4)1, 86HOU(13/9a)1⟩. Alkyl derivatives of the metal in the oxidation states (I) to (IV) are known.

2.11.4.5.1 Mn^I derivatives

Alkyl derivatives of manganese(I) have the formula $RMn(CO)_5$ or $RMn(CO)_4L$, L being a neutral ligand. The parent compound $MeMn(CO)_5$ was the first alkyl metal carbonyl complex to be prepared ⟨57JOC598A⟩. A wide range of methods can be employed for the synthesis of these species since both anionic or cationic forms of manganese pentacarbonyl may be prepared. Alkylation of $NaMn(CO)_5$ with alkyl halides, or transmetallation from $BrMn(CO)_5$ using alkyllithium, alkylmagnesium or alkylmercury reagents, both afford alkyl derivatives in good yields (Scheme 7) ⟨57JOC598A, 58LA(618)24⟩. Alternatively, a two-step sequence has been described involving acylation of $NaMn(CO)_5$ with acyl chlorides ⟨57JOC598B⟩, followed by thermal decarbonylation (Scheme 8) ⟨79JOM(172)405⟩. This method appears to be the most convenient for the preparation of alkyl derivatives since direct alkylation of the weakly nucleophilic $Mn(CO)_5^-$ requires the use of reactive alkyl halides. The stability of these complexes depends on the nature of the alkyl group. Although $MeMn(CO)_5$ is stable in air at room temperature, $EtMn(CO)_5$ is reported to decompose readily. The preparation of stable long-chain alkyl complexes ⟨92JOM(439)C25⟩ indicates that β-elimination is not, however, the decomposition pathway.

$$RMgX + BrMn(CO)_5 \xrightarrow{Et_2O,\text{ reflux}} RMn(CO)_5 \xleftarrow{\text{THF, RT}} RI + NaMn(CO)_5$$

R = Me, Bn (left); R = Me, Et, Bn (right)

Scheme 7

$$RCOCl + NaMn(CO)_5 \xrightarrow{\text{THF, RT}} RCOMn(CO)_5 \xrightarrow{\Delta,\ -CO} RMn(CO)_5$$

R = Me, Et, Pr^n, Bn, $c\text{-}C_6H_{11}\text{–}CH_2$, $n\text{-}C_6H_{13}$, $n\text{-}C_9H_{19}$, $n\text{-}C_{18}H_{37}$

Scheme 8

These compounds undergo some reactions which are useful for organic synthesis. For example, when treated with an alkene under high-pressure conditions, insertion of carbonyl and the alkene into the carbon–metal bond occurs, thus leading to a new organomanganese complex ⟨88JA2575⟩. An example is shown in Equation (37).

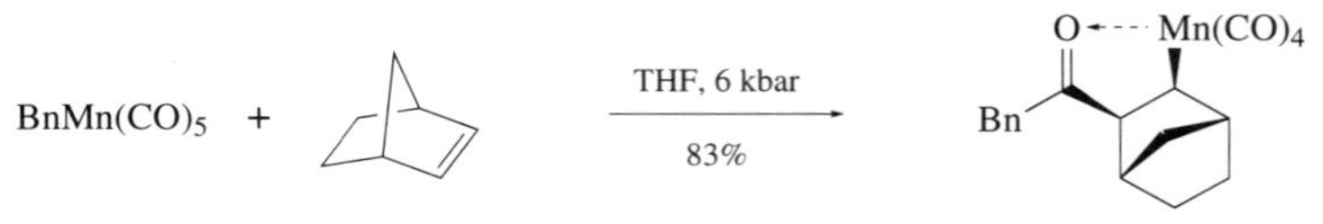

(37)

(43)

2.11.4.5.2 Mn^{II} derivatives

The preparation of alkylmanganese(II) derivatives and their use in organic synthesis has been extensively studied by Cahiez and co-workers ⟨B-83MI 211-02⟩. Their preparation involves oxidative addition (Grignard-type reaction) or transmetallation reactions.

Owing to the low reactivity of elemental manganese, preparation of alkylmanganese halides by oxidative addition is limited to the highly reactive allylic halides. They are prepared *in situ* in the presence of an electrophile (aldehyde or ketone) in order to avoid homocoupling ⟨89TL7373⟩. This reaction requires elevated temperatures with ethyl acetate as the solvent. However, the addition of a catalytic amount of zinc chloride also allows the reaction to be carried out in THF.

Since oxidative addition is restricted to reactive substrates, transmetallation is the most convenient method for the preparation of organomanganese(II) reagents. Three types of compounds may be obtained, RMnX, R_2Mn or R_3MnM (R = alkyl, M = MgX, Li), depending on the number of equivalents of reagent used. The order of stability is $R_3MnM > RMnX > R_2Mn$, the last being stable only for alkyl groups devoid of β-hydrogen atoms ⟨76JCS(D)2204⟩. Detailed experimental procedures for the preparation of these compounds and their precursors have been published ⟨89T4163⟩. Transmetallation may be achieved using both alkyllithium or alkylmagnesium reagents, and is suitable for nearly all primary, secondary or tertiary alkyl groups. Anhydrous manganese chloride (as a suspension in diethyl ether or THF) may be used as a precursor, although better results are obtained using freshly prepared manganese diiodide in ether. The use of the THF-soluble ate complex Li_2MnCl_4 (prepared from $MnCl_2$ and LiCl) allows transmetallation to occur instantaneously at 0 °C to give clear solutions. The reaction is performed under mild conditions in ether or THF (R_3MnMgX can be prepared only in THF) and yields are quantitative (Scheme 9 and Equations (38) and (39)).

RMgX or RLi $\xrightarrow{MnI_2,\ Et_2O,\ -10\ °C}$ RMnI

RMgX or RLi $\xrightarrow{MnCl_2,\ THF,\ -10\ °C}$ RMnCl

RMgX or RLi $\xrightarrow{Li_2MnCl_4,\ THF,\ 0\ °C}$ $RMnLi_2Cl_3$

R = primary, secondary, tertiary alkyl

Scheme 9

$$2RMgX \text{ or } RLi \xrightarrow{MnI_2,\ Et_2O \text{ or } MnCl_2,\ THF} R_2Mn \qquad (38)$$

$$3RMgX \text{ or } RLi \xrightarrow{MnCl_2,\ THF} R_3MnM \qquad (39)$$

M = MgX, Li

2.11.4.5.3 Mn^{III} and Mn^{IV} derivatives

These compounds are prepared by oxidation of lower oxidation state complexes. An Mn^{III} Schiff-base complex has been prepared by alkylation followed by oxidation ⟨77JINC153⟩, whereas two oxidations of alkylmanganese(II) to alkylmanganese(IV) compounds have been reported. Oxidation of dilithio(1-norbornyl)manganate affords tetra(1-norbornyl)manganese ⟨72JA2512⟩, whereas oxygen oxidation of various neopentylic dialkylmanganese compounds gives stable tetraalkylmanganese(IV) derivatives ⟨75JOM(99)C19⟩.

2.11.4.6 Alkyliron, -cobalt and -nickel Derivatives

Reviews are available covering the areas of organoiron chemistry ⟨82COMC-I(3)331, 86HOU(13/9a)175, B-87MI 211-04⟩, organocobalt chemistry ⟨82COMC-I(5)1, 84HOU(13/9b)1⟩, and organonickel chemistry ⟨82COMC-I(6)37, 84HOU(13/9b)628⟩. Alkyl derivatives of these metals in the oxidation states from (0) to (IV) are known.

2.11.4.6.1 M^0 derivatives

The M^0 oxidation state is accessible for iron and nickel only. The preparation of alkyl derivatives of iron(0) is of great importance in organic synthesis. Disodium tetracarbonylferrate ('Collman's reagent') was introduced as one of the first formyl anion equivalents ⟨75ACR342⟩. This reagent reacts with alkyl halides (order of reactivity: $I > Br > OTs > Cl \gg F$) to give monoalkylated derivatives in good yields. The reaction is of broad scope and tolerant of many functional groups in the alkyl chain. However, since this reagent is basic, tertiary halides cannot be used because of competing elimination, and secondary tosylates are preferred to the corresponding halides. The reaction is of S_N2-type at carbon, with inversion of configuration, and the resulting iron species may undergo a carbonyl migratory insertion reaction into the carbon–iron bond to form acyliron complexes. A second alkylation at iron followed by elimination affords ketones from two alkyl halides (Scheme 10).

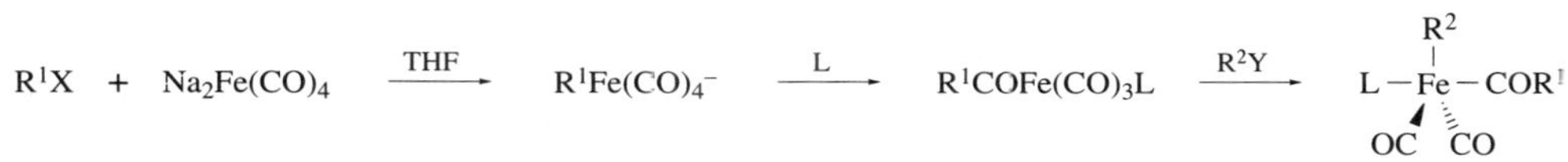

R^1, R^2 = primary, secondary alkyl; X, Y = I, Br, Cl, OTs; L = CO, PPh_3

Scheme 10

Attempts to prepare alkylnickel(0) complexes by reductive alkylation generally result in the elimination of elemental nickel, although the preparation of the anionic complex alkylnickel bis-(ethene) by reduction of $Ni(cod)_2$ with an alkyllithium reagent in the presence of TMEDA has been described ⟨76AG(E)621⟩.

2.11.4.6.2 M^I derivatives

The most important compounds of this type are alkylcobalt(I) derivatives.

(i) Preparation by alkylation

Alkylation of anionic cobalt species is the most convenient method for the preparation of complexes with formula $RCoL_4$; however, the nature of the ligand L has a great effect on the stability of the complex and the course of the reaction. Alkylcobalt tetracarbonyl compounds are very sensitive towards carbonyl insertion into the carbon–metal bond. Sequential exchange of carbonyl to phosphine (or phosphite) ligands brings further stability. Indeed, alkylation of $Co(PR_3)_4^-$ (R = Ph, OMe) proceeds easily to give exceptionally stable complexes. Owing to the instability of alkylcobalt tetracarbonyl derivatives, acylation of $Co(CO)_4^-$ followed by decarbonylation is not a useful method for their preparation.

(ii) Preparation by transmetallation

A few cobalt complexes have been prepared by transmetallation. These include pentacoordinated complexes $MeCo(PMe_3)_3L$ ($L = CO$, alkene, PMe_3), which are prepared by reaction of methyllithium with $ClCo(PMe_3)_3$ in the presence of the appropriate ligand.

(iii) Preparation by hydro- or carbometallation

The hydride complex $HCo(CO)_4$ reacts with alkenes to give alkylcobalt derivatives. This reaction is important as one of the key steps in the cobalt-catalysed hydroformylation of alkenes (the *oxo* synthesis) ⟨B-85MI 211-03⟩, when hydrometallation is followed by carbonyl insertion. Mechanistic studies have shown that this reaction, like most hydrometallation reactions, is a reversible process and double-bond isomerisation is therefore the main side reaction. The regioselectivity is rather poor; most of the substituted alkenes give mixtures, Markovnikov addition being predominant. A notable exception is styrene, which exclusively gives the branched compound. Catalyst modification, by exchange of two carbonyl ligands for two phosphine ligands, improves regioselectivity in favour of the primary alkyl derivative.

The hydride complex also cleaves the carbon–oxygen bond of epoxides and oxiranes to give α- or β-alkoxyalkyl derivatives. Mixtures are obtained with dissymmetric epoxides, with cleavage at the less-substituted carbon being predominant.

(iv) Preparation by oxidation or reduction

Oxidation or reduction are not general methods for the preparation of these complexes and only a few examples have been reported. Reduction of $M(acac)_3$ ($M = Fe$, Co) with alkylaluminum reagents in the presence of donor ligands affords alkyl metal(I) derivatives (Equation (40)). Methyl and benzyl derivatives have been synthesised with various phosphine ligands and various primary and secondary alkyl derivatives have been prepared in the presence of cod ligands.

$$R_3Al \text{ (or } R_2AlOEt) + M(acac)_3 \xrightarrow{L,\ Et_2O,\ -25\ ^\circ C} RML \qquad (40)$$

$M = Fe$; $R = Bn$; $L = 2PPh_3$
$M = Co$; $R = Me, Bn$; $L = 2PPh_3$; 70–80%
$M = Co$; $R = Me$; $L = dppe$; 82%
$M = Co$; $R = Et, Pr^i, Bu^n, Bu^i, n\text{-}C_8H_{17}$; $L = cod$

2.11.4.6.3 M^{II} derivatives

Alkyl compounds of Fe^{II}, Co^{II} and Ni^{II} are known. The methods for their preparation are similar to those employed for the preparation of M^{I} derivatives.

(i) Preparation by alkylation

The anionic complex $CpFe(CO)_2^-$ (also called 'Fp^-') has been introduced as a strongly nucleophilic equivalent to Collman's reagent ⟨70ACR417⟩. This complex may be prepared as its lithium, sodium or bromomagnesium salt. In contrast with Collman's reagent, Fp^- is more nucleophilic and may be alkylated under mild conditions with a wide range of primary, secondary, allylic or benzylic halides (or sulfonates) to give the easily isolable RFp complexes ⟨80JOC291⟩ (Equation (41)), which may be further transformed into carbonyl compounds.

$$RX \; + \; CpFe(CO)_2^- \xrightarrow{Et_2O,\ THF} RFeCp(CO)_2 \qquad (41)$$

R = primary, secondary, allylic, benzylic; X = I, Br, Cl, OTs

(ii) Preparation by transmetallation

Transmetallation reactions starting from iron(II), cobalt(II) or nickel(II) salts (generally chlorides) may be performed either with or without neutral ligands present, thus affording stabilised or nonstabilised alkyl derivatives.

Non-ligand-stabilised alkyliron or alkylcobalt derivatives have been studied extensively in their addition reactions to carbonyl compounds and cross-coupling reactions with vinyl halides. Mono- to tetraalkylated species RMX, R_2M, R_3MLi or R_4MLi_2 (R = alkyl, M = Fe, Co) may be prepared (generally *in situ*) by treatment of the anhydrous metal(II) chloride with alkyllithium reagents in THF at low temperature (Scheme 11) ⟨92CB157, 92CB163⟩. The order of stability is $R_4MLi_2 > R_3MLi > RMX > R_2M$ and the etherate complex of dilithium tetramethylferrate can even be isolated, albeit in low yields. Since anhydrous ferrous chloride is difficult to obtain, an alternative reductive transmetallation procedure has been developed ⟨92CB163⟩: anhydrous ferric chloride is reduced by treatment with one equivalent of alkyllithium, the resulting ferrous chloride suspension then undergoes transmetallation. The tetraalkylated compounds give the best results in reaction with electrophiles with the iron derivatives being superior to cobalt.

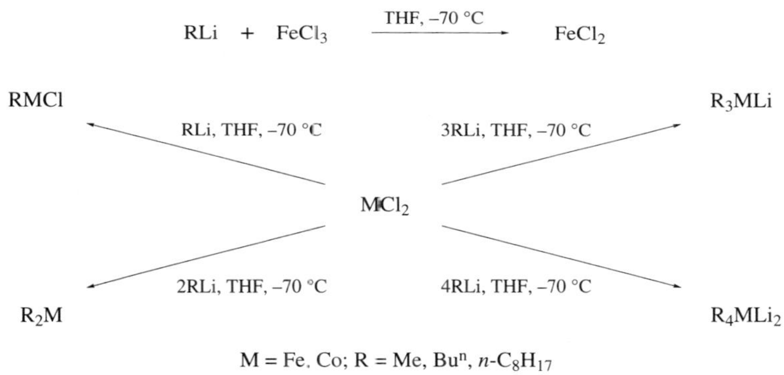

Scheme 11

A great variety of ligands give rise to stabilised complexes. Iron complexes with the formula $RFeCpL_2$ (L = CO, phosphine, bis(phosphine), isonitrile, etc.) ⟨85CB2407, 91JOM(419)245⟩ or nickel complexes with the formula R_2NiCpL or RNiCpL (L = phosphine, arsine) are prepared from the corresponding halide using alkyllithium or alkylmagnesium reagents. Cyclic nickel complexes are obtained using 1,4-dilithiobutane. Transmetallation reactions may also be achieved by reaction with simple metal salts in the presence of ligands: cobalt and nickel acetylacetonates are alkylated with alkylaluminum compounds in the presence of phosphine or bis(phosphine) ligands, during which one or two alkyl groups may be transferred to give stable complexes.

(iii) Preparation by hydrometallation, carbometallation or π to σ isomerisation

The π to σ isomerisation of alkene ligands has been developed as a valuable tool for organic synthesis. The cation $CpFe(CO)_2^+$ ('Fp^+') reacts with alkenes to give a π-complex. Upon treatment with a soft nucleophile, isomerisation and nucleophile addition occur to give an alkyliron complex (Equation (42)) ⟨77JA8426⟩. Many nucleophiles may be used including hydride, malonates, dialkylcuprates, enolates, enamines, or heteronucleophiles such as amines or thiols ⟨76JOM(108)93⟩. On the other hand, treatment with a base gives an allylic complex which undergoes [3 + 2] cycloadditions with electron-deficient alkenes ⟨92CRV97⟩. Cleavage of the iron–carbon bond is easily achieved by either oxidation or elimination. The overall process in this reaction is therefore an activation of alkenes towards nucleophiles by initial iron complexation.

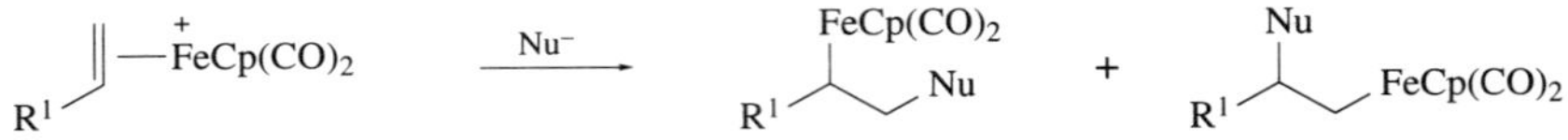

(42)

R^1 = SR, CO_2Me; Nu = H^-, $^-CH(CO_2Me)_2$, $R^2{}_2CuLi$, enolate, enamine, R^2SH, R^2NH_2

(iv) Preparation by oxidative addition

Direct addition of alkyl halides has been reported to be successful only with iron(0) or nickel(0) phosphine complexes.

2.11.4.6.4 M^{III} and M^{IV} derivatives

Porphyrin complexes of cobalt(III) or iron(III) play a vital role in numerous biological processes and their chemistry has been studied extensively. One natural alkylcobalt(III) porphyrin complex, coenzyme B_{12}, is involved in many biochemical reactions ⟨B-82MI 211-01⟩ and alkyliron(III) compounds are believed to be involved in the metabolism of various substrates with cytochrome P450 ⟨B-86MI 211-02⟩. In order to study the mechanism of these biological pathways, model alkyl metal porphyrin complexes have been prepared ⟨86ACR209, 88CRV1121⟩. These studies have underlined the weakness of the carbon–metal bond and its tendency, especially for Co, to undergo homolysis. Consequently, the use of alkylcobalt(III) complexes as radical precursors has been studied intensively, using simpler ligands such as Schiff bases ⟨90JCS(P1)2703, 90JCS(P1)2709⟩ or oximes ⟨89AG(E)325⟩.

Many chelate complexes such as porphyrins, precorrins, Schiff bases or oximes have been prepared for both metals and listing all the structures is beyond the scope of this chapter. The general methods for their preparation are transmetallation, oxidative addition (alkylation) or hydrometallation (Equations (43)–(45)). Whatever the method for their synthesis, these complexes are very air- and light-sensitive and reactions must be carried out under strictly anaerobic conditions.

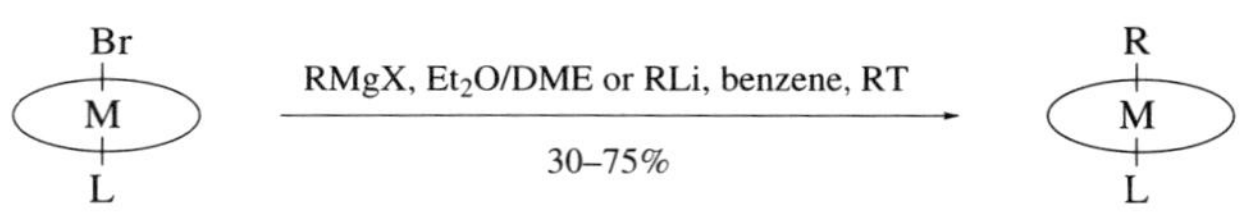

(43)

R = Me, Et, Pr^n, Bu^n; M = Fe, Co

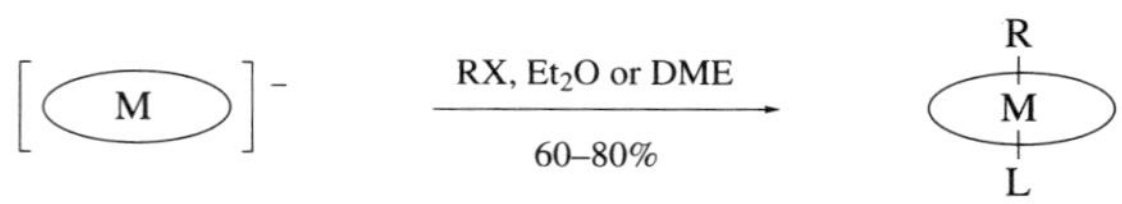

(44)

R = Me, Et, $(CH_2)_2OH$, $CH_2CH(OH)Me$, Bu^n, n-C_8H_{17}; M = Fe, Co

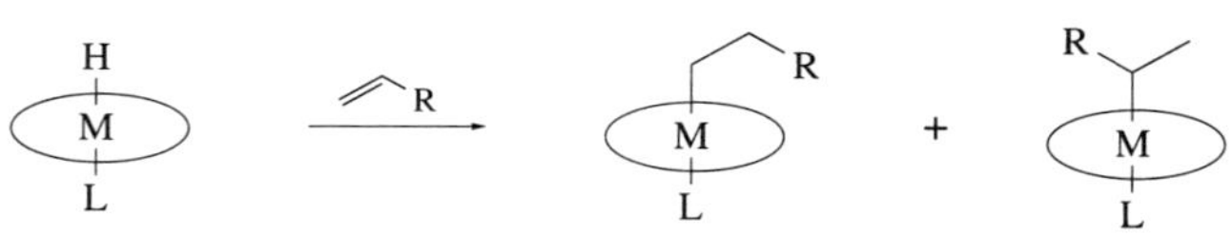

(45)

R = CH=CH$_2$, Ph, CN, CO_2Me; M = Fe, Co

 = porphyrin, precorrin, salen, salophen, bis(dimethylglyoximato); L = Py

Their preparation by transmetallation involves treatment of a chelate complex of a metal halide with an alkyllithium or alkylmagnesium reagent in the presence of a ligand (generally pyridine) and is suitable for Fe ⟨83JOM(253)65⟩ or Co ⟨84NJC759, 88JA5228⟩. Oxidative addition is also possible starting from M^I or M^{II} complexes. The most convenient method, however, is the alkylation of anionic iron or cobalt(I) complexes with alkyl halides (or sulfonates). These anionic complexes are prepared by chemical ⟨68IS65⟩ or electrochemical ⟨90JA1874⟩ methods. This method is particularly useful for the preparation of alkylcobalt(III) complexes as potential radical precursors ⟨88TL167, 89AG(E)325, 90JCS(P1)2691⟩. Treatment of an anionic cobalt(I) complex with an electron-deficient alkene under an atmosphere of hydrogen gives an alkylcobalt(III) complex via a hydrometallation reaction ⟨90JCS(P1)2709⟩. The regioselectivity of the reaction is pH dependent, with conjugate addition occurring in a neutral medium. Styrenes ⟨73JOM(55)375⟩ and 1,3-dienes ⟨90JCS(P1)2715⟩ also undergo hydrometallation to give conjugated complexes.

The iron(III) porphyrin complex-catalysed oxygenation of alkenes has been studied as a model for cytochrome P450. During this reaction, hydrometallation of the alkene in the presence of an external hydrogen source has been postulated to be the critical step and the presence of alkyliron(III) species has been demonstrated (Equation (46)) ⟨92CC1728, 92CL377, 93CPB292⟩.

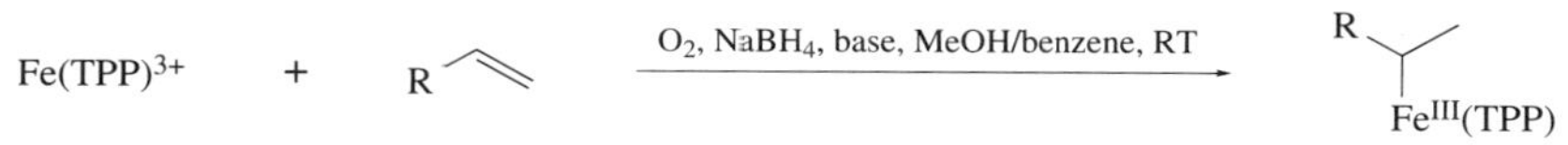

(46)

Alkyliron and alkylcobalt(III) compounds with ligands other than chelates have also been prepared, the most important being Cp-stabilised cobalt complexes. The dihalide $CpCoLI_2$ (L = phosphine, arsine, CO) may be mono- or dialkylated using Grignard reagents, while oxidative addition of alkyl iodides to $CpCo(PPh_3)_2$ affords cationic complexes.

The synthesis of alkyl metal(IV) derivatives is hampered by competitive reduction and the instability of the product. Stable tetrakis(1-norbornyl)iron or -cobalt complexes are prepared in the same way as for the manganese analogues. Oxidation of alkylcobalt(III) chelate complexes with various inorganic oxidants gives solutions of alkylcobalt(IV) derivatives which are stable only at low temperatures.

2.11.4.7 Alkylruthenium, -rhodium, -palladium, -rhenium, -osmium, -iridium and -platinum Derivatives

Late transition-metal complexes play an important role as catalysts for numerous organic reactions, although few of them involve σ-bonded alkyl derivatives. Alkylpalladium derivatives are found as intermediates in reactions such as the Heck reaction, the Wacker process, the metallo-ene reaction or cross-coupling reactions. These reactions are well documented and will not be discussed in this section. Depending on the metal, oxidation states of (I) to (VII) may be found. There are several reviews which document this chemistry: ⟨82COMC-I(4)691, 86HOU(13/9a)525⟩ (Ru); ⟨82COMC-I(5)277, 84HOU(13/9b)287⟩ (Rh); ⟨82COMC-I(6)279, 84HOU(13/9b)702, B-85MI 211-04⟩ (Pd); ⟨82COMC-I(4)161, 86HOU(13/9a)121⟩ (Re); ⟨82COMC-I(4)967, 86HOU(13/9a)613⟩ (Os); ⟨82COMC-I(5)541, 84HOU(13/9b)464⟩ (Ir); ⟨82COMC-I(6)471, 86HOU(13/9a)671⟩ (Pt). Comprehensive lists of structures and appropriate references can be found in these reviews.

2.11.4.7.1 M^I derivatives

(i) Preparation by transmetallation

Treatment of complexes of metal(I) chloride with methyllithium or methylmagnesium reagents produces the methyl derivatives, a process which is suitable for the preparation of rhodium(I) or iridium(I) phosphine complexes.

(ii) Preparation by alkylation

Simple alkylation is the most popular for the preparation of alkylrhenium complexes $RRe(CO)_5$; for example, the anion $Re(CO)_5^-$ is readily alkylated with active alkyl iodides or bromides. Acylation followed by decarbonylation produces the same products. Alkylation of the dianion $Os(CO)_4^{2-}$ (in analogy to Collman's reagent) fails to stop at the monoalkylation stage, owing to the very high nucleophilicity of the osmium anion. Finally, the preparation of alkylrhodium or alkyliridium phosphine complexes by electrochemical reduction of the corresponding chlorides in the presence of alkylating reagents has been described. This method compares well with the transmetallation in terms of yields and practicability.

(iii) Preparation by hydrometallation

The *oxo* synthesis (hydroformylation of alkenes) may also be performed in the presence of rhodium catalysts. The complex $HRh(CO)_4$ is more reactive than its cobalt counterpart; however, it does not react with better regioselectivity. As with cobalt, modified catalysts with general formula $HRh(CO)(PR_3)_2$ afford better selectivity in favour of the primary alkylrhodium intermediate. Various chiral phosphines or bis(phosphines) have been used for the asymmetric hydroformylation of styrene derivatives, although the best enantiomeric excess was a modest 50% ⟨81JOC4422⟩.

(iv) Other methods

Chiral rhenium nitrosyl complexes have been developed as potential chiral auxiliaries for the asymmetric formation of carbon–carbon bonds (Scheme 12) ⟨79JA1589, 83JA5804⟩. Treatment of the methyl complex (**44**) with trityl cation results in the unusual α-elimination to give a carbene complex (**45**) which then reacts with a nucleophile to give a homologated alkyl complex (**46**) ⟨82JA4865⟩. This sequence may be repeated to give a secondary alkyl derivative (**48**). Since carbene formation and addition of the nucleophile are stereoselective, a single diastereoisomer is obtained ⟨80JA3299⟩. Alternatively, treatment of the cationic complex (**49**) with an alkene affords a π-alkene complex (**50**) which upon treatment with a soft nucleophile (dialkyl cuprate reagent) undergoes π to σ isomerisation to give an alkyl complex as a single diastereoisomer ⟨90TL4417⟩. This latter reaction is analogous to the π to σ isomerisation of iron Fp complexes reported earlier.

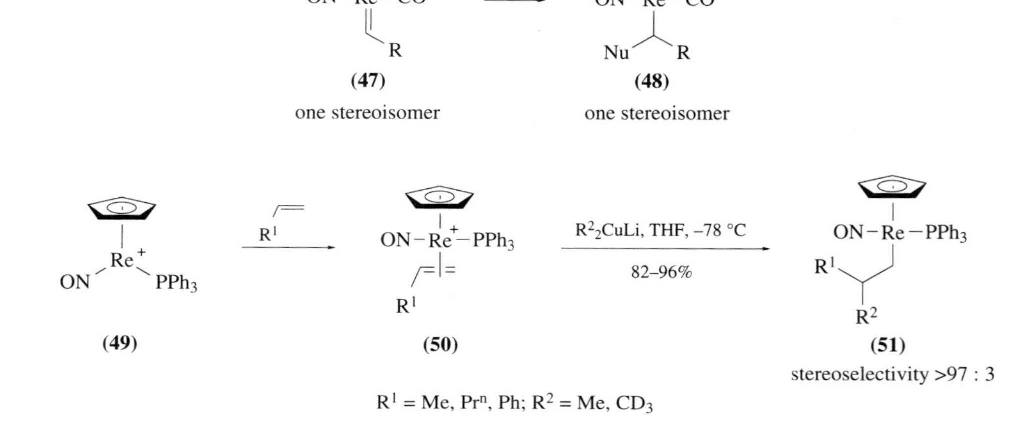

Scheme 12

2.11.4.7.2 M^{II} derivatives

(i) Preparation by transmetallation

Transmetallation techniques have been widely used for the preparation of palladium(II) and platinum(II) complexes. Reactions of organometallic reagents with simple palladium or platinum salts give highly unstable compounds which decompose into alkenes and alkanes. However, stabilised complexes can be produced and have the formula R_2ML_2 (M = Pd, Pt) where L is a neutral ligand (phosphine, phosphite, arsine, benzonitrile, cod, diphosphine or dipyridine). The transmetallation reaction is achieved using various alkyl metal reagents, depending on the ligand present, for example phosphine complexes are mono- or dialkylated using either alkyllithium or alkylmagnesium reagents (Scheme 13). Using 1,4-dilithiobutane, stable metallacycles are obtained ⟨80JCS(D)1633⟩. Transmetallation may also be achieved using alkyltin or alkylmercury compounds. Alternatively, alkyl complexes may be obtained by reaction of palladium bis(acetylacetonate) with alkylaluminum reagents in the presence of a ligand.

$$\text{RLi or RMgBr} + L_2MX_2 \xrightarrow[-60\ ^\circ\text{C to RT}]{\text{benzene}/Et_2O} RML_2X \longrightarrow R_2ML_2$$

R = Me, Et, Pr^n, Bn; X = I, Br, Cl; M = Pd, Pt; L = PEt_3, PPh_3, $AsPh_3$, PhCN, etc.

$$\text{RLi} + LMX_2 \xrightarrow[-60\ ^\circ\text{C to RT}]{\text{benzene}/Et_2O} R_2ML$$

L = dppe, bis(phospine), bis(amine), bipyridine, cod

Scheme 13

The stability of these complexes depends on the nature of the metal, the alkyl group and the ligand. Palladium complexes are less stable but also more reactive than their platinum counterparts. Complexes with triethylphosphine ligands are more stable than those with triphenylphosphine, probably for steric reasons. Alkyl complexes are also sensitive towards β-elimination, therefore methyl complexes are more stable than their higher analogues. In the early 1990s, the synthesis of exceptionally stable alkylpalladium and alkylplatinum complexes bearing a dithiocarbamate ligand was reported (Equation (47)) ⟨90OM16, 91OM902⟩. Stabilisation is believed to be due to π back-bonding of sulfur to the metal.

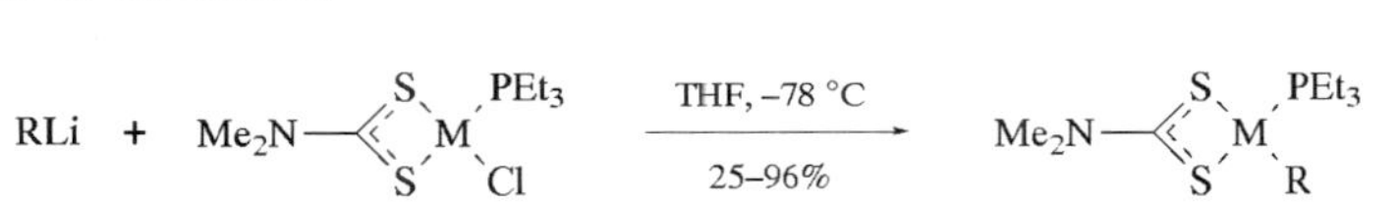

(47)

(52) M = Pt
(53) M = Pd

(54) M = Pt, R = Me, Pr^n, Pr^i, Bu^n, Bu^i, Bu^s, Bu^t
(55) M = Pd, R = Me, Pr^n, Pr^i, Bu^n, Bu^s

Osmium complexes with formula L_2OsCl_2 also react with alkyllithium or alkylaluminum reagents to give mono- or dialkylkated complexes, depending on the stoichiometry ⟨85JCS(D)573⟩, while π-arene osmium complexes $ArOs(L)Cl_2$ and ruthenium complexes $CpRuL_2Cl$ are monoalkylated using the same reagents ⟨83CB159, 83CB176⟩.

(ii) Preparation by orthometallation

Orthometallation reactions are observed with palladium or platinum complexes having bulky ligands, especially with tris(*o*-alkylphenylphosphine). Upon heating in a protic solvent, proton abstraction from the alkyl chain and extrusion of a ligand occurs, leading to the formation of a carbon–metal bond (Equation (48)). An isolated example of orthometallation of rhodium(III) phosphine complex has also been reported. Although triarylphosphine ligands are generally involved, orthometallation of a tris(*t*-butyl)phosphine palladium complex has been observed

⟨82OM654⟩. Nitrogen ligands may also promote this reaction, and an interesting example is the orthometallation of an 8-alkylquinoline ligand ⟨78CC344⟩.

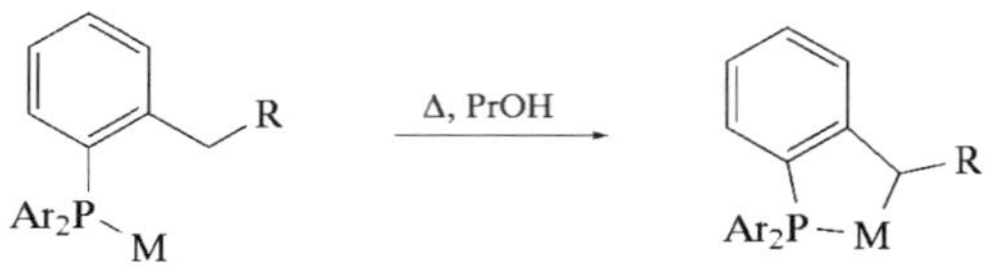

(48)

R = H, Me; M = Pt^{II}, Pd^{II}, Rh^{III}

(iii) Preparation by alkylation

Alkylation methods are useful for the synthesis of osmium(II) and ruthenium(II) complexes. The highly nucleophilic dianion $Os(CO)_4^{2-}$ is dialkylated using various alkyl halides or tosylates to give the extremely stable dialkyl derivatives in high yields. Similarly, alkyl hydride complexes are prepared by alkylation of the hydride anion $HOs(CO)_4^-$. The anionic ruthenium complex $CpRu(CO)_2^-$ reacts with alkyl halides in THF to give alkyl complexes in good yields. Compared to its iron analogue, the ruthenium complex is much less nucleophilic and only hard electrophiles (methyl, ethyl, allyl, benzyl) are reactive.

(iv) Preparation by oxidative addition

Addition of alkyl halides to metal(0) complexes is the most popular method for the preparation of monoalkylpalladium or -platinum complexes $RMXL_2$ ⟨76AOC345, 77ACR434⟩. Addition of alkyl halides (order of reactivity: primary > secondary and I > Br > Cl) on complexes ML_4 proceeds smoothly to give the *trans*-alkylchloro derivatives (or *cis* with bidentate ligands) in good yields (Scheme 14). The order of reactivity for the phosphine complexes is $M(PEt_3)_4 > M(PPh_3)_4 \geqslant M(CO)(PPh_3)_4$. When chiral secondary halides are used, the propensity for racemisation follows the same order, reaction with $M(CO)(PPh_3)_4$ giving complete inversion of configuration. These results are in favour of an S_N2 mechanism, although a radical mechanism has been proposed for some substrates ⟨80JCS(D)1448⟩. Direct addition of alkyl halides to palladium or platinum metal in Grignard-type reactions has been performed with some substrates ⟨85IS1961⟩. This reaction is of little practical use since the yields are low.

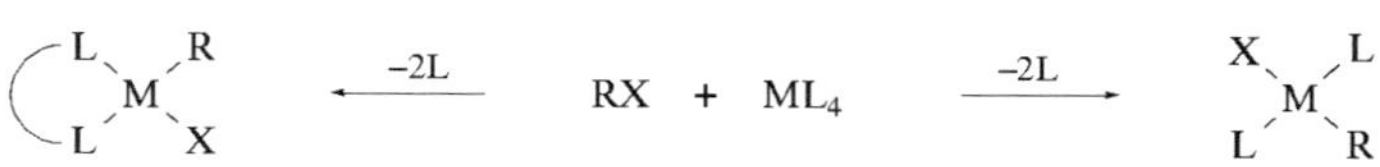

L_2 = dppe, bipyridine phenantrholine, cod

R = primary, secondary alkyl
M = Pt, Pd
X = I, Br, Cl

L = PEt_3, PBu_3, PPh_3, $AsPh_3$

Scheme 14

A variety of osmium(0) and ruthenium(0) complexes undergo oxidative addition reactions with alkyl halides. These include the nitrosyl ruthenium complex $RuCl(NO)(PPh_3)_2$, which reacts with methyl or benzyl halides, and π-arene complexes $ArM(PPh_3)_2$ (M = Ru, Os), which react with methyl iodide to give cationic complexes.

(v) Preparation by hydrometallation or π to σ isomerisation

Although osmium ⟨84JOM(276)C45⟩ or ruthenium ⟨85JCS(D)873⟩ hydride complexes add to carbon–carbon double bonds, platinum(II) compounds have received the most attention. The platinum complexes *trans*-$HPtCl(PR_3)_2$ reacts with ethene to give the ethylplatinum complex. However, the reaction is slow and requires high temperature and pressure. Substitution of the chloride in these reactions with other anions has been shown to increase the reaction rate, with the most reactive species being the cationic complex $[HPt(PEt_3)_2]^+$ X^- (X = ClO_4, BF_4, PF_6) and the stannate $HPt(SnCl_3)(PR_3)_2$. The use of platinum in the hydroformylation reaction is not very popular owing to its lower reactivity and poor regioselectivity, although platinum complexes with chiral phosphine ligands do give better enantiomeric excesses in asymmetric hydroformylation reactions. Systems giving enantiomeric excesses between 80% and 95% have been proposed ⟨91OM1183, 91OM2046⟩.

π to σ isomerisation reactions are analogous to the reaction of iron Fp-alkene complexes with nucleophiles. Reaction of cationic or neutral complexes of ruthenium, palladium, osmium or platinum with soft nucleophiles (hydride, amine, alcohol, phosphine or enolates) produces σ-bonded alkyl complexes. When the metal is palladium in these reactions, β-elimination occurs: the overall process is therefore a nucleophilic substitution on an alkene. Isomerisation may also occur upon addition of a neutral ligand on the metal, especially with π-allyl complexes of platinum and palladium.

2.11.4.7.3 M^{III} derivatives

Alkyl derivatives of rhenium, rhodium and iridium in the (III) oxidation state are known. Since methods for their preparation are similar to those for the preparation of M^{II} derivatives, they will be discussed only briefly here.

(i) Preparation by transmetallation

Bis- or trishalide complexes of rhenium, rhodium or iridium $XMCl_2(PPh_3)$ ($X = Cl$, Cp; $M = Re$, Rh, Ir) can be mono- or dialkylated using alkylmagnesium or alkylaluminum reagents. Other complexes which may be prepared by transmetallation are the rhenium clusters or bis(rhenium) derivatives. Alkyl porphyrin complexes of rhodium or iridium are similarly prepared from the corresponding chlorides by treatment with alkyllithium reagents.

(ii) Preparation by alkylation

Direct alkylation of an anionic metal(I) complexes is especially useful for the preparation of rhodium(III) and iridium(III) porphyrin complexes ⟨86ICA(121)27⟩.

(iii) Preparation by oxidative addition

Oxidative addition of alkyl halides on square planar rhodium(I) and iridium(I) halide complexes $MClX_3$ ($X =$ phosphine, CO, N_2) gives octahedral alkyl complexes via a *trans* addition (for a comprehensive list of structures, see references indicated at the beginning of the section). Important complexes in this series are Wilkinson's complex $Rh(PPh_3)_3Cl$ and the 'Vaska complex' $Ir(PPh_3)_2(CO)Cl$ and its rhodium analogue, which undergo oxidative addition with various alkyl halides.

(iv) Other methods

Rhodium(III) hydride complexes react with alkenes under mild conditions to give the alkyl derivatives in good yields. Two systems have been used, namely the neutral complex $RhH(PPh_3)_2Cl_2$ and the cationic complexes $RhH(NH_3)_4L^{2+}$ ($L = NH_3$, H_2O). Hydrometallation of electron-deficient alkenes with hydride or anionic iridium complexes has been described.

π-Alkene rhodium(I) complexes are isomerised to the σ-bonded alkyl derivatives upon treatment with an acid. Reaction with halo acids gives β-haloalkyl derivatives and with fluoroboric acid affords cationic complexes.

Finally, the formation of an alkylruthenium(III) porphyrin complex by reduction of a dialkylruthenium(IV) complex has been described ⟨87CC1110⟩. Reduction of $Me_2Ru(oep)$ (oep = octaethylporphyrin) to MeRu(oep) occurs simply by heating at 100°C. Homolysis of the carbon–ruthenium bond is involved in the decomposition pathway.

2.11.4.7.4 M^{IV} derivatives

Platinum derivatives are the most important and widespread compounds with this oxidation state. For example, trimethylplatinum iodide was prepared as early as 1907 and is one of the most stable alkyl metal compounds known. The principal methods for their preparation are, as expected, transmetallation and oxidative addition.

(i) Preparation by transmetallation

Various platinum(IV) complexes, such as $PtCl_4L_2$ (L = phosphine, arsine, pyridine) or $CpPtCl_3$ are alkylated using both alkyllithium or alkylmagnesium reagents. Substitution of all the halides appears to occur. Non-ligand-stabilised alkyl derivatives may be prepared by reaction of alkylmercury or alkylmagnesium reagents with $PtCl_4$, or better with K_2PtCl_6. This reaction does not proceed beyond trialkylation. Indeed, reaction of 15 equivalents of methylmagnesium iodide with platinum salts is the most convenient method for the synthesis of Me_3PtI. The more reactive alkyllithiums are reagents of choice for the preparation of peralkylated derivatives such as Li_2PtMe_6. Transmetallation techniques have also been applied to the synthesis of alkylrhenium(IV) clusters $R_{12}Re_4$ (R = Me, CH_2Bu^t, CH_2CMe_2Ph) using Grignard reagents, and for the preparation of a dimethylruthenium porphyrin complex with methyllithium ⟨86CC787⟩. Finally, a unique example of a transmetallation in aqueous medium has been described, where methylcobalamin reacts with K_2PtCl_6 to give $K_2MePtCl_5$ ⟨79JA1442⟩. This unique 'biomethylation' reaction is catalysed by platinum salts.

(ii) Preparation by oxidative addition

In principle, two oxidative addition reactions are possible which involve addition of either inorganic compounds or alkyl halides to alkylplatinum(II) complexes.

Although many inorganic compounds such as sulfides or metal halides add to platinum halides, halogens (chlorine, bromine and iodine) are the most used and add readily to give platinum(IV) complexes in excellent yields (Equation (49)). In some cases, elimination involving rupture of a carbon–platinum bond may occur. Alkyl halides readily react with alkylplatinum(II) complexes in polar solvents to give stable platinum(IV) derivatives (Equation (50)). This reaction is similar to the oxidative addition to platinum(0) derivatives in terms of reactivity, mechanism and stereoselectivity. Addition to bidentate complexes such as bipyridine or phenanthroline in these systems has been the subject of extensive mechanistic studies.

$$R_2PtL_2 \ + \ X_2 \xrightarrow{\text{benzene, RT}} R_2PtX_2L_2 \qquad (49)$$

X = I, Br, Cl; R = Me, $-(CH_2)_4-$; L = phosphine, arsine

$$R^2X \ + \ R^1{}_2PtL_2 \xrightarrow[\text{acetone, RT}]{\text{benzene or}} R^1{}_2R^2PtXL_2 \qquad (50)$$

R^1 = Me; R^2 = Me, CD_3, $Br(CH_2)_2Br$; X = I, Br, Cl; L = phosphine, arsine, bipyridine, phenanthroline

Another interesting oxidative addition is the reaction of cyclopropanes with platinum(II) salts, which leads to the formation of metallocyclobutanes by insertion of the metal into a carbon–carbon bond (Equation (51)). The reaction is regioselective for substituted cyclopropanes, but is inhibited by the presence of electron-withdrawing groups on the ring.

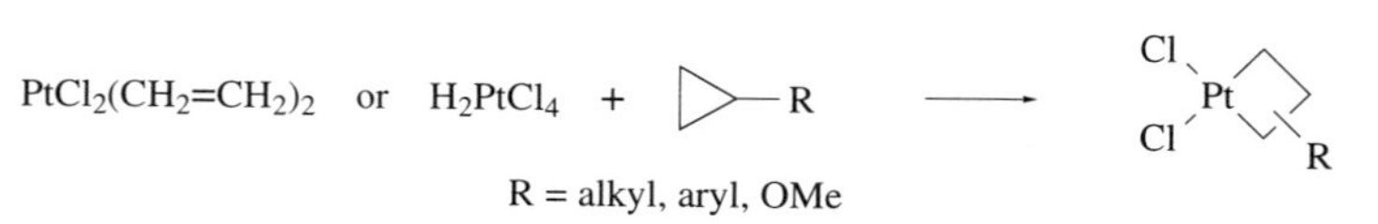

(51)

R = alkyl, aryl, OMe

2.11.4.7.5 M^V, M^{VI} and M^{VII} derivatives

Alkyl derivatives in these oxidation states are known for rhenium only, with the exception of the iridium(V) complex Cp^*IrMe_4, which is prepared by transmetallation from Cp^*IrCl_2 using trimethylaluminum, followed by oxidation ⟨81CC808⟩. A single alkylrhenium(V) derivative has been obtained by reductive alkylation of $CpReO_3$ with trimethylaluminum ⟨84AG(E)515⟩.

Alkylrhenium(VI) derivatives have been prepared from the *oxo* compound $ReOCl_4$ or $ReOCl_3(PPh_3)$. Reaction with methyllithium gives the tetramethylated compound Me_4ReO, which can be further alkylated with trimethylaluminum to give the explosive hexamethylated derivative Me_6Re.

The first report of the preparation of an alkylrhenium(VII) compound was achieved by oxidation of Me_4ReO with nitric oxide to give Me_3ReO_2. In the early 1990s a general method for the synthesis of monoalkyl trioxorhenium compounds was by reaction of dirhenium heptaoxide with alkylzinc reagents (Equations (52) and (53)) ⟨91AG(E)185, 93CB45⟩. Various alkyl groups are transferred in moderate to good yields and stabilisation may be provided by complexation with a quinuclidine ligand.

$$R_2Zn + Re_2O_7 \xrightarrow{\text{THF, }-40\ ^\circ\text{C}} 2RReO_3 \qquad (52)$$

R = Me, 95%; R = Et, 60%; R = Bu^n, 22%; n-C_5H_{11}, 40%; n-C_7H_{15}, 33%

$$R_2Zn + Re_2O_7 \xrightarrow{\text{Q, THF, }-40\ ^\circ\text{C}} 2R(Q)ReO_3 \qquad (53)$$

R = Pr^n, 40%; n-C_5H_{11}, 32%; i-C_5H_{11}, 31%; n-C_7H_{15}, 27%;
$CH_2CH(Me)Et$, 30%; $CH_2CH(Et)Bu^t$, 17%; cyclopropyl, 45%

Q =

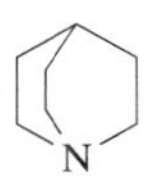

2.11.5 ALKYLCOPPER, -SILVER AND -GOLD DERIVATIVES

2.11.5.1 Alkylcopper Derivatives

The enormous development and importance of the chemistry of organocopper reagents has been stimulated by their use in fundamental carbon–carbon bond-forming reactions such as conjugate addition and halide displacement. Few organometallic reagents have played a greater role in organic synthesis. Their continual use in synthesis is still a major research activity and has been the subject of numerous reviews ⟨B-80MI 211-02, 82COMC-I(2)709, 91COS(1)107, 92OR(41)135⟩. Alkyl derivatives of copper are in the (I) oxidation state only and may be of three kinds—alkylcopper reagents RCu, and the ate complexes R_2CuM (low-order cuprates) and R_3CuM_2 (higher-order cuprates)—and are generally prepared using halogen–metal exchange or transmetallation reactions.

2.11.5.1.1 Preparation by halogen–metal exchange

Copper metal is not especially reactive towards alkyl halides and consequently the reaction requires very harsh conditions. However, the use of *activated* metal has proved to be very useful for the direct preparation of alkylcopper reagents. Rieke and co-workers have described the preparation of highly activated copper by reduction of copper(I) salts with lithium biphenylide, and its subsequent reactions with various alkyl halides (Scheme 15) ⟨92JA5110⟩. In the initial studies, $CuI \cdot PBu_3$ or $CuI \cdot PPh_3$ were used as substrates for reduction; however, further development showed that reduction of copper cyanide (solubilised by addition of LiCl or LiBr) gave more reactive species. Remarkably, the slurry of highly reactive copper reacts with alkyl or allylic halides (or allylic acetates) at temperatures as low as $-100\,^\circ C$, thus minimising side reactions such as Wurtz coupling or β-elimination. The conversion to alkylcopper reagents is essentially quantitative after a few minutes and metallation of allylic substrates is tolerant of many functional groups (e.g., esters, carbamates, nitriles, alkyl chlorides).

$$CuI{\bullet}PBu_3 \text{ or } CuCN{\bullet}LiCl \xrightarrow[\text{THF, }-100\ ^\circ\text{C}]{\text{lithium biphenylide}} Cu^* \xrightarrow{\text{RX, THF, }-100\ ^\circ\text{C}} RCu$$

RX = alkyl chloride, allylic chloride, allylic acetate

Scheme 15

2.11.5.1.2 Preparation by transmetallation

Despite the great potential of the halogen–metal exchange reaction with activated copper, transmetallation is the most convenient and general method for the preparation of alkyl derivatives of copper. Traditionally, organolithium or organomagnesium reagents have been used to initiate this reaction; however, alternative methods using other metals have been developed, involving synthesis of functionalised alkylcopper derivatives and preparation from substrates other than alkyl halides. These alternative aspects of transmetallation reactions are the subject of a review ⟨93S537⟩.

(i) Monoalkylcopper derivatives

Alkylcopper reagents are conveniently prepared by reaction of one equivalent of an organometallic reagent with a copper(I) salt in an ethereal solvent at low temperature (Equation (54)). Alkyllithium reagents are preferred to Grignard reagents for this reaction. As salt-free alkylcopper compounds are not very soluble in ethereal solvents, copper salts complexed with ligands are often used as starting materials, copper iodide tributylphosphine or copper bromide dimethyl sulfide being the most popular. Otherwise the presence of lithium salts ensures minimal solubility (the preparation of salt-free MeCu has been described ⟨85JA3197⟩). The reaction is generally quantitative and is suitable for primary, secondary, tertiary, allylic or benzylic derivatives. Alternatively, copper(II) salts may be used as starting materials with two equivalents of organometallic reagents, the first equivalent promoting reduction to copper(I). Dialkylzinc reagents are the most efficient for this procedure ⟨68JOM(12)225⟩. This reaction does not proceed cleanly, however, and therefore is not useful for synthetic purposes.

Apart from solubility, stability and reactivity are general problems associated with alkylcopper derivatives. They are thermally sensitive compounds and some of them have been reported to explode on attempted isolation. Decomposition reactions include β-elimination and homocoupling. Consequently, they are prepared at low temperature and used immediately. Alkylcopper reagents are also less reactive than the ate-complex species, even with complexation with a Lewis acid, which results in a dramatic enhancement of reactivity ⟨86AG(E)947⟩. It has been shown that preparation of alkylcopper reagents in neat dimethyl sulfide instead of ethereal solvent gives more stable and reactive compounds ⟨89T425⟩.

$$\mathrm{RM} + \mathrm{CuX} \xrightarrow{\mathrm{THF,\ Et_2O\ or\ DMS,\ -78\ ^\circ C}} \mathrm{RCu{\bullet}MX} \qquad (54)$$

R = alkyl; M = Li, MgBr; CuX = CuI, $\mathrm{CuI{\bullet}PBu_3}$, $\mathrm{CuI{\bullet}PEt_3}$, $\mathrm{CuBr{\bullet}SMe_2}$

(ii) Low-order cuprates

Low-order cuprates $\mathrm{R^1R^2CuM}$ have been the most widely used copper derivatives in organic synthesis since their introduction in 1952 ⟨52JOC1630⟩, both as stoichiometric reagents and as active species in the copper-catalysed reactions of organometallic reagents. Three different species may be prepared: homocuprates ($R^1 = R^2 =$ alkyl) mixed cuprates ($R^1 \neq R^2$) or heterocuprates ($R^1 =$ alkyl, $R^2 \neq$ alkyl).

(a) Homocuprates. These reagents are easily prepared by addition of two equivalents of an alkyllithium or alkylmagnesium reagent to a copper(I) salt (Equation (55)). They are more stable and soluble than the monoalkylcopper reagents and are generally prepared below 0 °C. Copper iodide is the most efficient precursor for stoichiometric preparation, while copper chloride is often used in copper-catalysed reactions. Compounds with secondary or tertiary alkyl groups are less stable than those with primary groups and are generally prepared with ligand-stabilised copper salts, especially copper bromide dimethyl sulfide complex. The reaction proceeds through a stepwise mechanism in which an alkylcopper is first formed, followed by ate complex formation.

$$2\mathrm{RM} + \mathrm{CuX} \xrightarrow{\mathrm{Et_2O\ or\ THF,\ 0\ ^\circ C}} \mathrm{R_2CuM} \qquad (55)$$

R = alkyl; M = Li, MgBr, MnCl, ZnI, ZnR; CuX = CuCl, $\mathrm{CuBr{\bullet}SMe_2}$, CuI, $\mathrm{CuI{\bullet}PR_3}$

Although lithium and magnesium are commonly used, transmetallation from other organometallic reagents has received attention. Alkylmanganese compounds have proved to be better starting materials in copper-catalysed reactions and show increased yields and selectivities in reaction with electrophiles ⟨89TL3541, 90TL7423⟩. Reactive species are prepared by addition of 1–5% of copper(I) chloride to a THF solution of the organometallic reagent at 0 °C. Alkyltitanium complexes have also been reacted with catalytic amounts of $CuBr \cdot SMe_2$ or $CuI \cdot LiCl$, although the exact nature of the active species is not known ⟨92T5709⟩.

The preparation of functionalised homocuprates has been achieved using various reagents; for example, by treatment of alkylsamarium(III) compounds (prepared from alkyl iodides and samarium diiodide) with copper iodide triethyl phosphite complex, which affords diiodosamarium dialkylcuprates ⟨92JOC1740, 93JOC3455⟩ (Scheme 16). Functional groups such as ethers, amides or chlorides are unaffected under these conditions. A more general protocol, however, is the transmetallation from dialkylzinc compounds which can contain a diverse array of functional groups such as chlorides, amides, esters, nitriles or protected hydroxyls. Reaction of the zinc compound with catalytic amounts of $CuBr \cdot SMe_2$ in THF/hexamethylphosphoramide affords functionalised cuprates, an example of which is shown in Equation (56).

$$2RI + 4SmI_2 \xrightarrow{\text{THF/HMPA, RT}} 2RSmI_2 \xrightarrow{\text{CuI•P(OEt)}_3} R_2CuSmI_3$$

HMPA = hexamethylphosphoramide

Scheme 16

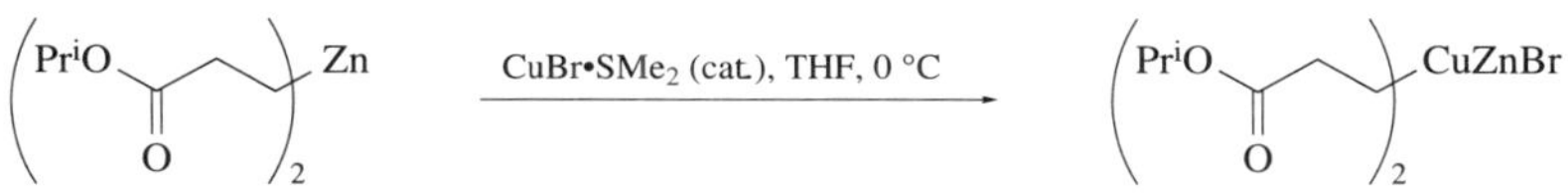

(56) (57) (56)

Another important aspect of this chemistry is the development of a one-pot hydrometallation–transmetallation procedure, which allows the direct preparation of alkyl cuprates from alkenes. Since hydrozirconation is one of the most efficient hydrometallation reactions, transmetallation from zirconium to copper has been studied extensively ⟨91JOC6494, 92SL718⟩. Primary, secondary and allylic groups are readily transferred to copper in THF solution, albeit relatively slowly. It is therefore necessary to raise the temperature above 0 °C and to use ligand-stabilised copper salts such as $CuBr \cdot SMe_2$ (Scheme 17).

Scheme 17

(b) Mixed cuprates. Mixed dialkylcuprates are prepared by sequential addition of different organometallic reagents to copper(I) salts. Their stability towards disproportionation into homocuprates depends on the structure and the mode of preparation ⟨78BSF(2)575⟩. Important mixed cuprate reagents are those containing one ligand tightly bound to copper which is not transferable. The discovery that alkynyl copper derivatives were unreactive ⟨69JOC3615⟩ prompted the use of mixed alkyl–alkynyl cuprate reagents for selective ligand transfer ⟨72JA7210⟩. These compounds are prepared by addition of one equivalent of an alkyllithium reagent to a solution of a copper acetylide in ether (or THF) at −78 °C. Various alkynes have been used as nontransferable ligands (also called 'residual' or 'dummy' ligands) (Scheme 18), the most popular being **(59)** because of its availability and the high solubility of its copper salts. Ligands other than alkynes may be used, the most common being listed in Scheme 18 ⟨72JA7210, 73JOC3893, 78JOC3418, 82TL3823⟩.

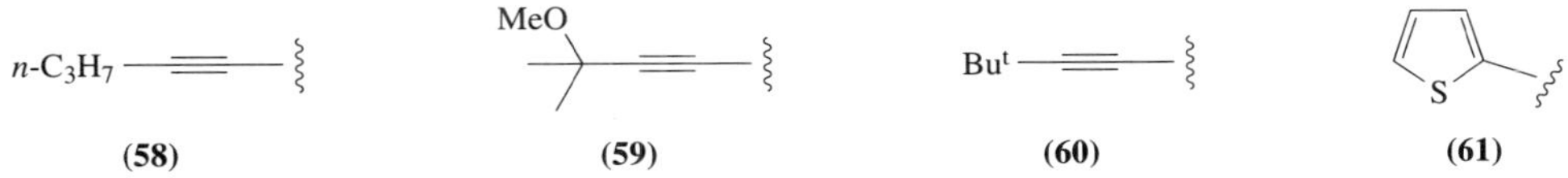

(58) **(59)** **(60)** **(61)**

Scheme 18

(c) Heterocuprates. Heterocuprates possess one alkyl ligand and one ligand other than halide which is bonded to copper through a heteroatom. Cyanide anion is also included in this group of ligands. As heteroligands are never transferred, they are used as residual ligands. They must, however, be tightly bound to copper in order to avoid disproportionation. The most important heterocuprates are cyanocuprates, which display remarkable stability and reactivity. These cuprates are readily prepared by addition of an alkyllithium reagent to copper cyanide in an ethereal solvent at 0°C, the reaction proceeding by direct ate complex formation. Other useful transmetallation procedures include the reaction of alkylzirconium derivatives with a catalytic amount of copper cyanide ⟨92TL5857⟩, or the preparation of stoichiometric zinc alkylcyanocuprates by reaction of alkylzinc iodides with copper cyanide (solubilised by addition of two equivalents of lithium chloride) ⟨88JOC2390, 89AG(E)351⟩. This latter method allows the preparation of reactive functionalised alkyl cuprate reagents in a mild and efficient fashion (Equations (57) and (58)). Other common residual heteroligands used in cuprate chemistry include bulky alkoxides or thioalkoxides ⟨73JA7788⟩ or diphenyl phosphide ⟨82JA5284⟩. These are prepared according to the following sequence: deprotonation of the heteroatom, complexation with copper by addition of a copper halide, then addition of the alkyllithium reagent. The main interest in the preparation of heterocuprates is the opportunity to use chiral ligands for enantioselective conjugate addition reactions ⟨92CRV771⟩. Some representative chiral residual ligands are shown in Scheme 19 ⟨86JA7115, 87JA2040, 91CC101⟩.

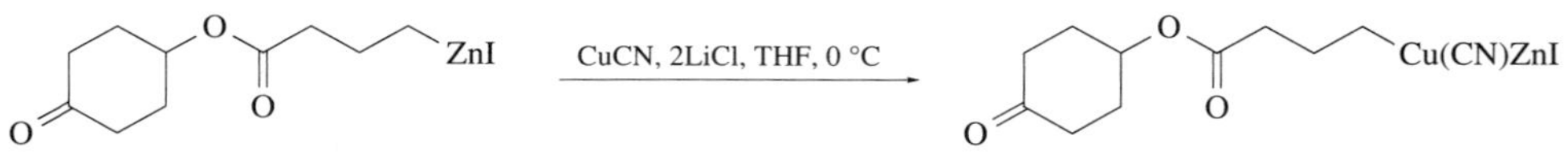

(62) **(63)** (57)

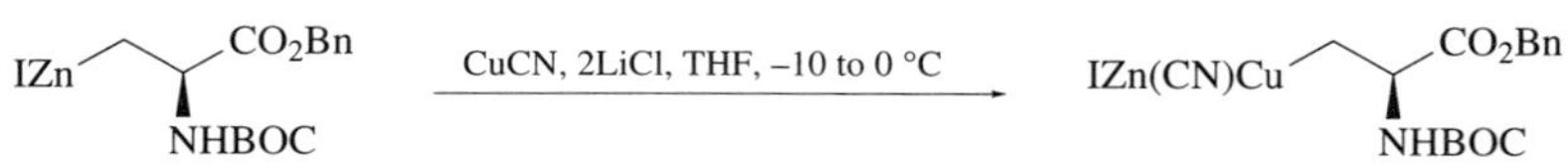

(64) **(65)** (58)

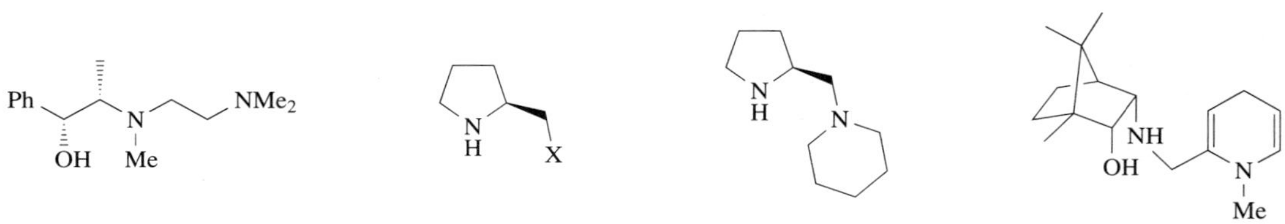

(66) **(67)** X = OMe **(68)** X = SMe **(69)** X = SPh **(70)** **(71)**

Scheme 19

(iii) Higher-order cuprates

The chemistry of higher-order cuprates has been the subject of intense study since their first preparation in 1981 and is the most rapidly expanding topic within organocopper chemistry. Their synthesis and properties have been detailed in reviews ⟨87S325, 90SL119⟩. Addition of two equivalents of an alkyllithium reagent to copper cyanide (or copper thiocyanate) at 0°C affords a dianionic species $R_2Cu(CN)Li_2$, which displays exceptional stability and reactivity. These cuprates are now the reagents of choice for the transfer of tertiary alkyl groups and reactions with hindered electrophiles. The π-acidic nature of the cyanide ligand is important for the formation of the triple-bonded copper species, since addition to copper halides never goes beyond the formation of low-order cuprates ⟨85JA3197⟩. However, it should be pointed out that the exact structure of higher-order cyanocuprates is still a matter of considerable controversy and arguments for and against their existence have been reported ⟨90JA4031, 90JA4032⟩.

The preparation of higher-order cuprates always involves the use of alkyllithium reagents, since addition of Grignard reagents, for example, leads to the formation of much less-reactive species ⟨86T2873⟩. As with low-order cuprates, residual ligands may be used for alkyl group economy, the most popular ligand being 2-thienyl ⟨85JOM(285)437⟩. Mixed higher-order cuprates are usually prepared according to the sequence shown in Scheme 20. The intermediate lithium 2-thienylcyanocuprate **(74)** is a stable reagent that is commercially available.

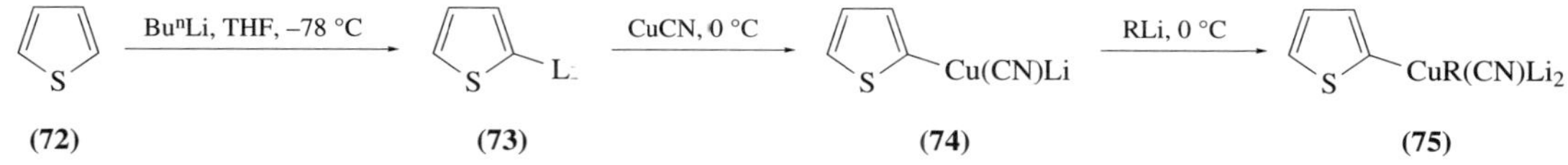

R = primary, secondary, tertiary alkyl

Scheme 20

Higher-order cyanocuprates are also able to promote alkyl group transfer from other organometallic reagents. A striking example is the preparation of higher-order allylic cuprates by reaction of $Me_2Cu(CN)Li_2$ with various allylic stannanes ⟨90JA4063⟩. This is the only successful route towards stable and reactive allylic cuprates.

2.11.5.1.3 Preparation by carbometallation

The only example of carbocupration of an alkene is the addition of cuprates (alkylcopper reagents are ineffective) to the carbon–carbon double bond of a cyclopropenone acetal ⟨88JA1297⟩. The addition is stereoselective (*syn* addition of an alkyl group and copper) and regioselective for dissymmetric substrates, delivering copper to the less-substituted carbon. Reaction with the chiral acetal **(76)** give excellent diastereoselectivities if the double bond is substituted (Equation (59)) ⟨90JA7428⟩.

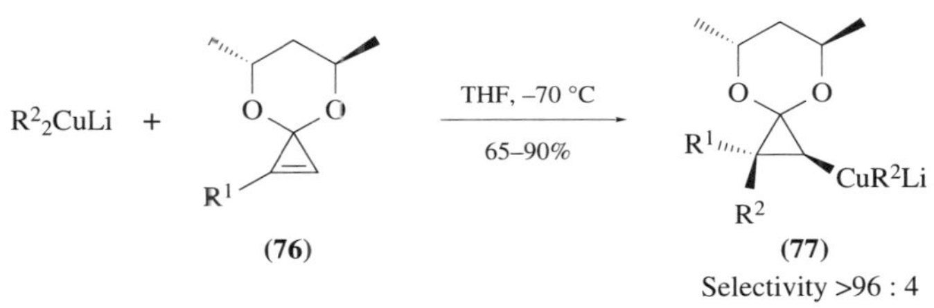

(59)

R^1 = Et, Ph; R^2 = Me, Et, Bu^n

2.11.5.2 Alkylsilver Derivatives

Most of the procedures developed for the synthesis of alkylcopper reagents may also be applied to silver ⟨82COMC-I(2)709⟩. However, the great thermal sensitivity and photosensitivity of alkylsilver derivatives has seriously limited their use in synthesis. The most available silver salt, silver nitrate, has been initially used for transmetallation using alkyllead species in methanol ⟨41CB1089⟩. Low reaction temperatures are required, since the products tend to decompose above −50°C. The discovery that silver compounds are stabilised by lithium bromide has led to the preparation of alkyl derivatives which are stable at temperatures up to −10°C ⟨79JOM(172)377⟩. Reaction of primary, secondary or tertiary alkyl Grignard reagents with AgBr·2LiBr gives the corresponding alkylsilver or dialkylargentate compounds, depending on the stoichiometry. Stabilisation with phosphine ligands has also been reported ⟨74JA2806⟩. Low- and higher-order cyanoargentates have been prepared by addition of one or two equivalents of methyllithium to silver cyanide ⟨92CB2409⟩. As with copper, both stability and reactivity were increased.

2.11.5.3 Alkylgold Derivatives

In contrast to copper and silver, alkylgold derivatives with the metal in oxidation state (I) and (III) are known (review on alkylgold chemistry: ⟨82COMC-I(2)765⟩). Monoalkylgold derivatives are very unstable species which decompose with precipitation of metallic gold. Consequently, they are always synthesised as phosphine complexes. A review containing a comprehensive list of synthesised alkylgold(I) complexes has been published ⟨94JOM(474)1⟩. Methods for their preparation include reaction of gold(I) chloride phosphine complex with alkyllithium or, preferably, alkylmagnesium reagents ⟨59CI(L)628, 73JA6487⟩. With an excess of the organometallic reagent, dialkyl aurates R_2Au^- are obtained ⟨73JCS(D)2620⟩.

Di-, tri- and tetraalkylgold(III) compounds are also known. Dialkylgold(III) halides are prepared by reaction of gold trichloride pyridine complex or digold hexabromide with Grignard reagents ⟨70AG(E)101⟩. These compounds are stable and are dimeric in structure. Trialkylgold derivatives are generally unstable species which decompose at low temperature unless they are stabilised by phosphine or arsine complexation as square planar structures. They are prepared either by transmetallation from gold(III) halides using alkyllithium reagents or, more conveniently, by alkylation of dialkyl aurates with alkyl iodides or bromides ⟨74JA6140⟩. In this case, different alkyl groups may be introduced. Addition of a further equivalent of alkyllithium reagent affords stable tetraalkyl aurate compounds ⟨75IC2402, 88CL63⟩.

2.11.6 ALKYLZINC, -CADMIUM AND -MERCURY DERIVATIVES

These elements have played a vital role in the history of organometallic chemistry. Dimethyl- and diethylzinc were the first alkyl metal compounds ever synthesised ⟨1849LA(79)171⟩, and organozinc and organomercury reagents have been the most convenient (if not the only) source for the preparation of many organometallic compounds. Since the development of the more reactive lithium or magnesium derivatives, their development and use in synthesis have been somewhat limited and the very high toxicity of mercury compounds has accelerated this decline in interest. However, the chemistry of zinc derivatives has been rejuvenated by discoveries since the early 1980s ⟨82COMC-I(2)823, 82COMC-I(2)863, 91COS(1)211⟩. Organic compounds of zinc, cadmium and mercury are stable and can be isolated. They are also far less reactive than their corresponding magnesium counterparts. However, their preparation is tolerant of many functional groups and alkylzinc compounds do not react with esters, nitriles, carbamates, amines, thiols, aliphatic chlorides or phosphates. It is therefore not surprising that most of the recent work has focused on the preparation of these functionalised alkylzinc reagents and their use in synthesis ⟨93CRV2117⟩.

2.11.6.1 Preparation by Oxidative Addition

Grignard-type reactions have been used as the main source of alkylzinc derivatives. Zinc is less reactive than magnesium and only alkyl iodides or activated (allylic or benzylic) bromides react. Simple alkylzinc iodides are metallated by heating with neat zinc dust, or by treatment with zinc dust or foil in THF ⟨62BSF974⟩. These methods are not suitable for the preparation of functionalised substrates, however—prior activation of the metal is necessary ⟨87T2203⟩. Chemical activation by addition of 1,2-dibromoethane or chlorotrimethylsilane results in an increase in the reactivity, and reproducible results for a wider range of substrates can be achieved ⟨88JOC2390, 92OS(70)195⟩. Further activation may be achieved if a catalytic amount of an alkali metal iodide is added and THF is replaced by a more polar aprotic solvent such as *N*,*N*′-dimethylpropyleneurea (DMPU) ⟨92JOC5425⟩. This method allows primary chlorides, bromides, mesylates or tosylates to react with zinc dust and benzylic or allylic mesylates or phosphates to be metallated without any homocoupling. This is undoubtedly the most general procedure for the preparation of alkylzinc derivatives reported to date (Scheme 21).

$$\mathrm{Zn} \xrightarrow{\mathrm{Br(CH_2)_2Br,\ TMS\text{-}Cl,\ DMPU}} \mathrm{Zn^*} \xrightarrow{\mathrm{RX,\ MBr,\ 40\text{–}80\ ^\circ C}} \mathrm{RZnX}$$

R = alkyl; X = I, Br, OMs, Cl, $OP(O)(OPh)_2$ (in order of reactivity); M = Li, Na, K, Cs

Scheme 21

As with magnesium, reduction of zinc chloride with lithium naphthalenide (LiN) ⟨81JOC4323⟩ or potassium graphite ⟨83JOC4108⟩ affords highly reactive zinc powder which inserts into the carbon–halogen bond of alkyl iodides or benzyl bromides at or just above room temperature. This latter method has also been applied to the synthesis of alkylcadmium derivatives by reduction of cadmium chloride with LiN ⟨85JOC416⟩. These results are remarkable since transmetallation has for a long time been the only route to alkylcadmium derivatives (Equation (60)).

$$RX \;+\; M^{*} \xrightarrow{\text{THF}} RMX \qquad (60)$$

RX = alkyl iodide, benzyl bromide; M = Zn, Cd

Another simple and inexpensive source of activated zinc is the zinc–copper couple ⟨73OSC(5)855⟩, which reacts with various functionalised alkyl iodides to give the zinc derivatives in excellent yields ⟨89OS(67)98⟩. The reaction is performed in benzene/dimethylacetamide at reflux or at room temperature under sonication conditions ⟨92JOC3397⟩. This mild and reproducible method is especially useful for the preparation of amino acid-derived zinc compounds (Equation (61)) ⟨89AG(E)351, 92JOC3397⟩. Ultrasound techniques are particularly suited for the preparation of alkylzinc compounds, especially in the Barbier reaction. The most remarkable feature in this reaction is that it can be performed in aqueous media. Thus, sonication of an alkyl iodide with zinc–copper couple in the presence of an electrophile affords the coupled products in good yield ⟨86TL3149⟩.

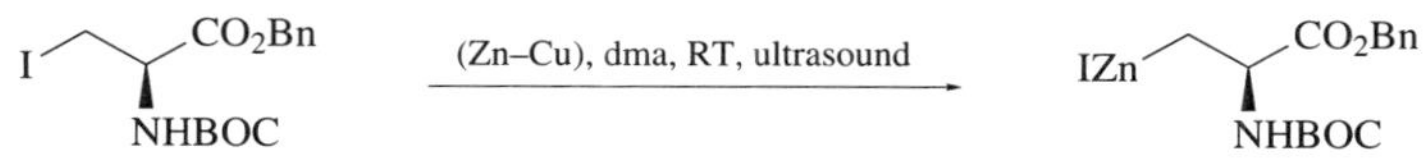

(78) **(64)** (61)

dma = dimethylacetamide

Not only is the preparation of alkylzinc halides tolerant of many functional groups but it is also remarkable for its mildness and lack of side reactions associated with the preparation of organometallic reagents such as Wurtz coupling or β-elimination; and even competitive deprotonation does not occur with acetylenic substrates (Equation (62)).

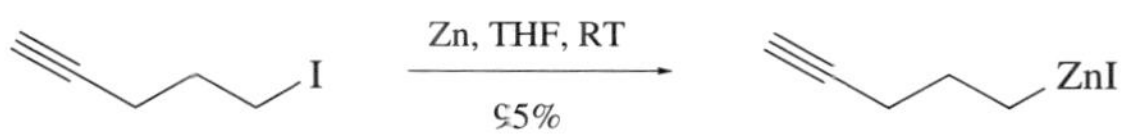

(79) **(80)** (62)

Simple alkylzinc iodides (from methyl to *n*-butyl) may be transformed into dialkylzinc derivatives upon heating and distillation, and this is the most convenient method for their isolation or purification.

2.11.6.2 Preparation by Halogen–Metal Interconversion

A general method for the synthesis of functionalised dialkylzinc reagents by halogen–metal interconversion has been reported (Scheme 22) ⟨92JOC1956⟩. Various primary alkyl iodides react with diethylzinc to give mixed alkylzinc species which are converted into homogeneous compounds by removal of diethylzinc under vacuum. The reaction may be catalysed with transition metals, although the nature of the catalyst has an influence on the outcome of the reaction: catalysis with copper(I) iodide gives dialkylzinc species ⟨93TL3115⟩, whereas catalysis with palladium complexes ⟨93JA7027, 93TL7911⟩ or manganese bromide/copper chloride ⟨94TL1177⟩ gives alkylzinc halides. This latter reaction is interesting since primary alkyl bromides may be used.

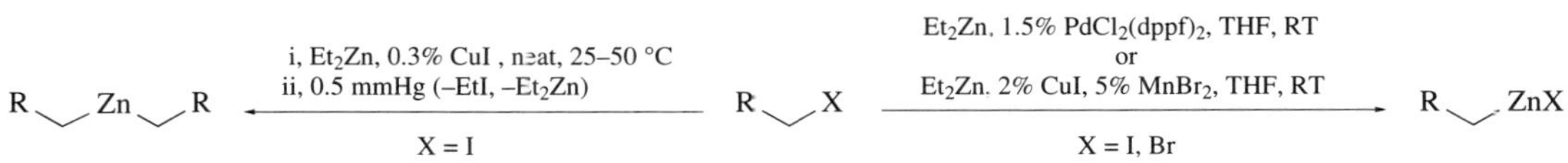

Scheme 22

2.11.6.3 Preparation by Transmetallation

Transmetallation is the most general procedure for the synthesis of mono- or dialkylated derivatives of zinc, cadmium or mercury (Scheme 23) ⟨73HOU(13/2a)553, 73HOU(13/2b)⟩. A great variety of metals have been used for the transfer of alkyl groups, and zinc derivatives are best prepared with alkyllithium, -magnesium, -mercury or -aluminum reagents ⟨79IS253⟩. Ate complexes of zinc, on the other hand, may be obtained with an added excess of alkyllithium reagents. Alkylcadmium derivatives are generally dialkylated and are prepared using alkyllithium or, better, with alkylmagnesium reagents. However, they exhibit lower thermal stability and are commonly prepared *in situ*. Monoalkyl-, homogeneous or mixed dialkylmercury compounds are prepared using alkyllithium or alkylmagnesium reagents and, to a lesser extent, alkylzinc, -lead, -tin, -thallium, -aluminum, -copper or -gold reagents. They are remarkably stable and may be isolated after aqueous workup procedures. Transmetallation reactions proceed easily and seem to be limited only by the availability of the starting organometallic reagents.

$$RM + ZnX_2 \xrightarrow[\text{DME or toluene}]{Et_2O,\ THF} RZnX \qquad X = I, Br, Cl$$

$$2RM + ZnX_2 \longrightarrow R_2Zn \qquad R = \text{alkyl}$$

$$4RLi + ZnX_2 \longrightarrow R_4ZnLi_2 \qquad M = Li, Mg, Al, Hg$$

$$2RM + CdCl_2 \xrightarrow{Et_2O \text{ or } THF} R_2Cd \qquad M = Li, Mg; R = \text{alkyl}$$

$$RM + HgX_2 \longrightarrow RHgX \qquad X = I, Br, Cl$$

$$R^1M + R^2HgX \longrightarrow R^1R^2Hg \qquad R = \text{alkyl}$$

$$2RM + HgX_2 \longrightarrow R_2Hg \qquad M = Li, Mg, Zn, Al, Pb, Au, Cu, Tl$$

Scheme 23

As for copper, hydrometallation–transmetallation procedures have received some attention as one-pot syntheses of alkyl metals from alkenes (Scheme 24). Transmetallation from boron to mercury is carried out by treatment of trialkylboranes with mercury acetate for primary alkyl groups, or mercury alkoxides for secondary substrates ⟨78AG(E)27⟩. A general method for transfer of alkyl groups from boron to zinc has been reported ⟨94SL410⟩. Hydroboration of terminal or internal alkenes with diethylborane, followed by treatment with neat diethylzinc, gives primary or secondary dialkylzinc compounds in excellent overall yield.

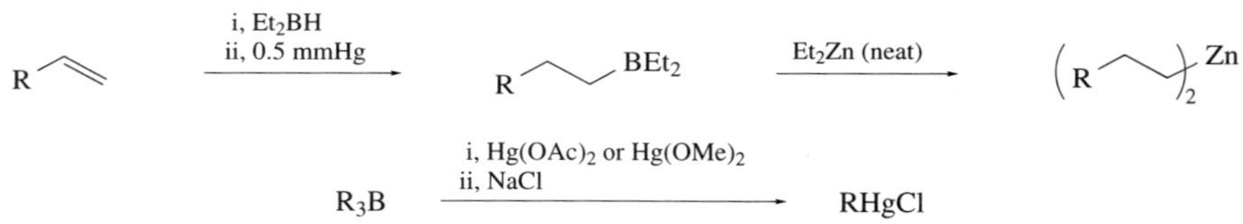

Scheme 24

2.11.6.4 Addition to Carbon–Carbon Double Bonds

2.11.6.4.1 Carbozincation

Studies of the carbozincation reaction have been restricted by the lack of accessibility and the poor reactivity of alkylzinc compounds towards carbon–carbon double bonds. However, alkylzinc derivatives show good reactivity in intramolecular carbometallation and 5-hexenylzinc iodide (prepared using activated zinc methods) smoothly cyclises at room temperature to give cyclopentylmethylzinc iodide in good yield ⟨93SL266⟩. Scope and selectivities in this reaction are similar to the reaction with the lithium analogue, the main advantages of zinc over lithium being the greater tolerance of functional groups and the possibility of the cyclisation of secondary alkylzinc compounds with excellent stereoselectivity (Scheme 25) ⟨91TL3317, 93TL6053⟩. This considerably widens the scope of intramolecular carbometallation reactions.

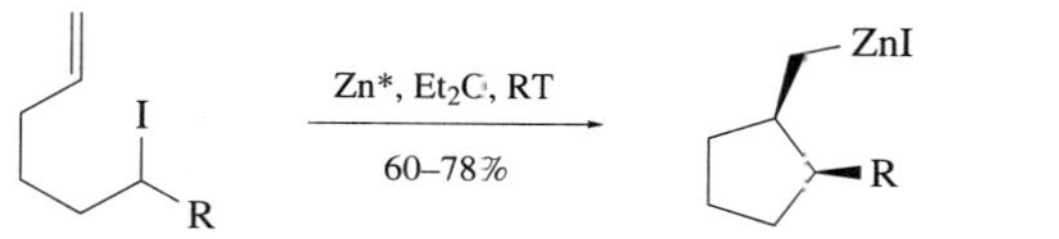

(81) **(82)**

cis : *trans* 66 : 34 to 81 : 19

R = H, Me, Pr^i, Bu^n, Bu^t, $SiMe_2Ph$

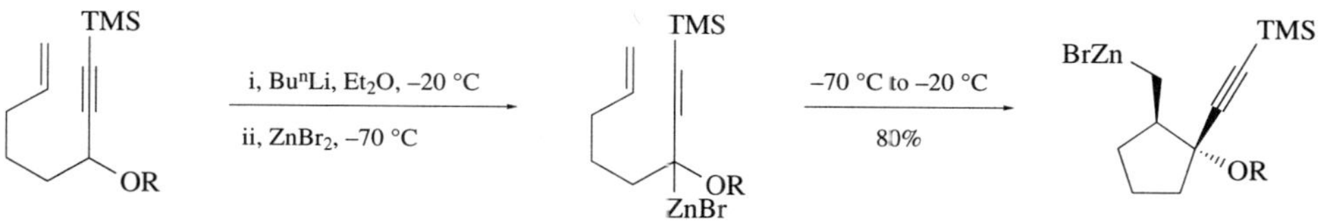

(83) **(84)** **(85)** single isomer

R = Me, methoxyethoxymethyl, $SiMe_2$Thex (Thex = 1,1,2-trimethylpropyl)

Scheme 25

2.11.6.4.2 Mercuration

The addition of mercury salts (acetate, trifluoroacetate or nitrate) to alkenes is a very well-known process and is one of the best methods for the preparation of alcohols. The reaction leads to the formation of a carbon–mercury bond which is stable to hydrolysis but can be reduced. This reaction has been used extensively in organic synthesis, especially when the intermediate mercuronium ion is trapped intramolecularly. Trapping with hydroxyl groups gives cyclic ethers, whereas trapping with electron-deficient alkenes gives carbocycles ⟨91COS(3)341⟩. Yields and stereoselectivities are generally excellent and this reaction is often a key step in the synthesis of natural products. Some examples are shown in Equations (63)–(65) ⟨75BSF770, 80JA1742, 88JA2506⟩.

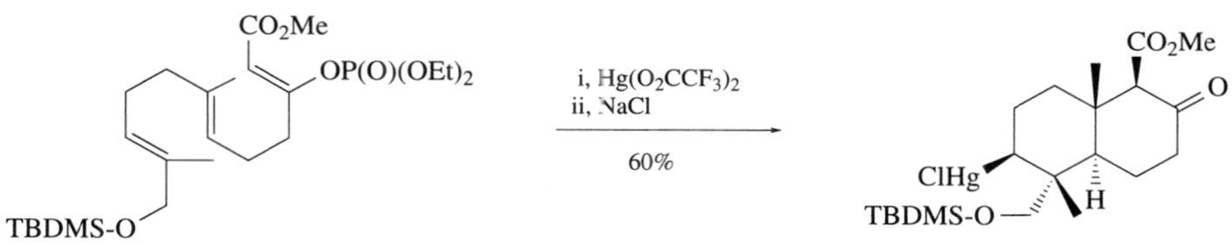

(86) **(87)** (63)

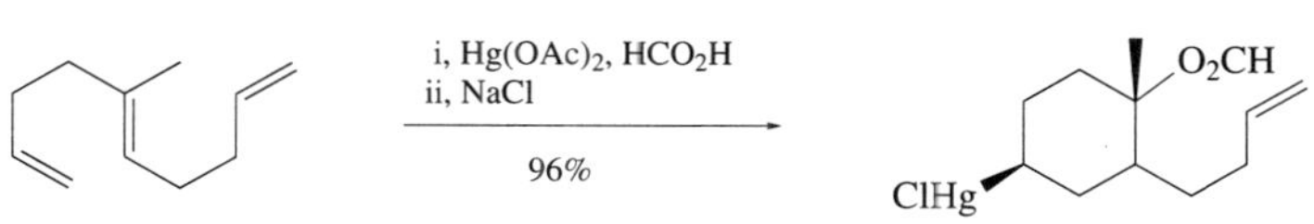

(88) **(89)** (64)

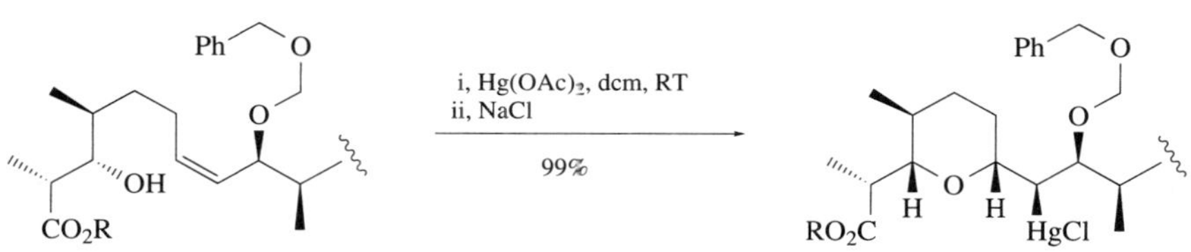

(90) dcm = dichloromethane **(91)** (65)

2.11.6.5 Miscellaneous Methods

2.11.6.5.1 Electrochemical synthesis

Electrochemical synthesis of alkylcadmium and alkylmercury compounds have been performed using both anodic and cathodic processes ⟨79PAC2005⟩. Anodic processes involve transmetallation from alkylmagnesium or alkylaluminum compounds, whereas cathodic processes involve oxidative addition of alkyl bromides. Reduction of ketones or alkenes to dialkylmercury compounds in acidic media is also possible ⟨75JOM(88)1⟩.

2.11.6.5.2 Decomposition reactions

Decarboxylation of mercury carboxylates to alkylmercury carboxylates is achieved by exposure to a radical initiator (UV light or a peroxide) ⟨80JOM(201)39⟩. Similarly, mercury salts of alkylsulfinic acids are desulfinated upon heating. This reaction may be performed by heating a sulfinic acid in the presence of mercury salts (halide or acetate) and is often referred to as the Peters reaction ⟨48OSC(1)519⟩.

2.11.6.5.3 Homologation with zinc reagents

Iodomethylzinc iodide (Simmons–Smith reagent) has been used for the one-carbon homologation of organometallic reagents in which allylic zinc derivatives are converted to homoallylic species in good yields ⟨93JOC2694⟩. Copper derivatives are also good substrates for this reaction, and an elegant example is the conversion of a vinylcopper into an allylic copper reagent (Equation (66)).

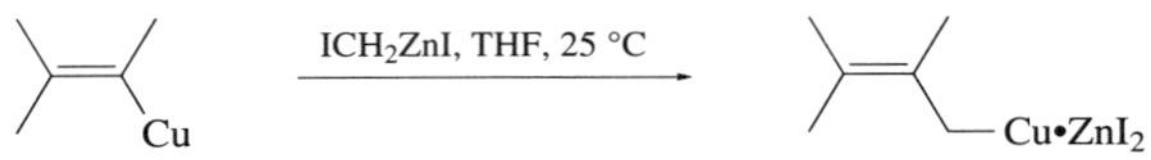

(66)

2.11.6.5.4 Reaction of bismetallic reagents

The synthesis of alkyl-1,1-bis(zinc) reagents via the 'metallo-Claisen' reaction has been developed as a powerful tool for acyclic stereoselection ⟨92PAC1857⟩. Some reactions of these bismetallic species have led to the formation of alkylzinc derivatives. For example, treatment with phenylsulfonyl chloride gives an α-chlorozinc compound, which is transformed into a secondary alkylzinc derivative upon treatment with an excess of Grignard or alkyllithium reagents ⟨93SL665⟩ (Scheme 26). Intramolecular delivery of an alkyl group from an intermediate trialkyzincate is believed to occur in this reaction. On the other hand, bismetallic compounds with an ether in the β position undergo elimination upon warming to give cyclopropylzinc halides (Equation (67)) ⟨93TL2303, 93TL7575⟩. The reaction is remarkable for its stereoselectivity in that a single isomer is obtained.

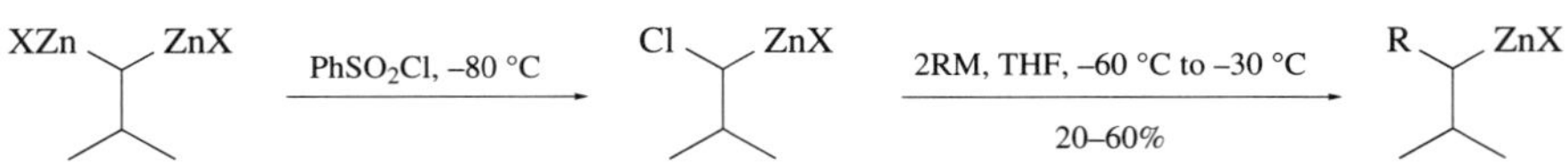

RM = MeLi, EtLi, Pr^iMgBr, Bu^nLi, Bu^sLi, Bu^tMgBr, allylMgBr

Scheme 26

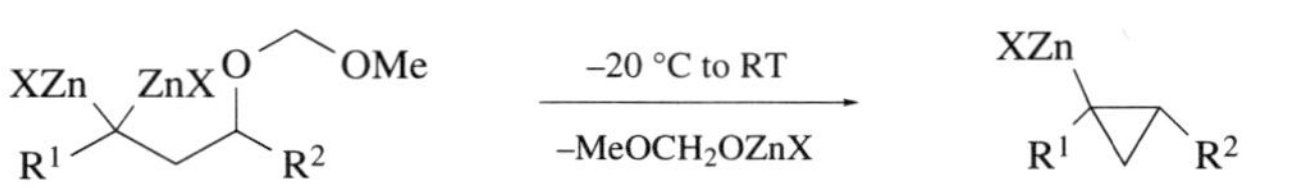

(67)

(92)
Stable at –20 °C

(93)
single isomer
$R^1 = CH_2CH{=}CH_2$
$R^2 = Pr^n, Bu^n$

2.11.7 ALKYLALUMINUM, -GALLIUM, -INDIUM AND -THALLIUM DERIVATIVES

2.11.7.1 Alkylaluminum Derivatives

Alkylaluminum reagents are widely used in organic synthesis as carbon nucleophiles, Lewis acids, alkyl radical precursors ⟨91AG(E)694⟩, or for the preparation of other alkyl metal compounds. Their preparation and properties have been reviewed extensively ⟨70HOU(13/4)9, B-72MI 211-01, 82COMC-I(1)555, B-87MI 211-05, 91COS(1)77⟩. In general, only derivatives with simple alkyl groups are used in synthesis. Most of the methods for their preparation were developed before 1970, and for industrial use.

2.11.7.1.1 Preparation by oxidative addition

Reaction of alkyl iodides or bromides with aluminum turnings affords alkylaluminum sesquichlorides $R_3Al_2X_3$ ⟨40JOC106, 51JA2854⟩. The reaction is suitable for primary halides from methyl to *n*-octyl (Scheme 27). Aluminum metal may be activated using various additives, or by using sonication techniques. This latter process allows reaction to proceed rapidly at room temperature ⟨83ZOB2519, 86JOM(307)273⟩. Aluminum sesquichlorides are converted into alkylaluminum dichlorides by addition of aluminum trichloride or into trialkylaluminum derivatives by addition of an alkyl metal reagent ⟨78USP4118409⟩. This latter reaction is also accelerated under sonication ⟨86JOM(307)273⟩. Alternatively, conversion to trialkylaluminum may be achieved by dehalogenation with sodium metal ⟨40JOC106⟩. These methods are the most convenient for the preparation of pure, unsolvated trialkylaluminum compounds on a laboratory scale. Direct synthesis of trialkylaluminum compounds from alkyl iodides is also possible using aluminum–magnesium alloy. Since diethyl ether is used as solvent, etherate complexes are obtained.

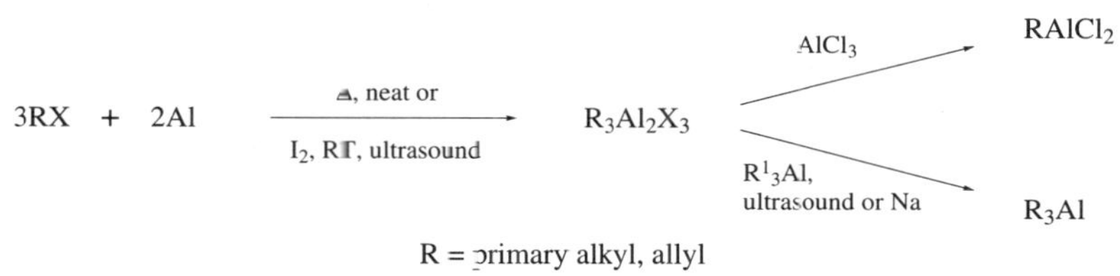

Scheme 27

2.11.7.1.2 Preparation by transmetallation

Reaction of aluminum trichloride with alkyllithium or alkylmagnesium reagents in diethyl ether gives trialkylaluminum compounds as their etherate complexes. Decomplexation may be achieved by distillation only when the product is thermally stable. This reaction is particularly useful for the preparation of compounds with secondary or tertiary groups which are stable as solvates ⟨64AG(E)640, 68LA(719)40⟩. Mixed trialkylaluminum derivatives are obtained by addition of an alkyllithium reagent to a dialkylaluminum chloride, and are stable if not heated. Transmetallation from alkylmercury compounds has the advantage of being performed in neat conditions; however, since high temperatures are required, this method is suitable only for thermally stable compounds.

2.11.7.1.3 Addition to carbon–carbon double bonds

Owing to the high reactivity of aluminum compounds towards carbon–carbon double bonds, hydro- and carboalumination are important methods for the industrial preparation of symmetrical or mixed trialkylaluminum compounds (Equations (68)–(70)).

(68)

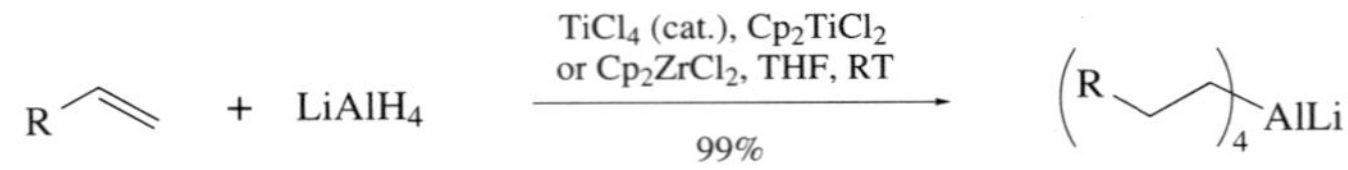

(69)

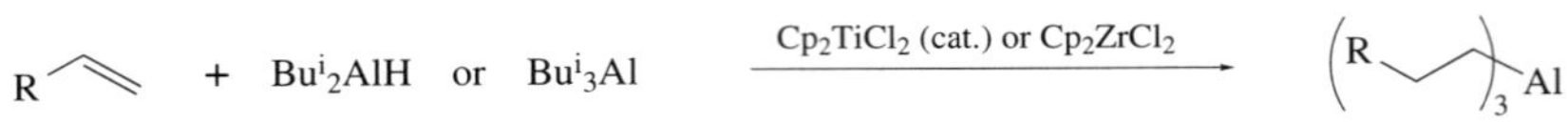

(70)

Aluminum hydride, AlH_3, reacts readily with various alkenes to give trialkylaluminum compounds. AlH_3 may be prepared *in situ* in the presence of an alkene by reaction of aluminum metal with hydrogen ⟨60LA(629)1⟩. This procedure is convenient if the alkene is not too reactive. Lithium aluminum hydride also reacts with alkenes to give tetraalkyl ate complexes $LiAlR_4$ ⟨54LA(589)91⟩; however, this reaction requires high temperatures unless transition-metal catalysts are added. Catalysis with titanium or zirconium salts allows the reaction to proceed under mild conditions ⟨77JOM(142)71⟩. Hydroalumination using dialkylaluminum hydrides affords symmetrical or mixed trialkylaluminum derivatives under milder conditions ⟨60LA(629)121⟩. The most important compound for this reaction is diisobutylaluminum hydride (dibal-H), which may be prepared prior to the reaction or *in situ* by thermal β-elimination of isobutene from triisobutylaluminum. Indeed, heating triisobutylaluminum with an alkene (of higher boiling point than isobutene) affords a trialkylaluminum by successive β-elimination and hydrometallation. Metallocycles may be obtained by hydrometallation of dienes. This reaction may also be catalysed with titanium or zirconium salts ⟨80TL1501⟩. The hydroalumination reaction follows the same patterns as other hydrometallation reactions: terminal double bonds only are reactive (with the exception of the strained norbornene derivatives) and the regioselectivity is high. The reaction is reversible and, with aluminum, the inverse reaction (β-elimination) is also very easy. Consequently, unsolvated secondary and tertiary alkylaluminum derivatives cannot be prepared by any method as they readily isomerise to the primary analogues (with the exception of cycloalkylaluminum compounds).

Unless very active alkenes are used, carboalumination reactions are catalysed with titanium or zirconium salts ⟨79JOC3603⟩, and the regioselectivity is identical to the hydroalumination process. The reaction is not reversible, but dehydroalumination often occurs. The overall process is therefore a nucleophilic substitution on an alkene.

2.11.7.1.4 Disproportionation and ligand exchange

Disproportionation methods are important for the preparation of aluminum derivatives which cannot be prepared by standard methods. Dialkylaluminum chloride is prepared by reaction of trialkylaluminum reagents, with aluminum chloride, zinc chloride or carbon tetrachloride. This reaction has also been applied to the synthesis of the bromides and to the fluorides by reaction with ZnF_2. Dialkylaluminum hydrides are prepared either by reaction between trialkylaluminum and aluminum hydride ⟨60LA(629)121⟩ or by that between a dialkylaluminum chloride and an alkali-metal hydride reagent ⟨54LA(589)91⟩. Finally, both di- and trialkylated compounds may be used for the preparation of dialkylaluminum cyanides, alkoxides, thioalkoxides or thiocyanates. An interesting development of this reaction is the preparation of bulky methylaluminum dialkoxides by reaction of trimethylaluminum with two equivalents of the corresponding alcohol at room temperature ⟨94T8983⟩. The very hindered reagents **(94)**–**(96)** induce remarkable selectivities in Lewis acid-catalysed reactions.

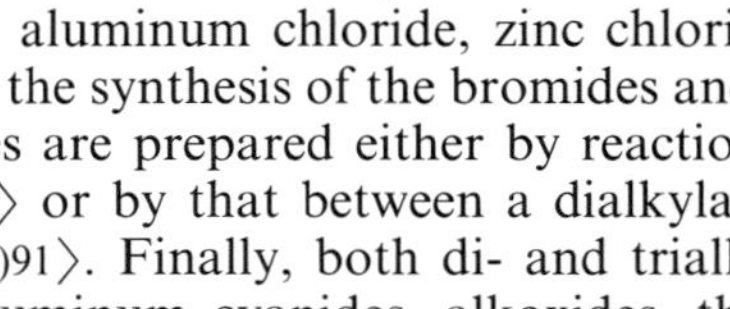

(94) MAD: $R^1 = Bu^t$, $R^2 = Me$
(95) MABR: $R^1 = Bu^t$, $R^2 = Br$
(96) MAPH: $R^1 = Ph$, $R^2 = H$

2.11.7.2 Alkylgallium and -indium Derivatives

Alkylgallium and -indium derivatives are less well studied organometallic reagents compared with those reported above, probably because of the very low availability of the metals ⟨70HOU(13/4)315, 70HOU(13/4)345, 82COMC-I(1)683⟩. Most of the methods for their preparation are similar to those for the preparation of alkylaluminum compounds. Oxidative addition of alkyl iodides to indium metal gives sesquiiodides which may undergo disproportionation ⟨74JOM(81)329⟩. Yields and reaction rate are increased using activated metal. Reduction of indium trichloride with potassium affords a highly reactive slurry of indium metal which reacts well with alkyl iodides to give dialkylated derivatives ⟨74JOM(67)C64⟩. Trialkylindium derivatives, on the other hand, are obtained by reaction of alkyl halides with indium–magnesium alloy (Equation (71)) ⟨63ZAAC(321)120⟩. Allylic bromides react rapidly with indium in DMF at room temperature to give the sesquibromide derivatives ⟨91JOC2538⟩. These reagents may also be prepared *in situ* by a Barbier-type reaction. Remarkably, this process may be performed in aqueous solvent under sonication conditions, and allylindium reagents prepared *in situ* in water or ethanol have been used in the synthesis of carbohydrate derivatives ⟨94JOC3714, 94T749⟩. Oxidative addition to gallium metal is not a satisfactory method, although alkylgallium diiodides may be prepared from the highly reactive gallium(I) iodide. The most convenient procedure consists of sonication of a mixture of gallium, iodine and an alkyl iodide in toluene ⟨90POL2763⟩. The use of indium(I) chloride in the Barbier reaction of allylic iodides has also been reported to work well ⟨89JOM(369)291⟩.

$$3RBr + In/Mg \xrightarrow{Et_2O,\ reflux} R_3In{\bullet}OEt_2 \qquad (71)$$

R = Me, 87%; Et, 72%; Pr^n, 75%; Pr^i, 70%; Bu^n, 71%; Bu^i, 85%; Bu^s, 79%

Transmetallation methods have been used for the preparation of mono-, di-, tri- and tetraalkylated derivatives of indium and gallium by reaction of organolithium reagents in benzene with $GaCl_3$ (Equation (72)) ⟨75IC2809⟩ and $InCl_3$ ⟨67JOM(8)427⟩, trialkylindium derivatives using Grignard compounds with $InCl_3$ in diethyl ether ⟨51ZAAC(267)39, 62JA3605⟩ or dialkylmercury compounds with indium metal without solvent ⟨74SRI373⟩. As with aluminum compounds, disproportionation reactions give mono- or dialkyl metal halides.

$$n\,RLi + GaCl_3 \xrightarrow{benzene,\ 70\ ^\circ C} R_nGaCl_{3-n} \qquad (72)$$

n = 3; R = Et, 96%; Pr^n, 74%; Bu^n, 97%; Bu^i, 67%; Bu^s, 95%; Bu^t, 84%
n = 2; R = Pr^n, 63%; Bu^n, yield not given
n = 1; R = Pr^n, 95%; Bu^n, yield not given

Although less reactive than aluminum, hydride compounds of gallium add to alkenes to give alkylgallium derivatives. Likewise, thermal decomposition of triisobutylgallium produces diisobutylgallium hydride, which adds to alkenes ⟨62JA3605⟩, whereas gallium hydride dihalide reacts with various terminal or cyclic alkenes with high regioselectivity ⟨67CB1129⟩. However, reaction with styrene or dienes leads to polymerisation.

2.11.7.3 Alkylthallium Derivatives

Mono-, di-, tri- or tetraalkylated derivatives of thallium have been prepared, although they differ in stability and reactivity. Mono- and trialkylthallium derivatives are very unstable and reactive species, but dialkylthallium halides are amongst the most stable and unreactive organometallic reagents known as of 1995. They are prepared by transmetallation or addition to carbon–carbon double bonds ⟨70HOU(13/4)363, B-71MI 211-01, 82COMC-I(1)725⟩.

2.11.7.3.1 Preparation by transmetallation

Transmetallation with thallium(III) halides is commonly used for the preparation of dialkylthallium halides. Monoalkylated species spontaneously disproportionate (although a few stable alkylthallium dichlorides are known ⟨70CC1037, 76BCJ2762⟩), whereas trialkyl derivatives are better prepared by stepwise procedures. Alkylmagnesium, -zinc or -aluminum reagents can be used to give variable yields of products (Equation (73)). Alternatively, reaction of a Grignard reagent with thallium(I) bromide has been used to give dialkyl products, via an unstable alkylthallium(I) intermediate (Equation (74)) ⟨68JOM(15)500⟩. The process is only suitable for primary alkyl compounds and yields are often moderate.

$$2RMgX \;+\; TlCl_3 \xrightarrow{Et_2O,\ -15\ ^\circ C} R_2TlCl \qquad (73)$$

R = Me, 62%; Pr^i, 31%; Bu^n, 80%; Bu^s, 30%; Bu^i, 83%; *i*-C_5H_{11}, 82%; *c*-C_6H_{11}, 33%

$$2RMgX \;+\; TlBr \xrightarrow{benzene/THF,\ reflux} R_2TlBr \qquad (74)$$

R = Me, 21%; Et, 40%; Pr^n, 55%; Bu^n, 56%; *n*-C_5H_{11}, 47%; *n*-C_6H_{13}, 48%; *n*-C_7H_{15}, 47%; *n*-C_8H_{17}, 32%; *n*-C_9H_{19}, 46%; *n*-$C_{10}H_{21}$, 33%

Trialkylthallium derivatives are best prepared from the dialkyl species by reaction with alkyllithium reagents in ether ⟨40JA2357, 90JOC3368⟩, or Grignard reagents in THF ⟨64JOM(2)281⟩, although direct preparation from thallium salts can be achieved by reaction of an organometallic reagent (generally alkyllithium) with thallium(I) halide and an alkyl iodide ⟨50JA1760⟩. This reaction, however, is used more commonly for the preparation of dialkylthallium halides, which are obtained in excellent yield by treating the reaction mixture with a halo acid (Scheme 28).

$$3RLi \;+\; 3RI \;+\; 2TlBr \xrightarrow{Et_2O,\ RT} 2R_3Tl \xrightarrow{2HX} 2R_2TlX$$

R = Me, X=I, 99%; R = Me, X = Cl, 81%; R = Et, X = I, 92%

Scheme 28

Monoalkylthallium carboxylates (generally acetates) are prepared by reduction of dialkyl thallium acetates with mercury salts; treatment of the products with alkyltin derivatives gives mixed dialkylthallium derivatives ⟨70JOM(25)353⟩.

2.11.7.3.2 Addition to carbon–carbon double bonds

The reaction of thallium(III) salts (acetate, trifluoroacetate, nitrate) with alkenes in protic solvents (oxythallation) is the most important application of organothallium chemistry for synthesis (Equation (75)) ⟨B-87MI 211-06⟩. The reaction leads to the formation of β-alkoxy alkylthallium derivatives, which do not disproportionate unless they are treated with trimethyl phosphite ⟨82OM440⟩. The reaction is analogous to the oxymercuration reaction but, unlike mercury compounds, alkylthallium adducts are not isolated but directly converted by reaction with nucleophiles, reduction or elimination. A few stable alkylthallium derivatives have been isolated and some examples are shown in Scheme 29 ⟨62LA(656)294, 64TL3393, 78CAR1⟩.

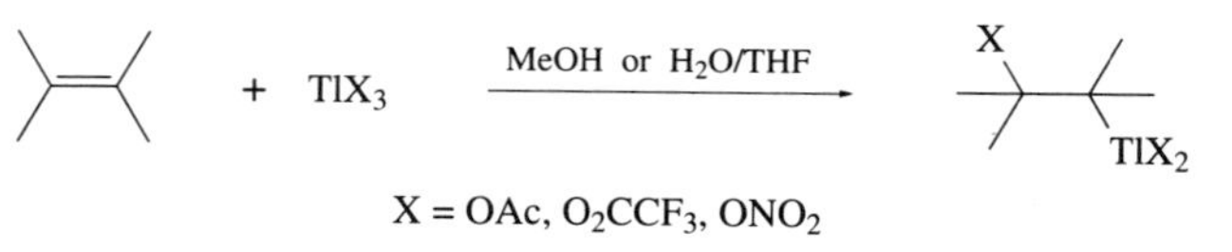

(75)

X = OAc, O_2CCF_3, ONO_2

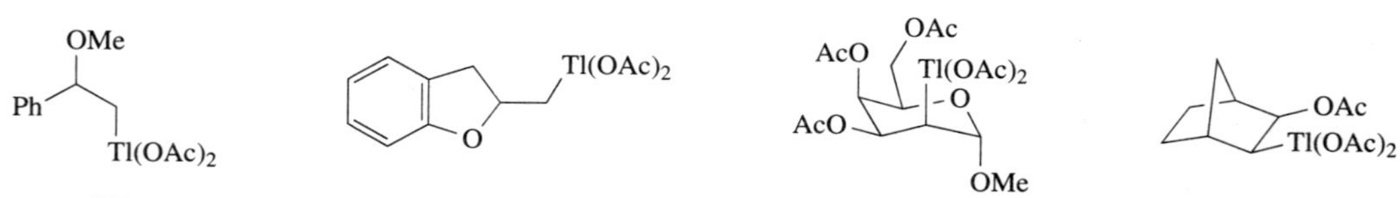

(97) **(98)** **(99)** **(100)**

Scheme 29

2.11.8 ALKYLTIN AND -LEAD DERIVATIVES

2.11.8.1 Alkyltin Derivatives

Alkyl derivatives of tin and lead were first prepared and used for industrial purposes, or for the preparation of other organometallic reagents. However, the development of modern organic methods such as carbanion or radical chemistry has highlighted the importance of alkyltin compounds in synthesis, and their preparation and reactions have been covered by reviews ⟨78HOU(13/6)181, 82COMC-I(2)519, B-87MI 211-07⟩. Mono, di-, tri- and tetraalkylated derivatives are known. Very important compounds are the mixed species $R^1_3SnR^2$, which are often used as carbanion and radical precursors (for R^2).

2.11.8.1.1 Preparation by transmetallation

The transmetallation process is the most common method for the preparation of simple, symmetrical tetraalkyltin derivatives, starting from tin tetrachloride. Grignard reagents are commonly used, although alkylaluminum reagents are recommended with larger alkyl groups (Equations (76) and (77)) ⟨62LA(653)157⟩. The production of contaminating di- or trialkylated species may be minimised if the organometallic reagent is used in excess and the reaction mixture is heated. Another method of preparation is the *in situ* preparation of alkylsodium derivatives by reacting alkyl chlorides with sodium in the presence of tin(IV) chloride (Wurtz method). This procedure is especially useful for large alkyl groups, when the Grignard reagent is not very reactive ⟨57USP2805234⟩. Tetraalkyltin derivatives are also useful starting materials for the synthesis of all other alkyltin derivatives (Scheme 30). Mono-, di- and trialkyltin chlorides (or bromides) are obtained by heating tetraalkyltin compounds with various amounts of tin tetrachloride (or bromide). The most important derivatives, trialkyltin chlorides, may be transformed into a variety of compounds by substitution (with amines, hydroxide, alkoxides, thioalkoxides, etc.) or reduction. Reduction with various metal hydrides gives trialkyltin hydrides, while reduction with sodium or magnesium gives hexaalkylditin derivatives. Substitution of an alkyl group on tin with a halide by reaction with bromine or iodine is also possible.

$$4RMgX + SnCl_4 \xrightarrow[86-96\%]{Et_2O,\ reflux} R_4Sn \qquad (76)$$

$$R = Me, Et, Pr^n, Bu^n$$

$$4R_3Al + 3SnCl_4 \xrightarrow{toluene,\ Et_2O} R_4Sn \qquad (77)$$

$$R = Me, Et, Pr^n, Bu^n$$

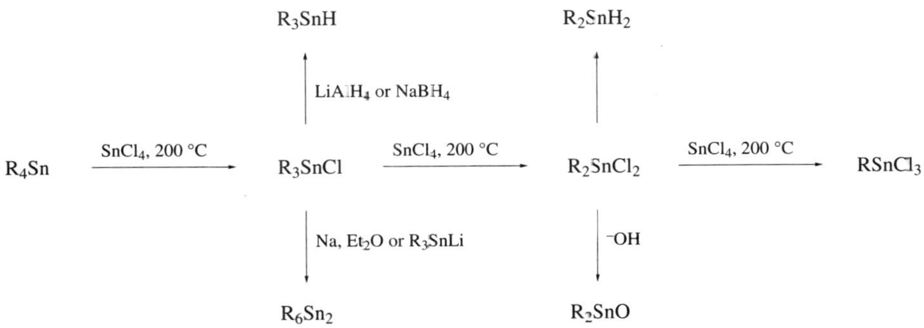

Scheme 30

The synthesis of mixed alkyltin species $R^1_nSnR^2_{4-n}$ ($n = 1, 2, 3$) by transmetallation is conveniently achieved by reaction of $R^1_nSnX_{4-n}$ with the appropriate amounts of an organometallic reagent. Various alkyl groups R^1 have been used, the most common being $R^1 = n$-butyl, since they offer more stability and less toxicity than their methyl or ethyl counterparts. The reaction is quite general and proceeds very easily. Grignard reagents are often used, although alkyllithium or alkylsodium reagents may also be employed for this reaction. The procedure is not limited to alkyltin halides, and transmetallation from alkyltin hydrides, oxides or alkoxides is similarly possible. An interesting application of this chemistry is the nucleophilic displacement of chiral alkoxides, which leads to the formation of an optically active, quadruply mixed tetraalkyltin derivative ⟨73JCS(P2)638⟩.

2.11.8.1.2 *Preparation by oxidative addition*

Addition of alkyl halides to tin powder or tin oxide affords dialkyltin dihalides ⟨53JA4103, 53JA4105⟩, whereas reaction with tin dichloride gives alkyltin trichlorides. Copper metal is used as a catalyst but the reaction is not very practical since high temperatures (above 300 °C) are required. Dialkyltin dichlorides have been converted into tetraalkyltin derivatives using the Wurtz method ⟨83JOM(255)473⟩.

2.11.8.1.3 *Preparation of $R^1_3SnR^2$*

The preparation of mixed tetraalkyltin derivatives by transmetallation has been described in Section 2.11.8.1.1.

(i) Alkylation

Despite being less popular than the transmetallation method, the preparation of mixed tetraalkyltin derivatives by alkylation of metallostannanes R^1_3SnM with alkyl halides is a convenient procedure, especially if organometallic reagents cannot be used. Metallostannanes may be prepared by three routes: by deprotonation of trialkyltin hydride with lithium diisopropylamide (M = Li), a Grignard reagent (M = MgX) or a metal hydride (M = Na, K); by halogen–metal exchange from R^1SnCl with lithium or sodium; or by cleavage of hexaalkylditin compounds with a metal (M = Li, Na) or an alkyllithium reagent.

Stannyl anions are strong nucleophiles which react easily with primary and secondary alkyl halides or tosylates (order of reactivity: I > Br > OTs > Cl and $Me_3Sn > Bu_3Sn > Ph_3Sn$) to give the alkylated compounds in good yields (Equation (78)) ⟨53JA2507, 63JOC237⟩. Elimination occurs with tertiary substrates. The mechanism of this reaction has been studied extensively ⟨78JA4834, 82JA2831, 82OM1496⟩ and seems to be very dependent on the reaction conditions. Under carefully controlled conditions, S_N2 reaction with complete inversion of configuration with chiral secondary tosylates or bromides occurs. The presence of a competing radical mechanism has been demonstrated by the partial cyclisation of 1-bromo-5-hexene ⟨80JOM(197)C19⟩ or fragmentation of cyclopropyl methyl halides.

$$R^1_3SnM + R^2X \xrightarrow{Et_2O \text{ or } THF} R^1_3SnR^2 \qquad (78)$$

R^1 = Me, Bu^n, Ph; R^2 = primary, secondary alkyl, allyl, benzyl
X = I, Br, OTs, Cl; M = Li, Na, K, MgBr

Other electrophiles react well with stannyl anions. For example, addition to epoxides occurs regioselectively on the less-substituted carbon to give β-hydroxyalkyl trialkylstannanes ⟨71JOM(33)C4⟩.

(ii) Conjugate addition

Trimethylstannyllithium ⟨77JA4836⟩, trimethylstannylsodium ⟨78JOC750⟩ or triphenylstannylpotassium ⟨80JOM(197)C19⟩ react rapidly with cyclic enones or α,β-unsaturated esters in THF at low temperature with high yields and 1,4-selectivity. However, more reactive species are obtained if copper salts are added. Trimethylstannylcopper dimethyl sulfide complex or stannyl homo- or heterocuprates react with a greater variety of conjugated ketones or esters, and their use is particularly recommended with sterically hindered and acyclic enones ⟨87CJC78⟩ (Equations (79) and (80)). Reaction of tributylstannylmagnesium bromide with enones in diethyl ether gives predominantly the alternative 1,2 addition products. However, these have been shown to rearrange into 1,4 addition products when heated ⟨71JOM(33)C7⟩.

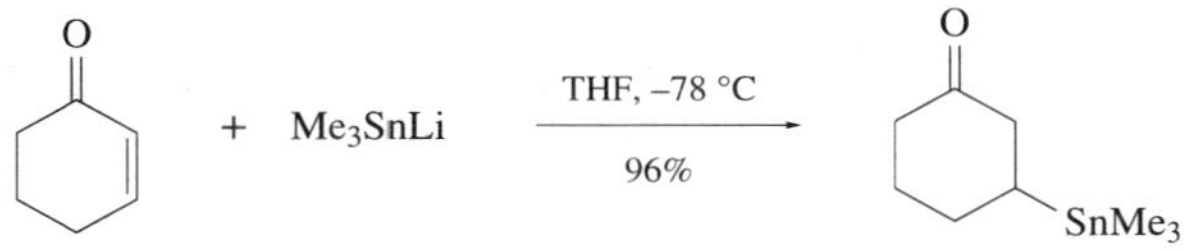

(79)

(101) **(102)** **(103)**

(80)

(104) **(105)** **(106)**

(iii) Hydrostannation

The hydrostannation reaction—for example, addition of R_3SnH or R_2SnH_2 to carbon–carbon double bonds—has been carried out with both simple and electron-deficient alkenes (Equation (81)). The reaction is catalysed with radical initiators, typically UV light irradiation or 2,2′-azobisisobutyronitrile (AIBN), and involves a reversible addition of the radical $R_3Sn\cdot$ on the double bond. Reaction with terminal alkenes gives the primary alkyltin compound regioselectively ⟨67LA(707)15⟩. Internal alkenes also react with tin hydride reagents under UV light irradiation with good regioselectivity, an example of which is shown in Equation (82) ⟨68JOC802⟩. Hydrostannation of α,β-unsaturated esters or nitriles proceeds under mild conditions and provides a convenient alternative to the conjugate addition of metallostannanes ⟨56CI(L)352, 64LA(705)10, 68BSF3385⟩. However, the reaction is not completely regioselective with hindered substrates and competitive reduction or anti-conjugate addition may occur, the best results for conjugate addition being the thermal addition of Ph_3SnH or the UV-catalysed addition of Bu_3SnH. Addition to a chiral ester, (−)-menthyl crotonate, leads to a diastereoisomeric excess of 20% ⟨75JOM(88)79⟩.

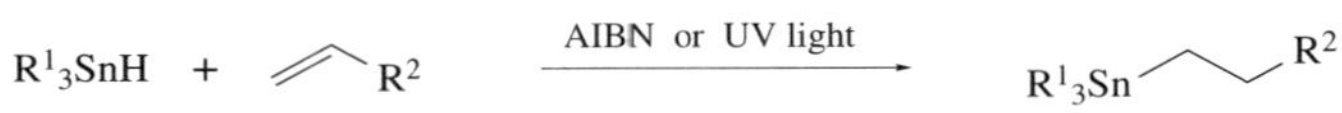

(81)

R^1 = Me, Bu^n, Ph; R^2 = alkyl, CN, CO_2Et
AIBN = 2,2′-azobisisobutyronitrile

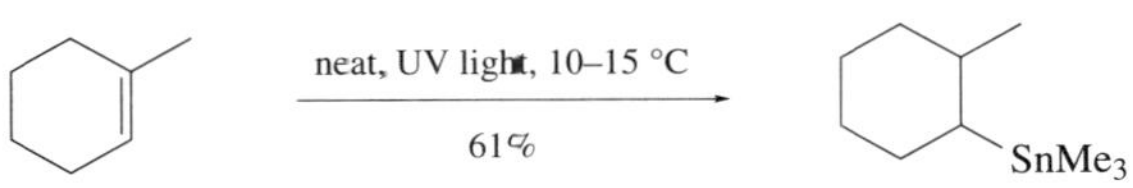

(82)

(107) **(108)**

stereochemistry not determined

2.11.8.2 Alkyllead Derivatives

The preparation of alkyllead derivatives is mostly employed for industrial purposes. The most important alkyllead compound is undoubtedly tetraethyllead, which is added to petrol to improve its octane rating. The use of alkyllead derivatives in organic synthesis has been largely hampered by the very high toxicity of lead compounds. Furthermore, lead derivatives are less thermally stable and photostable than their corresponding tin analogues and tend to decompose into radicals ⟨54CRV101, 75HOU(13/7)1, 82COMC-I(2)629⟩. Most of the methods for the preparation of organolead compounds were developed before 1950 and only few improvements have been reported since that date. The review published in 1954 therefore contains most of the crucial practical information.

The main industrial method for the preparation of tetraalkyllead compounds involves reaction of an alkyl halide with sodium–lead alloy in the presence of a catalyst. The mechanism of the reaction involves homolytic cleavage of the carbon–halide bond and subsequent reaction of the alkyl radical with lead. This method, which is important for large-scale preparation, is not very practical for laboratory purposes.

The preparation of tetraalkyllead compounds by transmetallation from lead(IV) halides using alkylmagnesium or alkyllithium reagents is complicated by competing reduction. Better results were obtained using lead tetraacetate as a starting material. Reaction with ethyl- or methylmagnesium bromide gives the tetraalkylated species in good yield (Equation (83)), while reaction with bulkier alkyl groups results in the transfer of only three alkyl groups and the formation or hexaalkyldilead compounds ⟨67JOC4062⟩. Alternatively, reaction of a Grignard or lithium reagent with lead(II) chloride affords a dialkyllead intermediate which disproportionates into tetraalkyllead and lead metal. A more convenient method consists of the reaction of an organometallic reagent with lead(II) chloride in the presence of an alkyl halide, by analogy with the synthesis of trialkylthallium compounds from Tl(I) salts (Equation (84)). Good yields of tetraalkyllead compounds are obtained and lead metal is not produced in this reaction. Alkyllithium reagents tend to give better results than their Grignard analogues. Conversion to alkyllead chlorides (R_3PbCl or R_2PbCl_2) is readily achieved by passing a stream of HCl gas into an ether solution of the tetraalkyllead reagent.

$$4RMgBr + Pb(OAc)_4 \xrightarrow[\text{R = Me, 89\%; R = Et, 94\%}]{\text{THF, 5 °C}} R_4Pb \qquad (83)$$

$$2RM + RX + PbCl_2 \xrightarrow{\text{Et}_2\text{O, reflux}} R_4Pb \qquad (84)$$

RM = MeLi, RX = MeI, 92%; RM = EtMgBr, RX = EtI, 85%

2.11.9 ALKYL ACTINIDE DERIVATIVES

Alkyl actinide derivatives have received considerable attention since the mid-1970s and several accounts or reviews are available ⟨76ACR217, 76ACR223, 82COMC-I(3)173⟩. Only alkyl derivatives of uranium and thorium in the (IV) oxidation state have been thoroughly investigated. In general, the chemistry of alkyl actinides is characterised by the low stability of the compounds, especially for the tetraalkyl derivatives, and all attempts to isolate tetraalkyluranium derivatives have been unsuccessful ⟨74JOM(67)61⟩ as these compounds undergo decomposition at room temperature. Thorium(IV) derivatives, on the other hand, are generally more stable, and their synthesis is less complicated by reduction. In the mid-1980s, the first polyalkylated derivative of an actinide in the (IV) oxidation state, heptamethylated thorium ate complex (**109**), has been synthesised and characterised by x-ray crystallography (Equation (85)) ⟨84JA6841⟩. This compound is stabilised by complexation with TMEDA and is stable for a few hours at room temperature. Another stable derivative is the uranium(V) complex Li_3UR_8 (R = Me, CH_2Bu^t) ⟨77JCS(D)812⟩.

$$7MeLi + ThCl_4 + 4TMEDA \xrightarrow[\text{45\%}]{\text{Et}_2\text{O, −78 °C}} Me_7Th(Li{\bullet}TMEDA)_3 \qquad (85)$$

(**109**)

Stable alkyl actinide complexes have the formula Cp_nMR_{4-n} (n = 2, 3) and they have been studied extensively. They are generally prepared by transmetallation using alkyllithium reagents. Reaction of Cp_3MCl (M = Th, U) produces the remarkably stable alkyl derivatives in moderate to good yield ⟨73ICA319, 82JA1860⟩, and alkylation of $Cp^*_2MCl_2$ (M = Th, U) affords in excellent yield mono- or dialkylated products (Equation (86)) ⟨81JA6650, 83JA6824⟩. Dialkyl complexes possessing neopentylic-type alkyl substitutents, such as (**110**), undergo *ortho*-metallation with loss of an alkyl group when heated above 50 °C (Equation (87)) ⟨86JA40⟩.

$$2RLi + Cp^*_2MCl_2 \xrightarrow{\text{Et}_2\text{O, −78 °C}} Cp^*_2MR_2 \qquad (86)$$

M = Th; R = Me, 60–70%; Bu^n, 56%; CH_2Bu^t, 56%; Bn, 67%
M = U; R = Me, 60–70%; Bn, 67%

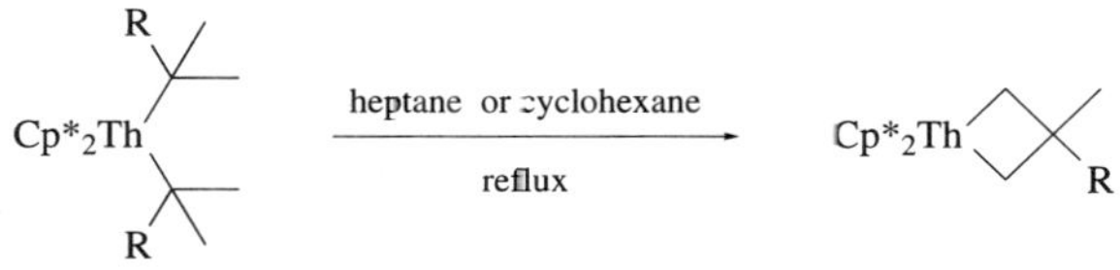

(87)

(110) R = Me, 65%; R = Et, 68% **(111)**

Another stabilising ligand is the hexamethyldisilazide (HMDS) anion and stable compounds of formula $(HMDS)_3MMe$ (M = U, Th) have been obtained by deprotonation of the hydride complex followed by alkylation ⟨81IC2991⟩. It is noteworthy that attempts to synthesise the same derivative by transmetallation from $(HMDS)_3MCl$ using alkyllithium reagents resulted only in ortho-metallation of one of the HMDS ligands.

2.12
Vinyl and Aryl Halides

CHRISTOPHER J. URCH
ZENECA Agrochemicals, Bracknell, UK

2.12.1 GENERAL METHODS FOR HALIDES

This first section on general methods for the synthesis of vinyl and aryl halides gives an overview of the major methods used. The subsequent sections on the synthesis of each of the individual halides then go into much greater detail on the available methodology and concentrate on how syntheses have been optimised for each class of compound. Methods which are unique to a particular vinyl or aryl halide are also considered separately in the section relating specifically to those compounds. A general review on the synthesis of organic halides was published in 1983 by Hudlicky and Hudlicky ⟨B-83MI 212-01⟩.

2.12.1.1 Vinyl Halides

Vinyl halides can be synthesised in a number of ways. Eliminations, particularly from dihalo compounds, to give vinyl halides are common. The elimination can be from either a geminal or vicinal dihalide. Usually it is a hydrogen halide that is eliminated (e.g., Equation (1)) ⟨51JA3329⟩, but decarboxylative eliminations (e.g., Equation (2)) ⟨84OS(62)39⟩ and the elimination of a silyl halide (e.g., Equation (3)) ⟨77SC475⟩ are also well known. The elimination of a dihalogen (either a molecule of a halogen or a mixed halogen) from a polyhalide is also a familiar route to vinyl halides (e.g., Equation (4)) ⟨85JOC1599⟩.

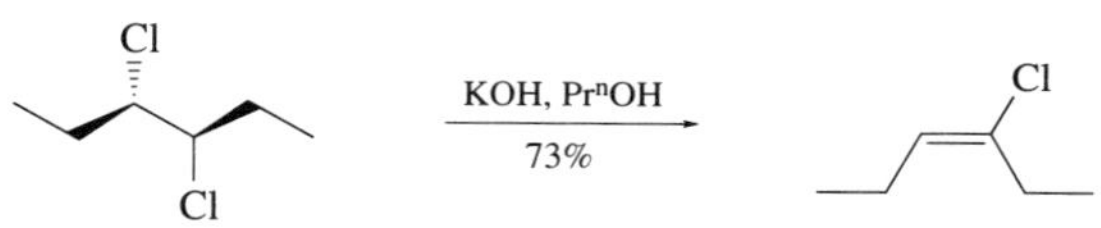

(1)

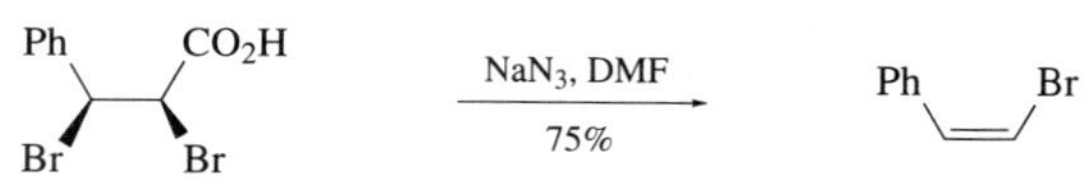

(2)

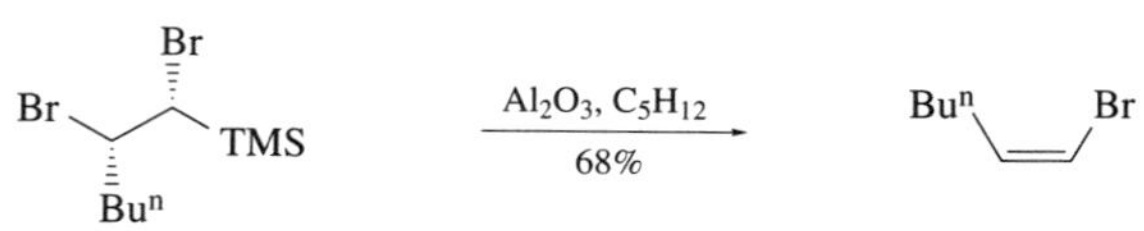

(3)

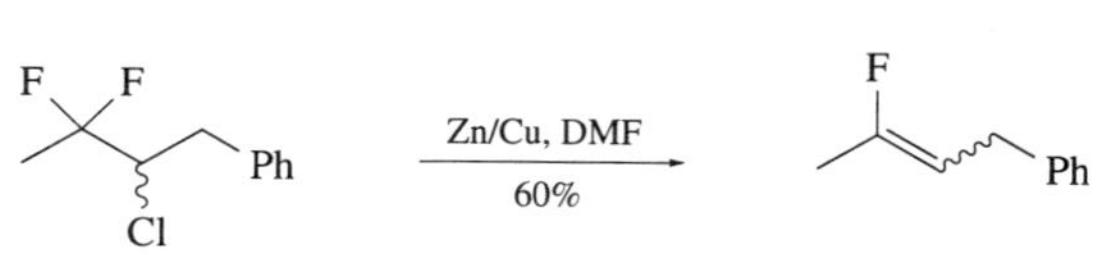

(4)

The net addition of a hydrogen halide to a triple bond is a common route to vinyl halides. This can be either the direct addition of a hydrogen halide (e.g., Equation (5)) ⟨36JA1806, 77JCS(P1)1797⟩ or, more usually, the addition of another species (frequently a borane) which is then converted into a halide, as shown in Equation (6) ⟨73JA5786⟩. This method can also be extended to the synthesis of vinyl halides by the formal addition of alkyl halides to triple bonds ⟨84JOM(268)5⟩. The halogenation of an alkyne leads directly to a 1,2-dihaloalkane as shown, for example, in Equation (7) ⟨66T2167, 80JOC4649⟩.

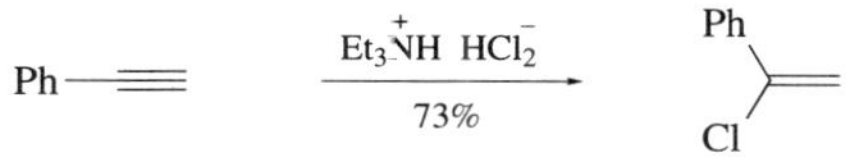

(5)

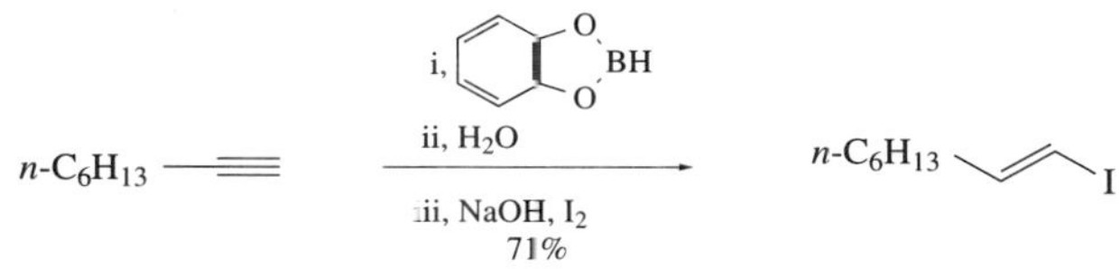

(6)

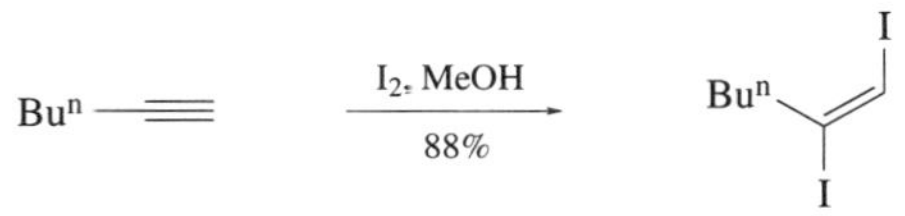

(7)

Related to these methods is the halogenation of a vinylic organometallic species, which is often, though not always, derived from an alkyne. Several metals are frequently used but perhaps the most common are boron (as mentioned previously), silicon and tin. Vinyl silanes ⟨78JOC4424⟩ and vinyl stannanes (e.g., Equation (8)) ⟨86JOC277, 89JOC5064⟩ react readily with some halogens or halogen equivalents, to give the corresponding vinyl halides, frequently via an addition–elimination reaction as mentioned above (e.g., Equation (3)). The reaction of a vinyl silane can proceed with retention or inversion of double-bond geometry depending on the conditions of the reaction ⟨74TL543⟩. Other vinyl metallic species such as vinyllithiums (e.g., Equation (9)) ⟨72T4883, 86JA2445⟩ and vinyl cuprates (e.g., Equation (10)) ⟨77TL3545⟩ can also be converted to vinyl halides.

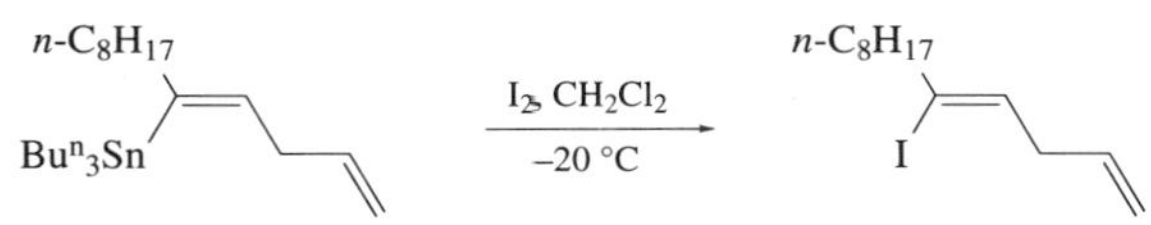

(8)

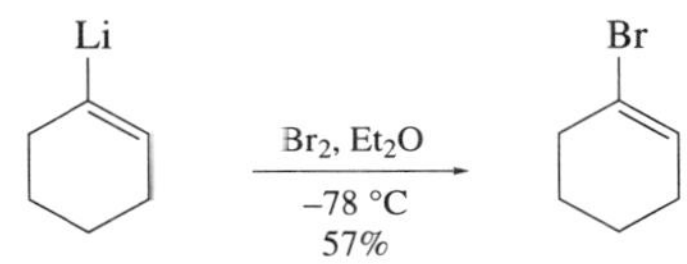

(9)

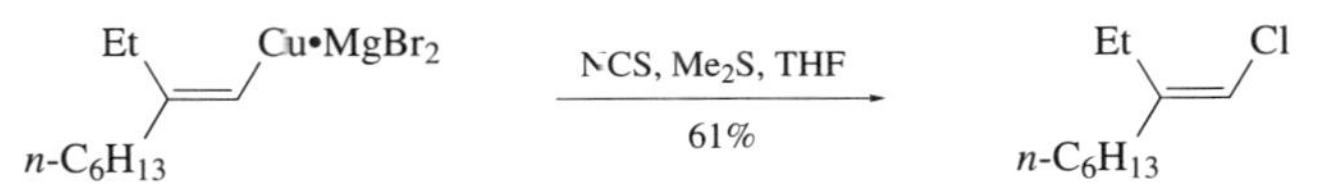

(10)

Carbonyl compounds can be readily transformed to vinyl halides in two distinct ways. First, this can be by treatment of an aldehyde or ketone with a halogenating reagent (such as phosphorus pentachloride) to give a vinyl halide with the same carbon skeleton, as illustrated in Equation (11) ⟨54JCS3659⟩. Second, a carbonyl compound (particularly an aldehyde) can be treated with a halogenated Wittig reagent (or the equivalent) to give a chain-extended vinyl halide. An example of this type of process is shown in Equation (12) ⟨59S75⟩.

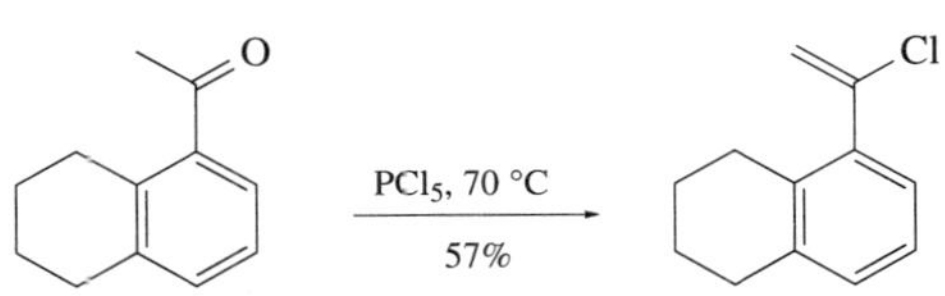

(11)

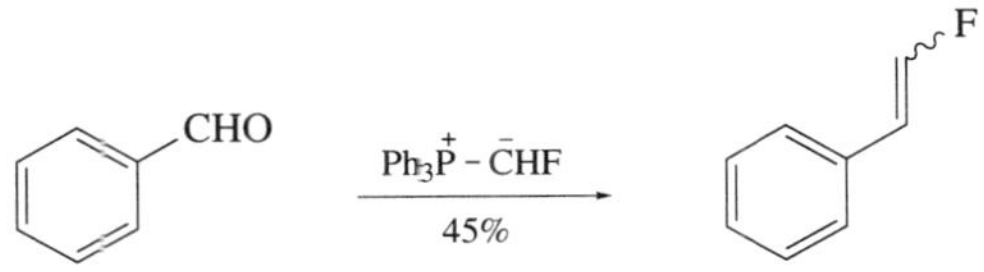

1:1 (*E*) : (*Z*)

(12)

2.12.1.2 Aryl Halides

Historically, one of the most common methods for the synthesis of aryl halides has been by the electrophilic substitution of an aromatic ring ⟨B-85MI 212-01⟩. The actual conditions employed in this process vary enormously depending on the substrate used and the halide to be introduced. Aryl rings with electron-donating substituents react readily to give the familiar *ortho/para* substitution pattern, as shown in Equation (13) ⟨61JA4605⟩. Conversely, electron-withdrawing substituents retard the rate of reaction and direct electrophilic halogenation occurs principally in the *meta* position, an example of which is illustrated in Equation (14) ⟨58JOC1412⟩. Often, for activated substrates undergoing chlorination or bromination, no catalyst is needed. Indeed, to control the reaction of anilines, these are frequently converted to anilides, particularly if monohalogenation is sought ⟨50JA2454⟩. A Lewis acid catalyst is often used for less active substrates. Ferric chloride is a common catalyst for chlorinations and is frequently prepared *in situ* by the addition of iron to the chlorine-containing reaction mixture ⟨57JCS1823⟩. The halogenating reagent is often the halogen itself or occasionally a mixed halogen. This is particularly so for chlorinations and brominations, though a host of other reagents such as NCS ⟨85S669⟩ and cupric bromide ⟨64JA427⟩ can also be used. Iodine is the least reactive of the halogens, but can be used with suitable catalysts or in the presence of an oxidizing agent which produces more electrophilic iodine species. Its use as an electrophilic reagent has been reviewed ⟨B-77MI 212-01⟩. Iodine monochloride is more reactive than iodine itself and can be used as the iodinating agent. For instance, tyrosine has been monoiodinated with iodine monochloride in aqueous acetic acid in 80–85% yield ⟨43JA1430⟩. Fluorine is usually too reactive to be of practical use for electrophilic aromatic substitutions and an alternative reagent such as fluoroxytrifluoromethane ⟨68CC806⟩ is used, though it is possible to use fluorine at low temperatures (e.g., Equation (15)) ⟨70JOC723⟩. Of key importance in all electrophilic aromatic substitution reactions is the control of the resultant substitution pattern. The methods used to achieve this for each halogen will be discussed in subsequent sections.

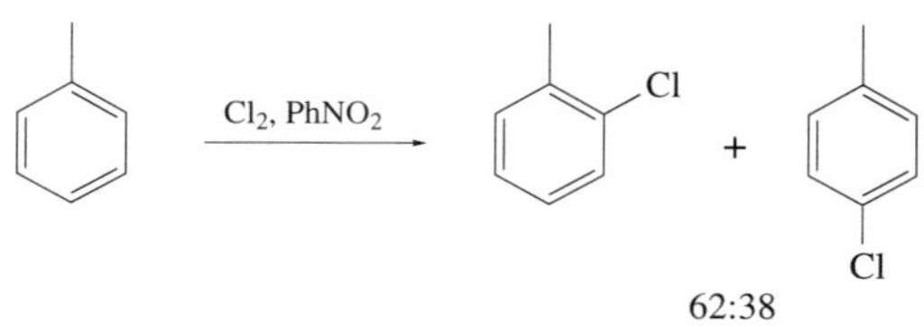

(13)

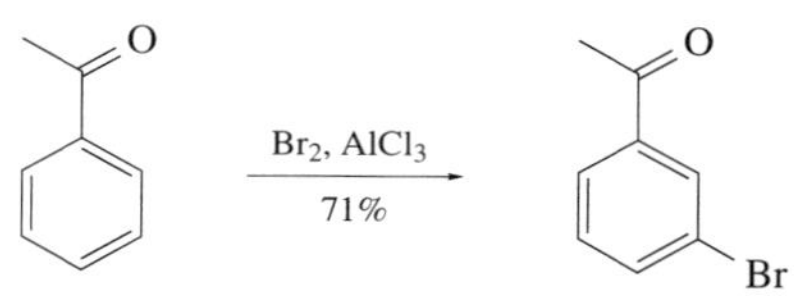

(14)

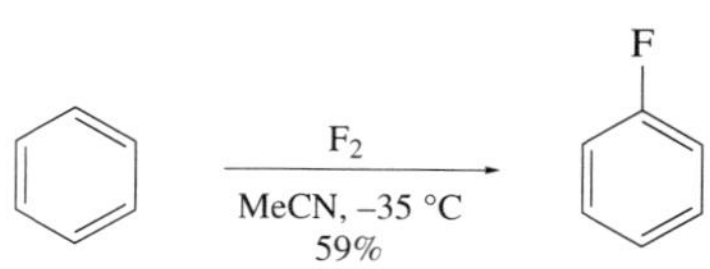

(15)

A variation of the electrophilic aromatic substitution reaction is the use of aryl metallic species with an electrophilic halogenating reagent. Several metals have been employed, with boron (e.g., Equation (16)) ⟨82OPP359⟩, silicon (e.g., Equation (17)) ⟨77AG(E)488, 79AG(E)402⟩ and tin (e.g., Equation (18)) ⟨72CC840⟩ being among the most commonly encountered. All of these methods have the advantage over electrophilic aromatic substitution of complete control of the substitution pattern of the aryl halide produced.

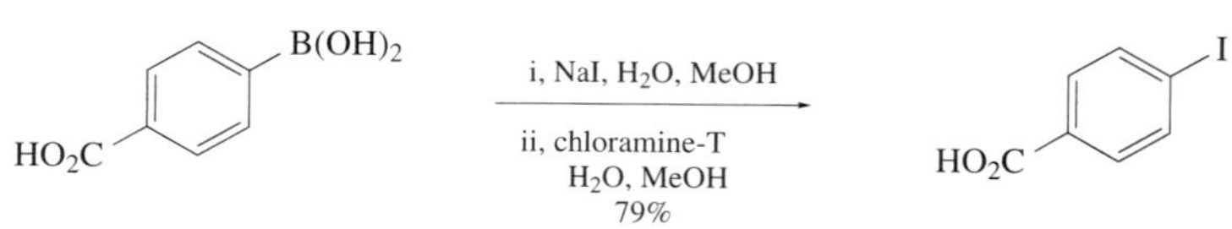

(16)

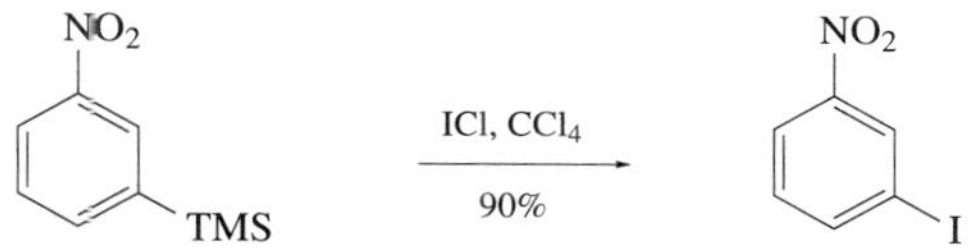

(17)

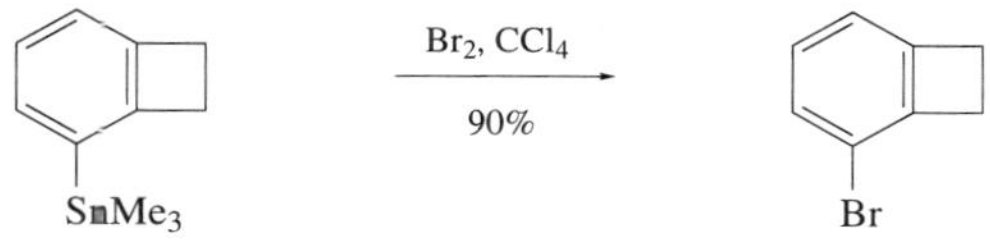

(18)

Nucleophilic aromatic substitution can also be used to prepare aryl halides. One application of this reaction is halide exchange, which is used to convert one aryl halide into another, less readily available, one. This reaction is particularly useful in the preparation of aryl fluorides as there are fewer methods available for their synthesis than for other aryl halides. An example is shown in Equation (19) ⟨56JA6034⟩. It is of course possible to utilise other leaving groups besides a halide. For instance, some phenols, particularly those with electron-withdrawing groups attached, can be converted to aryl chlorides by treatment with phosphorus pentachloride (e.g., Equation (20)) ⟨60DOK(133)855⟩; in suitably activated rings, a nitro group can be displaced by fluoride (e.g., Equation (21)) ⟨85TL2233⟩.

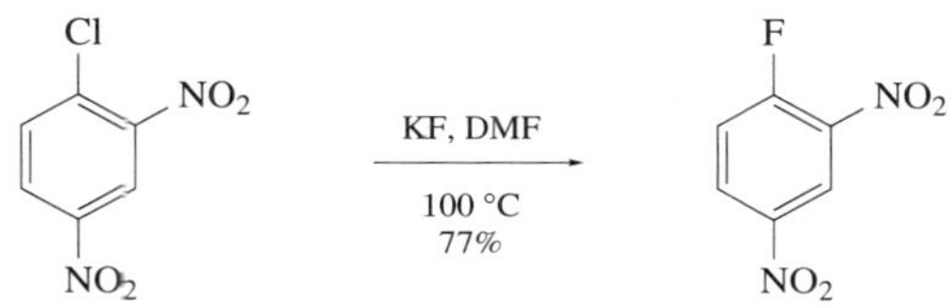

(19)

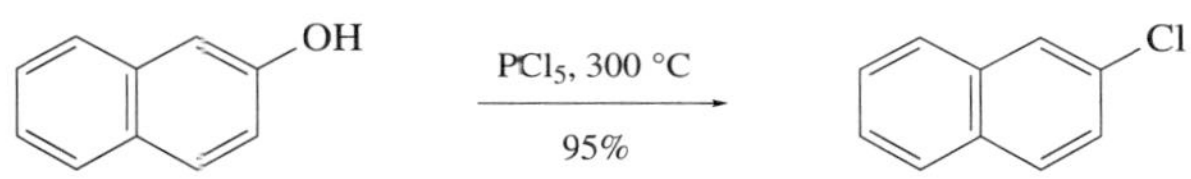

(20)

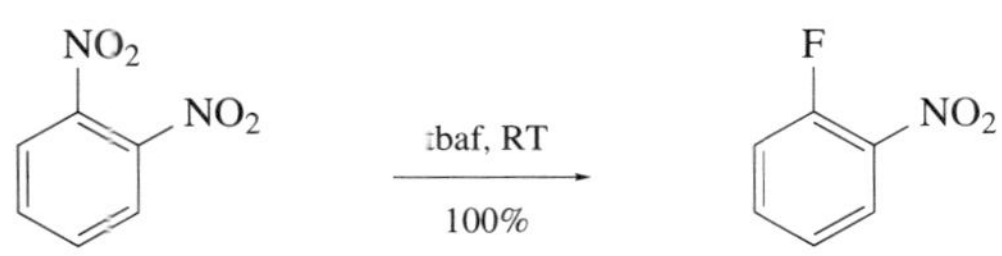

(21)

tbaf = tetra-*n*-butylammonium fluoride

One of the commonest methods of making aryl halides is the diazotization of an aniline followed by conversion to the corresponding aryl halide. All of the halides can be introduced to an aryl ring in this way. For aryl chlorides and bromides the Sandmeyer reaction is most commonly used. This involves treating an aryl diazonium salt with cuprous chloride or bromide to give the corresponding aryl chloride (e.g., Equation (22)) ⟨32OSC(1)170⟩ or bromide ⟨32OSC(1)136⟩, respectively. Copper in the presence of the appropriate hydrogen halide can also be used; the reaction is then known as the Gatterman reaction. Other chlorinating and brominating reagents are known, but their use is comparatively rare. Aryl iodides are readily prepared by simply treating an aryl diazonium salt with a source of iodide, as for example in Equation (23) ⟨60OS(40)105⟩. Aryl fluorides are most conveniently prepared by heating an aryl diazonium tetrafluoroborate salt, as the example in Equation (24) illustrates ⟨46JOC444⟩. This reaction is known as the Schiemann reaction, or sometimes as the Balz–Schiemann reaction ⟨31CB1340⟩.

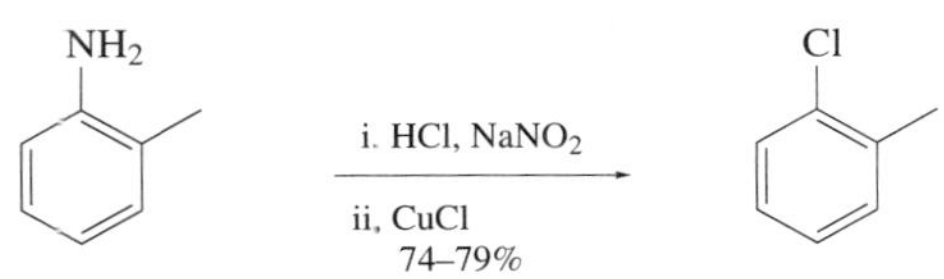

(22)

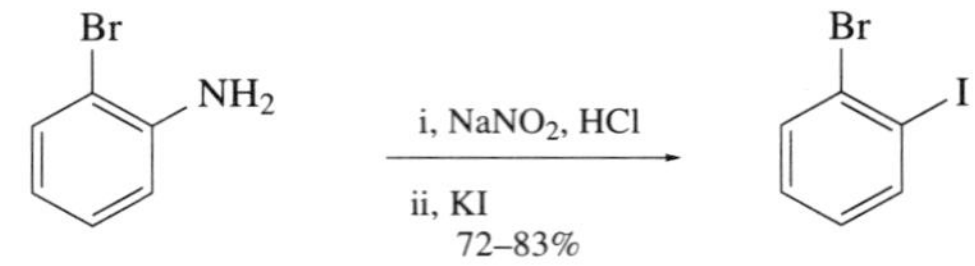

(23)

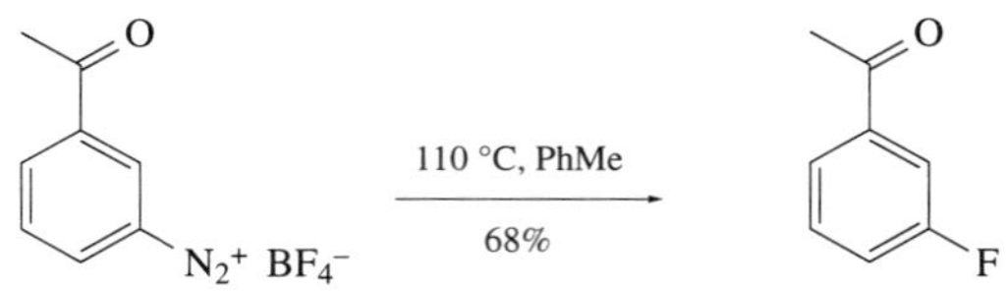

(24)

2.12.2 FLUORIDES

2.12.2.1 Vinyl Fluorides

2.12.2.1.1 Vinyl fluorides by elimination reactions

Vinyl fluorides can be made by the elimination of a hydrogen halide from either a vicinal or a geminal fluorohalide. A vicinal dihalide is often in turn made from an alkene by the addition of two halogen atoms across the double bond. Boche and Fahrmann have reported the synthesis of both (*E*)- and (*Z*)-vinyl fluorides by first adding bromine and fluorine atoms across an alkene and subsequently eliminating hydrogen bromide, as shown in Scheme 1 ⟨81CB4005⟩. The geometry of the initial double bond controls that of the final vinyl fluoride as both addition and elimination reactions are stereospecific. In a similar reaction, the method has been extended to the synthesis of trisubstituted vinyl fluorides in yields of 42–62% ⟨78S217⟩. The same overall reaction can also conveniently be achieved by adding iodine and fluorine atoms across the alkene and then eliminating hydrogen iodide ⟨91JOC2234⟩. In this case, starting with allylbenzene, (2-fluoroallyl)benzene was produced without contamination by an isomeric alkene, though this was the only example reported.

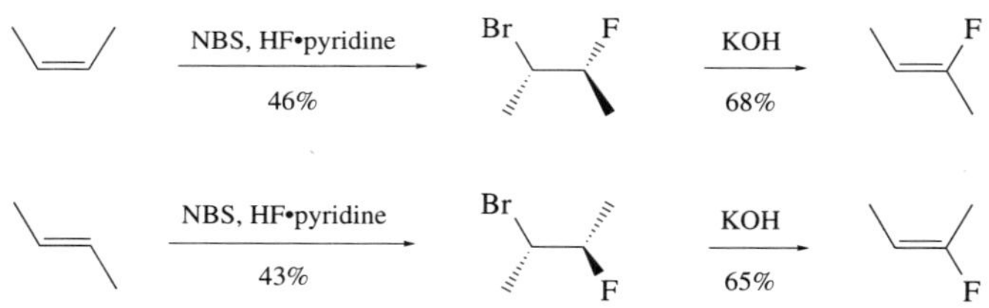

Scheme 1

Vicinal difluorides can also be eliminated to vinyl fluorides. In a two-step procedure, a cyclic ketone was first converted to the geminal difluorocycloalkane with sulfur tetrafluoride and then eliminated with alumina ⟨71JOC818⟩. The yield in both steps was variable, with hydrolysis back to the starting ketone sometimes being a problem. However, in favourable cases such as 1-fluorocyclohexene the overall yield was 44%. Posner *et al.* found that the problem of hydrolysis could be overcome by drying the alumina at 400 °C. In this manner a 20% yield of 1-fluorocycloheptene (**1**) was increased to 60% (Equation (25)) ⟨77JOC3173⟩.

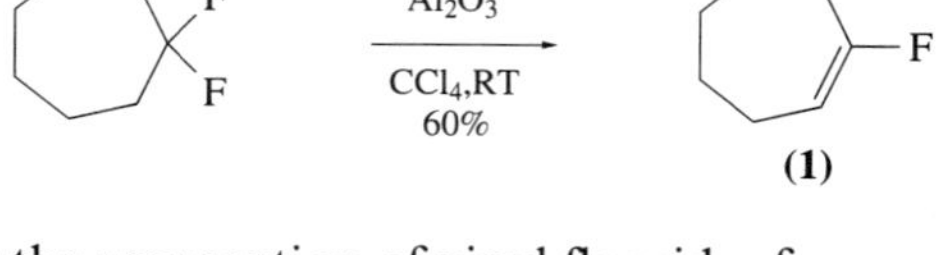

(**1**)

(25)

Related to these methods is the preparation of vinyl fluorides from epoxysilanes ⟨89TL967⟩. Ring opening of the epoxide with silicon tetrafluoride gave 2-fluoro-2-silyl alcohols which were then Peterson eliminated to give the vinyl fluoride with good control of stereochemistry. A decarboxylative elimination has also been used to prepare a vinyl fluoride ⟨79CL1499⟩. Another route to vinyl fluorides involving an elimination is the conversion of a thioacetal to an α-fluorothioether which is subsequently oxidized and eliminated ⟨87TL3901⟩. The elimination step gave a mixture of

(*E*)- and (*Z*)-fluoroalkenes in 40–80% yield. 2-Fluoroalkenes can be prepared by the elimination of 2-fluoroselenoxides which, in turn, can be prepared from a terminal alkene, as shown in Scheme 2 ⟨90TL973⟩. The reaction failed for 1,2-disubstituted double bonds, which gave allyl fluorides instead.

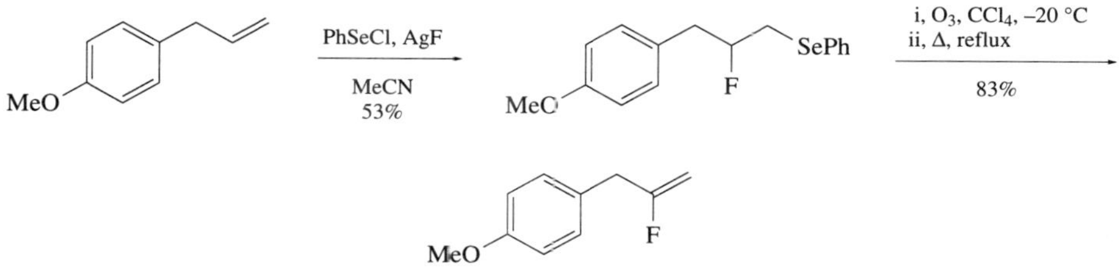

Scheme 2

2.12.2.1.2 *Vinyl fluorides from alkynes*

Fluoroalkenes are rarely made from alkynes. However, vinyl fluoride has been synthesised by the addition of hydrogen fluoride to alkyne ⟨46JA2467⟩. The reaction was performed using a mixture of mercuric chloride and barium chloride as catalyst to give vinyl fluoride in a yield of 82%. Hydrogen fluoride has also been added to electron-deficient alkynes ⟨86BSF910⟩. Tetrabutylammonium dihydrogentrifluoride was found to be an effective reagent with which to carry out this reaction, giving vinyl fluorides in yields in the range 57–90% for eight examples.

2.12.2.1.3 *Vinyl fluorides from vinyl metallic species*

The fluorination of a vinyl metallic species is a less common way of making vinyl fluorides than other vinyl halides as there are fewer readily available sources of electrophilic fluorine. However, a vinyllithium has been fluorinated with *N*-fluoro-*N*-norbornylsulfonamide ⟨84JA452⟩ to give the corresponding vinyl fluoride ⟨86JOC3561⟩. Vinylstannanes can be fluorinated with the dabco-based fluorinating agent (**2**) ⟨93TL3057⟩. The one example cited gave a yield of 71% (Equation (26)). Vinylstannanes can also be fluorinated with xenon difluoride in the presence of silver hexafluorophosphate ⟨92SC1461⟩ or silver triflate (Equation (27)) ⟨93SL207⟩. The latter was shown to be superior, giving yields of 50–80%.

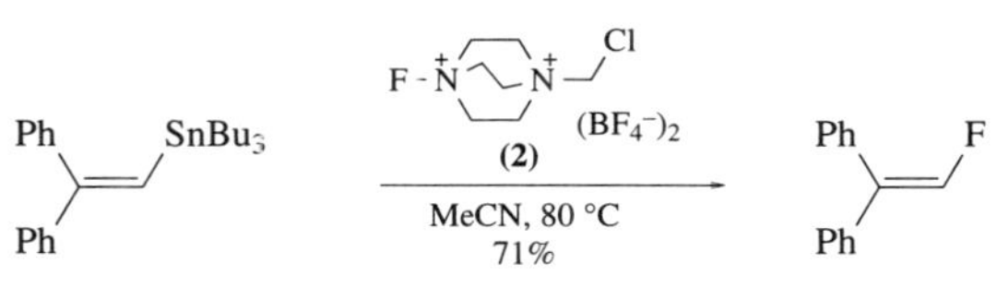

(26)

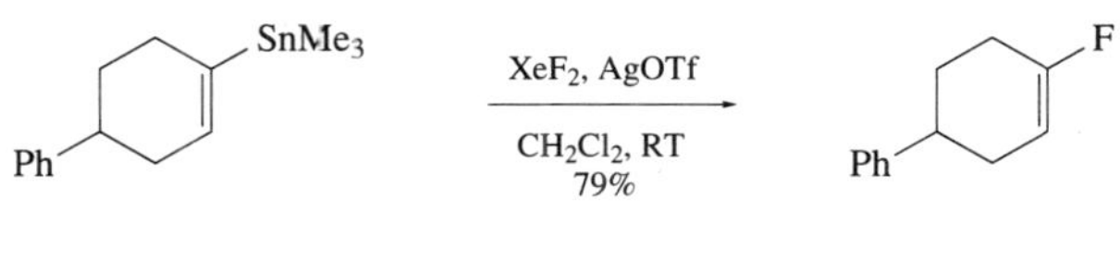

(27)

2.12.2.1.4 *Vinyl fluorides from carbonyl compounds*

Aldehydes can be converted into chain-extended vinyl fluorides by a Wittig reaction. Both aliphatic and aromatic aldehydes have been fluoromethylenated with a reagent derived from tributylphosphine and fluorotrichloromethane ⟨85JA2811⟩. For the 12 examples given, yields were in the range 40–60%. Aromatic aldehydes gave mainly (*E*)-vinyl fluorides, whereas aliphatic aldehydes gave the (*Z*)-geometry exclusively (Equation (28)). Schlosser and Zimmermann have described the conversion of aldehydes to vinyl fluorides in moderate yield via a Wittig reaction ⟨69S75⟩. A ketone was also fluoromethylenated, but the yield was only 16%. Trifluoroacetophenone can be fluoromethylenated by a Wittig procedure to give an approximately 50 : 50 (*E*) : (*Z*) mixture in 80% yield ⟨73JFC(3)447⟩. This reaction also worked with aldehydes, though the yields were somewhat lower. Schlosser and Christmann have reported the fluoromethylenation of an aldehyde by first making a betaine (**3**) at low temperature and then, before it could eliminate, deprotonating and fluorinating it *in situ* with perchloryl fluoride. Completion of the Wittig reaction afforded the vinyl fluoride (**4**) (Scheme 3) ⟨69S38⟩. In another Wittig reaction, perfluoroalkyl acyl fluorides have been converted to 1,2-difluoro-2-(perfluoroalkyl)alkenes ⟨83JA650⟩. The vinyl fluoride was produced in 52% yield with >95% (*E*)-geometry for the one example with experimental details.

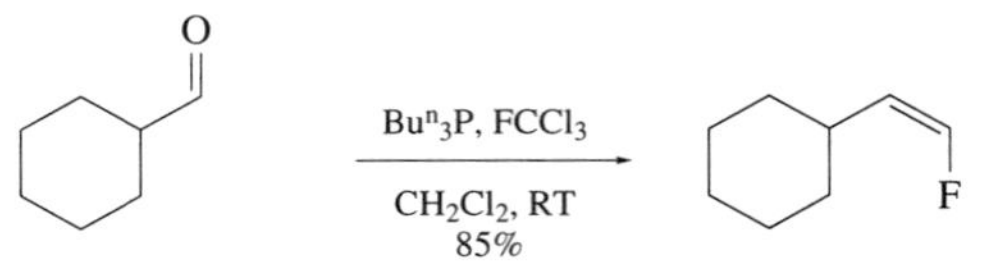

(28)

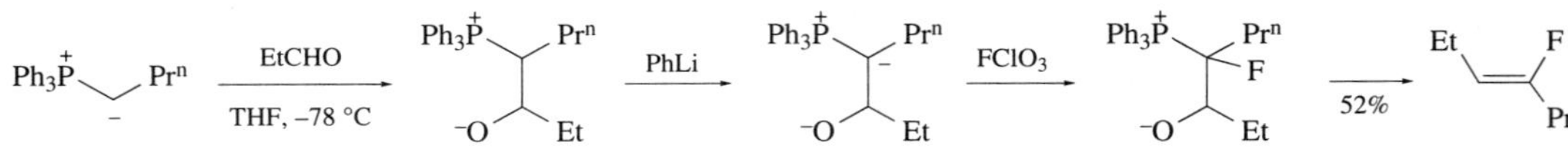

(**3**) (**4**)

Scheme 3

Aldehydes and ketones can be fluoromethylenated in a similar manner to the Wittig reaction but using a fluorinated sulfoximine rather than a phosphorus ylide. Thus, addition of the anion (**5**) to a carbonyl compound followed by elimination with aluminum amalgam gave 1-fluoroalkenes in good yield (Scheme 4). A 50 : 50 mixture of (*E*)- and (*Z*)-isomers was always obtained ⟨88TL3365⟩. Vinyl fluorides have also been synthesised by alkylating an α-fluoro-α-lithio sulfoxide followed by the thermal elimination of the resultant α-fluoro sulfoxide. Yields were in the range 50–80% ⟨83TL725⟩.

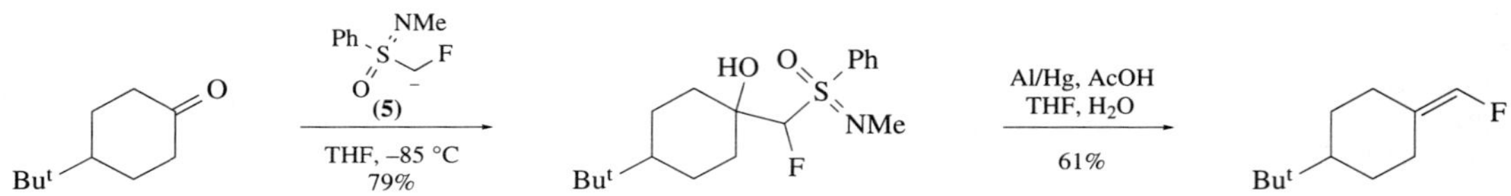

Scheme 4

2.12.2.1.5 Vinyl fluorides by miscellaneous methods

1-Fluoroalkenes, as predominantly the (*E*)-isomer, have been prepared by the selective mono-reduction of geminal difluoroalkenes ⟨79CL983⟩. Yields were in the range 78–96% for four examples. Fluorodienes have been prepared by Schlosser and co-workers by the fragmentation of 1-chloro-1-fluorocyclopropanes. This fragmentation can be initiated by treating an iodomethylcyclopropane with zinc ⟨80HCA1236⟩ or a silylmethylcyclopropane (**6**) with fluoride and is stereospecific as shown in Equation (29) ⟨84HCA284⟩. Yields were good in both reactions.

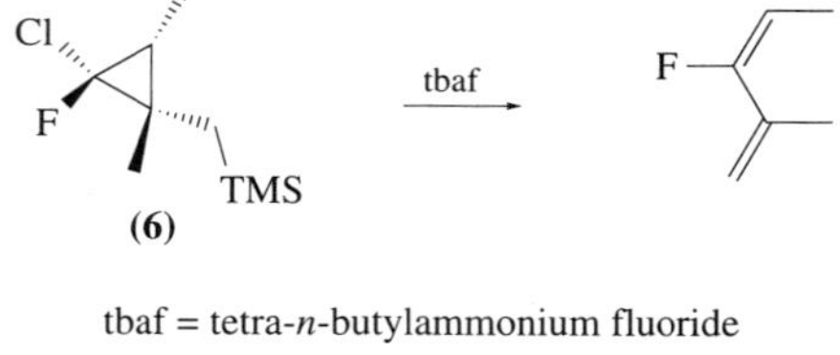

(**6**)

(29)

tbaf = tetra-*n*-butylammonium fluoride

2.12.2.2 Aryl Fluorides

2.12.2.2.1 Aryl fluorides by electrophilic aromatic substitution

As previously mentioned, electrophilic aromatic fluorinations usually use fluorinating agents other than fluorine itself. Fluoroxytrifluoromethane ⟨68CC806, 88CJC1479⟩ and bis(fluoroxy)difluoromethane have both been used as fluorinating reagents ⟨85JOC4576⟩. For example, the anilide (**7**) gave a high yield of the aryl fluoride (**8**) (Equation (30)). Caesium fluoroxysulfate has also been used in the fluorination of activated arenes ⟨81CC148, 81JA1964, 86JOC3242⟩. For the activated arenes studied in the last of these papers, it gave similar products to fluoroxytrifluoromethane. Another useful fluorinating agent is acetyl hypofluorite, which, as with the previous reagents, worked well with activated aromatic systems but less well when electron-withdrawing groups were attached to the aromatic ring ⟨81JOC4629, 86JOC1886⟩. With this reagent the major product is that of *ortho* rather than *para* substitution. For instance, anisole gave an 80% yield of a 10:1 *ortho*:*para* mixture ⟨84JOC806⟩. *N*-Fluoroperfluoroalkylsulfonamides have been reported to be good fluorinating reagents ⟨87JA7194, 91TL1631⟩. For example, the first of these papers reports that toluene gave an 80% yield of monofluorinated products in an *ortho*:*meta*:*para* ratio of 74:4:22 using $(CF_3SO_2)_2NF$. Xenon difluoride in the presence of hydrogen fluoride has been used as an aromatic fluorinating agent ⟨70JA6498, 71JOC2917⟩. Substituent effects were similar to those familiar in electrophilic aromatic reactions and yields were in the range 32–81%. Anisole gave 65% of a 46:4:50 *ortho*:*meta*:*para* mixture of monofluorinated products. With oxygen-substituted aromatic systems, xenon difluoride reacted directly without the presence of hydrogen fluoride to give the monofluorinated products in yields of 37–71% (Equation (31)) ⟨75JOC807⟩.

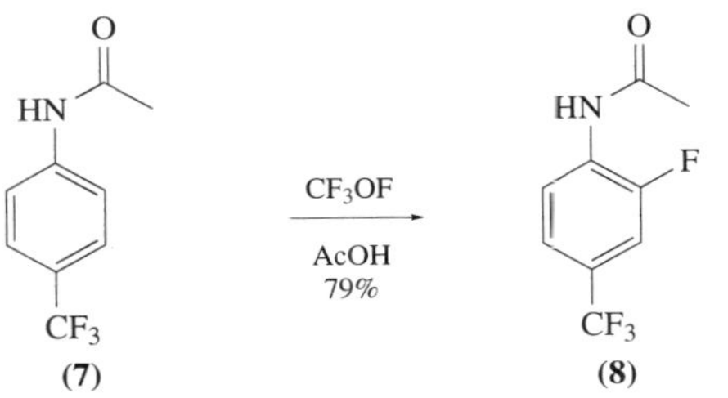

(30)

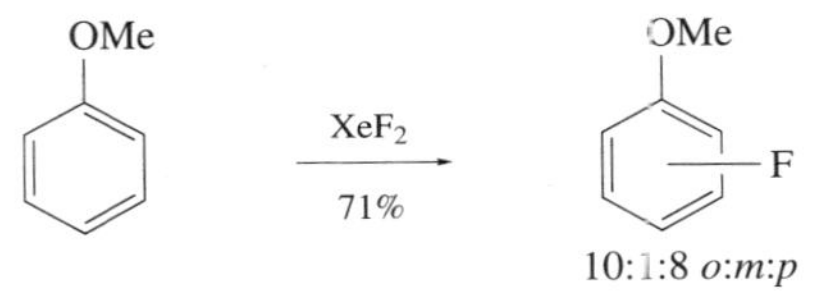

(31)

Photofluorination can also be used to prepare aryl fluorides. Fluorine itself has been used as the fluorinating agent ⟨81JMC93⟩ as has fluoroxytrifluoromethane ⟨70JA7494⟩. In the latter example, benzene was monofluorinated at −78 °C in a yield of 65% (with 10% trifluoromethoxybenzene as by-product).

2.12.2.2.2 *Aryl fluorides from aryl metallic species*

A wide variety of carbanions can be fluorinated with *N*-fluoro-*N*-alkylsulfonamides ⟨84JA452⟩. For instance, 1-napthol with potassium hydride gave 2-fluoro-1-napthol in 60% yield and phenylmagnesium bromide gave fluorobenzene in 50% yield. Aryllithiums were also fluorinated. Grignard reagents can also be fluorinated by *N*-fluoro-*O*-benzenedisulfonimide ⟨91TL1631⟩ and *N*-fluorolactams ⟨90JOC3373⟩. Arylstannanes have been fluorinated with fluorine, fluoroxytrifluoromethane and caesium fluoroxysulfate. Yields were particularly good (69–79%) when aryltrimethylstannanes were treated with caesium fluoroxysulfate (Equation (32)) ⟨86BSF930, 86CC1623⟩. Arylmercury compounds have been fluorinated with acetyl hypofluorite (Equation (33)) ⟨84CC655, 86JOC1886, 88TL1501⟩. Yields were typically in the range 50–60%, though the reaction failed if the arene was deactivated. Widdowson and co-workers have shown that *N*-methyldiethanolamine esters of aryl boronic acids (e.g., (**9**)) can be converted to the corresponding fluoroaromatics by treatment with caesium fluoroxysulfate, as shown in Equation (34) ⟨90SL761⟩. For the six examples given, yields ranged from 15% to 52%.

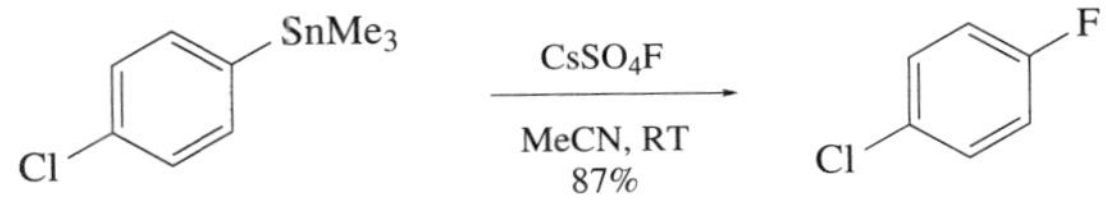

(32)

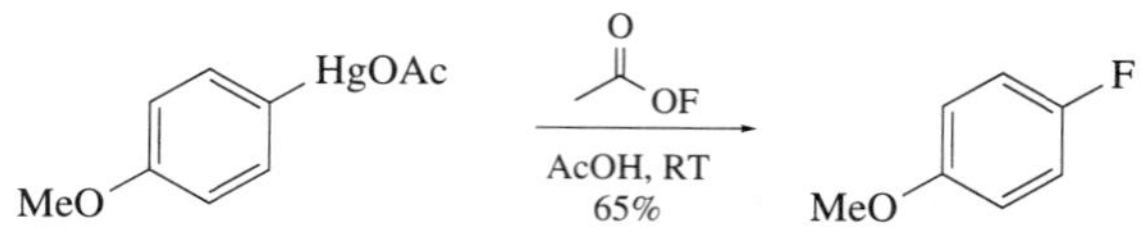

(33)

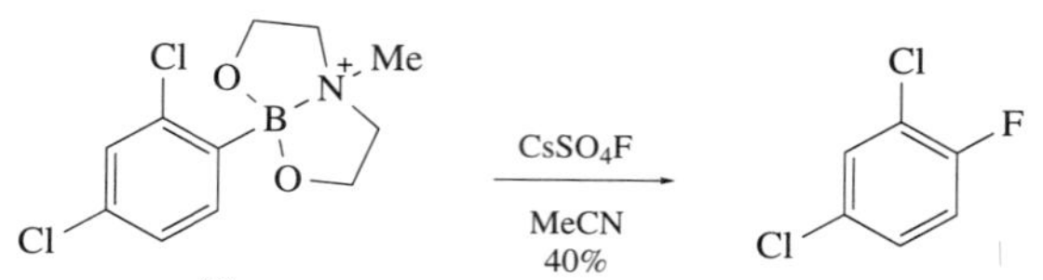

(9)

(34)

2.12.2.2.3 *Aryl fluorides by halogen exchange and related reactions*

The synthesis of fluoroaromatics by halogen exchange has been reviewed by Prescott ⟨78CI(L)56⟩ and de Nie-Sarink and Prillwitz ⟨84MI 212-01⟩. Aryl fluoride synthesis by chloride displacement from electron-deficient arenes has been reported using tetraphenylphosphonium hydrogen difluoride ⟨85CC672⟩ and a polymer-supported tetraphenylphosphonium salt catalyst ⟨89TL7199, 90CL769⟩. The polymer-supported catalyst has the advantage of being reusable. Other leaving groups besides halides can be displaced from an arene by fluoride in an analogous reaction to halogen exchange. Nitro groups have been displaced from electron-deficient aromatic rings, often in good yield ⟨83CC108⟩. This reaction has been catalysed by tetraphenylphosphorus bromide, when even a nitro leaving group with only a *meta* electron-withdrawing substituent on the aromatic ring was displaced by fluoride (Equation (35)) ⟨89CL2213⟩. Other leaving groups that have been reported to be displaced by fluoride include fluorosulfonyl ⟨88JOC4398⟩ and dimethylsulfonium ⟨85CPB1301⟩. Finally, diaryl iodonium salts have been converted to aryl fluorides, generally in good yields, by heating with potassium fluoride ⟨82JFC(21)385⟩.

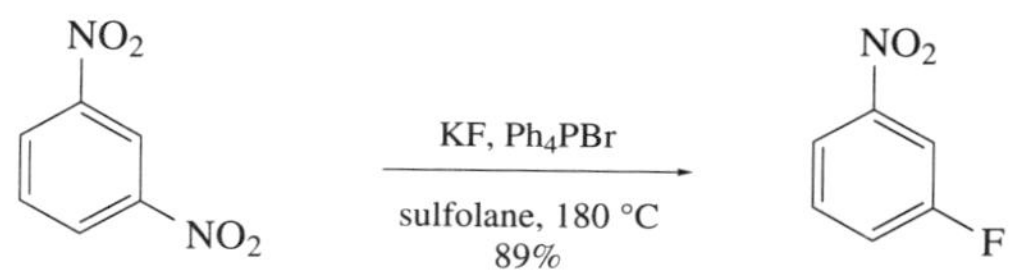

(35)

2.12.2.2.4 *Aryl fluorides from aryldiazonium salts*

The classical Schiemann reaction ⟨31CB1340, 47JA2022⟩ in which an aryldiazonium tetrafluoroborate salt is heated to produce an aryl fluoride has been much used and modified. It was reviewed by Suschitzky in 1965 ⟨65AFC(4)1⟩. One useful variation is that reported by Bergmann *et al.* in which the diazonium tetrafluoroborate was heated in the presence of copper powder or cuprous chloride ⟨56JA6037⟩. Both this procedure and the classical Schiemann reaction can give good yields, and which was better depended upon the substrate. Other salts such as hexafluorophosphate ⟨63OS(43)12⟩ and hexafluoroantimonate have also been used with success. Hexafluorophosphate salts have been shown to be superior to tetrafluoroborate in some, though not all, cases ⟨61JOC5149⟩. For instance, *p*-aminobenzoic acid failed to give the required aryl fluoride in the Schiemann procedure (tetrafluoroborate salt), but gave a 49% yield of *p*-fluorobenzoic acid using the hexafluorophosphate modification.

Another modification is the photolysis of aryldiazonium tetrafluoroborate or hexafluorophosphate salts rather than heating them ⟨71JOC631⟩. This modification worked well for some substrates, but was much poorer than pyrolysis for aromatic rings with electron-withdrawing groups. A further modification of the Schiemann reaction produced aryl fluorides by heating aryldiazonium tris(perfluoroalkyl)trifluorophosphates ⟨89JGU469⟩. In the one example in which a yield was quoted, *p*-nitrofluorobenzene was produced in 82%. Other methods have been reported for converting anilines into aryl fluorides. Olah and co-workers have described a one-pot procedure in which an aniline was converted into the corresponding aryl fluoride by pyridinium poly(hydrogen fluoride) and sodium nitrite ⟨79JOC3872⟩. Yoneda and co-workers have reported a similar procedure in which an aniline was treated with an organic base–hydrogen fluoride complex and sodium nitrite to give the aryl fluoride, usually in high yield (Equation (36)) ⟨87SC685⟩. It has the advantage over the Olah procedure of being experimentally easier to perform. In a further paper the same group reported that the fluorodediazotization step could be accelerated photochemically ⟨89SC865⟩. A further one-pot procedure for the synthesis of aryl fluorides from anilines has been reported by Milner ⟨92SC73⟩. Treatment of an aniline with nitrosonium tetrafluoroborate gave the aryl fluoride in consistently higher yields than the Schiemann reaction for the examples studied.

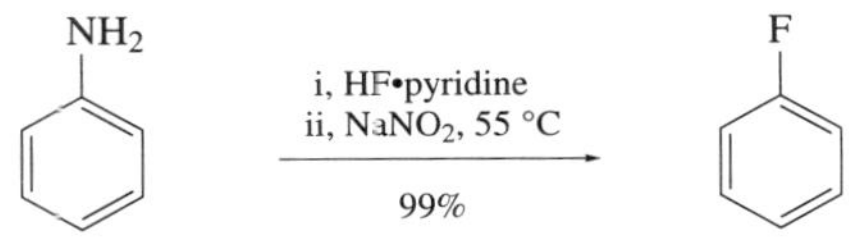

(36)

Aryl triazenes have been used as a source of aryldiazonium salts, avoiding the need to use nitrogen oxide species ⟨79CC914⟩. Thus, treatment of an aryl triazene (e.g., (**10**)) with 70% hydrogen fluoride in pyridine produced the corresponding aryl fluoride in a yield of up to 97% (Equation (37)). The reaction worked well for electron-donating and mildly electron-withdrawing aryl substituents.

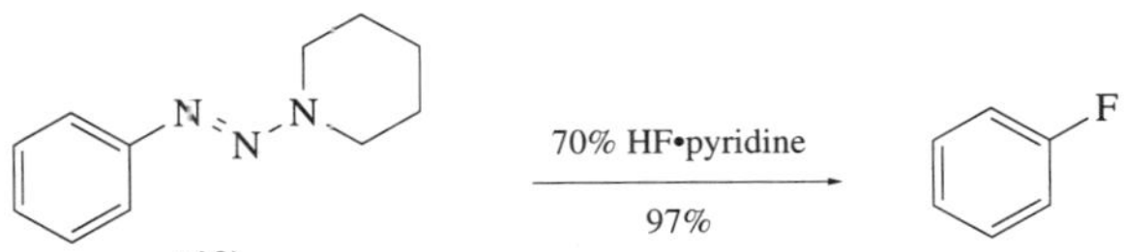

(**10**)

(37)

2.12.2.2.5 *Aryl fluorides by miscellaneous methods*

Aryl fluoroformates have been converted to the corresponding aryl fluorides by Ryan and co-workers ⟨85JFC(27)263⟩. The decarboxylative reaction occurred at 300 °C with a platinum on alumina catalyst. Stavber *et al.* have reported that hydroxyalkyl substituents on an aryl ring can be replaced by a fluorine simply by treatment with caesium fluoroxysulfate (Equation (38)) ⟨92CC274⟩. Seven examples were reported with yields in the range 60–85%. Finally, aryl fluorides have been synthesised by reaction of cyclopentadienes with fluorocarbenes under phase transfer conditions ⟨89IZV1935⟩. The aryl fluorides were produced in yields of 45–63%.

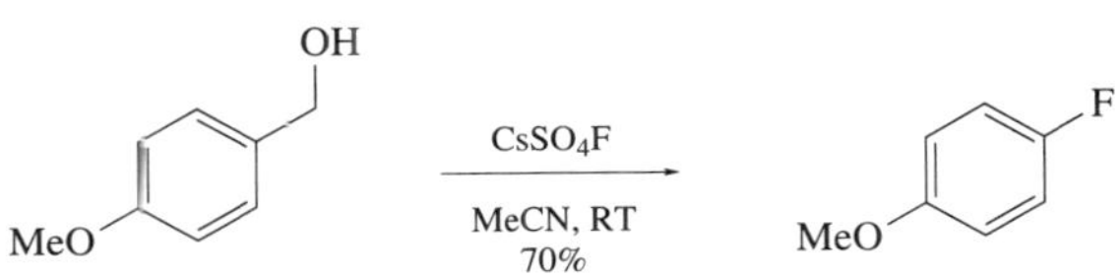

(38)

2.12.3 CHLORIDES

2.12.3.1 Vinyl Chlorides

2.12.3.1.1 *Vinyl chlorides by elimination reactions*

Vinyl chlorides have been synthesised by the elimination of vicinal dihalides under basic conditions ⟨51JA3329, 79HCA635⟩. The stereochemistry of the vinyl chloride derived by eliminating the sterically crowded dichloride (**11**) can be controlled to a large extent by the choice of solvent, varying from 95:5 to 7:93 (*E*):(*Z*) as shown in Scheme 5 ⟨79HCA1194⟩. The vinyl chloride (**13**) has been made by elimination of hydrogen chloride from the dichloride (**12**). Further treatment with potassium *t*-butoxide gave 1-chlorocyclopentene (**14**) (Scheme 6) ⟨71JOC1024⟩. Vinyl chlorides have also been synthesised by the elimination of a silyl halide. For instance, the strained chlorocyclopropene (**16**) has been made by the elimination of trimethylsilyl chloride from the trimethylsilylcyclopropane (**15**)

(Equation (39)) ⟨75TL3383⟩. The thermal elimination of sulfoxides and selenoxides has also been used to make vinyl chlorides ⟨89HCA1008⟩. For example, chloromethylphenyl sulfoxide (**17**) has been alkylated and then thermally eliminated to give a 1-chloroalkene as a mixture of (*E*)- and (*Z*)-isomers (Scheme 7) ⟨79TL617⟩. Engman has reported that phenylselenium trichloride can be added across an alkene, and the resulting phenylselenium dichloride hydrolysed to a selenoxide and then eliminated to give a vinyl chloride ⟨87JOC4086⟩.

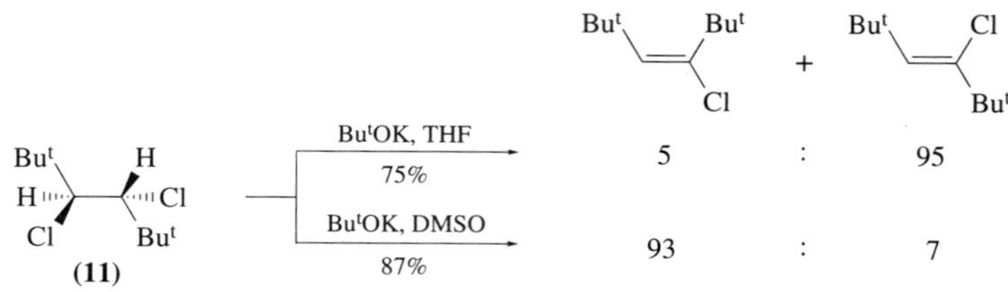

Scheme 5

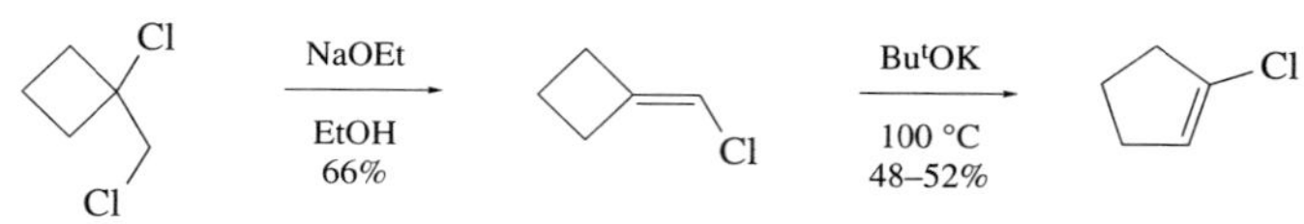

Scheme 6

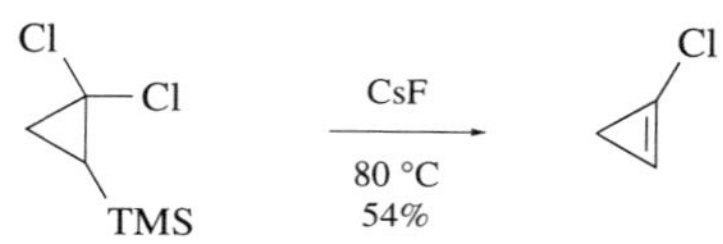

(39)

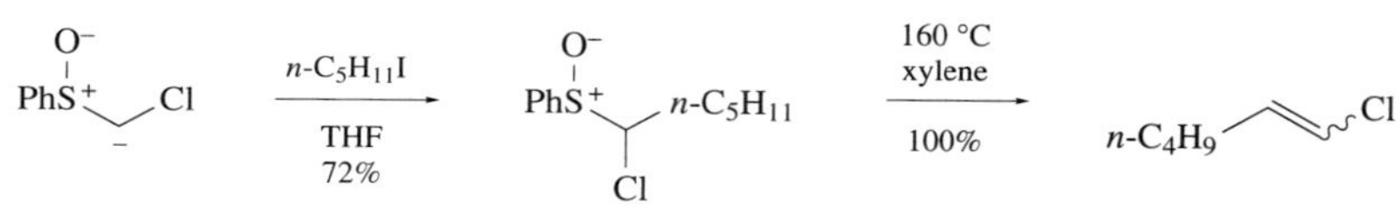

Scheme 7

2.12.3.1.2 *Vinyl chlorides from alkynes*

Hydrogen chloride can be added directly across an alkyne to give a vinyl chloride ⟨77JCS(P1)1797⟩. Alternatively, the alkyne can be hydroborated and then chlorinated. This reaction has been reported for terminal alkynes using chlorine as the chlorinating agent to afford (*Z*)-1-chloroalkenes in yields of 54–81% for the four examples given ⟨85TL279⟩. Alkynes have also been converted to vinyl chlorides by first being hydrozirconated to a vinylzirconium species followed by reaction with NCS ⟨75JA679⟩. Terminal alkynes gave 1-chloroalkenes with >98% regioselectivity. Disubstituted alkynes gave mixtures of the regioisomeric vinyl chlorides, though the selectivity could still be good if the two alkyne substituents were sufficiently sterically differentiated (Equation (40)). A formally related reaction added an alkyl cuprate across a terminal alkyne. The resultant vinyl cuprate was then treated with NCS to give a vinyl chloride in a yield of over 90%, with the alkyl and chloro groups added being *cis* to one another ⟨77RTC168⟩. A palladium catalyst has been used to effect the addition of an allyl chloride across an alkyne to give a 1-chloro-1,4-diene ⟨74TL1067, 76CL449, 78SC103⟩. Another approach to the synthesis of a vinyl chloride from an alkyne is to start with a 1-chloroalkyne which is then reduced. Lithium aluminum hydride has been used in this reduction to give (*E*)-vinyl chlorides in good yields (Equation (41)) ⟨79JA5101⟩.

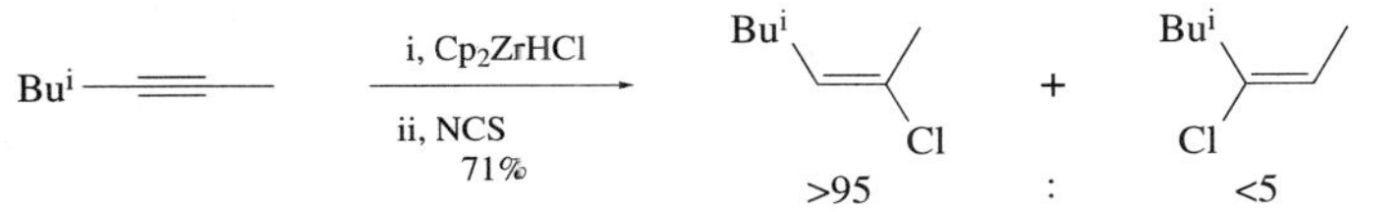

(40)

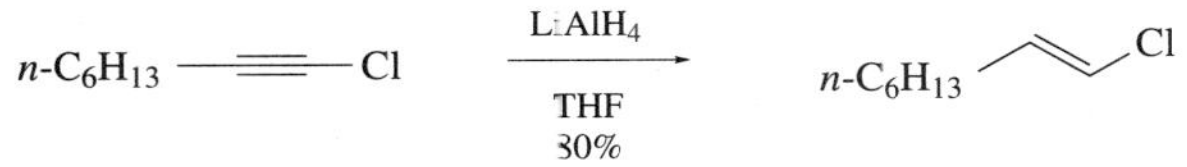

(41)

1,2-Dichloroalkenes can be prepared from alkynes by the addition of one equivalent of chlorine. This has been achieved using sulfuryl chloride ⟨81BCJ2843⟩ and iodobenzene dichloride ⟨75BSF2493⟩. Alkylphenyl alkynes have been dichlorinated with cupric chloride. The (*E*)-dichloroalkenes were obtained predominantly, in good yield (Equation (42)), except if a *t*-butyl alkyne was used when the (*Z*)-isomer was the major one formed ⟨75CC925, 77JCS(P1)676⟩.

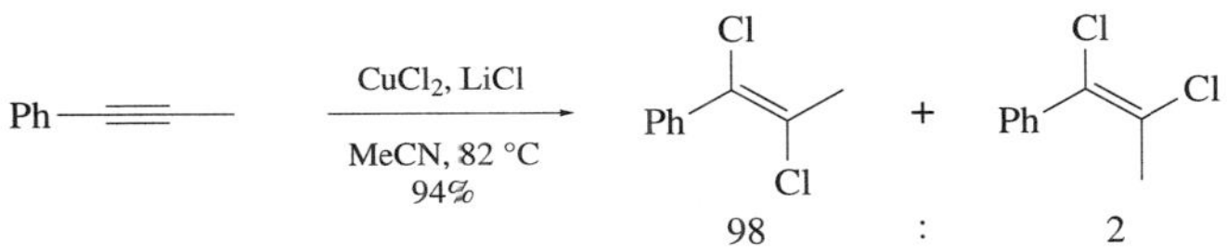

(42)

2.12.3.1.3 *Vinyl chlorides from vinyl metallic species*

Vinylsilanes can be converted to the corresponding vinyl chloride by treatment with chlorine followed by sodium methoxide-induced elimination ⟨74TL543⟩. Alternatively, the elimination step can be induced by fluoride ⟨82JA2659⟩. Vinylsilanes have also been converted to vinyl chlorides by the action of xenon difluoride, in the presence of lithium chloride (Equation (43)) ⟨90TL5633⟩. The chlorination of vinyl cuprates with NCS has been shown to yield vinyl chlorides, as illustrated in Equation (10) ⟨77TL3545⟩.

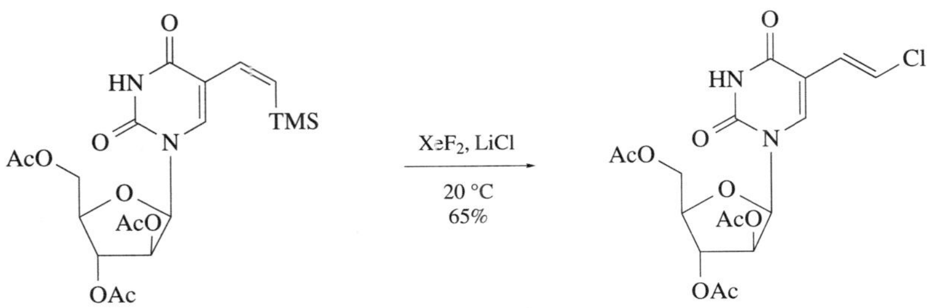

(43)

2.12.3.1.4 *Vinyl chlorides from carbonyl compounds*

Ketones can readily be converted to vinyl chlorides by a range of reagents. Phosphorus pentachloride ⟨54JCS3659⟩ and the Vilsmeier reagent ⟨79S533, 85JCS(P1)1127, 91JHC939⟩ have both been used successfully. Catechylphosphorus trichloride has also been used for this conversion ⟨63CB1387, 66CB1414⟩. Phosphorus pentachloride and catechylphosphorus trichloride give stereochemically complementary results with 2-methylcyclohexanone (**18**), allowing access to the two vinyl chlorides (**19**) and (**20**) (Scheme 8) ⟨85T1179⟩. Triphenylphosphine in carbon tetrachloride has also been used to prepare vinyl chlorides from acyclic and six-membered cyclic ketones ⟨72CC443⟩. Cyclopentanones and cyclobutanones, however, gave the products of dichloromethylenation. Other reagents for converting a ketone into a vinyl chloride include thionyl chloride ⟨88CCC110⟩ and dichloromethyl methyl ether in the presence of zinc chloride ⟨59CB83⟩. An α,β-unsaturated enone has been converted to a 1-chloro-1,3-diene with oxalyl chloride and oxalic acid in the presence of a saturated ketone ⟨62CJC818⟩.

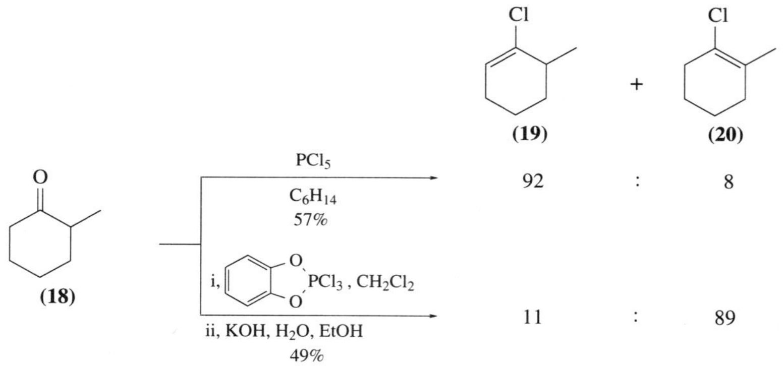

Scheme 8

Carbonyl compounds, particularly aldehydes, can also be converted to vinyl chlorides by a Wittig reaction ⟨61CB1373, 72CB1683⟩. A convenient way of achieving this transformation has been reported by Miyano *et al.* ⟨78CC446, 79BCJ1197⟩. They prepared the phosphonium salt by the reaction of triphenylphosphine with chloroiodomethane and then used potassium *t*-butoxide as the base to form the corresponding ylide. Excellent stereochemistry was achieved in some, but not all, cases. As can be seen in Equation (44), the reaction worked with ketones in addition to aldehydes. This transformation has also been carried out in a Wittig–Horner reaction, which also allowed the conversion of ketones to trisubstituted vinyl chlorides ⟨75S658, 83JOC2782⟩. Finally, vinyl chlorides have been synthesised by Corey *et al.* by chlorination of the betaine intermediate in a Wittig reaction and then allowing the reaction to proceed to completion ⟨70TL447⟩. The stereochemistry of the vinyl chloride was controlled by the choice of chlorinating agent, as is illustrated in Scheme 9.

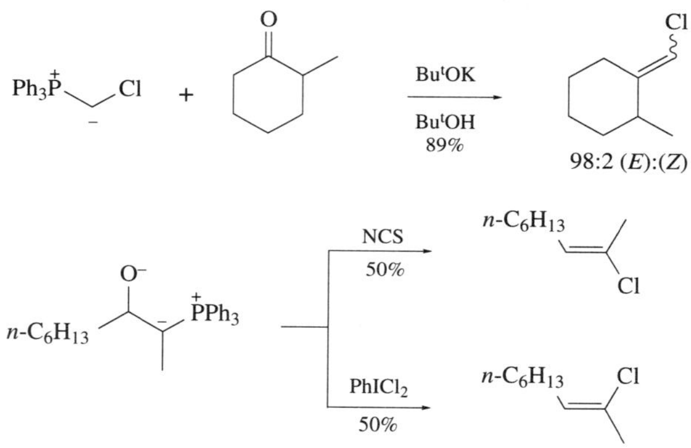

(44)

Scheme 9

Aldehydes can also be chloromethylenated in other ways. For instance, reaction with (bromochloromethyl)phenylmercury gave vinyl chlorides in good yield ⟨65JOM(3)337⟩. Takai *et al.* have achieved the transformation using chloroform and chromous chloride (Equation (45)) ⟨86JA7408⟩. The (*E*)-vinyl chlorides were the major product in all the reactions studied.

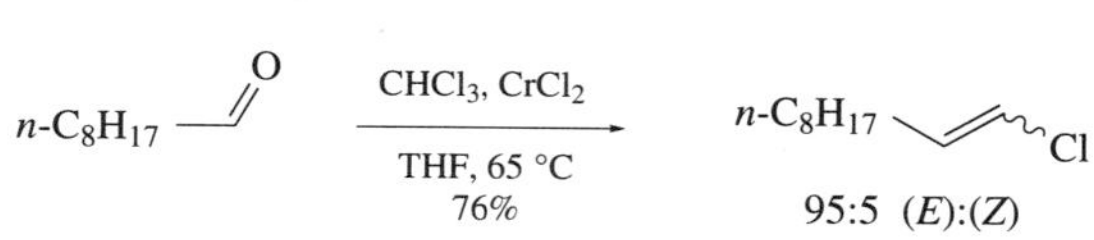

(45)

2.12.3.1.5 *Vinyl chlorides by miscellaneous methods*

Vinyl chlorides have been prepared from other vinyl halides. For example, vinyl bromides have been converted to the corresponding vinyl chlorides with retention of stereochemistry by cuprous chloride ⟨73BSF720, 81SC405⟩. It is also possible to make vinyl chlorides from chlorodiazirines either by rearrangement induced by photolysis ⟨85TL1651⟩, or by the coupling of unsymmetrical carbenes derived from the thermal decomposition of chlorodiazirines in the presence of diazo compounds ⟨87JOC1619⟩. The latter reaction, which is shown in Equation (46), has produced excellent yields of vinyl chlorides. Finally, the Ramberg–Backlund reaction has been used to prepare vinyl chlorides ⟨67JA4487⟩. The reaction only gave usable yields of the product when it was used to prepare 2,2-disubstituted vinyl chlorides (Equation (47)).

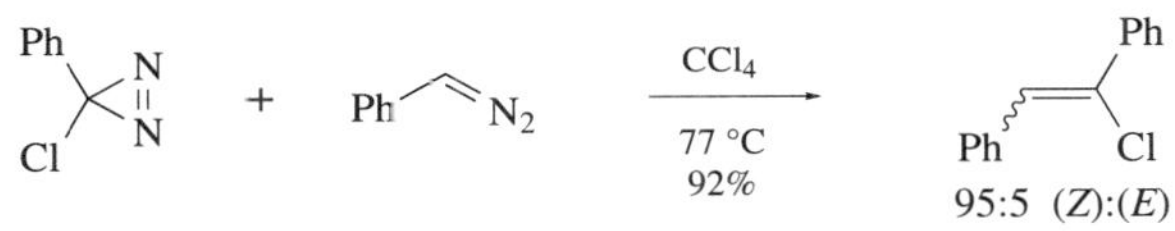

(46)

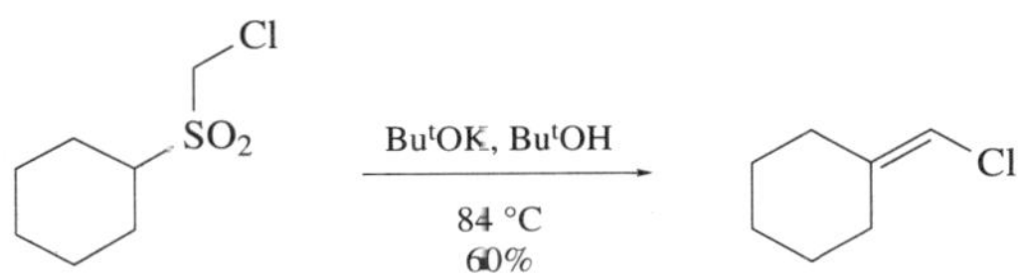

(47)

2.12.3.2 Aryl Chlorides

2.12.3.2.1 Aryl chlorides by electrophilic aromatic substitution

As described in the introductory section, electrophilic aromatic chlorinations can be carried out with chlorine itself and a suitable catalyst such as ferric chloride ⟨57JCS1823⟩. Chlorine has also been used with potassium carbonate and silica gel when somewhat greater *para* selectivity was achieved ⟨80CL1423⟩. Thus, reaction of an alkylbenzene gave the monochlorinated products with an *ortho*:*para* ratio in the range 46:54 to 16:84. The selective chlorination of the carbamate (**21**) has been achieved by *N*-chlorination followed by Orton rearrangement to give the aryl chloride (**22**) in excellent yield (Equation (48)) ⟨90JHC1549⟩. Chlorinations with chlorine gas can also be carried out in the solid phase ⟨74JOC1744⟩. Treatment of a thin film of phenol with chlorine for 10 minutes gave 2,4,6-trichlorophenol in 97% yield. Many other chlorinating agents have been used for electrophilic aromatic chlorinations. For instance, NCS has been used to chlorinate arenes and has been shown to give a mixture of regioisomers ⟨65JOC304⟩. Dichlorine monoxide has been reported to be a powerful chlorinating agent ⟨82JA4680⟩. With this reagent even the electron-deficient nitrobenzene (**23**) was monochlorinated in the presence of sulfuric acid to give the monochloride (**24**) (Equation (49)). A number of *N*-chlorinated reagents have been used to achieve aromatic chlorination. For example, trichloroisocyanuric acid chlorinated electron-rich aromatic rings in yields between 44% and 80% ⟨70JOC719⟩. *N*-Chlorobenzenedisulfonimide has been reported to be a more powerful chlorinating agent than chlorine ⟨69JOC3434⟩. It chlorinated toluene in less than 1 min at 25 °C to give a mixture of isomers. Iodobenzene bis(trifluoroacetate) and aluminum trichloride have been used to effect aromatic chlorinations ⟨82JCR(S)150⟩. Using this method, benzene was monochlorinated in 80% yield. Calcium hypochlorite has also been used to chlorinate electron-rich aromatic systems ⟨89SC799⟩. Although the reaction gave excellent yields for electron-rich arenes, benzene gave only a 22% yield of chlorobenzene and the reaction failed completely with nitrobenzene and other electron-poor arenes. Lastly, sulfuryl chloride has been used successfully for the monochlorination of phenols ⟨81JOC4486⟩. No dichlorination of the aryl ring was observed.

(48)

(**21**) (**22**)

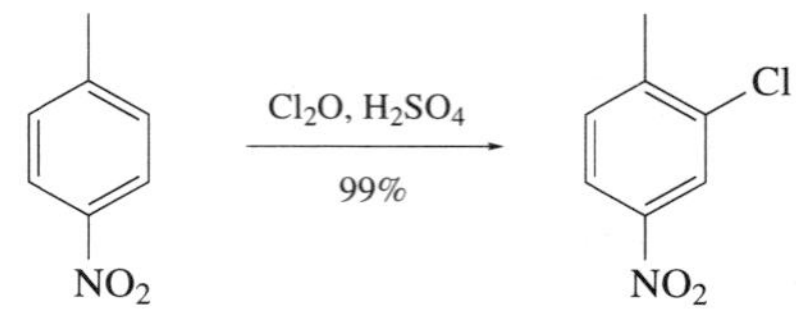

(23) **(24)** (49)

One of the key features in the electrophilic chlorination of aromatic rings is the control of the *ortho*:*para* ratio of products obtained. Selective *ortho* chlorination has been achieved, albeit in low yield, by the action of the chlorinating agent (**25**) on phenol (Equation (50)) ⟨82T2339⟩. Good *ortho* selectivities were also obtained when phenols were chlorinated with an *N*-chlorodialkylamine ⟨88TL1319⟩. However, yields of the monochlorinated products were not always good, with both unreacted starting material and dichlorinated products being present. Greater success has been achieved in regiospecific *para* chlorination. Yoshida and co-workers have shown that good *para* selectivities can be obtained by treatment of an arene with nitrobenzenesulfonyl peroxide in the presence of chloride ⟨88CL2017⟩. The *ortho*:*para* ratio was 96:4 with anisole, but only 55:45 with toluene. Broadly similar selectivities were seen when sulfuryl chloride and aluminum trichloride were used as the chlorinating agent ⟨85JOC2145⟩. *t*-Butyl hypochlorite with zeolites gave excellent yields of chlorinated products ⟨85S1157⟩. With toluene, 100% chlorination was achieved with an *ortho*:*para* ratio of 18:82, whereas with chlorobenzene the selectivity was 3:97 in a yield of 95%. Kodomori and co-workers have used cupric chloride supported on alumina to achieve excellent *para* selectivity in the chlorination of alkoxybenzenes ⟨87CL1901⟩. Anisole gave a 96% yield of monochlorinated products with an *ortho*:*para* ratio of 3:97 (Equation (51)). *N*-Chloroammonium salts have been used to chlorinate electron-rich arenes with excellent *para* selectivity ⟨83TL3117⟩. Anisole gave essentially 100% *para* selectivity (Equation (52)), though toluene reacted only slowly and without marked regioselectivity. Olah and co-workers have shown that chlorodimethylsulfonium chloride can be used to chlorinate aromatic rings with >98% *para* selectivity in yields in the range 75–87% ⟨86S868⟩. Finally, benzeneseleninyl chloride and aluminum trichloride have been used to chlorinate anisole with essentially 100% *para* selectivity ⟨88BCJ2226⟩. However, the regioselectivity of chlorination was poor with both toluene and phenol.

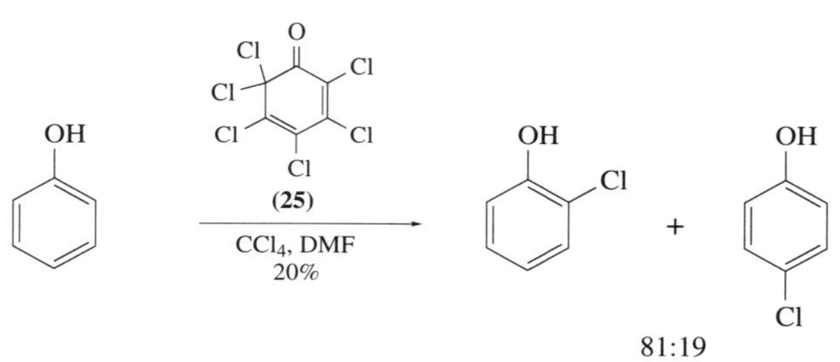

(50)

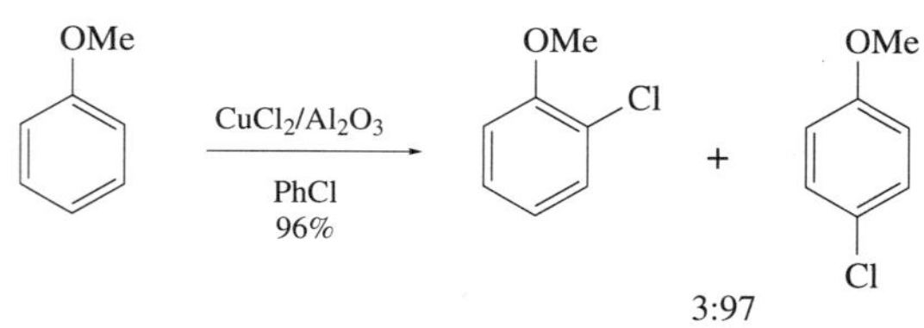

(51)

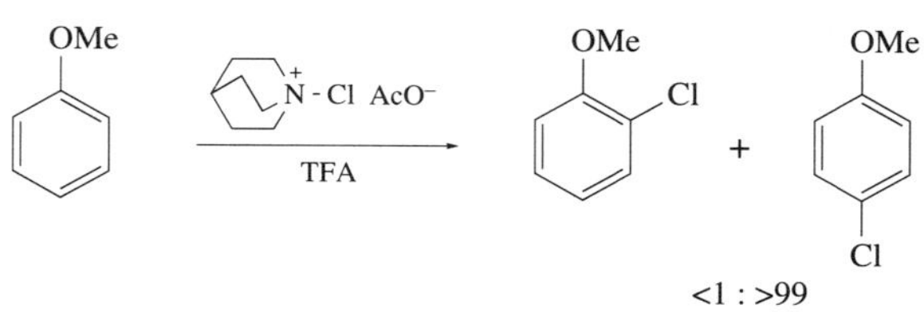

(52)

2.12.3.2.2 *Aryl chlorides from aryl metallic species*

The wealth of electrophilic aromatic chlorinations has led to there being comparatively few examples of the synthesis of aryl chlorides from aryl metallic species. However, aryl chlorides have been made by the chlorination of aryllithiums as in the following examples. Hexachloroacetone has been used as the chlorinating agent following a directed *ortho* metallation ⟨66CI(L)1635⟩. Katritzky *et al.* have used *p*-toluenesulfonyl chloride as the chlorinating agent to prepare 4-chlorodibenzothiophene ⟨90JHC1737⟩. Lastly, phenylacetylene has been *ortho* chlorinated by first doubly deprotonating it with *n*-butyllithium-potassium *t*-butoxide followed by chlorination with hexachloroethane ⟨81TL2495⟩.

2.12.3.2.3 *Aryl chlorides by halogen exchange and related reactions*

Halogen exchange is rarely used to make aryl chlorides, being more commonly encountered in the synthesis of less readily available aryl halides. Aryl chlorides have been synthesised from aryl bromides by treatment with cuprous chloride, though yields were moderate in most cases ⟨58JA1716⟩. Aryl chlorides can be made from aryl iodides in essentially the same way ⟨69JA6464⟩. Aryl iodides have also been converted to aryl chlorides by irradiation in carbon tetrachloride ⟨70JOC528⟩. For the 11 examples given, yields ranged from 51% to 96%.

Phenols have been converted to aryl chlorides with phosphorus pentachloride ⟨60DOK(133)855⟩, the Vilsmeier reagent ⟨86IJC328⟩ and phenylphosphorus tetrachloride ⟨90JOC3415⟩. For the last of these reagents yields of aryl chlorides were in the range 42–82%. The same reagent ($PhPCl_4$) has been used to transform nitrobenzenes into aryl chlorides (Equation (53)) ⟨88JOC2858⟩. Yields were generally good, varying from 67% to 94% for the 10 examples given.

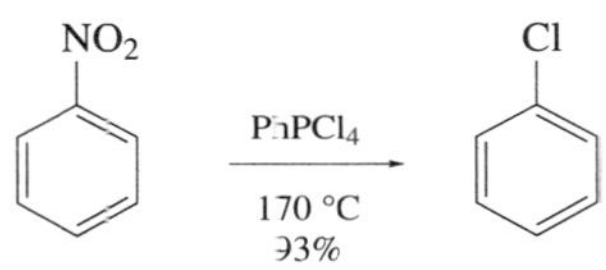

(53)

2.12.3.2.4 *Aryl chlorides from aryldiazonium salts*

Aryl chlorides are readily prepared by the Sandmeyer (or Gatterman) reaction ⟨32OSC(1)170⟩ and variations of it. Suzuki *et al.* have found that poly(ethylene glycol)–dichloromethane is an efficient solvent system for the Sandmeyer reaction ⟨84CC1523⟩. A one-step variation of the Sandmeyer reaction has been reported in which an aniline is treated with a cupric chloride–nitric oxide complex ⟨66RTC857⟩. Aryl chlorides were produced in yields of 30–90%. Another one-pot procedure involved treating an aniline with sodium nitrite and trimethylsilyl chloride in carbon tetrachloride, as shown in Equation (54) ⟨92TL3167⟩. For the 16 examples given yields were in the range 70–90%. An alternative diazotizing reagent to inorganic nitrite is *t*-butyl nitrite. Using this reagent Doyle *et al.* converted the aniline (**26**) to the aryl chloride (**27**) in 99% yield (Equation (55)) ⟨77JOC2426⟩. For other substrates yields ranged from 15% to 98%. *t*-Butyl thionitrite has also been used in the Sandmeyer reaction ⟨78TL4519, 80BCJ1065⟩. Yields were generally good.

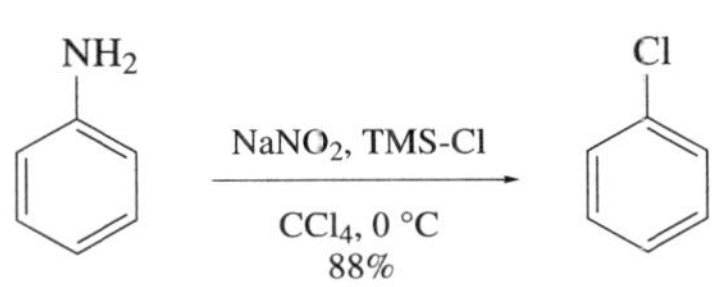

(54)

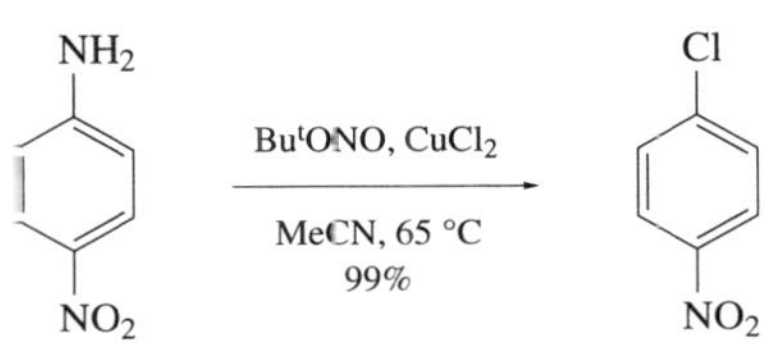

(55)

(26) **(27)**

2.12.3.2.5 Aryl chlorides by miscellaneous methods

Aryl chlorides have been made from benzoyl chlorides by a decarbonylation reaction. This has been achieved with catalytic $(Ph_3P)_3RhCl$ in generally excellent yields (Equation (56)) ⟨66TL1605⟩. The reaction has also been carried out with $(Ph_3P)_2Rh(CO)Cl$ ⟨66TL4713⟩ and palladium on charcoal or palladium chloride catalysts, again in excellent yields ⟨82TL371⟩. In a formally related reaction, arylsulfonyl chlorides have been converted to aryl chlorides with the loss of sulfur dioxide (Equation (57)) ⟨66TL3041⟩. Yields tended to be lower than in the related decarbonylation reaction.

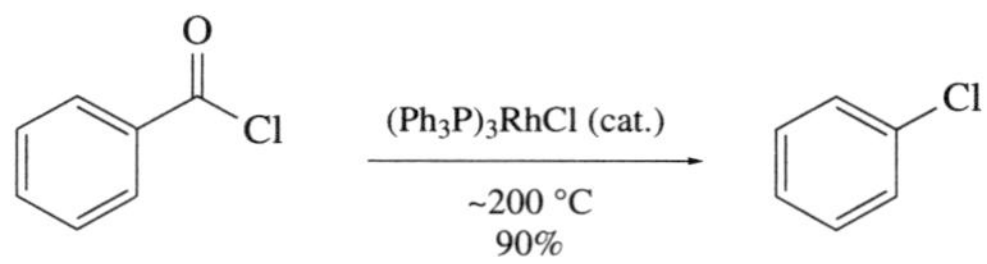

(56)

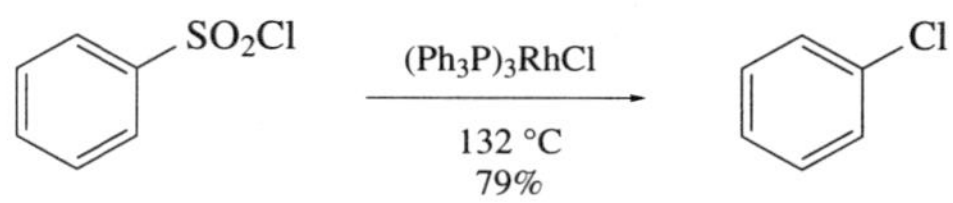

(57)

2.12.4 BROMIDES

2.12.4.1 Vinyl Bromides

2.12.4.1.1 Vinyl bromides by elimination reactions

Vinyl bromides are readily prepared by the elimination of vicinal dibromides. A 1,2-dibromoalkane has been eliminated to give the 1-bromoalkene in high yield ⟨66MI 212-01⟩. Vicinal dibromides within an alkyl chain have also been eliminated to vinyl bromides ⟨90JCS(P1)11⟩. The regiochemistry of the vinyl bromide formed can be difficult to control, though it is possible in some cases (Equation (58)). However, a mixture of double-bond isomers was formed. Elimination of 1-bromo-2-chlorocyclohexane has been shown to give predominantly 1-bromocyclohexene, though the corresponding vinyl chloride was also formed in significant amounts ⟨83JOC876⟩. Bases other than alkoxides can be used to effect the elimination. For instance, morpholine has been used to prepare cyclic vinyl bromides ⟨90S843⟩. The geometry of the elimination step has been studied by Tremmling *et al.* ⟨78JOC3076⟩.

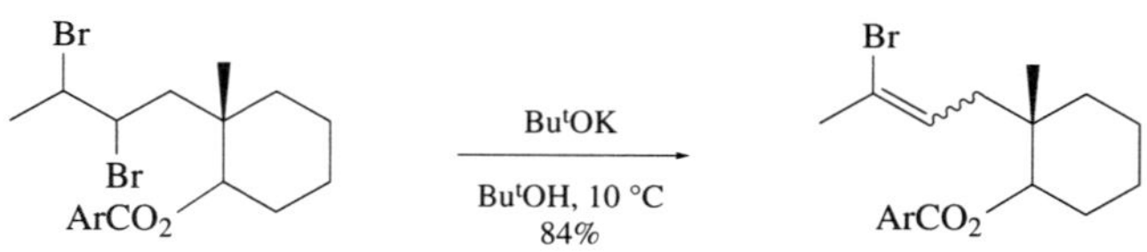

(58)

Decarboxylative eliminations can also be used to prepare vinyl bromides. The method has been used to prepare bromomethylenecycloalkanes in good yields ⟨65JOC2208⟩. It has also been used to prepare 2-bromopropene ⟨89CB2213⟩ and (*Z*)-1-bromopropene ⟨91JOC4066⟩. The latter example proceeded by a stereospecific addition of bromine to (*E*)-crotonic acid (**28**) to give the dibromide (**29**), followed by a stereospecific decarboxylative elimination to the vinyl bromide (**30**), as shown in Scheme 10.

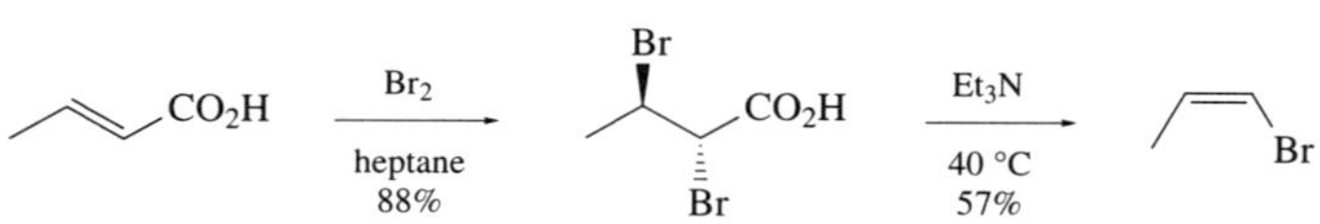

(28) **(29)** **(30)**

Scheme 10

The elimination of trimethylsilyl bromide has also been used in the stereospecific synthesis of vinyl bromides. Addition of bromine to a (*Z*)-vinyl silane followed by methoxide-induced elimination gave the (*E*)-vinyl bromide in 80% yield ⟨74TL543⟩. Hudrlik *et al.* have reported that an (*E*)- or (*Z*)-α,βepoxysilane can be converted stereospecifically to a (*Z*)- or (*E*)-vinyl bromide, respectively ⟨77JA1993⟩. The reaction proceeded by opening of the epoxide with HBr followed by the boron trifluoride-induced elimination of trimethylsilyl bromide.

Vinyl bromides have also been synthesised by the treatment of α-phenylsulfonyl carbanions with a bromolithiocarbenoid followed by elimination of the phenylsulfonyl group ⟨89JOM(379)201⟩. (*E*)-Vinyl bromides were produced selectively in good yields. The addition of phenylselenyl bromide to an alkene followed by oxidation to the selenoxide and thermal elimination has been shown to give vinyl bromides ⟨77TL3909⟩. As the initial addition of the phenylselenyl bromide to a terminal alkyne can be controlled to give either of the two regioisomers selectively, it is possible to make either a 2-bromoalkene (Scheme 11) or a 1-bromoalkene (providing, in this case, that elimination to give an allyl bromide is not possible).

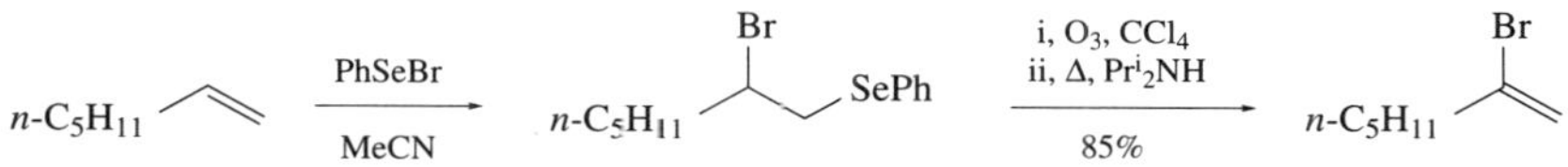

Scheme 11

2.12.4.1.2 *Vinyl bromides from alkynes*

Hydrogen bromide can be added across an alkyne directly to give a 2-bromoalkene. Grob and Cseh have added hydrogen bromide to phenylalkynes to give α-bromostyrenes in yields of 75–85% for five examples ⟨64HCA194⟩. The addition can also be effected using tetraethylammonium hydrogen bromide ⟨80S805⟩. In this reaction terminal alkynes were converted to 2-bromoalkenes in yields from 65% to 91% for the eight examples given. 2-Bromoalkenes can also be made by the addition of hydrogen bromide to trimethylsilyl alkynes ⟨74JOC3307⟩. 1-Bromoalkenes have been synthesised by the radical addition of hydrogen bromide to a terminal alkyne ⟨36JA1806, 63OR(13)150⟩. A few percent of the 2-bromoalkene was also formed. Alternatively, 2-bromoalkenes can be made by treating an alkyne with a *B*-bromoborane (e.g., *B*-Br-9-BBN), followed by protonolysis of the vinylborane (Equation (59)) ⟨83TL731⟩. Hydroboration can also be used to prepare (*Z*)-1-bromoalkenes ⟨83SC1027, 89JOC6068⟩. Brown *et al.* have reported that both (*E*)- and (*Z*)-vinyl bromides are available via hydroboration ⟨67JA4531, 73JA6456⟩. Hydroboration of 1-bromoalkynes has been used by Brown and Somayaji to make (*E*)-1-bromoalkenes ⟨84S919⟩. 1,2-Dialkylvinyl bromides have been synthesised by Masuda *et al.* by the hydroboration of 1-bromoalkenes ⟨78CL665⟩. Both (*Z*)- and (*E*)-isomers were obtained selectively by altering the conditions, as illustrated in Scheme 12. Brown *et al.* have also reported a stereoselective synthesis of (*E*)- and (*Z*)-disubstituted vinyl bromides using organoboron chemistry ⟨86S480⟩.

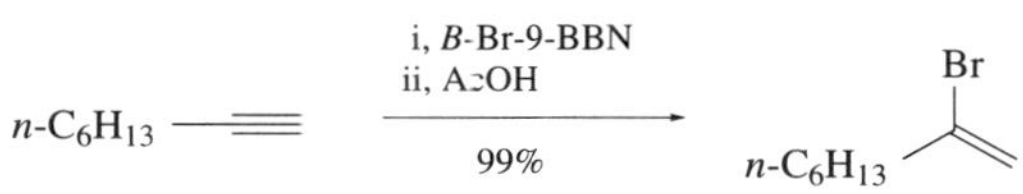

(59)

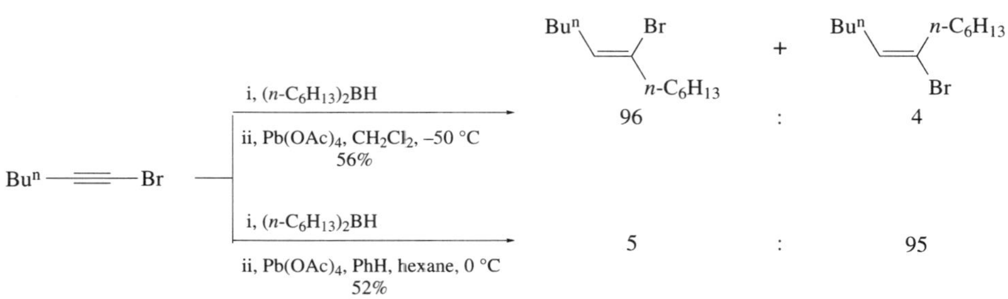

Scheme 12

Other metals besides boron have been used to effect additions to triple bonds which lead to vinyl bromides. Tributyltin hydride has been added to an alkyne and the resulting vinylstannane reacted with bromine to give an (*E*)-1-bromoalkene ⟨84JA5734, 86S453⟩. Vinyl bromides have also been synthesised by hydrozirconation of an alkyne followed by bromination with NBS ⟨75JA679⟩.

Normant *et al.* have shown that the vinyl cuprate (**31**), resulting from the addition of an alkyl cuprate to an alkyne, could be brominated via a vinylmercury species (Scheme 13) ⟨74S803⟩. A vinyl cuprate has also been brominated directly with NBS ⟨77RTC168⟩. It has been shown that it is possible to add allyl bromide across an alkyne with a palladium catalyst ⟨88TL5811⟩. The allyl bromide added *cis* to give 1-bromo-1,4-dienes in yields of 62–87%.

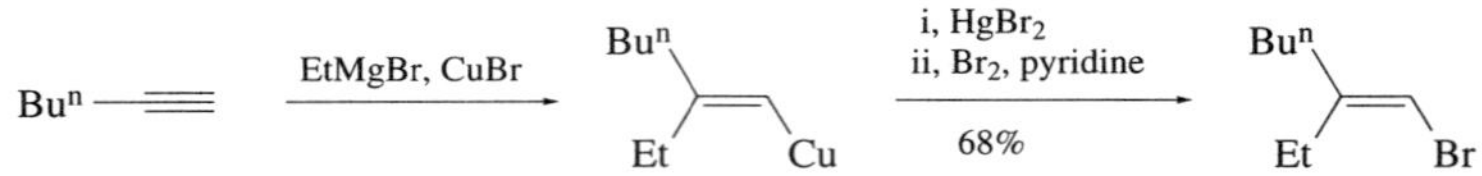

(31)

Scheme 13

The bromide-catalysed addition of bromine to phenylacetylene resulted in α,β-dibromostyrene ⟨70TL1629⟩. The reaction gave the (*E*)-stereoisomer as the major product. It is also possible to prepare other 1,2-dihaloalkenes. Thus, reaction of an alkyne with hydrogen fluoride and NBS gave a 1-bromo-2-fluoroalkene in which the bromine and fluorine had added *trans* ⟨70JOC1703⟩. The bromoboration of a 1-chloroalkyne has been used to prepare 1-chloro-2-bromoalkenes ⟨83S1005⟩. Moderate selectivity for either the (*Z*)- or (*E*)-isomers was achieved using this method. Finally, tellurium tetrachloride followed by NBS and aluminum trichloride has been used to prepare 1-bromo-2-chloroalkenes ⟨79CL1357⟩. The (*Z*)-dihaloalkenes were produced in reasonable yields (Equation (60)).

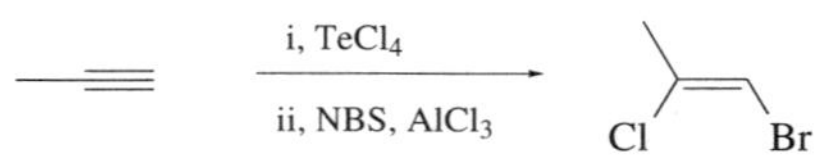

(60)

2.12.4.1.3 *Vinyl bromides from vinyl metallic species*

Several different vinyl metallic species have been used in the synthesis of vinyl bromides. Vinylsilanes have been converted to vinyl bromides with inversion of stereochemistry, as already described. They have also been reacted with cyanogen bromide and aluminum trichloride to give a vinyl bromide with retention of configuration ⟨77TL3317⟩. Furthermore, vinylsilanes have been transformed to vinyl bromides by the action of xenon difluoride and lithium bromide ⟨90TL5633⟩. Both vinylgermanes ⟨84TL3221, 85CL53⟩ and, as already mentioned, vinylstannanes ⟨84JA5734, 86S453⟩ have been converted to vinyl bromides by treatment with bromine. It is possible to prepare vinyl bromides from vinyllithiums by, for instance, treatment with bromine, as was illustrated in Equation (9) ⟨72T4883, 86JA2445⟩. Finally, a vinylselenide has been converted to the corresponding vinyl bromide by the action of bromine in benzene in a yield of 64% ⟨82TL3411⟩.

2.12.4.1.4 *Vinyl bromides from carbonyl compounds*

Ketones can be converted to vinyl bromides possessing the same carbon skeleton by a variety of different brominating agents. For instance, 1-tetralone gave 1-bromo-3,4-dihydronapthalene in 99% yield when treated with trichloroacetyl bromide ⟨71C59⟩. Phosphorus tribromide has also been used to synthesise vinyl bromides from ketones ⟨91TL827⟩. Another approach to this conversion, in which a ketone was first transformed to a ketal (with catechol), which was then treated with phosphorus tribromide to give the vinyl bromide in yields from 60% to 88%, has been reported ⟨86S122⟩.

Vinyl bromides can be synthesised from aldehydes by a Wittig reaction. The brominated ylide has been produced by bromine–lithium exchange from a 1,1-dibromoalkylphosphonium salt. Subsequent reaction of the ylide with aldehydes gave 1-bromoalkenes in moderate yields as a mixture of stereoisomers ⟨78JOC2833⟩. A brominated ylide has also been prepared by bromination of the boron-substituted ylide (**32**) and subsequent deprotonation with butyllithium ⟨86TL1995⟩. Reaction with aldehydes gave vinyl bromides in reasonable yields as a mixture of stereoisomers (Equation (61)). Aldehydes can also be converted to 1-bromoalkenes by the action of the ylide prepared from bromomethyltriphenylphosphonium bromide and potassium *t*-butoxide ⟨80TL4021⟩. Yields were in the range 47–81% for the five examples given. Both aldehydes and ketones have been bromomethylenated by the addition of dibromomethyllithium followed by reductive elimination on treatment with zinc in acetic acid ⟨81TL3745⟩. Yields were generally good, though a mixture of stereoisomers was produced (Equation (62)).

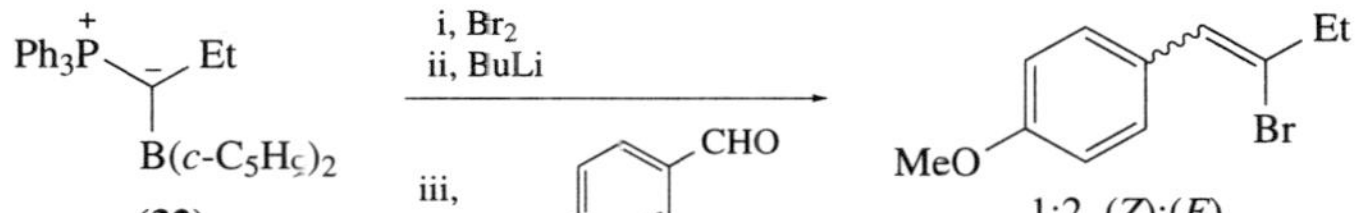

(32) (61)

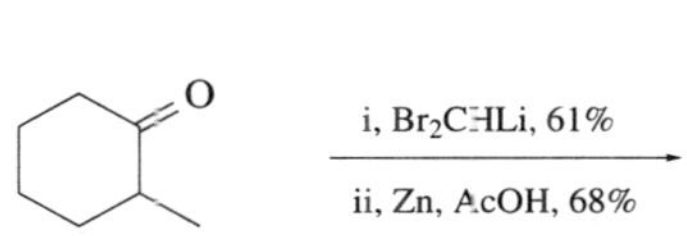

(62)

Aldehydes have also been converted to vinyl bromides by treatment with bromoform and chromous chloride in an analogous reaction to that described for the synthesis of vinyl chlorides (Equation (45)) ⟨86JA7408⟩. Generally good yields of the vinyl bromide were produced, with the (*E*)-isomer predominant.

2.12.4.1.5 *Vinyl bromides by miscellaneous methods*

Vinyl bromides have been prepared from α,β-unsaturated carboxylic acids by a brominative decarboxylation. (See Section 2.12.4.2.5 for details of the Hunsdiecker reaction.) The synthesis of bromomethylenecycloalkanes has been accomplished by treating the sodium salt of an acid (e.g., (**33**)) with bromine (Equation (63)) ⟨65JOC2208⟩. Yields of the resulting vinyl bromides were usually high. A stereoselective version of this reaction has been reported ⟨87TL3857⟩. The reaction has also been carried out using NBS as the brominating agent ⟨86JMC213⟩. A similar transformation has been achieved by Paquette *et al.*, but in a different manner ⟨91JOC6199⟩. An α,β-unsaturated acid was first transformed to its hydroxamate ester (**34**), which was then heated in bromotrichloromethane in the presence of 2,2′-azobisisobutyronitrile (AIBN) to give vinyl bromides in yields in the range 37–96% (Equation (64)).

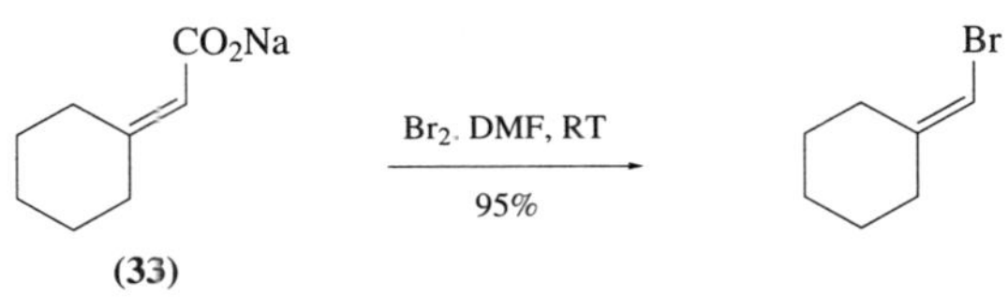

(33) (63)

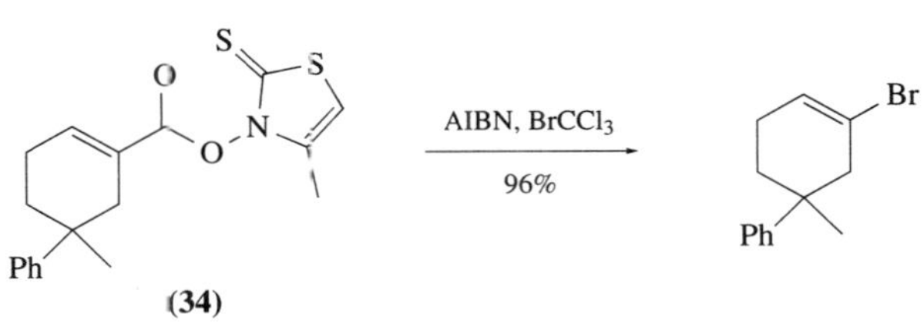

(34) (64)

2.12.4.2 Aryl Bromides

2.12.4.2.1 *Aryl bromides by electrophilic aromatic substitution*

Aryl bromides can be prepared by reaction of electron-rich aromatic systems with bromine without the aid of a catalyst ⟨53JA3596, 85HCA813⟩. With electron-rich arenes *ortho*/*para* selectivity is observed, as in the case of electrophilic aromatic chlorinations. Acid-sensitive arenes have been successfully brominated by bromine in the presence of trimethyl phosphate ⟨76S621⟩. 1,3,5-Tris(*t*-butyl)benzene (**35**) was brominated in a yield of 55% (Equation (65)).

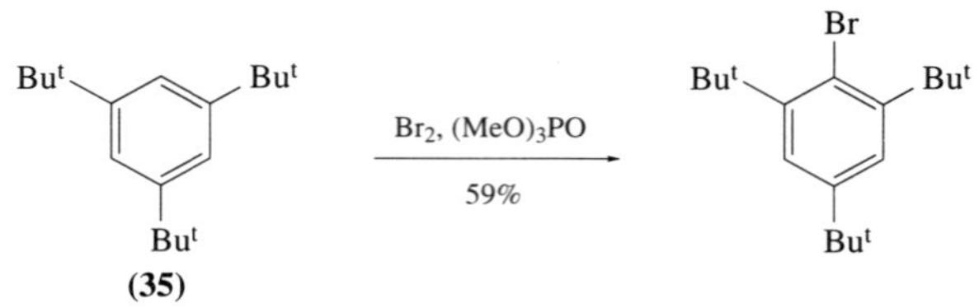

(65)

Bromine in the presence of silica gel has been used to achieve aromatic brominations in good yields ⟨74TL3395⟩. Many other brominating agents have been used in electrophilic aromatic brominations. Woodward used *N*-bromoacetamide to brominate α-oestradiol in high yield ⟨40JA1625⟩. *N,N*-Dibromophenylsulfonamide has been reported to be a good brominating agent for electron-rich arenes ⟨67BCJ418⟩. In a series of papers, Kajigaeshi *et al.* have reported on brominations using benzyltrimethylammonium tribromide ⟨87BCJ4187, 87CL627, 88BCJ597, 89BCJ439, 89BCJ3373⟩. Using this reagent, the bromination of phenols, alkyl benzenes and anilines was achieved in generally excellent yields, frequently in excess of 90%. Trifluoroacetyl bromide, produced from iodobenzene bis(trifluoroacetate) and bromine, has been used for the bromination of aromatic systems ⟨82JCR(S)150⟩. Bromine monofluoride (without a catalyst present) brominated a wide range of arenes ⟨88JOC5545⟩. Even electron-deficient substrates, such as ethyl benzoate, gave good yields of the brominated products (Equation (66)).

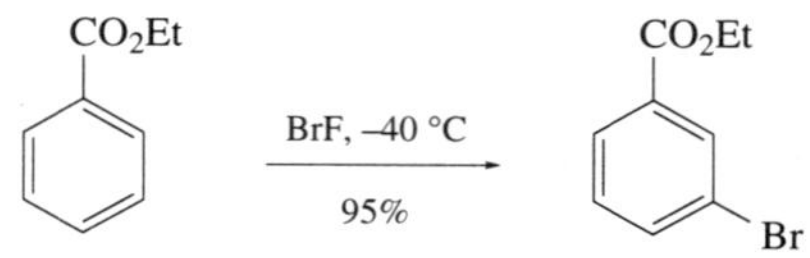

(66)

Cupric bromide supported on alumina has been used as a brominating agent for arenes ⟨88JOC2093⟩. Yields of the monobrominated products were usually excellent, though a mixture of *ortho* and *para* isomers was formed. McKillop *et al.* have shown that bromine in the presence of thallium triacetate is an effective aromatic brominating agent ⟨69TL1623, 72JOC88⟩. A wide range of activated aromatic systems were examined and aryl bromides were produced in yields from 60% to 93%. Normally, as has already been stated, phenols brominate *ortho*/*para*, but it has been shown that *meta* bromination can be accomplished if the reaction is performed in a super-acid (antimony pentafluoride–hydrogen fluoride) ⟨80CC110⟩. Yields were in the range 45–85% for the seven examples reported.

Electron-deficient arenes can also be brominated, but more forcing conditions are needed and *meta* products result. Acetophenone has been converted to *m*-bromoacetophenone by bromine with excess aluminum trichloride in 73% yield ⟨58JOC1412⟩. Bromination of the methyl group was not observed. Deactivated aromatic rings have also been brominated using bromine and mercuric oxide in the presence of sulfuric acid ⟨88JOC1799⟩. Other systems which have been used to brominate deactivated arenes include NBS in sulfuric acid ⟨65JOC304⟩, bromine and silver sulfate in sulfuric acid ⟨50JCS573⟩ and potassium bromate in sulfuric acid ⟨81JOC2169⟩.

Selective *ortho* bromination of a phenol has been reported by Pearson *et al.* ⟨67JOC2358⟩. Using *N*-bromo-*t*-butylamine at −70 °C, phenol was *ortho* brominated in 60% without contamination by the *para* isomer. Two equivalents of this reagent gave 2,6-dibromophenol in 87% yield. Another approach has been reported by Kim and Chang ⟨86CC1159⟩. They found that treatment of the diazo compound (**36**) with two equivalents of bromine gave *o*-bromophenol in 68% yield with only 2% of the *para* isomer (Equation (67)). It is also possible to prepare *p*-bromophenol selectively from the diazo compound (**36**) using 7 equivalents of bromine ⟨86CC1159⟩. No product of *ortho* bromination was observed in this case, in direct contrast to when only two equivalents of bromine were used. Phenol has been selectively *para* brominated in 93% yield by simply reacting with bromine in 1,2-dichloroethane ⟨58JOC280⟩. Both phenols and aromatic amines have been brominated exclusively *para* by tetrabutylammonium tribromide ⟨89CJC2061⟩. Yields were in the range 84–98% for 15 examples. Anilines have also been selectively *para* brominated using 2,4,4,6-tetrabromocyclohexa-2,5-dienone ⟨71JCS(C)3652⟩ and hexabromocyclopentadiene ⟨82CC778⟩. With the latter reagent, other electron-rich aromatic rings were also *para* brominated.

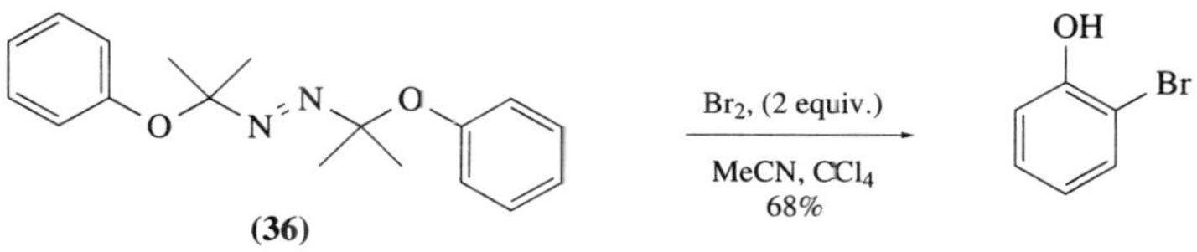

(36)

(67)

2.12.4.2.2 *Aryl bromides from aryl metallic species*

Arylsilanes are readily converted to aryl bromides in high yields by treatment with bromine ⟨75TL4539, 84TL483⟩. However, very activated arenes sometimes undergo normal electrophilic aromatic bromination rather than *ipso* substitution of the silyl substituent ⟨89JOC4372⟩. Arylstannanes have also been converted to aryl bromides ⟨82CC625⟩. Aryl bromides have been produced when Grignard reagents and phenyl aryllithiums were treated with bromine ⟨19JA287, 90JHC2151⟩.

2.12.4.2.3 *Aryl bromides by halogen exchange and related reactions*

Halogen exchange is rarely used to prepare aryl bromides. However, phenols have been converted to aryl bromides by reaction with triphenylphosphine dibromide ⟨64JA964, 67JOC1607⟩.

2.12.4.2.4 *Aryl bromides from aryldiazonium salts*

The Sandmeyer reaction has been used with great success to prepare aryl bromides ⟨32OSC(1)136, 45HCA628⟩. Variations include the use of poly(ethylene glycol)–dichloromethane ⟨84CC1523⟩ and dimethyl sulfoxide ⟨69OPP221⟩ as solvent and the use of mercuric bromide rather than cuprous bromide ⟨32CB1605, 41JA2847⟩. One-pot procedures for converting an aniline directly to an aryl bromide have also been reported. A cupric bromide–nitric oxide complex has been used to effect this transformation in yields in the range 6–96% ⟨66RTC857⟩. Alternatively, sodium nitrite and trimethylsilyl bromide in carbon tetrachloride gave excellent yields of aryl bromides ⟨92TL3167⟩. (See Equation (54) for the chloro analogue of this reaction.) *t*-Butyl nitrite and *t*-butyl thionitrite have both been used in place of sodium nitrite as described for the synthesis of aryl chlorides ⟨77JOC2426, 78TL4519, 80BCJ1065⟩.

2.12.4.2.5 *Aryl bromides by miscellaneous methods*

Carboxylic acids can be decarboxylatively brominated by treating their silver salts with bromine in the well-known Hunsdiecker reaction ⟨42CB291⟩, which has been reviewed by Wilson ⟨57OR(9)332⟩ and Johnson and Ingham ⟨57CRV219⟩. Electron-deficient aromatic acids undergo the reaction, though in variable yield, but benzoic acids with electron-donating substituents are brominated on the aromatic ring. The Cristol–Firth modification of the Hunsdiecker reaction using mercuric oxide ⟨61JOC280⟩ has been used to prepare aryl bromides in a more convenient manner than the original silver salt procedure ⟨65JOC415, 72JOC664⟩. A further improvement has been reported by Meyers and Fleming in which the Cristol-Firth modification was photoassisted ⟨79JOC3405⟩. This resulted in yields of aryl bromides in the range 75–96% for eight examples (Equation (68)). Barton and coworkers have achieved the same overall transformation, but by a different approach ⟨87T4321⟩. Benzoic acids were first converted to thiohydroxamic esters (**37**), which were then heated with AIBN in bromotrichloromethane to give aryl bromides in yields from 43% to 85% for the 11 examples given (Equation (69)). Finally, in an analogous reaction to that used to prepare aryl chlorides (Equation (56)), benzoyl bromides were decarbonylated with $(Ph_3P)_2Rh(CO)Cl$ to give aryl bromides ⟨66TL4713⟩.

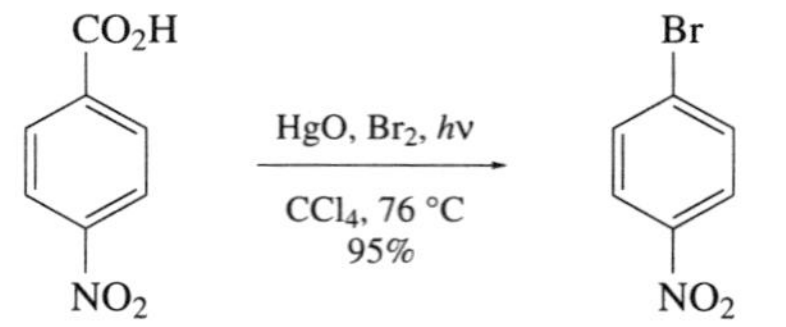

(68)

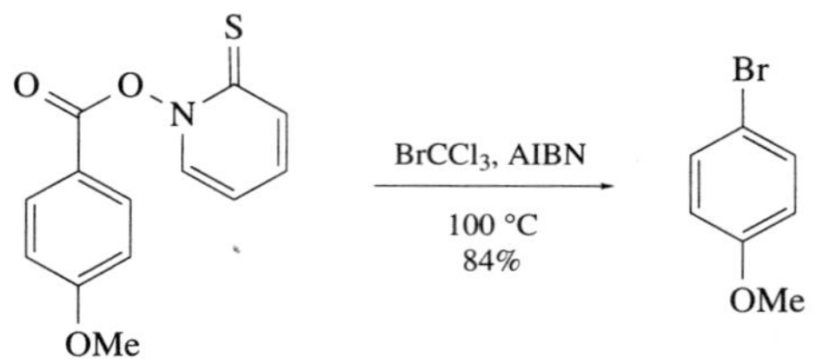

(37)

(69)

2.12.5 IODIDES

2.12.5.1 Vinyl Iodides

2.12.5.1.1 Vinyl iodides by elimination reactions

Vinyl iodides are rarely produced by elimination of vicinal diiodides as these are unstable and tend to decompose to an alkene and iodine. However, vinyl iodides have been made by a decarboxylative elimination ⟨90JCS(P1)47⟩. Thus, the diester (**38**) was hydrolysed and decarboxylatively eliminated to give the iodopropenoic acid (**39**) (Equation (70)). In addition, vinyl iodides have been prepared by the potassium iodide-induced elimination of trimethylsilyl trifluoroacetate ⟨74TL543⟩. Julia and co-workers have shown that vinyl iodides can be prepared by the action of lithium diiodomethane on an α-lithioalkylphenyl sulfone followed by elimination of the sulfone (Equation (71)) ⟨90BSF275⟩. The vinyl iodides were produced as a mixture of (*E*)- and (*Z*)-isomers.

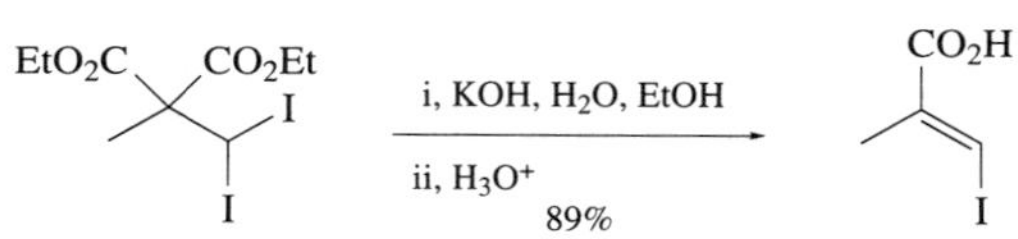

(38) (39)

(70)

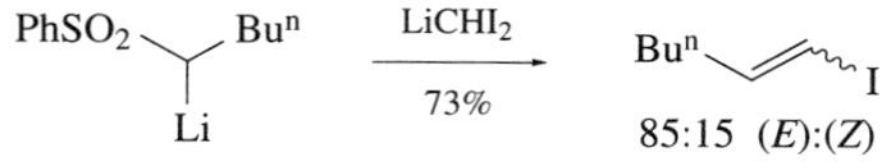

(71)

2.12.5.1.2 Vinyl iodides from alkynes

Vinyl iodides have been prepared by the addition of hydrogen iodide (prepared *in situ* from iodine and alumina) to an alkyne ⟨87TL4497⟩. The 2-iodoalkenes were prepared in yields from 36% to 85% for eight examples. 1-Iodoalkenes have been prepared by reaction of a terminal alkyne with a *B*-iodoborane (*B*-I-9-BBN) followed by protonolysis of the resulting vinylborane ⟨83TL731⟩. (See Equation (59) for the bromo analogue of this reaction.) 1-Iodoalkenes have also been synthesised by the reduction of 1-iodoalkynes with diimide ⟨75JOC1083, 88CC971⟩. The (*Z*)-stereoisomer was produced selectively. Another synthesis of (*Z*)-1-iodoalkenes has been reported by Brown *et al.* ⟨89JOC6068⟩. (*E*)-Vinylboranes were converted to (*Z*)-1-iodoalkenes in yields in the range 78–83% and virtually 100% stereospecificity. (*E*)-1-Iodoalkenes have been prepared by the hydrometallation of a terminal alkyne followed by iodination. Hydroboration (frequently using catechol borane) has been extensively used in this reaction with iodine ⟨73JA5786⟩ and iodine monochloride ⟨81SC247⟩ as the iodinating agents. Lithium borohydride in the presence of ethyl acetate has also been used as the hydroborating agent ⟨84JOC4822⟩. Other metals besides boron can be used in this reaction. Hydroalumination or hydrozirconation of an alkyne followed by iodination with iodine gave good yields of vinyl iodides ⟨67JA2753, 75JA679, 79JA4749⟩. Corey *et al.* have converted an internal alkyne into an (*E*)-vinyl iodide via hydrozirconation ⟨78JA4618⟩. In this example there was sufficient difference between the two substituents on the alkyne to give complete regiocontrol. A particularly common transformation of an alkyne into a vinyl iodide is in the case of a propargyl alcohol. Corey *et al.* showed that reaction with lithium aluminum hydride, in the presence of either aluminum trichloride or sodium methoxide, followed by the addition of iodine gave (*Z*)-2-iodo- and (*Z*)-3-iodoallyl alcohols, respectively (Scheme 14) ⟨67JA4245⟩. In a related reaction using hydromagnesiation of a propargyl alcohol, (*E*)-3-iodoallyl alcohols were produced ⟨81CC718⟩.

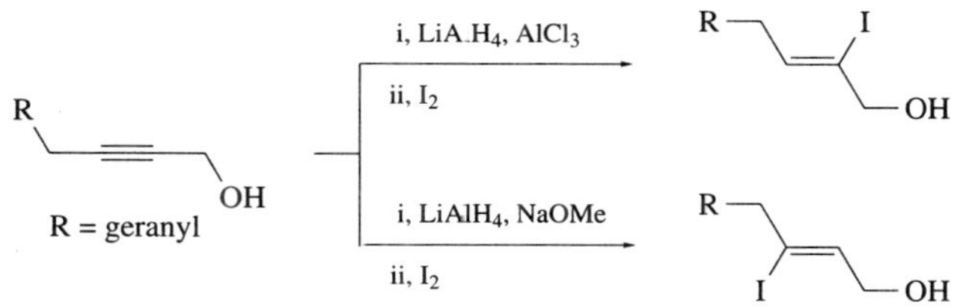

Scheme 14

As well as adding hydrogen iodide across an alkyne, many metallic species have been used to add an alkyl iodide. Organocuprates have been shown to add *cis* to terminal alkynes to give vinyl cuprates. Subsequent reaction with iodine gave 1-iodo-2-alkylalkenes (Equation (72)) ⟨71TL2583, 74JOM(77)269, 84OS(62)1⟩. Alkyne itself has been used in this reaction to give (*Z*)-1-iodoalkenes. Methylzirconation of an alkyne followed by iodination has been used to synthesise (*E*)-1-iodo-2-methylalkenes ⟨79S501, 80JA3298⟩. A zirconium complex has also been used to catalyse the allylalumination of an alkyne, which, following the addition of iodine, gave the vinyl iodide (**40**) (Equation (73)) ⟨84TL5863⟩. Finally, hydroalumination of a 1-chloroalkyne followed by the migration of an alkyl moiety from aluminum to the vinyl group with the displacement of chloride and iodination of the resulting vinylalane gave a vinyl iodide (**41**) with stereochemistry as shown in Equation (74) ⟨89JOC998⟩

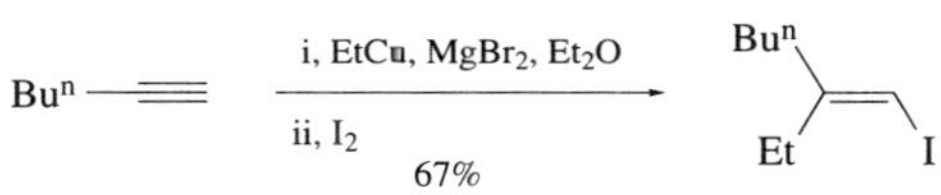

(72)

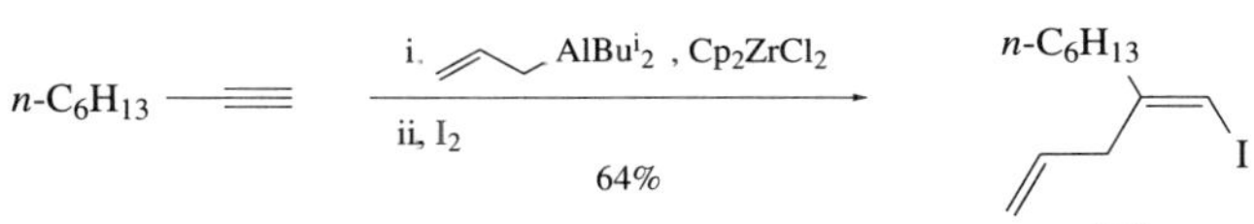

(40)

(73)

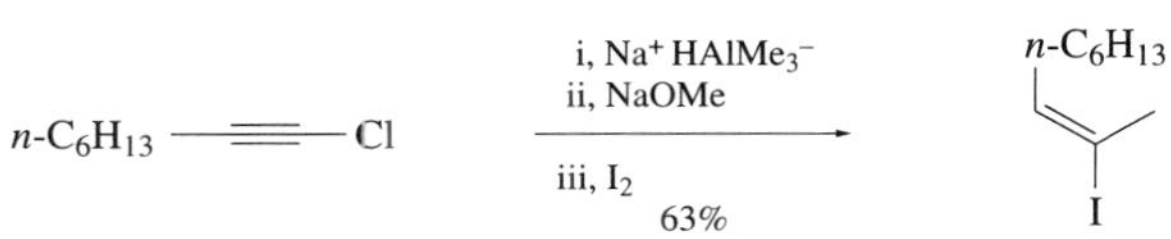

(41)

(74)

Vicinal 1,2-diiodoalkenes have been prepared by the addition of iodine to an alkyne, giving the (*E*)-stereoisomer selectively ⟨79JOC3931⟩. Iodine on alumina ⟨88JOC4477⟩ and iodine on florisil ⟨89JCR(S)313⟩ have both been shown to be convenient methods of performing this reaction. Other 1-iodo-2-haloalkenes have been made by a variety of methods. Treatment of a phenylalkyne with cuprous chloride and iodine gave good yields of the α-chloro-β-iodostyrene, predominantly as the (*E*)-isomer ⟨75CC925⟩. The same transformation has also been achieved using bis(pyridine)iodonium tetrafluoroborate in the presence of chloride ⟨90JOC3104⟩. Bromide was also used in this reaction to prepare the 1-iodo-2-bromoalkenes. 1-Iodo-2-chloroalkenes have been prepared by the reaction of an alkyne with tellurium tetrachloride followed by the addition of iodine, in the iodo equivalent of the reaction shown in Equation (60) ⟨79CL1357⟩.

2.12.5.1.3 *Vinyl iodides from vinyl metallic species*

Vinylsilanes have been selectively converted to vinyl iodides with both retention and inversion of configuration. Reaction with retention of configuration has been achieved in good yields by treatment with iodine in dichloromethane ⟨77TL3317⟩ or iodine monochloride in carbon tetrachloride ⟨89JOC868⟩. Inversion of stereochemistry resulted when iodine monochloride in DMF was used ⟨87JOC1100⟩ and when the vinylsilane was epoxidised and then treated with hydrogen iodide and boron trifluoride ⟨89JOC868⟩. Both vinylgermanes and vinylstannanes have been converted to vinyl iodides by reaction with iodine ⟨85CL53, 86JOC3561, 87JOC3687, 91TL3329⟩.

2.12.5.1.4 *Vinyl iodides from carbonyl compounds*

Barton and co-workers have converted a ketone to a vinyl iodide possessing the same carbon skeleton by first forming a hydrazone and then treating it with iodine in the presence of triethylamine ⟨62JCS470⟩. The same group later improved the reaction by using a guanidine base and inverse addition (adding the hydrazone to iodine) ⟨83TL1605, 88T147⟩. Vinyl iodides have also been prepared by a Wittig reaction. Stork and Zhao have reported a stereoselective synthesis of (*Z*)-1-iodoalkenes ⟨89TL2173⟩. Selectivity of up to 62 : 1 in favour of the (*Z*)-isomer was observed with benzaldehyde in a yield of 96%. Aliphatic aldehydes gave slightly lower (*Z*)-selectivity, but cyclohexanecarbaldehyde still gave the vinyl iodide (**42**) with a 15.5 : 1 (*Z*) : (*E*) selectivity in 71% (Equation (75)). In addition, cyclohexanone was iodomethylenated by this process in a yield of 61%. (*Z*)-Vinyl iodides have also been prepared by Corey *et al.* in the iodo equivalent of the reaction shown in Scheme 9 ⟨70TL447⟩. Finally, 1-iodoalkenes have been prepared by the reaction of aldehydes with iodoform and chromous chloride in the iodo version of the reaction shown in Equation (45) ⟨86JA7408⟩.

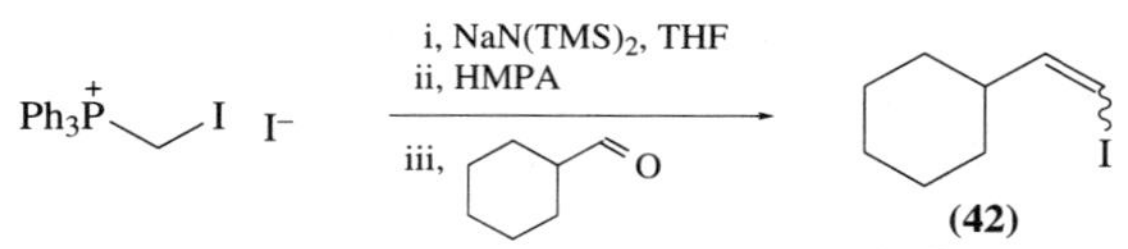

(**42**)
15.5:1 (*Z*):(*E*)

(75)

2.12.5.1.5 *Vinyl iodides by miscellaneous methods*

Vinyl iodides have been prepared from vinyl bromides by the action of potassium iodide and cuprous iodide in hexamethylphosphoramide (HMPA) ⟨88S236⟩. In a similar reaction, vinyl triflates have been converted to vinyl iodides by reaction with magnesium iodide in the presence of triethylamine ⟨86S222⟩. Lastly, vinyl iodides have been synthesised by treatment of 5,5-dialkyl-*N*-nitrosooxazolidinones (e.g., (**43**)) with sodium iodide and lithium 2-methoxyethoxide ⟨70JA4309⟩. The vinyl iodide (**44**) was produced in a yield of about 80%, as shown in Equation (76).

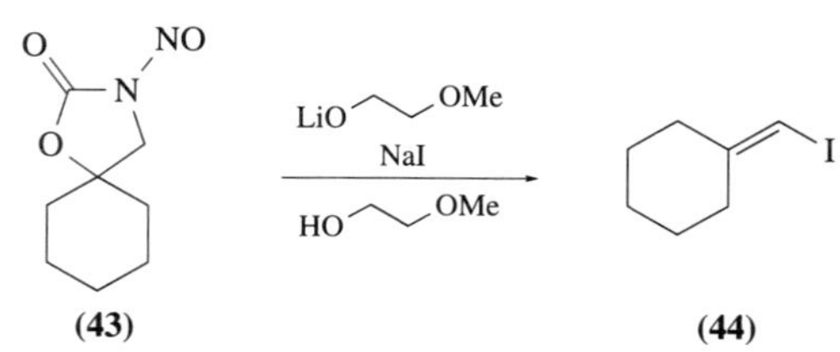

(**43**) (**44**)

(76)

2.12.5.2 Aryl Iodides

The synthesis of aryl iodides was reviewed by Merkushev in 1988 ⟨88S923⟩.

2.12.5.2.1 *Aryl iodides by electrophilic aromatic substitution*

Aryl iodides are readily prepared by the iodination of an aryl ring. With an electron-rich aromatic system the familiar *ortho*/*para* selectivity is seen. Iodine in the presence of nitric and sulfuric acids as the oxidant has been used to iodinate aromatic rings ⟨60MI 212-01⟩. This reaction, known as the Tronov–Novikov reaction, gave iodobenzene from benzene in a yield of 80%. Iodic acid has also been used to effect aromatic iodinations ⟨60LA(634)84⟩. Ethylbenzene gave *p*-iodoethylbenzene in 90% using this reaction. Several methods have been used to promote the iodination of an arene with iodine. Cornforth and Robinson used iodine with mercuric acetate in acetic acid to achieve iodinations in good yields ⟨42JCS684⟩. More recently, mercuric oxide with tetrafluoroboric acid has been shown to give good yields of a wide range of aryl iodides, including those with electron-withdrawing substituents ⟨84JCS(P1)2623⟩. Sugita *et al.* found that iodine with aluminum trichloride and cupric chloride gave aryl iodides in yields from 33% to 100% as mixtures of *ortho* and *para* isomers ⟨82CL1481⟩. Alternatively, iodine on alumina has been used to synthesise aryl iodides ⟨88JOC4477⟩. The yield of the product was very substrate dependent. Sy and co-workers have reported that alkyl benzenes can be iodinated with iodine and silver nitrate ⟨89TL3769⟩. Yields of a mixture of *ortho* and *para* products were in the range 43–95%. The same group have used iodine with silver sulfate to iodinate both alkyl and alkoxy benzenes ⟨90SC877⟩. Yields ranged from 71% to 97% for nine examples. Finally, iodobenzene bis(trifluoroacetate) with iodine has been used to prepare aryl iodides ⟨82JCR(S)150⟩.

Many other iodinating reagents besides iodine have been used. Ammonium hexanitratocerate with both iodide and iodine has been used to iodinate aryl systems ⟨81BCJ2847⟩. The reaction was particularly effective if tetrabutylammonium iodide was employed, when yields were in the range 86–100% for nine examples. Ferric iodide or aluminum triiodide in the presence of cupric salts have been shown to be effective iodine donors for the preparation of aryl iodides ⟨70JOC3436⟩. Arenes have also been iodinated by iodide and oxygen in the presence of catalytic nitrosonium tetrafluoroborate ⟨88JOC3548⟩. Phenols have been iodinated using sodium iodide with chloramine-T ⟨85TL2043⟩ and sodium iodide with *t*-butyl hypochlorite ⟨85JOC5384⟩. Kajigaeshi and co-workers have reported that aromatic rings can be iodinated with benzyltrimethylammonium dichloroiodate ⟨87CL2109, 88BCJ600, 89BCJ1349⟩. Yields were generally good for a large number of examples. McKillop *et al.* have shown that aryl iodides can be made easily and in high yield by the reaction of an aromatic substrate with thallic trifluoroacetate and potassium iodide ⟨69TL2427, 71JA4841⟩. The reaction proceeds via electrophilic substitution to give an arylthallium bis(trifluoroacetate). Aryl iodides were produced in yields of 70–100% for 17 examples (Equation (77)). The reaction has also been performed using thallic trifluoroacetate and iodine ⟨74BCJ1580⟩.

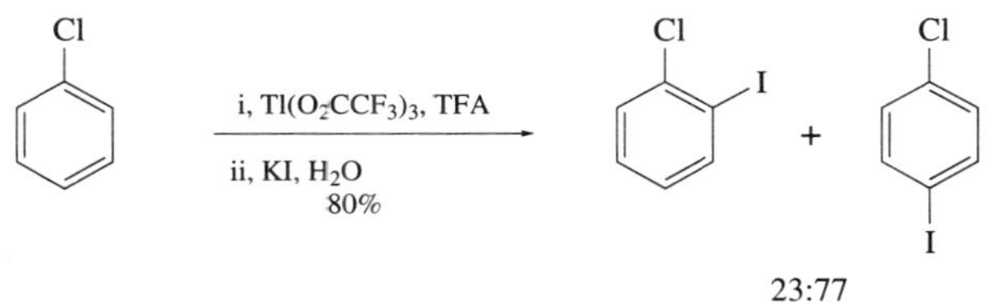

(77)

Electron-deficient aromatic rings have been iodinated in a number of ways to give the *meta* iodo products. For example, iodine and silver sulfate in sulfuric acid iodinated benzoic acid to give *m*-iodobenzoic acid in 75% yield ⟨50JCS3694⟩. Nitrobenzene was iodinated in 50% yield with iodine in oleum ⟨70JCS(C)1480⟩. Other iodinating systems that have been used include iodine with mercuric oxide and tetrafluoroboric acid ⟨84JCS(P1)2623⟩, iodine tris(trifluoroacetate) ⟨87BCJ4363⟩ and iodine monofluoride ⟨90JOC3552⟩.

Selectivity for the *ortho* iodination of phenols has been achieved by the use of iodine and thallous acetate ⟨76JCS(P1)1161⟩. Phenol was also iodinated predominantly *ortho* by iodine and cupric acetate ⟨84BCJ2691⟩. Selectivity for *para* iodination has been achieved in more ways. Aromatic ethers have been iodinated with high *para* selectivity with benzyltrimethylammonium dichloroiodate ⟨88CL795⟩. Anisole, for instance, gave *p*-iodoanisole in a yield of 92%. The same transformation has been effected by iodine and silver triflate, which gave a quantitative yield of *p*-iodoanisole ⟨77JCR(S)215⟩. Iodine with cupric acetate gave predominantly the *para* products when aniline and anisole (though not phenol) were iodinated ⟨84BCJ2691⟩. Antimony pentachloride with iodine has also been shown to lead to excellent *para* selectivity when used to iodinate anisole (Equation (78)) and halobenzenes ⟨74BCJ147⟩. Finally, the aromatic thallation reaction can be used to achieve a range of selectivities ⟨71JA4845⟩. Substitution occurred predominantly *meta* when the reaction was performed under thermodynamic control. If the reaction was performed under kinetic control, a substituent that allowed coordination (both electron-donating and electron-withdrawing) directed the iodination *ortho*, whereas if there was no possibility of coordination then iodination was principally *para*.

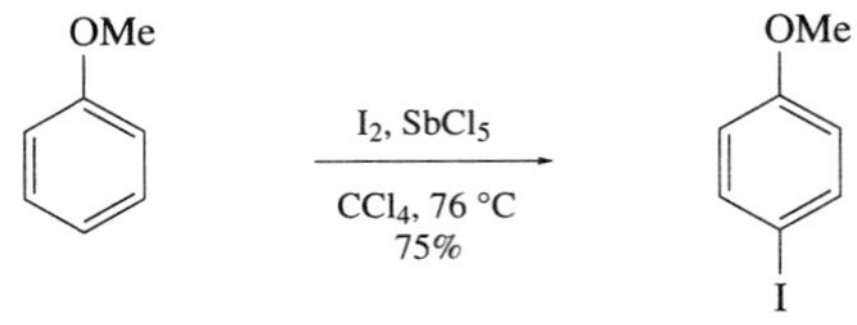

(78)

2.12.5.2.2 *Aryl iodides from aryl metallic species*

Arylboronic acids have been iodinated using iodine and chloramine-T, as was illustrated in Equation (16) ⟨82OPP359⟩. Aryl iodides have been prepared from arylsilanes by reaction with iodine monochloride ⟨76CC833, 83JOC1542⟩ and from an arylstannane by reaction with iodine ⟨80JOM(186)C33⟩. Aryl iodides have also been prepared from other aryl metallic species. Grignard reagents have been iodinated with both iodine ⟨19JA287⟩ and diiodoalkyne ⟨54CB1148⟩, and aryllithium reagents have been iodinated with iodine ⟨58JOC1161, 80CB1304⟩, as have arylcuprates ⟨74JIC1018⟩.

2.12.5.2.3 *Aryl iodides by halogen exchange and related reactions*

Aryl iodides have been prepared from aryl bromides by treatment with potassium iodide and low-valent nickel (prepared from nickel chloride and zinc) ⟨80BCJ3691⟩. Yields were good in most cases, though small amounts of biaryls were usually formed as a result of coupling. The reaction has also been performed using metallic nickel ⟨87JOC691⟩. In another reaction, aryl iodides have been prepared from aryl bromides by the action of potassium iodide in the presence of cuprous iodide in HMPA ⟨85CL411⟩. Yields were in the range 60–85% for nine examples. Both aryl chlorides and aryl bromides have been converted to aryl iodides by cuprous iodide supported on either alumina or charcoal ⟨87CC1409⟩. Yields were generally better starting with aryl bromides, being in the range 86–92% for three examples.

2.12.5.2.4 *Aryl iodides from aryldiazonium salts*

Aryl iodides can be prepared by treating an aryldiazonium salt with iodide, as was shown in Equation (23) ⟨60OS(40)105, 79TL2503⟩. Iodine itself has also been used in the reaction ⟨77TL3519, 81SC639⟩. A one-pot reaction using sodium nitrite and trimethylsilyl iodide has been reported ⟨92TL3167⟩. Other diazotizing agents besides sodium nitrite have been used. Amyl nitrite and iodine converted anilines to aryl iodides ⟨68JOC1636⟩. Small amounts of aryl diiodides were usually also formed. Isoamyl nitrite with diiodomethane as the iodinating reagent has been used to convert anilines to aryl iodides in generally good yields ⟨90JOC2543⟩.

Aryl triazenes can be used as a more stable equivalent of aryldiazonium salts in the preparation of aryl iodides. They have been converted to aryl iodides (via aryldiazonium salts) by reaction with an acid in the presence of iodide (Equation (79)) ⟨80S572, 87JCS(P1)1039, 90JOC4560⟩. Aryl triazenes have also been converted to aryl iodides by treatment with trimethylsilyl chloride and sodium iodide ⟨81JOC5239⟩ and an acid ion exchange resin with sodium iodide ⟨83JOC4394⟩. Katritzky has reported that anilines can be converted to aryl iodides via 2,4,6-triphenylpyridinium iodides (e.g., (**45**)) ⟨79JCS(P1)433⟩. Subsequent pyrolysis of the pyridinium iodides gave aryl iodides in good yields (Scheme 15). Arylarylazosulfones (available from aryldiazonium salts) have been converted to aryl iodides by potassium iodide with 18-crown-6 and trifluoroacetic acid ⟨85JOC1779⟩. Yields were in the range 61–84% for six examples.

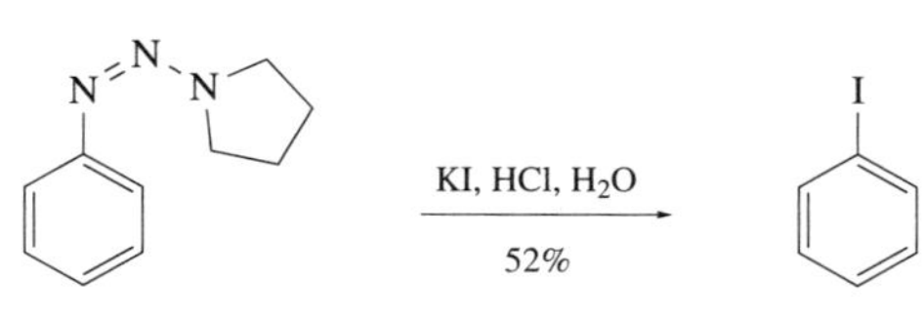

(79)

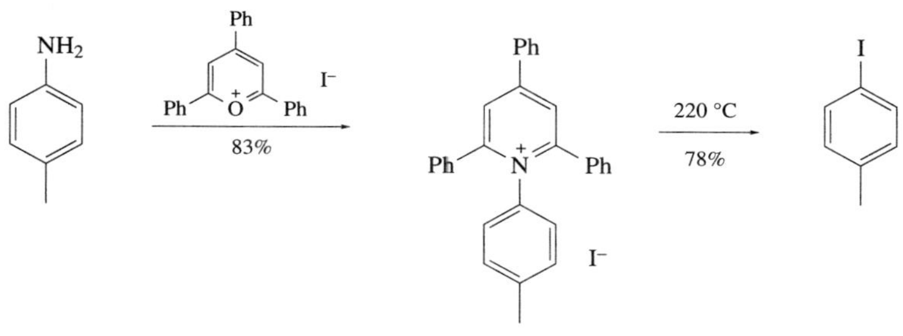

(45)

Scheme 15

2.12.5.2.5 Aryl iodides by miscellaneous methods

A modification of the Hunsdiecker reaction has been used to prepare aryl iodides ⟨83JOC3297⟩. Treatment of a benzoic acid with iodine and yellow mercuric oxide gave aryl iodides in yields of 50–75% for five examples. The same transformation has also been achieved with iodobenzene diacetate and iodine under irradiation to give aryl iodides in yields from 25% to 85% for six examples ⟨88SC1327⟩.

2.13
Alkenyl and Aryl Chalcogenides: Oxygen-based Functional Groups

CHARLES K.-F. CHIU
Pfizer Inc. Central Research, Groton, CT, USA

2.13.1 OXYGEN-BASED FUNCTIONAL GROUPS ATTACHED TO sp^2 CARBON

2.13.1.1 Enols and Phenols: ROH

2.13.1.1.1 Enols

An enol is the tautomer of the corresponding carbonyl compound, with the equilibrium in favor of the thermodynamically more stable keto form (Equation (1)). Under normal circumstances, enols exist in extremely low concentration, and yet they are important intermediates in many chemical transformations. In the past, enols were regarded as transient intermediates, but these elusive moieties can now be captured and characterized owing to advances in spectroscopic and experimental techniques. The term "simple enol" was coined by Hart for those enols which are not stabilized by other functionalities like electron-withdrawing groups (EWG). Rappoport defined "simple enols" as enols stabilized by steric demands. This section will deal only with these simple enols, and excludes those derived from 1,3-dicarbonyl compounds. Important reviews in this area are by Hart ⟨79CRV515⟩, Capon ⟨88ACR135⟩, Rappoport ⟨88ACR442⟩ and Kresge ⟨90ACR43⟩. The most recent and extensive review of the various aspects of enol chemistry was assembled by Rappoport ⟨B-90MI 213-01⟩.

enol keto (1)

(i) Simple aliphatic enols

Simple aliphatic enols are rather short-lived species. Although they can be generated in the gas phase under thermolytic conditions they are often generated in solution and not isolated.

(a) By thermolysis. The simplest enol is ethenol (vinyl alcohol) (**1**), the tautomer of acetaldehyde, which can be prepared in the gas phase by thermolytic dehydration of ethylene glycol at 940°C and 0.02–0.04 Torr (Equation (2)) ⟨76CPL399⟩. It has a half-life of approximately 30 min at room temperature in a pyrex flask.

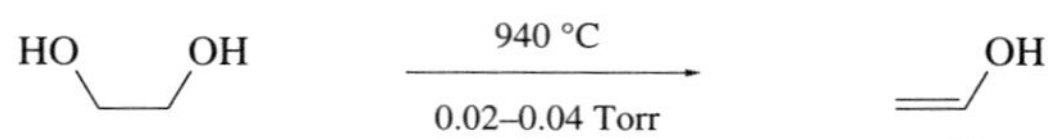

(**1**) (2)

The preparation of ethenol (**1**), as well as propen-2-ol (the enol of acetone) (**2**), can be performed more efficiently by thermolytic retrocycloaddition reaction of cyclobutanols (**3**) ⟨82JA2648⟩ or 3-thietanol (**4**) ⟨83JST(97)47⟩ (Equation (3)). This approach has been extended to the synthesis of higher homologs of ethenol by retro-Diels–Alder reaction of the norbornenols (**5**) ⟨88JA7984⟩ and the anthracene cycloadducts (**6**) ⟨79NJC195⟩ as shown in Scheme 1. In general, the enols generated are contaminated with 10–30% of their respective keto-tautomers.

(b) By photolysis. Similar to thermolysis, the enols generated photolytically in solution usually

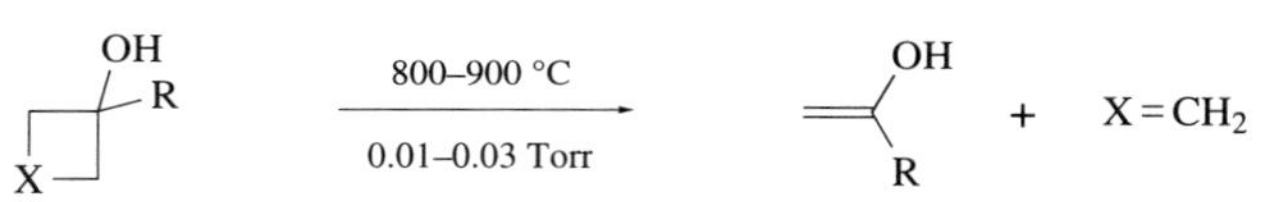

(3)

(**3**) X = CH_2, R = H, Me
(**4**) X = S, R = H

(**1**) R = H
(**2**) R = Me

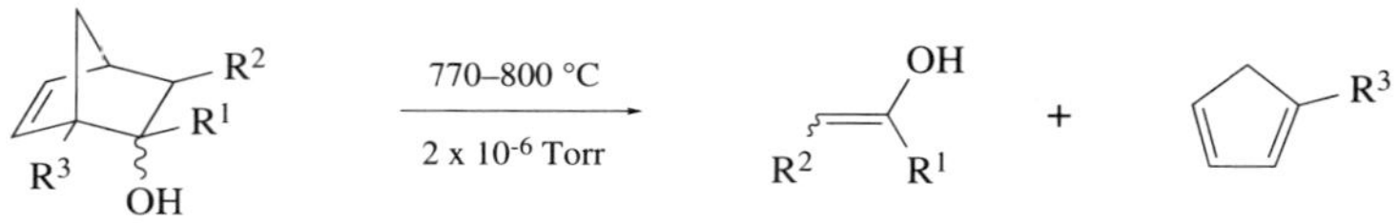

(**5**)

R^1 = H, Me, Ph, CH=CH_2, C≡CH
R^2 = H, Me, CH=CH_2
R^3 = H, Me

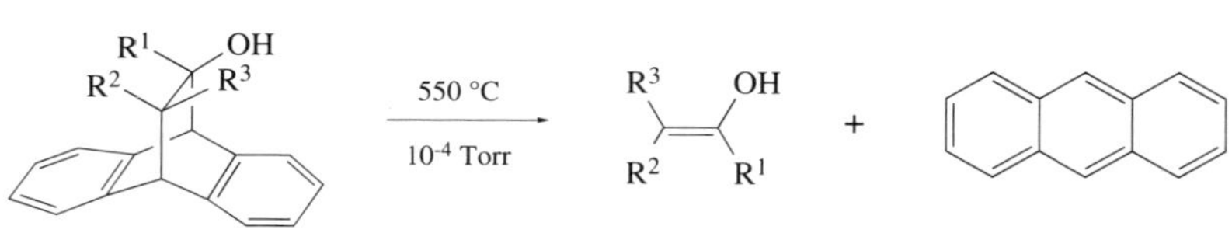

(**6**)

Scheme 1

have a limited lifetime. Nevertheless, the photolysis process provides a long enough time frame for extensive study of the resulting enols, often by UV spectroscopy ⟨90ACR43, 91SCI395⟩.

Flash photolysis of an aqueous solution of acyclic ketones results in the formation of enols through a Norrish type II process (Scheme 2) ⟨76AG(E)435, 84JA460⟩. In the case of the hydroxypentanone (**7**), both the enols of acetaldehyde and acetone are obtained (Equation (4)) ⟨87JA4000⟩. This technique has been applied to cyclic ketones; for example, irradiation of the indanone (**8**) in deoxygenated acetonitrile at 0 °C leads to the formation of the enol (**9**) in 98% conversion (Scheme 3) ⟨91AG(E)416⟩. Interestingly, in the presence of 10% of a chiral proton source, the optically active methylindanone (**10**) is obtained.

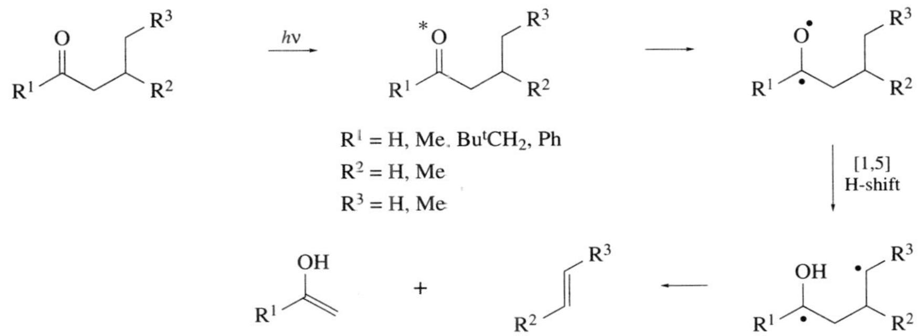

Scheme 2

(4)

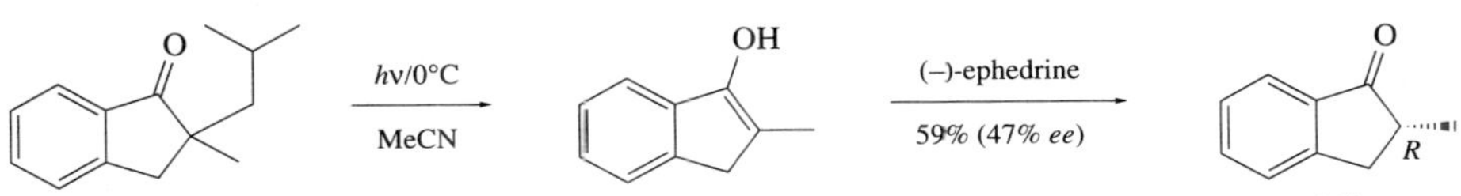

Scheme 3

In a "cross" Norrish type II process (intermolecular), acyclic and cyclic alcohols can be oxidized photolytically by photoexcited ketones. This results in the formation of two distinct enols of different reactivity, as illustrated by the reaction between cyclohexanol and acetone in Scheme 4. A variety of enols can be conveniently prepared by this protocol ⟨90JA4862⟩.

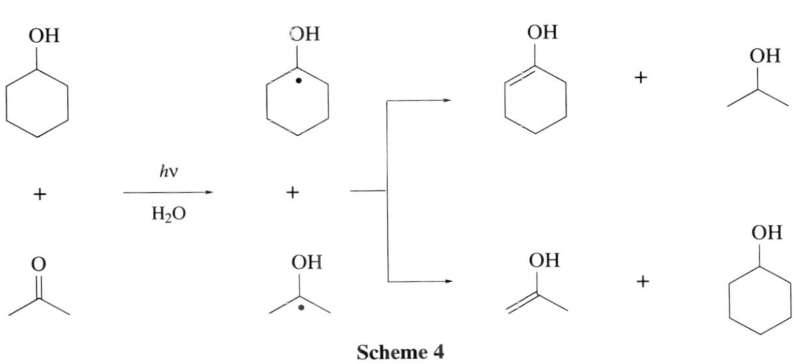

Scheme 4

The photohydration of alkynes provides another alternative to enol preparation (Scheme 5). Photoexcitation of aryl alkynes in aqueous HCl solution generates the vinyl cations (**11**), which undergo hydration to produce enols ⟨86HCA1331⟩. Enols are also accessible via the vinyl cations (**11**) by photohydrolysis of vinyl halides ⟨89CC1548⟩.

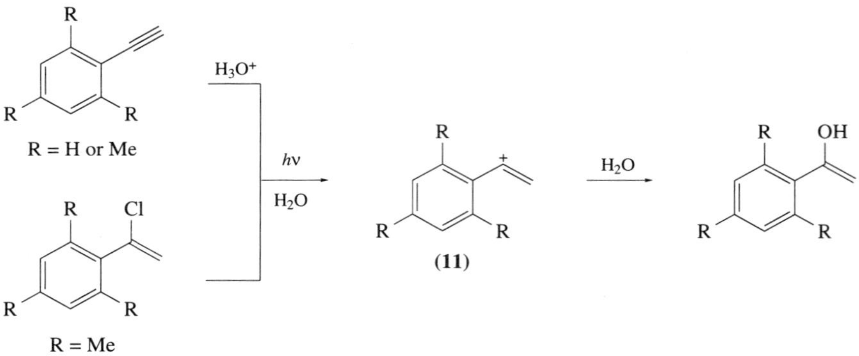

(**11**)

Scheme 5

Finally, enols can also be prepared by photoisomerization of α,β-unsaturated ketones ⟨76BCJ3673⟩. Irradiation of a vigorously deoxygenated acetonitrile solution of 1-acetyl-cyclooctene in a pyrex tube provides a 17 : 83 mixture of the (*E*)- and (*Z*)-dienols (**12**) in 80% yield (Equation (5)). Although the isomeric dienols cannot be separated and isolated in their pure forms, in the absence of oxygen, they are stable for at least 3 d as a dilute solution at room temperature.

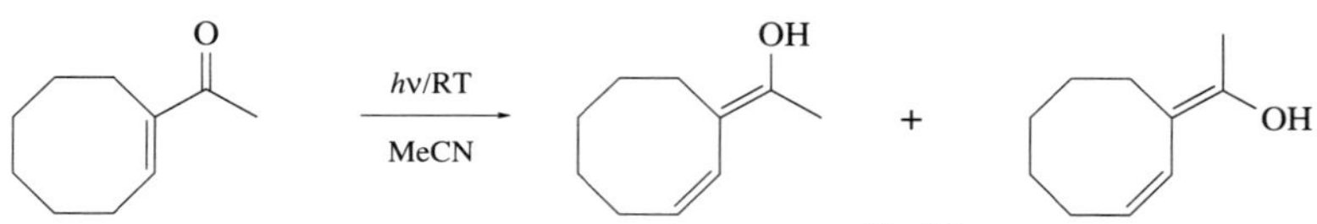

(**12-*E***) 17 : 83 (**12-*Z***) (5)

(c) Hydrolysis of O*-substituted enol ethers.* The preparation of enols in solution by means of hydrolysis of labile *O*-substituted enol substrates has been studied extensively by Capon and co-workers ⟨88ACR135⟩. At −20 °C, the labile *ortho*-ester (**13**) is hydrolyzed rapidly in a mixture of deuterated acetonitrile and water (90 : 10) to produce *O*-deuterated ethenol, along with chloroacetic acid and methyl formate (Equation (6)). The ethenol is stable in this acidic medium for several hours at −20 °C, but tautomerizes rapidly to acetaldehyde when neutralized by sodium deuterioxide, or upon warming to room temperature ($t_{1/2} \sim 10$ min) ⟨81JA1761⟩. Ethenol as well as a host of other enols have been made available by the hydrolysis of *ortho*-esters (**14**) or ketene acetals (**15**), in an acidic (10^{-4} M HCl) medium (99 : 1, deuterated acetone / water). In general, cyclic *ortho*-esters (**16**) are more reactive and are useful substrates for the generation of enols ⟨74JA4494⟩. These *ortho*-esters and ketene acetals are available readily (Scheme 6) and provide access to a wide range of simple enols in solution. The isolation of pure enol, however, is still technically challenging.

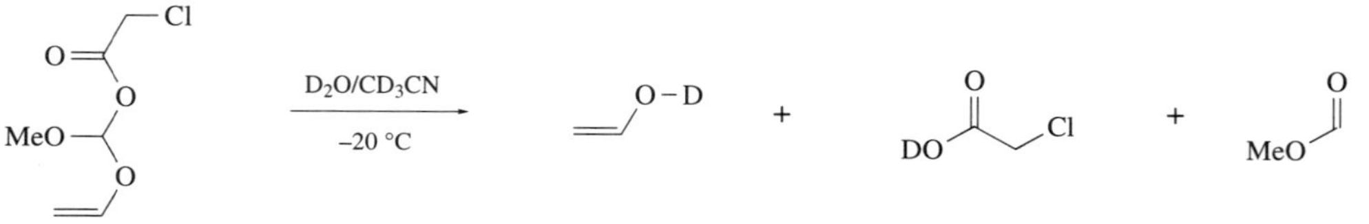

(**13**) (6)

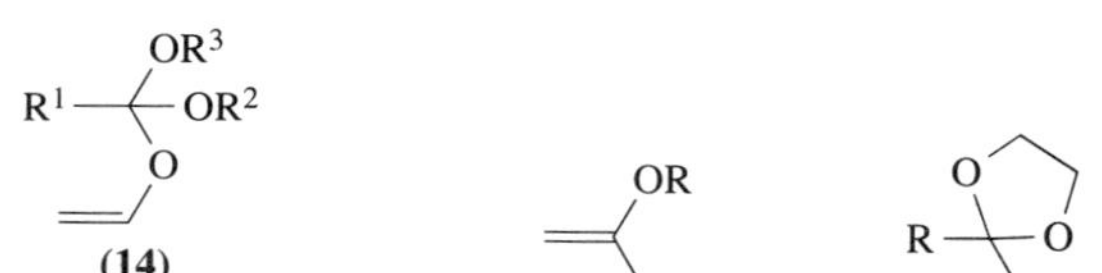

(**14**)
R^1 = H, Me
R^2 = Me, Ac
R^3 = Me, CH=CH$_2$

(**15**)
R = Me, CH=CH$_2$

(**16**)
R = Me, Ph

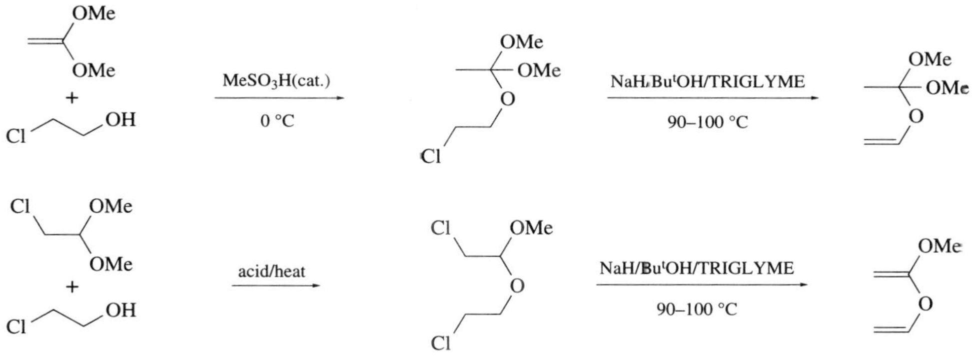

Scheme 6

Less labile enols can be generated by protonolysis of the corresponding trimethylsilyl enol ethers. Typically, the silyl enol ethers are treated with DCl (5×10^{-4} M) in a mixture of deuterated acetone and water at 32 °C, and the resulting enols are stable in the solution for several hours (Equation (7)) ⟨86TL3275, 87PAC1577⟩.

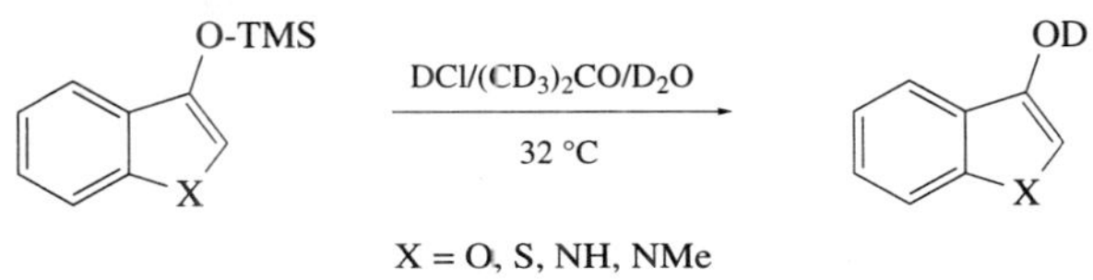

(7)

Hydrolysis of the amino dioxolane (**17**) led to the expulsion of DMF and the formation of the different enols (**18**)–(**20**) depending on the reaction conditions as shown in Scheme 7 ⟨72JA7832⟩. Although the lifetimes of these enols in protic solvent are rather short (5–15 min), in polar aprotic solvents like DMSO, the enol (**18**) is generated in quantitative yield and is stable for days. The lifetime of the enols can be extended further by replacing the enol proton with a deuterium.

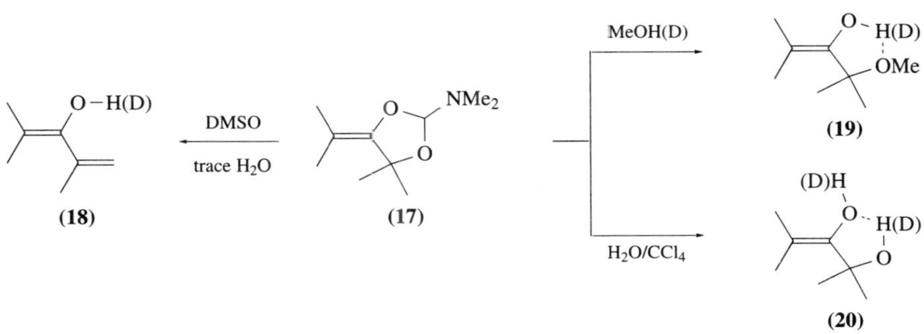

Scheme 7

(d) Isomerization of allylic alcohols. The ability of rhodium salts to induce C═C bond migration has been applied to the synthesis of enols (Scheme 8) ⟨87CC1213⟩. Isomerization of the alkenic bond in 2-methylprop-2-en-1-ol with $[Rh(CO)(PPh_3)_3]ClO_4$ (<0.2 mol%) in the absence of solvent at 0 °C leads to formation of the enol (**21**) in 95% yield along with a small amount of 2-methylpropanal ⟨88JA8244⟩. The *O*-deuterated enol was obtained in quantitative yield when the reaction was conducted in deuterated alcoholic solvent. The enol is stable at −78 °C (frozen) and remains unchanged for days, but tautomerizes and undergoes self aldol condensation within 24 h at 25 °C. The isomerization of 2-ethyl-prop-2-en-1-ol with the same rhodium complex leads to predominantly one isomeric enol (**22**) ⟨91JCS(D)443⟩. The geometry of the enolic double bond, however, could not be determined despite extensive spectroscopic analysis. The migration of the alkenic bond in allylic alcohols can also be effected by iridium complexes, for example, $[Ir(COD)(PhCN)(PPh_3)]ClO_4$ ⟨91JCS(D)443⟩. Prop-2-en-1-ol is isomerized rapidly to the (*Z*)-enol (**23-*Z***) as the kinetic product,

but is subsequently isomerized to the thermodynamically more stable (*E*)-isomer (**23-*E***) over time (Scheme 8).

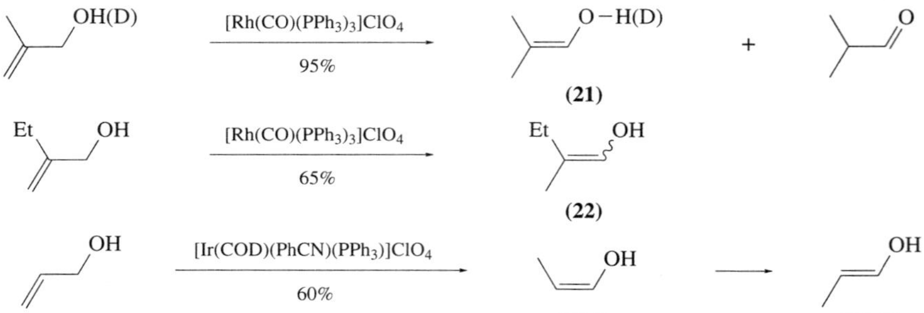

(21) **(22)** **(23-*Z*)** **(23-*E*)**

Scheme 8

This methodology is general and has been applied successfully to an array of allylic alcohols (Table 1) ⟨91CC1399, 91JA958⟩. Simple allylic alcohols are isomerized completely within 15 min. Phenyl substituents, however, retard the reaction and require hours to complete, presumably owing to complexation of the catalyst to the phenyl π-system. Surprisingly, the isomerization of homoallylic alcohols to enols is unproductive.

Table 1 Generation of enols by isomerization of allylic alcohols catalyzed by rhodium.

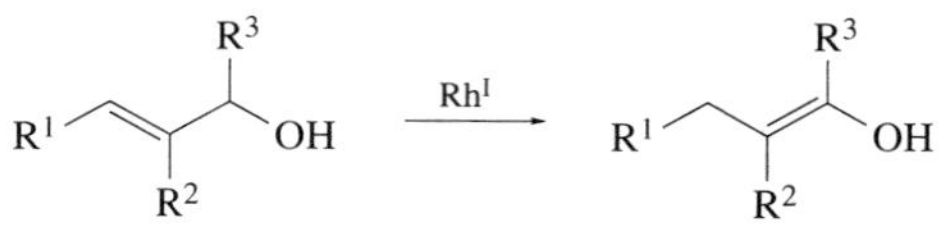

R¹	*R²*	*R³*	*Reaction time*	*Yield* (%)	(*Z*):(*E*)
H	H	H	14 min	89	48:52
H	Ph	H	5.5 h	86	50:50
H	H	Me	9 min	83	84:16
H	H	Ph	2.1 h	47	100:0
Ph	H	H	2.8 h	48	37:63
H	OEt	H	50 h	42	0:100
H	Me	H	16 min	96	n.a.
$(CH_2)_4$		H	27 h	71	n.a.

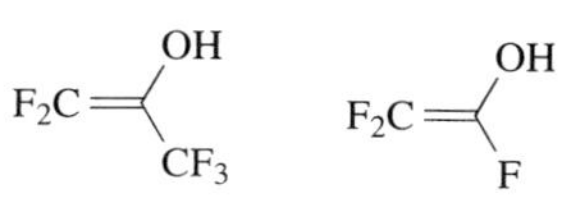

(24) **(25)**

(ii) Fluorinated simple enols

The enol of acetone (**2**) is less stable and tautomerizes about 100 times faster than that of acetaldehyde (**1**). Interestingly, perfluoropropen-2-ol (perfluoroacetone enol) (**24**) is a distillable liquid (b.p. = 54–55°C) while perfluoroethenol (perfluoroacetaldehyde enol) (**25**) is not known. Perfluoropropen-2-ol and analogs have been prepared by thermal decomposition of the corresponding enol phosphates in good yields (Scheme 9) ⟨75ZOR1370⟩. Alternatively, the volatile perfluoropropen-2-ol (**24**) can be generated in high yields by heating the enol phosphates (**26**), which are available readily from the reaction of chloropentafluoroacetone and trialkyl phosphites, in concentrated sulfuric acid (Scheme 10) ⟨74DOK(217)1320⟩. It should be noted that heating of the enol phosphate (**26**) in dilute sulfuric acid only leads to formation of the keto tautomer ⟨64JOC1876⟩.

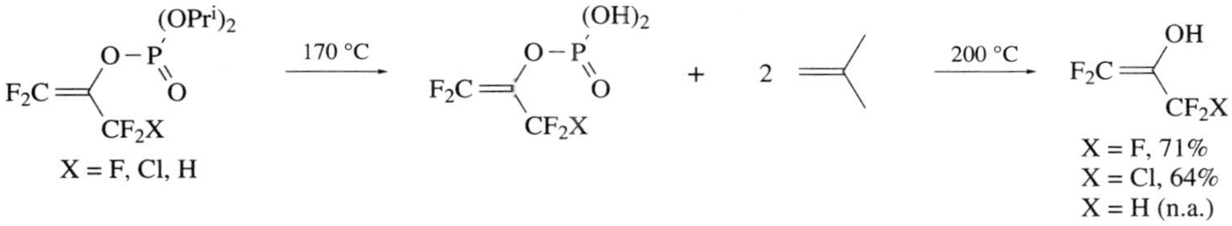

Scheme 9

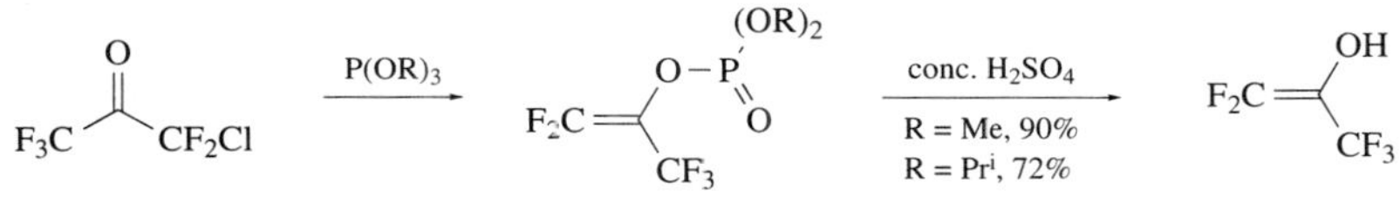

(26) **(24)**

Scheme 10

Most commonly, fluorinated enols are prepared by the acid hydrolysis of benzyl enol ethers. Hence, treating the benzyl enol ether (**27**) with concentrated sulfuric acid produces perfluoropropen-2-ol (**24**) in 82% yield (Equation (8)) ⟨75DOK(220)146⟩. This method is particularly effective in the preparation of cyclic fluorinated enols (Table 2).

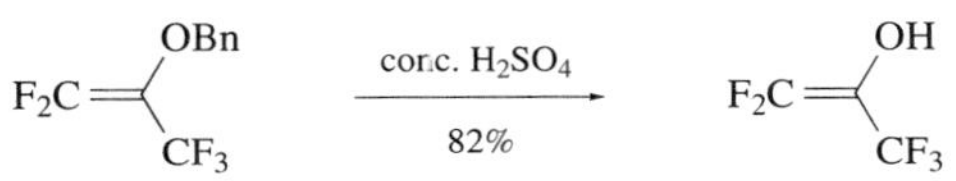

(27) **(24)** (8)

Table 2 Preparation of highly fluorinated cyclic enols.

Ring size	R^1	R^2	R^3	*Yield* (%)	*b.p.* (°C)	*Ref.*
$n = 1$	F	F	F	67	85–86	76DOK(229)870
$n = 1$	C_2F_5	F	F	73	91–92	78IZV1193
$n = 1$	CF_3	H	H	0[a]		78IZV1193
$n = 2$	F	F	F		101	78IZV493
$n = 3$	F	F	F	88	104–106	77DOK(233)591
$n = 3$	Cl	F	F		120	78IZV2775
$n = 3$	H	F	F			78IZV2775

[a] The enol tautomerizes rapidly and only the corresponding ketone is obtained

The 1,4-addition of hydrogen halide (HF or HCl) to an α,β-unsaturated perfluoroketone provides an alternative approach to fluorinated enols, although it is very substrate specific. The addition is successful only with highly reactive enones, such as perfluorocyclobutenone (**28**) (Equation (9)) ⟨76DOK(231)864⟩. Furthermore, the use of a basic solvent (ether, dioxane or THF) is essential.

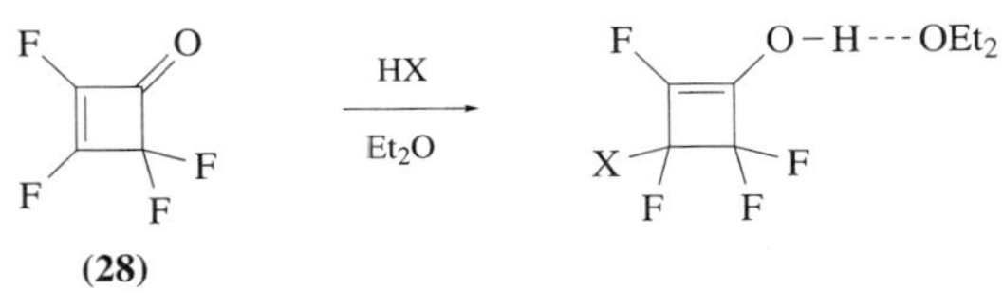

(28) (9)

X = F, 61%; X = Cl, 76%

(iii) Sterically congested stable enols

The simple aliphatic enols described earlier are rather unstable and have limited lifetimes because they tautomerize rapidly to their corresponding carbonyl compounds. Special handling is generally required for their preparation. In certain specific cases, however, the enol form is more stable owing to the steric environment, as exemplified by the tricyclic ketone (**29**). Attempts to epimerize the C-2 substituent of (**29**) with base only led to isolation of the crystalline enol (**30**) in quantitative yield (Equation (10)). X-ray structural analysis of both the ketone (**29**) and enol (**30**) suggests that the significant steric and dipole destabilization experienced by the ketone are relieved upon enolization ⟨87JA5553⟩. In other examples (Scheme 11), attempts to construct an oxetane ring for the taxol analogue (**31**) led to formation of the enol (**32**) predominantly ⟨91T9823⟩, and the enol (**33**) was obtained exclusively in the aziridination of 3-methylpentane-2,4-dione ⟨92CC1256⟩. These enols are stable owing to their unique structural makeup.

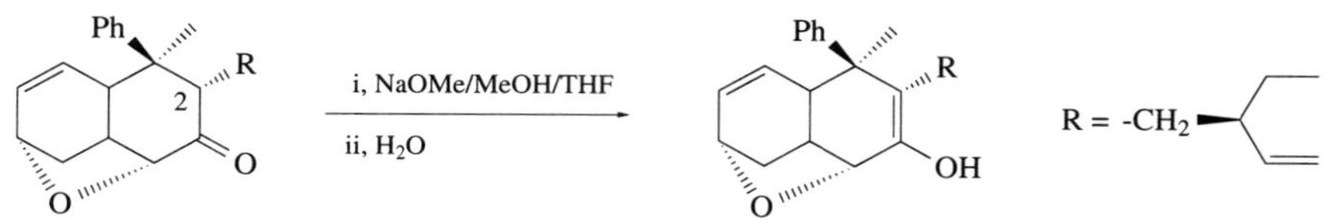

(**29**) (**30**) (10)

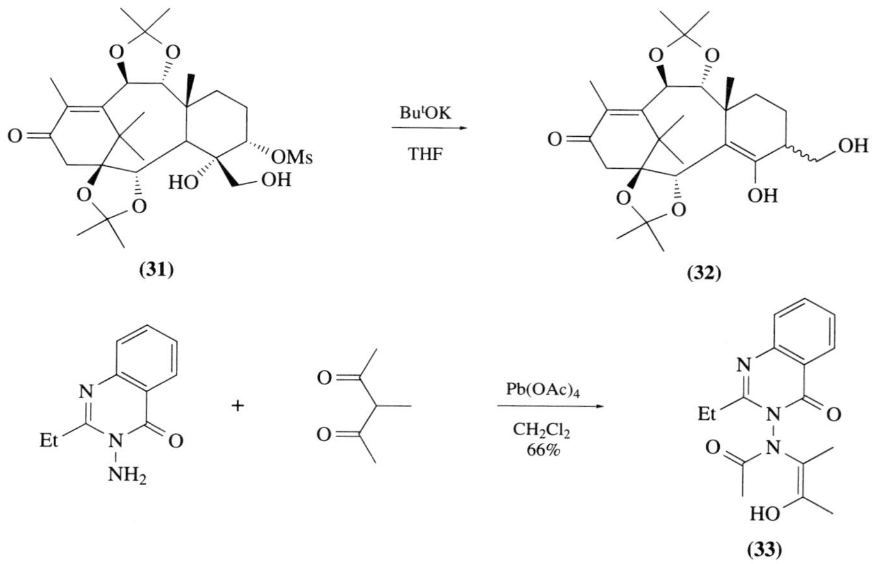

(**31**) (**32**) (**33**)

Scheme 11

In 1940, Fuson *et al.* ⟨40JA3250⟩ synthesized and isolated a stable simple enol (**34**) that existed as a crystalline solid. The generality of this phenomenon was demonstrated by subsequent preparation of a range of analogs commonly referred to as Fuson's enols, which were disclosed in a series of scientific papers spanning over a decade.

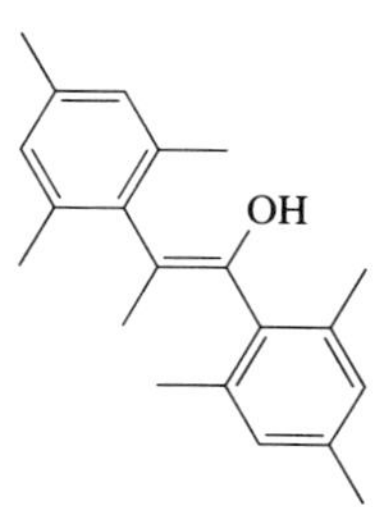

(**34**)

Fuson's enols are a family of ethenols substituted at C-1 and C-2 with bulky aryl groups, typically mesityl (2,4,6-trimethylphenyl, Mes), duryl (2,3,5,6-tetramethylphenyl, Dur), isoduryl (2,3,4,6-tetramethylphenyl, Dur^i) and tipyl (2,4,6-triisopropylphenyl, Tip). Other bulky aryl substituents such as 3,5-dibromomesityl (3,5-BrMes), pentamethylphenyl (PMP), 9-anthryl (Ant) and 9,10-dihydroanthryl (DHA), have since been introduced. The chemistry of Fuson's enols has been expanded by various research groups, particularly that of Rappoport ⟨88ACR442⟩. The aryl substituents contribute to the unusual stability of these enols by a combination of steric hindrance and electronic stabilization through conjugation. The criteria laid down by Fuson for stabilized ethenols, including enediols, are as follows: "Those which are stable have the following type of structure (**35**) in which A or B is an aryl group. If A is a radical of the mesityl type, B may be any hydrocarbon radical or a hydrogen atom. If B is a radical of the mesityl type, A may be hydroxyl or any hydrocarbon radical, but not hydrogen." ⟨45JA386⟩. The three basic classes of the Fuson's sterically crowded ethenols in descending order of relative stability towards tautomerization are: 1,2,2-triaryl > 2,2-diaryl > 1,2-diaryl. These polyaryl substituted ethenols are readily accessible by a variety of synthetic methods.

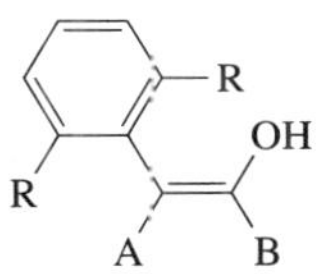

(35)

(a) From α,β-unsaturated ketones. The first stable enol synthesized in Fuson's laboratory was the 1,2-dimesityl ethenol (**34**), which was prepared by 1,4-hydrogenation of the α,β-unsaturated ketone (**36**) (Equation (11)) ⟨40JA3250⟩. The same reduction can also be achieved by lithium aluminum hydride ⟨90JOC2673⟩. It should be noted that the positioning of a bulky aryl substituent (e.g., Mes) at both C-1 and C-2 is essential for the stability of the enol.

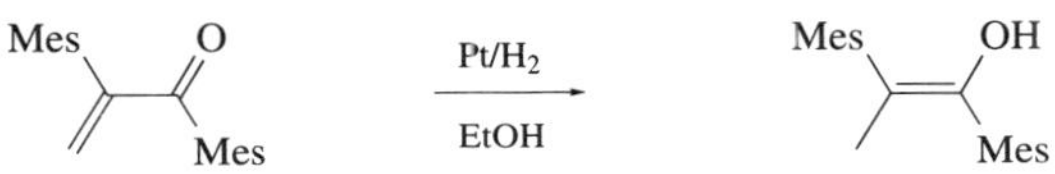

(36) **(34)** (11)

The conjugate addition of Grignard reagents to α,β-unsaturated ketones leads to the preparation of 2-alkyl-1,2-diaryl-ethenols (Equation (12)) ⟨44JA1873⟩. Oxidation of the enol (**37**) provides the α,β-unsaturated ketone which, in turn, can be subjected to further conjugate addition (Scheme 12). This addition/oxidation sequence has been manipulated efficiently for the synthesis of higher homologs, such as the isopropyl ethenol (**38**) ⟨90JOC2673⟩. Further elaboration of (**38**) to the *t*-butyl analog (**39**), however, fails to materialize. This methodology also provides an indirect route to the synthesis of 2,2-diaryl ethenols as outlined in Scheme 13 ⟨45JA1939⟩. The 1,4-addition of mesityl magnesium bromide to the enones (**40**) produces the enols (**41**) which are converted to the peroxides (**42**) by reaction with oxygen. Decomposition of the peroxide (**42**) by heating provides the 2,2-diaryl ethenols (**43**) and mesitoic acid. *Note, however, that extreme caution should be exercised due to the violent nature of this thermal decomposition.*

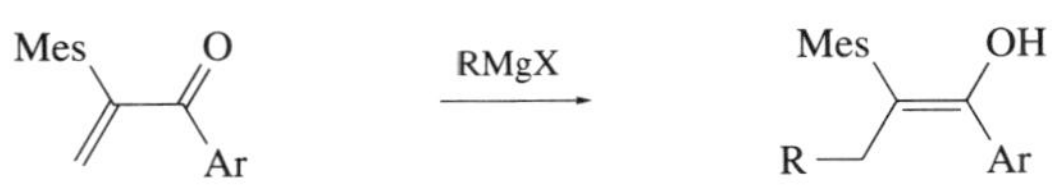

(12)

R = Me, Ph
Ar = Dur, Dur^i

(b) Dehydration of glycols. Dehydration of 1,2-diaryl-1,2-ethanediols under acidic conditions produces 2,2-diaryl ethenols ⟨43JA992⟩, and this method is applicable to a variety of substrates (Table 3). The reaction proceeds via a pinacol-type rearrangement to the carbonyl compounds, which readily tautomerize to give the enols as the isolated products. The dehydration can be effected by a variety of acids or simply by heating. The extreme stability of these enols is attested by the conversion of the enol (**44**) to the enol ether (**45**) exclusively in alcoholic HCl instead of hydrolysis to the aldehyde (**46**) (Scheme 14) ⟨44JA1109⟩.

(c) Ozonolysis of 1,1-*diarylethenes.* A novel synthesis of 2,2-diaryl ethenols is by ozonolysis of

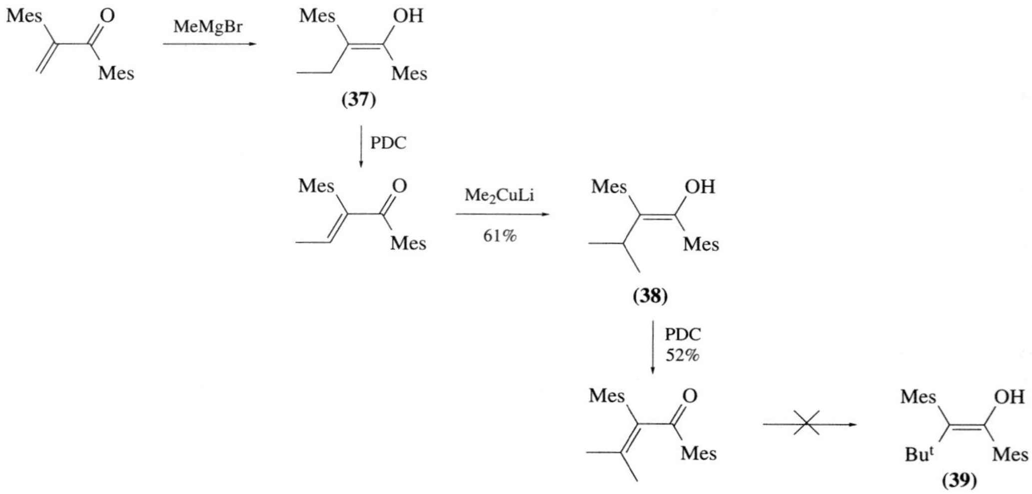

Scheme 12

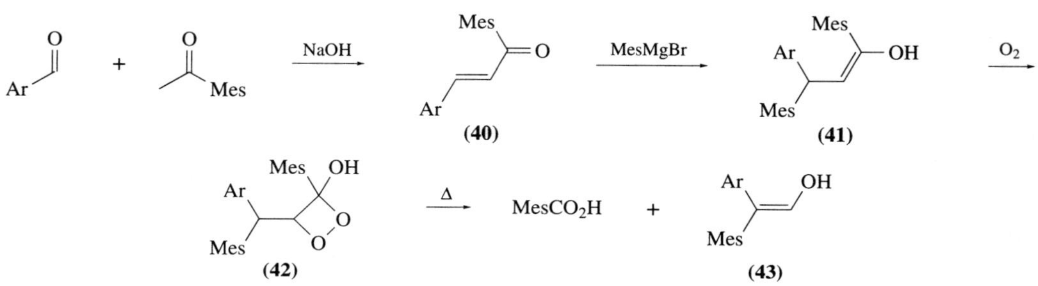

Ar = Ph, Mes, *p*-ClC_6H_4

Scheme 13

Table 3 Preparation of 2,2-diaryl substituted ethenol by acidic dehydration.

R[1]	*R[2]*	*Conditions*	*Yield* (%)	*m.p.* (°C)	*Ref.*
Mes	Mes	A, C, D	60	128–129	43JA992, 44JA1109
i-Dur	*i*-Dur	B		149.5–151.5	44JA1109
Ph	Mes	C	83	114–115	44JA1272
Ph	*i*-Dur	C	63	121–122	44JA1274
Tip	Tip	A, C, D, E		114–115	46JA389

Conditions: A. 55% H_2SO_4/Δ; B. 20:80 H_2SO_4/AcOH/Δ; C. 20:80 HCl/AcOH/Δ; D. P_2O_5/RT; E. Heating at 175–176 °C/6mm Hg

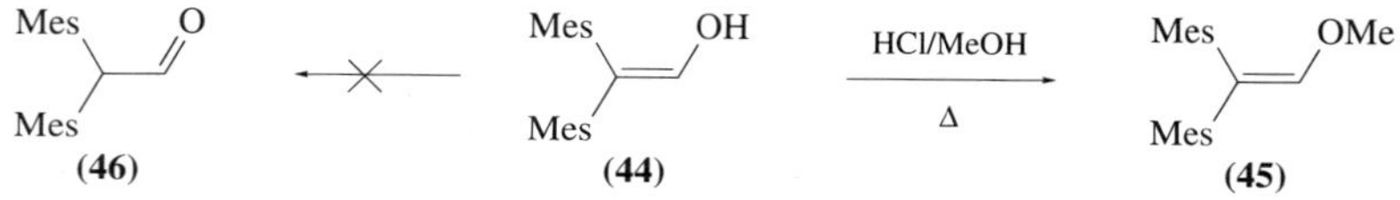

Scheme 14

1,1-diarylethenes. Instead of the usual oxidative cleavage of the C=C bond, the ethenols (**47**) are obtained (Equation (13)) ⟨44JA1274⟩.

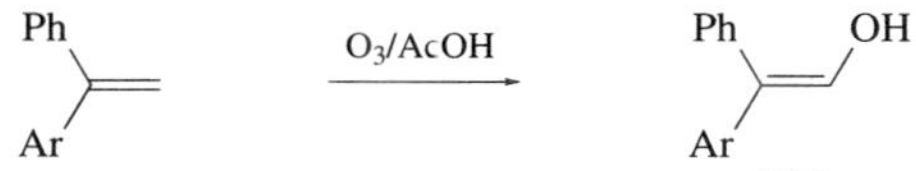

(**47**)

Ar = Mes, Dur[i]

(13)

(d) From ketenes. 2,2-Diaryl ethenols can also be prepared by the reduction of diaryl ketenes. The diaryl ketenes are readily available from the corresponding acids, and are reduced by *t*-butyl or cyclohexyl magnesium bromide to provide the 2,2-diaryl ethenols (**48**) in high yields (Scheme 15) ⟨44JA1464, 45JA1937⟩. The reduction by lithium aluminum hydride is equally effective ⟨84JA477⟩.

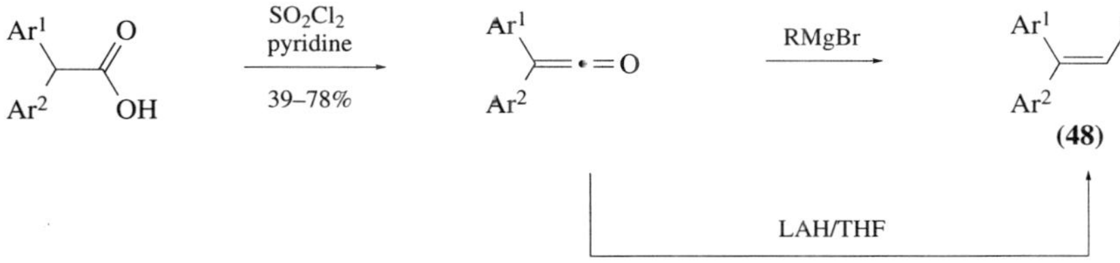

(**48**)

62%

$Ar^1 = Ar^2 =$ Mes

Ar[1]	*Ar[2]*	*Yield (%)*	*Grignard*
Mes	Ph	80	*c*-hexMgBr
Mes	Mes	86	Bu[t]MgBr
Dur	Ph	80	Bu[t]MgBr
3-BrMes	Ph	82	Bu[t]MgBr

Scheme 15

Nucleophilic addition of aryl magnesium bromide and aryllithium to diaryl ketenes provides a general approach towards the preparation of triaryl ethenols (Equation (14)) ⟨84JA477, 87JA2112⟩. Likewise, addition of alkyl metals provides access to 1-alkyl-2,2-diaryl ethenols (Scheme 16). However, the addition of alkyl Grignard reagents is often plagued by the aforementioned reduction side reaction. The undesired reduction process can be suppressed by the addition of $LiClO_4$, but only to a certain degree. Nonetheless, the use of alkyllithium reagents provides a better alternative ⟨85JA3669⟩.

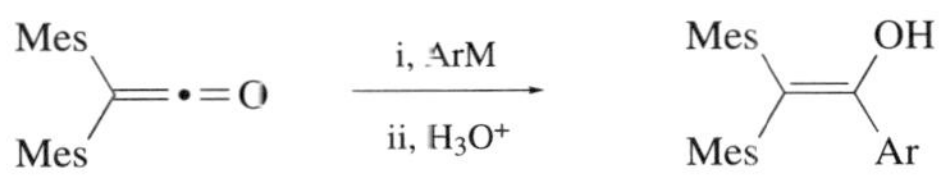

(14)

M = MgBr, Ar = Ph, 9-Ant, *m*-ClC_6H_4, *p*-Tol, *p*-$MeOC_6H_4$, *p*-$CF_3C_6H_4$
M = Li, Ar = *p*-$PhOC_6H_4$, 3,5-$Br_2C_6H_3$

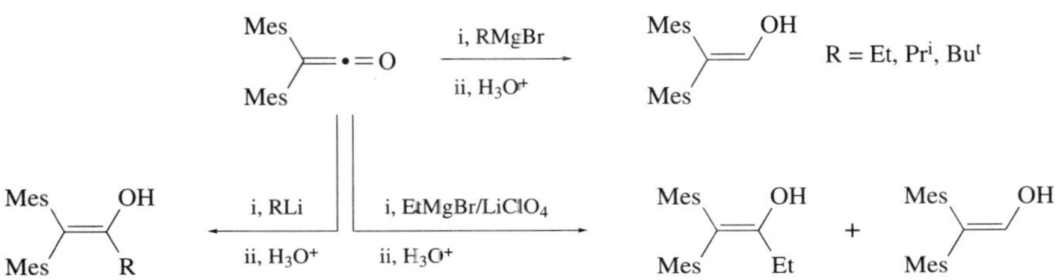

R = Pr[i], 43%
R = Bu[t], 55%

50 : 50

Scheme 16

(e) Enolization of carbonyl compounds. This is the most direct method as long as the substrates are suitably substituted with bulky aryl groups that push the equilibrium towards formation of

enols. Ironically, the preparation of the carbonyl compounds have sometimes proven to be more difficult than the respective enols ⟨44JA1109, 46JA389, 85JA1007⟩. Oxidative cleavage of glycols provides the aldehydes which tautomerize readily on treatment with base or heating, as demonstrated by the synthesis of the enol (**49**) (Scheme 17) ⟨48JA602⟩. Similarly, basic enolization of the ketone (**50**), which is easily obtained by reacting the acid chloride with mesityl magnesium bromide, delivers the triaryl ethenol (**51**) (Scheme 18) ⟨44JA1464⟩. The 1-tipyl-1-acenaphthyl ketone (**52**), an annulated version of a 1,2,2-triaryl ethenol, enolizes readily on exposure to base (Equation (15)) ⟨76JOC3599⟩. In the case of the enolization of the 9-anthryl ketone (**53**) under basic conditions (Bu^tOK/DMF), a 64 : 36 mixture of the (*E*)- and (*Z*)-enols (**54**) is obtained (Scheme 19). The (*E*)/(*Z*) mixture can be equilibrated under acidic conditions to furnish exclusively the thermodynamic (*Z*)-enol ⟨87JOC5205⟩.

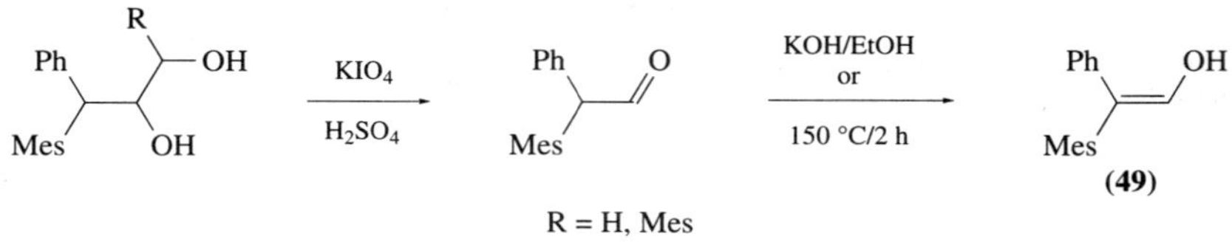

Scheme 17

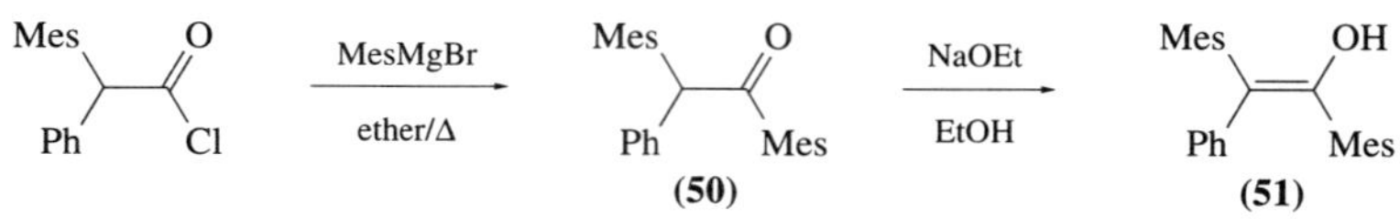

Scheme 18

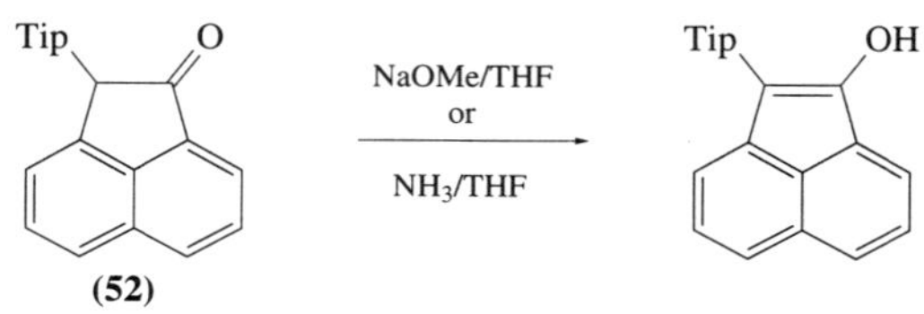

(15)

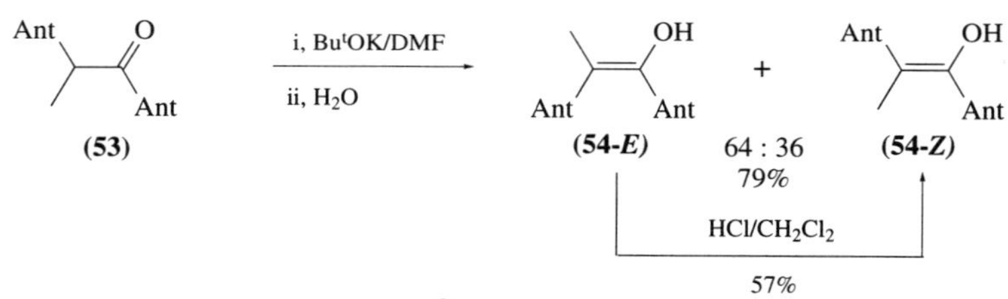

Scheme 19

2.13.1.1.2 *Phenols*

Phenols can be regarded as a class of special enols, stabilized by the aromaticity of the benzene ring. Owing to this stabilization, phenols exist in the enol form rather than the keto form (Equation (16)). In general, phenol can be prepared by elaboration of existing aromatic substrates. Alternatively, aliphatic fragments can be connected to form cyclohexanones which can be aromatized to phenols. This second approach allows more flexibility in the introduction of functionalities into the aromatic systems.

(16)

(*i*) *From aromatic substrates*

For the preparation of highly functionalized phenols, the classical aromatic electrophilic and nucleophilic substitution reactions play an important role in the introduction of various functional groups into an existing aromatic ring. The intrinsic electronic effects asserted by the existing substituents on the benzene ring, however, often dictate the manner in which other functionalities can be incorporated. It is not uncommon for a mixture of regioisomeric products to be obtained. In this section, the primary focus is on the formation of an aryl carbon–oxygen bond.

(a) Aromatic nucleophilic substitution. The fusion of benzene sulfonic acid (obtained from sulfonation of benzene) with sodium hydroxide at high temperatures is one of the oldest methods for the preparation of phenol, reported separately by Kekulé ⟨1866MI 213-01⟩ and Wurtz ⟨1867LA(144)121⟩. The reaction (Scheme 20) proceeds with direct displacement of the sulfonate group by a hydroxide anion ⟨66BCJ1212⟩. The harsh reaction conditions, however, may lead to decomposition of other incompatible functional groups.

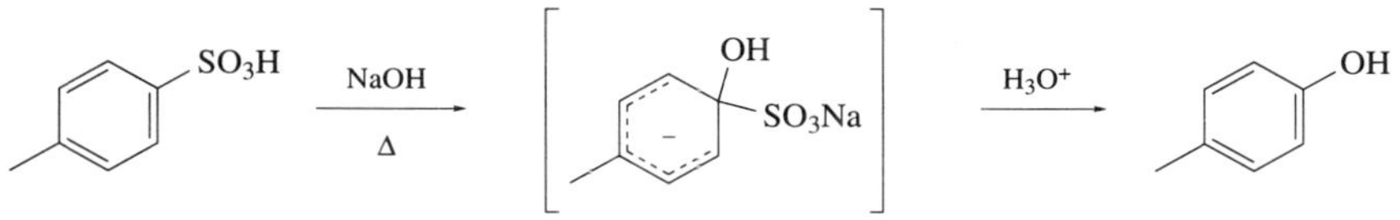

Scheme 20

Nucleophilic displacement of halides from halobenzenes by hydroxide is one of the most common methods used in the preparation of phenols and is widely practiced in industry. The reaction proceeds smoothly when the halides are activated by electron withdrawing groups (EWG) at the *ortho* and/or *para* positions. Vigorous reaction conditions (high temperature and pressure) are generally required for unactivated halides. The benzyne intermediate (**55**) ⟨57JA1458, 62CRV81⟩ is usually involved and results in a mixture of regioisomeric products (Scheme 21). Hydrolysis of aryl iodides can be catalyzed by copper sulfate under milder conditions, as in the conversion of iodovanillin (**56**) to the phenol (**57**) (Equation (17)) ⟨62CJC2175⟩.

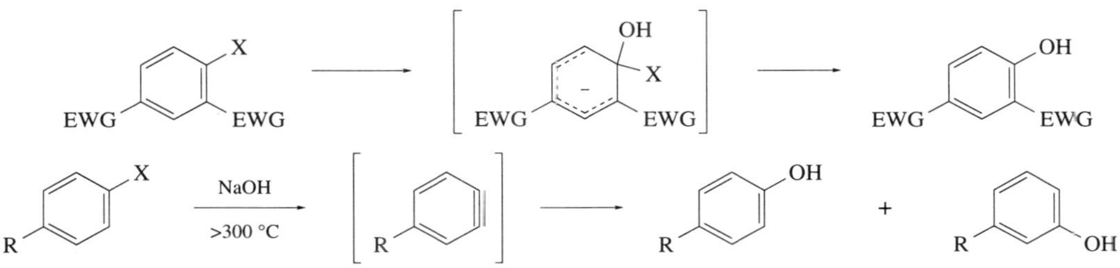

(**55**)

Scheme 21

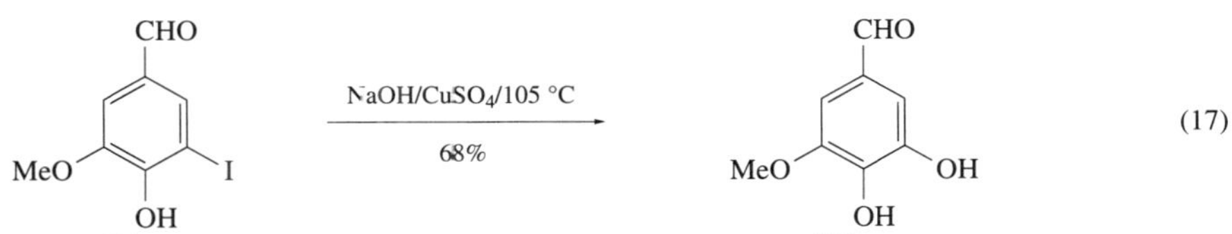

(**56**) (**57**)

The Bucherer reaction ⟨67AG(E)307⟩ has been used to convert aminonaphthalene to the naphthol (Scheme 22), although the transformation is restricted to the naphthalene system. Primary amino groups can be transformed into better leaving groups, such as a diazonium salt via diazotization, which can be hydrolyzed in turn to give phenols. Typically, the diazonium salt is thermally decomposed in nonnucleophilic acids such as aqueous sulfuric acid, as demonstrated in the preparation of the bromophenol (**58**) (Scheme 23) ⟨55OSC(3)130⟩. In extreme cases, heating with potassium trifluoroacetate in trifluoroacetic acid (TFA) is required, and the phenol (**59**) is obtained upon aqueous work up (Scheme 24) ⟨73CJC2347⟩. A milder procedure involves decomposition of a diazonium salt to the aryl radical (**60**) by copper(II) oxide, and subsequent oxidation by a large excess of copper(III) nitrate hydrate to afford the phenol in high yields (Scheme 25) ⟨77JOC2053⟩. This redox method is highly efficient and devoid of the side reactions commonly encountered in the thermal decomposition of diazonium salt, as illustrated by the preparation of the nitro phenol (**61**) in Scheme 26.

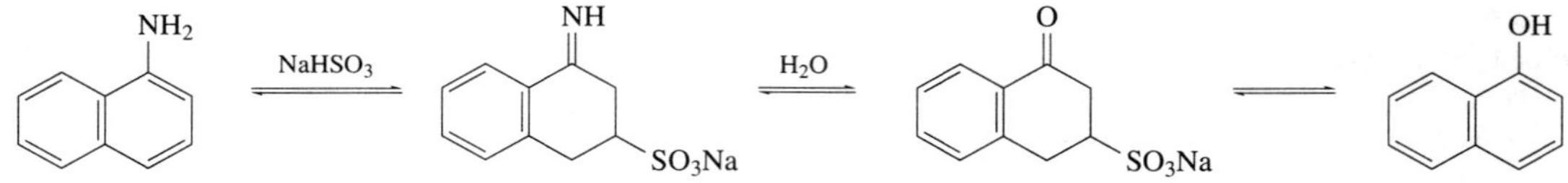

Scheme 22

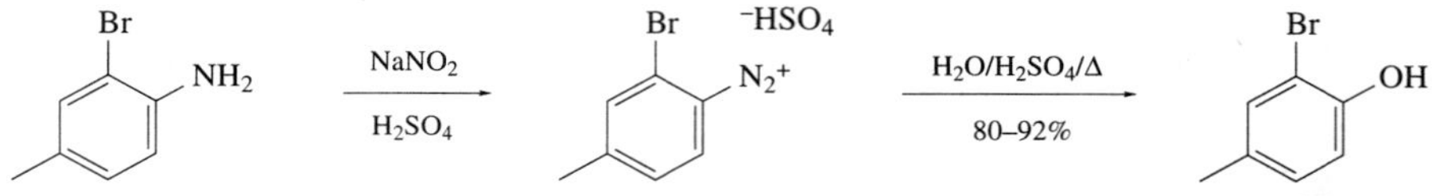

(58)

Scheme 23

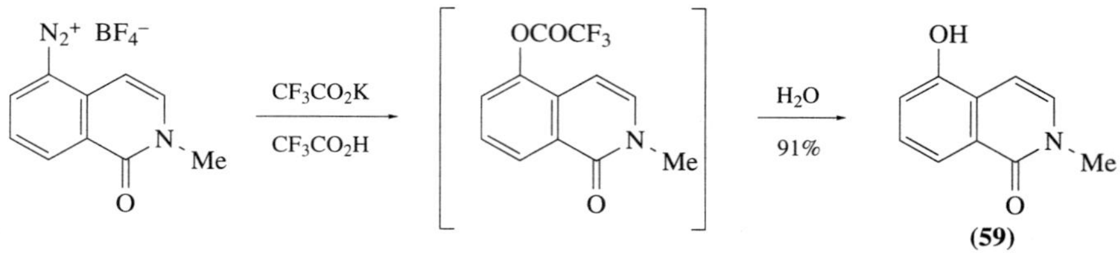

(59)

Scheme 24

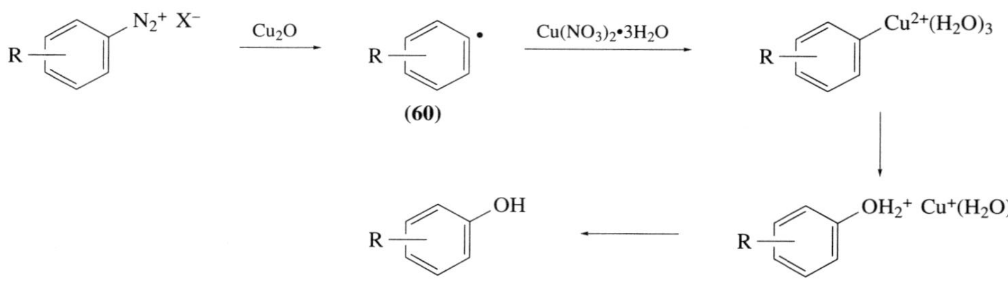

Scheme 25

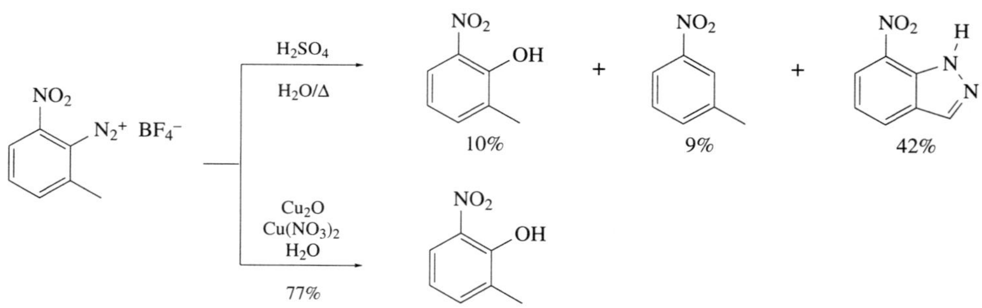

(61)

Scheme 26

When activated appropriately, an aryl nitro group is displaced just as easily. Treatment of the nitroarenes (**62**) with two equivalents of a sodium oximate in DMSO results in the formation of the phenols (**63**) in high yields (Equation (18)) ⟨74JOC3343⟩. 4-Nitrophenol is obtained in 90% yield when 1,4-dinitrobenzene is treated with potassium superoxide (Equation (19)) ⟨76TL2809⟩.

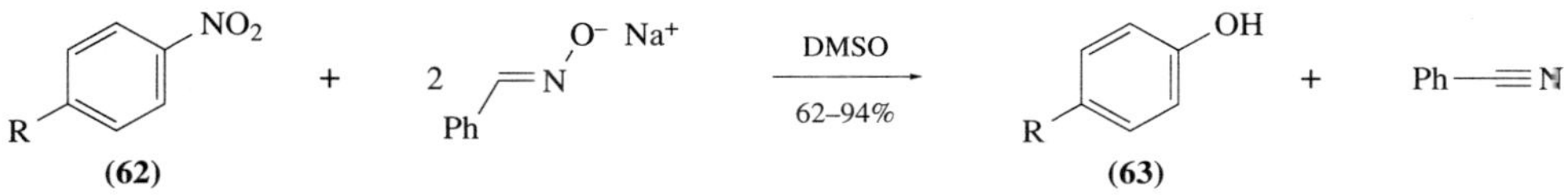 (18)

R = CO_2Me, NO_2, COPh, $CONH_2$

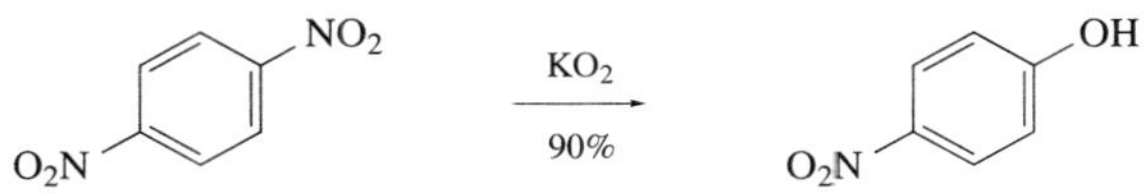 (19)

Upon exposure to aqueous acid, *N*-arylhydroxylamines undergo Bamberger rearrangement to provide *para*-aminophenols ⟨52QR34⟩. The reaction is actually an intermolecular nucleophilic addition of water to a nitrenium ion (**64**) ⟨84JOC4295⟩, as demonstrated by the degradation of the hydroxylamine (**65**) in aqueous perchloric acid (Scheme 27) ⟨87JA2824⟩.

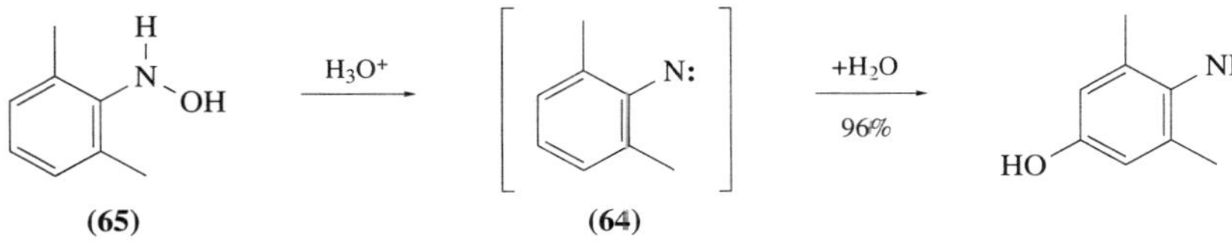

Scheme 27

Reaction of *N*-benzylamide with potassium superoxide provides an intriguing entry to *para*-substituted phenols (Scheme 28) ⟨81T2313⟩. The aryl cation intermediate (**66**) has been invoked, and the corresponding phenols (**67**) are isolated in reasonable yields (38–67%).

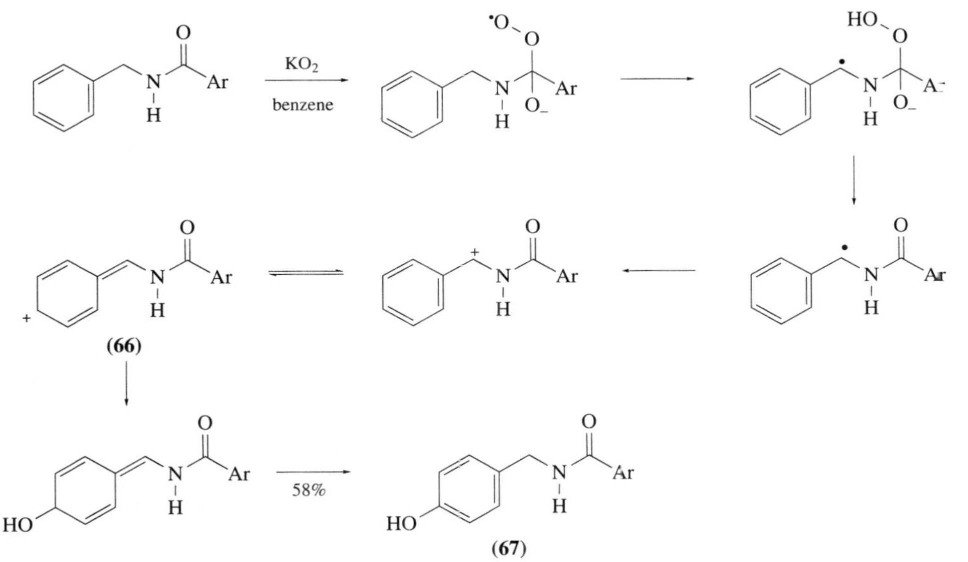

Scheme 28

(b) Electrophilic hydroxylation of arenes. Oxidation of the aryl C—H bond has been reviewed by Haines ⟨B-85MI 213-01⟩ and Dudfield ⟨91COS(7)329⟩. Direct hydroxylation of arenes can be effected by a wide array of oxidizing agents. Substituted arenes normally result in a mixture of regioisomeric phenols. For arenes substituted with electron donating groups (EDG), hydroxylation can occur at both the *ortho*- and *para*-positions, with the *ortho*-hydroxy product usually predominating. Furthermore, the introduction of the hydroxy group activates the aromatic nucleus towards further oxidation. The arenes are often used in large excess to the oxidizing agent in order to prevent over-oxidation of the products.

Electrophilic oxidation of arenes with hydrogen peroxide are achieved in the presence of either Lewis acids ($AlCl_3$, BF_3) or strong mineral acids (HF). The function of the acid is twofold: (1) to activate the peroxide to be more electrophilic, and (2) to deactivate the hydroxylated arene products from further oxidation via complexation or protonation of the oxygen. For substituted arenes, a mixture of regioisomeric phenols is normally obtained in moderate yield (Equation (20)) ⟨71JOC3184⟩. Olah and others have examined a variety of superacid systems to effect the oxidation, including SbF_5/HF ⟨86BSF625⟩, BF_3/HF ⟨81JOC4305⟩, pyridinium—$(HF)_x$ ⟨79S536⟩, FSO_3H ⟨78JOC865⟩, and HF ⟨70JOC4028⟩. The yields under these conditions are moderate, but the oxidation is not completely regioselective (Equation (20)).

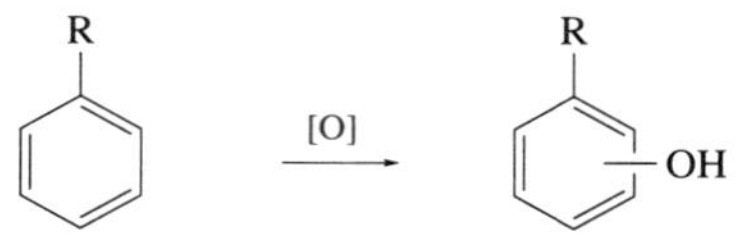

(20)

$H_2O_2/AlCl_3$

R	*(o)*	*(m)*	*(p)*	*Yield(%)*
OMe	44	1	55	70
Me	60	8	32	40
Cl	26	4	70	20
H	–	–	–	10

$H_2O_2/BF_3/HF/–78$ °C

R	*(o)*	*(m)*	*(p)*	*Yield(%)*
Me	70	9	21	52
Et	65	11	24	58
Pr^i	48	12	40	43
Bu^t	30	14	56	36
H	–	–	–	37

Alkyl peroxides (e.g., *t*-butyl hydroperoxide ⟨70BCJ293⟩, tetralin hydroperoxide ⟨69JOC3441⟩, etc.) work equally well in effecting the oxidation of arenes in the presence of acids with comparable yields. Notably, the combination of bis(trimethylsilyl) peroxide and trifluoromethanesulfonic acid (triflic acid) has been shown to provide both exceptional yields and regioselectivity (Equation (21)) ⟨89JOC1204⟩.

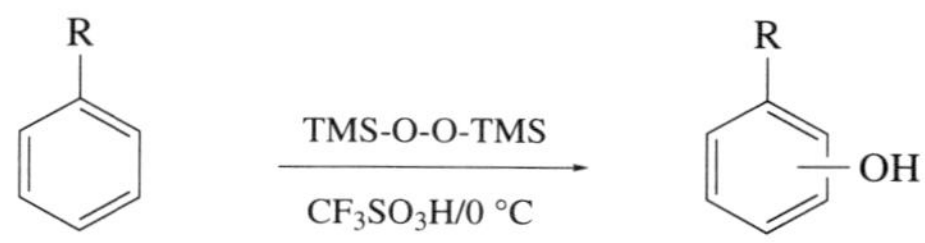

(21)

R	*(o)*	*(m)*	*(p)*	*Yield(%)*
Me	63	10	27	77
Cl	63	–	37	88
H	–	–	–	76

The oxidation of arenes can also be mediated by transition metals, as demonstrated by the classical Fenton's reagent ($H_2O_2/FeSO_4$) ⟨1894JCS899⟩, Udenfriend's reagent ($O_2/FeCl_3$/EDTA/ascorbic acid) ⟨54JBC(208)731, 60JA3472, 82JMOC333⟩ and Funabiki's reagent ($O_2/FeCl_3$/hydroquinone) ⟨89CL1267⟩. However, the yields are low for these oxidations, and they provide limited synthetic utility.

Peroxy acids are also highly effective in oxidizing arenes. Peroxytrifluoroacetic acid ⟨59JCS1804⟩, *meta*-chloroperoxybenzoic acid (mcpba) ⟨83S1046⟩, and peroxymonophosphoric acid (H_3PO_5 from P_2O_5/H_2O_2) ⟨74T3021⟩ are commonly used peroxy acids. The combining of a peroxy acid with a Lewis acid results in a highly potent oxidizing reagent. The oxidation of isodurene (**68**) with peroxytrifluoroacetic acid/BF_3 gives the phenol (**69**) exclusively in good yield ⟨71CRV337⟩. In contrast, similar oxidation of prehnitene (**70**) gives a variety of products (Scheme 29) ⟨64JOC2397⟩. For less substituted arenes (e.g., benzene, toluene, etc.) overoxidation usually occurs, resulting in tar formation. The inexpensive sodium perborate in conjunction with triflic acid has been exploited successfully in oxidizing arenes to their corresponding phenols in good yields with moderate regioselectivity (Equation (22)) ⟨91SL39⟩.

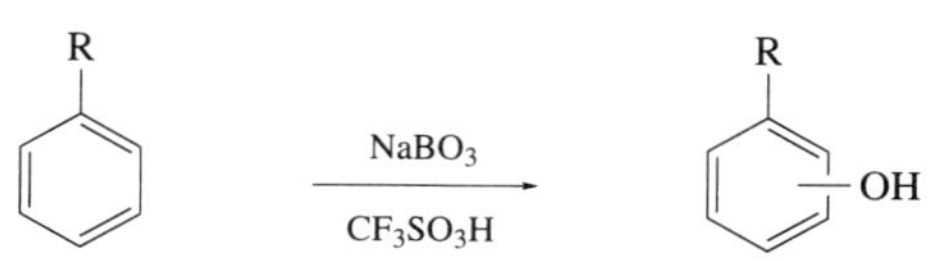

(22)

R	*(o)*	*(m)*	*(p)*	*Yield(%)*
Me	65	9	26	66
Et	65	11	24	68
Cl	32	9	59	71
H	–	–	–	62

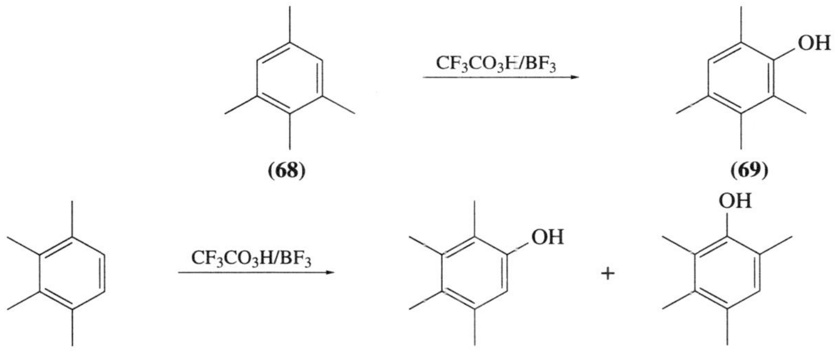

(68) **(69)** **(70)**

Scheme 29

The persulfate oxidation of phenol to hydroquinone is known as the Elbs oxidation ⟨1893JPR179, 88OR(35)421⟩. The water soluble sulfate salt (**71**) is the intermediate, and can be separated readily from the unchanged starting material and other organic products simply by washing with an organic solvent (Scheme 30). Subsequent acid hydrolysis of (**71**) provides the hydroquinone in high purity, and the yields are typically in the range of 20–50% ⟨48JCS2303⟩. This oxidation has been applied to highly functionalized substrates such as the naphthol (**72**) ⟨78CC242⟩. For *para*-substituted phenols, oxidation occurs at the *ortho*-position to give catechols in comparable yields. In contrast, oxidation of arylamines occurs predominantly at the *ortho*-position to provide the *ortho*-amino phenol (Scheme 31) ⟨53JCS3623⟩. The Elbs conditions have been modified (Et_4NOH/pyridine/$K_2S_2O_8$) and applied to more complex substrates, as in the synthesis of the flavone (**73**) (Equation (23)) ⟨84JOC645⟩.

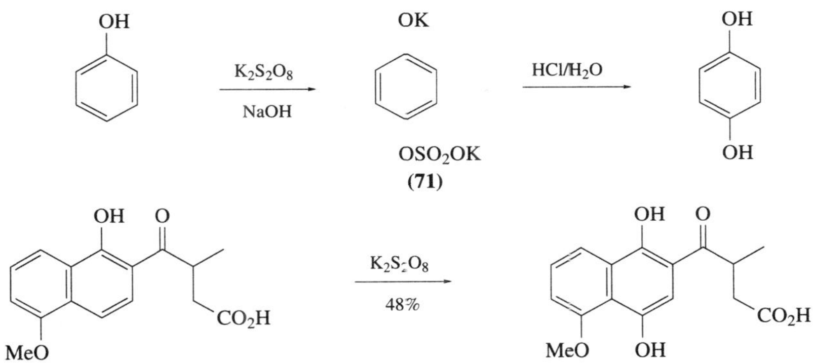

(71) **(72)**

Scheme 30

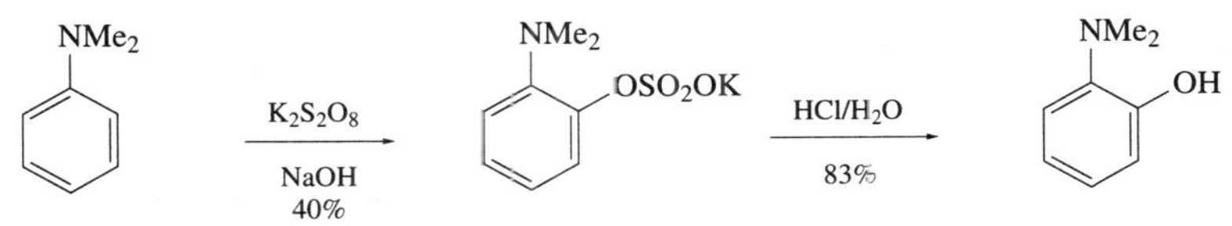

Scheme 31

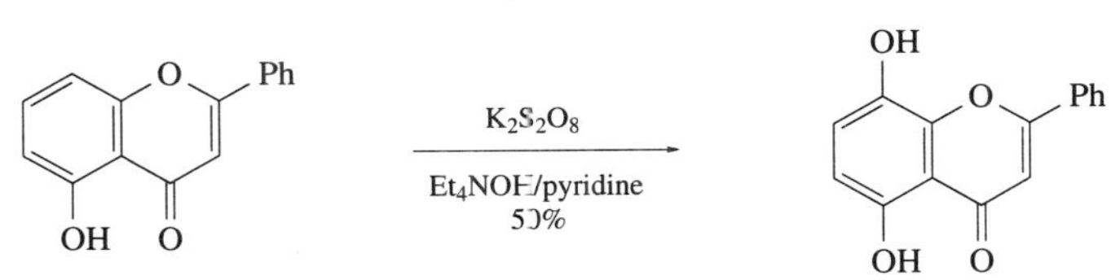

(73) (23)

Trifluoroacetyl nitrate, prepared *in situ* from ammonium nitrate and trifluoroacetic anhydride (TFAA), is also effective in oxidizing hydroxyarenes ⟨83JA4112⟩. In the synthetic approach to anthracycline analogues, the phenol (**74**) is oxidized readily to the hydroquinone (**75**) (Scheme 32) ⟨83JA4112⟩.

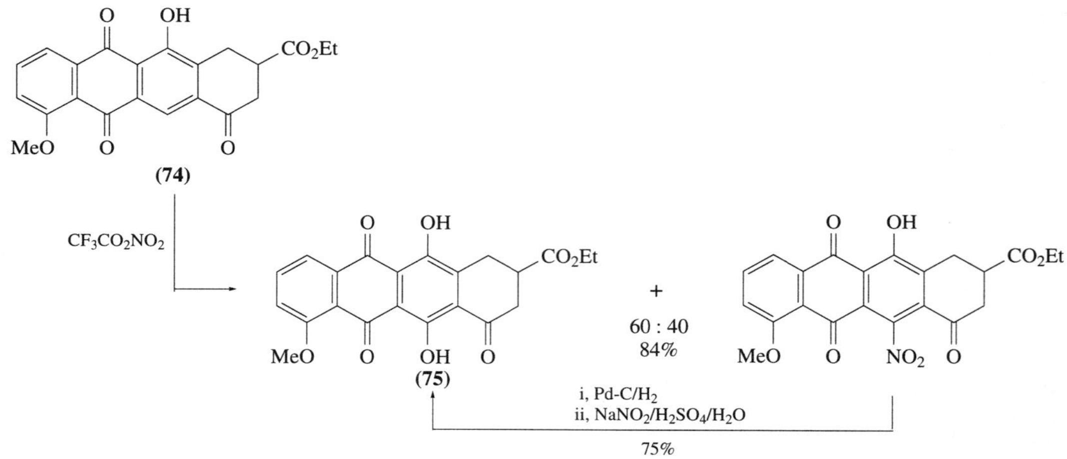

Scheme 32

Finally, oxidation of arenes, especially electron deficient systems, can be accomplished with cobalt(III) trifluoroacetate (one-electron oxidant) in high yields, via a cationic radical intermediate (Equation (24)) ⟨73JA7114⟩.

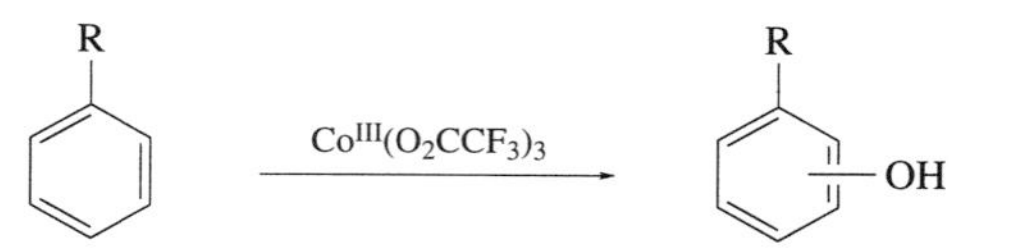

(24)

$R = OMe, Me, Et, Bu^t, I, Br, Cl, OAc, OPiv, O_2CCF_3, CF_3$

R	*(o)*	*(m)*	*(p)*	*Yield(%)*
Me	65	9	26	66
Et	65	11	24	68
Cl	32	9	59	71
H	–	–	–	62

(c) Oxidation of metallated arenes. The major disadvantage of the aforementioned electrophilic hydroxylation is the lack of regioselectivity. The oxidation of an aryl organometallic species will lead, however, to the formation of a phenol in a regioselective manner, because the oxygen is inserted into the existing carbon–metal bond. The classical Grignard reagent (arylmagnesium halides) and halogen–metal exchange processes provide ready access to regio-defined metallated arenes from widely available haloarenes (Scheme 33). The commonly used oxidants include *t*-butylperoxy-1,3,2-dioxaborolane ⟨83S107⟩, molybdenum peroxide–pyridine–hexamethylphosphoramide (MoOPH) ⟨77JOC1479⟩, the magnesium salt of *t*-butyl hydroperoxide ($Bu^tOOMgBr$) ⟨59JA4230⟩, bis(trimethylsilyl) peroxide ⟨86S633⟩, sulfonyloxaziridines ⟨87TL5115⟩, nitrobenzene ⟨77JOC2274, 88JMC1080⟩, and lead tetraacetate ⟨70JA3520⟩. The utility of these approaches is exemplified by the oxidation of the aryl magnesium bromides and aryl lithiums in Scheme 34.

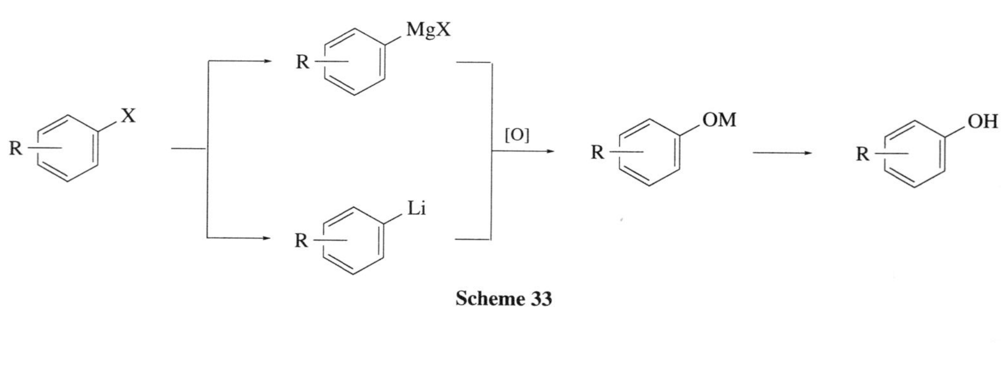

Scheme 33

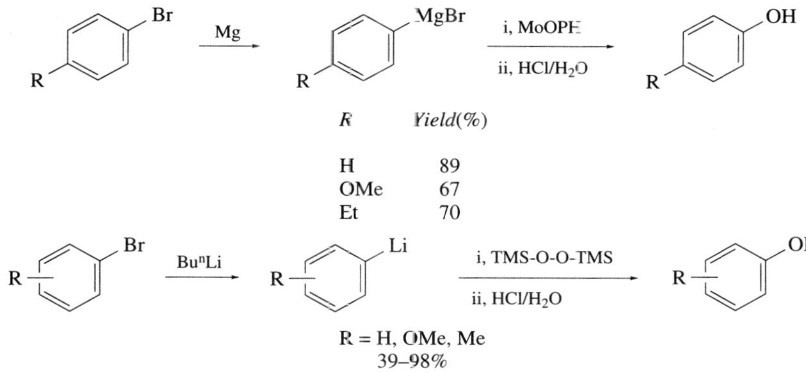

Scheme 34

Frequently, aryl magnesium, lithium, and even mercury ⟨71CC1475, 82CC1157⟩ and thallium ⟨75CC36⟩ derivatives, are transmetallated first with trimethyl borate ⟨57JOC1001, 69OS(49)90⟩ or diborane ⟨74JOM(76)C23⟩ to the boronate (**76**). Subsequent oxidative cleavage of the carbon–boron bond with hydrogen peroxide leads to the production of the corresponding phenols in high yields (Scheme 35).

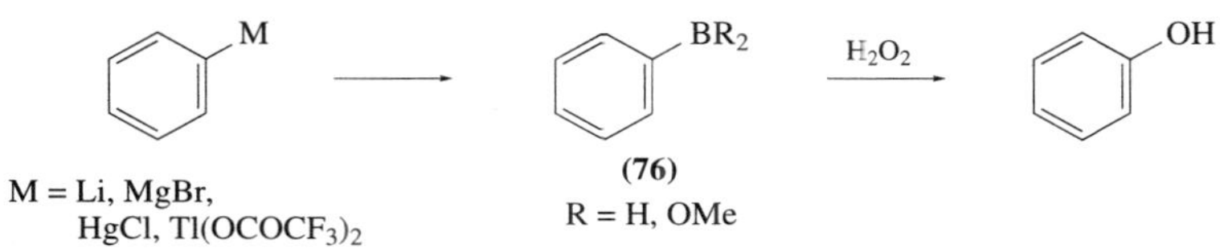

Scheme 35

Another way of generating regioselectively metallated arenes involves *ortho* directed metallation by existing functional groups on the aromatic ring (Scheme 36) ⟨B-74MI 213-01, 79OR(26)1⟩. The most widely used directing groups include alkoxy ⟨70JA2042, 83T2031⟩, tertiary amide ⟨87JOC674, 90CRV879⟩, oxazoline ⟨85T837⟩ and, interestingly, the formyl group ⟨88CC1350⟩ after derivatization *in situ* to the bidentate diamino alkoxide (**77**) (Scheme 37).

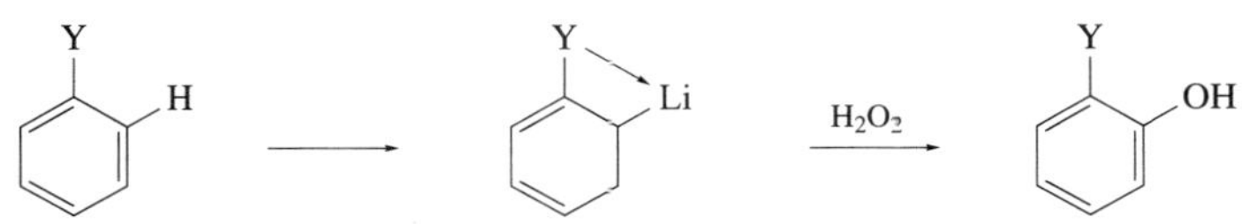

Scheme 36

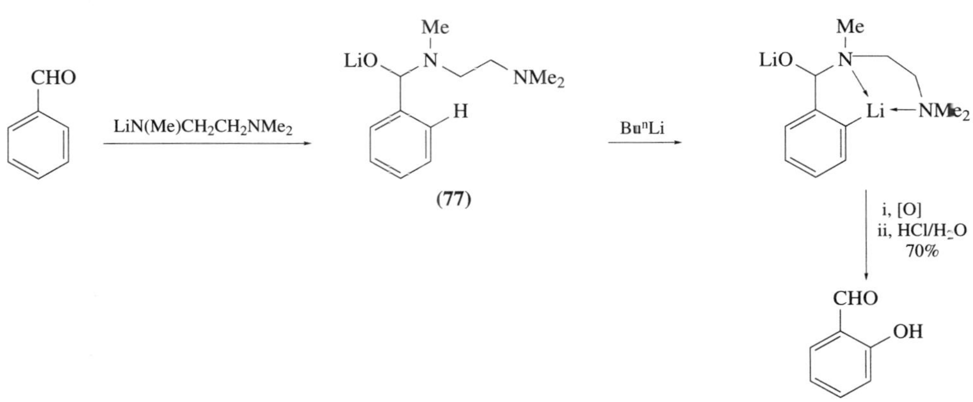

Scheme 37

(d) Baeyer–Villiger oxidation of acyl arenes. While Friedel–Crafts acylation and Vilsmeier–Haack formylation of arenes provides ready access to a variety of acyl and formyl arenes, the Baeyer–Villiger oxidation offers an efficient way to introduce oxygen into these aromatic ring systems. Subsequent hydrolysis of the resulting aryl esters provides an indirect yet reliable route to the synthesis of phenols (Scheme 38). Reviews on the Baeyer–Villiger oxidation have been conducted on numerous occasions and most recently by Krow ⟨93OR(43)251⟩.

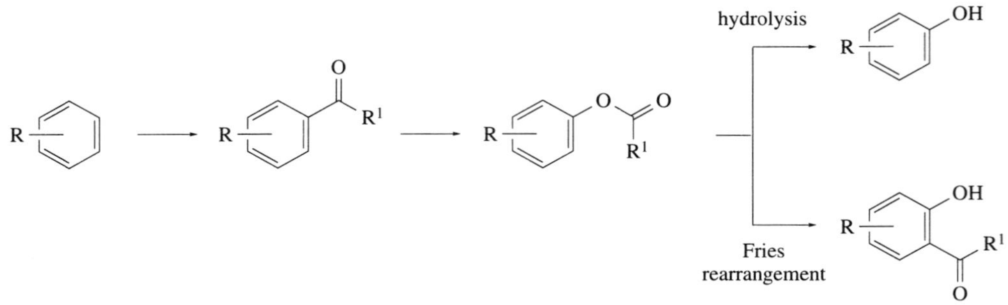

Scheme 38

Aryl aldehydes and ketones undergo Baeyer–Villiger oxidation with peroxyacetic acid ⟨61JOC1215⟩, mcpba ⟨74JCS(P1)1353⟩ and other peroxy acids, to produce aryl esters in high yields. Phenols can be isolated in excellent yield simply by hydrolysis of the corresponding esters as shown in Scheme 39. The use of urea–hydroperoxide (UHP) and TFA, a peroxytrifluoroacetic acid variant, is also highly effective ⟨90SL533⟩. Hydrogen peroxide, coupled with a catalytic amount of a selenium compound, typically selenium dioxide, provides an alternative to the peroxy acid system ⟨89S167⟩. This system is also found to be chemoselective in oxidizing an aryl aldehyde rather than ketone or other oxidizable functionalities, e.g., alkenes, as illustrated by the oxidation of the formyl chromanone (**78**) and chromene (**79**) (Scheme 40) ⟨92SC1555⟩. For unactivated aryl aldehydes, oxidation to the corresponding carboxylic acids is not uncommon. This undesired outcome is circumvented by using peroxymonophosphoric acid in acetonitrile, as in the conversion of the aryl aldehyde (**80**) to the phenol (**81**) (Scheme 41) ⟨91SC2335⟩. Molecular oxygen in the presence of iron(III) oxide ⟨92TL7557⟩ or clay ⟨94CC797⟩ is also highly effective (Equation (25)).

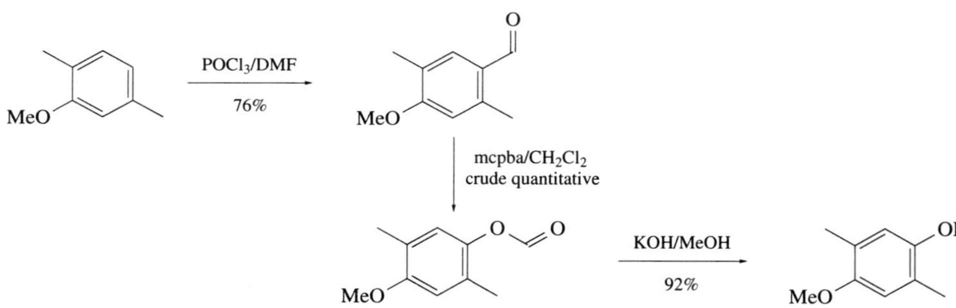

Scheme 39

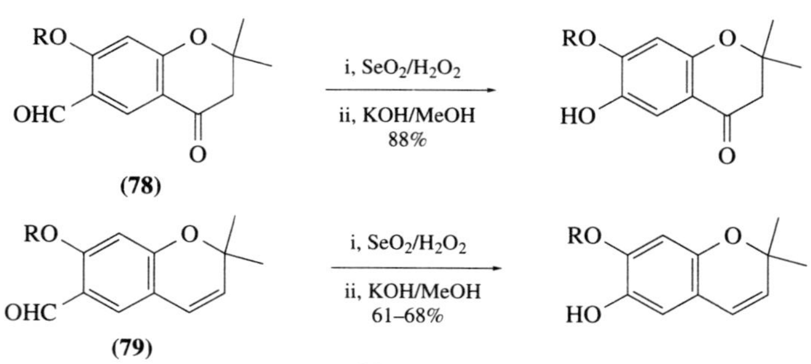

Scheme 40

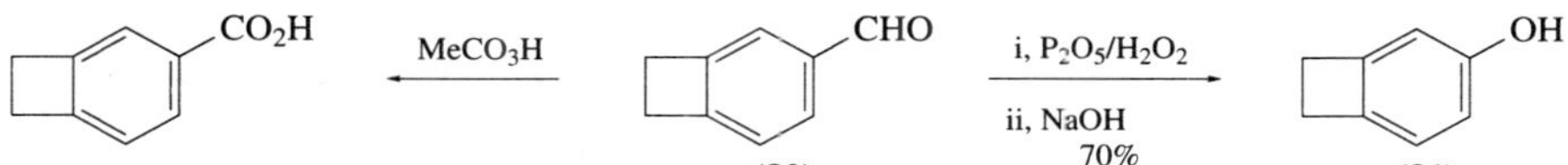

(80) **(81)**

Scheme 41

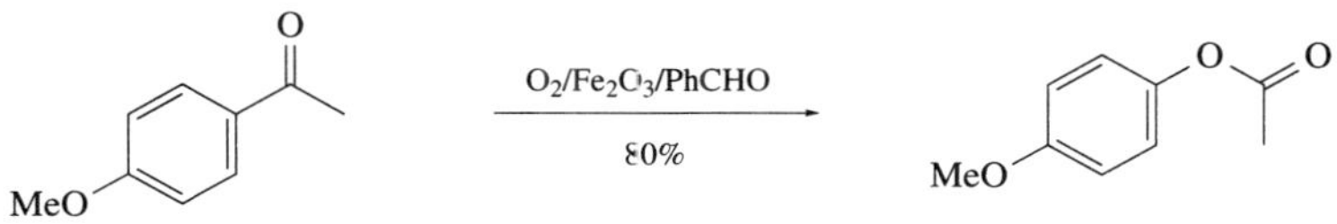

(25)

A similar oxidation of an *ortho*- or *para*- hydroxyaryl aldehyde by alkaline peroxide is known as the Dakin reaction ⟨09JA477⟩. This provides a route to the preparation of catechols and hydroquinones. Sodium percarbonate, an inexpensive reagent, has been applied successfully to the oxidation of a variety of hydroxybenzaldehydes and acetophenones (Equation (26)) ⟨92TL865⟩.

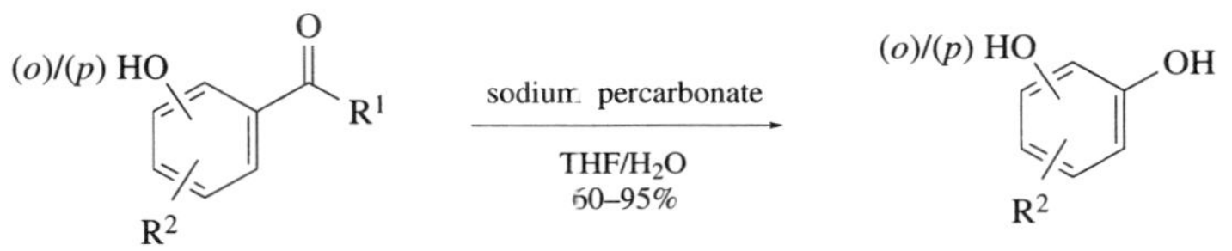

(26)

R^1 = H, Me
R^2 = H, MeO, Cl, NO_2

(e) Benzylic hydroperoxide rearrangement. An efficient industrial synthesis of phenol involves benzylic oxidation of cumene with oxygen or hydrogen peroxide ⟨68MI 213-01⟩. The resulting hydroperoxide rearranges, on treatment with mineral acid, to give phenol and acetone, two important industrial chemicals, in a single process (Scheme 42). In general, better yields are obtained with tertiary benzyl peroxides than with the secondary analogs. Addition of methyl magnesium bromide to acyl arenes provides a convenient route to benzyl carbinols. As an alternative to the Baeyer–Villiger protocol, oxidation and rearrangement of the benzyl carbinol to phenols (Scheme 43) can be achieved in one-pot with hydrogen peroxide promoted by H_2SO_4 ⟨81TL3451⟩, *p*-TsOH ⟨86JOC5436⟩ or BF_3 ⟨87TL1027⟩. Sodium perborate in the presence of excess BF_3 is also highly effective ⟨93TL7667⟩. This methodology has been skillfully utilized in the synthesis of selectively protected L-Dopa derivatives **(82)** ⟨87JOC5283⟩ and the various subunits **(83)** of the antitumor antibiotic CC-1065 ⟨87JA2717⟩. This approach has also been used successfully as an alternative to the Elbs oxidation in converting phenols to *para*-hydroquinones as depicted in Scheme 44 ⟨81TL2337⟩.

(f) Claisen rearrangement. Allyl aryl ethers undergo a [3,3]-sigmatropic allyl shift (aromatic

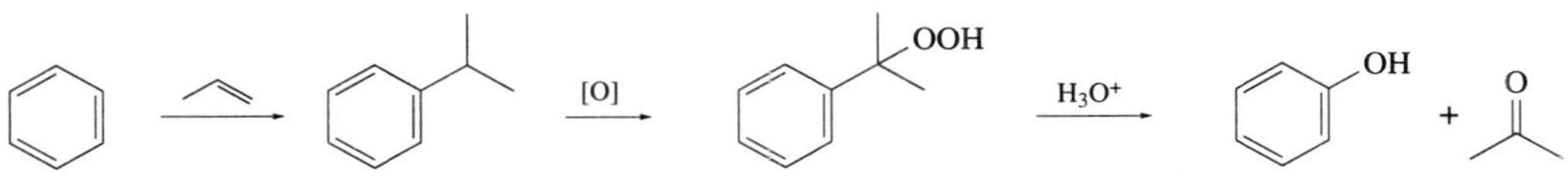

Scheme 42

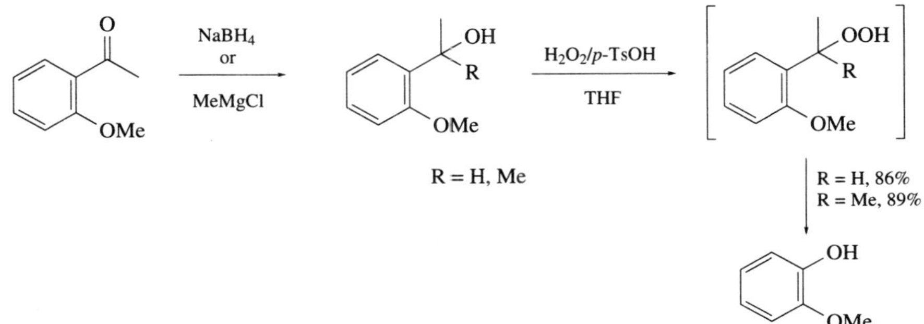

Scheme 43

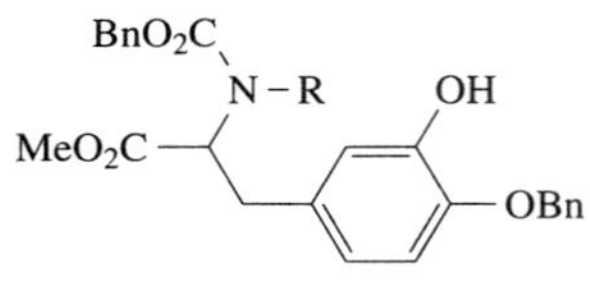

(82)

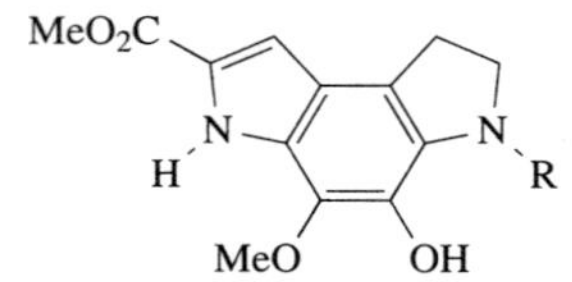

(83)

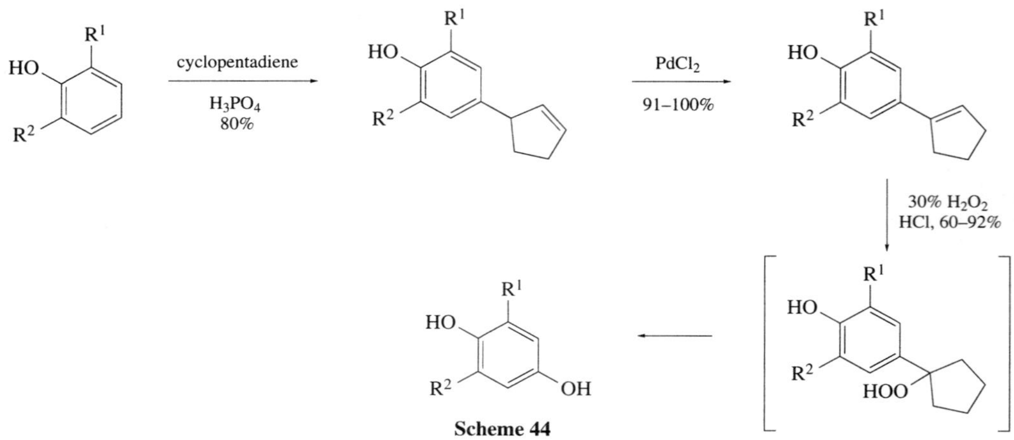

Scheme 44

Claisen rearrangement) to the *ortho*-dienone on heating, followed by rapid tautomerization to produce predominantly the *ortho*-allyl phenol (Scheme 45). Further migration of the allyl group via a second [3,3]-sigmatropic allyl shift to provide the *para*-allyl phenol is not uncommon. The subject of Claisen rearrangement has been reviewed by Rhoads and Raulins ⟨75OR(22)1⟩ and more recently by Wipf ⟨91COS(5)827⟩. The distribution of the *ortho*/*para*-products is influenced greatly by the reaction conditions, particularly by the polarity of the solvent ⟨63HCA2444, 68HCA1603⟩. The rearrangement has been performed under mild conditions with the aid of Lewis acids such as $TiCl_4$ ⟨75CL1041⟩, BCl_3 ⟨73HCA14⟩, Et_2AlCl ⟨70JOC3166⟩, acids such as acetic acid/sodium acetate ⟨76JOC3026⟩, or bases such as benzylamine ⟨79CJC3304⟩. The yields are generally satisfactory, as illustrated by the rearrangement of the allyloxyanthraquinone (**84**) in 88% yield (Equation (27)) ⟨79CJC3304⟩.

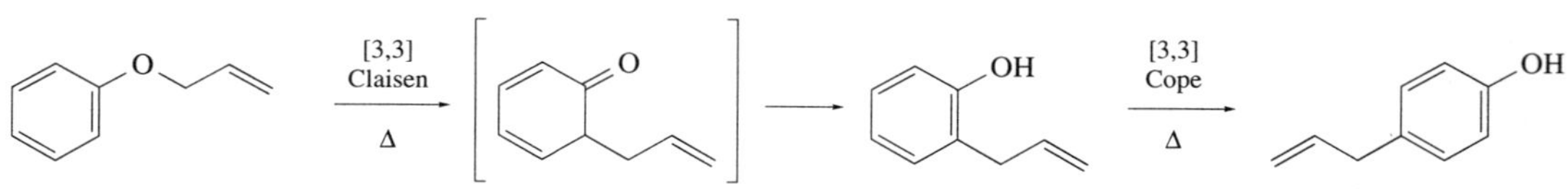

Scheme 45

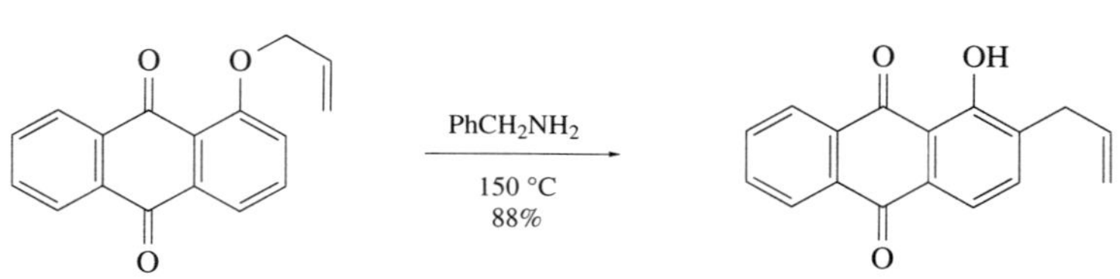

(84) (27)

(*g*) *Fries rearrangement.* The tandem Friedel–Crafts acylation/Baeyer–Villiger oxidation protocol (see above) provides access to a variety of phenyl esters. On heating with a Lewis acid, such as aluminum chloride, these esters undergo the Fries rearrangement ⟨08CB4271⟩ to give, predominantly, the *ortho*-acyl phenols in high yields (Scheme 38). Other widely used acid catalysts include $TiCl_4$ ⟨89S25⟩, BF_3 ⟨76TL1637⟩, methanesulfonic acid ⟨73AG(E)775⟩, anhydrous HF ⟨73JOC1924⟩ and boric acid ⟨39JCS398⟩. In contrast, the *para*-isomer predominates when the reaction is carried out in the presence of *β*-cyclodextrin in the photochemically induced rearrangement of phenyl acetate (Equation (28)) ⟨75AG(E)820⟩. An updated review in this area has been conducted by

Martin ⟨92OPP369⟩. Treatment of an *ortho*-bromophenyl ester with *s*-butyllithium results in a halogen–metal exchange followed by an intramolecular acyl transfer to give exclusively the *ortho*-acyl phenol (Equation (29)) ⟨87JOC322⟩ and is commonly regarded as the "metallo-Fries" rearrangement ⟨94TL1749⟩.

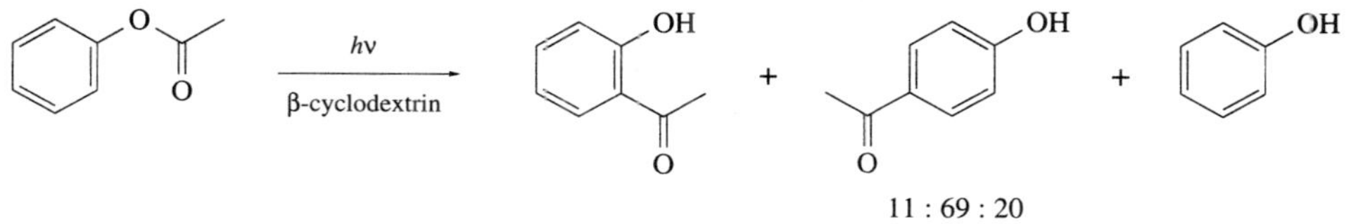

(28)

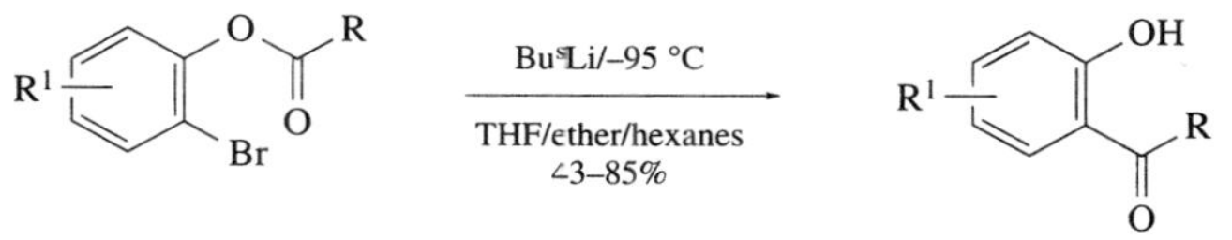

(29)

(h) Miscellaneous rearrangement. Upon treatment with four equivalents of Bu^{n}Li, the polyhaloethyl aryl ethers **(85)** are converted to the *ortho*-alkynyl phenols in good yields (70–85%) ⟨85JOC5430⟩. The benzofuran dianion intermediate **(86)** is invoked in the rearrangement (Scheme 46), with the polyhaloethyl group acting as a latent alkynyl group.

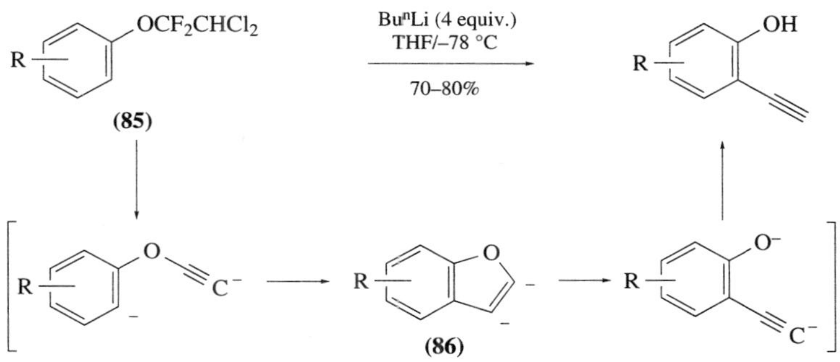

Scheme 46

(i) Reduction of quinones. Quinones can be transformed in high yields to the hydroquinones by a variety of reducing agents, such as sodium dithionite ⟨67MI 213-01⟩, chromium(II) ⟨69JCS(C)2349⟩, vanadium(II) ⟨76S815⟩, polymethylhydrosiloxane ⟨73JOC162⟩, diethylhydroxylamine ⟨75TL1695⟩, sodium borohydride ⟨76S528, 76JCS(P1)836⟩, trimethyl phosphite ⟨84HCA1406⟩, and zinc/acetic acid ⟨87H(26)65⟩. Halohydroquinones are obtained when *para*-benzoquinone is treated with a halosilane catalyzed by tetraethyl ammonium chloride (Scheme 47) ⟨78JOC3078⟩.

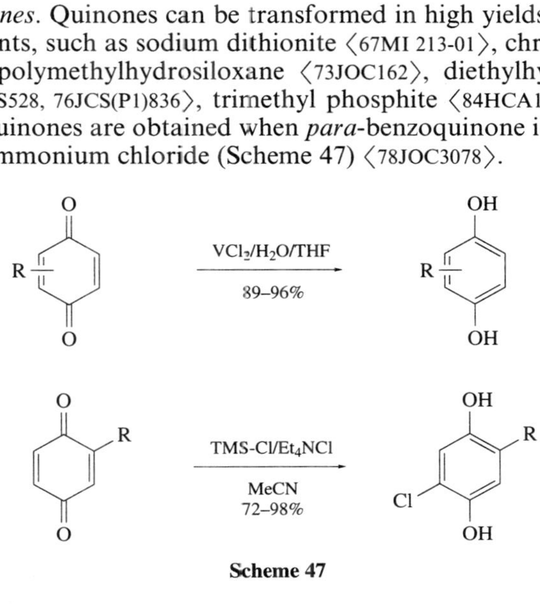

Scheme 47

(j) From O*-substituted phenolic ethers.* In daily laboratory research, phenols are often prepared (or rather liberated) by the cleavage of an oxygen–carbon bond in aryl alkyl ethers or aryl esters,

mostly due to the necessity for the temporary protection of the phenol moeity in a synthetic sequence. Various means of dealkylation of aryl alkyl ethers and hydrolysis of phenolic esters have been reviewed extensively by Bhatt and Kulkarni ⟨83S249⟩, Tiecco ⟨88S749⟩, and Green and Wuts ⟨B-91MI 213-01⟩. Generally, aryl alkyl ethers can be conveniently cleaved by treatment with hydrogen bromide ⟨71BCJ1986⟩, iodotrimethylsilane ⟨77JOC3761⟩, and Lewis acid in the presence of a soft nucleophile ⟨80JOC4275⟩. The hydrolysis of aryl esters, carbonates, carbamates or sulfonates is straightforward under either acidic or basic conditions.

(ii) Aromatization of alicyclic substrates

Cyclohexadienones undergo facile aromatization to the corresponding phenols, providing the basis for the preparation of aromatic compounds from alicyclic substrates.

(a) Aromatization of cyclohexanones, cyclohexenones and cyclohexadienones. Cyclohexadienones can be conveniently prepared from cyclohexenones via a two-step sequence, that is the introduction of a potential leaving group which is subsequently eliminated to give cyclohexadienones. The most commonly employed tactic in this regard is the bromination–dehydrobromination sequence. For this purpose, bromination has been effected by treatment with molecular bromine ⟨68CJC3625⟩, pyridinium hydrobromide perbromide ⟨75JOC3686⟩ or copper(II) bromide ⟨72JOC2901⟩ (Scheme 48). Dehydrobromination quite often is spontaneous, with concomitant aromatization to give the phenols. Occasionally, base is required to achieve the dehydrobromination. Similarly, cyclohexan-1,3-diones are elaborated to resorcinols ⟨91TL3441⟩ (Scheme 48). For substrates with other reactive functional groups, for example alkene double bonds, which are not compatible with the halogenation/dehydrohalogenation sequence, selenenylation followed by *syn* elimination of selenoxide offers a practical alternative (Scheme 49). The seleneno ketone is oxidized to the selenoxide which undergoes *syn* elimination followed by aromatization in one-pot to provide the phenols in moderate yields ⟨90TL6629, 82S683⟩. Sulfenylation of 2-methyl-cyclohexanone leads to the formation of the *ortho*-sulfenylated phenol (**87**) (Scheme 50) ⟨78TL1667⟩. In the case where the *ortho* substituent is lacking, the 2,6-bis(sulfenyl) phenol (**88**) is also obtained.

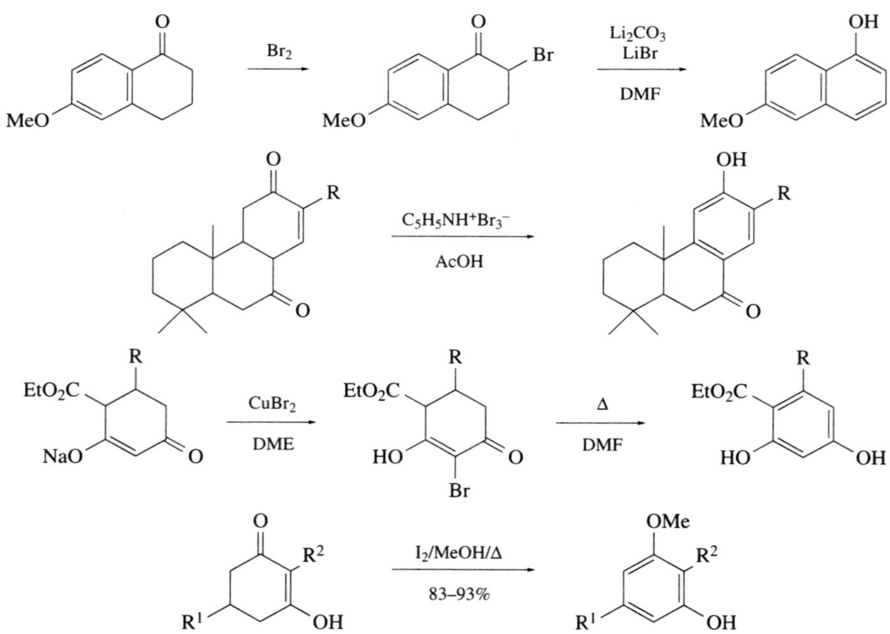

Scheme 48

Treatment of *ortho*-substituted cyclohexanones with sulfuric acid (two equivalents) in a mixture of AcOH and Ac_2O leads to the generation of the corresponding phenols in high yields (Equation (30)) ⟨74JOC2126⟩. The use of sulfuric acid and the stoichiometry are critical. In the absence of the *ortho*-substituent, only aldol condensation of the cyclohexanone is observed. Heating of

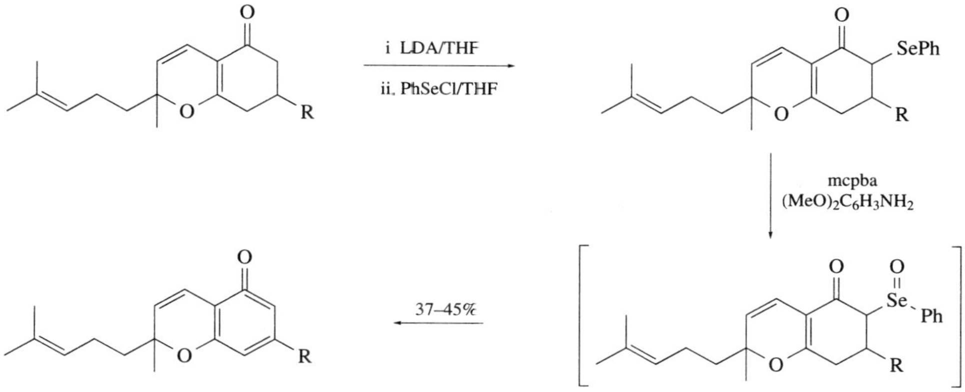

Scheme 49

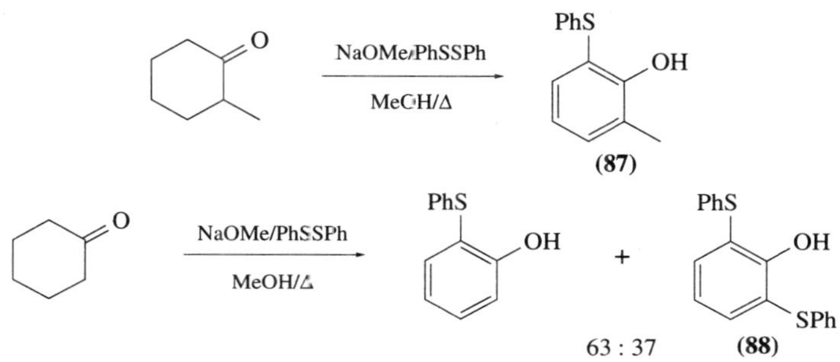

Scheme 50

disubstituted cyclohexan-1,4-diones in concentrated HCl or HBr provides the disubstituted phenols in good yields (70–80%) (Scheme 51) ⟨74CC584⟩. Once again, the substituents on the ring are essential for the success of the reaction.

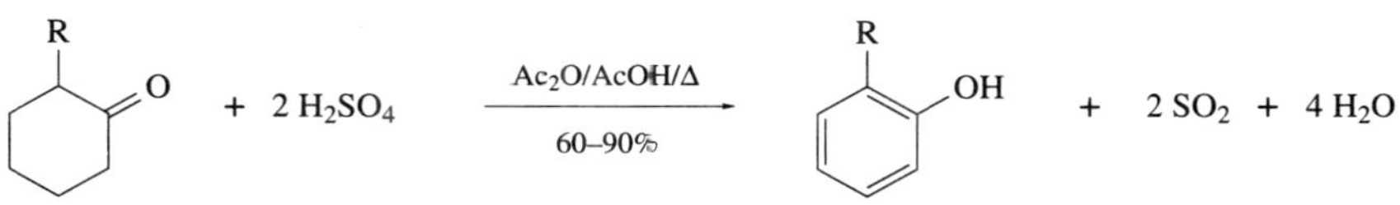

(30)

R = Me, Et, Pr^n, Bu^n, cyclohexyl

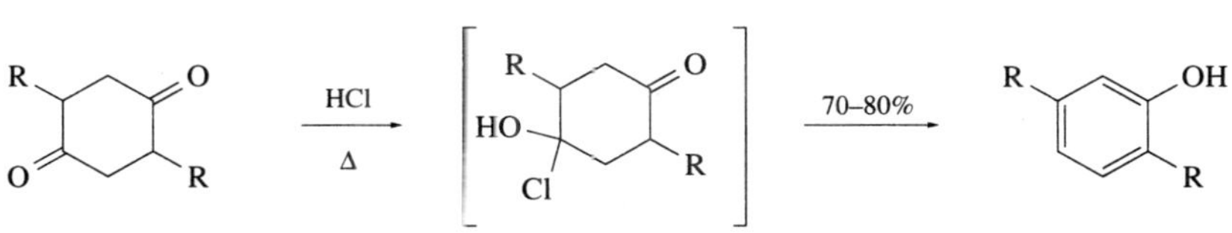

R = Me, Et

Scheme 51

Cyclohexanones and cyclohexenones can be aromatized oxidatively to phenols (Equation (31)). The dehydrogenation can be effected by a variety of reagents and catalysts such as palladium(0) ⟨76JOC977, 92TL4603⟩, palladium(II) ⟨72JPR170⟩, trityl tetrafluoroborate ⟨78CRV317⟩, *o*-chloranil ⟨79JOC4265⟩, ddq (2,3-dichloro-5,6-dicyano-1,4-benzoquinone, sulfur, selenium, and platinum ⟨B-48MI 213-01⟩. The common side reaction is dehydration that results in the formation of aromatic hydrocarbon instead of phenol.

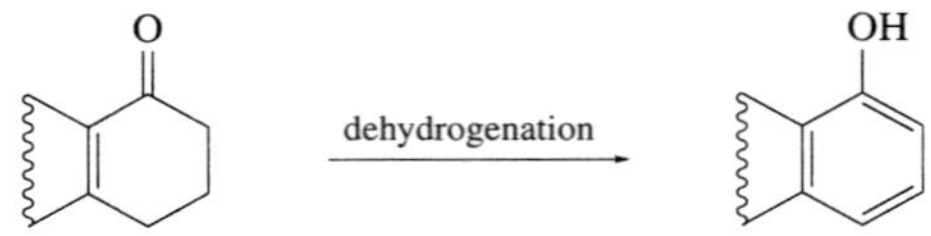

(31)

On the other hand, reductive aromatization of dienones has been achieved by treatment with lithium biphenyl ⟨64JA742⟩ or zinc ⟨63JOC783⟩ (Equation (32)). Dimethyl phosphite has been used in the aromatization of a bromo dienone in high yield (Scheme 52) ⟨63JOC345⟩.

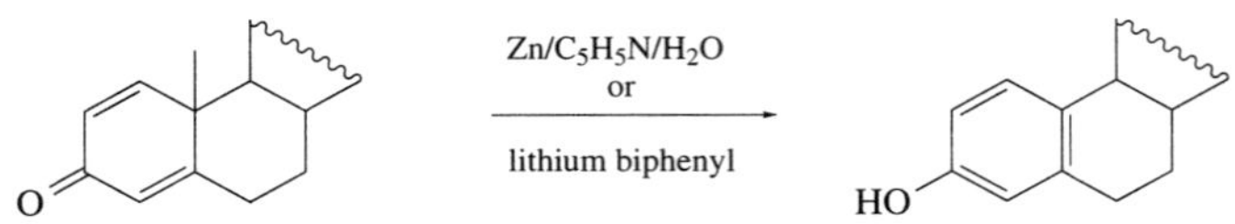

(32)

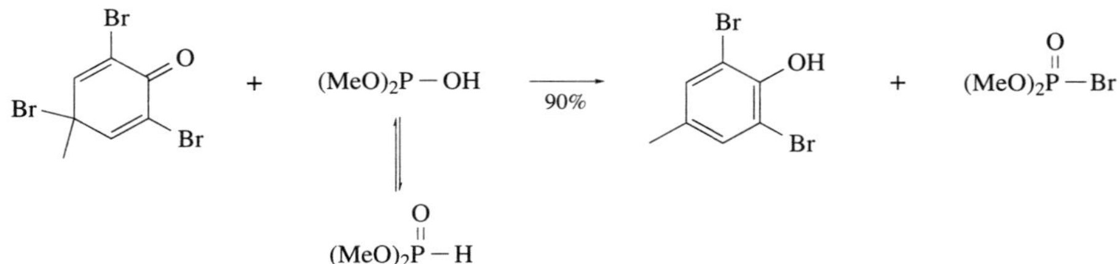

Scheme 52

Upon exposure to rhodium(III) chloride ⟨78TL2545, 82TL4843⟩ or *N*-lithioethylenediamine ⟨62JOC1430⟩, alkenyl cyclohexenones undergo remote olefinic bond migration and aromatize to the phenols. For example, carvone has been transformed to the phenol (**89**) in high yields (Equation (33)).

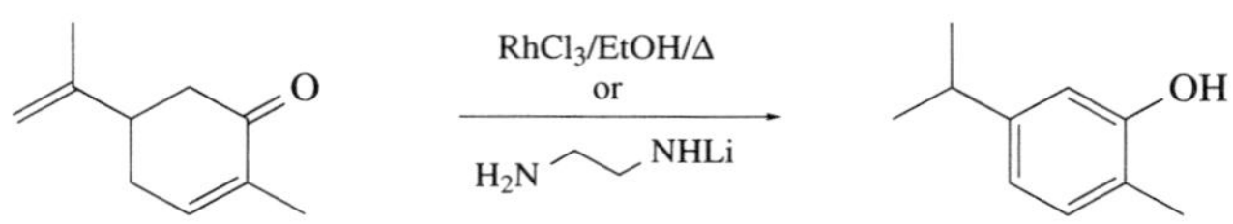

(33)

(**89**)

(b) Dienone–phenol rearrangement. The dienone–phenol rearrangement has been reviewed recently in detail by Whiting ⟨91COS(3)803⟩. The rearrangement normally occurs under acidic conditions and the yields are high. Trifluoroacetic acid (TFA) and trifluoroacetic anhydride (TFAA) are the most widely used acid catalysts. The steroidal dienone (**90**) rearranges in the presence of TFAA to yield predominantly the phenol (**91**) (1,3-shift product) and a small amount of the isomeric phenol (**92**) (1,2-shift product) (Equation (34)) ⟨64CB1940⟩. As with most cationic rearrangements, which group migrates is dependent on a combination of the migratory aptitude of the migrating group and the ability of the remaining group to stabilize the resulting cationic center. In a study of the rearrangement of the naphthoquinone diol (**93**), phenyl and higher homologs migrate in preference to the methyl group (Equation (35)) ⟨88TL4827⟩. Although the carboethoxy group has a low migratory aptitude, it migrates preferentially in the dienones (**94**) to give exclusively the phenols (**95**) (Equation (36)) ⟨74JA2121⟩. This simply reflects the fact that the carboalkoxy group is a relatively poor stabilizing group for cations. On treatment with TFA, the allyl 2,4-dienone (**96**) rearranges rapidly to the *para*-allyl phenol in quantitative yield, while the 2,5-dienone (**97**) rearranges to a mixture of regioisomeric phenols (Scheme 53) ⟨73HCA75⟩. The 1,3-shift of the allyl group can also be viewed as a Cope rearrangement. The regioselectivity is governed by the steric as well as electronic effects asserted by substituents on the ring. The presence of an electron donating group (EDG), for example a hydroxy or alkoxy group, has a profound effect on the regioselectivity as demonstrated by the rearrangement of the hydroxy dienones (**98**) (Scheme 54) ⟨94JOC1831⟩. Treatment of the dienone (**99**) with triethylamine leads to the formation of the phenol (**100**) in high yield via a Pummerer-type rearrangement (Equation (37)) ⟨79JOC4194⟩.

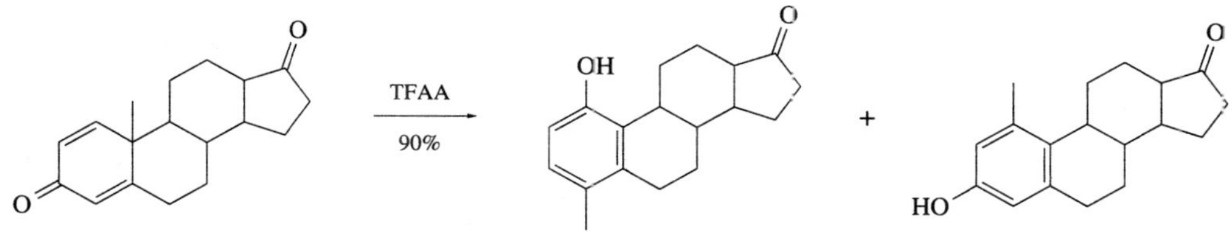

(90) (91) 89 : 11 (92) (34)

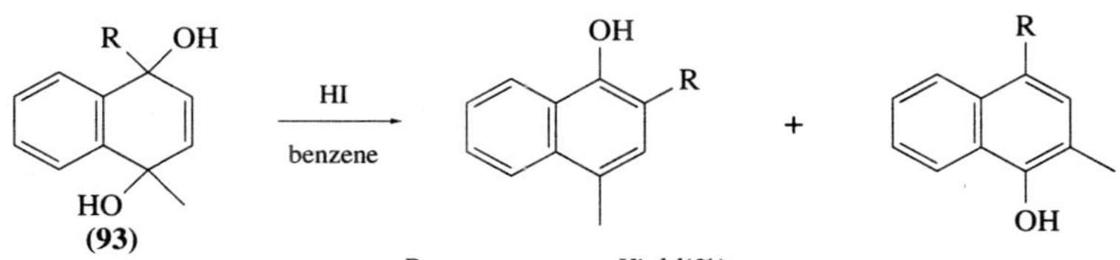

(93) (35)

R		*Yield(%)*
Bu^n	83 : 17	50
Bu^s	86 : 14	95
Ph	97 : 3	77

(94) (95) (36)

R = Me, Et, Ph

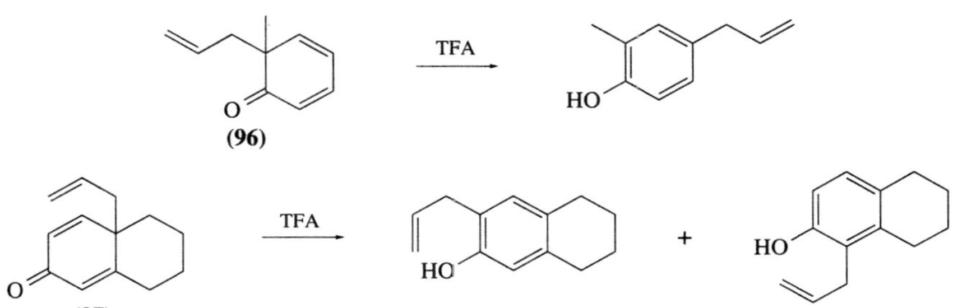

(96)

(97)

Scheme 53

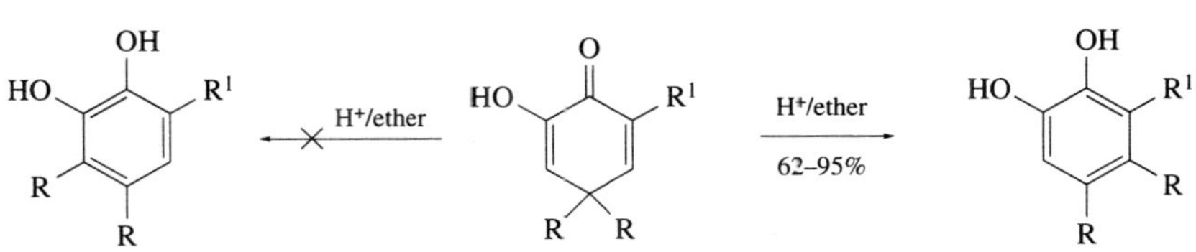

(98) R = Me, Ph; R^1 = H, Me, Ph, Bu^t

Scheme 54

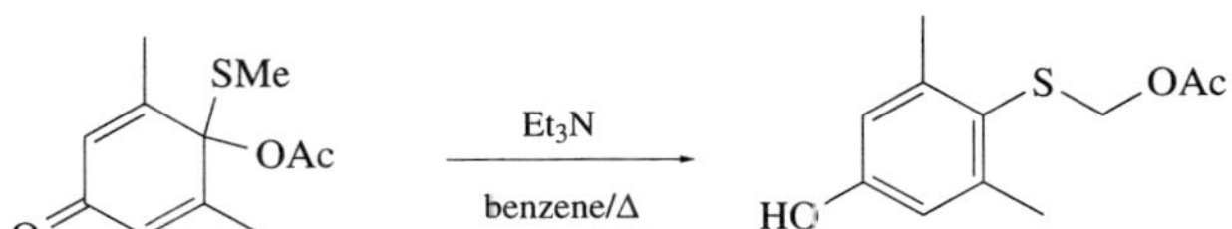

(99) (100) (37)

(iii) Annulation of acyclic fragments

The [4 + 2] Diels–Alder cycloaddition is a powerful tool in the synthesis of cyclic compounds and has been widely adopted for the synthesis of phenols. The application of electrocyclic reactions of dienyl ketenes, typically generated from ring-opening of alkenyl cyclobutenes, in the synthesis of phenols and polyhydroxy benzenes has also gained popularity in recent years. Aliphatic fragments can be connected, cyclized, and subsequently aromatized to highly substituted phenol systems. These annulation sequences typically involve the classical Aldol, Claisen, Dieckmann, Michael, etc., reactions of carbonyl compounds in a tandem fashion. The phenolic hydroxy group is normally derived from the enolization of the carbonyls in the precursors as a result of aromatization. An enormous wealth of synthetic approaches has been documented in the literature. An excellent review in this area was conducted earlier by Bamfield and Gordon ⟨84CSR441⟩.

(a) [3 + 3] Annulation. The Prelog condensation ⟨47HCA675, 47HCA1883⟩ of a ketone (nucleophilic 3-carbon component) possessing acidic α and α' methylenes with a 1,3-dicarbonyl compound (electrophilic 3-carbon component) provides an efficient synthesis of highly substituted phenols and polyhydroxy benzenes (Equation (38)). The reaction is typically conducted under basic conditions (NaOEt/EtOH) and is highly efficient. The switch to milder conditions (NaOH/MeOH/H_2O) has been found to improve the yields significantly ⟨80S708⟩. In cases where both the ketones and the 1,3-dicarbonyl compounds are unsymmetrical, mixtures of regioisomeric products are possible (Equation (39)). This can be easily circumvented by preferential enolization of the ketone component and reacting it with a 1,3-dicarbonyl compound that contains carbonyls of different reactivity as illustrated in Equation (40) ⟨74CC362⟩. Besides 1,3-dicarbonyl compounds, other variants, such as enaminones (**101**) ⟨80CC593⟩, β-amino carbonyl compounds (**102**) ⟨89CPB1434, 91CPB1132⟩, α,β-unsaturated carbonyl compounds (**103**) ⟨89CPB304⟩, and so on, have also been used extensively (Scheme 55). Interestingly, the cyclobutendione (**104**) has been employed as the electrophilic 3-carbon component in the efficient synthesis of the trihydroxybenzene (Scheme 56) ⟨68LA(717)54⟩. In a biomimetic synthesis of polyketides, dianion of acetoacetate (**105**) reacts with glutarates, and subsequent cyclization leads to the formation of a variety of tetrahydronaphthols (Scheme 57) ⟨86TL2401⟩. One of the unique features of this reaction is that the product is also a glutarate, which allows linear extension of the ring system simply by repeating the sequence.

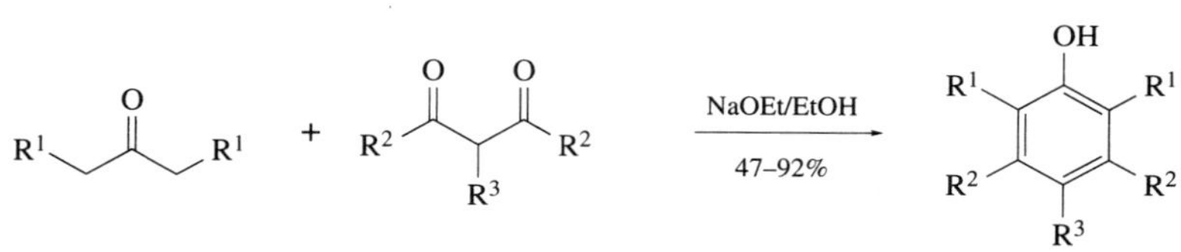

(38)

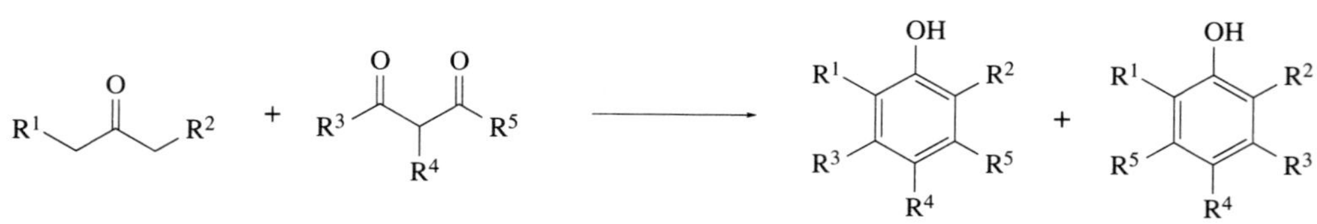

(39)

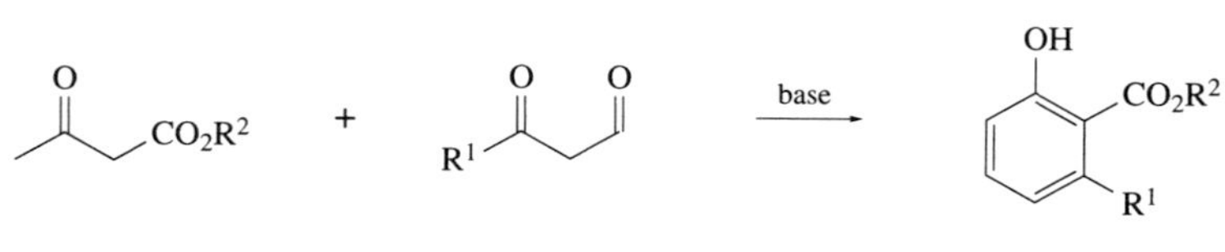

(40)

1,3-Bis(trimethylsilyloxy)-1-methoxy-1,3-butadiene (**106**) ⟨79CC578, 80JA3534⟩, a chemical equivalent of an acetoacetate dianion, condenses with 1,3-dicarbonyl compounds under the influence of $TiCl_4$ in a regioselective manner (Scheme 58) ⟨82TL2935⟩. On the other hand, the ethyl enol ether of acetoacetate (**107**) undergoes self-condensation in the presence of two equivalents of $TiCl_4$ to give the resorcinol derivative (**108**) (Equation (41)) ⟨75MI 213-01⟩.

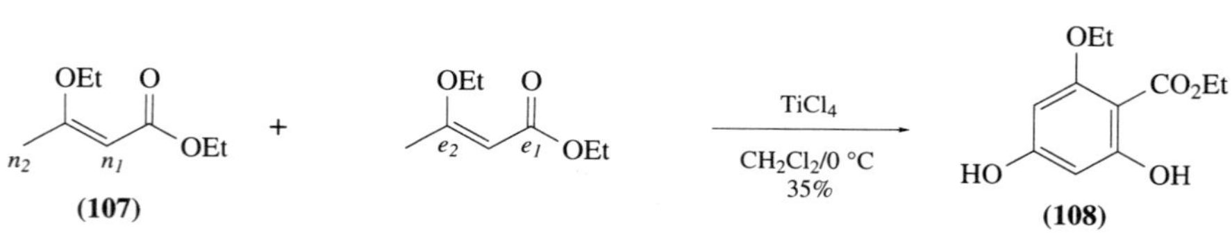

(107) (108)

(41)

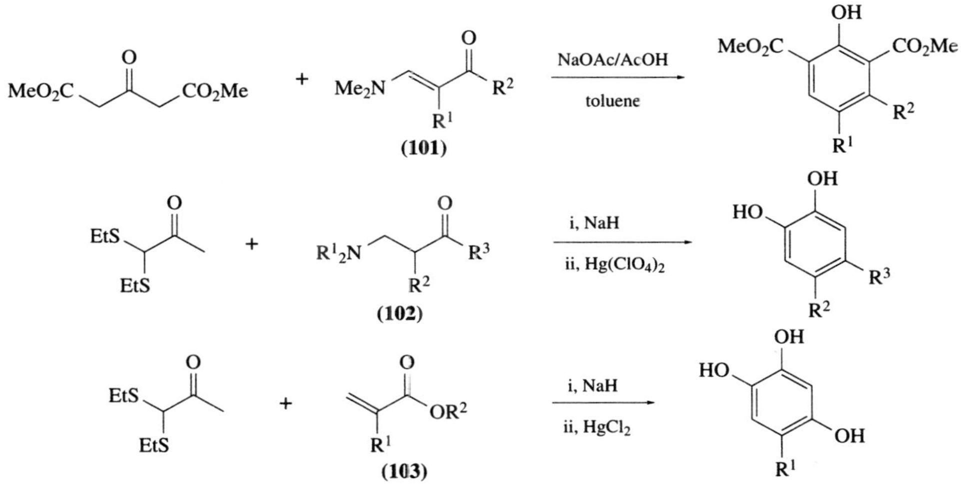

Scheme 55

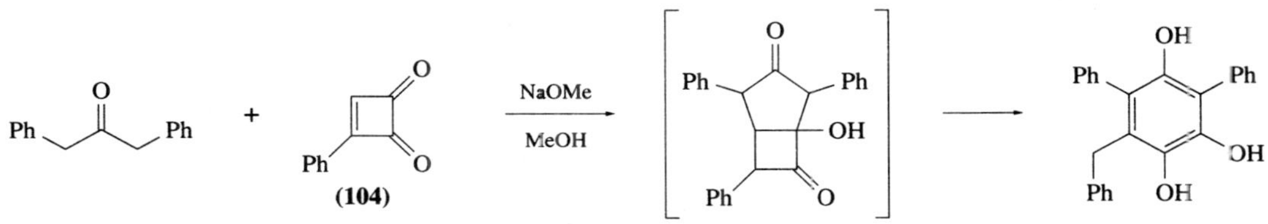

Scheme 56

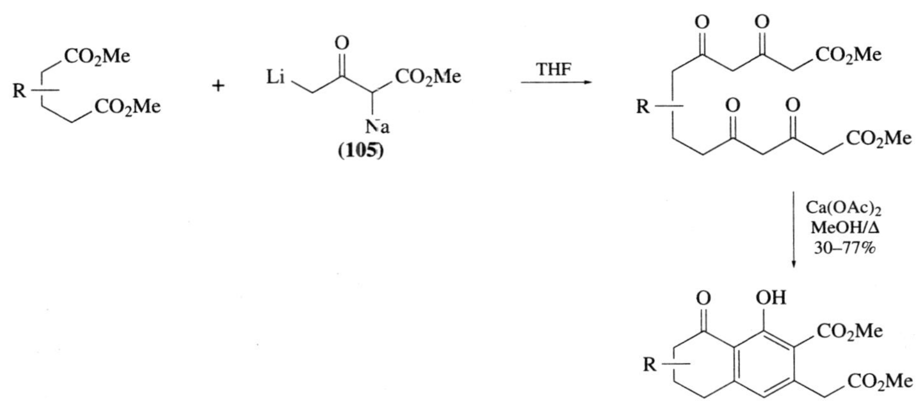

Scheme 57

Addition of isobutenylmagnesium chloride to the ketoketene thioacetal (**109**) leads to the formation of highly substituted phenols (Equation (42)) ⟨91MI 213-01⟩.

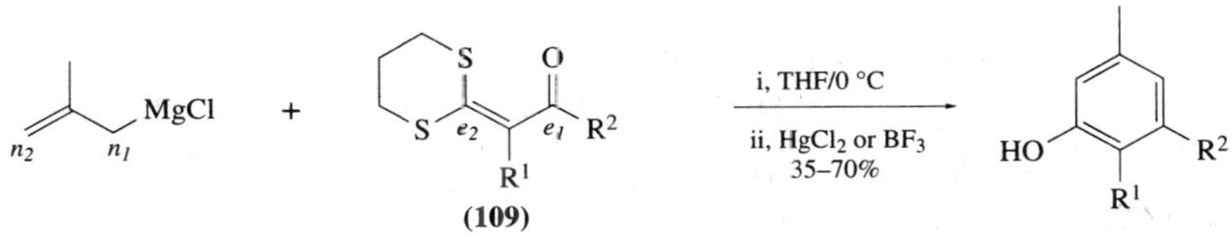

(42)

(b) [4 + 2] Annulation. Methyl vinyl ketone and its variants (**110**) (4-carbon component) have been widely utilized in the synthesis of phenols via the Robinson-type annulation (Scheme 59) ⟨75HCA104, 80JOC5002, 82JOC1200⟩. The built-in leaving group (X or Y) facilitates the conversion of

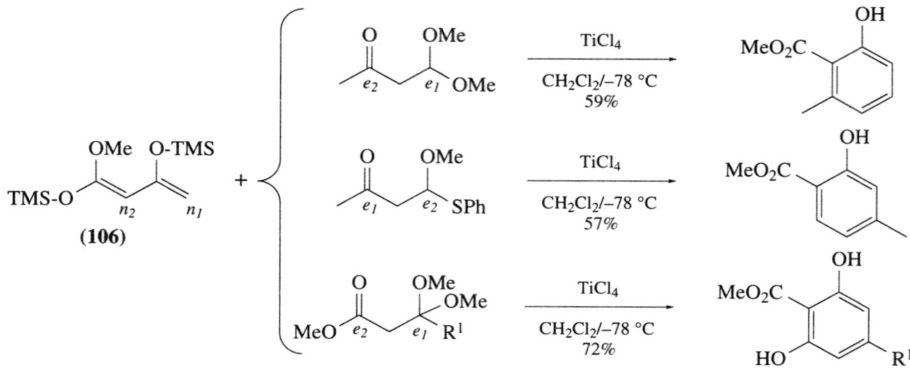

(106)

Scheme 58

the cyclohexenone (**111**) to the cyclohexadienone, which aromatizes readily to the phenol. On the other hand, the two-carbon component is typically a carbonyl compound, where the variation in its oxidation state is reflected in the resulting phenol. Hence, the resorcinol (**112**) ⟨78CC118⟩ and the trihydroxyarene (**113**) ⟨89CPB304⟩ are prepared from the acetate (**114**) and glyoxalate equivalent (**115**), respectively (Scheme 60).

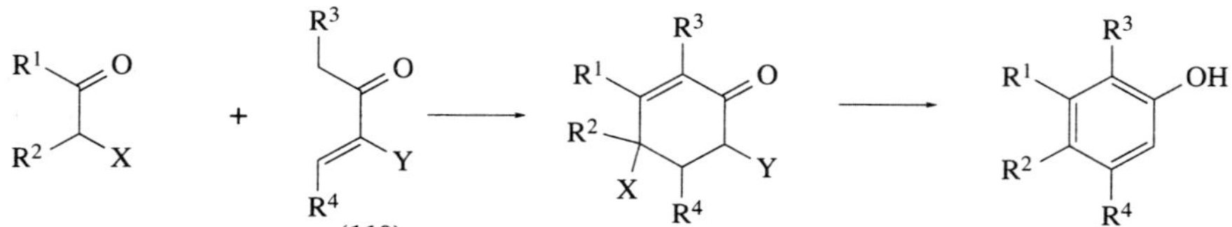

(110) **(111)**

X	*Y*
OH	H
S(O)Ph	H
H	SPh

Scheme 59

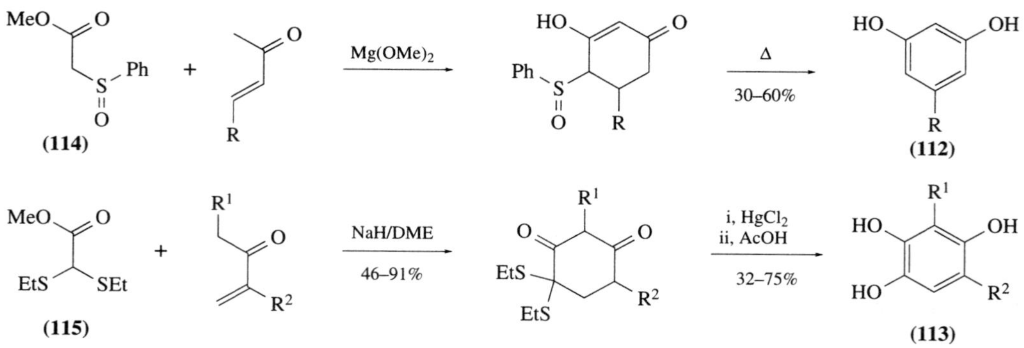

(114) **(112)** **(115)** **(113)**

Scheme 60

The annulation of *ortho*-toluic acid derivatives (4-carbon component) with activated alkenes (2-carbon component) in the synthesis of phenols is well documented (Scheme 61). Both the ester (**116**) ⟨78JOC178⟩ and lactone (**117**) ⟨78CC162, 79JA1628⟩ versions have been utilized successfully in the preparation of phenols. The corresponding resorcinol is obtained from the coupling of the toluic acid chloride (**118**) with a malonate ⟨79CC1165⟩.

Dianions of 1,3-dicarbonyl compounds (4-carbon component) undergo 1,4-addition with (*Z*)-3-bromoacrylate (2-carbon component) in moderate yields (38–55%). The resulting adducts (**119**) readily cyclize and aromatize to the phenols in good yields (57–80%) (Scheme 62) ⟨85TL4419⟩.

(c) [4+2] Cycloaddition. The application of the Diels–Alder cycloaddition towards the synthesis

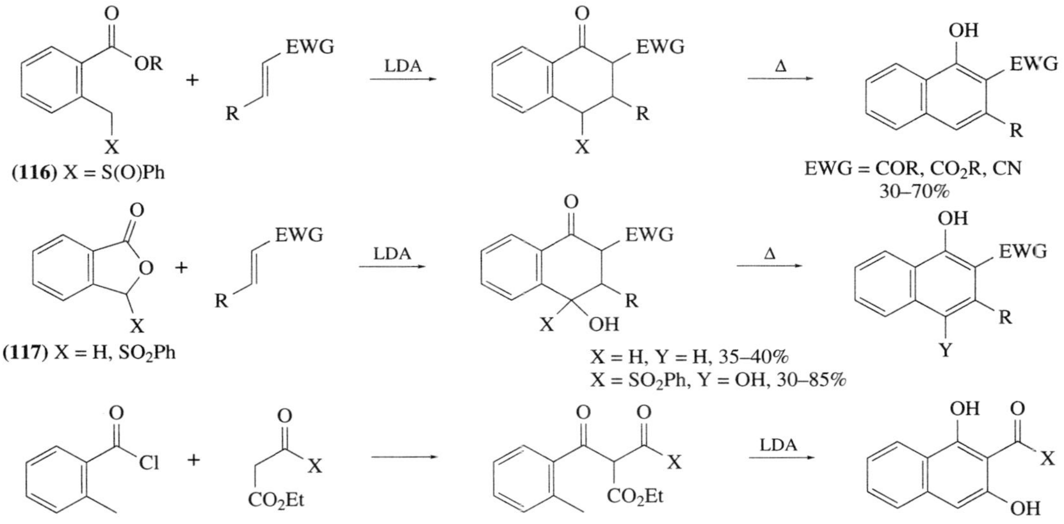

(116) X = S(O)Ph

(117) X = H, SO_2Ph

(118)

Scheme 61

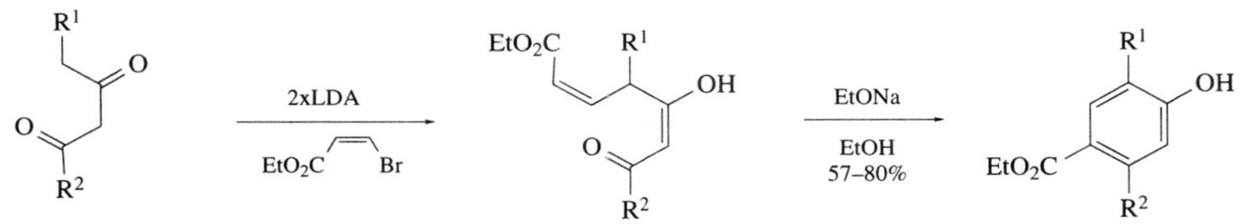

(119)

Scheme 62

of phenols and polyhydroxy benzenes is well documented. A review in this area has been conducted previously by Danishefsky ⟨81ACR400⟩.

Hydroxy-2H-pyran-2-ones are versatile dienic components in Diels–Alder reactions for the synthesis of phenols ⟨92T9111⟩. The reaction proceeds via a tandem [4 + 2] cycloaddition and cycloreversion with the extrusion of carbon dioxide in the process (Scheme 63). Both 3- and 4-hydroxy-pyran-2-ones (**120** and **121**) have been employed in reactions with a variety of dienophiles (**122**)–(**127**) ⟨75TL2389, 87CB1347⟩. Homophthalic anhydride (**128**), a chemical equivalent of 6-hydroxy-pyran-2-one, tautomerizes and reacts with dienophiles to furnish linear phenols (Scheme 64) ⟨81TL4283, 82JOC4376⟩. On the other hand, the readily available 3-carbomethoxy-pyran-2-ones (**129**) ⟨82TL4551⟩ undergo "inverse electron demand" Diels–Alder reactions with electron-rich dienophiles (e.g., 1,1,2-trimethoxyethylene, 1,1-dimethoxyethylene and vinylene carbonate) to give catechol and phenol derivatives in good to high yields (Scheme 65) ⟨87OS(65)98⟩.

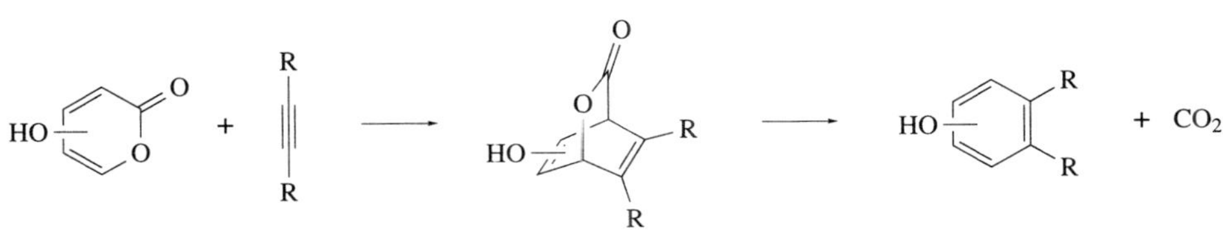

Scheme 63

Trimethylsilyloxy cyclohexadiene (**130**) undergoes [4 + 2] cycloaddition with electron-deficient alkynes (Scheme 66), and the extrusion of ethylene occurs only at higher temperatures (refluxing toluene) to produce the phenol (**131**) ⟨77TL611⟩.

The coupling of 1-methoxy-3-trimethylsilyloxy-1,3-butadiene (**132**, Danishefsky's diene) ⟨74JA7807⟩ with a variety of dienophiles, 3-nitro-cycloalkenones (**133**) ⟨77JA5810⟩ and 3-phenylsulfinyl-cycloalkenones (**134**) ⟨81TL603⟩ in particular, has also been utilized fully in the synthesis

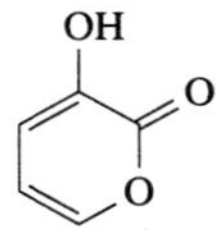

(120)

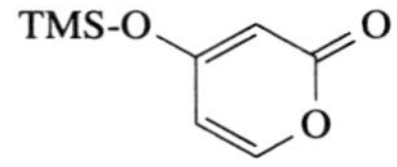

(121)

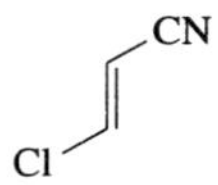

(122)

(123)

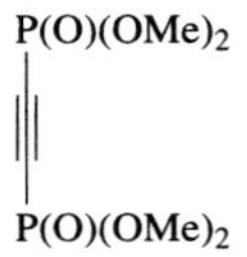

(124)

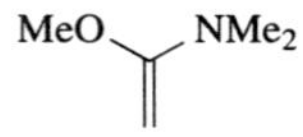

(125)

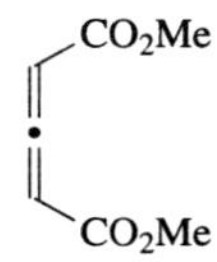

(126)

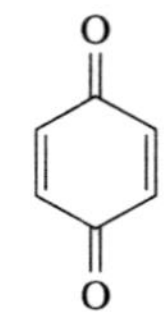

(127)

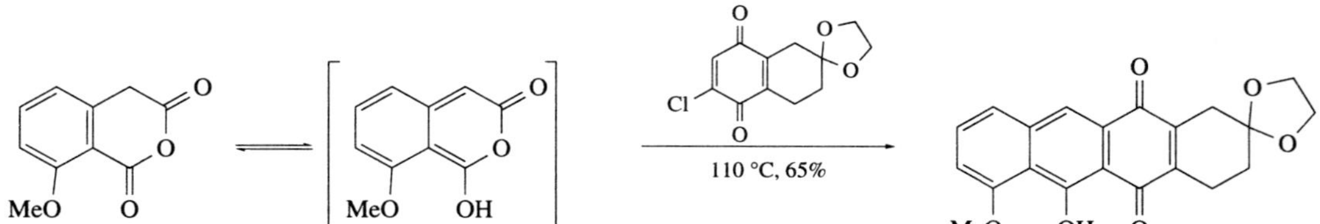

Scheme 64

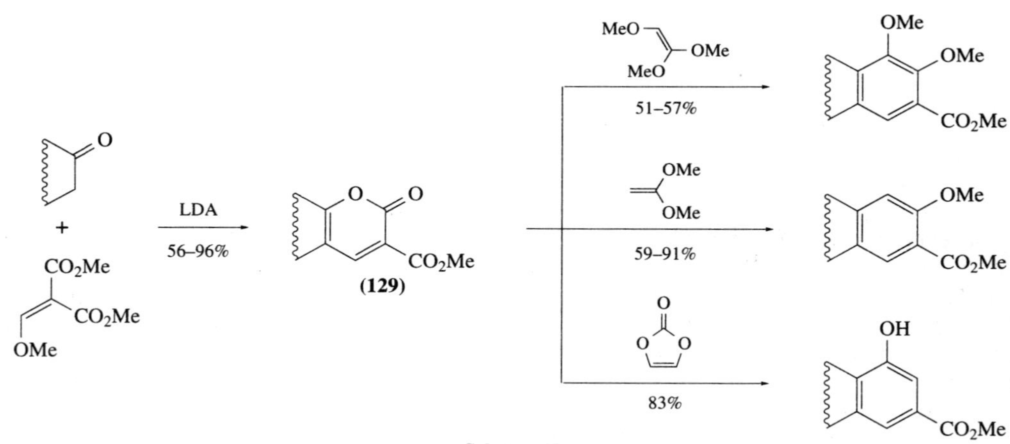

Scheme 65

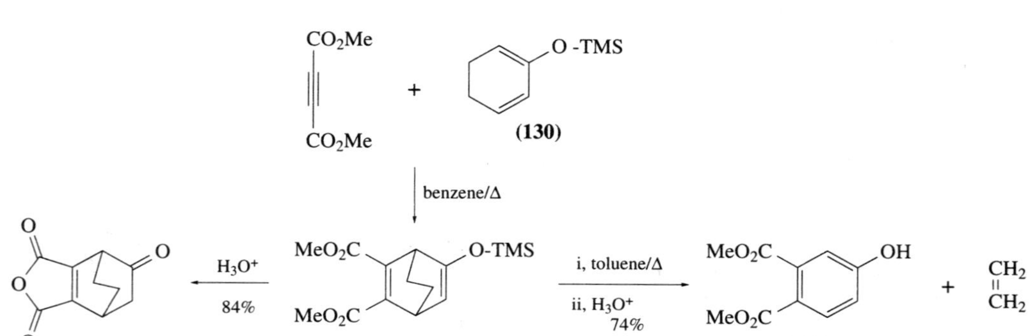

Scheme 66

of phenols (Scheme 67). Extensions of Danishefsky's diene, for example, 1,1-dimethoxy-3-trimethylsilyloxy-1,3-butadiene (**135**) ⟨76JCS(P1)1852, 78JOC379⟩, 1,1,4-trimethoxy-3-trimethylsilyloxy-1,3-butadiene (**136**) ⟨78JOC1435⟩, and 1,1,2,3-tetrakis(trimethylsilyloxy)-1,3-butadiene (**137**) ⟨81S381⟩, have been applied successfully to the synthesis of various polyhydroxy-benzenes (**138**–**140**). 3-Methoxymethylene-2,4-bis(trimethylsilyloxy)-1,4-pentadiene (**141**) has been employed in a double Diels–Alder reaction leading to the preparation of a highly functionalized phenol (**142**) (Scheme 68) ⟨84CL277⟩.

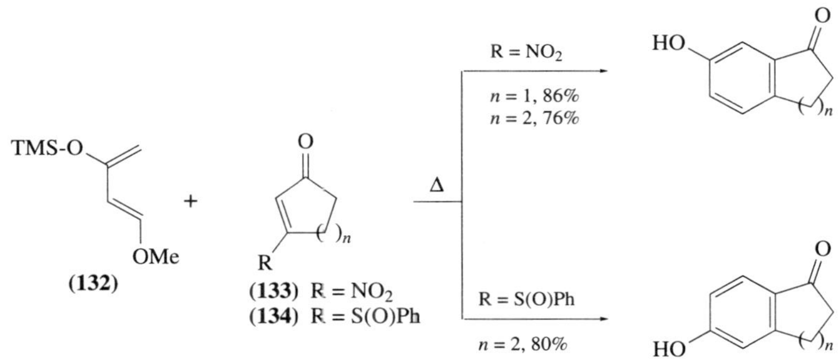

Scheme 67

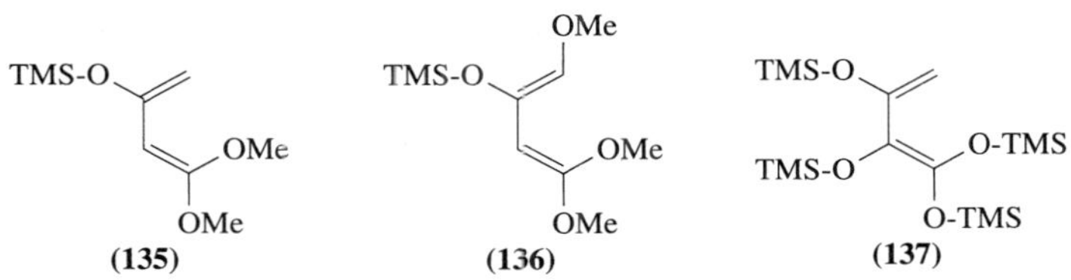

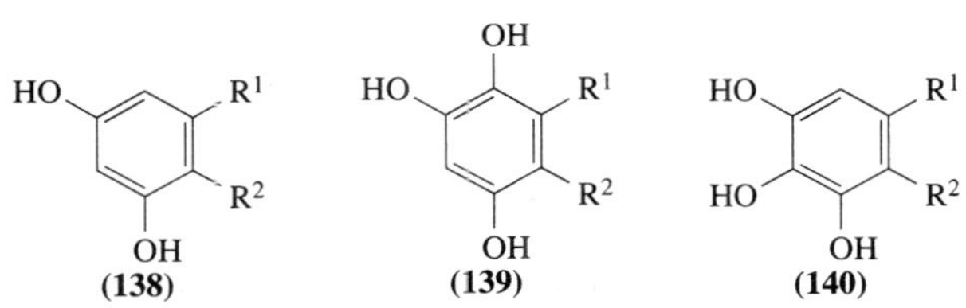

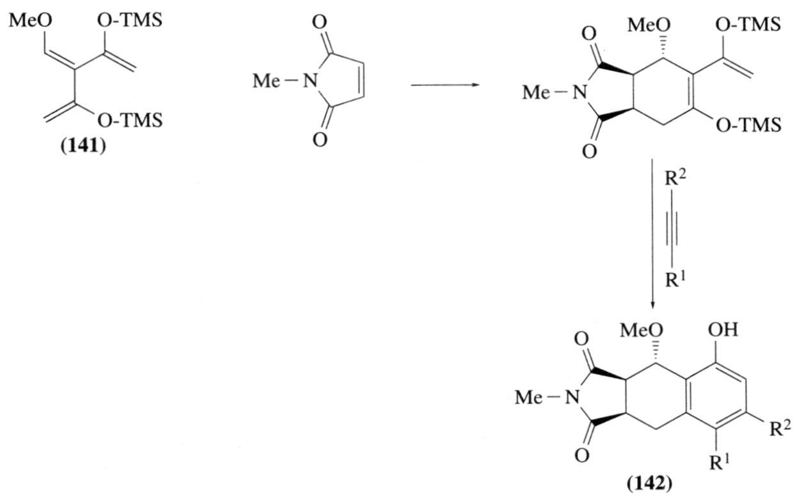

Scheme 68

Furan is a versatile diene for Diels–Alder reactions ⟨86CRV795⟩ and has been used in the synthesis of phenols. It reacts with benzyne ⟨56CB1334⟩ and 1-sulfonylallene ⟨83CC466⟩ to give the corresponding bicyclic adducts (**143** and **144**), which undergo ring opening to the phenols (Scheme 69). The 2,5-bis(trimethylsilyloxy)furan (**145**) has been used in the preparation of hydroquinones (Scheme 70) ⟨80TL3423⟩. An intramolecular Diels–Alder reaction of the alkynyl furan (**146**) provides an interesting entrance to the fused phenols (**147**) (Equation (43)) ⟨94TL729⟩.

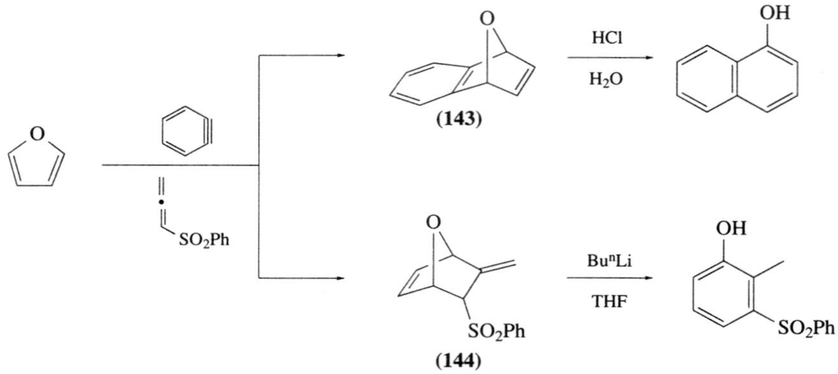

Scheme 69

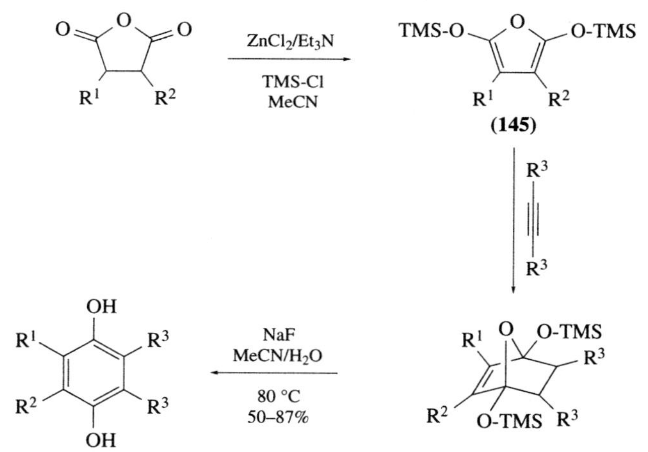

Scheme 70

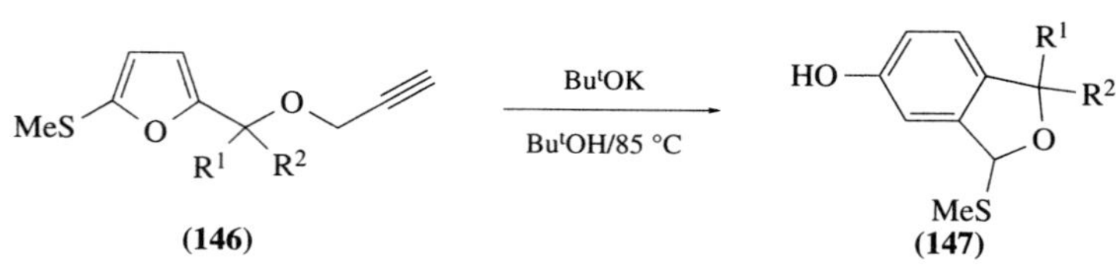

(43)

(d) [4 + 2] Annulation of alkenyl ketenes. The coupling of alkenyl ketenes with alkynes has been utilized in the synthesis of highly functionalized phenols. Alkenyl ketenes (**148**) can be prepared from α,β-unsaturated acid chlorides ⟨80JOC4810, 82TL4923⟩, the thermolysis of cyclobutenones ⟨84JOC1672⟩ or photochemical Wolff rearrangement of diazo enones ⟨90JA3093⟩. The coupling with alkynes proceeds with regio-control through a cascade of electrocyclic reactions to the alkenyl cyclobutenones (**149**), which undergo further rearrangement to the phenols (Scheme 71). The phenolic hydroxyl is derived from the ketene carbonyl. The use of the alkoxy- or silyloxy-alkynes ($Y = OR, OSiR_3$) leads to the synthesis of selectively protected resorcinol derivatives.

In an intramolecular fashion, the diazo dienone (**150**) is readily transformed to the dienyl ketene intermediate (**151**) via a thermal Wolff rearrangement (Scheme 72), which immediately cyclizes to the resorcinol derivative ⟨93TL7923⟩. The Wolff rearrangement can also be effected under photochemical conditions ⟨91TL5923⟩. The diazoketone (**152**) rearranges to the enyne-ketene (**153**) which readily

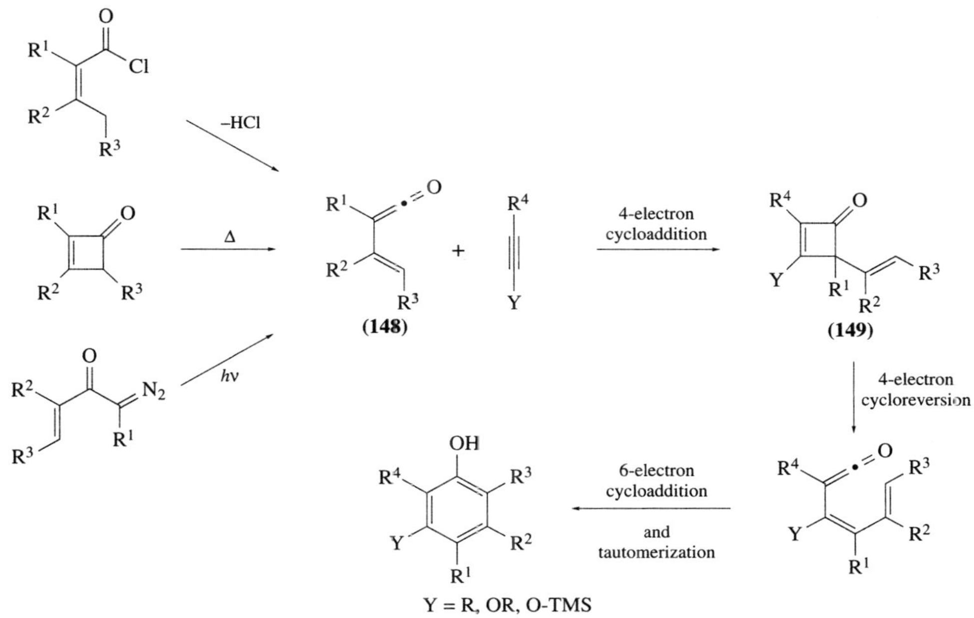

Scheme 71

cyclizes to the phenol (**154**) (Scheme 73). The diradical intermediate (**155**) is implicated and found to mimic the ene-diyne family of antibiotics in cleaving DNA by hydrogen abstraction ⟨94TL505⟩.

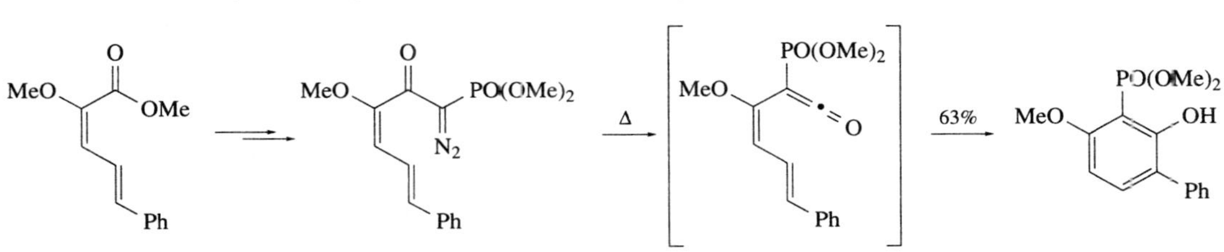

Scheme 72

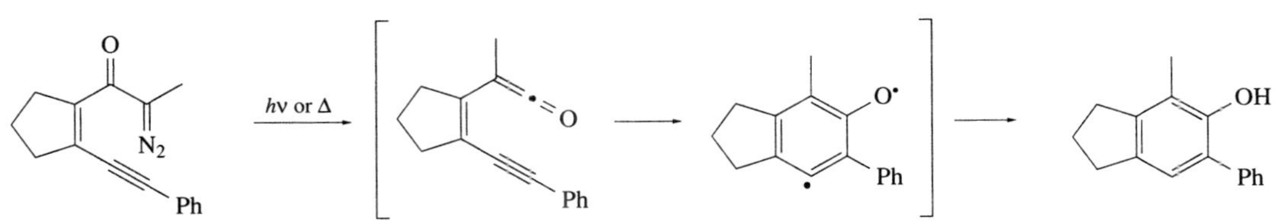

Scheme 73

Alternative routes to the 4-alkenyl-2-cyclobutenones include cross-coupling of 4-chloro-cyclobutenones with alkenyl metallics ⟨92JA1412⟩, 1,2-addition of alkenyl metallics ⟨91JOC4048⟩ and 1,4-addition of alkenyl cuprates ⟨93JA6101⟩ to cyclobuten-1,2-diones (Scheme 74). Thermal rearrangement of the resulting alkenyl cyclobutenones (**156** and **157**) leads to the preparation of a variety of highly substituted phenols, and the yields are generally good.

(e) Transition metal mediated benzannulation. The [3 + 2 + 1] benzannulation, first reported by Dötz ⟨75AG(E)644⟩, involves cycloaddition of a Fisher carbene with alkynes and insertion of carbon monoxide. The transition metal from the alkenyl carbene complex serves as a template to bring the various carbon fragments together (Scheme 75). The observed regioselectivity in the cyclization with unsymmetrical alkynes is governed by the steric differences of the substituents. The bulkier substituent is always *ortho*- to the phenolic hydroxyl in the products. The reaction proceeds through the dienyl ketene intermediate (**158**), wherein the phenolic hydroxyl originates from the inserted

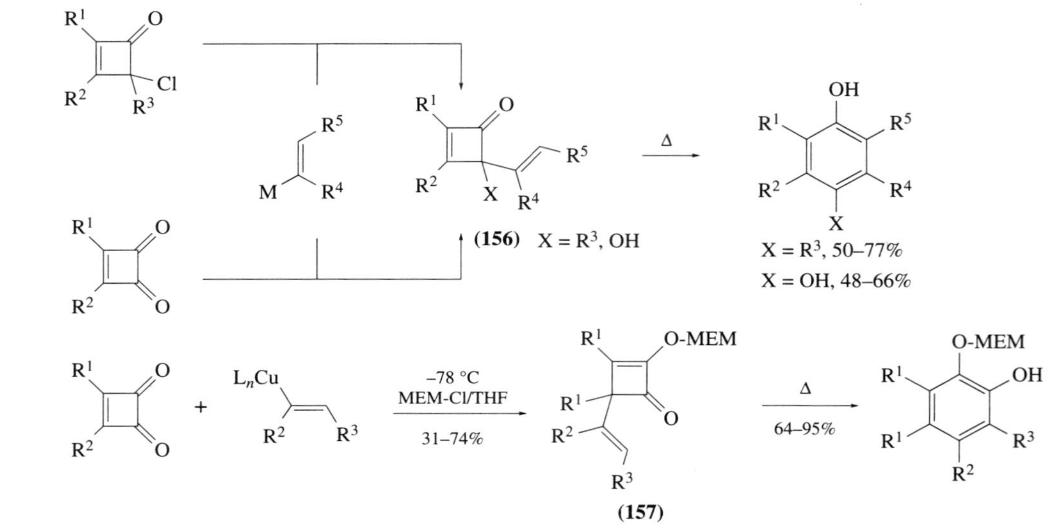

Scheme 74

carbon monoxide to give the 4-alkoxyphenols ⟨83JOM(252)C78, 84JOC2293⟩. The oxacycloalkylidene carbene complex (**159**) is a β-acyl alkenyl carbene equivalent ⟨93CC1811⟩, which undergoes benzannulation to furnish 2-acyl-4-alkoxyphenols (Scheme 76). Chromium is the most widely used transition metal, although molybdenum and tungsten are also encountered frequently. The transition metals are normally π-bound to the phenol products and are easily removed oxidatively by treatment with an iron(III) chloride–DMF complex ⟨81JA7677⟩. The [3 + 2 + 1] benzannulation has been reviewed by Dötz ⟨84AG(E)587⟩ and Wulff ⟨91COS(5)1065⟩.

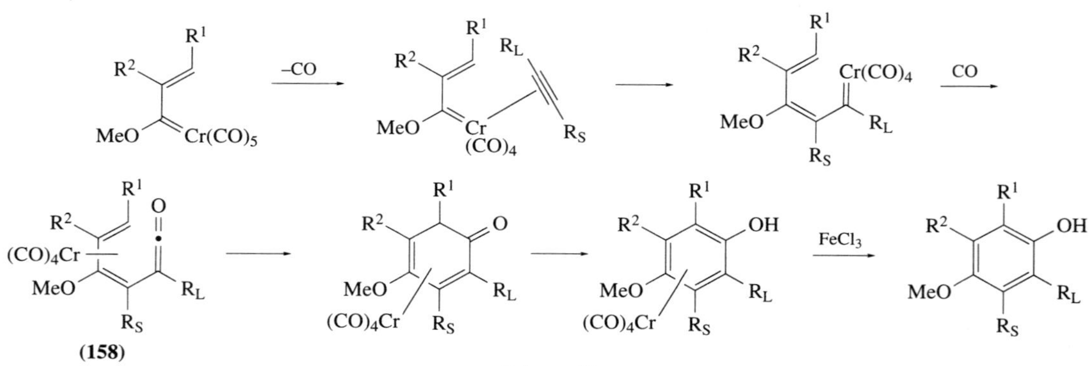

Scheme 75

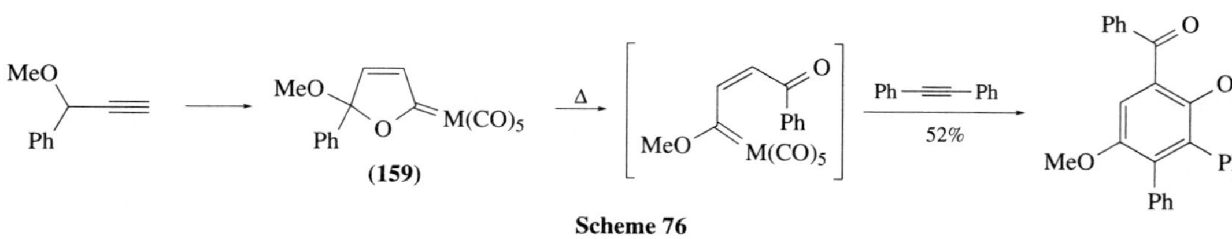

Scheme 76

Photolysis of the dienyl chromium carbenes (**160**) with carbon monoxide provides 2-alkoxyphenols in fair to high yields (Scheme 77) ⟨91JA7418⟩. Photo-carbonylation of 2-butyne with iron pentacarbonyl leads to the isolation of the *para*-quinone–iron complex (**161**), which is readily reduced to the 1,4-hydroquinone on treatment with HCl (Scheme 78) ⟨58JA1009⟩.

In a [1 + 4 + 1] annulation process, alkyl carbenes ⟨85JA1060⟩ and carbynes ⟨85TL2159⟩ carbonylate with diynes, and cyclize to provide annulated phenols in moderate yields (Scheme 79). When the

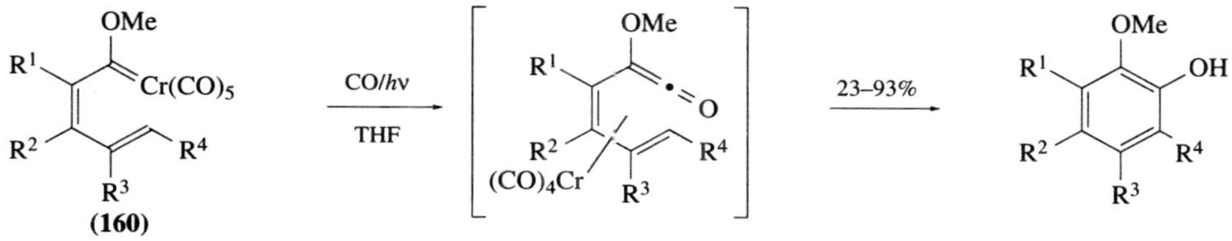

(160)

Scheme 77

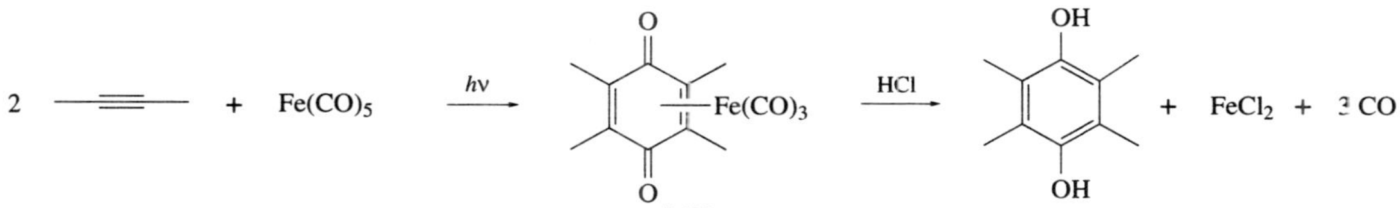

(161)

Scheme 78

diyne is tethered to the carbene complex, as in (**162**), a double intramolecular annulation leads to the isolation of bis-fused phenols (Equation (44)) ⟨91JA9873⟩.

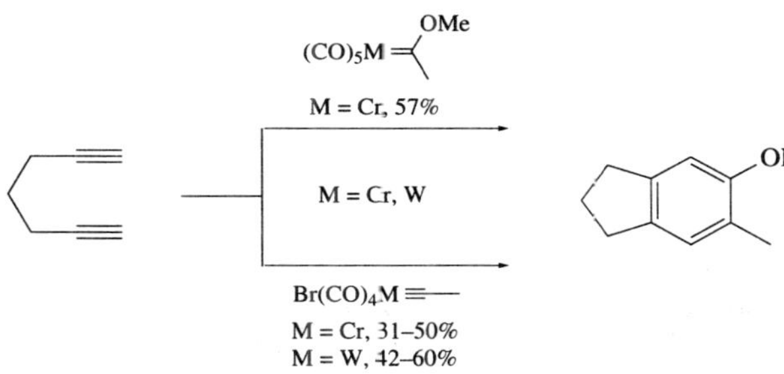

Scheme 79

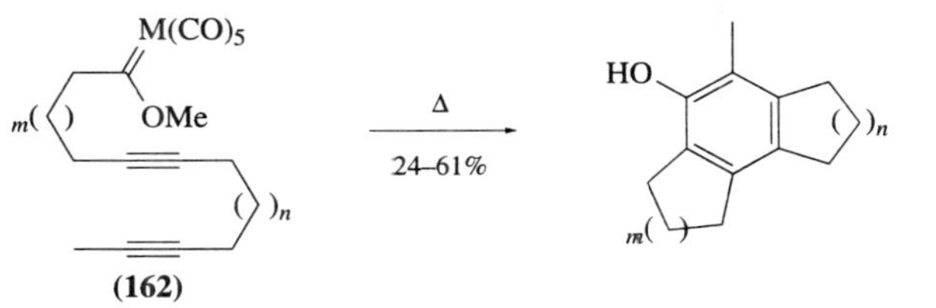

(162)

M = Cr, W
m, *n* = 0, 1, 2

(44)

Benzannulation, in tandem with the Diels–Alder reaction ⟨84JA7565⟩, is a powerful tool for the preparation of annulated phenols. The efficiency of this protocol is best illustrated in the construction of the tetracyclic steroidal ring system in one pot, via a tandem Diels–Alder/double benzannulation sequence of the triynyl carbene complex (**163**) (Equation (45)) ⟨91JA9873⟩.

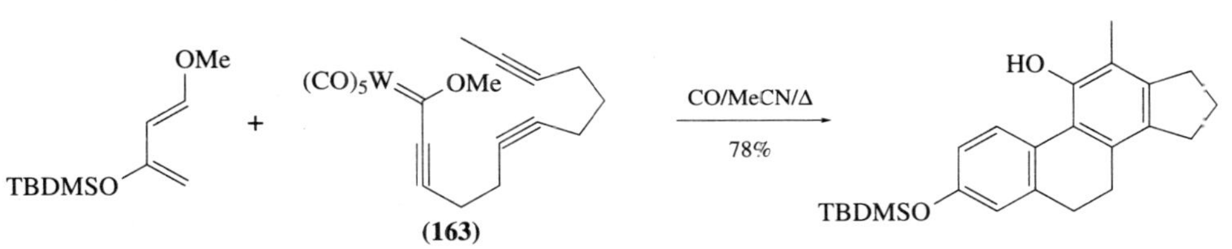

(163)

(45)

Ring-opening of acyl cyclopropenes with rhodium(I) produces the carbenoids, which insert readily into unactivated alkynes and cyclize to form substituted phenols (Scheme 80) ⟨92JA5881⟩. Alkenation of acyl cyclopropenes provides alkenyl cyclopropenes, which undergo carbonylation mediated by rhodium to produce phenols (Scheme 81) ⟨87JOC2631⟩.

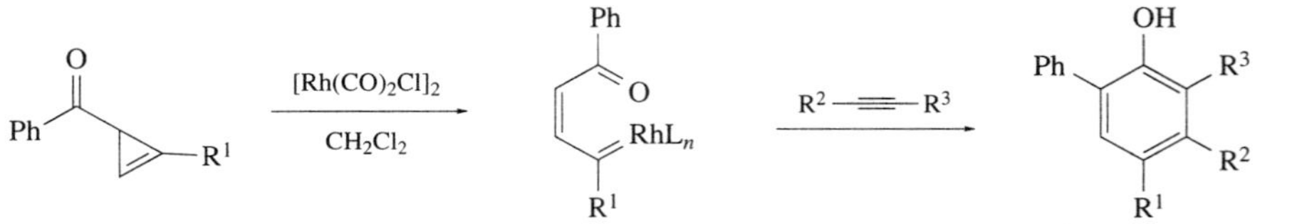

Scheme 80

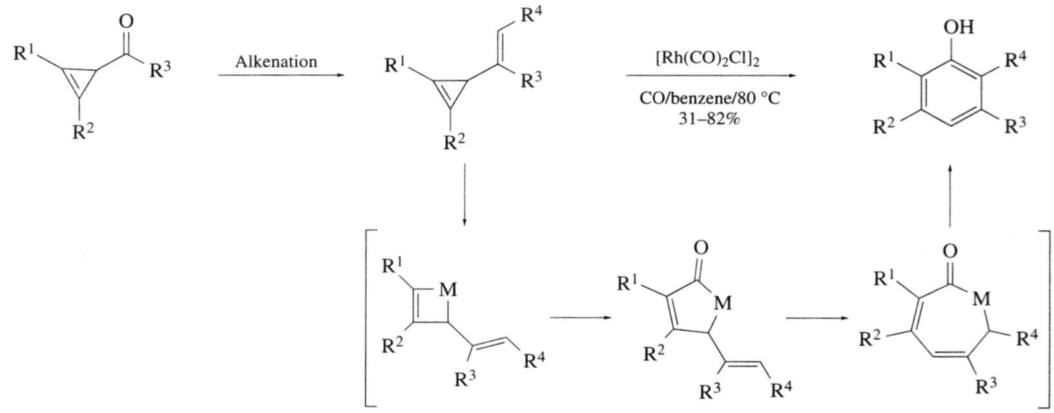

Scheme 81

Coupling of cyclobutenones with alkynes can be promoted by transition metals, in particular nickel ⟨91JA2771⟩ and cobalt ⟨92OM255⟩, via the alkenyl ketene metal complex (**164**) (Equation (46)). In general, the reaction proceeds under mild conditions with moderate regioselectivity.

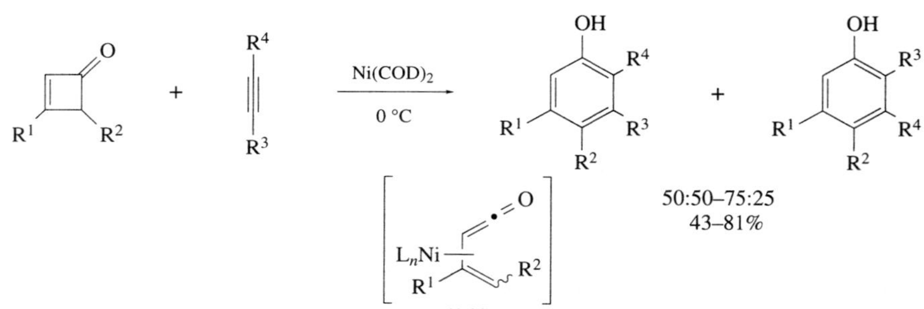

(46)

(164)

The activation of alkynes by niobium(III) chloride facilitates their addition to an aryl dialdehyde, leading predominantly to the formation of phenols (**165**) (Scheme 82) ⟨89JA5468⟩. The observed regioselectivity is attributed to the stabilization of the carbocation in the bicyclic intermediate (**166**) by the β-silyl group.

(f) Miscellaneous annulations. The enynone (**167**) cyclizes readily under acidic conditions (collidinium *p*-toluenesulfonate, CPTS) to the phenol (**168**), although the corresponding enol acetate provides the isomeric phenol (**169**) (Scheme 83) ⟨88TL6873⟩. The alkynyl acetophenones (**170**) also undergo facile cyclization on exposure to acid to give the naphthols (**171**) ⟨94TL1127⟩. On the other hand, base is used to induce cyclization of the enynone (**172**) to the phenol (**173**) (Scheme 84) ⟨89JA5472⟩.

Alkenyl β-keto esters (**174**) undergo oxidative cyclization in the presence of manganese(III) acetate and lithium chloride to produce the corresponding phenols (**175**) in moderate to good yields (Scheme 85) ⟨87TL6109, 89JOC38⟩.

Thermolysis of the readily available chlorodienic esters (**176**) leads to isolation of the phenols (**177**) in good yields ⟨88JOC880⟩. The chloro- substituent allows incorporation of other functionalities

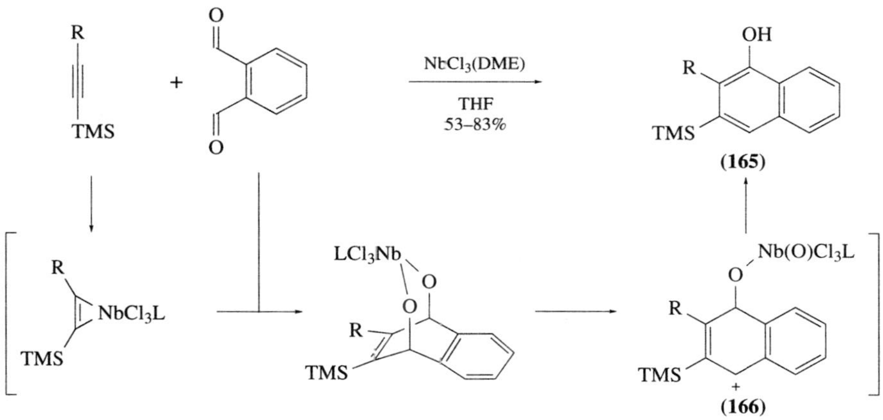

Scheme 82

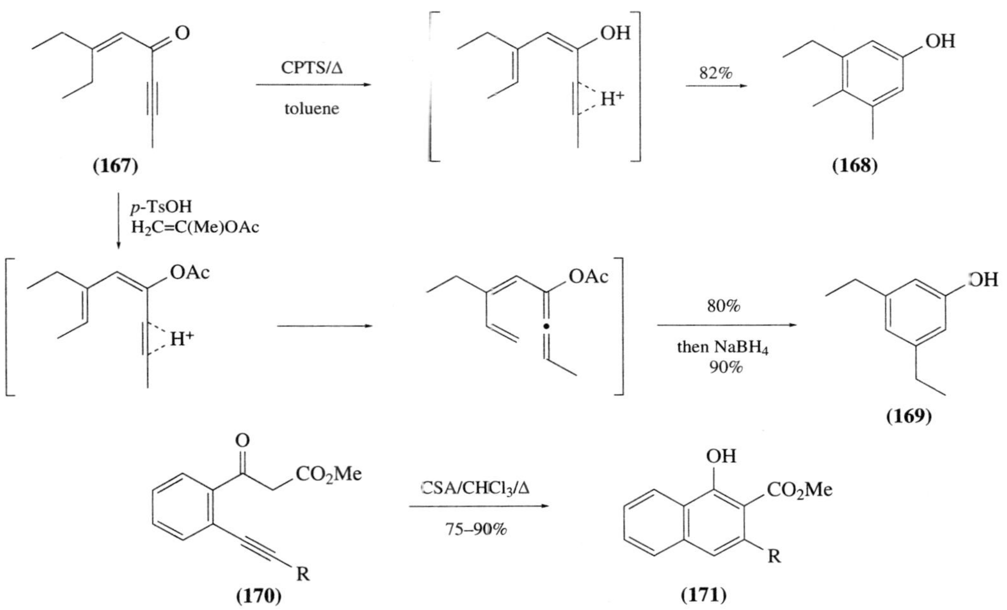

Scheme 83

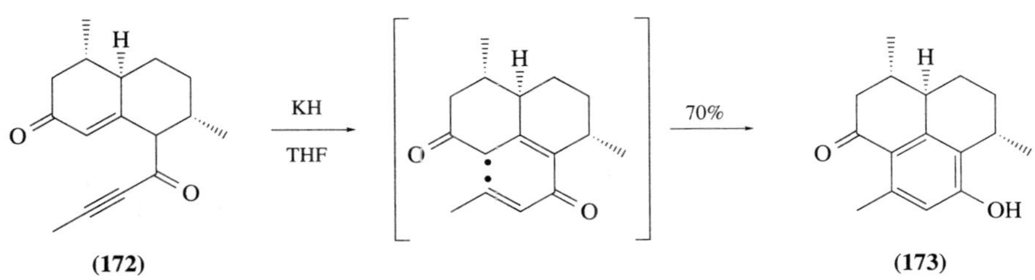

Scheme 84

for further elaboration, as shown by the synthesis of the hydroxy-xanthone acridinone (**178**) (Scheme 86) ⟨93TL6463⟩.

Finally, benzamides can be converted by a two-step sequence to the naphthols (**179**) in good overall yields (Scheme 87) ⟨86JOC271⟩.

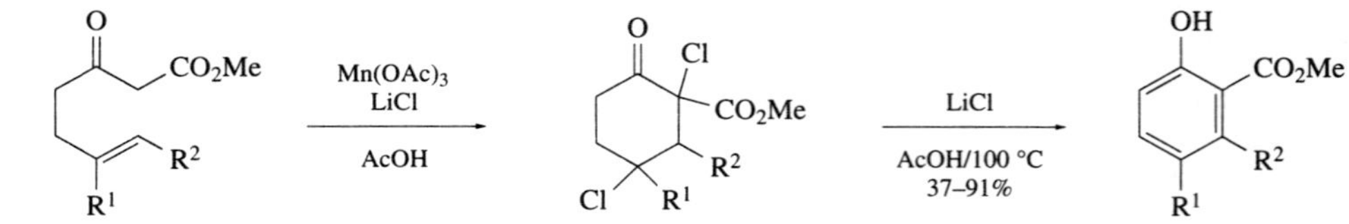

(174) (175)

Scheme 85

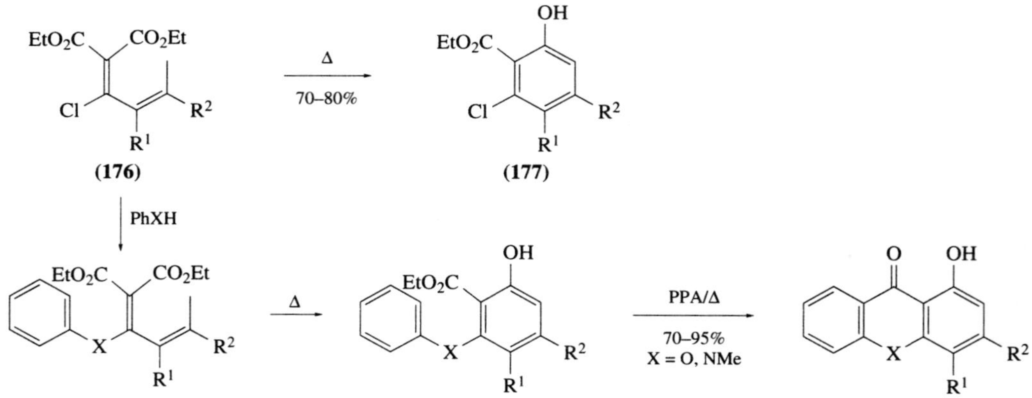

(176) (177)

(178)

Scheme 86

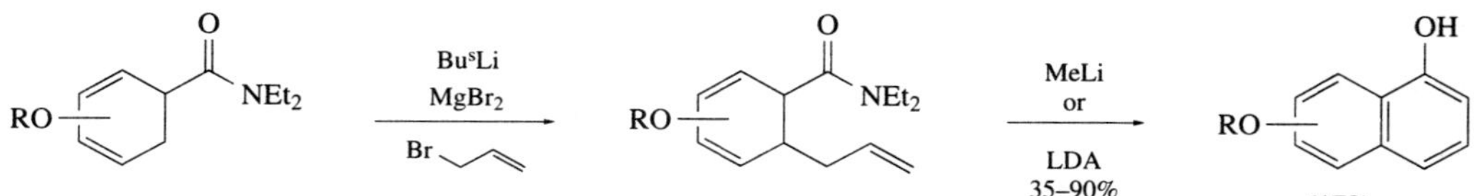

(179)

Scheme 87

2.13.1.2 Alkenyl and Aryl Ethers

2.13.1.2.1 Alkenyl ethers and esters

Generally, alkenyl ethers and esters are prepared by *O*-alkylation and acylation of enolates of carbonyl compounds. A two-step sequence involving acetalization of aldehydes and ketones followed by β-elimination of alcohol is also used widely. Alkenation of ketones and esters provides a practical alternative with a one-carbon homologation. The synthesis of alkenyl ethers has been reviewed previously by Shostakovskii *et al.* ⟨68RCR907⟩ and Chan ⟨91COS(2)595⟩.

(i) O-Alkylation/acylation of enols and enolates

An enolate is an ambient nucleophile on which alkylation can occur at either the oxygen or the carbon atom. In general, the choice of solvent, counterion and the nature of the alkylating agent, among other factors, are critical for the success of *O*-alkylation ⟨64AG(E)560, 70S1⟩ and hence in the preparation of alkenyl ethers from carbonyl compounds. Typically, polar, aprotic solvents (DMF, DMSO, HMPA) and large, soft cations (potassium, tetrabutylammonium) favor a free oxygen anion and render it more accessible to the alkylating agent. For example, cyclohexanone enolate undergoes predominantly *O*-methylation with dimethyl sulfate in DMSO, but mostly *C*-methylation in DME (Scheme 88) ⟨70RTC1153⟩. Furthermore, hard, reactive electrophiles favor attack by the more electronegative oxygen and hence selective *O*-alkylation. Alkenyl ethers are, therefore, obtained preferentially by reacting the enolates with reactive alkylating agents, for example, Meerwein's salt ⟨66CC51, 75CC644⟩, methyl fluorosulfonate ⟨72TL2677, 74S572⟩ and pentaethoxyphosphorane ⟨66JA1839⟩. In contrast, *O*-alkylation occurs in nonpolar solvents with less reactive alkylating agents for the thallium(I) salt of carbonyl compounds ⟨72CI(L)425⟩. The alkylation of an enolizable unsymmetrical ketone by carbenes, generated from the decomposition of diazoacetate by copper(I) salt, is not only oxygen specific but also regioselective, as demonstrated by the alkylation of 2-methylcyclohexanone (Equation (47)) ⟨78JOC1244⟩.

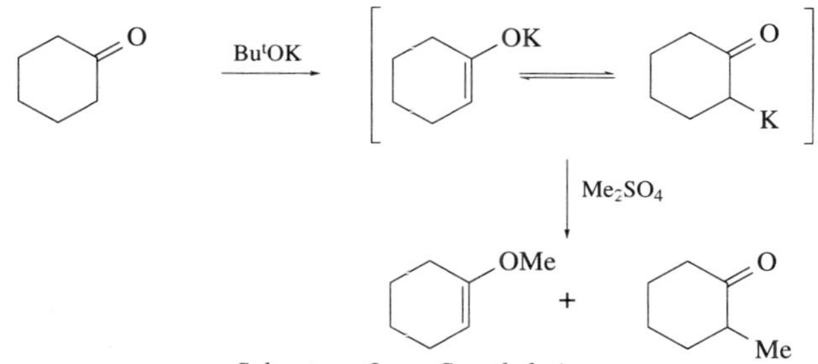

Solvent	*O vs C methylation*
DMSO	78 : 22
DME	35 : 65

Scheme 88

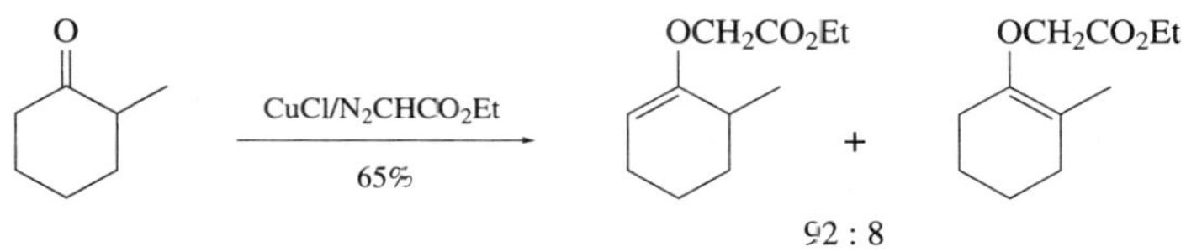

(47)

The preparation of phenyl alkenyl ethers can be achieved by treating an enol with triphenylbismuth bis-trifluoroacetate ⟨86TL3619⟩. The phenylation is accelerated by the use of copper(0) catalyst, as shown by the preparation of the alkenyl ether (**180**) in high yield (Equation (48)).

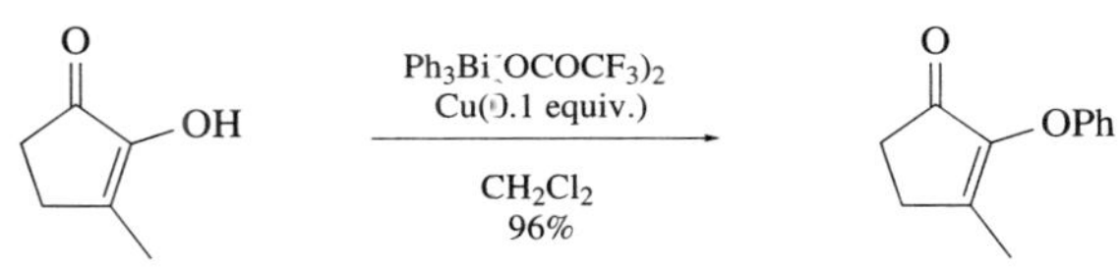

(48)

(180)

Besides the *O*-acylation of a metal enolate ⟨73JOC514⟩, alkenyl acetate is more commonly prepared by treating the parent carbonyl compounds with acetic anhydride or isopropenyl acetate ⟨64JCS1029⟩ under acid catalysis. The most frequently used acid catalysts include *p*-TsOH ⟨71OS(51)90⟩, perchloric acid ⟨64CJC212, 66JOC2105⟩ and hydrobromic acid ⟨70CC111⟩, and the reaction is normally under thermodynamic control. As a catalyst, dmap (4-dimethylaminopyridine) is particularly effective for aldehydes under basic conditions ⟨79SC157⟩. For fluoroalkyl ketones, exclusive *O*-acylation with an acid chloride in pyridine is observed ⟨93BSF664⟩. Acylation of metal enolates with chloroformates leads to the formation of alkenyl carbonates ⟨82TL703⟩. The use of HMPA as a cosolvent is essential for selective *O*-acylation ⟨78JOC2073⟩.

(ii) Isomerization of allyl ethers

The migration of the alkenic double bond in allyl ethers provides a convenient access to alkenyl ethers. The isomerization can be achieved by treatment with a strong base, such as potassium *t*-butoxide in DMSO ⟨61JA1773⟩, and the (*Z*)-isomer is formed preferentially under kinetic control ⟨93T11389⟩. Alternatively, the reaction can be catalyzed by transition metals under neutral conditions, and the yields are generally high. The most commonly used catalysts include palladium(0) ⟨80CC980⟩, rhodium(I) ⟨73JOC3224⟩, iridium(I) ⟨78CC694, 81S305⟩, palladium(II) ⟨73JCS(P1)2870⟩, and ruthenium(II) ⟨73JCS(P1)2870⟩. In the case of substituted allyl ethers, the (*Z*)-isomer normally predominates, with the exception of $[Ir(COD)(PMePh_2)_2]PF_6$, where the stereoselectivity is >97% in favor of the (*E*)-isomer (Equation (49) ⟨78CC694⟩.

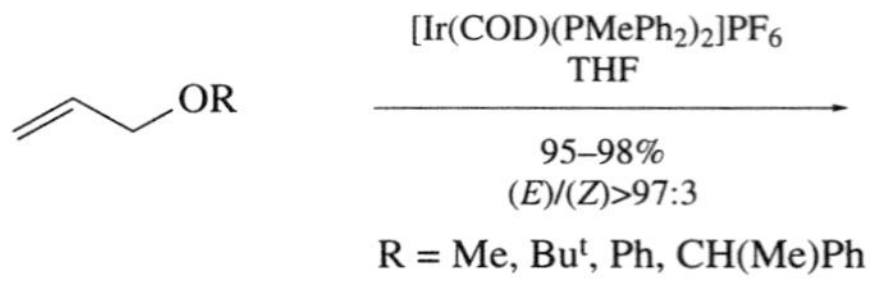

(49)

(iii) From alkynes

The classical alkoxymercuration ⟨73JOC4254⟩ of alkynes provides an easy access to alkenyl ethers (Scheme 89). Typically, alkynes are treated with an alcohol and a stoichiometric amount of mercuric salt in the presence of triethylamine or followed by sodium borohydride (reductive demercuration). Similarly, substituting the alcohol with a carboxylic acid results in the preparation of alkenyl ester. The reaction is regioselective in accordance with Markovnikov's rule. Instead of a reductive demercuration workup, the reaction can be terminated by addition of an electrophile (e.g., *N*-halosuccinimide) to introduce an extra functionality to the alkenyl ether product, as demonstrated by the intramolecular alkoxymercuration of the alkynol (**181**) (Equation (50)) ⟨82JA5842⟩. More conveniently, the reaction can be performed with a catalytic amount of mercuric chloride in good yield ⟨88S144⟩. Besides mercuric salts, the addition of carboxylic acids to alkynes can also be promoted by silver(I) oxide ⟨55OSC(3)853⟩, ruthenium(0) ⟨83OM1689⟩ and ruthenium(II) complexes ⟨86TL2125, 91SL755⟩. Interestingly, ruthenium(0) in the presence of bedentate bis-phosphine ligands promotes anti-Markovnikov addition of acid to alkyne and in a stereoselective fashion in favor of the (*Z*)-isomer (Equation (51)) ⟨93CC850⟩.

Scheme 89

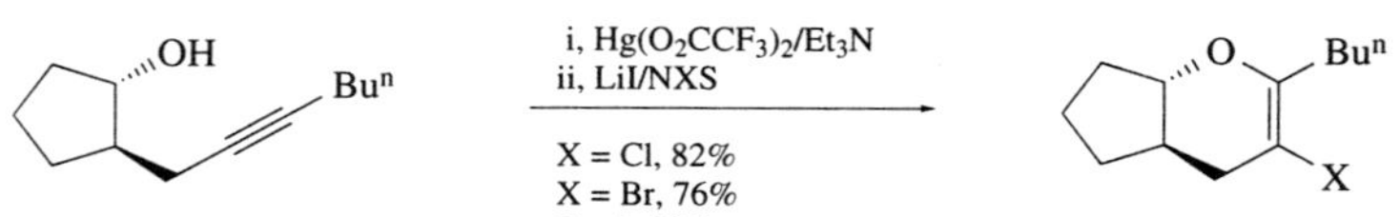

(**181**)

(50)

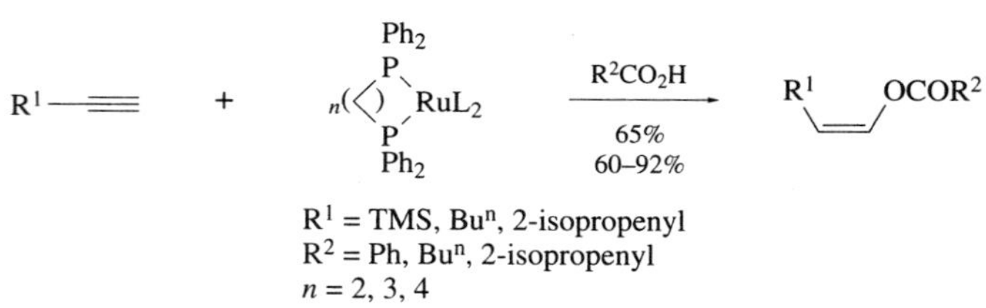

(51)

Reduction of alkoxyalkynes by lithium aluminum hydride or Birch reduction conditions is stereoselective to give the (*E*)-alkenyl ethers (Equation (52)) ⟨87JOC2919⟩. On the other hand, partial hydrogenation of alkoxyalkynes with Pd—$BaSO_4$ produces the (*Z*)-alkenyl ethers in high yields (Scheme 90) ⟨48RTC973⟩. The addition of organocopper reagents to alkoxyacetylenes is highly regioselective, and occurs exclusively at the substituted carbon to give α-substituted alkenyl ethers (**182**) in high yields ⟨77BSF693, 88JA6276⟩. For internal alkynes, the addition is not regioselective and reaction occurs at both carbons of the alkyne in equal proportion. The hydrostannation of ethoxyacetylene with trialkyltin hydride is both regio- and stereoselective, resulting in the formation of the (*Z*)-β-stannyl vinyl ethers (**183**) ⟨67JOM(9)285⟩. Transmetallation of (**183**) with *n*-butyllithium produces the (*Z*)-β-lithiated vinyl ether, which is highly stable in contrast to its (*E*)-counterpart ⟨77JA7365⟩. Its alkylation is stereospecific producing (*Z*)-β-substituted alkenyl ethers which are not readily accessible under normal circumstances.

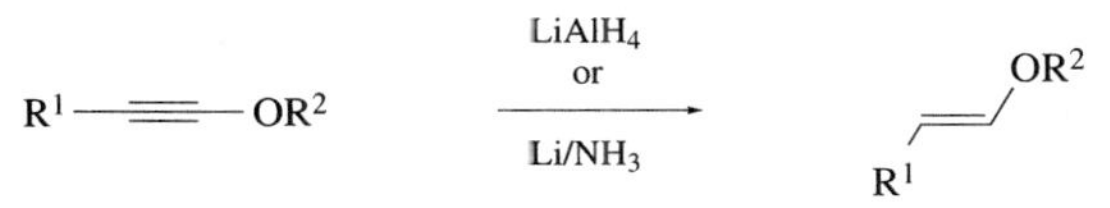

(52)

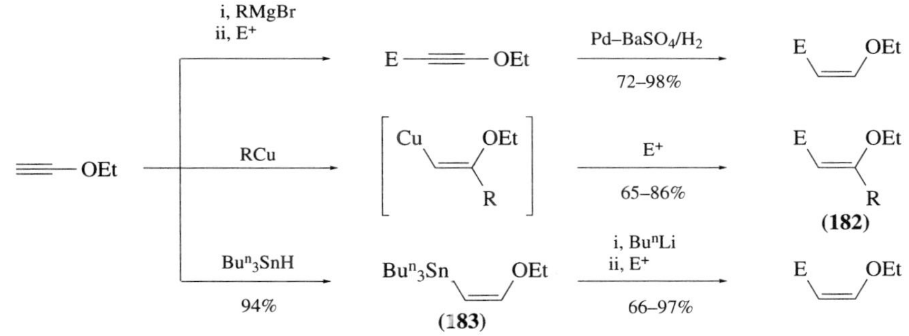

Scheme 90

Interestingly, the ethynyl group of certain nitrogen containing heterocycles undergoes addition of methanol to give alkenyl ethers with (*Z*)-geometry (Equation (53)) ⟨84S245⟩. The ethynyl group must be located at the unactivated positions, otherwise further addition of methanol leads to the formation of the dimethyl acetals.

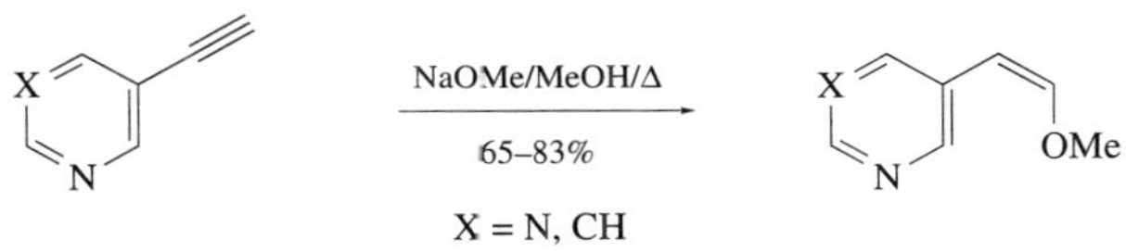

(53)

(iv) By 1,2-elimination of substituted alkyl ethers

Dehydrohalogenation of α- and β-haloalkyl ethers such as (**184**) and (**185**) with bases provides direct access to alkenyl ethers, although mixtures of (*E*) and (*Z*) geometric isomers are usually obtained. Tertiary amine bases ⟨86JOC3577⟩ are commonly used in the dehydrohalogenation (Scheme 91), for example, triethylamine ⟨65JCS5225⟩, ethyl diisopropylamine ⟨58CB380⟩, and dbu (1,5-diazobicyclo[5.4.0]undec-5-ene) ⟨93BCJ1156⟩. The combination of sodium hydroxide and a phase transfer catalyst (PTC) is also highly effective ⟨79S688⟩.

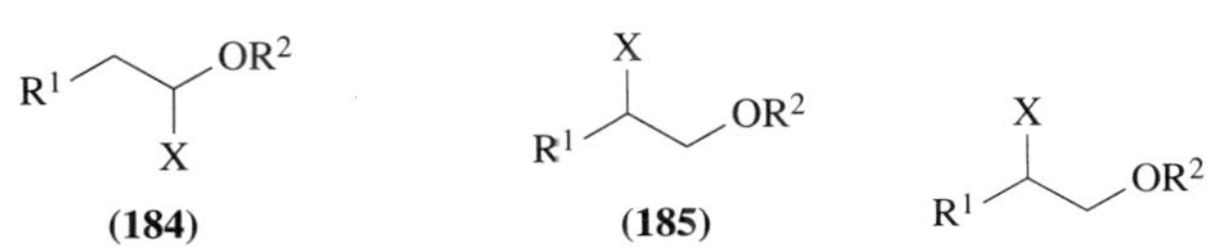

(184) **(185)** **(185)**

X = halogen

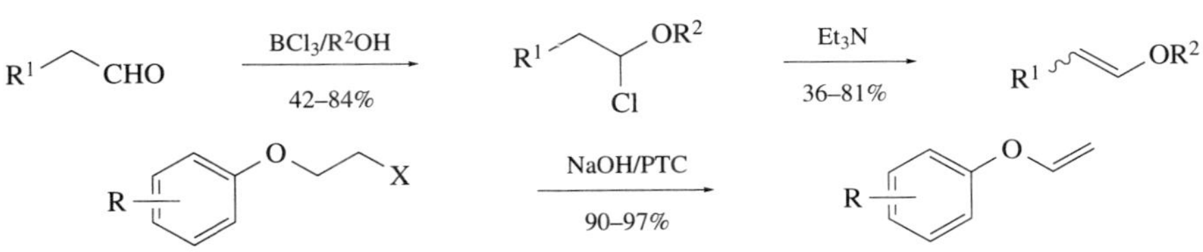

Scheme 91

The readily available *erythro* β-hydroxy acid (**186**) ⟨77TL4651⟩ undergoes either *syn*- or *anti*-elimination simply by manipulating the reaction conditions ⟨79CC52, 84TL4881⟩. Consequently, both

(*E*) and (*Z*) alkenyl ethers can be obtained stereospecifically in high yields as depicted in Scheme 92 (dead = diethyl azodicarboxylate).

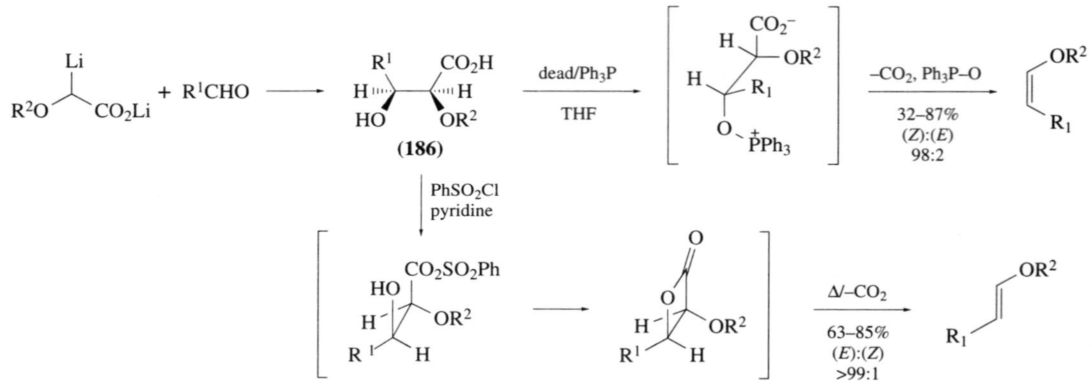

Scheme 92

Treatment of glycosyl halides with zinc/silver–graphite produces the glycal alkenyl ethers in high yields (Equation (54)). This methodology works well with both pyranose and furanose ⟨86CC1149⟩.

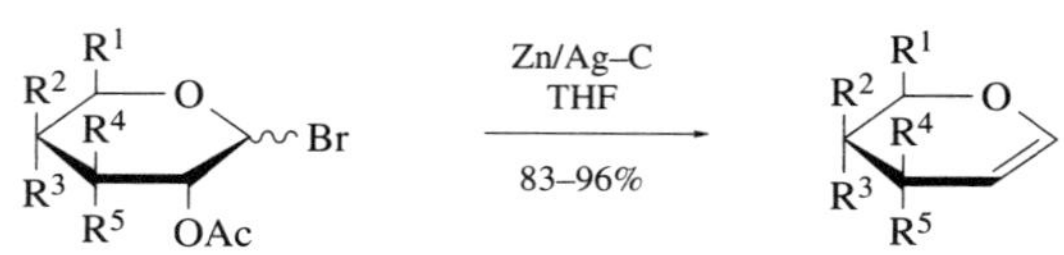

(54)

The lithiated sulfenyl (**187**) and sulfonyl (**188**) methyl ethers are versatile intermediates for the synthesis of alkenyl ethers (Scheme 93). Their alkylation with primary halides and subsequent elimination leads to the preparation of alkenyl ethers in good yields ⟨83TL4993, 91SL505⟩. Mixtures of geometric isomers are usually obtained, except in the case of dehydrosulfonylation which is highly stereospecific in favor of the (*E*)-isomer. The adducts between (**187**) and aldehydes have been elaborated to alkenyl ethers, but as an (*E*)/(*Z*) isomeric mixture ⟨84TL5997⟩. Both (**187**) and (**188**) also react with tosylhydrazones to provide an isomeric mixture of alkenyl ethers in a one-pot reaction (Scheme 93) ⟨79JA249, 83TL4993⟩.

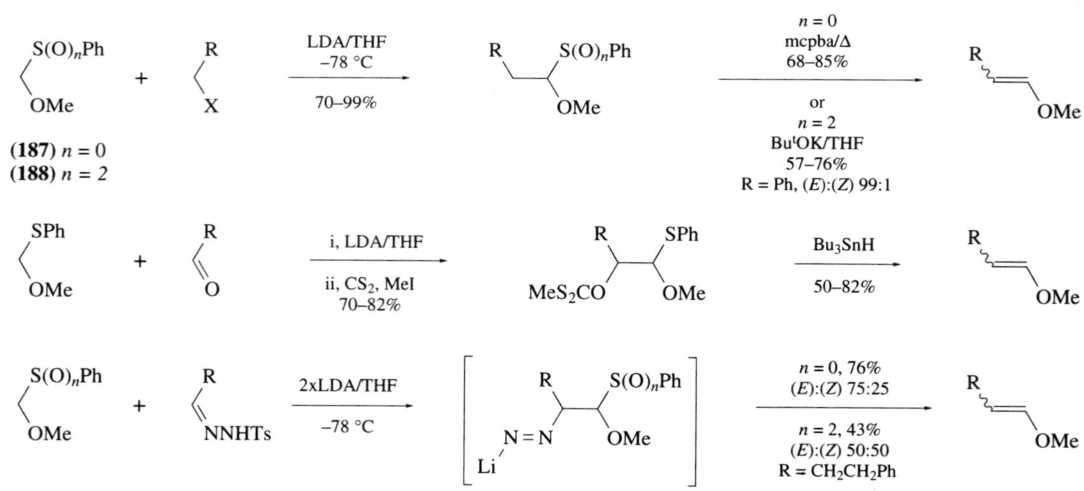

Scheme 93

(v) From acetals

Aldehydes and ketones react with alcohols under acidic conditions to give alkenyl ethers, which normally react further to result in the formation of the corresponding acetals (Scheme 94). Removal of the excess alcohol can push the equilibrium towards the formation of the alkenyl ethers. The use of orthoformates with removal of the alcohol byproduct provides a convenient one-pot synthesis of alkenyl ethers ⟨74S38, 74S348⟩. The preparation of alkenyl ethers by β-elimination of a molecule of alcohol from acetals is used more commonly. The reaction is normally achieved by heating an acetal in the presence of an acid catalyst ⟨65HOU(6/3)97, 73JOC2910⟩, as demonstrated by the preparation of the cyclic alkenyl ether **(189)** catalyzed by lutidinium *p*-toluenesulfonate (LPTS) (Equation (55)) ⟨94JOC311, 94JOC2910⟩. Other effective catalysts include pyridinium *p*-toluenesulfonate (PPTS) ⟨88JA649⟩, TMS-I/$(TMS)_2NH$ ⟨82TL323⟩, TMS—$Mn(CO)_5$ ⟨82TL631⟩, and triisobutylaluminum (TIBA) ⟨90JOC5814⟩. It is not uncommon that mixtures of regio- and geometric isomers are obtained (Equation (56)). It is worth noting that treatment of the meso-acetals **(190)**, derived from prochiral cyclohexanones, with TIBA leads to the synthesis of chiral alkenyl ethers with high d.e. (Scheme 95) ⟨86TL1363⟩.

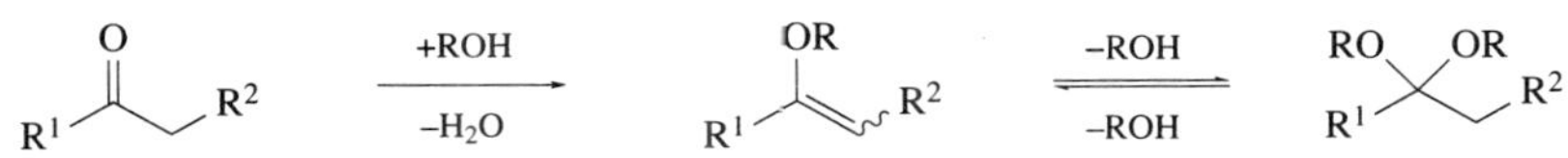

Scheme 94

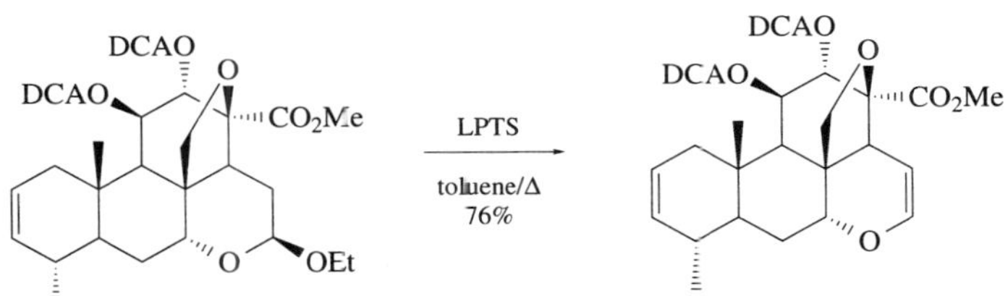

(189)

DCA = dichloroacetyl

(55)

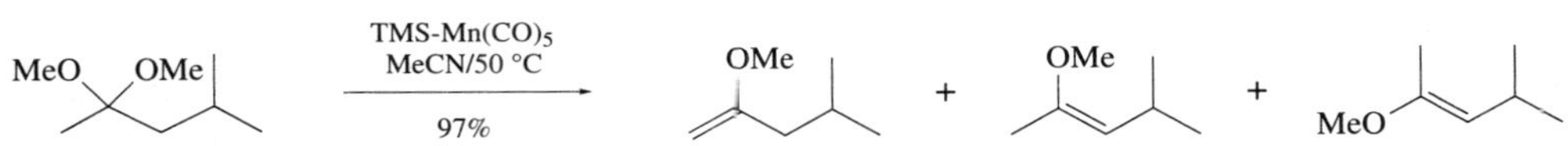

71 : 4 : 25

(56)

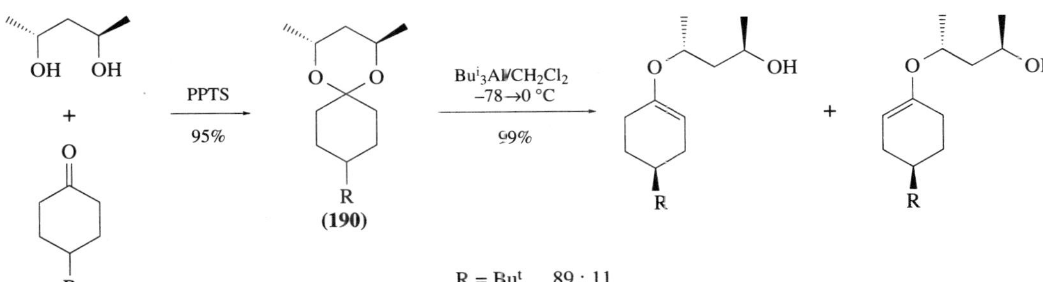

(190)

R = Bu^t 89 : 11
Et 86 : 14
Me 90 : 10

Scheme 95

The S_N2' addition of Grignard reagents to allylic acetals in the presence of a catalytic amount of copper(I) bromide provides alkenyl ethers with moderate stereoselectivity (Equation (57)) ⟨75TL3833, 79BSF(2)305⟩.

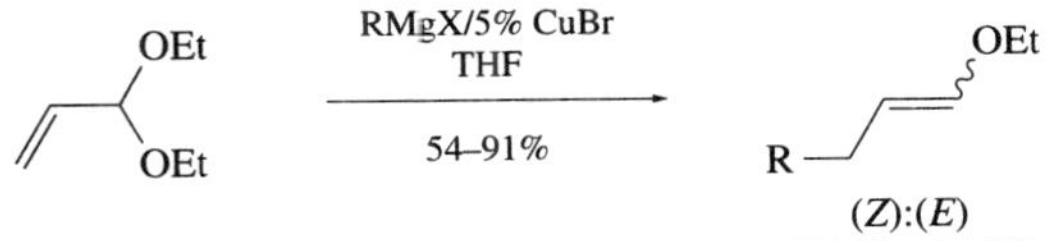

(*Z*):(*E*)
55:45 – 77:23

(57)

(vi) Classical alkenation of aldehydes and ketones

The popular Wittig and Wittig–Horner type alkenations of aldehydes and ketones provide a reliable approach to the synthesis of alkenyl ethers in good yields ⟨58JA6150, 62CB2514⟩. The phosphonium salt **(191)** ⟨75T89⟩, phosphine oxide **(192)** ⟨79JCS(P2)3099⟩, phosphonate **(193)** ⟨78TL3629⟩, and phosphonamide **(194)** ⟨78MI 213-01⟩ are commonly used. The phosphorus reagents are normally deprotonated by LDA (lithium diisopropylamide), and the subsequent reaction with carbonyl compounds is better conducted at low temperatures (−78 °C) to avoid undesired enolization. Although the addition is not stereoselective, the 50 : 50 mixture of diastereomeric adducts can be separated by chromatography or crystallization, as with the phosphine oxides **(195)** (Scheme 96) ⟨79JCS(P2)3099⟩. The adducts undergo stereospecific *syn*-elimination on treatment with base (sodium hydride or potassium *t*-butoxide), or simply by thermolysis ⟨78TL3629⟩ to give the alkenyl ethers in high yields.

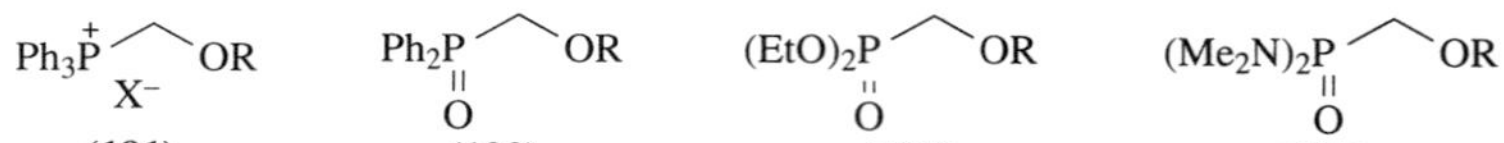

(191) **(192)** **(193)** **(194)**

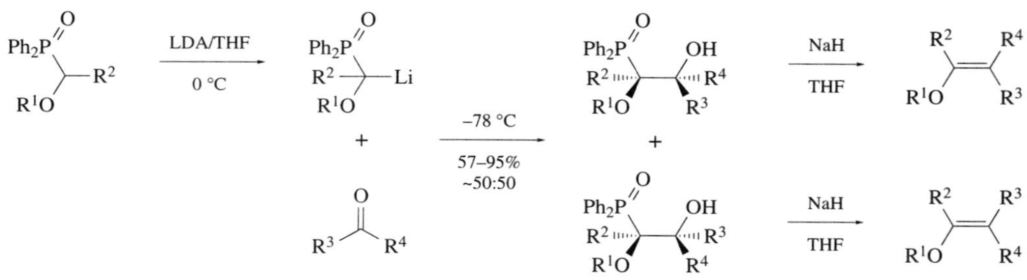

Scheme 96

A variation to the above ylide chemistry is the use of diazomethylphosphonate ⟨83JOC448, 89SC561⟩. Although this approach is limited to ketone substrates, it allows the option of introducing different *O*-alkyl groups in the products. In the presence of an alcohol and base, ketones are converted to alkenyl ethers in good yields (Equation (58)). In the case of unsymmetrical ketones, preferential formation of the (*E*)-isomer is observed.

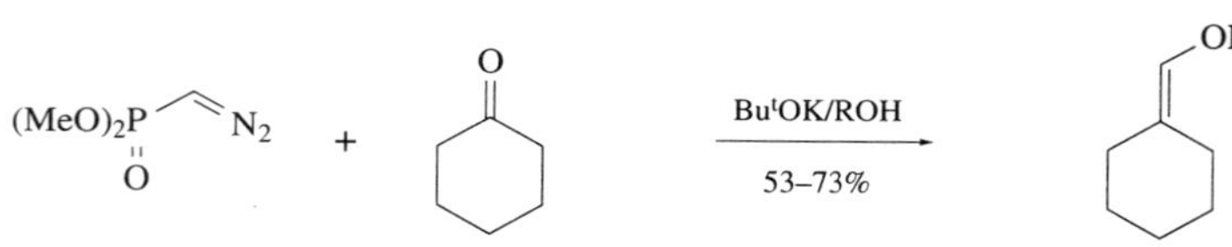 (58)

The Peterson type alkenation has also been applied to the synthesis of alkenyl ethers. The lithiated silylmethyl methyl ether reacts with aldehydes and ketones to give the hydroxysilane (Scheme 97). On treatment with potassium hydride *in situ*, alkenyl ethers are isolated in excellent yields ⟨79CC822⟩.

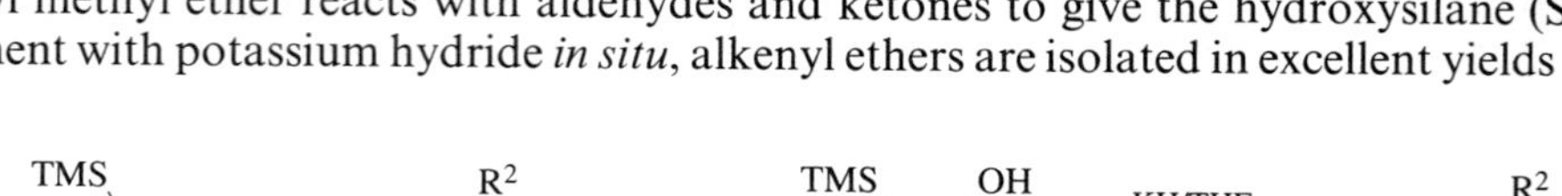

Scheme 97

(vii) Alkenation of esters, lactones and other carboxylic acid derivatives

The Tebbe reagent **(196)** ⟨78JA3611, 93OR(43)1⟩ is one of the most widely used organometallic reagents in the alkenation of carbonyl compounds. Its ability to alkenate ester and lactone carbonyls, which the Wittig reaction normally lacks, is valuable for the synthesis of alkenyl ethers ⟨80JA3270, 85JOC1212⟩. The Tebbe reagent can be prepared and used *in situ* without sacrificing any efficiency

⟨85JOC2386⟩. The utility of the Tebbe reagent is demonstrated in Scheme 98 in the transformation of the benzoate (**197**) ⟨90OS(69)72⟩ and the sugar lactone (**198**) ⟨84TL395, 86JOC5458⟩.

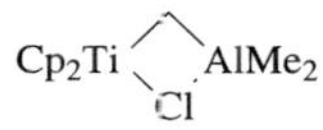

(196)

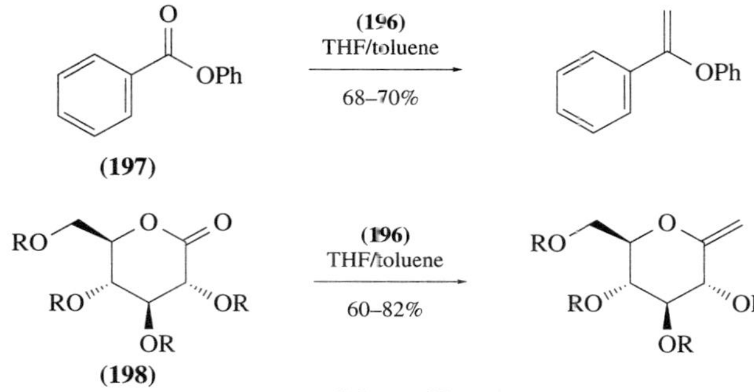

Scheme 98

While the Tebbe reagent only transfers an unsubstituted methylene group, other organometallic complexes have provided access to the preparation of tri- and tetra-substituted alkenyl ethers. For instance, the zirconium carbene complex (**199**) allows effective transfer of substituted methylenes to esters and lactones ⟨83JA640⟩ but lacks (*E*)/(*Z*) selectivity. Changing of the substrate from a lactone to an imidate, however, improves the stereoselectivity (Equation (59)) ⟨84JA8300⟩. Other transition metal alkylidene complexes such as tantalum ⟨76JA5399⟩ and tungsten ⟨72JA6543, 86OM398, 86CC531⟩ have also shown utility with reasonable stereoselectivity. Although the structure and composition of the active agent is not known, a mixture of $TiCl_4/Zn/TMEDA$ with gem dibromoalkanes is effective for the alkylidenation of esters and lactones with excellent (*Z*)-stereoselectivity (Equation (60)) ⟨87JOC4410⟩.

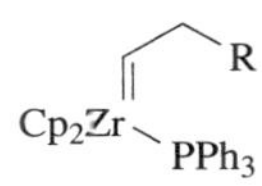

(199)

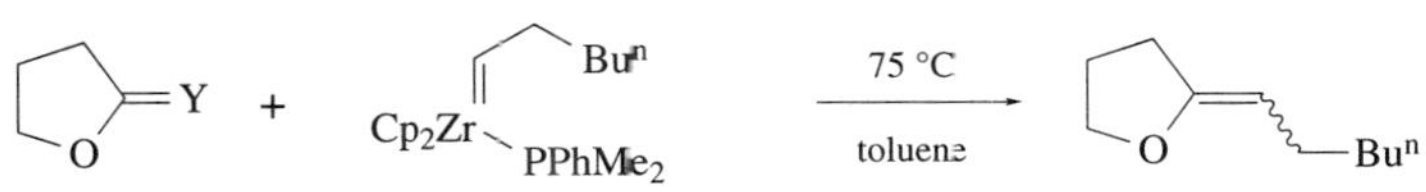 (59)

Y	*(E):(Z)*	*Yield(%)*
O	51:49	80
NPh	26:74	91

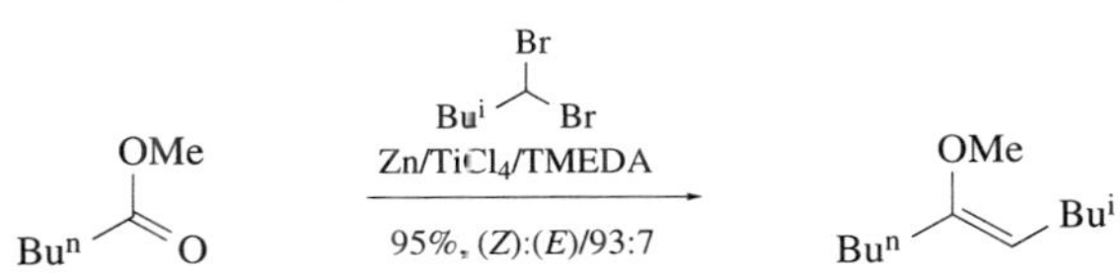 (60)

(viii) Homologation of ethenol ethers

Ethenol ether can be metallated easily at the α-position (Scheme 99), and its reaction with electrophiles leads to the formation of a variety of substituted alkenyl ethers ⟨74JA7125⟩. The advances in the transmetallation processes in conjunction with the transition metal catalyzed cross-coupling processes also expands the scope of alkylation, and allows the incorporation of alkenyl, alkynyl and aryl groups ⟨83JA943⟩. The coupling of ethenol ether with enol triflate can also be accomplished via a Heck type reaction mediated by palladium(II) ⟨89JOC1502⟩. The vinylation occurs regioselectively at the α-position to provide dienol ethers (Equation (61)).

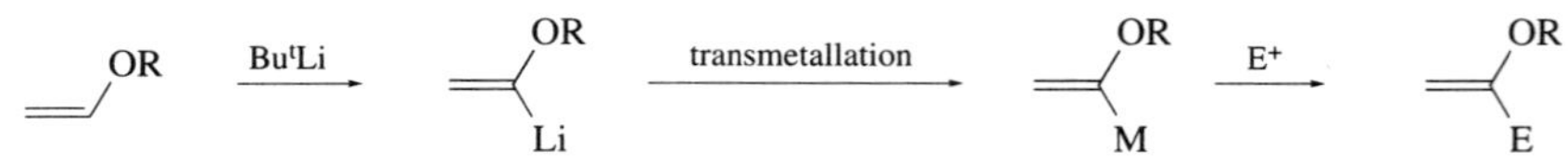

Scheme 99

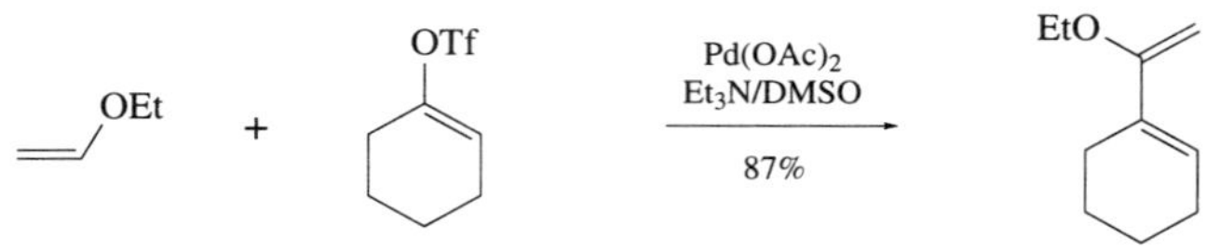

(61)

(ix) Miscellaneous

The Birch reduction of aryl alkyl ethers occurs regioselectively to give cyclic alkenyl ethers (Scheme 100) ⟨70S161, 92OR(42)1⟩. For an EDG such as methoxy, 2,5-reduction is observed, as illustrated by the reduction of anisole. In the presence of an EWG such as ester or keto-group, the ring is activated and 1,4-reduction is observed, as in the reduction of methyl salicylate.

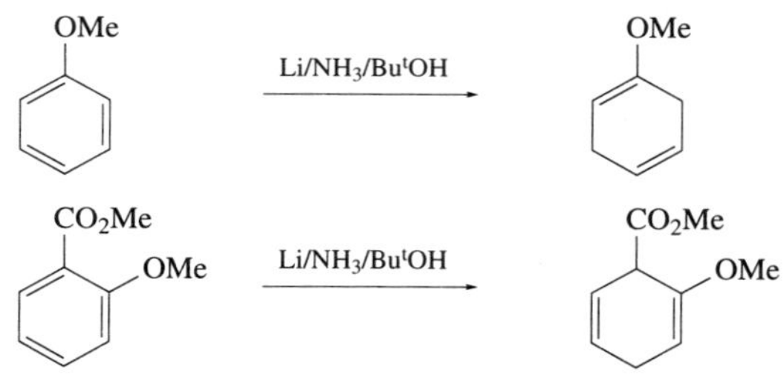

Scheme 100

A novel synthesis of alkenyl ethers involves phosphorylation of stereochemically defined ester enolates (Scheme 101). The stereochemical integrity is preserved during the reductive cleavage of the phosphate group to provide (*Z*)-alkenyl ethers in high yields ⟨87JOC2303⟩. The regioselective α-ring opening of stereochemically pure epoxysilanes produces the hydroxysilanes (**200**) (Scheme 102), which undergo *syn*-elimination on treatment with base to deliver (*E*)-alkenyl ethers stereospecifically in high overall yields ⟨77JA1993⟩.

Finally, alkenyl acetates can be obtained in good yields by heating alkenyl halides with copper(I) acetate with retention of the C=C bond configuration ⟨75JOM(93)415, 75TL3429⟩.

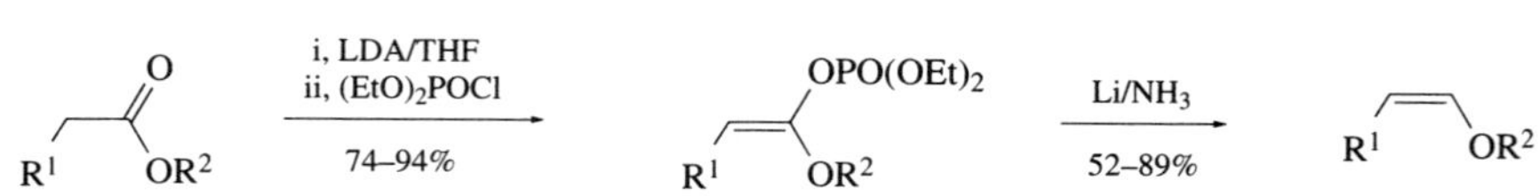

Scheme 101

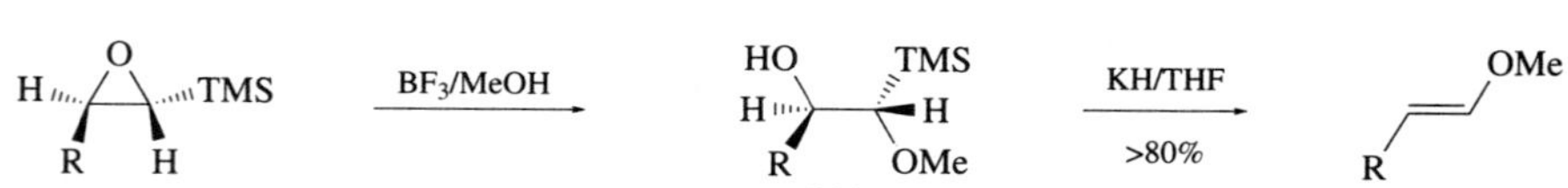

(200)

Scheme 102

2.13.1.2.2 *Aryl ethers and esters*

Alkyl aryl ethers and aryl esters are conveniently prepared by *O*-alkylation and acylation of phenols. The aromatic nucleophilic displacement of aryl halides by alkoxides has also been used widely. *O*-phenylation of alcohols is less common but provides a practical alternative.

(i) O-*Alkylation/acylation of phenols and phenolates*

In general, oxygen versus carbon alkylation is less of a problem for phenols as compared to enols. Typically, phenol is treated in the presence of a base with an alkylating agent (i.e., alkyl halides, sulfates or sulfonates) to produce alkyl aryl ethers in good yields (the Williamson etherification) ⟨B-91MI 213-01⟩. The alkylation can be greatly facilitated by the addition of phase transfer catalyst ⟨88T6677⟩ or calixarene ⟨88CL1773⟩. For "salt-free" conditions, other less widely used alkylating agents have been employed (e.g., alkyl orthocarbonate ⟨56ACS1006⟩, dialkylcarbonate ⟨79GEP(O)2807762⟩, pentaalkoxyphosphorane ⟨78JOC4672⟩, dialkyl oxalate ⟨72JOC3944⟩, trimethylsulfonium hydroxide ⟨79JOC638⟩, and methyl trichloroacetate ⟨84SC69⟩) as illustrated in Scheme 103.

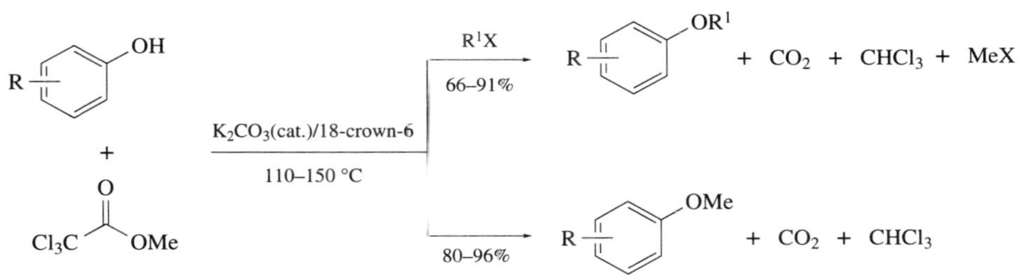

Scheme 103

In the case of dihydroxybenzenes, the more acidic phenolic hydroxyl is normally alkylated selectively, as demonstrated by the reaction of formyl catechol (Scheme 104) ⟨93H(36)13⟩. The combination of lithium carbonate in DMF is found to be particularly efficient for this purpose ⟨88SC1379⟩.

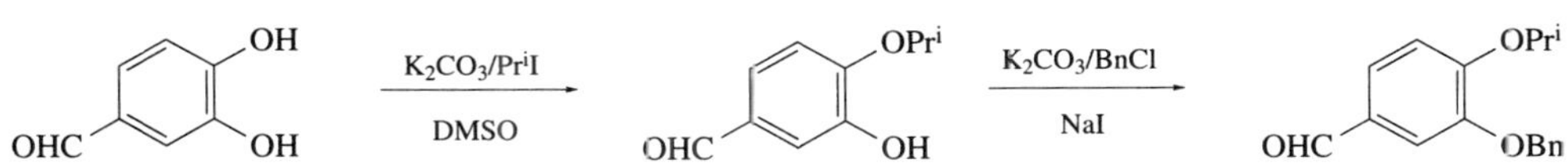

Scheme 104

Owing to the acidity of the phenolic hydroxyl, it can be condensed with alcohol under neutral conditions with dcc (dicyclohexyl carbodiimide) ⟨66CB1479⟩ or under the Mitsunobu condition (Ph_3P/dead) ⟨75CI(L)281, 75JCS(P1)461, 81S1⟩. Under the mild Mitsunobu conditions, the complication of acetylene–allene rearrangement was avoided in the coupling of the propargylic alcohols **(201)** with phenol (Equation (62)) ⟨89SC1255⟩.

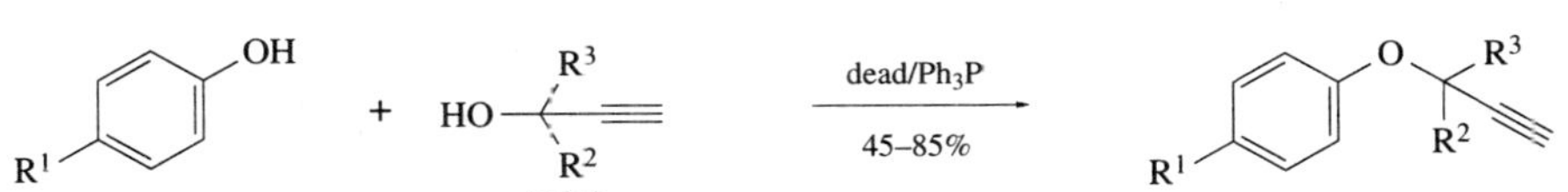

(201) (62)

R^1 = H, Me, NO_2, Cl, OMe
R^2, R^3 = H, Me, Ph

Despite the hazardous nature of diazomethane, it methylates phenols directly to methyl aryl ethers ⟨73TL1397⟩. Trimethylsilyldiazomethane, however, provides a safer alternative ⟨84CPB3759⟩. The alkylation of phenols by O—H insertion with rhodium carbenoids is highly efficient as illustrated by the diazophosphonoacetates in Equation (63) ⟨94T3177⟩.

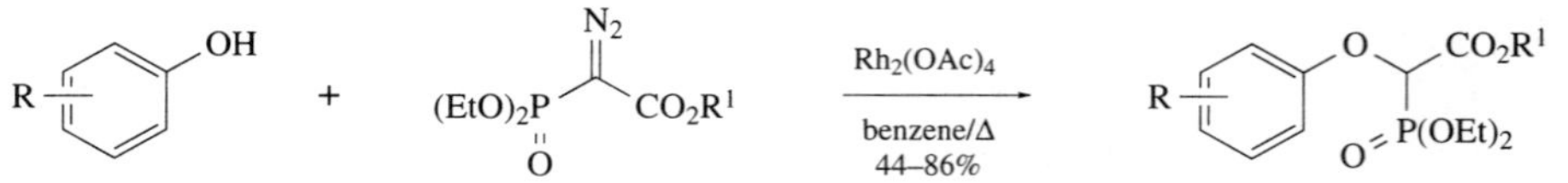

(63)

The preparation of *t*-butyl aryl ethers is typically conducted with isobutylene under acid catalysis ⟨55JOC1232, 86JOC111⟩. Alternatively, phenols are readily alkylated with *t*-butyl bromide in the presence of an amine ⟨78CL57⟩, or with the *t*-butyl acetal of DMF (**202**) ⟨83SC827⟩. Heating of phenols with a *t*-alkyl halide in the presence of $Ni(acac)_2$ provides a generic approach to *t*-alkyl aryl ethers ⟨82S186⟩.

The preparation of aryl esters is straightforward simply by heating the phenol with an acid chloride or anhydride ⟨86SC1043⟩.

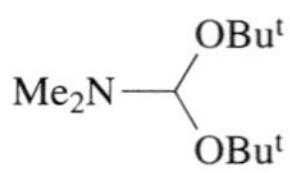

(202)

(ii) Aromatic nucleophilic substitution

The S_NAr displacement of activated aryl halides by alkoxide is commonly used in the preparation of alkyl aryl ethers. Even for unactivated substrates, reactions proceed smoothly with HMPA as the solvent ⟨76JOC732⟩. The reaction rate can also be enhanced by the addition of copper salts and has been extensively studied ⟨89T5565⟩. Transition metal carbonyls, for example, manganese ⟨86JOC2137⟩, chromium ⟨88JOM(347)365⟩, and iron ⟨89CC1363⟩, are effective in activating aryl halides in the reaction with alcohols and phenols. The removal of the transition metal complexes is easily achieved by treating with acetonitrile ⟨75JCS(D)1677⟩. The mildness of this protocol is demonstrated in the synthesis of the diaryl ether (**203**) (Scheme 105), where no racemization of the tyrosine moiety is observed ⟨86JOC2137⟩.

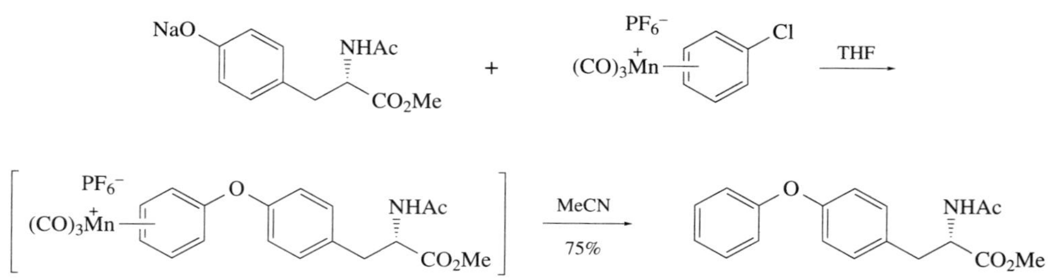

(203)

Scheme 105

Diaryl ethers can be prepared by the coupling of phenols with an aryl halide and it is generally referred to as the Ullmann reaction (Scheme 106). The coupling is typically conducted in pyridine at elevated temperatures and catalyzed by copper salts, as demonstrated by the synthesis of the diaryl ether (**204**) ⟨94SL49⟩. It is interesting to note that in 1,3-dimethyl-2-imidazolidinone (DMI), a solvent with a high dielectric constant and solvating power, the coupling of the hydroxybenzyl alcohol (**205**) with chlorobenzene occurs chemoselectively with the phenol moiety ⟨88CL899⟩. For activated aryl halides, coupling with phenol can be effectively carried out with potassium fluoride/alumina catalyzed by 18-crown-6 ether or PTC ⟨93JOC3229⟩.

(iii) Miscellaneous

Triphenylbismuth diacetate ⟨83JOC441⟩ is effective for the phenylation of alkyl alcohols and phenols ⟨86TL3615⟩. Furthermore, the reaction is accelerated greatly by the addition of a catalytic amount of copper(II) acetate ⟨86JCS(CC)65⟩.

t-Butyl aryl ethers can also be prepared by the oxidation of arylmagnesium bromide with *t*-butyl peroxybenzoate (**206**) ⟨61OS(41)91⟩ or *t*-butylperoxydialkylphosphates (**207**) ⟨72JOC2267⟩.

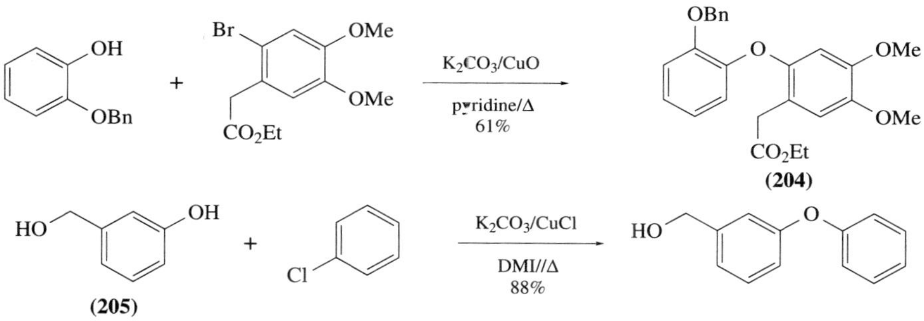

(204)

(205)

Scheme 106

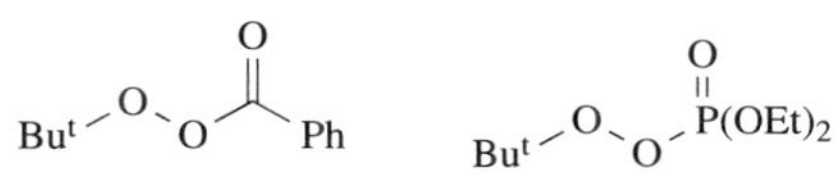

(206) (207)

2.13.1.3 Hypohalites (RO—Hal)

2.13.1.3.1 Alkenyl hypohalites

The preparation and isolation of alkenyl hypohalites (**208**) are not well documented in the literature because generally they are not stable. Nevertheless, their intermediacy in the halogenation of carbonyl compounds cannot be denied. In the NBS oxidation of an alcohol, the hypobromite intermediate is invoked. Since the enol tautomer is responsible for the reaction of carbonyl compounds, it is reasonable to assume the kinetic formation of the alkenyl hypohalite as a transient intermediate despite its elusiveness (Scheme 107).

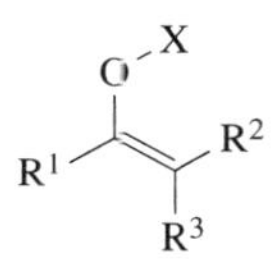

(208)

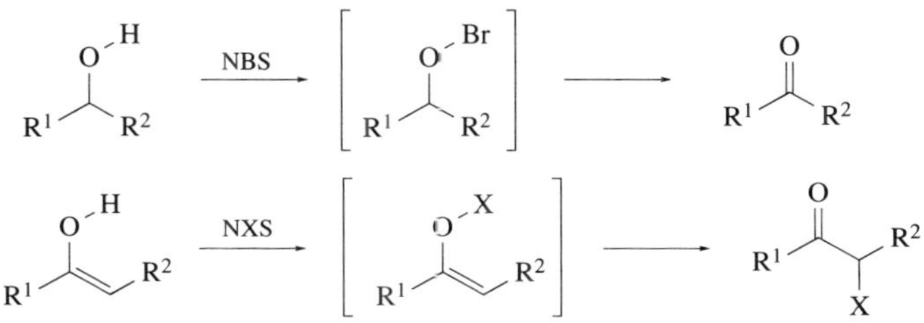

Scheme 107

2.13.1.3.2 Aryl hypohalites

Aryl hypohalites are believed to be intermediates in the halogenation of phenolic compounds ⟨67JOC2358⟩, which allows preferential *ortho*-halogenation (Scheme 108), particularly at higher pH ranges ⟨84JCS(P2)451⟩. It is suggested, however, that the iodination of phenol proceeds through a different pathway because phenyl hypoiodite is energetically too labile to be formed even as a transient intermediate ⟨91JCR(S)300⟩. There is a report of the isolation of hypobromite intermediates in the bromination of estradiol and its analog (Equation (64)) ⟨82JOC359⟩. The hypobromites (**209**) are solids and stable at ~7°C for several days. At room temperature and in the solid state, the hypobromites slowly rearrange to the thermodynamically more stable 2- and 4-bromoestradiols. In general, the isolation of simple phenyl hypohalites remains an unrealistic task.

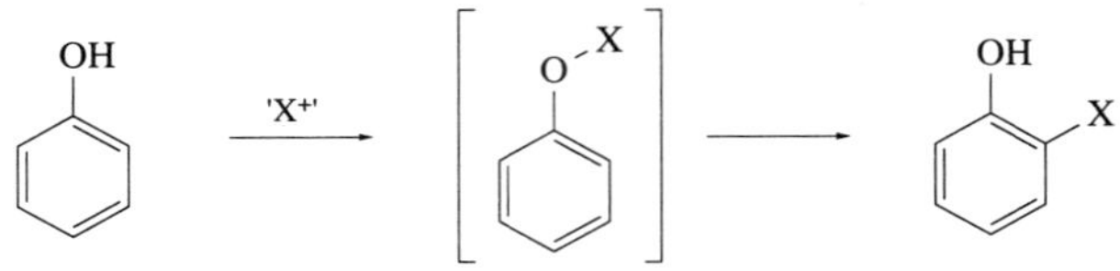

Scheme 108

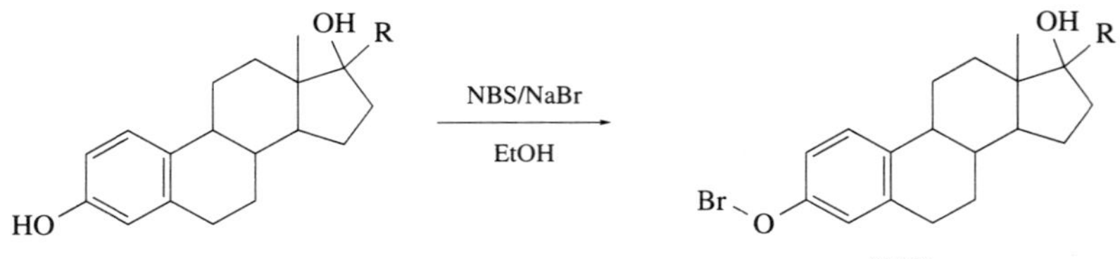

(64)

(209)

R = H, 96%
R = C≡CH, 79%

2.13.1.4 Peroxide Functions: RO—OH, RO—OR′

Although alkyl hydroperoxides and dialkyl peroxides are known and relatively stable, their alkenyl and aryl counterparts are, on the contrary, uncommon and unstable. There are scattered claims of the preparation of alkenyl and aryl peroxides in the older literature ⟨88HOU(E13)761⟩, but most of them have proved to be unfounded.

2.13.1.4.1 Alkenyl peroxides

Nature has the ability to synthesize these unstable compounds in an ingenious way, as exemplified by the marine natural product (**210**) ⟨74JA5580⟩, whereas attempts in the laboratory have not usually met with success ⟨63CB3044, 75S42⟩. In general, alkenyl peroxides are unstable and cannot be isolated ⟨92JCR(S)202⟩. The possible existence of the alkenyl peroxide may be limited to a transient intermediate which rapidly rearranges to the thermodynamically more stable α-oxygenated carbonyl compound. Both vinyl ether and acetate were reported to react with *t*-butyl hydroperoxide in the presence of mercuric salt to give the peroxy adduct (**211**) (Scheme 109) ⟨70JPR43⟩. Demercuration occurred on heating and led to the isolation of *t*-butoxyacetaldehyde, wherein the vinyl peroxide (**212**) was implicated as the intermediate.

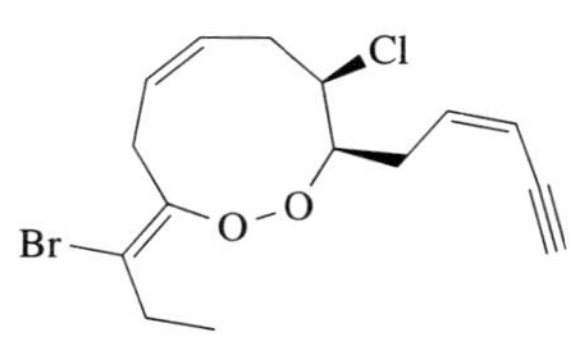

(210)

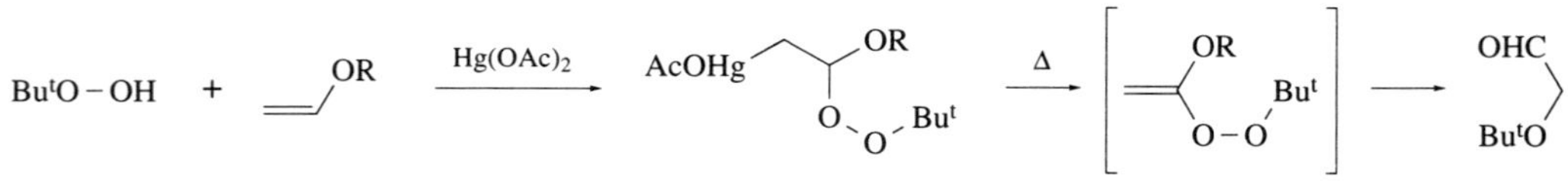

(211) **(212)**

R = Et, Ac

Scheme 109

2.13.1.4.2 Aryl peroxides

Owing to the stabilization of the aryloxy radical by the aromatic system, the dissociation energy of the peroxy bond in diphenyl peroxide is only ~16 kJ mol^{-1} and renders aryl peroxides highly unstable ⟨81JCS(P2)1512⟩. There have been reports of the preparation of aryl peroxides that were later found to be unsubstantiated ⟨72JOC3096, 79TL4363⟩. Nevertheless, the paper ⟨90M709⟩ describes the oxidation of an acridinone with mcpba leading to the formation of an aryl hydroperoxide (Equation (65)). The preparation of 2,4,6-trialkylbenzene monohydroperoxides by air oxidation of the corresponding benzenes is also the subject of some patents (Equation (66)) ⟨67USP3308164, 67USP3308165⟩.

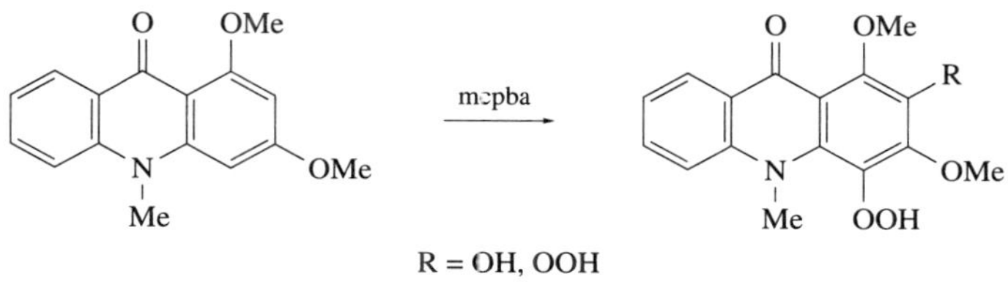

(65)

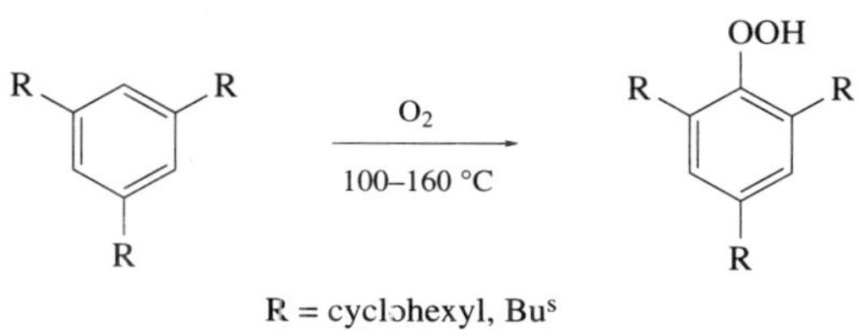

(66)

Despite the instability of aryl peroxides, some heteroaryl peroxides have been synthesized and found to be rather stable. In particular, trichloro-1,3,5-triazine reacts with *t*-butyl hydroperoxide, in the presence of alkaline or silver trifluoroacetate, to give the trisperoxytriazine (**213**) in 81% yield ⟨78T1231⟩. A series of the mono-, bis- and trisperoxy heterocyclic compounds has been prepared as crystalline solids (Scheme 110) ⟨76LA2331, 81JCS(P2)1512⟩. *Note that extreme caution should be exercised in handling these peroxy-compounds because they are potentially explosive.* Indeed, the monoperoxide (**214**) exploded during distillation while the trisperoxide (**213**) was found to decompose violently on heating.

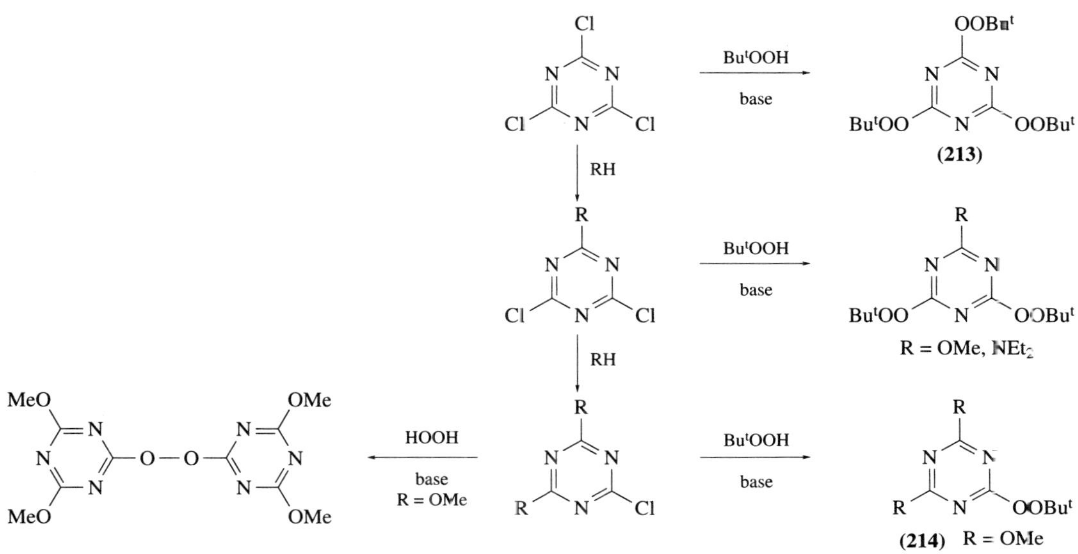

Scheme 110

2.13.1.5 Functions Based on RO—S, RO—Se and RO—Te Units

2.13.1.5.1 Alkenyl and aryl sulfenates

Sulfenylation of enols or enolates does not result in the formation of the alkenyl sulfenates, because reaction occurs only at the carbon atom ⟨46CRV(39)269, 92JOC1948⟩. Although phenols also undergo *C*-sulfenylation with sulfenyl chloride ⟨63JOC2807, 64JOC898⟩, phenolates are sulfenylated at the oxygen atom to give aryl sulfenates (Equation (67)), as described in the older literature ⟨12LA(391)55, 36JCS327⟩. The preparation of aryl sulfenate esters has been reviewed previously ⟨46CRV(39)269, B-73MI 213-01⟩. The most widely used sulfenylating reagents include sulfenyl halides, sulfenamide ⟨70TL5115⟩, and *N*-sulfenyl imides, for example, the phthalimide (**215**) ⟨70CC1466⟩ and succinimide (**216**) ⟨72CL441⟩.

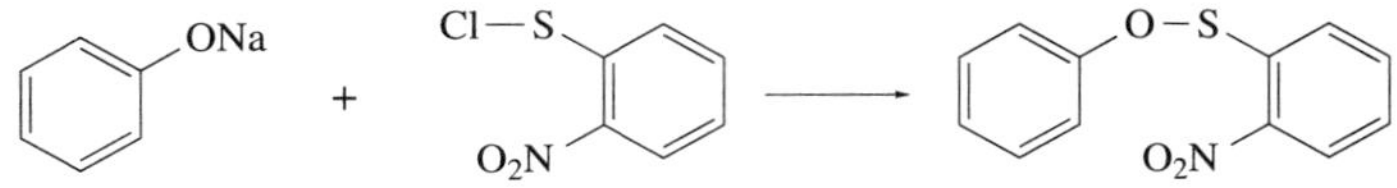

(67)

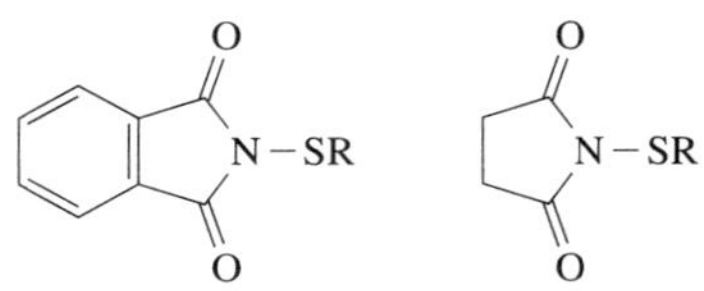

(215) **(216)**

2.13.1.5.2 Alkenyl and aryl sulfinates

Similar to the sulfenylation of enolizable carbonyl compounds, sulfinylation occurs only at the carbon atom, which prevents the formation of alkenyl sulfinates ⟨93SC1515⟩. However, aryl sulfinates are conveniently prepared by treating phenols with sulfinylating reagents (i.e. sulfinyl halides (**217**) and sulfinyl phthalimides (**218**) (Scheme 111) ⟨73JA2918, 76JOC3987⟩. The sulfinylation of *O*-silylated phenols is also a high yielding process ⟨80TL2067⟩. The most commonly used sulfinyl chlorides are readily available from the chlorination of the corresponding disulfides ⟨57JOC536⟩ or from the reaction between sodium sulfinate and thionyl chloride ⟨63OSC(4)937⟩. *N*-Alkyl and *N*-aryl sulfinyl phthalimides (**218**) are easily prepared from the mcpba oxidation of the corresponding sulfenyl phthalimides ⟨73JOC4328⟩.

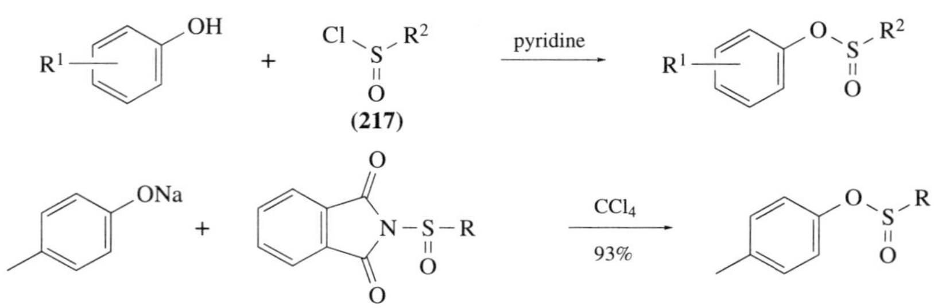

Scheme 111

2.13.1.5.3 Alkenyl and aryl sulfonates

Alkenyl sulfonates are generally prepared by sulfonylation of the enolates of carbonyl compounds with sulfonyl halides. Reaction can occur either at oxygen or carbon, and the selectivity is dependent largely on the cations of the enolates and the polarity of the solvents. Similar to the criteria set out for the alkylation of enolates (Section 2.13.1.2.1), large and soft cations (e.g., caesium and tetraalkylammonium versus lithium and potassium), and polar solvents generally promote *O*-sulfonylation (Equation (68)) ⟨82CB3687⟩. Addition of alkylsulfonic acid to alkynes mediated by mercuric oxide provides alternative access to alkenyl sulfonates (Equation (69)) ⟨82CB399⟩.

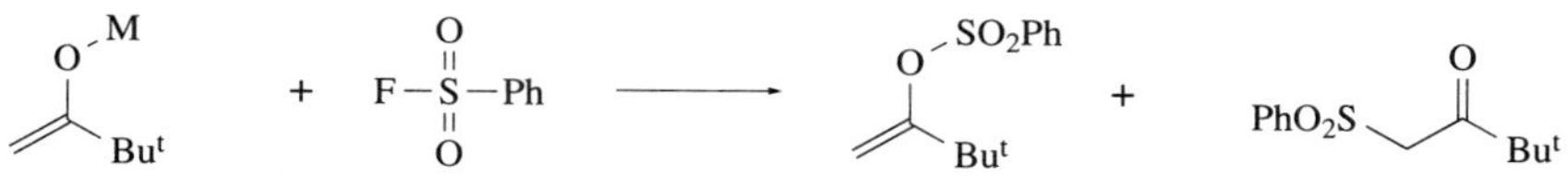

(68)

M	*Solvents*	*O vs C*	*Yield (%)*
Li	THF	0 : 100	40
Li	THF/HMPA	70 : 30	77
Cs	THF	100 : 0	99
R_4N	THF	100 : 0	90–96

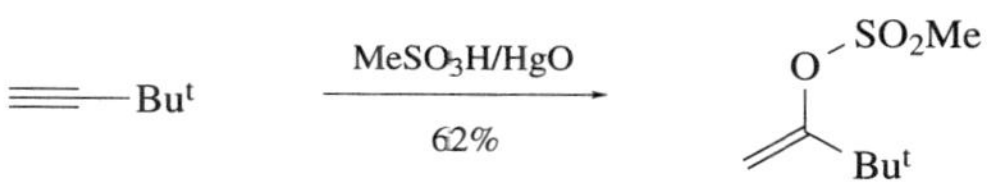

(69)

O-Sulfonylation of phenols is straightforward to furnish aryl sulfonates in high yields. Electrophilic oxidation of arenes with nitrobenzenesulfonyl peroxides has also been used to prepare aryl sulfonates ⟨66JOC153⟩.

Among the various alkenyl and aryl sulfonates, the fluorosulfonate ⟨94JOC6683⟩, perfluoroalkanesulfonates and trifluoromethanesulfonates (triflates) in particular ⟨82S85⟩ have received widespread attention in recent years due to their valuable synthetic utility, for instance, in Stille cross-coupling reactions ⟨84JA4630, 86AG(E)508, 93S735⟩. Alkenyl triflates can be prepared by trapping the metal enolates of carbonyl compounds with *N*-phenyl-trifluoromethanesulfonimide (Scheme 112) ⟨83TL979, 85S335⟩. A more convenient procedure is to treat an aldehyde or ketone with trifluoromethanesulfonic anhydride (triflic anhydride) in the presence of base ⟨70AG(E)521, 74OS(54)79⟩, especially sterically hindered pyridines ⟨80S283⟩. Alkenyl triflates have also been prepared by the addition of trifluoromethanesulfonic acid (triflic acid) to alkynes ⟨74JA1100⟩. Markovnikov's rule dictates the regiochemistry of the addition and mixtures of *cis* and *trans* isomers are usually obtained (Scheme 113).

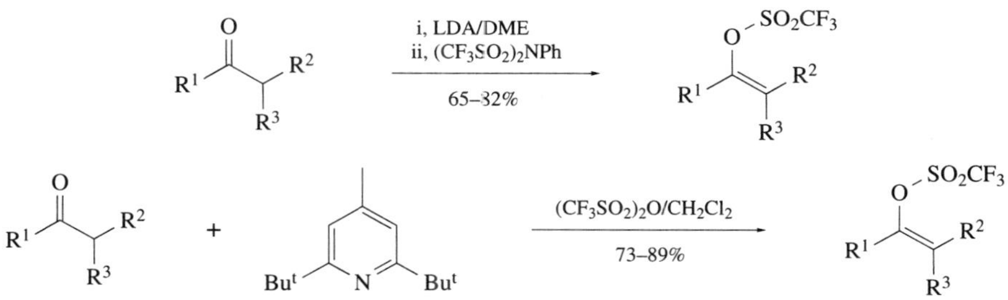

Scheme 112

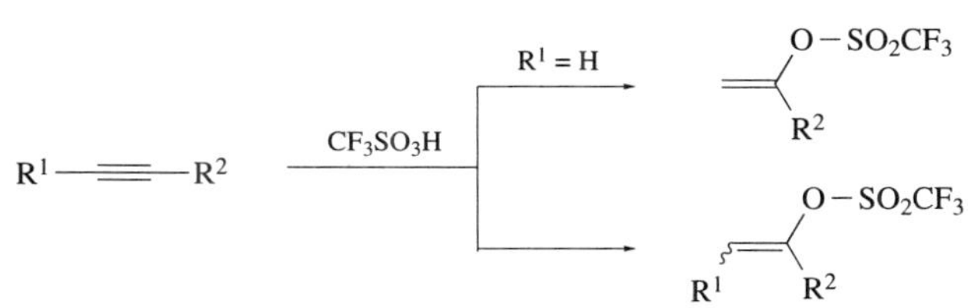

Scheme 113

Similarly, aryl triflates are easily prepared by reacting phenols with triflic anhydride in pyridine ⟨76JOC4099⟩ or by treating the metal phenolates with a triflating agent such as triflic anhydride, halide or imidazolide ⟨74S293⟩. Notably, *O*-silylated phenols are converted readily to aryl perfluoroalkanesulfonates with perfluorosulfonyl fluoride catalyzed by fluoride ion ⟨73LA20⟩. A novel synthesis of aryl triflates has been achieved by thermal or photochemical decomposition of aryldiazonium tetrafluoroborates in triflic acid (Equation (70)) ⟨91CL459⟩.

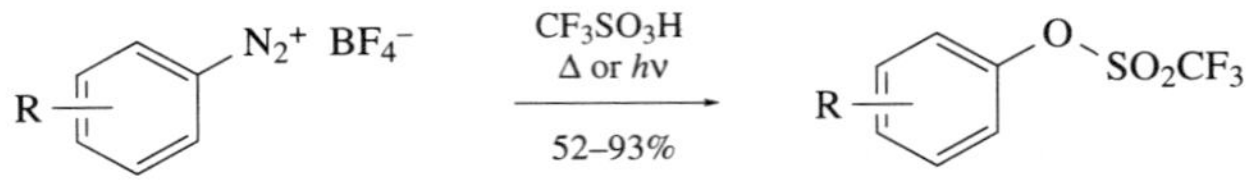

(70)

2.13.1.5.4 *Alkenyl and aryl sulfates*

Alkenyl and aryl sulfates are prepared by treating enolates and phenolates with chlorosulfonates (Equation (71)) ⟨48BSF156, 78MI 213-02⟩.

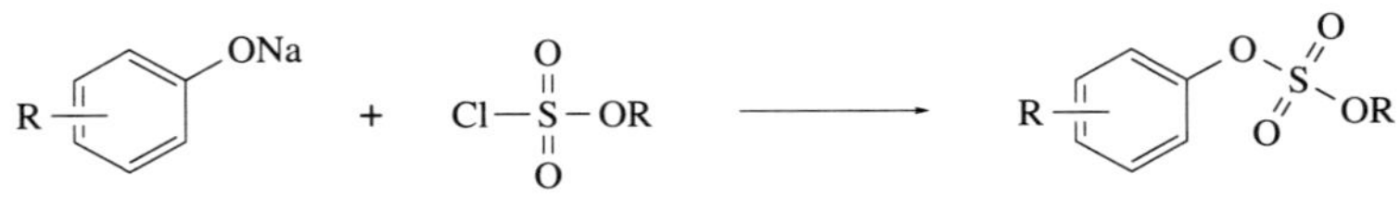

(71)

2.13.1.5.5 *Alkenyl and aryl O—Se and O—Te compounds*

Selenium and tellurium are not particularly oxygenophilic and reactions tend to occur at the carbon of enols and phenols. Nevertheless, the formation of the oxygen–selenium and oxygen–tellurium bonds is possible in transient intermediates. For instance, the oxidation of phenols with benzeneseleninic anhydride involves the formation of an aryl seleninate ester intermediate (Scheme 114) ⟨81JCS(P1)1473⟩.

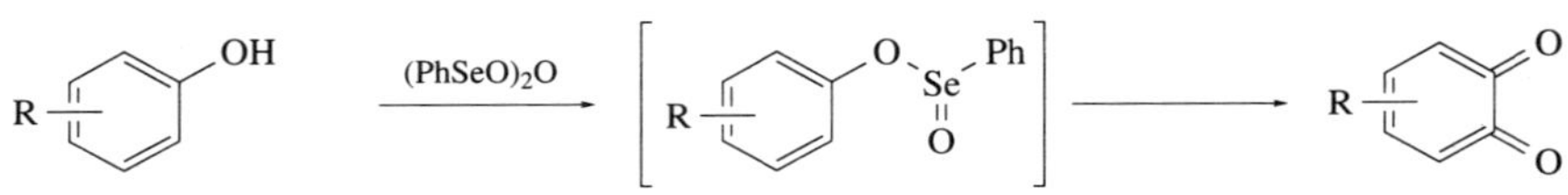

Scheme 114

Although reaction between acetylacetone and tellurium(IV) chloride leads to formation of the C—Te bonded 1-tellura-cyclohexan-3,5-dione ⟨76JCS(D)2307, 80JOM(188)255⟩, it has been reported ⟨90IJC(A)703⟩ that the reaction with dialkyl and diaryl tellurium(IV) methoxides or oxides provides the O—Te bonded enolates (Scheme 115). The sodium enolates of 1,3-diketones also undergo metal exchange with triphenyltellurium(IV) chloride. The corresponding tellurium enolates are highly stable and are isolated as solids from water ⟨85JOM(291)185⟩.

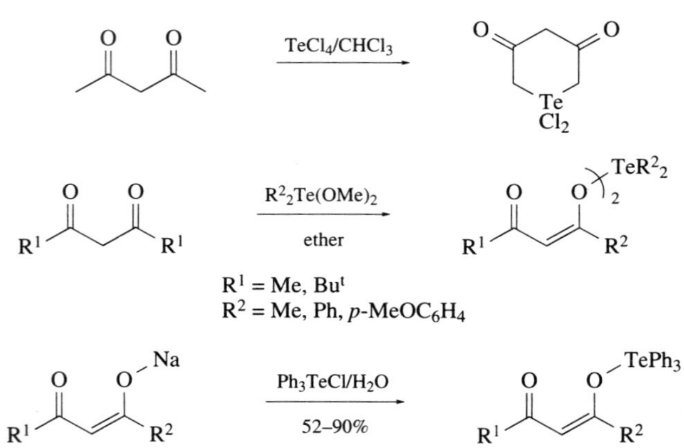

Scheme 115

2.13.1.6 Functions Based on the RO—N Unit

2.13.1.6.1 O-*Alkenyl and* O-*aryl hydroxylamines*

There are few precedents describing *O*-alkenyl hydroxylamines in the literature. Disclosures of *O*-aryl hydroxylamines are relatively more abundant.

The addition of *N*-*t*-butylhydroxylamine to a sulfonyl alkyne produces the (*Z*)-*O*-alkenyl hydroxylamine (86%) ⟨74JOC2641⟩. The substitution on the alkyne, however, has a significant impact on the outcome of the reaction (*O* vs *N* addition) (Equation (72)).

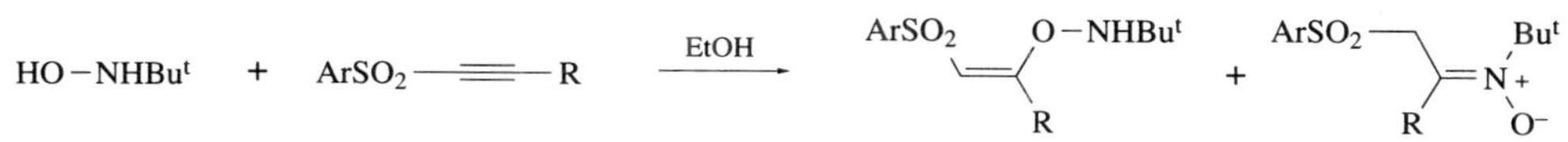
(72)

R	*O vs N*	*Yield*(%)
Me	100 : 0	86
H	0 : 100	63
Ph	40 : 60	94

O-Aryl hydroxylamines can be prepared by *O*-amination of phenol with an aminating reagent. Hydroxylamine-*O*-sulfonic acid ⟨57IS122⟩ is effective but the yield is usually low (15%) ⟨62CI(L)559⟩. Better yields (54–74%) can be achieved by using mesitylenesulfonyl hydroxylamine as shown in Equation (73) ⟨80S461⟩.

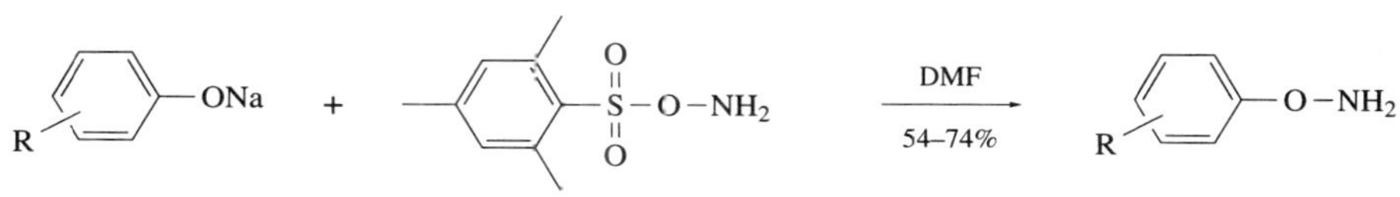
(73)

Alternatively, hydroxylamine can be *O*-arylated with diphenyliodonium bromide ⟨62CI(L)1244⟩ or activated aryl halides ⟨63HCA2009⟩. In order to ensure arylation of the oxygen, the nitrogen of the hydroxylamine is typically deactivated by derivatization as a carbamate. Both *N*-*t*-BOC hydroxylamine (*t*-BOC = *t*-butoxycarbonyl) ⟨72T3833⟩ and acetohydroxamate ⟨73JOC1239⟩ react readily with activated aryl halides (Scheme 116). The *N*-blocking groups can be easily removed by mild acid treatment to provide the *O*-aryl hydroxylamines, (**219**) for example. Like mesitylenesulfonyl hydroxylamine, the hydroxylamine (**219**) is an efficient amine transfer agent and reacts readily with a variety of phenols to provide the corresponding *O*-aryl hydroxylamines ⟨84JOC1348⟩. Most activated aryl halides are substituted with an EWG at the *ortho* and/or *para* position which consequently dictates the substitution of the aryl ring system. The use of chromium tricarbonyl complexes to activate the aryl halides adds a new dimension to the possible substitution and allows the inclusion of EDG as well as EWG on the aryl ring. *N*-*t*-BOC hydroxylamine reacts readily with a variety of aryl chlorides activated by complexation with chromium tricarbonyl (Scheme 117) ⟨88S344⟩. The chromium complexes can be easily removed by treatment with iodine.

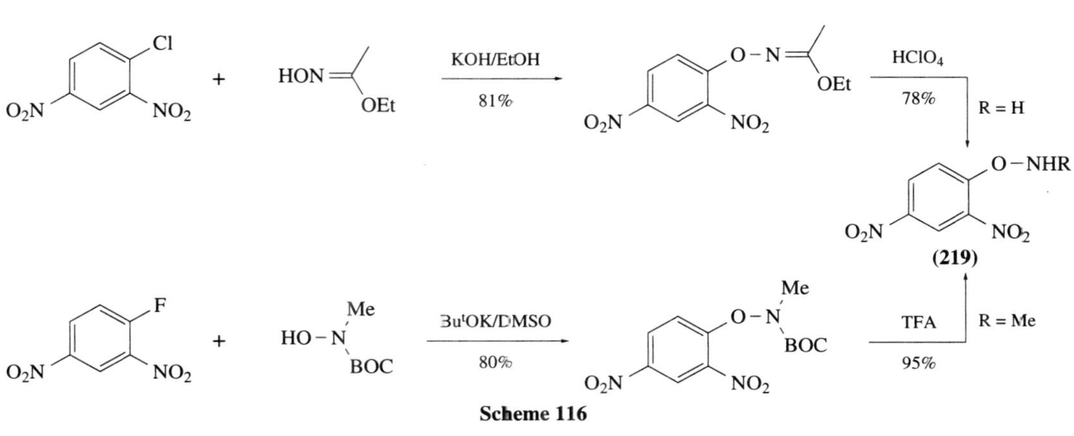

Scheme 116

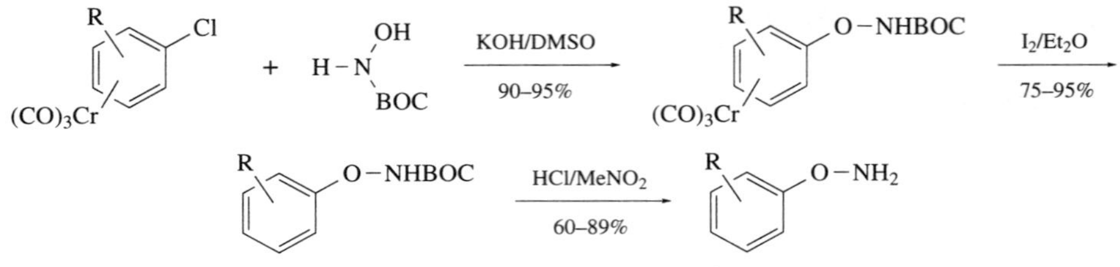

R = H, Cl, OMe, Me

Scheme 117

2.13.1.6.2 *O-Aryl oximes*

O-aryl hydroxylamines undergo facile condensation with aldehydes and ketones to give *O*-aryl oximes ⟨67JHC413, 75JOC126⟩. Alternatively, *O*-aryl oximes can be prepared directly from arylation of oximes with activated aryl halides (Equation (74)) ⟨67TL407, 67TL859, 75S782⟩. Arylation of oximes with haloarenes activated by chromium tricarbonyl, as described in the previous section, is also highly productive ⟨87S192⟩.

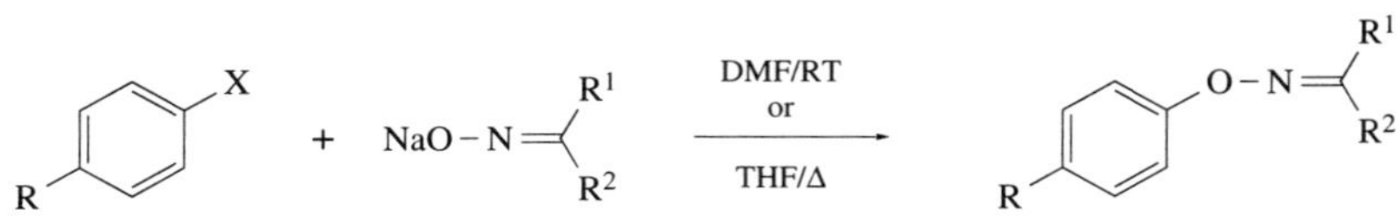

(74)

X = F, NO_2
R = NO_2, CF_3, CO_2Me, CN, CHO
R^1, R^2 = alkyl, aryl

2.13.1.6.3 *Miscellaneous RO—N compounds*

N-Hydroxyphthalimide reacts with activated aryl halides to give the *N*-aryloxyphthalimide (**220**) ⟨63HCA2009⟩. Similarly, phenylation of *N*-hydroxyphthalimide with diphenyliodonium chloride provides the *N*-phenoxyphthalimide (**221**), which undergoes hydrazinolysis to the *O*-phenyl hydroxylamine as shown in Scheme 118 ⟨77SC365⟩.

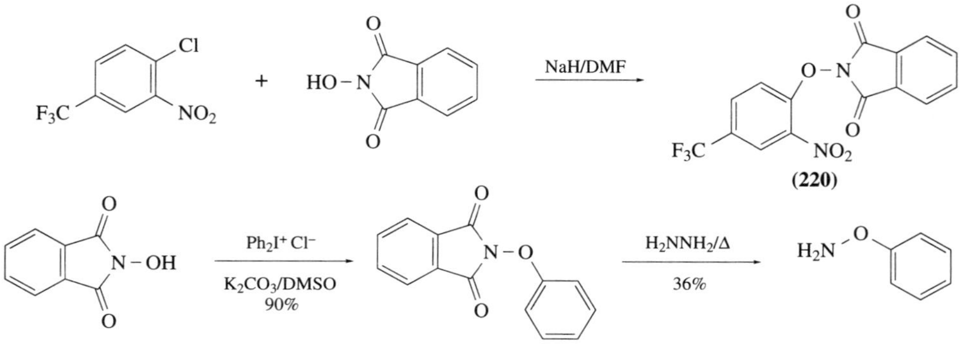

(220)

(221)

Scheme 118

2.13.1.7 Functions Based on RO—P, RO—As, RO—Sb and RO—Bi Units

Most attention on the phosphorus derivatives of enol and phenol is directed to the alkenyl and aryl phosphates, because they are useful synthetic intermediates. The chemistry of the enolates and phenolates of arsenic, antimony and bismuth is not studied and documented as widely.

2.13.1.7.1 Alkenyl and aryl phosphites

Alkenyl phosphites are generally prepared from the reaction of enolates and phosphorochloridites (**222**) as shown by the synthesis of the vinyl phosphites in Equation (75) ⟨59DOK(124)1251⟩. Triaryl phosphites are prepared by the reaction of phosphorus trichloride with 3 equivalents of phenols ⟨58MI 213-01⟩. The preparation of monoaryl phosphites is better achieved by treating sodium phenolates with phosphorochloridites (**222**).

(75)

(222)

R = Me, Et, Pr, Bu, Ph

2.13.1.7.2 Alkenyl and aryl phosphates/phosphoroamidates

The preparation and chemistry of alkenyl phosphates (also commonly known as enol phosphates) have been reviewed on separate occasions ⟨61CRV607, 64PAC307⟩. The most widely used procedure is the direct *O*-phosphorylation of enolates of carbonyl compounds. Phosphorus is highly oxygenophilic and *O*-phosphorylation occurs exclusively. Hence, treatment of enolates with dialkyl phosphorochloridate (**223**) and diamidophosphorochloridate (**224**) leads to the formation of the corresponding phosphates ⟨69TL2145⟩ and phosphoroamidates ⟨72JA5098, 83OS(61)116⟩ (Scheme 119). Alkenyl phosphates can also be prepared via the Perkow reaction ⟨52N353, 69CC112⟩ by warming α-halo aldehydes or ketones with trialkyl phosphites (Equation (76)).

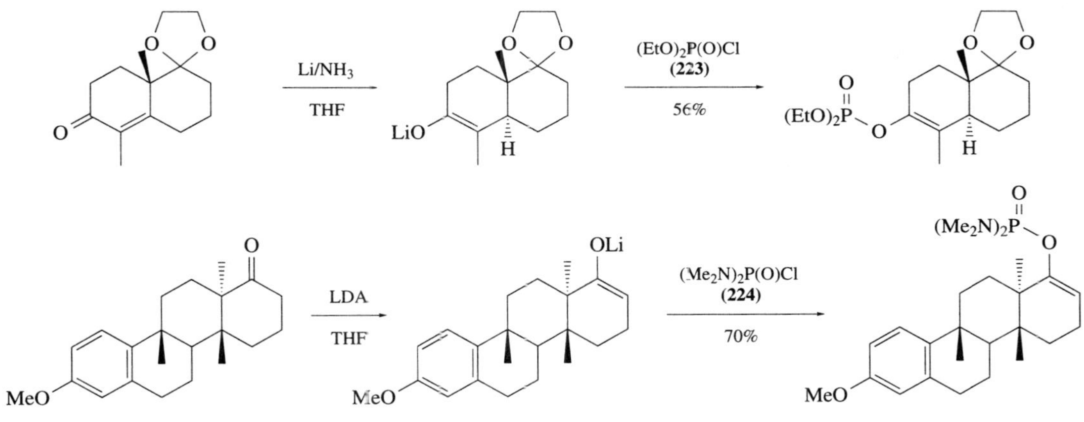

Scheme 119

(76)

X = Cl, Br, I
R = Me, Et, Bu, Bu[i]
R[1] = Me, Ph

Alternatively, alkenyl phosphates can be prepared by the addition of dialkyl hydrogen phosphates to alkynes, catalyzed by mercuric salts (Scheme 120) ⟨64JOC1817⟩. Mercuric salts have also been used to catalyze the reaction of vinyl acetate with phosphoric acid in the preparation of vinyl dihydrogen phosphate (**225**) ⟨59JBC(234)1⟩.

The oxidation of vinyl phosphites rarely provides vinyl phosphates, but results in extensive polymerization. Interestingly, heating of vinyl phosphites with sulfur leads to the isolation of vinyl thiophosphates (**226**) (50–79%) (Equation (77)) ⟨59DOK(124)1251⟩.

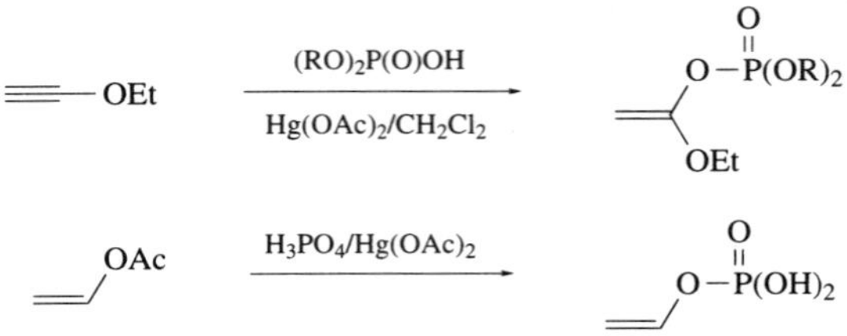

(225)

Scheme 120

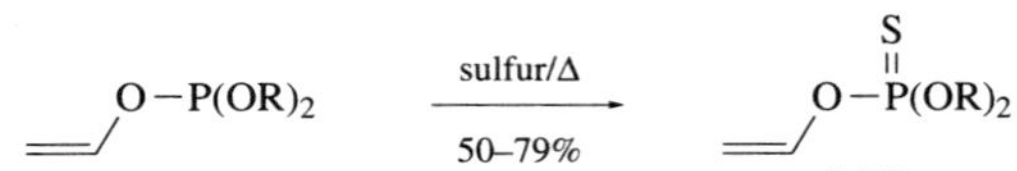

(226)

R = Et, Pr, Bu, Ph

(77)

Triaryl phosphates are prepared by the reaction of phosphorus oxychloride with phenols. Monoaryl phosphates can be obtained by controlling the stoichiometry of phenol, and the reaction is accelerated by a catalytic amount of sodium chloride (Scheme 121) ⟨56ZOB3060⟩. Phosphorylation of phenolates with dialkyl phosphorochloridate (**223**) offers a more direct route to monoaryl phosphates ⟨85SC411⟩. A novel approach involves the reaction of phenols with dialkyl phosphite and carbon tetrachloride where dialkyl phosphorochloridate is generated *in situ* (Scheme 121) ⟨55JCS522⟩. The oxidation of triaryl phosphites to the phosphates is quite general and can be effected by a variety of conditions, for example, dinitrogen tetroxide ⟨58JA5541⟩, ozone ⟨61JA845⟩, and chloramine T (sodium *p*-toluenesulfonchloroamide) ⟨61JCS3079⟩.

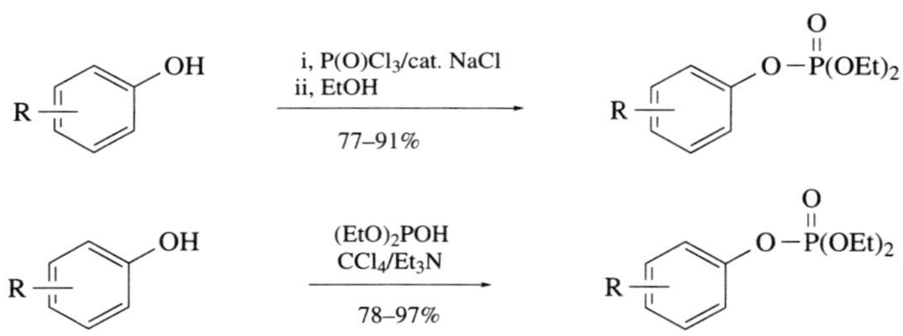

Scheme 121

In a manner analogous to the preparation of aryl phosphates, aryl phosphoroamidates are obtained by treating phenolates with diamidophosphorochloridate ⟨86JOC179, 90CPB2637⟩. Heating of phenols in HMPA in the presence of HCl provides a viable alternative (Equation (78)) ⟨74RTC252⟩.

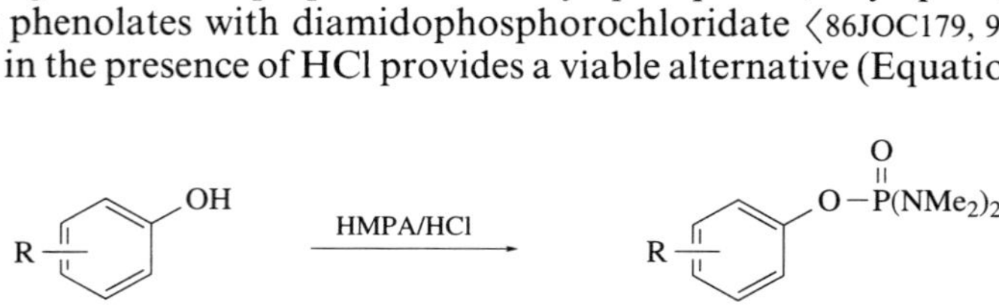

R = H, OMe, Cl, NO_2

(78)

2.13.1.7.3 Arsenic, antimony and bismuth enolates

In the enolates of arsenic, the metal tends to be carbon bound ⟨69ZOB2479⟩. The enolates of 1,3-dicarbonyl compounds, however, act as a bidentate ligand and bond via the oxygen as shown in the arsinidine complex (**227**) (Equation (79)) ⟨81CB1407⟩.

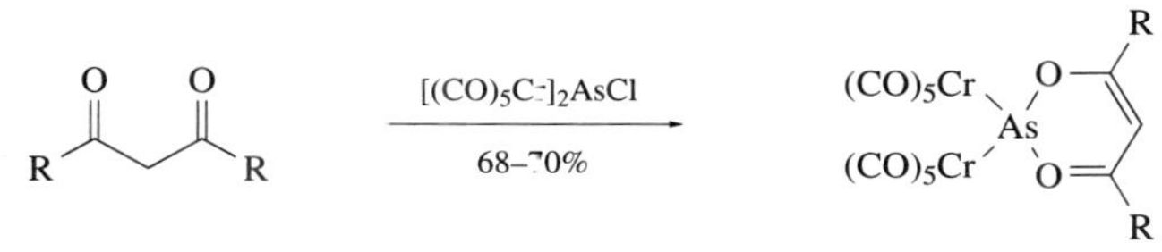

(227)

(79)

The antimony enolates, in contrast, exist in both oxygen and carbon bound isomeric forms. In general, increased steric bulk of the ligands on the antimony and/or at the α-carbon of the carbonyl compound promotes formation of the O—Sb bond ⟨74JOM(78)107⟩. Electronic effects also influence the ratio between the *O*- and *C*-isomers, but to a lesser extent ⟨74JOM(78)115⟩. The *O*-antimony(III) enolates are usually prepared by treating enol trifluoroacetates with alkoxystilbines (Scheme 122) ⟨69ZOB1174⟩, or enol acetates with aminostilbines ⟨73ZOB1264⟩. The reaction is accelerated by the presence of chlorostilbine as a catalyst. Metal exchange processes are also widely used in the preparation of antimony(III) enolates of 1,3-dicarbonyl compounds where the sodium enolates are treated with diphenylantimony(III) salt ⟨90MI 213-02⟩. Antimony(V) enolates have been prepared by direct enolization of acetylacetone with phenylstibonic acid ⟨67BCJ428⟩.

The chemistry of bismuth enolates is not widely documented. Bismuth(III) chloride has been used to catalyze an aldol reaction, however, the involvement of a bismuth enolate is not clear ⟨88TL4719⟩.

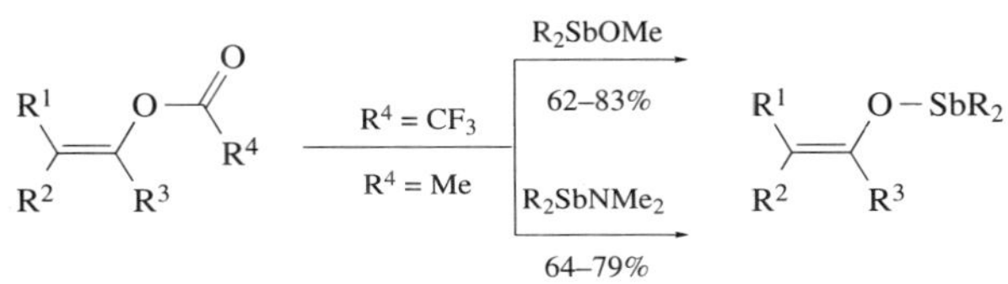

Scheme 122

2.13.1.7.4 *Arsenic, antimony and bismuth phenolates*

Phenols react readily with aminoarsine ⟨74ZAAC(406)52⟩ or aminostilbine ⟨70JOM(25)139⟩ to give the corresponding arsenic(III) and antimony(III) phenolates (Equation (80)). The esters of arsinic acid undergo transesterification with phenols to produce aryl arsinate (Equation (81)) ⟨72MI 213-01⟩. Transmetallation of sodium phenolates with antimony(III) and antimony(V) halides provides an easy access to the corresponding antimony phenolates ⟨75ZAAC(418)167⟩. The reduction of *ortho*-benzoquinone by an ester of arsonous acid ⟨73CB2738⟩ or stilbonous acid ⟨75ZAAC(418)279⟩ provides a novel entry to the corresponding aryl ester of *ortho*-arsonic acid and *ortho*-stilbonic acid (Equation (82)).

Bismuth phenolates can be accessed by treating sodium phenolates with bismuth(III) and bismuth(V) halides ⟨65FRP1397139, 72JOM(38)91⟩.

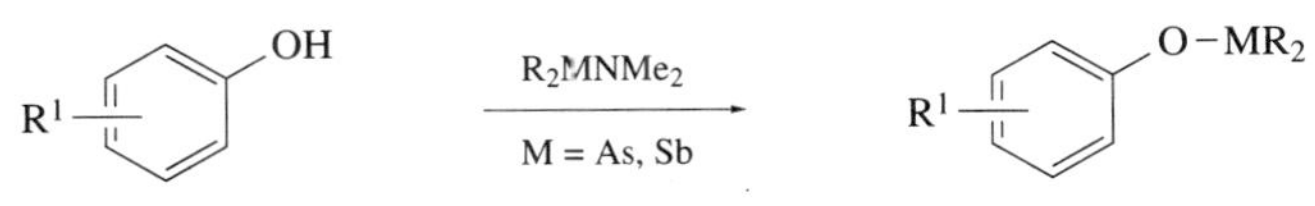

(80)

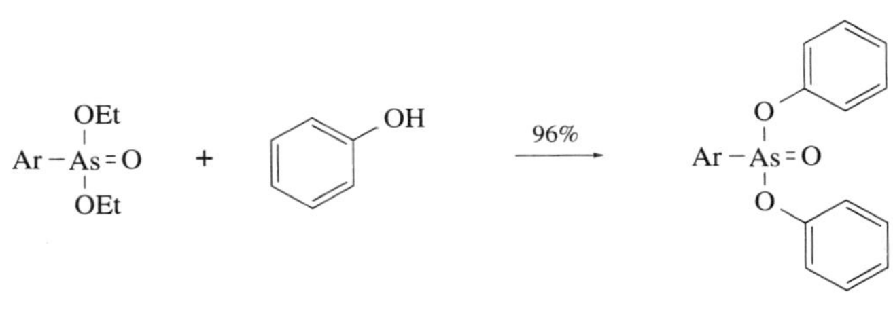

(81)

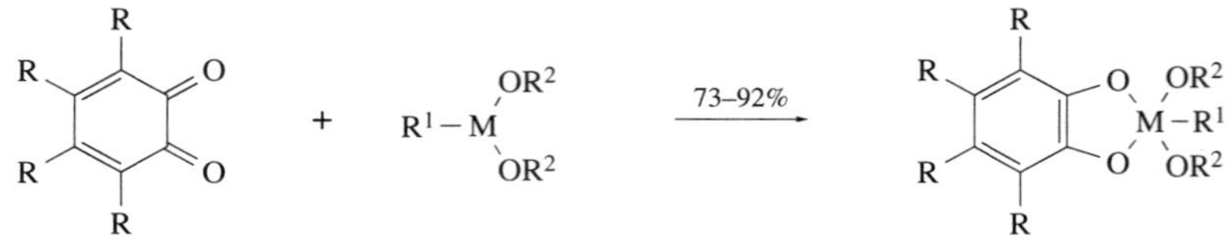

(82)

R = H, Cl
M = As, Sb; $R^1 = R^2$ = alkyl, aryl

2.13.1.8 RO—Metal and Metalloid Derivatives

2.13.1.8.1 Metal and metalloid enolates

Enolates play an important role in modern organic synthesis. The formation and synthetic application of metal enolates have been reviewed on many occasions, for example, by d'Angelo ⟨76T2979⟩, Evans ⟨B-84MI 213-01⟩ and Mekelburger and Wilcox ⟨91COS(2)99⟩. In general, metal enolates are highly reactive intermediates which are generated and used *in situ* and are not usually isolated. The preparation involves deprotonation of carbonyl compounds with strong bases. The enolization of ketones is fairly straightforward, but for the more reactive aldehydes it is often complicated by side reactions such as self-condensation. In the enolization of acyclic and macrocyclic carbonyl compounds, there are two possible geometries, (*E*) and (*Z*), for the resulting enolates (Scheme 123). The geometry of these enolates has a significant stereochemical impact on their subsequent reaction products such as in the aldol process ⟨91COS(2)181, 91COS(2)239⟩. The geometry of the enolates is governed by various factors, the solvent, reaction temperature, deprotonating base, and the participation of other functional groups on the carbonyl substrate (steric hindrance and chelation control). Normally, the (*E*)-enolates are preferentially generated under kinetic conditions (−78 °C in THF with bulky bases). Under thermodynamic control, where the enolates are allowed to equilibrate (0 °C in polar solvents), the (*Z*)-enolates predominate ⟨76JA2868, 80JA3959⟩. Furthermore, in the enolization of nonsymmetrical ketones, there are also issues of regioselective deprotonation. As a general rule, the most thermodynamically stable enolate is the one with the highest substitution.

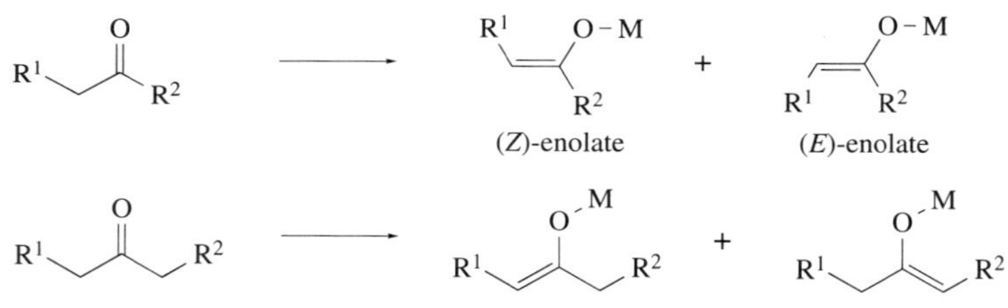

Scheme 123

(i) Enolates of group 1 metals

The bulky dialkyl amide bases of Group 1 metals, lithium in particular, are routinely used in the enolization of carbonyl compounds. The most commonly used amide bases include lithium diisopropylamide (LDA) ⟨69JOC2324⟩, lithium isopropylcyclohexylamide (LICA) ⟨71JA2318⟩, lithium tetramethylpiperidide (LTMP) ⟨73JA582⟩ and the hexamethyldisilazides (HMDS) of various Group 1 metals (LHMDS, NHMDS, KHMDS) ⟨63AG(E)617⟩. Apart from the amide bases, potassium hydride ⟨74JOC3913⟩, potassium *t*-butoxide ⟨74CRV45⟩ and alkaline metal triphenylmethide ⟨77SC137⟩ are also effective.

Stereo- and regioselective lithium enolates can be obtained from the treatment of geometrically defined silyl enol ethers ⟨68JA4462⟩ or alkenyl acetates ⟨65JOC2502⟩ with methyllithium (Equation (83)). Alkaline metal enolates can also be prepared from a Birch-type reduction of α,β-unsaturated carbonyl compounds ⟨76OR(23)1⟩. The reaction is regioselective as shown in the reduction of the decalone (**228**) in Equation (84) ⟨65JA275⟩.

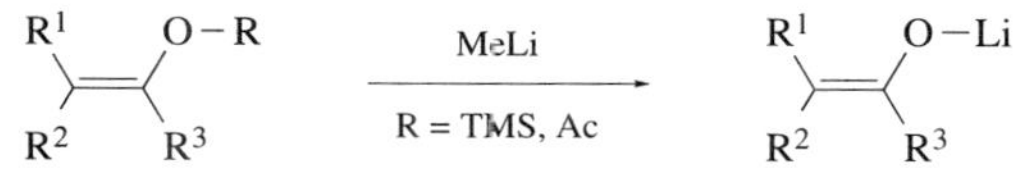

(83)

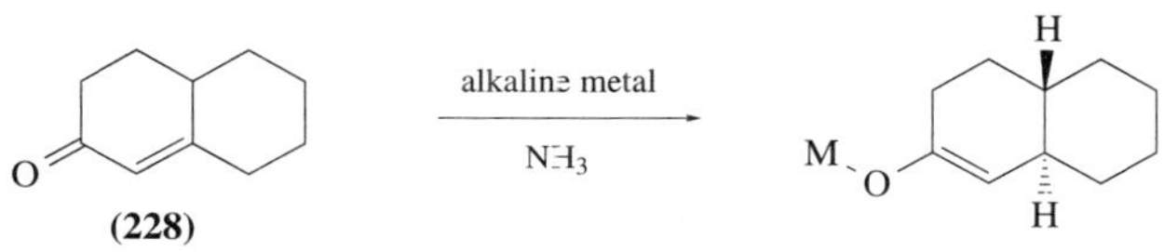

(228)

M = Li, Na, K

(84)

(ii) Enolates of group 2 metals

Magnesium bis-dialkylamides ⟨63AG(E)47⟩ and halomagnesium dialkylamides ⟨83TL1345⟩, readily derived from dialkylamines and Grignard reagents, are frequently used for the deprotonation of carbonyl compounds. The reaction is under thermodynamic control and results in the formation of the more substituted enolate (Equation (85)). Analogous to the preparation of Reformatsky reagents, treatment of α-bromoketones with magnesium metal leads to the formation of bromomagnesium enolates (Equation (86)) ⟨78T1349⟩. It is also common practice to exchange the metal of the readily available lithium enolates with magnesium halides to achieve different reactivity in reactions. However, the exact metal composition of the resulting enolates is not clear. Magnesium enolates are most frequently generated by the 1,4-addition of Grignard reagents to α,β-unsaturated carbonyl compounds catalyzed by copper(I) salts.

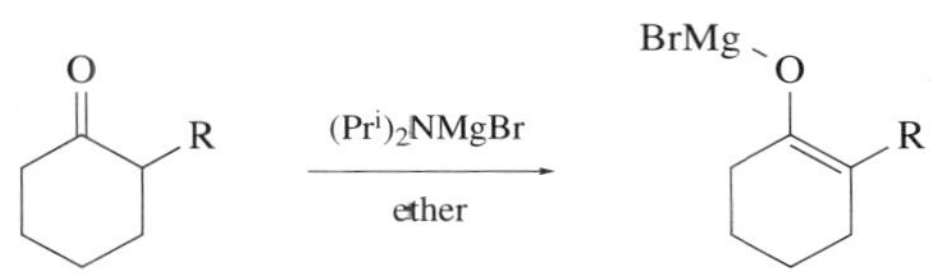

(85)

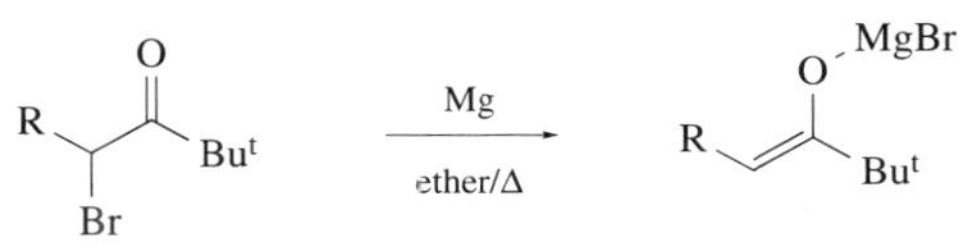

(86)

R = Me, Et, Pr^n, Pr^i, Bu^i, Bu^t, CH_2Bu^t

Calcium enolates are not encountered commonly in synthetic transformation. Nevertheless, they have been prepared by Benkeser reduction (calcium metal/ammonia reduction) of α,β-unsaturated ketones (Equation (87)) ⟨68BSF4265⟩.

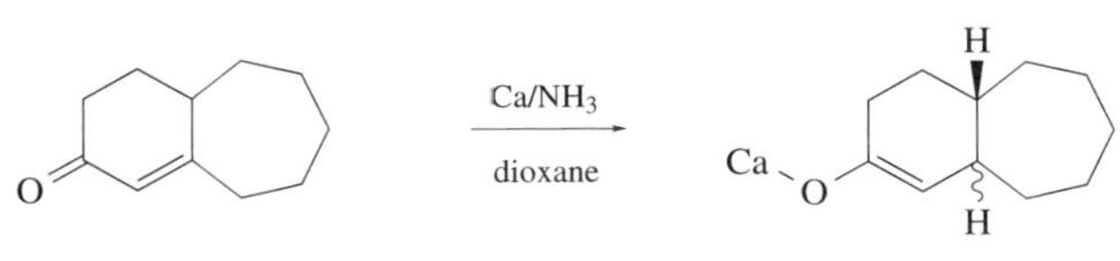

(87)

(iii) Enolates of group 13 metals and metalloids

Boron enolates, also known as enol borinates and vinyloxy boranes, have received enormous attention for their unique role in various carbon–carbon bond formation reactions, such as alkylation and aldol reactions. The stereochemical outcome of the reaction product correlates directly to the geometry of the boron enolate.

Organoboranes react regio- and stereoselectively with α-diazoketones to give (*E*)-boron enolates (Scheme 124) ⟨68JA5936, 87JOC1347⟩. Isomerization of the (*E*)-enolates to the thermodynamically more stable (*Z*)-enolates can be achieved with a catalytic amount of lithium phenoxide or pyridine ⟨79TL1665⟩. Heating of ketones with triethylborane catalyzed by diethylboron pivalate, however, leads to preferential formation of the (*Z*)-enolates with good regioselectivity (Equation (88) ⟨75LA1322⟩.

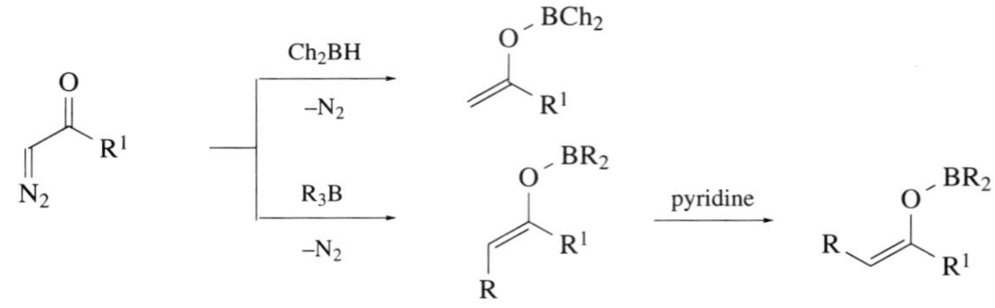

Scheme 124

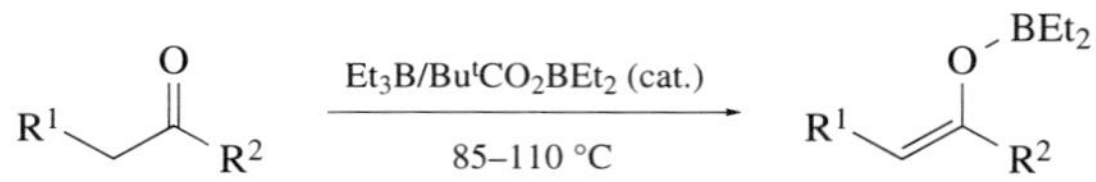

(88)

More conveniently, boron enolates are prepared by direct enolization of carbonyl compounds with reactive borylating agents, for example, dialkylboron triflate ⟨76CL559, 80BCJ174⟩ and halides ⟨89JA3441⟩, in the presence of a tertiary amine (Equation (89)). The geometry of the resulting boron enolate is found to be greatly influenced by the steric nature of the ligand (L) ⟨79TL2229, 79JA6120, 92JOC3767⟩ and the leaving group (X) ⟨93JOC147⟩ of the boron reagent (L_2BX), as well as the solvent and tertiary amine used in the reaction ⟨81JA3099, 93JOC7162⟩. The use of borylating reagents containing chiral ligands ⟨86JA8279, 86TL4721, 90T4663, 92JOC5173⟩ leads to the formation of chiral boron enolates which have been utilized successfully in enantio- and diastereoselective aldol reactions ⟨88CI(L)390⟩.

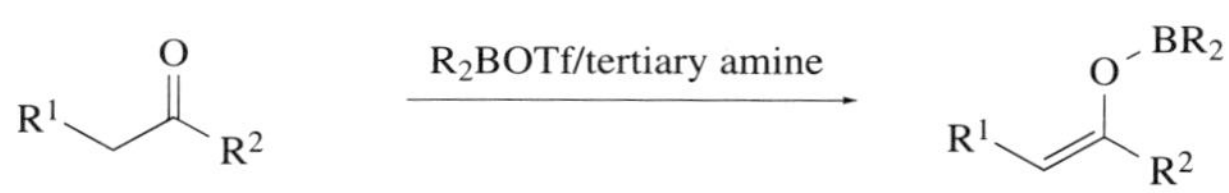

(89)

Another approach towards regio- and stereoselective formation of boron enolates is via 1,4-hydroboration of α,β-unsaturated ketones ⟨90CC1680, 91JOC5820⟩. Exclusive boron (*Z*)-enolates are obtained when (*E*)-α,β-unsaturated ketones are treated with bulky boranes (e.g., Ipc_2BH, Icr_2BH, Ch_2BH, Sia_2BH, and 9-BBN-H (9-borabicyclo[3.3.1]nonane) (Scheme 125)). Conjugate addition of trialkylborane to α,β-unsaturated ketones, however, is not stereoselective (Scheme 125) ⟨75LA2201⟩. The addition involves a free radical mechanism, and is greatly facilitated by radical initiators such as acyl peroxide ⟨70JA712⟩ and oxygen ⟨70JA714⟩. Transmetallation of the readily available lithium enolates with haloborane provides an alternate access to boron enolates, and the geometry of the enolates is generally preserved (Equation (90)) ⟨85TL1643⟩.

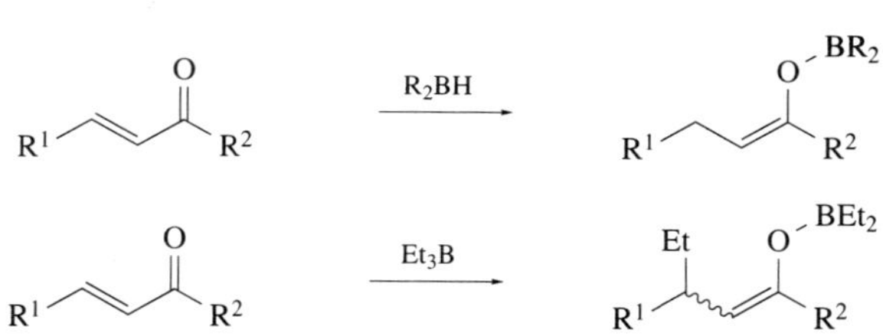

Scheme 125

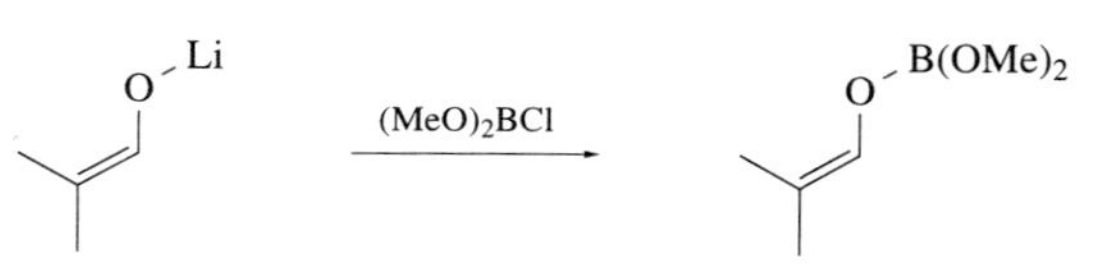

(90)

Deprotonation of carbonyl compounds with dialkylaluminum tetramethylpiperidide (R_2Al-TMP), generated from dialkylaluminum chloride and lithium TMP, leads to the formation of aluminum enolates (Equation (91)) ⟨79CL379⟩. Direct enolization of ketones can also be effected by treatment with trialkylaluminum ⟨85HCA961⟩. The transmetallation of preformed lithium enolates with alkylaluminum chloride $R_nAlCl_{(3-n)}$ or trialkylaluminum is common practice ⟨69TL1117⟩, although the true nature of the resulting enolates is not clear. The 1,4-addition of organoalane to α,β-unsaturated carbonyl compounds provides an alternate approach to the regioselective synthesis of aluminum enolates. Nickel acetoacetate catalyzes the addition of trimethylalane to an α,β-unsaturated ketone in a 1,4-fashion, and the (*Z*)-enolate is obtained preferentially (Equation (92)) ⟨74JOM365⟩. While most dialkylaluminum chloride reagents undergo polar conjugate addition to α,β-unsaturated carbonyl compounds, such as the *N*-acyl urethanes (**229**), dimethylaluminum chloride only proceeds via a radical pathway (Scheme 126) ⟨92SL343⟩.

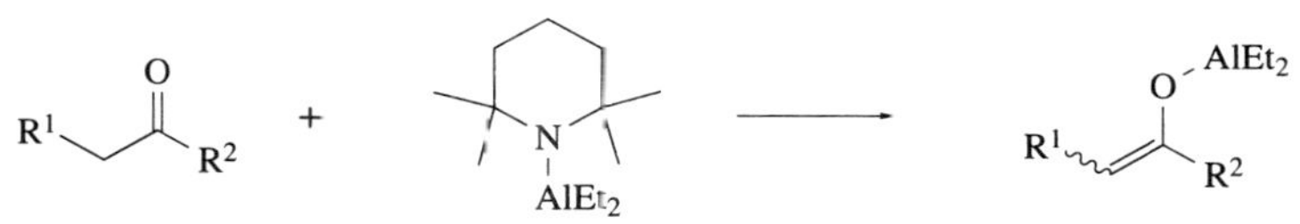

(91)

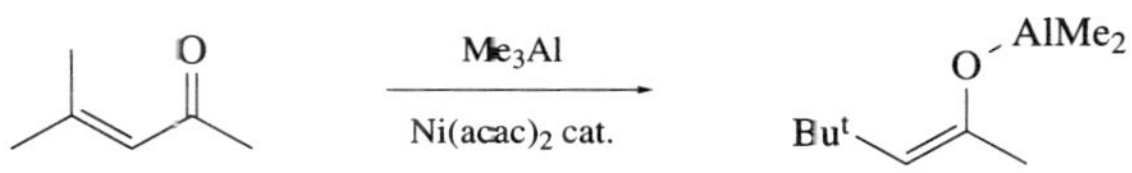

(92)

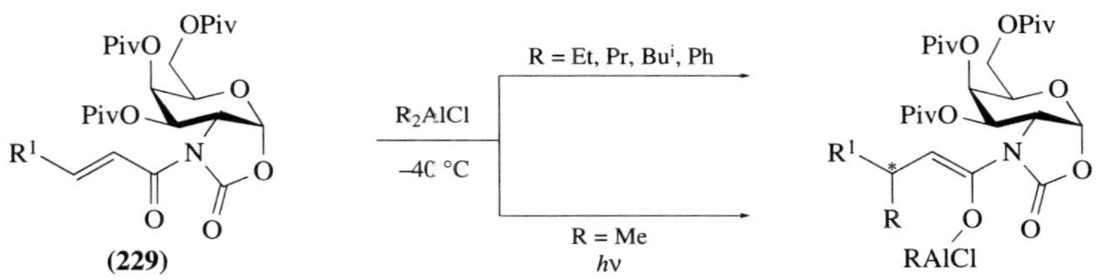

(**229**)

Scheme 126

(iv) Enolates of group 14 metals and metalloids

The synthesis of the *O*-silylated enolates of aldehydes and ketones, more commonly known as silyl enol ethers, has been reviewed by Brownbridge ⟨83S1⟩ and Chan ⟨91COS(2)599⟩ among others.

Silyl enol ethers are usually prepared by the silylation of metal enolates with silyl halides ⟨64JOM(1)476⟩. The geometry of the silyl enol ethers is inherited from the metal enolate precursors, as illustrated by the preparation of the kinetic ⟨69JOC2324⟩ and thermodynamic ⟨83TL1345⟩ silyl enol ethers in Scheme 127. Silyl chlorides are compatible with a variety of bases, especially at lower temperatures. Hence, enolization of ketones by potassium hydride in the presence of TBDMS-Cl (*t*-butyldimethylsilyl chloride) leads to thermodynamic silyl enol ethers ⟨84TL5099⟩, whereas the combination of LDA (lithium diisopropylamide) and trimethylsilyl chloride promotes the formation of kinetic products ⟨84TL495⟩ (Scheme 128). The use of silyl triflate and tertiary amine also provides an efficient one-pot synthesis of silyl enol ethers (Scheme 129) ⟨81TL3455, 84TL5953⟩. The combination of chlorosilane/sodium iodide/amine in acetonitrile provides a convenient alternative and is highly stereoselective even though the regioselectivity is moderate ⟨87T2075⟩. Trialkylsilyl perchlorates are effective for the stereoselective preparation of (*Z*)-silyl ketene acetals from esters (Equation (93)) ⟨84TL699⟩. The aziridine/TMS-Br/Ph_4SbBr system is not only regioselective, but also chemoselective for the enolization of cyclic ketones (Equation (94)) ⟨89CL1247⟩.

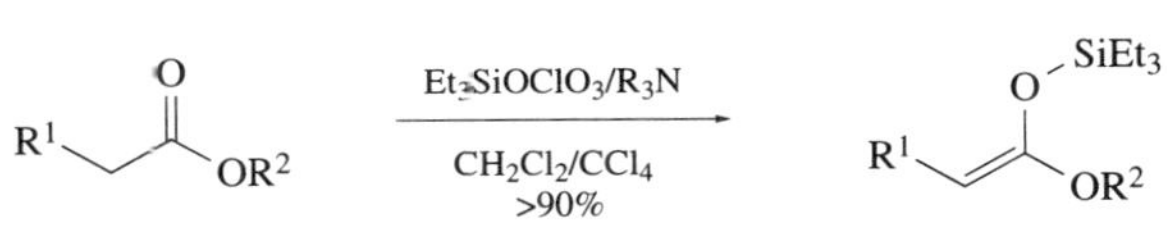

(93)

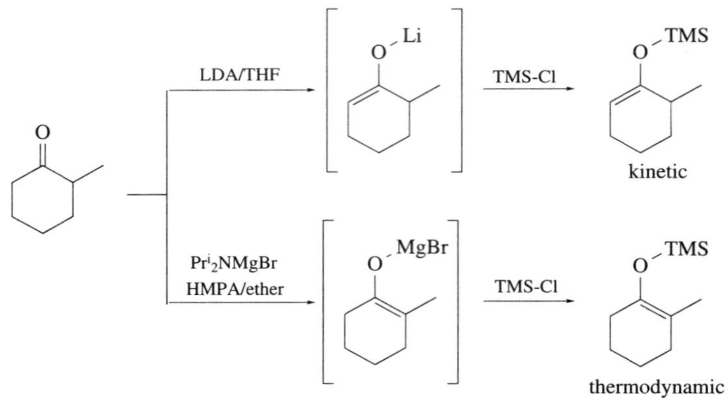

Scheme 127

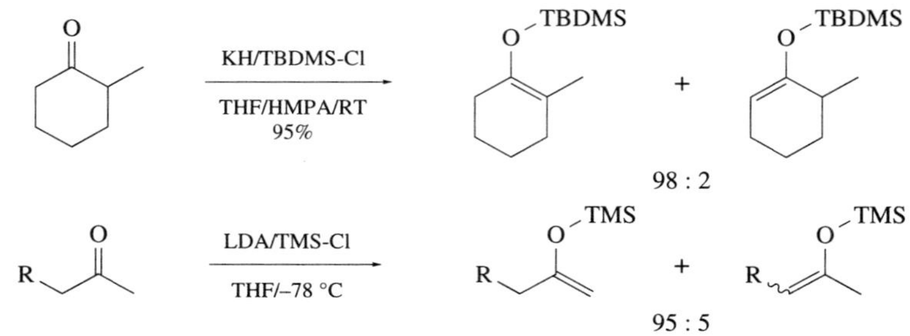

Scheme 128

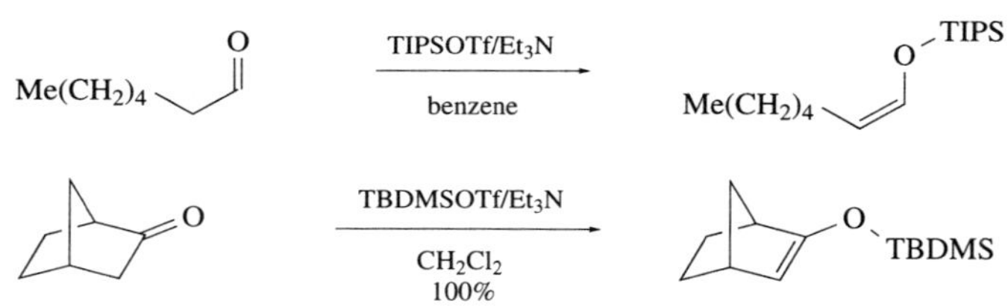

Scheme 129

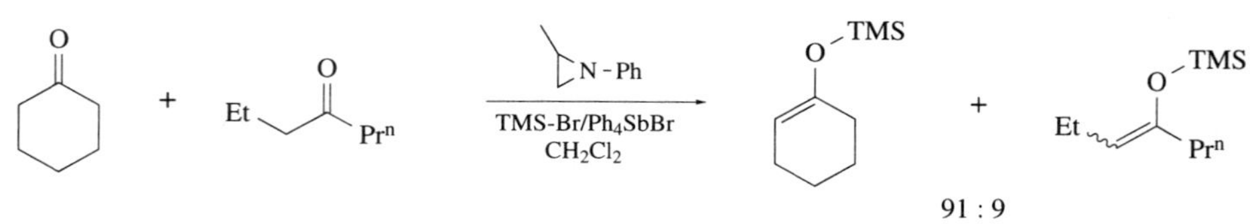

(94)

Other regio- and stereoselective syntheses of silyl enol ethers (Scheme 130) include alkylidenation of silyl esters ⟨88TL1065⟩, oxidation of alkenyl lithium with bis-TMS peroxide ⟨88TL4269⟩ and a novel rearrangement of β-silyloxy carbenoids ⟨93TL4985⟩. The rearrangement of acyl silanes also provides access to silyl enol ethers albeit without stereoselectivity ⟨75JOM(90)C6⟩.

While silicon is highly oxygenophilic, germanium bonds equally well with the oxygen and the carbon of an enolate ⟨72PAC409⟩. A mixture of the oxygen- and carbon-bonded isomer is obtained when stannylacetone is transmetallated with trimethylbromogermane ⟨68JOM(14)229⟩. Although the *C*-isomer predominates over the *O*-isomer, the equilibrium can be shifted by heating. Interestingly, pure *O*-germanium enolate of acetone has been prepared by treating an enol trifluoroacetate with aminogermane under mild conditions (Equation (95)), whereas more vigorous conditions are required with alkoxygermane ⟨69ZOB1174⟩. Transmetallation of lithium enolates with trialkylgermyl halides provides germanium enolates with the *C*-isomer being predominant ⟨87OM2568⟩.

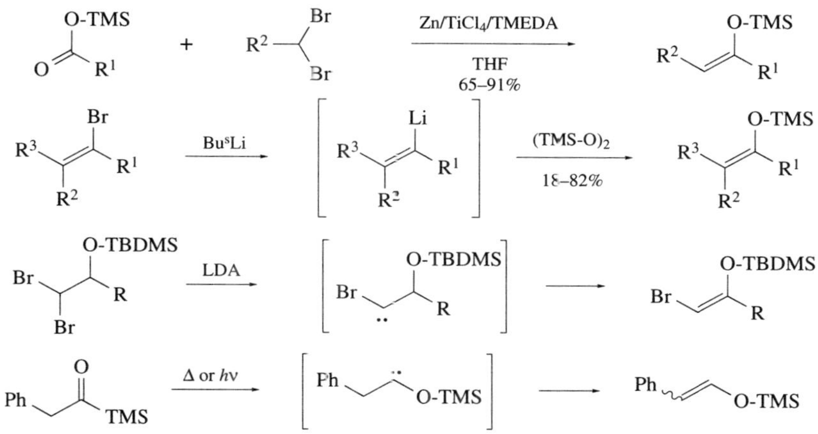

Scheme 130

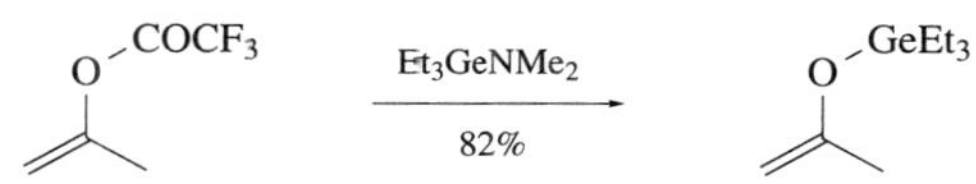

(95)

Tin(II) enolates display high diastereoselectivity in aldol reactions and receive wide acceptance. The ability of divalent tin to chelate further with chiral ligands, diamines in particular, consequently opens the door to various asymmetric transformations, aldol reaction for instance ⟨84T1381⟩. Tin(II) enolates can be generated by treating carbonyl compounds with tin(II) triflate in the presence of a tertiary amine (Equation (96)) ⟨82CL353, 92TL4233⟩. Aldehyde enolates are best prepared by oxidative addition of metallic tin to α-bromoaldehydes (Equation (97)) ⟨83CL1727⟩. Often, tin(II) enolates are obtained by transmetallation of lithium enolates with tin(II) salts ⟨87CL293⟩.

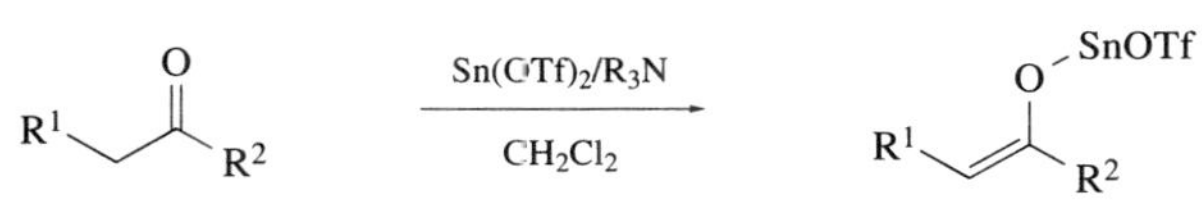

(96)

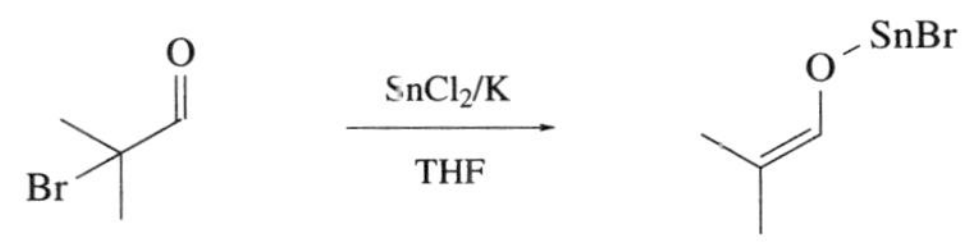

(97)

In contrast, tin(IV) enolates exist as a mixture of the oxygen- and carbon-bonded isomers ⟨72PAC409⟩. The equilibrium of the isomeric mixture is influenced by temperature, solvent, and substituents on the metal ⟨70JOM(24)359⟩. The tin(IV) enolates can be prepared by direct enolization of carbonyl compounds with aminostannanes ⟨70JOM(24)359⟩ or reaction between enol acetates and alkoxystannanes ⟨68JOM(11)97⟩ (Scheme 131). The 1,4-hydrostannation of α,β-unsaturated ketones also provides the oxygen-bound tin(IV) enolates with predominantly (*E*)-geometry (Equation (98)) ⟨67BSF1928⟩. A novel synthesis of tin(IV) enolates has been achieved by regioselective ring opening of diketene with tributyltin alkoxides (Scheme 132) ⟨93CC1067, 93CL1953⟩. Often, tin(IV) enolates are prepared by metal exchange of lithium enolates with tin(IV) salts and used *in situ* ⟨69TL111, 85JOC3846⟩.

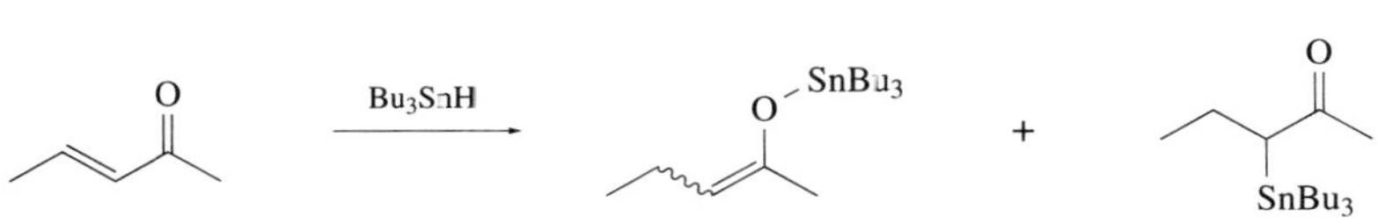

(98)

(*E*):(*Z*) 90:0 75:25

(v) Enolates of transition metals

The synthetic utility of enolates of the transition metals has received enormous attention in the search for unusual stereo- and regioselectivity in various synthetic transformations. Some of the

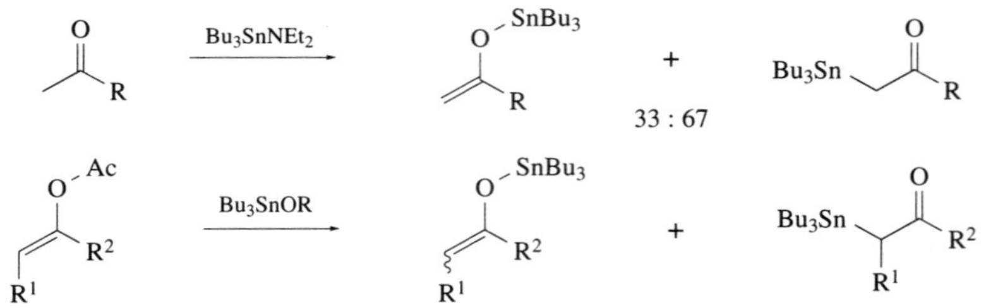

Scheme 131

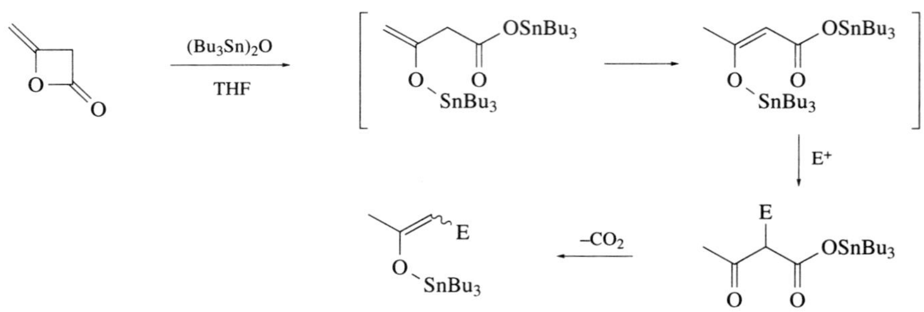

Scheme 132

transition metals are bonded to the oxygen of an enolate (alkaline metals), some are carbon bound e.g., mercury), while there are others that exist in both forms. Some of the more widely used transition metal enolates are included in this section for discussion.

(a) Copper. The 1,4-addition of organometallics to α,β-unsaturated carbonyl compounds can be efficiently mediated by a variety of copper(I) salts in a catalytic process. Extensive reviews in this area have been conducted on many occasions and in the 1990s by Wipf ⟨93S537⟩ and Nakamura ⟨91SL539⟩. Organocopper reagents undergo facile conjugate addition to α,β-unsaturated carbonyl compounds in the presence of a Lewis acid, such as BF_3 ⟨82JOC119, 86AG(E)947⟩ or $AlCl_3$ ⟨83CPB128⟩, leading to the formation of copper enolates (Equation (99)). Organocuprates undergo similar conjugate addition, however, the resulting adducts are not considered to be genuine copper enolates ⟨92OR(41)135⟩.

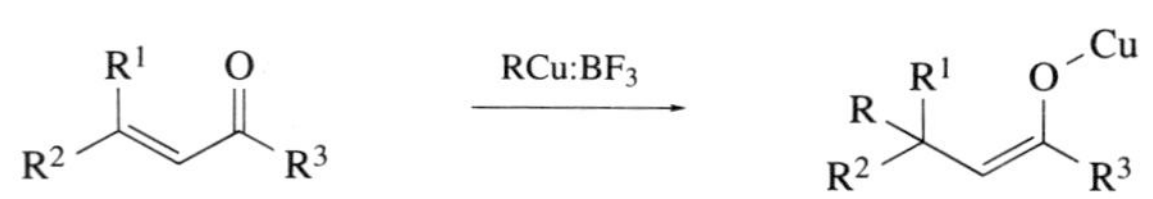

(99)

(b) Zinc. Zinc enolates are generally prepared from the action of activated zinc metal with α-halocarbonyl compounds, especially with α-haloester which is commonly known as the Reformatsky reaction ⟨1887CB1210, 89S571⟩. While it has been established that the zinc atom is carbon bonded in the Reformatski reagents (ester enolates), it is shown to be oxygen bonded by NMR spectroscopy in the ketone enolate. The oxygen-bonded zinc enolate is prepared by treating α-bromoketone with diethylzinc, however, the reaction is very substrate specific (Scheme 133) ⟨87OM2069⟩. Metal exchange of lithium enolates with zinc halides is often performed in synthetic operations, but the actual composition of the resulting metal enolates is not clear ⟨73JA3310⟩. The 1,4-addition of a mixed organozincate ⟨77CL679⟩ or Grignard reagents mediated by $ZnCl_2$—TMEDA ⟨86JOC3993⟩ to α,β-unsaturated ketones is highly efficient, even though the resulting adduct is not regarded as a *bona fide* zinc enolate.

(c) Titanium and zirconium. Titanium enolates are usually prepared by transmetallation of lithium enolates with organotitanium reagents, such as tris(isopropoxy)titanium chloride and tris-(diethylamino)titanium bromide (Equation (100)) ⟨81TL4691⟩. Treatment of a silyl enol ether with titanium tetrachloride leads to the formation of the trichlorotitanium enolate ⟨83TL3341⟩. Alternatively, trichlorotitanium (*Z*)-enolates can be obtained selectively from direct enolization of ketones with titanium tetrachloride in the presence of Hünig's base (Equation (101)) ⟨90JA866⟩. A novel synthesis of titanium enolates involves the reaction of the titanocene methylidene complex with acid chlorides (Equation (102)) ⟨83JA1664⟩ or acid anhydrides ⟨85JOC2316⟩.

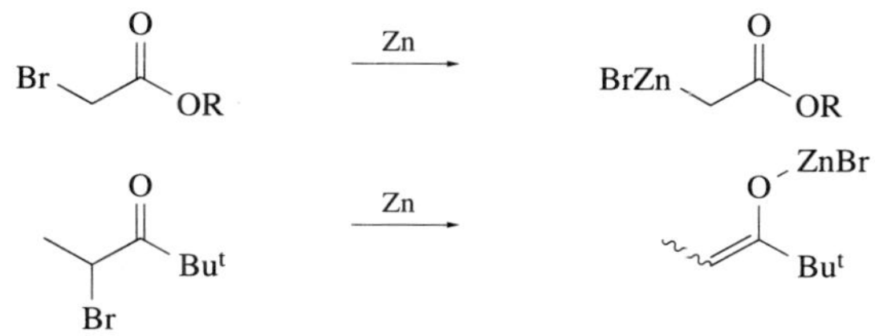

Scheme 133

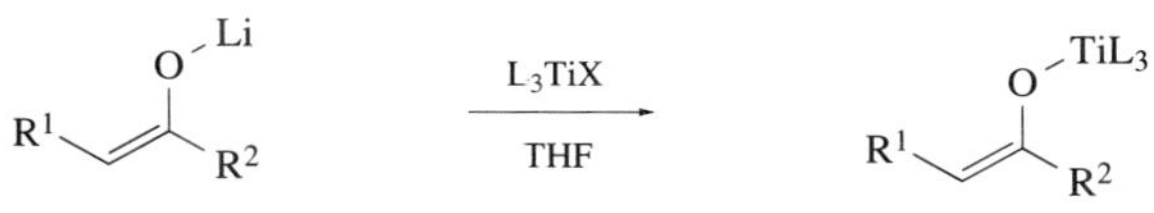

(100)

L = Pr^{i}O, Et$_2$N; X = Cl, Br

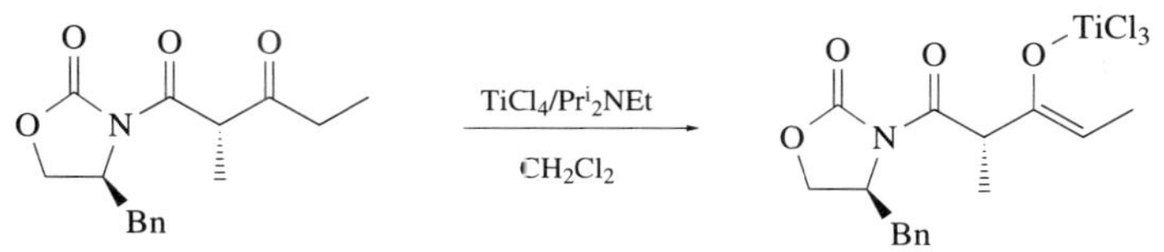

(101)

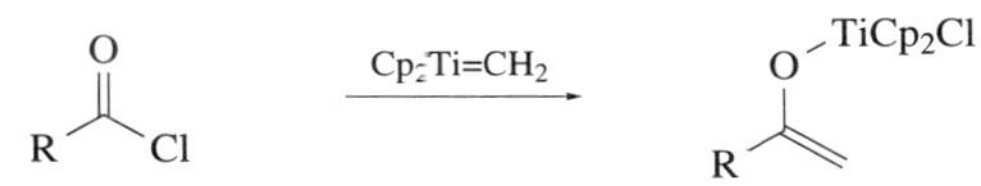

(102)

Zirconium enolates are often prepared by transmetallation of lithium enolates with dichlorozirconocene ⟨80TL3975, 80TL4607⟩. The addition of dimethylzirconocene to ketenes offers a viable alternative (Equation (103)) ⟨85IC654⟩.

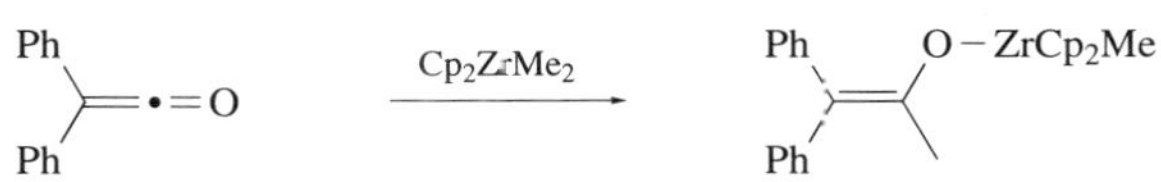

(103)

2.13.1.8.2 *Metal and metalloid phenolates*

Phenols are far more acidic than carbonyl compounds, with pK_a ranges between 7 and 11. Deprotonation of phenols is straightforward and can be achieved by treatment with metal hydroxides, alkoxides, hydrides and sometimes organometallics. Transmetallation provides a convenient alternative to the synthesis of phenolates of various metals and metalloids.

2.14
Vinyl and Aryl Chalcogenides: Sulfur-, Selenium- and Tellurium-based Functional Groups

PAUL C. TAYLOR

University of Warwick, Coventry, UK

2.14.1 INTRODUCTION

This chapter covers the synthesis of sulfur-, selenium- and tellurium-based functional groups attached to an sp^2 carbon. Clearly, however, many of the available synthetic methods are also applicable to sp^3-bound functional groups. These methods have been previously described (see Chapters 2.03 and 2.04). Therefore, the reader is advised to first consult Chapters 2.03 and 2.04, and to consider this chapter as a supplement which provides pointers towards functional group transformations that are particularly applicable to vinylic and arene substrates.

In keeping with the title of this work, the material in each section is primarily organised according to the functionality of the starting materials rather than the mechanism of the reaction. Nevertheless, wherever possible within this framework, mechanistically related reactions are discussed together. The coverage of material was not intended to be comprehensive. Instead an attempt has been made to direct the reader to readily available and informative primary literature, which thus provides a rapid introduction to each functional group.

2.14.2 SULFUR-BASED FUNCTIONAL GROUPS ATTACHED TO AN *sp*² CARBON

2.14.2.1 Arenethiols

2.14.2.1.1 Arenethiols from aryl halides

The transformation of aryl halides to arenethiols is an attractive synthetic route (Equation (1)). However, the one-step reaction proceeds via an aromatic nucleophilic substitution mechanism, and the process is often restricted to electron-deficient aromatic substrates. Clearly, a source of nucleophilic sulfur is required; one such example would be a metal sulfide. For example, the reduction of 1-bromo-4-nitrophenol to 4-aminothiophenol by sodium sulfide ⟨49JA1747⟩ is a very useful procedure (Equation (2)) ⟨86JOC3165⟩. Synthetically equivalent is the use of a metal polysulfide, followed by lithium aluminum hydride reduction ⟨68CB2956⟩.

$$\text{Ar—Hal} \xrightarrow{M_2S} \text{Ar—SH} \qquad (1)$$

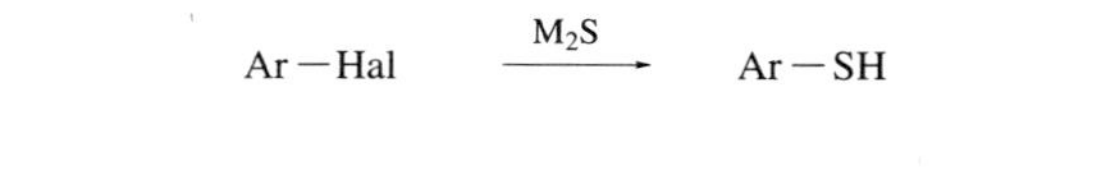

(2)

The reaction of aryl iodides with thiourea in the presence of nickel(0), followed by hydrolysis (Equation (3)), would seem to be an attractive route to arenethiols ⟨85CL1307⟩, but one which has yet to be widely used.

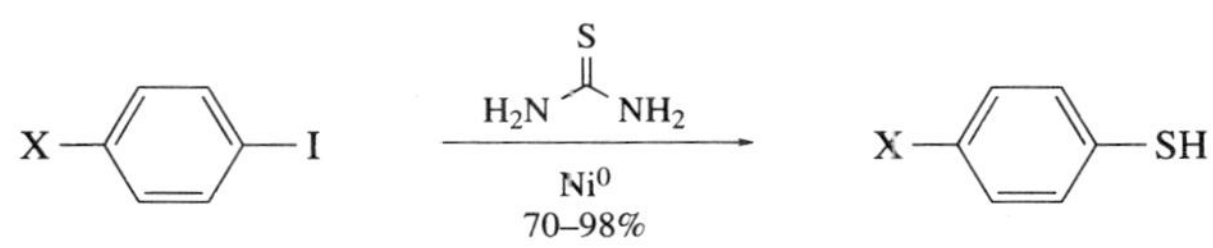

(3)

To circumvent the requirement for electron-deficient substrates, the aryl halide can be converted to the corresponding Grignard reagent, which will, in contrast, react with a source of electrophilic sulfur. Elemental sulfur is adequate for this purpose, and this reaction has been effectively used, after reductive work-up, to prepare sterically hindered thiols (Equation (4)) ⟨B-74MI 214-01⟩.

$$\text{Ar}-\text{MgHal} \xrightarrow[\text{ii, [H]}]{\text{i, } S_8} \text{Ar}-\text{SH} \quad (4)$$

2.14.2.1.2 *Arenethiols from phenols*

Another desirable synthetic approach to arenethiols is the transformation from the corresponding phenol. This process cannot be realised directly, but there are a number of multistep equivalents. Perhaps the most widely used of these methods uses the Newman–Kwart rearrangement (Scheme 1) ⟨66JOC410, 66JOC3980⟩. Although this route appears lengthy, the necessary evil of preparing intermediates has been elegantly turned into an advantage by using them as directing groups for metallation, and thus for further functionalisation (Scheme 2) ⟨92S112⟩.

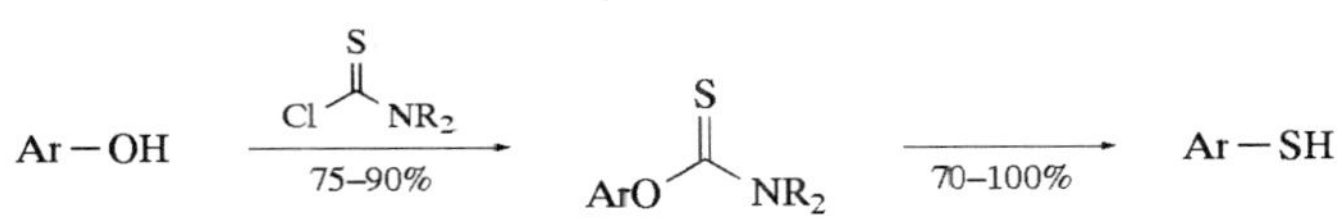

Scheme 1

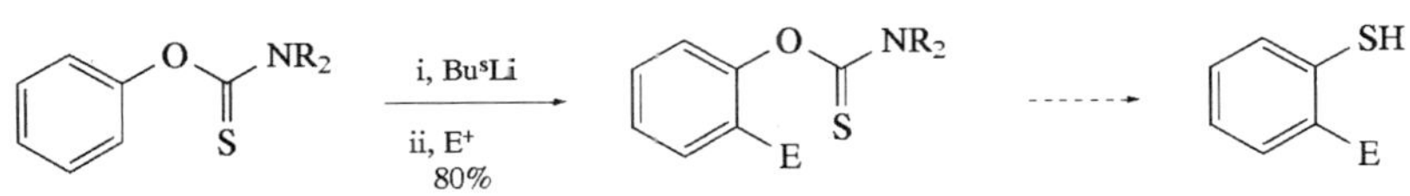

Scheme 2

2.14.2.1.3 *Arenethiols from arylamines*

The traditional nucleophilic aromatic substitution substrates are arenediazonium salts (**1**). However, their conversion to thiols has not been widely used. This may well change following the introduction of the apparently straightforward procedure shown in Scheme 3. A metal thiocarboxylate is used as the sulfur nucleophile, the ester being hydrolysed to provide the target thiols ⟨89T7411⟩.

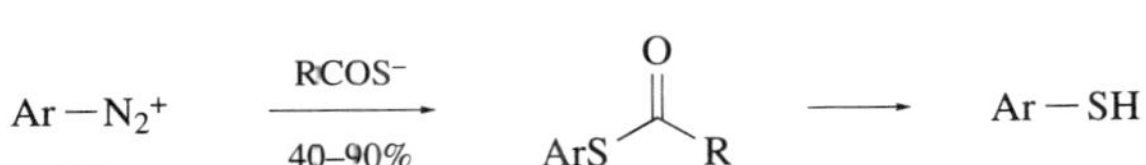

(**1**)

R = Me, Ph; Ar = Ph, 2,6-dialkylphenyl, *p*-IC_6H_4-, 1-naphthyl, *p*-arylphenyl

Scheme 3

2.14.2.1.4 Arenethiols by reduction

The conversion of disulfides to thiols has attracted a great deal of attention, and the various methods have been summarised in two papers ⟨84S498, 89JOC4458⟩ which describe methods of reduction that appear particularly useful for arene derivatives. The reagents $KHB(OPr^i)_3$ and $LiHAl(OBu^t)_3$ were shown to reduce diaryl disulfides in an efficient and very selective manner, even dialkyl disulfides being untouched (Equation (5)).

$$Ar{-}S{-}S{-}Ar \xrightarrow[90\text{–}100\%]{M^1HM^2(OR)_3} Ar{-}SH \qquad (5)$$

Ar = 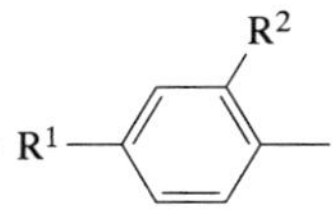

M^1 = Li, K; M^2 = B, Al; R^1 = Me, MeO, CN, Cl; R^2 = alkyl

The synthesis of arenethiols by electrophilic aromatic substitution cannot be achieved directly. However, a straightforward synthetic equivalent is offered by the electrophilic sulfonylation of arenes (see Sections 2.14.2.3 and 2.14.2.4) followed by the reduction of the sulfonyl group (usually a sulfonyl chloride) to the thiol (Scheme 4). A number of reagents have been used for the reduction ⟨84CL139⟩, but lithium aluminum hydride is usually the reagent of choice ⟨51JOC946⟩.

$$Ar{-}H \longrightarrow \underset{X\,=\,Cl,\ OH}{Ar{-}SO_2X} \xrightarrow[90\text{–}100\%]{LiAlH_4} Ar{-}SH$$

Scheme 4

The phosphine/iodine mixtures described by Oae provide an interesting alternative, as they permit reduction directly from sulfonic acids to thiols under rather mild conditions ⟨83BCJ3802, 84BCJ232⟩.

2.14.2.1.5 The Herz reaction

The Herz reaction provides an excellent method of introducing a thiol group *ortho* to an amino group ⟨57CRV1011⟩. A good example is the synthesis of the naphthalene derivative (**2**) (Scheme 5) ⟨90S539⟩.

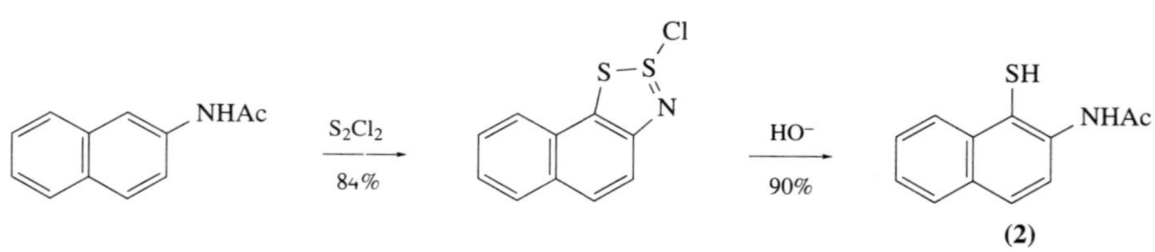

(2)

Scheme 5

2.14.2.2 Vinyl and Aryl Sulfides and their Higher-coordinated Derivatives

2.14.2.2.1 Vinyl sulfides

The available disconnections for vinyl sulfides are summarised in Scheme 6. Each of these possibilities is discussed below. The direct preparation of divinyl sulfide from acetylene and elemental sulfur has been reported and widely cited in the Russian literature ⟨83IZV230⟩.

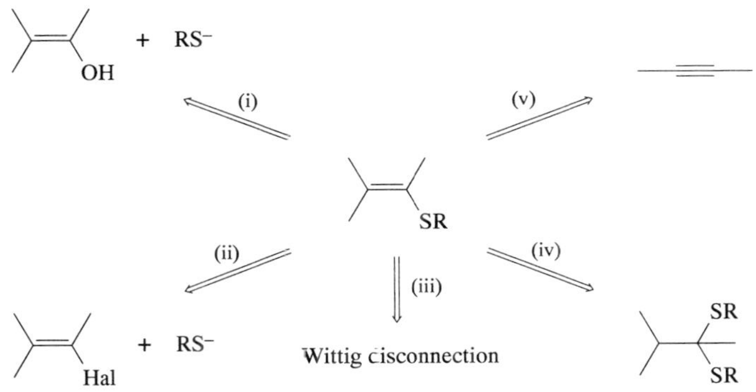

Scheme 6

(i) Vinyl sulfides by enolisation

Ketones react with thiols in the presence of various Lewis acids, mercuric triflate being perhaps the most useful, to provide vinyl sulfides directly (Equation (6)) ⟨68JOC780, 72JOC939, 83JA5075⟩. This process is generally applicable when the ketone is cyclic, but yields are often poor for other examples. A reaction sequence which is synthetically equivalent to the thiaenolisation of an aldehyde is shown in Scheme 7 ⟨85TL1795⟩.

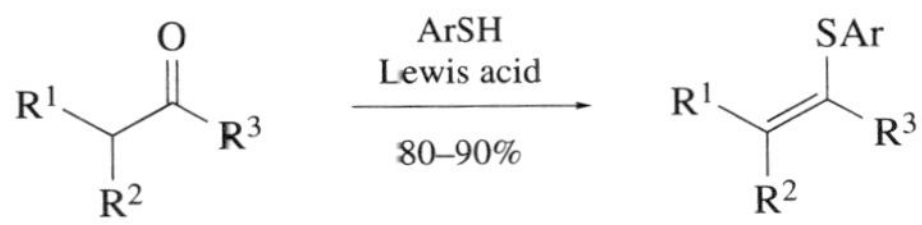

(6)

Ar = Ph, p-MeOC$_6$H$_4$–; R^1 = H, Me; R^2, R^3 = –(CH$_2$)$_n$–; n = 3, 4, 6, 10

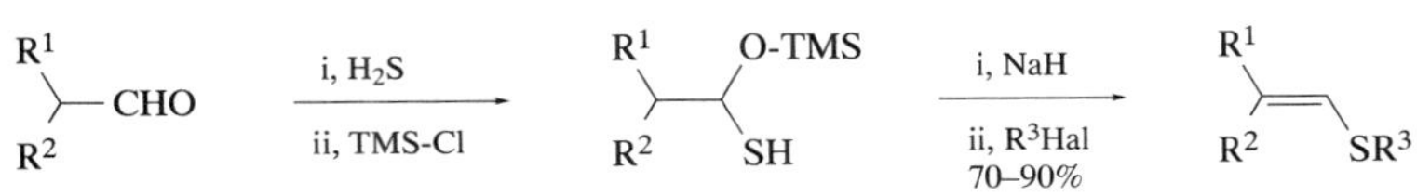

R^1, R^2 = H, Me, Et, –(CH$_2$)$_5$–; R^3 = PhCH$_2$, PhCH$_2$CH$_2$, Pr^n, Bu^i, cyclohexyl

Scheme 7

(ii) Vinyl sulfides from vinyl halides

It is possible, with careful control of the reaction conditions, to convert vinyl halides directly into vinyl sulfides (Equation (7)). The reaction was found to be completely stereospecific ⟨83JOC4795⟩.

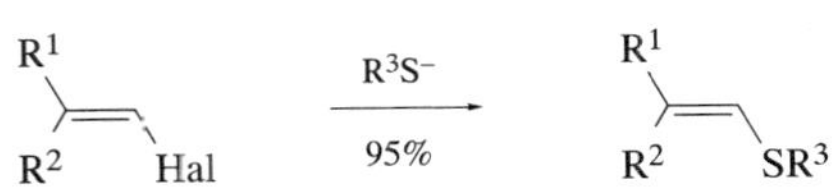

(7)

R^1, R^2 = H, Ph; R^3 = Me, Pr^i

(iii) Vinyl sulfides by Wittig-type reactions

As always, alkenes can be Wittig disconnected in two directions and both possibilities are indeed available for vinyl sulfide synthesis. Hence, sulfur-substituted Wittig–Horner reagents (**3**) react with a range of carbonyl compounds ⟨70JOC777⟩, and conventional Wittig reagents (**4**) react with thiolesters ⟨68BCJ2742⟩, both processes yielding vinyl sulfides (Scheme 8).

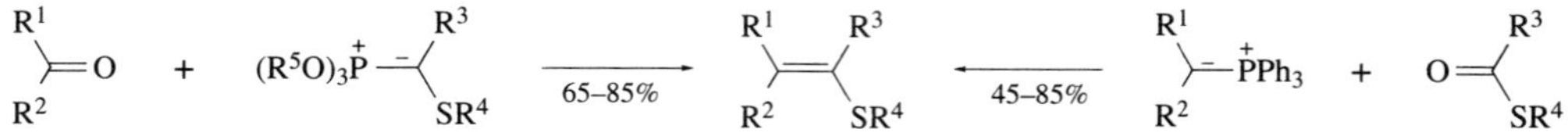

(3) (4)

$R^1, R^2 = H$, Me, pentyl, Ph; R^3 = Me, octyl; R^4 = Me

R^1, R^2 = Me, Ph, $-(CH_2)_5-$; R^3 = Ph, $p\text{-}ClC_6H_4$, $p\text{-}MeOC_6H_4$; R^4 = Me, Et

Scheme 8

(iv) *Vinyl sulfides by elimination*

Dithioacetals are useful precursors to vinyl sulfides. As they are readily prepared from carbonyl compounds, they offer a two-step equivalent of the enolisation strategy (i). Two of the more successful reagents for this transformation are zinc diiodomethane and cupric triflate ⟨85T4443, 88OSC(6)737⟩. The elimination of benzotriazole from a related precursor also yields vinyl sulfides ⟨91HCA1931⟩.

(v) *Vinyl sulfides from alkynes*

The addition of a thiol to an alkyne is a very attractive route to vinyl sulfides (Equation (8)). Indeed, radical addition, using one of a number of initiators, is an established process ⟨70AG(E)273⟩. More recently, triethylborane has been proposed as an initiator ⟨87CL1647⟩. However, mixtures of stereoisomers usually result.

(8)

R^1, R^2 = H, alkyl, aryl; R^3 = Me, Ph, Ac

If one of the alkyne substituents R^1 and R^2 is electron-withdrawing, base-catalysed addition of thiols can be useful ⟨72JOC2391⟩. In this case the thiol will often add with high *trans* selectivity. Yields and selectivity can be enhanced by the use of alumina ⟨91BCJ3485⟩.

The addition of thiols to alkynes in the alternative anti-Markovnikov sense can be induced by transition metal catalysts (Equation (9)). Palladium acetate was shown to be particularly effective in this respect ⟨92JA5902⟩.

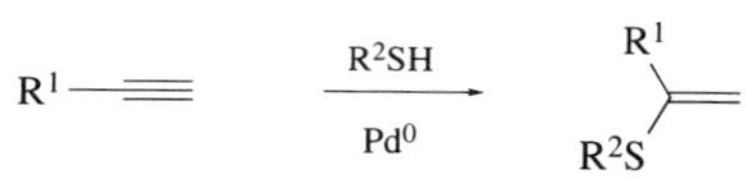

(9)

R^1 = CN, Ts, $p\text{-}NO_2C_6H_4$-, Ac; R^2 = alkyl, aryl

2.14.2.2.2 *Aryl sulfides*

Aryl sulfides can be disconnected in four ways (Scheme 9). The most obvious disconnection is to a thiophenolate and an alkyl halide (iv) ⟨74MI 214-01⟩, which corresponds to the general method of sulfide synthesis covered in Chapter 2.03. The other three disconnections, (i), (ii) and (iii), are discussed here, followed by some miscellaneous methods of aryl sulfide synthesis.

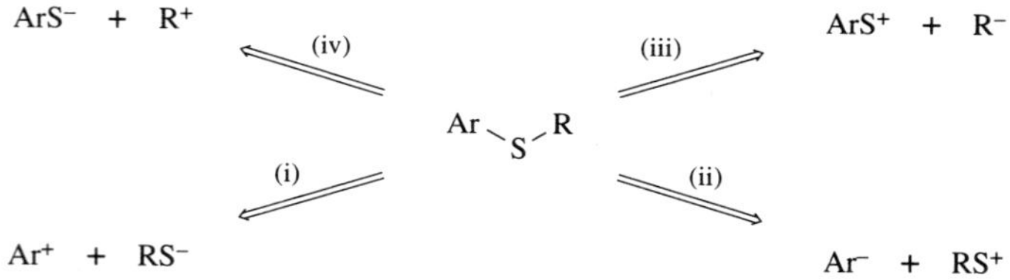

Scheme 9

(i) Aryl sulfides by nucleophilic aromatic substitution

It is convenient to distinguish three types of substrate for the nucleophilic aromatic substitution reaction: aryl halides, phenols and diazonium salts.

Aryl halides react with metal sulfides to yield symmetrical diaryl sulfides (Equation (10)), and nonsymmetrical diaryl sulfides are available from the reaction of Grignard reagents with diaryl disulfides ⟨48JCS528⟩.

$$Ar-Hal \xrightarrow{M_2S} Ar-S-Ar \qquad (10)$$

A very effective method of synthesising alkyl aryl sulfides from haloarenes was developed by Testaferri *et al.* ⟨83S751⟩, who discovered that, if a highly dipolar aprotic solvent is used, even relatively unactivated aryl halides undergo nucleophilic aromatic substitution (Equation (11)). The solvents of choice were dimethylformamide and dimethylacetamide, although polyglyme solvents and toluene under phase transfer conditions have also been proposed as alternatives ⟨85JOC4812, 84JOC1309⟩. Diaryl sulfides can also be prepared by this methodology ⟨91JOC862⟩.

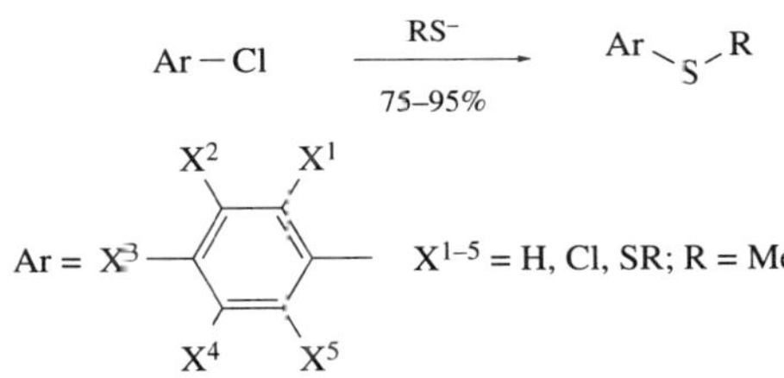

X^{1-5} = H, Cl, SR; R = Me, Et, Pr^i (11)

Transition metal-catalysed substitution of aryl halides by sulfur nucleophiles has also come to prominence. For example, nickel catalyses the reaction of arenethiols with simple aryl bromides (Equation (12)) ⟨81S892, 87CL2221⟩, and palladium facilitates the coupling of a functionalised stannylthiol and an aryl iodide (Equation (13)) ⟨91T8621⟩. Yields were excellent in both cases.

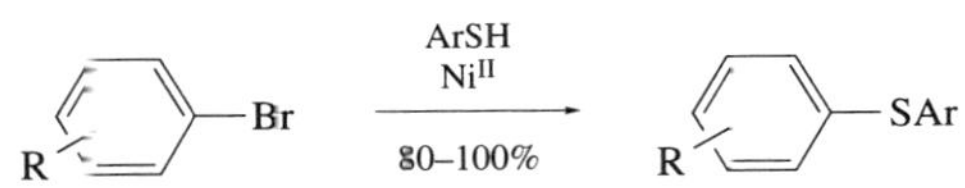

(12)

R = 2-Me, 3-Me, 4-Me, 4-CN, 4-NH_2, 4-MeO; Ar = Ph, p-ClC_6H_4-

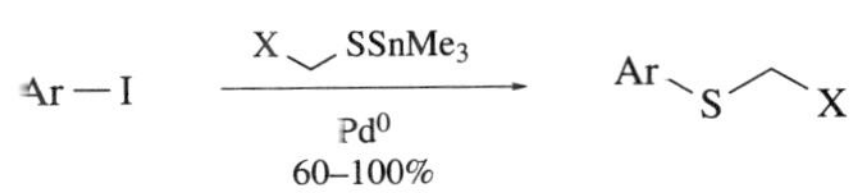

(13)

Ar = Ph, p-$MeOC_6H_4$-; X = Ph, vinyl, CO_2Me

In favourable cases, the nucleophilic aromatic substitution of phenols by thiolates is possible. The preparation of 2-naphthalenethiol (**5**) as shown in Equation (14) is representative ⟨89S955⟩.

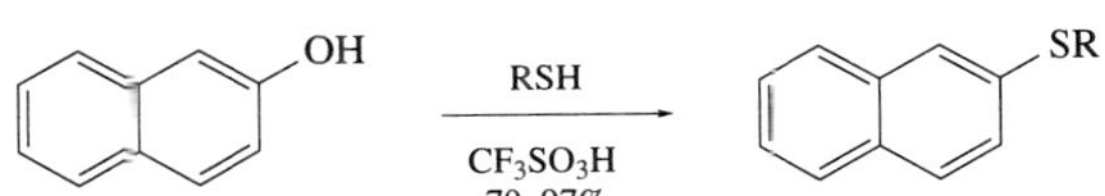

(**5**) (14)

R = n-alkyl, CH_2CO_2Et, aryl

Arenediazonium salts react as expected with sulfides to yield symmetrical diaryl sulfides (Equation (15)), but the more useful reaction with thiolates is complicated by formation of compounds such as (**6**). Although (**6**) can be converted to the target sulfides, a significant risk of explosion has made this procedure less attractive. Nevertheless, a satisfactory procedure has been developed which allows the preparation of a range of unsymmetrical diaryl sulfides (Equation (16)) ⟨86T4007⟩.

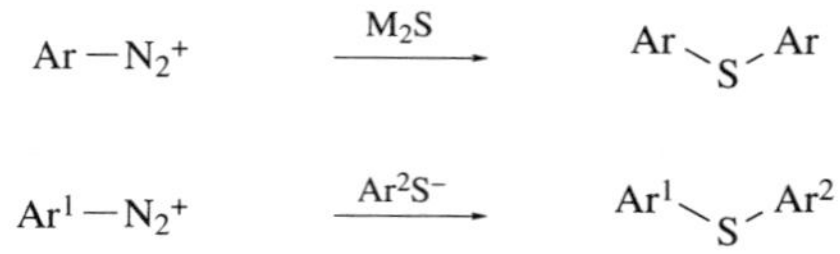

(15)

(16)

Ar^1 = Ph, naphthyl, 2-,3- or 4-halo- or nitrophenyl; Ar^2 = Ph, 2,4,6-trialkylphenyl, p-ClC_6H_4-

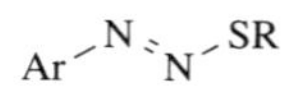

(6)

(ii) Aryl sulfides by electrophilic aromatic substitution

Arenes are known to react with sources of electrophilic sulfur. Thus, reaction with a mixture of elemental sulfur and disulfur dichloride yields symmetrical aryl sulfides (Equation (17)), whereas the use of arenesulfenyl chlorides permits access to unsymmetrical diaryl sulfides (Equation (17)) ⟨60JA5359, 68TL5071⟩.

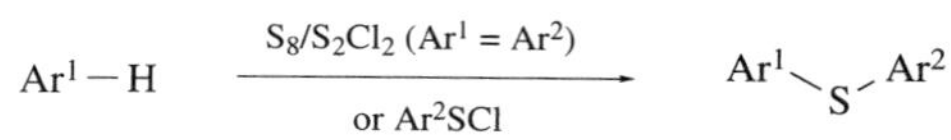

(17)

An important development in this area was the recent discovery that a mixed antimony halide catalyst is effective in rendering diaryl disulfides sufficiently electrophilic for aromatic substitution to occur smoothly and in high yield (Equation (18)) ⟨93CL1⟩

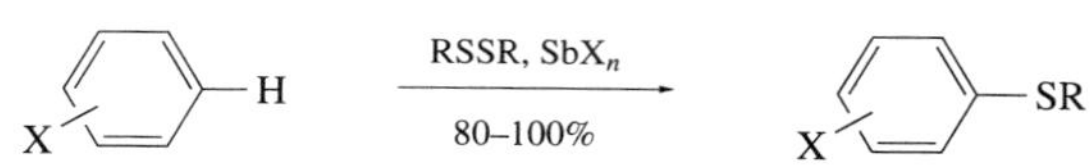

(18)

R = Me, Ph; X = H, Me, MeO

(iii) The reaction of R^- with a source of electrophilic sulfur

This is an uncommon method for the synthesis of aryl sulfides. One example is the reaction of a trialkylborane with diphenyl disulfide (Equation (19)) ⟨71JA3291⟩.

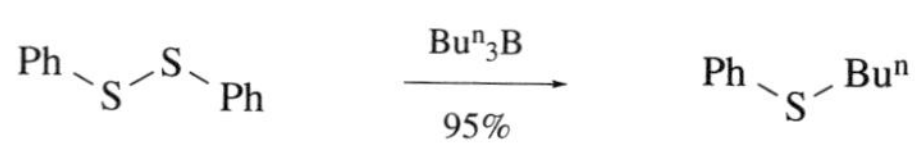

(19)

(iv) Aryl sulfides by addition reactions

Alkenes react with thiols, usually by a radical mechanism, to give the corresponding sulfides (Equation (20)) ⟨70AG(E)273⟩. An analogous reaction occurs with arenesulfenyl chlorides, to yield chloroalkyl derivatives ⟨77RCR548⟩.

(20)

X = H, Cl

Diaryl sulfides are also accessible by addition reactions, in this case the addition of ethyl sulfides to benzyne (Equation (21)). The process appears to be clean and high-yielding ⟨83CL249⟩.

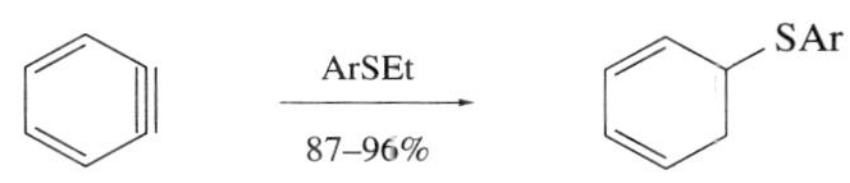

(21)

Ar = *p*-BrC_6H_4-, *p*-$MeOC_6H_4$-, 2,4,6-$Me_3C_6H_2$-, naphthyl

(v) Aryl sulfides by reduction

The preparation of sulfides from their corresponding sulfones and sulfoxides (Equation (22)), as well as their related imides, and from disulfides and thiocyanates has been discussed in a review ⟨91COS(8)403⟩.

$$\mathrm{Ar-SO}_n\mathrm{R} \xrightarrow[n\,=\,1 \text{ or } 2]{[\mathrm{H}]} \mathrm{Ar{-}S{-}R} \qquad (22)$$

2.14.2.2.3 *Vinyl and aryl sulfoxides*

(i) Vinyl sulfoxides

The use of vinyl sulfoxides is most important in asymmetric synthesis, and only syntheses of enantiomerically enriched products will be discussed here. Three disconnections are available, which will be discussed below (Scheme 10): (a) functional group interconversion to the corresponding sulfides; (b) disconnection of the carbon–sulfur bond; (c) disconnection of the alkene double bond. Methods of preparation of more highly functionalised vinyl sulfoxides and sulfinyldienes have been reviewed ⟨94MI 214-01⟩.

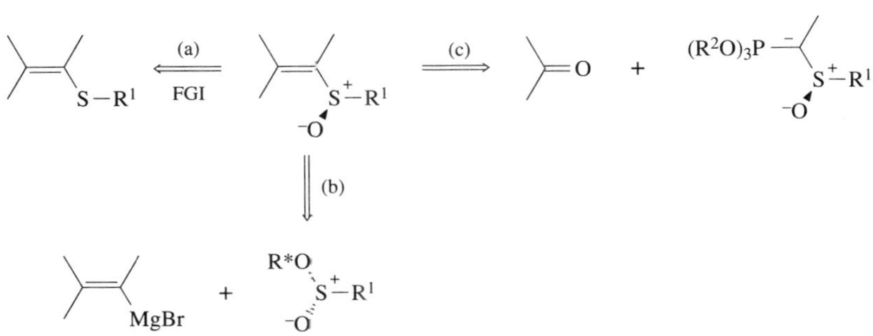

Scheme 10

(a) Three of the standard methods of asymmetric sulfoxidation (see Chapter 2.03) have been found to be useful for preparation of chiral vinyl sulfoxides, namely bioconversion, Kagan modified Sharpless oxidation and Davis chiral oxaziridine oxidation ⟨92TA629⟩. It appears that the choice of method is very dependent upon the substrate.

(b) The standard Anderson synthesis (see Chapter 2.03) can be adapted to the synthesis of chiral vinyl sulfoxides if vinyl Grignard reagents are employed ⟨78JOC4131⟩. However, it may well be more practical to use an alkynyl Grignard reagent as a synthetic equivalent (Scheme 11) ⟨85CL805, 87JOC1078⟩. The judicious choice of the reducing reagent thus leads to either *cis*- or *trans*-vinyl sulfoxides.

(c) The appropriate Wittig–Horner-type reagents (**7**) for the synthesis of chiral vinyl sulfoxides are simple to prepare ⟨78JOC473⟩. The reaction of (**7**) with carbonyl compounds proceeds smoothly but can result in mixtures of *cis* and *trans* products.

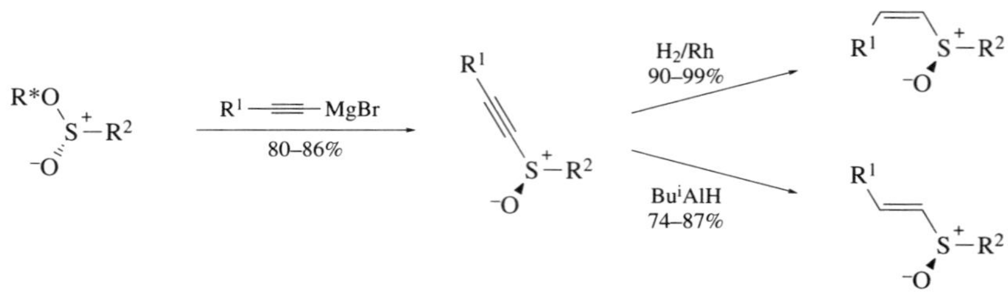

R* = menthyl; R^1 = *n*-alkyl, TMS; R^2 = *p*-tolyl

Scheme 11

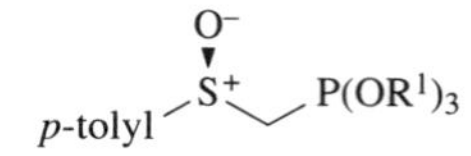

(7)

(ii) Aryl sulfoxides

Aryl sulfoxides are prominent among reagents for asymmetric synthesis. The most important synthetic methods are, therefore, ones which allow preparation of enantiomerically enriched sulfoxides. These methods can be divided into two categories: (a) asymmetric oxidation of sulfides (Equation (23)) and (b) nucleophilic substitution of chiral sulfinates (or analogues) (Equation (24)). Both have been reviewed ⟨94MI 214-01⟩ and are thus only briefly outlined here.

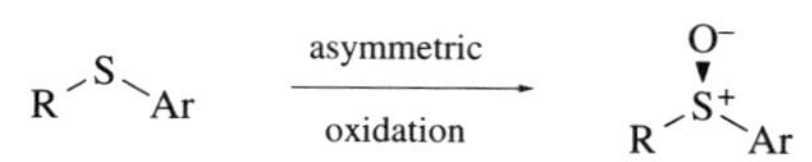

(23)

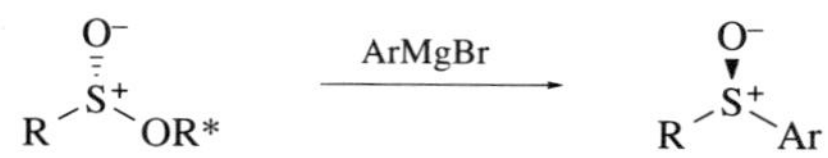

(24)

R* = chiral auxiliary

(a) The original Kagan method of asymmetric sulfoxidation has been supplemented by other catalytic methods including iron–porphyrin and manganese (salen) complexes as well as further titanium alkoxide-derived catalysts. Enzyme-catalysed oxidations also continue to be useful. The Davis oxaziridine provides an alternative, stoichiometric, approach. As remarked by Rayner, although high *ee* values are only observed in specific cases with most of these reagents, one of them will probably be effective for a given target sulfoxide ⟨94MI 214-01⟩.

(b) Chiral auxiliary routes to enantiomerically pure sulfoxides have also been reviewed ⟨94MI 214-01⟩. The use of diacetone-D-glucose as the auxiliary is particularly attractive.

2.14.2.2.4 Vinyl and aryl sulfones

(i) Vinyl sulfones

Methods for the preparation of vinyl sulfones have been reviewed ⟨90T6951⟩. Perhaps the most obvious route is via oxidation of the corresponding vinyl sulfides or sulfoxides using standard reagents (see Chapter 2.03 and ⟨91COS(8)403⟩) (Scheme 12). However, it is quite possible that the preparation of the appropriate sulfide or sulfoxide is not trivial. Thus, the direct synthesis of vinyl sulfones using one of the versatile methods described below is recommended.

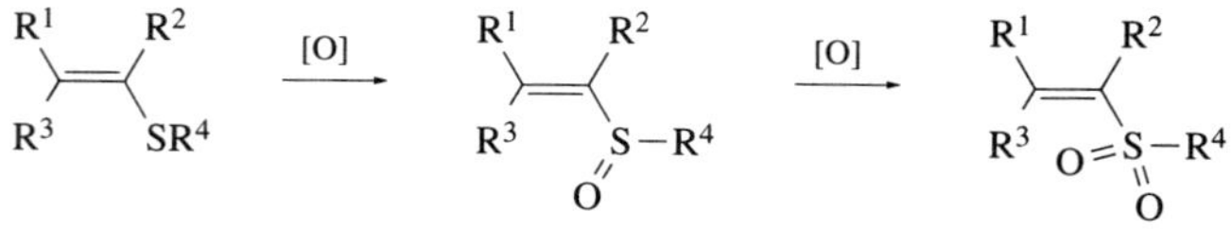

Scheme 12

(a) Vinyl sulfones by elimination. The high acidity of protons α to a sulfone group makes the transformation in Scheme 13 attractive as a disconnection of vinyl sulfones. Two disconnections of the precursor are clear, namely (i) to RSO_2X and an alkene or (ii) to a carbonyl compound (X = O) and a sulfone-stabilised carbanion (an aldol-type disconnection).

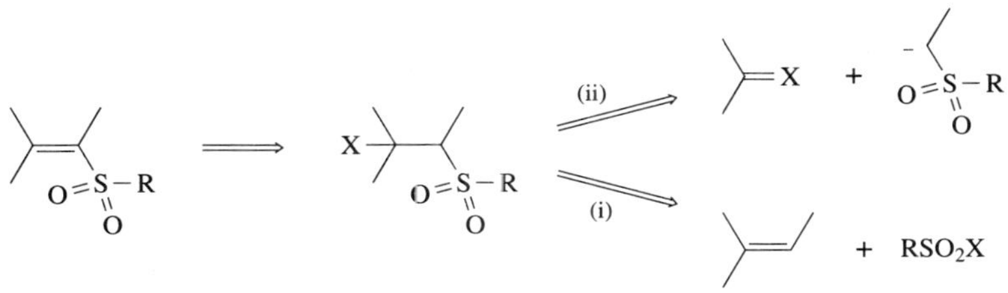

Scheme 13

The addition of RSO_2X to an alkene (cf. disconnection (i)) can be carried out directly, particularly when R is aryl and X is HgCl, SePh or I. The mercury reagents are prepared *in situ* by the reaction of sodium arenesulfinates and $HgCl_2$. A synthetically equivalent procedure is the addition of an arenesulfenyl halide to the alkene followed by the oxidation of the sulfide product to the sulfone (Scheme 14) ⟨90T6951⟩.

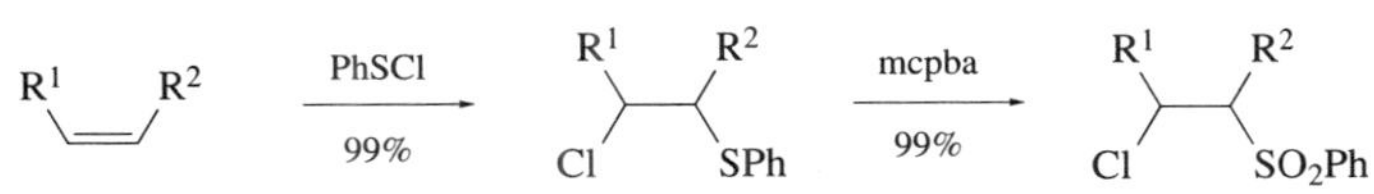

R^1, R^2 = cycloalkyl

Scheme 14

Sulfones are excellent carbanion-stabilising groups. Their aldol-type addition to carbonyl compounds (cf. disconnection (ii)) proceeds smoothly. Dehydration by one of a number of methods leads to the vinyl sulfone ⟨90T6951⟩. Mixtures of *cis* and *trans* isomers may result.

This methodology clearly invites extension to Peterson and Wittig analogues, as outlined below.

(b) Vinyl sulfones from Peterson and Wittig reactions. α-Silylsulfonates (**8a**) and α-sulfonylphosphonates (**8b**) are readily available. Compounds (**8a**) have been shown to react in a Peterson-type reaction, but often yield *cis/trans* isomeric mixtures. The Wittig–Horner equivalents (**8b**) have also been shown to react with aldehydes to produce, in this case exclusively, the *trans*-alkene (Equation (25)).

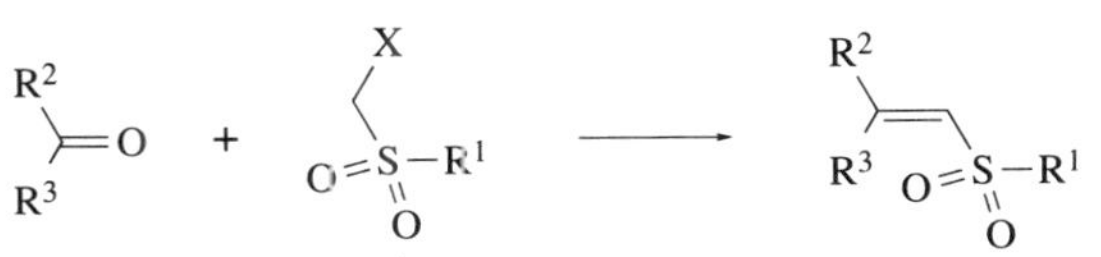

(25)

(**8a**) X = SiR^4_3, 65–92%
(**8b**) X = $P(OR^4)_3$, 72–97%

R^1 = Me, Ph, p-ClC_6H_4-; R^2, R^3 = H, alkyl, alkenyl, cycloalkyl, Ph

(ii) Aryl sulfones

Aryl sulfones can be prepared using many of the general methods described in Chapter 2.03, in particular via the oxidation of sulfides and sulfoxides. There are two strategies, however, which have proved particularly useful for the synthesis of both diaryl and aryl alkyl sulfones: the Friedel–Crafts sulfonylation of arenes and the reaction of arenesulfinate salts with aryl or alkylhalides.

The ready availability of arenesulfonyl chlorides and arenesulfonic acids makes Friedel–Crafts routes to unsymmetrical diaryl sulfones very attractive (Equation (26)) ⟨79COC(3)171⟩. The use of the latter can be recommended owing to improved procedures ⟨84S215, 84S323⟩. The preparation of alkyl aryl sulfones from the corresponding alkanesulfonyl chlorides is generally unsuccessful. However, the use of alkanesulfonyl fluorides has been proposed to overcome this problem (Equation (26)) ⟨84S215⟩.

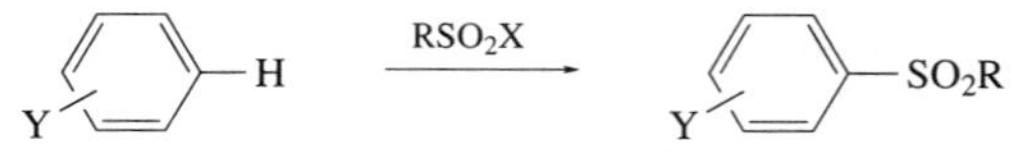

(26)

R = *n*-alkyl, vinyl; X = F; 60–90%
R = Ph, *p*-tolyl; X = Cl, OH; 45–95%
Y = H, Me, MeO, Cl

The use of phase transfer catalysts permits the use of arenesulfinates as nucleophiles. Hence, arenesulfonyl chlorides can be used to form the sulfinate *in situ*. Reaction with either electron-deficient aryl chlorides ⟨92S825⟩ or alkyl halides ⟨85JOC1327⟩ leads to diaryl or aryl alkyl sulfones, respectively (Scheme 15).

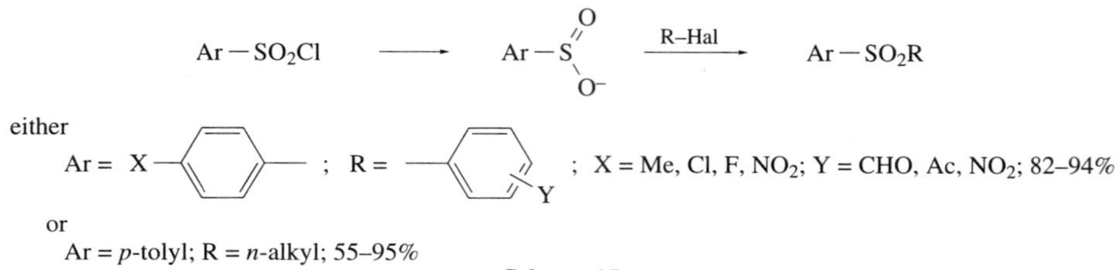

Scheme 15

2.14.2.3 Vinyl- and Arenesulfenyl Halides and their Higher-coordinated Analogues

2.14.2.3.1 *Arenesulfenyl halides*

Arenesulfenyl halides (**9**) have been increasing in use as reagents in organic synthesis. The fluorides are little known ⟨70S561⟩, and the iodides, although they can be prepared ⟨70JOC2006⟩, are rather too reactive to be synthetically useful. Arenesulfenyl bromides can be prepared and used in a very similar manner to their chloride analogues ⟨70S561⟩, but it is the latter compounds which have found most favour with synthetic chemists.

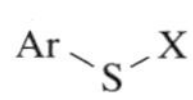

(**9**) X = F, Cl, Br, I

The most widely used method for the preparation of sulfenyl chlorides is the chlorinolysis of disulfides (Equation (27)) ⟨70S561⟩. Chlorine has been the most widely used reagent for this transformation, and for most aromatic substrates there is a limited risk of side reactions. However, as the use of chlorine on a small scale is not always convenient, one of the milder chlorinating reagents may be preferred, such as sulfuryl chloride or *N*-chlorosuccinimide.

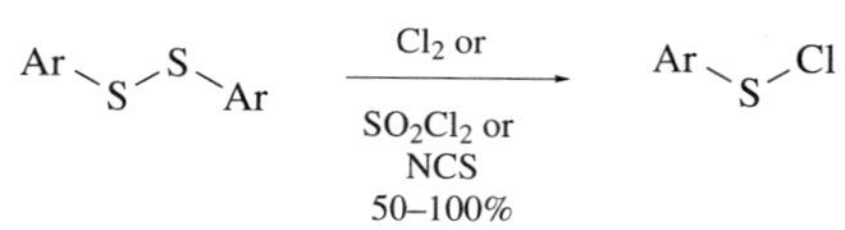

(27)

The second important method of arenesulfenyl chloride synthesis is the direct chlorination of an arenethiol (Equation (28)), which in fact generates the disulfide *in situ* ⟨70S561⟩. This route is obviously attractive if the thiol is readily available, but can be troublesome if the substrate has a functionality which is sensitive to chlorination.

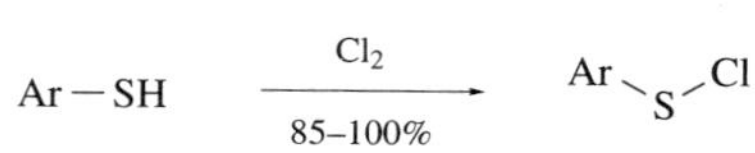

(28)

2.14.2.3.2 *Arenesulfinyl halides*

Chlorides are by far the most prominent of the arenesulfinyl halides (**10**). Arenesulfinyl bromides and iodides appear to be unknown, and arenesulfinyl fluorides are not widely used, although benzenesulfinyl fluoride can be prepared by a fluorine–oxygen exchange reaction ⟨64OS(44)39⟩.

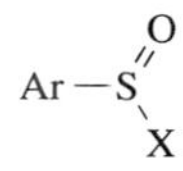

(10) X = F, Cl

The traditional methods of preparing sulfinyl chlorides have been reviewed ⟨70OPP235⟩. These routes proceed either from the reaction of sulfinic acids, which are not particularly accessible substrates, with thionyl chloride (Equation (29)), or from more convenient substrates using chlorine, which is not a practical reagent for small-scale synthesis. The use of sulfuryl chloride as the oxidising and chlorinating agent has therefore been proposed as an alternative (Equation (30)). Either arenethiols (**11a**) ⟨87S72⟩ or arenethiol esters (**11b**) ⟨87TL5193⟩ can be used as substrates.

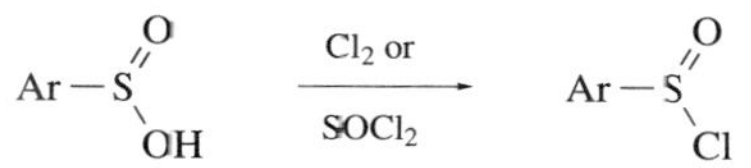

(29)

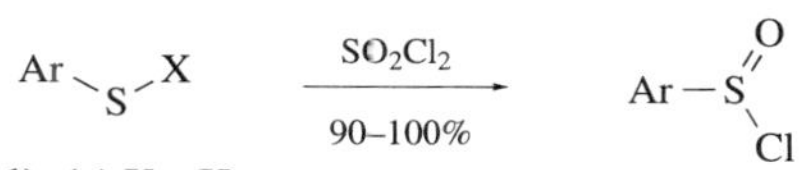

(30)

(11) (**a**) X = H
(**b**) X = Ac

Ar = Ph, *p*-tolyl

2.14.2.3.3 *Vinyl- and arenesulfonyl halides*

Vinylsulfonyl chlorides are generally prepared from the corresponding sulfonic acids (Equation (31)) ⟨72TL4319⟩. The unsubstituted parent compound (**12a**) can be synthesised by an elimination route from (**13a**) (Equation (32)) ⟨47JA1219⟩. Vinylsulfonyl fluoride (**12b**) has appeared regularly in the patent literature. The method of choice for its preparation appears to be from (**13b**) with magnesium oxide (Equation (32)) ⟨79JOC3847⟩.

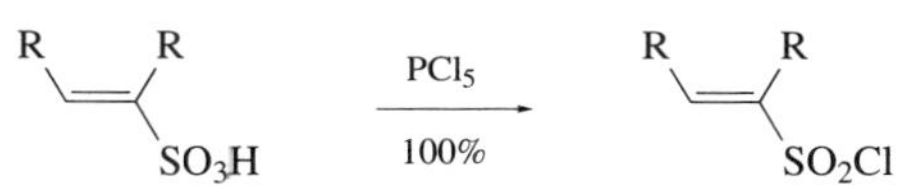

(31)

R = alkyl

(32)

(13) (**a**) X = Cl
(**b**) X = F

(12) (**a**) X = Cl
(**b**) X = F

Chlorides are by far the most widely used of the arenesulfonyl halides (**14**). Indeed, *p*-toluenesulfonyl chloride (tosyl chloride) is a standard laboratory chemical. The fluoride analogues can be prepared by halogen exchange from the chlorides ⟨77JOC2031⟩, and arenesulfonyl bromides and iodides are available from the corresponding hydrazides ⟨63JOC2766, 64MI 214-01⟩.

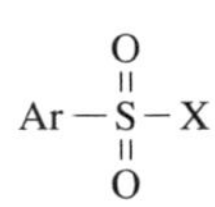

(14) X = F, Cl, Br, I

Arenesulfonyl chlorides are very easy to prepare, and are readily transformed into other aryl sulfur compounds ⟨92CL1483⟩. They can be synthesised from: arenes by electrophilic aromatic substitution with chlorosulfonic acid (Equation (33)) ⟨69S3⟩; from aryl halides via reaction of aryllithium ⟨86S852⟩, or aryl Grignard ⟨69S3⟩, reagents with sulfur dioxide and sulfuryl chloride (Equation (34)); from arenediazonium chlorides by reaction with sulfur dioxide and cuprous chloride (Equation (35)) ⟨69S3⟩; from arenethiols by reaction with sulfuryl chloride and potassium nitrate

(Equation (36)) ⟨92CL1483⟩; and from arenesulfonic acids by chlorination (Equation (37)) ⟨79COC(3)311⟩.

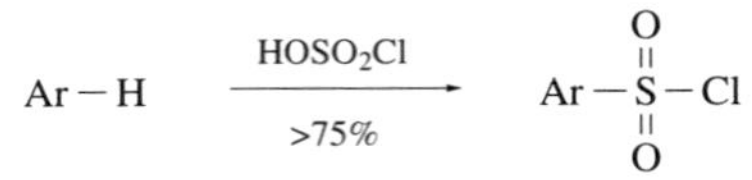

(33)

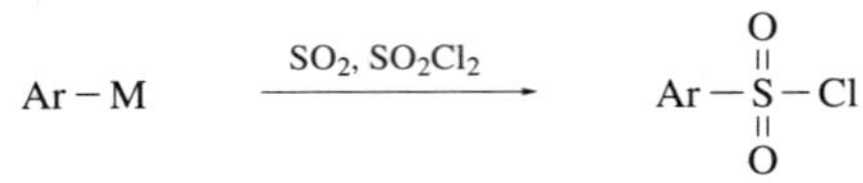

(34)

M = Li, 80–99%; MgBr, 53–64%

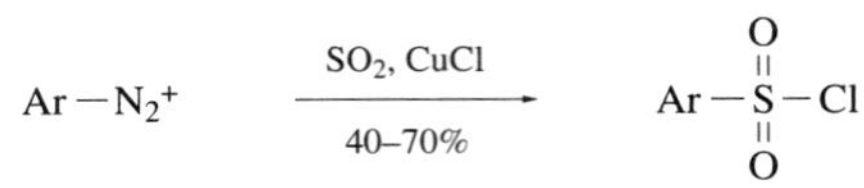

(35)

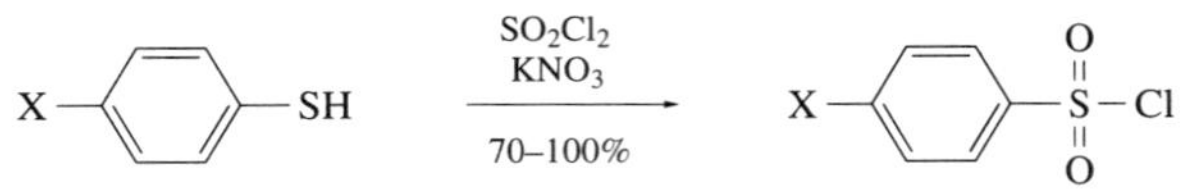

(36)

X = H, Me, Hal

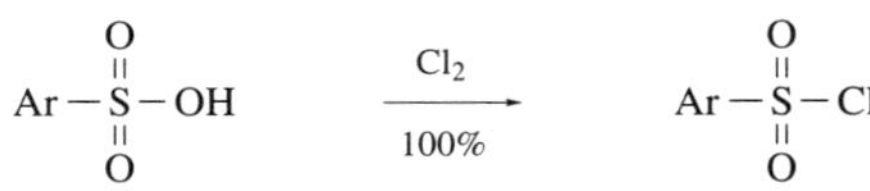

(37)

2.14.2.4 Arenesulfenic, Arenesulfinic and Vinyl- and Arenesulfonic Acids

2.14.2.4.1 Arenesulfenic acids

Arenesulfenic acids (**15**) are often proposed as intermediates in rearrangements of sulfoxides ⟨86JOC1033⟩. However, apart from a very few unusual examples where the acid is stabilized by intramolecular hydrogen bonding ⟨86JOC1033⟩, arenesulfenic acids are not isolable at room temperature. Their preparation, by flash vacuum pyrolysis of suitably substituted sulfoxides, and their isolation at low temperatures has, however, been reported ⟨81JOC3467, 83JOC3071⟩.

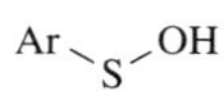

(15)

2.14.2.4.2 Arenesulfinic acids

The synthesis of arenesulfinic acids is a facile procedure from a number of different starting materials. However, the acids are not particularly stable. Storage as a salt is thus usually advisable ⟨79COC(3)261⟩. An important route to arenesulfinic acids is by reduction of readily available arenesulfonyl chlorides (Equation (38)), using a metal, for example zinc, or sodium sulfite as the reducing agent ⟨79COC(3)261⟩. One of the few more recent contributions to the area of sulfinic acid synthesis was the introduction of arenethiols as useful new reducing reagents for this transformation ⟨90S391⟩.

(38)

Other potential starting materials are: arenes, which undergo an aluminum trichloride-catalysed Friedel–Crafts reaction with sulfur dioxide, to yield the sulfinic acid directly (Equation (39)); aryl halides, via metallation to aryllithium or aryl Grignard reagents and subsequent reaction with sulfur dioxide (Equation (40)); arylamines, via reaction of arenediazonium salts with sulfur dioxide in the presence of copper (Equation (41)) ⟨79COC(3)261⟩; and arenethiols, by oxidation (overoxidation can be a problem) with mcpba, preferably (Equation (42)).

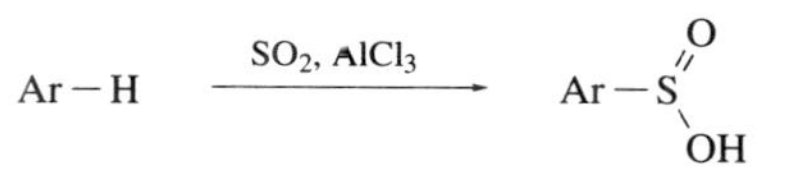
(39)

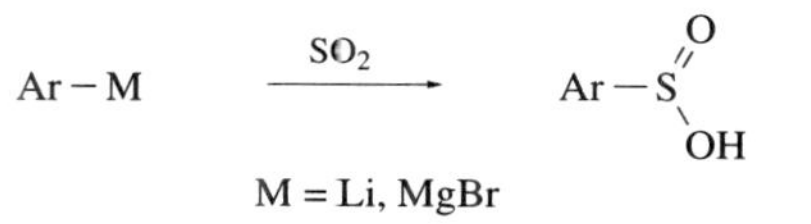
(40)

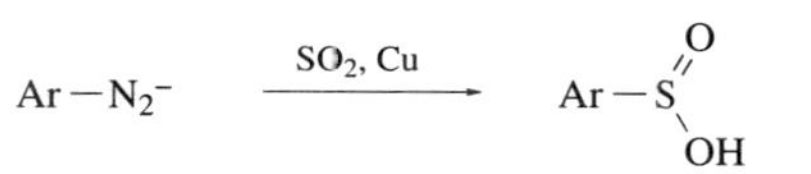
(41)

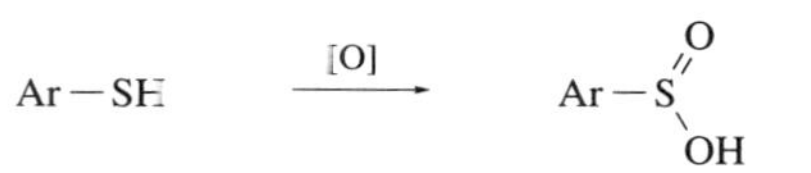
(42)

The manipulation of aryl sulfones can also provide an approach to arenesulfinic acids. The base-catalysed rearrangement of certain aryl alkyl sulfones permits an interesting three-step synthesis of arenesulfinic acids from arenethiols (Scheme 16) ⟨63JOC593⟩. Finally, the Smiles rearrangement provides a method of preparing *ortho*-substituted arenesulfinic acids from *ortho*-substituted sulfones, as shown in Scheme 17 ⟨70OR(18)99⟩.

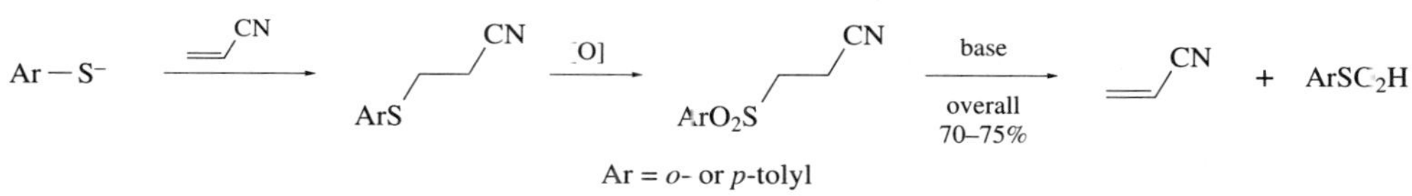

Scheme 16

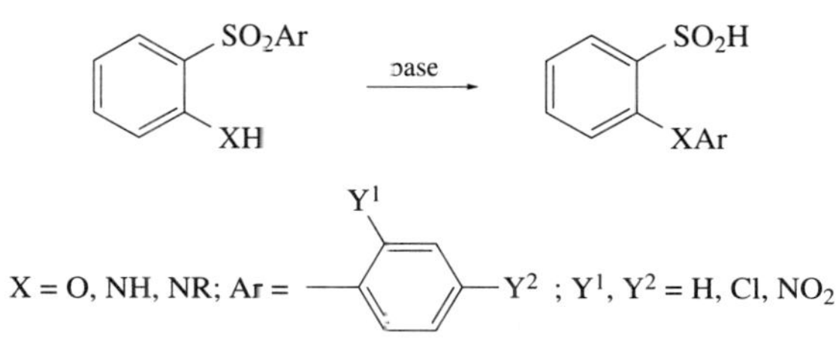

Scheme 17

2.14.2.4.3 *Vinyl- and arenesulfonic acids*

Vinylsulfonic acids are available from a potassium hydroxide-promoted rearrangement of dialkyl sulfones (Equation (43)) ⟨72TL4319⟩.

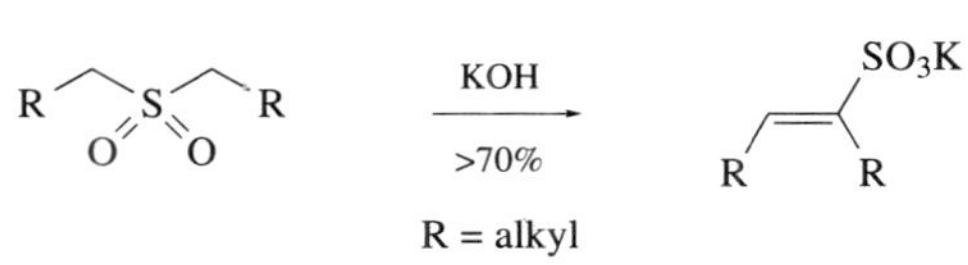
(43)

It is well known that arenesulfonic acids can be prepared by sulfonation of arenes, and indeed this process is one of the most important methods of forming an arene–sulfur bond (Equation (44)). The sulfonating agent is usually sulfuric acid, chlorosulfuric acid or sulfur trioxide. It can be difficult to handle pure sulfur trioxide, so its solution in sulfuric acid (oleum) is often preferred. If necessary, the reactivity of sulfur trioxide can be reduced by the formation of adducts with Lewis bases such as pyridine. The classic reviews in the field are by Gilbert ⟨62CRV549, B-65MI 214-01⟩. A more recent overview is provided by Andersen ⟨79COC(3)331⟩, and a useful list of references is given in a footnote to a paper discussing mechanistic aspects of the reaction ⟨85RTC153⟩.

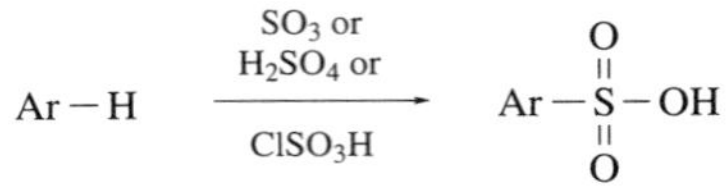

(44)

The main alternative approach to arenesulfonic acid synthesis is via oxidation. Although many arene sulfur compounds can be oxidised to the corresponding sulfonic acids ⟨79COC(3)331⟩, this is only likely to be a useful synthetic method if these compounds are readily available. The oxygen oxidation of arenethiols, under alkaline conditions (Equation (45)) ⟨63TL1131⟩, and the hydrogen peroxide oxidation of diaryl disulfides (Equation (46)), under acidic conditions ⟨65CI(L)81⟩, appear to be the most useful of these procedures.

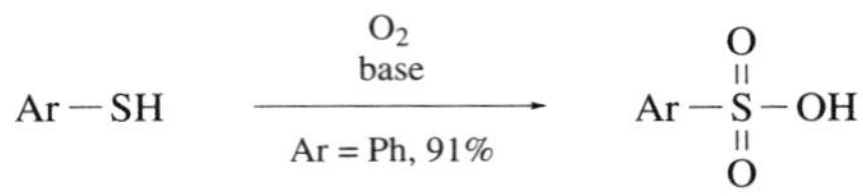

(45)

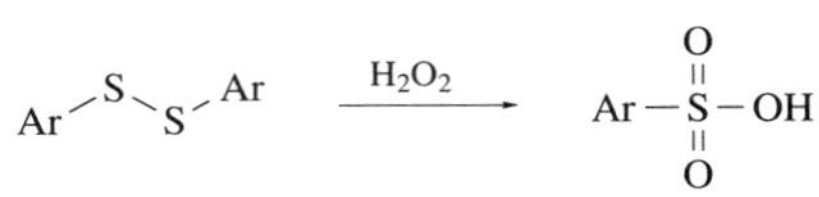

(46)

2.14.2.5 Arenesulfenates, Arenesulfinates, and Vinyl- and Arenesulfonates

2.14.2.5.1 Arenesulfenates

The classical method of preparing esters of arenesulfenic acids (arenesulfenates (**16**)) is by reaction of an alkoxide with an arenesulfenyl halide (Equation (47)) ⟨25CB409⟩. An application of this procedure was in the synthesis of the silyl derivative (**17**) ⟨91BCJ702⟩ (Equation (48)). However, an improved procedure which uses stannylated alcohols in place of alkoxides ⟨84S1042⟩, excludes further reaction of the product with the alkoxide.

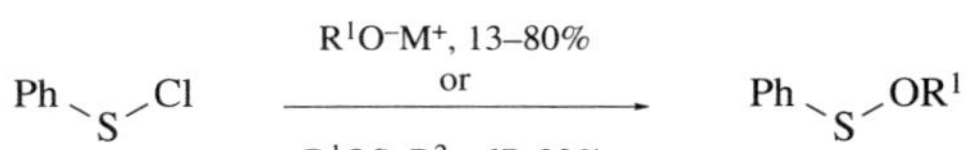

(**16**)
(47)

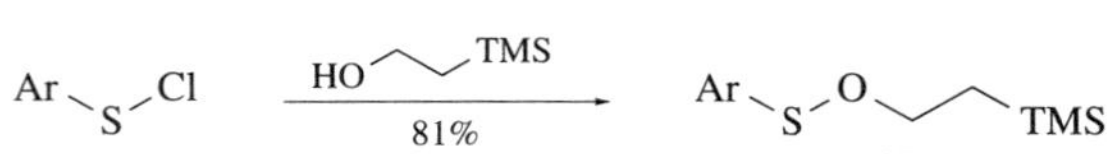

(**17**)
(48)

2.14.2.5.2 Arenesulfinates

Arenesulfinate esters are very important synthetic intermediates. In particular, the menthyl sulfinate (**18**) has been widely used for preparation of optically active sulfoxides. The emphasis here is therefore on methods of preparation of arenesulfinate esters which have applications in asymmetric synthesis.

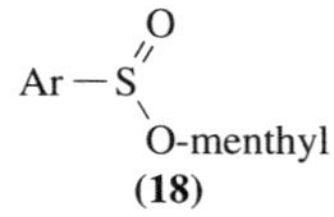

(**18**)

Sulfinate esters are most commonly prepared from sulfinyl chlorides (**19a**) and alcohols (Equation (49)) ⟨63JCS5741⟩, and indeed this method has been used for the synthesis of chiral sulfinates ⟨91JOC4552⟩. Esterification of sulfinic acids (**19b**) with alcohols is also possible (Equation (49)) ⟨79COC(3)317⟩, as is alkylation of the sulfinate (Equation (50)). The latter method seems excellent for the formation of arenesulfinate esters of secondary alcohols ⟨88T6687⟩. Some optically active highly functionalised sulfinates have been prepared from optically active arenesulfinamides (**19c**) ⟨83S1040⟩. Unfortunately, it was necessary to prepare the sulfinamides from sulfinates.

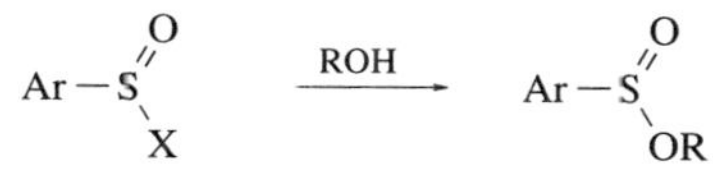

(49)

(19) **(a)**X = Cl
(b)X = OH
(c)X = NR_2

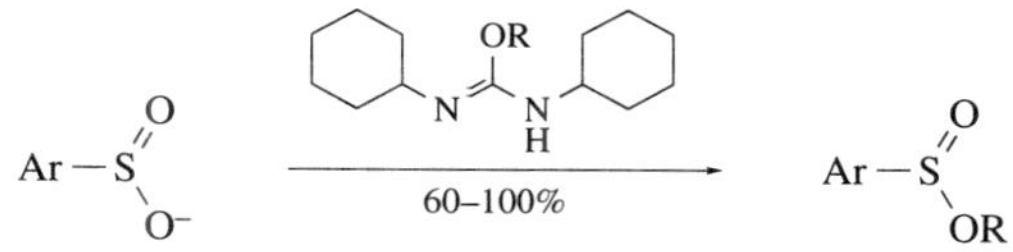

(50)

R = Me, Et, $PhCH_2$, Pr^i, Bu^s, 2-hexyl

Unfortunately, sulfinyl chlorides and sulfinic acids are not ideal starting materials as very few are readily available. For this reason, many groups have searched for approaches to sulfinate esters from more accessible sulfur compounds. The *N*-bromosuccinimide oxidation of diaryl disulfides in the presence of an alcohol, which can be menthol, is a good example of such an approach (Equation (51)) ⟨88S252⟩.

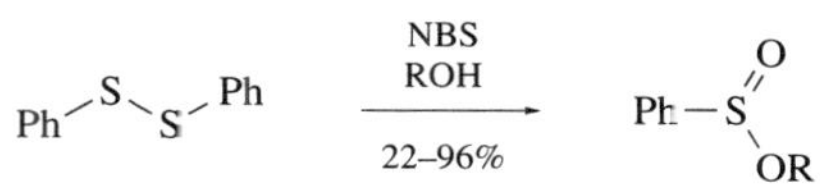

(51)

R = Me, Pr^i, cyclohexyl, menthyl

The current method of choice for the synthesis of the menthyl esters of a range of arenesulfinic acids uses readily available arenesulfonyl chlorides as precursors. Trimethyl phosphite reduction in the presence of menthol proceeds smoothly to give the target sulfinates (Equation (52)) ⟨87JOC2598⟩. An alternative approach from a sulfur(VI) compound is the reaction of tosyl cyanide with a variety of chiral alcohols ⟨91T9167⟩.

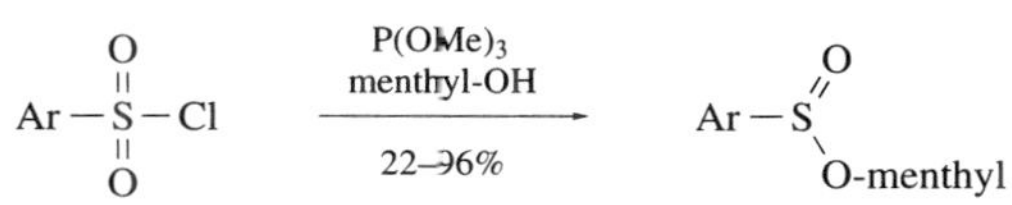

(52)

Ar = *p*-MeC_6H_4-, *p*-$MeOC_6H_4$-, 2,4,6-$Me_3C_6H_2$-, 2-naphthyl, 2-thiophenyl

2.14.2.5.3 *Vinyl- and arenesulfonates*

Vinylsulfonate esters are interesting dienophiles. Esters of the parent vinylsulfonic acid are best prepared by the reaction of vinylsulfonyl chloride with alcohols (Equation (53)) ⟨89AG(E)202⟩. However, the most general route makes use of Wittig–Horner-type reagents (**20**) ⟨87T5125⟩. This publication also includes citations of other, less generally applicable, methods.

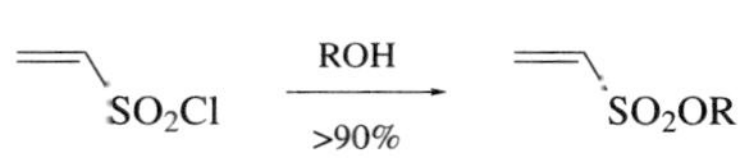

(53)

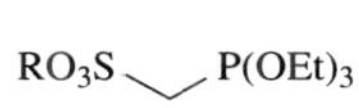

(20)
R = Et, Pr^i

The activation of alcohols as tosylates, or other arenesulfonates (**21**), is of immense importance to synthetic organic chemistry. A number of standard procedures for their preparation exist. These are essentially the same as those used for alkanesulfonates (see Chapter 2.03), are listed in Section 0–115 of March ⟨B-92MI 214-01⟩, and require no further comment here.

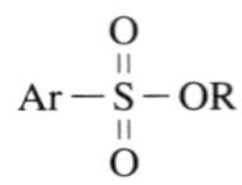

(21)

2.14.2.6 RSOX Functions

The reader is referred to Chapter 2.03.

2.14.2.7 RS(S)$_n$H, RS(S)$_n$R and RS(S)$_n$X Functions and Higher-coordinated Derivatives

2.14.2.7.1 Arylsulfanes (ArS(S)$_n$H)

Two methods of preparing arylsulfanes are available. The most general procedure involves conversion of an arenethiol via a sulfenyl chloride intermediate (Scheme 18) ⟨73CJC3403⟩. In this way, arylsulfanes with $n = 1$–3 could be characterised. Arylsulfanes where $n = 2$ can also be prepared from arenethiols via a hydrolysable disulfide ⟨71JOC3677⟩.

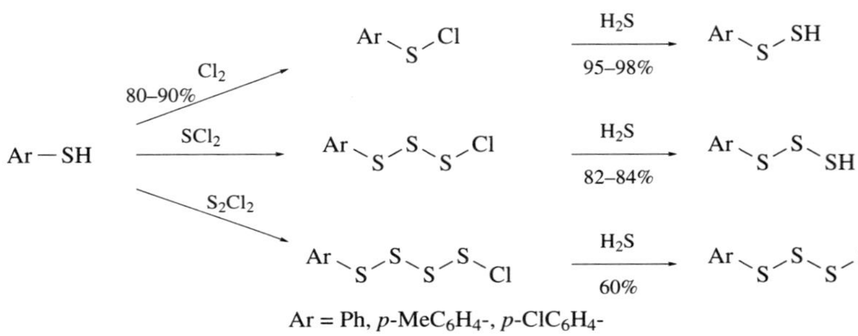

Scheme 18

2.14.2.7.2 Vinyl and aryl disulfides and aryl polysulfides (RS(S)$_n$R) and higher-coordinated derivatives

(i) Vinyl and aryl disulfides

Vinyl disulfides are best prepared by one of two methods which were published in 1969 ⟨69RTC519⟩. Both methods involve the reaction of vinylthiolates, generated *in situ*, with RSX compounds, where X is CN or SO_2R. The vinylthiolates were prepared either from vinyl Grignard reagents with elemental sulfur or from vinyl sulfides by reaction with lithium or sodium. Both routes appear synthetically useful (Scheme 19). The synthesis of divinyl disulfides has been reviewed ⟨92SR1⟩.

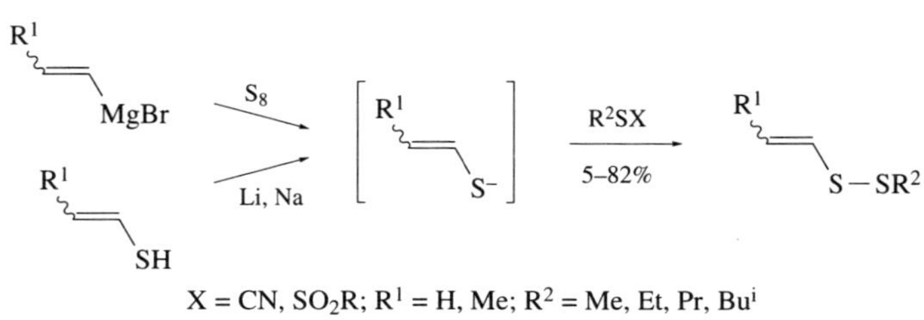

Scheme 19

Aryl disulfides, trisulfides and tetrasulfides are all known. In all three cases the greatest synthetic challenge is to produce unsymmetrical sulfides.

Symmetrical diaryl disulfides are well known. They are readily prepared by oxidation of thiols by the standard methods described in Chapter 2.03 and in a review (Equation (54)) ⟨91COS(7)757⟩. Reductive approaches to diaryl disulfides are also attractive due to the ready availability of arenesulfonyl chlorides (Equation (55)). As well as listing the many reagents which have been used for this transformation, a 1984 paper describes a useful alternative reagent, namely P_2I_4 ⟨84CL139⟩. Unsymmetrical aryl disulfides can also be prepared by most of the standard methods. A listing of the best of these methods is available ⟨94TL5381⟩.

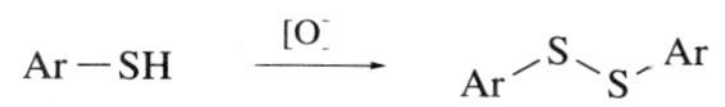

(54)

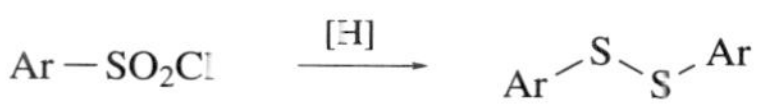

(55)

(ii) Aryl polysulfides

Symmetrical diaryl trisulfides are simple to prepare. Two methods from Harpp and co-workers appear useful ⟨76TL3001, 94TL5381⟩. The second of these (Scheme 20) is, in principle, applicable to the synthesis of unsymmetrical aryl trisulfides, but no examples are given. Otherwise, the reaction of aryl disulfanes with sulfenyl chlorides should be considered (Equation (56)) ⟨61JOC2483⟩.

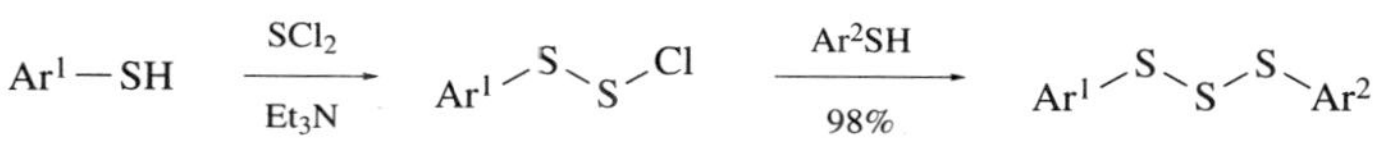

$Ar^1 = Ar^2 = p\text{-}Bu^tC_6H_4$

Scheme 20

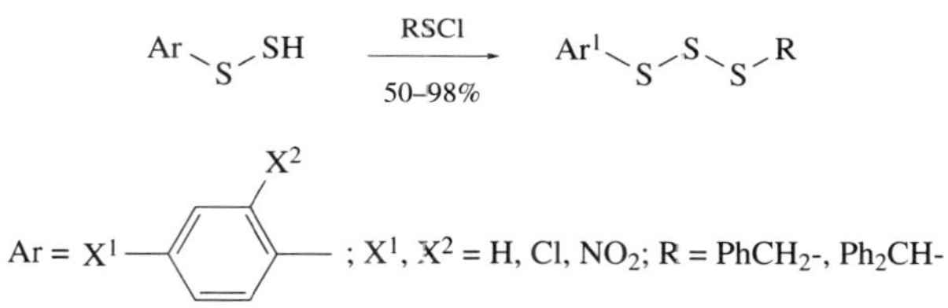

(56)

Symmetrical diaryl tetrasulfides are available from the reaction of arenethiols with disulfur dichloride ⟨1888JPR(2)207⟩. There appear to be no examples of unsymmetrical aryl tetrasulfides.

(iii) Higher-coordinated derivatives of aryl di- and polysulfides

Compounds which are formally the mono-, di-, tri- and tetraoxides (**22**)–(**25**) of aryl disulfides are all known, although the *S,S'*-dioxide isomers (**26**) of (**23**) remain elusive ⟨84CRV117⟩. They will be discussed in the listed order.

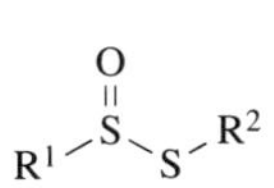

(22)

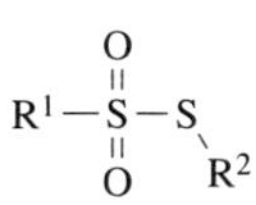

(23)

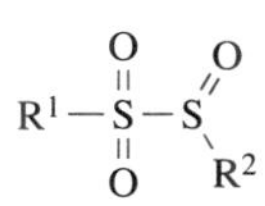

(24)

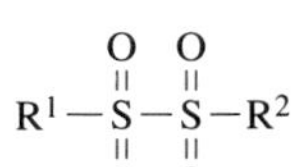

(25)

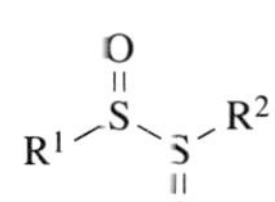

(26)

R^1 and/or R^2 = aryl

Arenethiosulfinate esters (**22**) are reasonably well known. The standard ester disconnection is valid for thiosulfinate esters, and, indeed, one of the most common methods for their preparation involves the reaction of thiols with arenesulfinyl chlorides (**27**) using pyridine as a base (Equation (57)) ⟨54RTC129⟩. An adaptation of this procedure, using stannylated thiols, is to be recommended if the older method fails ⟨83TL5173⟩. Other synthetic equivalents of $RS^+{=}O$ have been described.

Sulfinic acids can be activated to thiol attack ⟨80S937⟩, as can sulfinamides ⟨85TL5703⟩. The latter method permits the isolation of optically enriched thiosulfinates, as one sulfinamide is available in optically pure form. However, the reaction is not stereospecific, and some loss of optical purity is observed. Arenesulfinyl azides have also been converted to their thiosulfinates ⟨84JOC1931⟩.

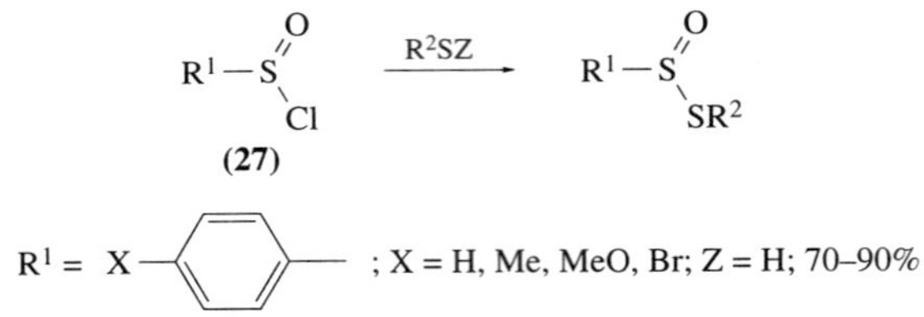

(57)

R^1, R^2 = Me, Bu^t, PhCH$_2$-, Ph; Z = SnR$_4$; 90-95%

An alternative disconnection exists with thiosulfinates that is not available for the carbon analogues, that is, carboxylic esters, namely functional group transformation of sulfoxide to sulfide. In fact, the oxidation of aryl disulfides has been extensively used to prepare arenethiosulfinate esters (Equation (58)) ⟨73IJS307⟩, but should be attempted with caution. Overoxidation is a significant problem and, for unsymmetrical examples, both sulfur centres are likely to be oxidised, producing regioisomers ⟨79COC(3)261⟩.

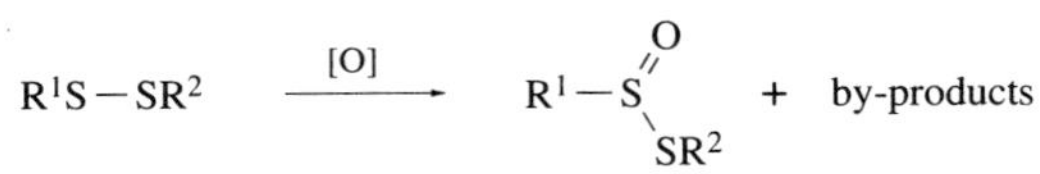

(58)

Esters of arenethiosulfonates (**23**) (R^1 = Ar), which have been proposed as useful sulfenylating agents ⟨77JA4405⟩, and aryl esters of thiosulfonates (**23**) (R^2 = Ar) will be discussed together. The traditional method of producing thiosulfonates is the oxidation of disulfides ⟨53BSB3⟩, but this procedure is usually only suitable for the preparation of symmetrical examples (Equation (59)). Otherwise, the sodium periodate oxidation of thiosulfinate esters (**22**) seems attractive (Equation (60)) ⟨81BCJ1443⟩.

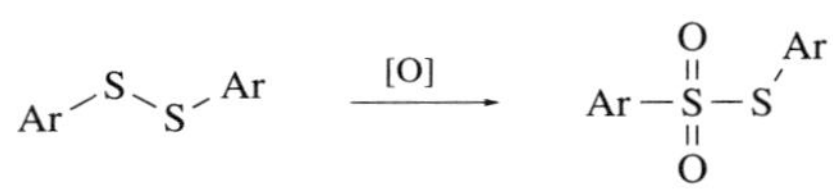

(59)

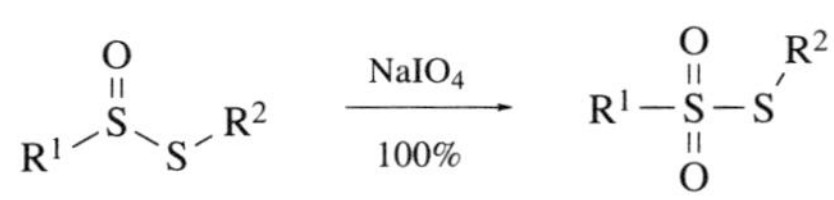

(60)

(**22**) (**23**)

R^1, R^2 = Me, Ph, *p*-tolyl

An entirely different approach, which is recommended for thiosulfonates with two identical aryl substituents, is the reaction of arenesulfonyl chlorides with potassium iodide (Equation (61)) ⟨81S888⟩.

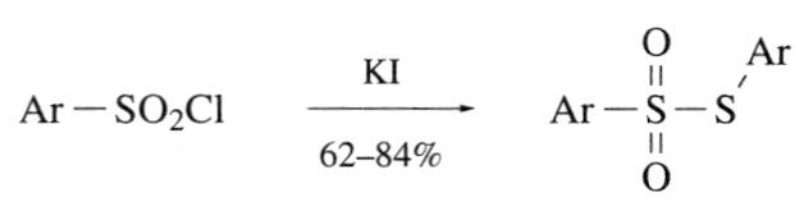

(61)

Ar = Ph, *p*-MeOC$_6$H$_4$-, *p*-MeC$_6$H$_4$-, *p*-ClC$_6$H$_4$-, 1,3,5-Me$_3$C$_6$H$_2$-

Arenesulfinyl sulfones (**24**) are still usually prepared by treating a sulfinate salt with a sulfonyl chloride ⟨60CB2736⟩.

Arenesulfonyl sulfonates (**25**) are reasonably well-known compounds. The various synthetic approaches have been summarised in a very helpful publication which also describes an elegant and simple new route to the target compounds ⟨93S490⟩. Previously, symmetrical arenesulfonyl arenesulfonates were prepared either by reduction of arenesulfonyl chlorides or oxidation of arenesulfinic acids (Scheme 21). Unsymmetrical examples were best approached from disulfides or sulfinyl sulfones.

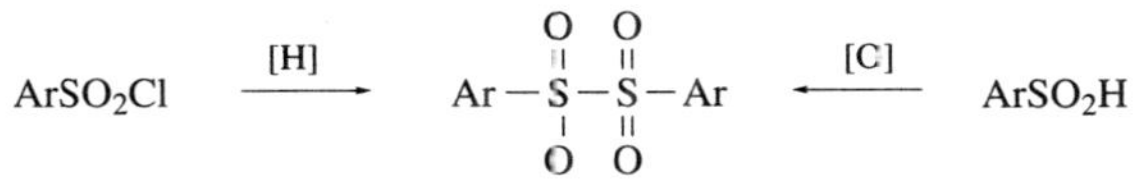

(25)

Scheme 21

The newer method permits the synthesis of a range of symmetrical and unsymmetrical sulfonylsulfones by the elimination of nitrogen from intermediates such as (**28**) derived from hydrazine and the appropriate sulfonyl chloride(s) (Scheme 22) ⟨93S490⟩.

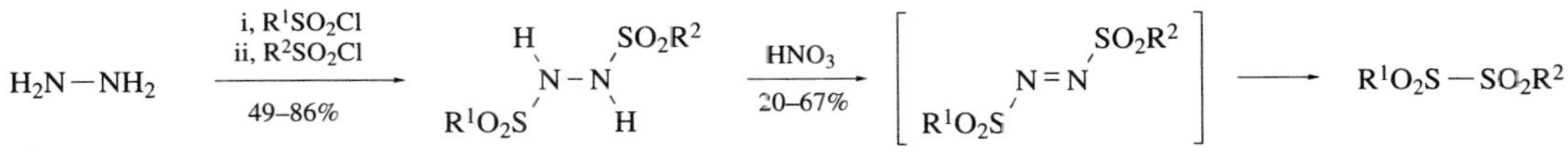

(28)

R^1, R^2 = alkyl, *p*-XC_6H_4- (X = H, alkyl, MeO, R_2N), naphthyl

Scheme 22

Dioxides (**29**) and tetroxides (**30**) of trisulfides have been prepared ⟨79JOC4135, 81JOC4894⟩. Further oxidation resulted in S—S bond cleavage ⟨81JOC4894⟩. The following paper should be consulted for both a new and older method of preparing a 'tetrasulfide tetroxide' (**31**) ⟨82CL2007⟩.

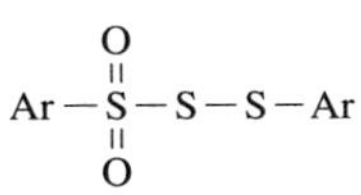

(29)

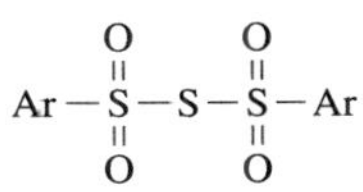

(30)

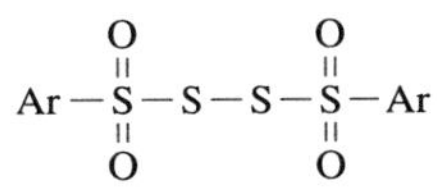

(31)

2.14.2.7.3 Arylchlorosulfanes (ArS(S)$_n$Cl)

Arylchlorosulfanes with $n = 1$ or 2 were prepared as precursors to arylsulfanes ⟨73CJC3403⟩ and were described in Section 2.14.2.7.1 (see Scheme 18).

2.14.2.8 Vinyl- and Arenesulfenamides, -sulfinamides, -sulfonamides and Derivatives

2.14.2.8.1 Vinyl- and arenesulfenamides and derivatives

The reader is referred to Chapters 2.03.9 and 2.15.5 for general references on sulfenamides. Isocyanate (**32**) and *N*-sulfinyl (**33**) derivatives are known ⟨70JAP7321096-R, 70JOM(24)629⟩.

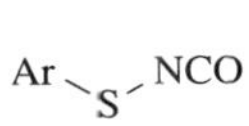

(32)

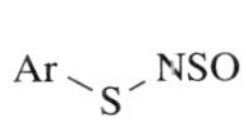

(33)

2.14.2.8.2 Vinyl- and arenesulfinamides and derivatives

The reader is referred to Chapters 2.03.9 and 2.15.5 for general references on sulfinamides. Arenesulfinyl isocyanates (**34**) can be prepared from the corresponding sulfinyl chloride with silver isocyanate ⟨69ZC305, 93JOC3469⟩.

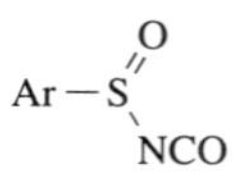

(34)

2.14.2.8.3 Vinyl- and arenesulfonamides and derivatives

Vinylsulfonamide (**35**) should be prepared by a definitive procedure ⟨81BSF(2)71⟩. Arenesulfonamides (**36**) are readily available from the standard reaction of sulfonyl chlorides with ammonia (Scheme 23) ⟨79COC(3)331⟩. Indeed, certain arenesulfonamides are commercially available (Ar = tolyl, for example), and will be discussed no further here.

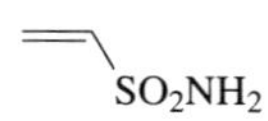

(35)

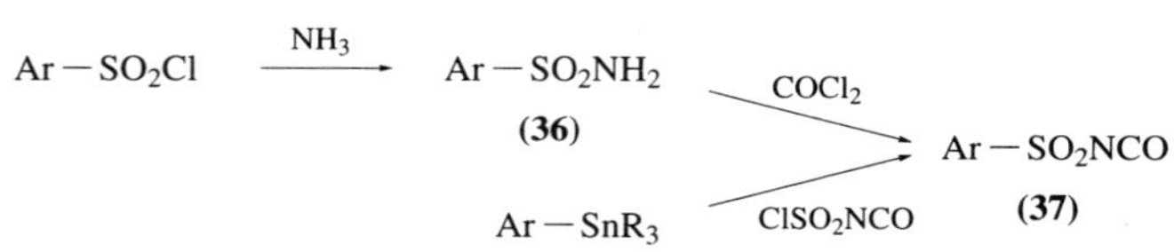

Ar = Ph, 2-, 3- or 4-alkylphenyl, 2-, 3- or 4-halophenyl, *p*-$MeOC_6H_4$; R = Me, Bu; >98%

Scheme 23

Arenesulfonyl isocyanates (**37**) were traditionally prepared by the reaction of the sulfonamide with either phosgene or oxalyl chloride (Scheme 23) ⟨64JOC2592, 66JOC2658⟩. An interesting alternative is the reaction of stannylated arenes with chlorosulfonyl isocyanate (Scheme 23). This rapid synthesis generates the targets *in situ*. Trapping to generate other sulfonamide derivatives was possible ⟨91CB1997⟩. Arenesulfonyl isothiocyanates (**38**) are also known ⟨65AG(E)430⟩. Using thionyl chloride in the place of phosgene yields the sulfinyl analogues (**39**) ⟨85TL3821⟩.

Ar — SO_2NCS **(38)**   Ar — SO_2NSO **(39)**

(Arenesulfonylimino)iodoarenes (**40**) are important as nitrenoid precursors ⟨93TL6105⟩. Arenesulfonylhydrazines ($ArSO_2NHNH_2$) and arenesulfonyl azides ($ArSO_2N_3$) and arenesulfonyl nitroso compounds ($ArSO_2NO$) are also known ⟨60OS(40)93, 61CB1067, 79TL3307⟩.

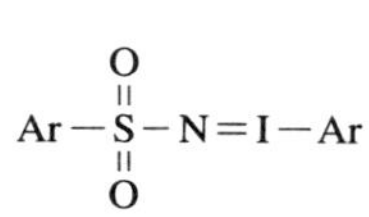

(40)

2.14.3 SELENIUM- AND TELLURIUM-BASED FUNCTIONAL GROUPS ATTACHED TO AN *sp*^{2} CARBON

2.14.3.1 Areneselenols and -tellurols

Areneselenols (**41**) are well-known compounds, and their oxidation to the corresponding diselenides (**42**) is easy (air is sufficient), making their storage difficult ⟨79COC(3)491⟩. In fact, by far the most practical way of generating an areneselenol is from its diselenide by one of the standard methods (Equation (62)) (see Chapter 2.04). Other potential starting materials are aryl halides, which lead to selenols either through direct nucleophilic substitution with a selenide dianion or indirectly via the reaction of the corresponding Grignard reagent with elemental selenium (Scheme 24), or arylamines via diazonium and selenocyanate intermediates (Equation (63)) ⟨79COC(3)491⟩.

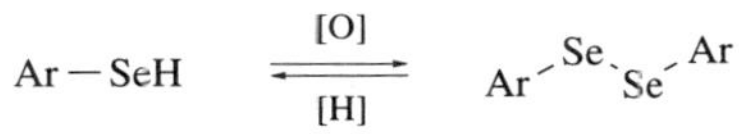
(41) (42) (62)

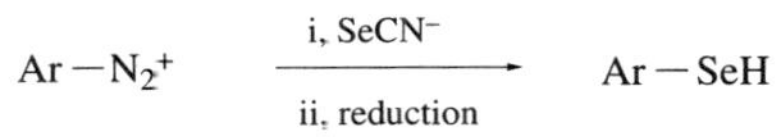
(63)

Ar = 8-quinolyl; 12% from $ArNH_2$

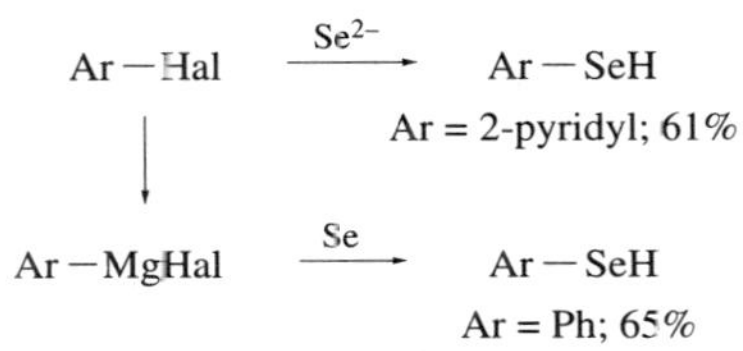

Scheme 24

Arenetellurols are even more prone to oxidation than their selenium analogues. In fact, they are probably best prepared *in situ* from the reaction of the corresponding ditelluride with a Grignard reagent (Scheme 25) ⟨63CB247⟩.

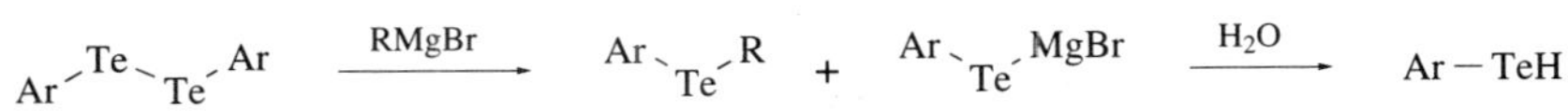

Ar = Ph, *p*-ROC_6H_4-, 1-naphthyl; quantitative yield

Scheme 25

2.14.3.2 Vinyl and Aryl Selenides and Tellurides and their Higher-coordinated Derivatives

2.14.3.2.1 Vinyl selenides

Four general approaches to the disconnection of vinyl selenides can be identified: (i) disconnection of the carbon–selenium bond; (ii) disconnection of the carbon–carbon double bond; (iii) functional group interconversion by addition of, for example, HX; (iv) functional group interconversion to alkynes. Available synthetic routes to vinyl selenides are thus classified below according to disconnection type.

(i) Vinyl selenides from vinyl halides

The most conceptually simple route to vinyl selenides is the direct reaction of vinyl halides with nucleophilic selenium species (Equation (64)). Suitable sources of nucleophilic selenium are selenates, in the presence of a nickel(II) catalyst, ⟨86JOC875⟩ or diselenides ⟨89CL769⟩. The opposite disconnection requires a vinyl nucleophile and a source of electrophilic selenium. Vinyl Grignard reagents, vinyllithium and vinylmercury reagents (Equation (65)) all react with selenenyl halides to fulfil these criteria ⟨81JOC2775⟩.

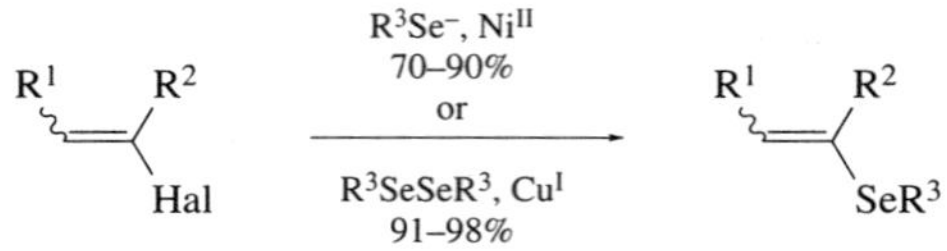

R^1, R^2 = H, alkyl, Ph (64)

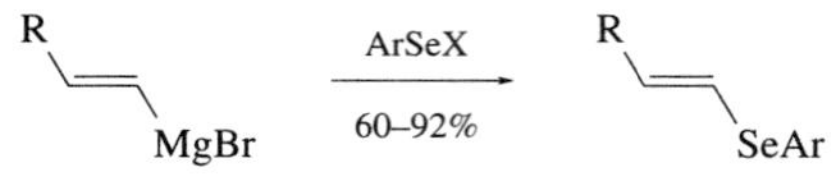

R = H, Me; Ar = Ph, *m*-$CF_3C_6H_4$-; X = Hal, SeAr (65)

(ii) *Vinyl selenides by Wittig- and Peterson-type reactions*

Conventional Wittig reagents react with aldehydes to produce vinyl selenides. However, Wittig–Horner reagents are preferred if high *trans* selectivity is desired (Equation (66)) ⟨78T1049⟩. Peterson-type organosilicon chemistry can also be used to access vinyl selenides ⟨79TL485⟩. A synthetically equivalent, and very useful, procedure is the condensation of a bis(arylseleno)methane anion with a carbonyl compound with subsequent elimination to yield the vinyl selenide (Equation (67)) ⟨81JOC2775⟩.

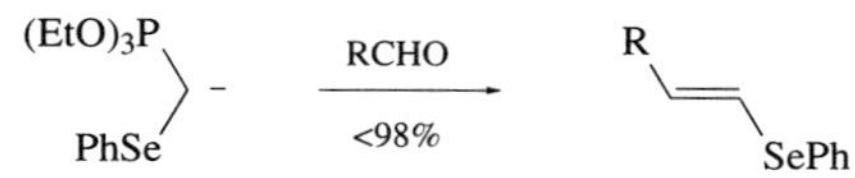

R = alkyl, aryl (66)

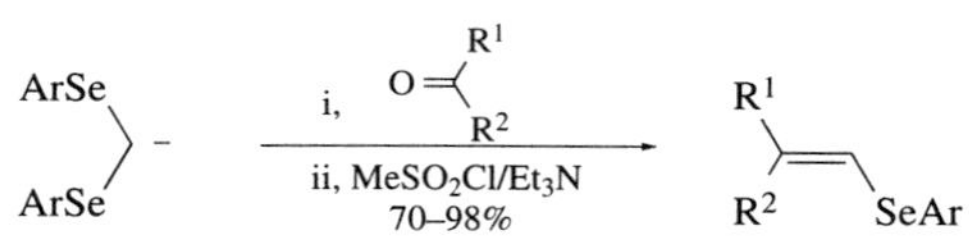

R^1, R^2 = H, Me; Ar = Ph, *m*-$CF_3C_6H_4$- (67)

(iii) *Vinyl selenides by elimination*

Selenides with a halide, or equivalent, in either the α or β position can be converted to vinyl selenides by reaction with a base ⟨81JOC2775⟩. A particularly useful variant of this procedure is the elimination of selenols from diselenoacetals, which are readily available from carbonyl compounds (Scheme 26) ⟨77TL3835⟩.

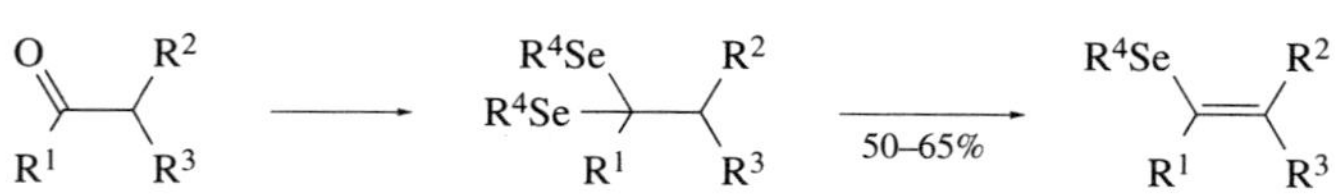

R^1, R^2 = H, *n*-alkyl, cycloalkyl; R^3 = H, Me; R^4 = Me, Ph

Scheme 26

(iv) *Vinyl selenides from alkynes*

Addition of selenols to terminal alkynes yields predominantly the *cis*-vinyl selenide (Equation (68)), whereas the lithium aluminum hydride reduction of selenoalkynes leads to the *trans* isomer (Equation (69)) ⟨76JOM(114)281⟩.

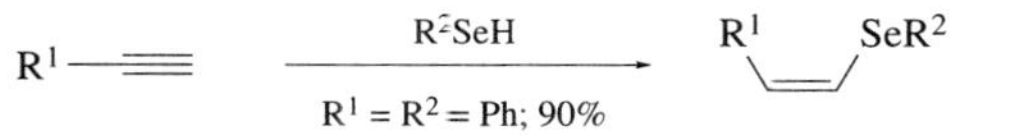

(68)

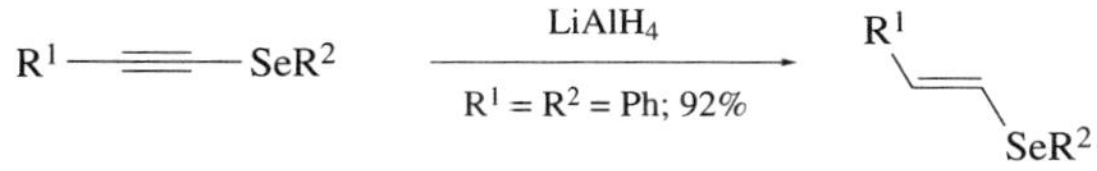

(69)

2.14.3.2.2 *Aryl selenides*

Diaryl selenides can be prepared from a number of starting materials. Arenes themselves can be functionalised by reaction with selenium tetrachloride or selenium oxychloride, followed by the zinc reduction of the intermediate dichloride (**43**) (Scheme 27) ⟨79COC(3)491⟩.

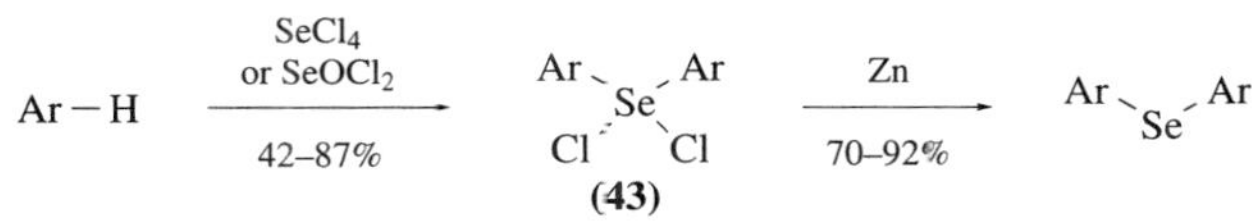

$Ar = p\text{-}ROC_6H_4\text{-}$; R = Me, Et, Pr, Bu, Ph

Scheme 27

Aryl halides are also useful precursors to diaryl selenides. If sufficiently activated to aromatic nucleophilic substitution they react directly with a selenide dianion, and, if converted to the corresponding arylmercury reagents, elemental selenium can be used as the electrophile (Scheme 28) ⟨79COC(3)491⟩. Metallated arenes also react with diaryl diselenides or aryl selenocyanates to yield unsymmetrical diaryl diselenides (Equation (70)) ⟨45JA1965, 56JA4028⟩.

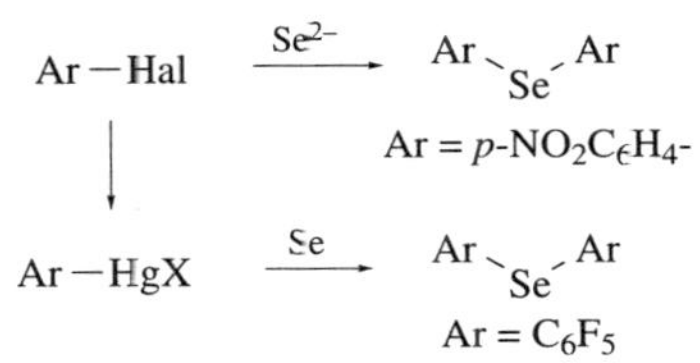

Scheme 28

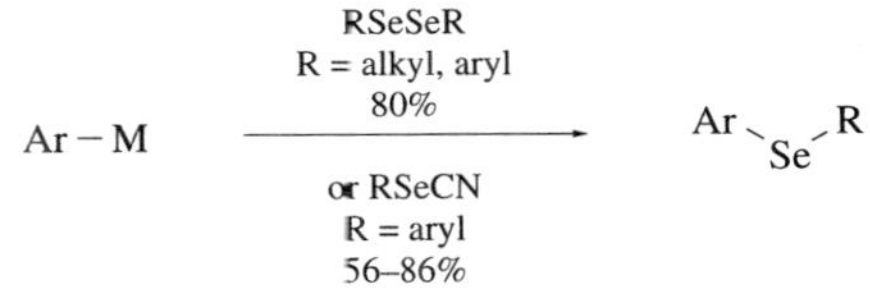

(70)

However, arylamines provide the most practical precursors to diaryl selenides. After diazotisation, reaction with a selenide dianion yields symmetrical diaryl selenides (Equation (71)) ⟨43OSC(2)238⟩.

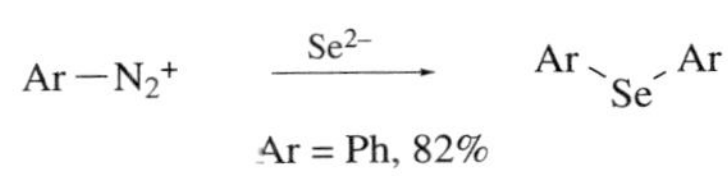

(71)

Alkyl aryl selenides are accessible from standard Williamson methodology (see Chapter 2.04). A particularly useful variant, which exploits readily available aryl halide starting materials, involves lithiation of the aryl halide followed by *in situ* reaction with selenium and alkylation of the so formed selenide (Scheme 29) ⟨85JOM(296)357, 91CB1315⟩. A very useful method for synthesis of aryl methyl selenides is the reaction of aryl halides, which need not be activated to nucleophilic substitution, with lithium methaneselenate (Equation (72)) ⟨83JOC4289⟩.

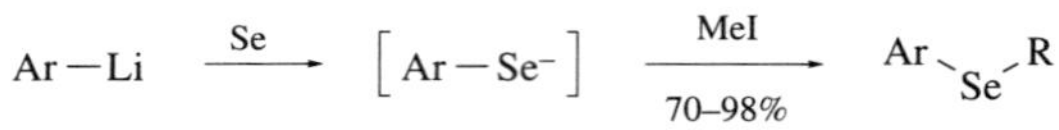

Ar = Ph, naphthyl, anthracenyl, 2-pyridyl

Scheme 29

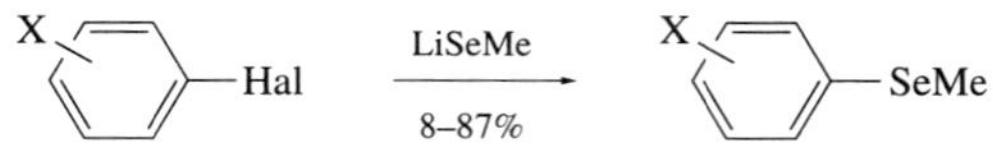

(72)

X = alkyl, aryl, Hal, RS, RSe

Finally, alkyl phenyl selenides are also available from the addition of benzeneselenyl bromide, or an equivalent, to alkenes ⟨76TL453⟩.

2.14.3.2.3 *Vinyl tellurides*

Vinyl tellurides are most easily prepared from alkynes. The addition of tellurols generated *in situ* from ditellurides proceeds with high regioselectivity to yield predominantly the (*Z*) isomer (Equation (73)) ⟨89OM1661⟩. A most interesting development in this area was the discovery that the carbon–tellurium bond will also add to alkynes ⟨92JA7591⟩. This reaction gives rise to highly substituted vinyl tellurides (Equation (74)). Unfortunately, mixtures of isomers were observed.

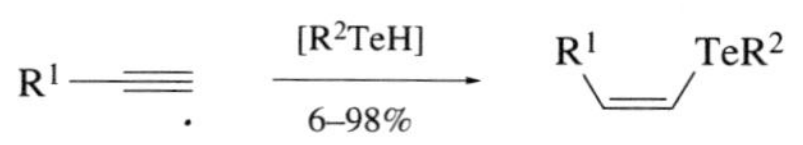

(73)

R^1, R^2 = alkyl, aryl, vinyl

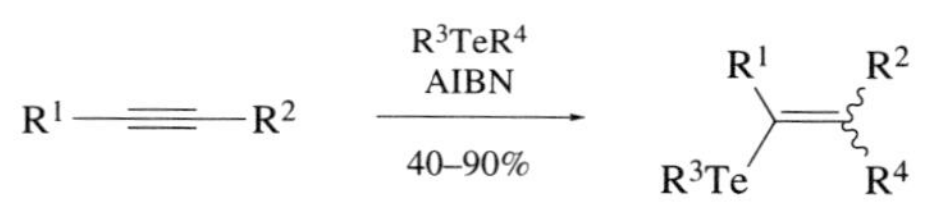

(74)

R^1, R^2 = alkyl, Ph; R^3, R^4 = H, alkyl, TMS, CO_2R, COR, Ph
AIBN = 2,2'-azobisisobutyronitrile

It has been shown that Wittig–Horner-type reagents can also be used to prepare vinyl tellurides (Equation (75)) ⟨94JCS(P1)717⟩. This method is complementary to the alkyne addition route in that it yields exclusively (*E*) products.

(75)

R = aryl, styryl

2.14.3.2.4 *Aryl tellurides*

Symmetrical diaryl tellurides can be prepared by a number of methods ⟨79COC(3)491⟩. Aryl halides are perhaps the most useful starting materials. Metallation and subsequent reaction with elemental tellurium or direct reaction with sodium telluride can be used to generate aryl tellurates *in situ*. Reaction with a second equivalent of aryl halide leads to the telluride (Scheme 30) ⟨92S281, 92S549⟩. Arylamines are also precursors to diaryl tellurides via diazotisation and reaction with sodium telluride or equivalent reagents (Equation (76)) ⟨92S281⟩. Diaryltellurium dichlorides (**44**) can be reduced to ditellurides.

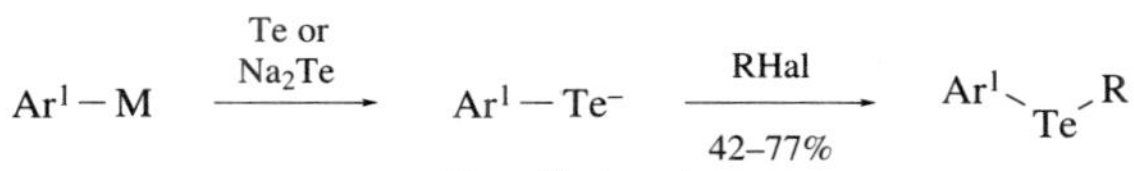

R = alkyl, aryl

Scheme 30

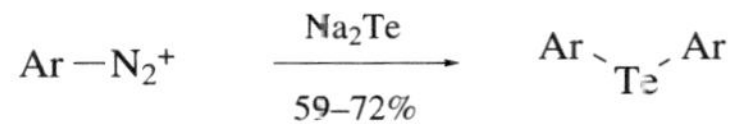 (76)

Ar = Ph, *p*-HalC$_6$H$_4$-, *p*-MeC$_6$H$_4$-, *p*-ROC$_6$H$_4$-

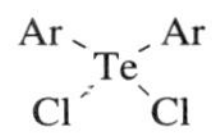

(44)

Unsymmetrical diaryl tellurides are probably most easily accessed by the reaction of diazonium salts with aryl tellurates, but the copper-catalysed reaction of aryl tellurates with aryl halides provides an alternative (Equation (77)) ⟨81CL1117⟩.

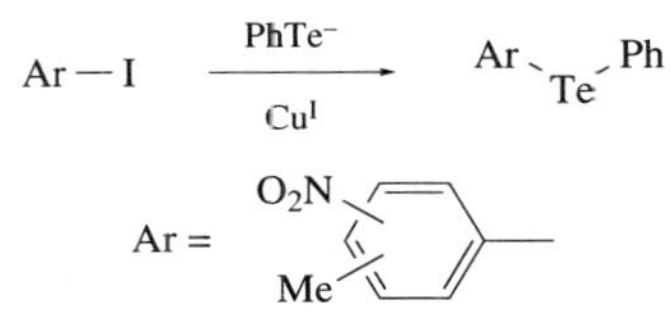 (77)

Alkyl aryl tellurides can be prepared by a standard Williamson-type reaction ⟨65ZOR2115⟩. Once again, *in situ* generation of the telluride, followed by trapping with an alkyl halide, is an attractive equivalent procedure (Scheme 30) ⟨92S549⟩. A useful alternative is the reaction of arenediazonium salts with ditellurides in the presence of potassium acetate (Equation (78)) ⟨82TL3905⟩.

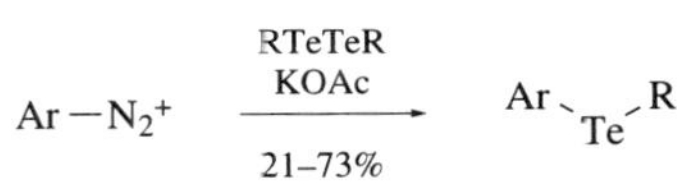 (78)

R = Me, Et; Ar = HalC$_6$H$_4$-, MeC$_6$H$_4$-

2.14.3.2.5 Selenoxides, telluroxides and related compounds

Selenoxides (**45**) are readily available via oxidation of selenides. There are a number of standard methods (see Chapter 2.04 and ⟨79COC(3)491⟩). Selenoxides with β hydrogen atoms are not usually isolable, as elimination to an alkene occurs spontaneously at room temperature. Indeed, this is a popular method of alkene synthesis. However, neither aryl nor vinyl substituents bear the necessary β hydrogen, and the compounds covered by this section are thus prominent among isolable selenoxides.

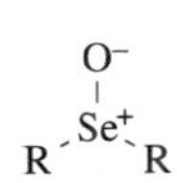

(45)

Aryl telluroxides (**46**) can be prepared from aryltellurium dihalides (**44**) by hydrolysis. The preference of tellurium(IV) compounds to be tetravalent is illustrated by the ready formation of the hydrate (**47**). Indeed, the tetravalent diaryltellurium dihalides (**44**) are much more important as synthetic intermediates than the telluroxides (**46**). Their preparation has been reviewed ⟨79RCR343⟩.

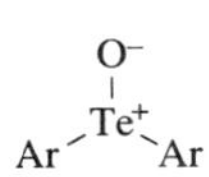

(46)

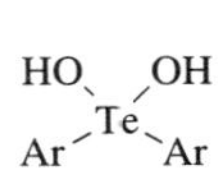

(47)

2.14.3.2.6 Selenones and tellurones

Selenones (**48**) are little-known compounds, but can be prepared by the oxidation of selenides (see Chapter 2.04). In particular, aryl-substituted selenones can be obtained using trifluoroperacetic acid as the oxidising agent ⟨68ZOB2509⟩. No tellurones have been isolated.

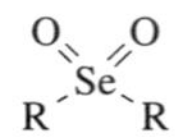

(48)

2.14.3.3 Vinyl- and Areneselenenyl and -telluryl Halides and their Higher-coordinated Analogues

2.14.3.3.1 Areneselenenyl and -tellurenyl halides

Areneselenenyl chlorides and bromides are readily available by the halogenolysis of diselenides or selenocyanates (Equation (79)) (see Chapter 2.04). Aryl alkyl selenides can also be used as starting materials (Equation (79)). Reaction with thionyl chloride leads to areneselenenyl chlorides ⟨28JCS2293⟩, and reaction with bromine to the corresponding bromides ⟨64BSB507⟩. However, this method has not found much application. Areneselenenyl iodides can be isolated from the reaction of sterically demanding diaryl diselenides with iodine (Equation (79)) ⟨87AG(E)780⟩.

$$\text{ArSeX} \xrightarrow[\text{SOCl}_2]{\text{Hal}_2 \text{ or}} \text{ArSeHal} \qquad (79)$$

X = SeAr, CN, alkyl

Arenetellurenyl chlorides and bromides have only been isolated as their thiourea adducts (**49**) ⟨66ACS132⟩. However, 2-naphthalenetellurenyl iodide can be isolated from the reaction of the appropriate ditelluride with iodine (Equation (80)) ⟨59CB40⟩. Characteristically, although arenetellurenyl halides are rare, aryltellurium trichlorides (**50**) are well known ⟨79RCR343⟩. They are easily prepared from deactivated arenes and tellurium trichloride, and are important starting materials for organotellurium chemistry.

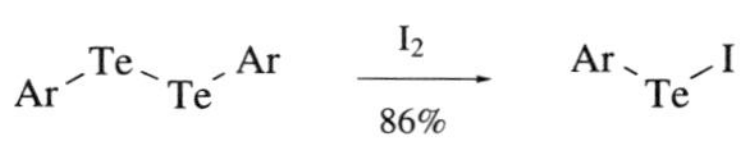

(80)

Ar = 2-naphthyl

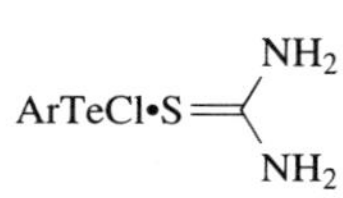

(49)

$ArTeCl_3$

(50)

2.14.3.3.2 Vinyl- and areneseleninyl and -tellurinyl halides

Vinylseleninyl halides are known. For example, the addition of selenium oxychloride to acetylene gives a high yield of vinyl derivative (**51**) ⟨71IJS(A)18⟩.

Areneseleninyl halides (**52**) can be prepared by standard methods (see Chapter 2.04), but an

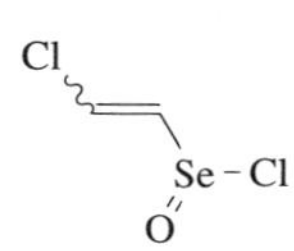

(51)

alternative route is available for areneseleninyl chlorides which may prove more useful, namely the reaction of the corresponding selenenyl chlorides with ozone (Equation (81)).

$$\text{ArSeHal} \xrightarrow{O_3} \textbf{(52)} \qquad (81)$$

Hal = Cl, Br

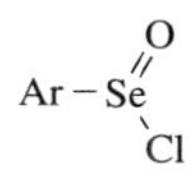

(52)

Arenetellurinic acids (**53**) can be prepared from the aryltellurium trihalides (**50**) by hydrolysis and by other standard methods (see Chapter 2.04).

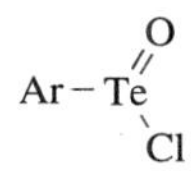

(53)

2.14.3.4 Functions Based on RSeOH Units and their Higher-coordinated Derivatives

2.14.3.4.1 Areneselenenic acids

Seleninic acids are not, in general, isolable compounds and are thus excluded from this text. Among the few stable examples is 2-nitrobenzeneseleninic acid (**54**), which is prepared by the hydrolysis of the corresponding seleninyl bromide ⟨35CB1540⟩. As with sulfenic acids, it appears that intramolecular hydrogen bonding is important to the anomalous stability of this compound.

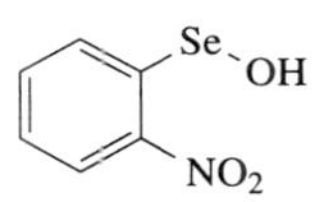

(54)

2.14.3.4.2 Areneseleninic acids

As well as the general methods for seleninic acid synthesis, of which oxidation of diselenides is perhaps the most useful (see Chapter 2.04 and ⟨91COS(7)757⟩), there are other procedures for particular cases. Hence, benzeneselenenyl bromide can be hydrolysed to yield benzeneseleninic acid via a selenenic acid intermediate (Scheme 31) ⟨32CB812⟩, and in favourable cases aryl selenides can be oxidised to the targets (Equation (82)) ⟨27JCS27⟩.

Scheme 31

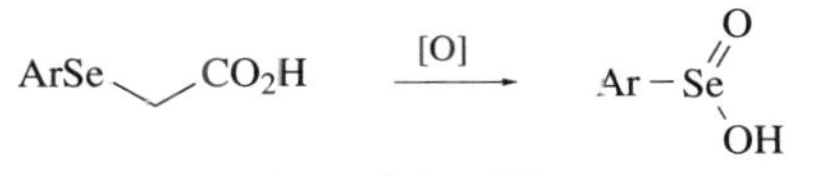

(82)

Ar = tolyl, xylyl

2.14.3.4.3 *Areneselenonic acids*

Selenonic acids (**55**) are extremely strong oxidising agents ⟨79COC(3)491⟩. A salt which approximates to benzeneselenonic acid can be prepared by the electrophilic aromatic substitution of benzene by selenium trioxide. The fluorene analogue can similarly be synthesised with H_2SeO_4 as the electrophile.

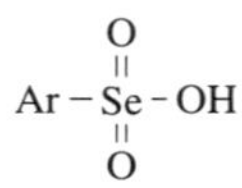

(55)

2.14.3.5 Functions Based on the R^1SeOR^2 Unit

Areneselenenic esters are, as expected, available by the reaction of alcohols with synthetic equivalents of $ArSe^+$ (Equation (83)). Areneselenenyl halides are suitable substrates, as are aryl selenocyanates ⟨79COC(3)491⟩, the reaction of the latter requiring silver acetate as a catalyst. The obvious condensation of alcohols with seleninic acids is restricted by the limited availability of the acids. However, the unstable acids can be trapped by esterification with alcohols (Equation (84)) ⟨86JOC295⟩.

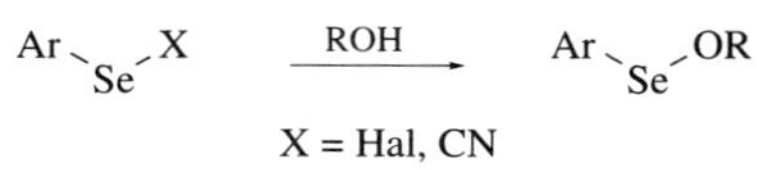

(83)

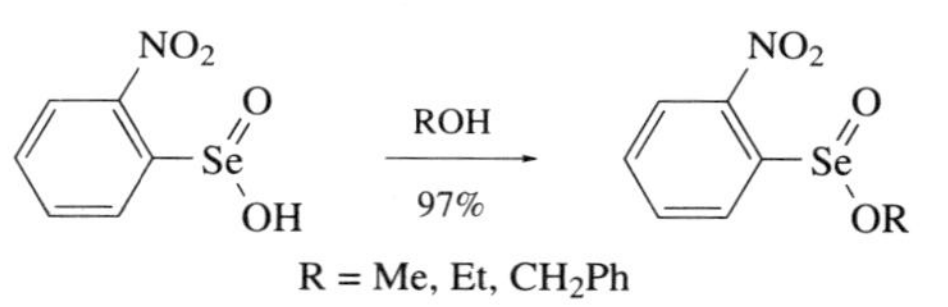

(84)

2.14.3.6 Aryl Di- and Triselenides and Aryl Ditellurides

2.14.3.6.1 *Aryl di- and triselenides*

Aryl diselenides are important starting materials for other arylselenium chemistry. Symmetrical derivatives are readily prepared by the oxidation of aryl selenols with a number of oxidising agents, including atmospheric oxygen (Equation (85)) ⟨91COS(7)757⟩. However, as aryl selenols are most conveniently prepared from the diselenides themselves, this method is not likely to be of preparative value.

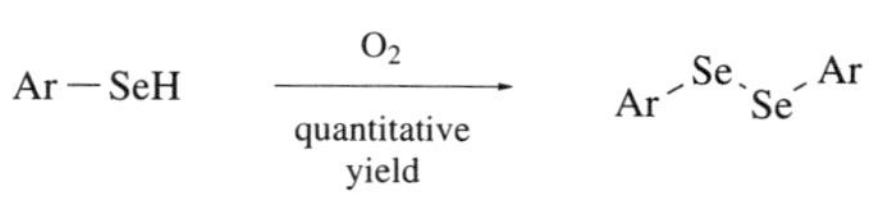

(85)

Electron-deficient aryl halides can be used as substrates, reaction with the diselenide dianion leading to the targets (Equation (86)) ⟨79COC(3)491⟩. Arylamines are also potential precursors of diaryl diselenides through the reaction of diazonium salts with sodium hydrogenselenide (Equation (87)) ⟨12CB1835⟩. Other arylseleno compounds can be converted to diselenides. In particular, aryl selenocyanates react with potassium hydroxide to yield symmetrical diaryl diselenides (Equation (88)) ⟨79COC(3)491⟩.

$$Ar{-}Hal \xrightarrow{Se_2^{2-}} Ar{-}Se{-}Se{-}Ar \qquad (86)$$

Ar = o-$NO_2C_6H_4$-, Hal = Cl, 57%

$$Ar{-}N_2^+ \xrightarrow{NaHSe} Ar{-}Se{-}Se{-}Ar \qquad (87)$$

Ar = o-$HO_2CC_6H_4$-

$$Ar{-}SeCN \xrightarrow{KOH} Ar{-}Se{-}Se{-}Ar \qquad (88)$$

One less obvious route to diaryl diselenides is through the reaction of arylaldehydes with sodium hydrogenselenide, which, after treatment with sodium borohydride and piperidine, leads to the target compounds (Equation (89)) ⟨76CC552⟩.

$$Ar{-}CHO \xrightarrow[67\text{–}92\%]{\text{i, NaHSe; ii, }NaBH_4\text{, piperidine}} Ar{-}Se{-}Se{-}Ar \qquad (89)$$

Ar = Ph, p-$Et_2NC_6H_4$-, naphthyl, anthryl

Symmetrical diaryl triselenides (**56**) are known. They can be prepared either by the reaction of selenocyanates with a thiol ⟨55CB1⟩ or from the disproportionation of diaryl diselenides ⟨62JOC3401⟩.

$$Ar{-}Se{-}Se{-}Se{-}Ar$$

(**56**)

2.14.3.6.2 *Aryl ditellurides*

Diaryl ditellurides can be prepared by a number of methods. The traditional route involves the reduction of aryltellurium trichlorides (Equation (90)), but these, as starting materials, are of somewhat limited availability ⟨90JOM(388)71⟩. Preferred routes exploit more readily available aryl halides as starting materials. Particularly recommended is the *in situ* generation of arene tellurates (**57**), followed by oxidative work-up (Equation (91)). The tellurates (**57**) can be generated by either the lithiation of aryl halides and subsequent reaction with elemental tellurium ⟨90JOM(388)71⟩ or, more simply, by the reaction of aryl iodides with sodium telluride in N-methylpyrrolidone ⟨92S549⟩.

$$Ar{-}TeCl_3 \xrightarrow{[H]} Ar{-}Te{-}Te{-}Ar \qquad (90)$$

$$\left[Ar{-}Te^-\right] \xrightarrow[51\text{–}84\%]{[O]} Ar{-}Te{-}Te{-}Ar \qquad (91)$$

(**57**)

Ar = Ph, naphthyl, 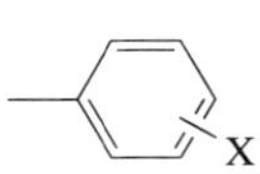 ; X = Hal, Me, MeO

2.14.3.7 RSeS and RTeS Functions

Arylselenyl sulfides are most conveniently prepared by the reaction of an aryl selenocyanate with a thiol (Equation (92)) ⟨50BSF245⟩. Areneselenenic acids can also be trapped with thiols to give selenylsulfides (Equation (92)) ⟨84JOC3106⟩.

$$Ar{-}SeX \xrightarrow{RSH} Ar{-}Se{-}S{-}R \qquad (92)$$

X = CN, OH; e.g., Ar = o-$NO_2C_6H_4$-, X = OH, R = Bu^t, 94%

2.14.3.8 RSeN Functions

Areneselenamides are available from areneseleninyl chlorides by reaction with ammonia (Equation (93)) ⟨40JA1859⟩.

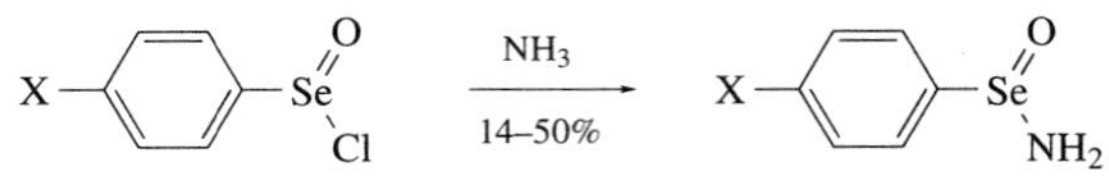

(93)

X = NO_2, NH_2, AcNH

2.15
Vinyl- and Arylnitrogen Compounds

GILLES SAUVÉ
Institut Armand-Frappier, Laval, Canada

and

VANGA S. RAO†
Bristol Myers Squibb Pharmaceutical Research Institute, Candiac, Canada

2.15.1 VINYL- AND ARYLNITROGEN COMPOUNDS

Vinylamines (enamines) generated most readily from aldehydes and ketones are widely used in the preparation of α-substituted aldehydes and ketones with high regio- and stereoselectivity. A number of reviews and books have been published on this subject ⟨63JA207, 63JCE194, B-63MI 215-01, 70S510, B-82MI 215-01, 82T405, 82T3363, B-88MI 215-01⟩. The developments in the preparation of vinylamines (enamines) and arylamines are discussed within this chapter.

2.15.2 ENAMINES AND ARYLAMINES

2.15.2.1 Enamines

2.15.2.1.1 From carbonyl compounds, acetals, ynamines, nitriles, and others

The classical preparation of simple enamines involves mixing an aldehyde (or a ketone) and an amine, with the removal of the water of condensation by azeotropic distillation with benzene or by a dehydrating agent such as calcium carbide, magnesium sulfate, calcium hydride, or a molecular sieve without the need for an acid or base catalyst. An acid catalyst, *p*-toluenesulfonic acid, Dowex-50, montmorillonite catalyst K10, boron trifluoride etherate, or even acetic acid can be employed for the condensation or when the uncatalyzed reaction is slow ⟨B-88MI 215-01⟩. The first general synthesis, derived from aldehydes, was discovered by Mannich and Davidsen ⟨36CB2106⟩. According to this method, aldehydes and secondary amines react in the cold in the presence of potassium carbonate to afford enamines which are isolated by distillation (Equation (1)).

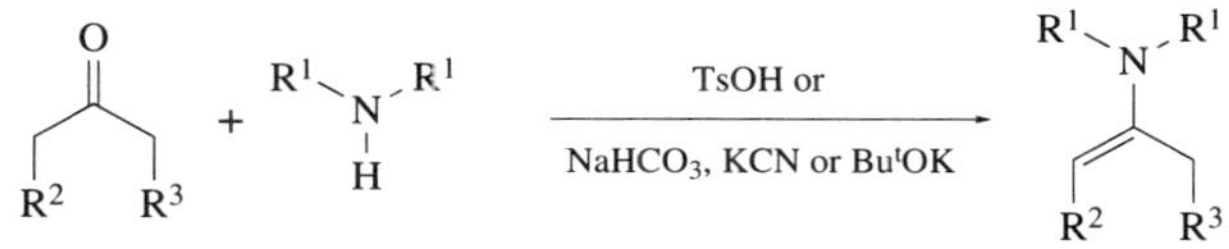

(1)

In the case of ketones, a wide range of dehydrating agents may be employed ⟨73OSC(5)808, 79TL2433, 80S320, 80TL2671, 84RTC305⟩. Hünig *et al.* ⟨73OSC(5)808⟩ indicated that an excess of amine is desirable due to some partitioning into the liberated water of condensation. All these methods are generally applicable to secondary amines and alkynones or cyclic ketones. The yields are typically in the range 72–80% with cyclohexanones, but are often lower for the reaction using sodium hydrogen carbonate as the dehydrating agent (41–64%) ⟨80S320⟩.

The "Wittig–Horner" reaction is an important method for the synthesis of enamines ⟨79TL2433, 80TL2671, 83JOC448, 84RTC305⟩. This versatile method uses an *N*,*N*-dialkyl or *N*-methyl-*N*-anilinomethyldiphenylphosphine oxide (**1**) as reagent. The relatively low basicity, but sufficient nucleophilicity of the ylide formed, is reflected in the high yields of product obtained even with highly enolisable ketones such as cyclopentanone and acetophenone (Scheme 1).

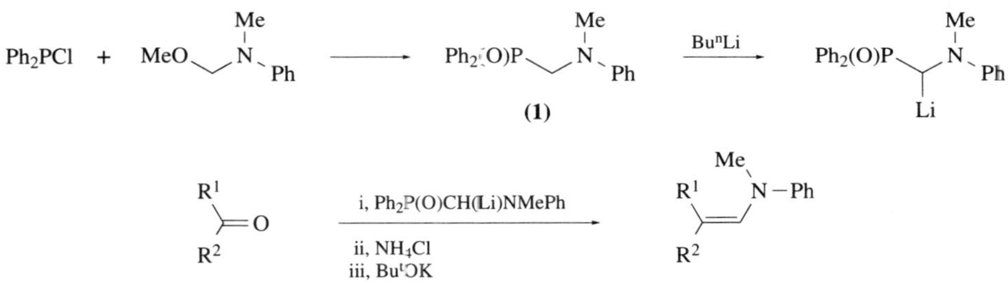

$R^1, R^2 = -(CH_2)_4-$, 86%
$R^1 = Me, R^2 = Ph$, 80%

Scheme 1

With unsymmetrical ketones, a mixture of geometric isomers has been observed. In the majority of cases the (*E*)-enamines are obtained. This is due to the isomerization of the (*Z*)-isomer to the thermodynamically more stable (*E*)-isomer under the reaction conditions.

A type of enamine phosphonate (**3**) ⟨71JOC2719, 73JOC2908⟩ was prepared from 1-alkynylphosphonate (**2**) and butylamine in fairly good yield (76–94%) (Equation (2)).

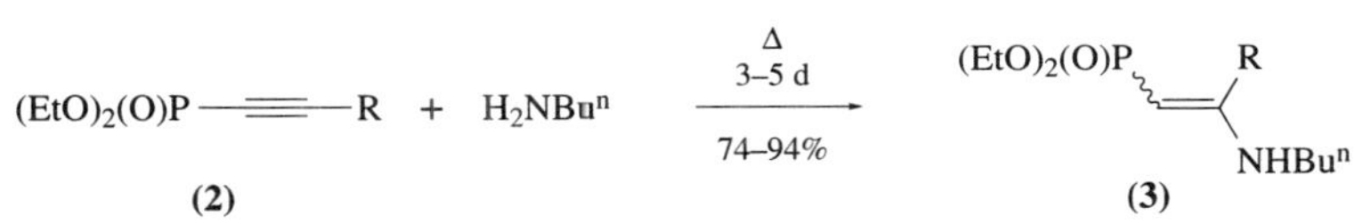

(2)

(2) **(3)**

$R = n\text{-}C_5H_{11}, n\text{-}C_6H_{13}, n\text{-}C_7H_{15}, c\text{-}C_5H_9, c\text{-}C_6H_{11}$, Ph, $Ph(CH_2)_2$

The reaction between an aryl aldehyde and hexamethylphosphorus triamide leads to the (aryl(dimethylamino)methyl)phosphonic bis(dimethylamide) (**4**) which can be deprotonated with *n*-butyllithium and condensed with aryl aldehydes to afford enamines in fairly good yield (Scheme 2) ⟨91S225⟩. The product enamines (**5**) represent useful precursors for the synthesis of more complex *β*-dicarbonyl derivatives ⟨82T1975⟩ and also for the stereoselective synthesis of (*E*)- or (*Z*)-alkenes ⟨89JA384⟩.

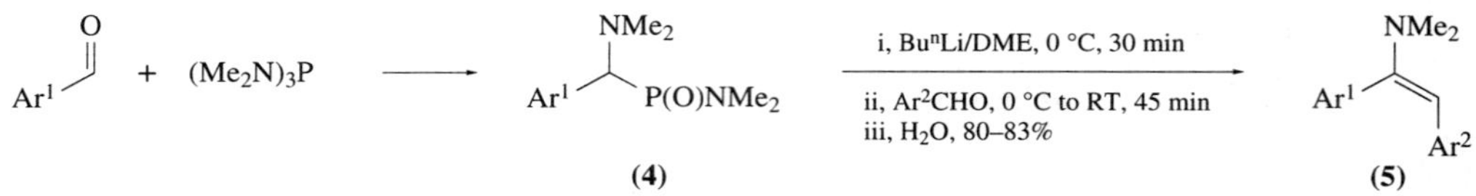

Scheme 2

The formation of enamine (**7**) from 2-nitro-5-carbomethoxytoluene (**6**) using *N*,*N*-dimethylformamide dimethyl acetal has been described as a precursor to the aldehyde (**8**) (Scheme 3) ⟨94TL219⟩.

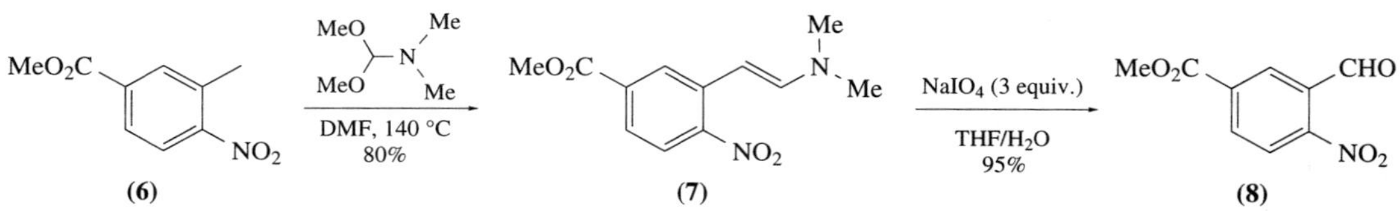

Scheme 3

Reaction of *N*,*N*-dimethylformamide diethyl acetal with a tertiary ethynyl alcohol (**9**) affords the dienamine (**10**) and/or the enamine *ortho*-formate (**11**) ⟨76JA7104⟩. The best yields are obtained with ethanol as solvent and a catalytic amount of pivalic acid (Equation (3)). Treatment of hydrocarbons with aminal ester or *N*,*N*-dimethylformamide diethylacetal results in the formation of enamines which can further be transformed into nitriles ⟨79TL1361⟩.

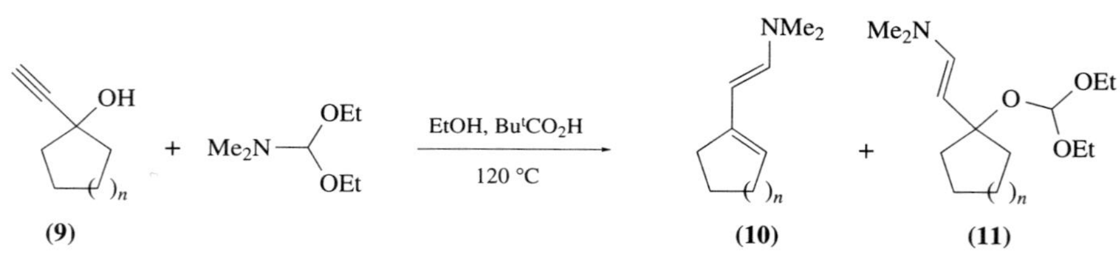

(3)

Butadiene undergoes cycloaddition with ynamines in the presence of an iron(0) catalyst to give the 1,4-cyclohexadienamine (**12**) ⟨79TL1499⟩. No isomerization to conjugated 1,3-cyclohexadienamine was observed (Equation (4)). Under similar conditions with other catalysts, such as copper(I) chloride, palladium(II) chloride, or the complex $H_2RuP(C_6H_5)_3$, the reaction gives rise to dimerization or polymerization of the ynamine.

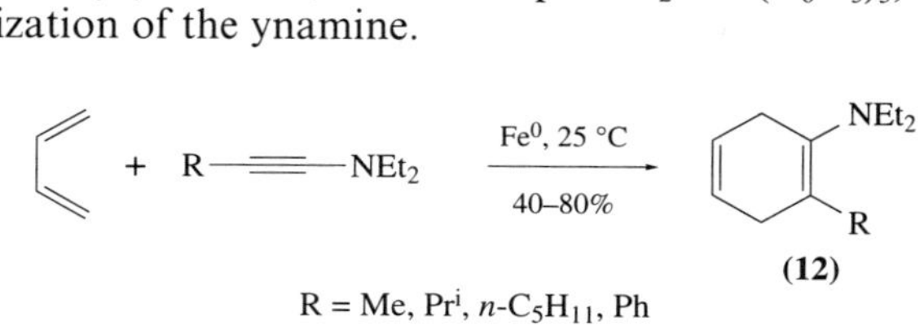

(4)

Stereoselective syntheses of (*Z*)-*γ*-amino enamines (**13**) have been reported by Barluenga *et al.* ⟨86S960⟩ via catalytic aminomercuration reactions (Scheme 4). The products are sensitive compounds which decompose on standing and must be handled under vigorously anhydrous conditions. Acid hydrolysis affords the corresponding *β*-aminoketones in almost quantitative yield. However, enediamines derived from aromatic amines are much more stable and in fact are not affected by sodium borohydride reduction in alkaline medium.

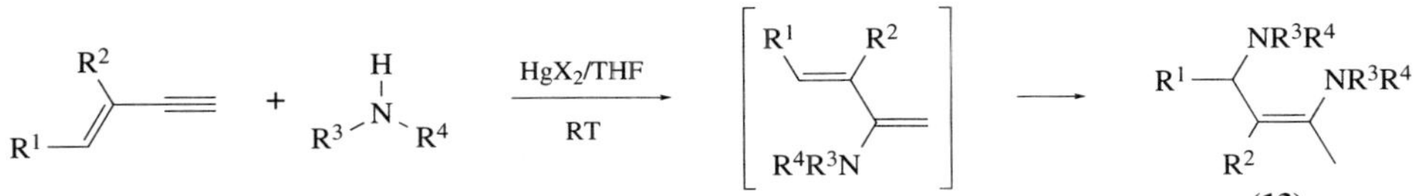

(13)

R^1, R^2 = H, Me, Et, Pr^n
R^3 = Ph, R^4 = H
R^3, R^4 = $-(CH_2)_5-$, $-(CH_2)_2O(CH_2)_2-$, Me, Et

Scheme 4

γ-Amino enamines (**14**) can also be prepared by the reaction of trimethylsilylamines ⟨84S493⟩ with α,β-unsaturated aldehydes (Equation (5)). The method provides a simple and efficient way to functionalized enamines. It is of preparative interest as the reaction occurs under aprotic conditions, gives nearly quantitative yields, and by-products are easily removed.

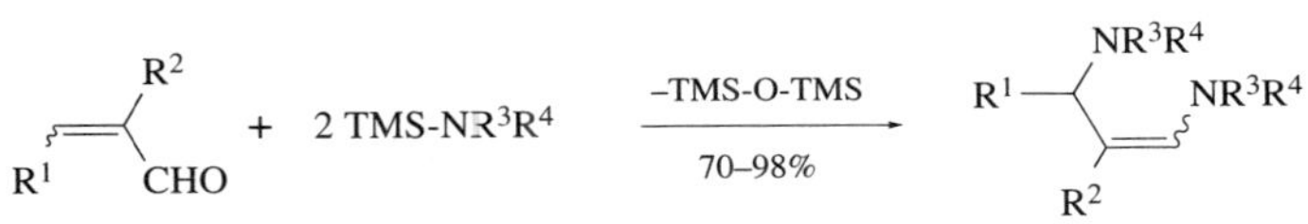

(5)

R^1, R^2 = alkyl, aryl **(14)**
R^3, R^4 = $-(CH_2)_2O(CH_2)_2-$, $-(CH_2)_4-$, $-(CH_2)_5-$, Me

The β-amino enamine (**15**) is readily prepared by the reaction of 2-halo-2-alkenals with secondary aliphatic amines (Equation (6)) ⟨83BSF(2)28, 89S446⟩.

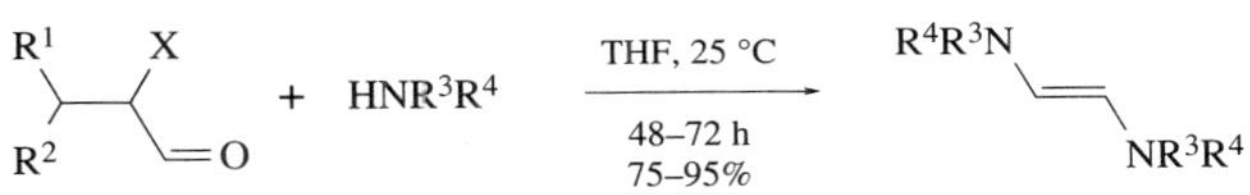

(6)

X = Cl, Br **(15)**

A convenient synthesis of enamines and dienamines was achieved by deprotonation of α-aminonitriles (**16**) with lithium diisopropylamide in tetrahydrofuran, followed by alkylation and dehydrocyanation (Scheme 5) ⟨79S127⟩. The significant advantage of this enamine synthesis is the high versatility resulting from the metallation–alkylation step, and high purity of the product enamines due to the absence of any acidic catalyst.

(16)

Scheme 5

Treatment of an ethereal solution of the methyl-substituted nitrogen heteroaryl (**17**) with two equivalents of *n*-butyllithium and subsequent addition of a secondary aminoacetonitrile (**18**) afforded the enamine (**19**) (Equation (7)) ⟨73JMC1113⟩. This reaction sequence usually gives moderate yields (67–88%), and requires two equivalents of *n*-butyllithium for the formation of the enamine due to the acidity of the secondary amino acetonitrile. The product enamine (**19**) was not isolated because of its instability.

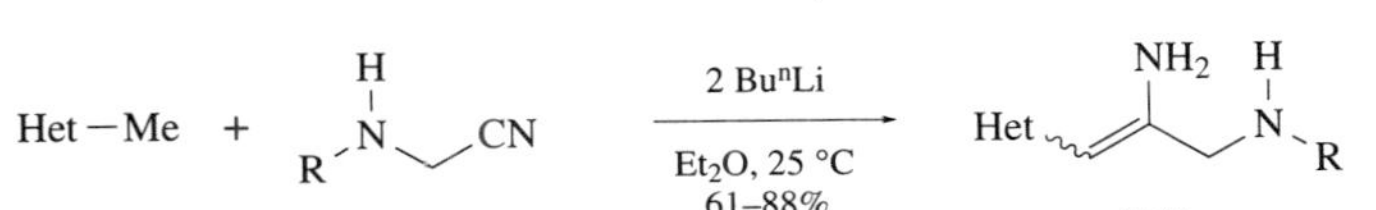

(7)

(17) **(18)** **(19)**

A one-pot synthesis of enamines such as (**21**) from substituted acetic acids (**20**) was reported by Knorr *et al.* (Scheme 6) ⟨83S785⟩. This method constitutes a convenient one-step decarboxylative formylation, leading directly to different types of enamines with yields between 36% and 89%.

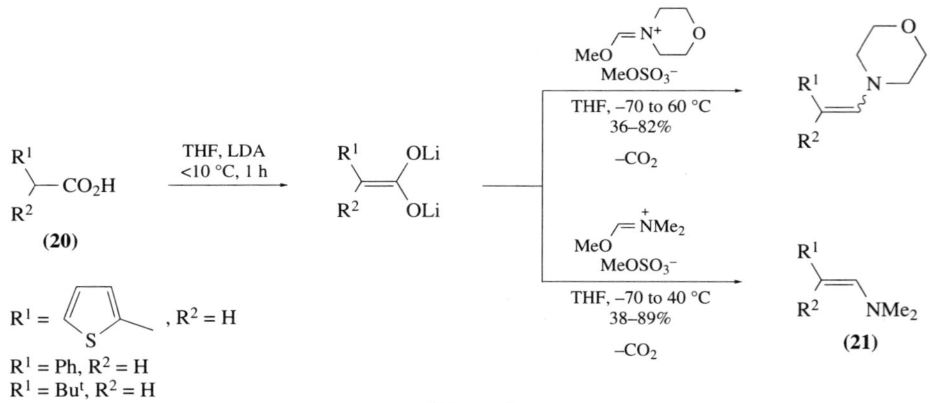

Scheme 6

An unexpected synthesis of (*E*)-enamines (**23**) was also reported ⟨87S727⟩ involving a rearrangement of ammonium ylides (**22**), especially the rearrangement of derivatives of benzazocine and benzazonine with sodium amide in liquid ammonia (Scheme 7). The starting material was obtained by quaternization of the amine with a suitable halide.

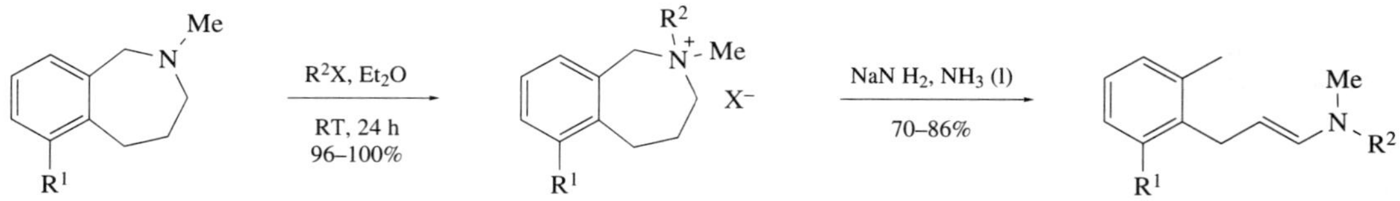

(**22**) (**23**)

R^1, R^2 = H, Me, Et
X = Br, I

Scheme 7

2.15.2.1.2 *Sulfonyl enamines and aminothiones*

β-Sulfonylenamines (**25**) have been prepared as follows: (a) by nucleophilic *β*-addition of a variety of amines to alkynic sulfones (**24**) (Equation (8)) ⟨64JCS5863, 66JCS217, 66JOC3543, 70JOC3275⟩; ethyleneimine adds to *p*-tolysulfonylacetylene (**24**; $R^1 = H$, $R^2 = p$-tolyl) giving >95% *cis*-1-ethylene-imino-2-(*p*-tolylsulfonyl)ethene while secondary and primary amines give the *trans* adduct via isomerization of the initially produced *cis* isomer; (b) by an unusual Knoevenagel-type condensation of bis(ethylsulfonyl)methane and benzaldehyde in the presence of an excess of piperidine ⟨72JOC1902⟩; (c) by reaction of an α-sulfonyl-acetaldehyde with aniline and *N*-methylaniline ⟨66CB1580⟩; and (d) by treatment of enolates of α-alkoxy- and α-alkylthio-arylsulfonyl acetaldehydes with aniline hydrochloride ⟨76S406⟩.

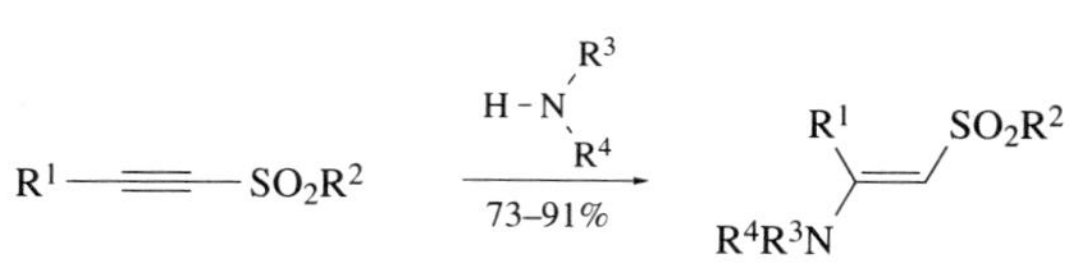

(8)

(**24**) (**25**)

R^1 = H, Me
R^2 = Tol
R^3, R^4 = Me, Pr^n

The preparation of β-sulfonylenamines (**27**) was also achieved from α,β-epoxy sulfones (**26**) (Scheme 8) ⟨83S628⟩, and metallation of 2,3,4,5-tetrahydro-6-methylpyridine with lithium diisopropylamide followed by treatment with a menthyl-(*S*)-sulfinate ⟨91S970⟩.

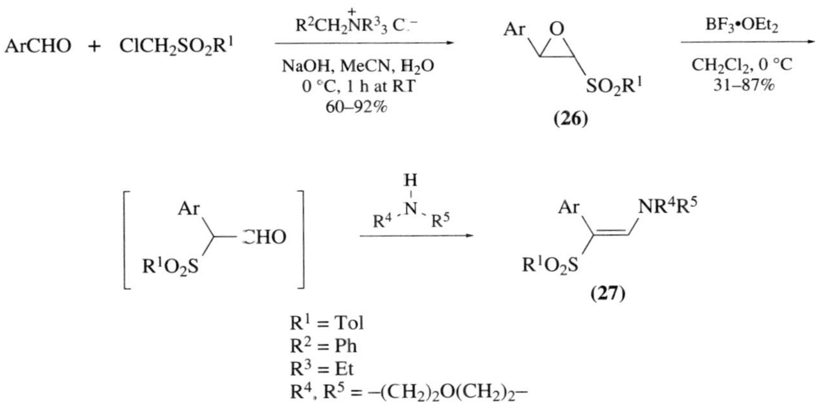

Scheme 8

β-Sulfonylenamines are almost inaccessible from the condensation of amines with β-ketosulfones as the latter undergo facile base-catalyzed C—C bond cleavage. However, they can efficiently be made from alkynes by the sequence of selenosulfonation, oxidation, and aminolysis ⟨85CJC2313⟩. When terminal alkynes are employed as precursors, the selenenylated enamines (**28**) generally dominate; however, their formation can be suppressed by employing selenic acid scavengers such as an enamine or ethyl vinyl ether (Scheme 9).

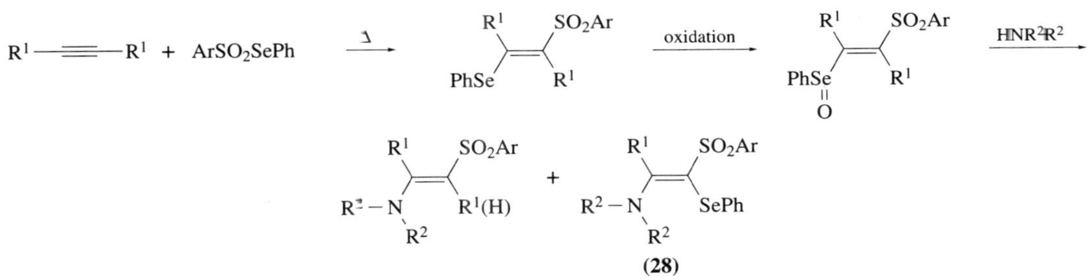

Scheme 9

The reaction of (*E*)-2-iodo-1-tosyl-1-alkenes with a primary or secondary amine in acetonitrile at room temperature produces the corresponding β-tosyl enamine (**29**) in almost quantitative yield (Equation (9)) ⟨92BCJ1379⟩. The enamine ($R = Ph(CH_2)_2$ or Pr^n) obtained with methylamine ($R^1 = Me$, $R^2 = H$) were a mixture of (*E*)- and (*Z*)-isomers, whereas pyrrolidine (R^1, $R^2 = —(CH_2)_4—$) gave only the (*E*)-isomer.

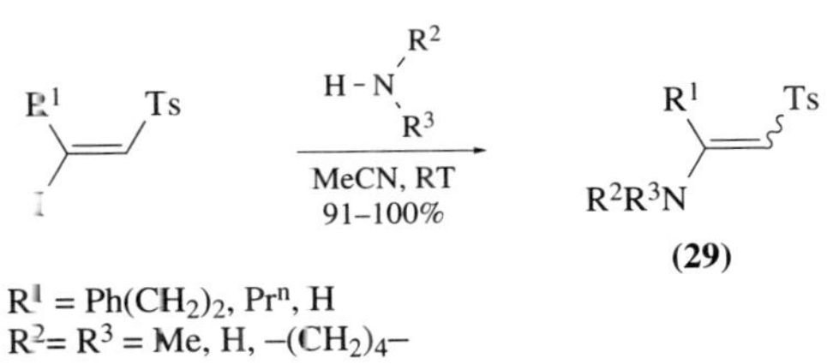

(9)

2,2-Bis(methylsulfonyl)vinylamines (**30**) are synthesized ⟨74JOC1432⟩ by the reaction of 2,2-bis(methylsulfonyl)vinyl ethylether with an amine (Equation (10)). The reaction conditions vary with the amine used.

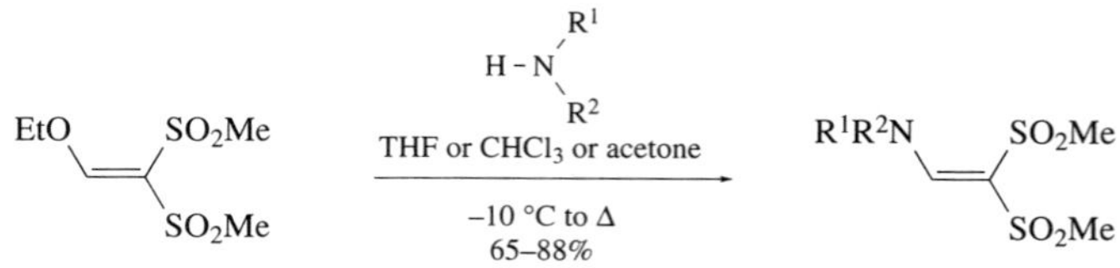

(10)

R^1 = H, Ph, Pr^n, PhNH
R^2 = H, Et, PhCO, MeNHCO, Ts

(30)

Enaminothiones (**31**) have been shown to be highly reactive synthons, they can be prepared by a variety of methods ⟨74CB2186, 74S298, 76S535, 89S631⟩. α-(**32**) and β-thioenamines (**33**) are also well known ⟨82TL2023, 84S304⟩.

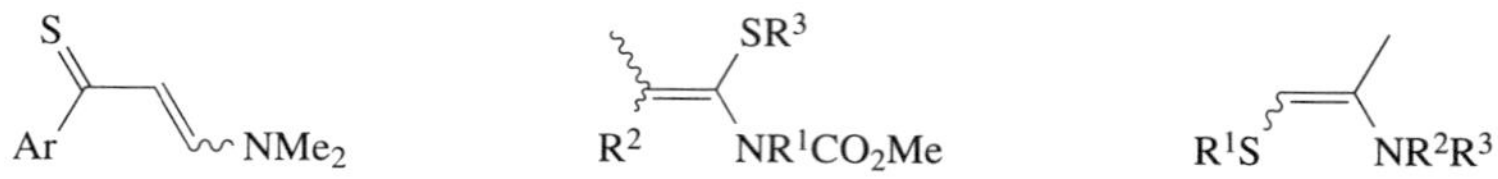

(31) **(32)** **(33)**

2.15.2.1.3 *Enaminones*

Enaminones are versatile intermediates in organic synthesis and their preparation involves the reaction of a primary or a secondary amine with a 1,3-diketone in an aromatic solvent with azeotropic removal of the water formed ⟨71JCS(C)2699, 81S880⟩. Baraldi *et al.* ⟨83S902⟩ reported a practical preparation of enaminones, characterized by the use of ammonium acetate as an active form of ammonia, and by employing the easily manipulable acetate form of low-boiling amines (Equation (11)). The yields are very high (70–98%). The main advantages of this method are that the reaction conditions are mild and inexpensive, and easy-to-handle reagents are involved. Enaminones can also be obtained from thioamides (**34**) (Scheme 10) ⟨76S535⟩.

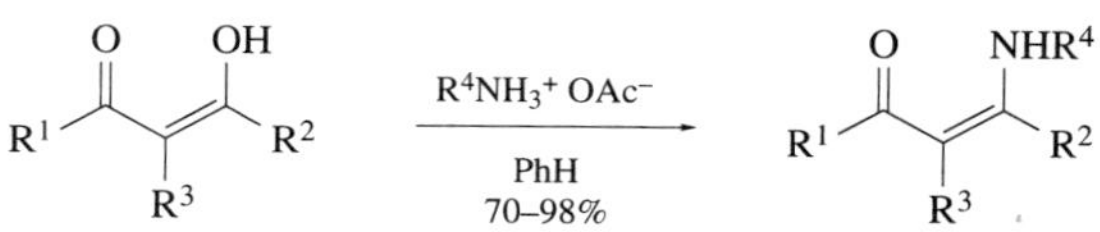

(11)

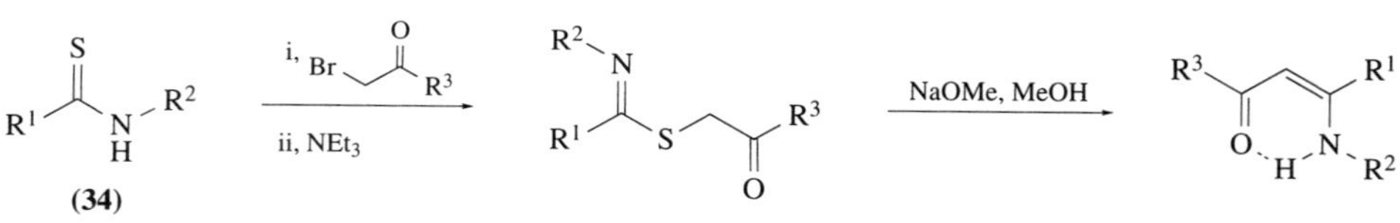

(34)

Scheme 10

René and El Cherif reported a one-pot synthesis of β-acylenamines (**35**) involving a Mannich reaction by heating a mixture of triethyl *ortho*-formate, ketone, and secondary amine in a pressure vessel (Equation (12)). References to several other methods can be found in this article ⟨88S138⟩.

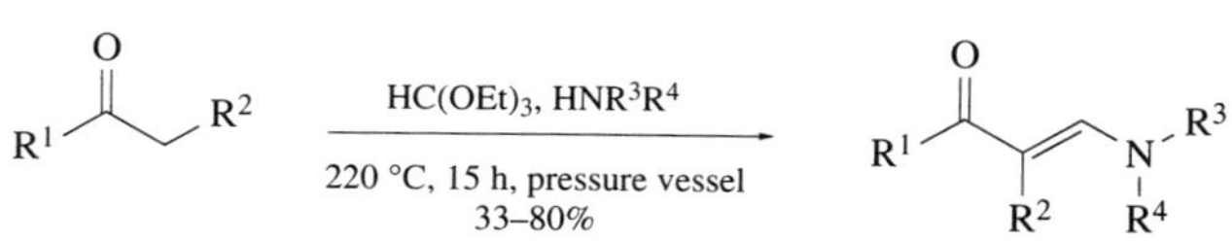

(12)

R^1 = Et, Ph, 2-furyl, etc.
R^2 = H, Me, Ph
R^3, R^4 = Et, $-(CH_2)_5-$, $-(CH_2)_2O(CH_2)_2-$

(35)

Enaminoketones (**37**) were obtained by the hydrogenation of isoxazoles (**36**) over either Raney nickel or platinum catalysts (Equation (13)) ⟨75JOC526, 75S20⟩.

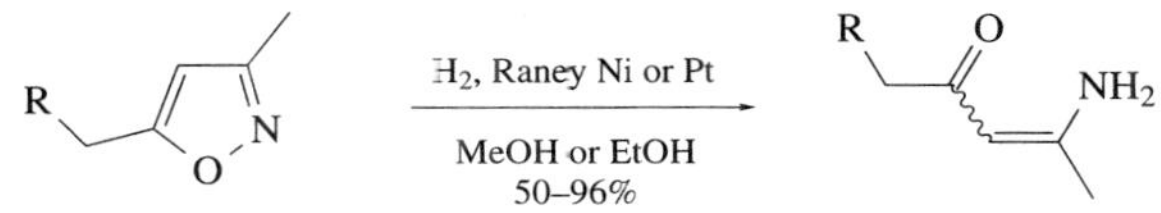

(36) R = H, Me, Et (37)

(13)

Treatment of diphenylsulfilimine (**38**) with *cis*- or *trans*-dibenzoylethene in benzene solution at room temperature afforded the primary *β*-carbonylenamine (**39**) in 50% yield (Equation (14)) ⟨76S30⟩.

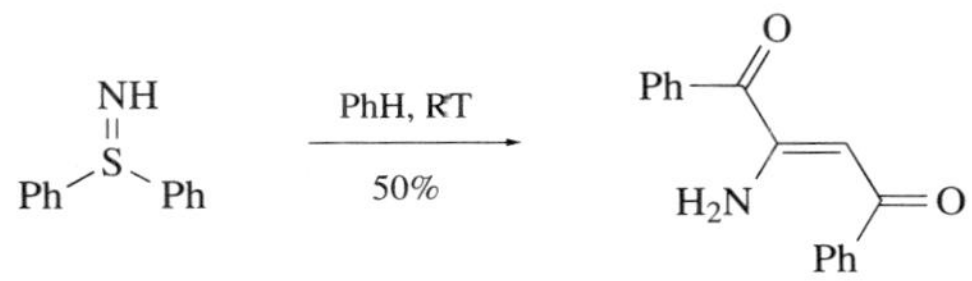

(38) (39)

(14)

Primary enaminoesters (**42**) were generated by the reaction of *N*-trimethylsilylimino-triphenylphosphorane (**41**) with *β*-dicarbonyl compounds (**40**) in the presence of a secondary alcohol and a catalytic amount of *p*-toluenesulfonic acid (Equation (15)) ⟨78JOC1460⟩.

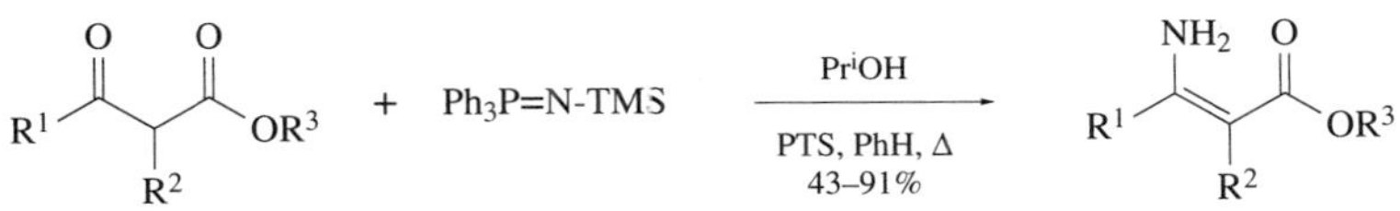

(40) (41) (42)

(15)

R^1 = Me
R^2 = H, Cl_3CCH_2–
R^3 = Et, Bu^t
R^1, R^2 = $-(CH_2)_3-$, $-(CH_2)_4-$

α-Acylenamines (**44**) are obtained by the reaction of alkyl pyruvates (**43**) with amines in the presence of arsenous chloride (Equation (16)) ⟨90S39⟩.

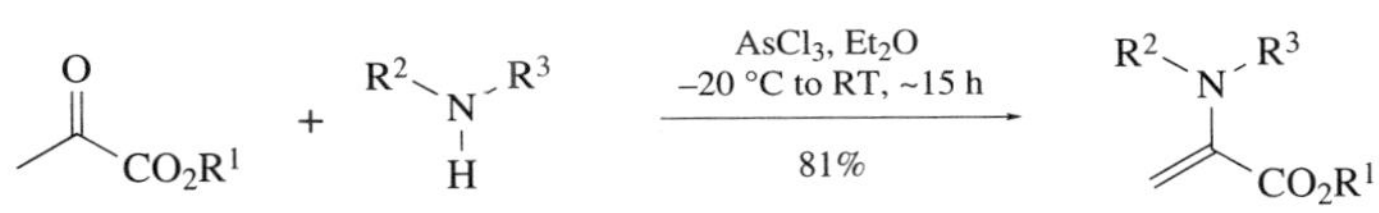

(16)

(43) R^1 = Me, Bu^t (44)
R^2, R^3 = Me, $-(CH_2)_5-$, $-(CH_2)_2O(CH_2)_2-$

Cyclic *α*-acylenamines (**46**) are prepared from 1-dialkylaminocyclohexenes (**45**) by reaction with ethyl or amyl nitrite without solvent or in an aprotic solvent in the presence of boron trifluoride etherate or aluminum chloride (Equation (17)) ⟨69JPR162⟩.

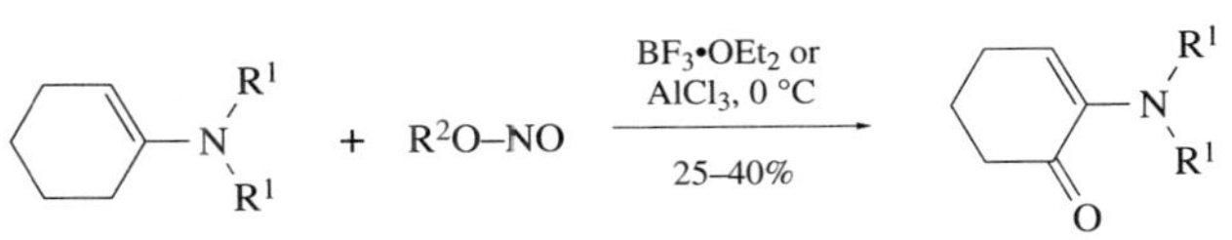

(17)

(45) R^1 = $-(CH_2)_4-$, $-(CH_2)_5-$, $-(CH_2)_2O(CH_2)_2-$ (46)
R^2 = Et, *i*-C_5H_{11}

An example in which an *α*- and *β*-acylenamine moiety may be found is provided by Bredereck *et al.* ⟨74CB1545⟩. Such a compound (**49**) was obtained by reacting dialkylamino derivatives possessing an active methylene group (**47**) with an excess of *t*-butoxy-bis(dimethylamino)methane (**48**) without solvent at temperatures between 120 °C and 170 °C (Equation (18)).

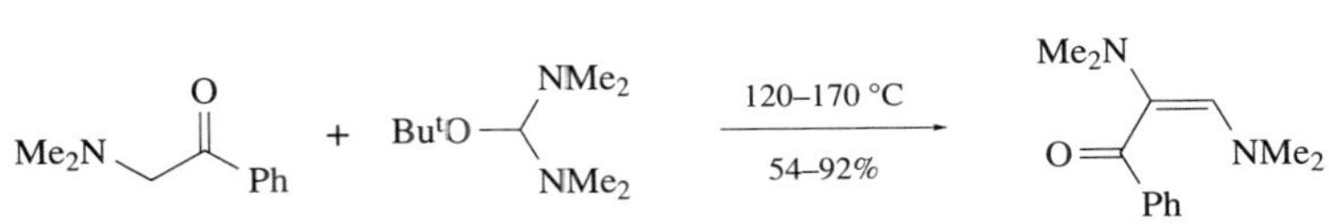

(18)

(47) (48) (49)

Enamino esters (**51**) were prepared from an epoxide (**50**) by reaction with sodium azide and ammonium chloride in 60% ethanol (Equation (19)) ⟨84S618⟩.

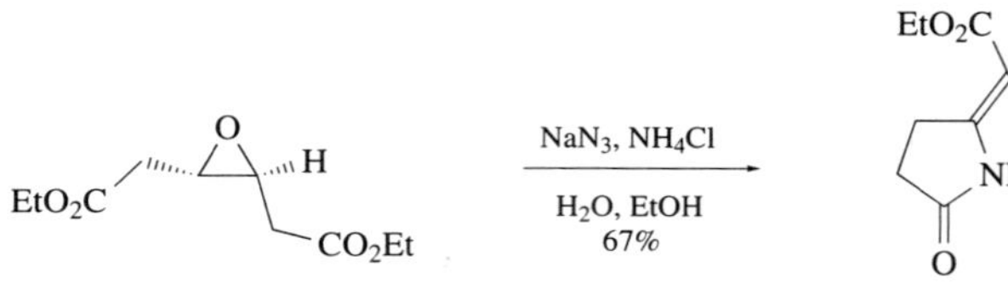

(19)

(50) **(51)**

Enamino esters (**54**) can be generated from iminium salts (**53**) (Gold's reagent) and esters or lactones (**52**) under basic conditions (Equation (20)) ⟨82SC939⟩.

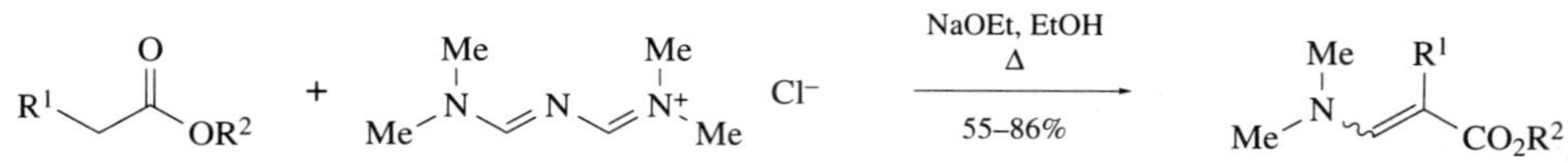

(20)

(52) **(53)** **(54)**

R^1 = H, Ph, CO_2Et, CN
R^2 = Et
R^1, R^2 = $-(CH_2)_3-$

Krogsgaard-Larsen *et al.* ⟨73ACS2802⟩, Wamhoff and Lichtenthäler ⟨75S426⟩, and McMurry ⟨88OSC(6)592⟩ have also reported the preparation of enamino esters.

Cyclic *β*,*β*-dicarbonyl enamines (**56**) were generated from the reaction of cyclic *β*-dicarbonyl compounds (**55**) with trimethylorthoformate and aniline or urea (Equation (21)) ⟨82ZN(B)105⟩.

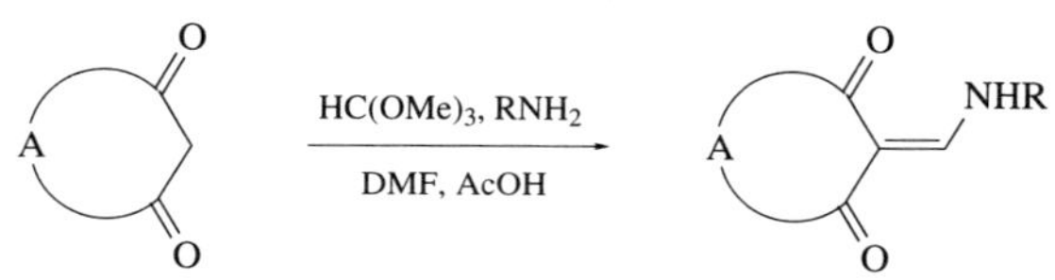

(21)

(55) **(56)**

R = Ph, $-CONH_2$
A = $-(CH_2)_3-$, etc.

In a similar manner L'Eplattenier and Vuitel ⟨76S543⟩ generated enamino lactones and enamino lactams by the reaction of a lactone or lactam, respectively, with trimethyl *ortho*-formate and aniline.

Buyle and Viehe ⟨68T4217⟩ reported the preparation of an *α*-chloro-*β*-acylenamine (**58**) from the reaction of amides (**57**) with phosgene at room temperature (Equation (22)).

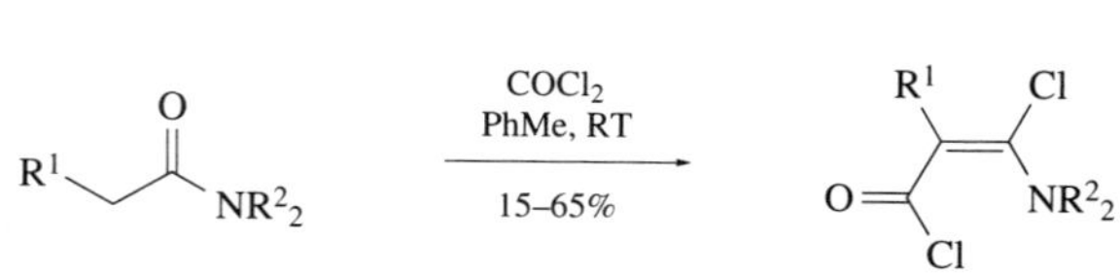

(22)

(57) **(58)**

Enaminones (**60**) were also generated by simple acylation of enamines (**59**) (Equation (23)), as described by Nilsson ⟨79ACS(B)547⟩ and Plath and Rohr ⟨82S318⟩.

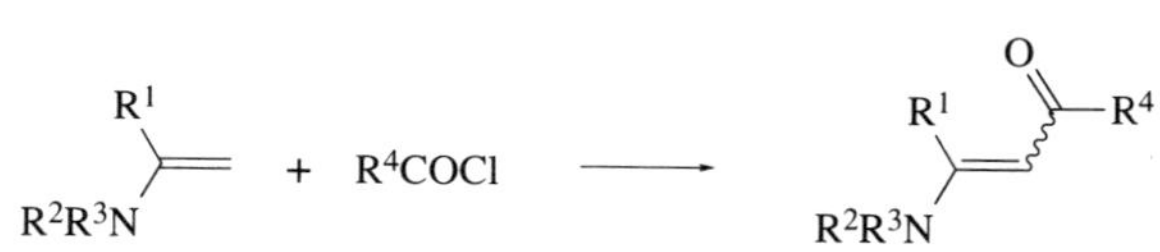

(23)

(59) **(60)**

R^1 = Et, Pr^i
R^2, R^3 = Et, $-(CH_2)_2O(CH_2)_2-$
R^4 = Pr^i, c-C_6H_{11}, Ph

2.15.2.1.4 N-*Acylenamines*

The enamides (or *N*-acyl enamines) are a related family of the enaminones. They are prepared from α,β-epoxysilanes **(61)** ⟨75JCS(P1)1237, 77JA1993⟩, by reaction with acetonitrile in the presence of boron trifluoride etherate at $-25\,^{\circ}C$ followed by acidification with a mixture of sulfuric acid/water/THF, and finally treatment with potassium hydride in THF at $20\,^{\circ}C$ to give *N*-acyl amides in an overall yield of 62% from epoxide (Equation (24)).

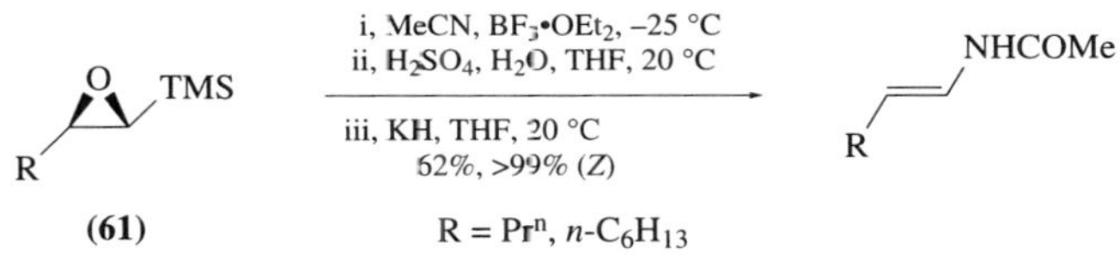

(24)

(61) R = Pr^n, n-C_6H_{13}

Ketoximes can be converted directly into α-*O*-acyl-*N*,*N*-diacyl enamines **(62)** ⟨76CC103⟩ by heating with acetic anhydride in the presence of either pyridine/acetylchloride, pyridinium hydrochloride or *p*-toluenesulfonic acid. The reaction can be extended to cyclic ketoximes and different acyl compounds (Equation (25)).

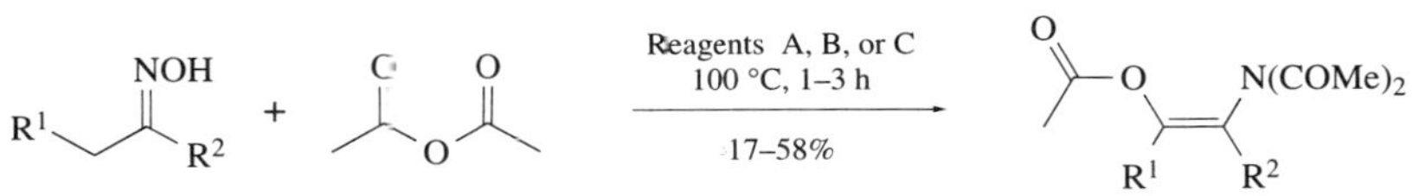

(25)

R^1 = Me
R^2 = Me, $Me_2C{=}CHCH_2$, Ph
R^1, R^2 = $-(CH_2)_3-$, $-(CH_2)_4-$

(62)

Reagents A: Pyridine/AcCl
Reagents B: Pyridinium hydrochloride
Reagents C: *p*-Toluenesulfonic acid

Reductive acetylation of ketoximes to enamides **(63)** ⟨75JCS(P1)1237⟩ can be carried out by the action of acetic anhydride in dry pyridine under reflux until all oxime acetate is consumed, followed by removal of the solvent and chromatography on alumina (Equation (26)).

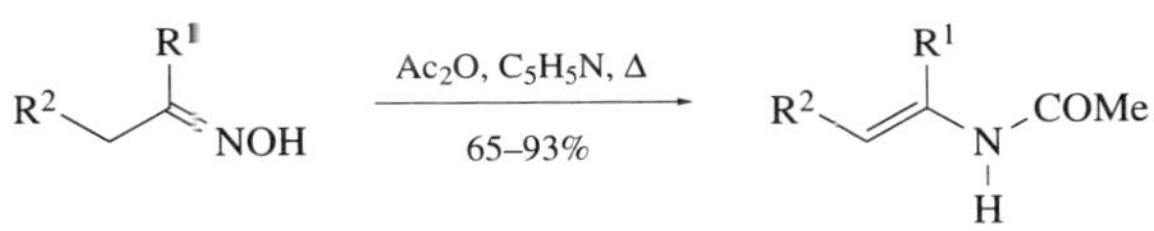

(26)

R^1, R^2 = $-(CH_2)_3-$, and other examples **(63)**

N-Acyl amino acids and *N*-acyl peptides ⟨81TL4263⟩, in a modified Dakin–West reaction give α-acylamino ketones **(64)**, the oximes of which smoothly undergo Beckmann fragmentation to give a mixture of (*E*)/(*Z*) enamides in ratios of 54 : 46 to 73 : 27 (Scheme 11).

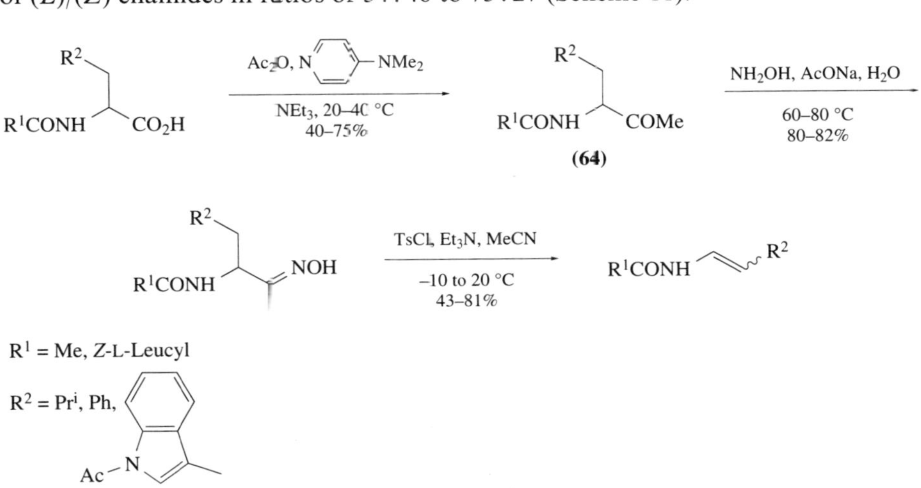

Scheme 11

Alkenyldialkylboranes (**65**), prepared from an alkyne and dialkylborane, react with potassium cyanide and trifluoroacetic anhydride to give (**66**), which when hydrolyzed with 2 N sodium hydroxide at 25 °C afford the *N*-acylenamine (**67**) (Scheme 12) ⟨74CC346⟩.

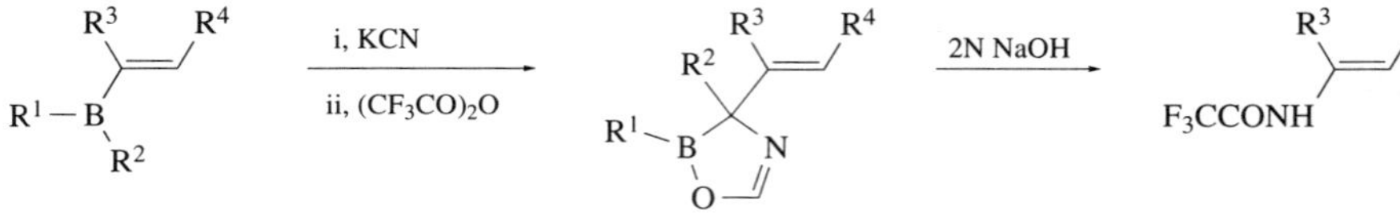

(65) **(66)** **(67)**

Scheme 12

The synthesis of cyclic *N*-acylenamines (**68**) was reported by Nyberg ⟨76S545⟩ from 1-methoxylated cyclic *N*-formylamines. However, the yield of enamides varied considerably due to difficulties encountered in removing the last traces of acid leading to polymerization products (Equation (27)).

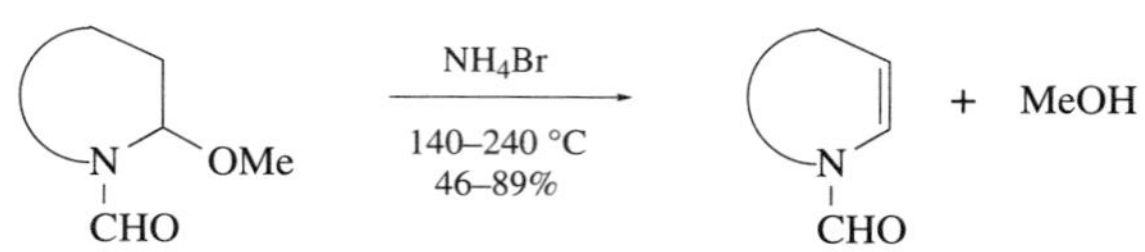

(68)

(27)

2.15.2.1.5 *α- and β-Cyanoenamines*

The *α*- and *β*-cyano enamines have received considerable attention as starting materials for the synthesis of a variety of compounds ⟨68T4677⟩. Tertiary *α*-cyano enamines have been synthesized from *α*,*β*-unsaturated aldehydes ⟨68T4677⟩, *α*-chloro enamines ⟨75JA2276⟩, *α*-halo aldehydes ⟨39BRP511430, 57JOC1714, 75S512, 77S495⟩, carboxamides ⟨71CB3983⟩, and from the addition of cyanogen bromide to enamines (Scheme 13) ⟨77S495⟩.

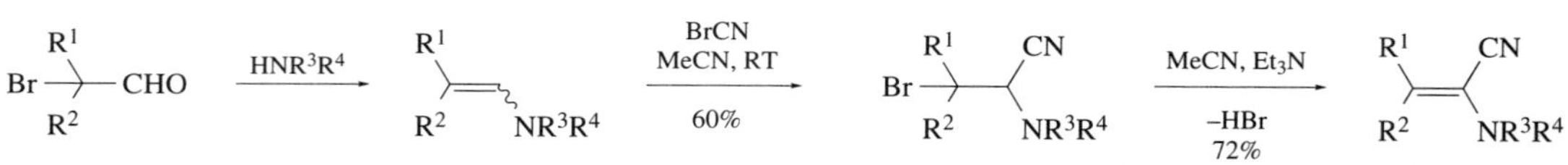

R^1 = Pr^i, R^2 = H
R^3, R^4 = –$(CH_2)_2O(CH_2)_2$–

Scheme 13

α-Cyano enamines (**70**) have been obtained via the Peterson reaction of *α*-(*N*-methylanilino)-*α*-(trimethylsilyl)acetonitrile (**69**) with formaldehyde (Equation (28)) ⟨78S897⟩.

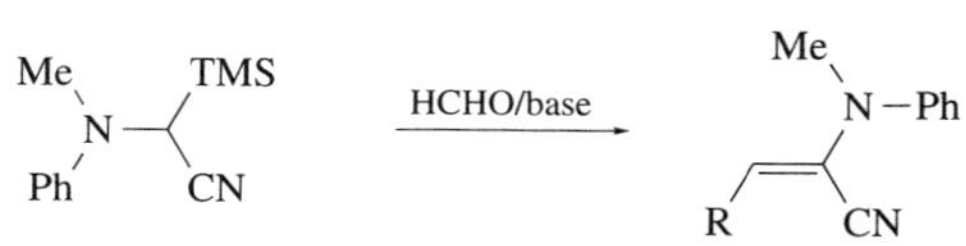

(69) **(70)**

(28)

Takahashi *et al.* ⟨83JOC3566⟩ prepared a variety of *α*-cyano enamines (**72**) by means of the Peterson reaction or via the condensation of carbanion of (**71**) with an aromatic aldehyde in dichloromethane in the presence of a phase-transfer catalyst, triethylbenzylammonium chloride (Equation (29)) ⟨84S1043⟩.

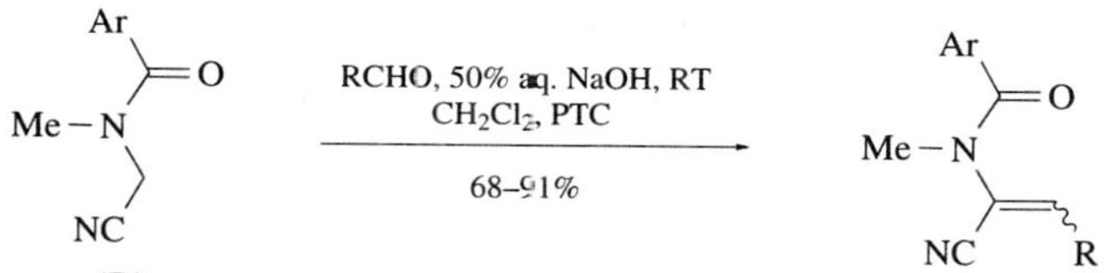

(71) R = aryl **(72)** (29)

Similarly, Jonczyk and Owczarczyk ⟨86S297⟩ prepared α-cyanoenamines by the condensation of the carbanion of α-(*N*-methylanilino)-acetonitrile with an aldehyde or ketone in the presence of a phase-transfer catalyst, tetrabutylammonium bromide. Secondary α-cyano enamines have been obtained from α-chloroaldimines and potassium cyanide in methanol ⟨76S741, 76T3063⟩. β-Cyano enamines have been prepared through reaction of an amine with, for example, β-chloroacrylonitrile ⟨64JOC1800⟩, cyanoacetylene ⟨69JCS(C)1086⟩, β-alkoxyacrylonitrile ⟨80GEP2912344⟩, 2-chloro-tetramethylguanidine ⟨66JOC1426⟩, dicyanoethylene ⟨77HCA2379⟩, and also from dimethylformamide acetals and acetonitrile ⟨70USP3542848⟩, and dineopentylformamide acetals and cyanoacetic acid ⟨65HCA1746⟩.

Novelli *et al.* ⟨68T2481⟩ obtained β-cyano enamines (**74**) by heating phenylcyanoacetaldehyde (**73**) with two to four equivalents of a dialkylformamide at 180 °C for 8 h (Equation (30)).

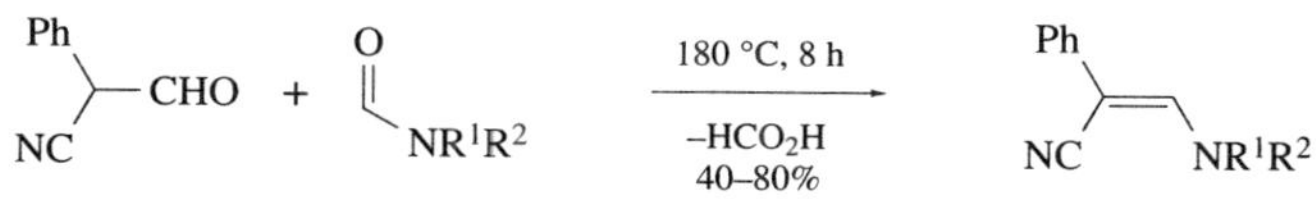

(73) **(74)** (30)

R^1R^2 = Me, Et, $-(CH_2)_4-$, $-(CH_2)_2-O-(CH_2)-$, etc.

René *et al.* reported an effective one-pot synthesis of β-cyano enamines (**75**) ⟨86S419⟩ through the reaction of ethyl *ortho*-formate, cyanoacetic acid and an amine (Equation (31)). Alkyl-substituted *ortho*-esters (e.g., R^2 = Me) also afforded good yields of the product, but with ethyl *ortho*-benzoate (R^2 = Ph) and morpholine, only the corresponding benzamide was recovered.

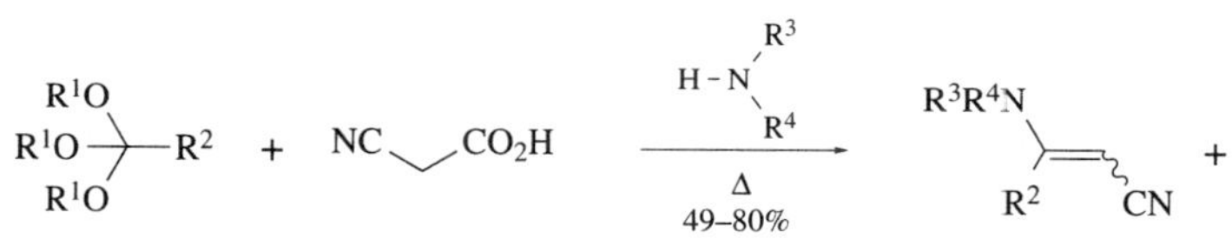

(31)

R^1 = Et **(75)**
R^2 = H, Me
R^3, R^4 = Me, Et, $-(CH_2)_4-$, $-(CH_2)_2O(CH_2)_2-$, etc.

Alkylation of a C α to a nitrile with another nitrile function, known as the Thorpe reaction, leads to β-cyanoenamines (**76**) in an analogous aldol-type condensation (Equation (32)) ⟨67OR(15)1, 88JCS(P1)269⟩. Thus, the intramolecular alkylation (Thorpe–Ziegler) is very useful for macrocycle synthesis.

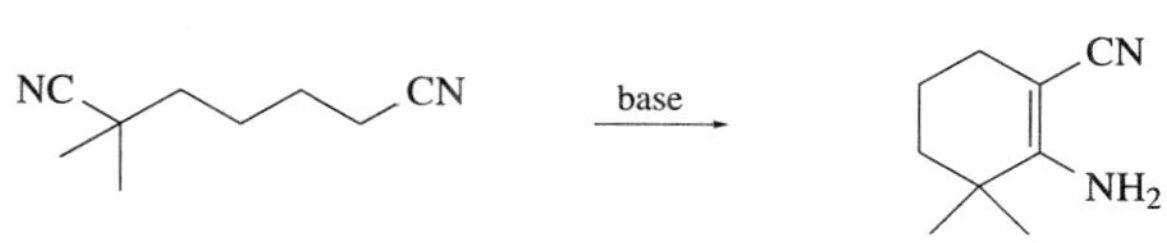

(76) (32)

2.15.2.1.6 *Miscellaneously substituted enamines*

In the course of a novel peptide backbone modification, Sauvé *et al.* reported the synthesis of β-mono ⟨88TL2295⟩ and β,β-difunctionalized ⟨88TL2299⟩ enamines using different strategies. In one method a thioamide is reacted with an α-halogenated methylene reagent to form the α-thioiminium salt which cyclizes to an episulfide in the presence of a base. Sulfur extrusion from this intermediate with a thiophilic reagent generates the functionalized enamine (**77**) according to the Eschenmoser sulfide ring contraction method (Scheme 14).

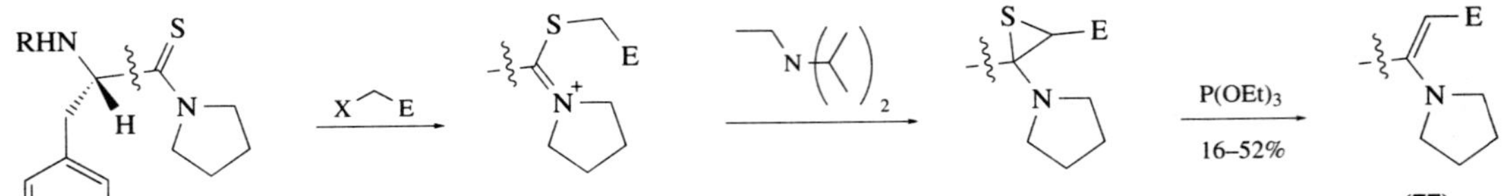

(77)

X = Cl, Br, I
E = CO_2Me, CN, COAr, etc.
R = Ac, CO_2Bu^t

Scheme 14

In a second procedure, Sauvé activated the *N*-protected amino acid/peptide carboxylic group with 1,1′-carbonyldiimidazole ⟨90JOC3002⟩. Nucleophilic attack by an active methylene reagent at the carbonyl group of the acylimidazole intermediate generated the enol, which in turn was activated as an enolphosphate and reacted with either a primary or secondary amine to yield the β,β-difunctionalized enamine (**78**) (Scheme 15).

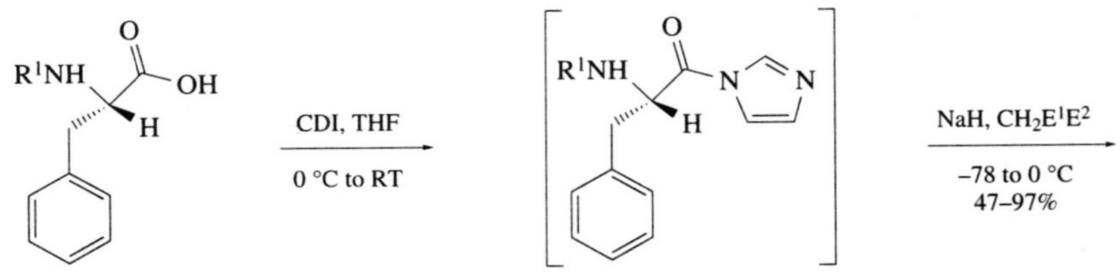

R^1 = Ac, CO_2Bu^t

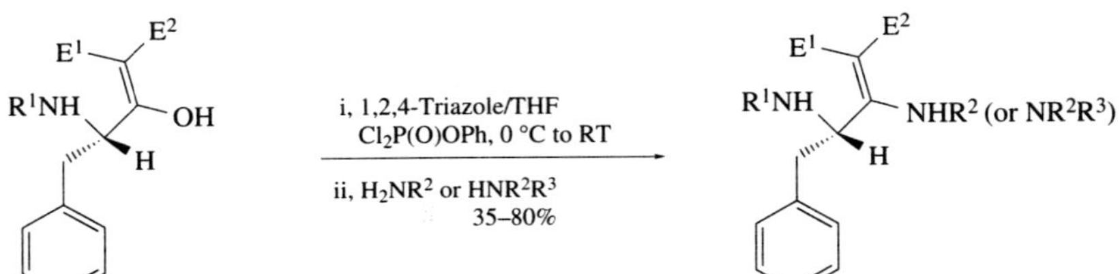

(78)

E^1 = CN, CO_2Me
E^2 = CN, CO_2Me, p-$NO_2C_6H_4$, $(EtO)_2PO$, SO_2Ph, 2-pyridyl
R^1 = CO_2Bu^t, *N*-Ac-Leu

E^1 = CN, CO_2Me, p-$NO_2C_6H_4$
E^2 = CN
R^2 = H
R^3 = H, Ph, CH_2Ph, Pr^i, $CHMeCO_2Me$
R^2, R^3 = $-(CH_2)_4-$
R^1 = CO_2Bu^t, *N*-Ac-Leu

Scheme 15

A third method employed thiophilic silver carbonate to generate the β,β-difunctionalized enamines (**79**) from thioamides and active methylene reagents ⟨90JOC2246⟩. It appears that when R = H, no reaction occurs (Equation (33)).

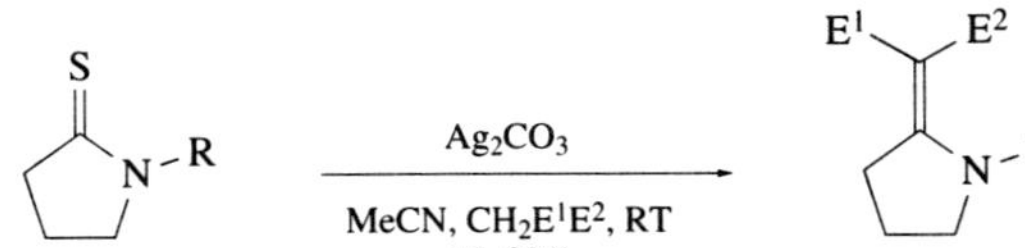

(33)

(79)

E^1 = CN, CO_2Me
E^2 = CN, CO_2Me, CO_2Bu^t, ArCO, NO_2, SO_2Ph, Ar, $(EtO)_2PO$
R = Me, H

The reaction between alkynic compounds and secondary amines is well known, and it has been postulated that this reaction proceeds via an enamine intermediate, which then reacts with more of the alkynic compound to give the adduct (**80**) (Scheme 16) ⟨63JCS1462⟩.

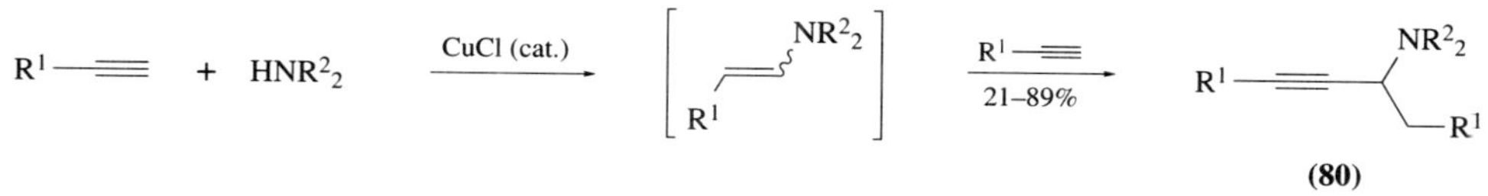

Scheme 16

Trichloro- ⟨90S750⟩, trifluoro-, and chlorodifluoroenamines ⟨74S816, 92JOC3807⟩, have been reported. The trifluoromethylated enamines (**83**, (*Z*)/(*E*) mixture) are synthesized by the condensation of various trifluoroacetamides with alkylidene phosphoranes (**82**), generated under salt-free conditions from the phosphonium salt (**81**) either with $NaNH_2$-hexamethyldisilazane or Bu^nLi (Scheme 17) ⟨89S309⟩.

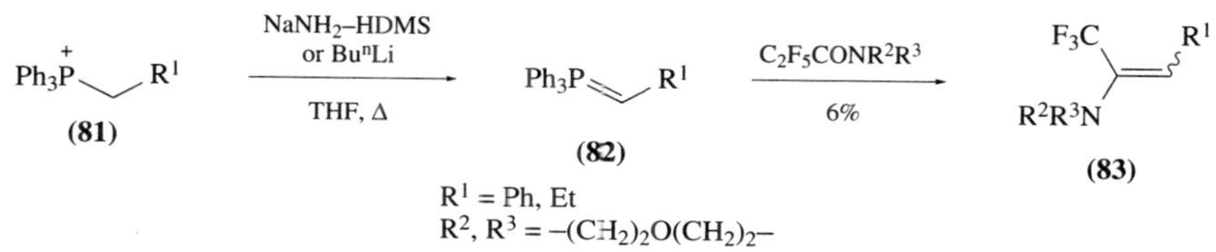

Scheme 17

Another group of enamines are the γ-metallated (silyl or stannyl) enamines ⟨B-63MI 215-01, 74S672, 82S717, 89S117, 92S1018⟩. These are prepared from α,β-unsaturated ketones (cyclic or acyclic), via β-stannylated carbonyl compounds, which react further with secondary amines, under standard reaction conditions, to give γ-stannylated enamines (**84**) (Scheme 18) ⟨89S117⟩.

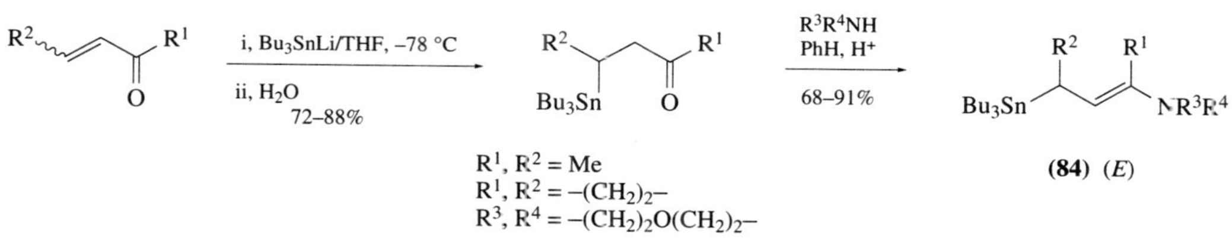

Scheme 18

With doubly enolizable ketones (**85**), the condensation occurs regioselectively (Scheme 19).

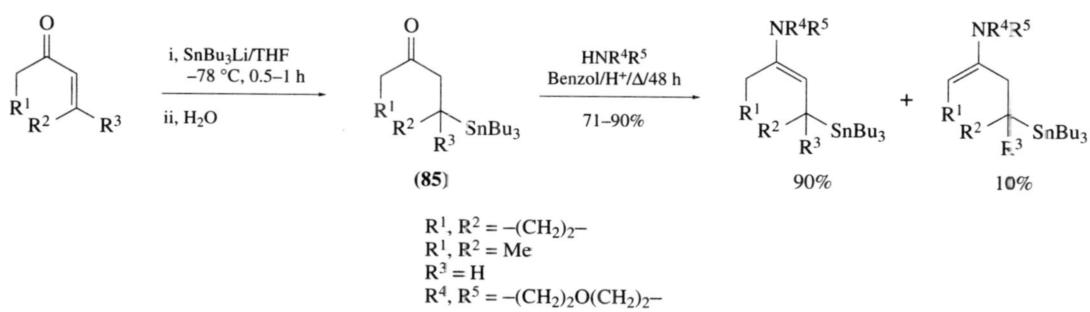

Scheme 19

More extensive reviews on the synthesis of enamines and arylamines, and their application in organic synthesis have been published (⟨63JCE194, B-63MI 215-01, B-70MI 215-01, 70S510, 79COC(2)131, 82T405, 82T3363, 83S517, B-88MI 215-01⟩ and references therein).

2.15.2.2 Arylamines

The various methods used for the preparation of aromatic amines have been reviewed by Lindsay ⟨79COC(2)131⟩.

2.15.2.2.1 *From hydrides (RH)*

The most widely used method for preparing aromatic amines involves nitration of the ring and subsequent reduction of the nitro group by a variety of reagents. A specific example is shown in Equation (34)) ⟨39JCS1299⟩.

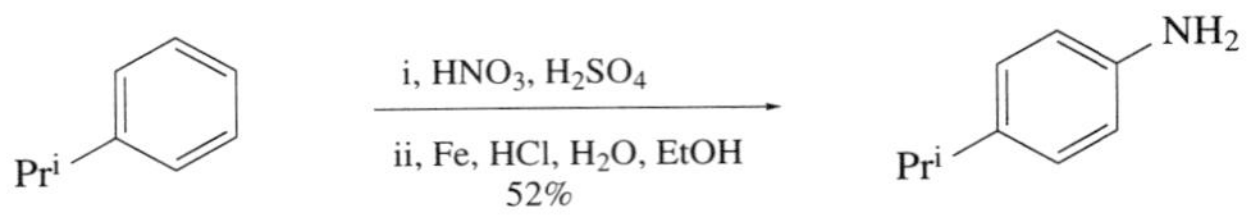

(34)

A variety of other less commonly used direct amination methods (Scheme 20) have been developed including the use of ammonia under high temperature and pressure and a catalyst consisting of a molybdenum, tungsten, or chromium salt with copper or nickel oxide ⟨60USP2948755⟩, NCl_3 ⟨66JA100⟩, Me_2NCl ⟨66CB1347⟩, $NH_2OSO_3H/AlCl_3$ ⟨61JA221, 61JA743⟩, $BuLi/NH_2OMe$ ⟨54OR(8)258⟩, Bu^nLi/TsN_3, Ni(Al) ⟨83TL4733⟩, and Pr^iNH_2/*m*-DCNB, *hν* ⟨87JOC753⟩.

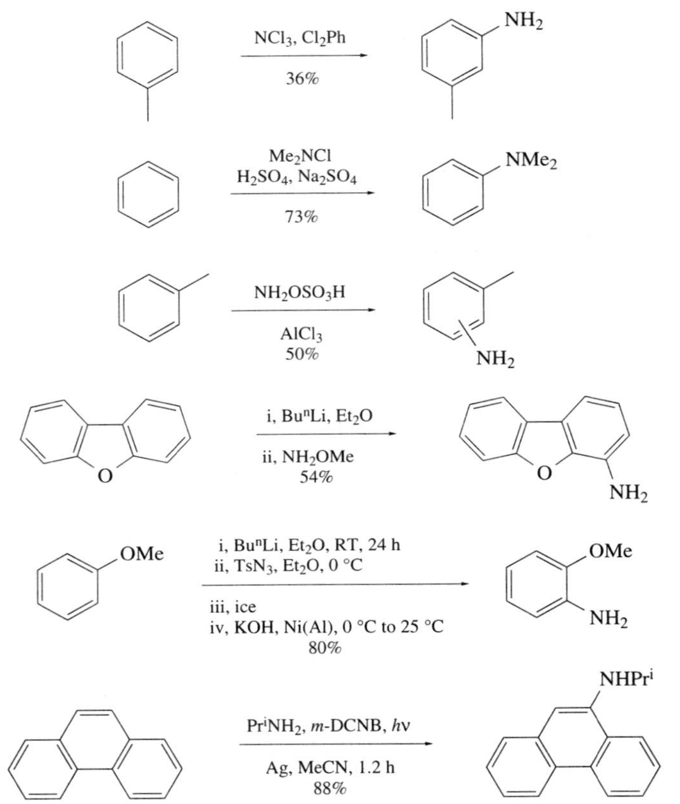

Scheme 20

2.15.2.2.2 *From nitro derivatives*

Nitro groups on aromatic rings have been reduced to the amine with a variety of common reducing agents and other reagents. An exhaustive list of these may be found in an excellent monograph by Larock ⟨B-89MI 215-01⟩.

Some examples employing more uncommon reagents and substrates containing other functional groups are given in Scheme 20. These include reduction of the nitro group on an anthraquinone derivative ⟨76S528⟩, reduction followed by substitution with a fluorine atom ⟨74JOC1758⟩, alkyne as an additional functionality in the molecule ⟨79JOC1233, 79JOC3671⟩, employing paraffin at 400 °C ⟨78S23⟩, and use of sodium formate in the presence of *N*-methylpyrrolidone ⟨81SC925⟩. Other examples include the use of tellurium ⟨80AG(E)1008, 80AG(E)1010, 84CL853, 86CL549⟩ and ruthenium complexes ⟨80TL2603, 84BCJ2440⟩. The nitro group on a phenyl ring carrying a chloro and a cyano group was also reduced to the amine employing tin chloride ⟨84TL839⟩ and the nitro group on an aromatic ring was reduced in preference to the one on an alkyl chain by the use of dry sodium sulfide in ethanol ⟨88TL635⟩. Some examples employing iron carbonyl complexes ⟨88TL5083⟩ and a polymer supported palladium reagents ⟨89TL251⟩ are also described (Scheme 21).

Di-*N*-methylated aromatic amines were obtained during the reduction of nitro group in the presence of formaldehyde (Scheme 22) ⟨73OSC(5)552, 76IJC(B)904⟩.

2.15.2.2.3 *From azides*

Aromatic amines were obtained by the reduction of the corresponding azide derivatives with a variety of reagents, a list of which may be found in Larock's monograph ⟨B-89MI 215-02⟩. An example employing vanadium chloride is shown in Equation (35) ⟨76S815⟩.

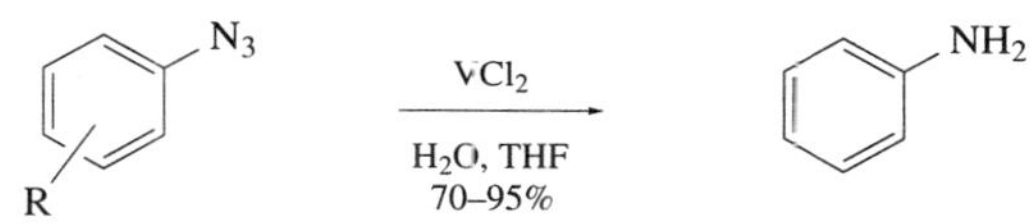

(35)

R = H, 2-Me, 4-Cl, etc.

2.15.2.2.4 *From halides*

Arylamines were prepared from aryl halides by employing magnesium metal in combination with azide reagents (Scheme 23), such as tosyl azide ⟨69JOC3430⟩, trimethylsilylmethyl azide ⟨83CC1322, 86TL6193⟩, phosphoryl azide ⟨84TL429⟩, and phenylthiomethyl azide ⟨81JA2483⟩.

Arylamines were also prepared from the corresponding halides by displacement with ammonia ⟨55OSC(3)307⟩ or a primary or secondary amine derivative in the presence of a metal (copper) ⟨87S383⟩, pressure ⟨87CL1187, 87CE551⟩, or heat ⟨49JA740⟩. Sodium amide in the presence of ammonia ⟨62CRV81⟩ and sodium or potassium salts of amines ⟨83JOC4397, 85JOC1334, 86JOC5157, 87JOC2619⟩ were also employed for this purpose. The last process is exemplified by the reaction of potassium phthalimide (**86**) with an aryl halide, a method known as the Gabriel synthesis (Equation (36)) ⟨69CC578⟩.

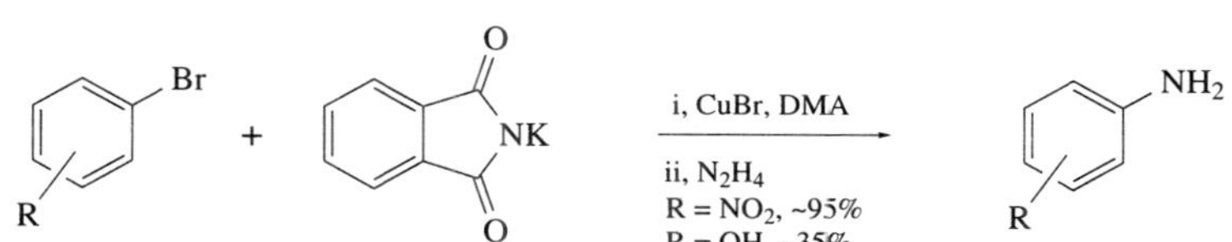

(36)

(**86**)

R = CN, NO_2, NH_2, OH

Aryl halides were also reacted with either an *N*-tin ⟨83CL927⟩ or *N*-copper ⟨74CC256⟩ derivative to yield the corresponding aryl amines (Scheme 24).

2.15.2.2.5 *From phenols*

Aryl amines were obtained from phenolic derivatives by the use of ammonia and sulfur dioxide ⟨42OR(1)105, 60JOC214⟩, $(EtO)_2P(O)Cl$ ⟨72JOC3571⟩, and a quinazoline (benzo[d]pyrimidine) reagent (Scheme 25) ⟨66JMC249, 68MI 215-01⟩.

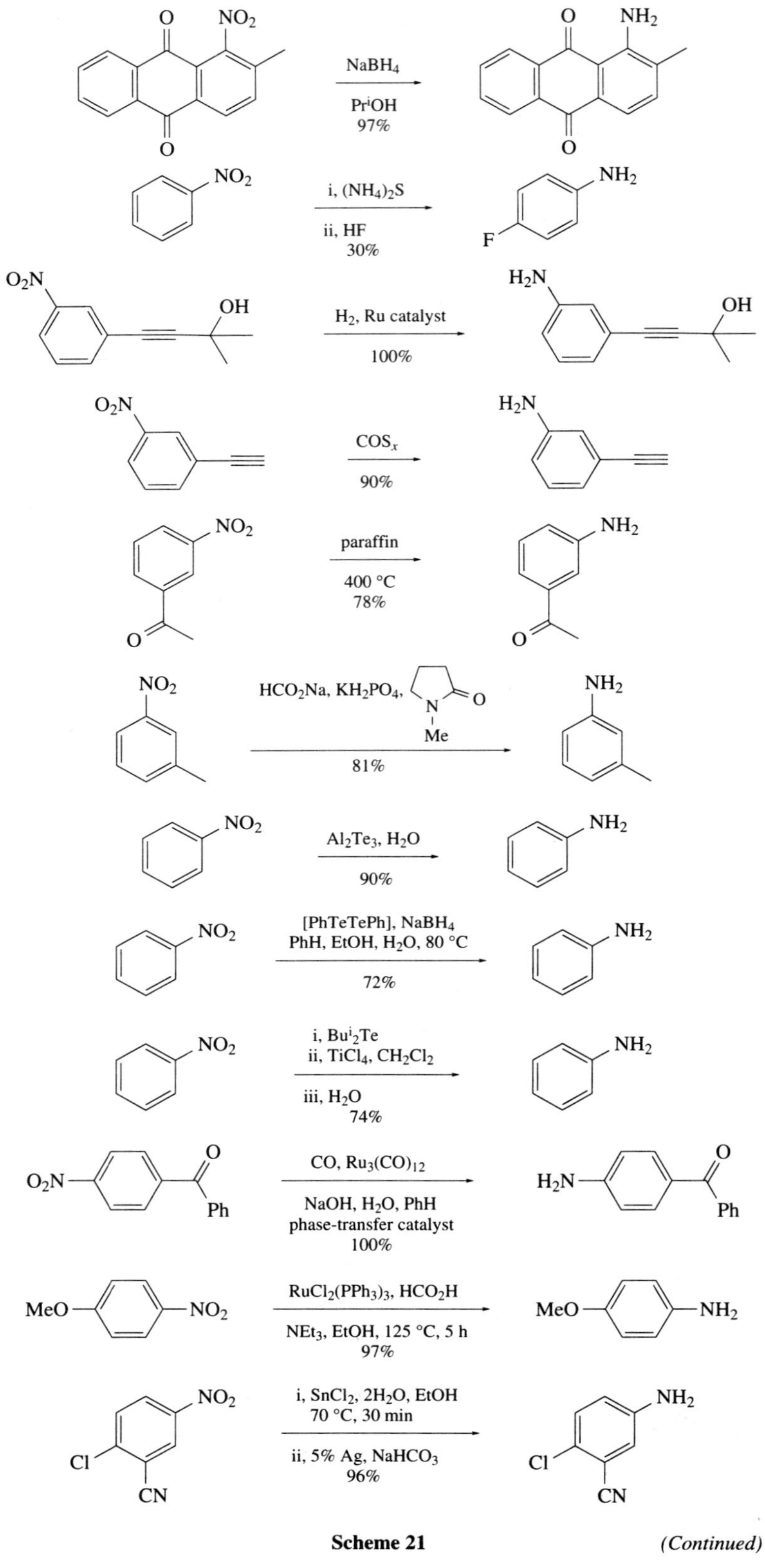

Scheme 21 *(Continued)*

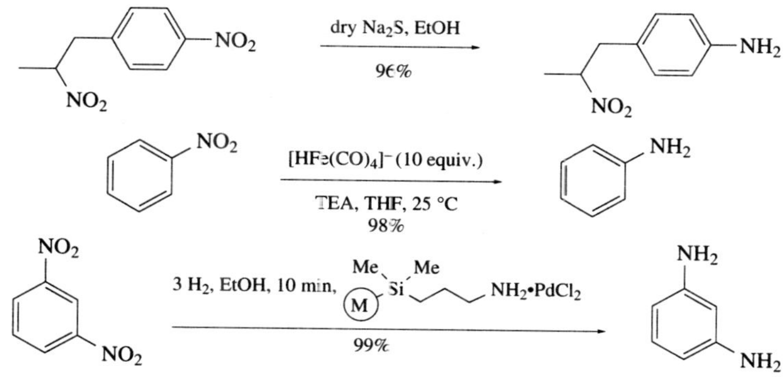

Scheme 21

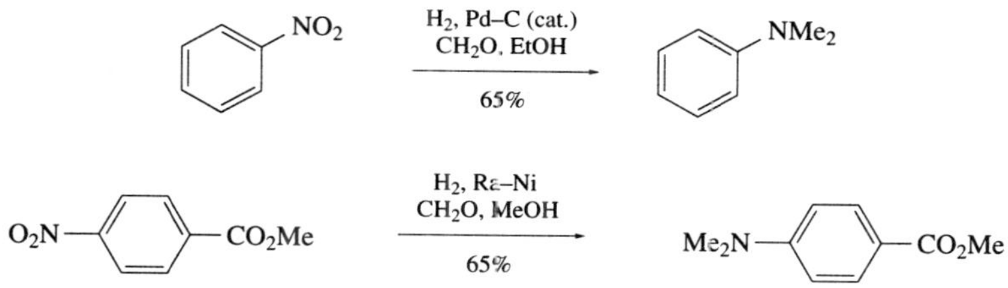

Scheme 22

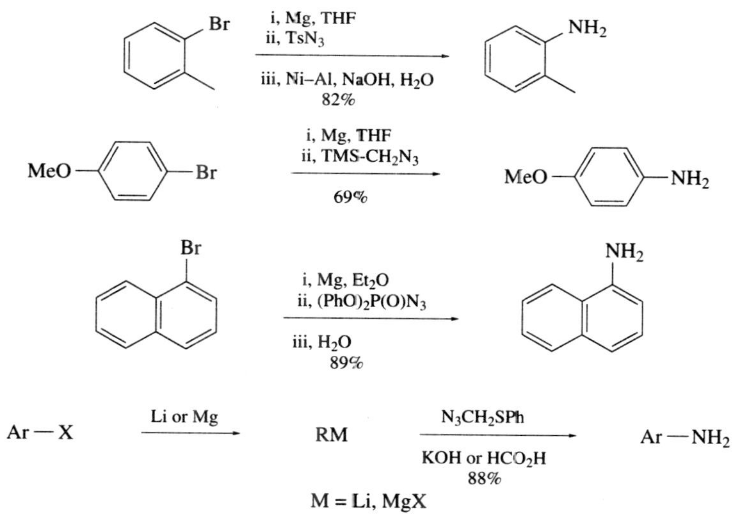

Scheme 23

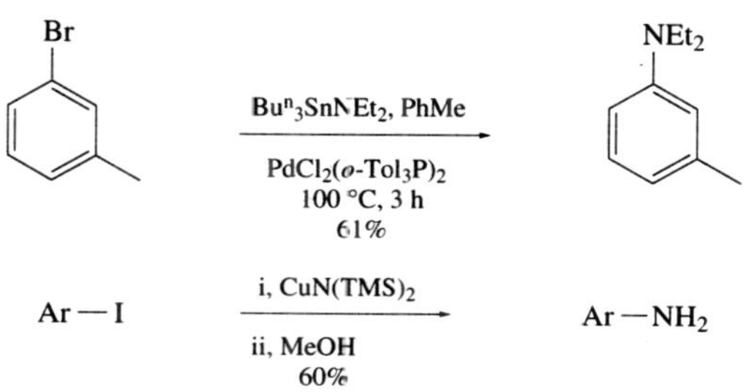

Scheme 24

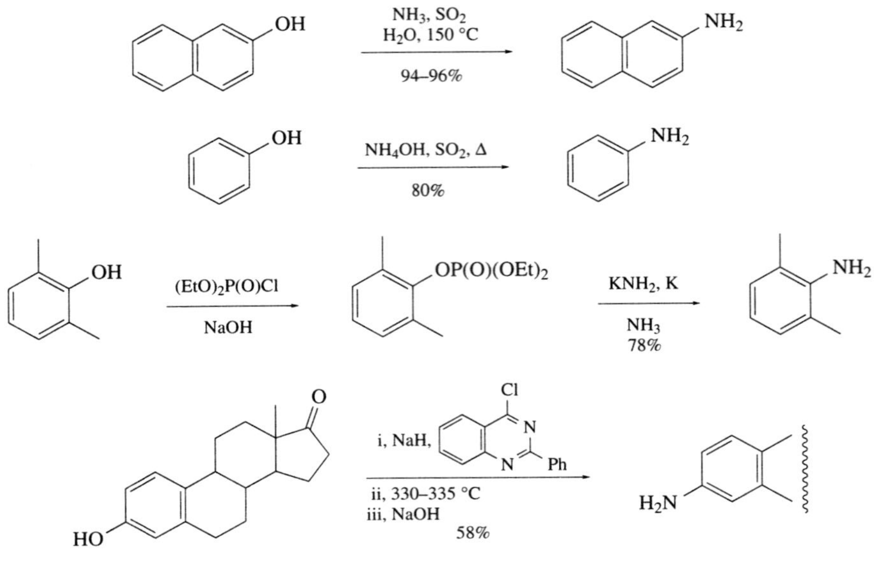

Scheme 25

2.15.2.2.6 *From ethers*

An *N*-methyl amine derivative was made from the methoxy compound under photochemical conditions (Equation (37)) ⟨66RTC56⟩.

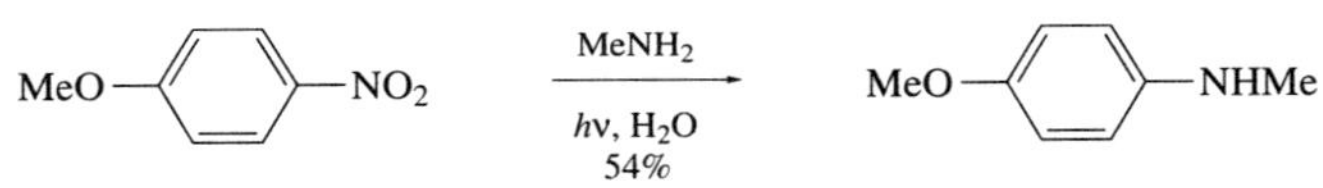

(37)

2.15.2.2.7 *From amines*

Aryl amines were also prepared from other amine derivatives (Scheme 26), such as *N*-tosyl ⟨88JOU2222⟩ and *N*-carboxy derivatives ⟨69SC1347⟩.

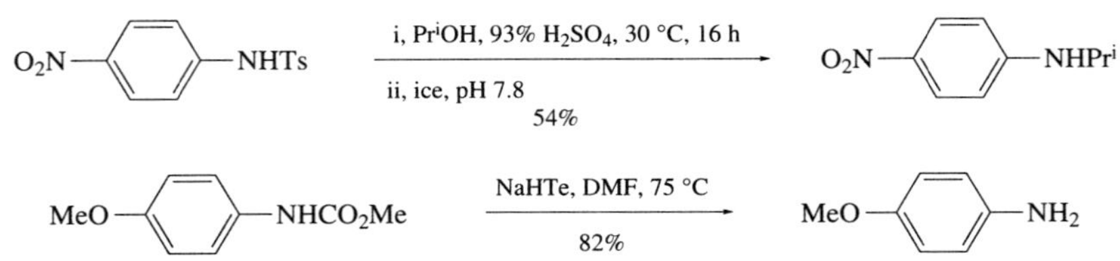

Scheme 26

2.15.2.2.8 *From carboxylic acids and derivatives*

Aromatic amides, generated from acids or acid halides, react with solutions of bromine or chlorine in sodium hydroxide to yield amines through a reaction known as the Hofmann rearrangement (Equation (38)) ⟨46OR(3)267, 49JA3929⟩.

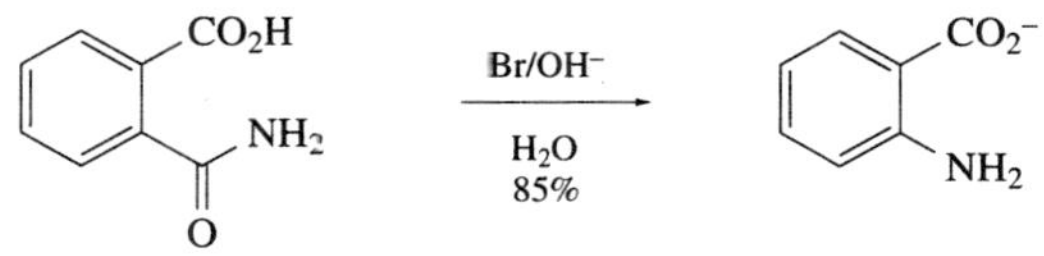

(38)

Aromatic acyl azides, prepared from the reaction of acyl chlorides with sodium azide, undergo Curtius rearrangement to isocyanates which hydrolyze on the addition of water to the aryl amines with the liberation of carbon dioxide (Scheme 27) ⟨21RTC285, 43CRV209, 46OR(3)337, 53JA2014, 63OSC(4)819, 73OSC(5)273, B-89MI 215-03⟩.

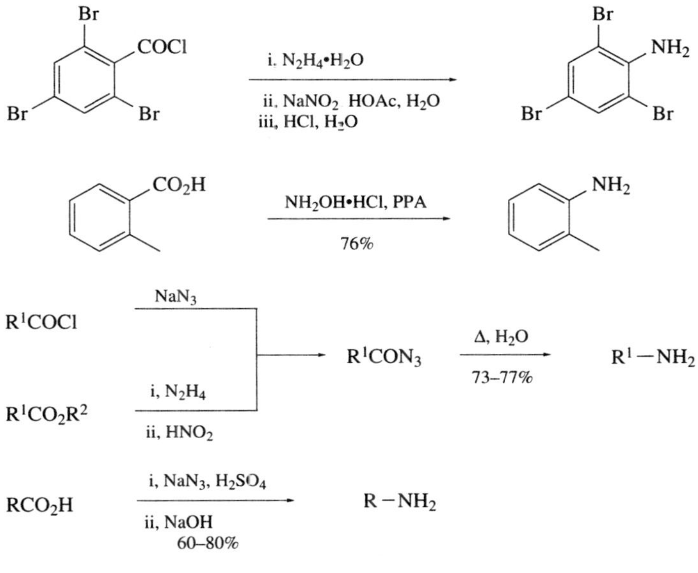

Scheme 27

2.15.3 SALTS OF ENAMINES AND ARYLAMINES

2.15.3.1 Salts of Enamines

Protonation of enamines first occurs at nitrogen ⟨63LA(665)101, B-69MI 215-02⟩ under kinetic control and the resulting enammonium salts have been isolated in some cases ⟨77BCJ1513⟩. Their rearrangement into the thermodynamically favored iminium salts has been followed by Matsushita *et al.* ⟨77BCJ1513⟩ by means of ^{1}H-NMR and IR techniques (Scheme 28). The subject of protonation, alkylation and acylation of enamines has been reviewed by Pitacco and Valentin ⟨B-82MI 215-01⟩. The vinylamine–imine tautomerism and conformational preferences have also been studied ⟨94JA642⟩.

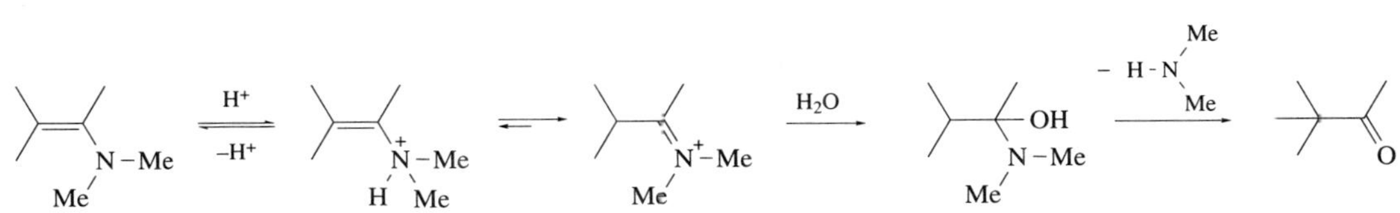

Scheme 28

When enamines are treated with alkyl halides, either *C*-alkylation or *N*-alkylation can occur. *N*-Alkylation was observed with unactivated halides as electrophiles and in particular with aldehydes ⟨68CR(C)623, 70CI(L)1188⟩. Some electrophilic alkenes, such as α-chloroacrylonitrile and methyl α-chloroacrylate react with morphilinocyclohexene (**87**) to give hexahydroindolium salts of the type

(**89**) in polar solvents. Cyclobutane adducts (**88**) were found in ether solution at low temperature; however, these are thermally unstable and easily converted into the salts (**89**) (Scheme 29) ⟨76BSB805⟩.

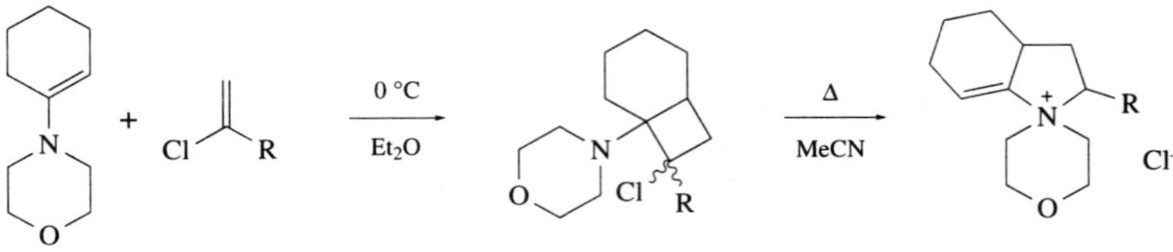

(**87**) (**88**) (**89**)

R = CN, CO_2Me

Scheme 29

C-Alkylated iminium salts, when subjected to hydrolysis, yield ketones. Since the enamine is normally formed from a ketone, the net result is alkylation of the ketone at the α-position. The method is known as the Stork enamine reaction (Scheme 30) ⟨63JA207, B-85MI 215-01⟩. The stereochemical implications in the formation of α-alkylated carbonyl compounds for which the iminium salts serve as intermediates have been well documented ⟨B-82MI 215-01⟩.

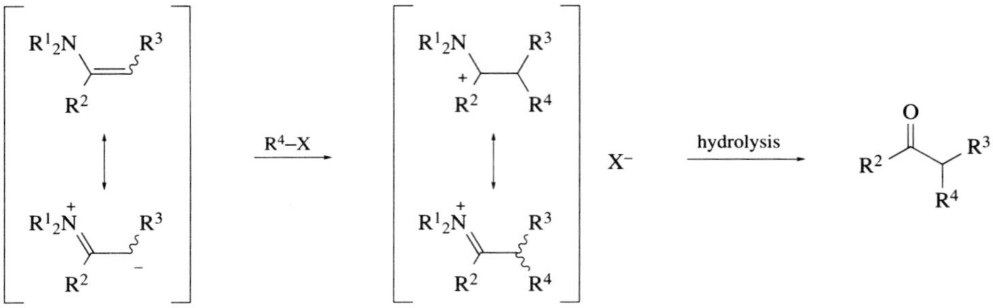

Scheme 30

Other examples of alkylation include α-bromocarbonyl compounds ⟨76ACS(B)1000⟩, enamines, and α-chloro enamines ⟨83CL1791⟩ leading to chloro alkylimidium chloride salts (**90**) (Equation (39)) ⟨69AG468, 76ACS(B)1000, 77S848⟩.

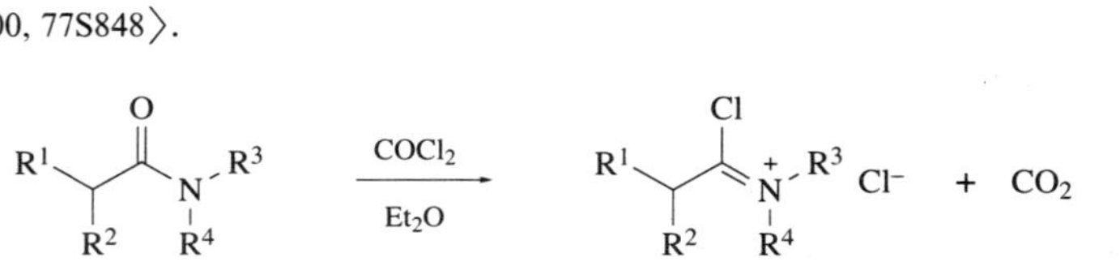

(**90**) (39)

α-Chloro enamines (**91**) ⟨B-76MI 215-01, 79CC1180, 80OS(59)26⟩, easily prepared by the treatment of chloroiminium salts with triethylamine, are known to react with carboxylic acids to form reactive carboxymethyleniminium chlorides (**92**) *in situ* without the liberation of hydrogen chloride. Accordingly, these can be reacted with one equivalent of Grignard reagent to obtain ketones in a one-pot operation. Thus, an α-chloro enamine may be used as a mild activating reagent for carboxylic acids (Scheme 31) ⟨83CL1791⟩.

Reaction of *N*,*N*-dimethyl-1-chloro-2-methylpropenylamine (**93**) with a slurry of freshly dried

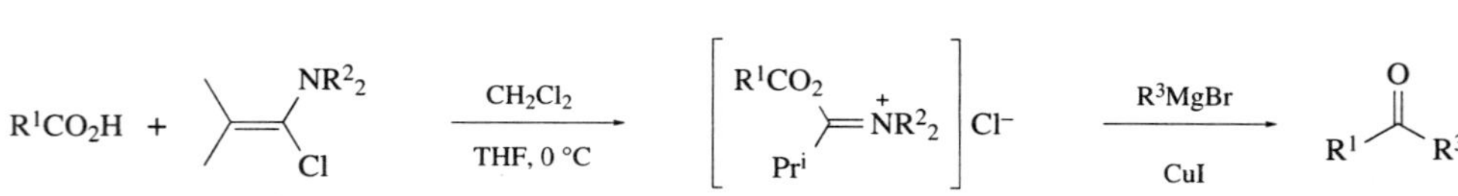

(**91**) (**92**)

Scheme 31

zinc chloride in refluxing dichloromethane affords a salt which reacts with an alkene to give the cyclobutylideneammonium salt (**94**) (Scheme 32) ⟨74AG272⟩.

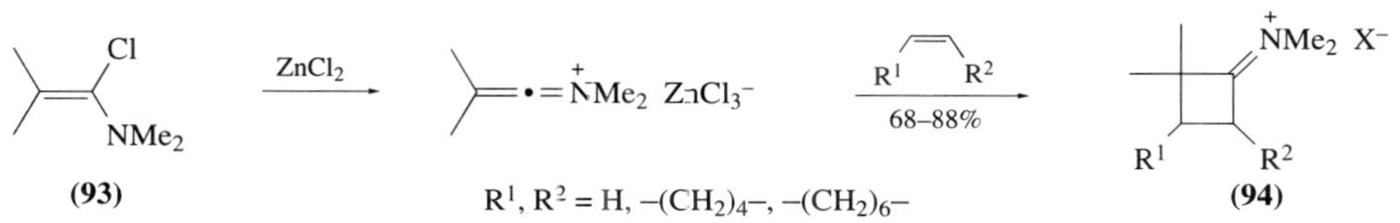

Scheme 32

Lewis acid-catalyzed electrophilic addition of 3-ethoxy-2-alkyl-acrolein to the nucleophilic carbon of an enamine gives attack at the aldehyde function, and 2-(3′-ethoxy-1′-hydroxy-2′propenyl)cycloalkanones (**95**) are obtained upon mild hydrolysis of the zwitterionic intermediate (Scheme 33) ⟨82TH1⟩.

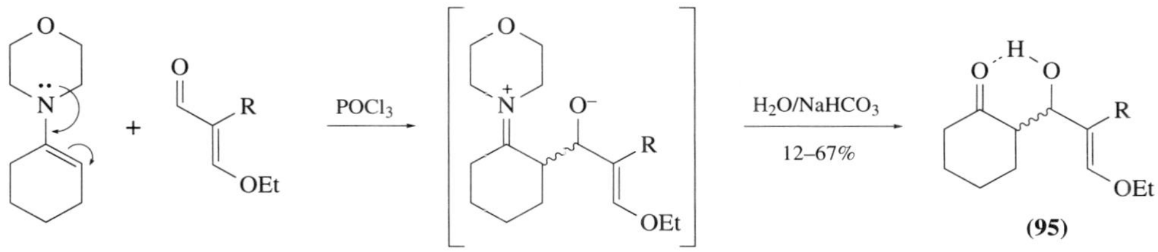

Scheme 33

Protonation of a mixture of isomeric enamines (**96**) and (**97**) by trifluoroacetic acid gives the iminium salt (**98**), which is regiospecifically deprotonated with *t*-butylamine, yielding the enamine isomer with a terminal double bond (**99**) (Scheme 34) ⟨78ACS(B)646⟩.

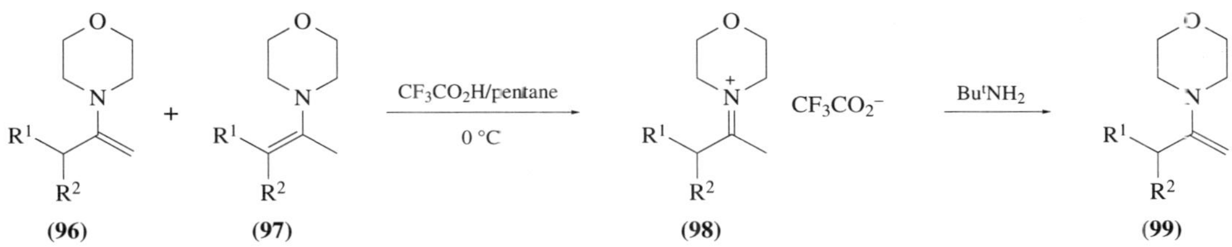

Scheme 34

β-Lithiated enamines, formed via the iminium salt, either in (*E*) or (*Z*) isomeric form, represent highly reactive carbon nucleophiles and thus are useful as enolate equivalents in organic synthesis ⟨70JOC1204, 77JA8356, 77TL3589, 79JOC3585, 82BSF(2)297, 82JCR(M)621, 82JCR(S)48, 83JCR(M)2101, 83JCR(S)222, 84T2989, 86JA304⟩. These nonambident "anions" react rapidly at low temperature with a wide range of electrophiles, such as deuterium oxide ⟨77JA8356, 79JOC3585, 86JA304⟩, alkyl halides ⟨77JA8356, 82JCR(M)621, 82JCR(S)48, 86JA304⟩, aldehydes ⟨77JA8356, 77TL3589, 82JCR(M)621, 82JCR(S)48⟩, and chloromethylsilanes ⟨81JOM(218)309, 83JCR(M)2101, 83JCR(S)222⟩, generating either β-substituted enamines or, after acid hydrolysis, the corresponding α-substituted carbonyl compounds. In 1991, Duhamel *et al.* ⟨91S649⟩ reported that β-lithiated enamines (**100**) undergo Michael addition with α,β-unsaturated esters and described a new and efficient synthesis of enamino esters and mono- or disubstituted-5-oxo esters upon acidic hydrolysis, with complete diastereoselectivity (de >98%) in a tandem Michael–alkylation process (Scheme 35).

2.15.3.2 Salts of Arylamines

Aromatic amine salts may be prepared simply by treating primary, secondary or tertiary amines with the common mono- or polyprotic mineral acids or organic acids capable of protonating the amine (Equation (40)).

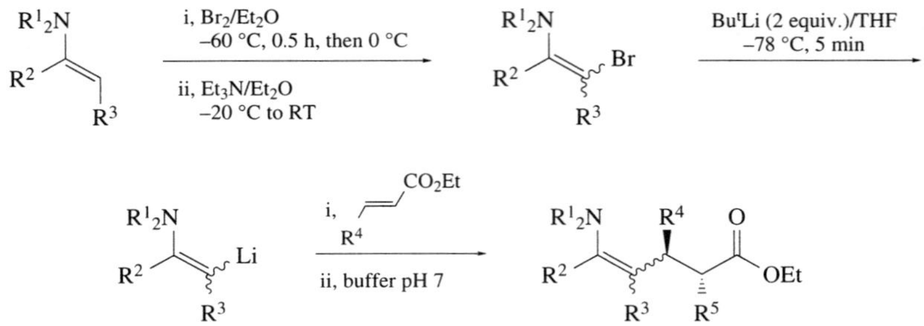

(100) $R^5 = H$
R^5 = Me; quenching enolate with MeI

Scheme 35

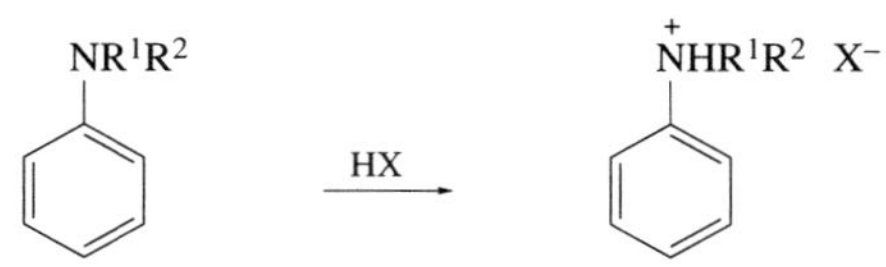

(40)

$X^- = F^-, Cl^-, Br^-, I^-, CF_3CO_2^-, CF_3SO_3^-$, etc.

$R^1 = R^2 = H$
$R^1 = H, R^2$ = substituent
R^1, R^2 = substituent

Quaternary ammonium salts are prepared in general by treating tertiary amines with any one of a variety of alkylating agents. An example in which methylbenzene sulfonate is used to alkylate dimethylaniline is shown in Scheme 36. If needed, the counteranion can be exchanged, in this case from sulfate (**101**) to ethoxide (**102**) (Scheme 36) ⟨53JA3632⟩.

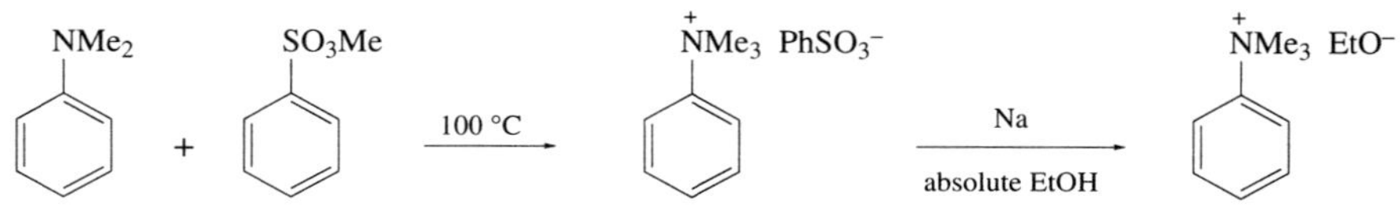

(101) **(102)**

Scheme 36

The use of methyl iodide to prepare the methiodide derivative (**104**) of (**103**) is shown in Equation (41) ⟨76IJC(B)904⟩.

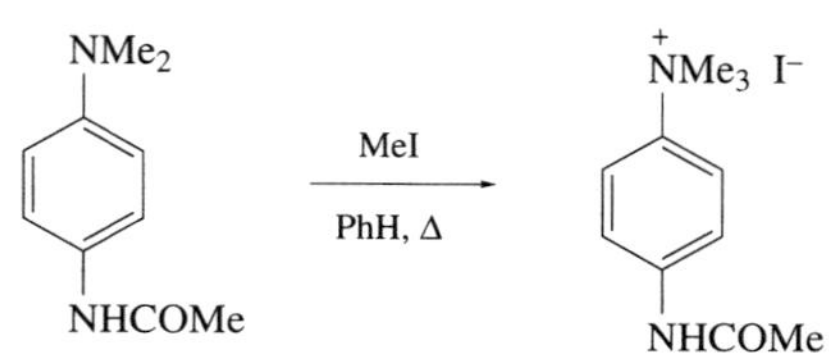

(41)

(103) **(104)**

Primary aromatic amines can be directly converted to the corresponding trimethyl quaternary salts by treating them simultaneously with dimethyl sulfate and sodium methoxide, and finally making the reaction mixture alkaline with 30% sodium hydroxide. The salts are crystallized from a mixture of acetone and methanol (Equation (42)) ⟨70TH134⟩.

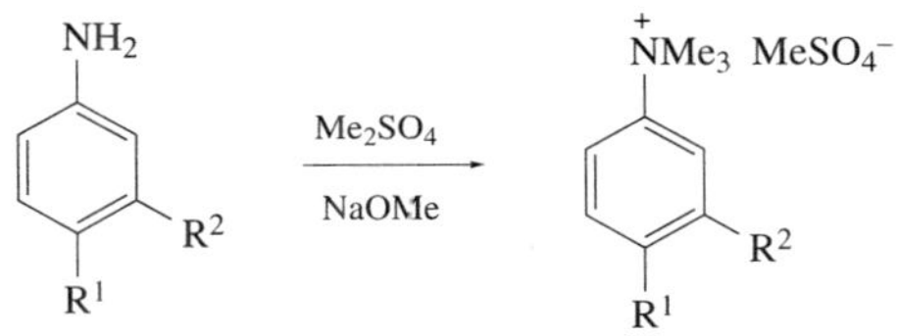

(42)

$R^1 = Me, R^2 = H$
$R^1 = CO_2Me, R^2 = H$
$R^1, R^2 = CO_2Me$

2.15.4 *N*-HALOENAMINES AND ARYLAMINES

Most methods of preparation of *N*-halo amines involve the halogenation of secondary amines ⟨70CR639⟩. *N*-Chloroamines have been the most widely studied, using aqueous hypochlorite or *t*-butyl hypochlorite as the source of positive chlorine. Monoalkylamines can be mono- or dichlorinated in a similar way, by employing appropriate amounts of the chlorinating agent (Equation (43)).

$$R\text{–}NH_2 + HOCl \xrightarrow{-H_2O} RNHCl \qquad (43)$$

Another method involves the cleavage of trimethylsilylamines with chlorine gas (Equation (44)) ⟨69ZN(B)774⟩.

$$R_2N\text{-TMS} + Cl_2 \longrightarrow R_2NCl + \text{TMS-Cl} \qquad (44)$$

Halogenation of an aromatic amine may occur either on carbon (double bond) or on nitrogen. However, hypochlorous acid and the hypochlorite ion are far better *N*- than *C*-chlorinating agents and react readily with secondary and tertiary amines. Thus, a solution of hypochlorous acid in ether reacts with aniline at −20°C to give *N,N*-dichloroaniline or *N*-chloroaniline ⟨72JA3884, 74JA5512, 76AJC367, 77TL2235⟩, depending on the reaction conditions (Equation (45)) ⟨B-68MI 215-02⟩. *N*-Bromo amines are prepared in a similar way, using sodium hypobromite.

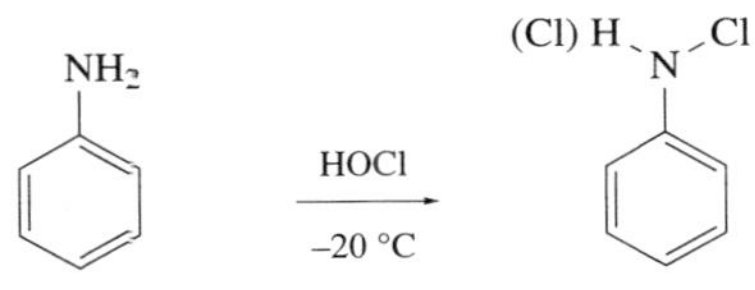

(45)

Indoles were transformed into 3-chloroindolenines (**106**) with sodium hypochlorite ⟨78JOC2639⟩, but it was shown that the reaction proceeded via the intermediacy of *N*-chloroindole (**105**) (Scheme 37).

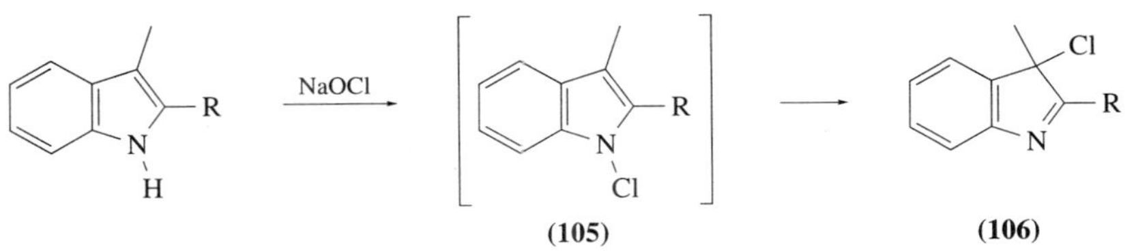

(**105**) (**106**)

Scheme 37

Thus, stirring a solution of indole in *n*-hexane, chloroform, or carbon tetrachloride with a freshly prepared solution of aqueous hypochlorite resulted in the formation of a light yellow solution containing an intermediate which could oxidize iodide ion to iodine. The structure of the titrimetrically observed species could be either *N*-chloroindole or 3-chloro-3H-indole. The yield varied between 80% and 92%, the highest yield obtained by using a concentration of starting indole of 0.01 M and with the aqueous sodium hypochlorite prepared from fresh calcium hypochlorite.

Reaction of (**107**) with sodium hypochlorite at −5°C to −10°C gave the very unstable 3-chloro-2,3-dimethylindolenine (**109**), which when treated with silver ion in methanol produced the

3-methoxyindolenine derivative (**110**). The possibility that the *N*-chloro derivative (**108**) might actually be the intermediate for (**109**) was supported by the conversion of the *N*-hydroxy compound (**111**) to (**113**) via the *p*-toluenesulfonate ester (**112**) (Scheme 38) ⟨72T2749⟩.

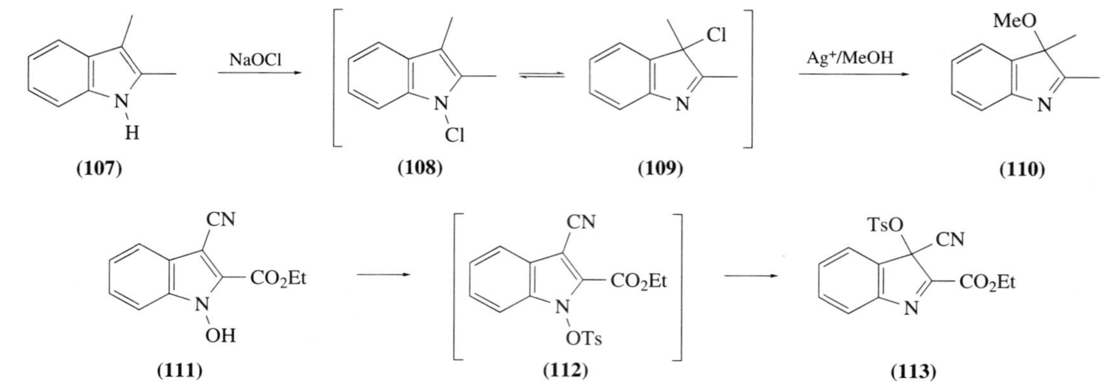

Scheme 38

The reaction of tetrafluorohydrazine with alkynic compounds is more complicated than that with alkenes, although the overall reaction appears to be similar ⟨67CRV655, 67PC-01⟩. The vinylbisdifluoroamine intermediate (**114**) is apparently not stable in most cases and undergoes rearrangement (Scheme 39).

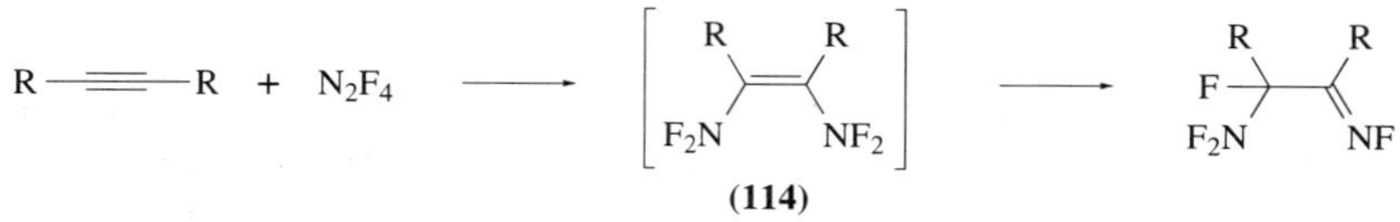

$R = Ph, CO_2Me, CN, Me$

Scheme 39

The product isolated from the reaction of hexafluoro-2-butyne ⟨67CRV665, 67PC1⟩ with tetrafluorohydrazine conducted at 170 °C lends strong support for this hypothesis. The intermediate vinyl difluoroamine derivative (**115**) is isolated in 90% yield as a 1 : 2 mixture of the *cis* and *trans* isomers (Equation (46)).

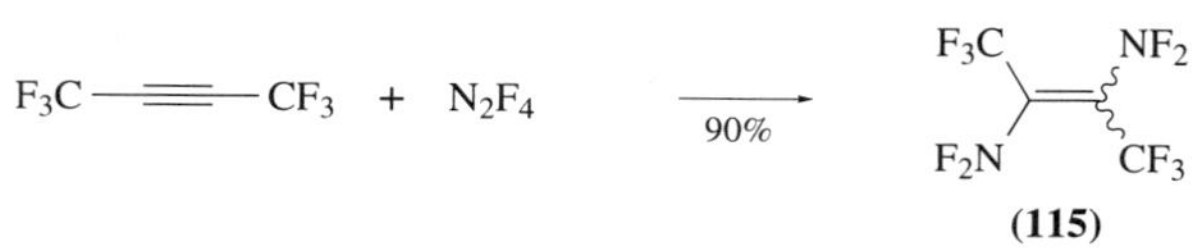

(46)

Allene (**116**) appears to undergo a rearrangement similar to that described above for the alkynic compounds, when treated with tetrafluorohydrazine (Scheme 40) ⟨67CRV665⟩.

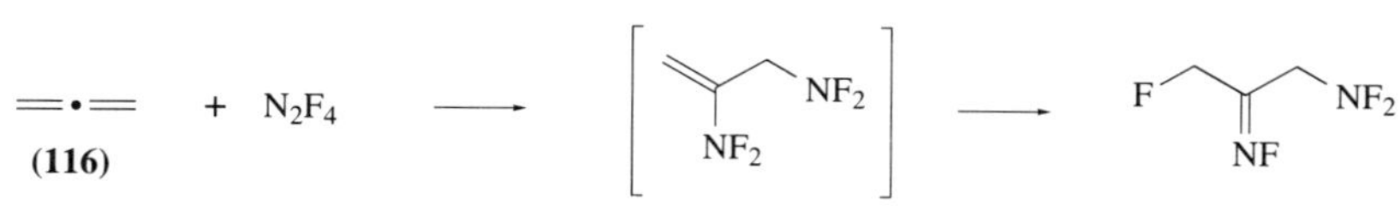

Scheme 40

Some evidence has been presented for the existence of *N*-fluoroammonium derivatives and related compounds ⟨66MI 215-01, 66MI 215-02, 67CRV665⟩. The complex which results from the treatment of dimethylfluoroamine with anhydrous hydrogen chloride may contain the $(CH_3)_2NHF^+$ cation ⟨66IC1629⟩. The evidence for this ion is based on the solubility characteristic of the complex and on its ^{19}F NMR spectrum. The salt, *N*-fluoro-*N*-methylisopropylidenammonium tetrafluoroborate

(**117**) can be obtained from the reaction of *t*-butyldifluoroamine with boron trifluoride (Equation (47)).

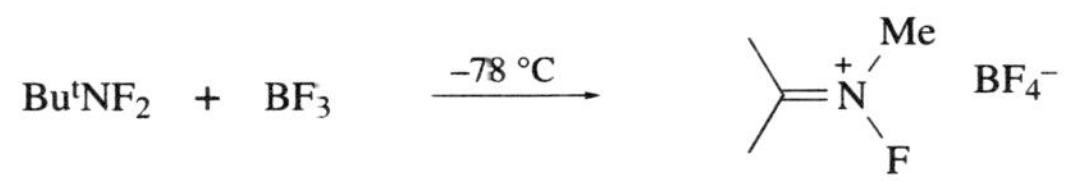

(**117**) (47)

The conversion of the difluoroamine (**118**) to the fluorosulfite (**120**) (Scheme 41), is an example of the migration of the phenyl group to nitrogen to produce *N*-phenyl-N-fluoroiminium salts (**119**) ⟨68JOC2664⟩, therefore this reaction also falls into the aromatic nucleophilic substitution class.

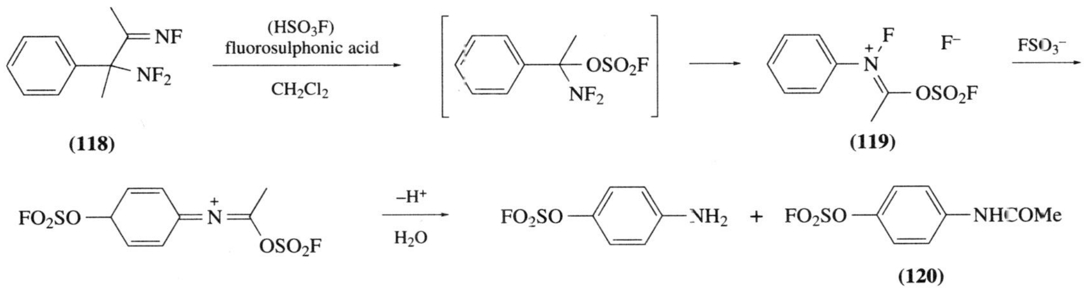

(**118**) (**119**) (**120**)

Scheme 41

2.15.5 HYDROXYLAMINES AND RELATED FUNCTIONS

The organic derivatives of hydroxylamine, NH_2OH, may be divided into the following classes:

(i) Alkyl and aryl substituted products in which one or more hydrogen atoms are replaced by organic groups.

(ii) *N*-Acyl substitution products: the hydroxamic acids and their derivatives.

(iii) Amine oxides that can be regarded as being derived from the hypothetical tautomeric form of hydroxylamine $H_3N^+—O^-$.

Acyclic hydroxylamines fall into three categories according to the substitution pattern on nitrogen and/or on oxygen. Basically, there are two types of monosubstituted hydroxylamines, either with the substituent on nitrogen (**121**) or with the substituent on the oxygen (**122**). Similarly, for disubstituted hydroxylamines, both substituents can be on nitrogen (**123**) or both nitrogen and oxygen can each possess one substituent (**124**). Obviously, only one type of acyclic trisubstituted hydroxylamine (**125**) is possible. In a similar pattern of substitution, cyclic hydroxylamines fall into the four subclasses—Structures (**126**)–(**129**). The methods of preparing these compounds are presented in the following pages, and only vinyl- and aryl-nitrogen compounds attached to one heteroatom by a single bond are treated.

RHNOH	H_2NOR	R_2NOH	RHNOR	R_2NOR
(**121**)	(**122**)	(**123**)	(**124**)	(**125**)

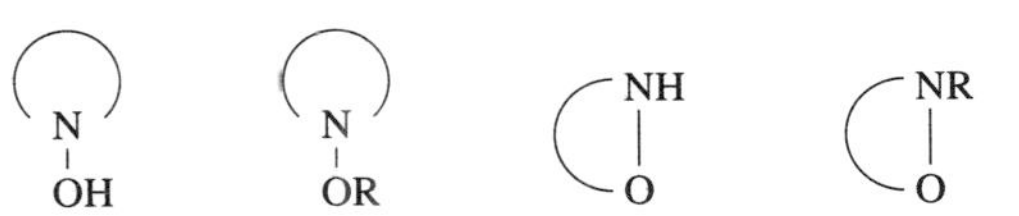

(**126**) (**127**) (**128**) (**129**)

2.15.5.1 Hydroxylamines

A general reductive technique to prepare *N*-monosubstituted acyclic hydroxylamines is the conversion of oximes to hydroxylamines with borane ⟨65JOC2877⟩ or sodium cyanoborohydride in an acidic medium ⟨71JA2897, 75S135, 76JOC863⟩. It has been demonstrated ⟨76TL233⟩ that unhindered oximes will react with excess organolithium reagents to produce *N*-alkylhydroxylamines (**130**) (Scheme 42).

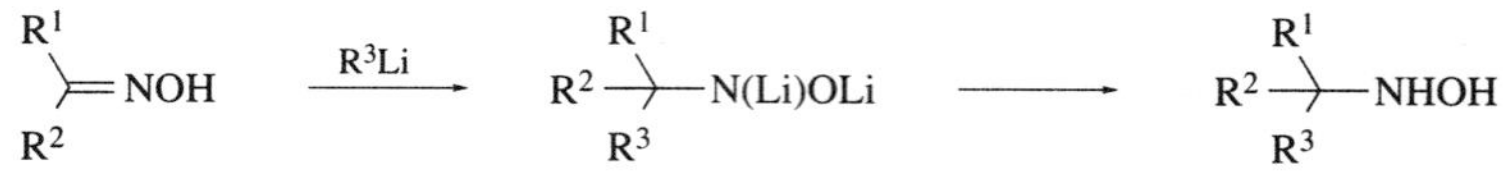

(130)

Scheme 42

Suitably activated benzene derivatives such as (**131**) (X = Cl, Br or OMe) can be directly converted into *N*-arylhydroxylamines by reaction with hydroxylamine (Equation (48)) ⟨57JA5739⟩.

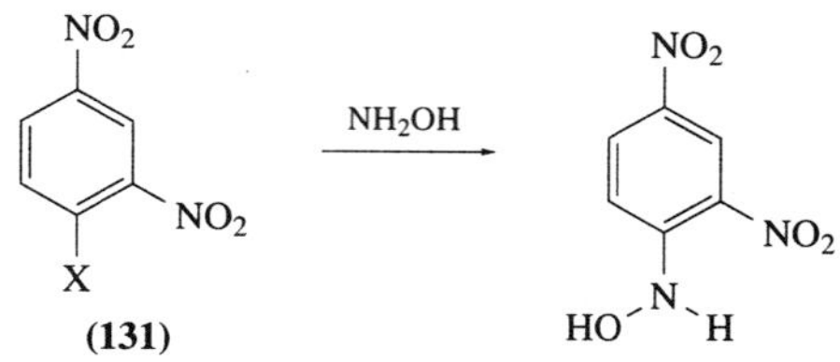

(131)

(48)

By far the most common and satisfactory method of preparing these compounds is by the reduction of a nitrogen-containing compound which is at a higher oxidation level than the hydroxylamine itself. For example, the reduction of nitro, oxime, and to a lesser extent, nitroso derivatives, has found extensive use. In the case of reduction of nitro compounds, the most commonly used procedure is zinc in the presence of ammonium chloride ⟨1894CB1548, 59JCS2094, 69JOC2269⟩. Yields in this reaction can be quite variable for both *N*-aliphatic and *N*-aromatic hydroxylamines and thus a large number of alternative reducing agents have been introduced over the years. These include the use of catalytic hydrogenation ⟨22CB875, 57LA(606)24, 70JOC2790⟩, electrochemical reduction ⟨51JA5675, 66BSF1867⟩, diborane ⟨65JOC2880⟩, aluminum amalgam ⟨64JA1139, 72OS(52)77⟩, and sodium hydrogen sulfide in the presence of calcium chloride (Scheme 43) ⟨21JCS768⟩.

$$R-NO_2 \xrightarrow{[H]} R-NO \xrightarrow{[H]} R-NHOH \xrightarrow{[H]} R-NH_2$$

Scheme 43

2.15.5.2 Nitrones

A discussion on the synthesis of nitrones is included here as they are known to exist in equilibrium with the tautomeric hydroxyenamines. The tautomerism depicted in Equation (49) between the nitrone (**132**) and the hydroxyenamine (**133**) was shown to exist completely towards the nitrone by NMR studies ⟨62CJC177⟩.

(132) **(133)**

(49)

However, indolenine *N*-oxides (**134**) have been known to exist in equilibrium with their *N*-hydroxyindole tautomers (**135**), the proportion of the latter can be increased by employing polar solvents (Equation (50)) ⟨65CR(260)2851, 67BSF1296, 78JCS(P1)1117⟩.

(134) **(135)**

(50)

The nitrone–hydroxyenamine equilibrium has been found to be influenced by the heteroatom as shown by the reaction of the oxime (**136**) with diazomethane (Scheme 44) ⟨76CB3462⟩.

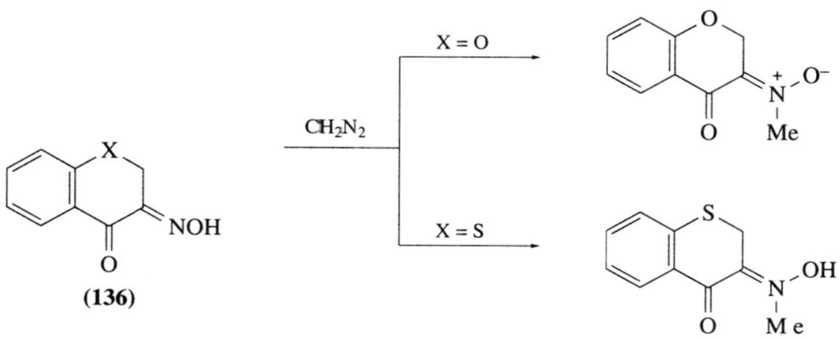

(136)

Scheme 44

Silver ion-catalyzed generation of cyclic nitrones (**137**) from allenic oximes has been used as 1,2-dipolarophiles for the synthesis of pyrrolizidine alkaloids (Scheme 45) ⟨68CC1017⟩.

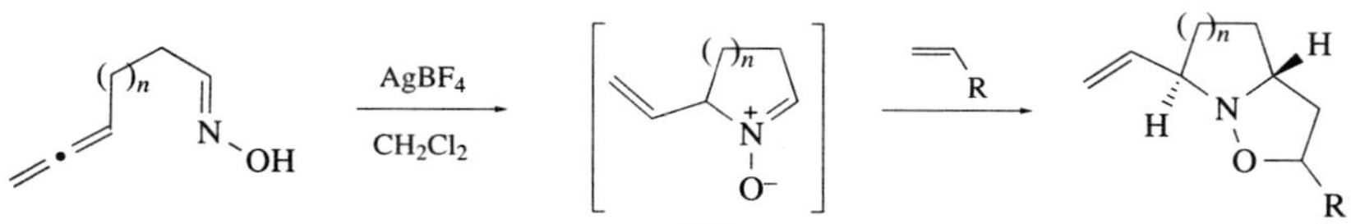

(137)

Scheme 45

Similarly oxime (nitroso) derivatives have been used in the synthesis of polycyclic nitrones ⟨75JCS(P1)1764, 76JCS(P1)2365⟩. Functionalized nitrones, such as acylnitrones, are synthesized in good yields by base-catalyzed condensation of nitrosoarenes with phenacylpyridinium bromide (**138**), an active methylene component in which the pyridinium moiety is a good leaving group, to give acylnitrones (**139**) (Equation (51)).

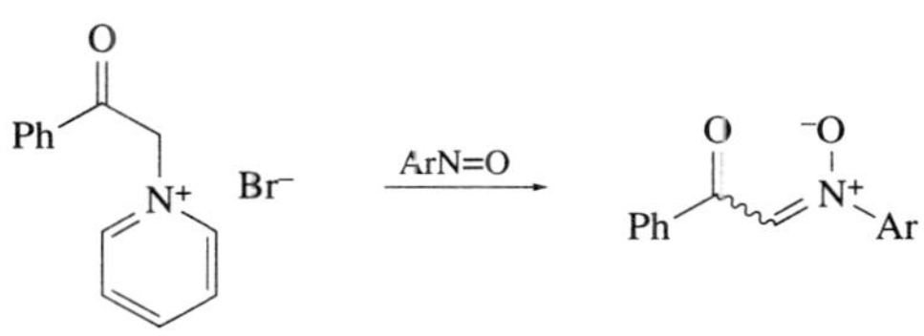 (51)

(138) **(139)**

Nitrones are also the end-products of the reactions of both alkyl and aryl nitroso compounds with diazoalkanes or sulfonium ylides ⟨64CRV473, 65QR329, B-66MI 215-03, 68HOU(10/4)311, 69JCS(C)1073, 71RTC548⟩. The reaction may involve oxaziridine intermediates (**140**) (Scheme 46).

2.15.5.3 *N*-Oxides

The most extensive synthetic investigation of dipole-stabilized carbanions from amine oxides has been carried out by Abromovitch *et al.* who have shown that a variety of 2-lithiopyridine *N*-oxides can be generated from the corresponding pyridine *N*-oxides and trapped in moderate yield by electrophiles ⟨67JA1537, 72JOC1690, 72JOC3584, B-74MI 215-01⟩. An example in which a cyclic *N*-hydroxy enamine (**142**) is generated from a pyridine *N*-oxide (**141**) ⟨69JHC989⟩ is shown in Scheme 47. Chloro, methyl, and ethoxy substituents in the 3- and 4-positions do not interfere with the metallation.

Tertiary amines can be converted to amine oxides by oxidation with hydrogen peroxide, but peracids are also important reagents for this purpose. An example in which benzoylperoxide is employed for this purpose is illustrated in Equation (52). Thus, addition of benzoylperoxide to

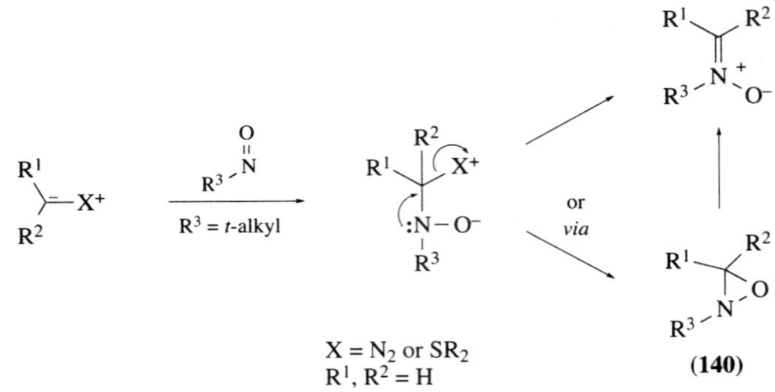

(140)

Scheme 46

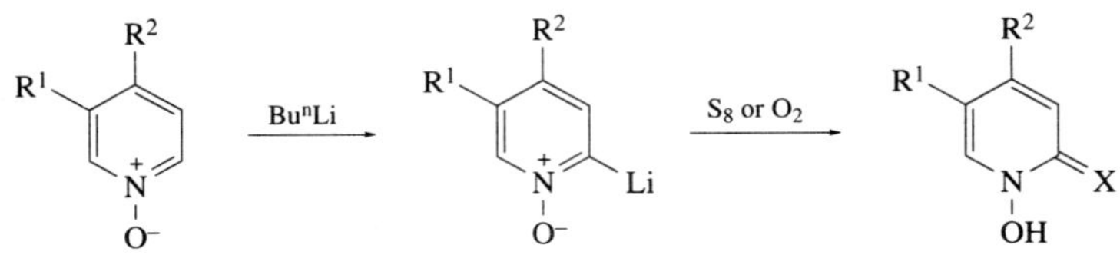

(141) **(142)**

R^1 = H, Me; R^2 = H, Me, Cl   X = O, S

Scheme 47

benzoquinuclidine (**143**) in $CHCl_3$ gave the *N*-oxide (**144**) in 80% yield, isolated as the picrate ⟨65TL783⟩.

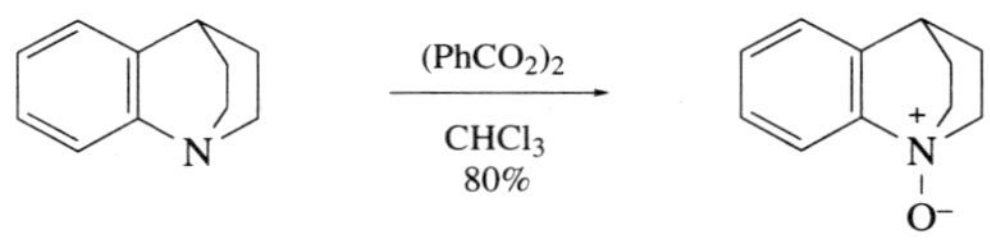

(52)

(143) **(144)**

2.15.6 THIOHYDROXYLAMINES AND RELATED FUNCTIONS

Thiohydroxylamines are compounds containing trivalent nitrogen bonded to divalent sulfur. They are derived formally from sulfenic acids RSOH, just as sulfinamides are derived from sulfinic acids RSO_2H, and sulfonamides are derived from sulfonic acids RSO_3H. They have been of interest to the chemical community over the years because of their industrial applications, their utility as synthetic reagents, and their interesting stereochemical properties. Thiohydroxylamines and their derivatives exist in a variety of forms ((**145**)–(**148**)). The nomenclature for these compounds is not thoroughly systematized.

(145) **(146)** **(147)** **(148)**

Compounds of the type ((**145**)–(**147**)) are referred to as sulfenamides by the Chemical Abstracts Authority. These compounds have also been termed thiohydroxylamines. In sulfenamides of the type (**145**), R^1 and R^2 can be H, alkyl, aryl, acyl, or X, where X is a heteroatom, and R^3 can be alkyl, aryl, acyl, or X. Cyclic sulfenamides (**147**), where nitrogen is part of an aliphatic or aromatic ring are also known. Sulfenamides (**145**) in which R^3 is NR_2, SNR_2 or $SSNR_2$ have been termed bisamine sulfides, bisamine disulfides, and bisamine trisulfides, respectively. Sulfenamides (**145**) in which R^3 is a halide are classified as sulfenyl halides, whereas when R^3 is OR or NR_2 they are considered also as *S*-amino derivatives of sulfenate esters and sulfenamides, respectively.

Compounds of the type (**146**) are called disulfenamides, bis(arylsulfen)imides, *N*,*N*-disulfenylamines, and occasionally *N*-(alkyl- or arylthio)sulfenamides. Finally, cyclic sulfenamides of the type (**148**) are correctly named as benzisothiazole derivatives.

The literature contains a number of reviews on sulfenic acid chemistry, the synthesis and reactions of sulfenamide derivatives, the stereochemistry, and the industrial application of sulfenamides. A large number of synthetic routes to sulfenamides have been developed over the years. There are several good reviews on the synthetic methods prior to 1979, which are cited in the review by Craine and Raban ⟨89CRV689⟩. Most of these methods can be categorized according to the oxidation state of sulfur in the sulfur-containing reagent: (i) use of sulfur reagent RSX (X = Cl, Br, OR, NR_2, SO_2Ar, SCN); (ii) use of the disulfide RSSR, in which the oxidation state of sulfur is lower by one unit; and (iii) use of thiols RSH, in which the oxidation state of sulfur is lower by two units.

2.15.6.1 Sulfenamides

2.15.6.1.1 From sulfenyl halides

The most common method employed in the synthesis of sulfenamides is the condensation of amines with sulfenyl halides ⟨89CRV689⟩. The reaction proceeds via nucleophilic attack of the amine on the sulfenyl halide (Scheme 48).

Scheme 48

The acid formed in the reaction is neutralized by excess amine or by another acid acceptor, such as triethylamine, alkali–metal hydroxide, and others. The aryl sulfenamides of 6-aminoquinoline *N*-oxide (**151**) are formed in 74–79% yield from the quinoline *N*-oxide (**149**) and the appropriate arenesulfenyl chloride (**150**) (Equation (53)) ⟨79MI 215-01⟩.

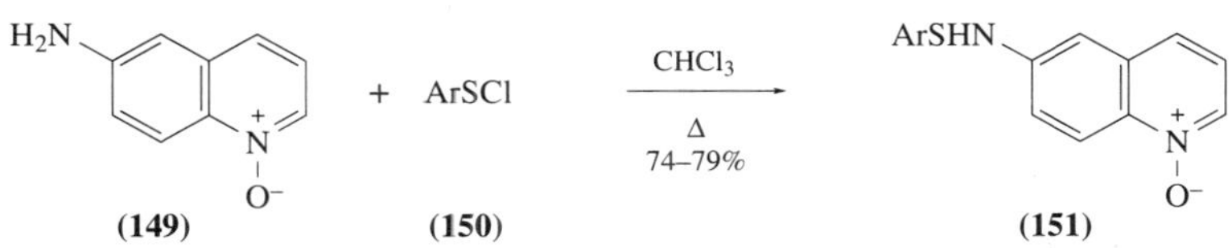

(**149**) (**150**) (**151**) (53)

The reaction of nitrobenzenesulfenyl chloride with indoles yields only 3-(arylthio)indoles under the usual reaction conditions ⟨80JOC1688⟩. In selected cases these can be thermally rearranged to the sulfenamides in a low yield ⟨82TL821⟩. Somewhat better yields of the sulfenamides are obtained at low reaction temperatures and by the use of the sodium salt of the indole ⟨85TL4443⟩.

Benzimidazoles react with 2,4-dinitrobenzenesulfenyl chloride (**152**) in the presence of triethylamine to produce sulfenamides (Equation (54)) ⟨84TL1337, 85JOC2205⟩.

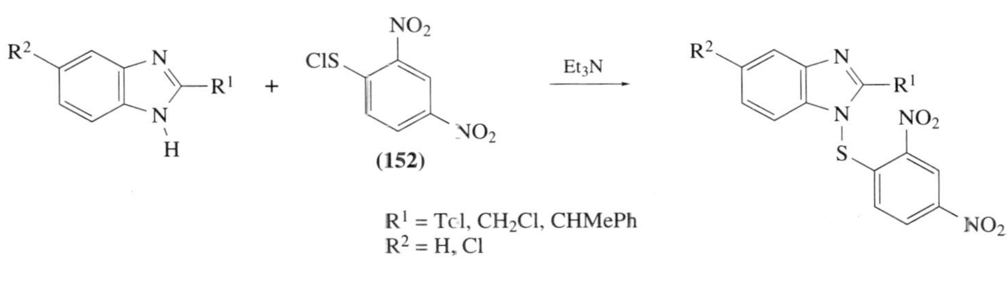

(**152**) (54)

R^1 = Tol, CH_2Cl, CHMePh
R^2 = H, Cl

The reaction of 2,4-dinitrobenzenesulfenyl chloride with diarylamines in the absence of base yields several products, with the product distribution dependent on the aryl substituents, thiohydroxylamine (**153**; Ar^1, Ar^2 = Ph) being the major product with diphenylamine (Scheme 49) ⟨89UP-01⟩.

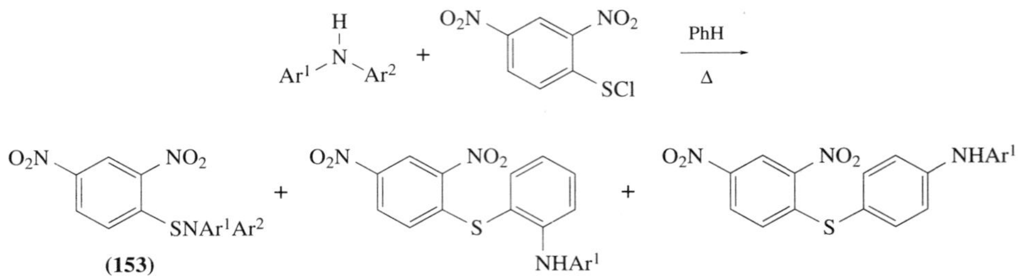

(153)

Scheme 49

3-Benzyl-4-methylcarbostyril reacts with 2,4-dinitrobenzene sulfenyl chloride and triethylamine to form the sulfenamide. However, under the same conditions, 8-ethyl-4-methylcarbostyril does not react and the by-product (enamine) is formed in small quantities. Similar results have been noted by Senning and Kelly ⟨72ACS2877⟩ and Traynelis and Rieck ⟨73JOC4339⟩.

2.15.6.1.2 From thiols and disulfides

Aromatic thiols and amines react to form sulfenamides in the presence of oxidizing reagents (Equation (55)).

$$ArSH + HNR_2 \xrightarrow{[O]} ArSNR_2 + H_2O \qquad (55)$$

Hydrogen peroxide, hypochlorites, halogens, potassium persulfate, and potassium ferrocyanide are reported to be useful oxidants ⟨79OPP33⟩. *t*-Butyl azidoacetate and *t*-butyl azidoformate react with *t*-butyl and benzyl thiols in the presence of copper(I) to give sulfenamides in moderate yields ⟨70JOC2979⟩.

2.15.7 SELENIUM AND TELLURIUM ANALOGUES OF THIOHYDROXYLAMINES AND THEIR DERIVATIVES

2.15.7.1 N—Se Compounds

The selenide chloride (**154**) reacts with a number of nucleophiles to give heterocycles typified by the structure (**155**) (Equation (56)) ⟨24CB1077⟩.

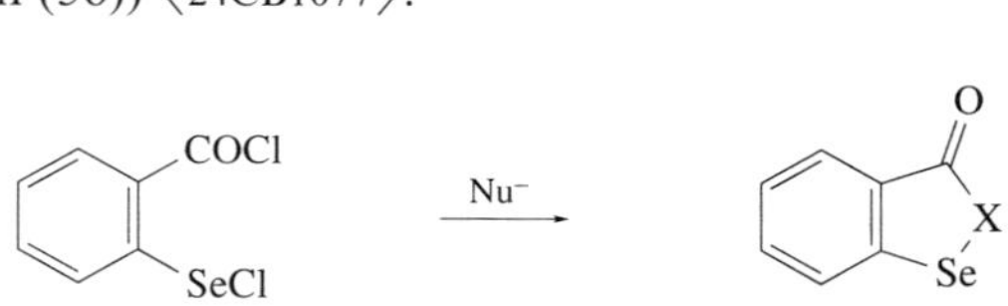

(56)

(154) X = NH, NPh, S, Se **(155)**

2.15.7.2 N—Te Compounds

Although the chemistry of organotellurium compounds has been widely studied since the early 1970s ⟨B-74MI 215-02, 85ACR274, 86S1⟩, relatively little attention has been paid to compounds containing N—Te bonds ⟨91OM1095⟩. Although organotellurium compounds bearing structures (**156**),

(**157**) and (**158**), where Te is tetravalent, have been extensively studied, only the cyclic tellurazole (**159**) has been reported containing a Te(II)—N-bond. In contrast, acyclic derivatives, i.e., tellurenamides, have not been reported so far, except for ditellurenylsulfonamide (**160**).

R_3TeN (**156**) R_2TeN_2 (**157**) RX_2TeN (**158**)

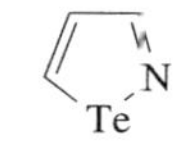

(**159**) $RSO_2N(TePh)_2$ (**160**)

2.15.8 HYDRAZINES AND RELATED FUNCTIONS

2.15.8.1 Hydrazines

The best-known organic derivative of hydrazine is phenyl hydrazine $PhNHNH_2$, which is also the simplest of the aryl hydrazines. It was first prepared in 1875 by Fischer ⟨20JCS1157, 21CB206⟩ and later proved to be one of the most valuable reagents in organic chemistry. An important method for the preparation of monoaryl hydrazines consists of the reduction of diazo compounds, and since primary aromatic amines can readily be diazotized, hydrazines containing variously substituted aryl groups can be obtained by this method ⟨89BCJ128⟩. The reduction can be carried out in several ways. Fischer's original method involved reducing the diazosulfonate with sulfite. Some diazo compounds such as those derived from *p*-nitroaniline, give a hydrazine disulfonate, and not a monosulfonate, but they are also easily reduced to the hydrazine, so that the method of preparation is unaffected ⟨22JCS715⟩.

In some cases it is better to reduce the diazosulfonate with zinc dust and acetic acid or with sodium hydrosulfite (Scheme 50) ⟨21MI 215-01, 79COC(2)224⟩.

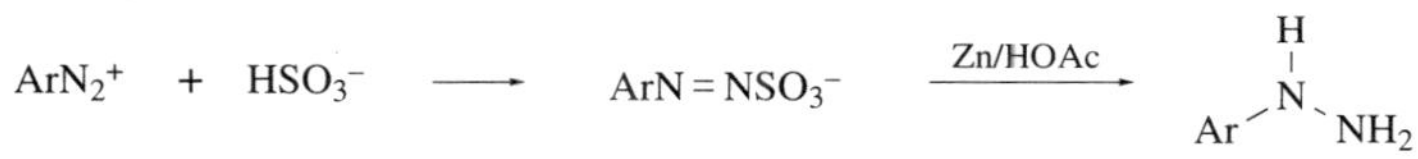

Scheme 50

Alternatively, the diazonium salt can be reduced directly to the hydrazine with stannous chloride in acidic solution ⟨B-66MI 215-04⟩. This method is often very convenient, because the hydrazine hydrochlorides are sparingly soluble in hydrochloric acid and crystallize out from the reaction mixture, it is however unsuitable if easily reducible groups, such as the nitro group, are present in the molecule. Sulfur dioxide ⟨42CB1096, 53CB1073⟩, triphenylphosphine or phosphites ⟨53CB1073⟩ may also be used as reducing agents.

King and Walton ⟨75S738⟩ designed the silylated copper reagent (**161**) for conversion of (non-activated) aromatic iodides into arylhydrazines (Scheme 51).

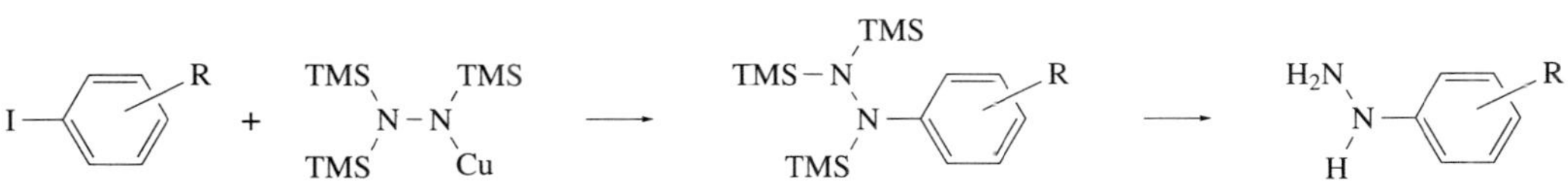

(**161**)

Scheme 51

Naphthyl hydrazines can be prepared by heating naphthols with hydrazine, but with a phenol the yield of the substituted hydrazine is poor. The aryl hydrazines exist commonly as liquids or low-melting solids and are colorless when pure, but like many organic bases, they quickly become brown on exposure to air. They are insoluble in water, but are soluble in dilute acids. Phenyl hydrazine decomposes slowly at its boiling point (242 °C) into ammonia, nitrogen, and benzene ⟨11JCS404⟩.

2.15.8.2 Azo Compounds

2.15.8.2.1 Azoalkenes

The chemistry of azoalkenes has received considerable attention since the early 1960s. The most important methods for the preparation of azoalkenes are based on the 1,4-elimination starting from suitably substituted hydrazones ⟨86OPP299⟩. Banert and Hagedorn ⟨92TL7331⟩ showed a new route to enazo compounds, which modifies the known [2.3] sigmatropic rearrangement of allylic 1,1-diazenes (**163**) to allyl azo compounds with the result that the azo function is attached to a vinylic position (Scheme 52).

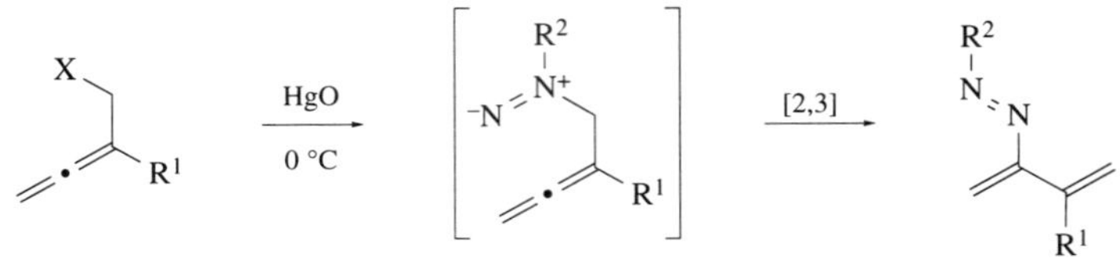

(162) **(163)** **(164)**

(165) R^1 = H, X = Br
(166) R^1 = H, X = $NMeNH_2$ (59%)
(167) R^1 = Me, X = Cl
(168) R^1 = Me, X = $NMeNH_2$ (59%)
(169) R^1 = Me, X = $N(CH_2CMe{=}CH_2)NH_2$ (43%)
(170) R^1 = Me, X = $N[(CH_2)_3CH{=}CH_2]NH_2$ (56%)
(171) R^1 = Me, X = $N[(CH_2)_3C{\equiv}CH]NH_2$ (47%)

(172) R^1 = H, R^2 = Me
(173) R^1 = R^2 = Me
(174) R^1 = Me, R^2 = $CH_2C(Me){=}CH_2$
(175) R^1 = Me, R^2 = $(CH_2)_3CH{=}CH_2$
(176) R^1 = Me, R^2 = $(CH_2)_3C{\equiv}CH$

Scheme 52

The 2-alkylazo substituted-1,3-butadienes were obtained in good yield (e.g., nearly 100%, (**173**)) by oxidation of 1-alkyl-1-(2,3-butadienyl)hydrazines ((**166**) and (**168**)–(**171**)) with yellow mercury(II) oxide or activated manganese dioxide. The precursors ((**166**) and (**168**)–(**171**)) were readily accessible from the reaction of halogen compounds (**165**) or (**167**) with methylhydrazine, methallylhydrazine, 4-pentenylhydrazine or 4-pentynylhydrazine (Scheme 52).

Most probably, the alkylazo substituted 1,3-butadiene (**164**) resulted from a short-lived intermediate (**163**) via a [2.3] sigmatropic rearrangement. The azoalkenes (**164**) could only be characterized and handled in dilute solution because they dimerized very rapidly even at room temperature. For the same reason a freshly prepared solution of (**164**) must be used for further reactions. Ludwig *et al.* ⟨80JOC4666⟩ reported a ^{13}C NMR study in which (*Z*)-(**177**) (>95% (*Z*)) was deprotonated, with low regioselectivity, on treatment with lithium diethylamide or lithium diisopropylamide at 0°C for 30 min, to give the *anti*- and *syn*-dimethylhydrazone lithio anions (**178**) and (**180**) (Scheme 53).

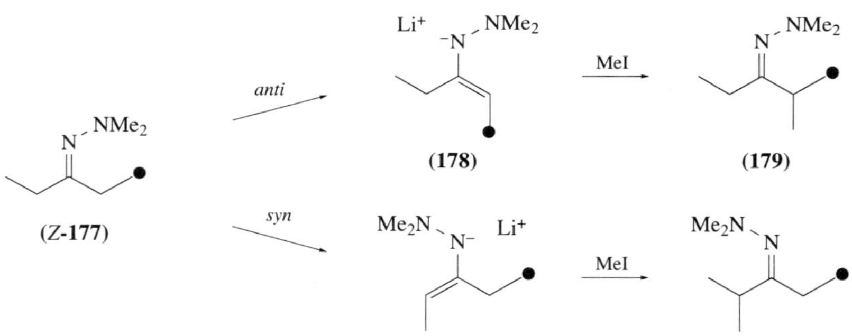

(*Z*-**177**) (**178**) (**179**) (**180**) (**181**)

Scheme 53

A carbon-13 labelled methyl group was employed for the assignment of ^{13}C NMR signals and for quantitative estimation of the amounts of (**178**) and (**180**) or (**179**) and (**181**).

The reaction of enamines with ω-bromoacetophenone semicarbazone in chloroform in the presence of triethylamine gives rise to 6-phenyl-2,5-dihydropyrazines (**182**) ⟨71AC(R)391⟩. Substitution in the 2-position depends on the enamine used. In some cases, 2-amino carbonyl-2,5-dihydropyridazines (R^3 = $CONH_2$) can be transformed into the 2,5-dihydropyridazines (R^3 = H) by acid hydrolysis (Equation (57)).

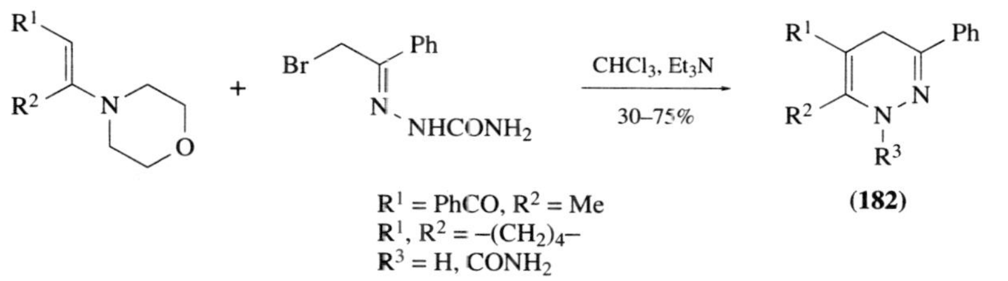

(57)

Several derivatives of 1-benzothiopyrano[3,4-*d*][1,2,3]-triazole were prepared by Rossi and Trimarco ⟨78S465⟩ through the cycloaddition of aryl azides to enamines of 4-oxo-4H-1-benzothiopyrans (Scheme 54). Enamines (**183**) were easily obtained by reacting the appropriate ketone with the secondary amine and titanium(IV) chloride. Their reactivity towards aryl azides is moderate and satisfactory results were obtained only on extended heating in acetonitrile or ethanol. The reductive deamination of the dihydrotriazole derivatives (**184**) was brought about by refluxing with sodium hydroxide in ethanol to give the corresponding triazole compounds.

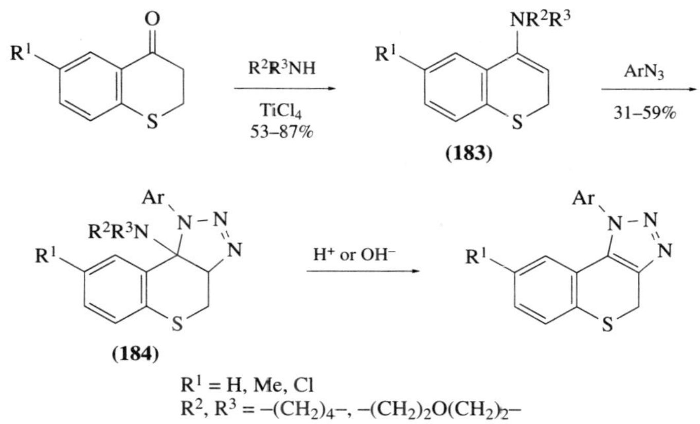

Scheme 54

An example of the synthesis of acyl triazoles using α-acyl vinyl phosphonium salt was discussed by Rasberger and Zbiral ⟨69M64⟩.

Gillis and Hagarty prepared the α,β-unsaturated azo derivative (**187**) from the α-halo hydrazone (**186**), obtained in turn by the reaction of α-chloroaldehyde (**185**) with methylhydrazine, by 1,4-dehydrohalogenation (Scheme 55) ⟨65JA4576⟩.

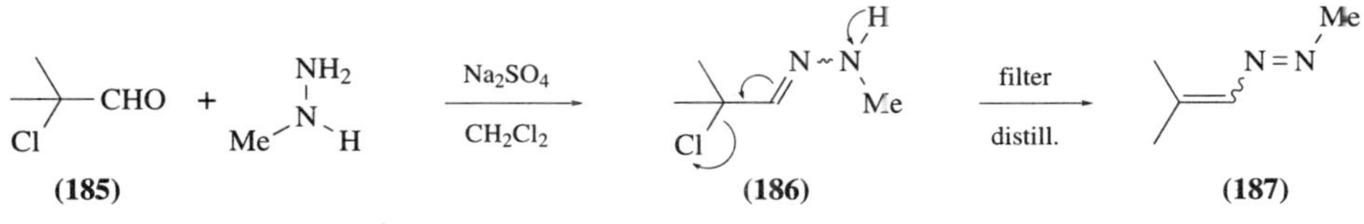

Scheme 55

In an analogous manner, advantage was taken of the epoxide ring opening (**188**) in the 1,4-elimination process to generate the azoalkene (**189**) (Scheme 56) ⟨B-72MI 215-01⟩.

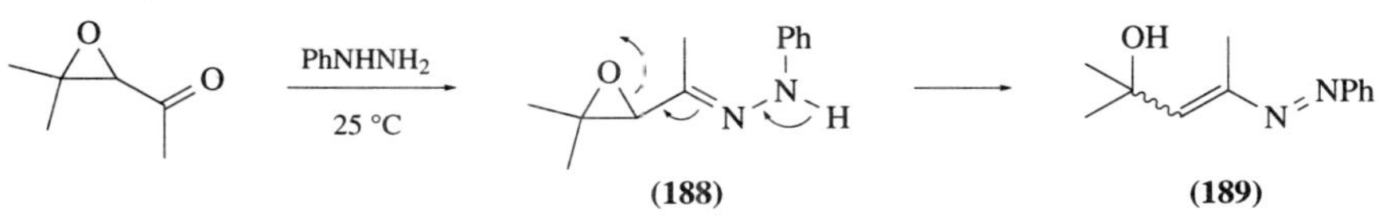

Scheme 56

2.15.8.2.2 *Aromatic azo compounds*

(i) From oxidation of primary aromatic amines

Aromatic azo compounds can be obtained by the oxidation of primary arylamines. The reagent most widely used for this purpose is activated manganese dioxide (Scheme 57) ⟨B-86MI 215-01⟩.

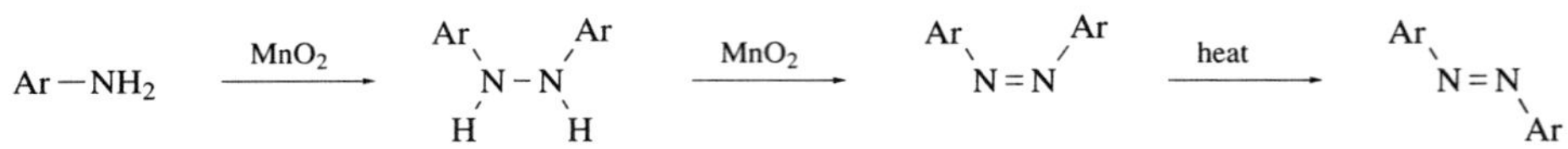

Scheme 57

Other reagents which bring about the oxidation of aromatic amines to the corresponding azo derivatives and more details on the subject may be found in an excellent review by Gilchrist ⟨91COS(7)738⟩. These include silver oxide on celite, lead tetraacetate, barium manganate, bispyridinesilver permanganate, sodium perborate, and sodium hypochlorite.

Potassium superoxide has been found to be capable of selectively oxidizing *ortho*- and *para*-substituted diamines and aminophenols to the corresponding azo compounds in good yield (Equation (58)) ⟨84AJC845⟩.

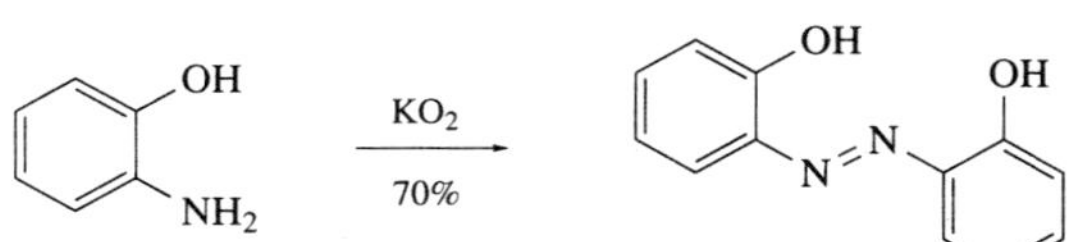

(58)

(ii) From oxidation of hydrazines

Syntheses of hydrazines can also be considered as syntheses of the corresponding azo compounds. The necessary oxidation is usually carried out by yellow mercuric oxide, bromine, hypochlorites, or cupric salts and air (Equation (59)) ⟨79COC(2)233⟩.

$$\text{ArHN}-\text{NHAr} \xrightarrow{\text{oxidant}} \text{ArN}=\text{NAr} \tag{59}$$

(iii) From reduction of aromatic nitro compounds

Reduction of aromatic nitro compounds often proceeds to generate mixtures of nitroso and hydroxylamine products which then condense to form azoxy and, eventually, symmetrically substituted azo compounds. A variety of reducing agents have been used to reduce nitroarenes to azo compounds. A mixture of zinc and sodium hydroxide is used most frequently ⟨91COS(8)364⟩ and is sufficiently mild that *p*-nitrostyrene (**190**) can be reduced to the azo compounds (**191**) without reduction of the vinyl group (Equation (60)) ⟨63JOC1232⟩.

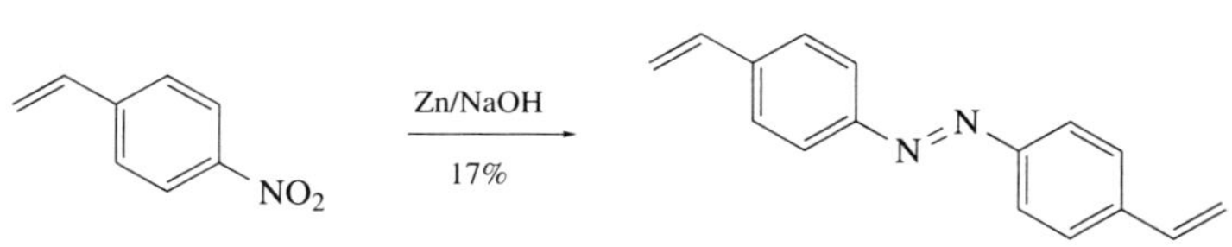

(60)

(190) **(191)**

Radell *et al.* ⟨56JOC1051⟩ described the formation of (**193**) possessing an intramolecular azo bridge by reduction of the dinitro compounds (**192**) with zinc and sodium hydroxide (Equation (61)).

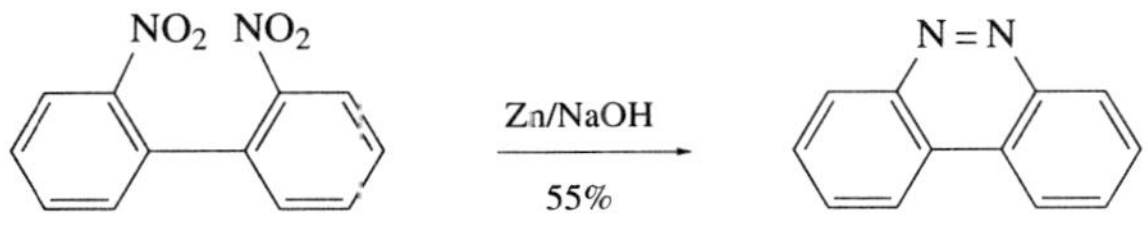

(192) (193) (61)

Sodium borohydride in methanol appears to effect the reductive coupling of 1,4-dinitrobenzene (**194**) to the corresponding 4,4′-dinitroazobenzene (**195**) (Equation (62)) ⟨88ACS(B)354⟩.

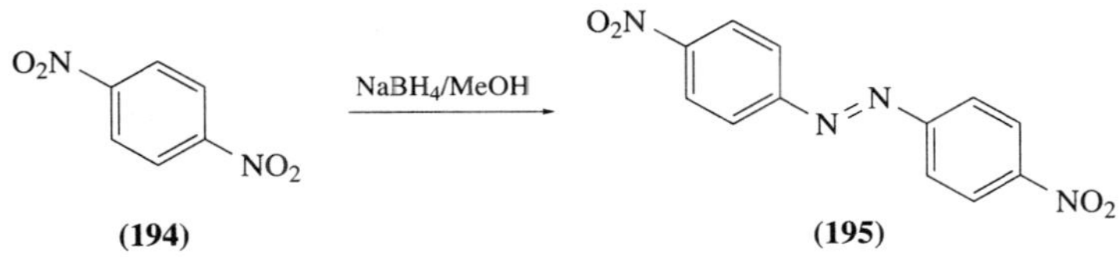

(194) (195) (62)

Some special methods for the preparation of aromatic azo compounds are summarized in the reactions shown below (Scheme 58) ⟨B-68MI 215-03, 79COC(2)236⟩. Some of these reactions are extremely important to the dye industry.

(iv) Other methods

Joucla and Rees ⟨84CC374⟩ reported the conversion of 3-amino-2-phenyl indazole (**196**) to the azide (**197**), which on boiling in tetrachloromethane or light petroleum gave a quantitative yield of 2-cyanoazobenzene (**198**) (Scheme 59).

Direct oxidation of the 3-amino compound (**196**) with lead tetraacetate in benzene at room temperature also gave the ring-opened product (**198**) quantitatively.

Diarylsulfur diimide (**199**) reacts with sodium to yield azobenzenes (Equation (63)) ⟨79COC(2)235⟩.

$$\underset{\textbf{(199)}}{\text{ArN=S=NAr}} \xrightarrow{\text{Na}} \text{ArN=NAr} \qquad (63)$$

2.15.8.3 Azoxy Compounds

2.15.8.3.1 α,β-Unsaturated azoxy compounds

A naturally occurring azoxy compound, elaiomycin (**200**), provides an example of α,β-unsaturated azoxy compound ⟨70JA1587⟩. It shows both carcinogenic and antibiotic properties ⟨69NAT765⟩. L-Dihydroelaiomycin (**201**) has been synthesized by the diazoate route in 12 steps with an 80% overall yield (Scheme 60) ⟨73JCS(P1)2778⟩.

2.15.8.3.2 Aromatic azoxy compounds

Azoxy compounds are of special interest because of their physiological activities and their utilization in liquid crystal systems ⟨74JA5158⟩. The chemistry of aromatic azoxy compounds has been reviewed ⟨31CRV117, B-75MI 215-01⟩. Many methods for preparing azoxy compounds have been investigated ⟨B-68MI 215-04, 73JA3070, 74ACR421, 74JOC407, 76JOC1751, 76S726⟩. Some of these methods suffer from a lack of regioselectivity where the *N*-substituents are not identical.

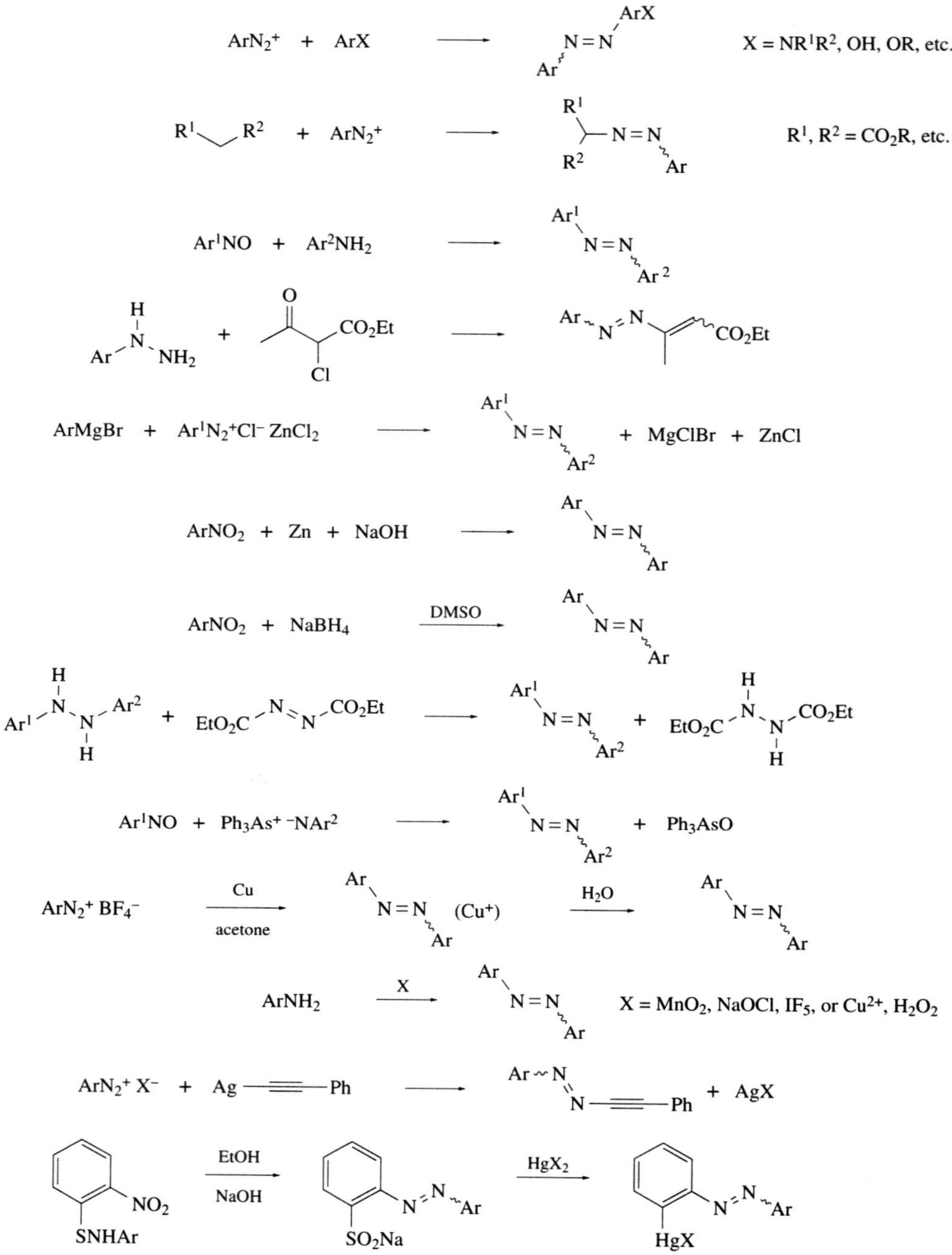

Scheme 58

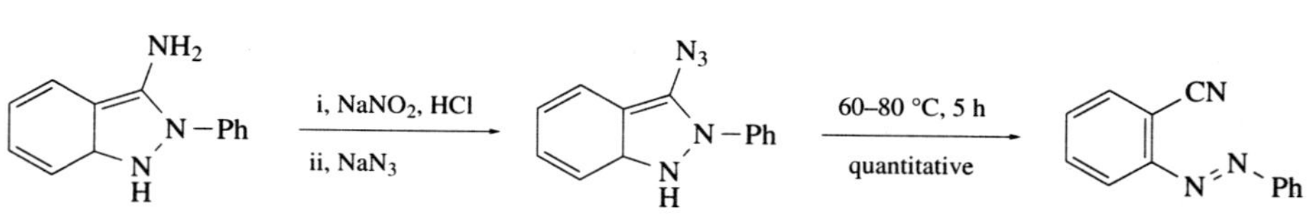

Scheme 59

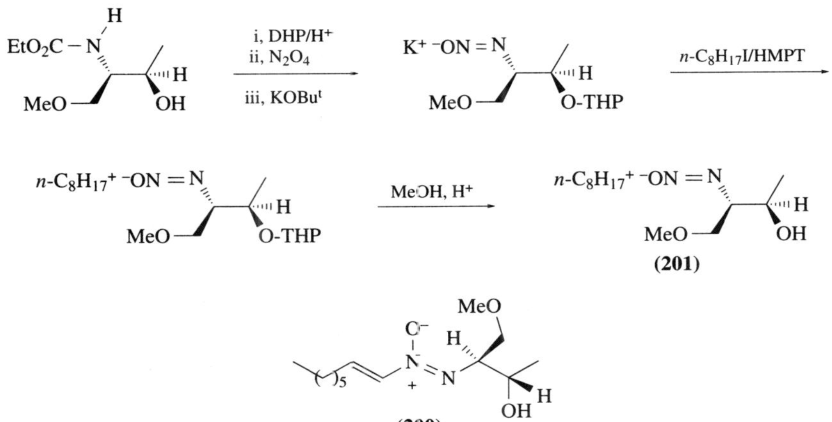

(201)

(200)

Scheme 60

(i) From aromatic nitroso compounds

The reaction of *N,N*-dichloroalkylamines with alkyl- or aryl-nitroso compounds provides a regioselective route to azoxy compounds (Equation (64)) ⟨76JOC1751⟩.

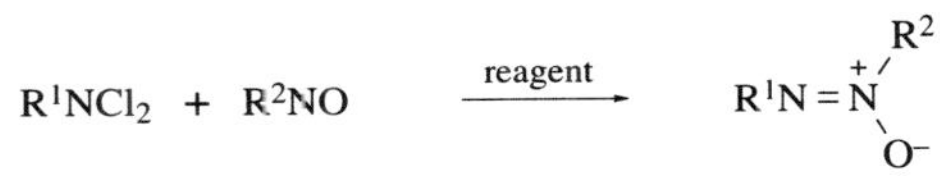

(64)

reagent = OH^-, CuCN, CuCl, KI

(ii) From aromatic amines

The peroxy acid oxidation of aromatic amines can give azodioxy compounds by reaction of nitroso compounds with hydroxylamine intermediates (Scheme 61) ⟨91COS(7)736⟩.

R = aromatic, aliphatic acid

Scheme 61

These products can also be obtained by peracetic acid oxidation of benzaldimines or by direct oxidation of amines (Scheme 62) ⟨79COC(2)245⟩.

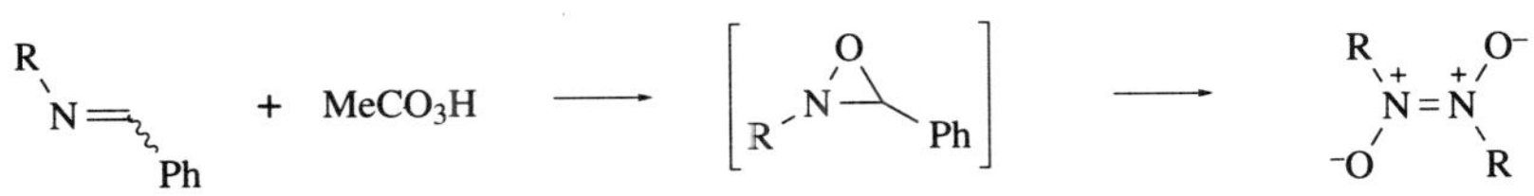

Scheme 62

(iii) From aromatic nitro compounds

Bimolecular reduction of aromatic nitro compounds may produce azoxy compounds. The preparation of *o,o′*-dicyanoazoxybenzene (**202**) is representative of the reduction of aromatic nitro compounds using zinc under acidic conditions (Equation (65)) ⟨61CJC862⟩.

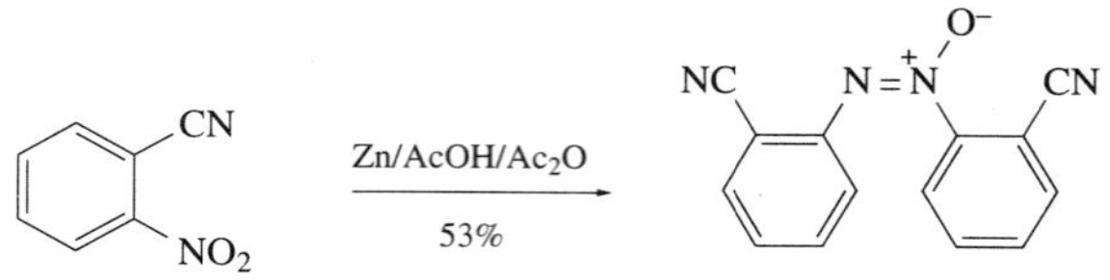

(202)

(65)

The reduction of aromatic nitro compounds to azoxy derivatives has also been accomplished with tin(II) chloride in a basic medium ⟨B-86MI 215-02⟩. Metallic reducing agents such as magnesium ⟨29LA(468)117⟩ and an alloy of sodium and lead (Drynap) ⟨67BCJ1538⟩ produce a mixture of azoxy and azo compounds (75%). Reduction with thallium metal in ethanol affords good yields (65–93%) of the corresponding azoxy compounds ⟨70JOC1670⟩.

Alcoholic potassium hydroxide and sodium alcoholate have also been used as reducing agents (Equation (66)) ⟨61JCS4260⟩.

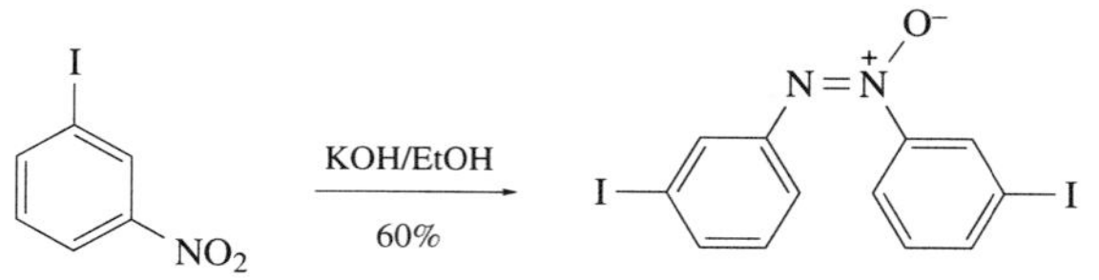

(66)

An interesting reduction of aromatic nitro compounds to azoxy derivatives using glucose in an alkaline medium offers the advantage of high yield, rapid rate and ease of product isolation ⟨51JA1323⟩. Other reducing agents which bring out the reduction of nitroarenes to azoxy compounds and more details on the subject are provided in a review by Kabalka and Varma ⟨91COS(8)366⟩. The reducing agents include potassium borohydride, sodium arsenate, phosphine, and yellow phosphorus.

2.15.8.4 Systems Containing *N,N,N*, *N,N,N,N* and *N,N,N,N,N* Functionalities

The presence of lone pairs of electrons on three or more contiguous sp^3 nitrogen atoms is destabilizing, and these compounds are only stable enough to be isolated when two or more of the nitrogens have sp^2 hybridization or are highly substituted. Triazenes (**204**) are the most widely-studied of this group and the most familiar route to them is the reaction of aromatic diazonium salts (**203**) with amines (Equation 67) ⟨67JOC1124⟩.

(203) **(204)**

(67)

Other methods of preparation include the reaction of diazonium salts (**205**) with amide anions (**206**) and the addition of Grignard reagents (**208**) or trialkylaluminums to azides (**207**) (Scheme 63) ⟨79COC(2)267⟩.

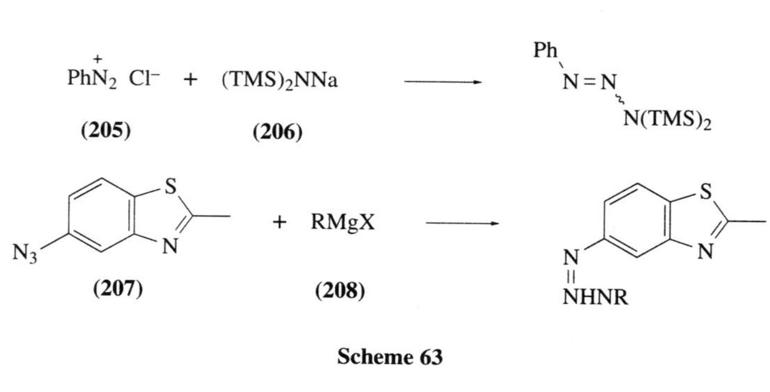

Scheme 63

Coupling of diazonium salts with hydroxylamines yields triazene *N*-oxides (**209**) (Equation (68)) ⟨79COC(2)267⟩.

(209) (68)

Only a few examples of substituted triazanes are known. The triazane (**211**) was prepared by the addition of aniline to azodicarboxylate (**210**) (Equation (69)) ⟨79COC(2)267⟩.

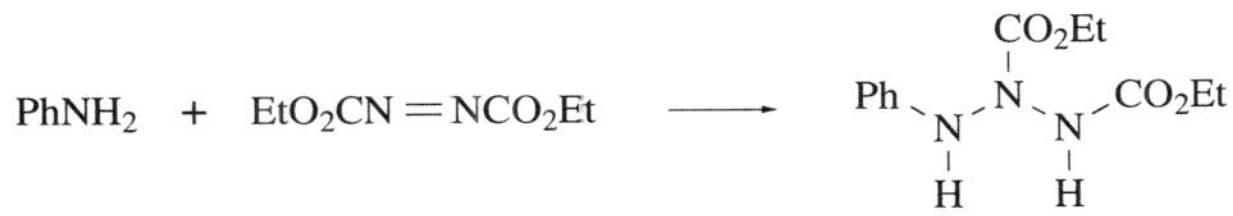

(210) (211) (69)

The instability of three nitrogens linked by σ-bonds is shown by the electrocyclic ring-opening of (**212**) to give the more stable ylide (**213**) (Scheme 64) ⟨75JCS(P1)26⟩.

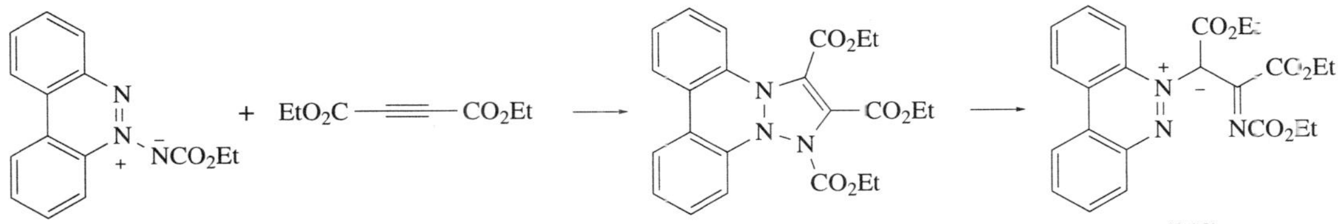

(212) (213)

Scheme 64

As in the case of triazenes, unsaturated tetrazenes are more abundant than tetrazanes. 1-Tetrazenes (**214**) are prepared by coupling of diazonium salts with hydrazines (Equation (70)) ⟨79COC(2)268⟩.

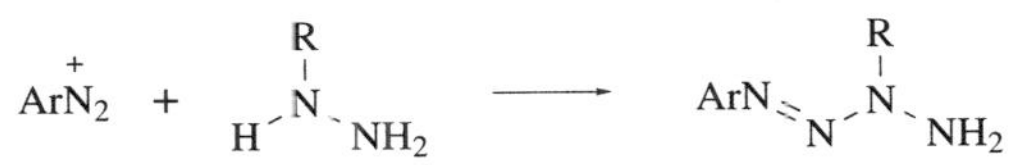

(214) (70)

Tetrazanes can be prepared in stable form only when they are highly substituted (Equation (71)) ⟨75JCS(P1)1065⟩.

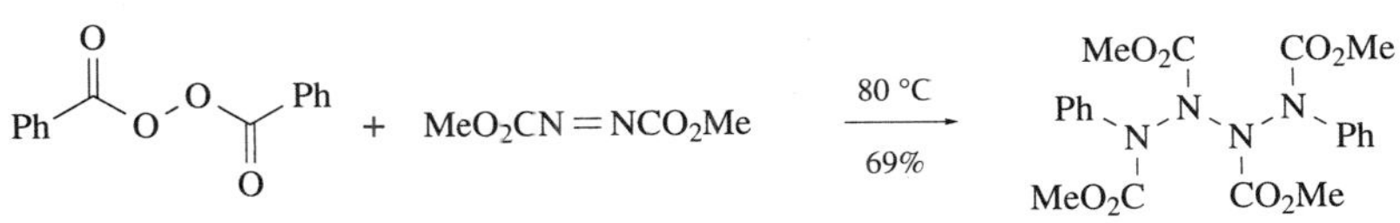

(71)

A few examples of pentazadienes (**215**) have been prepared by coupling of diazonium salts with primary amines (Equation (72)) ⟨73MI 215-01⟩.

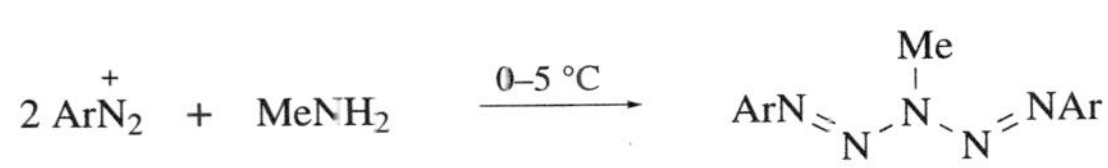

(215) (72)

2.15.9 *N*-VINYL AND *N*-ARYL COMPOUNDS OF THE TYPE RNY AND RN⁺Z (e.g., ArN_2^+)

2.15.9.1 Nitroso Compounds

2.15.9.1.1 α, β-Unsaturated nitroso compounds

The subject of conjugated nitroso compounds has been reviewed by Boyer ⟨B-69MI 215-01⟩ and Gilchrist ⟨83CSR53⟩. Only a few vinyl nitroso compounds have been isolated.

(i) From vinyl radicals and nitric oxide

The first stable compound, nitrosotrifluoroethylene (**217**), was reported in 1960. It was formed by the sensitized photolysis of iodotrifluoroethylene (**216**) in the presence of nitric oxide (Equation (73)) ⟨60JCS1398⟩.

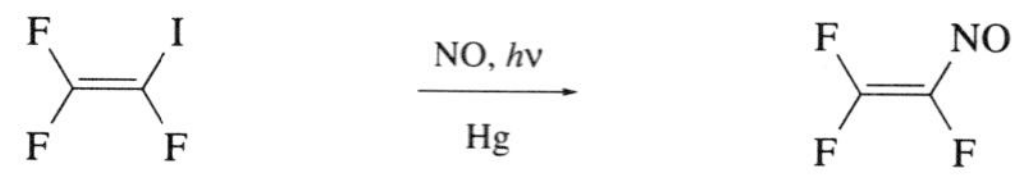

(216) (217) (73)

(ii) From nitrosohalides

Nitrosyl chloride alkene adducts undergo dehydrochlorination in the presence of base to form α, β-unsaturated nitroso compounds. Styrene nitrosochloride (**218**) on reaction with triethylamine in ether loses hydrogen chloride to form nitrosostyrene (**219**) (Equation (74)) ⟨65JPR123⟩.

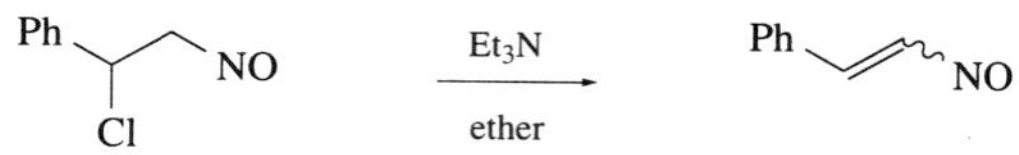

(218) (219) (74)

Similarly, nitrosoalkenes have been obtained from thujene nitrosochloride ⟨65MI 215-01⟩ and methyloleate nitrosochloride ⟨65MI 215-02⟩.

(iii) From α-halooximes

The most important method for the generation of vinyl-nitroso compounds involves the reaction of α-halooximes with bases. Thus, the first nitrosoalkene (**221**) stable at room temperature was obtained by elimination of hydrogen chloride from (**220**) with 1,5-diazabicyclo[4.3.0]non-5-ene (Equation (75)) ⟨75AG(E)70⟩.

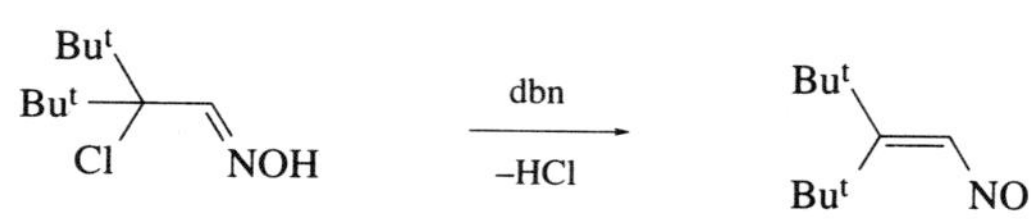

(220) (221) (75)

An interesting variation in which the leaving group is in the form of an epoxide is shown in (Equation (76)) ⟨75TL3117⟩.

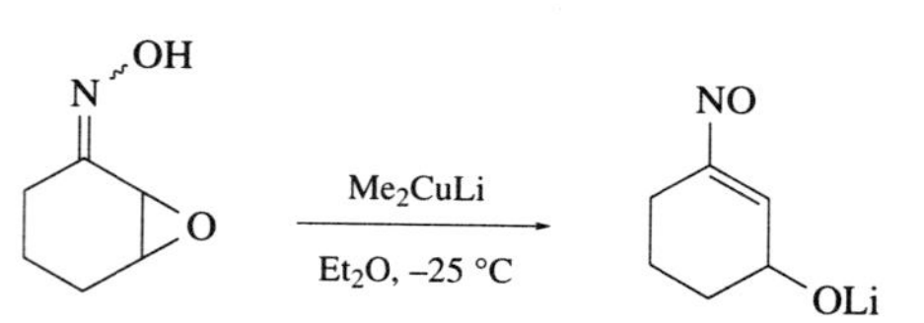

(76)

The formation of nitroso-alkene (**223**) has been postulated to arise from the acid-catalyzed dehydration of sulfoxide (**222**) (Equation (77)) ⟨80TL2025⟩.

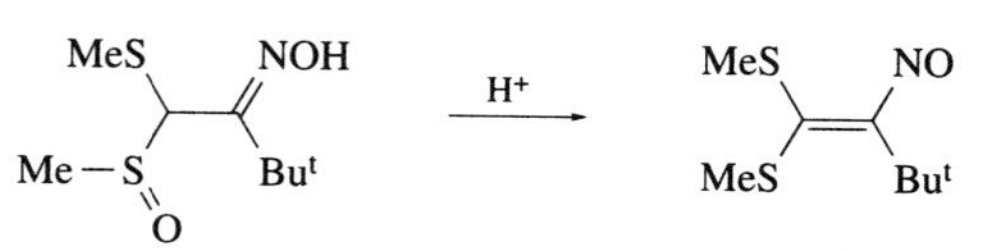

(222) (223) (77)

(iv) Other methods

Other methods are restricted to examples with special structural features, and the evidence for the formation of vinyl-nitroso intermediates is indirect, being based on the nature of the final product. The nucleophilic addition–elimination reactions of sulfur ylides with nitrile oxides probably also produce nitroso-alkenes as intermediates, but these then react further with the nucleophilic ylides to give 2:1 adducts as shown in Scheme 65 ⟨77JCS(P1)1196⟩.

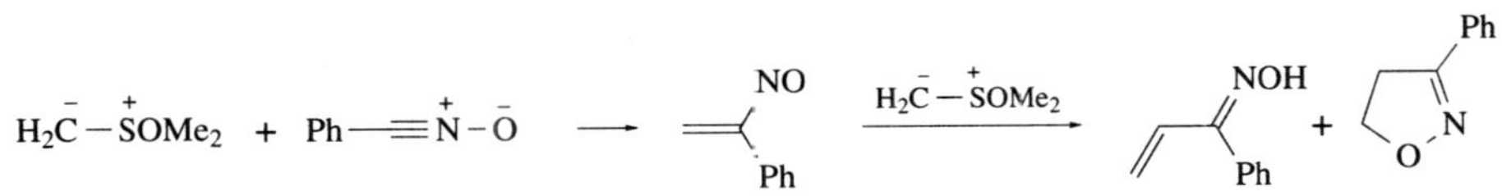

Scheme 65

2.15.9.1.2 *Aromatic nitroso compounds*

(i) From oxidative nitrosation

In the Baudisch reaction, oxidative nitrosation of aromatic compounds proceeds by the simultaneous introduction of the nitroso and hydroxyl groups into adjacent positions on the nucleus when treated with nitrosyl chloride and oxygen. Participation of copper salts appears necessary to stabilize the nitrosyl radical and to prevent the formation of *p*-nitrosophenols (Scheme 66) ⟨40SCI336, 47JOC1⟩.

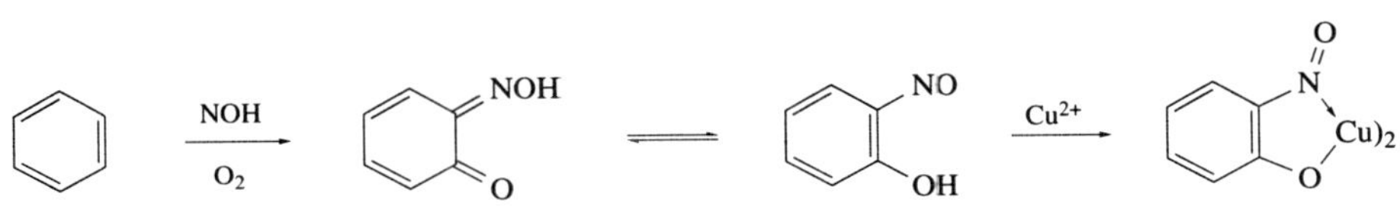

Scheme 66

(ii) From nitrosation of tertiary, secondary, and primary aromatic amines

The first known method for nitrosation at carbon consisted of the treatment of certain aromatic tertiary amines with nitrous acid ⟨1874CB809⟩. Thus, *p*-nitroso-*N,N*-dimethylaniline **(224)** was prepared as shown in Equation (78) ⟨47OSC(1)214⟩.

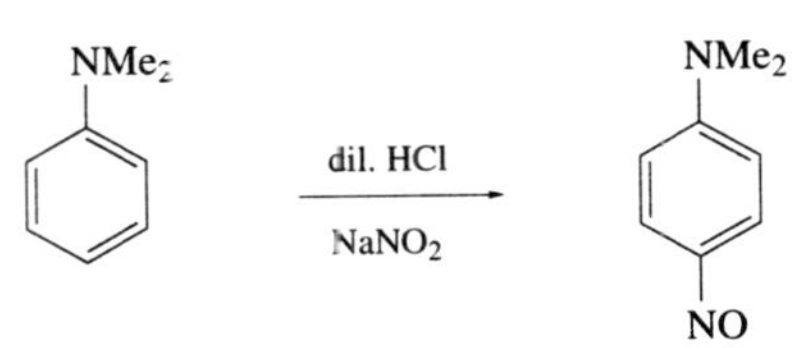

(224)

(78)

The nitroso group is not invariably introduced into the *para*-position and sometimes an *ortho*-position is selected (Equation (79)) ⟨64ZOB2094⟩.

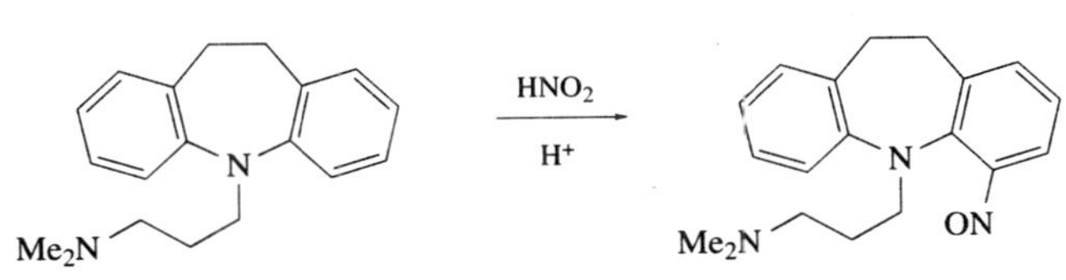

(79)

Nitrosation of secondary aromatic amines generally occurs initially at nitrogen and then an isomerization, known as the Fischer–Hepp reaction, proceeds to give the *para*-nitrosated derivative (Scheme 67) ⟨42LA(550)182, 50JCS2657⟩.

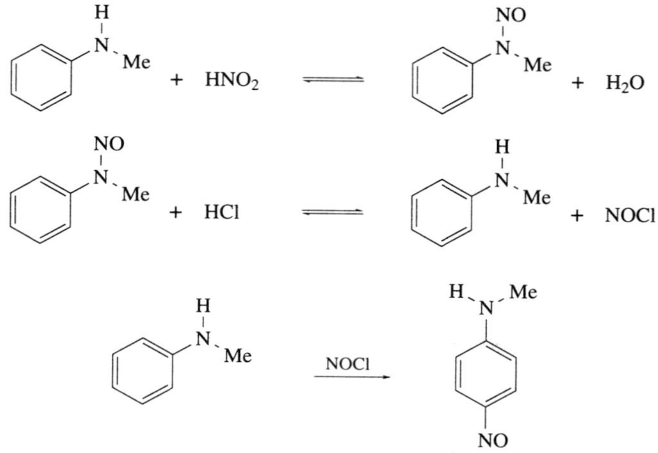

Scheme 67

A particularly interesting example of the Fischer–Hepp reaction is found in the isomerization of (**225**) to the ring-dinitrosated product (**226**) (Equation (80)) ⟨1895LA(286)145⟩.

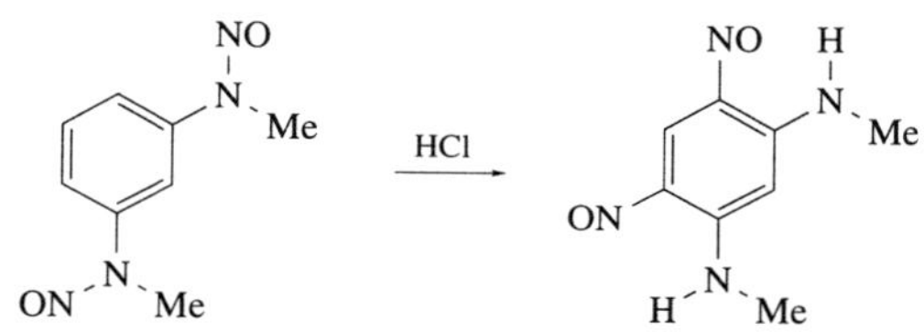

(**225**) (**226**) (80)

Under appropriate conditions, a variety of primary aromatic amines undergo nuclear nitrosation rather than diazotization. Nitrosyl sulfuric acid in concentrated sulfuric acid transforms α-naphthylamine into its 4-nitroso derivative ⟨38HCA1579⟩. A pyrimidine ring was nitrosated on the ring in the presence of both a primary and secondary amine function (Equation (81)) ⟨64LA(677)113⟩.

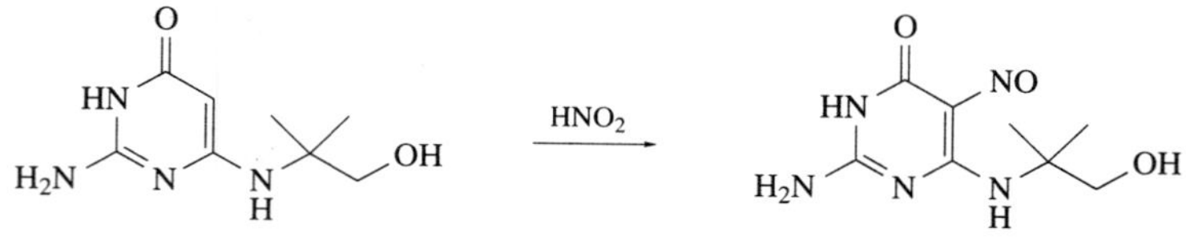

(81)

(*iii*) *Nitrosation of phenols*

Phenols readily undergo nitrosation at *ortho*- and *para*-positions. 2,6-Dibromophenol (**227**) reacts with sodium nitrite in aqueous alcohol to give (**228**) (Equation (82)) ⟨62JA4962⟩.

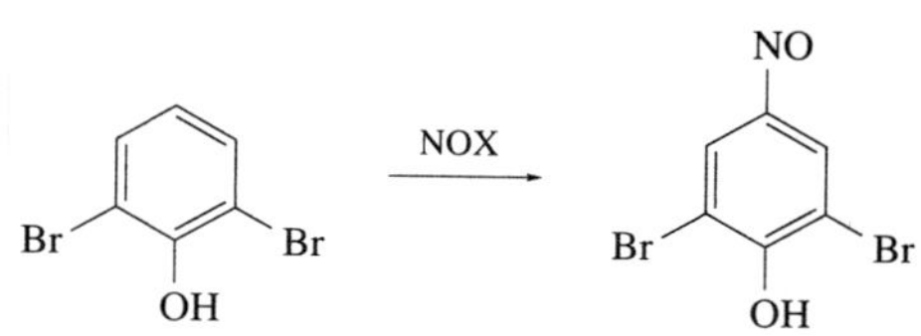

(**227**) (**228**) (82)

(iv) Nitrosation of aromatic ethers

Nitrosation of aromatic ethers is relatively unknown. Resorcinol diether (**229**) is partially transformed into a mixture of nitroso derivatives (Equation (83)) ⟨1881M371⟩.

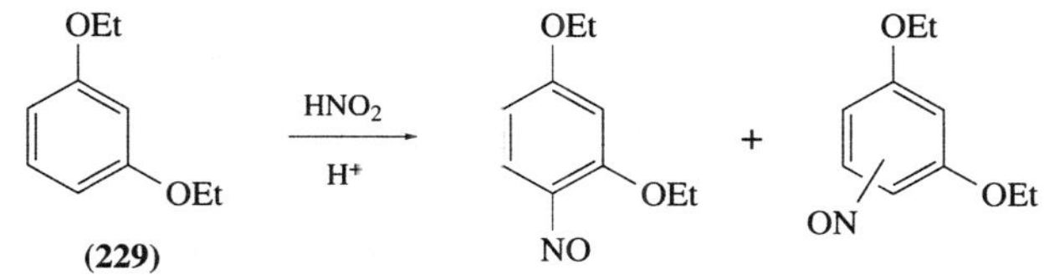

(83)

(v) Nitrosative decarboxylation

Nitrosation by displacement of substituents is rarely found. In a process known as nitrosodecarboxylation, 3,5-dibromo-4-hydroxybenzoic acid (**230**) was converted to the nitroso derivative (**231**) with sodium nitrite in an aqueous alcoholic solution (Equation (84)) ⟨58JOC648⟩.

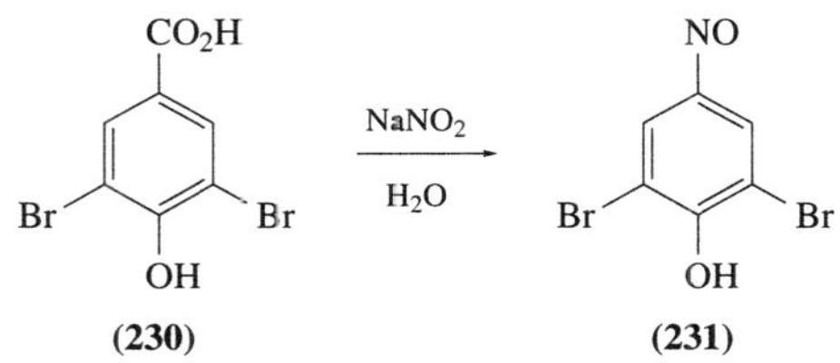

(84)

(vi) Oxidation of dioximes

The dioximes of both *ortho*- and *para*-benzoquinone undergo oxidations with alkaline ferricyanide, chlorine in ethanol or benzene, alkaline hypochloride, or dinitrogen trioxide to the dinitroso derivatives (Scheme 68) ⟨B-61MI 215-01⟩.

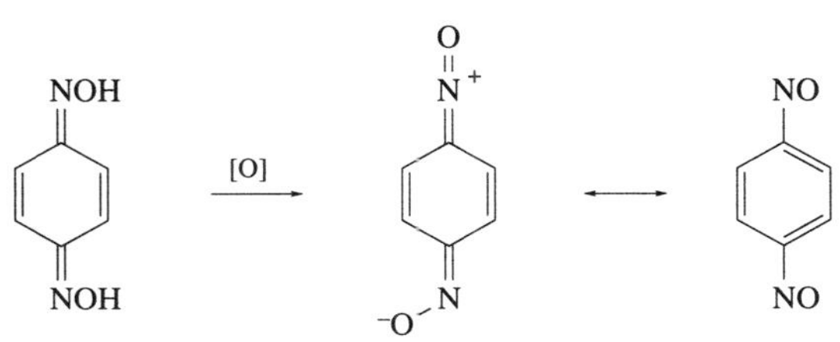

Scheme 68

(vii) Oxidation of hydroxylamines

In a two-step process nitrobenzene is reduced by zinc and ammonium chloride to phenylhydroxylamine (**232**) which is oxidized by sodium dichromate in sulfuric acid to nitrosobenzene (**233**) in an overall yield of 55% (Scheme 69) ⟨55OSC(3)668⟩.

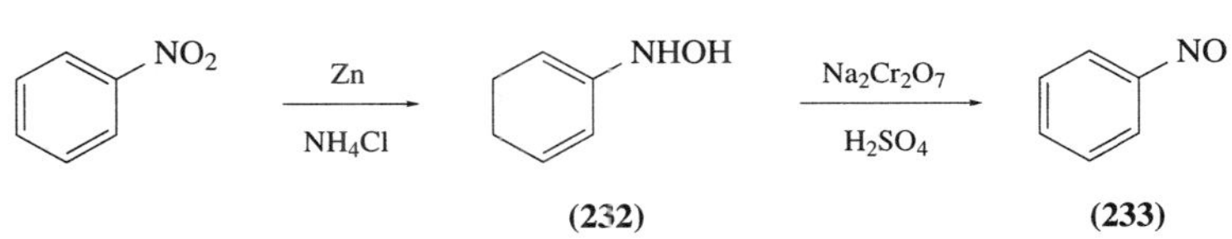

Scheme 69

Other reagents which have been used to oxidize hydroxylamines include mercuric oxide ⟨1897CB1656⟩, potassium ferricyanide ⟨64PCS230⟩, periodates in alkaline medium ⟨65CR(260)2512⟩, periodic acid ⟨62JOC1075⟩, chlorine ⟨59CCC1158⟩, air ⟨01LA(316)292⟩, and chromium trioxide ⟨26CB818, 65IZV1715⟩.

(viii) Oxidation of primary, secondary, and tertiary amines

Good yields have been reported for the oxidation of primary aromatic amines to corresponding nitroso derivatives by a peroxydisulfate in concentrated sulfuric acid (Equation (85)) ⟨02CB731, 55OSC(3)334, 68TL6155⟩.

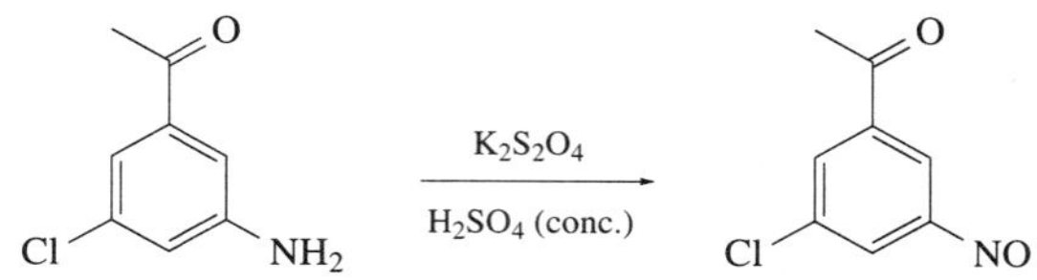

(85)

Other oxidizing reagents including Caro's acid (monoperoxy sulfuric acid) has been used as above ⟨03CB3803⟩. The elimination or suppression of further oxidation to the corresponding nitro compound is often a limiting factor to consider. Oxidation of *N*-benzylaniline with Caro's acid is known to produce nitrosobenzene along with other products ⟨02CB731⟩.

An indole derivative (**234**) was transformed to the nitroso compound (**235**) by a cold dilute permanganate solution (Equation (86)) ⟨1896CB2662⟩.

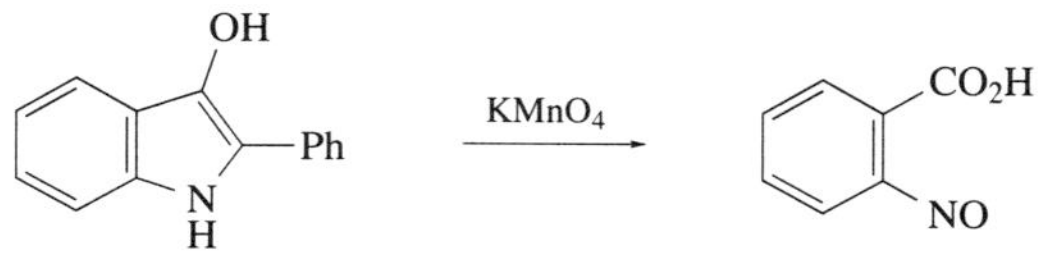

(86)

(234) **(235)**

ortho-Nitrosocinnamic acid (**236**) is a proposed intermediate in the oxidation of quinoline by hydrogen peroxide in acetic acid to *ortho*-nitrocinnamic acid (Scheme 70) ⟨64CPB326⟩.

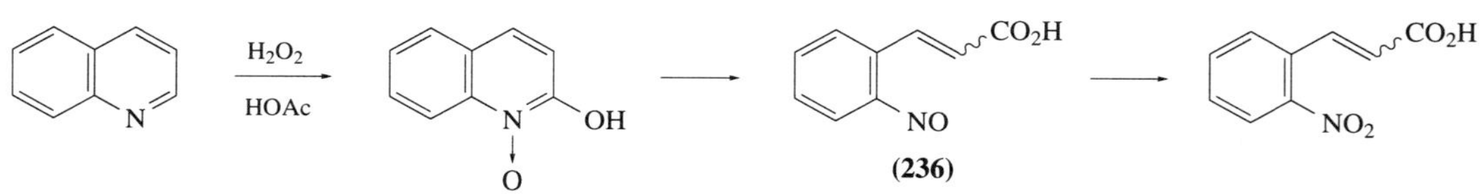

(236)

Scheme 70

Oxidation of 3-methyl-anthranil (**237**) occurs with opening of the isoxazole ring and the formation of *o*-nitrosoacetophenone (**238**) (Equation (87)) ⟨60JA1801⟩.

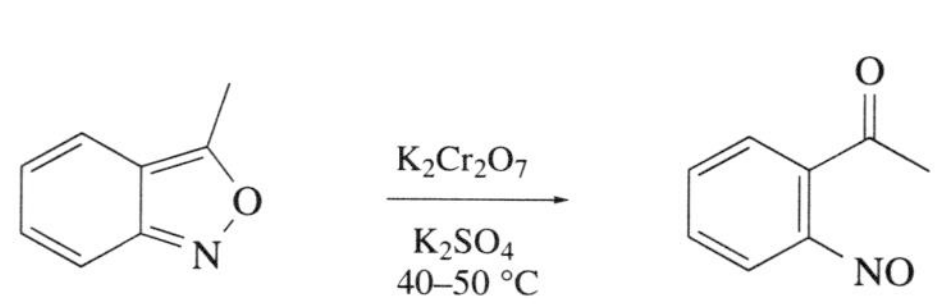

(87)

(237) **(238)**

(ix) Reduction and deoxygenation of nitro compounds

Reduction of a nitro compound leads first to the formation of a nitroso compound, with best yields if carried out in neutral media. An electrolytic reduction of nitrobenzene with a neutral electrolyte gave a good yield of nitrosobenzene ⟨08GEP1765⟩. The reduction of 3-nitro-4-dimethylaminotoluene (**239**) with sodium bisulfite was stopped at the nitroso stage with the formation of

1,5-dimethylbenzimidazole (**240**) as the end-product by an intramolecular condensation (Scheme 71) ⟨36JA225⟩.

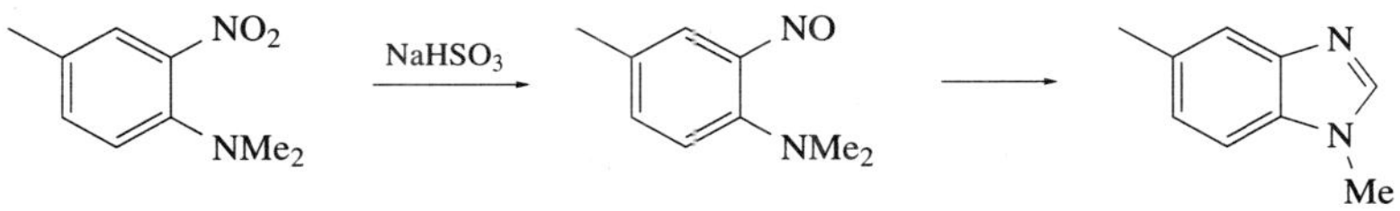

(239) **(240)**

Scheme 71

The transformation of *o*-nitromandelonitrile (**241**) by ammonia into *o*-nitrobenzoic acid (**242**) and of *o*-nitrobenzaldehyde by treatment with potassium cyanide into the same product (**242**) have been presumed to proceed as shown in Scheme 72 ⟨06CB2334, 06CB2339, 66JCE144⟩.

Other reducing and deoxygenation agents employed for this purpose have been described by Boyer ⟨B-69MI 215-01⟩.

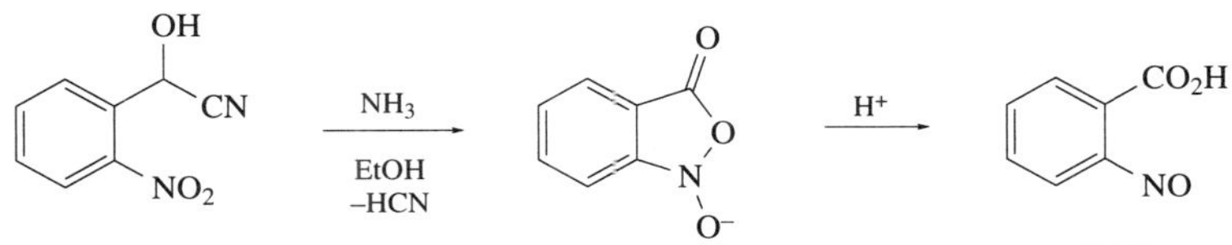

(241) **(242)**

Scheme 72

(x) Pyrolysis of heterocycles

Certain saturated and partially saturated rings containing a ring oxygen adjacent to a ring nitrogen bearing an exocyclic substituent undergo pyrolysis with the release of exocyclic substituent as the nitroso derivative (Scheme 73) ⟨63JA2683⟩.

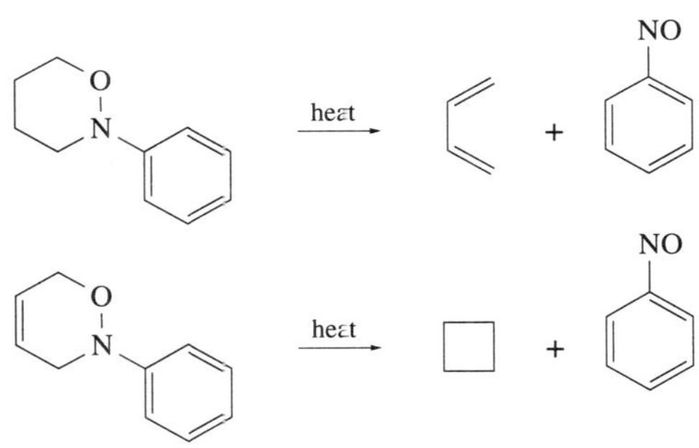

Scheme 73

2.15.9.2 Nitro Compounds

2.15.9.2.1 Nitroalkenes

Conjugated nitroalkenes are useful and versatile intermediates in organic synthesis ⟨78JA6294, 79C1, 86CRV751, 88S833, 91CSR95⟩. Due to the powerful electron-withdrawing property of the nitro group, they have found widespread application as Michael acceptors for the addition of stabilized carbon nucleophiles such as enamines ⟨66TL1821, 66TL6331, 68T1889, 73T3011, 74JCS(P1)660, 74T2741⟩, lithium enolates ⟨82CL1505⟩, silyl enol ethers, and silyl ketene acetals ⟨85ACR284⟩. Significant advances have been reported on the addition of nonstabilized organometallics derived from lithium ⟨85T4861⟩, magnesium ⟨47JCS1494, 47JCS1497, 88S379⟩, zinc ⟨92AG(E)1578⟩, aluminum ⟨88JOC45, 89JOC2391⟩, and copper ⟨82T389, 91TL441⟩. In addition, nitroalkenes have found notable utility as electron-deficient dienophiles in the Diels–Alder reaction ⟨80JOC1185, 81TL603⟩. Regardless of the application, the versatility of the chemistry of the nitro group stems from the ability to transform nitro-containing compounds into a variety of functionalized products such as aldehydes, ketones, oximes, nitrones, hydroxylamines, and amines ⟨B-70MI 215-02⟩.

Denmark *et al.* ⟨86JA1306⟩ have developed the use of conjugated nitroalkenes as electron-deficient heterodienes in Lewis acid-promoted [4 + 2] cycloaddition reactions. Most of these studies have involved 2-substituted nitroalkenes, as well as 1,2-disubstituted-1-nitroalkenes, and later they became interested in exploring [4 + 2] cycloadditions of 2,2-disubstituted-1-nitroalkenes.

Classically, nitroalkenes are prepared by the Henry reaction of a nitroalkane with a carbonyl compound ⟨71HOU(10/1)9, 87OPP283⟩, followed by dehydration of the resultant 2-nitro alcohol (**243**) (Scheme 74) ⟨47JCS1471⟩.

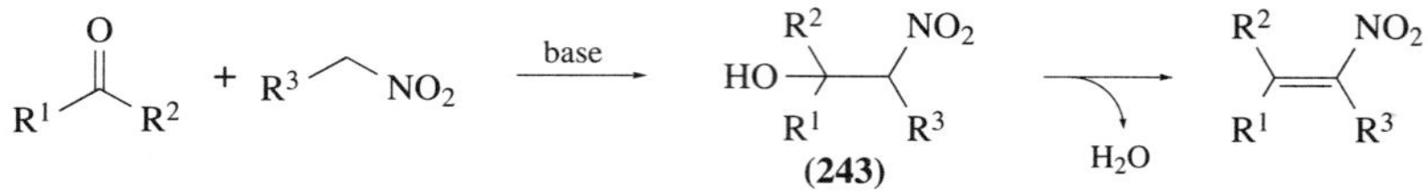

Scheme 74

Tamura and co-workers ⟨86JOC4368, 91COS(2)329⟩, have utilized *N*,*N*-dimethylethylenediamine to drive the condensation of ketones with nitroalkanes, but the β,γ-unsaturated tautomers are the major products (Scheme 75).

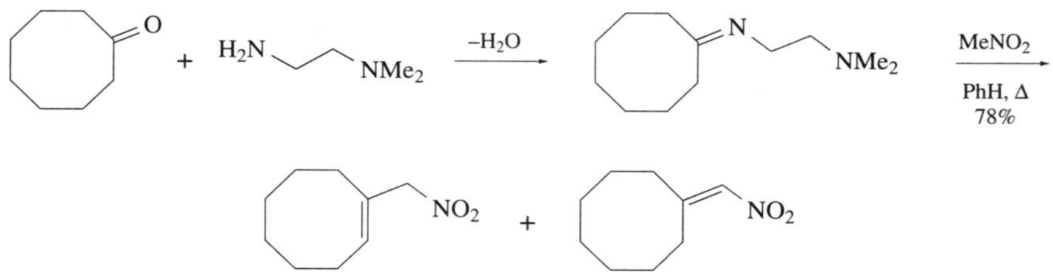

Scheme 75

In the early 1990s, Knochel and Rutherford ⟨91TL441⟩ described the preparation of 2,2-disubstituted 1-nitroalkenes from the addition–elimination of copper–zinc organometallic reagents to nitroalkenes substituted with leaving groups at the β-position (Equation (88)). This procedure allows for high-yield preparation of functionalized nitroalkenes, but is dependent upon the ease of preparation of the starting nitroalkene. A related example of addition–substitution of organolithium reagents to a β-sulfinyl nitroalkene has also been described ⟨92S1119⟩.

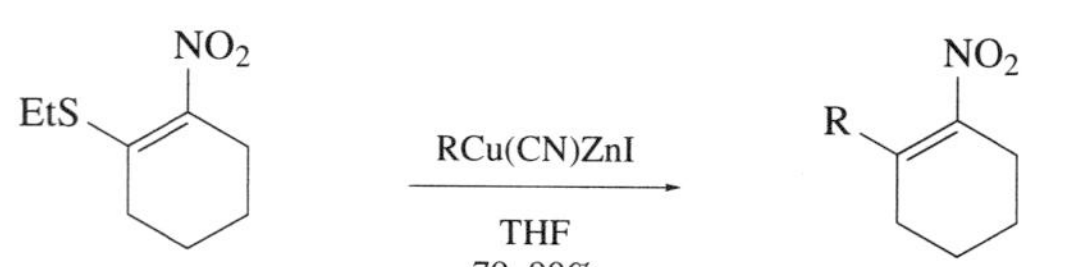

(88)

R = $(CH_2)_3CO_2Et$, $(CH_2)_3CN$, $(CH_2)_2PO(OEt)_2$, (*E*)-HC=CH-*n*C_6H_{13}

Sakakibara *et al.* ⟨81CC261, 82S261⟩ reported the preparation of 1-nitroalkenes from nitroalkanes utilizing selenium chemistry. It was found that nitronates, generated *in situ* from the treatment of nitroalkanes with *n*-butyllithium, were trapped upon quenching with phenyl selenenyl bromide to afford the corresponding nitroselenides (**244**). The selenoxides, obtained after oxidation of the selenides with hydrogen peroxide, spontaneously underwent elimination at room temperature to afford α,β-unsaturated-1-nitro-alkenes (**245**) in good yields (Scheme 76).

Utilizing a procedure developed by Rutherford and Knochel ⟨91TL441⟩, Michael-type addition of organometallics to the nitroalkene (**246**) afforded, after acidic work-up, the nitroalkane (**247**) in 75–95% yield. The purified nitroalkane was treated with *n*-butyllithium and the resulting lithium nitronate was quenched by the addition of phenyl selenenyl bromide. The resulting selenide (**248**), obtained in moderate yield (59–71%), was oxidized with hydrogen peroxide at room temperature

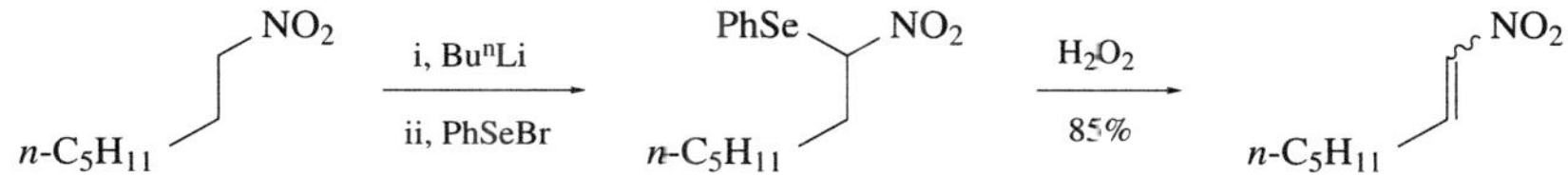

(244) (245)

Scheme 76

to afford the bright yellow nitroalkene (**249**) in good yield (77–93%). The nitroalkenes were obtained as (*Z*)/(*E*) isomeric mixtures, slightly enriched in the (*Z*) isomer (~ 1.5 : 1.0 (*Z*/*E*)) (Scheme 77).

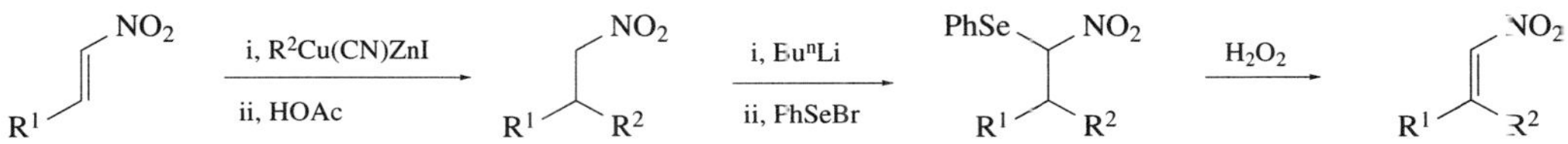

(246) (247) (248) (249)

R^1 = Ph, 3,4-di-OMe-Ph
R^2 = Et, Bu^n, Bu^i

Scheme 77

Similarly, Denmark and Marcin ⟨93JOC3850⟩ have developed a synthesis of 2,2-disubstituted-1-nitroalkenes (**250**), which is a combination of Knochel's and Sakakibara's work, involving the conjugate addition of copper–zinc organometallics to 1-nitroalkenes followed by phenylselenation and oxidative elimination (Scheme 78).

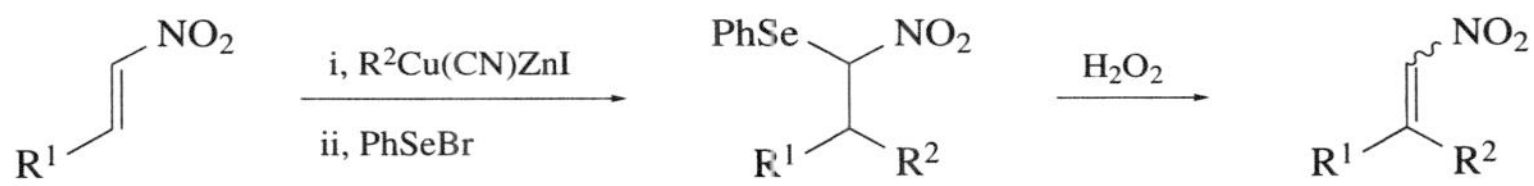

(250)

Scheme 78

The most versatile preparation of nitroalkenes involves the Henry condensation as mentioned earlier (see Scheme 75), it is routinely carried out under mildly basic conditions. Several reagents including dicyclohexylcarbodiimide (dcc) ⟨82S1017⟩ with copper(I) chloride as catalyst, pivaloyl chloride ⟨82TL3897, 84HCA261⟩, methanesulfonyl chloride, or phthalic anhydride ⟨B-70MI 215-02, 80JOC1185⟩ have been used for the dehydration step.

Barton *et al.* ⟨82CC551, 83BSF(2)61⟩ have reported that ketones, including 3*β*-hydroxyandrost-5-en-17-one reacted with nitromethane with ethylenediamine as a catalyst to directly provide the corresponding nitroalkenes.

Corey and Estreicher ⟨81TL603⟩ have described a convenient synthesis of the nitroenones from 2-cyclopentanone and 2-cyclohexanone.

Vankar and Bawa ⟨85SC1253⟩ adapted earlier Corey nitromercuration chemistry.

Dampawan and Zajac ⟨82TL135⟩ have reported a synthesis of cyclic nitroalkenes from *β*-nitroketones. This method is of particular interest in that the reaction intermediates, *β*-nitroalcohols, were not prepared via the Henry reaction, which in general, is not convenient for cyclic systems. In principle, nitroalkenes should be accessible from direct nitration of vinyl carbanions. However, this reaction is not practical due to potential anionic polymerization.

Corey and Estreicher ⟨80TL1113⟩ have modified the Shapiro reaction in an elegant synthesis of nitroalkenes from ketones. For example, cycloheptanone 2,4,6-triisopropylbenzenesulfonylhydrazone (**251**) was reacted with *s*-butyllithium in TMEDA-cyclohexane at −80°C to −10°C to produce the vinyl anion (**252**). This was trapped with chlorotrimethylstannane to provide (**253**) (84%), which reacted smoothly with tetranitromethane in dimethylsulfoxide to give 1-nitrocycloheptene (**254**) (87%) (Scheme 79). The method was also used to prepare alkyl-substituted-1-nitrocyclohexenes.

Hayama *et al.* ⟨82CL1109, 82TL4733⟩ and Seebach *et al.* ⟨85T4861⟩ have utilized selenium chemistry in the preparation of nitroalkenes. Hayama has developed nitroselenation as the most useful method for the conversion of unactivated alkenes into *β*-nitro selenides. On oxidation they obtained the selenoxide and thereby the nitroalkene. For example, cyclohexene (**255**) reacted with phenyl selenyl chloride, silver nitrate, and mercuric chloride in acetonitrile–THF at −78°C to 22°C to give the

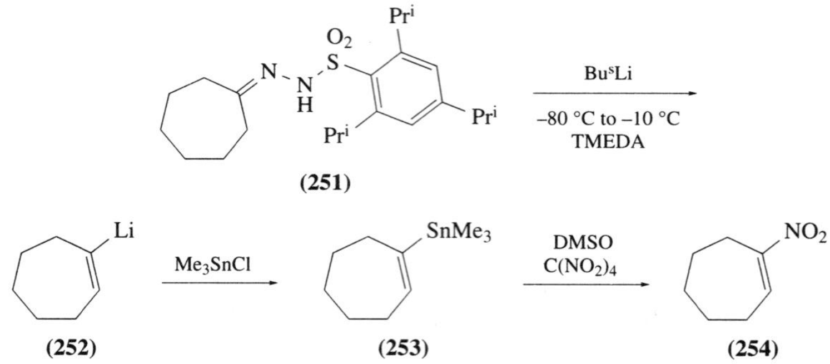

Scheme 79

trans-selenide (**256**) (81%). The mercuric chloride was added to suppress the formation of (**257**), although its mechanistic role is not clear. On oxidation with hydrogen peroxide, (**256**) was converted into (**258**) (93%) (Scheme 80). Both the nitroselenation (*trans*) and the selenoxide elimination (*syn*) reactions were stereospecific.

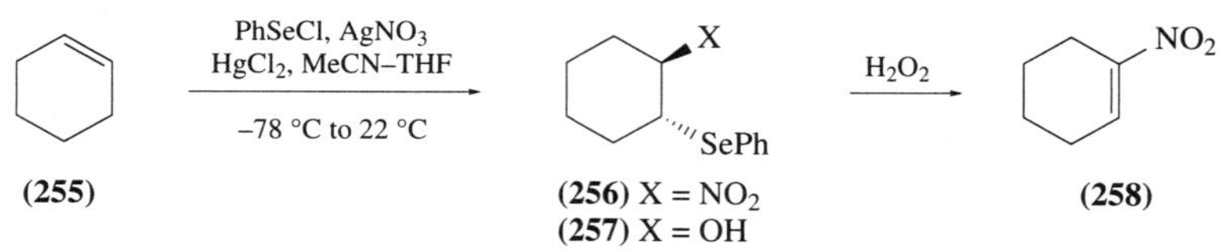

Scheme 80

Seebach *et al.* have employed phenyl selenyl trifluoroacetate for the functionalization of nitroalkenes via the *trans*-addition ⟨85T4861⟩. For example, (**259**) was reacted with phenyl selenyl chloride and silver trifluoroacetate in dichloromethane to give (**260**) (96%). Methanolysis in the presence of sodium hydrogen carbonate gave (**261**) (95%) and subsequent hydrogen peroxide oxidation regiospecifically gave (**262**) (Scheme 81).

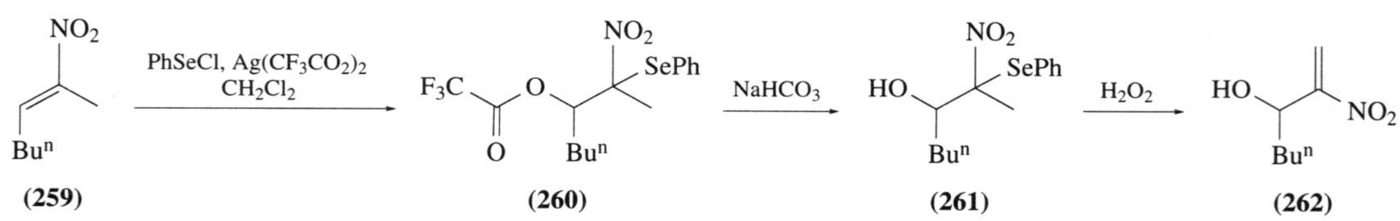

Scheme 81

Sakakibara *et al.* ⟨82BCJ635⟩ have reported that peroxytrifluoroacetic acid oxidation of α-chloro or α-bromo oximes in acetonitrile solution produced the corresponding nitroalkenes. The method was applied to the synthesis of several cyclic and acyclic nitroalkenes, which were obtained in modest yields (31–66%).

In a related protocol Yandovskii *et al.* ⟨81MI 215-01⟩ reported that two 1,1-disubstituted alkenes were oxidized by excess butyl or amyl nitrite in acetic acid to produce the nitroalkenes. In a further variation of the well-established alkene nitrosation protocol ⟨83TL4443⟩, Kunai *et al.* have developed an electrochemical method. Thus, electrolysis of aqueous sodium nitrite at a platinum anode in a divided cell in the presence of cyclohexene gave 1-nitro-1-cyclohexene in a yield of 25–57%, depending on the supporting electrolyte. The method was extended to the synthesis of 1-nitrocyclooctene (36%).

Simple nitroalkanes have been oxidized to produce nitroalkenes using selenium chemistry ⟨81CC261⟩. Phenyl selenyl bromide was added to the nitronate derived from (**263**) using *n*-butyllithium in THF to produce (**264**). Subsequent oxidation with hydrogen peroxide gave the base sensitive nitroalkene (**265**) (Scheme 82). The method is general for acyclic and cyclic systems.

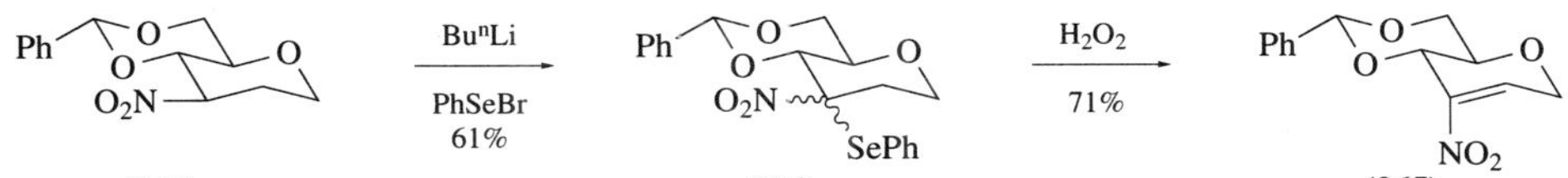

(263) (264) (265)

Scheme 82

2.15.9.2.2 β-Nitroenamines

β-Nitroenamines are useful synthetic intermediates ⟨B-70MI 215-02⟩. Royer and co-workers have described a method of preparation of these species by the condensation of triethyl *ortho*-formate, nitromethane, and secondary amines ⟨82S260⟩. Nitroacetic esters easily form the alkoxy methylene compounds (**266**) when heated with alkyl *ortho*-formates in the presence of acetic anhydride. The alkoxy group in (**266**) can be displaced by amines to form the nitrovinylamine (**267**) ⟨77CB2480, 79S666⟩. Nitroacetonitrile (**268**) leads to 3-amino-2-nitroacrylonitrile (**269**) by this procedure (Scheme 83).

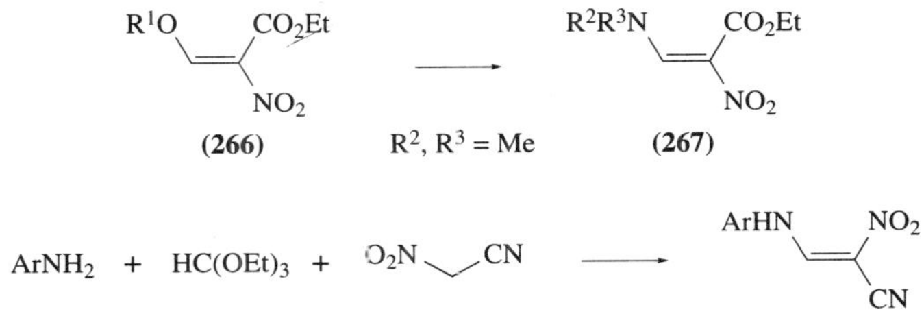

(266) R^2, R^3 = Me (267)

(268) (269)

Scheme 83

Some nitrovinylamines have been generated from enol-ethers of α-nitro ketones (Equation (89)) ⟨69JOC3279, 74JOC3351⟩. Alternatively treatment of 1-chloro-2-nitroethylenes (**270**) with amines leads to the nitrovinylamines (**271**) (Equation (90)) ⟨60JOC1877⟩.

(89)

(90)

(270) (271)

However, the difficulty of preparing the starting material makes this method of little preparative value. A related method is one proceeding via dichloronitroalkanes. These are produced when nitrosyl chloride is added to unsaturated compounds containing halogen. Treatment of (**272**) with aniline in ether leads to nitrovinylamines (**273**) (Scheme (84)) ⟨64ZOB1522⟩. Similar reactions with NaOBr have also been reported.

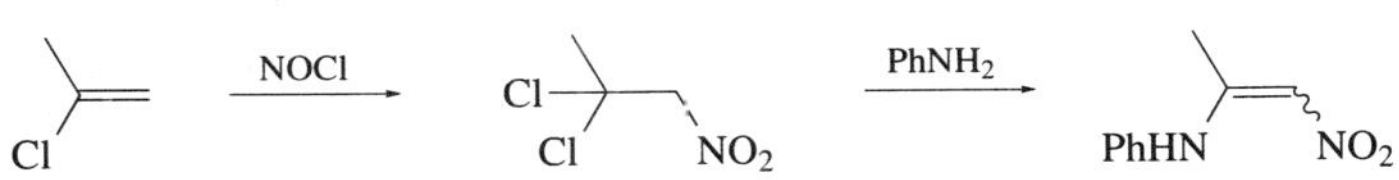

(272) (273)

Scheme 84

Transamination ⟨67AC(R)1266⟩ has also been employed as a preparative procedure; 1-dimethylamino-2-nitroethylene (**274**) is a very useful reagent for this purpose (Equation (91)).

(274) (91)

Fetell and Feuer ⟨78JOC1238⟩ have subjected 1-*t*-butylamino-2-methyl-2-nitroethylene (**275**) to transamination, but the poor reactivity of the substrate in this case illustrates the limited scope of this reaction (Equation (92)).

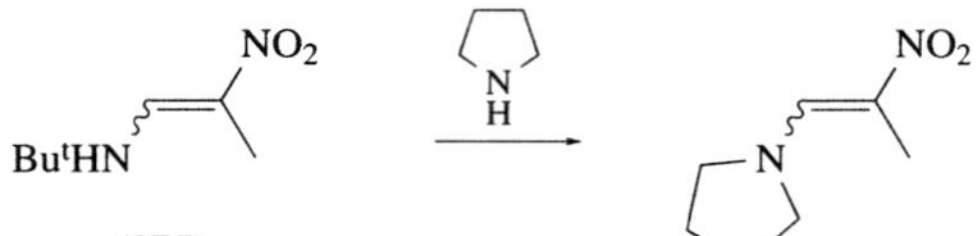

(275) (92)

One of the most common methods of preparing nitrovinylamines is by the reaction of amide acetals with nitromethane. Thus dimethylformamide diethylacetal gives 1-dimethylamino-2-nitroethylene (**276**) in 85% yield (Equation (93)) ⟨61LA(641)1⟩. A variety of substituted amideacetals have been reacted with nitromethane ⟨71CB3475⟩ and lactamacetals (**277**) have also been successfully used in this reaction (Equation (94)) ⟨71ZOR1146⟩. A simplified procedure, in some cases, makes use of the amide–dimethyl sulfate complex ⟨65CB3847, 71CB440⟩.

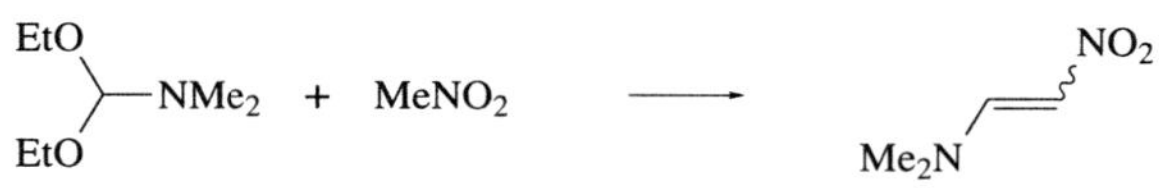

(276) (93)

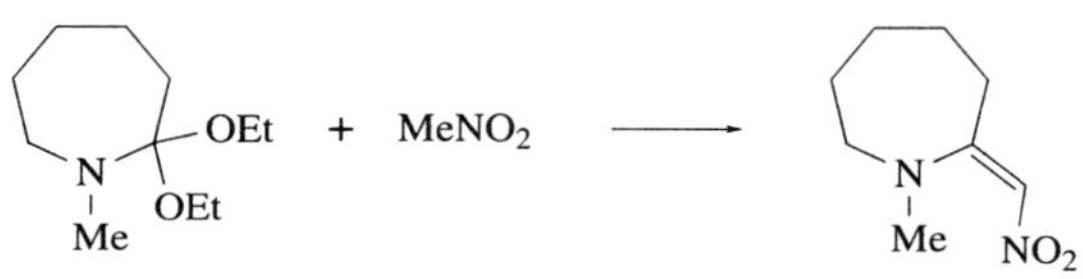

(277) (94)

Lactim ethers react easily with nitromethane to form the corresponding 2-nitromethylene derivatives ⟨60ZOB1855, 69BRP1167809, 69MIP156590⟩. An example is shown in Equation (95).

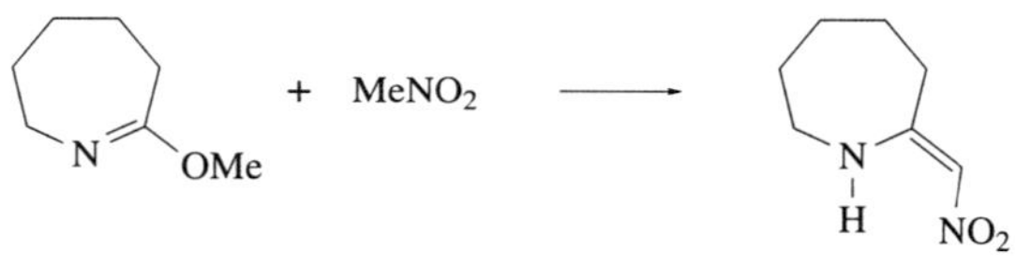

(95)

An interesting reaction ⟨79TL2525⟩ involves the addition of nitromethane to the nitrile group of ethyl cyanoformate, which is brought about with a combination of Lewis acid and a tertiary amine ($ZnCl_2$–Et_3N). The nitroenamine (**278**) is obtained in 18% yield (Equation (96)).

(278) (96)

Feuer and co-workers have discovered that aldimines can be nitrated by alkyl nitrate (e.g., *n*-propyl nitrate) in liquid ammonia in the presence of potassium amide to give the nitroenamines in about 50% yield (Equation (97)) ⟨78JOC497⟩. The reaction has been extended to alicyclic ketimines ⟨79JOC3410⟩.

(97)

The nucleophilic displacement of the halide from a 2-halogenoenamine by nitrite in DMSO to form the nitroenamine (**279**) in 27% yield has been reported (Equation (98)) ⟨73S109⟩.

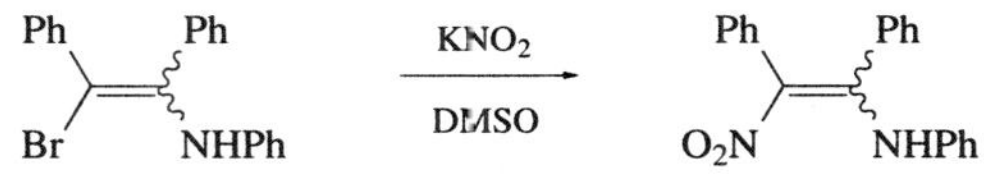

(**279**) (98)

Nitroenamines can be prepared from 1-nitro-2-phenylthioalkenes (**280**) and aliphatic amines in methanol or aromatic amines in acetonitrile at room temperature ⟨88S921⟩. The 1-nitro-2-phenylthioalkenes, in turn, are available from nitroalkanes, aldehydes, and thiols (Equation (99)) ⟨86JOC2139⟩.

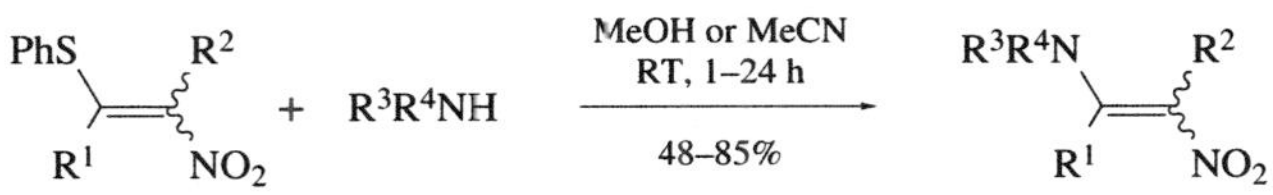

(**280**) (99)

$R^1, R^2 = H, Me$
$R^3, R^4 = -(CH_2)_4-$
$R^3 = Me, Bu^t, Ph, CH_2{=}CHCH_2$
$R^4 = H, Me, CH_2{=}CHCH_2$

A method ⟨82S260⟩ based on the reaction of primary nitroalkanes, triethyl *ortho*-formate, and (secondary) amines in the presence of *p*-toluenesulfonic acid as catalyst for the formation of 2-nitroenamines was reinvestigated by Krowczynski and Kozerski ⟨83S489⟩. They studied the reaction of nitrotoluene, 1-nitromethane, and 1-nitropropane with aniline, and primary and secondary heteroaromatic amines and found that the yields of 2-nitroenamines depend strongly on the type of substituents R^1, R^3, and R^4 (Scheme 85).

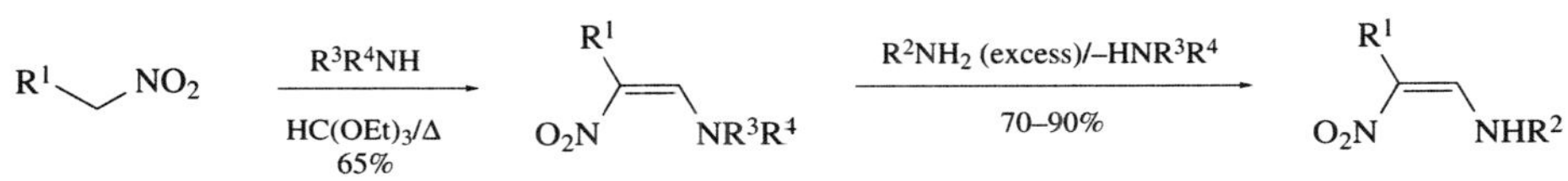

Scheme 85

The condensation of 1-nitroalkanes with triethyl *ortho*-formate and primary aliphatic amines gives only negligible amounts of the expected 2-nitroenamines, which can otherwise be conveniently prepared by transamination of the 2-nitroenamines with ammonia or primary aliphatic amines ⟨83S489⟩. A one-pot synthesis of nitroenamines by boiling a mixture of triethylformate, nitromethane and secondary amine has been reported by Royer and co-workers ⟨82S260⟩.

2.15.9.2.3 *Nitro-, di-, and polyenamines*

There are reports in the literature on the synthesis and further transformations of a few nitrodienamines and nitropolyenamines. In these compounds, the nitro and the amino groups are attached to the two termini of the conjugated alkenic system. The 4-*aci*-nitrocrotonic aldehyde derivatives (**281**), obtained from aldehydes and nitroenamines, on treatment with pyrrolidinium acetate give the nitrodienamines (**282**) (Scheme 86) ⟨76CB3541⟩. The nitrodienamines themselves can be used as reagents for the condensation with active methylene compounds leading to *aci*-nitro derivatives which can be converted to nitropolyenes, etc. ⟨78CB692⟩.

Nitrodienamines and nitropolyenamines undergo useful [4 + 2]- and [8 + 2]-cycloadditions with

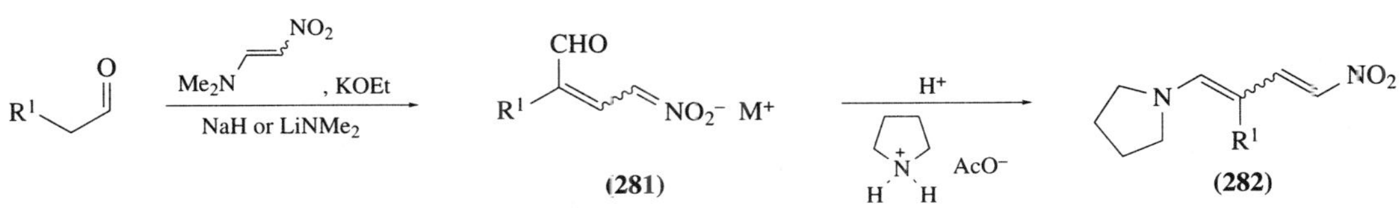

Scheme 86

substituted alkenes and acetylenes; loss of the amine from the adduct leads to aromatic products (Scheme 87) ⟨76CB3541, 78S592⟩.

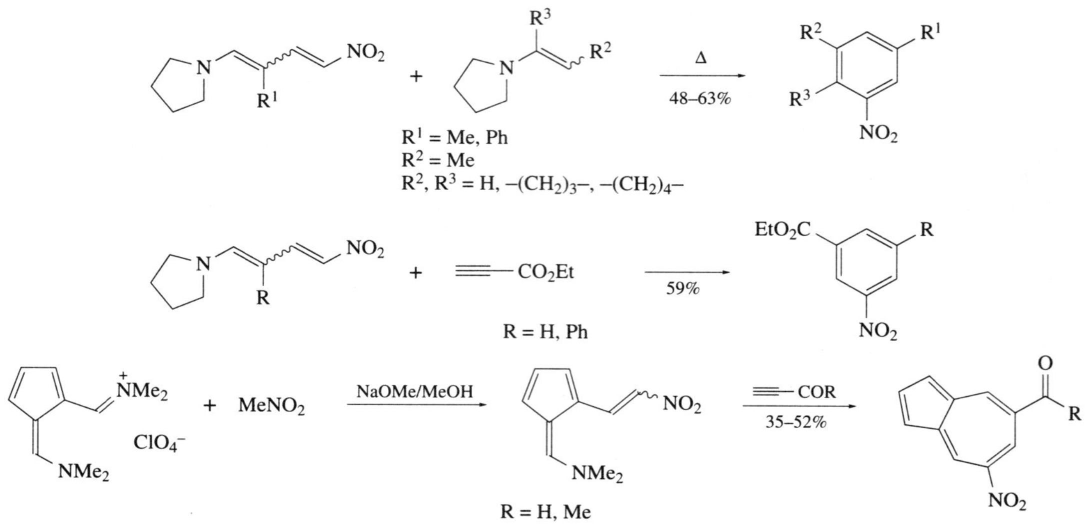

Scheme 87

The pentamethinium salt (**283**) reacts with nitromethane to give the 1-dimethylamino-6-nitrohexatriene (**284**). This undergoes thermal electrocyclization followed by elimination of dimethylamine to give nitrobenzene derivatives (**285**) (Scheme 88) ⟨72AG(E)315⟩.

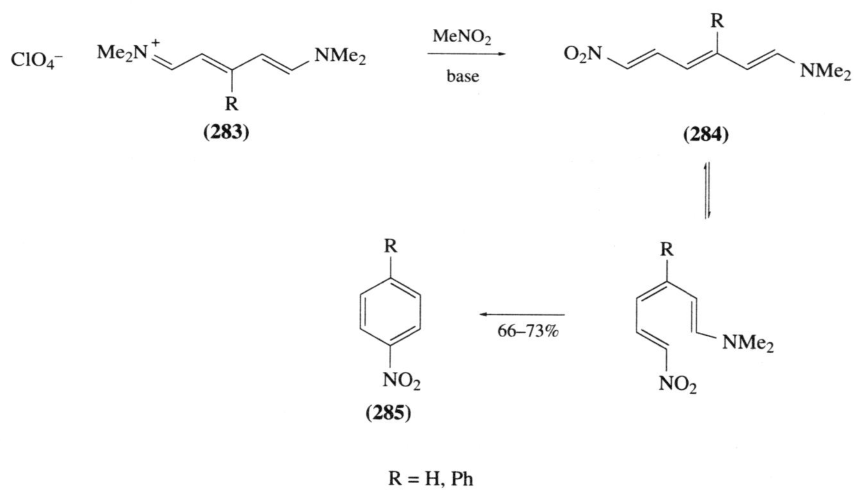

Scheme 88

A similar route involving electrocyclization and elimination of dimethylamine-containing intermediates has been used to prepare nitroazulenes (**286**) (Scheme 89) ⟨79S981⟩.

2.15.9.2.4 Aromatic nitro compounds

The field of aromatic nitration has been reviewed by Schofield in a monograph ⟨B-80MI 215-01⟩ and also briefly by Askani and Taber ⟨91COS(6)103⟩.

A nitro group is introduced in an aromatic ring usually by way of electrophilic aromatic substitution using nitric acid, its metal salts, mixed anhydrides, nitrate esters or nitronium salts ⟨77S217, 79MI 215-02, 83RCR284⟩. Alkyl benzenes have been nitrated with *n*-butyl nitrate, acetone cyanohydrin

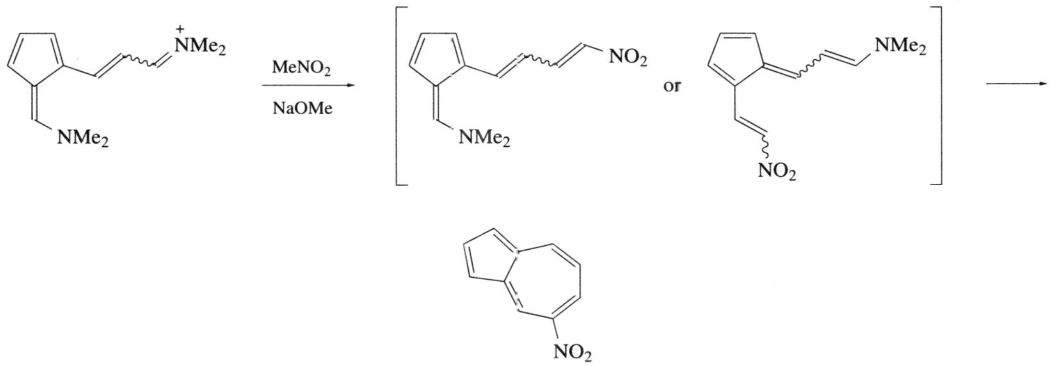

(286)

Scheme 89

nitrate or dinitrogen tetroxide over Nafion-H (perfluorinated sulfonic acid) catalyst, thus avoiding aqueous work-up (Equation (100)) ⟨82JOC596⟩.

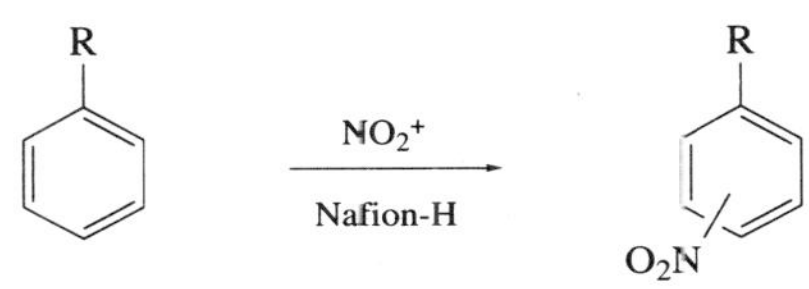

(100)

Trifluoroacetic nitrate, prepared *in situ* from ammonium nitrate and trifluoroacetic anhydride, has been used for nitration of aromatic compounds at room temperature with high yields (Equation (101)) ⟨81JOC3056⟩.

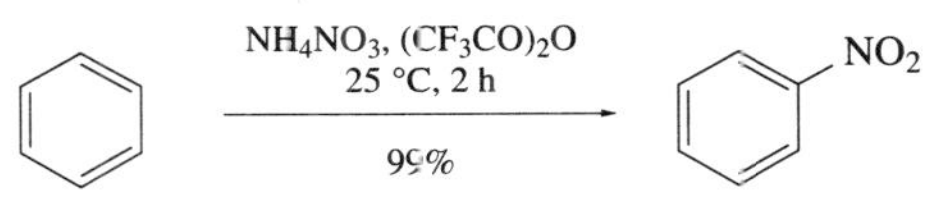

(101)

Phenols can be mononitrated in a two-phase system (ether–water) by sodium nitrate and aqueous hydrochloric acid in the presence of a catalytic amount of lanthanum(III) nitrate in yields generally above 80% (Equation (102)) ⟨86SC63, 89T191⟩.

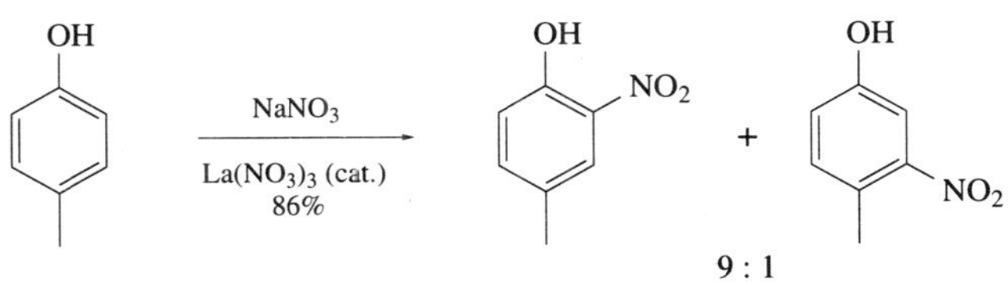

(102)

Aromatic amines are regioselectively nitrated with urea nitrate in sulfuric acid. *p*-Nitroanilines are formed exclusively. When the *para* position is occupied, nitration occurs to give the *m*-nitroaniline derivatives (Equation (103)) ⟨88SC2161⟩.

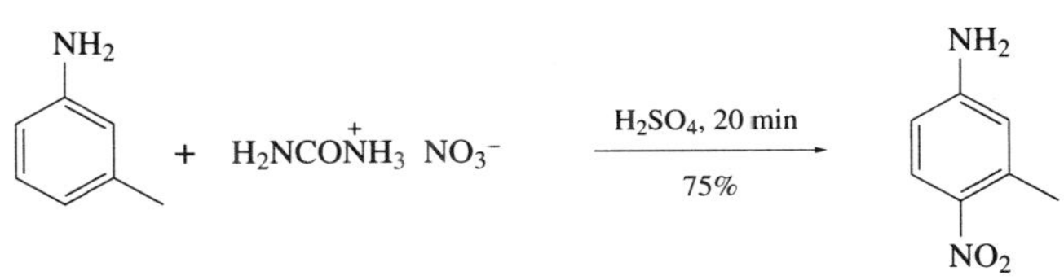

(103)

Silica-gel supported cerium(IV) ammonium nitrate (can) has been employed for controlled nitration of some naphthalene derivatives (Equation (104)) ⟨85S70⟩.

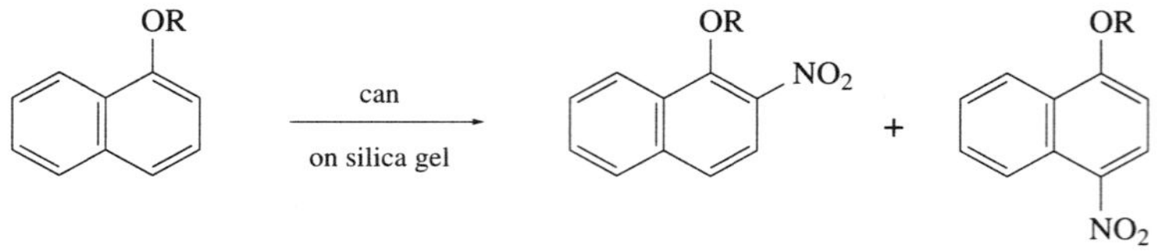

(104)

R = H, 80%	53%	47%
R = Me, 65%	0%	65%

Novel reagents consisting of metallic nitrates impregnated on montmorillonite have been introduced in aromatic nitration ⟨86ACR121, 88TL5657⟩. The effect of crown ethers on the selectivity of nitration of aromatic hydrocarbons and anisole has been studied with a mixture of trifluoroacetic anhydride and tetrabutylammonium nitrate as nitration agent ⟨85JOC4081⟩.

2.15.9.3 Aryl Thionitroso Compounds

The chemistry of thionitroso compounds remains elusive and unexplored, although thionitrosoarenes were first generated over twenty years ago from *N*,*N*-thiobis(arylamines) ⟨66AG(E)1048⟩. A few thionitrosoarenes have been generated and intercepted in Diels–Alder reactions, and unstable *N*-thionitrosoamines have been isolated ⟨66JA3842, 76JOC1333, 78TL4003⟩. Joucla and Rees ⟨84CC374⟩ employed the known opening of a five-membered heterocyclic ring with nitrene precursors in the α-position to generate the unstable thionitroso intermediate, which was then trapped as a cycloadduct. Thus, photolysis of (**287**) in diethyl ether at room temperature through pyrex with a 450 W high-pressure Hg lamp produced the thionitroso derivative (**288**). As this could not be isolated pure, the photolysis was repeated in the presence of an excess cyclopentadiene and the cycloadduct (**289**) was isolated in 60% yield (Scheme 90).

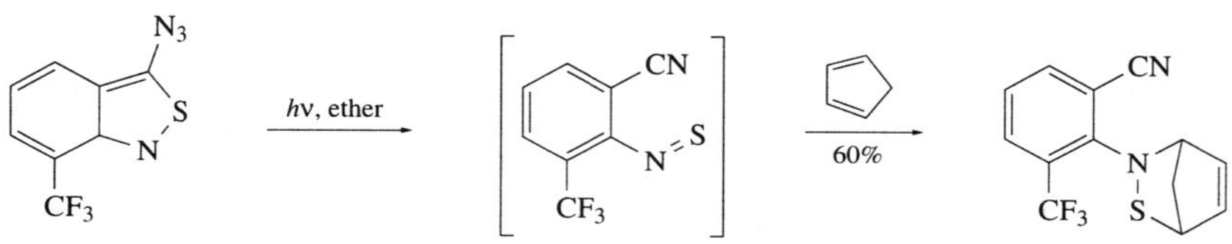

(**287**) (**288**) (**289**)

Scheme 90

Pedersen *et al.* ⟨78ACS(B)625⟩ showed that 2-thionitrosonitrobenzene (**291**) was formed as a reversible short-lived intermediate during flash photolysis of the thiadiazole derivative (**290**) (Equation (105)).

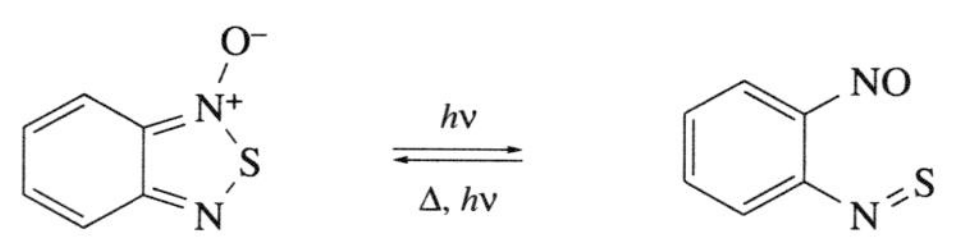

(105)

(**290**) (**291**)

Bryce and Taylor ⟨88CC950, 89TL3835, 94JCS(P1)1935⟩ developed a base-induced fragmentation of a series of *N*-(arylaminothio)phthalimides (**292**) to provide an efficient route to transient thionitrosoarenes (**293**), which were then trapped as Diels–Alder cycloadducts with butadiene and 2,3-dimethylbutadiene, or as ene adducts with alkenes (Scheme 91).

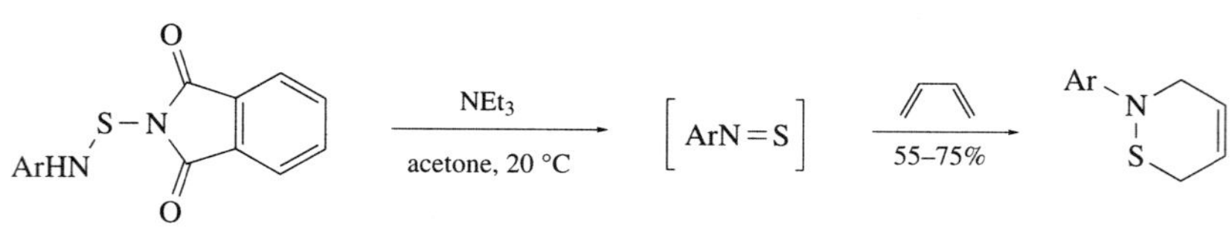

(**292**) (**293**)

Scheme 91

2.15.9.4 Sulfoximines

2.15.9.4.1 Vinyl sulfoximines

The synthesis and the use of *N*-sulfoximines in cycloaddition reactions have been reviewed by Weinreb ⟨88ACR313, 91COS(5)401⟩ and others ⟨62AG(E)89, 83SR215, 85RCR1706⟩. The vinyl sulfoximine (**295**), the vinyl being part of a ring system, was generated from the diamine (**294**) by reaction with PhNSO in the presence of a base (Equation (106)) ⟨94H(37)693⟩.

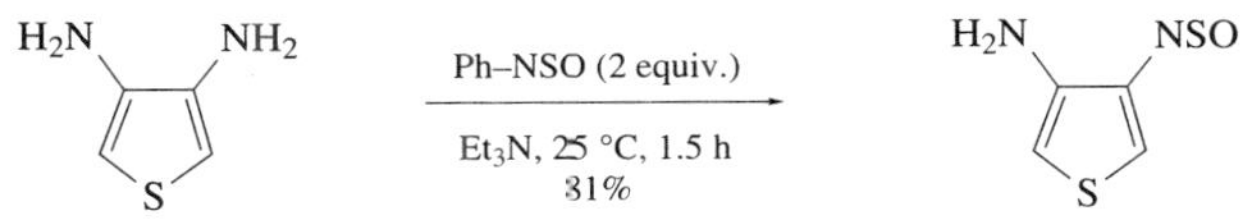

(106)

(**294**) (**295**)

2.15.9.4.2 Aromatic sulfoximines

N-Arylsulfoximines are usually prepared by reaction of thionyl chloride with aromatic amines or a "masked" amine group (Equation (107)) ⟨62AG(E)89⟩.

(107)

Sulfoximine (**298**) was obtained in 11% yield when a hexane solution of (**296**) was allowed to stand for 6 h at 40°C while oxygen was bubbled into the solution. Thionitrosoarene (**297**) was presumed to be the intermediate in this process (Scheme 92) ⟨89CL2087⟩.

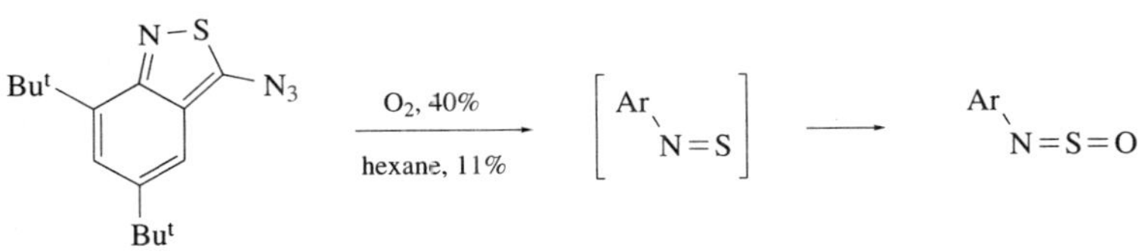

(**296**) (**297**) (**298**)

Scheme 92

Swern and co-workers ⟨76JOC2207⟩ have prepared a series of *N*-arylsulfoximines by a method involving the reaction of a complex, possibly a sulfurane (**299**), formed from *t*-butylhypochlorite and dimethylsulfoxide with arylamines (Scheme 93).

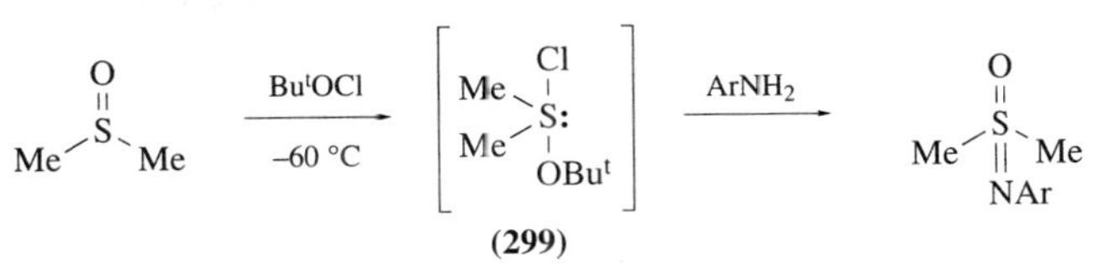

(**299**)

Scheme 93

2.15.9.5 *N*-Sulfonylamines

The most general approach to aromatic sulfonylamines (**300**) is the dehydrohalogenation of sulfamoyl chlorides; the *N*-sulfonylamines can be trapped by tertiary amines to give inner salts (**301**), which can serve as convenient sources of *N*-sulfonylamines (Scheme 94) ⟨71MI 215-01, 72JA6135⟩.

Lwowski and Scheiffele ⟨65JA4359⟩ inferred the interposition of *N*-sulfonylaniline (**304**) from the

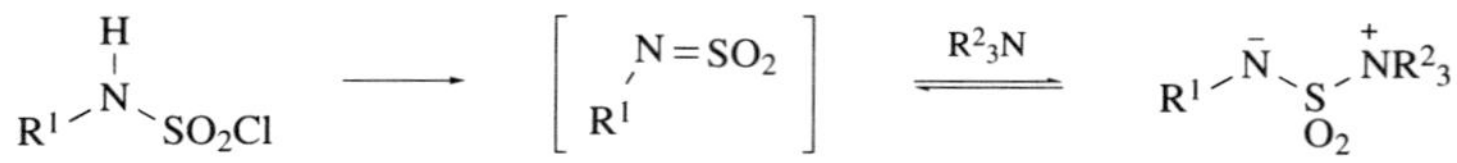

(300) (301)

Scheme 94

production of methyl *N*-phenylsulfamate (**305**) following the generation of benzene sulfonylnitrene (**303**) from (**302**) in methanol (Scheme 95).

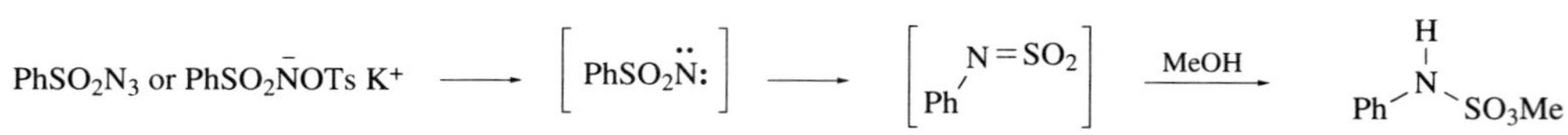

(302) (303) (304) (305)

Scheme 95

Similarly, Takahashi *et al.* ⟨89CL2087, 92JA1830⟩ inferred the formation of *N*-sulfonylamine (**306**) as an intermediate which then underwent an intramolecular 1,3-dipolar cycloaddition to give (**307**) (Scheme 96).

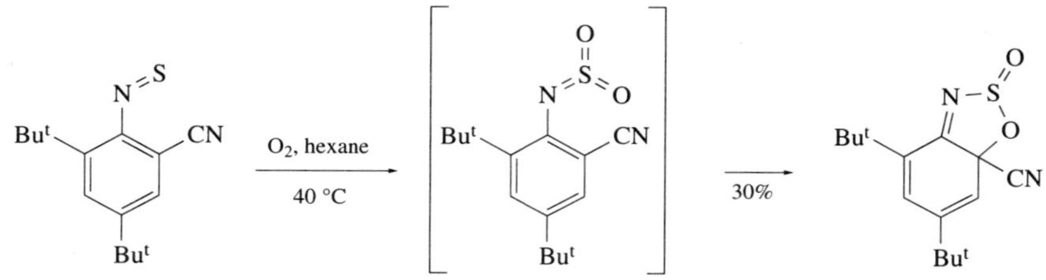

(306) (307)

Scheme 96

2.15.9.6 Sulfurdiimides

2.15.9.6.1 Vinyl sulfurdiimides

The following examples describe the synthesis of vinyl sulfurdiimides in which the unsaturation (vinyl part) is in a ring system. Tashiro and Mataka ⟨76H1243⟩ isolated the sulfurdiimide (**309**) in trace amounts during the reaction of the bishydrazole (**308**) with N_2S_4 (tetrasulfurtetranide) (Scheme 97).

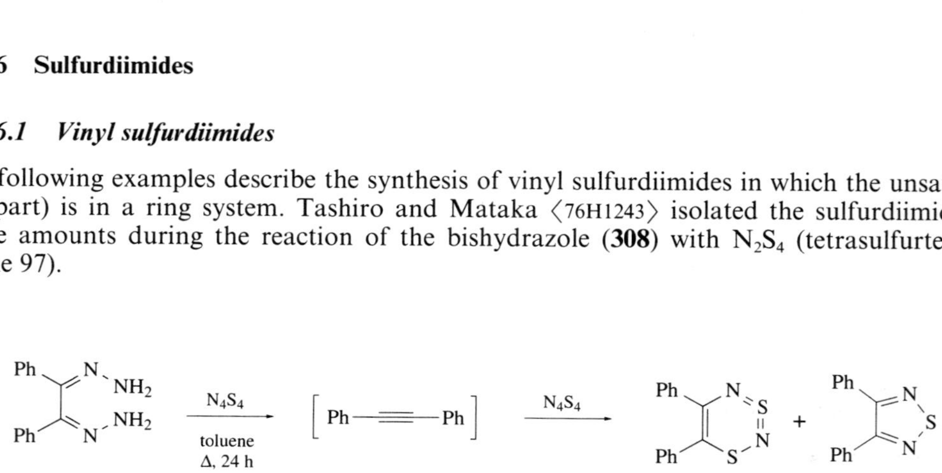

(308) (309) trace 22%

Scheme 97

Behforouz and Benrashid ⟨79TL4493⟩ reported the sulfurdiimide (**310**) as a transient intermediate *en route* to the preparation of (**311**) (Scheme 98).

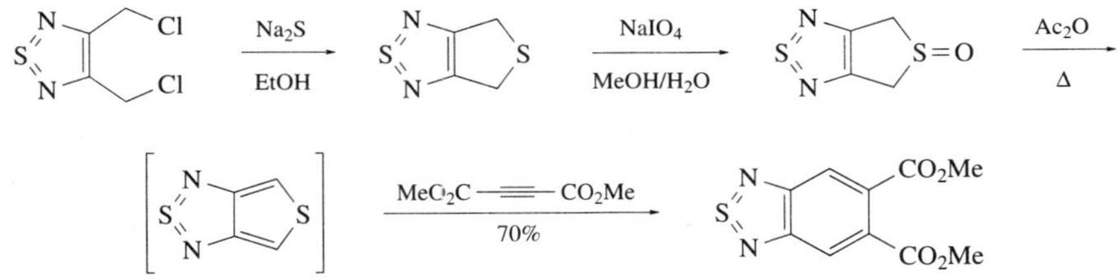

(**310**) (**311**)

Scheme 98

Oakley and co-workers ⟨83CJC1562⟩ reported the formation of a sulfurdiimide (**313**) consisting of a five-membered heterocyclic ring and a seven-membered heterocyclic ring fused to a benzene ring from a bifunctional sulfurdiimide (**312**) (Equation (108)).

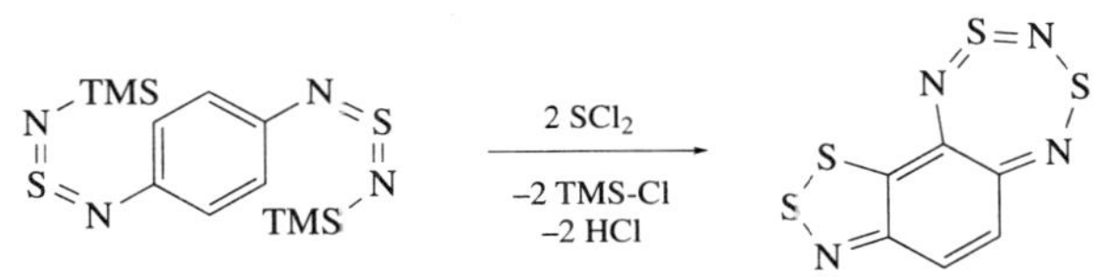

(108)

(**312**) (**313**)

Yamashita and co-workers ⟨94H(37)693⟩ isolated the vinylsulfurdiimide (**315**) from (**314**) by capturing it on a cold finger at −50 °C as yellow crystals during purification by sublimation. In the solid state or in concentrated solutions at room temperature, it decomposes to an insoluble and intractable colorless material. However, the *t*-butyl stabilized derivative (**316**) allowed its isolation and data collection for x-ray structural analysis at room temperature (Scheme 99).

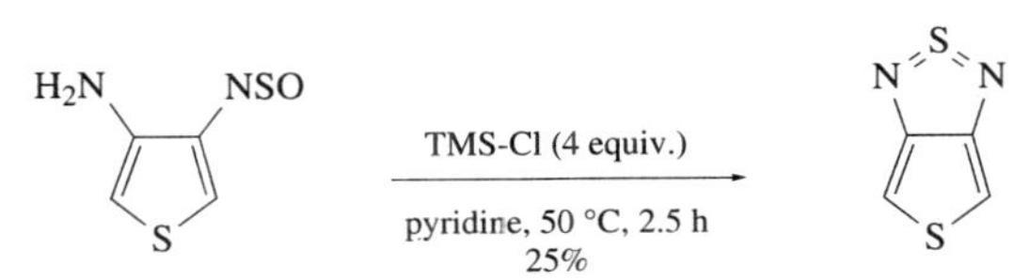

(**314**) (**315**)

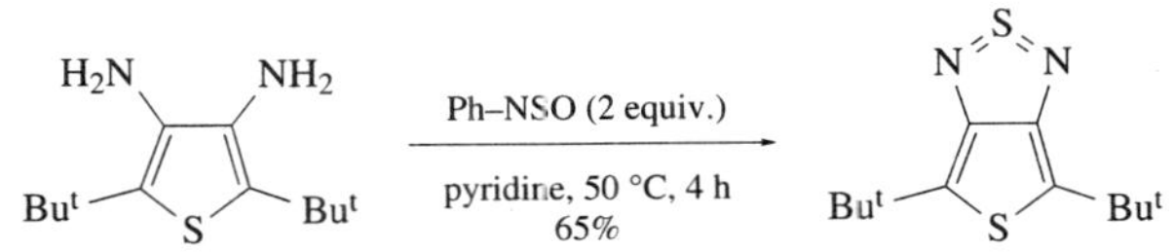

(**316**)

Scheme 99

2.15.9.6.2 *Aryl sulfurdiimides*

The aromatic sulfurdiimide (**318**) was synthesized in 58% yield as yellow crystals by heating the unstable thionitroso intermediate (**317**) at 180 °C in *o*-dichlorobenzene ⟨84CC374⟩. It is the product of dimerization of the thionitroso derivative (**317**) with loss of sulfur (Scheme 100); similar to the formation of dimethylsulfurdiimide from thionitrosomethane ⟨80JOC1691⟩. Similarly, Okazaki and co-workers ⟨89CL2087, 92JA1830⟩ reported the formation of aromatic sulfurdiimides under photolysis conditions. Diaryl sulfurdiimides were prepared by the reduction of primary amines with sulfur halides (Scheme 101).

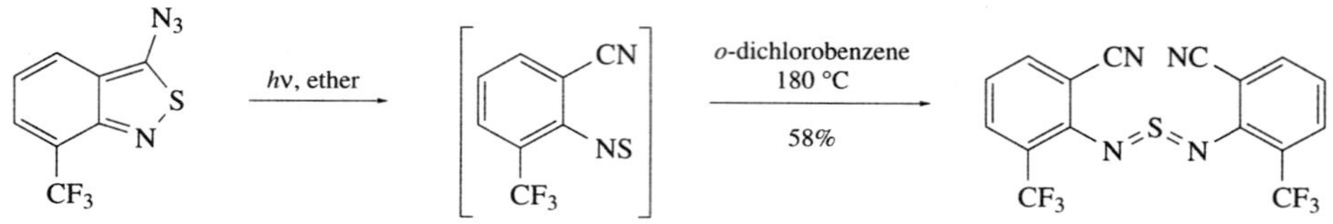

(317) **(318)**

Scheme 100

$$R^1NH_2 \xrightarrow[-2HX]{SX_4} R^1N{=}SX_3 \xrightarrow[-2HX]{R^2NH_2} R^2N{=}S{=}NR^1$$

X = Cl, F

Scheme 101

Diaryl sulfurdiimides were also prepared by the reduction of *N*-arylsulfoximes with sodium (Equation (109)) ⟨66TL1671, 67AG(E)149, B-68MI 215-05, 71BCJ1896⟩.

$$4\,ArN{=}S{=}O \;+\; 2\,Na \longrightarrow 2\,ArN{=}S{=}NAr \;+\; Na_2S_2O_4 \qquad (109)$$

Koenig and co-workers ⟨83CJC1562, 83CC73⟩ reported the reaction between *N*-phenyl-*N′*-trimethylsilylsulfurdiimide (**319**) and sulfur dichloride, which yielded the novel antiaromatic 12 π-electron heterocycle (**320**) (Scheme 102).

Heterocyclic examples of sulfurdiimides have been obtained from aromatic diamines by reaction with thionyl chloride or sulfur dioxide in the presence of triethylamine (Scheme 103) ⟨67AG(E)149⟩.

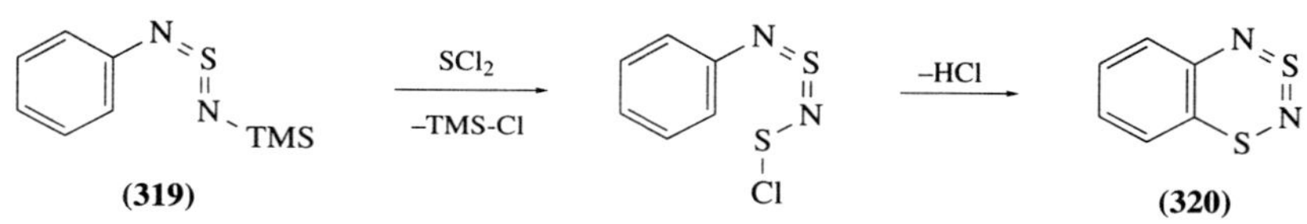

(319) **(320)**

Scheme 102

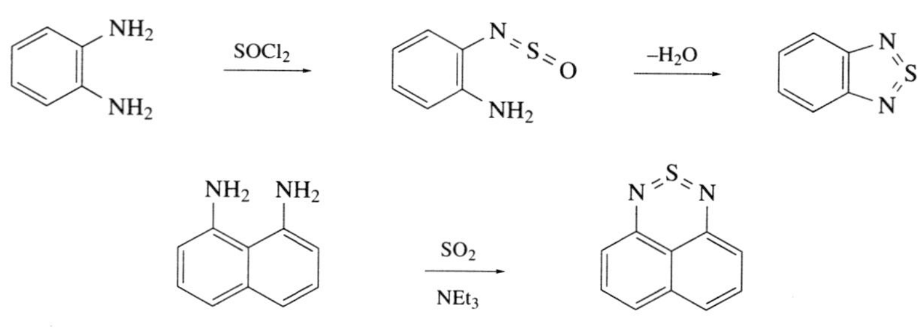

Scheme 103

2.15.9.7 Sulfinamides

Sulfinamides (**322**) are generally prepared by the treatment of sulfinylchlorides (**321**) with anilines (Equation (110) ⟨71IJS(B)277⟩.

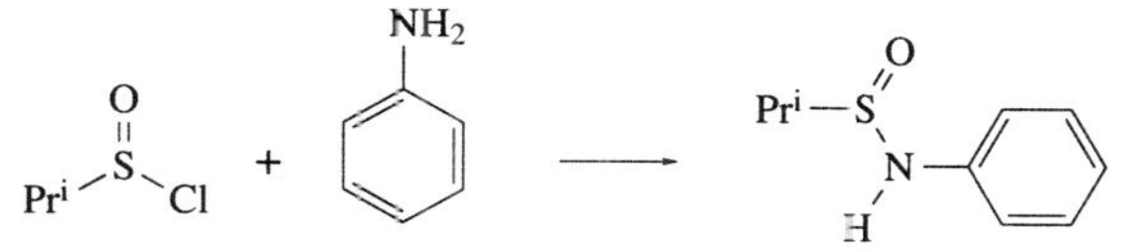

(321) **(322)** (110)

Another method involves the reaction of *N*-sulfinylaniline with Grignard reagents (Equation (111)) ⟨71IJS(B)277, 76SC129⟩.

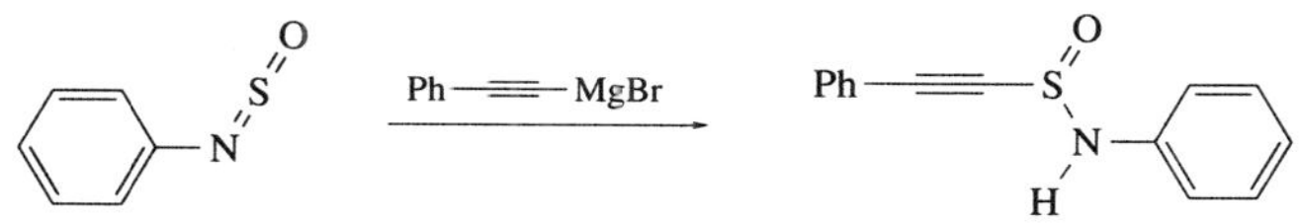

(111)

Tsuda *et al.* ⟨79CL149⟩ obtained sulfinamides in modest yields by coupling *p*-toluenesulfinic acid and aniline in the presence of dicyclohexylcarbodiimide (Equation (112)).

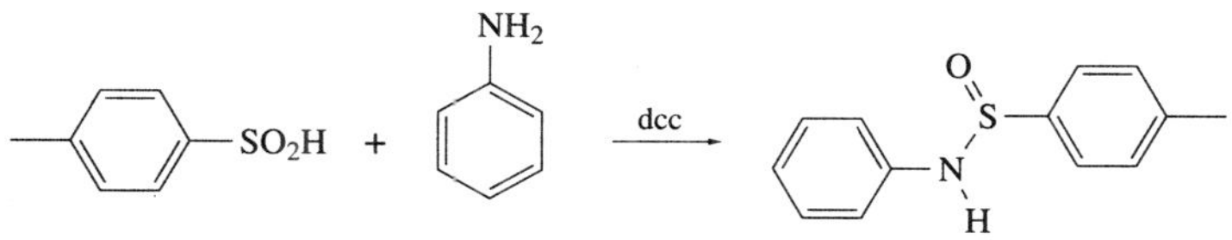

(112)

2.15.9.8 Sulfamides

Numerous examples of sulfamides $R^1R^2NSO_2NR^3R^4$ exist. Several review articles describe much of their chemistry ⟨58HOU(11/2)641, B-65MI 215-03, 73IJS469, 79COC(3)363⟩. Symmetrical sulfamides (**323**) were prepared by treating amines with sulfuryl chloride in pyridine ⟨58HOU(11/2)641, 60JCS4366⟩ or with Karl Fischer reagent (iodine and sulfur dioxide in pyridine (Equation (113)) ⟨73BCJ1254⟩.

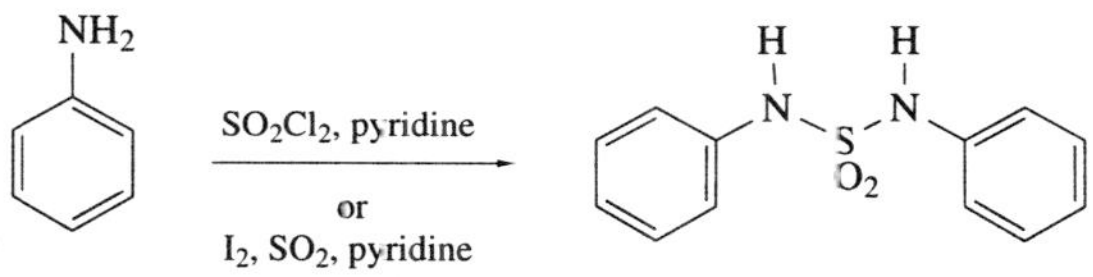

(323) (113)

2.15.9.9 Sulfonamides

Arenesulfonamides are generally stable, colorless solids. Many are of commercial importance, and some based on *p*-aminobenzenesulfonamides are widely used as antimicrobial agents. They are generally prepared by treating a primary or secondary amine with a sulfonyl chloride in the presence of a base, as shown in Equation (114) ⟨55HOU(9)285, 55HOU(9)343⟩.

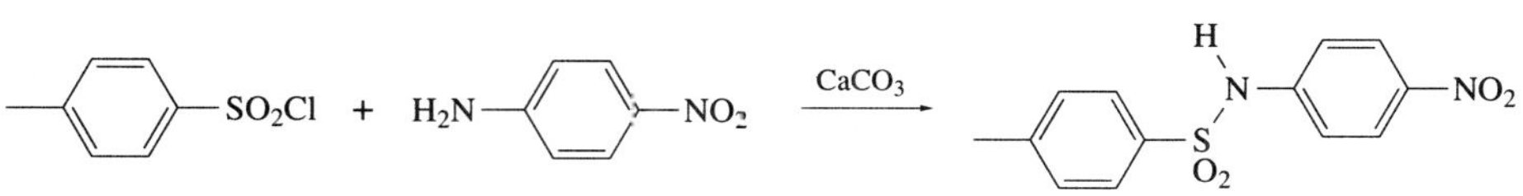

(114)

Cyclic sulfonamides or sultams (**324**) are formed by internal *N*-alkylation of linear sulfonamides (Equation (115)) ⟨60LA(635)91⟩.

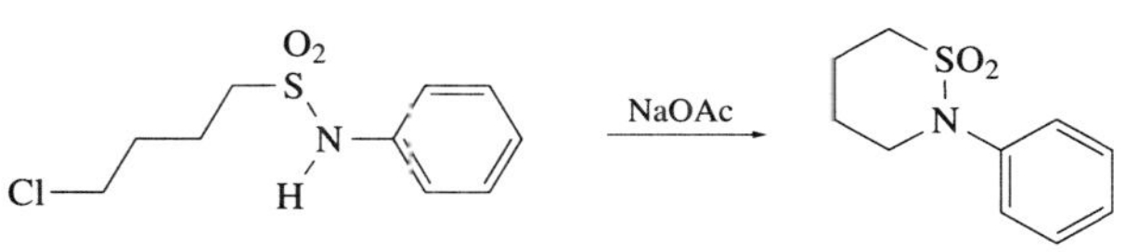

(324) (115)

2.15.9.10 *Aromatic nitrosamines*

Secondary aromatic nitrosamines are prepared, in general, by warming the corresponding amines with nitrous acid in aqueous or acetic acid solution (Equation (116)) ⟨79COC(2)363⟩.

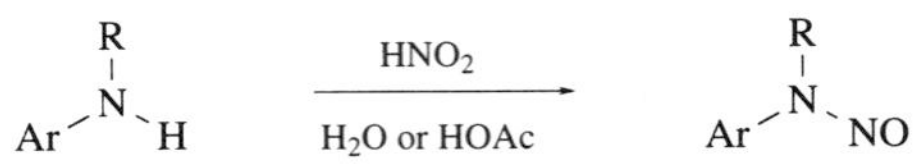

(116)

Nitrosyl chloride is known to give high yields of nitrosoamines ⟨76S462⟩. Other reagents such as, dinitrogen tri- and tetra-oxide, and nitrosonium tetrafluoroborate have also been used ⟨79COC(2)363⟩. *N*-Nitroso-3-nitrocarbazole has been found to act as a nitroso-group transfer reagent to secondary amines under nonacidic conditions ⟨79COC(2)363⟩.

Aromatic *N*-nitrosohydroxylamines may be prepared by the reaction of Grignard reagents with nitric oxide, and reaction with nitric oxide in the presence of a strong base is reported to be a successful procedure for the formation of a nitrosohydroxylamine from any compound having a hydrogen on carbon sufficiently activated to undergo nitrosation ⟨B-66MI 215-05⟩. Another preparation of *N*-nitrosohydroxylamines involves the reaction of an aryl diazonium compound with a solution of cupric chloride and ferrous sulfate saturated with nitric oxide ⟨64MI 215-01⟩.

2.15.9.11 Aromatic Nitramines

Direct nitration of weakly basic 2-nitro-, 4-nitro-, or 2,4-dinitroaniline may yield the nitramine along with the ring nitration products, and the latter may be formed via the nitramine (Equation (117)) ⟨72CC641, B-76MI 215-02⟩.

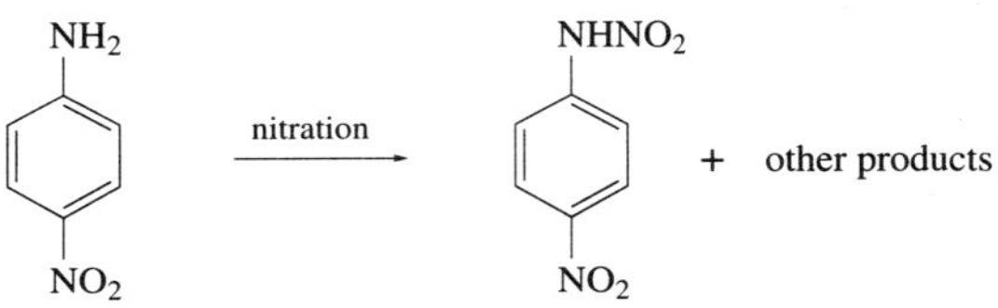

(117)

Nitration of *N*,*N*-dichloroamines in an excess of acetic anhydride gives *N*-chloronitramines (**325**), which may be reduced to primary nitramines (**326**) (Scheme (104)) ⟨48MI 215-01⟩.

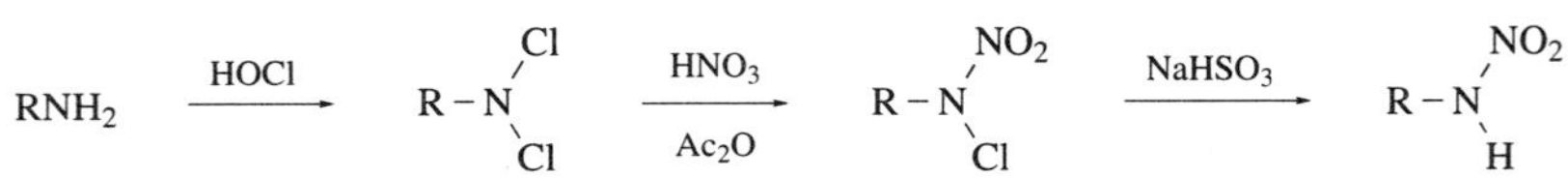

(325) **(326)**

Scheme 104

Conversion of a primary aromatic amine into its lithium derivative prior to its reaction with an alkylnitrate under alkaline conditions produces the corresponding nitramine ⟨65JOC2471⟩. Oxidation of arenediazotates, usually by potassium ferricyanide, provides a general method for the conversion of primary aromatic amines into nitramines (Scheme (105)) ⟨79COC(2)367⟩.

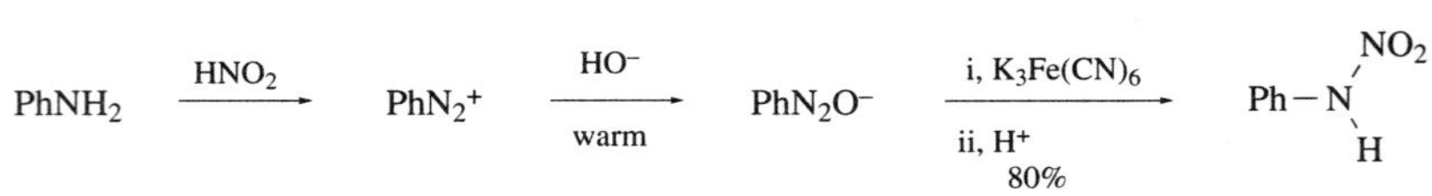

Scheme 105

2.15.9.12 Diazonium Salts

2.15.9.12.1 *Alkenyl (vinyl) diazonium salts*

(i) From α-diazocarbonyl compounds

Crystallization of hydrogen chloride, antimony pentachloride, and methyl benzoyldiazoacetate, or benzoyldiazoacetate, or benzoyldiazoacetophenone from dichloromethane affords crystalline salts (**328**) of hexachloro-antimonic acid from the diazocompounds (**327**) (Scheme 106) ⟨64AG(E)804, 66T1251⟩.

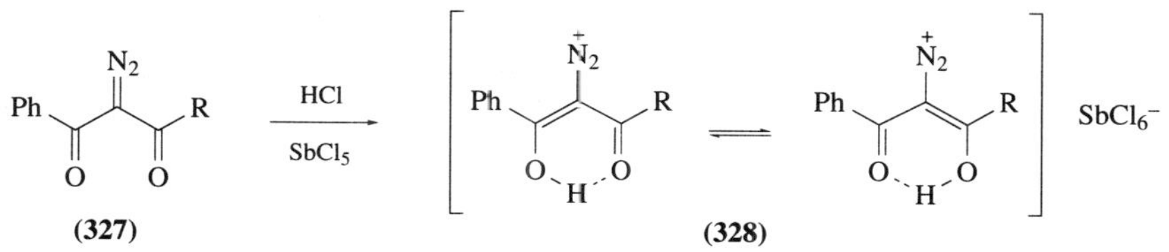

Scheme 106

The strongly basic imines (**329**) of 2-diazo-1,3-indanediones are protonated in ethanol–water by mineral acids to yield the salt (**330**) (Scheme 107) ⟨69LA(728)99⟩.

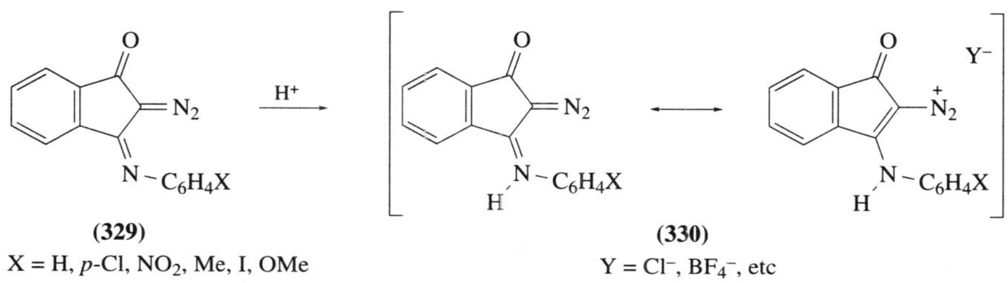

Scheme 107

α-Diazomonoketones were converted to the enoldiazonium ion only in strong acids such as $HF—SbF_5—SO_2$ or $FSO_3H—SbF_5—SO_2$ at −60 °C or less (Equation (118)) ⟨69CC1515, 70HCA1637⟩.

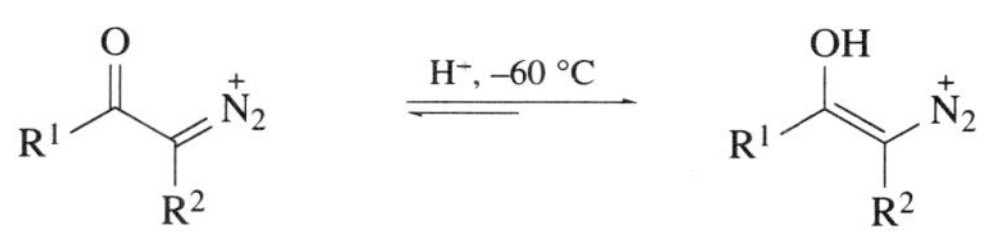

(118)

α-Diazoesters (**331**), in spite of their inability to undergo protonation, can be alkylated with triethyloxonium hexachloroantimonate to afford the isolable diazonium salts (Scheme 108) ⟨66T1251⟩.

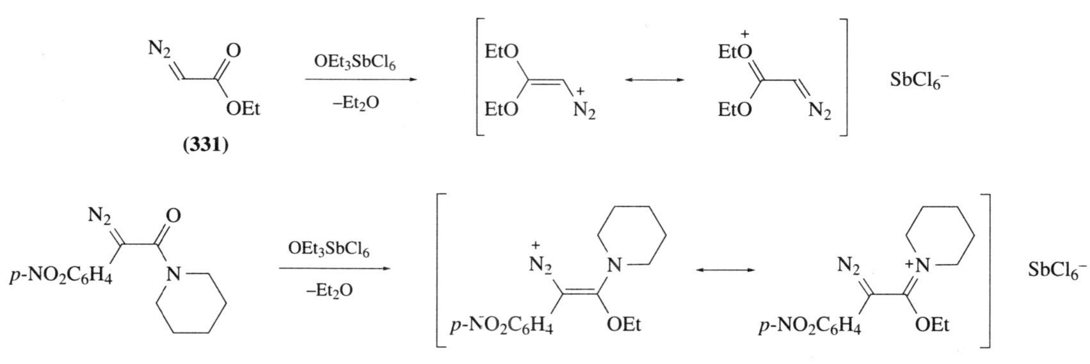

Scheme 108

(ii) From isocyanates and sulfinylamines

The synthesis of the 2,2-dichloroethylene diazonium hexachloroantimonate (**333**) was achieved by the action of nitrosyl hexachloroantimonate on the isocyanate (**332**) (Equation (119)) ⟨71TL2227⟩. Phenyl isocyanate gives benzene diazonium salts by the action of NOX ⟨58JOC1800⟩.

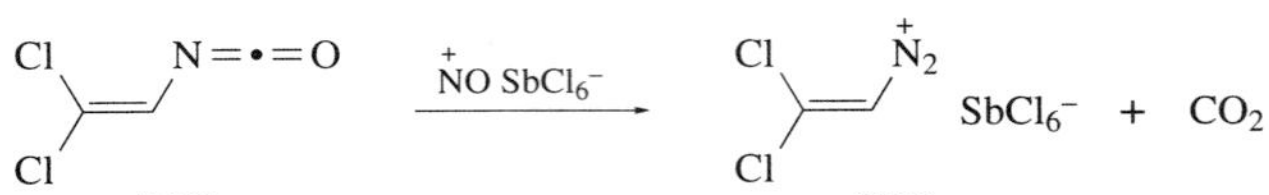

(**332**) (**333**) (119)

Aryl sulfinylamines (**334**) were converted to arenediazonium salts (**335**) (Equation (120)) ⟨65AG(E)148⟩. However, analogous preparations of alkenediazonium ions are unknown as the starting vinylsulfinylamines cannot be obtained.

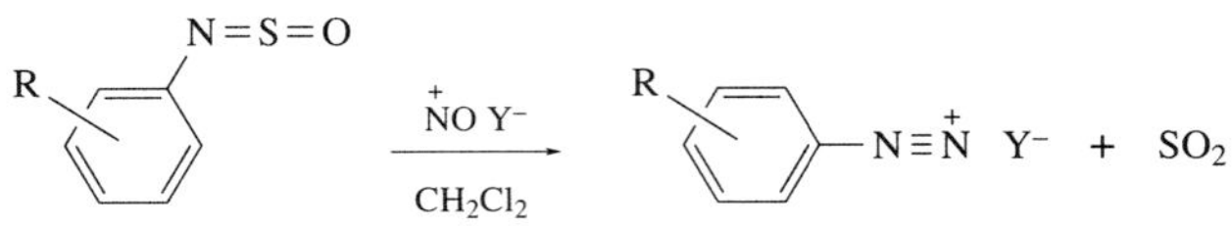

(**334**) (**335**) (120)

R = *p*-Cl, *p*-NO_2, *m*-NO_2
R = $SbCl_6$, ClO_4

(iii) From sulfonylhydrazones

The action of Lewis acids ($SbCl_5$, $SnCl_4$, $AlCl_3$) on the *p*-toluenesulfonylhydrazones of α-halo aldehydes or ketones (**336**) leads to alkene diazonium salts (**337**) (Scheme 109) ⟨70AG(E)954, 75CB402, 79AG(E)259⟩. The α-halogen can be replaced by the reaction of the alkoxy or acyloxy group, provided that removal is favored by geminal substituents R^1 and R^2. The rates of formation of (**337**) parallel the well-known order of Lewis acid strengths, $AlCl_3 > SbCl_5 > SnCl_4$.

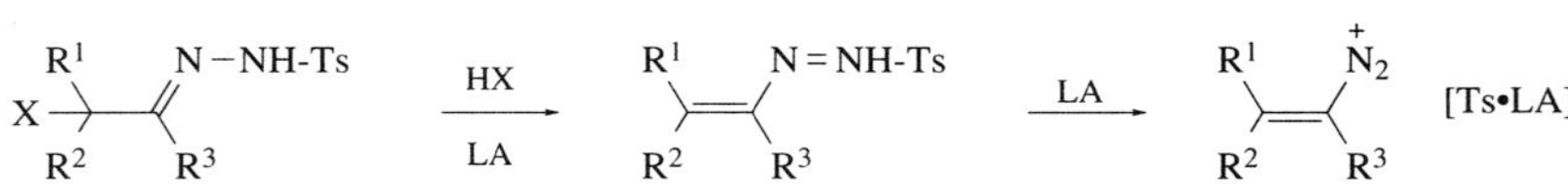

(**336**) (**337**)

LA = Lewis acid
X = F, Cl, Br

Scheme 109

Azofulvene (**338**) was converted to the salt (**339**), which was transformed into (**340**) by the action of hydrogen chloride. Furthermore, (**340**) was directly obtained from 9-fluorenecarbaldehyde tosylhydrazone (**341**) on treatment with antimony pentachloride, the latter also functioning as oxidizing agent. However, the related salt (**342**) is resistant to the attack of hydrogen chloride (Scheme 110) ⟨75CB402⟩.

(iv) From nitrosation of primary vinylamines

When diphenylvinylamine (**343**) is treated with *i*-amyl nitrite (NO^+-donating agent) in boiling benzene, diphenylalkyne is formed via the intermediacy of the alkenediazonium ion (**344**) (Scheme 111) ⟨65JA863, 65JA874⟩.

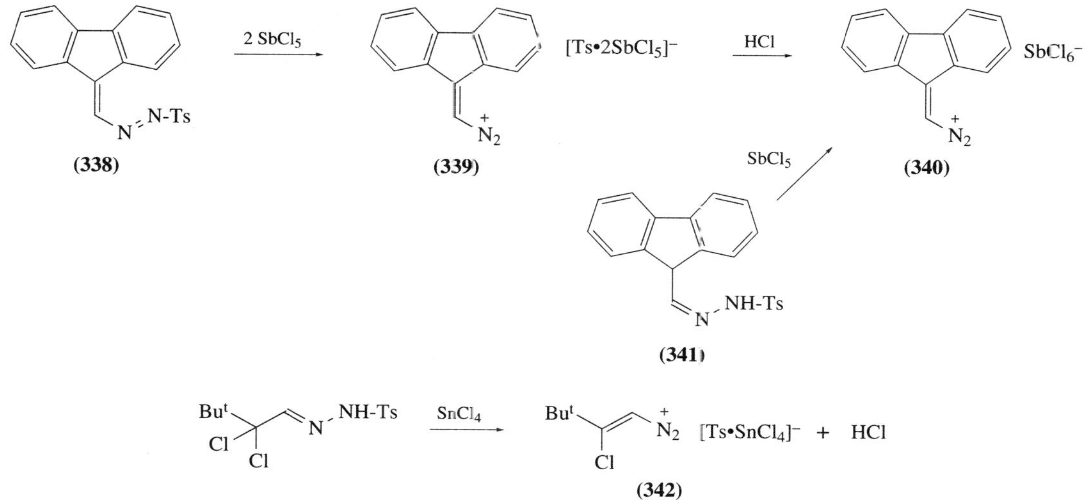

Scheme 110

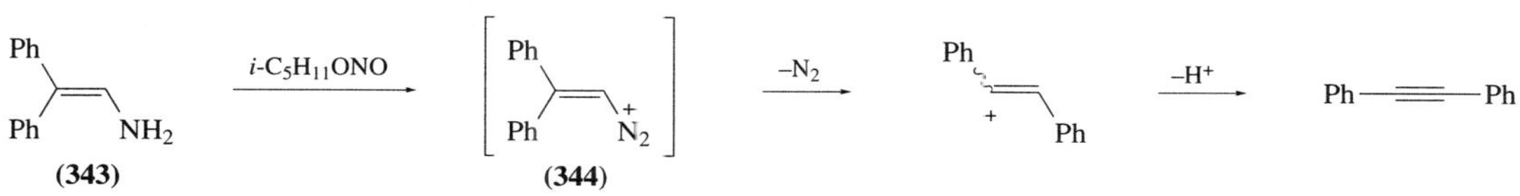

Scheme 111

The action of sodium nitrite in acetic acid on 3-amino-2-phenylindenone (**345**) generated the diazonium ion (**346**) which did not tend to eliminate nitrogen (Equation (121)) ⟨65JA874⟩.

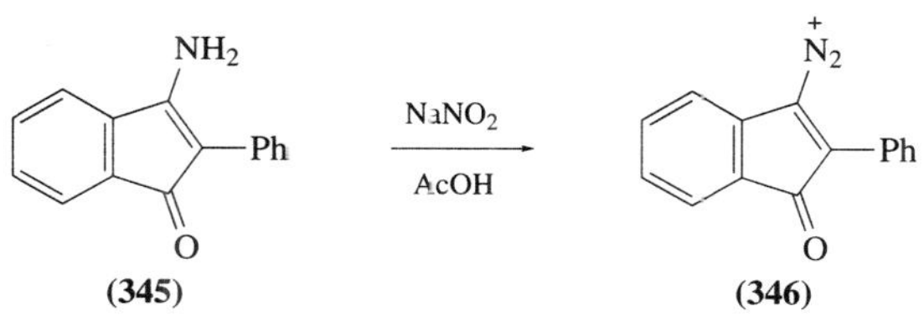

(121)

(v) *From acidolysis of vinyltriazenes*

The classical reaction of diaryl triazenes with acid to give aryldiazonium salts was extended to produce alkenediazonium ions from the analogous vinyltriazene intermediates (Equation (122)) ⟨67JA1960⟩.

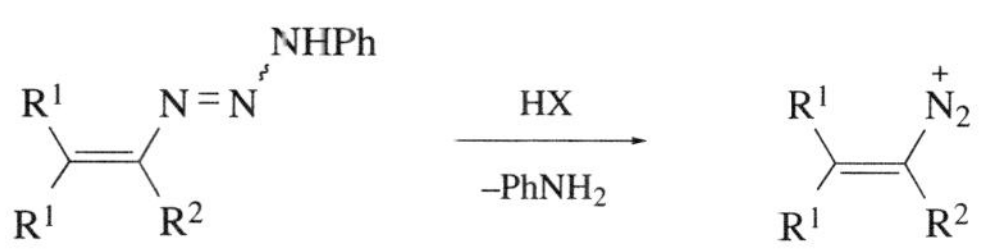

(122)

HX = AcOH, PhOH, TsOH
R^1 = Ph, R^2 = Ph
R^1 = Tol, R^2 = Ph
R^1 = Ph, R^2 = Me
R^1 = Ph, R^2 = H

(vi) *From base-promoted decomposition of* N*-nitrosooxazolidones*

The base-promoted decomposition of *N*-nitrosooxazolidones has been employed as a source of alkenediazonium ions ⟨51JA4199, 54JA1840, 69JA6461, 69JOC1220, 70JA4309, 70JA4312, 70JA7564, 70JOC2412,

72JOC3220, 73JOC2435, 73JOC2438, 74JOC553, 79CB2120⟩. Hassner and Reuss ⟨74JOC553⟩ proposed the mechanism involved in the generation of the alkenediazonium ion (**348**) by the action of sodium methoxide on (**347**) (Scheme 112).

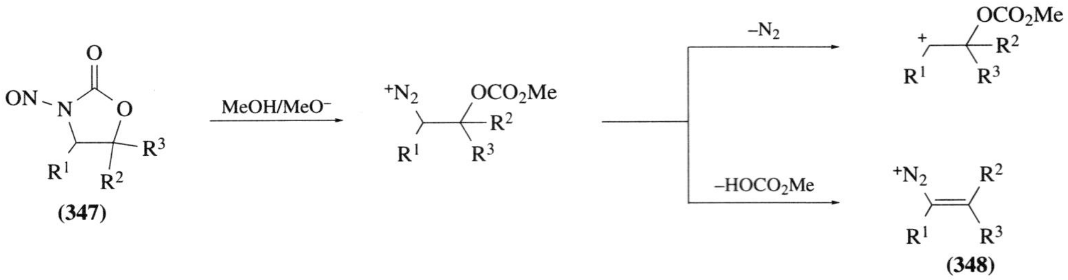

Scheme 112

(vii) From transfer of diazo groups

Reaction of bis(methanesulfonyl)diazomethane (**349**) with α-morpholinostyrene (**350**) in trichloromethane generated the intermediate diazonium ion (**351**) involving transfer of the diazo group from (**349**) to (**350**) (Equation (123)) ⟨68AG(E)138⟩.

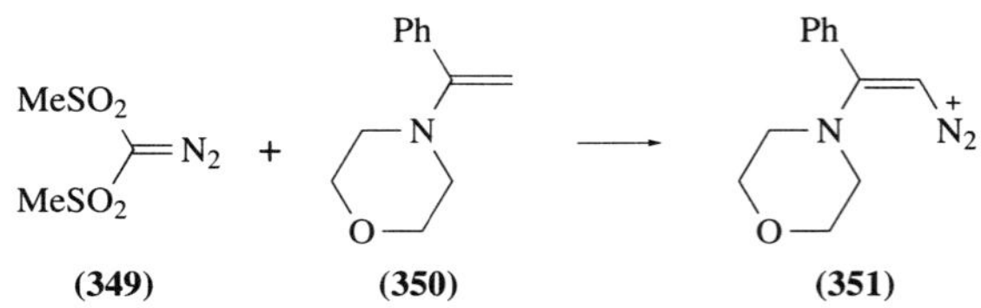

(123)

Generally, alkene diazonium salts may be considered stable enough to be isolated only if they form highly unstable vinyl cations on elimination of nitrogen.

2.15.9.12.2 *Aromatic diazonium salts*

(i) From diazotization of amines

The most widely used method of formation of an aromatic or heteroaromatic ⟨70RZC1131, 72KGS1690⟩ diazonium group is the diazotization of primary amines with NO^+ donors (Scheme 113).

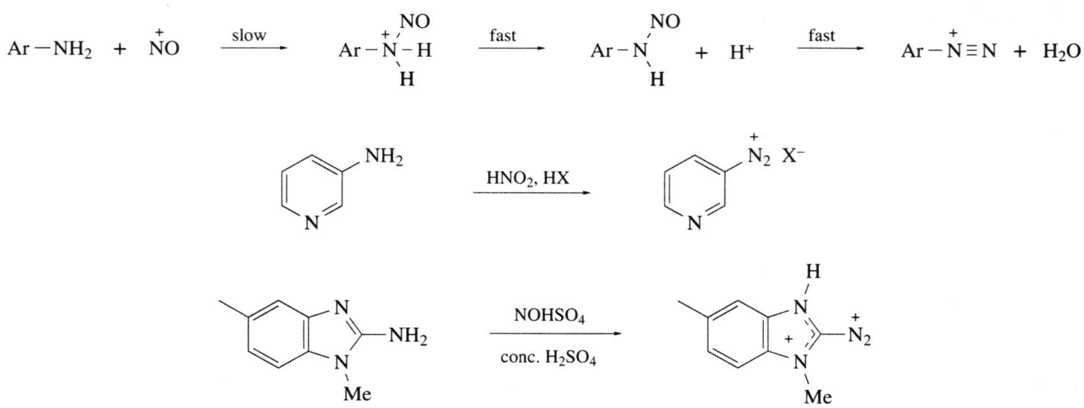

Scheme 113

A solution or suspension of the amine in dilute hydrochloric acid (2.5 equivs.) is treated with an equivalent of concentrated aqueous sodium nitrite at 0–10 °C. The reaction proceeds rapidly and quantitatively ⟨65HOU(10/3)16⟩. The excess nitrous acid can easily be removed by adding urea or sulfamic acid ⟨B-75MI 215-02⟩. Mineral acids such as sulfuric acid, phosphoric acid, etc., may be used if chloride ions are not desirable in the medium. In the case of amines which contain strongly acidic groups, and therefore exist as sparingly soluble betaines, a solution of the amine and sodium nitrite in dilute alkali is slowly added to the acid with vigorous stirring. Weakly basic amines bearing strongly electron-withdrawing groups on the aromatic or heteroaromatic nucleus are diazotized readily in concentrated acids ⟨65HOU(10/3)22⟩. These can be concentrated sulfuric acid alone ⟨B-52MI 215-01, 65HOU(10/3)23⟩ or in conjunction with glacial acetic acid or phosphoric acid, with concentrated fluoroboric acid, or with anhydrous hydrofluoric acid for the preparation of organofluorine compounds (see ⟨65HOU(10/3)22⟩) for references to the use of these acids). Solid diazonium salts are prepared by diazotization in organic solvents (glacial acetic acid ⟨72G527⟩, methanol, ethanol, formamide, DMF, acetone, etc., ⟨65HOU(10/3)28⟩) with nitrous acid esters (pentyl nitrite). Alternatively more stable BF_4^-, PF_6^-, and related stable salts may be precipitated by addition of the appropriate acid.

(ii) From other arenediazonium compounds

Diazotization of 7-amino-6-substituted 1,2,3-benzothiadiazole (**352**) gives an intermediate in which the diazonium group migrates to a different location from the original amine (Scheme 114) ⟨70JCS(C)2514⟩.

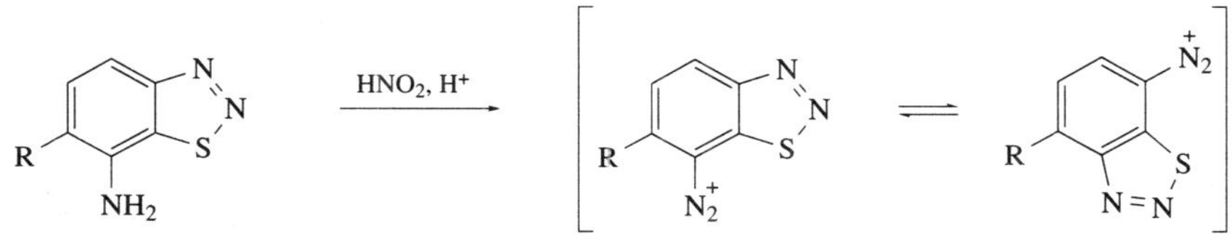

(352)

Scheme 114

Substituted hydroxy or substituted amino-aryl diazonium ions are prepared by the protonation of quinone diazides (**353**) ⟨65HOU(10/3)8, 65HOU(10/3)51⟩ or imines (**354**) ⟨65HOU(10/3)44⟩ by strong mineral acids (Scheme 115). Similarly, acylated diazonium ions (**356**) are prepared by acylation of quinone diazides (**355**) (Equation (124)) ⟨07GEP206455⟩, and the strong electron-donating nature of the quinone diazide oxygen enables further derivatization of the ring by electrophilic substitution (Scheme 116) ⟨65HOU(10/3)89⟩.

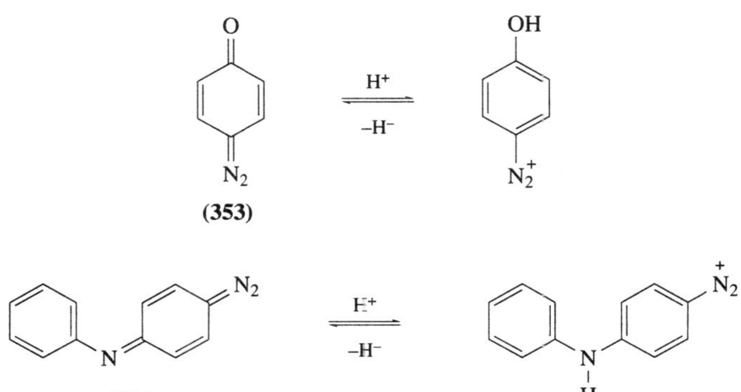

(353)

(354)

Scheme 115

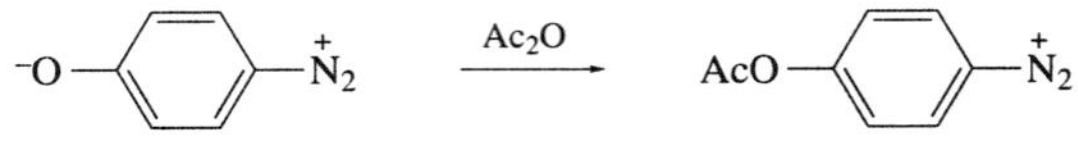

(355) (356)

(124)

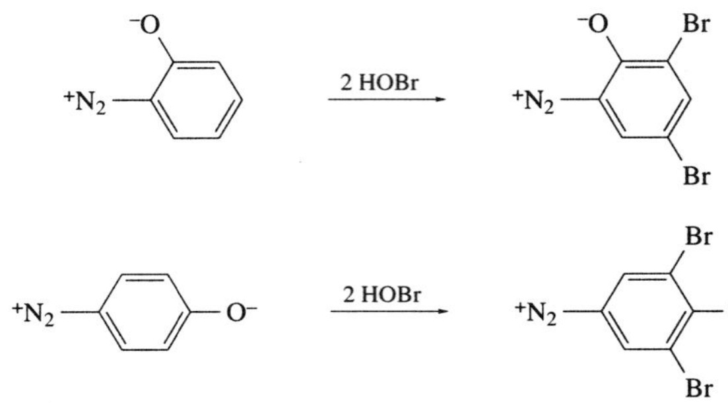

Scheme 116

Immediate oxidation of α-diazonaphthalene with alkaline potassium hexacyanoferrate(III) ⟨60AG303⟩ provided the β-oxygenated diazonium ion (Equation (125)).

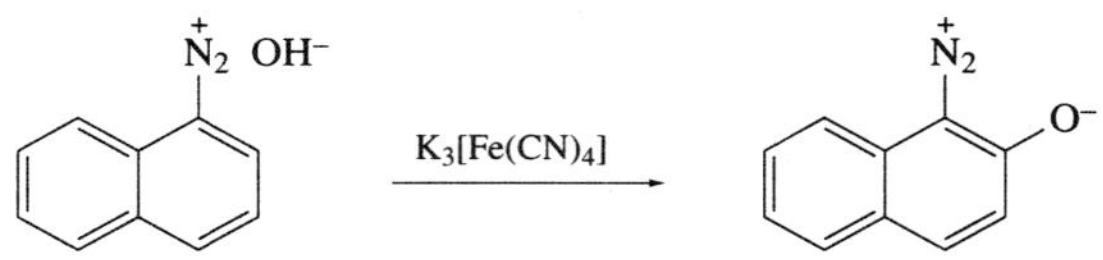

(125)

The electron-withdrawing nature of the diazonium group is used effectively to facilitate the substitution of good leaving groups from the *ortho* and *para* positions of aromatic diazonium compounds (Scheme (117)) ⟨69ACS2983⟩.

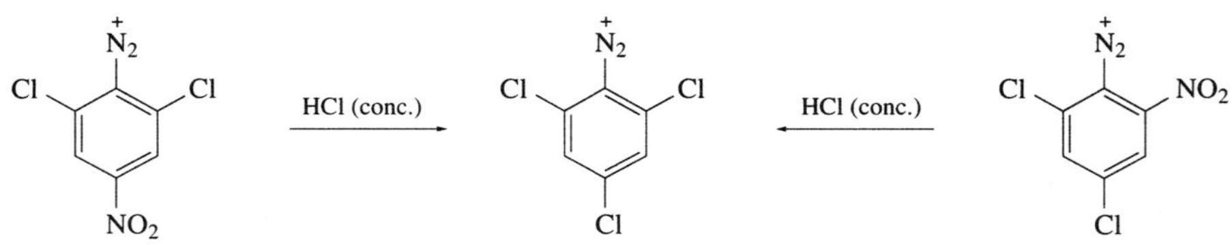

Scheme 117

(iii) From transdiazotization

Just as the diazonium ions from weakly basic amines are hydrolyzed easily, they can be aminolyzed by amines of high basicity providing the *trans*-diazotized products (Scheme 118) ⟨65HOU(10/3)46, B-75MI 215-03⟩.

(iv) From azo decoupling

The reversible cleavage of diazo/sulfonates and diazo/sulfones by mineral acids has been employed to generate diazonium ions ⟨65HOU(10/3)576, 65HOU(10/3)584⟩. The related alkali- or acid-catalyzed rearrangement of aryl azoxy sulfones is used to generate diazonium sulfates (Equation (126)).

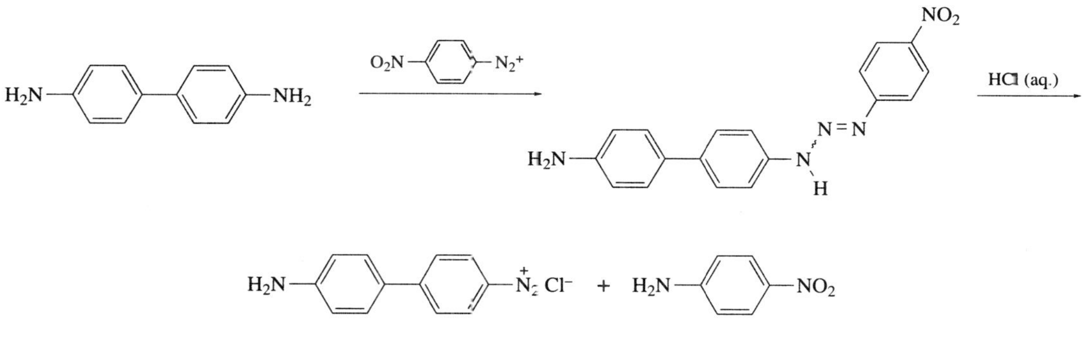

Scheme 118

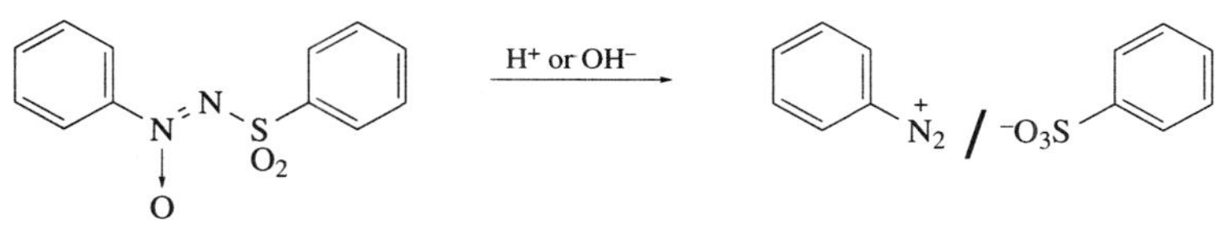 (126)

Cleavage of a C=N bond was observed during the reaction of chlorine with a phenylhydrazone to give the diazonium ion (Equation (127)) ⟨72JCC383⟩.

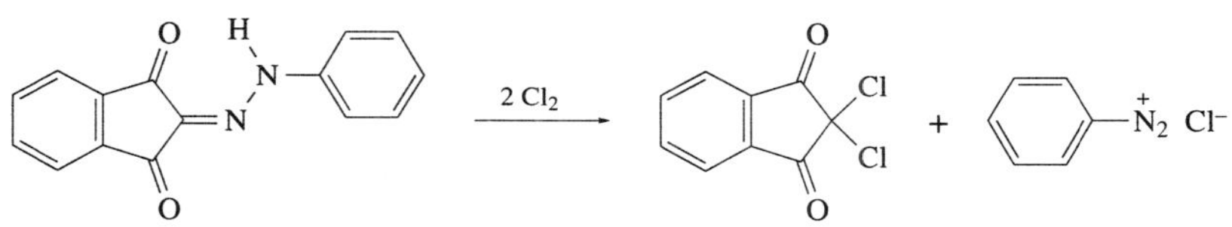 (127)

(v) From nitroso acylamines

Acylated aromatic amines Ar—NH—X, where X may be COR, SO_3H or NO_2 are readily nitrosated by NO^+ donors, in the two latter cases this leads to diazonium sulfates or nitrates. Diazotate anions are generated from nitroso- or acylamines by the action of a base ⟨1894CB914, 1897CB366, 51LA(573)163⟩. A pH-dependent equilibrium exists between diazotate anions and diazonium cations ⟨65HOU(10/3)552, 72CCC3467, 73CCC706⟩.

(vi) From nitrosation of imines

Crystalline diazonium salts are prepared by the treatment of azomethines with NOX (X = Cl ⟨35BSF627⟩, NO_3 ⟨64JOC3429⟩, BF_4 ⟨72JOC1597⟩) (Equation (128)).

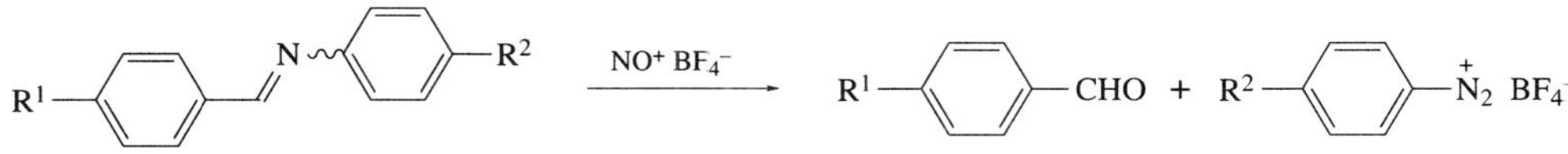 (128)

(vii) From hydroxylamines, oximes nitroso, and nitro compounds

N-Phenylhydroxylamine (**357**) is converted into benzene diazonium nitrate (**358**) with excess nitrous acid (Scheme 119) ⟨65HOU(10/3)71⟩.

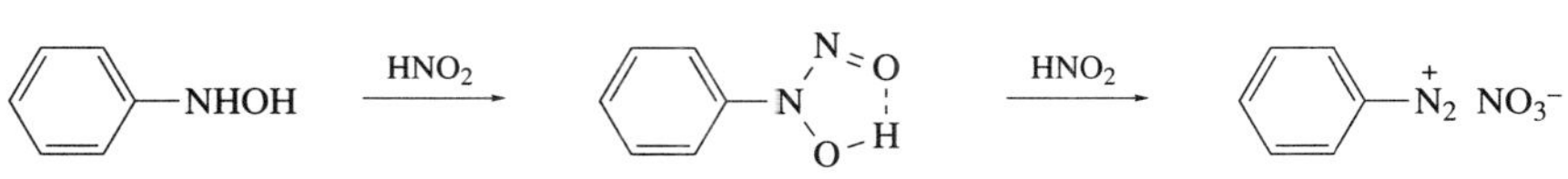

(**357**) (**358**)

Scheme 119

Similarly, the dioxime of anthraquinone (**359**) was converted to the diazonium salt (**360**) (Equation (129)).

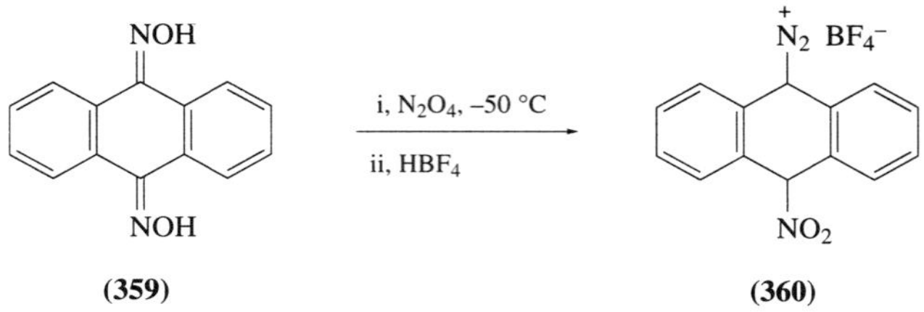

(129)

Aromatic nitroso compounds, in the higher oxidation state, undergo redox diazotization with excess nitrous acid (Equation (130)) ⟨65HOU(10/3)73⟩.

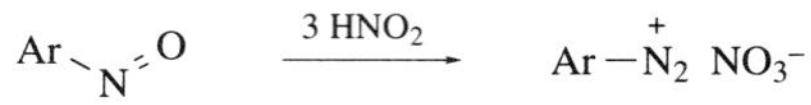

(130)

Electron-rich aromatic compounds are easily nitrosated to form nitroso compounds which in turn may be converted to diazonium nitrates in one step (Equation (131)) ⟨65JCS6186⟩.

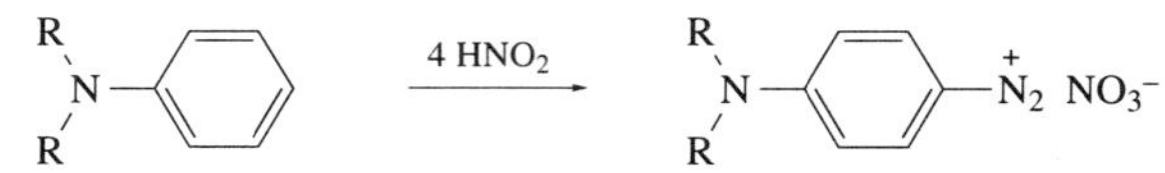

(131)

Less electron-rich aromatic compounds need more reactive nitrosating agents or metal catalysis ⟨B-75MI 215-04⟩, preferentially by Hg^{2+}. Reaction of nitrosobenzene and derivatives with hydroxylamine in the presence of sodium carbonate gives diazonium ions in an aldol-like condensation (Equation (132)) ⟨65HOU(10/3)86⟩.

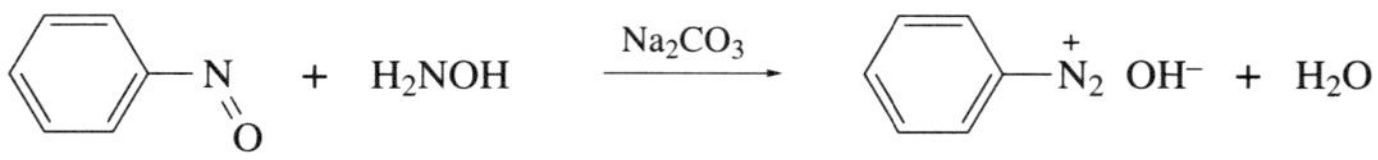

(132)

In a similar fashion, nitrobenzene in the next oxidation level, reacts with ammonia (in a lower oxidation level) in the presence of alkali to give the diazonium (diazotate) compound, however the yields are poor (Equation (133)) ⟨04CB629, 42JA19⟩.

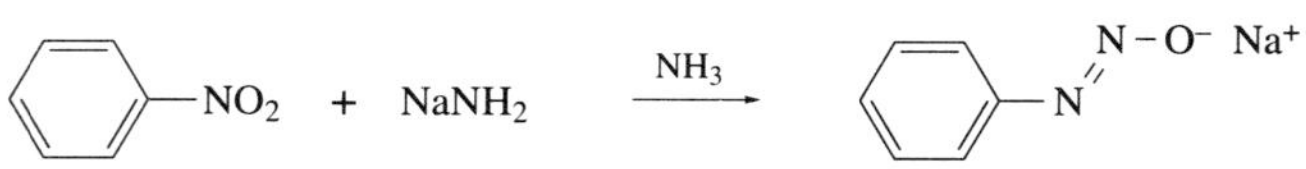

(133)

(viii) From arylhydrazines

The reaction of phenylhydrazine with nitrous acid, depending on reaction conditions, can yield a diazonium ion (Scheme 120) ⟨59HCA739⟩.

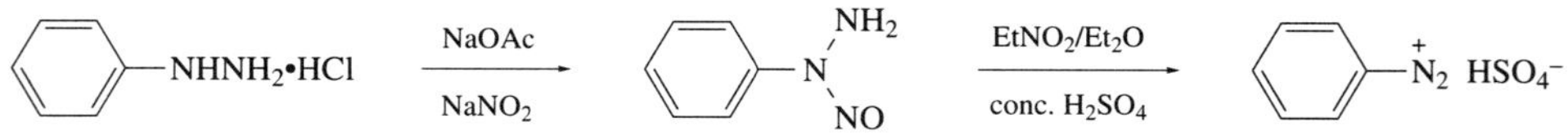

Scheme 120

An interesting fragmentation of *N*-chloro azocarbonamides ⟨71JPR642⟩ has been used to generate diazonium salts (Scheme 121).

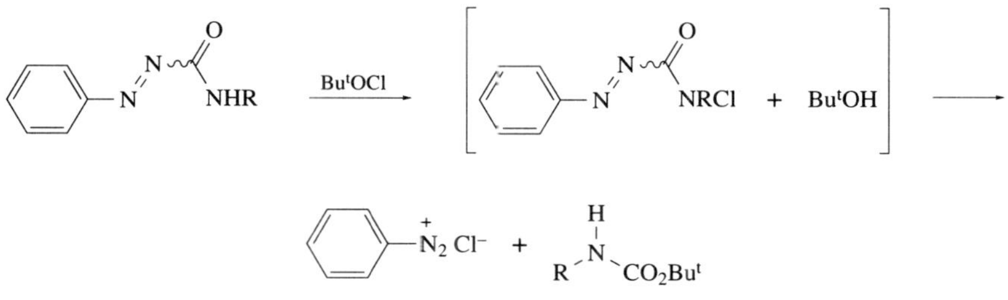

Scheme 121

2.15.10 N—P FUNCTIONS

There has been considerable interest in the chemistry of compounds containing the N—P bond ⟨58JA2686, B-65MI 215-04, B-66MI 215-06, B-67MI 215-01⟩. Phosphoramidates have been used in phosphorylation reactions for the synthesis of phosphate derivatives. Detailed studies of the chemical dynamics have demonstrated that the N—P bond of $RNHPO_3H^-$ is unstable and either generates the reactive *meta*phosphate ion, PO_3^-, in a unimolecular dissociative decomposition or undergoes rapid bimolecular transfer of the PO_3^- moiety to generate a new phosphate derivative ⟨58JA2686, 64JA1410, 65JA3199, B-65MI 215-04, 66JCS404, B-66MI 215-06, 67JA4714, 67JA6972, B-67MI 215-01, 71JA4009⟩. The lability of N—P bond plays an important role in biochemical systems for phosphorylation of various substrates. Therefore, studies of phosphate derivatives do not necessarily permit one to elucidate the fundamental properties of N—P bonds because of the perturbing effects of the other functional groups at phosphorus. In this context, Tyssee *et al.* ⟨73JA8066⟩ have examined the acid-catalyzed hydrolysis of phosphinanilides with particular attention to compound (**361**) (Equation (134)).

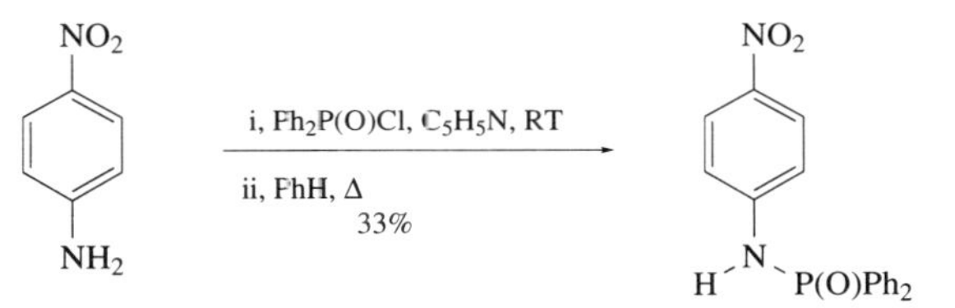

(361)

(134)

N-(*p*-Nitrophenyl)-*p*-methoxyphenylphenylphosphinamide (**363**) was prepared by the addition of the *p*-methoxyphenylphenylphosphinic acid (**362**) to an excess of thionyl chloride and the mixture was heated at 79 °C for 14 h. Excess thionyl chloride was removed by distillation. The acid chloride was refluxed with *p*-nitroaniline in pyridine for 48 h. The solid was purified by crystallization from ethanol to give the desired N—P compound (**363**) (Equation (135)) ⟨73JA8066⟩.

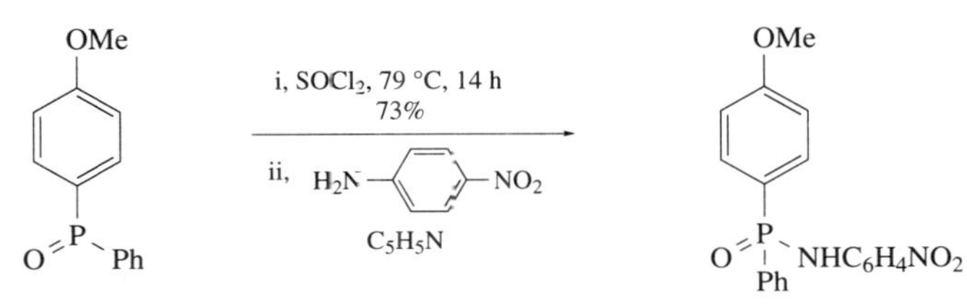

(362) **(363)**

(135)

The (−)-(*S*)-*N*-(phenyl)methylphenylphosphinamide (**365**) was prepared by refluxing (−)-(*S*)-methylmethylphenylphosphinate (**364**) with a tenfold excess of lithium anilide in benzene (Equation (136)) ⟨68JA4842⟩. The phosphinamide was purified by silica-gel chromatography followed by recrystallization.

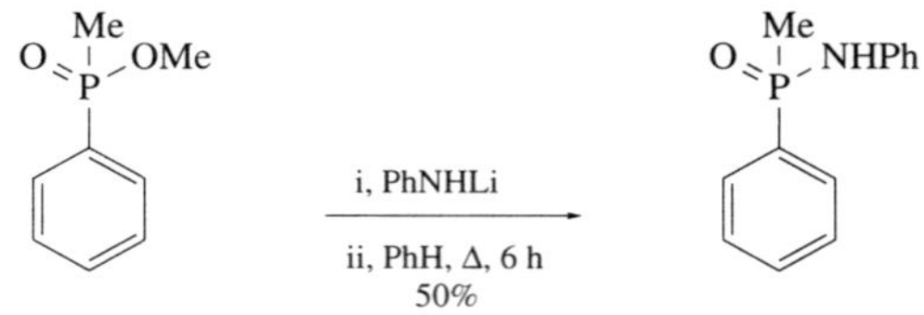

(364) **(365)** (136)

Enamides of phosphonamidates, phosphinamidates, and phosphinamidothioates are conveniently prepared by the reaction of the amides of the corresponding phosphorus(V) acid with substituted aldehydes, either in the presence of *p*-toluenesulfonic acid in chloroform or in the presence of *p*-toluenesulfonic acid and phosphorus pentoxide in boiling chloroform (Table 1) ⟨74AP(307)73⟩.

Table 1 Preparation of enamides of phosphonamidates, phosphinamidates, and phosphinamidothiotes.

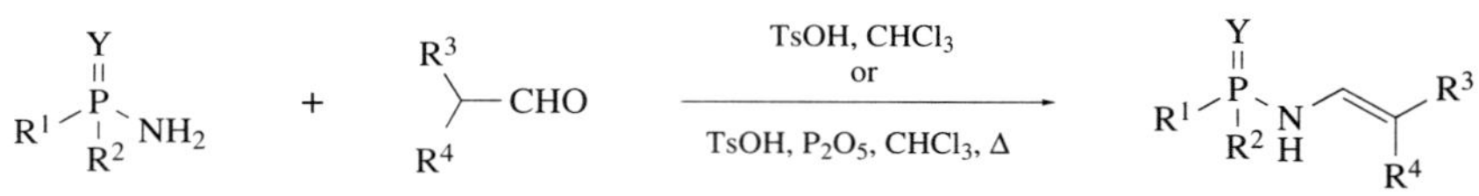

R^1	R^2	R^3	R^4	Y	*Yield* (%)
Ph	PhO	Ph	CN	O	47
Ph	PhO	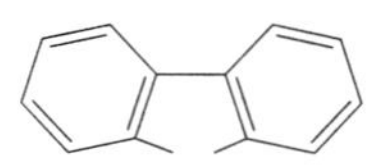		O	53
Ph	PhO	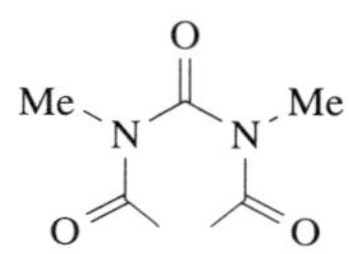		O	58
Ph	Ph	Ph	Ph	O	38
Me	Me	Ph	Ph	S	70
Me	Me	Ph	CN	S	28
Ph	Ph	Ph	Ph	S	45
Ph	Ph	Ph	CN	S	50
Me	Me	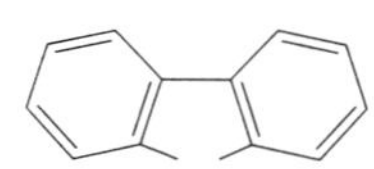		S	48
Ph	Ph	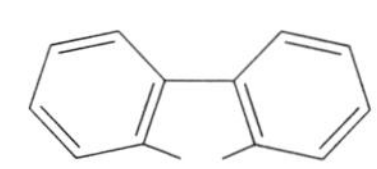		S	42
Ph	Ph	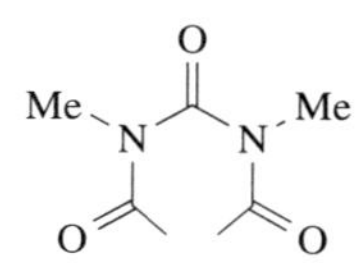		S	72

The methods most commonly employed for the transformation of primary amines into secondary amines involve the synthesis and subsequent catalytic hydrogenation of the corresponding Schiff bases. Procedures involving direct alkylation of the primary amines frequently afford mixtures of secondary and tertiary amines. Zwierzak *et al.* ⟨75S507, 76AG(E)302, 77AG(E)107⟩ found that the diethyl esters of *N*-alkyl or *N*-acylphosphoramidic acids, which are accessible from primary amines or their hydrochlorides, can readily be alkylated under the conditions of phase-transfer catalysis. Applicability of this procedure is limited, however, to the use of primary alkyl halides as alkylating agents. The protecting diethoxyphosphoryl group is removed from the alkylation product (**366**) by treatment of the latter with gaseous HCl in THF at room temperature to give the secondary amine (**367**) (Table 2).

Table 2 Preparation of diethyl esters of *N*-arylphosphoramidic acids by alkylation of diethyl phosphoramidates.

$$R^1NH_2 \xrightarrow{(EtO)_2P(O)H,\ CX_4} (EtO)_2P(O)NHR^1 \xrightarrow{R^2X,\ 50\%\ aq.\ NaOH,\ Bu^n_4N^+\ HSO_4^-}$$

$$\underset{\mathbf{(366)}}{(EtO)_2P(O)NR^1R^2} \xrightarrow{HCl} \underset{\mathbf{(367)}}{R^1R^2NH{\cdot}HCl}$$

R¹	*R²*	*Yield (%)* **(366)**	*Yield (%)* **(367)**
Et	Me	85	87
Et	Et	85	84
Et	Pr^n	67	91
Et	Bu^n	70	88
Et	$CH_2{=}CHCH_2$	93	90
Ph	Et	97	91
Ph	Bu^n	98	85
Bn	Bu^n	97	99
Bn	$CH_2{=}CHCH_2$	85	97
c-C_6H_{11}	Et	67	81
c-C_6H_{11}	Bu^n	35	83

2.15.11 N—As, N—Sb OR N—Bi FUNCTIONS

The organic chemistry of the group 15 elements has been a fascinating and fruitful area of study for chemists through several generations. Partly due to their historical role in the development of chemotherapy, the area has been somewhat dominated by organoarsenic compounds, but there is growing interest in the organic chemistry of antimony and bismuth, some aspects of which are relatively unexplored. A review of the organic chemistry of arsenic, antimony, and bismuth has been published by Poller 〈79COC(3)1114〉.

2.15.11.1 Arsinimines

Cadogan and Gosney have investigated the potential of the readily available triphenylarsine as a trapping agent for nitrenes 〈74JCS(P1)460〉. A few arsinimines were prepared by condensation of triarylarsine oxides with isocyanates 〈69ACS2935, 71ACS983〉, *N*-sulfinylamines 〈71ACS983〉 or *N*-sulfinylamides 〈65ACS1755, 71ACS983〉, by reaction of triphenylarsine dibromide with activated amides 〈67CA76112w〉, by treatment of triphenylarsine with sodium *N*-chlorotoluene-*p*-sulfonamide 〈37JCS527〉 (chloramine T) or *N,N′*-dichlorotoluene-*p*-sulfonamide 〈69CB2557〉, and in one case by the action of base on an aminoarsonium salt 〈60AG209〉. The simple approach of reaction of an azide with a tertiary arsine appears to have been used only once, and then only with partial success. In this case 〈69CB3731〉, a stannyl-substituted arsine on reaction with two moles of the azide gave an organostannylphenylaminoarsinimine (**368**) rather than the simple arsinimine, which was possibly an intermediate (Scheme 122).

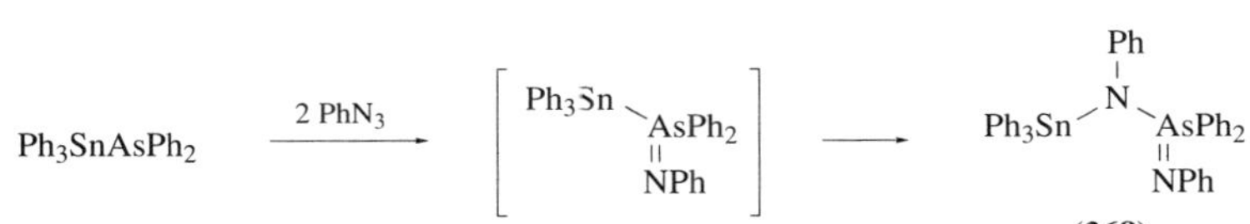

(368)

Scheme 122

Wittig and Hellwinkel ⟨62AG(E)53⟩ and Staudinger and Meyer ⟨19HCA635⟩, however, have reported that phenyl azide does not react with triphenylarsine. Thermolysis of pentafluorophenyl azide in molten triphenylarsine at 120°C, the decomposition point of the azide, gave *N*-pentafluorophenylarsinimine (**369**) as a colorless, crystalline solid which was stable in dry air but decomposed slowly in solution (Equation (137)). That no reaction occurred below the decomposition point of the azide is in accord with, but does not prove, nitrene participation.

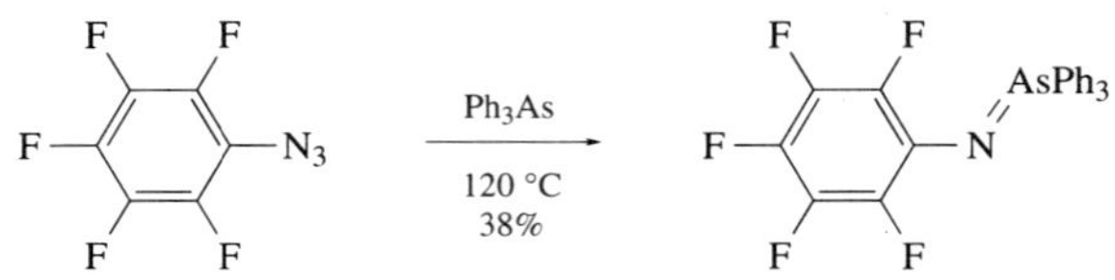

(**369**) (137)

Thermolysis of *p*-nitrophenyl azide in wet dioxane in the presence of triphenylarsine and copper bronze led to the isolation of a product corresponding to one mole of *N*-*p*-nitrophenyltriphenylarsinimine (**370**) containing one mole of water, as a yellow crystalline solid. Spectroscopic evidence (NMR, IR) led to its formulation as the hydrated arsinimine (**371**) (Scheme 123) similar in type to compounds such as $Ar_3As(OH)NHSO_2Ar$ which are readily formed from triarylarsine oxides and a weak acid such as toluene-*p*-sulfonamide ⟨32JCS958, 72JCS(P1)2793⟩. In accord with this, reaction of *p*-nitroaniline with triphenylarsine oxide gave a product (92%) identical in all respects with the hydrated arsinimine (**371**).

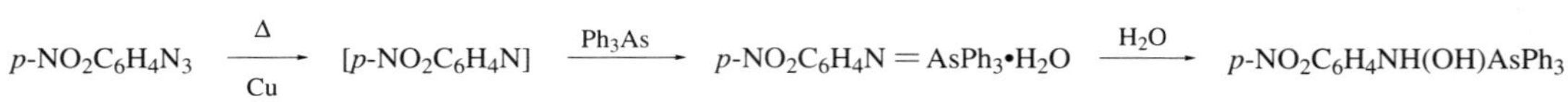

(**370**) (**371**)

Scheme 123

α-Elimination from *N*-*p*-nitrophenylsulfonyloxytoluene-*p*-sulfonamide (**373**) was used to convert triphenylarsine into triphenyl-*N*-*p*-tolylsulfonylarsinimine (**375**). This was identical to the product produced from *p*-tolylsulfonyl azide (**372**) and triphenylarsine, or the hydrated form of (**375**) obtained from chloramine T (**374**) ⟨37JCS527⟩ and triphenylarsine in the presence of water (Scheme 124).

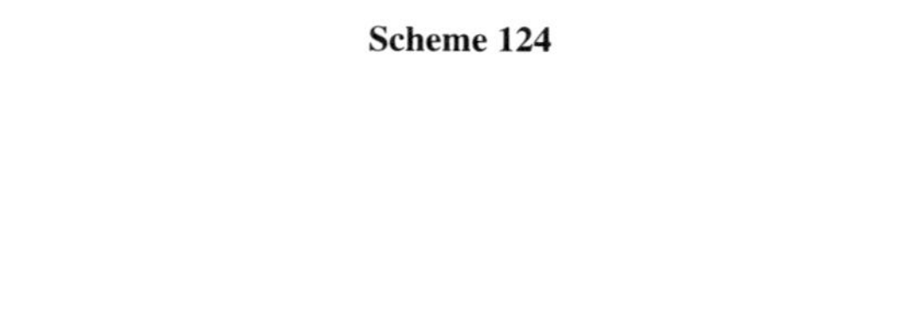

Scheme 124

2.15.12 COMPOUNDS WITH AN *N*—METALLOID BOND

2.15.12.1 Compounds with a Boron—Nitrogen Bond

Various syntheses of *N*-metalloid compounds like aminoboranes have been reported. When the imine (**376**) was treated with boron trichloride and triethylamine in dichloromethane, *N*-cyclohexyl-*N*-1-cyclohexenylaminodichloroborane (**377**), an aminohaloborane (Equation (138)) ⟨B-61MI 215-02, 79SC515⟩ was obtained.

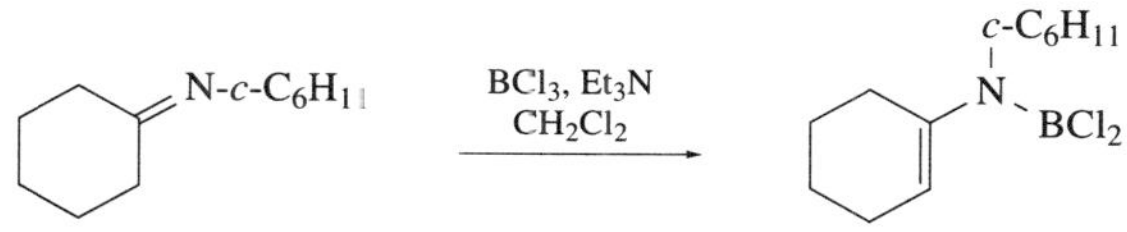

(376) (377) (138)

Similar treatment of the imine (**378**) gave *N*-cyclohexyl-*N*-2-butenylaminodichloroborane (**379**) as a mixture of *E* and *Z* isomers (Equation (139)) ⟨B-61MI 215-02, 79SC515⟩.

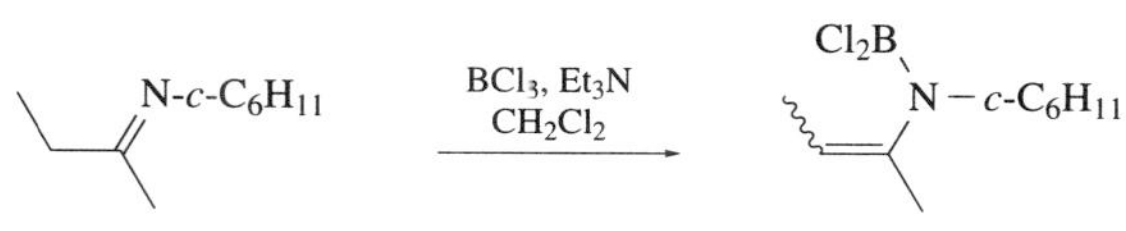

(378) (379) (139)

Reaction of (dimethylamino)organo haloboranes with AgCN yields Me_2N—B(R)CN ⟨77JOM(141)249⟩. Thermal treatment of the latter (R = Ph) leads to partial symmetrization, and the species with R = Me readily oligomerizes.

Bis(diphenylboryl)amine has been prepared by the ready symmetrization of TMS-$NHBPh_2$ and of $Me_2BNHBPh_2$ ⟨77CB2783⟩. In contrast, the species TMS-N(Me)BPh_2 is stable towards rearrangement, whereas the diborylamine $Me_2BN(Me)BPh_2$ rearranges near 50 °C to yield $PhBMe_2$ and $(—BPhN(Me)—)_3$.

Enaminodialkylboranes complex with Bu^tCN at 20 °C ⟨76BAU1374⟩ and upon heating the adducts, a rearrangement occurs with the formation of 1,2-azaborolidines (**380**).

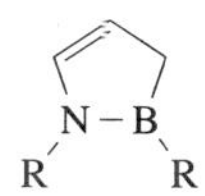

(380)

Synthesis of alkylarylketiminoboranes from ketimines and dialkylthioboric acid esters was also reported ⟨76IZV1823⟩. The systematic study of B—N compounds has its origins in the discovery of borazine by Stock and Pohland in 1926 ⟨26CB2215⟩, and since that time the field of B—N chemistry has become extensive and is rapidly expanding. The chemistry of B—N heterocycles, other than borazines, has been reviewed by Finch *et al.* ⟨69MI 215-03⟩. Paetzold ⟨67FCF437⟩ demonstrated that photolytic decomposition of a boron azide (**381**), gave high yields of the 1,3-diaza-2,4-diborane (**382**) involving a transfer of the amido group from the boron atom to the α-nitrogen atom of the azide (Equation (140)).

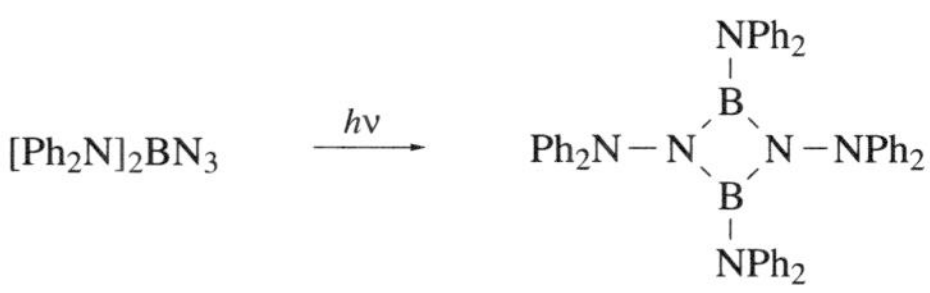

(381) (382) (140)

Paetzold ⟨67FCF437⟩ also showed that the same phenomenon occurs with boron azides (**383**) during their thermal decomposition. It was considered that the first step of the reaction was the formation of a triazoline (**384**) followed by evolution of nitrogen to yield an enamine (**385**), which then reacted in a manner typical of boron compounds with the loss of Me_2NH and subsequent dimerization to the diazadiborane ring (**386**) (Scheme 125).

2.15.12.2 Compounds with a Silicon—Nitrogen Bond

Several practical methods for the formation of Si—N bonds have been described in the literature ⟨61CRV361, 77JOM(137)56, 79JOM(8)1⟩. The most common practice is to treat a primary or secondary amine or amide with an appropriate chlorosilane. For weak amines or amines with bulky substituents, the formation of a Si—N bond usually requires more vigorous conditions (Equation (141)) ⟨81TL1787⟩.

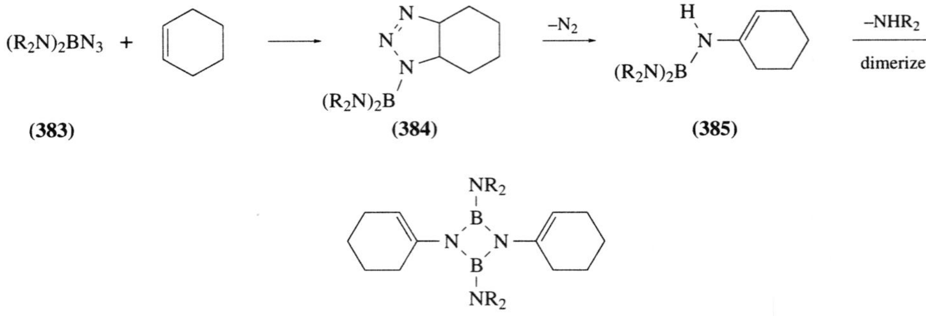

(386)

R = Me

Scheme 125

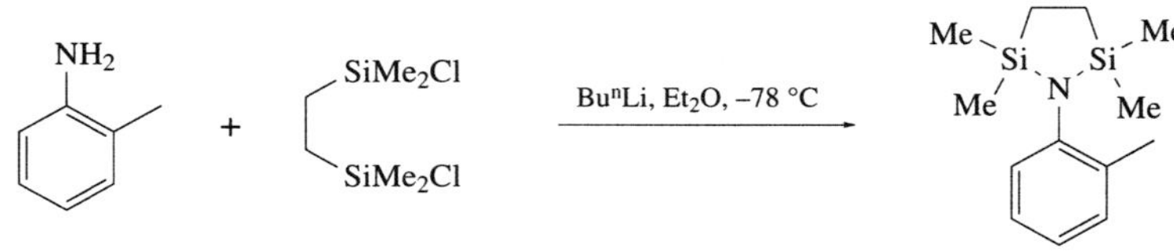

(141)

Another common technique is *trans*-silylation, the success of which depends on the requirement that either ammonia or a volatile amine be formed as the by-product (Scheme 126) ⟨78JOU1101⟩.

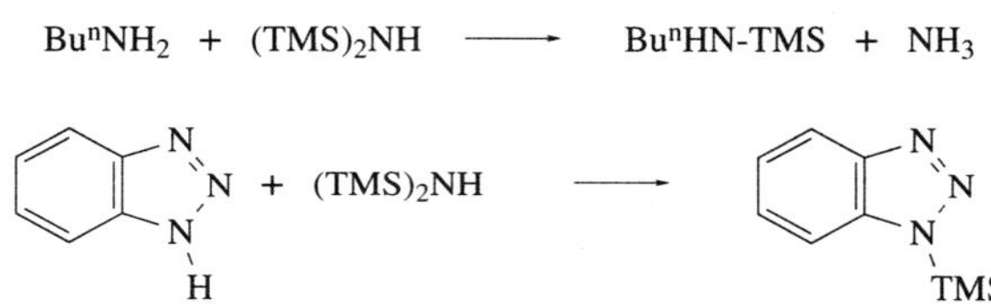

Scheme 126

A number of silylamines and silylimines have been prepared by the replacement of a nitrogen—metal bond by a N—Si bond, as shown in Scheme 127. It should be noted that the use of TMS-triflate leads to the exclusive formation of the Si—N bond ⟨80S630⟩.

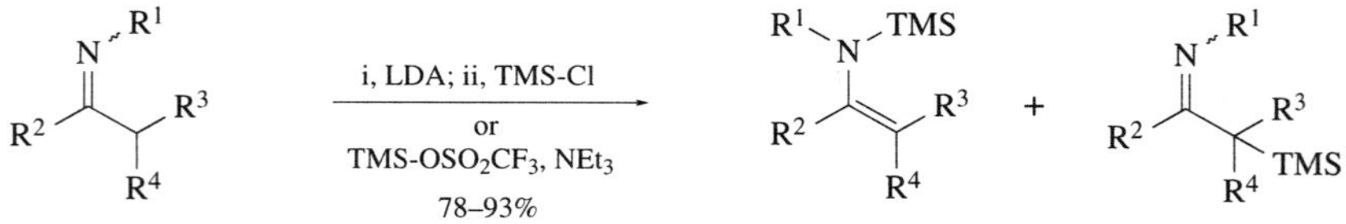

R^1 = Me, Pr^i, Ph
R^2 = H, Me, Et, Ph
R^3 = H, Me
R^4 = H, Et

Scheme 127

Several other methods for making Si—N bonds ⟨78ZN(B)1465⟩ have been developed although they have not been generally applicable. These include hydrosilylation (Equation (142)) and dehydrocondensation ⟨76S746, 79JOM(8)95, 79TL4311, 80S460, 82S273, 83S56⟩ of a silane and an amine (Scheme 128).

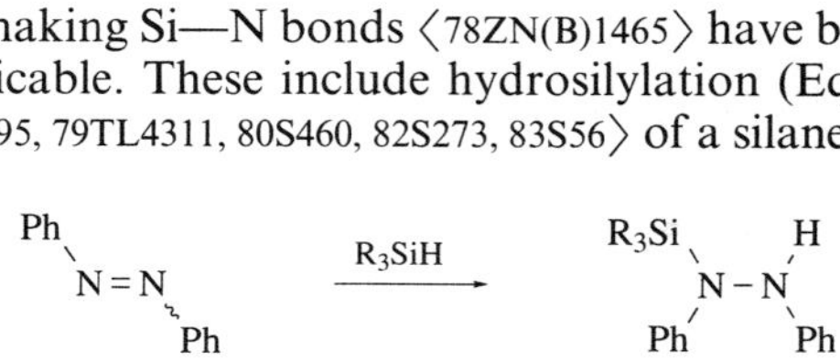

(142)

R = OEt, Et

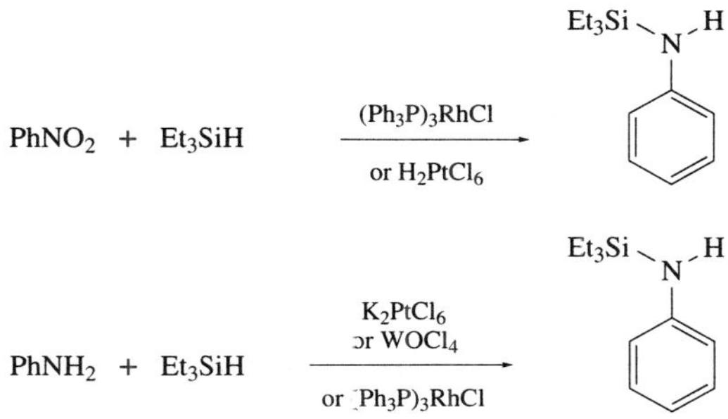

Scheme 128

Trimethylchlorosilane reacts with organotin imines or enamines leading mainly (or only) to the corresponding silicon imine or enamine (Equation 143) ⟨80JOM(193)165⟩.

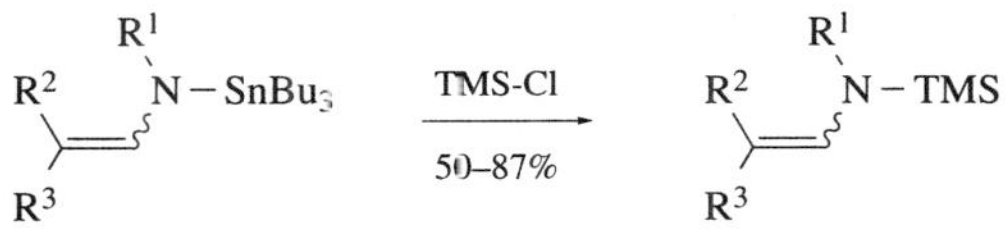

(143)

R^1 = Et, Bu^i,Pr^i, Bu^t
R^2 = Me, Et, C_5H_{11}
R^3 = Me, Et, H

The study of the Si—N versus Si—C equilibration shows clearly that in both cases the isomer obtained is the kinetic product of each reaction. The catalytic hydrosilylation ⟨86JOM(314)75⟩ of 1,4-diaza-1,3-dienes and 1-monoazadienes is performed with rhodium(I) complexes, carrying unsaturated *N*-containing control ligands as co-catalysts. Under mild conditions *N*-silylated enamines are formed as the main products. Reaction of 1-alkyl-4-aryl-1-aza-1,3-dienes with triethoxysilane proceeds quantitatively and gives, with good stereoselectivity, the corresponding *trans*-1-silyl-1-aza-2-butene (**387**) together with a *C*-silylated enamine (Equation (144)).

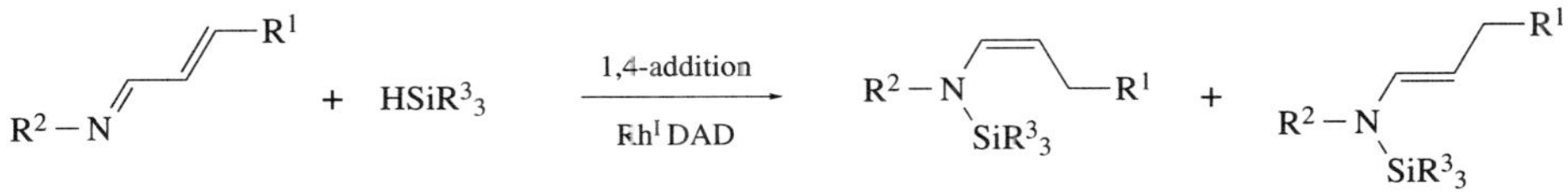

(144)

(387)

R^1 = Ph, Me
R^2 = Pr^i, Bu^t, $CHPr^i_2$, Tol
R^3 = OEt, Et

DAD = 1,4-diaza-1,3-diene

The reaction of vinyltriethylsilane with trimethylsilylazide solely gave the triethylsilyl-*N*-trimethylsilylaziridine ⟨68CB743⟩. However, on heating together (**388**) and trimethylsilyl azide under reflux for 5 days gave, not the expected aziridine, but instead the *trans*-*N*,*N*-bis(silyl)enamine (**389**) (Equation (145)) ⟨78JOM(152)C25⟩.

(145)

(388) **(389)**

p- or *m*-Bromo-*N*,*N*-bis(trimethylsilyl)aniline (**390**) ⟨75JOC1090⟩ was prepared by treating the corresponding bromoaniline with *n*-butyllithium (2.3 equiv.) in THF at room temperature, followed by chlorotrimethylsilane (2.3 equiv.). A maximum yield of the pure product was obtained when this excess of *n*-butyllithium chlorotrimethylsilane was utilized. Without this excess, an azeotrope of (**390**) and *p*- or *m*-bromo-*N*-trimethylsilylaniline was invariably formed (Equation (146)).

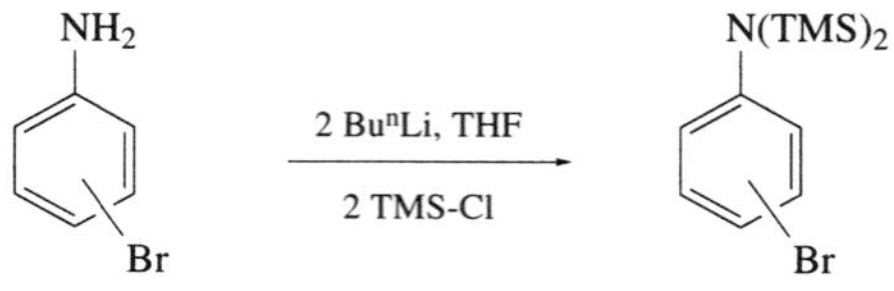

(390)

(146)

Exposing a hexane solution of equimolar amounts of the nitrile (**391**) and the complex (**392**) to UV light for several hours afforded the silylated enamines (**393**) in good yield (Equation (147)) ⟨81JOC3372, 83JOM(255)359, 85OM623⟩.

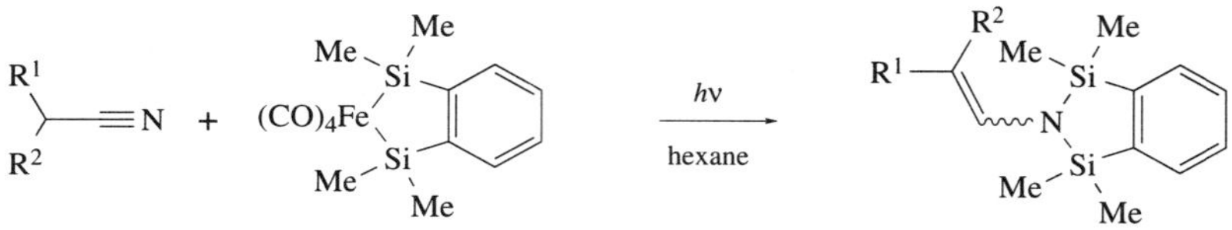

(147)

(391) **(392)** **(393)**

Burnell-Curty and Roskamp reported ⟨92JOC5063⟩ that tin(II) amides react with primary aldehydes to give the corresponding *trans*-enamines stereoselectively (Scheme 129).

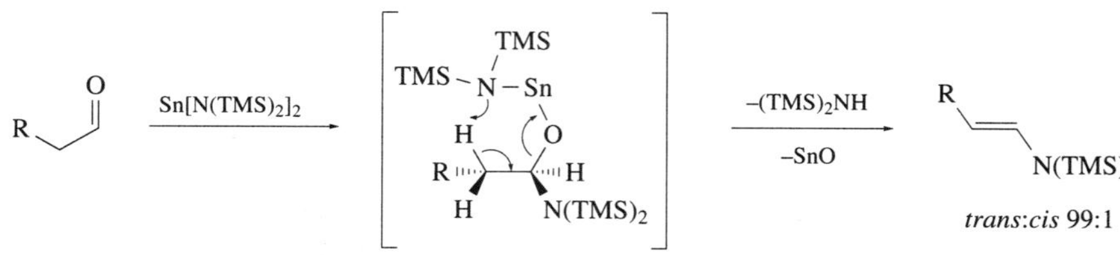

Scheme 129

Other syntheses of N—Si compounds include the reaction of pyridine with bis(trimethylsilyl)mercury, the product of which is generally an *N,N*-bis(trimethylsilyl)tetrahydro-4,4′- or -2,2′-bipyridyl ⟨76TL2173⟩. A cobalt carbonyl-catalyzed hydrosilylation of nitriles for the preparation of *N,N*-disilylarylamines was reported by Murai *et al.* ⟨90JOC449⟩. Cunico and Kuan described acetylsilane *O*-silylcyanohydrins as precursors to α-silyl ketones and β-siloxy-*N,N*-bissilylenamines ⟨90JOC4634⟩. The stereoselective synthesis of *N,N*-bis(trimethylsilyl)enamines from *N,N*-bis(trimethylsilyl)thioformamides was studied by Lüke and Walter ⟨77AG(E)535⟩.

2.15.12.3 Compounds with a Germanium—Nitrogen Bond

Germanium—nitrogen compounds are usually very sensitive to hydrolysis but can be easily synthesized according to the methods described below, provided moisture is excluded from the system. An addition reaction is shown in Equation (148) in which an *N*-aryl germanyl compound (**394**) is formed ⟨B-69MI 215-04, B-71MI 215-02⟩.

(148)

(394)

The best starting materials for the synthesis of a Ge—N bond are germanium halides. Ammonolysis or aminolysis of halogen—germanium bonds leads to the formation of the Ge—N bond with elimination of an ammonium halide (Scheme 130). The method is general, provided the metal halide is susceptible to nucleophilic attack ⟨B-80MI 215-02⟩.

$$R^1_3GeX + HNR^2_2 \longrightarrow [R^1_3GeX\ NHR^2_2] \xrightarrow{HNR^2_2} R^1_3GeNR^2_2 + HNR^2_2{\cdot}HX$$

Scheme 130

Another very useful synthetic method is transmetallation, which proceeds in good yield ⟨B-80MI 215-02, 83RTC65⟩. The best way to *N*-germylimines involves the reaction of halogermanes on iminolithium derivatives (**395**) (Scheme 131) ⟨67JOM(9)231, 73JCS(D)157, 76JCS(D)1⟩. *N*-Germylimines are obtained in good yield when R^1 and R^2 are both aryl groups. With aliphatic substituents, the hydrogen in the α-position of the ketimine enters into enamine tautomerism, as observed in the preparation of silicon analogues ⟨67JOM(9)231⟩.

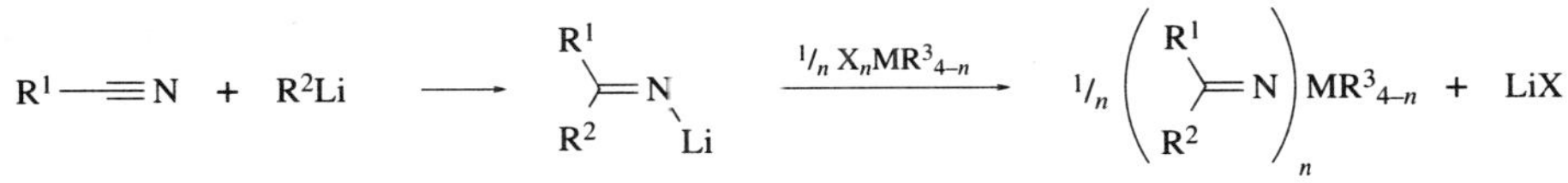

(**395**)

M = Si, Ge, Sn
R^1, R^2, R^3 = alkyl, aryl
n = 1–4

Scheme 131

Isocyanates react with the Ge—N bond to yield the C—N adduct while isothiocyanates (**396**) lead either to C—N or C—S adducts, the latter sometimes undergo rearrangement leading to the corresponding germanium sulfide (Scheme 132) ⟨B-80MI 215-02⟩.

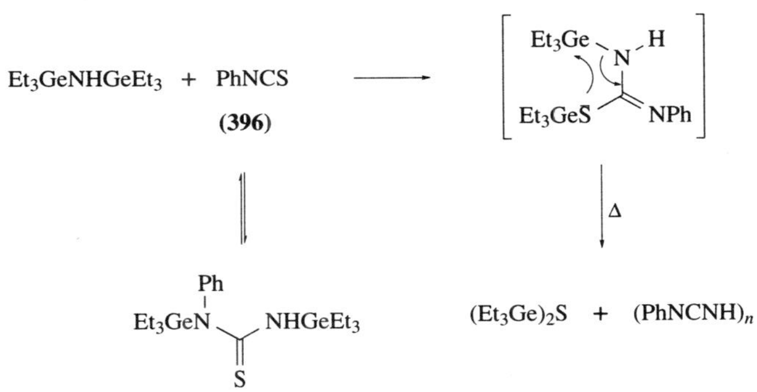

Scheme 132

For more details on the synthesis of Ge—N compounds, see Poller ⟨79COC(3)1073⟩, and Stagé and co-workers ⟨B-71MI 215-02⟩.

2.15.13 COMPOUNDS WITH AN *N*—METAL BOND

2.15.13.1 N—Sn, N—Pb, and N—Hg Bonds

The N—Sn bond is usually formed by substitution by a nitrogen nucleophile at a tin center. An organotin halide reacts with a free amine to give a five- or six-coordinate complex which merely dissociates on heating. Carboxamides and sulfonamides (**397**) are much more readily *N*-stannylated ⟨82COMC-I(1)599⟩ than are amines, and the reaction shown in Equation (149) occurs if the reagents are subjected to azeotropic dehydration.

$$(Bu_3Sn)_2O + 2\ PhNHSO_2Me \longrightarrow Bu_3SnNPhSO_2Me \qquad (149)$$

(**397**)

The N—Sn bond can also be formed by a variety of addition reactions of the general and specific types shown in Scheme 133.

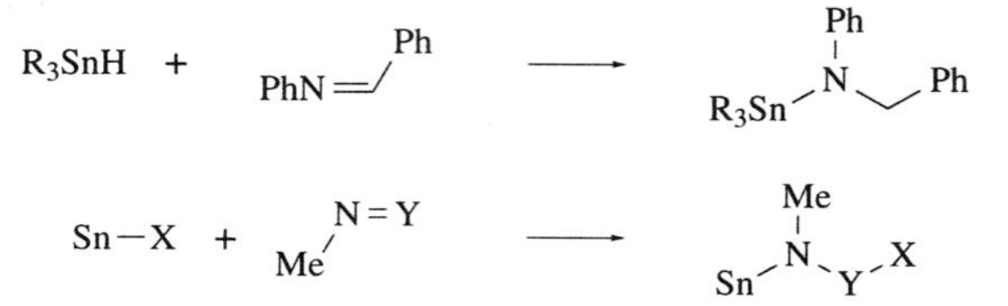

Scheme 133

The organotin derivatives ⟨76JOM(120)217, 80JOM(186)C9⟩ corresponding to the imine–enamine equilibrium have been obtained by two routes: (i) reaction of organotin enolates with a stannazane; (ii) reaction of tributyltin chloride with a magnesium derivative of an imine. The first method allows the preparation of less-hindered compounds, and the latter the more hindered ones.

Chiral organotin enamines ⟨85JOC3863⟩ are easily prepared from cyclic ketones, chiral amino alcohols (derived from amino acids), and an organotin precursor (Scheme 134).

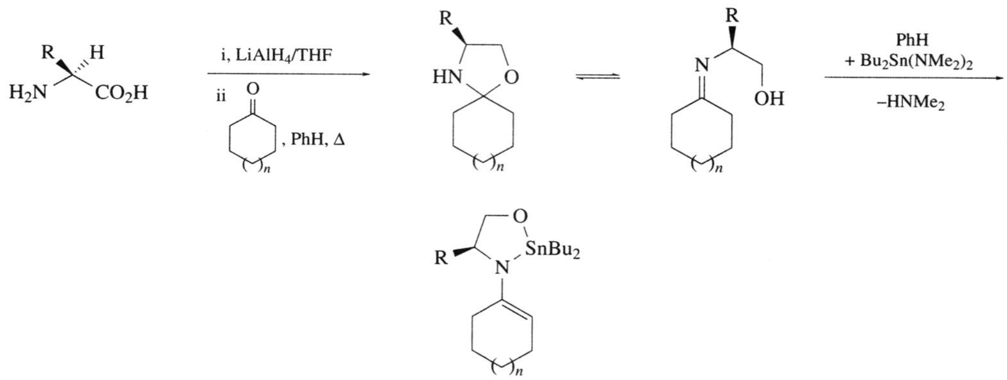

Scheme 134

The partial protolysis of organotin imine–enamines leads to organic secondary enamines which were found to be stable in completely aprotic media ⟨77JOM(137)23⟩. The metalloenamines are very often used for the synthesis of carbonyl compounds ⟨80JA5866⟩, hemiketal, and enol ethers ⟨75JOC2963⟩, asymmetric alkylation of acyclic ketones ⟨78JOC3245⟩ and enantioselective alkylation of cyclohexanone ⟨83TL511⟩. The asymmetric isomerization of allylamines to enamines catalyzed by 2,2′-bis(diphenylphosphino)-1,1′-binaphthyl-rhodium complexes was studied by Noyori and co-workers ⟨90JA4897⟩.

Amino–lead compounds, like the corresponding tin analogues, are generally obtained by the reaction of an organolead halide with the lithium or sodium derivative of the amine ⟨82COMC-I(2)662⟩. Like the triorganolead alkoxides, triorganolead amines (**398**) react with a wide variety of multiply-bonded reagents (Scheme 135).

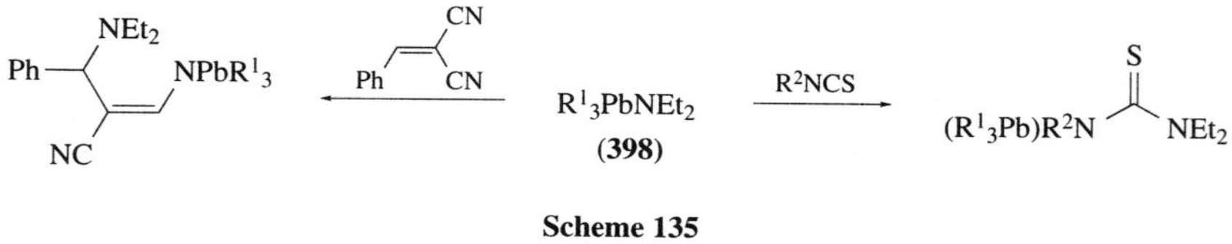

Scheme 135

By the elimination of benzene from various phenylaluminum derivatives, metallation can occur intramolecularly to yield aluminum heterocycles (**399**) (Equation (150)) ⟨64JA4221⟩.

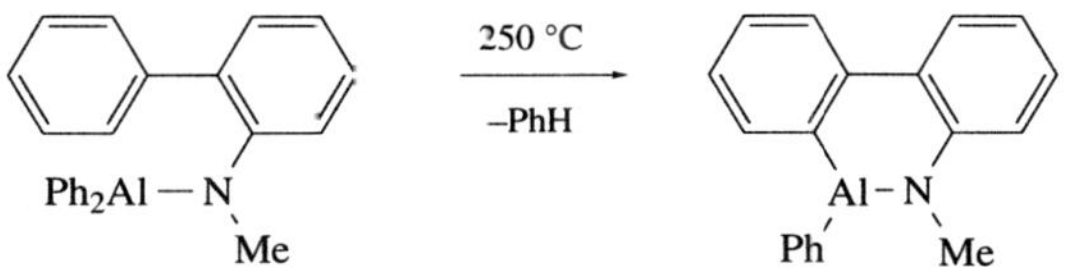

(150)

(399)

Examples of *N*-mercury enamines (**400**) were also reported (Equation (151)) ⟨64CB1155⟩.

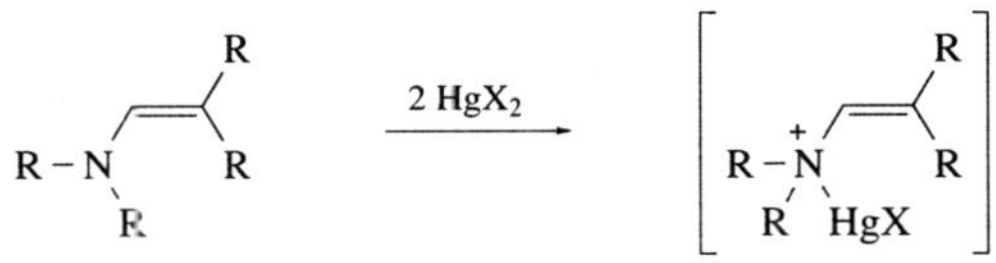

(151)

(400)

2.16
Vinyl- and Arylphosphorus Derivatives

TORU MINAMI
Kyushu Institute of Technology, Kitakyushu, Japan

and

KENTARO OKUMA
Fukuoka University, Japan

2.16.1 PHOSPHINES AND PHOSPHORANES

2.16.1.1 Phosphines

2.16.1.1.1 Primary phosphines (Table 1)

(i) Reduction of arylphosphonous ($ArP(OH)_2$) and arylphosphonic acids ($ArP(O)(OH)_2$) and their derivatives with $LiAlH_4$ (Method A)

Since methods for the synthesis of phosphines were reviewed by Maier ⟨B-72MI 216-01⟩ in 1972, there has been little development in their synthetic methods. That is, synthesis of primary phosphines has generally been achieved by the reduction of arylphosphonous, arylphosphonic, or arylphosphonothioic dichlorides, and corresponding esters such as diethyl arylphosphonites, arylphosphonodithioites or arylphosphonates with $LiAlH_4$ (or $LiBH_4$) in ether or THF (Equations (1)–(6)) ⟨52JA562, 52JA3414, 60CB1220, 74MI 216-01, 81TL4475, 82JA674, 84ZOB1917⟩. Vinyl- and ethynylphosphines are prepared in a similar manner ⟨63ZOB3774⟩.

$$2ArPCl_2 + LiAlH_4 \longrightarrow 2ArPH_2 + LiAlCl_4 \quad (1)$$

$$4ArP(O)Cl_2 + 3LiAlH_4 \longrightarrow 4ArPH_2 + 2LiAlCl_4 + LiAl(OH)_4 \quad (2)$$

$$4ArP(S)Cl_2 + 3LiAlH_4 \longrightarrow 4ArPH_2 + 2LiAlCl_4 + LiAl(SH)_4 \quad (3)$$

Table 1 Synthesis of primary phosphines ($ArPH_2$).

Entry	*Method*	*Ar*	*b.p.* (°C)(mmHg) or *m.p.* (°C)	*Ref.*
1	A	Ph	154–157(760)	52JA3414
2	A	Ph	72(50)	60CB1220
3	B	Ph	60(14)	65CB1681
4	A	$4\text{-}MeC_6H_4$	56–61(11)	74MI 216-01
5	A	$2\text{-}MeC_6H_4$	52–54(0.5)	67HCA1747
6	A	$4\text{-}MeOC_6H_4$	89–91(11)	74MI 216-01
7	A	$3\text{-}FC_6H_4$	52–53(21)	69CB2914
8	A	$4\text{-}FC_6H_4$	38–40(11)	74MI 216-01
9	A	$2\text{-}ClC_6H_4$	54–55(3)	67HCA1747
10	A	$3\text{-}ClC_6H_4$	77–78(12)	69ZOB841
11	A	$4\text{-}ClC_6H_4$	68–70(11)	74MI 216-01
12	A	$2,4,6\text{-}Bu^t_3C_6H_2$	144	82JA5820
13	A	$PhCH{=}CH$	82(2)	63ZOB3774
14	A	$2\text{-}H_2PC_6H_4$	71–73(3.5)	81TL4475
15	A	$4\text{-}H_2PC_6H_4$	69.0–70.5	62JOC2192
16	A	9-Anthracenyl	68–73	92CB2187

$$2ArP(OEt)_2 + LiAlH_4 \longrightarrow 2ArPH_2 + LiAl(OEt)_4 \quad (4)$$

$$4ArP(O)(OEt)_2 + 3LiAlH_4 \longrightarrow 4ArPH_2 + 2LiAl(OEt)_4 + LiAl(OH)_4 \quad (5)$$

$$2ArP(SEt)_2 + LiAlH_4 \longrightarrow 2ArPH_2 + LiAl(SEt)_4 \quad (6)$$

(ii) Reduction of arylphosphonous dichlorides ($ArPCl_2$) and arylphosphonates ($ArP(O)(OR)_2$) with hydrosilanes (Method B)

Reduction of arylphosphonous dichlorides and diethyl arylphosphonates with diphenylsilane, trichlorosilane, or polysilane also gives primary phosphines in good yields (Equations (7)–(9)) ⟨64CB1988⟩.

$$ArPCl_2 + Ph_2SiH_2 \longrightarrow ArPH_2 + Ph_2SiCl_2 \quad (7)$$

$$ArPCl_2 + 2HSiCl_3 \xrightarrow{2NEt_3} ArPH_2 + 2SiCl_4 \quad (8)$$

$$ArP(O)(OEt)_2 + 2Ph_2SiH_2 \longrightarrow ArPH_2 + Ph_2Si(OEt)_2 + (Ph_2SiO)_n \quad (9)$$

(iii) Hydrolysis of dialkali metal arylphosphides ($ArP(Metal)_2$) (Method C)

Although the preparation of arylphosphines using Methods C and D is not common, they have also been applied to the synthesis of primary phosphines. The primary arylphosphines can be obtained by hydrolysis of sodium (or lithium) phosphides, which are derived from arylphosphonous dichloride and sodium (or lithium) (Equations (10) and (11)) ⟨59CB2088⟩.

$$ArPCl_2 + 4Na \xrightarrow{\text{toluene, 100 °C}} ArPNa_2 + 2NaCl \quad (10)$$

$$ArPNa_2 \xrightarrow{H_2O} ArPH_2 \quad (11)$$

(iv) Disproportionation of arylphosphonous acids ($ArP(OH)_2$) (Method D)

Arylphosphonous acids on heating at high temperature undergo disproportionation to form arylphosphines and arylphosphonic acids (Equation (12)) ⟨52JCS3039⟩.

$$3ArP(OH)_2 \xrightarrow{250\ °C} ArPH_2 + 2ArP(O)(OH)_2 \quad (12)$$

2.16.1.1.2 *Secondary phosphines (Table 2)*

(i) Reduction of arylphosphinous (ArRPX) or arylphosphinic halides (ArRP(O)X), esters, and amides with $LiAlH_4$ (Method A)

Lithium aluminum hydride has also been found effective for reduction of arylphosphinous or arylphosphinic chlorides ⟨69CB2914⟩, esters ⟨58CB1583, 68CB2197⟩, and amides ⟨74JOC2296⟩ to form secondary arylphosphines (Equations (13)–(16)).

Table 2 Synthesis of secondary phosphines (ArRPH).

Entry	*Method*	*Ar*	*R*	*b.p.* (°C)(mmHg) *or m.p.* (°C)	*Ref.*
1	D	Ph	Bu^n	102(9)	59M792
2	C	Ph	Ph	112–113(1.5)	76IS161
3	B	$3\text{-}ClC_6H_4$	$3\text{-}ClC_6H_4$	No data	79JOC3741
4	A	$3\text{-}FC_6H_4$	$3\text{-}FC_6H_4$	144(13)	69CB2914
5	A	$4\text{-}FC_6H_4$	$4\text{-}FC_6H_4$	140–142(12)	69CB2914
6	B	$2\text{-}MeOC_6H_4$	$2\text{-}MeOC_6H_4$	No data	79JOC3741
7	B	$3\text{-}MeOC_6H_4$	$3\text{-}MeOC_6H_4$	No data	79JOC3741
8	C	$2\text{-}MeC_6H_4$	$2\text{-}MeC_6H_4$	104–106(0.1)	77JOC1236
9	C	$4\text{-}MeC_6H_4$	$4\text{-}MeC_6H_4$	121–122(2)	68CB2899
10	B	$2,4,6\text{-}Me_3C_6H_2$	$2,4,6\text{-}Me_3C_6H_2$	74	65CB1681
11	B	1-Naphthyl	1-Naphthyl	91–97	81JOC4422
12	B	2-Naphthyl	2-Naphthyl	94–96	81JOC4422
13	A	9-Anthracenyl	9-Anthracenyl	155–156	92CB2187

$$(m\text{-}FC_6H_4)_2PCl \xrightarrow[56\%]{LiAlH_4,\ ether,\ -10\ °C\ to\ reflux} (m\text{-}FC_6H_4)_2PH \quad (13)$$

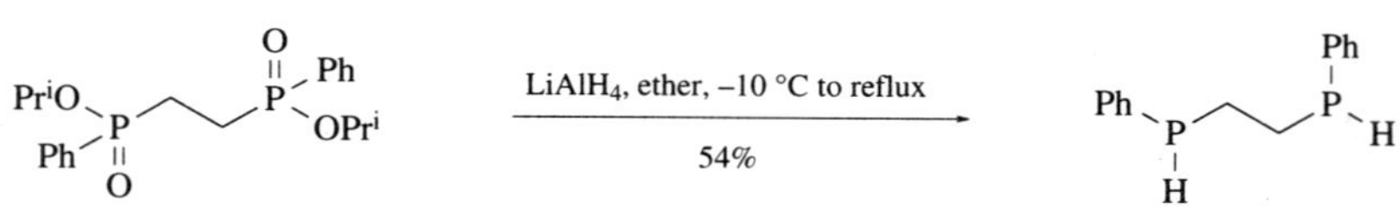

(14)

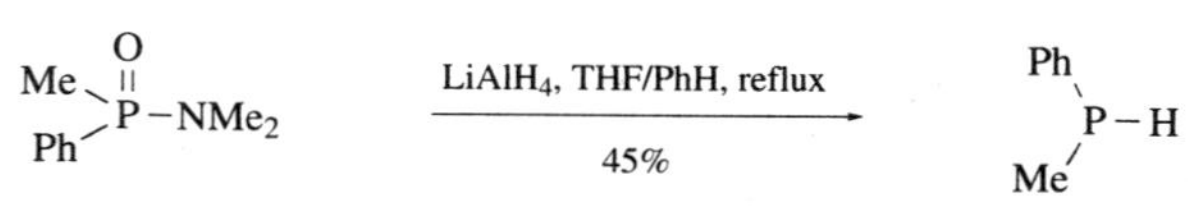

(15)

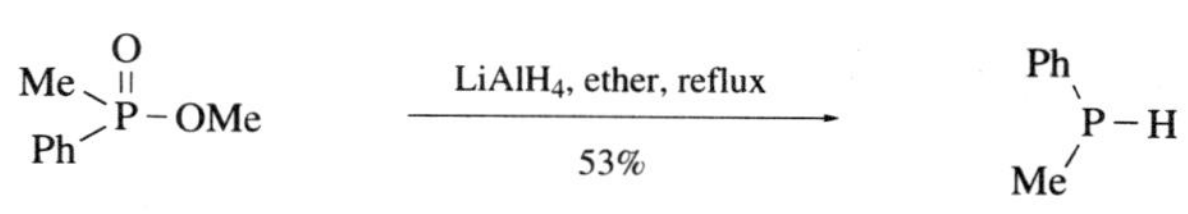

(16)

(ii) Reduction of arylphosphinous (ArRPOH) or arylphosphinic acids (ArRP(O)OH) and their derivatives with hydrosilanes (Method B)

The reduction of phosphinous or phosphinic acids and their chloride derivatives with hydrosilanes such as diphenylsilane ⟨81JOC4422⟩, phenylsilane, and trichlorosilane ⟨65CB1681⟩ is one of the most widely used methods for the synthesis of secondary phosphines (Equations (17)–(22)).

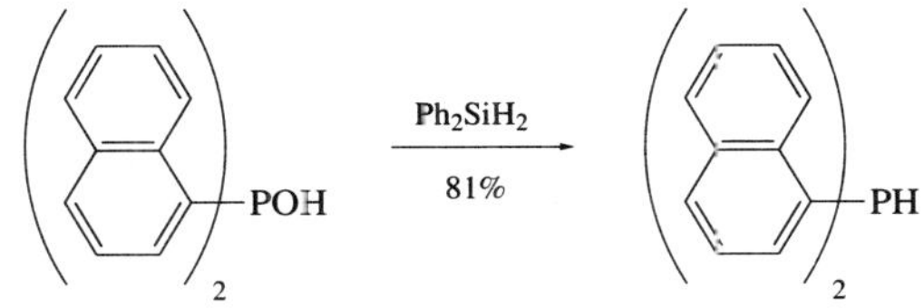
(17)

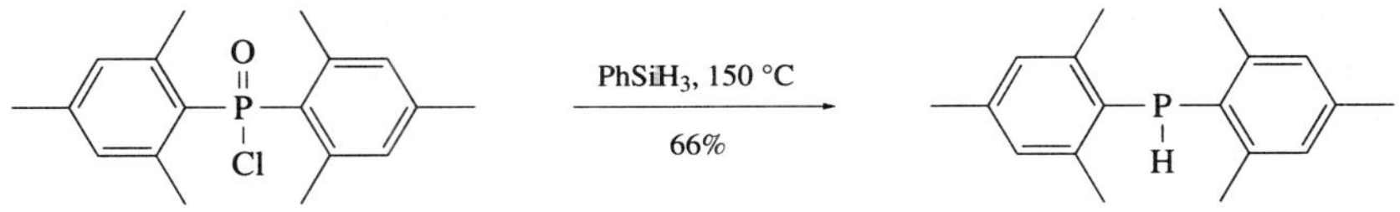
(18)

$$Ph_2P(O)OMe \xrightarrow[59\%]{Ph_2SiH_2,\ 190\ ^{\circ}C} Ph_2PH \qquad (19)$$

$$Ph_2P(O)Cl \xrightarrow[88\%]{Ph_2SiH_2,\ 200\ ^{\circ}C\ to\ 250\ ^{\circ}C} Ph_2PH \qquad (20)$$

$$Ph_2P(O)OH \xrightarrow[40\%]{HSiCl_3,\ PhH,\ reflux} Ph_2PH \qquad (21)$$

$$Ph_2PCl \xrightarrow[75\%]{i,\ HSiCl_3,\ Et_3N,\ PhH,\ reflux;\ ii,\ NaOH,\ H_2O} Ph_2PH \qquad (22)$$

(iii) Hydrolysis (or alcoholysis) of metal phosphides (Method C)

In the synthesis of symmetrical diarylphosphines, hydrolysis ⟨76IS161⟩ or alcoholysis ⟨68CB2899⟩ reactions of lithium (or sodium) diarylphosphides, which are prepared from triarylphosphines or diarylphosphinous chlorides and lithium (or sodium) metal, have frequently been employed, both for the availability of the starting materials and for the simplicity of working procedures (Schemes 1 and 2).

$$Ph_3P \xrightarrow{2Li,\ THF,\ RT} Ph_2PLi + PhLi \xrightarrow[81\%]{ice/H_2O} Ph_2PH + LiOH$$

Scheme 1

$$(p\text{-}MeC_6H_4)_2PCl \xrightarrow{2Na,\ toluene,\ reflux} (p\text{-}MeC_6H_4)_2PNa \xrightarrow[68\%]{MeOH/H_2O\ (1:1)} (p\text{-}MeC_6H_4)_2PH$$

Scheme 2

(iv) Alkylation of metal phosphides (Method D)

Synthesis of secondary phosphines via alkylation of monoalkali metal phosphides, prepared from primary phosphines and an alkali metal, with alkyl halides is not so widely utilized as the above-mentioned methods, since availability and handling of primary phosphines are problematic. However, this synthetic method has the advantage of ease of synthesis of unsymmetrical secondary phosphines (Scheme 3) ⟨59M792⟩.

$$PhPH_2 \xrightarrow{Na,\ ether,\ reflux} PhPHNa \xrightarrow[65\%]{MeI,\ RT\ to\ reflux} PhMePH$$

Scheme 3

2.16.1.1.3 Tertiary phosphines

(i) Reaction of arylphosphinous (ArRPX), or arylphosphonous Halides ($ArPX_2$), or phosphorus trihalides, or their ester derivatives with Grignard, organolithium, or other organometallic reagents (Method A)

The synthesis of tertiary phosphines via the reaction of phosphorus(III) halides such as phosphinous or phosphonous halides or phosphorus trichloride with Grignard or organolithium reagents is the most useful method from the point of view of generality and availability of the starting materials (Equations (23)–(25)) (Tables 3–5).

$$ArR^1PX^1 + R^2MgX^2 \text{ (or } R^2Li) \longrightarrow ArR^1R^2P + MgX^1X^2 \text{ (or } LiX^1) \quad (23)$$

$$ArPX^1_2 + 2RMgX^2 \text{ (or 2RLi)} \longrightarrow ArR_2P + 2MgX^1X^2 \text{ (or } 2LiX^1) \quad (24)$$

$$PX^1_3 + 3ArMgX^2 \text{ (or 3ArLi)} \longrightarrow Ar_3P + 3MgX^1X^2 \text{ (or } 3LiX^1) \quad (25)$$

R^1 = alkyl, aryl
X^1 = halogens, OR

Table 3 Preparation of tertiary phosphines (ArR^1R^2P) via Equation (23).

Entry	*Ar*	R^1	R^2	*b.p.* (°C)(mmHg) *or m.p.* (°C)	*Ref.*
1	Ph	Bu^t	Ph	118–122(0.2)	66CB1134
2	Ph	Ph	$4\text{-}ClC_6H_4$	155–157(0.02)	78JA7304
3	Ph	Ph	$2\text{-}ClC_6H_4$	106–108	78JA7304
4	Ph	Ph	CH_2CN	165–166(2)	87S938
5	Bu^t	Bu^t	Ph	97–100(0.6)	71JCS(A)2976
6	Ph	Ph	$2\text{-}OHCC_6H_4$	96–97	73LA1480
7	Ph	Ph	$3\text{-}OHCC_6H_4$	78–79	73LA1480
8	Ph	Ph	$3\text{-}BrC_6H_4$	160–182(0.1)	69CB1883
9	Ph	Ph	$2\text{-}NCC_6H_4$	146–147	81JOM(205)161
10	Ph	Ph	$2\text{-}Me_2NC_6H_4$	121	83PS165
11	Ph	Ph	$3\text{-}Me_2NC_6H_4$	78–80	76LA2126
12	Ph	Ph	$2\text{-}HOC_6H_4$	164–165	77IC2966
13	Ph	Ph	2-Pyridyl	85	83PS165
14	Ph	Ph	$2\text{-}CF_3C_6H_4$	86–87	77PS191
15	Ph	Ph	$3\text{-}CF_3C_6H_4$	44–45	76JCS(P1)2529
16	Ph	Ph	$3\text{-}FC_6H_4$	57	69CB2914
17	Ph	Ph	$2\text{-}FC_6H_4$	88–89	78JA7304
18	Ph	Ph	C_6F_5	68–70	68JCS(A)2149
19	Ph	1-Naphthyl	$4\text{-}C_6H_5C_6H_4$	192–193	71LA(751)17
20	Ph	Ph	—C≡C—C≡C—	105	66BSF1002
21	Ph	Ph	C≡C—Ph	43	66BSF1002
22	Ph	Ph	C≡CH	35	66BSF1002

Table 4 Synthesis of tertiary phosphines (ArR_2P) via Equation (24).

Entry	*Ar*	*R*	*b.p.* (°C)(mmHg) *or m.p.* (°C)	*Ref.*
1	Ph	$Me_2N(CH_2)_3$	102–105(0.005)	71HCA710
2	Ph	$2\text{-}MeC_6H_4$	73	78JA5130
3	$4\text{-}BrC_6H_4$	Ph	78–79	63LA(665)1
4	$3\text{-}ClC_6H_4$	Ph	195–198(4) 30–31	69ZOB841
5	Ph	$PhCH_2$	68–70	78IS172
6	Ph	Bu^n	104–108(1.7–2.0)	78IS171
7	Ph	$c\text{-}C_6H_{11}$	56–57	78IS171
8	Ph	Et	126–130(30)	78IS170
9	Ph	$n\text{-}C_6H_{13}$	236(50)	31JCS2109
10	$4\text{-}Me_2NC_6H_4$	Ph	153	63LA(665)1
11	Ph	CH_2=CH	55(0.5)	57JA5884

Table 5 Preparation of symmetrical tertiary phosphines (Ar_3P) via Equation (25).

Entry	*Ar*	*b.p.* (°C)(mmHg) *or m.p.* (°C)	*Ref.*
1	2-Pyridyl	115	80JCS(D)55
2	4-BrC_6H_4	163–164	74JCS(P2)1338
3	2-ClC_6H_4	185	37JCS527
4	3-ClC_6H_4	64–66	77PS191
5	4-ClC_6H_4	102	77PS191
6	$2,4\text{-(MeO)}_2C_6H_3$	187–188	64ZOB1446
7	$3,5\text{-(MeO)}_2C_6H_3$	131.5–132.5	64JPR294
8	$2\text{-Me}_2NC_6H_4$	106	83PS165
9	$4\text{-Me}_2NC_6H_4$	254	68LA(715)1
10	$2,4\text{-Me}_2C_6H_3$	154	74ZOB760
11	$2,5\text{-Me}_2C_6H_3$	155	01LA(315)43
12	$2,6\text{-Me}_2C_6H_3$	146.5–147.5	74ZOB760
13	2-FC_6H_4	147	82PS331
14	3-FC_6H_4	64	69CB2914
15	4-FC_6H_4	77–79	77PS191
16	2-MeOC_6H_4	203–204	61CB2514
17	3-MeOC_6H_4	112–114	64JPR294
18	4-MeOC_6H_4	131–132	61CB2514
19	2-MeC_6H_4	123–125	77PS191
20	3-MeC_6H_4	99–100	77PS191
21	4-MeC_6H_4	146	77PS191
22	2-MeSC_6H_4	No data	76IS168
23	$2\text{-CF}_3C_6H_4$	164–165	77PS191
24	$3\text{-CF}_3C_6H_4$	130–132(0.2)	77PS191
25	$4\text{-CF}_3C_6H_4$	78–80	80JFC239
26	$2,4,6\text{-(MeO)}_3C_6H_2$	146–147	63ZOB3050
27	$2,4,6\text{-Me}_3C_6H_2$	192–193	69ZOB1544
28	PhC≡C	92	68LA(714)1

As shown in Equations (26) ⟨87S938⟩, (27) ⟨92SL833⟩, and (28) ⟨78IS171⟩ and Scheme 4 ⟨65CB1134⟩, asymmetrical tertiary phosphines bearing different groups have been synthesized by the reaction of either arylphosphinous chlorides or arylphosphonous dichlorides with a variety of organolithium or Grignard reagents or vinyl copper compounds (Tables 3 and 4).

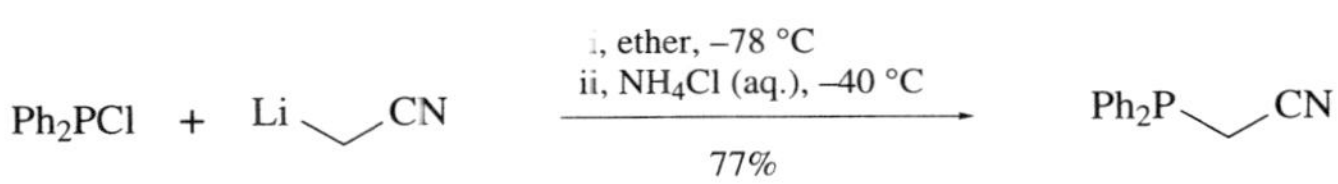

(26)

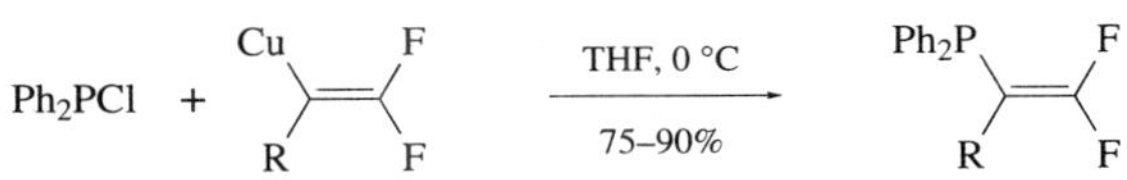

(27)

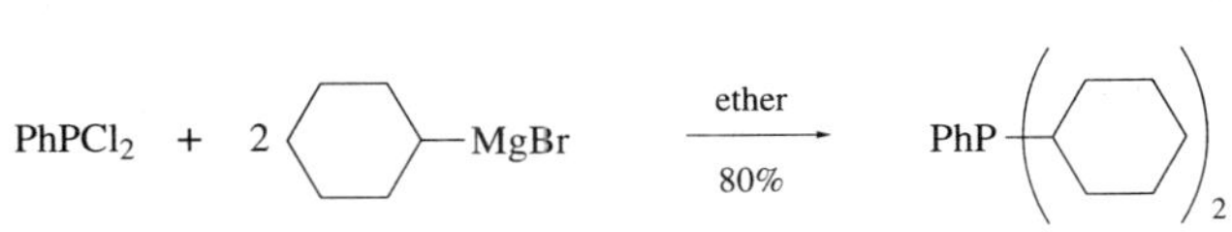

(28)

$$PhPCl_2 + Bu^tMgCl \xrightarrow{\text{ether, RT}} PhBu^tPCl \xrightarrow[37\%]{\text{PhMgBr, reflux}} Ph_2PBu^t$$

Scheme 4

Although symmetrical triarylphosphines have generally been synthesized from aryl Grignard or aryllithium reagents and phosphorus trichloride, the yields of the phosphines are influenced by the position and the number of substituents on the aryl group, that is, by steric and electronic factors. For example, tris(2-functionalized phenyl)phosphines such as tris(2-chlorophenyl)- ⟨37JCS527⟩ and tris(2-methylphenyl)phosphines ⟨77PS(3)191⟩ are obtained in rather low yields, compared with the corresponding 3- or 4-functionalized phosphines (Scheme 5).

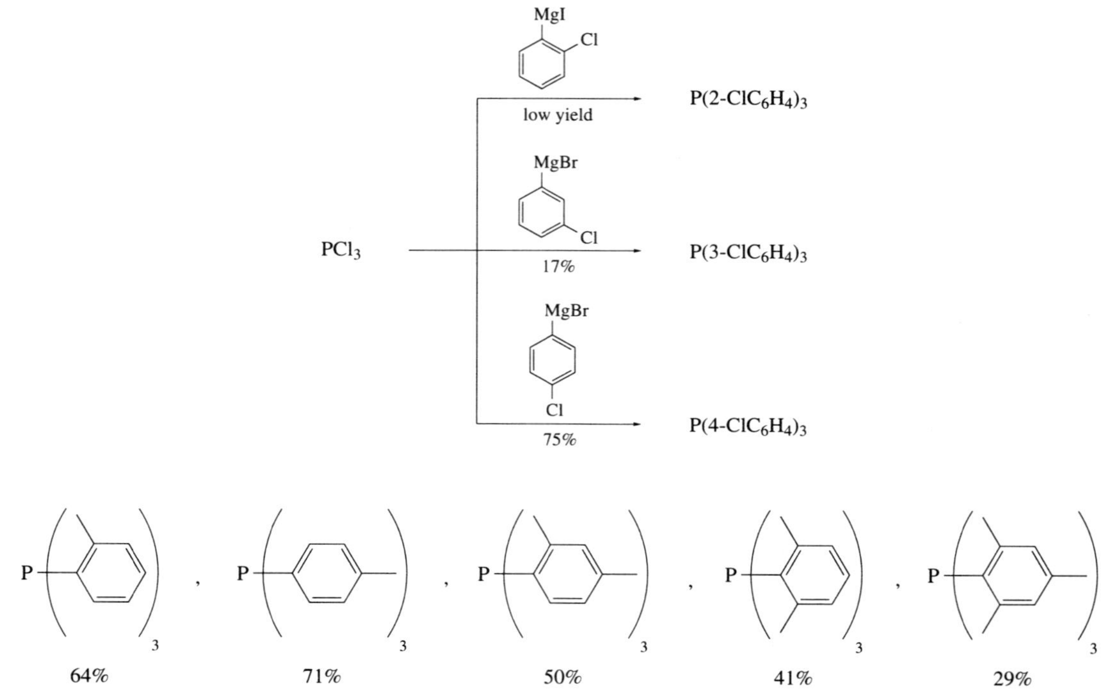

Scheme 5

Organotin compounds as well as Grignard or organolithium compounds have been used for the synthesis of trialkynylphosphines (Equation (29)) ⟨68LA(714)1⟩.

$$PCl_3 + 3Pr^n{}_3Sn{-}C{\equiv}C{-}Ph \xrightarrow[72\%]{THF,\ RT} P(C{\equiv}C{-}Ph)_3 + 3Pr^n{}_3SnCl \qquad (29)$$

(ii) Reaction of metal diarylphosphides with aryl or alkyl halides (Method B) (Tables 6 and 7)

Alkali metal diphenylphosphides are usually prepared from triphenylphosphine and lithium in THF or sodium in liquid ammonia, from diphenylphosphine and butyllithium, or from diphenylphosphinous chloride and alkali metals ⟨62ZC163⟩. In the preparative method for the diphenylphosphide anion from triphenylphosphine and lithium, which is more practical than those from other starting materials, the phenyllithium formed is selectively destroyed by addition of equivalent *t*-butyl chloride to avoid undesired side reactions (Equations (30) and (31)) ⟨77IS186⟩. The phosphide anion (lithium counterion) reacts with various alkyl, vinyl or aryl halides to produce a wide range of phosphines of the type Ph_2PR in high yields (Equation (32)) (Table 6).

$$Ph_3P + 2Li \xrightarrow{THF} Ph_2PLi + PhLi \qquad (30)$$

$$PhLi + Bu^tCl \xrightarrow{THF} PhH + LiCl + $$ 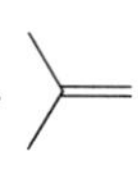 (31)

$$Ph_2PLi + RX \xrightarrow{THF} Ph_2PR + LiX \qquad (32)$$

R = alkyl, vinyl, aryl, etc.

Table 6 Preparation of tertiary phosphines via Method B.

Entry	*Ph_2PR*	*b.p.* (°C) (mmHg) *or m.p.* (°C)	*Ref.*
1	Ph_2PMe	120–122 (1.5)	76IS157
2	Ph_2PEt	138–141 (1)	76IS158
3	Ph_2PBu^n	80–90 (0.1)	76IS158
4	$Ph_2P\text{–}{}^cHex$	180 (2.5) 60–61	76IS159
5	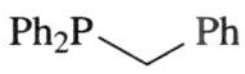	141–142	76IS159
6	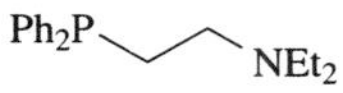	146 (1)	76IS160
7	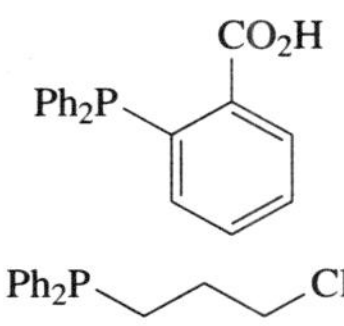	175–176	82IS175
8	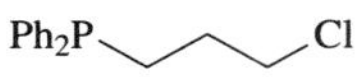	viscous oil	75JOM(94)327
9	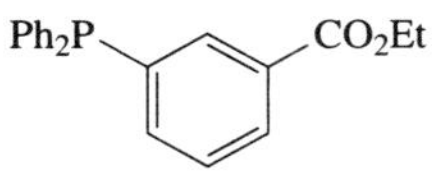	157–160	65M1021
10	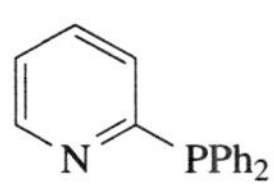	83–84	78JOC947
11	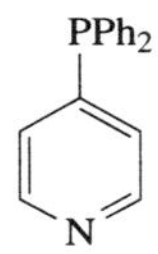	64–66	78JOC947

This method can be applied successfully to the synthesis of polyphosphines as well as monophosphines (Equation (33)) ⟨64JA2299⟩.

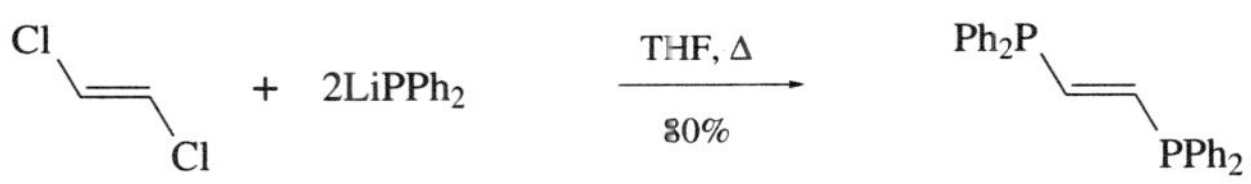

(33)

Sodium diphenylphosphide, as well as the lithium derivative, has been similarly utilized for the synthesis of tertiary phosphines (Scheme 6) ⟨81IS178⟩.

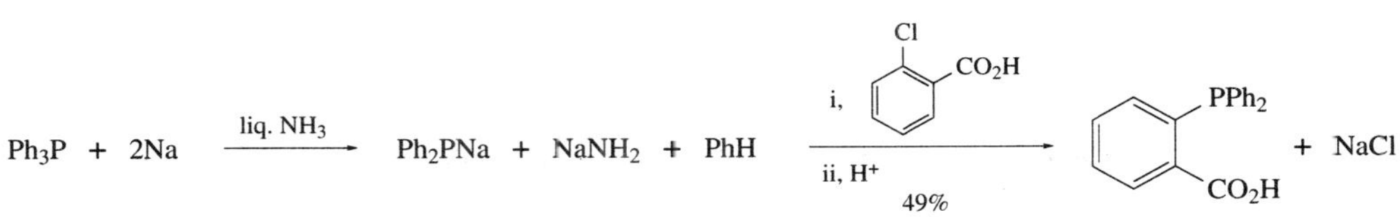

Scheme 6

(iii) Reduction of tertiary arylphosphine oxides (or sulfides) (Method C) (Table 8)

Tertiary phosphines have sometimes been prepared by reduction of the corresponding tertiary phosphine oxides or phosphine sulfides with a variety of reducing agents such as silanes ⟨64CB1988, 69CB1883⟩, $LiAlH_4$ ⟨58CB1583⟩, NaH ⟨58CB1583, 64CB1988⟩, LiH ⟨58CB1583⟩, CaH_2 ⟨64CB1988⟩, Na ⟨59CB2088⟩, and Raney nickel ⟨58CB1583⟩. Of these reducing agents, silanes and $LiAlH_4$ have normally been used for reduction of tertiary phosphine oxides, since drastic reaction conditions are necessary with other reducing agents (Equations (34)–(39)).

Table 7 Preparation of vinyl- and ethynyldiphosphines.

Entry	*Substrate*		*Phosphine*	*m.p.* (°C)	*Ref.*
1	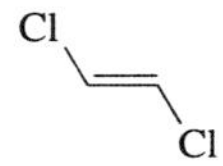	Ph_2PLi	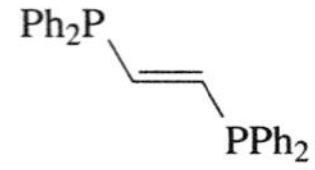	125–126	64JA2299
2	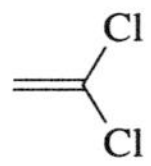	Ph_2PLi	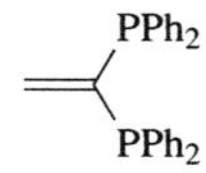	114	82JCS(D)1915
3	BrMg—≡—MgBr	Ph_2PCl	Ph_2P—≡—PPh_2	86	66BSF1002
4	BrMg—≡—≡—MgBr	Ph_2PCl	Ph_2P—≡—≡—PPh_2	105	66BSF1002
5	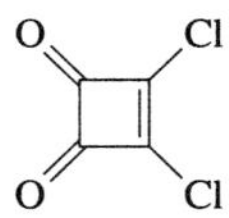	TMS-PPh_2	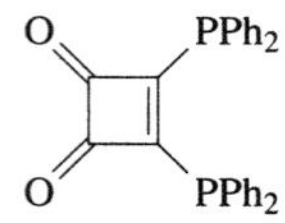	no data	73CB177

$$MePh(4\text{-}MeOC_6H_4)PO \xrightarrow[60\text{–}70\%]{LiAlH_4,\ Bu^nOBu^n,\ heat} MePh(4\text{-}MeOC_6H_4)P \quad (34)$$

$$PhP(O)(3\text{-}CNC_6H_4)_2 \xrightarrow[99\%]{HSiCl_3,\ PhH,\ reflux} PhP(3\text{-}CNC_6H_4)_2 \quad (35)$$

$$Ph_3PO \xrightarrow[43\%]{NaH,\ 350\ °C} Ph_3P \quad (36)$$

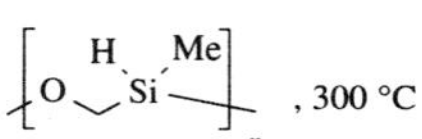

$$(2\text{-}MeC_6H_4)_3PO \xrightarrow[98\%]{[\text{polymer}]_n,\ 300\ °C} (2\text{-}MeC_6H_4)_3P \quad (37)$$

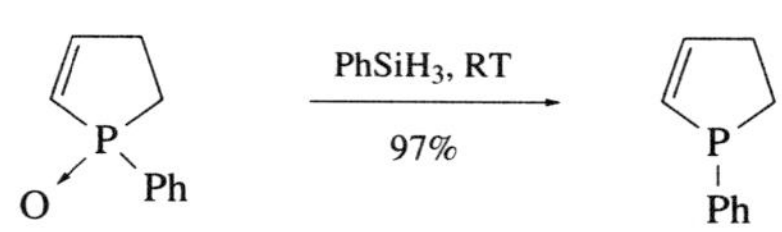

(38)

$$Ph_3PS \xrightarrow[90\%]{Na,\ naphthalene,\ THF,\ RT} Ph_3P \quad (39)$$

Table 8 Preparation of tertiary phosphines via Method C.

Entry	*Phosphine oxide*	*Phosphine*	*Reducing agent*	*m.p.* (°C)	*Ref.*
1	$(2\text{-}MeC_6H_4)_3PO$	$(2\text{-}MeC_6H_4)_3P$	MPS[a]	125	64CB1988
2	Ph_3PO	Ph_3P	MPS $PhSiH_3$ Ph_2SiH_2 Ph_3SiH	79	64CB1988
3	$(2\text{-}MeOC_6H_4)_3PO$	$(2\text{-}MeOC_6H_4)_3P$	MPS	204	64CB1988
4	$(3\text{-}NH_2C_6H_4)_3PO$	$(3\text{-}NH_2C_6H_4)_3P$	MPS	202	64CB1988
5	$(3\text{-}CNC_6H_4)_3PO$	$(3\text{-}CNC_6H_4)_3P$	Cl_3SiH	154	69CB1883
6	$(4\text{-}CNC_6H_4)_3PO$	$(4\text{-}CNC_6H_4)_3P$	Cl_3SiH	186	69CB1883
7	$(3\text{-}MeO_2CC_6H_4)_3PO$	$(3\text{-}HO_2CC_6H_4)_3P$	Cl_3SiH	276	69CB1883
8	$(4\text{-}MeO_2CC_6H_4)_3PO$	$(4\text{-}HO_2CC_6H_4)_3P$	Cl_3SiH	270	69CB1883
9	$Ph_2(2\text{-}NH_2C_6H_4)PO$	$Ph_2(2\text{-}NH_2C_6H_4)P$	MPS	82.5	78IC880

[a] MPS = methylpolysiloxane.

Since functional groups on phosphorus substituents do not interfere in reduction with silanes, this phosphine synthetic method, via reduction of tertiary phosphine oxides with silanes, has most often been applied to the synthesis of chiral phosphine ligands (Equations (40) and (41)) ⟨79AG655, 91TA667, 92CL547⟩.

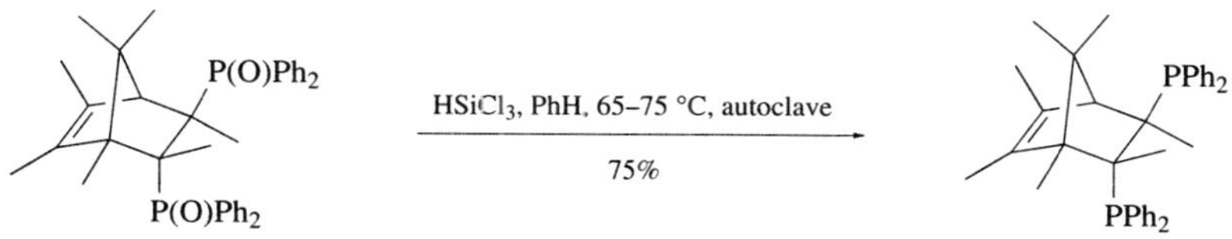

(40)

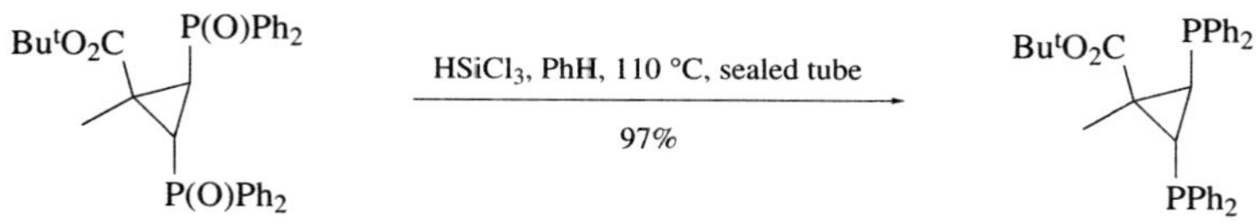

(41)

(iv) Miscellaneous methods

Although the methods are limited in scope, tertiary aromatic phosphines can also be prepared by the following reactions.

In a number of cases, aromatic compounds do not react with phosphorus trichloride to give tertiary aromatic phosphines. However, benzenes bearing strong electron-donating groups, such as *N,N*-dimethylaniline ⟨34LA(509)138⟩ and 1,3,5-trimethoxybenzene (Equation (42)) ⟨64ZOB1446⟩ or 1,3-dimethoxybenzene ⟨63ZOB3050⟩, react with phosphorus trichloride in the presence of pyridine and zinc chloride to provide tertiary aromatic phosphines in good yields.

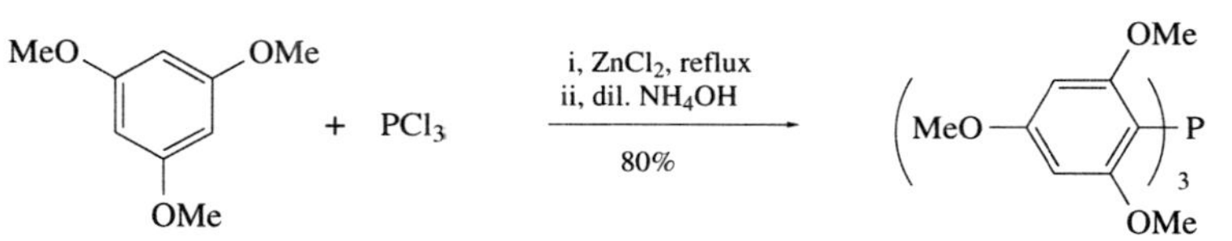

(42)

As an alternative approach to aromatic phosphines, heating alkali phosphides with arenesulfonic acid salts in ethers leads to aromatic phosphines in moderate to good yields (Equation (43)) ⟨65CB2431⟩.

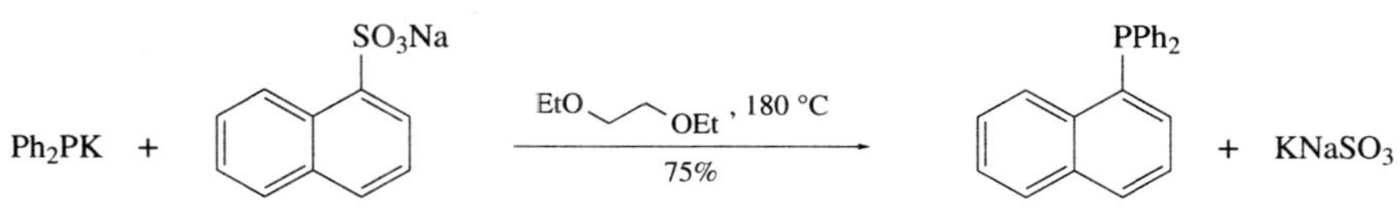

(43)

Michael addition of primary or secondary arylphosphines to vinyl cyanide proceeds without a catalyst, or with an acid or a basic catalyst, to furnish 2-cyanoethylphosphines (Equation (44)) ⟨52JCS4453⟩.

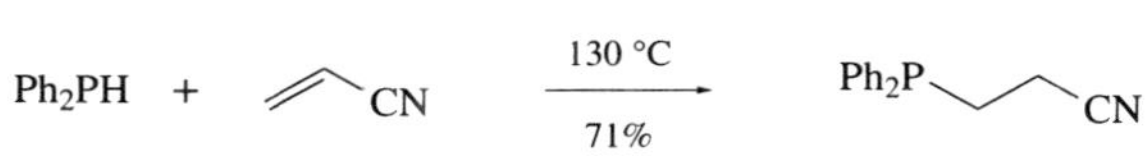

(44)

Reduction of amidophosphonium salts, prepared from phosphinous acid amides on alkylation, with $LiAlH_4$ readily causes P—N bond cleavage, converting them into tertiary phosphines (Equation (45)) ⟨80PS225⟩. This synthetic method can be a useful tool for the synthesis of optically active tertiary phosphines ⟨80PS(8)225⟩.

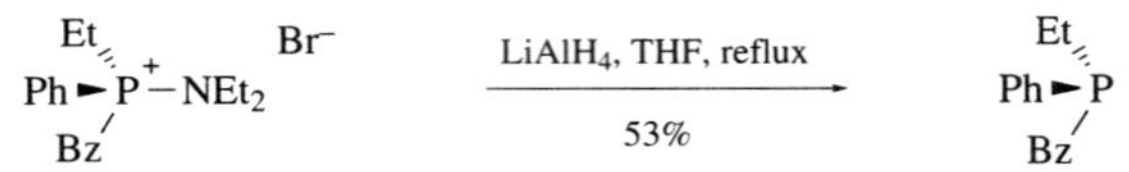

(45)

2.16.1.2 Phosphoranes

2.16.1.2.1 Pentaorganophosphoranes

Pentaorganophosphoranes are prepared by the reaction of tetraorganophosphonium salts with alkyl- or aryllithiums (Equations (46) and (47)) ⟨49LA(562)187, 65CB576⟩.

$$Ph_4P^+ I^- + PhLi \xrightarrow[60\%]{\text{ether, RT}} Ph_5P \quad (46)$$

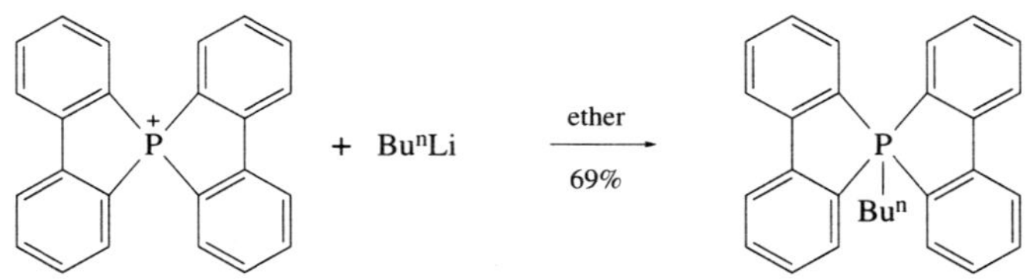

(47)

The reaction of an aminotriphenylphosphonium salt with two equivalents of phenyllithium also leads to pentaphenylphosphorane (Equation (48)) ⟨64CB741⟩. This phosphorane synthetic method, via aminophosphonium salts, can be applied to the synthesis of triphenylbiphenylenephosphorane (Equation (49)) ⟨64CB741⟩.

$$Ph_3\overset{+}{P}-N(Ph)(Me)\ I^- + 2PhLi \xrightarrow[84\%]{\text{ether, RT}} Ph_5P + Li-N(Ph)(Me) \quad (48)$$

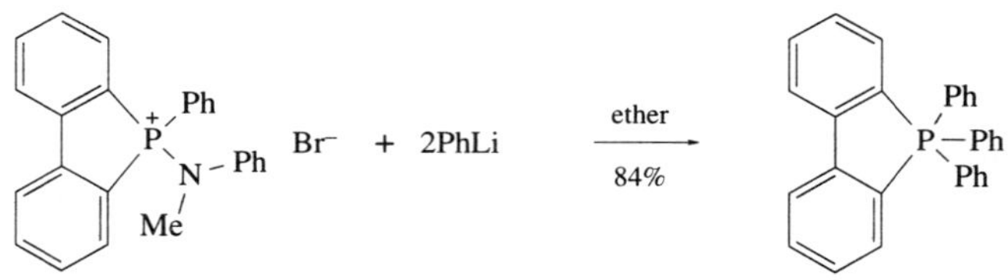

(49)

2.16.1.2.2 Dialkoxytriorgano- and alkoxytetraorganophosphoranes

Of phosphoranes, dialkoxytriorganophosphoranes have been well studied, owing to their structural interest and synthetic utility. Dialkoxyphosphoranes are readily prepared by oxidative addition of alkyl peroxides to triphenylphosphine (Equation (50)) ⟨69JA5243, 85JA5210⟩. However, the potential hazards associated with the preparation and use of peroxides serve to restrain extensive use of this methodology. Circumvention of this concern is achieved by using the synthetic method from dibromotriphenylphosphorane and alkoxides (Equations (51) and (52)) ⟨80JOC5052, 86JOC5490⟩.

$$Ph_3P + Et_2O_2 \xrightarrow{CH_2Cl_2 \text{ or toluene}} Ph_3P(OEt)_2 \quad (50)$$

$$Ph_3PBr_2 + 2CF_3CH_2O^-Na^+ \xrightarrow[95\%]{\text{ether, } CH_2Cl_2} Ph_3P(OCH_2CF_3)_2 \quad (51)$$

$$Ph_3PBr_2 + 2Bu^tCH_2O^-Li^+ \xrightarrow[48\%]{CH_2Cl_2/toluene} Ph_3P(OCH_2Bu^t)_2 \quad (52)$$

Diphenoxyphosphoranes are obtained by addition of two equivalents of phenols to a solution of triphenylphosphine and diethyl azodicarboxylate in chloroform (Equation (53)) ⟨82JA6876⟩.

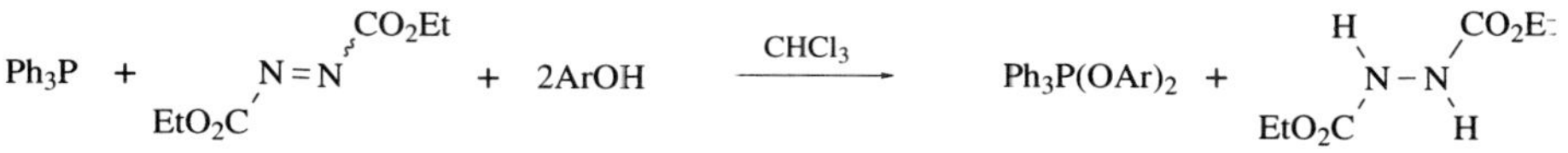
(53)

Although these phosphoranes are stable in solution, their isolation in pure form is difficult owing to their being hygroscopic and unstable to heat and acid. In contrast to acyclic dialkoxytriarylphosphoranes, polycyclic triaryldioxyphosphoranes of extreme stability have been synthesized (Equation (54)) ⟨71JA5236, 78CB13⟩.

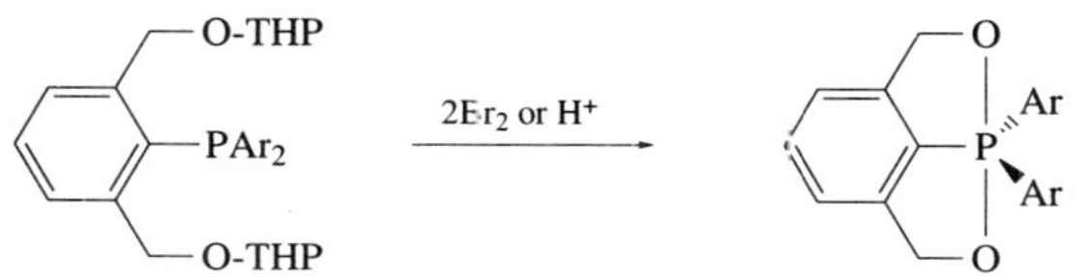
(54)

1,2-Oxaphosphetanes, members of the alkoxytetraorganophosphoranes, are usually unstable for isolation, but they are known to exist at low temperatures as intermediates of the Wittig reaction. In analogy with polycyclic triaryldioxyphosphoranes, as mentioned above, some isolable 1,2-oxaphosphetanes have been synthesized (Equations (55)–(57)) ⟨79AG945, 80AG740, 92JA4008⟩.

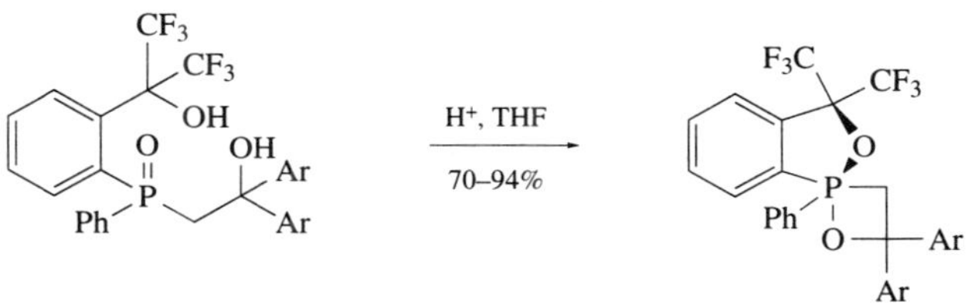
(55)

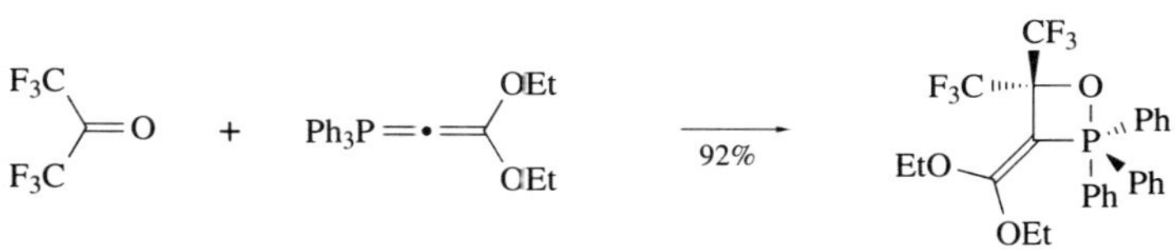
(56)

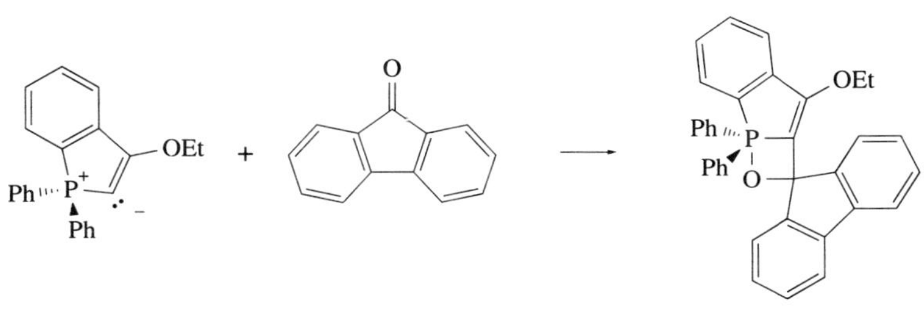
(57)

2,2,2-Triphenyl-1,2-oxaphospholane, obtained by treatment of a (3-hydroxypropyl)triphenylphosphonium salt with base, is rather unstable and hygroscopic (Equation (58)) ⟨76HCA695, 81T2855⟩.

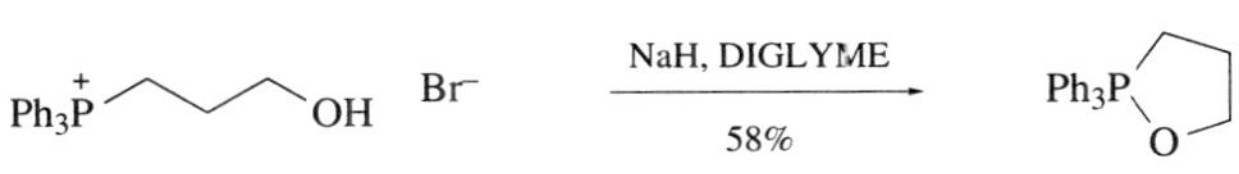
(58)

Pentaphenylphosphorane on treatment with phenol undergoes replacement of a single phenyl group with a phenoxy group, leading to phenoxytetraphenylphosphorane, which is relatively stable in air (Equation (59)) ⟨69IZV2234⟩.

$$Ph_5P + PhOH \xrightarrow[53\%]{\text{pyridine, RT}} Ph_4POPh + PhH \qquad (59)$$

2.16.2 PHOSPHORUS HALIDES

2.16.2.1 Arylphosphonous Dihalides (Aryldihalophosphines) ($ArPX_2$) (Table 9)

Phenylphosphonous dichloride (phenyldichlorophosphine), an important precursor in the synthesis of a large number of phenylphosphorus compounds, is readily synthesized by the Friedel–Crafts type of reaction of phosphorus trichloride with benzene in the presence of aluminum chloride, followed by removal of $AlCl_3$ as the insoluble $AlCl_3 \cdot P(O)Cl_3$ complex by addition of phosphorus oxychloride (Scheme 7) ⟨63OSC(4)784⟩.

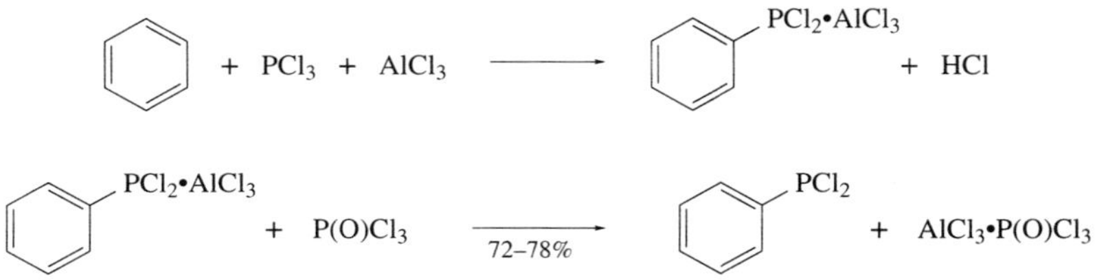

Scheme 7

Phenylphosphonous dibromide (phenyldibromophosphine) is prepared by similar Friedel–Crafts dibromophosphination of benzene with phosphorus tribromide in the presence of aluminum bromide, but the yield is not satisfactory (Equation (60)) ⟨66JCS(B)1162⟩.

$$PhH + PBr_3 \xrightarrow[17\%]{AlBr_3,\ \text{reflux}} PhPBr_2 \qquad (60)$$

The easiest and most efficient method for the synthesis of phenylphosphonous dibromide is via reorganization of halogen atoms between $PhPCl_2$ and PBr_3 (Equation (61)) ⟨83PS(15)93⟩, since phenylphosphonous dichloride is readily obtained.

$$PhPCl_2 + PBr_3 \xrightarrow[78\text{–}86\%]{100\text{–}200\ ^\circ C} PhPBr_2 + PCl_2Br \qquad (61)$$

A similar halogen exchange reaction between $PhPCl_2$ and sodium fluoride has been applied successfully to the synthesis of phenylphosphonous difluoride (Equation (62)) ⟨65CB552⟩.

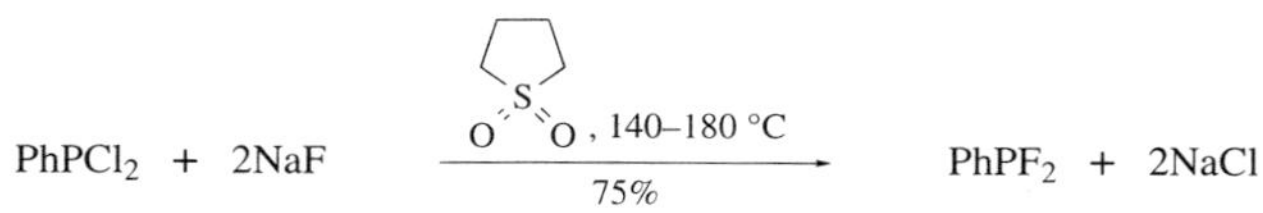

(62)

With alkylbenzenes instead of benzene in the Friedel–Crafts reaction of PCl_3, a mixture of *o*-, *m*-, and *p*-alkylbenzenephosphonous dichlorides is obtained (Equation (63)) ⟨61JOC3547, 65M1936⟩.

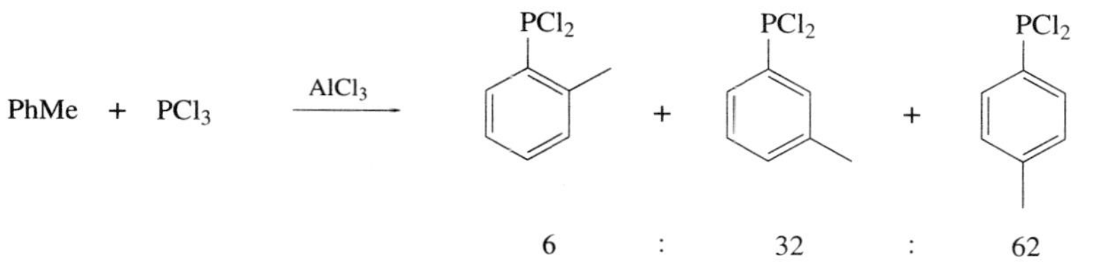 (63)

The synthesis of pure functionalized arylphosphonous dichlorides free from contamination of isomers has been achieved by the reaction of tetraethylphosphorodiamidous chloride (bis(diethylamino)chlorophosphine) with functionalized aryl Grignard reagents and subsequent treatment with dry HCl gas (Scheme 8) ⟨65M1936⟩.

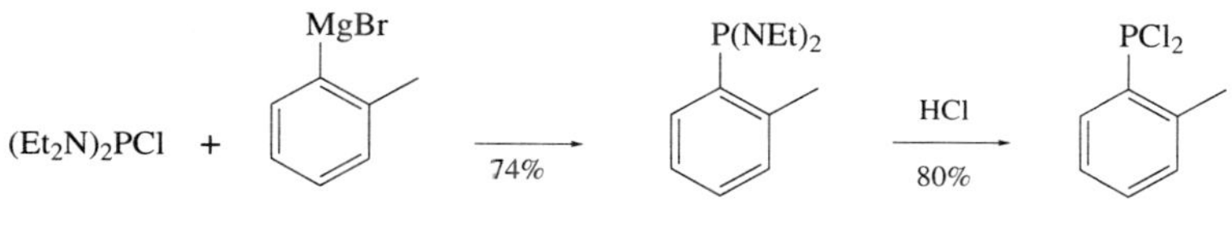

Scheme 8

Functionalized arylphosphonous dichlorides are produced by the magnesium reduction of aryl diazonium fluoroborate–phosphorus trichloride reaction products, although their yields are rather low (Scheme 9) ⟨61JA4124, 63JOC3315⟩.

$$ArN_2^+BF_4^- + PCl_3 \xrightarrow{AcOEt,\ CuBr} [Ar\overset{+}{P}Cl_3\ BF_4^-] \xrightarrow[21\text{–}33\%]{Mg} ArPCl_2$$

Scheme 9

Alkenylphosphonous dichlorides are easily prepared by the reduction of alkyltetrachlorophosphoranes, which are the reaction products of alkenes with phosphorus pentachloride, with elemental phosphorus in the presence of iodine ⟨55JA929⟩ or with phosphine followed by treatment with a tertiary amine (Scheme 10) ⟨78ZOB319⟩.

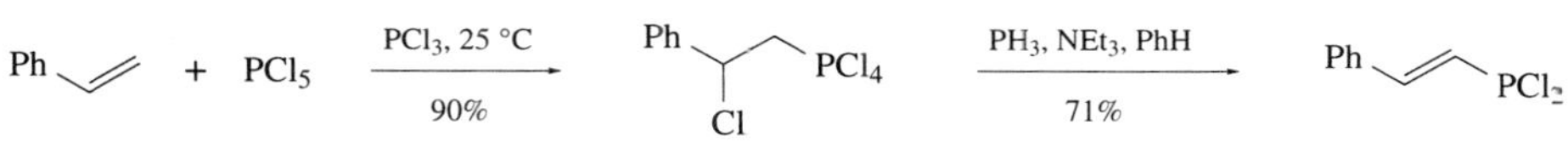

Scheme 10

Table 9 Synthesis of aryl- and alkenylphosphonous dihalides ($ArPX_2$).

Entry	*Ar*	*X*	*B.p.* (°C)(mmHg)	*Ref.*
1	Ph	Cl	68–70(1)	63OSC(4)784
2	2-MeC_6H_4	Cl	108–109(10)	65M1936
3	3-MeC_6H_4	Cl	102–104(9)	65M1936
4	4-MeC_6H_4	Cl	106–109(11)	65M1936
5	2-ClC_6H_4	Cl	113–117(9)	65M1936
6	3-ClC_6H_4	Cl	113–116(9)	65M1936
7	4-ClC_6H_4	Cl	115–118(9)	65M1936
8	4-FC_6H_4	Cl	82–83(10)	65M1936
9	Ph	Br	126–128(11)	66JCS(B)1162
10	Ph	F	31(11)	65CB552
11	PhCH=CH	Cl	135–138(8)	55JA929
12	Me_2C=CH	Cl	98–104(100)	55JA929
13	C_5H_{11}(Me)C=CH	Cl	70–72(3)	55JA929

2.16.2.2 Diarylphosphinous Halides (Diarylhalophosphines) (Ar_2PX) (Table 10)

The most convenient method for synthesizing diarylphosphinous chlorides (diarylchlorophosphines) is via disproportionation of the readily accessible arylphosphonous dichlorides (aryldichlorophosphines) in the presence of catalytic amounts of aluminum chloride (Equation (64)) ⟨61CB2122, 61CI(L)24⟩.

$$2PhPCl_2 \xrightarrow[67\text{–}70\%]{AlCl_3,\ reflux} Ph_2PCl + PCl_3 \qquad (64)$$

Synthesis of phosphinous chlorides via the reaction of phosphorus trichloride with Grignard reagents has not been popular, since isolation of the products of the partial replacement of chlorine atoms by aryl groups is impossible. However, the reactions of PCl_3 with *ortho*-substituted arylmagnesium bromides stop at the stage of the formation of diarylphosphinous chlorides at low temperatures (Equation (65)) ⟨78ZOB232⟩.

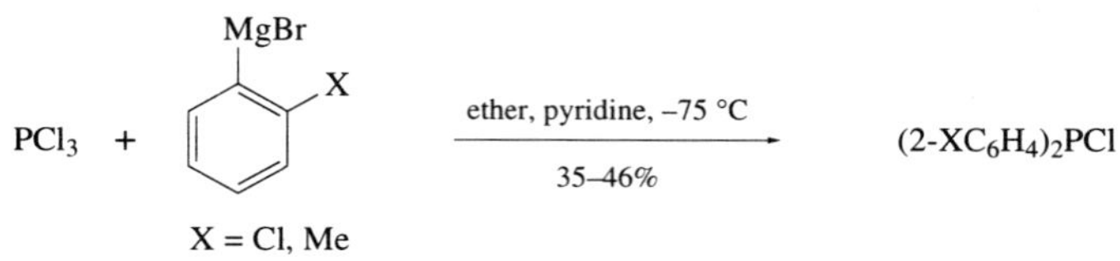

(65)

The limitations of this Grignard method can be solved by using diethylphosphoramidous dichloride ((*N*,*N*-diethylamino)dichlorophosphine) as a synthon equivalent of phosphorus trichloride. *P*,*P*-Diaryl-*N*,*N*-diethylphosphinous amides, prepared by the reaction of (*N*,*N*-diethylamino) dichlorophosphine with Grignard reagents, are treated with dry hydrogen chloride to convert them into diarylphosphinous chlorides in good yields (Scheme 11) ⟨78JA7304⟩.

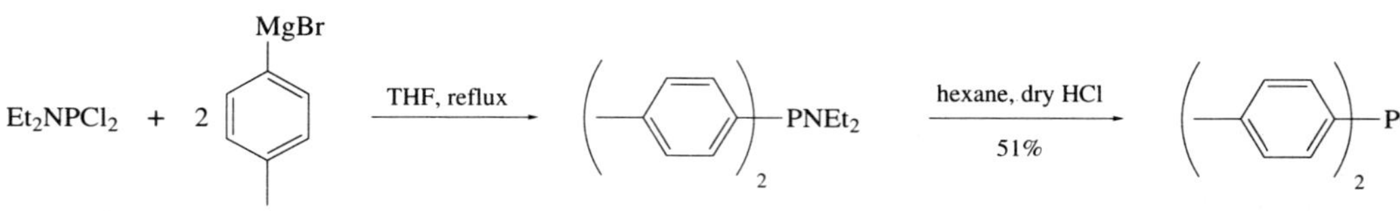

Scheme 11

From the industrial perspective, the one-step synthesis of arylphosphonous chlorides and diarylphosphinous chlorides via the reaction of elemental white phosphorus with aryl halides, in the presence of Lewis acid catalysts, is useful as a simple synthetic method, notwithstanding the moderate yields (Equation (66)) ⟨75JOC3586⟩.

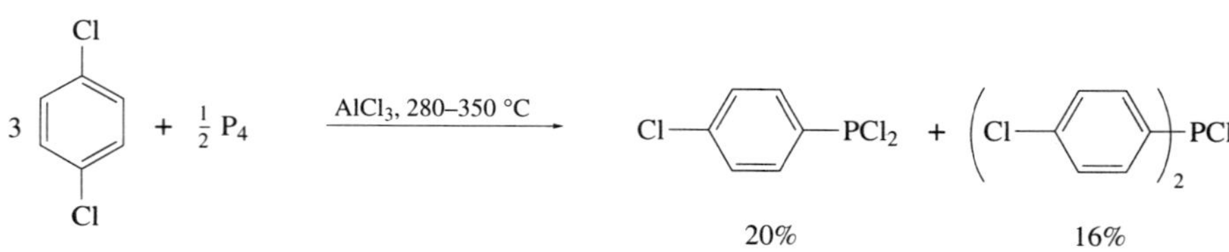

(66)

In analogy with the reaction of aryldiazonium fluoroborate with PCl_3 to yield arylphosphonous dichlorides (see Section 2.16.2.1), unsymmetrical diarylphosphinous chlorides are prepared by the reduction of the reaction products of substituted aryldiazonium fluoroborate with arylphosphonous dichlorides using aluminum at 30–50 °C (Scheme 12) ⟨63JOC3315⟩. The yields lie in the 30–50% range, but nevertheless, the synthesis is useful as a means of obtaining unsymmetrical diarylphosphinous chlorides.

$$Ar^1\overset{+}{N}_2\,BF_4^- + Ar^2PCl_2 \xrightarrow{AcOPr^i,\ CuBr} [Ar^1Ar^2\overset{+}{P}Cl_2\,BF_4^-] \xrightarrow[30\text{–}50\%]{Al} Ar^1Ar^2PCl$$

Scheme 12

Treatment of diarylphosphinous chlorides with iodotrimethylsilane leads to the formation of the corresponding diarylphosphinous iodides in high yields (Equation (67)) ⟨80S823⟩.

$$Ph_2PCl + TMS\text{-}I \xrightarrow[77\%]{PhH,\ 20\ °C} Ph_2PI + TMS\text{-}Cl \qquad (67)$$

Diphenylphosphinous fluoride (fluorodiphenylphosphine) is prepared by the reaction of benzoyl fluoride with diphenyl-*N*-propylphosphinous amide or by the action of sodium fluoride on diphenylphosphinous chloride. However, the fluorophosphine readily disproportionates to give trifluorodiphenylphosphorane and tetraphenyldiphosphine (Scheme 13) ⟨70JCS(C)878⟩.

$$Ph_2PNHPr^n + PhCOF \longrightarrow Ph_2PF + PhCONHPr^n$$

$$3Ph_2PF \longrightarrow Ph_2PF_3 + Ph_2P-PPh_2$$

Scheme 13

Diarylphosphinous bromides are easily prepared via rearrangement of halogen atoms between diarylphosphinous chloride and phosphorus tribromide in a similar manner as for arylphosphonous dibromides (Equation (68)) ⟨83PS(15)93⟩.

$$3Ph_2PCl + PBr_3 \xrightarrow[87\%]{220\ °C} 3Ph_2PBr + PCl_3 \qquad (68)$$

Table 10 Synthesis of arylphosphinous halides (Ar_2PX).

Entry	*Ar*	*X*	*b.p.* (°C)(mmHg) *and/or m.p.* (°C)	*Ref.*
1	Ph	Cl	140(0.6)	61CB2122, 61CI(L)24
2	2-MeC_6H_4	Cl	179–183(7) 36	78ZOB232
3	2-MeC_6H_4	Cl	120–122(0.03)	78JA7304
4	2-ClC_6H_4	Cl	208–210(7) 38	78ZOB232
5	4-MeC_6H_4	Cl	125–127(0.03)	78JA7304
6	1-Naphthyl	Cl	130–132	75JOC3586
7	4-ClC_6H_4	Cl	67–69(0.3)	75JOC3586
8	3-Cl-4-MeC_6H_3	Cl	176–182(1.1)	75JOC3586
9	Ph	I	120–123(0.05)	80S823

2.16.2.3 Arylhalophosphoranes (Table 11)

In general, dihalotriphenylphosphoranes (Ph_3PX_2: X = Cl, Br, I) are prepared by adding equimolar halogen to Ph_3P in acetonitrile (Equation (69)) ⟨66JCS(A)520⟩.

$$Ph_3P + X_2 \xrightarrow{MeCN} Ph_3PX_2 \qquad (69)$$

$$X = Cl, Br, I$$

For the synthesis of dichlorotriphenylphosphorane, the use of hexachloroethane as a chlorinating agent is more practical for purposes of laboratory handling (Equation (70)) ⟨77CB2382, B-79MI 216-01, B-79MI 216-02⟩.

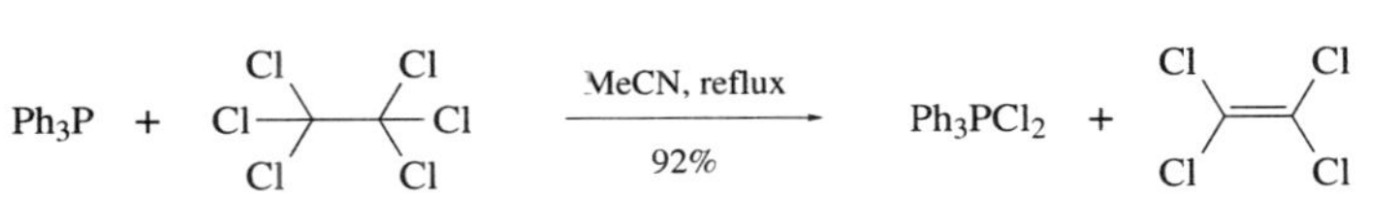

(70)

Difluorotriphenylphosphorane is readily synthesized by the fluorination of triphenylphosphine with tetrafluorohydrazine (Equation (71)) ⟨65IC764⟩ or sulfur tetrafluoride (Equation (72)) ⟨60JA6176, 68CPB1009⟩.

$$2Ph_3P + N_2F_4 \xrightarrow[45\%]{C_6H_5Cl,\ -196\ °C\ \text{to RT}} 2Ph_3PF_2 + N_2 \quad (71)$$

$$2Ph_3P + SF_4 \xrightarrow[69\%]{PhH,\ 50\ °C\ \text{to}\ 150\ °C} 2Ph_3PF_2 + S \quad (72)$$

Trifluorodiphenylphosphoranes are, analogously to difluorotriphenylphosphoranes, obtained by the fluorination of diphenylphosphorus compounds as follows: diphenylphosphine with tetrafluorohydrazine ⟨65IC764⟩, diphenylphosphinic acid with sulfur tetrafluoride (Equation (73)) ⟨60JA6176⟩, or diphenylphosphinous chloride with arsenic(III) fluoride (Equation (74)) ⟨67IS69⟩, etc.

$$Ph_2P(O)H + SF_4 \xrightarrow[42\%]{50\ °C\ \text{to}\ 120\ °C} Ph_2PF_3 + SO + HF \quad (73)$$

$$3Ph_2PCl + 3AsF_3 \xrightarrow[65\text{–}72\%]{\sim 50\ °C} 3Ph_2PF_3 + 2As + AsCl_3 \quad (74)$$

A wide variety of methods for the preparation of trichlorodiphenylphosphorane has been reported; for example, the action of dry chlorine gas on diphenylphosphinodithioic acid (Equation (75)) ⟨81CB3485⟩ or on diphenylphosphinous chloride (Equation (76)) ⟨64IC428⟩ in carbon tetrachloride, or thermolysis of trichloroacetyldiphenylphosphine at 0–40 °C ⟨82CB2181⟩ give trichlorodiphenylphosphorane in excellent yields.

$$Ph_2P(S)SH + 3Cl_2 \xrightarrow[93\%]{CCl_4,\ RT} Ph_2PCl_3 + HCl + S_2Cl_2 \quad (75)$$

$$Ph_2PCl + Cl_2 \xrightarrow[97\%]{CCl_4,\ -196\ °C\ \text{to}\ 20\ °C} Ph_2PCl_3 \quad (76)$$

In analogy with the synthesis of trichlorodiphenylphosphorane, tribromo-, or triiododiphenylphosphorane can be obtained by reacting Ph_2PBr or Ph_2PI with an equimolar amount of bromine or iodine in a suitable solvent (Equations (77) and (78)) ⟨76ZOB252, 82POL123⟩.

$$Ph_2PBr + Br_2 \xrightarrow{CCl_4,\ RT} Ph_2PBr_3 \quad (77)$$

$$Ph_2PI + I_2 \xrightarrow[80\%]{PhH,\ \text{reflux}} Ph_2PI_3 \quad (78)$$

The preparation of aryltetrafluorophosphoranes is achieved by fluorination of arylphosphonic acids or arylphosphonous dichlorides with sulfur(IV) fluoride (Equation (79)) ⟨60JA6176⟩, antimony(III) fluoride (Equation (80)) ⟨67IS69, 68JA5236⟩, or arsenic(III) fluoride ⟨64IC410⟩.

$$PhPO(OH)_2 + 3SF_4 \xrightarrow[58\%]{100\ °C\ \text{to}\ 150\ °C} PhPF_4 + 3SOF_2 + 2HF \quad (79)$$

$$3PhPCl_2 + 4SbF_3 \xrightarrow[82-87\%]{} 3PhPF_4 + 2Sb + 2SbCl_3 \quad (80)$$

Tetrachlorophenylphosphorane is easily synthesized by the action of dry chlorine gas on phenylphosphonous dichloride in carbon tetrachloride in a sealed stainless steel tube (Equation (81)) ⟨64IC428⟩.

$$PhPCl_2 + Cl_2 \xrightarrow[95\%]{CCl_4,\ -196\ °C\ \text{to}\ 25\ °C} PhPCl_4 \quad (81)$$

With excess bromine, phenylphosphonous diiodide reacts with the replacement of iodine by bromine to give tetrabromophenylphosphorane (Equation (82)) ⟨79ZOB546⟩.

$$PhPI_2 + 3Br_2 \xrightarrow[80\text{–}90\%]{PhH, RT} PhPBr_4 \quad (82)$$

Various aryliodophosphoranes, analogous to diarylphosphinous iodides, are easily prepared by the reaction of the corresponding arylchlorophosphoranes with iodotrimethylsilane in high yields (Equation (83)) ⟨80S823⟩.

$$PhPCl_4 + 4TMS\text{-}I \xrightarrow[91\%]{PhH,\ 20\ ^\circ C} PhPI_4 \quad (83)$$

Table 11 Preparation of arylhalophosphoranes ($Ar_{5-n}PX_n$).

Entry	$Ar_{5-n}PX_n$	*m.p.* (°C)(mmHg)	*Ref.*
1	Ph_3PF_2	160–162	65IC765
2	Ph_3PCl_2	207–210	77CB2382
3	Ph_3PBr_2	No data	73OSC(5)142, 73OSC(5)249
4	Ph_3PI_2	172–175 (dec.)[a]	80S823
5	Ph_2PF_3	92–93(0.4)	67IS69
6	Ph_2PCl_3	199–201	81CB3485
7	Ph_2PBr_3	No data	82POL123
8	Ph_2PI_3	132–135	76ZOB252
9	$PhPF_4$	133–134(760) 58(60)	67IS64
10	$PhPCl_4$	75–76	64IC428
11	$PhPBr_4$	134–136	79ZOB546
12	$PhPI_4$	90–91	80S823

[a] Decomposes.

2.16.3 COMPOUNDS WITH A P—O BOND

2.16.3.1 Arylphosphonous ($ArP(OH)_2$) and Unsaturated Phosphonous Acids, and their Derivatives (Table 12)

Since arylphosphonous acids (**1**) normally exist as the phosphoryl tautomers, arylphosphinic acids (**2**) (Scheme 14), free acids containing a single carbon–phosphorus bond are termed phosphinic acids in this chapter.

$$\underset{\substack{\textbf{(1)}\\ \text{Arylphosphonous acids}}}{ArP(OH)_2} \rightleftharpoons \underset{\substack{\textbf{(2)}\\ \text{Arylphosphinic acids}}}{ArPH(O)OH}$$

Scheme 14

As a wide variety of arylphosphonous dichlorides is readily accessible, described in Section 2.16.2.1, the synthesis of arylphosphinic acids and their derivatives, by the use of the corresponding phosphonous dichlorides is most convenient. Hydrolysis of arylphosphonous or alkenylphosphonous dichlorides with aqueous hydrochloric acid (or ethanol–water) leads to arylphosphinic acids (Equation (84)) ⟨50JA4291, 61JOC850⟩ or alkenylphosphinic acids (Equation (85)) ⟨55JA929⟩.

$$PhPCl_2 \xrightarrow{10\%\ HCl\ \text{or}\ EtOH,\ H_2O} PhP(OH)_2 \quad (84)$$

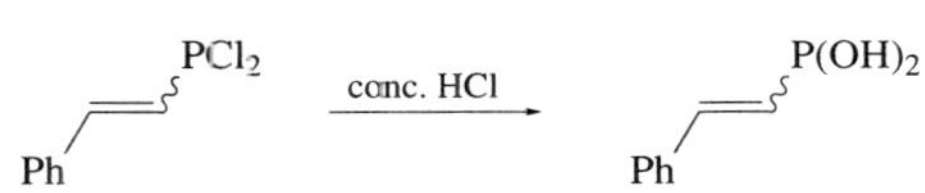

(85)

The reaction of the phosphonous dichlorides with two equivalents of alcohol or one equivalent of glycol, in the presence of tertiary amines, under mild conditions gives dialkyl (Equations (86) and (89)) ⟨60JA423, 67JOC2172, 78ZOB319⟩ or cyclic phosphonites (Equations (87) and (88)) ⟨64JOC2572, 70M776, 83CC857⟩ in good yields.

$$PhPCl_2 + 2MeOH \xrightarrow[68\%]{\text{hexane, pyridine, 0 °C to 10 °C}} PhP(OMe)_2 \qquad (86)$$

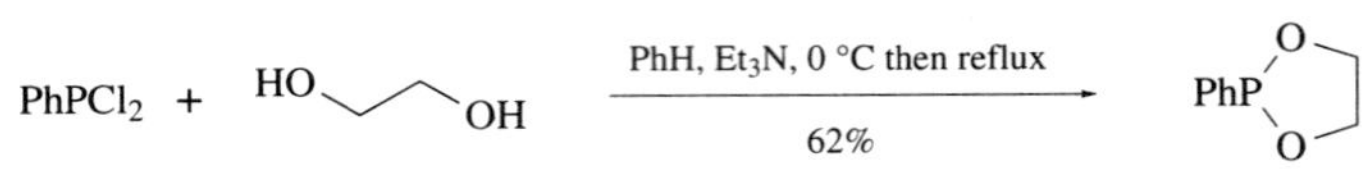

(87)

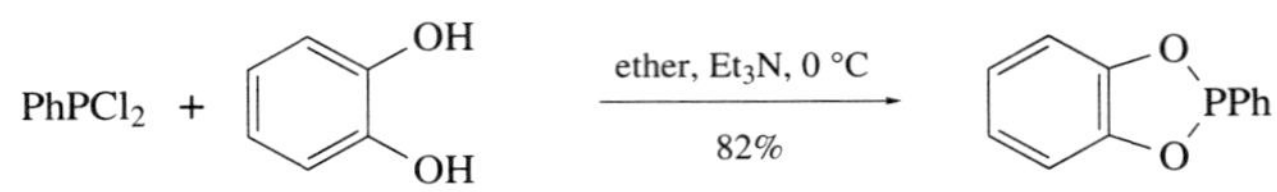

(88)

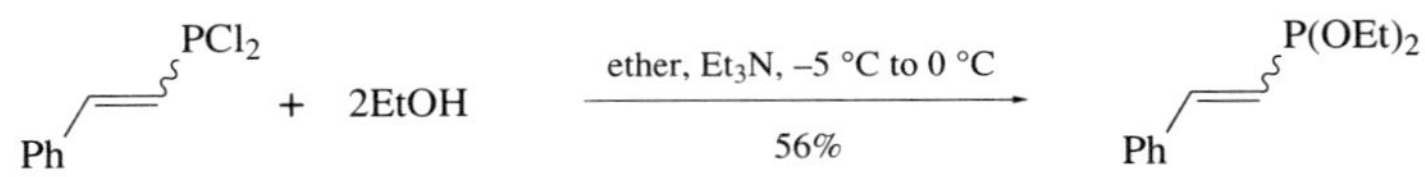

(89)

Alternatively, the dialkyl arylphosphonites are also prepared in satisfactory yields by the action of arylmagnesium halides on dialkyl phosphorochloridites at low temperature (Equation (90)) ⟨50MI 216-01, 57DOK(117)996⟩.

$$PhMgBr + ClP(OR)_2 \xrightarrow[54\text{–}75\%]{\text{ether, –60 °C to –65 °C}} PhP(OR)_2 \qquad (90)$$

Dialkyl vinyl- and ethynylphosphonites are similarly prepared in good yields, via the reaction of the corresponding Grignard reagents with dialkyl phosphorochloridites (Equations (91) and (92)) ⟨62ZOB3351, 73ZOB2178⟩.

(91)

(92)

The preparation of esters of unsaturated phosphonous acids via the reaction of Grignard reagents with dialkyl phosphorochloridites, is more frequently used than that via the reaction of phosphonous dichlorides with alcohols, since vinyl- or ethynylmagnesium bromides are easily prepared or commercially available.

When the reactions of phosphonous chlorides with alcohols are performed in the absence of tertiary amines, monoalkyl phosphonites are formed (Equations (93)–(95)) ⟨50JA4292, 64JCS4750, 78ZOB319⟩.

$$PhPCl_2 + Bu^nOH \xrightarrow[70\text{–}75\%]{\text{5 °C to 10 °C}} PhPH(O)OBu^n \qquad (93)$$

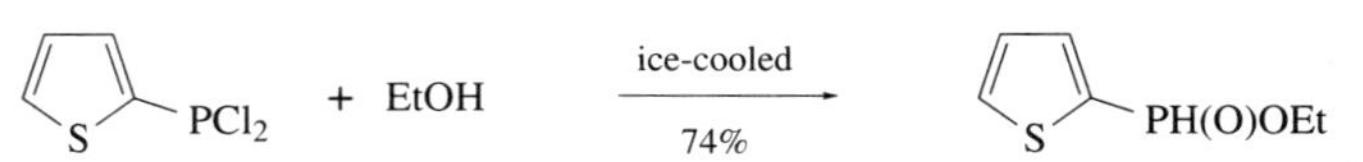

(94)

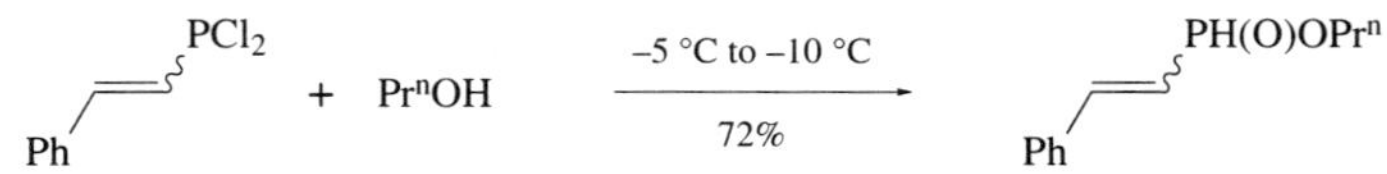

(95)

The monoalkyl phosphonites are alternatively synthesized via the reaction of the phosphinic acids produced above with diazoalkanes (Equation (96)) ⟨62JOC1012⟩.

$$p\text{-}ClC_6H_4PH(O)OH + CH_2N_2 \xrightarrow[40\text{–}50\%]{\text{ether}} p\text{-}ClC_6H_4PH(O)OMe \qquad (96)$$

Arylphosphonous acid monoester monochlorides (arylphosphonochloridous acid esters), are prepared according to Equations (97) and (98) ⟨62CB2993, 80ZOB803⟩.

Table 12 Synthesis of dialkyl arylphosphonites ($ArP(OR)_2$).

Entry	*ArP(OR)₂*	*b.p.* (°C) (mmHg)	*Ref.*
1	$PhP(OMe)_2$	98 (17)	60JA423
2	$PhP(OEt)_2$	62.5–63 (1)	57DOK996
3	$PhP(OPr^n)_2$	73–74 (1)	57DOK996
4	$PhP(OBu^n)_2$	97.5–98.5 (1)	57DOK996
5	$PhP(OPh)_2$	223–224 (10)	50MI 216-01
6	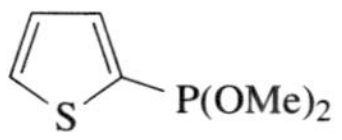	57 (0.9)	78ZOB319
7	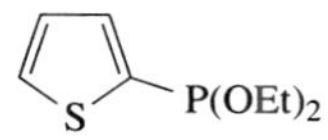	81–83 (3)	78ZOB319
8	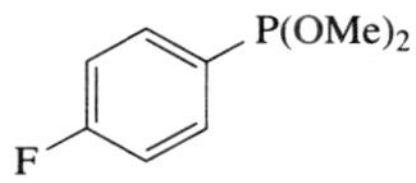	75 (2.7)	68JA5236
9	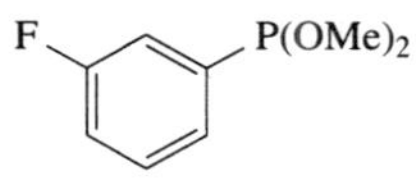	45 (0.7)	68JA5236
10		49–51.5 (2)	62ZOB3351
11	≡—$P(OEt)_2$	58 (9)	73ZOB2178
12	≡—$P(OBu^n)_2$	58.8–60 (2)	62ZOB3351

$$PhPCl_2 + PhP(OEt)_2 \xrightarrow[74\%]{Et_2O,\ \text{ice-cooled}} 2PhP(OEt)Cl \quad (97)$$

$$PhP(NEt_2)OEt \xrightarrow[31\%]{Et_2\overset{+}{N}H_2Cl^-,\ 100\ °C\ \text{to}\ 150\ °C} PhP(OEt)Cl \quad (98)$$

Their homologous phenylphosphonofluoridous acid esters can be produced by mixing phenylphosphonous difluoride and an alcohol in a 2 : 1 molar ratio at temperatures above 10°C (Equation (99)) ⟨79CJC1903⟩.

$$2PhPF_2 + ROH \xrightarrow{<10\ °C} Ph{-}P(F)(OR) + PhPF_3H \quad (99)$$

2.16.3.2 Arylphosphinous Acids (ArRPOH) and their Derivatives (Table 13)

Diarylphosphinous acids (**3**) and diarylphosphine oxides (**4**) (Scheme 15) are in tautomeric equilibrium with each other analogous to that between arylphosphonous acids and arylphosphinic acids.

$$\underset{\text{Diarylphosphinous acids}}{\underset{(\mathbf{3})}{Ar_2POH}} \rightleftharpoons \underset{\text{Diarylphosphine oxides}}{\underset{(\mathbf{4})}{Ar_2P(O)H}}$$

Scheme 15

2.16.3.2.1 Diarylphosphinous halides with water ⟨59JOC2013, 63JOC3315⟩ or alcohols (Method A)

Diarylphosphinous halides show similar chemical properties to those of arylphosphonous dihalides toward water (Equations (100) and (101)) ⟨78CB13⟩, or alcohols (Equation (102)) ⟨78ZOB232⟩.

$$Ar_2PCl \xrightarrow{H_2O,\ PhH\ or\ CCl_4} Ar_2P(O)H \qquad (100)$$

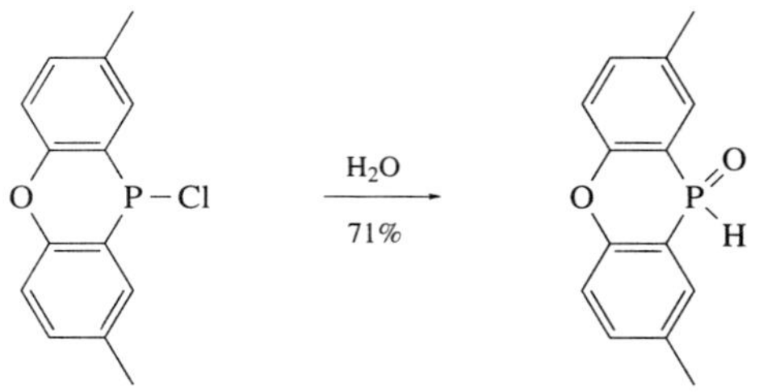

(101)

$$(o\text{-}MeC_6H_4)_2PCl + Bu^nOH \xrightarrow[62\%]{Et_3N,\ -5\ °C} (o\text{-}MeC_6H_4)_2POBu^n \qquad (102)$$

The preparation of diarylphosphine oxides via hydrolysis of diarylphosphinous halides is less useful than that of arylphosphinic acids from the corresponding hydrolysis of arylphosphonous dihalides (cf. Section 2.16.3.1) since many diarylphosphinous halides are relatively hard to access.

2.16.3.2.2 Dialkyl phosphonates with Grignard reagents (Method B)

The simplest way of preparing symmetrical secondary phosphine oxides is via the reaction of dialkyl phosphonates, such as diethyl or dibutyl phosphonates, with Grignard reagents, followed by acidic hydrolysis (Scheme 16 and Equation (103)) ⟨67T1065, 69BCJ2975, 80JCS(P2)154⟩.

$$3p\text{-}ClC_6H_4MgBr + (EtO)_2P(O)H \xrightarrow[-PhCl]{THF} (p\text{-}ClC_6H_4)_2POMgBr + 2EtOMgBr \xrightarrow[73\%]{20\%\ H_2SO_4} (p\text{-}ClC_6H_4)_2P(O)H$$

Scheme 16

$$3o\text{-}MeC_6H_4MgBr + (Bu^nO)_2P(O)H \xrightarrow{ether,\ H^+} (o\text{-}MeC_6H_4)_2P(O)H \qquad (103)$$

The advantage of this method is the inclusion of a wide variety of functional groups on the Grignard reagents, available from the corresponding halides. Organolithium compounds are occasionally used in the reaction with dialkyl phosphonates instead of Grignard reagents ⟨57CI(L)235⟩. Similar preparations of unsymmetrical secondary phosphine oxides are possible using esters of monosubstituted phosphinic acids and Grignard reagents, which give unsymmetrical secondary phosphine oxides in moderate to good yields (Equation (104)) ⟨68JA3459⟩.

$$ArPH(O)OR^2\ or\ R^1PH(O)OR^2 \xrightarrow[30\text{–}75\%]{R^1MgX\ or\ ArMgX} Ar-P(=O)(H)-R^1 \qquad (104)$$

2.16.3.3 Tertiary Arylphosphine Oxides

2.16.3.3.1 Oxidation of tertiary phosphines (Method A)

The oxidation of tertiary phosphines with a wide variety of oxidizing agents is one of the most useful methods of synthesizing tertiary phosphine oxides. Although oxygen, peroxy compounds,

Table 13 Preparation of arylphosphine oxides (ArRP(O)H).

Entry	*Method*	*Ar*	*R*	*m.p.* (°C) or *b.p.* (°C)(mmHg)	*Ref.*
1	A	Ph	4-ClC_6H_4	48–50	63JOC3315
2	A	Ph	4-BrC_6H_4	53–55	63JOC3315
3	A	Ph	3-ClC_6H_4	35–36.5	63JOC3315
4	A	Ph	4-CNC_6H_4	Oil	63JOC3315
5	A	Ph	3-$CF_3C_6H_4$	Oil	63JOC3315
6	A	4-C_6H_4	3-ClC_6H_4	Oil	63JOC3315
7	B	Ph	Ph	53–55	67T1065
8	B	4-ClC_6H_4	4-ClC_6H_4	131–133	67T1065
9	B	3-ClC_6H_4	3-ClC_6H_4	160–180(0.02)	79JOC3741
10	B	2-MeC_6H_4	2-MeC_6H_4	93–95	80JCS(P2)154
11	B	4-MeC_6H_4	4-MeC_6H_4	102–103	67T1065
12	B	1-Naphthyl	1-Naphthyl	79	73ZOB997
13	B	4-$CF_3C_6H_4$	4-$CF_3C_6H_4$	65–67	67T1065
14	B	3-$CF_3C_6H_4$	3-$CF_3C_6H_4$	53–54	67T1065
15	B	Ph	Bu^n	136–138(2)	68JA3459
16	B	Ph	$PhCH_2$	111–113	68JA3459
17	B	Ph	4-$BrC_6H_4CH_2$	160.5–161.5	68JA3459

sulfur oxides, inorganic oxidants, etc., have been used as oxidizing agents, hydrogen peroxide is most widely employed owing to the advantages of availability, ease in handling, formation of inert water as a by-product, and efficiency as an oxidant (Equations (105)–(109)) ⟨64JPR294, 65JOC2826, 66BSF1002, 76ZOB1701, 77IS183⟩.

$$Ar_3P + H_2O_2 \longrightarrow Ar_3P{=}O + H_2O \qquad (105)$$

$$[3,5\text{-}(MeO)_2C_6H_3]_3P + 30\%\ H_2O_2 \xrightarrow{\text{acetone, RT}} [3,5\text{-}(MeO)_2C_6H_3]_3P{=}O \qquad (106)$$

$$(p\text{-}MeC_6H_4)_2(p\text{-}Me_2NC_6H_4)P + 3\%\ H_2O_2 \xrightarrow{\text{acetone, RT}} (p\text{-}MeC_6H_4)_2(p\text{-}Me_2NC_6H_4)P{=}O \qquad (107)$$

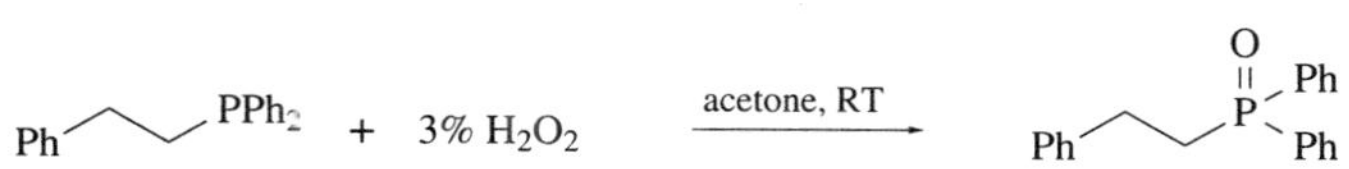

(108)

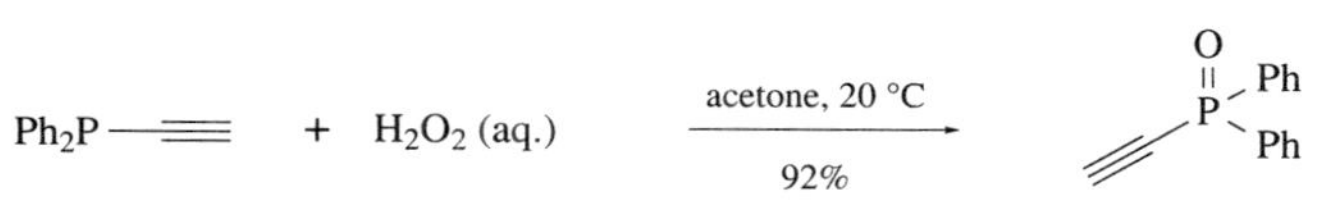

(109)

2.16.3.3.2 *Phosphorus halides or esters with organometallic reagents (Method B)*

In analogy with the synthesis of tertiary phosphines described in Section 2.16.1.1.3, tertiary phosphine oxides are prepared from the reaction of phosphorus oxyhalides (Equation (110)) ⟨65JOC97⟩, phosphinyl halides (Equations (111) and (112)) ⟨50JA4820, 76ZOB1701, 78ZOB91, 88OS(67)20⟩ or phosphonic dihalides (Equation (113)) ⟨52JA4526, 65JOM(4)291⟩ or their esters ⟨55JCS5771, 65ZOB1574⟩ with organometallic reagents, such as Grignard and organolithium reagents.

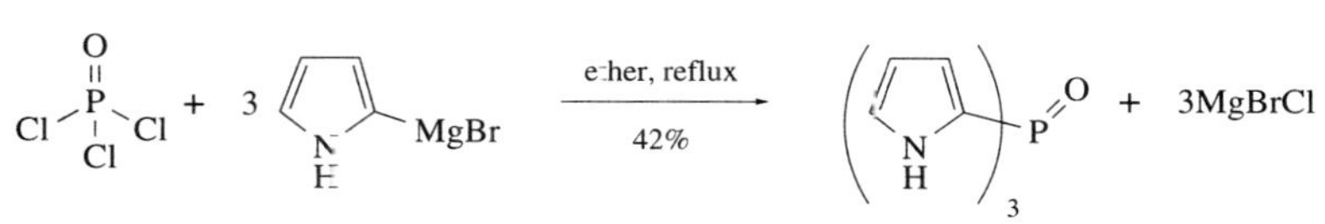

(110)

$$(o\text{-}MeC_6H_4)_2P(O)Cl + PhMgBr \xrightarrow[72\%]{\text{ether, reflux}} (o\text{-}MeC_6H_4)_2PhP{=}O + MgBrCl \qquad (111)$$

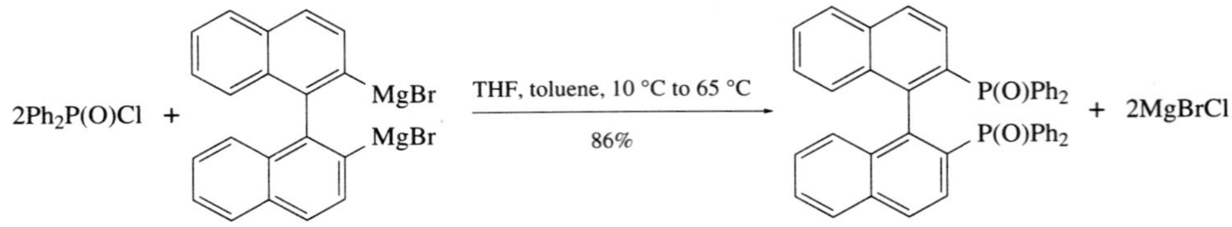
(112)

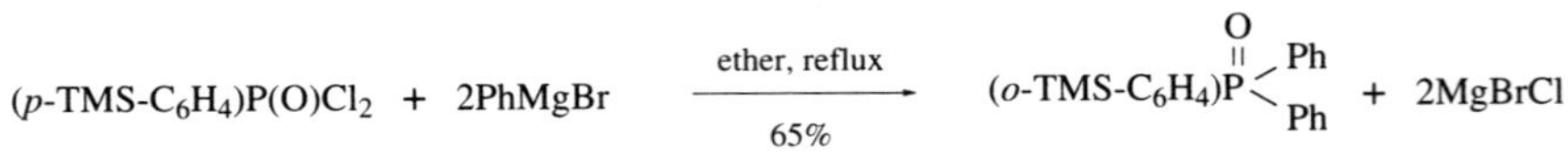
(113)

The synthesis of tertiary phosphine oxides of the type $Ar_2RP(O)$ is via the reaction of diethyl phosphonate with arylmagnesium bromide. This method has the advantage of simplicity, since the whole sequence is carried out in a single reaction vessel (Scheme 17) ⟨65JCS5771⟩.

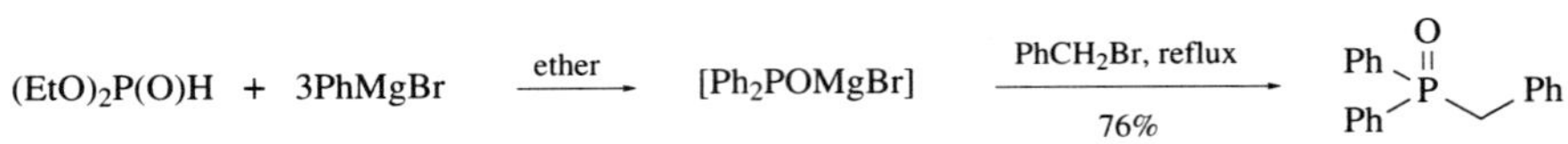

Scheme 17

2.16.3.3.3 *Michaelis–Arbuzov reaction of phosphinic acid esters with alkyl halides (Method C)*

Tertiary phosphine oxides are also obtained by the Michaelis–Arbuzov reaction, which involves the reaction of phosphinic acid esters with alkyl halides to give intermediate phosphonium salts, followed by rapid decomposition (Scheme 18) ⟨55JA3526⟩.

$$ArR^1POR^2 + R^3X \longrightarrow ArR^1\overset{+}{P}(OR^2)R^3\ X^- \longrightarrow ArR^1P(O)R^3 + R^2X$$

Scheme 18

The use of the Michaelis–Arbuzov reaction to prepare tertiary phosphine oxides has the disadvantage that the alkyl halides (R^2X) generated frequently react with the starting esters as alkylating agents.

An optically active phosphine oxide has been prepared by the Michaelis–Arbuzov reaction of asymmetric methyl ethylphenylphosphinite with methyl iodide, which proceeds with retention of configuration at phosphorus (Equation (114)) ⟨75CC382⟩.

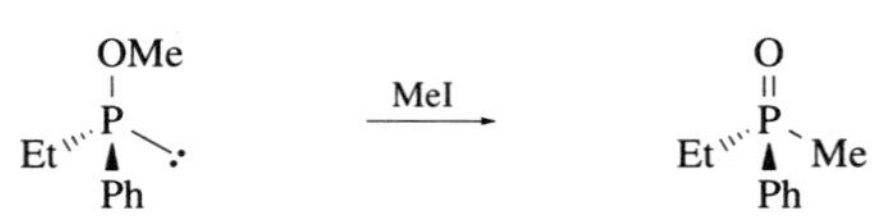
(114)

2.16.3.3.4 *Hydrolysis of phosphonium salts (Method D)*

The hydrolysis of phosphonium salts in the presence of hydroxide ion gives phosphine oxides and a hydrocarbon (Equation (115)), and numerous examples have been reported to date. The rate of hydrolysis depends upon the ability of R^4 to depart as a carbanion. Thus, the ease of displacement of R^4 is in the order $CH_3COCH_2 > CH_2Ph >$ aryl > alkyl ⟨29JCS2342⟩.

(115)

The hydrolysis of acyclic phosphonium salts takes place with inversion of configuration at phosphorus (Equation (116)) ⟨59JA3805⟩.

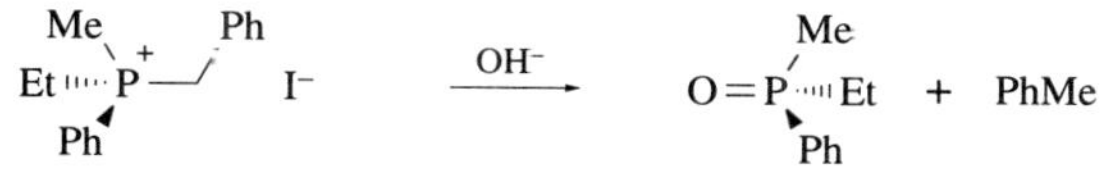

(116)

The alkaline hydrolysis of 1-cycloalkenyltriphenylphosphonium salts gives 1-cycloalkenyldiphenylphosphine oxides in quantitative yields (Equation (117)) ⟨79CB355, 83JOC2569⟩, while similar treatment of vinyltriphenylphosphonium bromide leads to a mixture of diphenyl(α-phenylethyl)phosphine oxide, 1,2-bis(diphenylphosphinyl)ethane, and styrene ⟨69JCS(C)2038⟩.

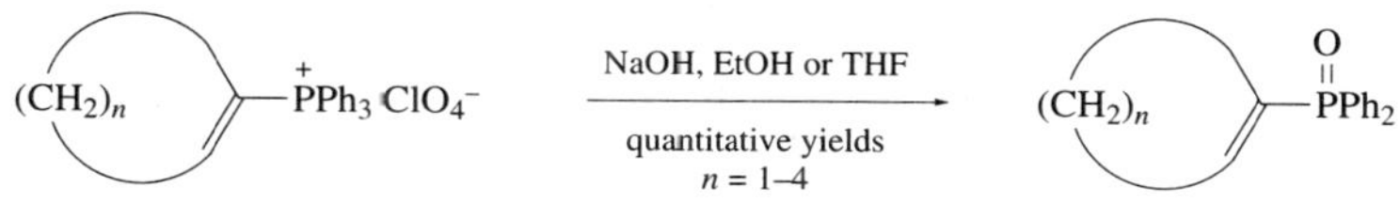

(117)

2.16.3.3.5 *Palladium-catalyzed reaction of alkenyl or aryl bromides with secondary phosphine oxides (Method E)*

The reaction of diphenylphosphine oxide or alkylphenylphosphine oxides with alkenyl or aryl bromides in the presence of triethylamine and a catalytic amount of tetrakis(triphenylphosphine)palladium leads to alkenyldiphenyl- and alkenylalkylphenylphosphine oxides or alkylarylphenylphosphine oxides in good yields (Equations (118) and (119)) ⟨84S781, 86S691⟩.

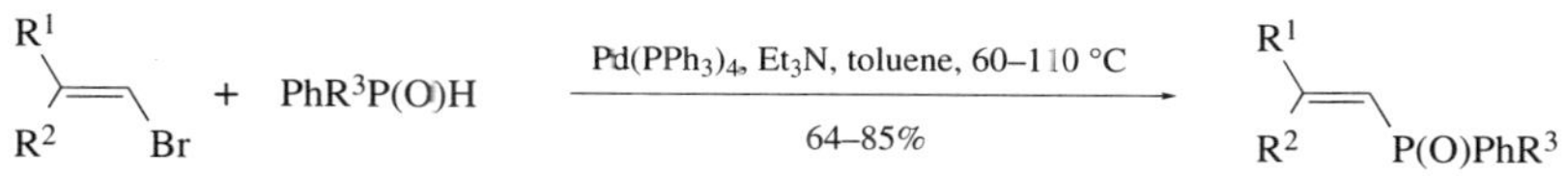

(118)

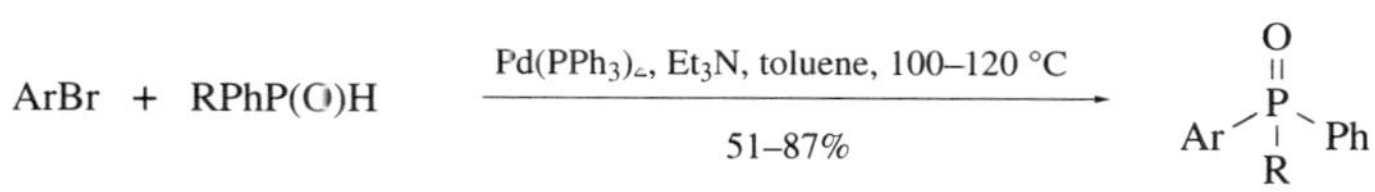

(119)

2.16.3.4 Arylphosphinic Acids (ArRP(O)OH) and their Derivatives

2.16.3.4.1 *Oxidation of secondary phosphines or phosphine oxides (Method A)*

This method has been widely used for the synthesis of diarylphosphinic acids, since diarylphosphine oxides are easily obtained from the reaction of dialkyl phosphonates with Grignard reagents, as described in Section 2.16.3.2 (Equation (120)) ⟨82T1511⟩.

$$(4\text{-ClC}_6\text{H}_4)_2\text{P(O)H} \xrightarrow[64\%]{30\%\ \text{H}_2\text{O}_2,\ \text{CH}_2\text{Cl}_2,\ \text{RT}} (4\text{-ClC}_6\text{H}_4)_2\text{P(O)OH} \qquad (120)$$

Secondary phosphines, instead of secondary phosphine oxides, are also in general use (Equation (121)) ⟨78CB13, 82JCS(P1)2289⟩.

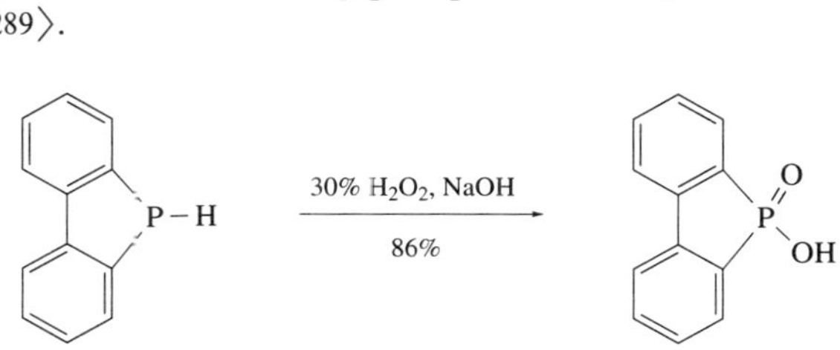

(121)

2.16.3.4.2 *Hydrolysis of arylphosphinyl halides ($Ar_2P(O)Cl$) or arylphosphinic amides ($Ar_2P(O)NR_2$) (Method B)*

The hydrolysis of phosphinyl halides generated *in situ*, either from oxidation of phosphinous chlorides or from halogenation of secondary phosphine oxides, is one of the most useful syntheses of phosphinic acids (Equations (122)–(124)) ⟨66IS71, 76JCS(P1)1739, 79JOC2226, 80JCS(P2)154⟩.

$$(o\text{-MeC}_6\text{H}_4)_2\text{P(O)Cl} \xrightarrow[28\%]{\text{ice, H}_2\text{O}} (o\text{-MeC}_6\text{H}_4)_2\text{P(O)OH} \qquad (122)$$

$$\text{Ph}_2\text{PCl} \xrightarrow[90\text{–}96\%]{\text{H}_2\text{O, H}_2\text{O}_2\text{, NaOH}} \text{Ph}_2\text{P(O)OH} \qquad (123)$$

$$(p\text{-MeOC}_6\text{H}_4)_2\text{P(O)H} \xrightarrow[66\%]{\text{NaOH, Br}_2} (p\text{-MeOC}_6\text{H}_4)_2\text{P(O)OH} \qquad (124)$$

Diarylphosphinic amides, prepared from arylmagnesium halides and phosphoramidic dichlorides ($Cl_2P(O)NR_2$) easily undergo acidic hydrolysis to give the corresponding phosphinic acids (Equations (125) and (126)) ⟨67JOC1572, 69CB2922, 82JCS(P1)2299⟩.

$$(m\text{-MeOC}_6\text{H}_4)_2\text{P(O)NEt}_2 \xrightarrow[59\%]{\text{HCl, H}_2\text{O, heat}} (m\text{-MeOC}_6\text{H}_4)_2\text{P(O)OH} \qquad (125)$$

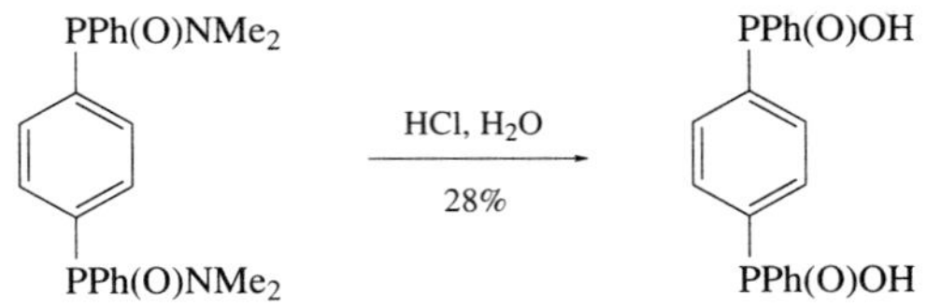

(126)

2.16.3.4.3 *Phosphorus trichloride with aryldiazonium salts (Method C)*

Reaction of phosphorus trichloride with aryldiazonium tetrafluoroborates in the presence of copper(I) bromide, followed by hydrolysis, gives a mixture of arylphosphonic and diarylphosphinic acids in 20–50% and 10–30% yields, respectively (Scheme 19) ⟨51JA5658, 53JA683, 53JA1379, 61JA4124⟩.

$$\text{PCl}_3 + \text{Ar}\overset{+}{\text{N}}_2\,\text{BF}_4^- \xrightarrow{\text{CuBr, EtOAc}} \left[\text{Ar}\overset{+}{\text{P}}\text{Cl}_3 + \text{Ar}_2\overset{+}{\text{P}}\text{Cl}_2\right]\text{BF}_4^- \xrightarrow{\text{H}_2\text{O}} \text{ArP(O)(OH)}_2 + \text{Ar}_2\text{P(O)OH}$$

Scheme 19

2.16.3.4.4 *Alkaline hydrolysis of tertiary phosphine oxides (Method D)*

The synthesis of diarylphosphinic acids via alkaline hydrolysis of triarylphosphine oxides has no practical value, since the reaction requires vigorous conditions (heating above 250 °C for a long period under alkali fusion) ⟨58CB64⟩. However, triheteroarylphosphine oxides undergo alkaline hydrolysis under rather mild conditions to produce diheteroarylphosphinic acids in good to moderate yields (Equation (127)) ⟨77JCS(P2)1705⟩.

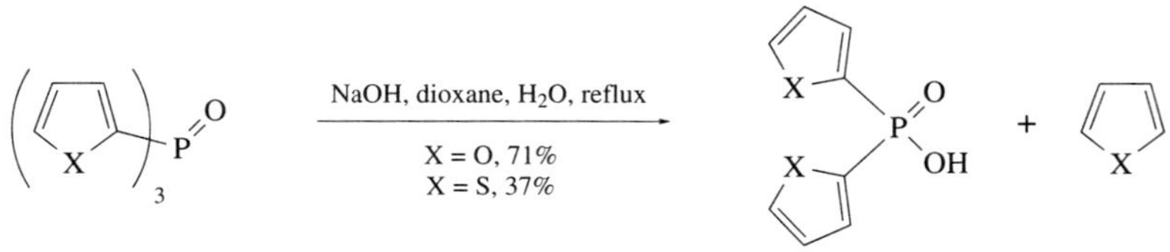

(127)

The reactivity of the triheteroarylphosphine oxides to hydrolysis decreases in the order 2-furyl > 2-thienyl > 1-methylpyrrol-2-yl. From the synthetic point of view, this route to heteroarylphosphinic acids is useful, since heteroarylphosphine oxides are readily available, either by direct synthesis from an organometallic derivative with phosphoryl chloride (Section 2.16.3.3.2) or via oxidation of tertiary phosphines (Section 2.16.3.3.1).

2.16.3.4.5 *Arylphosphinic acid derivatives*

Arylphosphinic chlorides are prepared by various methods: (1) chlorination of diarylphosphinic acids or diarylphosphine oxides with phosphorus pentachloride (Equation (128)) ⟨55JA1864⟩, thionyl chloride (Equations (129) and (130)) ⟨57JCS2413, 69CB2914, 82T1511⟩, trichlorodiarylphosphoranes ⟨55JA1864⟩, *N*-chlorosuccinimide ⟨57JCS2413⟩, etc.; (2) the reaction of trichlorodiarylphosphoranes with phosphorus pentoxide (Equation (131)) ⟨55JA1864⟩; and (3) the reaction of phosphorus oxychloride with activated aromatic compounds (Equation (132)) ⟨76JCS(P1)1739⟩ or *o*-substituted aromatic Grignard reagents (Equation (133)) ⟨79JOC2226⟩.

$$Ph_2P(O)OH + PCl_5 \xrightarrow[76\%]{70\text{–}80\ ^\circ C} Ph_2P(O)Cl \quad (128)$$

$$(p\text{-}ClC_6H_4)_2P(O)OH + SOCl_2 \xrightarrow{\text{reflux}} (p\text{-}ClC_6H_4)_2P(O)Cl + SO_2 + HCl \quad (129)$$

$$Ph_2P(O)H + SOCl_2 \xrightarrow[32\%]{CHCl_3,\ \text{heat}} Ph_2P(O)Cl + SO + HCl \quad (130)$$

$$3Ph_2PCl_3 + P_2O_5 \xrightarrow[79\%]{PhH,\ 30\text{–}35\ ^\circ C} 3Ph_2P(O)Cl + 2P(O)Cl_3 \quad (131)$$

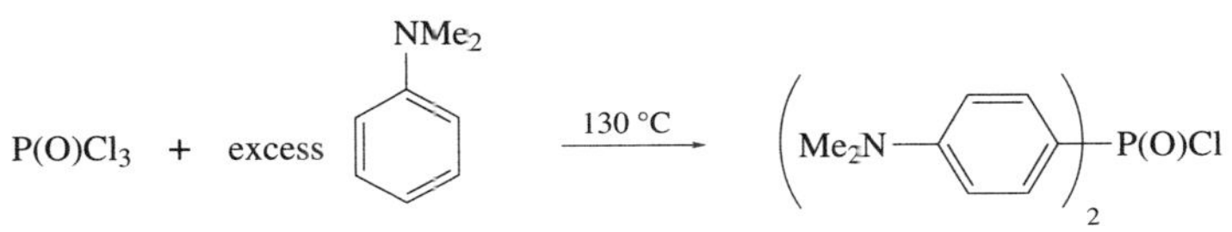

(132)

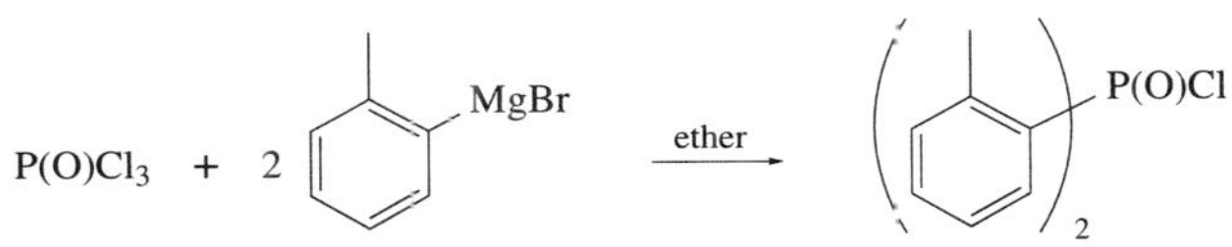

(133)

Method 3 is limited to the synthesis of diarylphosphinic chlorides having a strongly electron-donating substituent or with a bulky group at the 2-position and is therefore not general. Arylphosphinic fluorides are prepared by electrochemical fluorination of secondary phosphines in the presence of Et_4NBF_4 ⟨82ZOB2721⟩, or by halogen exchange reaction between a phosphinic chloride and KF (Equation (134)) ⟨75JOC2059⟩.

$$PhRP(O)Cl + KF \xrightarrow{MeCN,\ \text{reflux}} PhRP(O)F + KCl \quad (134)$$

A wide variety of methods for the preparation of arylphosphinic acid esters has been developed as follows: (1) the Arbuzov reaction of dialkyl arylphosphonites with alkyl halides ⟨68JA4842, 73JOC2703⟩; (2) alcoholysis of arylphosphinic chlorides ⟨58JOC1298, 64JCS1240, 80JCS(P2)1505⟩; (3) alkylation of arylphosphinic acids ⟨50JA4292, 68JA3459, 80JCS(P2)1505, 82T1511⟩; (4) metallation-induced rearrangement of diaryl ethylphosphates ⟨86JOC179⟩; (5) addition of phenylphosphinic acid esters to alkenes ⟨71CC146⟩; and (6) palladium-catalyzed reaction of alkenyl or aryl bromides with monoalkyl phosphonites. Method 1 is normally used for the preparation of phosphinic acid esters (Equation (135)) ⟨68JA4842, 73JOC2703⟩. Organohalo compounds used in this method are mostly limited to aliphatic halides, and therefore unsymmetrical alkyl alkylarylphosphinates are formed. However, the nickel salt-catalyzed reaction of phosphonic acid esters, with aryl or heteroaryl halides, leads to diarylphosphinic acid esters in good yields (Equation (136)) ⟨79CB2428⟩.

$$ArP(OR^1)_2 + R^2X \xrightarrow{\text{heat}} ArR^2P(O)OR^1 + R^1X \quad (135)$$

$$PhP(OEt)_2 + ArBr \xrightarrow[50\text{–}80\%]{NiX_2,\ 160\ ^\circ C} ArPhP(O)OEt + EtBr \quad (136)$$

The Arbuzov reaction is also applied to the synthesis of diphosphinates from diphosphonites and alkyl halides (Equation (137)) ⟨67JOC2172⟩.

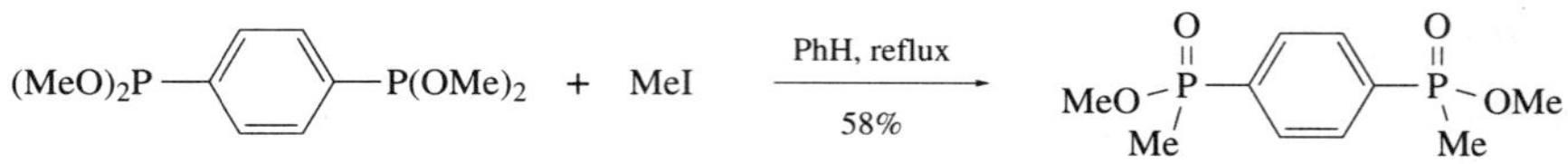 (137)

The synthesis of phosphinic acid esters via alcoholysis of arylphosphinic chlorides (Method 2) is less useful than Method 1. The alcoholysis of phosphinic chlorides is analogous to that of phosphinous and phosphonous acid chlorides.

The reaction of asymmetric phosphinic chlorides with optically active alcohols leads to a mixture of diastereomeric esters, which are separable as optically active phosphinic acid esters (Equation (138)) ⟨68JA3869, 80JCS(P2)1505⟩. Optically active phosphinates are also obtained by acid-catalyzed alcoholysis of optically active phosphinic amides ⟨79S110⟩, or by base-catalyzed hydrolysis of optically active dialkoxyphosphonium salts ⟨69JA7393, 72JOC2272⟩.

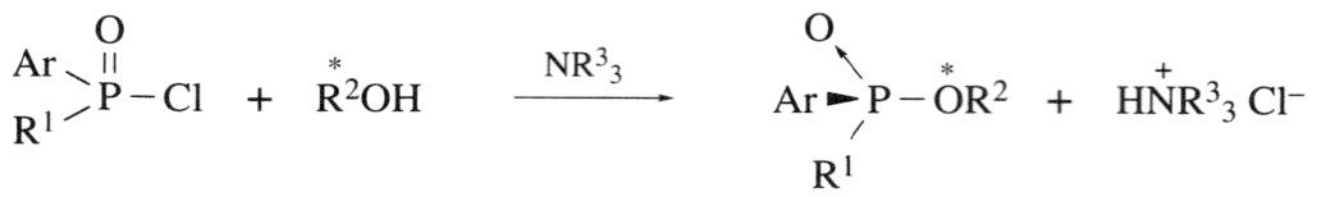 (138)

Method 3 is only significantly practical for the reaction of phosphinic acids with diazoalkanes ⟨50JA4292, 77JCS(P2)1705, 80JCS(P2)1505, 82T1511⟩.

Method 4 has become a useful route to diarylphosphinic acid esters; for example, treatment of diaryl ethylphosphates with lithium diisopropylamide, yields ethyl bis(2-hydroxyaryl)phosphinates. The reaction involves the double migration of phosphorus from oxygen to *o*-positioned carbon (Scheme 20) ⟨86JOC179⟩.

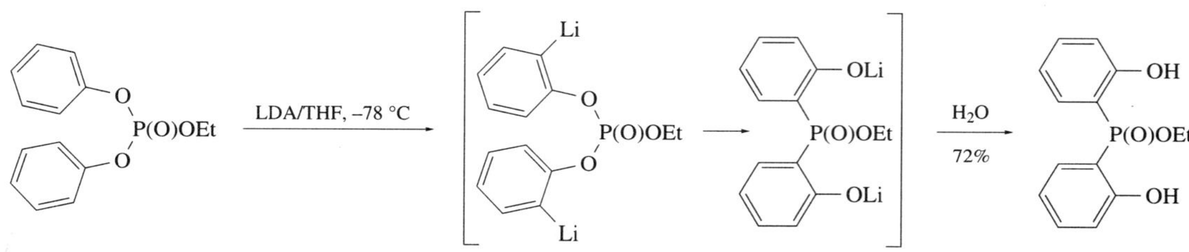

Scheme 20

Addition of primary phosphinic acid esters to activated or simple alkenes occurs under base- ⟨51NAT1035⟩ or peroxide-catalyzed conditions (Equation (139)) ⟨71CC146⟩ to give secondary phosphinic acid esters (Method 5).

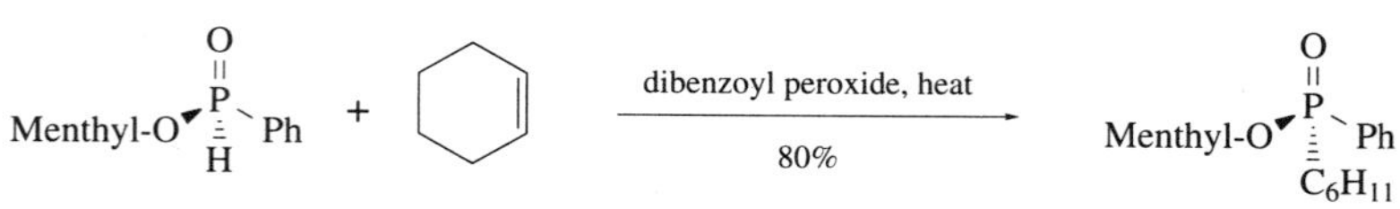 (139)

Analogously to the palladium-catalyzed synthesis of tertiary alkenylphosphine oxides (cf. Section 2.16.3.3.5), alkyl alkenylphenylphosphinates are synthesized stereospecifically by the palladium-catalyzed reaction of monoalkyl phosphonites with alkenyl or aryl bromides in the presence of triethylamine, with moderate to good yields (Equations (140) and (141)) ⟨84S778, 86S240⟩.

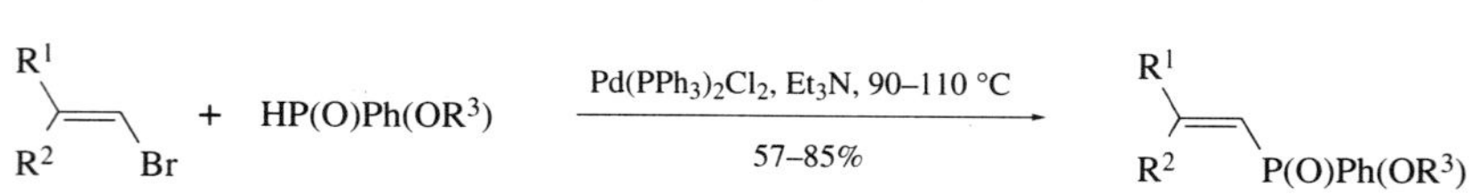 (140)

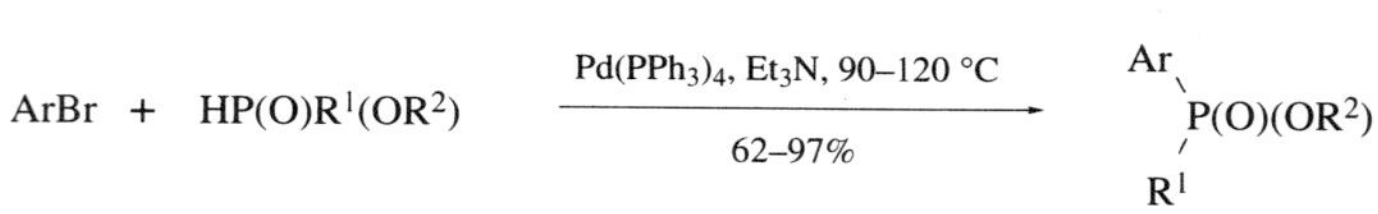 (141)

2.16.3.5 Arylphosphonic ($ArP(O)(OH)_2$) and Unsaturated Phosphonic Acids, and their Derivatives ⟨57CRV479, 64HOU(12/1)338, B-76MI 216-01⟩

2.16.3.5.1 Hydrolysis of phosphonic halides, esters, and amides, and aryltrichlorophosphonium salts (Method A)

Warming an aqueous solution of arylphosphonic chlorides to reflux leads to arylphosphonic acids (Equation (142)) ⟨56CB688⟩. Similar treatment of alkenylphosphonic dichlorides ⟨74ZOB85⟩ or alkenyltetrachlorophosphoranes ⟨33CB278⟩ affords the corresponding phosphonic acids.

$$ArP(O)Cl_2 + 2H_2O \xrightarrow{\text{heat}} ArP(O)(OH)_2 + 2HCl \quad (142)$$

A more general method for the preparation of arylphosphonic acids is acid- ⟨48JA3465, 70CB2428, 76JCS(P2)1444⟩ or base- ⟨77JCS(P2)789⟩ catalyzed hydrolysis of dialkyl arylphosphonates (Equation (143)).

$$ArP(O)(OR)_2 \xrightarrow{H^+, \text{ or } OH^-, H_2O, \text{ reflux, 9–12 h}} ArP(O)(OH)_2 + 2ROH \quad (143)$$

In comparison with phosphonic acid esters, arylphosphonic diamides undergo acidic hydrolysis under rather mild conditions to lead to the free acids (Equation (144)) ⟨51DOK(78)709, 51JA5896⟩.

$$ArP(O)(NR_2)_2 \xrightarrow[\text{60–90\%}]{H^+, H_2O, \text{ reflux, 30 min}} ArP(O)(OH)_2 + 2R_2NH \quad (144)$$

Hydrolysis of aryltrichlorophosphonium salts [$ArP^+Cl_3\ BF_4^-$] generated from the reaction of aryldiazonium salts with PCl_3 in the presence of CuBr also produces phosphonic acids in moderate yields ⟨51JA5658, 52JA753⟩.

2.16.3.5.2 Dealkylation of phosphonic acid esters with iodotrimethylsilane (Method B)

Phosphonic acid dialkyl esters are transformed in high yields into the corresponding bistrimethylsilyl esters on treatment with iodotrimethylsilane, and subsequent hydrolysis of the silyl esters selectively affords phosphonic acids, without modification of a variety of other functionalities (Equation (145)) ⟨78CC870, 78TL2523, 80JCS(P1)1150, 91JOC833⟩.

$$RP(O)(OEt)_2 + 2TMS\text{-}I \xrightarrow[\text{quantitative yields}]{\text{i, } CHCl_3, 0\ ^\circ C \text{ to RT; ii, } H_2O \text{ or MeOH}} RP(O)(OH)_2 + 2EtI + TMS\text{-}O\text{-}TMS \quad (145)$$

$R = CH_2{=}CH, HC{\equiv}C$, aryl

Since the synthesis of phosphonic acids via the hydrolysis of phosphonate diesters usually requires vigorous acidic or basic conditions, treatment with iodotrimethylsilane is preferable to the hydrolysis method in terms of conditions, yields, applicability to functionalized phosphonates, etc.

2.16.3.5.3 Miscellaneous methods

Pyridine-2- or -4-phosphonic acids are conveniently prepared by the addition of phosphorous acid to 1-(2-pyridyl)- or 1-(4-pyridyl)pyridinium salts ⟨79S452⟩.

2.16.3.5.4 Phosphonic acid derivatives

Treatment of arylphosphonic acids with phosphorus pentachloride or thionyl chloride leads to arylphosphonic dichlorides (aryldichlorophosphine oxides) ⟨64HOU(12/1)338⟩. The more important preparations of arylphosphonic dichlorides are achieved by the action of chlorine on a mixture of phosphorus pentoxide and arylphosphonous dichlorides (Equation (146)) ⟨48JA186⟩, or by the oxidation of arylphosphonous dichlorides with oxidizing agents (Equation (147)) ⟨69ZOB841⟩.

$$3ArPCl_2 + P_2O_5 + 3Cl_2 \xrightarrow[80\text{–}90\%]{} 3Ar(O)Cl_2 + 2POCl_3 \quad (146)$$

$$ArPCl_2 + SO_2Cl_2 \longrightarrow ArP(O)Cl_2 + SOCl_2 \quad (147)$$

Alkenylphosphonic dichlorides are provided by the addition of phosphorus pentachloride to alkenes, with subsequent reaction of the adducts with sulfur dioxide ⟨73OSC(5)1005⟩ or with triethylamine (Scheme 21) ⟨65ZOB1577⟩.

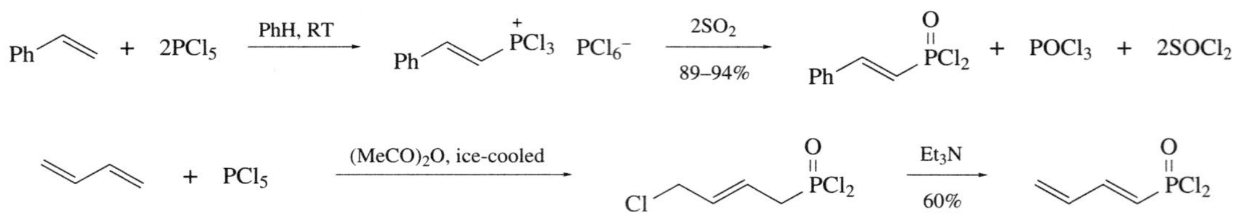

Scheme 21

Arylphosphonic difluorides are obtained by fluorination of phosphonic acids with sulfur tetrafluoride ⟨57USP2950306⟩ or of phosphonic acid chlorides with antimony trifluoride ⟨64JCS4551⟩ or zinc fluoride ⟨59ZOB3766⟩.

The reaction of phosphonic dichlorides with alcohols or phenols in the absence of tertiary amines gives phosphonic acid monoalkyl esters (Equation (148)) ⟨64HOU(12/1)338⟩ or phosphonochloridates (Equation (149)) ⟨56JA3063⟩.

$$ArP(O)Cl_2 + 2ROH \xrightarrow{\text{heat}} ArP(O)(OH)(OR) + RCl + HCl \quad (148)$$

$$PhP(O)Cl_2 + PhOH \xrightarrow[59\%]{\text{heat}} PhP(O)(OPh)Cl + HCl \quad (149)$$

Monoalkyl arylphosphonates are conveniently prepared by the mono-*O*-alkylation of bis(tetramethylammonium) salts of arylphosphonic acids (Equation (150)) ⟨90S140⟩.

$$ArP(O)(OH)_2 \xrightarrow[44\text{–}67\%]{\text{i, } Me_4NOH\text{, EtOH; ii, RBr, MeCN, EtOH, reflux}} ArP(O)(OH)(OR) \quad (150)$$

Phosphonic acid esters are practically prepared according to the following methods: (1) alcoholysis of phosphonic dichlorides ⟨68ZOB2285, 69CB2922, 77JCS(P2)789, 88S912⟩; (2) the Arbuzov reaction or the Arbuzov type reaction of aryl halides with trialkyl phosphites; (3) the reaction of dialkyl phosphorochloridates with Grignard or organolithium reagents ⟨79S452⟩; (4) the Michaelis–Becker reaction or the Michaelis–Becker type reaction of dialkyl phosphonate ions with aryl halides; (5) the palladium-catalyzed reaction of aryl halides with dialkyl phosphonates; and (6) oxidation of phosphonous acid diesters.

A classical, but still useful method for the preparation of dialkyl aryl-, vinyl-, or alkynylphosphonates is esterification (Method 1) of the corresponding phosphonic dichlorides with alcohols or phenols (2 molar equivalents) in the presence of tertiary amines (2 molar equivalents) (Equation (151)) ⟨68ZOB2285, 69CB2922, 77JCS(P2)789, 88S912⟩.

$$ArP(O)Cl_2 + 2ROH \xrightarrow{C_5H_5N} ArP(O)(OR)_2 \quad (151)$$

Dialkyl arylphosphonates or vinylphosphonates are not produced via simple Arbuzov or Michaelis–Becker reaction of aryl or vinyl halides with trialkyl phosphites or dialkyl phosphonate anions. However, the reaction of aryl, or vinyl, or thienyl halides with trialkyl phosphites in the presence of nickel chloride or bromide (5–10 mol%) takes place to give dialkyl arylphosphonates in good yields (Method 2) (Equations (152)–(155)) ⟨70CB2428, 70T5529, 90S140⟩.

$$ArBr + P(OEt)_3 \xrightarrow[70\text{–}90\%]{NiX_2,\ 150\text{–}160\ ^\circ C} ArP(O)(OEt)_2 \qquad (152)$$

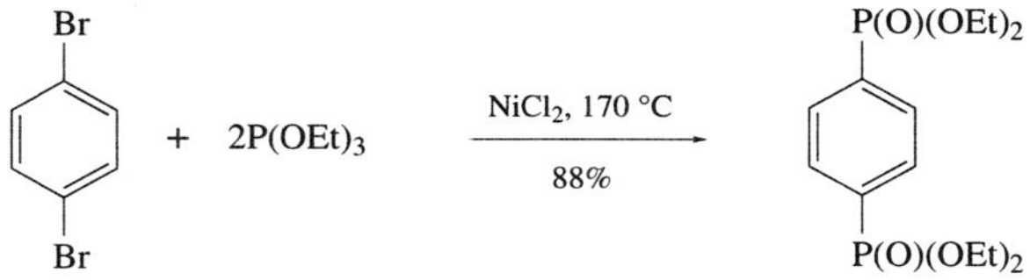

(153)

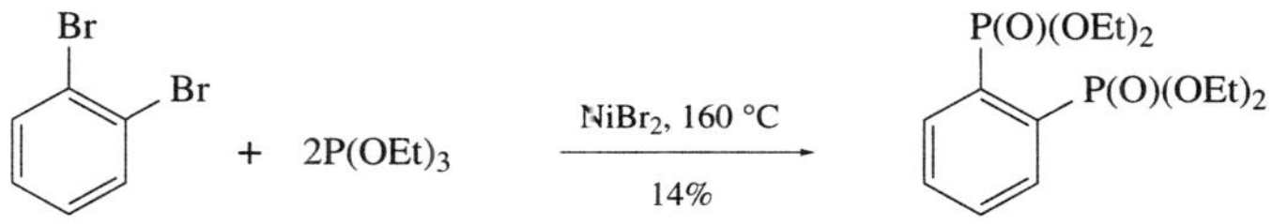

(154)

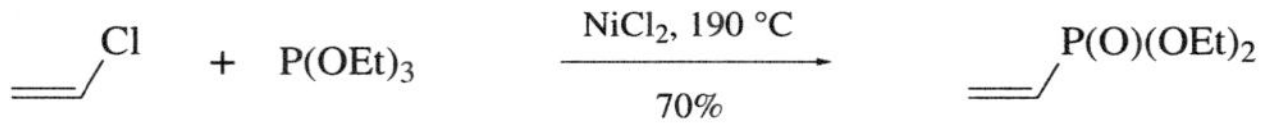

(155)

o-Substituted aryl halides give rather low yields of arylphosphonates compared with the corresponding *m*- or *p*-substituted isomers.

A variety of substituted phenylphosphonates and heteroarylphosphonates are synthesized in good yields by the low-temperature ultraviolet photolysis of the corresponding aryl iodides in excess trialkyl phosphites (Scheme 22) ⟨66JOC2455, 66TL5049, 68JOC632, 79S615⟩.

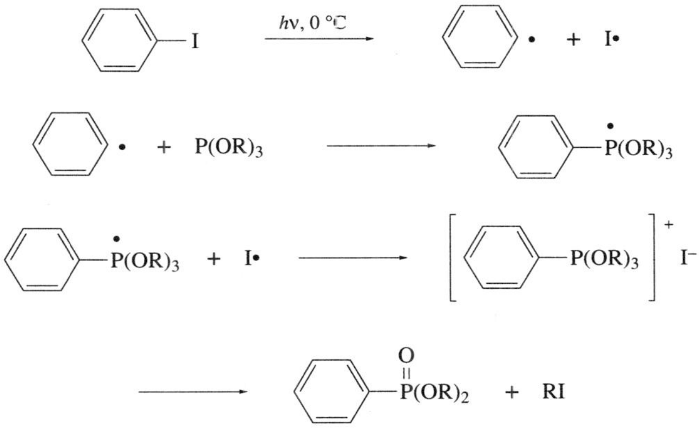

Scheme 22

The reactions in Scheme 22 constitute a catalytic or photolytic modification of the Arbuzov reaction. In contrast, alkenyl- and alkynylphosphonates are readily produced by the Arbuzov reaction of trialkyl phosphites with activated bromoalkenes (Scheme 23) ⟨82TL2603⟩ and bromoalkynes (Equations (156) and (157)) ⟨63ZOB2863, 71JA3694⟩.

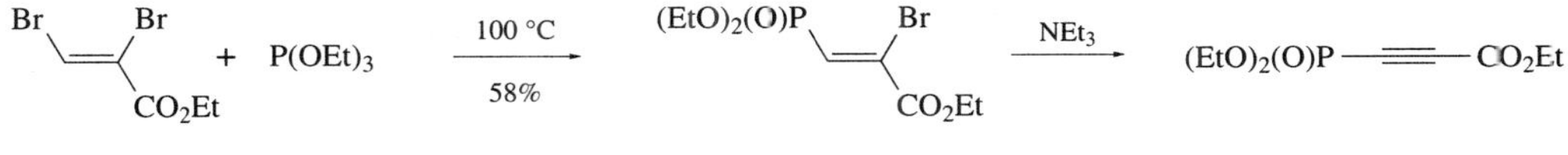

Scheme 23

$$Ph{-}C{\equiv}C{-}Br + P(OEt)_3 \xrightarrow[75\%]{THF,\ heat} Ph{-}C{\equiv}C{-}P(O)(OEt)_2 + EtBr \qquad (156)$$

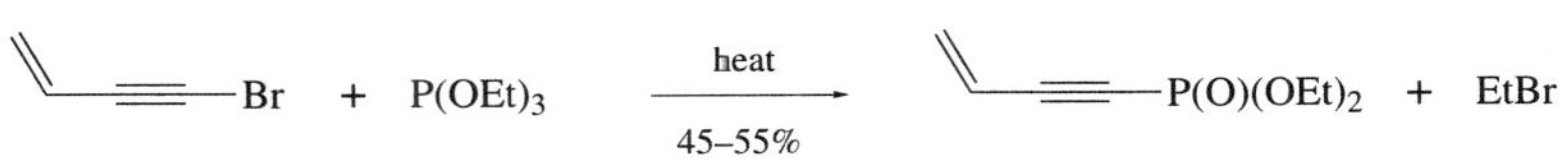

(157)

Method 3 is occasionally used for the synthesis of aryl- and vinylphosphonates, although it does tend to give low to moderate yields (Equations (158) and (159)) ⟨79S452⟩.

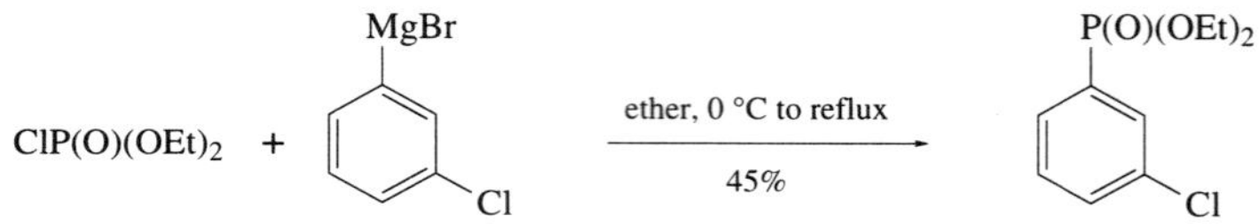

(158)

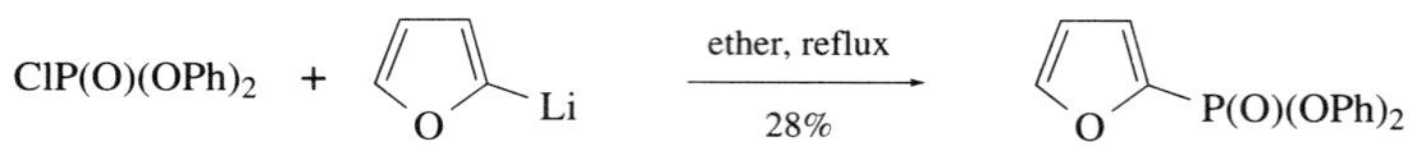

(159)

The reaction of alkyl halides with alkali metal salts of dialkyl phosphonates (the Michaelis–Becker reaction) is an important method for the preparation of dialkyl alkylphosphonates, but this reaction along with the Arbuzov reaction is not applicable to aryl halides. However, various aryl iodides react with potassium dialkyl phosphonates in liquid ammonia under irradiation (350 nm) to form dialkyl arylphosphonates in good yields (Equation (160)) ⟨74JOC3612, 79JOC4918⟩. In this photostimulated reaction, *o*-fluoroiodobenzene suffers replacement of iodine to afford only the monosubstitution product, while *o*-chloroiodo-, *o*-bromoiodo-, and *o*-diiodobenzene give the disubstitution products (Equations (161) and (162)) ⟨78JOC1867, 79JOC4918⟩.

$$\mathrm{ArI} + \mathrm{(RO)_2PO^- K^+} \xrightarrow[87\text{–}96\%]{\text{liq. } \mathrm{NH_3},\ h\nu} \mathrm{ArP(O)(OR)_2} + \mathrm{KI} \qquad (160)$$

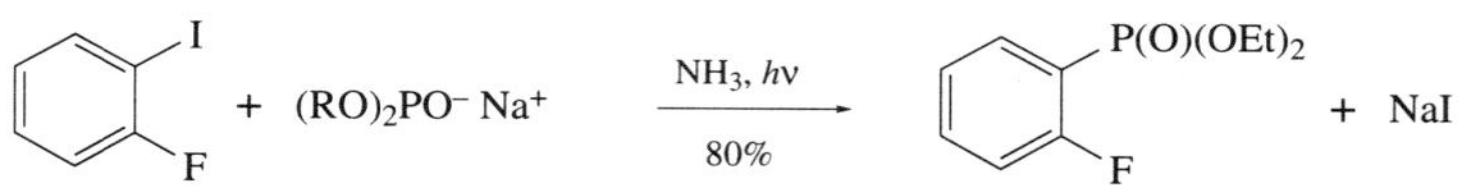

(161)

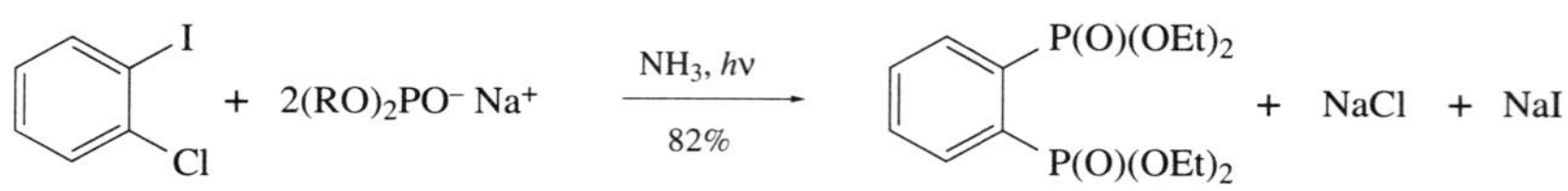

(162)

The reaction of alkynyl halides with sodium diethyl phosphonate under mild conditions gives diethyl alkynylphosphonates (Equation (163)) ⟨71JA3694⟩.

$$\mathrm{Ph{-}C{\equiv}C{-}Br} + \mathrm{(RO)_2PO^- Na^+} \xrightarrow[61\%]{\text{THF, 0 °C}} \mathrm{Ph{-}C{\equiv}C{-}P(O)(OEt)_2} \qquad (163)$$

The palladium-catalyzed synthesis of dialkyl aryl- and vinylphosphonates is important, since this synthetic method has the advantages of wide applicability, good yields, selectivity, availability of starting reagents and simple experimental operation (Equations (164)–(167)) ⟨82BCJ909⟩.

$$\mathrm{ArX} + \mathrm{(RO)_2P(O)H} \xrightarrow{\mathrm{Pd(PPh_3)_4}\text{ (cat.), } \mathrm{Et_3N}\text{, 90 °C}} \mathrm{ArP(O)(OR)_2} + \mathrm{Et_3N{\bullet}HX} \qquad (164)$$

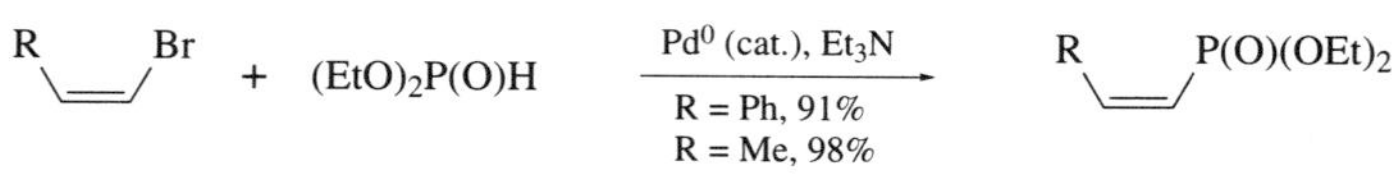

(165)

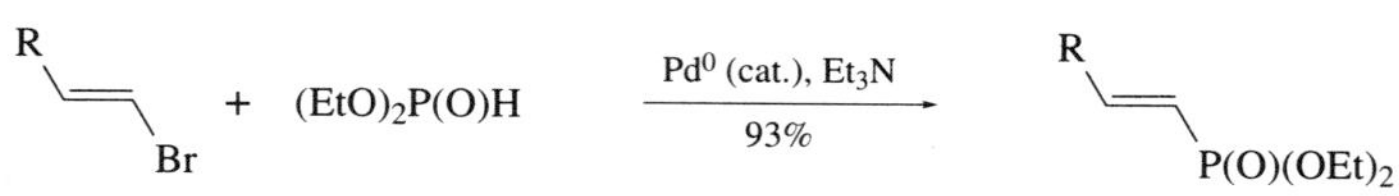

(166)

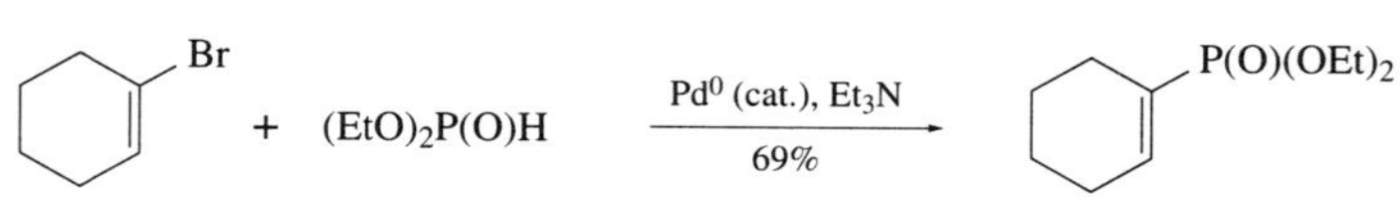

(167)

For preparation of dialkyl aryl- or alkenylphosphonates via the oxidation of the corresponding phosphonites (Method 6), oxidants used on other trivalent phosphorus compounds are also applicable here. For example, the oxidation of dialkyl vinylphosphonites by active manganese dioxide gives dialkyl vinylphosphonates in good yields with no side reactions (Equation (168)) ⟨62ZOB3351⟩.

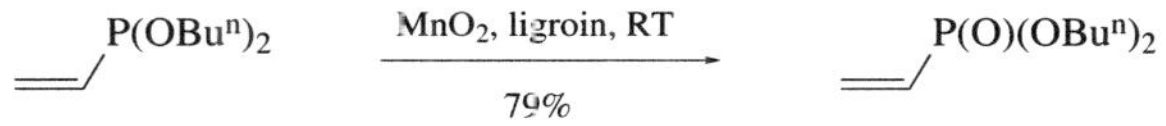

(168)

The synthesis and synthetic applications of functionalized vinylphosphonates have been reviewed by Minami and Motoyoshiya ⟨92S333⟩.

2.16.4 COMPOUNDS WITH A P—S BOND

2.16.4.1 Arylthiophosphinous Acids (Ar_2PSH) and their Derivatives

Arylthiophosphinous acid (diphenylthiophosphinous acid) probably exists as diphenylphosphine sulfide. Arylthiophosphinous acids are prepared by phenylation of thiophosphonic acid esters with phenylmagnesium halides or by the reaction of aminophosphine with hydrogen sulfide (Equations (169) and (170)) ⟨66HCA1249⟩.

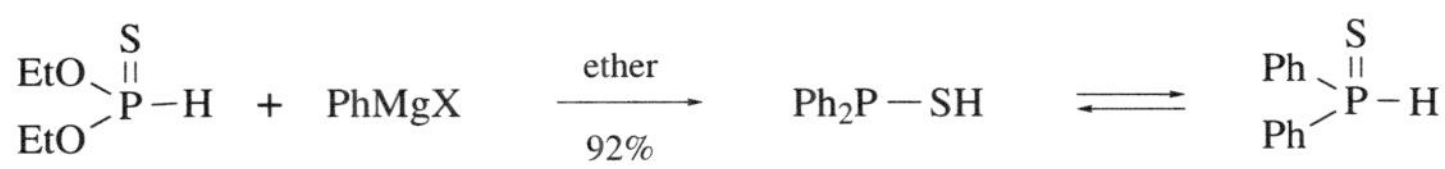

(169)

$$Ar_2PNR_2 + 2H_2S \longrightarrow Ar_2PSH + R_2NH_2SH \quad (170)$$

When tetraphenyldiphosphine disulfide is irradiated in methanol for 6 h with a low-pressure mercury lamp, diphenylphosphine sulfide and *O*-methyldiphenylphosphine sulfide are obtained (Equation (171)) ⟨73BCJ898⟩. Their alkyl esters are usually obtained by dehydrochlorination of diphenylphosphine with sulfenyl chlorides ⟨62ZOB4017⟩, or by heating diphenylphosphinous chloride with alkyl xanthates ⟨67T3049⟩.

$$Ph_2P(S)P(S)Ph_2 + MeOH \xrightarrow{h\nu} Ph_2P(S)H + Ph_2P(S)OMe \quad (171)$$

Diphenylphosphinous chloride also reacts with hexamethylcyclotrisilathiane to afford bisdiphenylthiophosphino sulfide (Equation (172)) ⟨64JCS5585⟩.

$$6Ph_2PCl + (Me_2SiS)_3 \longrightarrow 3Ph_2PSPPh_2 + 3Me_2SiCl_2 \quad (172)$$

When phenylthiophosphinic dichloride is dechlorinated with magnesium in THF in the presence of 2,3-dimethylbutadiene at 50 °C, unstable cyclic phenylthiophosphinous acids are formed, which further react with sulfur or oxygen to afford 4,5-dimethyl-2-phenyl-3,6-dihydro-1,2-thiaphosphorin 2-sulfide or the corresponding 2-oxide in 44% or 21% yields, respectively ⟨75BCJ546⟩. A thiophosphinic acid thioester reacts with tributylphosphine to give a phenylthiophosphinous acid arylthioester (Equation (173)) ⟨80PS(8)221⟩.

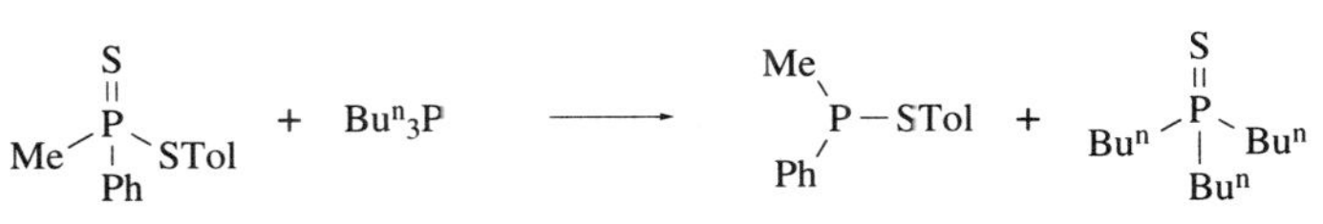

(173)

Optically active methyl *t*-butylphenylthiophosphinite is formed by the reaction of optically active *t*-butyl(phenyl)(methylthio)chlorophosphonium trifluoromethanesulfonate with triphenylphosphine with inversion of configuration, followed by oxidation (Scheme 24) ⟨92CC1718⟩.

2.16.4.2 Arylthiophosphonous Acids (ArP(SH)$_2$) and their Derivatives (ArP(SH)(Hal), etc.)

Phenylthiophosphonous acid *O*-esters are prepared by the reaction of dialkyl phenylphosphonites with hydrogen sulfide (Scheme 25) ⟨70ZN111⟩. Chloroethoxyphenylphosphine reacts with ethanethiol to give diethyl arylthiophosphonite (Equation (174)) ⟨62CB2993⟩.

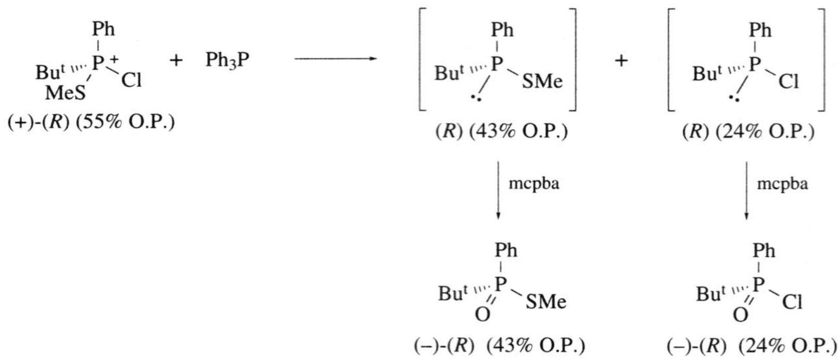

Scheme 24

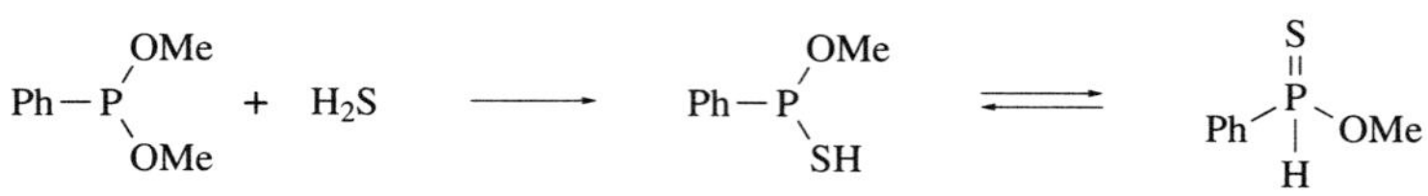

Scheme 25

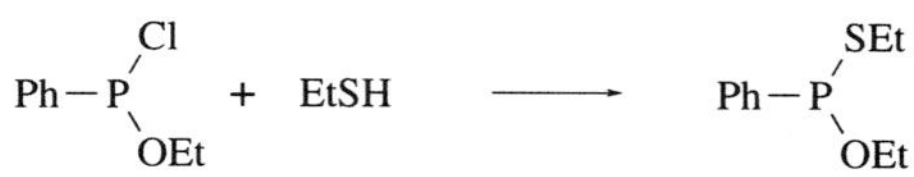

(174)

Phenyl dithiophosphonous acid diesters are prepared by the reaction of phenylphosphonous dichloride with alkylthiotrimethylsilanes, disulfides, or alkane dithiols (Equations (175)–(177) and Scheme 26) ⟨64JCS5585, 68CB1381, 76JPC2417, 82CC691⟩.

$$PhPCl_2 + 2EtS\text{-}TMS \longrightarrow PhP(SEt)_2 + 2TMS\text{-}Cl \qquad (175)$$

$$PhPCl_2 + EtSSEt \xrightarrow{Zn,\ THF} PhP(SEt)_2 + ZnCl_2 \qquad (176)$$

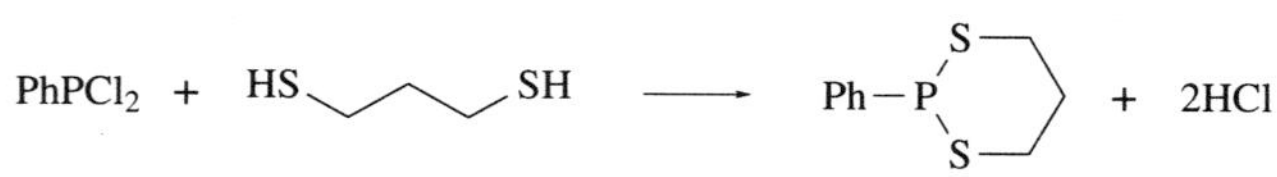

(177)

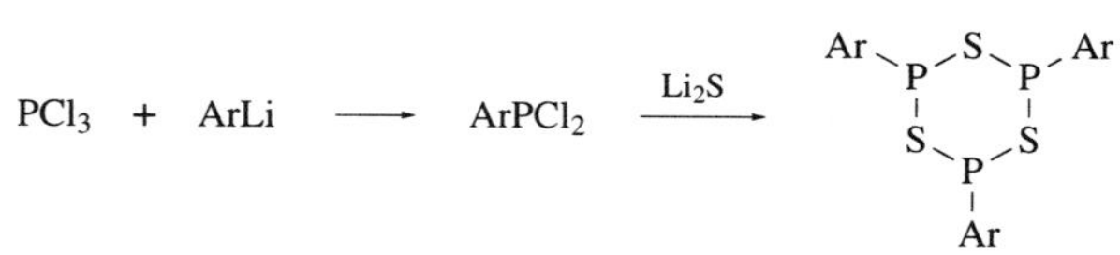

Scheme 26

Phenylthiophosphonous acid fluoride is obtained by the reaction of $PhPF_2$ with hydrogen sulfide in the presence of triethylamine (Scheme 27) ⟨81ZAAC(480)90⟩.

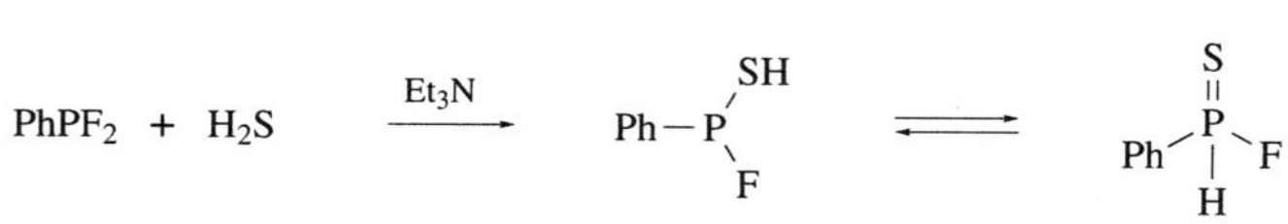

Scheme 27

2.16.4.3 Arylphosphine Sulfides ($ArR^1R^2P{=}S$)

Tertiary alkylphosphine sulfides were originally prepared by the direct reaction of trialkylphosphine with elemental sulfur ⟨1857LA(101)1⟩. There are now numerous examples of the use of elemental sulfur as a reagent for the direct sulfurization of phosphines to the corresponding phosphine sulfides. This method is applicable to the synthesis of arylphosphine sulfides and alkylarylphosphine sulfides and is still one of the best preparations of these types of compounds (Equation (178)). Other syntheses of triarylphosphine sulfides include the reaction of phosphines with disulfides, trisulfides or sulfenamides (Equations (179) and (180)) ⟨68ZN1030⟩. Many other sulfur-containing compounds react with triarylphosphine to afford triarylphosphine sulfides.

(178)

(179)

(180)

Thiiranes react with triphenylphosphine or triphenylphosphonium ylides to yield the corresponding alkenes and triphenylphosphine sulfide ⟨59JA6330, 87CL357⟩. 1-Phenylphosphole has been found to react with sulfur to give the corresponding phosphine sulfide dimer (Equation (181)) ⟨68TL1755⟩.

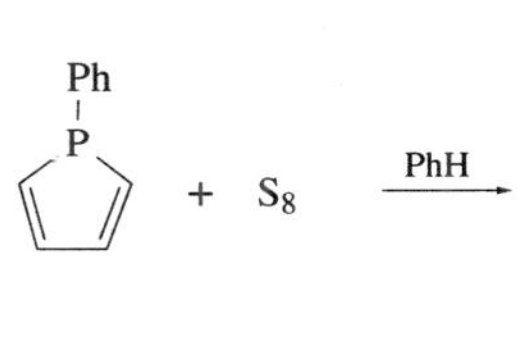
(181)

Reaction of triphenylphosphonium ylides with elemental sulfur is a good method for the synthesis of thioketones, 1,2,3,4,6-pentathiepane, and alkenes, in which triphenylphosphine sulfide is obtained as a side product ⟨19HCA635, 72BCJ506, 88BCJ4323⟩.

Vinylphosphine sulfides are important reagents for the synthesis of polyphosphines. Dimethylthiophosphinyl bromide, methylthiophosphoryl chloride, and thiophosphoryl chloride are converted to dimethylvinylphosphine sulfide, methyldivinylphosphine sulfide, and trivinylphosphine sulfide, respectively, by reaction with vinylmagnesium bromide. This method can be applied to large-scale synthesis. These vinylphosphine sulfides react with diphenylphosphine to afford poly(*t*-phosphine) monosulfides, which undergo desulfurization to give poly (*t*-phosphines) (Scheme 28) ⟨75JA53⟩. The synthesis of fluorinated vinylphosphine sulfide has been reported ⟨92SL833⟩.

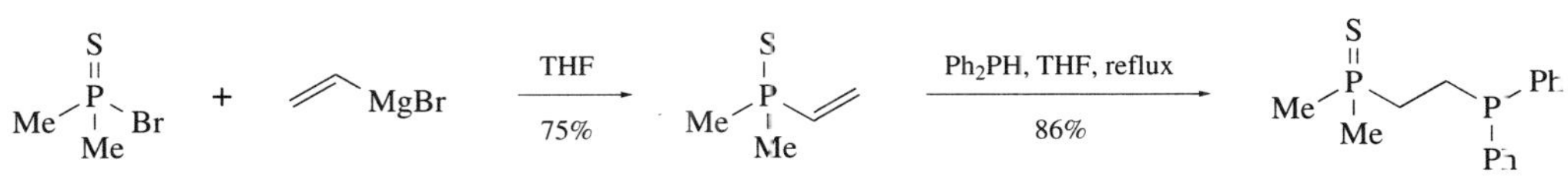

Scheme 28

2.16.4.4 Arylthiophosphinic Acids and their Derivatives ($Ar_2P(S)(OH)$, $Ar_2P(O)SH$, $Ar_2P(S)SH$, $Ar_2P(S)Hal$)

t-Butylphenylphosphinic acid chloride is converted to thiophosphinic acid by treatment with NaSH followed by the addition of acid ⟨77ZN(B)584⟩.

Many methods for the preparation of aromatic thiophosphinic acid *S*-esters have been reported, and include the sulfenylation of diarylphosphine oxides (Equation (182)) ⟨67T1065⟩ or phosphinous acid esters (Equation (183)) ⟨68JCS(C)2131⟩, reaction of diphenylphosphinous acid chloride with sulfinic acid (Equation (184)) ⟨67M1792⟩, oxidation of phosphinous acid *S*-esters, thiation of phosphinic acid chlorides (Equations (185) and (186)) ⟨69ZOB341⟩, and alkylation of thiophosphinic acid ⟨76PS(1)211⟩.

$$Ph_2P(O)H + PhSSPh \longrightarrow Ph_2P(O)SPh + PhSH \quad (182)$$

$$Ph_2POMe + PhSSPh \longrightarrow Ph_2P(O)SPh + PhSMe \quad (183)$$

$$2Ph_2PCl + PhSO_2Na \longrightarrow Ph_2P(O)SPh + Ph_2P(O)Cl \quad (184)$$

$$Ph_2PSR \xrightarrow{[O]} Ph_2P(O)SR \quad (185)$$

$$Ph_2P(O)Cl + NaSR \longrightarrow Ph_2P(O)SR + NaCl \quad (186)$$

A Friedel–Crafts type reaction of benzene with $P(S)Cl_3$ in the presence of Lewis acids affords diphenylthiophosphinic acid chloride (Equations (187) and (188)) ⟨62JINC1073, 64HCA120⟩.

$$Ph(S)Cl_3 + AlCl_3 + PhH \longrightarrow Ph_2P(S)Cl \quad (187)$$

$$PhP(S)Cl_2 + Me_4Pb \longrightarrow PhMeP(S)Cl \quad (188)$$

Chlorination of diphosphine disulfides is another method for the preparation of arylthiophosphinic acid chloride ⟨62JOC4444⟩.

Bifunctional dithiophosphinic acids are obtained by nucleophilic fission of perthiophosphonic anhydrides with di-Grignard reagents (Equation (189)) ⟨78CB629⟩. Direct thiation of ethylenebisphosphines is an alternative method for the preparation of bifunctional dithiophosphinic acids ⟨68CB2197⟩.

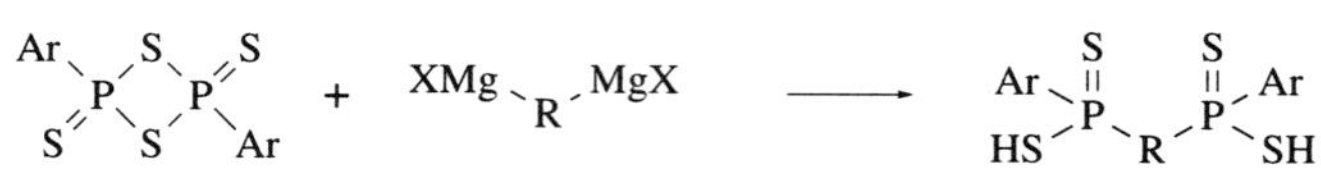
(189)

Dechlorination of phenylthiophosphonic acid dichloride with magnesium in the presence of 2,3-dimethylbutadiene affords cyclic dithiophosphinic acid or thiophosphinic acid derivatives (Scheme 29) ⟨71CC1186⟩.

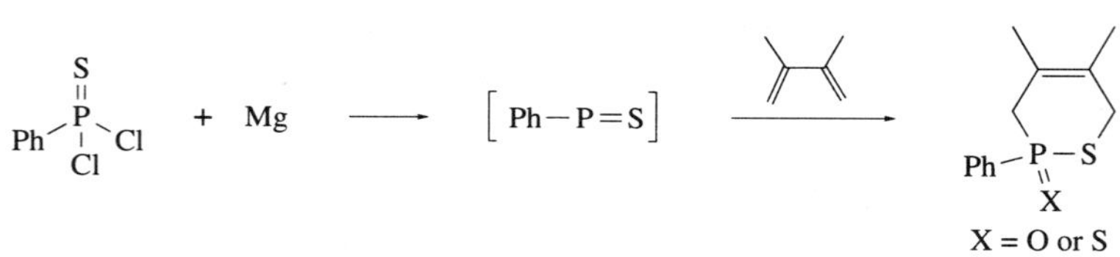

Scheme 29

2.16.4.5 Arylthiophosphonic Acids and their Derivatives ($ArP(S)(OH)_2$, $ArP(O)(OH)SH$, $ArP(O)(SH)_2$, $ArP(S)OH(SH)$, $ArP(S)(Hal)_2$)

Arylthiophosphonic acid chlorides are converted to thiophosphonic acid *S*-esters by hydrolysis (Equation (190)) ⟨66ZOB500⟩.

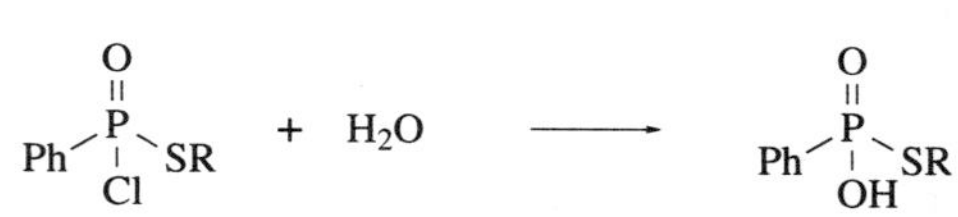
(190)

Menthyl phenylphosphinate reacts with sulfur and methyl iodide to afford menthyl *S*-methylphenylthiophosphonate with retention of configuration at phosphorus (Scheme 30) ⟨71JA3792⟩.

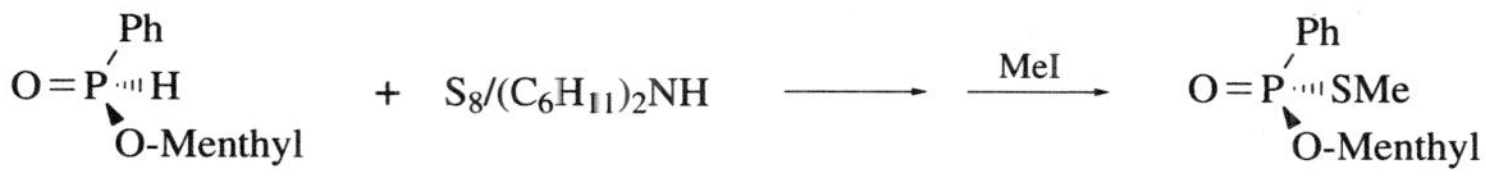

Scheme 30

There are many methods for the synthesis of arylthiophosphonic acid *O*-,*S*-diesters, which include alkylation of arylthiophosphonic acid *O*-esters (Equation (191)) ⟨74JAP7401534⟩, sulfenylation of phenylphosphinic acid *O*-esters (Equation (192)) ⟨67JAP6703120⟩ or phenylphosphonous acids (Equation (193)) ⟨63JA1618⟩, and isomerization of phenylphosphonothioic acid *O*-,*O*-diesters (Equation (194)) ⟨68TL5015⟩.

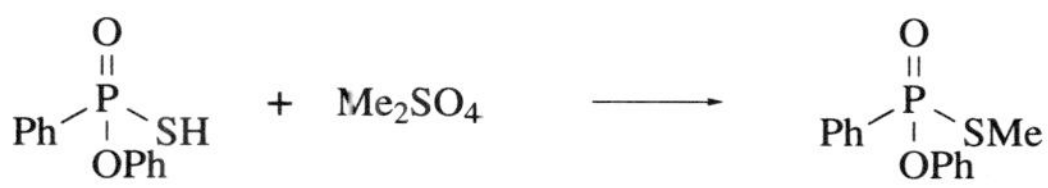

(191)

(192)

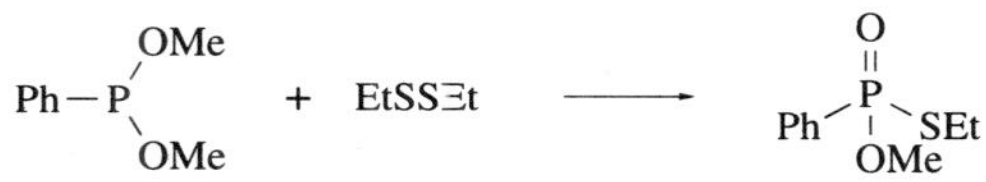

(193)

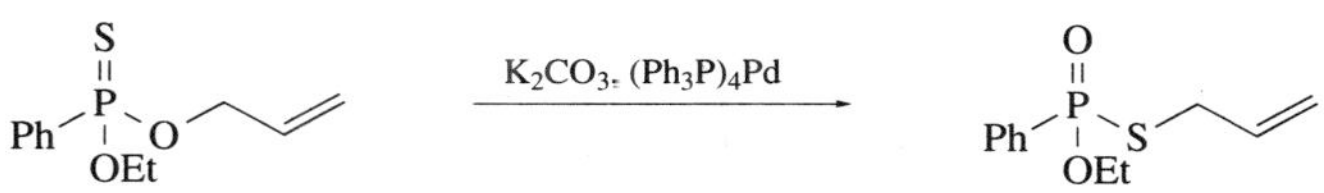

(194)

Peake *et al.* prepared cyclic trithiophosphonic acid diesters by thiation of dithiaphospholanes (Equation (195)) ⟨72JCS(P2)380⟩.

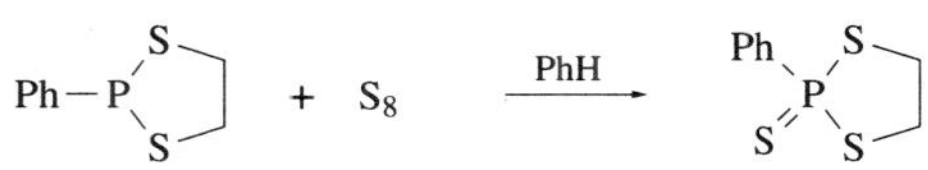

(195)

Phenyltrithiophosphonic acid bis(trimethylsilyl) ester is formed from phenylbis(trimethylsilyl)phosphine by the addition of sulfur ⟨86ZAAC(543)7⟩. Interaction of phenylphosphonic dichloride with hydrogen sulfide in the presence of triethylamine in benzene yields trimeric thionophosphine oxide (phenylmonothiophosphonic anhydride) (Equation (196)) ⟨72HCA1991⟩.

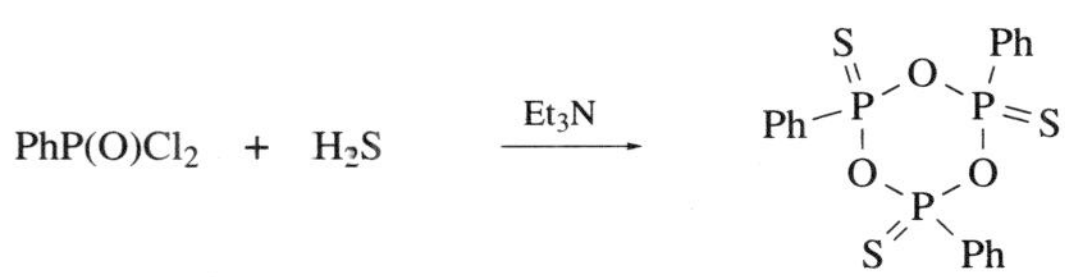

(196)

Sodium fluoride suspended in sulfolane is found to react with phenylphosphonothioic dichloride, to give phenylphosphonothioic difluoride (Equation (197)) ⟨60JOC2015⟩.

$$PhP(S)Cl_2 + 2NaF \longrightarrow PhP(S)F_2 + 2NaCl \qquad (197)$$

Perthiophosphonic anhydrides react with phenyltetrafluorophosphorane, to give bis(fluorophenylthiophosphinyl) sulfides in ~60% yields (Equation (198)) ⟨72JCS(D)1590⟩.

(198)

Perthiophosphinic anhydrides react with ethylene glycol to afford *O*-(2-hydroxyethyl) hydrogenarylphosphonodithioates and *O,O*-(1,2-ethylene)bis(arylphosphonodithioic) acid. However, it has been reported that the products are 1,3,2-dioxaphosphorinane-2-sulfide, cyclic trithiopyrophosphate and thioacetamide (Scheme 31) 〈93T1271〉.

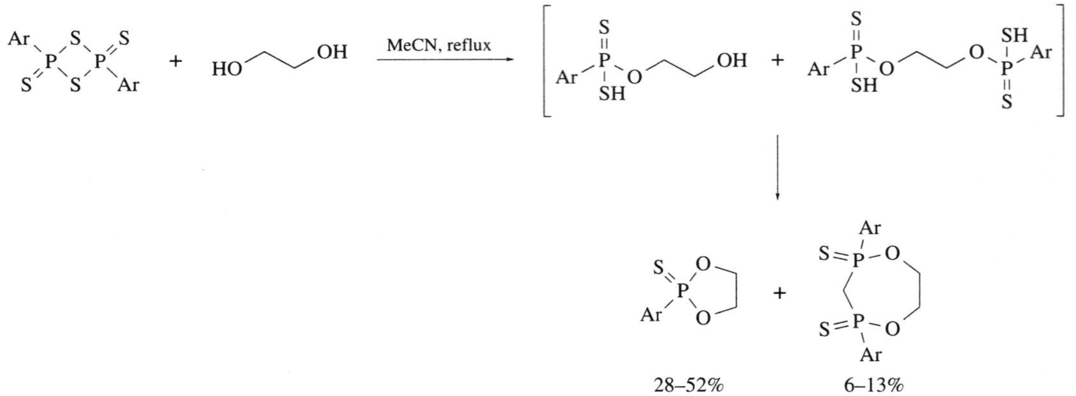

Scheme 31

2.16.4.6 Arylphosphorane Compounds with Sulfur Function

Sulfur-containing spirocyclic phosphoranes are prepared by the reaction of dithiocatechol phosphoranes with $PhPF_5^-$ or by the reaction of phenyltetrafluorophosphorane with *o*-phenyleneoxathia- or *o*-phenylenedithiabis(trimethylsilane) in 64–91% yields. The crystal structures of these spirocyclic phosphoranes suggest that the geometries around the phosphorus atoms are 31–93% displaced from the trigonal bipyramid toward the rectangular pyramid (Scheme 32 and Equation (199)) 〈73CC144, 79JA3790, 83IC1771〉.

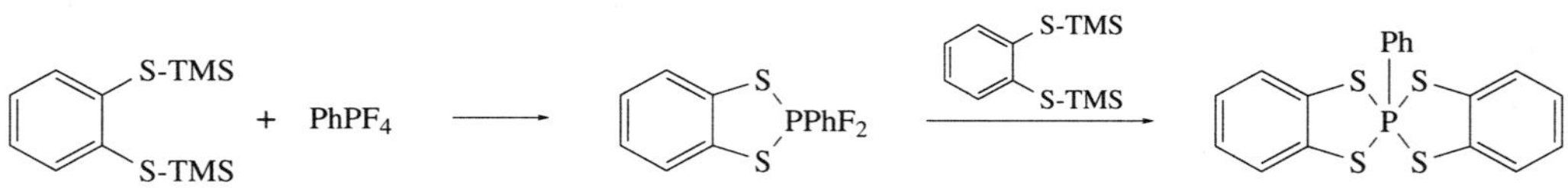

Scheme 32

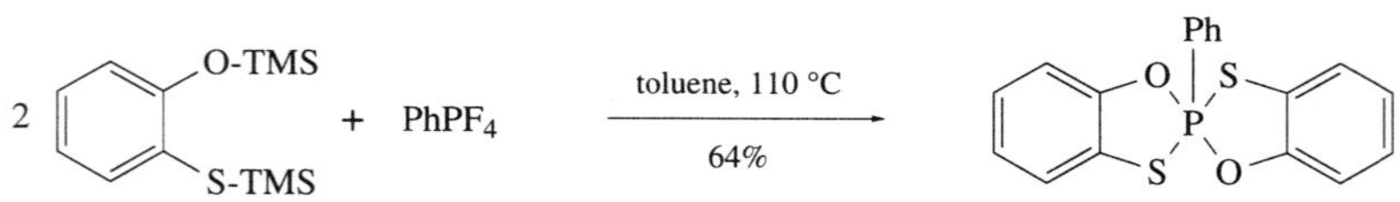

(199)

Aryl 2-nitrophenyl sulfides are transformed into benzothiazaphoranes by the reaction with 2-phenyl-1,3,2-dioxaphosphorane in approximately 70% yield; the x-ray crystallographic structure shows a distorted rectangular pyramid.

2.16.5 COMPOUNDS WITH A P—Se AND/OR A P—Te BOND

The chemistry of P—Se and P—Te bonds is not well documented in comparison with that of P—N and P—S bonds, because of their relative instability. However, interest in heteroatomic chemistry has included compounds containing selenium or tellurium because of the progress of NMR spectroscopy. The NMR study of ^{77}Se and ^{125}Te nuclei, in addition to the ^{31}P nucleus, gives much information on the structural analysis of Se- and Te-containing compounds 〈B-92MI 216-01〉.

Diphenylphosphinous chloride is transformed to diphenylselenophosphinous acid *Se*-phenyl ester by reaction with phenylselenomagnesium chloride (Equation (200)) 〈65CI(L)1076〉.

$$Ph_2PCl + PhSeMgCl \longrightarrow Ph_2PSePh \qquad (200)$$

Phenyldihalophosphines react with methylselenylsilane or mercury diselenides to give phenyldiselenaphosphonous acid esters (Equations (201) and (202)) ⟨75INC233, 79ZAAC(448)35⟩.

$$PhPCl_2 + 2TMS\text{-}SeMe \longrightarrow PhP(SeMe)_2 + 2TMS\text{-}Cl \qquad (201)$$

$$PhPI_2 - Hg(SeCF_3)_2 \longrightarrow PhP(SeCF_3)_2 + HgI_2 \qquad (202)$$

Phosphine selenides react with oxygen to afford the corresponding selenophosphinic acids ⟨62MI 216-01⟩. Selenium reacts with phosphinous acid chlorides or esters to give selenophosphinic acid esters (Equations (203) and (204)) ⟨78ZOB2673⟩. Tetraphenyldiphosphine dioxide reacts with phenylselenenyl bromide to give diphenylselenophosphinic acid *Se*-phenyl ester ⟨68CB3070⟩.

$$Ph_2PCl + Se_n + Pr^iOH \longrightarrow Ph_2P(=Se)\text{-}OPr^i \qquad (203)$$

$$Ph_2POBu^n + Se_n \longrightarrow Ph_2P(=Se)\text{-}OBu^n \qquad (204)$$

Petragnani *et al.* have developed six ways to synthesize diphenylselenophosphinic acid *Se*-phenyl ester (Equations (205)–(207) and Schemes 33–35) ⟨68CB3070⟩.

$$Ph_2P(O)Cl + PhSeMgBr \longrightarrow Ph_2P(O)SePh \qquad (205)$$

$$Ph_2PSePh + PhSeCl \longrightarrow \xrightarrow{H_2O} Ph_2P(O)SePh \qquad (206)$$

$$Ph_2P(O)P(O)Ph_2 + PhSeBr \longrightarrow Ph_2P(O)SePh \qquad (207)$$

$$Ph_2PCl + PhSeH \longrightarrow Ph_2PSePh \xrightarrow{[O]} Ph_2P(O)SePh$$

Scheme 33

$$Ph_2PMgCl + PhSeBr \longrightarrow Ph_2PSePh \xrightarrow{[O]} Ph_2P(O)SePh$$

Scheme 34

$$Ph_2PMgCl + PhSeSePh \longrightarrow Ph_2PSePh \xrightarrow{[O]} Ph_2P(O)SePh$$

Scheme 35

Diphenylselenophosphinic acid chloride reacts with ethanethiol to give diphenylselenothiophosphinic acid *S*-ethyl ester (Equation (208)) ⟨78ZOB2673⟩.

$$Ph_2P(Se)Cl + EtSH \xrightarrow{Et_3N} Ph_2P(Se)SEt \qquad (208)$$

Unusual three-membered cyclic selenadiphosphiranes are formed by the reaction of sterically hindered diphosphene with selenium along with diselenoxophosphoranes (Equation (209)) ⟨84CL603⟩.

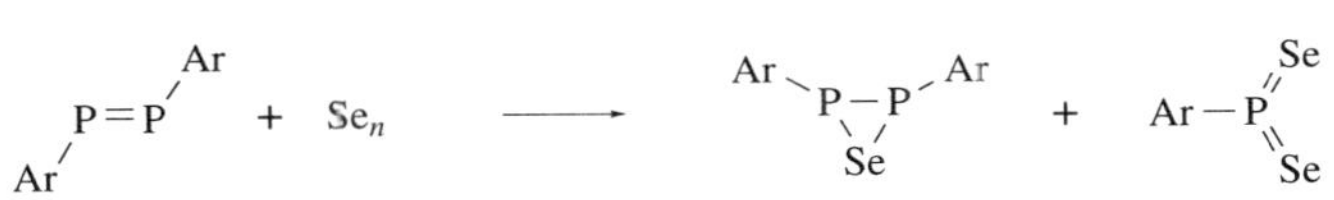

(209)

2,4-Di-*t*-butyl-6-(dimethylamino)phenyldiselenoxyphosphorane has been found to be deselenated with hexamethylphosphorus triamide to give the corresponding metastable selenoxophosphine, which disproportionates to afford the corresponding diselenoxophosphorane and diphosphene (Equation (210)) ⟨93TL1043⟩.

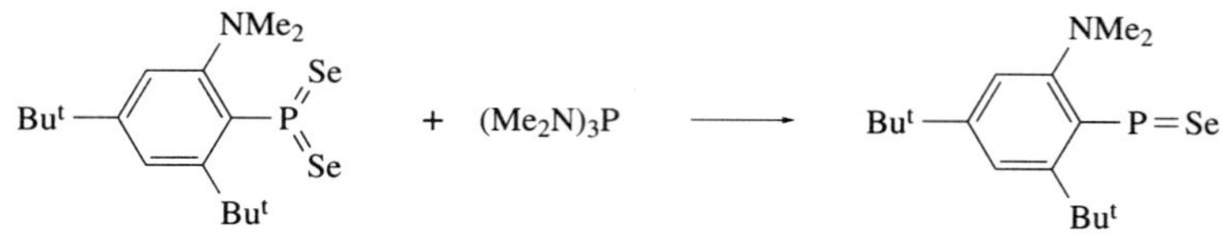

(210)

Triarylphosphine selenides have been produced by fusing triaryl phosphine with elemental selenium (Equation (211)) ⟨01LA(315)43⟩. This is a typical and general method for the synthesis of aryl- and alkylphosphine selenides, and diphenylphosphine selenide is also prepared this way (Equation (212)) ⟨66HCA1000⟩. Arylphosphines have been found less reactive toward selenium or tellurium than alkylphosphines. In addition, arylphosphine selenides are also prepared by the reaction of phosphine with diselenide (Equation (213)) ⟨77BCJ3007⟩.

$$Ar_3P + Se_n \longrightarrow Ar_3P{=}Se \quad (211)$$

$$Ph_2PH + Se_n \longrightarrow Ph_2HP{=}Se \quad (212)$$

(213)

Similar treatment of triphenylphosphoranylidenehydrazones with elemental selenium gives the corresponding selones and triphenylphosphine selenide (Equation (214)) ⟨75CC539, 76JCS(P1)2079⟩. Reaction of Wittig reagents with elemental selenium forms alkenes and triphenylphosphine selenide (Scheme 36) ⟨87TL6649, 88JA624⟩.

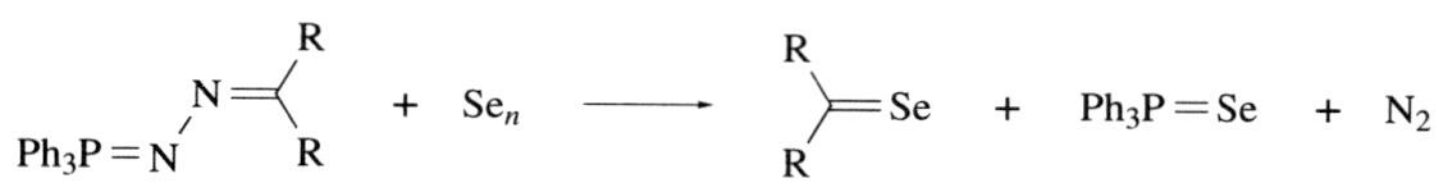

(214)

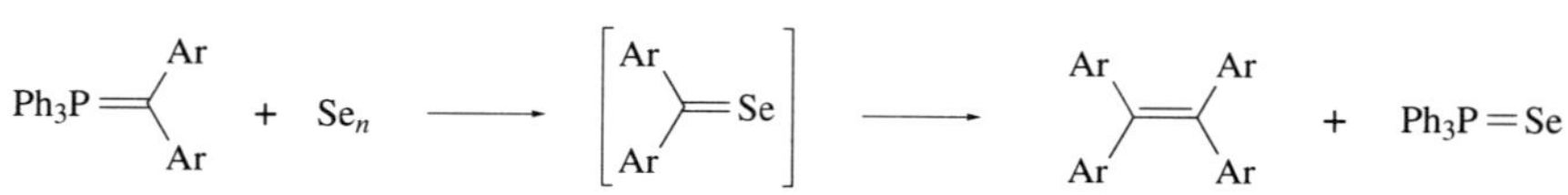

Scheme 36

Diphenylselenophosphinous acid phenyl ester upon heating undergoes isomerization to give triphenylphosphine selenide ⟨69MI 216-01⟩.

Campbell *et al.* have reported that a phosphine oxide is transformed into the corresponding phosphine selenide by heating with selenium in potassium dihydrogen phosphate at 270–370 °C (Equation (215)) ⟨65JCS2184⟩.

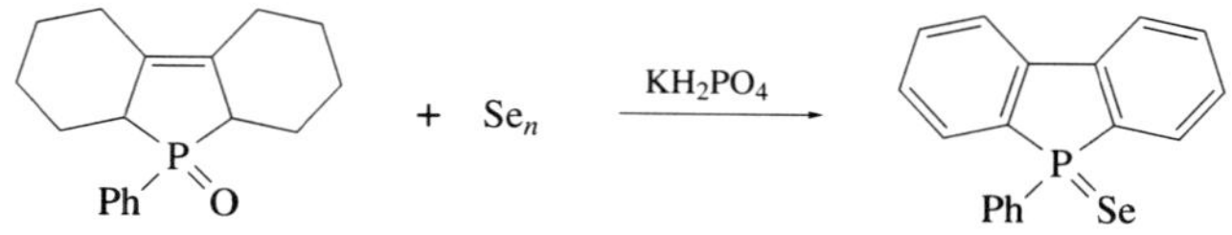

(215)

Triphenylphosphine telluride is found to be a relatively unstable solid and to exist as $Ar_3P{=}Te \cdot PAr_3$ (Scheme 37) ⟨73ACS1939⟩. Monophenylphosphine tellurides are prepared by direct telluriation of phosphine in refluxing toluene (Equation (216)) ⟨65JOM(4)320⟩.

$$Ar_3P + Te \longrightarrow Ar_3P{=}Te{\bullet}PAr_3 \rightleftharpoons Ar_3P + Ar_3P{=}Te$$

Scheme 37

$$PhR_2P + Te \longrightarrow PhR_2P{=}Te \quad (216)$$

2.16.6 COMPOUNDS WITH A P—N BOND

2.16.6.1 Amides of Arylphosphinous Acid (Ar_2PNR_2, $ArPHNR_2$)

A typical method for the preparation of arylphosphinous amide is amination of arylphosphinous halides (Equations (217) and (218)) ⟨61JOC611, 66JCS(C)1592⟩.

$$Ph_2PCl + R_2NH \longrightarrow Ph_2PNR_2 \quad (217)$$

$$(C_6F_5)_2PCl + TMS\text{-}NMe_2 \longrightarrow (C_6F_5)_2PNMe_2 \quad (218)$$

This method is applied to the synthesis of atropisomeric chiral bisaminophosphine derivatives, which are useful reagents for asymmetric hydrogenation ⟨80CL729, 83CL441⟩. Three-membered cyclic phosphinous amides are prepared by the reaction of 1-chloro-1H-phosphirene with lithium dialkylamides (Equation (219)) ⟨89AG(E)225⟩.

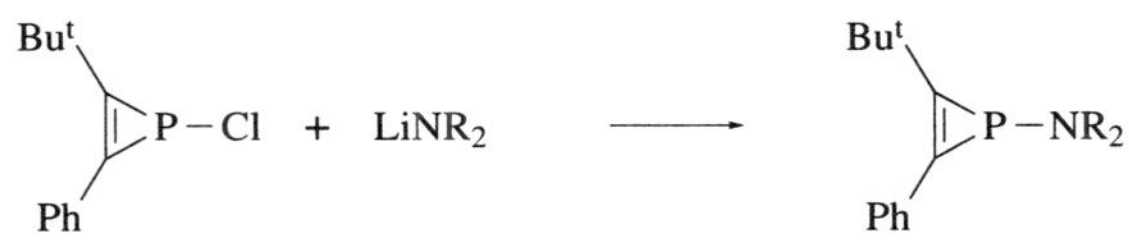

(219)

2.16.6.2 Amides of Arylphosphonous Acid ($ArP(NR)_2$, $ArP(SR)NR_2$, $ArP(NR_2)Hal$)

The typical methods for the synthesis of arylphosphonous acid diamides ($ArP(NR^1R^2)_2$) include the amination of aryldichlorophosphines and the reaction of aryl Grignard reagents with diaminochlorophosphine ⟨66JCS1592, 67JCS1028⟩. Cyclic arylphosphonous diamides are synthesized by this method (Equation (220)) ⟨71IC1028⟩. Cyclic aminoethylphosphinites (1,3,2-oxazaphospholidines) are synthesized by a similar method ⟨67BCJ147⟩.

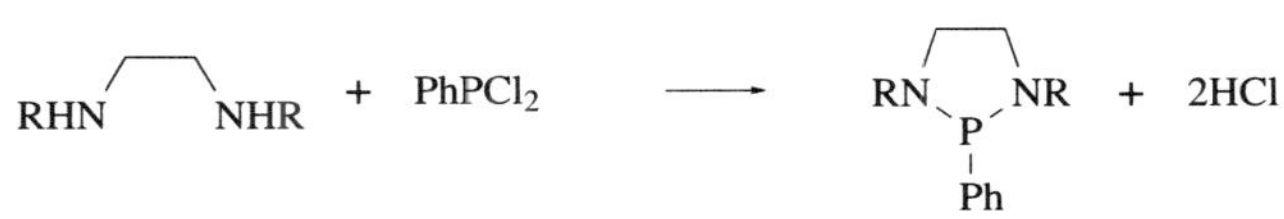

(220)

Transformation of aminodichlorophosphines into arylphosphonous amide chlorides is accomplished by the reaction with equivalent aryl magnesium bromide ⟨66JCS1592⟩. Phosphonousamide chloride is also obtained by equimolecular amounts of dichlorophosphine and arylphosphonous diamides ⟨79ZOB2230⟩.

2.16.6.3 Arylphosphine Imides ($Ar_3P{=}NR$)

Triarylphosphine imides are well known as aza-Wittig reagents, which are synthetically versatile for the preparation of imines, amines, and nitrogen heterocycles. Their synthesis may be achieved by direct amination of triarylphosphine, or by reaction of triarylphosphine dihalides with amines. The synthesis of phosphine imides (phosphazenes) was developed by Staudinger in 1919, and is via the addition of an organic azide to a phosphorus(III) derivative, as shown in Scheme 38 ⟨19HCA635⟩. Decomposition of the intermediate phosphazides, which can be isolated in some cases, leads to the phosphine imides with evolution of nitrogen.

$$Ph_3P + PhN_3 \longrightarrow Ph_3P{=}N{-}N{=}NPh \longrightarrow Ph_3P{=}NPh$$

Scheme 38

Many azides such as aryl, acyl, and sulfonyl azides have been used in this reaction; trimethylsilyl azide is particularly useful, since the silyl group from the resulting phosphine imides may easily be removed. Cyclodextrin phosphine imide was prepared from the reaction of cyclodextrin azides with triphenylphosphine (Scheme 39) ⟨78HCA2190⟩. This is a good method for the preparation of aminocyclodextrin.

$$\alpha\text{-CD}(N_3)_6 \xrightarrow{Ph_3P} \alpha\text{-CD}(N{=}PPh_3)_6 \xrightarrow{NH_4OH} \alpha\text{-CD}(NH_2)_6$$

Scheme 39

N-Acylphosphine imides are easily prepared by the reaction of *N*-trimethylsilylphosphine imide with acyl chlorides or acid anhydrides (Equation (221)) ⟨72S695⟩.

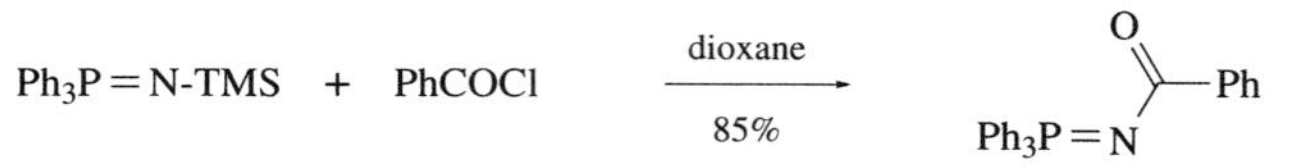

(221)

Another method, described by Kirsanov and Horner, consists of the reaction of triphenylphosphine dihalides with amines. This method is widely applied to the synthesis of the phosphine imides containing various substituents (Scheme 40) ⟨56ZOB907, 59LA(627)142⟩.

$$Ar_3PCl_2 + RNH_2 \longrightarrow Ar_3P(Cl)NHR \longrightarrow Ar_3P{=}NR + RNH_3^+\,Cl^-$$

Scheme 40

Treatment of phosphorus(III) compounds with *N*-chloroamines produces phosphonium salts, which can be converted into monophosphazenes with base. The use of sodium salts of chloroamines, for example, chloramine T, requires no base.

2-Hydroxy- or 2-amino-azides react with triphenylphosphine to afford 5-membered cyclic oxaaza- or diazaphosphoranes (Scheme 41) ⟨84HCA1238⟩.

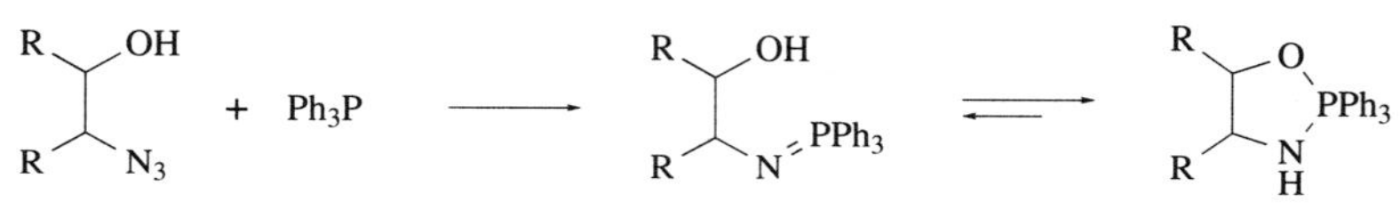

Scheme 41

Amino-substituted monophosphazenes are transformed into polyaminophosphazenes by reaction with tris(dimethylamino)phosphonium chloride followed by demethylation ⟨87AG(E)709⟩.

Phosphorus ylides react with Schiff bases or nitriles to afford phosphine imides.

2.16.6.4 Amides of Arylphosphinic Acids ($Ar_2P(O)NR_2$, $Ar_2P(Se)NR_2$)

Relatively few reports on the synthesis of arylphosphinic amides and related compounds have been made. The most convenient method for the synthesis of arylphosphinic amides is via the reaction of arylphosphinic halides with amines (Equation (222)) ⟨57JCS2413⟩. Arylthiophosphinic amides are prepared in a similar manner (Equation (223)) ⟨72JCS(P1)1184⟩.

$$Ph_2P(O)Cl + PhNH_2 \longrightarrow Ph_2P(O)NHPh \tag{222}$$

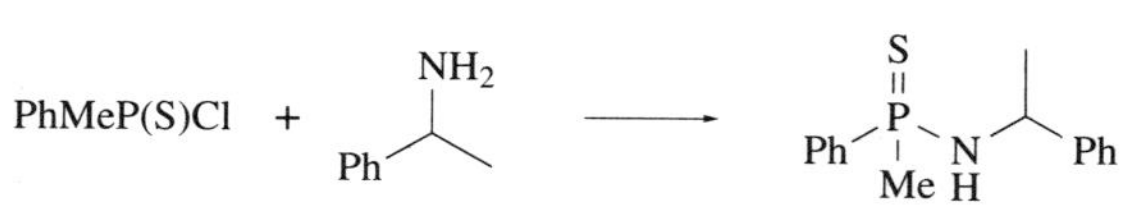

(223)

Felcht and Regitz have reported a novel synthesis of phosphinic pyrazolides via 1,5-sigmatropic rearrangement of the reaction product of a phosphinyldiazomethane with methyl propiolate; they undergo aminolysis to afford other phosphinic amides (Scheme 42) ⟨77LA1309⟩.

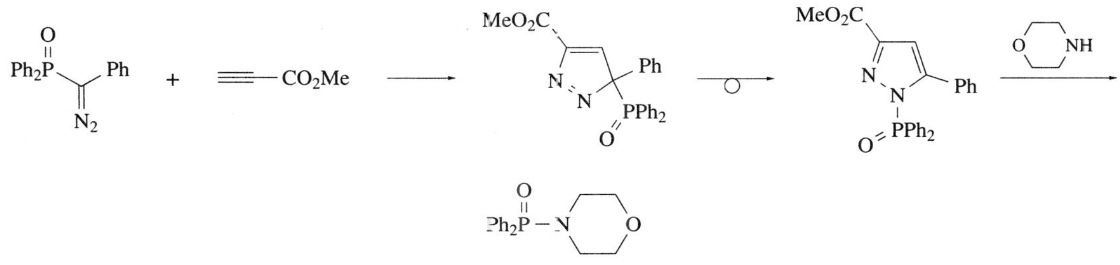

Scheme 42

An optically active phosphinic amide is prepared by the reaction of an asymmetric 1,3,2-oxaazaphospholane with alkyl halides, which proceeds with retention of configuration at phosphorus (Equation (224)) ⟨89ZN(B)481⟩.

(224)

Diphenyltrifluorophosphorane is converted to trimethylsilylfluorodiphenylphosphinimidate by treatment with lithium bis(trimethylsilyl)amide at −78 °C in 71% yield (Equation (225)) ⟨77IC1460⟩.

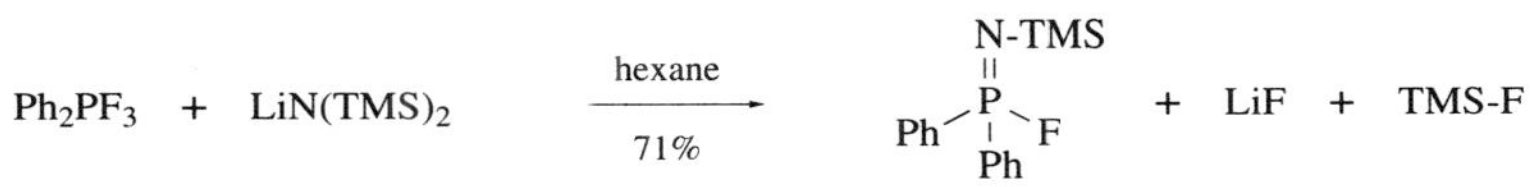

(225)

Phenylthiophosphinic amides are treated with $NiCl_2$ or $CoCl_2$ to afford the corresponding thiophosphinic *N*-amidate chelates, whose x-ray crystallographic and magnetic analyses have shown that these complexes are planar in the solid state ⟨91ZAAC(606)119⟩.

2.16.6.5 Amides of Arylphosphonic Acids ($ArP(O)(NR_2)_2$, $ArP(O)(NR_2)Hal$, $ArP(S)(NR_2)_2$, etc.)

There have been many methods reported for the synthesis of arylphosphonic amides. The simplest and most convenient method is the reaction of arylphosphoryl chlorides with amines. Since the first introduction, many researchers have developed and applied this method (Equation (226)) ⟨1896LA(293)193, 57JOC265⟩. Thio-analogues of this class of compounds are prepared in a similar manner (Equation (227)) ⟨1896LA(293)215, 60JOC666⟩.

$$PhP(O)Cl_2 + NH_3\,(\text{liq.}) \xrightarrow[94\%]{} PhP(O)(NH_2)_2 \quad (226)$$

$$PhP(S)Cl_2 + 2PhNH_2 \xrightarrow[61\%]{\text{ether, RT}} PhP(S)(NHPh)_2 \quad (227)$$

Aryl Grignard reagents react with dianilido phosphorochloridate to give arylphosphonic acid dianilides (Scheme 43) ⟨51JA5896⟩.

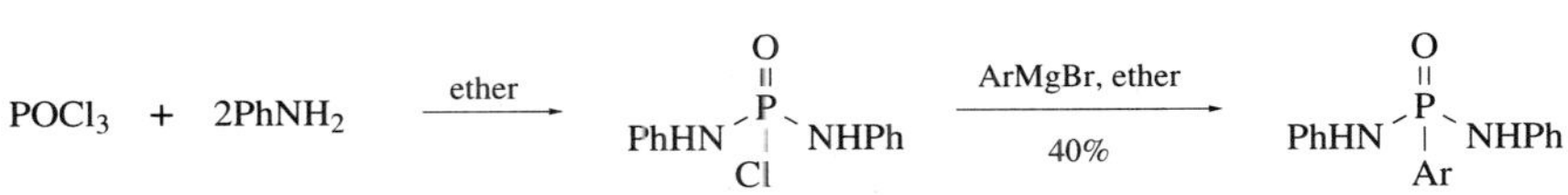

Scheme 43

Oxidation of arylphosphonous amides is another method for the synthesis of arylphosphonic amides. Arylthiophosphonic amides and arylselenophosphonic amides are synthesized in similar manners (Equations (228)–(230)) ⟨67JCS(A)1492⟩.

$$PhP(NHPr^i)_2 + MgO_2 \xrightarrow[33\%]{PhH} PhP(O)(NHPr^i)_2 \quad (228)$$

$$PhP(NHPr^i)_2 + S_8 \xrightarrow[100\%]{PhH} PhP(S)(NHPr^i)_2 \quad (229)$$

$$PhP(NHBu^t)_2 + Se_n \xrightarrow[87\%]{PhH} PhP(Se)(NHBu^t)_2 \quad (230)$$

Dimethyl phenylphosphonite and phenylphosphonous bis(dimethylamide) are transformed into the corresponding phenylphosphorimidates in good yields via the Staudinger reaction. Dimethyl phenylphosphorimidate is thermally (140 °C) rearranged to give methyl *N*,*N*-dimethylphenylphosphoramidate, together with many side products (Scheme 44) ⟨75JCS(D)12⟩.

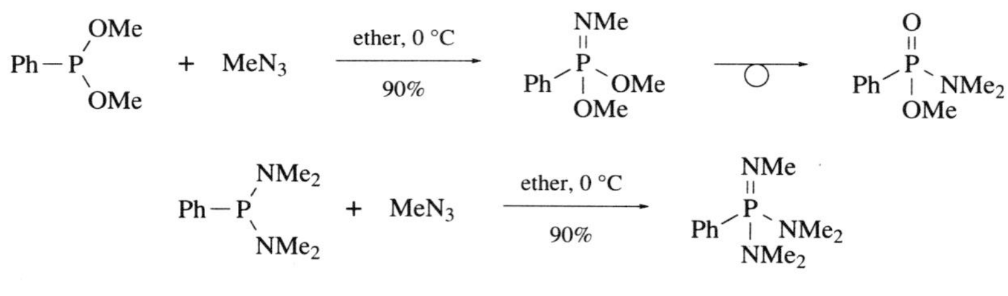

Scheme 44

In the reaction with amines, arylphosphonic or arylthiophosphonic difluorides suffer replacement of a fluoride atom with the amino group to furnish arylphosphonic or arylthiophosphonic fluoroamides (Equations (231)–(233)) ⟨69ZN(B)1250, 70ZAAC(375)140, 77IC1460⟩.

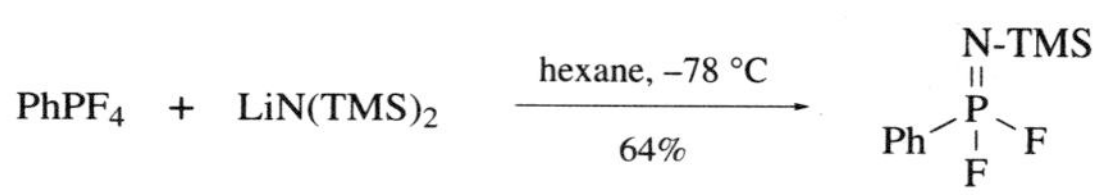

(231)

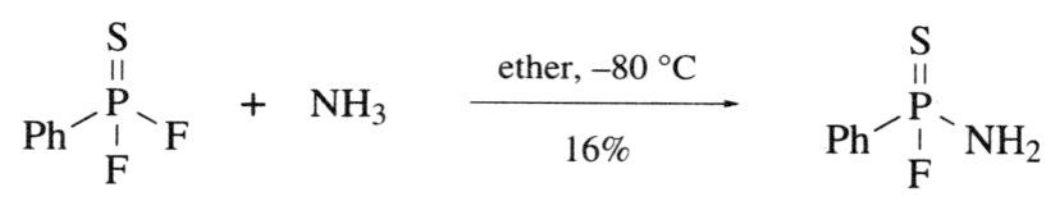

(232)

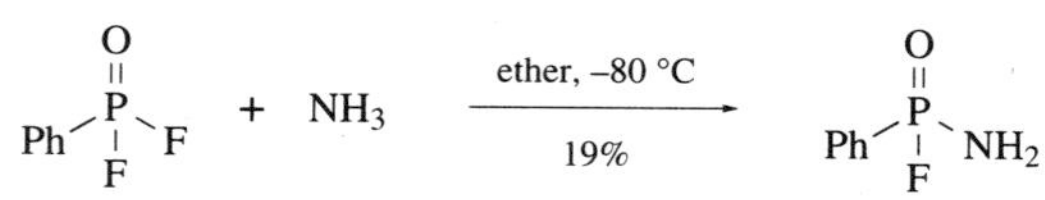

(233)

N,*N*-Dimethyltrimethylsilylamine reacts with diphenyldithiophosphonic anhydride to give trimethylsilyl phenyldithiophosphonic dimethylamide (Equation (234)) ⟨77ZAAC(431)221⟩.

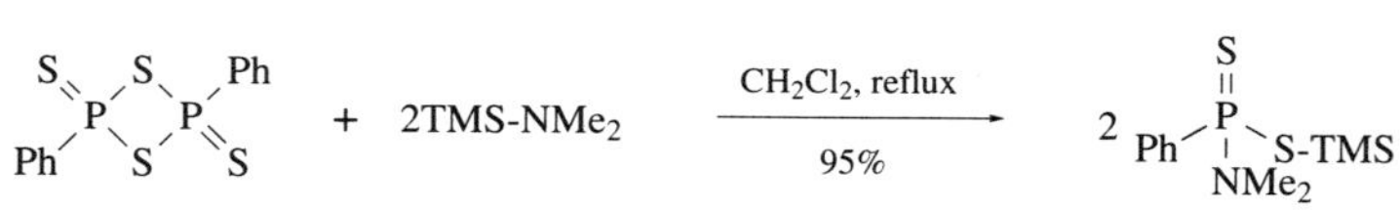

(234)

Photolysis of methylphenylphosphinic azide in methanol gives methyl *N*-methylphenylphosphonamidate and methyl *N*-phenylmethylphosphonamidate, which are produced via migration of the phenyl or methyl group from P to N in a Curtius-like rearrangement. Intermediacy of monomeric metaphosphonimidates is proposed (Scheme 45) ⟨82T3073⟩.

Cyclic arylphosphonic or arylthiophosphonic amides (1,3,2-oxazaphospholidine-2-one and 1,3,2-oxazaphospholidine-2-thione) are obtained by the reaction of ephedrine with phenylphosphonous dichloride, followed by oxidation. The crystal structure of phosphonic amides exhibits a C(4) envelope, with C(4) above the C(5)–O(1)–P(2)–N(3) plane (Scheme 46) ⟨91JCS(P2)2081⟩.

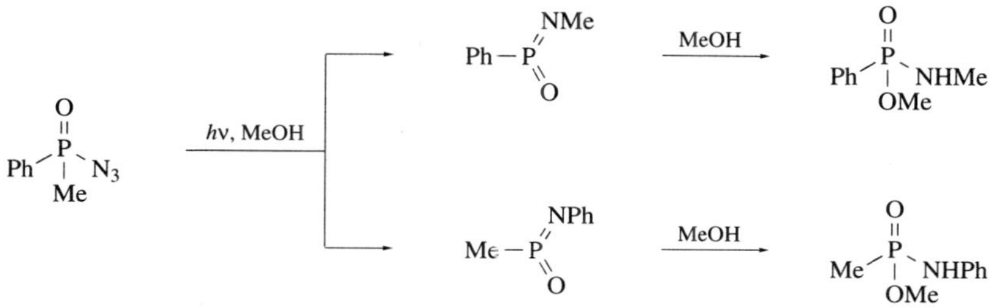

Scheme 45

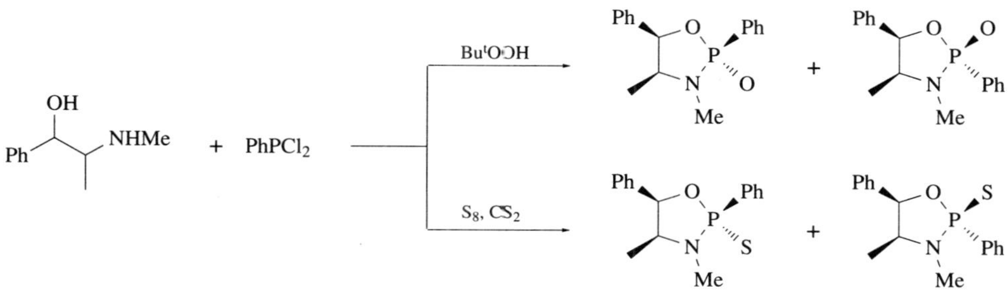

Scheme 46

2.16.6.6 Arylphosphorus Compounds Incorporating an NY Function

2.16.6.6.1 Y = S; Ar_2P—N═S, Ar_2P═N—S, etc.

Wiegräbbe and Bock reported the synthesis of phosphine imides with an N—S bond, which includes the reaction of triphenylphosphine dibromide with sulfonamides, or of triphenylphosphine with sulfonyl azides or chloramine-T (Equations (235)–(237)) ⟨68CB1414⟩.

$$Ph_3PBr_2 + RSO_2NH_2 \longrightarrow Ph_3P{=}N{-}SO_2R \qquad (235)$$

$$Ph_3P + RSO_2N_3 \longrightarrow Ph_3P{=}N{-}SO_2R \qquad (236)$$

$$Ph_3P + MeC_6H_4SO_2NNaCl \longrightarrow Ph_3P{=}N{-}SO_2C_6H_4Me \qquad (237)$$

2.16.6.6.2 Y = N; Ar_2P—N═N—R, Ar_2P—N—N═N

Diarylphosphinic azides are prepared by the reaction of diarylphosphinyl chlorides with sodium azide. By this method, phenylphosphonic and phenylthiophosphonic dichlorides are transformed to the corresponding azides by the reaction with sodium azide (Equations (238)–(240)) ⟨61JA4466, 65JOC3866⟩.

$$Ph_2P(O)Cl + NaN_3 \xrightarrow{\text{pyridine}} Ph_2P(O)N_3 \qquad (238)$$

$$PhP(O)Cl_2 + 2NaN_3 \xrightarrow{\text{pyridine}}$$ 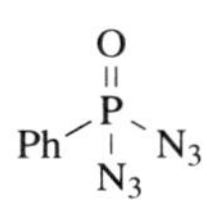 (239)

$$PhP(S)Cl_2 + 2NaN_3 \xrightarrow{\text{pyridine}}$$ 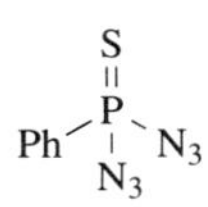 (240)

These diazides react stepwise with triphenylphosphine to afford intermediate azidophosphine imides, which further react with triphenylphosphine to yield bisphosphine imides (Scheme 47) ⟨65JOC3866⟩.

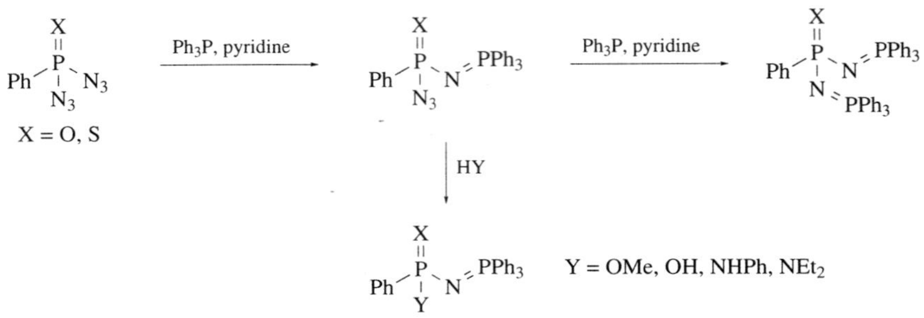

Scheme 47

Analogous to the reaction of diamine with $Ph_3P—CCl_4$, treatment of hydrazine with $Ph_3P—CCl_4$ results in the formation of *N,N′*-bis(triphenylphosphonio)hydrazine chloride, followed by dehydrochlorination to give bis(triphenylphosphoranediyl)hydrazine (Scheme 48) ⟨68ZAAC(363)176⟩.

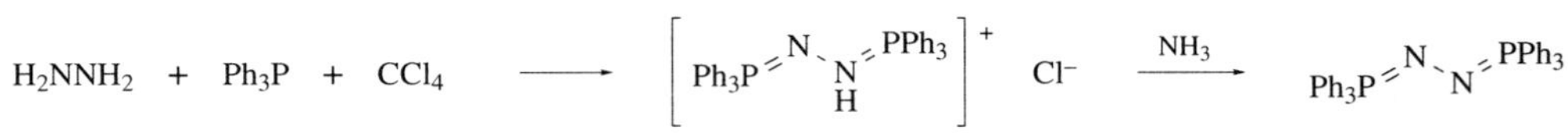

Scheme 48

2.16.6.6.3 Y = P—; $Ar_2P—N{=}P$, etc.

The best-known compounds having P—N=P bonds are cyclophosphazenes, the synthesis and chemistry of which have been reviewed by Keat and Shaw ⟨B-73MI 216-01, 79COC(2)1301, B-90MI 216-01, 91CRV119⟩.

Diphenylphosphinous chloride reacts with ammonium chloride, followed by the addition of chlorine in tetrachloroethane, to afford imide-tetraphenyldiphosphinic acid trichloride, which is treated with formic acid to give imide-tetraphenylphosphinic acid (Scheme 49) ⟨63CB3091⟩.

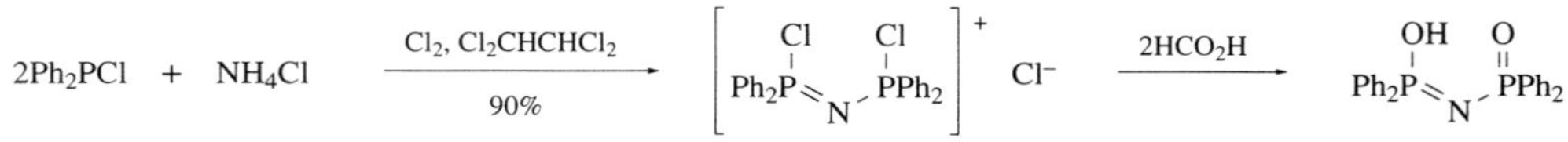

Scheme 49

2.16.7 COMPOUNDS WITH A P—P, P—As, P—Bi, OR P—Sb BOND

2.16.7.1 Arylphosphorus Compounds with a P—P Bond

The best-known compounds containing a P—P bond are diphosphines ($Ar_2P—PAr_2$). The first reported aryl diphosphine was tetraphenyldiphosphine, prepared by the reaction of diphenylphosphine with diphenylphosphinous chloride (Equation (241)) ⟨1888CB1505⟩. These compounds are easily obtained by direct phosphorylation of phosphines, reduction of diphosphine disulfides, Würtz–Fittig-like reactions, and cyclization reactions (Equations (242)–(244)).

$$Ph_2PH + ClPPh_2 \longrightarrow Ph_2P—PPh_2 + HCl \qquad (241)$$

$$R_2PNa + ClPR_2 \longrightarrow R_2P{-}PR_2 + NaCl \qquad (242)$$

(243)

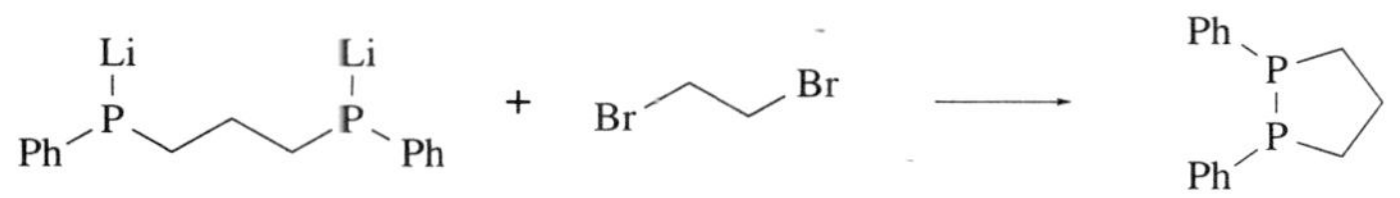

(244)

The reaction of 1,2-diphenyl-1,2-bis(trimethylsilyl)diphosphine with phenylphosphonous dichloride produces triphenylcyclotriphosphine, together with the corresponding cyclotetra-, cyclopenta-, and cyclohexaphosphines, although the relative concentration of triphenylcyclotriphosphine in the reaction mixture decreases with time ⟨77ZAAC(432)67⟩. A three-membered diphosphine has been similarly prepared by Wagner *et al.* (Equation (245)) ⟨89AG(E)225⟩.

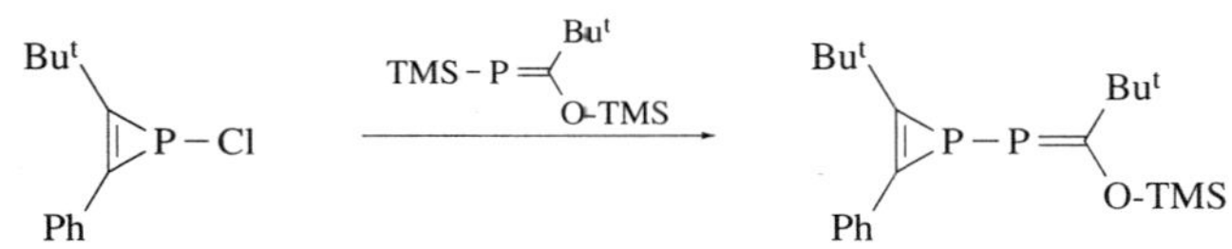

(245)

Yoshifuji *et al.* have developed bis[2,4-di-*t*-butyl-6-(dimethylamino)phenyl]diphosphene, which is more stable than the corresponding 2,4,6-tri-*t*-butyl derivative ⟨93TL1043⟩.

o-Phenylenebis(phenylphosphine) is converted to 1,2,3-benzotriphosphole, by reaction with alkylphosphonous dichlorides, and reacts with elemental sulfur to afford the corresponding 1,3-disulfides (Scheme 50) ⟨72JCS(P1)2548⟩.

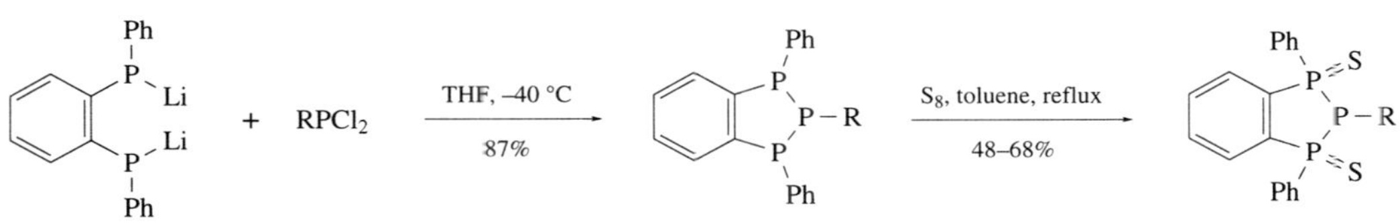

Scheme 50

2.16.7.2 Arylphosphorus Compounds with a P—As Bond

o-Phenylenebis(phenylphosphine) is transformed to 1,3,2-benzodiphospharsoles by reaction with methyl- and phenylarsonous dichloride (Scheme 51) ⟨72JCS(P1)2548⟩.

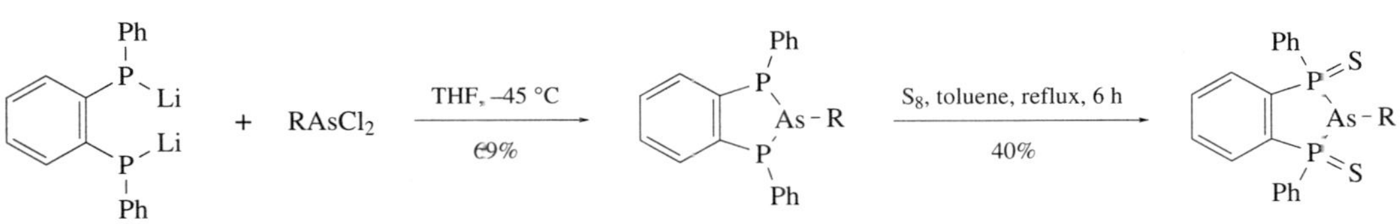

Scheme 51

The cycloaddition reaction of phosphaalkynes is a good method for the preparation of phosphorus-containing heterocycles. By this method, P—As-containing heterocycles are also obtained. The arsenic atom is located in a bridgehead position (Equation (246)) ⟨88AG(E)709⟩.

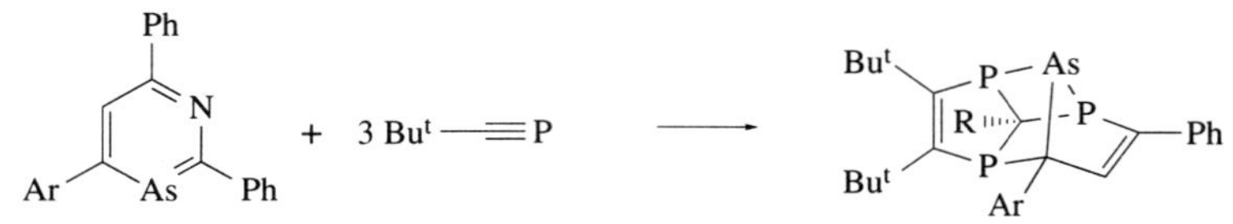

(246)

The compounds containing a P═As double bond, phosphaarsenes, are synthesized by reaction of 2,4,6-tri-*t*-butylphenylphosphine with $RAsCl_2$ or by reaction of 2,4,6-tri-*t*-butylphenylarsine with phosphonous dichlorides (Equations (247) and (248)) ⟨83CC881, 84IC2582, 87JOM(326)C6⟩.

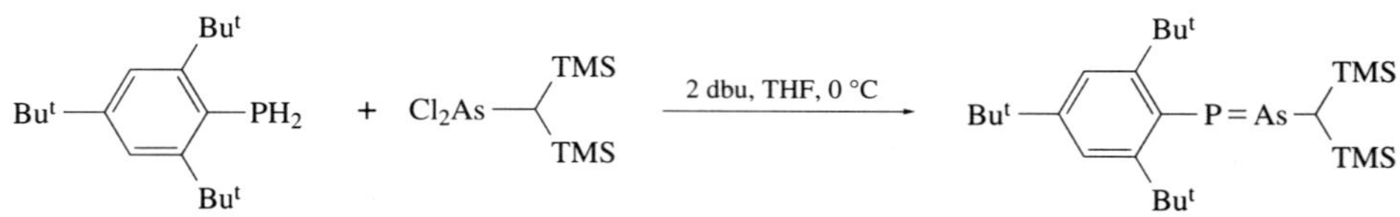

(247)

dbu = 1,5-diazabicyclo[5.4.0]undec-5-ene

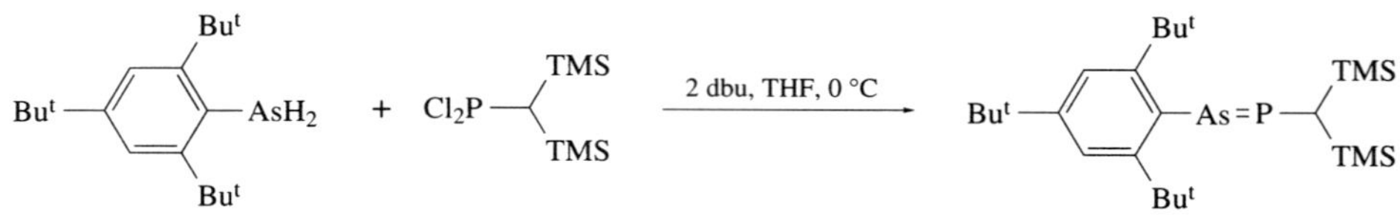

(248)

Phosphaarsenyl complexes react with sulfur to give the corresponding metal-functionalized phosphathiaarsiranes (Equation (249)) ⟨91CB725⟩.

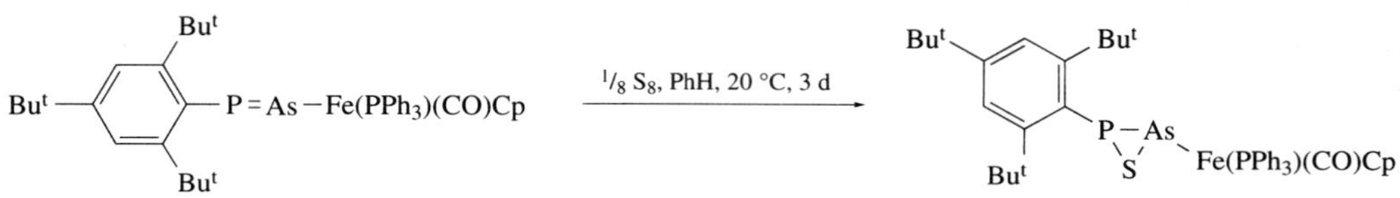

(249)

2.16.7.3 Arylphosphorus Compounds with a P—Sb Bond

Preparations of compounds with a P—Sb bond were not reported until the early 1980s. The synthesis of aromatic phosphastibenes was successfully carried out by the reaction of the lithium salt of 2,4,6-tri-*t*-butylphenyltrimethylsilylphosphine with aminodichlorostibane ⟨83TL3625⟩. Romanenko *et al.* have reported the synthesis of phosphastibene by the reaction of lithium aryl phosphine with $RNSbCl_2$ (Equation (250)) ⟨85ZOB2141⟩.

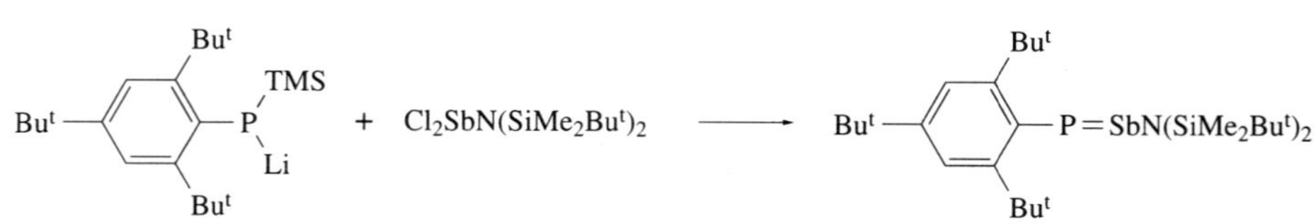

(250)

2.16.8 COMPOUNDS WITH A P—METALLOID BOND

2.16.8.1 Arylphosphorus Compounds with a P—Si Bond

Arylbis(trimethylsilyl)-, or aryl(trimethylsilyl)phosphines, or diaryl(trimethylsilyl)phosphines are readily synthesized by the reaction of the corresponding dialkali metal or monoalkali metal arylphosphides with chlorotrimethylsilane (Equations (251)–(253)) ⟨73CB177, 77IS186, 78ZAAC(443)42⟩. Aryl(trimethylsilyl)phosphines are alternatively obtained by the treatment of arylbis(trimethylsilyl)phosphines with methanol containing a catalytic amount of hydrochloric acid (Equation (254)) ⟨78ZAAC(443)42⟩.

$$PhPLi_2 + 2TMS\text{-}Cl \xrightarrow[91\%]{\text{ether, RT}} PhP(TMS)_2 \quad (251)$$

$$PhPHLi + TMS\text{-}Cl \xrightarrow[65\%]{\text{ether, 20 °C}} PhPH(TMS) \quad (252)$$

$$Ph_2PLi + TMS\text{-}Cl \xrightarrow[80\%]{\text{THF}} Ph_2H(TMS) \quad (253)$$

$$PhP(TMS)_2 \xrightarrow[78\%]{\text{HCl (cat.), THF, MeOH, reflux}} PhPH(TMS) \quad (254)$$

2.16.8.2 Arylphosphorus Compounds with a P—Ge Bond

Alkylgermyl- or arylgermylphosphines are prepared in a similar way to silylphosphines as described above from the reaction of lithium phosphides with alkylgermyl or arylgermyl halides (Equations (255)–(257)) ⟨65JCS4283, 69CB2900⟩.

$$PhPH_2 + 2Ph_3GeCl \xrightarrow[34\%]{\text{PhH, Et}_3\text{N, RT}} PhP(GePh_3)_2 \quad (255)$$

$$Ph_2PLi + Et_3GeBr \xrightarrow[79\%]{\text{THF, RT}} Ph_2PGeEt_3 \quad (256)$$

$$2Ph_2PLi + Ph_2GeBr_2 \xrightarrow[34\%]{\text{THF, RT}} (Ph_2P)_2GePh_2 \quad (257)$$

Mixed metallaphosphines, having an asymmetric phosphorus center prepared from lithium aryl(trimethylsilyl)phosphides and trimethylchlorogermane, undergo scrambling to form arylbis-(trimethylgermyl)phosphines and arylbis(trimethylsilyl)phosphines (Scheme 52) ⟨77JCR(M)3101, 77JCR(S)272⟩.

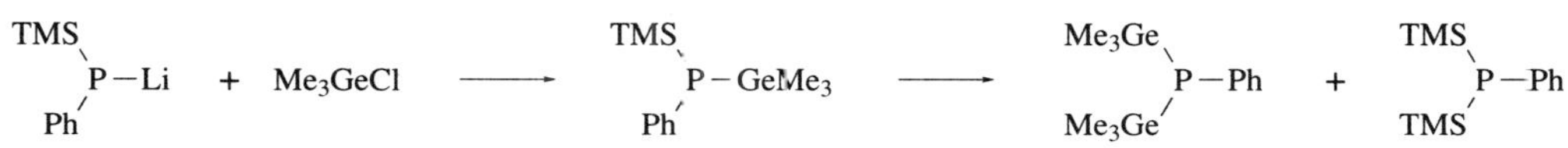

Scheme 52

2.16.8.3 Arylphosphorus Compounds with a P—B Bond

Phosphinodiarylboranes are usually prepared by addition of chlorodiphenylborane to a mixture of secondary phosphines and triethylamine (Equation (258)), or by addition of sodium phosphide to chlorodiarylborane (Equation (259)) ⟨61JCS1000⟩.

$$Ph_2PH + Ph_2BCl \xrightarrow[61\%]{\text{PhH, Et}_3\text{N}} Ph_2P{-}BPh_2 \quad (258)$$

$$Ar^1{}_2PNa + Ar^2{}_2BCl \xrightarrow[66\text{–}67\%]{\text{THF}} Ar^1{}_2P{-}BAr^2{}_2 \quad (259)$$

These monomeric phosphinodiarylboranes are relatively resistant to hydrolysis and air oxidation. Their analogues, arylphosphinidenebis(diphenylborane) ($ArP(BPh_2)_2$) and bis(diarylphosphino) phenylborane ($(Ar_2P)_2BPh$), are similarly prepared from chlorodiphenylborane and arylphosphine, and dichlorophenylborane and diarylphosphine (Equations (260) and (261)) ⟨61JCS5053⟩.

In contrast to phosphinodiarylboranes, both these P–B compounds are sensitive to atmospheric oxidation.

$$PhPH_2 + 2Ph_2BCl \xrightarrow{\text{xylene, NEt}_3} PhP(BPh_2)_2 \qquad (260)$$

$$2Ph_2PH + PhBCl_2 \xrightarrow[78\%]{\text{PhH, NEt}_3} (Ph_2P)_2PBPh \qquad (261)$$

2.16.9 COMPOUNDS WITH A P—METAL BOND

2.16.9.1 Arylphosphorus Compounds with a P—Sn Bond

Most stannylphosphines are prepared either by the reaction of primary or secondary phosphines with organotin halides in the presence of a base (Equation (262)) ⟨69CB2900⟩, or by the reaction of halophosphines with organotin lithium (Equation (263)) ⟨64CB2395⟩.

$$PhPH_2 + 2Ph_3SnCl \xrightarrow[90\%]{\text{PhH, NEt}_3} PhP(SnPh_3)_2 \qquad (262)$$

$$PhPCl_2 + 2Ph_3SnLi \xrightarrow[41\%]{\text{THF}} PhP(SnPh_3)_2 \qquad (263)$$

Diaryl(trimethylsilyl)phosphines undergo transmetallation reactions with triorganotin halides to give the corresponding diaryl(triorganostannyl)phosphines and chlorotrimethylsilane (Equation (264)) ⟨80JOM(190)53⟩.

$$Ph_2P\text{-TMS} + Ph_3SnCl \xrightarrow[75\%]{\text{toluene}} Ph_2PSnPh_3 + \text{TMS-Cl} \qquad (264)$$

The Sn—N or Sn—O bonded compounds react with a P—H-containing compound providing phosphinotin compounds (Equations (265) and (266)) ⟨65JOM(3)295, 66LA(692)1⟩.

$$Ph_2PH + R_3SnNMe_2 \xrightarrow[\substack{R = Me, 80\% \\ R = Et, 70\%}]{} Ph_2PSnR_3 + HNMe_2 \qquad (265)$$

$$Ph_2PH + (Bu^n{}_3Sn)_2O \xrightarrow{\text{CaH, PhH, 80 °C}} Ph_2PSnBu^n{}_3 \qquad (266)$$

2.16.9.2 Arylphosphorus Compounds with a P—Pb Bond

Preparations of organolead phosphines usually involve treatment of organolead chlorides with phosphines in the presence of triethylamine (Equations (267) and (268)) ⟨69CB2900⟩. The organolead phosphines are unstable on moderate heating, compared with other organoelement phosphines ($R^1{}_2PMR^2{}_3$ and $(R^1{}_2P)_2MR^2{}_2$: M = Si, Ge, Sn).

$$Ph_2PH + Ph_3PbCl \xrightarrow[77\%]{\text{Et}_3\text{N, PhH}} Ph_2P{-}PbPh_3 \qquad (267)$$

$$PhPH_2 + 2Ph_3PbCl \xrightarrow[75\%]{\text{Et}_3\text{N, PhH}} PhP{-}(PbPh_3)_2 \qquad (268)$$

2.16.10 HEXAVALENT PHOSPHORUS COMPOUNDS

As representative examples of hexacoordinated phosphorus compounds, hexafluorophosphate salts are well known and are usually stable and solid substances having melting points and octahedral structures. However, relatively little is known of aryl-substituted hexacoordinated phosphate compounds.

Arylpentahalophosphates are generally prepared by the reaction of aryltetrafluoro- or aryltetrachlorophosphoranes with halogeno anions or silyl compounds. For example, $PhPF_4$ is converted into $PhPF_5^-$ by reaction with dimethyl sulfoxide ⟨65IC119⟩. Another route to phenylpentafluorophosphate involves reaction with *N,N,N′,N′*-tetramethylspiro[silamidazolidine] (Equation (269)) ⟨68IC1327⟩.

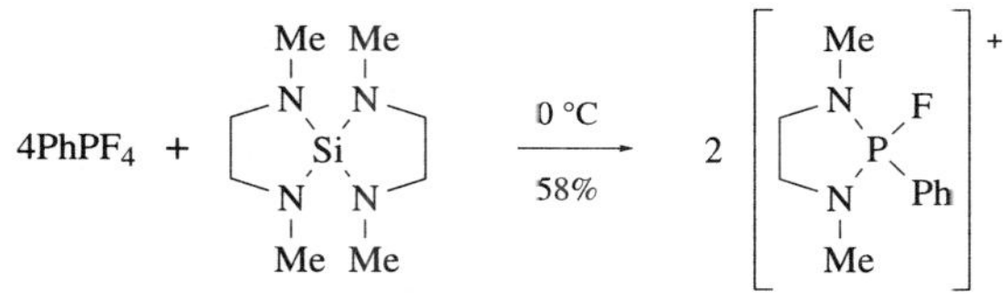

(269)

Phenyldimethylaminotrifluorophosphorane disproportionates to afford phenylbis(dimethylamino)fluorophosphonium phenylpentafluorophosphate upon standing for several weeks (Equation (270)) ⟨64JA4500⟩.

(270)

Hellwinkel showed many examples of spirocyclic hexacoordinated phosphorus compounds containing biphenyl groups. For example, the reaction of PCl_5 with 2,2′-dilithium-biaryl affords the corresponding spirocyclic tetraarylphosphonium hexaarylphosphates. Interestingly, optically active potassium trisbiphenylenephosphates are prepared by the reaction of lithium trisbiphenylenephosphate with *N*-methylbrucinium iodide followed by resolution and addition of potassium iodide, and have remarkably high specific rotations (Scheme 53) ⟨65AG(E)356, 65CB576⟩.

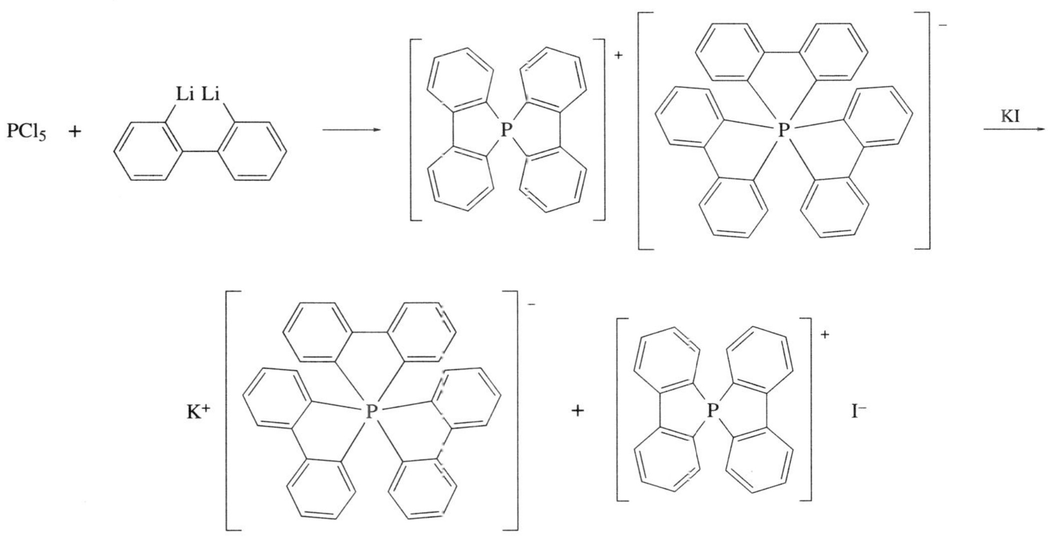

Scheme 53

Another type of hexacoordinated phosphorus compound is the substance containing charge distribution in a molecule which has been prepared by Ramirez and co-workers. Thus, the spiropentaoxyphosphorane reacts with pyridine to afford the intramolecular zwitterionic hexacoordinated adduct in 80% yield. The characteristic feature of this compound appears in the ^{31}P chemical shift, which is much higher than the shifts of the related pentacoordinated phosphorus structure (Equation (271)) ⟨74JA7269⟩.

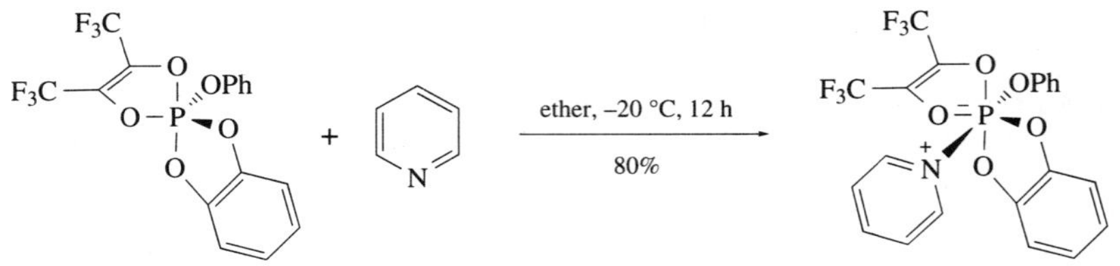

(271)

Bis(*o*-phenylenedioxy)chlorophosphorane is transformed to a cationic hexacoordinated bis(pyridine)bis(benzcatechoyl)phosphonium chloride by the reaction with pyridine. When 2,2′-bipyridine is used as a base, the more stable hexacoordinated phosphonium chloride is obtained. ^{1}H NMR study of the bis(pyridine) adduct has shown that the adduct has a *cis* structure (Equation (272)) ⟨76ZN(B)1058⟩.

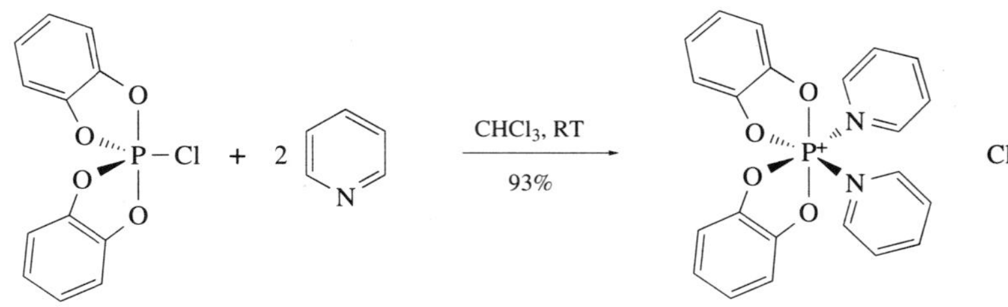

(272)

Simple acyclic hexacoordinated phosphorus compounds are not as stable as the cyclic ones. Denney *et al.* have reported the formation of phenyl- or diphenylphosphate species containing trifluoroethoxy groups in solution. The reaction of phenyltetrakis(2,2,2-trifluoroethoxy)phosphorane or diphenyltris(2,2,2-trifluoroethoxy)phosphorane with sodium 2,2,2-trifluoroethoxide resulted in the formation of the corresponding monophenyl- or diphenylphosphate. In this reaction, diphenylphosphate is initially produced as the *cis* isomer, which rapidly isomerizes to a mixture of *trans* (82%) and *cis* (18%) (Equation (273)) ⟨81JA1785⟩.

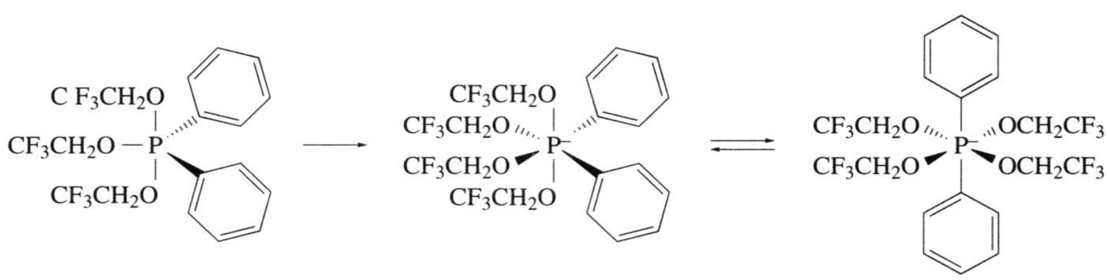

(273)

2.17
Vinyl- and Arylarsenic, -antimony and -bismuth Compounds

ROGER W. READ
The University of New South Wales, Sydney, NSW, Australia

2.17.1 ARSINES AND ARSORANES AND THEIR ANTIMONY AND BISMUTH ANALOGUES

The preparation of organometallic compounds derived from arsenic, antimony and bismuth is described in monographs by Patai ⟨B-94MI 217-01⟩, Aylett ⟨B-79MI 217-01⟩, Samaan ⟨78HOU(13)1⟩ and Doak and Freedman ⟨B-70MI 217-01⟩, and in reviews by Wardell ⟨82COMC-I(2)⟩ and Smith ⟨B-75MI 217-01⟩. In addition, there have appeared reviews on organoarsenic ⟨66AOC(4)146⟩, -antimony ⟨B-90MI 217-01⟩, -antimony and -bismuth compounds ⟨79MI 217-02⟩, organobismuth species ⟨B-77MI 217-01, 82CRV15⟩, pentaarylbismuth compounds ⟨92AOC(34)207⟩, compounds of dicoordinate arsenic ⟨91RCR162⟩, group 15 heterobenzenes ⟨82TCC125⟩, and other heterocyclic derivatives ⟨B-70MI 217-02, 84CHEC(1)539⟩, and of organoantimony and organobismuth compounds in organic synthesis ⟨B-89MI 217-01⟩ all of which refer to methods of synthesis. Annual series of reports of the literature have continued for antimony ⟨92JOM(442)1⟩ and bismuth ⟨92JOM(442)61⟩, but have fallen behind or stopped for arsenic ⟨83JOM(257)67⟩.

2.17.1.1 Organometal(III) and (V) Compounds through Carbon–Metal Formation

2.17.1.1.1 Oxidative metallation

Vinyl- and phenylarsines and -stibines can be prepared most directly by heating elemental As or Sb with the corresponding organo halides, particularly in the presence of Cu, however, mixtures of mono-, di- and triorganometal(III) halides result, and the method is of mostly historical interest ⟨78HOU(13)45⟩. Products derived from Bi decompose at the temperatures at which they are formed ⟨60JINC213⟩. Reactions of highly reduced metal species are discussed in 2.17.1.1.3.

2.17.1.1.2 Transmetallation

The most widely used synthetic method for the preparation of organometal(III) derivatives is the treatment of arsenic, antimony and bismuth halides with organo-Mg or organo-Li reagents ⟨30JCS463, 41OSC(1)550⟩. The trihalides are difficult to limit to single substitution and therefore trialkenyl- or triarylmetal derivatives normally result (Equation (1)), but sterically hindered organics and less reactive halides can selectively yield the monosubstituted compounds ⟨89JOM(366)73⟩.

$$3\,PhLi + MCl_3 \longrightarrow Ph_3M \qquad (1)$$

$$M = As,\ Sb\ or\ Bi$$

Mixed triorganometallic (Equation (2)) ⟨80JOM(186)39⟩ and -dimetallic compounds (Equation (3)) ⟨85JOM(287)57⟩ are accessible by this route. Notably, the As—Cl bond is more labile than alkenyl and aryl C—Cl and aryl C—Br bonds (Equation (4)) ⟨75JOM(88)171, 75SC393⟩. In addition to halides, alkoxy groups have been displaced by alkenyl and aryl-Mg and -Li reagents and use of structure **(1)** provides a convenient route to high purity triorganoarsine derivatives for metal vapour deposition (Equation (5)) ⟨90IC3502⟩.

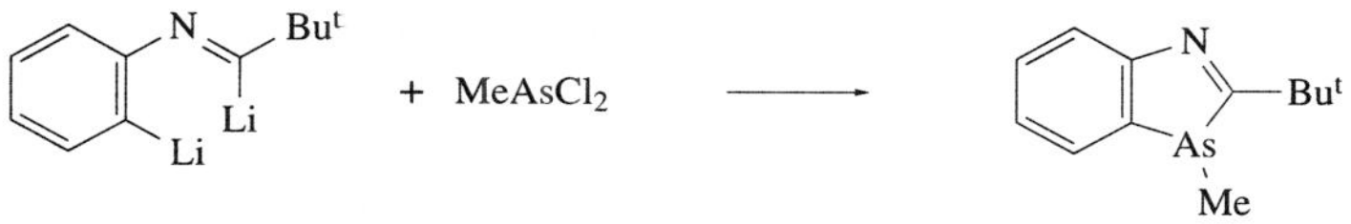

(2)

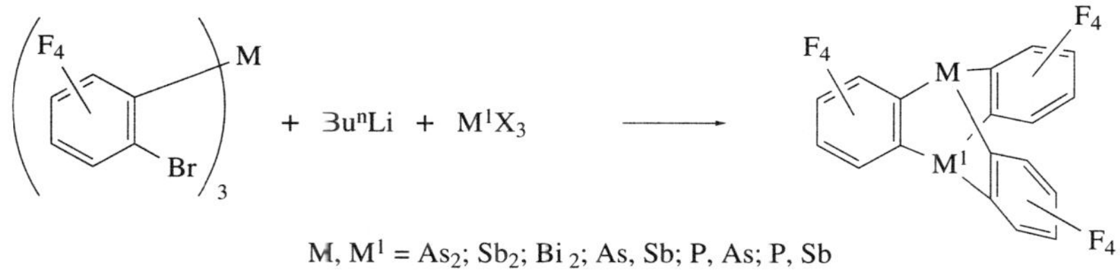

(3)

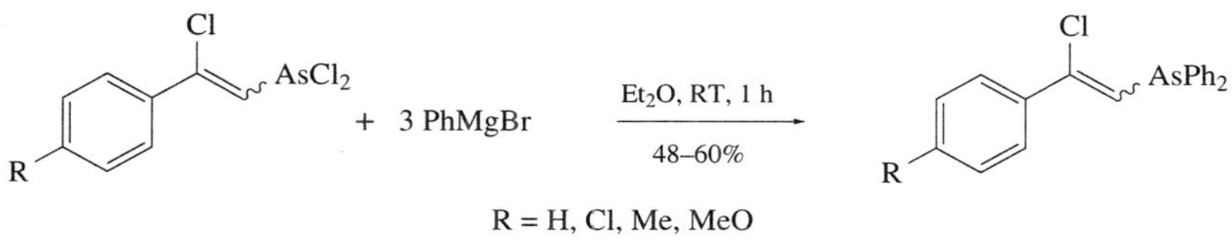

(4)

(5)

(1) R = aryl, vinyl, alkyl
M = MgBr or Li

Organometallic derivatives of Hg, Cd, Sn, Si and Zr have been used to deliver aryl and alkenyl groups to As, Sb and Bi (Equations (6)–(10)) ⟨67MI 217-01, 85TL4067, B-86MI 217-01, 87JOM(334)323, 90AG(E)771⟩. The reactions proceed less readily than with their organo-Mg and -Li counterparts (see Equation (7)), but in all cases the alkenyl group is transferred stereospecifically with retention of the carbon configuration.

(6)

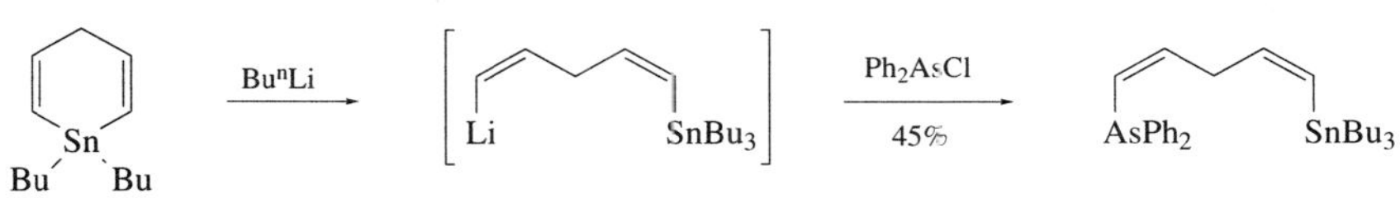

(7)

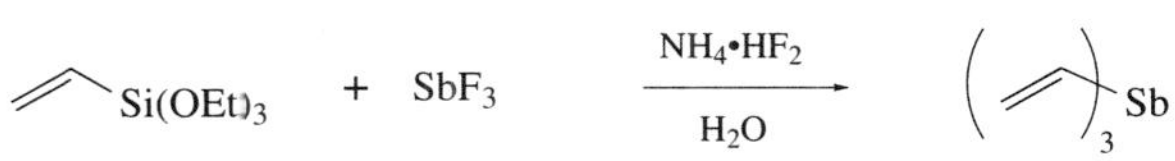

(8)

$$(C_6F_5)_2Cd + BiX_3 \xrightarrow[\text{50 °C, 12 h; 71\%}]{\text{DIGLYME, MeCN}} (C_6F_5)_3Bi \qquad (9)$$

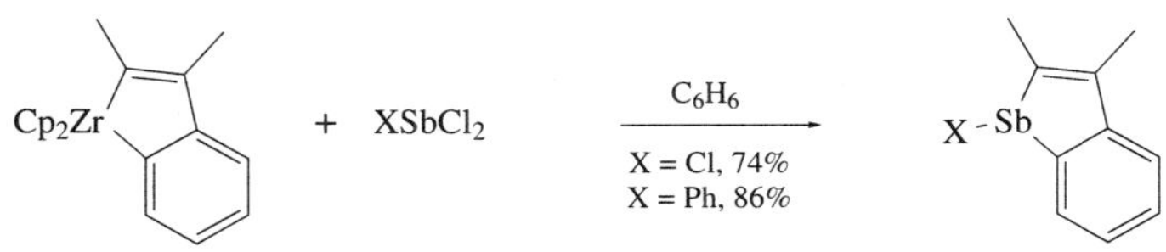

(10)

Pentaorganometal(V) compounds have been prepared, and in some cases (Sb and Bi) hexa-organometalate complexes obtained, through reaction of appropriate ratios of organo-Li reagents with R_3BiX_2 (R = aryl; X = F, Cl, Br, OAc) Ph_4MBr (M = As, Sb) and $Ph_3M{=}NSO_2Ph$ (M = Sb, Bi) ⟨52LA(577)26, 52LA(578)136, 64CB789, 89CB803⟩; reaction of R_3MX_2 with only 1 equivalent of organo-Li affords R_3M.

Metathesis between group 15 elements also provides a synthesis of Ph_5Sb (Equation (11)) ⟨88JGU2050⟩.

$$Ph_3Sb + Ph_5Bi \xrightarrow[\text{20 °C, 48 h; 43\%}]{\text{toluene}} Ph_5Sb \; (+ \; Ph_3Bi) \qquad (11)$$

2.17.1.1.3 *Halogen–metal exchange*

The reactions of elemental As and Sb with Na or K metal in liquid NH_3 yield highly reduced species, which react with aryl halides under irradiation by light to form symmetrical triaryl derivatives (Equation (12)) ⟨84JOM(270)177⟩.

$$\underset{M = \text{As or Sb}}{M} \xrightarrow[\text{ii, 4 PhX(X = Cl, Br)/}h\nu\text{; 40–90\%}]{\text{i, Na(or K)/NH}_3\text{(l)}} Ph_3M \qquad (12)$$

Alkali metal diorganoarsines and -stibines have been used to displace halide ion from alkenyl and aryl halides to provide mixed triorganyl derivatives. The most versatile of these reagents $HAsPh_2$ is capable of reacting with the least activated of substrates, organochlorine compounds, and is tolerant of other functionalities (Equation (13)) ⟨71JOM(33)347⟩. Sodium salts, $NaAsMe_2$ and $NaSbPh_2$, successfully displace aryl bromides (Equations (14) and (15)) ⟨67CJC675, 75JOM(88)171⟩ but fail to react, or give a less satisfactory reaction with, *o*-dichlorobenzene ⟨71JOM(33)347, 75JOM(88)171⟩. Aromatic substrates show no indication of isomerization during treatment with arsenide or stibnide reagents, indicating that benzyne formation is not involved. However, scrambling of aryl groups does occur, especially with Na or K salts in liquid NH_3 in the presence of light, and an $S_{RN}1$ mechanism operates ⟨82JOC77⟩.

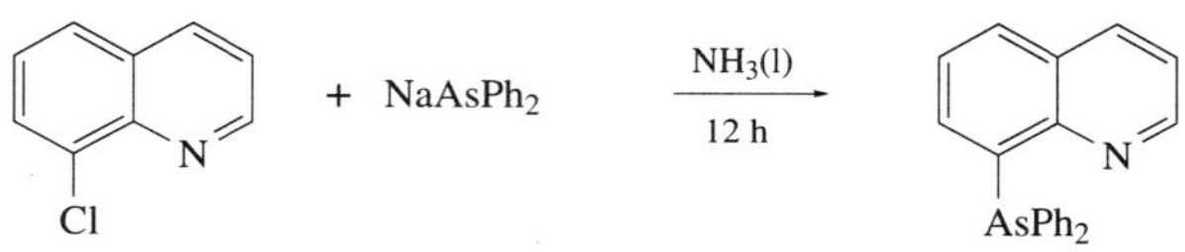

(13)

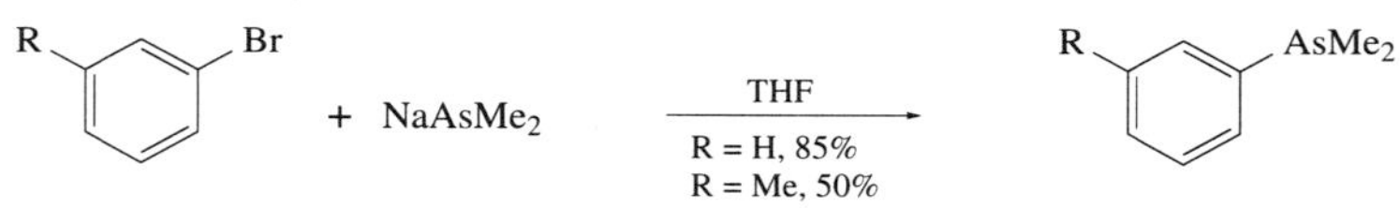

(14)

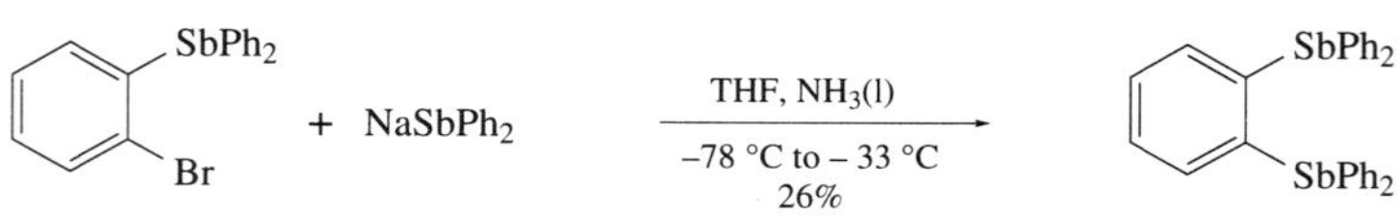

(15)

Alkenyl halides undergo substitution with $LiAsPh_2$ in high yield and with retention of configuration at carbon (Equations (16) and (17)) ⟨67JOC2627, 68JOC1681, 68MI 217-01⟩. With dihalides, monosubstitution can be made to predominate ⟨76JCS(D)1429⟩, but in reaction with $NaAsMe_2$ an elimination–addition process occurs with loss of stereochemical integrity (Equation (18)) ⟨68IC2003, 68ICA379, 71JOM(33)347⟩.

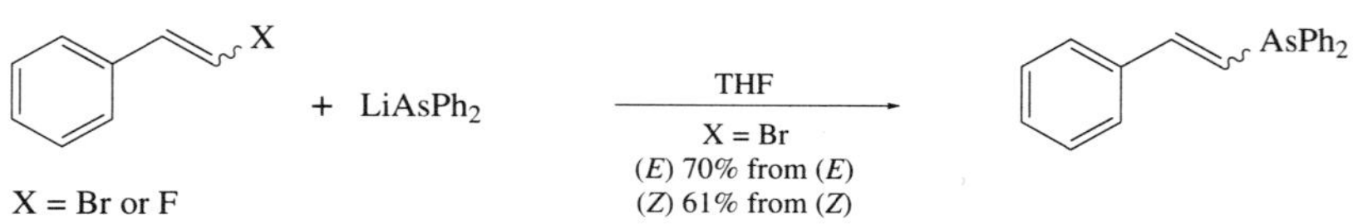

(16)

(17)

$n = 2$	(*E*), X = $AsPh_2$	58% from (*E*)
$n = 2$	(*Z*), X = $AsPh_2$	61% from (*Z*)
$n = 0.5$	(*E*), X = Cl	58% from (*E*)
$n = 0.5$	(*Z*), X = Cl	72% from (*Z*)

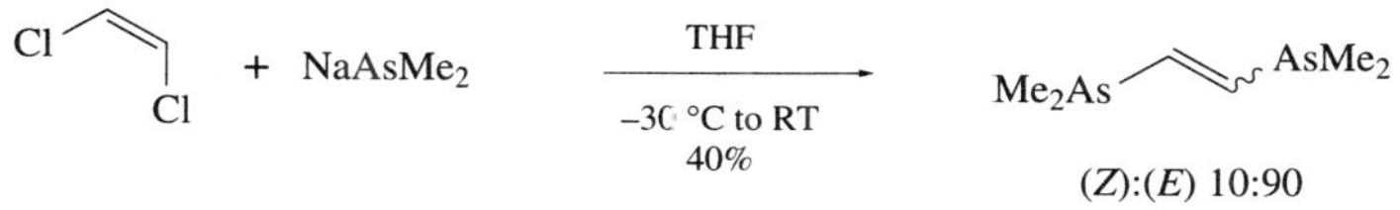

(18)

Direct displacement of aryl and alkenyl halides by neutral group 15 reagents is unknown, but an S_N2' displacement involving (**2**) by Me_2AsH followed by thermolysis and a second S_N2' displacement has given (**3**) in good overall yield (Scheme 1) ⟨68JOM(12)133⟩.

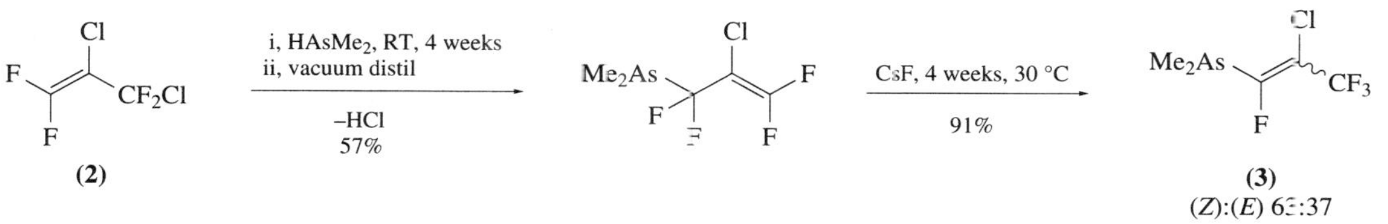

Scheme 1

2.17.1.1.4 *Reactions of organometal reagents with arenes and alkynes*

Electrophilic substitution takes place readily between arsenic halide and oxygen derivatives and alkene and aromatic substrates. In the absence of activating groups, as in (**4**) ⟨23JCS2489⟩ and (**5**) ⟨58JCS1719⟩, Lewis or other acid catalysis is necessary (Scheme 2).

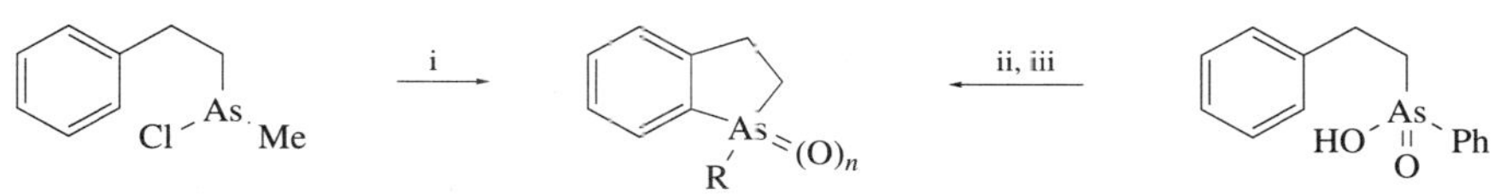

(**4**) R = Me, Ph (**5**)

i, $AlCl_3$, CS_2, gentle heat; ii, conc, H_2SO_4 (n = 1); iii, SO_2 (n = 0)

Scheme 2

The addition of aryl and alkenyl secondary arsines and stibines to alkynes has been reviewed ⟨85CZ53⟩. Reaction occurs with alkynes bearing varied functionality to yield (*E*) and (*Z*) 1,2-disubstituted alkenes (Equation (19)); where ratios of isomers have been determined, the (*Z*) alkene predominates significantly. The primary arsine complex (**6**) undergoes addition to terminal and nonterminal alkynes. Products of single or double *trans*-addition are obtained depending upon the bulk of the alkyne, and highest yields are obtained using toluene as solvent ⟨87JOM(336)337, 92JOM(441)389⟩ (Equation (20)) (in which AIBN is 2,2′-azobisisobutyronitrile). The reaction can be made to yield heterocycles ⟨93JOM(444)53⟩.

(19)

R^1 = Ph, alkyl, alkenyl, hydroxyalkyl, ferrocenyl, carboxylate ester, nitrile
R^2, R^3 = dialkyl, diaryl, alkyl/aryl, alkyl/aminoalkyl, alkyl/aminoaryl

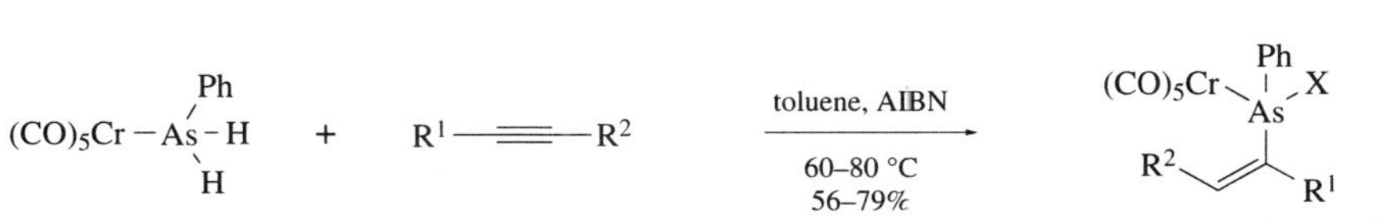

(20)

(**6**) X = H or $C(R^1)=CHR^2$

R^1, R^2 = Ph, H; Ph_2MeSi, H; Ph, Ph; Me, Me

Base-catalysed additions to alkynes also occur, and the stereochemical course of the reaction can be influenced by the addition of an appropriate amine (Equation (21)) ⟨67TL4447⟩. Organoarsenic halides undergo thermal, Lewis acid and photochemically initiated addition to alkynes (Equations

(22)–(24)) ⟨65JOM(3)406, 72IZV213, 84TL1357⟩. Addition to alkynyl ethers is followed by an eliminative fragmentation leading to arsinylketenes (Equation (22)).

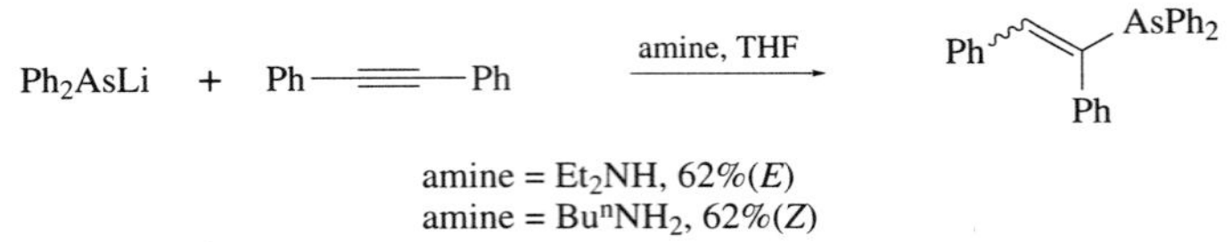
(21)

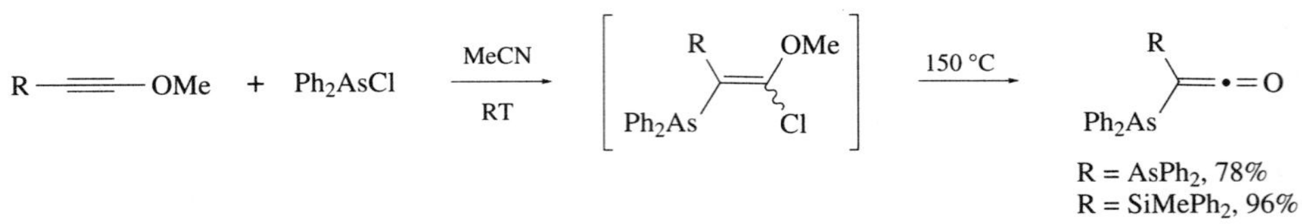
(22)

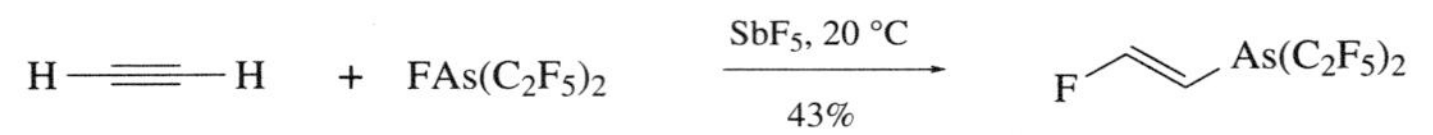
(23)

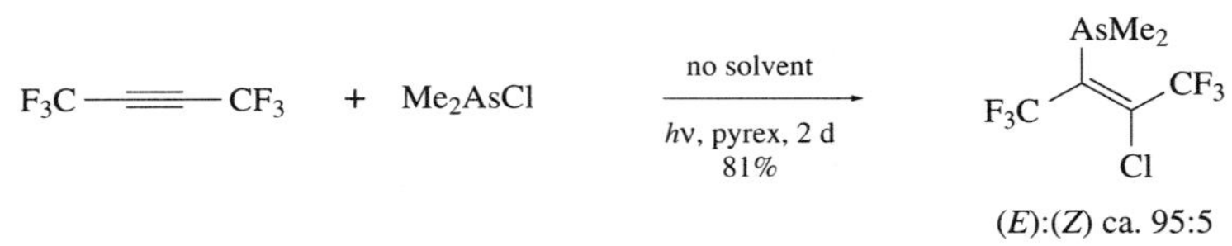
(24)

Addition of $Sb(NMe_2)_3$ to esters of acetylenedicarboxylic acid gives products of three-fold addition as a mixture of (*E*) and (*Z*) isomers ⟨75BSF(2)1185⟩.

Tetraorganodiarsines add to hexafluorobut-2-yne in a nonstereoselective manner under prolonged standing, but in the presence of $(CF_3)_2CO$ there is a strong preference for the (*Z*) isomeric products (Equation (25)) ⟨65JOM(4)151, 73CJC936⟩. Photochemical and radical induced addition of diarsines to phenylacetylene gives predominantly the (*E*) isomer ⟨71JPR254⟩.

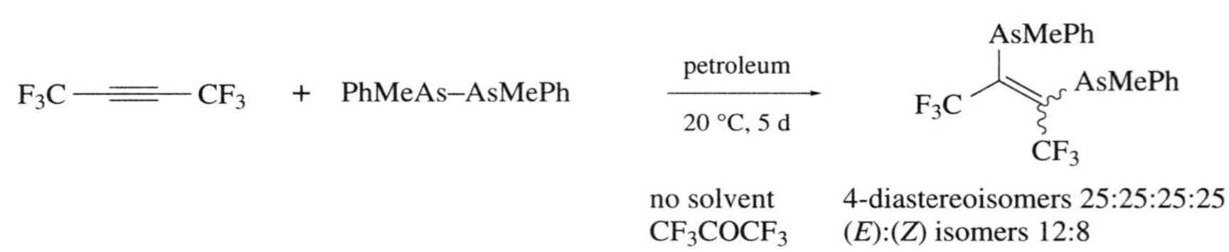
(25)

2.17.1.1.5 *Formation from diazonium salts*

The reaction of aryldiazonium tetrafluoroborates with elemental Bi is useful for the synthesis of Ar_3Bi ⟨58DOK(122)614⟩. The method is less successful with As and Sb but has been used to prepare tetraorganostibonium salts from Cu catalysed reactions with Ar_3Sb ⟨75JOM(85)179⟩.

2.17.1.2 Organometal(III) and (V) Compounds through Modification of Existing Organic Groups

2.17.1.2.1 *Alkenylmetal compounds from addition to alkyne derivatives*

The addition of $HSbPh_2$ and $HSnPh_3$ to (**7**) proceeds without complication to give alkenes (**8**) and (**9**) (Equation (26)) ⟨67DOK(172)1329⟩. Dialkylethynylstibines undergo more complex addition as does di-*n*-butylethynylarsine with $Bu^n{}_3SnH$ ⟨75IZV897⟩. Hydroboration of (**10**), followed by hydrolytic cleavage, affords ligand (**11**) (1,2-bis(dimethylarsino)ethylene (edas)) in good yield, but the alkene is unstable. It is best stored as its Fe^{III} complex (**12**) from which it can be regenerated (Scheme 3) ⟨71JOM(33)347⟩. Addition of HCl to dialkylarsinoacetylides gives only low to moderate yields of 1-alkylarsino-2-chloroalkenes ⟨82MI 217-01⟩. The addition of Ph_2AsCl and Ph_3GeBr to arsino ynethers is an efficient route to arsinoketenes (Equation (22)) ⟨84TL1357⟩. In an isolated process, (**13**) undergoes reaction with the carbonyl group of ketene to give an oxetene (**14**) that rearranges spontaneously to (**15**) and can be made to undergo intramolecular cycloaddition (Scheme 4) ⟨86CB1953⟩.

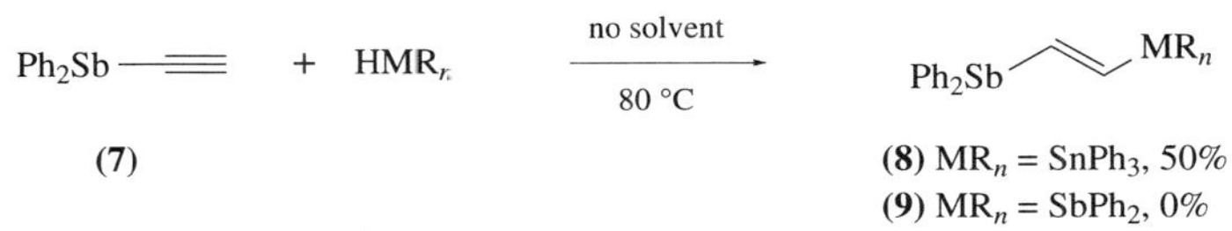

(26)

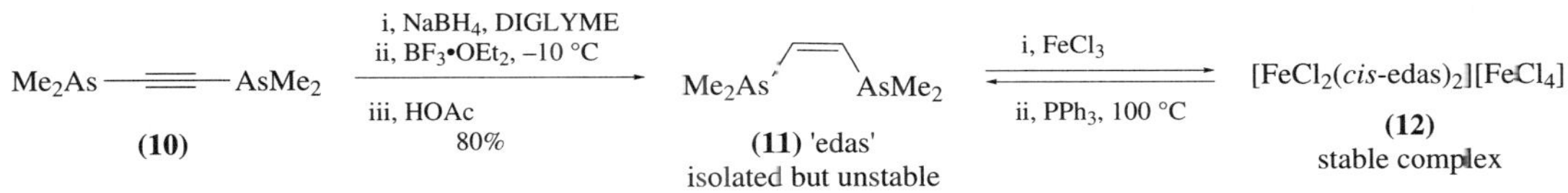

Scheme 3

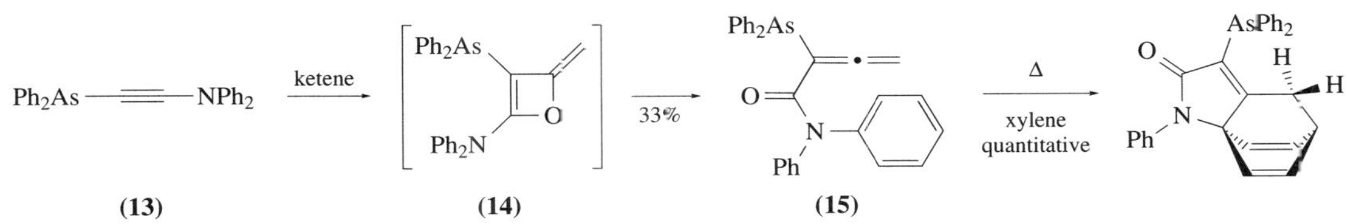

Scheme 4

2.17.1.2.2 Alkenylmetal compounds through β-elimination

Vinylarsonium bromide salts are prepared through β-elimination of a halide (Equation (27)) and can undergo subsequent addition with a range of carbon and heteroatom nucleophilic reagents ⟨75T2121⟩. Diphenylvinylarsane oxide (**16**) is prepared in high yield through fragmentation of monolithiated (**17**) or (**18**) (Equation (28)) ⟨82CB654⟩. The method via (**18**) gives an overall yield almost twice as high as the alternative synthesis via oxidation of diphenylvinylarsane.

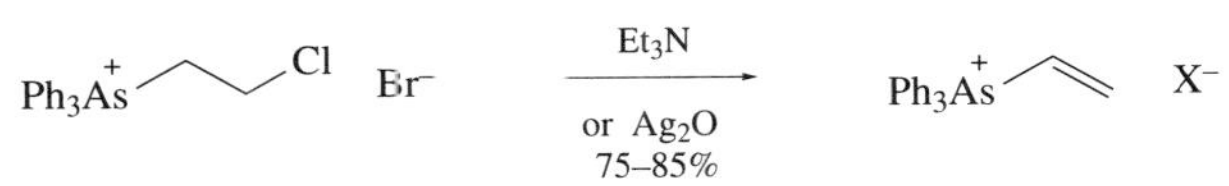

(27)

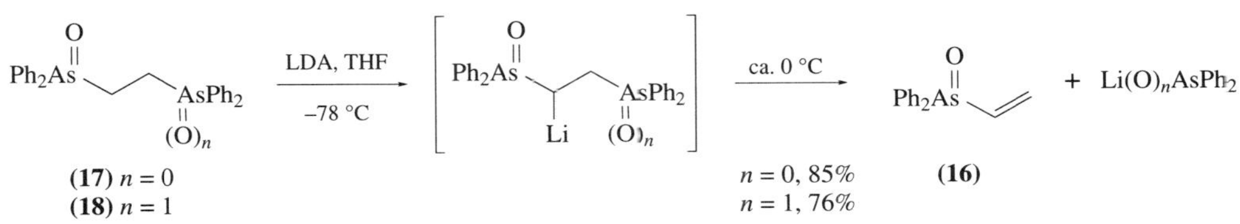

(28)

2.17.1.2.3 Alkenylmetal salts from corresponding ylides

Alkenylarsonium salts can be prepared through trapping of carbonyl-stabilised arsorane ylides. Reaction of the phenacylide with alkyl halides and $Me_3O^+BF_4^-$ affords the *O*-alkyl product, whereas the path of acylation depends upon the reagent used ⟨70JOC2678, 84DOK(274)606, 84IZV700, 86IZV181⟩. Similarly, acylation of the α-thenoyl ylide (**19**) using acid anhydrides gave *C*-acylation whereas reaction of (**19**) with acid chlorides gave products of *O*-acylation (Scheme 5) ⟨88MI 217-01⟩. A comparative study of alkylation of both phosphorus and arsenic β-ketoylides has shown that the arsenic ylides are more susceptible to *C*-alkylation than the phosphorus ylides ⟨87JOU1417⟩.

An unusual variant of these reactions enables preparation of Ph_2As-substituted As and P ylides (Scheme 6) ⟨87CB1281⟩.

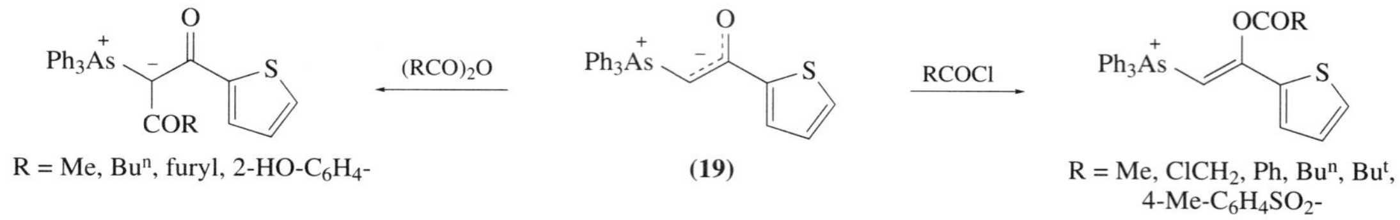

Scheme 5

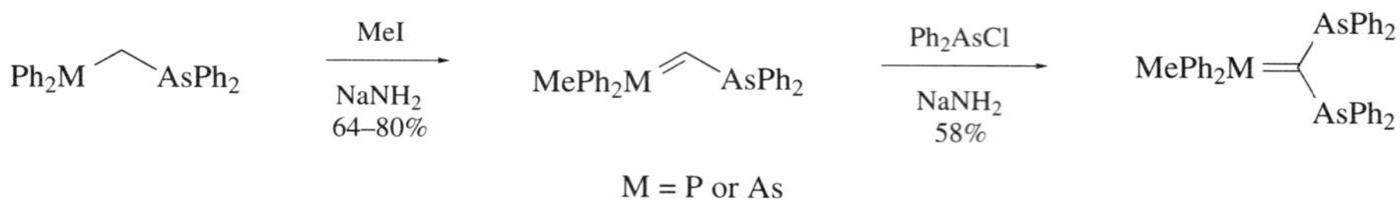

Scheme 6

2.17.1.2.4 *Arsinoketenes*

Free arsinoketenes (**20**; $R^1 = 4\text{-}MeC_6H_4$) can be prepared in almost quantitative yield through thermal decomposition of their tungsten coordination complex (**21**) following treatment with PMe_3 ⟨83AG(E)1001⟩. Tungsten complexes of such ketenes are prepared from η^2-ketenyl complexes (**22**) by reaction with alkyl- or arylhalogenarsanes (Scheme 7) ⟨84CB427, 86JOM(317)175⟩.

This sequence has seen no further development, and there appear to be no examples of stibaketenes or bismaketenes reported in the literature.

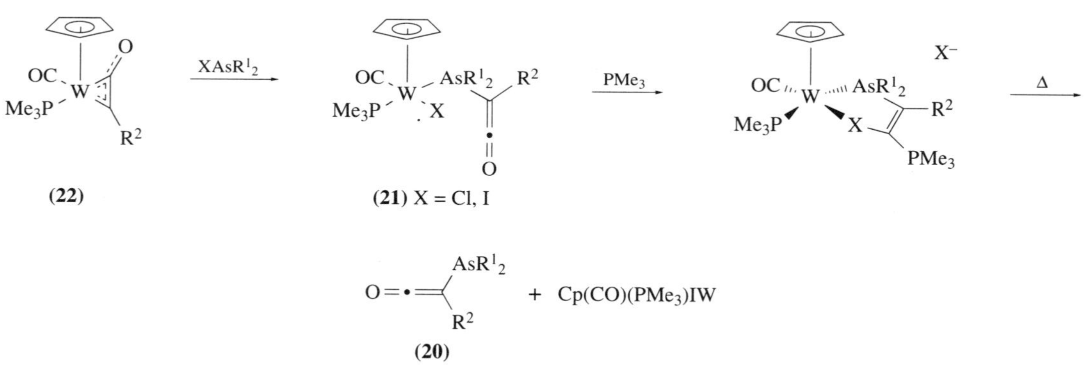

Scheme 7

2.17.1.2.5 *Thermal processes: cycloaddition and thermal rearrangements*

The group 15 heterobenzenes (**23**) (Scheme 8) (Section 2.17.1.4) are thermally unstable and give cycloaddition dimers in the absence of other reagents, or can be trapped as Diels–Alder adducts, both products of which contain alkenylmetal functions. Arsacumulenes also dimerize spontaneously to give dialkenyl-1,3-diarsetanes (Equation (29)) ⟨90TL6331⟩.

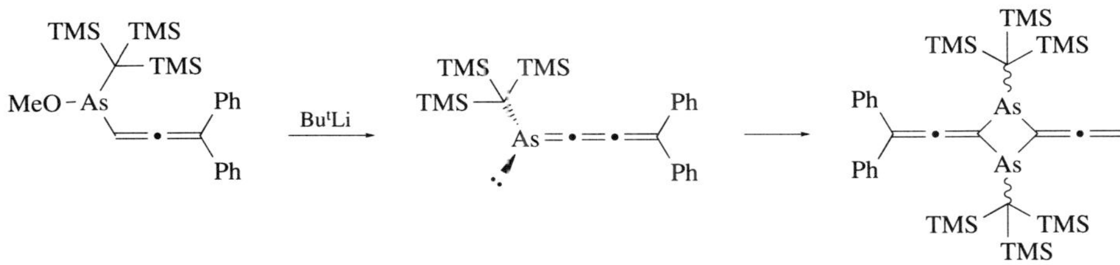

(29)

Thermal isomerization of prop-2-ynyl arsenite esters takes place above 160 °C in the presence of catalytic amounts of prop-2-yn-1-ol to give 3-arsinoacroleins (Equation (30)) ⟨88JGU1173⟩.

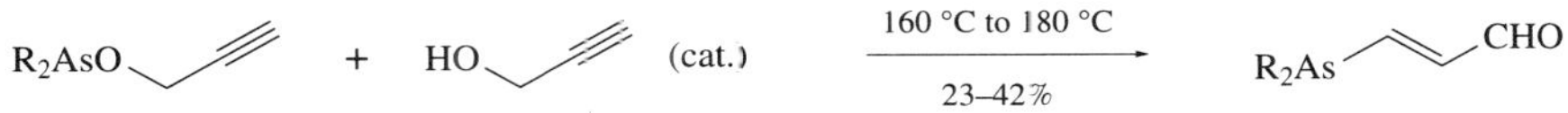

(30)

2.17.1.3 Organometal(III) Compounds through Reductive Processes

Reduction of R_4MX by hydride reagents ⟨56ZN(B)751, 56ZN(B)753⟩, of R_3MX_2 by zinc, mercury, aluminium or boron hydrides, bisulfite or hydrazine ⟨78HOU(13)465⟩, and of R_3MO ⟨85MI 217-01⟩, all provide R_3M compounds.

Covalent and ionic Ph_3RSbX (R = Ph or Me; X = Br, I, OR^1, SR^1 (R^1 = alkyl or aryl)) are thermolysed to produce organometal(III) derivatives (Equation (31)) ⟨80JOM(188)345⟩. Similarly, Ar_5M (e.g., Ar = Ph; M = As, Sb or Bi) upon heating ⟨52LA(577)26; 60JGU3203⟩ or in the presence of $Cu(OAc)_2$ ⟨88JGU629⟩ provides Ar_3M (Equation (32)). Mixed aryl systems behave similarly ⟨68CB121, 71CB1761⟩. Trivinylantimony undergoes loss of butadiene at 450–500 °C to yield vinylantimony(I) which can be trapped as its dideuteride in a deuterium atmosphere ⟨91CM96⟩.

$$Ph_4Sb{-}SC_6H_4X \xrightarrow[140\ °C,\ 3\ h]{\text{no solvent}} \underset{93\text{–}99\%}{Ph_3Sb} + [PhSC_6H_4X + (XC_6H_4S)_2 + Ph_2 + PhH] \qquad (31)$$

X = H, *p*-Br, *p*-OMe, *o*-OMe, *p*-Me

$$Ph_5Sb \xrightarrow[\text{or } Cu(OAc)_2\text{ (cat.), solvent, RT to 80 °C}]{\text{no cat., solvent, 160 °C to 200 °C}} Ph_3Sb + \text{other products dependent upon solvent} \qquad (32)$$

e.g. toluene gives Ph_2 + PhH
BuOH gives BuOPh + PhH

2.17.1.4 Heteroaromatic Compounds of Arsenic, Antimony and Bismuth

The chemistry of group 15 heteroaromatic compounds has received considerable attention. Heterobenzenes (**23**; M = As, Sb or Bi; R^1,R^2 = H) have been prepared from (**24**) through the intermediacy of dihydrostannin (**25**) (Scheme 8) ⟨71JA3293, 71JA6690, 72JA7596⟩. Use of (**26**; R^3 = OH or OMe but not OEt) as a starting material is particularly useful for the preparation of 4-substituted derivatives ⟨73AG(E)931, 76TL3125⟩. In this process HOCl or MeOCl is eliminated from (**26**). The method is preferred when a 4-carboxaldehyde derivative is sought. The metal–halide bond of (**26**) can also be first reduced with trialkyltin hydride, and then MeOH or EtOH is eliminated ⟨74AG(E)668⟩. A hetero-Diels–Alder reaction involving an arsaalkene has been used to provide an arsenin with mixed functionality (Equation (33)) ⟨89TL349⟩.

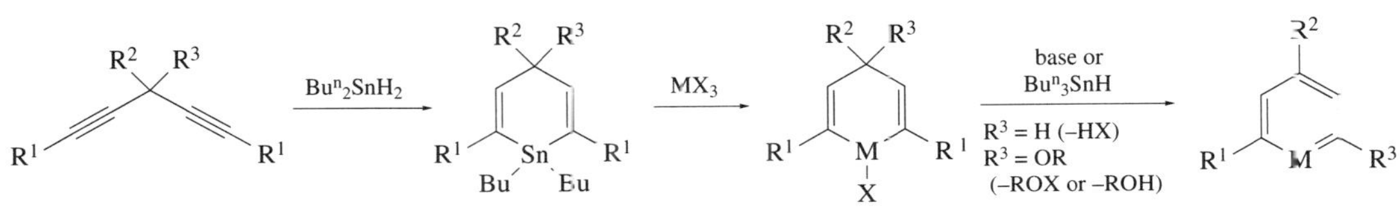

(**24**) (**25**) M = As, Sb or Bi (**26**) (**23**)
X = Cl or Br

Scheme 8

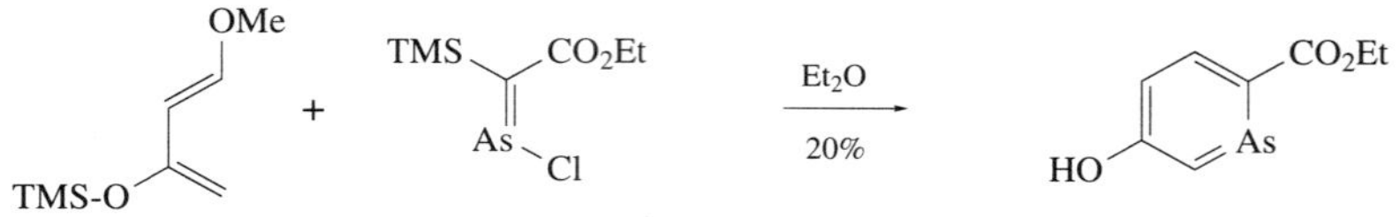

(33)

Nonbenzenoid heteroaromatic compounds, condensed ring systems bearing group 15 hetero-benzene rings, and heterocycles bearing arsenic, and in fewer cases antimony and bismuth, and at least one other heteroatom (As, P, N, O, S) have been synthesised ⟨84CHEC(1)539⟩. Approaches to these compounds are illustrated in Schemes 9–11 ⟨79CC880, 80JOM(202)C95, 83OM1005, 84OM495, 87AG(E)229, 88TL535, 88TL539⟩.

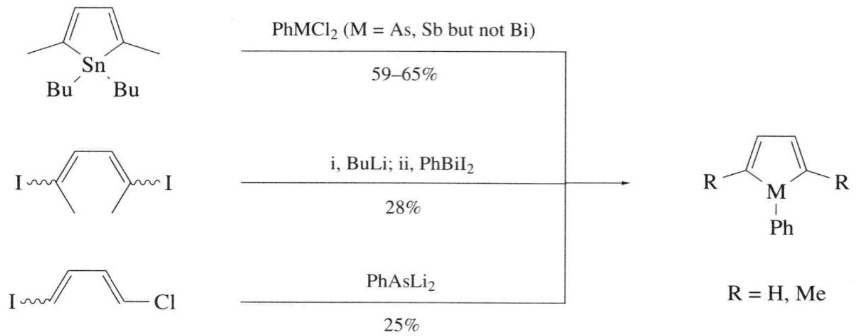

Scheme 9

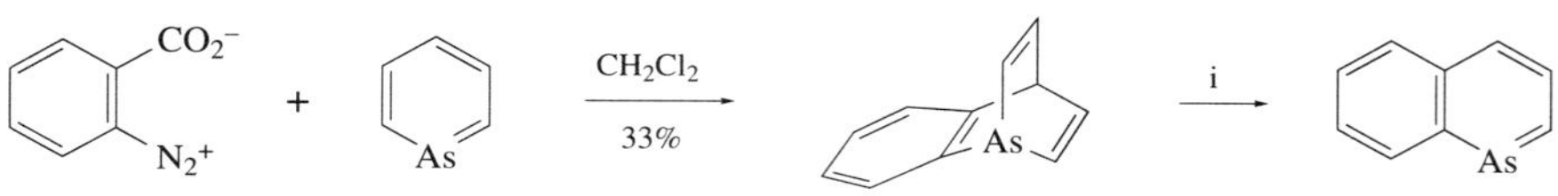

i, 3,6-di(2-pyridyl)pyridazine, $CDCl_3$, 25 °C

Scheme 10

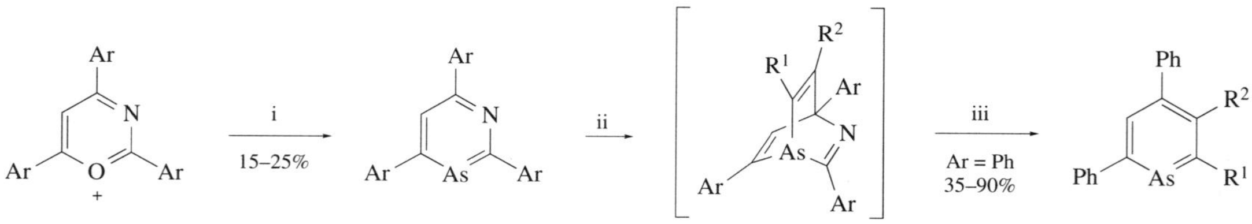

i, $As(TMS)_3$, KF, 18-crown-6, MeCN; ii, R^1—≡—R^2
iii, toluene, reflux, 3–96 h (–PhCN)

R^1, R^2 = H, Ph; CHO, Ph; Ph, Ac; H, CO_2Me

Scheme 11

2.17.2 ARSENIC HALIDES AND THEIR ANTIMONY AND BISMUTH ANALOGUES

Many of the methods described in Section 2.17.1, such as oxidative metallation (As and Sb) and electrophilic aromatic substitution (As) can be applied to the synthesis of metal(III) halides. Transmetallation using alkyllithium and Grignard reagents is difficult to stop at single substitution and is therefore impractical ⟨91JOM(402)55⟩. Reagents that employ less electropositive metals, e.g., Hg and Sn can be made selective and do provide access to mono- and diorgano As, Sb and Bi^{III} halides, e.g., Equations (6) and (10), and conversion (**25**) to (**26**).

The reaction of Grignard reagents with organometal(V) halides and oxides (at least in the cases of As and Sb) (Equations (34) and (35)) does permit synthesis of tetraphenylmetal(V) halides.

$$Ph_3MCl_2 + PhMgBr \longrightarrow Ph_4MX \qquad (34)$$

$$M = As\ Sb;\ X = Cl, Br$$

$$Ph_3AsO \xrightarrow[\text{ii, HX}]{\text{i, PhMgBr}} Ph_4AsX \qquad (35)$$

$$X = Cl, Br, I$$

The addition of $AsCl_3$ to acetylene (Equation (36); R = H) is an efficient route to the chemical warfare agent, Lewisite, wherein it yields only the (*E*)-isomer (Section 2.17.1.2.1). A similar reaction occurs between $(C_2F_5)_2AsF$ and acetylene ⟨72IZV213⟩.

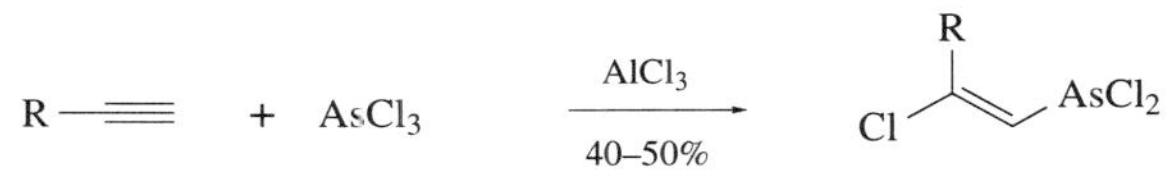

(36)

R = H, 4-ClC$_6$H$_4$; 4-MeC$_6$H$_4$; 4-MeOC$_6$H$_4$

Similar addition to arylacetylenes (Equation (36); R = aryl) affords analogous chloroalkenylarsine dichlorides ⟨75SC393⟩. Meanwhile, addition of $SbCl_5$ to acetylene proceeds in the presence or absence of $HgCl_2$ catalyst and the degree of reaction is temperature dependent (Scheme 12) ⟨45BAU251, 71IZV2103⟩.

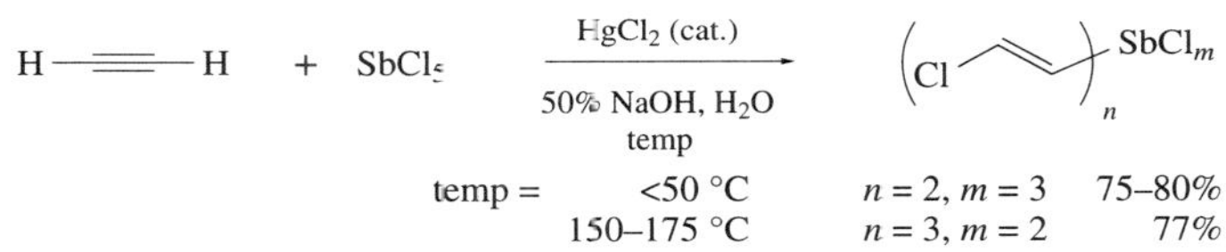

Scheme 12

The following methods apply more specifically to the preparation of halometal derivatives through introduction of the halometal bond.

2.17.2.1 Halogenation of Arsines, Stibines and Bismuthines

The preparation of R_3MX_2 (R = alkyl, alkenyl, aryl; M = As, Sb, Bi; X = F, Cl, Br, I) is most readily achieved through addition of halogens to R_3M, including low temperature direct fluorination and electrochemical fluorination (R = Ph, C_6F_5; M = As) ⟨85JGU1332, 86AX(C)1875⟩. The thermal stability of triarylmetal halides decreases in the order F > Cl > Br > I so that $Ph_3As \cdot I_2$ exists as a charge transfer complex ⟨87AG(E)264⟩.

Other halogenating agents have included $PhIF_2$, XeF_2 ⟨76JGU2477⟩, SF_4 ⟨60JA6176⟩, $PhICl_2$ ⟨83JOM(248)269⟩, SO_2Cl_2, $PbCl_4$, $CuCl_2$ and $TlCl_3$ ⟨78JOM(149)315, 78JOM(154)13⟩. Interhalogens and halopseudohalogens IX (X = Cl, Br, N_3, NCO) also react ⟨85POL251⟩.

Diarylmetal(V) trihalides, R_2MX_3 (R = aryl; M = As, Sb; X = F, Cl, Br; X ≠ I), are prepared by addition of halogen or equivalent reagents to, e.g., Ph_2AsCl, Ph_2SbF or Ph_2AsH.

Organometal(III) halides are provided by direct halogenation of $R^1_2R^2M$ (R^1 = aryl; R^2 = aryl, H, halide, SR^3, MR^1_2; M = As, Sb, Bi; R^3 = lower alkyl) followed by spontaneous or thermal loss of R^2X (Equations (37)–(41) and Scheme 13) ⟨22JCS(121)91, 53JCS16, 56N59, 64CB789, 72IZV1612, 75MI 217-02, 83JOM(248)269, 84JCS(D)2365, 89JGU2500⟩. Thermal stability of the intermediate dihalide is lowest when the metal is Bi and decreases in the order of halide = F > Cl > Br > I. Iodination can therefore yield two-fold substitution (Equation (41)). The leaving ability of R groups is in the order: benzyl > Me > CF_3 > alkyl > Ph. Treatment of $R^1_2R^2_3Sb$ with X_2 (X = Br, I) can give $R^1_2R^2_2SbX$ or $R^1_2R^2SbX_2$ through exothermic cleavage of alkenyl (R^2) in the presence of alkyl or Ph (R^1) groups ⟨64IZV1197, 64IZV1202⟩.

$$PhAsH_2 + 2\,SO_2Cl_2 \xrightarrow[90\%]{Et_2O} PhAsCl_2 \quad (37)$$

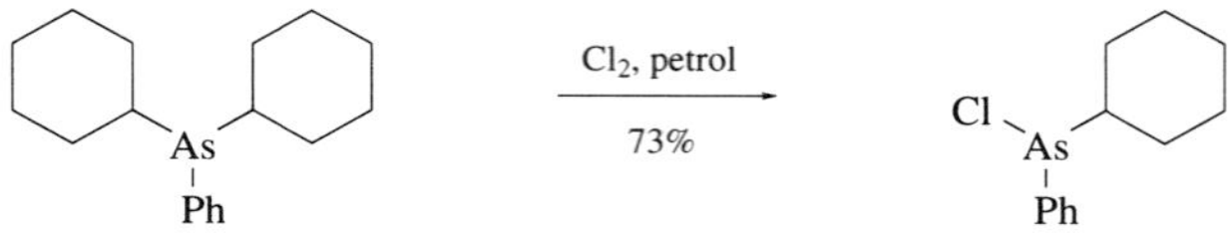

(38)

$$Ph_3SbBr_2 \xrightarrow[92\%]{220–230\ °C} Ph_2SbBr \quad (39)$$

$$Ph_3Bi + IX \xrightarrow[X = Cl,\ 94\%;\ X = Br,\ 91\%]{Et_2O} Ph_2BiX \quad (40)$$

$$Ph_3Bi + I_2 \xrightarrow[48\%]{Et_2O} PhBiI_2 \quad (41)$$

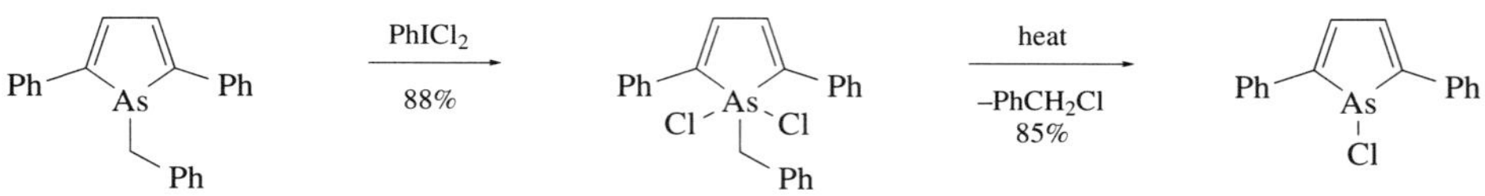

Scheme 13

2.17.2.2 Disproportionation and Transmetallation Reactions

A convenient and often high yielding source of RMX_2 and R_2MX (M = As, Sb, Bi; R = aryl and vinyl) is through disproportionation reactions of R_3M and MX_3. The method is most often used to prepare Bi derivatives (Equations (42) and (43)) ⟨41JA207, 57JA5884⟩.

$$2\,Ar_3Bi + BiCl_3 \xrightarrow[Ar = Ph,\ 99\%;\ Ar = 4\text{-}MeC_6H_4,\ 93\%]{Et_2O} Ar_2BiCl \quad (42)$$

$$Ar_3Bi + 2\,BiBr_3 \xrightarrow[Ar = 2\text{-}MeC_6H_4,\ 95\%;\ Ar = 1\text{-naphthyl},\ 89\%]{Et_2O\ or\ CHCl_3} ArBiBr_2 \quad (43)$$

In similar transmetallation processes, Bi^{III} derivatives can be prepared from Ar_3Bi by treatment with Lewis acids, e.g., PCl_3, $AsCl_3$ ⟨22JCS104, 60CB1417⟩, $HgCl_2$, $TlCl_3$, $SbCl_3$ ⟨57MI 217-01⟩, $AlCl_3$ ⟨51ZOB93⟩, $SiCl_4$ ⟨60MI 217-01⟩.

2.17.2.3 Displacement Reactions

Halide exchange is used to prepare $ArMX_2$ and ArRMX (Ar = aryl, alkenyl; R = alkyl or aryl; M = As, Sb, Bi; X = F, Cl, Br, I) (Equation (44)) ⟨39MI 217-01, 41JA575, 65JCS5132⟩ and less so the organometal(V) halides (Equation (45)) ⟨53MI 217-01⟩. Acetone is not used as a solvent in the latter reactions because it participates in the reaction ⟨71JCS(A)562⟩.

$$(R)(Ar)As\text{–}Cl + NaI \xrightarrow{acetone} (R)(Ar)As\text{–}I \quad (44)$$

Ar = Ph, Me, 90%
Ar, R = Ph, Ph, 85%
Ar, R = $(3\text{-}NO_2\text{-}4\text{-}HO\text{-}C_6H_4)_2$, 70%

$$Ar_3SbCl_2 + KF \xrightarrow[75\text{–}90\%]{EtOH\text{-}H_2O} Ar_3SbF_2 \quad (45)$$

Ar = Ph, 2-MeC_6H_4, 3-MeC_6H_4, 4-MeC_6H_4, 4-ClC_6H_4, 4-BrC_6H_4, 1-naphthyl

Displacement of halide equivalents, e.g., NMe_2, OEt, OAc, SEt, $OSbPh_2$, $BiPh_2$, either directly or following Lewis or Brönsted acid catalysis, is another versatile approach to organometal(III) halides (Equations (46)–(48)) ⟨65JGU1253, 68JCS(A)483, 72IZV1612, 84JCS(D)2365⟩.

$$Ar_2As{-}NR_2 \xrightarrow[\text{reflux, 1 h}]{\text{conc. HX}} Ar_2As{-}X \quad (46)$$

Ar = Ph, X = Cl, 83%
Ar = 4-MeC_6H_4, X = Br, 61%
Ar = 4-F-C_6H_4, X = Cl, 65%

$$(C_6F_5As{=}S)_4 + 4\,HgCl_2 \xrightarrow[89\%]{Et_2O\text{, reflux, 30 min}} 4\,C_6F_5AsCl_2 \quad (47)$$

$$Ph(Et)As{-}SEt + Br_2 \xrightarrow{CCl_4} Ph(Et)As{-}Br \quad (48)$$

Organometal(V) chlorides (As, Sb) are also prepared from organometal(V) oxides and acids by treatment with HCl (Equation (49)) ⟨20LA(421)174, 22LA(429)144, 35JA913⟩.

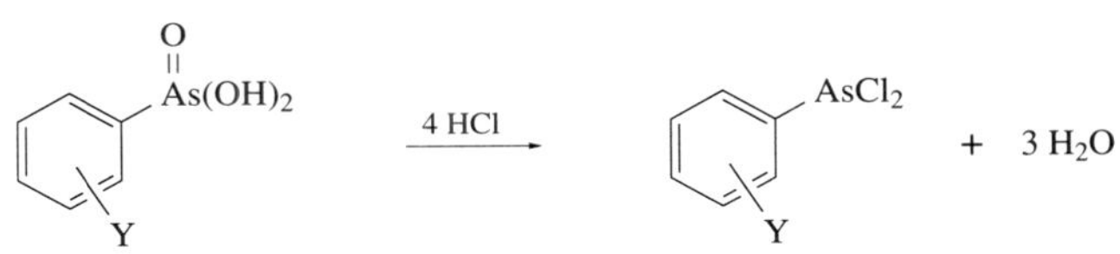

(49)

Y = H, 2-Cl, 3-NO_2, 4-NH_2, 2-Cl-5-NO_2, 4-Cl-3-NO_2

Eliminative substitution of chlorovinyl groups is limited in its applications but has given aryl and alkenylarsinic dichlorides (Equation (50)) ⟨55IZV472⟩.

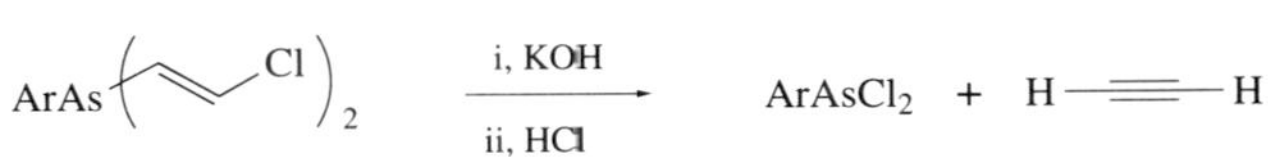

(50)

Ar = Ph, 79%
Ar = 4-MeC_6H_4, 92%
Ar = CH=CHPh, 42%

2.17.2.4 Reductive Halogenation of Arsonic and Arsinic Acids and their Antimony Analogues

Another useful approach to RMX_2 and R^1R^2MX (M = As, Sb; R, R^1 = aryl, R^2 = aryl or alkyl; X = Cl, Br) is the reductive halogenation of RMO_3H_2 and $R^1R^2MO_2H$ (Equation (51)) ⟨41JA575, 71IC2676, 83JGU1832, 83JGU2449⟩. Common reducing agents include SO_2, $SnCl_2$ (gives lower yields than SO_2), PCl_3 and PBr_3.

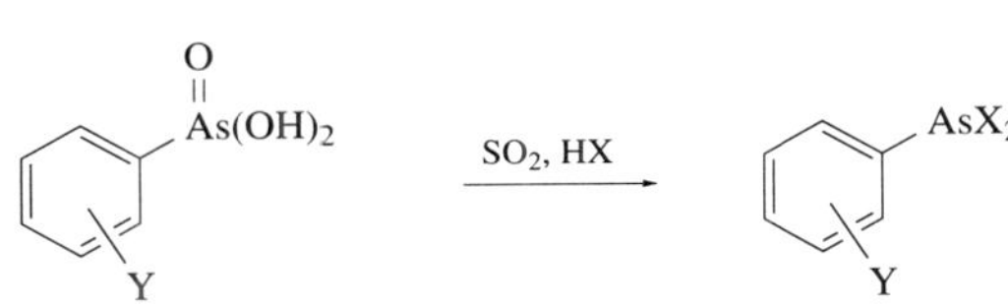

(51)

X = Cl; Y = H, 2-Me, 3-Me, 4-Me, 4-Cl, 4-Br, 4-I, 4-OMe, 4-NO_2, 56–84%
X = Br; Y = H, 2-Br, 2-NO_2, 4-Br, 4-NO_2, 77–92%

2.17.3 ARSENIC COMPOUNDS WITH AN As—O BOND AND THEIR ANTIMONY AND BISMUTH ANALOGUES

2.17.3.1 Preparation through Metal–Carbon Bond Formation

The preparation of arylarsonic and stibonic acids, $ArMO_3H_2$ (Ar = substituted phenyl, polycyclic aromatic and heteroaromatic), is achieved through treatment of $ArN_2 \cdot X$ (X = Cl, HSO_4, BF_4) with $M(ONa)_3$ (M = As,Sb), usually in the presence of Cu^I (Bart reaction) (Equation (52)) ⟨78HOU(13)295⟩. Highest yields are achieved with $ArN_2 \cdot BF_4$ salts. Treatment of haloaromatics with $As(ONa)_3$ (Rosenmund reaction) offers an alternative route to arsonic acids (Equation (53)) ⟨21CB438⟩.

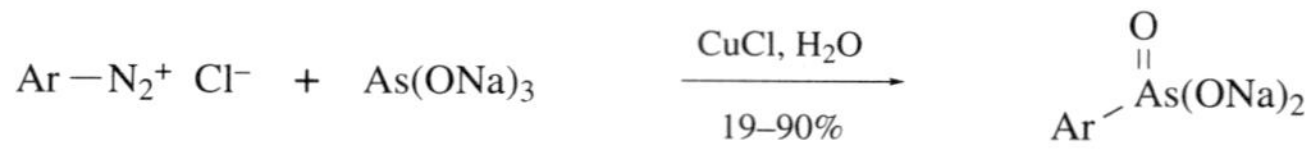

(52)

Ar = substituted phenyl, biphenyl, naphthyl, benzoquinonyl, anthraquinonyl, pyridyl, furanyl, fluorenyl
50 examples

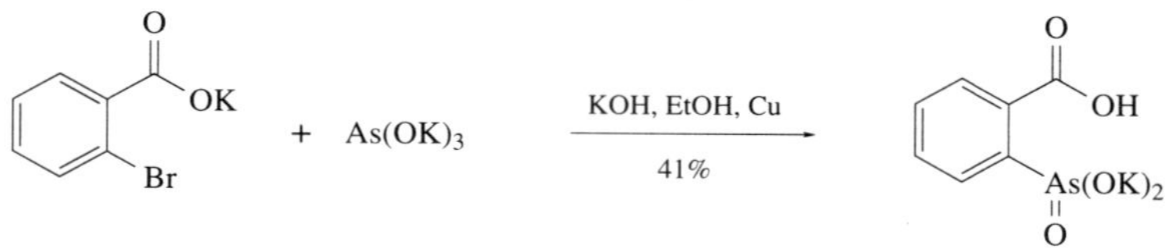

(53)

High yields of $ArAsO_3H_2$ can also be obtained from treatment of electron-rich aromatics with H_3AsO_4 under conditions of water removal (Béchamp reaction) ⟨40JA3142⟩, but Ar_2AsO_2H is also formed. Intramolecular reactions of As and Sb derivatives are more practical (Equations (54) and (55)) ⟨36JA1573, 52JCS4448⟩.

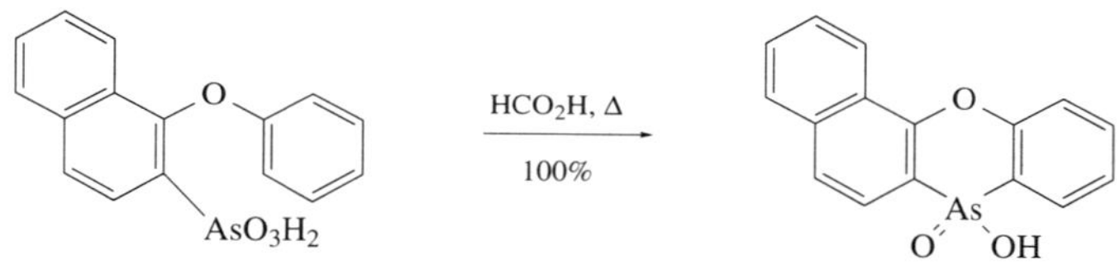

(54)

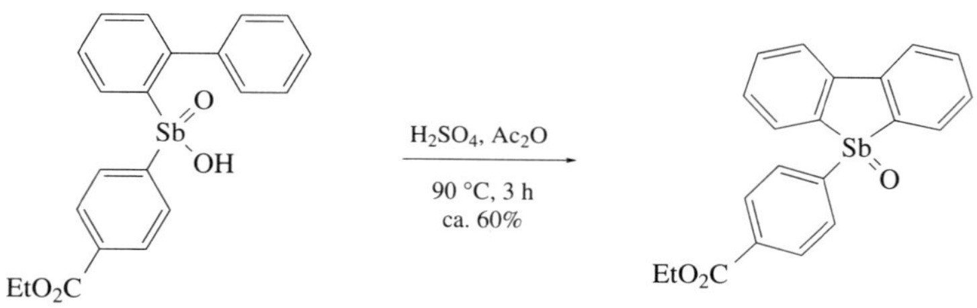

(55)

A hydrolytic method that avoids the problem of diarylation involves the reaction of MCl_3 with aryldiazonium salts (Scheller reaction) (Scheme 14) ⟨40JA167, 46JA1987⟩. Modifications of this approach have used hexachloroantimonate salts ⟨54IZV410⟩, while reaction of $ArMCl_2$ intentionally gives diaryl derivatives (Scheme 15) ⟨52DOK(87)991, 53DOK(91)1341, 54ZOB1259⟩. Fluorinated derivatives, e.g., (**27**), have also been prepared by a tandem fluorination process (Scheme 16) ⟨68JCS(C)2317⟩.

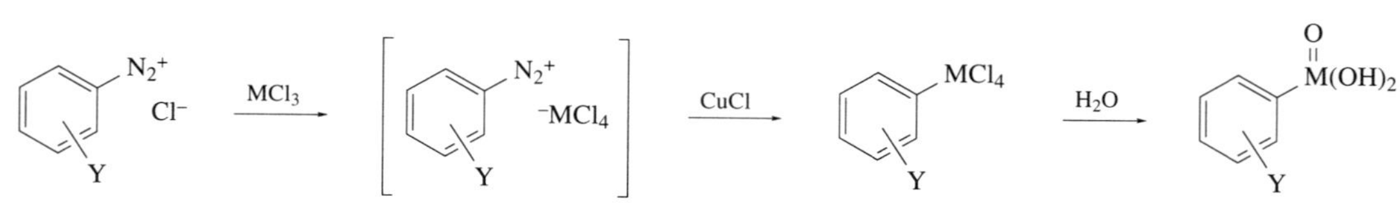

M = As Y = 3-CO$_2$H, 3-NO$_2$, 4-SO$_2$NH$_2$, 3-NO$_2$-4-Me overall yield 40–76%
M = Sb Y = H, 3-Me, 4-Me, 4-Cl, 4-NO$_2$, 3-CO$_2$H, 4-CO$_2$H overall yield 24–78%

Scheme 14

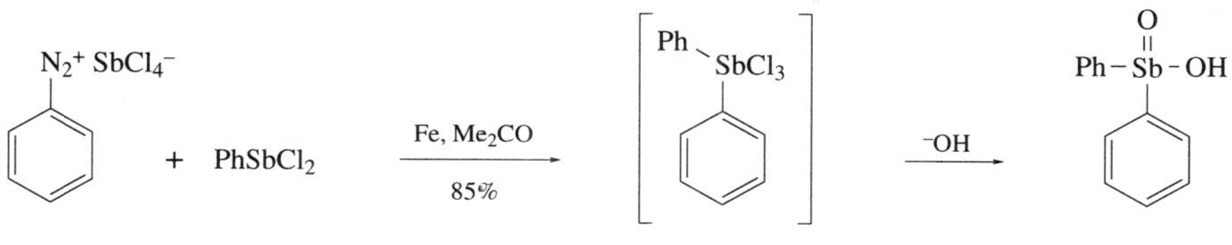

Scheme 15

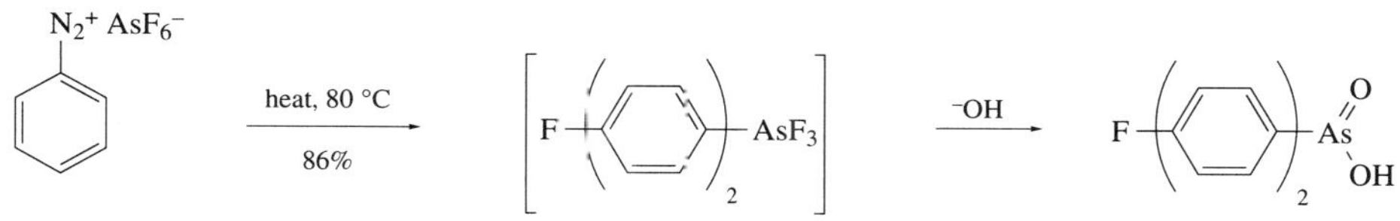

Scheme 16

Like and mixed diarylarsinic acids are available in low to moderate yield by a modification of the Bart reaction (Equation (56)). The preparation of $Ar_3As{=}O$ through reaction of diazonium salts with As_2O_3 has also been described ⟨78HOU(13)376⟩.

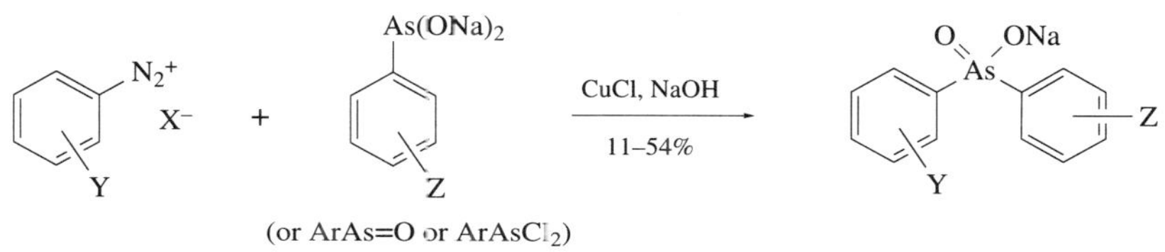 (56)

2.17.3.2 Formation of Metal–Oxygen Bonds by Substitution Reactions

A general approach to metal(III) alkoxide derivatives is the displacement of halide ions and other leaving groups from RMX_2 and R_2MX (R = aryl, alkenyl and alkyl; M = As, Sb, Bi; X = halide, SCN, OR^1, NR^1_2; R^1 = lower alkyl) by oxygen nucleophiles under base-promoted or dehydrating conditions (Equations (57)–(59)) ⟨37JIC349, 64NEP6405309, 70JOM(25)139, 72JGU536, 72JGU1960, 82ZN(B)815⟩. Exchange reactions between alkoxide or acyloxy groups can be achieved when the reaction conditions enable removal of one component from participation ⟨82ZN(B)815⟩.

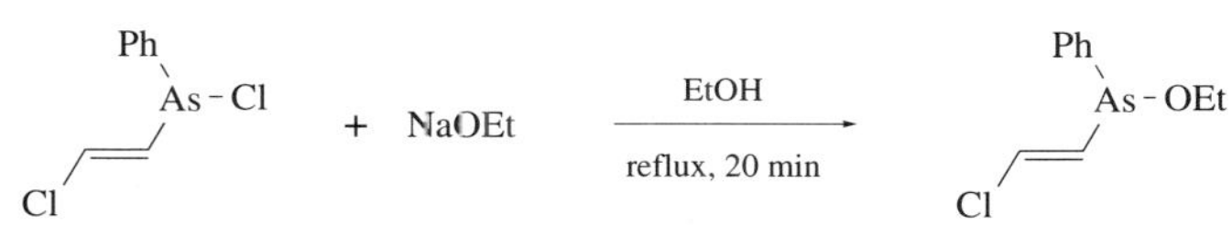 (57)

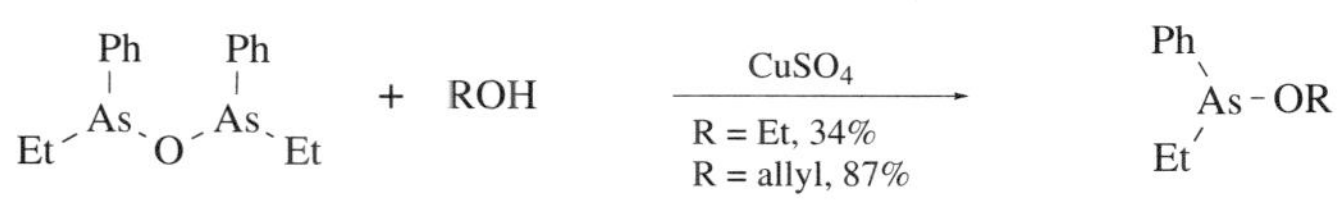 (58)

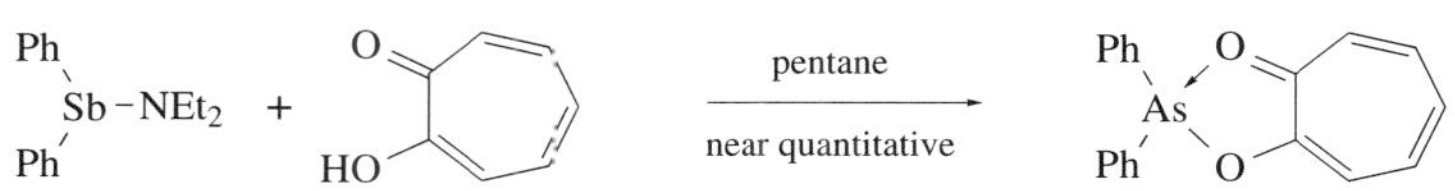 (59)

Bis(diorganometal) oxides $(R_2M)_2O$ (R = aryl or alkyl; M = As, Sb) are readily prepared by alkaline hydrolysis of diorganometal(III) halide or pseudohalide derivatives. Alkaline hydrolysis of $ArMX_2$ (Ar = aryl, heteroaryl; M = As, Sb, Bi; X = halogen, OR, O_2CR, NR_2; R = lower alkyl) yields monoorganometal(III) oxides $(ArM{=}O)_n$, which are frequently cyclooligomeric or polymeric in solution ⟨78HOU(13)167⟩. Without an alkali, arylarsenic derivatives bearing electron-withdrawing substituents on the aryl ring are hydrolysed to $ArAs(OH)_2$. The ease of hydrolysis increases with the electronegativity of the X groups (e.g., for halogens, F > I).

The arylarsenic anhydrides, $ArAs{=}O$, and their imine derivatives, $ArAs{=}NPh$, react with hydroxylic reagents to give arsenic esters, $ArAs(OR)_2$ (Ar = aryl; R = alkyl) ⟨72JGU1960, 87JGU2496⟩.

Organometal(V) derivatives of As, Sb and Bi bearing OR, OOR, $OCOR^1$ (R = alkyl; R^1 = alkyl or aryl) dibasic acid and (O) groups can be prepared by base-catalysed substitution reactions (Equations (60)–(62)) ⟨51JA4470, 64LA(678)167, 71CJC2529, 72JOM(36)323, 75ACS(B)362, 85JGU63⟩. Intermediate products of single substitution are commonly obtained (Equations (63) and (64)) ⟨72JOM(38)91, 73JOM(55)143⟩.

$$PhAsX_4 + 4\,NaOMe \xrightarrow[\text{near quantitative}]{MeOH} PhAs(OMe)_4 \qquad (60)$$

X = Cl, Br, I

$$Ph_3SbBr_2 + HO_2Bu^t \xrightarrow[\text{benzene or benzene-}Et_2O]{\text{base}} Ph_3Sb(O_2Bu^t)_2 \qquad (61)$$

base = NH_3, 75%
base = $NaNH_2$, 77%
base = NaOMe, 85%

$$Ph_3BiCO_3 + 2\,RCO_2H \longrightarrow Ph_3Bi(O_2CR)_2 \qquad (62)$$

R = CH_2Cl, 97%
R = 2-$HO_2CC_6H_4$, 63%
R = 2-$NH_2C_6H_4$, 90%
R = 4-OHC_6H_4, 82%
R = 4-$NH_2C_6H_4$, 70%

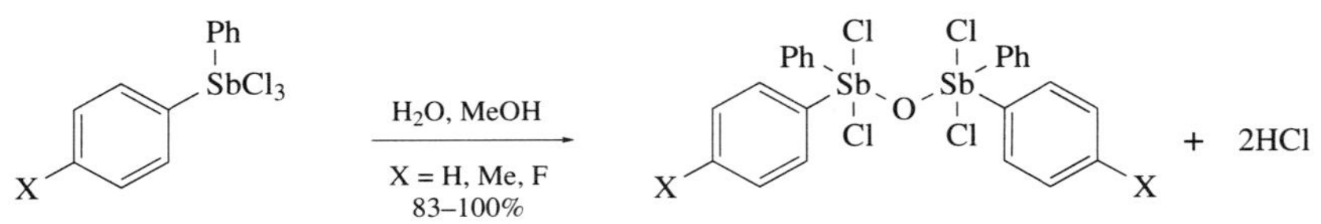

(63)

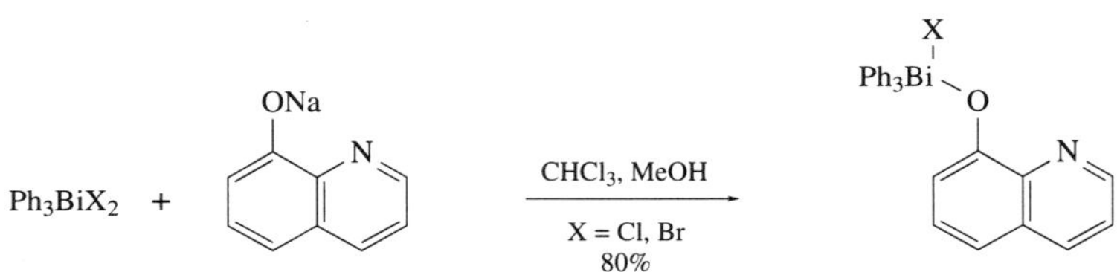

(64)

2.17.3.3 Formation of Metal–Oxygen Bonds through Esterification Processes

Esters have been prepared directly from arsonic and arsinic acids by esterification processes using azeotropic or extractive removal of water with $CuSO_4$ or Ca_2C ⟨73JGU1921⟩. With aromatic acids and higher alcohols the yields are nearly quantitative ⟨67JGU855⟩.

Transesterification using phenols is also possible when the leaving group can be easily expelled from the reaction mixture ⟨72MI 217-01⟩.

Insertion of CO_2 into M—X bonds (M = Sb, Bi; X = NR_2 or OR) has been used a number of times to prepare metal–oxygen derivatives ⟨1889LA(251)323, 70JOM(25)139⟩.

2.17.3.4 Formation through Oxidative Processes

Metal(III) and metal(V) oxides have been prepared directly by oxidative processes, e.g., Equations (65)–(68) ⟨81JGU1056, 84JCS(D)2365, 85JGU63, 88JGU1934, 89MI 217-02⟩. Common reagents include: O_2 ⟨84ZN(B)887⟩, H_2O_2, Bu^tOOH, SeO_2, HgO, $KMnO_4$, $Pb(OAc)_4$, $Hg(OAc)_2$, N_2O_4, O_3, benzoquinone and $K_2H_2P_2O_8$ ⟨85CJC2285⟩. Solvent effects have been noted (Scheme 17). Reactions with alkylaryl- and triarylmetal compounds proceed without complication with H_2O_2, HgO, $KMnO_4$. Mixtures of oxide and selenoxide are often obtained with SeO_2. Chiral arsines do not racemize during oxidation provided the reactions proceed without formation of dihydroxides ⟨64TL3271, 72JGU538, 73JGU1123⟩.

$$\mathrm{Ph_3Sb + Bu^tO_2H \xrightarrow[\text{5–10 °C, 2 h then RT, 2 h; 80\%}]{C_6H_6} Ph_3Sb(O_2Bu^t)_2} \quad (65)$$

$$\mathrm{Ph_3Bi + Bu^tO_2H + 2\,RCO_2H \xrightarrow[\text{87–100\%}]{Et_2O,\ RT} Ph_3Bi(OCOR)_2} \quad (66)$$

R = Me, Bu, Ph, CH=CH$_2$

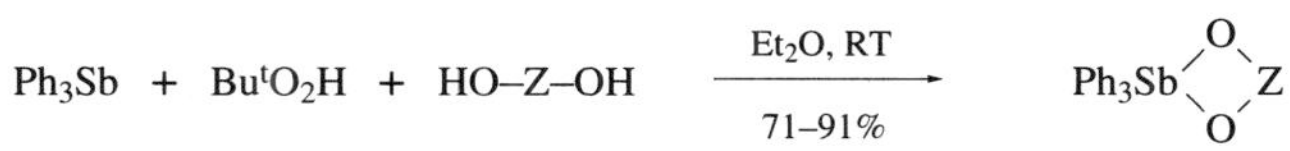

(67)

7 examples, including Z = –CH$_2$CH$_2$– and *o*-C$_6$H$_4$

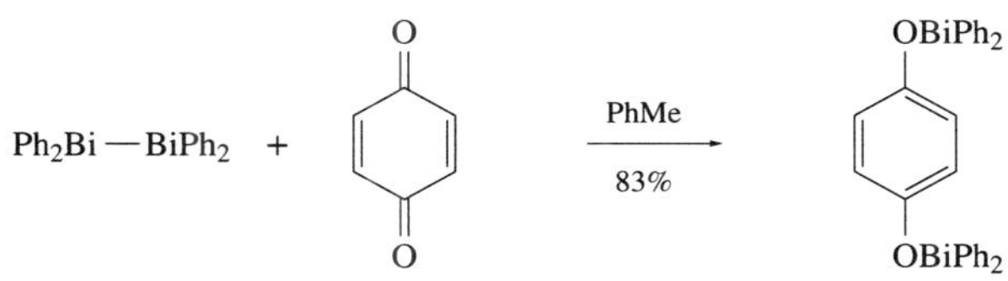

(68)

$$\mathrm{Ph_3Sb + Bu^tO_2H + (RCO)_2O \xrightarrow[\text{50 °C, 3 h}]{solvent} Ph_3Sb(OCOR)_2}$$

R	*Solvent*	*Yield(%)*
Me	Et$_2$O	59
Et	Et$_2$O	52
Ph	Et$_2$O	50
Ph	THF	48
Ph	PhMe	32
Ph	CCl$_4$	7

Scheme 17

Derivatives with allyl groups or alkenyl groups undergo competing oxidative reactions with H_2O_2 and $KMnO_4$. These problems are overcome through use of MnO_2 ⟨67JOC3707⟩.

Oxidation, especially with halogenating reagents, can be accompanied by hydrolysis to give metal oxides indirectly. Silyloxy compounds, e.g., (**28**), have been formed in a similar process (Equation (69)) ⟨1894CB263, 24JCS2275, 70JPR10⟩. It can also proceed with concomitant cleavage of metal–alkyl or metal–aryl bonds (Equations (70)–(72)) ⟨45CR(221)562, 53MI 217-02, 64NEP6405309, 73JOM(55)143, 86ICA(113)43, 86JGU287⟩.

$$\mathrm{Ph_3M + R_3SiOX \xrightarrow{CCl_4,\ -30\ °C} Ph_3M(OSiR_3)X} \quad (69)$$

M = As; R = Ph; X = Cl or Br
M = Sb; R = Ph or Me; X = Cl or Br

(28)

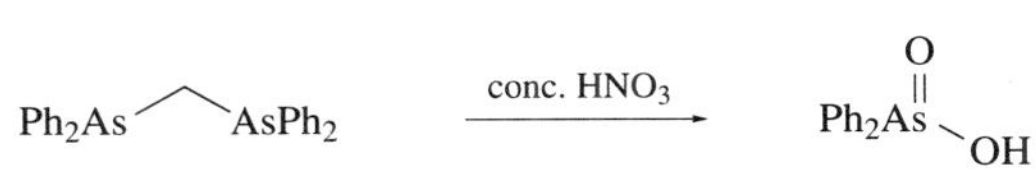

(70)

$$\mathrm{Ph_3Sb \xrightarrow[\text{H_2O–acetone, reflux, 8 h; 90\%}]{NaOH,\ H_2O_2} Ph_2Sb(=O)OH} \quad (71)$$

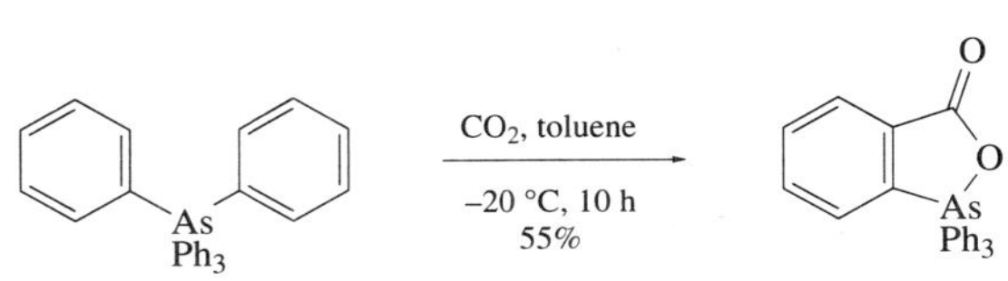

(72)

The preparation of arsine oxides is also achieved in high yield by oxidation of the corresponding sulfides with N_2O_4 or DMSO/H^+ ⟨72JGU538, 73S307⟩ The reaction with N_2O_4 proceeds, in at least one example, with partial inversion of configuration.

2.17.3.5 Preparation by Reductive Methods

Metal(III) oxide derivatives can also be prepared in high yield by reduction of compounds of higher oxidation state using SO_2 or SO_2/HCl (the reaction is catalysed by iodide and appears to proceed through the dihalide), PCl_3 and $PhNHNH_2$ as reducing agents (Equations (73) and (74)) ⟨40JA3142, 44JA191, 48JOC438⟩. In many cases involving Sb derivatives and SO_2/HCl, the intermediate dichloride can be isolated and hydrolysed separately ⟨20LA(421)174, 30JA2046, 32JCS1815⟩.

$$\mathrm{ArAsO_3H_2} \xrightarrow[\text{72–100\%}]{\mathrm{SO_2}} (\mathrm{ArAs{=}O})_n \text{ or } \mathrm{ArAs(OH)_2} \qquad (73)$$

Ar = Ph, 3-$NO_2C_6H_4$, 4-HO_2C-C_6H_4, 4-NC-C_6H_4, 2-NO_2-3-HOC_6H_3, 2-Me-5-$NO_2C_6H_3$, 9-oxofluoren-1-yl, 9-oxofluoren-4-yl

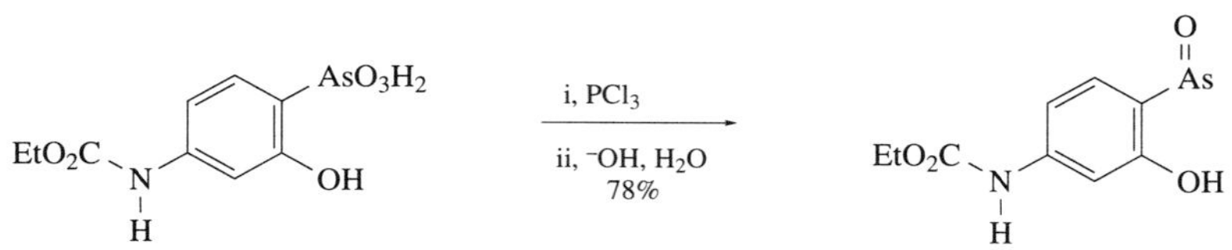

(74)

Modification of the type of oxygen derivative can be achieved by a variety of eliminative processes that are usually thermally induced ⟨67CB1544, 72AG(E)441⟩. In particular, the preparation of arsinous esters from chiral arsine oxides proceeds with inversion of configuration (Equation (75)) ⟨72JGU1952⟩.

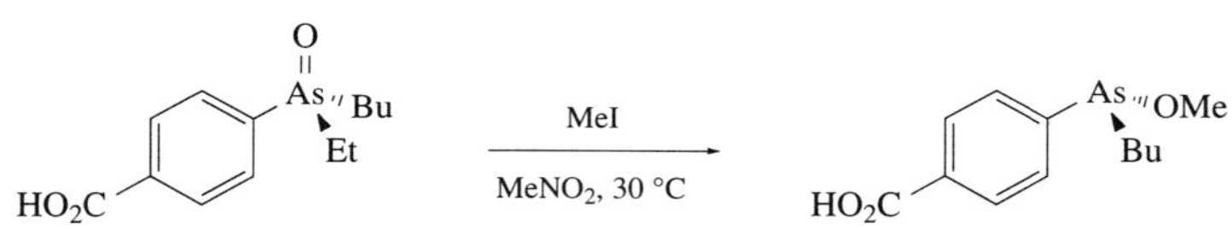

inversion of configuration

(75)

Similar processes can be used to introduce acyloxy substituents into triaryl- and penta-arylantimony and triarylbismuth (Equations (76) and (77)) ⟨64NEP6405309, 71ZAAC(386)147, 82ZN(B)815, 84AJC527⟩. In the cleavage of (4-X—$C_6H_4)_3$Bi (X = H, Me, OMe, OEt) with RCO_2H (R = Me, CF_3), high yields of diacyloxy and monoacyloxy derivatives can be obtained selectively ⟨84AJC527⟩.

$$\mathrm{Ph_5Sb} + \mathrm{RCO_2H} \xrightarrow[\text{R = H or Me, 77–79\%}]{\mathrm{C_6H_6},\ 50\text{–}60\ ^\circ\mathrm{C},\ 5\ \mathrm{h}} \mathrm{Ph_4Sb{-}OCOR} \qquad (76)$$

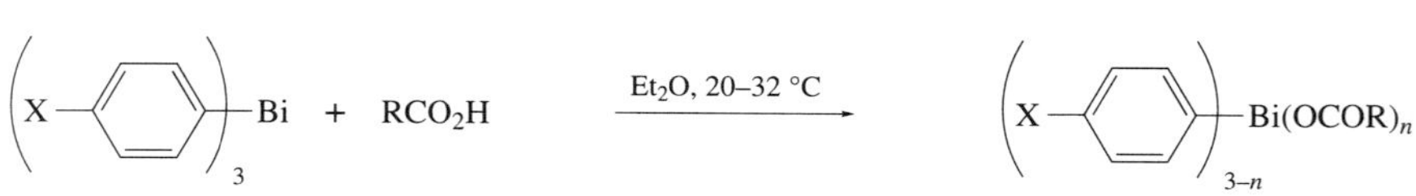

(77)

X = H, Me, OMe, OEt

R = Me $n = 2$ 47–82%
R = CF_3 $n = 1$ 25–76%

2.17.4 ARSENIC, ANTIMONY AND BISMUTH COMPOUNDS WITH THE HETEROATOM BONDED TO OTHER CHALCOGENS

Syntheses of chalcogen derivatives other than those of oxygen involve introduction of the chalcogen by direct or indirect means rather than through carbon–metal bond formation. One recent exception is the preparation of stibathiolanes (Equation (78)) ⟨91JA165⟩.

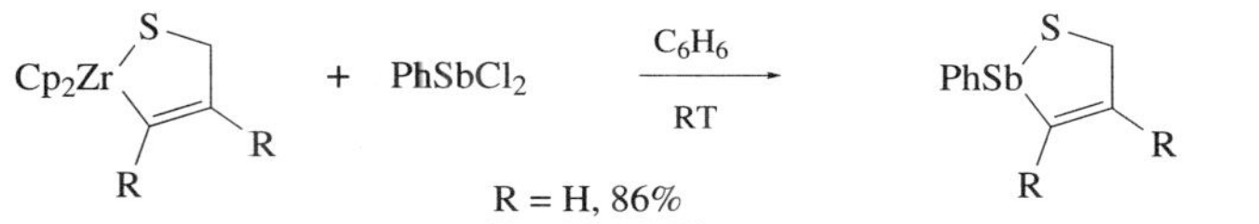

R = H, 86%
R = Me, 84%
R = Ph, 85%

(78)

2.17.4.1 Preparation of Organometal(III) Derivatives through Nucleophilic Displacement Reactions

The synthesis of thioarsenic esters and thioarsinic esters is most commonly achieved by nucleophilic substitution reactions on R_2AsX and $RAsX_2$ (R = alkenyl, aryl; X = halogen, OR^1, NR^1_2; R^1 = lower alkyl) (Scheme 18) ⟨50JCS18, 61JGU3315, 64DOK(157)371, 67JGU1578⟩. In a few cases the reaction has been extended to Se and Te derivatives (Equation (79)) ⟨70JCS(C)810⟩. The reaction with dithiols is the basis of the action of British anti-Lewisite ⟨47JCS592, 73ACS1337, 73ACS2135⟩.

$$R^1AsCl_2 \xrightarrow{R^2SH,\ Et_3N\ \text{or}\ NaSR^2} R^1As(SR^2)_2$$

R¹	*R²*	*Yield(%)*
CH=CHCl	Ph	–
Ph	Ph	64
Ph	Et	40
Ph	Pr^n	57
Ph	Bu^n	57
Ph	Hex^n	10
4-Me-C_6H_4	Pr^n	68
2,4-Cl_2-C_6H_3	Pr^n	67
2,4-Cl_2-C_6H_3	Bu^n	72

Scheme 18

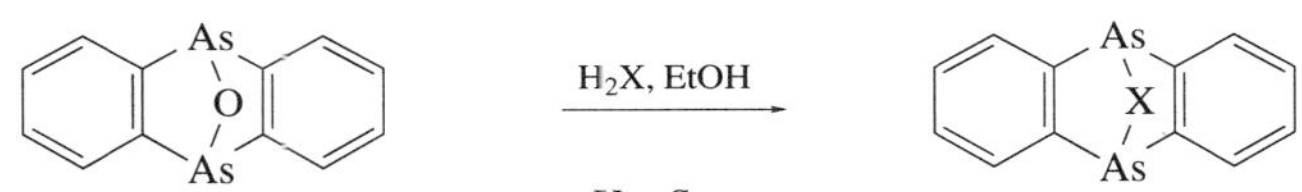

X = Se
X = Te

(79)

Antimony, and in a few cases bismuth analogues, can be prepared similarly ⟨51JA2880, 68JOM(11)299⟩, although various attempts to prepare $R_3Sb(SR^1)_2$ (R = aryl; R^1 = alkyl) have failed, probably due to their thermal instability. In all situations, care must be taken over temperature because exchange of the chalcogen ligands can take place ⟨71IZV1726⟩.

Nucleophilic reagents other than simple thiols and thiolates (e.g., $NaSC(S)NR^1_2$ (NR^1_2 = NMe_2, NEt_2, aziridine, piperidine, morpholine), $KSCO_2Me$, KSC(S)OEt, $NaSP(O)(OR^2)_2$, $HSP(S)(OR^2)_2$ (R^2 = Bu, iPr)) have also been used successfully (Equation (80)) ⟨32JA2993, 50JCS18, 65IC973, 66JGU1986, 68JOM(11)627, 90PS(47)357⟩.

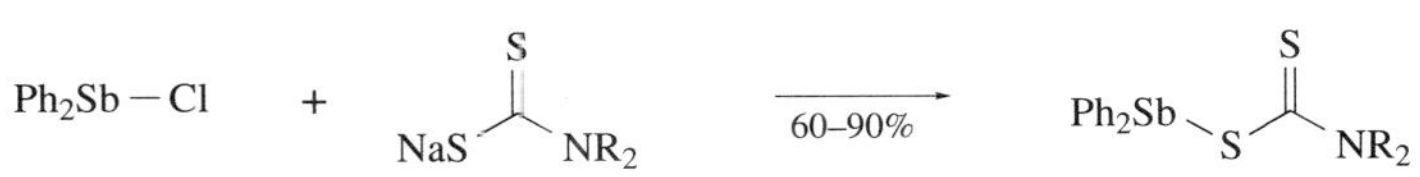

R = Me, Et, Ph

(80)

Displacement reactions to yield dithiocarbonate and dithiocarbamate derivatives can occur with reagents generated *in situ* (Equation (80)) or through treatment of esters and amides with CS_2 (Equation (81)) ⟨69ZAAC(369)178, 70JOM(25)139, 89JGU2254⟩.

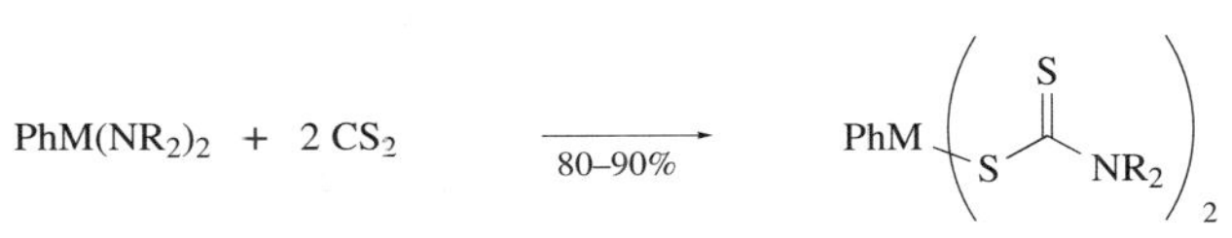

M = As or Sb; NR_2 = NEt_2, piperidino or morpholino

(81)

Thioepoxides have also been used as nucleophilic reagents to displace SCN ligands from alkylarsenic derivatives, but there has been no report of analogous reactions with aryl derivatives. However, addition of S nucleophiles, including thiols and thioethers, to RM═X (R = aryl, M = As or Sb, X = O, S or NPh) can give rise to thioesters or sulfides (Equation (82)) ⟨32JCS593, 32JCS2505, 68JOM(11)627, 69ZAAC(370)227, 87JGU2496⟩.

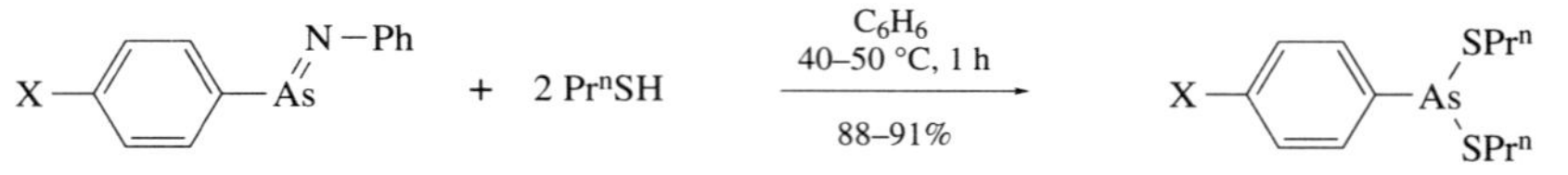
(82)

X = H, Br, Me

Treatment of $RAsX_2$ or RAs═O (R = aryl) with H_2S and its equivalents under alkaline or thermal conditions affords $(RAs{=}S)_n$ ⟨66JOM(5)326, 68JCS(A)483, 78HOU(13)206⟩. Such reactions have not been reported with Sb or Bi derivatives, nor have they been extended to examples using H_2Se or H_2Te.

2.17.4.2 Preparation of Organometal(V) Derivatives through Displacement Reactions

The synthesis of As and Sb^V derivatives from oxide and halide derivatives can be achieved through treatment with H_2S under alkaline conditions (Equations (83)–(85)) ⟨63ACS1875, 71CB3782, 75JGU1476, 75JOM(96)57⟩.

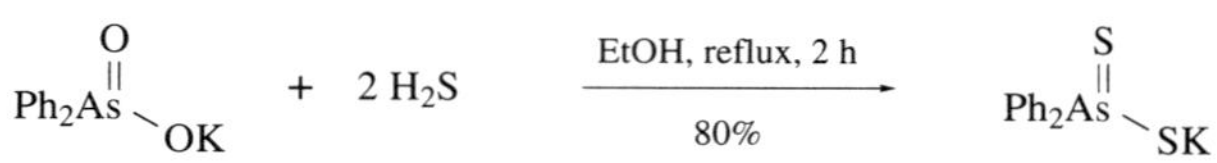
(83)

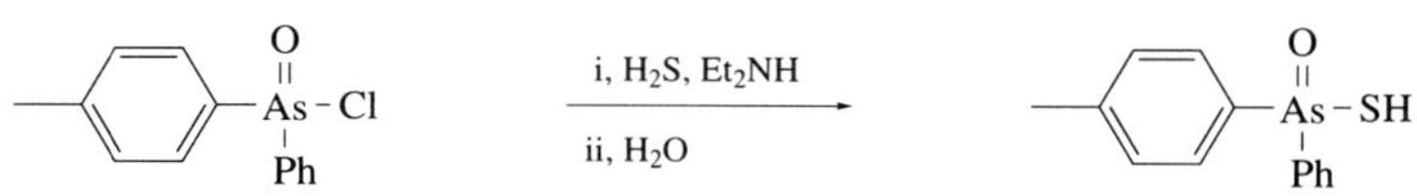
(84)

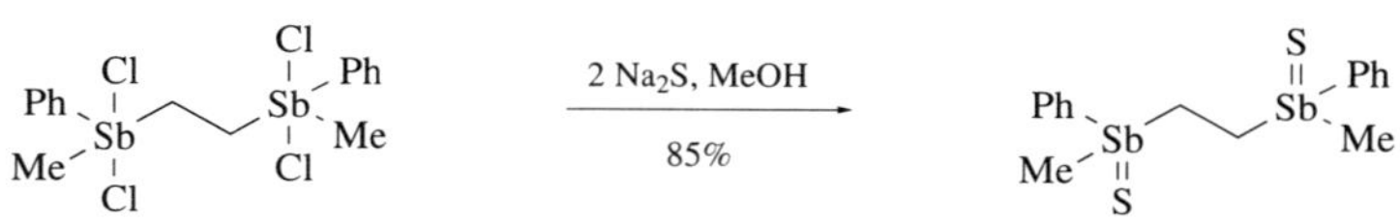
(85)

The reaction of H_2Se with Ph_3AsCl_2 in alkaline solution has been used to provide $Ph_3As{=}Se$ (Equation (86)) but purely gaseous H_2Se causes reduction of Ph_3AsX_2 (X = Cl, Br) ⟨63ACS1875, 64IC580⟩.

$$Ph_3AsCl_2 + H_2Se \xrightarrow[0\ °C]{NH_3,\ EtOH} Ph_3As{=}Se \qquad (86)$$

2.17.4.3 Preparation of Sulfur, Selenium and Tellurium Derivatives by Oxidative Processes

Sulfur derivatives of organoarsenic(III) and organoarsenic(V) compounds are prepared from a wide variety of precursors, including mono-, di- and triorganoarsines $(PhAs)_6$, tetraaryldiarsenides, through oxidative processes ⟨1882CB1952, 02LA(321)141, 33JA1161, 70JOM(21)231⟩. Elemental sulfur is commonly employed with some tolerance of other functional groups (Equation (87)) ⟨70JOM(21)231⟩. Chiral arsines react with retention of configuration ⟨68JGU370⟩. The corresponding selenation is only known for aliphatic arsines ⟨64IC580⟩.

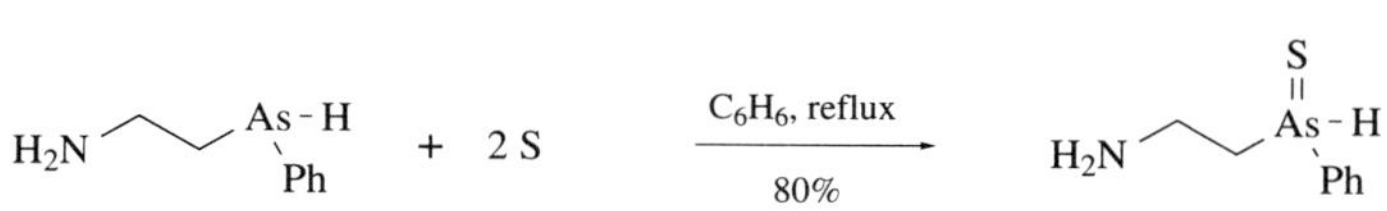
(87)

$(Ph_2Sb)_2Se$ and $(Ar_2Bi)_2X$ (Ar = Ph, 4-MeC_6H_4, X = S and Se) have similarly been prepared from $(Ar_2M)_2$ by treatment with elemental X (Equation (88)), but in the latter case Te is unreactive ⟨84JCS(D)2365, 84JOM(277)257, 84ZN(B)887, 88IC3730⟩. Monometal chalcogen derivatives, Ar_2MXPh

(Ar = Ph, 4-MeC_6H_4; M = Sb or Bi; X = S, Se or Te) have been prepared from $(Ar_2M)_2$ by reaction with $(PhX)_2$ (Equation (89)) ⟨84JOM(277)257, 84ZN(B)1668, 87ZN(B)695, 88IC3730⟩.

$$Ar_2M{-}MAr_2 \quad + \quad X \longrightarrow Ar_2M{\overset{X}{\frown}}MAr_2 \qquad (88)$$

M = Sb; Ar = Ph; X = Se or Te
M = Bi; Ar = Ph; 4-MeC_6H_4; X = S or Se

$$Ar_2M{-}MAr_2 \quad + \quad PhX{-}XPh \longrightarrow 2 \quad Ar_2M{\frown}XPh \qquad (89)$$

M = Sb, Bi; Ar = Ph. 4-C_6H_4; X = S, Se or Te

Xanthate derivatives, $ArAs(S_2CNR_2)_2$ can be prepared by insertion of CS_2 into the As—N bonds of $ArAs(NR_2)_2$ (see Section 2.17.4.1). Preparation of $Ph_2AsSP(S)Ph_2$ through an interesting reductive process from $Ph_2As(S)SNa$ has also been described ⟨83JOM(252)295⟩.

2.17.5 ARSENIC COMPOUNDS WITH AN As—N BOND AND THEIR ANTIMONY AND BISMUTH ANALOGUES

2.17.5.1 Preparation by Displacement of Halide and Halide Equivalents from the Metal

Aryl- and alkenylarsenic(III) amides can be prepared by treatment of $Cl_nAs(NR_2)_{3-n}$ (R = lower alkyl or $(CH_2)_5$; n = 1,2) with organometallic reagents, but such approaches have not been used to prepare Sb or Bi derivatives ⟨64ZAAC(330)233⟩.

A more versatile approach, applicable to organometal(III) and (V) amides, has been through the reaction of nucleophilic nitrogen reagents with organometal halides ⟨69ZAAC(369)178, B-70MI 217-01⟩. The reactions can occur with primary or secondary amines, and are made more efficient through use of alkali metal, often Li, amide salts. Another advance has been the use of silylated amines which react with elimination of R_3SiX (Equation (90)) ⟨91JGU405⟩. The TMS group can be retained if N-silylated Li or Na amides are employed. Phase transfer catalysis has been used to prepare $R_3Sb(NCSe)_2$ (R = Me, aryl or C_6F_5) and $Ph_3RSb(NCSe)$ (R = Ph or allyl) by this route ⟨86POL1581⟩.

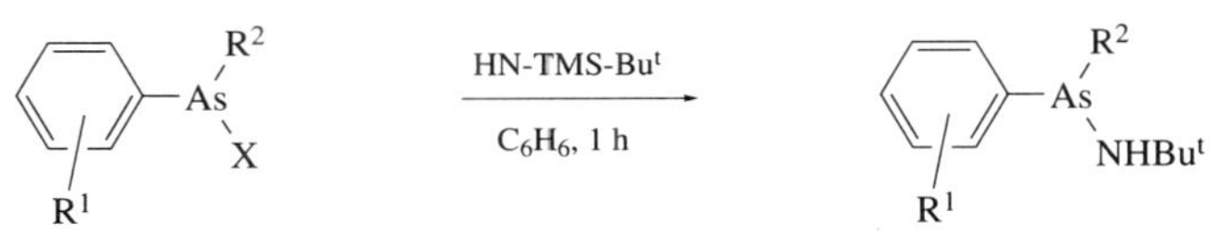

(90)

R[1] = H, *o*-OMe, *m*-Me; R[2] = X = Cl, Br, I, 64–83%
R[1] = H, *p*-Me, *p*-Br; R[2] = aryl; X = Cl, 77–84%

The range of amines used with arylmetal halides has been more limited than that reported with alkyl derivatives, i.e., with alkyl, aryl, TMS, $P(alkyl)_2$, $=P(alkyl)_3$, =C(Ph)N(Me)TMS, $AsMe_2$, $AsBu^t_2$ substituents ⟨78HOU(13)284⟩, but a study in the late 1980s expanded the range of Ar_2SbNR_2 (Ar = Ph, C_6F_5; NR_2 = heterocyclic, e.g., succinimide and benzotriazolide) that were available ⟨89POL581⟩, and thiodiimine $Ph_2AsN{=}S{=}NAsPh_2$ has been prepared in this way ⟨84CB1999⟩. Chloride is less easily displaced by amines than is bromide, and this enables selective formation of monosubstitution products ⟨70JGU127⟩. Enantiomers of $Bu^tPhAsNMe_2$ have been prepared from $Bu^tPhAsCl$ by successive treatment with 1-menthol and $LiNMe_2$ ⟨88JGU1040⟩.

Metal(III) and metal(V) azide derivatives of As, Sb and Bi are readily prepared by this method ⟨68JOM(13)529, 73JOM(50)129, 74IC1252⟩. Solvent significantly influences the rate of the reaction, with hydroxylic solvents including H_2O, conferring the highest rates.

Arsonic amides may be obtained from the corresponding esters. Strong heating is required to achieve reaction with amines, and products of single substitution can be obtained in good yield (Equation (91)) ⟨74JGU1506⟩. Amide exchange is also feasible when the displaced amine is volatile ⟨75ZAAC(412)202⟩.

$$PhAs(=O)(OR)_2 + HNBu_2 \xrightarrow[\substack{R = Et,\ 77\% \\ R = Pr^i,\ 58\% \\ R = Pr^n,\ 78\%}]{100\ ^\circ C\ to\ 155\ ^\circ C,\ 3\text{–}6\ h} PhAs(=O)(OR)(NBu_2) \quad (91)$$

Primary amines, and sulfur and phosphorus amides commonly yield metal imines with As and Sb dihalides (Equation (92)) ⟨66AOC(4)146, 71ACS983, 74JCS(P1)466⟩. Similar arsenic derivatives are prepared by treatment of 10-chloro- and 10-methyl-5,10-dihydrophenarsazines with chloramine ⟨72IC1226⟩ (see Section 2.17.5.2). An interesting outcome where this is not the case is the formation of (**29**) (Equation (93)) ⟨70JPR751⟩.

$$Ph_3AsBr_2 + H_2N{-}SO_2Ph \xrightarrow{2\ Et_3N} Ph_3As{=}N{-}SO_2Ph \quad (92)$$

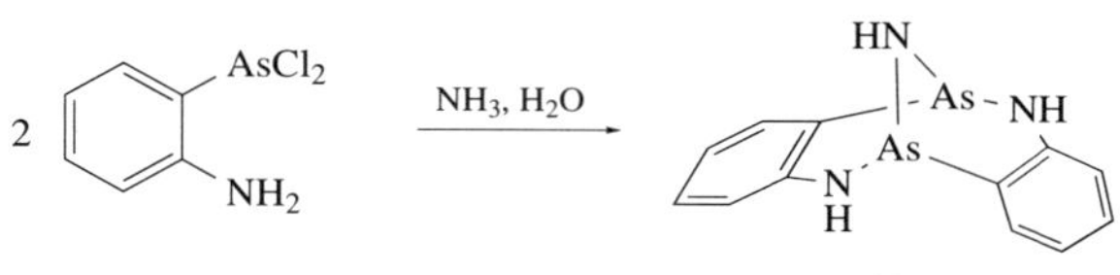

(93)

(29)

2.17.5.2 Preparation by Formation of Metal–Nitrogen Bonds through Oxidative Processes

A versatile synthesis of stable metal imines of As, Sb and Bi compounds is by treatment of the organometal with azido, chloramine or other nitrene generating reagent (Equation (94)) ⟨72IC1226, 74JCS(P1)460⟩.

$$Ph_3As + RN_3 \xrightarrow{heat,\ no\ solvent} Ph_3As{=}NR \quad (94)$$

R = CN, 40%
R = CO_2Et, 66%
R = SO_2Me, 74%
R = C_6F_5, 38%

At low temperatures it is possible to trap chloro amines which are probably intermediates to the imines ⟨67IC1178⟩. However, the reaction of Ph_3As with TosNClNa (Chloramine-T) is suggested to proceed through $Ph_3As{=}O$ ⟨72JCS(P1)2793, 74JCS(P1)460⟩. Certainly, there are other general approaches to arsinimines from $Ph_3As{=}O$ (Scheme 19) ⟨65ACS1755, 71ACS983, 74JCS(P1)466⟩.

$Ph_3As{=}O$ —i→ $Ph_3As{=}NR^1$
R^1 = Bz, $PO(OEt)_2$, Ts, SO_2Ph, SO_2Me, C_6H_4-4-NO_2

$Ph_3As{=}O$ —ii→ $Ph_3As{=}NR^2$
R^2 = Ph, Ac, Bz, $POPh_2$, Ts

$Ph_3As{=}O$ —iii (65–90%)→ $Ph_3As{=}NR^3$
R^3 = SO_2Me, SO_2Ph, $SO_2C_6H_4$-4-Me, C_6H_4-4-NO_2, Bz, $P(O)Ph_2$

i, $R^1N{=}S{=}O$ or $R^1N{=}S{=}NR^1$, C_6H_6, reflux; ii, $R^2N{=}\bullet{=}O$; iii, R^3NH_2, Ac_2O, reflux

Scheme 19

2.17.6 ARSENIC, ANTIMONY AND BISMUTH COMPOUNDS WITH THE HETEROATOM BONDED TO OTHER GROUP 15 ELEMENTS

2.17.6.1 Symmetrical Diarsines, Distibines and Dibismuthines

Interest in this group of compounds is derived largely from the thermochromic behaviour of many distibine and dibismuthine derivatives ⟨90AOC(30)77⟩. The synthesis of As, Sb and Bi derivatives is most readily achieved by treatment of diorganometal(III) alkali metal salts with an oxidant, e.g., I_2 ⟨83OM1005⟩, O_2/EtOH, Ph_3CClO_4, Bu^tOCl ⟨83JOM(248)269⟩ and $CuCl_2$ ⟨90AX(C)774⟩, the corresponding diorganometal(III) halides ⟨83JOM(248)269⟩, or more generally a 1,2-dihaloethane (Scheme 20) ⟨83JOM(253)C21, 83OM1859, 83OM1573, 87OM1185⟩. Symmetrical distibines are obtained directly from diarylstibine halides by reaction with Mg, but under these conditions monoorganostibine dihalides give tetrameric products ⟨86ZN(B)321⟩. In the late 1980s, treatment of Ph_2BiI with Cp_2Co was reported as a new route to $Ph_2BiBiPh_2$ in 80% yield ⟨88IC3730⟩. It has also been noted that treatment of 1-chloro-2,3,4,5-tetramethylarsole with alumina gives the dimer in 40% yield ⟨83OM1859⟩.

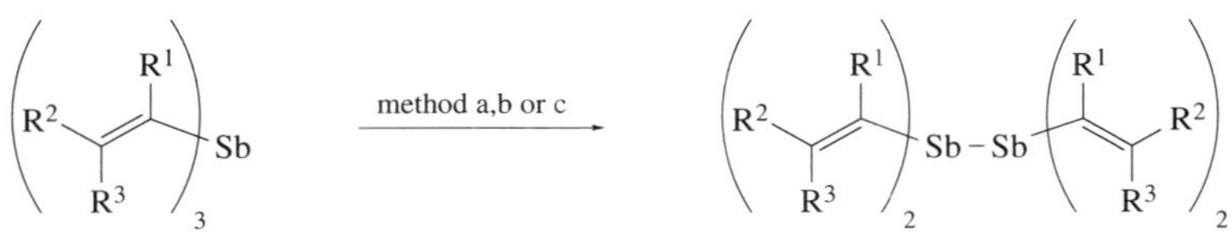

R¹	*R²*	*R³*	*Method*	*Yield(%)*
H	H	H	a	57
H	H	H	b	86
Me	H	H	b	86
H	Me	H	b	77
H	H	Me	b	80
H	Me	H	c	76
Me	Me	H	c	69

method: a, i, Li, THF, 25 °C, 2 h; ii, $BrCH_2CH_2Br$, THF, 30 min;
b, i, Na, NH_3, 1 h; ii, $ClCH_2CH_2Cl$, 30 min; c, i, K, NH_3, 1 h; ii, NH_4Br, 30 min; iii, $BrCH_2CH_2Br$

Scheme 20

2.17.6.2 Mixed Organodiarsines, -distibines and -arsinophosphines

Unsymmetrical tetraorganodiarsines $Ph_2AsAsMe_2$ are formed in solution through treatment of Ph_2AsH and Me_2AsNMe_2 at −60 °C but they cannot be isolated; above −40 °C metathesis takes place giving only the symmetrical products ⟨88ICA(150)51⟩. Similar exchange has been reported between Ph_2PPPh_2 and $Ph_2AsAsPh_2$, but when As is replaced by Sb the exchange occurs to a lesser extent ⟨84IC1504⟩. Between different distibines, the degree of exchange depends upon the nature of the substituents; again mixed distibines have not been isolated ⟨88POL2601⟩.

Mixed compounds containing doubly bound As—As and As—P have been prepared and are stable, but bulky groups attached to As appear to be essential for stability (Equation (95)) ⟨84IC2582⟩.

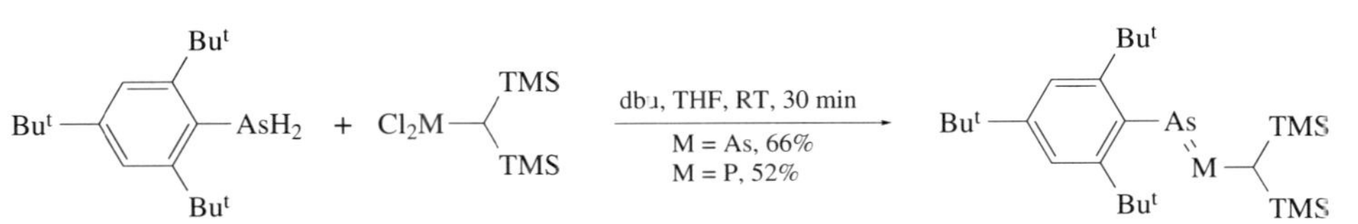

(95)

2.17.6.3 Cyclopolyarsines and their Antimony and Bismuth Analogues

Another major group of compounds are the cyclopolyarsines $(RAs)_n$ (R = alkyl or aryl). Most aryl derivatives are formed as hexamers ($n = 6$), and are available by reduction of the arsonic acids with H_3PO_2, reaction of arsenic dihalides with alkali metals, Hg (preferred when the halide = I), or H_3PO_2, and oxidation of primary arylarsines ⟨75JOM(84)1⟩. Fewer Sb—Sb bonded compounds exist, but they are prepared by similar methods. An exchange reaction has been developed that gives polyarsines cleanly and in high yield (Equation (96)) ⟨87IC1638⟩. Condensation methods have been used to prepare the triphosphatriarsine (**30**) (Equation (97)) ⟨83IZV1152⟩. The synthesis of dihydroarsaphosphetes (**31**), through ring expansion of phosphirenes (**32**), and their chemistry has been described (Scheme 21) ⟨93BSF521⟩.

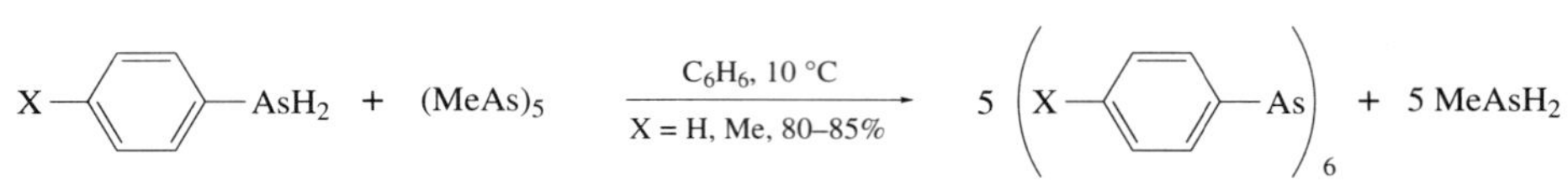

(96)

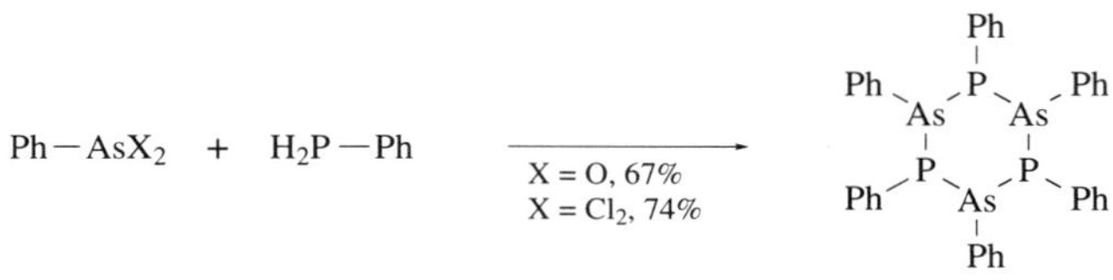

(97)

(**30**)

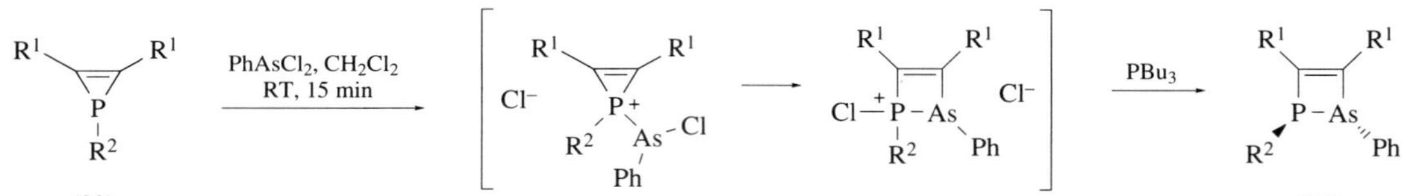

(**32**) (**31**)

R[1]	*R[2]*	*Yield(%)*
Ph	Ph	45
Me	Ph	30
Ph	Et	15

Scheme 21

2.17.7 ARSENIC, ANTIMONY AND BISMUTH COMPOUNDS WITH THE HETEROATOM BONDED TO A METALLOID

There are few known examples of aryl or alkenyl metal compounds of this type other than those mentioned in Sections 2.17.4 and 2.17.6, although alkyl substituted As, Sb and Bi derivatives bound to Si and Ge are reported ⟨68JCS(A)2623, 81ZN(B)48⟩. The method of synthesis that has been used with aryl derivatives involves halide displacement from organostibine halides (Equation (98)) ⟨86ZN(B)321, 89PS(44)129⟩.

$$n\ \text{TMS-MgCl} + \text{Ph}_{n-3}\text{SbCl}_n \xrightarrow[\text{RT, 2–4 h}]{n\ \text{Mg, THF}} \text{Ph}_{n-3}\text{Sb(TMS)}_n \qquad (98)$$

$n = 1$, 63%
$n = 2$, 58%

2.17.8 ARSENIC, ANTIMONY AND BISMUTH COMPOUNDS WITH THE HETEROATOM BONDED TO A METAL

2.17.8.1 Alkali Metal Derivatives

Synthesis can take place through a variety of methods, many of which have been described in earlier sections (Equations (99)–(102)).

$$Ph_3M + Na \longrightarrow Ph_2MNa \quad M = As, Sb \text{ or } Bi \tag{99}$$

$$Ph_2MCl + Na \longrightarrow Ph_2MNa \quad M = As, Sb \text{ or } Bi \tag{100}$$

$$Ph_2MH + Na \longrightarrow Ph_2MNa \quad M = As \text{ or } Sb \tag{101}$$

$$PhMH_2 + 2\,BuLi \longrightarrow PhMLi_2 \quad M = As \text{ or } Sb \tag{102}$$

Coordinating solvents can influence the σ or π nature of bonding of lithium arsolide species (Scheme 22) ⟨91CB2453⟩.

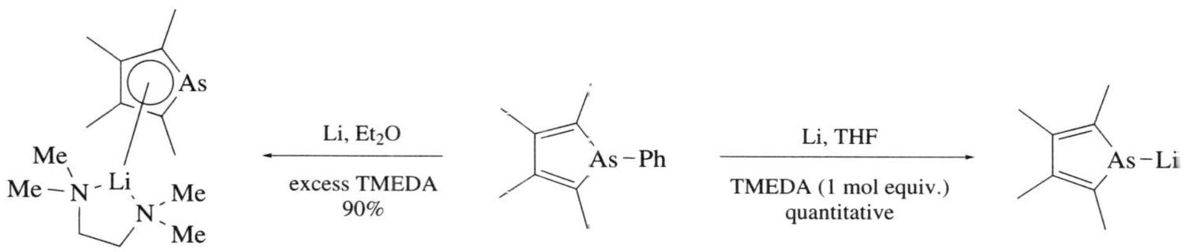

(33)

Scheme 22

2.17.8.2 Other Main Group Derivatives

Apart from the derivatives mentioned in earlier parts of this chapter, arylstibines containing an Sb—Sn bond can be prepared from halostibines or alkali metal stibides by displacement reactions (Equations (103) and (104)) ⟨67JOM(8)105⟩.

$$n\,Ph_3SnLi + Ph_{3-n}SbCl_n \xrightarrow{\text{THF, ambient}} Ph_{3-n}Sb(SnPh_3)_n \quad n = 1, 64\%;\ n = 2, 40\% \tag{103}$$

$$n\,Ph_2SbNa + Ph_{4-n}SnCl_n \xrightarrow{NH_3} Ph_{4-n}Sn(SbPh_2)_n \quad n = 1, 34\%;\ n = 2, 29\%;\ n = 3, 35\%;\ n = 4, 68\% \tag{104}$$

Synthesis of the first diarsadigalletane by a similar route has also been described ⟨92JOM(439)C33⟩. The π-bonded lithium arsolide complex (**33**) also reacts to give acceptable yields of σ-bonded Al and Ga derivatives (Scheme 23) ⟨91CB2453⟩.

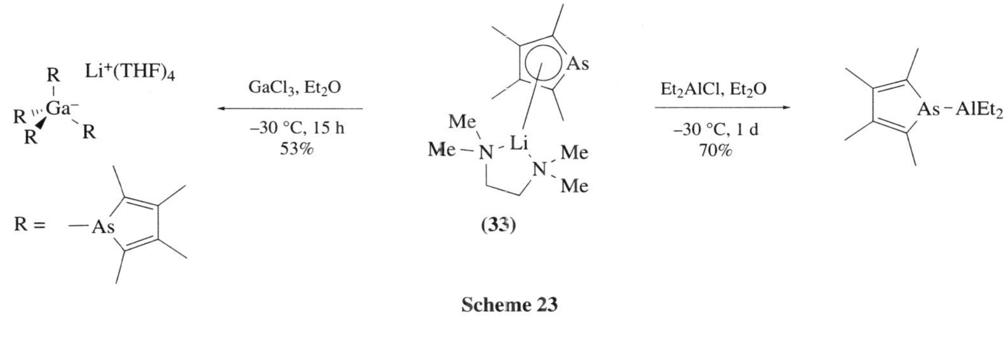

Scheme 23

2.17.8.3 Transition Metal Derivatives

In this review, π-bonded complexes involving triorganometal(III) compounds as a simple ligand are not considered ⟨B-79MI 217-02⟩. However, heterocyclopentadienides of As, Sb and Bi are versatile ligands towards transition metals. η^1-Coordinated, e.g., $Mn(CO)_5$ complexes **(34)** and η^5-coordinated, e.g., $Mn(CO)_3$ complexes **(35)** (the heterocymantrenes), and Fe^{II} complexes **(36)** (the heteroferrocenes), react by halide displacement (Schemes 24 and 25) ⟨91CB2453, 91OM2068, 93JOM(447)197⟩.

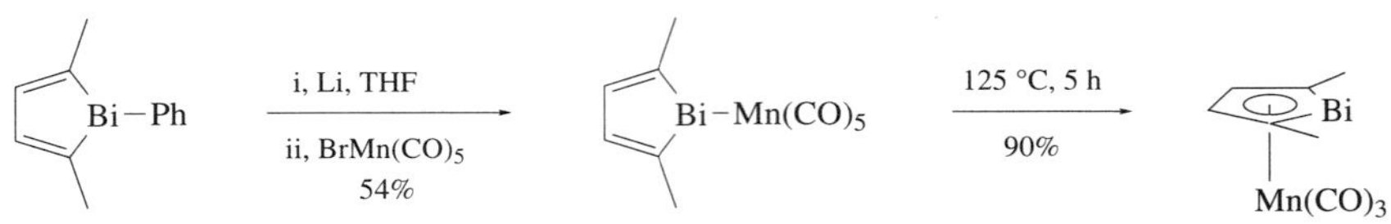

(34) **(35)**

Scheme 24

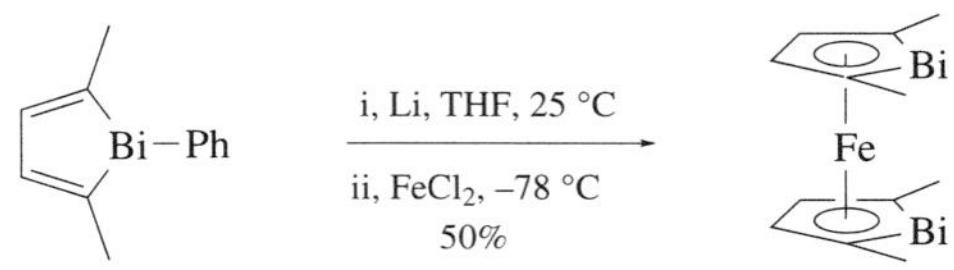

(36)

Scheme 25

In contrast, in an alternative synthetic approach, the chloroarsinidene compound **(37)** reacts with carbon nucleophiles to give neutral **(38)** or anionic adduct complexes **(39)** in the presence of excess reagents (Scheme 26) ⟨93JOM(448)83⟩.

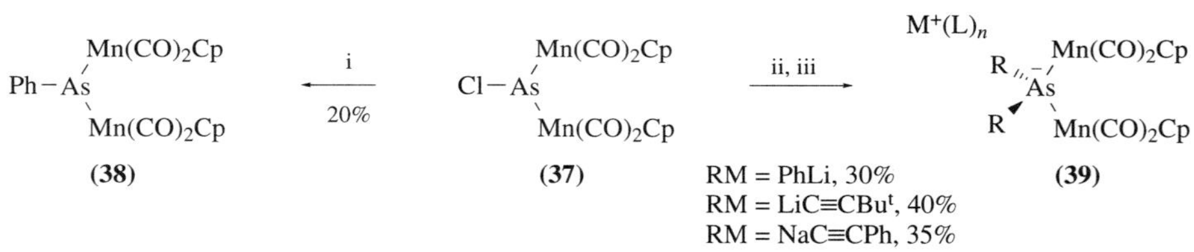

i, PhLi, THF–Et_2O, –50 °C to 20 °C; ii, RM (2.5 mol equiv.), THF–Et_2O, 20 °C; iii, 12-crown-4 or 222-Krypt

Scheme 26

Taking their lead from known phosphorus chemistry, two groups have reported the preparation of chromium complexed vinyl substituted arsanes **(40)** and **(41)** through reaction of $Cr(CO)_5AsPhH_2$ with alkynes (Equations (105) and (106)) ⟨87JOM(336)337, 92JOM(441)389⟩. Steric hindrance plays a role in the degree of substitution that occurs. This approach has also been used in a synthesis of 1-phospha-4-arsacyclohexa-2,5-dienes ⟨93JOM(444)53⟩. Another group has synthesised a series of arsinoketenes stabilised as ligands in tungsten complexes ⟨84CB427⟩, and demonstrated their conversion to novel arsenic–tungsten heterocycles through addition of trialkylphosphines (Scheme 7, Section 2.17.1.2.4) ⟨83OM1266⟩. Thermolysis of these heterocycles affords the free arsenoketenes ⟨86JOM(317)175⟩.

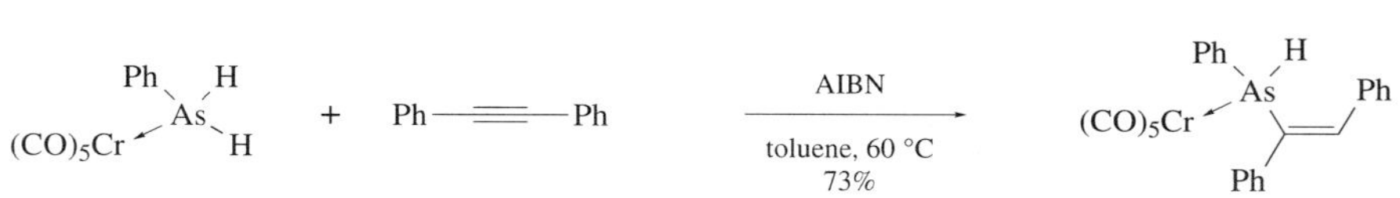

(105)

(40)

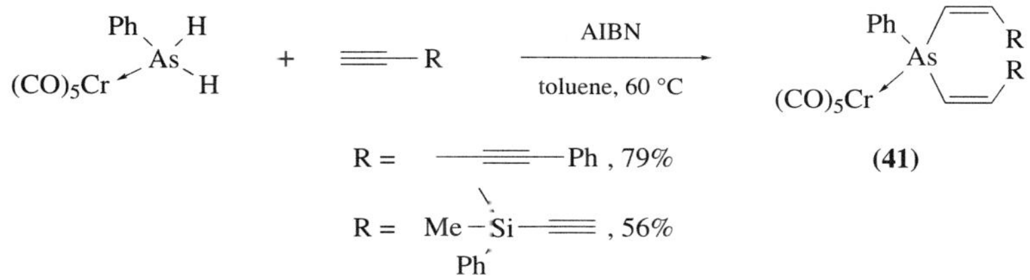

(41)

(106)

The role of bis(tertiary arsines) as ligands for transition metals can be complicated by ligand rearrangement or cleavage during reactions. These processes can lead to novel products, e.g., **(42)** and **(43)** (Equations (107) and (108)) ⟨75JCS(D)172, 82IC3731⟩.

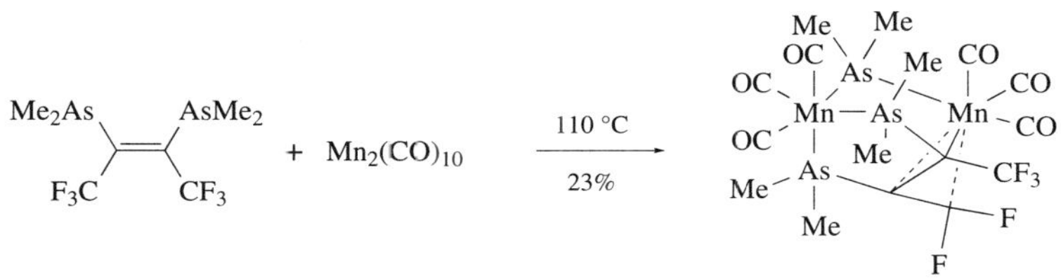

(42)

(107)

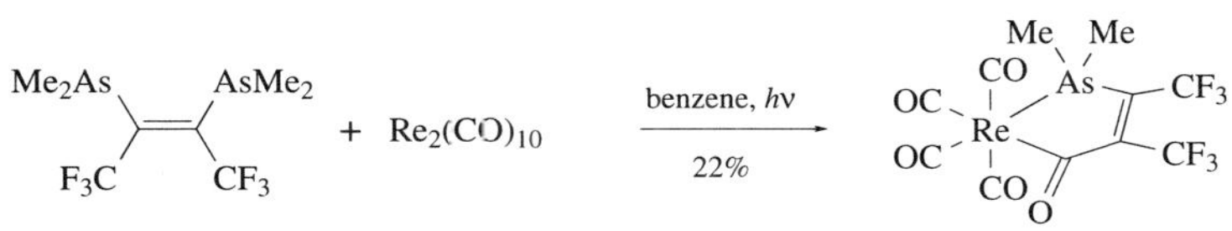

(43)

(108)

2.18
Vinyl- and Arylsilicon, -germanium, and Boron Compounds

LÀSZLÒ HEVESI

Facultés Universitaires N.-D. de la Paix, Namur, Belgium

2.18.1 BORON DERIVATIVES

The rapid development of organoboron chemistry is clear from the number of books ⟨B-61MI 218-01, B-72MI 218-01, B-73MI 218-01, B-75MI 218-01, B-88MI 218-01⟩ and review articles ⟨79COC(3)687, B-81MI 218-01, 82COMC-I(7)111, 82COMC-I(7)143, 82COMC-I(7)161, 82COMC-I(7)303, 82COMC-I(7)323, 74S77, 85OR(33)1, 89T1859, 89CRV1535, 91COS(1)487, 91COS(7)593, 91COS(8)703⟩ that have been published over the years.

It follows that the synthetic methods allowing the preparation of various types of organic boron compounds have been constantly updated and expanded. One of the latest and possibly the most helpful compilation is *Borane Reagents* by A. Pelter, K. Smith, and H. C. Brown ⟨B-88MI 218-01⟩: in addition to giving a critical review of the reagents and methods, this book also provides representative experimental procedures for the preparation of boron reagents, as well as typical syntheses that make use of these reagents.

Because of this, classical methods will not be presented in detail, but will be briefly summarized and newer methods will then be discussed.

2.18.1.1 Alkenyl Boron Compounds

2.18.1.1.1 Addition of boron reagents to alkynes

This is certainly the most widely applied reaction type for the preparation of alkenyl boron derivatives, and according to the boron reagent that is added to the alkyne, it is possible to distinguish: (i) hydroboration; (ii) halo- and thioboration; (iii) carboboration; and (iv) metalloboration reactions. Among these, hydroboration is by far the most popular.

(i) Hydroboration of alkynes

Figure 1 features frequently used hydroboration reagents (**1**)–(**20**), as well as halo- and thioboration reagents (**21**) and (**22a, b, c**). Complexing Lewis bases such as Et_2O, THF, Me_2S, or TMEDA are added to some of these reagents in order to stabilize them, for example against disproportionation in the case of haloboranes ⟨77IC2938, 80CC1169⟩. On the other hand, incorporation of one or two branched alkyl, cycloalkyl, or substituted aryl groups into the borane serves essentially to increase the steric bulk of the reagent in order to suppress double hydroboration of alkynes.

Double hydroboration and controlling the regiochemistry of the hydroboration are indeed the two major problems in the synthesis of alkenyl boranes.

Although quite regioselective in the case of terminal alkynes, hydroboration using borane-THF (**1**) cannot be stopped at the monohydroboration stage ⟨61JA3834, 67JA291⟩. Compounds such as $BH_2Cl \cdot OEt_2$ (**13**) ⟨73JOC1617, 76JA1785⟩ and $BHCl_2 \cdot OEt_2$ (**15**) ⟨76JA1798⟩ yield mixtures unless 10–40% excess of the alkyne is used, however $BHBr_2 \cdot SMe_2$ (**16**) gives (*E*)-1-alkenyldibromoboranes in good yield even in stoichiometric amounts ⟨80JOC389, 89JOC6075⟩. Uncomplexed $BHCl_2$, generated from trimethylsilane and boron trichloride, yields only 1-hexenyldichloroborane from 1-hexyne at $-78\,^{\circ}C$ ⟨90JOC2274⟩.

Alkenylboranes are obtained from terminal alkynes in good to excellent yields using thexylborane (**3**) ⟨68JA6243, 71JA3395⟩, thexylchloroborane (**17**) ⟨80JA5919, 86S674⟩, disiamyl-7, and dicyclohexylboranes (**8**) ⟨65ZOB1669, 67JA3652, 71JA3395, 73S672, 88S103⟩, 9-borabicyclo[3.3.1]nonane (9-BBN-H) (**10**) ⟨76JA7832, 77JOC579, 79JA96, 80JOC550, 86JOC3398, 92JOC4017⟩, diarylboranes (**11**) ⟨83TL1433, 91TL6239⟩, catecholborane (**18**) ⟨72JA4370, 75JA5249, 79JOC3374⟩, pinacolborane (**19**) ⟨92JOC3482⟩, and 2-bora-1,3-dithiolane (**20**) ⟨77JOC3243⟩.

Some of these boranes have been found to react with conjugated terminal enynes in a highly chemo-, regio- and stereoselective fashion to give dienylboranes ⟨65ZOR1669, 71JA3395, 87TL4169, 90JOC5324, 93JA2268⟩. It is interesting, however, that the Pd(0)-catalyzed hydroboration of 3-methylbutenyne with catecholborane gave an allenic boronate ⟨89TL3789⟩.

Internal alkynes, unless having a symmetrically substituted triple bond ⟨61JA3834, 76JA1785, 76JA1798⟩, give mixtures of regioisomeric alkenylboranes ⟨71JA3395, 73JOC1617, 80JOC389⟩. The ratio of isomers in these mixtures will change with the relative steric bulk of the two substituents in favor of the one where boron is attached to the alkenic carbon bearing the smaller alkyl group, so that preparatively useful regioselectivities can be observed ⟨70JA1427, 71JA3395, 75JA5249, 76JA7832, 77JOC579, 79JA96, 80JOC389⟩. Of course, the steric bulk of the borane acts in the same direction ⟨71JA3395,

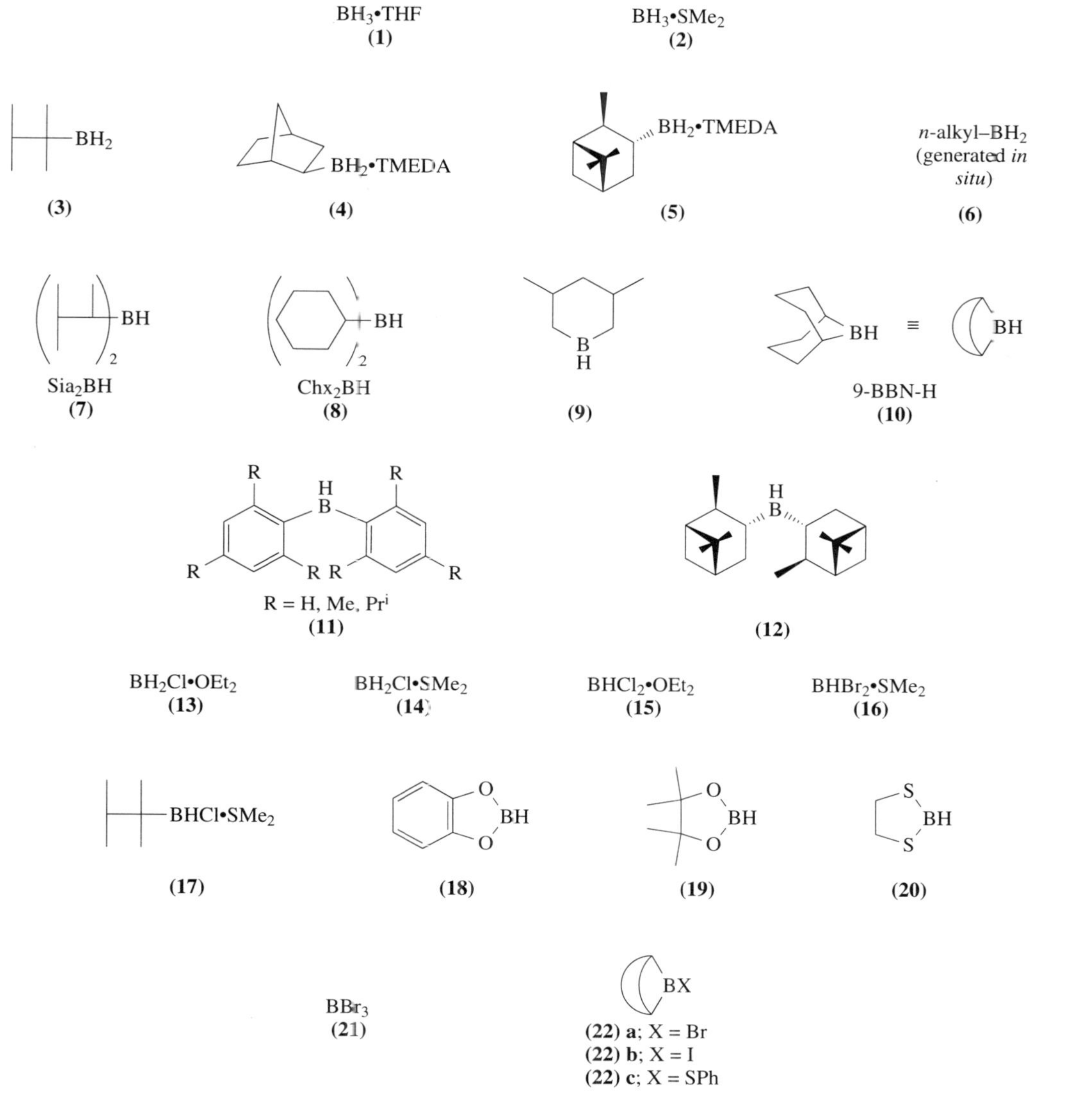

Figure 1 Common hydroboration reagents

83TL1433, 91TL6239⟩, but complications due to electronic effects from the alkyne can arise ⟨75JA5249, 79JA96, 80JOC389⟩.

Heteroatom-substituted alkynes represent a class of internal alkynes whose hydroboration products are of particular interest. Thus, hydroboration of 1-halo-1-alkynes gives 1-halo-1-alkenylboron derivatives (boranes, boronic esters and acids) which can be transformed in a variety of ways to give (i) (*Z*)-1-halo-1-alkenes ⟨67JA5086⟩; (ii) (*Z*)-1-alkenylboranes ⟨84OM1392, 88JOC239, 89JOC6075⟩; (iii) internal (*E*)-alkenylboranes ⟨67JA5086, 82JOC754, 82JOC3808, 84S920, 86JOC5277, 88JOC6009⟩, and (*E*)-alkenes, free of regioisomeric impurity and of opposite stereochemistry to those obtained by hydroboration of internal alkynes; (iv) 1,3-dienylboranes of well defined regio- and stereochemistry ⟨77JA5192, 88S130, 91T343⟩, as well as dienes ⟨80JOC549, 91T343⟩, α,β-epoxyketones ⟨88S130⟩, α,β-unsaturated ketones ⟨88S130, 91T343⟩, enynes, and 1,2,3-trienes ⟨91T343⟩.

Hydroboration of 1-ethoxy- and 1-alkylthio-1-alkynes occurs regiospecifically to attach the boron moiety at position 2; palladium- and/or nickel-catalyzed coupling of the 2-alkenylboranes produced from aryl or vinyl halides leads to a variety of vinylic ethers ⟨87S373⟩ and vinyl or dienyl sulfides ⟨93JOC5351⟩.

The regiochemistry of the hydroboration of 1-silyl-1-alkynes can be variable depending on the combination of steric effects on the silyl alkynes and on the borane ⟨89JA4873, 90TL4545, 91TL7365⟩, but in common cases 1-silyl-1-boryl-1-alkenes are obtained, which can be deboronated to give acylsilanes ⟨81S288, 90T2573⟩ or vinylic silanes (see Section 2.18.2.1.3(ii)(a)).

Finally, it is worth noting that some alkyne hydroborations are subject to transition metal (Ni, Pd, Rh) catalysis ⟨87JA5867, 89TL3789, 93JOC5351⟩.

(ii) Halo- and thioboration of alkynes

B-Bromo- and *B*-iodo derivatives of 9-borabicyclo [3.3.1]nonane (9-BBN) (**22a,b**) add to terminal alkynes with high chemo-, regio-, and stereoselectivity to form 2-halo-1-alkenylboranes which can be converted into 2-halo-1-alkenes or into haloenynes ⟨83TL731, 83TL735, 89TL5153⟩ (Scheme 1). Interestingly, *B*-bromo-9-BBN and *n*-hexyldibromoborane react with 1-halo-1-alkynes with opposite stereoselectivity, the former leading to (*Z*)-1,2-dihalo-1-alkenes and the latter to (*E*)-1,2-dihalo-1-alkenes as major products after protonolysis ⟨83S1005⟩.

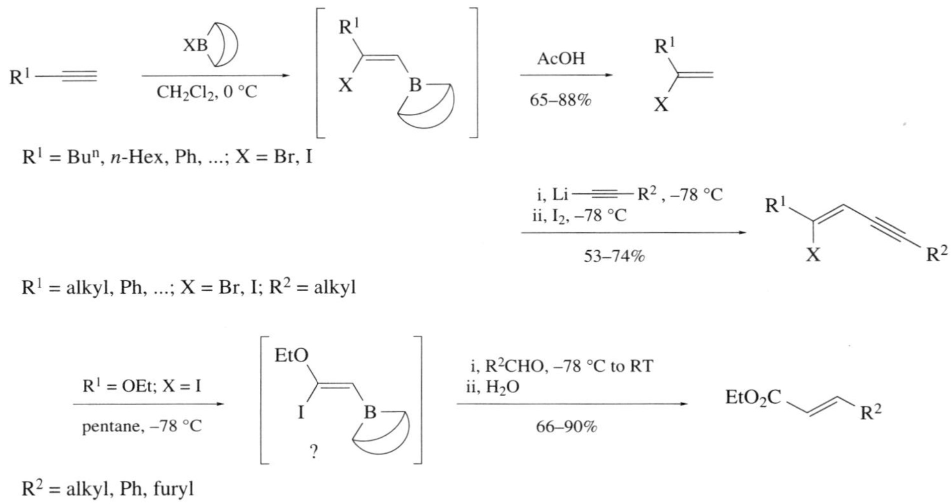

Scheme 1

Iodoboration of ethoxyethyne affords the same regiochemistry as does its hydroboration and the ethoxyiodovinylborane intermediate leads to α,β-unsaturated esters on reaction with aldehydes (Scheme 1) ⟨89TL5153⟩.

Addition of the much more reactive tribromoborane to terminal alkynes occurs readily at $-78\,^{\circ}$C and the 2-bromo-1-alkenyldibromoboranes undergo various oxidation ⟨85S406⟩, 1,4-addition ⟨90TL247⟩, and transition metal catalyzed coupling reactions (Scheme 2) ⟨88TL1811, 90TL247⟩.

Scheme 2

The palladium-catalyzed thioboration of terminal alkynes using (**22c**) or other *B*-alkylthio-9-BBN derivatives (Scheme 3) and coupling of the (*Z*)-2-thio-1-alkenylboranes formed with aryl-, alkenyl-, and terminal alkynyl halides can be carried out in 'one pot' to produce the corresponding unsaturated sulfides with high stereoselectivity and in good to excellent yield ⟨93JA7219⟩.

(iii) Carboboration of alkynes

In contrast to the hydroboration reagents of Figure 1, trialkylboranes do not react with alkynes. The 1,2-addition reactions that occur between trimethylsilylethyne or ethoxyethyne and triallyl-

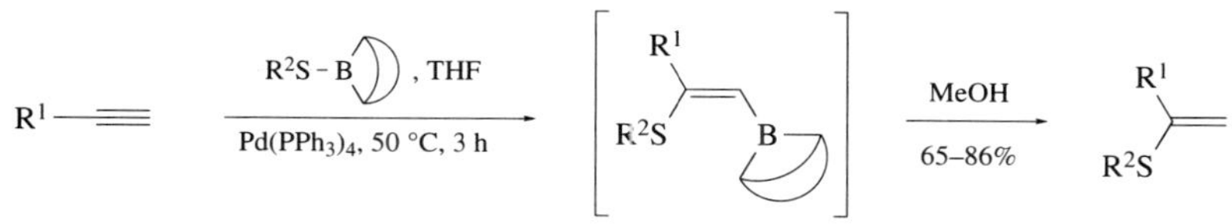

R^1 = alkyl, R^2 = Ph, alkyl

Scheme 3

and tribenzylboranes ⟨81S841, B-88MI 218-01⟩ are rather exceptional reactions in that both partners are very reactive.

Carboboration reactions also occur easily between trialkylboranes and 1-stannyl-1-alkynes leading to 1-boryl-2-stannyl-1-alkenes (Equation (1)) ⟨77ZN(B)1400, 81ZN(B)1265, 85CC1199, 86CC403, 86CC1170, 89T1119⟩. This reaction, however, is not a conventional 1,2-addition of the borane to the C≡C triple bond; rather it is reminiscent of the well-known reaction of 1-alkynyltrialkyl borates with electrophiles (Section 2.18.1.1.1(v)). Indeed, the transformation has been interpreted as starting with a tin to boron transmetallation of the starting alkyne to produce a borate-like intermediate. This then rearranges with B→C migration of one of the alkyl groups under the influence of the electrophilic tin moiety attacking position 2 of the triple bond ⟨89T1119⟩.

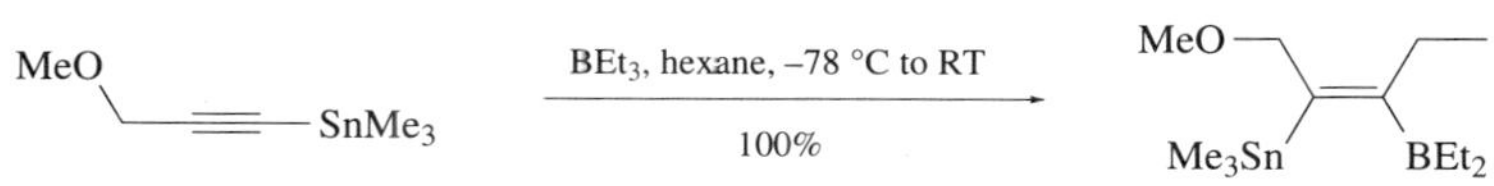

(1)

(iv) Metalloboration of alkynes

Among the various boron halides ⟨91CRV35⟩, diboron tetrachloride and di-*t*-butyl diboron dichloride have been the most widely studied in their reactions with acetylenic derivatives. Diboron tetrachloride adds easily to terminal and internal alkynes in a 1,2-manner to form (*Z*)-1,2-bis(dichloroboryl) alkenes ⟨80CB3352, 89ZN(B)1179⟩ (Equation (2)), which can be transformed further into diboraheterocycles ⟨80CB3352⟩, 1,3-dihydro-1,3-diborafulvenes ⟨88AG(E)299, 89CB401⟩, and 1,4-dihydro-1,4-diboranaphthalenes ⟨90CB2273⟩.

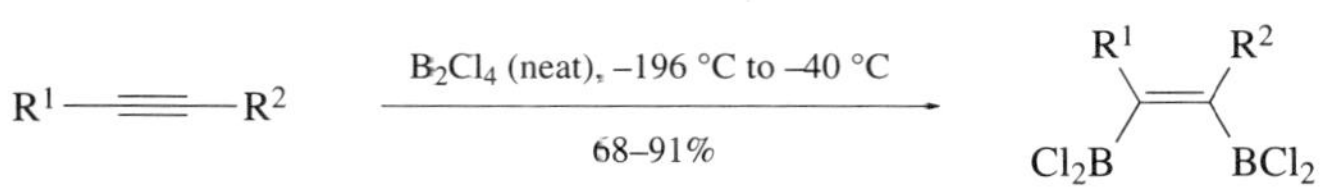

(2)

R^1, R^2 = H, Me, Et, Bu^t

Silyl and germyl alkynes react with diboron chlorides; as in the case of stannyl alkynes, 1,1-addition of the two boryl moieties occurs accompanied by the migration of the initially present metal moiety (stannyl, silyl, germyl groups) (Equations (3), (4)) ⟨84ZN(B)1042, 93ZN(B)719⟩.

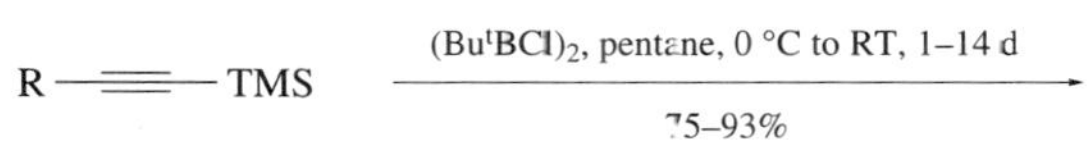

(3)

R = TMS, Me, CH_2Bu^t, Ph

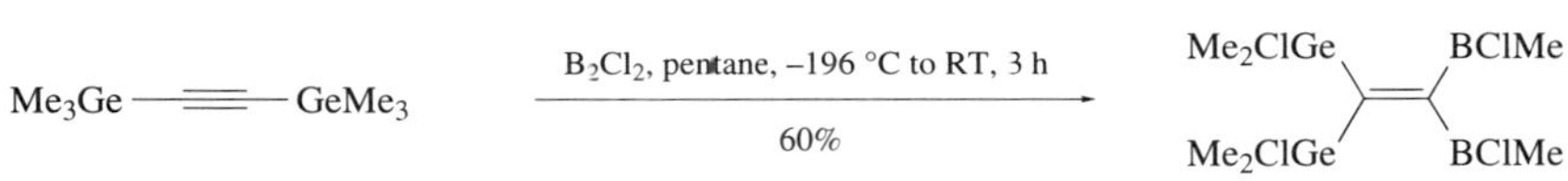

(4)

In addition to these double borations, terminal alkynes and terminal enynes also undergo silylboration ⟨86TL2007⟩ (see Section 2.18.2.1.2(v)(c), Equation (66)) and stannylboration reactions (Scheme 4) ⟨88TL261, 91T1163⟩. In all of these reactions the regiochemistry of the metallometallation step is greatly dependent on the catalyst and the alkyne structure.

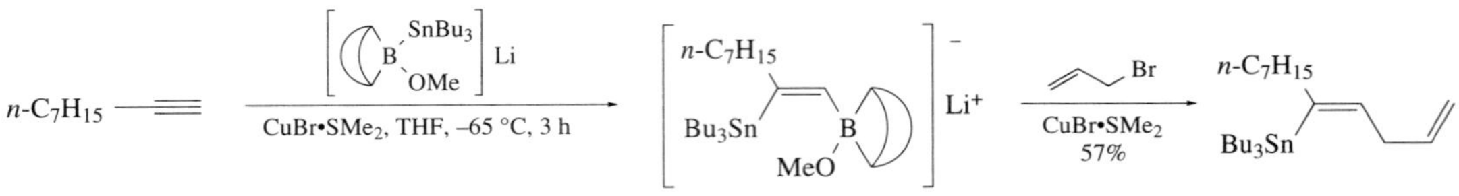

Scheme 4

(v) From alkynylborates with electrophiles

The reaction of alkynylborates (easily prepared from 1-lithioalkynes and trialkylboranes) with various electrophiles was reviewed in the late 1980s ⟨B-88MI 218-01⟩. Initial electrophilic attack usually occurs on the β-carbon atom of the alkynyl portion followed by migration of one of the alkyl groups of the boron moiety: the whole process gives an alkenylborane, which is not isolated but further transformed *in situ*. For example, protonation of this intermediate ⟨73CC544, 74TL2961, 75S376, 76JCS(P1)2435⟩ leads to disubstituted alkenes, alkylation to trisubstituted alkenes ⟨76JCS(P1)2419, 76JCS(P1)2428⟩; reaction with iodine gives another alkyne by sodium hydroxide mediated deiodoborylation of the 2-iodo-1-alkenylborane intermediate ⟨91T343⟩, and reaction with trialkylstannyl chlorides produces 2-stannyl-1-alkenylboranes ⟨86JOC767, 93JOC1949⟩.

2.18.1.1.2 *Alkenylboranes from other alkenylmetals*

Examples of early syntheses of limited synthetic value are the reactions of vinylzinc, -mercury and -tin derivatives with boron trifluoride or tribromide ⟨59ZN(B)809, 60JA6218, 60JA6238⟩. However, *B*-bromo-9-BBN produces vinyl-9-BBN in good yield on reaction with tributylvinylstannane (Equation (5)) ⟨90JA7423, 92JOC5768⟩.

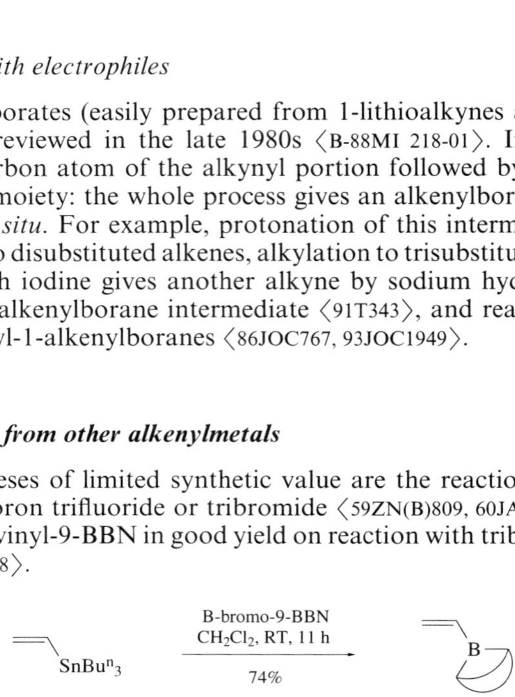

(5)

As for the synthesis of alkenylsilanes by a similar route (see Section 2.18.2.1.1), the most useful starting alkenylmetals are alkenyllithiums and alkenylmagnesiums. Thus, alkenylboronic esters can be prepared efficiently by reacting alkenylmagnesium chlorides or bromides with trialkylboric esters ⟨59JA5004, 60JA4228, 65JA1526, 89CRV1535, 89T1859⟩ (Equation (6)). In the case of mixed boric esters (prepared from boric acid and an equimolar mixture of a 1,4-diol and an alcohol) cyclic alkenylboronic esters are obtained exclusively ⟨66JOC2766⟩.

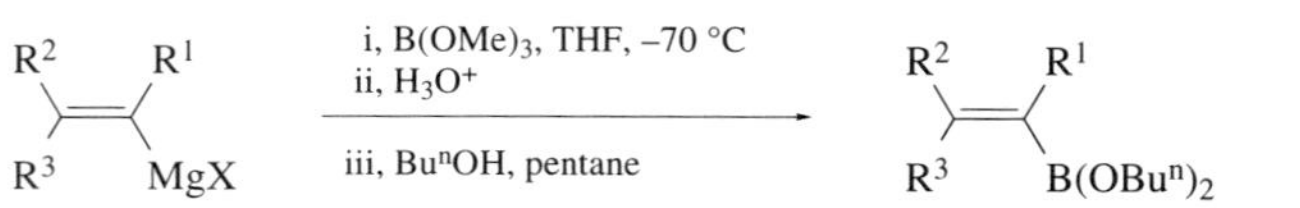

(6)

$R^1 = R^2 = R^3 = H$, 72%; $R^1 = R^2 = H$, $R^3 = Me$, 70%; $R^2 = H$, $R^1 = R^3 = Me$, 56%

Due to the high degree of retention of alkene configuration in halogen/lithium exchange reactions, alkenyllithiums are even better suited for the synthesis of alkenylboronic esters from alkenyl bromides and iodides (Equation (7)) ⟨75JOC1083, 80JOC389, 88TL21⟩. It is worth noting that the boronic esters so obtained can easily be hydrolyzed into alkenylboronic acids and esterified to form other alkenylboronic esters ⟨88TL21, 88JOC6009, 90JOC4986⟩; however, careful control of the hydrolysis conditions may be required ⟨93JA2268⟩.

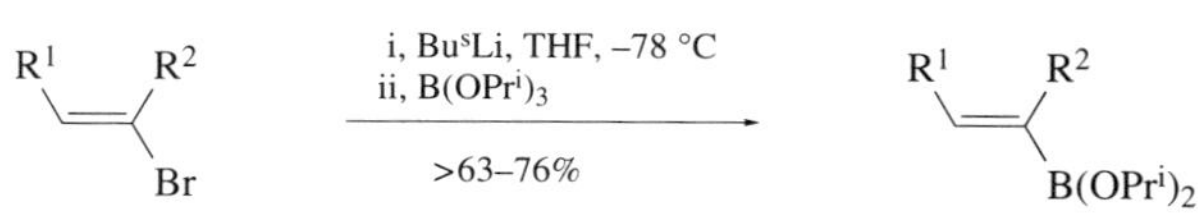

(7)

$R^1 = n$-Hex, Bu^n, $(CH_2)_3Cl$; $R^2 = Me$, Bu^n, Et, Pr^i

Allenylmetals (see Section 2.18.2.3.1) can also be borylated with trialkoxyboranes ⟨76IZV1580, 76IZV2062⟩ or with *B*-methoxy-9-BBN ⟨92JOC794⟩ to give allenylboronic esters (Scheme 5). Allenylboronic acid is also accessible from the reaction of propargylmagnesium bromide with trimethoxyborane (Equation (8)) ⟨86JA483⟩.

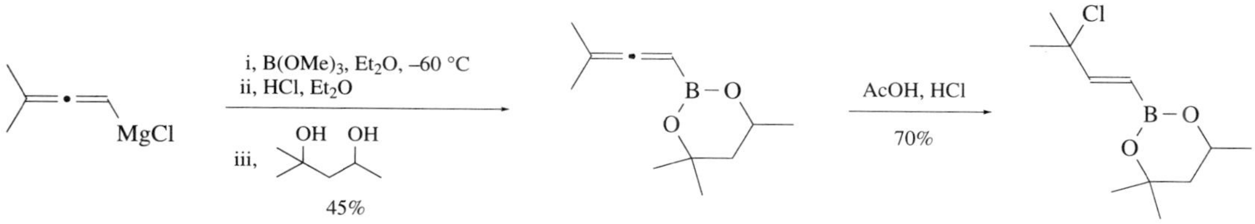

Scheme 5

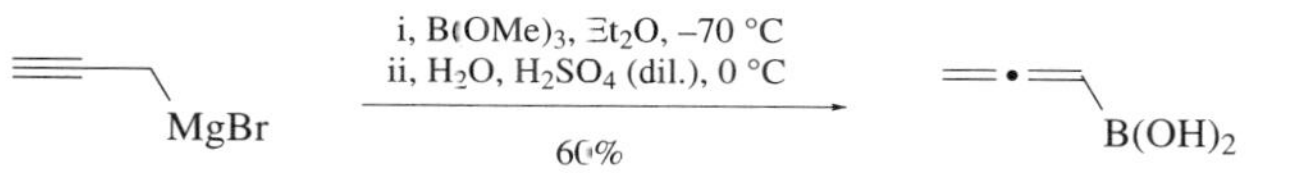

(8)

2.18.1.1.3 *Alkenylboranes from boron stabilized carbanions*

Tris(ethylenedioxyboryl)methane, on reaction with methyl- or *n*-butyllithium, gives bis(ethylenedioxyboryl)methyllithium which condenses with aldehydes and ketones in a boron–Wittig type reaction ⟨83TL635, 89TL5643, 89TL5647⟩. The products, alkenylboronic esters or acids are obtained in high yield with (*E*) selectivity in the case of aldehydes, and as a mixture of stereoisomers in the case of ketones (Equation (9)) ⟨75JA5608⟩. The same bis(boryl)methyllithium can also be obtained by lithium tetramethylpiperide (LITMP) deprotonation of bis(ethylenedioxyboryl)methane, and tris(boryl)methyllithium can be generated by treatment of tetrakis(ethylenedioxyboryl)methane with *n*-butyllithium ⟨89T1859⟩.

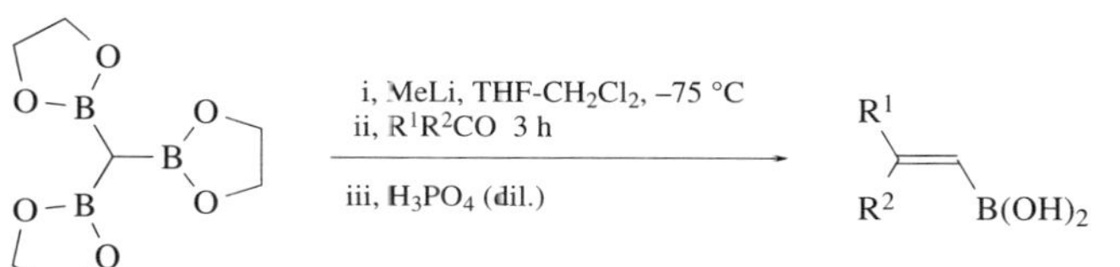

(9)

R^1 = Pr^n, R^2 = H, 68%; R^1 = Ph, R^2 = H, 87%; R^1 = Ph, R^2 = Me, 80% (*E*+*Z*)

In the case of the trimethylsilylborylmethyllithiums, the outcome of the reaction with carbonyl compounds depends on the structure of the boryl moiety: trimethylsilylpinacolborylmethyllithium exclusively produces alkenylpinacolboronates ⟨80CC39⟩ (Equation (10) in a Peterson-type reaction, whereas the more hindered trimethylsilyldimesitylborylmethyllithium gave with benzophenone a mixture of vinylsilane and vinylborane ⟨83TL637⟩.

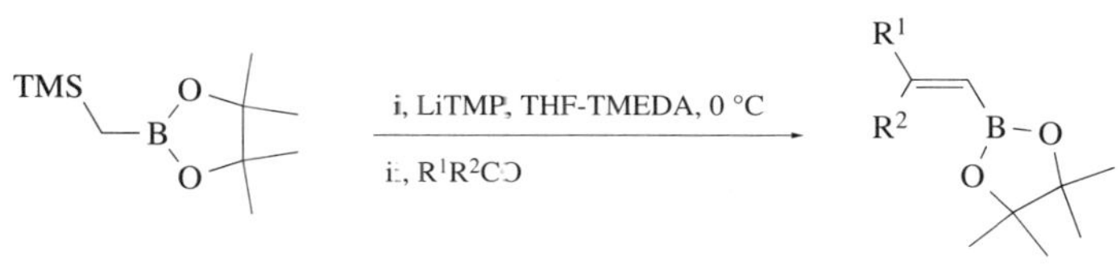

(10)

R^1 = *n*-C_6H_{13}, Ph, R^2 = H, 73, 34% (*E*/*Z* = 30/70). R^1 = R^2 = Bu^n, 74%

2.18.1.1.4 *Miscellaneous reactions*

As mentioned in Section 2.18.1.1.1, a more economic and highly (*E*) selective way of converting 1-bromo-1-alkenylboranes into internal (*E*)-alkenylboranes is to treat 1-bromo-1-alkenylboronic esters successively with alkyllithiums or alkyl Grignard reagents followed by methanol or sodium methoxide ⟨86JOC5277⟩. Although this is a two-step synthesis, it allows the preparation of the otherwise inaccessible internal alkenylboronic esters in high yield. When stereodefined alkenyllithiums are used instead of alkyllithiums, the same procedure leads to 2-(1,3-dienyl) boronic esters in good yield and in >97% stereochemical purity (Equation (11)) ⟨91TL3655⟩.

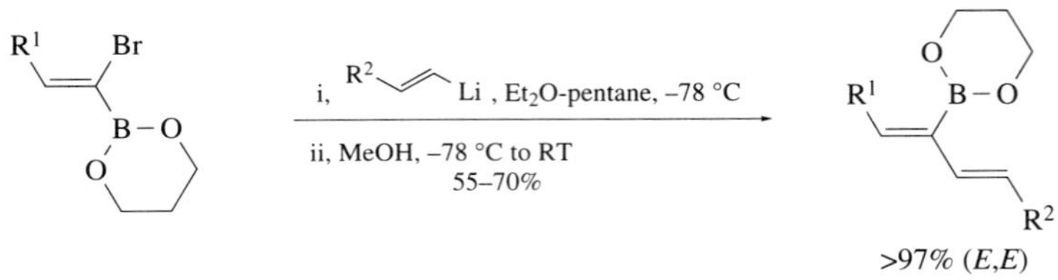

(11)

R^1 = Bu^n, Bu^t, $Cl(CH_2)_3$; R^2 = Bu^n, *n*-Hex, *c*-Pent, Bu^t

The problem arising from the competitive mono- and dihydroboration of terminal alkynes ⟨79JA96, B-88MI 218-01⟩ can be avoided by converting terminal alkynes into 1,1-bis(boryl)alkanes by the use of a twofold excess of 9-BBN-H, and effecting a dehydroborylation reaction by means of an appropriate aldehyde (Equation (12)) ⟨93JA6065⟩.

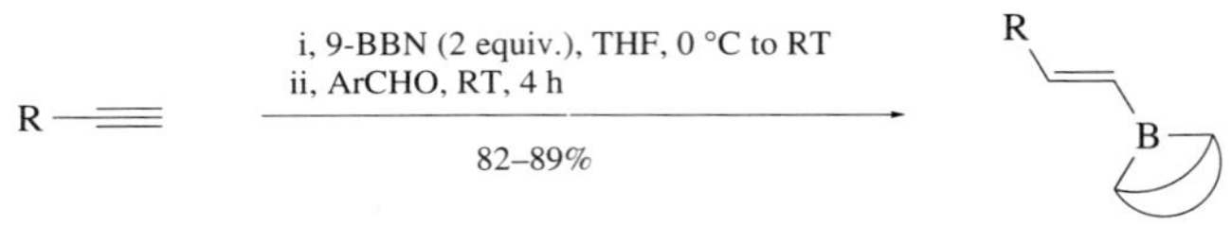

(12)

R = Me, Pr^i, Bu^n, Ph, TMS; Ar = Ph, 1-Naph

In close analogy with other known vinyl carbenoids ⟨67CB3371, 76TL2201⟩, 2,2-difluoro-1-lithio-1-vinyltosylate reacts with trialkylboranes. The product 1,1-difluoro-2-alkenyldialkylboranes can be transformed into alkenes ⟨89TL1641⟩, ketones ⟨89TL5437⟩, iodoalkenes ⟨89TL6379⟩, or coupled efficiently with acyl chlorides or aryl iodides to produce *β*,*β*-difluoroenones or 1,1-difluoro-2-aryl-1-alkenes (Scheme 6) ⟨92TL337, 92TL3779⟩.

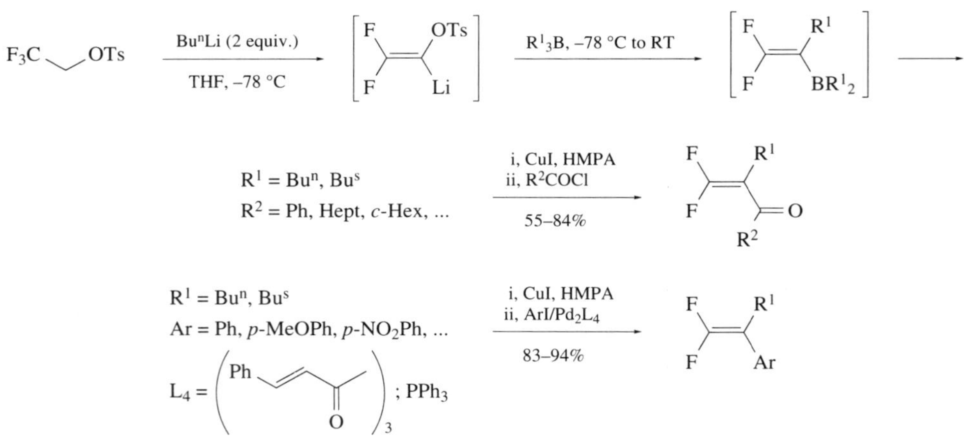

Scheme 6

2.18.1.2 Aryl Boron Compounds

As a consequence of the restricted synthetic utility of this type of boron derivative, as compared to that of alkenylboranes, there are few novel methods of preparation.

The rather inconvenient method used at the beginning of the twentieth century for the preparation of arylboronic acids (consisting of the reaction of arylmercurys with boron trihalides ⟨1882CB180, 1894CB244, 01LA(315)19, 01LA(315)26⟩ followed by hydrolysis) has been the subject of major improvement by the use of aryl Grignard reagents in conjunction with trialkoxyboranes ⟨09CB3091⟩ such as trimethoxyborane ⟨38JA105, 63OSC(4)68, 88S950⟩ or tributoxyborane ⟨90JOC4581⟩. The reaction of arylmagnesiums with borane-THF complex is another interesting modification ⟨83JOM(259)269⟩. Also, vinylarylboronic esters are formed by reacting arylmagnesiums with vinylboronic esters ⟨63JOC2171⟩. Finally, aromatic electrophilic substitution of phenyl boron compounds can be a valuable source of some otherwise not accessible aryl boron derivatives ⟨90JOC46⟩.

2.18.2 SILICON DERIVATIVES

2.18.2.1 Alkenylsilanes

Several reviews have appeared on both the reactivity ⟨79S761, 82COMC-I(7)515, B-89MI 218-01, B-89MI 218-02, 89OR57⟩ and the preparations ⟨79S761, 82COMC-I(2)1, B-87MI 218-01, B-89MI 218-01⟩ of this important class of intermediates in organic synthesis. Therefore, this section will not provide a comprehensive review of the subject but will summarize the most efficient synthetic methods, including the newest developments.

The most widely used methods for the preparation of alkenylsilanes from suitably functionalized unsaturated (alkenyl, alkynyl, or allenyl) derivatives are outlined in Scheme 7. In addition to these, a number of other methods will be discussed, some being less general but quite efficient in particular cases.

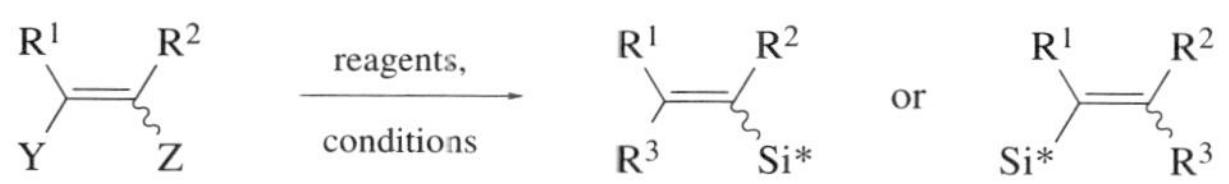

(a) R^1, R^2, Y or Z = H, alkyl, aryl, ... R^1, R^2, Z or Y = metal	⟶	2.18.2.1.1	From alkenyl metal and Si*–X
(b) R^1, R^2 = H, alkyl, aryl, ... Y, Z = C–C bond	⟶	2.18.2.1.2	From alkynes and Si*–H or Si*–metal
(c) R^1 = H, alkyl, aryl, ... R^2 = Si* Y, Z = C–C bond	⟶	2.18.2.1.3	From alkynyl silanes and R^3–metal
(d) R^1, Y = H, alkyl, aryl, ... R^2, Z = C=C double bond	⟶	2.18.2.1.4	From allenes and Si*–metal
(e) R^1, Y = H, alkyl, aryl, ... R^2 = Si* Z = H, Halogen	⟶	2.18.2.1.5	From existing alkenyl silanes and R^3–X

Scheme 7

2.18.2.1.1 From alkenyl metals (Li, Na, K, Mg) and electrophilic silicon reagents

(i) Alkenyl metals by hydrogen/metal exchange

A large variety of alkenes, cycloalkenes, vinyl halides, ethers, sulfides, sulfoxides, selenides, tellurides, and isocyanides can be deprotonated with strong bases ⟨78RTC69, 84TL741, 84TL745, B-87MI 218-01⟩ under suitable conditions ⟨79OR(26)1, B-88MI 218-02⟩. Silylation of the resulting alkenyl metals (most often lithium) has not been carried out in all cases, but representative examples are given in Equations (13) and (14), and Scheme 8 ⟨76JOM(105)C29, 87BCJ777⟩. Although vinylpotassium can be prepared efficiently from ethene (Bu^nLi/Bu^tOK/TMEDA/hexane/$-25\,°C$; $\geqslant 90\%$ yield) ⟨86CC260⟩, the simple ethenylsilanes have not been prepared by this route. Deprotonation of higher alkenes or unstrained cycloalkenes is more problematic because of the competitive formation of alkenyl and allyl metals. Nevertheless, in some cases the method does have preparative value, for example trimethylsilylnorbornadiene was prepared in 92% yield ⟨86RTC66⟩ and phenyldimethylsilylbornene was obtained as a 7 : 1 mixture of the two regioisomers (Equation (13)) ⟨91JOC1951⟩.

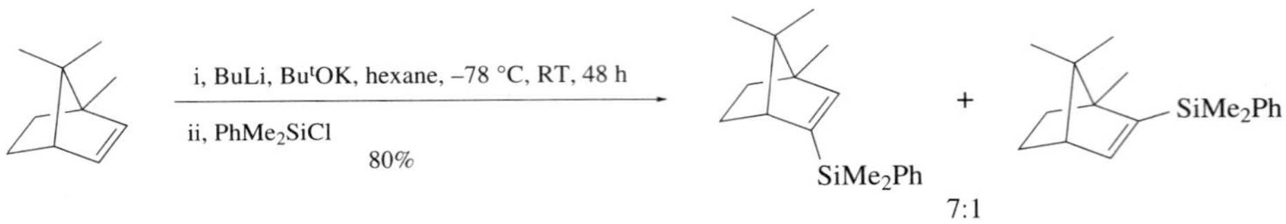

(13)

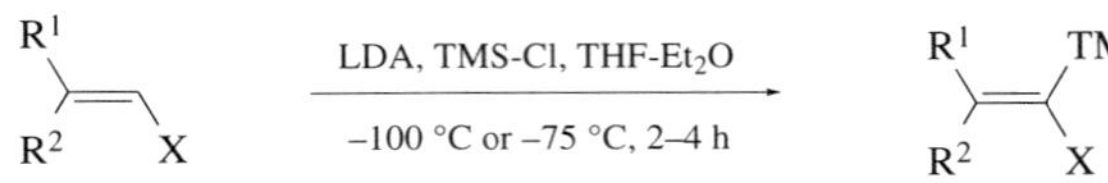

(14)

$R^1 = R^2 = H$, X = Br, 60%
$R^1 = R^2 = Me$, X = Br, 54%
$R^1 = H$, $R^2 = Me$, X = Cl, 23% (*E*+*Z*)
$R^1 = Ph$, $R^2 = H$, X = Br, 55%
$R^1 = H$, $R^2 = CH_2Cl$, X = Cl, 59% (*E*+*Z*)

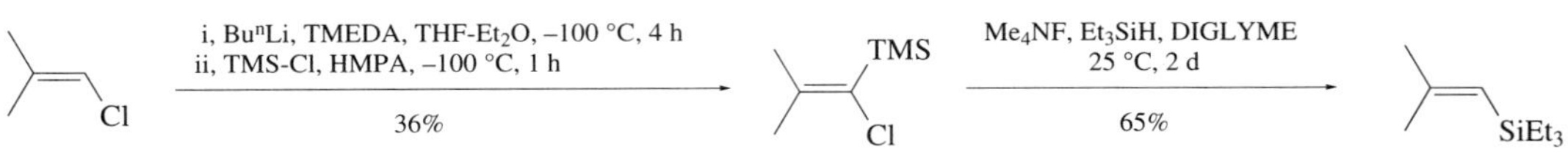

Scheme 8

When heteroatomic substituents are attached to the C═C double bond of alkenes, they strongly activate the remaining hydrogens at the α or β positions towards hydrogen/metal exchange ⟨79OR(26)1, B-87MI 218-01, B-88MI 218-02⟩. Thus, α-halovinyllithiums can be generated and silylated in reasonable yields (Scheme 8, Equation (14)) ⟨76JOM(105)C29, 87BCJ777⟩. The products α-halovinylsilanes are efficient sources of vinyl carbenes which are capable of insertion into Si—H bonds leading to another vinylsilane (Scheme 8) ⟨76JOM(105)C29⟩.

Using the same (deprotonation/silylation) methodology, various heterosubstituted vinylsilanes have been prepared in good to excellent yields (Scheme 9, Equations (15)–(18)). Deprotonation of vinyl ethers at the α position is also readily performed ⟨74JA7125⟩; however, the α-alkoxyvinyllithiums are silylated only in moderate yields (Equation (15)) ⟨76JOM(107)229, 85JOC5410, 89S179, 89S647⟩. Lithiation followed by silylation of (i) phenylvinyl sulfide (Scheme 9) ⟨77CC522, 81TL587, 83JCS(P1)1131, 88BCJ3957⟩, (ii) arylvinyl selenides (Equation (16)) ⟨78AG(E)526, 78JOC3794, 81JOC2775⟩, and (iii) phenylvinyl telluride (Equation (17)) ⟨82AG(E)410, 83CB1001⟩ has been carried out according to a very similar protocol. It is interesting to note for phenylthiovinylsilanes that the corresponding sulfoxides can be produced in high yield by mcpba oxidation ⟨77CC522, 88BCJ3957⟩.

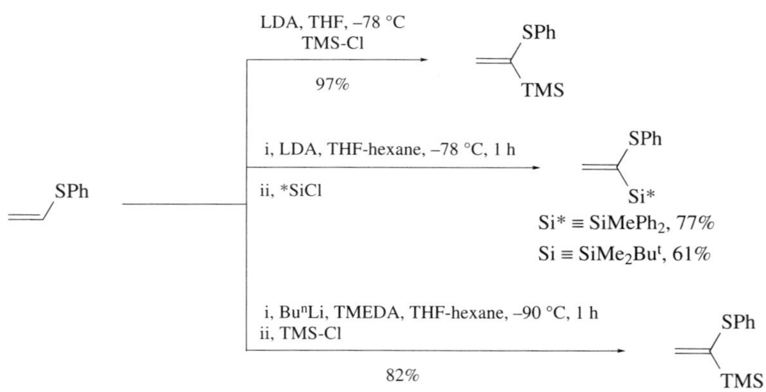

Scheme 9

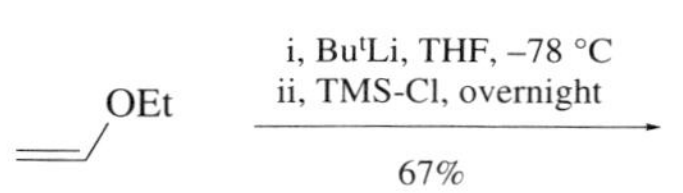

(15)

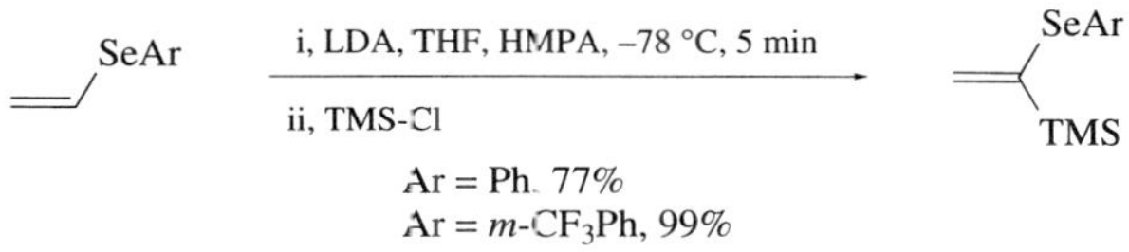

(16)

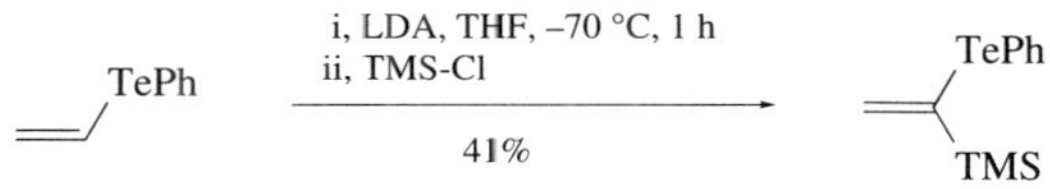

(17)

Many other functionalized alkenes have been deprotonated and reacted efficiently with various electrophiles ⟨B-87MI 218-01, B-88MI 218-02⟩. α-Lithiovinyl isocyanides have also been silylated in good yields (Equation (18)) ⟨77LA1167⟩.

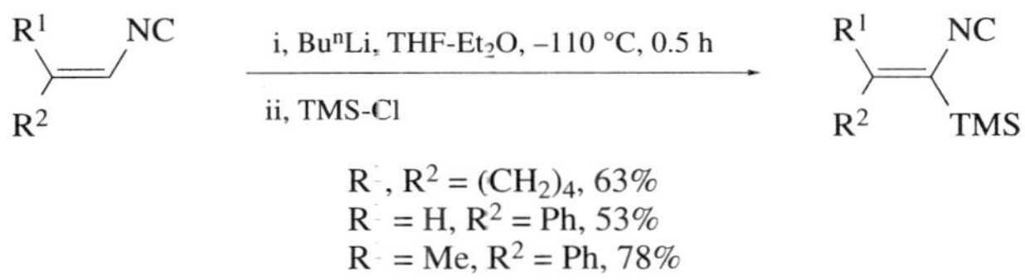

(18)

(ii) *Alkenyl metals by halogen/metal exchange*

Halogen/metal exchange is the most widespread and most general method for the generation of alkenyl metals including alkenyllithiums, -sodiums, -potassiums, and bromomagnesiums ⟨B-87MI 218-01, B-88MI 218-02, B-88MI 218-03⟩. In addition to its simplicity, the method gained popularity because the vinylsilanes are usually formed in a stereospecific manner and the starting halides can be prepared by a variety of highly stereoselective/specific reactions or reaction sequences ⟨70TL447, 81S841, 83T877⟩. It has been shown ⟨64JA883⟩ that the stereochemistry of the vinylsilanes is identical to that of the precursor vinyl metals (Equations (19)–(27)). Monometalloalkenyl derivatives display a high degree of configurational stability ⟨64JA883, 76TL4839, 78CB2785, 81S841, 83T877⟩; however, 1,1-dimetalloalkenes are usually configurationally unstable ⟨86JA3402⟩.

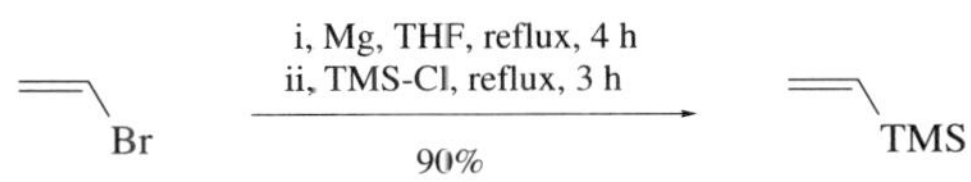

(19)

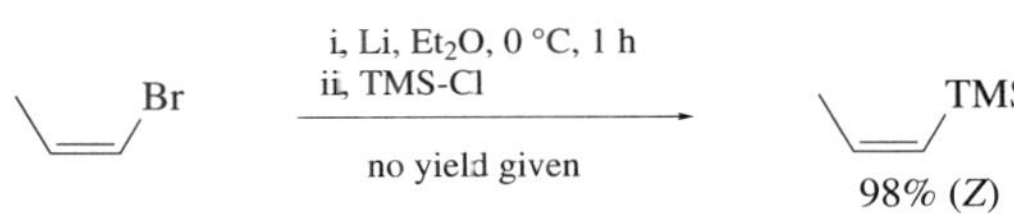

(20)

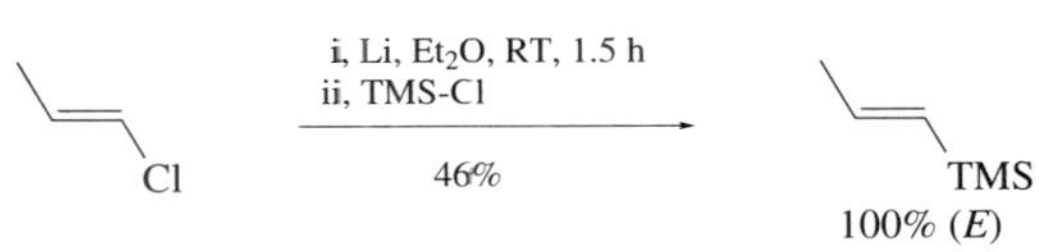

(21)

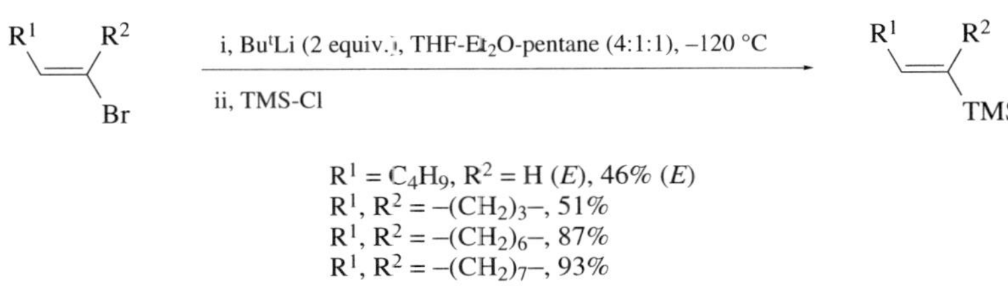

(22)

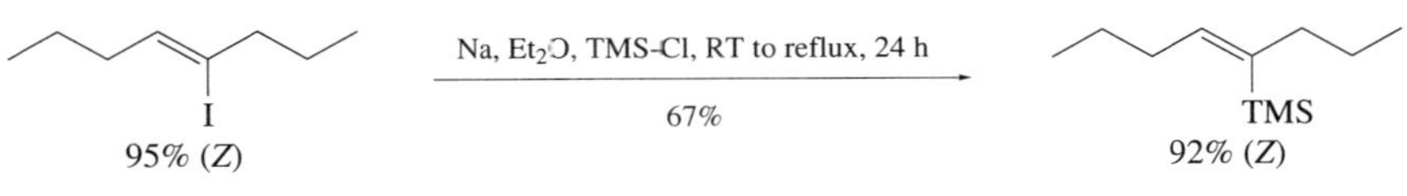

(23)

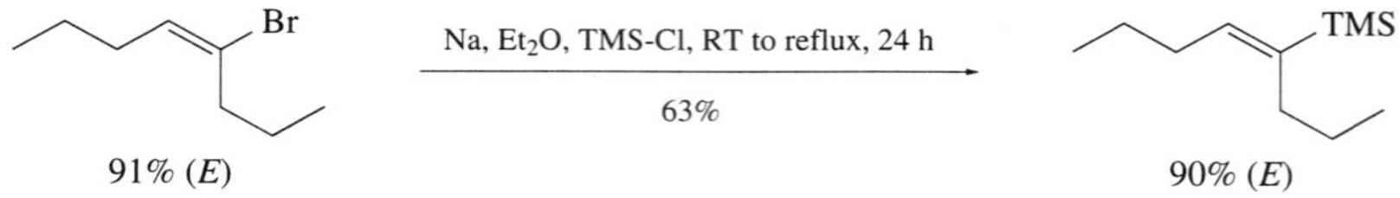

(24)

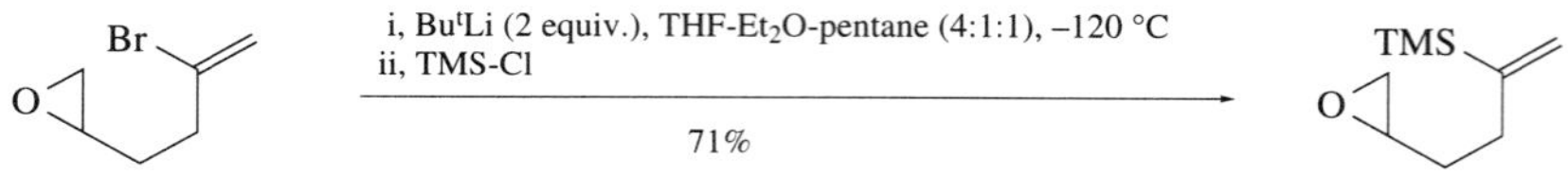

(25)

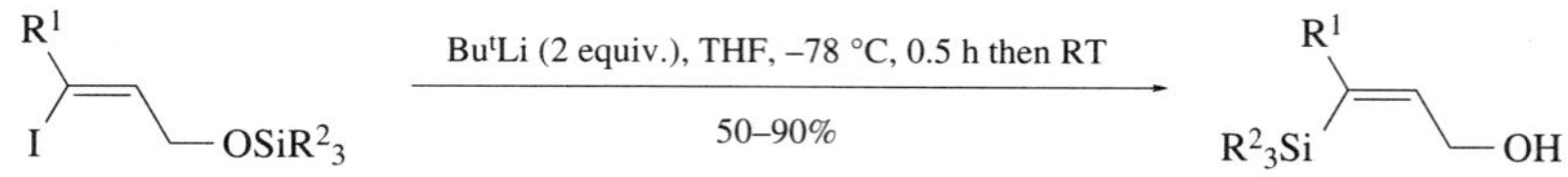

(26)

R^1 = alkyl, Ph, PhS, PhSe, TMS; $R^2{}_3Si$ = TMS, TBDMS, tips

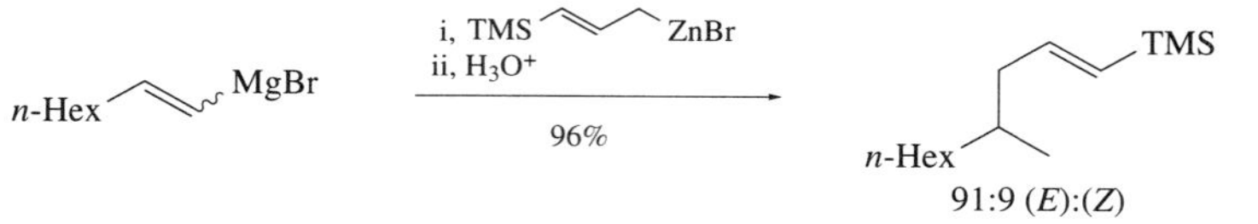

(27)

Since the formation and reaction of vinyl Grignard reagents requires higher temperatures (Equation (19)) ⟨63CJC2977⟩, these are configurationally less stable. Thus, on reaction with chlorosilanes vinyl Grignards usually produce mixtures of (*E*) and (*Z*) vinylsilanes ⟨B-69MI 218-01, 83T877⟩. Silylation of vinyllithiums (Equations (20)–(22)) ⟨64JA883, 76TL4839, 78CB2785⟩ or vinylsodiums (Equations (23) and (24)) ⟨83T877⟩ proceeds with high retention of configuration which can be further increased if the vinyllithium is produced by treatment with Bu^tLi at low temperature (Equations (22) and (26)) ⟨76TL4839, 78CB2785, 90TL6137⟩. At $-100\,^{\circ}C$ even a sensitive epoxide survives metallation (Equation (25)) ⟨86JOC2381⟩. Stereoselective addition of an allylic zinc bromide across the C=C double bond of a vinylic Grignard reagent produces a 1,1-bis(metallo) derivative which can then be protonated (Equation (27)) ⟨86TL5727⟩.

(iii) Alkenyl metals by transmetallation

Metal/metal exchange reactions represent an extremely powerful tool in organic synthesis ⟨81S841, 82AG(E)410, 84OR375, B-88MI 218-02, 92OR135, 93S537⟩. This chapter will also consider metal/silicon exchange as a form of transmetallation, as this reaction is important for the synthesis of vinylsilanes (see Sections 2.18.2.1.2. and 2.18.2.1.3). However, only those transmetallations involving stable, pre-existing alkenyl metals will be considered herein (Equations (28)–(33)) ⟨64JA883, 76JOC1480, 77CB867, 82AG(E)410, 82TL3411, 87JOC4868⟩.

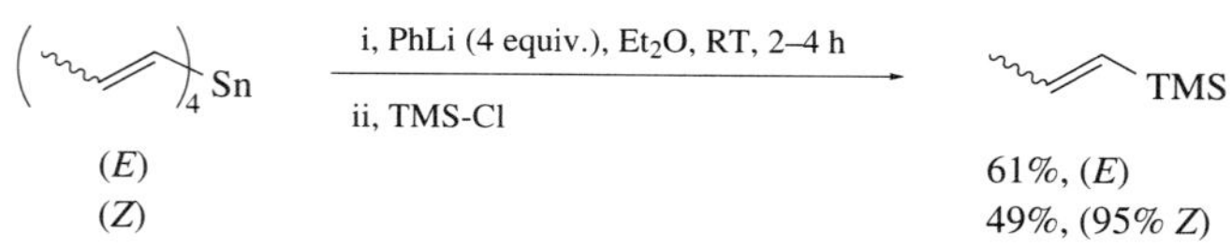

(28)

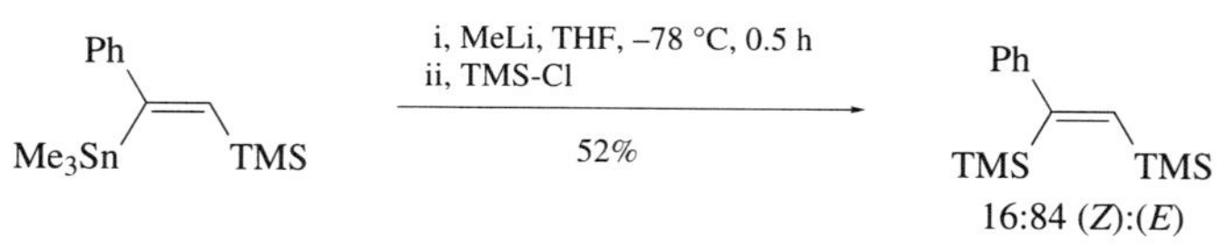

(29)

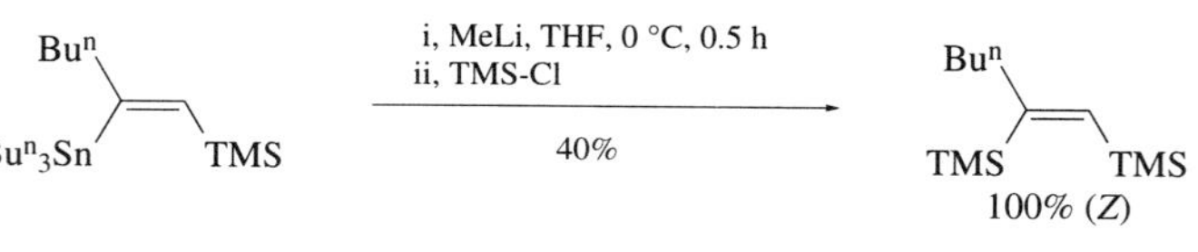

(30)

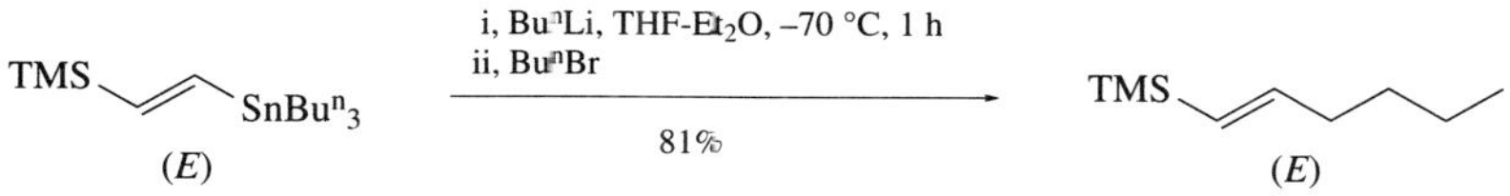

(31)

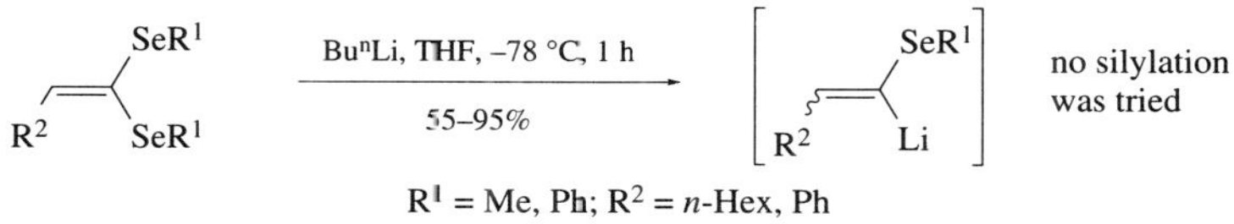

(32)

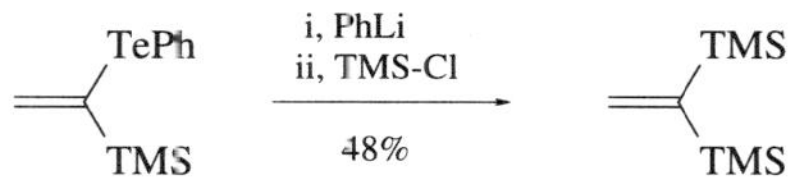

(33)

Tin/lithium exchange is well-known to occur on reaction of a vinylstannane with phenyl- or alkyllithium. The process is essentially stereospecific (Equations (28)–(31)) ⟨64JA883, 76JOC1480, 87JOC4868⟩, unless alpha to an aryl group (Equation (29)) ⟨87JOC4868⟩. Even though α-lithiovinylselenides are formed efficiently from 1,1-bis(seleno)alkenes (Equation (32)) ⟨77CB867, 82TL3411⟩, their silylation has not been attempted. The cleavage and silylation of a vinyltelluride was of moderate efficiency (Equation (33)) ⟨82AG(E)410⟩, presumably due to β-elimination leading to silylacetylene.

(iv) Alkenyllithiums from ketones through arenesulfonylhydrazones

This is a modification of the Shapiro reaction ⟨76OR(23)405⟩ where vinyllithiums, generated from arenesulfonylhydrazones and alkyllithiums, are quenched by chlorosilanes to produce alkenylsilanes ⟨76S801, 77TL159, 78JOC147, 78JOC1620⟩. Formation of the key intermediate vinyllithiums (−30 to 0 °C, monitored by the formation of N_2 bubbles) is largely controlled by steric and chelation effects ⟨80JOC3017, 90OR(39)1⟩ to give the less substituted (*Z*)-double bond with the lithium atom attached to the ex-carbonyl carbon; thus (*E*)-vinylsilanes are usually produced with high regio- and stereoselectivity (Equations (34)–(38)) ⟨76S801, 78JOC147, 80JOC3017, 84CJC2415, 90OR(39)1⟩. Optimal reaction conditions involve the use of hexane-TMEDA ⟨78JOC147⟩ or pure TMEDA ⟨76S801, 80JOC3017⟩ (recyclable) as solvent with an excess of base (3–4 equiv. of Bu^{n}Li) at −78 °C to 0 °C or room temperature.

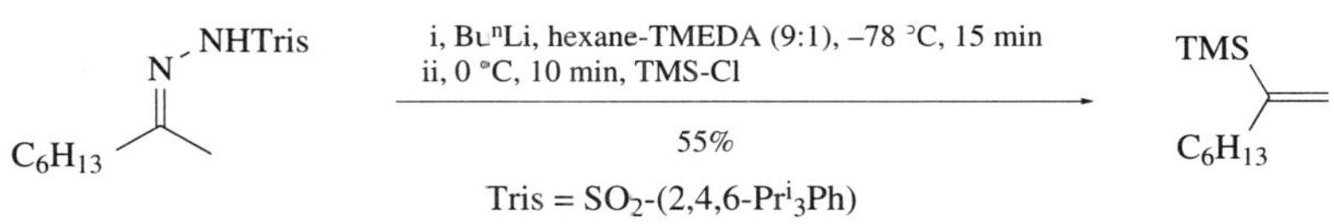

(34)

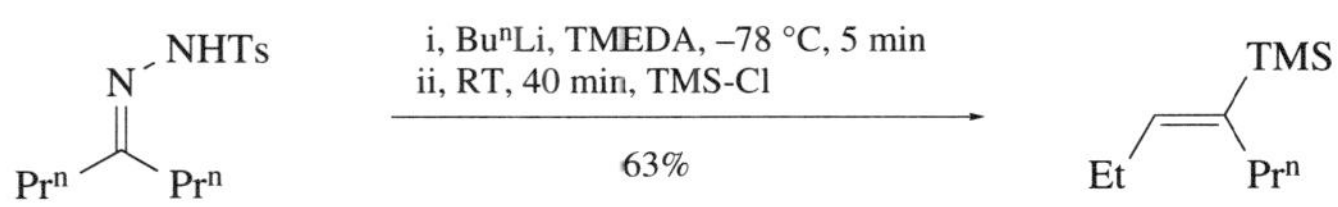

(35)

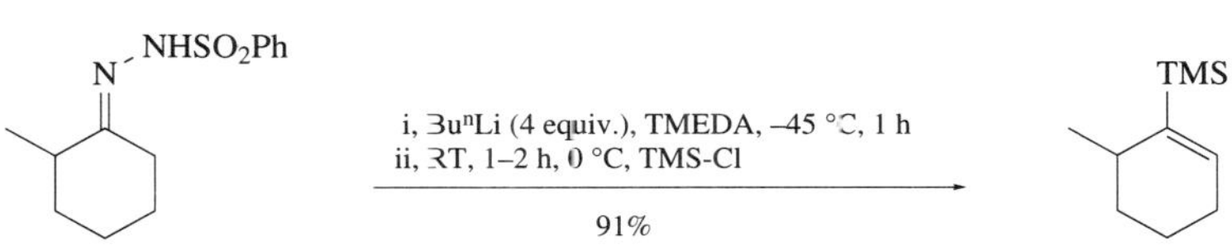

(36)

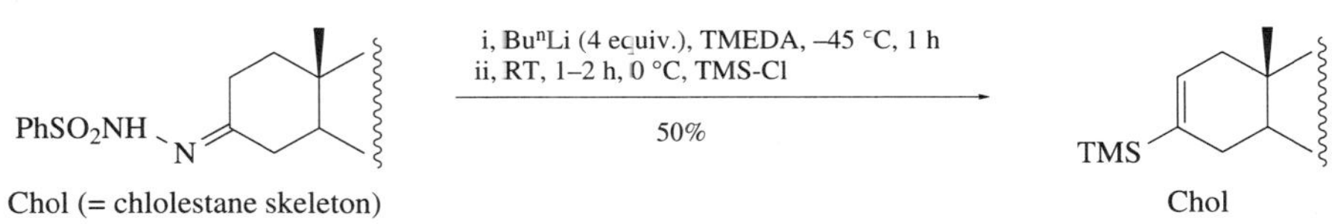

(37)

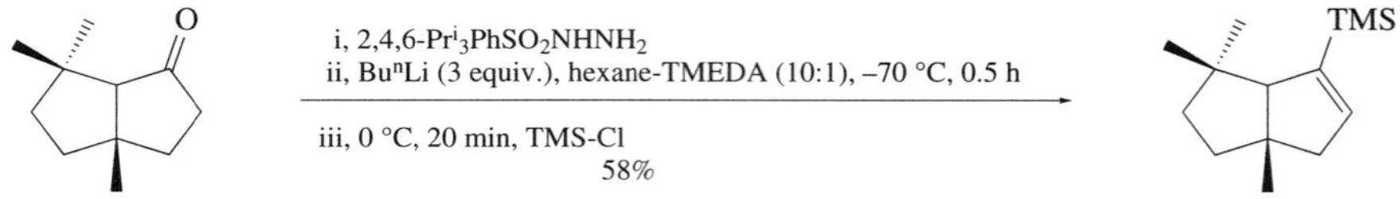

(38)

(v) Transition metal catalyzed coupling of alkenyl and silyl moieties

The increasing significance of transition metal catalysis in the synthesis of vinyl-(aryl) silanes is demonstrated by the results presented here together with those discussed in later sections.

The coupling partners suitable for use in these reactions include terminal alkenes possessing no allylic hydrogen ⟨83CC1193, 86JOC3890, 87JOC4864⟩, terminal vinyl bromides or iodides ⟨78AG(E)279, 83TL4895, 87TL4715, 91SL845⟩ and 4-tosyl-1-cyclohexenyl triflate ⟨92OM1095⟩ as the alkenic moieties together with hydrosilanes ⟨83CC1193, 86JOC3890, 87JOC4864⟩, di- or polychlorodisilanes ⟨78AG(E)279⟩, silyl anion equivalents ⟨83TL4895, 87TL4715⟩ and Bu$^n_3$SnTMS ⟨91JOC3846⟩ as silyl sources. The catalysts used include Ti, Zr, Hf ⟨92OM1095⟩, Co ⟨83CC1193, 87JOC4864, 93JA2151⟩, Ru ⟨86JOC3890⟩, Ni ⟨83TL4895⟩, and Pd based complexes ⟨78AG(E)279, 87TL4715, 91JOC3846⟩ (Equations (39)–(44)).

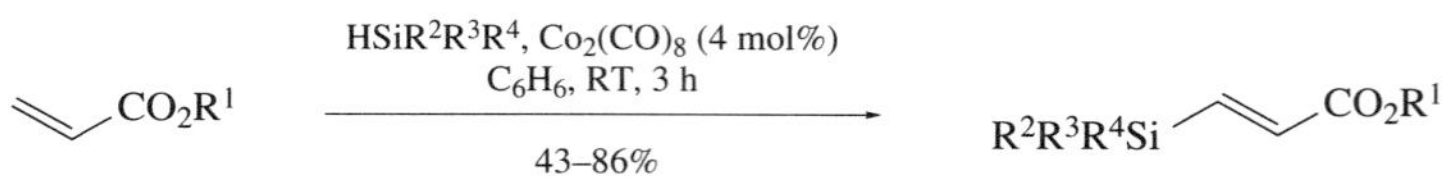

(39)

R^1 = Me, Et, Bu^n; R^2, R^3, R^4 = Me, Et, OEt, Ph

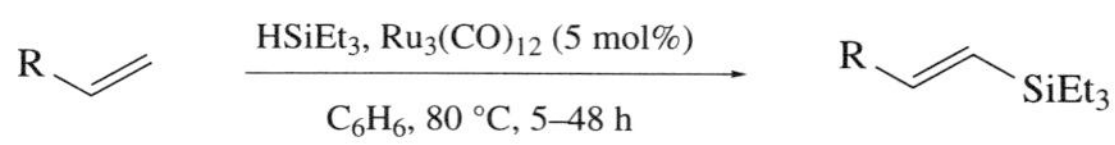

(40)

R =		
	p-X-Ph (X = H, Me, OMe, Cl)	83–100%
	Bu^t	88%
	Bu^{n}O	58%
	Et$_3$Si	94%

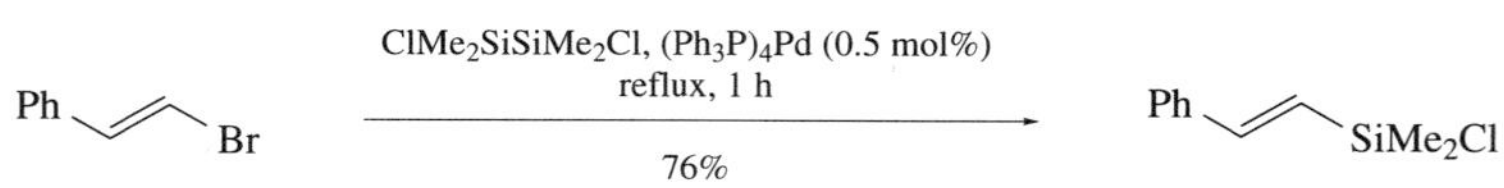

(41)

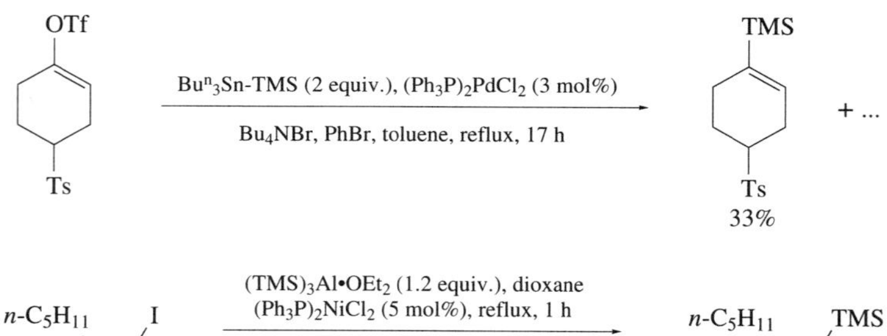

(42)

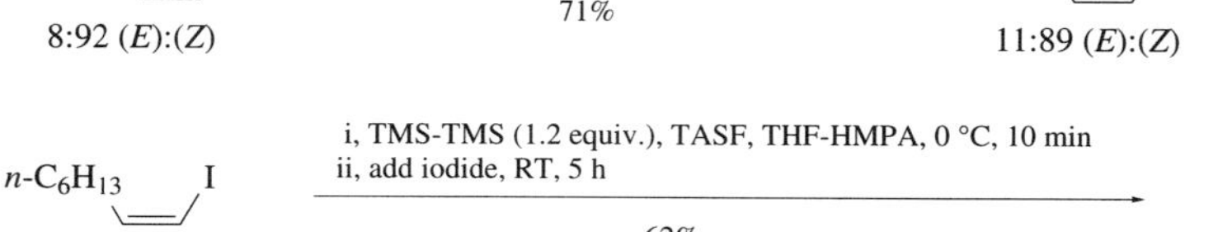

(43)

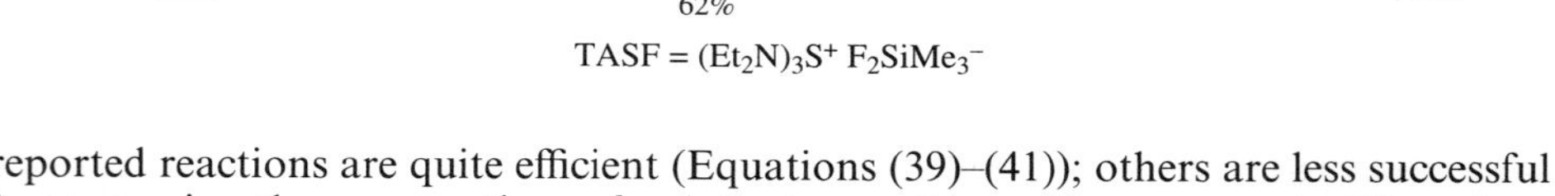

TASF = $(Et_2N)_3S^+\ F_2SiMe_3^-$

Some of the reported reactions are quite efficient (Equations (39)–(41)); others are less successful (Equation (42)) or require the preparation of relatively sensitive silicon reagents (Equations (43),(44)).

2.18.2.1.2 Vinylsilanes from alkynes

The reactions presented in this section constitute the most useful methods for preparing vinylsilanes, usually with high regio- and stereocontrol either without any further transformation of the adduct, or followed by various transmetallation, demetallation, alkylation, and silylation (or a combination of these) steps.

(i) Radical addition of hydrosilanes to alkynes

Peroxide-initiated addition of $HSiCl_3$ to straight chain terminal alkynes is a (*Z*)-selective process ($Z/E \sim 3/1$) leading to 1-trichlorosilyl-1-alkenes (Equation (45)) in moderate yield, which can be converted to 1-trimethylsilyl-1-alkenes on reaction with MeMgBr ⟨61JA4385⟩. Higher (*Z*)/(*E*) ratios have been observed in the Et_3B/O_2 induced additions of Ph_3SiH ⟨93BCJ2356⟩, TMS-$SiPh_2H$ ⟨93BCJ2356⟩, and $(TMS)_3SiH$ (Equation (46)) ⟨92JOC3994, 93BCJ2356⟩. The amount of (*E*)-alkenylsilane formed was observed to increase with reaction temperature, or when 2,2′-azobisisobutyronitrile (AIBN) was used as initiator ⟨92JOC3994⟩. Steric interactions also favor the formation of the (*E*)-isomer ⟨61JA4385, 92JOC3994, 93BCJ2356⟩. 1,2-Disubstituted phenylethynes gave, on reaction with $(TMS)_3SiH$, mixtures of stereoisomeric vinylsilanes, the silyl group being attached exclusively to the carbon atom β to the phenyl group ⟨92JOC3994⟩.

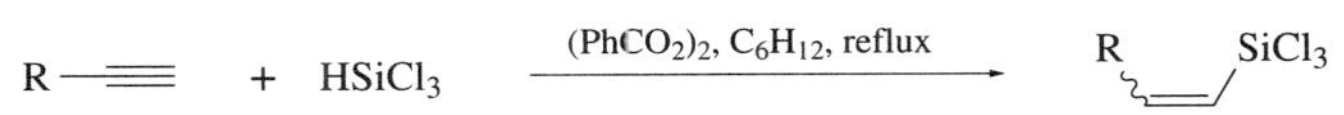

(45)

R	*Yield (%)*	*(Z):(E)*
Pr^n	35	4:1
Bu^n	36	3:1
$Pent^n$	43	3:1
Pr^i	37	2.6:1
Bu^t	39	(*E*) only

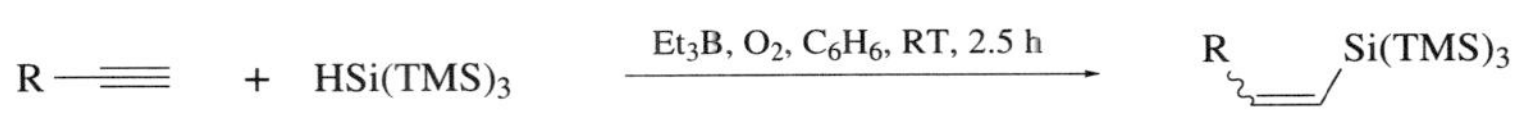

(46)

R = Ph, CO_2Et, CH_2OH, Dec^n	50–98%	>17:1 (*Z*):(*E*)
R = Bu^t	88%	<1:100 (*Z*):(*E*)

In the case of a 1,6-enyne bearing no substituent on the terminal carbon radical addition of $(TMS)_3SiH$ produced a mixture of mono- and bicyclic vinylsilanes, whereas 1,1-dimethyl-1,6-enynes gave monocyclic compounds in high yield (Equation (47)) ⟨93BCJ2348⟩. For similar but nonradical cyclizations, see also Section 2.18.2.1.2(vii).

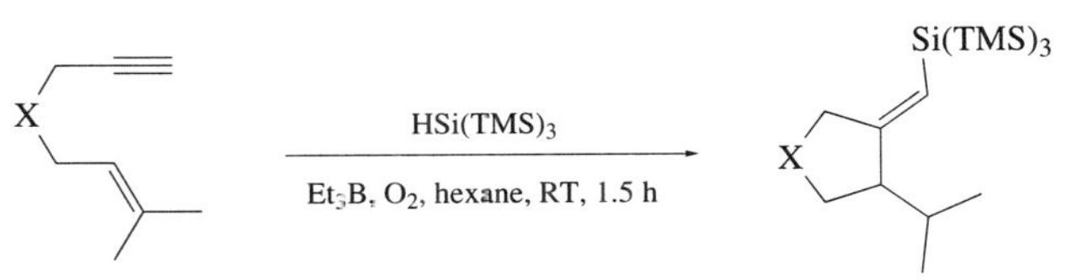

(47)

X = $C(CO_2Me)_2$ 94%
X = O 83%

(ii) Transition metal catalyzed hydrosilylation of alkynes

The additions of hydrosilanes to unsaturated compounds in the presence of transition metal catalysts has attracted great interest over the years ⟨B-79MI 218-01⟩ and a considerable amount of knowledge on the mechanism ⟨65JA16, 67JA1640, 93OM988⟩ and stereochemistry ⟨61JA4385, B-66MI 218-01, B-69MI 218-01⟩ of these transformations has been accumulated. As a result, this type of reaction is now a widely used method for the synthesis of vinylsilanes. However, it is important to note that the regio- and stereochemical outcome of the hydrosilylation of alkynes can be greatly affected by the alkyne structure, the nature of the catalyst and the silane reagent, as well as by the reaction conditions (Equations (48)–(51); Schemes 10–12).

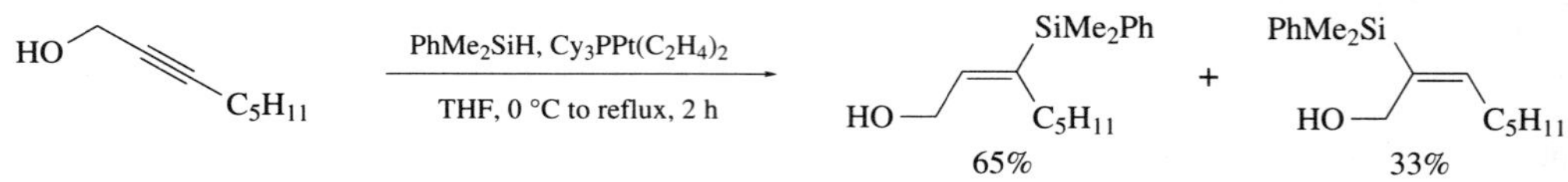

(48)

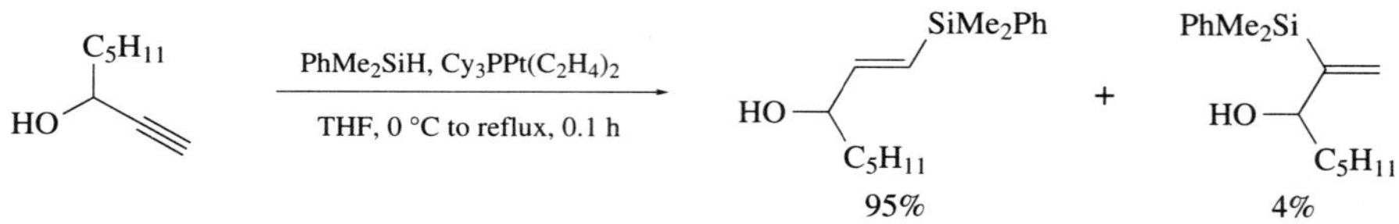

(49)

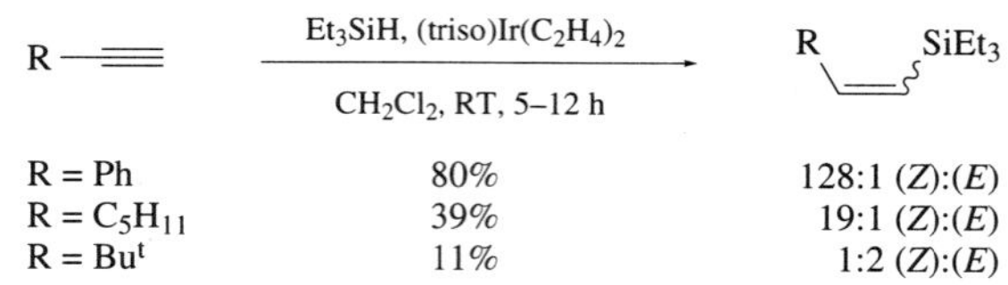

(50)

triso = tris(diphenyloxophosphoranyl)methanide

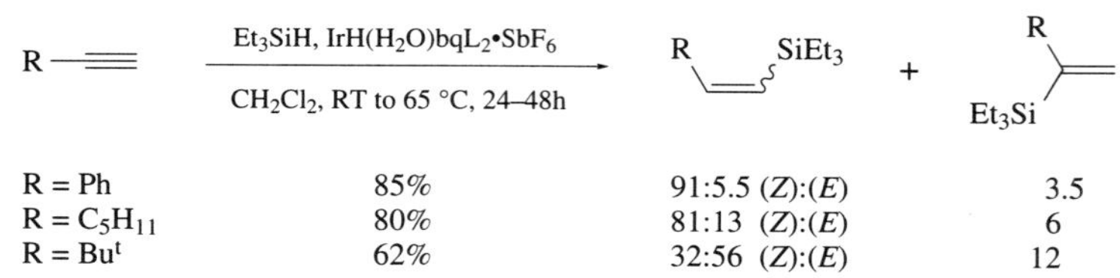

(51)

bq = 7,8-benzoquinolinate; L = PPh_3

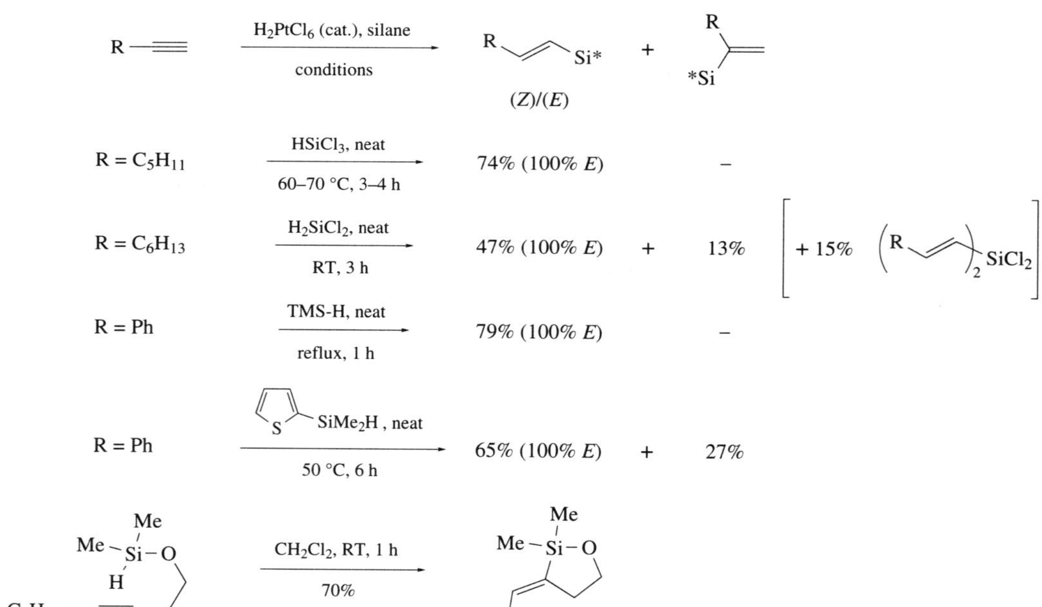

Scheme 10

In the presence of platinum-based catalysts only *syn*-addition occurs, leading to (*E*)-vinylsilanes (Schemes 10, 11; Equations (48) and (49)) ⟨61JA4385, 74JOM(69)193, 76JOM(121)293, 78JOC4424, 86ZOR140, 88TL6955⟩, but the regiochemistry of the reaction can be reversed to some extent, depending on the nature of the silane used ⟨74JOM(69)193, 86ZOR140⟩ (Scheme 10). The nature of the catalyst and the reaction conditions can also be of considerable importance (Scheme 11) ⟨88TL6719, 90TL1051⟩ in

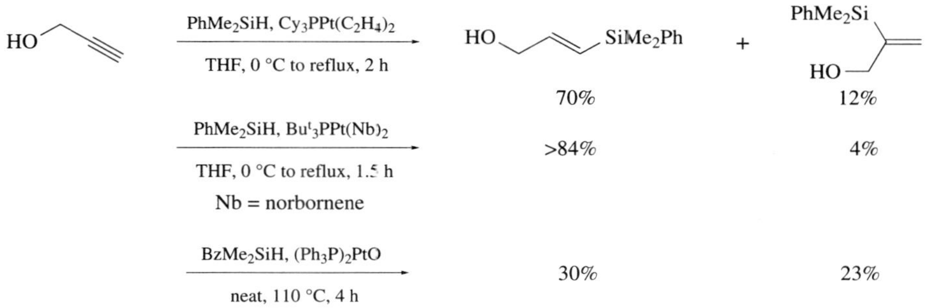

Scheme 11

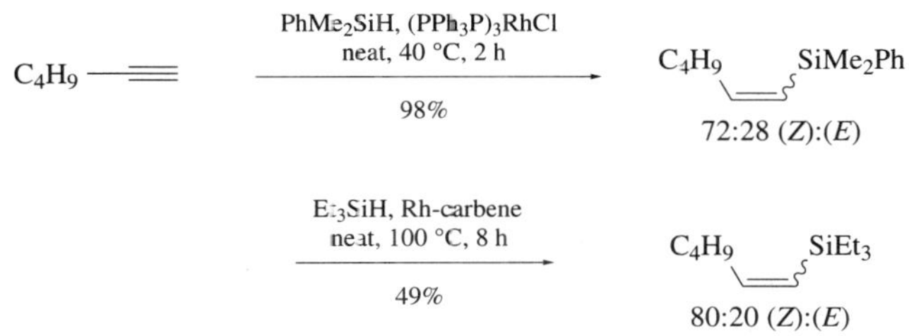

Scheme 12

determining the orientation of hydrosilylation and the substitution pattern of the alkyne (Equations (48) and (49)) ⟨90TL1051⟩.

Although quite efficient, rhodium ⟨74JOM(66)C14, 77IZV2389, 77JOM(137)293⟩ and iridium catalysts ⟨90CC1056, 90JA7984, 93JOM(447)177⟩ cause a reversal in the stereochemistry of alkyne hydrosilylation (Scheme 12; Equations (50) and (51)). Thus, rhodium catalysts transform terminal alkynes exclusively into 1-silyl-1-alkenes having (*Z*)/(*E*) ratios of 65/35 to 80/20 (Scheme 12) and these ratios can go up to 128/1 using an iridium catalyst (Equation (50)) ⟨90JA7984⟩. Using another iridium catalyst, small amounts 1-alkyl-1-silylethenes appear in the product mixture ⟨93JOM(447)177⟩ (Equation (51)). Also, bulky (Bu^t) substituents favor the formation of (*E*)-vinylsilanes (Equations (50) and (51)). It is worth mentioning that vinylstannanes, also obtainable by Pd(0)-catalyzed hydrostannylation of alkynes ⟨90JOC1857, 90TL6621, 91TL5047, 92SL886⟩, can be transformed into vinylsilanes according to transmetallation/silylation protocols (Section 2.18.2.1.1(iii)).

(iii) Bis-silylation of alkynes

The first examples of transition metal mediated "double silylation" of alkynes were reported by Kumada and co-workers; the reactions were stoichiometric between $Ni(bipy)(SiX_3)_2$ type complexes (where X = Cl or Me) and excess of the alkyne, and gave (*E* + *Z*)-1,2-bis(trimethylsilyl)alkenes after methylation with MeMgBr in 40–50% yield ⟨74JOM(76)95, 74JOM(76)105⟩.

The palladium-catalyzed addition of disilanes appears more attractive from the preparative point of view, since the reactions are often highly (*Z*)-selective, and the products are formed in higher yields ⟨75JOM(86)C27, 76JOM(114)C19, 80JOM(186)51, 81JOM(216)149⟩ (Equations (52) and (53)). One drawback of these reactions is that they proceed only in low yield with TMS-TMS, and the (*Z*)-selectivity seems to depend also on the presence of electronegative substituents in the disilane.

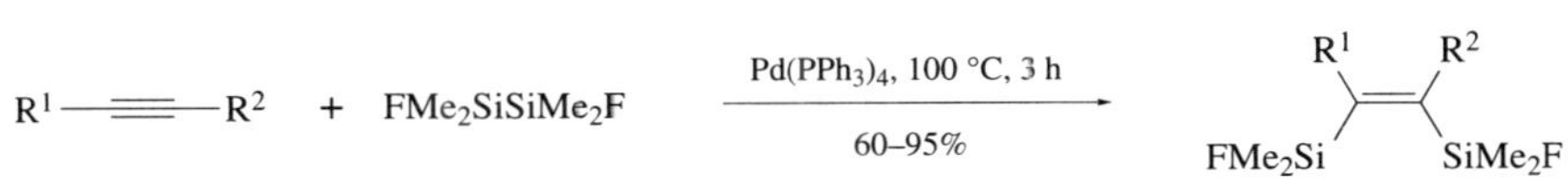

(52)

R^1 = Ph, R^2 = H, Ph; R^1 = R^2 = Et

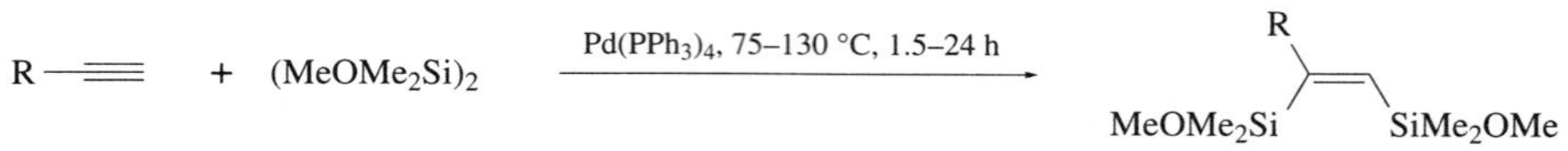 (53)

R = Ph 78% 92:8 (Z):(E)
R = Bu^n 76% 84:16 (Z):(E)

Intramolecular variants of the above bis-silylations have also been reported ⟨75JA931, 91JOC1948⟩ (Equations (54) and (55)), and the products can be of synthetic value.

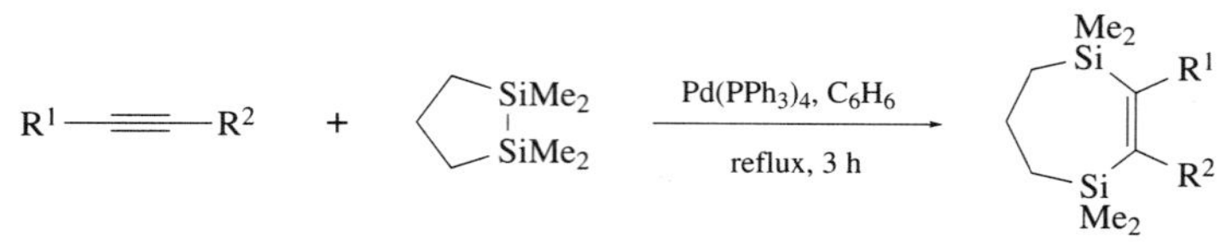 (54)

$R^1 = R^2 = CO_2Me$, 83%
R^1 = Ph, R^2 = H, 85%

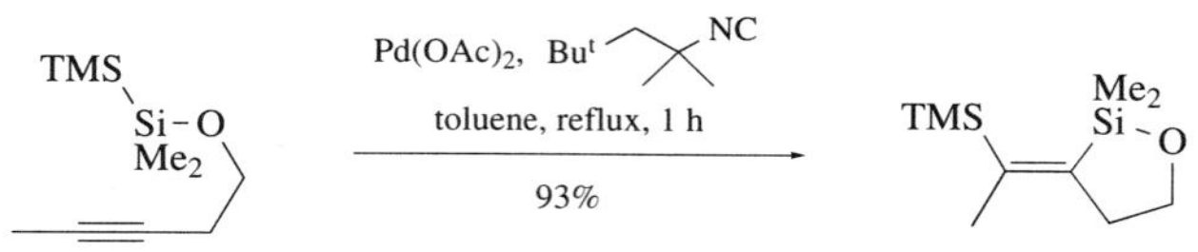 (55)

Under harsher conditions alkynes react with TMS-Cl/Mg/HMPA to give mixtures of bis- and/or polysilylated alkenic products in low to moderate yield ⟨75JOM(87)169⟩ (Equation (56)).

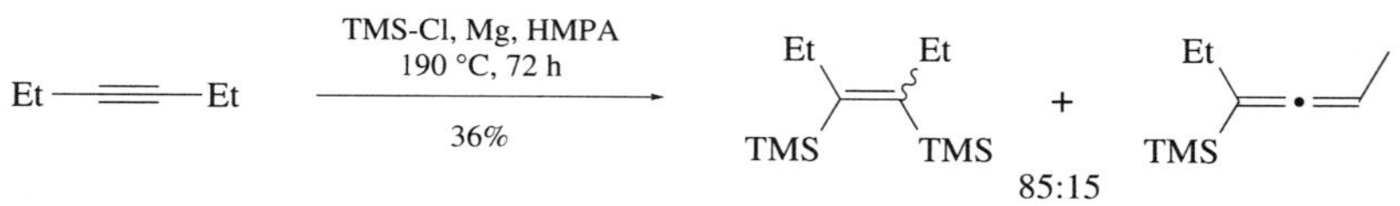 (56)

(iv) Stannylsilylation of alkynes

This is in fact a specific example of metallometallation of alkynes (see also the next section), which occurs under palladium catalysis and is most efficient for terminal alkynes reacting with trialkylstannyltrialkylsilanes ⟨85CC354, 85JOC3666, 86JOC3561, 87JOC4868, 94SL77⟩. In most cases the additions take place with high regio- and stereoselectivities to produce (Z)-1-silyl-2-stannyl alkenes generally in good yield (Equations (57) and (58)). The products undergo Sn/Li exchange (Equation (59)) with retention of the double bond configuration or tin/halogen exchange (Equation (60)) with predominant inversion of the double bond geometry. Palladium-catalyzed coupling reactions with allyl, vinyl, and aryl halides ⟨87JOC4868⟩, as well as with acid chlorides ⟨86JOC3561, 86TL2801, 87JOC4868⟩, can also be carried out efficiently on 1-silyl-2-stannylalkenes (see Section 2.18.2.1.5(iii)). Interestingly, trialkylstannyldimethylsilanes produce silacyclopentadienes (silols) on Pd-catalyzed reaction with excess alkynes ⟨93JOC5045⟩ (Equation (61)).

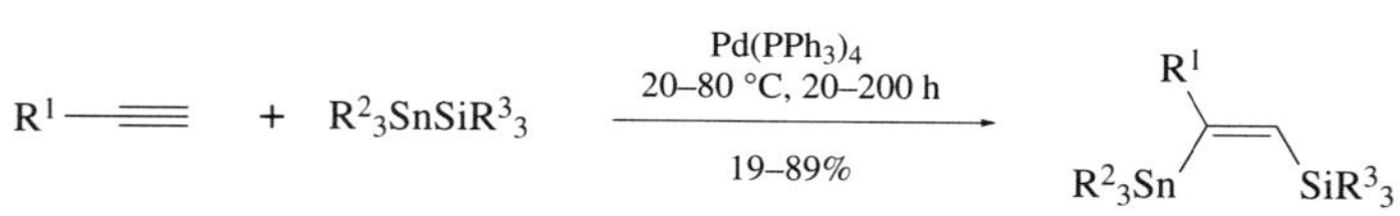 (57)

R^1 = Bu^n, $PhCH_2$, $HOCH_2$, HOCH(R'), ...; R^2 = Me, Bu^n; R^3 = Me

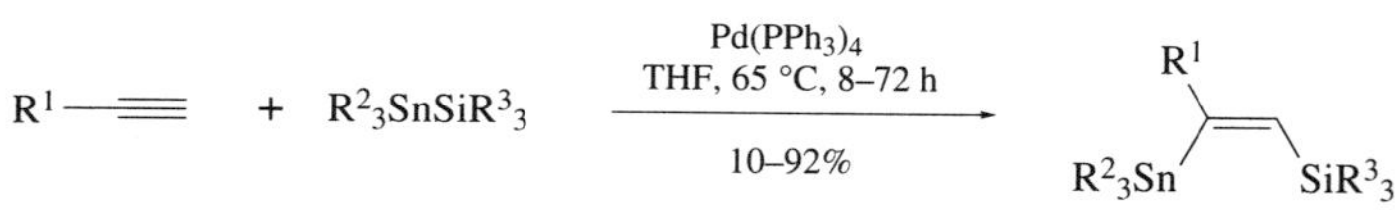 (58)

R^1 = Ph, H, Bu^n, Bu^t, Pr^i, $Cl(CH_2)_3$, $CN(CH_2)_3$ ...; R^2 = Me, Bu^n; $R^3{}_3$ = Me_3, Bu^nMe_2, Bu^tMe_2, $PhMe_2$

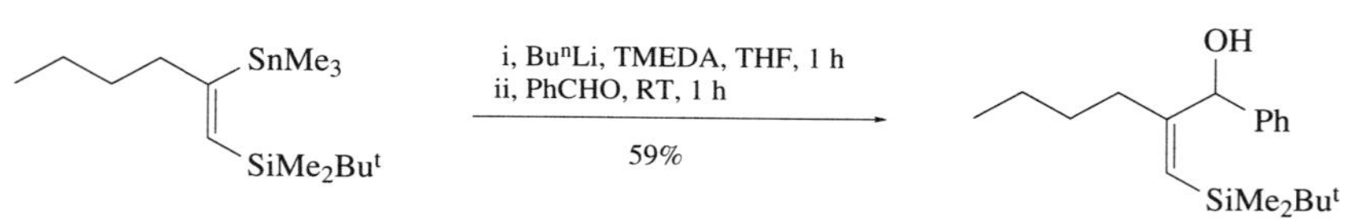 (59)

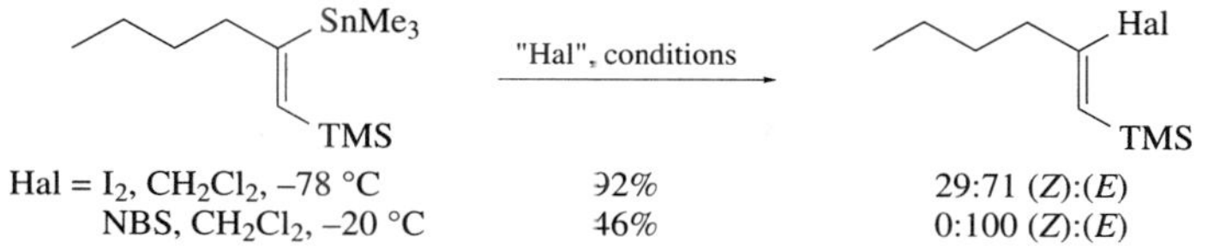

(60)

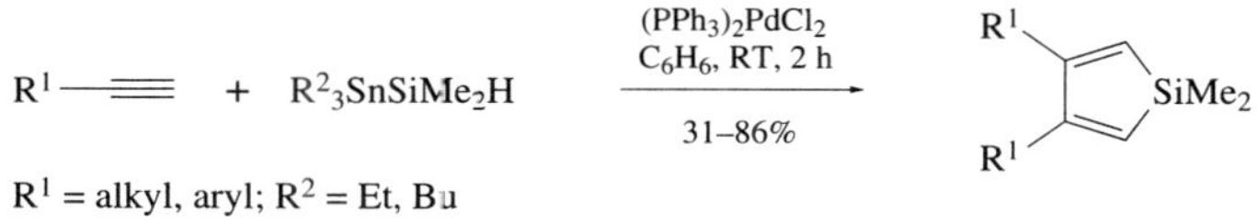

(61)

As for vicinal double stannylation ⟨86CC626⟩, stannylsilylation of di-substituted alkynes produces mixtures of isomers ⟨87JOC4868⟩.

(v) Silylmetallation of alkynes

Metallometallation of unsaturated substrates, intensely investigated during the last 15 years, has appeared as an extremely powerful and versatile tool in organic synthesis, primarily because of the high regio-, stereo-, and chemoselectivities that can be obtained. This results from the large number of ways of achieving appropriate 'tuning' of the behavior of the two metallic components.

For the synthesis of vinylsilanes, one of the metal components has to be silicon-based; however, the other one can be variable. In addition to Si/Sn the presently known combinations are Si/Cu, Si/Mg, Si/Zn, Si/B, Si/A1, Si/Mn, and Si/Ti.

(a) Silylcupration. Pioneered by Fleming and co-workers, the stereospecific *syn* addition of silyl copper(I) reagents to terminal alkynes followed by the reaction of the 1,2-bis-metalloalkene intermediates with various electrophiles allows for the synthesis of a number of functionalized vinylsilanes (Scheme 13) ⟨80CC276, 81JCS(P1)2527, 84JCS(P1)1805, 85S899, 87S325⟩. Reagents such as $PhMe_2SiCu \cdot LiCN$ ⟨80CC276⟩ and $(PhMe_2Si)_2CuLi \cdot LiCN$ ⟨81JCS(P1)2527, 88JA6565, 91JOC770, 92G65⟩, or $PhMe_2Si(Me)CuLi \cdot LiCN$ ⟨84JCS(P1)1805, 91JOC770⟩ and $(PhMe_2Si)_3 \cdot CuLi_2$ ⟨91JOC770⟩ give significantly different regioselectivities due to the variation of the structure of the copper reagents.

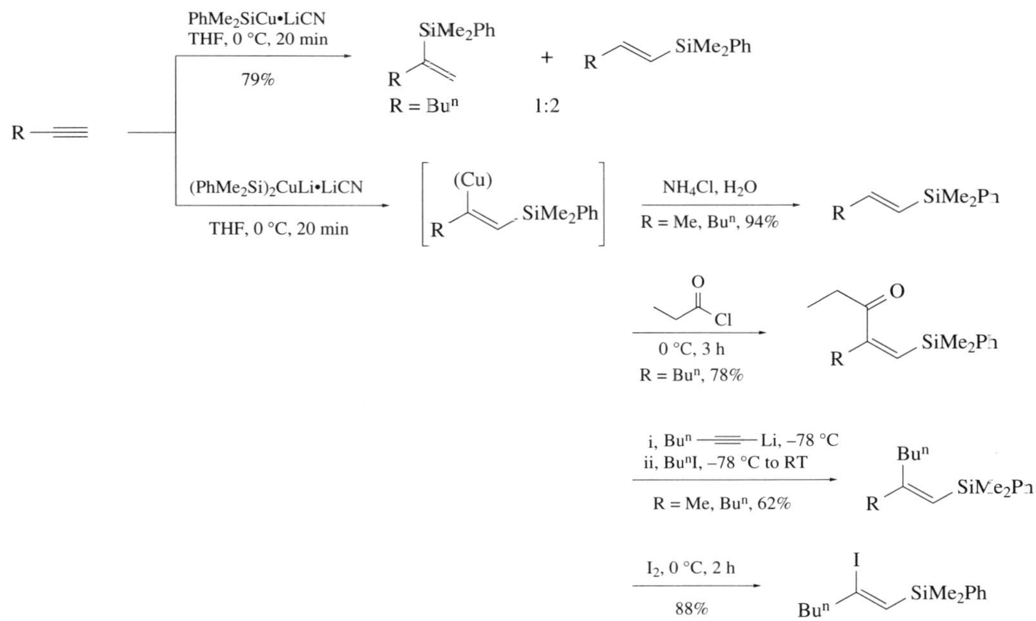

Scheme 13

The second step in these sequences (i.e. reaction of the intermediate vinylcoppers with electrophiles) may be a delicate one because of the variable reactivity of different electrophiles ⟨75OR(22)253, B-80MI 218-01, 81S841⟩. The more reactive ones (H^+, RCOCl, I_2, MeI) ⟨81JCS(P1)2527⟩ can be used directly, but the vinylcopper must be activated ⟨92OR135⟩ by adding hexamethylphosphoramide (HMPA) or *n*-hexynyllithium in the case of less reactive electrophiles (Scheme 13) ⟨81JCS(P1)2527, 88JA6565⟩.

It has been shown that a number of functional groups are compatible with silylcupration. *N*-Protected propargylamines ⟨89JOC1473, 91S1201, 92PAC439⟩ (Scheme 14), propargyl sulfides (Equation (62)) ⟨92SL981⟩ and an ynamine (Equation (63)) ⟨93TL3311⟩ have been transformed into functionalized vinylsilanes using this protocol.

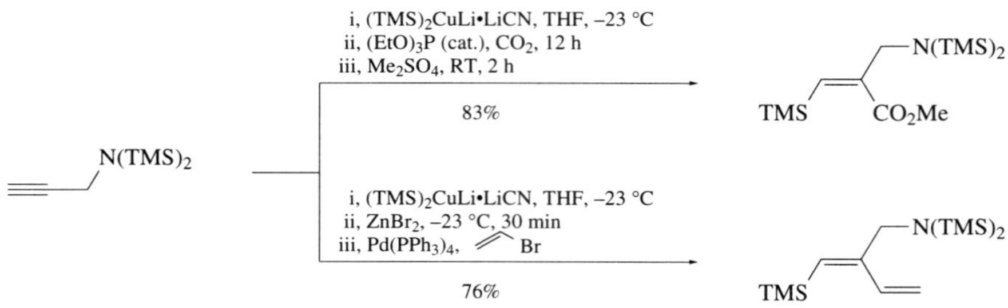

Scheme 14

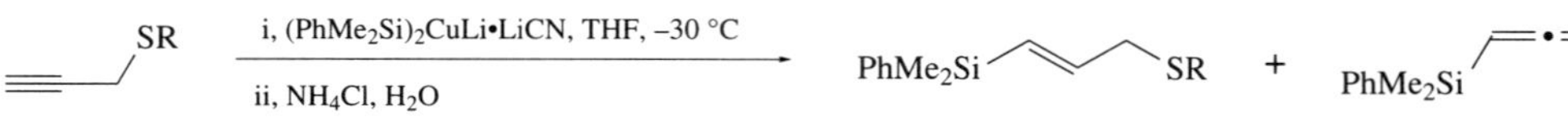

(62)

R = Bu^t	79%	21%
R = Ph	0%	80%

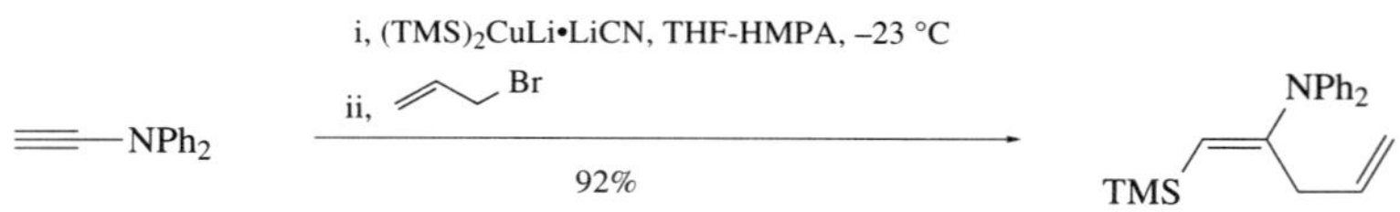

(63)

(*b*) *Addition of silyl zinc, magnesium, and aluminum reagents.* Tris(trimethylsilyl)aluminum reacts with alkynes by transfer of all three silyl moieties to produce vinylsilanes after protonolysis of the intermediate vinylalanes. The reactions proceed by clean *syn* addition and with high regioselectivity in the case of terminal alkynes (Equation (64)) ⟨80TL4069, 82CZ441⟩.

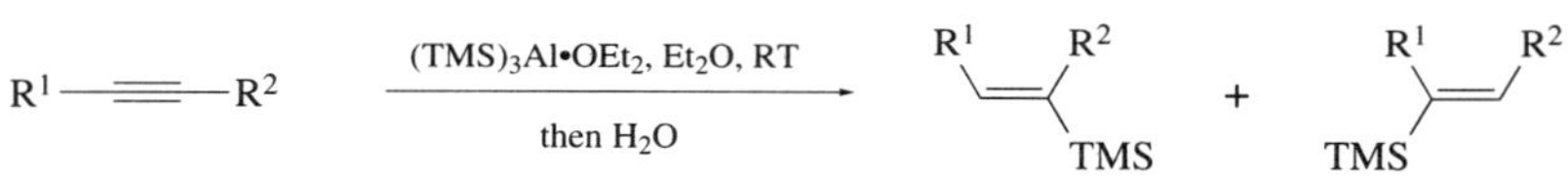

(64)

R^1 = Bu^n, R^2 = H	95%	95:5
R^1 = Pr^n, R^2 = Me	88%	70:30
R^1 = Ph, R^2 = H	92%	100:0

Monosilylzinc, -magnesium and -aluminum reagents also add to alkynes in the presence of (Cu(I), Pd, Pt based) catalysts and produce vinylsilanes with excellent selectivities (Scheme 15; Equation (65)) ⟨83JA4491, 85TL4629, 86CB1581, 86T4427, 88T4277⟩. Both zincate or alanate ⟨85TL4629, 86T4427⟩ (Scheme 15) and simple Si–Zn, Si–Mg, Si–Al reagents (Equation (65)) ⟨83JA4491, 88T4277⟩ have been used. The stereochemistry of the products always results from *syn* addition; however, regiochemistry is influenced by the structure of the reagent and catalyst. Furthermore, in the cases of Si–Mg and Si–Zn addition the intermediate vinylmagnesiums and vinylzincs can be reacted with electrophiles other than a proton, such as MeI, RCHO, TMS-Cl ⟨83JA4491, 85TL4629⟩. This greatly extends the usefulness of these silylmetallation reactions. When coupled with the fluoride ion mediated protodesilylation of the product vinylsilanes ⟨85T3257⟩, the whole transformation can be regarded as the stereospecific and highly regioselective alkylation of alkynes.

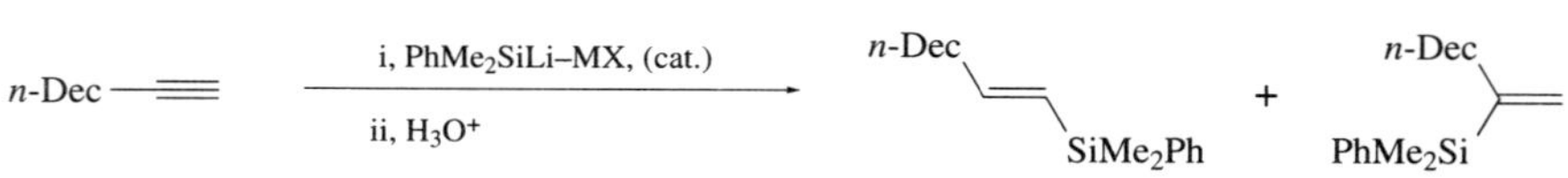

(65)

MX = MeMgI	cat = *cis*-$PtCl_2(PBu^n{}_3)_2$	90%	>99 : <1
	$PdCl_2(PPh_3)_2$	76%	60 : 40
MX = Et_2AlCl	cat = $RhCl(PPh_3)_3$	70%	91 : 9
	$PdCl_2(P(o\text{-Tol})_3)_2$	85%	15 : 85

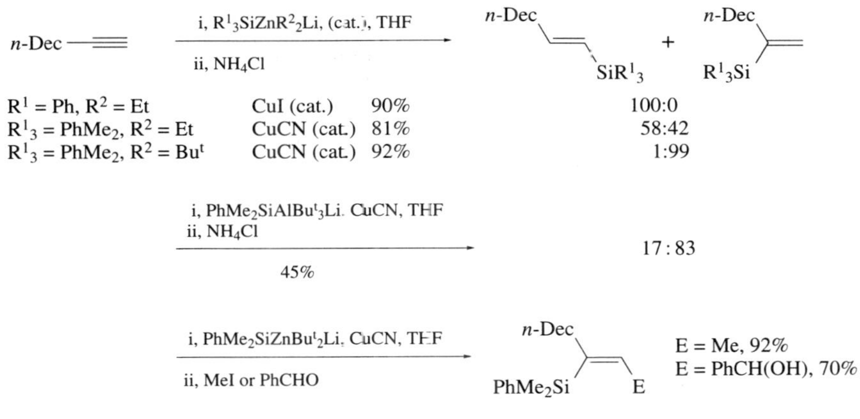

Scheme 15

(c) Addition of silyl boron, manganese, and titanium reagents. These reactions have been investigated less thoroughly than those described in Sections 2.18.2.1.2(v)(a) and 2.18.2.1.2(v)(b), but in all the cases the results are consistent with *syn* additions of the Si–B ⟨78JOM(156)C20⟩, Si–Mn ⟨88T4277⟩, and Si–Ti ⟨82JOM(232)C9⟩ couples. For the silylborate reagents the regiochemistry of addition greatly depends on the catalyst (Equation (66)) ⟨86TL2007⟩. The Si–Mn ⟨90OR(39)1⟩ reagent delivers two silyl moities leading to net bis-silylation with ill-defined stereochemistry of the starting alkyne (Equation (67)) ⟨88T4277⟩. The Si–Ti reagent is believed to have the structure Cp_2TiSiR_3 resulting from one electron reduction of Ti(IV) by R_3SiLi (Equation (68)) ⟨82JOM(232)C9⟩.

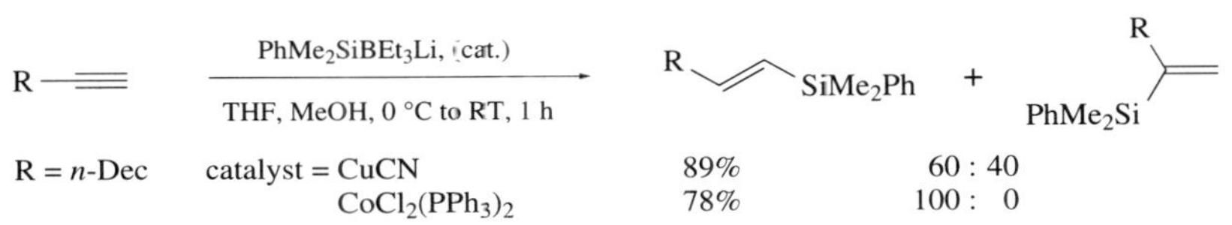

(66)

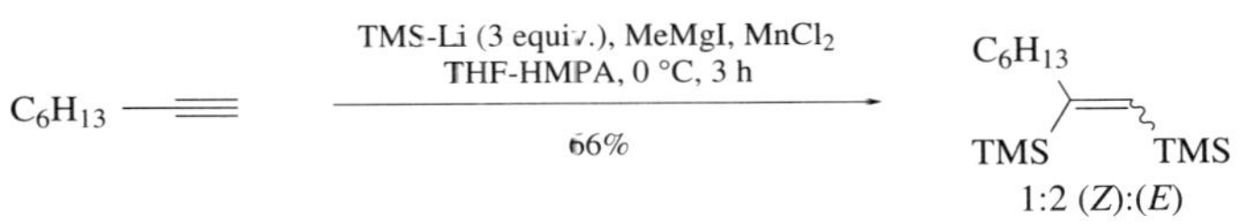

(67)

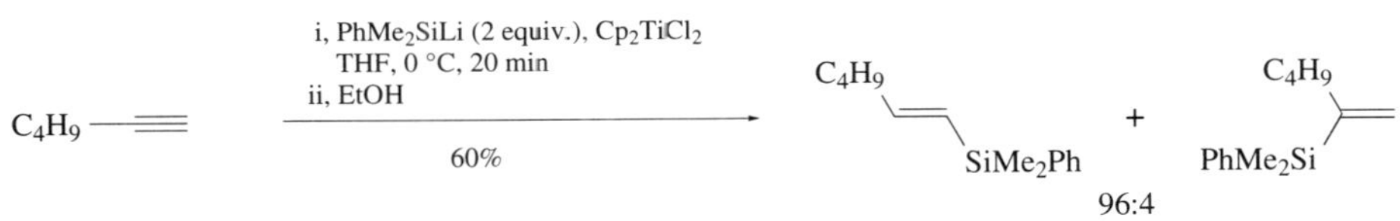

(68)

(vi) Carbometallation of alkynes

Carbometallation ⟨81S841, 93S537⟩ as well as hydrometallation ⟨75JOM(86)143, 84OR375, 88S1⟩ of alkynes is not a practical route to vinylsilanes. The reasons for this are twofold: either the necessary intermediate vinyl metal is difficult or impossible to form (metal = Li, Mg), or when easy to form (metal = B, Al, Cu, Zr), it does not react with common silyl electrophiles such as TMS-Cl, $PhMe_2SiCl$, etc. ⟨79S826, 81S841, 92OR(41)135⟩. Transmetallation might be a remedy ⟨92HCA170, 93S537⟩; however alternative methods are often more straightforward.

(vii) Carbosilylation of alkynes

In this transition metal (Pd, Ni, Rh) catalyzed reaction the carbon and silicon moieties are attached to the C≡C triple bond in a single-step operation. In the intermolecular version, trimethylsilylcyanide can be added to alkynes and arylethynes to produce β-silylacrylonitriles with fair to good (*Z*) stereocontrol (Equation (69)) ⟨88JOC3539⟩. When diynes ⟨89JA6478⟩ or enynes ⟨92JA6580⟩ are reacted with hydrosilanes in the presence of Ni or Rh catalysts, cyclic vinylsilanes are formed in good yields and with high stereocontrol (Equations (70) and (71)). Vinyldipropargylamine reacts with triethylsilane and carbon monoxide with incorporation of one or two molecules of CO depending on the pressure and the catalyst used (Scheme 16) ⟨92JA6580⟩. Palladium-catalyzed insertion of some ethynes into the C—Si bond of sila-cyclobutanes leading to vinylsilane type products was reported in 1991 ⟨91BCJ1461⟩.

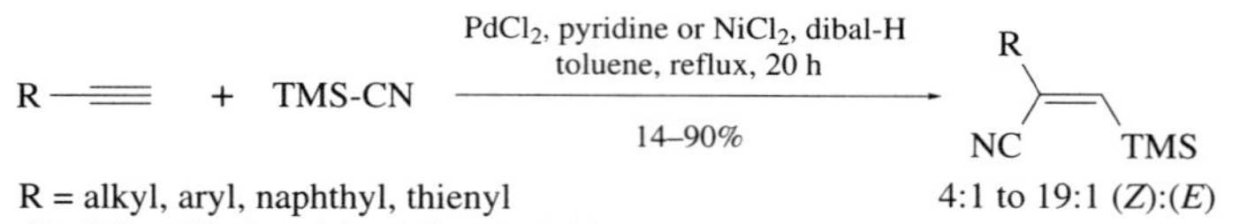

(69)

R = alkyl, aryl, naphthyl, thienyl
dibal-H = diisobutylaluminium hydride

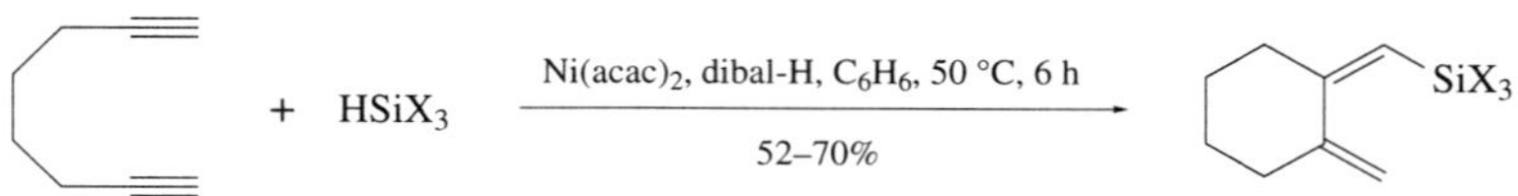

(70)

X_3 = $MeEt_2$, Me_2NEt_2, Me_2OPr^i, $Me(OEt)_2$, $(OEt)_3$ 94:6 (*Z*):(*E*)

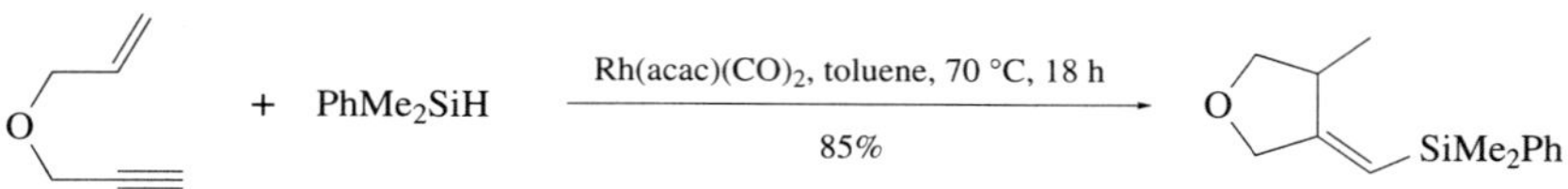

(71)

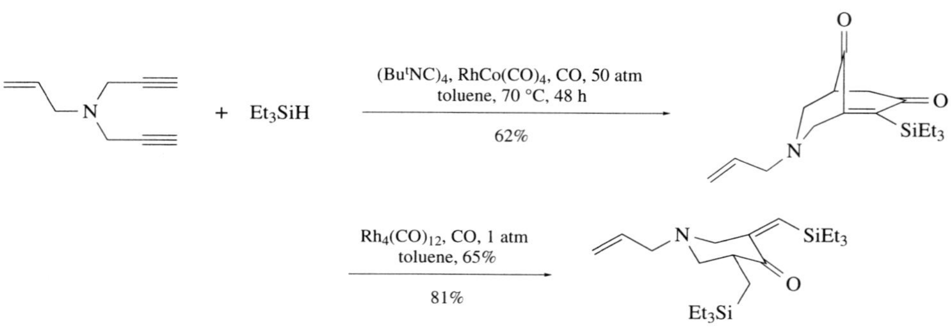

Scheme 16

(viii) *Aminosilylation of acetylenedicarboxylate*

Acetylenedicarboxylate is the only alkyne reported to undergo clean addition reactions with *N*-silyl amines ⟨68JOM(14)327, 78JOM(152)39⟩ or imines (Equations (72)–(74)) ⟨87CC1195⟩. Although one stereoisomer is formed in all cases, the stereochemistry of the additions has not been established with certainty.

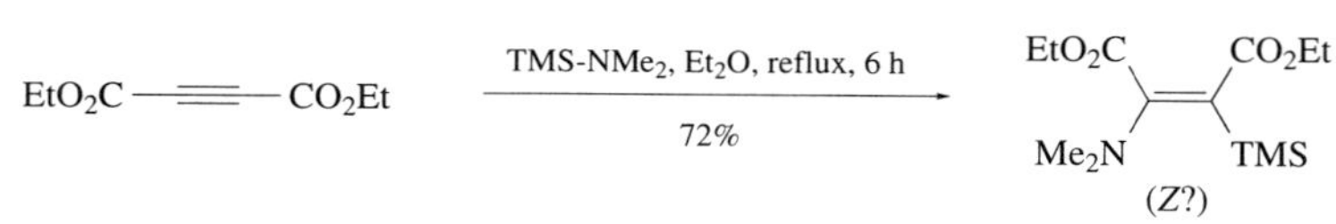

(72)

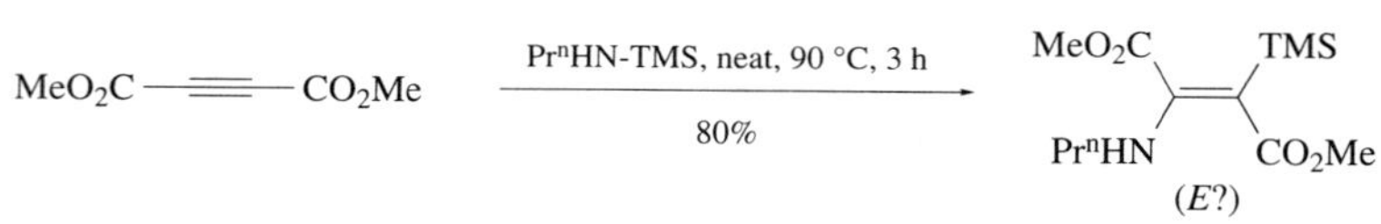

(73)

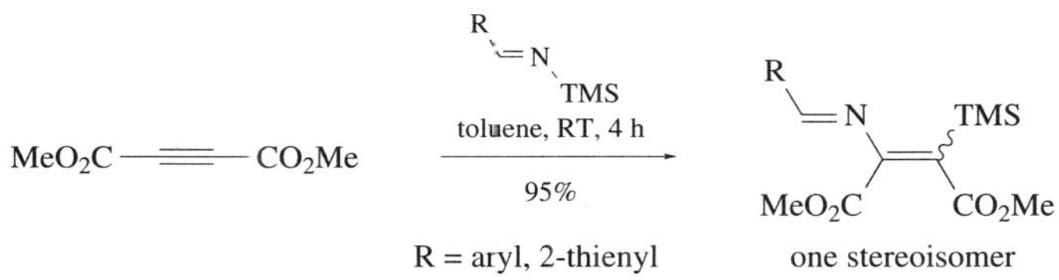

(74)

2.18.2.1.3 *Vinylsilanes from silyl alkynes*

Alkynylsilanes can be transformed into alkenylsilanes by various methods including semihydrogenation, hydride reduction, hydrometallation, carbometallation, cycloadditions, hydrocarbonylation, and halogenation. Many of these reactions are facilitated by transition metal catalysts.

(i) Semihydrogenation and hydride reduction

Raney nickel ⟨76JOM(121)293, 77TL2001⟩ as well as P-2Ni ⟨78JOC4424⟩ catalysts appear to effect semihydrogenation of alkynylsilanes with high (>94%) (*Z*) selectivity (Equations (75), (76)).

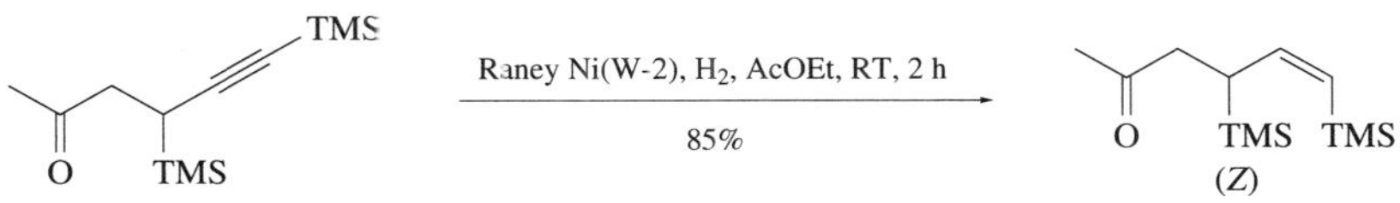

(75)

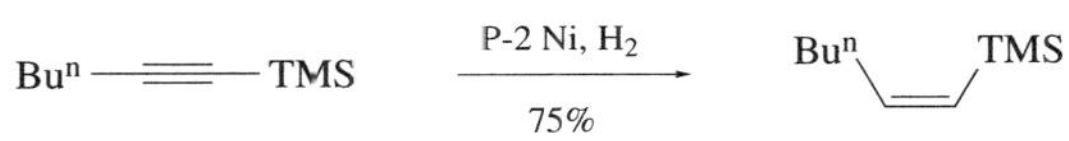

(76)

Partial reductions on palladium/charcoal ⟨71JA2080⟩ or on Pd/$BaSO_4$ ⟨77JA7613, 93JOC4823⟩ also proceed with (*Z*) selectivity, but in some cases the (*Z*)/(*E*) ratios were found to depend on both concentration ⟨93JOC4823⟩ and substrate structure (Equation (77) ⟨77JA7613, 93JOC4823⟩.

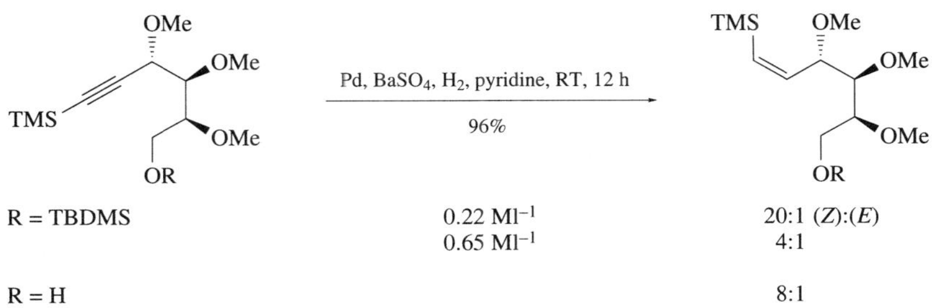

(77)

Lithium aluminum hydride (LAH) reduction of 3-hydroxypropynylsilanes gave (*E*)-vinylsilanes ⟨74JA3684, 81JCS(P1)2415⟩. On closer scrutiny, however, the stereochemical outcome of these reactions depends strongly on the solvent as well as on the heterogeneity of the reaction mixtures ⟨82JOC4595, 83TL4295⟩. An (*E*)-specific and high yielding procedure has been described using sodium bis(2-methoxyethoxy)aluminum hydride in ether (Equation (78)) ⟨82JOC4595, 87TL2041, 88T3953, 88T4073, 93JOC2714⟩.

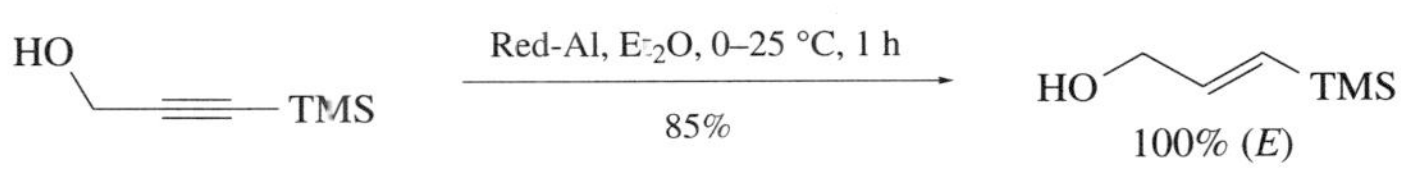

(78)

(ii) Hydrometallation of alkynylsilanes

Combined with subsequent proto-demetallation, the title reaction is conceptually equivalent to semihydrogenation or hydride reduction of alkynylsilanes, but occasionally it can lead to differing stereochemistry in the product vinylsilanes. The presently known hydrometallations are hydroboration, hydroalumination, hydrosilylation (and double silylation), hydromagnesiation, and hydrozirconation.

(a) Hydroboration. Application of the well-known hydroboration of alkynes (see Section 2.18.1.1.1(i)) to alkynylsilanes leads, after protonolysis of the intermediate vinylborane–vinylsilanes, to (*Z*) vinylsilanes in high yield (Equation (79)) ⟨74TL543, 78JOC4424, 84S113, 89JA4873, 88T3919, 90TL5113, 93JOC4823⟩.

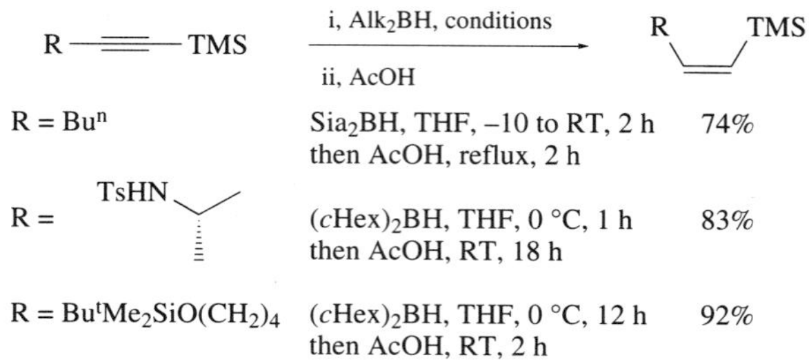

(79)

The regiochemistry of the hydroboration can be completely reversed by an appropriate choice of the boron alkyl substituents and the alkyne and silyl moieties ⟨89JA4873, 91TL7365⟩. The most widely used boranes (dicyclohexyl ⟨74TL543, 88T3919, 90TL5113, 93JOC4823⟩, disiamyl ⟨78JOC4424⟩, 1,2-dimethylpropyl ⟨84S113⟩) give 1-boryl-1-silyl alkene addition intermediates with TMS-alkynes. Low and variable yields of (*Z*)-vinylsilane were obtained from 3-TMS-propynoate ⟨77JA7613⟩.

In addition to proto-deboronation, the 1-boryl-1-silyl alkene intermediates can also undergo carbo-deboronation after suitable activation of the boron moiety by MeLi and cuprous iodide (Scheme 17) ⟨76JOC2941, 77T2987, 82TL3387⟩. Efficient activation was also found using NaOH/CuI (Equation (80)) ⟨86BCJ659⟩. In the particular case of 3-chloro-1-TMS-propyne, hydroboration followed by aqueous NaOH treatment led to (*E*)-vinylsilanes ⟨92BCJ685⟩.

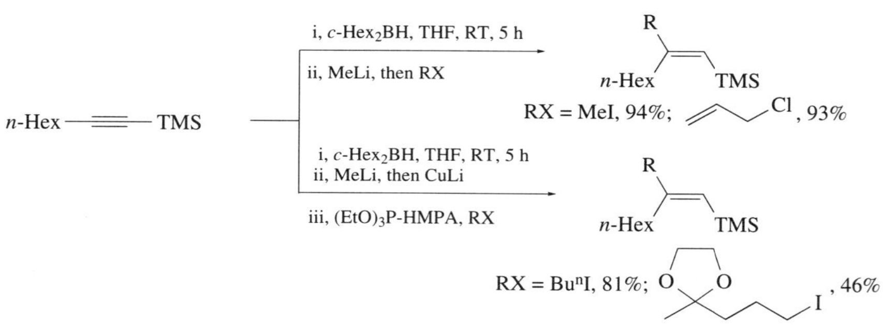

Scheme 17

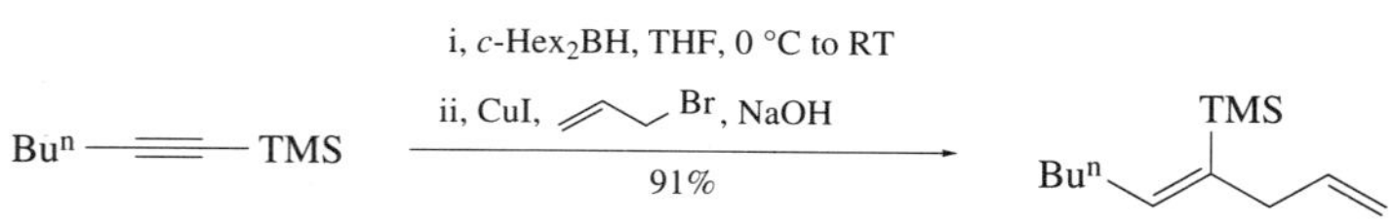

(80)

(b) Hydroalumination. In close analogy with hydroboration, hydroalumination of alkynylsilanes is another versatile method for obtaining alkenylsilane derivatives, since the intermediate 1-silyl-1-aluminoalkenes can be dealuminated in a variety of ways ⟨84OR375⟩. It is important to note however, that while the regiochemistry of hydroalumination is the same as that of hydroboration, the stereochemistry is solvent dependent ⟨76JOC2214, 76JOC2215⟩, with hydrocarbon solvents (hexane, heptane) giving (*E*)-vinylsilanes and ether (or ether–hydrocarbon mixtures) affording (*Z*)-vinylsilanes after proto-dealumination. Treatment of the aluminosilylalkene intermediates with aqueous acid ⟨76JOM(121)293, 78JOC4424, 80S803, 84S113, 88T4009, 92G65⟩ affords terminal vinylsilanes in high yields and reaction with halogens (I_2 or BrCN, NBS, NCS) leads to highly stereodefined 1-halo-1-silyl alkenes ⟨78JOC2739, 81S999, 82TL3387, 83TL2055, 84JOC1838⟩. Alkylative dealumination can be effected either through the 'ate' complex (produced by addition of MeLi) ⟨76JOC2214, 76JOC2215, 88T4009⟩, or through transmetallation with MeLi ⟨88T4009, 91JOC2468⟩ or with MeLi/CuI/$P(OEt)_3$

⟨84TL131⟩. Transmetallation with MeLi alone gives the 1-lithio-1-silyl alkene which usually isomerizes ⟨86JA3402⟩, with subsequent alkylation leading to internal alkenylsilanes of opposite stereochemistry ⟨88T4009⟩.

(c) Other hydrometallations. Hydromagnesiations ⟨83TL1041, 90TL6399, 92SL849⟩ and hydrozirconations ⟨91JOC2590⟩ of the C≡C triple bond of alkynylsilanes is a highly *cis*-selective process (Equations (81) and (83)). In the case of 1-silyl-3-propynols however, hydromagnesiation occurs to give the *trans*-product (Equation (82)) ⟨90TL6399⟩. The intermediate vinylmagnesiums can be alkylated ⟨83TL1041⟩, carbonated ⟨90TL6399⟩, silylated ⟨84JOC1838⟩, or stannylated ⟨90TL3105, 90TL5829⟩, allowing for the syntheses of a wide range of elaborated vinylsilanes. While hydromagnesiation is usually effected in ether, alkylation and stannylation of the vinylmagnesiums are carried out in THF and silylation requries HMPA.

Radical addition of triphenyltinhydride to TMS-alkyne has also been reported ⟨87JA2547⟩.

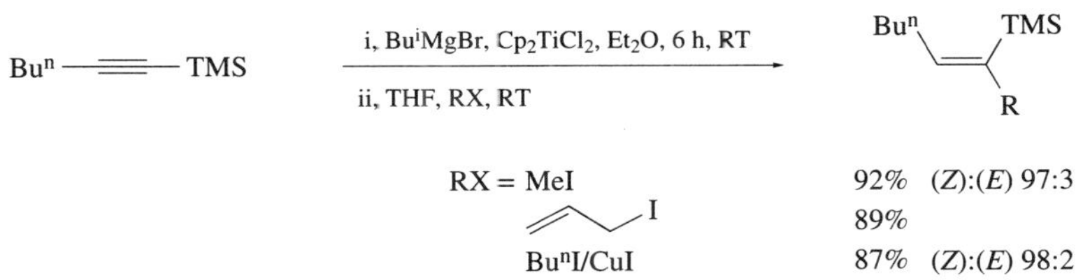

(81)

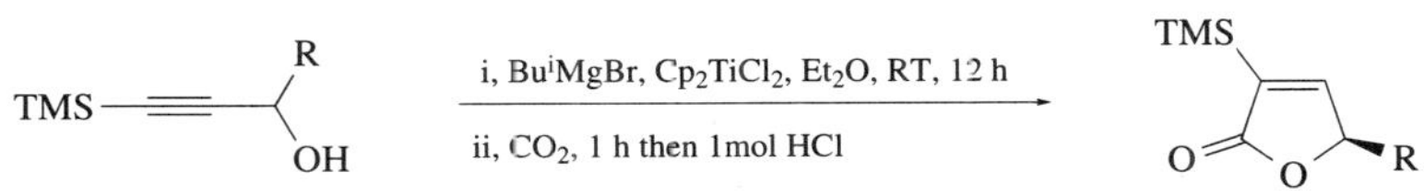

(82)

R = Me, 78%; *n*-Pent, 92%; *c*-Hex, 86%; CH_2OCH_2Ph, 84%

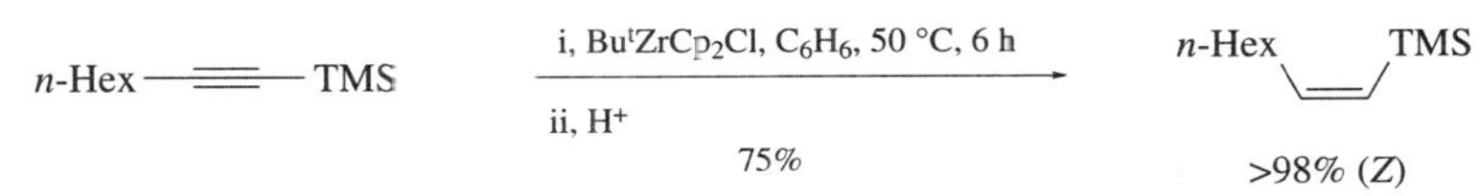

(83)

(iii) Hydrosilylation, double silylation, and silylmetallation

These reactions, basically identical to those of Sections 2.18.2.1.2(ii)–(v), lead to alkenes possessing two or three silyl moieties attached to the double bond. Therefore, such alkenes can be considered as polymetalloalkenes and usually exhibit distinguishable reactivities at the various metallic (i.e., silicon) centers.

Hydrosilylation of alkynylsilanes (using chlorohydrosilanes and chloroplatinic acid, for example) proceeds predominantly with regiochemistry opposite to hydrosilylation of terminal alkynes ⟨B-69MI 218-01, 79JOC155⟩. Moreover, protodesilylation of the so-obtained 1,2-bis(silyl)alkene gives essentially the 2-silyl-1-alkene ⟨79JOC155⟩ (Equation (84)); compare with Schemes 10–12, Equations (48)–(51)).

Double silylations of alkynylsilanes either using TMS-Cl/Mg in HMPA ⟨75JOM(87)169⟩, or with methoxymethyldisilanes or methylchlorodisilanes under Pd(0) catalysis ⟨81JOM(216)149, 90BCJ3103, 93AG(E)1473⟩, are known; however, these reactions are sometimes low yielding and not general.

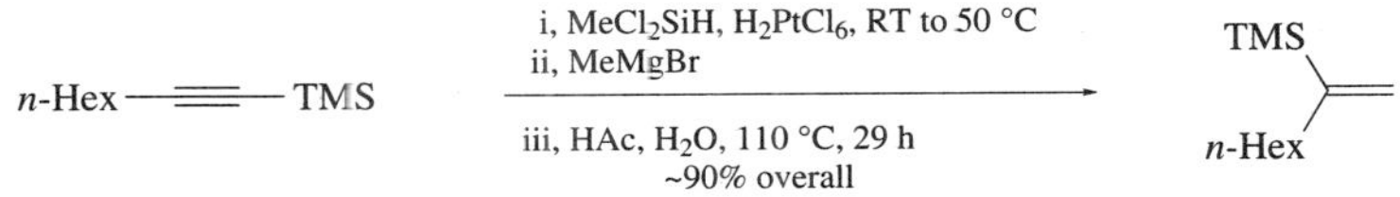

(84)

Silylmetallations (i.e. Si/Al ⟨82CZ441⟩, Si/Zn, ⟨86T4427⟩, Si/Ti ⟨82JOM(226)C9⟩) of alkynylsilanes are also closely related to those of alkynes (Section 2.18.2.1.2(v), Scheme 15; Equations (64)–(65)); however, the presence of a silyl group in both the starting alkyne and the reagent can produce more subtle and variable regio- and stereochemical outcomes. Thus, silylalumination of TMS-1-hexyne gives 1,1-bis(TMS)-1-hexene (Equation (85)) ⟨82CZ441⟩, whereas silylzincation invariably produces 1,2-bis(silyl)-1-alkenes of (*E*) or (*Z*) stereochemistry, depending on the absence or presence of a free hydroxy group in the vicinity of the triple bond (Equations (86), (87)) ⟨86T4427⟩.

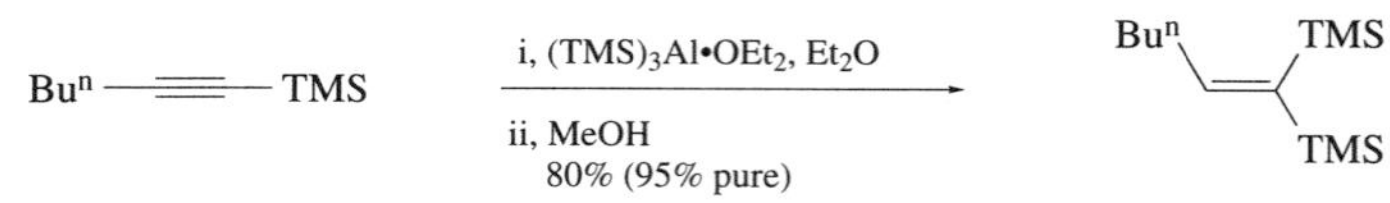

(85)

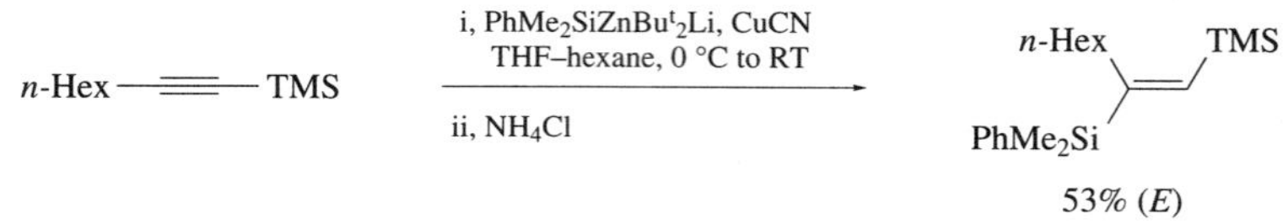

(86)

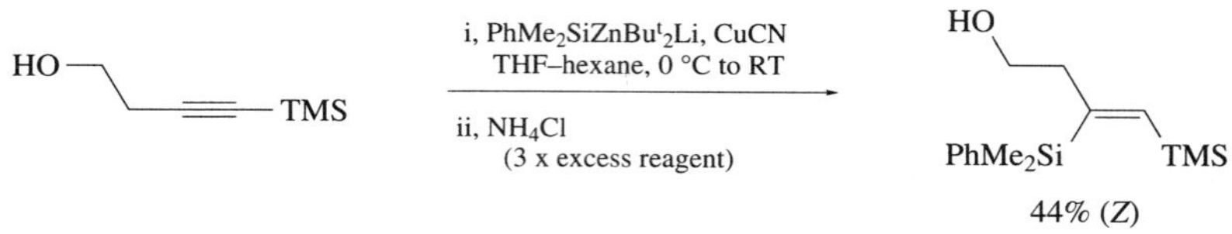

(87)

Silyltitanation of TMS–phenylacetylene gives a low yield of a mixture of isomers with β,β-bis(silyl)styrene predominating (Equation (68)) ⟨82JOM(226)C9⟩.

(iv) Carbometallation of alkynylsilanes

Carbometallation of alkynes is not a flexible route to vinylsilanes (Section 2.18.2.1.2(vi) ⟨81S841, 92OR135⟩). By contrast, this type of reaction on alkynylsilanes has emerged during the last decade as an extremely versatile method for the syntheses of both vinylsilane and arylsilane type compounds.

In the broad spectrum of carbometallations of interest here, one can distinguish those using main group organometallic reagents (metal = Mg, Cu, Zn, Al), and those using transition metal based reagents (Ti, Zr, Hf, Ta, Cr, Mo, W, Ru, Co, Rh, Ni, Pd, Pt).

(a) Carbomagnesiation and carbocupration. Since Grignard reagents do not add easily to alkynes ⟨81S841⟩, their utility in vinylsilane synthesis from alkynylsilanes is limited to MeMgBr, because the addition reaction requires a transition metal catalysis. Quenching the intermediary 1-magnesio-1-silyl alkene with various electrophiles gives tetrasubstituted alkenes with 100% regioselectivity and with stereoselectivities of 80 : 20 to 90 : 10, corresponding to predominant "*cis*-1,2-dialkylation" of the starting silyl alkyne (Equation (88)) ⟨78JA4624⟩. The limitation of this transformation to MeMgBr arises from the prevalent hydromagnesiation reaction of higher Grignard reagents in the presence of transition metal catalysts ⟨78JA4624, 83TL1041⟩ (Section 2.18.2.1.3(ii)(c)).

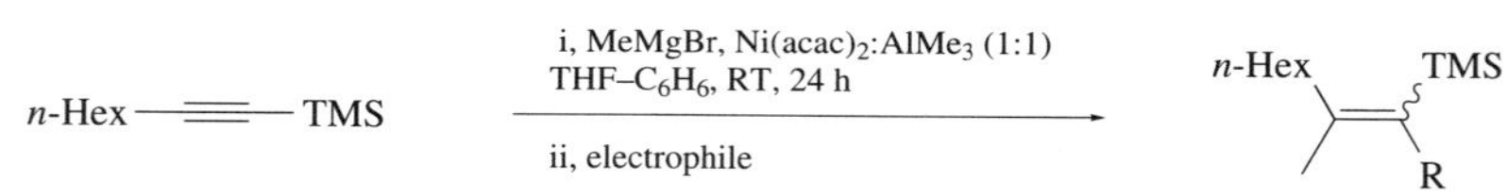

(88)

R = H	80%	(*E*):(*Z*) 1:9
R = CO_2	61%	(*E*):(*Z*) 1:4
R = I	71%	(*E*):(*Z*) 9:1
R = Me	50%	(*E*):(*Z*) 1:4

Presumably due to steric factors, carbocuprations are limited to silyl acetylene ⟨79JOM(177)145⟩. Nevertheless, this single precursor permits the synthesis of various internal vinylsilanes due to the *cis*-1,2-dialkylation proceeding with a regioselectivity as high as that of carbomagnesiation, but with much higher stereoselectivity (Equation (89)) ⟨77TL1805, 77TL1823, 79JOM(177)145, 90TL4937⟩.

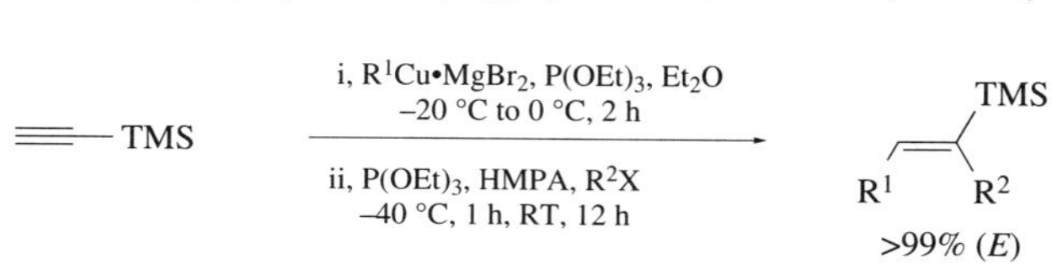

(89)

$R^1 = C_3H_7$, $R^2X = n\text{-}C_5H_{11}I$	64%
$R^1 = C_3H_7$, $R^2X = n\text{-}C_6H_{13}I$	69%
$R^1 = n\text{-}C_6H_{13}$, $R^2X = n\text{-}C_3H_7I$	58%
$R^1 = n\text{-}C_6H_{13}$, R^2X = 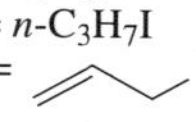 Br	57%

As well as the 1,2-dialkylation product, the alkenyl copper intermediate can also give 1-halo-1-silyl alkenes on reaction with various halogenating reagents ⟨77TL1823, 90TL4937⟩, further extending the usefulness of this method.

One example of an intramolecular carbocupration leading to a silyl alkylidenecyclopentane has been reported (Equation (90)) ⟨91JA5735⟩.

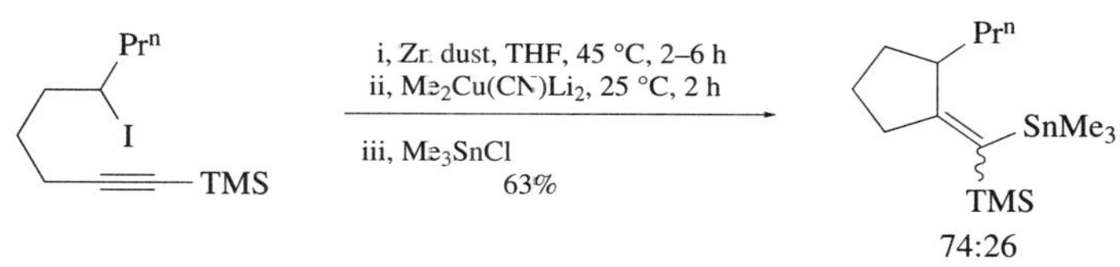

(90)

(b) Allylzincation. In contrast to Grignard ⟨81S841⟩ reagents, organozincs are able to add to alkynes, although internal alkynes do not react ⟨83JA6761⟩. On the other hand, alkynylsilanes have been found to react readily with allylic zinc reagents in THF to produce, after quenching with iodine ⟨83JA6761⟩ or a proton ⟨83JOC5409⟩, the corresponding vinylsilanes (Equation (91)). The reactions are completely regiodefined placing the zinc moiety on the carbon atom bearing the silyl group; however, the predominant *cis* addition products are often contaminated by their stereoisomers. This is thought to be due to isomerization of the intermediary vinyl zinc compounds. Increasing the bulk of the alkyl substituent can invert the isomeric ratio in the product (Equation (91)).

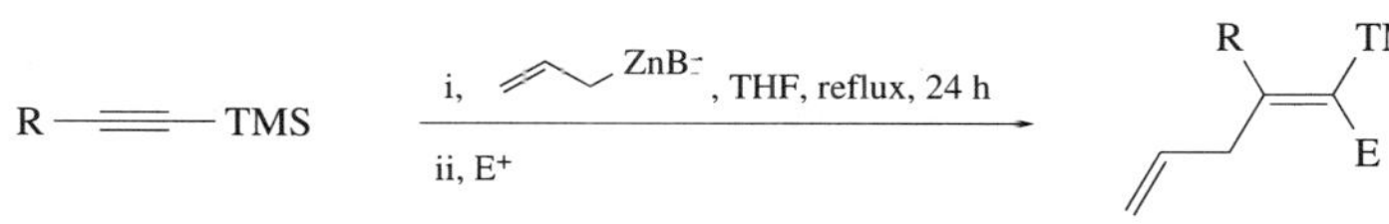

(91)

R = C_6H_{13}, E = I	83%	85:15 (*Z*):(*E*)
R = C_4H_9, E = H	70%	85:15 (*Z*):(*E*)
R = Pr^n(PhCH$_2$O)CH, E = H	74%	<5:95 (*Z*):(*E*)

The intermediate alkenyl zinc derivative can be alkylated intramolecularly using a Pd(0)-catalyzed substitution reaction of the zinc nucleophile with an allyl ether moiety. This tandem allylzincation-Pd(0)-catalyzed cyclization produces 4-methylenecyclopentenes (Equation (92)) ⟨87TL2889, 92T6087⟩. Further sophistication of this strategy has been achieved by also performing the allylzincation step in an intramolecular fashion to afford fused bicyclic 4-methylenecyclopentenylsilanes (Equation (93)) ⟨92T6105⟩. Interestingly, allylaluminum reagents do not react with 1-silyl-1-alkynes ⟨85TL5863⟩.

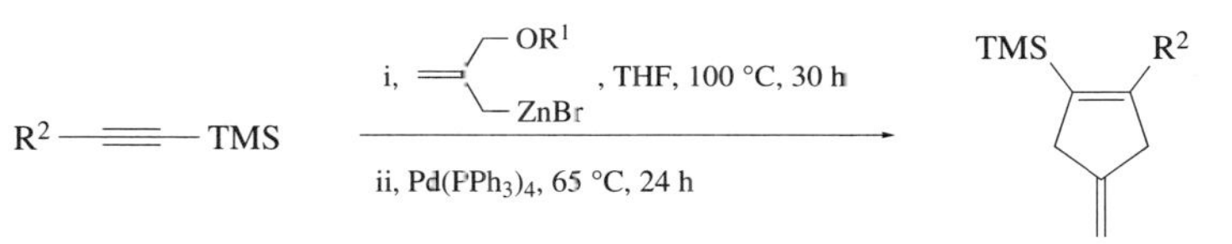

(92)

R^1 = Ph, R^2 = Bu^n, CH_2OMe, CH_2CH_2O-TMS — 72%, 51%, 78%

R^1 = CH_2Ph, R^2 = CH_2CH_2CH, 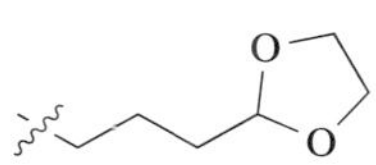 — 67%, 91%

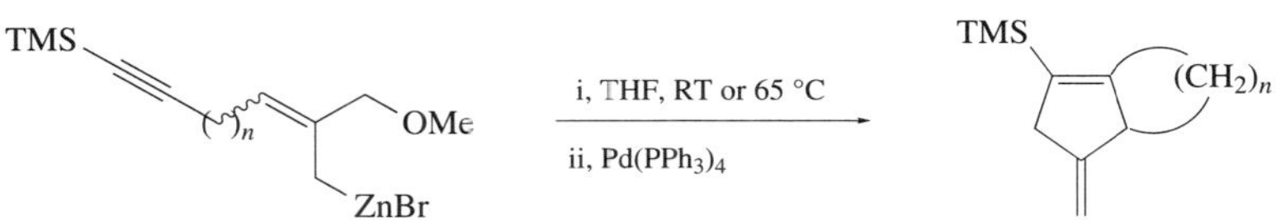

(93)

n = 3; i, RT, 2 h; ii, RT, 3.5 h	84%
n = 4; i, 65 °C, 2 h; ii, 65 °C, 2 h	67%
n = 5; i, 65 °C, 24 h; ii, 65 °C, 4 h	44%

(c) Carboalumination. In close analogy with the zirconocene dichloride mediated carboalumination of alkynes ⟨78JA2252, 78JA2254, 85JA6639, 88S1⟩, trimethylaluminum ⟨78JOM(156)C20⟩ or alkyl aluminum chlorides ⟨78JOM(159)C13, 79JOM(179)C37, 84JOC725⟩ have been found to react with alkynylsilanes in the presence of a stoichiometric amount of titanocene dichloride. When quenching of the alkenyl metal intermediates was carefully controlled excellent yields of 1-alkenyl silanes

and 1-halo-1-silyl alkenes were obtained with high regio-and stereoselectivities (Equation (94)) ⟨79JOM(179)C37, 84JOC725⟩.

(94)

i, Me_2AlCl, Cp_2TiCl_2 (1:1)			
ii, 25% NaOH (aq.)	R = Me, X = H	93%	95:5
	R = Et, X = H	85%	95:5
i, Me_2AlCl, Cp_2TiCl_2 (1:1)			
ii, Add to I_2, −78 °C	R = Me, X = I	65%	88:12
i, Me_2AlCl, Cp_2TiCl_2 (1:1)			
ii, Add to NBS, −78 °C	R = Et, X = Br	80%	95:5

By an unusual cycloalkylation of the intermediate butenyl metals, 1-silyl-4-halo-1-butynes led, under similar conditions but in the presence of zirconocene dichloride, to 1-silyl-2-methylcyclobutene (Equation (95)) ⟨83JA6344, 84JA6105⟩.

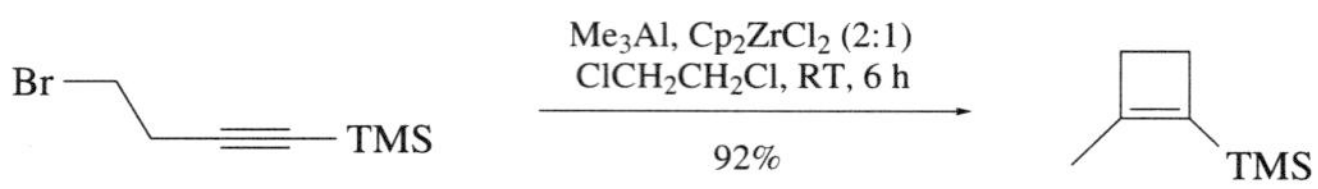

(95)

Transmetallation of the alkenylalane intermediate by means of dilithiodi(1-hexynyl) cyanocuprate allowed Michael additions to cyclohexenone to be carried out efficiently (Equation (96)) ⟨92JOC3178⟩.

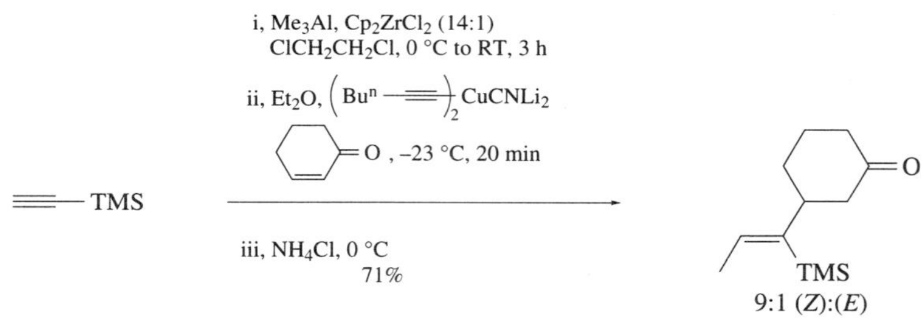

(96)

(d) Cyclizations involving transition metals. The most well known of these cyclizations is the $Co_2(CO)_8$ mediated (in stoichiometric amounts) Pauson–Khand [2 + 2 + 1] cycloaddition of alkynes (or 1-silyl-1-alkynes) with alkenes and carbon monoxide to give cyclopentenones (or 2-silylcyclopentenones) ⟨91OR(40)1⟩. The intramolecular version of this reaction has been applied to the construction of various fused-ring systems (Equation (97)) ⟨87JA7495, 87JOC1483⟩. Irradiation (Hg medium pressure lamp) of TMS-ethyne or α,ω-silyldiynes in the presence of $CpCo(CO)_2$ produced bis(silyl)cyclopentadienones ⟨80AG(E)1023⟩ which were also formed in high yield in Ni(0)-promoted intramolecular cycloadditions with isocyanide incorporation (Equation (98)) ⟨89JOC3517, 92SL539⟩.

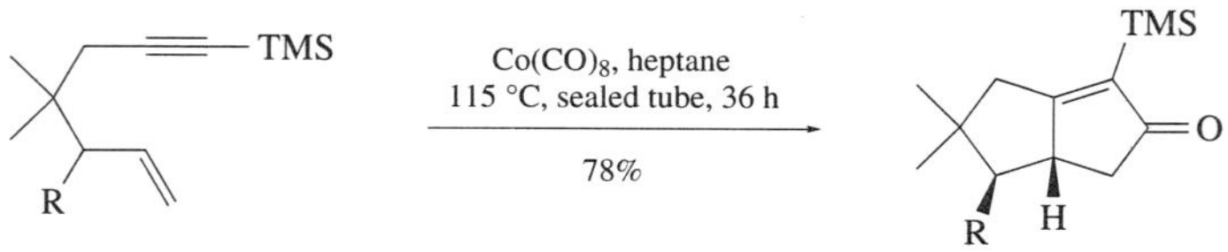

(97)

R = $MeOCH_2OCH_2CH_2$

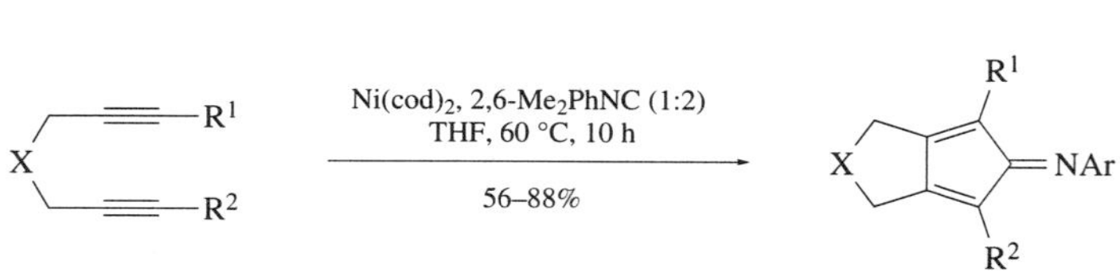

(98)

X = O, $(CH_2)_{1-3}$; R^1, R^2 = Et, Ph, TMS

Conceptually similar reactions have been found to occur when titanocene ⟨84JA6422⟩ and more importantly zirconocene ⟨87JA2788, 88S1⟩ reagents were used in conjunction with α,ω-diynes or α,ωenynes. The major difference lies in the fact that the zirconocene-induced reactions proceed by discrete zirconacarbocyclic intermediates which, if desired, can be isolated and characterized, or decomposed in different ways (Equation (99); Scheme 18) ⟨87JA2788, 89JA3336, 93JA10394⟩. It has been reported that in favorable cases two different alkynes can be coupled in a regioselective manner by insertion of one alkyne into the zirconacyclopropene intermediate obtained from hydrozirconation (using Schwartz's reagent, Cp_2ZrHCl) of a second alkyne. This process led to an isomeric mixture of zirconacyclopentadienes which could be equilibrated (at 80 °C) into a single isomer and decomposed by treatment with acid or iodine ⟨89JA2870⟩.

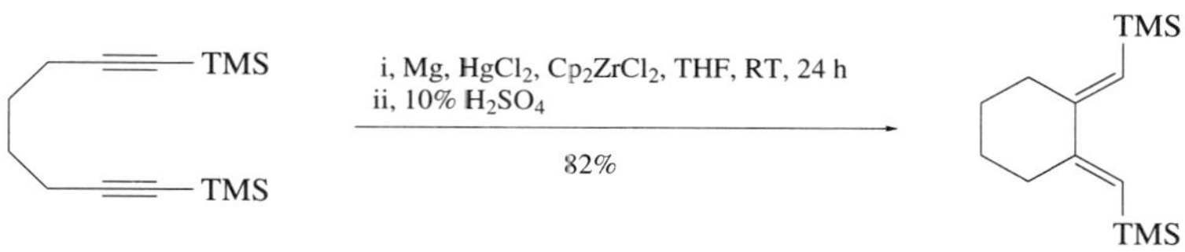

(99)

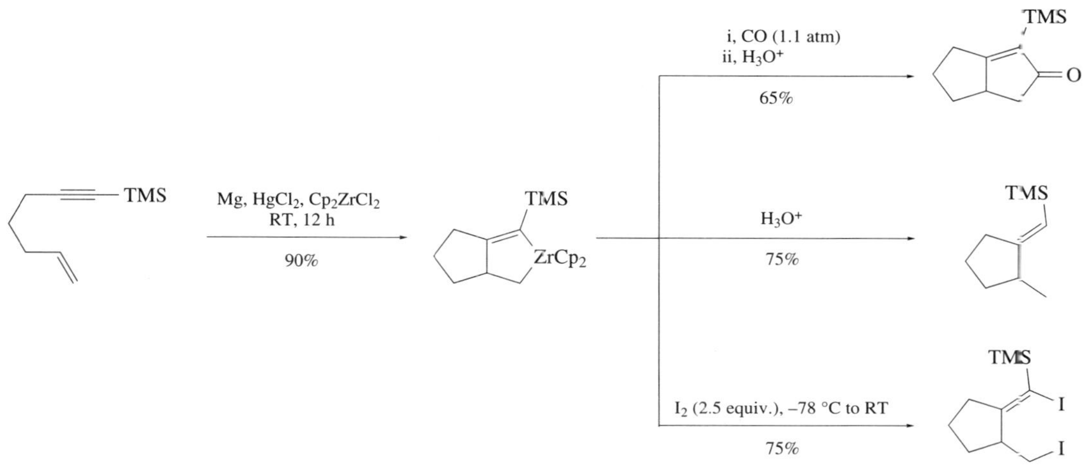

Scheme 18

Further developments in the early 1990s include the chemoselective step by step decomposition of five-membered zirconacarbocyclic intermediates ⟨92JOC3519, 93CC1042⟩, as well as the generation of novel zirconacarbocyclic and zirconaheterocyclic intermediates from benzynes, imines, hydrazones, nitriles, isocyanides, and aldehydes. These intermediates give rise to silyl substituted benzothiophenes ⟨89JOC2793⟩, vinylsilanes possessing allylic amino ⟨89JA4486, 91TL5465, 92SL143⟩, hydrazino ⟨88TL1811⟩, and alcohol ⟨93TL687⟩ functionalities and β-silyl enones ⟨93JOC5595⟩.

Low valent niobium and tantalum are also thought to form complexes with acetylenes (and silyl alkynes), possibly metallacyclopropenes, which can (i) be hydrolyzed ⟨90JOC1707, 90TL365, 90TL369, 92JOC1615⟩ to give alkenes or vinylsilanes; (ii) be reacted with aldehydes and ketones to give allyl alcohols in a highly regio- and stereoselective manner ⟨90JOC1707, 91JOC5980⟩; (iii) be combined with aldehydes in the presence of isonitriles to yield furans ⟨90JOC5310⟩; or (iv) give insertion products with isocyanates and acyl cyanides, the decomposition of which leads to β-silyl enamides or to silylalkenyl cyanohydrins ⟨92BCJ1543⟩.

Metal carbene complexes involving chromium, molybdenum, tungsten ⟨93SL363⟩, rhodium ⟨93JOC21⟩, and cobalt ⟨93JA9846⟩ on reaction with silyl alkynes can also produce various silylated cyclization products.

Finally, the Pd(0)-catalyzed cycloisomerizations of enynes containing silylalkyne moieties can provide elegant routes to polyfunctionalized *exo*-cyclic vinylsilanes ⟨91MI 218-01, 93TL4735⟩. *Exo*-alkylidene silacyclobutenyl species can be formed by reaction of dialkynylsilanes with cationic platinum hydride complexes ⟨89JA7615, 90JA7229, 92SL681⟩.

(v) Cycloadditions of alkynylsilanes

(a) Diels–Alder cycloadditions. Alkynylsilanes are poor dienophiles. However, in favorable cases [4 + 2] cycloadditions can give cyclic vinylsilanes in acceptable yields. Thus, while trimethyl-

silylethyne reacts poorly with cyclopentadiene even under forcing conditions, its trichlorosilyl analogue gives a high yield of the cycloadduct (70 °C/2 h/93%) ⟨71JOC929⟩. Also, in the case of intramolecular [4 + 2] cycloadditions, TMS–alkynyl termini were found to react efficiently with diene moieties (140–180 °C/60%) ⟨89JOC2271, 92JOC5279⟩.

(b) Other cycloadditions. Dichloroketene, when generated from trichloroacetyl chloride and activated zinc, was found to add cleanly to alkynylsilanes to form silyldichlorocyclobutenones ⟨83TL23, 93JA10097⟩ which could be hydrolyzed rapidly to the corresponding silylcyclobutenediones (Equations (100) and (101)) ⟨93JA10097⟩.

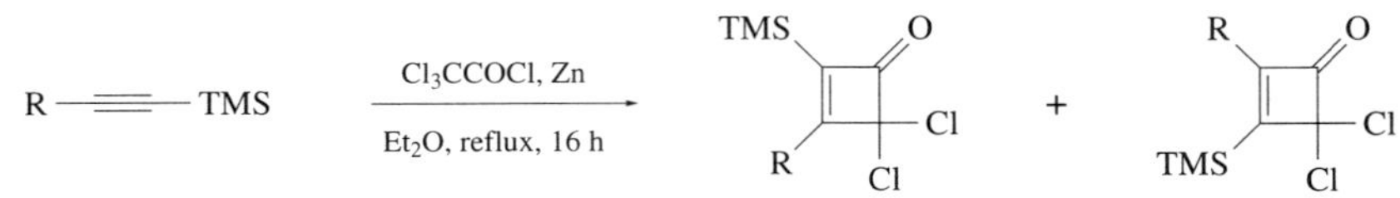

(100)

R = H	67%	1:99
R = Me	74%	64:36
R = Ph	93%	100:0
R = OEt	86%	100:0

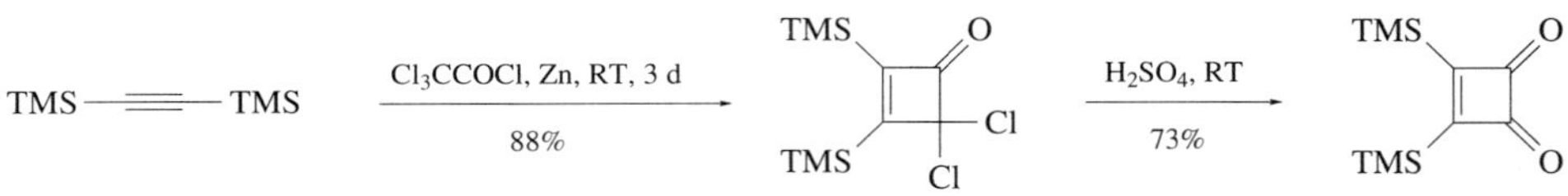

(101)

p-Tolyl-2-trimethylsilylethynyl sulfone proved to be an excellent acceptor of the 1,3-dipolar 2-diazopropane, the resulting 3,3-dimethyl-4-(*p*-tolylsulfonyl)-5-trimethylsilyl-3H-pyrazole ⟨90T1145⟩ gave the corresponding TMS-cyclopropene by light induced nitrogen extrusion in nearly quantitative yields (Scheme 19) ⟨87JOC4760⟩.

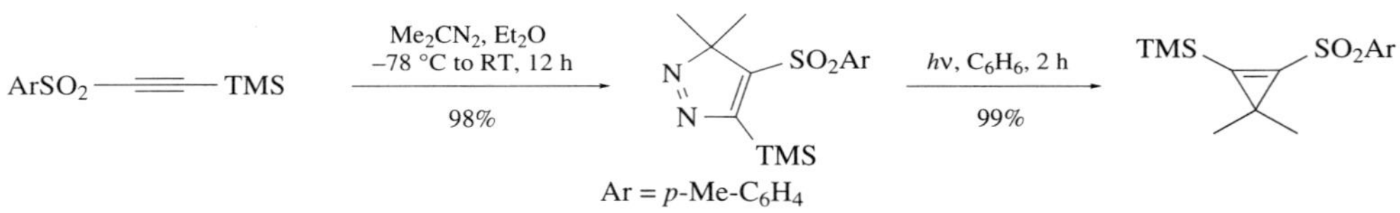

Scheme 19

Cyclization of alkynylsilanes with *o*-halobenzaldehydes could be effected under Pd(0) catalysis to give silyl indenones with complete regiocontrol (Equation (102)) ⟨93JOC4579⟩.

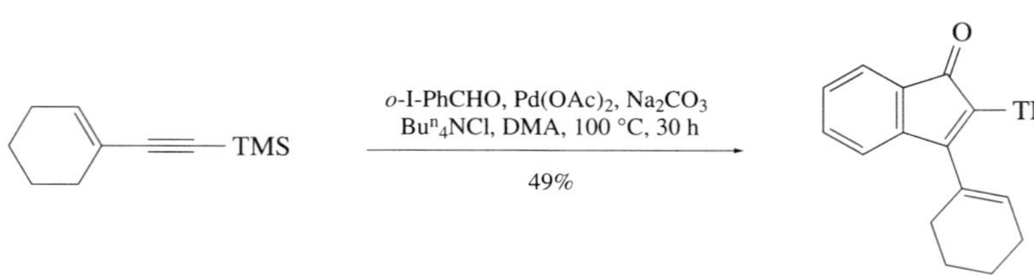

(102)

(vi) Hydrocyanation and hydroformylation

The Ni(0)-catalyzed hydrocyanation of alkynylsilanes ⟨87S1032⟩ (Equation (103)) and the Ni(0)- or Pd(0)-catalyzed silylcyanation of alkynes (see Equation (69)) give β-silylacrylonitriles having complementary stereochemistry arising from the predominant *syn* addition in both cases. Further reduction of the nitriles by diisobutylaluminum hydride (dibal-H)/ – 78 °C) ⟨87S1032⟩ affords β-silyl-α,β-unsaturated aldehydes.

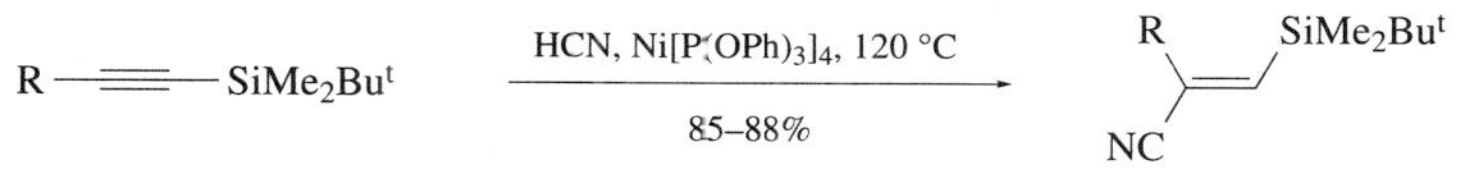

(103)

R = Me, Bu^n

Palladium-catalyzed hydroformylation of various alkynylsilanes has also been reported. When carried out in the presence of $SnCl_2$ in ethanol solvent, the reaction leads to (*E*)-*β*-ethoxycarbonylvinylsilanes in a highly regio- and stereoselective manner with excellent yields (Equation (104)) ⟨92CC1358, 93JCS(P1)1031⟩. In turn, this method appears complementary in regioselectivity to the hydroalumination (followed by carbonation and esterification) of the same alkynylsilanes (see Section 2.18.2.1.3(ii)(b)) ⟨91JOC2468⟩.

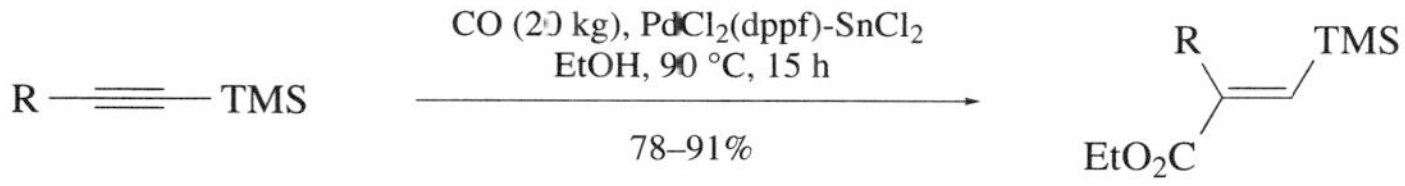

(104)

R = H, Bu^n, Oct, *c*-Hex, Ph, $Cl(CH_2)_3$

(vii) Halogenation and hydrohalogenation

Bromination of trimethylsilylethyne and of 2-bromoethynylsilane gives the di- and tribromovinylsilanes, respectively ⟨88TL1833⟩. On the other hand, controlled hydrobromination of 1-TMS-alkynes affords good yields of 1-TMS-2-bromoalkenes ⟨79JOC4781⟩.

2.18.2.1.4 *Vinylsilanes from allenes and ketenes*

(i) From allenes by silylmetallation

(a) Silyllithiation and silylmagnesiation. Although the mechanism has not been ascertained, the reaction of 1,2-cyclononadiene with TMS-Cl/Li gives 1,9-bis(TMS)-1-cyclononene in good yield ⟨78TL57⟩. The yield is much lower when Me_2HSiCl is used instead of TMS-Cl and the transformation seems to be limited to cyclononadiene as the cyclic allene component.

In the presence of transition metal catalysts, the same allene has been reported to undergo silylmetallations (metal = Mg, Zn, Al) with a regiochemistry highly dependent on both the silylmetal used and on the catalyst. The CuI-catalyzed silylmagnesiation appears to be best suited for the synthesis of cyclic and open chain 2-silyl-1-alkenes alkylated in the allylic position (Equation (105)) ⟨84TL1163⟩.

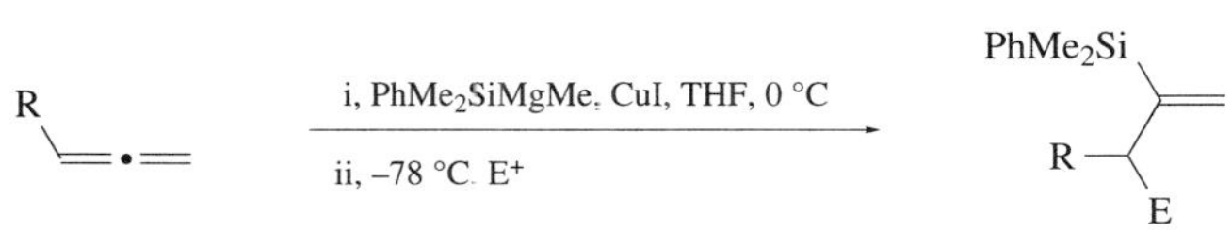

(105)

R = H, E = H, $H_2C{=}CHCH_2$, PhCH(OH), TMS 30–62%
R = $C_{10}H_{21}$, E = H 68%

(b) Silylcupration. As in the case of alkynes (Section 2.18.2.1.2(v)(a)), silylcuprate reagents appear to add readily to allenes; however, the regioselectivity of the addition strongly depends on the allenes, substitution pattern and on temperature.

Thus, at −78 °C, bis(phenyldimethylsilyl)cuprate reacted with allene and phenylallenes to give vinylsilanes. However, monoalkyl allenes gave mixtures of vinyl- and allylsilanes, and di- or trialkylallenes gave predominantly allylsilanes ⟨86CC1010, 88TL1825, 89T413⟩. At −78 °C the regiochemistry of addition to allene of bis(*t*-butyldiphenylsilyl)cuprate was reversed ⟨91JCS(P1)2811⟩.

Silylcupration of allenes giving vinylsilanes gains in flexibility from the possibility of scavenging the vinylsilane–allyl copper intermediates with a variety of electrophiles (Scheme 20) ⟨89T413⟩.

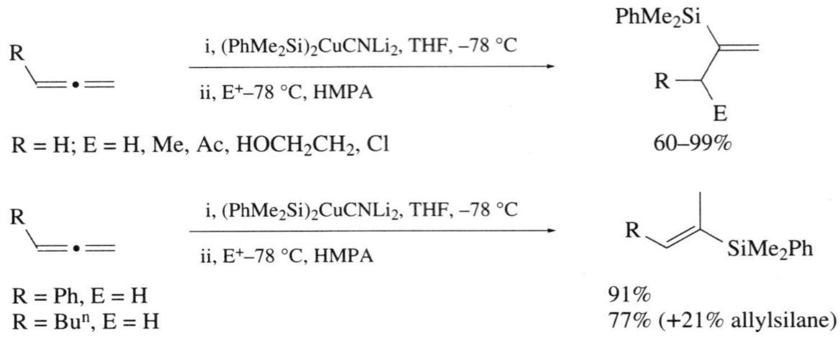

Scheme 20

Brief investigation of silymagnesiation and silylcupration of 1-phenylthio-1-TMS allene indicated that reactions similar to the above had occurred giving 1-phenylthio-1,2-bis(TMS) alkenes as products ⟨90BCJ51⟩.

It has also been shown that stannylcupration of allenes could be a valuable source of vinylstannanes ⟨90CC1030, 92JCS(P1)327⟩.

(c) Silylstannylation of allenes. The limited available data shows that this Pd(0)-catalyzed reaction of allenes takes place such that the silyl group is attached to the central carbon atom of the 1,2-diene fragment, leading to vinylsilanes in all cases. However, the site of attachment of the stannyl group depends on the allene's substituents and thus stereoisomeric mixtures of vinylsilanes–allylstannanes are usually obtained ⟨85CC354, 91TL1187⟩.

(ii) From allenylsilanes

While both the radical-induced and the Pd(0)-catalyzed additions of hydrostannanes and hydrogermanes to allenylsilanes result in regioisomeric product mixtures ⟨88BCJ2693⟩, the $TiCl_4$-mediated cationic [3 + 2] cyclization (called "TMS–cyclopentene annulation") leads to mono- and bicyclic TMS-cyclopentenes with remarkable regio- and stereoselectivity (Equation (106)) ⟨83T935⟩. This reaction has been extended with success to cyclizations involving tropylium ions ⟨89JA389⟩ (leading to azulenes; see Section 2.18.2.4.4) and benzyl cations ⟨91TL2327⟩ (leading to silylated dihydronaphthalenes through formal [3 + 3] cyclization).

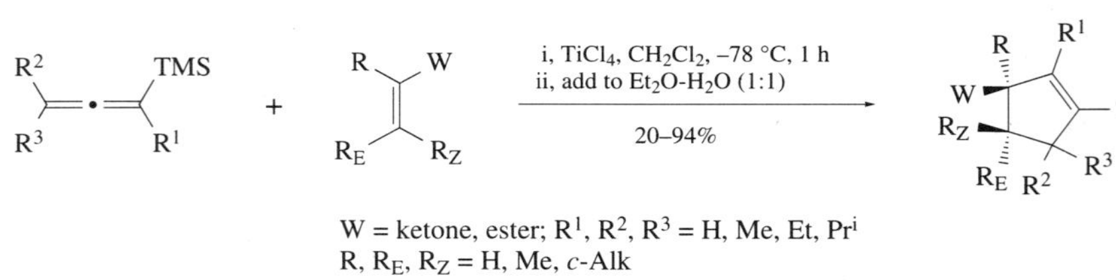

(106)

On the other hand, intramolecular cyclization producing TMS-dihydrofurans can be induced by the action of silver(I) ions on TMS-substituted α-allenic alcohols ⟨86JOC745⟩.

(iii) From ketenes

Some trimethylsilyl enol derivatives have been prepared either by addition of TMS–Li to ketenes ⟨90JOC3634, 90TL555⟩, or by reaction of alkoxyketene acetals with TMS–ketene (Equations (107) and (108)) ⟨90JOC4215⟩. The latter reaction, however, was unsuccessful with simple alkenes or silyl enol ethers.

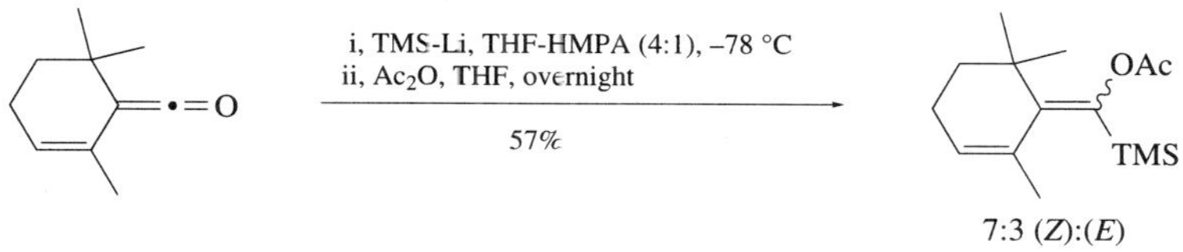

(107)

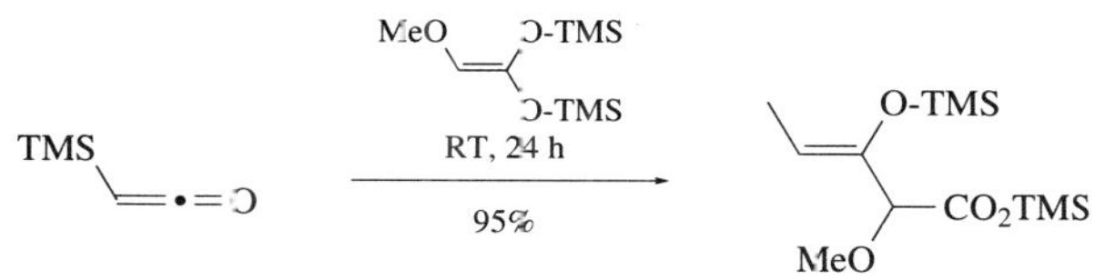

(108)

2.18.2.1.5 Alkenylsilanes from simple vinyl- or allylsilanes

More complex alkenylsilanes can be elaborated from relatively simple and readily available functionalized vinylsilanes ⟨63CJC2977, 77CB867, 78JOC2739, 79JOC4781, 81S999⟩. These include α- or β-bromovinylsilanes without or with an (*E* or *Z*) alkyl substituent in the β position; 1-silyl-2-stannyl- or 1,2-bis(silyl)ethenes; and vinyl- or allylsilanes bearing a leaving group in the allylic position.

Some of the most frequently used strategies as applied to these precursors will be illustrated in this section.

(i) Vinylsilanes as anion sources

Metallation (Equations (19)–(27)) or transmetallation (Equations (28)–(33)) of vinyl halides or stannanes can also be performed on α- and β-halovinyl- or α- and β-stannylvinylsilanes. The so-obtained silyl vinylmagnesiums ⟨63CJC2977, 73JA6152, 73JA6867, 78OS(58)152, 79TL1073, 82BSF395, 83HCA2377, 83JA4491, 83T925, 88TL6541, 92JOC5610, 92SL849⟩ and silyl vinyllithiums ⟨74AG(E)83, 74TL3365, 76JOC1480, 76JOM(107)C1, 76TL4439, 77CB867, 78JOC2739, 79JOC4781, 82TL3387, 83JA6344, 83TL2055, 84JA6105, 90TL2917, 92JOC3270, 92SL129, 93SL189⟩ have been reacted directly, or after transmetallation to give silyl vinylcopper(I) ⟨79TL1073, 83JA4491⟩, bis(silylvinyl)cuprate ⟨79JOC4781⟩ or silyl vinylcerium ⟨88TL6541⟩ reagents, with a variety of electrophiles leading to more elaborate alkenylsilanes (Schemes 21–24). A number of these structures can be synthesized by the methods outlined in the previous sections, while for others the reactions shown here represent unique methods of access.

(ii) Bromovinylsilanes as vinyl cation equivalents

A few examples of the well-known substitution reactions of alkenyl substrates with lower order cuprates ⟨92OR135⟩ have been reported to occur on bromovinylsilanes. Thus, the tetrahydropyranyl ether of (*E*)-1-bromo-1-trimethylsilyl-1-propen-3-ol reacts with the mixed cuprate (PhS)Bu^sCuLi to give, after Bu^s transfer, the corresponding substitution product with retention of double bond configuration ⟨83TL2055⟩. Similarly, 1,3-dibromo-1-trimethylsilyl-1-propene gives 5-trimethylsilyl-5-undecene on reaction with butylcuprate ⟨93SL189⟩. The substitution reaction also occurs with methyllithium in the presence of a catalytic amount of copper(I) iodide ⟨83TL2055⟩.

(iii) Transition metal catalyzed couplings

α- and β-bromovinylsilanes have served as building blocks for the syntheses of alkenylsilanes in Pd(0)-catalyzed coupling reactions. They have been reacted with alkyl ⟨79TL1073⟩ or aryl ⟨84TL83⟩ Grignard and zinc reagents, as well as with arylstannanes ⟨93SL771⟩. Conversely, β-bromomagnesiovinylsilane was coupled with vinyl bromides with high stereoselectivity (Schemes 25 and 26) ⟨88TL3705⟩. Most remarkable are the carbonylative cyclizations of a 1-iodo-1-silyl-1,4-diene leading to α-silyl cyclopentenones (Scheme 27; see also Scheme 16) ⟨83JA6761, 85JA8289⟩.

In a similar manner, 1-silyl-2-stannyl- as well as 1,2-bis(silyl)- ⟨89JOC268, 90TL2719, 91SL845⟩ ethenes have been coupled under Pd(0) catalysis to vinyl and aryl triflates ⟨84JA7500, 86JA3033, 87JA5478,

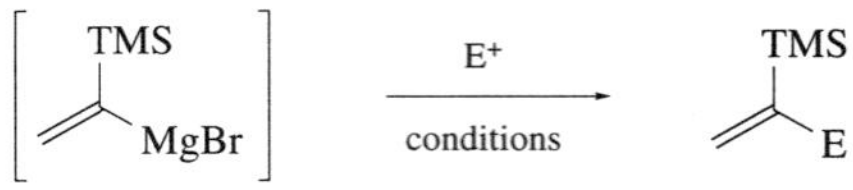

E$^+$	*Conditions*	*Yield (%)*
CO_2	THF	not given
MeCHO	THF	80
	THF, 0 °C to RT, 15 h	70
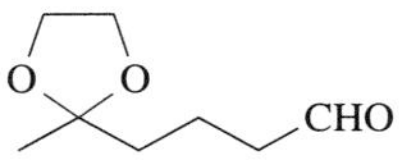	THF, –5 °C to reflux	>54
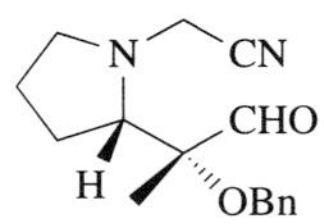	$CeCl_3$, THF, –78 °C	71
n-Octyl-I, (Br, OTs)	CuI, THF, 0 °C to RT	97 (91, 83)
PhSeBr	THF	51–60
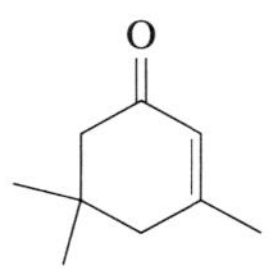	CuI, THF	80 (1,4-addition)

Scheme 21

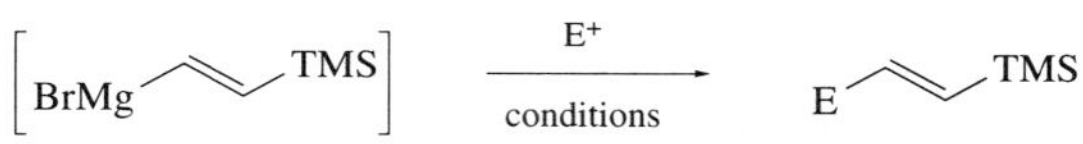

E$^+$	*Conditions*	*Yield (%)*
n-Octyl-I (Br, OTs)	CuI, THF, 0 °C to RT	77 (38, 40)
isophorone	CuI, THF	62 (1,4-addition)
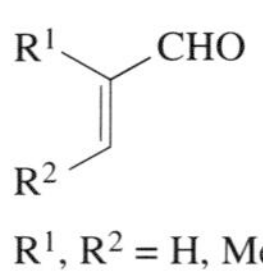 R^1, R^2 = H, Me, Ph, $(CH_2)_n$	THF, –30 °C to 0 °C	51–92 (1,2-addition)

Scheme 22

90TL2719⟩, to vinyl and arylhalides ⟨87JA813, 89JOC268⟩, to acid chlorides ⟨86TL2801⟩, and to an allyl bromide ⟨87JOC4868⟩ (Schemes 28, 29; Equation (109)). It is of interest to note that aryl iodides ⟨86JOC5286⟩, phenylpalladium acetate ⟨84TL5789, 87JOC1276⟩ and vinyl bromides ⟨90SL405⟩ can be coupled to simple otherwise unfunctionalized vinylsilanes (Scheme 30).

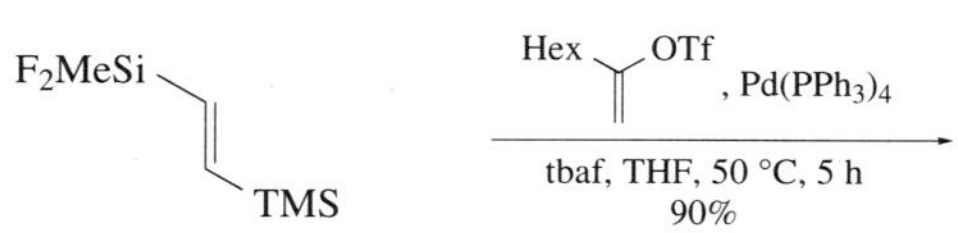

(109)

tbaf = tetra-*n*-butylammonium fluoride

(iv) Alkenylsilanes from vinyl- and allylsilanes

When the hydroxyl group of an allyl alcohol–vinylsilane ⟨76JOM(107)C1, 76TL4439, 92SL189⟩ (such as those of Schemes 21–24) or that of an allyl alcohol–allylsilane ⟨81TL3079, 91TL855⟩ is converted

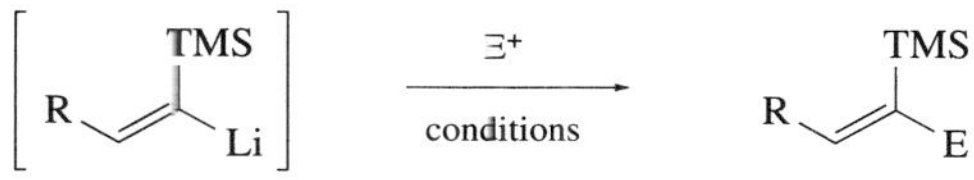

E⁺ (R = H)	*Conditions*	*Yield (%)*
R^1CHO	THF, –78 °C	75–80
$(CH_2)_5CO$, Ph_2CO	THF, –78 °C, 2 h	60, 69
n-Pent-I	THF, –78 °C, 1 h	83
Cyclohexenone	CuI (0.5 equiv.), Et_2O, –78 °C (1,4-addition)	71
		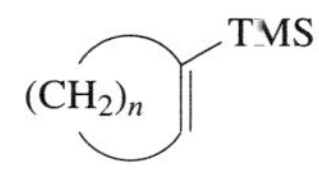
R = $(CH_2)_n$Br (*n* = 2, 3, 4)	Et_2O, –78 °C to RT	81, 84, 64
R = Pr^n; E^+ = cyclohexenone	CuI (0.5 equiv.), Et_2O, –78 °C (1,4-addition)	82

Scheme 23

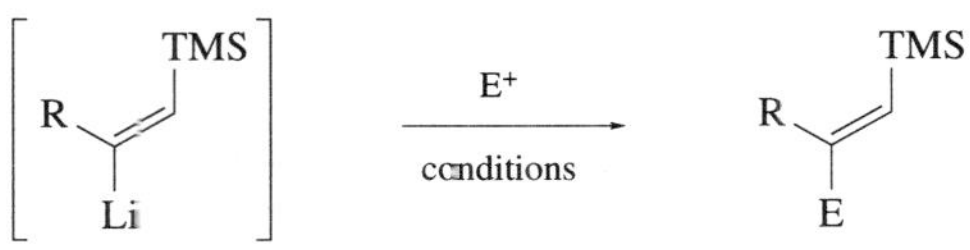

E⁺ (R = H)	*Conditions*	*Yield (%)*
CO_2, Bu^nBr, TMS-Cl	THF-Et_2O, –70 to –100 °C	86, 81, 84
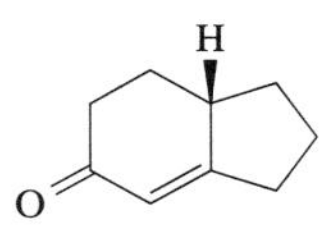	CuI (0.5 equiv.), THF, –78 °C	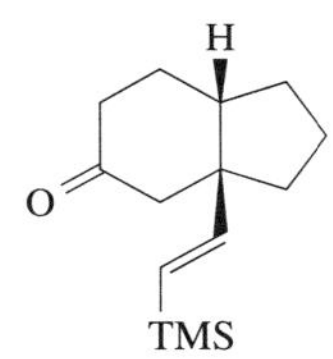 64
$(PhSe)_2$	THF, –78 °C to RT	64
R = Pr^n, E^+ = cyclohexenone	CuI (0.5 equiv.), THF, –78 °C (1,4-addition)	70

Scheme 24

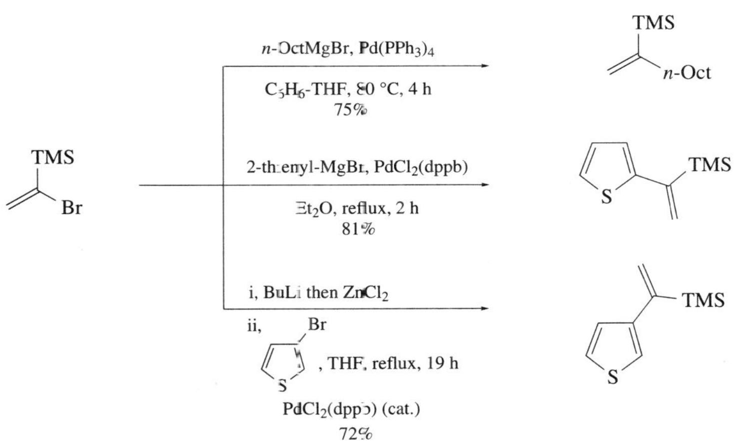

Scheme 25

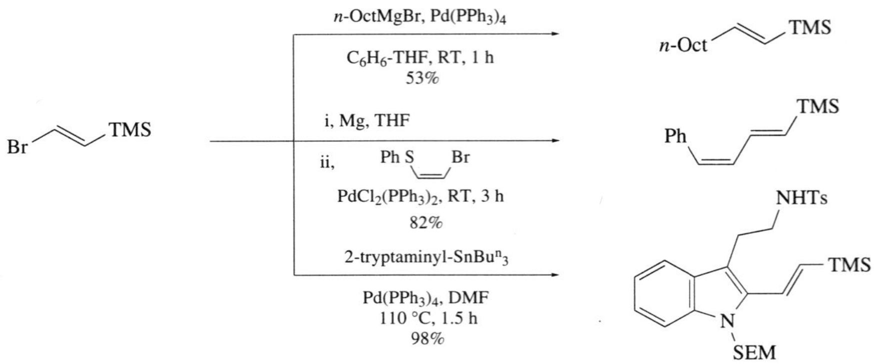

Scheme 26

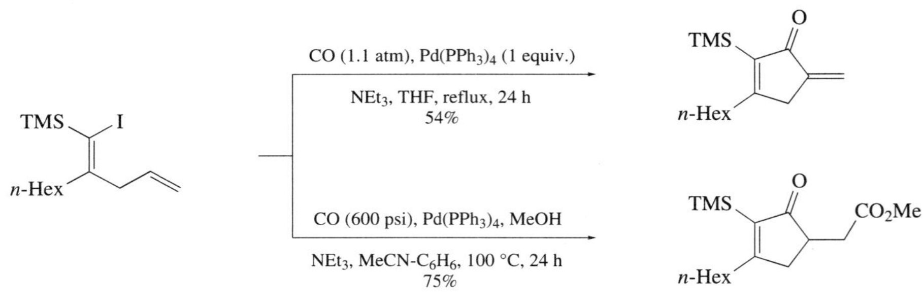

Scheme 27

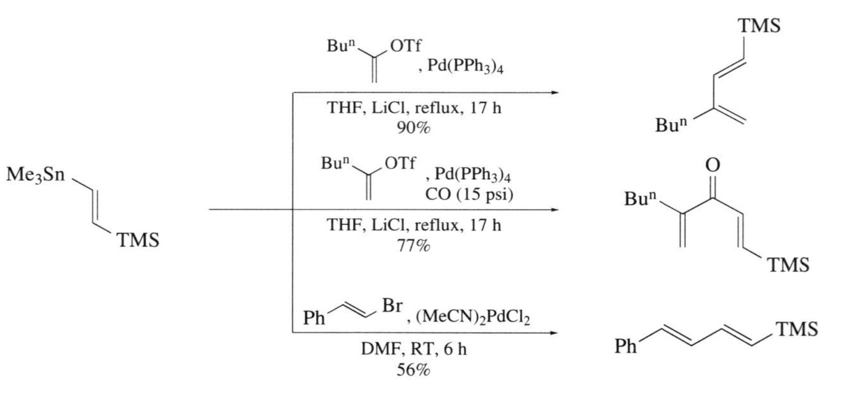

Scheme 28

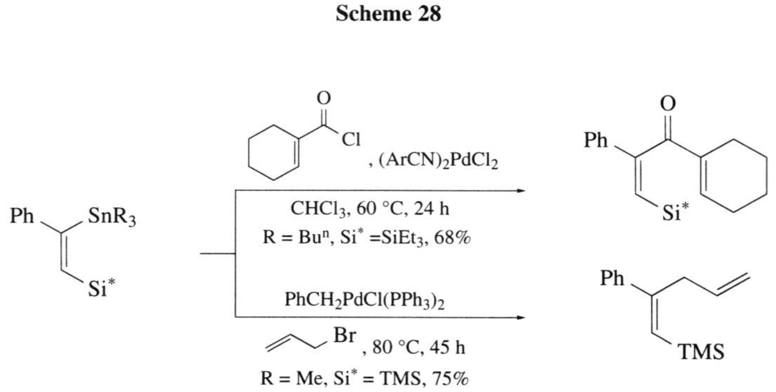

Scheme 29

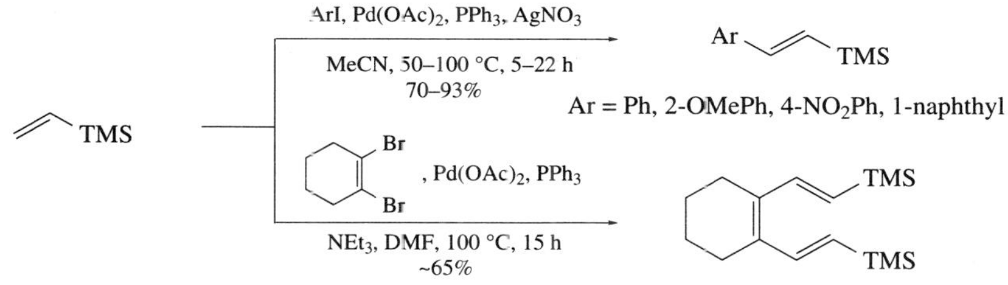

Scheme 30

into a better leaving group, displacement in an S_N2' type reaction by a suitable carbon nucleophile generates a new alkenylsilane (Scheme 31; (Equation (110)). However, with internal vinylsilane–allyl chlorides, S_N2 reaction is observed ⟨76TL4439⟩.

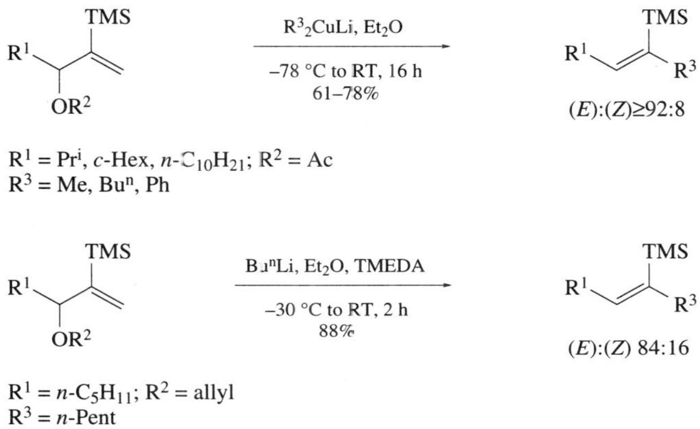

Scheme 31

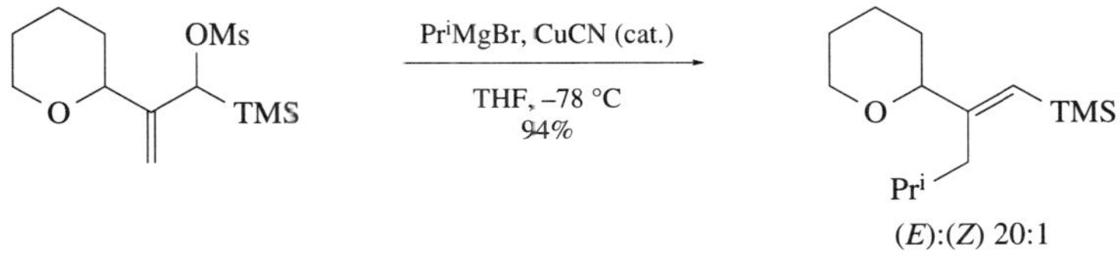

(110)

Optically pure allylsilane–allyl acetates ⟨90TA801⟩ undergo palladium-catalyzed [3,3] allylic rearrangement to give (*E*)-alkenylsilanes with complete chirality transfer (Equation (111)) ⟨90JOC5564⟩.

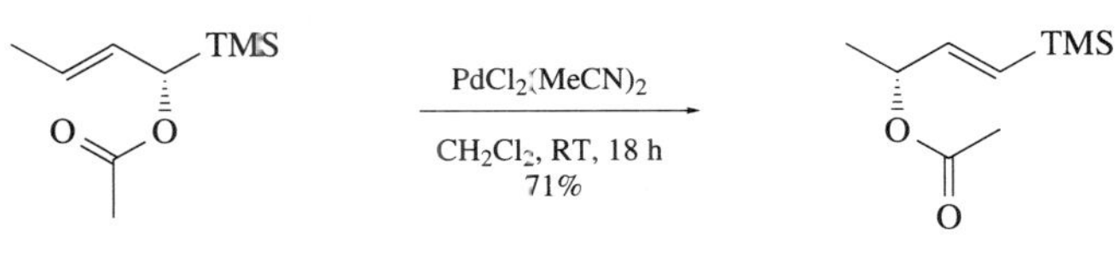

(111)

A great deal of work has been devoted to the study of the regioselectivity of reactions of various silicon-substituted allylmetals with electrophiles. It was found that the α/γ (with respect to silicon) orientation depended on the metal, the substitution pattern of the allylmetal, the nature of the electrophile and the reaction conditions. Thus, γ-products were given on reaction of silyl allyllithiums with alkyl halides when the silyl group was bulky ⟨84TL717, 85TL5375⟩, with aldehydes and ketones ⟨80JA5004⟩ and with trimethoxyborane ⟨81TL2751⟩. In the case of 1-silyl-1,3-bis(phenylthio)-3-propenyllithium ⟨92SL883⟩ γ-products were given with all electrophiles. α-Products predominated in the reactions of silylallylmagnesium bromide with carbonyl compounds, whereas silylallyl zinc and cadmium derivatives gave γ-products ⟨78TL2383⟩. Additions of silylallylcopper reagents to carbonyls gave γ-products ⟨81TL2985, 88T3997⟩, unsubstituted silylallyltitaniums led to α-products ⟨82TL4589,

86BCJ657⟩, while 1,3-disubstituted allyltitanium gave γ-products with aldehydes ⟨83CC921⟩. Vinylsilane products were also obtained from 2-silylpropenyl zinc ⟨84TL4383, 93SL277, 93SL279⟩ and tin ⟨93JA9305⟩ derivatives.

2.18.2.1.6 Alkenylsilanes from carbonyl compounds by silylalkenation

(i) Wittig-type alkenation

The reaction of the readily available silicon stabilized phosphorus ylides ⟨B-66MI 218-02, 75CB1732, 81TL4705, 86CB535, 93ZN719⟩, on reaction with carbonyl compounds generally does not lead to alkenylsilanes, due to the fact that the betaïne intermediate decomposes following the Peterson pathway rather than the Wittig pathway ⟨73JOC2664, 75CB1732, 81TL4705⟩. One vinylogous silyl substituted phosphorus ylide has been reported, however, to undergo Wittig alkenation with a series of aliphatic and aromatic aldehydes to give 1-silyl-1,3-dienes (Equation (112)) ⟨90TL543⟩.

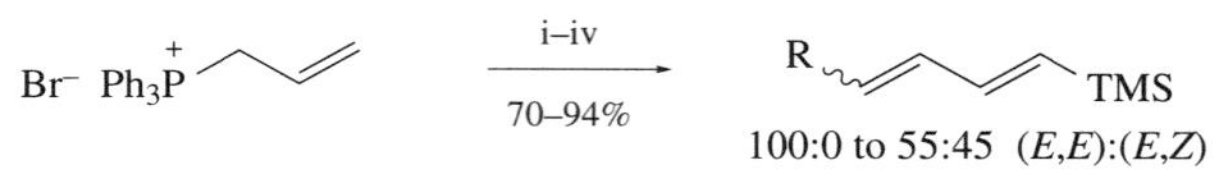

(112)

i, Bu^nLi, THF, 0 °C; ii, TMS-Cl; iii, –70 °C, Bu^nLi, 0 °C; iv, RCHO, 2–4 h

On the other hand, while the betaïnes formed from phosphorus ylides and aromatic acylsilanes ⟨90CSR147, 92OPP553⟩ usually undergo Brook rearrangement ⟨B-66MI 218-03⟩, those derived from aliphatic acylsilanes lead to alkenylsilanes in good yield and with high (*Z*) selectivity (Equation (113)) ⟨88TL2425, 88TL2777, 92TL6915⟩. Benzoyl silanes also gave α-silylstyrenes in high yield ⟨79JOC413⟩.

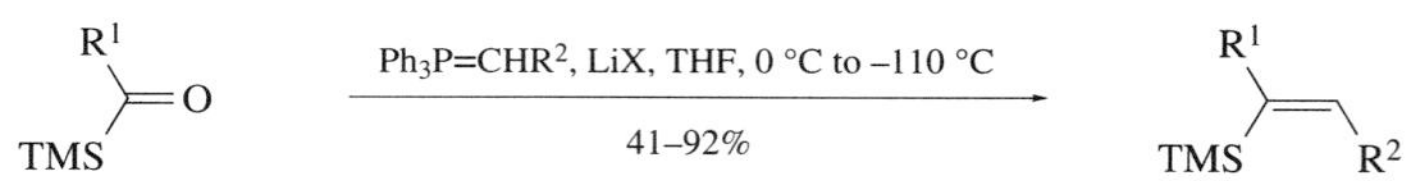

(113)

R^1 = Me, Bu^n, $Cl(CH_2)_4$, Pr^i; R^2 = Me, Pr^n

Although not a Wittig reaction, cyclization of ω-halogenoacylsilanes to give 2-silyldihydrofurans and pyrans is of interest (Equation (114)) ⟨92JOC7010⟩.

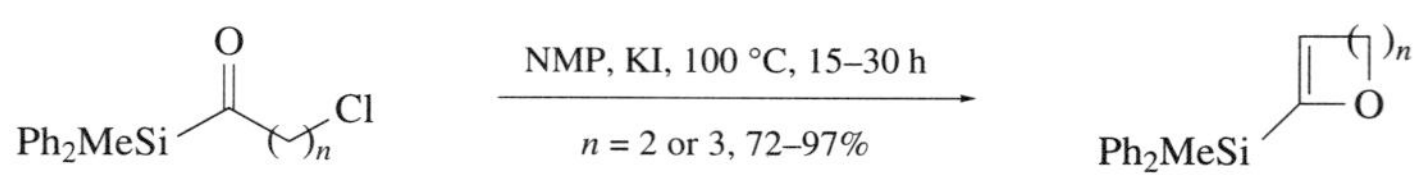

(114)

(ii) Peterson-type alkenation

Addition of reagents containing at least two silyl groups geminally linked to a carbanionic center to carbonyl compounds leads to alkenylsilanes ⟨73TL4193, 74AG(E)83, 76TL4041, 77CB852, 80JCS(P1)2485, 93SL863⟩ through a Peterson-type reaction ⟨84S384⟩. Because of steric hindrance by the bulky silyl groups, only nonenolizable aldehydes give high yields of alkenylsilanes (Scheme 32). Incorporation of heteroatomic or functionalized substituents into the starting α,α-bis(silyl)carbanion allows for the synthesis of the corresponding substituted alkenylsilanes ⟨76TL4041, 77CB852, 93SL863⟩.

(iii) Other alkenations

Reactions of heterosubstituted α-monosilyl carbanions with aldehydes and ketones can be conducted in such a way as to avoid Peterson alkenation and the resulting intermediates may be subjected to other types of alkenations to produce alkenylsilanes (Scheme 33) ⟨79TL485, 88S234⟩.

Thioacetals react with trimethylsilylmethylmagnesium chloride (or α-silyl thioacetals with Grignard reagents) under $NiCl_2(PPh_3)_2$ or $CpTi(CH_2TMS)_3$ catalysis to produce alkenyl-, dienyl- or trienylsilanes (Equation (115)) ⟨90JA9356⟩. Stoichiometric amounts of $Cp_2Ti(CH_2TMS)_2$ can be

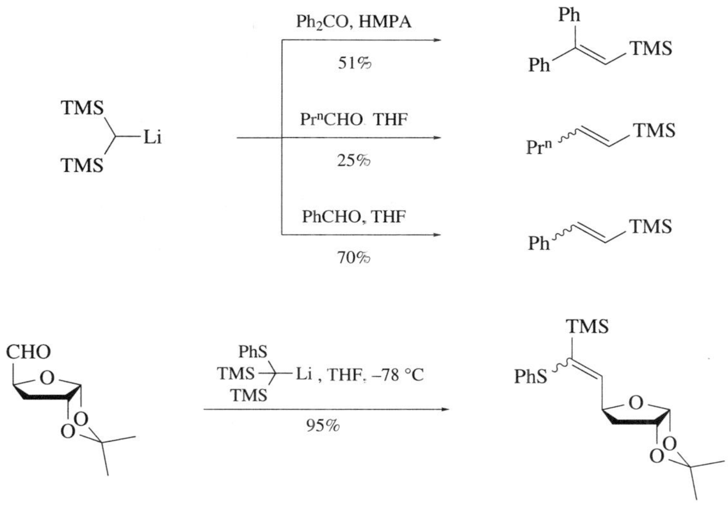

Scheme 32

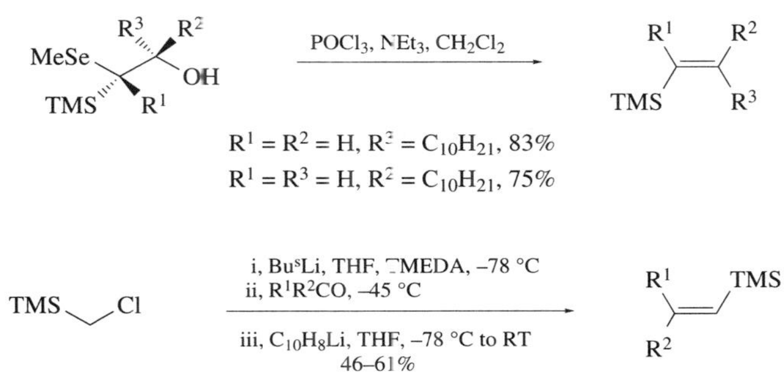

Scheme 33

used to transform aldehydes, ketones, esters, and lactones into alkenylsilanes in a remarkable silylalkenation reaction (Equation (116)) ⟨92SL665⟩.

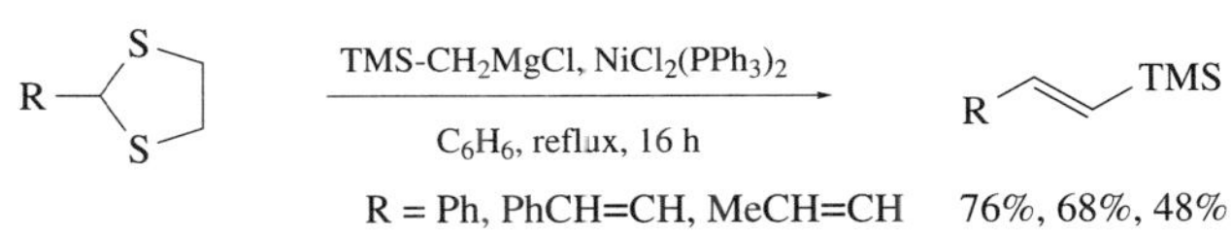

(115)

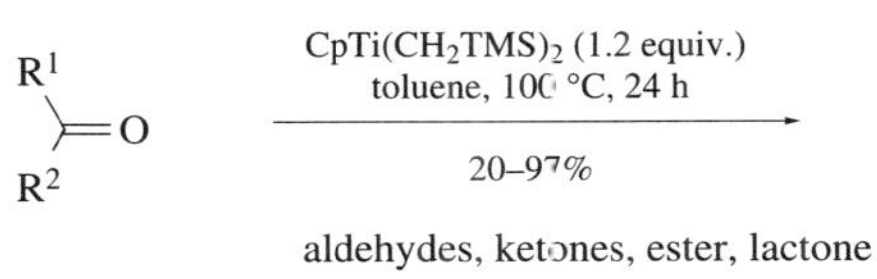

(116)

2.18.2.1.7 *Miscellaneous methods*

(i) Reduction of aromatic compounds

The electrochemical ⟨74JCS(P1)2055⟩, the Birch type ⟨75JCS(P1)470, 92OR1⟩ and the Benkeser type ⟨75JCS(P1)475, 92OR1⟩ reductions of aromatic silanes have been studied. Although a complex mixture of products is often produced, these methods do have preparative value in selected cases. For example, electrolysis of 1,3-bis(trimethylsilyl)benzene ($MeNH_2$, $LiCl/-50\,°C/4$ h) gave a 75% yield of 1,3-bis(trimethylsilyl)-1,4-cyclohexadiene ⟨74JCS(P1)2055⟩ and Birch reduction of the same arylsilane (Li/liq. NH_3/EtOH/-70°C) gave 2-trimethylsilyl-1,4-cyclohexadiene in 85% yield ⟨75JCS(P1)470⟩. Similarly, reductive silylation of TMS–benzene (TMS-Cl/Li/THF) gave 1,3,4,5,6-pentakis(trimethylsilyl)-1-cyclohexene quantitatively ⟨86T669⟩ and electrochemical reductive silylation of *o*-chlorotoluene (Al anode/TMS-Cl/4.4 F.M^{-1}/THF–HMPA) gave a mixture of two geometrical isomers of 1-methyl-2,5,6-tris(TMS)-1,3-cyclohexadiene in 76% yield ⟨92JOC4705⟩.

The great efficiency and high diastereoselectivity of the reductive alkylation of chiral *N*-carbalkoxy-3-silylpyridium salts with metal enolates ⟨93JOC5035⟩ is being exploited in asymmetric syntheses of piperidine alkaloids (Equation (117)) ⟨93JA8851⟩.

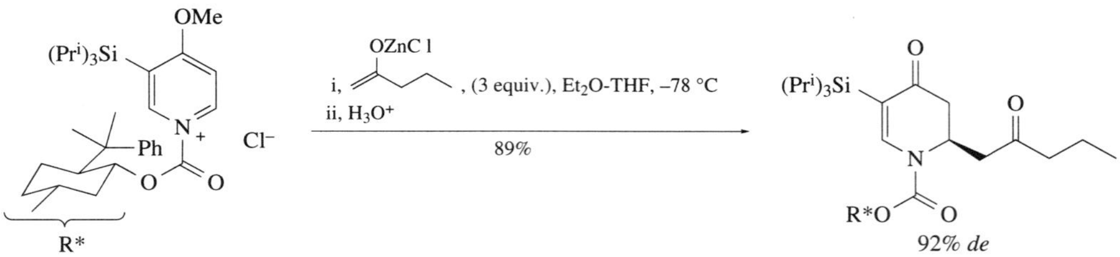

(117)

(ii) Diels–Alder cycloadditions

Diels–Alder reactions using alkynylsilanes as dienophiles have been mentioned (see Section 2.18.2.1.3(v)(a)) as possible routes to cycloalkenylsilanes. These compounds can also be formed from the same type of cyclization process when the silyl group is attached to the 2 and/or 3 positions of a 1,3-diene or heterodiene ⟨83T949, 85S899, 91TL2971⟩. Nevertheless, in spite of some remarkable results (Equation (118)) ⟨85JOC3421, 86TL2761, 87TL397, 88TL1833⟩, the internally substituted silyl dienes often show low Diels–Alder reactivity ⟨90JOC2282, 93JA6625, 93JOC967⟩ as well as low regioselectivity ⟨86JOC3435⟩.

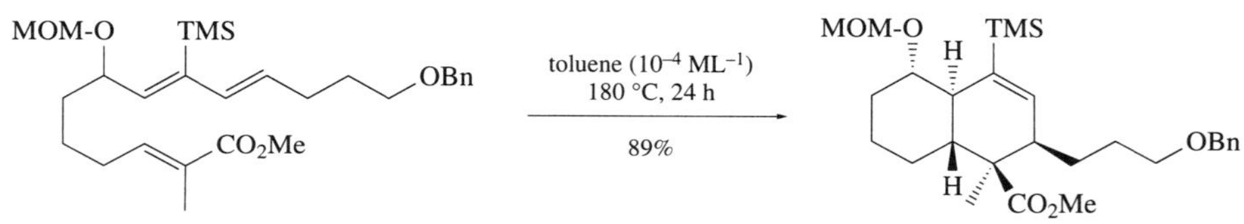

(118)

(iii) Metathesis and co-metathesis of vinylsilanes

Among the various possibilities, the most interesting combination is the co-metathesis of simple vinylsilanes with alkenes. While in some cases good yields in alkenylsilane and high (*E*) selectivities have been observed using $RuCl(CO)(PPh_3)_2$ ⟨91CC703⟩ and $RuCl_2(PPh_3)_2$ ⟨91JMOC113, 92JMOC307, 93PJC1361⟩ as catalysts, the co-metathesis reaction seems to be limited at present to terminal alkenes ⟨91JMOC113⟩.

(iv) Other methods

On treatment with *n*-butyllithium, silylpropargyl silyl ethers undergo O→C silyl migration (retro-Brook rearrangement) to give α-silylsilylpropargyl alcohols which can be transformed further with Lewis acids into α,β-bis(silyl)acroleins or methyl vinyl ketones (Equation (119)) ⟨91AG(E)1687, 93JOC2714⟩.

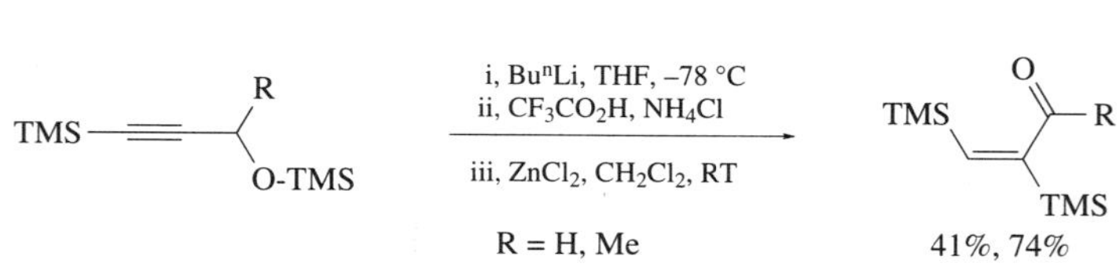

(119)

(*E*)- and/or (*Z*)-alkenylsilanes are obtained in good to high yield when silyl epoxides are treated with an excess of alkyllithium (Equation (120)) ⟨91TL3457⟩.

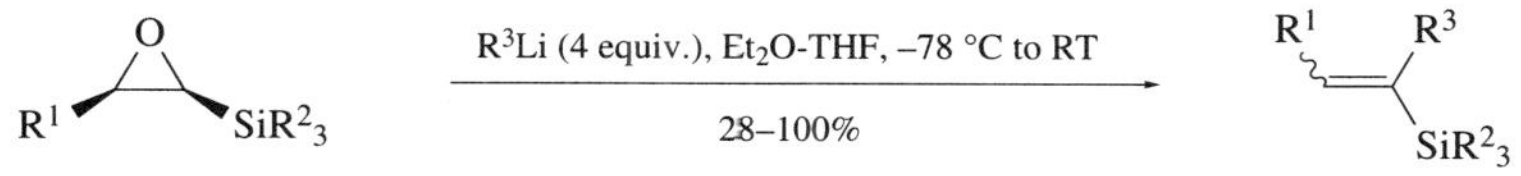

(120)

R^1 = Me, Pr^n, Pr^i, Bu^n; R^2 = Me, Et, Pr^i; R^3 = Me, Bu^n, Bu^t

In addition of palladium-catalyzed coupling reactions (Schemes 25–27), mixed silyl–manganese–magnesium reagents react with variously substituted alkenyl halides, sulfides, phosphates or triflates to give the corresponding alkenylsilanes in excellent yields ⟨88T4277⟩.

2.18.2.2 Conjugated Dienyl- and Enynylsilanes

This topic was thoroughly reviewed in the early 1990s ⟨93S349⟩. For this reason, only a few reactions will be considered here, which are interesting variants of those already reviewed, or which have appeared in the mid 1990s.

In an alternative Wittig-type reaction phenylthio(diphenylphosphinoxido) methyllithium reacted with *β*-silylacroleine to give a 3 : 1 mixture of (*E*,*E*) and (*E*,*Z*)-1-silyl-4-phenylthio-1,4-butadiene in 95% yield ⟨88TL6719⟩.

2-Bromo-1-alkenyldibromoboranes (obtained from terminal alkynes and BBr_3) were coupled to 1-silyl-1-ethenylzinc chloride under Pd catalysis, and the resulting silyldienylboranes were deboronated to produce 2-silyl-1,3-butadienes in good yield ⟨88TL1811⟩. Similarly, the 1-silyl-2-ethenyl zinc derivative could be transformed into 1-silyl-2-aminomethyl-1,3-butadiene (Scheme 14) ⟨89JOC1473⟩.

Besides disilanes and 1-silyl-2-stannyl ethenes ⟨93S349⟩, unsymmetrical 1,2-disilyl ethenes can also be coupled with vinyl iodides ⟨89JOC268⟩ or with vinyl triflates ⟨90TL2719⟩ in the presence of palladium catalysts to give 1-silyl-1,3-dienes.

In a remarkable three-component coupling reaction a terminal alkyne, vinylstannane, TMS-I, Pd(0) system produced the 1-silyl-1,3-diene with moderate stereoselectivity, and the terminal alkyne, 1-alkynylstannane, TMS-I, Pd(0) systems led to 1-silyl enynes in excellent yields and with high stereoselectivities ⟨91JA7778⟩. Interesting selectivities in the products have been observed in transition metal-catalyzed hydrosilylation of 1,4-bis(trimethylsilyl)butadiyne as a function of the catalyst and the hydrosilane reagent ⟨92BCJ1280⟩.

Finally, dienylsilanes have been obtained by the nickel-catalyzed cycloaddition of terminally silyl-substituted diynes with carbon dioxide ⟨90JOC2978⟩ and of a terminally 2-thienyl-substituted diyne with 1-trimethyl-2-diphenylhydrodisilane ⟨92JA5867⟩.

2.18.2.3 Cumulenylsilanes

2.18.2.3.1 Allenylsilanes

(i) Metallation of allenic and propargylic derivatives

Hydrogen–lithium exchange on allene and its alkylated derivatives ⟨B-87MI 218-01⟩ followed by silylation is a straightforward route to allenylsilanes (Equation (121)) ⟨81S875⟩. Similarly, alkoxyallenes ⟨80TL3987, 82JA1119, 83T949, 93S165⟩, silylallenes ⟨80TL3987, 92JOC794, 93JA3794⟩, and allenylphenylselenide ⟨92JOC5610⟩ have been lithiated and silylated, alkylated or boronated.

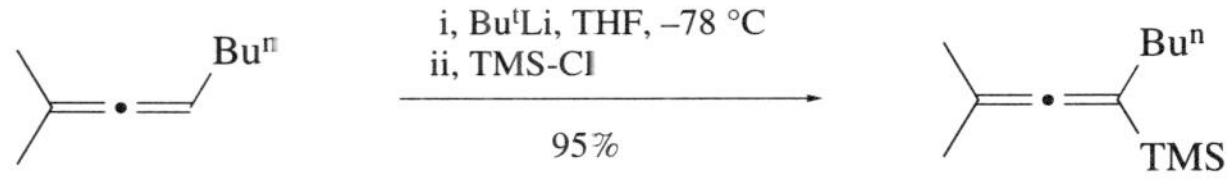

(121)

Propargyl ethers ⟨68RTC916, 71JOM(30)295, 71TL591, 71TL593, 73TL2585, 82JOM(232)C1⟩, propargyl sulfide ⟨90BCJ51⟩, and 1-silyl-1-alkynes ⟨83JOC5376⟩ have also been lithiated, zincated, or aluminated

⟨81TL1579⟩. Subsequent silylation in the first two cases gave the allenylsilanes, while in the latter case, the regiochemistry (allenic or propargylic product) depended on the metal and the electrophile ⟨81TL1579, 83JOC5376⟩.

(ii) From 1-silyl-1-alkynes

Compounds of this type possessing a leaving group in position 3 such as acetate, methanesulfinate, methanesulfonate (MsO) or tosylate (TsO) undergo 1,3-substitution reactions (S_N2') to produce allenylsilanes when treated with certain organocopper nucleophiles (Equation (122)) ⟨79S390, 80JOC4536, 80JOC4740, 80TL163, 83T935⟩.

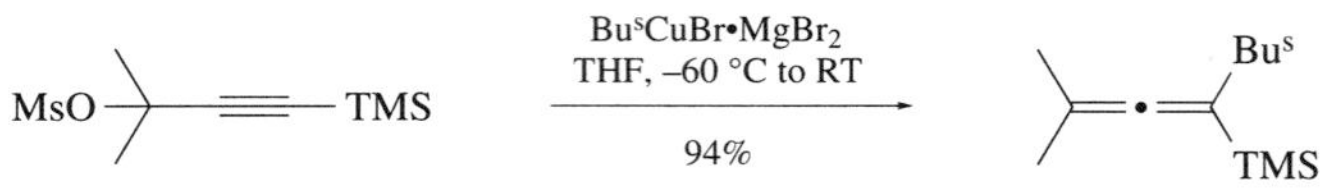

(122)

(iii) 1,6-Addition to enynes

Organocuprates ⟨90CB2173⟩ and organolithium/copper(I) 2-(dimethylaminomethyl)phenylthiolate reagents ⟨93JOC5849⟩ have been observed to add to conjugated (silylated) enynyl ketones and esters in a 1,6-manner leading to (silylated) allenic derivatives (Equation (123)).

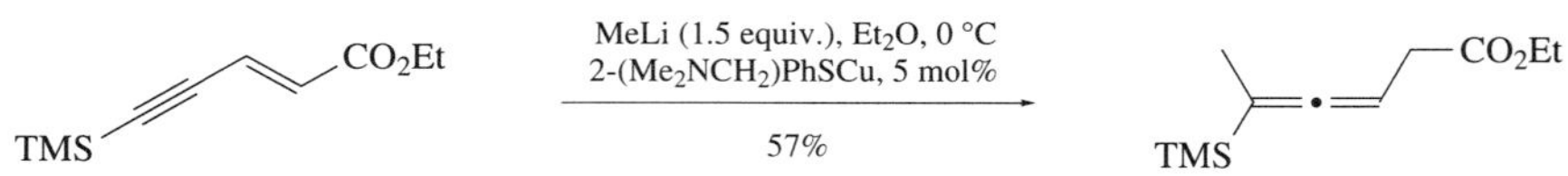

(123)

(iv) Miscellaneous reactions

At low temperature, Friedel–Crafts reaction of 1,3-bis(trimethylsilyl)propyne gave 1-TMS-1-acetylallene in good yield ⟨81TL3401⟩.

A number of conjugated enynes produced allenylsilanes or mixtures thereof when treated with TMS-Cl/Mg/HMPA ⟨75JOM(87)169, 81TL3179⟩, or with TMS-Cl/Li/Et_2O-THF/-20°C ⟨81TL3179⟩.

Silylcupration of a propargylic acetate ⟨81JCS(P1)2527⟩ and of propargyl phenyl sulfide ⟨92SL981⟩ gave allenylsilanes in variable to good yields.

1-Silyl-2-stannyl-3-hydroxy-1-alkenes were found to decompose into allenylsilanes when heated with an equimolar amount of Me_2SnBr_2 ⟨87JOC4868⟩.

Pentamethyl(3′-trimethylsilyloxy-1-alkynyl)disilanes were transformed into 1,1-bis(silyl)allenes on treatment with an excess of TMS-Cl in the presence of catalytic amounts of $FeCl_3$ or stoichiometric amounts of $TiCl_4$ ⟨90TL5607⟩.

Vacuum pyrolysis (retro-ene cleavage) of trimethylsilylpropargyl ethers led to allenylsilanes in good to high yields and free of acetylenic isomers ⟨88TL609⟩.

Under appropriate conditions the platinum- and/or rhodium-catalyzed hydrosilylation of 1,4-bis(trimethylsilyl)butadiyne gave a polysilylated allene in high yields and free of other isomers ⟨92BCJ1280⟩.

2.18.2.3.2 Cumulenylsilanes

Only a few examples of this class of compounds have been described. For example, treatment of 1-dimethylamino-4-methoxy-4-methyl-2-pentyne with two equivalents of butyllithium (THF/-30 to -100°C) followed by silylation at -30°C produced 1-dimethylamino-1-trimethylsilyl-4-methyl-1,2,3-pentatriene in 64% yield ⟨82JOM(233)C25⟩. 1,4-Bis(trimethylsilyl)butatriene and 1,1,4,4-tetrakis(trimethylsilyl)butatriene have been prepared by ruthenium-catalyzed dimerization of TMS–ethyne ⟨93BCJ987⟩ and by palladium-catalyzed double silylation of 1,4-bis(trimethylsilyl)butadiyne, respectively ⟨90BCJ3103⟩.

2.18.2.3.3 *Silyl ketenes*

Silyl ketenes appear to be the thermodynamic silylation products of lithium ynolates ⟨86JA7127⟩. Unsubstituted silyl ketenes can be prepared in good to excellent yields by thermolysis of 1-silyl-2-alkoxyethynes (Equation (124)) ⟨74JOC3607, 90JOC395, 90JOC1108⟩ or by dehydration of silylacetic acid (DCC/NEt_3/Et_2O/0 °C) ⟨89S568⟩.

(124)

Si* = TMS	R = Et	65%
$SiEt_3$, $SiBu^tMe_2$	R = Et	84%, 86%
TMS, $SiBu^tMe_2$, $SiBu^tPh_2$	R = Bu^t	63%, 100%, 91%

The reaction of iodotrimethylsilane with ethoxyalkynes such as 1-ethoxy-1-hexyne or 1-ethoxy-2-trimethylsilylethyne resulted in the formation of *n*-butyl trimethylsilylketene and bis(trimethylsilyl)ketene in 57 and 50% yield, respectively ⟨79S740⟩. The latter compound is also obtained in high yield by silylation of TMS–ketene enolate ⟨78JOC376⟩. Bromotrimethylsilylketene in hexane solution was prepared from TMS–ketene after bromination followed by dehydrobromination ⟨76TL1553⟩. Trimethylsilyl vinyl ketene can be produced from (*Z*)-2-TMS-2-butenoic acid in 39–50% overall yield ⟨80JOC4810⟩ and photolysis or thermolysis of 3,4-bis(silyl)cyclobut-3-ene-1,2-diones gives stable silyl-substituted bis(ketenes) ⟨93JA10097⟩.

In contrast with lithium ynolates for which *O*-silylation takes place under kinetic (i.e. low temperature) conditions ⟨86JA7127⟩, lithium thio- ⟨76CC1008⟩ and lithium selenoynolates ⟨79RTC55⟩ undergo kinetically controlled *C*-silylation thereby producing silylthio- or silylselenoketenes in high yield (Equation (125)).

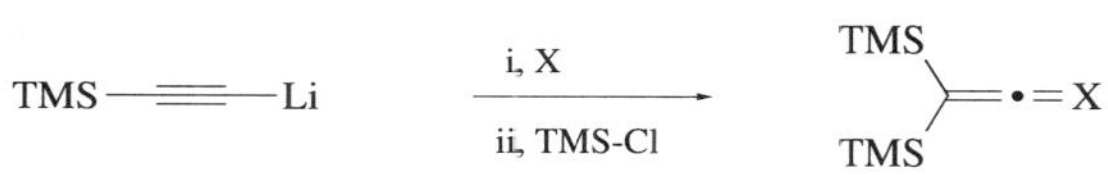

(125)

X = S	Et_2O, –45 °C	90%
X = Se	hexane-Et_2O, –60 °C	80%

Allyl substituted silylthio- and silylselenoketenes ⟨93T1189⟩ could be prepared conveniently by [3,3] rearrangements of trimethylsilylethynylallyl sulfides ⟨77TL4307⟩ and selenides (Equation (126)) ⟨80TL4251⟩.

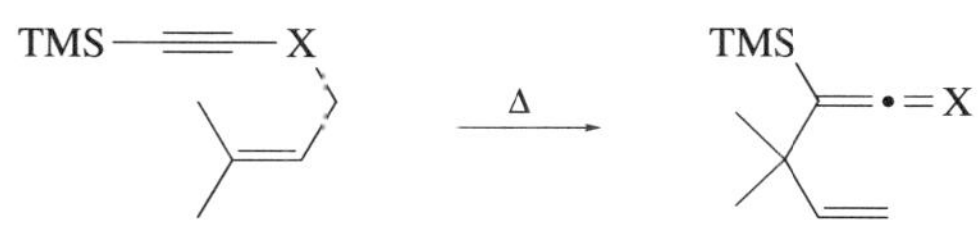

(126)

X = S	50 °C	80%
X = Se	50 °C, 5 h	46%

2.18.2.4 Arylsilanes

2.18.2.4.1 *By Würtz–Fittig type coupling*

A variety of simple arylsilanes have been prepared by heating a mixture of an aryl halide (chloride, bromide), chlorosilane and of a metal such as lithium, sodium, or magnesium, usually in stoichiometric amounts or with a small excess of the metal/chlorosilane. For example, TMS–benzene (90%) from chlorobenzene ⟨53JA2821⟩, TMS-*p*-toluene (87%) from *p*-chlorotoluene ⟨53JA2821⟩, tetra-*p*-biphenylylsilane (90%) from *p*-bromobiphenyl ⟨55JA6227⟩, 1,4-bis (TMS)benzene (84%) from 1,4-dichlorobenzene ⟨B-69MI 218-02⟩, 1,3,5-tris(TMS)benzene (74%) from 1,3,5-trichlorobenzene ⟨86T669⟩, have been obtained by this route. It was shown in the early 1990s that, even for preparative purposes, electrolysis can replace the metal in these reactions ⟨92JOC4705⟩.

2.18.2.4.2 *By silylation of arylmetals*

As in the preparation of vinylsilanes (Section 2.18.2.1.1), arylsilanes can be made by silylation of the corresponding aryllithiums or -magnesiums.

Simple examples include tetraphenylsilane (99%) ⟨46JA1675⟩, tetra-2-naphthylsilane (81%) ⟨50JA4280⟩, diphenyl-di-*p*-tolylsilane (78%) ⟨46JA1675⟩, 1,1-diphenyldibenzosilol (49%) ⟨55JA6380⟩, and dimethyl-bis(3,4-dimethylphenyl)silane (68%) ⟨73JOC4271⟩. Chlorosilanes are routinely used as silylating agents, but alkoxysilanes ⟨46JA1675⟩ and even hydrosilanes ⟨50JA4702, 57JA4560⟩ react efficiently with phenyllithium.

Access to the necessary aryl- and heteroarylmetals is provided mainly by hydrogen/metal and halogen/metal exchange reactions. While halogen/metal exchange is restricted by the availability of the appropriate aryl halide, the site selective or directed metallations exhibit a remarkable flexibility due to the continuous development of new directed metallation groups (DMGs) in both the homo- and heteroaromatic series ⟨78JOC731, 79OR(26)1, 83T2009, B-87MI 218-01, B-88MI 218-02, 90CRV879⟩. Since one can find in these reviews a great number of arylsilane syntheses by this route, only a few illustrative and newer examples are mentioned here.

1,3-Dichlorobenzene is metallated in position 2 ⟨88S803⟩, whereas proton abstraction on *t*-butyl (3-trifluoromethyl)phenylcarbamate ⟨87TL1851⟩ occurs in position 6 (Equations (127) and (128)). Silylation can also take place by O → C silyl migration (Equation (129)) ⟨76AG(E)428, 83TL4515, 87TL5965, 90TL4267⟩.

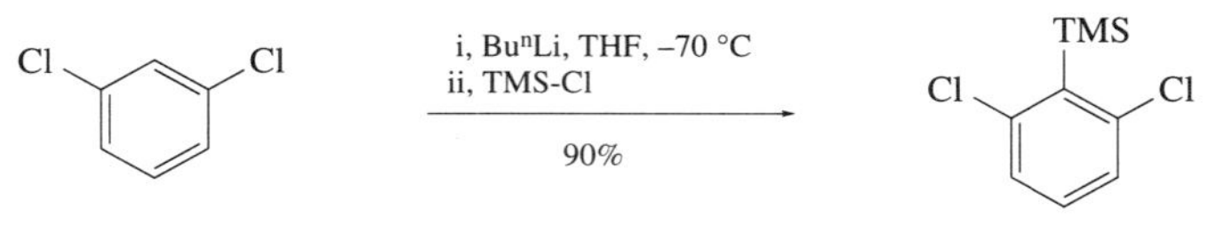

(127)

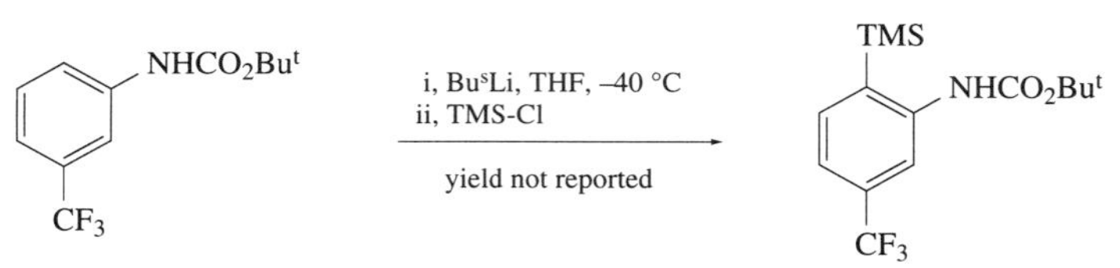

(128)

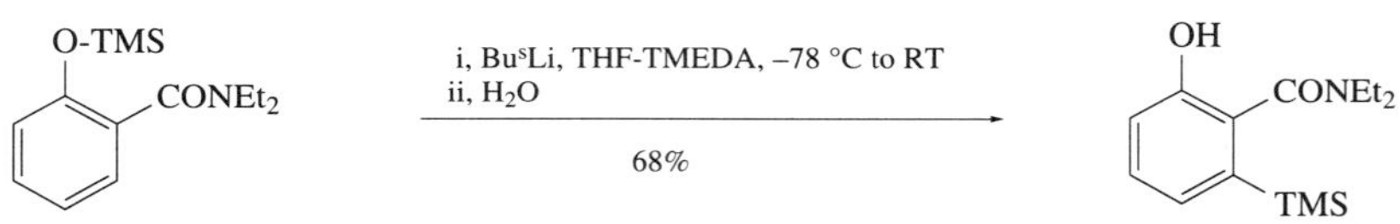

(129)

Chromium tricarbonyl arene complexes undergo asymmetric directed metallation with high selectivity in some cases (Equation (130)) ⟨93JOC6182⟩. Interesting transformations can occur when more than one DMG is present simultaneously ⟨90TL4267, 92SL112, 92SL360⟩.

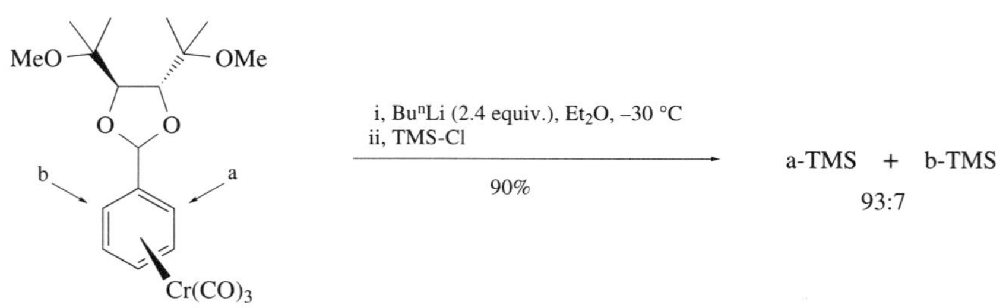

(130)

Considerable work has been devoted to directed metallation of heterocyclic compounds ⟨79OR1, 83T2009, 90CRV879⟩. While the heteroatom controlled direct metallation of pyridines often results in complex product mixtures ⟨83JOC4156, 83T2009, 84CC257⟩, the presence of an additional heteroatomic DMG leads to excellent selectivities (Scheme 34) ⟨90JOC292, 90TL4267, 92SL112⟩.

Metallation of five-membered ring heterocycles can be effected rather unambiguously in most cases ⟨79OR(26)1, 84CC258, 85S633, 85S945, 86S757, B-87MI 218-01, 92S954, 92SL327⟩, and the corresponding silanes are obtained in good to high yields. Although metallation of *N*-alkylpyrroles occurs at the α-position, a β-silylpyrrole was made from *N*-triisopropylsilylpyrrole (Scheme 35) ⟨90TL6785⟩.

2.18.2.4.3 By electrophilic substitution

Due to the greater ionic character of the carbon–silicon bond, as compared to that of the carbon–hydrogen bond, many electrophilic reagents preferentially attack arylsilanes at the *ipso* position

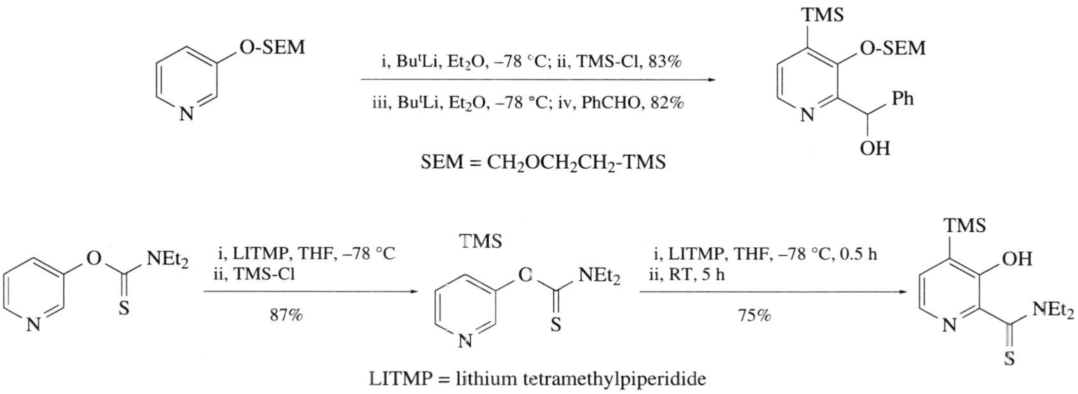

Scheme 34

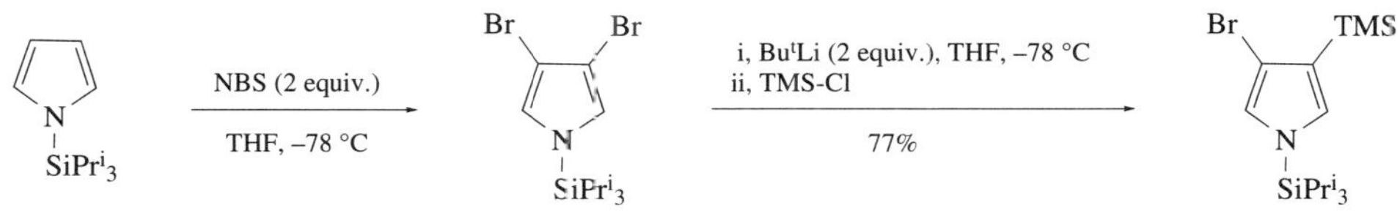

Scheme 35

⟨79S761, B-89MI 218-03⟩. This reactivity can be used to advantage for the preparation of substituted arylsilanes where aromatic silanes bearing two or more silyl groups are available from other routes. For example, selective monobromo- ⟨67CB2681, 82JA2613, 84JOC4687, 88JOC1815⟩, monoiodo- ⟨84JOC4687, 93S611⟩, monoprotio- ⟨85CR(301)1289⟩, and monoacetylative ⟨85CR(301)1289⟩ desilylations of bis(silyl)benzene and pyrazole derivatives have been described with the reactions proceeding in excellent yields.

When strongly *ortho-*, *para-*directing groups are present in addition to silyl groups, the silyl substituents may not be affected by the incoming electrophile, with hydrogen being displaced instead of silicon ⟨79S761, B-89MI 218-03, 93SL171⟩. Occasionally both types of substitutions can take place simultaneously ⟨88JOC1815⟩, and in rare cases silyl groups can be introduced into heteroaromatic rings by way of electrophilic substitution ⟨86T1299⟩.

2.18.2.4.4 By cycloaddition reactions

(i) [2 + 3] Dipolar cycloadditions

1,3-Dipoles such as diazoalkanes and nitrile oxides add to various silyl alkynes to form silyl pyrazoles ⟨67CB2681, 72CB470, 72CB1759⟩ and silyl isoxazoles (Equations (131) and (132)) ⟨74CB3717⟩. Diazomethane also adds to 1,2-bis(trimethylsilyl)ethenes to form pyrazolines which can be aromatized into bis(trimethylsilyl)pyrazines ⟨81CB2293⟩. Silyl-substituted isoxazoles, pyrazoles, and triazoles can be produced by stepwise cyclization of conjugated 1-silyl-1-alkynones with hydroxylamine ⟨79CB2829⟩, *N*-tosylhydrazine ⟨83T935⟩, and hydrazoic acid ⟨79CB2829⟩, respectively.

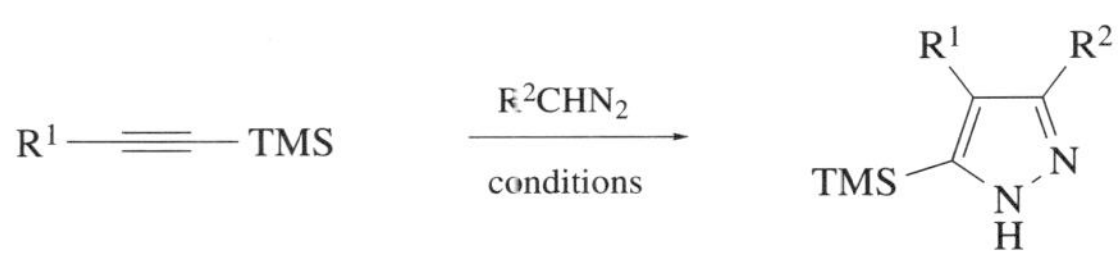

(131)

R^1 = Ph; R^2 = H	Et$_2$O, RT, 14 d	35%
R^1 = TMS; R^2 = CO$_2$Et	neat, 110 °C	92%
R^1 = TMS; R^2 = H	Et$_2$O, RT, 4 d	95%

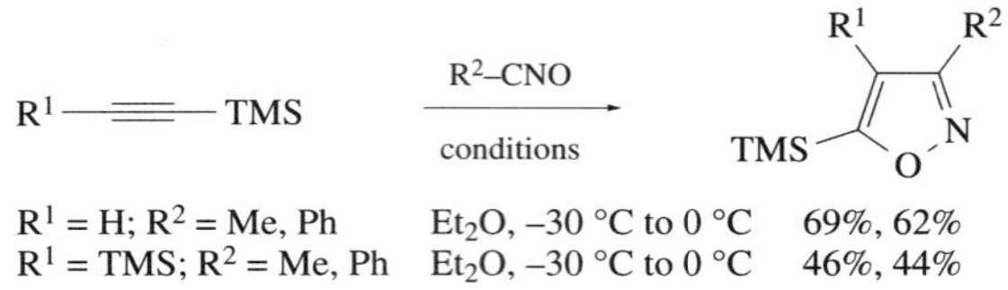

(132)

Similar polyazaheterocyclic silanes are also obtained on reaction of lithium trimethylsilyldiazomethanide with substituted ketenimines ⟨91S1163⟩.

(ii) Diels–Alder and other cyclizations

Silyl vinyl ketene and ketenimines undergo [4 + 2] cycloadditions to give silylated phenol ⟨80JOC4810⟩ and anilines ⟨87TL397⟩ respectively, in reasonable to high yields (Equation (133)).

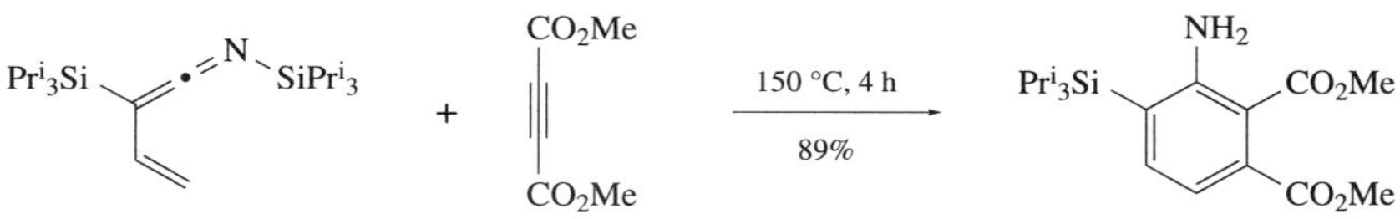

(133)

Diels–Alder addition of silylalkynes with dimethyl-1,2,4,5-tetrazine-3,6-dicarboxylate followed by retro-Diels–Alder nitrogen elimination leads to silyl pyridazines (Equation (134)) ⟨81CB3154⟩. [2 + 2 + 2] Cyclization of silyl pyridines with dimethyl acetylenedicarboxylate gives silyl quinolizines (Equation (135)) ⟨83CB2564⟩ and cationic cyclization of the tropylium ion with allenylsilanes produces silyl azulenes (Equation (136)) ⟨89JA389⟩.

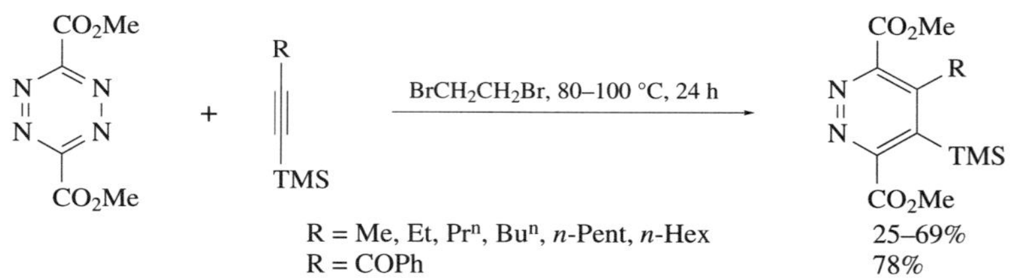

(134)

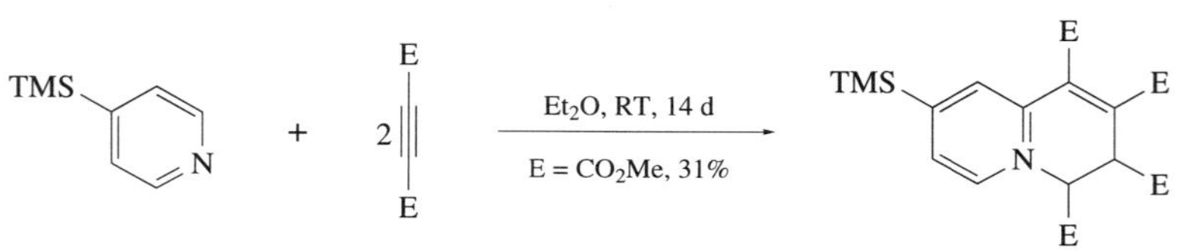

(135)

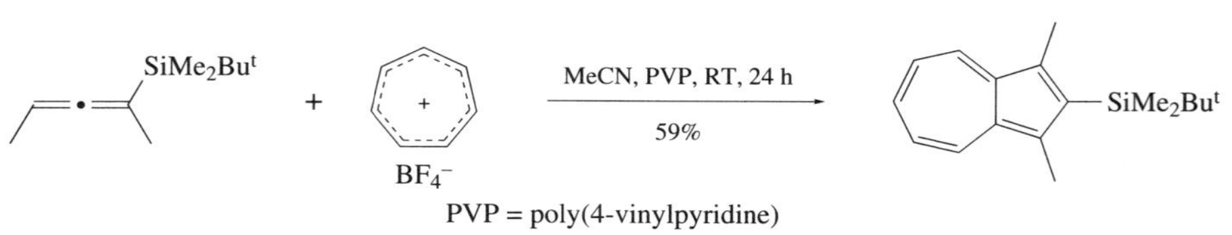

(136)

2.18.2.4.5 Transition metal-catalyzed reactions

The cobalt-catalyzed co-oligomerization of bis(trimethylsilyl)acetylene with 1,5-diynes ⟨75JA5600, 77JA5483, 84AG(E)539, 94SL92⟩ into bis(trimethylsilyl)benzocyclobutenes (Equation (137)) has been applied in various synthetic strategies ⟨85JA5670, 93JOC4298, 93S579⟩.

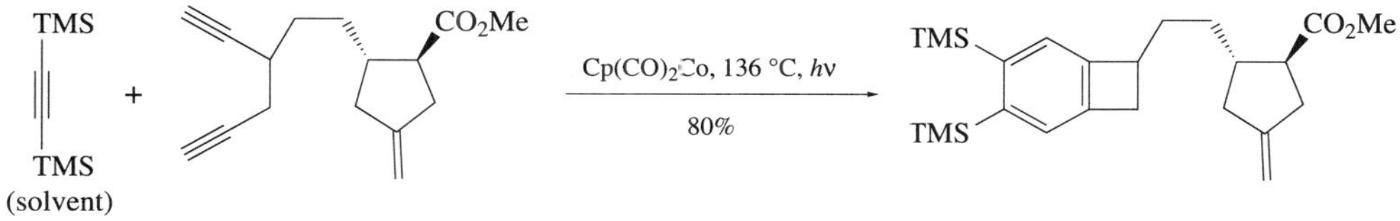

(137)

2-Trimethylsilylbenzothiophenes have been formed by coupling zirconocene complexes of benzynes with silylalkynes, followed by sulfur dichloride treatment of the intermediate zirconabenzocyclopentadienes (Equation (138)) ⟨89JOC2793⟩. Similarly, tantalum- ⟨90JOC5310⟩, rhodium- ⟨93JOC21⟩, and cobalt- ⟨93JA1586, 93JA9846⟩ mediated cyclizations of 1-silylalkynes to give silyl furans have been described.

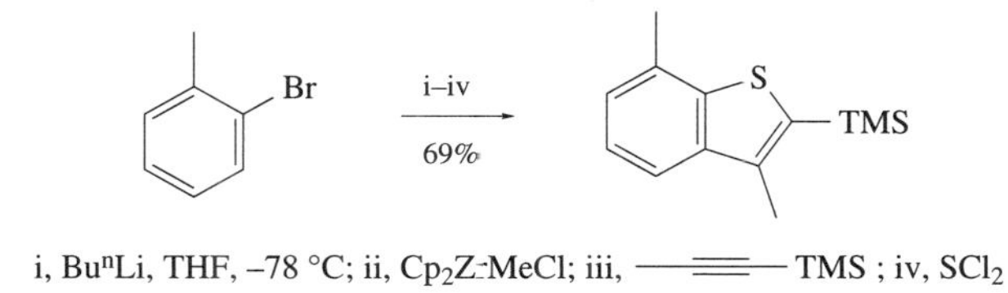

(138)

Benzenoid arylsilanes have been obtained by nickel, palladium, and platinum catalyzed coupling of various silicon reagents with aryl halides ⟨83TL4895, 91JOC3846, 91SL845, 93OM2065⟩, aromatic acid chlorides ⟨89JA5886, 89OM2609⟩, and aromatic hydrocarbons (Equations (139)–(141)) ⟨93OM2065⟩.

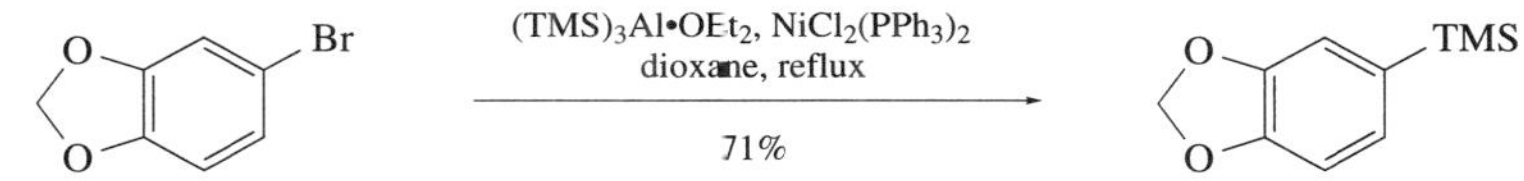

(139)

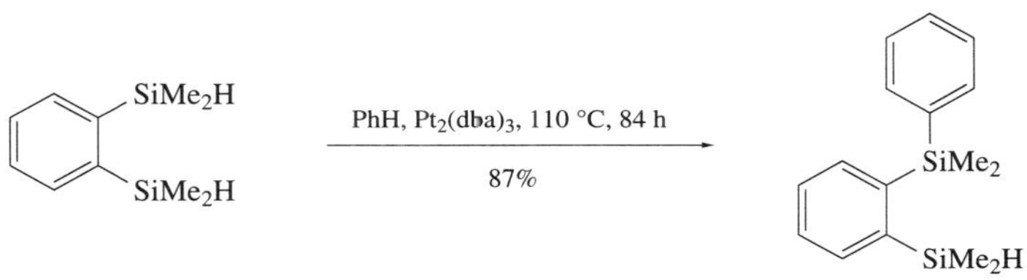

(140)

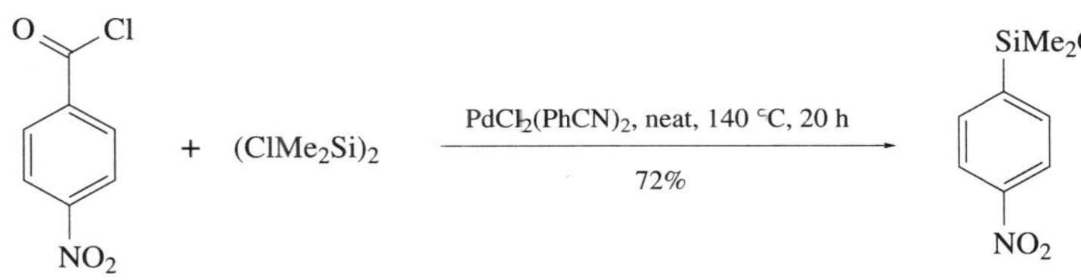

(141)

2.18.2.4.6 Miscellaneous reactions

Reductive silylation of aromatic compounds, as well as metallation–silylation of dihydroaromatics leads to silyl-dihydro derivatives which can be aromatized with cadmium chloride ⟨75JOC3097, 79JOC4275⟩, by hydride transfer ⟨78AG(E)278⟩ or air oxidation ⟨90TL6179⟩.

2-Alkyl-4-trimethylsilyl furans are obtained by cyclization–dehydration of the LDA-generated enolates of β-oxo-silyl epoxides ⟨84TL4383⟩.

Some silylmethyldialkylamines react with benzyne generated *in situ*, to give *ortho*-silyl-*N*-dialkyl anilines after addition and rearrangement ⟨76JOC3559⟩.

2.18.3 VINYL- AND ARYLGERMANIUM COMPOUNDS

2.18.3.1 Alkenylgermanes

In close analogy to the synthesis of alkenylsilanes, the most commonly used methods of preparation of alkenylgermanes are (i) from alkenyl metals, (ii) from alkynes, and (iii) from alkynylgermanes.

2.18.3.1.1 *From alkenylmetals*

Alkenylmagnesiums or -lithiums ⟨B-87MI 218-01⟩ prepared from the corresponding alkenyl halides react easily with trialkylgermanium halides (chlorides or bromides) to produce alkenylgermanes ⟨B-71MI 218-01, 79COC(2)1061⟩. Grignard reagents are known to give somewhat better yields ⟨B-71MI 218-01⟩. In the absence of alkenyl halides, other sources of alkenyllithiums can be used with comparable efficiency. For example, lithiation of (*E*)-1-methoxy-1,3-butadiene with Bu^tLi followed by reaction with chlorodimethylgermane leads to the corresponding dienylgermane in good yield (Equation (142)) ⟨80JOC541⟩.

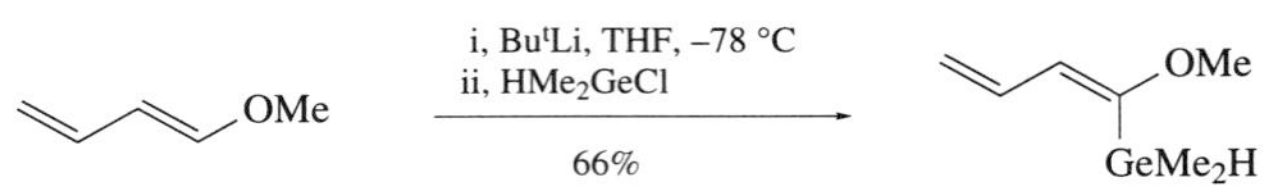

(142)

Alkenyllithiums generated from *N*-tosyl hydrazones with *n*-butyllithium (Equation (143)) ⟨77TL159⟩ or from 1-silyl-2-stannyl alkenes by transmetallation (Equation (144)) ⟨87JOC4868⟩ react with Me_3GeCl to produce alkenylgermanes in reasonable to good yields (see also Equations (28)–(38)).

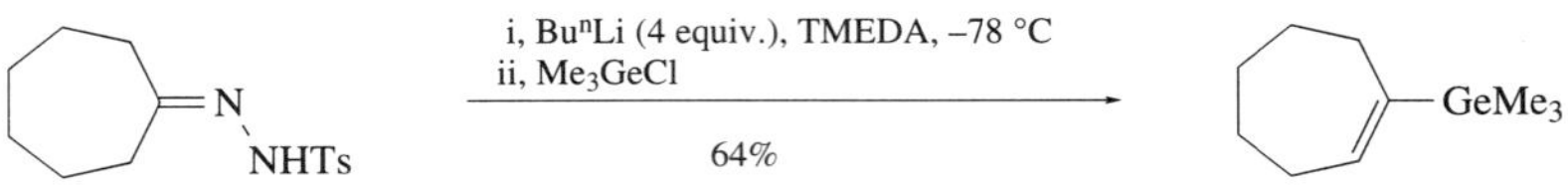

(143)

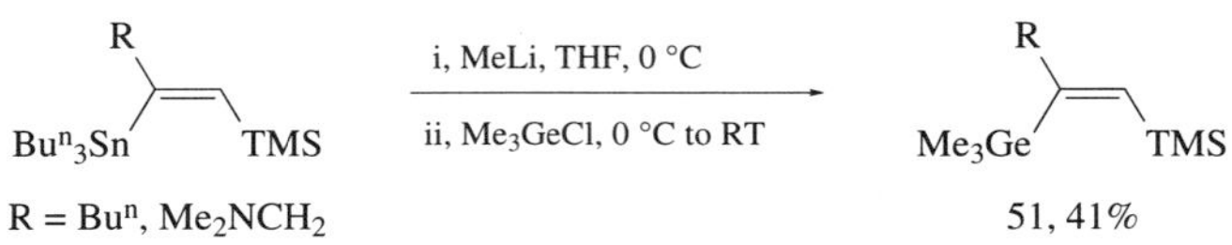

(144)

2.18.3.1.2 *From alkynes*

In close analogy with alkenylsilanes (see Section 2.18.2.1.2), alkynes represent a major source of alkenylgermanes. However, considerably less effort has been invested in the synthesis of alkenylgermanes and thus little information has appeared concerning hydrogermylation, carbogermylation, metallogermylation or double germylation of alkynes. By contrast, the reaction of diorganylgermylenes with alkynes has emerged as a novel method of increasing applicability.

(i) Addition of germyl reagents

Radical addition of hydrogermanes to various unsaturated organic substrates has been described ⟨82COMC-I(2)399⟩. In the presence of radical initiators such as 2,2′-azobisisobutyronitrile (AIBN) or Et_3B, predominant *trans* addition to alkynes leads to (*Z*)-alkenylgermanes. However, using Et_3B, the stereochemistry of the product is greatly affected by the reaction temperature and the nature of the hydrogermane ⟨87TL3709⟩.

Radical reaction of 2-benzothiazolylpropargyl sulfide with Ph_3GeH gives triphenylallenylgermane in high yield ⟨86S776⟩.

Transition metal catalyzed hydrogermylation of alkynes also proceeds with excellent yields, giving (*E*)-alkenylgermanes predominantly ⟨72JOM(40)73, 84TL3221, 87BCJ3468⟩. As expected, the nature of the catalyst strongly influences both the regio- and the stereochemistry of the reaction; among those tried $Pd(PPh_3)_4$ gives the highest (*E*) selectivity ⟨87BCJ3468⟩. However, for several terminal alkynes, even with this catalyst, formation of the 2-germyl-1-alkene regioisomer cannot be suppressed. Hydrogermylation of allenes gives mixtures of allyl- and alkenylgermanes ⟨88BCJ2693⟩.

The sole example of carbogermylation reported is the palladium- or nickel-catalyzed cyanogermylation of terminal alkynes and arylalkynes ⟨90JOC3393⟩, which occurs in the same way as the addition of TMS-CN ⟨88JOC3539⟩ (see Equation (69)), producing (*Z*)-β-germyl-α-alkyl/aryl acrylonitriles in high yield. In contrast to silylcupration (Section 2.18.2.1.2(v)(a); Schemes 13 and 14; Equations (62) and (63)), germylcupration of terminal alkynes is only effective when a proton donor (Bu^nOH or Bu^nCHO) is present in the reaction mixture. Using $(Et_3Ge)_2Cu(SMe_2)Li$ as the germylcupration reagent, it is possible to capture the vinylcopper intermediates with electrophiles other than protons (methyl iodide, allyl bromide) ⟨84TL3217⟩. In all these reactions, the nature of the copper(I) salt and its ligands strongly influences the regioisomeric ratio of the product alkenylgermanes.

Palladium(0)-catalyzed stannylgermylation of conjugated acetylenic esters affords mixtures of easily separated regio- and stereoisomers in which the (*E*)-α-stannyl-β-germylalkenyl esters predominate ⟨87CC1025⟩.

When tetrakis(2,6-diethylphenyl)digermene is refluxed in benzene solution containing a large excess of phenylacetylene the corresponding 1,2-digermacyclobut-3-ene is formed in a [2 + 2] cycloaddition type reaction ⟨88TL3383⟩.

(ii) Addition of germylenes to alkynes

The heavy carbene analogues germylenes ⟨73JOM(56)1, 91CRV311, B-91MI 218-01⟩ are best generated by thermal or photochemical decomposition of 7-germabenzonorbornadienes ⟨83JA897, 83TL4083, 85TL4523, 86TL3251⟩, which in turn are obtained in good yield from substituted germoles and benzyne ⟨83JA897⟩.

The first germirenes have been isolated from the reaction of a thiacycloheptyne with 7-germabenzonorbornadienes (Scheme 36) ⟨83TL4083⟩. Other alkynes do not lead to germirenes; rather, in the absence ⟨88TL5245⟩ or in the presence of $Pd(PPh_3)_4$ catalyst ⟨91SL431⟩, substituted germacyclopentadienes (germoles) or 1,4-digermacyclohexadienes are formed (Scheme 37). Under Pd(0) catalysis, conjugated alkynones lead to oxagermamethylenecyclopentenes which on treatment with acetic acid give β-germylalkenones of (*Z*) configuration ⟨91SL431⟩. In the presence of proton sources germylenes add to terminal alkynes as well as to cyanoallenes to form substituted germyl alkenes (Equations (145) and (146)) ⟨90SL113⟩.

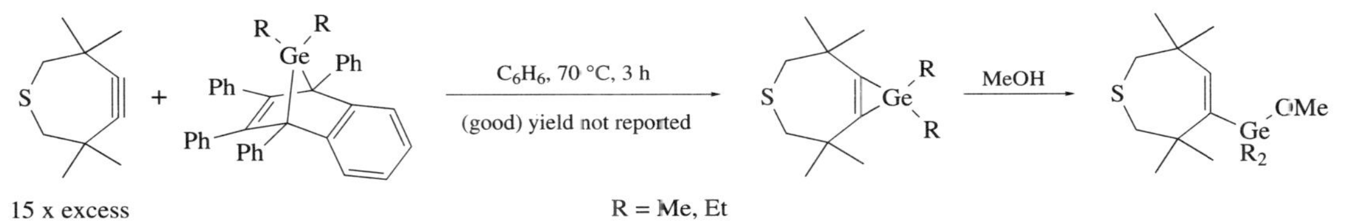

Scheme 36

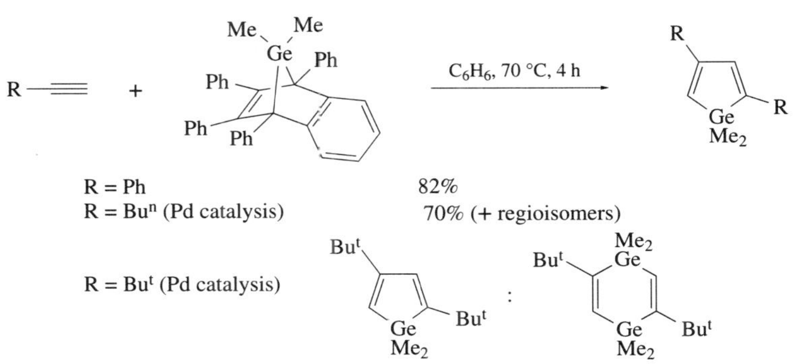

Scheme 37

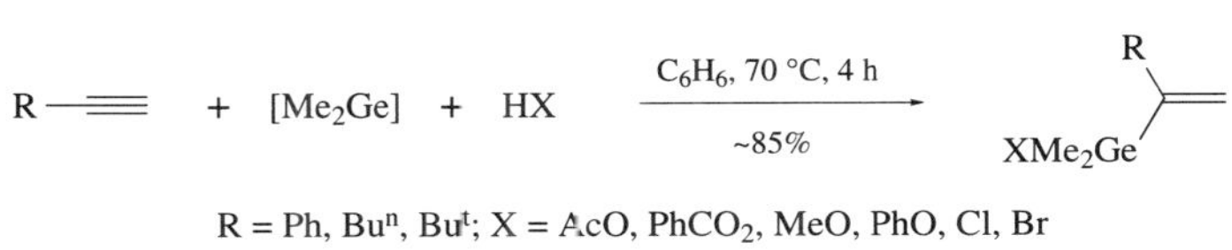

(145)

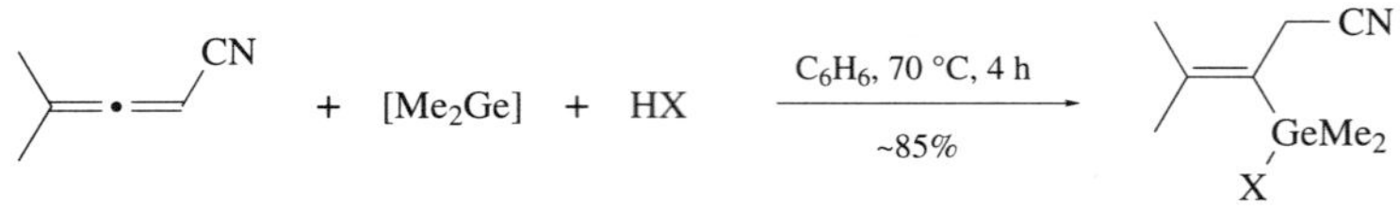

(146)

X = F, PhO

2.18.3.1.3 *From alkynylgermanes*

Alkynylgermanes are readily accessible from the reaction of alkynylmagnesiums or -lithiums with trialkylgermyl halides.

Partial hydrogenation of alkynylgermanes is possible over Lindlar's catalyst, but is unsuccessful using Raney nickel ⟨61CR(253)673⟩.

Hydroalumination followed by protonolysis is a more powerful method because the stereochemistry of the reaction can be controlled almost completely. Using Bu^i_2AlH in hydrocarbon solvents (pentane, heptane), (*E*)-alkenylgermanes are obtained, whereas a 1 : 1 mixture of Bu^i_2AlH and *N*-methylpyrrolidine in the same solvent produces the (*Z*) isomers (Scheme 38) ⟨71JOC3520⟩. In the case of the reduction of 1-octynylgermanes with Bu^i_2AlH the use of diethylether as cosolvent was reported to produce pure (*Z*)-alkenylgermane ⟨84TL3221⟩.

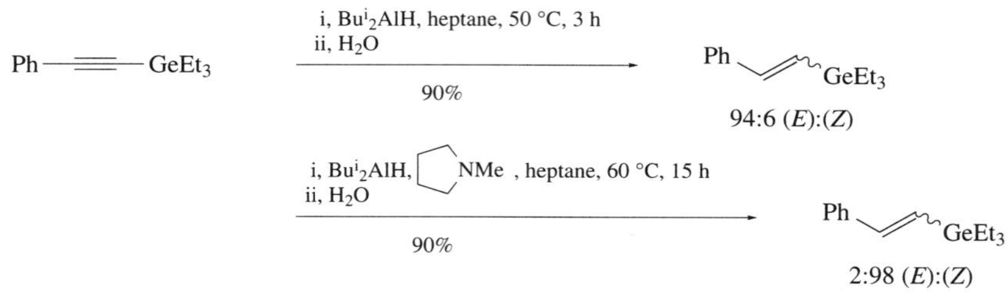

Scheme 38

Hydroboration of alkynylgermanes followed by alkylation of the germylalkenyl borane intermediate leads to internal (*Z*)-alkenylgermanes of definite regiochemistry (Equation (147)). The corresponding (*E*) isomer (Scheme 39) ⟨84TL3221⟩ can be prepared by a longer route providing that the bromine adduct is stable enough to avoid base-promoted dehydrobromination such that spontaneous bromodegermylation can occur.

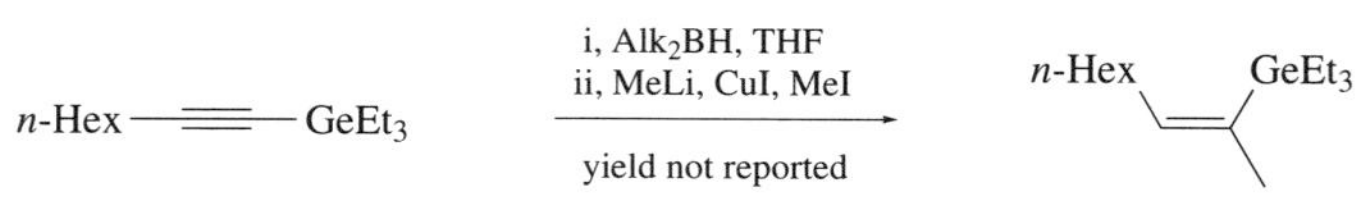

(147)

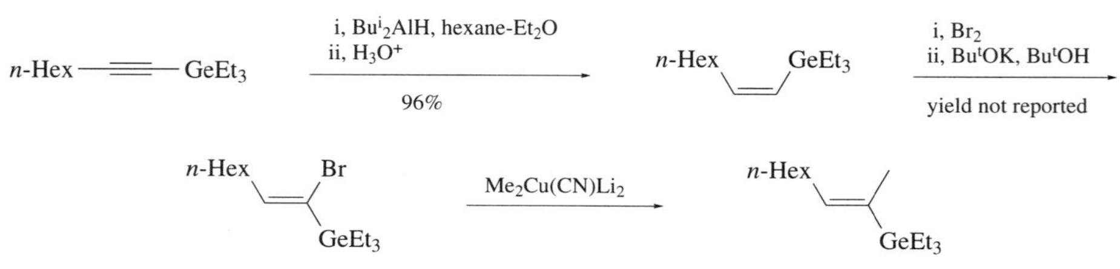

Scheme 39

2.18.3.2 Arylgermanes

The reactions of aromatic Grignard and lithium reagents with trialkylgermyl halides constitute the most widely used methods for the preparation of arylgermanes ⟨B-71MI 218-01, 82COMC-I(2)399⟩.

Where available, Grignard reagents are the reagents of choice due to lower cost and higher product yields. For example, *m*-chloro- and *m*-methoxyphenyltrimethylgermanes have been prepared in high yield by the Grignard route ⟨87JOC664⟩.

Arylation of $GeCl_4$ stops at the Ar_3GeCl stage when the aryl Grignard reagent is a sterically hindered one ⟨B-71MI 218-01, 80JOM(202)5⟩. Nevertheless, tris(2,5-dimethylphenyl) germanium bromide was alkylated in good yield using a 10-fold excess of various alkylmagnesium bromides in toluene at reflux ⟨78ZOB639⟩. In similar proportions, tricyclohexylgermanium bromide and phenyllithium were found to react efficiently (cyclohexane/Rfx/1 h/75%); however, no reaction was observed using *o*-tolyllithium ⟨46JA1720⟩. Furyl- and thienylgermanes have also been obtained in good yields from the corresponding aryllithiums ⟨92S954⟩.

The preparation of hexakis(trimethylgermyl)benzene from magnesium, chlorotrimethylgermanium and hexabromobenzene (THF/12 h/7.6% yield) illustrates that large amounts of steric strain can be tolerated in some cases ⟨86JA6664⟩.

2.18.3.3 Germoles

The syntheses and the chemistry of various elementacyclopentadienes, that is metalloles, were reviewed at the beginning of the 1990s ⟨90CRV215, 90CRV265⟩. The spontaneous or palladium-catalyzed cyclization of germylenes with alkynes has been mentioned (Schemes 36 and 37; Equations (145) and (146)). Germylenes also undergo [1 + 4] cycloadditions with conjugated dienes to give 1-germa-3-cyclopentenes ⟨66BSF327, 73JOM(56)1, 91CRV311⟩ which can be oxidized to give germoles ⟨83JA897⟩.

A valuable alternative method is the efficient transmetallation of zirconacyclopentadienes by germanium tetrachloride, since the so-formed 1,1-dichlorogermoles ⟨88JA2310⟩ can be readily alkylated ⟨66BSF327⟩.

Germoles bearing two different groups on the germanium atom can be prepared from the 1,1-dimethyl derivatives by ligand exchange with *n*-butyllithium or phenyllithium ⟨85TL1315⟩.

1,1-Dimethyl-2-*t*-butyl-benzo-3-azagermole can be prepared in good yield on reaction of dichlorodimethylgermanium with the dilithio compound obtained from phenylisocyanide following α-addition of Bu^tLi and ortholithiation ⟨78JOC731⟩.

2.18.3.4 Miscellaneous Germanium Derivatives

As well as structures such as cyclic, polycyclic, and polyhedral oligogermanes ⟨91AG(E)902, B-92MI 218-01⟩, a number of structurally simple, previously unknown or uncommon germanium compounds were reported in the early 1990s.

Unlike thiones which can be isolated when bulky alkyl or aryl groups are attached to the C=S double bond, even sterically crowded germanethiones are too unstable to be isolated. They generally dimerize into 1,3-dithiagermetanes ⟨91TL6877⟩.

Analogous to silenes ⟨B-91MI 218-01⟩, a few sterically hindered germenes (i.e. germaalkenes) could be prepared and isolated when stabilized by an extra ligand such as Et_2O or Et_3N (Equation (148)) ⟨87JA4411⟩.

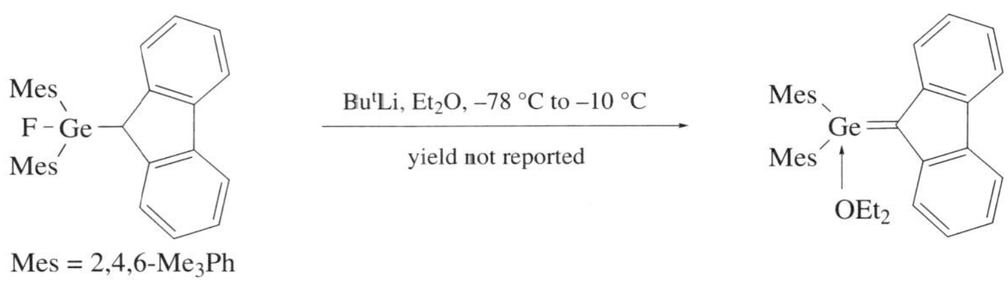

(148)

Digermenes can be produced by photolysis of suitably substituted cyclotrigermanes ⟨82JA6136⟩ or bis(silyl)germanes (Equation (149)) ⟨85TL1281⟩, and further transformed into various addition products ⟨88TL3383, 89JA1888⟩.

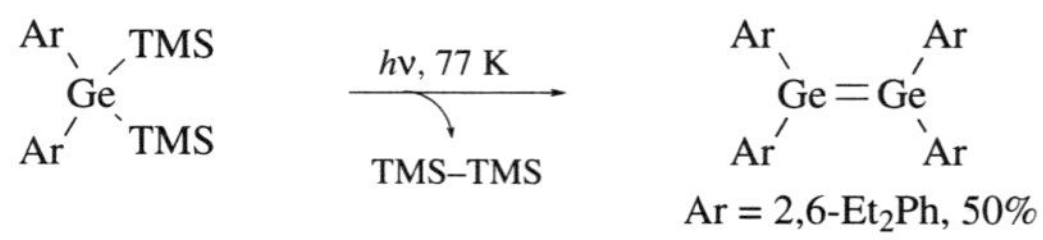

(149)

A new synthesis of acylgermanes consists of the Peterson-type alkenation of an aldehyde with lithiomethoxytrimethylsilyltrimethylgermylmethane producing an (*E*,*Z*) mixture of 1-methoxy-1-germyl alkenes which are hydrolyzed in high yield into the acylgermanes (Equation (150)) ⟨92SL843⟩.

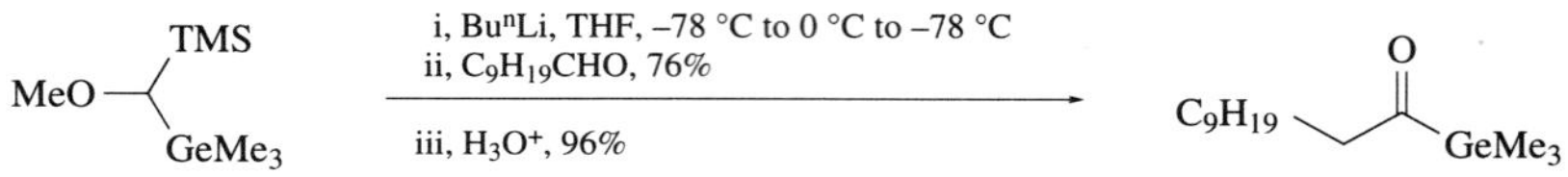

(150)

2.19
Vinyl- and Arylmetals

EI-ICHI NEGISHI and DANIELE CHOUEIRY
Purdue University, West Lafayette, IN, USA

2.19.1 INTRODUCTION AND GENERAL DISCUSSION

2.19.1.1 Definition and Classification of Vinyl- and Arylmetals and the Scope of Discussion

Vinyl- and arylmetals may be defined as those organometals in which metal atoms are σ-bonded (as opposed to π-bonded) to vinyl and aryl groups, respectively. This definition is, in most cases, straightforward and unambiguous. Inasmuch as σ- and π-bonds as well as sp^n hybridization, where $n = 1$, 2, or 3, are artificial notions with which chemists try to best approximate and visualize reality, ambiguities are inevitable. For example, the products of complexation or oxidative coupling of alkynes with certain transition metal compounds may be viewed as either alkyne–metal π-complexes (**1a**) or metallacyclopropenes (**1b**), in which a metal atom is simultaneously σ-bonded to the two sp^2 C atoms (Equation (1)). Some such compounds are better approximated by (**1a**), others by (**1b**). However, no known compound of a given molecular formula and a given atom-linking sequence has existed as both (**1a**) and (**1b**), which are identifiable as two discrete species. It is therefore reasonable to view them as the resonance hybrids of (**1a**) and (**1b**), and either of the two may be interchangeably used to represent them. In this chapter, they are arbitrarily viewed as alkyne–metal complexes, and their synthesis is therefore not discussed. Certain organotransition metal complexes which are normally considered as π-complexes may simultaneously possess σ-carbon–metal bond character. Representative examples include η^3-allylmetals and η^5-cyclopentadienylmetals which may also be viewed as σ, π-complexes, where the σ-bonded C atom is sp^2 hybridized. Regardless of detailed bonding schemes these compounds may be formally viewed as allylmetal derivatives and their synthesis is not discussed here. However, in cases where η^5-cyclopentadienylmetals are σ-metallated as in (**2**), they should be viewed as arylmetals. These dimetallic compounds are discussed in Chapter 4.16. Other special classes of vinyl- and arylmetals discussed in other chapters include alkenylidenemetals (**3**) (Chapter 3.12), 1,1-dimetalloalkenes (**4**) (Chapter 4.23), and acylmetals (**5a**), which often possess metal–carbene (**5b**) and/or vinylmetal (**5c**) characters (Scheme 1) (Chapters 5.10, 5.25 and 5.29).

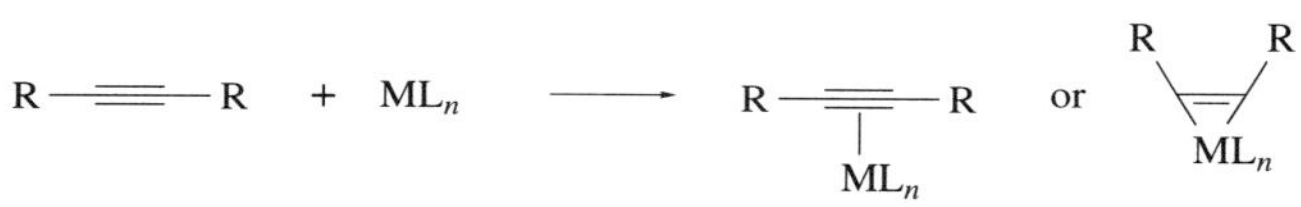

(1)

(**1a**) (**1b**)

(**2**)

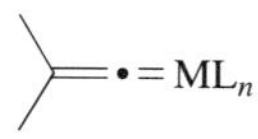

(**3**)

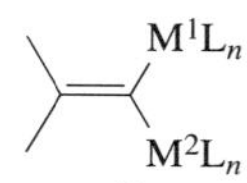

(**4**)

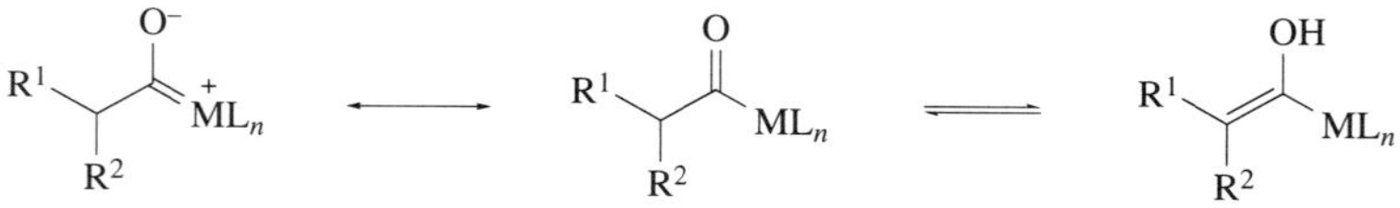

(**5b**) (**5a**) (**5c**)

Scheme 1

Aryl groups may include phenyl, naphthyl, and other higher homologs containing benzene rings, such as biphenyl, anthranyl, phenanthryl, and so on, as well as a variety of other aromatic rings, such as furyl, thienyl, imidazoyl, pyridyl, pyrimidyl, and so on. Their metal derivatives may contain one or more metal atoms and one or more carbon and other substituents. In many cases, structural variations may not call for the use of different synthetic methods. In other special cases, however, some reactions applicable only to certain specific structural types have proved to be the methods of choice. *Ortho*-lithiation of arenes with heteroatom-containing substituents is a representative example of such reactions, which deserve special attention.

Vinylmetals may also contain one or more metal atoms and one or more substituents. Various monoalkenes containing different numbers of metals and substituents are classified in Table 1. Here again certain methods, such as oxidative addition and metal–halogen exchange reactions of vinyl halides are relatively less affected by structural variations than some other methods. Since vinylmetals can be not only regiodefined but also stereodefined, a large number of selective methods producing certain specifically regio- and stereodefined vinylmetals have been developed, as discussed later.

Table 1 Vinylmetals containing one vinyl group.

Classification	*Structure*		
Monometallated			
Non-substituted	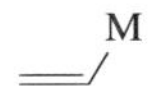		
Monosubstituted	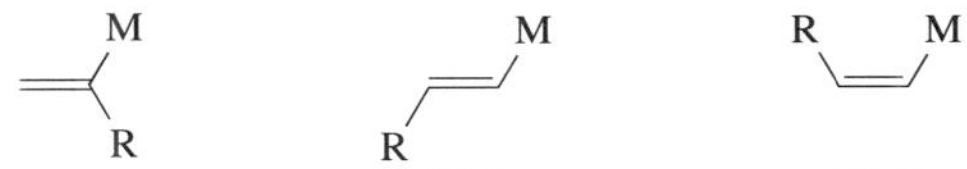 α-	(*E*)-β	(*Z*)-β
Disubstituted	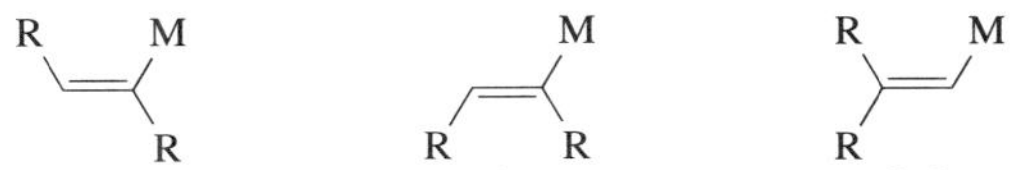 α, β-*trans*	α, β-*cis*	β, β-
Trisubstituted	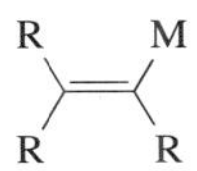		
Dimetallated			
Non-substituted	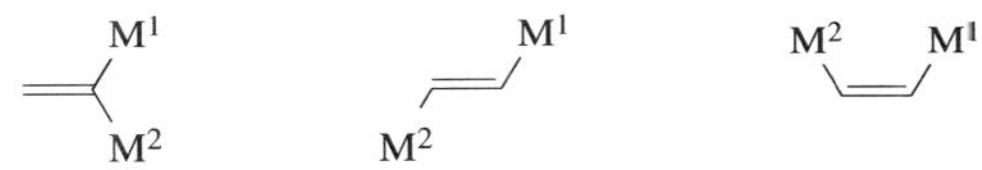 1,1-	(*E*)-1,2-	(*Z*)-1,2-
Monosubstituted	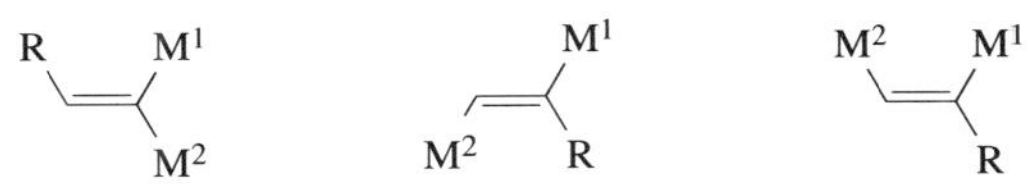		
Disubstituted	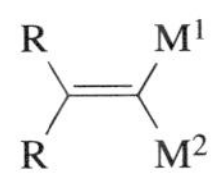		
Trimetallated			
Tetrametallated	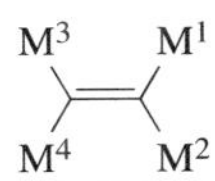		

R = C and other substituents.

Vinyl groups may contain more than one carbon–carbon double bond. Such dienyl and polyenyl groups may be conjugated, cumulated, or isolated. Furthermore, vinylmetals may also be cyclic. Some select three- to seven-membered metallacycles consisting of carbon, and metal atoms are shown in Table 2. Similar lists may be compiled for di- and multimetallated cyclic compounds as well as heteroatom-containing derivatives.

Vinyl- and arylmetals of various types mentioned above may exist as either monomers or oligo- and polymeric aggregates. For example, alkenylalanes are known to exist, in many cases, as alkenyl-bridged dimers. In this chapter, however, no special attention is paid to the degree of aggregation, and all organometals are treated as monomers unless otherwise stated. Organometals discussed here may be prepared as (a) thermally stable, discrete species that can be characterized and even isolated for full identification or use in a separate step; (b) stoichiometrically *in situ* generated, unstable species; or (c) catalytically *in situ* generated transient species, which may or may not be isolated and/or characterized. Distinction among these species is not generally noted here, and due attention must be paid along this line when applying the methods discussed to specific cases.

Table 2 Some representative cyclic vinylmetals.

Ring size	*Structure*
Three-membered	
Four-membered	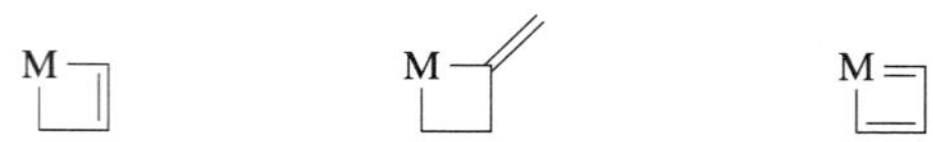

Five-membered	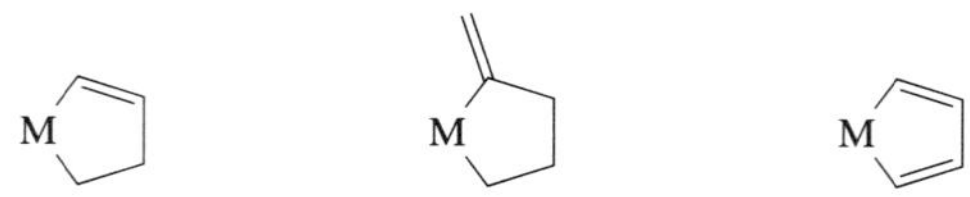
	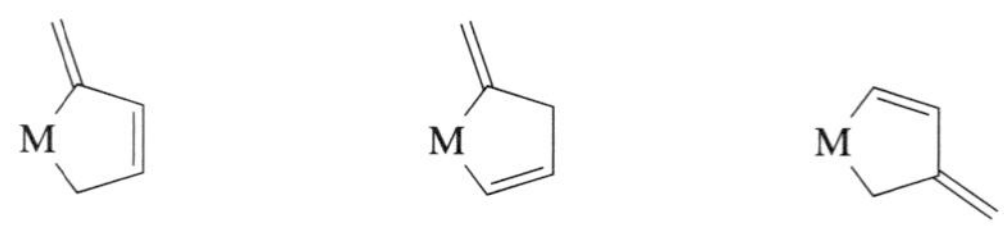
	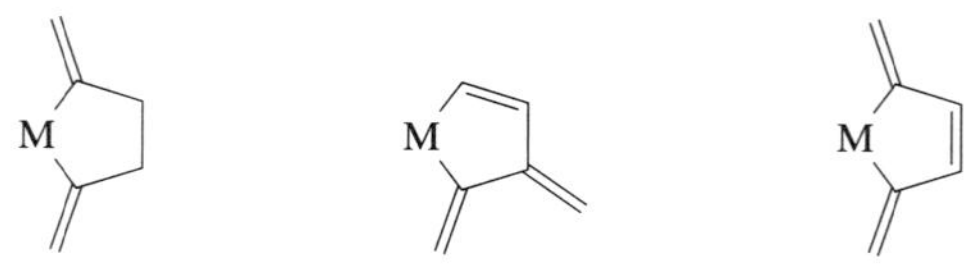
Six-membered	

Seven-membered	

[a] Viewed here as π-complexes.

2.19.1.2 General Discussion of the Synthesis of Vinyl- and Arylmetals

Organometals are, in most cases, prepared from (a) organic halides and related derivatives; (b) active C—H compounds; (c) unsaturated compounds, such as alkenes, alkynes, and arenes; and (d) organometals containing the same or different metals ⟨B-80MI 219-02⟩. The great majority of the currently known methods can be classified into a dozen categorically different methods summarized in Table 3. As indicated by the specific examples shown in Table 3, all of these methods with the exception of π-complexation (or oxidative coupling) are applicable to the preparation of aryl- and alkenylmetals. A semantic issue of whether or not the products of oxidative coupling of alkynes with metal complexes should be viewed as alkenylmetals is discussed above. It should also be noted that alkyne–metal π-complexes can react further with a variety of nucleophiles and electrophiles to produce unequivocal alkenylmetals, as exemplified by the transformation shown in Equation (2) ⟨83OM101⟩.

Table 3 Methods of preparation of organometallic compounds.

Method and description	*General equations and specific examples*
A. From halides and related compounds	
Method I	
Oxidative metallation of organic halides and related compounds (oxidative addition to RX)	$RX + ML_n \longrightarrow RML_nX$ or $RX + 2ML_n \longrightarrow RML_nX + ML_nX$ ex. $PhI + Pd(PPh_3)_4 \longrightarrow PhPd(PPh_3)_2Br$ $H_2C{=}CHBr + 2Li \longrightarrow H_2C{=}CHLi + LiBr$
Method II	
Oxidative displacement	$RX + {}^-ML_n \longrightarrow RML_n + X^-$ ex. $PhBr + {}^-CuMe_2 \longrightarrow PhCuMe_2 + Br^-$
Method III	
Metal–halogen exchange	$RX + R^1ML_n \longrightarrow RML_n + R^1X$ ex. $PhBr + Bu^nLi \longrightarrow PhLi + Bu^nBr$
B. From active C–H compounds	
Method IV	
Oxidative C–H activation (oxidative addition to C–H)	$RH + ML_n \longrightarrow RML_nH$ or $RH + ML_n \longrightarrow RML_n + 1/2H_2$ ex. $PhH + RhCp^*(PMe_3) \longrightarrow Ph(H)RhCp^*(PMe_3)$
Method V	
Metal–hydrogen exchange (nonredox)	$RH + ML_nX \longrightarrow RML_n + HX$ ex. $PhH + TlX_3 \longrightarrow PhTlX_2 + TlX$ $X = O_2CCF_3$
C. From unsaturated compounds	
Method VI	
Hydrometallation	 ex. $Bu^nC{\equiv}CH + HAl(Bu^i)_2 \longrightarrow (E)\text{-}Bu^nCH{=}CHAl(Bu^i)_2$
Method VII	
Carbometallation	 ex. $PhC{\equiv}CH + Me_3Al \xrightarrow{Cl_2ZrCp_2\ (cat.)} (E)\text{-}PhMeC{=}CH(AlMe_2)$
Method VIII	
Halo-, oxy-, and other heterometallation	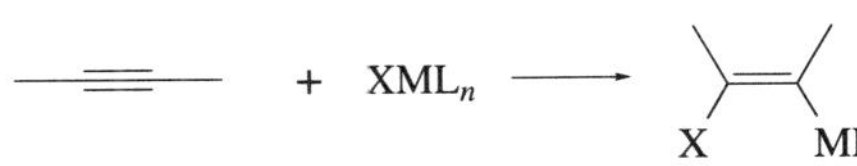 ex. $MeC{\equiv}CMe + Hg(OAc)_2 \xrightarrow{NaCl} (E)\text{-}AcO(Me)C{=}C(Me)HgCl$
Method IX	
Complexation or oxidative coupling	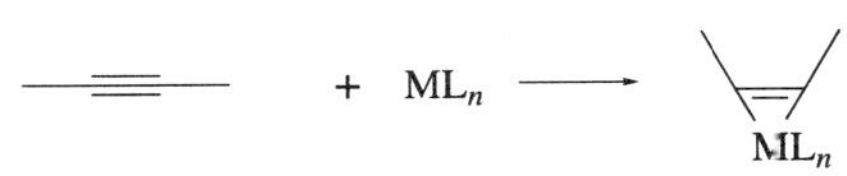 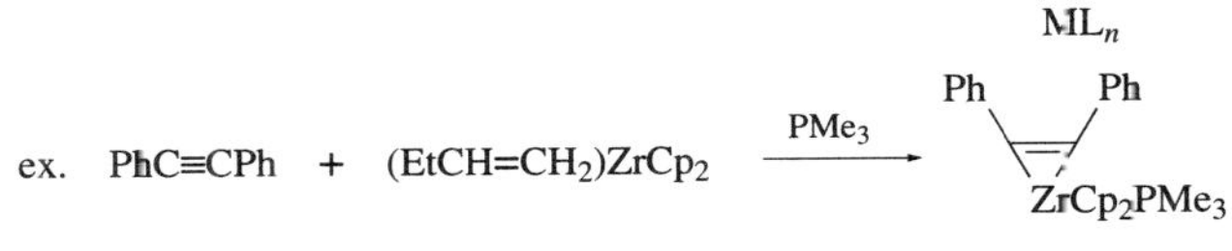
Method X	
Oxidative metallation of unsaturated compounds	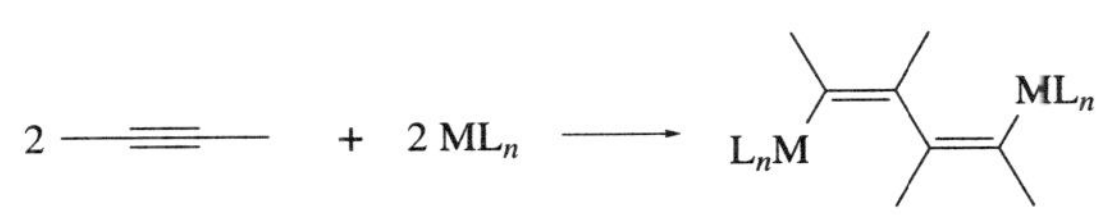 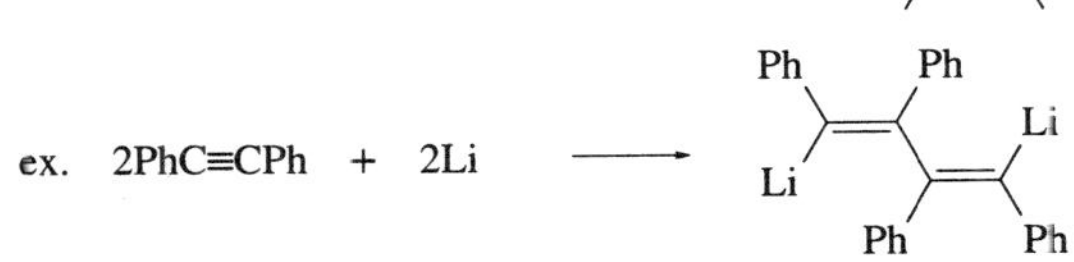

(continued)

Table 3 (continued)

Method and description	*General equations and specific examples*
D. From organometals (cf. Method VII)	
Method XI	
Transmetallation (nonredox)	$RM + M^1L_nX \longrightarrow RM^1L_n + MX$
	ex. $PhMgBr + BrSnBu_3 \longrightarrow PhSnBu_3 + MgBr_2$
Method XII	
Redox transmetallation	$RML_n + M^1L_n \longrightarrow RM^1L_n + ML_n$
	ex. $(H_2C{=}CH)_4Pb + 4Li \longrightarrow 4\,H_2C{=}CHLi + Pb$

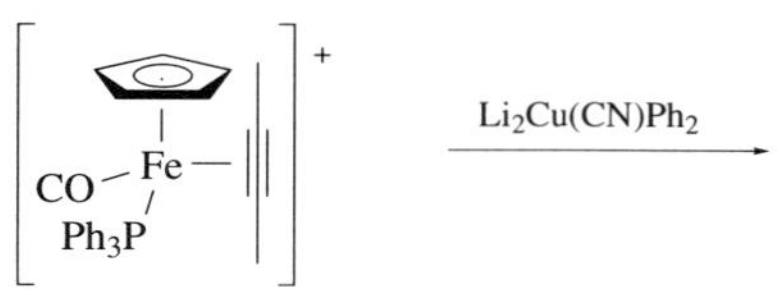

(2)

Of the dozen methods shown in Table 3, oxidative metallation of organic halides (Method I) and metal–halogen exchange (Method III) provide two of the most common methods of forming aryl- and alkenylmetals, especially those containing magnesium and lithium. In this sense, organic halides currently represent the most widely used precursors to these classes of organometals. Once these organometals are obtained, a wide variety of organometals containing other main group and transition metals can be prepared via transmetallation, especially Method XI. Thus, the majority of the currently known examples of the synthesis of vinyl- and arylmetals may be represented by Scheme 2.

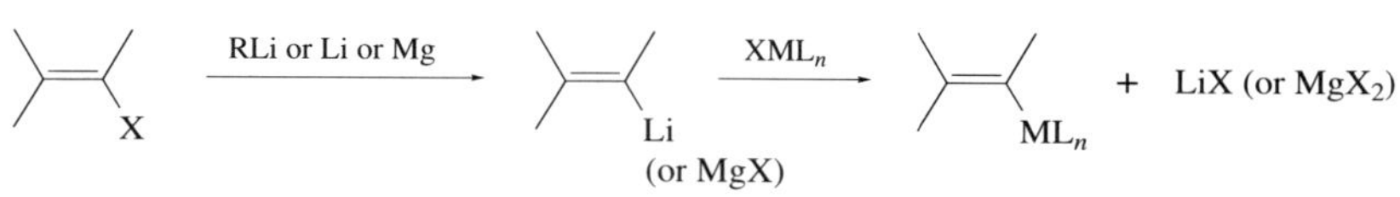

$$ArX \xrightarrow{\text{RLi or Li or Mg}} ArLi\ (\text{or } MgX) \xrightarrow{XML_n} ArML_n + LiX\ (\text{or } MgX_2)$$

Scheme 2

In addition to these methods, various addition reactions of HML_n (hydrometallation), RML_n (carbometallation), and XML_n (halo- and oxymetallation as well as other heterometallation reactions) provide convenient and attractive routes to vinylmetals, which are often highly regio- and stereoselective (Scheme 3). There also are some methods that are specific to the synthesis of arylmetals. The reaction of arenediazonium salts with metals to give arylmetals ⟨29CB1010⟩ is a representative example (Equation (3)). Although formally classified as a metal–hydrogen exchange reaction (Method V), the reaction of arenes with electrophilic metal compounds, such as $Hg(OAc)_2$, $AlCl_3$, $Tl(NO_3)_3$, and $Pd(OAc)_2$, via electrophilic aromatic substitution is also a process specific to the synthesis of arylmetals. In addition to all these methods there are a number of miscellaneous methods involving the use of other types of starting compounds. For example, carbenoids and metal compounds can produce metal–carbene complexes ⟨B-87MI 219-02⟩, some of which, for example (**3**), represent vinylmetals. Other important classes of starting compounds include metal carboxylates, which can decompose to give organometals ⟨66JCS(C)106⟩, as exemplified by Equation (4) and tosylhydrazones which can be treated with an appropriate base, for example Bu^nLi, to prepare vinyllithiums (the Shapiro reaction) ⟨76OR(23)405, 90OR(39)1⟩.

$$ArN_2^+X^- + HgX_2 \longrightarrow ArN_2^+HgX_3^- \xrightarrow{2Cu} ArHgX + N_2 + 2CuX \qquad (3)$$

$$2HO_2CC_6F_5 \xrightarrow{Hg(OAc)_2} Hg(O_2CC_6F_5)_2 \xrightarrow{\Delta} Hg(C_6F_5)_2 + 2CO_2 \qquad (4)$$

In carbometallation (Method VII) and transmetallation (Methods XI and XII) both starting

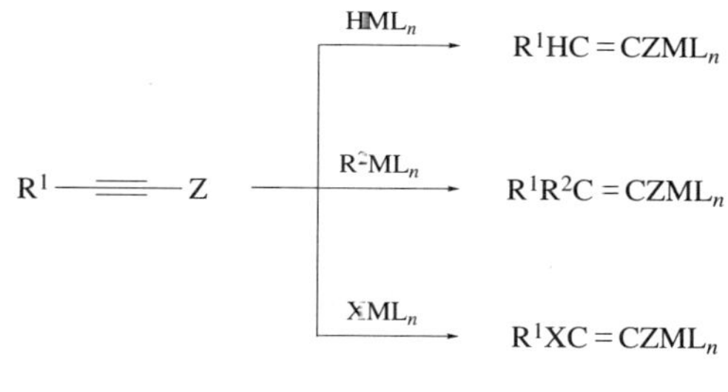

Scheme 3

compounds and products are organometals. Strictly speaking, they should therefore be considered as organometallic interconversion reactions rather than organometallic preparations. In fact, metal-containing reagents can be organometallic in most of the other methods as well. Aside from this semantic issue, however, it is important to recognize that a large number of organometallic interconversion reactions that are not usually considered as preparative methods do provide routes to organometals, many of which have proved to be the methods of choice for specific cases. For example, oxidative addition of aroyl chlorides to rhodium complexes gives aroylrhodium chlorides which can decompose to give arylrhodium derivatives via decarbonylation (Equation (5)) ⟨81JA1847⟩. These reactions are too numerous to be systematically covered in this chapter. Only select examples will therefore be presented.

$$\mathrm{ArCOCl} \longrightarrow \mathrm{ArCORhL}_n\mathrm{Cl} \longrightarrow \mathrm{ArRhL}_n\mathrm{Cl} \qquad (5)$$

2.19.1.3 Characterization, Identification, and Quantitative Analysis of Vinyl- and Arylmetals

In cases where vinyl- and arylmetals are thermally stable and isolable in pure forms, their characterization, identification, and quantitative analysis are essentially the same as those of organic compounds. Commonly used spectroscopic methods include ^{1}H, ^{13}C and other NMR, IR, UV-visible, and mass spectroscopies as well as x-ray crystallography. Together with elemental analysis and molecular weight measurements, these spectroscopic methods usually provide detailed structural information. In reality, many vinyl- and arylmetals are thermally labile. Even in such cases, some spectroscopic methods, such as NMR spectroscopy, have proved to be indispensable for characterization and/or quantitative analysis. Furthermore, those organometals which contain metals of relatively low electronegativity, usually $\leqslant 1.8$–1.9, are readily protonated or deuterated. The organic products thus obtained may be characterized and identified. Highly electronegative metals, such as Hg, B, Si, Sn, and some late transition metals, tend to form hydrolytically stable carbon–metal bonds, and they are usually characterized and identified as such. Alternatively, vinyl- and arylmetals may be treated with iodine to produce the corresponding iodides, often in nearly quantitative yields. Incorporation of D (but not H) and/or I is, in many cases, a clear indication of the formation of the corresponding organometals, and high-yield formation of deuterated and/or iodinated compounds often permits quantitative analysis of the parent organometals (Scheme 4).

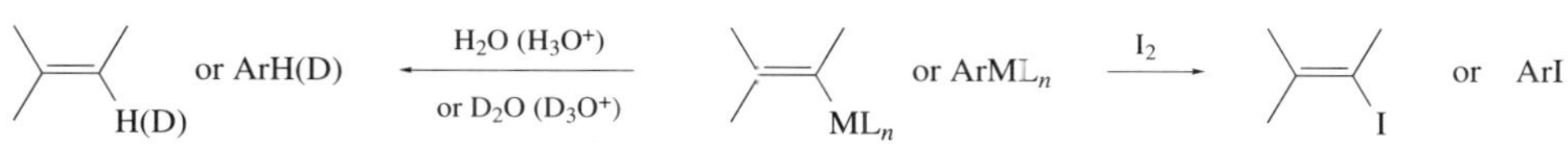

Scheme 4

2.19.2 OXIDATIVE METALLATION OF VINYL AND ARYL HALIDES (METHODS I AND II)

2.19.2.1 Overview

The reaction between a metal or a metal compound of low oxidation state and an organic halide provides one of the most direct and general methods for preparing organometals of various classes. In the oxidative addition reaction shown in Equation (6) and the oxidative displacement reaction shown in Equation (8) the metal suffers from two-electron oxidation, while the "oxidative addition" process shown in Equation (7) involves oxidation by one electron per metal. In oxidative addition the number of bonds increases by one. Furthermore, the M—X bonds are often stronger than the R—X bonds. Consequently, the process is, in general, thermodynamically favorable, unless some other factors such as aggregation, ligation, and solvation offset thermodynamic equilibria. Kinetic factors, especially in the dissociation, deligation, and desolvation stages, also play crucial roles. In general, highly electropositive and readily ionizable (or oxidizable) metals, that is, those of low ionization energies, readily participate in oxidative addition. Among the main group metals lithium and other alkali metals and magnesium and some other alkaline earth metals most readily participate in the reaction. In cases where vinyl and aryl halides are used the Wurtz coupling does not represent a threatening side reaction. However, β-elimination producing alkynes and benzynes can be a serious side reaction in cases where strongly basic organometals, such as those containing sodium and potassium, and/or organic halides having a strong tendency to undergo β-elimination are used. On balance, lithium and magnesium represent the two most widely employed main group metals. Oxidative addition reactions of lithium and magnesium are believed to involve radical processes ⟨80JA217, 80JA226, 80JA231, 80JA239, 80JA2493⟩ proceeding via (i) one electron transfer from a metal to a halide producing a metal cation and the radical anion of the halide; and (ii) the second electron transfer to give a carbanionic species and a metal cationic species, that is, M^+ or MX^+.

$$\text{R-X} + \text{ML}_n \longrightarrow \text{R-ML}_n\text{-X} \qquad (6)$$

$$\text{or} \quad \text{R-X} + 2\text{ML}_n \longrightarrow \text{R-ML}_n + \text{X-ML}_n \qquad (7)$$

$$\text{or} \quad \text{R-X} + {}^{-}\text{ML}_n \longrightarrow \text{R-ML}_n + \text{X}^- \qquad (8)$$

Bulk metals are multiple metal–metal bonded polymeric species which must break down to provide atomic or oligomeric metals to be incorporated in the oxidative addition product. Furthermore, the surfaces of bulk metals are generally coated, to varying degrees, with the corresponding metal oxides. These represent kinetic barriers to be overcome. To cope with the latter problem various techniques ranging from mechanically scratching, shaving, and grinding bulk metals to chemically exposing uncoated metal surfaces through sacrificial use of highly reactive species, such as CH_3I, 1,2-dibromoethane and iodine have been devised ⟨B-54MI 219-01, 81S585, 93AG(E)164⟩. A more recent development of sonication ⟨80JA7927, 89S787⟩ is also noteworthy.

In addition to these devices there are other more substantial modifications including (i) chemical generation of active metals by pretreatment of metal salts with highly electropositive metals, for example potassium; (ii) use of metals activated by arenes and graphite; (iii) use of alloys and amalgams; (iv) use of ligated metals; and (v) physical activation by metal vaporization.

2.19.2.2 Generation of Active Metals by Treatment of Metal Salts with Alkali Metals

A potentially useful method for generating active metals is to treat metal salts with an alkali metal in an ethereal or hydrocarbon solvent ⟨75TCC1⟩. Although potassium was extensively used in early studies, more recent reports suggest that lithium naphthalide is a superior reagent ⟨89SCI1260⟩. Highly reactive metallic species containing Mg, Zn, Cd, Al, In, Tl, Sn, Cr, Ni, Pd, Rt, Cu, U, Th, and Pu have been prepared by this method. Whereas all previous efforts to prepare the Grignard reagent from fluorobenzene and magnesium had failed, magnesium powder generated by the reaction of $MgCl_2$ with potassium in the presence of KI reacted with *p*-fluorotoluene to give *p*-tolylmagnesium fluoride (Scheme 5) ⟨72JA7178, 73CC879⟩. Similarly, active zinc powder generated from $ZnBr_2$ and potassium in THF reacts with phenyl bromide to give the corresponding phenylzinc bromide ⟨73CC269⟩ (Scheme 6). Direct preparation of stereo- and regiodefined vinylcoppers ⟨88JOC4482⟩ is a

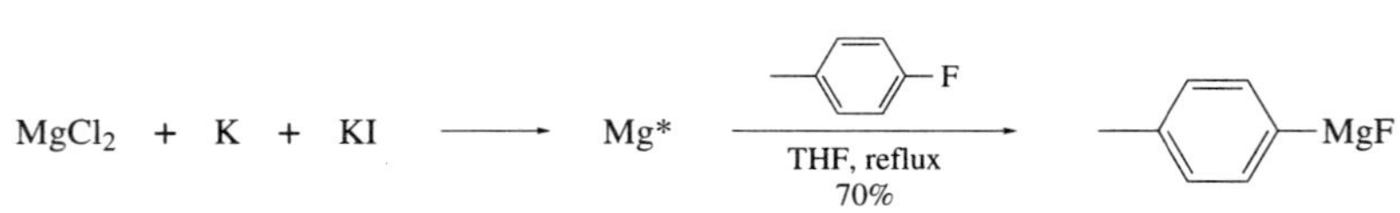

Scheme 5

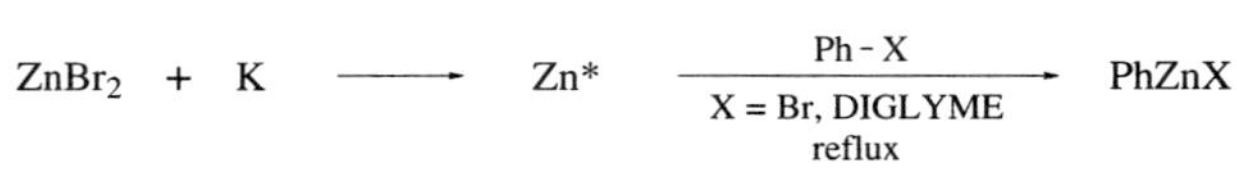

Scheme 6

potentially superior alternative to many indirect procedures developed over the past few decades (Equation (9)). Some representative examples of vinyl- and arylmetals obtained by this method are shown in Table 4.

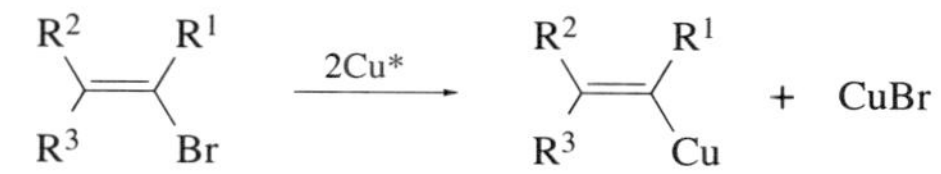
(9)

Table 4 Oxidative metallation of vinyl and aryl halides using activated metals in THF.

M	*RX*	*Temp.* (°C)	*Time* (min)	*Yield of RM* (%)	*Ref.*
Mg	*p*-ClC_6H_4Br	25	120	100[a]	74JA1775
Mg	$CH_2{=}CMeBr$	25	5	100	74JA1775
Ca	PhF	25	b	≥85	90JOC5045
Ca	*p*-MeC_6H_4Cl	−20	30	≥86	90JOC5045
Zn	*p*-BrC_6H_4COOEt	reflux	120	100	91JOC1445
Zn	$PhCBr{=}CH_2$	b	b	≥91	91JOC1445
Cu	PhI	25	10	98	84JOC5280
Cu	$PhCH{=}CHBr$	0	b	82	88JOC4482

[a] Yield of the di-Grignard reagent. [b] Information not available.

2.19.2.3 Oxidative Metallation with Metals Activated by Arenes and Graphite

Alkali and alkaline earth metals readily interact with fused arenes and graphite to form complexes of discrete compositions. These complexes have been shown to be excellent sources of active metals ⟨93AG(E)164⟩. One of the most extensively studied complexes is magnesium–anthracene readily formed by the reaction of magnesium with anthracene in THF at 0–60 °C (Equation (10)) ⟨88ACR261, 90CB1507⟩. This appears to be a superior source of magnesium; thus, for example, propargyl chloride is smoothly converted to propargylmagnesium chloride. In a more general sense, treatment of various metal salts such as those containing V, Mo, Fe, Co, Ni, Pd, and Pt with magnesium–anthracene can give "naked" and ligated metal species (Equation (11)) ⟨83AG(E)728⟩. Graphite also serves as a polyfused aromatic species. Thus, for example, potassium and graphite interact at 150 °C to give a discrete species represented by C_8K (Equation (12)) ⟨26ZAAC246⟩. This reagent can serve as a potassium source displaying some attractive features such as high reactivity, which can be used in generating other active metals (Equation (13)) ⟨76PAC177, 78CC927, 85PAC1887, B-87MI 219-01, 88AOC(28)85⟩.

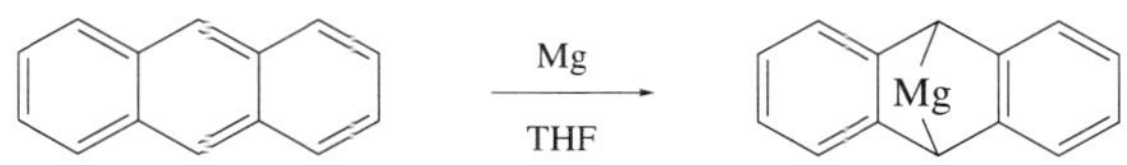
(10)

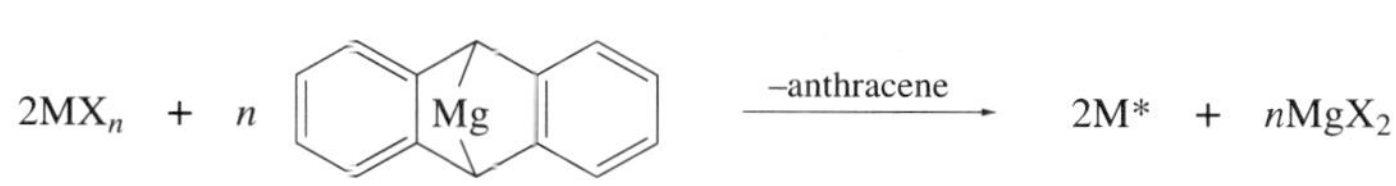
(11)

$$8C + K \xrightarrow[\text{Ar}]{150\,°C,\ 15\ \text{min}} C_8K \qquad (12)$$

$$nC_8K + MX_n \xrightarrow{\text{THF, reflux}} M^*\ (\text{graphite}) + nKX \qquad (13)$$

2.19.2.4 Oxidative Metallation with Alloys and Amalgams

Alloys and amalgams consisting of highly electropositive and electronegative metals can react with organic halides to give organometals containing the more electronegative metals of the two (Equation (14)) ⟨77JOM(127)127⟩.

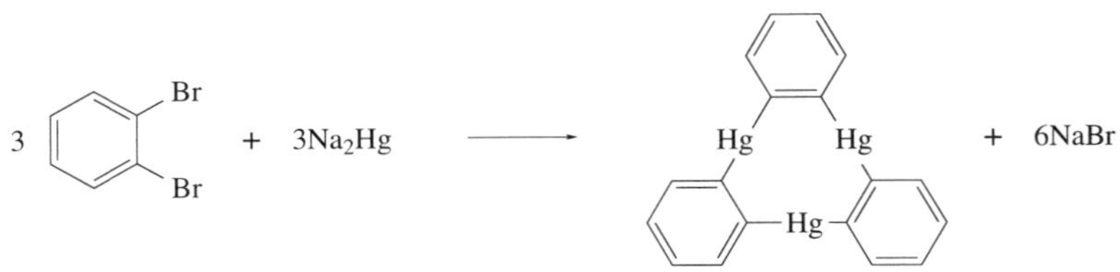

(14)

2.19.2.5 Oxidative Metallation with Transition Metal Complexes

Certain transition metals and their complexes readily undergo oxidative addition reactions with organic halides including vinyl and aryl halides ⟨77ACR434, B-87MI 219-03⟩. Such reactions may be either discrete stoichiometric processes or part of catalytic processes which might even be thermodynamically unfavorable, if they are carried out independently as isolated reactions. Many of these reactions appear to be concerted processes proceeding with retention of configuration, which must require transition metal species capable of providing at least one pair of filled nonbonding and empty orbitals. Those species that contain late transition metals most readily satisfy this requirement, and many of them, for example, $ClRh(PPh_3)_3$, are isolable as stable species. With two pairs of filled nonbonding and empty orbitals a metal species may interact either simultaneously or stepwise with a π-bond and the carbon–halogen bond of a vinyl or aryl halide to produce the corresponding organometal. With just one pair a metal species may initially interact with a π-bond to give a metallacyclopropane which may then undergo β-elimination to give the oxidative addition product (Scheme 7).

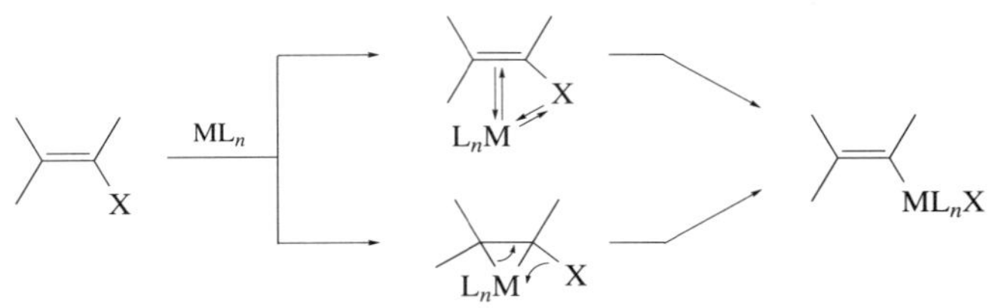

Scheme 7

In the oxidative displacement reaction metallic anions displace halides from organic halides. Such displacement reactions of vinyl and aryl halides involve substitution at sp^2 carbon centers. Some radical and concerted processes are conceivable. In cases where retention of alkenyl configuration is observed as in many known reactions, such reactions may share common mechanistic features with the concerted oxidative addition reaction discussed above. Thus, at least one pair of filled nonbonding and empty orbitals may be required. Certain cuprates ($^-CuR_2$) and aurates ($^-AuR_2$) satisfy this requirement. Indeed, the cross-coupling reaction of diorganylcuprates with vinyl halides proceeds with retention, and one plausible mechanism involves the intermediacy of a vinylcopper(III) species (Scheme 8) ⟨75OR(22)253⟩. Triorganylcoppers are generally very unstable and difficult to detect. However, some triorganylgolds have been characterized, and their reductive elimination reaction has been investigated ⟨72JOM(40)C81, 73JA6487⟩. Rhodium(I) porphyrin anions react with vinyl halides to give oxidative displacement products with retention of configuration (Equation (15)) ⟨75JA6461⟩.

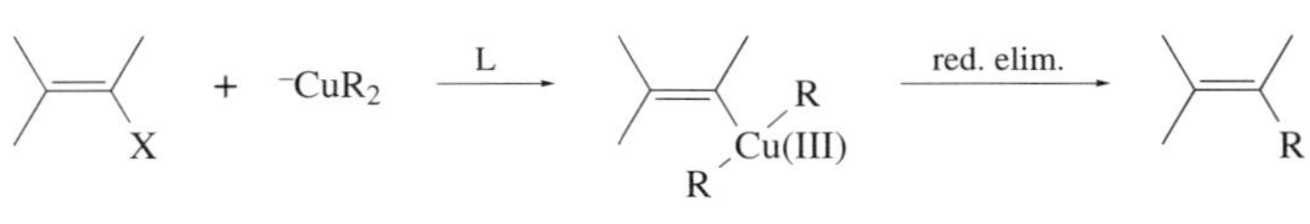

Scheme 8

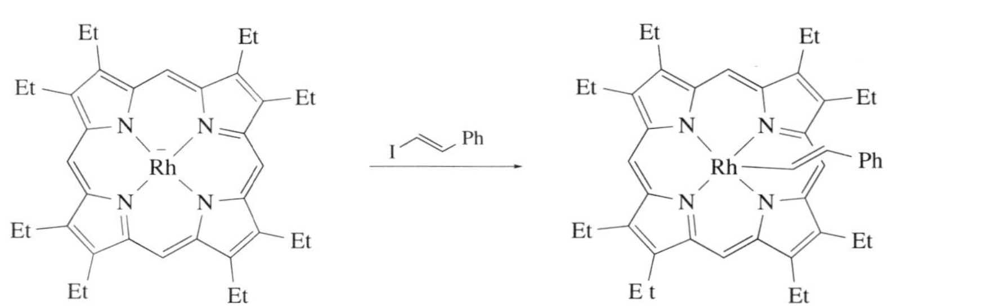

(15)

2.19.2.6 Oxidative Metallation with Metal Vapors

Metals can be vaporized as monoatomic and oligomeric clusters. This vaporization technique has been widely used for metal plating in industry. Some pioneering works over the past few decades have shown that metal vapors are highly reactive and potentially useful reagents for the preparation of organometals, providing routes to even those that are otherwise inaccessible ⟨72MI 219-01, 73ACR97, 75ACR393, 75AG(E)218⟩. In short, metals are vaporized at high temperatures using heaters, electron guns, lasers, or arcs in a vacuum chamber (10^{-3} to 10^{-6} torr). Metal vapors deposit on the walls of the reactor kept at very low temperatures, while vapors of organic substrates are directed through a shower head, codeposited in the same area, and allowed to react with the metals. Metals that have been utilized include both main group metals, such as Li, Mg, Ca, Zn, Cd, B, Al, In, Si, Ge, and Sn, and transition metals, such as Ti, Zr, V, Nb, Ta, Cr, Mo, W, Mn, Fe, Ru, Co, Ir, Ni, Pd, Pt, Ag, Au, U, and various lanthanide metals. Although widely applicable, there are some serious difficulties and limitations. Competitive polymerization of metals is a serious side-reaction. Organic halides must be sufficiently volatile so that they may be introduced to the reactor as vapors.

2.19.3 METAL–HALOGEN EXCHANGE REACTIONS OF VINYL AND ARYL HALIDES (METHOD III)

$$R^1{-}X \quad + \quad R^2{-}ML_n \longrightarrow R^1{-}ML_n \quad + \quad R^2{-}X \qquad (16)$$

This reaction can proceed forward in cases where R^1 is more electronegative or acidic than R^2, that is pK_a of $R^1H < pK_a$ of R^2H. It provides one of the most widely used methods for preparing organolithiums (Equation (16)). The Wurtz coupling reaction is one of the serious side-reactions. For the metal–halogen reaction to serve as a satisfactory preparative method it must be substantially faster than the competing Wurtz reaction. With alkali metals other than lithium the metal–halogen exchange process may be facile, but so is the Wurtz reaction. So, this metal–halogen exchange reaction does not appear to be generally useful. With magnesium and other more electronegative metals, on the other hand, the metal–halogen exchange process itself appears to be too slow to be of practical utility. At present, lithium appears to be the only metal with which differentiation of metal–halogen exchange and coupling reaction can be readily and generally achieved ⟨51OR(6)339⟩. Even so, its successful use requires some precautionary measures. In relatively nonpolar solvents including diethyl ether, the Wurtz coupling reaction is sufficiently slow. In more polar solvents, such as THF, however, the lithium–halogen exchange process is rapidly followed by the coupling reaction at room temperature (Scheme 9) ⟨74JOC3452⟩. The unwanted coupling reaction can be avoided through the use of tertiary alkyllithiums, for example, Bu^tLi. In this case the by-product is a tertiary alkyl halide which does not readily couple with the desired organolithium. It does, however, readily react with another molecule of the tertiary alkyllithium, necessitating the use of two equivalents of the tertiary alkyllithium. Significantly, tertiary alkyllithiums are considerably more reactive towards tertiary alkyl halides than the desired vinyl- and aryllithiums (Scheme 10) ⟨74CB847⟩.

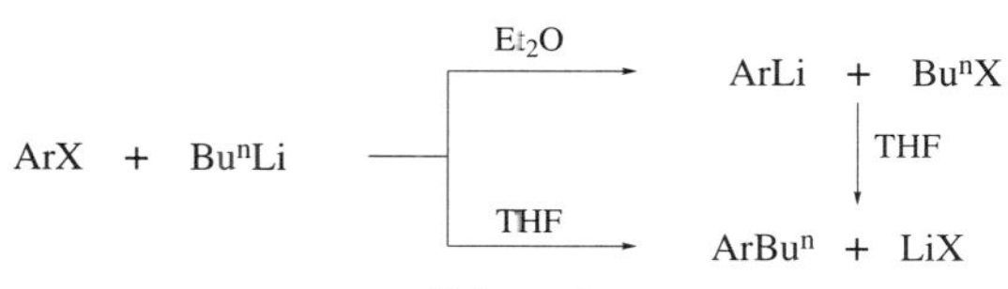

Scheme 9

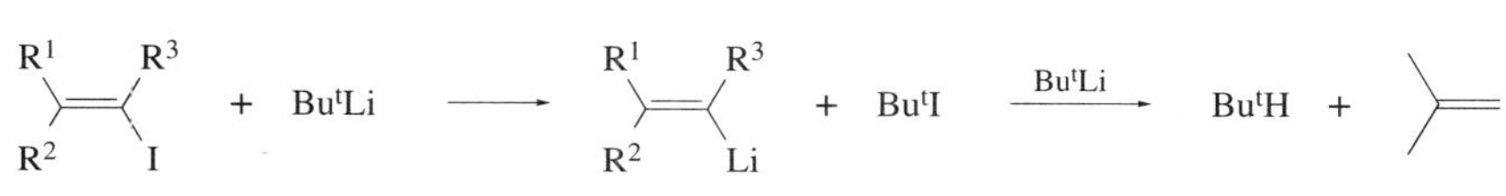

Scheme 10

The order of reactivity of halogens is I > Br > Cl ≫ F. With organic iodides and bromides the reaction is often very fast even at −78 °C. The configuration of vinyl groups is generally retained unless there are some groups, such as Ph and silyl ⟨77TL3969, 86JA3402⟩, that can induce facile stereoisomerization in the α position.

Although a few different mechanisms including radical and "ate" complexation paths have been proposed, the retention of vinyl configuration suggests that the "ate" complexation mechanism (Scheme 11) ⟨85JA4101⟩ is highly plausible for those cases that involve vinyl and even aryl groups.

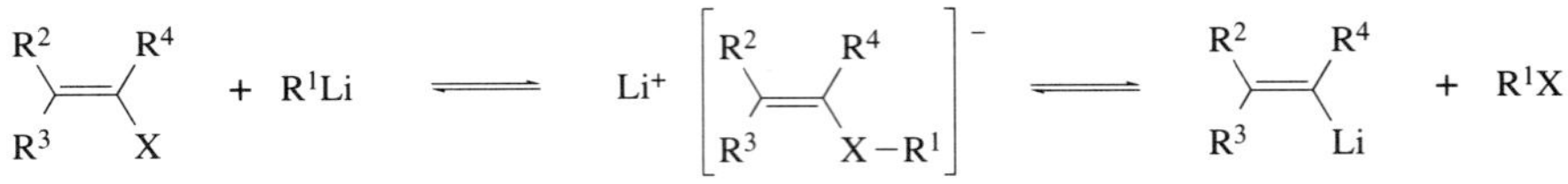

Scheme 11

Synthetically useful examples of halogen–metal exchange reactions involving metals other than lithium are rare. In the following cyclization reaction dialkylcuprates presumably undergo Cu–I exchange, which is then followed by carbonyl addition (Scheme 12) ⟨70JA395⟩.

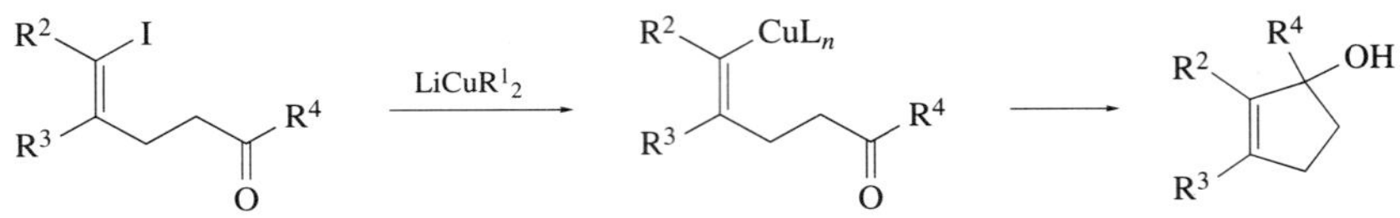

Scheme 12

2.19.4 OXIDATIVE C—H ACTIVATION (METHOD IV)

$$R-H \quad + \quad ML_n \longrightarrow R-ML_n-H \qquad (17)$$

The process shown in Equation (17) is fundamentally similar to the oxidative addition reaction of organic halides discussed earlier. However, it is intrinsically less favorable than the corresponding reaction of organic halides. In some cases, it may even be endothermic. The unfavorable thermodynamics in turn influence the kinetics as well. For concerted processes the requirement for the metal complexes appears to be the same as in the oxidative addition reaction of organic halides, that is, the availability of at least one pair of filled nonbonding and empty orbitals. In the absence of a proximal π-bond, σ and σ^* C—H orbitals must directly interact with the metal frontier orbitals. In the cases of alkenes and arenes, a 16-electron transition metal species may undergo a sequence of reaction consisting of π-complexation and β-elimination (Scheme 13). In cases where two each of filled nonbonding and empty orbitals are available, C=C and C—H bonds may simultaneously interact with metal. These aspects are, however, not yet fully established, and further clarification is necessary. Another reason for the relative paucity of C—H oxidative addition is that, in many instances, substrates contain more reactive functional groups, and their reactions tend to overshadow otherwise observable C—H oxidative addition. Mainly for these reasons the current scope of the synthesis of vinyl- and arylmetals by this reaction is rather limited ⟨85CRV245⟩. A few representative examples of intermolecular C—H activation are shown in Equation (18) and Schemes 14–16 ⟨65JCS843, 72CC1114, 82JA352, 83JA3929, 84JA1650⟩.

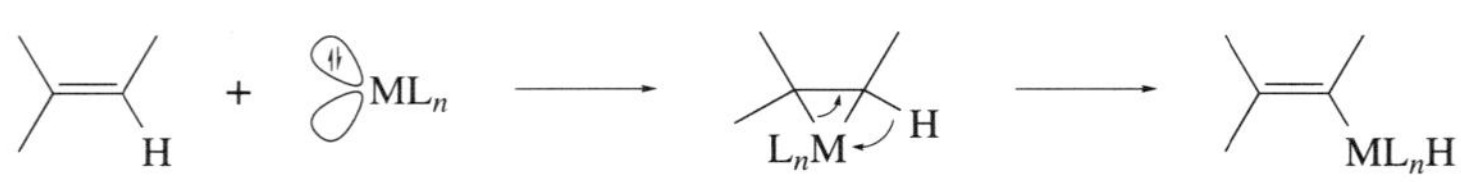

Scheme 13

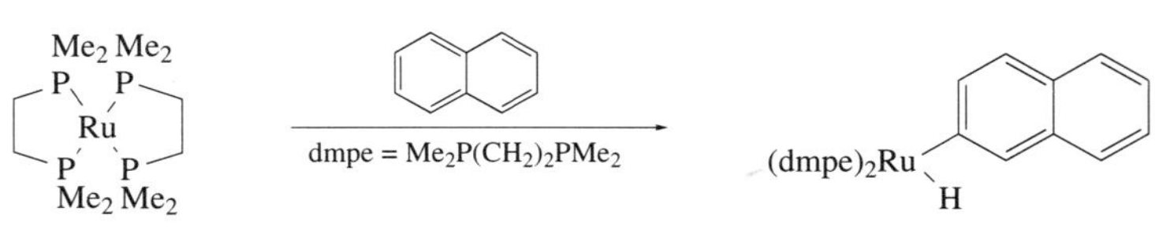

(18)

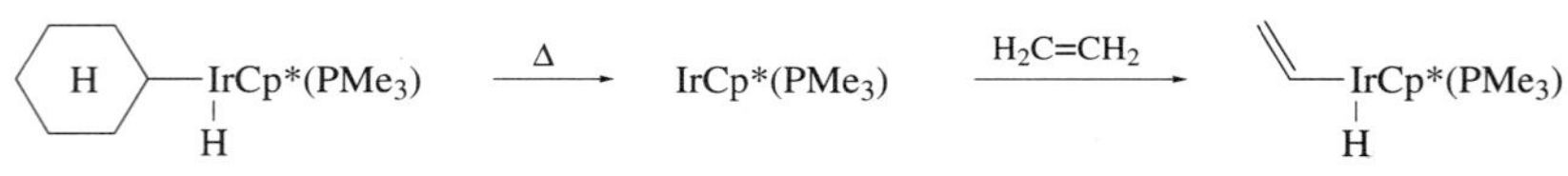

Scheme 14

$$\mathrm{Me{-}RhCp^*(PMe_3)(H)} \xrightarrow{-17\,^\circ\mathrm{C}} \mathrm{RhCp^*(PMe_3)} \xrightarrow{\mathrm{PhH}} \mathrm{Ph{-}RhCp^*(PMe_3)(H)}$$

Scheme 15

$$\mathrm{Cp^*_2WH_2} \xrightarrow{h\nu} \mathrm{Cp^*_2W} \xrightarrow{\mathrm{PhH}} \mathrm{Cp^*_2W(H){-}Ph}$$

Scheme 16

In the intermolecular oxidative addition reaction two molecules give one product molecule. On the other hand, the number of molecules does not change in the intramolecular version of the reaction. In the absence of excessive conformational strain associated with cyclization, the intramolecular process should therefore be both thermodynamically and kinetically more favorable than the corresponding intermolecular reaction. Indeed, a large number of cyclic C—H activation reactions are known, as exemplified by those shown in Equations (19) ⟨86JOM(310)27⟩ and (20) ⟨69JA6983⟩. These reactions can offer some attractive possibilities. More often than not, however, these reactions represent unwanted side-reactions destroying metal reagents and catalysts through ligand C—H activation and some efforts may have to be made to avoid them in order to favor other processes of interest.

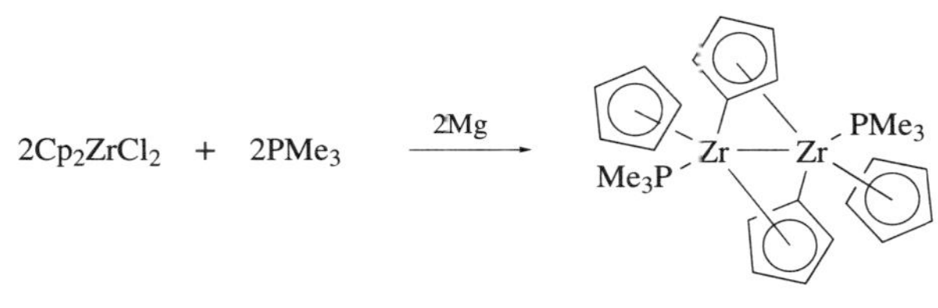

(19)

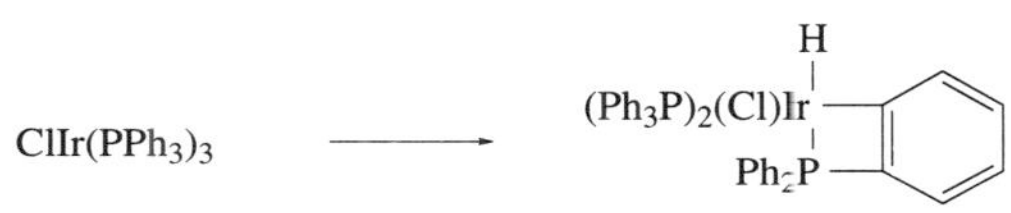

(20)

2.19.5 NONREDOX METAL–HYDROGEN EXCHANGE (METHOD V)

Unlike oxidative C—H activation the metathetical process shown in Equation (21) does not involve a redox process, which represents a fundamental difference between the two. In cases where X in Equation (21) is a carbon group, the critical requirement is the same as that for metal–halogen exchange, that is pK_a of HR < pK_a of HX. If X is a heteroatom or heteroatom group, however, other factors may also be involved, and the overall thermodynamics must be carefully evaluated. From a more practical viewpoint there are two different types of reactions that are of synthetic importance.

$$\mathrm{R{-}H} \;+\; \mathrm{X{-}ML}_n \longrightarrow \mathrm{R{-}ML}_n \;+\; \mathrm{H{-}X} \qquad (21)$$

2.19.5.1 Metallation of Arenes via Electrophilic Substitution

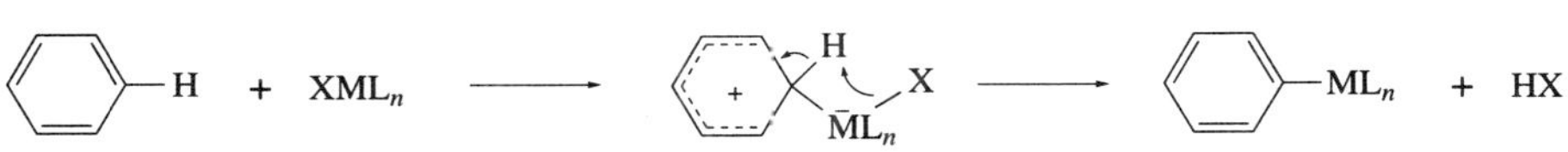

Scheme 17

As indicated in Scheme 17, there is no fundamental difference between the reactions discussed here and various known electrophilic aromatic substitution reactions. As such, the reaction is applicable only to the synthesis of arylmetals, and the corresponding reactions of alkenes and alkynes tend to give addition products, as discussed in Section 2.19.8. Electrophilic metal compounds, such as those containing B, Al, Sn, Hg, Tl, and Pd, have participated in the reaction. In general, mixtures of *o*-, *m*-, and *p*-isomers are obtained. However, certain heteroatom substitutents strongly direct metallation to the *ortho* position, as shown by some results in Table 5 as well as in Equations (22) ⟨68JA909, B-87MI 219-05⟩ and (23) ⟨81JOC4416⟩.

Table 5 Regiochemistry of the metallation of substituted benzenes.

XML$_n$	*Z of PhZ*	*Isomer distribution* *o*	*m*	*p*	*Ref.*
$Hg(OAc)_2$	Me	29	12	59	62JA1650
$Hg(OAc)_2$	OMe	14		86	60JA1939
BCl_3	Me		18	82	59JA2597
$Tl(O_2CCF_3)_3$	Me	9	4	87	71JA4845
$Tl(O_2CCF_3)_3$	CH_2OH	>99			71JA4845
$Tl(O_2CCF_3)_3$	CH_2OAc	50	21	29	71JA4845

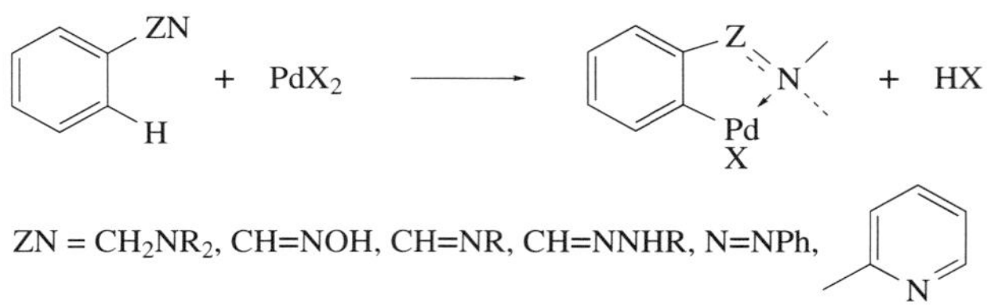

(22)

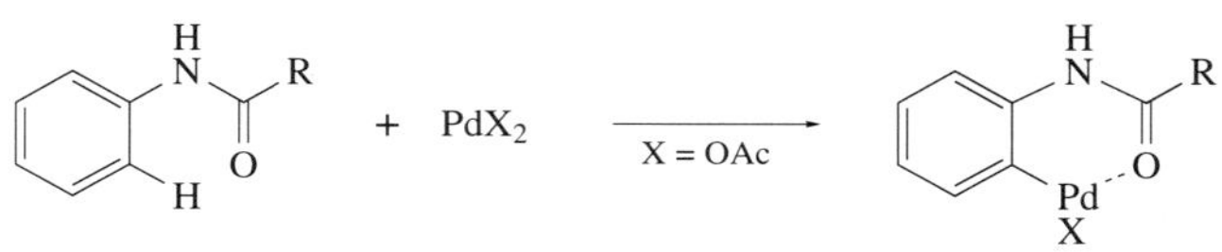

(23)

2.19.5.2 Metallation of Alkenes and Arenes via Deprotonation with Bases

(24)

$$\text{Ph—H} + \text{LiR} \longrightarrow \text{Ph—Li} + \text{HR} \qquad (25)$$

The reactions shown in Equations (24) and (25) should be thermodynamically favorable in cases where the starting alkenes and arenes are more acidic than HR. The reaction must also be kinetically favorable. To satisfy the latter requirement, the pK_a differential should generally be considerably larger than 2–3, which necessitates the use of highly basic reagents, such as alkylmetals containing Li, Na, and K. One serious limitation of this method is that allylic and benzylic C—H bonds, which are often present in alkenes and arenes, respectively, can be competitively abstracted. In many cases, these C—H bonds tend to preferentially react to give allylic and benzylic organometals (Equations (26) and (27)) ⟨70C109, 76JA4674⟩. Selection of the reaction site among competing C—H groups of the same class is another difficult synthetic task. Recent developments, however, permit considerable regio- and/or stereochemical control. It is generally difficult to predict accurately the point of attack in cases where two or more competing C—H bonds are present. However, strongly basic reagents, such as KCH_2SiMe_3, tend to attack the most acidic site, whereas reactions with weaker bases, such as Bu^{n}Li-KOBu^t, can be strongly influenced by other factors, such as steric hindrance (Scheme 18) ⟨76HCA453⟩.

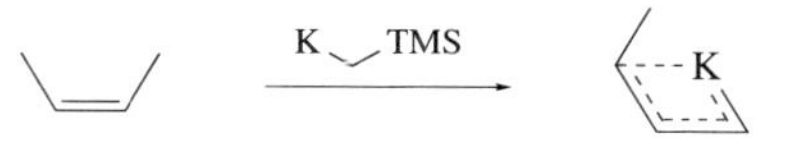

(26)

(27)

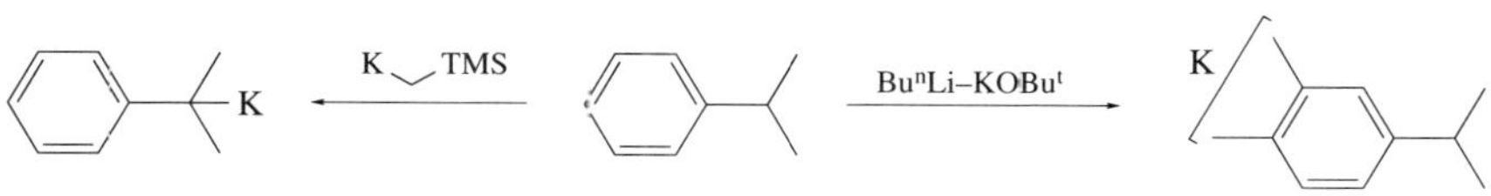

Scheme 18

Proximal heteroatom containing groups and other metal cation and/or carbanion stabilizing groups accelerate and direct metallation often in a highly regioselective manner. Lithiation with alkyllithiums, for example Bu^{n}Li, Bu^{s}Li, and Bu^{t}Li, and lithium amides, for example LiN(Pr^i)$_2$ (LDA) and lithium 2,2,6,6-tetramethylpiperidide (LTMP), is by far the most widely employed reaction of this type ⟨79OR(26)1, 82ACR306⟩. Although both α- and β-lithiations of alkenes and arenes have been observed, α-lithiation appears to be generally favored in cases where both are possible. Some of the most commonly observed regioselectivity patterns are shown in Scheme 19. In the absence of α-hydrogens, vinyl derivatives may undergo β-metallation as shown in Equation (28) ⟨86JA304⟩.

(28)

2.19.6 HYDROMETALLATION (METHOD VI)

Hydrometallation of alkynes producing vinylmetals can be *syn*-stereoselective, *anti*-stereoselective, or nonstereoselective. All three types of disubstituted vinylmetals, that is α-, (E)-β-, and (Z)-β-substituted vinylmetals, are, in principle, accessible via regio- and stereoselective hydrometallation. Of the three possible trisubstituted derivatives, *cis*- and *trans*-α,β-disubstituted vinylmetals are similarly accessible via hydrometallation (Scheme 20). (Here, the *cis* and *trans* geometry refers to the stereochemistry of the two nonmetallic substituents. To minimize possible confusion the less precise *cis–trans* nomenclature is deliberately chosen over the more precise and hence inflexible (E)–(Z) nomenclature.) On the other hand, β,β'-disubstituted vinylmetals are not accessible via hydrometallation (cf. Section 2.19.7). Hydrometallation can also be applicable to the synthesis of the parent vinylmetals. Its synthesis does not require any stereoselective process unless labelling with D or T is involved. Furthermore, its synthesis must avoid a general tendency of hydrometallation reactions to undergo dihydrometallation to give dimetalloalkanes.

2.19.6.1 *syn*-Hydrometallation of Alkynes

Addition of metal–hydrogen bonds to alkenes and alkynes in a *syn* manner is referred to as *syn*-hydrometallation. In principle, any metal hydrides capable of providing a low-lying empty orbital should be able to participate in the concerted *syn*-hydrometallation reaction as suggested by the HOMO–LUMO interaction scheme shown in Scheme 21. In the absence of any other factors, the sum of C—H and M—C bonds is expected to be stronger than the sum of C—C π-bond and M—H bond. In reality, other factors do play significant roles, which affect not only thermodynamics but also kinetics. Some noteworthy factors include π-complexation, solvation, and aggregation. Other competitive processes, including *anti*-hydrometallation, may also limit the scope of *syn*-hydrometallation. Specifically the high tendency of group 1 and 2 metal hydrides to form stable and insoluble aggregates severely limits the scope of *syn*-hydrometallation involving these metals, and nearly the same statement is applicable to group 11 and 12 metals as well. The most versatile and

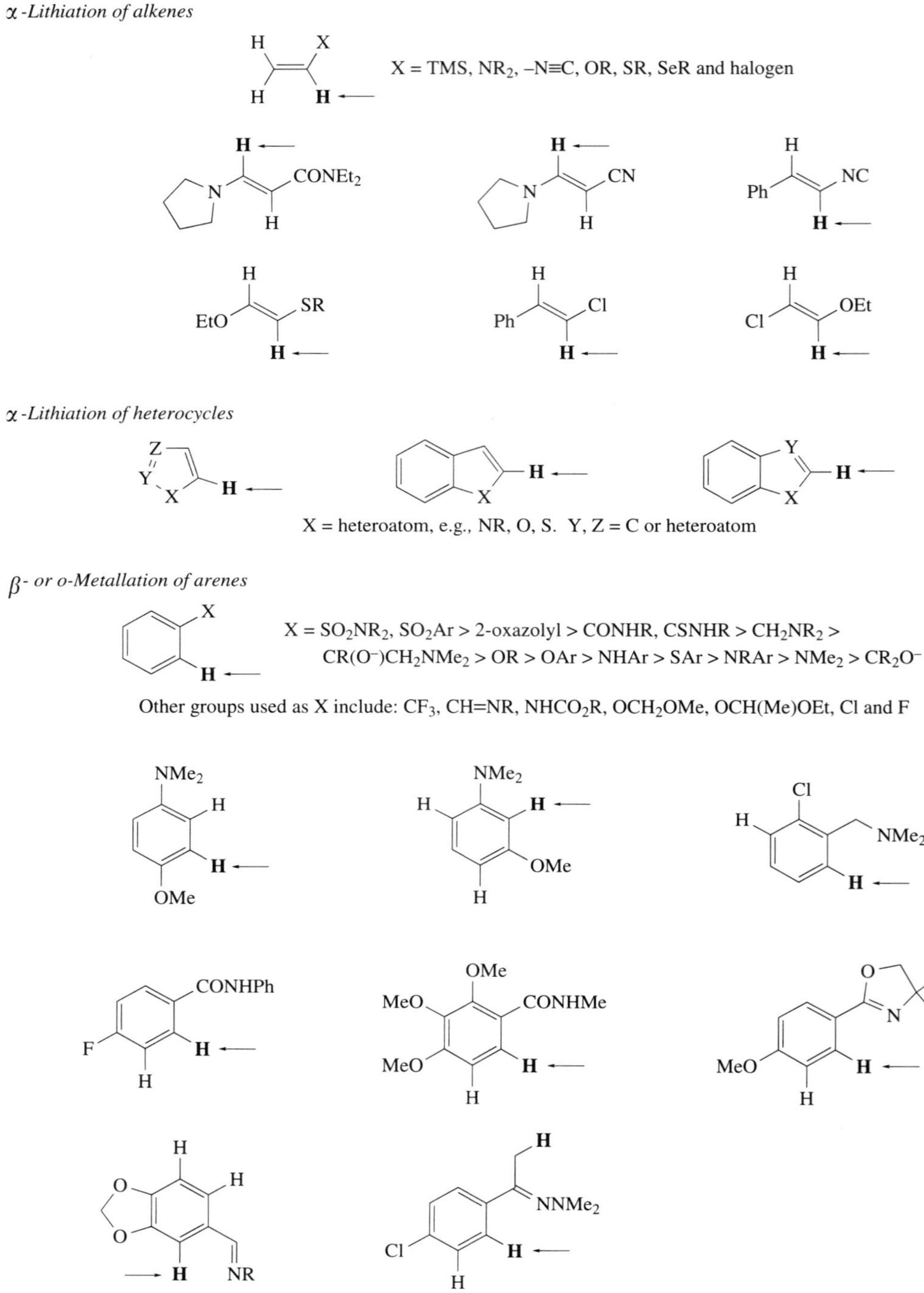

Scheme 19

synthetically useful *syn*-hydrometallation reaction involving a main group metal is hydroboration ⟨B-75MI 219-01, B-88MI 219-01⟩ discussed in Chapter 2.18. Aluminum and other heavier group 13 metals also undergo *syn*-hydrometallation with alkynes. Although hydrostannation of alkynes can readily take place, its stereochemistry is very much dependent on substrates and reaction conditions, as discussed later. Essentially all d-block transition metals must be capable of participating in *syn*-hydrometallation, and many have been observed and/or suggested as part of catalytic cycles. And

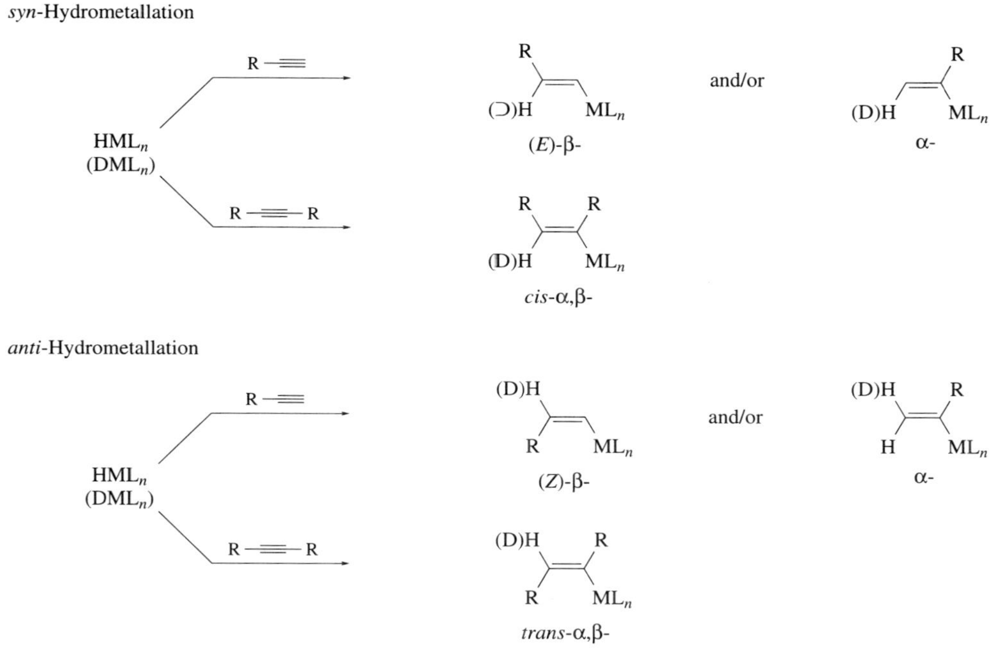

Scheme 20

yet, the number of examples of well-documented stoichiometric *syn*-hydrometallation reactions producing vinylmetals as discrete and identifiable species is relatively small. The most widely investigated case is hydrozirconation. In summary, hydroalumination and hydrozirconation, in addition to hydroboration, dominate the current scope of *syn*-hydrometallation of alkynes.

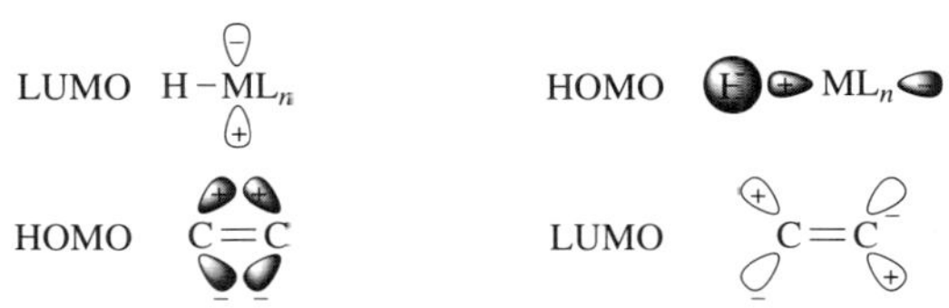

Scheme 21

2.19.6.1.1 **syn-*hydroalumination***

Dialkylaluminum hydrides, such as diisobutylaluminum hydride (dibal-H), react with both terminal and internal alkynes, usually at about 50 °C, to give the corresponding *syn*-hydroalumination products ⟨56CB444, 60LA(629)222⟩. This reaction has been extensively reviewed ⟨70HOU(13/4)23, 79COC(3)1013, 82COMC-I(1)555, 91COS(8)733⟩. Although the reaction of acetylene itself with dibal-H is not clean, alkyl-substituted terminal alkynes react with dibal-H either neat or in hydrocarbon solvents, to give (*E*)-1-alkenylalanes typically in ⩾90% yields along with a ⩽5% yield each of alkynylalanes and 1,1-dialuminoalkanes (Equation (29)). Both stereoselectivity and regioselectivity appear to be essentially 100%. The corresponding reaction run in THF and NEt_3 lead to di- and trihydroalumination, respectively. The reaction of other types of alkynes can be complicated by various undesirable side-reactions. Arylated ethynes and conjugated enynes usually give alkynylalanes to the extents of 20–30%. More remote aryl and alkenyl substituents, however, do not seriously affect the course of reaction. The regioselectivity of hydroalumination of internal alkynes depends on the substituents. Unless the two substituents are radically different, mixtures of the two possible regioisomers result. With internal alkynes competitive carbometallation involving the alkenylalane products can also be a serious side-reaction. The stereochemistry of hydroalumination of alkynylsilanes and alkynlgermanes is condition-dependent. In ether (40 °C) or tertiary amines (60 °C) the reaction is >95% *syn*-selective, whereas it is ca. 96% *anti*-selective in hydrocarbons. Alkynlstannanes and 1-haloalkynes react with dibal-H only to undergo displacement of these hetero-substituents with aluminum. These facts are summarized in Table 6. The regiochemistry of hydro-alumination tends to be reversed with alkynes terminally substituted with electron-donating N, P, O, and S groups. These heteroatoms in other locations of alkynes can also cause complications, such as deactivation of dibal-H through complexation and cleavage of carbon–heteroatom bonds, especially in the propargylic position. The use of alkenylalanes in organic synthesis has been extensively reviewed ⟨B-80MI 219-03, 84OR375⟩.

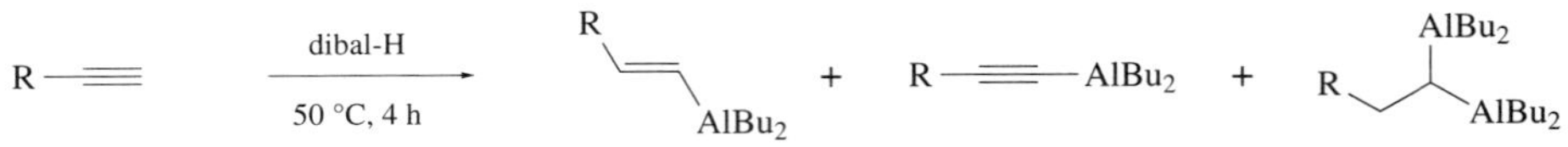
(29)

R = Bu^n	90%	6%	4%
R = cyclohexyl	94%	2%	4%
R = Bu^t	97%	trace	trace

Table 6 Hydroalumination of alkynes with diisobutylalane (dibal-H).

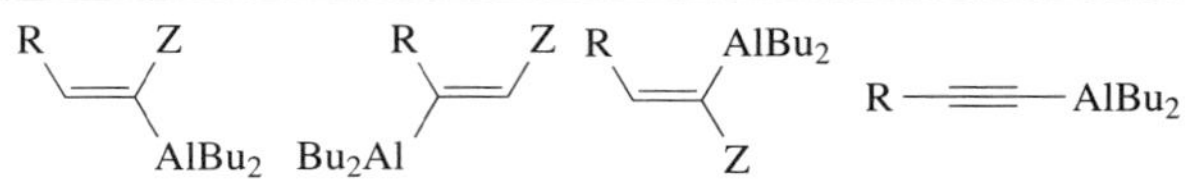

Alkyne (RC≡CZ)	*Temp.* (°C)	*Product* (%)				*Ref.*
Bu^{n}C≡CH	50	88	2		6	B-78MI 219-01
PhC≡CBu^t	50	0	100			71JOC3520
PhC≡CBr	100				100	71JOC3520
PhC≡C–TMS	60	96		4		71JOC3520
PhC≡CSnMe$_3$	20				100	71JOC3520

2.19.6.1.2 syn-*hydrozirconation*

The reaction of terminal alkynes with $HZrCp_2Cl$ gives stereo- and regioselectively (*E*)-1-alkenylzirconium derivatives (Equation (30)) ⟨71JOM(27)373, 75JA679⟩. Acetylene itself reacts similarly ⟨82TL63, 83OM1555⟩. The corresponding reaction of internal alkynes generally is also essentially 100% stereoselective, but gives mixtures of two possible regioisomers, which can be equilibrated in the presence of a catalytic amount of $HZrCp_2Cl$ to significantly improve the regioselectivity, as indicated by the results summarized in Table 7. A plausible mechanism of this regioisomerization reaction is shown in Equation (31) ⟨75JA679⟩.

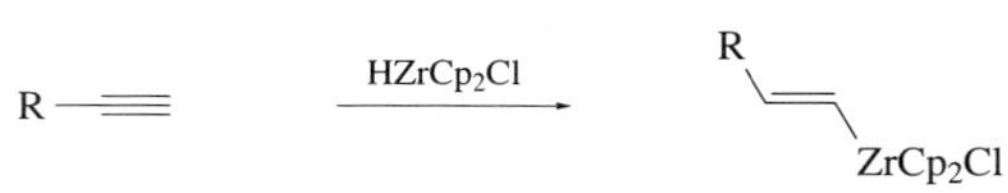
(30)

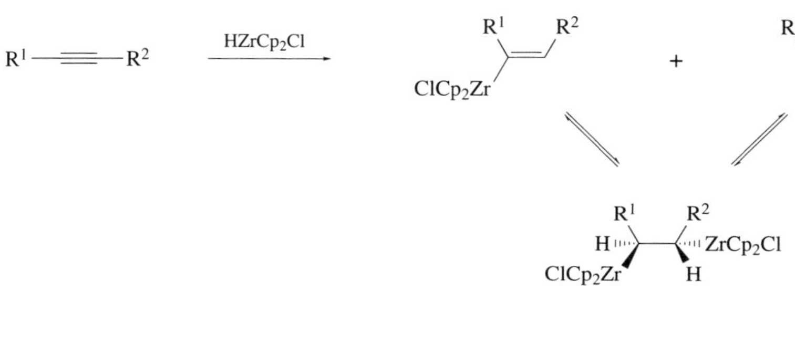
(31)

Table 7 Hydrozirconation of internal alkynes and regiochemistry of the products ⟨75JA679⟩.

$R^1C{\equiv}CR^2$	$ClCp_2Zr(R^1)C{=}CHR^2/R^1HC{=}C(R^2)ZrCp_2Cl$	
	Initial ratio	*Final ratio*
$MeC{\equiv}CEt$	55/45	89/11
$MeC{\equiv}CPr^n$	69/31	91/9
$MeC{\equiv}CPr^i$	84/16	>98/2
$MeC{\equiv}CBu^t$	>98/2	>98/2

Although $HZrCp_2Cl$ is a reasonably satisfactory hydrozirconating agent, its preparation involving filtration under an inert atmosphere is not only cumbersome but also can be complicated by the formation of undesirable species such as Cp_2ZrH_2. To alleviate the former difficulty some procedures for *in situ* generation of "$HZrCp_2Cl$" involving (i) mixing $LiAlH_4$ and Cl_2ZrCp_2 or $NaAlH_2(OR)_2$ and Cl_2ZrCp_2 ⟨84TL3407, 90TL7257⟩; and (ii) treatment of Cl_2ZrCp_2 with one equivalent of Bu^tMgCl ⟨91JOC2590⟩ have been developed. To eliminate Cp_2ZrH_2 a modified procedure for generating $HZrCp_2Cl$ involving the use of CH_2Cl_2 has been developed ⟨87TL3895⟩. Generally speaking, the scope of *syn*-hydrozirconation of alkynes lies somewhere between those of hydroboration and hydroalumination, hydroboration being most widely applicable. Thus, for example, alkoxyethynes ⟨77JA3168⟩ and propargyl alcohols ⟨77JA8045, 78TL4381, 80JA1333⟩ are readily hydrozirconated without complications observed in the corresponding hydroalumination reactions.

Similarly to alkenylalanes, alkenylzirconium derivatives are useful for the selective preparation of a wide variety of alkenes via displacement of Zr with H, D, halogen, other heteroatoms, carbon, and other metals. These aspects have been extensively reviewed ⟨76AG(E)333, 85MI 219-01, 88S1⟩. Of various carbon–carbon bond-forming reactions of alkenylzirconium derivatives carbonylation with CO and isonitriles ⟨87TL295, 88TL1631, 88TL3445⟩, Al-promoted acylation ⟨77JA638, 79JA3521⟩, Ni- or Pd-catalyzed cross-coupling ⟨77JA3168, 78JA2254, 78TL1027⟩, and Ni-catalyzed or Cu-promoted conjugate addition ⟨77JA8045, 78TL4381, 80JA1333⟩ are particularly noteworthy. In the Al-promoted acylation, Ni- or Pd-catalyzed cross-coupling, and Ni-catalyzed conjugate addition reactions mentioned above, the alkenyl group of alkenylzirconium derivatives must be transferred from Zr to Al, Ni, or Pd, respectively. Transfer of alkenyl groups should, in principle, be favorable in cases where metals that are more electronegative than Zr, such as Cu, Hg, Si, and Sn, are involved, as discussed in Section 2.19.10.

2.19.6.1.3 Other syn-*hydrometallation reactions of alkynes*

No other *syn*-hydrometallation reactions involving just one metal have been extensively developed as generally applicable and synthetically useful stoichiometric routes to alkenylmetals. The reaction of 5-decyne with a reagent generated *in situ* from KBu^s_3BH (two equivalents) and CuI in THF gives after hydrolysis, an 88 : 12 mixture of (*Z*)- and (*E*)-5-decenes ⟨74CC762⟩. Although the (*Z*)-5-decene is thought to be formed via hydrocupration, no firm data supporting this notion have been obtained. Similarly, it has been reported that $CuH\text{-}MgX_2$ complexes, prepared *in situ* from either $MgH_2\text{-}CuX$ ⟨78JOC757⟩ or $NaH\text{-}CuX\text{-}MgX_2$ ⟨87CC495⟩, react with terminal alkynes to give alkenylcoppers. Hydrometallation reactions of alkenes involving some other transition metals, such as Sc ⟨87JA203, 90SL74⟩, Y ⟨92JOC3266⟩, and various lanthanides ⟨85JA8103, 85JA8111⟩, for example La, Nd, Sm, and Lu, have been reported. However, very few have been applied to hydrometallation of alkynes. One rare example of such reactions is shown in Equation (32) ⟨84JA1291⟩.

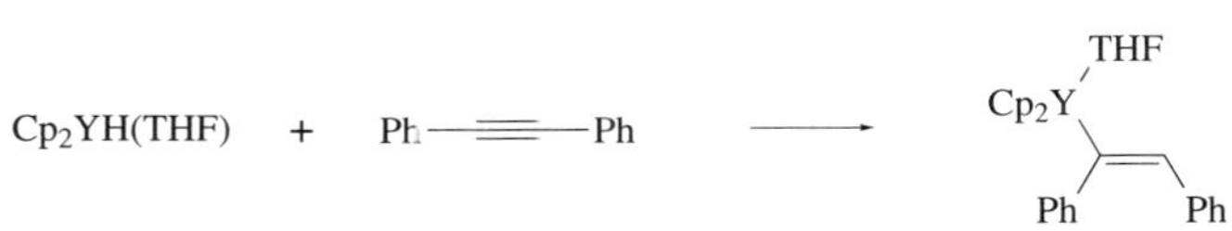

(32)

Much more promising are transition metal-catalyzed *syn*-hydrometallation reactions, of which the Pt-catalyzed hydrosilation reaction of alkynes (Chapter 2.18) is currently the most significant example. Although limited in scope, the reaction of Bu^iMgBr with silylated propargyl alcohols catalyzed by Cp_2TiCl_2 proceeds regio- and stereoselectivity (Equation (33)) ⟨83TL1809, 85JOM(285)53⟩.

Hydroboration and hydroalumination of alkenes can be catalyzed by various transition metal complexes such as those containing Rh ⟨91CR1179⟩, Ni ⟨68JOM(12)P33⟩, and Zr ⟨80TL1501⟩. However, little is known about the corresponding reactions of alkynes.

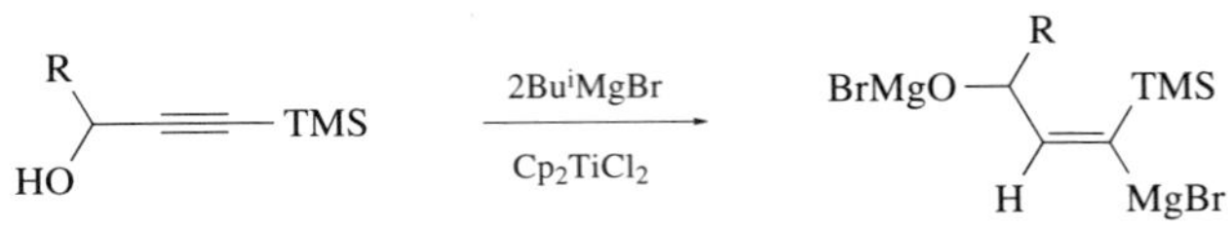

(33)

2.19.6.2 *anti*-Hydrometallation Reactions of Alkynes

Certain metal hydrides undergo *anti*-addition of the metal–hydrogen bond to alkynes. Some representative examples are shown in Equations (34) and (35), and Scheme 22. A few notable features common to these reactions are as follows. The metal hydride reagents are coordinatively saturated 8- or 18-electron species. Electron-deficient alkynes appear to be more reactive than their electron-rich counterparts. Although their mechanisms are not very clear, the features mentioned above suggest some one-electron transfer mechanisms involving alkenyl radical anionic species. Concerted $[_{\pi}2_a + {}_{\sigma}2_s]$ and other nonradical mechanisms cannot be ruled out, however. With the exception of the *anti*-hydroalumination reaction ⟨66T1741⟩ involving $LiAlH_4$, $NaAlH_2(OCH_2CH_2OMe)_2$ (Red-Al), and other hydridoaluminate reagents, the *anti*-hydrometallation reactions of alkynes such as those involving Mo and W ⟨76JMOC285⟩ have not been widely used for synthetic purposes.

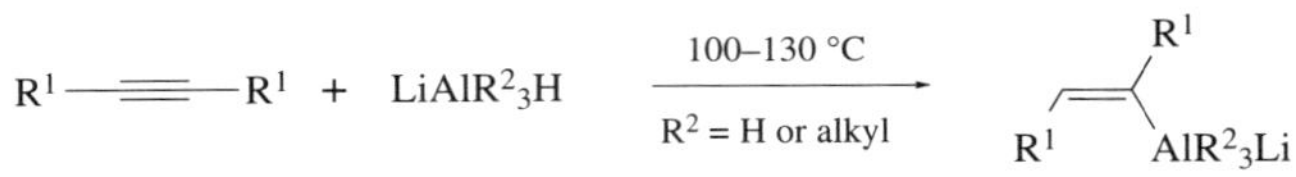

(34)

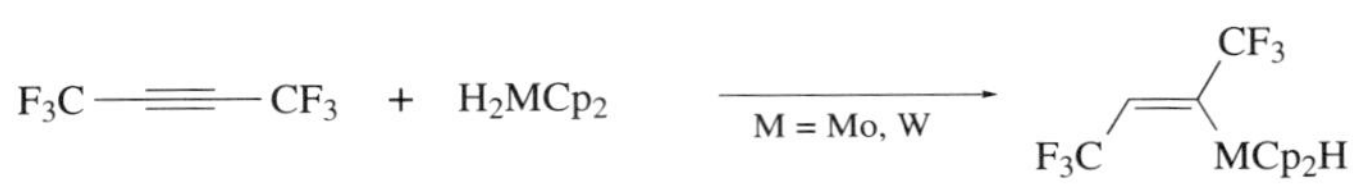

(35)

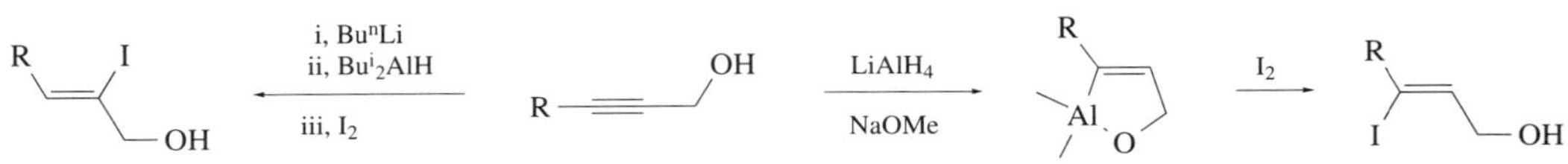

Scheme 22

The reaction of propargylic alcohols with hydridoaluminate reagents is particularly important in the preparation of various trisubstituted allylic alcohols. The regiochemistry of the reaction is very much dependent on the choice of reagents. The use of $LiAlH_4$ and NaOMe leads to the formation of 1-oxa-2-alumina-3-cyclopentene derivatives, which, on iodinolysis, are converted to (*Z*)-3-iodo-2-alken-1-ols ⟨67JA4245⟩, whereas sequential treatment of propargyl alcohols with Bu^nLi, dibal-H, and I_2 affords (*Z*)-2-iodo-2-alken-1-ols (Scheme 22) ⟨70JA6314⟩. These allyl alcohol derivatives can be further converted to a variety of trisubstituted alkenes of defined regio- and stereochemistry, thereby providing a selective general route to natural products of terpenoid origin.

2.19.6.3 Stereorandom Hydrometallation Reactions of Alkynes

There are some hydrometallation reactions the stereochemistry of which is strongly substrate-dependent and often unpredictable. In some special cases, however, they can be highly stereoselective. The currently most representative example of this class of hydrometallation reactions is hydrostannation. The hydrostannation reaction of alkynes is normally carried out at 50–100 °C either in the presence or in the absence of radical initiators ⟨67JOM(9)285, 67JOM(9)295⟩. Hydrostannation most likely proceeds via radical and/or polar mechanisms. As the results summarized in Table 8 indicate, the reaction usually gives a mixture of regio- and stereoisomers ⟨67JOM(9)285, 67JOM(9)295⟩. Highly regio- and stereoselective hydrostannation reactions, such as those cases where alkynes containing CH_2OTHP ⟨75JOC2265⟩, CN ⟨67JOM(9)285, 67JOM(9)295⟩, and $SnBu_3$ ⟨74JA5582⟩

are involved, provide selective routes to heterosubstituted alkenes (Equations (36)–(38)). On the other hand, the inherently stereo-random nature of hydrostannation also offers some advantages which can be synthetically exploited, as in the conversion of 1,4-pentadiyne into a 3-stanna-1,4-cyclohexadiene derivative (Equation (39)) ⟨71JA1804⟩.

Table 8 Hydrostannation of terminal alkynes.

R of R_3SnH	*R^1 of $R^1C{\equiv}CH$*	*Temp. (°C)*	*Time (h)*	*Total yield (%)*	*Product composition (%)*			*Ref.*
					I	*II*	*III*	
β-*syn*-selective								
Bu	CH_2OTHP	80	2	89	100	0	0	75JOC2265
Bu	$SnBu_3$			88	100	0	0	74JA5582
								67DOK96
β-*anti*-selective (>90%)								
Et	OEt	50	1.5	100	5	91	4	67JOM(9)285
								67JOM(9)285
α-selective								
Et	CN	20	1	100	0	0	100	67JOM(9)285
								67JOM(9)285
β-selective but stereorandom								
Me	Bu	60	7	85	69	29	2	67JOM(9)285
								67JOM(9)285

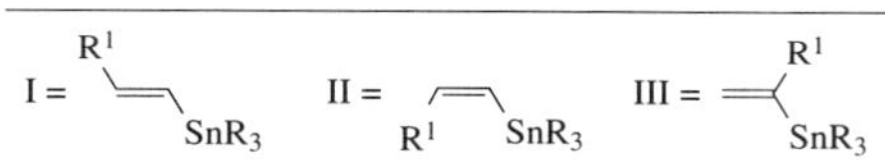

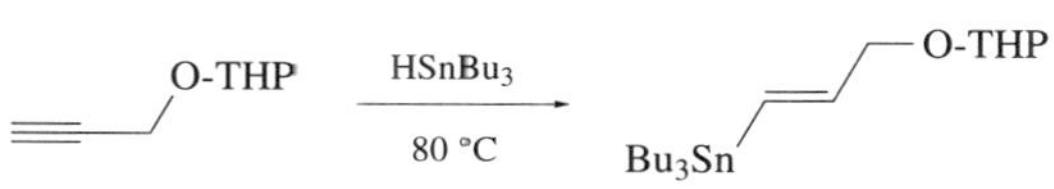

(36)

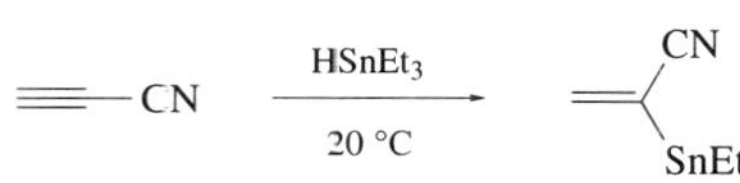

(37)

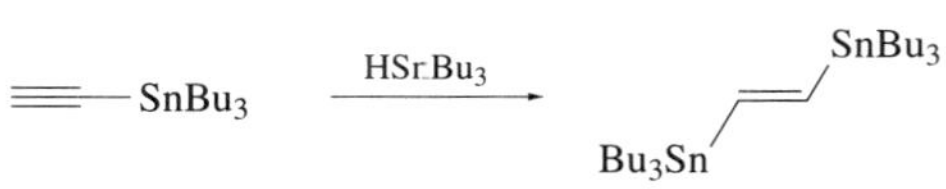

(38)

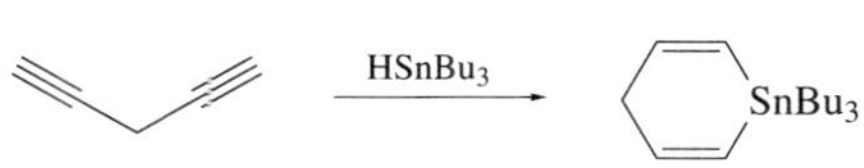

(39)

2.19.7 CONTROLLED CARBOMETALLATION (METHOD VII)

Carbometallation of alkynes can be *syn*-stereoselective, *anti*-stereoselective, or non-stereoselective. With unsymmetrical alkynes, two different regioisomers can be obtained. Since the products of carbometallation are organometals, they may undergo further carbometallation reactions to give oligomeric and polymeric products. In this section, however, emphasis is placed on controlled *syn*-monocarbometallation reactions (Scheme 23). Such monocarbometallation processes can also be cyclic, as detailed later in this section.

Carbometallation may proceed by a variety of mechanisms. However, the concerted four-center *syn*-addition (Scheme 24) and the six-centered cyclic addition involving allylic, allenic, and pro-

syn-Monocarbometallation

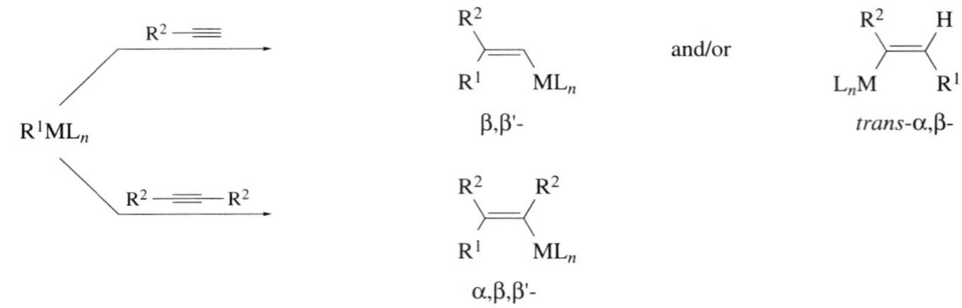

Scheme 23

pargylic organometals (Scheme 25) appear to be the two most representative processes. The former reaction is closely analogous to *syn*-hydrometallation, for which the presence or ready availability of a low-lying empty metal orbital is one crucial requirement. Some other significant mechanisms have also been found to be operative in carbometallation. Various main group and transition metals participate in carbometallation reactions. Of current synthetic importance are, however, those involving Li, Mg, Zn, B, Al, Cu, Ti, Zr, and Pd, even though it may be anticipated that those involving various other transition metals including lanthanides are likely to play increasingly important roles in organic synthesis.

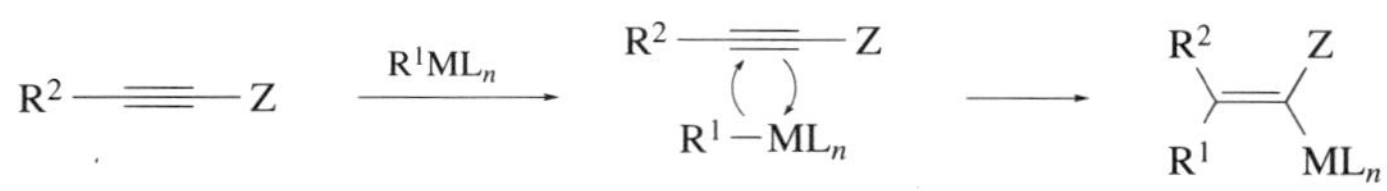

Scheme 24

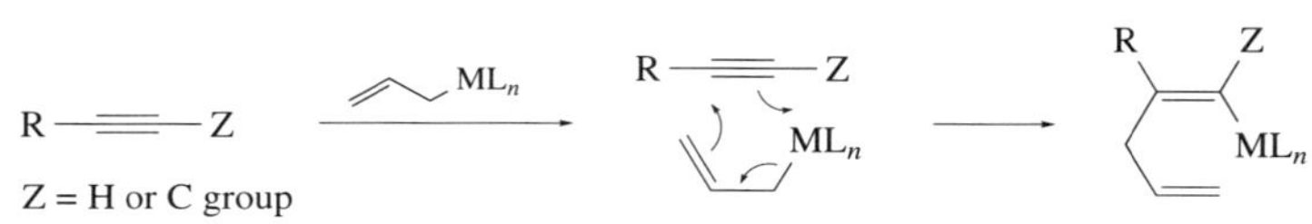

Scheme 25

2.19.7.1 Carbometallation with Organometals Containing Group 1 and 2 Metals

The intermolecular reactions of alkynes with organolithiums, organomagnesiums, and organozincs are generally sluggish and are of limited synthetic utility. With terminal alkynes, abstraction of the terminal hydrogen usually occurs preferentially. Nonetheless, there are a number of specific reactions of synthetic interest, of which the following are representative. In cases where alkynes are highly reactive and/or not readily prone to competing side reactions, *syn*-carbolithiation may be observed. In Scheme 26 such a *syn*-carbolithiation is, however, followed by stereoisomerization and Li–H exchange ⟨63JA3897⟩. Although still limited, cyclic carbolithiation of alkynes, especially alkynylsilanes, has recently been shown to be of considerable synthetic utility ⟨90TL493, 90TL627⟩. The formation of five-membered rings can be high-yielding, but six-membered rings are generally obtained in low yields (Equation (40)). A related reaction of Grignard reagents displays a similar scope (Scheme 27) ⟨84TL1999⟩.

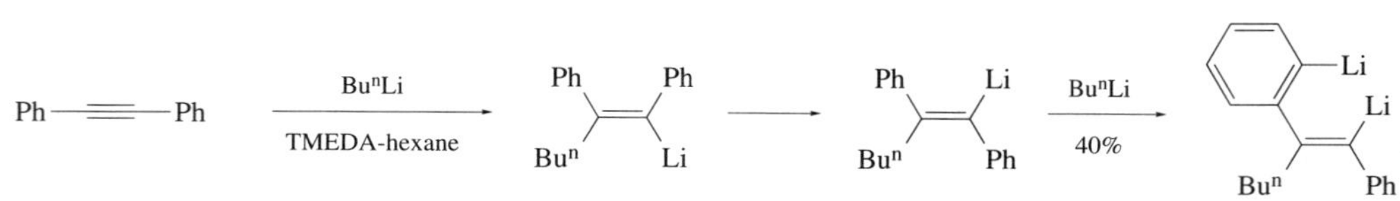

Scheme 26

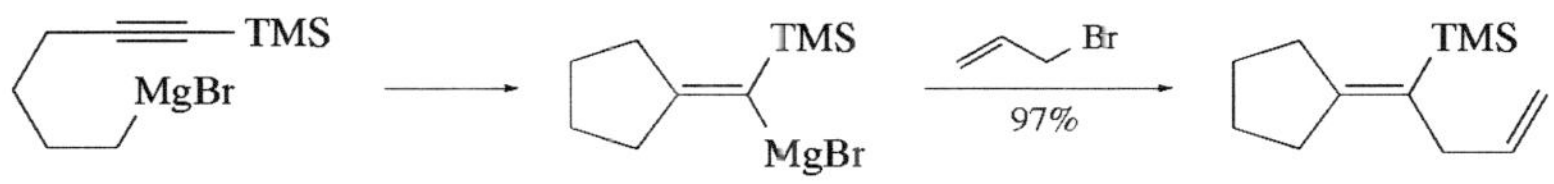

Scheme 27

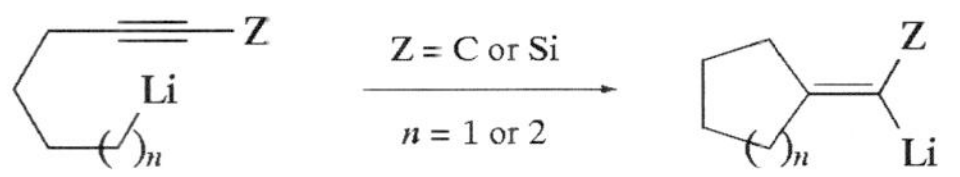

(40)

90% 95% 85% 50%

Allylic, propargylic, and allenic organometals containing magnesium and zinc are not only readily available but also react more readily with alkynes than other types of derivatives containing these metals. Metallated terminal alkynes appear to be particularly reactive ⟨79JOM(168)227, 79JOM(168)233, 83JA6761, 83JOC5409⟩, as shown in Schemes 28 and 29.

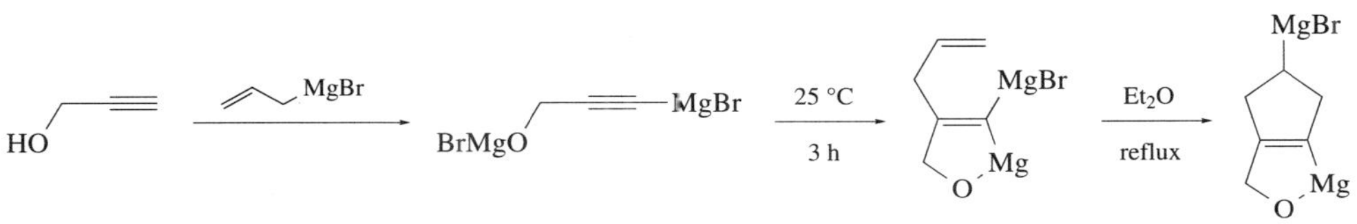

Scheme 28

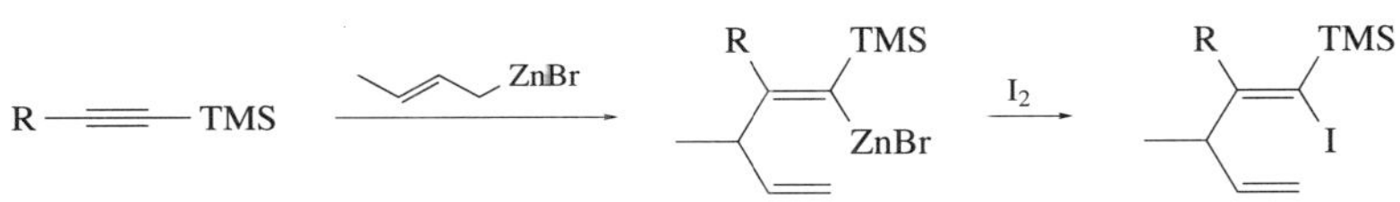

Scheme 29

One of the most promising ways of using polar organometals containing magnesium and zinc is to promote their carbometallation reactions with transition metal catalysts. Carbozincation of simple alkynes with alkylzinc (Equation (41)) ⟨83OM563⟩ and allylzinc (Scheme 30) ⟨83JA6761, 83JOC5409⟩ derivatives can be significantly catalyzed by I_2ZrCp_2. The regiochemistry of the Zr-catalyzed crotylzincation is opposite to that observed in Scheme 29, suggesting that the Zr-catalyzed reaction may involve a four-centered process, whereas the uncatalyzed reaction may proceed via a six-centered transition state. The Zr-catalyzed reaction has been applied to the synthesis of cycloalkenylzinc derivatives containing four- to six-membered cycloalkenes (Scheme 31) ⟨84JA6105, 88JOC913⟩.

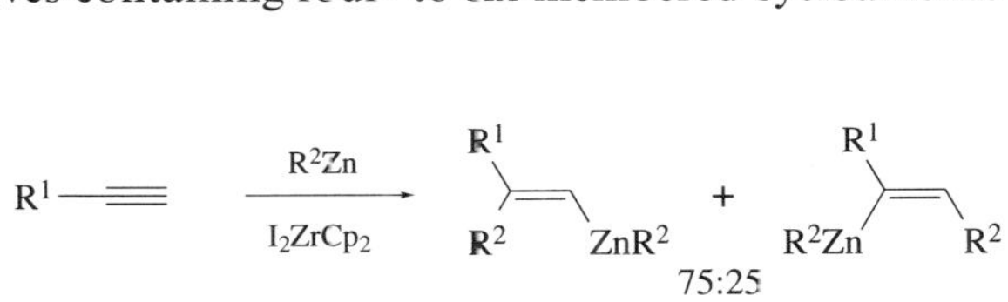

(41)

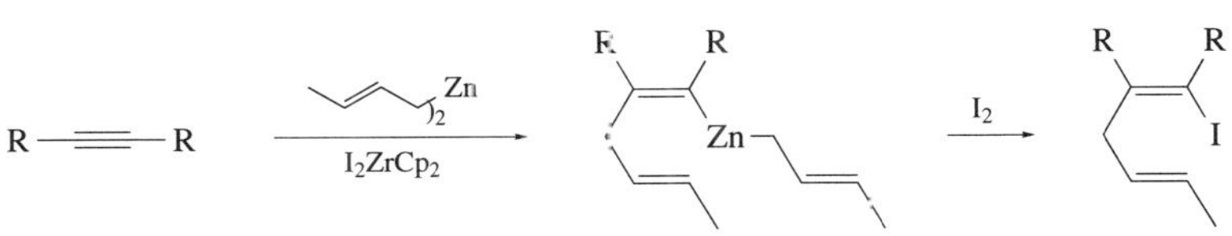

Scheme 30

Another synthetically useful carbometallation reaction of Grignard reagents involves catalysis by CuI (Scheme 32) ⟨79JOM(168)227, 79JOM(168)233⟩. This reaction is an *anti*-addition process similar to the *anti*-hydroalumination of propargyl alcohols. As such, this reaction represents rare examples of the preparation of stereo- and regio-defined tetrasubstituted alkenes with four different substituents.

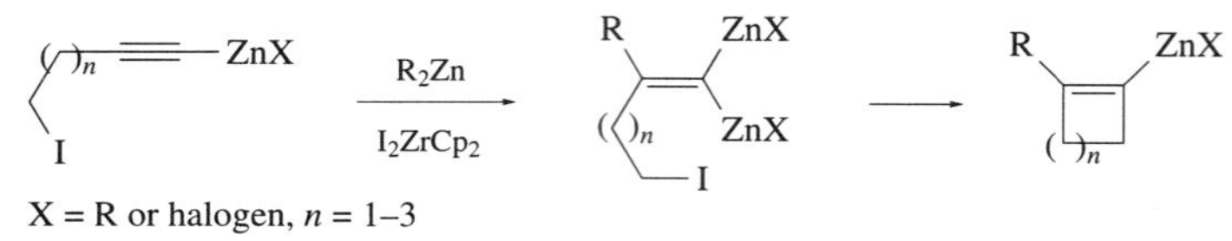

Scheme 31

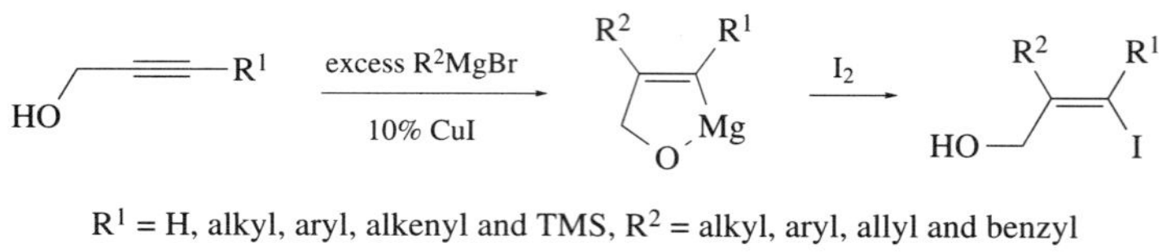

Scheme 32

Finally, it is noteworthy that some reactions of halogenated arenes with organolithiums and Grignard reagents lead to the formation of arylmetals that can arise via carbometallation of benzyne intermediates, as shown in Scheme 33 ⟨83TL4935, 85CC690⟩ and Scheme 34 ⟨85JOC3104, 85JOC5524⟩.

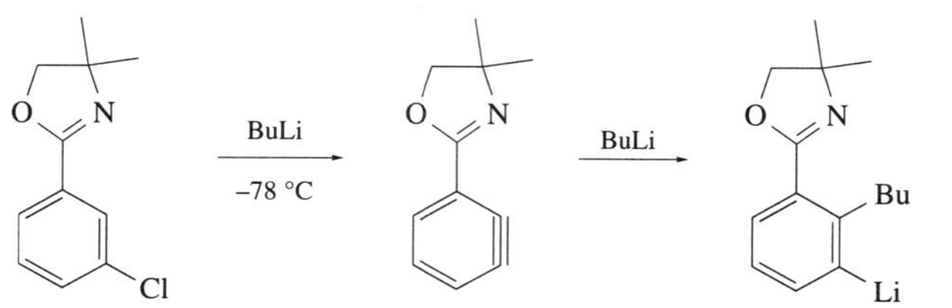

Scheme 33

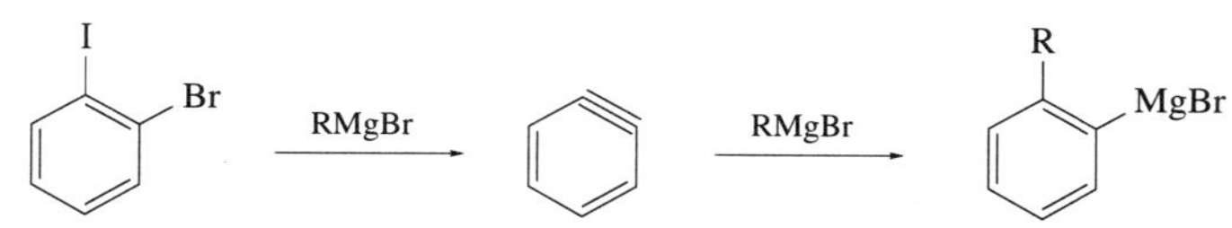

Scheme 34

2.19.7.2 Zirconium-catalyzed Carboalumination and Related Reactions

The reactions of alkynes with organoalanes have long been known. However, uncatalyzed carboalumination of alkynes is generally of low synthetic utility. Acetylene itself reacts with trialkylalanes to give (*Z*)-1-alkenylalanes in good yields (Equation (42)) ⟨60LA(629)222⟩. However, the corresponding reaction of terminal alkynes gives mainly alkynylalanes, and that of internal alkynes has a tendency to produce dimeric and oligomeric products (Equation (43)) ⟨74JA6815⟩. Some of these difficulties can be readily overcome by using Cl_2ZrCp_2 and other zirconocene derivatives as catalysts. The Zr-catalyzed carboalumination proceeds well with Me_3Al ⟨78JA2252, 81PAC2333, 85JA6639, 86OCS467⟩, allyldialkylalanes ⟨84TL5863⟩, and benzylalanes ⟨84TL5863⟩. With ethyl and higher alkyl derivatives, other reactions including hydroalumination can significantly compete with the desired carboalumination, and the carbocupration reaction discussed in Section 2.19.7.3 is generally superior.

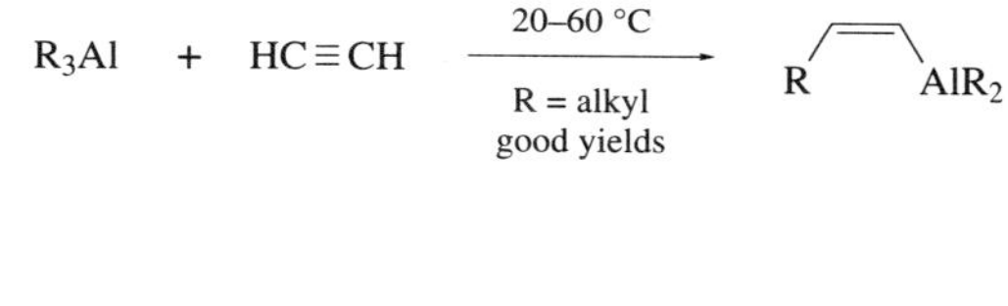 (42)

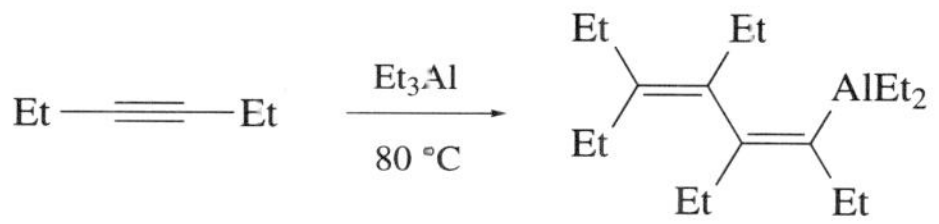

(43)

Specifically, 2-deuterio-1-phenylethyne reacts with two equivalents of Me_3Al and 0.05 to 1 equivalent of Cl_2ZrCp_2 to give in essentially quantitative yield a 95 : 5 regioisomeric mixture of two *syn*-addition products (Scheme 35) ⟨78JA2252, 81PAC2333, 85JA6639, 86OCS467⟩. Thus, the reaction is >98% *syn*-selective and 95% regioselective. Little or no abstraction of the terminal hydrogen takes place. The reaction of alkyl-substituted terminal alkynes proceeds similarly (Equation (44)). Internal alkynes also react (Equation (45)). In cases where the two substituents are different, mixtures of the two regioisomers usually result. A recent modification of the reaction conditions through the use of H_2O leads to a considerable acceleration of the reaction and somewhat more favorable results in some cases ⟨93AG(E)1068⟩. The Zr-catalyzed methylalumination can accommodate various heterofunctional groups such as OH, $OSiMe_2Bu^t$, SPh, I, and $SiMe_3$ ⟨81JOC4093, 82TL27⟩. Neither conjugated ⟨81JA2882⟩ nor isolated ⟨79S501⟩ alkenyl groups interfere with the reaction. Of particular synthetic utility are those alkynes containing heterofunctional groups in the propargylic or homopropargylic position. Interestingly and in marked contrast to some other known carbometallation reactions of propargyl and homopropargyl derivatives, this reaction is very highly stereoselective (>98%) and regioselective (generally 92–100%) (Equation (46)). Table 9 illustrates the scope of this methylalumination reaction.

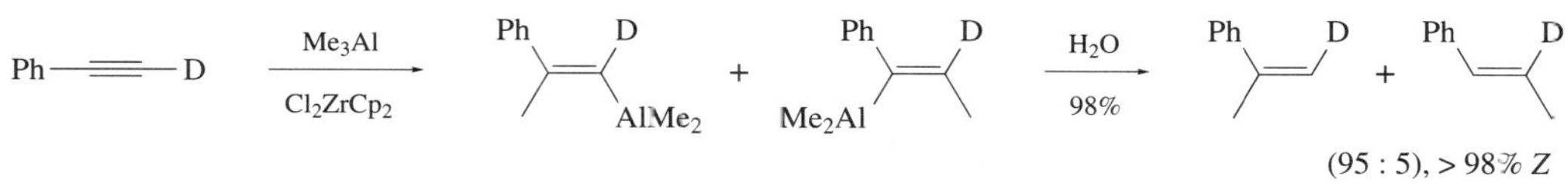

Scheme 35

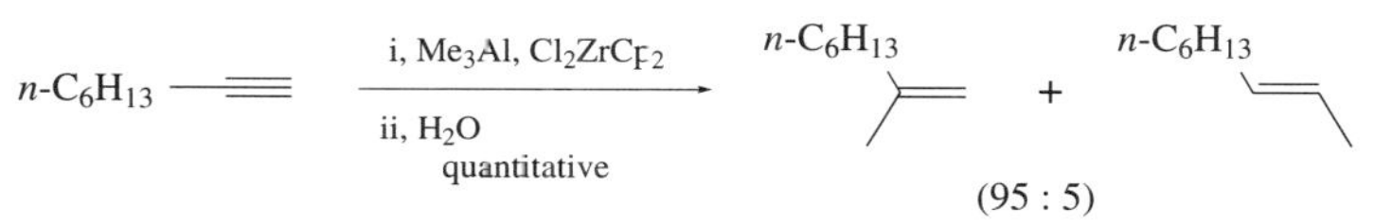

(44)

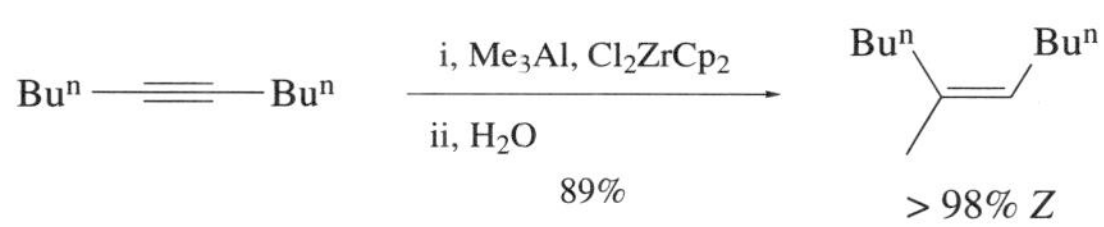

(45)

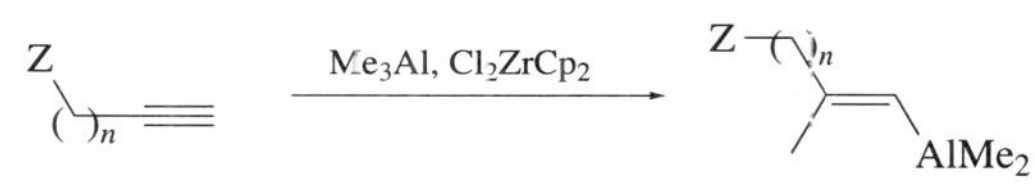

(46)

Z = OH, $OSiMe_2Bu^t$, SPh, I, TMS, etc. n = 1 or 2

Table 9 Carboalumination of terminal alkynes catalyzed by zirconocene dichloride.

Alkyne	*Organoalane*	*Major product*		*Ref.*
		Yield (%)	*Regioselectivity*[a] (%)	
$HC{\equiv}CHex^n$	Me_3Al	≥98	≥95	78JA2252, 85JA6639, 86OCS467
$HC{\equiv}CPh$	Me_3Al	≥98	≥95	78JA2252, 85JA6639, 86OCS467
$HC{\equiv}CCH{=}CH_2$	Me_3Al	≥86	≥95	81JA2882, 84OS31
$HC{\equiv}C(CH_2)_2CH{=}CMe_2$	Me_3Al	72	≥95	79S501
$HC{\equiv}CCH(OH)Pent^n$	Me_3Al	60	>98	81JOC4093
$HC{\equiv}CCH_2SiMe_3$	Me_3Al	63	≥95	82TL27
$HC{\equiv}C(CH_2)_2I$	Me_3Al	74	>98	81JOC4093
$HC{\equiv}C(CH_2)_4CH{=}CH_2$	$CH_2{=}CHCH_2AlMe_2$	98	75	84TL5863
$HC{\equiv}CHex^n$	$(PhCH_2)_3Al$	94	70	84TL5863

[a] The major regioisomer is terminally aluminated, and the stereoisomeric purity in each case is ≥98%.

From the viewpoint of synthetic applications, alkenylalanes are very useful reagents. They react with a wide variety of electrophiles, and the C—Al bond can be readily converted into various C—X bonds (X = H, D, halogen, Hg, B, Zr, Cu, C, etc.) with essentially complete retention of stereochemistry via protonolysis, deuterolysis, halogenolysis ⟨79S501⟩, transmetallation (see Section 2.19.10), reactions with various carbon electrophiles such as ClCOOEt, CO_2, $(CH_2O)_n$, $ClCH_2OMe$ ⟨78TL2357⟩, and epoxides ⟨80S1034⟩, preferably after "ate" complexation, and Pd- or Ni-catalyzed cross-coupling reactions ⟨78JA2254, 81JA2882⟩. Applications include the syntheses of allylic or homoallylic alcohols ⟨78TL2357, 80S1034⟩, conjugated dienes or enynes ⟨78JA2254⟩, 1,4-dienes ⟨81JA2882⟩ and 1,5-dienes or enynes ⟨80JA3298⟩. This methodology has been applied to the synthesis of various natural products such as geraniol ⟨78TL2357⟩, farnesol ⟨80JA3298⟩, monocyclofarnesol ⟨80JOC2526⟩, α-farnesene ⟨81JA2882⟩, dendrolasin ⟨80JOC5223⟩, mokupalide ⟨80JOC5223⟩, vitamin A ⟨91TL6683⟩, brassinolide ⟨80JA6580⟩, milbemycin ⟨82JA4708⟩, verrucarin ⟨84JOC1772, 84JOC4332⟩, udoteatrial ⟨83JOC1556⟩, and zoapatanol ⟨85JCS(P1)1589⟩.

The methylalumination of alkynes with Me_3Al-Cl_2ZrCp_2 appears to involve direct addition of a Me—Al bond, rather than a Me—Zr bond, to alkynes via a four-centered process which is facilitated by a $ZrCp_2$ species, as depicted in Scheme 36 ⟨81JA4985⟩. However, the reaction of 1-pentynyldimethylalane with either this reagent system or preformed $Cl(Me)ZrCp_2$ must involve direct addition of the Me—Zr bond, as depicted in Scheme 37 ⟨81JA1276⟩.

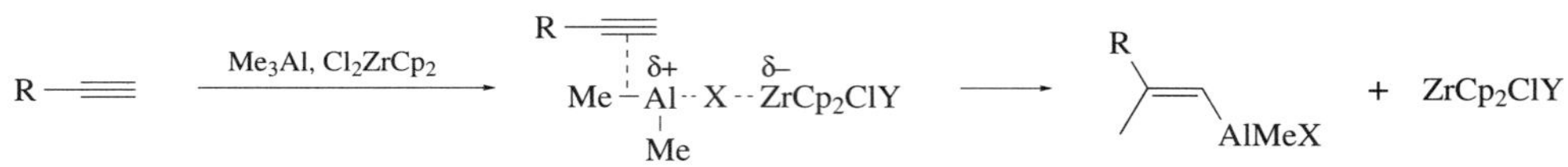

X, Y = Me or Cl

Scheme 36

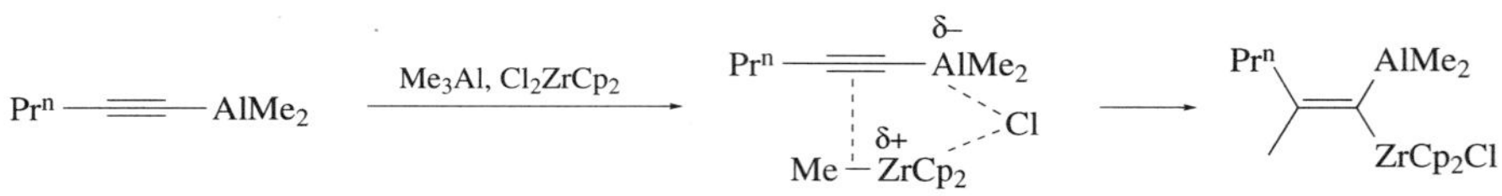

Scheme 37

The reaction of 4-bromo-1-(trimethylsilyl)-1-butyne and the corresponding 1-dimethylalumino derivative with Me_3Al-Cl_2ZrCp_2 gives the respective 2-methylcyclobutenylmetals. Whereas the alkynylalane reaction proceeds via regiospecific σ-cyclization ⟨84JA6105, 88JOC913⟩, the alkynylsilane reaction involves a novel nonregiospecific π-cyclization (Equations (47) and (48)) ⟨83JA6344⟩.

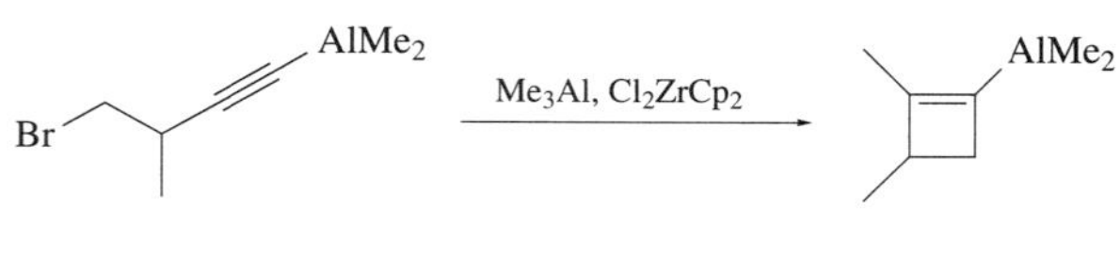

(47)

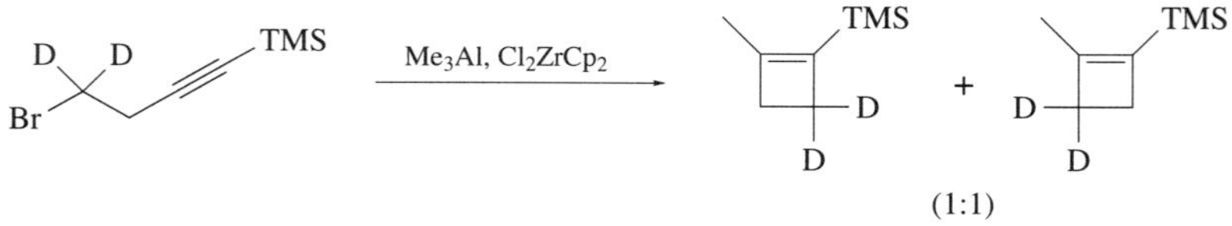

(48)

2.19.7.3 Carbocupration of Alkynes

Various types of organocoppers react with alkynes to give *syn*-addition products (Equation (49)) ⟨76MI 219-01, 81S841, 91COS(4)865⟩. Those organocoppers that have been used for this purpose include $RCu \cdot MgX_2$, R_2CuMgX, R_2CuLi, RCu(X)Li, where X = OBu^t, SPh, CN, etc. The choice of reagents depends on the alkyne structure, and the reaction conditions must be carefully selected for a given combination of reactants. Carbocupration is currently the most general controlled *syn*-

carbometallation reaction of alkynes. However, those cases where Me and allyl derivatives of copper are used, the reaction is too sluggish to be practically useful. In this sense, this reaction and the zirconium-catalyzed carboalumination are complementary to each other. Acetylene itself is very reactive, and all types of organocopper reagents add to acetylene (Equation (50)) ⟨84OS(62)1⟩. In addition to (*Z*)-alkenylcoppers, (*Z*,*Z*)-dienylcoppers can also be prepared using an excess of acetylene (4 equivalents) (Equation (51)) ⟨82TL5151⟩. Carbocupration of terminal alkynes is usually carried out using RCu · MgX at −35 to −15 °C (Equation (52)) ⟨71TL2583⟩. While nonfunctionalized internal alkynes do not usually undergo carbocupration, some intramolecular examples are known (Equation (53)) ⟨75JA7171⟩. Proximal heteroatoms in alkynes strongly affect the course of carbocupration ⟨76MI 219-01, 81S841, 91COS(4)865⟩. This is in sharp contrast with the predictable regioselectivity observed with the zirconium-catalyzed carboalumination of propargyl and homopropargyl derivatives (see Section 2.19.7.2). Some representative results are summarized in Table 10.

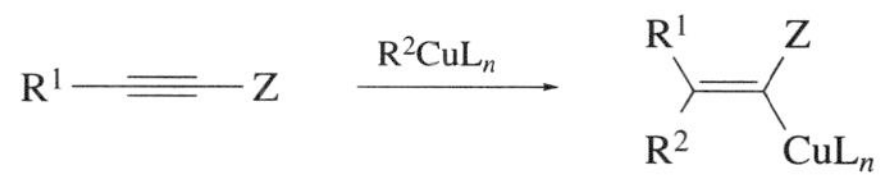

(49)

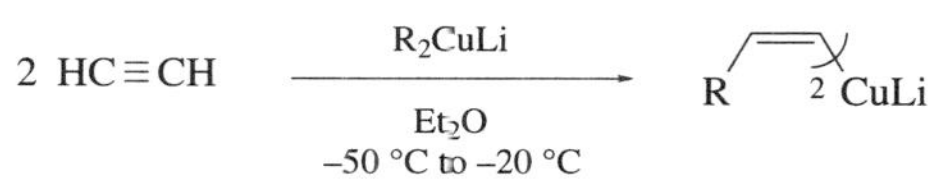

(50)

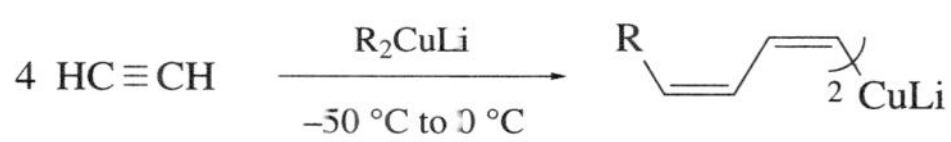

(51)

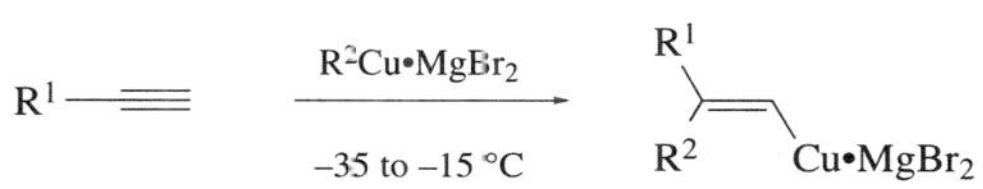

(52)

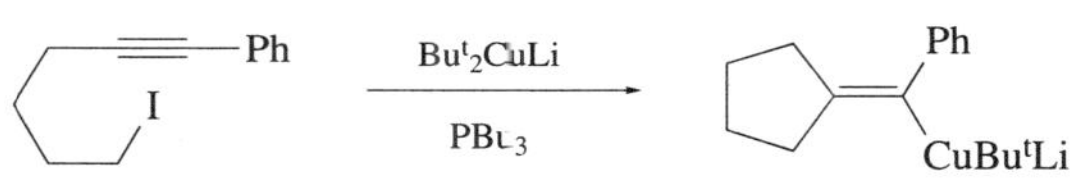

(53)

Table 10 Carbocupration of heterosubstituted alkynes ⟨76MI 219-01⟩.

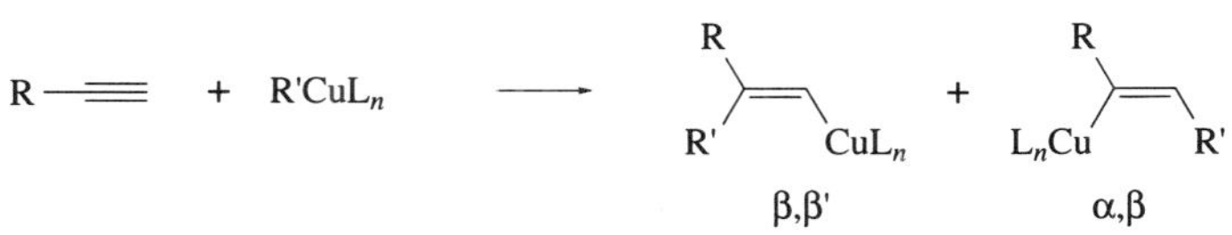

R of RC″CH	*R'CuL$_n$*	*Solvent*	*Product*	
			Total Yield (%)	β,β'/α,β
NPh_2	EtCu•$MgBr_2$	THF	80	100/0
OEt	EtCu•$MgBr_2$	THF	80	100/0
SEt	EtCu•$MgBr_2$	THF	95	0/100
CH_2OEt	BuCu•$MgBr_2$	Et_2O	65	97/3
CH_2OEt	BuCu•$MgBr_2$	THF	42	7/93
CH_2SEt	BuCu•$MgBr_2$	Et_2O	44	7/93
CH_2SEt	BuCu•$MgBr_2$	THF	32	98/2

Addition of Si—Cu ⟨81JCS(P1)2527, 84JCS(P1)1805⟩, Ge—Cu ⟨84TL2221, 84TL3217⟩, and Sn—Cu ⟨84TL2151, 85JOM163, 85S898, 85S899⟩ bonds to alkynes provides a highly stereoselective route to alkenylmetals containing Si, Ge, and Sn, respectively (Equations (54)–(56)). The regioselectivities of these reactions are, however, in the 70–90% range.

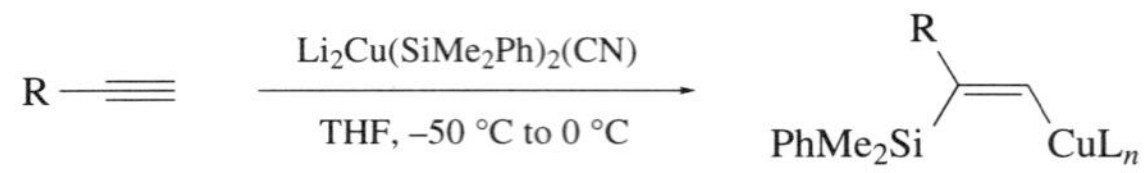

(54)

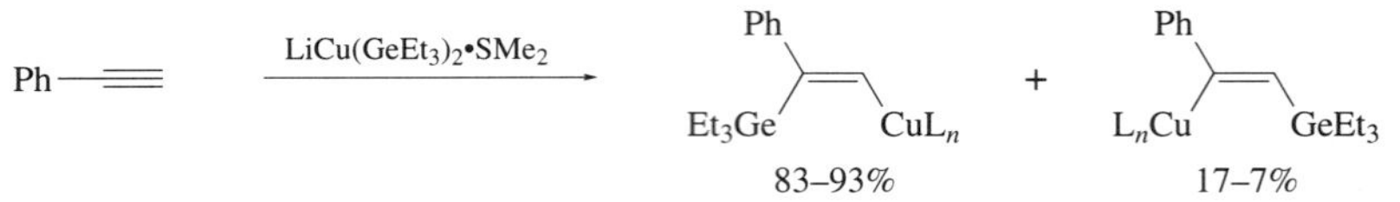

(55)

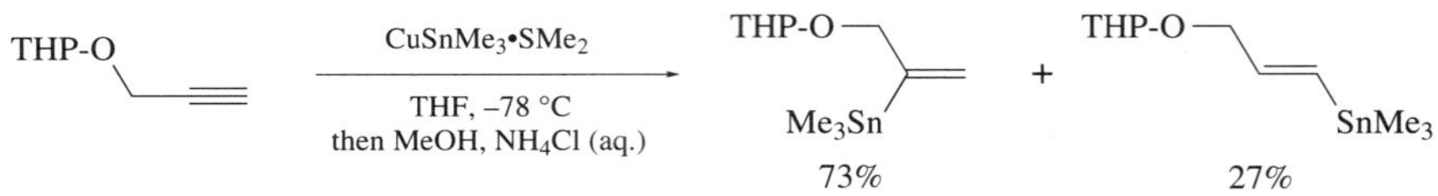

(56)

Alkenylcoppers are versatile intermediates. In addition to protonolysis, deuterolysis, and halogenolysis, the copper-containing group can be substituted with those containing Si, Sn, P, S, and so on ⟨76MI 219-01, 81S841, 91COS(4)865⟩. Substitution of copper with a carbon group can be achieved by direct cross-coupling with allylic, benzylic, and primary alkyl halides in the presence of hexamethylphosphoramide (HMPA), Pd-catalyzed cross-coupling with alkenyl, aryl, alkynyl, and acyl halides, as well as addition to aldehydes, ketones, CO_2, CS_2, ArNCO, epoxides, and enones ⟨91COS(4)865⟩.

2.19.7.4 Carbometallation with Other Organotransition Metals

Carbometallation of organotransition metals is a ubiquitous process which is, in principle, observable with all transition metals, since the only key requirement for concerted *syn*-carbometallation is the presence or ready availability of an empty metal orbital. In reality, however, controlled mono- or oligo-carbometallation of alkynes producing alkenylmetals as discrete products is relatively rare. For one thing, the second and third transition series metals with the exception of a few such as zirconium and molybdenum are quite expensive. Consequently, most of their synthetically attractive reactions are catalytic. In the following two subsections, some representative stoichiometric and catalytic carbometallation reactions with organotransition metals are discussed.

2.19.7.4.1 Stoichiometric carbometallation with organotransition metals

There are scattered examples of stoichiometric carbometallation with organotransition metals, as shown in Equations (57) and (58), and Schemes 38 and 39. The reaction shown in Equation (57) ⟨81JA1276⟩ is analogous to that shown in Scheme 37. The reaction shown in Scheme 38 ⟨88JA2575⟩ is an example of acylmetallation. The reaction of $CpCo(PPh_3)Me_2$ with PhC≡CPh ⟨80JA7003⟩ is accompanied by reductive elimination (Scheme 39). Nucleophilic attack with carbanions on alkynes complexed with transition metals can lead to an overall *anti*-carbometallation, as in Equation (58) ⟨84OM134⟩.

(57)

Scheme 38

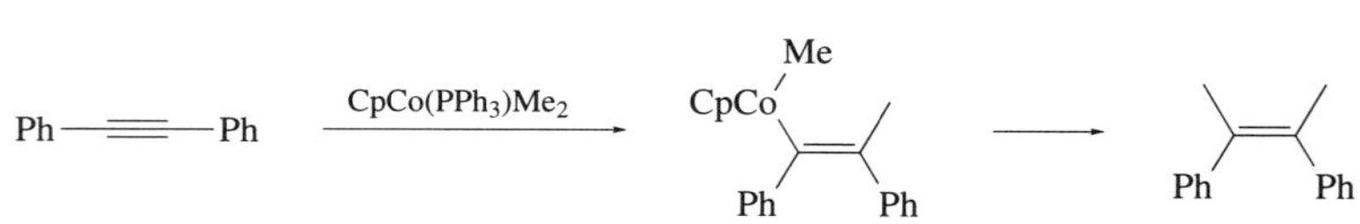

Scheme 39

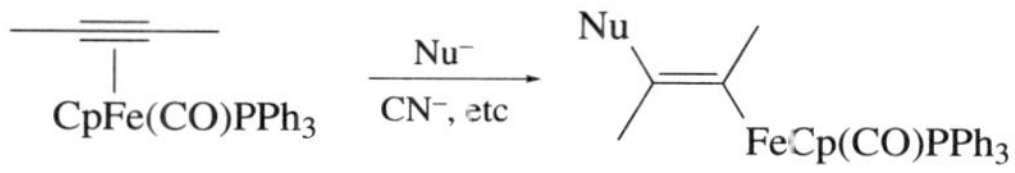

(58)

The reactions of alkynes with metal–carbene and metal–carbyne complexes lead to the formation of metallacyclobutenes and metallacyclobutadienes, respectively (Scheme 40). Some specific and stoichiometric examples follow. Diphenylacetylene reacts with either the Tebbe reagent, that is $Cp_2Ti(Cl)CH_2AlMe_2$, or titanacyclobutane derivatives to give the corresponding titanacyclobutene derivative (Equation (59)) ⟨80JA6149, 82JA7491⟩. Allenes react similarly (Equation (60)) ⟨83JA5490⟩. Methylenetitanocene is believed to be the active titanium species. Another example involving a late transition metal, iridium is shown in Scheme 41 ⟨84OM1223⟩. The reaction shown in Equation (61) ⟨82JA6808⟩ is an example of the metal–carbyne reactions. Although organometallic intermediates are usually not detected and characterized, the reactions of various Fischer-type carbene complexes containing chromium, molbydenum, and tungsten with alkynes are thought to give the corresponding metallacyclobutenes as key intermediates ⟨91COS(5)1065⟩. These intermediates are further converted under the reaction conditions into various types of products, such as cyclobutenones, indenes, and naphthol derivatives. A plausible mechanistic scheme is shown in Scheme 42, although some alternative mechanisms have also been proposed.

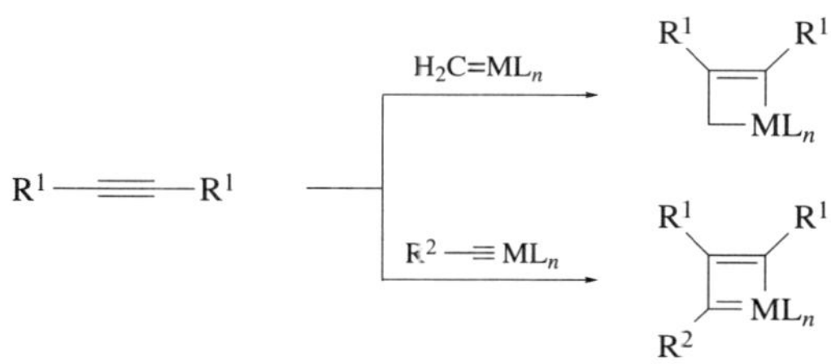

Scheme 40

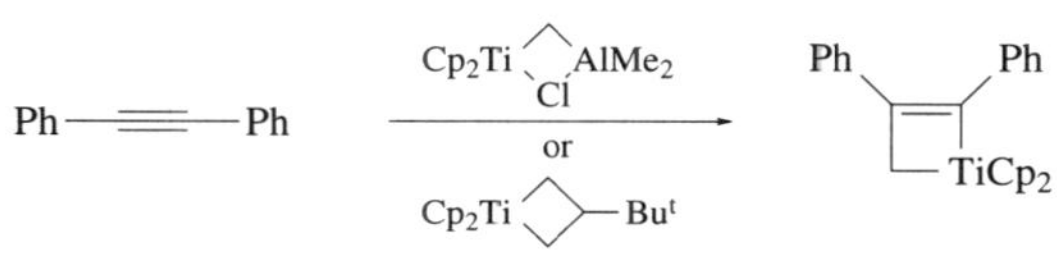

(59)

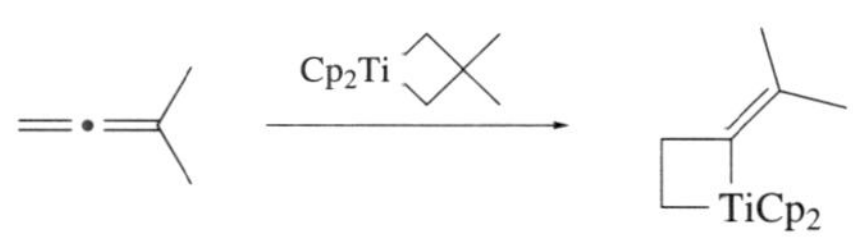

(60)

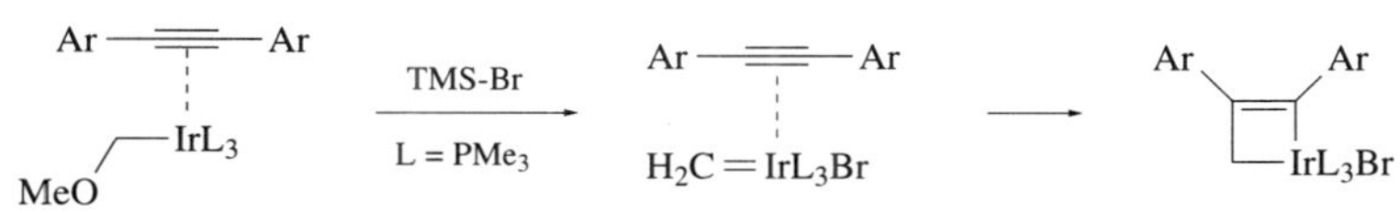

Scheme 41

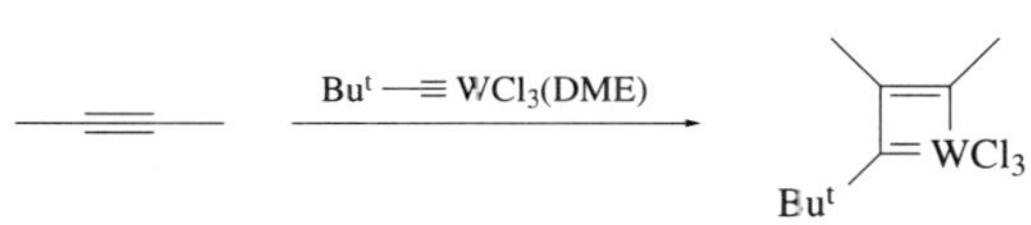

(61)

2.19.7.4.2 *Carbometallation with organotransition metals in catalytic reactions*

Although less well defined and characterized, catalytic carbometallation of alkynes with organotransition metals is synthetically very attractive. Carbopalladation of alkynes, which had until

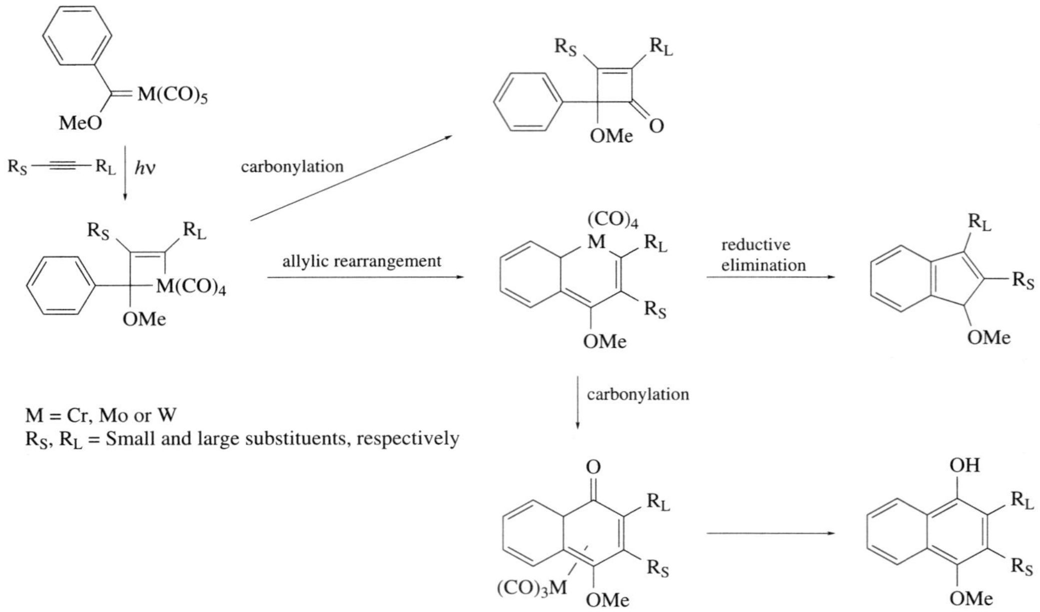

Scheme 42

recently been applied mainly to induce symmetrical cyclooligomerization of alkynes ⟨76ACR93⟩, has been shown to be particularly useful for the preparation of complex organic molecules ⟨92PAC323⟩. Although the carbopalladation products have not been characterized in these reactions, they are undoubtedly formed as indicated in the following reactions, judging from various types of organic products obtained via subsequent addition–elimination with alkenes (Schemes 43 and 44) ⟨89JA3454⟩, cross-coupling (Scheme 45) ⟨90TL4393⟩, and carbonylation (Scheme 46) ⟨89JA3454, 92PAC323⟩. Unlike the corresponding carbopalladation of alkenes, the alkyne reaction is "living," and the alkenylpalladium products can undergo further carbometallation leading to a cascade of carbometallation reactions (Equation (62) and Scheme 47) ⟨88JA7255, 90JA8590⟩. It is noteworthy that the regiochemistry of carbopalladation of simple alkynes is opposite to that of carboalumination and carbocupration ⟨93T5471⟩. The regiochemistry of cyclic carbopalladation however is "normal," and it strongly favors the *exo*-mode cyclization ⟨93T5471⟩. Along a similar line an intramolecular alkoxycarbonylpalladation reaction is also known (Scheme 48) ⟨79JA4107⟩.

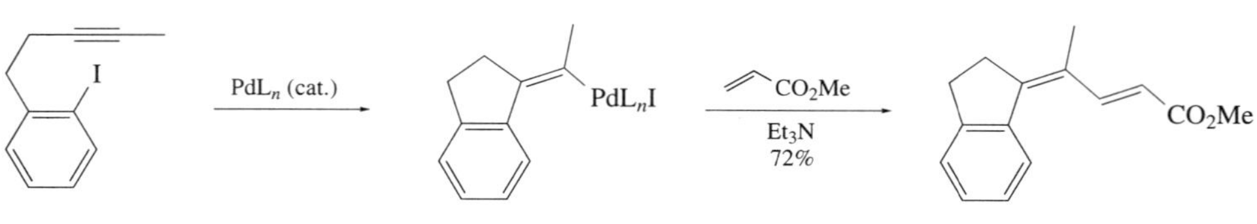

Scheme 43

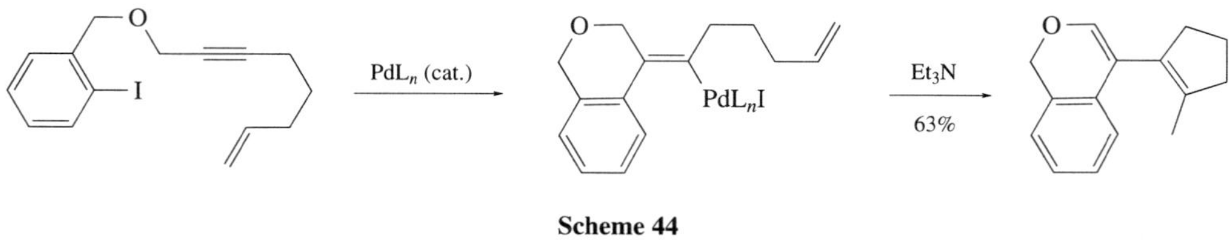

Scheme 44

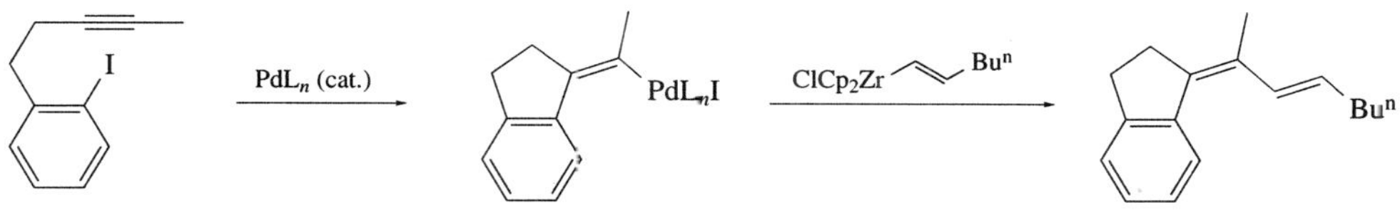

Scheme 45

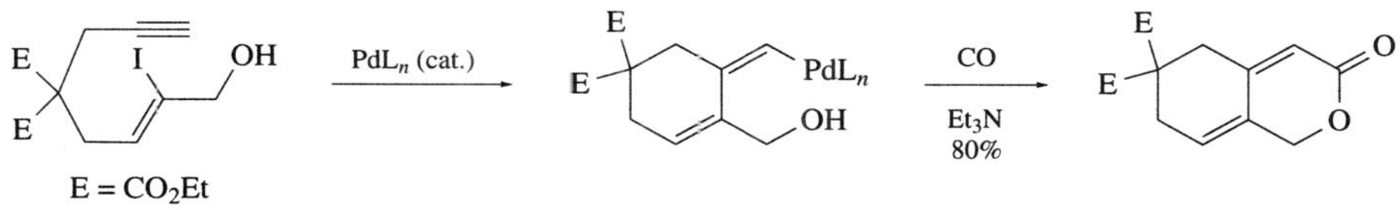

Scheme 46

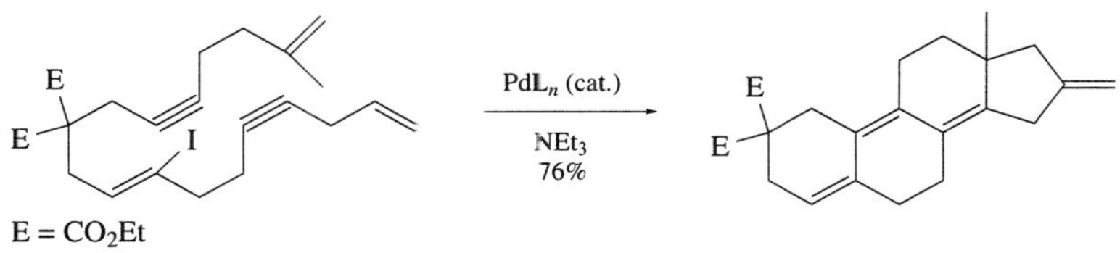

(62)

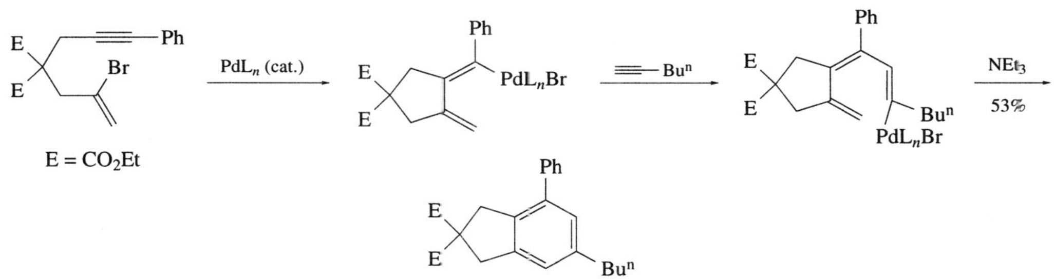

Scheme 47

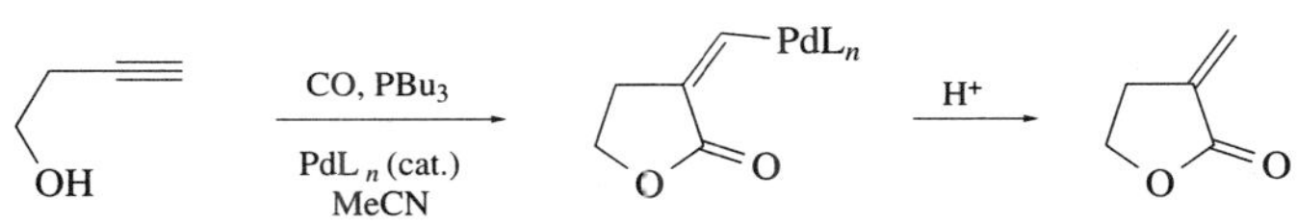

Scheme 48

Catalytic reactions of metal–carbene complexes with alkynes have recently been developed into synthetically useful reactions. The alkyne polymerization reactions producing poly(acetylenes) are representative examples (Scheme 49) ⟨80JA422⟩. Many other catalytic carbometallation reactions of alkynes with organotransition metals are known. However, few have found widespread synthetic applications. One novel *anti*-carbometallation involving yttrium is shown in Scheme 50 ⟨87OM2053⟩.

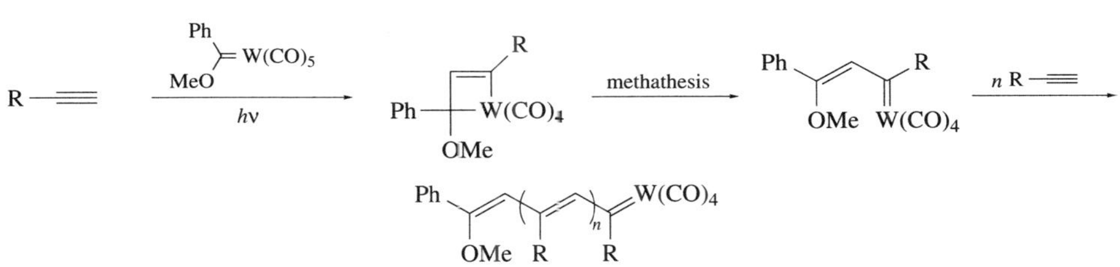

Scheme 49

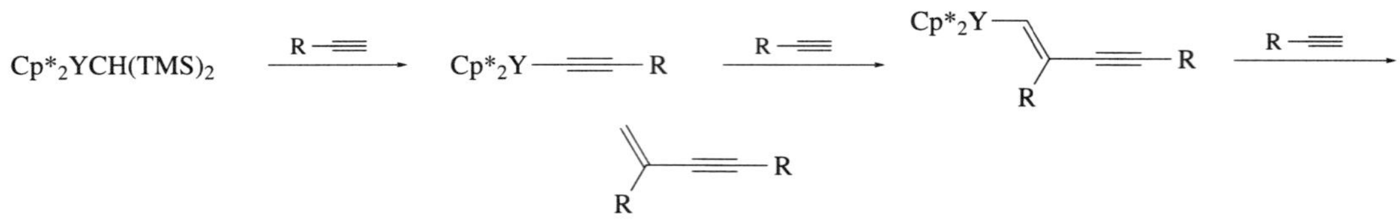

Scheme 50

2.19.8 OXYMETALLATION, HALOMETALLATION, AND OTHER HETEROMETALLATION REACTIONS (METHOD VIII)

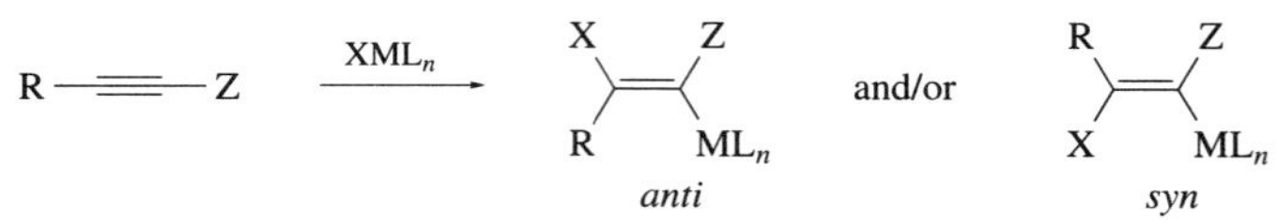

Z = H or C group; X = O, halogen, etc.

Scheme 51

β-Oxy-, β-halo-, and other β-heterosubstituted organometals are often unstable, and they undergo β-elimination to give alkenes and alkynes. Thus, Scheme 51 is thermodynamically unfavorable with most metals. With relatively electronegative metals, however, conversion of alkynes into alkenes may provide a sufficient driving force to make the heteroatom-metal bond addition thermodynamically favorable. Indeed, relatively electronegative, heavy metals, for example Hg, Tl, Pd, and Pt, readily undergo oxymetallation, halometallation, and related heterometallation reactions. Both *syn* and *anti* additions have been observed, and most of such reactions are stereoselective. The *anti* addition is generally thought to proceed via a stepwise polar mechanism, while the *syn* addition must be concerted. The polar reactions must be mechanistically related to the electrophilic metallation reaction of arenes (see Section 2.19.5.1). Whereas oxymetallation appears to be essentially limited to heavy metals, halometallation has been observed even with some light metals, for example boron.

Oxymercuration ⟨B-86MI 219-01⟩ is a well-documented subject and is therefore only briefly discussed here. Conversion of acetylene into acetaldehyde via mercury-catalyzed hydration must involve oxymercuration. A presumed β-hydroxyvinylmercury intermediate decomposes to give acetaldehyde under the reaction conditions. On the other hand, the reaction of alkynes with $Hg(OAc)_2$ gives β-acetoxyalkenylmercury derivatives as discrete products. The stereochemistry depends on both alkynes and reaction conditions (Equations (63) and (64)) ⟨54IZV1008, 58AG166⟩. Halomercuration ⟨49IZV305⟩ proceeds similarly (Equations (65) and (66)). The intramolecular version of oxymercuration reactions of alkynes has been used to prepare α-alkylidene furan derivatives (Equations (67) and (68)) ⟨81JA5459, 82JA5842⟩. Other heavy main group metals such as thallium can also participate in oxymetallation and halometallation. However, the addition products tend to decompose under the reaction conditions, and alkenylmetals may not be obtained (Equation (69)) ⟨71JA7331, 73JA1296⟩.

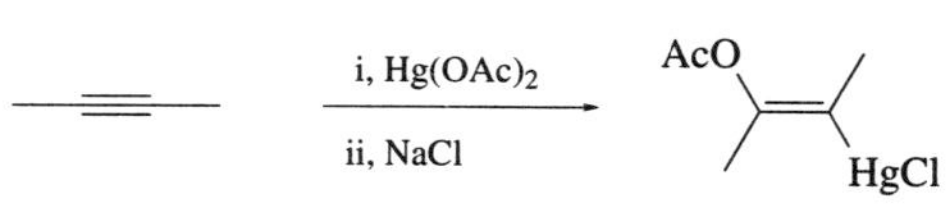

(63)

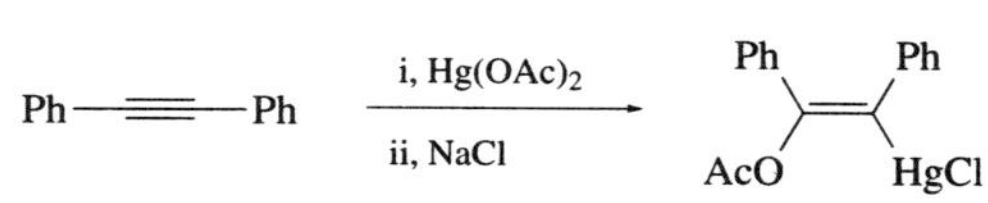

(64)

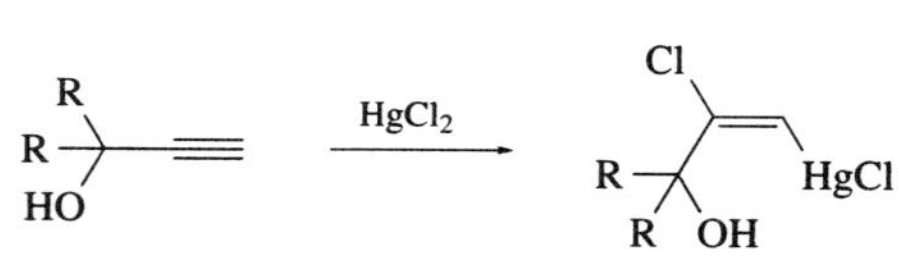

(65)

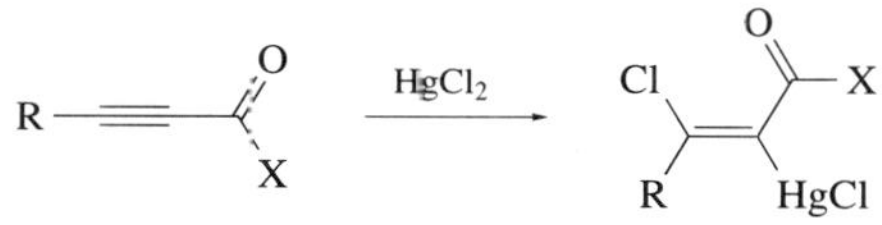

(66)

X = Me, OH, OMe, OEt, *etc.*

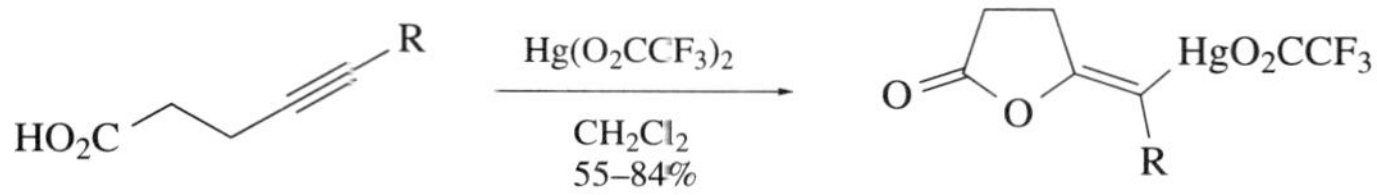

(67)

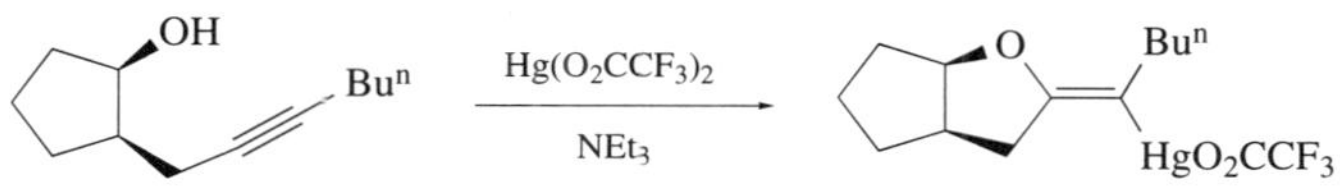

(68)

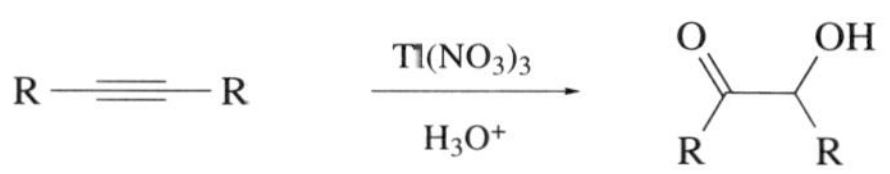

(69)

Oxymetallation and related reactions of palladium and other heavy transition metal compounds are widely known. Oxypalladation of alkenes is the basis for the Wacker oxidation of ethylene to produce acetaldehyde. This and related subjects have been extensively reviewed ⟨B-80MI 219-01, B-85MI 219-01⟩. Although much less is known about oxymetallation and other heterometallation reactions of alkynes involving transition metals, intramolecular oxypalladation of alkynoic acids ⟨86JA2753⟩ appears to be very promising (Scheme 52). Particularly attractive is the use of Pd(0) catalysts in conjunction with aryl and alkenyl halides (Scheme 53) ⟨92JOC976⟩.

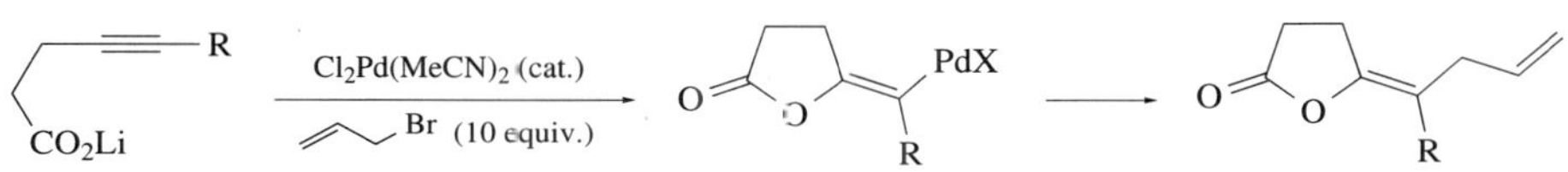

Scheme 52

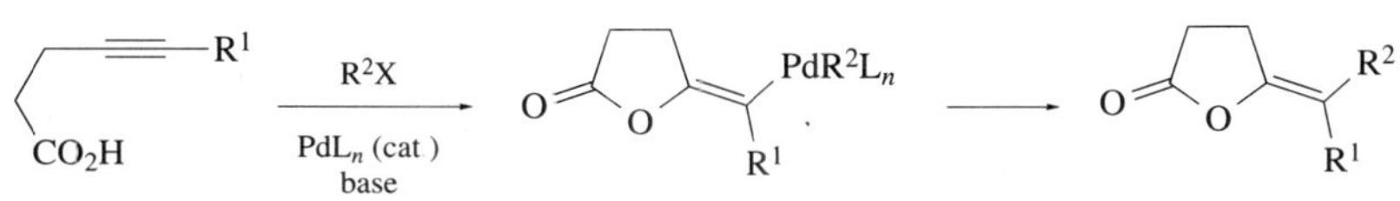

Scheme 53

2.19.9 OXIDATIVE COUPLING (METHOD IX) AND OXIDATIVE METALLATION (METHOD X) OF ALKYNES

Oxidation coupling

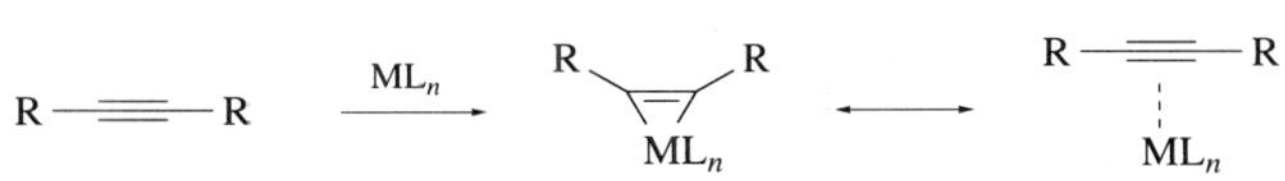

Scheme 54

Oxidative metallation

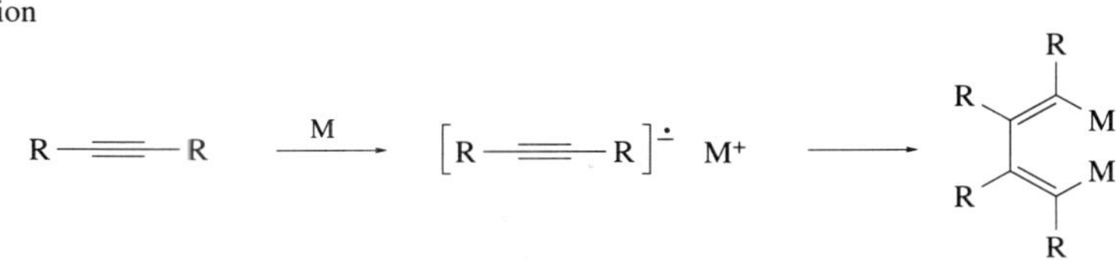

Scheme 55

2.19.9.1 Oxidative Coupling–Ring Expansion Tandem Processes

The reaction of alkynes with transition metal species having at least one each of empty and filled nonbonding orbitals can produce complexes which are best viewed as resonance hybrids of metallacyclopropenes and alkyne–metal π-complexes (Scheme 54) (cf. Section 2.19.1.1). Likewise, the corresponding reactions of alkenes give resonance hybrids of metallacyclopropanes and alkene–metal complexes. A number of such compounds have been prepared as discrete species and fully characterized and identified by various analytical methods including x-ray crystallography. Since metallacyclopropenes are not considered to be alkenylmetals here, oxidative coupling itself does not usually serve as a method for preparing such organometals. There are some exceptions, however. Thus, for example, the oxidative coupling reaction of cyclopropenes produces, depending on the cyclopropenes, metal complexes including ligands, and reaction conditions, 2-metallabicyclo[1.1.0]-butanes, 2-propenylidene-metal complexes, and/or 3-metallacyclobutenes (Scheme 56). The reaction of cyclopropenes with '$ZrCp_2$' (Scheme 57) is a representative example ⟨89AG(E)610, 91CB2165⟩. Similar reactions of benzocyclopropenes give metallabenzocyclobutene derivatives, for example Equation (70) ⟨86AG(E)640⟩.

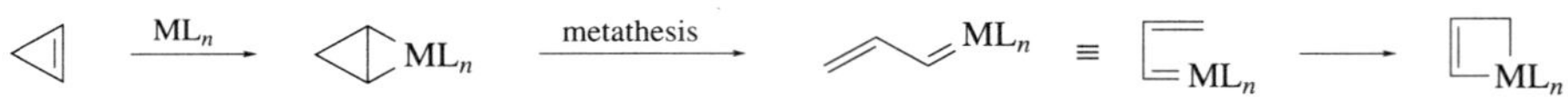

Scheme 56

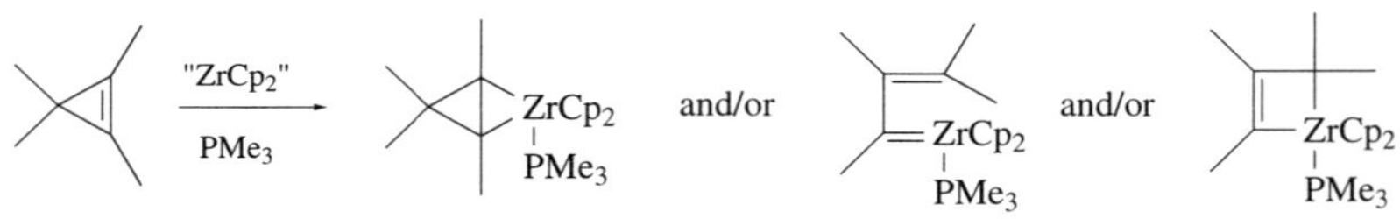

Scheme 57

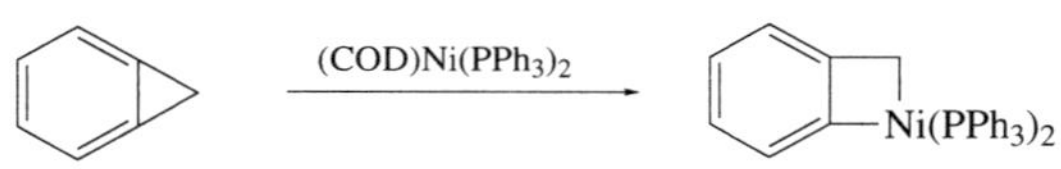

(70)

Much more important as routes to vinylmetals are various combinations of oxidative coupling and reactions of the resultant metallacyclopropanes and metallacyclopropenes with a variety of π-compounds, such as alkenes, alkynes, carbonyl compounds, and nitriles. Some fundamental patterns for the preparation of vinylmetals by these reactions are shown in Scheme 58. All of these processes are known. Some other heterofunctional groups, such as imino and thiocarbonyl, have also been used. The reactions of metallacyclopropanes and metallacyclopropenes with alkenes and alkynes may formally be viewed as carbometallation reactions, and the presence or ready availability of a metal empty orbital appears to be the crucial requirement. The same factor should facilitate the reactions with carbonyl compounds and nitriles as well. These oxidative coupling–ring expansion tandem sequences may also be observed with difunctional organic compounds containing alkenyl, alkynyl, carbonyl, and cyano groups. There are 10 different binary combinations involving these four functional groups, and most of these types of difunctional compounds have been shown to participate in the reactions shown in Scheme 58. With some transition metals, zirconium in particular, both three-membered and five-membered metallacycles have been isolated and/or spectroscopically characterized. In the other cases, the formation of the three- and/or five-membered metallacycles is assumed. In this section, some representative examples of these reactions are discussed with emphasis on those organometals that have been characterized and identified.

Cyclic oligomerization of alkynes and related reactions promoted or catalyzed by transition metal complexes to give benzene derivatives, cyclobutadiene–metal complexes and other cyclic compounds have been known for several decades ⟨B-87MI 219-04⟩. Earlier reactions are, in general, suitable only for the preparation of highly symmetrical molecules and are of limited synthetic utility. Some of the more recent procedures, most notably alkyne cocyclotrimerization reactions catalyzed by $CpCoL_2$, where L is CO or ethylene ⟨84AG(E)539⟩, are applicable to the synthesis of complex organic compounds. In many of these reactions the formation of both benzene derivatives and cyclobutadiene–metal complexes are assumed to take place via metallacyclopentadienes without firm

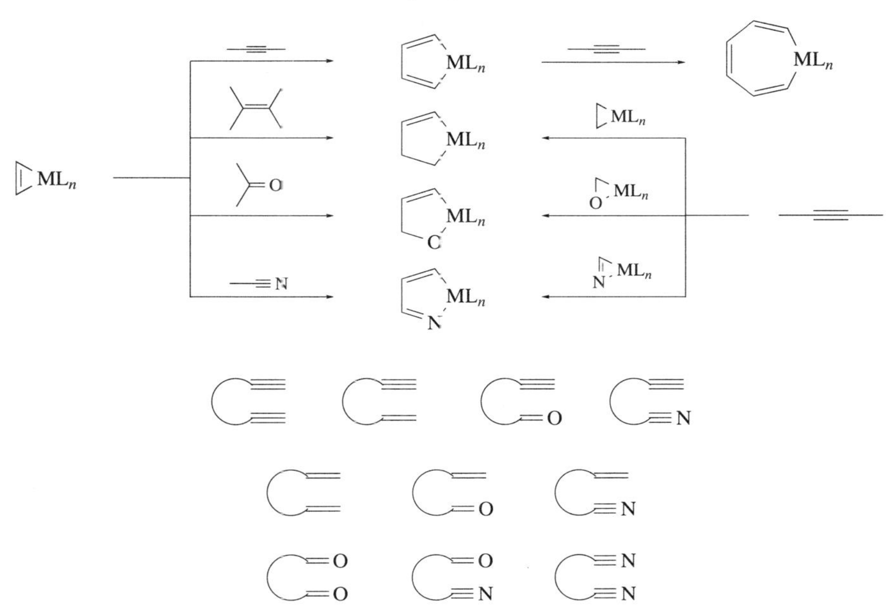

Scheme 58

experimental supports (Scheme 59). In a limited number of cases, for example the reaction shown in Scheme 60 ⟨68IC1298⟩, the intermediacy of metallacyclopentadienes has been firmly demonstrated.

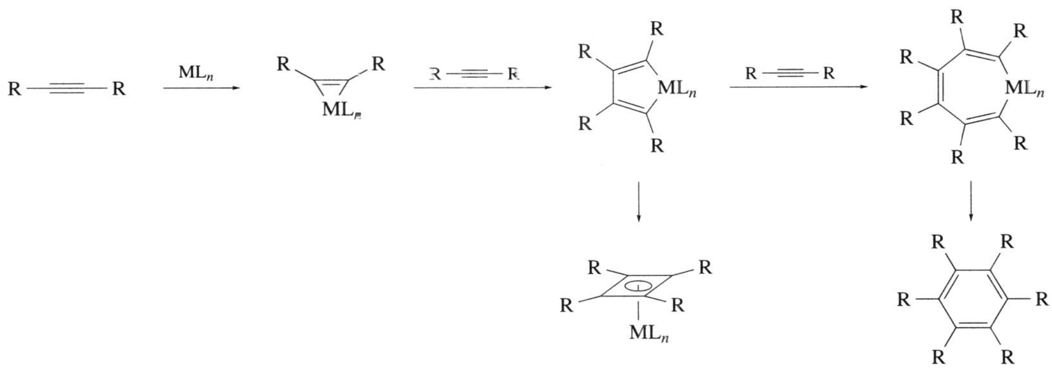

Scheme 59

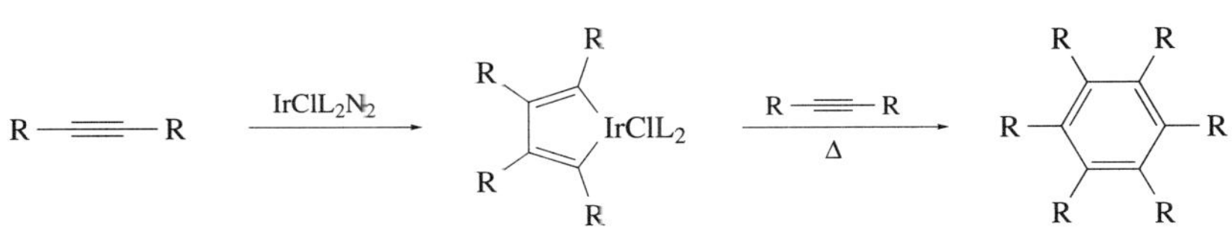

Scheme 60

Of more recent origin are the preparation of various five-membered carbocycles and heterocycles. The currently most widely used reaction of this type is the cobalt-promoted cyclopentenone synthesis ⟨73JCS(P1)977, 77ANY(295)2, 81JOC5357, 81JOC5436, 83JA2477⟩ represented by Scheme 61. This reaction has been extensively reviewed in recent years ⟨91COS(5)1037, 91OR(40)1⟩. The organocobalt intermediates shown in Scheme 61 do not appear to have been isolated and/or characterized, and it is not even clear if the reaction can be stopped at these intermediary stages. Closely related to this reaction are a carbonylative enyne bicyclization promoted by $W(CO)_6$ (Equation (71)) ⟨93JA1154⟩, an enyne–isonitrile cobicyclization promoted by $Ni(COD)_2$ (Equation (72)) ⟨92SL539⟩, and a carbonylative diyne bicyclization promoted by $Fe(CO)_5$ (Equation (73)) ⟨92OM4096⟩. In Equation (73) the products isolated are cyclopentadienone–iron complexes. In none of these reactions, however, have vinylmetal intermediates been isolated and/or characterized. In the reaction shown in Equation (72), both metallacyclopentene and metallacyclobutene derivatives have been considered as possible intermediates.

Scheme 61

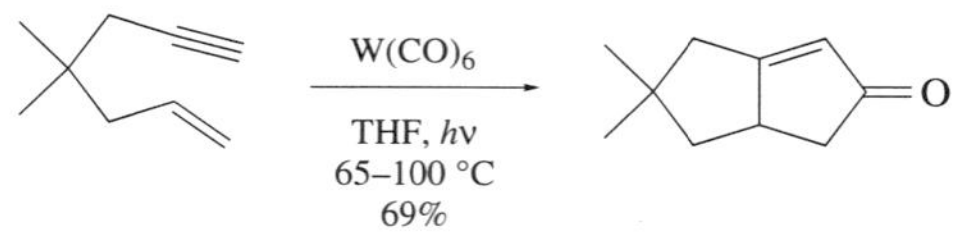

(71)

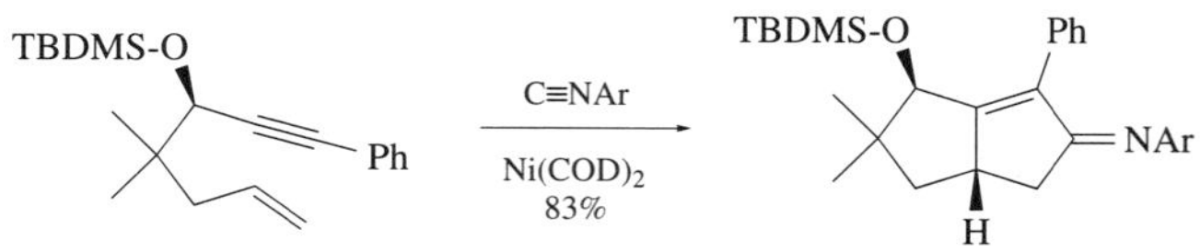

(72)

TBDMS = *t*-butyldimethylsilyl

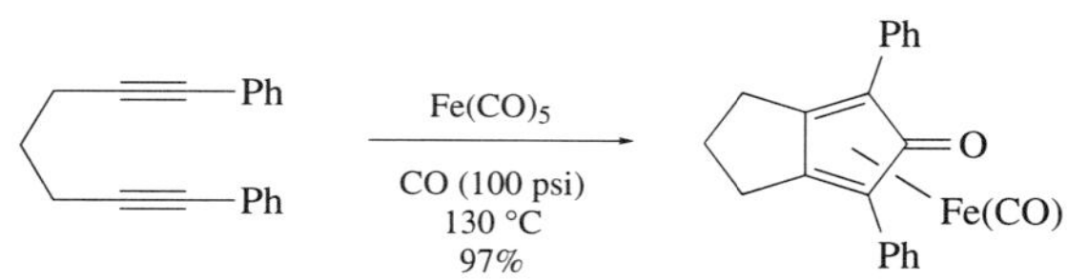

(73)

In contrast with those reactions discussed above which involve late transition metals, related reactions of early transition metals can proceed stepwise, producing both three-membered and five-membered metallacycles, which can be further transformed into organic products. All three metals of the titanium triad, especially zirconium, have been shown to undergo a wide variety of oxidative coupling–ring expansion tandem processes which have been extensively reviewed ⟨88CRV1047, 88S1, 91COS(5)1163⟩. Thermal decomposition of dialkylzirconocenes containing one or more β-hydrogens leads to the formation of alkene-zirconocenes ⟨86TL2829, 89CL761, 89JA3336⟩. Weak alkene ligands, for example 1-butene ⟨87CL623⟩ and isobutylene ⟨91OM825⟩, can be readily displaced with alkynes and other π-compounds (Scheme 62). On the other hand, if the alkene ligands are tightly bound to zirconium, for example ethylene-zirconocene, the reaction of such alkene-zirconocenes with alkynes and other π-compounds may now lead to the formation of the corresponding five-membered zirconacycles (Scheme 63). The exact courses of these reactions are often difficult to predict, and they may have to be experimentally determined.

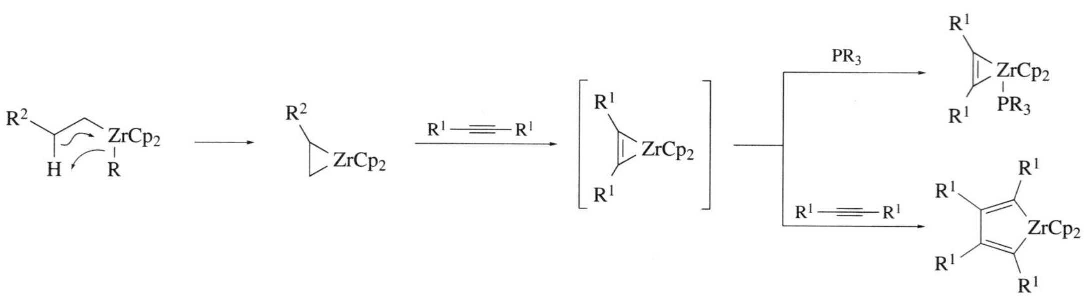

Scheme 62

Alkyne-zirconocenes ⟨86JA7441, 87JA2544, 88CRV1047⟩ and benzyne-zirconocenes ⟨79JA3659, 86JA7411, 88CRV1047⟩ have been most widely prepared by related thermal decomposition reactions of alkenyl- and arylzirconocene derivatives containing alkenyl and aryl β-hydrogens, respectively (Schemes 64 and 65). In these reactions, no free alkynes and benzynes appear to be generated, which prevents the formation of often-undesirable zirconacyclopentadienes and zirconafluorenes. On the other

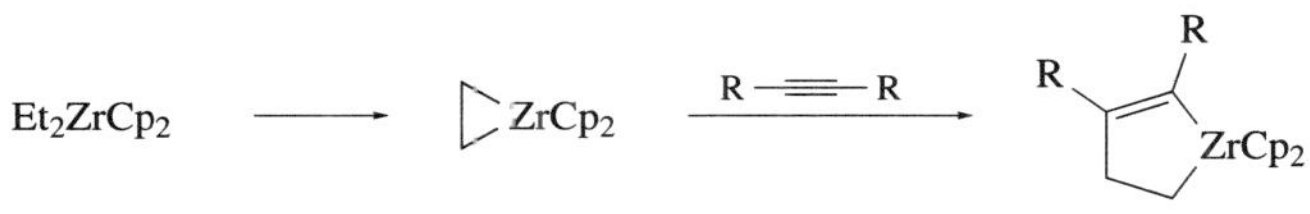

Scheme 63

hand, the alkene displacement reaction with alkynes (Scheme 62) is carried out in the presence of free alkynes, and this reaction tends to proceed further to give the zirconacyclopentadienes unless the intermediary zirconacyclopropenes are trapped by phosphines or other suitable reagents ⟨87CL623⟩.

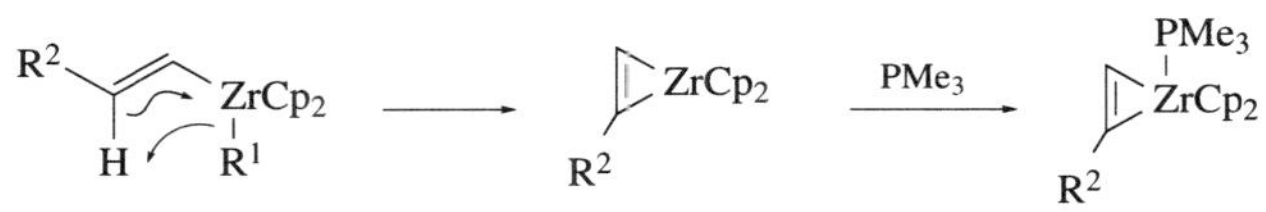

Scheme 64

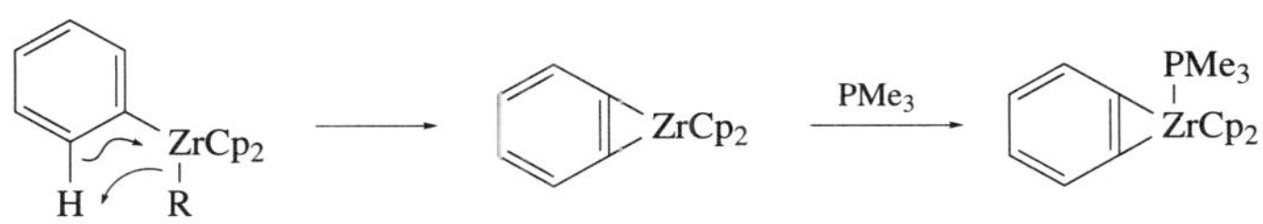

Scheme 65

Alkyne-zirconocenes react with alkenes, alkynes, aldehydes, ketones, and nitriles to give the corresponding five-membered zirconacycles in a predictable manner, presumably because the original alkyne ligands are tightly bound to zirconium (Scheme 66) ⟨86JA7441, 87CL623, 87JA2544, 88CRV1047⟩. Benzyne-zirconocenes react similarly ⟨79JA3659, 86JA7411, 88CRV1047⟩. Yet another route to five-membered zirconacycles is to generate zirconacyclopropanes containing nitrogen ⟨89JA776⟩ and other heteroatoms in the three-membered ring and allow them to react with alkynes and other π-compounds (Scheme 67).

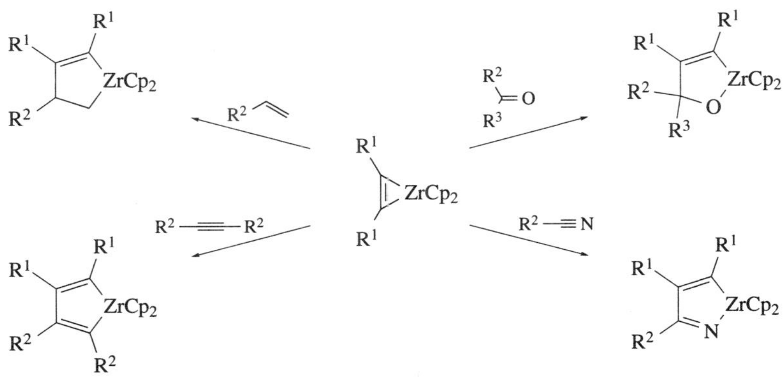

Scheme 66

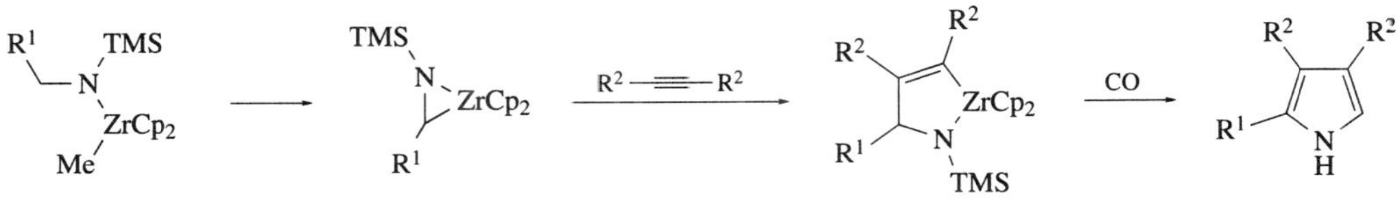

Scheme 67

Finally, many of the five-membered zirconacycles, such as zirconacyclopentanes ⟨89CC852, 90CC182⟩ and zirconacyclopentenes ⟨93TL687, 94TL695⟩ are prone to ring contraction with concomitant release of an alkene. In the presence of stronger π-ligands, such as alkynes, aldehydes, and nitriles, the ring contraction–expansion sequence leads to the formation of new five-membered zirconacycles, even though some of these reactions may be complicated by competitive formation of seven-membered zirconacycles (Scheme 68) ⟨93TL687, 94TL695⟩.

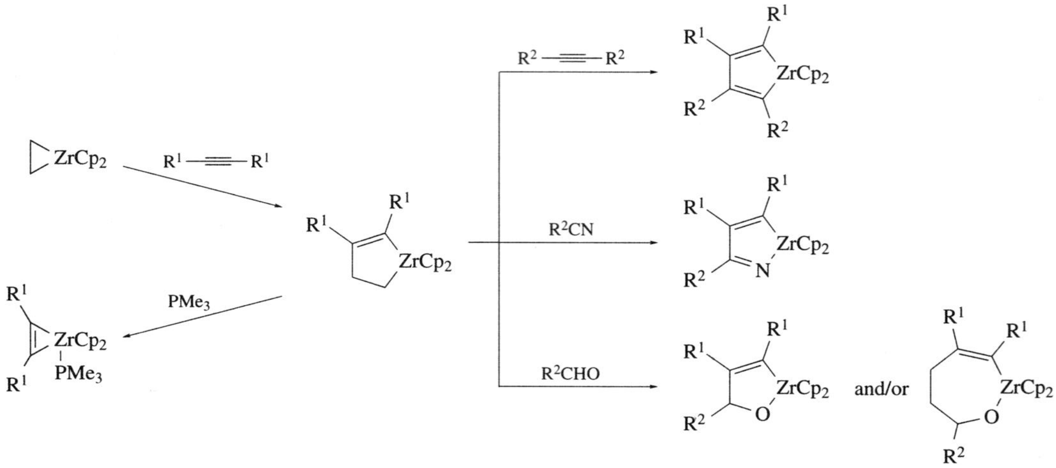

Scheme 68

In many of these reactions both three- and five-membered zirconacycles can be isolated and/or characterized. The structures of some representative members of these compounds have been fully established by chemical and spectroscopic means including x-ray analysis. The zirconium-promoted bicyclization reactions of dienes ⟨89JA6435, 89TL5105⟩, enynes ⟨86TL2829, 88JA7128, 89CL761, 89JA3336⟩, and diynes ⟨86TL2829, 87JA2788, 89CL761, 89JA3336⟩ are particularly attractive from the viewpoint of the synthesis of complex natural products, such as phorbol ⟨90JA4956⟩, pentalenic acid ⟨91JA7424⟩, and dendrobine (Scheme 69) ⟨92JOC3519⟩. Since the five-membered zirconacycles can be prepared as discrete intermediates, they can be subsequently converted into a variety of organic products. This feature makes the zirconium-based methodology potentially more versatile than those involving late transition metals. Much less is known about related reactions of titanium and hafnium, even though they appear to be capable of undergoing many related reactions, as suggested by the examples shown in Scheme 70 ⟨84JA6422⟩ and Scheme 71 ⟨89CS457⟩.

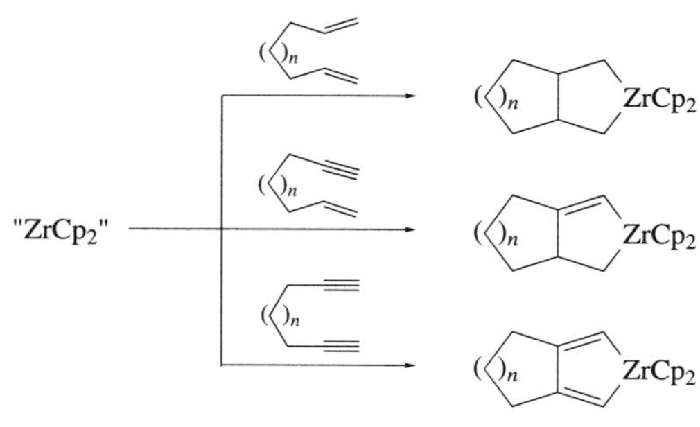

Scheme 69

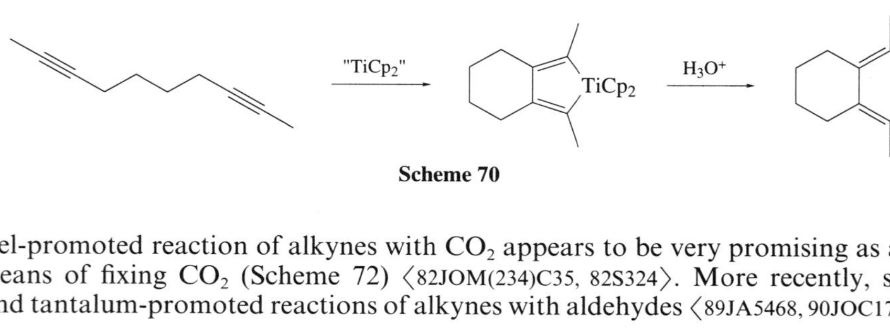

Scheme 70

The nickel-promoted reaction of alkynes with CO_2 appears to be very promising as a potentially catalytic means of fixing CO_2 (Scheme 72) ⟨82JOM(234)C35, 82S324⟩. More recently, some related niobium- and tantalum-promoted reactions of alkynes with aldehydes ⟨89JA5468, 90JOC1707⟩, ketones

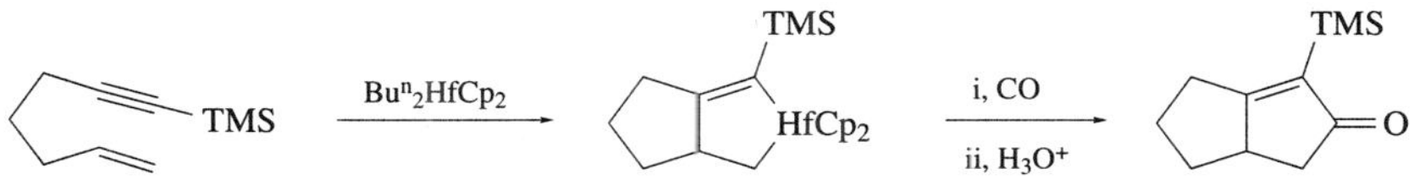

Scheme 71

⟨89JA5468, 90JOC1707⟩, isocyanates ⟨92BCJ1543⟩, and hydrazones ⟨92CL99⟩ which are thought to proceed via alkyne–metal complexes and five-membered metallacycles have been reported. Their merits and demerits relative to those involving zirconium and other metals are not yet very clear.

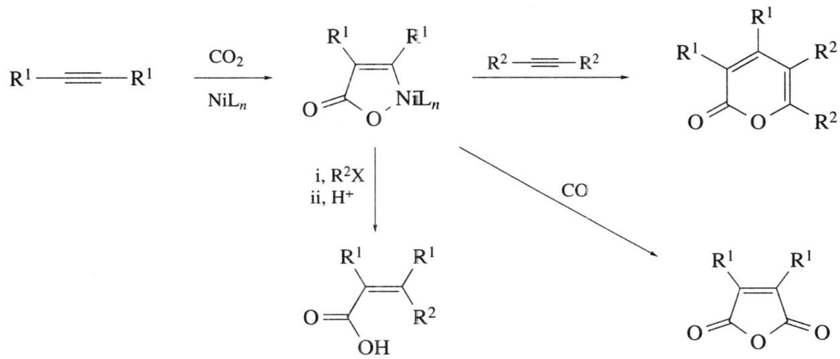

Scheme 72

2.19.9.2 Oxidative Metallation of Alkynes

Metals of low ionization potentials can undergo one-electron transfer reactions with alkynes with low-lying LUMOs to form ion pairs consisting of metal cations and anion radicals derived from alkynes. Coupling of such radical anions gives dimeric alkenylmetals (Scheme 55). In reality, only alkali metals, for example lithium, seem to readily participate in this reaction (Scheme 73) ⟨73JOM(47)225⟩.

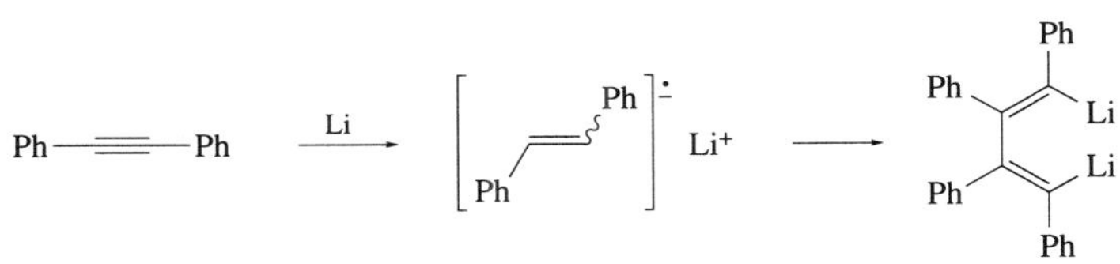

Scheme 73

Although the synthetic utility of the reaction is severely limited by the fact that it is applicable mainly to the synthesis of symmetrical dimeric alkenylmetals, this reaction has been used to prepare via transmetallation metallacyclopentadienes containing various metals such as boron ⟨69JA4575⟩, aluminum ⟨77JOM(127)C29⟩, silicon ⟨61JA4406⟩, germanium ⟨59JA3163⟩, and tin ⟨59JA3163⟩ (Equation (74)) that cannot be directly prepared by the oxidative coupling–carbometallation route.

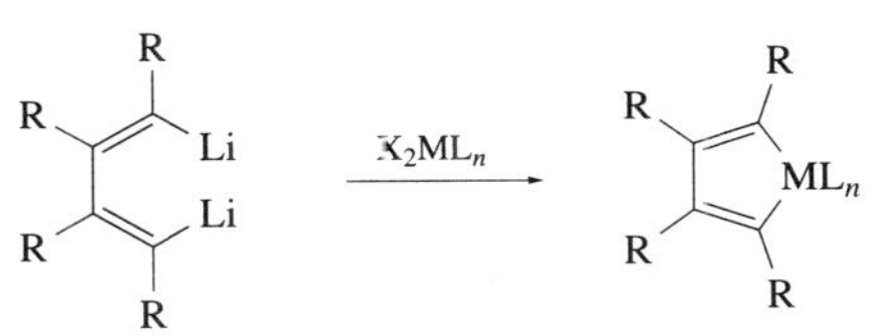

(74)

M = B, Al, Si, Ge, Sn, etc.

2.19.10 NONREDOX AND REDOX TRANSMETALLATION REACTIONS (METHODS XI AND XII)

Nonredox transmetallation

$$L_nM^1-R \quad + \quad X-M^2L_n \longrightarrow L_nM^1-X \quad + \quad R-M^2L_n \tag{75}$$

Redox transmetallation

$$L_nM^1-R \quad + \quad M^2L_n \longrightarrow L_nM^1 \quad + \quad R-M^2L_n \tag{76}$$

Transfer of a carbon group from one metal to another is defined as transmetallation, more precisely organometallic transmetallation. Such processes may be of either nonredox (Equation (75)) or redox (Equation (76)) type. Since the nonredox transmetallation is synthetically far more important than its redox counterpart, it is simply referred to as transmetallation, while the latter will be specifically referred to as redox transmetallation. It should be noted that, in many cases, interaction of two organometals leads to the formation of complexes consisting of both metals. In cases where the electronegativity differential between two metals is sufficiently large, the bimetallic complexes can be viewed as "ate" complexes. For example, the reaction of an organolithium (R^1Li) with a chlorodialkylalane ($ClAlR_2$) might be generally thought to give R^1AlR_2 and LiCl. In reality, however, the product is, in most instances, better represented by $LiAlR_2R^1Cl$, unless AlR_2R^1 is removed from LiCl by distillation or some other means. It is generally safe to assume, unless otherwise proven, that even alkali and alkaline earth metal halides and alkoxides may not exist as separate species in the presence of coordinatively unsaturated organometals. In the present discussion, however, this factor is not rigorously treated, and such metal salts may be shown separately.

Transmetallation provides a versatile method for the preparation of essentially all types of organometals including aryl- and vinylmetals. Since it is such a ubiquitous process, discussion of specific examples would be impractical and inappropriate. The following discussion is to provide some general knowledge, underlying principles, and reliable predictability. A wide variety of organometals participates in transmetallation reactions between two (or more) metal species. It is kinetically favored by the presence or ready availability of empty orbitals at one or both metal centers, suggesting that, in such cases, transmetallation is most likely to proceed via either synchronous or stepwise three- or multicenter bonding involving the two participating metals (Scheme 74). As in any other reactions of coordinatively unsaturated organometals, solvents may play significant roles in transmetallation, although their precise effects are often difficult to predict. In cases where stabilization of polar intermediates is crucial, solvents of appropriate polarity can promote transmetallation. However, strongly coordinating solvents may hinder transmetallation by plugging the crucial empty coordinating site. On the other hand, at the nonpolar end of the polarity scale, coordinatively unsaturated organometals tend to form tight and unreactive clusters. As a consequence the ease and rate of transmetallation often peak at an optimum point on the solvent polarity scale. Such an optimum point must however be experimentally determined.

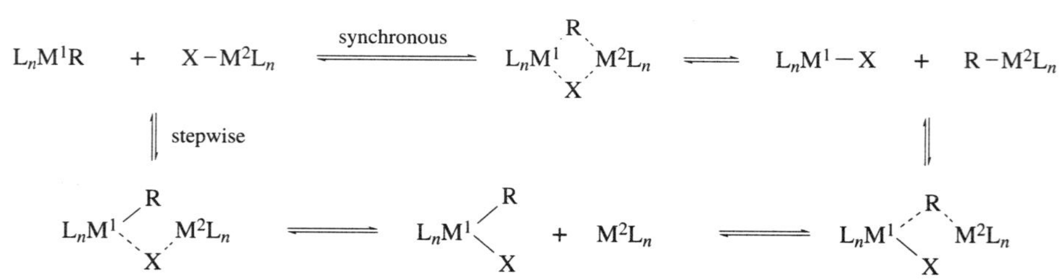

Scheme 74

The direction of transmetallation is readily predicted in most cases on the basis of the electronegativities of two participating metals. In a ligand exchange process via transmetallation, the formation of a bond between the more electropositive metal and the more electronegative ligand provides a major driving force. So in the reaction between an organometal and a metal halide, for example, the alkylating (arylating and alkenylating in this chapter) ability of organometals decreases as the electronegativity of metals increases. Among some representative main group metals, the following order is usually observed: $K \geqslant Na \geqslant Li > Mg > Zn \geqslant Al > Sn > Hg$, B, and Si. Coupled with the ready availability via oxidative metallation and metal–halogen exchange, organolithiums and Grignard reagents are the two most important classes of compounds for transmetallation (Equations (77) and (78)). Transfer of aryl and alkenyl groups to essentially all metals including even alkali and alkaline earth metals can be achieved using these organometals.

$$\text{ArLi(MgX)} \xrightarrow{\text{XML}_n} \text{ArML}_n \tag{77}$$

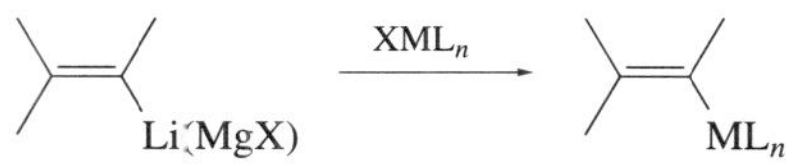

(78)

Use of organolithiums and Grignard reagents may however be associated with some undesirable consequences. First, they generally lack chemoselectivity because of their high reactivity towards various functional groups. With respect to transmetallation itself, these organometals have a strong tendency to undergo multiple and exhaustive transmetallation as well as "ate" complexation. It is therefore desirable in some cases to avoid their use and seek less reactive organometals providing an optimum electronegativity differential. This point is illustrated below using the preparation of organozirconiums as a representative example. The reaction of RLi and RMgX with Cl_2ZrCp_2 tends to give R_2ZrCp_2 regardless of the molar ratio of the two reactants. Thus, the reaction of vinyllithiums or the corresponding Grignard reagents with Cl_2ZrCp_2 cannot be stopped cleanly at the monovinylation stage, producing mainly divinylzirconocenes, which may decompose under the reaction conditions. On the other hand, alkenylaluminate complexes (but not alkenylalanes) react with Cl_2ZrCp_2 to give the monoalkenylated product (Scheme 75) ⟨82TL3327⟩. As might be expected, steric and electronic factors also play significant roles. Although the following examples do not pertain to the preparation of aryl- or vinylmetals, they illustrate the significance of these other factors (Scheme 76) ⟨91OM825⟩.

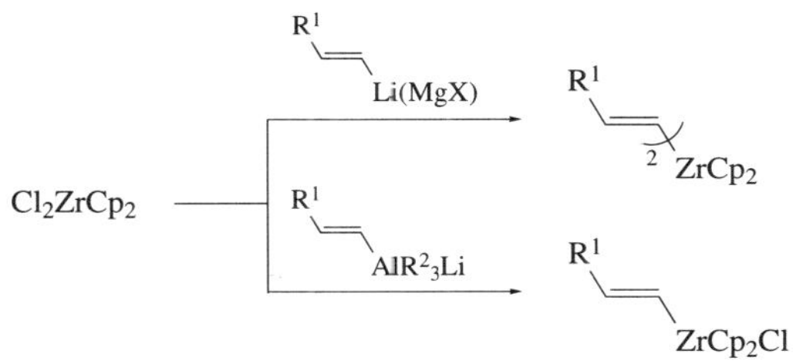

Scheme 75

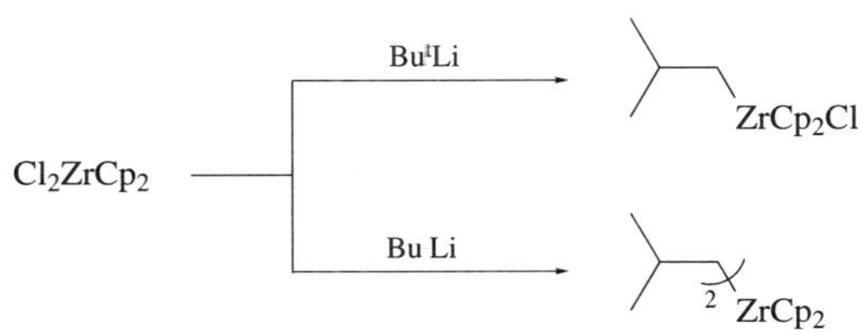

Scheme 76

In cases where the electronegativities of two participating metals are similar, transmetallation may be reversible, often producing equilibrium mixtures. It may also proceed cleanly in either direction depending on ligands and other factors. For example, in sharp contrast with the reaction shown in Scheme 75, in which an alkenyl group is transferred from aluminum to zirconium, treatment of alkenylzirconocene chlorides with $AlCl_3$ gives cleanly alkenyldichloroalanes and Cl_2ZrCp_2 via transfer of an alkenyl group from zirconium to aluminum (Scheme 77) ⟨77JA638, 79JA3521⟩. In the transmetallation reaction involving an exchange between a carbon group and a more electronegative ligand, such as halogen and oxygen groups, it is not practical to transfer the carbon group from the more electronegative metal to the more electropositive one. Thus, for example, treatment of alkenylzirconocene chlorides with magnesium halides will not give alkenylmagnesiums in significant yields, because it is a contrathermodynamic process. On the other hand, transfer of the same alkenyl group from zirconium to magnesium may be feasible in the reaction of alkenylzirconocene chlorides with Grignard reagents, because exchange between two carbon groups should be nearly thermoneutral (Scheme 78).

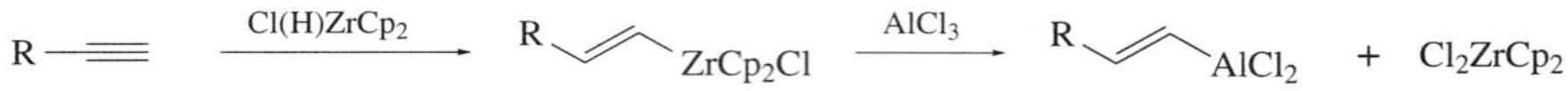

Scheme 77

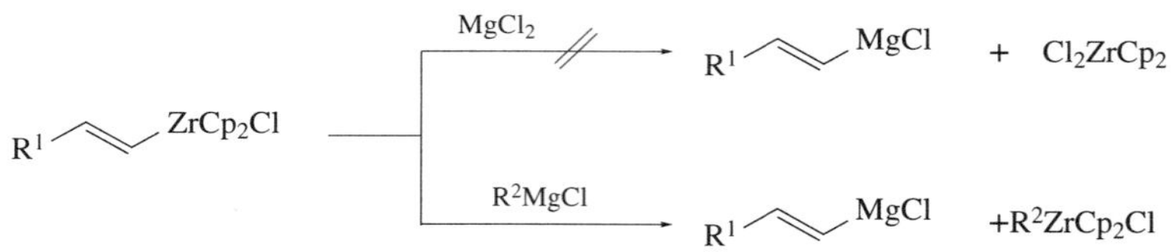

Scheme 78

In catalytic reactions even contrathermodynamic transmetallation may occur as elementary steps in a catalytic cycle provided that such processes are kinetically feasible and that the overall catalytic process is thermodynamically favorable. For example, the palladium-catalyzed cross-coupling reaction of a wide variety of organometals containing such metals as Mg, Zn, B, Al, Sn, Zr, and Cu has been shown to be effective ⟨82ACR340⟩. In many cases, the transmetallation step is rate-determining ⟨87JA2393⟩. In cases where highly electronegative metals, such as B, Sn, Cu, and even Si, are involved, it is not clear if the transmetallation step is thermodynamically favorable. This however may not be critically important, since the overall process is thermodynamically favorable due mainly to the formation of metal salts as by-products (Scheme 79). Similar situations may be encountered in the transmetallation reaction of alkenylmetals containing relatively electronegative metals, such as Sn and B, with Cu salts ⟨88JA2641⟩.

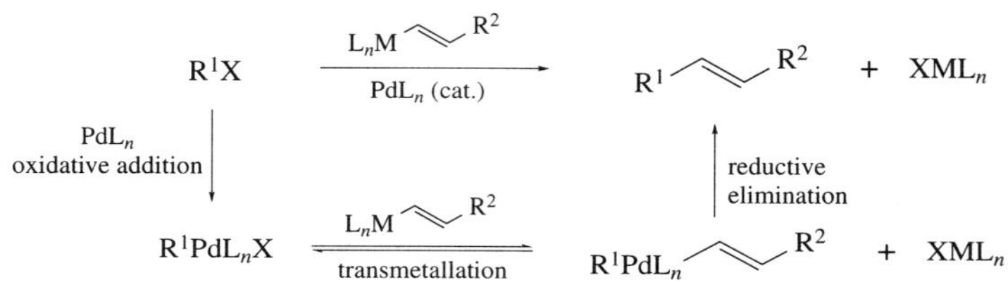

Scheme 79

Within the restrictions discussed above transmetallation permits interconversions among organometals containing a wide variety of metals, thereby significantly elevating the synthetic utility of a given class of organometallic compounds. This point may be further illustrated by the interconversion scheme showing various interconversions to and from zirconium which is used as a representative metal (Scheme 80) ⟨77JA638, 77JA8045, 79JA3521, 82TL3327, 86TL1537, 87JA2393, 88JA2310, 90JA7440⟩.

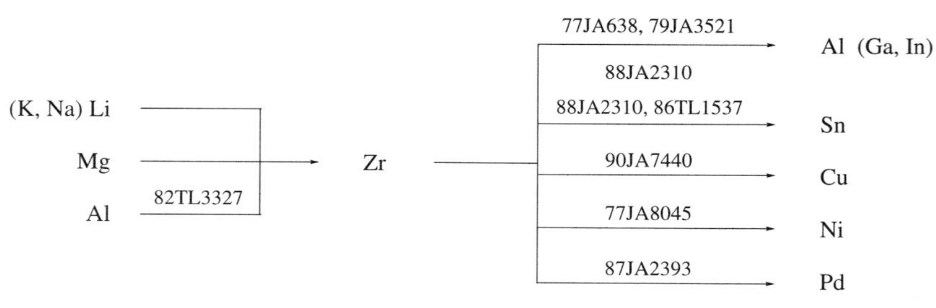

Scheme 80

The redox transmetallation reaction can proceed forward in cases where the free energy of formation of RM^1 is more positive (or less negative) than that of RM^2 (Scheme 81). Organometals containing some heavy metals, such as cadmium and mercury, can serve as RM^1, and alkali and alkaline earth metals readily serve as M^2. Since organometals containing alkali and alkaline metals are readily available from other methods discussed earlier, the synthetic need and value of redox transmetallation is not very high. However, it provides a valuable route to salt-free organometals containing alkali and alkaline earth metals.

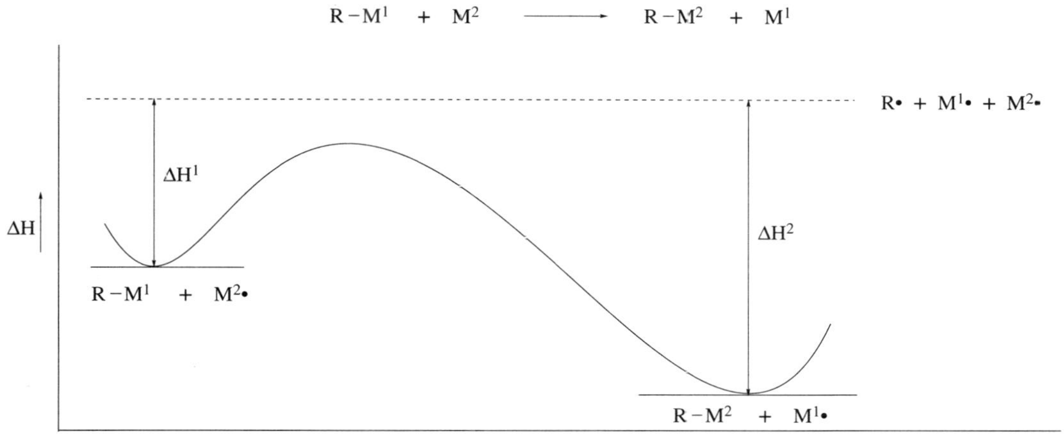

Scheme 81

2.19.11 SUMMARY

The results presented in this chapter may be summarized as shown in Table 11. Undoubtedly, this table will have to be modified as new data are obtained.

Table 11 Summary of the synthesis of vinyl- and arylmetals.

Metal	*Method*[a]											
	I Oxid. M of RX	*II Oxid. disp.*	*III M—X exchange*	*IV Oxid. C—H act.*	*V Nonredox M—H ex.*	*VI Hydro-metal.*	*VII Carbo-metal.*	*VIII Hetero-metal.*	*IX Oxid. coupl.*	*X Oxid. M of alkyne*	*XI Trans-metal.*	*XII Redox transmet.*
Group 1												
Li	general[b]		general		considerable[b]		moderate[b]			some[b]		general
Na, K	some				some							some
(Rb, Cs, Fr)												
Group 2												
Be	some										considerable	
Mg	general				some	some	moderate				considerable	
Ca, Sr, Ba, Ra	some										considerable	
Group 3												
Sc, Y	some					some	some				general	
Lanthanides	some					some					general	
Actinides	some										general	
Group 4												
Ti, Zr, Hf	some			some		general	considerable		general		general	
Group 5												
V, Nb, Ta	some			some		some	some		considerable		general	
Group 6												
Cr, Mo, W	some			some		some	considerable		some		general	
Group 7												
Mn, (Tc), Re	some						some				general	

Group 8											
Fe, Ru, Os	some			some							general
Group 9											
Co, Rh, Ir	some	some		some							general
Group 10											
Ni, Pd, Pt	general	some		some	some (Ar)[c]	general	general	general	some		general
Group 11											
Cu	moderate	general	some			some	general				general
Ag, Au	some	some									general
Group 12											
Zn (Cd)	considerable						moderate				general
Hg	some				considerable (Ar)[c]			general			general
Group 13											
Al	moderate				some (Ar)[c]	general	general				general
Ga, In	some					some					general
Tl	some				considerable (Ar)[c]			considerable			general
Group 14											
Sn	some				some (Ar)[c]	general					general
Pb						some					general

[a] For definition of each method, see Section 2.19.1.2. [b] The terms "general," "considerable," and "moderate" indicate methods of general, considerable, and moderate applicabilities, respectively, while the term "some" indicates that some examples are known but that the scope is not well delineated. [c] The Ar in parentheses indicates that the method is applicable to arylmetals only.

2.20
Stabilized Substituted Ions and Radicals Bearing One Heteroatom ($R^1R^2C^-X$, $R^1R^2C^+X$, $R^1R^2C^{\bullet}X$)

STEPHEN D. LINDELL
AgrEvo UK Ltd, Saffron Walden, UK

2.20.1 STABILIZED CARBANIONS BEARING ONE HETEROATOM ($R^1R^2C^-X$)

Stabilized substituted carbanions bearing a single heteroatom are of great importance in modern synthetic organic chemistry ⟨79OR(26)1, 80T2531, 81S185, 82OR(27)1, 83TCC85, 84AG(E)932, 90OR(38)1, 91COS(1)729, 91COS(2)55⟩. Consequently, there has been a keen interest in isolating these compounds and in determining their structures in order to gain a greater insight into their properties. In most cases, the α-heteroatom stabilized carbanions are formed from appropriate precursors via deprotonation or metal–halogen exchange and then crystallized in the presence of a strong chelating agent (e.g., TMEDA). Many x-ray crystal structures have now been determined and in several cases these have enabled a better understanding of reactivity patterns, stereoselectivity or reaction mechanism ⟨85JOM(285)1, 89OM2308, 91JA4002, 92CB2265, 93AG(E)1032⟩. In addition to providing structural information, in some cases isolation can yield anions with new properties ⟨89AG(E)69, 92JCS(P1)1721⟩.

2.20.1.1 Carbanions Adjacent to Halogen

α-Haloalkyllithiums are well known in the literature, where they are commonly referred to as carbenoids due to their carbene-like reactivity ⟨72AG(E)473⟩. They are most often prepared by metal–

halogen exchange between an alkyllithium reagent and a dihaloalkane at $-100\,°C$, as exemplified by Equations (1) ⟨83HCA308⟩ and (2) ⟨84TL835⟩. α-Chloroalkylzincs have been made by treatment of organo-*gem*-biszinc with phenylsulfonyl chloride ⟨93SL665⟩. Iodomethylzinc iodide can be prepared by reaction of diiodomethane with zinc metal at $40\,°C$ ⟨71JOM(30)151⟩ and this in turn can be converted to the much less stable α-iodomethylcopper by addition of CuI ⟨89JOC5202⟩. The bis(iodomethyl)zinc complex (**1**) has been isolated as a crystalline solid and its x-ray crystal structure has been determined ⟨91JA723⟩. However, in general these organometallics are rather unstable and readily decompose to carbenes, so that for synthetic use they must be formed at low temperatures in the presence of a trapping electrophile.

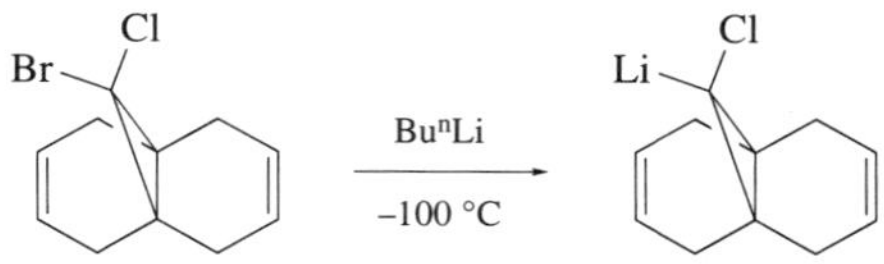

(1)

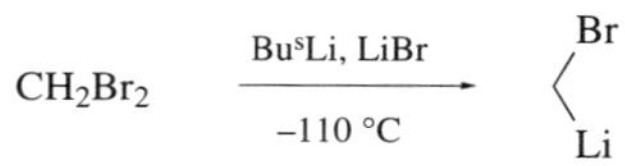

(2)

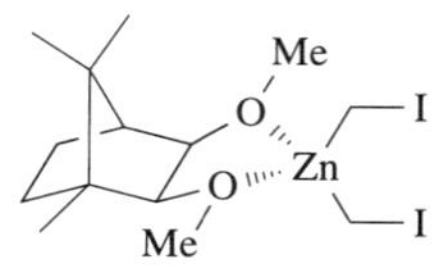

(1)

α-Haloalkenyllithiums are also known and can be prepared either via deprotonation ⟨64TL1137⟩ or metal–halogen exchange ⟨90JOM(400)19, 93JOC546⟩. The chlorolithioethene (**2**), which was formed by deprotonation of the chloroethene with *n*-butyllithium at $-78\,°C$, has been isolated as a crystalline solid in 39% yield ⟨93AG(E)1032⟩. The x-ray crystal structure was determined at $-90\,°C$ and showed a lengthening of the C—Cl bond consistent with a carbenoid character for (**2**).

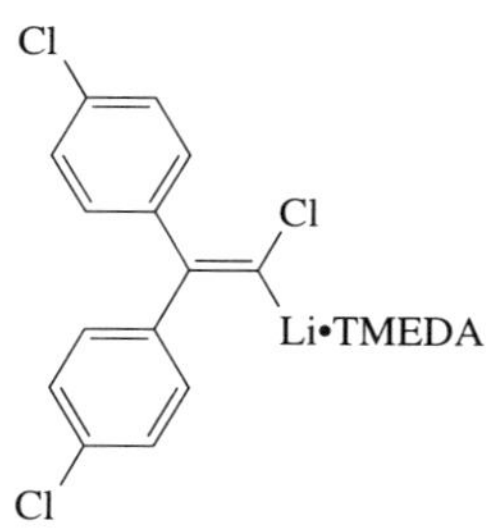

(2)

Stable hypervalent iodonium ylides are easily prepared by the reaction of stabilized enolates with $PhI(OAc)_2$, as shown in Equation (3) ⟨89LA167⟩, and several x-ray crystal structures have been published ⟨90AX(C)1300, 92CC1487⟩.

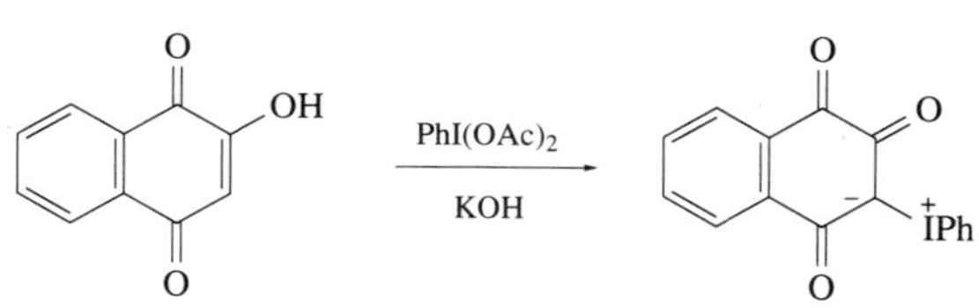

(3)

Further details on the preparation of carbanions adjacent to halogens can be found in Chapters 4.03.5.1 and 4.18.5.1.

2.20.1.2 Carbanions Adjacent to Chalcogen

α-Alkoxycarbanions are calculated to be 5–10 kcal mol^{-1} more stable than the corresponding alkyl anions, due to the σ inductive effect of oxygen ⟨84JA6467⟩. Treatment of ethers with butylpotassium yields α-alkoxypotassium anions directly ⟨83TL3165, 84TL745⟩. The α-alkoxylithium anions can be formed by treating the corresponding ethers with a strong lithium base (e.g., *t*-BuLi), provided there is an additional anion stabilizing feature also present, for example, carbonyl ⟨81JOC2363⟩, cyano ⟨87JOC564⟩ or alkene ⟨79TL833⟩. In a remarkable experiment, Hoppe *et al.* ⟨89AG(E)69⟩ isolated the chiral lithium compound (**3**), by deprotonating an achiral allyl ether with *s*-butyllithium in the presence of (−)-sparteine (Equation (4)). The x-ray crystal structure of the related alkoxylithium species (**4**) has been solved and shows that the lithium is bonded only to C-1 of the allyl system ⟨91AG(E)321⟩. The crystal structure of the silyloxymethyllithium (**5**) has also been determined and indicates that the anionic carbon is hybridized between sp^2 and sp^3 due to delocalization of the charge into the phenyl rings ⟨92CB2265⟩. In both (**4**) and (**5**), the bond between the anionic carbon and oxygen is elongated, suggesting a carbenoid rather than a normal carbanion structure ⟨93CB1873⟩.

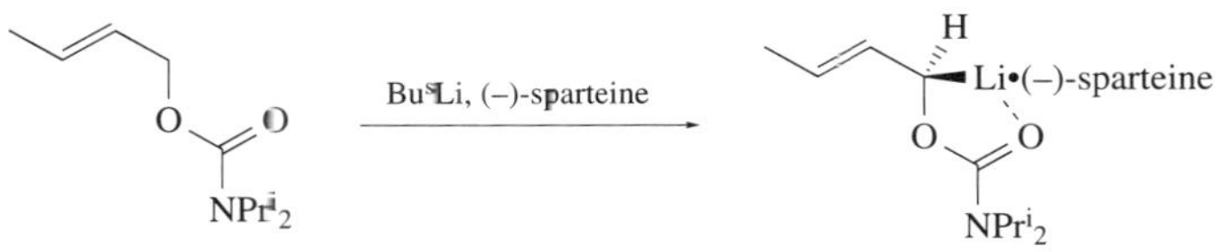

(**3**)

(4)

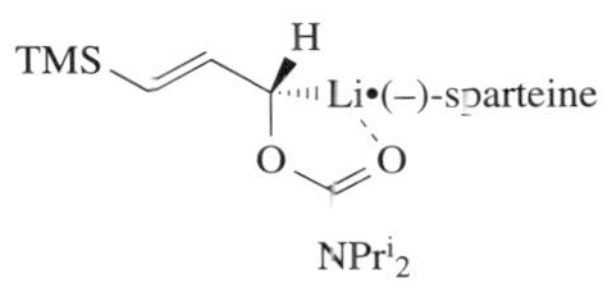

(**4**)

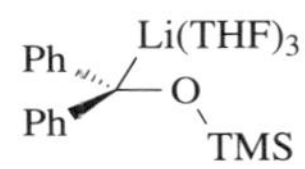

(**5**)

In the absence of stabilizing features, the lithium anions are best formed by reductive lithiation of α-(phenylthio) ethers ⟨84JA1130⟩, or transmetallation of α-alkoxy stannanes ⟨88JA842⟩. The reaction of α-haloalkyl carboxylates with zinc metal yields the organozinc derivatives which can be converted to the copper reagents by addition of cuprous salts ⟨93JOC588⟩. Vinyl ethers ⟨74JA7125⟩ and furans ⟨82AHC(30)167⟩ are readily deprotonated with lithium bases to form α-oxygen-substituted alkenyllithiums. The 2-lithiobenzofuran: TMEDA complex (**6**) has been isolated and its crystal structure determined ⟨89OM1688⟩.

(**6**) X = O
(**7**) X = S

Further details on the preparation of α-alkoxycarbanions can be found in Chapters 4.08.2.1 and 4.20.4.1.

α-Sulfenyl carbanions are more stable than their oxygen counterparts by about 15 kcal mol^{-1} ⟨84JA6467⟩. As a consequence, it has been possible to isolate several as crystalline solids and to determine their crystal structures. Addition of *n*-butyllithium to solutions of dimethyl and phenylmethyl sulfides in undiluted TMEDA gave crystals of the corresponding α-lithiosulfides (**8**) and (**9**) as TMEDA complexes ⟨84HCA224⟩. Treatment of phenylbenzyl sulfide in THF/hexane with *n*-butyllithium at 0 °C gave crystals of (phenylthio)benzyllithium (**10**) as a THF solvate ⟨91CB543⟩. In all three cases the anionic carbon was sp^3 hybridized and the bond distance between this carbon and the sulfur was shortened relative to the nonlithiated precursors. The crystal structure of lithiated 2-butenyl *t*-butyl sulfide (**11**) indicates that the lithium atom is bound to all three carbons of the allyl moiety ⟨88HCA299⟩. In general, these α-lithiosulfides are unstable as solids above about −20 °C.

α-Sulfenyl Grignard reagents can be formed by reaction of α-chlorosulfides with magnesium ⟨82CL1697⟩.

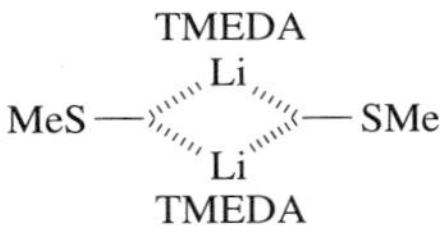

(8)

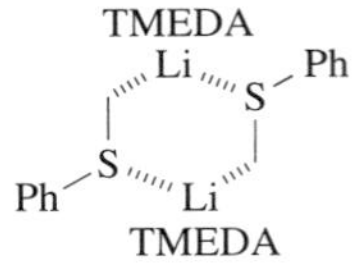

(9)

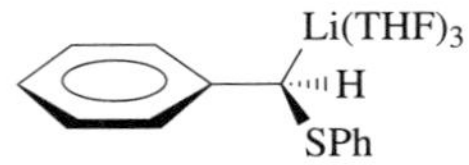

(10)

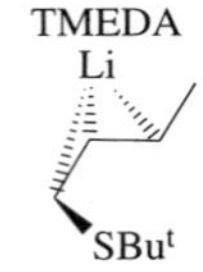

(11)

α-Sulfenylalkenyllithiums are formed when vinyl sulfides are treated with *n*-butyllithium in the presence of TMEDA ⟨83JCS(P1)1131⟩. The 2-lithiobenzothiophene (**7**) has been isolated as a crystalline solid and its x-ray crystal structure determined ⟨89OM1688⟩.

Only a single example of an α-sulfinyl carbanion has been crystallized. This was formed by treating α-methylbenzyl(phenyl)sulfoxide with *n*-butyllithium in the presence of TMEDA ⟨86AG(E)1011⟩. The crystal structure of (**12**) indicates that the lithium is bonded to the sulfoxide oxygen and not to the anionic carbon, which is hybridized between sp^2 and sp^3.

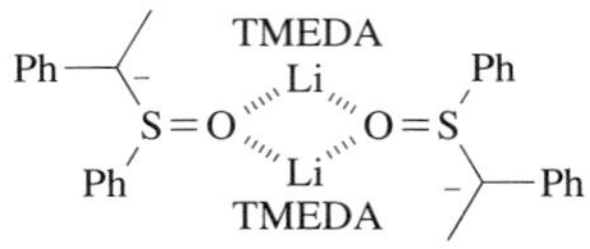

(12)

α-Sulfonyl carbanions are easily formed by deprotonation of sulfones and several have been isolated as crystalline solids, most of which are stable under an inert atmosphere at 20 °C. In all cases they formed dimeric complexes with the metal, sulfur, and oxygens forming an eight-membered ring. The anionic carbon of the lithiosulfones (**13**) ⟨85AG(E)859⟩, (**14**) ⟨88TL1259, 91JA4002⟩, (**15**) ⟨86AG(E)939⟩ and (**20**) ⟨88AG(E)846⟩ is sp^3 hybridized, whereas that of the α-phenyl substituted compounds (**16**) ⟨85AG(E)573⟩, (**17**) ⟨88TL1259⟩, (**18**) ⟨89AG(E)277⟩, and (**19**) ⟨89AG(E)1025⟩ is sp^2 hybridized. The lithium is bonded to the sulfone oxygen atoms in structures (**13**)–(**19**) and only in the cyclopropyl derivative (**20**) is it bonded to carbon. The tetraalkylammonium salt of (**19**) has been prepared in 84% yield by deprotonation of the sulfone with tetrabutylammonium hydroxide ⟨90AG(E)100⟩. In contrast to the sulfonyl lithiums described above, the x-ray crystal structure of the α-sulfonimidoyl lithium (**21**) indicates that the metal is bonded directly to the anionic carbon ⟨86AG(E)938⟩.

$[PhSO_2CH_2Li{\bullet}TMEDA]_2$

(13)

$[PhSO_2C(Me)_2Li{\bullet}DIGLYME]_2$

(14)

$[PhSO_2CH(CH{=}CH_2)Li{\bullet}DIGLYME]_2$

(15)

$[PhSO_2CHPhLi{\bullet}TMEDA]_2$

(16)

$[PhSO_2C(Me)PhLi{\bullet}TMEDA]_2$

(17)

$[MeSO_2CHPhLi{\bullet}TMEDA]_2$

(18)

$[F_3CSO_2CHPhLi{\bullet}TMEDA]_2$

(19)

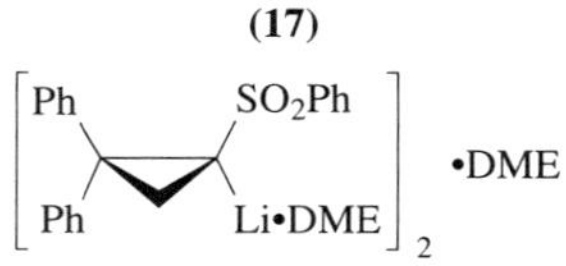

(20)

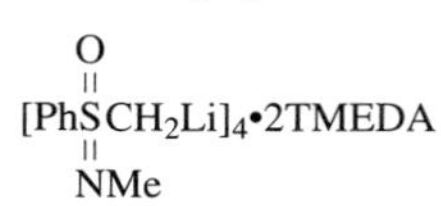

(21)

Sulfonium ⟨65JA1353, 69JOC1240⟩ and sulfoxonium ⟨87T2609⟩ ylides are commonly prepared from the corresponding salts by deprotonation with a base (Equation (5)), or by the (often intramolecular) rhodium catalyzed reaction of sulfides ⟨87JA3010, 91T8611⟩ or sulfoxides ⟨88TL6009⟩ with diazo-compounds (Equation (6)). Sulfonium ylides can also be prepared by the condensation of sulfoxides with active methylene compounds in the presence of acetic anhydride ⟨67T4279⟩. In cases when extra stabilizing groups are also present on the anionic carbon (e.g., β-carbonyl or cyano) these ylides can be isolated as solids and several x-ray crystal structures have been determined ⟨69AX(B)73, 81JCS(P1)212, 83JHC399, 84AX(C)842, 87JA3010, 88TL6009, 88JCS(P1)1023, 90CPB329, 91T8611⟩. In most cases the anionic carbon is sp^2 hybridized, though in the case of the cyclic sulfoxonium ylide (**22**) it was pyramidal ⟨88TL6009⟩.

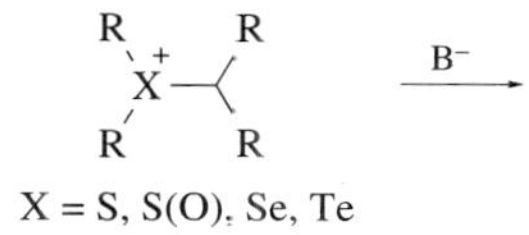

(5)

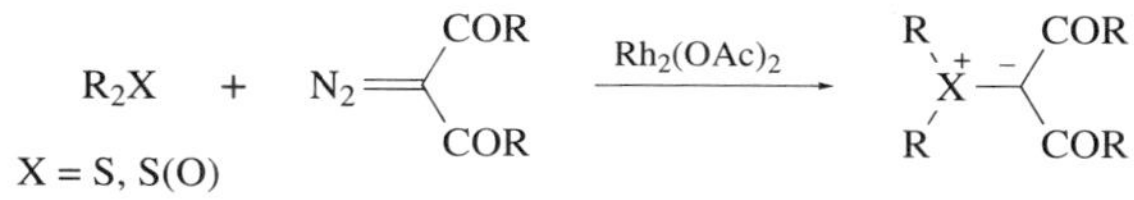

(6)

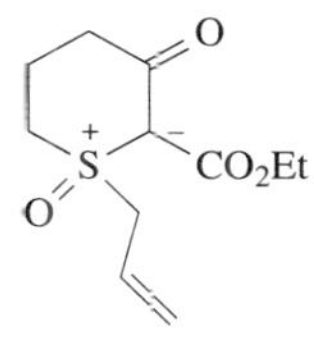

(22)

The structure, synthesis, and reactions of sulfur stabilized carbanions have been recently reviewed ⟨89AG(E)277, 91COS(1)505⟩. Further details on their synthesis can be found in Chapters 4.08.2.2 and 4.20.4.2.

α-Selenyl, selenoxyl, and selenonyl carbanions have all been prepared, usually by deprotonation of the corresponding carbon acid with a lithium or potassium base. Their synthesis and reactions have been extensively reviewed ⟨80T2531, 91COS(1)629⟩ and further preparative details can be found in Chapters 4.08.2.3 and 4.20.4.3. Selenonium ⟨74AG(E)274⟩ and telluronium ⟨83TL5109⟩ ylides can be similarly prepared from the corresponding salts by reaction with a strong base (Equation (5)). When extra stabilizing groups are also present on the anionic carbon atom (e.g., β-carbonyl) the selenonium ylides can be isolated as solids ⟨73T917, 74JA4099⟩ and x-ray crystal structures have been determined including that of the chiral ylide (**23**) ⟨92JCS(P1)1721⟩.

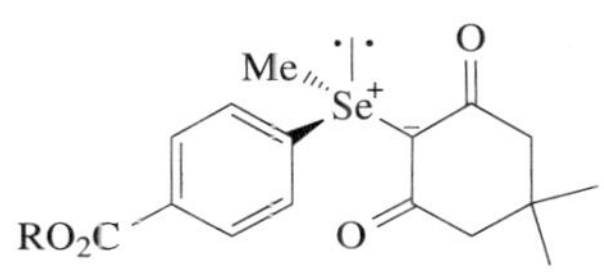

(23)

2.20.1.3 Carbanions Adjacent to Nitrogen

Carbanions can be formed adjacent to tertiary amines by deprotonation with strong bases (e.g., *s*-BuK) ⟨84TL1353⟩ or by transmetallation of α-aminostannanes with *n*-butyllithium ⟨71JA4027⟩. Such anions are more stable than the corresponding alkyl species, but have only been isolated when other stabilizing features are also present. For example, α-dimethylaminobenzyllithium (**24**) was isolated in 50% yield, following treatment of the corresponding α-aminostannane with *n*-butyllithium ⟨93CB1887⟩. The x-ray crystal structure indicates that the anionic carbon is sp^3 hybridized and that the lithium is bound to this carbon and the nitrogen. The lithioallylamine (**25**) was obtained as a stable crystalline solid by deprotonation of the corresponding amine ⟨90CB1853⟩. In this case, the x-ray crystal structure showed that the carbanionic carbon was sp^2 hybridized (indicating that the anion was strongly delocalized) and the lithium atom was bound to the γ-allylic carbon atom. The crystal structure of an α-lithioaminonitrile has also been published ⟨89AG(E)277⟩.

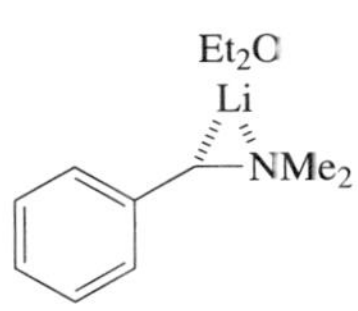

(24)

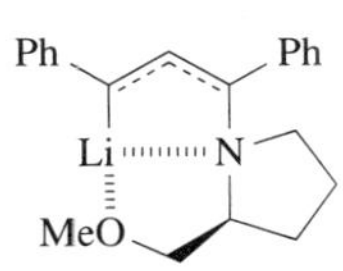

(25)

In general, it is easier to deprotonate adjacent to nitrogen, and the resulting anions are more stable, if an activating group is present on the nitrogen. Many such groups are known including amide, thioamide, urea, carbamate, amidine, nitrosamine and isonitrile ⟨84CRV471⟩. They work by stabilizing the transition state for deprotonation via resonance delocalization (e.g., nitrosamines, isonitriles) or by dipole stabilization (e.g., amides, amidines). When such an activating group is attached to a piperidine, then interesting differences in the structure of the α-lithio species can be observed. Thus, cyclic nitrosamines normally lose the axial proton (Equation (7)) whereas cyclic amides lose the equatorial proton (Equation (8)) ⟨78HCA3100⟩. By attaching an activating group to an enamine nitrogen, it is possible to form α-aminoalkenyllithiums via deprotonation with a strong base (e.g., *t*-BuLi) ⟨85JOC1019⟩. The reaction of α-haloalkyl imides with zinc metal yields the organozinc derivatives which can be easily transmetallated to give the corresponding copper organometallics ⟨93JOC588⟩.

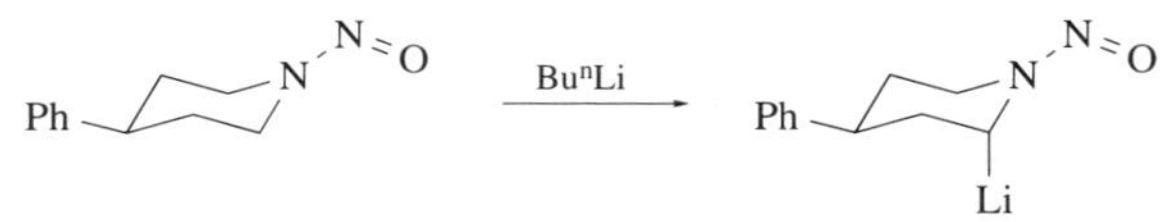

(7)

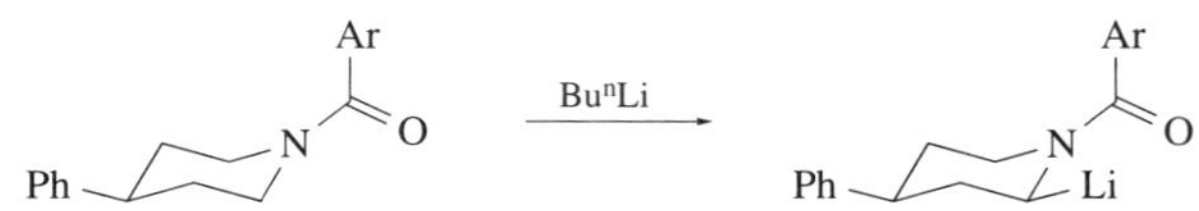

(8)

Deprotonation of *N*-benzyl-*N*-methylpivalamide with *t*-butyllithium in the presence of (−)-sparteine yielded the α-lithiated amide (**26**) in 63% yield ⟨93CB1887⟩. The x-ray crystal structure shows that the lithium is bonded directly to the anionic carbon which is sp^3 hybridized. The magnesiopivaloyltetrahydroisoquinoline : THF complex (**27**) was prepared by deprotonation with *t*-butyllithium followed by addition of magnesium bromide and then crystallized at −78 °C from a mixture of THF and pentane ⟨85JOM(285)1⟩. The crystal structure indicates that the magnesium is bonded directly to the anionic carbon and occupies a pseudoequatorial position on the ring system. The crystal structure has also been determined for the lithiodiphenyl-methylisonitrile : sparteine complex (**28**), which crystallized from THF at −28 °C ⟨92AG(E)79⟩. The lithium is bound to the carbon of the isonitrile in a linear arrangement and the anionic carbon appears sp^2 hybridized due to delocalization of the negative charge into the two phenyl rings.

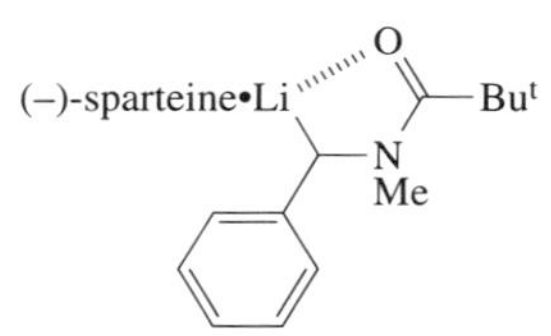

(26)

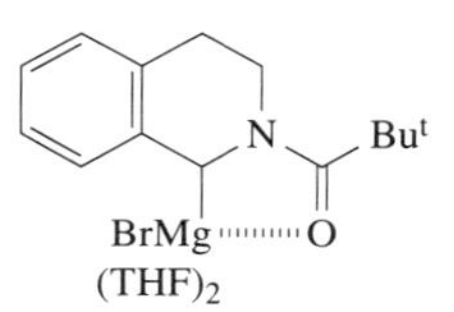

(27)

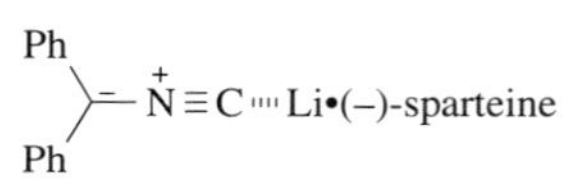

(28)

Anions are readily formed adjacent to nitrogen in heteroaromatic rings (e.g., pyrroles, pyrimidines, imidazoles etc.), usually either via deprotonation or metal–halogen exchange. The synthesis of these species has been extensively reviewed ⟨93AHC(56)155⟩ and are not be covered further here. An x-ray crystal structure has been published for the lithiopyrrole (**29**) ⟨86AG(E)1103⟩.

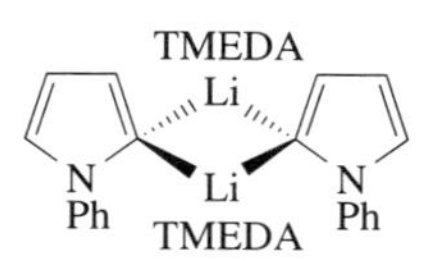

(29)

The hydrogens adjacent to a nitro group are easily removed, and synthetically useful concentrations of the anions can often be formed with relatively weak bases (e.g., Et_3N, K_2CO_3) ⟨91COS(2)321⟩. Several such nitronate anions have been crystallized and their x-ray crystal structures have been determined ⟨54AX145, 74AX(B)147, 81AX(B)344, 82AX(B)1124, 87AG(E)78⟩. In all cases, the anionic carbon was sp^2 hybridized and the metal was bound to the nitro group. As an example to illustrate their preparation, treatment of phenylnitromethane with lithium ethoxide in ethanol yielded crystals of α-nitrobenzyllithium (**30**), which were stable at 20°C ⟨87AG(E)78⟩ The isolation and crystallization of a nitrosonate anion has also been reported ⟨79UKZ811⟩.

$[PhCHNO_2Li{\bullet}EtOH]_n$

(30)

Stabilized nitrogen ylides are easily prepared by deprotonation of the corresponding quaternary salts. For example, the trimethylammonium ylide (**31**) ⟨82M1451⟩ and the pyridinium ylide (**32**) ⟨62CB1108⟩ were prepared from the salts by treatment with silver oxide and potassium carbonate, respectively. Due to their high stability, they are easily isolated and a number of x-ray crystal structures have been published ⟨79AX(B)2443, 83M783, 83M1097, 88CB843, 91T5277, 91AX(C)328⟩.

(31)

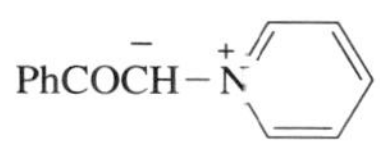

(32)

Diazoalkanes can also be viewed as having an ylide type structure (**33**) and hence as α-azacarbanions. The compounds are well known, are stable to air and water and have been isolated as solids when further stabilized by an aryl ring ⟨82S419, 83JA6096⟩ or a carbonyl group ⟨78AX(B)828, 81CJC1463, 86IZV134, 86JST141, 87CAR(168)7, 91JOC1453⟩. In the case of the diazocarbonyl compounds, x-ray crystal structures have also been determined, but do not definitively differentiate between the different canonical structures. The synthesis and reactions of diazo compounds have been extensively reviewed ⟨91COS(1)843, 91COS(6)103⟩.

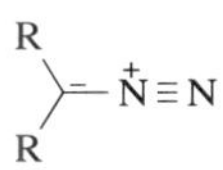

(33)

Further details covering the preparation of carbanions adjacent to nitrogen can be found in a review ⟨91COS(1)459⟩ and in Chapters 3.12.2, 4.11.2.1, and 4.21.5.1.

2.20.1.4 Carbanions Adjacent to Other Group 15 Elements

Addition of trimethylphosphine to a mixture of *n*-butyllithium and TMEDA yielded high melting crystals (>310°C) of the lithio(dimethylphosphino)methane : TMEDA complex (**34**) in 89% yield ⟨89JCS(D)105⟩. Similar treatment of dimethylphenyl and methyldiphenylphosphine gave the α-phosphinoalkyl lithiums (**35**) ⟨89JCS(D)105⟩ and (**36**) ⟨89OM2308⟩. The x-ray crystal structures indicate that the anionic carbon is sp^3 hybridized. The bond between this carbon and phosphorus is much shortened, approaching the length of a P=C bond (1.69 Å) ⟨84T765⟩, reflecting the stabilization of the carbanion by phosphorus. The crystal structures for the dimethylaluminium analogue of (**34**) ⟨85OM231⟩ and the dicyclopentadienyl nickel analogue of (**36**) ⟨82ZN(B)1146⟩ have both been reported.

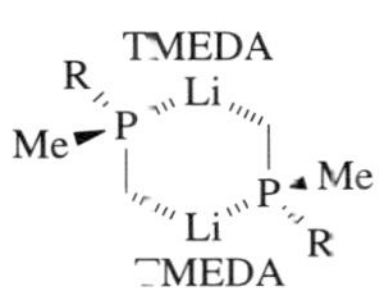

(34) R = Me
(35) R = Ph

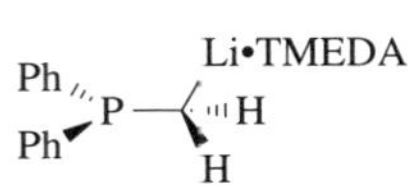

(36)

Phosphoryl-stabilized carbanions are more easily formed and more stable than those adjacent to trivalent phosphorus, due to the increased acidity of the hydrogens and the greater polarizability of the phosphoryl group. Such carbanions are of great synthetic importance for alkene synthesis and their synthesis and use has been extensively reviewed ⟨77OR(25)73, 89CRV863, 91COS(1)729⟩. Several have been isolated as crystalline solids and have had their x-ray crystal structures determined. For example, treatment of diethylbenzylphosphonate with LDA at 20°C, followed by addition of dabco, gave crystals of the diethyllithiobenzylphosphonate : dabco complex (**37**) ⟨91CB861⟩. The lithio diazaphosphorinanes (**38**) ⟨90JA864⟩ and (**39**) ⟨91JA1468⟩ have also been isolated, this time using *s*-butyllithium as the base. In all three cases, the x-ray crystal structures showed that the lithium was bound to the phosphonyl oxygen and not the carbanionic carbon, which was slightly pyramidalized, but predominantly sp^2 hybridized. The anionic carbon–phosphorus bond distances were shorter than observed for lithio derivatives (**34**)–(**36**), reflecting the increased stabilization of the carbanion.

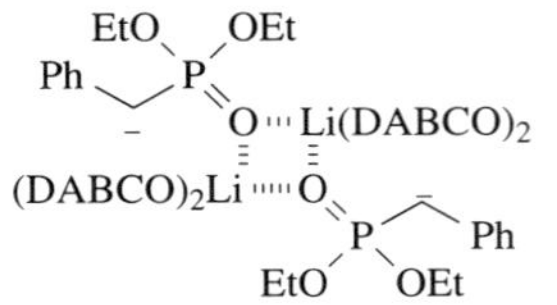

(**37**)

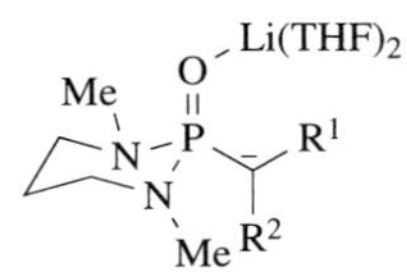

(**38**) R^1 =H; R^2 = Ph
(**39**) R^1 = R^2 = Me

The bidentate carbanions of type (**40**; M = Li, Na, K, Mg, Cu, Co) are well known as stable solids ⟨86PAC1133⟩ and x-ray crystal structures of the magnesium ⟨85CB3529⟩, copper ⟨88AX(C)626⟩ and cobalt ⟨68IC18⟩ derivatives have been published.

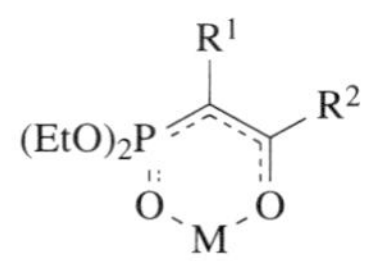

(**40**)

Phosphonium ylides, better known as Wittig reagents, are well characterized in the literature and both their preparation and use in synthesis have been thoroughly reviewed ⟨65OR(14)270, 77AG(E)349, 89CRV863, 91COS(6)171⟩. Many have been isolated as solids and several x-ray crystal structures have been determined ⟨69JCS(B)350, 73JA6158, 77CB3168, 91JCS(P2)1077, 93AG(E)1746⟩ including that of the carboxylic acid (**41**) ⟨75JCS(P2)1030⟩ which serves to illustrate the high degree of stability achievable with stabilized ylides. Diethyoxyvinylidinetriphenylphosphorane (**42**), synthesized via deprotonation of the corresponding vinylphosphonium salt, has a nonlinear structure with a P—C_α—C_β angle of 125.6° and P—C_α bond length of 1.84 Å, both of which suggest a noncumulene structure ⟨74CB1949⟩. In contrast, the comparative values for the phosphorane (**43**) are 168° and 1.68 Å ⟨67JCS(A)1913⟩ which are more consistent with the depicted cumulene structure.

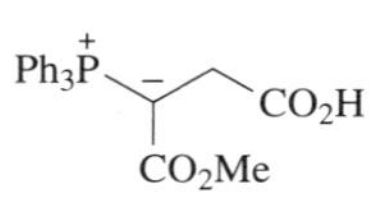

(**41**)

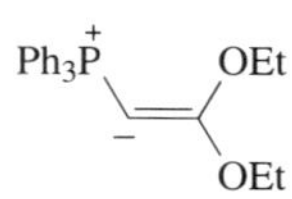

(**42**)

$Ph_3P{=}\bullet{=}\bullet{=}S$

(**43**)

Methylenetriphenylarsorane (**44**) was prepared by treating methyltriphenylarsonium bromide with sodium amide ⟨75CC668⟩. In contrast to the equivalent phosphonium ylide, where the carbanionic carbon was sp^2 hybridized, in this case the ylidic carbon is sp^3 hybridized. As expected, in stabilized arsonium ylides, the anionic carbon is also sp^2 hybridized ⟨75JCS(D)1284⟩. Stibonium and bismuthonium ylides have been made, though they are not common ⟨67CC1042⟩.

$Ph_3\overset{+}{As}-\overset{-}{C}H_2$

(**44**)

Further information describing the preparation of carbanions adjacent to phosphorus can be found in Chapters 3.13, 4.12.4.1 and 4.22.4.1. Further details on the equivalent arsenic, antimony, and bismuth carbanions can be found in Chapters 4.13.3 and 4.22.8.

2.20.1.5 Carbanions Adjacent to Metalloids

Silicon is able to stabilize adjacent carbanions, even though it is more electropositive than hydrogen or carbon, via hyperconjugative interactions and *p*—*d* π-bonding. The α-silylalkyl carbanions can be formed by deprotonation of silanes with a strong base (e.g., *n*-BuLi), or by the substitution of an α-heteroatom (chlorine, sulfur, selenium, silicon, or tin) for a metal ⟨84S384⟩. Treatment of vinyl silanes with lithium or magnesium bases does not give α-silylalkenyl anions and instead, the reagents add to the alkene to give α-silylalkyl anions in high yields ⟨83JCS(P1)1131, 84S384⟩.

Several α-silylbenzyllithiums have been isolated as crystalline solids and their x-ray crystal structures have been determined ⟨82CC14, 84JCS(D)321, 84JCS(D)1801, 85JCS(D)337, 89JCS(D)105, 91CB543, 91OM1282⟩. As an example of their preparation, treatment of benzyltrimethylsilane with *n*-butyllithium in the presence of TMEDA yielded α-(trimethylsilyl)benzyllithium : TMEDA (**45**), which was crystallized from *n*-hexane at −22 °C ⟨91CB543⟩. The anionic carbon of (**45**) is predominantly sp^3 hybridized and the lithium is bound both to this carbon and to the adjacent aromatic carbon. In some other cases, the lithium forms a third bond with a second aromatic carbon, as shown by (**46**) where one lithium atom is η^2 bonded while the other is η^3 bonded ⟨85JCS(D)337⟩.

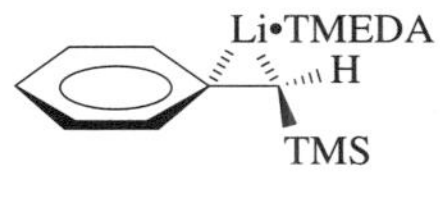

(**45**)

(**46**)

The synthesis and reactions of α-silyl carbanions have been reviewed ⟨84S384, 89JOM(360)39, 90OR(38)1, 91COS(1)579⟩ and further preparative details can be found in Chapters 4.14.2.1 and 4.23.2.1.

Boron is an efficient π-acceptor and stabilizes adjacent carbanions to the same degree as formyl or cyano groups ⟨81JOC1693⟩. Despite this very high degree of stabilization, formation of boron stabilized anions by deprotonation is complicated by the predilection of boranes to react with bases, to form coordinatively saturated borates. This tendency can be overcome by attaching bulky groups on to the boron, so that, for example, dimesitylmethylborane can be deprotonated in THF at 25 °C with lithium dicyclohexylamide ⟨83TL627⟩. Upon addition of two equivalents of 12-crown-4, the boron methylidenide ion (**47**) crystallizes from solution ⟨87JA2541⟩. The x-ray crystal structure shows that the anionic carbon is sp^2 hybridized and that the B—CH_2 bond length is much shortened, which is consistent with significant double bond character. This latter observation is further supported by the high barrier to rotation observed for the related anion (**48**) ⟨83TL623⟩.

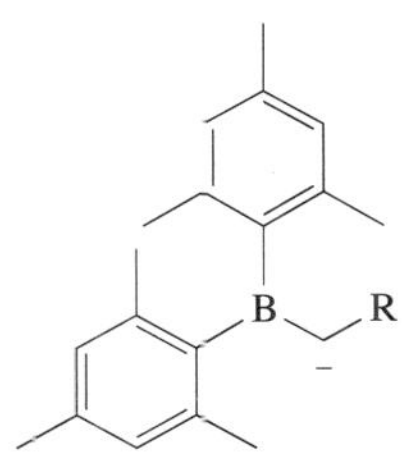

(**47**) R = H
(**48**) R = Ph

Boron stabilized organometallics are also readily prepared by base treatment of 1,1-diboryl alkanes ⟨66TL2535, 73S37⟩. Boronic esters are less easily deprotonated than boranes and the α-lithio species are best prepared by transmetallation of the α-stannyl species which are in turn prepared from the α-halo compound ⟨85OM1690⟩. The reaction of bases with borylstannylalkenes yields α-borylalkenyl compounds ⟨91PAC403⟩. Treatment of α-halo boronic esters with zinc metal yields (dialkoxyboryl)alkyl zincs which can be readily converted to cuprates ⟨90JA7431⟩.

The synthesis and reactions of boron stabilized carbanions has been reviewed ⟨91COS(1)437⟩ and further details covering their preparation can be found in Chapters 4.14.2.3.1 and 4.23.2.1.

2.20.1.6 Carbanions Adjacent to Metals

gem-Dimetalloalkanes are well known in the literature, though to date no x-ray crystal structures appear to have been determined. 1,1-Dilithio and 1,1-dipotassio compounds can be prepared via thermal decomposition of the dimetallo salts of hydrazones (Equation (9)) ⟨68JA7287⟩. Pyrolysis of 1-lithio-2,2,3,3-tetramethylcyclopropane at 170 °C gave a 40% isolated yield of the dilithio compound (**49**) ⟨85JA5313⟩. Methylene magnesium and zinc halides have been prepared by the reaction of magnesium and zinc metal with diodomethane ⟨70T1281, 83TL2043⟩. However, by far the most general method for preparing 1,1-dimetalloalkanes is via the metalla-Claisen carbozincation of an alkenyl organometallic (Scheme 1) ⟨86TL1039, 92SL633⟩. The resulting metallozinc compounds are formed in good yields and are stable in solution at room temperature. The lithio and magnesiozincs can be converted to cuprozincs by addition of copper(I) cyanide ⟨86TL4427⟩. *gem*-Dimetalloalkenes, including the 1,1-dilithio compound **(50)** ⟨84AG(E)224⟩, are also known.

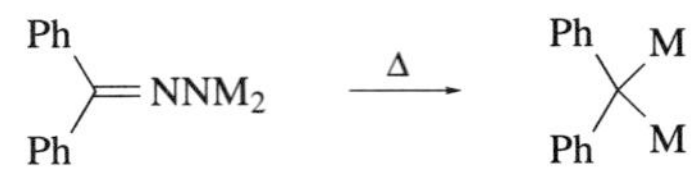

M = Li, K (9)

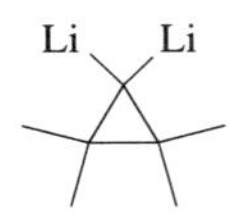

(49)

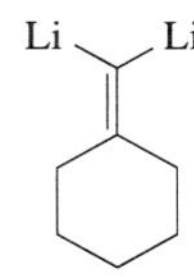

(50)

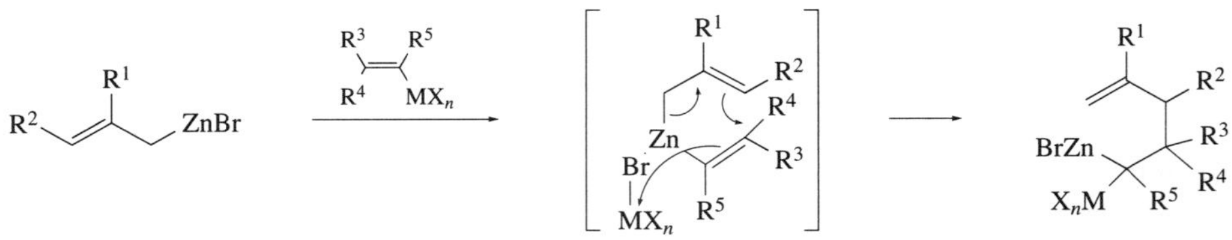

M = Li, Mg, Al

Scheme 1

Further details on the preparation of gem-dimetallo compounds can be found in Chapters 4.15, 4.16, and 4.23.

2.20.2 STABILIZED CARBOCATIONS BEARING ONE HETEROATOM ($R^1R^2C^+X$)

In addition to stabilizing carbanions ⟨84JA6467⟩, many heteroatoms can also stabilize carbocations ⟨65JA2490⟩. Such heteroatom stabilized carbocations are of real importance in synthetic organic chemistry ⟨70SCI1298, 91OR(40)157, 91COS(2)707, 91COS(2)733, 91COS(2)893, 91COS(2)953, 91COS(6)261, 93JA497⟩ and several have now been isolated. In general, they are formed by protonation of an appropriate precursor with a super acid or by treatment of a halide with a strong Lewis acid. A number of x-ray crystal structures have been determined and just as with their carbocation counterparts, these can provide an insight into reaction mechanism ⟨74AG(E)1, 87JA882⟩.

2.20.2.1 Carbocations Adjacent to Halogen

Carbocations adjacent to a halogen atom are easily generated by treatment of dihaloalkanes with antimony pentafluoride (Equation (10)) ⟨69JA2955⟩. Dialkylhalomethyl cations are only stable at low temperatures in solution, but diarylhalomethyl cations are more stable and several have been isolated as solids ⟨72S307, 81LA1415⟩. For example, chlorodiphenylcarbenium hexachloroantimonate **(51)** was obtained in 95% yield when SbF_5 was added to a solution of dichlorodiphenylmethane. The trifluoro- and trichlorocyclopropenium salts **(53)** ⟨64JA1459, 69JA415⟩ and the chlorovinyl dication **(54)** ⟨62TL983⟩ have also been isolated.

$$R_2CX_2 \xrightarrow{SbF_5} R_2\overset{+}{C}X \qquad (10)$$

X = F, Cl, Br, I

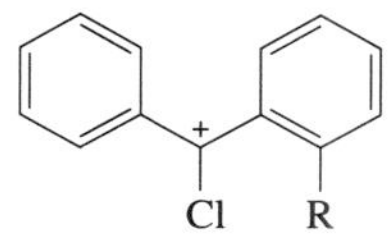

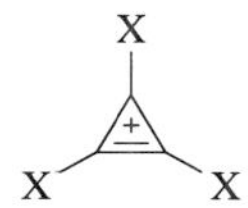

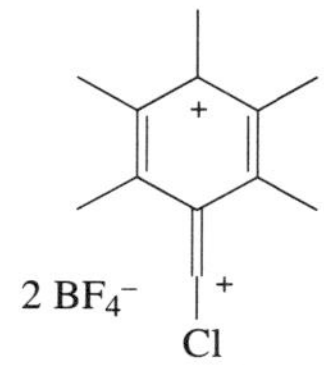

(51) R = H; $SbCl_6^-$
(52) R = Cl; SbF_6^-

(53) X = F; SbF_6^-
X = Cl; $AlCl_4^-$

(54)

The crystal structure of the diarylchloromethyl cation **(52)** has been determined by x-ray crystallography and shows significant shortening of the C—Cl bond ⟨93JA1731⟩. This was interpreted as support for the stabilization of the cation, via back-donation from the lone pairs on chlorine. The crystal structure for the highly delocalized cation **(55)** has also been determined and shows a lengthening of the C—Br bond, suggesting this is not a bromine stabilized cation ⟨91JCS(P2)1563⟩.

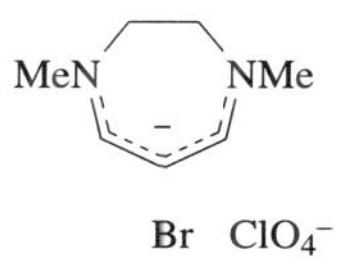

(55)

2.20.2.2 Carbocations Adjacent to Chalcogen

Protonation of aliphatic ketones with super acids (e.g., SbF_5/FSO_3H) at $-60\,^\circ C$ leads to formation of α-hydroxy carbocations (Scheme 2). NMR studies indicate that the ions exist predominantly in the oxonium form and that the oxygen stabilizes the positive charge to the same extent as an α-phenyl ring ⟨68JA938, 69JA5801⟩. When extra stabilizing groups are present in addition to oxygen (e.g., phenyl, cyclopropyl) then it is possible to isolate these cations and several x-ray crystal structures have been determined ⟨65JCS1097, 86JA1692, 86JA1694, 89AX(C)1520, 90JA8912⟩. For example, the α-hydroxy carbocations **(56)** ⟨90JA8912⟩ and **(57)** ⟨86JA1692⟩ were isolated as moisture and temperature sensitive solids upon addition of SbF_5/HF to a solution of the corresponding ketones at $-78\,^\circ C$. In both cases, the x-ray crystal structures showed that the cationic carbon was sp^2 hybridized and that the C—O bond length was lengthened relative to the ketones, but shorter than the C—O bond length found in ethers. This suggests these compounds have some degree of carbocationic structure.

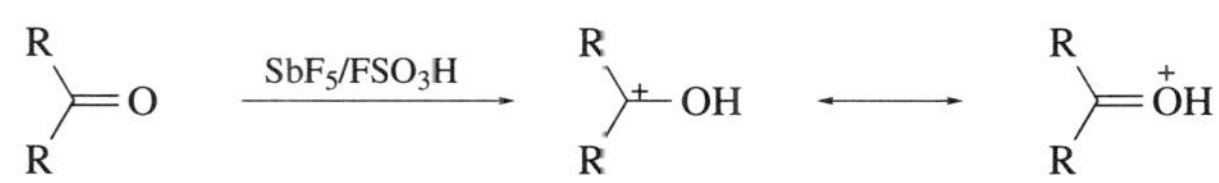

Scheme 2

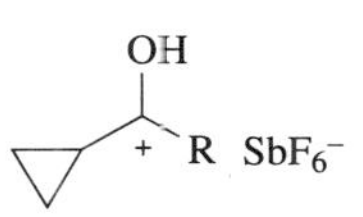

(56) R = Me
(57) R = Ph

α-Alkoxy carbocations can be prepared by the reaction of Lewis acids ⟨66JA4934⟩ and Brönsted acids ⟨79PJC861⟩ with acetals (Equation (11)). They are also formed by treatment of α-haloethers with Lewis acids ⟨67JA2993⟩. In general, these compounds are rather unstable and additional stabilizing factors need to be present for isolation, as exemplified in the dihydrofurylium salt **(58)** ⟨79PJC861⟩ and norbornyl cation **(59)** ⟨66JA4934⟩. The x-ray crystal structure of **(59)** was most consistent with an α-alkoxy carbocation as shown, rather than a nonclassical bridged ion ⟨87JA4749⟩.

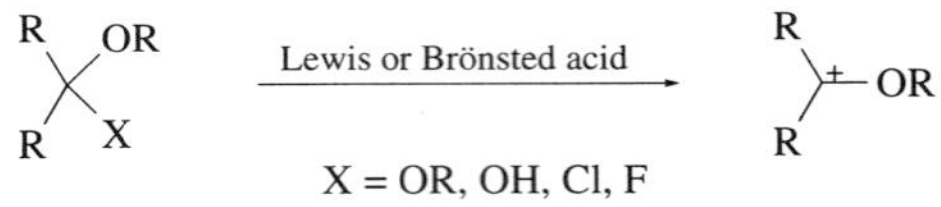

(11)

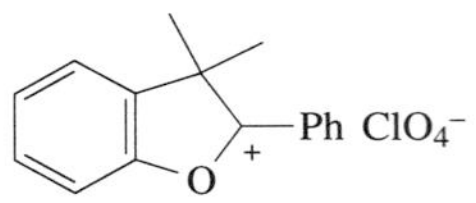

(58)

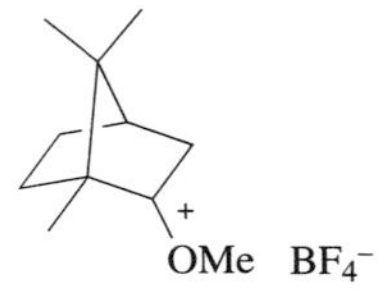

(59)

Acyl carbocations (**60**) have been isolated in high yield upon treatment of aliphatic acyl chlorides with SbF_5/HF ⟨72S306⟩ and aromatic acyl fluorides with just SbF_5 ⟨75LA128⟩. Crystallographic studies on the methyloxocarbonium ion confirm the expected linear structure of these ions ⟨66JA1572⟩.

$$R-\overset{+}{C}=O \quad SbF_6^-$$

(60)

Other oxygen-containing carbocations which have been successfully isolated include cyclopropenium ⟨83JOC3038⟩, pyrylium ⟨73OSC(5)1135, 85H(23)2709⟩, and tropylium ⟨83JOC3043, 86JA3613⟩ salts. The stability of these compounds is much enhanced by their aromatic nature. The preparation and chemistry of pyrylium salts has been extensively reviewed ⟨82AHC(2)1, 90AHC(50)157⟩.

Carbocations can also be generated adjacent to sulfur and selenium; however, in general these species are less well characterized than their oxygen counterparts. Treatment of 2,2-dimethyl-1,3-oxathiolan with SbF_5/FSO_3H at $-60\,°C$ yielded exclusively the dication (**61**) which was characterized by NMR ⟨71BSF541⟩. In a similar way, the reaction of selenoacetals with $SnCl_4$ at $-40\,°C$ generated selenium stabilized carbocations which were characterized by their reaction with enol ethers ⟨84TL4441⟩. The sulfur, selenium, and tellurium containing cyclopropenium ions (**62**) have all been isolated as solids ⟨80S227⟩, and an x-ray crystal structure has been published for the dithiolium salt (**63**) ⟨66ACS1874⟩.

The preparation and chemistry of carbocations adjacent to a chalcogen has been reviewed ⟨92PSS155⟩.

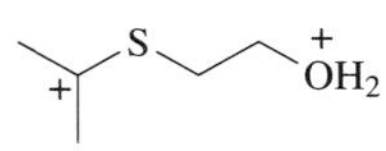

(61)

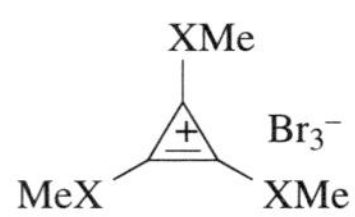

(62) X = S, Se, Te

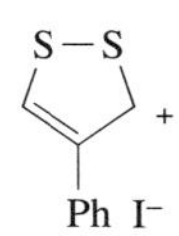

(63)

2.20.2.3 Carbocations Adjacent to Nitrogen

Dialkyliminium salts (**64**) can be conveniently prepared and isolated in high yields by reacting methylenediamines with a reactive halide (Equation (12)) ⟨60CB1305, 69AP(302)354, 76AG(E)239, 80JOC524⟩ or by the combination of secondary amine salts with carbonyl compounds (Equation (13)) ⟨63JOC3021, 85JOC4553⟩. The x-ray crystal structures of the salts (**64**) ⟨86MI 220-1, 89IC1295, 90AG(E)188, 92CJC2040⟩ and of protonated imines ⟨75CSC135, 77AX(B)3782, 87CJC2000, 92POL2351⟩ typically exhibit C—N bond lengths of 1.26–1.29 Å, suggesting these compounds exist predominantly in the iminium form with the charge localized on nitrogen. However, in some cases where the positive charge is also stabilized by delocalization, then it is possible to form and isolate α-aminocarbocations in which significant charge appears to be localized on the carbon. For example, the x-ray crystal structure of the protonated imine (**65**) has been determined and compared with that of the corresponding imine ⟨85JA2067⟩. There is a lengthening of the C(9)—N(9) bond and shortening of the C(9)—C(1a) and C(9)—C(8a) bonds consistent with a carbocationic structure as shown. Even so, the C(9)—N(9) bond length of 1.32 Å is significantly shorter than a typical aromatic C—N single bond length of 1.40 Å ⟨87JA882⟩. The crystal structure of the stabilized iminium salt **(66)** ⟨89JA5424⟩ exhibits a similar C—N bond length to that found for structure **(65)**.

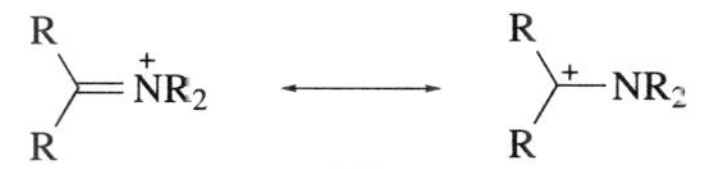

(64)

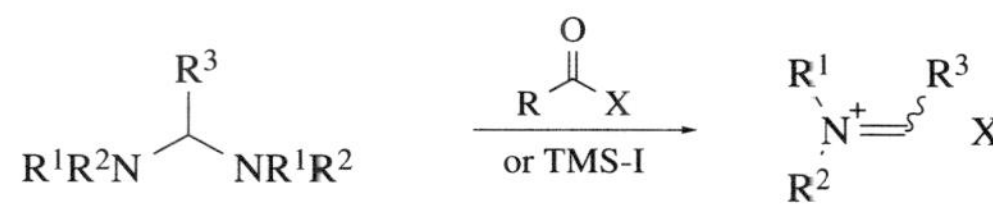

(12)

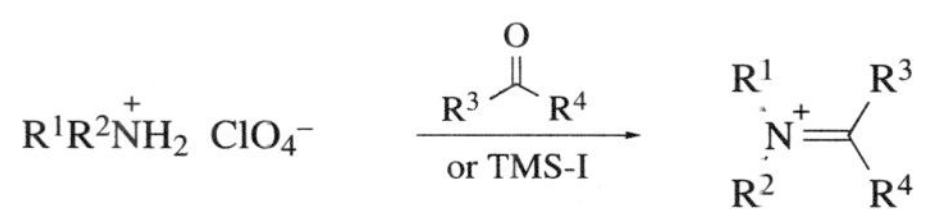

(13)

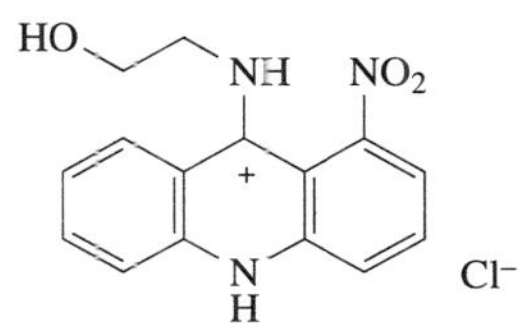

(65)

The reaction of 2,4,6-tripyrrolidinobenzene with alkyl iodides gives the σ-complexes **(67)** (Wheland intermediates) in generally good yields ⟨68TL4265⟩. The x-ray crystal structures of several such complexes have been determined and the bond between the carbocationic carbon and nitrogen is a similar length to that found in structures **(65)** and **(66)** ⟨87JA882, 88JOC4379⟩.

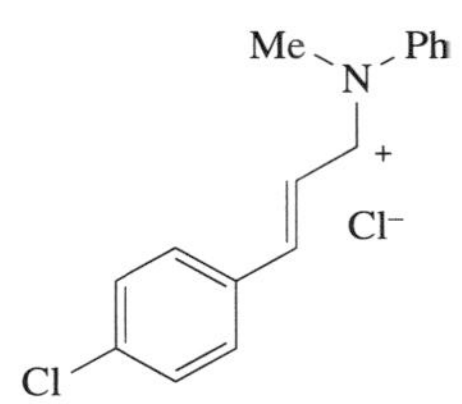

(66)

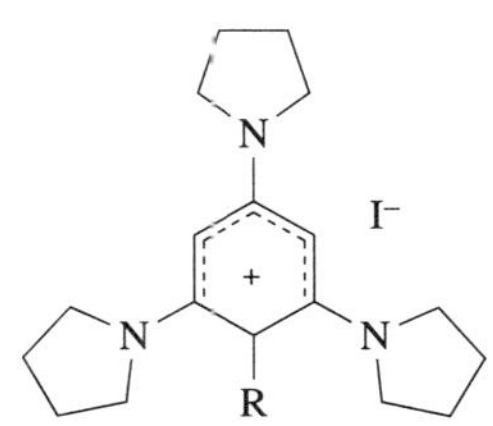

(67)

Nitrogen-substituted cyclopropenium cations ⟨72JA1688, 92AX(C)912⟩ and a wide variety of heteroaromatic iminium salts, which are stabilized by their aromatic nature, have also been isolated. The preparation and chemistry of heteroaromatic iminium salts has been reviewed ⟨85H(23)1513, 87AHC(41)275⟩ and several x-ray crystal structures have been determined ⟨85CB4259, 87AX(C)313, 87T1273, 89JA1299, 90K668⟩.

The preparation and chemistry of iminium salts has been extensively reviewed ⟨91COS(2)893, 91COS(2)953, 91COS(2)1007, 91COS(2)1047⟩.

Both NMR ⟨68JA4666⟩ and x-ray crystal ⟨77ACS(B)56⟩ data indicate that the charge on nitrilium salts **(68)**, which in principle can exist in a carbocationic form, is in fact localized almost entirely on nitrogen. The salts are readily prepared from nitriles by protonation ⟨59LA(626)60, 68JA4666⟩ or by treatment with a Lewis acid and then an alkyl halide or aryl azide ⟨56CB209, 84CB1900⟩. The reaction of nitrilium salts with aromatic aldehydes yields *N*-acyliminium salts ⟨91T205⟩.

(68)

2.20.3 STABILIZED RADICALS BEARING ONE HETEROATOM ($R^1R^2C^{\bullet}X$)

Carbon-centered radicals bearing one heteroatomic substituent are calculated to be more stable than the corresponding unsubstituted radicals ⟨87JOC3062⟩. The degree of stabilization increases in the order halogen < phosphorus < chalcogen < nitrogen < boron. However, unlike heteroatom-stabilized carbanions and carbocations, the radicals have not, in general, been isolated. This is due to the tendency of radicals to dimerize and to be quenched by even minor impurities in the reaction medium, so that most radicals have only a transitory existence. These processes are slowed in sterically-hindered structures and where steric bulk is sufficiently great, then persistent carbon-centered radicals can be formed and in some cases can be isolated ⟨76ACR13, 85ACR380⟩. A good example is the perchlorodiphenylmethyl radical **(69)**, which was prepared by oxidation of the corresponding carbanion with iodine and was isolated as a solid in 82% yield ⟨71JA2215⟩. This compound was stable enough to enable its x-ray crystal structure to be determined ⟨71JPC1246⟩. Reduction of benzophenone with lithium, and several other metals, gives the stable ketyl radical **(70)** ⟨64JA2538⟩. A wide variety of persistent heteroatom-stabilized radicals can be formed by addition of transient radicals to C—C or C—heteroatom multiple bonds (Equations (14)–(17)). These include α-aminoalkyl radicals ⟨74JA6202⟩, α-phosphinoalkyl radicals ⟨75JA5526⟩, α-silylalkyl radicals ⟨74JA6203, 75JA4269⟩, and α-borylalkyl radicals ⟨81JOC179⟩. The preparation and structure of transient heteroatom-stabilized radicals has been reviewed ⟨B-73MI 220-1⟩.

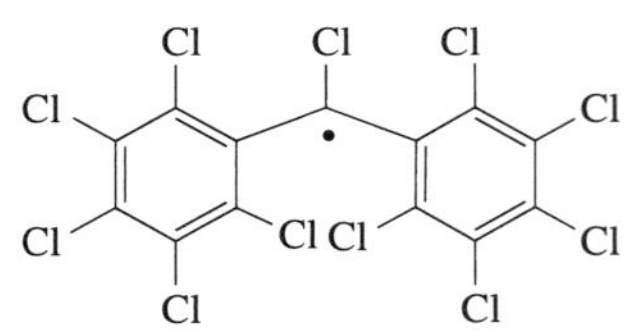

(69)

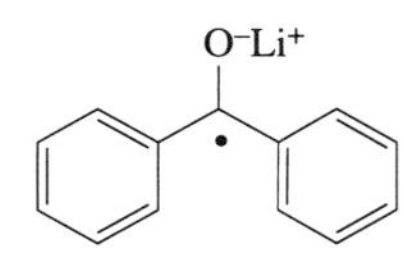

(70)

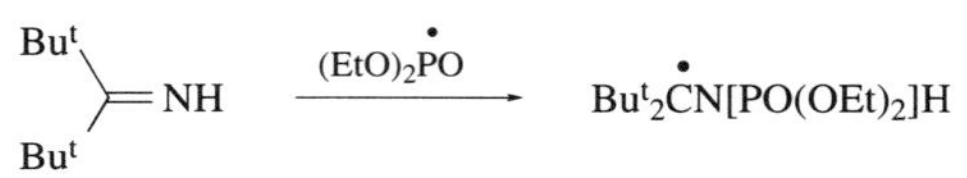

(14)

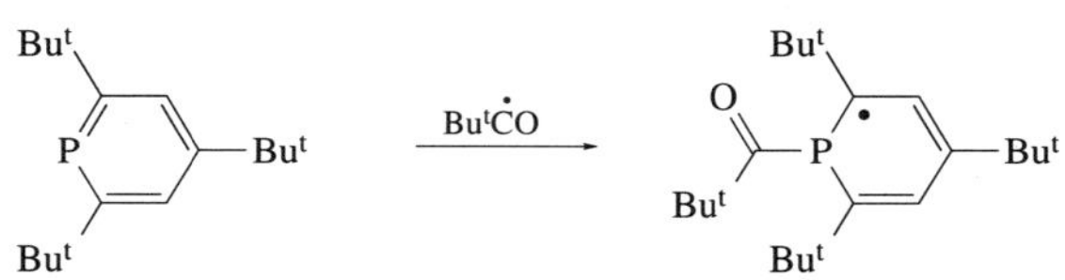

(15)

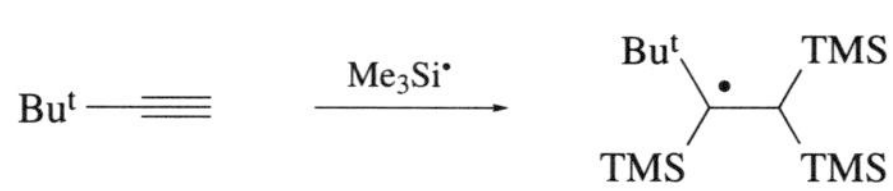

(16)

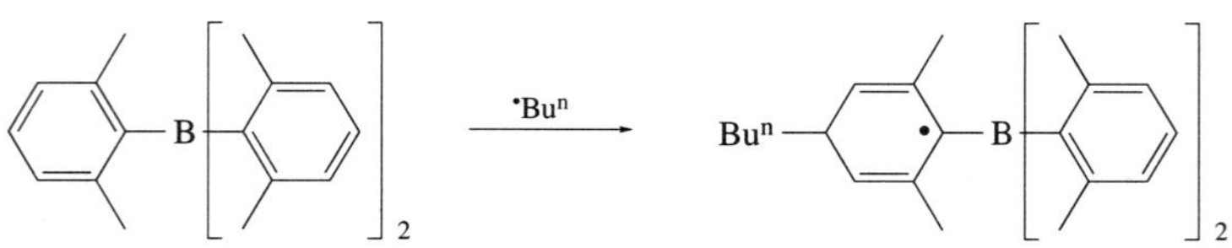

(17)

2.21
Alkynyl Halides and Chalcogenides

PETER J. STANG
University of Utah, Salt Lake City, UT, USA

and

VIKTOR V. ZHDANKIN
University of Minnesota-Duluth, MN, USA

2.21.1 ALKYNYL HALIDES

2.21.1.1 General Methods

Alkynyl halides belong to one of the most useful and well-investigated classes of substituted alkynes. The general principles of the preparation and chemistry of these compounds have been discussed in several comprehensive reviews ⟨B-69MI 221-01, 77HOU(5/2a)110, B-88MI 221-01⟩. By analogy with alkyl and alkenyl halides these compounds can be expected to be valuable precursors in the syntheses of various alkynes by acetylenic nucleophilic substitution which may be realized by a variety of mechanisms ⟨B-78MI 221-02, 76ACR358⟩. In view of this property there is a substantial interest in the synthetic application of alkynyl halides ⟨B-69MI 221-01, 84JOC3429, 86JOC5320, 85SC463⟩

and alkynyl iodonium salts ⟨92AG(E)274⟩, the latter being especially reactive toward nucleophilic reagents due to the very high leaving group ability of the iodonium moiety.

The stability and physical and chemical properties of alkynyl halides vary in a broad range depending on the nature of the halogen ⟨B-69MI 221-01⟩. Alkynyl fluorides are the least stable, and the chemistry of these compounds is quite different from other alkynyl halides. The first synthesis of alkynyl fluorides was only reported in 1959 ⟨59JA803⟩, and there were only a few representatives of this class known even in 1994. In contrast, alkynyl chlorides, bromides, and iodides are relatively stable and have been known for over 120 years ⟨B-69MI 221-01⟩. The newest member of the family of alkynyl halides, alkynyliodonium salts ($R^1C{\equiv}CI^+R^2{\cdot}X^-$), became readily available only in recent years ⟨92AG(E)274⟩.

A vast variety of synthetic procedures for the preparation of alkynyl halides have been reported in the literature. However, most of these procedures can be generalized as three major methods: (i) elimination (mainly dehydrohalogenation or dehalogenation) reactions, (ii) alkylation of haloethynylides, XC≡CM, and (iii) reactions of terminal alkynes, RC≡CH, or alkynylides, RC≡CM, with halogens or electrophilic halogenating reagents.

The first approach—elimination—is the most general and has been widely used in the synthesis of alkynyl fluorides, chlorides, and bromides. The best procedures are based on dehalogenation or dehydrohalogenation of various polyhaloalkenes or alkanes. Dehalogenation is usually achieved either by use of metals, such as zinc or magnesium, or with organolithium compounds under mild conditions. The reactivity of halogen atoms in the dehalogenation reaction reduces sharply from iodine to fluorine ($I > Br > Cl > F$), which allows the selective preparation of fluoro- or chloroalkynes from the corresponding mixed haloalkenes. A typical example of such a reaction is illustrated by the preparation of fluoroethyne (**2**) by treatment of 1,2-dibromo-1-fluoroethene (**1**) with magnesium in THF (Equation (1)) ⟨62CB319, 62M952⟩.

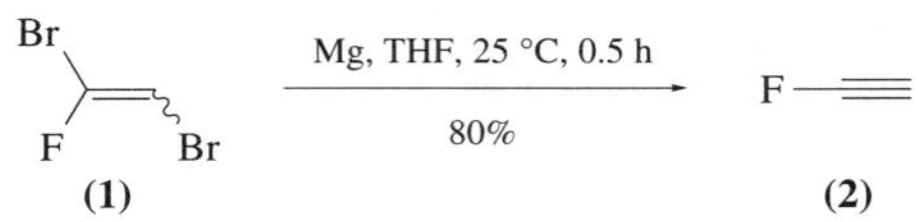

(1)

The dehalogenation method has, however, limited application because of the possibility of further metal exchange in the originally formed alkynyl halides with the formation of the corresponding metal alkynylides. The dehydrohalogenation procedures are more commonly applied due to the better selectivity of the elimination reaction. The best results are given by reactions of polyhaloalkenes with oxygen bases such as KOH, NaOH, or NaOR. The configuration of the starting alkenes is especially important; the *trans* arrangement of the hydrogen and the halogen facilitates elimination ⟨52JA629⟩. The selective preparation of alkynyl chlorides (**4**) by dehydrohalogenation of *cis*-1,2-dichloro-1-alkenes (**3**) (Equation (2)) is a representative example of such an elimination ⟨54CJC500⟩.

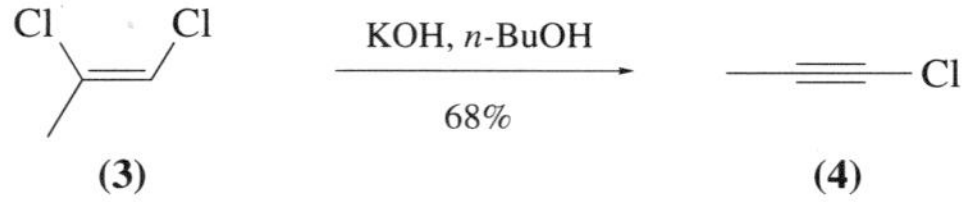

(2)

The dehydrohalogenation approach is usually complicated by various side-reactions due to nucleophilic addition or addition–elimination processes. More convenient methods are based on substitution at the acetylenic carbon, namely alkylation of haloethynylides, XC≡CM (Equation (3)), and reactions of terminal alkynes, RC≡CH, or alkynylides, RC≡CM, with halogens or various electrophilic halogenating reagents (Equation (4)).

$$X{-}{\equiv}{-}M \quad + \quad RX \longrightarrow R{-}{\equiv}{-}X \qquad (3)$$

M = Li, Na; X = Cl, Br, I

$$R{-}{\equiv}{-}M \quad + \quad X_2 \longrightarrow R{-}{\equiv}{-}X \qquad (4)$$

M = Li, Na; X = Cl, Br, I

Both methods are widely used in the synthesis of alkynyl chlorides and bromides. The first approach, the so-called haloethynylation reaction (Equation (3)), is especially useful for the preparation of various substituted alkynyl chlorides. The second method (Equation (4)) and its variations represent the major entry into alkynyl iodides and alkynyl iodonium salts.

2.21.1.2 Fluorides

Only a few examples of alkynyl fluorides are known, namely fluoroethyne, difluoroethyne, fluorochloroethyne, perfluoropropyne, *t*-butylfluoroethyne, and fluoropropiolyl fluoride ⟨B-69MI 221-01⟩. All of these compounds are highly unstable and require special caution in their preparation and handling. The only known synthetic approach to alkynyl fluorides is based on various elimination reactions ⟨B-69MI 221-01⟩. The parent fluoroethyne (**2**) was first prepared in 1959 in essentially quantitative yield by the pyrolysis of fluoromaleic anhydride (**5**) in vacuum (Equation (5)) ⟨59JA803⟩.

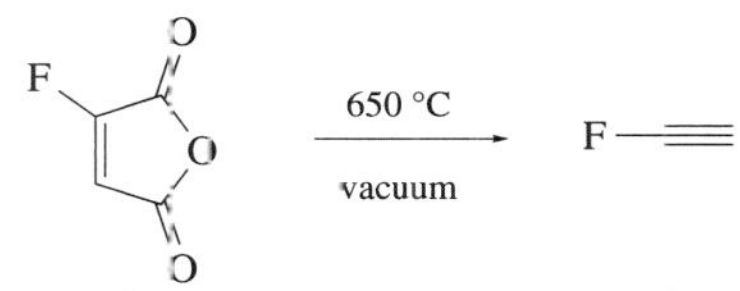

(5) **(2)** (5)

A simpler synthesis of fluoroethyne (**2**) consists of the dehalogenation of 1,2-dibromo-1-fluoroethene (**1**) with magnesium in THF (Equation (1)) ⟨62CB319, 62M952⟩. Fluoroethyne is a colorless gas that condenses below −100 °C to a mobile, highly explosive liquid. It does not explode as a gas at atmospheric pressure, but slowly condenses with the formation of 1,2,4-trifluorobenzene and some polymeric material ⟨59JA803⟩.

The formation of difluoroethyne (**7**) and fluoropropiolyl fluoride (**8**) was first detected by the mass-spectroscopic analysis of products of thermal decomposition of difluoromaleic anhydride (**6**) (Equation (6)) ⟨58USP2831835, 65JPC2484⟩. However, neither of these two fluoroalkynes were isolated in individual form in this work ⟨58USP2831835⟩.

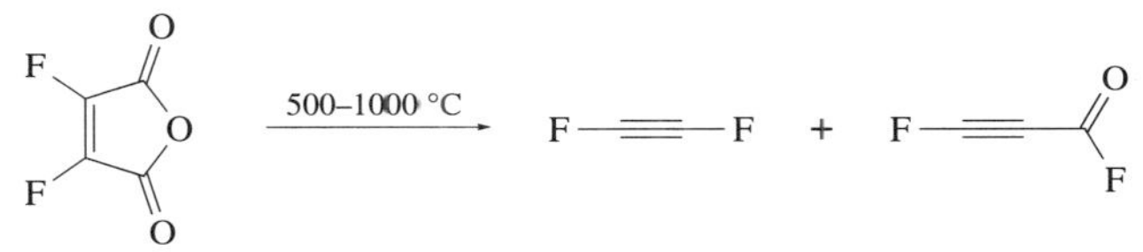

(6) **(7)** **(8)** (6)

Two other approaches to difluoroethyne (**7**) have been reported in the literature, namely the photolysis of tetrafluoroethylene ⟨65JPC2484⟩ and the radiolysis of hexafluoroethane ⟨65JCP(42)2255⟩. In both cases, difluoroethyne (**7**) was obtained in a low yield besides a complex mixture of other products. It was separated by gas chromatography as a very unstable gas, and identified by IR spectroscopy.

Fluorochloroethyne (**10**) can be prepared by dehydrochlorination of dichlorofluoroethene (**9**) in the gas phase with solid potassium hydroxide (Equation (7)) ⟨70CB1198⟩. This extremely explosive compound was purified by distillation in high vacuum and identified by IR and NMR spectroscopy ⟨B-69MI 221-01⟩.

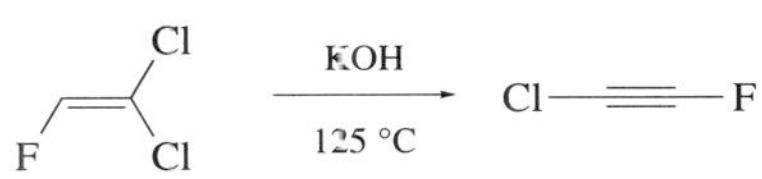

(9) **(10)** (7)

Perfluoropropyne (**11**) was prepared by two different elimination reactions (Equations (8) and (9)) and identified by IR and ^{19}F NMR spectroscopy ⟨68TL3909⟩. It is a colorless gas (boiling point −50 °C) which can be stored at 25 °C under atmospheric pressure for several days; however, it slowly polymerizes to a waxy solid when kept at higher pressures.

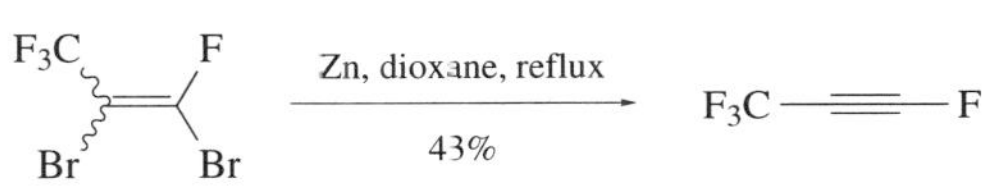

cis:*trans* = 3:2 **(11)** (8)

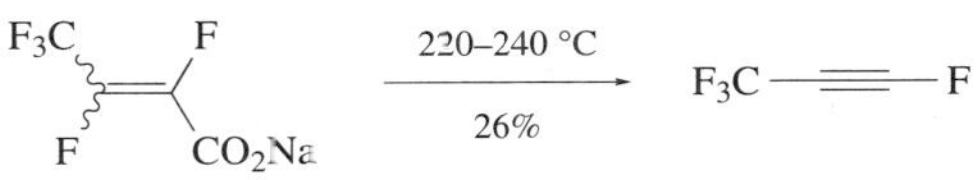

cis:*trans* = 1:1 **(11)** (9)

The only known example of an alkyl-substituted fluoroalkyne is *t*-butylfluoroethyne (**13**) ⟨64AG(E)747, 64AG(E)755, 65AG(E)746⟩. This compound can be prepared by the dehydrohalogenation of the fluorobromoalkene (**12**) with potassium hydroxide at about 120 °C under reduced pressure (Equation (10)) ⟨64AG(E)747⟩.

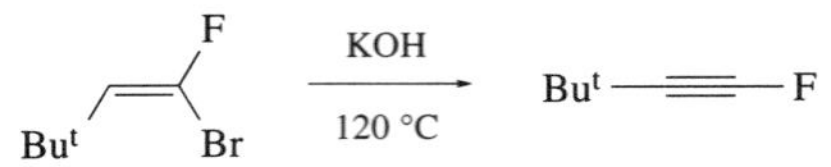

(12) **(13)** (10)

t-Butylfluoroethyne (**13**) is a highly unstable compound. As a liquid it polymerizes spontaneously and exothermally even below 0 °C with the formation of various oligomers, mainly trimers ⟨64AG(E)755, 65AG(E)746⟩. However, it can be handled unchanged for a short time at room temperature as a gas or a dilute solution in ether.

Alkynyl fluorides attracted significant attention due to their high reactivity toward nucleophiles and as partners in cycloaddition reactions ⟨B-69MI 221-01⟩. However, the value of these compounds as reagents in organic synthesis is sharply reduced because of their thermal instability.

2.21.1.3 Chlorides

A large variety of alkynyl chlorides are known ⟨B-69MI 221-01⟩. The first preparation of the parent representative of the class, monochloroethyne (**15**), was reported over 100 years ago in 1880 by the pyrolysis of dichloroacrylic acid ⟨1880LA(203)83⟩. A number of more convenient synthetic procedures based on the dehydrochlorination of polychlorinated ethenes with different bases have been developed more recently ⟨38JCS1358, 52JA629, 43CB80, 08CB314, 09CB4232, 38JCS1218, 59CB1270, 59CB1950⟩. Reactions of 1,2-dichloroethene (preferably the *cis* isomer (**14**) with sodium methoxide or hydroxide lead to clean dehydrochlorination with the formation of monochloroethyne (**15**) (Equation (11)) ⟨38JCS1358, 52JA629, 43CB80, 08CB314, 09CB4232, 38JCS1218⟩.

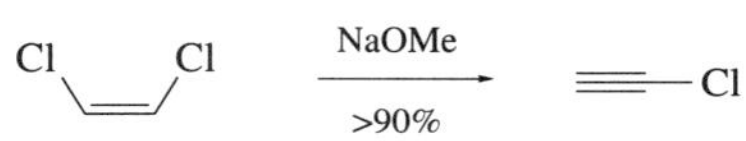

(14) **(15)** (11)

Stronger bases, such as sodium amide and alkyllithiums, react with dichloroethenes, forming chloroethynylides (**16**), which can be hydrolyzed to monochloroethyne (**15**) by careful addition of water (Scheme 1) ⟨59CB1270, 59CB1950⟩.

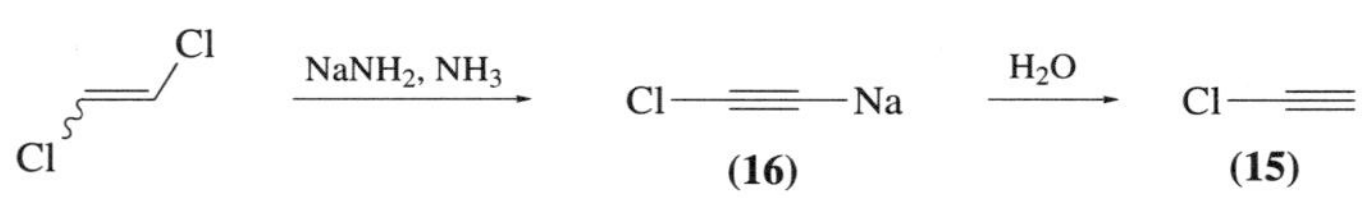

(16) **(15)**

Scheme 1

Neat monochloroethyne (**15**) is a highly unstable compound ⟨B-69MI 221-01⟩. It inflames spontaneously in air and may explode violently ⟨38JCS1358⟩. However, generated *in situ* in the form of sodium or lithium alkynylide it can be conveniently used for further transformations without isolation.

Dichloroethyne (**18**) is best obtained by passing a mixture of trichloroethene (**17**) and ether over heated potassium hydroxide (Equation (12)) ⟨31CB1324, 30CB1941, 60JOC1626, 61LA(640)5⟩. The azeotropic mixture of (**17**) with ether (boiling point 32 °C) is stable, and can be stored indefinitely in the dark under an inert atmosphere. Pure dichloroethyne (**18**) was found to detonate violently and to react explosively with air ⟨30CB1941⟩.

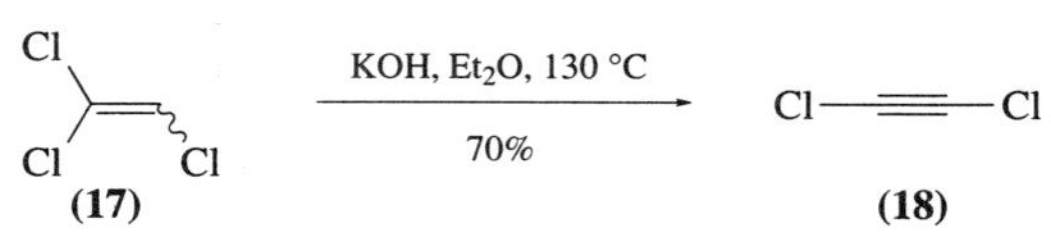

(17) **(18)** (12)

Several other syntheses of dichloroethyne (**18**) have been reported in the literature, such as pyrolysis of carbon tetrachloride or of trichloroacrylic acid salts and the reaction of ethyne with potassium hypochlorite ⟨B-69MI 221-01⟩. However, these methods are hazardous and give only low yields of the product ⟨B-69MI 221-01, B-88MI 221-01⟩.

The dehydrochlorination approach can be also used for the synthesis of some substituted alkynyl chlorides. For example, chloropropyne (**20**) is conveniently prepared in moderate yield by treatment of *cis*-1,2-dichloropropyne (**19**) with potassium hydroxide in boiling butanol (Equation (13)) ⟨54CJC500⟩.

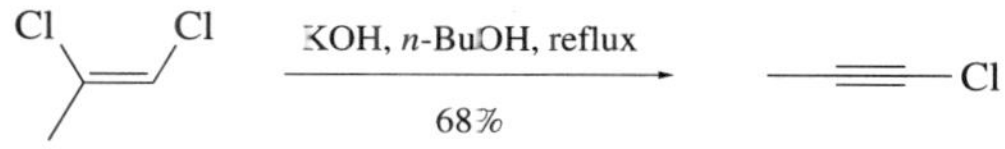

(**19**) (**20**) (13)

A variety of alkynyl chlorides (**22**) have been prepared by similar elimination techniques from 1,1-dichloroalkenes (**21**) and potassium hydroxide (Equation (14)) ⟨58IZV846, 70CB1216, 92TL683⟩.

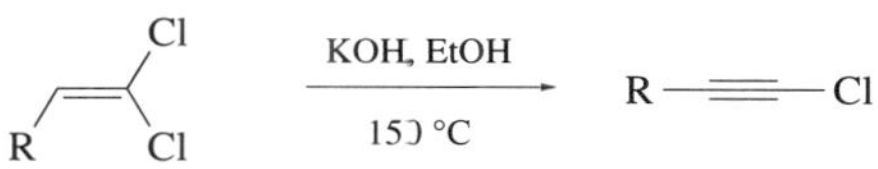

(**21**) R = Bu^t, $R^1CH(OR^2)$– (**22**) (14)

In a modification of this elimination procedure, lithium diethylamide is used as the base instead of potassium hydroxide, and the starting *gem*-dichloroalkenes (**21**) are prepared from aldehydes by a Wittig–Horner reaction ⟨75S458⟩.

The elimination approach has only a limited practical application for the synthesis of alkynyl chlorides due to various side-reactions. More general are the methods which are based on substitution at the acetylenic carbon. Especially useful is the chloroethynylation approach, in which the metal derivatives of chloroethyne (**16**) and (**23**) are condensed with alkyl halides or carbonyl compounds. Metal chloroethynylides can be generated in nearly quantitative yields by a reaction of dichloroethenes with sodium amide in liquid ammonia (Scheme 1) ⟨59CB1270⟩ or butyllithium in ether (Equation (15)) ⟨59CB1950⟩.

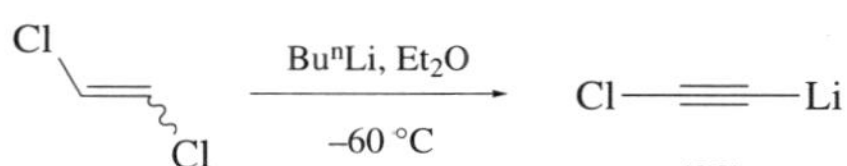

(**23**) (15)

Metal chloroethynylides (**16**) and (**23**) are relatively stable in solution and have been used in many syntheses of alkynyl chlorides. Several examples of chloroethynylation reactions are shown in Equations (16)–(22). Sodium chloroethynylide (**16**) in liquid ammonia reacts smoothly with alkyl halides (Equation (16)), ketones (Equation (17)), and triphenyltin chloride (Equation (18)), forming the corresponding substituted alkynyl chlorides (**24**), (**25**) and (**26**) in high yield ⟨59CB1270, 68BSF193⟩.

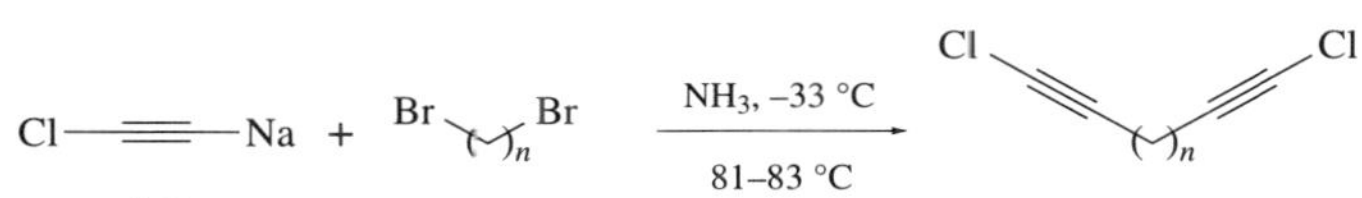

(**16**) (**24**) n = 4, 6 (16)

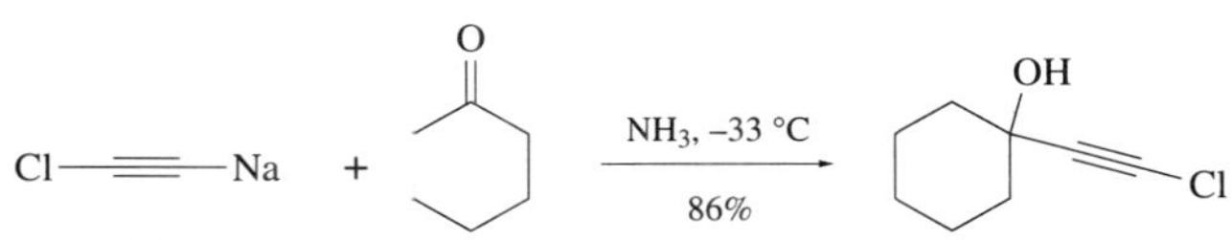

(**16**) (**25**) (17)

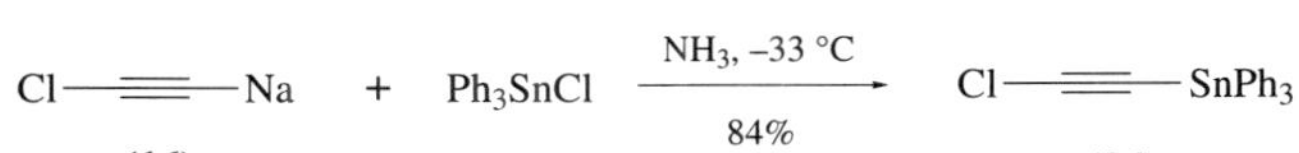

(**16**) (**26**) (18)

Lithium chloroethynylide (**23**) in ether solution can be condensed with more sensitive substrates, which are unstable in liquid ammonia. For example, aldehydes can be chloroethynylated with the formation of the corresponding 3-hydroxyalkynyl chlorides (**27**) in high yields (Equation (19)) ⟨59CB1950⟩.

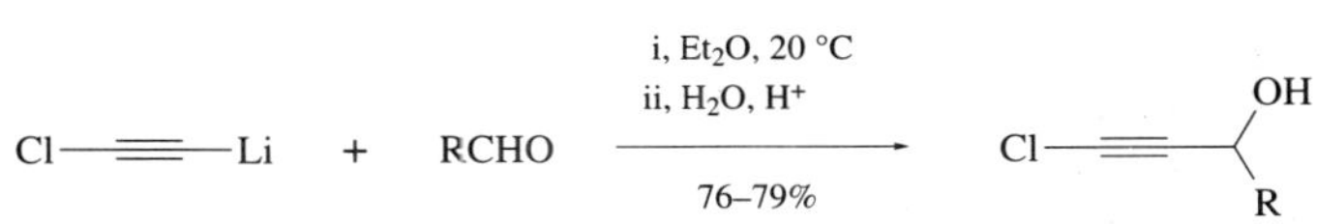

(**23**) (**27**) R = H, Et, C_5H_{11} (19)

An analogous approach was used for the chloroethynylation of some sensitive ketones, such as hexafluoroacetone (Equation (20)) ⟨68JOC819⟩, for the preparation of chloropropiolic acid (**29**) from carbon dioxide (Equation (21)) ⟨59CB1950, 65AG(E)49⟩, and for the preparation of silylated chloroethynes ⟨62CB3064⟩ and their group 14 analogs (**30**) (Equation (22) ⟨66JOM(6)109⟩.

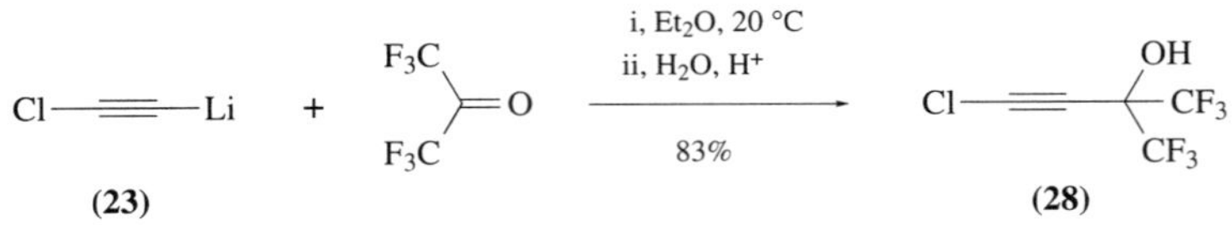

(20)

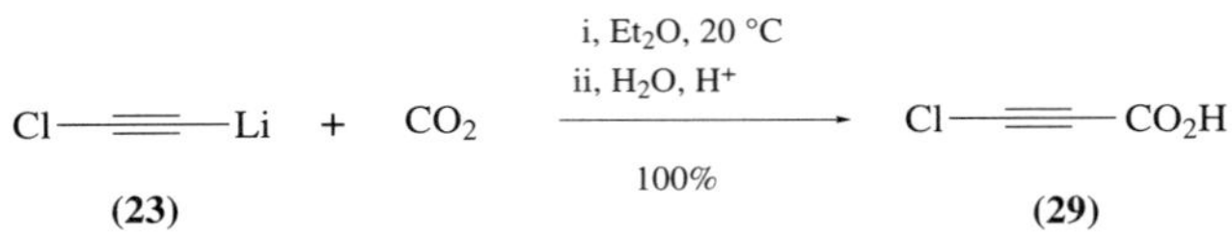

(21)

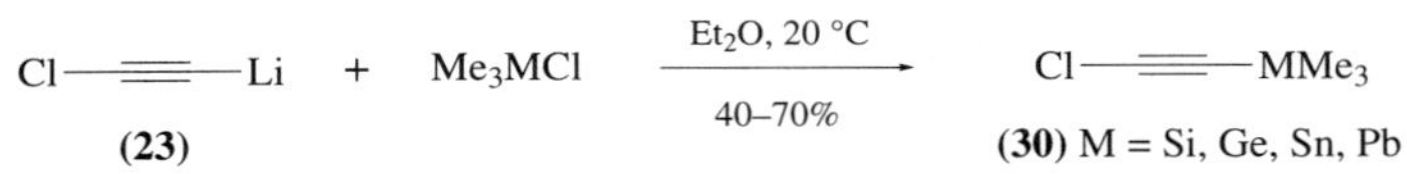

(22)

One important application of the chloroethynylation reaction is the synthesis of various chloroethynyl steroidal derivatives with strong antifertility activity ⟨61JA4663, 62JCS4995, 66T2829, 65BSF71⟩.

The reactions of terminal alkynes or the corresponding alkynylides with chlorinating reagents such as chlorine, hypochlorites, sulfuryl chloride, and arylsulfonyl chlorides represent the most general and widely used approach to alkynyl chlorides. Direct preparation of alkynyl chlorides from terminal alkynes is best achieved by using metal hypochlorites. Both monochloroethyne and dichloroethyne can be prepared by passing ethyne through a sodium or potassium hypochlorite solution ⟨30CB1863, 30CB1886⟩. A similar procedure has been applied to the synthesis of a variety of substituted chloroethynes (Equation (23)) ⟨B-88MI 221-01, 30CB1863, 63JA1648, 56BSF1615, 47BSF841⟩.

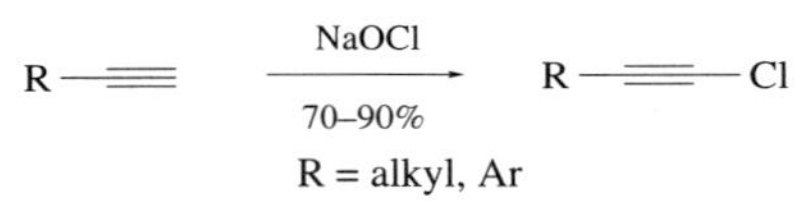

(23)

Equation (23) gives best results for the synthesis of alkynyl chlorides from relatively "acidic" alkynes (such as arylethynes) or from ethynyl carbinols which are soluble in water ⟨B-88MI 221-01⟩.

Employment of metal alkynylides instead of alkynes as starting materials in reactions with chlorinating reagents usually improves the yields and makes the procedure more mild and general. Chlorination of alkynylides with elemental chlorine at low temperature has been used in a number of syntheses of alkynyl chlorides ⟨37JA1307, 33JA4667, 64ACS1629, 66T965⟩. Typical examples of this approach are shown in Equations (24)–(26).

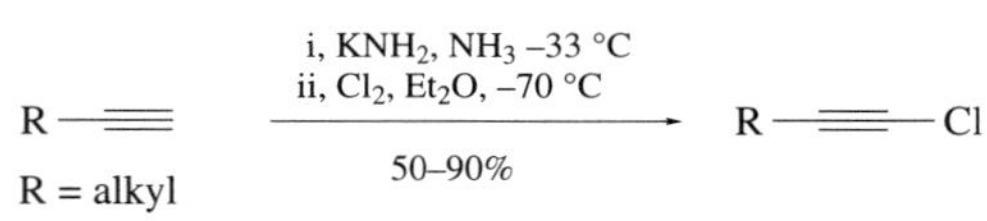

(24)

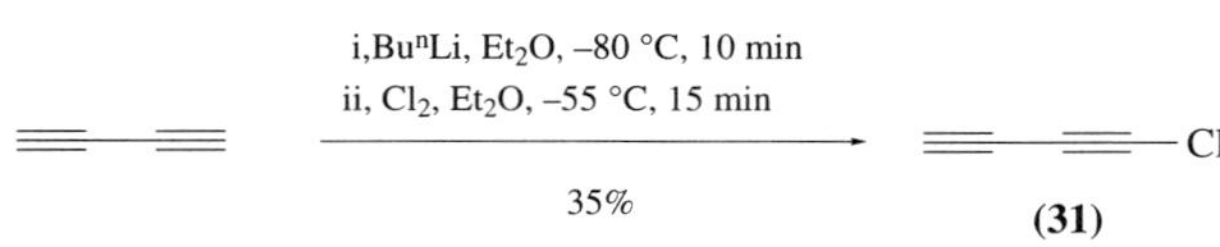

(25)

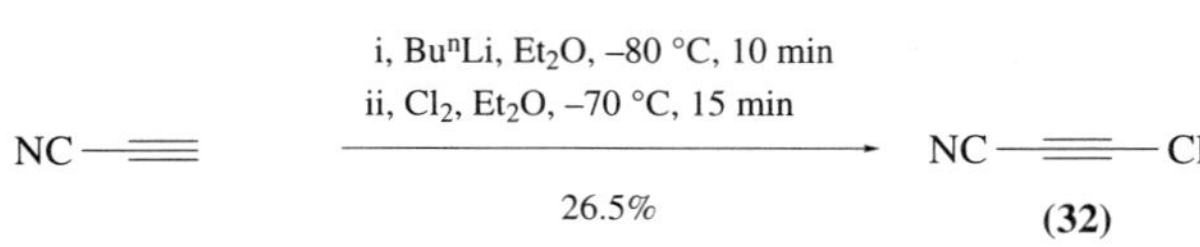

(26)

Reaction conditions for the chlorination are very mild and allow the preparation of alkynyl chlorides (**31**) and (**32**) even from such unstable and sensitive alkynes as butadiyne (Equation (25)) ⟨66T965⟩ and cyanoethyne (Equation (26)) ⟨64ACS1629⟩.

The safest and most convenient laboratory procedure for the synthesis of alkynyl chlorides involves the reaction of alkynylides with arylsulfonyl chlorides ⟨B-88MI 221-01, 31AC(P)309, 58LA(614)37⟩.

Phenylsulfonyl and *p*-tolylsulfonyl chlorides can be used as the reagents in the reaction with alkali metal derivatives of both alkyl- and aryl-substituted alkynes to give the corresponding alkynyl chlorides in good yield (Equation (27)).

$$R{-}{\equiv}{-}M \;+\; ArSO_2Cl \xrightarrow[60\text{–}76\%]{} R{-}{\equiv}{-}Cl \;+\; ArSO_2M \qquad \textbf{(27)}$$

R = alkyl, aryl; M = Li, Na, K; Ar = Ph, 4-MeC_6H_4

In a typical procedure ⟨B-88MI 221-01⟩, phenylsulfonyl chloride is slowly added to a suspension of the metal alkynylide in ether or THF at −20 °C and then the reaction mixture is gradually warmed to room temperature. Aqueous work up and careful distillation give the pure alkynyl chloride ⟨B-88MI 221-01⟩.

2.21.1.4 Bromides

Alkynyl bromides can usually be prepared by the same general approaches as chlorides. The elimination approach has been employed in the synthesis of monobromoethyne (**33**) ⟨38JCS1358, 64IJ121⟩ and dibromoethyne (**34**) ⟨1861LA(119)182⟩. Monobromoethyne was first prepared by dehydrohalogenation of *cis*-dibromoethene with sodium hydroxide in ethanol (Equation (28)) in the form of a relatively stable gas ⟨38JCS1358, 1861LA(119)182⟩. Dehydrobromination of tribromoethene with sodium hydroxide under similar conditions can be used for the preparation of dibromoethyne (Equation (29)).

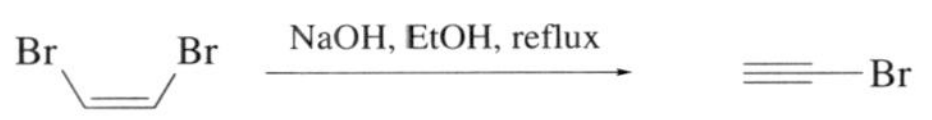

(33) (28)

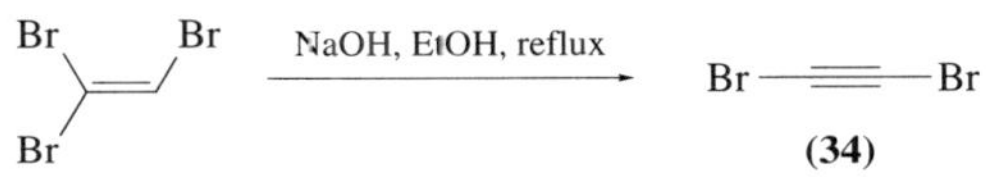

(34) (29)

The elimination method has no practical importance for the preparation of substituted alkynyl bromides, and only a few examples of such reactions have been reported in the literature ⟨B-69MI 221-01⟩.

The haloethynylation approach, which represents one of the most important methods for the preparation of alkynyl chlorides, has only very limited practical application in the synthesis of alkynyl bromides. The principal reagents in this case are sodium or lithium bromoethynylides (**35**) and (**36**), which can be generated *in situ* by dehydrobromination of the corresponding precursors with sodium amide (Equation (30)) ⟨B-69MI 221-01, 66USP3249630⟩ or phenyllithium (Scheme 2) ⟨61CI(L)1619⟩. The consequent reaction of lithium bromoethynylide with triphenylchlorosilane gives the respective silylated bromoethyne (**37**) in moderate yield (Scheme 2).

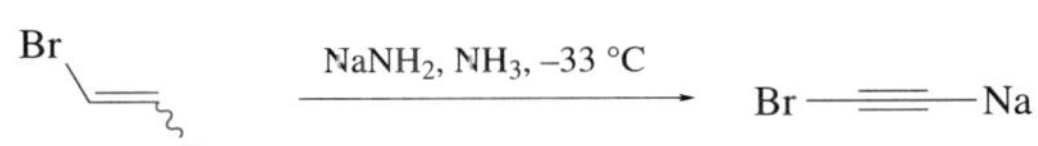

(35) (30)

(36) (37)

Scheme 2

Direct functionalization of terminal alkynes or the corresponding alkynylides with brominating reagents is the most general and widely used approach to synthesize alkynyl bromides. The best laboratory procedure involves the reaction of terminal alkynes with metal hypobromides (Equation (31)) ⟨B-88MI 221-01, B-69MI 221-01⟩. This method was first found by Strauss *et al.* in 1930 ⟨30CB1863⟩ and has been significantly improved in more recent procedures ⟨73OSC(5)921, 90S984⟩. In a typical procedure the alkyne is added to a mixture of bromine, potassium hydroxide, and water at 0 °C, and then the reaction mixture is vigorously agitated for 0.5–6 hours at room temperature. This method gives high yields with arylethynes ⟨B-88MI 221-01, 73OSC921⟩, lower terminal alkynes (propyne–hexyne) ⟨90S984⟩, enynes, and diynes ⟨B-88MI 221-01⟩.

$$R{-}{\equiv} \xrightarrow[80\text{–}99\%]{KOBr,\ H_2O} R{-}{\equiv}{-}Br \qquad (31)$$

R = alkyl, aryl

Terminal alkynes can be directly converted into bromides by several more recent methods ⟨84AG(E)727, 91TL2169, 90TL3141⟩. The reaction of terminal alkynes with bis(trimethylsilyl)peroxide and copper or zinc bromide gives alkyl-, aryl- and trimethylsilyl-substituted bromoalkynes in moderate yields (Equation (32)) ⟨91TL2169⟩.

$$R{-}{\equiv} + \text{TMS-O-O-TMS} + MBr_n \xrightarrow[45\text{–}75\%]{THF,\ -15\ \text{to}\ 25\ ^\circ C} R{-}{\equiv}{-}Br \qquad (32)$$

$R = n\text{-}C_8H_{17}$, Ph, TMS, $(TMS)_2NCH_2$; M = Cu, Zn; n = 1, 2

Bromination of alkynes with *N*-bromosuccinimide and silver nitrate (Equation (33)) is especially convenient for the preparation of biologically active bromoethynyl-substituted steroids ⟨84AG(E)727⟩.

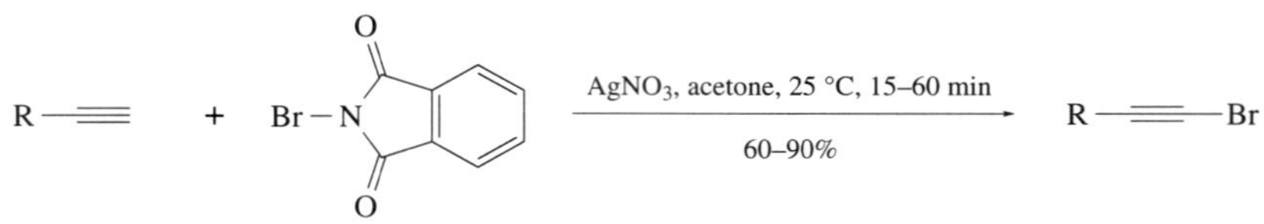

(33)

R = 17α or 17β steroids

A very mild and efficient procedure is the bromination of terminal alkynes with the PPh_3/CBr_4 reagent ⟨90TL3141⟩. This procedure gives an almost quantitative yield of alkynyl bromides and works well for very sensitive alkynes, such as benzylic, propargylic, and allylic ethers and acetals.

The reaction of various brominating reagents with alkali metal alkynylides represents another general entry into alkynyl bromides. The best method to prepare bromoalkynes in which the substituent is a higher alkyl or cycloalkyl group is a low-temperature reaction of bromine with the lithiated alkyne in ether or THF (Equation (34)) ⟨B-88MI 221-01⟩.

$$R{-}{\equiv}{-}Li \xrightarrow[\text{over } 75\%]{Br_2,\ Et_2O,\ \text{hexane},\ -60\ \text{to}\ -40\ ^\circ C} R{-}{\equiv}{-}Br \qquad (34)$$

R = alkyl, cycloalkyl

In this procedure various brominating reagents have been applied instead of bromine, and other metal derivatives can also be used instead of lithium alkynylides ⟨B-69MI 221-01⟩. Bromination of potassium or lithium alkynylides in inert solvents can be achieved with *N*-bromosuccinimide ⟨60LA(638)33, 66T2829⟩, arylsulfonyl bromides ⟨65USP3172912, 52JA398⟩, and some other brominating reagents ⟨B-69MI 221-01⟩. In some cases, bromination of other metal derivatives, such as alkynyl magnesium bromides ⟨05BSF184, 37JA1307⟩, mercury alkynylides ⟨63JCS2295, 26JA469, 51JCS893⟩, and copper alkynylides ⟨65ZOR415, 63IZV2213, 64IZV392⟩, gives better yields of alkynyl bromides and has some advantages over more common procedures.

2.21.1.5 Iodides and Alkynyliodonium Salts

2.21.1.5.1 The synthesis of alkynyl iodides

Alkynyl iodides are best prepared by the direct iodination of terminal alkynes or the corresponding alkynylides. Iodine in liquid ammonia has been found to be very effective for the direct iodination of free alkynes (Equation (35)) ⟨33JA2150, 32JA787, 33JCS741⟩.

$$R{-}{\equiv} \xrightarrow[70\text{–}99\%]{I_2,\ NH_3,\ -33\ ^\circ C} R{-}{\equiv}{-}I \qquad (35)$$

R = aryl, alkyl, alkenyl

In a modification of this procedure, iodination of the terminal alkyne is performed in a DMF solution in the presence of catalytic amounts of copper(I) iodide with sodium carbonate as the base ⟨88CC909⟩.

The direct iodination of ethyne is best achieved by passing it through the solution of sodium hypoiodide in water (Equation (36)) ⟨04CB4415, 1897CB290, 11JA1598⟩. Diiodoethyne (**38**) is the main product of this reaction, and can be separated as a precipitate.

$$H{-}C{\equiv}C{-}H \xrightarrow[80\text{–}90\%]{NaOI,\ H_2O} I{-}C{\equiv}C{-}I \quad \textbf{(38)} \qquad (36)$$

Hypoiodides can also be used for the iodination of other terminal alkynes; however, the yields of alkynyl iodides in this reaction usually are relatively low (27–50%) ⟨B-69MI 221-01⟩.

Iodination of metal alkynylides represents another general entry into alkynyl iodides. Reactions of acetylenic Grignard reagents with iodine was one of the first general methods reported for the preparation of alkynyl iodides (Equation (37)) ⟨26AC(P)3⟩.

$$R{-}C{\equiv}C{-}MgBr \xrightarrow[70\text{–}90\%]{I_2,\ Et_2O,\ 0\ ^\circ C} R{-}C{\equiv}C{-}I \qquad (37)$$

R = aryl, alkyl

Alkynylides of sodium, copper, mercury, and silver can also be iodinated with iodine, and, in particular, iodination of sodium derivatives in refluxing liquid ammonia is reported as a good synthesis of iodoalkynes ⟨34JA1106, 33JA2150⟩. In a more recent procedure, sodium alkynylides prepared *in situ* from alkynes and MeONa in methanol are selectively iodinated with bis(pyridine)iodine(I) tetrafluoroborate (Equation (38)) ⟨87S661⟩.

$$R{-}C{\equiv}C{-}Na \ + \ I(Py)_2BF_4 \xrightarrow[50\text{–}90\%]{MeOH,\ 25\ ^\circ C,\ 30\ min} R{-}C{\equiv}C{-}I \qquad (38)$$

R = Ph, Bu^n, $HOCH_2$, $CH_2{=}CHCH_2$, $MeOCH_2$, $BrCH_2$, etc.

Another procedure employs the reaction of lithium alkynylides with bis(trimethylsilyl) peroxide and zinc iodide (Equation (39)) ⟨89S461⟩.

$$R{-}C{\equiv}C{-}Li \ + \ TMS\text{-}O\text{-}O\text{-}TMS \ + \ ZnI_2 \xrightarrow[60\text{–}85\%]{Et_2O,\ -78\ to\ 25\ ^\circ C,\ 2h} R{-}C{\equiv}C{-}I \qquad (39)$$

R = n-C_6H_{13}, Bu^t, Ph, TMS, $(TMS)_2NCH_2$, $HOMe_2C$, etc.

Various iodoalkynes have been synthesized by the reaction of an etherial suspension of copper alkynylides with iodine ⟨65ZOR415, 63IZV2213, 64IZV392⟩.

2.21.1.5.2 *The synthesis of alkynyliodonium salts*

Alkynyl(phenyl)iodonium salts are the newest member of the family of alkynyl halides. In recent years these compounds have attracted substantial interest due to their effective alkynylating properties and biological activity ⟨91ACR304⟩. The first preparation of an alkynyliodonium species, namely the chloride (**39**), involved the reaction of lithium phenylethynylide and (dichloro) iodobenzene (Equation (40)) as reported by Beringer and Galton in 1965 ⟨65JOC1930⟩. However, the yield of the compound (**39**) was very low, and it was highly unstable at room temperature.

$$Ph{-}C{\equiv}C{-}Li \ + \ PhICl_2 \xrightarrow[12\text{–}20\%]{Et_2O,\ 0\text{–}5\ ^\circ C} Ph{-}C{\equiv}C{-}I(Cl){-}Ph \quad \textbf{(39)} \qquad (40)$$

The next synthesis of an alkynyliodonium species from the interaction of *p*-diethynylbenzene with (bistrifluoroacetoxy)iodobenzene in chloroform was reported by Merkushev *et al.* (Equation (41)) ⟨79DOK(245)607⟩. The iodonium salt (**40**) was isolated in good yield as a crystalline compound but was characterized only by IR spectroscopy and by identification of its hydrolysis product.

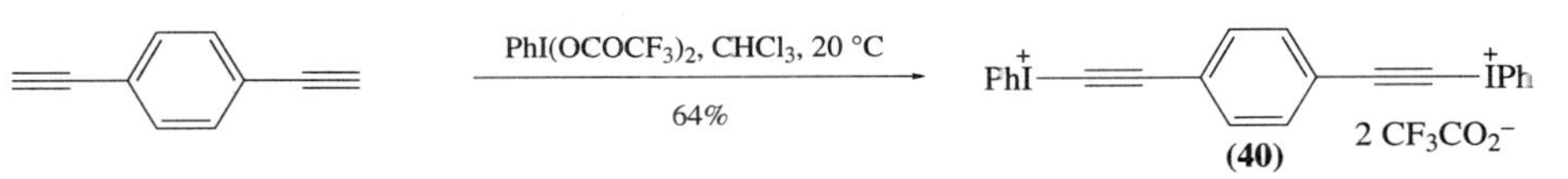

(41)

A more general approach to stabilize alkynyl(phenyl)iodonium compounds as tosylates was developed by Koser *et al.* in the early 1980s ⟨81JOC4324, 84JOC4700⟩. According to this method terminal alkynes are allowed to react with [hydroxy(tosyloxy)iodo]benzene (**41**) ⟨70JOU889, 90SL365⟩ in refluxing chloroform (Equation (42)). Unfortunately, this reaction is not selective, which causes purification problems and greatly diminishes the yields of the desired alkynyliodonium tosylates (**42**).

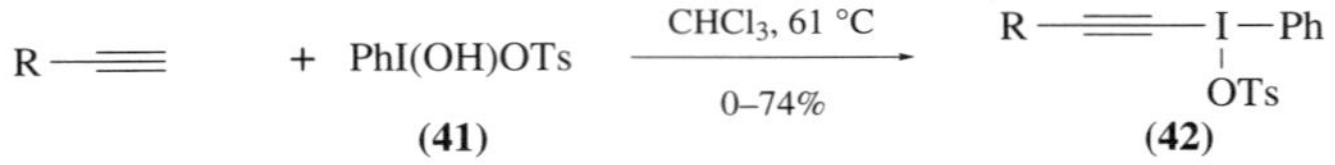
(42)

R = Bu^t, Bu^s, Ph, cyclo-C_6H_{11}, etc.

A number of modifications and improvements for the ready preparation of alkynyl(phenyl) iodonium tosylates (**42**) were developed ⟨85JA1452, 87JA228, 88JOC4105⟩. At the same time, the preparation of some other alkynyliodonium salts were reported in the literature. Fujita and co-workers have synthesized a variety of alkynyl(phenyl)iodonium tetrafluoroborates (**43**) in good yield by the reaction of trimethylsilyl-substituted alkynes with iodosobenzene and $BF_3 \cdot Et_2O$ in dichloromethane, followed by treatment with aqueous $NaBF_4$ (Equation (43)) ⟨85TL4501⟩.

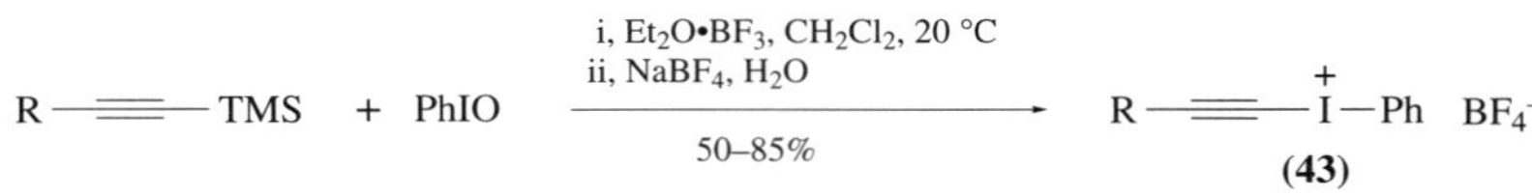
(43)

R = alkyl, aryl

Treatment of bis(trimethylsilyl)ethyne (**44**) with iodosobenzene and $BF_3 \cdot Et_2O$ under similar conditions and subsequent desilylation with HF/H_2O gives the parent ethynyl(phenyl)iodonium tetrafluoroborate (**45**) (Equation (44)) ⟨90CC118⟩.

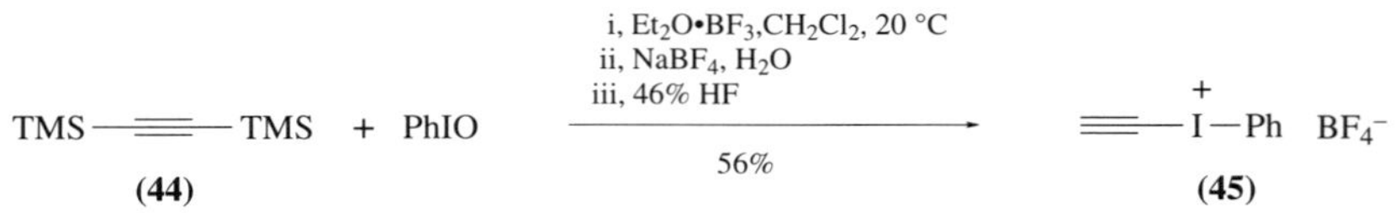
(44)

The most stable and well-investigated alkynyliodonium species are represented by the trifluoromethanesulfonate (triflate) salts (**47**). These compounds can be prepared by two general methods. In the first procedure (Equation (45)), a trimethylsilylalkyne is treated with Zefirov's reagent (**46**) generated *in situ* from iodosobenzene and trifluoromethanesulfonic anhydride ⟨91JOC3912⟩. A variety of alkyl- and aryl-substituted ethynyl(phenyl)iodonium triflates (**47**) can be obtained by this method as stable, microcrystalline solids in good yields.

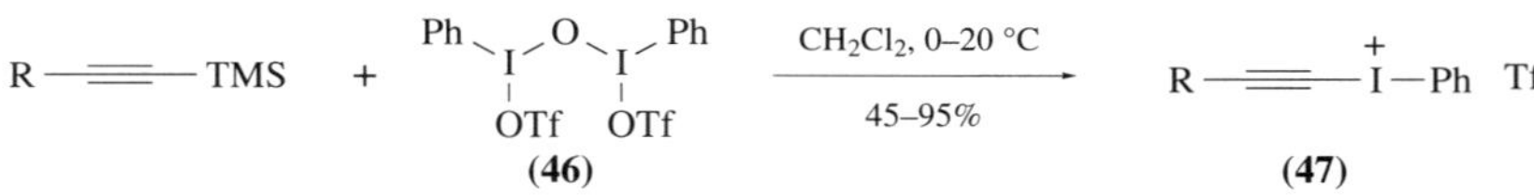
(45)

R = Me Bu^t, Bu^n, TMS, Ph, etc.; Tf = CF_3SO_2

The parent ethynyl(phenyl)iodonium triflate, HC≡CIPh·OTf, can be prepared from tributyltinethyne and reagent (**46**) by a similar procedure ⟨90AG(E)287⟩.

The second general procedure involves the use of tin-substituted alkynes and the cyanoiodonium triflate (**48**) ⟨90TL4828⟩ for the preparation of a wide variety of β-alkyl- and β-arylethynyl (phenyl)iodonium triflates. The previously inaccessible β-functionalized ethynyl(phenyl)iodonium triflates (**49**) may also be synthesized in this way (Equation (46)) ⟨91JA5870⟩.

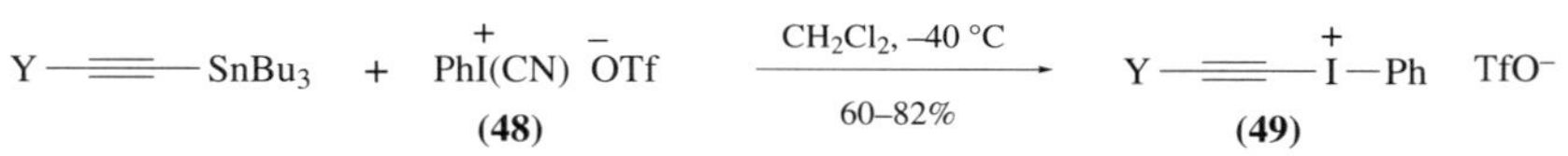
(46)

Y = PhC(O), CN, Cl, MeOC(O), $MeOCH_2$, etc.

Likewise, this procedure allows formation of the novel bis(iodonium)ethyne (**50**) (Equation (47)), which can serve as an effective C_2 transfer reagent in reactions with various nucleophiles and is also a highly reactive dienophile in the [4 + 2] cycloaddition with dienes ⟨90JA6437, 91JA4571⟩.

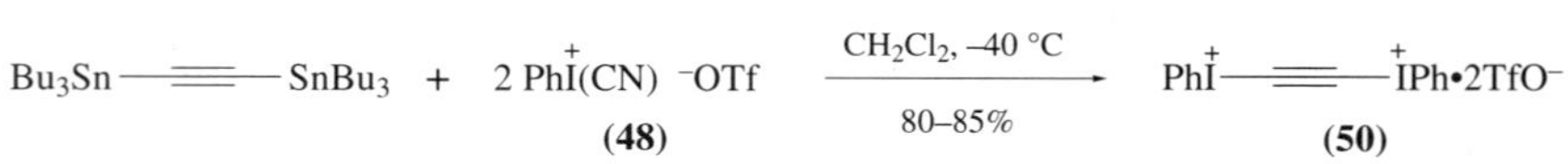
(47)

Until the early 1990s, the iodonium transfer reaction between phenyl(cyano)iodonium triflate (**48**) and tinalkynes was the most general and simple method for the preparation of alkynyliodonium species. Due to the very mild reaction conditions it can even be applied to such sensitive substrates as diynes (Equations (48)–(50)) ⟨90JA6437, 92JOC1861, 91S1073⟩. The products of these reactions (Equations (48)–(50)), bis(iodonium)diynes (**51**) ⟨90JA6437⟩, (**52**) ⟨92JOC1861⟩, and mono(iodonium) butadiynes (**53**) ⟨91S1073⟩, are isolated as microcrystalline solids at low temperatures; however, they decompose at room temperature in a few hours.

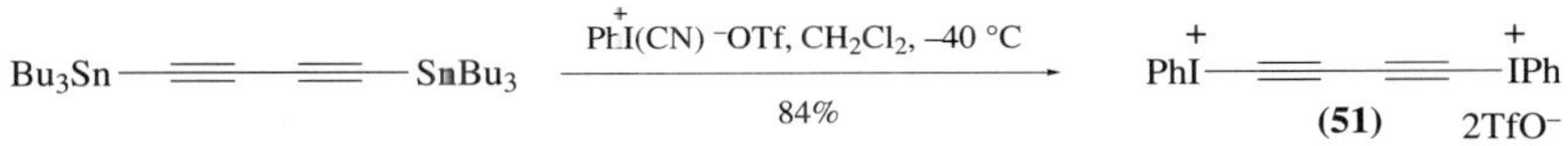
(48)

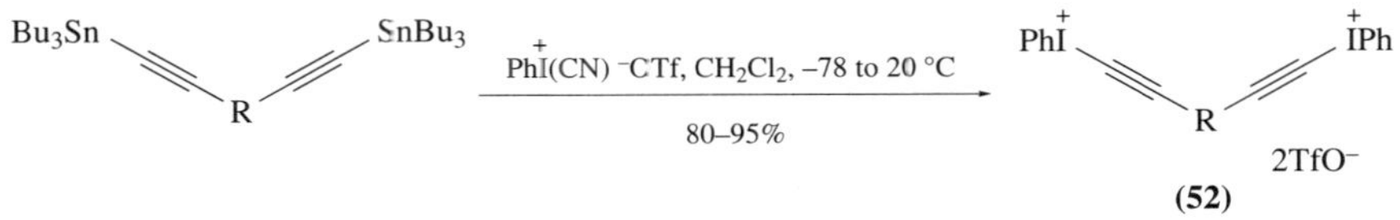
(49)

$R = (CH_2)_n$, n = 1–5, or $(C_6H_4)_m$, m = 1, 2

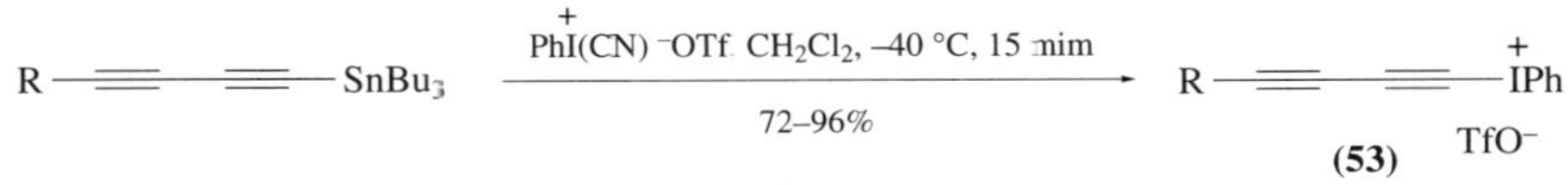
(50)

R = TMS, Bu^t, Bu^n, Me, Ph

Several other structural types of alkynyliodonium salts have been synthesized. The reaction of bis(trimethylsilyl)ethyne with *p*-bis(iodosyl)benzene (**54**) in the presence of trimethylsilyl triflate affords a diiodonium salt (**55**) (Equation (51)) which has pronounced biological activity due to its highly effective methoxatin-inhibiting property ⟨92MI 221-01, 93JA11702⟩.

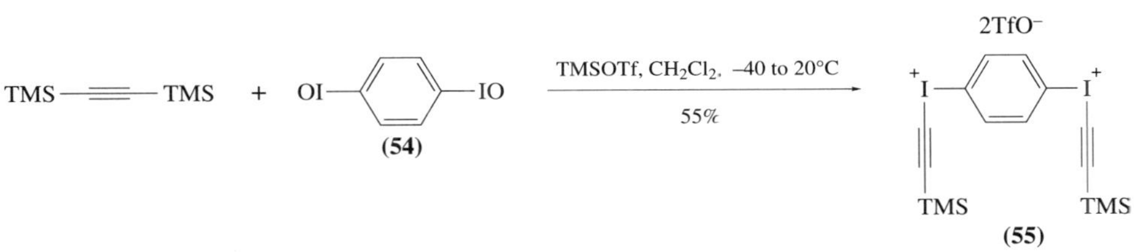
(51)

Novel dialkynyliodonium triflates (**57**) were first prepared by the reaction of silylated alkynes with iodosyl triflate (**56**) (Equation (52)) ⟨91JA8997⟩. A more mild and efficient method for the synthesis of compounds (**57**) involves the iodonium transfer between (dicyano)iodonium triflate and tin-substituted alkynes ⟨92TL1419⟩.

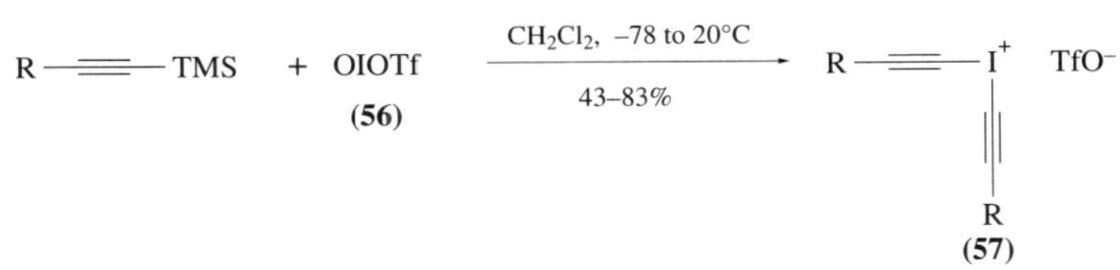
(52)

R = Bu^t, TMS, Pr$^i{}_3$Si

2.21.2 ALKYNYL CHALCOGENIDES

2.21.2.1 The Synthesis of Oxygen-based Functional Groups Attached to an *sp*-Carbon

Alkynyl compounds with the O—C*sp* bonding are represented by ynol ethers and esters which are derivatives of the hypothetical alkynols RC≡COH. The chemistry of ynol ethers has been well developed since their first preparation at the end of the nineteenth century ⟨B-69MI 221-02, B-60MI 221-01⟩. Several exhaustive reviews ⟨B-69MI 221-02, B-60MI 221-01, 65HOU(6/3)116, B-78MI 221-01, 89RCR948⟩ covering the literature on ynol ethers and their analogs up to 1985 have been published since the 1960s. In contrast to ynol ethers, esters of alkynols were unknown until the mid-1980s, when the first preparation of alkynyl tosylates was reported ⟨85JA1452⟩. In the following years a wide variety of alkynyl carboxylate, phosphate, and sulfonate esters have been prepared from alkynyliodonium salts. The chemistry of these novel derivatives of alkynols has been summarized in a review ⟨91ACR304⟩. Since the mid-1980s, considerable interest and research activity has arisen toward alkynols themselves and such derivatives as alkynolate salts and silyl ynol ethers.

2.21.2.1.1 *Ynols and ynolates*

Ynols or hydroxyalkynes (**58**) are the triple bond analogs of enols. Like enols they are tautomers of the corresponding carbonyl species—ketenes (**59**) (Equation (53)) ⟨90ACR43⟩.

(58) **(59)** (53)

This tautomeric equilibrium is almost completely on the side of ketenes (**59**), which according to high-level *ab initio* calculations are some 37 kcal mol^{-1} (155 kJ mol^{-1}) more stable than the corresponding ynols ⟨82JOC1869, 89JA8297⟩. However, in the gas phase, alkynols may exist as relatively stable molecules due to a very high energy barrier for the isomerization of (**58**) to (**59**). For example, the activation energy for the isomerization of hydroxyethyne to ketene is calculated to be 73 kcal mol^{-1} (306 kJ mol^{-1}) ⟨86AG(E)282⟩. Due to its stability in the form of isolated molecules, hydroxyethyne has been considered as a possible constituent of interstellar clouds, planetary atmospheres, and flames ⟨82JPC2835⟩. Moreover, two direct observations of hydroxyethyne have been reported ⟨86AG(E)282, 89AG(E)181⟩. In 1986 Schwarz and co-workers reported the generation and identification of hydroxyethyne in a tandem mass spectrometer ⟨86AG(E)282⟩. More recently, hydroxyethyne (**61**) has been generated photochemically from semisquaric acid (**60**) in an argon matrix at a temperature of 12 K (Equation (54)), and characterized by IR spectroscopy ⟨89AG(E)181⟩.

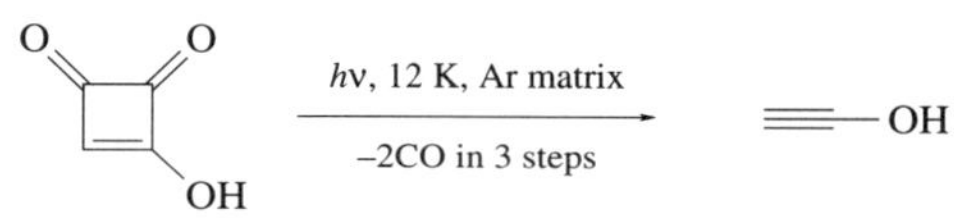

(60) **(61)** (54)

Interestingly, the IR spectrum of hydroxyethyne (**61**) is typical of any unsymmetrically substituted alkyne, with the stretching band of the triple bond at 2198 cm^{-1}, the terminal C—H at 3340 cm^{-1}, and the O—H at 3501 cm^{-1}.

In a series of studies by Kresge and co-workers the chemistry of ynols was examined in aqueous solution ⟨90ACR43, 89JA2355, 87CJC1719, 87CJC441⟩. In this research, phenylhydroxyethyne (**63**) was generated as a transient intermediate by flash photolysis of phenylhydroxycyclopropenone (**62**) (Scheme 3) and its subsequent transformations were investigated kinetically ⟨89JA2355, 87CJC1719⟩.

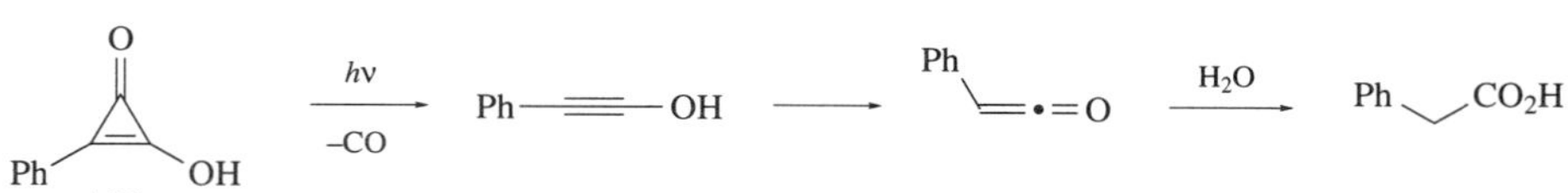

(62) **(63)**

Scheme 3

One of the most remarkable results of this research ⟨90ACR43⟩ was the determination of the unusually high acidity of ynol (**63**) with an estimated pK_a of less than 2.8. This indicates that ynols are at least seven orders of magnitude more acidic than their double bond analogs (enols) and more acidic than even carboxylic acids. The results of *ab initio* calculations ⟨89JA8297⟩ confirm and provide insight into the high acidity of hydroxyalkynes.

Ynolate anions are substantially more stable than ynols. Moreover, in 1994, several convenient procedures for the preparation of ynolate salts were known, and ynolates have been found to have important synthetic applications.

Ethynediolate salts, MOC≡COM (M = alkali or alkali earth metal), are well-known products of oxo-carbon chemistry and can be prepared by the interaction of the corresponding metal with carbon monoxide ⟨63HCA1121, 66HCA907, 65HCA1229, 88JPC37⟩. For example, the colorless potassium ethynediolate salt (**64**) can be obtained as a principal product by passing carbon monoxide into a solution of potassium in liquid ammonia (Equation (55)) ⟨63HCA1121⟩.

$$2\,K + 2\,CO \xrightarrow{\text{liquid NH}_3} KO{-}{\equiv}{-}OK \quad \textbf{(64)} \qquad (55)$$

The structure of ethynediolate salts was investigated by x-ray diffraction ⟨63HCA1121⟩ and matrix IR spectroscopy ⟨88JPC37⟩. The C—C bond length in compound (**64**) is 0.121 nm, which is typical of an acetylenic triple bond. The C—O distance of 0.128 nm has an intermediate value between a single and a double bond. However, the IR spectroscopy indicates single bond character for the C—O bond in ethynediolate anion ⟨88JPC37⟩. Further support of structure (**64**) as the ynolate salt was obtained from its chemistry: one of the products of the protonolysis of this salt with anhydrous hydrogen chloride is hexahydroxybenzene, which may result from the trimerization of the initially formed dihydroxyethyne, HOC≡COH ⟨65HCA1229⟩.

There has been substantial recent interest in the preparation and chemistry of alkynolate anions. In 1982, Kowalski *et al.* found a general approach to alkynolate anions (**68**) starting from α-haloketones (**65**) by a sequence of reactions represented in Scheme 4 ⟨82JA321⟩.

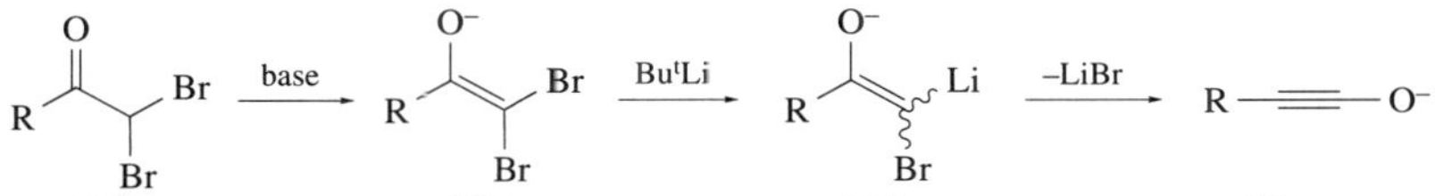

(**65**) (**66**) (**67**) (**68**)

Scheme 4

The key step in this scheme is the rearrangement of the carbene generated from the intermediate (**67**), obtained by metal–halogen exchange in the haloenolate anion (**66**). The reaction is usually performed in a THF solution at low temperature; however, the resulting solutions of lithium alkynolate salts are relatively stable even at room temperature. Instead of α-dibromoketones (**65**), monobromo- or monochloroketones can be used as the starting material; such bases as lithium hexamethyldisilazide or methyllithium are applied in the enolization step (**65**)→(**66**) ⟨82JA321⟩. Kowalski *et al.* further modified this preparation of alkynolates (Scheme 4) using ethyl esters of carboxylic acids and dibromomethyllithium as the starting material for the generation of enolates (**66**) ⟨92JOC7194, 85JA1429, 87TL2463, 86JA7127⟩.

Several less general approaches to alkynolate anions have also been reported in the literature ⟨78JOC376, 79LA219, 87JA7568, 86JA7125⟩. Trimethylsilylethynolate (**70**) can be generated by deprotonation of the ketene (**69**) in THF solution (Equation (56)) ⟨78JOC376⟩. Phenylethynolate (**72**) has been prepared via isoxazole (**71**) (Scheme 5) ⟨79LA219⟩, and lithiumethynolate (**74**) has been reported as an intermediate product in the reaction of dihydrofuran (**73**) with butyllithium (Scheme 6) ⟨87JA7568⟩.

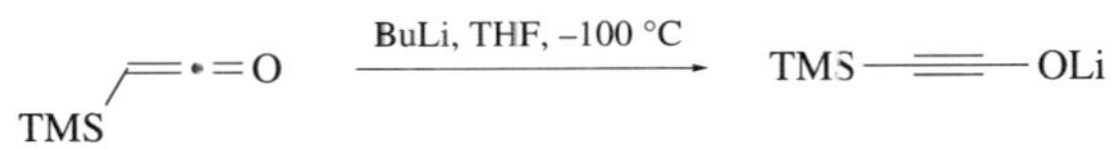

(**69**) (**70**) (56)

(**71**) (**72**)

Scheme 5

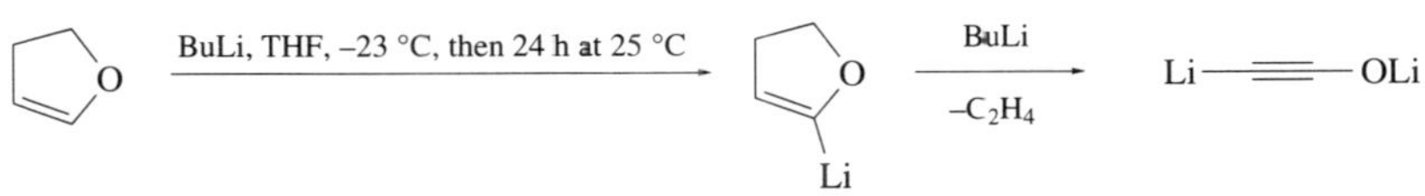

(**73**) (**74**)

Scheme 6

Recent approaches to lithium alkynolates involve cleavage of some ynol ethers and esters with methyllithium ⟨86JA7127, 86JA7125⟩. Desilylation of silyl ynol ethers with methyllithium at room temperature in THF solution leads to the selective formation of the corresponding alkynolates

(Equation (57)) ⟨86JA7127⟩. Similarly, alkynyl tosylates can be selectively cleaved with methyllithium at low temperatures (Equation (58)) ⟨86JA7125⟩.

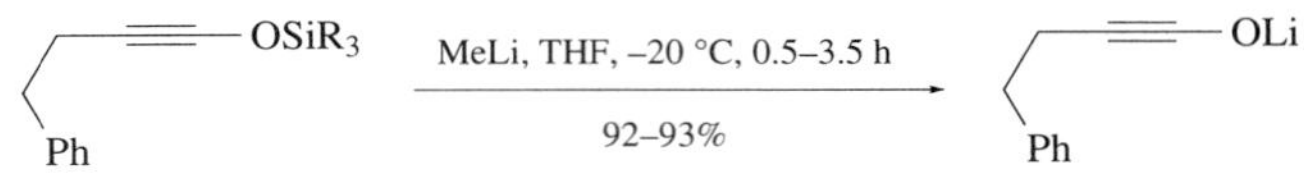

(57)

$SiR_3 = Si(Pr^i)_3$ or $Si(Bu^t)Me_2$

$$R{-}{\equiv}{-}OTs \xrightarrow[R = Bu^t \text{ or } Bu^s]{\text{MeLi, THF or DME, }-20\,^\circ\text{C}} R{-}{\equiv}{-}OLi \qquad (58)$$

2.21.2.1.2 *Silyl ynol ethers*

The first reliable synthesis of silyl ynol ethers (**77**) was reported in 1985 by Maas and Brückmann ⟨85JOC2801⟩. This method employs α-diazocarbonyl compounds (**75**) and their silylated derivatives (**76**) as the starting material (Scheme 7), and is based on a carbene rearrangement ⟨85JOC2801⟩.

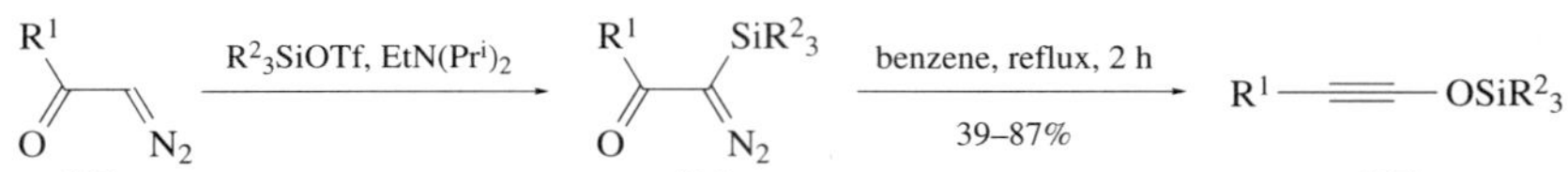

(**75**) (**76**) (**77**)

R = aryl, heteroaryl

$SiR^2_3 = SiEt_3, Si(Pr^i)_3, SiMe_2(Bu^t)$

Scheme 7

In the early 1990s, the simplest and most general procedure for the preparation of silyl ynol ethers is the reaction of lithium alkynolates, generated *in situ* from ethyl esters, with silyl chlorides, according to Equation (59) ⟨86JA7127⟩.

$$RCO_2Et \xrightarrow[50\text{–}75\%]{\text{i, 2.2 equiv. LiTMP, 2.2 equiv. }CH_2Br_2;\ \text{ii, 5 equiv. BuLi, THF, }-78\text{ to }30\,^\circ\text{C};\ \text{iii, 5 equiv. }Pr^i_3SiCl\text{, THF, }-78\text{ to }0\,^\circ\text{C}} R{-}{\equiv}{-}OSiPr^i_3 \qquad (59)$$

R = cyclohexyl, $PhCH_2$, Ph, etc.

TMP = 2,2,6,6-tetramethylpiperidine

However, this method (Equation (59)) does not work for the preparation of siloxyalkynes in which R^1 is a lower alkyl group or hydrogen ⟨88TL4917⟩. Danheiser *et al.* ⟨88TL4917⟩ have found a more conventional approach to silyl ynol ethers by the base-promoted dehydrohalogenation of (*Z*)-2-bromovinyl ethers (**78**) (Equations (60) and (61)). This method can be applied even for the synthesis of the parent siloxyethyne (**79**), as well as other silyl ynol ethers.

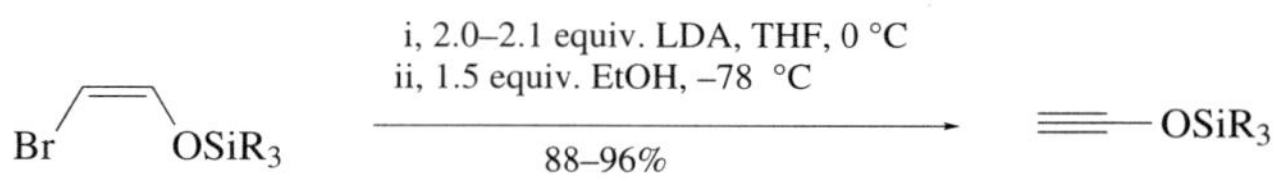

(**78**) (**79**) (60)

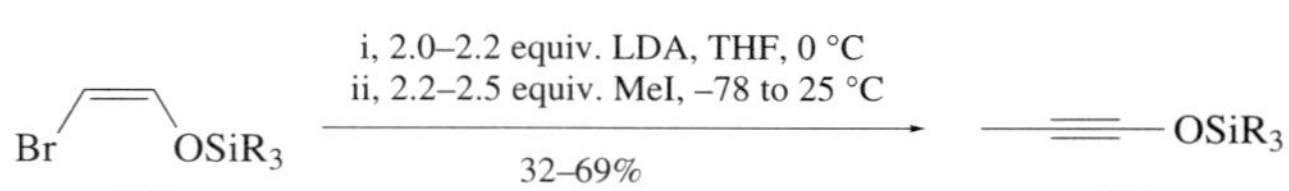

(**78**) (**80**) (61)

$SiR_3 = SiPr^i_3, SiMe_2Bu^t, SiPh_2Bu^t$

Silyl ynol ethers bearing bulky silyl groups (Pr^i_3Si, Bu^tMe_2Si) are thermally stable oils which can be distilled without decomposition or purified by chromatography ⟨86JA7127, 86JA7125, 85JOC2801, 88TL4917⟩. Products containing small groups are less stable: triethylsiloxyethynes partially decompose on column chromatography or on distillation ⟨85JOC2801⟩, and trimethylsilyl derivatives are unstable even at room temperature ⟨85JA7127⟩. All siloxyalkynes are moisture-sensitive; their hydrolysis affords the corresponding carboxylic acids ⟨85JOC2801⟩.

Silyl ynol ethers have highly characteristic spectral properties. The triple bond stretching in the IR spectra of siloxyalkynes is at 2260–2270 cm^{-1}, which compares with that of ynol esters and alkoxyalkynes ⟨86JA7125, 85JOC2801⟩. Particularly noteworthy and characteristic are the alkyne carbon signals in the ^{13}C NMR spectra. The signals of the α carbons are in the normal alkyne region of 85–89 ppm whereas the β carbons are at 27–40 ppm ⟨85JOC2801, 88TL4917⟩. This considerable upfield shift of the β carbons can be explained by the electron-rich nature of the siloxyalkynes due to resonance donation according to Equation (62).

(62)

2.21.2.1.3 *The synthesis of alkynyl ethers*

(i) By dehydrohalogenation of haloacetals and haloalkenyl ethers

The most general and well-developed procedures for the preparation of alkynyl ethers are based on dehydrohalogenation or dehalogenation of various haloacetals (e.g., acetals of chloroacetaldehyde (**81**)) or haloalkenyl ethers (**82**) (Scheme 8) ⟨B-69MI 221-02, B-60MI 221-01, 65HOU(6/3)116, B-78MI 221-01, 89RCR948⟩. The first product in these eliminations is usually the alkoxyethynylide (**83**), which can be subsequently quenched with a variety of electrophiles such as water, alkylhalides, ketones, etc., to give the final product (**84**).

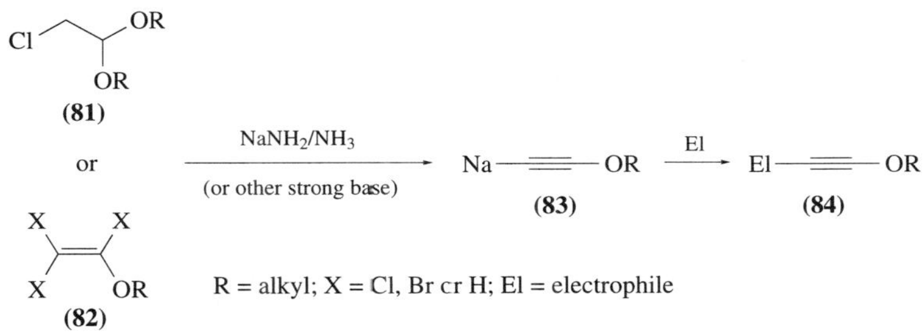

Scheme 8

Numerous examples of such reactions (Scheme 8) have been listed in earlier reviews ⟨B-69MI 221-02, B-60MI 221-01, 65HOU(6/3)116, B-78MI 221-01, 89RCR948⟩. In the well-known standard synthesis of alkoxyalkynes ⟨72OPP89, 77OS(57)65, 88S287, B-88MI 221-01⟩ the reaction of acetals of chloroacetaldehyde (**81**) with sodium amide in liquid ammonia has been used to generate sodium alkoxyethynylides (**83**). In the early 1990s, this elimination procedure has been further improved by using lithium diethylamide in THF solution (Scheme 9) ⟨87JOC2332⟩. In this work ⟨87JOC2332⟩, *in situ* generated lithium alkoxyethynylides (**85**) were subsequently converted to either ethoxyethyne (**86**) by quenching with saturated aqueous sodium chloride, or to ethoxyethynyl carbinols (**87**) by subsequent reaction with ketones or aldehydes.

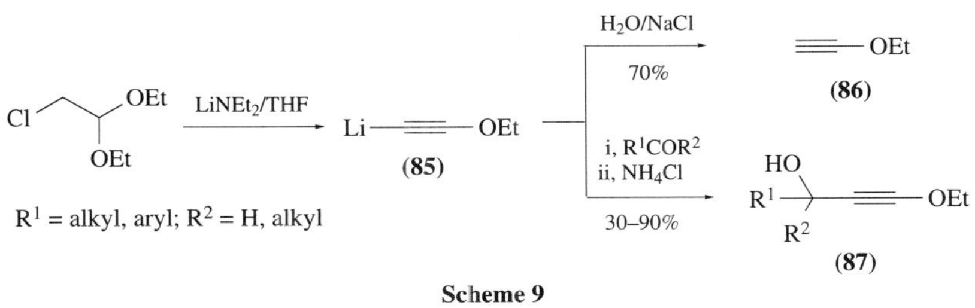

Scheme 9

Dehydrohalogenation or dehalogenation of haloalkenyl ethers probably is the most universal approach to alkoxyalkynes ⟨B-69MI 221-02, B-60MI 221-01, 65HOU(6/3)116, B-78MI 221-01, 89RCR948⟩. This approach has been significantly improved and applied to the preparation of various alkynyl ethers since the mid-1980s. In contrast to older procedures, in which sodium hydroxide or sodium amide is used as a base, these new methods usually employ alkyllithium or potassium hydride. In 1985, Smithers ⟨85S556⟩ developed a procedure based on the reaction of α-chloro-β,β-dibromovinyl ethers (**88**) with butyllithium (Scheme 10). In contrast to previous methods the first step of this reaction is metallation of the vinylic bromine bond and subsequent elimination of LiBr, resulting in bromoalkynes (**89**). The consequent halogen–metal exchange leads to lithium alkoxyethynylides (**90**), which can be further functionalized *in situ* by reaction with such electrophiles as water, ketones, and alkyl bromides ⟨85S556⟩.

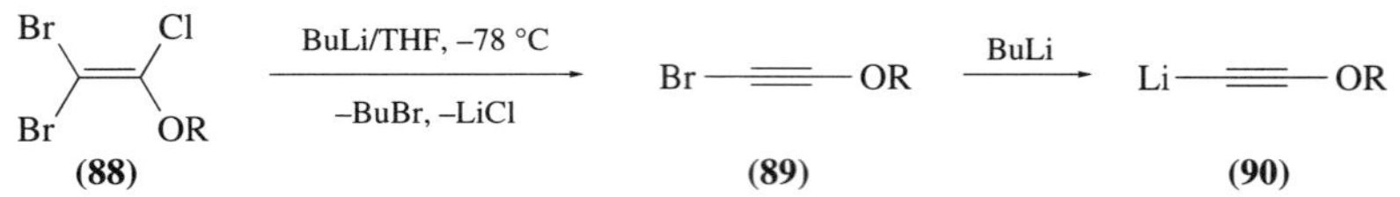

R = Me, Et, *n*-C_8H_{17}, etc.

Scheme 10

In the early 1990s, the dehydrohalogenation method was further modified by using 1,2-dichlorovinyl ethers (**91**) as starting compounds ⟨92S495, 87JOC2919⟩. Löffler and Himbert applied the reaction of 1,2-dichlorovinyl ethers (**91**) with butyllithium to generate lithium alkoxyethynylides (**90**), which were subsequently quenched with ketones or arylated by the palladium-catalyzed cross-coupling with aryl halides to yield alkoxyarylethynes (**92**) (Scheme 11) ⟨92S495⟩.

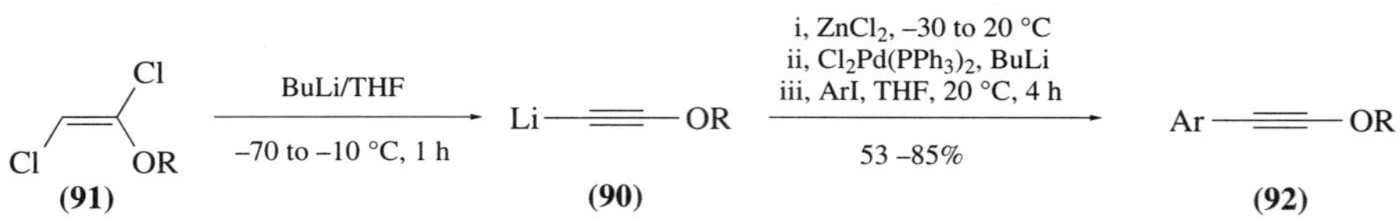

R = Et, Pr^i, Bu^n

Ar = Ph, 4-ClC_6H_4, 4-MeC_6H_4, 3-ClC_6H_4.

Scheme 11

This procedure (Scheme 11) is especially convenient since 1,2-dichlorovinyl ethers (**91**) can be easily prepared from readily available and inexpensive trichloroethylene and sodium alkoxides ⟨92S495⟩. A similar method has been used by Greene and co-workers ⟨87JOC2919⟩ for the preparation of chiral alkynyl ethers (**94**) and (**95**) from the corresponding alcohols (**93**) and trichloroethylene (Scheme 12).

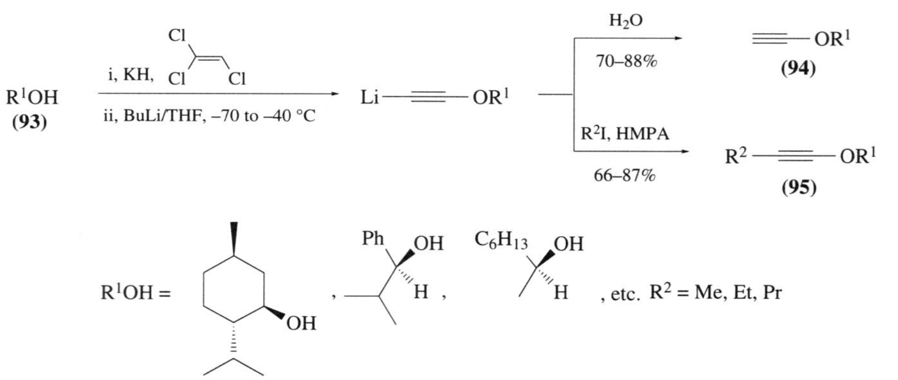

Scheme 12

Based on the elimination reaction, Pericàs *et al.* ⟨87T2311⟩ have developed a method for the synthesis of alkynyl ethers (**97**), derived from tertiary alcohols with a bulky alkyl group such as *t*-butyl and adamantyl. The key step in this synthesis is the dehydrobromination of 1-bromo-2-alkoxyethylene (**96**) with sodium amide in ammonia or lithium diisopropylamide in a hexane–THF solution (Equation (63)) ⟨87T2311⟩.

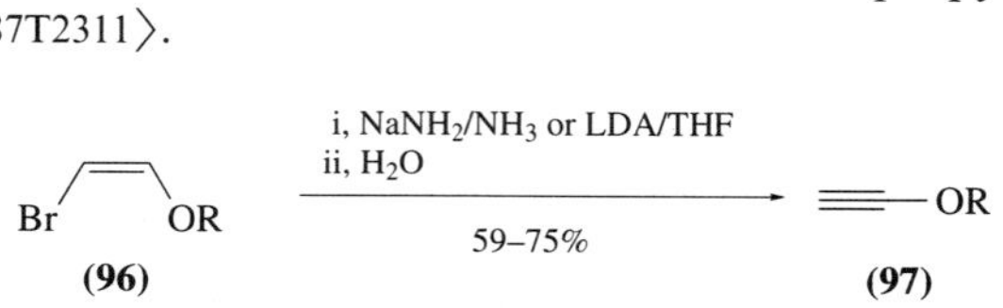

(63)

R = Bu^t, 1-adamantyl

A similar elimination procedure has been used for the preparation of relatively stable, sterically hindered dialkoxyalkynes (**99**) (Equation (64)) ⟨81T1441, 87OS(65)68, 82T1505⟩. In this case the precursors are 1,2-diethers (**98**), which can be prepared from 1,4-dioxane in several steps.

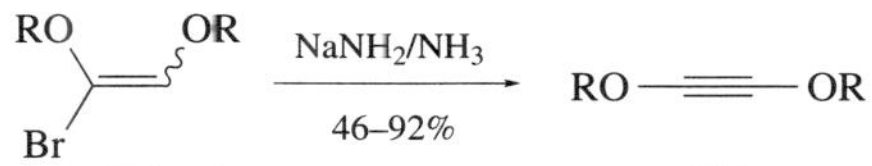

(98) (*Z*) + (*E*) mixture **(99)**

R = Pr^i, Bu^t, -CH_2Bu^t (64)

In the late 1980s, a dehydrohalogenation approach was applied to the generation of 1,4-dioxacyclohexyne (*p*-dioxyne) (**101**) (Scheme 13) ⟨88CC942⟩. This highly unstable cyclic alkynyl diether was formed as a transient molecule in the reaction of dibromide (**100**) with Bu^tLi at −78 °C. Consequent cyclotrimerization of *p*-dioxyne (**101**) gives the final isolated product (**102**).

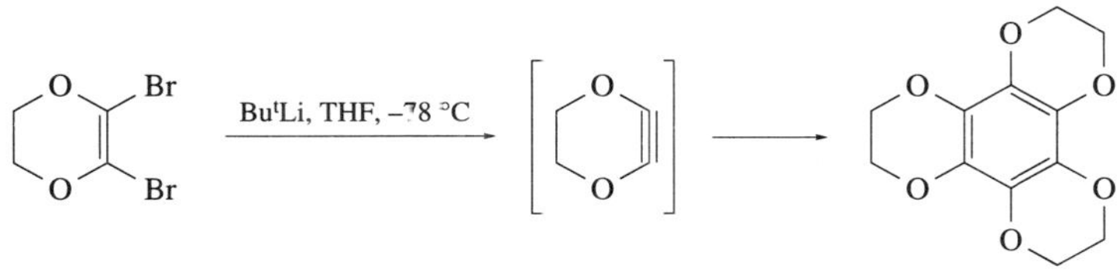

(100) **(101)** **(102)**

Scheme 13

(ii) By functionalization of terminal alkoxyethynes

The parent alkoxyalkynes, HC≡COR (**103**), are the most readily available representatives of alkynyl ethers. Moreover, ethoxyethyne is even commercially available, or can be easily prepared by standard, well-developed procedures ⟨B-69MI 221-02, B-60MI 221-01, 65HOU(6/3)116, B-78MI 221-01, 89RCR948⟩. These compounds can usually be further functionalized by standard methods employed for terminal alkynes.

The most general synthetic approach to alkoxyalkynes (**105**) is the reaction of alkoxyethynylides (**104**) with electrophiles according to Scheme 14.

$$\equiv\!-OR \xrightarrow{\text{base}} M\!-\!\equiv\!-OR \xrightarrow{\text{El}} El\!-\!\equiv\!-OR$$

(103) **(104)** M = Li, Na, etc. **(105)**

El = electrophile

Scheme 14

Reactions of alkoxyethynylides with carbonyl compounds (Scheme 9) or their coupling with aryl or alkyl halides (Schemes 11 and 12) have been widely applied in the synthesis of a variety of alkoxyalkynes (see the literature for recent examples ⟨93T2159, 90TL7597, 92SL502⟩). Also, this method is especially useful for the preparation of various element containing alkoxyalkynes. For example, Lukashev and Kazankova ⟨90MI 221-01⟩ synthesized a number of phosphorylated alkoxyethynes (**106**) by the reaction of lithium alkoxyethynylides with phosphine halides (Equation (65)).

$$X_2PCl \;+\; Li\!-\!\equiv\!-OR \longrightarrow X_2P\!-\!\equiv\!-OR \qquad (65)$$

(106)

X = Pr^i, Bu^t, Ph, C_6F_5, OEt, NPr^i_2; R = Me, Et, Bu, Ph

A similar approach has been employed in the syntheses of silicon ⟨83MI 221-01, 84LA1358⟩, germanium ⟨87MI 221-01⟩, tin ⟨87MI 221-01⟩, and boron ⟨77IZV1456⟩ substituted alkoxyethynes.

Another well-known transformation of terminal alkynes is an oxidative dimerization leading to butadiynes. Pericàs and co-workers ⟨90JA7405⟩ found that alkoxyethynes also can be oxidatively dimerized in the presence of a copper(I) catalyst, yielding relatively stable 1,4-dialkoxy-1,3-butadiynes (**107**) (Equation (66)).

$$2\;\equiv\!-OR \xrightarrow[\text{65–95\%}]{O_2\text{, acetone, CuI•2TMEDA (cat.)}} RO\!-\!\equiv\!-\!\equiv\!-OR \qquad (66)$$

(107)

R = Bu^t, 1-adamantyl, cyclohexyl, etc.

(iii) By reactions of alkynyl halides with alkoxide anions

Nucleophilic substitution of halogen at an alkyne carbon in principle represents another general approach to alkynyl ethers (Equation (67)).

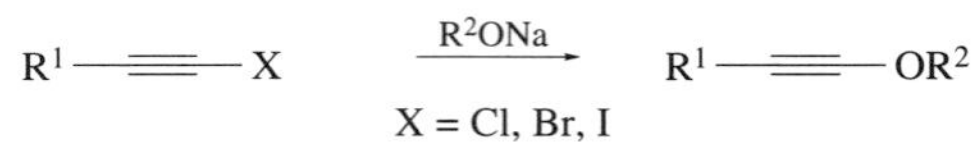

(67)

A study of the reaction of phenylethynyl halides with the methoxide anion revealed a complex addition–elimination mechanism for this substitution ⟨71T2651⟩. Such reactions require application of DMSO as a solvent and afford alkynyl ethers only in 42–46% yield ⟨71TL1753⟩. Reactions of phosphorylated chloroalkynes (**108**) with alkoxide anions are even less selective and afford products of nucleophilic addition to the triple bond as major products ⟨85JGU1771⟩. However, the analogous reaction of compounds (**108**) with the phenoxide anion gives phenoxyethynes (**109**) in moderate yield (Equation (68)) ⟨85JGU1771⟩.

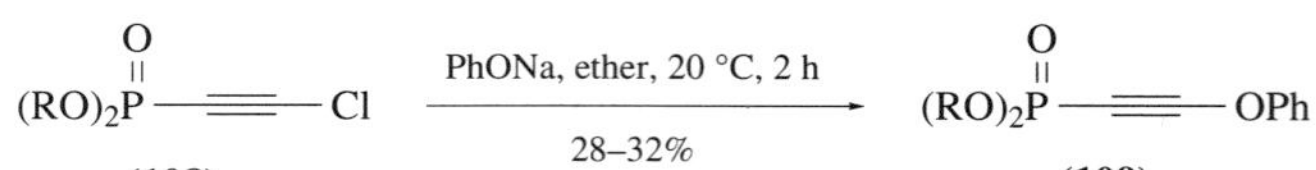

(68)

(**108**) (**109**)

R = Me, Et

The use of a better leaving group in the starting ethyne in principle should make this reaction more selective and improve the yield of the product. Indeed, alkynyliodonium salts which have a very good leaving group, iodobenzene, react with alkoxide anions under mild conditions affording alkynyl ethers as major products ⟨91JA4571⟩. For example, the bis-iodonium ethyne (**50**) reacts with lithium phenolate even at low temperatures to give diphenoxyethyne (**110**) in 57% isolated yield (Equation (69)) ⟨91JA4571⟩.

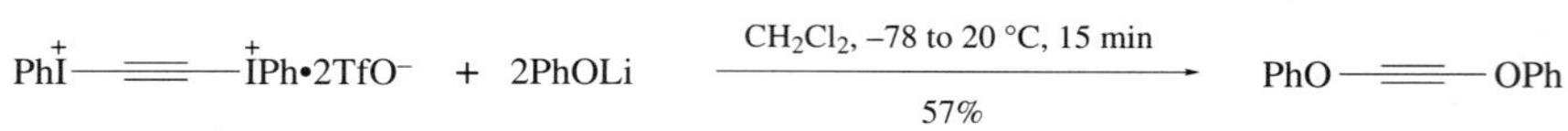

(69)

(**50**) (**110**)

2.21.2.1.4 The synthesis of alkynyl esters

The most general approach to the preparation of functionalized alkynes including alkynyl ethers involves elimination techniques, mostly dehydrohalogenation of appropriate alkene precursors. However, these and related procedures do not work for the preparation of alkynol esters (**111**)–(**113**), and this undoubtedly is one of the major reasons why these compounds remained unknown until the mid-1980s ⟨85JA1452⟩.

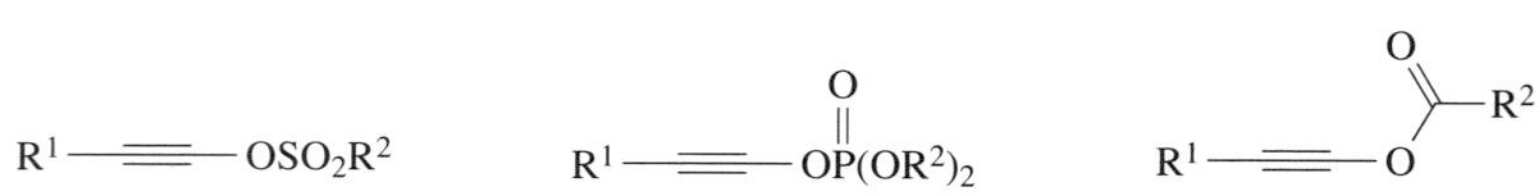

(**111**) (**112**) (**113**)

The first successful preparation of alkynyl esters involved the use of alkynyl(phenyl)iodonium salts (**114**)–(**116**), which are now readily available through a variety of methods ⟨92AG(E)274⟩. The first reported members of the alkynyl esters—alkynyl sulfonates (**111**)—have been prepared by treatment of alkynyl(phenyl)iodonium tosylates or mesylates (**114**) with catalytic CuOTf or AgOTf in dry acetonitrile (Equation (70)) ⟨85JA1452, 87JA228⟩.

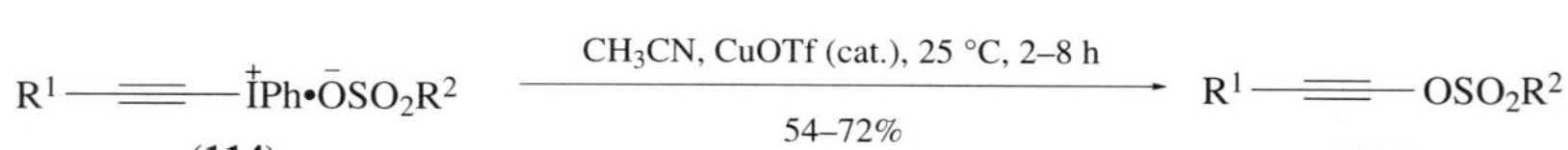

(70)

(**114**) (**111**)

R^1 = Me, Bu, Bu^s, Bu^t, Ph
R^2 = Me, *p*-MeC_6H_4

In a similar manner, decomposition of alkynyl(phenyl)iodonium phosphates (**115**) affords alkynyl phosphate esters (Equation (71)) ⟨86JA7832, 89JA2225⟩. This reaction does not even require any catalyst, since alkynyl(phenyl)iodonium phosphates (**115**) are substantially less stable than tosylates

(**114**), and smoothly decompose upon standing at room temperature for a few hours in CH_2Cl_2 or $CHCl_3$, yielding alkynyl phosphates (**112**) and iodobenzene as the by-product.

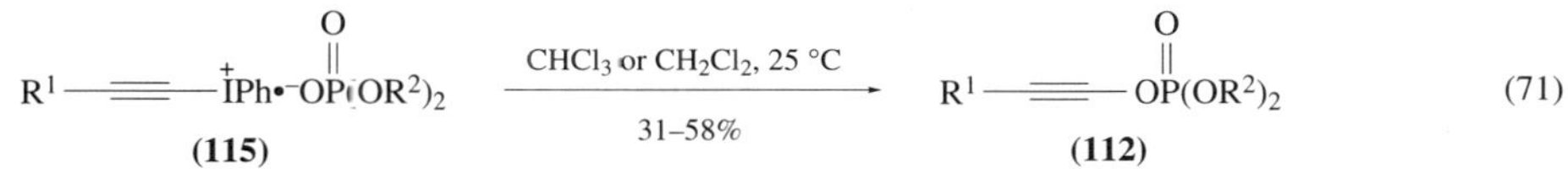

R^1 = Me, Bu^n, Bu^s, Bu^t, Pr^n, n-C_6H_{13}
R^1 = Me, Et, CH_2Ph

Alkynyl(phenyl)iodonium carboxylates (**116**) are even less stable than phosphates (**115**) and decompose immediately upon generation from iodonium triflates (**117**) or benzoates (**118**), with the formation of the desired alkynyl carboxylates (**113**) (Scheme 15) ⟨88JA3272⟩.

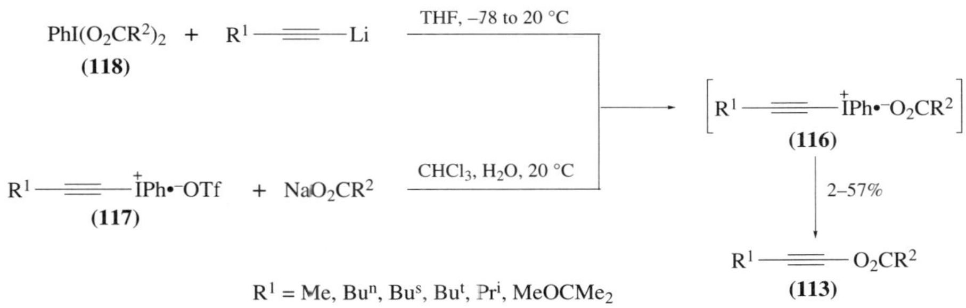

R^1 = Me, Bu^n, Bu^s, Bu^t, Pr^i, $MeOCMe_2$
R^1 = Me, CMe_3, Ph, 4-$MeOC_6H_4$,
4-$NO_2C_6H_4$, 3,5-$(MeO)_2C_6H_3$, CH_2CH_2Ph, $CHPh_2$

Scheme 15

Formalistically, these reactions (Equations (70) and (71), Scheme 15) are nucleophilic alkynyl substitutions (S_NA) with the corresponding anions (sulfonate, phosphate, or carboxylate) acting as nucleophiles and alkynyliodonium species (**114**)–(**116**) as the electrophilic substrates. However, the actual details of the mechanism are considerably more complex (Scheme 16).

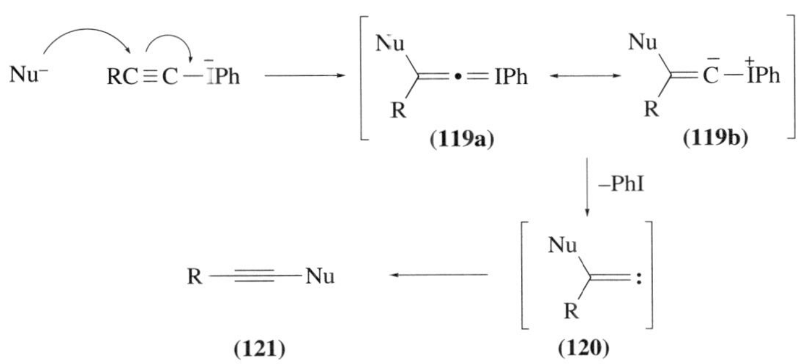

Scheme 16

The first step in this scheme is a Michael addition of the nucleophile to the β carbon of the alkynyliodonium salt to give the ylide (**119**). Loss of iodobenzene from the ylide (**119**) gives the alkylidenecarbene (**120**) which rearranges to the alkyne (**121**) in the absence of external traps. This mechanism is experimentally supported by the isolation of the cyclic by-products (**125**) besides the major products, the alkynyl esters (**124**) in the reaction of the alkynyliodonium salt (**122**) with nucleophiles (Scheme 17) ⟨92AG(E)274⟩. These cyclic enol esters are the result of the insertion of the intermediate carbene (**123**) into the tertiary-ε-carbon–hydrogen bond.

In the early 1990s, the reaction of alkynyliodonium salts with the corresponding anionic nucleophiles was the most general and selective approach to alkynyl esters. However, two more conventional but less general syntheses have also been reported in the literature ⟨93TL1461, 87TL785⟩. Hoppe and Gonschorrek ⟨87TL785⟩ have prepared ethynyl carbamates (**127**), which are also members of the family of alkynyl esters, by treatment of the allene (**126**) with butyllithium (Equation (72)).

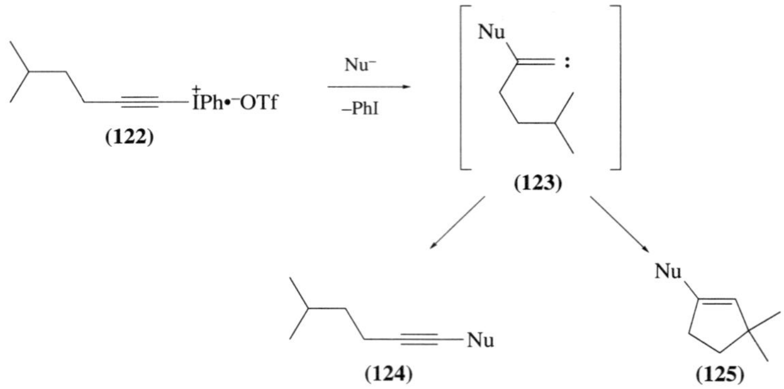

Nu = $PhCO_2$, $(EtO)_2PO_2$, etc.

Scheme 17

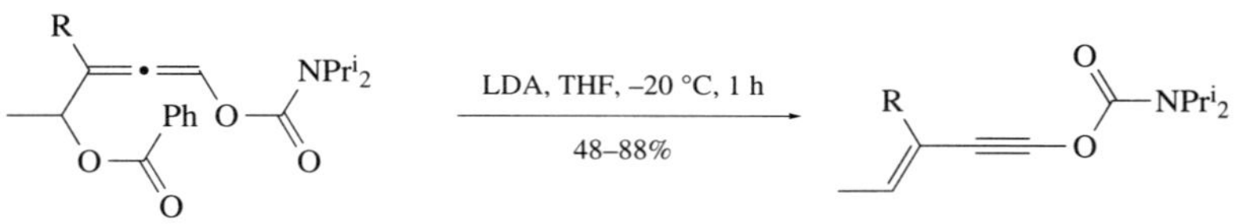

(126) **(127)**

R = H, Me (72)

In 1993, Zhdankin and Stang have reported the preparation of alkynyl esters (**112**) and (**113**) by the reaction of lithium alkynolates with the corresponding acid chlorides (Equations (73) and (74)) ⟨93TL1461⟩. This method works well for the synthesis of alkynyl phosphates (**112**); however, it has only limited applicability in the preparation of alkynylcarboxylates (**113**) because of low selectivity ⟨93TL1461⟩

(112)

R = Bu^t, Bu^n, Me, Ph (73)

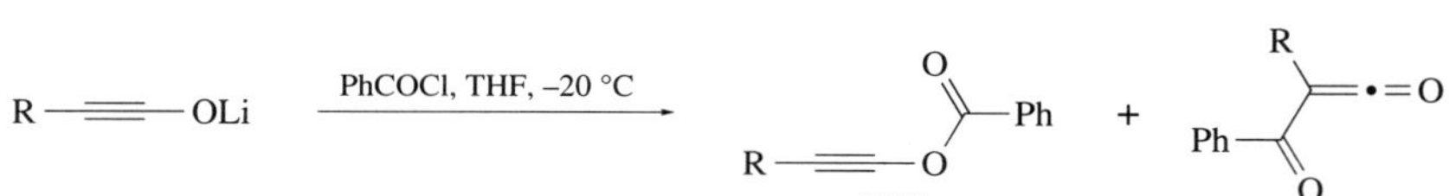

(113)

R = Bu^t, Ph (74)

2.21.2.2 The Synthesis of Sulfur-based Functional Groups Attached to an *sp*-Carbon

2.21.2.2.1 The synthesis of alkynyl thioethers and thioesters

(i) By elimination reactions

A broad variety of alkynyl thioethers are known ⟨B-69MI 221-02, 89RCR948⟩. The general principles of the preparation of these compounds are similar to the analogous *O*-derivatives. Especially useful are elimination reactions. Dehydrohalogenation of halothioethers is usually achieved by the action of potassium hydroxide ⟨56JA4783, 56JA2760, 56RTC481, 61RTC237, 56RTC1459, 59RTC252, 67JOC2063⟩ or sodium amide in liquid ammonia ⟨62RTC510, 63RTC1119, 66BSF3016⟩. For example, ethynyl thioether (**129**) can be prepared in moderate yield by dehydrobromination of bromothioether (**128**) with an alcoholic solution of potassium hydroxide (Equation (75)) ⟨56JA2760⟩.

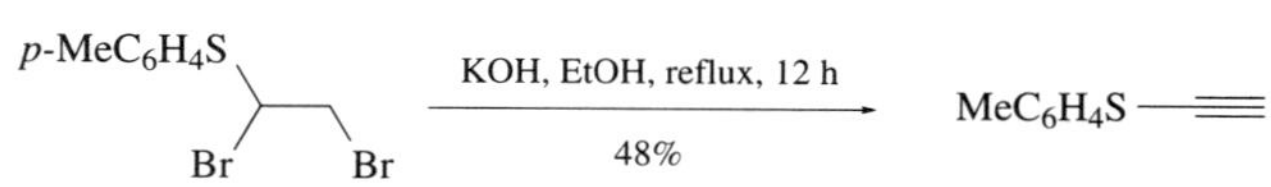

(128) **(129)** (75)

The rate of the dehydrobromination with potassium hydroxide and the yield of alkynyl thioether depends significantly on the structure of the starting substrate. In many cases this reaction is very slow and is complicated by nucleophilic addition of the solvent to the product ⟨B-69MI 221-02⟩.

Dehydrobromination with sodium amide in liquid ammonia is a method of choice for the preparation of thermally unstable and volatile thioethers **(130)** (Equation (76)) ⟨B-88MI 221-01⟩.

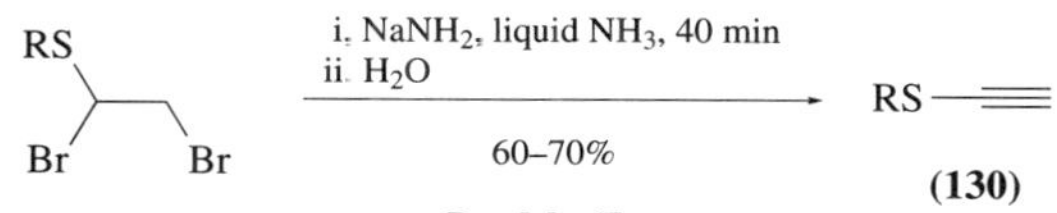

(130)

R = Me, Et (76)

An analogous procedure has been successfully applied in the synthesis of various ethynyl thioethers ⟨66BSF3016⟩ and diethynyl sulfide **(131)** (Equation (77)) ⟨62RTC510, 63RTC1119⟩.

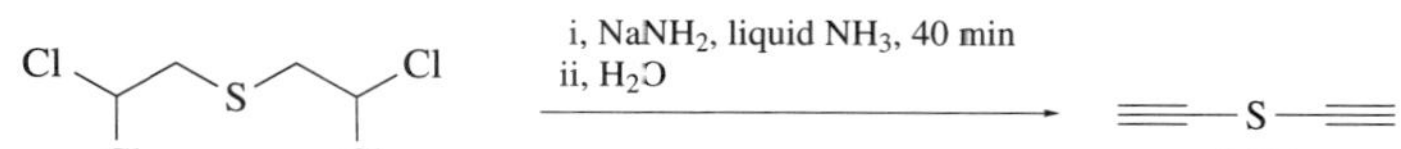

(131) (77)

Some other strong bases have been successfully applied as dehydrohalogenation reagents in the synthesis of alkynylsulfides. Vinylethynyl sulfide **(134)** has been prepared in good yield by the opening of the ring in the cyclopropane **(132)** with subsequent dehydrochlorination of the intermediate diene **(133)** (Scheme 18) ⟨72BCJ509, 73JOC1361⟩.

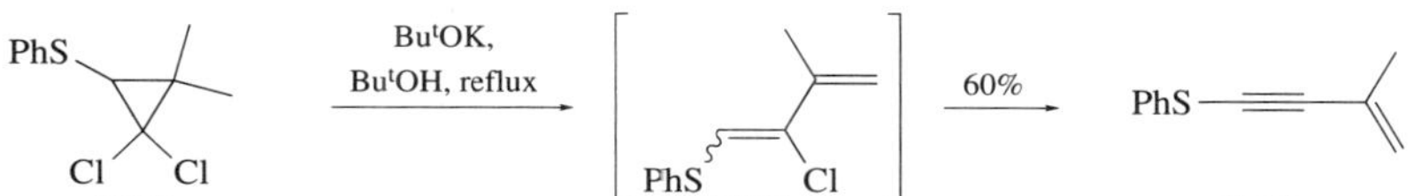

(132) **(133)** **(134)**

Scheme 18

The etherial solution of ethylthioethynyllithium **(136)** can be conveniently prepared by dehydrochlorination of the vinylic thioether **(135)** with butyllithium (Equation (78)) ⟨63BSF1876⟩. This reaction is especially important since the thioethynylide **(136)** is suitable for further transformation by reactions with carbonyl compounds or other electrophiles.

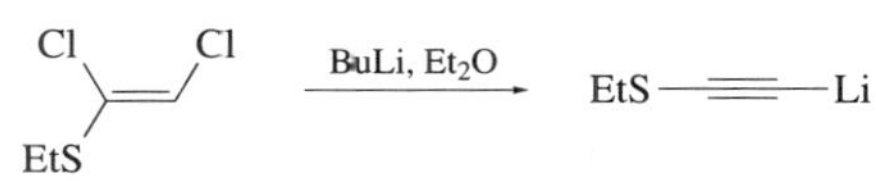

(135) **(136)** (78)

Analogously, thiobutynylides **(138)** have been generated *in situ* by dechlorination of the diene **(137)** (Equation (79)) ⟨85ZN(B)645⟩. A series of butadiynyl sulfides **(139)** has been synthesized by the subsequent reaction of alkynylides **(138)** with various electrophiles (Equation (80)) ⟨85ZN(B)645⟩.

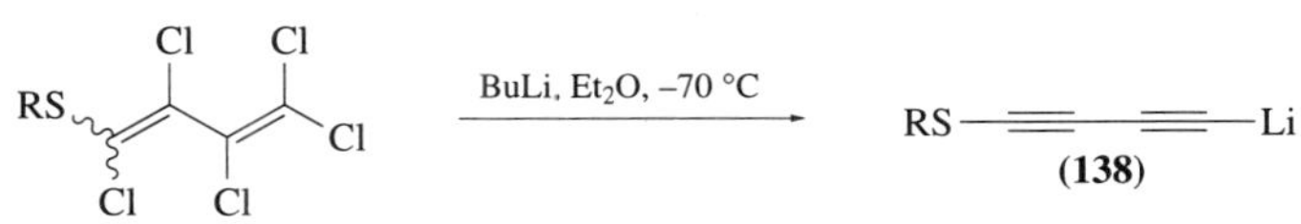

(138)

(137) R = Ph, cycloalkyl (79)

(138) **(139)** (80)

R^1 = cycloalkyl, aryl

R^2 = H, TMS, Ph_3Sn, etc.

Not only halothioethers can serve as starting compounds in the elimination procedures; readily available bis(thio)ethenes react with alkali amides in liquid ammonia or with butyllithium to form the corresponding metal thioethynylides as the result of 1,2 elimination of the thiol moiety ⟨56JA4783, 59JA3386, 60RTC866⟩. For example, phenylthioethyne **(141)** was first obtained by the reaction of bis(phenylthio)ethene **(140)** with butyllithium (Equation (81)) ⟨56JA4783⟩.

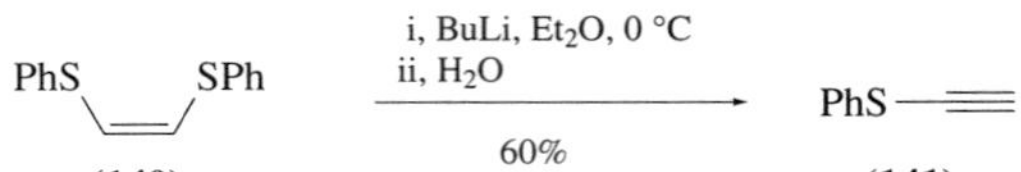

(140) **(141)** (81)

In a more recent procedure ⟨93S1067⟩, phenylthioethyne (**141**) was prepared by the pyrolysis of the readily available phenylthio derivative of Meldrum's acid (**142**) (Equation (82)).

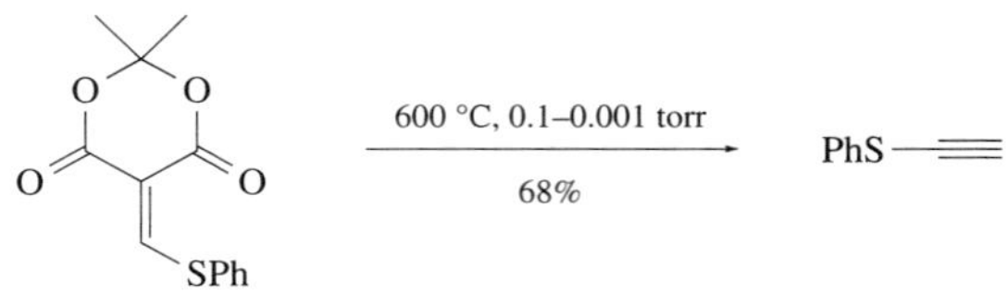

(142) **(141)** (82)

(ii) By reactions of alkynylides with thiolating reagents

A wide variety of alkynyl thioethers can be prepared in high yields by reactions of metal alkynylides with thiolating agents, such as sulfenyl halides, disulfides, thiocyanates, or elemental sulfur ⟨B-88MI 221-01, B-69MI 221-02⟩. A convenient and quick method consists of adding a disulfide to a lithium or sodium alkynylide in liquid ammonia (Equation (83)) ⟨63RTC375, 62RTC533, 61RTC244, 66RTC580, 68RTC97, 66RTC952⟩.

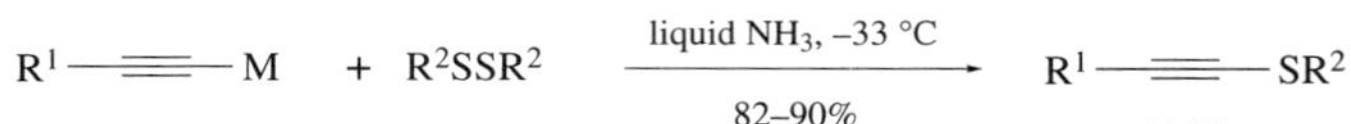

(143) (83)

R^1, R^2 = alkyl, aryl

M = Li, Na

Ether or THF can be used as a solvent instead of liquid ammonia in this reaction, and thiocyanates can serve as efficient thiolating reagents ⟨B-88MI 221-01, B-69MI 221-02⟩.

The buteneyne thioester (**146**) is best prepared by the reaction of the sodium alkynylide (**144**) with powdered sulfur in cold liquid ammonia and consequent alkylation of the intermediate sodium thiolate (**145**) with bromoethane (Scheme 19) ⟨B-88MI 221-01⟩. This procedure has general application in the synthesis of various alkynyl thioethers, selenoethers, and telluroethers ⟨62RTC583, 64RTC208, 64ZOB1899, 67ZOR2102, 65ZOR2157⟩.

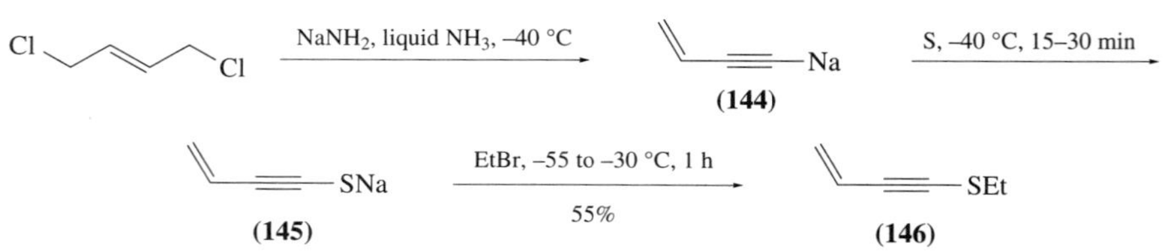

Scheme 19

The reaction of lithium alkynylides with sulfur in a THF–hexane solution provides a general entry into alkynyl lithium thiolates (**147**) (Equation (84)), which can be subsequently alkylated with various alkyl halides and chloromethyl ethers ⟨B-88MI 221-01, 63N302⟩, acylated with acetyl bromide or ethyl chloroformate (Equation (85)) ⟨B-88MI 221-01⟩, silylated with chlorotrimethylsilane (Equation (86)) ⟨77RTC179, 76CC1008⟩, or converted into alkynyl thiocyanates (**148**) by the reaction with cyanogen chloride (Equation (87)) ⟨B-88MI 221-01, 71RTC1098⟩.

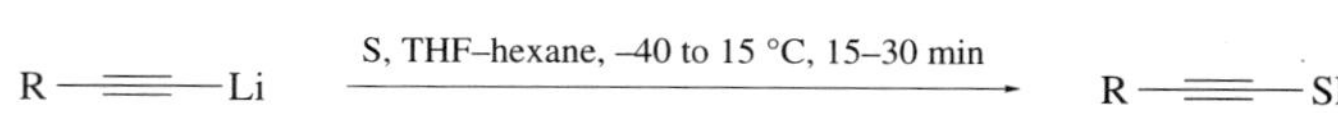

(147) (84)

R = Me, Et, Bu^t, TMS, etc.

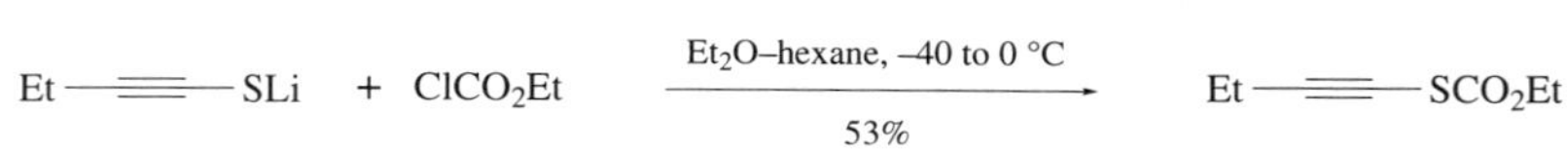

(85)

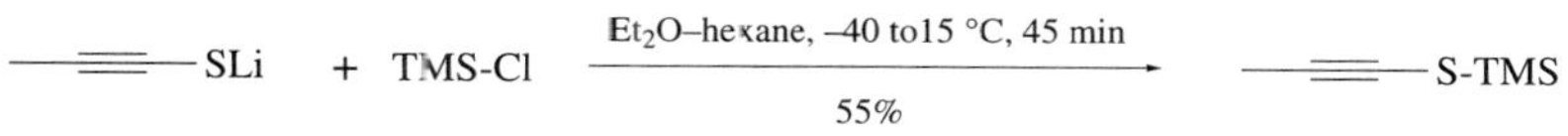

(86)

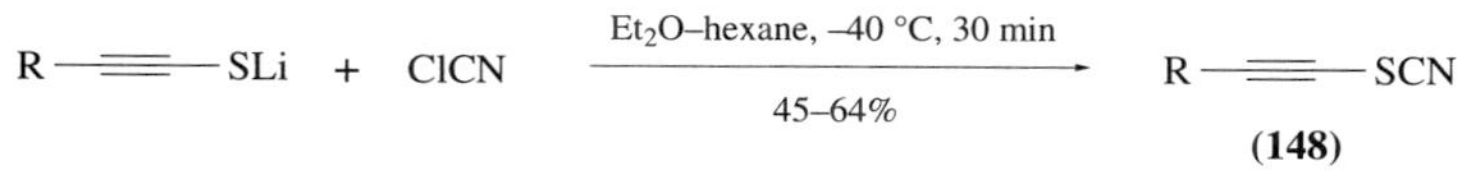

(87)

R = Me, Bu^t

(iii) By reactions of alkynyl halides with thiolates

Nucleophilic substitution of halogen at the *sp*-carbon of an alkynyl halide represents another general approach to alkynyl thioethers and analogous compounds. Alkylthiolates and arylthiolates smoothly react with alkynyl chlorides or bromides in aprotic polar solvents to afford alkynyl thioethers in moderate to good yields (Equation (88)) ⟨B-69MI 221-02, 63JA1648, 69RTC179⟩.

(88)

R^1, R^2 = alkyl, aryl; X = Cl, Br

Alkynyl(phenyl)iodonium salts are even better substrates in these reactions, because of the excellent leaving group abilities of the iodobenzene moiety ⟨91JA4571⟩. For example, the bis(iodonium)ethyne (**50**) reacts with the thiophenolate anion under very mild conditions, affording bis(phenylthio)ethyne (**149**) in good yield (Equation (89)) ⟨91JA4571⟩.

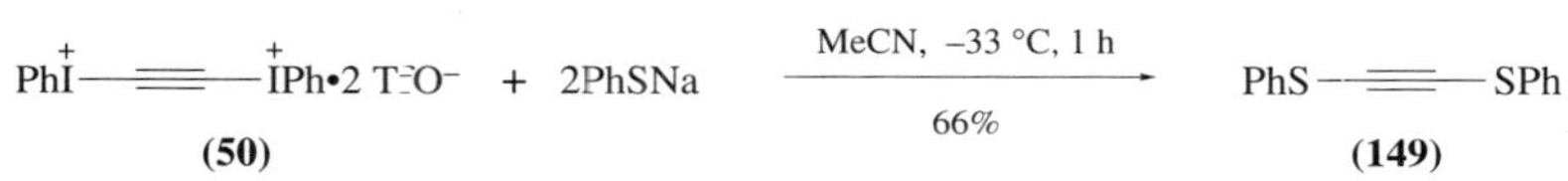

(89)

Reactions of this type are especially useful for the preparation of alkynyl thioesters and alkynyl thiocyanates. There has been substantial interest in the preparation and study of *S*-alkynyl esters of diethyl dithiophosphoric acid (**150**) due to their exceptional insecticidal activity ⟨91RCR885⟩. Both alkynyl thiophosphates (**150**; X = O) and alkynyl dithiophosphates (**150**; X = S) can be synthesized by the reaction of alkynyl bromides with the corresponding copper salts (Equation (90)) ⟨83IZV1662, 84DOK(274)822, 91RCR885⟩.

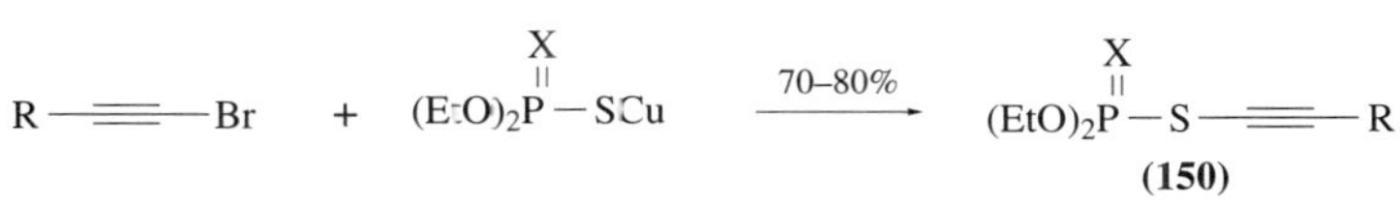

(90)

R = Bu^n, Bu^t, CH_2CH_2SEt, Ph, cyclo-C_6H_{11}, etc.
X = O, S

Even better results can be achieved by the employment of alkynyliodonium salts (**42**) instead of alkynyl bromides in this reaction. In this case, copper catalysis is not needed, and the yields of alkynyl dithiophosphates (**151**) are even better (Equation (91)) ⟨93JOC1924⟩.

(91)

R^1 = Bu^t, Ph; R^2 = Me, Et, Pr^n, Bu^n, $PhCH_2$, Ph
TEBA = $PhCH_2N^+Et_3{\bullet}Cl^-$

Alkynyliodonium salts are also useful precursors in the synthesis of alkynyl thiocyanates (**148**) by the reaction with the thiocyanate anion ⟨92SL535, 93T3043⟩. Various thiocyanates (**148**) can be prepared by shaking of a solution of the appropriate iodonium triflate (**47**) in CH_2Cl_2 with an aqueous solution of sodium thiocyanate (Equation (92)) ⟨92SL535⟩

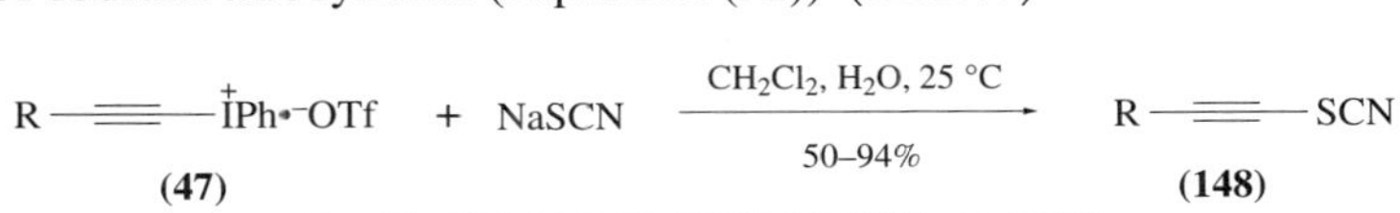

(92)

R = Me, Bu^t, Bu^n, Ph, TMS, $MeOCH_2$, Bu^{t}C(O)

A similar procedure has been applied in the synthesis of bis(alkynyl) dithiocyanates (**152**) from bis(alkynyl) bis(iodonium) salts (**52**) (Equation (93)) ⟨93T3043⟩

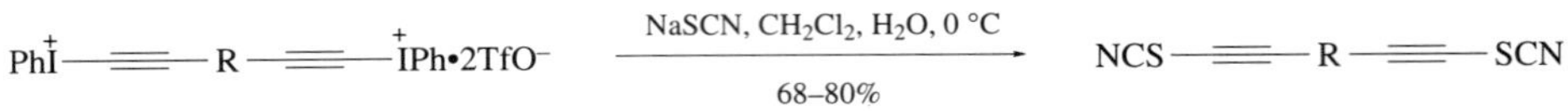

(93)

(52) **(152)**

$R = (CH_2)_n$, n = 1–5 or $(C_6H_4)_m$, m = 1, 2

2.21.2.2.2 *The synthesis of alkynyl sulfoxides and sulfones*

(i) By oxidation of alkynyl thioethers

There is substantial interest in alkynyl sulfones because of their extensive use in organic synthesis as activated ethyne equivalents. Typical applications of alkynyl sulfones include cycloadditions, and addition reactions with amines, cuprates, alkoxides, and organolithium reagents ⟨B-93MI 221-01⟩.

The traditional, well-developed approach to alkynyl sulfones and sulfoxides includes oxidation of sulfur in alkynyl thioethers ⟨B-69MI 221-02, B-88MI 221-01, 56JA4783, 56JA2760, 12LA(394)325, 54RTC565, 23CB2290, 60JA3799, 66BSF3016, 66JOC3545, 66JCS(B)1188⟩. Oxidation of thioethers to sulfones is best achieved by organic peracids. In a typical procedure, peracetic acid is used to oxidize an alkynyl thioether to afford the sulfone (**153**) in an excellent yield (Equation (94) ⟨B-88MI 221-01⟩.

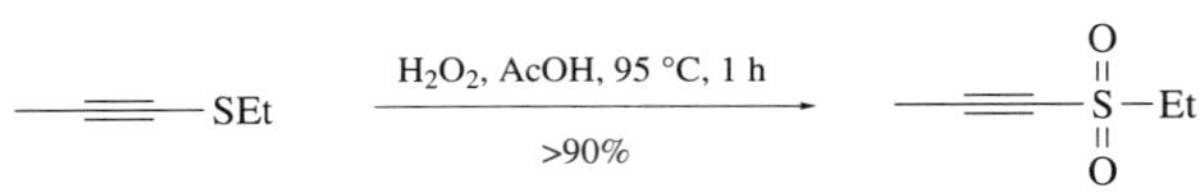

(94)

(153)

The analogous reaction with *m*-chloroperbenzoic acid in chloroform proceeds under milder conditions and can be recommended for the oxidation of more sensitive sulfides ⟨87JCS(P2)1275, 81TL3347, 86JOM(304)115⟩. For example, the thioethynylphosphonate ester (**154**) has been selectively oxidized to the alkynyl sulfone (**155**) with an excess of *m*-chloroperbenzoic acid at 0°C in a chloroform solution (Equation (95)) ⟨87JCS(P2)1275⟩. Use of an equimolecular amount of peracid under the same conditions (Equation (95)) results in the formation of the analogous sulfoxide in moderate yield ⟨87JCS(P2)1275⟩.

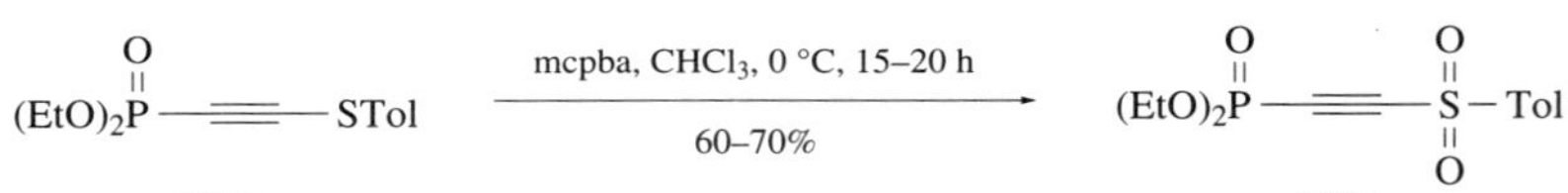

(95)

(154) **(155)**

Employment of sodium periodate instead of peracids in the oxidation of alkynyl thioethers leads to the selective formation of alkynyl sulfoxides (Equation (96)) ⟨B-88MI 221-01⟩.

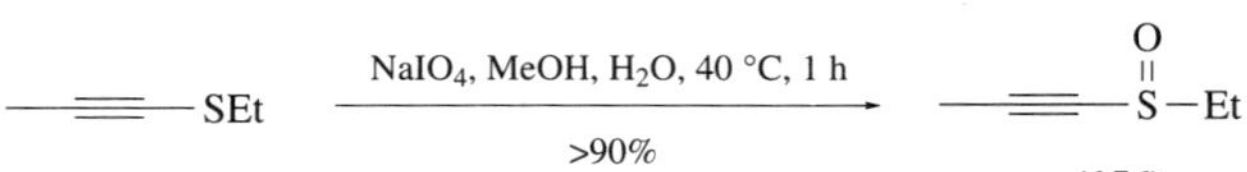

(96)

(156)

(ii) By elimination reactions

A variety of elimination procedures have been applied to the synthesis of alkynyl sulfones ⟨71JOC1727, 89CC154, 90SL290, 92SL825, 83JOC3077, 85JOC2372⟩. A mild and efficient procedure involves dehydrohalogenation of β-iodoalkenyl sulfones with potassium carbonate in aqueous methanol (Equation (97)) ⟨71JOC1727⟩. Yields of products (**157**) in this reaction are excellent for alkynyl toluene sulfones, but low in the case of alkynyl alkyl sulfones.

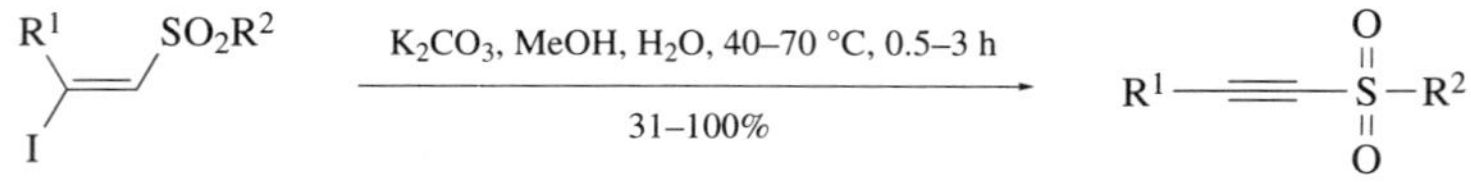

(97)

(157)

R = Ph, Bu^t, Pr^i, Bu^n; R^2 = Tol, Et, Me, Pr^i

1-(Phenylsulfonyl)alkynes (**159**) are conveniently prepared from readily available β-ketosulfones (**158**) by treatment with triflic anhydride and diisopropylethylamine (Equation (98)) ⟨92SL825⟩.

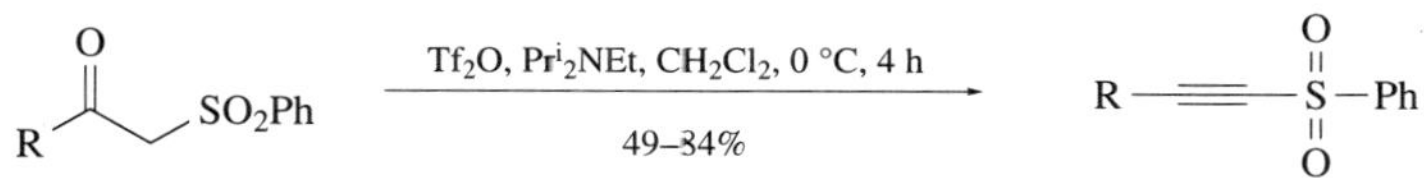

(98)

(158) **(159)**

R = Ph, Bu^t, Pr^i, Me_2CHCH_2, $Me(CH_2)_{10}$, etc.

Vinyl triflates are plausible intermediates in this reaction (Equation (98)). In an analogous procedure (Equation (99)), alkynyl sulfones are prepared by elimination from vinyl phosphates (**160**), which can be synthesized from β-ketosulfones (**158**) and diethyl chlorophosphate ⟨90SL290⟩.

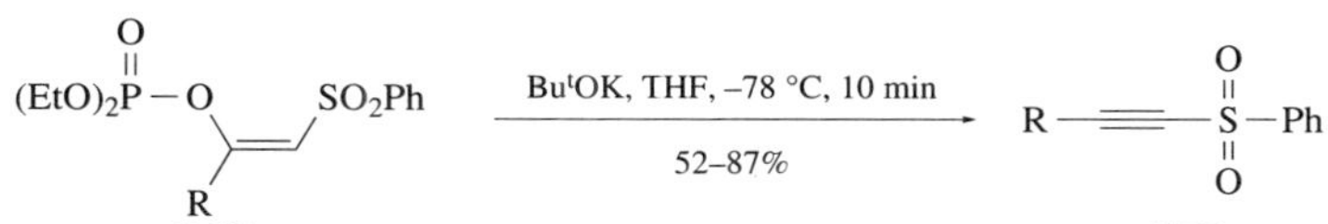

(99)

(160) **(159)**

R = Ph, Pr^n, Pr^i, cyclo-Pr, 4-$MeOC_6H_4$, 4-ClC_6H_4

Similarly, oxidative elimination in selenoalkenes (**161**) affords alkynyl sulfones in good yield (Equation (100)) ⟨83JOC3077⟩.

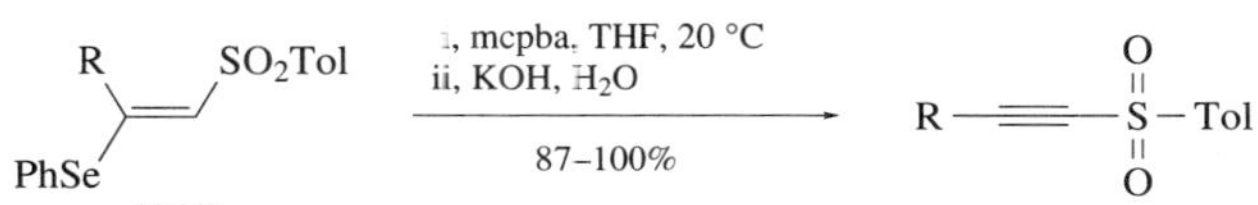

(100)

(161)

R = TMS, $Me(CH_2)_7$, $HOCH_2CH_2$, Ph

Two reported elimination procedures for the synthesis of alkynyl sulfones involve diazotization of sulfonylated 5-aminoisoxazoles ⟨85JOC2372⟩ and decarboxylative demercuration of (sulfonyl)ethenyl mercury precursors ⟨89CC154⟩.

(iii) By substitution at an sp-*carbon*

Several procedures exist for the preparation of various alkynyl sulfones by substitution at the acetylenic carbon. Parent arylsulfonylethynes can be further functionalized by deprotonation and subsequent reaction with electrophiles. For example, *p*-toluenesulfonylethyne (**162**) reacts with butyllithium at $-100\,^{\circ}C$, affording the thermally unstable lithium ethynylide (**163**), which can be converted into the carboxylic acid (**164**) by treatment with excess carbon dioxide gas (Scheme 20) ⟨88TL6409⟩.

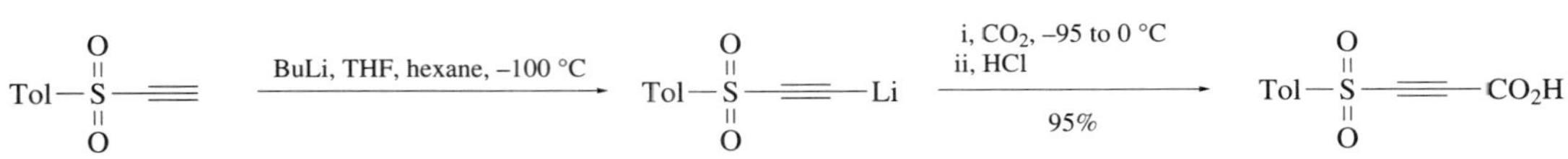

(162) **(163)** **(164)**

Scheme 20

Direct one-step functionalization of *p*-toluenesulfonylethyne (**162**) into the stannyl derivatives (**165**) is achieved by the reaction with an alkoxytrialkyltin (Equation (101)) ⟨86JOM(304)115⟩.

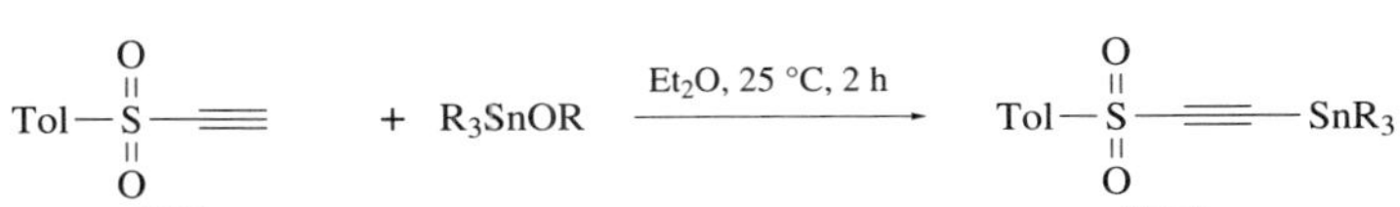

(101)

(162) R = alkyl **(165)**

Alkynyl sulfoxides (**166**) are conveniently prepared by the reaction of a lithiated alkyne with an alkanesulfinyl chloride in THF at low temperature (Equation (102)) ⟨B-88MI 221-01⟩. A similar reaction with thionyl chloride affords symmetrical bis(alkynyl) sulfoxides (**167**) (Equation (103)) ⟨B-88MI 221-01⟩.

(102)

(166)

R^1 = alkyl; R^2 = Ar, Me

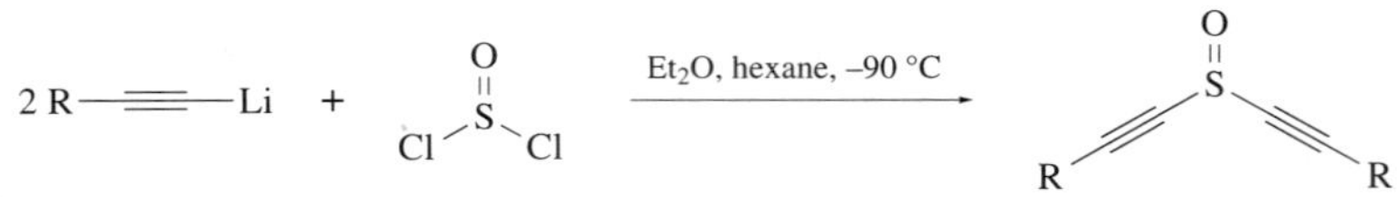

(103)

(167)

R = alkyl

The newest synthetic procedure for the synthesis of alkynyl sulfones employs the reaction of alkynyliodonium salts with sodium sulfinate ⟨93JOC5235, 92SC1997⟩. Two modifications of this procedure have been reported in the literature. In the first procedure, alkynyl(phenyl)iodonium tosylates (**42**) are treated with sodium sulfinates in a two-phase system of chloroform and water in the presence of a catalytic amount of TEBA (Equation (104)) ⟨92SC1997⟩.

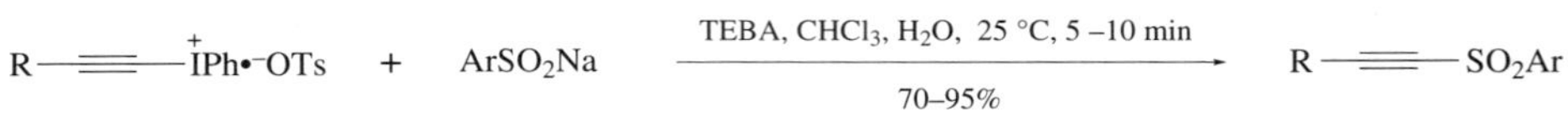

(104)

(42)

R = Ph, Bu^t; Ar = Ph, *p*-Tol, *p*-ClC_6H_4, *m*-$NO_2C_6H_4$

TEBA = $PhCH_2N^+Et_3{\bullet}Cl^-$

In the second procedure, alkynyl(phenyl)iodonium triflates (**47**) are used as the starting materials in the reaction with anhydrous sodium *p*-toluenesulfinate in dichloromethane (Equation (105)). This procedure works well for the preparation of the parent (tolylsulfonyl)ethyne, some functionalized ethynyl sulfones (**168**) (Equation (105)), and bis(sulfonylethynyl)benzene (**169**) (Equation (106)) ⟨93JOC5235⟩.

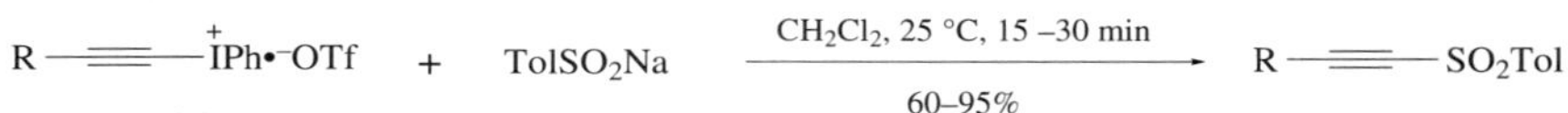

(105)

(47) **(168)**

R = H, Me, Et, Ph, Pr^i_3, $ClCH_2$, $BrCH_2$, $HOCH_2$

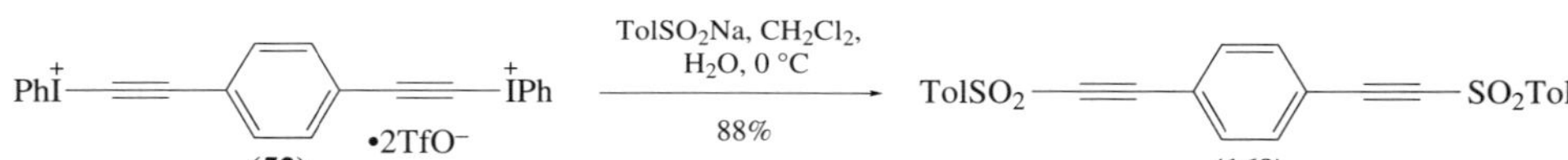

(106)

(52) **(169)**

2.21.2.3 The Synthesis of Selenium- and Tellurium-based Functional Groups Attached to an *sp*-Carbon

There is substantial interest and research activity in alkynyl selenides and tellurides due to their increasing importance as reagents and intermediates in organic synthesis ⟨89RCR948, 93SL211, 81JOM(216)287, 84S240, 86SC283, 86CC1067, 88SC1979, 90CB417, 91JCS(P1)2049, 90ZOR2635, 90SC2181, 90IZV1865, 88OM84, 82CL1244⟩. Both alkynyl selenides and tellurides are usually prepared by procedures based on the substitution at alkynyl carbon; however, the elimination approach is also known. In a well-developed and efficient procedure ⟨B-88MI 221-01, B-69MI 221-02⟩, alkynyl selenolates or tellurolates (**170**) are prepared by the reaction of sodium alkynylides with elemental selenium or tellurium in liquid ammonia (Equation (107)). Subsequent alkylation of the sodium salt (**170**) affords the respective alkynyl selenoethers or telluroethers (**171**) in good yield (Equation (108)). Besides ordinary alkynes, this procedure (Equations (107) and (108)) can be used for the synthesis of selenoethers and telluroethers from various diynes and enynes ⟨B-69MI 221-02⟩.

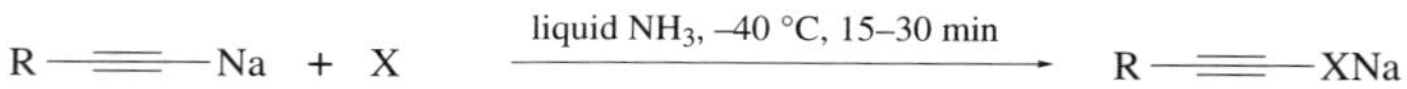

(107)

(170)

X = Se, Te

$$\underset{\textbf{(170)}}{R^1{-}C{\equiv}C{-}XNa} + R^2X \xrightarrow[70\text{–}85\%]{\text{liquid } NH_3\text{, 1 h}} \underset{\textbf{(171)}}{R^1{-}C{\equiv}C{-}XR^2} \quad (108)$$

R^1 = H, alkyl, $CH_2{=}CH$; R^2 = Me ; X = Se, Te

In a recent modification of this method, THF is used as the solvent for this reaction ⟨88OM84, 90CB417⟩. For example, the alkynyl tellurides (**172**) can be prepared in excellent yields by the reaction of the lithium alkynylides with tellurium, followed by trapping of the intermediate lithium tellurolates with alkyl bromides (Equation (109)) ⟨88OM84⟩.

$$Ph{-}C{\equiv}C{-}Li \xrightarrow[73\text{–}97\%]{\text{i, Te, THF-hexane, 0 °C to reflux, 30 min; ii, RBr, 25 °C, 40 min}} \underset{\textbf{(172)}}{Ph{-}C{\equiv}C{-}TeR} \quad (109)$$

R = Et, Bu^n, n-$C_{12}H_{25}$, Pr^i, Pr^nCHMe, CH_2CH_2Ph

This procedure (Equation (109)) works very well for the synthesis of alkynyl (alkyl)tellurides or -selenides. The aryl derivatives (**173**) (Equation (110)) are best obtained by the reaction of lithium alkynylides with an arylselenium bromide ⟨81JOM(216)287⟩ or an aryltellurium bromide generated *in situ* from a diaryl ditelluride and bromine ⟨88OM84⟩.

$$R{-}C{\equiv}C{-}Li + ArXBr \xrightarrow[73\text{–}94\%]{\text{THF–hexane, 0 to 25 °C, 1 h}} \underset{\textbf{(173)}}{R{-}C{\equiv}C{-}XAr} \quad (110)$$

R = Pr^n, Bu^n, Ph; Ar = Ph, 4-$MeOC_6H_4$; X = Se, Te

Several further modifications of this method (Equation (110)) have been reported in the literature ⟨90CB417, 91JCS(P1)2049, 93SL211⟩. In one variation, alkynyl stannanes or alkynyl magnesium bromides were used instead of lithium alkynylides in the reaction with arylselenium halides ⟨90CB417⟩. The second method involves diphenyl diselenide–copper(I) triflate as the selenylating agent in the reaction with the trimethylsilyl-substituted alkynes (Equation (111)) ⟨91JCS(P1)2049⟩. This procedure gives especially good results in the phenyselenylation of some complex alkynes, such as steroidal derivatives ⟨91JCS(P1)2049⟩.

$$R{-}C{\equiv}C{-}TMS + Ph_2Se_2 \xrightarrow[68\text{–}92\%]{\text{CuOTf, PhH, }CaCO_3\text{, dioxane, reflux, 2.5 h}} \underset{\textbf{(174)}}{R{-}C{\equiv}C{-}SePh} \quad (111)$$

R = Me, $Ph(CH_2)_2$, etc.

A variety of alkynes were directly converted into the corresponding phenyl alkynyl selenides (**174**) by treatment with diphenyl diselenide and iodobenzene diacetate (Equation (112)) ⟨93SL211⟩.

$$R{-}C{\equiv}CH + Ph_2Se_2 \xrightarrow[15\text{–}81\%]{PhI(OAc)_2\text{, }CH_2Cl_2\text{, 40 °C, 10–15 h}} \underset{\textbf{(174)}}{R{-}C{\equiv}C{-}SePh} \quad (112)$$

R = Ph, Bu^n, Bu^t, TMS, n-C_6H_{13}, n-C_8H_{17}, $Me_2C(OH)$, MeO_2C, etc.

Acetylenic nucleophilic substitution of the halogen in alkynyl halides represents another possible approach to alkynyl selenides and tellurides. Lithium selenoates and telluroates smoothly react with alkynyl bromides in DMF or THF to afford products of substitution (**175**) in good yield (Equation (113)) ⟨88SC1979⟩.

$$Ph{-}C{\equiv}C{-}Br + RXLi \xrightarrow[75\text{–}90\%]{\text{DMF or THF, 25 °C, 1 h}} \underset{\textbf{(175)}}{Ph{-}C{\equiv}C{-}XR} \quad (113)$$

R = Bu^n, Ph, $PhC{\equiv}C$, n-$C_5H_{11}C{\equiv}C$; X = Se, Te

A similar reaction of (chloroethynyl)thioethers (**176**) with selenols in DMSO under alkaline conditions was used to synthesize (thioethynyl) selenides (**177**) (Equation (114)) ⟨90IZV1865⟩. These products (**177**) can also be obtained by dehydrohalogenation of alkenes (**178**) with strong bases (Equation (115)) ⟨90IZV1865⟩.

$$\underset{(176)}{R^1S{-}{\equiv}{-}Cl} + R^2SeH \xrightarrow[42\text{–}82\%]{\text{KOH, DMSO, 20 to 65 °C, 7 h}} \underset{(177)}{R^1S{-}{\equiv}{-}SeR^2} \qquad (114)$$

R^1, R^2 = Et, Pr^n, Pr^i, Bu^n, Ph

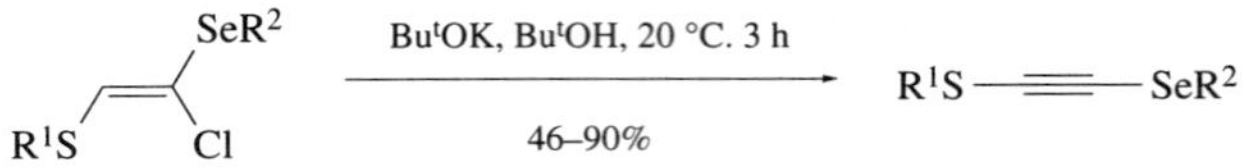

(178) **(177)** (115)

R^1, R^2 = Et, Pr^n, Pr^i, Bu^n, Ph

Several other elimination reactions can be applied to the synthesis of alkynyl selenides ⟨84S240, 87RCR267, 81CB673⟩. The vacuum pyrolysis of selenophosphoranes (**179**) leads to the elimination of triphenylphosphine oxide and affords (alkynyl)aryl selenides (**180**) (Equation (116)) ⟨84S240⟩.

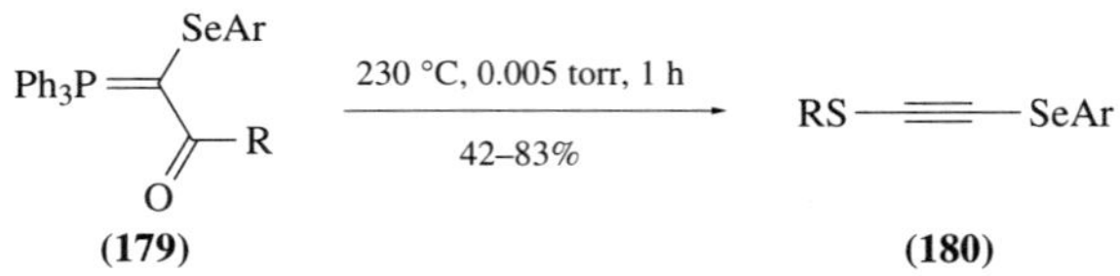

(179) **(180)** (116)

R = Me, CF_3, Ph, Tol, 4-ClC_6H_4, 4-$MEOC_6H_4$, 2-thienyl

Ar = Ph, 3-$CF_3C_6H_4$, (2-NO_2)(4-Cl)C_6H_3

Di(methylseleno)ethyne (**182**) can be obtained by the thermal decarbonylation of 3,4-di(methylseleno)cyclobutene-1,2-dione (**181**) (Equation (117)) ⟨81CB673⟩.

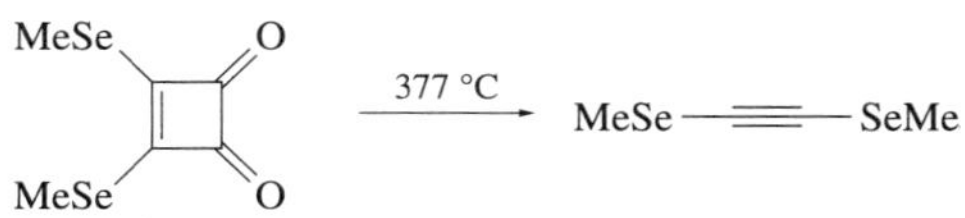

(181) **(182)** (117)

Reaction of substituted selenodiazoles (**183**) with butyllithium results in the loss of nitrogen and the formation of alkynyl selenolates (**184**), which can be alkylated *in situ* to afford (alkynyl)alkyl selenides (Equation (118)) ⟨87RCR267⟩.

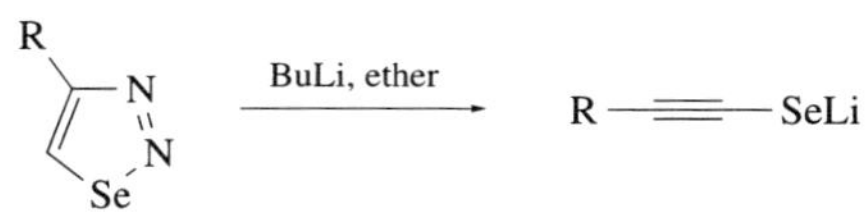

(183) **(184)** (118)

R = alkyl, aryl

In contrast to the alkynyl compounds of sulfur, stable alkynyl selenoxides and telluroxides are unknown. Oxidation of alkynyl selenides with peracids results in the elimination of the selenium moiety and the formation of diynes ⟨86CC1067, 82CL1244⟩ or terminal alkynes ⟨82CL1244⟩. However, alkynyl selenoxides are considered as the intermediates in these oxidations ⟨86SC283⟩. The only known stable alkynyl derivatives of polyvalent tellurium are dichlorotelluroalkynes (**186**), which can be prepared in high yield by the reaction of alkynyl tellurides (**185**) with an excess of sulfuryl chloride (Equation (119)) ⟨90SC2181⟩.

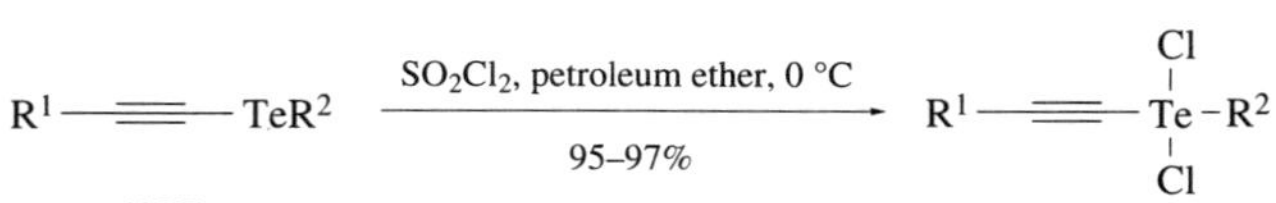

(185) **(186)** (119)

R^1 = Ph, n-C_5H_{11}; R^2 = Bu^n

These products (**186**) can serve as useful intermediates in the synthesis of trihalo-substituted alkenes ⟨90SC2181⟩.

2.22
Alkynylnitrogen and -phosphorus Compounds

KEVIN I. BOOKER-MILBURN
University of East Anglia, Norwich, UK

2.22.1 ALKYNYLNITROGEN COMPOUNDS

2.22.1.1 Amines and their Salts

The synthesis and chemistry of *N*,*N*-disubstituted-1-amino-1-alkynes (**1**) or ynamines was extensively reviewed in 1969 ⟨B-69MI 222-01⟩, 1976 ⟨76T1448⟩, and 1986 ⟨86TCC89⟩. Unless otherwise stated they will be referred to as ynamines.

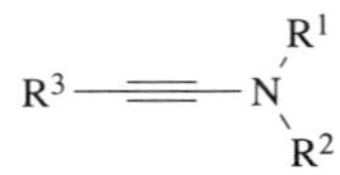

(1)

R^1, R^2 and R^3 = H, alkyl, vinyl or aryl

2.22.1.1.1 *Primary and secondary ynamines*

Historically, because of their facile tautomerization to ketenimines (Equation (1)), 1° and 2° ynamines were postulated as unisolable, fleeting intermediates. Early synthetic routes, which would have produced 1° and 2° ynamines, invariably led to amides derived from the corresponding tautomeric ketenimines ⟨28RTC960⟩. These species have, however, been generated and studied both at low temperature in a matrix and in solution.

(1)

For example, Wentrup and Winter found that flash vacuum pyrolysis of a variety of substituted 4-(anilinomethylene)-3-methylisoxazol-5(4*H*)-ones (**2**) allowed the generation of the 2° ynamines (**3**), which could be trapped at −196 °C and observed by IR spectroscopy ⟨80AG(E)720⟩. On warming to ca. −70 °C to −60 °C, the ynamines were observed to undergo tautomerization to the corresponding ketenimines (**4**). This would imply that it may be possible to study the chemical reactions of these ynamines at intermediate temperatures, e.g., −100 °C to −90 °C (Scheme 1).

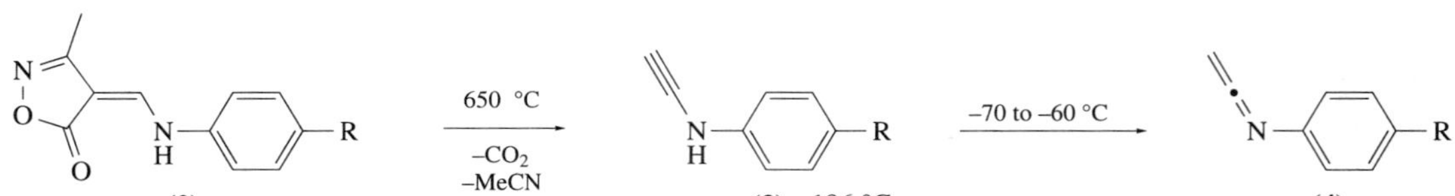

(2) R = H, Me, OMe, OEt, NO_2 (3) −196 °C (4)

Scheme 1

More recently, Kresge *et al.* found that flash photolysis of phenylaminocyclopropenones (**5**) in aqueous solution gives carboxylic acid amides (**7**) as the ultimate reaction products via short-lived, transient intermediates, which, from their behavior, were concluded to be the corresponding 1° and 2° ynamines (**6**) (Scheme 2) ⟨91AG(E)1356⟩. Although this method still does not enable the isolation of 1° and 2° ynamines, it may allow their generation in neutral solution, via conventional photolysis at low temperatures (i.e., −100 °C), and their subsequent chemical reactions with *in situ* reagents to be studied.

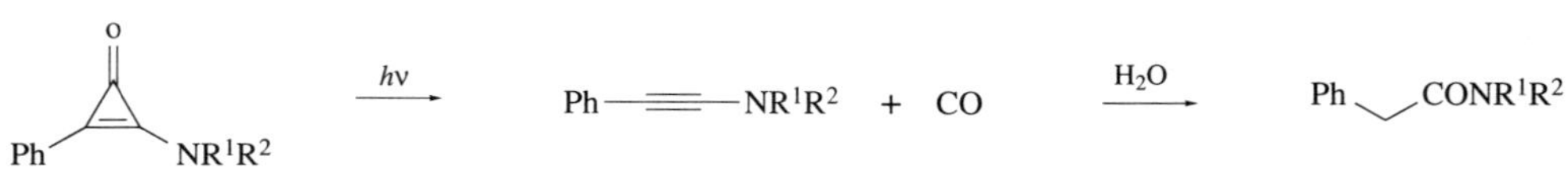

(5) (6) (7)

$R^1 = R^2 = H$
$R^1 = H, R^2 = C_6F_5$
$R^1 = H, R^2 = C_6H_{11}$
$R^1 = H, R^2 = CHMe_2$

Scheme 2

2.22.1.1.2 *Tertiary ynamines*

Unlike the corresponding 1° and 2° counterparts, 3° ynamines cannot undergo tautomerization to ketenimines, and although they are isolable, thermally stable compounds, they are readily hydrolyzed to amides. They are by far the most studied of the alkynylnitrogen compounds and, as such, a large portion of this chapter will be devoted to them. Their synthesis can be divided into three main preparative methods: (i) by nucleophilic substitution reactions, (ii) by elimination reactions, and (iii) by isomerization reactions.

(i) Substitution reactions

(a) From acetylide anions and nitrogen electrophiles. One of the earliest approaches to ynamines involved the attractive prospect of nucleophilic substitution of a suitable nitrogen electrophile with an alkynic anion. Initially, however, this apparently simple approach proved to be highly unsuitable as a general method for ynamine synthesis. For example, in 1960 it was reported that treatment of diethylchloramine with phenylethyne magnesium bromide gave the corresponding ynamine in a meagre 1.7% yield (Equation (2)) ⟨60LA(639)33⟩.

$$Ph{-}{\equiv}{-}MgBr \xrightarrow[1.7\%]{ClNEt_2,\ Et_2O} Ph{-}{\equiv}{-}NEt_2 + MgClBr \qquad (2)$$

By the use of more sophisticated acetylides and aminating reagents, these earlier problems have largely been overcome. For example, it was found that the reaction of electrophilic reagents prepared from dimethylhydroxylamine with a variety of alkynic cuprates (**8**) gave rise to the corresponding ynamines in moderate to good yields (Scheme 3) ⟨83AG(E)53⟩. The authors reported that success was highly dependent on the method used to prepare the alkynic cuprate and optimized yields were obtained using three different methods. Although the generality of this method needs to be tested with substituted aminating reagents other than dimethyl, it does offer the most direct approach to alkynyl-substituted *N*,*N*-dimethylynamines. Table 1 summarizes the optimized results obtained from the three cuprate methods.

$$\left(R{-}{\equiv}\right)_3 CuLi_2 \xrightarrow[Et_2O,\ -78\ ^\circ C]{MeSO_3NMe_2 \text{ or } Ph_2PO_2NMe_2} R{-}{\equiv}{-}NMe_2$$

(8)

$$2\ R{-}{\equiv}{-}Li + R{-}{\equiv}{-}Cu \xrightarrow{\text{method A}} (\mathbf{8})$$

$$3\ R{-}{\equiv}{-}Li + CuCN \xrightarrow{\text{method B}} (\mathbf{8})$$

$$3\ R{-}{\equiv}{-}Li + CuBr/Me_2S \xrightarrow{\text{method C}} (\mathbf{8})$$

Scheme 3

Table 1 Ynamines by electrophilic amination of alkynyl cuprates (Scheme 3).

Entry	*R*	*Method*	*X—NMe₂*	*Yield (%)*
1	Ph	A	$MeSO_3NMe_2$	52
2	Pr^n	C	$Ph_2PO_2NMe_2$	87 (NMR)
3	Bu^n	B	$Ph_2PO_2NMe_2$	78
4	Hex^n	B	$Ph_2PO_2NMe_2$	82
5	Cyclohexyl	B	$Ph_2PO_2NMe_2$	44
6	Bu^t	C	$Ph_2PO_2NMe_2$	60
7	Me_3Si	C	$Ph_2PO_2NMe_2$	67
8	PhS	C	$Ph_2PO_2NMe_2$	45

(b) From alkynyl halides and metal amides. This has proved to be a fairly general method for the synthesis of 3° ynamines and involves the nucleophilic displacement of halogen from an alkynyl halide (see Chapter 2.21) using metal amides. The most commonly used system involves the reaction of alkynyl chlorides with alkali metal amides (usually lithium amides). The general reaction is shown in Equation (3). In general, the displacement of alkynyl chlorides with lithium amides would appear to give the best results and has been the most extensively used. The reactions are usually carried out at low temperature (−80 °C) in ether, although with sodium amides higher temperatures are usually required (Scheme 4) ⟨64AG(E)506⟩.

(3)

$R^1 = R^2$ = alkyl, aryl; R^3 = H, alkyl, aryl; M = Na or Li; X = Cl, Br, F

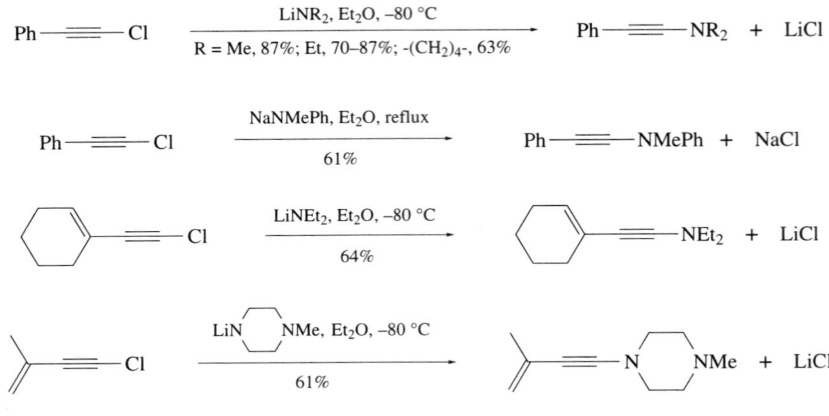

Scheme 4

The *in situ* synthesis of the more reactive 1-fluoroalkynes from 1,1-difluoroalkenes allowed a more practical approach to ynamines to be developed ⟨71JOC1438⟩. The attractive feature of this route is that the difluoroalkenes are readily available from simple aromatic or aliphatic aldehydes and overall yields are commonly 30–40% (Scheme 5) ⟨65JOC1027⟩.

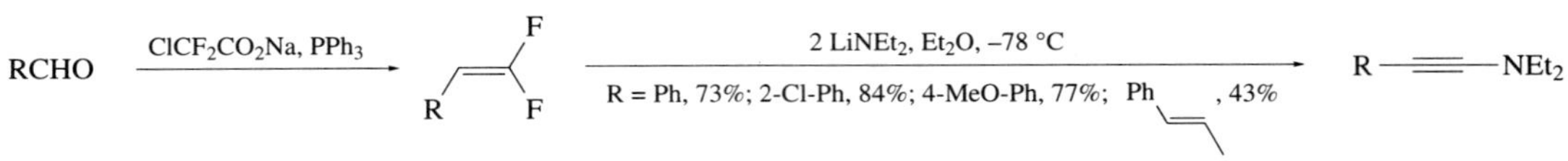

Scheme 5

The difluoroalkene route also allows easy access to alkynyl sulfur ⟨77BCJ3069⟩ and selenium ⟨82S1083⟩ substituted ynamines in high yields. The starting trifluoroalkanes are readily available from trifluoroethanol. In this route, three equivalents of lithium amide are used; the first equivalent eliminates fluoride to give the difluoroalkene; this then undergoes further elimination and substitution as above (Scheme 6).

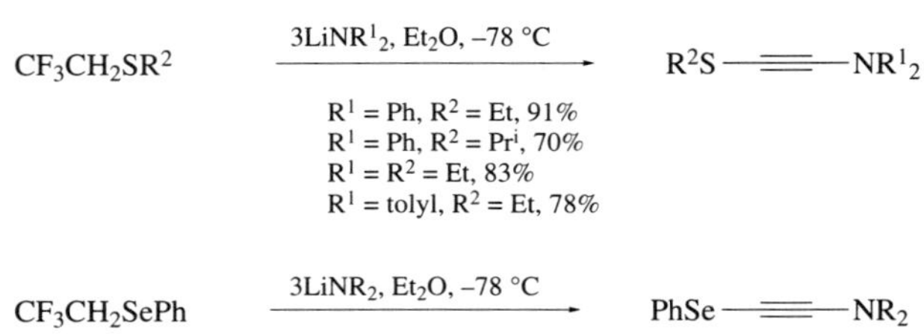

R = Me, 95%; Pr^i, 75%; Et, 85%; $-(CH_2)_5-$, 78%

Scheme 6

(c) From alkynyl halides and tertiary amines. This method involves the treatment of alkynyl halides with tertiary amines and proceeds via the ammonium salts (**9**). Since these salts can often be isolated in quantitative yield, this provides a convenient method for the synthesis of the quaternary ammonium salts of ynamines. Further reaction of (**9**) with excess base leads to the respective ynamines in reasonable overall yield (Scheme 7). Successful conversion of the initial ammonium salt to the ynamine is dependent on the use of an excess of the tertiary amine; this helps to shift the equilibrium via removal of RX by transalkylation to R_4NX. Scheme 8 summarizes the more general examples of this reaction ⟨65CR(260)1984, 66ZOR1994, 70CB564⟩.

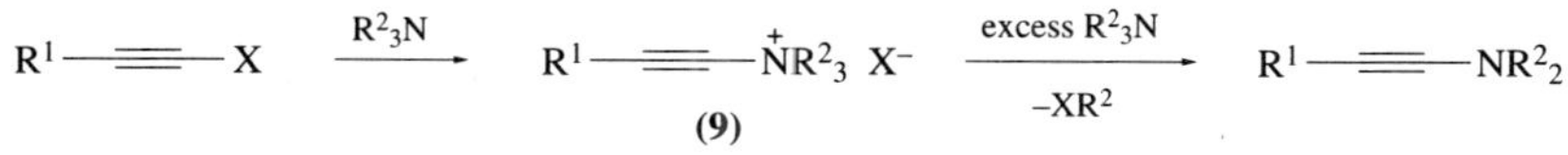

Scheme 7

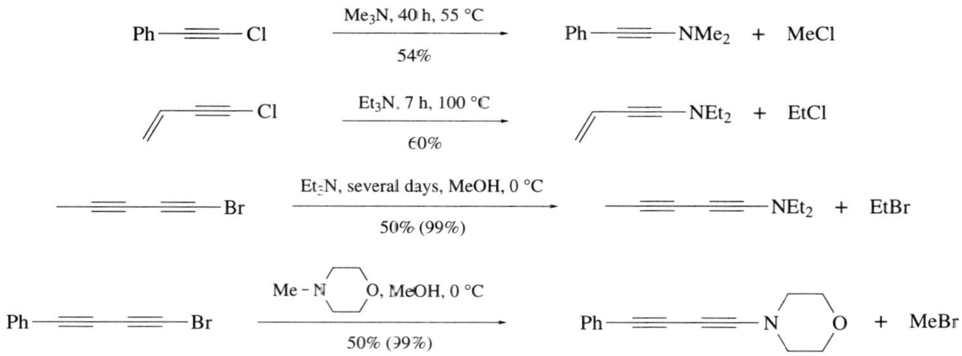

Yields in brackets refer to isolated ammonium salts

Scheme 8

(d) From alkynyl ethers and metal amides. The displacement of the ether group in alkynyl ethers (or ynethers: see Chapter 2.21) with lithium amides can be achieved in generally good yields, as demonstrated by Brandsma *et al.* ⟨64RTC1211⟩. The desired conversion is usually achieved by heating a mixture of the alkynylether and lithium amide to 120°C and distilling off the product from the reaction mixture (Scheme 9).

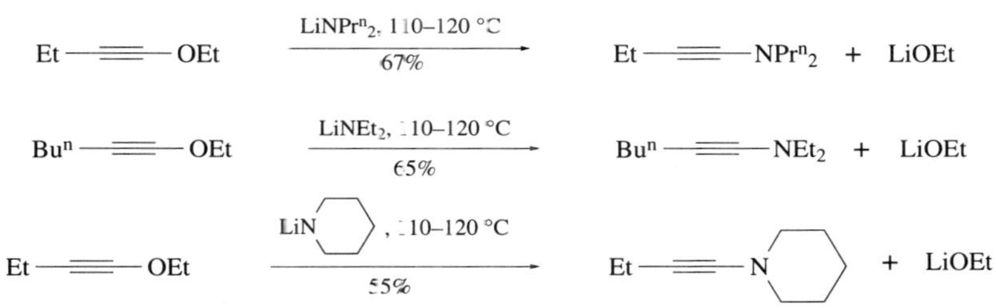

Scheme 9

The above nucleophilic substitutions are thought to proceed via two generally accepted mechanisms: (i) direct attack at the *sp* carbon bearing the halogen atom (C-1) followed by β-elimination to give the ynamine, or (ii) attack at C-2 followed by α-elimination and onium rearrangement via a carbene to the ynamine. Pathway (i) is preferred when X = F or ether since, especially with fluorine, these groups would direct substitution *ipso*. Pathway (ii) is considered more likely when X is an electron-stabilizing group such as chlorine (Scheme 10).

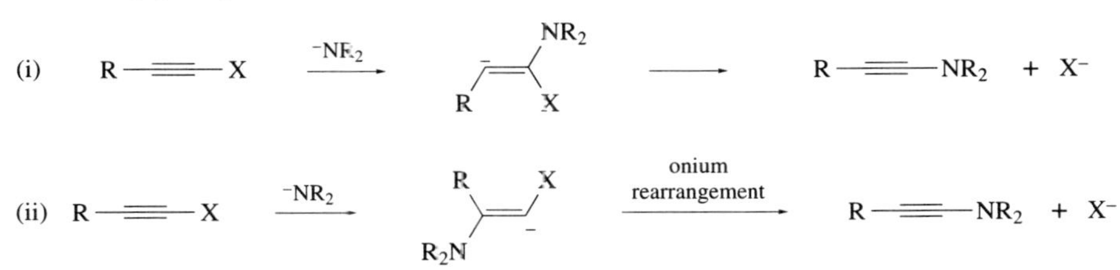

Scheme 10

(e) By functionalization of ynamines. The rich and varied chemistry of organosilanes and -stannanes has been used to good effect for the synthesis of a variety of alkynyl substituted ynamines.

The basic principle involves the electrophilic substitution of ynamines substituted at the alkynyl position with a trialkylstannyl or trialkylsilyl group. A huge contribution to this field has been made by Himbert, who has found that these stannylated and silylated ynamines are highly reactive species that selectively react with a large number of electrophiles to give a range of diversely functionalized ynamines that would in general be inaccessible by other methods. For example, both species react readily with acid chlorides, giving rise to a useful push–pull alkyne synthesis ⟨81LA1907⟩. In general, stannanes tend to give better results, with the silanes requiring more reactive acylating agents such as trichloroacetyl chloride (Scheme 11).

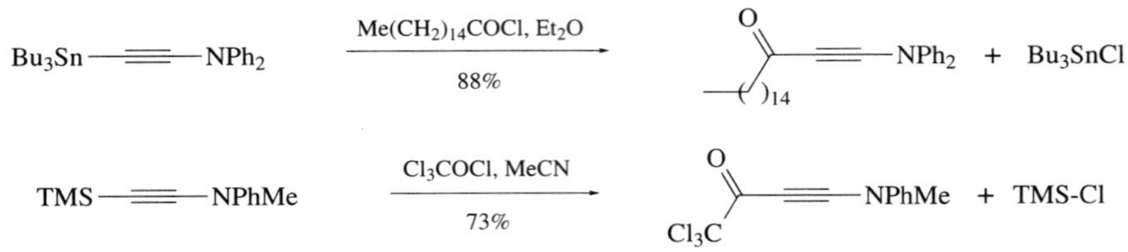

Scheme 11

Reaction of the silane (**10**) with diphenylketene gives rise to the silyl enol ether derivative (**11**) in high yield ⟨79LA1828⟩. Use of simpler ketenes is complicated by competing 2 + 2 cycloadditions. Similarly, reaction of the stannylated ynamine (**12**) with 4-chlorophenyl isocyanate yields the imine (**13**), which upon mild hydrolysis yields the push–pull alkynic amide (**14**) in excellent overall yield (Scheme 12). This reaction is of wide scope and gives a variety of ynamine amides in generally good yields ⟨81LA1844⟩.

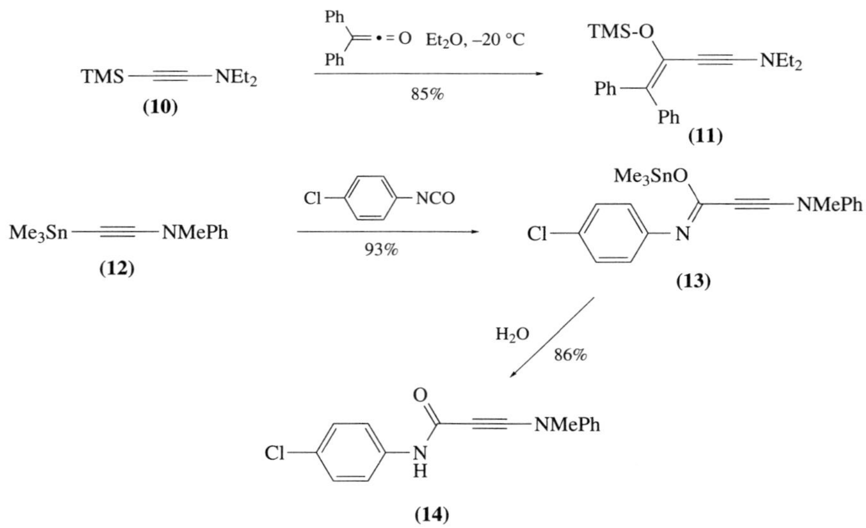

Scheme 12

Isothiocyanates ⟨78TL1951⟩ and carbodiimides ⟨83LA1185⟩ also react with stannylated ynamines to give the corresponding ynamine imines in moderate to good yields (Scheme 13). In both cases, the reactions are general and a range of ynamines, isothiocyanates and carbodiimides were investigated.

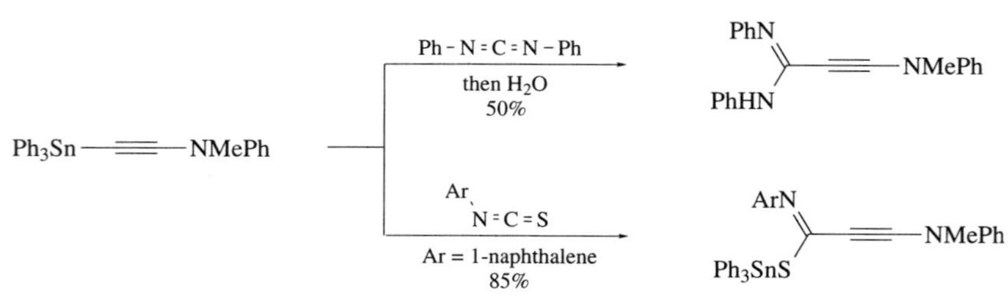

Scheme 13

A further interesting reaction in this class is the reaction of silylated ynamines with dimethylethyne dicarboxylate, which gives rise to the highly functionalized ynamine silyl maleates (**15**) in good yields ⟨78JOC199⟩. Propiolates also react in a similar manner (Equation (4)).

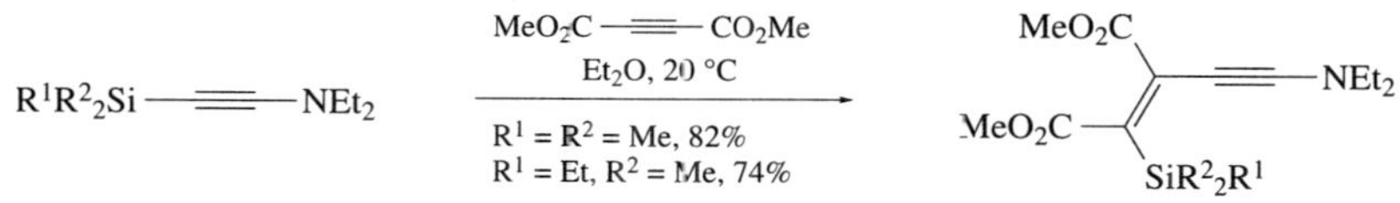

(4)

(15)

The propargyl protons of alkynyl substituted ynamines are acidic enough to be deprotonated with a strong base; this leads to a resonance-stabilized anion (**16**) that can react with electrophiles at either the α- or γ-positions. Lithiation of the diethylynamine (**17**) and alkylation with chlorotrimethylsilane gave the silylated ynamine (**18**) in good yield via γ-alkylation ⟨70JOC3405⟩. However, attempts to react the lithiated ynamine with other electrophiles, such as methyl and allyl bromide, resulted in failure. From IR analysis of the crude reaction mixture, it was thought that α-alkylation had occurred, which resulted in the formation of unstable allenic amines. Use of the more hindered diisopropylynamine suppressed α-alkylation and allowed access to the methyl and allyl derivatives in moderate yield (Scheme 14). Use of even bulkier substituents on the nitrogen allows for preferential γ-alkylation and hence a useful ynamine extension sequence. For example, the lithiated tetramethylpiperidino substituted ynamine (**19**) underwent γ-alkylation with a variety of electrophiles in moderate yield (Equation (5)) ⟨70JOC3405⟩.

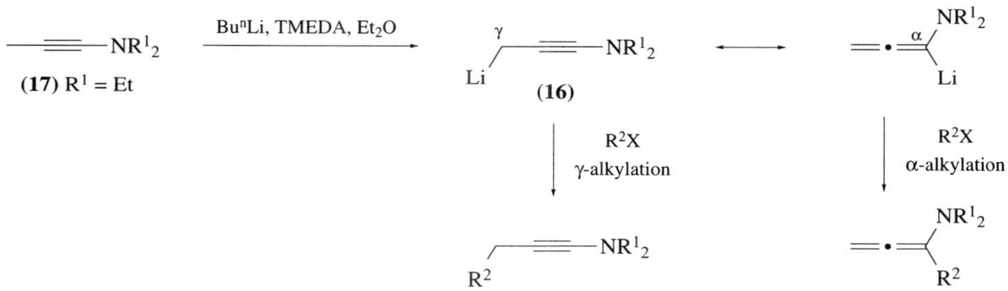

(18) R^1 = Et, R^2X = TMS-Cl, 86%

Scheme 14

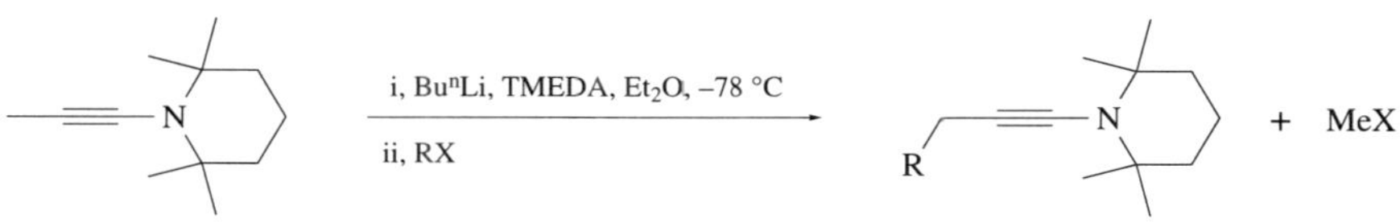

(5)

(19)

RX	*Yield* (%)
TMS-Cl	60
allyl iodide	45
MeI	35
Bu^nI	50

(ii) Elimination reactions

This section will deal with the synthesis of ynamines where the nitrogen atom is already present in the starting material and the triple bond is introduced via an elimination reaction. Many useful ynamine syntheses have been developed over the years using elimination strategies, and most methods rely upon the use of halogenated enamines, which are readily available from substituted amides.

(a) From α-chloroenamines. Treatment of amides with a variety of inorganic acid chlorides gives, initially, the corresponding amide chloride (**20**), which on treatment with base yields the α-chloroenamines (**21**). In an early ynamine synthesis, Viehe found that direct treatment of amide

chlorides with lithium amides produced the ynamines (**22**) in good yield. The reaction is thought to proceed via the α-chloroenamine (**21**) (Scheme 15) ⟨66AG(E)584⟩.

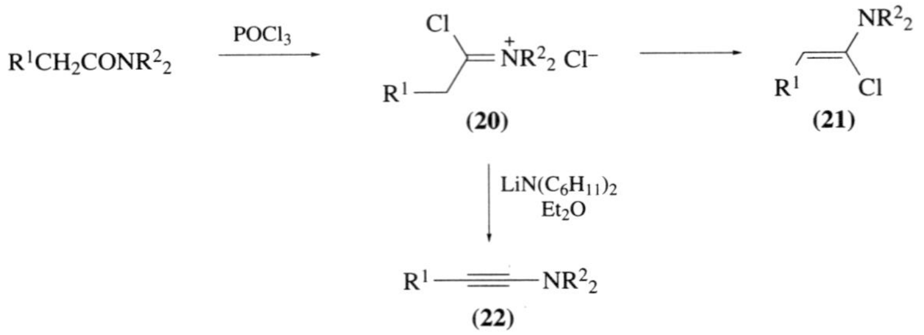

R^1 = Me, R^2 = Et, 58%
$R^1 = R^2$ = Et, 78%

Scheme 15

One of the advantages of the α-chloroenamine route is that it is not just limited to the synthesis of *N,N*-dialkyl ynamines and, indeed, it allowed the synthesis of the previously unknown ynamides as well as the carbamate and urea derivatives ⟨77JCR(S)105⟩. For example, treatment of a variety of secondary amides with phosgene led to the α-chloroenamines (**23**) containing a carbamoyl chloride moiety. Nucleophilic substitution allowed the synthesis of three different substituted α-chloroenamines which upon elimination with potassium *t*-butoxide yielded the corresponding *N*-substituted ynamines in moderate to good yield (Scheme 16). It is interesting to note that the ynamides (entry 1) are much more stable to hydrolysis and can survive acidic conditions that would normally hydrolyze *N,N*-dialkyl ynamines to amides.

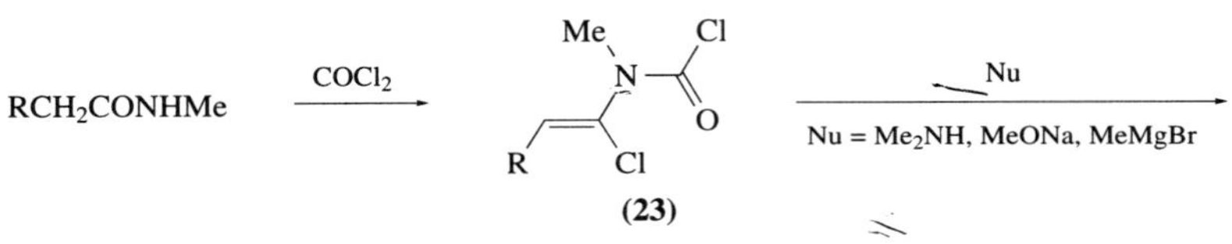

Entry	*R*	*Nu*	*Yield* (%)
1	Bu^t	Me	52
2	Bu^t	NMe$_2$	40
3	Ph	NMe$_2$	70
4	Ph	OMe	90

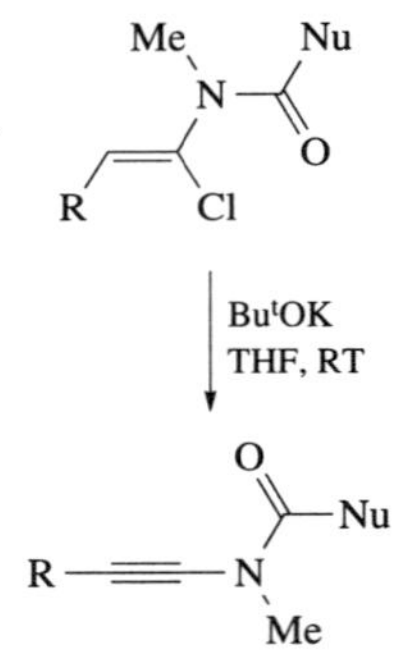

Scheme 16

The related cyanine salts (**24**) have also proved useful for the synthesis of ynamine amidines (**25**) via an amination/elimination sequence ⟨76C387⟩. For example, treatment of (**24**) with a variety of primary amines gave the corresponding amidines in reasonable overall yield. Similarly, treatment of the cyanine iodide (**26**) with triethylamine gave the remarkably stable ynamine iminium salt (**27**) in high yield (Scheme 17) ⟨81JA6142⟩.

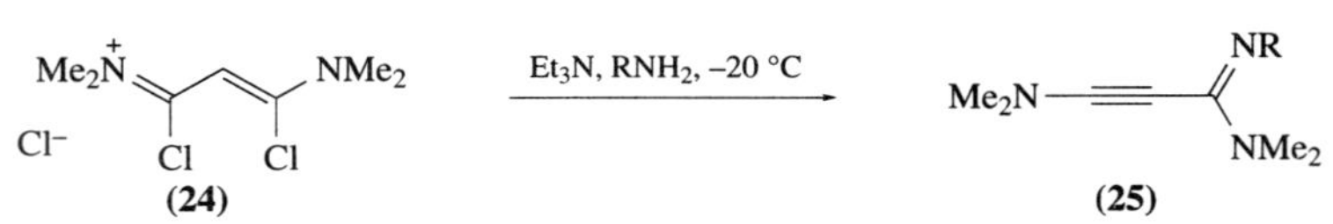

R = H, 53%; Me, 50%; Pr^i, 63%; Bu^t, 49%; C_6H_{11}, 54%

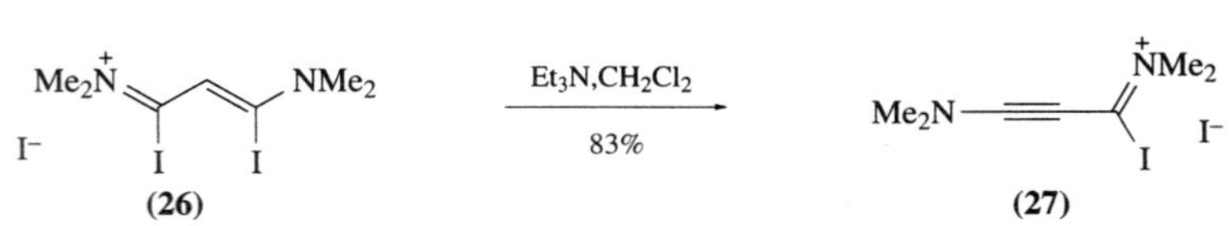

Scheme 17

(b) From α,β-di- and trichloroenamines. It had been known for some time that deoxygenation of trichloroacetamides with either phosphines or phosphites leads to the corresponding α,β-trichloroenamines (**28**) in good yields ⟨62JA1868⟩. Following pioneering early studies by Ficini *et al.* ⟨64BSF871, 65BSF2787⟩, it was found that treatment of these enamines with *n*-butyllithium gave, after a metallation/elimination sequence, the lithiated ynamines (**29**), which upon protonation with water yielded the corresponding ynamines (**30**) in good overall yield ⟨72CB2963⟩. The overall process represents an operationally simple method for the synthesis of ynamines from readily available trichloroacetamides (Scheme 18).

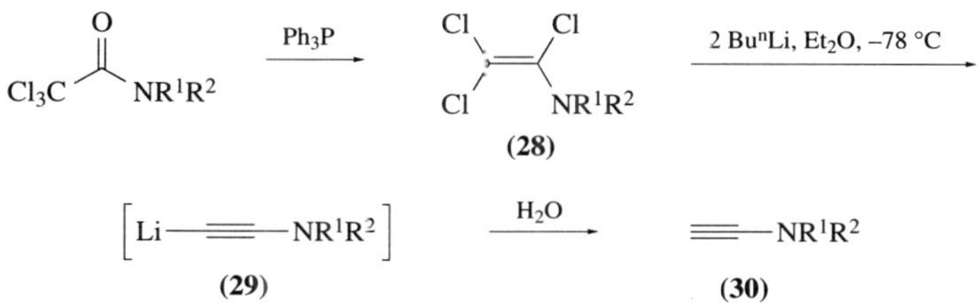

		Yield (%)	
R¹	*R²*	*(29)*	*(30)*
Et	Ph	85	51
Pr^i	Ph	72	77
Bz	Ph	77	43

Scheme 18

The very electron-rich ynediamines (**33**) have been known for some time, but their synthesis has involved the use of hazardous and complicated procedures involving the toxic and explosive dichloroethyne ⟨70CB1209⟩ or fluorodichloroethylene, which has to be synthesized by a multistep procedure ⟨64AG(E)506, 70CB1198⟩. Modification of the above methodology has recently led to a delightfully simple, one-pot synthesis of these relatively inaccessible ynediamines from a variety of substituted trichloroenamines ⟨91SL117⟩. Thus, treatment of a variety of trichloroenamines (**28**) with a mixture of one equivalent each of *n*-butyllithium and a range of lithium diorganoamides gives initially the eliminated chloroynamine (**31**), which reacts with the lithium amide to give the lithiated ketene-*N*,*N*-acetal (**32**). This then undergoes an onium rearrangement to furnish the ynediamines (**33**) in good overall yields (Scheme 19).

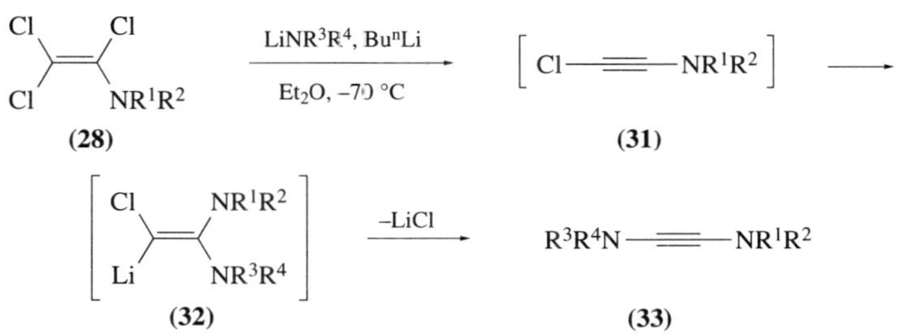

R¹	*R²*	*R³*	*R⁴*	*Yield (%)*
Me	Ph	$-(CH_2)_2-O-(CH_2)_2-$	$-(CH_2)_2-O-(CH_2)_2-$	37
Et	Et	Et	Ph	59
Me	Ph	Et	Ph	55
Et	Et	Et	Et	51

Scheme 19

α,β-Dichloroenamines of the general structure (**34**) also readily undergo elimination to ynamines on treatment with *n*-butyllithium. The α,β-dichloroenamines themselves are easily synthesized by deoxygenation of the corresponding dichloroacetamides with trialkylphosphines (Scheme 20).

As mentioned above, the elimination of α,β-di- and trichloroenamines with *n*-butyllithium initially leads to the lithiated ynamines (**35**), which can be protonated to give the corresponding unsubstituted ynamines. If, however, an alkylating agent is added prior to work up, then the corresponding

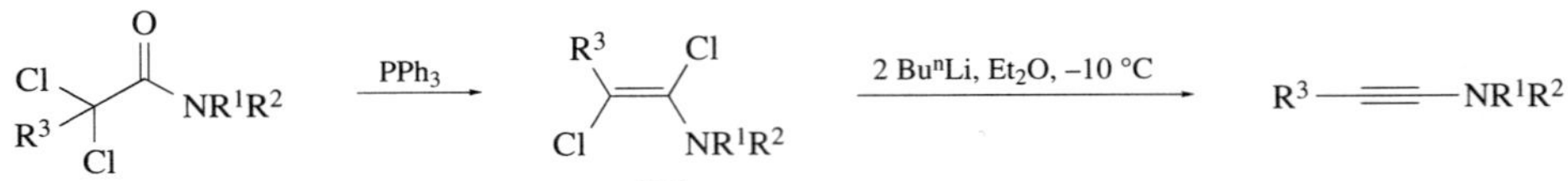

(34)

$R^1 = R^2 = Et, R^3 = Ph$, 70%
$R^1 = R^2 = Ph, R^3 = H$, 65%
$R^1 = Me, R^2 = R^3 = Ph$, 73%

Scheme 20

alkylated ynamines (**36**) can be obtained in good yield (Scheme 21). When the alkylating agent is an alkyl halide or tosylate then it is usually necessary to add a cosolvent such as hexamethylphosphoramide (HMPA) ⟨65BSF2187, 67CR(C)1496⟩. The use of ketones as alkylating agents allows access to the corresponding alkynols ⟨68JOC4406⟩. For the *N*,*N*-diethylynamine, it was found that the best results were obtained using alkyl tosylates as alkylating agents (Scheme 22).

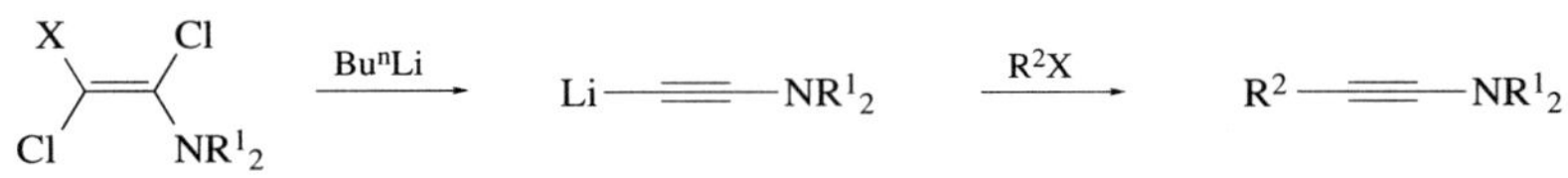

X = H or Cl **(35)** **(36)**

Scheme 21

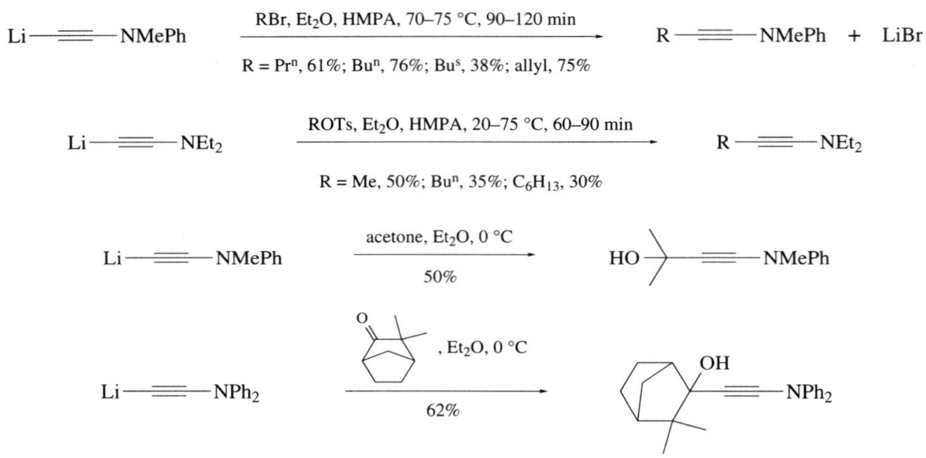

Scheme 22

This is probably the method of choice for the synthesis of alkynyl-substituted ynamines because of the huge variety of derivatives that can be prepared from the simple and readily available lithiated ynamines (**35**). It is also the most valuable method for the synthesis of alkynyl-substituted ynamines containing silicon ⟨71CR(C)289⟩, tin ⟨81LA1907⟩, germanium ⟨74CB2513⟩, phosphorus ⟨74CB2513⟩ and boron ⟨78JOM(145)17⟩. Equation (6) illustrates the versatility of this method for the preparation of these heterosubstituted ynamines.

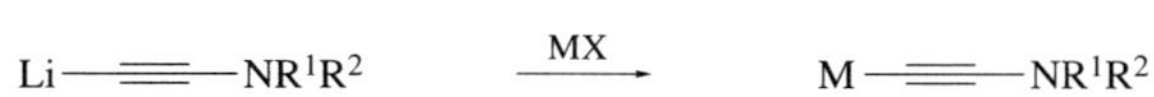

(6)

$R^1 = Ph, R^2 = Me$, MX = TMS-Cl, 85%
$R^1 = R^2 = Et$, MX = Bu_3SnCl, 70%
$R^1 = R^2 = Ph$, MX = Ph_3GeBr, 80%
$R^1 = Ph, R^2 = Bz$, MX = Ph_2PCl, 57%
$R^1 = R^2 = Et$, MX = 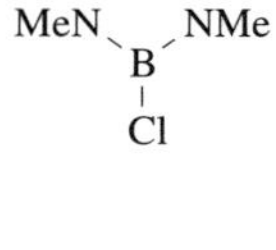, 85%

(c) From β-chloroenamines. Unlike α,β-di- and trihaloenamines, the corresponding β-halo-enamines do not undergo elimination easily on treatment with alkyllithiums. The usual products that are observed are a result of the formation of the β-lithioenamine obtained via halogen–metal exchange. β-Bromoenamines with an electron-withdrawing group on the β-carbon, however, readily undergo elimination to the corresponding ynamines in good yield ⟨69HCA2641, 78HCA1609⟩. This method provides convenient access to a number of ynamines bearing amide, carbamate and urea substituents at the alkynic carbon (Equation (7)). The elimination is usually carried out at room temperature in THF using potassium *t*-butoxide as base, and yields are commonly in the 70% range. The interesting electronic nature of this class of ynamines has led them to be termed "push–pull alkynes" ⟨84TL273⟩. Table 2 summarizes some of the more general examples of this ynamine route.

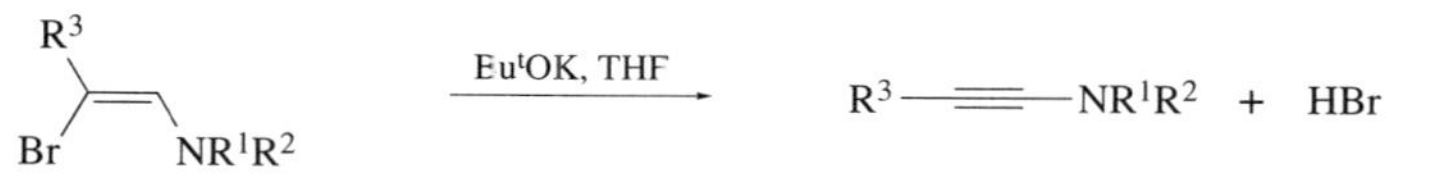

(7)

Table 2 Ynamine synthesis from β-chloroenamines (Equation (7)).

Entry	R^3	R^1	R^2	*Yield* (%)
1	Me	Et	Et	71
2	Ph	Et	Et	72
3	4-ClPh	Me	Me	62
4	Me	Me	Me	75
5	H	Me	Me	70
6	OMe	Me	Me	72
7	NMe_2	Me	Me	55

(d) From thioamides and ketene N,S-*acetals.* Thioamides (**37**) and the corresponding ketene *N,S*-acetals (**38**) are easily synthesized from amides in high yield by treatment with phosphorus pentasulfide alone or with alkylation, respectively. Both undergo elimination to ynamines in moderate to good yields on heating with an excess of sodium or potassium amide ⟨66AG(E)584, 70TL3141⟩. Although this would appear to be a versatile and economic ynamine synthesis, it has not received the same level of attention as other methods. Scheme 23 summarizes some of the more general examples of this type of elimination. In terms of preparative merits, this method allowed the synthesis of 15–16 g batches of the simplest *N,N*-dimethylynamine in 55% yield ⟨71RTC265⟩.

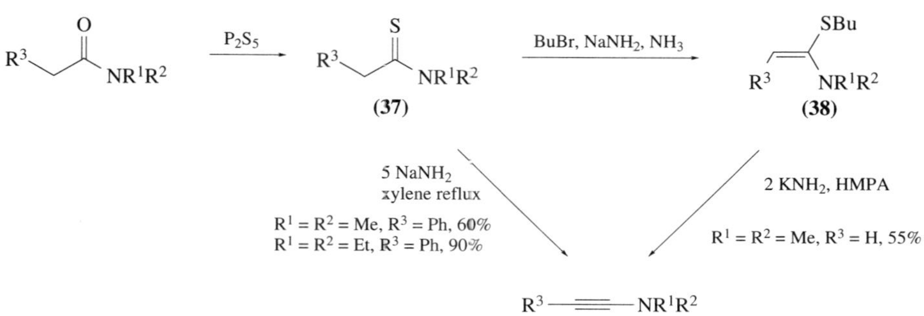

Scheme 23

(iii) Isomerization reactions

It is interesting to note that the first successful ynamine synthesis was a result of serendipity rather than rational design ⟨58JOC1389⟩. An attempt to react propargyl bromide with phenothiazine to produce the propargyl amine (**39**) using sodium hydride as base led to the isolation of the ynamine (**40**) in 70% yield. Reinvestigation of this reaction in DMSO using potassium hydroxide as base led to the isolation of (**40**) in almost quantitative yield ⟨65CR(260)1984⟩. The reaction proceeds by the initial formation of the propargyl amine followed by base-catalyzed isomerization to the ynamine (Scheme 24).

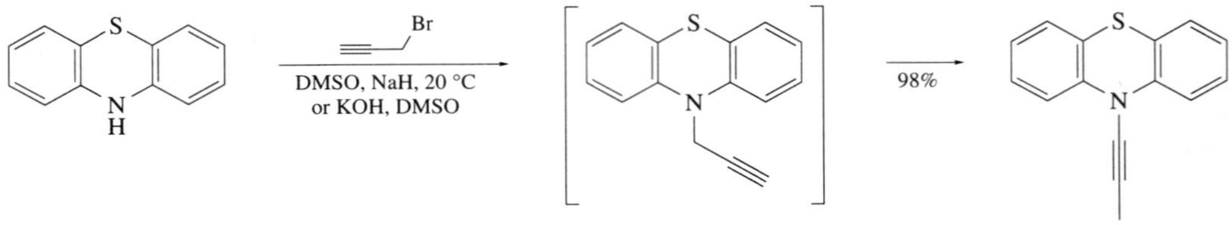

(**39**) (**40**)

Scheme 24

This would appear to be an ideal and economic method for the synthesis of ynamines from amines and a number of systems have been investigated. However, with many of the examples investigated, the reaction is not quite as simple as it would first appear and often equilibrium mixtures of ynamines and aminoallenes are formed, which seriously limits this method as a general ynamine synthesis. These problems are discussed in detail in the excellent ynamine review by Motte and Janousek ⟨86TCC89⟩. Nevertheless, many ynamines have been prepared by this method, often in good yield, and this is probably the method of choice for the preparation of the simple *N*,*N*-diphenyl- and *N*,*N*-diethylynamines. In fact, the latter are prepared commercially via this route. Equation (8) summarizes some of the more general examples of this method.

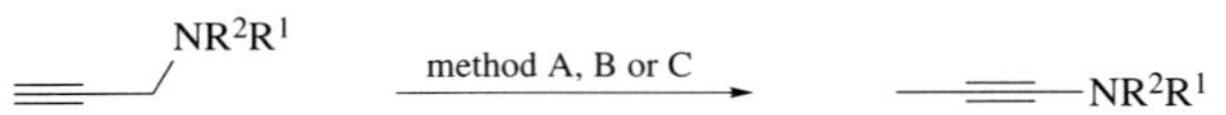

(8)

$R^1 = R^2 = Ph$, A, 50%
$R^1 = R^2 = Et$, B, 65–75%
$R^1 = Ph$, $R^2 = Me$, C, 60%
$R^1 = R^2 = -(CH_2)_5-$, B, 38%

method A: DMSO, KOH, 20 °C; method B: hexane, Al_2O_3, KNH_2 cat., 65 °C;
method C: molecular distillation at 200 °C, KOH

(iv) Miscellaneous methods

There have been a number of "non-classical" methods of ynamine synthesis reported in the literature and although some of them are of purely mechanistic interest, a few of them have proved useful for the synthesis of ynamines that are difficult to access by other methods. For example, Breslow reported that pyrolysis of the diaminocyclopropenone (**41**) yielded the ynediamines (**42**) in good yield by the thermal extrusion of carbon monoxide ⟨80TL3241⟩. Similarly, pyrolysis of the fluorinated triazene (**43**) gave the bis(trifluoromethyl)ynediamine (**44**), which would probably be inaccessible from existing routes (Scheme 25).

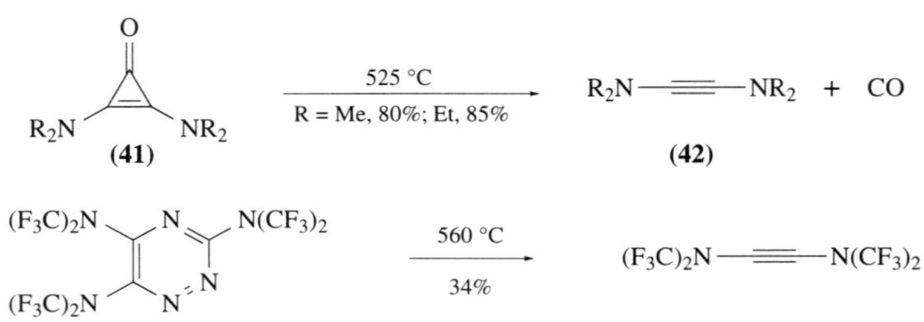

(**41**) (**42**) (**43**) (**44**)

Scheme 25

A very interesting oxidative dimerization was reported by Ficini in which treatment of the ynamine (**45**) under an atmosphere of oxygen in the presence of cuprous chloride gave the bis(dialkylamino)butadiynes (**46**) in relatively good yield ⟨74BSF1535⟩. In a related reaction, Peterson found that a 3:2 mixture of *N*,*N*-dimethyl-2-phenylethynylamine and 1,4-diphenylbutadiyne could be synthesized by the oxidative coupling of phenylethyne with dimethylamine using cupric acetate ⟨68TL5357⟩. The isolated yields of the ynamine were reported to be greater than 40%; however, substantial amounts of phenylacetamide were obtained because of unavoidable *in situ* hydrolysis. The reaction also proceeds with piperidine or diethylamine, although no yields were given (Scheme 26).

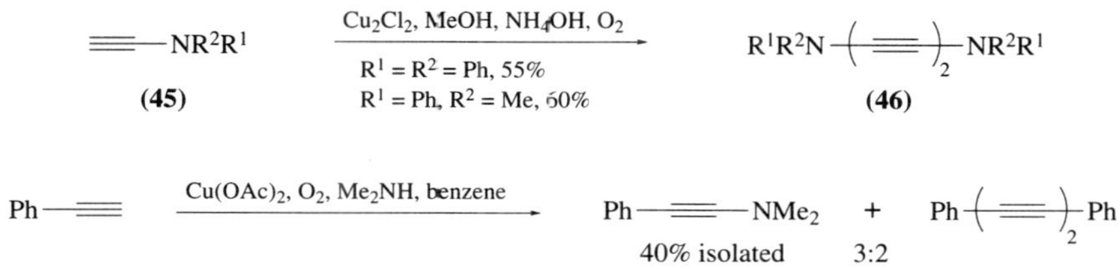

Scheme 26

A very useful method for the preparation of triphenylphosphonium ynamine salts involves the treatment of stabilized phosphonium ylides (**47**) with triphenylphosphonium dibromide followed by treatment of the resulting adduct (**48**) with triethylamine, which gives the ynamine salts (**49**) in excellent yields (Scheme 27) ⟨83CB1320⟩.

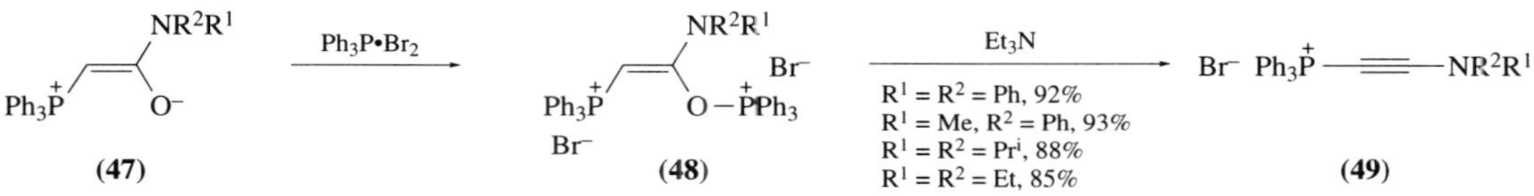

Scheme 27

2.22.1.1.3 *Summary*

While it is desirable in organic synthesis to have one or two general and reliable methods for the synthesis of a particular functional group, this is not the case with ynamines and, therefore, it is difficult to recommend a single, standard synthetic procedure. Each method has its own merits in terms of accessibility of starting materials, ease of operation and functionality introduced, and it is these factors that have to be considered carefully when choosing a particular method.

2.22.1.2 N=Y Functions

2.22.1.2.1 *Nitrosoalkynes*

Nitrosoalkynes were first reported in 1968 and were found to be extremely unstable compounds that rearranged to acyl cyanides at room temperature ⟨68JCS(C)1324⟩. They can, however, be prepared and studied (albeit in solution) by the reaction of mercury bis(acetylides) with nitrosyl chloride at −40 °C in chloroform. This gives rise to an intense green solution of the nitrosoalkyne, which is characterized by the strong alkyne (2290 cm^{-1}) and NO (1580 cm^{-1}) stretching frequencies in the infrared. On warming to room temperature, these are replaced by the weak nitrile (2210 cm^{-1}) and a strong carbonyl stretch (1710 cm^{-1}) as the acyl nitrile is formed. An alternative procedure involves the reaction of *t*-butyltrimethylstannylethyne with dinitrogen tetroxide at low temperature ⟨75C515⟩.

This gives yields of at least 40% which are based on further *in situ* reactions of the *t*-butylnitrosoethyne. This method avoids the hypersensitivity problems associated with handling mercury acetylides (Scheme 28).

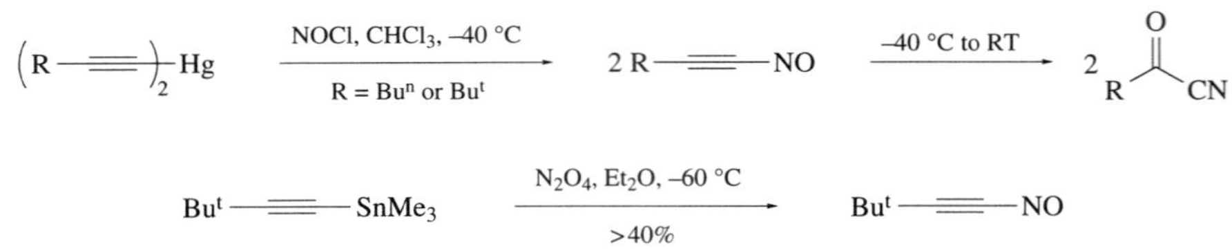

Scheme 28

2.22.1.2.2 *Nitroalkynes*

Nitroalkynes where the nitro group is attached directly to the *sp* carbon are a highly reactive species, many of which are unstable to isolation and can explode on purification. The first reported synthesis involved dehydrobromination of 1-bromo-1-nitro-2-phenylethylene with aniline and gave a very low yield of phenyl nitroethyne which exploded on attempted vacuum distillation ⟨30CB1707⟩. A later and related approach involved the elimination of iodide from *β*-iodo-nitroalkene derivatives, which themselves are readily obtained from alkylalkynes ⟨69AG(E)273⟩. Use of standard amine bases to effect the elimination led to side reactions, and poor product yields were obtained. The best results were obtained by passing the nitroalkene over potassium hydroxide under reduced pressure. For example, multigram quantities of the relatively stable *t*-butylnitroethyne could be obtained in almost quantitative yield by this method. Spectral evidence suggested that both the *n*-propyl and isopropyl analogues could also be synthesized, although they could not be isolated. Oxidation of *t*-butylnitrosoalkynes also provides a route towards the corresponding nitroalkyne; however, the poor yields and inaccessibility of the highly unstable nitroso compounds mean that this method is of limited synthetic value (Scheme 29) ⟨75C515⟩.

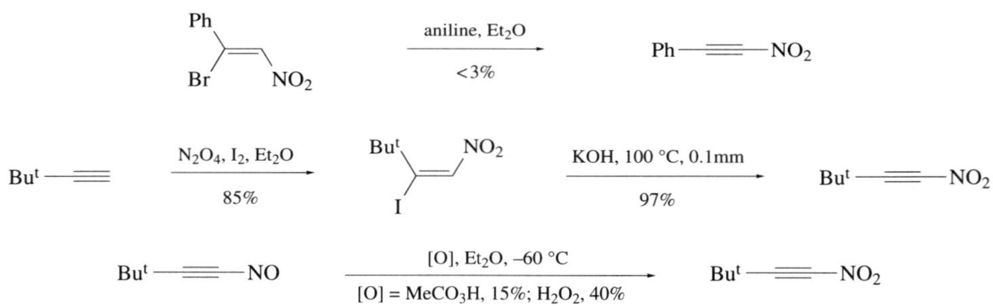

Scheme 29

A more practical and general approach involves the direct nitration of substituted trialkylstannylalkynes with either dinitrogen pentoxide or nitronium tetrafluoroborate ⟨75C516⟩. This gives good yields of the *t*-butyl derivative and also allows access to the less stable Pr^i and Pr^n derivatives in moderate yield (Equation (9)).

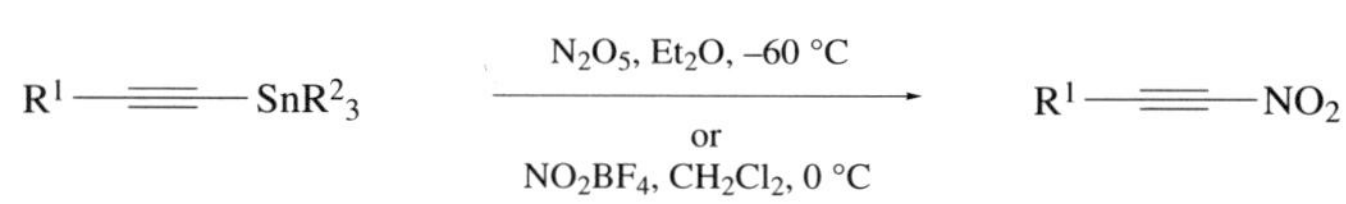

(9)

		Yield (%)	
R^1	R^2	N_2O_5	NO_2BF_4
Bu^t	Me	40	70
Bu^t	Bu	35	40
Pr^i	Me	30	50
Pr^i	Bu		10
Bu^n	Me	10	

Trialkylsilyl substituted nitroalkynes are stable in dilute solution for extended periods of time. A number of methods for their synthesis have been developed and most rely upon the nitration of a silyl- or stannylalkyne with a suitable nitronium source. The method of choice involves the reaction of bis(silylalkynes) with either nitronium tetrafluoroborate or hexafluorophosphate ⟨87JOC2294⟩. The purity of the nitronium source is crucial and the authors found that freshly triturated samples gave the best results (Scheme 30).

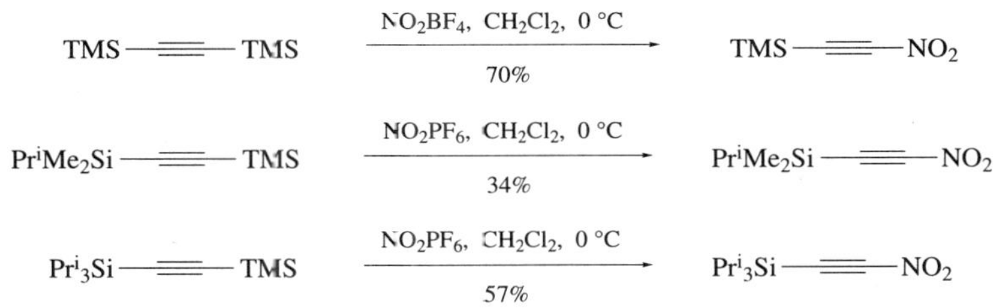

Scheme 30

With unsymmetrical (bis)silanes, the smaller trimethylsilyl group usually undergoes substitution in preference to more bulky groups, although small amounts of mixed products are occasionally observed. The authors found these nitroalkynes to be highly reactive species that underwent a number of useful cycloaddition reactions with a variety of dienes and dipoles ⟨90JOC1916⟩.

2.22.1.2.3 N-*Methylene-ynamines*

This elusive class of compounds was first reported in 1987 and are stable to isolation at room temperature, although they undergo decomposition in air after a few hours ⟨87AG(E)923⟩. They can be prepared via nucleophilic substitution of oxime sulfonate esters by alkynic cuprate reagents. For example, treatment of the cuprate derived from phenylethyne with a diphenyl or phenylmethyl oxime sulfonate led to the corresponding *N*-methylene-ynamines in moderate yield (Equation (10)).

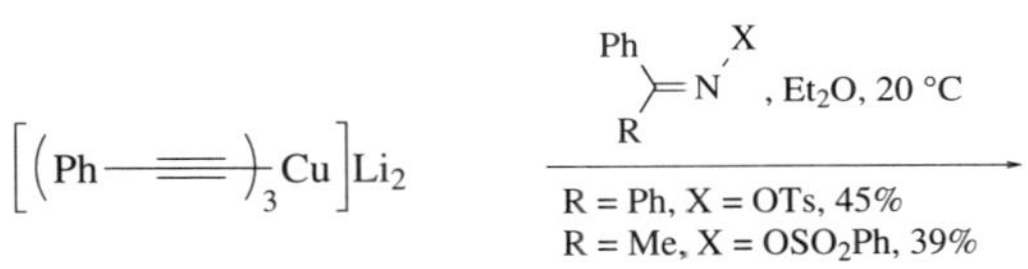

(10)

A more conventional approach involved the substitution of an alkynyl chloride with a suitably functionalized imine; this leads to the bis(dimethylamino) and diphenyl derivatives in poor to moderate yields, respectively ⟨88TL5355⟩. In this particular example, the trisubstituted alkene could be treated further with *n*-butyllithium to give the lithium acetylide and reacted with chlorodiphenylmethylsilane to give a complex *N*-methylene-ynamine containing a diynyl system (Scheme 31).

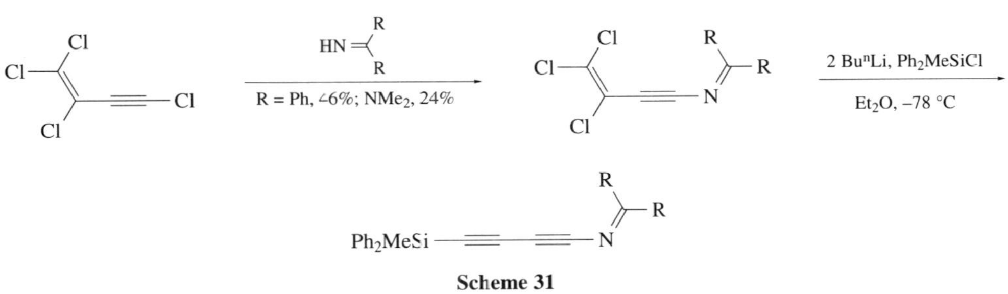

Scheme 31

Another approach to *N*-methylene-ynamines involves the treatment of the *N,N*-bis(trimethylsilyl)ynamines (**50**) with simple aldehydes under fluoride catalysis ⟨89TL6147⟩. This gives the phenyl- and *t*-butylynamines in reasonable yield (Equation (11)).

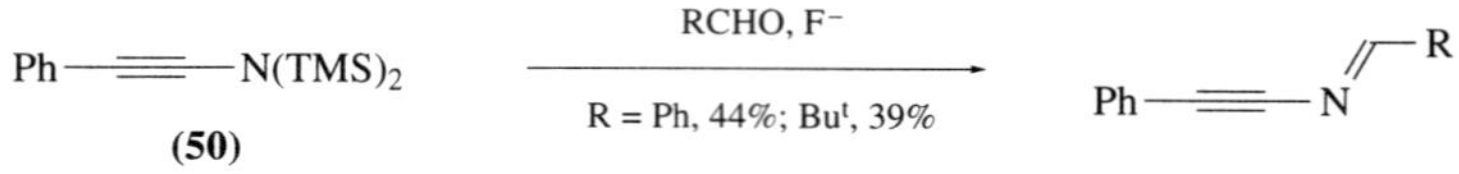

(50) (11)

2.22.1.2.4 *Azoalkynes*

The synthesis of azoalkynes was first reported by Sladkov and co-workers, who found that such compounds bearing electron-withdrawing aryl groups could be prepared by the reaction of silver acetylides with arenediazonium salts in aqueous solution ⟨66JOU1440⟩. However, yields were only poor to moderate and the products were often difficult to isolate from the reaction mixtures. Modification of Sladkov's original procedure by Huang led to a general synthesis of this class of compound (Equation (12)) ⟨75JOC124⟩. It was found that reaction of purified arenediazonium salts (either as the chlorides or bromides) with silver phenylacetylide in a chloroform–ethanol solvent system gave good yields of the corresponding azoalkynes, which could be purified by column chromatography. The results are summarized in Table 3. Apart from entry 2, all the azoalkynes were reported to be stable at room temperature. At elevated temperatures, however, these compounds undergo an interesting dimerization to yield a new heteroaromatic system ⟨69CC1393, 74BCJ946⟩. For example, prolonged heating of the azoalkyne (**51**) in cyclohexane gave the tetra-azapentalene (**52**) in 60% yield (Equation (13)).

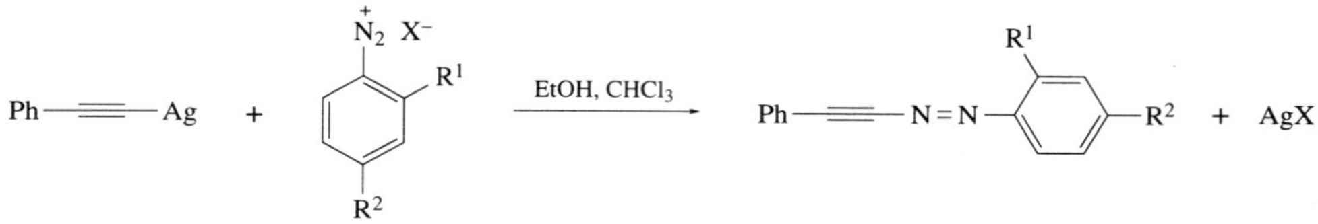

(12)

Table 3 Azoalkyne synthesis from arenediazonium salts (Equation (12)).

Entry	R^1	R^2	*Yield* (%)
1	H	H	65
2	MeO	H	33
3	Cl	H	70
4	Br	H	70
5	I	H	60
6	NO_2	H	72
7	CO_2Et	H	81
8	CN	H	55
9	Cl	CO_2Me	60
10	H	PhCO	87

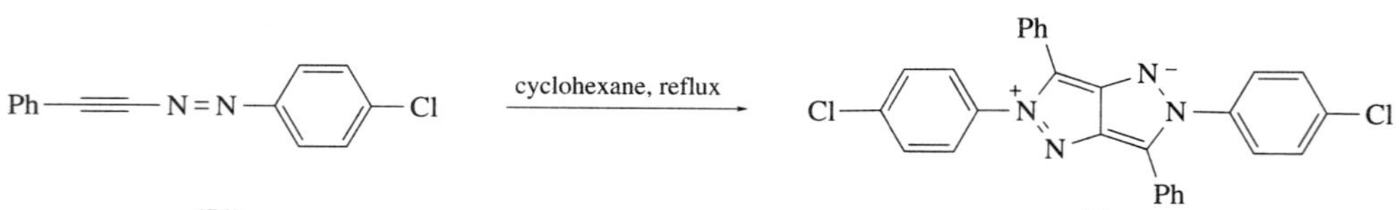

(51) **(52)** (13)

2.22.1.2.5 *Alkynyl azides*

The synthesis of alkynyl azides has long been the goal of a number of groups, but, despite numerous attempts, so far no member of this class of alkynyl nitrogen compound has been successfully isolated. An early attempt to prepare phenylazidoethyne involved the treatment of sodium phenylacetylide with tosyl azide; however, the only product isolated was the triazole (**53**), which was thought to arise from cyclization of the initial adduct (**54**) ⟨58JOC1051⟩. Later investigations involved an attempted synthesis of phenylazidoethyne by the elimination of halogen from the dihaloazidoalkene (**55**) using zinc metal, which resulted in the formation of *cis*- and *trans*-dicyanostilbene (**56**) in 67% yield ⟨69JA6122⟩. This is thought to arise by formation of the phenylazidoethyne followed by loss of nitrogen and further reactions of the resulting alkynylnitrene leading to (**56**) (Scheme 32).

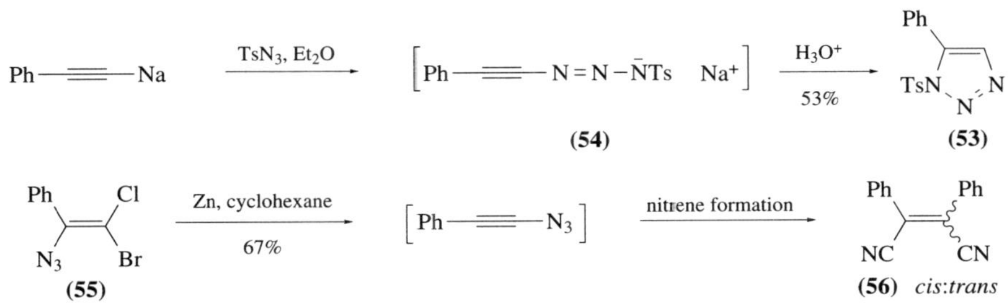

Scheme 32

The most compelling evidence so far for the existence of alkynyl azides has come from investigations made by Tanaka of the nucleophilic attack of an azide ion on alkynyl halides ⟨83CC329⟩. It was found that the reaction between phenylchloroethyne with sodium azide in DMSO gave rise to a complex mixture of products (Scheme 33). From this mixture, the following compounds were isolated: the vinyl azide (**57**) arising from addition to the alkyne (15–20%), benzonitrile (4–6%), benzoic acid (15–17%) and the rather unusual *N*-(phenylethynyl)-*S*,*S*-dimethylsulfoximine (**58**) (5–8%). The sulfoximine (which is the first example of this type of alkynyl nitrogen compound) is thought to arise from the initial formation of the putative alkynyl azide followed by rapid loss of nitrogen to give the nitrene (**59**) which undergoes trapping with DMSO to give the product (**58**). This trapping of nitrenes with sulfoxides is well documented in the literature ⟨81S165⟩.

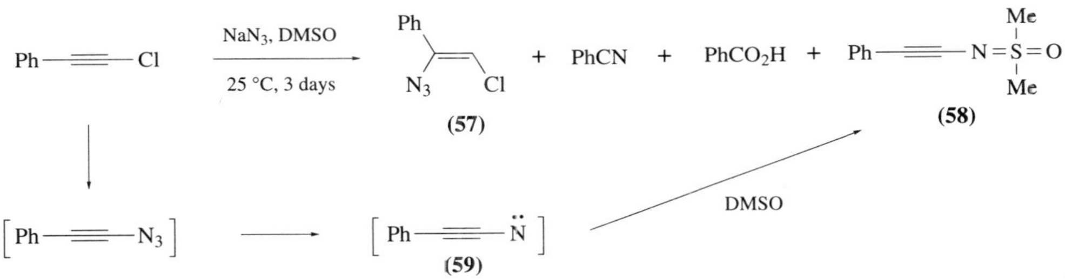

Scheme 33

2.22.1.2.6 *Alkynyl isocyanates and isothiocyanates*

(i) Alkynyl isocyanates

One of the first synthetic studies of alkynyl isocyanates involved the synthesis ⟨26JFR314⟩ of phenylpropioyl azide (**60**) and its decomposition via a photoinduced Curtius rearrangement ⟨69JA6122⟩. The alkynyl isocyanate (**61**) was not isolated but its presence was, however, inferred from the rearrangement products, phenylacetonitrile and 9,10-dicyanophenanthrene, that were obtained (Scheme 34). Later investigations by Walton, King and co-workers allowed the first authentic synthesis of isocyanatoethyne (**62**) (Scheme 35) ⟨78CC351⟩. It was found that a variety of silylated hydroxamic acid derivatives undergo fragmentation and rearrangement on heating to give the corresponding isocyanates in high yield. More specifically, it was found that treatment of propioloyl chloride with TMSONLiTMS in ether at −40 °C gave the corresponding hydroxamic acid derivative, which upon heating in mesitylene gave the unstable isocyanatoethyne (**62**) in quantitative yield. The mechanism is thought to proceed by loss of $(Me_3Si)_2O$ and formation of the acyl nitrene followed by rearrangement to the isocyanate (Scheme 34).

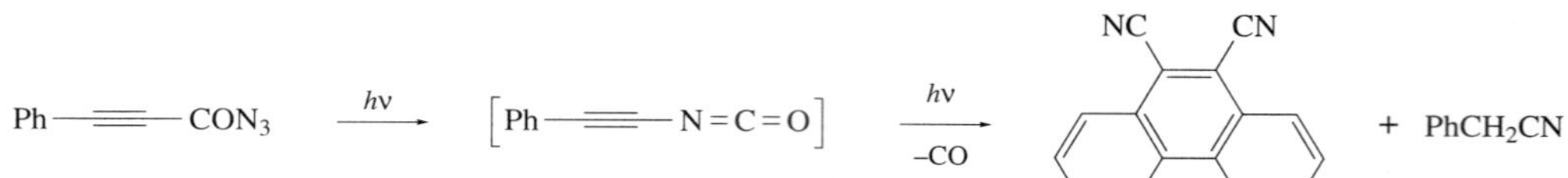

(**60**) (**61**)

Scheme 34

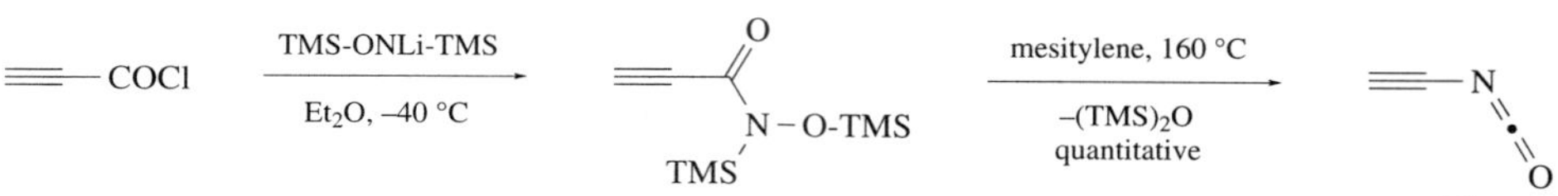

(**62**)

Scheme 35

(ii) Alkynyl isothiocyanates

Several reports of the synthesis of alkynyl isothiocyanates have appeared in the Russian literature over the years and all rely on the nucleophilic displacement of an alkynyl halide with thiocyanate ion (Equation (14)). The most recent disclosure describes a high-yielding method for the synthesis of aromatic propargylic derivatives by the reaction of a variety of aryl substituted 1-bromo- or 1-iodopropargylic ethers with sodium or potassium thiocyanate ⟨87UKZ34⟩. In general, it was found that bromides reacted faster than iodides. Table 4 summarizes the results.

(14)

Table 4 Aromatic propargylic isothiocyanate synthesis from 1-bromo or 1-iodo propargylic ethers with varying aryl substitutions.

Aryl substitution (Ar in Equation (14))	2-*Chloro-phenyl*	2,4-*Di-chloro-phenyl*	2,4,5-*Trichloro-phenyl*	2,4,6-*Trichloro-phenyl*	*Penta-chloro-phenyl*	2-*Bromo-phenyl*	2,4-*Di-bromo-phenyl*	2,4,6-*Tri-bromo-phenyl*
Yield (%)	90	93	95	96	96	68	82	87

2.22.1.3 N≡X Functions

Of the two possible representatives in this class (namely diazonium salts and isonitriles) only diazonium salts have been synthesized to date. In their studies on nitrosoalkynes, Tedder and Robson found that treatment of di-1-hexynyl mercury with nitrosyl chloride at low temperature gave rise to a blue–green solution of 1-nitroso-1-hexyne ⟨63AG1033⟩. Treatment of this solution with nitric oxide gave a deep gold–yellow solution of the diazonium nitrate salt (**63**). This salt was not isolated; however, its existence was confirmed by a number of *in situ* reactions. For example, treatment with 2-naphthol gave the usual azo coupling product, which on reaction with base yielded 1-hexyne and the corresponding diazoketone (**64**). Treatment with aniline gives the amino-azo compound (**65**), which upon hydrolysis yields benzenediazonium nitrate and hexanenitrile. Alternatively, addition of dimethylaminobenzene gives the azoalkyne (**66**), which upon acidic hydrolysis yields 1-hexyne and dimethylaminobenzene-diazonium nitrate (Scheme 36).

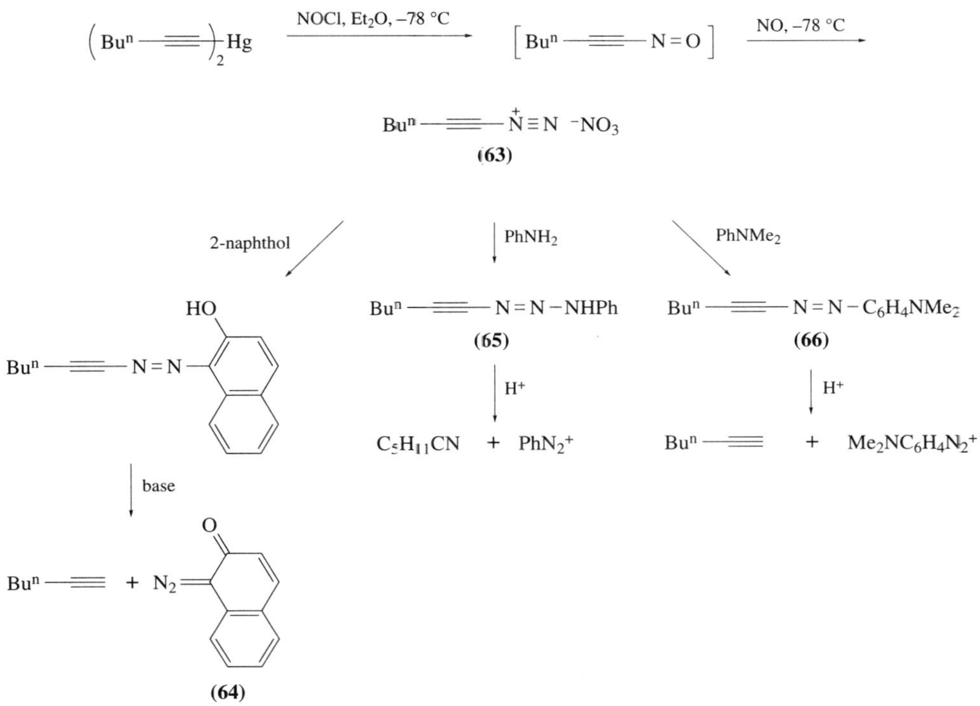

Scheme 36

2.22.1.4 N—P Functions

The synthesis of alkynylnitrogen compounds containing an N—P bond has been reported once in the literature ⟨76TL835⟩. In a study of the synthesis and isomerization reactions of *N*-propargylphosphoramides, it was found that treatment of the phosphoramide (**67**) with sodium hydride in THF gave rise to a mixture of the allene (**68**) and the alkynylnitrogen isomer (**69**). The ratio of products was dependent on the reaction temperature, although no yields were reported (Scheme 37).

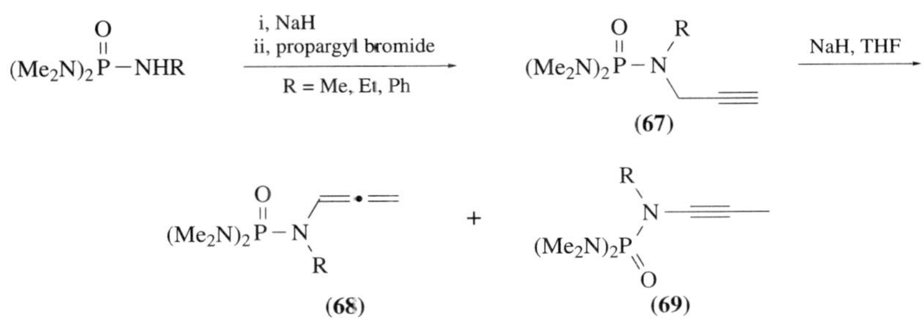

Scheme 37

2.22.1.5 N—Si Functions

It is remarkable, given the reactivity of *N*,*N*-dialkylynamines towards hydrolysis, that the synthesis of the potentially much more labile *N*,*N*-disilyl derivatives was ever considered. Despite these potential problems, however, two syntheses of this intriguing class of ynamine have been reported, both being quite different in their approach. For example, in 1971 West reported that treatment of acetonitrile with an excess of *t*-butyllithium led to the highly reactive dimetallated species Li_2C_2HN, which upon treatment with a number of chlorosilane derivatives yielded a mixture of the trisilylynamines (**70**) and the trisilylketenimines (**71**) in high yield ⟨71JA1714⟩. The proportion of ynamine present was dependent on the nature of the chlorosilane and the reaction solvent used. These ynamines are stable at room temperature but rearrange to the ketenimines (**71**) in quantitative yield upon heating (Scheme 38).

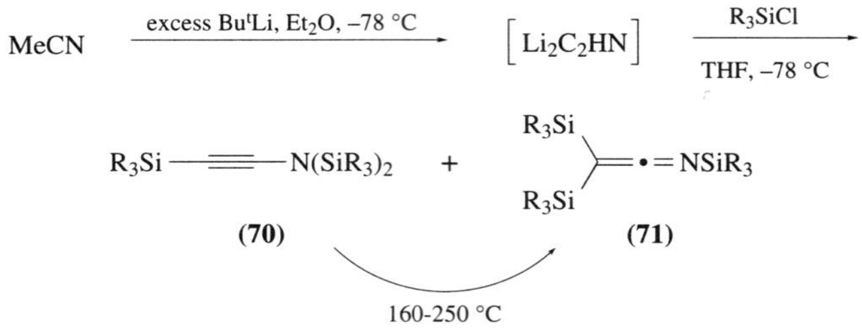

R_3Si = TMS, 85%, (**70**)/(**71**) 1/4

R_3Si = Bu^{t}Me$_2$Si, 83%, (**70**)/(**71**) 2/1

R_3Si = HMe$_2$Si, 80%, (**70**)/(**71**) 2/1

Scheme 38

More recently Würthwein reported that treatment of the silylated hydroxylamine sulfonate (**72**) with four different lithium acetylides gave rise to the corresponding bis(silyl)ynamines (**73**) (Equation (15)) ⟨89TL6147⟩. In two of the examples, the yields were less than 20%; however, given the potential instability of such compounds it is remarkable that they can be synthesized in the first place. These authors also reported that the silylated ynamines (**73**) were stable at room temperature but underwent isomerization to the corresponding ketenimines on warming to 100°C. Hydrolysis yields the corresponding nitriles as final products. It is the author's intention to use these ynamines as valuable intermediates for the synthesis of compounds such as alkynyl diazonium salts and alkynyl isocyanides.

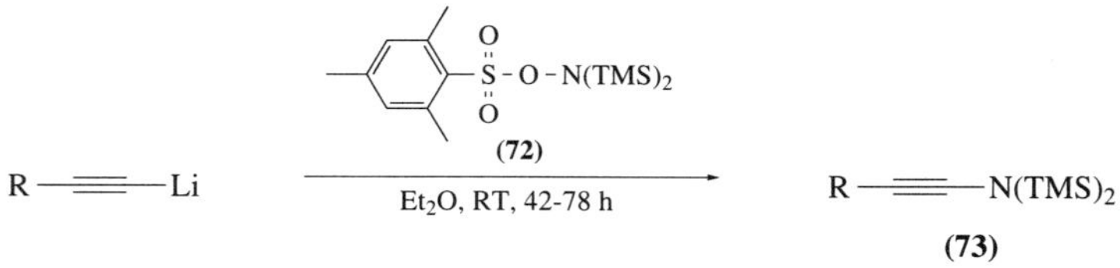

(15)

R = Ph, 63%; Bu^t, 35%; TMS, 12%; Me, 18%

2.22.2 ALKYNYLPHOSPHORUS COMPOUNDS

In marked contrast to alkynylnitrogen compounds, the synthesis of most alkynylphosphorus compounds has been relatively straightforward. This has been mainly because many of the phosphorus derivatives are directly accessible via the nucleophilic substitution of electrophilic phosphorus with metal acetylides.

2.22.2.1 Alkynylphosphines

The preparation of alkynylphosphines was first reported in 1921, but it was not until the studies of Hartmann that general and synthetically useful methods were developed. In general, alkynylphosphines are readily available by the nucleophilic substitution of phosphorus halides by metal acetylides ⟨56ZAAC(287)261, 61AG173⟩. The use of the readily available alkynyl Grignard reagents gives particularly good results and allows easy access to a large range of functionalized alkynylphosphines.

In general, this methodology would appear to be limited only by the availability of the requisite metal acetylides (Equation (16)).

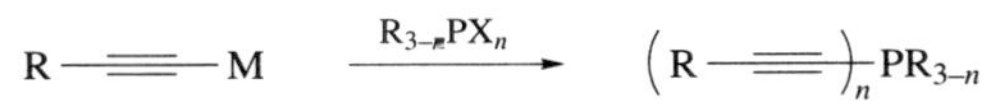

(16)

R = H, alkyl, alkenyl, alkynyl, aryl
M = MgX, Na, Li, HgX

Simple dialkylalkynylphosphines tend to be reactive species that undergo oxidation and polymerization in air. The corresponding diphenylalkynyl derivatives, however, are usually solids that are stable to aerial oxidation and are inert to hydrogenation using palladium catalysis, although they do undergo reduction under more vigorous conditions with Raney nickel. Scheme 39 summarizes just a few of the many examples of alkynylphosphines synthesized by the alkynic Grignard route ⟨66BSF1002⟩.

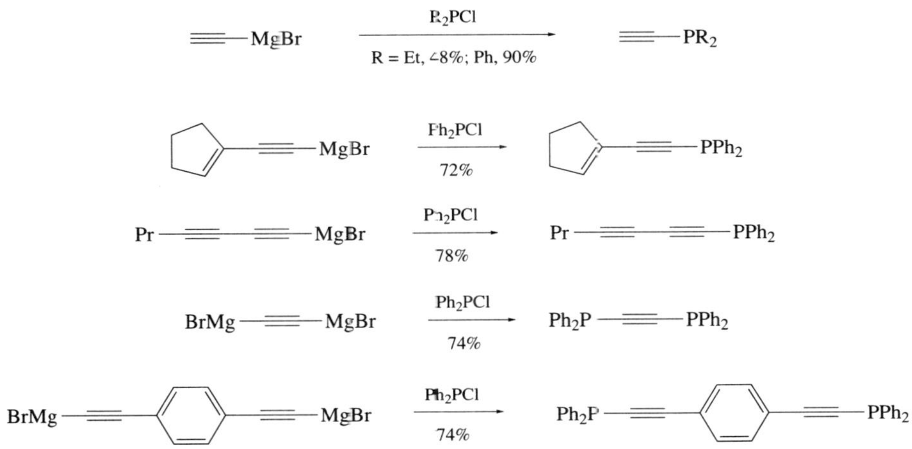

Scheme 39

Furthermore, use of the appropriate phosphorus halides allows access to both di- and trialkynylphosphines in high yields by the Grignard method (Scheme 40) ⟨60CR(250)866, 64RTC1301⟩.

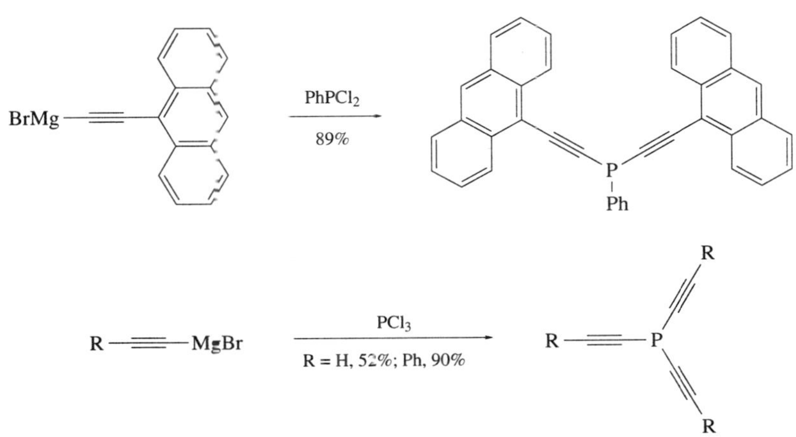

Scheme 40

Although alkynyl Grignard reagents provide an ideal method for the synthesis of alkynylphosphines, the reaction of many other metal acetylides with phosphorus halides has been investigated and should not be overlooked (in certain cases better yields can be obtained). For example, the reaction of a number of readily available lithium acetylides (by deprotonation with *n*-butyllithium) with a range of dialkylchlorophosphines gave high yields of the corresponding alkynylphosphines (Scheme 41) ⟨68JOM(15)69, 71CJC2706, 73JOC1611⟩.

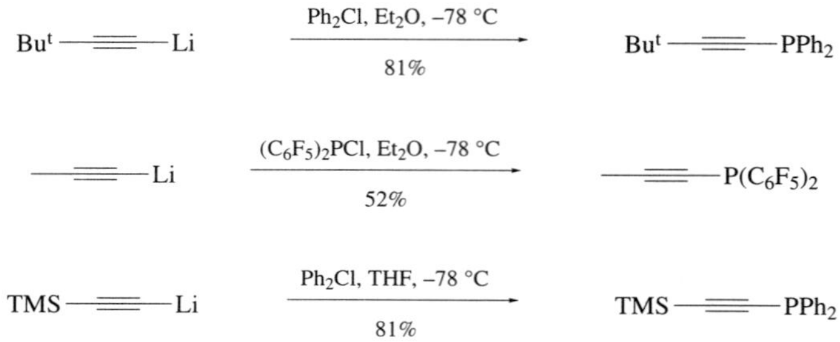

Scheme 41

In a modification of the above procedure, the lithium acetylide is added to a solution of dichloro(phenyl)phosphine to give the *in situ* chlorophosphine (**74**), which on further treatment with *n*-butyllithium yields the alkynylphosphine (**75**) ⟨73JOC1611⟩. This method is useful for the synthesis of unsymmetrically substituted alkynylphosphines, as it would be difficult or inconvenient to synthesize relatively inaccessible starting materials such as phenyl(butyl)chlorophosphine (Scheme 42).

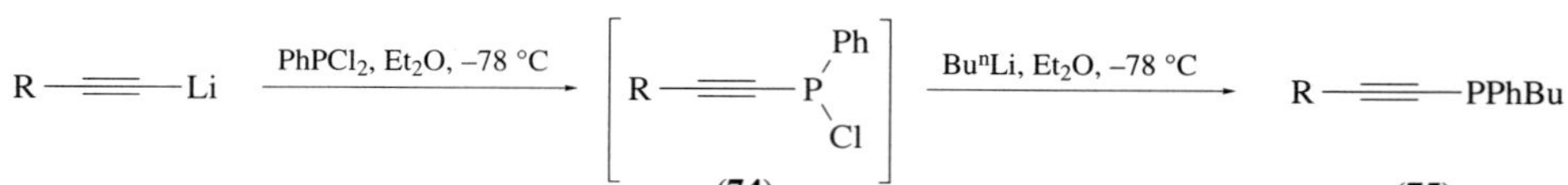

(74) **(75)**

R = Ph, 53%; Pr^n, 61%; Bu^n, 60%

Scheme 42

Mercury (bis)acetylides yield trialkynylphosphines on treatment with phosphorus triiodide ⟨64N213⟩. However, this method is less attractive than the Grignard route, as other groups have reported that these types of organomercurial can cause severe hypersensitivity problems and must be handled with extreme caution (Equation (17)).

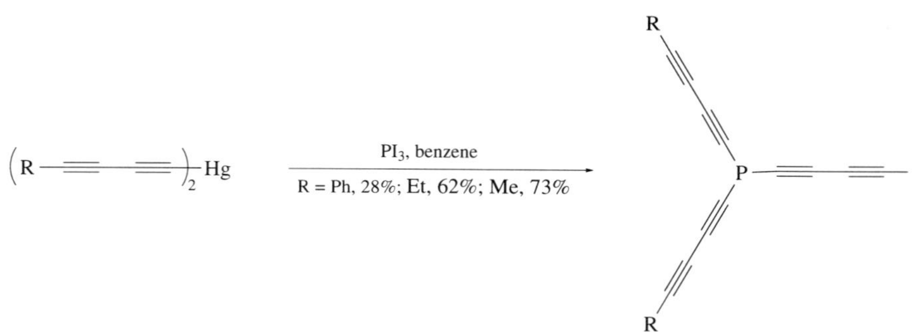

(17)

Another useful method involves the nucleophilic substitution of alkynylphosphonites. For example, Aguiar *et al.* found that a range of alkynylphosphines could be prepared by the reaction of alkynyllithiums with dialkylphosphinous chlorides and further *in situ* reaction of the resulting alkynylphosphonites (**76**) with a variety of Grignard reagents ⟨69JOC2684⟩. Similarly, the alkynylphosphonite (**77**) was found to undergo nucleophilic substitution on phosphorus with a variety of substituted aryl Grignard reagents to provide the diarylalkynylphosphines in reasonable yield ⟨76MI 222-01⟩. This method would appear to provide a more convenient alternative to the alkynyl Grignard substitution of electrophilic phosphines, as the dialkyl- or diarylchlorophosphines required for the synthesis of (**78**) would be fairly inaccessible (Scheme 43).

In a reversal of the above strategy, it has been reported that the reaction of anionic phosphines with alkynyl chlorides gives reasonable yields of the corresponding alkynylphosphines ⟨62CB268⟩. For example, reaction of phenylchloroethyne with lithium diphenyl- or dicyclohexylphosphine gave the alkynylphosphines in 35% and 50% yield, respectively (Equation (18)).

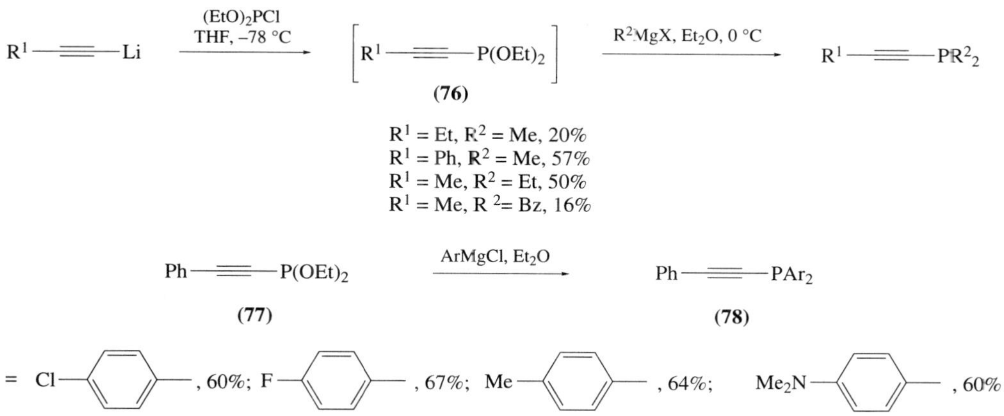

Scheme 43

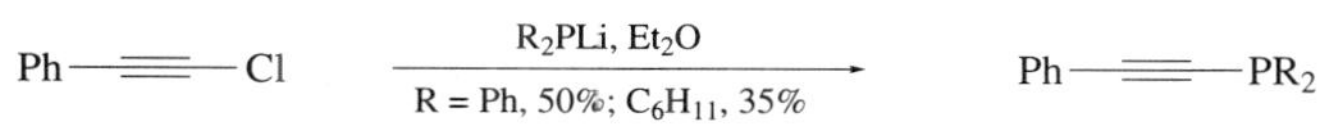

(18)

Another very useful route involves the deprotonation of an existing alkynylphosphine followed by reaction with an electrophile to give a new substituted alkynylphosphine. This method is of wide scope and allows access to a large range of complex alkynylphosphines from simple and readily available starting materials. Deprotonation with lithium amide and reaction with either alkyl halides or ketones gives good yields of the corresponding alkyl or carbinol derivatives, respectively ⟨62RTC993⟩. Use of *n*-butyllithium as base allowed a range of phosphine, arsine, silyl, stannyl, germyl and plumbyl derivatives to be synthesized from the resulting (diphenylphosphino)ethynyllithium in moderate to good yield by this method (Scheme 44) ⟨68JOM(15)69, 71CJC2706⟩.

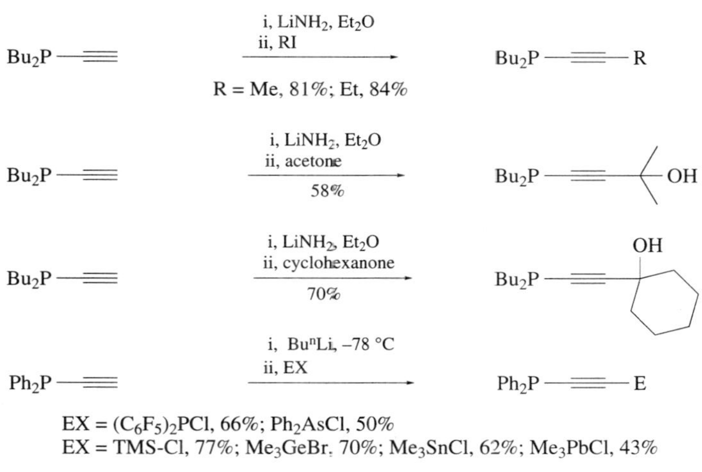

Scheme 44

Other reactions of this class include the preparation of propiolic acid derivatives (**79**) by the reaction of (diphenylphosphino)ethynyl magnesium bromide with carbon dioxide (Equation (19)) and an operationally simple and high-yielding synthesis of propargyl alcohol derivatives (**81**) by the reaction of the phosphine (**80**) with a range of ketones under basic conditions using potassium hydroxide in THF (Equation (20)) ⟨66BSF1002⟩.

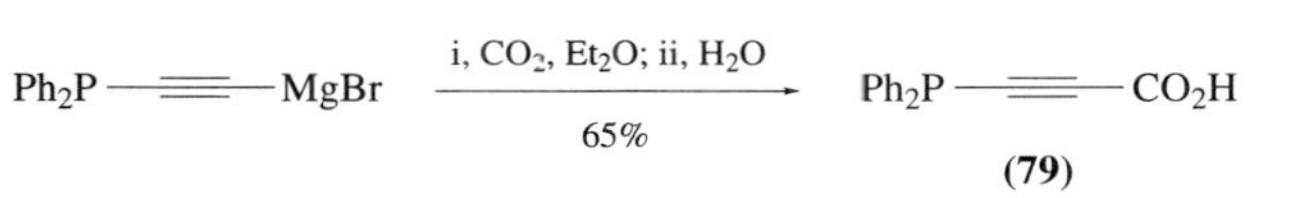

(19)

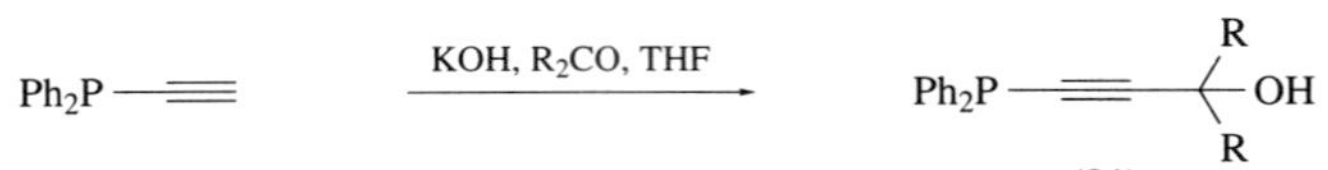

(80) **(81)** (20)

R_2CO = acetone, 63%; cyclohexanone, 74%; benzophenone, 85%; fluorenone, 85%, anthraquinone, 90%

2.22.2.2 Alkynylphosphonium Salts

Alkynylphosphonium salts are readily prepared by two main synthetic routes: (i) by the reaction of alkynyl halides with trialkyl- or triarylphosphines or (ii) by quaternization of alkynylphosphines with a suitable alkyl halide.

(i) From alkynyl halides

The synthesis of alkynylphosphonium salts by the nucleophilic displacement of halide from an alkynic carbon was first reported in 1962 by independent studies from the groups of Miller ⟨62JA2020⟩ and Viehe ⟨62CB319⟩ and later by Hoffmann ⟨64TL983⟩. In general, the salts are best prepared by reaction of the appropriate phosphine with an alkynyl halide (chloride, bromide or fluoride) in an aprotic solvent (usually ether). Although reaction times were slow, often requiring several days, good yields of the salts could be obtained in this way ⟨72JOC2168⟩. The hygroscopic product usually precipitates from the reaction mixture in a high state of purity (Scheme 45). Reaction of triphenylphosphine with dichloroethyne at low temperature in ether gave the triphenylphosphonium chloroethyne (**82**) in excellent yield ⟨76MI 222-02⟩. Previous attempts at this reaction resulted in low yields and mixtures of products ⟨62CB319⟩. Similarly, the reaction of dimethoxyphosphoryl-chloroethyne (**83**) with triphenylphosphine gave the unusual zwitterionic species (**84**) by loss of methyl chloride from the intermediate adduct (**85**). Although an almost quantitative yield of (**84**) was obtained, only one example of this interesting reaction has been reported (Scheme 45) ⟨76ZN(B)172⟩.

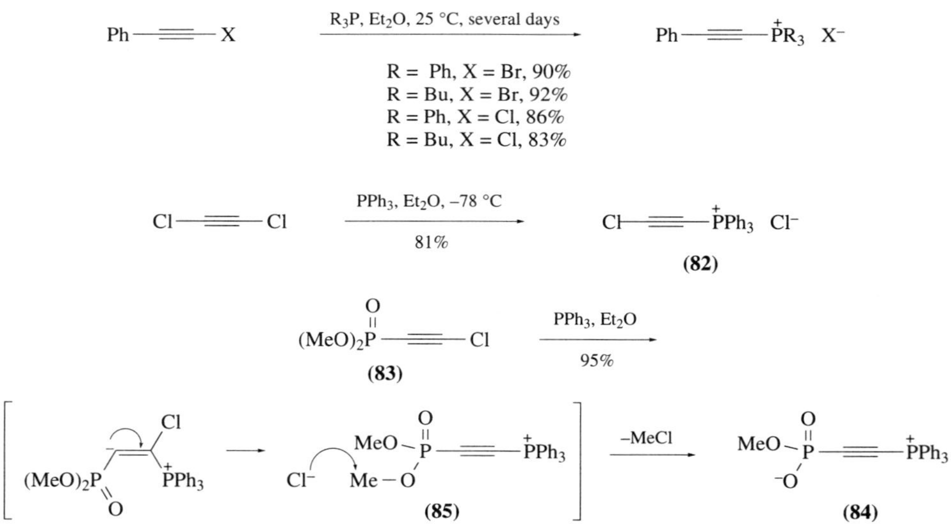

Scheme 45

Unfortunately the above methodology is not suitable for the synthesis of alkyl-substituted alkynylphosphines, as reaction of phosphines with alkyl alkynyl halides results in mixtures of products arising from nucleophilic attack at the halogen as well as at the α- and β-carbon atoms. More recent studies have led to an ingenious solution to this problem using alkynyl(phenyl)iodonium salts ⟨87CC1708⟩. It was found that treatment of alkynyliodonium tetrafluoroborates of the general structure (**86**) in THF with triphenylphosphine in sunlight at $-78\,^{\circ}C$ for only 2 h gave the corresponding phosphonium tetrafluoroborates in quantitative yields (Equation (21)). This is in marked contrast to the prolonged reaction times (2–3 days) required with alkynyl halides. It is thought to proceed via a free-radical chain mechanism as there is no reaction observed in the dark. The iodonium salts are readily available from alkynylsilanes in good yield ⟨85TL4501⟩. Although the generality of this procedure needs to be investigated further, it would appear to be the method of choice for the synthesis of alkynylphosphonium salts, via this type of nucleophilic displacement strategy, as it avoids the use of the hazardous (toxic, explosive) alkynyl halides, which can often be difficult or tedious to prepare.

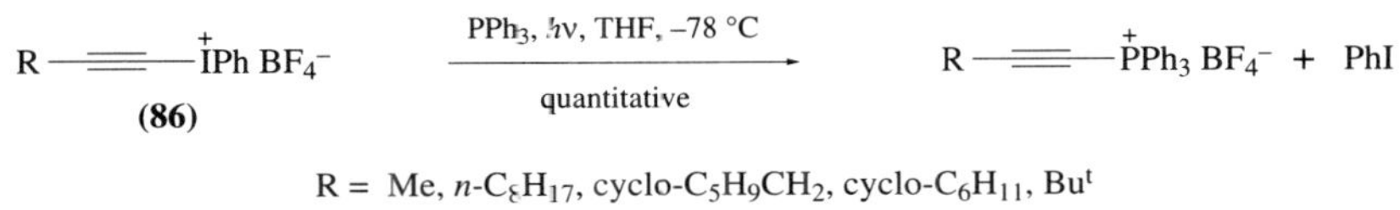

(21)

(ii) Quaternization of alkynylphosphines

This route has not received the same level of attention as the above methods; however, it does provide a simple and convenient method for the synthesis of phosphonium salts in which the groups on phosphorus are different. For example, Cadiot and co-workers found that treatment of a variety of phosphines with methyl iodide gave the corresponding phosphonium iodides in acceptable yields ⟨66BSF1002⟩. Alkynylbis(phosphines) also give good yields of the corresponding alkynylphosphonium salts on treatment with either methyl or ethyl iodide ⟨56ZAAC(287)261⟩. The reaction of alkynylphosphines with allyl bromide ⟨79LA43⟩ allows access to (allyl)arylalkynylphosphonium salts (Scheme 46) ⟨56ZAAC(287)261, 66BSF1002, 67CR(C)995, 72ZAAC(394)74, 79LA43⟩.

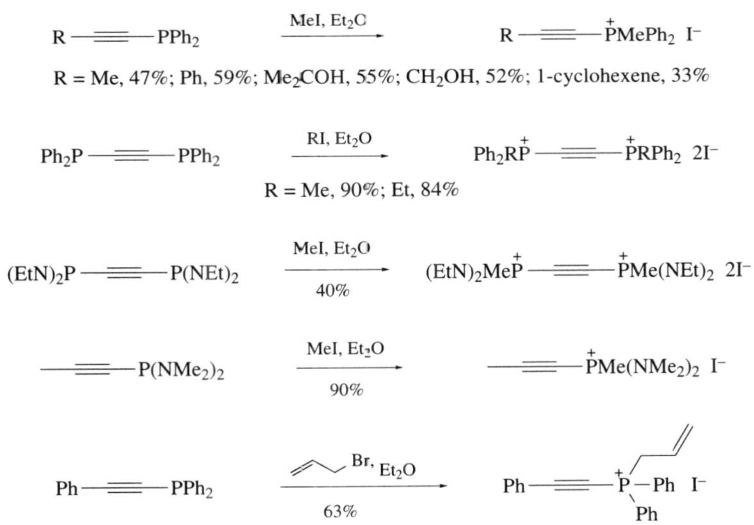

Scheme 46

(iii) Miscellaneous methods

Another very useful method involves the reaction of stabilized phosphonium ylides (**87**) with triphenylphosphonium dibromide followed by treatment of the resulting adduct (**88**) with triethylamine ⟨83CB1320⟩. This results in the elimination of triphenylphosphine oxide to give the alkynylphosphonium salts (**89**) in excellent yields (Scheme 47).

A further intriguing and related method involves the reaction of aromatic acid chlorides with the stable crystalline hexaphenylcarbodiphosphorane (**90**), which leads initially to the unusual phosphonium salts (**91**). Heating in toluene under reflux for 12–18 h furnishes the alkynylphosphonium salts (**92**) by thermal extrusion of triphenylphosphine oxide. When phthaloyl dichloride was used as the starting acyl halide, the bisphosphonium salt (**93**) was obtained in excellent overall

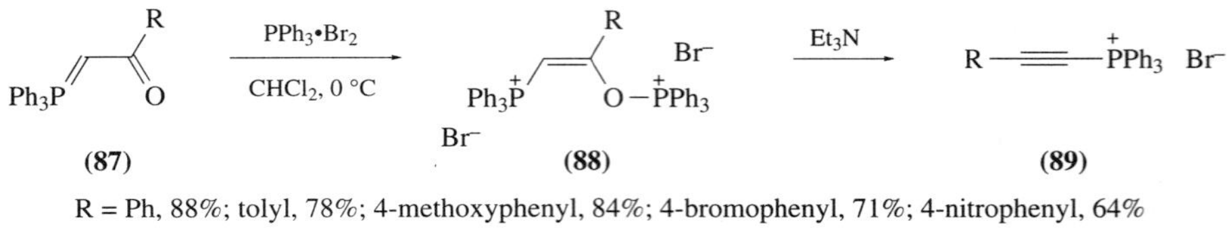

Scheme 47

yield. This method leads to an economic synthesis of alkynylphosphonium salts in good overall yields from readily available starting materials (Scheme 48) ⟨77AG(E)45⟩.

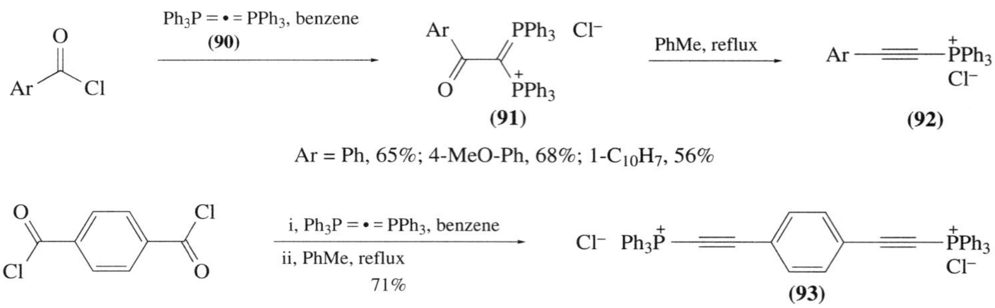

Scheme 48

Treatment of phthalic anhydride with the phosphorane **(90)** gave the zwitterionic species **(95)** which upon treatment with methyl iodide yielded the interesting alkynylphosphonium salt **(96)** in 60% yield by a mechanism similar to that described above (Scheme 49) ⟨78TL3343⟩.

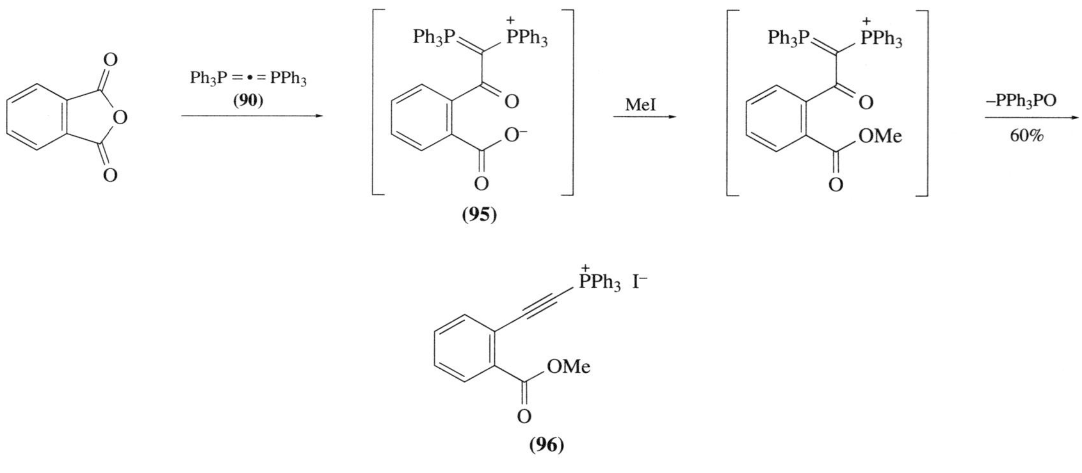

Scheme 49

2.22.2.3 Alkynyl—PX_2 Functions

2.22.2.3.1 Alkynylphosphonites (X = OR)

This class of alkynylphosphorus compound is readily accessible by the reaction of lithium acetylides with dialkyl phosphorochloridites (Equation (22)). They tend to be reactive compounds that polymerize when heated ⟨69JOC2684⟩. As mentioned in Section 2.22.2.1, they are often synthesized *in situ* and used without isolation, and so in many cases no yields are quoted. Scheme 50 illustrates a number of examples of alkynylphosphonites synthesized by this method ⟨69JOC2684, 75JGU1832, 84CC200⟩.

(22)

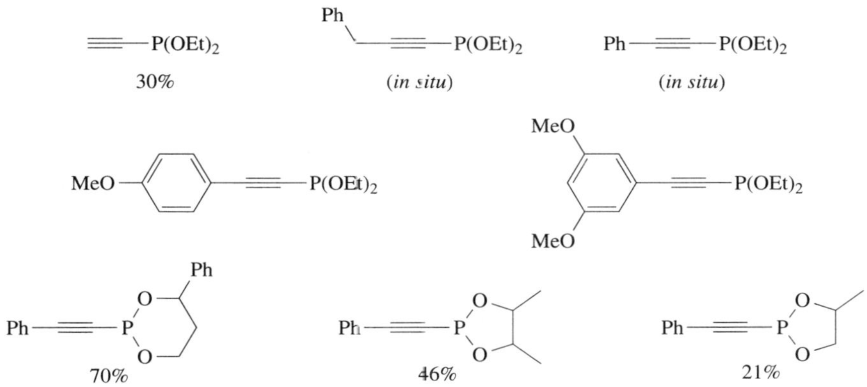

Scheme 50

2.22.2.3.2 Alkynylaminophosphines (X = NR)

These compounds are readily available through the reaction of alkynic Grignards with bis(amino)chlorophosphites (Equation (23)). The reaction is of wide scope and good yields of both mono- and dialkynylaminophosphines, as well as mixed alkylaminophosphines, can be obtained via this procedure. One interesting observation is that treatment of the bis(aminophosphine) derivative (**97**) with dry hydrogen chloride gas allows access to the bis(dichlorophosphinyl)alkyne (**98**) in 49% yield (Equation (24)). However, this compound was reported to be extremely air and moisture sensitive and exploded on distillation. Scheme 51 summarizes a few of the many examples of alkynylaminophosphines ⟨67CR(C)995, 72ZAAC(394)74⟩.

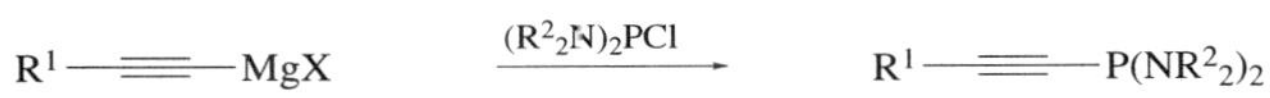

(23)

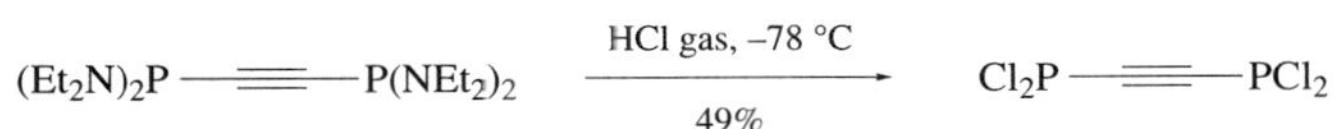

(24)

(97) **(98)**

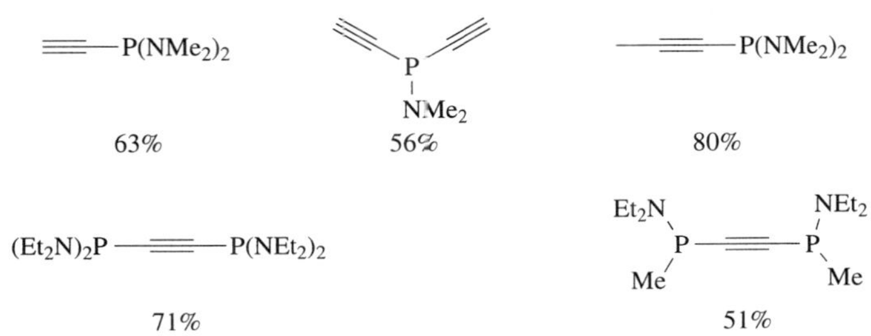

Scheme 51

2.22.2.4 Alkynylphosphine Chalcogenides

2.22.2.4.1 Alkynylphosphine oxides (X = O)

Alkynylphosphine oxides have been studied in great detail and can be synthesized by two main preparative methods: (i) by the oxidation of an existing alkynylphosphine or (ii) by the reaction of a metal acetylide with a chlorophosphine oxide. Other methods have also been investigated, such as a number of interesting preparations using functionalized diazophosphine derivatives, although the generality of these methods has not been proved to the extent of (i) and (ii).

(i) From alkynylphosphines

Although alkynylphosphines are reactive enough to undergo oxidation to the phosphine oxides in air ⟨63N403⟩, the oxidation is usually best achieved using 30% aqueous hydrogen peroxide solution (Equation (25)). This provides a practical, high yielding and economic synthesis of alkynylphosphine oxides from readily available starting materials. Unlike the corresponding phosphines, alkynylphosphine oxides are usually stable crystalline materials. This method also gives high yields of the phosphine oxides of alkynylaminophosphines. Scheme 52 illustrates a number of examples of compounds prepared by the hydrogen peroxide oxidation of alkynylphosphines ⟨56ZAAC(287)261, 66BSF1002, 72BSF1600, 72ZAAC(394)74⟩.

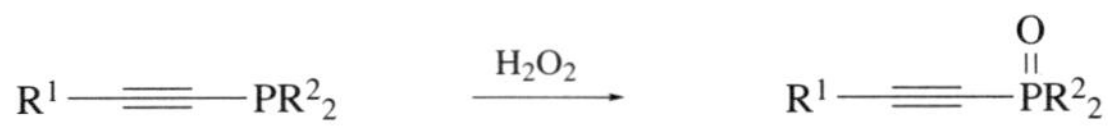

(25)

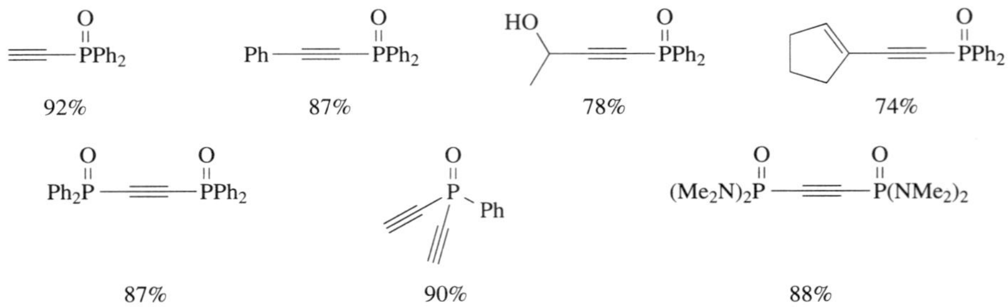

Scheme 52

(ii) From chlorophosphine oxides

The addition of alkynyl Grignard reagents to chlorophosphine oxides provides a very convenient method for the synthesis of mono-, di- and tri(alkynyl)phosphine oxides in one synthetic operation (Equation (26)). Alkynylaminophosphine oxides can also be obtained by this method using chloro(amino)phosphine oxides and alkynyl Grignard reagents. In general, yields are usually good and the reaction would appear to be limited only by the availability of the requisite chlorophosphines. It is interesting to note that, in one case, use of an alkynyllithium reagent and dimethylmethoxyphosphine oxide gave none of the expected product; the only products observed were those arising from nucleophilic attack of the alkynyllithium at the alkoxy carbon of the chlorophosphine oxide. Scheme 53 illustrates a number of examples of alkynylphosphine oxides prepared by the above method ⟨60CR886, 67CR(C)995, 69RTC71, 73RTC762⟩.

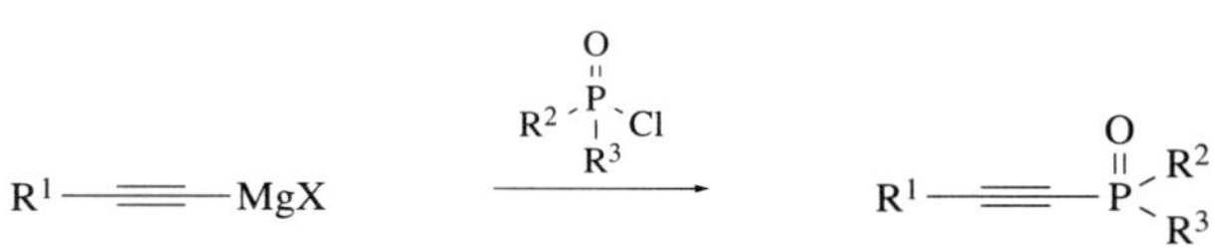

(26)

R^1 = H, alkyl, aryl, etc.
R^2 = R^3 = alkyl, aryl, Cl

(iii) Other methods

Disteldorf and Regitz reported an interesting alternative to the above methods, which utilizes the (diazomethyl)diphenylphosphine oxide (**99**) ⟨76CB546⟩. It was found that (**99**) added to a variety of aldehydes, under base catalysis, in an aldol-like manner to give the hydroxydiazo compounds (**100**)

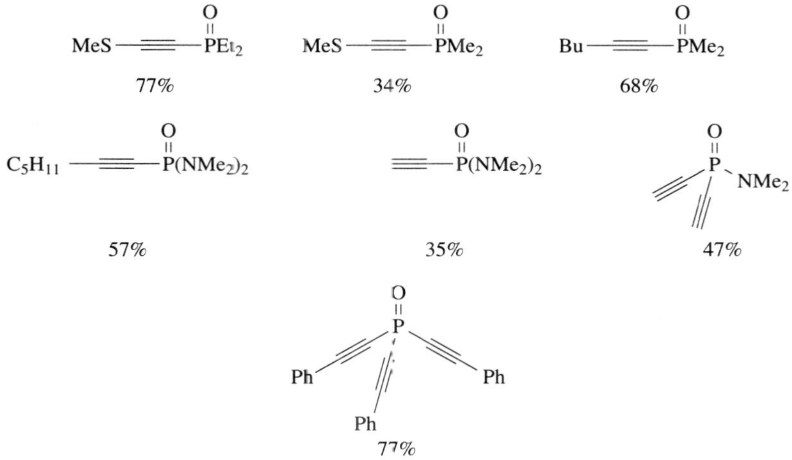

Scheme 53

in good yields. Treatment of these compounds with boron trifluoride induced a sequence of events that involved loss of hydroxide to give the vinyl diazonium species (**101**) which after elimination of nitrogen yielded the phosphine oxides (**102**) in good overall yields. The whole sequence provides an attractive and synthetically useful two-step sequence for the synthesis of alkynylphosphine oxides from aldehydes. This method is particularly suitable for the synthesis of heteroaromatic substituted alkynylphosphine oxides (Table 5, entries 6 and 7), as well as the alkenyl and alkynyl substituted derivatives (Table 5, entries 8 and 9), as the starting materials required for the synthesis of these oxides by methods (i) and (ii) above would be relatively inaccessible (Scheme 54).

Table 5 Synthesis of alkynylphosphine oxides from various aldehydes (Scheme 54).

Entry	*1*	*2*	*3*	*4*	*5*	*6*	*7*	*8*	*9*
R	Me	tolyl	Ph	4-CN-Ph	2-naph.	2-thienyl	2-furyl	MeCH=CHPh	MeC≡CPh
Yield (**100**) (%)	85	48	63	85	68	68	58	55	66
Yield (**102**) (%)	71	70	58	62	45	58	55	57	57

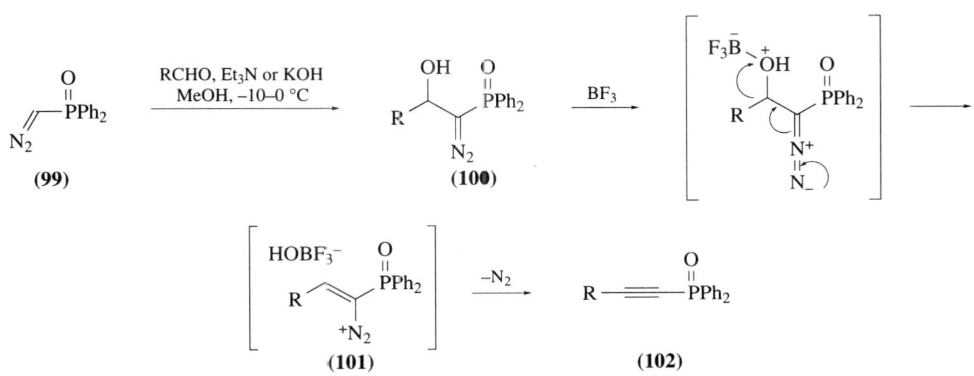

Scheme 54

Another method has been reported whereby either (trimethylsilyl)[bis(diisopropylamino)-phosphino]diazomethane (**103**) or the lithio derivative (**104**) was reacted with an acid chloride to give good yields of the corresponding alkynylphosphonium oxides ⟨88JA6463⟩. The reaction is thought to proceed via the initial formation of the diazo ketone (**105**), which then loses nitrogen to form the ylide (**106**). This then undergoes an intramolecular Wittig-like reaction to generate the phosphine oxide (**107**). Although only three examples were investigated, this method is attractive in that it uses simple starting materials such as acid chlorides (Scheme 55).

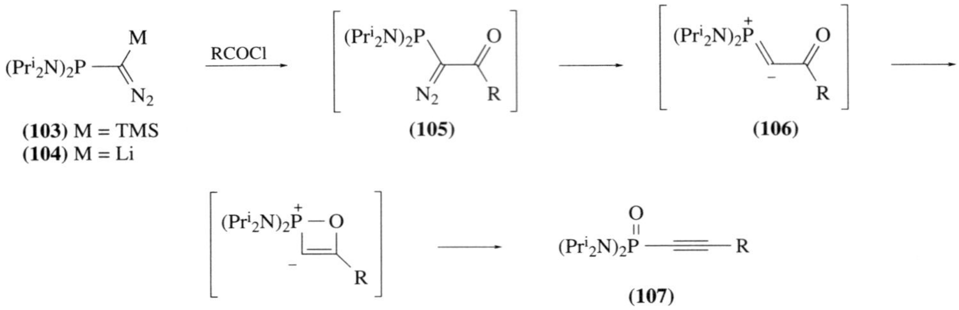

M = TMS, R = tolyl, 85%
M = Li, R = Bu^t, 70%
M = Li, R = tolyl, 70%

Scheme 55

The same authors also reported that the diazomethylenephosphorane (**108**) reacted with benzaldehyde to give the alkynylphosphine oxide (**110**) ⟨92BSF367⟩. It is thought that the reaction proceeds by the initial formation of the diazo-oxaphosphetane (**109**), which after loss of nitrogen and HCl yields the phosphine oxide. Although only one example of this interesting reaction has been disclosed, this method may, after further investigation, provide an operationally simple method for alkynylphosphine oxide synthesis from simple aldehydes (Scheme 56). Regitz *et al.* also reported an interesting alkynylphosphine oxide synthesis in their study of carbenes generated from (diazomethyl)cyclopropenes ⟨79AG(E)531⟩. It was found that irradiation of the cyclopropene (**111**) gave the carbene (**112**), which underwent a cheletropic reaction to give a mixture of products from which the alkynylphosphine oxide (**113**) could be isolated in 49% yield (Scheme 57).

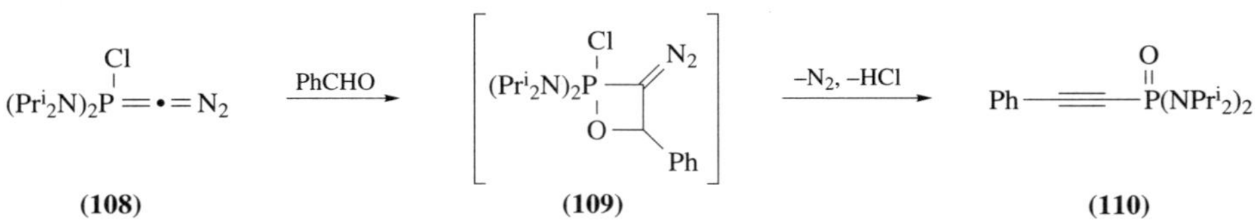

Scheme 56

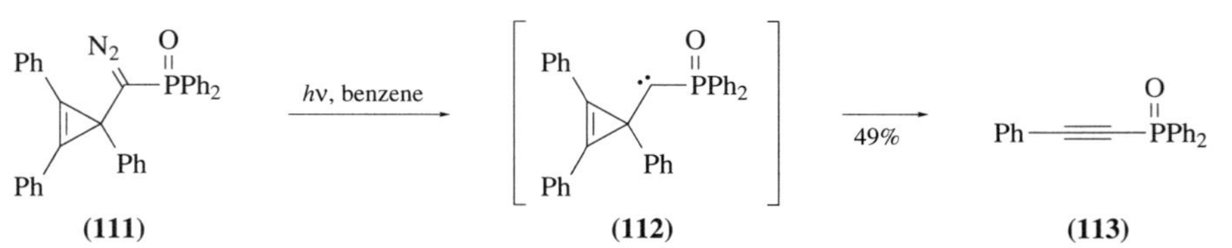

Scheme 57

2.22.2.4.2 *Alkynylphosphine imides (X=NR)*

This class of alkynylphosphorus compound is readily accessible from the reaction of the corresponding phosphine and an azide via the well-known Staudinger reaction (Equation (27)). The reaction is of wide scope and a variety of azides have been studied, including aryl, alkylsulfonyl and arylsulfonyl azides. Yields are generally high and the products are usually stable solids, although they can be easily hydrolyzed to the corresponding alkynylphosphine oxides on heating in water. Two groups have investigated this reaction in detail and in both cases no products arising from 1,3-dipolar addition reactions to the triple bond were observed. Scheme 58 summarizes the results of the two investigations ⟨74CB2513, 74JGU1461, 77LA1766⟩.

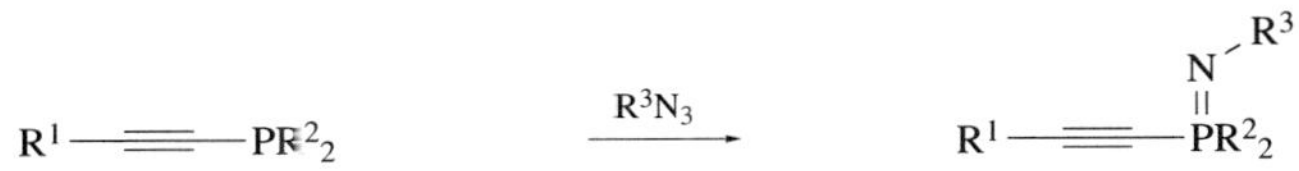

(27)

$R^1 = R^2 = H$, alkyl, aryl; R^3 = aryl, alkylsulfonyl, arylsulfonyl

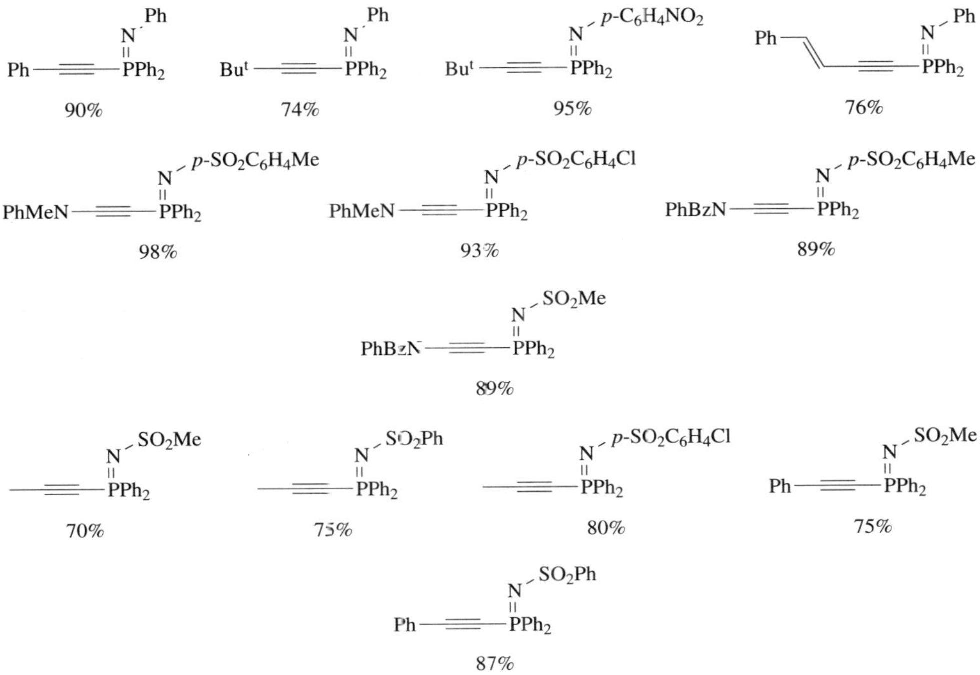

Scheme 58

2.22.2.4.3 *Alkynylphosphine sulfides*

The synthesis of alkynylphosphine sulfides can be achieved in two ways: (i) by the reaction of an existing phosphine with sulfur or (ii) by the reaction of an alkynyllithium with a dialkylhalophosphine sulfide. Although both methods give excellent yields, (i) is the method of choice simply because it has been studied more extensively, the starting alkynylphosphines are readily available and it is very economic in that elemental sulfur is used.

(i) From phosphines

The conversion of alkynylphosphines to the corresponding sulfides is usually carried out by heating the phosphine in a suitable solvent (usually carbon disulfide or benzene) with sulfur powder. This furnishes the usually crystalline sulfides in a high state of purity, often in quantitative yield (Equation (28)). This method is also suitable for the synthesis of bis(dialkylamino)alkynylphosphine sulfides. Scheme 59 illustrates just a few of the many examples of this method ⟨56ZAAC(287)261, 62CB268, 66BSF1002, 72ZAAC(394)74, 74CB2513⟩.

(28)

$R^1 = R^2 = H$, alkyl, aryl, vinyl, etc.

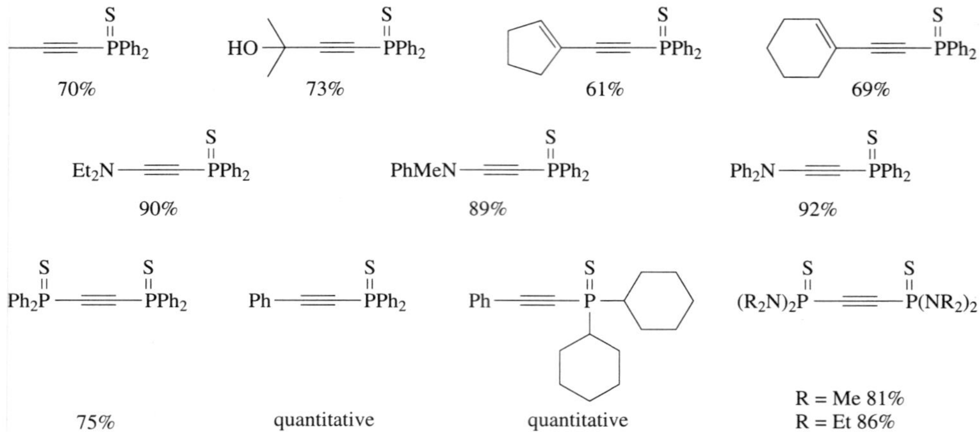

Scheme 59

(ii) From dialkylhalophosphine sulfides

An alternative method to the above involves the reaction of alkynyllithiums with dialkylhalophosphine sulfides. For example, Aguiar *et al.* found that treatment of tetramethylbiphosphine sulfide (**114**) with bromine gave dimethylbromophosphine sulfide (**115**), which upon treatment with 1-butynyllithium yielded dimethyl-1-butynylphosphine sulfide (**116**) in 74% yield ⟨69JOC3349⟩. 1-Phenylethynyllithium reacted similarly to give dimethylphenylethynylphosphine sulfide (Scheme 60).

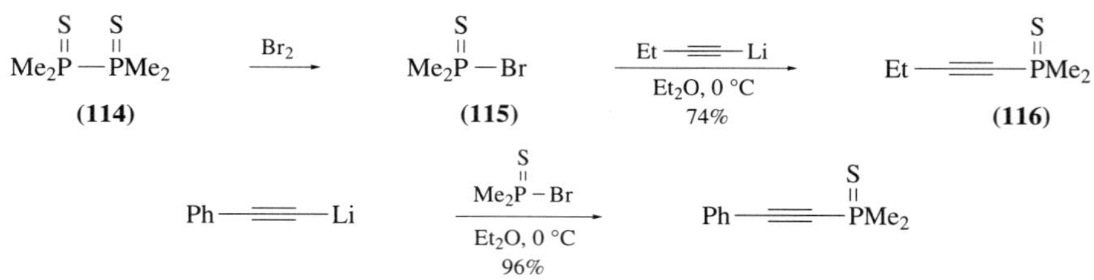

Scheme 60

2.22.2.4.4 *Alkynylphosphine selenides*

Like the corresponding sulfides, these alkynylphosphine derivatives are readily synthesized by the reaction of an existing phosphine with elemental selenium (Equation (29)). The reaction is conveniently carried out by the portionwise addition of selenium powder to a solution of the phosphine in an inert solvent. Reaction times are short (usually of the order 5–30 min) and the crystalline products are often isolated in near quantitative yield. A large range of selenides can be accessed by this method from mono-, di- and trialkylaminophosphine starting materials. A selection of selenides prepared by this method are illustrated in Scheme 61 ⟨72ZAAC(394)74, 74CB2513⟩.

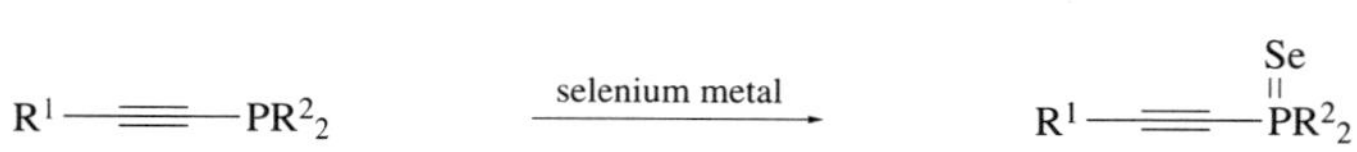

(29)

$R^1 = R^2$ = H, alkyl, aryl, vinyl, etc.

2.22.2.5 Alkynylphosphonates

These alkynylphosphorus derivatives are best prepared via two different synthetic methods involving either (i) an Arbusov-type reaction of an alkynyl halide with a trialkyl phosphite or (ii) the

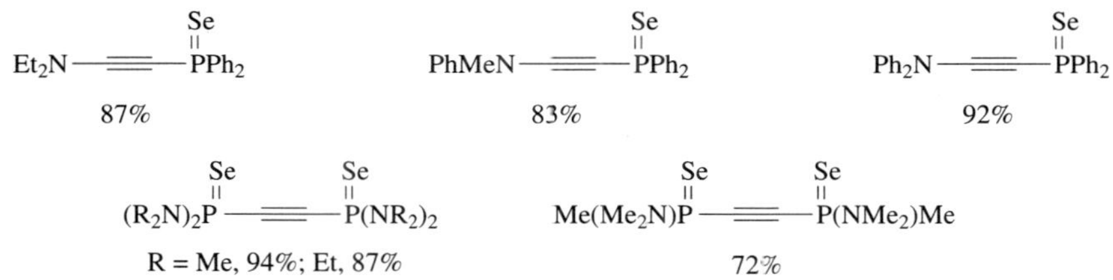

Scheme 61

reaction of alkynyl Grignard reagents with dialkyl or diaryl halophosphates. Both methods give good yields and although (i) has been investigated more extensively, method (ii) is more attractive from a practical point of view as it avoids the use of the hazardous alkynyl halides.

(i) From alkynyl halides

The synthesis of alkynylphosphonate esters by the reaction of trialkyl phosphites with alkynyl halides (Equation (30)) was first reported by Simpson and co-workers ⟨60CI(L)47, 63JCS3351⟩ and has been investigated since by a number of groups worldwide ⟨69CR(268)358, 69JOC1483⟩ with notable contributions from the Russian workers Petrov and Ionin ⟨62JGU2355, 63JGU2791, 67JGU898⟩. A wide range of alkynylphosphonates are available in good yield by this method, which is limited only by the availability of the alkynyl halide. The reaction is usually carried out by heating the alkynyl halide with the trialkyl phosphite in an inert solvent until the evolution of the alkyl halide is complete. In many cases the reaction proceeds at or below room temperature. One interesting observation is that treatment of the alkynyl phosphonate (**117**) with phosphorus pentachloride in benzene gave the alkynylphosphonic dichloride (**118**) in excellent yield (Equation (31)) ⟨65JGU1582⟩. Scheme 62 illustrates a number of examples of alkynylphosphonate synthesis from the literature ⟨63JGU2791, 65JGU1910, 67JGU898, 69CR(268)358, 69JCS(C)2273, 81TL1875⟩.

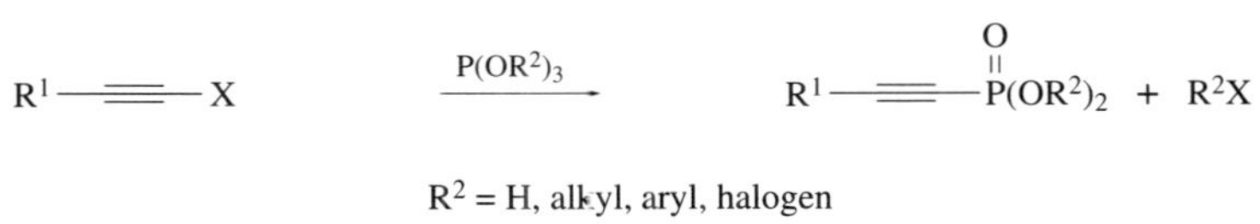 (30)

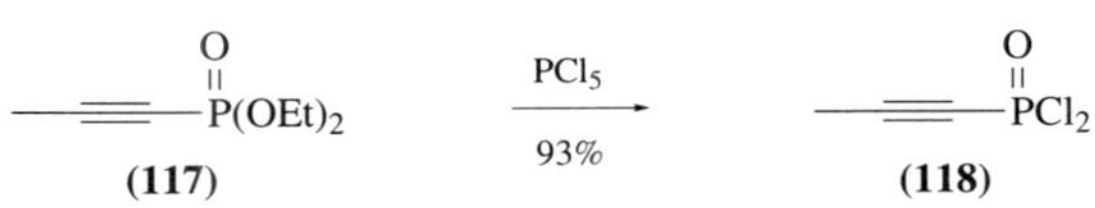 (31)

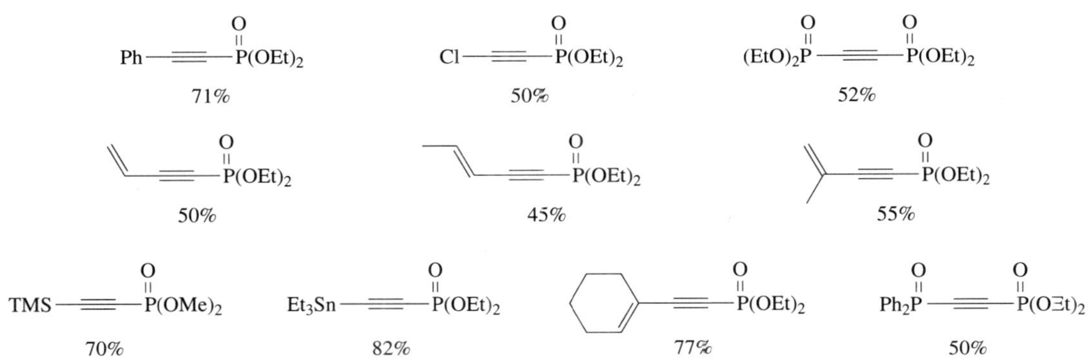

Scheme 62

For a number of years, the mechanism of the above reactions was assumed to be of the Arbusov type. This, however, has largely been disproved through the studies of Burt and Simpson who, through a series of clever experiments, deduced that the reaction proceeds via initial abstraction of halogen by the phosphorus to give an acetylide anion which then reacts with the resulting phosphonium halide to give the alkynylphosphonium species (**119**) ⟨71JCS(C)2872⟩. Loss of alkyl halide then gives the product (Scheme 63). By using hydroxylic solvents it was possible to protonate the intermediate acetylide anion.

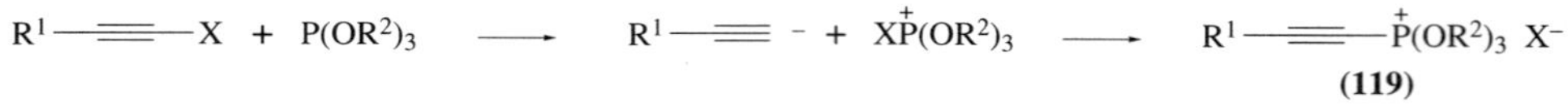

Scheme 63

In a related procedure, sodium diethyl phosphite reacts with a variety of alkynyl bromides in ether to give moderate to good yields of the resulting phosphate esters (Scheme 64) ⟨65CR(261)1019, 66BSF1707⟩. The reaction is thought to proceed by abstraction of bromide to give the alkynic anion and the phosphoryl bromide followed by nucleophilic attack at phosphorus. This is supported by the fact that use of the alkynyl alcohol derivative (**120**) gives rise to the phosphate ester (**122**) by equilibration to the intermediate alkoxide (**121**).

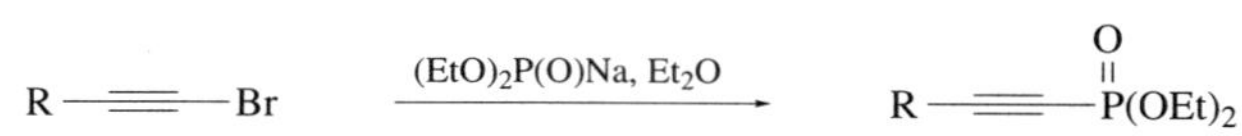

$R = C_5H_{11}$, 73%; Ph, 50%; Me_2COH, 68%; 2-propenyl, 65%

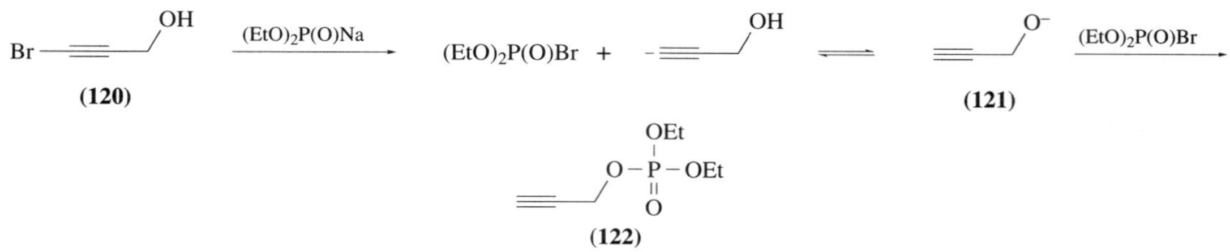

Scheme 64

(ii) From alkynyl Grignard reagents

Early investigations of the reaction of acetylide ions with dialkyl halophosphates as a route towards alkynylphosphonates gave poor yields ⟨60CI(L)47, 63JCS3351⟩. This was thought to be caused by the polymerization of the products from reaction of the acetylide with the newly formed alkynylphosphonate. It was later shown by Chattha and Aguiar, however, that this strategy (Equation (32)) provided an ideal method for the synthesis of phosphonates from readily available alkynyl Grignard reagents ⟨71JOC2719⟩. Furthermore, it is probably the method of choice as most of the starting materials are commercially available and it also avoids the use of the hazardous alkynyl halides. The reaction is conveniently carried out by converting the alkyne to the corresponding Grignard reagent by treatment with ethyl magnesium bromide followed by the addition of either the dialkyl or diphenyl chlorophosphate. This method is also amenable to large-scale work, e.g., 50 g batches of the dibutylphosphonate of THP-protected propargyl alcohol could be prepared in one operation using this methodology ⟨79JOC865⟩. Table 6 summarizes the results of Chattha and Aguiar's investigations.

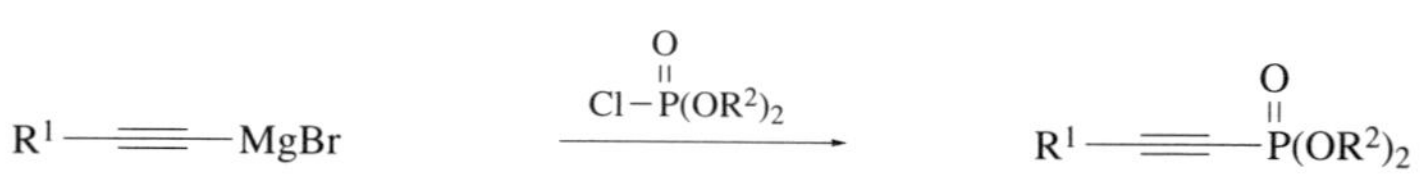

(32)

R^1 = alkyl, cycloalkyl, aryl
R^2 = alkyl, aryl

The same authors also reported that this method provides a high yielding entry to the analogous alkynyl-1-thiophosphonates (**123**) by the reaction of alkynyllithiums with dialkyl chlorothiophosphates (Equation (33)) ⟨71JOC2720⟩. Best results were obtained using a catalytic amount of pyridine, which is thought to enhance the nucleophilicity of the alkynyllithium by coordinating to

Table 6 Synthesis of phosphonates from alkynyl Grignard reagents (Equation (32)).

Entry	*1*	*2*	*3*	*4*	*5*	*6*	*7*	*8*	*9*	*10*
R^1	Me	Pr^n	Bu^n	n-C_6H_{13}	Ph	$PhCH_2CH_2$	n-C_6H_{13}	n-C_5H_9	Me	Pr^n
R^2	Et	Et	Et	Et	Et	Et	Et	Et	Ph	Me
Yield (%)	76	59	64	52	53	60	51	70	74	57

the lithium atom thus labilizing the carbon–lithium bond. Low yields of products were obtained when the alkynyl Grignard reagents were used. The results are summarized in Table 7.

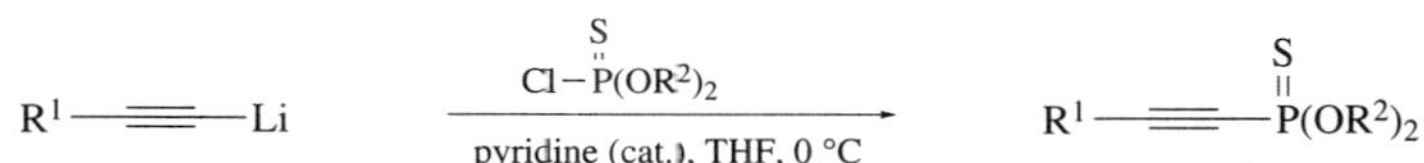

(123) (33)

R^1 = alkyl, cycloalkyl, phenyl
R^2 = alkyl

Table 7 Synthesis of alkynyl-1-thiophosphonates (Equation (33)).

Entry	*1*	*2*	*3*	*4*	*5*	*6*	*7*	*8*	*9*
R^1	Me	Pr^n	Bu^n	n-C_6H_{13}	$PhCH_2CH_2$	C_6H_{11}	C_5H_9	Ph	Ph
R^2	Et	Et	Et	Et	Et	Et	Et	Et	Me
Yield (%)	76	61	66	79	52	74	81	52	35

(iii) Miscellaneous methods

Although there are a number of alternative methods for the synthesis of alkynylphosphonates they cannot compete, in terms of generality, with the methods above. For example, one of the first alkynylphosphonate syntheses involved dehydrohalogenation of substituted vinyl halides with potassium hydroxide ⟨33CB278⟩. Other methods include the prototropic rearrangement of allene derivatives which themselves are prepared by sigmatropic rearrangement of propargyl phosphites ⟨62JOC1828⟩. Elimination and isomerization of a bromo(allyl)phosphonate to the corresponding alkynylphosphonate has also been reported ⟨65CR(260)1984⟩.

2.22.2.6 Alkynylphosphoranes

The synthesis of pentavalent phosphorus compounds has been an area of intense activity since the early 1960s. There have been a number of reports on the synthesis of alkynylphosphoranes starting from a trivalent alkynylphosphorus compound. For example, Fluck *et al.* found that treatment of the alkynylaminophosphine (**124**) with sulfur tetrafluoride gave the phosphorane (**125**) in 44% yield. Similarly, reaction of the fluoro(amino)alkynylphosphine (**126**) with sulfur tetrafluoride gave the phosphorane (**127**) in 38% yield (Scheme 65) ⟨87ZAAC(548)63⟩.

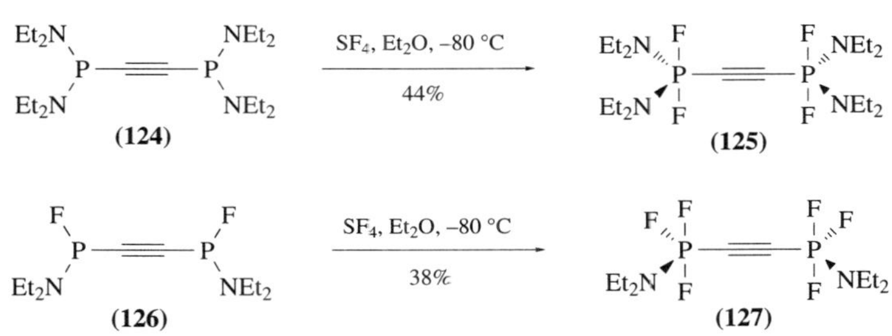

Scheme 65

The same authors also reported an ingenious synthesis of the phosphorane (**129**) by treating the phosphine (**128**) with tetrachloro-*o*-benzoquinone at $-80\,^\circ C$ ⟨88ZAAC(567)39⟩. The crystalline product was isolated in almost quantitative yield (Equation (34)).

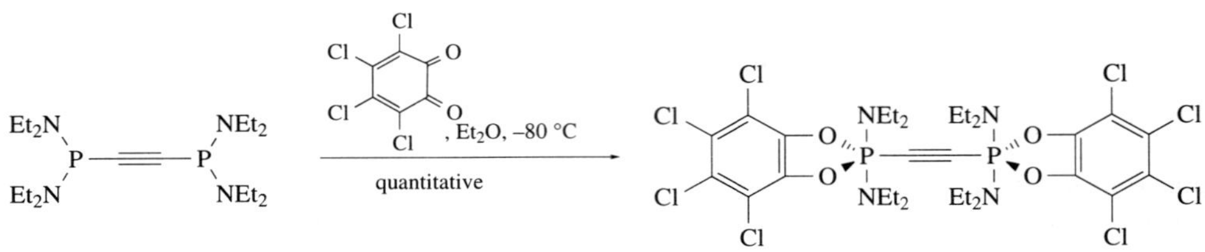

(128) (129)

(34)

A Russian group also active in this area reported that a variety of substituted alkynylphosphonite derivatives (**130**) reacted with pyruvic acid esters at $0\,^\circ C$ in ether to give the alkynylphosphoranes (**131**) in moderate to excellent yields ⟨89TL577, 89ZOB1979, 91ZOB611⟩. These reactive phosphoranes underwent rearrangement to the bicyclic vinylphosphoranes (**132**), which were the only products when the reaction was carried out at reflux in benzene. Use of acyl cyanides gives the cyano-substituted alkynylphosphoranes (**133**) (Scheme 66).

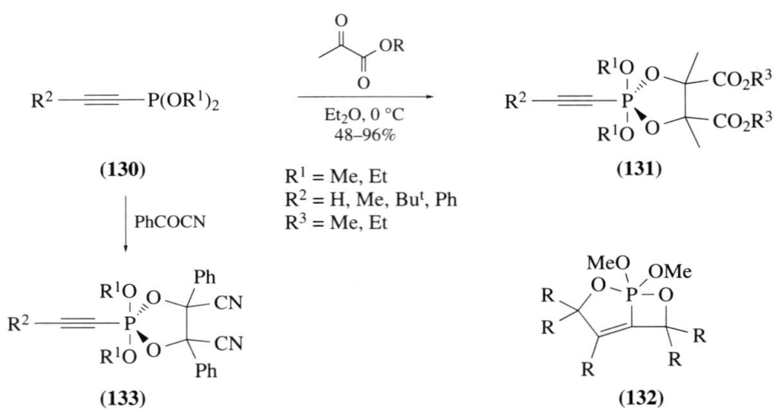

Scheme 66

2.23
Alkynylarsenic, -antimony, -bismuth, -boron, -silicon, -germanium and -metal Compounds

WILLIAM KITCHING
The University of Queensland, Brisbane, Qld, Australia

and

KLAUS KWETKAT
HÜLS AG, Marl, Germany

2.23.1 ALKYNYLARSENIC, -ANTIMONY AND -BISMUTH DERIVATIVES

2.23.1.1 Toxicity, Stability and Manipulation

It should be noted that considerable caution is required in any procedure involving these derivatives which incorporate, in addition to the very reactive alkyne moiety, the metalloid centre and ancillary ligands (organic groups, halide, etc.). Certain combinations are potentially hazardous, and known to be biologically active. The available, but somewhat scattered literature on arsenic, antimony and bismuth compounds has been summarised by Thayer ⟨74JOM(76)265⟩ and the biological effects and toxicology of arsenic and antimony compounds, and to a lesser extent, bismuth has been discussed. These derivatives are of variable stability, with the bismuth compounds generally being the least stable, and showing sensitivity to oxygen and moisture. Some alkynyl compounds are reported to be explosive, and a precautionary note is to be found in the reviews by Davidsohn and Henry ⟨67CRV73⟩ and Azerbaev *et al.* ⟨74RCR657⟩. For example, triethynyl derivatives explode in air whereas *t*-butyl substitution, which is often the method used, may increase stability. Stability with respect to heat and light varies considerably with the metalloid and its oxidation state.

Laboratory procedures preventing skin contact, inhalation and ingestion of these compounds should be rigorously adopted.

2.23.1.2 Alkynylarsenic Compounds

The first reports of such compounds appeared in ⟨21BSF1024⟩ and ⟨23LA(432)39⟩. The first reviews of this subject were by Davidsohn and Henry ⟨67CRV73⟩, followed by Azerbaev *et al.* ⟨74RCR657⟩. The most recent summary is to be found in ⟨78HOU(13/8)33⟩. Coverage of the literature for this entire chapter extends up to early 1993. Examination of the literature indicates that four general methods of preparation for the alkynyl arsine moiety can be identified, and certain compounds (e.g., As(V)) are accessible by oxidation of the arsenic centre following formation of the arsenic–alkynyl bond. There is inevitably some overlap between these categories, which do, however, form a useful basis for discussion. Additional listings of alkynylarsenicals are found in the cited reviews and are not retabulated.

Method 1: Reaction of an arsenic halide with a metal acetylide

This procedure is well illustrated in the key paper by Reiff and Pant ⟨69JOM(17)165⟩ who describe the synthesis of alkynyl compounds of phosphorus, arsenic and antimony, and demonstrate the stabilising effect of *t*-butyl groups in the alkyne moiety on these derivatives. Thus the tris(3,3-dimethyl-1-butynyl)arsine was obtained as shown in Equation (1). This arsine can be stored at room temperature, is not light sensitive, is remarkably thermally stable, melting without decomposition at 62 °C, and can be sublimed. Peroxide oxidation to yield the arsine oxide was partially successful ($\approx$50%). Preparation of the precursor alkynylithium reagents, from 1-alkynes, is easily achieved by acid–base reaction with *n*-butyllithium in tetrahydrofuran. Similarly, alkynyl derivatives of the type R—C≡C—MgX are routinely made by reaction of the 1-alkyne with, for example, ethylmagnesium bromide. For more extensive discussions, the books by Wakefield ⟨B-74MI 223-01⟩ and Brandsma and Verkruijsse ⟨B-90MI 223-01⟩ should be consulted. These approaches have been explored by Lukehart and co-workers ⟨92OM3993⟩ and full details are given for the synthesis shown in Equation (2). A very interesting variant, that provided a disubstituted acetylene incorporating both arsenic and phosphorus, was reported by Carty *et al.* ⟨71CJC2706⟩ (Equation (3)). Alkynylsodium intermediates have also been employed but less generally than the alkynylithium reagents easily generated by hydrogen–lithium exchange. A tandem elimination (HCl)-deprotonation sequence is described in a German patent ⟨62GEP1126388⟩ and summarised in Equation (4). Hartmann *et al.* ⟨59N321⟩ and Gavrilov ⟨72JGU1957⟩ have also reported the use of sodium intermediates (Equation (5)). A wide selection of alkynylarsenicals has been made using both magnesium and sodium alkynes, with further derivatisation (oxidation) being described. The procedures are summarised in Equations (6) and (7) and moderate to excellent yields are reported ⟨65CR(261)1996, 68CR(C)937⟩. Variations of these procedures employing ether as solvent have been published ⟨59N321, 88JGU2136⟩, but in certain cases replacement of ether by chloroform is claimed to be efficacious ⟨57ZAAC(290)348⟩, as shown in Equation (8). There is one report ⟨82IZV933⟩ that lithium tetraalkynylaluminate R—C≡C)$_4$AlLi (in THF) efficiently alkynylates arsenic when provided as PhAsCl$_2$, and with manipulation of the conditions, either product shown in Equation (9) can be obtained. The precursor aluminate is generated by reaction of the alkyne (R—C≡CH) with $LiAlH_4$, but R in Equation (9) is unusual (e.g., 2,2-dimethyl-2,3-dihydropyran) and the generality of the procedure is unclear. Phenylethynylsilver has also been employed by Gol'ding *et al.* ⟨72IZV529⟩ as an alkynyl transfer reagent for phosphorus, arsenic and antimony, as shown in Equation (10). Alkynylstannanes have also been employed for transfer to arsenic ⟨65N59, 68LA(717)1, 68IZV2287⟩ but the applications are limited, although more promising than the use of alkynyl silanes or germanes ⟨66IC1004⟩.

$$AsCl_3 + 3\,Bu^t{-}C{\equiv}C{-}Li \xrightarrow{THF} (Bu^t{-}C{\equiv}C)_3As \tag{1}$$

$$Bu^t{-}C{\equiv}CH \xrightarrow[\text{ii, }PhAsCl_2]{\text{i, }Bu^nLi\ (THF)} (Bu^t{-}C{\equiv}C)_2AsPh \tag{2}$$

$$Ph_2P{-}C{\equiv}CH \xrightarrow[\text{ii, }Ph_2AsCl]{\text{i, }Bu^nLi\ (THF)} Ph_2P{-}C{\equiv}C{-}AsPh_2 \tag{3}$$

$$Ph_2AsBr + ClCH{=}CHCl + 2\,Na \xrightarrow[82\%]{NH_3\ (l)} Ph_2As{-}C{\equiv}C{-}Cl \tag{4}$$

$$R^1{}_2AsCl + R^2{-}C{\equiv}C{-}Na \xrightarrow{THF} R^2{-}C{\equiv}C{-}AsR^1{}_2 \tag{5}$$

$$R^1{}_{3-n}AsX_n + R^2{-}(C{\equiv}C)_m{-}MgX \xrightarrow{THF} R^1{}_{3-n}As\left[(C{\equiv}C)_m{-}R^2\right]_n \tag{6}$$

$$Ph_2AsCl + Na{-}C{\equiv}C{-}CH_3 \xrightarrow[96\%]{hexane} Ph_2As{-}C{\equiv}C{-}CH_3 \tag{7}$$

$$R_2AsI + BrMg{-}C{\equiv}C{-}MgBr \xrightarrow{CHCl_3} R_2As{-}C{\equiv}C{-}AsR_2 \tag{8}$$

R = Et, *n*-hexyl, 1-naphthyl

$$(R{-}C{\equiv}C)_4AlLi \xrightarrow[85\%]{PhAsCl_2} R{-}C{\equiv}C{-}AsClPh$$
$$(R{-}C{\equiv}C)_4AlLi \xrightarrow[90\%]{PhAsCl_2,\ THF} (R{-}C{\equiv}C)_2AsPh \tag{9}$$

$$AsCl_3 + 3\,Ph{-}C{\equiv}C{-}Ag \xrightarrow[50\,^\circ C,\ 1\ h]{2\ PhH : 1\ THF} (Ph{-}C{\equiv}C)_3As \tag{10}$$

Dimetallic derivatives of alkynes, for example, M(—C≡C)—M, have also been employed in alkynylarsine syntheses. The conditions used are similar to those required for R—C≡C—M derivatives, and are illustrated in Equation (11). Reaction of R_2AsCl (R = phenyl, 1-naphthyl) with two molar equivalents of sodium acetylide (in NH_3) proceeds in moderate yield (≈60%) to provide bisarsines (Equation (12)). The method is difficult to control with respect to stoichiometry and appears to be useful for aryl derivatives only.

$$2\,R_2AsX + M{-}(C{\equiv}C)_n{-}M \longrightarrow R_2As{-}(C{\equiv}C)_n{-}AsR_2 \tag{11}$$

n = 1; M = MgBr, Cl, I; R = Pr^n, Bu^n, Bu^i, *n*-Pent, *n*-Hex
n = 1; M = $SnBu_3$; R = Ph
n = 2; M = $SnMe_3$; R = Ph
n = 2; M = MgBr, Cl, I; R = Et, Pr^n, Pr^i, Bu^i

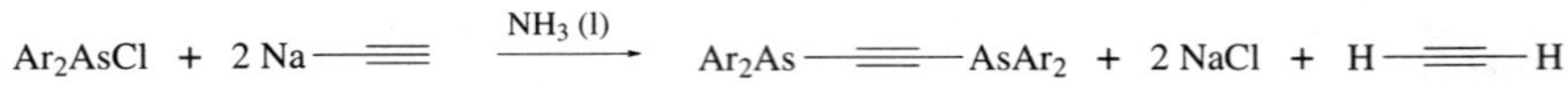
(12)

Method 2: Condensation of sufficiently acidic terminal alkynes

Cullen and Leeder ⟨66IC1004⟩ have investigated methods for direct use of the alkyne. In most cases addition occurs but the necessary subsequent elimination is generally difficult, as for example in the addition of dimethylarsine (Equation (13)). Use of dimethylarsenic chloride appeared more attractive, as formation of the alkynylarsine would now require, in the final step, elimination of hydrogen chloride. Some success resulted, but only at high temperatures (Equation (14)). The propynylarsine is presumably formed by elimination of HCl from $(CH_3)_2AsCH{=}C(CF_3)Cl$. Considering that Sn—N, Sn—O and Sn—S bonds react readily with acidic C—H bonds, analogous arsenic derivatives would be expected to react similarly. This was demonstrated, although the As—S system is less reactive than As—N, with the As—O method being most economical, as water eliminated in the final step does not readd to the starting alkyne. Examples are shown in Scheme 1. Essentially the same approach has been described by Kuz'min *et al.* ⟨70JGU1100⟩, but utilising phenylacetylene (Equation (15)). Arsinic acids were also expected to enter into condensation reactions with phenylethyne to provide alkynylarsines. However, tertiary arsines of the type $R_2AsC{\equiv}CPh$ were isolated, and the sequence of events as shown in Scheme 2 was suggested ⟨68JGU1295⟩. The mechanisms of these transformations, which provided tertiary arsines in moderate to low yields, were unclear.

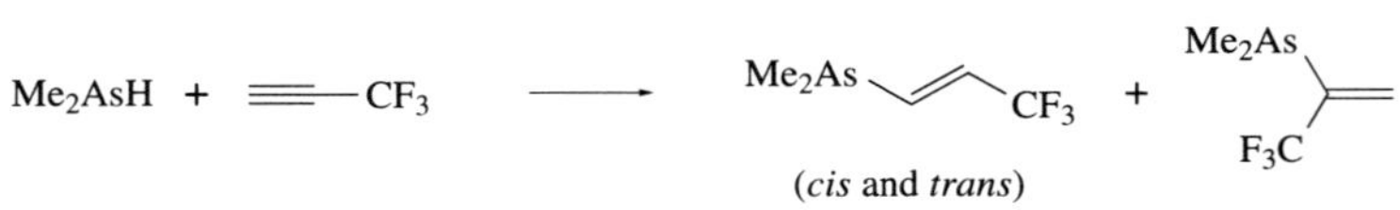
(13)

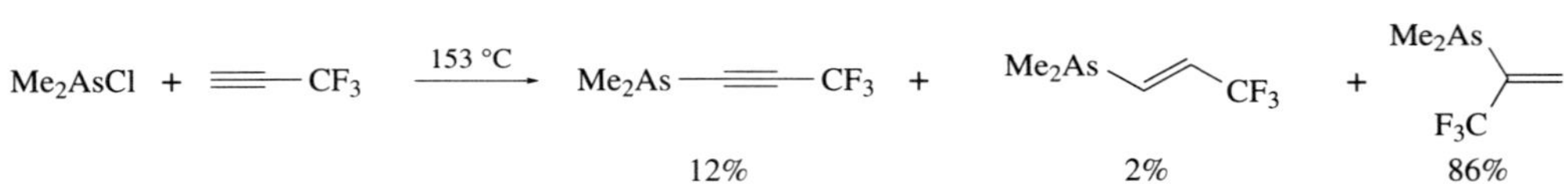
(14)

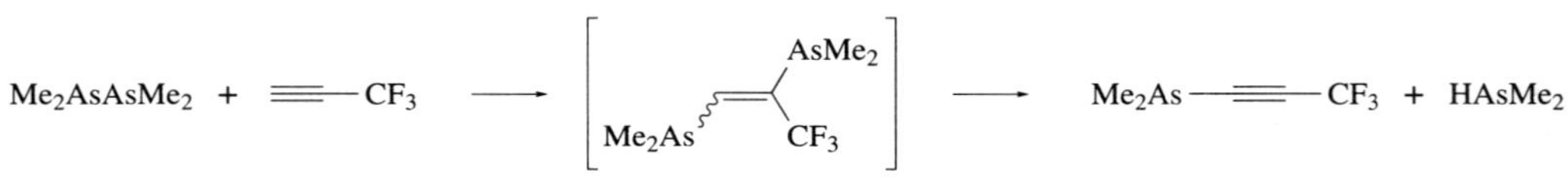

Scheme 1

(15)

R = alkyl; X = I, OR, OC(O)Me

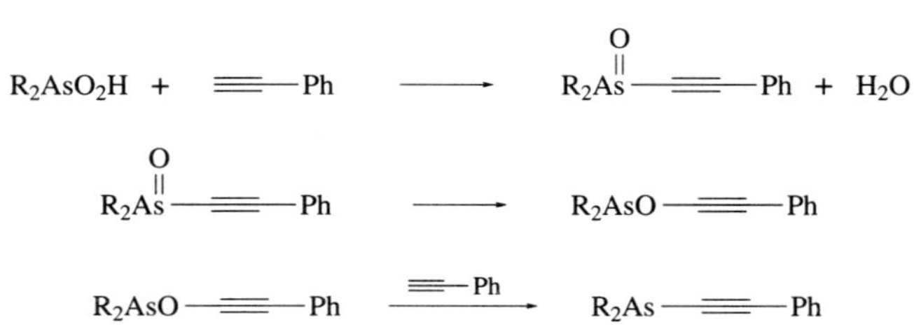

Scheme 2

Method 3: Modification of preformed alkynylarsines

Formation of metallic derivatives of alkynyl derivatives (e.g., $R_2AsC{\equiv}CH$) creates possibilities for synthesis of more functionalised alkynyl arsines. These reports date to the early 1960s, and Luijten *et al.* ⟨64RTC1301⟩ reported that the Grignard reagent from $R_2AsC{\equiv}CH$ formed easily and could be converted to carboxyl derivatives. The same method has been discussed by Kuz'min and Pavlova in a number of reports ⟨69JGU1068, 69JGU1797⟩ and the magnesium derivatives have been reacted with anhydrides, aldehydes, ketones and carbon dioxide to provide the expected addition products in moderate yields. These are illustrated in Scheme 3. Similarly, treatment of the arsine with either Et_2NH or KOH ⟨87JGU2497⟩, generated a reactive form of the arsine which reacts with formaldehyde or acetone. The procedure using potassium hydroxide was previously utilised by Benaim ⟨65CR(261)1996⟩. Certain alkynylarsines form organometallic complexes on reaction with dicobaltoctacarbonyl under mild conditions, the products being sterically hindered carbon–cobalt tetrahedrane cluster units ⟨90ZN(B)1529⟩. This chemistry is summarised in Equation (16), with addition occurring to all three alkynyl moieties under the correct stoichiometric conditions. Heating the above complex, where $n = 1$, in toluene provides the structurally interesting dimer (**1**).

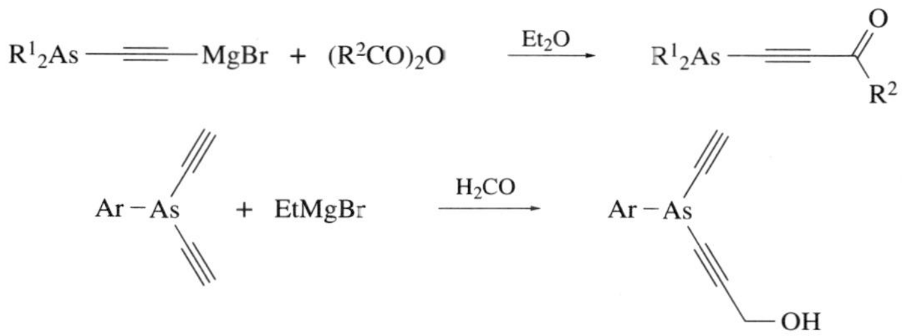

Scheme 3

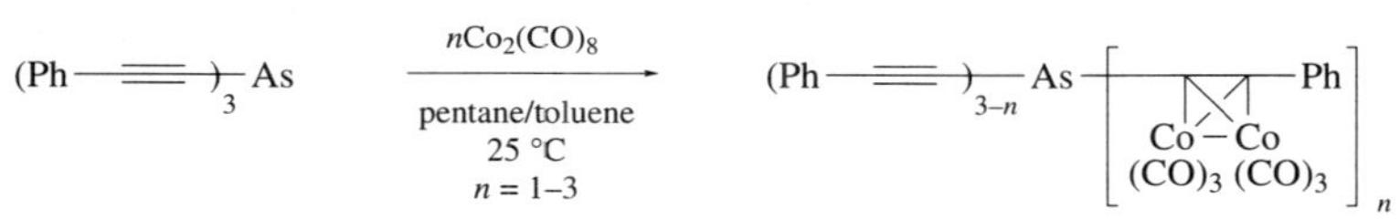 (16)

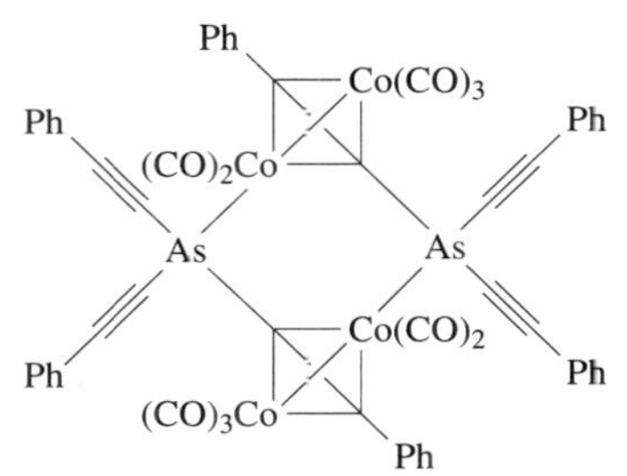

(**1**)

Method 4: Quaternisation and related reactions

Quaternisation using methyl iodide was reported in 1957 by Hartmann and Nowak ⟨57ZAAC(290)348⟩ as shown in Equation (17). A more extensive study of methylation was reported by Benaim ⟨65CR(261)1996⟩ on a variety of more complex arsines as shown in Equation (18), and both methyl iodide and tosylate were used with acetonitrile being a suitable solvent. Direct addition of organylbromides to triethylarsine also proceeded in almost quantitative yields to provide the compounds (**2**) and (**3**).

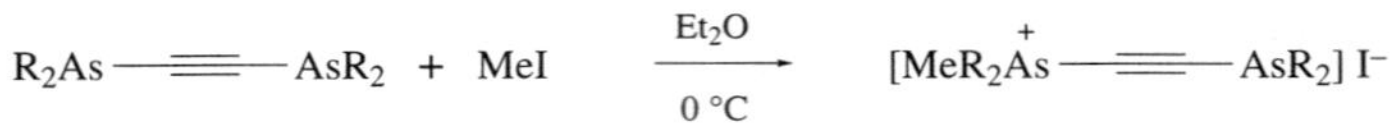
(17)

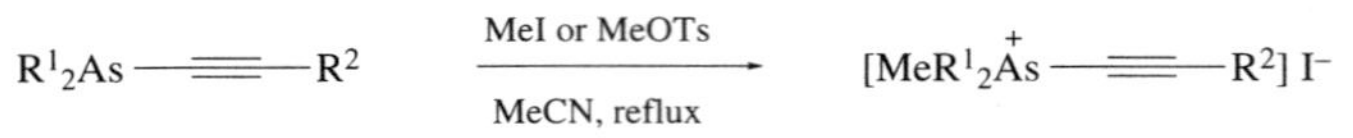
(18)

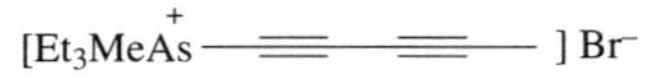
(2)

(3)

Method 5: Oxidation at arsenic

Oxidation of alkynylarsines is more difficult than that of the corresponding phosphines, and use of hydrogen peroxide in acetic acid or acetone provides the oxide or hydroxide in high yield, with the bishydroxide undergoing dehydration on heating, or in a desiccator ⟨65CR(261)1996⟩ (Equation (19)). Lukehart and co-workers ⟨92OM3993⟩ have also reported that oxidation is efficient with *t*-butylhydroperoxide, to obtain a fair yield of the arsine oxide based on the *t*-butylethyne derivative (Equation (20)). Reaction of this oxide with a cationic platinum hydride reagent resulted in the unexpected elimination of $(PhAsO)_n$ and a platinum complex containing both alkynyl ligands. Perbenzoic acid has been employed also for formation of arsenic oxides ⟨62ZOB1349⟩. Alkyl- and alkynylarsenic acids have been acquired ⟨71JINC3177⟩ from aminoarsenic chlorides which involves an interesting hydrolytic substitution of the amine followed by peroxide oxidation in the second step (Equation (21)). These liquid arsinic acids were difficult to purify and underwent slow decomposition to produce arsenic trioxide.

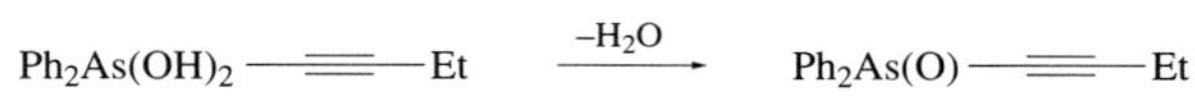
(19)

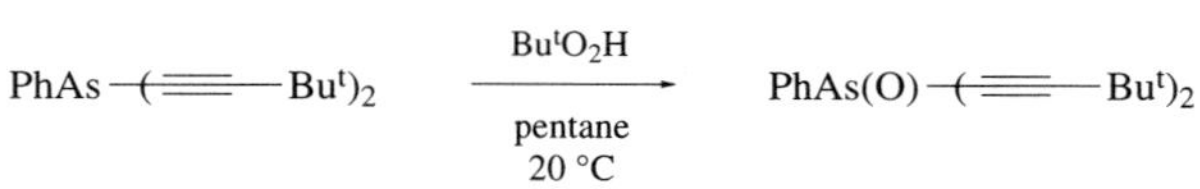
(20)

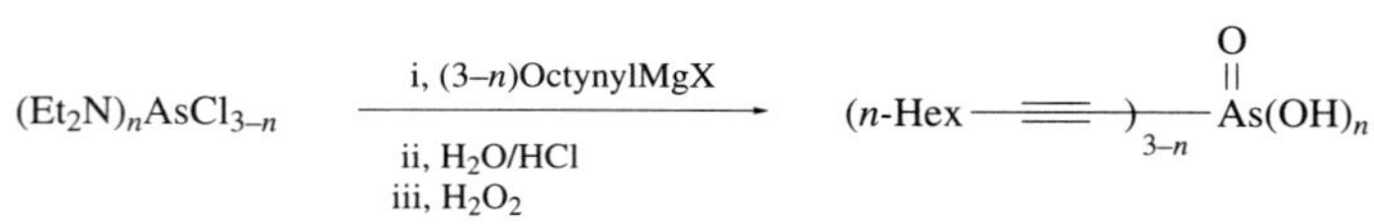
(21)

2.23.1.3 Alkynylantimony Compounds

Alkynylantimony derivatives have not been exhaustively reviewed as a group, but information is available within the general review of alkynyl derivatives of arsenic, antimony and bismuth ⟨67CRV73⟩, and additional information is located in the handbooks *Houben-Weyl* ⟨78HOU(13/8)33⟩ and *Gmelin* ⟨B-82MI 223-01⟩. The reported procedures for acquiring alkynyl antimony systems are discussed under the headings used in presenting the alkynylarsenic derivatives.

Method 1: Reaction of an antimony halide with a metal acetylide

Alkynyllithium reagents have been key precursors for alkynylantimony compounds, for both the +3 and +5 oxidation states. In the latter case, the triorganoantimony was oxidised with halogen to generate R_3SbX_2 which was treated with two molar equivalents of the alkynyllithium. These procedures are summarised in Scheme 4. Introduction of the simple ethynyl group (HC≡C) has not been performed with HC≡CLi, whereas the more general use of ethynylsodium requires care in the choice of solvent. For example, if the ethynyl derivative is the target compound, THF should be used, as HC≡CNa is less soluble in this solvent and causes a lower level of product deprotonation, and hence a lower level of a disubstituted alkyne. Alternatively, use of liquid NH_3 for HC≡CNa leads to a higher level of the disubstituted compound ⟨61ZAAC(312)186⟩. These aspects are shown below in Equations (22)–(24). Acetylides other than the parent, for example, RC≡CNa, have also been utilised in THF solvent ⟨59N321⟩. Acetylenic Grignard reagents are the most common source of the alkynyl group for preparing alkynylantimony compounds. For example, $(HC{\equiv}C)_3Sb$ is prepared in this way using ether–THF as solvent (Equation (25)) ⟨64RTC1301⟩. This procedure works well for alkynylating mono- and diorganoantimony halides (Equations (26) and (27)). Certain acetylides of tin and silver have also found application in this area, but their use is restricted to a few examples. Mono- and bis(tributylstannyl)alkynes react in benzene with diphenylantimony halides as shown in Equation (28) ⟨68IZV2287⟩. Tri(phenylethynyl)antimony has been prepared directly from the trichloride and phenylethynylsilver (Equation (29)).

$$R^1{}_2SbX + R^2{-}C{\equiv}C{-}Li \longrightarrow R^1{}_2Sb{-}C{\equiv}C{-}R^2$$

R^1 = 1-Naphthyl; R^2 = Ph

$$SbCl_3 + 3\,R^1{-}C{\equiv}C{-}Li \longrightarrow (R^1{-}C{\equiv}C)_3Sb$$

R^1 = Bu^t, Ph

$$R^1{}_3Sb + X_2 \longrightarrow R^1{}_3SbX_2 \longrightarrow R^1{}_3Sb(C{\equiv}C{-}R^2)_2$$

R^1 = Me; R^2 = Me; X_2 = Cl_2
R^1 = MeC≡C–; R^2 = Me; X_2 = Cl_2
R^1 = *p*-Tol; R^2 = Ph; X_2 = Br_2

Scheme 4

$$R_2SbX + HC{\equiv}C{-}Na \longrightarrow R_2Sb{-}C{\equiv}CH \qquad (22)$$

$$R_2Sb{-}C{\equiv}CH + HC{\equiv}C{-}Na \longrightarrow R_2Sb{-}C{\equiv}C{-}Na + H{-}C{\equiv}C{-}H \qquad (23)$$

$$R_2Sb{-}C{\equiv}C{-}Na + R_2SbX \longrightarrow R_2Sb{-}C{\equiv}C{-}SbR_2 + NaX \qquad (24)$$

$$SbCl_3 + 3\,HC{\equiv}C{-}MgBr \xrightarrow[\text{ii, } NH_4Cl,\ -30\ ^\circ C]{\text{i, } Et_2O/THF,\ -30\ ^\circ C} (HC{\equiv}C)_3Sb \qquad (25)$$

$$R^1SbX_2 + 2\,R^2{-}C{\equiv}C{-}MgBr \longrightarrow R^1Sb(C{\equiv}C{-}R^2)_2 \qquad (26)$$

$$R^1{}_2SbX + R^2{-}C{\equiv}C{-}MgBr \longrightarrow R^1{}_2Sb{-}C{\equiv}C{-}R^2 \qquad (27)$$

$$2\,Ph_2SbCl + Bu^n{}_3Sn{-}C{\equiv}C{-}SnBu^n{}_3 \xrightarrow[-2Bu^n{}_3SnCl]{PhH} Ph_2Sb{-}C{\equiv}C{-}SbPh_2 \qquad (28)$$

$$SbCl_3 + 3\,Ph{-}C{\equiv}C{-}Ag \xrightarrow{PhH/THF} (Ph{-}C{\equiv}C)_3Sb \qquad (29)$$

Method 2: Condensation of sufficiently acidic terminal alkynes

This procedure was shown to be effective for a number of alkynylarsines, but in 1994 has not found application in antimony chemistry.

Method 3: Modification of preformed alkynyl derivatives

Deprotonation of ethynyl derivatives, generally to form the magnesium derivative, permits efficient condensation with aldehydes and ketones, for example, to yield the hydroxyalkyl compounds. This is exemplified in Equation (30) and a listing of such compounds is available ⟨B-81MI 223-01, B-81MI 223-02⟩. The metal carbonyl complex formed by reaction of tri(phenylethynyl) antimony with dicobaltoctacarbonyl as described in the arsine case, has been fully characterised, but the interesting dimer (**1**) could not be obtained for the antimony case.

(30)

Method 4: Quaternisation and related reactions

There are few examples in the antimony series, reported by Hartmann and Kuhl, formed by methylation of the alkynylantimony compounds with methyl iodide ⟨61ZAAC(312)186⟩ (Equation (31)). These compounds are easily hydrolysed, but show good stability under anhydrous conditions.

$$Me_2Sb{-}C{\equiv}C{-}SbMe_2 + MeI \longrightarrow [Me_3Sb{-}C{\equiv}C{-}SbMe_2]^+\ I^- \quad (31)$$

Method 5: Oxidation at antimony

Two basic reaction types fall into this category, viz. halogen addition to generate antimony(V) derivatives, and oxidation by peroxide. Halogen addition was unimportant for arsenic, but there are several examples for antimony, and in fact, it is the first step in converting R_3Sb to R_5Sb (see Scheme 4) ⟨81ZAAC(474)157⟩. There are many examples of such halogen addition. Oxidation at antimony, in the controlled way described for arsenic, has not been achieved, and there are no well characterised alkynylantimony oxides. Hartmann and Kuhl ⟨61ZAAC(312)186⟩ examined oxidation (with H_2O_2 in acetone) of a number of alkynyl compounds, but could not isolate alkynyl containing antimony oxides. An attempt to generate antimony sulphides also failed.

2.23.1.4 Alkynylbismuth Compounds

Well characterised alkynylbismuth compounds are few in number, and thus useful synthetic methods are not extensive. Despite efforts, trialkynyl derivatives are unknown, and a report in 1991 suggests such compounds undergo rapid polymerisation ⟨91ZN(B)1319⟩. This feature is part of the difficulty in purifying dialkynylbismuth derivatives, which are thought to undergo dissociation to the trialkynylbismuth compounds which then polymerise. The monoalkynyl compounds are well characterised and are generally acquired by reaction of the magnesium or sodium alkynes with diarylbismuthhalides, for example Equation (32) ⟨62ZAAC(317)54, 91ZN(B)1319⟩. *Ortho*-substituted arylbismuth compounds are generally more stable and use of methyl and 2-propyl groups for this purpose is effective (Scheme 5) ⟨92BCJ3504⟩. As far as the authors are aware, there are no examples of pentaorganobismuth compounds, and attempts to make such derivatives with at least one alkynyl group result in cleavage of this group.

$$MeBiBr_2 + 2\ HC{\equiv}C{-}Na \longrightarrow MeBi(C{\equiv}CH)_2 \quad (32)$$

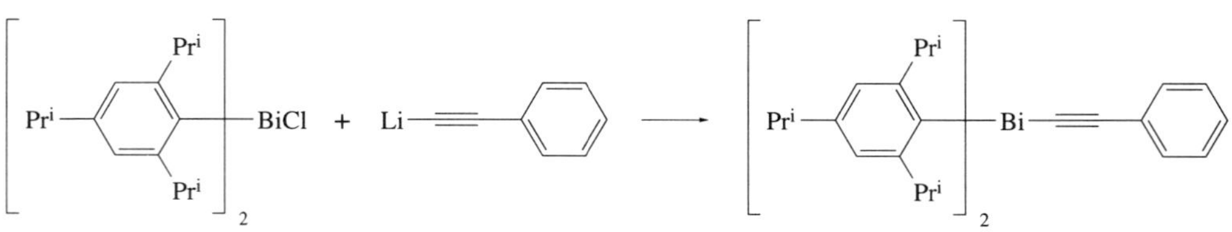

Scheme 5

2.23.2 ALKYNYLBORON, -SILICON AND -GERMANIUM DERIVATIVES

2.23.2.1 Alkynylboron Compounds

2.23.2.1.1 Triorganoboranes

The early review by Davidsohn and Henry ⟨67CRV73⟩ discussed alkynyl derivatives generally, but the position as of 1984 with respect to alkynylboranes is contained within an exhaustive summary of organoboron chemistry, which includes a very useful presentation of ^{11}B NMR data ⟨82HOU(13/3b)1, 84HOU(13/3c)1⟩. The approach to be adopted here will utilise selected examples to illustrate important general methods. The most abundant compounds in this general category are those that contain one alkynyl group, and several methods are generally applicable.

Method 1: Reaction of a boron halide with a metal acetylide

Use of sodium or lithium acetylides to effect 1-alkynylation of diorganoboron halides (mostly chlorides) is best conducted using pentane, cyclohexane or benzene as solvent. This procedure was utilised in 1993 to obtain diphenylphosphinoethynyldimesitylborane and (ferrocenylethynyl)-dimesitylborane in good yields ⟨93JOM(449)27⟩, and is summarised in Scheme 6. Molecular structures of the products have been reported.

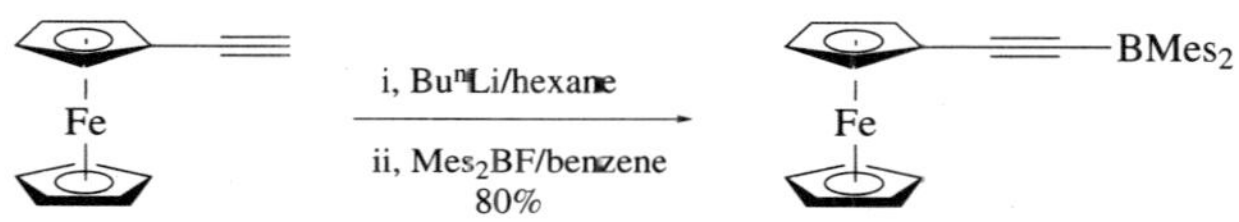

Scheme 6

Soulié and Cadiot ⟨66BSF3846⟩ have developed procedures employing alkynylsodium reagents and emphasised the necessity for working with an inert solvent (hydrocarbon) to avoid the formation of Lewis base adducts. Subsequently, Köster and co-workers ⟨68LA(717)21⟩ reported on the ease of borate formation and further irreversible reactions that result in lower yields. Hydrocarbons and not ethers should be employed as solvents to minimise those complications. The products shown in Equation (34) are unstable and inflame on exposure to air and this behaviour is typical of the more volatile alkynylboranes.

$$\text{Mes}{-}{\equiv}{-}\text{MgBr} + \text{Mes}_2\text{BF} \xrightarrow[67\%]{\text{THF}} \text{Mes}{-}{\equiv}{-}\text{BMes}_2 \qquad (33)$$

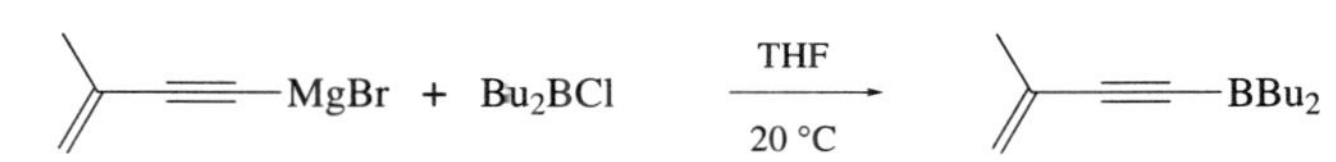

(34)

Eisch *et al.* ⟨90JA1847⟩ have employed 2,4,6-trimethyl- and 2,6-dimethylphenylethynyl Grignard reagents to alkynylate dimesitylboron fluoride as shown in Equation (33), with THF being employed as the solvent. An ethynylmagnesium bromide has been reported to behave well in reaction with Bu$^n{}_2$BCl, utilising ether as solvent (Equation (34)) ⟨62ZAAC(317)54⟩. The product is unstable and has a boiling point of 22–25°C (0.1 mm). A somewhat unusual procedure has been reported which involves loss of iodine from the boron iodide and an 1-iodoalkyne (Equation (35)) ⟨75ZAAC(418)273⟩. This method proceeds in low yield (28%) and is not general.

$$\text{Ph}{-}{\equiv}{-}\text{I} + \text{Me}_2\text{BI} \xrightarrow[28\%]{\text{CS}_2} \text{Ph}{-}{\equiv}{-}\text{BMe}_2 + \text{I}_2 \qquad (35)$$

Proper adaptations of the procedures exemplified in Scheme 6 and Equations (33)–(35) above led to di- and trialkynylboranes ⟨64JOM(1)384, 72CJC1005⟩ respectively, with the latter being of varying stability.

Method 2: Reaction of a Lewis base with a boron trifluoride diethyletherate

A method regarded as being the most versatile for the formation of mono- and dialkynylboranes was described by Köster and co-workers ⟨68LA(717)21⟩, and involves treatment of certain Lewis-base adducts with the diethyletherate of boron trifluoride. The Lewis bases routinely employed were trimethylamine or pyridine (Equation (36)), with the latter being used for dialkynylboranes. The starting boranes were acquired from the generally solid Lewis base adducts as shown in Equation (37), and have a much diminished reactivity towards oxygen.

$$Y{\bullet}R^1{}_2B{-}{\equiv}{-}R^2 + Et_2O{\bullet}BF_3 \longrightarrow R^1{}_2B{-}{\equiv}{-}R^2 + F_3BY + Et_2O \quad (36)$$

$$Y = NEt_3,\ NC_5H_5$$

$$R^2{-}{\equiv}{-}Li + RBX_2{\bullet}Y \longrightarrow R^1B{-}({\equiv}{-}R^2)_2{\bullet}Y + 2LiX \quad (37)$$

$$X = F,\ Cl$$
$$Y = NEt_3,\ NC_5H_5$$

Method 3: Reaction of methoxydialkylborane with alkynylithium reagents

A procedure for obtaining alkynyldialkylboranes that involves a borate intermediate has been reported by Brown and Sinclair ⟨77JOM(131)163⟩. The reaction of methoxy-dialkylborane with alkynylithium reagents in THF yields lithium methoxyalkynyldialkylborinates which on treatment with 1.3 equiv. of boron trifluoride diethyletherate provide alkynyldialkylboranes (isolated as THF adducts) in essentially quantitative yields (Equations (38 and 39)).

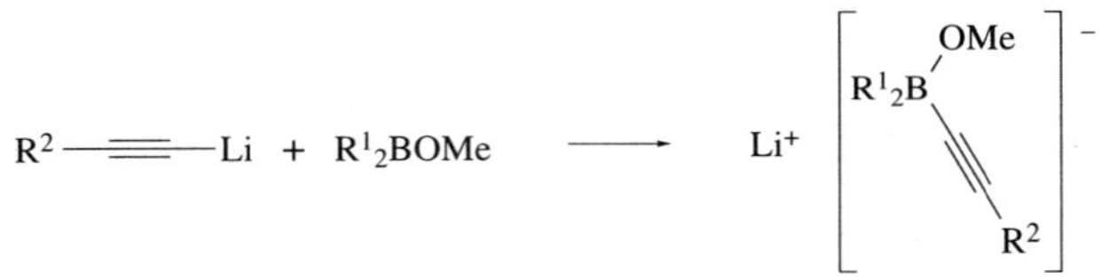

(38)

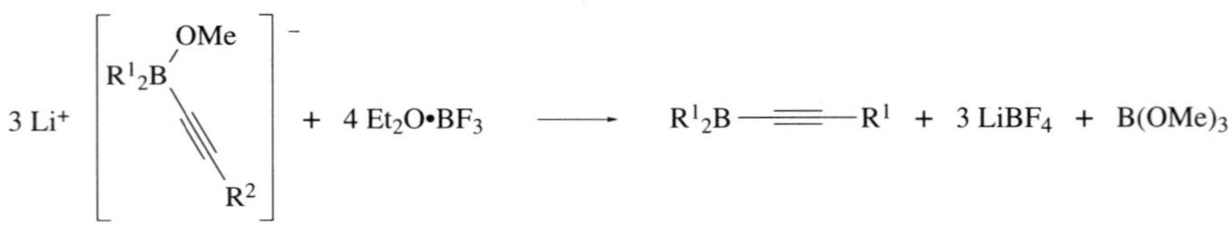

(39)

In the latter two methods, comment on the Lewis-base adducts of the alkynylboranes has been made, and these derivatives are of importance generally because of their increased stability compared with the uncomplexed boranes.

Method 4: Reaction of an organoborane with a Lewis base

Formation of Lewis-base adducts is straightforward from the organoborane and the Lewis base such as amines or ethers ⟨66BSF1981⟩. Displacement of an unusual leaving group (2-iodoethoxy) from boron by an alkynyl Grignard reagent, in the presence of pyridine, has been employed by Soulié and Cadiot ⟨66BSF1981⟩ and others ⟨57IZV812, 60CR(251)727⟩ to obtain the adducts directly (Equation (40)). This is a procedure that probably lacks generality and involves hydrolysis of a sodium tetraalkynylborate ⟨56ZN(B)676, 64JOC3225⟩.

(40)

2.23.2.1.2 Organoboranes with heteroatoms σ-bonded to boron

Method 5: Reaction of an alkynyl sodium with chloroaminoborane

Alkynylaminoboranes were first reported in 1966 ⟨66BSF3846⟩ and were obtained by reacting an alkynyl sodium with chloroaminoborane (Equation (41)) which in turn was obtained from borontrichloride and the amine ⟨57JCS381⟩. Yields of the distillable alkynylaminoboranes were low to moderate. As part of this general programme, Soulié and Cadiot ⟨66BSF3846⟩ also synthesised a number of alkynylamino-chloroboranes (Equation (42)) and reported detailed spectroscopic data.

$$R^2\text{—}\equiv\text{—}Na \; + \; ClB(NR^1{}_2)_2 \xrightarrow[N_2]{-60\,^\circ C} R^2\text{—}\equiv\text{—}B(NR^1{}_2)_2 \qquad (41)$$

R^1 = Ph, alkyl, alkenyl
R^2 = Et, Me

$$2 \;\text{—}\equiv\text{—}MgBr \; + \; Me_2NBCl_2 \longrightarrow (\text{—}\equiv\text{—})_2B\text{—}NMe_2 \qquad (42)$$

Alkynyl Grignard reagents can also be employed for alkynylation of boron in these systems (Equation (43)) ⟨80CB1235⟩. Alkynyl 2,4,6-trichloroborazines have been reported ⟨65IC657⟩ from the reaction of alkynyl Grignard reagents with 2,4,6-trichloroborazine (Equation (43)).

$$3\,R^2\text{—}\equiv\text{—}MgBr \; + \; Cl_3B_3N_3R^1{}_3 \longrightarrow (R^2\text{—}\equiv\text{—})_3B_3N_3R^1{}_3 \qquad (43)$$

R^1 = H; R^2 = Me
R^1 = H, Ph; R^2 = Ph

Aminochloroboranes on reaction with a 1-alkyne in the presence of a base (Et_3N) provide alkynylaminoboranes in good yield (Equation (44)) ⟨82HOU(13/3b)69⟩. A very unusual alkynylborane resulted from the reaction shown in Equation (45) ⟨87CB1437⟩ and carbenoidal intermediates were considered to be involved. The product of interest here was formed in a low yield (17%) and its structure proven by an x-ray determination.

$$R^2\text{—}\equiv \; + \; (R^1{}_2N)(R^3)BCl \xrightarrow{Et_3N} R^2\text{—}\equiv\text{—}B(NR^1{}_2)R^3 \qquad (44)$$

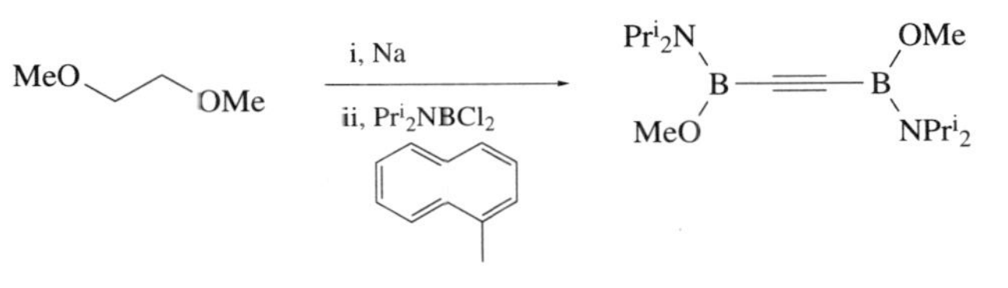

(45)

Alkynylalkoxyboranes have attracted considerable attention since the mid 1960s, mainly from a synthetic viewpoint, but this class of alkynylboranes was discussed in 1966 by Soulié and Cadiot ⟨66BSF3846⟩, with a summary of earlier examples and synthetic methods.

Method 6: Reaction of an alkynylmetal with a borate, and an alkynyl sodium with a chloroborate

Cadiot developed two additional procedures, the first involving the reaction of an alkynylmetal derivative with a borate to provide the alkylborate salt which is redistributed with boron trichloride (Scheme 7) ⟨66BSF3848⟩. The second method (Equation (46)) is the direct reaction between an alkynyl sodium and a chloroborate. These methods afford monoalkynyldialkoxyboranes in moderate yields for a range of alkynyl groups and all are mobile, distillable liquids which are sensitive to hydrolysis.

$$R\text{—}\equiv\text{—}Na \; + \; ClB(OBu)_2 \xrightarrow{-40\,^\circ C} R\text{—}\equiv\text{—}B(OBu)_2 \qquad (46)$$

In the late 1980s, Brown *et al.* described the preparation and characterisation of organyl-1-alkynylalkoxyboranes which were previously not well characterised ⟨87OM629, 88JOC1391⟩. Brown *et al.* discovered that these compounds were formed in high yields in the reaction of a 1-alkynylithium

$$\equiv\!\!-M + B(OR)_3 \longrightarrow M[(RO)_3B-\!\!\equiv]$$

$$M[(RO)_3B-\!\!\equiv] + {}^1/_3\,BCl_3 \longrightarrow \equiv\!\!-B(OR)_3 + MCl + {}^1/_3B(OR)_3$$

M = Na, Li, K, MgBr or MgCl
R = Me, Bu^n

Scheme 7

with a boronic ester, followed by treatment with dry HCl in ether (Scheme 8). Lower temperatures are used to favour the addition complex, as shown in Scheme 8. This methodology is applicable to a wide variety of boronic esters and is not restricted to hindered alkyl systems. The same is true for the alkynylithium moiety. These organyl-1-alkynylalkoxyboranes are colourless, stable, distillable liquids that store well in the cold under N_2. Alkyl(phenylalkynyl)alkoxy-boranes are thermally unstable and prone to polymerisation. Brown *et al.* reported an extension of this general approach to 1-alkynyldiisopropoxyboranes ⟨83OM1316, 88TL2631⟩ and this procedure is shown in Scheme 9.

$$R-\!\!\equiv\!\!-Li + Bu^sB(OPr^i)_2 \xrightarrow[-78\,^\circ C]{} Li[Bu^s(OPr^i)_2B-\!\!\equiv\!\!-R]$$

$$Li[Bu^s(OPr^i)_2B-\!\!\equiv\!\!-R] \xrightarrow[-78\,^\circ C]{+\,HCl} Bu^s(OPr^i)B-\!\!\equiv\!\!-R + LiCl + Pr^iOH$$

Scheme 8

$$R-\!\!\equiv\!\!-Li + B(OPr^i)_3 \longrightarrow Li[R-\!\!\equiv\!\!-B(OPr^i)_3] \xrightarrow[Et_2O]{dry\ HCl} R-\!\!\equiv\!\!-B(OPr^i)_2 + Pr^iOH + LiCl$$

Scheme 9

Most of the boranes are also stable distillable liquids and can be stored indefinitely under N_2 at room temperature. The procedure is a high yielding one and substitution in the acetylene moiety does not noticeably affect yields. Halide groups are also tolerated. For both series of compounds, tabulations of ^{13}C NMR and IR data are presented.

There is one example of an alkynyl(aminocarboxy)borane formed by direct carboxylation of an alkynyl(bisamino)borane ⟨67CB2377⟩.

2.23.2.1.3 *Tetraorganoborates*

Method 7: Reaction of an alkynyl metal with a triorganoborane

This is the most direct method for acquiring tetraorganoborates. The triorganoborane may already incorporate an alkynyl moiety. This approach has been well described by Pelter and Brown and co-workers ⟨75CC857, 76JOC1078, 76TL4385⟩ and its versatility is clear from examples in Scheme 10. Further applications of these procedures are to be found in the reports by Brown and co-workers ⟨86JOC4507, 86JOC4521⟩. The unusual sodium reagent, alk-3-en-1-ynyl sodium adds very efficiently to triphenylborane, in either THF or ether to form the borate (Equation (47)).

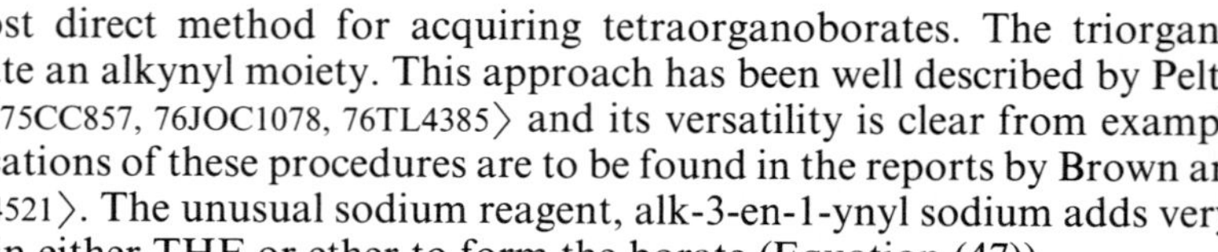

(47)

Borates are also accessible directly from boron trihalides on reaction with four molar equivalents of an alkynyl metal derivative using an alkane as solvent at $-60\,^\circ C$. The yields are poor (20%) and accompanied by red polymeric material (Equation (48)). However, this can be circumvented by using $BF_3\cdot Et_2O$ as the boron source (Equation (49)).

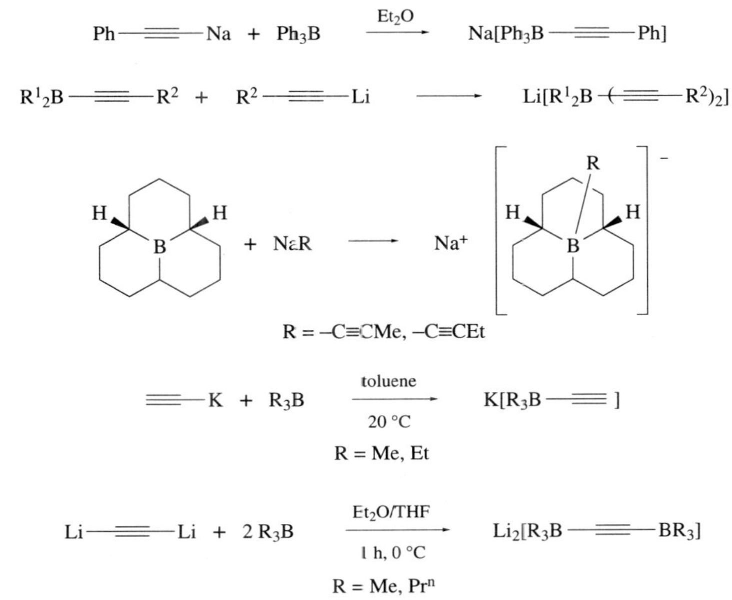

Scheme 10

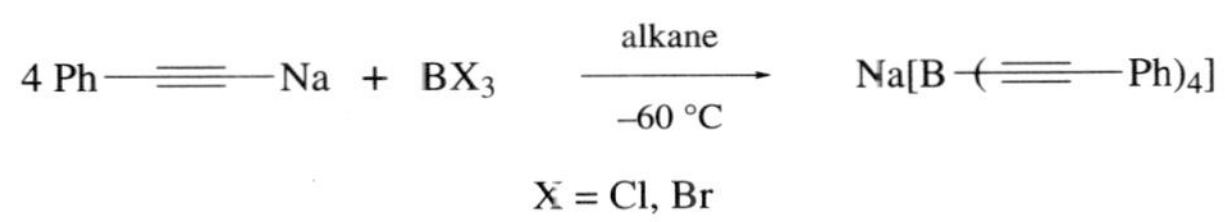
(48)

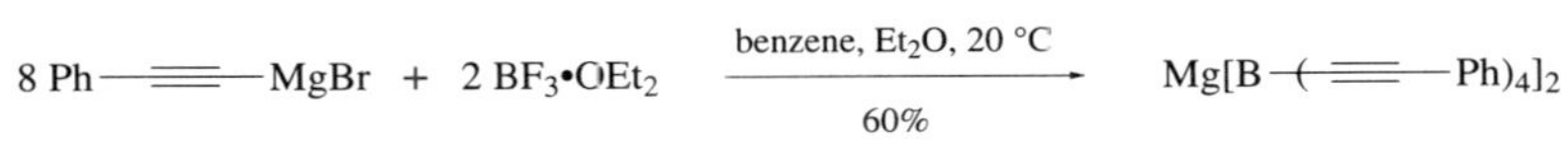
(49)

Method 8: Reaction of Lewis-base adducts with alkynylmetal reagents

Lewis-base adducts are useful boron sources for reaction with alkynylmetal reagents, and very high yields have been reported, as shown in Equation (50) ⟨68LA(717)21⟩. A variation of the approach of Equation (50), is the use of a Lewis-base adduct of a dialkylborane as shown in Scheme 11.

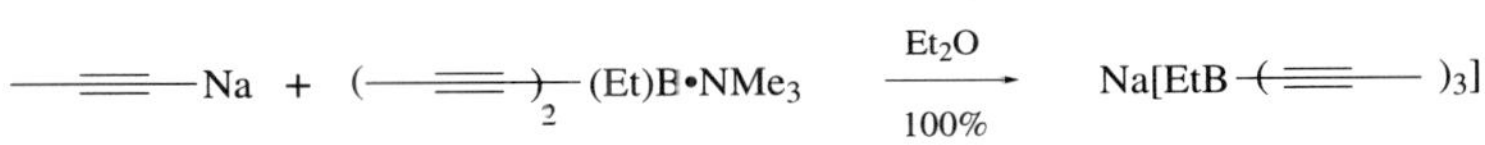
(50)

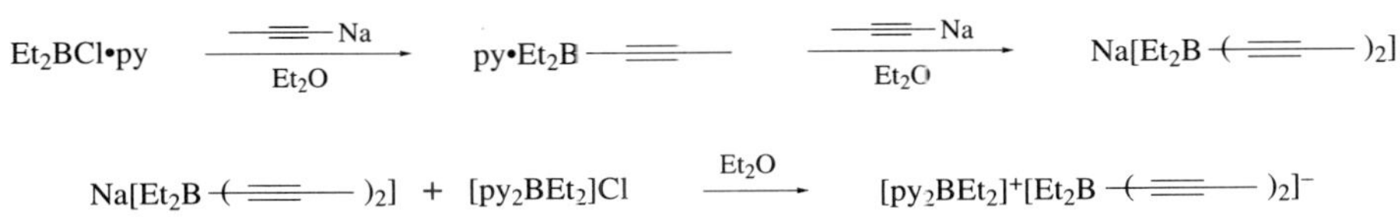

Scheme 11

Method 9: Reaction of borohydrides with alkynes

Certain borohydrides react directly with 1-alkynes at 20 °C in hydrocarbon solvents to provide borates in quantitative yields (Equation (51)) and both active positions of ethyne can be utilised ⟨68LA(717)21, 75LA2130⟩. The use of ether as solvent promotes side-reactions ⟨65TL1901, 74MI 223-01⟩.

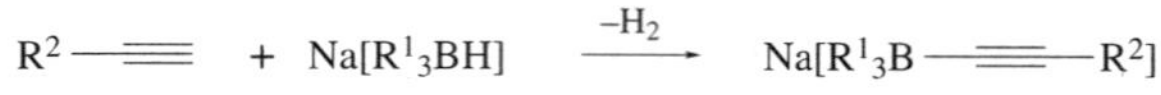

(51)

R^1 = Me, Et, Pr^n
R^2 = alkyl, aryl, vinyl

A novel series of organotin cations incorporating intramolecular stabilisation from a 1-alkynylboron fragment has been reported by Wrackmeyer *et al.* ⟨91AG(E)1370, 92CB1597⟩. These unusual cations made from tetraalkynyl stannanes and triethyl borane have been characterised spectroscopically and in one case by x-ray crystallography (**4**).

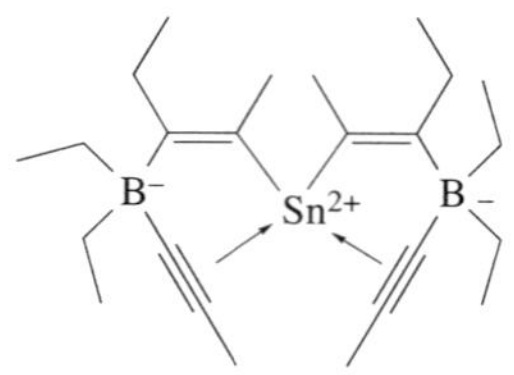

(4)

Several examples of interesting zwitterionic borates have been described, with quaternisation of an (aminoalkyl)alkynylborate being reported by Binger and Köster ⟨75CB395⟩ (Equation (52)). Oxidation of a Lewis base adduct of a borane with trimethylamine oxide proceeds in very high yield (≈98%) as shown in Equation (53) ⟨68LA(717)21⟩.

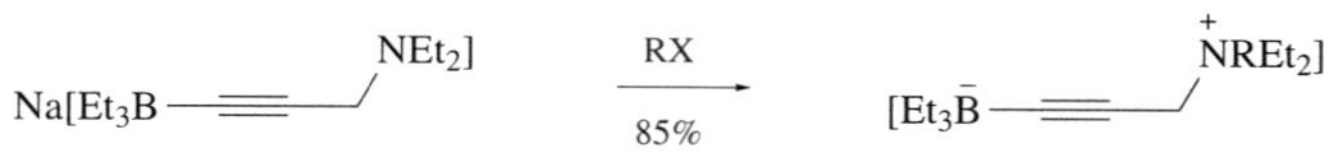

(52)

RX = triethyloxonium tetrafluoroborate

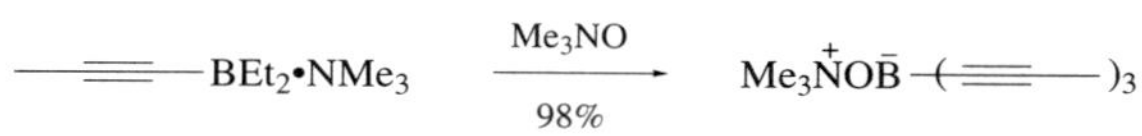

(53)

An arrangement of theoretical interest with respect to stabilisation of C_2 by donor and acceptor units has been synthesised by Bestmann *et al.* ⟨89AG(E)1219⟩ starting with ethynyldiphenylphosphane, as shown in Equation (54). The crystal structure showed a nonlinear arrangement for B—C≡C—P in contrast to predictions from calculations.

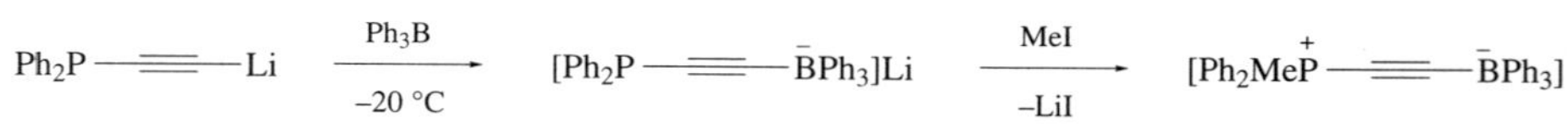

(54)

Two interesting examples of metal substituted 1-alkynylborates have been reported. Deprotonation of lithium ethynyltrihexylborate proceeds at −78°C with *n*-butyllithium as shown in Equation (55) ⟨76TL3969⟩. Ethynyltrimethylsilane reacts with sodium triethylborohydride to form a silylsubstituted ethynylborate (Equation (56)) ⟨76S118⟩.

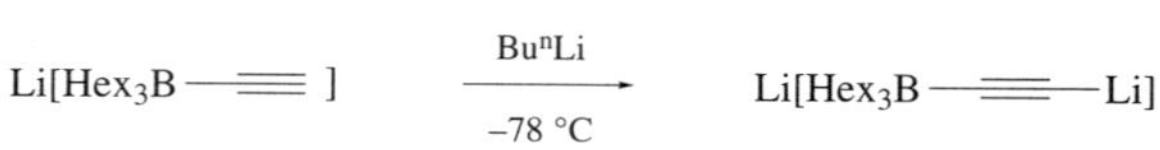

(55)

(56)

Lithium 1-alkynyldialkylmethoxyborates have been mentioned previously (see Equations (38) and (39)).

A disproportionation reaction has been employed in the synthesis of sodium dimethylhydro-1-propynyl borate, with release of the volatile triethylborane providing much of the driving force (Equation (57)). During the reaction, about 25% of the starting material undergoes hydroboration ⟨65TL1901⟩.

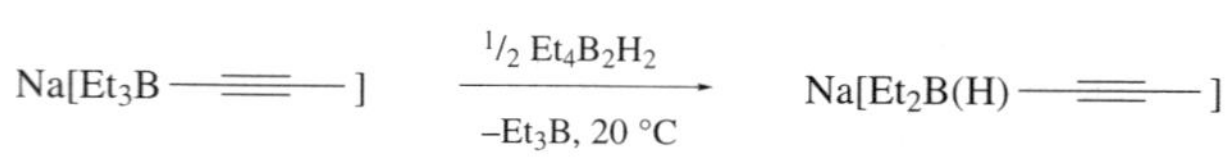

(57)

2.23.2.2 Alkynylsilicon Compounds

Reviews of the literature up to 1967 ⟨67CRV73⟩ and 1980 ⟨80HOU(13/5)33⟩ are available on this expanding subject with a later review ⟨92RCR1091⟩ being restricted to silicon containing 1,3-alkenynes and related unsaturated derivatives. In this chapter examples of general methods are given, together with developments made in the early 1990s.

Method 1: The reaction of an organosilylhalide with an alkynyllithium or magnesium derivative

This reaction has been very widely employed, encompassing large structural variations in the organic groups on silicon or as part of the alkynyl unit (Scheme 12, Equation (58)) ⟨52DOK(85)1297, 64JOM(2)97, 70JOM(21)837⟩.

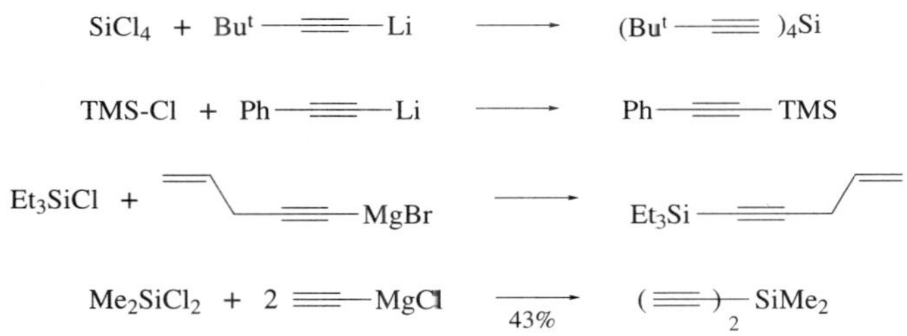

Scheme 12

 (58)

X = Cl; n = 1–4; M = Li, MgX

The triflate group, because it confers considerable electroplicity on Si, is now commonly employed as a leaving group, and alkynylation of silyltriflates represents a minor change to a well-established procedure (Equation (59)). This approach has been used to generate a polymeric alkynylsilane system (Equation (60)), which may be further functionalised by phenyl group cleavage ⟨91ZAAC(603)109⟩. The ready displacement of triflate from silicon permits access to a wide range of alkynylsilyl systems as shown in Scheme 13. An interesting series of polysilacycloalkynes has been prepared from a dimagnesioalkyne and dichloropolysilanes using THF as solvent ⟨93OM803⟩.

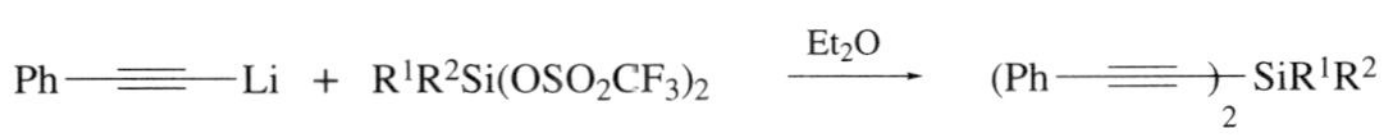 (59)

$R^1 = R^2$ = Me, Bu^t
R^1 = Me; R^2 = H

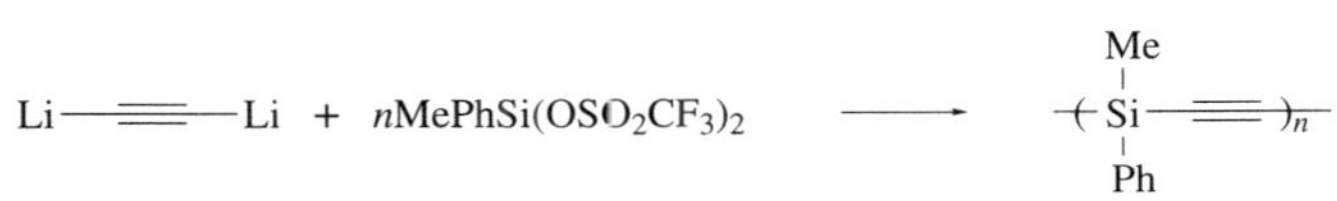 (60)

For polysilacycloalkynes ($n = 6$), photolysis results in low yields of ring contracted silanes, and in the presence of acetone, acetone adducts of the bicyclic silacyclopropenes were characterised ⟨93T6359⟩. *Trans*-bis-1-alkynyl(pthalocyaninato)silanes required as models for one-dimensional conductors, have been obtained by treating the corresponding dichlorides with 1-alkynyl Grignard reagents in THF.

For these substitutions at silicon to proceed in good yield it was found that Grignard reagents were more effective and provided purer products than the corresponding lithium or sodium derivatives. In addition, solvents other than THF gave complex mixtures and overall erratic results ⟨81JOM(204)315⟩.

With respect to Method 1, the use of alkynyl sodium derivatives for halide displacement at silicon has not been popular, and this may relate to the ease and rapidity of alkynylithium generation from the alkynes and readily available organolithium reagents, for example, n-butyllithium.

1-Alkynylstannanes, although of lower reactivity than the lithium or magnesium derivatives, nevertheless react with bromosilanes, in a stepwise fashion if required, to yield alkynyl or dialkynyl

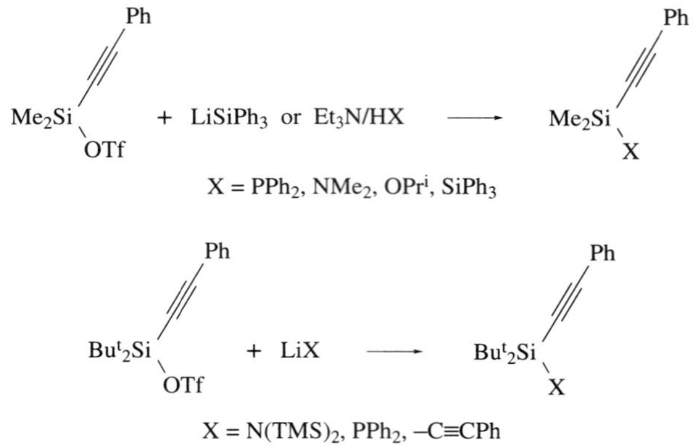

Scheme 13

silanes as shown in Equations (61) and (62). The neat reagents are heated at ca. 150–180 °C for 1–5 h and the reaction, although presently lacking generality, could be rendered more useful, if the effect of Lewis acids were clarified ⟨92JGU787⟩.

$$R^2{-}C{\equiv}C{-}SnR^1_3 + Me_2SiBr_2 \xrightarrow{\text{heat}} R^2{-}C{\equiv}C{-}SiBrMe_2 + R^1_3SnBr \quad (61)$$

$$R^2{-}C{\equiv}C{-}SnR^1_3 + R^2{-}C{\equiv}C{-}SiBrMe_2 \xrightarrow{\text{heat}} (R^2{-}C{\equiv}C)_2SiMe_2 + R^1_3SnBr \quad (62)$$

$R^1 = Me$; $R^2 = CH_2$-TMS, Ph
$R^1 = Et$; $R^2 = Bu^t$

A propargyl stannane, on conversion to the Grignard reagent with ethylmagnesium bromide, reacts with bromotrimethylsilane to yield the alkynylsilane (Equation (63)) ⟨87JGU2482⟩ in good yield (≈80%). Propargyl-allene isomerisation did not occur during this sequence.

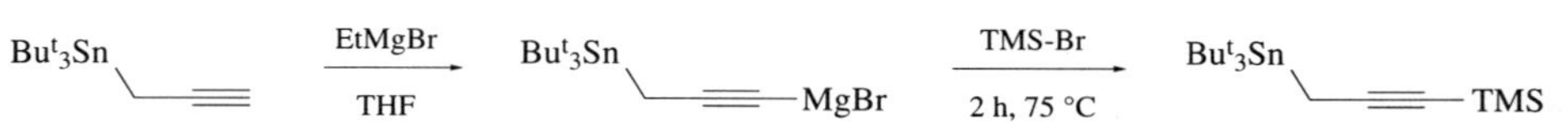

(63)

Chlorotrimethoxysilanes have been used as starting materials for the synthesis of ethynylmethoxysilanes and moderate yields were obtained when either alkynyl sodium or magnesium reagents were employed. These methods are exemplified in Equation (64). Polymeric organosiloxanes react with alkynyl Grignard reagents to form alkynylsilanols which can be further transformed ⟨63IZV97⟩.

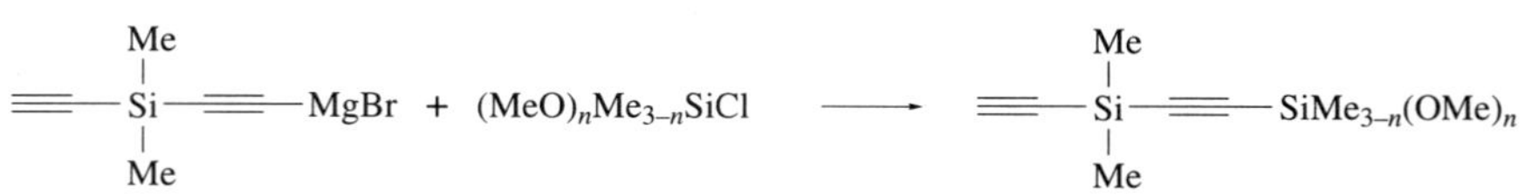

(64)

An unusual procedure that involves the use of catalytic amounts of sodium (5% mole) and of hexamethylphosphoramide (HMPA) or DMSO (10% mole) is the reaction of a 1-alkyne with a triorganosilane or hexaalkyldisilane. The reaction is presumed to commence with the formation of the alkynyl anions and the production of H_2 may constitute the main driving force for the reaction ⟨69CR(C)72⟩. Moderate to good yields of isolated products were reported (Equation (65)).

$$R^2{-}C{\equiv}CH + YSiR^1_3 \xrightarrow{-HY} R^2{-}C{\equiv}C{-}SiR^1_3 \quad (65)$$

Y = H; $R^1 = Et$, Pr^n, (Et_2Me); $R^2 = Ph$, Bu^n, Et_3Si, Et_2MeSi
Y = TMS; $R^1 = Me$; $R^2 = Bu^n$

The TMS-Cl/Mg/HMPA reagent, developed extensively by the French group of Calas and Dunoguès ⟨76MI 223-01⟩, enters into reaction at ca. 50–80 °C with propargylic and allenic bromides to yield isomeric mixtures of the unsaturated silanes and disilanes (Equation (66)) ⟨72CR(C)1857⟩.

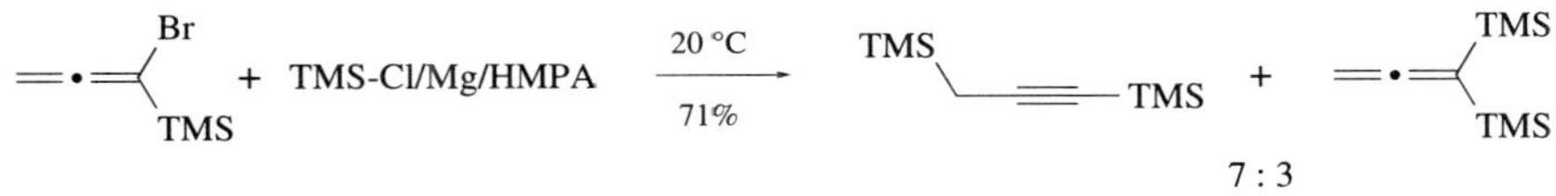
(66)

The ditosylate of 2,4-hexadiyne-1,6-diol has been used as a starting material for silyl-substituted triynes, and high yields were obtained by adding *n*-butyllithium to a mixture of the ditosylate and chlorotrimethylsilane at −78 °C (Equation (67)) ⟨91JA6943⟩. In a second step, more relevant groups, for example *i*-Pr_3Si and *t*-$BuPh_2Si$ were readily substituted for TMS.

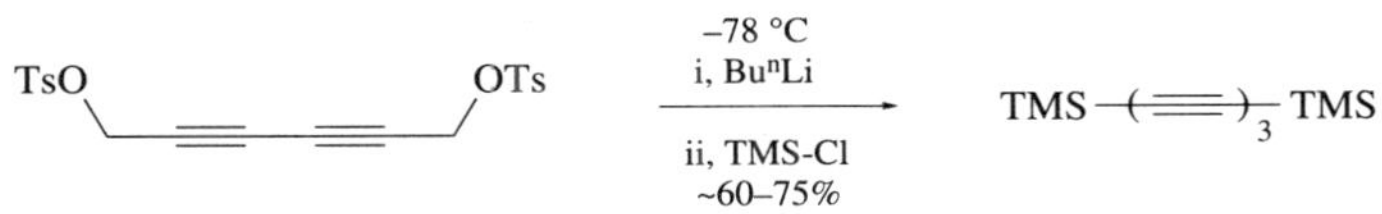
(67)

A series of novel cross-conjugated ethynylsilanes of the general type (**5**) has been reported. For R = TMS, a crystal structure determination has been reported ⟨92AG(E)1101, 92AG(E)1241⟩.

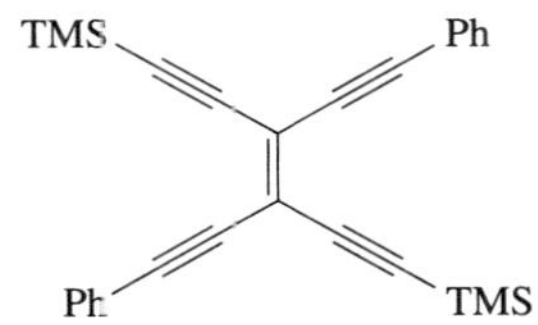
(**5**)
m.p. 150 °C
yellow

Method 2: Ethynyl derivatives of silatranes

These compounds have received attention mainly from Russian workers Voronkov *et al.*, and a synthetic method was developed that provides them in good yield (Equation (68)) ⟨86JGU849⟩. Maintaining an equimolar mixture of the reactants (30 °C) in a vacuum for 4 h was required. More examples of alkynylsilatranes have been acquired by substituting the alkoxy groups on silicon with a tris(2-hydroxy)amine ⟨79JGU536⟩.

(68)

Method 3: Alkynylsilanes from furans

Furans have served as starting materials for the synthesis of a variety of alkynylsilanes. Reductive silylation of ethyl 2-furoates leads to the silyl enol ether of 3,5-bis(trimethylsilyl)pent-4-yn-1-al (Equation (69)) ⟨77CC76⟩; whereas low temperature photolysis in pentane of 2,5-bis(trimethylsilyl)furan led ultimately to the alkynylsilanes shown in Scheme 14 ⟨83JA6316⟩.

(69)

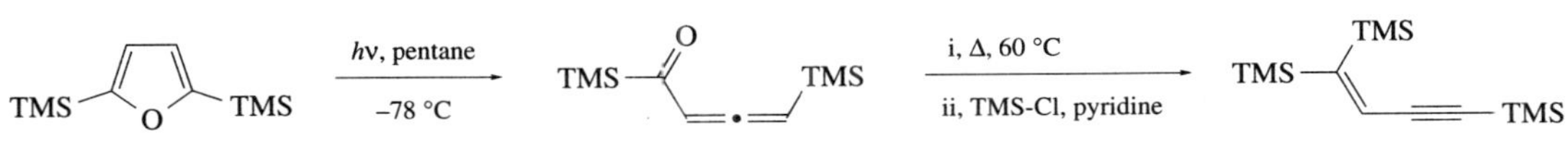

Scheme 14

Method 4: Modifications of alkynylsilanes

There are a number of reports describing modifications of existing alkynylsilanes. For example, reduction of alkynylsilyl chlorides generates new alkynyl silanes (Equation (70)) ⟨70ZAAC(377)48⟩. A standard form of modification of certain alkynylsilanes involves replacement of the active hydrogen with metallic groups ('carbanion' generation) followed by reaction with electrophiles. In this general way, lithium, magnesium, tin and copper derivatives have been first formed and then reacted with a variety of metallic ⟨93JOM(449)53⟩ or carbon electrophiles. This is shown in general form in Equation (71) and Scheme 15.

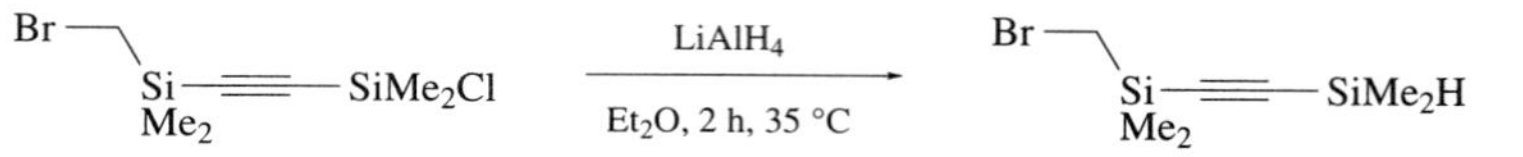

(70)

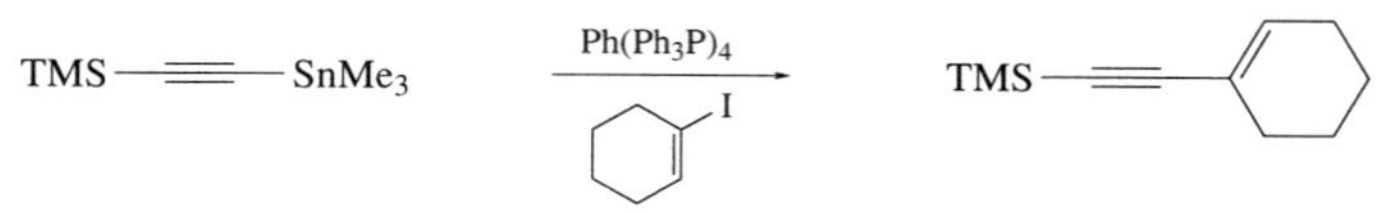

(71)

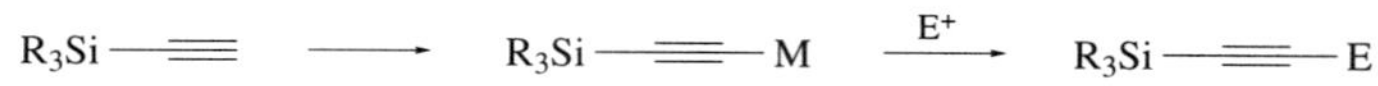

Scheme 15

Reaction of aminostannanes with silylalkynes is also general and will be discussed in detail later (Section 2.23.3.5), but these silyl stannylacetylenes are valuable precursors in palladium(0) catalysed cross-coupling reactions ⟨86AG(E)508, 92S803⟩.

2-(Trimethylsilyl)ethynyltitanium triisopropoxide has found useful applications in stereoselective addition to certain aldehydes as shown in Equation (72). The silicon moiety can be altered as shown by the example in Equation (73) ⟨90TL2739⟩ and in general, a variety of transformations can be executed in organic molecules without affecting the alkynylsilane grouping.

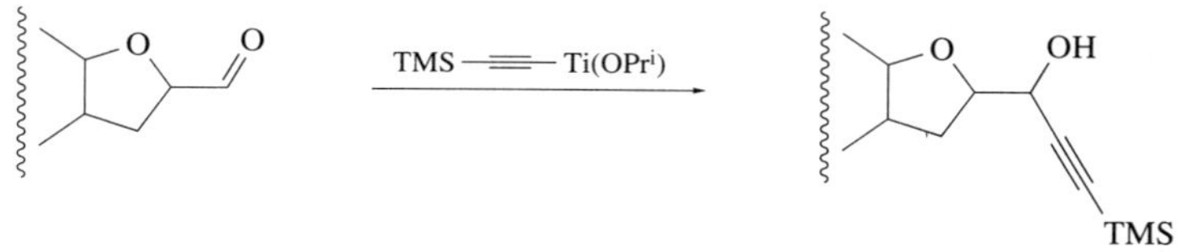

(72)

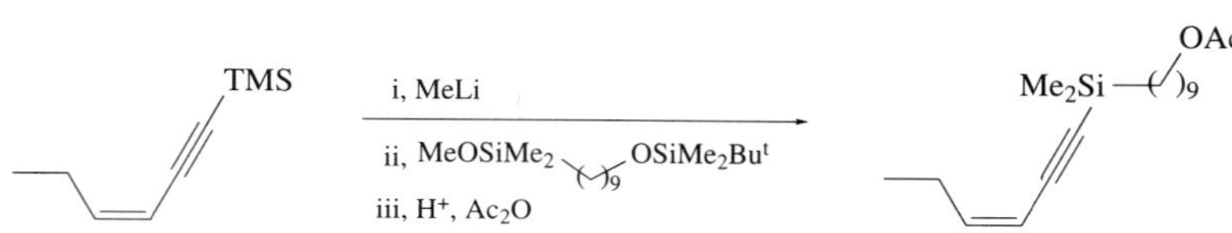

(73)

2.23.2.3 Alkynylgermanium Compounds

Alkynylgermanes were thoroughly reviewed by Davidsohn and Henry ⟨67CRV73⟩ and briefly by Glockling in his monograph ⟨B-69MI 223-01⟩. Since 1968, a number of reviews of organogermanium chemistry generally, but which include extensive discussion of alkynylgermanes, have appeared ⟨B-88MI 223-01, B-89MI 223-01, B-90MI 223-02⟩. These later reviews (1988–1990) contain extensive tabulations of compounds, their properties and methods of formation.

A wide range of akynylgermanes are known and in general these present no particular difficulties in synthesis or subsequent manipulation. Examination of the literature shows that preparative procedures can be organised into three groups.

Method 1: Reaction of organogermanium halide with metal acetylide

A wide range of alkynyl metal derivatives react with organogermanium halides as shown in Equation (74). The tetraalkynylgermanes are a well characterised class and have been acquired by alkynylation of $GeCl_4$ with alkynylmagnesium bromides, using THF or ether as solvent (Equation (74)). Tetraethynylgermane explodes on rapid heating and is shock sensitive ⟨66JOM(5)29, 67SA(A)1169⟩. Use of the Grignard reagent from propargyl bromide promotes a mixture of the propargyl and allenyl derivatives ⟨67BSF777⟩. Sodium and lithium acetylides also react efficiently with $GeCl_4$ to form the tetraalkynyl germanes, with ether or THF being the solvents ⟨B-88MI 223-01⟩. Further examples of Method 1 are shown in Equations (75)–(77) ⟨66JGU570, 71JGU122, 79MI 223-01⟩.

$$R^1_{4-n}GeX_n + nR^2{-}C{\equiv}C{-}M \longrightarrow R^1_{4-n}Ge{-}(C{\equiv}C{-}R^2)_n \qquad (74)$$

$n = 1$–4
X = Cl, Br, I
M = MgBr, MgI, Li, Na, Bu_3Sn, Al

$$EtGeCl_3 + Ph{-}C{\equiv}C{-}MgCl \xrightarrow{\text{THF}} EtGe{-}(C{\equiv}C{-}Ph)_3 \qquad (75)$$

m.p. = 71–72 °C

$$Et_3GeBr + Li{-}C{\equiv}C{-}OEt \xrightarrow[\text{3 h, 40 °C; 40\%}]{\text{ether}} Et_3Ge{-}C{\equiv}C{-}OEt \qquad (76)$$

$$Bu_3GeCl + Na[Al{-}(C{\equiv}C{-}Bu)_4] \xrightarrow[\text{120 °C, 40 min; 81\%}]{\text{THF, 60 °C, 1 h}} Bu_3Ge{-}C{\equiv}C{-}Bu \qquad (77)$$

Transfer of alkynyl groups from aluminum to germanium has been employed, and proceeds in high yield, although the generality of the method is unclear ⟨67JGU515⟩. In a similar way, trialkylgermanium halides have been alkynylated by treatment with appropriate stannanes and HMPA in THF but yields and procedures are sketchy ⟨81IZV1280⟩.

Aminoalkynylstannanes also engage in this reaction as shown in Equation (78) ⟨79AG(E)405⟩. Aminogermanes and -stannanes have been utilised in sequential reactions to provide alkynylgermanes bearing tin or silicon groups. These are illustrated in Equation (79) ⟨67JOM(9)435⟩.

$$Ph_3GeBr + R_3Sn{-}C{\equiv}C{-}N(Me)Ph \xrightarrow[\text{55\%}]{} Ph_3Ge{-}C{\equiv}C{-}N(Me)Ph \qquad (78)$$

$$Ph_3GeNEt_2 + Na{-}C{\equiv}CH \xrightarrow[\text{20 min}]{\text{THF, 65 °C}} \xrightarrow[\text{65 °C, 10 min}]{Ph_3SiCl\text{, THF}} Ph_3Ge{-}C{\equiv}C{-}SiPh_3 \qquad (79)$$

Displacement of bromide from bromogermane ($BrGeH_3$) has been reported to occur with methyl and ethylnylmagnesium bromides, as well as with indenyl potassium. Various deuterium isotopomers were also acquired (Equation (80)) ⟨67JCP(46)658⟩.

$$(D_3)H_3GeBr + MR \longrightarrow (D_3)H_3GeR \qquad (80)$$

Method 2: Condensation of sufficiently acidic terminal alkynes

1-Alkynes functioning as protic acids react with a range of organogermanium derivatives to yield alkynylgermanes, as shown generally in Equation (81).

$$R^1_3GeY + HC{\equiv}C{-}R^2 \longrightarrow R^1_3Ge{-}C{\equiv}C{-}R^2 + HY \qquad (81)$$

This method has been utilised for y = amino, or organocadmium groups (Equations (82) and (83)) ⟨67BSF777, 73JGU626⟩. Radical induced alkynylations ⟨73JGU626⟩ may be regarded as a variant of this general procedure (Equation (84)), although products, mainly of addition, are also formed ⟨72JOM(34)83⟩.

$$(Et_3Ge)_2NH + Ph{-}{\equiv} \xrightarrow[150\ ^\circ C,\ 6.5\ h;\ 53\%]{(NH_4)_2SO_4} Ph{-}{\equiv}{-}GeEt_3 \qquad (82)$$

$$(Et_3Ge)_2Cd + Ph{-}{\equiv} \xrightarrow[58\%]{\text{toluene, reflux, 5 h}} Ph{-}{\equiv}{-}GeEt_3 \qquad (83)$$

$$Et_3GePEt_2 + Ph{-}{\equiv} \xrightarrow{70\ ^\circ C,\ 6\ h,\ AIBN} Ph{-}{\equiv}{-}GeEt_3 \qquad (84)$$

An unusual reaction of a vinyl triflate incorporated as part of an enediyne system attached to adamantane leads in poor yield to the alkynylgermane (Equation (85)) ⟨81JA6437⟩.

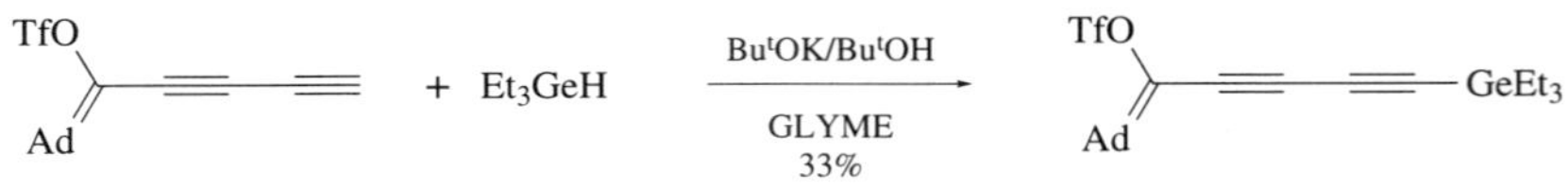

(85)

Ad = adamantylidene

Method 3: Modifications of preformed alkynylgermanes

Modifications of systems incorporating the alkynyl germanium substructure or a carbon–germanium bond that may be transformed to the alkynyl grouping, have been described and are quite abundant. Details of these reactions are located in the reviews cited at the beginning of this section, but it is useful to note the range of transformations reported, for example: 1-alkynyl metallation followed by electrophile trapping; eliminations (e.g., H_2O, HBr) in the side-chain; hydrostannylation; halogenations; complexation, for example, with transition metal carbonyls; destannylation; substitution reactions (e.g., of OH); oxidation (e.g., with peroxides or chromium(VI)); hydrazone formation within keto side-chains; cycloadditions.

2.23.3 ALKYNYLMETAL DERIVATIVES

2.23.3.1 Alkynylcopper Compounds

Alkynylcopper compounds have been known since the beginning of the twentieth century and there has been considerable discussion concerning their structures. More recently, the increased focus on organocopper chemistry generally, encouraged by important applications in organic synthesis, has resulted in alkynylcopper derivatives receiving closer scrutiny as part of this general renaissance. The synthetic applications of organocuprates have been presented in a number of reviews ⟨75ZAAC(418)273, B-80MI 223-01, 89T349⟩ and organocopper compounds generally have been discussed in the *Handbook of Inorganic Chemistry* ⟨B-83MI 223-01, B-83MI 223-02, B-86MI 223-01, B-87MI 223-01, B-87MI 223-02, B-87MI 223-03, B-87MI 223-04⟩. It is necessary to distinguish several categories of organocopper compounds that contain an alkynyl–Cu linkage. Those of the cuprate type are rarely characterised, being formed at reduced temperatures generally, and engaged directly in *in situ* reaction at reduced temperatures. The approach here is to discuss systems of the type R–Cu(L) (where R = alkynyl and L = ligand) initially, followed by 'cuprates' of various orders.

2.23.3.1.1 Alkynylcopper(I) systems, RC≡C—Cu(L)

An excellent in-depth presentation of preparative methods for compounds of this type is available ⟨B-86MI 223-01⟩. The information is organised into eight generalised preparative methods and important variations are also noted. The most important general method is the reaction of an alkyne with a copper(I) salt as an aqueous ammonia solution $[Cu(NH_3)_x]^+$ (Equation (86)) (Ilosvay's reagent ⟨1899CB2697, 01ZAAC(40)123⟩); or with solutions of the alkyne in alcohols, acetone, dioxan, etc. The RC≡CCu are generally of very low solubility in water and frequently precipitate. Use of excess Ilosvay's reagent may 'solubilise' the desired product, presumably because of complexation.

$$R{-}C{\equiv}CH + [Cu(NH_3)_x]^+_{organic} \xrightarrow{H_2O \text{ or solvent}} R{-}C{\equiv}C{-}Cu \quad (86)$$

Another general method of importance is reminiscent of those discussed previously for other metals viz. the reaction of alkynyl lithium or potassium reagents with salts of copper(I) (Equation (87)). Solvents employed have been liquid NH_3, ether or ether/alcohol mixtures. However, reaction of RC≡CH with copper(II) salt leads to a mixture of RC≡C—Cu and $Cu(C{\equiv}CR)_2$ which are very easily separated. A summary of the structural properties of RC≡CCu has been presented ⟨B-86MI 223-01⟩.

$$R{-}C{\equiv}C{-}Li + Cu^I{}_x \longrightarrow R{-}C{\equiv}C{-}Cu \quad (87)$$

2.23.3.1.2 *Alkynylcuprate systems (e.g., $[R^1C{\equiv}C{-}Cu{-}R^2]^-M^+$)*

Another category of alkynylcopper(I) compounds are the 'cuprates', the simplest form being $[R^1C{\equiv}C{-}Cu{-}R^2]^-\ M^+$ where, for example, M = Li, MgX, etc. Nothing is implied here concerning the details of the solution structure of these derivatives.

Higher order members of this class ⟨90SL119⟩ are of increasing importance as cuprates generally are of major utility in carbon–carbon bond forming reactions. The alkynyl cuprates when used in this way generally transfer the organic group R more efficiently than the alkynyl group, the latter then being referred to as a 'dummy group'.

Tabulations of preparative procedures for alkynyl cuprates are available ⟨B-83MI 223-01, B-83MI 223-02⟩ but these are of two basic types. With lithium as the counterion, the procedure involves addition of an equivalent of the first organolithium reagent (R^1Li) to copper(I) halide, generally stabilised with a ligand, for example, $Bu^n{}_3P$, followed by addition of an equivalent of the second organolithium reagent (R^2Li). This solution then contains the mixed cuprate R^1R^2CuLi. When R^1 is 1-alkynyl, this R^1Li is added first. For example, using THF at −78 °C, a colourless to pale yellow solution of $(H_2C{=}CMe)(MeOC(Me)C{\equiv}C)CuLi$ has been prepared ⟨78JOC3418⟩.

The second general method is relevant when R^2Cu, typically aryl or alkyl, is relatively stable and can be treated in ether or THF, often in the presence of a solubilising liquid or solvent (e.g., HMPA), with the appropriate R^1Li. In this way (Bu^n) (*n*-PentC≡C)CuLi was prepared in ether at 20 °C ⟨73BSF2137⟩. This general procedure is normally applied when Grignard reagents are used instead of organolithium solutions. The alkynyl group can be incorporated into higher order cuprates, for example, of the type $R^1{}_2R^2CuLi_2$. Presumably (**6**) is formed in the sequence shown in Scheme 16 and transfers the *n*-butyl group in the reaction with iodocyclohexane ⟨81JA7672⟩. Alkynyl groups are bound to copper in a variety of other structural types which are of less general interest ⟨B-87MI 223-01⟩.

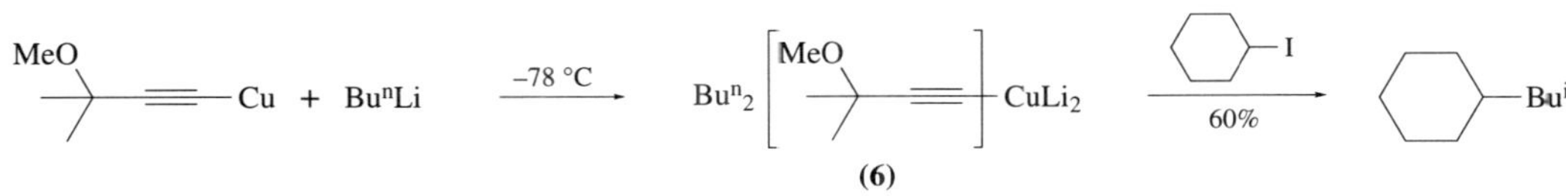

(**6**)

Scheme 16

2.23.3.2 Alkynylzinc Compounds

Organozinc compounds were reviewed as a general class in 1973 ⟨73HOU(13/2a)608, 73HOU(13/2a)622⟩ but the entries on alkynylzinc compounds were few. In 1992, the palladium-catalysed cross-coupling of organozinc compounds with organic systems has been reviewed ⟨92T9577⟩ with some examples based on alkynylzinc reagents. Much of the focus on alkynylzinc compounds is from a synthetic perspective and very often these reagents are generated in solution and used directly.

A distinction between the types of organozinc compounds is worthwhile for the purpose of discussion of synthetic methods. Three basic structural types are identified: (i) alkynylzinc halides (RC≡C—ZnX), (ii) dialkynyl zinc derivatives $(R^1C{\equiv}C)_2Zn$ or $R^1C{\equiv}CZnR^2$ and (iii) zincates, $M_n[Zn(C{\equiv}CR^1)_{2+n}]$.

2.23.3.2.1 *Alkynylzinc halides*

The traditional method of formation of alkynylmetal derivatives using alkynyl lithium or Grignard reagents ⟨62BSF974⟩ has been utilised for zinc compounds also (Equation (88)). Full descriptions of the procedures are available ⟨77CC683, 78JOC358, 93TL4623⟩. Based on the yield of the product subsequently derived from the alkynylzinc halide, it is clear that this reagent forms very efficiently. Direct reaction of zinc, either as dust, turnings or couples (e.g., with copper), with haloalkynes has also been utilised and a procedure described ⟨87TL2387⟩.

$$R{-}C{\equiv}C{-}Li + ZnX_2 \xrightarrow{\text{THF, hexane, }-78\ ^\circ\text{C}} R{-}C{\equiv}C{-}ZnCl \qquad (88)$$

R = H, Me, *n*-Pent, Ph, Bu^n, *n*-Hex, *c*-Hex, $(CH_2)_3Cl$, $H_2C{=}C(Me){-}CH_2{-}C{\equiv}C{-}Pr^n$

2.23.3.2.2 *Dialkynylzinc compounds*

Cleavage of diethylzinc with a terminal alkyne acting as an acid was reported by Durand (Equation (89)) ⟨23CR992, 24BSF166⟩ and was subsequently developed. For example, phenylacetylene may be used in a one- or two-molar stoichiometry to yield the unsymmetrical alkynyl, or the dialkynyl derivatives. The former product slowly disproportionates as shown in Scheme 17. Complexation with NH_3 was used as part of the purification procedure. The reaction of zinc amide with acetylene was also employed ⟨62CB2155⟩. Similar cleavage of dimethylzinc with 1-alkynes has been reported, but the conditions required are more vigorous and require heating (ca. 80 °C) of the neat reactants for extended periods in sealed ampoules (Equation (90)) ⟨68JOM(11)393⟩.

$$Et_2Zn + H{-}C{\equiv}C{-}H \longrightarrow Zn(C{\equiv}CH)_2 \qquad (89)$$

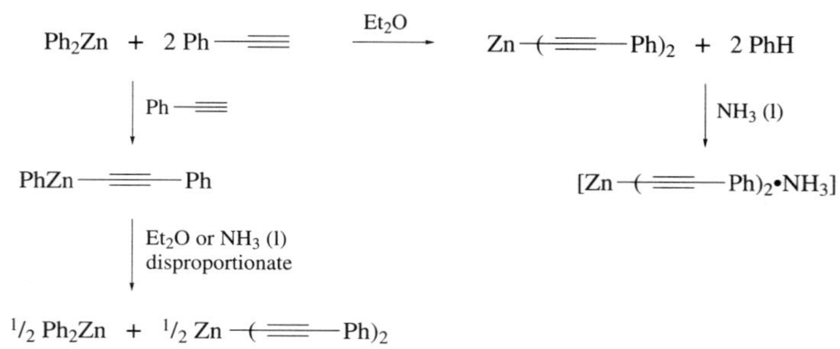

Scheme 17

$$Me_2Zn + R{-}C{\equiv}CH \longrightarrow Zn(C{\equiv}C{-}R)_2 \qquad (90)$$

R = Ph, *n*-Pent

The relative rates of acid cleavage of phenyl vs. alkyl groups from zinc are demonstrated by reactions of diethyl and diphenyl zinc compounds with aminoalkynes (the amino group complexes internally with zinc in the products). With diethylzinc, replacement of only one ethyl group is possible, whereas both phenyl groups are cleaved from diphenylzinc (Equations (91) and (92)) ⟨78JOM(153)C37, 79JOM(174)129⟩. Solvent effects in 1-alkyne cleavages of a variety of diorganozinc compounds have been examined and dipolar aprotic solvents result in great acceleration (Equation (93)) ⟨65JOM(3)257, 90JCS(P1)937⟩. Examples of alkynyl zincates are uncommon, but a preparative procedure and structural information is available for a potassium derivative. The method of formation is shown in Scheme 18, and the complex forms tetragonal lattices containing four formula units with isolated tetrahedral $[M(C{\equiv}CH)_4]$ groups. This complex is not explosive or pyrophoric ⟨68JOM(14)21⟩.

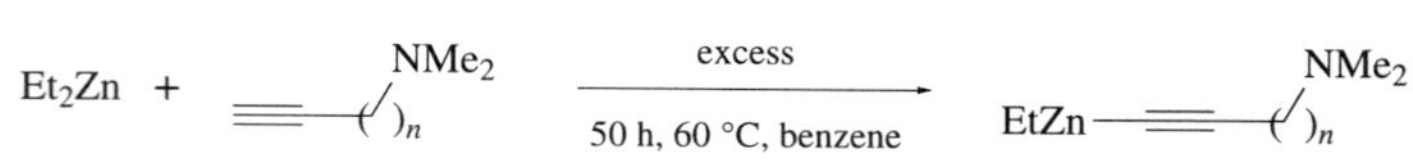

(91)

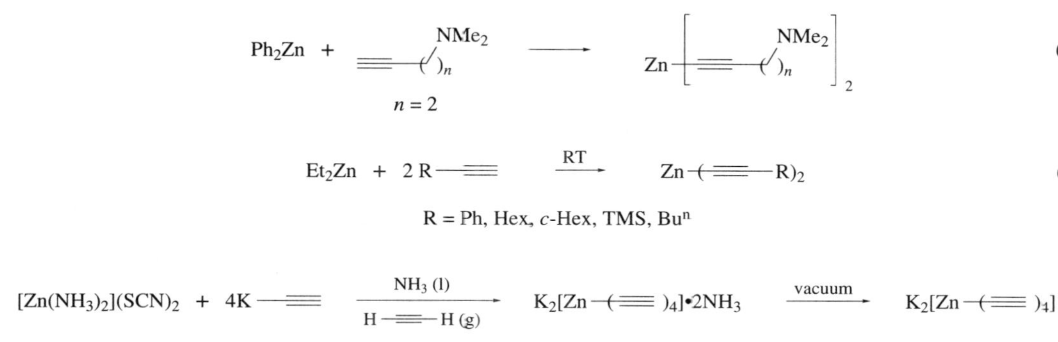

(92)

(93)

Scheme 18

2.23.3.3 Alkynylmercury Compounds

Note: All organomercury compounds should be regarded as being extremely toxic, and manipulations with such derivatives should be conducted with great care. The preferred method for destruction is by oxidation, for example with bromine in water or nitric acid ⟨68MI 223-01, 74HOU(13/2b)9, 74HOU(13/2b)26⟩.

As part of a general discussion of organomercury chemistry, alkynyl derivatives were reviewed in 1974 ⟨B-74HOU(13/2b)9, B-74HOU(13/2b)26⟩ following the general review by Makarova and Nesmeyanov ⟨B-67MI 223-01⟩. The range of alkynylmercury compounds is not extensive, and is not subdivided in this discussion.

Alkynylmercury compounds of the type $R^1Hg{-}C{\equiv}CR^2$ are readily prepared from the appropriate organomercury halide and alkynyl metal derivatives—the method emphasised throughout this whole chapter (Equation (94))—and yields are invariably good.

$$R^1Hg + R^2{-}{\equiv}{-}M \longrightarrow R^1Hg{-}{\equiv}{-}R^2 \qquad (94)$$

R^1 = various alkyl, phenyl
R^2 = H, Me, Ph, TMS
M = MgBr, K, $SnBu_3$

Redistribution methods have also found application in this area and dialkyl mercurials, for example, react with dialkynyl mercurials to provide the unsymmetrical mercurial. Certain silylmercurials of this type are unstable (Equations (95)–(97)) ⟨86SA(A)1107⟩. Alkynylmercuric acetates have been generated and used *in situ* (Equation (98)) ⟨83JCS(P1)1087, 88CC121⟩.

$$Hg(TMS)_2 + Hg({\equiv}{-}R)_2 \longrightarrow 2\ TMS\text{-}Hg{-}{\equiv}{-}R \qquad (95)$$

$$HgMe_2 + Hg({\equiv}{-}R)_2 \longrightarrow 2\ MeHg{-}{\equiv}{-}R \qquad (96)$$

R = Bu^n, Ph, TMS

$$Hg(OAc)_2 + Hg({\equiv}{-}R)_2 \longrightarrow 2\ R{-}{\equiv}{-}HgOAc \qquad (97)$$

$$2\ Ph{-}{\equiv}{-}MgBr + HgCl_2 \xrightarrow[\sim 85\%]{THF} (Ph{-}{\equiv}{-})_2Hg \qquad (98)$$

There are several additional routes of varying generality to dialkynylmercurials. The reaction of phenylethynyl Grignard reagent (>2 equiv.) with $HgCl_2$ has been employed on occasions (Equation (99)) ⟨63JCS2295⟩, but a more popular procedure utilises the acidity of a terminal alkyne. Reaction of such alkynes with mercuric halides in the presence of a base is one variant of this approach (Equation (100)) ⟨68LA(717)21⟩. When $Hg(OAc)_2$ is utilised, no base is required and methanol is used as solvent ⟨68IZV2287⟩.

$$2\,R{-}{\equiv} + HgCl_2 \xrightarrow[79\%]{Et_3N} 2\,Hg({\equiv}{-}R)_2 \quad (99)$$

$R = Bu^n, Ph$

$$2\,R{-}{\equiv} + K_2HgI_4 + 2\,KOH \longrightarrow Hg({\equiv}{-}R)_2 + 4\,KI + 2\,H_2O \quad (100)$$

The most popular application of this approach uses a complexed mercury(II) iodide or cyanide as the mercury source, again in a basic medium (Equation (101)) ⟨26JA469⟩. A listing of examples is available ⟨B-74MI 223-01⟩. Haloalkynes and alkenes have been reported to be useful in this procedure (Equation (101)) ⟨08CB316, 09CB4233, 38JCS1218⟩. Peroxyalkynes have been applied also in this way (Equation (102)) ⟨85JGU1691, 86JGU1812⟩. Mercuric thiocyanate has been employed as a mercury source in this general approach (Equation (103)) ⟨64CB3317⟩, using $NH_3(l)$ as solvent. When R = H, the product is stable only as the ammonia complex.

$$Cl{-}{\equiv} + Hg(CN)_2 + 2\,KOH \longrightarrow Hg({\equiv}{-}Cl)_2 \quad (101)$$

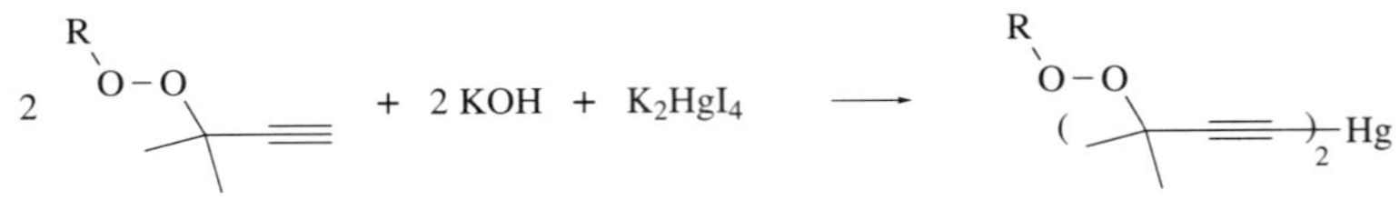

(102)

$$Hg(SCN)_2 + Ba(SCN)_2 + 4\,K{-}{\equiv}{-}R \longrightarrow Ba[Hg({\equiv}{-}R)_4] \quad (103)$$

R = Ph, H

2.23.3.4 Alkynylaluminum Compounds

The review of Davidsohn and Henry summarised the relatively limited information on alkynylaluminum compounds in 1967 ⟨67CRV73⟩. As part of a general review of organoaluminum chemistry, alkynyl derivatives were discussed further in 1970 ⟨70(HOU(13/4)154⟩, but there have been no major innovations in preparative methods since this time. For convenience of discussions, two types of alkynylaluminum compounds are identified; alkynylalanes, $R^1_{3-n}Al(C{\equiv}CR^2)_n$ and complex alkali–aluminum alkynes $M[R^1_nAl(C{\equiv}CR^2)_{4-n}]$. Of interest is the paucity of data on alkynylaluminum halides ⟨62MI 223-01⟩.

2.23.3.4.1 *Alkynylalanes $R^1_{3-n}Al(C{\equiv}CR^2)_n$*

Halide displacement from R^1_2AlX by an alkynyl useful reagent (e.g., sodium) is a commonly employed procedure as shown in (Equation (104)) ⟨60LA222, 65CR(261)1992, 78JOM(159)373⟩. These reactions (Equation (104)) were conducted at room temperature or below, and proceeded in moderate yield, except when $R^2 = H$ where polymerisation was a complication. Alkynyllithium reagents were employed in the mid 1980s to effect the same result but in higher yields ⟨84BCJ108, 85BCJ2425, 85JOC3923⟩. The dialkylalkynylalanes ($n = 1$) do not readily disproportionate and are weak Lewis acids.

$$R^1_2AlX + Na{-}{\equiv}{-}R^2 \xrightarrow{\text{hydrocarbon solvent}} R^1_2Al{-}{\equiv}{-}R^2 + NaX \quad (104)$$

R^1 = Me, Et, Ph
R^2 = Me, Et, $CH_2{=}CMe$, Ph, *n*-Hex
X = Br, I, Cl

Alkynylalanes where $n = 3$ are also accessible by a simple adaptation of the methods shown in Equation (104). For example, alkynylsodium reagents, generated in liquid NH_3, react with aluminum trichloride (as its etherate) to form the alkynylalane (Equation (105)). The products are usually isolated as the etherate complexes, and although uncomplexed trialkylalanes are known ⟨70HOU(13/4)154⟩, they are less stable. When pyridine is the Lewis base these complexed alanes are stable to distillation. However, attempts to remove ether from $(HC{\equiv}C)_3Al \cdot Et_2O$ by warming results in explosive decomposition ⟨62MI 223-01, 63CB1462⟩. Mixed alanes of the type $R^1Al(C{\equiv}CR^2)_2$ are accessible by a redistribution reaction, shown in Equation (106) ⟨62MI 223-01⟩.

$$R^2{-}{\equiv}{-}Na \xrightarrow[AlCl_3 \cdot OR^1_2]{NH_3\ (l)} Al({\equiv}{-}R^2)_3 \cdot OR^1_2 \qquad (105)$$

R^1 = H, Me, Et, Ph

$$R^1_3Al + 2\ Al({\equiv}{-}R^2)_3 \longrightarrow 3\ R^1Al({\equiv}{-}R^2)_2 \qquad (106)$$

1-Alkynes have been employed as protic acids to cleave Al—C and Al—H bonds and hence form alkynylalanes ⟨62JA3820, 63AG918, 67CRV73⟩. Phenylalkyne preferentially cleaves a phenyl group, rather than hydride from aluminum (Equation (107)) ⟨65AJC14⟩.

An unusual route to alkynylalanes ($n = 3$) involves the transfer of alkynyl groups from mercury to aluminum metal, or alane as its amine complex (see Equations (108) and (109)). In the second method (Equation (110)) some reduction of the alkynylalane product occurs.

$$Ph_2AlH + Ph{-}{\equiv} \longrightarrow Ph(H)Al{-}{\equiv}{-}Ph \qquad (107)$$

$$Hg({\equiv}{-}R)_2 + Al \xrightarrow[60\text{–}80\%]{ether} Al({\equiv}{-}R)_3 \qquad (108)$$

R = alkyl, aryl

$$AlH_3 \cdot NMe_3 + 3\ Hg({\equiv}{-}Pr)_2 \xrightarrow{-Hg^0} Al({\equiv}{-}Pr)_3 + H_2\uparrow \qquad (109)$$

$$M[AlH_4] + 4\ {\equiv}{-}R \xrightarrow{solvent} M[Al({\equiv}{-}R)_4] \qquad (110)$$

M = Li, Na, K
R = H, alkyl, aryl

2.23.3.4.2 *Complex alkali–aluminum alkynes, $M[R^1_nAl(C{\equiv}CR^2)_{4-n}]$*

Alkali metal alanates $M(AlH_4)$, where M = Li, Na, K, react with active alkynes in various solvents to provide the title compounds (Equation (111)). The basicity of the solvent should be increased with decreasing acidity of the alkyne. Solvents employed are, for example, diethylether, DIGLYME, GLYME, THF and pyridine ⟨55JOC1337, 60JA3560, 63AG918, 63IZV1146, 64IZV1989⟩.

Mixed alkylalkynylaluminates may be obtained in a number of ways, and some of these are outlined in the review of Davidsohn and Henry ⟨67CRV73⟩. In 1990, the reaction of alkynylithium reagents with Me_3Al was employed to generate alkynylaluminates, which were used in epoxide ring openings of anhydrosugars ⟨90S285⟩. Hydridoalkylaluminates are accessible by a redistribution procedure (Equations (111) and (112)) ⟨65JGU1677⟩.

$$R{-}{\equiv}{-}Li + Me_3Al \xrightarrow[0\ ^\circ C\ to\ RT]{hexane} [R{-}{\equiv}{-}AlMe_3]Li \qquad (111)$$

R = TMS, $CH_2OSiBu^tMe_2$, Bu^n, *n*-Hex, $(CH_2)_2OSiBu^tPh_2$

$$NaAlH_4 + NaAl({\equiv}{-}Bu)_4 \xrightarrow[xylene,\ 60\%]{reflux} NaH_2Al({\equiv}{-}Bu)_2 \qquad (112)$$

2.23.3.5 Alkynyltin Compounds

The first definitive review of alkynylstannes is due to Davidsohn and Henry ⟨67CRV73⟩ and since that time many reviews and monographs discussing organotin chemistry generally have appeared.

These include some data on alkynylstannanes, but this coverage is limited ⟨B-70MI 223-01, B-70MI 223-02, B-87MI 223-05, B-88MI 223-02, 89MI 223-02⟩.

The general method for alkynylstannane synthesis is the reaction of an alkynylmetal derivative with a halotin as shown in Equations (113) and (114)). A list of stannanes made by this method is available ⟨88G1⟩, and it is considered that the Grignard variation of this general method is reliable for moderate to large scale preparations for common alkynes. When limited amounts of 1-alkynes are available, exchange with Bu^nLi to generate the alkynyl lithium is preferred ⟨88G1⟩.

$$R_{4-n}SnX_n + nR{-}C{\equiv}C{-}M \longrightarrow R_{4-n}Sn{-}(C{\equiv}C{-}R)_n + MX \qquad (113)$$

M = Na, Li, MgX

$$R^1{-}C{\equiv}CH + Bu^nLi \xrightarrow[\text{hexane}]{\text{ether}} R^1{-}C{\equiv}C{-}Li \xrightarrow{R^2{}_3SnCl} R^1{-}C{\equiv}C{-}SnR^2{}_3 \qquad (114)$$

Symmetrical distannylalkynes, which can be made from dilithioethyne are important precursors of other alkynylstannanes, by Sn–Li exchange, followed by reactions with electrophiles (Equation (115)). Sodium acetylides are considered to have advantages over the lithium derivatives ⟨88G1⟩, and the reader's attention is directed to a report of an improved method for preparing $NaC{\equiv}CH$ ⟨82S211⟩.

$$R^1{}_3Sn{-}C{\equiv}C{-}SnR^1{}_3 \xrightarrow[\text{ii, } R^2X]{\text{i, } Bu^nLi} R^1{}_3Sn{-}C{\equiv}C{-}R^2 \qquad (115)$$

A second method that proceeds quantitatively is the reaction of a terminal alkyne with an aminostannane, and it is likely this procedure will be more widely used. Alkynylstannanes, difficult to obtain by more orthodox procedures, are readily acquired by the aminostannane route (Equation (116)) ⟨88G1⟩. A full critique of this and other methods for making alkynylstannanes is available ⟨88G1⟩.

$$R_{4-n}Sn(NR^2{}_2)_n \xrightarrow{n\ HC{\equiv}C{-}R^2} R_{4-n}Sn{-}(C{\equiv}C{-}R^2)_n \qquad (116)$$

Condensation reactions also find important application in alkynylstannane synthesis. Organotin oxides or hydroxides react with active alkynes to provide the alkynylstannanes (Equation (117)). Removal of water may be achieved by azeotropic distillation with benzene, addition of CaH_2 or molecular sieves, depending on the hydrolytic and thermal stability of the components ⟨89T1145⟩. A listing of alkynylstannanes obtained in this way is available ⟨88G1⟩.

$$(R_3Sn)_2O + 2\ HC{\equiv}C{-}R \longrightarrow 2\ R_3Sn{-}C{\equiv}C{-}R + H_2O \qquad (117)$$

Thermal decarboxylation is often a route to carbon–metal bonds and this has been utilised in the synthesis of alkynylstannanes. Tin salts of acetylene mono- or dicarboxylic acids have been employed, as shown in Equation (118) ⟨64RTC295⟩. Alkynyltrichlorostannanes are considered to be formed from 1-alkynes, stannic chloride and tributylamine and are effective for the alkynylation of aldehydes, metals and enones under mild conditions, and in high yields (Scheme 19) ⟨92CL2479⟩.

$$Bu_3SnO_2C{-}C{\equiv}C{-}Ph \xrightarrow{\sim 60\ ^\circ C} Bu_3Sn{-}C{\equiv}C{-}Ph + CO_2 \qquad (118)$$

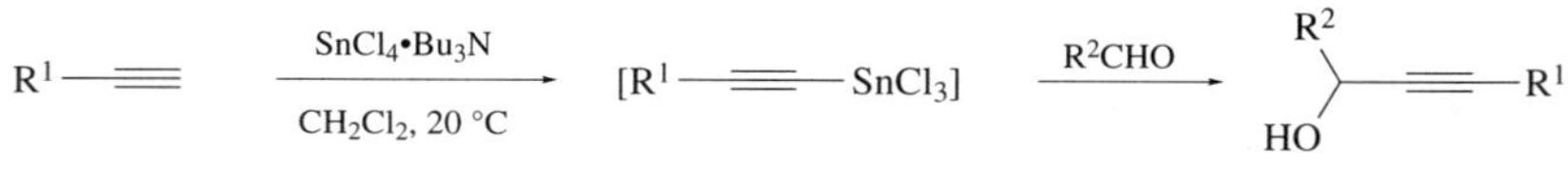

Scheme 19

The bulky stannylene, bis(trimethylsilyl)methylstannylene inserts into ethyne at room temperature to afford (**7**) as one of a number of products (Equation (119)). This reaction does not appear to have any obvious synthetic generality ⟨93JOM(453)C11⟩.

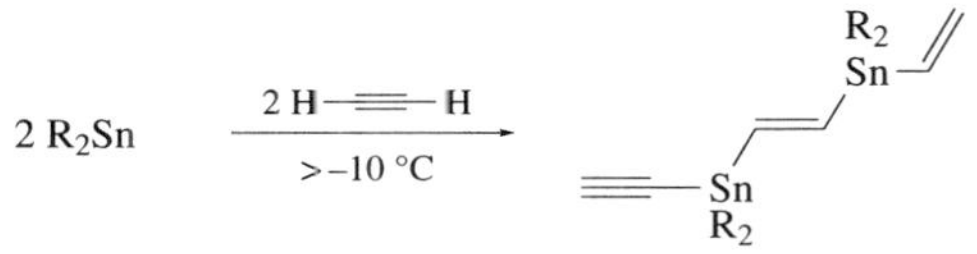

(119)

(**7**)

R = CH(TMS)$_2$

2.23.3.6 Alkynyllead Compounds

The much cited review of Davidsohn and Henry ⟨67CRV73⟩ includes information on alkynylplumbanes, and the position at 1968 is further discussed in the monograph of Shapiro and Frey ⟨B-68MI 223-02⟩. As a general rule the plumbanes, which are more reactive to a variety of reagents, appear to be prone to disproportionation, and explosive decomposition on heating in some cases. With respect to alkynylation at lead, the general methods outlined for the stannanes are applicable, as shown in Equation (120).

$$R^1{}_3Pb + R^2-C\equiv C-M \longrightarrow R^1{}_3Pb-C\equiv C-R^2 \quad (120)$$

R^1 = Et, Ph, Me, *c*-Hex
M = MgBr, Li, Na

Variations of the method shown in Equation (120), for example using R_2PbX_2 and $K_2[PbCl_6]$ ⟨69JOM(16)413⟩, have been reported, with the latter lead(IV) salt affording the tetraalkynylplumbane (Equations (121) and (122)).

$$K_2PbCl_6 + 4\,Bu^t-C\equiv C-Li \longrightarrow Pb(C\equiv C-Bu^t)_4 \quad (121)$$

$$Ph_2PbX_2 + 2\,Li-C\equiv C-R \longrightarrow Ph_2Pb(C\equiv C-R)_2 \quad (122)$$

Enynylplumbanes have been acquired by triorganolead–sodium displacement of enynylbromides (Equation (123)), with these compounds being explosive when heated, and readily hydrolysed in moist air ⟨62DOK(143)855⟩. The elimination of methanol or water from organolead hydroxides or alkoxides and 1-alkynes has also been utilised and good yields were reported (Equations (124) and (125)) ⟨67DOK(175)1079, 68LA(716)19⟩. The method of Equation (125) was adapted to produce 1,2-bis(trimethylphenyl)ethyne ⟨68JCS(C)317⟩.

$$R_3PbNa + Br-C\equiv C-CH=CH_2 \xrightarrow[80\%]{NH_3} R_3Pb-C\equiv C-CH=CH_2 \quad (123)$$

R = Et

$$Bu^n{}_3PbY + HC\equiv C-Ph \xrightarrow[\text{azeotrope}]{\text{benzene}} Bu^n{}_3Pb-C\equiv C-Ph + H_2O \quad (124)$$

Y = OH

$$Et_3PbOMe + HC\equiv C-R \xrightarrow[-MeOH,\ RT]{\text{benzene}} Et_3Pb-C\equiv C-R \quad (125)$$

R = Ph, Bu^n

Decarboxylation ⟨68JCS(C)317⟩ and the use of aminotrialkylplumbanes ⟨72JOM(40)143⟩ have also been reported (Equations (126) and (127)). Much of the information summarised in Equations (120)–(127) is discussed elsewhere ⟨75HOU(13/7)73⟩.

$$Ph_3PbOCO{-}{\equiv}{-}Ph \xrightarrow[-CO_2]{180\,°C} Ph_3Pb{-}{\equiv}{-}Ph \qquad (126)$$

$$Me_3PbNEt_2 + F_3C{-}{\equiv} \longrightarrow Me_3Pb{-}{\equiv}{-}CF_3 \qquad (127)$$

A class of alkynyllead compounds with interesting synthetic applications are alk-1-ynyllead triacetates ⟨91AJC1353⟩. Pinhey reports that reactions of bis(alk-1-ynyl)mercurials or trialkyl(alk-1-ynyl)stannanes with lead tetraacetate produce species behaving as expected for alk-1-ynyllead triacetates (Equations (128) and (129)). These compounds are labile and evidence for their existence is based upon their decomposition products and reactions. Disproportionation is the major decomposition route, although quickly after formation alk-1-ynylation of β-dicarbonyl compounds was possible. Disproportionation provides the tetraalkylplumbane (Scheme 20). The use of alkynyltrimethylstannanes in the preparation of the alk-1-ynyllead triacetates has been reported and again successful alkylation of β-dicarbonyl compounds was achieved (Equation (130)).

$$(R{-}{\equiv}{-})_2Hg + Pb(OAc)_4 \xrightarrow{CHCl_3} R{-}{\equiv}{-}HgOAc + R{-}{\equiv}{-}Pb(OAc)_3 \qquad (128)$$

$$R^1{-}{\equiv}{-}SnR^2{}_3 + Pb(OAc)_4 \xrightarrow{CHCl_3} R^1{-}{\equiv}{-}Pb(OAc)_3 + R^2{}_3SnOAc \qquad (129)$$

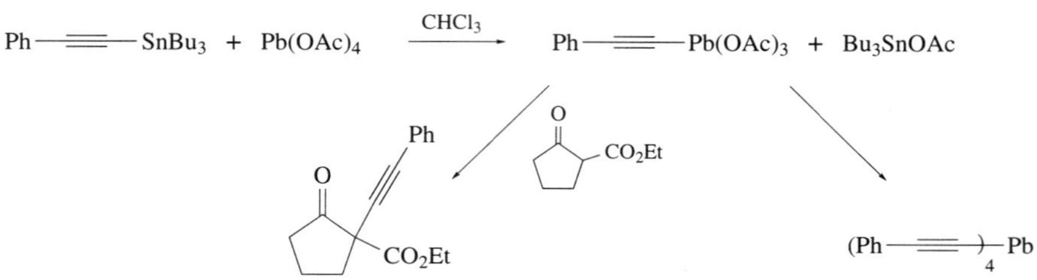

Scheme 20

$$R{-}{\equiv}{-}SnMe_3 + Pb(OAc)_4 \longrightarrow R{-}{\equiv}{-}Pb(OAc)_3 + Me_3SnOAc \qquad (130)$$

Wrackmeyer *et al.* have prepared and studied the reactions of bis(trimethylsilylethynyl) dimethylplumbane and proposed some interesting intermediates for reactions with trialkyl-organoboranes ⟨90JOM(399)1⟩.

References

EXPLANATION OF THE REFERENCE SYSTEM

Throughout this work, references are designated by a number–lettering coding of which the first two numbers denote tens and units of the year of publication, the next one to three letters denote the journal, and the final numbers denote the page. This code appears in the text each time a reference is quoted; the advantages of this system are outlined in the Introduction. The system has been used previously in "Comprehensive Heterocyclic Chemistry," eds A. R. Katritzky and C. W. Rees, Pergamon, Oxford, 1984 and is based on that used in the following two monographs: (a) A. R. Katritzky and J. M. Lagowski, "Chemistry of the Heterocyclic N-Oxides," Academic Press, New York, 1971; (b) J. Elguero, C. Marzin, A. R. Katritzky and P. Linda, "The Tautomerism of Heterocycles," in "Advances in Heterocyclic Chemistry," Supplement 1, Academic Press, New York, 1976.

The following additional notes apply:

1. A list of journal codes in alphabetical order, together with the journals to which they refer, is given immediately following these notes. Journal names are abbreviated throughout using the CASSI (Chemical Abstracts Service Source Index) system.
2. Each volume contains all the references cited *in that volume*; no separate lists are given for individual chapters.
3. The list of references is arranged in order of (a) year, (b) journal in alphabetical order of journal code, (c) part letter or number if relevant, (d) volume number if relevant, (e) page number.
4. In the reference list the code is followed by (a) the complete literature citation in the conventional manner and (b) the number(s) of the page(s) on which the reference appears, whether in the text or in tables, schemes, etc.
5. For nontwentieth-century references the year is given in full in the code.
6. For journals which are published in separate parts, the part letter or number is given (when necessary) in parentheses immediately after the journal code letters.
7. Journal volume numbers are *not* included in the code numbers unless more than one volume was published in the year in question, in which case the volume number is included in parentheses immediately after the journal code letters.
8. Patents are assigned appropriate three-letter codes.
9. Frequently cited books are assigned codes.
10. Less common journals and books are given the code "MI" for miscellaneous with the whole code for books prefixed by the letter "B-".
11. Where journals have changed names, the same code is used throughout, e.g. CB refers to both *Chem. Ber.* and to *Ber. Dtsch. Chem. Ges.*

Journal Codes

AAC	Antimicrob. Agents Chemother.
ABC	Agric. Biol. Chem.
AC	Appl. Catal.
AC(P)	Ann. Chim. (Paris)
AC(R)	Ann. Chim. (Rome)
ACH	Acta Chim. Acad. Sci. Hung.

ACR	Acc. Chem. Res.
ACS	Acta Chem. Scand.
ACS(A)	Acta Chem. Scand., Ser. A
ACS(B)	Acta Chem. Scand., Ser. B
AF	Arzneim.-Forsch.
AFC	Adv. Fluorine Chem.
AG	Angew. Chem.
AG(E)	Angew. Chem., Int. Ed. Engl.
AHC	Adv. Heterocycl. Chem.
AHCS	Adv. Heterocycl. Chem. Supplement
AI	Anal. Instrum.
AJC	Aust. J. Chem.
AK	Ark. Kemi
AKZ	Arm. Khim. Zh.
AM	Adv. Mater. (Weinheim, Ger.)
AMLS	Adv. Mol. Spectrosc.
AMS	Adv. Mass. Spectrom.
ANC	Anal. Chem.
ANL	Acad. Naz. Lncei
ANY	Ann. N. Y. Acad. Sci.
AOC	Adv. Organomet. Chem.
AP	Arch. Pharm. (Weinheim, Ger.)
APO	Adv. Phys. Org. Chem.
AQ	An. Quim.
AR	Annu. Rep. Prog. Chem.
AR(A)	Annu. Rep. Prog. Chem., Sect. A
AR(B)	Annu. Rep. Prog. Chem., Sect. B
ARP	Annu. Rev. Phys. Chem.
ASI	Acta Chim. Sin. Engl. Ed.
ASIN	Acta Chim. Sin.
AX	Acta Crystallogr.
AX(A)	Acta Crystallogr., Part A
AX(B)	Acta Crystallogr., Part B
B	Biochemistry
BAP	Bull. Acad. Pol. Sci., Ser. Sci. Chim.
BAU	Bull. Acad. Sci. USSR, Div. Chim. Sci.
BBA	Biochim. Biophys. Acta
BBR	Biochim. Biophys. Res. Commun.
BCJ	Bull. Chem. Soc. Jpn.
BEP	Belg. Pat.
BJ	Biochem. J.
BJP	Br. J. Pharmacol.
BMC	Bioorg. Med. Chem. Lett.
BP	Biochem. Biopharmacol.
BPJ	Br. Polym. J.
BRP	Br. Pat.
BSB	Bull. Soc. Chim. Belg.
BSF	Bull. Soc. Chim. Fr.
BSF(2)	Bull. Soc. Chim. Fr., Part 2
C	Chimia
CA	Chem. Abstr.
CAN	Cancer
CAR	Carbohydr. Res.
CAT	Chim. Acta Turc.
CB	Chem. Ber.

CBR	Chem. Br.
CC	J. Chem. Soc., Chem. Commun.
CCA	Croat. Chem. Acta
CCC	Collect. Czech. Chem. Commun.
CCR	Coord. Chem. Rev.
CE	Chem. Express
CEN	Chem. Eng. News
CHE	Chem. Heterocycl. Compd. (Engl. Transl.)
CHEC	Comp. Heterocycl. Chem.
CI(L)	Chem. Ind. (London)
CI(M)	Chem. Ind. (Milan)
CJC	Can. J. Chem.
CJS	Can. J. Spectrosc.
CL	Chem. Lett.
CLY	Chem. Listy
CM	Chem. Mater.
CMC	Comp. Med. Chem.
COC	Comp. Org. Chem.
COMC-I	Comp. Organomet. Chem., 1st edn.
COS	Comp. Org. Synth.
CP	Can. Pat.
CPB	Chem. Pharm. Bull.
CPH	Chem. Phys.
CPL	Chem. Phys. Lett.
CR	C. R. Hebd. Seances Acad. Sci.
CR(A)	C. R. Hebd. Seances Acad. Sci., Ser. A
CR(B)	C. R. Hebd. Seances Acad. Sci., Ser. B
CR(C)	C. R. Hebd. Seances Acad. Sci., Ser. C
CRAC	Crit. Rev. Anal. Chem.
CRV	Chem. Rev.
CS	Chem. Scr.
CSC	Cryst. Struct. Commun.
CSR	Chem. Soc. Rev.
CT	Chem. Tech.
CZ	Chem.-Ztg.
CZP	Czech. Pat.
DIS	Diss. Abstr.
DIS(B)	Diss. Abstr. Int. B.
DOK	Dokl. Akad. Nauk SSSR
DP	Dyes Pigm.
E	Experientia
EC	Educ. Chem.
EF	Energy Fuels
EGP	Ger. (East) Pat.
EJM	Eur. J. Med. Chem.
EUP	Eur. Pat.
FCF	Forschr. Chem. Forsch.
FCR	Fluorine Chem. Rev.
FES	Farmaco Ed. Sci.
FOR	Forschr. Chem. Org. Naturst.
FRP	Fr. Pat.
G	Gazz. Chim. Ital.
GAK	Gummi Asbest Kunstst.
GEP	Ger. Pat.
GEP(O)	Ger. Pat. Offen.

GSM	Gen. Synth. Methods
H	Heterocycles
HAC	Heteroatom Chem.
HC	Chem. Heterocycl. Compd.
HCA	Helv. Chim. Acta
HOU	Methoden Org. Chem. (Houben-Weyl)
HP	Hydrocarbon Process
IC	Inorg. Chem.
ICA	Inorg. Chim. Acta
IEC	Ind. Eng. Chem. Res.
IJ	Isr. J. Chem.
IJC	Indian J. Chem.
IJC(A)	Indian J. Chem., Sect. A
IJC(B)	Indian J. Chem., Sect. B
IJM	Int. J. Mass Spectrom. Ion Phys.
IJQ	Int. J. Quantum Chem.
IJS	Int. J. Sulfur Chem.
IJS(A)	Int. J. Sulfur Chem., Part A
IJS(B)	Int. J. Sulfur Chem., Part B
IS	Inorg. Synth
IZV	Izv. Akad. Nauk SSSR Ser. Khim.
JA	J. Am. Chem. Soc.
JAN	J. Antibiot.
JAP	Jpn. Pat.
JAP(K)	Jpn. Kokai
JBC	J. Biol. Chem.
JC	J. Chromatogr.
JCC	J. Coord. Chem.
JCE	J. Chem. Ed.
JCED	J. Chem. Eng. Data
JCI	J. Chem. Inf. Comput. Sci.
JCP	J. Chem. Phys.
JCPB	J. Chim. Phys. Physico-Chim. Biol.
JCR(M)	J. Chem. Res. (M)
JCR(S)	J. Chem. Res. (S)
JCS	J. Chem. Soc.
JCS(A)	J. Chem. Soc. (A)
JCS(B)	J. Chem. Soc. (B)
JCS(C)	J. Chem. Soc. (C)
JCS(D)	J. Chem. Soc., Dalton Trans.
JCS(F1)	J. Chem. Soc., Faraday Trans. 1
JCS(F2)	J. Chem. Soc., Faraday Trans. 2
JCS(P1)	J. Chem. Soc., Perkin Trans. 1
JCS(P2)	J. Chem. Soc., Perkin Trans. 2
JCS(S2)	J. Chem. Soc. (Suppl. 2)
JEC	J. Electroanal. Chem. Interfacial Electrochem.
JEM	J. Energy Mater.
JES	J. Electron. Spectrosc.
JFA	J. Sci. Food. Agri.
JFC	J. Fluorine Chem.
JGU	J. Gen. Chem. USSR (Engl. Transl.)
JHC	J. Heterocycl. Chem.
JIC	J. Indian Chem. Soc.
JINC	J. Inorg. Nucl. Chem.
JLC	J. Liq. Chromatogr.

JMAS	J. Mat. Sci.
JMC	J. Med. Chem.
JMOC	J. Mol. Catal.
JMR	J. Magn. Reson.
JMS	J. Mol. Sci.
JOC	J. Org. Chem.
JOM	J. Organomet. Chem.
JOU	J. Org. Chem. USSR (Engl. Transl.)
JPC	J. Phys. Chem.
JPJ	J. Pharm. Soc. Jpn.
JPO	J. Phys. Org. Chem.
JPP	J. Pharm. Pharmacol.
JPR	J. Prakt. Chem.
JPS	J. Pharm. Sci.
JPS(A)	J. Polym. Sci., Polym. Chem., Part A
JPU	J. Phys. Chem. USSR (Engl. Transl.)
JSC	J. Serbochem. Soc.
JSP	J. Mol. Spectrosc.
JST	J. Mol. Struct.
K	Kristallografiya
KFZ	Khim. Farm. Zh.
KGS	Khim. Geterotsikl. Soedin.
KO	Kirk-Othmer Encyc.
KPS	Khim. Prir. Soedin.
L	Langmuir
LA	Liebigs Ann. Chem.
LC	Liq. Cryst.
LS	Life Sci.
M	Monatsh. Chem.
MAC	Macromol. Chem.
MC	Mendeleev Chem. J. (Engl. Transl.)
MCLC	Mol. Cryst. Liq. Cryst.
MI	Miscellaneous [book/journal]
MIP	Miscellaneous Pat.
MM	Macromolecules
MP	Mol. Phys.
MRC	Magn. Reson. Chem.
N	Naturwissenschaften
NAT	Nat.
NEP	Neth. Pat.
NJC	Nouv. J. Chim.
NKK	Nippon Kagaku Kaishi (J. Chem. Soc. Jpn.)
NKZ	Nippon Kagaku Zasshi
NZJ	N. Z. J. Sci. Technol.
OCS	Organomet. Synth.
OM	Organometallics
OMR	Org. Magn. Reson.
OMS	Org. Mass Spectrom.
OPP	Org. Prep. Proced. Int.
OR	Org. React.
OS	Org. Synth.
OSC	Org. Synth., Coll. Vol.
P	Phytochemistry
PA	Polym. Age
PAC	Pure Appl. Chem.

PAS	Pol. Acad. Sci.
PB	Polym. Bull.
PC	Personal Communication
PCS	Proc. Chem. Soc.
PHA	Pharmazi
PHC	Prog. Heterocycl. Chem.
PIA	Proc. Indian Acad. Sci.
PIA(A)	Proc. Indian Acad. Sci., Sect. A
PJC	Pol. J. Chem.
PJS	Pak. J. Sci. Ind. Res.
PMH	Phys. Methods Heterocycl. Chem.
PNA	Proc. Natl. Acad. Sci. USA
POL	Polyhedron
PP	Polym. Prepr.
PRS	Proceed. Roy. Soc.
PS	Phosphorus Sulfur
QR	Q. Rev., Chem. Soc.
QRS	Quart. Rep. Sulfur. Chem.
QSAR	Quant. Struct. Act. Relat. Pharmacol. Chem. Biol.
RC	Rubber Chem. Technol.
RCM	Rapid Commun. Mass Spectrom.
RCP	Rec. Chem. Prog.
RCR	Russ. Chem. Rev. (Engl. Transl.)
RHA	Rev. Heteroatom Chem.
RJ	Rubber J.
RP	Rev. Polarogr.
RRC	Rev. Roum. Chim.
RS	Ric. Sci.
RTC	Recl. Trav. Chim. Pays-Bas
RZC	Rocz. Chem.
S	Synthesis
SA	Spectrochim. Acta
SA(A)	Spectrochim. Acta, Part A
SAP	S. Afr. Pat.
SC	Synth. Commun.
SCI	Science
SL	Synlett
SM	Synth. Met.
SR	Sulfur Reports
SRI	Synth. React. Inorg. Metal-Org. Chem.
SS	Sch. Sci. Rev.
SST	Org. Compd. Sulphur, Selenium, Tellurium [R. Soc. Chem. series]
SUL	Sulfur Letters
SZP	Swiss Pat.
T	Tetrahedron
T(S)	Tetrahedron, Suppl.
TA	Tetrahedron Asymmetry
TAL	Talanta
TCA	Theor. Chim. Acta
TCC	Top. Curr. Chem.
TCM	Tetrahedron, Comp. Method
TFS	Trans. Faraday Soc.
TH	Thesis
TL	Tetrahedron Lett.
TS	Top. Stereochem.

UK	Usp. Khim.
UKZ	Ukr. Khim. Zh. (Russ. Ed.)
UP	Unpublished Results
URP	USSR Pat.
USP	US Pat.
WCH	Wiadom. Chem.
YGK	Yuki Gosei Kagaku Kyokaishi
YZ	Yakugaku Zasshi (J. Pharm. Soc. Jpn.)
ZAAC	Z. Anorg. Allg. Chem.
ZAK	Zh. Anal. Khim.
ZC	Z. Chem.
ZN	Z. Naturforsch.
ZN(A)	Z. Naturforsch., Teil A
ZN(B)	Z. Naturforsch., Teil B
ZOB	Zh. Obshch. Khim.
ZOR	Zh. Org. Khim.
ZPC	Hoppe–Seyler's Z. Physiol. Chem.
ZPK	Zh. Prikl. Khim.

VOLUME 2 REFERENCES

1760MI 209-01 L. C. Cadet de Gassincourt; *Mem. Math. Phys. Savants Etrangers*, 1760, **3**, 363. 494, 505, 506

1834LA(11)1 W. C. Zeise; *Justus Liebigs Ann. Chem.*, 1834, **11**, 1. 119

1849LA(79)171 E. Frankland; *Justus Liebigs Ann. Chem.*, 1849, **79**, 171. 68, 590

1852JCS106 W. Williamson; *J. Chem. Soc.*, 1852, 106. 88
1852JCS229 W. Williamson; *J. Chem. Soc.*, 1852, 229. 88

1857LA(101)1 A. Cahours and A. W. Hofmann; *Justus Liebigs Ann. Chem.*, 1857, **101**, 1. 461, 853

1858LA(106)334 L. Carius; *Justus Liebigs Ann. Chem.*, 1858, **106**, 334. 100

1859LA(111)93 L. Carius; *Justus Liebigs Ann. Chem.*, 1859, **111**, 93. 100
1859PRS568 E. Frankland and B. F. Duppa; *Proc. Roy. Soc.* (*London*), 1859, **10**, 568. 520

1861LA(119)182 V. Sawitsch; *Justus Liebigs Ann. Chem.*, 1861, **119**, 182. 1017
1861LA(119)313 L. Carius; *Justus Liebigs Ann. Chem.*, 1861, **119**, 313. 135

1865LA(136)891 C. Marker; *Justus Liebigs Ann. Chem.*, 1865, **136**, 891. 150

1866MI 213-01 A. Kekulé; *Compt. Rendu.*, 1866, **64**, 753. 647

1867LA(143)118 C. Friedel and A. Ladenburg; *Justus Liebigs Ann. Chem*, 1867, **143**, 118. 108
1867LA(144)121 C. Wurtz; *Justus Liebigs Ann. Chem.*, 1867, **144**, 121. 647

1868CB177 A. W. Hofmann; *Ber. Dtsch. Chem. Ges.*, 1868, **1**, 177. 127

1870CB858 B. Rathke; *Ber. Dtsch. Chem. Ges.*, 1870, **3**, 858. 166

1871CB430 A. W. Hofmann; *Ber. Dtsch. Chem. Ges.*, 1871, **4**, 430. 427
1871CB605 A. W. Hofmann; *Ber. Dtsch. Chem. Ges.*, 1871, **4**, 605. 427

1872LA(161)178 E. Linneman; *Justus Liebigs Ann. Chem.*, 1872, **161**, 178. 54

1873CB292 A. W. Hofmann; *Ber. Dtsch. Chem. Ges.*, 1873, **6**, 292. 427
1873LA(167)212 B. Rathke; *Justus Liebigs Ann. Chem.*, 1873, **167**, 212. 239

1874CB646 H. Bunte; *Ber. Dtsch. Chem. Ges.*, 1874, **7**, 646. 122
1874CB809 A. Baeyer and H. Caro; *Ber. Dtsch. Chem. Ges.*, 1874, **7**, 809. 779
1874CB1688 A. Michaelis and J. Ananoff; *Ber. Dtsch. Chem. Ges.*, 1874, **7**, 1688. 451

1879BSF(2)551	M. Hanriot; *Bull. Soc. Chim. Fr.*, Part 2, 1879, **32**, 551.	105
1880LA(203)83	O. Wallach; *Justus Liebigs Ann. Chem.*, 1880, **203**, 83.	1014
1881CB891	C. Hell; *Ber. Dtsch. Chem. Ges.*, 1881, **14**, 891.	28
1881M371	A. Kraus; *Monatsh. Chem.*, 1881, **12**, 371.	781
1882CB180	A. Michaelis and P. Becker; *Ber. Dtsch. Chem. Ges.*, 1882, **15**, 180.	906
1882CB1952	A. Michaelis and C. Schulte; *Ber. Dtsch. Chem. Ges.*, 1882, **15**, 1952.	890
1883CB175	A. Janny; *Ber. Dtsch. Chem. Ges.*, 1883, **16**, 175.	103
1883CB1439	G. Meyer; *Ber. Dtsch. Chem. Ges.*, 1883, **16**, 1439.	497
1884JPR280	G. McGowan; *J. Prakt. Chem.*, 1884, **30**, 280.	181
1885CB2109	A. Michaelis and W. LaCoste; *Ber. Dtsch. Chem. Ges.*, 1885, **18**, 2109.	465
1886LA(233)60	A. Michaelies and U. Paetow; *Justus Liebigs Ann. Chem.*, 1886, **233**, 60.	490
1887CB224	S. Gabriel; *Ber. Dtsch. Chem. Ges.*, 1887, **20**, 224.	299
1887CB1210	S. N. Reformatsky; *Ber. Dtsch. Chem. Ges.*, 1887, **20**, 1210.	702
1887CB1516	A. Marquardt; *Ber. Dtsch. Chem. Ges.*, 1887, **20**, 1516.	490
1887CB2376	P. Klason; *Ber. Dtsch. Chem. Ges.*, 1887, **20**, 2376.	169
1887LA(241)150	A. Michaelis and B. Landmann; *Justus Liebigs Ann. Chem.*, 1887, **241**, 150.	101
1888CB857	T. Sandmeyer; *Ber. Dtsch. Chem. Ges.*, 1888, **19**, 857.	102
1888CB1505	C. Dörken; *Ber. Dtsch. Chem. Ges.*, 1888, **21**, 1505.	864
1888CB2035	A. Marquardt; *Ber. Dtsch. Chem. Ges.*, 1888, **21**, 2035.	490
1888G434	G. Bertoni; *Gazz. Chim. Ital.*, 1888, **18**, 434.	104
1888JCS636	N. Colie; *J. Chem. Soc.*, 1888, **53**, 636.	452
1888JPR(2)207	R. Otto and A. Milch; *J. Prakt. Chem. Part 2*, 1888, **37**, 207.	723
1889LA(251)323	A. Michaelis and A. Marquardt; *Justus Liebigs Ann. Chem.*, 1889, **251**, 323.	886
1890CB203	R. Behrend and K. Leuchs; *Ber. Dtsch. Chem. Ges.*, 1890, **21**, 203.	103
1890LA(260)40	O. Hinsberg; *Justus Liebigs Ann. Chem.*, 1890, **260**, 40.	101
1891JCS1931	B. K. Blount; *J. Chem. Soc.*, 1891, 1931.	452
1893CB310	A. Michaelis and O. Storbeck; *Ber. Dtsch. Chem. Ges.*, 1893, **26**, 310.	386
1893CB1567	A. Werner; *Ber. Dtsch. Chem. Ges.*, 1893, **26**, 1567.	103
1893JPR179	K. Elbs; *J. Prakt. Chem.*, 1893, **48**, 179.	651
1894CB244	A. Michaelis and M. Behrens; *Ber. Dtsch. Chem. Ges.*, 1894, 244.	906
1894CB263	A. Michaelis and A. Lösner; *Ber. Dtsch. Chem. Ges.*, 1894, **27**, 263.	887
1894CB914	E. Bamberger; *Ber. Dtsch. Chem. Ges.*, 1894, **27**, 914.	805
1894CB1548	E. Bamberger; *Ber. Dtsch. Chem. Ges.*, 1894, **27**, 1548.	764
1894CB3350	A. Werner and E. Sonnenfeld; *Ber. Dtsch. Chem. Ges.*, 1894, **27**, 3350.	103
1894JCS899	H. S. H. Fenton; *J. Chem. Soc.*, 1894, 899.	650
1894MI 202-01	V. Gasselin; *Ann. Chim. Phys.*, 1894, **3**, 5.	110
1895CB449	F. Lengfeld; *Ber. Dtsch. Chem. Ges.*, 1895, **28**, 449.	100
1895LA(286)145	O. Fischer and E. Diepolder; *Justus Liebigs Ann. Chem.*, 1895, **286**, 145.	780
1896CB710	A. Michaelis and K. Luxembourg; *Ber. Dtsch. Chem. Ges.*, 1896, **29**, 710.	414, 416, 418
1896CB2662	E. Fischer; *Ber. Dtsch. Chem. Ges.*, 1896, **29**, 2662.	782
1896LA(293)193	A. Michaelis; *Justus Liebigs Ann. Chem.*, 1896, **293**, 193.	861
1896LA(293)215	A. Michaelis; *Justus Liebigs Ann. Chem.*, 1896, **293**, 215.	861
1897CB290	H. Biltz; *Ber. Dtsch. Chem. Ges.*, 1897, **30**, 290.	1019
1897CB366	E. Bamberger; *Ber. Dtsch. Chem. Ges.*, 1897, **30**, 366.	805
1897CB1656	O. Piloty and I. Ruff; *Ber. Dtsch. Chem. Ges.*, 1897, **30**, 1656.	782
1898BSF120	A. Reychler; *Bull. Soc. Chim. Fr.*, 1898, **3**, 120.	180
1898CB1048	A. Michaelis and R. Kaehne; *Ber. Dtsch. Chem. Ges.*, 1898, **31**, 1048.	456
1899CB2697	L. Ilosvay von Nagyilosva; *Ber. Dtsch. Chem. Ges.*, 1899, **26**, 2697.	1094
1899CR(128)511	M. Guerbet; *C. R. Hebd. Seances Acad. Sci.*, 1899, **128**, 511.	87
00CR(130)1322	V. Grignard; *C. R. Hebd. Seances Acad. Sci.*, 1900, **130**, 1322.	557
01CB3246	W. Markovnikoff and P. Zuboff; *Ber. Dtsch. Chem. Ges.*, 1901, **34**, 3246.	87
01LA(315)19	A. Michaelis and G. Thévenot; *Justus Liebigs Ann. Chem.*, 1901, **315**, 19.	906
01LA(315)26	A. Michaelis and E. Richter; *Justus Liebigs Ann. Chem.*, 1901, **315**, 26.	906

01LA(315)43	A. Michaelis; *Justus Liebigs Ann. Chem.*, 1901, **315**, 43.	455
01LA(316)292	E. Bamberger and A. Rising; *Justus Liebigs Ann. Chem.*, 1901, **316**, 292.	782
01LA(318)157	J. U. Nef; *Justus Liebigs Ann. Chem.*, 1901, **318**, 157.	87
01ZAAC(40)123	L. Ilosvay von Nagyilosva; *Z. Anorg. Allg. Chem.*, 1901, **40**, 123.	1094
02CB731	R. Hubner; *Ber. Dtsch. Chem. Ges.*, 1902, **35**, 731.	782
02JCS79	H. P. Stevens; *J. Chem. Soc.*, 1902, **81**, 79.	120
02LA(321)141	A. Michaelis; *Justus Liebigs Ann. Chem.*, 1902, **321**, 141.	890
02MI 202-01	S. Weber; *Chem. Ztbl.*, 1902, **I**, 364.	102
03CB3803	W. D. Langley, E. Bamberger and R. Hubner; *Ber. Dtsch. Chem. Ges.*, 1903, **36**, 3803.	782
03CR(136)1676	L. Bouveault and G. Blanc; *C. R. Hebd. Seances Acad. Sci.*, 1903, **136**, 1676.	54
03CR(137)60	L. Bouveault and G. Blanc; *C. R. Hebd. Seances Acad. Sci.*, 1903, **137**, 60.	54
03CR(137)925	V. Auger; *C. R. Hebd. Seances Acad. Sci.*, 1903, **137**, 925.	495
03JCS1021	T. Purdie and J. C. Irvine; *J. Chem. Soc.*, 1903, 1021.	89
03LA(326)149	A. Michaelis; *Justus Liebigs Ann. Chem.*, 1903, **326**, 149.	409, 410
03LA(326)160	A. Michaelis; *Justus Liebigs Ann. Chem.*, 1903, **326**, 160.	410
03M773	A. Hochsfetter and M. Cohn; *Monatsch. Chem.*, 1903, **24**, 773.	314
04BSF748	L. Bouveault and G. Blanc; *Bull. Soc. Chim. Fr.*, 1904, **31**, 748.	54
04CB629	E. Bamberger and A. Wetter; *Ber. Dtsch. Chem. Ges.*, 1904, **37**, 629.	806
04CB690	O. C. Billeter; *Ber. Dtsch. Chem. Ges.*, 1904, **37**, 690.	249
04CB4415	H. Biltz and E. Kuppers; *Ber. Dtsch. Chem. Ges.*, 1904, **37**, 4415.	1019
04CR(138)148	L. Bouveault and G. Blanc; *C. R. Hebd. Seances Acad. Sci.*, 1904, **138**, 148.	54
04CR(138)982	F. Taboury; *C. R. Hebd. Seances Acad. Sci.*, 1904, **138**, 982.	125
04CR(138)1705	V. Auger; *C. R. Hebd. Seances Acad. Sci.*, 1904, **138**, 1705.	507
05BSF184	Z. Iotschich; *Bull. Soc. Chim. Fr.*, 1905, **34**, 184.	1018
05CB53	O. Ruff and C. Albert; *Ber. Dtsch. Chem. Ges.*, 1905, **38**, 53.	108
05CB646	T. Posner; *Ber. Dtsch. Chem. Ges.*, 1905, **38**, 646.	193
05CB670	O. Dimroth; *Ber. Dtsch. Chem. Ges.*, 1905, **38**, 670.	387
05CB1661	I. Taurke; *Ber. Dtsch. Chem. Ges.*, 1905, **38**, 1661.	108
05MI 209-01	W. M. Dehn; *Am. Chem. J.*, 1905, **33**, 101.	495, 498, 500
06CB2334	G. Heller and H. S. Meyer; *Ber. Dtsch. Chem. Ges.*, 1906, **39**, 2334.	783
06CB2339	C. H. Heller; *Ber. Dtsch. Chem. Ges.*, 1906, **39**, 2339.	783
06CB3905	O. Dimroth; *Ber. Dtsch. Chem. Ges.*, 1906, **39**, 3905.	387
06CR(142)1151	V. Auger; *C. R. Hebd. Seances Acad. Sci.*, 1906, **142**, 1151.	506
06JCS1	F. E. Francis; *J. Chem. Soc.*, 1906, 1.	104
06MI 209-01	W. M. Dehn and B. B. Wilcox; *Am. Chem. J.*, 1906, **35**, 1.	498
07GEP206455	Farbenf. Bayer; *Ger. Pat.* 206 445 (1907) (*Fortschr. Teerfabrik. Verwandt. Industriezweige*, 1911, **9**, 300).	803
08CB314	K. A. Hofmann and H. Kirmreuther; *Ber. Dtsch. Chem. Ges.*, 1908, **41**, 314.	1014
08CB316	K. A. Hofmann and H. Kirmreuther; *Ber. Dtsch. Chem. Ges.*, 1908, **41**, 316.	1098
08CB4115	T. P. Hilditch and S. Smiles; *Ber. Dtsch. Chem. Ges.*, 1908, **41**, 4115.	172
08CB4271	K. Fries and G. Finck; *Ber. Dtsch. Chem. Ges.*, 1908, **41**, 4271.	656
08GEP1765	O. Dieffenbach; *Ger. Pat.* 192 519 (1908) (*Chem. Abstr.*, 1908, **2**, 1765).	782
08JCS1833	M. Gazdar and S. Smiles; *J. Chem. Soc.*, 1908, **93**, 1833.	144
08MI 209-01	W. M. Dehn; *Amer. Chem. J.*, 1908, **30**, 88.	507
09BSF405	A. Mailhe and M. Murat; *Bull. Soc. Chim. Fr.*, 1909, **5**, 405.	125
09CB49	L. Tschugaeff; *Ber. Dtsch. Chem. Ges.*, 1909, **42**, 49.	279
09CB3091	E. Khotinsky and M. Melamed; *Ber. Dtsch. Chem. Ges.*, 1909, 3091.	906
09CB4232	K. A. Hofmann and H. Kirmreuther; *Ber. Dtsch. Chem. Ges.*, 1909, **42**, 4232.	1014
09CB4233	K. A. Hofmann and H. Kirmreuther; *Ber. Dtsch. Chem. Ges.*, 1909, **42**, 4233.	1098
09CB4811	N. Prilezhaev; *Ber. Dtsch. Chem. Ges.*, 1909, **42**, 4811.	90
09JA477	H. D. Dakin; *J. Am. Chem. Soc.*, 1909, **42**, 477.	655
09JCS1725	T. S. Price and D. F. Twiss; *J. Chem. Soc.*, 1909, 1725.	122
10JCS2585	T. P. Hilditch and S. Smiles; *J. Chem. Soc.*, 1910, 2585.	172
10LA(376)239	J. Thiele; *Justus Liebigs Ann. Chem.*, 1910, **376**, 239.	386, 388 389, 403
11JA1598	W. H. Dehn, *J. Am. Chem. Soc.*. 1911, **33**, 1598.	1019
11JCS404	F. D. Chattaway and M. Aldridge; *J. Chem. Soc.*, 1911, **99**, 404.	769
12CB1835	R. Lesser and R. Weiss; *Ber. Dtsch. Chem. Ges.*, 1912, **45**, 1835.	734
12LA(391)55	T. Zincke and F. Farr; *Justus Liebigs Ann. Chem.*, 1912, **391**, 55.	688
12LA(392)133	H. Wieland and H. Fressel; *Justus Liebigs Ann. Chem.*, 1912, **392**, 133.	388
12LA(394)325	E. Fromm, H. Benzinger and F. Shäfer; *Justus Liebigs Ann. Chem.*, 1912, **394**, 325.	1034
14CB1269	L. Chugaev and W. Chlopin; *Ber. Dtsch. Chem. Ges.*, 1914, **47**, 1269.	279, 292

14JA2202	L. W. Jones and L. Neuffer; *J. Am. Chem. Soc.*, 1914, **36**, 2202.	103
14JCS1672	D. F. Twiss; *J. Chem. Soc.*, 1914, **105**, 1672.	237
15CB524	H. Lecher; *Ber. Dtsch. Chem. Ges.*, 1915, **48**, 524.	133
15CB1473	G. Gruttner and M. Wiernik; *Ber. Dtsch. Chem. Ges.*, 1915, **48**, 1473.	489
16CB239	F. Fichter and E. Elkind; *Ber. Dtsch. Chem. Ges.*, 1916, **49**, 239.	506
17CB262	W. Schlenk and J. Hotlz; *Ber. Dtsch. Chem. Ges.*, 1917, **50**, 262.	555, 556
17MI 202-01	T. Milobendzki and K. Szulgin; *Chem. Polsk.*, 1917, **15**, 66.	105
18JA822	B. T. Brooks and I. Humphrey; *J. Am. Chem. Soc.*, 1918, **40**, 822.	40
19CB413	D. Vorländer and E. Mittag; *Ber. Dtsch. Chem. Ges.*, 1919, **52**, 413.	194, 239
19CR(164)795	J. Guyot and L. J. Simon; *C. R. Hebd. Seances Acad. Sci.*, 1919, **164**, 795.	102
19HCA635	H. Staudinger and J. Meyer; *Helv. Chem. Acta*, 1919, **2**, 635.	810, 853, 859
19IEC105	R. H. Uhlinger and R. V. Cook; *Ind. Eng. Chem. Res.*, 1919, **11**, 105.	491, 495
19IEC817	J. F. Norris; *Ind. Eng. Chem. Res.*, 1919, **11**, 817.	495
19JA287	R. L. Datto and H. K. Miller; *J. Am. Chem. Soc.*, 1919, **41**, 287.	627, 632
20CB1013	W. Steinkopf and W. Mieg; *Ber. Dtsch. Chem. Ges.*, 1920, **53**, 1013.	495
20CB1329	H. Wieland and E. Schamberg; *Ber. Dtsch. Chem. Ges.*, 1920, **53**, 1329.	381, 382, 383
20JA830	J. B. Conant and A. A. Cook; *J. Am. Chem. Soc.*, 1920, **42**, 830.	456
20JA856	H. L. Gray and G. O. Gutekunst; *J. Am. Chem. Soc.*, 1920, **42**, 856.	134
20JCS889	R. H. Vernon; *J. Chem. Soc.*, 1920, 889.	288
20JCS1157	M. O. Forster; *J. Chem. Soc.*, 1920, **117**, 1157.	769
20LA(421)174	H. Schmidt; *Justus Liebigs Ann. Chem.*, 1920, **421**, 174.	883, 888
21BSF1024	E. de Mahler; *Bull. Soc. Chim. Fr.*, 1921, **29**, 1024.	1076
21CB206	K. Hoesch; *Chem. Ber.*, 1921, **54**, 206.	76
21CB438	K. W. Rosenmund; *Chem. Ber.*, 1921, **54**, 438.	884
21JA591	O. B. Helfrich and E. E. Reid; *J. Am. Chem. Soc.*, 1921, **43**, 591.	165
21JA880	R. L. Kramer and E. E. Reid; *J. Am. Chem. Soc.*, 1921, **43**, 880.	118, 135
21JCS768	R. D. Haworth and A. Lapworth; *J. Chem. Soc.*, 1921, 768.	764
21MI 202-01	C. Neuberg and J. Hirsch; *Biochem. Z.*, 1921, **115**, 282.	38
21MI 202-02	G. P. Plant and N. V. Sidgwick; *J. Soc. Chem. Ind.*, 1921, **40**, 14.	102
21MI 215-01	L. Thompson; *J. Soc. Dyers Colourists*, 1921, **37**, 7.	769
21RTC285	L. C. Janse; *Recl. Trav. Chim. Pays-Bas.*, 1921, **40**, 285.	757
22BSF545	C. Volmar; *Bull. Soc. Chim. Fr.*, 1922, **31**, 545.	489
22CB875	K. Brand and J. Steiner; *Chem. Ber.*, 1922, **55**, 875.	764
22CB1424	H. Lecher and M. Wittwer; *Ber. Dtsch. Chem. Ges.*, 1922, **55**, 1474.	219
22CB2423	H. Lecher and K. Simon; *Ber. Dtsch. Chem. Ges.*, 1922, **55**, 2423.	127
22JA805	A. J. Quick and R. Adams; *J. Am. Chem. Soc.*, 1922, **44**, 805.	498
22JA2556	H. L. Lochte, W. A. Noyes and J. R. Bailey; *J. Am. Chem. Soc.*, 1922, **44**, 2556.	388, 403
22JCS91	F. Challenger and J. F. Wilkinson; *J. Chem. Soc.*, 1922, **121**, 91.	882
22JCS104	F. Challenger and L. R. Ridgway; *J. Chem. Soc.*, 1922, **121**, 104.	882
22JCS594	F. G. Mann and W. J. Pope; *J. Chem. Soc.*, 1922, **121**, 594.	170
22JCS715	W. Davies; *J. Chem. Soc.*, 1922, **121**, 715.	769
22JCS1052	F. G. Mann and W. J. Pope; *J. Chem. Soc.*, 1922, **121**, 1052.	160
22JCS1161	C. F. Ward; *J. Chem. Soc.*, 1922, **121**, 1161.	29
22LA(429)144	H. Schmidt; *Justus Liebigs Ann. Chem.*, 1922, **429**, 144.	883
22MI 207-01	K. A. Taipale; *J. Russ. Phys-Chem. Soc.*, 1922–3, **54**, 638 (*Chem. Abstr.*, 1924, **18**, 3049).	388
23CB553	J. von Braun and W. Kaiser; *Ber. Dtsch. Chem. Ges.*, 1923, **56**, 553.	172
23CB679	M. Bergmann, R. Ulpts and C. Witte; *Ber. Dtsch. Chem. Ges.*, 1923, **56**, 679.	387
23CB1933	O. Diels; *Ber. Dtsch. Chem. Ges.*, 1923, **56**, 1933.	388
23CB2290	E. Fromm and H. Landmann; *Ber. Dtsch. Chem. Ges.*, 1923, **56**, 2290.	1034
23CR(176)992	J. F. Durand; *C. R. Hebd. Seances Acad. Sci.*, 1923, **176**, 992.	1096
23HCA249	E. Cherbuliez and O. Schrauder; *Helv. Chim. Acta*, 1923, **6**, 249.	176
23HCA256	E. Cherbuliez and O. Schrauder; *Helv. Chim. Acta*, 1923, **6**, 256.	182
23JA1071	W. H. Carruthers and R. Adams; *J. Am. Chem. Soc.*, 1923, **45**, 1071.	52
23JCS2489	E. E. Turner and F. W. Bury; *J. Chem. Soc.*, 1923, **123**, 2489.	875
23JCS2999	F. D. Chattaway and O. G. Backeberg; *J. Chem. Soc.*, 1923, **123**, 2999.	94
23LA(431)231	B. K. Mereshkowsky; *Justus Liebigs Ann. Chem.*, 1923, **431**, 231.	87
23LA(432)39	M. Wieland and H. Wesche; *Justus Liebigs Ann. Chem.*, 1923, **431**, 39.	1076
23RTC1050	H. Prins; *Recl. Trav. Chim. Pays-Bas*, 1923, **42**, 1050.	54
24BSF166	J. F. Durand; *Bull. Soc. Chim. Fr.*, 1924, **35**, 166.	1096
24CB795	B. Helferich and J. Hausen; *Ber. Dtsch. Chem. Ges.*, 1924, **57**, 795.	108
24CB1077	R. Lesser and R. Weiss; *Ber. Dtsch. Chem. Ges.*, 1924, **57**, 1077.	768
24CR(178)1827	A. Bert; *C. R. Hebd. Seances Acad. Sci.*, 1924, **178**, 1827.	100
24JA1675	W. H. Carruthers and R. Adams; *J. Am. Chem. Soc.*, 1924, **46**, 1675.	52

24JCS2275	H. Burton and C. S. Gibson; *J. Chem. Soc.*, 1924, **125**, 2275.	887
24JCS1463	P. K. Dutt; *J. Chem. Soc.*, 1924, **125**, 1463.	181
25CB409	H. Lecher, F. Holschneider, K. Köberle, W. Speer and P. Stöcklin; *Ber. Dtsch. Chem. Ges.*, 1925, **58**, 409.	720
25CRV43	G. Calingaert; *Chem. Rev.*, 1925, **2**, 43.	
25G973	G. Levi and G. Watta; *Gazz. Chim. Ital.*, 1925, **54**, 973.	135
25JA395	M. C. Taylor, R. B. MacMullin and C. A. Gammal; *J. Am. Chem. Soc.*, 1925, **47**, 395.	94
25JA1449	H. Gilman and N. J. Beaber; *J. Am. Chem. Soc.*, 1925, **47**, 1449.	135
25ZAAC(142)83	P. H. Hermans; *Z. Anorg. Allg. Chem.*, 1925, **142**, 83.	111
26AC(P)3	V. Grignard and H. Perrichon; *Ann. Chim. (Paris)*, 1926, **5**, 3.	1019
26CB818	W. Borsche and E. Feske; *Ber. Dtsch. Chem. Ges.*, 1926, **59**, 818.	782
26CB2215	A. Stock and E. Pohland; *Ber. Dtsch. Chem. Ges.*, 1926, **59**, 2215.	811
26CR(182)612	J. B. Senderens; *C. R. Hebd. Seances Acad. Sci.*, 1926, **182**, 612.	89
26JA469	J. R. Johnson and W. L. McEwen; *J. Am. Chem. Soc.*, 1926, **48**, 469.	1018, 1098
26JA776	C. G. Moses and E. E. Reid; *J. Am. Chem. Soc.*, 1926, **48**, 776.	126
26JPR314	Th. Curtius and E. Kenngott; *J. Prakt. Chem.*, 1926, **112**, 314.	1055
26PRS523	G. T. Morgan and G. R. Davies; *Proc. Roy. Soc.*, 1926, **110**, 523.	489, 493, 498
26ZAAC246	K. Fredenhagen and G. Cadenbach; *Z. Anorg. Allg. Chem.*, 1926, **158**, 246.	959
27CR(185)779	M. Valeur and P. Gaillot; *C. R. Hebd. Seances Acad. Sci.*, 1927, **185**, 779.	494, 505, 506
27CR(185)956	M. Valeur and P. Gaillot; *C. R. Hebd. Seances Acad. Sci.*, 1927, **185**, 956.	494
27JA1017	W. C. Harden and F. Dunning; *J. Am. Chem. Soc.*, 1927, **49**, 1017.	261
27JA2011	M. T. Bogert and A. Stull; *J. Am. Chem. Soc.*, 1927, **49**, 2011.	279, 292
27JA2517	W. H. Hartung and H. Adkins; *J. Am. Chem. Soc.*, 1927, **49**, 2517.	29
27JCS27	W. H. Porritt; *J. Chem. Soc.*, 1927, 27.	733
27JCS1798	E. V. Bell and G. M. Bennett; *J. Chem. Soc.*, 1927, 1798.	147
28CB1439	E. Larsson; *Ber. Dtsch. Chem. Ges.*, 1928, **61**, 1439.	126
28CB1906	W. Steinkopf and H. Dudek; *Ber. Dtsch. Chem. Ges.*, 1928, **61**, 1906.	507
28CB1911	W. Steinkopf, H. Dudek and S. Schmidt; *Ber. Dtsch. Chem. Ges.*, 1928, **61**, 1911.	484, 489, 492
28JA536	C. S. Palmer and A. B. Scott; *J. Am. Chem. Soc.*, 1928, **50**, 536.	507
28JA2739	G. H. Coleman; *J. Am. Chem. Soc.*, 1928, **50**, 2739.	338
28JCS321	G. T. Morgan and H. Burgess; *J. Chem. Soc.*, 1928, 321.	288
28JCS2293	O. J. K. Edwards, W. R. Gaythwaite, J. Kenyon and H. Phillips; *J. Chem. Soc.*, 1928, 2293.	286, 732
28JCS3179	F. L. Gilbert and T. M. Lowry; *J. Chem. Soc.*, 1928, 3179.	288
28JPR198	H. Biltz and A. Beck; *J. Prakt. Chem.*, 1928, **118**, 198.	134
28LA(462)186	W. Strecker and W. Daniel; *Justus Liebigs Ann. Chem.*, 1928, **462**, 186.	101, 102
28RTC942	H. J. Backer; *Recl. Trav. Chem. Pays-Bas*, 1928, **47**, 942.	364
28RTC960	I. J. Rinkes; *Recl. Trav. Chim. Pays-Bas*, 1928, **47**, 960.	1040
29CB598	V. Ipatieff, G. Rasuwajev and V. Stromski; *Ber. Dtsch. Chem. Ges.*, 1929, **62**, 598.	408
29CB1010	A. N. Nesmeyanov; *Ber. Dtsch. Chem. Ges.*, 1929, **62**, 1010.	956
29JA1576	H. Gilman, E. A. Zoellner and J. B. Dickey; *J. Am. Chem. Soc.*, 1929, **51**, 1576.	560
29JA3131	W. R. Kirner and G. H. Richter; *J. Am. Chem. Soc.*, 1929, **51**, 3131.	133
29JCS1048	H. V. A. Briscoe, J. B. Peel and P. L. Robinson; *J. Chem. Soc.*, 1929, 1048.	171
29JCS1262	W. C. Davies, P. L. Pearse and W. J. Jones; *J. Chem. Soc.*, 1929, 1262.	433, 434, 435
29JCS2076	T. M. Lowry and F. L. Gilbert; *J. Chem. Soc.*, 1929, 2076.	289
29JCS2342	G. W. Fenton and C. K. Ingold; *J. Chem. Soc.*, 1929, 2342.	842
29LA(468)117	L. Zechmeisten and P. Rom; *Justus Leibigs Ann. Chem.*, 1929, **468**, 117.	776
29MI 208-01	A. E. Arbuzov and G. Kamai; *J. Russ. Phys. Chem. Soc.*, 1929, **61**, 2037 (*Chem. Abstr.*, 1930, **24**, 5736).	461
30CB1136	H. Ramser and E. Wilberg; *Ber. Dtsch. Chem. Ges.*, 1930, **63**, 1136.	110
30CB1707	J. Loevenich and H. Gerber; *Ber. Dtsch. Chem. Ges.*, 1930, 1707.	1052
30CB1863	F. Strauss, L. Kollek and W. Heyn; *Ber. Dtsch. Chem. Ges.*, 1930, **63**, 1863.	1016, 1017
30CB1886	F. Strauss, L. Kollek and H. Hauptmann; *Ber. Dtsch. Chem. Ges.*, 1930, **63**, 1886.	1016
30CB1941	E. Ott, W. Ottmeyer and K. Packendorff; *Ber. Dtsch. Chem. Ges.*, 1930, **63**, 1941.	1014
30CB2856	J. von Braun and K. Weissbach; *Ber. Dtsch. Chem. Ges.*, 1930, **63**, 2856.	172
30JA376	M. E. P. Friedrich and C. S. Marvel; *J. Am. Chem. Soc.*, 1930, **52**, 376.	483, 487
30JA651	L. C. Swallen and C. E. Boord; *J. Am. Chem. Soc.*, 1930, **52**, 651.	128, 132
30JA2046	D. E. Worrell; *J. Am. Chem. Soc.*, 1930, **52**, 2046.	888
30JA4500	du Vigneaud, L. F. Audrieth and H. S. Loring; *J. Am. Chem. Soc.*, 1930, **52**, 4500.	126
30JCS463	W. J. C. Dyke, W. C. Davies and W. J. Jones; *J. Chem. Soc.*, 1930, 463.	483, 872
30JCS1920	W. J. C. Dyke and W. J. Jones; *J. Chem. Soc.*, 1930, 1920.	492, 497
30JCS2426	W. J. C. Dyke and W. J. Jones; *J. Chem. Soc.*, 1930, 2426.	492
30JPR223	N. Wigren; *J. Prakt. Chem.*, 1930, **126**, 223.	498
31AC(P)309	R. Truchet; *Ann. Chim. (Paris)*, 1931, **16**, 309.	1016
31CB1324	E. Ott and K. Packendorff; *Ber. Dtsch. Chem. Ges.*, 1931, **64**, 1324.	1014
31CB1340	G. Schiemann and R. Pillarsky; *Ber. Dtsch. Chem. Ges.*, 1931, **64**, 1340.	609, 614
31CRV117	H. E. Bigelow; *Chem. Rev.*, 1931, **9**, 117.	773

31JA3407 F. C. Wagner and E. E. Reid; *J. Am. Chem. Soc.*, 1931, **53**, 3407. 191
31JCS185 W. J. C. Dyke, G. Davies and W. J. Jones; *J. Chem. Soc.*, 1931, 185. 487, 492, 497, 502
31JCS2109 I. K. Jackson, W. C. Davies and W. J. Jones; *J. Chem. Soc.*, 1931, 2109. 824
31JPR357 B. A. Arbuzov; *J. Prakt. Chem.*, 1931, **131**, 357. 452, 452
31LA(485)258 W. Voss and E. Blanke; *Justus Liebigs Ann. Chem.*, 1931, **485**, 258. 100
31ZAAC(202)1 E. Wiberg and W. Sutterlin; *Z. Anorg. Allg. Chem.*, 1931, **202**, 1. 110
31ZAAC(202)22 E. Wiberg and W. Sutterlin; *Z. Anorg. Allg. Chem.*, 1931, **202**, 22. 110
31ZOB190 M. J. Kratt and F. W. Ljutkina; *Z. Obshch. Khim.*, 1931, **1**, 190. 102

32CB409 W. Steinkopf, I. Schubart and J. Roch; *Ber. Dtsch. Chem. Ges.*, 1932, **65**, 409. 484
32CB812 O. Behagel and H. Seibert; *Ber. Dtsch. Chem. Ges.*, 1932, **65**, 812. 733
32CB1605 H.-W. Schwechten; *Ber. Dtsch. Chem. Ges.*, 1932, **65**, 1605. 627
32JA787 T. H. Vaughn and J. A. Nieuwland; *J. Am. Chem. Soc.*, 1932, **54**, 787. 1018
32JA977 G. K. Rolefson and C. F. Oldershaw; *J. Am. Chem. Soc.*, 1932, **54**, 977. 103
32JA1674 L. M. Ellis and E. E. Reid; *J. Am. Chem. Soc.*, 1932, **54**, 1674. 119
32JA2088 J. F. Norris and G. W. Rigby; *J. Am. Chem. Soc.*, 1932, **54**, 2088. 89
32JA2441 A. H. Kohlhase; *J. Am. Chem. Soc.*, 1932, **54**, 2441. 181
32JA2993 F. F. Blicke and U. O. Oakdale; *J. Am. Chem. Soc.*, 1932, **54**, 2993. 889
32JA3364 W. H. Hunter and E. B. Sorenson; *J. Am. Chem. Soc.*, 1932, **54**, 3364. 181
32JA3714 F. C. Whitmore and A. L. Houk; *J. Am. Chem. Soc.*, 1932, **54**, 3714. 560
32JCS483 W. Davies and J. H. Dick; *J. Chem. Soc.*, 1932, 483. 183
32JCS593 A. Cohen; *J. Chem. Soc.*, 1932, 593. 890
32JCS958 F. G. Mann; *J. Chem. Soc.*, 1932, 958. 810
32JCS1815 M. E. Brinnard, W. J. C. Dyke, W. H. Jones and W. J. Jones; *J. Chem. Soc.*, 1932, 1815. 484, 492, 497, 888
32JCS2284 W. J. Jones, W. J. C. Dyke, G. Davies, D. C. Griffiths; *J. Chem. Soc.*, 1932, 2284. 484
32JCS2505 A. Cohen, H. King and W. I. Strangeways; *J. Chem. Soc.*, 1932, 2505. 890
32JCS2880 W. C. Davies and C. J. O. R. Morris; *J. Chem. Soc.*, 1932, 2880. 456
32JGU524 G. H. Kamai; *J. Gen. Chem. USSR* (*Engl. Transl.*), 1932, **2**, 524 (*Chem. Abstr.*, 1933, **27**, 966). 442
32JPR257 H. Rheinboldt, F. Mott and E. Motzkus; *J. Prakt. Chem.*, 1932, **134**, 257. 134
32OSC(1)136 L. A. Bigelow; *Org. Synth., Coll. Vol.*, 1932, **1**, 136. 609, 627
32OSC(1)170 C. S. Marvel and S. M. McElvain; *Org. Synth., Coll. Vol.*, 1932, **1**, 170. 609, 621

33BSF1070 P. Carr and D. Libermann; *Bull. Soc. Chim. Fr.*, 1933, **53**, 1070. 100
33CB278 E. Bergmann and A. Bondi; *Ber. Dtsch. Chem. Ges.*, 1933, **66**, 278. 847, 1073
33JA766 R. L. Shriner and E. A. Parker; *J. Am. Chem. Soc.*, 1933, **55**, 766. 103
33JA1161 F. F. Blicke and L. D. Powers; *J. Am. Chem. Soc.*, 1933, **55**, 1161. 890
33JA1242 R. E. Stutz and R. L. Shriner; *J. Am. Chem. Soc.*, 1933, **55**, 1242. 126
33JA2150 T. H. Vaughn and J. A. Nieuwland; *J. Am. Chem. Soc.*, 1933, **55**, 2150. 1018, 1019
33JA2932 W. E. Bachmann and R. F. Cockerill; *J. Am. Chem. Soc.*, 1933, **55**, 2932. 125
33JA3888 W. A. Noyes; *J. Am. Chem. Soc.*, 1933, **55**, 3888. 104
33JA4020 A. B. Burg and H. Schlesinger; *J. Am. Chem. Soc.*, 1933, **55**, 4020. 109, 111
33JA4667 R. A. Jacobson and W. H. Carothers; *J. Am. Chem. Soc.*, 1933, **55**, 4667. 1016
33JCS739 R. C. Murrary; *J. Chem. Soc.*, 1933, 739. 192
33JCS741 T. H. Vaughn and J. A. Nieuwland; *J. Chem. Soc.*, 1933, 741. 1018
33JCS776 H. H. Hart; *J. Chem. Soc.*, 1933, 776. 451
33JCS989 G. F. Fenton and C. K. Ingold; *J. Chem. Soc.*, 1933, 989. 442

34JA1106 D. J. Pflaum and H. H. Wenzke; *J. Am. Chem. Soc.*, 1934, **56**, 1106. 1019
34JA1204 C. Barkenbus and J. J. Owen; *J. Am. Chem. Soc.*, 1934, **56**, 1204. 100
34JA1628 W. M. Laver and C. M. Langkammerev; *J. Am. Chem. Soc.*, 1934, **56**, 1628. 190
34LA(509)138 L. Koenigs and H. Friedrich; *Justus Liebigs Ann. Chem.*, 1934, **509**, 138. 829

35CB1011 A. Meuwsen and H. Gebhardt; *Ber. Dtsch. Chem. Ges.*, 1935, **68**, 1011. 99
35CB1540 O. Behagel and W. Müller; *Ber. Dtsch. Chem. Ges.*, 1935, **68**, 1540. 733
35JA913 C. B. Biswell and C. S. Hamilton; *J. Am. Chem. Soc.*, 1935, **57**, 913. 883
35JA1098 B. H. Nicolet; *J. Am. Chem. Soc.*, 1935, **57**, 1098. 117
35JA2172 L. C. Raiford and S. E. Hazlet; *J. Am. Chem. Soc.*, 1935, **57**, 2172. 173
35JCS366 F. A. Paneth and H. Loleit; *J. Chem. Soc.*, 1935, 366. 491, 505
35JCS901 S. Astin, L. de. V. Moulds and H. L. Riley; *J. Chem. Soc.*, 1935, 901. 101
35ZAAC(222)92 E. Wiberg and W. Sutterlin; *Z. Anorg. Allg. Chem.*, 1935, **222**, 92. 111

36AC(P)560 R. Levaillant; *Ann. Chim.* (*Paris*), 1936, **6**, 560. 102
36BSF2310 E. Billmann and K. A. Jensen; *Bull. Soc. Chim. Fr.*, 1936, **3**(5), 2310. 288
36CB937 A. Meuwsen and H. Gebhardt; *Ber. Dtsch. Chem. Ges.*, 1936, **69**, 937. 99, 172
36CB2753 R. Criegee and H. Stanger; *Ber. Dtsch. Chem. Ges.*, 1936, **69**, 2753. 203
36JA225 W. M. Lauer, M. M. Sprung and C. M. Langkamerer; *J. Am. Chem. Soc.*, 1936, **58**, 225. 783
36JA1348 T. B. Johnson and J. M. Sprague; *J. Am. Chem. Soc.*, 1936, **58**, 1348. 120, 125
36JA1573 G. W. Bowers and C. S. Hamilton; *J. Am. Chem. Soc.*, 1936, **58**, 1573. 884
36JA1806 C. A. Young, R. R. Vogt and J. A. Nieuwland; *J. Am. Chem. Soc.*, 1936, **58**, 1806. 606, 623
36JCS327 E. K. Learmonth and S. Smiles; *J. Chem. Soc.*, 1936, 327. 688
36JCS688 W. Gerrard; *J. Chem. Soc.*, 1936, 688. 101

36JPR251	H. Meerwein and H. Sönke; *J. Prakt. Chem.*, 1936, **147**, 251.	112
36MI 202-01	M. Kitagawa; *J. Biochem. Jpn.*, 1936, **24**, 107.	103
36ZAAC(228)341	J. Meyer and W. Sporman; *Z. Anorg. Allg. Chem.*, 1936, **228**, 341.	94
37CB675	H. Rheinboldt and C. Tetsch; *Ber. Dtsch. Chem. Ges.*, 1937, **70**, 675.	134
37JA811	H. A. Young; *J. Am. Chem. Soc.*, 1937, **59**, 811.	213
37JA1307	P. A. McCusker and R. R. Vogt; *J. Am. Chem. Soc.*, 1937, **59**, 1307.	1016, 1018
37JA1837	J. M. Sprague and T. B. Johnson; *J. Am. Chem. Soc.*, 1937, **59**, 1837.	120, 125
37JA2068	W. C. Johnson and A. Pechukas; *J. Am. Chem. Soc.*, 1937, **59**, 2068.	482, 483
37JA2439	J. M. Sprague and T. B. Johnson; *J. Am. Chem. Soc.*, 1937, **59**, 2439.	177
37JA2473	W. F. H. Jackman and J. Kenyon; *J. Am. Chem. Soc.*, 1937, **59**, 2473.	127
37JCS527	F. G. Mann and E. J. Chaplin; *J. Chem. Soc.*, 1937, 527.	809, 810, 825
37JIC349	H. N. Das-Gupta; *J. Indian Chem. Soc.*, 1937, **14**, 349 (*Chem. Abstr.*, 1937, **31**, 8532).	885
38CB712	B. Helferich and A. Gnüchtel; *Ber. Dtsch. Chem. Ges.*, 1938, **71**, 712.	181
38CB1873	G. Sheibe and O. Stoll; *Ber. Dtsch. Chem. Ges.*, 1938, **71**, 1873.	99
38HCA1579	L. Blangey; *Helv. Chim. Acta*, 1938, **21**, 1579.	780
38JA105	H. R. Snyder, J. A. Kuck and J. R. Johnson; *J. Am. Chem. Soc.*, 1938, **60**, 105.	906
38JA121	J. R. Johnson and M. G. Van Campen, Jr.; *J. Am. Chem. Soc.*, 1938, **60**, 121.	112
38JA538	C. M. Suter, P. B. Evans and J. M. Kiefere; *J. Am. Chem. Soc.*, 1938, **60**, 538.	101
38JA544	H. G. Haulton and H. V. Tartar; *J. Am. Chem. Soc.*, 1938, **60**, 544.	186
38JA1486	I. B. Douglass and T. B. Johnson; *J. Am. Chem. Soc.*, 1938, **60**, 1486.	177
38JA1930	J. G. Aston and D. E. Ailman; *J. Am. Chem. Soc.*, 1938, **60**, 1930.	405
38JA2370	A. Lippert and E. E. Reid; *J. Am. Chem. Soc.*, 1938, **60**, 2370.	261
38JA2452	S. O. Johns and E. E. Reid; *J. Am. Chem. Soc.*, 1938, **60**, 2452.	116, 117, 137, 138
38JA2810	W. W. Binkley and E. F. Degering; *J. Am. Chem. Soc.*, 1938, **60**, 2810.	102
38JCS341	M. P. Balfe, C. A. Chaplin and H. Phillips; *J. Chem. Soc.*, 1938, 341.	288, 291
38JCS1218	M. Fitzgibbon; *J. Chem. Soc.*, 1938, 1218.	1014, 1098
38JCS1358	L. A. Bashford, H. J. Emeleus and H. V. A. Briscoe; *J. Chem. Soc.*, 1938, 1358.	1014, 1017
38JGU1825	N. J. Sheerdina and B. J. Kokko; *J. Gen. Chem. USSR.*, 1938, **8**, 1825.	305
38JOC175	M. S. Karasch, E. M. May and F. R. Mayo; *J. Org. Chem.*, 1938, **3**, 175.	190
38USP2123082	W. A. Schultze; *US Pat.* 2 123 082 (1938) (*Chem. Abstr.*, 1938, **32**, 6667).	166
39BRP511430	I. G. Farben; *Br. Pat.* 511 430 (1939) (*Chem. Abstr.*, 1940, **34**, 5860).	748
39CB657	H. Reinboldt and E. Motzkus; *Ber. Dtsch. Chem. Ges.*, 1939, **72**, 657.	172, 194
39JA71	V. N. Ipatieff and B. S. Friedman; *J. Am. Chem. Soc.*, 1939, **61**, 71.	478
39JA673	H. C. Brown, A. B. Burg and H. Schlesinger; *J. Am. Chem. Soc.*, 1939, **61**, 673.	109, 111
39JA940	L. Spiegler and J. M. Tinker; *J. Am. Chem. Soc.*, 1939, **61**, 940.	23
39JA2548	T. B. Johnson and I. B. Douglass; *J. Am. Chem. Soc.*, 1939, **61**, 2548.	177, 178
39JA2971	E. S. Cook and G. W. Kreke; *J. Am. Chem. Soc.*, 1939, **61**, 2971.	132
39JA3093	M. S. Karasch, R. T. E. Schenek and F. R. Mayo; *J. Am. Chem. Soc.*, 1939, **61**, 3093.	190
39JA3250	W. W. Binkley and E. P. Degering; *J. Am. Chem. Soc.*, 1939, **61**, 3250.	366
39JA3432	H. S. Kharasch and H. C. Brown; *J. Am. Chem. Soc.*, 1939, **61**, 3432.	23
39JCS99	W. Gerrard; *J. Chem. Soc.*, 1939, 99.	100
39JCS398	Ch. Weizmann, L. Haskelberg and T. Berlin; *J. Chem. Soc.*, 1939, 398.	656
39JCS1299	R. D. Haworth and R. L. Barker; *J. Chem. Soc.*, 1939, 1299.	752
39JOC162	H. Gilman and L. D. Apperson; *J. Org. Chem.*; 1939, **4**, 162.	489
39MI 217-01	M. Sartori and E. Recchi; *Ann. Chim. Applicata*, 1939, **29**, 128 (*Chem. Abstr.*, 1939, **33**, 9175).	882
40CRV351	F. R. Mayo and C. Walling; *Chem. Rev.*, 1940, **27**, 351.	4, 116, 117, 439
40JA167	G. O. Doak; *J. Am. Chem. Soc.*, 1940, **62**, 167.	884
40JA1044	S. Zuffanti; *J. Am. Chem. Soc.*, 1940, **62**, 1044.	191
40JA1173	E. J. Mills and M. T. Bogert; *J. Am. Chem. Soc.*, 1940, **62**, 1173.	129
40JA1625	R. B. Woodward; *J. Am. Chem. Soc.*, 1940, **62**, 1625.	626
40JA1859	C. K. Banks and C. S. Hamilton; *J. Am. Chem. Soc.*, 1940, **62**, 1859.	736
40JA2357	H. Gilman and R. G. Jones; *J. Am. Chem. Soc.*, 1940, **62**, 2357.	598
40JA3098	R. B. Sandin and L. F. Fieser; *J. Am. Chem. Soc.*, 1940, **62**, 3098.	13
40JA3142	C. K. Banks and C. S. Hamilton; *J. Am. Chem. Soc.*, 1940, **62**, 3142.	884, 888
40JA3250	R. C. Fuson, J. Corse and C. H. McKeever; *J. Am. Chem. Soc.*, 1940, **62**, 3250.	642, 643
40JOC24	P. H. Latimer and R. W. Bost; *J. Org. Chem.*, 1940, **5**, 24.	191
40JOC106	A. V. Grosse and J. M. Mavity; *J. Org. Chem.*, 1940, **5**, 106.	595
40SCI336	O. Baudisch; *Science*, 1940, **92**, 336.	779
40USP2223948	C. Bremner; *US Pat.* 2 223 948 (1940) (*Chem. Abstr.*, 1941, **35**, 1805).	111
41CB759	O. Westphal; *Ber. Dtsch. Chem. Ges.*, 1941, **74**, 759.	373, 378, 380, 382
41CB774	O. Westphal; *Ber. Dtsch. Chem. Ges.*, 1941, **74**, 774.	388
41CB1089	G. Semerano and L. Riccoboni; *Ber. Dtsch. Chem. Ges.*, 1941, **74**, 1089.	589
41CB1365	O. Westphal; *Ber. Dtsch. Chem. Ges.*, 1941, **74**, 1365.	378, 380, 383
41JA207	H. Gilman and H. L. Yablunky; *J. Am. Chem. Soc.*, 1941, **63**, 207.	882
41JA575	F. T. Blicke and S. R. Safir; *J. Am. Chem. Soc.*, 1941, **63**, 575.	882, 883
41JA658	T. S. Price and D. F. Twiss; *J. Am. Chem. Soc.*, 1941, **63**, 658.	227
41JA1118	F. C. Whitmore and R. S. Thorpe; *J. Am. Chem. Soc.*, 1941, **63**, 1118.	88

41JA1764	M. S. Schechter and H. L. Haller; *J. Am. Chem. Soc.*, 1941, **63**, 1764.	178
41JA2361	A. H. Nathan and M. T. Bogert; *J. Am. Chem. Soc.*, 1941, **63**, 2361.	136
41JA2847	M. S. Newman and P. H. Wise; *J. Am. Chem. Soc.*, 1941, **63**, 2847.	627
41JA2938	D. S. Tarbell and C. Weaver; *J. Am. Chem. Soc.*, 1941, **63**, 2938.	162
41JA3530	K. Folkers, A. Russell and R. W. Bost; *J. Am. Chem. Soc.*, 1941, **63**, 3530.	177
41JCS70	M. P. Balfe and K. N. Nandi; *J. Chem. Soc.*, 1941, 70.	291
41LA(547)1	F. Klages; *Justus Liebigs Ann. Chem.*, 1941, **547**, 1.	380
41OSC(1)25	O. Kamm and C. S. Marvel; *Org. Synth., Coll. Vol.*, 1941, **1**, 25.	31
41OSC(1)36	R. H. Goshorn, T. Boyd and E. F. Degering; *Org. Synth., Coll. Vol.*, 1941, **1**, 36.	31
41OSC(1)84	R. Adams and C. S. Marvel; *Org. Synth., Coll. Vol.*, 1941, **1**, 84.	192
41OSC(1)115	H. T. Clarke and E. R. Taylor; *Org. Synth., Coll. Vol.*, 1941, **1**, 115.	29
41OSC(1)142	J. E. Copenhaver and A. M. Whaley; *Org. Synth., Coll. Vol.*, 1941, **1**, 142.	25
41OSC(1)144	J. F. Norris and A. W. Olmsted; *Org. Synth., Coll. Vol.*, 1941, **1**, 144.	25
41OSC(1)401	F. C. Whitmore and M. C. Whitmore; *Org. Synth., Coll. Vol.*, 1941, **1**, 401.	393
41OSC(1)428	J. von Braun; *Org. Synth., Coll. Vol.*, 1941, **1**, 428.	33
41OSC(1)521	E. C. Kendall and B. F. McKenzie; *Org. Synth., Coll. Vol.*, 1941, **1**, 521.	30
41OSC(1)550	G. S. Hiers; *Org. Synth., Coll. Vol.*, 1941, **1**, 550.	872
42CB34	F. Asinger, W. Schmidt and F. Ebeneder; *Ber. Dtsch. Chem. Ges.*, 1942, **75**, 34.	175
42CB291	H. Hunsdiecker and C. Hunsdiecker; *Ber. Dtsch. Chem. Ges.*, 1942, **75**, 291.	627
42CB344	F. Asinger and F. Ebeneder; *Ber. Dtsch. Chem. Ges.*, 1942, **75**, 344.	178
42CB1096	K. Pfannstiel and J. Janecke; *Ber. Dtsch. Chem. Ges.*, 1942, **75**, 1096.	769
42JA19	F. W. Bergstrom and J. S. Buehler; *J. Am. Chem. Soc.*, 1942, **64**, 19.	806
42JA149	H. E. Westlake and G. Dougherty; *J. Am. Chem. Soc.*, 1942, **64**, 149.	227
42JA392	G. Calingaert, H. Soroos and V. Hnizda; *J. Am. Chem. Soc.*, 1942, **64**, 392.	494
42JA1128	F. L. Benton and T. E. Dillon; *J. Am. Chem. Soc.*, 1942, **64**, 1128.	33
42JA1337	L. F. Audrieth and A. D. F. Toy; *J. Am. Chem. Soc.*, 1942, **64**, 1337.	412, 413
42JA1553	L. F. Audrieth and A. D. F. Toy; *J. Am. Chem. Soc.*, 1942, **64**, 1553.	413
42JA1963	S. M. McElvain and P. M. Walters; *J. Am. Chem. Soc.*, 1942, **64**, 1963.	29
42JA1966	S. M. McElvain, R. L. Clarke and G. D. Jones; *J. Am. Chem. Soc.*, 1942, **64**, 1966.	29
42JCS570	M. L. Bird and F. Challenger; *J. Chem. Soc.*, 1942, 570.	286, 288, 291
42JCS684	J. W. Cornforth and R. Robinson; *J. Chem. Soc.*, 1942, 684.	631
42JOC23	P. Allen; *J. Org. Chem.*, 1942, **7**, 23.	186, 193
42JOC81	S. W. Lee and G. Dougherty; *J. Org. Chem.*, 1942, **5**, 81.	177, 178, 179
42JOC472	W. E. Vaughan and F. F. Rust; *J. Org. Chem.*, 1942, **5**, 472.	116
42LA(550)182	P. W. Neber and H. Rauscher; *Justus Liebigs Ann. Chem.*, 1942, **550**, 182.	780
42OR(1)105	N. L. Drake; *Org. React.*, 1942, **1**, 105.	753
42OR(1)63	R. C. Fuson and C. H. McKeever; *Org. React.*, 1942, **1**, 63.	13
43CB80	E. Ott and G. Dittus; *Ber. Dtsch. Chem. Ges.*, 1943, **76**, 80.	1014
43CRV209	H. L. Yale; *Chem. Rev.*, 1943, **33**, 209.	757
43JA992	R. C. Fuson and S. P. Rowland; *J. Am. Chem. Soc.*, 1943, **65**, 992.	643, 645
43JA1430	P. Block, Jr. and G. Powell; *J. Am. Chem. Soc.*, 1943, **65**, 1430.	608
43JCS4	A. A. Goldberg; *J. Chem. Soc.*, 1943, 4.	191
43JCS636	A. I. Vogel, *J. Chem. Soc.*, 1943, 636.	25
43OSC(2)31	J. R. Byers and J. B. Dickey; *Org. Synth., Coll. Vol.*, 1943, **2**, 31.	121
43OSC(2)74	J. C. Eck and C. S. Marvel; *Org. Synth., Coll. Vol.*, 1943, **2**, 74.	29
43OSC(2)108	W. A. Noyes; *Org. Synth., Coll. Vol.*, 1943, **2**, 108.	104
43OSC(2)109	G. R. Dutton and C. R. Noller; *Org. Synth., Coll. Vol.*, 1943, **2**, 109.	107
43OSC(2)111	C. M. Suter and H. L. Gerhardt; *Org. Synth., Coll. Vol.*, 1943, **2**, 111.	102
43OSC(2)112	C. M. Suter and H. L. Gerhart; *Org. Synth., Coll. Vol.*, 1943, **2**, 112.	100
43OSC(2)171	H. R. Snyder and L. A. Brooks; *Org. Synth., Coll. Vol.*, 1943, **2**, 171.	30
43OSC(2)177	H. S. Rhinesmith; *Org. Synth., Coll. Vol.*, 1943, **2**, 177.	30
43OSC(2)208	H. H. Hatt; *Org. Synth., Coll. Vol.*, 1943, **2**, 208.	376
43OSC(2)211	H. H. Hatt; *Org. Synth., Coll. Vol.*, 1943, **2**, 211.	379, 385
43OSC(2)238	H. M. Leicester; *Org. Synth., Coll. Vol.*, 1943, **2**, 238.	729
43OSC(2)270	T. W. Abbott and D. Althousen; *Org. Synth., Coll. Vol.*, 1943, **2**, 270.	30
43OSC(2)308	G. Braun; *Org. Synth., Coll. Vol.*, 1943, **2**, 308.	31
43OSC(2)322	W. W. Hartman, J. R. Byers and J. B. Dickey; *Org. Synth., Coll. Vol.*, 1943, **2**, 322.	36
43OSC(2)336	R. A. Pacaud and C. F. H. Allen; *Org. Synth., Coll. Vol.*, 1943, **2**, 336.	23
43OSC(2)358	C. R. Noller and R. Dinsmore; *Org. Synth., Coll. Vol.*, 1943, **2**, 358.	31
43OSC(2)395	H. H. Hatt; *Org. Synth., Coll. Vol.*, 1943, **2**, 395.	373
43OSC(2)512	A. P. Black and F. H. Babers; *Org. Synth., Coll. Vol.*, 1943, **2**, 512.	393
43OSC(2)547	R. W. Bost and M. W. Conn; *Org. Synth., Coll. Vol.*, 1943, **2**, 547.	132
43OSC(2)558	C. S. Marvel and M. S. Sparberg; *Org. Synth., Coll. Vol.*, 1943, **2**, 558.	191
43OSC(2)576	E. M. Faber and G. E. Miller; *Org. Synth., Coll. Vol.*, 1943, **2**, 576.	132
44IEC610	E. E. Dreger; *Ind. Eng. Chem. Res.*, 1944, **36**, 610.	102
44JA82	K. Campbell, A. Sonmers and B. Campbell; *J. Am. Chem. Soc.*, 1944, **66**, 82.	449
44JA191	F. Linsker and M. T. Bogert; *J. Am. Chem. Soc.*, 1944, **66**, 191.	888
44JA1109	R. C. Fuson, P. L. Southwick and S. P. Rowland; *J. Am. Chem. Soc.*, 1944, **66**, 1109.	643, 645, 646
44JA1272	R. C. Fuson, N. Rabjohn and D. J. Byers; *J. Am. Chem. Soc.*, 1944, **66**, 1272.	645

44JA1274 R. C. Fuson, L. J. Armstrong, W. E. Wallace and J. W. Kneisley; *J. Am. Chem. Soc.*, 1944, **66**, 1274. 645
44JA1464 R. C. Fuson, L. J. Armstrong, J. W. Kneisley and W. J. Shenk, Jr.; *J. Am. Chem. Soc.*, 1944, **66**, 1464. 645, 646
44JA1873 R. C. Fuson, D. J. Byers, S. P. Rowland, P. L. Southwick and C. A. Sperati; *J. Am. Chem. Soc.*, 1944, **66**, 1873. 643
44JOC89 J. F. Audrieth and M. Sveda; *J. Org. Chem.*, 1944, **9**, 89. 366
44JOC211 B. N. Meals; *J. Org. Chem.*, 1944, **9**, 211. 483
44OR(2)49 A. L. Henne; *Org. React.*, 1944, **2**, 49 18
44ZPC(281)156 G. Ivanovics and L. Vargha; *Hoppe-Seyler's Z. Physiol. Chem.*, 1944, **281**, 156. 177

45BAU251 A. N. Nesmeyanov and A. E. Borisov; *Bull. Acad. Sci. USSR. Div. Chim. Sci.*, 1945, 251 (*Chem. Abstr.*, 1946, **40**, 2123). 881
45CR(221)562 A. Etienne; *C. R. Hebd. Seances Acad. Sci.*, 1945, **221**, 562. 887
45HCA628 E. Hardegger, D. Redlich and A. Gal; *Helv. Chim. Acta.*, 1945, **28**, 628. 627
45JA386 R. C. Fuson, L. J. Armstrong, D. H. Chadwick, J. W. Kneisley, S. P. Rowland, W. J. Shenk, Jr. and Q. F. Soper; *J. Am. Chem. Soc.*, 1945, **67**, 386. 643
45JA594 R. O. Clinton, C. M. Suter, S. C. Laskowski, M. Jackman and W. Huber; *J. Am. Chem. Soc.*, 1945, **67**, 594. 128
45JA1180 G. M. Kosolapoff; *J. Am. Chem. Soc.*, 1945, **67**, 1180. 456
45JA1578 S. M. McElvain, A. Jelinek and K. Rorig; *J. Am. Chem. Soc.*, 1945, **67**, 1578. 191
45JA1844 H. Gilman and L. A. Woods; *J. Am. Chem. Soc.*, 1945, **67**, 1844. 129
45JA1937 R. C. Fuson, R. E. Foster, W. J. Shenk, Jr. and E. W. Maynert; *J. Am. Chem. Soc.*, 1945, **67**, 1937. 645
45JA1939 R. C. Fuson, E. W. Maynert and W. J. Shenk, Jr.; *J. Am. Chem. Soc.*, 1945, **67**, 1939. 643
45JA1965 T. W. Campbell and J. D. McCullough; *J. Am. Chem. Soc.*, 1945, **67**, 1965. 729
45JA2014 J. Waser and V. Schomaker; *J. Am. Chem. Soc.*, 1945, **67**, 2014. 507
45JCS14 H. Burton and E. Hoggarth; *J. Chem. Soc.*, 1945, 14. 203
45JCS30 D. R. Lyon and F. G. Mann; *J. Chem. Soc.*, 1945, 30. 488
45JCS37 F. G. Holliman and F. G. Mann; *J. Chem. Soc.*, 1945, 37. 288
45JCS45 F. G. Holliman and F. G. Mann; *J. Chem. Soc.*, 1945, 45. 488
45JCS380 H. McCombie, B. C. Saunders and G. J. Stacey; *J. Chem. Soc.*, 1945, 380. 104
45JCS382 F. R. Atherton, H. T. Howard and A. R. Todd; *J. Chem. Soc.*, 1945, 382. 105
45JCS660 F. R. Atherton, H. T. Openshaw and A. R. Todd; *J. Chem. Soc.*, 1945, 660. 412
45JCS666 G. Bulmer and F. G. Mann; *J. Chem. Soc.*, 1945, 666. 123
45MI 208-01 A. E. Arbuzov and A. I. Razumov; *Izv. Akad. Nauk SSSR Otd. Khim. Nauk*, 1945, 167 (*Chem. Abstr.*, 1964, **40**, 341). 451
45USP2471472 H. Woodstock; *US Pat.* 2 471 472 (1945) (*Chem. Abstr.*, 1949, **43**, 7499). 456

46CRV269 N. Kharasch, S. J. Potemp and H. L. Wehrmeister; *Chem. Rev.*, 1946, **39**, 269. 357, 688
46JA389 R. C. Fuson, D. H. Chadwick and M. L. Ward; *J. Am. Chem. Soc.*, 1946, **68**, 389. 645, 646
46JA1161 W. F. Gilliam, R. N. Meals and R. O. Sauer; *J. Am. Chem. Soc.*, 1946, **68**, 1161. 538
46JA1675 H. Gilman and R. N. Clark; *J. Am. Chem. Soc.*, 1946, **71**, 1675. 942
46JA1720 O. H. Johnson and W. H. Nebergall, *J. Am. Chem. Soc.*, 1946, **71**, 1720. 949
46JA1860 W. H. C. Rueggeberg, A. Ginsburg and W. A. Cooke; *J. Am. Chem. Soc.*, 1946, **68**, 1860. 259, 260
46JA1864 T. H. Maren; *J. Am. Chem. Soc.*, 1946, **68**, 1864. 260
46JA1987 G. O. Doak and H. G. Steinman; *J. Am. Chem. Soc.*, 1946, **68**, 1987. 884
46JA2103 R. L. Frank and P. V. Smith; *J. Am. Chem. Soc.*, 1946, **68**, 2103. 119, 120
46JA2402 N. J. Leonard and Z. W. Wicks; *J. Am. Chem. Soc.*, 1946, **68**, 2402. 33
46JA2467 A. E. Newkirk; *J. Am. Chem. Soc.*, 1946, **68**, 2467. 611
46JA2671 H. Rheinboldt and E. Giesbrecht; *J. Am. Chem. Soc.*, 1946, **68**, 2671. 287
46JCS1050 C. J. Culvenor, W. Davies and K. H. Pausacker; *J. Chem. Soc.*, 1946, 1050. 136
46JCS1126 H. J. Emeléus and H. G. Heal; *J. Chem. Soc.*, 1946, 1126. 165
46JCS1138 C. W. Shoppee; *J. Chem. Soc.*, 1946, 1138. 25
46JOC444 B. L. Zenitz and W. H. Hartung; *J. Org. Chem.*, 1946, **11**, 444. 609
46JOC469 R. C. Fuson, C. C. Price, R. A. Bauman, O. H. Bullitt, W. R. Hatchard and W. Everett; *J. Org. Chem.*, 1946, **11**, 469. 166
46JOC475 R. C. Fuson, C. C. Price and D. M. Burness; *J. Org. Chem.*, 1946, **11**, 475. 138
46JOC499 R. C. Fuson, D. M. Burness, R. E. Foster and R. D. Lipscomb; *J. Org. Chem.*, 1946, **11**, 499. 166
46MI 203-01 I. M. Kolthoff, D. R. May, P. Morgan, H. A. Laitinen and A. S. O'Brien; *Anal. Chem.*, 1946, **18**, 442 (*Chem. Abstr.*, 1946, **40**, 5359). 126
46OR(3)267 E. S. Wallis and J. F. Lane; *Org. React.*, 1946, **3**, 267. 320, 756
46OR(3)337 P. A. S. Smith; *Org. React.*, 1946, **3**, 337. 757

47ACS307 O. Foss; *Acta Chem. Scand.*, 1947, **1**, 307. 216
47BSF351 R. Vieillefosse; *Bull. Soc. Chim. Fr.*, 1947, 351. 182
47BSF841 J. Colonge and L. Cumet; *Bull. Soc. Chim. Fr.*, 1947, 841. 1016
47CB219 M. Goehring and A. Stamm; *Chem. Ber.*, 1947, **80**, 219. 99
47CR(224)570 G. Darzens; *C. R. Hebd. Seances Acad. Sci.*, 1947, **224**, 570. 54
47CRV381 J. Kleinberg; *Chem. Rev.*, 1947, **40**, 381. 12
47HCA675 V. Prelog, O. Metzler and O. Jeger; *Helv. Chim. Acta*, 1947, **30**, 675. 662
47HCA1650 G. Schetty; *Helv. Chim. Acta*, 1947, **30**, 1650. 248

47HCA1883 V. Prelog, L. Ruzicka and O. Metzler; *Helv. Chim. Acta*, 1947, **30**, 1883. 662
47JA559 R. C. Fuson and W. Shive; *J. Am. Chem. Soc.*, 1947, **69**, 559. 494, 506
47JA927 C. K. Banks, J. F. Morgan, R. L. Clark, E. B. Hatlelid, F. H. Kahler, H. W. Paxton, E. J. Cragoe, R. J. Anders, B. Elpern, R. F. Coles, J. Lawhead and C. S. Hamilton; *J. Am. Chem. Soc.*, 1947, **69**, 927. 491, 496, 498
47JA950 A. A. Morton, E. E. Majat and R. L. Letsinger; *J. Am. Chem. Soc.*, 1947, **69**, 950. 556
47JA1219 E. F. Landau; *J. Am. Chem. Soc.*, 1947, **69**, 1219. 717
47JA1675 A. A. Morton and M. E. Holden; *J. Am. Chem. Soc.*, 1947, **69**, 1675. 556
47JA1710 D. S. Laverne, J. H. Bailey and C. J. Cavallito; *J. Am. Chem. Soc.*, 1947, **69**, 1710. 212, 232
47JA1976 F. C. Whitmore, L. H. Sommer and J. Gold; *J. Am. Chem. Soc.*, 1947, **69**, 1976. 537
47JA2022 G. B. Bachmann and L. L. Lewis; *J. Am. Chem. Soc.*, 1947, **69**, 2022. 614
47JA2158 W. E. Bessinger and F. E. Kung; *J. Am. Chem. Soc.*, 1947, **69**, 2158. 100
47JA2236 O. Foss; *J. Am. Chem. Soc.*, 1947, **69**, 2236. 237
47JA2328 C. D. Hurd and L. L. Gershheim; *J. Am. Chem. Soc.*, 1947, **69**, 2328. 117, 138
47JA2672 H. R. Snyder, J. M. Stewart and J. B. Ziegler; *J. Am. Chem. Soc.*, 1947, **69**, 2672. 129, 136
47JA2675 H. R. Snyder, J. M. Stewart and J. B. Ziegler; *J. Am. Chem. Soc.*, 1947, **69**, 2675. 128, 136
47JCS134 J. I. Cunneen; *J. Chem. Soc.*, 1947, 134. 117
47JCS592 L. A. Stocken; *J. Chem. Soc.*, 1947, 592. 119, 259, 260, 889
47JCS674 F. R. Atherton and A. R. Todd; *J. Chem. Soc.*, 1947, 674. 412
47JCS963 A. T. Fuller and H. King; *J. Chem. Soc.*, 1947, 963. 103
47JCS1106 R. F. Naylor; *J. Chem. Soc.*, 1947, 1106. 132
47JCS1471 G. D. Buckley and C. W. Scaife; *J. Chem. Soc*, 1947, 1471. 784
47JCS1477 R. L. Heath and A. Lambert; *J. Chem. Soc.*, 1947, 1477. 117
47JCS1494 G. D. Buckley; *J. Chem. Soc*, 1947, 1494. 783
47JCS1497 G. D. Buckley and E. Ellery; *J. Chem. Soc.*, 1947, 1497. 783
47JCS1532 R. F. Naylor; *J. Chem. Soc.*, 1947, 1532. 117
47JOC1 G. Cronheim; *J. Org. Chem.*, 1947, **12**, 1. 779
47OR(4)337 P. A. Smith; *Org. React.*, 1947, **4**, 337. 319
47OSC(1)214 R. Adams and G. H. Coleman; *Org. Synth.*, *Coll. Vol.*, 2nd edn, 1947, **1**, 214. 779
47QR358 N. Levy and J. D. Rose; *Q. Rev.*, *Chem. Soc.*, 1947, **1**, 358. 393

48BSF156 L. Bollinger; *Bull. Soc. Chim. Fr.*, 1948, 156. 690
48CB210 A. Schöberl and R. Hamm; *Chem. Ber.*, 1948, **81**, 210. 213
48CRV271 C. Djerassi; *Chem. Rev.*, 1948, **43**, 271. 27
48IEC1129 W. A. Proell, C. E. Adams and B. H. Shoemaker; *Ind. Eng. Chem. Res.*, 1948, **40**, 1129 (*Chem. Abstr.*, 1948, **42**, 7239d). 177
48JA165 C. L. Stevens; *J. Am. Chem. Soc.*, 1948, **70**, 165. 3
48JA186 A. D. F. Toy; *J. Am. Chem. Soc.*, 1948, **70**, 186. 847
48JA550 H. Gilman and L. A. Woods; *J. Am. Chem. Soc.*, 1948, **70**, 550. 274
48JA602 R. C. Fuson and T.-L. Tan; *J. Am. Chem. Soc.*, 1948, **70**, 602. 646
48JA695 H. Adkins and H. R. Billica; *J. Am. Chem. Soc.*, 1948, **70**, 695. 52
48JA1115 R. N. Lewis; *J. Am. Chem. Soc.*, 1948, **70**, 1115. 237
48JA1176 L. C. King, R. M. Dodson and L. A. Subluskey; *J. Am. Chem. Soc.*, 1948, **70**, 1176. 125
48JA1450 E. Pierson, M. Giella and M. Tishler; *J. Am. Chem. Soc.*, 1948, **70**, 1450. 138
48JA1762 C. K. Banks, J. Controulis, D. F. Walker, E. W. Tillotson, L. A. Sweet and O. M. Gruhzit; *J. Am. Chem. Soc.*, 1948, **70**, 1762. 260
48JA2053 D. T. Hurd; *J. Am. Chem. Soc.*, 1948, **70**, 2053. 40
48JA2295 G. W. Watt and R. C. Thompson; *J. Am. Chem. Soc.*, 1948, **70**, 2295. 427
48JA2434 L. W. Clemence and M. T. Leffler; *J. Am. Chem. Soc.*, 1948, **70**, 2434. 261
48JA3465 G. M. Kosolapoff; *J. Am. Chem. Soc.*, 1948, **70**, 3465. 847
48JA3526 M. S. Kharasch and A. F. Zavist; *J. Am. Chem. Soc.*, 1948, **70**, 3526. 176
48JA3667 J. L. Szabo and E. T. Stiller; *J. Am. Chem. Soc.*, 1948, **70**, 3667. 138
48JA3882 A. D. F. Toy; *J. Am. Chem. Soc.*, 1948, **70**, 3882. 108
48JA3945 H. Gilman and R. H. Kyle; *J. Am. Chem. Soc.*, 1948, **70**, 3945. 420
48JA3955 C. R. Noller and P. J. Hearst; *J. Am. Chem. Soc.*, 1948, **70**, 3955. 182
48JA4045 J. R. Ritter and P. P. Minieri; *J. Am. Chem. Soc.*, 1948, **70**, 4045. 308
48JCS147 P. R. Carter and D. H. Hey; *J. Chem. Soc.*, 1948, 147. 250
48JCS528 H. Burton and W. A. Davy; *J. Chem. Soc.*, 1948, 528. 711
48JCS699 B. C. Saunders, G. J. Stacey, F. Wild and I. G. E. Wilding; *J. Chem. Soc.*, 1948, 699. 412
48JCS849 C. Djerassi; *J. Chem. Soc.*, 1948, 849. 203
48JCS1313 R. Heap and B. C. Saunders; *J. Chem. Soc.*, 1948, 1313. 413
48JCS1683 Z. Földi and J. Kollonitsch; *J. Chem. Soc.*, 1948, 1683. 117
48JCS1892 F. N. Woodward; *J. Chem. Soc.*, 1948, 1892. 129
48JCS2177 L. J. Goldsworthy, G. F. Harding, W. L. Norris, S. G. P. Plant and B. Selton; *J. Chem. Soc.*, 1948, 2177. 128
48JCS2303 W. Baker and N. C. Brown; *J. Chem. Soc.*, 1948, 2303. 651
48JGU729 A. G. Kostsova; *J. Gen. Chem. USSR*, 1948, **18**, 729 (*Chem. Abstr.*, 1949, **43**, 6568h). 178
48JOC97 M. S. Kharasch and C. F. Fuchs; *J. Org. Chem.*, 1948, **13**, 97. 117
48JOC280 M. E. Havill, I. Joffe and H. W. Post; *J. Org. Chem.*, 1948, **13**, 280. 108
48JOC438 H. H. Fox; *J. Org. Chem.*, 1948, **13**, 438. 888
48JOC822 C. S. Marvel and R. S. Johnson; *J. Org. Chem.*, 1948, **13**, 822. 187, 273
48MI 203-01 S. Siggia and R. L. Edserg; *Anal. Chem.*, 1948, **20**, 938 (*Chem. Abstr.*, 1949, **43**, 1289d). 181

B-48MI 213-01	P. A. Plattner; "Newer Methods of Preparative Organic Chemistry," Interscience, New York, 1948, vol. 1 p. 21.	659
48MI 215-01	G. N. R. Smart and G. F. Wright; *Can. J. Res.* (*B*), 1948, **26**, 284.	798
48OSC(1)492	F. C. Whitmore and F. H. Hamilton; *Org. Synth., Coll. Vol.*, 1948, **1**, 492.	187
48OSC(1)519	F. C. Whitmore, F. G. Hamilton and N. Thurman; *Org. Synth., Coll. Vol.*, 1948, **1**, 519.	594
48RTC973	J. F. Arens and D. A. van Dorp; *Recl. Trav. Chim. Pays-Bas*, 1948, **67**, 973.	676
48USP2437795	C. Walling; *US Pat.* 2 437 795 (1948) (*Chem. Abstr.*, 1948, **42**, 4198).	429
48USP2437797	C. Walling; *US Pat.* 2 437 797 (1948) (*Chem. Abstr.*, 1948, **42**, 4198).	429
48ZOB2008	A. Arbuzov and K. V. Nikonorov; *Zh. Obshch. Khim.*, 1948, **18**, 2008.	450
49CB594	M. Seyhan; *Chem. Ber.*, 1949, **72**, 594.	125
49CRV34	D. Swern; *Chem. Rev.*, 1949, **45**, 34.	348
49IZV305	A. N. Nesmeyanov and N. K. Kochetkov; *Izv. Akad. Nauk. SSSR, Ser. Khim.*, 1949, **74**, 305.	982
49JA84	E. A. Fehnel and M. Carmack; *J. Am. Chem. Soc.*, 1949, **71**, 84.	133, 135
49JA369	G. M. Kosolapoff; *J. Am. Chem. Soc.*, 1949, **71**, 369.	454
49JA740	T. W. Campbell; *J. Am. Chem. Soc.*, 1949, **71**, 740.	753
49JA1478	H. Gilman and L. Fullhart; *J. Am. Chem. Soc.*, 1949, **71**, 1478.	125, 135
49JA1747	H. Gilman and G. C. Gainer; *J. Am. Chem. Soc.*, 1949, **71**, 1747.	706
49JA2221	W. V. E. Doering and E. M. Beringer; *J. Am. Chem. Soc.*, 1949, **71**, 2221.	193
49JA2270	F. Kipnis, I. Levy and J. Ornfelt; *J. Am. Chem. Soc.*, 1949, **71**, 2270.	126, 223
49JA3079	D. I. Weisblat and D. A. Lyttle; *J. Am. Chem. Soc.*, 1949, **71**, 3079.	392
49JA3107	J. R. Dice and J. N. Bowden; *J. Am. Chem. Soc.*, 1949, **71**, 3107.	3
49JA3451	A. B. Burg and C. L. Randolph, Jr.; *J. Am. Chem. Soc.*, 1949, **71**, 3451.	418, 419
49JA3929	A. C. Cope, T. T. Foster and P. H. Towle; *J. Am. Chem. Soc.*, 1949, **71**, 3929.	756
49JCS278	C. C. J. Culvenor, W. Davies and N. J. Heath; *J. Chem. Soc.*, 1949, 278.	129
49JCS282	C. C. J. Culvenor, W. Davies and N. J. Heath; *J. Chem. Soc.*, 1949, 282.	128
49JCS315	M. F. Clarke and L. N. Owen; *J. Chem. Soc.*, 1949, 315.	203
49JCS1883	M. I. Gillibrand and A. H. Lamberton; *J. Chem. Soc.*, 1949, 1883.	386
49JCS2126	A. H. Ford-Moore; *J. Chem. Soc.*, 1949, 2126.	163
49JCS2723	R. H. Hall and B. K. Howe; *J. Chem. Soc.*, 1949, 2723.	138
49JCS2921	H. G. Cook, J. D. Ilett, B. C. Saunders, G. J. Stacey, H. G. Watson, I. G. E. Wilding and G. J. Woodcock; *J. Chem. Soc.*, 1949, 2921.	409, 410, 412, 413
49JOC272	I. B. Douglass, V. C. Simpson and A. K. Sawyer; *J. Org. Chem.*, 1949, **14**, 272.	168
49JOC310	C. R. Hauser and M. J. Weiss; *J. Org. Chem.*, 1949, **14**, 310.	421
49JOC429	M. S. Kharasch, E. V. Jensen and S. Weinhouse; *J. Org. Chem.*, 1949, **14**, 429.	448, 489
49JOC813	G. Gever and K. Hayes; *J. Org. Chem.*, 1949, **14**, 813.	374
49JOC921	E. J. Carr, G. P. Smith and G. Allinger; *J. Org. Chem.*, 1949, **14**, 921.	358
49LA(562)187	G. Wittig and M. Rieber; *Justus Liebigs Ann. Chem.*, 1949, **562**, 187.	830
49MI 203-01	I. M. Kolthoff and W. E. Harris; *Anal. Chem.*, 1949, **21**, 963 (*Chem. Abstr.*, 1950, **44**, 76a).	172
49OR(5)301	M. L. Moore; *Org. React.*, 1949, **5**, 301.	310
49USP2472471	L. T. Elby; *US Pat.* 2 472 471 (1949) (*Chem. Abstr.*, 1949, **43**, 6646c).	133
50BSF245	H. Rheinboldt and M. Perrier; *Bull. Soc. Chim. Fr.*, 1950, 245.	735
50CB87	H. Brintzinger, K. Pfannstiel, H. Koddebusch and K. E. Kling; *Chem. Ber.*, 1950, **83**, 87.	167
50CR(230)304	H. Felkin; *C. R. Hebd. Seances Acad. Sci., Ser. A.*, 1950, **230**, 304.	314
50DOK303	A. Ya. Yakubovich, V. A. Ginsburg and S. P. Makarov; *Dokl. Akad. Nauk SSSR*, 1950, **71**, 303 (*Chem. Abstr.*, 1950, **44**, 8320a).	498
50G180	S. Bezzi and P. Lanza; *Gazz. Chim. Ital.*, 1950, **80**, 180.	213
50IS50	E. G. Rochow; *Inorg. Synth.*, 1950, **3**, 50.	538
50JA731	R. C. Arnold, A. P. Lien and R. M. Alm; *J. Am. Chem. Soc.*, 1950, **72**, 731.	126
50JA1033	C. S. Marvel and P. D. Caesar; *J. Am. Chem. Soc.*, 1950, **72**, 1033.	125
50JA1687	U. Weiss and S. Sokol; *J. Am. Chem. Soc.*, 1950, **72**, 1687.	122
50JA1753	H. E. Herr and F. W. Heyl; *J. Am. Chem. Soc.*, 1950, **72**, 1753.	31
50JA1760	H. Gilman and R. G. Jones; *J. Am. Chem. Soc.*, 1950, **72**, 1760.	598
50JA2454	R. Adams and J. R. Gordon; *J. Am. Chem. Soc.*, 1950, **72**, 2454.	608
50JA2735	T. W. Campbell, W. Burney and T. L. Jacobs; *J. Am. Chem. Soc.*, 1950, **72**, 2735.	89
50JA3595	R. L. Rowland, W. L. Perry, E. L. Foreman and H. L. Friedman; *J. Am. Chem. Soc.*, 1950, **72**, 3595.	274
50JA4147	R. B. Fox; *J. Am. Chem. Soc.*, 1950, **72**, 4147.	448
50JA4280	H. Gilman and C. G. Brannen; *J. Am. Chem. Soc.*, 1950, **72**, 4280.	942
50JA4291	G. M. Kosolapoff and J. S. Powell; *J. Am. Chem. Soc.*, 1950, **72**, 4291.	837
50JA4292	G. M. Kosolapoff; *J. Am. Chem. Soc.*, 1950, **72**, 4292.	838, 845, 846
50JA4702	W. H. Nebergall; *J. Am. Chem. Soc.*, 1950, **72**, 4702.	942
50JA4820	D. C. Morrison; *J. Am. Chem. Soc.*, 1950, **72**, 4820.	841
50JA4890	R. O. Roblin and J. W. Clapp; *J. Am. Chem. Soc.*, 1950, **72**, 4890.	177
50JA5491	H. J. Lucas, F. W. Mitchell and C. N. Scully; *J. Am. Chem. Soc.*, 1950, **72**, 5491.	410
50JCS18	W. A. Waters and J. H. Williams; *J. Chem. Soc.*, 1950, 18.	889
50JCS406	D. E. Ames and R. E. Bowman; *J. Chem. Soc.*, 1950, 406.	31
50JCS573	D. H. Derbyshire and W. A. Waters; *J. Chem. Soc.*, 1950, 573.	626
50JCS579	J. H. Chapman and L. N. Owen; *J. Chem. Soc.*, 1950, 579.	128
50JCS1371	H. J. Bridger and E. W. Pittman; *J. Chem. Soc.*, 1950, 1371.	289
50JCS1769	J. E. Gardiner and B. A. Kilby; *J. Chem. Soc.*, 1950, 1769.	413

50JCS2088	W. Gerrard, A. Nechvatal and B. M. Wilson; *J. Chem. Soc.*, 1950, 2088.	101
50JCS2657	J. Glazer, E. D. Hughes, C. K. Ingold, A. T. James, G. T. Jones and E. Robers; *J. Chem. Soc.*, 1950, 2657.	780
50JCS2938	L. W. C. Miles and L. N. Owen; *J. Chem. Soc.*, 1950, 2938.	128
50JCS3125	H. G. Cook, J. D. Ilett, B. C. Saunders and G. J. Stacey; *J. Chem. Soc.*, 1950, 3125.	101
50JCS3309	M. P. Balfe, J. Kenyon and C. E. Searle; *J. Chem. Soc.*, 1950, 3309.	118
50JCS3694	D. H. Derbyshire and W. A. Waters; *J. Chem. Soc.*, 1959, 3694.	631
50JOC162	M. Hamell and R. Levine; *J. Org. Chem.*, 1950, **15**, 162.	
50JOC795	I. B. Douglass and F. T. Martin; *J. Org. Chem.*, 1950, **15**, 795.	168
50JOC865	J. G. Affleck and G. Dougherty; *J. Org. Chem.*, 1950, **15**, 865.	227
50LA(566)210	H. Bestian; *Justus Liebigs Ann. Chem.*, 1950, **566**, 210.	129, 413
50LA(567)195	K. Ziegler and H. G. Gellert; *Justus Liebigs Ann. Chem.*, 1950, **567**, 195.	555
50MI 216-01	G. Kamai and E. A. Gerasimova; *Trudy Kazan. Khim. Tekhnol. Inst. im. S. M. Kirova*, 1950, **15**, 26.	838, 839
50OS(30)35	A. J. Speziale; *Org. Synth.*, 1950, **30**, 35.	121
50OS(36)86	G. J. M Van der Kerk and J. G. A. Luitjen; *Org. Synth.*, 1956, **36**, 86.	478
50OS(30)58	P. J. Hearst and C. R. Noller; *Org. Synth.*, 1950, **30**, 58.	182
50RTC638	J. Stratling and H. J. Backer; *Recl. Trav. Chim. Pays-Bas*, 1950, **69**, 638.	126, 127
50USP2580695	W. D. Niederhauser; *US Pat.* 2 580 695 (1950) (*Chem. Abstr.*, 1952, **46**, 7116).	167

51ACS115	O. Foss; *Acta Chem. Scand.*, 1951, **5**, 115.	238
51ACS690	R. Dahlbom; *Acta Chem. Scand.*, 1951, **5**, 690.	117
51CB170	H. Bohme and W. Krause; *Chem. Ber.*, 1951, **84**, 170.	339
51CB911	E. Schneider; *Chem. Ber.*, 1951, **84**, 911.	166, 171, 172
51DOK(78)709	B. M. Mikhailov and N. F. Kucherova; *Dokl. Akad. Nauk SSSR*, 1951, **78**, 709.	847
51IEC1596	R. D. Morin and A. E. Bearse; *Ind. Eng. Chem. Res.*, 1951, **43**, 1596.	89
51JA459	F. Misani, T. W. Fair and L. Reiner; *J. Am. Chem. Soc.*, 1951, **73**, 459.	163
51JA855	G. M. Kosolapoff and J. F. MaCollough; *J. Am. Chem. Soc.*, 1951, **73**, 855.	458
51JA998	R. E. Buckles and J. W. Long; *J. Am. Chem. Soc.*, 1951, **73**, 998.	24
51JA1323	H. W. Galbraith, E. F. Degering and E. F. Hitch; *J. Am. Chem. Soc.*, 1951, **73**, 1323.	776
51JA2251	F. G. Bordwell and W. H. McKellin; *J. Am. Chem. Soc.*, 1951, **73**, 2251.	143
51JA2509	J. D. Roberts and R. H. Mazur; *J. Am. Chem. Soc.*, 1951, **73**, 2509.	11
51JA2647	L. A. Subluskey and L. C. King; *J. Am. Chem. Soc.*, 1951, **73**, 2647.	212
51JA2854	H. Adkins and C. Scanley; *J. Am. Chem. Soc.*, 1951, **73**, 2854.	595
51JA2880	H. Gilman and H. L. Yale; *J. Am. Chem. Soc.*, 1951, **73**, 2880.	889
51JA3196	E. S. Gould and J. D. McCullough; *J. Am. Chem. Soc.*, 1951, **73**, 3196.	287
51JA3329	M. C. Hoff, K. W. Greenlee and C. E. Boord; *J. Am. Chem. Soc.*, 1951, **73**, 3329.	606, 615
51JA3627	D. T. McAllan, T. V. Cullum, R. A. Dean and F. A. Fidler; *J. Am. Chem. Soc.*, 1951, **73**, 3627.	212, 219
51JA4043	K. Mislow; *J. Am. Chem. Soc.*, 1951, **73**, 4043.	89
51JA4101	G. M. Kosolapoff and R. M. Watson; *J. Am. Chem. Soc.*, 1951, **73**, 4101.	455
51JA4199	M. S. Newman and A. Kutner; *J. Am. Chem. Soc.*, 1951, **73**, 4199.	801
51JA4464	R. Duschinsky and H. Gainer; *J. Am. Chem. Soc.*, 1951, **73**, 4464.	212
51JA4470	H. Gilman and H. L. Yale; *J. Am. Chem. Soc.*, 1951, **73**, 4470.	886
51JA4590	A. B. Burg and W. E. McKee; *J. Am. Chem. Soc.*, 1951, **73**, 4590.	452
51JA4953	K. C. Brannock; *J. Am. Chem. Soc.*, 1951, **73**, 4953.	259
51JA5041	H. M. Curry and J. P. Mason; *J. Am. Chem. Soc.*, 1951, **73**, 5041.	386
51JA5202	T. Wagner-Jauregg, J. J. O'Neill and W. H. Summerson; *J. Am. Chem. Soc.*, 1951, **73**, 5202.	412
51JA5392	G. M. Kosolapoff and J. F. McCulough; *J. Am. Chem. Soc.*, 1951, **73**, 5392.	454
51JA5466	G. M. Kosolapoff and R. M. Watson; *J. Am. Chem. Soc.*, 1951, **73**, 5466.	455
51JA5658	G. O. Doak and L. D. Freedman; *J. Am. Chem. Soc.*, 1951, **73**, 5658.	844, 847
51JA5675	A. I. Ryer and G. B. L. Smith; *J. Am. Chem. Soc.*, 1951, **73**, 5675.	764
51JA5896	D. C. Morrison; *J. Am. Chem. Soc.*, 1951, **73**, 5896.	847, 861
51JCS893	B. B. Elsner and P. F. M. Paul; *J. Chem. Soc.*, 1951, 893.	1018
51JCS2123	R. Brown, W. E. Jones and A. R. Pinder; *J. Chem. Soc.*, 1951, 2123.	117
51JCS3292	D. A. Peak and T. I. Watkins; *J. Chem. Soc.*, 1951, 3292.	176, 177
51JOC395	L. Wolinski, H. Tieckelmann and H. W. Post; *J. Org. Chem.*, 1951, **16**, 395.	262
51JOC524	M. S. Kharasch, W. Nudenberg and G. J. Mantell; *J. Org. Chem.*, 1951, **16**, 524.	116
51JOC621	C. Ziegler and J. M. Sprague; *J. Org. Chem.*, 1951, **16**, 621.	176, 177
51JOC767	P. Allen, L. Karger, J. D. Haygood and J. Schrensel; *J. Org. Chem.*, 1951, **16**, 767.	193
51JOC892	J. P. Clay; *J. Org. Chem.*, 1951, **16**, 892.	478
51JOC946	L. Field and F. A. Grunwald; *J. Org. Chem.*, 1951, **16**, 946.	708
51JOC964	L. Field and F. A. Grunwald; *J. Org. Chem.*, 1951, **16**, 964.	187
51LA(573)163	R. Huisgen; *Justus Liebigs Ann. Chem.*, 1951, **573**, 163.	805
51M970	F. Drahowzal and D. Klamann; *Monatsh. Chem.*, 1951, **82**, 970.	225
51NAT1035	B. Bochwic and J. Michalski; *Nature*, 1951, **167**, 1035.	846
51OR(6)339	R. G. Jones and H. Gilman; *Org. React.*, 1951, **6**, 339.	554, 961
51QR75	A. H. Lamberton; *Q. Rev., Chem. Soc.*, 1951, **5**, 75.	385
51RTC254	H. J. Backer; *Recl. Trav. Chim. Pays-Bas*, 1951, **70**, 254.	364
51USP2529493	W. D. Harris, H. D. Tate and J. W. Zukel; *US Pat.* 2 529 493 (1951) (*Chem. Abstr.*, 1951, **45**, 1293).	101
51USP2722538	H. L. Jackson; *US Pat.* 2 722 538 (1951) (*Chem. Abstr.*, 1956, **50**, 4218).	478

51ZAAC(267)39 F. Runge, W. Zimmermann, H. Pfeiffer and I. Pfeiffer; *Z. Anorg. Allg. Chem.*, 1951, **267**, 39. 597
51ZOB93 Z. M. Manulkin and A. N. Tatarenko; *Zh. Obshch. Khim. Lett.*, 1951, **21**, 93 (*Chem. Abstr.*, 1951, **45**, 7038). 882

52BRP673451 Bataafsce Petroleum Maatschappij; *Br. Pat.* 673 451 (1952) (*Chem. Abstr.*, 1953, **47**, 5426). 439
52DOK(85)1297 A. D. Petrov and S. I. Sadukzade; *Dokl. Akad. Nauk SSSR*, 1952, **85**, 1297 (*Chem. Abstr.*, 1953, **47**, 4281). 1089
52DOK(87)991 O. A. Reutov; *Dokl. Akad. Nauk SSSR*, 1952, **87**, 991 (*Chem. Abstr.*, 1954, **48**, 143). 884
52IZV894 A. I. Razumov, O. A. Mukhacheva and S. Do-Khen; *Izv. Akad. Nauk SSSR Ser. Khim.*, 1952, 894 (*Chem. Abstr.*, 1953, **47**, 10 466). 466
52JA161 G. R. Norman, W. M. Leuser and T. N. Mastin; *J. Am. Chem. Soc.*, 1952, **74**, 161. 255, 256
52JA308 E. S. Lewis and C. E. Boozer; *J. Am. Chem. Soc.*, 1952, **74**, 308. 25
52JA398 L. Field; *J. Am. Chem. Soc.*, 1952, **74**, 398. 1018
52JA562 R. J. Horvat and A. Furst; *J. Am. Chem. Soc.*, 1952, **74**, 562. 820
52JA629 S. I. Miller and R. M. Noyes; *J. Am. Chem. Soc.*, 1952, **74**, 629. 1012, 1014
52JA753 G. O. Doak and L. D. Freedman; *J. Am. Chem. Soc.*, 1952, **74**, 753. 847
52JA1421 H. H. Anderson; *J. Am. Chem. Soc.*, 1952, **74**, 1421. 417, 418
52JA2006 H. R. Snyder and R. E. Heckert; *J. Am. Chem. Soc.*, 1952, **74**, 2006. 194
52JA2526 J. K. Weil, L. P. Witnauer and A. J. Stirton; *J. Am. Chem. Soc.*, 1952, **75**, 2526. 190
52JA3282 A. R. Stiles, F. F. Rust and W. E. Vaughan; *J. Am. Chem. Soc.*, 1952, **74**, 3282. 439
52JA3414 L. D. Freedman and G. O. Doak; *J. Am. Chem. Soc.*, 1952, **74**, 3414. 820, 821
52JA3693 C. Hanna and F. W. Schueler; *J. Am. Chem. Soc.*, 1952, **74**, 3693. 379
52JA3982 T. L. Cairns, G. L. Evans, A. W. Larchar and B. C. McKusick; *J. Am. Chem. Soc.*, 1952, **74**, 3982. 130
52JA4526 P. W. Morgan and B. C. Herr; *J. Am. Chem. Soc.*, 1952, **74**, 4526. 452, 841
52JA4891 J. B. Ziegler and A. C. Shabica; *J. Am. Chem. Soc.*, 1952, **74**, 4891. 24
52JA5418 R. H. Williams and L. A. Hamilton; *J. Am. Chem. Soc.*, 1952, **74**, 5418. 450
52JA5880 J. M. Stewart and H. P. Cordts; *J. Am. Chem. Soc.*, 1952, **74**, 5880. 179, 222
52JCS11 A. K. Holliday and J. Sowler; *J. Chem. Soc.*, 1952, 11. 110
52JCS800 J. Baddiley and E. M. Thain; *J. Chem. Soc.*, 1952, 800. 127
52JCS1193 J. W. Baker and D. M. Easty; *J. Chem. Soc.*, 1952, 1193. 103
52JCS2552 G. R. A. Brandt, H. J. Emeleus and R. N. Haszeldine; *J. Chem. Soc.*, 1952, 2552. 486, 505
52JCS2669 C. L. Arcus and H. E. Strauss; *J. Chem. Soc.*, 1952, **3**, 2669. 30
52JCS3039 F. G. Mann and I. T. Millar; *J. Chem. Soc.*, 1952, 3039. 437, 821
52JCS3334 W. Griffin and D. H. Hey; *J. Chem. Soc.*, 1952, 3334. 177
52JCS4191 B. W. Langley, B. Lythgoe and L. S. Rayner; *J. Chem. Soc.*, 1952, 4191. 403, 404
52JCS4331 C. T. Mortimer and H. A. Skinner; *J. Chem. Soc.*, 1952, 4331. 491
52JCS4448 I. G. M. Campbell; *J. Chem. Soc.*, 1952, 4448. 884
52JCS4453 F. G. Mann and I. T. Millar; *J. Chem. Soc.*, 1952, 4453. 829
52JCS4887 R. E. Dabby, J. Kenyon and R. E. Mason; *J. Chem. Soc.*, 1952, 4887. 188
52JOC1630 H. Gilman and R. G. Jones; *J. Org. Chem.*, 1952, **17**, 1630. 586
52LA(577)26 G. Wittig and K. Clauss; *Justus Liebigs Ann. Chem.*, 1952, **577**, 26. 873, 879
52LA(578)136 G. Wittig and K. Clauss; *Justus Liebigs Ann. Chem.*, 1952, **578**, 136. 873
52M1398 D. Klamann; *Monatsh. Chem.*, 1952, **83**, 1398. 9
52MI 207-01 H. Breederveld and H. I. Waterman; *Research* (*London*), 1952, **5**, 537 (*Chem. Abstr.*, 1953, **47**, 3169g). 416, 417
52MI 208-01 B. A. Arbuzov and N. I. Rizpolozhenskii; *Izv. Akad. Nauk SSSR, Otd. Khim. Nauk*, 1952, 956 (*Chem. Abstr.*, 1953, **47**, 9904). 451
B-52MI 215-01 H. E. Fierz-David and L. Blangley; in "Grundlegende Operationen der Farbenchemie," Springer-Verlag, Wien, 1952, p. 244. 803
52N353 W. Perkow, K. Ullerich and F. Meyer; *Naturwissenschaften*, 1952, **39**, 353 (*Chem. Abstr.*, 1953, **47**, 8248e). 693
52QR34 E. D. Hughes and C. K. Ingold; *Q. Rev., Chem. Soc.*, 1952, **6**, 34. 649
52RTC1228 H. Kloosterziel, M. H. Deinema and H. J. Backer; *Recl. Trav. Chim. Pays-Bas*, 1952, **71**, 1228. 251
52USP2584112 H. C. Brown; *US Pat.* 2 584 112 (1952) (*Chem. Abstr.*, 1952, **46**, 9580). 428
52ZOB1528 A. Ya. Yakubovich and S. P. Makarov; *Zh. Obshch. Khim.*, 1952, **22**, 1528 (*Chem. Abstr.*, 1953, **47**, 8010c). 498

53ACS227 O. Foss; *Acta Chem. Scand.*, 1953, **7**, 227. 238
53ACS1293 G. Wittig and K. Torssell; *Acta Chem. Scand.*, 1953, **7**, 1293. 487, 488
53BSB3 H. J. Backer; *Bull. Soc. Chim. Belg.*, 1953, **62**, 3. 724
53CB272 E. G. Walaschewski; *Chem. Ber.*, 1953, **86**, 272. 486, 491
53CB557 H. Brintzinger and M. Langheck; *Chem. Ber.*, 1953, **86**, 557. 167
53CB1073 L. Horner and H. Stöhr; *Chem. Ber.*, 1953, **86**, 1073. 769
53CB1483 R. Meier; *Chem. Ber.*, 1953, **86**, 1483. 420
53CJC976 A. B. Bannard, J. R. Gilpin, G. R. Vavasourn and A. F. McKay; *Can. J. Chem.*, 1953, **31**, 976. 456
53DOK(92)69 L. A. Khismatullina; *Dokl. Akad. Nauk SSSR*, 1953, **92**, 69 (*Chem. Abstr.*, 1954, **48**, 9945). 442
53DOK(91)1341 A. N. Nesmeyanov, O. A. Reutov and O. A. Ptitsyna; *Dokl. Akad. Nauk SSSR*, 1953, **91**, 1341 (*Chem. Abstr.*, 1954, **48**, 11 375). 884
53HCA325 W. G. Dauben, D. F. Dickel, O. Jeger and V. Prelog; *Helv. Chim. Acta*, 1953, **36**, 325. 71

53JA134 A. A. Morton and E. F. Cluff; *J. Am. Chem. Soc.*, 1953, **75**, 134. 556
53JA192 H. C. Brown, H. Schlesinger, I. Sheft and D. M. Ritter; *J. Am. Chem. Soc.*, 1953, **75**, 192. 109
53JA199 H. Schlesinger, H. C. Brown, H. R. Hockstra and L. R. Rapp; *J. Am. Chem. Soc.*, 1953, **75**, 199. 109
53JA205 H. Schlesinger, H. C. Brown and A. E. Finholt; *J. Am. Chem. Soc.*, 1953, **75**, 205. 109
53JA228 A. B. Burg and F. G. A. Stone; *J. Am. Chem. Soc.*, 1953, **75**, 228. 109
53JA243 J. M. Stewart and C. H. Burnside; *J. Am. Chem. Soc.*, 1953, **75**, 243. 167
53JA683 L. D. Freedman, H. Tauber, G. O. Doak and H. J. Magnuson; *J. Am. Chem. Soc.*, 1953, **75**, 683. 844
53JA1379 L. D. Freedman, H. Tauber, G. O. Doak and H. J. Magnuson; *J. Am. Chem. Soc.*, 1953, **75**, 1379. 844
53JA2014 H. R. Snyder, C. T. Elston and D. B. Kellom; *J. Am. Chem. Soc.*, 1953, **75**, 2014. 757
53JA2118 G. L. O'Connor and H. R. Nace; *J. Am. Chem. Soc.*, 1953, **75**, 2118. 125, 127
53JA2507 H. Gilman and S. D. Rosenberg; *J. Am. Chem. Soc.*, 1953, **75**, 2507. 600
53JA2821 H. Freiser, M. V. Eagle and J. Speier; *J. Am. Chem. Soc.*, 1953, **75**, 2821. 941
53JA3379 P. C. Croft and G. M. Kosolapoff; *J. Am. Chem. Soc.*, 1953, **75**, 3379. 454
53JA3493 C. Djerassi and C. T. Lenk; *J. Am. Chem. Soc.*, 1953, **75**, 3493. 34
53JA3513 M. Rubin and B. H. Armbrecht; *J. Am. Chem. Soc.*, 1953, **75**, 3513. 28
53JA3596 G. M. Kosolapoff; *J. Am. Chem. Soc.*, 1953, **75**, 3596. 625
53JA3632 C. Djerassi, N. Frick and L. E. Geller; *J. Am. Chem. Soc.*, 1953, **75**, 3632. 760
53JA3869 R. I. Wagner and A. B. Burg; *J. Am. Chem. Soc.*, 1953, **75**, 3869. 427, 429
53JA3872 A. B. Burg and R. I. Wagner; *J. Am. Chem. Soc.*, 1953, **75**, 3872. 474
53JA4103 A. C. Smith, Jr. and E. G. Rochow; *J. Am. Chem. Soc.*, 1953, **75**, 4103. 600
53JA4105 A. C. Smith, Jr. and E. G. Rochow; *J. Am. Chem. Soc.*, 1953, **75**, 4105. 600
53JA4372 C. Walling and S. A. Buckler; *J. Am. Chem. Soc.*, 1953, **75**, 4372. 96
53JA4520 C. G. Overberger and R. W. Cummins; *J. Am. Chem. Soc.*, 1953, **75**, 4520. 478
53JA4582 I. B. Douglass and C. E. Osborne; *J. Am. Chem. Soc.*, 1953, **75**, 4582. 168
53JA6019 F. G. Bordwell and H. M. Andersen; *J. Am. Chem. Soc.*, 1953, **75**, 6019. 127
53JA6263 H. C. Brown and E. J. Mead; *J. Am. Chem. Soc.*, 1953, **75**, 6263. 59
53JCS549 F. Kurzer; *J. Chem. Soc.*, 1953, 549. 172
53JCS16 F. Challenger and C. F. Allpress; *J. Chem. Soc.*, 1953, 16. 881
53JCS1541 A. G. Davies, R. V. Foster and A. M. White; *J. Chem. Soc.*, 1953, 1541. 95, 96
53JCS1552 H. J. Emeleus, R. N. Haszeldine and E. G. Walaschewski; *J. Chem. Soc.*, 1953, 1552. 481, 482, 486, 491, 492, 498
53JCS1709 R. J. Bridgewater and C. W. Shoppee; *J. Chem. Soc.*, 1953, 1709. 25
53JCS2224 S. R. Landauer and H. N. Rydon; *J. Chem. Soc.*, 1953, 2224. 8
53JCS2257 H. G. Khorana and A. R. Todd; *J. Chem. Soc.*, 1953, 2257. 108
53JCS3219 R. N. Haszeldine and J. M. Kidd; *J. Chem. Soc.*, 1953, 3219. 166, 167, 172
53JCS3623 E. Boyland, D. Mason and P. Sims; *J. Chem. Soc.*, 1953, 3623. 651
53JOC36 D. C. Gregg, K. Hazelton and T. F. McKeon; *J. Org. Chem.*, 1953, **18**, 36. 193
53JOC868 K. Balenovic, D. Cerar and L. Filipovic; *J. Org. Chem.*, 1953, **18**, 868. 12
53JOC1112 C. L. Stevens and W. Holland; *J. Org. Chem.*, 1953, **18**, 1112. 29
53JOC1591 D. A. Shirley and J. R. Zeitz; *J. Org. Chem.*, 1953, 1591. 119, 225
53LA(579)23 S. Hunig and S. Boes; *Justus Liebigs Ann. Chem.*, 1953, **579**, 23. 144
53LA(581)182 D. Klamann and G. Hofbauer; *Justus Liebigs Ann. Chem.*, 1953, **581**, 182. 364
53LA(582)133 W. Reppe and H. Vetter; *Justus Liebigs Ann. Chem.*, 1953, **582**, 133. 44
53MI 207-01 H. Breederveld and H. I. Waterman; *Research* (*London*), 1953, **6**, 43S (*Chem. Abstr.*, 1954, **48**, 9314a). 416, 417
53MI 207-02 H. Breederveld and H. I. Waterman; *Research* (*London*), 1953, **6**, 18 (*Chem. Abstr.*, 1954, **48**, 6374h). 417, 421
53MI 207-03 H. Breederveld and H. I. Waterman; *Research* (*London*), 1953, **6**, 48S (*Chem. Abstr.*, 1954, **48**, 10545a). 417
53MI 207-04 H. Breederveld and H. I. Waterman; *Research* (*London*), 1953, **6**, 41S (*Chem. Abstr.*, 1954, **48**, 9313e). 417
53MI 217-01 V. P. Glushkova, T. V. Talalaeva, Z. P. Razmanova, G. S. Izhdanov and K. A. Kocheshkov; *Sbornik Statei Obshch. Khim.*, 1953, **2**, 992 (*Chem. Abstr.*, 1955, **49**, 6859i). 882
53MI 217-02 A. I. Titov and B. B. Levin; *Sbornik Statei Obshch. Khim.*, 1953, **2**, 1478 (*Chem. Abstr.*, 1955, **49**, 4504g). 887
53OR(7)99 J. H. Brewster and E. L. Eliel; *Org. React.*, 1953, **7**, 99. 302
53OR(7)198 H. A. Hageman; *Org. React.*, 1953, **7**, 198. 33
53OR(7)378 D. Swern; *Org. React.*, 1953, **7**, 378. 90
53RTC612 H. Kloosterziel, J. S. Boerema and H. J. Backer; *Recl. Trav. Chim. Pays-Bas*, 1953, **72**, 612. 226
53RTC993 E. Farenhorst and E. C. Kooyman; *Recl. Trav. Chim. Pays-Bas*, 1953, **72**, 993. 403
53USP2642461 R. C. Morris and J. L. VanWinkle; *US Pat.* 2 642 461 (1953) (*Chem. Abstr.*, 1954, **48**, 8814). 462
53USP2659714 P. Harman and A. R. Stiles; *US Pat.* 2 659 714 (1953) (*Chem. Abstr.*, 1954, **48**, 12 168). 451
53ZAAC(273)161 H. Siebert; *Z. Anorg. Allg. Chem.*, 1953, **273**, 161. 488
53ZN608 E. Wiberg and U. Krerke; *Z. Naturforsch.*, 1953, **8b**, 608. 110

54ACS135 G. Forsberg; *Acta Chem. Scand.*, 1954, **8**, 135. 93
54AX145 D. J. Sutor, F. J. Llewellyn and H. S. Maslen; *Acta Crystallog.*, 1954, **7**, 145. 1002
54BRP810930 A. F. Childs and L. T. D. Williams; *Br. Pat.* 810 930 (1954) (*Chem. Abstr.*, 1959, **53**, 15 978). 458

54BSF1109 R. Fort and L. Denivelle; *Bull. Soc. Chim. Fr.*, 1954, 1109. 94
54CB325 H. Brintzinger and M. Langheck; *Chem. Ber.*, 1954, **87**, 325. 167
54CB919 N. Kreutzkamp; *Chem. Ber.*, 1954, **87**, 919. 427
54CB1148 V. Franzen; *Chem. Ber.*, 1954, **87**, 1148. 632
54CB1661 H. Bohme, F. Freimuth and E. Mundles; *Chem. Ber.*, 1954, **87**, 1661. 212
54CJC500 A. T. Morse and L. C. Leitch; *Can. J. Chem.*, 1954, **32**, 500. 1012, 1015
54CJC545 R. Renaud and L. C. Leitch; *Can. J. Chem.*, 1954, **32**, 545. 377
54CRV101 R. W. Leeper, L. Summers and H. Gilman; *Chem. Rev.*, 1954, **54**, 101. 601
54HCA375 M. Viscontini, K. Adank, N. Merckling, K. Ehrhardt and P. Karrer; *Helv. Chim. Acta*, 1954, **37**, 375. 227
54IZV410 A. N. Nesmeyanov, O. A. Reutov and P. G. Knol; *Izv. Akad. Nauk SSSR, Ser. Khim.*, 1954, 410 (*Chem. Abstr.*, 1955, **49**, 9651). 884
54IZV1008 A. E. Borisov, V. D. Vil'chevskaya, and A. N. Nesmeyanov; *Izv. Akad. Nauk. SSSR, Ser. Khim.*, 1954, 1008. 982
54JA386 F. G. A. Stone and A. B. Burg; *J. Am. Chem. Soc.*, 1954, **76**, 386. 481, 482, 509
54JA585 A. H. Schlesinger and D. T. Mowry; *J. Am. Chem. Soc.*, 1954, **76**, 585. 127
54JA655 O. M. Friedman and A. M. Seligman; *J. Am. Chem. Soc.*, 1954, **76**, 655. 414
54JA1030 L. H. Sommer and L. J. Tyler; *J. Am. Chem. Soc.*, 1954, **76**, 1030. 538
54JA1082 F. G. Bordwell, H. M. Andersen and B. H. Pitt; *J. Am. Chem. Soc.*, 1954, **76**, 1082. 125
54JA1828 M. W. Bullock, J. A. Brockman, E. L. Patterson, J. V. Pierce, M. H. von Saltza, F. Sanders and E. L. R. Stokstad; *J. Am. Chem. Soc.*, 1954, **76**, 1828. 214
54JA1840 M. S. Newman and W. M. Edwards; *J. Am. Chem. Soc.*, 1954, **76**, 1840. 801
54JA2984 H. R. Williams and H. S. Mosher; *J. Am. Chem. Soc.*, 1954, **76**, 2984. 95
54JA2987 H. R. Williams and H. S. Mosher; *J. Am. Chem. Soc.*, 1954, **76**, 2987. 95
54JA3468 W. D. Emmons; *J. Am. Chem. Soc.*, 1954, **76**, 3468. 386
54JA3945 F. G. Bordwell, M. L. Peterson and C. S. Rondesgveet; *J. Am. Chem. Soc.*, 1954, **76**, 3945. 190
54JA3957 F. G. Bordwell and M. L. Peterson; *J. Am. Chem. Soc.*, 1954, **76**, 3957. 191
54JA4869 L. F. Audrieth and L. H. Diamond; *J. Am. Chem. Soc.*, 1954, **76**, 4869. 374
54JA5011 A. J. Speziale, J. A. Stephens and Q. E. Thompson; *J. Am. Chem. Soc.*, 1954, **76**, 5011. 202
54JA5357 W. E. Truce and F. D. Hoerger; *J. Am. Chem. Soc.*, 1954, **76**, 5357. 203
54JBC(208)731 S. Udenfriend, C. T. Clark, J. Axelrod and B. B. Brodie; *J. Biol. Chem.*, 1954, **208**, 731. 650
54JCS22 G. E. Coates and F. Glockling; *J. Chem. Soc.*, 1954, 22. 421
54JCS881 H. J. Emeleus, R. N. Haszeldine and R. C. Paul; *J. Chem. Soc.*, 1954, 881. 498
54JCS1212 J. Grimley and A. K. Holliday; *J. Chem. Soc.*, 1954, 1212. 110
54JCS1750 A. J. Ellis and F. G. Soper; *J. Chem. Soc.*, 1954, 1750. 338
54JCS2200 A. G. Davies, R. V. Foster and A. M. White; *J. Chem. Soc.*, 1954, 2200. 95, 96
54JCS2281 D. G. Coe, S. R. Landauer and H. N. Rydon; *J. Chem. Soc.*, 1954, 2281. 8
54JCS2526 G. E. Coates and F. Glockling; *J. Chem. Soc.*, 1954, 2526. 421
54JCS2819 F. G. Mann and F. H. C. Stewart; *J. Chem. Soc.*, 1954, 2819. 26
54JCS2832 F. G. Mann, I. T. Millar and F. H. C. Stewart; *J. Chem. Soc.*, 1954, 2832. 445
54JCS3659 A. D. Campbell; *J. Chem. Soc.*, 1954, 3659. 607, 617
54JOC1165 R. B. Scott, J. B. Gayle, M. S. Heller and R. E. Lutz; *J. Org. Chem.*, 1954, **20**, 1165. 273
54LA(585)142 H. Bohme and G. Zinner; *Justus Liebigs Ann. Chem.*, 1954, **585**, 142. 211
54LA(589)91 K. Ziegler, H. G. Gellert, H. Martin, K. Nagel and J. Schneider; *Justus Liebigs Ann. Chem.*, 1954, **589**, 91. 596
54LA(590)1 R. Huisgen, F. Jakob, W. Siegel and A. Cadus; *Justus Liebigs Ann. Chem.*, 1954, **590**, 1. 375
B-54MI 202-01 M. S. Kharasch and O. Reinmuth; "Grignard Reactions of Nonmetallic Substances," Prentice-Hall, New York, 1954. 67
54MI 203-01 T. Nomura, M. Yokoi and K. Yamasaki; *Proc. Japan Acad.*, 1954, **29**, 342 (*Chem. Abstr.*, 1955, **49**, 12 274g). 264
54MI 207-01 H. Breederveld and H. I. Waterman; *Research* (*London*), 1954, **7**, 5S (*Chem. Abstr.*, 1955, **49**, 2307b). 416, 417
B-54MI 211-01 M. S. Kharash and O. Reinmuth; "Grignard Reactions of Nonmetallic Substances," Prentice-Hall, New York, 1954. 557
B-54MI 219-01 M. S. Karasch and O. Reinmuth; "Grignard Reactions of Nonmetallic Substances," Prentice-Hall, New York, 1954. 958
54MIP1032247 G. Schrader; *DAS* 1 032 247 (1954) (*Chem. Abstr.*, 1960, **54**, 22 363a). 464
54NAT125 J. Kollonitsch, O. Fuchs and V. Gábor; *Nature* (*London*), 1954, **173**, 125. 59, 60
54OR(8)258 H. Gilman and J. W. Morton; *Org. React.*, 1954, **8**, 258. 752
54OR(8)1 H. Adkins; *Org. React.*, 1954, **8**, 1. 52
54PRS(A)389 L. Bateman and K. R. Hargrave; *Proc. Roy. Soc., Ser A*, 1954, **224**, 389. 146
54RTC129 H. J. Backer and H. Kloosterziel; *Recl. Trav. Chim. Pays-Bas*, 1954, **73**, 129. 723
54RTC565 H. J. Backer and J. Strating; *Recl. Trav. Chim. Pays-Bas*, 1954, **73**, 565. 1034
54USP2686803 R. D. Stayner; *US Pat.* 2 686 803 (1954) (*Chem. Abstr.*, 1955, **49**, 11 000). 451
54USP2693482 R. D. Stayner; *US Pat* 2 693 482 (1954) (*Chem. Abstr.*, 1955, **49**, 13 287). 451
54USP2882313 M. W. Kellogg Co., J. Kwiatek and J. W. Copenhaver; *US Pat.* 2 882 313 (1954) (*Chem. Abstr.*, 1959, **53**, 16 965). 478
54ZAAC(277)258 K. Issleib and A. Brack; *Z. Anorg. Allg. Chem.*, 1954, **277**, 258. 433
54ZOB1259 O. A. Reutov and V. V. Kondrateva; *Zh. Obshch. Chim.*, 1954, **24**, 1259 (*Chem. Abstr.*, 1955, **49**, 12 339h). 884

55AG32 F. Seel, H. Jones, L. Riehl and J. Langer; *Angew. Chem.*, 1955, **67**, 32. 183
55AG304 A. Luttringhaus and K. Hagele; *Angew. Chem.*, 1955, **67**, 304. 214

55AG424	K. Ziegler; *Angew. Chem.*, 1955, **67**, 424.	60
55AG425	K. Ziegler; *Angew. Chem.*, 1955, **67**, 425.	60
55CB1	H. Rheinboldt and E. Giesbrecht; *Chem. Ber.*, 1955, **88**, 1.	735
55CB201	D. Klamann and H. Bertsch; *Chem. Ber.*, 1955, **88**, 201.	176, 181
55CB1544	H. Hock and H. Kropf; *Chem. Ber.*, 1955, **88**, 1544.	97
55CRV485	R. Boschan, R. T. Merrow and R. W. van Dolah; *Chem. Rev.*, 1955, **55**, 485.	103
55DOK(105)489	G. Kamai and Z. L. Khisamova; *Dokl. Akad. Nauk SSSR*, 1955, **105**, 489 (*Chem. Abstr.*, 1956, **50**, 11234h).	414, 415
55HOU(9)1	M. Quaedlieg; *Methoden Org. Chem.* (*Houben-Weyl*), 1955, **9**, 1.	175
55HOU(9)2	B. F. Muth; *Methoden Org. Chem.* (*Houben-Weyl*), 1955, **9**, 2.	175
55HOU(9)3	H. Eckoldt; *Methoden Org. Chem.* (*Houben-Weyl*), 1955, **9**, 3.	175
55HOU(9)4	H. Bohme; *Methoden Org. Chem.* (*Houben-Weyl*), 1955, **9**, 4.	211
55HOU(9)5	A. Schöberl and A. Wagner; *Methoden Org. Chem.* (*Houben-Weyl*), 1955, **9**, 5.	216, 225
55HOU(9)260	A. Schöberl and A. Wagner; *Methoden Org. Chem.* (*Houben-Weyl*), 1955, **9**, 260.	402
55HOU(9)285	M. Quaedvlieg; *Methoden Org. Chem.* (*Houben-Weyl*), 4th edn, 1955, **9**, 285.	797
55HOU(9)343	F. Muth; *Methoden Org. Chem.* (*Houben-Weyl*), 4th edn, 1955, **9**, 343.	797
55IZV472	I. L. Knunyants and V. Yu. Pilskaya; *Izv. Akad. Nauk SSSR, Ser. Khim.*, 1955, 472 (*Chem. Abstr.*, 1956, **50**, 6298).	883
55IZV823	K. N. Anisimov, N. E. Kolobova and A. N. Nesmejanov; *Izv. Akad. Nauk SSSR, Ser. Khim.*, 1955, 823 (*Chem. Abstr.*, 1956, **50**, 9319).	464
55JA80	O. C. Dermer and M. T. Edmison; *J. Am. Chem. Soc.*, 1955, **77**, 80.	249
55JA181	D. C. Morrison; *J. Am. Chem. Soc.*, 1955, **77**, 181.	167, 257
55JA416	L. J. Reed and C.-I. Niu; *J. Am. Chem. Soc.*, 1955, **77**, 416.	127, 214
55JA568	C. Djerassi, M. Gorman, F. X. Markey and E. B. Oldenberg; *J. Am. Chem. Soc.*, 1955, **77**, 568.	123
55JA704	H. Hauptmann and H. Berl; *J. Am. Chem. Soc.*, 1955, **77**, 704.	212
55JA790	H. Zimmer, L. F. Audrieth, M. Zimmer and R. A. Rowe; *J. Am. Chem. Soc.*, 1955, **77**, 790.	379
55JA929	E. N. Walsh, T. M. Beck and W. H. Woodstock; *J. Am. Chem. Soc.*, 1955, **77**, 929.	833, 837
55JA1041	D. L. Chamberlain and M. S. Kharasch; *J. Am. Chem. Soc.*, 1955, **77**, 1041.	165
55JA1578	G. L. O'Connor and H. R. Nace; *J. Am. Chem. Soc.*, 1955, **77**, 1578.	111
55JA1796	S. Wawzonek and R. C. Nagler; *J. Am. Chem. Soc.*, 1955, **77**, 1796.	421
55JA1843	J. Fugger, J. M. Tien and I. M. Hunsberger; *J. Am. Chem. Soc.*, 1955, **77**, 1843.	374
55JA1864	W. A. Higgins, P. W. Vogel and W. G. Craig; *J. Am. Chem. Soc.*, 1955, **77**, 1864.	845
55JA1866	S. O. Li and R. E. Eakin; *J. Am. Chem. Soc.*, 1955, **77**, 1866.	412
55JA2453	J. H. Robson and J. Reinhart; *J. Am. Chem. Soc.*, 1955, **77**, 2453.	386
55JA2457	S. G. Cohen and C. H. Wang; *J. Am. Chem. Soc.*, 1955, **77**, 2457.	403
55JA2871	J. F. Allen and O. H. Johnson; *J. Am. Chem. Soc.*, 1955, **77**, 2871.	412
55JA3131	L. H. Diamond and L. F. Audrieth; *J. Am. Chem. Soc.*, 1955, **77**, 3131.	374, 379
55JA3411	R. H. Williams and L. A. Hamilton; *J. Am. Chem. Soc.*, 1955, **77**, 3411.	454
55JA3526	C. Stuebe, W. M. LeSuer and G. R. Norman; *J. Am. Chem. Soc.*, 1955, **77**, 3526.	462, 842
55JA3826	C. Djerassi, J. Grossman and G. H. Thomas; *J. Am. Chem. Soc.*, 1955, **77**, 3826.	34
55JA4100	C. G. Overberger, L. C. Palmer, B. S. Marks and N. R. Byrd; *J. Am. Chem. Soc.*, 1955, **77**, 4100.	388
55JA4387	W. D. Emmons and J. P. Freeman; *J. Am. Chem. Soc.*, 1955, **77**, 4387.	385
55JA4557	W. D. Emmons and A. S. Pagano; *J. Am. Chem. Soc.*, 1955, **77**, 4557.	397
55JA5351	L. Zervas and P. G. Katsoyannis; *J. Am. Chem. Soc.*, 1955, **77**, 5351.	412
55JA5354	L. Zervas and P. G. Katsoyannis; *J. Am. Chem. Soc.*, 1955, **77**, 5354.	414
55JA5446	R. C. L. J. Thomas; *J. Am. Chem. Soc.*, 1955, **77**, 5446.	214
55JA6061	W. D. Emmons and J. P. Freeman; *J. Am. Chem. Soc.*, 1955, **77**, 6061.	386
55JA6064	H. I. Jacobson, R. G. Harvey and E. V. Jensen; *J. Am. Chem. Soc.*, 1955, **77**, 6064.	256
55JA6227	L. Spialter, D. C. Priest and C. W. Harris; *J. Am. Chem. Soc.*, 1955, **77**, 6227.	941
55JA6231	C. H. Schramm, H. Lemaire and R. H. Carlson; *J. Am. Chem. Soc.*, 1955, **77**, 6231.	191
55JA6380	H. Gilman and R. D. Gorsich; *J. Am. Chem. Soc.*, 1955, **77**, 6380.	942
55JCS522	G. W. Kenner and N. R. Williams; *J. Chem. Soc.*, 1955, 522.	694
55JCS1307	N. S. Johary and L. N. Owen; *J. Chem. Soc.*, 1955, 1307.	203
55JCS1470	J. D. Edwards, W. Gerrard and M. F. Lappert; *J. Chem. Soc.*, 1955, 1470.	110
55JCS2039	M. H. Maguire and G. Shaw; *J. Chem. Soc.*, 1955, 2039.	451
55JCS2901	R. N. Haszeldine and J. M. Kidd; *J. Chem. Soc.*, 1955, 2901.	194
55JCS3173	E. D. Hughes, C. K. Ingold and J. D. H. Mackie; *J. Chem. Soc.*, 1955, 3173.	25
55JCS3997	P. Bruck and A. H. Lamberton; *J. Chem. Soc.*, 1955, 3997.	386, 405
55JOC9	C. R. Rondestvedt, Jr., R. M. Scribner and C. E. Wulfman; *J. Org. Chem.*, 1955, **20**, 9.	112
55JOC475	H. Z. Lecher and E. M. Hardy; *J. Org. Chem.*, 1955, **20**, 475.	122
55JOC981	C. E. Claff and A. A. Morton; *J. Org. Chem.*, 1955, **20**, 981.	557
55JOC1019	D. W. Goheen and W. R. Vaughan; *J. Org. Chem.*, 1955, **20**, 1019.	278
55JOC1165	R. B. Scott, J. B. Gayle, M. S. Heller and R. E. Lutz; *J. Org. Chem.*, 1955, **20**, 1165.	175
55JOC1232	D. R. Stevens; *J. Org. Chem.*, 1955, **20**, 1232.	684
55JOC1337	W. J. Bailey and C. R. Pfeifer; *J. Org. Chem.*, 1955, **20**, 1337.	1099
55MI 202-01	W. Gerrard, G. Machell and P. Tolcher; *Res. Correspondence*, 1955, **8**, 57.	101
55MI 202-02	T. Sakaguchi; *Pharm. Bull.* (*Tokyo*), 1955, **3**, 170.	111
55NAT346	J. Kollonitsch, O. Fuchs and V. Gábor; *Nature* (*London*), 1955, **174**, 346.	59, 60
55OSC(3)105	N. H. Cromwell and R. Benson; *Org. Synth., Coll. Vol.*, 1955, **3**, 105.	30
55OSC(3)119	W. D. McPhee and E. Klingsberg; *Org. Synth., Coll. Vol.*, 1955, **3**, 119.	12
55OSC(3)127	P. Z. Bedoukian; *Org. Synth., Coll. Vol.*, 1955, **3**, 127.	28

55OSC(3)130	H. E. Ungnade and E. F. Orwoll; *Org. Synth., Coll. Vol.*, 1955, **3**, 130.	647
55OSC(3)188	M. S. Newman, M. D. Farbman and H. Hipsher; *Org. Synth., Coll. Vol.*, 1955, **3**, 188.	21
55OSC(3)195	O. Grummitt and A. Buck; *Org. Synth., Coll. Vol.*, 1955, **3**, 195.	13
55OSC(3)209	R. C. Fuson, J. W. Kneisley and E. W. Kaiser; *Org. Synth., Coll. Vol.*, 1955, **3**, 209.	30
55OSC(3)217	A. Roebuck and H. Adkins; *Org. Synth., Coll. Vol.*, 1955, **3**, 217.	45
55OSC(3)244	C. E. Redemann, F. O. Rice, R. Roberts and H. P. Ward; *Org. Synth., Coll. Vol.*, 1955, **3**, 244.	385
55OSC(3)307	W. A. Wisansky and S. Ansbacher; *Org. Synth., Coll. Vol.*, 1955, **3**, 307.	753
55OSC(3)334	W. D. Langley; *Org. Synth., Coll. Vol.*, 1955, **3**, 334.	782
55OSC(3)350	L. I. Smith and M. M. Falkof; *Org. Synth., Coll. Vol.*, 1955, **3**, 350.	30
55OSC(3)523	C. S. Marvel; *Org. Synth., Coll. Vol.*, 1955, **3**, 523.	29
55OSC(3)668	G. H. Coleman, C. M. McCloskey and F. Stuart; *Org. Synth., Coll. Vol.*, 1955, **3**, 668.	781
55OSC(3)692	D. W. Andrus; *Org. Synth., Coll. Vol.*, 1955, **3**, 692.	32
55OSC(3)785	H. Adkins and R. E. Burks, Jr.; *Org. Synth., Coll. Vol.*, 1955, **3**, 785.	30
55OSC(3)793	L. H. Smith; *Org. Synth., Coll. Vol.*, 1955, **3**, 793.	31
55OSC(3)848	C. S. Marvel; *Org. Synth., Coll. Vol.*, 1955, **3**, 848.	29
55OSC(3)853	R. H. Wiley; *Org. Synth., Coll Vol.*, 1955, **3**, 853.	676
55OSC(4)363	G. G. Urquhart, J. N. Gates and R. Connor; *Org. Synth., Coll. Vol.*, 1955, **4**, 363.	120
55RZC960	J. Michalski; *Rocz. Chem.*, 1955, **29**, 960 (*Chem. Abstr.*, 1956, **50**, 10 641).	451
55USP2821554	H. Johnston; *US Pat* 2 821 544 (1955) (*Chem. Abstr.*, 1958, **52**, 10 145).	171
55ZAAC(282)86	J. Goubeau, H. J. Becher and F. Griffel; *Z. Anorg. Allg. Chem.*, 1955, **282**, 86.	110
55ZOB1943	A. V. Dombrovskii and G. M. Prilutskii; *Zh. Obshch. Khim.*, 1955, **25**, 1943 (*Chem. Abstr.*, 1956, **50**, 8450b).	190
56ACS1006	B. Smith; *Acta Chem. Scand.*, 1956, **10**, 1006.	683
56AG678	M. Goehring and G. Weis; *Angew. Chem.*, 1956, **68**, 678.	399
56BSF1615	M. Julia and J. M. Surzur; *Bull. Soc. Chim. Fr.*, 1956, 1615.	1016
56CB209	H. Meerwein, P. Laasch, R. Mersch and J. Spille; *Chem. Ber.*, 1956, **89**, 209.	1009
56CB444	G. Wilke and H. Müller; *Chem. Ber.*, 1956, **89**, 444.	967
56CB685	H. Esser, K. Rastädter and G. Reuter; *Chem. Ber.*, 1956, **89**, 685.	403
56CB688	L. Anschütz and H. Wirth; *Chem. Ber.*, 1956, **89**, 688.	847
56CB1334	G. Wittig and L. Pohmer; *Chem. Ber.*, 1956, **89**, 1334.	667
56CI(L)352	G. J. M. Van der Kerk, J. G. A. Luitjen and J. G. Noltes; *Chem. Ind. (London)*, 1956, 352.	601
56CJC1093	M. Kulka; *Can. J. Chem.*, 1956, 1093.	124
56CJC1819	R. A. McIvor, G. D. McCarthy and G. A. Grant; *Can. J. Chem.*, 1956, **34**, 1819.	257
56CRV219	R. G. Johnson and F. K. Ingham; *Chem. Rev.*, 1956, **56**, 219.	12
56CRV959	M. F. Lappert; *Chem. Rev.*, 1956, **56**, 959.	111
56GEP1023033	K. Schimmelschmidt and W. Denk; *Ger. Pat.* 1 023 033 (1956) (*Chem. Abstr.*, 1960, **54**, 5466).	456
56GEP948330	H. Freytag, F. Lober and G. Domagk; *German Pat.*, 948 330 (1956) (*Chem. Abstr.*, 1959, **53**, 76 267).	361
56IZV927	K. N. Anisimov and N. E. Kolobova; *Izv. Akad, Nauk SSSR, Ser. Khim.*, 1956, 927 (*Chem. Abstr.*, 1957, **51**, 4934).	478
56JA348	H. L. Goering, D. I. Relyea and D. W. Larsen; *J. Am. Chem. Soc.*, 1956, **78**, 348.	117
56JA976	C. Stuebe and H. P. Lankelma; *J. Am Chem. Soc.*, 1956, **78**, 976.	253
56JA1211	G. M. Omietanski and H. H. Sisler; *J. Am. Chem. Soc.*, 1956, **78**, 1211.	384
56JA1218	L. Fishbein and J. A. Gallaghan; *J. Am. Chem. Soc.*, 1956, **78**, 1218.	30
56JA1645	R. L. Hinman; *J. Am. Chem. Soc.*, 1956, **78**, 1645.	377, 381, 382
56JA2467	R. T. Beltrami and E. R. Bissell; *J. Am. Chem. Soc.*, 1956, **78**, 2467.	381, 382, 383
56JA2469	M. S. Newman and W. B. Lutz; *J. Am. Chem. Soc.*, 1956, **78**, 2469.	223
56JA2582	H. C. Brown and B. C. Subba Rao; *J. Am. Chem. Soc.*, 1956, **78**, 2582.	126
56JA2760	W. E. Truce, H. E. Hill and M. M. Boudakian; *J. Am. Chem. Soc.*, 1956, **78**, 2760.	1030, 1034
56JA3063	K. L. Marsi, C. A. VanderWerf and W. E. McEwen; *J. Am. Chem. Soc.*, 1956, **78**, 3063.	848
56JA3135	B. Harrell and A. H. Corwin; *J. Am. Chem. Soc.*, 1956, **78**, 3135.	32
56JA3613	H. C. Brown, E. J. Mead and C. J. Shoaf; *J. Am. Chem. Soc.*, 1956, **78**, 3613.	109, 532
56JA3874	G. Omietanski, A. D. Kelmers, R. W. Shellman and H. H. Sisler; *J. Am. Chem. Soc.*, 1956, **78**, 3874.	374
56JA4028	B. Greenberg, E. S. Gould and W. Burlant; *J. Am. Chem. Soc.*, 1956, **78**, 4028.	729
56JA4285	R. G. Jones, G. Karmas, G. A. Morton, Jr. and H. Gilman; *J. Am. Chem. Soc.*, 1956, **78**, 4285.	421
56JA4447	R. T. Hopkins and P. W. Vogel; *J. Am. Chem. Soc.*, 1956, **78**, 4447.	258
56JA4783	W. E. Parham and F. L. Stright; *J. Am. Chem. Soc.*, 1956, **78**, 4783.	1030, 1031, 1034
56JA5428	M. B. Frankel and K. Klager; *J. Am. Chem. Soc.*, 1956, **78**, 5428.	386
56JA5580	A. J. Speziale and P. C. Hamm; *J. Am. Chem. Soc.*, 1956, **78**, 5580.	176, 177
56JA5694	H. C. Brown and B. C. Subba Rao; *J. Am. Chem. Soc.*, 1956, **78**, 5694.	40
56JA6034	G. C. Finger and C. W. Kruse; *J. Am. Chem. Soc.*, 1956, **78**, 6034.	609
56JA6037	E. D. Bergmann, S. Berkovic and R. Ikan; *J. Am. Chem. Soc.*, 1956, **78**, 6037.	614
56JA6181	R. T. Major and H. Peterson; *J. Am. Chem. Soc.*, 1956, **78**, 6181.	243, 360
56JCS173	T. Gromstadt and R. N. Haszeldine; *J. Chem. Soc.*, 1956, 173.	181, 182, 183, 194
56JCS437	S. E. Livingstone; *J. Chem. Soc.*, 1956, 437.	128
56JCS824	P. B. Brindley, W. Gerrard and M. F. Lappert; *J. Chem. Soc.*, 1956, 824.	111
56JCS1755	P. A. Briscoe, F. Challenger and P. S. Duckworth; *J. Chem. Soc.*, 1956, 1755.	366
56JCS2705	C. E. Addison and J. C. Sheldon; *J. Chem. Soc.*, 1956, 2705.	150

56JCS3006	T. Colclough, W. Gerrard and M. F. Lappert; *J. Chem. Soc.*, 1956, 3006.	99
56JCS3139	G. Sosnovsky; *J. Chem. Soc.*, 1956, 3139.	194
56JCS3631	R. N. Haszeldine and B. O. West; *J. Chem. Soc.*, 1956, 3631.	486
56JCS3677	J. S. Watson; *J. Chem. Soc.*, 1956, 3677.	383, 388
56JCS3933	D. A. Burton and P. F. G. Praill; *J. Chem. Soc.*, 1956, 3933.	94
56JCS4066	J. C. Crawhall, D. F. Elliott and K. C. Hooper; *J. Chem. Soc.*, 1956, 4066.	214
56JCS4669	A. G. Davies and R. Feld; *J. Chem. Soc.*, 1956, 4669.	95
56JOC1051	J. Radell, L. Spialter and J. Hollander; *J. Org. Chem.*, 1956, **21**, 1051.	772
56JOC1217	M. Cecchini and E. Giesbrecht; *J. Org. Chem.*, 1956, **21**, 1217.	279
56JOC4003	N. Kornblum, R. J. Clutter and W. J. Jones; *J. Org. Chem.*, 1956, **78**, 4003.	397
56N59	L. Anschütz and H. Wirth; *Naturwissenschaften*, 1956, **43**, 59	881
56RTC481	J. F. Arens and T. Doornbos; *Recl. Trav. Chim. Pays-Bas*, 1956, **75**, 481.	1030
56RTC1459	J. F. Arens, H. C. Volger, T. Doornbos, J. Bonnema, J. W. Greidanus and J. H. van den Hende; *Recl. Trav. Chim. Pays-Bas*, 1956, **75**, 1459.	1030
56ZAAC(287)208	F. Hein, K. Issleib and H. Rabold; *Z. Anorg. Allg. Chem.*, 1954, **287**, 208.	441
56ZAAC(287)261	H. Hartmann, C. Beermann and H. Czempik; *Z. Anorg. Allg. Chem.*, 1956, **287**, 261.	1058, 1063, 1066, 1069
56ZAAC(288)201	K. Issleib and W. Seidel; *Z. Anorg. Allg. Chem.*, 1956, **288**, 201.	452
56ZN(B)676	U. Küerke; *Z. Naturforsch., Teil B*, 1956, **11**, 676.	1084
56ZN(B)750	E. Wiberg and K. Modritzer; *Z. Naturforsch., Teil B*, 1956, **11**, 750.	487
56ZN(B)751	E. Wiberg and K. Mödritzer; *Z. Naturforsch., Teil B*, 1956, **11**, 751.	879
56ZN(B)753	E. Wiberg and K. Mödritzer; *Z. Naturforsch., Teil B*, 1956, **11**, 753.	879
56ZOB126	G. Kamai and Z. L. Khisamova; *Zh. Obshch. Khim.*, 1956, **26**, 126 (*Chem. Abstr.*, 1956, **50**, 13730a).	414, 415
56ZOB907	A. V. Kirsanov and R. G. Makitra; *Zh. Obshch. Khim.*, 1956, **26**, 907.	860
56ZOB2463	A. I. Razumov and A. Mukhacheva; *Zh. Obshch. Khim.*, 1956, **26**, 2463 (*Chem. Abstr.*, 1957, **51**, 1822).	453
56ZOB3060	V. V. Katyshkina and M. Ya. Kraft; *Zh. Obshch. Khim.*, 1956, **26**, 3060 (*Chem. Abstr.*, 1957, **51**, 8028a).	694
56ZOB3378	K. A. Petrov and G. A. Sokol'skii; *Zh. Obshch. Khim.*, 1956, **26**, 3378 (*Chem. Abstr.*, 1957, **51**, 8029b).	407
56ZOB3426	G. Kamai and L. A. Khismatullina; *Zh. Obshch. Khim.*, 1956, **26**, 3426 (*Chem. Abstr.*, 1957, **51**, 9512).	442
57ACS571	G. Bergson; *Acta Chem. Scand.*, 1957, **11**, 571.	292
57AG677	P. McCusker; *Angew. Chem.*, 1957, **69**, 677.	111
57AG713	A. Scholerl and H. Grafje; *Angew. Chem.*, 1957, **69**, 713.	214
57ANC75	C. R. Stahl and S. Siggia; *Anal. Chem.*, 1957, 154 (*Chem. Abstr.*, 1957, **51**, 17611a).	478
57BEP560503	Société Belge du Titane, S. A., *Belg. Pat.* 560 503 (1957) (*Chem. Abstr.*, 1960, **54**, 1923).	565
57CB2075	M. Beck-Goehring and J. Sambeth; *Chem. Ber.*, 1957, **90**, 2075.	414
57CI(L)235	J. L. Willans; *Chem. Ind.* (*London*), 1957, 235.	840
57CI(L)1622	A. G. Davies and R. B. Moodie; *Chem. Ind.* (*London*), 1957, 1622.	99
57CRV219	R. G. Johnson and R. K. Ingham; *Chem. Rev.*, 1956, **57**, 219.	627
57CRV1011	W. K. Warburton; *Chem. Rev.*, 1957, **57**, 1011.	708, 847
57CRV1021	H. H. Sisler, G. M. Omietanski and B. Rudner; *Chem. Rev.*, 1957, **57**, 1021.	383
57DOK(115)305	G. Kamai and N. A. Chadaeva; *Dokl. Akad. Nauk SSSR*, 1957, **115**, 305 (*Chem. Abstr.*, 1958, **52**, 6161).	500
57DOK(117)817	M. I. Kabachnik and E. N. Tsvetkov; *Dokl. Akad. Nauk SSSR*, 1957, **117**, 817 (*Chem. Abstr.*, 1958, **52**, 8070).	451
57DOK(117)996	A. N. Nesmeyanov, T. P. Tolstaya and L. S. Isaeva; *Dokl. Akad. Nauk SSSR*, 1957, **117**, 996.	838, 839
57GEP1101407	E. Kühle and R. Wegler; *German. Pat.* 1 101 407 (1957) (*Chem. Abstr.*, 1962, **56**, 12 805).	365
57HCA918	H. U. Daeniker and J. Druey; *Helv. Chim. Acta*, 1957, **40**, 918.	374
57HCA2148	H. U. Daeniker and J. Druey; *Helv. Chim. Acta*, 1957, **40**, 2148.	176, 180
57HOU(11/1)1	E. Moeller; *Methoden Org. Chem.* (*Houben-Weyl*), 1957, **11/1**, 1.	298, 303
57HOU(11/1)961	E. Moeller; *Methoden Org. Chem.* (*Houben-Weyl*), 1957, **11/1**, 961.	302
57IS122	H. J. Matsuguma and L. F. Audrieth; *Inorg. Synth.*, 1957, **5**, 122.	691
57IZV69	G. Kamai and N. A. Chadaeva; *Izv. Akad. Nauk SSSR, Ser. Khim.*, 1957, 69 (*Chem. Abstr.*, 1960, **54**, 6521).	500
57IZV812	B. M. Mikhailov and V. A. Vaver; *Izv. Akad. Nauk SSSR, Ser. Khim.*, 1957, 812 (*Chem. Abstr.*, 1958, **52**, 3667).	1084
57JA572	W. R. McBride and H. W. Kruse; *J. Am. Chem. Soc.*, 1957, **79**, 572.	388
57JA717	F. G. Bordwell and P. J. Boutom; *J. Am. Chem. Soc.*, 1957, **79**, 717.	150
57JA974	J. L. Speier, J. A. Webster and G. H. Barnes; *J. Am. Chem. Soc.*, 1957, **79**, 974.	536
57JA1222	D. D. Nyberg and B. E. Christensen; *J. Am. Chem. Soc.*, 1957, **79**, 1222.	103
57JA1458	A. T. Bottini and J. D. Roberts; *J. Am. Chem. Soc.*, 1957, **79**, 1458.	647
57JA1970	M. E. Wolff and A. Burger; *J. Am. Chem. Soc.*, 1957, **79**, 1970.	412
57JA2311	F. L. M. Pattison and J. J. Norman; *J. Am. Chem. Soc.*, 1957, **79**, 2311.	19
57JA2553	C. Djerassi and J. Grossman; *J. Am. Chem. Soc.*, 1957, **79**, 2553.	126
57JA3474	S. Dykstra and H. S. Mosher; *J. Am. Chem. Soc.*, 1957, **79**, 3474.	95
57JA3493	F. G. Bordwell and W. A. Hewett; *J. Am. Chem. Soc.*, 1957, **79**, 3493.	117
57JA3503	A. Segre, R. Viterbo and G. Parisi; *J. Am. Chem. Soc.*, 1957, **79**, 3503.	214
57JA3567	W. J. Bailey and S. A. Buckler; *J. Am. Chem. Soc.*, 1957, **79**, 3567.	442

57JA3570	F. W. Hoffmann, T. C. Simmons and L. J. Clunz; *J. Am. Chem. Soc.*, 1957, **79**, 3570.	458
57JA4560	H. Gilman and E. A. Zuech; *J. Am. Chem. Soc.*, 1957, **79**, 4560.	942
57JA5000	D. S. Breslow and R. R. Hough; *J. Am. Chem. Soc.*, 1957, **79**, 5000.	208
57JA5190	G. F. Hennson, P. A. McClusker, E. C. Ashby and A. J. Rutkowski; *J. Am. Chem. Soc.*, 1957, **79**, 5190.	521
57JA5628	R. S. Porter and G. H. Cady; *J. Am. Chem. Soc.*, 1957, **79**, 5628.	98
57JA5739	W. D. Emmons; *J. Am. Chem. Soc.*, 1957, **79**, 5739.	764
57JA5884	L. Maier, D. Seyferth, F. G. A. Stone and E. G. Rochow; *J. Am. Chem. Soc.*, 1957, **79**, 5884.	444, 824, 882
57JA6248	M. Hauptschein, M. Braid and F. E. Lawlor; *J. Am. Chem. Soc.*, 1957, **79**, 6248.	166
57JA6483	D. S. Acker and W. J. Wayne; *J. Am. Chem. Soc.*, 1957, **79**, 6483.	224
57JA6522	W. D. Emmons; *J. Am. Chem. Soc.*, 1957, **79**, 6522.	390
57JCS348	J. D. Edwards, W. Gerrard and M. F. Lappert; *J. Chem. Soc.*, 1957, 348.	110
57JCS381	M. F. Lappert and C. A. Pearce; *J. Chem. Soc.*, 1957, 381.	418, 419, 1085
57JCS739	M. J. Frazer, W. Gerrard and M. F. Lappert; *J. Chem. Soc.*, 1957, 739.	110
57JCS799	D. S. Margolis and R. W. Pittman; *J. Chem. Soc.*, 1957, 799.	278, 292
57JCS1823	G. Olah, A. Pavlath and G. Varsanyi; *J. Chem. Soc.*, 1957, 1823.	608, 619
57JCS2078	G. W. A. Fowles and C. M. Pleass; *J. Chem. Soc.*, 1957, 2078.	421
57JCS2413	B. B. Hunt and B. C. Saunders; *J. Chem. Soc.*, 1957, 2413.	845, 860
57JCS2574	J. Burdon, I. Farazmand, M. Stacey and J. C. Tatlow; *J. Chem. Soc.*, 1957, 2574.	182, 364
57JCS2640	T. Gramstadt and R. N. Haszeldine; *J. Chem. Soc.*, 1957, 2640.	181, 182
57JCS3708	J. W. Dale, H. J. Emeleus, R. N. Haszeldine and J. H. Moss; *J. Chem. Soc.*, 1957, 3708.	486, 492, 505
57JCS4547	D. Barnard; *J. Chem. Soc.*, 1957, 4547.	192, 287
57JOC145	Q. W. Decker and H. W. Post; *J. Org. Chem.*, 1957, **22**, 145.	227
57JOC265	W. C. Smith and L. F. Audrieth; *J. Org. Chem.*, 1957, **22**, 265.	861
57JOC358	P. A. S. Smith and E. E. Most, Jr.; *J. Org. Chem.*, 1957, **22**, 358.	384
57JOC452	L. M. Werbel, T. P. Dawson, J. R. Hooton and T. E. Dalbey; *J. Org. Chem.*, 1957, **22**, 452.	496, 498
57JOC536	I. B. Douglass and D. R. Poole; *J. Org. Chem.*, 1957, **22**, 536.	174, 688
57JOC561	E. Campaigne and S. W. Osburn; *J. Org. Chem.*, 1957, **22**, 561.	123
57JOC972	A. F. Torralba and T. C. Myers; *J. Org. Chem.*, 1957, **22**, 972.	456
57JOC980	F. G. Bordwell and W. A. Hewett; *J. Org. Chem.*, 1957, **22**, 980.	176, 180
57JOC1001	M. F. Hawthorne; *J. Org. Chem.*, 1957, **22**, 1001.	653
57JOC1136	H. C. Brown and B. C. Subba Rao; *J. Org. Chem.*, 1957, **22**, 1136.	40
57JOC1186	B. Loev and J. T. Massengale; *J. Org. Chem.*, 1957, **22**, 1186.	413
57JOC1714	S. C. Temin; *J. Org. Chem.*, 1957, **22**, 1714.	748
57JOC598A	R. D. Closson, J. Kozikowski and T. H. Coffield; *J. Org. Chem.*, 1957, **22**, 598.	573
57JOC598B	T. H. Coffield, J. Kozikowski and R. D. Closson; *J. Org. Chem.*, 1957, **22**, 598.	573
57LA(606)24	L. Horner and H. Steppan; *Justus Liebigs Ann. Chem.*, 1957, **606**, 24.	764
57LA(606)75	H. Bohme and H.-D. Stachel; *Justus Liebigs Ann. Chem.*, 1957, **606**, 75.	211
57MI 203-01	V. Duvieneaud and G. B. Brown; *Biochem. Prep.*, 1957, **5**, 93.	126, 214
57MI 208-01	B. A. Arbuzov, N. I. Rizpolozhenskii and M. A. Zvereva; *Izv. Akad. Nauk SSSR, Otd. Khim. Nauk*, 1957, 179 (*Chem. Abstr.*, 1957, **5**, 11 237).	462
57MI 217-01	F. Kh. Solomakhina; *Trudy Tashkentsk. Farmatseut. Inst.*, 1957, **1**, 321 (*Chem. Abstr.*, 1961, **55**, 15 389).	882
57MIP1057112	H. Schlor, E. Schegk and G. Schrader; *DAS* 1 057 112 (1957) (*Chem. Abstr.*, 1960, **54**, 12 988b).	463
57OR(9)332	C. V. Wilson; *Org. React.*, 1957, **9**, 332.	12, 627
57OR(9)350	C. V. Wilson; *Org. React.*, 1957, **9**, 350.	45
57RTC129	J. H. Uhlenbroeck; *Recl. Trav. Chim. Pays-Bas*, 1957, **76**, 129.	176, 180, 271
57RZC715	J. Michalski and Cz. Krawiecki; *Rocz. Chem.*, 1957, **31**, 715 (*Chem. Abstr.*, 1958, **52**, 5284).	466
57RZC879	J. Michalski, H. Strzelecka and J. Wieczorkowski; *Rocz. Chem.*, 1957, **31**, 879 (*Chem. Abstr.*, 1958, **52**, 8037h).	412
57USP2805234	C. R. Glosky; *US Pat.* 2 805 234 (1957) (*Chem. Abstr.*, 1958, **52**, 2030).	599
57USP2950306	W. C. Smith; *US Pat.* 2 950 306 (1957) (*Chem. Abstr.*, 1961, **55**, 2569).	848
57ZAAC(290)348	H. Hartmann and G. Nowak; *Z. Anorg. Allg. Chem.*, 1957, **290**, 348.	1076, 1079
58IZV846	L. I. Zakharkin; *Izv. Akad. Nauk SSSR, Ser. Khim.*, 1958, 846.	1015
58AC(P)108	L. Horner and O. H. Brasedow; *Ann. Chim. (Paris)*, 1958, **612**, 108.	193
58AG166	G. Drefahl, G. Heublein and A. Wintzer; *Angew. Chem.*, 1958, **70**, 166.	982
58AG251	A. Rieche; *Angew. Chem.*, 1958, **70**, 251.	96
58CB64	L. Horner, H. Hoffmann and H. G. Wippel; *Chem. Ber.*, 1958, **91**, 64.	454, 844
58CB380	S. Hünig and M. Kiessel; *Chem. Ber.*, 1958, **91**, 380.	677
58CB1057	H. Krimm; *Chem. Ber.*, 1958, **91**, 1057.	390
58CB1181	F. Cramer and H. Hettler; *Chem. Ber.*, 1958, **91**, 1181.	106
58CB1212	A. Rieche, E. Schmitz and E. Beyer; *Chem. Ber.*, 1959, **92**, 1212.	96
58CB1583	L. Horner, H. Hoffmann and P. Beck; *Chem. Ber.*, 1958, **91**, 1583.	428, 430, 431, 441, 822, 827
58CB2130	F. Asinger, G. Geiseler and M. Hoppe; *Chem. Ber.*, 1958, **91**, 2130.	175
58CB2345	J. Gosselck; *Chem. Ber.*, 1958, **91**, 2345.	288
58CB2528	F. Feher and W. Kruse; *Chem. Ber.*, 1958, **91**, 2528.	360
58CRV439	B. S. Thyagarajan; *Chem. Rev.*, 1958, **58**, 439.	214
58CRV925	P. S. Bailey; *Chem. Rev.*, 1958, **58**, 925.	348

58DOK(122)614	A. N. Nesmeyanov, T. P. Tolstaya and L. S. Isaeva; *Dokl. Akad. Nauk SSSR*, 1958, **122**, 614 (*Chem. Abstr.*, 1959, **53**, 4178a).	876
58HCA244	H. Keller, H. Netter and B. Niemann; *Helv. Chim. Acta*, 1958, **41**, 244.	414
58HCA593	G. Holze and W. Jenny; *Helv. Chim. Acta*, 1958, **41**, 593.	282
58HOU(11)190	H. Freytag, F. Moller, G. Pieper and H. Soll; *Methoden Org. Chem.* (*Houben-Weyl*), 1958, **11**, 190.	347, 348
58HOU(11/2)99	G. Pieper; *Methoden Org. Chem.* (*Houben-Weyl*), 1958, **11/2**, 99.	385
58HOU(11/2)178	H. Söll; *Methoden Org. Chem.* (*Houben-Weyl*), 1958, **11/2**, 178.	385
58HOU(11/2)641	A. Dorlars; *Methoden Org. Chem.* (*Houben-Weyl*), 4th edn, 1958, **11/2**, 641.	399
58HOU(11/2)738	A. Dorlars; *Methoden Org. Chem.* (*Houben-Weyl*), 1958, **11/2**, 738.	399
58JA617	G. F. Hennson, P. A. McClusker and A. J. Rutkowski; *J. Am. Chem. Soc.*, 1958, **80**, 617.	521
58JA838	L. Field and J. E. Lawson; *J. Am. Chem. Soc.*, 1958, **80**, 838.	215
58JA1009	H. W. Sternberg, R. Markby and I. Wender; *J. Am. Chem. Soc.*, 1958, **80**, 1009.	670
58JA1107	A. B. Burg and P. J. Slota; *J. Am. Chem. Soc.*, 1958, **80**, 1107.	409
58JA1336	D. Seyferth; *J. Am. Chem. Soc.*, 1958, **80**, 1336.	483, 487
58JA1680	L. Goodman, A. Benitez and B. R. Baker; *J. Am. Chem. Soc.*, 1958, **80**, 1680.	129
58JA1716	W. B. Hardy and R. B. Fortenbaugh; *J. Am. Chem. Soc.*, 1958, **80**, 1716.	621
58JA2516	W. E. Bull, J. A. Seaton and L. F. Audrieth; *J. Am. Chem. Soc.*, 1958, **80**, 2516.	388
58JA2584	M. C. Caserio, J. D. Roberts, M. Neeman and W. S. Johnson; *J. Am. Chem. Soc.*, 1958, **80**, 2584.	89
58JA2686	J. D. Chanley and E. Feageson; *J. Am. Chem. Soc.*, 1958, **80**, 2686.	807
58JA3945	F. W. Hoffmann, D. H. Wadsworth and H. D. Weiss; *J. Am. Chem. Soc.*, 1958, **80**, 3945.	463
58JA4546	C. Walling, F. R. Stacey, S. E. Jamison and E. S. Huyser; *J. Am. Chem. Soc.*, 1958, **80**, 4546.	104
58JA5541	J. R. Cox, Jr. and F. H. Westheimer; *J. Am. Chem. Soc.*, 1958, **80**, 5541.	694
58JA6150	S. G. Levine; *J. Am. Chem. Soc.*, 1958, **80**, 6150.	680
58JA6212	P. T. Gilham and H. G. Khorana; *J. Am. Chem. Soc.*, 1958, **80**, 6212.	106
58JA6562	C. G. Overberger and A. V. DiGiulio; *J. Am. Chem. Soc.*, 1958, **80**, 6562.	377, 403
58JCS692	G. R. Knox and P. L. Pauson; *J. Chem. Soc.*, 1958, 692.	182
58JCS1451	E. W. Abel, J. D. Edwards, W. Gerrard and M. F. Lappert; *J. Chem. Soc.*, 1958, 1451.	111, 112
58JCS1719	E. R. H. Jones and F. G. Mann; *J. Chem. Soc.*, 1958, 1719.	875
58JCS1963	H. A. C. Montgomery and J. H. Turnbull; *J. Chem. Soc.*, 1958, 1963.	413
58JCS2372	A. G. Davies and R. B. Moodie; *J. Chem. Soc.*, 1958, 2372.	99
58JCS2834	L. Bateman and R. W. Glazebrook; *J. Chem. Soc.*, 1958, 2834.	180
58JCS2939	J. W. Dale, H. J. Emeleus and R. N. Haszeldine; *J. Chem. Soc.*, 1958, 2939.	288, 291
58JCS3039	V. M. Clark, G. W. Kirby and A. R. Todd; *J. Chem. Soc.*, 1958, 3039.	414
58JCS3746	E. W. Abel; *J. Chem. Soc.*, 1957, 3746.	418
58JGU2963	I. P. Komkov, K. V. Karavanov and S. Z. Ivin; *J. Gen. Chem. USSR* (*Engl. Transl.*), 1958, **28**, 2963 (*Chem. Abstr.*, 1959, **53**, 9035).	478
58JOC50	S. H. Langer, S. Connell and I. Wender; *J. Org. Chem.*, 1958, **23**, 50.	539
58JOC280	H. E. Podall and W. E. Foster; *J. Org. Chem.*, 1958, **23**, 280.	626
58JOC311	W. D. Emmons, A. S. Pagano and T. E. Stevens; *J. Org. Chem.*, 1958, **23**, 311.	386
58JOC330	I. B. Douglass and B. S. Farah; *J. Org. Chem.*, 1958, **23**, 330.	174, 187
58JOC363	A. H. Weinstein, R. M. Pierson, B. Wargotz and T. F. Yen; *J. Org. Chem.*, 1958, **23**, 363.	175
58JOC381	D. L. Chamberlain, D. Peters and M. S. Kharasch; *J. Org. Chem.*, 1958, **23**, 381.	165
58JOC529	R. L. Hinman and K. L. Hamm; *J. Org. Chem.*, 1958, **23**, 529.	379
58JOC636	F. G. Bordwell and W. A. Hewett; *J. Org. Chem.*, 1958, **23**, 636.	118
58JOC648	R. A. Henry; *J. Org. Chem.*, 1958, **23**, 648.	781
58JOC1051	J. H. Boyer, C. H. Mack, N. Goebel and L. R. Morgan; *J. Org. Chem.*, 1958, **23**, 1051.	1055
58JOC1133	D. T. Warner and L. L. Coleman; *J. Org. Chem.*, 1958, **23**, 1133.	366
58JOC1161	R. C. Fuson and B. Freedman; *J. Org. Chem.*, 1958, **23**, 1161.	632
58JOC1230	A. F. Graefe; *J. Org. Chem.*, 1958, **23**, 1230.	388
58JOC1298	J. M. Denham and R. K. Ingham; *J. Org. Chem.*, 1958, **23**, 1298.	845
58JOC1322	M. S. Kharasch and G. N. Sosnovsky; *J. Org. Chem.*, 1958, **23**, 1322.	96
58JOC1389	H. E. Zaugg, L. R. Swett and G. R. Stone; *J. Org. Chem.*, 1958, **23**, 1389.	1049
58JOC1412	D. E. Pearson, H. W. Pope, W. W. Hargrove and W. E. Stamper; *J. Org. Chem.*, 1958, **23**, 1412.	608, 626
58JOC1449	G. Wendt and W. F. Brice; *J. Org. Chem.*, 1958, **23**, 1449.	274
58JOC1525	N. H. Koenig, G. S. Sasin and D. Swern; *J. Org. Chem.*, 1958, **23**, 1525.	120, 180
58JOC1595	P. A. S. Smith, J. M. Clegg and J. Lakritz; *J. Org. Chem.*, 1958, **23**, 1595.	376
58JOC1694	N. Kharasch and R. B. Langford; *J. Org. Chem.*, 1958, **23**, 1694.	194
58JOC1800	G. B. Bachmann and W. Michalowicz; *J. Org. Chem.*, 1958, **23**, 1800.	800
58JOC2044	H. Gilman, H. A. McNich and D. Wittenberg; *J. Org. Chem.*, 1958, **23**, 2044.	553
58LA(614)37	H. H. Schlubach and R. Repenning; *Justus Liebigs Ann. Chem.*, 1958, **614**, 37.	1016
58LA(614)66	A. Schöberl and H. Gräfje; *Justus Liebigs Ann. Chem.*, 1958, **614**, 66.	214, 215
58LA(617)62	H. Bohme and G. van Ham; *Justus Liebigs Ann. Chem.*, 1958, **617**, 62.	211
58LA(617)71	A. Schöberl and H. Gräfje; *Justus Liebigs Ann. Chem.*, 1958, **617**, 71.	219
58LA(617)257	A. Baeyer; *Justus Liebigs Ann. Chem.*, 1958, **617**, 257.	492
58LA(618)24	W. Hieber and G. Wagner; *Justus Liebigs Ann. Chem.*, 1958, **618**, 24.	573
58LA(618)173	S. Goldschmidt and B. Acksteiner; *Justus Liebigs Ann. Chem.*, 1958, **618**, 173.	404
B-58MI 203-01	E. E. Reid; in "Organic Chemistry of Bivalent Sulfur," Chemical Publishing Co., New York, 1958, vol. 1.	132, 213

58MI 208-01	O. Y. Okhlobystin and L. I. Zakharkin; *Izv. Akad. Nauk SSSR, Otd. Khim. Nauk*, 1958, 1006 (*Chem. Abstr.*, 1959, **53**, 1122).	434
58MI 213-01	D. Houalla and R. Wolf; *Compt. Rendu.*, 1958, **247**, 482 (*Chem. Abstr.*, 1959, **53**, 7072d).	693
58MIP1067011	W. Lorenz and G. Schrader; *DAS* 1 067 011 (1958) (*Chem. Abstr.*, 1962, **56**, 1482).	464
58USP2831835	W. J. Middleton, *US Pat.* 2 831 835 (1958) (*Chem. Abstr.*, 1958, **52**, 14 658f).	1013
58USP2853518	W. J. Balon; *US Pat* 2 853 518 (1958) (*Chem. Abstr.* 1959, **53**, 5202).	435
58ZAAC(294)81	A. Dornow and H. H. Gehrt; *Z. Anorg. Allg. Chem.*, 1958, **294**, 81.	419
59ACS35	L. Tansjö; *Acta Chem. Scand.*, 1959, **13**, 35.	417
59ACS2155	O. Foss and S. Hauge; *Acta Chem. Scand.*, 1959, **13**, 2155.	289
59AG321	E. Jucker; *Angew. Chem.*, 1959, **71**, 321.	373
59AG349	L. Horner and E. H. Winkelmann; *Angew. Chem.*, 1959, **71**, 349.	27
59AG574	L. Maier; *Angew. Chem.*, 1959, **71**, 574.	428
59BRP823483	Kali-Chemie Akt. Ges.; *Br. Pat.* 823 483 (1959) (*Chem. Abstr.*, 1960, **54**, 14 125).	431
59BSF1544	R. Emiliozzi, L. Pichat and M. Herbert; *Bull. Soc. Chim. Fr.*, 1959, 1544.	126
59BSF1887	R. Emiliozzi and L. Pichat; *Bull. Soc. Chim. Fr.*, 1959, 1887.	232
59CB40	G. Vicentini, E. Giesbrecht and L. R. M. Pitombo; *Chem. Ber.*, 1959, **92**, 40.	732
59CB83	A. Rieche and H. Gross; *Chem. Ber.*, 1959, **92**, 83.	182, 617
59CB704	K. Issleib and A. Tzschach; *Chem. Ber.*, 1959, **92**, 704.	431
59CB850	H. Grunze; *Chem. Ber.*, 1959, **92**, 850.	105
59CB1048	G. Kresze and U. Uhlich; *Chem. Ber.*, 1959, **92**, 1048.	240
59CB1118	K. Issleib and A. Tzschach; *Chem. Ber.*, 1959, **92**, 1118.	427, 437
59CB1149	M. Becke-Goehring and H. Jenne; *Chem. Ber.*, 1959, **92**, 1149.	361
59CB1177	U. Schmidt and P. Grafen; *Chem. Ber.*, 1959, **92**, 1177.	214
59CB1270	H. G. Viehe; *Chem. Ber.*, 1959, **92**, 1270	1014, 1015
59CB1842	F. Klages and H. Wolf; *Chem. Ber.*, 1959, **92**, 1842.	383, 384
59CB1910	D. Klamann, C. Sars and M. Zelenka; *Chem. Ber.*, 1959, **92**, 1910.	363, 399
59CB1950	H. G. Viehe; *Chem. Ber.*, 1959, **92**, 1950	1014, 1015, 1016
59CB2088	L. Horner, P. Beck and H. Hoffmann; *Chem. Ber.*, 1959, **92**, 2088.	462, 821, 827
59CB2521	R. Gösl and A. Meuser; *Chem. Ber.*, 1959, **92**, 2521.	380, 383, 384
59CB2637	K. Modritzer; *Chem. Ber.*, 1959, **92**, 2637.	415, 492
59CB2716	H. Hock and F. Ernst; *Chem. Ber.*, 1959, **92**, 2716.	112
59CB3175	K. Issleib and D. W. Müller; *Chem. Ber.*, 1959, **92**, 3175.	437
59CCC1158	O. Wichterle, V. Gregor, A. Dubansky and V. Seidl; *Collect. Czech. Chem. Commun.*, 1959, **24**, 1158 (*Chem. Abstr.*, 1959, **53**, 11 384).	782
59CI(L)541	P. R. Bloomfield and K. Parvin; *Chem. Ind.* (*London*), 1959, 541.	437
59CI(L)542	P. T. Gilham and G. M. Tener; *Chem. Ind.* (*London*), 1959, 542.	105, 106
59CI(L)628	G. Galvin, G. E. Coates and P. S. Dixon; *Chem. Ind.* (*London*), 1959, 628.	590
59CI(L)849	S. Grim and D. Seyferth; *Chem. Ind.* (*London*), 1959, 849.	447
59CJC1498	A. A. Oswald; *Can. J. Chem.*, 1959, **37**, 1498.	105
59DOK(124)1251	A. N. Nesmeyanov, I. F. Lutsenko, Z. S. Kraits and A. P. Bokovoi; *Dokl. Akad. Nauk SSSR*, 1959, **124**, 1251 (Engl. Transl. p. 155; *Chem. Abstr.*, 1959, **53**, 16931g)	693
59DOK(127)1217	B. A. Arbuzov, G. M. Vinokurova and I. A. Perfileva; *Dokl. Akad. Nauk SSSR*, 1959, **127**, 1217 (*Chem. Abstr.*, 1960, **54**, 1377).	439
59DOK(128)299	G. Kamai and B. D. Chernokal'skii; *Dokl. Akad. Nauk SSSR*, 1959, **128**, 299 (*Chem. Abstr.*, 1960, **54**, 7538c).	498
59HCA739	K. Clusius and K. Schwarzenbach; *Helv. Chim. Acta*, 1959, **42**, 739.	806
59HCA1653	H. H. Bosshard, R. Mory, M. Schmid and H. Zollinger; *Helv. Chim. Acta*, 1959, **42**, 1653.	182
59IS69	E. N. Walsh and A. D. F. Toy; *Inorg. Synth.*, 1959, **7**, 69.	411
59IS73	A. D. F. Toy and E. N. Walsh; *Inorg. Synth.*, 1959, **7**, 73.	413
59JA1	A. B. Burg and L. R. Grant; *J. Am. Chem. Soc.*, 1959, **81**, 1	481, 491, 510
59JA148	F. W. Hoffmann, B. Kagan and J. H. Canfield; *J. Am. Chem. Soc.*, 1959, **81**, 148.	463
59JA248	K. Kumli, W. E. McEwan and C. A. VanderWerf; *J. Am. Chem. Soc.*, 1959, **81**, 248.	443
59JA714	H. Tilles; *J. Am. Chem. Soc.*, 1959, **81**, 714.	137
59JA803	W. J. Middleton and W. H. Sharkey; *J. Am. Chem. Soc.*, 1959, **81**, 803.	1012, 1013
59JA1103	M. M. Rauhut, I. Hechenbleikner, H. A. Currier, F. C. Schaefer and V. P. Wystrach; *J. Am. Chem. Soc.*, 1959, **81**, 1103.	428
59JA1137	C. Walling and R. Rabinowitz; *J. Am. Chem. Soc.*, 1959, **81**, 1137.	127
59JA2002	F. G. Bordwell, R. D. Chapman and C. E. Osborne; *J. Am. Chem. Soc.*, 1959, **81**, 2002.	206
59JA2191	C. H. Robinson, L. Finckenor, E. P. Oliveto and D. Gould; *J. Am. Chem. Soc.*, 1959, **81**, 2191.	24
59JA2597	E. L. Muetterties; *J. Am. Chem. Soc.*, 1959, **81**, 2597.	964
59JA2805	J. H. Biel, A. F. Drukker, T. F. Mitchell, E. P. Sprengeler, P. A. Nuhfer, A. C. Conway and A. Horita; *J. Am. Chem. Soc.*, 1959, **81**, 2805.	373, 374, 379, 381
59JA3163	F. C. Leavitt, T. A. Manuel, and F. Johnson; *J. Am. Chem. Soc.*, 1959, **81**, 3163.	989
59JA3294	R. L. Hinman and K. L. Hamm; *J. Am. Chem. Soc.*, 1959, **81**, 3294.	388
59JA3386	W. E. Parham, R. F. Motter and G. L. Mayo; *J. Am. Chem. Soc.*, 1959, **81**, 3386.	1031
59JA3805	K. W. Kumli, W. E. McEwen and C. A. Vander Werf; *J. Am. Chem. Soc.*, 1959, **81**, 3805.	842
59JA4108	H. C. Brown and B. C. Subba Rao; *J. Am. Chem. Soc.*, 1959, **81**, 4108.	40
59JA4230	S.-O. Lawesson and N. C. Yang; *J. Am. Chem. Soc.*, 1959, **81**, 4230.	89, 652
59JA4803	M. Grayson, P. T. Keough and G. A. Johnson; *J. Am. Chem. Soc.*, 1959, **81**, 4803.	442
59JA4810	E. Kober; *J. Am. Chem. Soc.*, 1959, **81**, 4810.	166
59JA5004	D. S. Matteson; *J. Am. Chem. Soc.*, 1959, **81**, 5004.	904

59JA5430 M. T. Beachem, J. T. Shaw, G. D. Sargent, R. B. Forrenbaugh and J. M. Salsburg; *J. Am. Chem. Soc.*, 1959, **81**, 5430. 271, 273

59JA5546 W. R. McBride and E. M. Bens; *J. Am. Chem. Soc.*, 1959, **81**, 5546. 388

59JA5800 J. Meinwald and O. L. Chapman; *J. Am. Chem. Soc.*, 1959, **81**, 5800. 350

59JA5836 M. F. Hawthorne; *J. Am. Chem. Soc.*, 1959, **81**, 5836. 269, 522

59JA6177 J. C. Sheldon and S. Y. Tyree, Jr.; *J. Am. Chem. Soc.*, 1959, **81**, 6177. 460

59JA6273 G. W. Parshall and R. V. Lindsey, Jr.; *J. Am. Chem. Soc.*, 1959, **81**, 6273. 427

59JA6330 D. B. Denney and M. J. Boskin; *J. Am. Chem. Soc.*, 1959, **81**, 6330. 853

59JBC(234)1 E. Baer, L. J. Ciplijauskas and T. Visser; *J. Biol. Chem.*, 1959, **234**, 1. 693

59JCP(30)1222 W. J. Lehmann, H. G. Weiss and I. Shapiro; *J. Chem. Phys.*, 1959, **30**, 1222. 109

59JCS372 W. R. Cullen and H. J. Emeleus; *J. Chem. Soc.*, 1959, 372. 503

59JCS429 M. H. Abraham and A. G. Davies; *J. Chem. Soc.*, 1959, 429. 112

59JCS1529 W. Gerrard, M. F. Lappert and B. A. Mountfield; *J. Chem. Soc.*, 1959, 1529. 111, 267

59JCS1804 R. D. Chambers, P. Goggin and W. K. R. Musgrave; *J. Chem. Soc.*, 1959, 1804. 650

59JCS2094 R. Bonnett, R. F. C. Brown, V. M. Clark, I. O. Sutherland and A. Todd; *J. Chem. Soc.*, 1959, 2094. 764

59JCS2835 S. Hinton and F. G. Mann; *J. Chem. Soc.*, 1959, 2835. 461

59JCS2927 D. W. Aubery and M. F. Lappert; *J. Chem. Soc.*, 1959, 2927. 419, 533

59JCS2969 B. Milligan and J. M. Swan; *J. Chem. Soc.*, 1959, 2969. 176, 177

59JGU2125 N. N. Yarovenko and M. A. Raksha; *J. Gen. Chem. USSR* (*Engl. Transl.*), 1959, 2125. 19

59JOC356 M. C. Hoff and P. Hill; *J. Org. Chem.*, 1959, **24**, 356. 428

59JOC807 J. Tsurugi and T. Nakabayashi; *J. Org. Chem.*, 1959, **24**, 807. 211

59JOC973 I. B. Douglass and B. S. Farah; *J. Org. Chem.*, 1959, **24**, 973. 166, 173, 232

59JOC1029 N. Kharasch and A. J. Parker; *J. Org. Chem.*, 1959, **24**, 1029. 126

59JOC1419 R. L. Arceneaux, J. G. Frick, Jr., E. K. Leonard and J. D. Reid; *J. Org. Chem.*, 1959, **24**, 1419. 413

59JOC2004 I. B. Douglass; *J. Org. Chem.*, 1959, **24**, 2004. 185

59JOC2013 R. C. Miller; *J. Org. Chem.*, 1959, **24**, 2013. 430, 840

59JPC1319 I. Shapiro and H. G. Weiss; *J. Phys. Chem.*, 1959, **63**, 1319. 109

59LA(625)88 G. A. Olah, A. Oswald and S. Kuhn; *Justus Liebigs Ann. Chem.*, 1959, **625**, 88. 411

59LA(626)60 F. Klages, R. Ruhnau and W. Hauser; *Justus Liebigs Ann. Chem.*, 1959, **626**, 60. 1009

59LA(627)142 L. Horner and H. Oediger; *Justus Liebigs Ann. Chem.*, 1959, **627**, 142. 860

59LA(627)195 F. Asinger, M. Thiel and G. Lipfert; *Justus Liebigs Ann. Chem.*, 1959, **627**, 195. 233

59M148 F. Pass and H. Schindlbaur; *Monatsh. Chem.*, 1959, **90**, 148. 428

59M792 F. Pass, E. Steininger and H. Schindlbauer; *Monatsh. Chem.*, 1959, **90**, 792. 429, 822, 823

59N321 H. Hartmann, W. Reiss and B. Karbstein; *Naturwissenschaften*, 1959, **46**, 321. 1076, 1081

59RTC252 H. J. Boonstra, L. Brandsma, A. M. Wiegman and J. F. Arens; *Recl. Trav. Chim. Pays-Bas*, 1959, **78**, 252. 1030

59T36 M. C. Caserio, J. D. Roberts, M. Neeman and W. S. Johnson; *Tetrahedron*, 1959, **6**, 36. 89

59TL24 A. Marquet and J. Jacques; *Tetrahedron Lett.*, 1959, **9**, 24. 24

59USP2912465 H. E. Ramsden; *US Pat.* 2 912 465 (1959) (*Chem. Abstr.* 1960, **54**, 2170). 433

59ZN(B)809 B. Bartocha, C. M. Douglas and Y. M. Gray; *Z. Naturforsch., Teil B*, 1959, **14**, 809. 904

59ZOB942 N. N. Yarovenko, V. N. Shemanina and G. B. Gazieva; *Zh. Obsch. Khim.*, 1959, **29**, 942. 288

59ZOB1450 T. A. Mastruikova, A. E. Shipov and M. I. Kabachnik; *Zh. Obshch. Khim.*, 1959, **29**, 1450 (*Chem. Abstr.*, 1960, **54**, 9729). 454

59ZOB1671 M. I. Kabachnik, E. I. Golubeva, D. M. Paikin, M. P. Shabanova, N. M. Gamper and L. F. Efimova; *Zh. Obshch. Khim.*, 1959, **29**, 1671. 464

59ZOB3154 B. V. Ioffe and L. E. Poroshin; *Zh. Obshch. Khim.*, 1959, **29**, 3154 (*Chem. Abstr.*, 1960, **54**, 11975e). 380

59ZOB3766 L. M. Yagupolskii and Z. M. Ivanova; *Zh. Obshch. Khim.*, 1959, **29**, 3766. 848

59ZOB3792 N. N. Yarovenko and A. S. Vasil'eva; *Zh. Obshch. Khim.*, 1959, **29**, 3792 (*Chem. Abstr.*, 1960, **54**, 19 479e). 171

60ACS1980 S. Åkerfeldt; *Acta Chem. Scand.*, 1960, 1980. 120

60AG209 R. Appel and D. Wagner; *Angew. Chem.*, 1960, **72**, 209. 809

60AG210 H. Niebergall; *Angew. Chem.*, 1960, **72**, 210. 431

60AG303 J. Sauer and R. Huisgen; *Angew. Chem.*, 1960, **72**, 303. 804

60BCJ1743 Y. Nomura and Y. Takeuchi; *Bull. Chem. Soc. Japan*, 1960, **33**, 1743. 130

60BRP916751 Continental Oil Co.; *Br. Pat.* 916 751 (1960) (*Chem. Abstr.*, 1963, **58**, 12 422). 186

60BSF1794 E. Gryszkiewicz-Trochimowski, J. Quinchon and O. Gryszkiewicz-Trochimowski; *Bull. Soc. Chim. Fr.*, 1960, 1794. 465

60CB782 H. Baganz and G. Dransch; *Chem. Ber.*, 1960, **93**, 782. 166, 167

60CB1078 H. Nöth and H. Beyer; *Chem. Ber.*, 1960, **93**, 1078. 109

60CB1220 M. Sander; *Chem. Ber.*, 1960, **93**, 1220. 451, 820, 821

60CB1249 G. Hesse and M. Förderreuther; *Chem. Ber.*, 1960, **93**, 1249. 100

60CB1305 H. Böhme and K. Hartke; *Chem. Ber.*, 1960, **93**, 1305. 1008

60CB1323 C. Stölzer and A. Simon; *Chem. Ber.*, 1960, **93**, 1323. 411, 412, 413

60CB1417 E. O. Fischer and S. Schreiner; *Chem. Ber.*, 1960, **93**, 1417. 882

60CB1852 K. Issleib and A. Tzschach; *Chem. Ber.*, 1960, **93**, 1852. 431

60CB2736 H. Bredereck, A. Wagner, H. Beck and R. J. Klein; *Chem. Ber.*, 1960, **93**, 2736. 724

60CB2902 H. A. Staab and K. Wendel; *Chem. Ber.*, 1960, **93**, 2902. 205

60CCC2191 V. Ettel and J. Horák; *Collect. Czech Chem. Commun.*, 1960, 2191 (*Chem. Abstr.*, 1961, **55**, 5402). 440

60CI(L)47 G. Hunt, B. C. Saunders and P. Simpson; *Chem. Ind.* (*London*), 1960, 47. 1071, 1072
60CJC343 A. F. McKay, D. L. Garmaise, G. Y. Paris and S. Gelblum; *Can. J. Chem.*, 1960, **38**, 343. 102
60CJC1428 G. A. Olah and A. A. Oswald; *Can. J. Chem.*, 1960, **38**, 1428. 408, 414
60CJC1431 G. A. Olah and A. A. Oswald; *Can. J. Chem.*, 1960, **38**, 1431. 408, 414
60CR(250)866 W. Chodkiewicz, P. Cadiot and A. Willemart; *C. R. Hebd. Seances Acad. Sci.*, 1960, **250**, 866. 1059
60CR(251)727 J. Soulié and A. Willemart; *C. R. Hebd. Séances Acad. Sci.*, 1960, **251**, 727. 1084
60CRV243 K. D. Berlin and G. B. Butler; *Chem. Rev.*, 1960, **60**, 243. 462
60CRV448 C. H. Depuy and R. W. King; *Chem. Rev.*, 1960, **60**, 448. 346
60DOK(133)855 O. M. Nefedov, Y. L. Levkov and A. D. Petrov; *Dokl. Acad. Nauk. SSSR*, 1960, **133**, 855 (*Chem. Abstr.*, 1960, **54**, 24 567). 609, 621
60DOK(135)323 M. I. Kabachnik and E. N. Tsvetkov; *Dokl. Akad. Nauk SSSR*, 1960, **135**, 323. 450
60DOK(135)1150 A. I. Razumov and B. G. Liorber; *Dokl. Akad. Nauk SSSR*, 1960, **135**, 1150 (*Chem. Abstr.*, 1961, **55**, 13 298). 465
60G383 A. Cerniani, G. Modena and P. E. Todesco; *Gazz. Chim. Ital.*, 1960, **90**, 383. 145
60HOU(5/4)1 E. Muller (ed.); *Methoden Org. Chem.* (*Houben-Weyl*), 1960, **5/4** (bromine and iodine), 1. 2, 6
60IS113 I. T. Millar, H. Heaney, D. M. Heinekey and W. C. Fernelius; *Inorg. Synth.*, 1960, **6**, 113. 491
60IS116 I. T. Millar, H. Heaney, D. M. Heinekey and W. C. Fernelius; *Inorg. Synth.*, 1960, **6**, 116. 491
60IZV386 B. M. Mikhailov and Y. N. Bubnov; *Izv. Akad. Nauk. SSSR, Ser Khim.*, 1960, 386 (*Chem. Abstr.*, **54**, 20 934). 269
60IZV651 M. I. Kabachnik, T. Ya. Medved, G. K. Kozlova, V. S. Balabukha, E. A. Mironova and L. I. Tikhonova; *Izv. Akad. Nauk SSSR, Ser. Khim.*, 1960, 651 (*Chem. Abstr.*, 1960, **54**, 22 329). 464
60JA403 M. V. George, D. J. Peterson and H. Gilman; *J. Am. Chem. Soc.*, 1960, **82**, 403. 545
60JA423 H. J. Harwood and J. D. W. Grisley; *J. Am. Chem. Soc.*, 1960, **82**, 423. 837, 839
60JA472 S. Raynolds and R. Levine; *J. Am. Chem. Soc.*, 1960, **82**, 472. 422
60JA748 M. F. Hawthorne; *J. Am. Chem. Soc.*, 1960, **82**, 748. 269
60JA909 A. J. Speziale and R. C. Freeman; *J. Am. Chem. Soc.*, 1960, **82**, 909. 8
60JA1801 A. H. Riebel, R. E. Erickson, C. J. Abshire and P. S. Bailey; *J. Am. Chem. Soc.*, 1960, **82**, 1801. 782
60JA1883 W. A. Piccoli, G. G. Haberland and R. L. Merker; *J. Am. Chem. Soc.*, 1960, **82**, 1883. 538
60JA1939 H. C. Brown and M. Dubeck; *J. Am. Chem. Soc.*, 1960, **82**, 1939. 964
60JA2145 A. B. Burg and P. J. Slota, Jr.; *J. Am. Chem. Soc.*, 1960, **82**, 2145. 431
60JA2148 A. B. Burg and P. J. Slota, Jr.; *J. Am. Chem. Soc.*, 1960, **82**, 2148. 431
60JA2640 D. H. R. Barton, J. M. Beaton, L. E. Geller and M. M. Pechet; *J. Am. Chem. Soc.*, 1960, **82**, 2640. 391
60JA2728 L. A. Carpino, A. A. Santilli and R. W. Murray; *J. Am. Chem. Soc.*, 1969, **82**, 2728. 377, 379
60JA2872 W. E. Truce, D. P. Tate and D. N. Burdge; *J. Am. Chem. Soc.*, 1960, **82**, 2872. 127, 188
60JA3133 L. A. Carpino; *J. Am. Chem. Soc.*, 1960, **82**, 3133. 379
60JA3222 H. C. Brown and G. Zweifel; *J. Am. Chem. Soc.*, 1960, **82**, 3222. 40
60JA3472 R. R. Grinstead; *J. Am. Chem. Soc.*, 1960, **82**, 3472. 650
60JA3560 G. B. Smith; *J. Am. Chem. Soc.*, 1960, **82**, 3560. 1099
60JA3799 W. E. Truce, W. Bannister, B. Groten, H. Klein, R. Cruse, A. Levy and E. Roberts; *J. Am. Chem. Soc.*, 1960, **82**, 3799. 1034
60JA3816 T. A. Gover and J. E. Willard; *J. Am. Chem. Soc.*, 1960, **82**, 3816. 33
60JA3866 H. C. Brown and W. Korytnyk; *J. Am. Chem. Soc.*, 1960, **82**, 3866. 109
60JA3919 M. Grayson and P. Keough; *J. Am. Chem. Soc.*, 1960, **82**, 3919. 444
60JA4001 A. Bowers, L. C. Ibanez, E. Denot and R. Becerra; *J. Am. Chem. Soc.*, 1960, **82**, 4001. 18
60JA4007 A. Bowers, E. Denot and R. Becerra; *J. Am. Chem. Soc.*, 1960, **82**, 4007. 18
60JA4228 D. S. Matteson; *J. Am. Chem. Soc.*, 1960, **82**, 4228. 904
60JA4708 H. C. Brown and G. Zweifel; *J. Am. Chem. Soc.*, 1960, **82**, 4708. 514
60JA4903 T. Emery and J. B. Neilands; *J. Am. Chem. Soc.*, 1960, **82**, 4903. 390
60JA5354 N. P. Neureiter and F. G. Bordwell; *J. Am. Chem. Soc.*, 1960, **82**, 5354. 118
60JA5359 S. Oae and C. Zalut; *J. Am. Chem. Soc.*, 1960, **82**, 5359. 712
60JA5794 W. A. Henderson, Jr. and S. A. Buckler; *J. Am. Chem. Soc.*, 1960, **82**, 5794. 433, 444
60JA6176 W. C. Smith; *J. Am. Chem. Soc.*, 1960, **82**, 6176. 476, 835, 836, 881
60JA6218 F. E. Brinckman and F. G. A. Stone; *J. Am. Chem. Soc.*, 1960, **82**, 6218. 904
60JA6238 S. L. Stafford and F. G. A. Stone; *J. Am. Chem. Soc.*, 1960, **82**, 6238. 904
60JA6242 R. J. Brotherton, A. L. McCloskey, L. L. Petterson and H. Steinberg; *J. Am. Chem. Soc.*, 1960, **82**, 6242. 418, 419
60JCS225 D. McHale, J. Green and P. Mammalis. *J. Chem. Soc.*, 1960, 225. 103
60JCS1024 A. M. Creighton and L. N. Owen; *J. Chem. Soc.*, 1960, 1024. 123
60JCS1030 S. M. Iqbal and L. N. Owen; *J. Chem. Soc.*, 1960, 1030. 123, 124
60JCS1103 H. J. Emeleus and S. N. Nabi; *J. Chem. Soc.*, 1960, 1103. 239
60JCS1398 C. E. Griffin and R. N. Haszeldine; *J. Chem. Soc.*, 1960, 1398. 778
60JCS1665 J. Michalski, J. M. Modro and J. Wieczorkowski; *J. Chem. Soc.*, 1960, 1665. 187, 199
60JCS4130 M. H. Abraham; *J. Chem. Soc.*, 1960, 4130. 112
60JCS4366 E. W. Parnell; *J. Chem. Soc.*, 1960, 4366. 797
60JCS4406 E. W. Abel; *J. Chem. Soc.*, 1960, 4406. 264
60JCS5079 R. G. Bacon and R. S. Irvin; *J. Chem. Soc.*, 1960, 5079. 361
60JGU3203 G. A. Razuvaev, N. A. Osanova, N. P. Shulaev and B. M. Tsyzin; *J. Gen. Chem. USSR* (*Engl. Transl.*), 1960, **30**, 3203. 879
60JINC213 L. Maier, E. J. Rochow and W. C. Fernelius; *J. Inorg. Nucl. Chem.*, 1960, **16**, 213. 872

60JOC20	R. E. Buckles and D. F. Knaack; *J. Org. Chem.*, 1960, **25**, 20.	23
60JOC24	R. E. Buckles, J. L. Forrester, R. L. Burham and T. W. McGee; *J. Org. Chem.*, 1960, **25**, 24.	24
60JOC214	M. S. Newman and R. H. B. Galt; *J. Org. Chem.*, 1960, **25**, 214.	753
60JOC352	C. King; *J. Org. Chem.*, 1960, **25**, 352.	162
60JOC621	S. Wawzonek, P. D. Klimstra and R. E. Kallio; *J. Org. Chem.*, 1960, **25**, 621.	95
60JOC666	E. J. Reist, I. G. Junga and B. R. Baker; *J. Org. Chem.*, 1960, **25**, 666.	861
60JOC1626	J. H. Wotiz, F. Huba and R. Vendley; *J. Org. Chem.*, 1960, **25**, 1626.	1014
60JOC1877	R. D. Campbell and F. J. Schultz; *J. Org. Chem.*, 1960, **25**, 1877.	787
60JOC1901	R. J. Kennedy and A. M. Stock; *J. Org. Chem.*, 1960, **25**, 1901.	192
60JOC1996	W. J. Bailey, S. A. Buckler and F. Marktscheffel; *J. Org. Chem.*, 1960, **25**, 1996.	442, 444
60JOC2015	C. W. Tullock and D. D. Coffman; *J. Org. Chem.*, 1960, **25**, 2015.	855
60JOC2016	C. W. Tullock and D. D. Coffman; *J. Org. Chem.*, 1960, **25**, 2016.	169
60JPC682	W. J. Cooper and J. F. Masi; *J. Phys. Chem.*, 1960, **64**, 682.	109
60LA(629)1	K. Ziegler, H. G. Gellert, H. Lehmkuhl, W. Pfohl and K. Zosel; *Justus Liebigs Ann. Chem.*, 1960, **629**, 1.	596
60LA(629)121	K. Ziegler, H. G. Gellert, K. Zosel, E. Holzkamp, J. Schneider, M. Söll and W. R. Kroll; *Justus Liebigs Ann. Chem.*, 1960, **629**, 121.	596
60LA(629)222	G. Wilke and H. Müller; *Justus Liebigs Ann. Chem.*, 1960, **629**, 222.	967, 974
60LA(629)241	K. Ziegler, F. Krupp and K. Zosel; *Justus Liebigs Ann. Chem.*, 1960, **629**, 241.	43, 112
60LA(631)1	E. Benzing; *Justus Liebigs Ann. Chem.*, 1960, **631**, 1.	404
60LA(634)84	H. O. Wirth, O. Konigstein and W. Kern; *Justus Liebigs Ann. Chem.*, 1960, **634**, 84.	631
60LA(635)91	B. Helferich and K. G. Kleb; *Justus Liebigs Ann. Chem.*, 1960, **635**, 91.	797
60LA(637)146	F. Asinger, M. Thiel and W. Schäfer; *Justus Liebigs Ann. Chem.*, 1960, **637**, 146.	167
60LA(638)33	V. Wolf and E. Kowitz; *Liebigs Ann. Chem.*, 1960, **638**, 33.	1018
60LA(639)33	K. Ziegler, R. Köster, H. Lehmkuhl and K. Reinert; *Justus Liebigs Ann. Chem.*, 1960, 33.	1041
60MI 202-01	F. D. Gunstone; *Adv. Org. Chem.*, 1960, **1**, 103.	44
60MI 202-02	B. M. Mikhailov and V. A. Dorokhov; *Proc. Acad. Sci., USSR*, 1960, **130**, 137.	112
60MI 202-03	E. L. Schumann, L. A. Paquette, R. V. Heinzelmann, D. P. Wallach, J. P. DaVanzo and M. E. Greig; *J. Med. Pharm. Chem.*, 1960, **5**, 343.	103
B-60MI 203-01	E. E. Reid; in "Organic Chemistry of Bivalent Sulfur," Chemical Publishing Co. Inc., New York, 1960, vol. 2.	132, 137
B-60MI 203-02	E. E. Reid; in "Organic Chemistry of Bivalent Sulfur," Chemical Publishing Co., New York, 1960, vol. 3.	217
60MI 212-01	B. V. Tronov and A. N. Novikov; *Izvest. Vysshikh Ucheb. Zavedenii, Khim. i. Khim. Tekhnol.*, 1960, **3**, 872 (*Chem. Abstr.*, 1961, **55**, 8348).	631
60MI 217-01	F. Kh. Solomakhina; *Trudy Tashkentsk. Farmatseut. Inst.*, 1960, **2**, 317 (*Chem. Abstr.*, 1962, **57**, 11 230c).	882
B-60MI 221-01	J. F. Arens; in "Advances in Organic Chemistry," eds. R. A. Rafael, E. C. Taylor and H. Wynberg, Interscience, New York, 1960, vol. 2, pp. 117–212.	1021, 1025, 1027
60NAT1086	A. T. Austin; *Nature* (*London*), 1960, **188**, 1086.	88
60OR(11)317	A. C. Cope and E. R. Trumbull; *Org. React.*, 1960, **11**, 317.	346
60OS(40)62	I. B. Douglass and B. S. Farah; *Org. Synth.*, 1960, **40**, 62.	166, 173
60OS(40)93	L. Friedman, R. L. Little and W. R. Reichle; *Org. Synth.*, 1960, **40**, 93.	726
60OS(40)105	H. Heaney and I. T. Millar; *Org. Synth.*, 1960, **40**, 105.	609, 632
60RCR331	E. N. Zil'berman; *Russ. Chem. Rev.* (*Engl. Trans.*), 1960, **29**, 331.	308
60RTC866	H. J. Boonstra and J. F. Arens; *Recl. Trav. Chim. Pays-Bas*, 1960, **79**, 866.	1031
60RZC1675	A. Markowska and J. Michalski; *Rocz. Chem.*, 1960, **34**, 1675 (*Chem. Abstr.*, 1962, **56**, 7345).	465
60T76	Y. Schaapsma, A. F. Bickel and E. C. Kooyman; *Tetrahedron*, 1960, **16**, 76.	127
60TL20	K. Balenovic, N. Brebant and D. Francetic; *Tetrahedron Lett.*, 1960, 20.	147
60USP2948755	Universal Oil Products Co.; *US Pat.* 2 948 755 (1960) (*Chem. Abstr.*, 1960, **55**, 1527).	752
60USP2957931	L. A. Hamilton and R. H. Williams; *US Pat.* 2 957 931 (1960) (*Chem. Abstr.*, 1961, **55**, 10 317h).	451
60USP2959621	H. Niebergalland and B. Langenfeld; *US Pat.* 2 959 621 (1960) (*Chem. Abstr.*, 1961, **55**, 7289).	431
60ZAAC(303)53	A. Simon and R. Paetzold; *Z. Anorg. Allg. Chem.*, 1960, **303**, 53.	101
60ZAAC(305)88	R. Paetzold, H. D. Schumann and A. Simon; *Z. Anorg. Allg. Chem.*, 1960, **305**, 88.	291
60ZOB784	F. Y. Perveev; *Zh. Obshch. Khim.*, 1960, **30**, 784 (*Chem. Abstr.*, 1961, **55**, 1580).	427
60ZOB1431	N. P. Bednyagina and I. Ya. Postovskii; *Zh. Obshch. Khim.*, 1960, **30**, 1431.	361
60ZOB1855	R. G. Glushkov and O. Yu. Magidson; *Zh. Obshch. Khim.*, 1960, **30**, 1855 (*Chem. Abstr.*, 1961, **55**, 7430).	788
60ZOB2348	A. N. Pudovik and I. V. Konovalova; *Zh. Obshch. Khim.*, 1960, **30**, 2348 (*Chem. Abstr.*, 1961, **55**, 8326).	454
60ZOB2995	K. A. Petrov, N. K. Blyznyuk and V. P. Korotkova; *Zh. Obshch. Khim.*, 1960, **30**, 2995 (*Chem. Abstr.*, 1961, **55**, 18 561).	454
60ZOB4069	N. N. Yarovenko, M. A. Raksha and V. N. Shemanina; *Zh. Obshch. Khim.*, 1960, **30**, 4069.	288
61ACS438	G. Aksnes; *Acta Chem. Scand.*, 1961, **15**, 438.	444
61ACS938	A. Fredga and A. Lindgren; *Acta Chem. Scand.*, 1961, **15**, 938.	278
61AG26	K. Issleib and A. Tzschach; *Angew. Chem.*, 1961, **73**, 26.	486, 506, 510
61AG173	H. Hartmann; *Angew. Chem.*, 1961, **73**, 173.	1058
61AK(18)1143	G. Bergson and A. Biezais; *Ark. Kemi*, 1961, **18**, 1143.	237

61CB113	K. Issleib and S. Häusler; *Chem. Ber.*, 1961, **94**, 113.	438
61CB186	H. Hoffmann and R. Gruenewald; *Chem. Ber.*, 1961, **94**, 186.	446
61CB392	K. Issleib and H. Völker; *Chem. Ber.*, 1961, **94**, 392.	432
61CB634	H. Noth and G. Mikulaschek; *Chem. Ber.*, 1961, **94**, 634.	358
61CB712	L. Horner and H. Fernekess; *Chem. Ber*, 1961, **94**, 712.	388
61CB1067	J. Goerdeler and H. Ullmann; *Chem. Ber.*, 1961, **94**, 1067.	726
61CB1263	L. Birkofer, W. Konkel and A. Ritter; *Chem. Ber.*, 1961, **94**, 1263.	262
61CB1317	L. Horner, P. Beck and V. G. Toscano; *Chem. Ber.*, 1961, **94**, 1317.	454
61CB1331	H. Hoffmann; *Chem. Ber.*, 1961, **94**, 1331.	446
61CB1373	G. Wittig and M. Schlosser; *Chem. Ber.*, 1961, **94**, 1373.	618
61CB1447	E. Amberger; *Chem. Ber.*, 1961, **94**, 1447.	481, 490
61CB1505	H. Nöth and H.-J. Vetter; *Chem. Ber.*, 1961, **94**, 1505.	410
61CB1682	J. Goerdelier and M. Budnowski; *Chem. Ber.*, 1961, **94**, 1682.	168
61CB1876	K. Rühlmann; *Chem. Ber.*, 1961, **94**, 1876.	263
61CB2122	L. Horner, P. Beck and V. G. Toscano; *Chem. Ber.*, 1961, **94**, 2122.	834, 835
61CB2166	E. Schmitz and R. Ohme; *Chem. Ber.*, 1961, **94**, 2166.	374
61CB2244	K. Issleib and G. Thomas; *Chem. Ber.*, 1961, **94**, 2244.	433
61CB2514	O. Neunhoeffer and L. Lamza; *Chem. Ber.*, 1961, **94**, 2514.	825
61CB2644	A. Cosmatos, I. Photaki and L. Zervas; *Chem. Ber.*, 1961, **94**, 2644.	412
61CB2656	K. Issleib and K. Krech; *Chem. Ber.*, 1961, **94**, 2656.	429, 461
61CB2664	K. Issleib and G. Döll; *Chem. Ber.*, 1961, **94**, 2664.	429
61CB3051	L. Maier; *Chem. Ber.*, 1961, **94**, 3051.	431
61CB3056	L. Maier; *Chem. Ber.*, 1961, **94**, 3056.	431
61CI(L)24	M. P. Brown and H. B. Silver; *Chem. Ind.* (*London*), 1961, 24.	834, 835
61CI(L)469	K. Balenovic, I. Bregovic, D. Francetic, I. Monkovic and V. Tomasic; *Chem. Ind.* (*London*), 1961, 469.	478
61CI(L)1619	H. Gilman and D. Aoki; *Chem. Ind.* (*London*), 1961, 1619.	1017
61CJC862	E. Cullen and Ph. L'Ecuyer; *Can. J. Chem.*, 1961, **39**, 862.	775
61CJC1386	I. M. Thomas; *Can. J. Chem.*, 1961, **39**, 1386.	112, 421
61CJC1855	W. R. Cullen; *Can. J. Chem.*, 1961, **39**, 1855.	482
61CR(252)4007	A. Guillemonat and J. L. Chevalier; *Compt. Rend.*, 1961, **252**, 4007.	361
61CR(253)673	P. Mazerolles, M. Lesbre and D. Huy-Giao; *C. R. Hebd. Seances Acad. Sci.*, 1961, **253**, 673.	948
61CRV361	R. Fessenden and J. S. Fessenden; *Chem. Rev.*, 1961, **61**, 361.	543, 811
61CRV607	F. W. Lichtenthaler; *Chem. Rev.*, 1961, **61**, 607.	693
61DOK(139)385	B. M. Mikailov and L. S. Vasil'ev; *Dokl. Akad. Nauk SSSR*, 1961, **139**, 385 (*Chem. Abstr.*, 1975, **82**, 111 344).	533
61DOK(139)1359	E. I. Grinstein, A. B. Bruker and L. Z. Soborovskii; *Dokl. Akad. Nauk SSSR*, 1961, **139**, 1359 (*Chem. Abstr.*, 1962, **56**, 1475).	440
61E475	C. Meystre, K. Heusler, J. Kalvoda, P. Wieland, G. Anner and A. Wettstein; *Experientia*, 1961, **17**, 475.	95
61GEP1104520	G. Schrader; *Ger. Pat.* 1 104 520 (1961) (*Chem. Abstr.*, 1962, **56**, 14 328e).	464
61GEP1114190	L. Horner and A. Mentrup; *Ger. Pat.* 1 114 190 (1961) (*Chem. Abstr.*, 1962, **57**, 2256).	441
61GEP1156403	G. Schulze and G. Weiss; *German Pat.* 1 156 403 (1961) (*Chem. Abstr.*, 1964, **60**, 4061).	365
61HCA61	B. Iselin; *Helv. Chim. Acta*, 1961, **44**, 61.	171
61HCA1812	E. Cherbuliez, B. Baehler, F. Hunkeler and J. Rabinowitz; *Helv. Chim. Acta*, 1961, **44**, 1812.	456
61IZV604	M. I. Kabachnik and T. Ya. Medved; *Izv. Akad. Nauk SSSR, Ser. Khim.*, 1961, 604 (*Chem. Abstr.*, 1961, **55**, 23 319).	463
61JA159	G. M. Tener; *J. Am. Chem. Soc.*, 1961, **83**, 159.	105, 106
61JA168	S. A. Buckler and V. P. Wystrach; *J. Am. Chem. Soc.*, 1961, **83**, 168.	440, 446
61JA221	P. Kovacic and R. P. Bennett; *J. Am. Chem. Soc.*, 1961, **83**, 221.	752
61JA367	M. F. Hawthorne; *J. Am. Chem. Soc.*, 1961, **83**, 367.	524
61JA535	J. K. Ruff; *J. Am. Chem. Soc.*, 1961, **83**, 535.	421
61JA649	J. G. Moffat and H. G. Khorana; *J. Am. Chem. Soc.*, 1961, **83**, 649.	414
61JA743	P. Kovacic and J. L. Foote; *J. Am. Chem. Soc.*, 1961, **83**, 743.	752
61JA747	D. C. Iffland, L. Salisbury and W. R. Schafer; *J. Am. Chem. Soc.*, 1961, **83**, 747.	403
61JA831	M. F. Hawthorne; *J. Am. Chem. Soc.*, 1961, **83**, 831.	515
61JA845	Q. E. Thompson; *J. Am. Chem. Soc.*, 1961, **83**, 845.	452, 694
61JA996	C. L. Frye, G. E. Vogel and J. A. Hall; *J. Am. Chem. Soc.*, 1961, **83**, 996.	547
61JA1241	H. C. Brown and G. Zweifel; *J. Am. Chem. Soc.*, 1961, **83**, 1241.	518
61JA1256	L. Field, C. B. Hoelzel, J. M. Locke and J. E. Lawson; *J. Am. Chem. Soc.*, 1961, **83**, 1256.	198
61JA1345	M. J. Hawthorne; *J. Am. Chem. Soc.*, 1961, **83**, 1345.	267
61JA1515	J. Kollonitsch; *J. Am. Chem. Soc.*, 1961, **83**, 1515.	111
61JA1613	D. Seyferth and S. O. Grim; *J. Am. Chem. Soc.*, 1961, **83**, 1613.	524
61JA1773	C. C. Price and W. H. Snyder; *J. Am. Chem. Soc.*, 1961, **83**, 1773.	675
61JA2122	W. D. English, A. L. McCloskey and H. Steinberg; *J. Am. Chem. Soc.*, 1961, **83**, 2122.	419
61JA2541	M. F. Hawthorne; *J. Am. Chem. Soc.*, 1961, **83**, 2541.	515, 522
61JA2835	J. K. Ruff; *J. Am. Chem. Soc.*, 1961, **83**, 2835.	418, 421
61JA3384	B. Sternbach and A. G. MacDiarmid; *J. Am. Chem. Soc.*, 1961, **83**, 3384.	110
61JA3417	H. C. Brown and A. W. Moerikofer; *J. Am. Chem. Soc.*, 1961, **83**, 3417.	518
61JA3834	H. C. Brown and G. Zweifel; *J. Am. Chem. Soc.*, 1961, **83**, 3834.	900
61JA3985	J. Radell, J. W. Connolly and A. J. Raymond; *J. Am. Chem. Soc.*, 1961, **83**, 3985.	94
61JA4124	L. D. Quin and J. J. S. Humphrey; *J. Am. Chem. Soc.*, 1961, **83**, 4124.	833, 844
61JA4357	G. Claeson, G. Androes and M. Calvin; *J. Am. Chem. Soc.*, 1961, **83**, 4357.	212

61JA4381 K. Moedritzer; *J. Am. Chem. Soc.*, 1961, **83**, 4381. 448

61JA4385 R. A. Robert, A. Benkeser, M. L. Burrous, L. E. Nelson and J. V. Swisher; *J. Am. Chem. Soc.*, 1961, **83**, 4385. 913, 914

61JA4406 E. H. Braye, W. Hubel and I. Caplier; *J. Am. Chem. Soc.*, 1961, **83**, 4406. 465, 989

61JA4466 R. A. Baldwin and R. M. Washburn; *J. Am. Chem. Soc.*, 1961, **83**, 4466. 863

61JA4605 L. M. Stock and A. Himoe; *J. Am. Chem. Soc.*, 1961, **83**, 4605. 608

61JA4663 J. H. Fried, T. S. Bry, A. G. Oberster, R. E. Beyler, T. B. Windholz, J. Hannah, L. H. Sarett and S. L. Steelman; *J. Am. Chem. Soc.*, 1961, **83**, 4663. 1016

61JCS238 Z. Pelchowicz; *J. Chem. Soc.*, 1961, 238. 458

61JCS896 J. Chatt and R. G. Hayler; *J. Chem. Soc.*, 1961, 896. 437

61JCS922 J. Dale; *J. Chem. Soc.*, 1961, 922. 111

61JCS1000 G. E. Coates and J. G. Livingstone; *J. Chem. Soc.*, 1961, 1000. 867

61JCS1398 D. Hamer and R. G. Leckey; *J. Chem. Soc.*, 1961, 1398. 491

61JCS1825 A. C. Chapman, W. S. Holmes, N. L. Paddock and H. T. Searle; *J. Chem. Soc.*, 1961, 1825. 407

61JCS2066 J. R. Case, N. H. Ray and H. L. Roberts; *J. Chem. Soc.*, 1961, 2066. 165

61JCS2070 J. R. Case, N. H. Ray and H. L. Roberts; *J. Chem. Soc.*, 1961, 2070. 165

61JCS3079 J. I. G. Cadogan and H. N. Moulden; *J. Chem. Soc.*, 1961, 3079. 694

61JCS3208 J. M. F. Braddock and G. E. Coates; *J. Chem. Soc.*, 1961, 3208. 494

61JCS4260 B. T. Newbold; *J. Chem. Soc.*, 1961, 4260. 776

61JCS4263 S. T. D. Gough and S. Trippett; *J. Chem. Soc.*, 1961, 4263. 442

61JCS4933 E. W. Abel; *J. Chem. Soc.*, 1961, 4933. 263

61JCS5053 G. E. Coates and J. G. Livingstone; *J. Chem. Soc.*, 1961, 5053. 867

61JCS5339 D. Barnard, L. Bateman, M. E. Cain, T. Colcough and J. I. Cunneen; *J. Chem. Soc.*, 1961, 5339. 144, 173

61JCS5545 S. Cooper, M. J. Frazer and W. Gerrard; *J. Chem. Soc.*, 1961, 5545. 110

61JGU3315 G. Kamai and N. A. Chadaeva; *J. Gen. Chem. USSR* (*Engl. Transl.*), 1961, **31**, 3315. 889

61JOC280 S. J. Cristol and W. C. Firth, Jr.; *J. Org. Chem.*, 1961, **26**, 280. 12, 627

61JOC611 H. H. Sisler and N. L. Smith; *J. Org. Chem.*, 1961, **26**, 611. 859

61JOC630 F. Sondheiemer and M. Nussim; *J. Org. Chem.*, 1961, **26**, 630. 40

61JOC850 A. W. Frank; *J. Org. Chem.*, 1961, **26**, 850. 837

61JOC1152 R. G. Hiskey, F. I. Carroll, R. M. Babb, J. O. Bledsoe, R. T. Puckett and B. W. Roberts; *J. Org. Chem.*, 1961, **26**, 1152. 219

61JOC1215 D. G. Crosby; *J. Org. Chem.*, 1961, **26**, 1215. 654

61JOC1357 H. Feuer, R. Miller and C. B. Lawyer; *J. Org. Chem.*, 1961, **26**, 1357. 396

61JOC1533 H. E. Baumgarten and R. E. Allen; *J. Org. Chem.*, 1961, **26**, 1533. 412

61JOC1819 H. H. Sisler, H. S. Ahuja and N. L. Smith; *J. Org. Chem.*, 1961, **26**, 1819. 384

61JOC1996 I. B. Douglass, B. S. Farah and E. G. Thomas; *J. Org. Chem.*, 1961, **26**, 1996. 166, 173, 174

61JOC2478 K. C. Schreiber and V. P. Fernandes; *J. Org. Chem.*, 1961, **26**, 2478. 168

61JOC2483 T. Nakabayashi and J. Tsurugi; *J. Org. Chem.*, 1961, **26**, 2483. 723

61JOC2531 T. E. Stevens; *J. Org. Chem.*, 1961, **26**, 2531. 403

61JOC2740 G. N. Walker, M. A. Moore and B. M. Weaver; *J. Org. Chem.*, 1961, **26**, 2740. 381

61JOC3024 C. S. Rondestvedt; *J. Org. Chem.*, 1961, **26**, 3024. 121

61JOC3250 A. Finch, J. C. Lockhart and J. Pearn; *J. Org. Chem.*, 1961, **26**, 3250. 110

61JOC3401 I. L. Doerr, I. Wempen, D. A. Clarke and J. J. Fox; *J. Org. Chem.*, 1961, **26**, 3401. 212

61JOC3547 R. A. Baldwin, K. A. Smitheman and R. M. Washburn; *J. Org. Chem.*, 1961, **26**, 3547. 832

61JOC3780 T. P. Johnston and A. Gallagher; *J. Org. Chem.*, 1961, **26**, 3780. 214

61JOC4010 R. Cramer and D. D. Coffman; *J. Org. Chem.*, 1961, **26**, 4010. 401

61JOC4190 J. P. Freeman; *J. Org. Chem.*, 1961, **26**, 4190. 386

61JOC4626 M. M. Rauhut and H. A. Currier; *J. Org. Chem.*, 1961, **26**, 4626. 450

61JOC4628 M. M. Rauhut and H. A. Currier; *J. Org. Chem.*, 1961, **26**, 4628. 454

61JOC4770 W. A. Henderson, S. A. Buckler, N. E. Day and M. Grayson; *J. Org. Chem.*, 1961, **26**, 4770. 448

61JOC5138 M. M. Rauhut, H. A. Curner, A. M. Semsel and V. P. Wystrach; *J. Org. Chem.*, 1961, **26**, 5138. 428

61JOC5149 K. G. Rutherford, W. Redmond and J. Rigamonti; *J. Org. Chem.*, 1961, **25**, 5149. 614

61JOC5205 R. A. Zingaro and R. E. McGlothlin; *J. Org. Chem.*, 1961, **26**, 5205. 461

61JPS193 L. J. Leeson, J. A. Lowery, G. M. Sieger and S. Muller; *J. Pharm. Sci.*, 1961, **50**, 193. 111

61LA(640)5 R. Riemschneider and K. Brender; *Justus Liebigs Ann. Chem.*, 1961, **640**, 5. 1014

61LA(640)79 H. Hellmann and O. Schumacher; *Justus Liebigs Ann. Chem.*, 1961, **640**, 79. 447

61LA(641)1 H. Meerwein, W. Florian, N. Schön and G. Stopp; *Justus Liebigs Ann. Chem.*, 1961, **641**, 1. 788

61LA(646)65 L. Horner and A. Mentrup; *Justus Liebigs Ann. Chem.*, 1961, **646**, 65. 441

B-61MI 203-01 N. Kharasch (ed); "Organic Sulfur Compounds," ed. Pergamon, New York, 1961, vol. 1. 172, 216, 217

61MI 203-02 J. M. Zgliczynski and T. Stelmaszynska; *Acta. Biochim. Polon.*, 1961, **8**, 123 (*Chem. Abstr.*, 1962, **57**, 4991h). 213

B-61MI 203-03 I. B. Douglass; in "Organic Sulfur Compounds," N. Kharasch, Pergamon, New York, 1961, vol. 1, chap. 30. 213

61MI 209-01 R. Pietsch and E. Pichler; *Mikrochim. Acta*, 1961, 582. 498

B-61MI 215-01 J. H. Boyer; in "Heterocyclic Compounds," ed. R. C. Elderfield, Wiley, 1961, vol. 7, pp. 463–522. 781

B-61MI 215-02 G. Williams; in "The Organic Chemistry of Boron," Academic Press, New York, 1961, p. 308. 810, 811

B-61MI 218-01 W. Gerrard; "The Organic Chemistry of Boron," Academic Press, New York, 1961. 900

61OS(41)91 C. Frisell and S.-O. Lawesson; *Org. Synth.*, 1961, **41**, 91. 684

61RTC237 L. Brandsma and J. F. Arens; *Recl. Trav. Chim. Pays-Bas*, 1961, **80**, 237. 1030
61RTC244 J. R. Nooi and J. F. Arens; *Recl. Trav. Chim. Pays-Bas*, 1961, **80**, 244. 1032
61T1 L. Pichat, M. Herbert and M. Thiers; *Tetrahedron*, 1961, **12**, 1. 291
61TL607 A. Maccioni, F. Montanari, M. Secci and M. Tramonti; *Tetrahedron Lett.*, 1961, 607. 147
61USP2969390 S. A. Buckler; *US Pat.* 2 969 390 (1961) (*Chem. Abstr.*, 1961, **55**, 14 381). 440
61USP2992072 G. F. Huff and W. H. Schechter; *US Pat.* 2 992 072 (1961) (*Chem. Abstr.*, 1961, **55**, 23 953d). 110
61USP2992266 A. D. McElroy and R. M. Adams; *US Pat.* 2 992 266 (1961) (*Chem. Abstr.*, 1961, **55**, 24 795b). 110
61USP3014060 J. D. Bush; *US Pat.* 3 014 060 (1961) (*Chem. Abstr.*, 1962, **56**, 8563e). 110
61ZAAC(312)186 H. Hartmann and G. Kuhl; *Z. Anorg. Allg. Chem.*, 1961, **312**, 186. 489, 1081, 1082
61ZOB217 A. P. Terent'ev and R. A. Gracheva; *Zh. Obshch. Khim.*, 1961, **55**, 1869 (*Chem. Abstr.*, 1961, **55**, 22 205g). 478
61ZOB507 T. A. Mastryukova, A. E. Shipov and M. I. Kabachnik; *Zh. Obshch. Khim.*, 1961, **31**, 507 (*Chem. Abstr.*, 1961, **55**, 22 101). 462
61ZOB949 E. L. Gefter; *Zh. Obshch. Khim.*, 1961, **31**, 949 (*Chem. Abstr.*, 1959, **53**, 14 937). 451
61ZOB3411 K. A. Petrov, V. A. Parshina and V. A. Gaidamak; *Zh. Obshch. Khim.*, 1961, **31**, 3411 (*Chem. Abstr.*, 1962, **57**, 4692). 440
61ZOB3417 K. A. Petrov and V. A. Parshina; *Zh. Obshch. Khim.*, 1961, **31**, 3417 (*Chem. Abstr.*, 1962, **57**, 4692). 440

62ACS772 B. Magnusson; *Acta Chem. Scand.*, 1962, 772. 130
62ACS1536 B. Magnusson; *Acta Chem. Scand.*, 1962, 1536. 130
62AG(E)53 G. Wittig and H. Hellwinkel; *Angew. Chem., Int. Ed. Engl.*, 1962, **1**, 53. 810
62AG(E)89 G. Kresze, A. Maschke, R. Albrecht, K. Bederke, H. P. Patachke, H. Smalla and A. Trede; *Angew. Chem., Int. Ed. Engl.*, 1962, **1**, 89. 793
62AG(E)97 G. Kresze, A. Maschke, R. Albrecht, K. Bederke, H. P. Patzschke, H. Smalla and A. Trede; *Angew. Chem., Int. Ed. Engl.*, 1962, **1**, 97. 100
62AG(E)405 R. Gösl; *Angew. Chem., Int. Ed. Engl.*, 1962, **1**, 405. 387
62AJC570 I. W. Stapleton and J. M. Swan; *Aust. J. Chem.*, 1962, **15**, 570. 220
62AJC710 R. S. Dickson and B. O. West; *Aust. J. Chem.*, 1962, **15**, 710. 483
62AP(295)615 K. Winterfeld and W. Haring; *Arch. Pharm.* (*Weinheim, Ger.*), 1962, **295**, 615. 176, 177
62BEP609076 W. Lorenz and G. Schrader; *Belg. Pat.* 609 076 (1962) (*Chem. Abstr.*, 1962, **57**, 16 661). 461
62BSF974 M. Gaudemar; *Bull. Soc. Chim. Fr.*, 1962, **175**, 974. 590, 1096
62BSF2126 M. Demuynck and J. Vialle; *Bull. Soc. Chim. Fr.*, 1962, 2126. 130
62CB64 H. Niebergall and B. Laudenfeld; *Chem. Ber.*, 1962, **95**, 64. 431, 461
62CB268 K. Issleib and G. Harzfeld; *Chem. Ber.*, 1962, **95**, 268. 437, 1060, 1069
62CB319 H. G. Viehe and E. Franchimont; *Chem. Ber.*, 1962, **95**, 319. 1012, 1013, 1062
62CB473 M. Sander; *Chem. Ber.*, 1962, **95**, 473. 428
62CB680 E. Schmitz and D. Habisch; *Chem. Ber.*, 1962, **95**, 680. 374
62CB803 H. Bredereck, F. Effenberger and E. H. Schweizer; *Chem. Ber.*, 1962, **95**, 803. 29
62CB1108 F. Krönhnke and K. Gerlach; *Chem. Ber.*, 1962, **95**, 1108. 1003
62CB1664 F. Cramer and H. Nuenhoeffer; *Chem. Ber.*, 1962, **95**, 1664. 106, 107
62CB1764 J. Jentzsch, J. Fabian and R. Mayer; *Chem. Ber.*, 1962, **95**, 1764. 130
62CB2012 E. Schmitz and R. Ohme; *Chem. Ber.*, 1962, **95**, 2012. 377
62CB2155 R. Nast, O. Künzel and R. Müller; *Chem. Ber.*, 1962, **95**, 2155. 1096
62CB2514 G. Wittig, W. Böll and K.-H. Krück; *Chem. Ber.*, 1962, **95**, 2514. 680
62CB2976 W. Ziegenbein and K.-H. Hornung; *Chem. Ber.*, 1962, **95**, 2976. 26
62CB2993 E. Steininger; *Chem. Ber.*, 1962, **95**, 2993. 451, 461, 838, 851
62CB3064 H. G. Viehe; *Chem. Ber.*, 1962, **95**, 3064. 1016
62CI(L)559 C. L. Bumgardner and R. L. Lilly; *Chem. Ind.* (*London*), 1962, 559. 691
62CI(L)1244 J. S. Nicholson and D. A. Peak; *Chem. Ind.* (*London*), 1962, 1244. 691
62CJC177 R. Bonnett and D. E. McGreer; *Can. J. Chem.*, 1962, **40**, 177. 764
62CJC342 N. R. Fetter and B. Bartocha; *Can. J. Chem.*, 1962, **40**, 342. 421
62CJC449 D. C. Bradley and I. M. Thomas; *Can. J. Chem.*, 1962, **40**, 449. 409, 420, 421
62CJC818 R. Deghengi and R. Gaudry; *Can. J. Chem.*, 1962, **40**, 818. 618
62CJC1355 D. C. Bradley and I. M. Thomas; *Can. J. Chem.*, 1962, **40**, 1355. 409, 421
62CJC2175 S. K. Banerjee, M. Manolopoulo and J. M. Pepper; *Can. J. Chem.*, 1962, **40**, 2175. 647
62CRV81 H. Heaney; *Chem. Rev.*, 1962, **62**, 81 647, 753
62CRV549 E. E. Gilbert; *Chem. Rev.*, 1962, **62**, 549. 720
62CRV599 B. A. Bohm and P. I. Abell; *Chem. Rev.*, 1962, **62**, 599. 4, 29
62DOK(143)855 S. V. Zavgorodnii and A. A. Petrov; *Dokl. Akad. Nauk SSSR* 1962, **143**, 855 (*Chem. Abstr.*, 1962, **57**, 3466). 1101
62GEP1126388 H. G. Viehe; *Ger. Pat.* 1 126 388, (1962), (*Chem. Abstr.*, 1963, **58**, 6860). 1076
62HCA358 B. J. R. Nicolaus, G. Pagini and E. Testa; *Helv. Chim. Acta*, 1962, **45**, 358. 103
62HCA717 B. J. R. Nicolaus, E. Bellasio and E. Testa; *Helv. Chim. Acta*, 1962, 717. 182, 248
62HOU(5/3)1 E. Muller (ed.); *Methoden Org. Chem.* (*Houben-Weyl*), 1962, **5/3**, (fluorine and chlorine) 1. 2, 6, 18
62HOU(5)796 R. Stroh; *Methoden Org. Chem.* (*Houben-Weyl*), 1962, **5**, 796. 334, 337
62IC756 R. M. Rosenberg and E. L. Muetterties; *Inorg. Chem.*, 1962, **1**, 756. 166
62JA190 G. Zweifel, K. Nagase and H. C. Brown; *J. Am. Chem. Soc.*, 1962, **84**, 190. 518
62JA769 W. R. Vaughan and R. D. Carlson; *J. Am. Chem. Soc.*, 1962, **84**, 769. 27
62JA847 L. Field, C. B. Hoelzel and J. M. Locke; *J. Am. Chem. Soc.*, 1962, **84**, 847. 215
62JA1621 W. E. Truce and J. J. Breiter; *J. Am. Chem. Soc.*, 1962, 1621. 127
62JA1650 H. C. Brown and G. Goldman; *J. Am. Chem. Soc.*, 1962, **84**, 1650. 964

62JA1868 A. J. Speziale and L. R. Smith; *J. Am. Chem. Soc.*, 1962, **84**, 1868. 1047
62JA2020 S. I. Miller, C. E. Orzech, C. A. Welch, G. R. Ziegler and J. I. Dickstein; *J. Am. Chem. Soc.*, 1962, **84**, 2020. 1062
62JA2785 J. K. Kochi; *J. Am. Chem. Soc.*, 1962, **84**, 2785. 97
62JA2922 A. U. Blackham and N. L. Eatough; *J. Am. Chem. Soc.*, 1962, **84**, 2922. 403
62JA2938 E. J. Corey and R. B. Mitra; *J. Am. Chem. Soc.*, 1962, **84**, 2938. 225
62JA3442 E. J. Griffiths and A. B. Burg; *J. Am. Chem. Soc.*, 1962, **84**, 3442. 452
62JA3605 J. J. Eisch; *J. Am. Chem. Soc.*, 1962, **84**, 3605. 597
62JA3820 J. J. Eisch; *J. Am. Chem. Soc.*, 1962, **84**, 3820. 1099
62JA3887 L. Zervas and I. Photaki; *J. Am. Chem. Soc.*, 1962, **84**, 3887. 273, 275
62JA4962 K. M. Ibne-Rasa; *J. Am. Chem. Soc.*, 1962, **84**, 4962. 780
62JCS470 D. H. R. Barton, R. E. O'Brien and S. Sternhell; *J. Chem. Soc.*, 1962, 470. 630
62JCS1330 B. J. Brisdon, G. W. A. Fowles and B. P. Osborne; *J. Chem. Soc.*, 1962, 1330. 421
62JCS1786 D. W. Aubrey, W. Gerrard and E. F. Mooney; *J. Chem. Soc.*, 1962, 1786. 418, 419
62JCS1967 D. H. R. Barton, M. V. George and M. Tomoeda; *J. Chem. Soc.*, 1962, 1967. 131
62JCS2089 G. Ayrey, D. Barnard and D. T. Woodbridge; *J. Chem. Soc.*, 1962, 2089. 287, 291
62JCS2172 B. Milligan and J. M. Swan; *J. Chem. Soc.*, 1962, 2172. 172, 220
62JCS2529 H. J. Emeleus, K. J. Packer and N. Welcman; *J. Chem. Soc.*, 1962, 2529. 499, 500
62JCS2953 P. J. Banyon and J. I. G. Cadogon; *J. Chem. Soc.*, 1962, 2953. 254
62JCS3818 R. P. Nielsen and H. H. Sisler; *J. Chem. Soc.*, 1962, 3818. 387
62JCS4088 D. W. Aubery, M. F. Lappert and M. K. Majumdar; *J. Chem. Soc.*, 1962, 4088. 533
62JCS4288 A. G. Davies; *J. Chem. Soc.*, 1962, 4288. 95
62JCS4361 A. J. Downs; *J. Chem. Soc.*, 1962, 4361. 169
62JCS4995 C. Burgess, D. Burn, J. W. Ducker, B. Ellis, P. Feather, A. K. Hiscock, A. P. Leftwick, J. S. Mills and V. Petrow; *J. Chem. Soc.*, 1962, 4995. 1016
62JGU2355 B. I. Ionin and A. A. Petrov; *J. Gen. Chem. USSR* (*Engl. Transl.*), 1962, **32**, 2355. 1071
62JINC275 L. Maier; *J. Inorg. Nucl. Chem.*, 1962, **24**, 275. 449
62JINC1073 L. Maier; *J. Inorg. Nucl. Chem.*, 1962, **24**, 1073. 854
62JOC1 D. O. Cowan and H. S. Mosher; *J. Org. Chem.*, 1962, **27**, 1. 67
62JOC93 B. C. Cossar, J. O. Fournier and D. L. Fields and D. D. Reynolds; *J. Org. Chem.*, 1962, 93. 121
62JOC282 N. J. Leonard and C. R. Johnson; *J. Org. Chem.*, 1962, **27**, 282. 149
62JOC1012 L. D. Quin and M. R. Dysart; *J. Org. Chem.*, 1962, **27**, 1012. 838
62JOC1020 J. K. Ruff; *J. Org. Chem.*, 1962, **27**, 1020. 419
62JOC1041 C. Djerassi and B. Tursch; *J. Org. Chem.*, 1962, **27**, 1041. 130
62JOC1075 T. Emery and J. B. Neilands; *J. Org. Chem.*, 1962, **27**, 1075. 782
62JOC1398 I. B. Douglass and D. A. Koop; *J. Org. Chem.*, 1962, **27**, 1398. 194
62JOC1430 B. S. Tyagi, B. B. Ghatge and S. C. Bhattacharyya; *J. Org. Chem.*, 1962, **27**, 1430. 660
62JOC1493 G. D. Cooper and H. L. Finkbeiner; *J. Org. Chem.*, 1962, **27**, 1493. 561
62JOC1828 A. P. Boisselle and N. A. Meinhardt; *J. Org. Chem.*, 1962, **27**, 1828. 1073
62JOC1933 P. T. Lansbury and V. A. Pattison; *J. Org. Chem.*, 1962, **27**, 1933. 478
62JOC2192 J. E. M. Evleth, L. D. Freedman and R. I. Wagner; *J. Org. Chem.*, 1962, **27**, 2192. 821
62JOC2198 G. Peters; *J. Org. Chem.*, 1962, **27**, 2198. 460
62JOC2573 C. Screttas and A. F. Isbell; *J. Org. Chem.*, 1962, **27**, 2573. 434
62JOC2667 G. Olah and S. J. Kuhn; *J. Org. Chem.*, 1962, **27**, 2667. 182
62JOC2853 J. S. Showell, J. R. Russell and D. Swern; *J. Org. Chem.*, 1962, **27**, 2853. 192
62JOC2976 C. Walling and A. Padwa; *J. Org. Chem.*, 1962, **27**, 2976. 95
62JOC3401 H. A. Silverwood and M. Orchin; *J. Org. Chem.*, 1962, **27**, 3401. 735
62JOC4397 E. R. Glazier; *J. Org. Chem.*, 1962, **27**, 4397.
62JOC4444 K. A. Pollart and H. J. Harwood; *J. Org. Chem.*, 1962, **27**, 4444. 854
62LA(652)7 G. Wilke and P. Heimbach; *Justus Liebigs Ann. Chem.*, 1962, **652**, 7. 96, 97
62LA(653)157 W. P. Neumann; *Justus Liebigs Ann. Chem.*, 1962, **653**, 157. 599
62LA(656)204 H. J. Kabbe; *Justus Liebigs Ann. Chem.*, 1962, **656**, 204. 478
62LA(657)86 B. Helferich and R. Hoffmann; *Justus Liebigs Ann. Chem.*, 1962, **657**, 86. 249, 365
62LA(659)49 H. J. Hellmann, J. Bader, H. Birkner and O. Schumacher; *Justus Liebigs Ann. Chem.*, 1962, **659**, 49. 440, 442, 446
62M230 F. Pass, E. Steininger and H. Zorn; *Monatsh. Chem.*, 1962, **93**, 230. 427
62M952 R. Riemschneider, L. Weil and K. Nolde; *Monatsh. Chem.*, 1962, **93**, 952. 1012, 1013
B-62MI 202-01 H. C. Brown; "Hydroboration," W. A. Benjamin, New York, 1962. 40
62MI 208-01 B. A. Arbuzov, L. A. Shapshinskaya and V. M. Erokina; *Izv. Akad. Nauk SSSR Otd. Khim. Nauk*, 1962, 2074 (*Chem. Abstr*., 1963, **58**, 11 396). 454
B-62MI 210-01 H. C. Brown; "Hydroboration," Benjamin, New York, 1962. 514
62MI 216-01 B. Krawieka, Z. Skyrzypczynski and J. Michalski; *Phosphorus*, 1962, **3**, 177. 857
62MI 223-01 P. Chine; *Chimica Ind.*, 1962, **44**, 1220. 1098, 1099
62OR(12)101 N. Kornblum; *Org. React.*, 1962, **12**, 101. 392
62RTC510 L. Brandsma and J. F. Arens; *Recl. Trav. Chim. Pays-Bas*, 1962, **81**, 510. 1030, 1031
62RTC533 J. R. Nooi, J. F. Arens; *Recl. Trav. Chim. Pays-Bas*, 1962, **81**, 533. 1032
62RTC583 L. Brandsma, H. E. Wijers and J. F. Arens; *Recl. Trav. Chim. Pays-Bas*, 1962, **81**, 583. 1032
62RTC993 W. Voskuil and J. F. Arens; *Recl. Trav. Chim. Pays-Bas*, 1962, **81**, 993. 1061
62RZC1781 J. Michalski and Z. Tulimowski; *Rocz. Chem.*, 1962, **36**, 1781 (*Chem. Abstr.*, 1963, **59**, 10 109). 461
62T1221 S. A. Buckler and M. Epstein; *Tetrahedron*, 1962, **18**, 1211, 1221. 440, 450
62T1231 M. Epstein and S. A. Buckler; *Tetrahedron*, 1962, **18**, 1231. 440
62T1527 H. Beecken and F. Korte; *Tetrahedron*, 1962, **18**, 1527. 245

62TL93 K. K. Andersen; *Tetrahedron Lett.*, 1962, 93. 151

62TL983 H. Hart and J. S. Fleming; *Tetrahedron Lett.*, 1962, 983. 1006

62USP3031509 M. D. Marshall and H. J. Harwood; *US Pat.* 3 031 509 (1962) (*Chem. Abstr.*, 1962, **57**, 8618). 431

62ZAAC(317)54 H. Hartmann, G. Habenicht and W. Reiss; *Z. Anorg. Allg. Chem.*, 1962, **317**, 54. 1082, 1083

62ZC163 K. Issleib; *Z. Chem.*, 1962, **2**, 163. 826

62ZN(B)807 F. Weygand and G. Zumach; *Z. Naturforsch., Teil B*, 1962, **17**, 807. 270

62ZOB553 K. A. Petrov, V. A. Parshin and M. B. Luzanova; *Zh. Obshch. Khim.*, 1962, **32**, 553 (*Chem. Abstr.*, 1963, **58**, 5714). 442

62ZOB1349 A. A. Petrov, S. V. Zavgorodnii and A. Kormer; *Zh. Obshch. Khim.*, 1962, **32**, 1349, (*Chem. Abstr.*, 1963, **58**, 1481). 1080

62ZOB1974 K. A. Petrov, V. P. Evdakov, K. A. Bilevich and Yu. S. Kosarev; *Zh. Obshch. Khim.*, 1962, **32**, 1974. 451

62ZOB2848 G. Kamai and G. M. Rusetskaya; *Zh. Obshch. Khim.*, 1962, **32**, 2848 (*Chem. Abstr.*, 1963, **58**, 7965). 436

62ZOB3062 K. A. Petrov, V. P. Evdakov, L. I. Mizrakh and V. P. Romodin; *Zh. Obshch. Khim.*, 1962, **32**, 3062. 461

62ZOB3070 K. A. Petrov, V. P. Evdakov, G. I. Abramtsev and A. K. Strautman; *Zh. Obshch. Khim.*, 1962, **32**, 3070 (*Chem. Abstr.*, 1963, **58**, 11 395h). 411

62ZOB3351 M. I. Kabachnik, C. Jung-yu and E. N. Tsvetkov; *Zh. Obshch. Khim.*, 1962, **32**, 3351. 838, 839, 851

62ZOB3579 T. A. Mastryukova, A. E. Shipov and M. I. Kabachnik; *Zh. Obshch. Khim.*, 1962, **32**, 3579 (*Chem. Abstr.*, 1963, **58**, 11 394). 462

62ZOB4017 K. A. Petrov, V. A. Parshina, B. A. Orlov and G. M. Tsypina; *Zh. Obshch. Khim.*, 1962, **32**, 4017. 460, 851

63AC(R)1869 M. Fioretti and M. Portell; *Ann. Chim. (Rome)*, 1963, **53**, 1869. 498

63ACS273 B. Magnusson; *Acta Chem. Scand.*, 1963, **17**, 273. 130

63ACS1806 O. Foss, S. Husebye and K. Maroy; *Acta Chem. Scand.*, 1963, **17**, 1806. 289

63ACS1875 K. A. Jensen and P. H. Nielson; *Acta Chem. Scand.*, 1963, **17**, 1875. 890

63ACS2141 N. C. Hansen; *Acta. Chem. Scand.*, 1963, **17**, 2141. 368

63AG918 P. Binger; *Angew. Chem.*, 1963, 918. 1099

63AG1033 J. M. Tedder and E. Robson; *Angew. Chem.*, 1963, **75**, 1033. 1056

63AG(E)47 U. Wannagat and H. Kuckertz; *Angew. Chem., Int. Ed. Engl.*, 1963, **2**, 47. 697

63AG(E)157 E. Schmitz, R. Ohme and S. Schramm; *Angew. Chem., Int. Ed. Engl.*, 1963, **2**, 157. 403

63AG(E)218 H.-J. Vetter, H. Strametz and H. Nöth; *Angew. Chem., Int. Ed. Engl.*, 1963, **2**, 218. 408, 415

63AG(E)328 I. Ruidisch and M. Schmidt; *Angew. Chem., Int. Ed. Engl.*, 1963, **2**, 328. 264

63AG(E)370 R. Mayer, G. Hiller, M. Nitzschke and J. Jentzsch; *Angew. Chem., Int. Ed. Engl.*, 1963, **2**, 370. 130

63AG(E)617 C. R. Krüger and E. G. Rochow; *Angew. Chem., Int. Ed. Engl.*, 1963, **2**, 617. 696

63AJC636 T. N. Bell, B. J. Pullman and B. O. West; *Aust. J. Chem.*, 1963, **16**, 636. 486

63BRP917921 H. Coates and P. A. T. Hoye; *Br. Pat.* 917 921 (1963) (*Chem. Abstr.*, 1963, **58**, 13 795f). 119

63BRP921463 E. Hofman; *Br. Pat.* 921 463 (1963) (*Chem. Abstr.*, 1963, **59**, 1682). 429

63BSF470 J. M. Geneste and A. Kergomard; *Bull. Soc. Chim. Fr.*, 1963, 470. 95

63BSF1876 J. Normant; *Bull. Soc. Chim. Fr.*, 1963, 1876. 1031

63CB247 N. Petragnani; *Chem. Ber.*, 1963, **96**, 247. 727

63CB1019 M. Wieber and M. Schmidt; *Chem. Ber.*, 1963, **96**, 1019. 262

63CB1131 F. Feher; *Chem. Ber.*, 1963, **96**, 1131. 361

63CB1387 H. Gross and J. Gloede; *Chem. Ber.*, 1963, **96**, 1387. 617

63CB1462 H. Teichmann and G. Hilgetag; *Chem. Ber.*, 1963, **96**, 1462. 1099

63CB1561 M. Wieber and M. Schmidt; *Chem. Ber.*, 1963, **96**, 1561. 262

63CB2186 K. Issleib, K. Krech and K. Gruber; *Chem. Ber.*, 1963, **96**, 2186. 433

63CB2822 M. Wieber and M. Schmidt; *Chem. Ber.*, 1963, **96**, 2822. 262

63CB3044 A. Rieche and P. Dietrich; *Chem. Ber.*, 1963, **96**, 3044. 686

63CB3050 E. Müller and E. W. Schmidt; *Chem. Ber.*, 1963, **96**, 3050. 170

63CB3068 H. Feichtinger; *Chem. Ber.*, 1963, **96**, 3068. 175

63CB3091 E. Fluck and F. L. Goldmann; *Chem. Ber.*, 1963, **96**, 3091. 864

63CB3184 E. Steininger; *Chem. Ber.*, 1963, **96**, 3184. 431, 453

63CB3289 L. Birkofer, A. Ritter and H. Galler; *Chem. Ber.*, 1963, **96**, 3289. 541

63CI(L)609 R. D. Youssefyeh and Y. Mazur; *Chem. Ind. (London)*, 1963, 609. 26, 33

63CI(L)1877 G. W. Kirby; *Chem. Ind. (London)*, 1963, 1877. 105

63CI(L)1954 E. S. Huyser and D. N. DeMott; *Chem. Ind. (London)*, 1963, 1954. 27

63CJC322 W. R. Cullen; *Can. J. Chem.*, 1963, **41**, 322. 505, 506

63CJC2424 W. R. Cullen; *Can. J. Chem.*, 1963, **41**, 2424. 499

63CJC2671 B. E. Leggetter and R. K. Brown; *Can. J. Chem.*, 1963, **41**, 2671. 128

63CJC2977 A. Ottolenghi, M. Fridkin and A. Zilkha; *Can. J. Chem.*, 1963, **41**, 2977. 910, 931

63CRV557 H. F. Van Woerden; *Chem. Rev.*, 1963, **63**, 557. 100, 101

63HCA1121 E. Weiss and W. Büchner; *Helv. Chim. Acta*, 1963, **46**, 1121 and references therein. 1022, 1023

63HCA2009 A. O. Ilvespää and A. Marxer; *Helv. Chim. Acta*, 1963, **46**, 2009. 691, 692

63HCA2444 J. Borgulya, H.-J. Hansen, R. Barner and H. Schmid; *Helv. Chim. Acta*, 1963, **46**, 2444. 656

63HCA2464 E. Cherbuliez, G. Weber and J. Rabinowith; *Helv. Chim. Acta*, 1963, **46**, 2464. 451

63HCA2667 L. Maier; *Helv. Chim. Acta*, 1963, **46**, 2667. 451

63HOU(6/2)325 A. Berthmann and H. Ratz; *Methoden Org. Chem. (Houben-Weyl)*, **6/2**, 325. 104

63HOU(6/3)1 H. Meerwein; *Methoden Org. Chem. (Houben-Weyl)*, 1965, **6/3**, 1. 89

63HOU(12/1)1	K. Sasse, *Methoden Org. Chem.* (*Houben-Weyl*), 1963, **12/1**, 1.	409
63IZV97	N. V. Komarov, Yu. V. Maroshin, A. D. Lebedeva and L. I. Astafeva; *Izv. Akad. Nauk SSSR, Ser. Khim.*, 1963, 97.	1090
63IZV1146	L. I. Zakharkin and V. V. Gavrilenko; *Izv. Akad. Nauk SSSR, Ser. Khim.*, 1963, 1146 (*Chem. Abstr.*, 1963, **59**, 8772).	1099
63IZV1227	M. I. Kabachnik and E. N. Tsvetkov; *Izv. Akad. Nauk SSSR, Ser. Khim.*, 1963, **7**, 1227.	450
63IZV2213	A. M. Sladkov, L. Y. Ukhin and V. V. Korshak; *Izv. Akad. Nauk SSSR, Ser. Khim.*, 1963, 2213.	1018, 1019
63JA79	M. Grayson; *J. Am. Chem. Soc.*, 1963, **85**, 79.	442
63JA207	G. Stork, A. Brizzolara, H. Landesman, J. Szmuszkovicz and R. Terrel; *J. Am. Chem. Soc.*, 1963, **85**, 207.	738, 758
63JA749	J. F. Harris and F. W. Stacey; *J. Am. Chem. Soc.*, 1963, **85**, 749.	166
63JA1200	W. Lwowski, T. J. Maricich and T. W. Mattingly; *J. Am. Chem. Soc.*, 1963, **85**, 1200.	298
63JA1618	R. G. Harvey, H. I. J. Jacobson and E. V. Jensen; *J. Am. Chem. Soc.*, 1963, **85**, 1618.	855
63JA1648	G. R. Ziegler, C. A. Welch, C. E. Orzech, S. Kikkawa and S. I. Miller; *J. Am. Chem. Soc.*, 1963, **85**, 1648.	1016, 1033
63JA2063	H. C. Brown and A. W. Moerikofer; *J. Am. Chem. Soc.*, 1963, **85**, 2063.	518
63JA2066	G. Zweifel and H. C. Brown; *J. Am. Chem. Soc.*, 1963, **85**, 2066.	518
63JA2072	G. Zweifel, N. R. Ayyangar and H. C. Brown; *J. Am. Chem. Soc.*, 1963, **85**, 2072.	516, 517
63JA2144	L. A. Carpino; *J. Am. Chem. Soc.*, 1963, **85**, 2144.	377, 379
63JA2462	W. A. Henderson, M. Epstein and F. S. Seichter; *J. Am. Chem. Soc.*, 1963, **85**, 2462.	471
63JA2683	C. E. Griffin, N. F. Hepfinger and B. L. Shapiro; *J. Am. Chem. Soc.*, 1963, **85**, 2683.	783
63JA3897	J. E. Mulvaney, Z. G. Gardlund, and S. L. Gardlund; *J. Am. Chem. Soc.*, 1963, **85**, 3897.	972
63JA3992	A. R. Donmaux, J. E. McKeon and D. J. Trecker; *J. Am. Chem. Soc.*, 1993, **91**, 3992.	1566
63JCE194	J. A. West; *J. Chem. Ed.*, 1963, **40**, 194.	738, 751
63JCS1268	H. J. Emeleus and N. Welcman; *J. Chem. Soc.*, 1963, 1268.	292
63JCS1462	K. C. Brannock, R. D. Burpitt and J. G. Thweatt; *J. Chem. Soc.*, 1963, **28**, 1462.	751
63JCS1582	W. Gerrard, M. Howarth, E. F. Mooney and D. E. Pratt; *J. Chem. Soc.*, 1963, 1582.	520
63JCS2101	D. C. Bradley and D. A. W. Hill; *J. Chem. Soc.*, 1963, 2101.	108
63JCS2295	B. Eglinton and W. McCrae; *J. Chem. Soc.*, 1963, 2295.	1018, 1097
63JCS2636	S. N. Nabi, S. Ahmad and S. Ahmad; *J. Chem. Soc.*, 1963, 2636.	172
63JCS3351	B. C. Saunders and P. Simpson; *J. Chem. Soc.*, 1963, 3351.	1071, 1072
63JCS3692	P. Robson, T. A. Smith, R. Stephens and J. C. Talow; *J. Chem. Soc.*, 1963, 3692.	194
63JCS4864	H. Hogeveen and F. Montanari; *J. Chem. Soc.*, 1963, 4864.	147
63JCS5741	C. J. M. Stirling; *J. Chem. Soc.*, 1963, 5741.	721
63JCS(C)798	R. A. Zingaro, R. E. McGlothin and R. M. Hedges; *J. Chem. Soc. (C)*, 1963, **59**, 798.	500, 502
63JGU2791	B. I. Ionin and A. A. Petrov; *J. Gen. Chem. USSR* (*Engl. Transl.*), 1963, **33**, 2791.	1071
63JGU2831	G. Kamai and R. G. Miftakhova; *J. Gen. Chem. USSR* (*Engl. Transl.*), 1963, **33**, 2831.	496, 497
63JOC237	C. Tamborski, F. E. Ford and E. J. Soloski; *J. Org. Chem.*, 1963, **28**, 237.	600
63JOC345	B. Miller; *J. Org. Chem.*, 1963, **28**, 345.	660
63JOC481	T. Mukaiyama, T. Hata and O. Mitsunobu; *J. Org. Chem.*, 1963, **28**, 481.	107
63JOC593	W. E. Truce and F. E. Roberts; *J. Org. Chem.*, 1963, **28**, 593.	188, 271, 719
63JOC783	K. Tsuda, E. Ohki and S. Nozoe; *J. Org. Chem.*, 1963, **28**, 783.	660
63JOC870	B. Boduszek and J. L. Kice; *J. Org. Chem.*, 1963, **28**, 870.	187
63JOC957	J. L. Kice; *J. Org. Chem.*, 1963, **28**, 957.	122
63JOC1140	H. S. Schultz, H. B. Fryermuth and S. R. Buc; *J. Org. Chem.*, 1963, **28**, 1140.	156
63JOC1232	H. J. Shine and J. T. Chamness; *J. Org. Chem.*, 1963, **28**, 1232.	772
63JOC1244	G. V. D. Tiers; *J. Org. Chem.*, 1963, **28**, 1244.	181, 182
63JOC1901	N. Kharasch and R. B. Langford; *J. Org. Chem.*, 1963, **28**, 1901.	169
63JOC1945	E. E. Gilbert; *J. Org. Chem.*, 1963, **28**, 1945.	157
63JOC1952	K. Griesbaum, A. A. Oswald, E. R. Quiram and W. Naegele; *J. Org. Chem.*, 1963, **28**, 1952.	137
63JOC2171	D. S. Matteson and R. W. H. Mah; *J. Org. Chem.*, 1963, **28**, 2171.	907
63JOC2485	C. F. Bennett, D. W. Goheen and N. S. MacGregor; *J. Org. Chem.*, 1963, **28**, 2485.	180
63JOC2508	J. P. Freeman; *J. Org. Chem.*, 1963, **28**, 2508.	404, 405
63JOC2766	A. C. Poshkus, J. E. Herweh and F. A. Magnotta; *J. Org. Chem.*, 1963, **28**, 2766.	717
63JOC2775	O. F. Beumel, Jr. and R. F. Harris; *J. Org. Chem.*, 1963, **28**, 2775.	420
63JOC2807	B. S. Farah and E. E. Gilbert; *J. Org. Chem.*, 1963, **28**, 2807.	688
63JOC2992	D. Horton, M. L. Wolfrom and H. G. Garg; *J. Org. Chem.*, 1963, **28**, 2992.	171
63JOC3021	N. J. Leonard and J. V. Paukstelis; *J. Org. Chem.*, 1963, **28**, 3021.	1008
63JOC3246	C. N. Yiannios and J. P. Karabinos; *J. Org. Chem.*, 1963, **28**, 3246.	214
63JOC3315	L. D. Quin and R. E. Montgomery; *J. Org. Chem.*, 1963, **28**, 3315.	833, 834, 840, 841
63JOC3430	H. J. Harwood and K. A. Pollart; *J. Org. Chem.*, 1963, **28**, 3430.	461
63JOC3537	A. D. Bliss, W. K. Cline, C. E. Hamilton and O. J. Sweeting; *J. Org. Chem.*, 1963, **28**, 3537.	176
63JOC3608	B. Milligan, B. Saville and J. M. Swan; *J. Org. Chem.*, 1963, **28**, 3608.	122
63LA(665)1	H. Goety, F. Nerdel and K.-H. Wiechel; *Justus Liebigs Ann. Chem.*, 1963, **665**, 1.	824
63LA(665)101	G. Opitz and A. Griesinger; *Justus Liebigs Ann. Chem.*, 1963, **665**, 101.	757
B-63MI 202-01	D. M. Brown; in "Advances in Organic Chemistry," Interscience, New York, 1963, vol. 3, p. 75.	105
63MI 203-01	D. Cavallini, B. Mondovi and C. Demarco; *Biochem. Prep.*, 1963, **10**, 72.	211
63MI 204-01	S. H. Chu, W. H. H. Gunther and H. G. Mautner; *Biochem. Prep.*, 1963, **10**, 153.	278
63MI 208-01	L. Maier; *Prog. Inorg. Chem.*, 1963, **5**, 27.	428
B-63MI 215-01	J. Szmuszkovicz; in "Advances in Organic Chemistry," ed. R. A. Raphael, E. C. Taylor and H. Wynberg, Interscience, New York, 1963, vol. 4, pp. 1–113.	738, 751

63N302	M. Schmidt and V. Potschka; *Naturwissenschaften*, 1963, **50**, 302.	1032
63N403	H. Hartmann and A. Meixner; *Naturwissenschaften*, 1963, **50**, 403.	1066
63OR(13)1	G. Zweifel and H. C. Brown; *Org. React.*, 1963, **13**, 1.	40
63OR(13)150	F. W. Stacey and J. F. Harri, Jr.; *Org. React.*, 1963, **13**, 150.	4, 29, 392, 623
63OS(43)12	K. G. Rutherford and W. Redmond; *Org. Synth.*, 1963, **43**, 12.	614
63OSC(4)795	A. N. Bourns, H. W. Embleton and M. K. Hansuld; *Organic Synthesis Coll. Vol.*, 1963, **4**, 795.	320
63OSC(4)66	C. G. Overberger, P.-T. Huang and M. B. Berenbaum; *Org. Synth., Coll. Vol.*, 1963, **4**, 66.	403
63OSC(4)68	R. M. Wasburn, E. Levens, C. F. Albright and F. A. Billig; *Org. Synth., Coll. Vol.*, 1963, **4**, 68.	906
63OSC(4)108	F. L. Greenwood, M. D. Kellert and J. Sedlak; *Org. Synth., Coll. Vol.*, 1963, **4**, 108.	27
63OSC(4)125	H. M. Teeter and E. W. Bell; *Org. Synth., Coll. Vol.*, 1963, **4**, 125.	94
63OSC(4)162	E. W. Warnhoff, D. G. Martin and W. S. Johnson; *Org. Synth., Coll. Vol.*, 1963, **4**, 162.	21
63OSC(4)169	E. L. Eliel, M. T. Fisk and T. Prosser; *Org. Synth., Coll. Vol.*, 1963, **4**, 169.	25
63OSC(4)195	L. F. Fieser; *Org. Synth., Coll. Vol.*, 1963, **4**, 195.	30
63OSC(4)274	C. G. Overberger, P.-T. Huang and M. B. Berenbaum; *Org. Synth., Coll. Vol.*, 1963, **4**, 274.	378
63OSC(4)321	H. Stone and H. Shechter; *Org. Synth., Coll. Vol.*, 1963, **4**, 321.	36
63OSC(4)323	H. Stone and H. Shechter; *Org. Synth., Coll. Vol.*, 1963, **4**, 323.	36
63OSC(4)368	H. Feuer and G. Leston; *Org. Syn., Coll. Vol.*, 1963, **4**, 368.	393
63OSC(4)398	C. F. Allen and M. J. Kalm; *Org. Synth., Coll. Vol.*, 1963, **4**, 398.	29
63OSC(4)401	J. Speziale; *Org. Synth , Coll. Vol.*, 1963. **4**, 401.	120
63OSC(4)454	N. Kornblum and R. K. Blackwood; *Org. Synth., Coll. Vol.*, 1963, **4**, 454.	393
63OSC(4)491	H. Kofod; *Org. Synth. Coll. Vol.*, 1963, **4**, 491.	120
63OSC(4)525	A. I. Vogel, J. Leicester and W. A. T. Macey; *Org. Synth., Coll. Vol.*, 1963, **4**, 525.	19
63OSC(4)529	A. O. Snoddy; *Org. Synth., Coll. Vol.*, 1963, **4**, 529.	204
63OSC(4)543	H. Stone and H. Schechter; *Org. Synth., Coll. Vol.*, 1963, **4**, 543.	34
63OSC(4)608	C. F. Allen and M. J. Kalm; *Org. Synth., Coll. Vol.*, 1963, **4**, 608.	29
63OSC(4)612	A. C. Cope; *Org. Synth., Coll. Vol.*, 1963, **4**, 612.	347
63OSC(4)638	H. M. Foster and H. R. Snyder; *Org. Synth., Coll. Vol.*, 1963, **4**, 638.	120
63OSC(4)652	R. B. Moffett; *Org. Synth., Coll. Vol.*, 1963, **4**, 652.	395
63OSC(4)674	L. Field and R. D. Clark; *Org. Synth., Coll. Vol.*, 1963, **4**, 674.	156
63OSC(4)724	N. Kornblum and H. Ungnade; *Org. Synth., Coll. Vol.*, 1963, **4**, 724.	393
63OSC(4)753	H. L. Herzog; *Org. Synth., Coll. Vol.*, 1963, **4**, 753.	9
63OSC(4)784	B. Bruchner and J. L. B. Lockhart; *Org. Synth., Coll. Vol.*, 1963, **4**, 784.	832, 833
63OSC(4)819	P. A. S. Smith; *Org. Synth., Coll. Vol.*, 1963, **4**, 819.	757
63OSC(4)844	P. E. Fanta; *Org. Synth., Coll. Vol.*, 1963, **4**, 844.	398
63OSC(4)851	N. A. Khan, F. E. Deatherage and J. B. Brown; *Org. Synth., Coll. Vol.*, 1963, **4**, 851.	30
63OSC(4)8623	J. K. Weil, R. G. Bistline and A. J. Stirton; *Org. Synth., Coll. Vol.*, 1963, **4**, 862.	190
63OSC(4)921	E. Campaigne and B. F. Tullar; *Org. Synth., Coll. Vol.*, 1963, **4**, 921.	27
63OSC(4)937	F. Kurzer; *Org. Synth., Coll. Vol.*, 1963, **4**, 937.	688
63OSC(4)969	N. A. Khan; *Org. Synth., Coll. Vol.*, 1963, **4**, 969.	30
63OSC(4)984	E. F. M. Stephenson; *Org. Synth., Coll. Vol.*, 1963, **4**, 984.	27
63RTC68	L. Brandsma and H. Wijers; *Rec. Trav. Chim. Pays Bas*, 1963, **82**, 68.	278, 279
63RTC302	W. Voskuil and J. F. Arens; *Rec. Trav. Chim. Pays-Bas*, 1963, **82**, 302.	434
63RTC375	H. Hogeveen and W. Drenth; *Recl. Trav. Chim. Pays-Bas*, 1963, **82**, 375.	1032
63RTC773	H. Berger; *Recl. Trav. Chim. Pays-Bas*, 1963, **82**, 773.	192
63RTC1119	L. Brandsma; *Recl. Trav. Chim. Pays-Bas*, 1963, **82**, 1119.	1030, 1031
63T1563	U. Hasserodt, K. Hunger and F. Korte; *Tetrahedron*, 1963, **19**, 1563.	253, 462
63TL1131	T. J. Wallace and A. Schrielsheim; *Tetrahedron Lett.*, 1963, 1131.	192, 720
63USP3081350	S. Andreades; *US Pat.* 3 081 350 (1963) (*Chem. Abstr.*, 1963, **59**, 5024).	194
63USP3116334	S. A. Buckler and M. Epstein; *US Pat.* 3 116 334 (1963).	450
63ZAAC(319)230	H. J. Berthold and G. Groh; *Z. Anorg. Allg. Chem.*, 1963, **319**, 230.	566
63ZAAC(321)120	E. Todt and R. Dötzer; *Z. Anorg. Allg. Chem.*, 1963, **321**, 120.	597
63ZOB1246	K. A. Petrov, V. P. Evdakov and L. I. Mizrakh; *Zh. Obshch. Khim.*, 1963, **33**, 1246.	451
63ZOB1919	A. B. Bruker, M. K. Baranaev, E. I. Grinshtein, R. I. Novoselova, V. V. Prokhorova and L. Z. Soborovskii; *Zh. Obshch. Khim.*, 1963, **33**, 1919 (*Chem. Abstr.*, 1963, **59**, 11 207).	440
63ZOB2631	E. V. Kuznetsov, T. V. Sorokina and R. K. Valetdinov; *Zh. Obshch. Khim.*, 1963, **33**, 2631 (*Chem. Abstr.*, 1964 **60**, 542).	442
63ZOB2863	B. E. Ionin and A. A. Petrov; *Zh. Obshch. Khim.*, 1963, **33**, 2863.	849
63ZOB3050	I. S. Protopopov and M. Y. Kraft; *Zh. Obshch. Khim.*, 1963, **33**, 3050.	825, 829
63ZOB3774	G. M. Bogolyubov and A. A. Petrov; *Zh. Obshch. Khim.*, 1963, **33**, 3774.	820, 821
64ACS815	A. Biezais and G. Bergson; *Acta Chem. Scand.*, 1964, **18**, 815.	238
64ACS954	N. H. Christensen; *Acta Chem. Scand.*, 1964, **18**, 954.	157
64ACS1629	E. Kloster-Jensen; *Acta Chem. Scand.*, 1964, **18**, 1629.	1016
64ACS2199	C. T. Pedersen; *Acta Chem. Scand.*, 1964, **18**, 2199.	388
64AG(E)66	W. Stamm and A. Breindel; *Angew. Chem., Int. Ed. Engl.*, 1964, **3**, 66.	485
64AG(E)326	M. Schlosser; *Angew. Chem., Int. Ed. Engl.*, 1964, **3**, 326.	556
64AG(E)333	E. Schmitz; *Angew. Chem., Int. Ed. Engl.*, 1964, **3**, 333.	374
64AG(E)506	H. G. Viehe and M. Reinstein; *Angew. Chem., Int. Ed. Engl.*, 1964, **3**, 506.	1041, 1047
64AG(E)560	R. Gompper; *Angew. Chem., Int. Ed. Engl.*, 1964, **3**, 560.	674
64AG(E)640	H. Lehmkuhl; *Angew. Chem., Int. Ed. Engl.*, 1964, **3**, 640.	595
64AG(E)702	O. J. Scherer and M. Schmidt; *Angew. Chem., Int. Ed. Engl.*, 1964, **3**, 702.	415

64AG(E)747	H. G. Viehe, R. Merenyi, J. F. M. Oth and P. Valange; *Angew. Chem., Int. Ed. Engl.*, 1964, **3**, 747.	1014
64AG(E)755	H. G. Viehe, R. Merenyi, J. F. M. Oth, J. R. Senders and P. Valange; *Angew. Chem., Int. Ed. Engl.*, 1964, **3**, 755.	1014
64AG(E)804	K. Bott; *Angew. Chem., Int. Ed. Engl.*, 1964, **3**, 804.	799
64ANC1768	E. Sekido, Q. Fernando and H. Freiser; *Anal. Chem.*, 1964, **36**, 1768.	278
64BBR(17)146	K. E. Pfitzner and J. G. Moffatt; *Biochem. Biophys. Res. Commun.*, 1964, **17**, 146.	105
64BSB507	M. Renson and J.-L. Piette; *Bull. Soc. Chim. Belg.*, 1964, **73**, 507.	732
64BSF871	J. Ficini and C. Barbara; *Bull. Soc. Chim. Fr.*, 1964, 871.	1047
64BSF952	J. Huet; *Bull. Soc. Chim. Fr.*, 1964, 952.	312
64CB49	E. Schmitz and K. Schinkowski; *Chem. Ber.*, 1964, **97**, 49.	378
64CB449	H. Schmidbaur, H.-S. Arnold and E. Beinhofer; *Chem. Ber.*, 1964, **97**, 449.	492, 494
64CB741	G. Wittig and E. Kochendoerfer; *Chem. Ber.*, 1964, **97**, 741.	830
64CB789	G. Wittig and D. Hellwinkel; *Chem. Ber.*, 1964, **97**, 789.	873, 881
64CB1155	K. Brodersen, G. Opitz, D. Breitinger and D. Menzel; *Chem. Ber.*, 1964, **97**, 1155.	817
64CB1940	E. Hecker; *Chem. Ber.*, 1964, **97**, 1940.	660
64CB1988	H. Fritzsche, U. Hasserodt and F. Korte; *Chem. Ber.*, 1964, **97**, 1988.	821, 827, 828
64CB2395	H. Schumann, H. Köpe and M. Schmidt; *Chem. Ber.*, 1964, **97**, 2395.	868
64CB2614	E. Müller and E. W. Schmidt; *Chem. Ber.*, 1964, **97**, 2614.	169, 170
64CB2903	F. Puschel and C. Kaiser; *Chem. Ber.*, 1964, **97**, 2903.	191
64CB3317	R. Nast and C. Richers; *Chem. Ber.*, 1964, **97**, 3317.	1098
64CB3430	K. Issleib and G. Harzfeld; *Chem, Chem. Ber.*, 1964, **97**, 3430.	440
64CI(L)1582	G. F. Freeguard and L. H. Long; *Chem. Ind.* (*London*), 1964, 1582.	36
64CJC212	J. Champagne, H. Favre, D. Vocelle and I. Zbikowski; *Can. J. Chem.*, 1964, **42**, 212.	675
64CJC2710	G. Sumrell, B. M. Wyman, R. G. Howell and M. C. Harvey; *Can. J. Chem.*, 1964, **42**, 2710.	35
64CJC2903	J. Brault and J. M. Lalancette; *Can. J. Chem.*, 1964, **42**, 2903.	533
64CPB326	H. Ito; *Chem. Pharm. Bull.*, 1964, **12**, 326 (*Chem. Abstr.*, 1964, **60**, 15 703).	782
64CR(259)3737	C. Belinski, G. Français, C. Horny and F. X. L. Keraly; *C. R. Hebd. Seances Acad. Sci.*, 1964, **259**, 3737.	387
64CRV473	J. Hamer and A. Macaluso; *Chem. Rev.*, 1964, **64**, 473.	765
64DOK(157)139	A. J. Titov and A. N. Barishnikova; *Dokl. Akad. Nauk SSSR*, 1964, **157**, 139 (*Chem. Abstr.*, 1964, **61**, 9396).	188
64DOK(157)371	N. A. Chadaeva, G. Kamai and K. A. Mamakov; *Dokl. Akad. Nauk SSSR Ser. Khim.*, 1964, **157**, 371 (*Chem. Abstr.*, 1964, **61**, 9524h).	889
64HCA27	L. Maier; *Helv. Chim. Acta*, 1964, **47**, 27.	461
64HCA120	L. Maier; *Helv. Chim. Acta*, 1964, **47**, 120.	854
64HCA194	C. A. Grob and G. Cseh; *Helv. Chim. Acta*, 1964, **47**, 194.	623
64HCA1448	L. Maier; *Helv. Chim. Acta*, 1964, **47**, 1448.	463
64HCA1647	E. Cherbuliez, F. Hunkeler, G. Weber and J. Rabinowitz; *Helv. Chim. Acta*, 1964, **47**, 1647.	104
64HCA2137	L. Maier; *Helv. Chim. Acta*, 1964, **47**, 2137.	441
64HOU(12/1)338	K. Sasse; *Methoden Org. Chem.* (*Houben-Weyl*), 1964, **12/1**, 338.	847, 848
64HOU(12/2)97	G. Schrader; *Methoden Org. Chem.* (*Houben-Weyl*), 1964, **12/2**, 97.	409
64HOU(12/2)104	G. Schrader; *Methoden Org. Chem.* (*Houben-Weyl*), 1964, **12/2**, 104.	410
64HOU(12/2)393	G. Schrader; *Methoden Org. Chem.* (*Houben-Weyl*), 1964, **12/2**, 393.	412
64IC410	R. Schmutzler; *Inorg. Chem.*, 1964, **3**, 410.	836
64IC428	D. L. Herring and C. M. Douglas; *Inorg. Chem.*, 1964, **3**, 428.	836, 837
64IC580	R. A. Zingaro and A. Merijanian; *Inorg. Chem.*, 1964, **3**, 580.	483, 502, 890
64IC609	K. Moedritzer; *Inorg. Chem.*, 1964, **3**, 609.	416
64IJ121	W. Oppenheim and L. M. Shorr; *Isr. J. Chem.*, 1964, **2**, 121.	1017
64IZV392	A. M. Sladkov and L. Y. Ukhin; *Izv. Akad. Nauk SSSR, Ser. Khim.*, 1964, 392.	1018, 1019
64IZV644	I. L. Knunyants, M. G. Linkova and N. D. Kuleshova; *Izv. Akad. Nauk SSSR, Ser. Khim.*, 1964, 644 (*Chem. Abstr.*, 1961, 2966).	214
64IZV1197	A. N. Nesmeyanov, A. E. Borisov and N. V. Norikova; *Izv. Akad. Nauk SSSR, Ser. Khim.*, 1964, 1197 (*Chem. Abstr.*, 1964, **61**, 12 032a).	881
64IZV1202	A. N. Nesmeyanov, A. E. Borisov and N. V. Norikova; *Izv. Akad. Nauk SSSR, Ser. Khim.*, 1964, 1202 (*Chem. Abstr.*, 1964, **61**, 12 032e).	881
64IZV1989	L. L. Ivanov, V. V. Gavrilenko and L. I. Zakharkin; *Izv. Akad. Nauk SSSR, Ser. Khim*, 1964, 1989 (*Chem. Abstr.*, 1965, **62**, 7660).	1099
64JA209	G. G. Long, G. O. Doak and L. D. Freedman; *J. Am. Chem. Soc.*, 1964, **86**, 209.	492
64JA393	G. Zweifel and H. C. Brown; *J. Am. Chem. Soc.*, 1964, **86**, 393.	42, 57
64JA397	H. C. Brown, N. R. Ayyangar and G. Zweifel; *J. Am. Chem. Soc.*, 1964, **86**, 397.	42, 57, 519
64JA427	P. Kovacic and K. E. Davis; *J. Am. Chem. Soc.*, 1964, **86**, 427.	608
64JA742	H. L. Dryden, Jr., G. M. Webber and J. J. Wieczorek; *J. Am. Chem. Soc.*, 1964, **86**, 742.	660
64JA883	D. Seyferth and L. G. Vaughan; *J. Am. Chem. Soc.*, 1964, **86**, 883.	909, 910, 911
64JA964	G. A. Wiley, R. L. Hershkowitz, B. M. Rein and B. C. Chung; *J. Am. Chem. Soc.*, 1964, **86**, 964.	7, 627
64JA1030	W. R. Cullen and R. G. Hayter; *J. Am. Chem. Soc.*, 1964, **86**, 1030.	512
64JA1139	P. A. S. Smith, H. R. Alul and R. L. Baumgarten; *J. Am. Chem. Soc.*, 1964, **86**, 1139.	764
64JA1410	W. P. Jencks and M. Gilchrist; *J. Am. Chem. Soc.*, 1964, **86**, 1410.	807
64JA1459	S. W. Tobey and R. West; *J. Am. Chem. Soc.*, 1964, **86**, 1459.	1006
64JA1630	T. M. Jacob and H. G. Khorana; *J. Am. Chem. Soc.*, 1964, **86**, 1630.	143
64JA2299	A. M. Aguiar and D. Daigle; *J. Am. Chem. Soc.*, 1964, **86**, 2299.	827, 828

64JA2378 W. E. McEwan, K. F. Kumli, A. Blade-Font, M. Zanger and C. A. VanderWerf; *J. Am. Chem. Soc.*, 1964, **86**, 2378. 436, 442, 443

64JA2395 D. M. Lemal, F. Menger and E. Coats; *J. Am. Chem. Soc.*, 1964, **86**, 2395. 379, 388

64JA2538 N. Hirota and S. I. Weissman; *J. Am. Chem. Soc.*, 1964, **86**, 2538. 1010

64JA2956 P. S. Skell and R. R. Pavlis; *J. Am. Chem. Soc.*, 1964, **86**, 2956. 35

64JA3283 H. M. Walborsky, F. J. Impastato and A. E. Young; *J. Am. Chem. Soc.*, 1964, **86**, 3283. 554

64JA3565 H. C. Brown, W. R. Heydkamp, E. Breuer and W. S. Murphy; *J. Am. Chem. Soc.*, 1964, **86**, 3565. 305

64JA3660 L. Zervas, D. Borovas and E. Gazis; *J. Am. Chem. Soc.*, 1964, **86**, 3660. 358

64JA4074 J. Meinwald, Y. C. Meinwald and T. N. Baker, III; *J. Am. Chem. Soc.*, 1964, **86**, 4074. 391

64JA4221 J. J. Eisch and M. E. Healy; *J. Am. Chem. Soc.*, 1964, **86**, 4221. 816

64JA4500 R. Schmutzler; *J. Am. Chem. Soc.*, 1964, **86**, 4500. 869

64JA5037 A. G. Anderson, Jr. and F. J. Freenor; *J. Am. Chem. Soc.*, 1964, **86**, 5037. 32

64JA5326 S. G. Cohen and S. Y. Weinstein; *J. Am. Chem. Soc.*, 1964, **86**, 5326. 88

64JA5359 A. R. Young, II and R. Ehrlich; *J. Am. Chem. Soc.*, 1964, **86**, 5359. 421

64JCS1020 W. Hewerton, R. A. Shaw and B. C. Smith; *J. Chem. Soc.*, 1964, 1020. 465

64JCS1029 T. G. Halsall, D. W. Theobald and K. B. Walshaw; *J. Chem. Soc.*, 1964, 1029. 675

64JCS1240 P. C. Crofts, I. M. Downie and K. Williamson; *J. Chem. Soc.*, 1964, 1240. 845

64JCS2100 A. A. Watson; *J. Chem. Soc.*, 1964, 2100. 218

64JCS3122 E. W. Abel and D. A. Armitage; *J. Chem. Soc.*, 1964, 3122. 399

64JCS3832 P. J. E. Brownlee, M. E. Cox, B. O. Handord, J. C. Marsden and G. T. Young; *J. Chem. Soc.*, 1964, 3832. 223

64JCS4551 R. Schmutzler; *J. Chem. Soc.*, 1964, 4551. 848

64JCS4750 M. Bentov, L. David and E. D. Bergmann; *J. Chem. Soc.*, 1964, 4750. 838

64JCS4962 M. Asscher and D. Vofski; *J. Chem. Soc.*, 1964, 4962. 157

64JCS5482 R. T. Wragg; *J. Chem. Soc.*, 1964, 5482. 478

64JCS5585 E. W. Abel, D. A. Armitage and R. P. Bush; *J. Chem. Soc.*, 1964, 5585. 851, 852

64JCS5863 C. J. M. Stirling; *J. Chem. Soc.*, 1964, 5863. 742

64JCS5892 H. J. Emeleus and B. W. Tattershall; *J. Chem. Soc.*, 1964, 5892. 360

64JCS5975 E. W. Abel and D. A. Armitage; *J. Chem. Soc.*, 1964, 5975. 231

64JMC229 W. H. H. Günther and H. G. Mautner; *J. Med. Chem.*, 1964, **7**, 229. 279, 292

64JMC493 E. F. Elslager, E. A. Weinstein and D. F. Worth; *J. Med. Chem.*, 1964, **7**, 493. 373, 378, 381

64JOC360 C. G. Overberger and A. Drucker; *J. Org. Chem.*, 1964, **29**, 360. 123

64JOC898 S. Andreades, J. F. Harris and W. A. Sheppard; *J. Org. Chem.*, 1964, **29**, 898. 688

64JOC918 L. Field, T. F. Parsons and R. R. Crenshaw; *J. Org. Chem.*, 1964, **29**, 918. 178

64JOC951 I. B. Douglass and D. A. Koop; *J. Org. Chem.*, 1964, **29**, 951. 172

64JOC986 S. Herbstman; *J. Org. Chem.*, 1964, **29**, 986. 485, 508, 510

64JOC1350 A. Hassner and C. Heathcock; *J. Org. Chem.*, 1964, **29**, 1350. 392

64JOC1688 G. R. Collins; *J. Org. Chem.*, 1964, **29**, 1688. 245

64JOC1800 F. Scotti and E. J. Frazza; *J. Org. Chem.*, 1964, **29**, 1800. 749

64JOC1817 H. H. Wasserman and D. Cohen; *J. Org. Chem.*, 1964, **29**, 1817. 693

64JOC1876 D. W. Wiley and H. E. Simmons; *J. Org. Chem.*, 1964, **29**, 1876. 640

64JOC2025 K. B. Doifode and M. G. Marathey; *J. Org. Chem.*, 1964, **29**, 2025. 28

64JOC2397 H. Hart and C. A. Buehler; *J. Org. Chem.*, 1964, **29**, 2397. 650

64JOC2572 T. Mukaiyama, T. Fujisawa, Y. Tamura and Y. Yokota; *J. Org. Chem.*, 1964, **29**, 2572. 837

64JOC2592 J. E. Franz and C. Osuch; *J. Org. Chem.*, 1964, **29**, 2592. 726

64JOC3055 W. J. Baumann and H. K. Mangold; *J. Org. Chem.*, 1964, **29**, 3055. 272

64JOC3225 E. C. Ashby and W. E. Foster; *J. Org. Chem.*, 1964, **29**, 3225. 1084

64JOC3429 R. M. Scriber; *J. Org. Chem.*, 1964, **29**, 3429. 805

64JOC3459 L. King and G. K. Ostrum; *J. Org. Chem.*, 1964, **29**, 3459. 28

64JOC3561 E. Galangay, H. Engel, A. Szabo and J. Fried; *J. Org. Chem.*, 1964, **29**, 3561. 274

64JOM(1)369 R. A. Zingaro and A. Merijanian; *J. Organomet. Chem.*, 1964, **1**, 369. 502

64JOM(1)384 M. F. Lappert and B. Prokai; *J. Organomet. Chem.*, 1964, **1**, 384. 1083

64JOM(1)476 C. R. Krüger and E. G. Rochow; *J. Organomet. Chem.*, 1964, **1**, 476. 699

64JOM(2)97 C. Eaborn and D. R. M. Walton; *J. Organomet. Chem.*, 1964, **2**, 97. 1089

64JOM(2)281 O. Yu. Okhlobystin, K. A. Bilevitch and L. I. Zakharkin; *J. Organomet. Chem.*, 1964, **2**, 281. 598

64JPR128 R. Sowda; *J. Prakt. Chem.*, 1964, **23**, 128. 367

64JPR294 L. Lamza; *J. Prakt. Chem.*, 1964, **25**, 294. 825, 841

64LA(667)113 W. Pfleiderer and H. Walter; *Justus Liebigs Ann. Chem.*, 1964, **677**, 113. 780

64LA(678)167 A. Rieche, J. Dahlmann and D. List; *Justus Liebigs Ann. Chem.*, 1964, **678**, 167. 886

64LA(679)83 H. Meister; *Justus Liebigs Ann. Chem.*, 1964, **679**, 83. 403, 404, 405

64LA(705)10 W. P. Neumann and R. Sommer; *Justus Liebigs Ann. Chem.*, 1964, **705**, 10. 601

64M33 M. Pailer and H. Huemer; *Monatsh Chem.*, 1964, **95**, 33. 268

B-64MI 201-01 G. A. Olah and W. S. Tolgyesi; in "Friedel–Crafts and Related Reactions," ed. G. A. Olah, Wiley, New York, 1964, vol. 2, p. 659. 13

B-64MI 202-01 H. Steinberg; "Organoboron Chemistry," Wiley, New York, 1964. 109, 111, 112

B-64MI 207-01 W. Theilacker and E. Wegner; in "W. Foerst, Academic Press, London, 1964, vol. 3, p. 303. 374

64MI 209-01 R. Pietsch and P. Ludwig; *Mikrochim. Acta*, 1964, 1082. 498

B-64MI 210-01 H. Steinberg; "Organoboron Chemistry," Wiley-Interscience, New York, 1964, vol. 1. 531

64MI 210-02 J. J. Zuckerman; *Adv. Inorg. Chem. Radiochem.*, 1964, **6**, 383. 538

64MI 214-01 L. M. Litvinenko; *Zh. Obsch. Khim. Lett.*, 1964, **34**, 3739 (Engl. 3780). 717

64MI 215-01 F. Miniski and R. Galli; *Chimica et Industrie*, 1964, **46**, 423. 798

64N213 H. Hartmann and H. Fratzscher; *Naturwissenschaften*, 1964, **51**, 213. 1060
64NEP6405309 M. and T. Chemicals Inc.; *Neth. Pat. Appl.*, 1964, 6 405 309 (*Chem. Abstr.*, 1965, **62**, 16 300). 885, 887, 888
64OS(44)39 W. A. Sheppard; *Org. Synth.*, 1964, **44**, 39. 716
64PAC307 B. A. Arbusow; *Pure Appl. Chem.*, 1964, **9**, 307. 693
64PAC225 L. Horner; *Pure Appl. Chem.*, 1964, **9**, 225. 441
64PCS230 C. J. W. Gutch and W. A. Waters; *Proc. Chem. Soc.*, 1964, 230. 782
64RTC208 L. Brandsma, H. E. Wijers and C. Jonker; *Recl. Trav. Chim. Pays-Bas*, 1964, **83**, 208. 1032
64RTC295 J. G. A. Luijten and G. J. M. Van der Kerk; *Rec. Trav. Chim. Pays-Bas*, 1964, **83**, 295. 1100
64RTC1211 P. P. Montijn, E. Harryvan and L. Brandsma; *Recl. Trav. Chim. Pays-Bas*, 1964, **83**, 1211. 1043
64RTC1301 W. Voskuil and J. F. Arens; *Rec. Trav. Chim. Pays-Bas*, 1964, **83**, 1301. 1059, 1079, 1081
64T173 A. Finch and J. Peran; *Tetrahedron*, 1964, **20**, 173. 268
64T1265 D. A. Swann and J. H. Turnbull; *Tetrahedron*, 1964, **20**, 1265. 125
64TL175 L. Horner and H. Winkler; *Tetrahedron Lett.*, 1964, 175. 461
64TL983 H. Hoffmann and H. Förster; *Tetrahedron Lett.*, 1964, 983. 1062
64TL1137 G. Köbrich and K. Flory; *Tetrahedron Lett.*, 1964, 1137. 998
64TL3271 L. Horner and H. Winkler; *Tetrahedron Lett.*, 1964, 3271. 886
64TL3393 K. C. Pande and S. Winstein; *Tetrahedron Lett.*, 1964, 3393. 598
64ZAAC(326)280 A. Tzschach and W. Lange; *Z. Anorg. Allg. Chem.*, 1964, **326**, 280. 414, 482, 503
64ZAAC(328)21 K. Issleib and F. Krech; *Z. Anorg. Allg. Chem.*, 1964, **328**, 21. 444
64ZAAC(330)233 H.-J. Vetter and H. Nöth; *Z. Anorg. Allg. Chem.*, 1964, **330**, 233. 414, 415, 891
64ZAAC(331)35 H.-J. Vetter, H. Nöth and U. Hayduk; *Z. Anorg. Allg. Chem.*, 1964, **331**, 35. 415
64ZAAC(332)179 K. Issleib and B. Hamann; *Z. Anorg. Allg. Chem.*; 1964, **332**, 179. 481, 486, 505, 510
64ZAAC(332)238 H. Bock and K. L. Kompa; *Z. Anorg. Allg. Chem.*, 1964, **332**, 238. 388
64ZOB1441 I. N. Zhmurova and B. S. Drach; *Zh. Obshch. Khim.*, 1964, **34**, 1441 (*Chem. Abstr.*, 1964, **61**, 5490f). 407
64ZOB1446 I. S. Protopopov and M. Y. Kraft; *Zh. Obshch. Khim.*, 1964, **34**, 1446. 825, 829
64ZOB1522 K. A. Ogloblin and V. P. Semenov; *Zh. Obshch. Khim.*, 1964, **34**, 1522 (*Chem. Abstr.*, 1964, **61**, 5500). 787
64ZOB1899 A. A. Petrov, S. I. Radchenko, K. S. Mingaleva, I. G. Savich and V. B. Lebedev; *Zh. Obshch. Khim.*, 1964, **34**, 1899. 1032
64ZOB1948 G. Y. Epshtein, I. A. Usov and S. Z. Ivin; *Zh. Obshch. Khim.*, 1964, **34**, 1948 (*Chem. Abstr.*, 1964, **61**, 8178c). 167
64ZOB2094 B. A. Porai-Koshits, E. N. Sof'ina and I. Ya. Kvitko; *Zh. Obshch. Khim.*, 1964, **34**, 2094 (*Chem. Abstr.*, 1964, **61**, 6990). 779
64ZOB3055 I. N. Zhmurova and B. S. Drach; *Zh. Obshch. Khim.*, 1964, **34**, 3055 (*Chem. Abstr.*, 1965, **62**, 14480d). 407

65ACS1490 G. Fuchs; *Acta Chem. Scand.*, 1965, **19**, 1490. 117, 124
65ACS1755 A. Senning; *Acta Chem. Scand.*, 1965, **19**, 1755. 809, 892
65AFC(4)1 H. Suschitzky; *Adv. Fluorine Chem.*, 1965, **4**, 1. 614
65AG172 K. Issleib, R. Kümmel and H. Zimmermann; *Angew. Chem.*, 1965, **77**, 172. 430
65AG(E)49 G. Köbrich; *Angew. Chem., Int. Ed. Engl.*, 1965, **4**, 49. 1016
65AG(E)148 K. Bott; *Angew. Chem., Int. Ed. Engl.*, 1965, **4**, 148. 800
65AG(E)244 W. Kuchen and B. Knop; *Angew. Chem., Int. Ed. Engl.*, 1965, **4**, 244. 465
65AG(E)356 D. Hellwinkel; *Angew. Chem., Int. Ed. Engl.*, 1965, **4**, 356. 869
65AG(E)430 K. Dickore, E. Kühle and B. Anders; *Angew. Chem., Int. Ed. Engl.*, 1965, **4**, 430. 726
65AG(E)524 E. Höft and A. Rieche; *Angew. Chem., Int. Ed. Engl.*, 1965, **4**, 524. 96
65AG(E)746 H. G. Viehe; *Angew. Chem., Int. Ed. Engl.*, 1965, **4**, 746. 1014
65AJC14 J. R. Surtees; *Aust. J. Chem.*, 1965, **18**, 14. 1099
65AJC903 A. Jones and M. Talmet; *Aust. J. Chem.*, 1965, **18**, 903. 447
65ANC164 R. E. Humphrey and J. L. Potter; *Anal. Chem.*, 1965, **37**, 164 (*Chem. Abstr.*, 1965, **62**, 7669d). 127
65BCJ1989 H. Takei, I. Watanabe and T. Mukaiyama; *Bull. Chem. Soc. Jpn.*, 1965, **38**, 1989. 401
65BSB21 E. J. Goethals and M. Verzele; *Bull. Soc. Chim. Belg.*, 1965, **74**, 21. 177
65BSF71 M. Minssen and J. Jacques; *Bull. Soc. Chim. Fr.*, 1965, 71. 1016
65BSF787 G. H. Weinreich and M. Jufresa; *Bull. Soc. Chim. Fr.*, 1965, 787. 190, 208
65BSF2187 Y. J. Israéli; *Bull. Soc. Chim. Fr.*, 1965, 2187. 1048
65BSF2368 D. Houalla, M. Sanchez and R. Wolf; *Bull. Soc. Chim. Fr.*, 1965, 2368. 411
65BSF2787 J. Ficini and C. Barbara; *Bull. Soc. Chim. Fr.*, 1965, 2787. 1047
65CB171 H. Fritzche, U. Hasserodt and F. Korte; *Chem. Ber.*, 1965, **98**, 171. 440
65CB363 H. Hoffmann and H. Diehr; *Chem. Ber.*, 1965, **98**, 363. 446
65CB552 R. Schmutzler; *Chem. Ber.*, 1965, **98**, 552. 832, 833
65CB576 D. Hellwinkel; *Chem. Ber.*, 1965, **98**, 576. 830, 869
65CB977 A. Tzschach and W. Deylig; *Chem. Ber.*; 1965, **98**, 977 485, 510
65CB998 F. O. Bobbio and P. A. Bobbio; *Chem. Ber.*, 1965, **98**, 998. 130
65CB1455 H. Böhme and O. Müller; *Chem. Ber.*, 1965, **98**, 1455. 169
65CB1588 H.-H. Stroh and H.-G. Scharnow; *Chem. Ber.*, 1965, **98**, 1588. 373
65CB1681 H. Fritzsche, U. Hasserodt and F. Korte; *Chem. Ber.*, 1965, **98**, 1681. 428, 431
65CB1715 L. Horner and H. Neumann; *Chem. Ber.*, 1965, **98**, 1715. 188
65CB2091 K. Issleib and H. R. Roloff; *Chem. Ber.*, 1965, **98**, 2091. 438
65CB2429 H. W. Roesky and A. Hoff; *Chem. Ber.*, 1965, **98**, 2429. 368
65CB2431 H. Zorn, H. Schindlbauer and H. Hagen; *Chem. Ber.*, 1965, **98**, 2431. 829

65CB3377	H. Dorn, A. Zubek and E. Hilgetag; *Chem. Ber.*, 1965, **98**, 3377.	380
65CB3847	T. Severin and B. Brück; *Chem. Ber.*, 1965, **98**, 3847.	788
65CI(L)81	T. Chao and W. B. Hardy; *Chem. Ind.* (*London*), 1965, 81.	720
65CI(L)223	L. H. Long and G. F. Freeguard; *Chem. Ind.* (*London*), 1965, 223.	32, 36
65CI(L)680	J. E. Hesse and F. K. Truby; *Chem. Ind.* (*London*), 1965, 680.	213
65CI(L)1076	N. Petragnani and M. D. M. Campos; *Chem. Ind.* (*London*), 1965, 1076.	856
65CJC840	H. H. Baer and T. Nielson; *Can. J. Chem.*, 1965, **43**, 840.	398
65CJC1884	N. R. Fetter and F. M. Peters; *Can. J. Chem.*, 1965, **43**, 1884.	421
65CJC3074	H. H. Baer and F. Kienzle; *Can. J. Chem.*, 1965, **43**, 3074.	398
65CJC3407	R. A. Abramovich, J. Roy and V. Uma, *Can. J. Chem.*, 1965, **43**, 3407.	249
65CR(260)1984	H. Normant and G. Sturtz; *C. R. Hebd. Seances Acad. Sci.*, 1965, **260**, 1984.	1042, 1050, 1073
65CR(260)2512	J. Armand and R. M. Minville; *C. R. Hebd. Seances Acad. Sci.*, 1965, **260**, 2512.	782
65CR(260)2851	M. Mousseron-Canet and J.-P. Boca; *C. R. Hebd. Seances Acad. Sci.*, 1965, **260**, 2851.	764
65CR(261)1019	G. Sturtz and C. Charrier; *C. R. Hebd. Seances Acad. Sci.*, 1965, **261**, 1019.	1072
65CR(261)1992	H. Demarne; *C. R. Hebd. Seances Acad. Sci.*, 1965, **261**, 1992.	1098
65CR(261)1996	J. Benaim; *C. R. Hebd. Seances Acad. Sci.*, 1965, **261**, 1996.	1076, 1079, 1080
65CRV385	A. Senning; *Chem. Rev.*, 1965, **65**, 385.	204
65FRP1397139	J. R. Leebrick; *Fr. Pat.* 1 397 139 (1965) (*Chem. Abstr.*, 1965, **63**, 5449a).	695
65FRP1401930	J. R. Geigy; *Fr. Pat.* 1 401 930 (1965) (*Chem. Abstr.*, 1965, **63**, 11 615).	427, 430, 437
65HCA1229	W. Büchner; *Helv. Chim. Acta*, 1965, **48**, 1229.	1022, 1023
65HCA1746	H. Brechbühler, H. Büchi, E. Hatz, J. Schreiber and A. Erdenmoser; *Helv. Chim. Acta*, 1965, **48**, 1746.	749
65HOU(6/3)97	H. Meerwein; *Methoden Org. Chem.* (*Houben-Weyl*), 1965, **6**, 97.	679
65HOU(6/3)116	H. Meerwein; *Methoden. Org. Chem.* (*Houben-Weyl*), 1965, **6/3**, 116–118.	1021, 1025, 1027
65HOU(10/3)8	R. Pütter; *Methoden Org. Chem.*, (*Houben-Weyl*), 1965, **10/3**, 8.	803
65HOU(10/3)16	R. Pütter; *Methoden Org. Chem.*, (*Houben-Weyl*), 1965, **10/3**, 16.	803
65HOU(10/3)22	R. Pütter; *Methoden Org. Chem.*, (*Houben-Weyl*), 1965, **10/3**, 22.	803
65HOU(10/3)23	R. Pütter; *Methoden Org. Chem.*, (*Houben-Weyl*), 1965, **10/3**, 23.	803
65HOU(10/3)28	R. Pütter; *Methoden Org. Chem.*, (*Houben-Weyl*), 1965, **10/3**, 28.	803
65HOU(10/3)44	R. Pütter; *Methoden Org. Chem.*, (*Houben-Weyl*), 1965, **10/3**, 44.	803
65HOU(10/3)46	R. Pütter; *Methoden Org. Chem.*, (*Houben-Weyl*), 1965, **10/3**, 46.	804
65HOU(10/3)51	R. Pütter; *Methoden Org. Chem.*, (*Houben-Weyl*), 1965, **10/3**, 51.	803
65HOU(10/3)71	R. Pütter; *Methoden Org. Chem.*, (*Houben-Weyl*), 1965, **10/3**, 71.	805
65HOU(10/3)73	R. Pütter; *Methoden Org. Chem.*, (*Houben-Weyl*), 1965, **10/3**, 73.	806
65HOU(10/3)86	R. Pütter; *Methoden Org. Chem.*, (*Houben-Weyl*), 1965, **10/3**, 86.	806
65HOU(10/3)89	R. Pütter; *Methoden Org. Chem.*, (*Houben-Weyl*), 1965, **10/3**, 89.	803
65HOU(10/3)552	R. Pütter; *Methoden Org. Chem.*, (*Houben-Weyl*), 1965, **10/3**, 552.	805
65HOU(10/3)576	R. Pütter; *Methoden Org. Chem.*, (*Houben-Weyl*), 1965, **10/3**, 576.	804
65HOU(10/3)584	R. Pütter; *Methoden Org. Chem.*, (*Houben-Weyl*), 1965, **10/3**, 584.	804
65IC119	E. L. Muetterties and W. Mahler; *Inorg. Chem.*, 1965, **4**, 119.	477, 869
65IC657	H. Watanabe, T. Totani and T. Yoshizaki; *Inorg. Chem.*, 1965, **4**, 657.	1085
65IC764	W. C. Firth, S. Frank, M. Garber and V. P. Wystrach; *Inorg. Chem.*, 1965, **4**, 764.	835, 836
65IC765	W. C. Firth, S. Frank, M. Garber and V. P. Wystrach; *Inorg. Chem.*, 1965, **4**, 765.	837
65IC973	E. J. Kupchik and P. J. Calabretta; *Inorg. Chem.*, 1965, **4**, 973.	889
65IC1458	N. E. Miller; *Inorg. Chem.*, 1965, **4**, 1458.	487
65IZV1715	B. L. Dyatkin, E. P. Mochalina and I. L. Knunyants; *Izv. Akad. Nauk SSSR, Ser. Khim.*, 1965, 1715 (*Chem. Abstr.*, 1965, **63**, 17 882).	782
65JA16	A. J. Chalk and J. F. Harrod; *J. Am. Chem. Soc.*, 1965, **87**, 16.	913
65JA230	J. H. Prager and P. G. Thompson; *J. Am. Chem. Soc.*, 1965, **87**, 230.	94
65JA275	G. Stork, P. Rosen, N. Goldman, R. V. Coombs and J. Tsuji; *J. Am. Chem. Soc.*, 1965, **87**, 275.	696
65JA863	D. Y. Curtin, J. A. Kampmeier and B. R. O'Connor; *J. Am. Chem. Soc.*, 1965, **87**, 863.	800
65JA874	D. Y. Curtin, J. A. Kampmeier and M. L. Farmer; *J. Am. Chem. Soc.*, 1965, **87**, 874.	800, 801
65JA1109	C. R. Johnson and D. McCants; *J. Am. Chem. Soc.*, 1965, **87**, 1109.	144, 149
65JA1213	A. B. Burg and J. Singh; *J. Am. Chem. Soc.*, 1965, **87**, 1213.	492
65JA1345	E. J. Corey and M. Chaykovsky; *J. Am. Chem. Soc.*, 1965, **87**, 1345.	153
65JA1353	E. J. Corey and M. Chaykovsky; *J. Am. Chem. Soc.*, 1965, **87**, 1353.	1000
65JA1526	D. S. Matteson and J. D. Liedtke; *J. Am. Chem. Soc.*, 1965, **87**, 1526.	904
65JA2113	A. B. Burg and K. Gosling; *J. Am. Chem. Soc.*, 1965, **87**, 2113.	461
65JA2490	R. W. Taft, R. H. Martin and F. W. Lampe; *J. Am. Chem. Soc.*, 1965, **87**, 2490.	1006
65JA3199	W. P. Jencks and M. Gilchrist; *J. Am. Chem. Soc.*, 1965, **87**, 3199.	807
65JA3286	W. F. Brill; *J. Am. Chem. Soc.*, 1965, **87**, 3286.	96
65JA3965	R. G. Hiskey and D. N. Harpp; *J. Am. Chem. Soc.*, 1965, **87**, 3965.	220
65JA4194	J. J. Eisch and G. R. Hiesk; *J. Am. Chem. Soc.*, 1965, **87**, 4194.	561
65JA4359	W. Lwowski and E. Scheiffele; *J. Am. Chem. Soc.*, 1965, **87**, 4359.	793
65JA4447	T. R. Hopkins and P. W. Vogel; *J. Am. Chem. Soc.*, 1965, **87**, 4447.	255
65JA4576	B. T. Gillis and J. D. Hagarty; *J. Am. Chem. Soc.*, 1965, **87**, 4576.	771
65JA4922	L. Zervas, I. Photaki, A. Cosmatos and D. Borovas; *J. Am. Chem. Soc.*, 1965, **87**, 4922.	274
65JCP(42)2255	L. Kevan and P. Hamlet; *J. Chem. Phys.*, 1965, **42**, 2255.	1013
65JCS541	M. Green; *J. Chem. Soc.*, 1965, 541.	448
65JCS843	J. Chatt and J. M. Davidson, *J. Chem. Soc.*, 1965, 843.	962
65JCS1097	I. C. Paul and G. A. Sim; *J. Chem. Soc.*, 1965, 1097.	1007
65JCS1539	S. Brewis, W. Dent and R. Smith; *J. Chem. Soc.*, 1965, 1539.	445

65JCS1870	G. E. Coates and D. Ridley; *J. Chem. Soc.*, 1965, 1870.	421
65JCS1944	K. Jones and M. F. Lappert; *J. Chem. Soc.*, 1965, 1944.	421
65JCS2184	I. G. M. Campbell, R. C. Cookson, M. B. Hocking and A. N. Hughes; *J. Chem. Soc.*, 1965, 2184.	858
65JCS2438	D. H. R. Barton, H. P. Faro, E. P. Serebryakov and N. F. Woolsey; *J. Chem. Soc.*, 1965, 2438.	13
65JCS2901	B. Milligan and J. M. Swan; *J. Chem. Soc.*, 1965, 2901.	228
65JCS3241	O. T. Beachley and G. E. Coates; *J. Chem. Soc.*, 1965, 3241.	511
65JCS3354	W. Gee, R. Shaw and B. Smith; *J. Chem. Soc.*, 1965, 3354.	444
65JCS4256	M. F. Lappert, H. Pyszora and M. Rieker; *J. Chem. Soc.*, 1965, 4256.	419
65JCS4283	E. H. Brooks, F. Glockling and K. A. Hooton; *J. Chem. Soc.*, 1965, 4283.	867
65JCS5132	H. Heaney and I. T. Millar; *J. Chem. Soc.*, 1965, 5132.	882
65JCS5225	D. K. Black and S. R. Landor; *J. Chem. Soc.*, 1965, 5225.	677
65JCS5516	B. C. Smith and G. H. Smith; *J. Chem. Soc.*, 1965, 5516.	452
65JCS5771	I. M. Downie and G. Morris; *J. Chem. Soc.*, 1965, 5771.	841, 842
65JCS6186	C. Sellers and H. Suschitzky; *J. Chem. Soc.*, 1965, 6186.	806
65JCS6638	G. M. Kosolapoff; *J. Chem. Soc.*, 1965, 6638.	452
65JCS7098	E. W. Abel, D. A. Armitage and R. P. Bush; *J. Chem. Soc.*, 1965, 7098.	263
65JCS7511	N. Welcman and H. Regev; *J. Chem. Soc.*, 1965, 7511.	292
65JGU1253	N. K. Bliznyuk, G. S. Levskaya and E. N. Matyukhina; *J. Gen. Chem. USSR* (*Engl. Transl.*), 1965, **35**, 1253.	883
65JGU1582	L. N. Mashlyakovskii and B. I. Ionin; *J. Gen. Chem. USSR* (*Engl. Transl.*), 1965, **35**, 1582.	1071
65JGU1677	L. I. Zakharkin, V. V. Gavrilenko and L. L. Ivanov; *J. Gen. Chem. USSR* (*Engl. Transl.*), 1965, **35**, 1677.	1099
65JGU1910	B. I. Ionin and A. A. Petrov; *J. Gen. Chem. USSR* (*Engl. Transl.*), 1965, **35**, 1910.	1071
65JHC152	M. A. Misch and J. E. Robertson; *J. Heterocycl. Chem.*, 1965, **2**, 152.	89
65JMC845	W. H. H. Gunther and H. G. Mautner; *J. Med. Chem.*, 1965, **8**, 845.	278, 279, 292
65JOC33	F. I. Carroll, H. M. Dickson and M. E. Wall; *J. Org. Chem.*, 1965, **30**, 33.	274
65JOC38	R. P. Pinnell, E. S. Huyser and J. Kleinberg; *J. Org. Chem.*, 1965, **30**, 38.	183, 272
65JOC97	C. E. Griffin, R. P. Peller, K. R. Martin and J. A. Peters; *J. Org. Chem.*, 1965, **30**, 97.	841
65JOC175	R. R. Crenshaw and L. Field; *J. Org. Chem.*, 1965, **30**, 175.	166, 173, 232
65JOC304	F. L. Lambert, W. D. Ellis and R. J. Parry; *J. Org. Chem.*, 1965, **30**, 304.	619, 626
65JOC415	J. A. Davis, J. Herynk, S. Carroll, J. Bunds and D. Johnson; *J. Org. Chem.*, 1965, **30**, 415.	12, 627
65JOC633	I. B. Douglass; *J. Org. Chem.*, 1965, **30**, 633.	178, 187, 196
65JOC933	G. K. Helmkamp, H. N. Cassey, B. A. Olsen and D. J. Pettitt; *J. Org. Chem.*, 1965, **30**, 933.	172
65JOC1027	S. A. Fuqua, W. G. Duncan and R. M. Silverstein; *J. Org. Chem.*, 1965, **30**, 1027.	1042
65JOC1384	W. J. Middleton and W. H. Sharkey; *J. Org. Chem.*, 1965, **30**, 1384.	166, 170
65JOC1488	J. E. Franz, J. F. Herber and W. S. Knowles; *J. Org. Chem.*, 1965, **30**, 1488.	391
65JOC1930	F. M. Beringer and S. A. Galton; *J. Org. Chem.*, 1965, **30**, 1930.	1019
65JOC2208	J. L. Wolinsky and K. L. Erickson; *J. Org. Chem.*, 1965, **30**, 2208.	622, 625
65JOC2454	D. L. Klayman; *J. Org. Chem.*, 1965, **30**, 2454.	237, 278, 291, 292
65JOC2471	L. J. Winters, D. B. Learn and S. C. Desai; *J. Org. Chem.*, 1965, **30**, 2471.	386, 798
65JOC2502	H. O. House and B. M. Trost; *J. Org. Chem.*, 1965, **30**, 2502.	696
65JOC2635	J. P. Schaefer and D. S. Weinberg; *J. Org. Chem.*, 1965, **30**, 2635.	7
65JOC2692	Q. E. Thompson, M. M. Crutchfield, M. W. Dietrich and E. Pierron; *J. Org. Chem.*, 1965, **30**, 2692.	100
65JOC2703	Q. E. Thompson; *J. Org. Chem.*, 1965, **30**, 2703.	99
65JOC2826	A. M. Aguiar and D. Daigle; *J. Org. Chem.*, 1965, **30**, 2826.	841
65JOC2877	H. Feuer, B. F. Vincent and R. S. Bartlett; *J. Org. Chem.*, 1965, **30**, 2877.	763
65JOC2880	H. Feuer, R. S. Bartlett, B. F. Vincent and R. S. Anderson; *J. Org. Chem.*, 1965, **30**, 2880.	341, 764
65JOC3278	L. Spialter, D. H. O'Brien, G. L. Untereiner and W. A. Rush; *J. Org. Chem.*, 1965, **30**, 3278.	377, 403
65JOC3689	D. Rosenthal, G. Brandrup, K. H. Davis and M. Wall; *J. Org. Chem.*, 1965, **30**, 3689.	129
65JOC3866	R. A. Baldwin; *J. Org. Chem.*, 1965, **30**, 3866.	863, 864
65JOC3985	J. W. Hamersma and E. I. Snyder; *J. Org. Chem.*, 1965, **30**, 3985.	14
65JOM(3)257	O. Yu. Okhlobystin and L. I. Zakharkin; *J. Organomet. Chem.*, 1965, **3**, 257.	1096
65JOM(3)295	K. Jones and M. F. Lappert; *J. Organomet. Chem.*, 1965, **3**, 295.	421, 868
65JOM(3)337	D. Seyferth, H. D. Simmons, Jr. and G. Singh; *J. Organomet. Chem.*, 1965, **3**, 337.	618
65JOM(3)406	W. R. Cullen, D. S. Dawson and G. E. Styan; *J. Organomet. Chem.*, 1965, **3**, 406.	876
65JOM(4)151	W. R. Cullen and G. E. Styan; *J. Organomet. Chem.*, 1965, **4**, 151.	482, 487, 876
65JOM(4)291	R. W. Bott, B. F. Dowden and C. Eaborn; *J. Organomet. Chem.*, 1965, **4**, 291.	841
65JOM(4)320	R. A. Zingaro, B. H. Steeves and K. Irgolic; *J. Organomet. Chem.*, 1965, **4**, 320.	465, 858
65JPC2484	J. Heicklen and V. Knight; *J. Phys. Chem.*, 1965, **69**, 2484.	1013
65JPR123	W. Pritzkow, H. Schaefer, P. Pabst, A. Ebenroth and J. Beger; *J. Prakt. Chem.*, 1965, **29**, 123.	778
65LA(682)21	R. Koster, W. Larbig and G. W. Rotemund; *Justus Liebigs Ann. Chem.*, 1965, **682**, 21.	523
65LA(687)214	M. Regitz and D. Stadler; *Justus Liebigs Ann. Chem.*, 1965, **687**, 214.	404
65LA(687)236	W. Lüttke and V. Schabaker; *Justus Liebigs Ann. Chem.*, 1965, **687**, 236.	403, 404
65M1021	H. Schindlbauer; *Monatsh. Chem.*, 1965, **96**, 1021.	
65M1936	H. Schindlbauer; *Monatsh. Chem.*, 1965, **96**, 1936.	832, 833
B-65MI 202-01	J. G. Buchanan, N. A. Hughes and G. A. Swan; in "Rodd's Chemistry of Carbon Compounds," ed. S. Coffey, Elsevier, Amsterdam, 1965, vol. 1B, and supplements therein.	38, 52, 86, 89
65MI 202-02	A. G. Davies; *Progress in Boron Chemistry*, 1965, **1**, 265.	96, 99, 109

65MI 203-01 R. E. Humphrey, A. L. McCrary and R. M. Webb; *Talanta*, 1965, **12**, 727 (*Chem. Abstr.*, 1965, **63**, 6846h). 127

B-65MI 203-02 E. E. Gilbert; in "Sulfonation and Related Reactions," Wiley-Interscience, New York, 1965. 175, 191

B-65MI 203-03 E. E. Reid; in "Organic Chemistry of Bivalent Sulfur," Chemical Publishing Co. Ltd., New York, 1965, vol. 1, chap. 1. 192

B-65MI 205-01 W. A. L. Spialter and J. A. Pappalardo; "The Acyclic Aliphatic Tertiary Amines," Macmillan, New York, 1965, pp. 44–52. 298, 320

B-65MI 205-02 P. A. S. Smith; "The Chemistry of Open-Chain Organic Nitrogen Compounds," Benjamin, New York, 1965, vol. 1, pp. 60–78 and 115–122. 298

B-65MI 207-01 H. Goldwhite; in "Rodd's Chemistry of Carbon Compounds," ed. S. Coffey, Elsevier, Amsterdam, 1965, vol. 1B, p. 93. 391

B-65MI 210-01 V. Bazant, V. Chvalovsky and J. Rathousky; "Organosilicon Compounds," Academic Press, New York, 1965, vols. 1 and 2. 535

B-65MI 214-01 E. E. Gilbert; "Sulfonation and Related Reactions," Wiley, New York, 1965. 720

65MI 215-01 R. D. Shrivastava and S. S. Deshapande; *Proc. Natl. Inst. Sci. India, Pt. A.*, 1965, **31**, 180 (*Chem. Abstr.*, 1966 **64**, 6696, 9770). 778

65MI 215-02 W. R. Miller, E. H. Pryde, J. C. Cowan and H. M. Teeter; *J. Am. Oil Chem. Soc.*, 1965, **42**, 713. 778

B-65MI 215-03 E. E. Gilbert; "Sulfation and Related Reactions," Interscience, New York, 1965, ch. 7. 797

B-65MI 215-04 R. F. Hudson; "Organo-phosphorus Chemistry," Academic Press, New York, 1965, ch. 8. 807

65N59 H. Hartmann, B. Karbstein and W. Reiss; *Naturwissenschaften*, 1965, **52**, 59. 1077

65OR(14)270 A. Maercker; *Org. React.*, 1965, **14**, 270. 1004

65QR329 G. R. Delpierre and M. Lamchen; *Q. Rev., Chem. Soc.*, 1965, **19**, 329. 765

65TL783 R. Huisgen and W. Kolbeck; *Tetrahedron Lett.*, 1965, 783. 766

65TL1157 L. Horner and W. D. Balzer; *Tetrahedron Lett.*, 1965, 1157. 441

65TL1487 D. H. Clemens, A. J. Belle and J. L. O'Brien; *Tetrahedron Lett.*, 1965, 1487. 399

65TL1901 P. Binger and R. Köster; *Tetrahedron Lett.*, 1965, 1901. 1087, 1088

65TL2271 T. J. Wallace and A. Schrielsheim; *Tetrahedron Lett.*, 1965, **21**, 2271. 192

65URP173230 V. S. Tsivunin, G. Kh. Kamai and R. Sh. Khisamutdinova; *USSR Pat.* 173 230 (1965) (*Chem. Abstr.*, 1966 **64**, 755). 461

65USP3172912 N. R. Easton and C. W. Ryan; *US Pat.* 3172912 (1965) (*Chem. Abstr.*, 1965, **62**, 14527e). 1018

65USP3454631 H. P. C. Lee; *US Pat.* 3 454 631 (1965) (*Chem. Abstr.*, 1967, **67**, 63 783). 170

65ZAAC(335)289 R. Paetzold and D. Lienig; *Z. Anorg. Allg. Chem.*, 1965, **335**, 289. 291

65ZAAC(336)36 A. Tzschach and W. Deylig; *Z. Anorg. Allg. Chem.*, 1965, **336**, 36. 482, 489

65ZAAC(336)234 K. Issleib and R. D. Bleck; *Z. Anorg. Allg. Chem.*, 1966, **336**, 234. 440

65ZAAC(338)22 V. R. Paetzold and E. Rönsh; *Z. Anorg. Allg. Chem.*, 1965, **338**, 22. 369

65ZAAC(338)195 R. Paetzold and E. Roensch; *Z. Anorg. Allg. Chem.*, 1965, **338**, 195. 291, 292

65ZAAC(339)289 K. Issleib and B. Hamann; *Z. Anorg. Allg. Chem.*, 1965, **339**, 289. 481, 486, 487, 502, 505, 510

65ZAAC(339)298 K. Issleib, B. Hamann and L. Schmidt; *Z. Anorg. Allg. Chem.*, 1965, **339**, 298. 481, 486, 489, 505, 506, 510

65ZN(B)181 K. Issleib and L. Brüsehaber; *Z. Naturforsch., Teil B*, 1965, **20**, 181. 439

65ZN(B)916 K. Issleib and D. Haferburg; *Z. Naturforsch., Teil B*, 1965, **20**, 916. 438

65ZOB344 I. N. Zhmurova, B. S. Drach and A. V. Kirsanov; *Zh. Obshch. Khim.*, 1965, **35**, 344 (*Chem. Abstr.*, 1965, **62**, 13030b). 407

65ZOB732 K. A. Petrov, R. A. Baksova and L. V. Khorkhoyanu; *Zh. Obshch. Khim.*, 1965, **35**, 732. 458

65ZOB1412 M. A. Englin, S. P. Makarov, S. S. DuBov and A. Y. Yakubovich; *Zh. Obshch. Khim.*, 1965, **35**, 1412 (*Chem. Abstr.*, 1965, **63**, 14 344e). 166, 170

65ZOB1574 M. I. Kabachnik, T. A. Mastryukova and A. E. Shipov; *Zh. Obshch. Khim.*, 1965, **35**, 1574. 841

65ZOB1577 L. N. Mashlyakovskii and B. I. Ionin; *Zh. Obshch. Khim.*, 1965, **35**, 1577. 848

65ZOB1669 V. U. Markova, V. A. Korner and A. A. Petrov; *Zh. Obshch. Khim.*, 1965, **35**, 1669. 900

65ZOR415 A. M. Sladkov, L. Y. Ukhin, G. N. Gorshkova, M. A. Chubarova, A. G. Makhsumov and V. I. Kasatochkin; *Zh. Org. Khim.*, 1965, **1**, 415. 1018, 1019

65ZOR1256 G. P. Balabanov and Y. I. Dergunov; *Zh. Org. Khim.*, 1965, **1**, 1256 (*Chem. Abstr.*, 1967, **66**, 55 169e). 478

65ZOR2115 S. I. Radchenko and A. A. Petrov; *Zh. Org. Khim.*, 1965, **1**, 2115. 731

65ZOR2157 S. I. Radchenko and A. A. Petrov; *Zh. Org. Khim.*, 1965, **1**, 2157. 1032

66ACS123 O. Foss and K. Marov; *Acta Chem. Scand.*, 1966, **20**, 123. 289

66ACS132 O. Foss and K. Husebye; *Acta Chem. Scand.*, 1966, **20**, 132. 289, 732

66ACS1874 A. Hordvik and E. Sletten; *Acta Chem. Scand.*, 1966, **20**, 1874. 1008

66ACS2508 G. Aksnes and K. Bergesen; *Acta Chem. Scand.*, 1966, **20**, 2508. 453

66AG(E)418 C. R. Russ and A. G. MacDiarmid; *Angew. Chem., Int. Ed. Engl.*, 1966, **5**, 418. 508

66AG(E)580 R. Koster and Y. Morita; *Angew. Chem., Int. Ed. Engl.*, 1966, **5**, 580. 112

66AG(E)584 R. Buijle, A. Halleux and H. G. Viehe; *Angew. Chem., Int. Ed. Engl.*, 1966, **5**, 584. 1046, 1049

66AG(E)937 L. Kuhnen; *Angew. Chem., Int. Ed. Engl.*, 1966, **5**, 893. 156

66AG(E)1048 P. Tavs; *Angew. Chem., Int. Ed. Engl.*, 1966, **5**, 1048. 792

66AHC(6)95 F. Johnson and R. Madroñero; *Adv. Heterocycl. Chem.*, 1966, **6**, 95. 308

66AOC(4)146 W. R. Cullen; *Adv. Organomet. Chem.*, 1966, **4**, 146. 872, 892

66BCJ364 S. Oae, Y. Onishi, S. Kozuka and W. Takakai; *Bull. Chem. Soc. Jpn.*, 1966, **39**, 364. 148

66BCJ1212 S. Oae, N. Furukawa, M. Kise and M. Kawanishi; *Bull. Chem. Soc. Jpn.*, 1966, **39**, 1212. 647

66BCJ1309 S. Oae and K. Ikura; *Bull. Chem. Soc. Japan*, 1966, **39**, 1309. 187

66BCJ1778 M. Kobayashi and K. Honda; *Bull. Chem. Soc. Jpn.*, 1966, **39**, 1778. 399

66BSF327 P. Mazerolles and G. Manuel; *Bull. Soc. Chim. Fr.*, 1966, 327. 949

66BSF1002 C. Charrier, W. Chodkiewicz and P. Cadiot; *Bull. Soc. Chim. Fr.*, 1966, 1002. 824, 828, 841, 1059, 1061, 1063, 1066, 1069

66BSF1707 G. Sturtz, C. Charrier and H. Normant; *Bull. Soc. Chim. Fr.*, 1966, 1707. 1072

66BSF1867 M. Le Guyader; *Bull. Soc. Chim. Fr.*, 1966, 1867. 764

66BSF1981 J. Soulié and P. Cadiot; *Bull. Soc. Chim. Fr.*, 1966, 1981. 1084

66BSF3016 G. Pourcelot and P. Cadiot; *Bull. Soc. Chim. Fr.*, 1966, 3016. 1030, 1031, 1034

66BSF3846 J. Soulié and P. Cadiot; *Bull. Soc. Chim. Fr.*, 1966, 3846. 1083, 1085

66CA-P2297 N. K. Bliznyuk, Z. N. Kvasha and A. F. Kolomiets; *Chem. Abstr.*, 1966, **65**, P2297. 448

66CB1134 H. Hoffmann and P. Schellenbeck; *Chem. Ber.*, 1966, **99**, 1134. 433, 434, 462, 824, 825

66CB1347 H. Bock and K. L. Kompa; *Chem. Ber.*, 1966, **99**, 1347. 752

66CB1414 K. Schank, B. Eistert and J. H. Felzmann; *Chem. Ber.*, 1966, **99**, 1414. 617

66CB1479 E. Vowinkel; *Chem. Ber.*, 1966, **99**, 1479. 683

66CB1580 H. W. Wanzlick and H. Ahrens; *Chem. Ber.*, 1966, **99**, 1580. 742

66CB1609 R. Muller and C. Dathe; *Chem. Ber.*, 1966, **99**, 1609. 487

66CB1663 W. Kuchen and B. Knop; *Chem. Ber.*, 1966, **99**, 1663. 465

66CB1718 H. Reinheckel and D. Jahnke; *Chem. Ber.*, 1966, **99**, 1718. 187

66CB3118 R. Appel, H. Heinen and R. Schöllhorn; *Chem. Ber.*, 1966, **99**, 3118. 384

66CB3384 L. Khunen; *Chem. Ber.*, 1966, **99**, 3384. 348

66CC51 G. J. Heiszwolf and H. Kloosterziel; *J. Chem. Soc., Chem. Commun.*, 1966, 51. 674

66CI(L)379 E. E. Aynsley, N. N. Greenwood and J. B. Leach; *Chem. Ind. (London)*, 1966, 379. 278

66CI(L)1635 R. L. Gay, T. F. Crimmins and C. R. Hauser; *Chem. Ind. (London)*, 1966, 1635. 621

66CJC2315 C. F. H. Allen and W. J. Humphlett; *Can. J. Chem.*, 1966, **44**, 2315. 117

66CJC2339 J. R. Campbell, J. K. N. Jones and S. Wolfe; *Can. J. Chem.*, 1966, **44**, 2339. 23, 26

66CRV529 G. Sosnovsky and J. H. Brown; *Chem. Rev.*, 1966, **66**, 529. 98

66DOK(168)807 V. D. Akamsin and N. I. Rizpolozhenskii; *Dokl. Akad. Nauk SSSR*, 1966, **168**, 807. 460

66DOK(170)96 M. G. Lin'kova, A. M. Orlov and I. L. Knunyants; *Dokl. Akad. Nauk SSSR*, 1966, **170**, 96 (*Chem. Abstr.*, 1967, **66**, 54 992f). 167

66HCA842 L. Maier; *Helv. Chim. Acta*, 1966, **49**, 842. 428, 440

66HCA907 W. Büchner; *Helv. Chim. Acta*, 1966, **49**, 907. 1022

66HCA1000 L. Maier; *Helv. Chim. Acta*, 1966, **49**, 1000. 465, 858

66HCA1249 L. Maier; *Helv. Chim. Acta*, 1966, **49**, 1249. L. Maier; *Br. Pat.*, 1968, 1 101 334 (*Chem. Abstr.*, 1968, **69**, 3004). 459, 460, 462, 851

66HCA2344 M. Harris and J. Feisst; *Helv. Chim. Acta*, 1966, **49**, 2344. 170

66HCA2488 L. Maier; *Helv. Chim. Acta*, 1966, **49**, 2488. 444

66IC164 G. S. Reddy and R. Schmutzler; *Inorg. Chem.*, 1966, **5**, 164. 477

66IC187 A. Merijanian and R. A. Zingaro; *Inorg. Chem.*, 1966, **5**, 187. 497, 502

66IC1004 W. R. Cullen and W. R. Leeder; *Inorg. Chem.*, 1966, **5**, 1004. 1077, 1078

66IC1297 P. Nicpon and D. W. Meek; *Inorg. Chem.*, 1966, **5**, 1297. 465

66IC1629 R. A. Wiesboeck and J. K. Ruff; *Inorg. Chem.*, 1966, **5**, 1629. 762

66IC1835 K. Utvary and H. H. Sisler; *Inorg. Chem.*, 1966, **5**, 1835. 387

66IS71 L. R. Ocone, C. W. Schaumann and B. P. Block; *Inorg. Synth.*, 1966, **8**, 71. 843

66IZV493 V. D. Akamsin and N. I. Rizpolozhenskii; *Izv. Akad. Nauk SSSR, Ser. Khim.*, 1966, 493 (*Chem. Abstr.*, 1966, **65**, 5480). 462

66JA100 P. Kovacic, J. A. Levisky and C. T. Goralski; *J. Am. Chem. Soc.*, 1966, **88**, 100. 752

66JA137 S. F. Nelsen and P. D. Bartlett; *J. Am. Chem. Soc.*, 1966, **88**, 137. 403

66JA1078 P. S. Forgioue, G. S. Sprague and H. J. Troffkin; *J. Am. Chem. Soc.*, 1966, **88**, 1078. 388

66JA1458 H. C. Brown, P. M. Weissman and N. M. Yoon; *J. Am. Chem. Soc.*, 1966, **88**, 1458. 126

66JA1464 H. C. Brown and N. M. Yoon; *J. Am. Chem. Soc.*, 1966, **88**, 1464. 60

66JA1572 F. P. Boer; *J. Am. Chem. Soc.*, 1966, **88**, 1572. 1008

66JA1828 E. L. Eliel, T. W. Doyle, R. A. Daignault and B. C. Newman; *J. Am. Chem. Soc.*, 1966, **88**, 1828. 128

66JA1839 D. B. Denney and L. Saferstein; *J. Am. Chem. Soc.*, 1966, **88**, 1839. 674

66JA2201 D. J. Pasto, C. C. Cumbo and J. Hickman; *J. Am. Chem. Soc.*, 1966, **88**, 2201. 83

66JA2328 F. Yoneda, K. Suzuki and Y. Nitta; *J. Am. Chem. Soc.*, 1966, **88**, 2328. 213

66JA3178 A. H. Cowley, A. B. Burg and W. R. Cullen; *J. Am. Chem. Soc.*, 1966, **88**, 3178. 507

66JA3842 W. J. Middleton; *J. Am. Chem. Soc.*, 1966, **88**, 3842. 385, 386, 792

66JA3874 T. Higuchi and K. H. Genoch; *J. Am. Chem. Soc.*, 1966, **88**, 3874. 148

66JA4531 J. K. Ruff, A. R. Pitochelli and M. Lustig; *J. Am. Chem. Soc.*, 1966, **88**, 4531. 94

66JA4757 J. J. Tufariello and L. T. C. Lee; *J. Am. Chem. Soc.*, 1966, **88**, 4757. 524

66JA4934 T. G. Traylor and C. L. Perrin; *J. Am. Chem. Soc.*, 1966, **88**, 4934. 1007

66JA5245 J. L. Kice, J. M. Anderson and N. E. Pawlowski; *J. Am. Chem. Soc.*, 1966, **88**, 5245. 122

66JCE144 R. F. Stockel, J. C. Oppelt and V. Arendt; *J. Chem. Ed.*, 1966, **43**, 144. 783

66JCS217 C. H. McMullen and C. J. M. Stirling; *J. Chem. Soc.*, 1966, 217. 742

66JCS404 N. K. Hamer, *J. Chem. Soc.*, 1966, 404. 807

66JCS(A)520 A. D. Beveridge, G. S. Harris and F. Inglis; *J. Chem. Soc. (A)*, 1966, 520. 835

66JCS(A)706 N. N. Greenwood, E. J. F. Ross and A. Storr; *J. Chem. Soc. (A)*, 1966, 706. 421

66JCS(A)1780 B. Cohen and A. G. MacDiarmid; *J. Chem. Soc. (A)*, 1966, 1780. 399, 400

66JCS(B)1162 A. Finch, P. J. Gardner and K. K. S. Gupta; *J. Chem. Soc. (B)*, 1966, 1162. 832, 833

66JCS(B)1188 L. Du Nunno, G. Modena and G. Scorrano; *J. Chem. Soc. (B)*, 1966, 1188. 1034

66JCS(C)106 J. E. Connet, A. G. Davies, G. B. Deacon, and J. H. S. Green; *J. Chem. Soc. C*, 1966, 106. 956

66JCS(C)245 J. H. Davies, J. D. Downer and P. Kirby; *J. Chem. Soc. (C)*, 1966, 245. 439

66JCS(C)415 E. D. Brown, S. M. Iqbal and L. N. Owen; *J. Chem. Soc. (C)*, 1966, 415. 128

66JCS(C)813 D. MacKay and W. A. Waters; *J. Chem. Soc. (C)*, 1966, 813. 379, 388

66JCS(C)1171	R. E. Banks, R. N. Haszeldine and A. Peppin; *J. Chem. Soc.* (*C*), 1966, 1171.	183, 194
66JCS(C)1592	M. G. Barlow, M. Green, R. N. Haszeldine and H. G. Higson; *J. Chem. Soc.* (*C*), 1966, 1592.	859
66JCS(C)2176	M. J. Gallagher and I. D. Jenkins; *J. Chem. Soc.*, (*C*), 1966, 2176.	414
66JGU570	A. M. Sladkov and L. K. Luneva; *J. Gen. Chem. USSR*, 1966, **36**, 570 (*Zh. Obshch. Khim.*, 1966, **36**, 553).	1093
66JGU1986	N. A. Chadaeva, G. Kh. Kamai and K. A. Mamakov; *J. Gen. Chem. USSR* (*Engl. Transl.*), 1966, **36**, 1986.	889
66JHC14	R. Ratz and A. D. Bliss; *J. Heterocycl. Chem.*, 1966, **3**, 14.	259
66JMC249	D. F. Morrow and R. M. Hoffer; *J. Med. Chem.*, 1966, **9**, 249.	753
66JOC153	R. L. Dannley and G. E. Corbett; *J. Org. Chem.*, 1966, **31**, 153.	689
66JOC221	T. J. Bardos, N. Datta-Gupta and P. Hebborn; *J. Org. Chem.*, 1966, **31**, 221.	260
66JOC410	H. Kwart and E. R. Evans; *J. Org. Chem.*, 1966, **31**, 410.	707
66JOC524	C. H. Robinson, L. Milewich and P. Hofer; *J. Org. Chem.*, 1966, **31**, 524.	389
66JOC588	G. P. Papp and S. A. Buckler; *J. Org. Chem.*, 1966, **31**, 588.	440
66JOC615	E. P. Papadopoulos, A. Jarrar and C. H. Issidorides; *J. Org. Chem.*, 1966, **31**, 615.	213
66JOC776	E.-A. I. Haiba; *J. Org. Chem.*, 1966, **31**, 766.	117
66JOC842	J. E. Dunbar and J. H. Rogers; *J. Org. Chem.*, 1966, **31**, 842.	358
66JOC957	D. P. Graham and W. Weinmayr; *J. Org. Chem.*, 1966, **31**, 957.	122
66JOC1202	W. H. H. Gunther; *J. Org. Chem.*, 1966, **31**, 1202.	127, 278, 292
66JOC1217	T. J. Wallace; *J. Org. Chem.*, 1966, **31**, 1217.	213
66JOC1426	A. J. Papa; *J. Org. Chem.*, 1966, **31**, 1426.	749
66JOC2105	A. J. Liston; *J. Org. Chem.*, 1966, **31**, 2105.	675
66JOC2179	R. Hiskey and J. B. Adams; *J. Org. Chem.*, 1966, **31**, 2179.	275
66JOC2357	R. T. von Aller, R. B. Scott and E. L. Brockelbank; *J. Org. Chem.*, 1966, **31**, 2357.	179
66JOC2455	J. B. Plumb, R. Obrycki and C. E. Griffin; *J. Org. Chem.*, 1966, **31**, 2455.	849
66JOC2658	H. Ulrich, B. Tucker and A. A. R. Sayigh; *J. Org. Chem.*, 1966, **31**, 2658.	726
66JOC2766	W. G. Woods, I. S. Bengelsdorf and D. L. Hunter; *J. Org. Chem.*, 1966, **31**, 2766.	904
66JOC3071	T. J. Wallace; *J. Org. Chem.*, 1966, **31**, 3071.	214, 215
66JOC3391	H. R. Hays and T. J. Logan; *J. Org. Chem.*, 1966, **31**, 3391.	428
66JOC3396	D. L. Klayman and J. W. Lown; *J. Org. Chem.*, 1966, **31**, 3396.	290
66JOC3543	W. E. Truce and D. G. Brady; *J. Org. Chem.*, 1966, **31**, 3543.	742
66JOC3545	W. E. Truce and D. G. Brady, *J. Org. Chem.*, 1966, **31**, 3545.	1034
66JOC3817	H. R. Hays; *J. Org. Chem.*, 1966, **31**, 3817.	427, 429, 432, 444
66JOC3980	M. S. Newman and E. A. Karnes; *J. Org. Chem.*, 1966, **31**, 3980.	707
66JOC4178	G. Zweifel and C. C. Whitney; *J. Org. Chem.*, 1966, **31**, 4178.	518
66JOM(5)29	W. Davidsohn and M. C. Henry; *J. Organomet. Chem.*, 1966, **5**, 29.	1093
66JOM(5)326	E. W. Abel and D. A. Armitage; *J. Organomet. Chem.*, 1966, **5**, 326.	890
66JOM(5)446	E. G. Claeys; *J. Organomet. Chem.*, 1966, **5**, 446.	491
66JOM(6)17	K. Irgolic, R. A. Zingaro and M. R. Smith; *J. Organomet. Chem.*, 1966, **6**, 17.	498, 504
66JOM(6)109	W. Steingross and W. Zeil; *J. Organomet. Chem.*, 1966, **6**, 109.	1016
66JOM(6)259	O. J. Scherer, P. Hornig and M. Schmidt; *J. Organomet. Chem.*, 1966, **6**, 259.	415, 416, 489, 503
66JOM(6)364	W. R. Cullen, P. S. Dhaliwal and G. E. Styan; *J. Organomet. Chem.*, 1966, **6**, 364.	487
66JOU1440	A. M. Sladkov, L. Yu. Ukhin and G. N. Gorshkova; *J. Org. Chem. USSR* (*Engl. Transl.*), 1966, **2**, 1440.	1054
66JPR96	W. Schulze, G. Letsch and H. Fritzsche; *J. Prakt. Chem.*, 1966, **33**, 96.	381
66LA(692)1	W. P. Neumann, B. Schneider and R. Sommer; *Justus Liebigs Ann. Chem.*, 1966, **692**, 1.	868
B-66MI 202-01	T. Eicher; "The Chemistry of the Carbonyl Group," Wiley, New York, 1966, pt. 1, p. 621.	67
66MI 202-02	R. O. Mumma; *Lipids*, 1966, **1**, 221.	102
B-66MI 203-01	A. A. Oswald and T. J. Wallace; in "The Chemistry of Organic Sulfur Compounds," Pergamon, New York, 1966, vol. 2.	213
66MI 212-01	W. G. Knipprath and R. A. Stein; *Lipids*, 1966, **1**, 81.	622
66MI 215-01	W. E. Tolberg, R. T. Rewick, R. S. Stringham and M. E. Hill; *Inorg. Nucl. Chem. Lett.*, 1966, **2**, 79.	762
66MI 215-02	K. O. Criste, J. P. Guertin and A. E. Pavlath; *Inorg. Nucl. Chem. Lett.*, 1966, **2**, 83.	762
B-66MI 215-03	P. A. S. Smith; "Open-Chain Nitrogen Compounds," Benzamin, New York, 1966, vol. 2, ch. 8, pp. 29–63.	765
B-66MI 215-04	I. T. Millar and H. D. Springall; "The Organic Chemistry of Nitrogen," Clarendon Press, Oxford, 1966, p. 504.	769
B-66MI 215-05	P. A. S. Smith; "Open-Chain Nitrogen Compounds," Benzamin, New York, 1966, vol. 2, p. 501.	798
B-66MI 215-06	T. C. Bruice and S. J. Benkovic; "Bioorganic Mechanisms," W. A. Benjamin, New York, 1966, vol. 2.	807
B-66MI 218-01	R. A. Benkeser; in "Organosilicon Chemistry," Butterworths, London, 1966, vol. 1, p. 133.	913
B-66MI 218-02	D. Seyferth, G. Singh and R. Suzuki; in "Organosilicon Chemistry," Butterworths, London, 1966, vol. 1, p. 159.	936
B-66MI 218-03	A. G. Brook; in "Organosilicon Chemistry," Butterworths, London, 1966, vol. 1, p. 215.	936
66PAC35	M. G. Voronkov; *Pure Appl. Chem.*, 1966, **13**, 35.	547
66QRS285	N. Kharasch, Z. S. Ariyan and A. J. Havlik; *Quart. Rep. Sulfur. Chem.*, 1966, **1**, 285.	216
66RTC56	M. E. Kronenberg, A. Van Der Heyden and E. Havinga; *Recl. Trav. Chim. Pays-Bas.*, 1966, **85**, 56.	756
66RTC580	J. H. van Boom, L. Brandsma and J. F. Arens; *Recl. Trav. Chim. Pays-Bas*, 1966, **85**, 580.	1032
66RTC857	W. Brackman and P. J. Smit; *Recl. Trav. Chim. Pays-Bas*, 1966, **85**, 857.	621, 627

66RTC952 J. H. van Boom, L. Brandsma and J. F. Arens; *Recl. Trav. Chim. Pays-Bas*, 1966, **85**, 952. 1032
66T399 A. M. Schoffstall and H. Tieckelmann; *Tetrahedron*, 1967, **22**, 399. 105
66T965 E. Kloster-Jensen; *Tetrahedron*, 1966, **22**, 965. 1016
66T1251 K. Bott; *Tetrahedron*, 1966, **22**, 1251. 799
66T1741 L. H. Slaugh; *Tetrahedron*, 1966, **22**, 1741. 970
66T2167 M. L. Poutsma and J. L. Karten; *Tetrahedron*, 1966, **22**, 2167. 606
66T2829 C. Burgess, D. Burn, P. Feather, M. Howarth and V. Petrow; *Tetrahedron*, 1966, **22**, 2829. 1016, 1018
66TL875 M. Cherest, H. Felkin, C. Frajerman, C. Leon, G. Roussi and G. Swerczewski; *Tetrahedron Lett.*, 1966, 875. 561
66TL1605 J. Blum; *Tetrahedron Lett.*, 1966, 1605. 622
66TL1671 W. Wucherpfenning and G. Kresze; *Tetrahedron Lett.*, 1966, 1671. 796
66TL1821 A. Risaliti, S. Fatutta, M. Forchiassin and E. Valentin; *Tetrahedron Lett.*, 1966, **7**, 1821. 783
66TL2535 G. Zweifel and H. Arzoumanian; *Tetrahedron Lett.*, 1966, 2535. 1005
66TL3041 J. Blum; *Tetrahedron Lett.*, 1966, 3041. 622
66TL4713 J. Tsuji and K. Ohno; *Tetrahedron Lett.*, 1966, 4713. 622, 627
66TL5049 R. Obrycki and C. E. Griffin; *Tetrahedron Lett.*, 1966, 5049. 849
66TL5493 H. Sakurai, K. Tominaga, T. Watanabe and M. Kumada; *Tetrahedron Lett.*, 1966, 5493. 537
66TL6331 A. Risaliti, M. Forchiassin and E. Valentin; *Tetrahedron Lett.*, 1966, **7**, 6331. 783
66USP3243450 M. Grayson; *US Pat.* 3 243 450 (1966) (*Chem. Abstr.*, 1966, **64**, 19 678). 442
66USP3249630 H. G. Viehe; *US Pat.* 3249630 (1966) (*Chem. Abstr.*, 1966, **65**, 747g). 1017
66USP3261743 J. H. Wilson, Jr.; *US Pat.* 3 261 743 (1966) (*Chem. Abstr.*, 1966, **65**, 9662). 461
66ZAAC(343)268 K. Issleib and H. Oehme; *Z. Anorg. Allg. Chem.*, 1966, **343**, 268. 462
66ZAAC(346)241 K. Issleib and O. Löw; *Z. Anorg. Allg. Chem.*, 1966, **346**, 241. 438
66ZOB484 A. B. Bruker, E. I. Grinshtein and L. Z. Soborovskii; *Zh. Obshch. Khim.*, 1966, **36**, 484 (*Chem. Abstr.*, 1966, **65**, 740). 430
66ZOB500 A. A. Neimysheva, I. V. Savchuk and I. L. Knunjants; *Zh. Obshch. Khim.*, 1966, **36**, 500. 464, 854
66ZOB564 N. G. Feshchenko and A. V. Kirsanov; *Zh. Obshch. Khim.*, 1966, **36**, 564 (*Chem. Abstr.*, 1966, **65**, 742). 453
66ZOB1436 V. S. Tsivunin, G. Kh. Kamai, V. V. Kormachev and G. S. Ukader; *Zh. Obshch. Khim.*, 1966, **36**, 1436 (*Chem. Abstr.*, 1967, **66**, 11 007). 461
66ZOR1994 I. A. Maretina and A. A. Petrov; *Zh. Org. Khim.*, 1966, **2**, 1994. 1042

67AC(R)1266 L. Marchetti and V. Passalacqua; *Ann. Chim.* (*Rome*), 1967, **57**, 1266 (*Chem. Abstr.*, 1968, **68**, 95 420). 787
67ACS652 E. Rebane; *Acta Chem. Scand.*, 1967, **21**, 652. 291
67ACS1257 B. Östman; *Acta Chem. Scand.*, 1967, **21**, 1257. 104
67AG109 G. Kresze and W. Wucherpfennig; *Angew. Chem.*, 1967, **79**, 1079. 245
67AG(E)107 G. Opitz; *Angew. Chem., Int. Ed. Engl.*, 1967, **6**, 107. 204
67AG(E)149 G. Kresze and W. Wucherpfenning; *Angew. Chem., Int. Ed. Engl.*, 1967, **6**, 149. 796
67AG(E)307 H. Seeboth; *Angew. Chem., Int. Ed. Engl.*, 1967, **6**, 307. 647
67AG(E)355 J. Ellermann and F. Poersch; *Angew. Chem., Int. Ed. Engl.*, 1967, **6**, 355. 432
67AG(E)362 F. Eckstein and K. H. Scheit; *Angew. Chem., Int. Ed. Engl.*, 1967, **6**, 362. 105
67AG(E)544 H. Distler; *Angew. Chem., Int. Ed. Engl.*, 1967, **6**, 544. 122
67AG(E)633 T. Kauffmann, J. Albrecht, D. Berger and J. Legler; *Angew. Chem., Int. Ed. Engl.*, 1967, **6**, 633. 421
67AG(E)703 F. Feher and D. Degen; *Angew. Chem., Int. Ed. Engl.*, 1967, **6**, 703. 212
67AG(E)706 F. Seel, W. Gombler and R. Budenz; *Angew. Chem., Int. Ed. Engl.*, 1967, **6**, 706. 166
67AG(E)951 F. Effenberger and G. Kkiefer; *Angew. Chem., Int. Ed. Engl.*, 1967, **6**, 951. 245
67BCJ147 T. Fujisawa, Y. Yokota and T. Mukaiyama; *Bull. Chem. Soc. Jpn.*, 1967, **40**, 147. 859
67BCJ418 S. Ito and T. Oya; *Bull. Chem. Soc. Jpn.*, 1967, **40**, 418. 626
67BCJ428 Y. Kawasaki and R. Okawara; *Bull. Chem. Soc. Jpn.*, 1967, **40**, 428. 695
67BCJ1538 K. Tabei and M. Yamaguchi; *Bull. Chem. Soc. Jpn.*, 1967, **40**, 1538. 776
67BCJ2380 O. Mitsunobu and M. Yamada; *Bull. Chem. Soc. Jpn.*, 1967, **40**, 2380. 118, 124
67BCJ2388 T. Mukaiyama and T. Endo; *Bull. Chem. Soc. Jpn.*, 1967, **40**, 2388. 215
67BCJ2913 N. Inukai, K. Nakano and M. Murakami; *Bull. Chem. Soc. Jpn.*, 1967, **40**, 2913. 219
67BSF777 J. C. Mason, M. le Quan and P. Cadiot; *Bull. Soc. Chim. Fr.*, 1967, 777. 1093
67BSF1213 M. Demuynck and J. Vialle; *Bull. Soc. Chim. Fr.*, 1967, 1213. 130
67BSF1296 M. Mousseron-Canet and J.-P. Boca; *Bull. Soc. Chim. Fr.*, 1967, 1296. 764
67BSF1928 M. Pereyre and J. Valade; *Bull. Soc. Chim. Fr.*, 1967, 1928. 701
67CA76112w B. D. Chernokal'Skii, S. S. Nasybullina, R. R. Shagidullin, I. A. Lamanova and G. Kamai; *Chem. Abstr.*, 1967, **66**, 76 112w. 809
67CB412 K. Issleib and K. Jasche; *Chem. Ber.*, 1967, **100**, 412. 439
67CB692 H. Hoffmann and P. Schellenbeck; *Chem. Ber.*, 1967, **100**, 692. 434
67CB1000 H. Schmidbauer and W. Wolfsberger; *Chem. Ber.*, 1967, **100**, 1000. 470
67CB1129 H. Schmidbaur and H. F. Klein; *Chem. Ber.*, 1967, **100**, 1129. 597
67CB1482 M. Hasserodt; *Chem. Ber.*, 1967, **100**, 1482. 170
67CB1544 J. Dahlmann and A. Rieche; *Chem. Ber.*, 1967, **100**, 1544. 888
67CB2159 H. Beecken; *Chem. Ber.*, 1967, **100**, 2159. 245
67CB2377 R. H. Cragg, M. F. Lappert, H. Nöth, P. Schweizer and B. P. Tilley; *Chem. Ber.*, 1967, **100**, 2377. 1086
67CB2681 L. Birkofer and M. Franz; *Chem. Ber.*, 1967, **100**, 2681. 943
67CB2685 K. Issleib and H. Oehme; *Chem. Ber.*, 1967, **100**, 2685. 430, 454

67CB3331 K. Issleib and R. Kümmel; *Chem. Ber.*, 1967, **100**, 3331. 427
67CB3371 G. Köbrich and H. R. Merkle; *Chem. Ber.*, 1967, **100**, 3371. 906
67CC131 C. J. M. Stirling; *J. Chem. Soc., Chem. Commun.*, 1967, 131. 159
67CC270 S. Braverman and Y. Stabinsky; *J. Chem. Soc., Chem. Commun.*, 1967, 270. 154
67CC985 L. Sagramora, P. Koch, A. Garbesi and A. Fava; *J. Chem. Soc., Chem. Commun.*, 1967, 985. 362
67CC1042 D. Lloyd and M. I. C Singer; *J. Chem. Soc., Chem. Commun.*, 1967, 1042. 1004
67CC1189 R. M. Dodson and T. P. Nelson; *J. Chem. Soc., Chem. Commun.*, 1967, 1189. 153
67CJC379 W. R. Cullen and P. S. Dhaliwal; *Can. J. Chem.*, 1967, **45**, 379. 499
67CJC495 R. Greenhalgh and M. A. Weinberger; *Can. J. Chem.*, 1967, **45**, 495. 108
67CJC675 J. R. Philips and J. H. Vis; *Can. J. Chem.*, 1967, **45**, 675. 485, 505, 510, 874
67CR(C)995 M. C. Charrier and M.-P. Simonnin; *C. R. Hebd. Seances Acad. Sci., Ser. C.*, 1967, **264**, 995. 1063, 1065, 1066
67CR(C)1496 J. Ficini, C. Barbara and A. Duréault; *C. R. Hebd. Seances Acad. Sci., Ser. C*, 1967, **265**, 1496. 1048
67CRV73 W. E. Davidsohn and M. C. Henry; *Chem. Rev.*, 1967, **67**, 73. 1076, 1080, 1083, 1089, 1092, 1098, 1099, 1101
67CRV665 J. K. Ruff; *Chem. Rev.*, 1967, **67**, 665. 334, 335, 762
67DOK(172)1329 A. N. Nesmeyanov, A. E. Borisov and N. V. Novikova; *Dokl. Akad. Nauk SSSR*, 1967, **172**, 1329 (*Chem. Abstr.*, 1967, **67**, 3127r). 876
67DOK(174)96 A. N. Nesmeyanov and A. E. Borison; *Dokl. Akad. Nauk SSSR*, 1967, **174**, 96. 971
67DOK(175)1079 M. F. Shostakovskii, N. V. Komarov and T. I. Ermolova; *Dokl. Akad. Nauk SSSR*, 1967, **175**, 1079 (*Chem. Abstr.*, 1968, **68**, 2970). 1101
67FCF437 P. Paetzold; *Fortschr. Chem. Forsch.*, 1967, **8**, 437. 811
67FRP1493679 R. Aries; *Fr. Pat.* 1 493 679 (1967) (*Chem. Abstr.*, 1968, **69**, 35 667z). 194
67HCA1747 L. Maier and J. J. Daly; *Helv. Chim. Acta*, 1967, **50**, 1747. 821
67HOU(10/2)169 E. Enders; *Methoden Org. Chem.* (*Houben-Weyl*), 1967, **10/2**, 169. 373
67HOU(10/2)757 E. Müller; *Methoden Org. Chem.* (*Houben-Weyl*), 1967, **10/2**, 757. 402, 404
67HOU(10/2)827 E. Müller; *Methoden Org. Chem.* (*Houben-Weyl*), 1967, **10/2**, 827. 387, 388
67IC1178 R. L. McKenny and H. H. Sisler; *Inorg. Chem.*, 1967, **6**, 1178. 892
67ICA34 M. H. O'Brien, G. O Doak and G. G. Long; *Inorg. Chim. Acta*, 1967, **1**, 34. 492
67IS63 R. Schmutzler; *Inorg Synth.*, 1967, **9**, 63. 477
67IS64 R. Schmutzler; *Inorg Synth.*, 1967, **9**, 64. 837
67IS69 R. Schmutzler; *Inorg Synth.*, 1967, **9**, 69. 836, 837
67IS92 G. O. Doak, G. G. Long and M. E. Key; *Inorg. Synth.*, 1967, **9**, 92. 492
67IS160 R. D. Feltham and W. Silverthorn; *Inorg. Synth.*, 1967, **10**, 159. 482
67IZV1606 N. A. Buina, I. A. Nuretdinov and N. P. Grechkin; *Izv. Akad. Nauk SSSR, Ser. Khim.*, 1967, 1606 (*Chem. Abstr.*, 1968, **68**, 12594p). 410
67IZV1976 V. D. Akamsin and N. I. Rizpolozhenskii; *Izv. Akad. Nauk SSSR, Ser. Khim.*, 1967, 1976 (*Chem. Abstr.*, 1968, **68**, 29 783). 463
67JA291 G. Zweifel, H. Arzoumanian; *J. Am. Chem. Soc.*, 1967, **89**, 291. 41, 900
67JA561 H. C. Brown and G. Zweifel; *J. Am. Chem. Soc.*, 1967, **89**, 561. 523
67JA716 W. H. Graham and C. P. Freeman; *J. Am. Chem. Soc.*, 1967, **89**, 716. 335
67JA718 J. R. Shetton and K. E. Davis; *J. Am. Chem. Soc.*, 1967, **89**, 718. 153, 184
67JA725 N. Kornblum, T. M. Davies, G. W. Earl, N. C. Holy, R. C. Kerber, M. T. Musser and D. H. Snow; *J. Am. Chem Soc.*, 1967, **89**, 725. 393
67JA1040 A. P. Lane and A. B. Burg; *J. Am. Chem. Soc.*; 1967, **89**, 1040. 509
67JA1102 R. I. Wagner, L. V. Freeman, H. Goldwhite and D. G. Rowsell; *J. Am. Chem. Soc.*, 1967, **89**, 1102. 430
67JA1417 C. D. Hurd and F. F. Cesark; *J. Am. Chem. Soc.*, 1967, **89**, 1417. 376
67JA1522 H. C. Brown and P. Geoghan, Jr.; *J. Am. Chem. Soc.*, 1967, **89**, 1522. 40
67JA1537 R. A. Abramovitch, M. Saha, E. M. Smith and R. T. Coutts; *J. Am. Chem. Soc.*, 1967, **89**, 1537. 765
67JA1640 A. J. Chalk and J. F. Harrod; *J. Am. Chem. Soc.*, 1967, **89**, 1640. 913
67JA1960 W. M. Jones and F. W. Miller; *J. Am. Chem. Soc.*, 1967, **89**, 1960. 801
67JA2435 R. G. Laughlin and W. Yellin; *J. Am. Chem. Soc.*, 1967, **89**, 2435. 164
67JA2502 G. M. Atkins Jr. and E. M. Burgess; *J. Am. Chem. Soc.*, 1967, **89**, 2502. 367
67JA2753 G. Zweifel and C. C. Whitney; *J. Am. Chem. Soc.*, 1967, **89**, 2753. 629
67JA2782 R. Koster and B. Rickborn; *J. Am. Chem. Soc.*, 1967, **89**, 2782. 524
67JA2993 G. A. Olah and J. M. Bollinger; *J. Am. Chem. Soc.*, 1967, **89**, 2993. 1007
67JA3652 G. Zweifel, H. Arzoumanian and C. C. Whitney; *J. Am. Chem. Soc.*, 1967, **89**, 3652. 900
67JA3662 R. E. Partch; *J. Am Chem. Soc.*, 1967, **89**, 3662. 86
67JA4099 F. Wudl, D. A. Lightner and D. Cram; *J. Am. Chem. Soc.*, 1967, **89**, 4099. 187
67JA4245 E. J. Corey, J. A. Katzenellenbogen and G. H. Posner; *J. Am. Chem. Soc.*, 1967, **89**, 4245. 629, 970
67JA4313 L. R. Anderson and W. B. Fox; *J. Am. Chem. Soc.*, 1967, **89**, 4313. 99
67JA4316 P. G. Thompson; *J. Am. Chem. Soc.*, 1967, **89**, 4316. 99
67JA4487 L. A. Paquette, L. S. Wittenbrook and V. V. Kane; *J. Am. Chem. Soc.*, 1967, **89**, 4487. 619
67JA4531 H. C. Brown, D. W. Bowman, S. Misumi and M. K. Unni; *J. Am. Chem. Soc.*, 1967, **89**, 4531. 623
67JA4714 S. J. Benkovic and P. A. Benkovic; *J. Am. Chem. Soc.*, 1967, **89**, 4714. 807
67JA4784 O. Korpiun and K. Mislow, *J. Am. Chem. Soc.*, 1967, **89**, 4784. 441
67JA5086 G. Zweifel and H. Arzoumanian; *J. Am. Chem. Soc.*, 1967, **89**, 5086. 901
67JA5478 H. C. Brown and E.-I. Negishi; *J. Am. Chem. Soc.*, 1967, **89**, 5478. 518

67JA5838 M. S. Simon, J. B. Rogers, W. Saenger and J. Z. Gougoutas; *J. Am. Chem. Soc.*, 1967, **89**, 5838. 170

67JA6972 I. Oney and M. Caplow; *J. Am. Chem. Soc.*, 1967, **89**, 6972. 807

67JAP6703120 M. Kado, T. Maeda and E. Yoshinaga; *Jpn. Pat.* 6 703 120 (*Chem. Abstr.*, 1967, **66**, 115 445d). 855

67JCP(46)658 R. W. Lovejoy and D. R. Baker; *J. Chem. Phys.*, 1967, **46**, 658. 1093

67JCS1028 A. P. Lane, D. A. Morton-Blake and D. S. Payne; *J. Chem. Soc.*, 1967, 1028. 859

67JCS(A)56 G. E. Coates and D. Ridley; *J. Chem. Soc.* (*A*), 1967, 56. 421

67JCS(A)652 B. J. Aylett and J. Emsley; *J. Chem. Soc.* (*A*), 1967, 652. 416

67JCS(A)1308 R. G. Cavell and R. C. Dobbie; *J. Chem. Soc.* (*A*), 1967, 1308. 482

67JCS(A)1347 M. C. Baird, J. T. Mague, J. A. Osborne and G. Wilkinson; *J. Chem. Soc.* (*A*), 1967, 1347. 478

67JCS(A)1492 A. P. Lane, D. A. Morton-Blake and D. S. Payne; *J. Chem. Soc.* (*A*), 1967, 1492. 862

67JCS(A)1913 J. J. Daly; *J. Chem. Soc.* (*A*), 1967, 1913. 1004

67JCS(A)1918 B. J. Aylett and J. Emsley; *J. Chem. Soc.* (*A*), 1967, 1918. 416

67JCS(A)2014 E. W. Abel, B. C. Crosse and G. V. Hutson; *J. Chem. Soc.* (*A*), 1967, 2014. 278

67JCS(C)252 D. M. Brown and G. H. Jones; *J. Chem. Soc.* (*C*), 1967, 252. 12

67JCS(C)2015 J. D. Hobson and J. G. McCluskey; *J. Chem. Soc.* (*C*), 1967, 2015. 303

67JCS(C)2131 R. S. Davidson; *J. Chem. Soc.* (*C*), 1967, 2131. 460

67JGU515 V. V. Gavrilenko, L. L. Ivanov and L. I. Zakharkin; *J. Gen. Chem. USSR* (*Engl. Transl.*), 1967, **37**, 515 (*Zh. Obshch. Khim.*, 1967, **37**, 550). 1093

67JGU855 G. S. Levskaya, A. F. Kolomiets; *J. Gen. Chem. USSR* (*Engl. Transl.*), 1967, **37**, 855. 886

67JGU898 V. S. Zavgorodnii, B. I. Ionin and A. A. Petrov; *J. Gen. Chem. USSR* (*Engl. Transl.*), 1967, **37**, 898. 1071

67JGU1578 M. P. Osipova, G. Kh. Kamai and N. A. Chabaeva; *J. Gen. Chem. USSR* (*Engl. Transl.*), 1967, **37**, 1578. 889

67JHC413 T. Sheradsky, *J. Heterocycl. Chem.*, 1967, **4**, 413. 692

67JOC31 Z. Suzuki and K.-I. Morita; *J. Org. Chem.*, 1967, **32**, 31. 26

67JOC97 R. G. Hiskey, T. Mizoguchi and E. L. Smithwick; *J. Org. Chem.*, 1967, **32**, 97. 219

67JOC236 M. Grayson and C. E. Farley; *J. Org. Chem.*, 1967, **32**, 236. 460, 463

67JOC282 J. S. Dix and C. R. Bresson; *J. Org. Chem.*, 1967, **32**, 282. 130

67JOC416 R. F. Merritt and F. A. Johnson; *J. Org. Chem.*, 1967, **32**, 416. 335

67JOC510 J. G. Traynham, D. B. Stone and J. L. Couvillio; *J. Org. Chem.*, 1967, **32**, 510. 420

67JOC677 N. J. Leonard and R. Y. Ning; *J. Org. Chem.*, 1967, **32**, 677. 129

67JOC727 F. Yoneda, K. Suzuki and Y. Nitta; *J. Org. Chem.*, 1967, **32**, 727. 213

67JOC781 S. O. Grim, W. McFarlane and E. F. Davidoff; *J. Org. Chem.*, 1967, **32**, 781. 435

67JOC1060 H. R. Hays and R. G. Laughlin; *J. Org. Chem.*, 1967, **32**, 1060. 452

67JOC1063 T. Obata and T. Mukaiyama; *J. Org. Chem.*, 1967, **32**, 1063. 106

67JOC1115 D. M. Gardiner, R. Helitzer and D. H. Rosenblatt; *J. Org. Chem.*, 1967, **32**, 1115. 336, 339

67JOC1124 E. V. Brown and J. J. Duffy; *J. Org. Chem.*, 1967, **32**, 1124. 776

67JOC1261 J. R. Piper and T. P. Johnston; *J. Org. Chem.*, 1967, **32**, 1261. 214

67JOC1572 R. A. Baldwin and M. T. Cheng; *J. Org. Chem.*, 1967, **32**, 1572. 844

67JOC1607 J. P. Schaefer and J. Higgins; *J. Org. Chem.*, 1967, **32**, 1607. 627

67JOC1717 D. J. Peterson; *J. Org. Chem.*, 1967, **32**, 1717. 434

67JOC2063 J. F. Harris; *J. Org. Chem.*, 1967, **32**, 2063. 1030

67JOC2172 R. A. Baldwin, C. O. Wilson, Jr. and R. I. Wagner; *J. Org. Chem.*, 1967, **32**, 2172. 453, 837, 845

67JOC2358 D. E. Pearson, R. D. Wysong and C. V. Breder; *J. Org. Chem.*, 1967, **32**, 2358. 626, 685

67JOC2360 J. C. Stowell; *J. Org. Chem.*, 1967, **32**, 2360. 377, 403

67JOC2445 D. B. Denney and F. C. Gross; *J. Org. Chem.*, 1967, **32**, 2445. 434

67JOC2465 A. L. Ternay, D. W. Chasar and M. Sax; *J. Org. Chem.*, 1967, **32**, 2465. 478

67JOC2580 H. M. Gilow and G. L. Walker; *J. Org. Chem.*, 1967, **32**, 2580. 288

67JOC2627 A. M. Aguiar and T. G. Archibald; *J. Org. Chem.*, 1967, **32**, 2627. 874

67JOC3032 A. Yamazaki, I. Kumashivo and T. Takenish; *J. Org. Chem.*, 1967, **32**, 3032. 192

67JOC3111 L. S. Skattebøl, B. Boulette and S. Solomon; *J. Org. Chem.*, 1967, **32**, 3111. 149

67JOC3645 R. V. Norton, G. M. Beverly and I. B. Douglass; *J. Org. Chem.*, 1967, **32**, 3645. 478

67JOC3707 M. A. Weiner and G. Pasternack; *J. Org. Chem.*, 1967, **32**, 3707. 887

67JOC4034 R. C. Petry and J. P. Freeman; *J. Org. Chem.*, 1967, **32**, 4034. 335, 336

67JOC4062 K. C. Williams; *J. Org. Chem.*, 1967, **32**, 4062. 602

67JOC4066 S. E. Cremer and R. J. Chorvat; *J. Org. Chem.*, 1967, **32**, 4066. 441

67JOC4134 A. I. Meyers and J. C. Sircar; *J. Org. Chem.*, 1967, **32**, 4134. 398

67JOM(8)105 H. Schumann, T. Östermann and M. Schmidt; *J. Organomet. Chem.*, 1967, **8**, 105. 895

67JOM(8)225 Y. Takashi; *J. Organomet. Chem.*, 1967, **8**, 225. 488

67JOM(8)427 H. C. Clark and A. L. Pickard; *J. Organomet. Chem.*, 1967, **8**, 427. 597

67JOM(9)231 L. H. Chan and E. G. Rochow; *J. Organomet. Chem.*, 1967, **9**, 231. 815

67JOM(9)285 A. J. Leusink, H. A. Budding, and J. W. Marman; *J. Organomet. Chem.* 1967, **9**, 285. 676, 970, 971

67JOM(9)295 A. J. Leusink, H. A. Budding, and W. Drenth; *J. Organomet. Chem.*, 1967, **9**, 295. 970, 971

67JOM(9)435 W. Findeiss, W. E. Davidsohn and M. C. Henry; *J. Organomet. Chem.*, 1967, **9**, 435. 1093

67JOM(10)P9 K. Matsuzaki and T. Yasukawa; *J. Organomet. Chem.*, 1967, **10**, P9. 566

67JPS292 W. O. Foye, A. M. Hebb and J. Micles; *J. Pharm. Sci.*, 1967, **56**, 292. 214

67LA(707)15 W. P. Neumann, H. Niermann and B. Schneider; *Justus Liebigs Ann. Chem.*, 1967, **707**, 15. 601

67LA(709)248 A. Tzschach and R. Schwarzer; *Justus Liebigs Ann. Chem.*, 1967, **709**, 248. 482, 510

67M1518 A. Gandini, P. Schenone and G. Bignardi; *Monatsh. Chem.*, 1967, **98**, 1518. 204

67M1792 H. Schindlbauer and W. Priloszowich; *Monatsh. Chem.*, 1967, **99**, 1792. 854

67M1944 H. Reinheckel and G. Tauber; *Monatsh. Chem.*, 1967, **98**, 1944. 393

B-67MI 205-01 Rylander; "Catalytic Hydrogenation over Platinum Metals," Academic Press, New York, 1967, pp. 139–159. 314

67MI 209-01 Y. L. Baay and A. G. MacDiarmid; *Inorg. Nucl. Chem. Lett.*, 1967, **3**, 159. 512

67MI 209-02 Y. Matsumura, M. Shindo and R. Okawara; *Inorg. Nucl. Chem. Lett.*, 1967, **3**, 219. 502

B-67MI 210-01 A. N. Nesmeyanov and K. A. Kocheshkov; "Methods of Elemento-Organic Chemistry," North Holland Publishing Company, Amsterdam, 1967, vol. 1. 520

B-67MI 210-02 M. F. Lappert; "Chemistry of Boron and its Compounds," ed. E. L. Muetterties, Wiley, New York, 1967, p. 443. 520

B-67MI 210-03 R. J. H. Voorhoeve; "Organohalosilanes: Precursors to Silicones," Elsevier, Amsterdam, 1967. 538

B-67MI 211-01 S. T. Ioffe and A. Nesmeyanov; "The Organic Chemistry of Beryllium, Magnesium, Calcium, Strontium and Barium," North-Holland, Amsterdam, 1967. 557

B-67MI 213-01 L. F. Fieser and M. Fieser; "Reagents for Organic Synthesis," Wiley, New York, 1967, vol. 1, p. 1081. 657

B-67MI 215-01 A. J. Kirby and S. G. Warren; in "The Organic Chemistry of Phosphorus," Elsevier, New York, 1967, ch. 10. 807

67MI 217-01 R. Müller, S. Reichel and C. Dathe; *Inorg. Nucl. Chem. Lett.*, 1967, **3**, 125. 873

B-67MI 223-01 L. G. Makarova and N. Nesmeyanov; "The Organic Compounds of Mercury," North-Holland, 1967, p. 1 1097

67NEP6614945 Deutsche Advance Produktion GmbH; *Neth. Pat.*, 6 614 945 (1967) (*Chem. Abstr.*, 1967, **67**, 108 757). 433

67NKK991 T. Hashimoto, M. Sugita, H. Kitano and K. Fukui; *Nippon Kagaku Kaishi*, 1967, **88**, 991. 279, 288

67OR(15)1 J. P. Schaefer and J. J. Bloomfield; *Org. React.*, 1967, **15**, 1. 749

67OR(15)204 G. Jones; *Org. React.* 1967, **15**, 204. 394

67PC-01 G. N. Sausen and A. L. Logothetis; *Personal Communication.* 7622

67SA(A)1169 R. E. Sacher, D. H. Lemmon and F. A. Miller; *Spectrochim. Acta, Part A*, 1967, 1169. 1093

67T1065 M. Grayson, C. E. Farley and C. A. Strenli; *Tetrahedron*, 1967, **23**, 1065. 840, 841, 854

67T2243 A. Zwierzak and A. Koziara; *Tetrahedron*, 1967, **23**, 2243. 414

67T3049 W. H. Dietsche; *Tetrahedron*, 1967, **23**, 3049. 851

67T4279 H. Nozaki, D. Tunemoto, Z. Morita, K. Nakamura, K. Watanabe, M. Takuku and K. Kondó; *Tetrahedron*, 1967, **23**, 4279. 1000

67TL407 A. Mooradian; *Tetrahedron Lett.*, 1967, 407. 692

67TL859 D. Kaminsky, J. Shavel, Jr. and R. I. Meltzer; *Tetrahedron Lett.*, 1967, 859. 692

67TL4131 T. P. Povlock; *Tetrahedron Lett.*, 1967, 4131. 36

67TL4447 A. M. Aguiar, T. G. Archibald and L. A. Kapicak; *Tetrahedron Lett.*, 1967, 4447. 875

67USP3308164 A. F. Shepard; *US Pat.* 3 308 164 (1967) (*Chem. Abstr.*, 1967, **67**, P2892z). 687

67USP3308165 A. F. Shepard; *US Pat.* 3 308 165 (1967) (*Chem. Abstr.*, 1967, **67**, P32479n). 687

67ZAAC(351)18 K. Issleib and G. Harzfeld; *Z. Anorg. Allg. Chem.*, 1967, **351**, 18. 440

67ZAAC(352)167 R. Paetzold and E. Wolfram; *Z. Anorg. Allg. Chem.*, 1967, **352**, 167. 288, 289, 292

67ZC151 H. Dorn and K. Walter; *Z. Chem.*, 1967, **7**, 151. 380

67ZN(B)1238 K. Issleib and F. Ungváry; *Z. Naturforsch., Teil B*, 1967, **22**, 1238. 430

67ZOB852 A. M. Pinchuk, L. N. Markovskii, E. S. Levchenko and V. I. Shevchenko; *Zh. Obshch. Khim.*, 1967, **37**, 852 (*Chem. Abstr.*, 1967, **67**, 108407k). 407

67ZOB856 A. M. Pinchuk, I. M. Kosinskaya and V. I. Shevchenko; *Zh. Obshch. Khim.*, 1967, **37**, 856 (*Chem. Abstr.*, 1967, **67**, 108169j). 407

67ZOB896 I. N. Zhmurova and A. P. Martynyuk; *Zh. Obshch. Khim.*, 1967, **37**, 896 (*Chem. Abstr.*, 1968, **68**, 29350d). 407

67ZOB1173 E. A. Krasilnikova, A. M. Potapov and A. I. Razumov; *Zh. Obshch. Khim.*, 1967, **37**, 1173. 461

67ZOB1692 E. E. Nifant'ev and L. V. Matveeva; *Zh. Obshch. Khim.*, 1967, **37**, 1692. 451

67ZOB2080 Z. S. Novikova, M. V. Proskurina, L. I. Petrovskaya, I. V. Bogdanova, N. P. Galitskova and I. F. Lutsenko; *Zh. Obshch. Khim.*, 1967, **37**, 2080 (*Chem. Abstr.*, 1968, **68**, 78 392). 436

67ZOB2522 R. K. Valetdinov, E. V. Kuznetsov, M. R. Rakhimova and M. K. Khasanov; *Zh. Obshch. Khim.*, 1967, **37**, 2522 (*Chem. Abstr.*, 1968, **68**, 87 353). 442

67ZOB2693 A. M. Pinchuk, I. M. Kosinskaya and V. I. Shevchenko; *Zh. Obshch. Khim.*, 1967, **37**, 2693 (*Chem. Abstr.*, 1968, **69**, 67069x). 407

67ZOB2787 P. M. Zavlin and T. F. Kondyurina; *Zh. Obshch. Khim.*, 1967, **37**, 2787. 469

67ZOR1282 E. S. Levchenko, L. N. Markovskii and A. V. Kirsanov; *Zh. Org. Khim.*, 1967, **3**, 1282 (*Chem. Abstr.*, 1967, **67**, 99792c). 399, 400

67ZOR2102 Yu. A. Boiko, B. S. Kupin and A. A. Petrov; *Zh. Org. Khim.*, 1967, **3**, 2102. 1032

68AG970 H. Stetter, M. Krause and W. D. Last, *Angew. Chem.*, 1968, **80**, 970. 174

68AG986 M. S. Gibson and R. W. Bradshaw; *Angew. Chem.*, 1968, **80**, 986. 299

68AG(E)112 V. Bazant, J. Joklik and J. Rathousky, *Angew. Chem., Int. Ed. Engl.*, 1968, **7**, 112. 478

68AG(E)138 U. Schöllkopf, E. Wiskott and K. Riedel; *Angew. Chem., Int. Ed. Engl.*, 1968, **7**, 138. 802

68AG(E)735 R. Koster and W. Fenzl; *Angew., Chem. Int. Ed. Engl.*, 1968, **7**, 735. 532

68AG(E)919 M. S. Gibson and R. W. Bradshaw; *Angew. Chem., Int. Ed. Engl.*, 1968, **7**, 919. 299

68B2898 H. E. Ganther; *Biochemistry*, 1968, **7**, 2898. 236

68BAP567 M. Michalska; *Bull. Acad. Pol. Sci. Ser. Sci. Chim.*, 1968, **16**, 567 (*Chem. Abstr.*, 1969, **70**, 87466y). 397

68BCJ1724 Z. Suzuki and K. Morita; *Bull. Chem. Soc. Jpn.*, 1968, **41**, 1724. 10

68BCJ2742 T. Kumamoto, K. Hosoi and T. Mukaiyama; *Bull. Chem. Soc. Jpn.*, 1968, 2742. 709

68BCJ3015 N. Kuneda, K. Sakai and S. Oae; *Bull. Chem. Soc. Jpn.*, 1968, **41**, 3015. 187

68BSB153 L. Christiaens and M. Renson; *Bull. Soc. Chim. Belg.*, 1968, **77**, 153. 288

68BSF193 J. Rauss-Godineau, J. Barrabis, W. Chodkiewickz and P. Cadiot; *Bull. Soc. Chim. Fr.*, 1968, 193. 1015

68BSF3385 M. Pereyre, J. Colin and J. Valade; *Bull. Soc. Chim. Fr.*, 1968, 3385. 601

68BSF4265 R. Granger, J.-P. Chapat, J. Crassous and F. Simon; *Bull. Soc. Chim. Fr.*, 1968, 4265. 697

68CA-P106877 N. K. Bliznyuk, L. D. Protasova, Z. N. Kvasha and S. L. Varshavskii; *Chem. Abstr.* 1968, **69**, P106 877. 448

68CAR(6)253 A. K. Chatterjee and D. L. MacDonald; *Carbohydr. Res.*, 1968, **6**, 253. 105

68CB121 D. Hellwinkel and G. Kilthau; *Chem. Ber.*, 1968, **101**, 121. 879

68CB743 E. Ettenhuber and K. Rühlmann; *Chem. Ber.*, 1968, **101**, 743. 813

68CB1271 H. Schmidbaur and G. Jones; *Chem. Ber.*, 1968, **101**, 1271. 504

68CB1381 U. Schmidt, I. Boie, C. Osterroht, R. Schroer and H.-F. Grutzmacher; *Chem. Ber.*, 1968, **101**, 1381. 506, 852

68CB1414 W. Wiegräbbe and H. Bock; *Chem. Ber.*, 1968, **101**, 1414. 863

68CB2197 K. Issleib and H. Weichmann; *Chem. Ber.*, 1968, **101**, 2197. 431, 454, 822, 854

68CB2899 L. Horner, P. Beck and R. Luckenbach; *Chem. Ber.*, 1968, **101**, 2899. 822, 823

68CB2956 W. Rundel; *Chem. Ber.*, 1968, **101**, 2956. 706

68CB3070 N. Petragnani, V. G. Toscano and M. D. M. Campos; *Chem. Ber.*, 1968, **101**, 3070. 857

68CB3438 E. Lindner and H. Kranz; *Chem. Ber.*, 1968, **101**, 3438. 438

68CB3454 W. Kuchen and H. Mayatepek; *Chem. Ber.*, 1968, **101**, 3454. 462

68CB3567 H.-H. Horhold and H. Eibisch; *Chem. Ber.*, 1968, **101**, 3567. 245, 250

68CB3612 K. Issleib, R. Kummel, H. Oehme and I. Meissner; *Chem. Ber.*, 1968, **101**, 3612. 427, 430, 462

68CB3883 A. Schmidpeter and J. Ebling; *Chem. Ber.*, 1968, **101**, 3883. 468

68CB3976 A. Schmidt; *Chem. Ber.*, 1968, **101**, 3976. 504

68CB4148 O. J. Scherer and G. Shieder; *Chem. Ber.*, 1968, **101**, 4148. 470

68CC9 R. M. Dodson, P. D. Hammen and R. A. Davis; *J. Chem. Soc., Chem. Commun.*, 1968, 9. 200

68CC495 E. C. Alyea, J. S. Basi, D. C. Bradley and M. H. Chisholm; *J. Chem. Soc., Chem. Commun.*, 1968, 495. 421

68CC806 D. H. R. Barton, A. K. Ganguly, R. H. Hesse, S. N. Loo and M. M. Pechet; *J. Chem. Soc., Chem. Commun.*, 1968, 806. 608, 612

68CC976 Y. Iskander, E. M. E. Mansour, A. G. A. Rehiem, Y. Riad and A. A. Youssef; *J. Chem. Soc., Chem. Commun.*, 1968, 976. 26

68CC1017 T. Gallagher and D. Lathbury; *J. Chem. Soc., Chem. Commun.*, 1968, 1017. 765

68CC1290 A. W. Downs; *J. Chem. Soc., Chem. Commun.*, 1968, 1290. 290

68CC1350 I. M. Downie, J. B. Lee and M. F. S. Matough; *J. Chem. Soc., Chem. Commun.*, 1968, 1350. 7

68CJC1149 J. R. Shelton, J. F. Garmish, C. K. Liang, P. L. Samuel, P. Kovacic and L. W. Haynes; *Can. J. Chem.*, 1968, **46**, 1149. 403

68CJC3625 T. R. Kasturi and T. Arunachalam; *Can. J. Chem.*, 1968, **46**, 3625. 658

68CPB1009 Y. Kobayashi and C. Akashi; *Chem. Pharm. Bull.*, 1968, **16**, 1009. 7, 835

68CPB1784 Y. Kobayashi, C. Akashi and K. Morinaga; *Chem. Pharm. Bull.*, 1968, **16**, 1784. 20

68CR(C)270 R. Burgada and H. Germa; *C. R. Hebd. Seances Acad. Sci., Ser. C*, 1968, **267**, 270. 476

68CR(C)623 A. Kirrmann and E. Elkik; *C. R. Hebd. Seances Acad. Sci., Ser. C.*, 1968, 623. 757

68CR(C)937 J. Benaim; *C. R. Hebd. Seances Acad. Sci., Ser. C*, 1968, **262**, 937. 1076

68DOK(181)877 Zh. L. Evtikov, N. A. Razumova and A. A. Petrov; *Dokl. Akad. Nauk SSSR*, 1968, **181**, 877 (*Chem. Abstr.*, 1969, **70**, 78 084). 454

68DOK(182)337 Yu. I. Baranov, O. F. Filippov, S. L. Varshavskii and M. I. Kabachnik; *Dokl. Akad. Nauk SSSR*, 1968, **182**, 337. 449

68HCA405 L. Maier; *Helv. Chim. Acta*, 1968, **51**, 405. 452

68HCA1603 F. Scheinmann, R. Barner and H. Schmid; *Helv. Chim. Acta*, 1968, **51**, 1603. 656

68HOU(10/4)311 W. Rundel; *Methoden Org. Chem., (Houben-Weyl)*, 1968, **10/4**, 311. 765

68IC18 F. A. Cotton, R. Hügel and R. Eiss; *Inorg. Chem.*, 1968, **7**, 18. 1004

68IC391 E. W. Lawless and L. D. Harman; *Inorg. Chem.*, 1968, **7**, 391. 175

68IC698 K. Utvary and H. H. Sisler; *Inorg. Chem.*, 1968, **7**, 698. 387

68IC834 C. G. Moreland, M. H. O'Brien, C. E. Douthit and G. G. Long; *Inorg. Chem.*, 1968, **7**, 834. 492

68IC1253 E. Louis and G. Urry; *Inorg. Chem.*, 1968, **7**, 1253. 5411

68IC1298 J. P. Collman, J. W. Kang, W. F. Little and M. P. Sullivan; *Inorg. Chem.*, 1968, **7**, 1298. 985

68IC1327 R. Schmutzler; *Inorg. Chem.*, 1968, **7**, 1327. 869

68IC2003 R. D. Feltham, H. G. Metzger and W. Silverthorn; *Inorg. Chem.*, 1968, **7**, 2003. 874

68ICA379 M. A. Bennett, G. J. Erskine and J. D. Wild; *Inorg. Chem. Acta*, 1968, **2**, 379. 874

68IS65 G. N. Schrauzer; *Inorg. Synth.*, 1968, **11**, 65. 578

68IS124 W. L. Jolly; *Inorg. Synth.*, 1968, **11**, 124. 427

68IS126 W. L. Jolly; *Inorg. Synth.*, 1968, **11**, 126. 430

68IS157 G. W. Parahall; *Inorg. Synth.*, 1968, **11**, 157. 431

68IZV190 R. G. Kostyanovskii, V. V. Yakshin and S. L. Zimont; *Izv. Akad. Nauk SSSR, Ser. Khim.*, 1968, 190 (*Chem. Abstr.*, 1968, **69**, 43 987). 439

68IZV1889 M. G. Lin'kova and I. L. Knunyants; *Izv. Akad. Nauk SSSR, Ser. Khim.*, 1968, 1889 (*Chem. Abstr.*, 1969, **70**, 3975c). 167

68IZV2287 A. E. Borisov, A. I. Borisova and L. V. Kudryavtseva; *Izv. Akad. Nauk SSSR, Ser. Khim.*, 1968, 2287 (*Chem. Abstr.*, 1969, **70**, 29 023r). 1077, 1081, 1097

68JA808 D. D. Tanner and G. C. Gidley; *J. Am. Chem. Soc.*, 1968, **90**, 808. 33, 95

68JA909 A. C. Cope and E. C. Friedrich; *J. Am. Chem. Soc.*, 1968, **90**, 909. 964

68JA938 G. A. Olah and M. Calin; *J. Am. Chem. Soc.*, 1968, **90**, 938. 1007

68JA2459 T. L. Emmick and R. L. Letsinger; *J. Am. Chem. Soc.*, 1968, **90**, 2459. 450

68JA2651 R. G. Laughlin; *J. Am. Chem. Soc.*, 1968, **90**, 2651. 402

68JA2677	R. G. Hiskey and R. L. Smith; *J. Am. Chem. Soc.*, 1968, **90**, 2677.	220
68JA2686	H. C. Brown and N. M. Yoon; *J. Am. Chem. Soc.*, 1968, **90**, 2686.	83
68JA3459	T. L. Emmick and R. L. Letsinger; *J. Am. Chem. Soc.*, 1968, **90**, 3459.	840, 841, 845
68JA3869	A. Nudelman and D. J. Cram; *J. Am. Chem. Soc.*, 1968, **90**, 3869.	362, 846
68JA4096	M. T. Reagen and A. Nickon; *J. Am. Chem. Soc.*, 1968, **90**, 4096.	249
68JA4462	G. Stork and P. F. Hudrlik; *J. Am. Chem. Soc.*, 1968, **90**, 4462.	539, 696
68JA4666	G. A. Olah and T. E. Kiovsky; *J. Am. Chem. Soc.*, 1968, **90**, 4666.	1009
68JA4842	O. Korpiun, R. A. Lewis, J. Chickos and K. Mislow; *J. Am. Chem. Soc.*, 1968, **90**, 4842.	807, 845
68JA5038	H. C. Brown, M. W. Rathke and M. M. Rogic; *J. Am. Chem. Soc.*, 1968, **90**, 5038.	34
68JA5236	J. W. Rakshys, R. W. Taft and W. A. Sheppard; *J. Am. Chem. Soc.*, 1968, **90**, 5236.	835, 839
68JA5280	E. F. Knights and H. C. Brown; *J. Am. Chem. Soc.*, 1968, **90**, 5280.	516
68JA5403	C. T. Ratcliffe and J. M. Shreeve; *J. Am. Chem. Soc.*, 1968, **90**, 5403.	166
68JA5936	J. Hooz and S. Linke; *J. Am. Chem. Soc.*, 1968, **90**, 5936.	697
68JA6225	W. S. Johnson, T. Li, D. J. Faulkner and S. F. Campbell; *J. Am. Chem. Soc.*, 1968, **90**, 6225.	35
68JA6243	G. Zweifel, N. L. Polston and C. C. Whitney; *J. Am. Chem. Soc.*, 1968, **90**, 6243.	900
68JA6556	A. D. Norman; *J. Am. Chem. Soc.*, 1968, **90**, 6556.	473
68JA6726	F. Ramirez, C. P. Smith and J. F. Pilot; *J. Am. Chem. Soc.*, 1968, **90**, 6726.	476
68JA7173	M. N. Ackermann, J. L. Ellenson and D. H. Robison; *J. Am. Chem. Soc.*, 1968, **90**, 7173.	402
68JA7287	E. M. Kaiser, F. E. Henoch and C. R. Hauser; *J. Am. Chem. Soc.*, 1968, **90**, 7287.	1005
68JA7378	J. L. Kice and K. Ikura; *J. Am. Chem. Soc.*, 1968, **90**, 7378.	208
68JCS839	A. D. Brown, Jr. and G. M. Kosolapoff; *J. Chem. Soc.*, 1968, 839.	450
68JCS1133	A. Davison, J. A. McClevarty and G. Wilkinson; *J. Chem. Soc.*, 1963, 1133.	478
68JCS(A)483	M. Green and D. Kirkpatrick; *J. Chem. Soc. (A)*, 1968, 483.	833, 890
68JCS(A)1406	R. G. Cavell and R. C. Dobbie; *J. Chem. Soc. (A)*, 1968, 1406.	472, 508
68JCS(A)1909	K. Gosling, D. J. Holman, J. D. Smith and B. N. Ghose; *J. Chem. Soc. (A)*, 1968, 1909.	435
68JCS(A)2149	R. D. W. Kemmitt, D. I. Nichols and R. D. Peacock; *J. Chem. Soc. (A)*, 1968, 2149.	824
68JCS(A)2605	A. Storr; *J. Chem. Soc. (A)*, 1968, 2605.	421
68JCS(A)2623	E. W. Abel, R. Honigschmidt-Grossich and S. M. Illingworth; *J. Chem. Soc. (A)*, 1968, 2623.	508, 509, 894
68JCS(A)2962	R. H. Cragg; *J. Chem. Soc. (A)*, 1968, 2962.	267
68JCS(B)1416	P. S. Crofts and I. S. Fox; *J. Chem. Soc. (B)*, 1968, 1416.	461
68JCS(C)317	A. G. Davies and R. J. Puddephatt; *J. Chem. Soc. (C)*, 1968, 317.	1101
68JCS(C)416	R. W. Kelly and P. J. Sykes; *J. Chem. Soc. (C)*, 1968, 416.	23
68JCS(C)796	E. S. Alexander, R. N. Hazeldine, M. J. Newlands and A. E. Tipping; *J. Chem. Soc.(C)*, 1968, 796.	339
68JCS(C)922	B. M. Culbertson and S. Dietz; *J. Chem. Soc. (C)*, 1968, 922.	478
68JCS(C)1265	S. N. Bhattacharya, C. Eaborn and D. R. M. Walton; *J. Chem. Soc. (C)*, 1968, 1265.	176
68JCS(C)1317	U. Folli, D. Iarossi, F. Montanari and G. Torre; *J. Chem. Soc. (C)*, 1968, 1317.	147
68JCS(C)1324	E. Robson, J. M. Tedder and D. J. Woodcock; *J. Chem. Soc. (C)*, 1968, 1324.	1051
68JCS(C)1704	K. Blazevic, R. P. Houghton and C. S. Williams; *J. Chem. Soc., (C)*, 1968, 1704.	421
68JCS(C)2131	R. S. Davidson; *J. Chem. Soc. (C)*, 1968, 2131.	854
68JCS(C)2317	C. Sellers and H. Suschitzky; *J. Chem. Soc. (C)*, 1968, 2317.	884
68JCS(C)2374	B. J. Auret, D. R. Boyd and H. B. Henbest; *J. Chem. Soc. (C)*, 1968, 2374.	151
68JCS(C)2713	D. R. Hogg, J. H. Smith and P. W. Vipond; *J. Chem. Soc. (C)*, 1968, 2713.	169
68JGU370	Yu. F. Gatilov, L. B. Ionov and G. Kh. Kamai; *J. Gen. Chem. USSR*, 1968, **38**, 370.	890
68JGU1295	K. I. Kuz'min and Z. U. Panfilovich; *J. Gen. Chem. USSR (Engl. Transl.)*, 1968, **38**, 1295.	1078
68JHC397	M. J. Kornet and S. I. Tan; *J. Heterocycl. Chem.*, 1968, **5**, 397.	377, 381, 383
68JINC1715	H. G. Ang, G. Manoussakis and Y. O. El-Nigumi; *J. Inorg. Nucl. Chem.*, 1968, **30**, 1717.	504
68JMC446	S.-H. Chu and H. G. Mautner; *J. Med. Chem.*, 1968, **11**, 446.	278
68JMC504	W. T. Smith, Jr and W. Y. Chen; *J. Med. Chem.*, 1968, **11**, 504.	378, 386
68JOC38	D. D. Tanner and G. C. Gidley; *J. Org. Chem.*, 1968, **33**, 38.	23
68JOC313	G. B. Bachmann and K. G. Strawn; *J. Org. Chem.*, 1968, **33**, 313.	397
68JOC344	S. Temple; *J. Org. Chem.*, 1968, **33**, 344.	176
68JOC579	A. M. Aguiar and M. G. Raghanan Nair; *J. Org. Chem.*, 1968, **33**, 579.	452
68JOC632	R. Obrycki and C. E. Griffin; *J. Org. Chem.*, 1968, **33**, 632.	849
68JOC780	D. J. Peterson; *J. Org. Chem.*, 1968, **33**, 780.	709
68JOC802	R. Sommer and H. G. Kuivila; *J. Org. Chem.*, 1968, **33**, 802.	601
68JOC819	R. E. Dear and E. E. Gilbert; *J. Org. Chem.*, 1968, **33**, 819.	1016
68JOC1008	C. M. Sharts; *J. Org. Chem.*, 1968, **33**, 1008.	335
68JOC1275	D. J. Martin and C. C. Greco; *J. Org. Chem.*, 1968, **33**, 1275.	123
68JOC1374	R. C. Slagel; *J. Org. Chem.*, 1968, **33**, 1374.	384
68JOC1636	L. Friedman and J. F. Chlebowski; *J. Org. Chem.*, 1968, **33**, 1636.	632
68JOC1681	A. M. Aguiar, J. T. Mague, H. J. Aguiar, T. G. Archibald and G. W. Prejean; *J. Org. Chem.*, 1968, **33**, 1681.	874
68JOC2095	R. L. Talbott; *J. Org. Chem.*, 1968, **33**, 2095.	98
68JOC2104	I. B. Douglass and R. V. Norton; *J. Org. Chem.*, 1968, **33**, 2104.	172, 173, 174
68JOC2111	W. H. Mueller and P. E. Butler; *J. Org. Chem.*, 1968, **33**, 2111.	359
68JOC2407	I. Bhatnagar and M. V. George; *J. Org. Chem.*, 1968, **33**, 2407.	388
68JOC2590	E. Schweizer, M. El Bakoush, K. Light and K. Oberle; *J. Org. Chem.*, 1968, **33**, 2590.	445
68JOC2664	T. E. Stevens; *J. Org. Chem.*, 1968, **33**, 2664.	763
68JOC3088	L. G. Vaughan and R. V. Lindsey, Jr.; *J. Org. Chem.*, 1968, **33**, 3088.	440
68JOC3422	J. S. Meek and J. S. Fowler; *J. Org. Chem.*, 1968, **33**, 3422.	156

68JOC3690 H. R. Hays; *J. Org. Chem.*, 1968, **33**, 3690. 450, 452

68JOC3976 R. Harville and S. F. Reed; *J. Org. Chem.*, 1968, **33**, 3976. 148

68JOC4406 M. E. Kuehne and P. J. Sheeran; *J. Org. Chem.*, 1968, **33**, 4406. 1048

68JOM(11)97 M. Pereyre, B. Bellegarde, J. Mendelsohn and J. Valade; *J. Organomet. Chem.*, 1968, **11**, 97. 701

68JOM(11)299 M. Shindo, Y. Matsumura and R. Okawara; *J. Organomet. Chem.*, 1968, **11**, 299. 502, 889

68JOM(11)393 E. A. Jeffery and T. Mole; *J. Organomet. Chem.*, 1968, **11**, 393. 1096

68JOM(11)615 A. J. Chalk and T. J. Hoogeboom; *J. Organomet. Chem.*, 1968, **11**, 615. 554

68JOM(11)627 E. J. Kupchik and C. Th. Theisen; *J. Organomet. Chem.*, 1968, **11**, 627. 889, 890

68JOM(12)133 H. Goldwhite, D. G. Rowsell and C. Valdez; *J. Organomet. Chem.*, 1968, **12**, 133. 487, 875

68JOM(12)225 K.-H. Thiele and J. Köhler; *J. Organomet. Chem.*, 1968, **12**, 225. 586

68JOM(12)377 C. F. McBrearty, Jr., K. Irgolic and R. A. Zingaro; *J. Organomet. Chem.*, 1968, **12**, 377. 414, 504

68JOM(12)401 M. T. Chen and J. W. George; *J. Organomet. Chem.*, 1968, **12**, 401. 292

68JOM(12)P33 J. J. Eisch and M. W. Foxton; *J. Organomet. Chem.*, 1968, **12**, P33. 970

68JOM(13)529 W. T. Reichle; *J. Organomet. Chem.*, 1968, **13**, 529. 891

68JOM(14)21 E. Weiss and H. Plass; *J. Organomet. Chem.*, 1968, **14**, 21. 1096

68JOM(14)229 I. F. Lutsenko, Yu. I. Baukov, I. Yu. Belavin and A. N. Tvorogov; *J. Organomet. Chem.*, 1968, **14**, 229. 700

68JOM(14)327 T. A. George and M. F. Lappert; *J. Organomet. Chem.*, 1968, **14**, 327. 920

68JOM(15)69 W. Siebert, W. E. Davidsohn and M. C. Henry; *J. Organomet. Chem.*, 1968, **15**, 69. 1060, 1061

68JOM(15)500 A. McKillop, L. F. Elson and E. C. Taylor; *J. Organomet. Chem.*, 1968, **15**, 500. 598

68JPR257 R. Ohme, E. Schmitz and L. Sterk; *J. Prakt. Chem.*, 1968, **37**, 257. 374

68LA(713)74 R. Ohme and H. Preuschhof; *Justus Liebigs Ann. Chem.*, 1968, **713**, 74. 374, 378, 380

68LA(714)1 H. Hartmann; *Justus Liebigs Ann. Chem.*, 1968, **714**, 1. 436, 825, 826

68LA(715)1 H. Goetz and D. Probst; *Justus Liebigs Ann. Chem.*, 1968, **715**, 1. 825

68LA(716)19 W. P. Neumann and F. G. Kleiner; *Justus Liebigs Ann. Chem.*, 1968, **716**, 19. 1101

68LA(717)1 R. Köster, H.-J. Horstschäfer and P. Binger; *Justus Liebigs Ann. Chem.*, 1968, **717**, 1. 1077

68LA(717)21 P. Binger, G. Benedikt, G. W. Rotermund and R. Köster; *Justus Liebigs Ann. Chem.*, 1968, **717**, 21. 1083, 1084, 1087, 1088, 1097

68LA(717)54 W. Ried and W. Kunkel; *Justus Liebigs Ann. Chem.*, 1968, **717**, 54. 662

68LA(719)40 H. Lehmkuhl; *Justus Liebigs Ann. Chem.*, 1968, **719**, 40. 595

68LA(719)169 R. Koster and M. A. Grassberger; *Justus Liebigs Ann. Chem.*, 1968, **719**, 169. 526

68LA(720)1 R. Köster, H. Bellut and S. Hattori; *Justus Liebigs Ann. Chem.*, 1968, **720**, 1. 418

B-68MI 202-01 J. A. Monick; "Alcohols, Their Chemistry, Properties and Manufacture," Reinhold, New York, 1968. 38, 87

68MI 202-02 N. J. Stevens and J. R. Livingston; *Chem. Eng. Prog.*, 1968, **64**, 61. 112

68MI 202-03 D. R. Arnold; *Adv. Photochem.*, 1968, **6**, 301. 92

B-68MI 203-01 S. Karo (ed); "Organic Functional Group Preparations," Academic Press, London, 1968. 116, 132

B-68MI 203-02 G. A. Russell and G. J. Mikol; in "Mechanisms of Molecular Migrations," ed. B. S. Thyagarajan, Wiley, New York, 1968, vol. 1. 174

B-68MI 203-03 J. Tsurugi; in "Mechanisms of Reactions of Sulfur Compounds," 1968, vol. 2. 211

B-68MI 205-01 "The Chemistry of the Amino Group," ed. S. Patai, Interscience, London, 1968. 298, 300, 303

B-68MI 206-01 B. C. Challis and A. R. Butler; in "The Chemistry of the Amino Group," ed. S. Patai, Interscience, London, 1968, p. 302. 335, 337, 339, 340, 344, 348

68MI 208-01 M. V. Proskurina, I. F. Lutsenko, Z. S. Novikova and N. P. Voronova; *Khim. Org. Soedin. Fosfora, Akad. Nauk SSSR, Otd. Obshch. Tekh. Khim.*, 1967, 8 (*Chem. Abstr.*, 1968, **69**, 52 217). 436

B-68MI 210-01 R. W. Bott and C. Eaborn; in "Organometallic Compounds of the Group IV Elements," ed. A. G. MacDiarmid, Dekker, New York, 1968, vol. 1, part 1, p. 105. 536, 538

68MI 213-01 G. P. Armstrong; "Mod. Chem. Ind., Proc. Int. Union Pure Appl. Chem. Symp.," 1968, 165 (*Chem. Abstr.*, 1970, **73**, 45007f). 655

68MI 215-01 R. B. Conrow and S. Bernstein; *Steroids*, 1968, **11**, 151. 753

B-68MI 215-02 B. C. Challis and A. R. Butler; in "The Chemistry of the Amino Group," ed. S. Patai, Interscience, London, 1968, p. 302. 761

B-68MI 215-03 S. R. Sandler and W. Karo; "Organic Functional Group Preparations," Academic Press, New York, 1968, vol. 1. 773

B-68MI 215-04 P. Y. Sollenberger and R. B. Martin; in "The Chemistry of the Amino Group," ed. S. Patai, Interscience, London, 1968, pp. 395–397. 773

B-68MI 215-05 W. Wucherpfenning and G. Kresze; "Mechanisms of Reactions of Sulfur Compounds," 1968, vol. 2, p. 89. 796

68MI 217-01 G. Marchese and F. Naso; *Boll. Sci. Fac. Chim. Ind. Bologna*, 1968, **26**, 283 (*Chem. Abstr.*, 1968, **70**, 106 074n). 874

68MI 223-01 J. M. Barnes and L. Magoos; *Organomet. Chem. Rev.*, 1968, **3**, 137. 1097

B-68MI 223-02 H. Shapiro and F. W. Frey; "The Organic Compounds of Lead," Wiley, New York, 1968, p. 207. 1101

68NKK83 T. Hashimoto, H. Kitano and K. Fukui; *Nippon Kagaku Kaishi*, 1968, **89**, 83. 288

68OS(48)4 P. Kovacic and S. S. Chaudhary; *Org. Synth.*, 1968, **48**, 4. 338

68OSC(6)670 N. A. LaBel and D. Hwang; *Org. Synth., Coll. Vol.*, 1968, **6**, 184. 354

68QRS189 D. L. Klayman and R. J. Shine; *Quart. Rep. Sulfur Chem.*, 1968, **3**, 189. 220

68RCR907 M. F. Shostakovskii, B. A. Trofimov, A. S. Atavin and V. I. Lavrov; *Russ. Chem. Rev. (Engl. Transl.)*, 1968, **37**, 907. 674

68RTC97 J. H. van Boom, L. Brandsma and J. F. Arens; *Recl. Trav. Chim. Pays-Bas*, 1968, **87**, 97. 1032

68RTC916 S. Hoff, L. Brandsma and J. F. Arens; *Recl. Trav. Chim. Pays-Bas*, 1968, **87**, 916. 939

68SCI3836 J. P. Ferris; *Science*, 1968, **161**, 3836. 106

68T1889 A. Risaliti, M. Forchiassin and E. Valentin; *Tetrahedron*, 1968, **24**, 1889. 783

68T2481 A. Novelli, A. P. G. V. de Varela and J. D. Bonafede; *Tetrahedron*, 1968, **24**, 2481. 749

68T4217 R. Buyle and H. G. Viehe; *Tetrahedron*, 1968, **24**, 4217. 746

68T4677 L. A. Yanovskaya, Ch. Shachidayatov, E. P. Prokofiev, G. M. Andrianova and V. F. Kucherov; *Tetrahedron*, 1968, **24**, 4677. 748

68TL85 D. B. Denny, D. Z. Denny and L. A. Wilson; *Tetrahedron Lett.*, 1968, 85. 476

68TL319 J. Sauer and K. K. Mayer; *Tetrahedron Lett.*, 1968, 319. 163

68TL1755 G. Märkl and R. Potthast; *Tetrahedron Lett.*, 1968, 1755. 853

68TL3057 D. F. Veber, J. D. Milkowski, R. G. Denkewalter and R. Hirschmann; *Tetrahedron Lett.*, 1968, 3057. 223

68TL3607 P. Claus and W. Vycudiliak; *Tetrahedron Lett.*, 1968, 3607. 162

68TL3909 R. E. Banks, M. G. Barlow, W. D. Davies, R. N. Haszeldine, K. Mullen and D. R. Taylor; *Tetrahedron Lett.*, 1968, 3909. 1013

68TL4265 R. Niess, K. Nagel and P. Effenberger; *Tetrahedron Lett.*, 1968, 4265. 1009

68TL4479 D. Wagner, D. Gertner and A. Zilka; *Tetrahedron Lett.*, 1968, 4479. 190

68TL5015 Y. Yamada, K. Mukai and H. Yoshioka; *Tetrahedron Lett.*, 1968, 5015. 855

68TL5071 T. Fujisawa, T. Kobori, N. Ohtsuka and G. Tsuchihashi; *Tetrahedron Lett.*, 1968, 5071. 712

68TL5149 A. J. Krubsack and T. Higa; *Tetrahedron Lett.*, 1968, 5149. 170

68TL5333 M. Wieber and W. R. Hoos; *Tetrahedron Lett.*, 1968, 5333. 476

68TL5357 L. I. Peterson; *Tetrahedron Lett.*, 1968, 5357. 1051

68TL5629 J. F. Biellman and J. B. Ducep; *Tetrahedron Lett.*, 1968, 5629. 143

68TL5907 T. Mukaiyama and K. Takahashi; *Tetrahedron Lett.*, 1968, 5907. 118, 124, 213, 217

68TL6155 T. W. Gorrod; *Tetrahedron Lett.*, 1968, **59**, 6155. 782

68USP3392180 L. A. Hamilton; *US Pat.* 3 392 180 (1968) (*Chem. Abstr.*, 1968, **69**, 77 475). 434

68ZAAC(356)195 K. H. Thiele and K. Jakob; *Z. Anorg. Allg. Chem.*, 1968, **356**, 195. 566

68ZAAC(358)44 H. Nöth and H. Suchy; *Z. Anorg. Allg. Chem.*, 1968, **358**, 44. 112

68ZAAC(360)293 R. Paetzold and G. Bochmann; *Z. Anorg. Allgem. Chem.*, 1968, **360**, 293. 287

68ZAAC(361)296 O. J. Scherer and J. Wokulat; *Z. Anorg. Allg. Chem.*, 1968, **361**, 296. 415

68ZAAC(363)84 H. Schmidbaur and M. Bergfeld; *Z. Anorg. Allg. Chem.*, 1968, **363**, 84. 494

68ZAAC(363)176 R. Appel, B. Blaser and G. Siegemund; *Z. Anorg. Allg. Chem.*, 1968, **363**, 176. 864

68ZC41 R. Ohme and A. Zubek; *Z. Chem.*, 1968, **8**, 41. 373

68ZN1030 F. Fehér and D. Kurz; *Z. Naturforsch.*, 1968, **23b**, 1030. 853

68ZOB609 E. A. Krasil'nikova, A. M. Potapov and A. I. Razumov; *Zh. Obshch. Khim.*, 1968, **38**, 609 (*Chem. Abstr.*, 1968, **69**, 43 993). 462

68ZOB1551 E. A. Smirnov, Yu. M. Zinov'ev and V. A. Petrunin; *Zh. Obshch. Khim.*, 1968, **38**, 1551 (*Chem. Abstr.*, 1968, **69**, 96 827y). 448

68ZOB1906 G. I. Drozd, S. Z. Ivin and V. V. Sheluchenko; *Zh. Obshch. Khim.*, 1968, **38**, 1906. 476

68ZOB2285 E. N. Tsvetkov, D. I. Lobanov and M. I. Kabachnik; *Zh. Obshch. Khim.*, 1968, **38**, 2285. 848

68ZOB2509 L. M. Yagupol'skii and V. G. Voloshchuk; *Zh. Obshch. Khim.*, 1968, **38**, 2509. 287, 732

69ACS2935 P. Froyen; *Acta Chem. Scand.*, 1969, **23**, 2935. 809

69ACS2983 B. Andersson and B. Lamm; *Acta Chem. Scand.*, 1969, **23**, 2983. 804

69AG468 L. Ghosez, B. Haveaux and H. G. Viehe; *Angew. Chem.*, 1969, **81**, 468. 758

69AG(E)273 V. Jäger and H. G. Viehe; *Angew. Chem., Int. Ed. Engl.*, 1969, **8**, 273. 1052

69AG(E)450 D. Seebach and N. Peleties; *Angew. Chem., Int. Ed. Engl.*, 1969, **8**, 450. 281

69AG(E)677 H. Alper; *Angew. Chem., Int. Ed. Engl.*, 1969, **8**, 677. 125, 221, 228

69AG(E)773 F. Seel, W. Gombler; *Angew. Chem., Int. Ed. Engl.*, 1969, **8**, 773. 166

69AJC765 W. D. Crow and I. Gosney; *Aust. J. Chem.*, 1969, **22**, 765. 235

69AP(302)354 J. Gloede, J. Freiberg, W. Bürger, G. Ollmann and H. Gross; *Arch. Pharm. (Weinheim, Ger.)*, 1969, **302**, 354. 1008

69AX(B)73 A. T. Christensen and W. G. Whitmore; *Acta Crystallogr., Part B*, 1969, **25**, 73. 1000

69B2557 H. E. Ganther and C. Corcoran; *Biochemistry*, 1969, **8**, 2557. 478

69BBR(37)593 N. S. Kosower, E. M. Kosower, B. Wertheim and W. S. Correa; *Biochem. Biophys. Res. Commun.*, 1969, **37**, 593. 216

69BCJ2975 I. Ojima, K. Akiba and N. Inamoto; *Bull. Chem. Soc. Jpn.*, 1969, **42**, 2975. 840

69BCJ3606 K. Suga, S. Watanbe, T. Fiyita and T. P. Pan; *Bull. Chem. Soc. Jpn.*, 1969, **42**, 3606. 308

69BRP1167809 A. Etienne and Y. Correia; *Br. Pat.* 1 167 809 (1969) (*Chem. Abstr.*, 1970, **72**, 31 602p). 788

69BSB191 E. J. Goethals, J. Huylebroeck and W. Smolders; *Bull. Soc. Chim. Belg.*, 1969, **78**, 191. 232

69BSF3609 G. Rio and J. Berthelot; *Bull. Soc. Chim. Fr.*, 1969, **10**, 3609. 209

69BSF4169 M. Gaudry and A. Marquet; *Bull. Soc. Chim. Fr.*, 1969, 4169. 29

69CAR(9)201 R. H. Bell, D. Horton and M. J. Miller; *Carbohydr. Res.*, 1969, **9**, 201 (*Chem. Abstr.*, 1969, **70**, 78 267g). 171

69CB1489 L. Almasi and L. Paskucz; *Chem. Ber.*, 1969, **102**, 1489. 361

69CB1883 G. P. Schiemenz and H.-U. Siebeneick; *Chem. Ber.*, 1969, **102**, 1883. 824, 827, 828

69CB2557 A. Schonberg and E. Singer; *Chem. Ber.*, 1969, **102**, 2557. 809

69CB2900 H. Schumann, P. Schwabe and O. Stelzer; *Chem. Ber.*, 1969, **102**, 2900. 867, 868

69CB2914 H. Schindlbauer and W. Prikoszovich; *Chem. Ber.*, 1969, **102**, 2914. 821, 822, 824, 825, 845

69CB2922 W. Prikoszovich and H. Schindlbauer; *Chem. Ber.*, 1969, **102**, 2922. 844, 848

69CB3357 H. Stetter, M. Krause and W.-D. Last; *Chem. Ber.*, 1969, **102**, 3357. 174

69CB3731 H. Schumann and A. Roth; *Chem. Ber.*, 1969, **102**, 3731. 809

69CB3955 R. Appel and D. Rebban; *Chem. Ber.*, 1969, **102**, 3955. 483

69CB4136 H. Schmidbaur, J. Weidlein and K.-H. Mitschke; *Chem. Ber.*, 1969, **102**, 4136. 488

69CC112 M. Fetizon, M. Jurion and N. T. Anh; *J. Chem. Soc., Chem. Commun.*, 1969, 112. 693
69CC365 W. D. Kingsbury and C. R. Johnson; *J. Chem. Soc., Chem. Commun.*, 1969, 365. 148
69CC578 R. G. R. Bacon and A. Karim; *J. Chem. Soc., Chem. Commun.*, 1969, 578. 753
69CC832 R. H. Cragg; *J. Chem. Soc., Chem. Commun.*, 1969, 832. 267
69CC1064 F. Lappert and A. R. Sanger; *J. Chem. Soc., Chem. Commun.*, 1969, 1064. 421
69CC1159 R. M. Dodson and T. P. Nelson; *J. Chem. Soc., Chem. Commun.*, 1969, 1159. 153
69CC1205 W. H. Mueller and M. B. Dines; *J. Chem. Soc., Chem. Commun.*, 1969, 1205. 208
69CC1393 J. H. Lee, A. Matsumoto, O. Simamura and M. Yoshida; *J. Chem. Soc., Chem. Commun.*, 1969, 1393. 1054
69CC1515 M. Allard, J. Levisalles and J. M. Sommer; *J. Chem. Soc., Chem. Commun.*, 1969, 1515. 799
69CCC4010 F. Hrabák; *Collect. Czech. Chem. Commun.*, 1969, **34**, 4010. 376
69CI(L)200 A. P. Marchand and W. R. Weimar, Jr; *Chem. Ind. (London)*, 1969, 200. 32
69CJC2371 K. R. Kopecky and T. Gilban; *Can. J. Chem.*, 1969, **47**, 2371. 403
69CJC3271 H. M. Kolenbrander; *Can. J. Chem.*, 1969, **47**, 3271. 270
69CJC4281 L. K. Peterson and G. L. Wilson; *Canad. J. Chem.*, 1969, **47**, 4281. 470
69CR(268)358 G. Peiffer, A. Guillemonat, J.-C. Traynard and M. Faure; *C. R. Hebd. Seances Acad. Sci.*, 1969, **268**, 358. 1071
69CR(C)72 R. Calas and P. Bourgeois; *C. R. Hebd. Seances Acad. Sci., Ser. C.*, 1969, **269**, 72. 1090
69CR(C)834 T. Cuvigny and H. Normant; *C. R. Hebd. Seances Acad. Sci., Ser. C*, 1969, **268**, 834. 421
69HCA2641 H.-J. Gais, K. Hafner and M. Neuenschwander; *Helv. Chim. Acta*, 1969, **52**, 2641. 1049
69IC255 W. K. Musker and R. R. Stevens; *Inorg. Chem.*, 1969, **8**, 255. 524
69IC279 J. Mack and C. H. Yoder; *Inorg. Chem.*, 1969, **8**, 279. 417, 418
69IC1006 J. Summers and H. H. Sisler; *Inorg. Chem.*, 1969, **8**, 1006. 471
69IC1032 L. K. Krannich and H. H. Sisler; *Inorg. Chem.*, 1969, **8**, 1032. 506
69IZV2234 G. A. Razuvaev, N. A. Osanova and I. K. Grigor'eva; *Izv. Akad. Nauk. SSSR, Ser. Khim.*, 1969, 2234. 832
69JA415 P. B. Sargeant and C. G. Krespan; *J. Am. Chem. Soc.*, 1969, **91**, 415. 1006
69JA1224 H. C. Brown and E.-I. Negishi; *J. Am. Chem. Soc.*, 1969, **91**, 1224. 518
69JA1226 H. C. Brown and W. C. Dickinson; *J. Am. Chem. Soc.*, 1969, **91**, 1226. 518
69JA1310 D. E. Gould, L. R. Anderson, D. E. Young and W. B. Fox; *J. Am. Chem. Soc.*, 1969, **91**, 1310. 94
69JA1649 K. J. Wynne and J. W. George; *J. Am. Chem. Soc.*, 1969, **91**, 1649. 288
69JA1679 V. Grahauskas and K. Baum; *J. Am. Chem. Soc.*, 1969, **91**, 1679. 335
69JA2788 K. Naumann, G. Zon and K. Mislow; *J. Am. Chem. Soc.*, 1969, **91**, 2788. 440
69JA2915 A. H. Cowley and J. L. Mills; *J. Am. Chem. Soc.*, 1969, **91**, 2915. 434
69JA2955 G. A. Olah and M. B. Comisarow; *J. Am. Chem. Soc.*, 1969, **91**, 2955. 1006
69JA4273 C. P. Hoiberg and R. O. Mumma; *J. Am. Chem. Soc.*, 1969, **91**, 4273. 102
69JA4575 J. J. Eisch, N. K. Hota, and S. Kozima; *J. Am. Chem. Soc.*, 1969, **91**, 4575. 989
69JA4724 K. L. Marsi; *J. Am. Chem. Soc.*, 1969, **91**, 4724. 441
69JA4986 F. Knoll, H. C. Marsmann and J. R. Wazer; *J. Am. Chem. Soc.*, 1969, **91**, 4986. 507
69JA5107 J. M. Simson and W. Lwowski; *J. Am. Chem. Soc.*, 1969, **91**, 5107. 298
69JA5243 D. B. Denney, D. Z. Denney, B. C. Chang and K. L. Marsi; *J. Am. Chem. Soc.*, 1969, **91**, 5243. 830
69JA5415 T. C. Shields and A. N. Kurtz; *J. Am. Chem. Soc.*, 1969, **91**, 5415. 131
69JA5645 P. D. Henson, K. Naumann and K. Mislow; *J. Am. Chem. Soc.*, 1969, **91**, 5645. 441
69JA5646 H. C. Brown and M.-H. Rei; *J. Am. Chem. Soc.*, 1969, **91**, 5646. 89
69JA5801 G. A. Olah and A. M. White; *J. Am. Chem. Soc.*, 1969, **91**, 5801. 1007
69JA6122 J. H. Boyer and R. Selvarajan; *J. Am. Chem. Soc.*, 1969, **91**, 6122. 1055
69JA6461 M. S. Newman and T. B. Patrick; *J. Am. Chem. Soc.*, 1969, **91**, 6461. 801
69JA6464 C. E. Castro, R. Havlin, V. K. Honwad, A. Malte and S. Moje; *J. Am. Chem. Soc.*, 1969, **91**, 6464. 621
69JA6609 A. H. Cowley and D. S. Dierdorf; *J. Am. Chem. Soc.*, 1969, **91**, 6609. 461, 508
69JA6983 M. A. Bennett and D. L. Milner; *J. Am. Chem. Soc.*, 1969, **91**, 6983. 963
69JA7012 K. Naumann, G. Zon and K. Mislow; *J. Am. Chem. Soc.*, 1969, **91**, 7012. 441
69JA7023 G. Zon, K. E. DeBruin, K. Naumann and K. Mislow; *J. Am. Chem. Soc.*, 1969, **91**, 7023. 441
69JA7027 K. E. DeBruin, G. Zon, K. Naumann and K. Mislow; *J. Am. Chem. Soc.*, 1969, **91**, 7027. 440, 441
69JA7393 K. E. DeBruin and K. Mislow; *J. Am. Chem. Soc.*, 1969, **91**, 7393. 846
69JA7525 R. G. Hiskey, A. M. Thomas, R. L. Smith and W. C. Jones; *J. Am. Chem. Soc.*, 1969, **91**, 7525. 219, 220
69JCS(A)890 D. C. Goodall; *J. Chem. Soc. (A)*, 1969, 890. 278, 279
69JCS(A)980 D. C. Bradley and M. N. Gitlitz; *J. Chem. Soc. (A)*, 1969, 980. 409, 421
69JCS(A)992 T. A. George and M. F. Lappert; *J. Chem. Soc. (A)*, 1969, 992. 415, 416, 418, 419
69JCS(A)2704 J. E. Drake, N. Goddard and C. Riddle; *J. Chem. Soc. (A)*, 1969, 2704. 474
69JCS(B)350 J. C. J. Bart; *J. Chem. Soc. (B)*, 1969, 350. 1004
69JCS(B)749 D. Bethell, A. R. Newall, G. Stevens and D. Whittaker; *J. Chem. Soc. (B)*, 1969, 749. 89
69JCS(C)397 A. H. Lamberton and H. M. Yusuf; *J. Chem. Soc. (C)*, 1969, 397. 405
69JCS(C)626 T. Higuchi, A. Hussain and I. H. Pitman; *J. Chem. Soc.(C)*, 1969, 626 338
69JCS(C)986 W. L. F. Armarego; *J. Chem. Soc. (C)*, 1969, 986. 395
69JCS(C)1046 R. Jones and J. B. Pattison; *J. Chem. Soc. (C)*, 1969, 1046. 36
69JCS(C)1073 J. E. Baldwin, A. K. Qureshi and B. Sklarz; *J. Chem. Soc. (C)*, 1969, 1073. 389, 765
69JCS(C)1086 T. Sasaki, T. Yoshioka and K. J. Shoji; *J. Chem. Soc. (C)*, 1969, 1086. 749
69JCS(C)1605 R. Baker, J. Hudec and K. L. Rabone; *J. Chem. Soc. (C)*, 1969, 1605. 87
69JCS(C)2038 J. R. Shutt and S. Trippett; *J. Chem. Soc. (C)*, 1969, 2038. 843

69JCS(C)2273 D. W. Burt and P. Simpson; *J. Chem. Soc.* (*C*), 1969, 2273. 1071

69JCS(C)2334 F. E. Hardy, P. R. H. Speakman and P. Robson; *J. Chem. Soc.* (*C*), 1969, 2334. 145

69JCS(C)2349 J. R. Hanson and S. Mehta; *J. Chem. Soc.* (*C*), 1969, 2349. 657

69JCS(D)1023 C. W. Sink and A. B. Harvey; *J. Chem. Soc., Dalton Trans.*, 1969, 1023. 278

69JGU1068 K. I. Kuz'min and L. A. Pavlova; *J. Gen. Chem. USSR* (*Engl. Transl.*), 1969, **39**, 1068. 1079

69JGU1797 K. I. Kuz'min and L. A. Pavlova; *J. Gen. Chem. USSR* (*Engl. Transl.*), 1969, **39**, 1797. 1079

69JGU2161 A. Ya. Yakubovich, I. M. Filatova, E. L. Zaitseva and A. P. Simonov; *J. Gen. Chem. USSR* (*Engl. Transl.*), 1969, **39**, 2161. 470

69JHC627 W. H. Mueller and M. Dines; *J. Heterocycl. Chem.*, 1969, **6**, 627. 165, 169

69JHC989 R. A. Abramovitch and E. E. Knaus; *J. Heterocycl. Chem.*, 1969, **6**, 989. 765

69JINC271 M. Bermann and K. Utvary; *J. Inorg. Nucl. Chem.*, 1969, **31**, 271. 468

69JINC3684 K. D. Crosbie and G. M. Sheldrick; *J. Inorg. Nucl. Chem.*, 1969, **31**, 3684. 428, 429, 432

69JOC36 L. Field and R. B. Barbee; *J. Org. Chem.*, 1969, **34**, 36. 213, 215

69JOC97 U. G. Nayak and R. L. Whistler; *J. Org. Chem.*, 1969, **34**, 97. 129

69JOC911 P. Kovacic and M. K. Lowery; *J. Org. Chem.*, 1969, **34**, 911. 338

69JOC940 A. L. Ternay, L. Ens, J. Herrman and S. Evans; *J. Org. Chem.*, 1969, **34**, 940. 478

69JOC1220 M. S. Newman and A. O. M. Okorodudu; *J. Org. Chem.*, 1969, **34**, 1220. 801

69JOC1240 A. W. Johnson and R. T. Amel; *J. Org. Chem.*, 1969, **34**, 1240. 1000

69JOC1483 D. Seyferth and J. D. H. Paetsch; *J. Org. Chem.*, 1969, **34**, 1483. 1071

69JOC1792 L. Field and R. B. Barbee; *J. Org. Chem.*, 1969, **34**, 1792. 216, 218

69JOC1817 H. Feuer and D. M. Braunstein; *J. Org. Chem.*, 1969, **34**, 1817. 314

69JOC2163 S. Hanessian and N. R. Plessas; *J. Org. Chem.*, 1969, **34**, 2163. 25

69JOC2254 F. D. Greene, J. C. Stowell and W. R. Bergmark; *J. Org. Chem.*, 1969, **34**, 2254. 378, 381

69JOC2263 F. D. Greene, W. R. Bergmark and J. G. Pacifici; *J. Org. Chem.*, 1969, **34**, 2263. 403

69JOC2269 F. D. Greene and J. F. Pazos; *J. Org. Chem.*, 1969, **34**, 2269. 764

69JOC2324 H. O. House, L. J. Czuba, M. Gall and H. D. Olmstead; *J. Org. Chem.*, 1969, **34**, 2324. 696, 699

69JOC2397 Y. H. Chiang, J. S. Luloff and E. Schipper; *J. Org. Chem.*, 1969, **34**, 2397. 352, 363

69JOC2446 V. Grakauskas; *J. Org. Chem.*, 1969, **34**, 2446. 13

69JOC2684 A. M. Aguiar, J. R. S. Irelan, C. J. Morrow, J. P. John and G. W. Prejean; *J. Org. Chem.*, 1969, **34**, 2684. 435, 435, 1060, 1064

69JOC2720 P. Aeberli and W. J. Houlihan; *J. Org. Chem.*, 1969, **34**, 2720. 382, 384

69JOC2955 W. H. Mueller; *J. Org. Chem.*, 1969, **34**, 2955. 117

69JOC2997 W. E. Thien and W. R. McBride; *J. Org. Chem.*, 1969, **34**, 2997. 388, 389

69JOC3097 W. E. Truce, J. W. Fieldhouse, D. J. Vrencur, J. R. Novell, R. W. Campbell and D. G. Brady; *J. Org. Chem.*, 1969, **34**, 3097. 204

69JOC3112 N. A. Lebel, R. F. Czaja and A. De Boer; *J. Org. Chem.*, 1969, **34**, 3112. 137

69JOC3279 P. L. Southwick, J. A. Fitzgerald, R. Madhav and D. A. Welsh; *J. Org. Chem.*, 1969, **34**, 3279. 787

69JOC3349 A. M. Aguiar, J. R. Smiley Irelan and N. S. Bhacca; *J. Org. Chem.*, 1969, **34**, 3349. 1070

69JOC3430 P. A. S. Smith, C. D. Rowe and L. B. Bruner; *J. Org. Chem.*, 1969, **34**, 3430. 753

69JOC3434 J. B. Hendrickson, S. Okano and R. K. Bloom; *J. Org. Chem.*, 1969, **34**, 3434. 619

69JOC3441 P. F. Wolf and R. K. Barnes; *J. Org. Chem.*, 1969, **34**, 3441. 650

69JOC3519 E. L. Albano and D. Horton; *J. Org. Chem.*, 1969, **34**, 3519. 35

69JOC3615 H. O. House and W. Fisher, Jr.; *J. Org. Chem.*, 1969, **34**, 3615. 587

69JOC3991 F. K. Lautenschlaeger and N. V. Schwartz; *J. Org. Chem.*, 1969, **34**, 3991. 221, 226

69JOC4121 G. B. Bachmann and N. W. Connon; *J. Org. Chem.*, 1969, **34**, 4121. 397

69JOC4170 K. Mori and Y. Nakamura; *J. Org. Chem.*, 1969, **34**, 4170. 123

69JOM(16)413 B. C. Pant, W. E. Davidsohn and M. C. Henry; *J. Organomet. Chem.*, 1969, **16**, 413. 1101

69JOM(16)P69 O. J. Sherer and W. Janssen; *J. Organomet. Chem.*, 1969, **16**, P69. 503

69JOM(17)165 H. F. Reiff and B. C. Pant; *J. Organomet. Chem.*, 1969, **17**, 165. 1076

69JOM(17)201 V. V. Korshak, V. A. Zamyatina, A. I. Solomatina and P. V. Petrovskii; *J. Organomet. Chem.*, 1969, **17**, 201. 430

69JOM(18)299 H. Bürger and W. Kilian; *J. Organometal. Chem.*, 1969, **18**, 299. 545

69JOM(19)367 U. Wannagat, E. Bogush and R. Braun; *J. Organomet. Chem.*, 1969, **19**, 367. 503

69JOM(20)129 H. Bürger and H. J. Neese; *J. Organomet. Chem.*, 1969, **20**, 129. 566

69JOM(121)123 G. Cahiez, D. Bernard, J. F. Normant and J. Villiras; *J. Organomet. Chem.*, 1969, **121**, 123. 186

69JOU971 N. V. Svetlakov, I. E. Moisak and I. G. Averko-Antonovich; *J. Org. Chem. USSR* (*Engl. Transl.*), 1969, 971. 25, 31

69JPR162 G. Drefahl, G. Heublein and G. Tetzlaff; *J. Prakt. Chem.*, 1969, **311**, 162. 745

69LA(725)1 E. Schmitz, H. Fechner-Schramm and S. Schramm; *Justus Liebigs Ann. Chem.*, 1969, **725**, 1. 387

69LA(728)99 M. Regitz and H. Schwall; *Justus Liebigs Ann. Chem.*, 1969, **728**, 99. 799

69LA(729)40 G. Weiss and G. Schulze; *Justus Liebigs Ann. Chem.*, 1969, **729**, 40. 367, 368

69M64 M. Rasberger and E. Zbiral, *Monatsh. Chem.*, 1969, **100**, 64. 771

69M311 H. Hoffmann, H. Forster and G. Tor-Poghassian; *Monatsh*, 1969, **100**, 311. 470

69M959 H. Wittmann, E. Ziegler, H. Stark and G. Dworak; *Monatsh. Chem.*, 1969, **100**, 959. 400

B-69MI 201-01 M. L. Poutsma; in "Methods in Free Radical Chemistry," ed. E. S. Huyser, Dekker, New York, 1969, vol. 1, p. 79. 2

B-69MI 201-02 M. L. Poutsma; in "Methods in Free Radical Chemistry," ed. E. S. Huyser, Dekker, New York, 1969, vol. 2, p. 121. 2

B-69MI 203-01 R. M. Kellogg; in "Methods in Free Radical Chemistry," ed. E. J. Huyser, Dekker, New York, 1969, vol. 2. 117

B-69MI 206-01 W. A. Sheppard and C. M. Sharts; "Organic Fluorine Chemistry," W. A. Benjamin Inc., New York, 1969, p. 205. 335

B-69MI 207-01 J. H. Boyer; in "The Chemistry of the Nitro and Nitroso Groups," ed. H. Feuer, Interscience, New York, 1969, p. 215. 389

B-69MI 207-02 H. O. Larson; in "The Chemistry of the Nitro and Nitroso Groups," ed. H. Feuer, Interscience, New York, 1969, p. 301. 391, 392

69MI 207-03 S. S. Washburne and W. R. Peterson; *Inorg. Nuc. Chem. Lett.*, 1969, **5**, 17. 416, 417

69MI 209-01 Y. Matsumura and R. Okawara; *Inorg. Nucl. Chem. Lett.*, 1969, **5**, 449. 486, 510

B-69MI 215-01 J. H. Boyer; in "The Chemistry of the Nitro and Nitroso Groups," ed. H. Feuer, Interscience, New York, 1969, ch. 5, pp. 216–299. 777, 783

B-69MI 215-02 A. G. Cook (ed.); "Enamines, Synthesis, Structure and Reactions," Marcel Dekker, New York, 1969, p. 101. 757

69MI 215-03 A. Finch, J. B. Leach and J. H. Morris; *Organomet. Chem. Rev.*, 1969, **A4**, 1. 811

B-69MI 215-04 F. Glockling; "The Chemistry of Germanium," Academic Press, New York, 1969, pp. 1–234. 814

69MI 216-01 R. A. N. McLean; *Inorg. Nucl. Chem. Lett.*, 1969, **5**, 745. 858

B-69MI 218-01 R. A. Benkeser; in "Organosilicon Chemistry," Butterworths, London, 1969, vol. 2, p. 389. 910, 913, 923

B-69MI 218-02 D. Ballard, T. Brennan, F. W. G. Fearon, K. Shiina, I. Haiduc and H. Gilman; in "Organosilicon Chemistry," Butterworths, London, 1969, vol. 2, p. 449. 941

B-69MI 221-01 S. Y. Delavarenne and H. G. Viehe; in "Chemistry of Acetylenes," ed. H. G. Viehe, Dekker, New York, 1969, chap. 10, pp. 651–750. 1011, 1013, 1014, 1017, 1018, 1019

B-69MI 221-02 L. Brandsma, H. J. T. Bos and J. F. Arens; in "Chemistry of Acetylenes," ed. H. G. Viehe, Dekker, New York, 1969, chap. 11, pp. 751–860. 1021, 1025, 1027, 1030, 1031, 1032, 1033, 1034, 1036

B-69MI 222-01 H. G. Viehe; "Chemistry of Acetylenes," Dekker, New York, 1969, p. 861. 1039

B-69MI 223-01 F. Glockling; "The Chemistry of Germanium," Academic Press, London, 1969, p. 168. 1092

69MIP156590 J. Rakoczi and I. Beck; *Hung. Pat.* 156 590 (1969) (*Chem. Abstr.*, 1970, **72**, 43 497w). 788

69NAT765 R. Schoental; *Nat*, 1969, **221**, 765. 773

69OPP221 M. Kobayashi, E. Yamada, M. Matsui and N. Nobumasa; *Org. Prep. Proced. Int.*, 1969, **1**, 221. 627

69OR(17)213 L. I. Krimen and D. J. Cota; *Org. React.*, 1969, **17**, 213. 308

69OS(49)90 R. L. Kidwell, M. Murphy and S. D. Darling; *Org. Synth.* 1969, **49**, 90. 653

69RCR640 A. L. Fridman, V. P. Ivshin and S. S. Novikov; *Russ. Chem. Rev.* (*Engl. Transl.*), 1969, **38**, 640. 385

69RTC71 W. Hagens, H. J. T. Bos, W. Voskuil and J. F. Arens; *Recl. Trav. Chim. Pays-Bas*, 1969, **88**, 71. 1066

69RTC179 J. G. A. Kooyman, H. J. T. Bos, L. Brandsma and J. F. Arens; *Recl. Trav. Chim. Pays-Bas*, 1969, **88**, 179. 1033

69RTC513 L. Brandsma; *Recl. Trav. Chim. Pays-Bas*, 1969, **88**, 513. 220

69RTC519 H. E. Wijers, H. Boelens, A. Van der Gren and L. Brandsma; *Recl. Trav. Chim. Pays-Bas*, 1969, **88**, 519. 722

69S3 E. E. Gilbert; *Synthesis*, 1969, 3. 175, 176, 717

69S38 M. Schlosser and K.-F. Christmann; *Synthesis*, 1969, 38. 612

69S39 H. J. Ziegler, L. Walgraeve and F. Binon; *Synthesis*, 1969, 39. 29

69S75 M. Schlosser and M. Zimmermann; *Synthesis*, 1969, 75. 607, 611

69S112 H. R. Hudson; *Synthesis*, 1969, 112. 6

69S135 K. W. Field and P. Kovacic; *Synthesis*, 1969, 135. 23

69SC1347 X. J. Zhou and Z. Z. Huang; *Synth. Commun.*, 1969, **19**, 1347. 756

69T761 G. Snatzke, H. Laurent and R. Wiechert; *Tetrahedron*, 1969, **25**, 761. 103

69T1097 S. Chan, H. Goldwhite, H. Keyzer, D. G. Rowsell and R. Tang; *Tetrahedron*, 1969, **25**, 1097. 430, 438

69T4283 F. G. Weber; *Tetrahedron*, 1969, **25**, 4283. 24

69TL447 T. Minami, O. Aoki, H. Miki, Y. Onshiro and T. Agawa; *Tetrahedron Lett.*, 1969, 447. 244

69TL477 D. Carr, T. P. Seden and R. W. Turner; *Tetrahedron Lett.*, 1969, 477. 163

69TL1117 P. A. Tardella; *Tetrahedron Lett.*, 1969, 1117. 699

69TL1447 D. N. Harpp and J. G. Gleason; *Tetrahedron Lett.*, 1969, 1447. 199

69TL1623 A. McKillop, D. Bromley and E. C. Taylor; *Tetrahedron Lett.*, 1969, 1623. 626

69TL2145 R. E. Ireland and G. Pfister; *Tetrahedron Lett.*, 1969, 2145. 693

69TL2427 A. McKillop, J. S. Fowler, M. J. Zelesko, J. D. Hunt, E. C. Taylor and G. McGillivray; *Tetrahedron Lett.*, 1969, 2427. 631

69TL2743 K. Kondo, A. Negishi and G. I. Tsuchihashi; *Tetrahedron Lett.*, 1969, 2743. 200, 222

69TL3631 T. Taguchi, Y. Kiyoshima, O. Komori and M. Mori; *Tetrahedron Lett.*, 1969, 3631. 123

69TL3707 J. F. Biellman and J. B. Ducep; *Tetrahedron Lett.*, 1969, 3707. 143

69TL3901 M. J. Gibian and D. V. Winkelman; *Tetrahedron Lett.*, 1969, 3901. 216

69TL4117 G. Kresze and H. Grill; *Tetrahedron Lett.*, 1969, 4117. 399

69TL4643 V. T. Bandurco and J. P. Snyder; *Tetrahedron Lett.*, 1969, 4643. 404

69ZAAC(364)316 R. Masthoff, G. Krieg and C. Vieroth; *Z. Anorg. Allg. Chem.*, 1969, **364**, 316. 421

69ZAAC(369)178 E. Fluck and G. Jakobson; *Z. Anorg. Allg. Chem.*, 1969, **369**, 178. 889, 891

69ZAAC(370)171 O. J. Scherer and P. Klussmann; *Z. Anorg. Allg. Chem.*, 1969, **370**, 171. 472

69ZAAC(370)227 K. Sommer; *Z. Anorg. Allg. Chem.*, 1969, **370**, 227. 890

69ZC105 A. Zubek and R. Liebig; *Z. Chem.*, 1969, **9**, 105. 380

69ZC305 G. Jänchen and G. Westphal; *Z. Chem.*, 1969, **9**, 305. 726

69ZN(B)179 A. Schmidpeter and H. Brecht; *Z. Naturforsch., Teil B*, 1969, **24**, 179. 293

69ZN(B)774 K. Seppelt and W. Sundermeyer; *Z. Naturforsch., Teil B.*, 1969, **24**, 774. 761

69ZN(B)1250 H. W. Roesky and W. G. Böwing; *Z. Naturforsch., Teil B*, 1969, **24**, 1250. 862
69ZOB341 V. K. Khairullin, M. A. Vasyanina and A. N. Pudovik; *Zh. Obshch. Khim.*, 1969, **39**, 341. 854
69ZOB841 D. I. Lobanov, E. N. Tsuetkov and M. I. Kabachnik; *Zh. Obshch. Khim.*, 1969, **39**, 841. 821, 824, 847
69ZOB1174 I. F. Lutsenko, V. L. Foss and N. M. Semenenko; *Zh. Obshch. Khim.*, 1969, **39**, 1174 (Plenum Engl. Transl. p. 1141). 695, 700
69ZOB1543 V. S. Abramov, V. I. Barabanov and Z. Ya. Sazonova; *Zh. Obshch. Khim.*, 1969, **39**, 1543 (*Chem. Abstr.*, 1969, **71**, 113 045). 463
69ZOB1544 B. I. Stepanov, E. N. Karpova and A. I. Bokanov; *Zh. Obshch. Khim.*, 1969, **39**, 1544. 825
69ZOB1804 G. M. Bogolyubov, Yu. N. Shlyk and A. A. Petrov; *Zh. Obshch. Khim.*, 1969, **39**, 1804. 292
69ZOB2139 N. G. Feschenko, E. A. Melnichuk and A. V. Kirsanov; *Zh. Obshch. Khim.*, 1969, **39**, 2139. 449
69ZOB2367 Zh. L. Evtikov, N. A. Razumova, and A. A. Petrov; *Zh. Obshch. Khim.*, 1969, **39**, 2367 (*Chem. Abstr.*, 1970, **72**, 43 791). 454
69ZOB2418 A. N. Pudovik and I. V. Gur'yanova and G. V. Romanov; *Zh. Obshch. Khim.*, 1969, **39**, 2418 (*Chem. Abstr.* 1970, **72**, 79 164). 454
69ZOB2479 V. V. Kudinova, Yu. A. Veits, V. L. Foss and I. F. Lutsenko; *Zh. Obshch. Khim.*, 1969, **39**, 2479. 694
69ZOR268 B. V. Ioffe and L. M. Gershtein; *Zh. Org. Khim.*, 1969, **5**, 268. 403
69ZOR1088 B. G. Boldyrev and S. A. Kolesnikova; *Zh. Org. Khim.*, 1969, **5**, 1088. 358
69ZOR1735 B. V. Ioffe, Z. I. Sergeva and Y. Y. Dumpis; *Zh. Org. Khim.*, 1969, **5**, 1735. 377

70ACR417 R. B. King; *Acc. Chem. Res.*, 1970, **3**, 417. 576
70AG(E)101 B. Armer and H. Schmidbauer; *Angew. Chem., Int. Ed. Engl.*, 1970, **9**, 101. 590
70AG(E)273 K. Griesbaum; *Angew. Chem., Int. Ed. Engl.*, 1970, 273. 137, 710, 712
70AG(E)521 T. E. Dueber, P. J. Stang, W. D. Pfeifer, R. H. Summerville, M. A. Imhoff, P. v. R. Schleyer, K. Hummel, S. Bocher, C. E. Harding and M. Hanack; *Angew. Chem., Int. Ed. Engl.*, 1979, **9**, 521. 689
70AG(E)954 K. Bott; *Angew. Chem., Int. Ed. Engl.*, 1970, **9**, 954. 800
70AJC1559 P. S. Elmes, S. Middleton and B. O. West; *Aust. J. Chem.*, 1970, **23**, 1559. 507
70B1626 D. T. Gibson, M. Hensley, H. Yoshika and R. J. Mabry; *Biochemistry*, 1970, **9**, 1626. 39
70BCJ293 S. Hashimoto and W. Koike; *Bull. Chem. Soc. Jpn.*, 1970, **43**, 293. 650
70BCJ1890 J. Tsurugi, Y. Abe and S. Kawamura; *Bull. Chem. Soc. Jpn.*, 1970, **43**, 1890. 211
70BCJ3232 J. Koketsu, S. Kojima and Y. Ishi; *Bull. Chem. Soc. Jpn.*, 1970, **43**, 3232. 469
70BCJ3299 H. Suzuki; *Bull. Chem. Soc. Jpn.*, 1970, **43**, 3299. 13
70BCJ3543 O. Tsuge and S. Iwanami; *Bull. Chem. Soc. Jpn.*, 1970, **43**, 3543. 249
70BSB353 J. L. Piette and M. Renson; *Bull. Soc. Chim. Belg.*, 1970, **79**, 353. 280
70BSB367 J. L. Piette and M. Renson; *Bull. Soc. Chim. Belg.*, 1970, **79**, 367. 280
70BSB383 J. L. Piette and M. Renson; *Bull. Soc. Chim. Belg.*, 1970, **79**, 383. 280
70BSF192 R. Burgada and J. Roussel; *Bull. Soc. Chim. Fr.*, 1970, 192. 414
70C109 C. G. Screttas, J. F. Estham, and C. W. Kaminski; *Chimia*, 1970, **24**, 109. 964
70CA88061 T. J. Logan and T. W. Rave; *Chem. Abstr.*, 1970, **73**, 88 061. 470
70CB318 L. Horner and M. Ernst; *Chem. Ber.*, 1970, **103**, 318. 441, 442
70CB564 R. Fuks and H. G. Viehe; *Chem. Ber.*, 1970, **103**, 564. 1042
70CB799 B. Deubzer, M. Elian, E. O. Fischer and H. P. Fritz; *Chem. Ber.*, 1970, **103**, 799. 483
70CB1198 S. Y. Delavarenne and H. G. Viehe; *Chem. Ber.*, 1970, **103**, 1198. 1013, 1047
70CB1209 S. Y. Delavarenne and H. G. Viehe; *Chem. Ber.*, 1970, **103**, 1209. 1047
70CB1216 H. G. Viehe and S. Y. Delavarenne; *Chem. Ber.*, 1970, **103**, 1216. 1015
70CB2052 K-H. König and B. Zeeh; *Chem. Ber.*, 1970, **103**, 2052. 382, 383
70CB2152 R. Appel and J. Kohnke; *Chem. Ber.*, 1970, **103**, 2152. 399
70CB2428 P. Tavs; *Chem. Ber.*, 1970, **103**, 2428. 845, 847, 848
70CB3733 R. Appel and D. Hänssgen; *Chem. Ber.*, 1970, **103**, 3733. 384
70CC111 P. Toft and A. J. Liston; *J. Chem. Soc., Chem. Commun.*, 1970, 111. 675
70CC216 D. E. Gould, C. T. Ratcliffe, L. R. Anderson and W. B. Fox; *J. Chem. Soc., Chem. Commun.*, 1970, 216. 98
70CC305 J. S. Grossert; *J. Chem. Soc., Chem. Commun.*, 1970, 305. 249
70CC425 R. H. Anderson and R. H. Cragg; *J. Chem. Soc., Chem. Commun.*, 1970, 425. 260
70CC777 T. Oishi, K. Kamata and Y. Bau; *J. Chem. Soc., Chem. Commun.*, 1970, 777. 365
70CC1037 M. D. Johnson; *J. Chem. Soc., Chem. Commun.*, 1970, 1037. 597
70CC1466 D. H. R. Barton, G. Page and D. A. Widdowson; *J. Chem. Soc., Chem. Commun.*, 1970, 1466. 195, 688
70CC1495 J. Michalski, A. Okruszek and W. Stec; *J. Chem. Soc., Chem. Commun.*, 1970, 1495. 452
70CI(L)499 B. J. Aylett, I. A. Ellis and C. J. Porritt; *Chem. Ind. (London)*, 1970, 499. 416
70CI(L)1188 P. W. Hickmott and H. Suschitzky; *Chem. Ind. (London)*, 1970, 1188. 757
70CJC845 T. Durst and K.-C. Tin; *Can. J. Chem.*, 1970, **48**, 845. 202, 207
70CJC1885 W. R. Cullen and M. C. Waldman; *Can. J. Chem.*, 1970, **48**, 1885. 487
70CJC2745 B. H. Al-Sader and R. J. Crawford, *Can. J. Chem.*, 1970, **48**, 2745. 376, 403
70CPB235 K. Sirakawa, O. Aki, T. Tsujikawa and T. Tsuda; *Chem. Pharm. Bull.*, 1970, **18**, 235. 228
70CR639 P. Kovacic, M. K. Lowery and K. W. Field; *Chem. Rev.*, 1970, **70**, 639. 298, 305, 761
70CR(C)589 J. Blais, J. Soulie and P. Cadiot; *C. R. Hebd. Seances Acad. Sci., Ser. C*, 1970, **271**, 589. 521
70CR(C)2086 P. Savignac, J. Chenault and P. Chabriev; *C. R. Hebd. Seances Acad. Sci., Ser. C*, 1970, **270**, 2086. 255
70CRV639 P. Kovacic, M. K. Lowery and K. W. Field; *Chem. Rev.*, 1970, **70**, 639. 20, 337, 338, 339
70HCA1637 C. Wentrup and H. Dahn; *Helv. Chim. Acta*, 1970, **53**, 1637. 799

70HOU(13/1)1 U. Schöllkopf; *Methoden Org. Chem.* (*Houben-Weyl*), 1970, **13/1**, 1. 551

70HOU(13/1)261 H. F. Ebel; *Methoden Org. Chem.* (*Houben Weyl*), 1970, **13/1**, 261. 556

70HOU(13/4)9 H. Lehmkhul, K. Ziegler and H. G. Gellert; *Methoden Org. Chem.* (*Houben-Weyl*), 1970, **13/4**, 9. 595

70HOU(13/4)154 G. Bähr, P. Burba, H. Lehmkuhl and K. Ziegler; *Methoden Org. Chem.* (*Houben-Weyl*), 1970, **13/4**, 154. 1098

70HOU(13/4)315 G. Bähr and P. Burba; *Methoden Org. Chem.* (*Houben-Weyl*), 1970, **13/4**, 315. 597

70HOU(13/4)345 G. Bähr and P. Burba; *Methoden Org. Chem.* (*Houben-Weyl*), 1970, **13/4**, 345. 597

70HOU(13/4)363 G. Bähr and P. Burba; *Methoden Org. Chem.* (*Houben-Weyl*), 1970, **13/4**, 363. 597

70HOU(13/4)23 H. Lehmkuhl, K. Ziegler, and H. G. Gellert; *Methoden Org. Chem.* (*Houben-Weyl*), 1970, XIII/4, 23. 967

70IC106 K. J. Wyne and P. S. Pearson; *Inorg. Chem.*, 1970, **9**, 106. 292

70IC146 R. Ehrlich; *Inorg. Chem.*, 1970, **9**, 146. 421

70IC862 J. C. Summers and H. H. Sisler; *Inorg. Chem.*, 1970, **9**, 862. 490

70IC870 A. D. Norman; *Inorg. Chem.*, 1970, **9**, 870. 47, 474

70JA190 A. F. Cook; *J. Am. Chem. Soc.*, 1970, **92**, 190. 255

70JA395 E. J. Corey and I. Kuwajima; *J. Am. Chem. Soc.*, 1970, **92**, 395. 962

70JA712 H. C. Brown and G. W. Kabalka; *J. Am. Chem. Soc.*, 1970, **92**, 712. 698

70JA714 H. C. Brown and G. W. Kabalka; *J. Am. Chem. Soc.*, 1970, **92**, 714. 698

70JA1427 G. Zweifel, A. Horng and J. T. Snow; *J. Am. Chem. Soc.*, 1970, **92**, 1427. 900

70JA1587 W. J. McGahren and P. Kunstmann; *J. Am. Chem. Soc.*, 1970, **92**, 1587. 773

70JA2042 J. L. Webb, C. K. Mann and H. M. Walborsky; *J. Am. Chem. Soc.*, 1970, **92**, 2042. 653

70JA2467 R. C. Larock and H. C. Brown; *J. Am. Chem. Soc.*, 1970, **92**, 2467. 520

70JA3035 A. Fava, L. Senatore and E. Ciuffarin; *J. Am. Chem. Soc.*, 1970, **92**, 3035. 195

70JA3223 P. D. Bartlett and A. P. Schaap; *J. Am. Chem. Soc.*, 1970, **92**, 3223. 98

70JA3520 E. C. Taylor, H. W. Altland, R. H. Danforth, G. McGillivray and A. McKillop; *J. Am. Chem. Soc.*, 1970, **92**, 3520. 652

70JA3825 C. R. Johnson and E. V. Jonsson; *J. Am. Chem. Soc.*, 1970, **92**, 3825. 365

70JA3969 H. C. Marsmann and J. R. Wazer; *J. Am. Chem. Soc.*, 1970, **92**, 3969. 493, 495, 496

70JA4309 M. S. Newman and C. D. Beard; *J. Am. Chem. Soc.*, 1970, **92**, 4309. 630, 801

70JA4312 M. S. Newman and T. B. Patrick; *J. Am. Chem. Soc.*, 1970, **92**, 4312. 801

70JA4586 D. S. Malament and J. M. McBride; *J. Am. Chem. Soc.*, 1970, **92**, 4586. 404

70JA5457 W. M. Jones and D. D. Maness; *J. Am. Chem. Soc.*, 1970, **92**, 5457. 204

70JA6314 E. J. Corey, H. A. Kirst, and J. A. Katzenellenbogen; *J. Am. Chem. Soc.*, 1970, **92**, 6314. 970

70JA6498 M. J. Shaw, H. H. Hyman and R. Filler; *J. Am. Chem. Soc.*, 1970, **92**, 6498. 613

70JA6660 H. C. Brown and C. F. Lane; *J. Am. Chem. Soc.*, 1970, **92**, 6660. 30

70JA6894 B. Rickborn and M. T. Wuehsthoff; *J. Am. Chem. Soc.*, 1970, **92**, 6894. 59

70JA6983 H. C. Brown and S. K. Gupta; *J. Am. Chem. Soc.*, 1970, **92**, 6983. 531

70JA7224 T. H. Chan, J. P. Montillier, W. F. Van Horn and D. N. Harpp; *J. Am. Chem. Soc.*, 1970, **92**, 7224. 221

70JA7494 J. Kollonitsch, L. Barash and G. A. Doldouras; *J. Am. Chem. Soc.*, 1970, **92**, 7494. 15, 613

70JA7564 M. S. Newman and C. D. Beard; *J. Am. Chem. Soc.*, 1970, **92**, 7564. 801

70JA7629 S. J. Brois, J. F. Pilot and H. W. Barnum; *J. Am. Chem. Soc.*, 1970, **92**, 7629. 220

70JAP7321096-R I. Takahashi, S. Ozaki and A. Tamaoki (Mitsui Toatsu Chemical Inc.); *Jpn. Pat.*, 7321096-R (1970) (*Chem. Abstr.*, 1973, **79**, 115 309). 725

70JCED356 J. M. Kanamueller; *J. Chem. Eng. Data*, 1970, **15**, 356. 386

70JCS(A)1049 S. C. Peake and R. Schmutzler; *J. Chem. Soc.* (*A*), 1970, 1049. 476

70JCS(A)2359 M. Fild and R. Schmutzler; *J. Chem. Soc.* (*A*), 1970, 2359. 448

70JCS(B)735 A. G. Davies and S. C. W. Hook; *J. Chem. Soc.* (*B*), 1970, 735. 501

70JCS(C)270 D. B. Hope, C. D. Morgan and M. Wälki; *J. Chem. Soc.* (*C*), 1970, 270. 188

70JCS(C)810 D. W. Allen, J. C. Coppola, O. Kennard, F. G. Mann, W. D. S. Motherwell and D. G. Watson; *J. Chem. Soc.* (*C*), 1970, 810. 889

70JCS(C)878 C. Brown, M. Murray and R. Schmutzler; *J. Chem. Soc.* (*C*), 1970, 878. 835

70JCS(C)1088 B. T. Hayes and T. S. Stevens; *J. Chem. Soc.* (*C*), 1970, 1088. 379

70JCS(C)1480 J. Arotsky, R. Butler and A. C. Darby; *J. Chem. Soc.* (*C*), 1970, 1480. 631

70JCS(C)1578 I. El-Sayed El-Kholy, F. K. Rafta and M. M. Mishrikey; *J. Chem. Soc.*, 1970, 1578. 121

70JCS(C)1714 J. A. Blair and R. J. Gardner; *J. Chem. Soc.* (*C*), 1970, 1714. 377

70JCS(C)2195 R. J. Holman and M. J. Perkins; *J. Chem. Soc.* (*C*), 1970, 2195. 389

70JCS(C)2409 I. T. Kay and N. Punja; *J. Chem. Soc.* (*C*), 1970, 2409. 271

70JCS(C)2514 E. Haddock, P. Kirby and A. W. Johnson; *J. Chem. Soc.* (*C*), 1970, 2514. 803

70JCS(C)2529 P. C. Crofts and D. M. Parker; *J. Chem. Soc.* (*C*), 1970, 2529. 448

70JGU127 Yu. F. Gatilov and L. B. Ionov; *J. Gen. Chem. USSR* (*Engl. Transl.*), 1970, **40**, 127. 891

70JGU1100 K. I. Kuz'min and L. A. Pavlova; *J. Gen. Chem. USSR* (*Engl. Transl.*), 1970, **40**, 1100. 1078

70JOC10 C. D. Broaddus; *J. Org. Chem.*, 1970, **35**, 10. 554

70JOC279 J. E. Dunbar and J. H. Rogers; *J. Org. Chem.*, 1970, **35**, 279. 358

70JOC528 F. Kienzle and E. C. Taylor; *J. Org. Chem.*, 1970, **35**, 528. 621

70JOC719 E. C. Jeunge, D. A. Beal and W. P. Duncan; *J. Org. Chem.*, 1970, **35**, 719. 619

70JOC723 V. Grakauskas; *J. Org. Chem.*, 1970, **35**, 723. 608

70JOC749 L. Field and C. H. Foster; *J. Org. Chem.*, 1970, **35**, 749. 215

70JOC777 E. J. Corey and J. I. Shulman; *J. Org. Chem.*, 1970, **35**, 777. 709

70JOC905 L. A. Paquette, R. W. Houser and M. Rosen; *J. Org. Chem.*, 1970, **35**, 905. 158

70JOC960 G. Koga and J.-P. Anselme; *J. Org. Chem.*, 1970, **35**, 960. 420

70JOC1118 R. G. Hiskey and B. F. Ward; *J. Org. Chem.*, 1970, **35**, 1118. 220

70JOC1204	S. J. Huang and M. V. Lessard; *J. Org. Chem.*, 1970, **35**, 1204.	759
70JOC1226	W. E. Truce and D. J. Vrencur; *J. Org. Chem.*, 1970, **35**, 1226.	207
70JOC1468	H. Feuer and F. Brown, Jr.; *J. Org. Chem.*, 1970, **35**, 1468.	377, 382
70JOC1545	V. Grahauskas and K. Baum; *J. Org. Chem.*, 1970, **35**, 1545.	335
70JOC1670	A. Mckillop, R. A. Raphael and E. C. Taylor; *J. Org. Chem.*, 1970, **35**, 1670.	776
70JOC1703	R. E. A. Dear; *J. Org. Chem.*, 1970, **35**, 1703.	624
70JOC1839	M. N. Sheng and J. G. Zajacek; *J. Org. Chem.*, 1970, **35**, 1839.	90
70JOC2006	E. Ciuffarin and G. Guaraldi; *J. Org. Chem.*, 1970, **35**, 2006.	171, 172, 716
70JOC2319	J. P. H. Verheyden and J. G. Moffatt; *J. Org. Chem.*, 1970, **35**, 2319.	8
70JOC2412	M. S. Newman and C. D. Beard; *J. Org. Chem.*, 1970, **35**, 2412.	801
70JOC2482	F. D. Greene and S. S. Hecht; *J. Org. Chem.*, 1970, **35**, 2482.	403, 404
70JOC2678	A. W. Johnson and H. Schubert; *J. Org. Chem.*, 1970, **35**, 2678.	877
70JOC2716	E. L. Eliel and T. W. Doyle; *J. Org. Chem.*, 1970, **35**, 2716.	128
70JOC2785	M. L. Scheinbaum; *J. Org. Chem.*, 1970, **35**, 2785.	391
70JOC2790	M. L. Scheinbaum; *J. Org. Chem.*, 1970, **35**, 2790.	764
70JOC2826	H. Zimmer, M. Jayawant and P. Gubsch; *J. Org. Chem.*, 1970, **35**, 2826.	304
70JOC2979	T. Saegusa, Y. Ito and T. Shimizu; *J. Org. Chem.*, 1970, **35**, 2979.	240, 359, 768
70JOC3012	N. E. Heimer and L. Field; *J. Org. Chem.*, 1970, **35**, 3012.	358
70JOC3086	J. L. Kurz and J. C. Harris; *J. Org. Chem.*, 1970, **35**, 3086.	278
70JOC3166	F. M. Sonnenberg; *J. Org. Chem.*, 1970, **35**, 3166.	656
70JOC3259	D. N. Harpp and J. G. Gleason; *J. Org. Chem.*, 1970, **35**, 3259.	212
70JOC3267	L. Field, J. L. Vanhorne and L. W. Cunningham; *J. Org. Chem.*, 1970, **35**, 3267.	172
70JOC3275	W. E. Truce and L. D. Markley; *J. Org. Chem.*, 1970, **35**, 3275.	742
70JOC3401	R. C. Neuman, Jr. and E. S. Alhadeff; *J. Org. Chem.*, 1970, **35**, 3401.	403
70JOC3405	E. J. Corey and D. E. Cane; *J. Org. Chem.*, 1970, **35**, 3405.	1045
70JOC3436	W. C. Baird, Jr., and J. H. Surridge; *J. Org. Chem.*, 1970, **35**, 3436.	631
70JOC3641	B. C. Newman and E. L. Eliel; *J. Org. Chem.*, 1970, **35**, 3641.	128
70JOC3967	S. R. Sandler; *J. Org. Chem.*, 1970, **35**, 3967.	26
70JOC4028	J. A. Vesely and L. Schmerling; *J. Org Chem.*, 1970, **35**, 4028.	650
70JOC4148	R. G. Hiskey, G. W. Davis, M. E. Safdy, T. Inui, R. A. Upham and W. C. Jones; *J. Org. Chem.*, 1970, **35**, 4148.	219
70JOC4209	D. T. Witiak and M. C. Lu; *J. Org. Chem.*, 1970, **35**, 4209.	129
70JOC4217	W. E. Truce, C. T. Goralski, L. W. Christensen and R. H. Bavry; *J. Org. Chem.*, 1970, **35**, 4217.	157
70JOC4227	K. Kato and O. Mitsunobu; *J. Org. Chem.*, 1970, **35**, 4227.	213
70JOC4229	G. B. Bachmann and T. F. Biermann; *J. Org. Chem.*, 1970, **35**, 4229.	397
70JOM(21)231	A. Tzschach and D. Drohne; *J. Organomet. Chem.*, 1970, **21**, 231.	890
70JOM(21)351	H. Schumann and U. Arbenz; *J. Organomet. Chem.*, 1970, **21**, 351.	4744
70JOM(21)411	H. Schumann, L. Rosch and O. Stelzer; *J. Organomet. Chem.*, 1970, **21**, 411.	474
70JOM(21)837	U. Küerke; *J. Organomet. Chem.*, 1970, **21**, 83.	1077
70JOM(22)637	G. N. Chremos and R. A. Zingaro; *J. Organomet. Chem.*, 1970, **22**, 637.	497, 502
70JOM(22)653	H. A. Meinema and J. G. Noltes; *J. Organomet. Chem.*, 1970, **22**, 653.	488
70JOM(24)359	I. F. Lutsenko, Yu. I. Baukov and I. Yu. Belavin; *J. Organomet. Chem.*, 1970, **24**, 359.	701
70JOM(24)629	D. A. Armitage and M. J. Clark; *J. Organomet. Chem.*, 1970, **24**, 629.	725
70JOM(25)17	M. Schmidt and H. D. Block; *J. Organomet. Chem.*, 1970, **25**, 17.	294
70JOM(25)139	H. A. Meinema and J. G. Noltes; *J. Organomet. Chem.*, 1970, **25**, 139.	695, 885, 886, 889
70JOM(25)353	T. Abe, H. Kurosawa and R. Okawara; *J. Organomet. Chem.*, 1970, **25**, 353.	598
70JOM(25)389	A. R. Bassindale and D. R. M. Walton; *J. Organomet. Chem.*, 1970, **25**, 389.	132
70JOU889	O. Neiland and B. Karele; *J. Org. Chem. USSR* (*Engl. Transl.*), 1970, **6**, 889.	1020
70JPR10	J. Dahlman and L. Austenat; *J. Prakt. Chem.*, 1970, **312**, 10.	887
70JPR129	E. Wenschuh and B. Fritzsche; *J. Prakt. Chem.*, 1970, **312**, 129.	363
70JPR349	R. Ohme and H. Preuschhof; *J. Prakt. Chem.*, 1970, **312**, 349.	374, 380
70JPR751	B. Chiswell and K. A. Verrall; *J. Prakt. Chem.*, 1970, **312**, 751.	892
70JPR43	E. Schmitz and O. Brede; *J. Prakt. Chem.*, 1970, **312**, 43 (*Chem. Abstr.*, 1970, **73**, 14082x).	686
70M776	M. Wieber and W. R. Hoos; *Monatsh. Chem.*, 1970, **101**, 776.	837
B-70MI 205-01	K. Haada; "The Chemistry of the Carbon–Nitrogen Double Bond," Wiley, New York, 1970, pp. 266–272.	312
B-70MI 205-02	B. C. Challis and J. A. Challis; "The Chemistry of the Amides," Wiley, New York, 1970, pp. 795–801.	320
70MI 208-01	"Organophosphorus Chemistry," Specialist Periodical Reports, Royal Society of Chemistry, London, 1970–93.	427
B-70MI 209-01	G. O. Doak and L. D. Freedman; in "Organometallic Compounds of Arsenic, Antimony, and Bismuth," Wiley, New York, 1970.	481
70MI 209-02	H. Burger and H. J. Neese; *Inorg. Nucl. Chem. Lett.*, 1970, **6**, 299.	512
70MI 210-01	E. W. Abel and S. M. Illingworth; *Organometal. Chem. Rev. A*, 1970, **5**, 143.	544
B-70MI 215-01	S. Dayagi and Y. Degani; in "The Chemistry of Carbon–Nitrogen Double Bond," ed. S. Patai, Interscience, London, 1970, pp. 61–148.	751
B-70MI 215-02	H. H. Baer and L. Urbas; in "The Chemistry of the Nitro and Nitroso Groups," ed. H. Feuer, Interscience, New York, 1970, vol. 2, pp. 75–200.	784, 785, 787
B-70MI 217-01	G. O. Doak and L. D. Freedman; "Organometallic Compounds of Arsenic, Antimony and Bismuth," Wiley, New York, 1970.	872, 891
B-70MI 217-02	F. G. Mann; "The Heterocyclic Derivatives of Phosphorus, Arsenic, Antimony and Bismuth," 2nd Edition, Wiley, New York, 1970.	872

B-70MI 223-01	R. C. Poller; "The Chemistry of Organotin Compounds," Logos Press, London, 1970, p. 25.	1100
B-70MI 223-02	W. P. Neumann; "The Organic Chemistry of Tin," Wiley, London, 1970, p. 27.	1100
70OPP235	M. L. Kee and I. B. Douglass, *Org. Prep. Proced. Int.*, 1970, **2**, 235.	174, 717
70OR(18)99	W. E. Truce, E. M. Kreider and W. W. Brand; *Org. React.*, 1970, **18**, 99.	719
70OR(18)403	S. H. Pine; *Org. React.*, 1970, **18**, 403.	303
70OS(50)56	M. N. Sheng and J. G. Zajacek; *Org. Synth.*, 1970, **50**, 56.	348
70PAC285	D. H. R. Barton; *Pure Appl. Chem.*, 1970, 285.	15
70QRS245	Q. E. Thompson; *Quarterly Reports of Sulfur Chemistry*, 1970, **5**, 245.	361
70RTC159	A. Mackor and T. J. de Boer; *Recl. Trav. Chim. Pays-Bas*, 1970, **89**, 159.	391
70RTC151	A. Mackor and T. J. de Boer; *Recl. Trav. Chim. Pays-Bas*, 1970, **89**, 151.	391
70RTC164	A. Mackor and T. J. de Boer; *Recl. Trav. Chim. Pays-Bas*, 1970, **89**, 164.	389
70RTC1153	G. J. Heiszwolf and H. Kloosterziel; *Recl. Trav. Chim. Pays-Bas*, 1970, **89**, 1153.	674
70RZC1131	P. Tomasik, E. Kucharzewska-Rusek and A. Thomas; *Rocz. Chem.*, 1970, **44**, 1131.	802
70S1	W. J. LeNoble; *Synthesis*, 1970, 1.	674
70S7	E. S. Huyser; *Synthesis*, 1970, 7.	21, 27
70S161	R. G. Harvey; *Synthesis*, 1970, 161.	682
70S209	M. L. Milhailovic and Z. Cekovic; *Synthesis*, 1970, 209.	93
70S510	M. E. Kuehne; *Synthesis*, 1970, 510.	738, 751
70S545	M. S. Heller and A. M. Adelman; *Synthesis*, 1970, 545.	249
70S561	E. Kühle; *Synthesis*, 1970, 561.	165, 716
70SCI1298	G. A. Olah; *Science*, 1970, **168**, 1298.	1006
70T1281	F. Bertini, P. Grasselli, G. Zubiani and G. Cainelli; *Tetrahedron*, 1970, **26**, 1281.	1006
70T1463	W. D. Crow and I. Gosney; *Tetrahedron*, 1970, **26**, 1463.	235
70T3279	H. Kropf, C.-R. Bernert and L. Dahlenburg; *Tetrahedron*, 1970, **26**, 3279.	97
70T3527	A. Zwierzak and A. Koziara; *Tetrahedron*, 1970, **26**, 3527.	412
70T5529	P. Tavs and H. Weitkamp; *Tetrahedron*, 1970, **26**, 5529.	848
70TH134	V. S. Rao; "Applications of Nuclear Magnetic Resonance in the Study of Conjugated Systems," PhD Thesis, Andhra University, 1970, p. 134.	760
70TL215	D. J. Pasto and P. W. Wojtkowski; *Tetrahedron Lett.*, 1970, 215.	524
70TL441	A. Takamizawa, K. Hirai and T. Ishiba; *Tetrahedron Lett.*, 1970, 441.	224
70TL447	E. J. Corey, J. I. Shulman and H. Yamamoto; *Tetrahedron Lett.*, 1970, 447.	618, 630, 909
70TL699	P. E. Pfeffer and L. S. Silbert; *Tetrahedron Lett.*, 1970, 699.	393
70TL1203	V. Boekelheide and J. L. Mondt; *Tetrahedron Lett.*, 1970, 1203.	213
70TL1313	D. Sianesi, G. C. Bernardi and G. Moggi; *Tetrahedron Lett.*, 1970, 1313.	175
70TL1427	H. Grill and G. Kresze; *Tetrahedron Lett.*, 1970, 1427.	244
70TL1629	J. Konig and V. Wolf; *Tetrahedron Lett.*, 1970, 1629.	624
70TL1687	P. M. Draper, T. H. Chan and D. N. Harpp; *Tetrahedron Lett.*, 1970, 1687.	4366
70TL2517	M. L. Shulman, V. N. Yoldikov and A. Y. Khorlin; *Tetrahedron Lett.*, 1970, 2517.	32
70TL3141	A. Halleux, H. Reimlinger and H. G. Viehe; *Tetrahedron Lett.*, 1970, **36**, 3141.	1049
70TL3411	T. Mukaiyama and T. Taguchi; *Tetrahedron Lett.*, 1970, 3411.	361
70TL3447	J. A. Vida; *Tetrahedron Lett.*, 1970, 3447.	25
70TL3547	K. S. Boustany and A. B. Sullivan; *Tetrahedron Lett.*, 1970, 3547.	217
70TL3551	D. N. Harpp, D. K. Ash, T. G. Back, J. G. Gleason, B. A. Orwig, W. F. Vanhorn and J. P. Snyder; *Tetrahedron Lett.*, 1970, 3551.	217
70TL4449	M. Haake; *Tetrahedron Lett.*, 1970, 4449.	164
70TL4983	K. Boustany and J. P. Van der Kooi; *Tetrahedron Lett.*, 1970, 4983.	241, 358
70TL5115	T. Mukaiyama, S. Kobayashi and T. Kumamoto; *Tetrahedron Lett.*, 1970, 5115.	688
70USP3542848	W. Leimgruber and M. Weigele; *US Pat.* 3 542 848 (1970) (*Chem. Abstr.*, 1971, **74**, 41 945h).	749
70ZAAC(375)140	H. W. Roesky and W. Kloker; *Z. Anorg. Allg. Chem.*, 1970, **375**, 140.	862
70ZAAC(372)171	C. Fritz, G. Becker and D. Kummer; *Z. Anorg. Allg. Chem.*, 1970, **372**, 171.	473, 474
70ZAAC(375)55	K. Sommer; *Z. Anorg. Allg. Chem.*, 1970, **375**, 55.	494, 497
70ZAAC(376)37	K. Sommer; *Z. Anorg. Allg. Chem.*, 1970, **376**, 37.	432
70ZAAC(376)150	K. Sommer; *Z. Anorg. Allg. Chem.*, 1970, **376**, 150.	490
70ZAAC(377)48	G. Fritz and M. Hähnke; *Z. Anorg. Allg. Chem.*, 1970, **377**, 48.	1092
70ZAAC(377)120	K. Sommer; *Z. Anorg. Allg. Chem.*, 1970, **337**, 120.	493
70ZAAC(377)128	K. Sommer; *Z. Anorg. Allg. Chem.*, 1970, **337**, 128.	489
70ZAAC(377)278	K. Sommer; *Z. Anorg. Allg. Chem.*, 1970, **337**, 278.	486, 490, 510
70ZAAC(379)56	K. Sommer; *Z. Anorg. Allg. Chem.*, 1970, **379**, 56.	430
70ZC33	D. Hass and D. Cech; *Z. Chem.*, 1970, **10**, 33.	416
70ZC118	A. Tzschach and H. Nindel; *Z. Chem.*, 1970, **10**, 118.	486, 487, 510
70ZN111	R. Schliebs; *Z. Naturforsch., Teil B*, 1970, **25**, 111.	851
70ZOB1692	V. A. Shokol, V. V. Doroshenko and G. J. Derkach; *Zh. Obshch. Khim.*, 1970, **40**, 1692 (*Chem. Abstr.*, 1971, **74**, 75 967g).	169
71ACR193	E. M. Kosower; *Acc. Chem. Res.*, 1971, **4**, 193.	402
71AC(R)391	V. Sprio and S. Plescia; *Ann. Chim.* (*Rome*), 1971, **61**, 391.	770
71ACS983	P. Froyen; *Acta Chem. Scand.*, 1971, **25**, 983.	809, 892
71ACS2337	P. E. Iverson; *Acta Chem. Scand.*, 1971, **25**, 2337.	379
71AG(E)494	W. Stec, A. Okruszek and J. Michalski; *Angew. Chem., Int. Ed. Engl.*, 1971, **10**, 494.	441
71AG(E)813	K.-H. Link and R. Turley; *Angew. Chem., Int. Ed. Engl.*, 1971, **10**, 813.	387
71B4089	H. E. Gonther; *Biochemistry*, 1971, **10**, 4089.	236

71BCJ1896	O. Tsung and S. Mataka; *Bull. Chem. Soc. Jpn.*, 1971, **44**, 1896.	796
71BCJ1986	I. Kawasaki, K. Matsuda and T. Kaneko; *Bull. Chem. Soc. Jpn.*, 1971, **44**, 1986.	658
71BCJ2750	O. Tsuge and S. Iwanami; *Bull. Chem. Soc. Jpn.*, 1971, **44**, 2750.	249
71BCJ3427	O. Mitsunobu and M. Eguchi; *Bull. Chem. Soc. Jpn.*, 1971, 3427.	87, 118, 124
71BSF136	R. Burgada; *Bull. Soc. Chim. Fr.*, 1971, 136.	410, 411
71BSF541	F. Guinot, G. Lamaty and H. Munsch; *Bull. Soc. Chim. Fr.*, 1971, 541.	1008
71BSF1649	J.-J. Barieux and J. Gore; *Bull. Soc. Chim. Fr.*, 1971, 1649.	43
71BSF1921	G. Wittig; *Bull. Soc. Chim. Fr.*, 1971, 1921.	88
71BSF3571	A. Horeau, L. Ménager and H. Kagan; *Bull. Soc. Chim. Fr.*, 1971, 3571.	561
71C59	H. Gessels, J. Hanuise, R. R. Smolders, C. Brouyere, B. Lemaire and A. Wins; *Chimia*, 1971, **25**, 59.	624
71CB440	T. Severin and H. Kullmer; *Chem. Ber.*, 1971, **104**, 440.	788
71CB917	H. Stetter and E. Smulders; *Chem. Ber.*, 1971, **104**, 917.	390, 397, 403, 405
71CB1088	G. Büttner and S. Hünig; *Chem. Ber.*, 1971, **104**, 1088.	374, 385, 404
71CB1104	G. Büttner and S. Hünig; *Chem. Ber.*, 1971, **104**, 1104.	404
71CB1247	S. P. von Halasz and O. Glemser; *Chem. Ber.*, 1971, **104**, 1247.	400
71CB1761	D. Hellwinkel and B. Knabe; *Chem. Ber.*, 1971, **104**, 1761.	879
71CB1842	H. Schmidbaur and K.-H. Mitschke; *Chem. Ber.*, 1971, **104**, 1842.	502
71CB2023	R. Appel and J. Kohnke; *Chem. Ber.*, 1971, **104**, 2023.	165, 402
71CB2648	R. Appel and J. Kohnke; *Chem. Ber.*, 1971, **104**, 2648.	399
71CB3234	H. Schmidbauer and G. Kammel; *Chem. Ber.*, 1971, **104**, 3234.	163
71CB3475	H. Bredereck, W. Kantlehner and D. Schweizer; *Chem. Ber.*, 1971, **104**, 3475.	788
71CB3740	H. Timmler and J. Kurz; *Chem. Ber.*, 1971, **104**, 3740.	258
71CB3782	A. Müller and P. Werle; *Chem. Ber.*, 1971, **104**, 3782.	890
71CB3875	R. Appel and J. Kohnke; *Chem. Ber.*, 1971, **104**, 3875.	402
71CB3983	H. Plieninger, R. El-Berins and H. Mah; *Chem. Ber.*, 1971, **104**, 3983.	748
71CC146	W. B. Farnham, R. K. Murray and K. Mislow; *J. Chem. Soc., Chem. Commun.*, 1971, 146.	845, 846
71CC474	N. M. Roscher; *J. Chem. Soc., Chem. Commun.*, 1971, 474.	94
71CC979	A. G. Anastassiou and B. Chao; *J. Chem. Soc., Chem. Commun.*, 1971, 979.	153
71CC1186	S. Nakayama, M. Yoshifuji, R. Okazaki and N. Inamoto; *J. Chem. Soc., Chem. Commun.*, 1971, 1186.	854
71CC1413	D. A. Armitage and C. C. Tso; *J. Chem. Soc., Chem. Commun.*, 1971, 1413.	360, 361
71CC1475	S. W. Breuer, M. J. Leatham and F. G. Thorpe; *J. Chem. Soc., Chem. Commun.*, 1971, 1475.	653
71CC1614	R. J. Hargrove, T. E. Deuber and P. J. Stang; *J. Chem. Soc., Chem. Commun.*, 1971, 1614.	204
71CJC943	J. F. King, A. Hawson, B. L. Huston, L. J. Danks and J. Komery; *Can. J. Chem.*, 1971, **49**, 943.	179
71CJC2529	R. G. Goel and H. S. Prasad; *Can. J. Chem.*, 1971, **49**, 2529.	886
71CJC2586	C. Freppel, R. Favier, J.-C. Richer and M. Zador; *Can. J. Chem.*, 1971, **49**, 2586.	40
71CJC2706	A. J. Carty, N. K. Hota, T. W. Ng, H. A. Patel and T. J. O'Connor; *Can. J. Chem.*, 1971, **49**, 2706.	1060, 1061, 1076
71CJC4047	J. M. Lalancette and A. Freche; *Can. J. Chem.*, 1971, **49**, 4047.	226
71COC(5)123	D. W. Hutchinson; *Comp. Org. Chem.*, 1971, **5**, 123.	82
71CPB1022	A. Takamizawa, K. Hirai and T. Ishiba; *Chem. Pharm. Bull.*, 1971, **19**, 1022.	224
71CR(C)289	J. Ficini and A. Dureault; *C. R. Hebd. Seances Acad. Sci., Ser. C*, 1971, **273**, 289.	1048
71CRV337	H. Hart; *Chem. Rev.*, 1971, **71**, 337.	650
71FCF61	G. Zon and K. Mislow; *Fortschr. Chem. Forsch.*, 1971, **19**, 61.	441, 443
71HCA710	M. Roth, P. Dubs, E. Götschi and A. Eschonmoss; *Helv. Chim. Acta*, 1971, **54**, 710.	824
71HOU(10/1)9	G. Schickh, H. G. Apel, H. H. Padeker, A. Schwarttz and A. Segnitz; *Methoden Org. Chem.* (Hoube-Weyl), 1971, **10/1**, 9	784
71HOU(10/1)12	O. von Schickh, G. Apel, H. G. Padeken, H. H. Schwarz and A. Segnitz; *Methoden Org. Chem.* (*Houben-Weyl*), 1971, **10/1**, 12.	391, 392, 394
71HOU(10/1)891	H. Metzger and H. Meier; *Methoden Org. Chem.* (*Houben-Weyl*), 1971, **10/1**, 891.	389
71HOU(10/1)1091	B. Zeeh and H. Metzer; *Methoden Org. Chem.* (*Houben-Weyl*), 1971, **10/1**, 1091.	340, 343, 344, 346, 350, 353, 354, 355, 356
71IC893	R. A. Kovar and E. C. Ashby; *Inorg. Chem.*, 1971, **10**, 893.	421
71IC900	R. A. Kovar, R. Culbertson and E. C. Ashby; *Inorg. Chem.*, 1971, **10**, 900.	418, 419
71IC906	R. G. Beach and E. C. Ashby; *Inorg. Chem.*, 1971, **10**, 906.	421
71IC1028	M. K. Das and J. J. Zuckerman; *Inorg. Chem.*, 1971, **10**, 1028.	859
71IC1888	R. G. Beach and E. C. Ashby; *Inorg. Chem.*, 1971, **10**, 1888.	421
71IC2362	F. A. Cotton and A. Shaver; *Inorg. Chem.*, 1971, **10**, 2362.	468
71IC2676	B. R. Cook, C. A. McAuliffe and D. W. Meek; *Inorg. Chem.*, 1971, **10**, 2676.	883
71ICA311	L. M. Zaboronski and J. M. Shreeve; *Inorg. Chim. Acta*, 1971, **5**, 311.	166, 170
71IJS215	T. Namata and S. Oae; *Int. J. Sulfur Chem.*, 1971, **A1**, 215.	360
71IJS(A)18	C. D. Hurd, and O. Faucher; *Int. J. Sulfur Chem., Part A*, 1971, **1**, 18.	732
71IJS(A)130	V. Calo, F. Ciminale, G. Lopez and P. E. Todesco; *Int. J. Sulfur Chem.* (*A*), 1971, **1**, 130.	148
71IJS(B)277	C. J. M. Stirling; *Int. J. Sulfur Chem.* (*B*), 1971, **6**, 277.	172, 796, 797
71IZV1726	N. A. Chadaeva, G. Kh. Kamai and K. A. Mamakov; *Izv. Akad. Nauk SSSR, Ser. Khim.*, 1971, 1726 (*Chem. Abstr.*, 1972, **76**, 25 387p).	889
71IZV2103	A. N. Nesmeyanov and A. E. Borisov; *Izv. Akad. Nauk SSSR, Ser. Khim.*, 1971, 2103 (*Chem. Abstr.*, 1972, **76**, 3976c).	881
71JA438	N. C. Deno, W. E. Billups, R. Fishbein, C. Pierson, R. Whalen and J. C. Wyckoff; *J. Am. Chem. Soc.*, 1971, **93**, 438.	20
71JA1389	R. L. Ayres, C. J. Michejda and E. P. Rack; *J. Am. Chem. Soc.*, 1971, **93**, 1389.	35

71JA1508 A. Suzuki, S. Nozawa, M. Harada, M. Itoh, H. C. Brown and M. M. Midland; *J. Am. Chem. Soc.*, 1971, **93**, 1508. 34

71JA1714 G. A. Gornowicz and R. West; *J. Am. Chem. Soc.*, 1971, **93**, 1714. 1058

71JA1804 A. J. Ashe; *J. Am. Chem. Soc.*, 1971, **93**, 1804. 971

71JA1816 H. C. Brown and S. K. Gupta; *J. Am. Chem. Soc.*, 1971, **93**, 1816. 527, 528

71JA1992 T. Tsuji and E. M. Kosower; *J. Am. Chem. Soc.*, 1971, **93**, 1992. 402

71JA2080 G. Stork and E. Colvin; *J. Am. Chem. Soc.*, 1971, **93**, 2080. 921

71JA2150 A.. H. Cowley, J. L. Mills, T. M. Loehr and T. V. Long, II; *J. Am. Chem. Soc.*, 1971, **93**, 2150. 488

71JA2215 M. Ballester, J. Riera, J. Castaner, C. Badia and J. M. Monso; *J. Am. Chem. Soc.*, 1971, **93**, 2215. 1010

71JA2318 M. W. Rathke and A. Lindert; *J. Am. Chem. Soc.*, 1971, **93**, 2318. 420, 696

71JA2437 D. N. Harpp and J. G. Gleason; *J. Am. Chem. Soc.*, 1971, **93**, 2437. 143

71JA2692 M. Raban and F. B. Jones; *J. Am. Chem. Soc.*, 1971, **93**, 2692. 365

71JA2802 H. C. Brown and S. K. Gupta; *J. Am. Chem. Soc.*, 1971, **93**, 2802. 531

71JA2870 J. E. Baldwin, G. Höfle and S. C. Choi; *J. Am. Chem. Soc.*, 1971, **93**, 2870. 223

71JA2897 R. P. Borch, M. D. Bernstein and H. D. Durst; *J. Am. Chem. Soc.*, 1971, **93**, 2897. 763

71JA3291 H. C. Brown and M. M. Midland; *J. Am. Chem. Soc.*, 1971, **93**, 3291. 138, 712

71JA3293 A. J. Ashe, III; *J. Am. Chem. Soc.*, 1971, **93**, 3293. 879

71JA3395 G. Zweifel, G. M. Clark and N. L. Polston; *J. Am. Chem. Soc.*, 1971, **93**, 3395. 900

71JA3694 A. Fujii and S. I. Miller; *J. Am. Chem. Soc.*, 1971, **93**, 3694. 849, 850

71JA3792 J. Donohue, N. Mandel, W. B. Farnham, R. K. Murray, Jr., K. Mislow and H. P. Benschop; *J. Am. Chem. Soc.*, 1971, **93**, 3792. 855

71JA3816 J. P. Snyder, L. Lee and D. G. Farnum; *J. Am. Chem. Soc.*, 1971, **93**, 3816. 404

71JA4009 S. J. Benkovic and E. J. Sampson; *J. Am. Chem. Soc.*, 1971, **93**, 4009. 807

71JA4027 D. J. Peterson; *J. Am. Chem. Soc.*, 1971, **93**, 4027. 1001

71JA4062 H. C. Brown and S. K. Gupta; *J. Am. Chem. Soc.*, 1971, **93**, 4062. 522

71JA4065 E. W. Turnbloom and T. J. Katz; *J. Am. Chem. Soc.*, 1971, **93**, 4065. 475

71JA4158 R. B. King and P. N. Kapoor; *J. Am. Chem. Soc.*, 1971, **93**, 4158. 439

71JA4301 S. Otsuka, A. Nakamura, T. Kano and K. Tani; *J. Am. Chem. Soc.*, 1971, **93**, 4301. 442

71JA4841 A. McKillop, J. D. Hunt, M. J. Zelesko, J. S. Fowler, E. C. Taylor, G. McGillivray and F. Kienzle; *J. Am. Chem. Soc.*, 1971, **93**, 4841. 631

71JA4845 E. C. Taylor, F. Kienzle, R. L. Robey, A. McKillop and J. D. Hunt; *J. Am. Chem. Soc.*, 1971, **93**, 4845. 631, 964

71JA4907 A. J. Kresge, Y. Chiang, P. H. Fitzgerald, R. S. McDonald and G. H. Schmid; *J. Am. Chem. Soc.*, 1971, **93**, 4907. 40

71JA5236 D. D. Swank, C. N. Canghlan, F. Ramirez and J. F. Pilot; *J. Am. Chem. Soc.*, 1971, **93**, 5236. 831

71JA5305 C. W. Schroeck and C. R. Johnson; *J. Am. Chem. Soc.*, 1971, **93**, 5305. 362

71JA5677 R. A. Zingaro, K. J. Irgolic, D. H. O'Brien and L. J. Edmonson, Jr.; *J. Am. Chem. Soc.*, 1971, **93**, 5677. 499

71JA6267 S. Kukolfa; *J. Am. Chem. Soc.*, 1971, **93**, 6267. 196

71JA6690 A. J. Ashe, III; *J. Am. Chem. Soc.*, 1971, **93**, 6690. 879

71JA7331 A. McKillop, O. H. Oldenziel, B. O. Swann, E. C. Taylor, and R. L. Robey; *J. Am. Chem. Soc.*, 1971, **93**, 7331. 982

71JA7337 J. Swigert and K. G. Taylor; *J. Am. Chem. Soc.*, 1971, **93**, 7337. 404, 405

71JCS(A)562 R. G. Goel and H. S. Prasad; *J. Chem. Soc. (A)*, 1971, 562. 882

71JCS(A)1125 G. W. Fraser, R. D. Riddick and P. M. Watkins; *J. Chem. Soc. (A)*, 1971, 1125. 370

71JCS(A)1237 A. Storr and A. D. Penland; *J. Chem. Soc. (A)*, 1971, 1237. 421

71JCS(A)1433 J. S. Basi, D. C. Bradley and M. H. Chisholm; *J. Chem. Soc. (A)*, 1971, 1433. 421

71JCS(A)1931 D. F. Evans, F. G. Fazakerly and R. F. Philips; *J. Chem. Soc. (A)*, 1971, 1931. 564

71JCS(A)2976 B. E. Mann, B. L. Shaw and R. M. Slade; *J. Chem. Soc. (A)*, 1971, 2976. 824

71JCS(B)1330 M. J. Cook, A. R. Katritzky and M. M. Manas; *J. Chem. Soc. (B)*, 1971, 1330. 347

71JCS(C)945 D. H. Ballard and A. J. Bloodworth; *J. Chem. Soc. (C)*, 1971, 945. 98

71JCS(C)1660 A. G. Davies and S. C. W. Hook; *J. Chem. Soc. (C)*, 1971, 1660. 492, 497

71JCS(C)1812 W. L. F. Armarego; *J. Chem. Soc. (C)*, 1971, 1812. 395

71JCS(C)2404 T. P. Forshaw and A. E. Tipping; *J. Chem. Soc. (C)*, 1971, 2404. 383

71JCS(C)2699 J. V. Greenhill; *J. Chem. Soc. (C)*, 1971, 2699. 744

71JCS(C)2741 D. C. Bradley and M. H. Chisholm; *J. Chem. Soc.*, *(C)*, 1971, 2741. 421

71JCS(C)2840 D. A. Armitage and M. J. Clark; *J. Chem. Soc. (C)*, 1971, 2840. 358

71JCS(C)2872 D. W. Burt and P. Simpson; *J. Chem. Soc. (C)*, 1971, 2872. 1072

71JCS(C)3652 V. Calo, F. Ciminale, L. Lopez and P. E. Todesco; *J. Chem. Soc. (C)*, 1971, 3652. 626

71JCS(C)3867 D. A. Armitage, M. J. Clark and A. C. Kinsey; *J. Chem. Soc. (C)*, 1971, 3867. 358, 361

71JGU122 S. V. Ponomarev, M. B. Erman, S. A. Lebedev, S. Ya. Pechurina and I. F. Lutsenko; *J. Gen. Chem. USSR (Engl. Transl.)*, 1971, **41**, 122 (*Zh. Obshch. Khim.*, 1971, **41**, 127). 1093

71JGU475 V. N. Prons, M. P. Grinblat and A. L. Klebanskii; *J. Gen. Chem. USSR (Engl. Trans.)*, 1971, **41**, 475. 468

71JINC3177 K. J. Irgolic, L. R. Kallenbach and R. A. Zingaro; *J. Inorg. Nucl. Chem.*, 1971, **33**, 3177. 498, 1080

71JMC982 L. G. Humber, F. Herr and M-P. Charest; *J. Med. Chem.*, 1971, **14**, 982. 27

71JMC995 L. Field, W. S. Hanley and I. McVeigh; *J. Med. Chem.*, 1971, **14**, 995. 218

71JMC1052 D. E. Butler, S. M. Alexander, J. W. McLean and L. B. Strand; *J. Med. Chem.*, 1971, **14**, 1052. 378, 379

71JOC309	L. Field and P. M. Giles; *J. Org. Chem.*, 1971, **36**, 309.	218
71JOC322	D. N. Harpp, J. G. Gleason and D. K. Ash; *J. Org. Chem.*, 1971, **36**, 322.	199
71JOC488	R. G. Hiskey, L. M. Beacham, V. G. Matl, J. N. Smith, E. B. Williams, A. M. Thomas and E. T. Wolters; *J. Org. Chem.*, 1971, **36**, 488.	219
71JOC532	M. H. Karger and Y. Mazur; *J. Org. Chem.*, 1971, **36**, 532.	208
71JOC540	M. H. Karger and Y. Mazur; *J. Org. Chem.*, 1971, **36**, 540.	208
71JOC631	R. C. Petterson, A. DiMaggio, III, A. L. Herbert, T. J. Haley, J. P. Mykytka and I. M. Sarkar; *J. Org. Chem.*, 1971, **36**, 631.	614
71JOC818	D. R. Strobach and G. A. Boswell, Jr.; *J. Org. Chem.*, 1971, **36**, 818.	610
71JOC829	S. H. Pine and B. L. Sanchez; *J. Org. Chem.*, 1971, **36**, 829.	310
71JOC929	R. F. Cunico; *J. Org. Chem.*, 1971, **36**, 929.	928
71JOC1024	K. L. Erickson, J. Markstein and K. Kim; *J. Org. Chem.*, 1971, **36**, 1024.	615
71JOC1285	D. K. Myers and L. D. Quin; *J. Org. Chem.*, 1971, **36**, 1285.	449
71JOC1314	D. N. Harpp and J. G. Gleason; *J. Org. Chem.*, 1971, **36**, 1314.	207, 213
71JOC1438	D. R. Strobach; *J. Org. Chem.*, 1971, **36**, 1438.	1042
71JOC1700	H. H. Sisler and N. K. Kotia; *J. Org. Chem.*, 1971, **36**, 1700.	278
71JOC1727	W. E. Truce and G. C. Wolf; *J. Org. Chem.*, 1971, **36**, 1727.	1034
71JOC1790	D. J. Pasto and P. W. Wojtkowski; *J. Org. Chem.*, 1971, **36**, 1790.	532
71JOC2088	W. C. Baird, Jr., J. H. Surridge and M. Buza; *J. Org. Chem.*, 1971, **36**, 2088.	24
71JOC2207	R. W. Heintzelman, R. B. Bailey and D. Swern; *J. Org. Chem.*, 1971, **36**, 2207.	163
71JOC2530	J. Danehy, C. P. Egan and J. Switalski; *J. Org. Chem.*, 1971, **36**, 2530.	172
71JOC2561	R. Walter and J. Roy; *J. Org. Chem.*, 1971, **36**, 2561.	287
71JOC2574	J. M. Larkin and K. L. Kreuz; *J. Org. Chem.*, 1971, **36**, 2574.	398
71JOC2693	R. M. Dodson, P. D. Hammen and R. A. Davis; *J. Org. Chem.*, 1971, **36**, 2693.	200
71JOC2698	R. M. Dodson, P. D. Hammen, E. H. Jancis and G. Klose; *J. Org. Chem.*, 1971, **36**, 2698.	188
71JOC2703	R. M. Dodson, P. D. Hammen and J. Y. Fan; *J. Org. Chem.*, 1971, **36**, 2703.	200
71JOC2719	M. S. Chattha and A. M. Aguiar; *J. Org. Chem.*, 1971, **36**, 2719.	739, 1072
71JOC2720	M. S. Chattha and A. M. Aguiar; *J. Org. Chem.*, 1971, **36**, 2720.	1072
71JOC2917	M. J. Shaw, H. H. Hyman and R. Filler; *J. Org. Chem.*, 1971, **36**, 2917.	613
71JOC3055	J. C. Stowell; *J. Org. Chem.*, 1971, **36**, 3055.	390
71JOC3145	W. E. Phillips and K. W. Ratts; *J. Org. Chem.*, 1971, **36**, 3145.	168
71JOC3184	M. E. Kurz and G. J. Johnson; *J. Org. Chem.*, 1971, **36**, 3184.	650
71JOC3311	P. R. Jones and E. J. Goller; *J. Org. Chem.*, 1971, **36**, 3311.	69
71JOC3324	W. C. Baird, Jr., J. H. Surridge and M. Buza; *J. Org. Chem.*, 1971, **36**, 3324.	23, 24, 30
71JOC3520	J. J. Eisch and M. W. Foxton; *J. Org. Chem.*, 1971, **36**, 3520.	948, 968
71JOC3677	J. Tsurugi and T. Nakabayashi; *J. Org. Chem.*, 1971, **36**, 3677.	211, 722
71JOC3687	U. Lerch and J. G. Moffat; *J. Org. Chem.*, 1971, **36**, 3687.	162
71JOC3828	D. N. Harpp and T. G. Back; *J. Org. Chem.*, 1971, **36**, 3828.	217
71JOC3861	U. Lerch and J. G. Moffat; *J. Org. Chem.*, 1971, **36**, 3861.	162
71JOM(27)373	P. C. Wailes, H. Weigold, and A. P. Bell; *J. Organomet. Chem.*, 1971, **27**, 373.	968
71JOM(28)81	A. D. Norman; *J. Organomet. Chem.*, 1971, **28**, 81.	474
71JOM(28)C1	H. C. Brown and E.-I. Negishi; *J. Organomet. Chem.*, 1971, **28**, C1.	516
71JOM(30)151	D. Seyferth and S. B. Andrews; *J. Organomet. Chem.*, 1971, **30**, 151.	998
71JOM(30)295	Y. Leroux and R. Mantione; *J. Organomet. Chem.*, 1971, **30**, 295.	939
71JOM(32)313	H.-J. Neese and H. Bürger; *J. Organomet. Chem.*, 1971, **32**, 313.	566
71JOM(32)335	W. Brüser, K. H. Thiele, P. Zdunneck and F. Brune; *J. Organomet. Chem.*, 1971, **32**, 335.	566
71JOM(33)347	R. D. Feltham and H. G. Metzger; *J. Organomet. Chem.*, 1971, **33**, 347.	874, 876
71JOM(33)C4	J.-C. Lahournère and J. Valade; *J. Organomet. Chem.*, 1971, **33**, C4.	600
71JOM(33)C7	J.-C. Lahournère and J. Valade; *J. Organomet. Chem.*, 1971, **33**, C7.	601
71JPC1246	J. Silverman, L. J. Soltzberg, N. F. Yannoni and A. P. Krukonis; *J. Phys. Chem.*, 1971, **75**, 1246.	1010
71JPR254	A. Tzschach and S. Bänsch; *J. Prakt. Chem.*, 1971, **313**, 254.	876
71JPR259	A. Tzschach and V. Kiesel; *J. Prakt. Chem.*, 1971, **313**, 259.	507
71JPR642	R. Ohme and H. Preuschhof; *J. Prakt. Chem.*, 1971, **313**, 642.	806
71LA(744)15	H. Bohme, R. Braun and L. Hafner; *Justus Liebigs Ann. Chem.*, 1971, **744**, 15.	157
71LA(746)149	G. Wittig and A. Hesse; *Justus Liebigs Ann. Chem.*, 1971, **746**, 149.	420
71LA(751)17	G. Wittig, H. Braun and H.-J. Cristau; *Justus Liebigs Ann. Chem.*, 1971, **751**, 17.	824
B-71MI 201-01	G. W. Brown; in "The Chemistry of Functional Groups," ed. S. Patai, Wiley, New York, 1971, p. 593.	6
B-71MI 202-01	S. Patai (ed.); "The Chemistry of the Hydroxyl Group," Wiley, New York, 1971.	38
B-71MI 202-02	D. Swern (ed.); "Organic Peroxides," Wiley, New York, 1971.	95, 97, 98
71MI 209-01	R. G. Goel and D. R. Ridley; *Inorg. Nucl. Chem. Lett.*; 1971, **7**, 21.	504
71MI 210-01	H. Noth and W. Storch; *Synth. Inorg. Metal-Org. Chem.*, 1971, **1**, 197.	520
B-71MI 211-01	A. G. Lee; "The Chemistry of Thallium," Elsevier, Amsterdam, 1971.	597
71MI 215-01	O. Tsuge and S. Mataka; *J. Synth. Org. Chem. Jpn.*, 1971, **29**, 933.	793
B-71MI 215-02	M. Lesbre, P. Mazerolles and J. Stagé; in "The Organic Compounds of Germanium," ed. Z. Rappoport, Wiley, London, 1971.	814, 815
B-71MI 218-01	M. Lesbre, P. Mazerolles and J. Satgé; "The Organic Compounds of Germanium," ed. Z. Rappoport, Wiley, London, 1971.	946, 948, 949
71OS(51)90	C.-L. Mao and C. R. Hauser; *Org. Synth.*, 1971, **51**, 90.	675
71PS(1)71	L. Maier; *Phosphorus Sulfur*, 1971, **1**, 71.	462
71RTC265	L. Brandsma; *Recl. Trav. Chim. Pays-Bas*, 1971, **90**, 265.	1049
71RTC373	J. A. Maassen and T. J. de Boer; *Recl. Trav. Chim. Pays-Bas*, 1971, **90**, 373.	390

71RTC548	A. H. M. Kayen and T. J. de Boer; *Recl. Trav. Chim. Pays-Bas*, 1971, 548.	765
71RTC1098	J. Meijer and L. Brandsma; *Recl. Trav. Chim. Pays-Bas*, 1971, **90**, 1098.	1032
71S89	Y. Ohshiro, M. Komatsu and T. Agawa; *Synthesis*, 1971, 89.	303
71S204	J. R. Shelton and C. K. Liang; *Synthesis*, 1971, 204.	403
71S501	J. Kalvoda and K. Heusler; *Synthesis*, 1971, 501.	94
71S527	H. Arouzmanian and J. Metzger; *Synthesis*, 1971, 527.	40
71S563	E. Kühle; *Synthesis*, 1971, 563.	138, 165
71S590	E. Vilsmaier and B. Hlock; *Synthesis*, 1971, 590.	195
71S617	E. Kühle; *Synthesis*, 1971, 617.	165
71T2617	A. N. Thakore, P. Pope and A. C. Oehlschlager; *Tetrahedron*, 1971, **27**, 2617.	26, 32
71T2651	R. Tanaka, M. Rodgers, R. Simonaitis and S. I. Miller, *Tetrahedron*, 1971, **27**, 2651.	1028
71T3163	A. Zwierak and K. Kluba; *Tetrahedron*, 1971, **27**, 3163.	107
71T4245	J. B. Lambert and W. L. Oliver; *Tetrahedron*, 1971, **27**, 4245.	431
71T5753	U. Eisner and T. Krishnamurthy; *Tetrahedron*, 1971, **27**, 5753.	229
71T5861	J. F. Biellmann and J. B. Ducep; *Tetrahedron*, 1971, **27**, 5861.	143
71TL87	K. Kikugawa and M. Ichino; *Tetrahedron Lett.*, 1971, 87.	31
71TL591	Y. Leroux and R. Mantione; *Tetrahedron Lett.*, 1971, **12**, 591.	939
71TL593	R. Mantione and Y. Leroux; *Tetrahedron Lett.*, 1971, **12**, 593.	939
71TL691	J. M. Bachhawat and N. K. Mathur; *Tetrahedron Lett.*, 1971, 691.	45
71TL1511	H. Sakurai, A. Okada, M. Kira and K. Yonezawa; *Tetrahedron Lett.*, 1971, 1511.	546
71TL1753	R. Tanaka and S. I. Miller; *Tetrahedron Lett.*, 1971, 1753.	1028
71TL1761	H. Kise, G. F. Whitfield and D. Swern; *Tetrahedron Lett.*, 1971, 1761.	478
71TL2053	T. Takamoto, R. Sudoh and T. Nagakawa; *Tetrahedron Lett.*, 1971, 2053.	397
71TL2177	R. Luckenbach; *Tetrahedron Lett.*, 1971, 2177.	441
71TL2183	H. G. Richey, Jr., W. R. Erickson and A. S. Heyn; *Tetrahedron Lett.*, 1971, 2183.	561
71TL2227	K. Bott; *Tetrahedron Lett.*, 1971, 2227.	800
71TL2583	J. F. Normant and M. Bourgain; *Tetrahedron Lett.*, 1971, 2583.	629, 977
71TL3361	R. M. Carlson and A. H. Funk; *Tetrahedron Lett.*, 1971, 3361.	40
71TL3995	M. W. Rathke and A. Lindert; *Tetrahedron Lett.*, 1971, 3995.	29, 34
71TL4953	D. N. Harpp and T. G. Back; *Tetrahedron Lett.*, 1971, 4953.	241, 358
71USP3544593	E. A. Friedheim; *US Pat.* 3 544 593 (1971) (*Chem. Abstr.*, 1971, **74**, 87 991).	237
71USP3600417	A. J. Lundeen and J. E. Yates; *US Pat.* 3 600 417 (1971) (*Chem. Abstr.*, 1971, **75**, 151 349).	112
71ZAAC(381)103	J. Muller; *Z. Anorg. Allg. Chem.*; 1971, **381**, 103.	504
71ZAAC(383)136	K. Sommer; *Z. Anorg. Allg. Chem.*, 1971, **383**, 136.	493
71ZAAC(386)139	H. Schmidbaur, K.-H. Mitschke, J. Weidlein and St. Cradock; *Z. Anorg. Allg. Chem.*, 1971, **383**, 139.	488, 504
71ZAAC(386)147	H. Schmidbaur, K.-H. Mitschke and J. Weidlein; *Z. Anorg. Allg. Chem.*, 1971, **386**, 147.	888
71ZN(B)73	R. Appel and E. Lassmann; *Z. Naturforsch. Teil B*, 1971, **26**, 73.	400
71ZN(B)81	A. Schmidpeter and H. Rossnecht; *Z. Naturforsch., Teil B*, 1971, **26**, 81.	469
71ZN(B)386	A. Zwierzak and R. Gramze; *Z. Naturforsch., Teil B*, 1971, **26**, 386.	255
71ZOB1644	A. F. Grapov, O. B. Mikhailova, L. V. Razvodovskaya and L. L. Melnikov; *Zh. Obshch. Khim.*, 1971, **41**, 1644.	468, 469, 470
71ZOB2372	E. E. Nifant'ev and I. V. Shilov; *Zh. Obshch. Khim.*, 1971, **41**, 2372.	469
71ZOR258	G. S. Gol'din, K. O. Averbakh and M. A. Novak; *Zh. Org. Khim.*, 1971, **7**, 258.	383
71ZOR1146	V. G. Granik and R. G. Glushkov; *Zh. Org. Khim.*, 1971, **7**, 1146.	788
72ACS279	P. Meurling, K. Sjöberg and B. Sjöberg; *Acta Chem. Scand.*, 1972, **26**, 279.	123
72ACS2877	A. Senning and J. Kelly; *Acta Chem. Scand.*, 1972, **26**, 2877.	768
72AG(E)229	R. Scheffold and E. Saladin; *Angew. Chem., Int. Ed. Engl.*, 1972, **11**, 229.	8, 36
72AG(E)315	C. Jutz and R. M. Wagner; *Angew. Chem., Int. Ed. Engl.*, 1972, **11**, 315.	790
72AG(E)441	G. Märkl and H. Hauptmann; *Angew. Chem., Int. Ed. Engl.*, 1972, **11**, 441.	888
72AG(E)473	G. Köbrich; *Angew. Chem. Int. Ed. Engl.*, 1972, **11**, 473.	997
72BCJ506	H. Tokunaga, K. Akiba and N. Inamoto; *Bull. Chem. Soc. Jpn.*, 1972, **45**, 506.	853
72BCJ509	W. E. Parham, S. Kajigaeshi and S. H. Groen; *Bull. Chem. Soc. Jpn.*, 1972, **45**, 509.	1031
72BCJ2244	T. Mukaiyama, S. Yamamoto and M. Shiono; *Bull. Chem. Soc. Jpn.*, 1972, **45**, 2244.	524
72BJ723	J. Carlsson, H. Drevin and R. Axen; *Biochem. J.*, 1972, **173**, 723.	219
72BSF1600	M. Maumy; *Bull. Soc. Chim. Fr.*, 1972, 1600.	1066
72CB406	F. Seel and K. D. Velleman; *Chem. Ber.*, 1972, **195**, 406.	471
72CB454	W. Siebert and A. Ospica; *Chem. Ber.*, 1972, **105**, 454.	294
72CB470	L. Birkofer and M. Franz; *Chem. Ber.*, 1972, **105**, 470.	943
72CB1683	G. Kobrich, D. Merkel and K.-W. Theim; *Chem. Ber.*, 1972, **105**, 1683.	618
72CB1759	L. Birkofer and M. Franz; *Chem. Ber.*, 1972, **105**, 1759.	943
72CB1962	W. Becker and H. Noth; *Chem. Ber.*, 1972, **105**, 1962.	474, 509
72CB2963	G. Himbert and M. Regitz; *Chem. Ber.*, 1972, **105**, 2963.	1047
72CC71	H. C. Brown and Y. Yamamoto; *J. Chem. Soc., Chem. Commun.*, 1972, 71.	524
72CC122	D. H. R. Barton, R. H. Hesse, M. M. Pechet, G. Tarzia, H. T. Toh and N. D. Westcott; *J. Chem. Soc., Chem. Commun.*, 1972, 122.	16
72CC443	N. S. Isaacs and D. Kirkpatrick; *J. Chem. Soc., Chem. Commun.*, 1972, 443.	617
72CC641	J. H. Ridd and E. F. V. Scriven; *J. Chem. Soc., Chem. Commun.*, 1972, 641.	798
72CC840	C. Eaborn, A. A. Najam and D. R. M. Walton; *J. Chem. Soc., Chem. Commun.*, 1972, 840.	608
72CC1114	C. Ciannotti and M. L. H. Green; *J. Chem. Soc., Chem. Commun.*, 1972, 1114.	962
72CCC3467	O. Machackova and V. Sterba; *Collect. Czech. Chem. Commun.*, 1972, **37**, 3467.	805

72CI(L)425	M. T. Pizzorno and S. M. Albonico; *Chem. Ind.* (*London*), 1972, 425.	675
72CJC1005	J. Hooz and R. B. Layton; *Can. J. Chem.*, 1972, **50**, 1005.	1083
72CJC1167	O. E. Edwards, D. Vocelle and J. W. ApSimon; *Can. J. Chem.*, 1972, **50**, 1167.	337
72CJC3109	N. J. Bunce; *Can. J. Chem.*, 1972, **50**, 3109.	27
72CJC3502	J. T. Edward and J. Whiting; *Can. J. Chem.*, 1972, **50**, 3502.	366
72CL259	K. Narasaka, M. Hayashi and T. Mukaiyama; *Chem. Lett.*, 1972, 259.	143
72CL441	Y. Abe and J. Tsurugi; *Chem. Lett.*, 1972, 441.	688
72CL935	T. Fujisawa and T. Kobori; *Chem. Lett.*, 1972, 935.	226
72CR(C)1857	G. Mérault, P. Bourgeois and J. Dunoguès; *C. R. Hebd. Seances Acad. Sci., Ser. C*, 1972, **274**, 1857.	1090
72G527	M. Colonna, L. Greci and P. Bruni; *Gazz. Chim. Ital.*, 1972, **102**, 527.	803
72HCA481	C. Ganter and N. Wigger; *Helv. Chim. Acta*, 1972, **55**, 481.	124
72HCA1991	J. J. Daly, L. Maier and F. Sanz; *Helv. Chim. Acta*, 1972, **55**, 1991.	855
72IC1150	J. Rathke and R. Schaeffer; *Inorg. Chem.*, 1972, **11**, 1150.	524
72IC1226	L. A. Krannich and H. H. Sisler; *Inorg. Chem.*, 1972, **11**, 1226.	892
72IJ101	P. D. Bartlett and M. Lahav; *Isr. J. Chem.*, 1972, **10**, 101.	98
72IJS199	J. Tsurugi and T. Nakabayashi; *Int. J. Sulfur Chem.*, 1972, **7**, 199.	211
72IZV213	Yu. L. Kopaevich, G. G. Belen'kii, Z. A. Stumbrevichute, L. S. German and I. L. Knunyants; *Izv. Akad. Nauk SSSR, Ser. Khim.*, 1972, 213 (*Chem. Abstr.*, 1972, **77**, 19 757p).	876, 881
72IZV529	I. R. Gol'ding and A. M. Sladkov; *Izv. Akad. Nauk SSSR, Ser. Khim.*, 1972, 529 (*Chem. Abstr.*, 1972, **77**, 88 613a).	1077
72IZV1612	N. A. Chandaeva, G. Kamai and K. A. Mamakov; *Izv. Akad. Nauk SSSR, Ser. Khim.*, 1972, 1612 (*Chem. Abstr.*, 1972, **77**, 140 234u).	881, 883
72IZV2510	M. A. Belavenster, M. L. Mikheer, V. M. Pavlow and G. A. Sokolskii; *Izv. Akad. Nauk. SSSR, Ser. Khim.*, 1972, 2510 (*Chem. Abstr.*, 1973, **78**, 83 757).	208
72JA208	J. B. Lambert, C. E. Mixan and D. S. Bailey; *J. Am. Chem. Soc.*, 1972, **94**, 208.	161
72JA642	E. Block; *J. Am. Chem. Soc.*, 1972, **94**, 642.	186
72JA679	O. Mitsunobu, M. Wada and T. Sano; *J. Am. Chem. Soc.*, 1972, **94**, 679.	87, 118, 124
72JA1668	A. T. Ku and M. Sundaralingam; *J. Am. Chem. Soc.*, 1972, **94**, 1688.	1009
72JA2512	B. K. Bower and H. G. Tennent; *J. Am. Chem. Soc.*, 1972, **94**, 2512.	575
72JA3561	H. C. Brown, E.-I. Negishi and P. L. Burke; *J. Am. Chem. Soc.*, 1972, **94**, 3561.	516, 522
72JA3884	P. G. Gassman, G. A. Campbell and R. C. Frederick; *J. Am. Chem. Soc.*, 1972, **94**, 3884.	761
72JA4184	B. Stephenson, G. Solladie and H. S. Mosher; *J. Am. Chem. Soc.*, 1972, **94**, 4184.	9
72JA4370	H. C. Brown and S. K. Gupta; *J. Am. Chem. Soc.*, 1972, **94**, 4370.	527, 900
72JA4392	R. A. Moss, M. J. Landon, K. M. Luchter and A. Mamantov; *J. Am. Chem. Soc.*, 1972, **94**, 4392.	405
72JA4628	J. L. Fry, E. M. Engler and P. v. R. Schleyer; *J. Am. Chem. Soc.*, 1972, **94**, 4628.	68
72JA4784	A. Wright, D. Ling, P. Boudjouk and R. West; *J. Am. Chem. Soc.*, 1972, **94**, 4784.	131
72JA5077	E. F. Ullmann and P. Singh; *J. Am. Chem. Soc.*, 1972, **94**, 5077.	405
72JA5098	R. E. Ireland, D. C. Muchmore and U. Hengartner; *J. Am. Chem. Soc.*, 1972, **94**, 5098.	693
72JA5786	K. Kondo, A. Negishi and I. Ojima; *J. Am. Chem. Soc.*, 1972, **94**, 5786.	222
72JA6135	G. M. Atkins, Jr. and E. M. Burgess; *J. Am. Chem. Soc.*, 1972, **94**, 6135.	401, 793
72JA6190	E. J. Corey and A. Venkateswarlu; *J. Am. Chem. Soc.*, 1972, **94**, 6190.	539
72JA6543	C. P. Casey and T. J. Burkhardt; *J. Am. Chem. Soc.*, 1972, **94**, 6543.	681
72JA6854	M. W. Rathke and R. Kow; *J. Am. Chem. Soc.*, 1972, **94**, 6854.	525
72JA7178	R. D. Rieke and P. M. Hudnall; *J. Am. Chem. Soc.*, 1972, **94**, 7178.	558, 958
72JA7210	E. J. Corey and D. J. Beames; *J. Am. Chem. Soc.*, 1972, **94**, 7210.	587
72JA7406	R. J. Crawford and K. Takagi; *J. Am. Chem. Soc.*, 1972, **94**, 7406.	403
72JA7596	A. J. Ashe, III and M. D. Gordon; *J. Am. Chem. Soc.*, 1972, **94**, 7596.	879
72JA7832	E. A. Schmidt and H. M. R. Hoffmann; *J. Am. Chem. Soc.*, 1972, **94**, 7832.	639
72JCP(57)4434	C. W. Sink and A. B. Harvey; *J. Chem. Phys.*, 1972, **57**, 4434.	278
72JCS(D)326	A. Storr, B. S. Thomas and A. D. Penland; *J. Chem. Soc., Dalton Trans.*, 1972, 326.	421
72JCS(D)847	D. M. Revitt and D. B. Sowerby; *J. Chem. Soc., Dalton Trans.*, 1972, 847.	504
72JCS(D)1590	R. K. Harris, J. R. Woplin, M. Murray and R. Schmutzler; *J. Chem. Soc., Dalton Trans.*, 1972, 1590.	855
72JCS(D)1766	J. C. Fuggle, D. W. A. Sharp and J. M. Winfield; *J. Chem. Soc., Dalton Trans.*, 1972, 1766.	421
72JCS(D)2313	G. W. A. Fowles, D. A. Rice and J. D. Wilkins; *J. Chem. Soc., Dalton Trans.*, 1972, 2313.	570
72JCS(P1)1184	K. Ellis, D. J. H. Smith and S. Trippett; *J. Chem. Soc., Perkin Trans. 1*, 1972, 1184.	860
72JCS(P1)1543	J. Sinnreich and M. Asscher; *J. Chem. Soc., Perkin Trans. 1*, 1972, 1543.	157
72JCS(P1)1692	R. O. C. Norman, R. Purchase, C. B. Thomas and J. B. Aylward; *J. Chem. Soc., Perkin Trans. 1*, 1972, 1692.	388
72JCS(P1)2548	E. G. Mann and A. J. H. Mercer; *J. Chem. Soc., Perkin Trans. 1*, 1972, 2548.	508, 865
72JCS(P1)2793	D. W. Allen, F. G. Mann and J. C. Tebby; *J. Chem. Soc., Perkin Trans. 1*, 1972, 2793.	810, 892
72JCS(P2)380	S. C. Peake, M. Fild, R. Schmutzler, R. K. Harris, J. M. Nichols and R. G. Rees; *J. Chem. Soc., Perkin Trans. 2*, 1972, 380.	855
72JGU536	Yu. F. Gatilov and M. G. Kralichkina; *J. Gen. Chem. USSR* (*Engl. Transl.*), 1972, **42**, 536.	885
72JGU538	Yu. F. Gatilov, L. B. Ionov and V. M. Gavrilov; *J. Gen. Chem. USSR* (*Engl. Transl.*), 1972, **42**, 538.	885, 888
72JGU1952	Yu. F. Gatilov, L. B. Ionov, S. S. Molodtsov and V. P. Kovyrzina; *J. Gen. Chem. USSR* (*Engl. Transl.*), 1972, **42**, 1952.	888
72JGU1957	V. I. Gavrilov; *J. Gen. Chem. USSR* (*Engl. Transl.*), 1972, **42**, 1957.	1076
72JGU1960	R. G. Miftakova, F. V. Akhmetova and R. Kh. Sultanova; *J. Gen. Chem. USSR* (*Engl. Transl.*), 1972, **42**, 1960.	885

72JMC968	R. D. Westland, M. L. Mouk, J. L. Holmes, R. A. Cooley, J. S. Hong and M. M. Grenan; *J. Med. Chem.*, 1972, **15**, 968.	129
72JOC88	A. McKillop, D. Bromley and E. C. Taylor; *J. Org. Chem.*, 1972, **37**, 88.	626
72JOC333	M. D. Bentley, I. B. Douglass and J. A. Lacadie; *J. Org. Chem.*, 1972, **37**, 333.	232
72JOC383	W. M. Moon; *J. Org. Chem.*, 1972, **37**, 383.	805
72JOC626	A. G. Anderson, Jr. and F. J. Freenor; *J. Org. Chem.*, 1972, **37**, 626.	32
72JOC664	N. J. Bunce; *J. Org. Chem.*, 1972, **37**, 664.	12, 627
72JOC887	C. M. Leir; *J. Org. Chem.*, 1972, **37**, 887.	166
72JOC939	F. A. Carey and A. S. Court; *J. Org. Chem.*, 1972, **37**, 939.	709
72JOC1121	H. Kise, G. F. Whitfield and D. Swern; *J. Org. Chem.*, 1972, **37**, 1121.	121
72JOC1466	E. I. Snyder; *J. Org. Chem.*, 1972, **37**, 1466.	7
72JOC1597	M. P. Doyle, W. Wierenga and M. A. Zaleta; *J. Org. Chem.*, 1972, **37**, 1597.	805
72JOC1690	R. A. Abramovitch, E. M. Smith, E. E. Knaus and M. Saha; *J. Org. Chem.*, 1972, **37**, 1690.	765
72JOC1902	A. R. Friedman and D. R. Graber; *J. Org. Chem.*, 1972, **37**, 1902.	742
72JOC2168	J. I. Dickstein and S. I. Miller; *J. Org. Chem.*, 1972, **37**, 2168.	1062
72JOC2259	T. J. Curphey and K. S. Prasad; *J. Org. Chem.*, 1972, **37**, 2259.	373, 376
72JOC2267	G. Sosnovsky, E. H. Zaret and M. Konieczny; *J. Org. Chem.*, 1972, **37**, 2267.	684
72JOC2272	K. E. DeBruin and J. R. Peterson; *J. Org. Chem.*, 1972, **37**, 2272.	846
72JOC2367	R. M. Dodson, K. S. Srinivasan, R. F. Sauers and K. S. Sharma; *J. Org. Chem.*, 1972, **37**, 2367.	221, 225
72JOC2391	W. E. Truce and G. J. W. Tichenor; *J. Org. Chem.*, 1972, **37**, 2391.	710
72JOC2710	L. Field and Y. H. Khim; *J. Org. Chem.*, 1972, **37**, 2710.	212, 218
72JOC2810	G. B. Bachmann and R. J. Maleski; *J. Org. Chem.*, 1972, **37**, 2810.	398
72JOC2901	R. S. Marmor; *J. Org. Chem.*, 1972, **37**, 2901.	658
72JOC3096	R. E. Schwerzel and J. E. Leffler; *J. Org. Chem.*, 1972, **37**, 3096.	687
72JOC3220	M. S. Newman and S. J. Gromelski; *J. Org. Chem.*, 1972, **37**, 3220.	802
72JOC3516	R. W. Murray and S. L. Jindal; *J. Org. Chem.*, 1972, **37**, 3516.	232
72JOC3549	C. L. Borders, D. L. MacDonell and J. L. Chambers; *J. Org. Chem.*, 1972, **37**, 3549.	183
72JOC3550	M. D. Bachi and K. J. Ross-Petersen; *J. Org. Chem.*, 1972, **37**, 3550.	230
72JOC3571	J. De Jong and J. H. Boyer; *J. Org. Chem.*, 1972, **37**, 3571.	753
72JOC3584	R. A. Abramovitch, R. T. Coutts and E. M. Smith; *J. Org. Chem.*, 1972, **37**, 3584.	765
72JOC3615	F. E. Condon; *J. Org. Chem.*, 1972, **37**, 3615.	377
72JOC3810	T. Minami, K. Yamataka, Y. Ohshiro, T. Agawa, N. Yasuoka and N. Kasai; *J. Org. Chem.*, 1972, **37**, 3810.	244
72JOC3944	E. E. Smissman, M. D. Corbett, S. El-Antably and K. C. Kroboth; *J. Org. Chem.*, 1972, **37**, 3944.	683
72JOC4396	J. P. Berry, A. F. Isbell and G. Hunt; *J. Org. Chem.*, 1972, **37**, 4396.	457
72JOM(32)97	W. S. Haller and K. J. Irgolic; *J. Organomet. Chem.*, 1972, **32**, 97.	292
72JOM(34)83	J. Satgé, C. Couret and J. Escudie; *J. Organomet. Chem.*, 1972, **34**, 83.	1093
72JOM(35)C20	C. Servens and M. Peyreye; *J. Organomet. Chem.*, 1972, **35**, C20.	78
72JOM(36)323	R. G. Goel and H. S. Prasad; *J. Organomet. Chem.*, 1972, **36**, 323.	886
72JOM(36)C13	H. Sakurai and A. Okada; *J. Organomet. Chem.*, 1972, **36**, C13.	544
72JOM(38)91	G. Faraglia, E. Rivarda and F. Di Bianca; *J. Organomet. Chem.*, 1972, **38**, 91.	695, 886
72JOM(39)301	L. S. Sagan, R. A. Zingaro and K. J. Irgolic; *J. Organomet. Chem.*, 1972, **39**, 301.	494, 500
72JOM(40)73	R. J. P. Corriu and J. J. E. Moreau; *J. Organomet. Chem.*, 1972, **40**, 73.	946
72JOM(40)143	R. J. Puddephatt and G. H. Thistlethwaite; *J. Organomet. Chem.*, 1972, **40**, 143.	1101
72JOM(40)C81	A. Tamaki and J. K. Kochi; *J. Organomet. Chem.*, 1972, **40**, C81.	960
72JOM(44)37	R. H. Cragg, J. P. N. Husband, G. C. H. Jones and A. F. Weston; *J. Organomet. Chem.*, 1972, **44**, 2C 37.	358
72JOM(44)233	H. C. Brown and A. B. Levy; *J. Organomet. Chem.*, 1972, **44**, 233.	526
72JPR891	C. Bischoff and E. Schröder; *J. Prakt. Chem.*, 1972, **314**, 891.	392
72JPR170	V. B. Bierling, K. Kirschke, H. Oberender and M. Schulz; *J. Prakt. Chem.*, 1972, **314**, 170.	659
72KGS1690	S. N. Kolodyazhnaya, A. M. Simonov and I. G. Uryukina; *Khim. Geterotsikl. Soedin.*, 1972, 1690.	802
72LA(762)93	G. Kresze and U. Wagner; *Justus Liebigs Ann. Chem.*, 1972, **762**, 93.	366
B-72MI 201-01	E. Nechvatal; "Advances in Free Radical Chemistry," ed. G. H. Williams, Wiley, New York, 1972, vol. 4, p. 175.	3
B-72MI 202-01	B. J. Walker; "Organophosphorus Chemistry," Penguin, London, 1972.	105
B-72MI 203-01	P. C. Jocelyn; in "Biochemistry of the Thiol Group," Academic Press, London, 1972.	127
B-72MI 208-01	L. Maier; in "Organic Phosphorus Compounds," ed. G. M. Kosolapoff and L. Maier, Wiley-Interscience, New York, 1972, vol. 1, chap. 1, p. 1–77.	426, 428, 440
B-72MI 208-02	L. Maier; in "Organic Phosphorus Compounds," ed. G. M. Kosolapoff and L. Maier, Wiley-Interscience, New York, 1972, vol. 1, chap. 2, p. 326–332.	431
B-72MI 208-03	P. Beck; in "Organic Phosphorus Compounds," ed. G. M. Kosolapoff and L. Maier, Wiley-Interscience, New York, 1972, vol. 2, chap. 4, p. 190–508.	427, 444
B-72MI 208-04	L. Maier; in "Organic Phosphorus Compounds," ed. G. M. Kosolopoff and L. Maier, Wiley-Interscience, New York, 1972, vol. 1, chap. 2, p. 293–431.	475
72MI 208-05	M. A. Ward and R. S. Shaw; *Topics Phos. Chem.*, 1972, **7**, 1.	446
B-72MI 210-01	H. C. Brown; "Boranes in Organic Chemistry," Cornell University Press, Ithaca, New York, 1972.	514, 524
B-72MI 210-02	V. Chvalovsky; in "Organometallic Reactions," ed. E. I. Becker and M. Tsutsui, Wiley, New York, 1972, vol. 3, p. 191.	536

B-72MI 211-01 T. Mole and E. Jeffery; "Organoaluminum Compounds," Elsevier, Amsterdam, 1972. 595

72MI 213-01 V. S. Gamayurova, M. G. Kralichkina and B. D. Chernokal'skii; *Izv. Vyssh. Ucheb. Zaved., Khim. Tekhnol.*, 1972, **15**, 1023 (*Chem. Abstr.*, 1972, **77**, 152294b). 695

B-72MI 215-01 H. O. House; "Modern Synthetic Reactions," Benzamin, New York, 1972, 2nd edn, p. 238. 771

B-72MI 216-01 L. Maier; in "Organic Phosphorus Compounds," ed. G. M. Kosolapoff and L. Maier, Wiley, New York, 1972, vol. 1, chap. 1. 820

72MI 217-01 V. S. Gamayurova, M. G. Kralichkina and B. D. Chernokalskii; *Izv. Vyss. Ucheb. Zaved. Khim. Khim. Tekhnol.*, 1972, **15**, 1023 (*Chem. Abstr.*, 1972, **77**, 152 294). 886

B-72MI 218-01 H. C. Brown; "Boranes in Organic Chemistry," Cornell University Press, Ithaca, NY, 1972. 900

72MI 219-01 P. L. Timms; *Adv. Inorg. Radiochem.*, 1972, **14**, 121. 961

72OPP89 M. S. Newman, J. R. Geib and W. M. Stalick; *Org. Prep. Proceed. Int.*, 1972, **4**, 89. 1025

72OR(19)279 R. A. Sheldon and J. K. Kochi; *Org. React.*, 1972, **19**, 279. 12

72OS(52)77 A. Calder, A. R. Forrester and S. P. Hepburn; *Org. Synth.*, 1972, **52**, 77. 764

72PAC409 I. F. Lutsenko; *Pure Appl. Chem.*, 1972, **30**, 409. 700, 701

72PAC575 K. H. Thiele; *Pure Applied Chem.*, 1972, **30**, 575. 569

72RCR740 L. S. Boguslauskaya; *Russ. Chem. Rev. (Engl. Transl.)*, 1972, **41**, 740. 5, 18

72S44 G. P. Pollini, A. Barco and G. De Guilli; *Synthesis*, 1972, 44. 395

72S203 M. Okahara, K. Matsunaga and S. Komori; *Synthesis*, 1972, 203. 162

72S306 G. A. Olah and J. J. Svoboda; *Synthesis*, 1972, 306. 1008

72S307 G. A. Olah and J. J. Svoboda; *Synthesis*, 1972, 307. 1006

72S483 G. L. Grady and S. K. Chokshi; *Synthesis*, 1972, 483. 89

72S632 J. W. Timberlake, M. L. Hodge and K. Betterton; *Synthesis*, 1972, 632. 403

72S695 H. R. Kriecheldorf; *Synthesis*, 1972, 695. 860

72S699 H. C. Brown and Y. Yamamoto; *Synthesis*, 1972, 699. 524

72SRI33 G. A. Razuvaev, V. N. Latyaeva, G. A. Vasil'eva and L. I. Vyshenskaya; *Synth. React. Inorg. Metal-Org. Chem.*, 1972, **2**, 33. 566

72SRI53 L. Fournier and J. G. Riess; *Synth. React. Inorg. Metal.-Org. Chem.*, 1972, **2**, 53. 483

72T2749 P. G. Gassman, G. A. Campbell and G. Mehta; *Tetrahedron*, 1972, **28**, 2749. 762

72T3833 T. Sheradsky, G. Salemnick and Z. Nir; *Tetrahedron*, 1972, **28**, 3833. 691

72T4883 A. T. Bottini, F. P. Corson, R. Fitzgerald and K. A. Frost, II; *Tetrahedron*, 1972, **28**, 4883. 607, 624

72TL181 K. Hovius and J. B. F. N. Engberts; *Tetrahedron Lett.*, 1972, 181. 250, 364, 365

72TL411 H. Takaku and Y. Shimada; *Tetrahedron Lett.*, 1972, 411. 106

72TL847 D. U. Robert and J. G. Riess; *Tetrahedron Lett.*, 1972, 847. 10

72TL2285 H. Felkin, J. D. Umpleby, E. Hagaman and E. Wenkert; *Tetrahedron Lett.*, 1972, 2285. 562

72TL2677 J. B. Press and H. Shechter; *Tetrahedron Lett.*, 1972, **28**, 2677. 674

72TL3567 J. C. Hinshaw; *Tetrahedron Lett.*, 1972, 3567. 226

72TL3869 N. S. Isaacs and D. Kirkpatrick; *Tetrahedron Lett.*, 1972, 3869. 26

72TL4067 P. L. Stotter and K. A. Hill; *Tetrahedron Lett.*, 1972, 4067. 3

72TL4137 Y. Tamura, K. Sumoto, J. Minikawa and M. Ikeda; *Tetrahedron Lett.*, 1972, 4137. 161

72TL4319 C. Y. Meyers and L. L. Ho; *Tetrahedron Lett.*, 1972, 4319. 717, 719

72TL4339 E. J. Corey, C. U. Kim and M. Takeda; *Tetrahedron Lett.*, 1972, 4339. 25, 32

72TL4907 W. P. Weber and J. P. Shepherd; *Tetrahedron Lett.*, 1972, 4907. 44

72TL5313 D. N. Harpp and T. G. Back; *Tetrahedron Lett.*, 1972, 5313. 363

72ZAAC(394)74 W. von Kuchen and K. Koch; *Z. Anorg. Allg. Chem.*, 1972, **394**, 74. 1063, 1065, 1066, 1069, 1070

72ZAAC(394)171 A. Schmidpeter, J. Ebling, H. Stary and C. Weingand; *Z. Anorg. Allg. Chem.*, 1972, **394**, 171. 468

72ZAAC(394)293 K. H. Thiele and U. Dieckmann; *Z. Anorg. Allg. Chem.*, 1972, **394**, 293. 572

72ZC129 H. Dorn and A. Zubek; *Z. Chem.*, 1972, **12**, 129. 377

72ZN(B)769 A. Schmidpeter, W. Zeiss and H. Eckert; *Z. Naturforsch., Teil B*, 1972, **27**, 769. 468

72ZOB1285 E. I. Babkina and I. V. Vershchurnski; *Zh. Obshch. Khim.*, 1972, **42**, 1285. 448

72ZOB1936 E. E. Nifantev and I. V. Shilov; *Zh. Obshch. Khim.*, 1972, **42**, 1936. 414

72ZOR2057 L. N. Markovskii, G. S. Fedyuk and E. S. Levchenko; *Zh. Org. Khim.*, 1972, **8**, 2057. 399, 400

73ACR97 P. S. Skell, J. J. Havel, and M. J. McGlinchey; *Acc. Chem. Res.*, 1973, **6**, 97. 961

73ACR341 C. R. Johnson; *Acc. Chem. Res.*, 1976, **6**, 341. 363

73ACS1337 W. Aksenes and O. Vikane; *Acta Chem. Scand.*, 1973, **27**, 1337. 889

73ACS1451 T. Kempe and T. Norin; *Acta Chem. Scand.*, 1973, **27**, 1451. 177

73ACS1939 T. Austad, T. Rod, K. Ase, J. Songstad and A. H. Norbury; *Acta Chem. Scand.*, 1973, **27**, 1939. 858

73ACS2135 W. Aksenes and O. Vikane; *Acta Chem. Scand.*, 1973, **27**, 2135. 889

73ACS2802 P. Krogsgaard-Larsen, S. B. Christensen and H. Hjeds; *Acta Chem. Scand.*, 1973, **27**, 2802. 746

73ACS3180 B. Bjellqvist; *Acta Chem. Scand.*, 1973, **27**, 3180. 191

73AG(E)163 J. Buddrus; *Angew. Chem., Int. Ed. Engl.*, 1973, **12**, 163. 45

73AG(E)775 F. Effenberger, H. Klenk and P. L. Reiter; *Angew. Chem., Int. Ed. Engl.*, 1973, **12**, 775. 656

73AG(E)877 H. Staendeke and H. J. Kleiner; *Angew. Chem., Int. Ed. Engl.*, 1973, **12**, 877. 448

73AG(E)931 G. Märkl and F. Kneidl; *Angew. Chem., Int. Ed. Engl.*, 1973, **12**, 931. 879

73BCJ898 T. Emoto, R. Okazaki and N. Inamoto; *Bull. Chem. Soc. Jpn.*, 1973, **46**, 898. 851

73BCJ1254 M. Nojima, S. Hasegawa and N. Tokura; *Bull. Chem. Soc. Jpn.*, 1973, **46**, 1254. 797

73BCJ2156 T. Minami, M. Fukuda, M. Abe and T. Agawa; *Bull. Chem. Soc. Jpn.*, 1973, **46**, 2156. 399

73BCJ2253 T. Nishio, M. Yoshioka, H. Oayama and N. Sugiyama; *Bull. Chem. Soc. Jpn.*, 1973, **46**, 2253. 228

73BSF210 A. Le Berre, A. Étienne, J. Coquelin and C. Jacquot; *Bull. Soc. Chim. Fr.*, 1973, 210. 176

73BSF720 R. Lapouyade, M. Daney, M. Lapenue and H. Bouas-Laurent; *Bull. Soc. Chim. Fr.*, 1973, 720. 619

73BSF1465 P. Duhamel, L. Duhamel and J.-Y. Valnot; *Bull. Soc. Chim. Fr.*, 1973, 1465. 28

73BSF2137 M. Bourgain and J. F. Normant; *Bull. Soc. Chim. Fr.*, 1973, 2137. 1095

73CB177 H. J. Becher, D. Fenske and E. Langer; *Chem. Ber.*, 1973, **106**, 177. 828, 866

73CB1052 K.-H. Linke, R. Turley and E. Flaskamp; *Chem. Ber.*, 1973, **106**, 1052. 382

73CB1226 H. Schmidbaur, K.-H. Mitschke, W. Buchner, H. Stuhler and J. Weidlein; *Chem. Ber.*, 1973, **106**, 1226. 488

73CB2738 M. Wieber, B. Eichhorn and J. Götz; *Chem. Ber.*, 1973, **106**, 2738. 695

73CB3097 R. Kreher and W. Wißmann; *Chem. Ber.*, 1973, **106**, 3097. 388

73CB3645 K. H. Mitschke and H. Schmidbaur; *Chem. Ber.*, 1973, **106**, 3645. 488

73CC4 F. Jung and T. Durst; *J. Chem. Soc., Chem. Commun.*, 1973, 4. 200

73CC110 R. Aneja, A. P. Davies and J. A. Knaggs; *J. Chem. Soc., Chem. Commun.*, 1973, 110. 7

73CC144 M. Eisenhut, R. Schmutzler and W. S. Sheldrick; *J. Chem. Soc., Chem. Commun.*, 1973, 144. 856

73CC212 N. Furukawa, T. Inoue, T. Aida and S. Oae; *J. Chem. Soc., Chem. Commun.*, 1973, 212. 32

73CC269 R. D. Rieke, P. M. Hudnall, and S. Uhm; *J. Chem. Soc., Chem. Commun.*, 1973, 269. 958

73CC273 B. E. Maryanoff, R. Tang and K. Mislow; *J. Chem. Soc., Chem. Commun.*, 1973, 273. 462

73CC544 A. Pelter, C. R. Harrison and D. Kirkpatrick; *J. Chem. Soc., Chem. Commun.*, 1973, 544. 904

73CC695 D. L. J. Clive; *J. Chem. Soc., Chem. Commun.*, 1973, 695. 282, 291

73CC813 J. W. Gilje, R. W. Brown and A. H. Cowley; *J. Chem. Soc., Chem. Commun.*, 1973, 813. 476

73CC821 J. Gore, P. Place and M. L. Roumestant; *J. Chem. Soc., Chem. Commun.*, 1973, 821. 9

73CC879 R. D. Rieke and S. F. Bales; *J. Chem. Soc., Chem. Commun.*, 1973, 879. 958

73CCC706 J. Jahelka, O. Machackova and V. Sterba; *Collect. Czech. Chem. Commun.*, 1973, **38**, 706. 805

73CJC936 W. R. Cullen and L. Mihichuk; *Can. J. Chem.*, 1973, **51**, 936. 876

73CJC2347 D. E. Horning, D. A. Ross and J. M. Muchowski; *Can. J. Chem.*, 1973, **51**, 2347. 647

73CJC2504 A. L. MacDonald and J. Trotter; *Can. J. Chem.*, 1973, **51**, 2504. 361

73CJC3403 H. J. Langer and J. B. Hyne; *Can. J. Chem.*, 1973, **51**, 3403. 722, 725

73CL599 J. Okamoto and H. Sakurai; *Chem. Lett.*, 1973, 599. 459

73CPB867 Y. Kobayashi, I. Kumadaki, A. Ohsawa and M. Honda; *Chem. Pharm. Bull.*, 1973, **21**, 867. 20

73E389 M. Walti and D. B. Hope; *Experientia*, 1973, **29**, 389. 214

73HCA14 J. Borgulya, R. Madeja, H.-J. Hansen, H. Schmid and R. Barner; *Helv. Chim. Acta*, 1973, **56**, 14. 656

73HCA75 U. Widmer, J. Zsindely, H. J. Hansen and H. Schmid; *Helv. Chim. Acta*, 1973, **56**, 75. 660

73HOU(13/2a)4 G. Bähr and H. O. Kalinowski; *Methoden Org. Chem.* (*Houben Weyl*), 1973, **13/2a**, 4. 563

73HOU(13/2a)47 K. Nützel; *Methoden Org. Chem.* (*Houben-Weyl*), 1973, **13/2a**, 47. 557

73HOU(13/2a)533 G. Bähr and H. O. Kalinowski; *Methoden. Org. Chem.* (*Houben-Weyl*), 1973, **13/2a**, 533. 563

73HOU(13/2a)553 K. Nützel; *Methoden Org. Chem.* (*Houben-Weyl*), 1973, **13/2a**, 553. 591

73HOU(13/2a)608 K. Nützel; *Methoden Org. Chem.* (*Houben-Weyl*), 1973, **13/2a**, 608. 1095

73HOU(13/2a)622 K. Nützel; *Methoden Org. Chem.* (*Houben-Weyl*), 1973, **13/2a**, 622. 1095

73HOU(13/2a)859 K. Nützel; *Methoden Org. Chem.* (*Houben-Weyl*), 1973, **13/2a**, 859. 1095

73HOU(13/2b) H. Straub, K. P. Zeller and H. Leditscke; *Methoden Org. Chem.* (*Houben-Weyl*), 1973, **13/2b**. 591

73IC2119 J. G. Morse and K. W. Morse; *Inorg. Chem.*, 1973, **12**, 2119. 414

73IC2482 R. G. Montemayor and R. W. Parry; *Inorg. Chem.*, 1973, **12**, 2482. 409

73ICA319 G. Brandi, M. Brunelli, G. Lugli and A. Mazzei; *Inorg. Chim. Acta*, 1973, **7**, 319. 602

73IJS71 F. A. Davis; *Int. J. Sulfur Chem.*, 1973, **8**, 71. 357

73IJS307 N. Isenberg and M. Grdinic; *Int. J. Sulfur Chem.*, 1973, **8**, 307. 232, 724

73IJS469 W. J. Spillane; *Int. J. Sulfur Chem.*, 1973, **8**, 469. 366, 797

73JA197 D. L. Klayman and T. S. Griffin; *J. Am. Chem. Soc.*, 1973, **95**, 197. 279, 292

73JA581 R. A. Olofson and C. M. Dougharty; *J. Am. Chem. Soc.*, 1973, **95**, 581. 420

73JA582 R. A. Olofson and C. M. Dougherty; *J. Am. Chem. Soc.*, 1973, **95**, 582. 696

73JA920 P. Chao and D. M. Lemal; *J. Am. Chem. Soc.*, 1973, **95**, 920. 153

73JA922 D. M. Lemal and P. Chao; *J. Am. Chem. Soc.*, 1973, **95**, 922. 153

73JA982 C. A. Brown; *J. Am. Chem. Soc.*, 1973, **95**, 982. 423

73JA1296 A. McKillop, O. H. Oldenziel, B. O. Swann, E. C. Taylor, and R. L. Robey; *J. Am. Chem. Soc.*, 1973, **95**, 1296. 982

73JA1562 M. J. S. Dewar and W. B. Jennings; *J. Am. Chem. Soc.*, 1973, **95**, 1562. 378, 381, 382

73JA1677 E. F. Ullman, L. Call and S. S. Tseng; *J. Am. Chem. Soc.*, 1973, **95**, 1677. 405

73JA2697 K. B. Sharpless and R. F. Lauer; *J. Am. Chem. Soc.*, 1973, **95**, 2697. 85

73JA2719 S. F. Nelsen and R. T. Landis, II; *J. Am. Chem. Soc.*, 1973, **95**, 2719. 382

73JA2918 L. Senatore, E. Ciuffarin, A. Fava and G. Levita; *J. Am. Chem. Soc.*, 1973, **95**, 2918. 688

73JA3070 R. A. Moss and E. M. Love; *J. Am. Chem. Soc.*, 1973, **95**, 3070. 405, 773

73JA3228 J. R. Roberts and K. U. Ingold; *J. Am. Chem. Soc.*, 1973, **95**, 3228. 388

73JA3310 H. O. House, D. S. Crumrine, A. Y. Teranishi and H. D. Olmstead; *J. Am. Chem. Soc.*, 1973, **95**, 3310. 702

73JA4064 D. B. Denney, D. Z. Denney and Y. F. Hsu; *J. Am. Chem. Soc.*, 1973, **95**, 4064. 165

73JA4194 T. E. Nappier, D. W. Meek, R. M. Kirchner and J. A. Ibers; *J. Am. Chem. Soc.*, 1973, **95**, 4194. 438

73JA4634 J. B. Lambert, C. E. Mixan and D. H. Johnson; *J. Am. Chem. Soc.*, 1973, **95**, 4634. 287

73JA5786 H. C. Brown, T. Hamaoka and N. Ravindran; *J. Am. Chem. Soc.*, 1973, **95**, 5786. 606, 628

73JA5831 T. M. Maricich, R. A. Jourdenais and T. A. Ulbight; *J. Am. Chem. Soc.*, 1973, **95**, 5831. 364

73JA6137 K. B. Sharpless, R. F. Lauer and A. Y. Teranishi; *J. Am. Chem. Soc.*, 1973, **95**, 6137. 282, 287

73JA6152	G. Stork and B. Ganem; *J. Am. Chem. Soc.*, 1973, **95**, 6152.	931
73JA6158	H. L. Ammon, G. L. Wheeler and P. H. Watts, Jr.; *J. Am. Chem. Soc.*, 1973, **95**, 6158.	1004
73JA6263	E. Samuel and M. D. Rausch; *J. Am. Chem. Soc.*, 1973, **95**, 6263.	566
73JA6456	H. C. Brown, T. Hamaoka and N. Ravindran; *J. Am. Chem. Soc.*, 1973, **95**, 6456.	623
73JA6487	A. Tamaki, S. A. Magennis, and J. K. Kochi; *J. Am. Chem. Soc.*, 1973, **95**, 6487.	590, 960
73JA6757	E.-I. Negishi and H. C. Brown; *J. Am. Chem. Soc.*, 1973, **95**, 6757.	516
73JA6840	B. M. Trost and T. N. Salzmann; *J. Am. Chem. Soc.*, 1973, **95**, 6840.	139, 140
73JA6867	R. K. Boeckman Jr.; *J. Am. Chem. Soc.*, 1973, **95**, 6867.	931
73JA7114	J. K. Kochi, R. T. Tang and T. Bernath; *J. Am. Chem. Soc.*, 1973, **95**, 7114.	652
73JA7171	J. J. Partridge, N. K. Chadhu and W. R. Uskokovic; *J. Am. Chem. Soc.*, 1973, **95**, 7171.	519
73JA7788	G. H. Posner, C. E. Whitten and J. J. Sterling; *J. Am. Chem. Soc.*, 1973, **95**, 7788.	588
73JA7917	D. Arigoni, A. Vasella, K. B. Sharpless, H. P. Jensen; *J. Am. Chem. Soc.*, 1973, **95**, 7917.	292
73JA8066	D. A. Tyssee, L. P. Bausher and P. Haake; *J. Am. Chem. Soc.*, 1973, **95**, 8066.	807
73JA8191	D. B. Denney, D. Z. Denney and Y. F. Hsu; *J. Am. Chem. Soc.*, 1973, **95**, 8191.	165
73JA8486	S. Krishnamurthy, R. M. Schubert and H. C. Brown; *J. Am. Chem. Soc.*, 1973, **95**, 8486.	83
73JCS(D)157	M. F. Lappert and D. E. Palmer; *J. Chem. Soc., Dalton Trans.*, 1973, 157.	815
73JCS(D)961	G. W. A. Fowles, D. A. Rice and J. D. Wilkins; *J. Chem. Soc., Dalton Trans.*, 1973, 961.	570
73JCS(D)1716	J. W. Anderson, G. K. Barker, J. E. Drake and M. Rodger; *J. Chem. Soc., Dalton Trans.*, 1973, 1716.	278, 294
73JCS(D)1876	A. Majid, D. W. A. Sharp, J. M. Winfield and I. Hanley; *J. Chem. Soc., Dalton Trans.*, 1973, 1876.	421
73JCS(D)2436	G. Santini-Scampucci and J. G. Reiss; *J. Chem. Soc., Dalton Trans.*, 1973, 2436.	570
73JCS(D)2620	A. Tamaki and J. K. Kochi; *J. Chem. Soc., Dalton Trans.*, 1973, 2620.	590
73JCS(P1)676	R. N. Haszeldine, R. B. Rigby and A. E. Tipping; *J. Chem. Soc., Perkin Trans. 1*, 1973, 676.	193
73JCS(P1)977	I. U. Khand, G. R. Knox, P. L. Pauson, W. E. Watts, and M. I. Foreman; *J. Chem. Soc., Perkin Trans. 1*,1973, 977.	985
73JCS(P1)2778	R. A. Moss and T. B. K. Lee, *J. Chem. Soc., Perkin Trans. 1*, 1973, 2778.	773
73JCS(P1)2870	P. Golborn and F. Scheinmann; *J. Chem. Soc., Perkin Trans. 1*, 1973, 2870.	675
73JCS(P2)638	U. Folli, D. Iarossi and F. Taddei; *J. Chem. Soc., Perkin Trans. 2*, 1973, 638.	600
73JFC(3)447	D. J. Burton and P. E. Greenlimb; *J. Fluorine Chem.*, 1973, **3**, 447.	611
73JGU626	V. T. Bychkov, N. S. Vyazankin, N. F. Cherepinnikova and G. A. Razuvaev; *J. Gen. Chem. USSR* (*Engl. Transl.*), 1973, **43**, 626 (*Zh. Obshch. Khim.*, 1973, **43**, 629).	1093
73JGU825	N. A. Chadaeva, K. A. Mamakov, R. R. Shagidullin and G. K. Kamai; *J. Gen. Chem. USSR* (*Engl. Transl.*), 1973, **43**, 825.	501
73JGU1123	Yu. F. Gatilov and F. D. Yambushev; *J. Gen. Chem. USSR* (*Engl. Transl.*), 1973, **43**, 1123.	886
73JGU1921	V. S. Gamayurova, V. K. Kuz'min, B. D. Chernokalskii and R. R. Shagidullin; *J. Gen. Chem. USSR* (*Engl. Transl.*), 1973, **43**, 1921.	886
73JINC311	B. L. Kalsotra, R. K. Multani and B. D. Jain; *J. Inorg. Nucl. Chem.*, 1973, **35**, 311.	565
73JINC2659	R. T. Markham, E. A. Dietz, Jr. and D. R. Martin; *J. Inorg. Nucl. Chem.*, 1973, **35**, 2659.	434
73JMC736	D. F. Morrow, P. C. Johnson, H. Torabi, D. Williams, D. L. Wedding, J. W. Craig, R. F. Majewski, J. P. Braselton and D. G. Gallo; *J. Med. Chem.*, 1973, **16**, 736.	381
73JMC1113	R. F. Meyer, C. D. Stratton, S. G. Hastings and R. M. Corey; *J. Med. Chem.*, 1973, **16**, 1113.	741
73JOC158	R. H. Fish; *J. Org. Chem.*, 1973, **38**, 158.	527
73JOC162	J. Lipowitz and S. A. Bowman; *J. Org. Chem.*, 1973, **38**, 162.	657
73JOC182	H. C. Brown and N. Ravindran; *J. Org. Chem.*, 1973, **38**, 182.	41
73JOC489	J. H. Wotiz, P. M. Barelski and D. F. Koster; *J. Org. Chem.*, 1973, **38**, 489.	420, 422
73JOC514	H. O. House, R. A. Auerbach, M. Gall and N. P. Peet; *J. Org. Chem.*, 1973, **38**, 514.	675
73JOC1239	Y. Tamura, J. Minamikawa, K. Sumoto, S. Fuji and M. Ikeda; *J. Org. Chem.*, 1973, **38**, 1239.	357, 691
73JOC1361	W. E. Parham, W. D. McKown, V. Nelson, S. Kajigaeshi and N. Ishikawa; *J. Org. Chem.*, 1973, **38**, 1361.	1031
73JOC1418	N. Kornblum and P. A. Wade; *J. Org. Chem.*, 1973, **38**, 1418.	393, 395
73JOC1611	M. S. Chattha and A. M. Aguiar; *J. Org. Chem.*, 1973, **38**, 1611.	1060
73JOC1617	H. C. Brown and N. Ravindran; *J. Org. Chem.*, 1973, **38**, 1617.	900
73JOC1813	J. P. Lorand, R. W. Grant, P. A. Samuel, E. M. O'Connell, J. Zaro, J. Pilote and R. W. Wallace;*J. Org. Chem.*, 1973, **38**, 1813.	355
73JOC1870	S. Yamaguchi and H. S. Mosher; *J. Org. Chem.*, 1973, **38**, 1870.	64
73JOC1924	J. B. Norell; *J. Org. Chem.*, 1973, **38**, 1924.	656
73JOC2412	C. J. Michejda and R. W. Schluenz; *J. Org. Chem.*, 1973, **38**, 2412.	381
73JOC2422	H. C. Brown and B. A. Carlson; *J. Org. Chem.*, 1973, **38**, 2422.	530
73JOC2435	M. S. Newman and V. Lee; *J. Org. Chem.*, 1973, **38**, 2435.	802
73JOC2438	M. S. Newman and W. C. Liang; *J. Org. Chem.*, 1973, **38**, 2438.	802
73JOC2576	P. L. Stotter and K. A. Hill; *J. Org. Chem.*, 1973, **38**, 2576.	28, 29
73JOC2664	C. Trindle, J.-T. Hwang and F. A. Carey; *J. Org. Chem.*, 1973, **38**, 2664.	936
73JOC2703	J. F. Bunnett, J. O. Edwards, D. V. Wells, H. J. Brass and R. Curci; *J. Org. Chem.*, 1973, **38**, 2703.	845
73JOC2908	M. S. Chattha and A. M. Aguiar; *J. Org. Chem.*, 1973, **38**, 2908.	739
73JOC2910	M. S. Newman and M. C. Vander Zwan; *J. Org. Chem.*, 1973, **38**, 2910.	679
73JOC2965	D. A. Jones, R. A. Mikiulec and R. H. Mazur; *J. Org. Chem.*, 1973, **38**, 2965.	214
73JOC3224	E. J. Corey and J. W. Suggs; *J. Org. Chem.*, 1973, **38**, 3224.	675
73JOC3893	H. O. House and M. J. Umen; *J. Org. Chem.*, 1973, **38**, 3893.	587

73JOC3968	H. C. Brown, J.-J. Katz and B. A. Carlson; *J. Org. Chem.*, 1973, **38**, 3968.	530
73JOC4070	W. G. Filby, K. Gunther and R. D. Penzhorn; *J. Org. Chem.*, 1973, **38**, 4070.	186
73JOC4092	C. G. Scouten and H. C. Brown; *J. Org. Chem.*, 1973, **38**, 4092.	41
73JOC4254	P. F. Hudrlik and A. M. Hudrlik; *J. Org. Chem.*, 1973, **38**, 4254.	676
73JOC4271	J. R. Pratt and S. F. Thames; *J. Org. Chem.*, 1973, **38**, 4271.	942
73JOC4328	D. N. Harpp and T. G. Back; *J. Org. Chem.*, 1973, **38**, 4328.	363, 688
73JOC4339	V. J. Traynelis and J. N. Rieck; *J. Org. Chem.*, 1973, **38**, 4339.	768
73JOM(47)33	R. van Veen and F. Bickelhaupt; *J. Organomet. Chem.*, 1973, **47**, 33.	523
73JOM(47)225	V. Kalyanaraman and M. V. George; *J. Organomet. Chem.*, 1973, **47**, 225.	989
73JOM(50)26	I. Ojima, M. Nihonyana and Y. Nagai; *J. Organomet. Chem.*, 1973, **50**, C26.	264
73JOM(50)129	R. G. Goel and H. S. Prasad; *J. Organomet. Chem.*, 1973, **50**, 129.	891
73JOM(50)C26	I. Ojima, M. Nihonyanagi and Y. Nagai; *J. Organomet. Chem.*, 1973, **50**, C26.	540
73JOM(51)223	H. A. Meinema, H. F. Martens and J. G. Noltes; *J. Organomet. Chem.*, 1973, **51**, 223.	485, 486, 487, 505
73JOM(54)61	H. C. Brown and B. A. Carlson; *J. Organomet. Chem.*, 1973, **54**, 61.	530
73JOM(55)143	G. O. Doak and J. M. Summy; *J. Organomet. Chem.*, 1973, **55**, 143.	886, 887
73JOM(55)375	K. N. V. Duong, A. Ahond, C. Merienne and A. Gaudemer; *J. Organomet. Chem.*, 1973, **55**, 375.	579
73JOM(56)1	J. Satgé, M. Massol and P. Rivière; *J. Organomet. Chem.*, 1973, **56**, 1.	947, 949
73JOM(57)269	P. J. Davidson, M. F. Lappert and R. Pearce; *J. Organomet. Chem.*, 1973, **57**, 269.	566
73JOM(61)83	R. S. Glass; *J. Organomet. Chem.*, 1973, **61**, 83.	263, 541
73LA20	H. Niederprüm, P. Voss and V. Beyl; *Justus Liebigs Ann. Chem.*, 1973, 20.	689
73LA1480	G. P. Schiemenz and H. Kaack; *Justus Liebigs Ann. Chem.*, 1973, 1480.	824
73LA1644	K. Hartke and G. Golz; *Justus Liebigs Ann. Chem.*, 1973, 1644.	240
73M503	A. Ecke, I. Boie and U. Schmidt; *Monatsh. Chem.*, 1973, **104**, 503.	256
73MI 201-01	J. F. Mills and J. A. Schneider; *Ind. Eng. Chem. Prod. Res. Dev.*, 1973, **12**, 160.	24
B-73MI 201-02	M. L. Poutsma; in "Free Radicals," ed. J. K. Kochi, Wiley, New York, 1973, p. 159.	2, 21
B-73MI 201-03	E. S. Huyser; in "The Chemistry of Functional Groups," ed. S. Patai, Wiley, New York, 1973, p. 550.	2
B-73MI 201-04	W. G. Nigh; in "Oxidation in Organic Chemistry," Part B, ed. W. S. Trahanovsky, Academic Press, New York, 1973, p. 67.	28
B-73MI 202-01	W. Gerrard and H. R. Hudson; "Organic Phosphorus Compounds," Wiley, New York, 1973, vol. 5, p. 21.	104, 105
B-73MI 202-02	E. Cherbuliez; "Organic Phosphorus Compounds," Wiley, New York, 1973, vol. 6, p. 211.	105
B-73MI 202-03	E. G. Rochow; "Comprehensive Inorganic Chemistry," Pergamon, Oxford, 1973, vol. 1, p. 1425.	108
B-73MI 206-01	E. Kühle; "The Chemistry of the Sulfenic Acids," Georg Thieme, Stuttgart, 1973.	357
B-73MI 207-01	J. Mathieu and J. Weill-Raynal; "Formation of C—C Bonds," Thieme, Stuttgart, 1973.	393, 394, 396
73MI 208-01	R. Luckenbach; *Phosphorus*; 1973, **3**, 77.	441, 442
B-73MI 210-01	G. M. L. Cragg; "Organoboranes in Organic Synthesis," Dekker, New York, 1973.	514
B-73MI 210-02	V. Bazant; "Handbook of Organosilicon Compounds," Marcel Dekker, New York, 1973, vols. 1–4.	535
B-73MI 211-01	M. Schlosser; "Polar Organometalle," Springer, Berlin, 1973.	551, 557
B-73MI 213-01	E. Kühle; "The Chemistry of the Sulfenic Acids," Georg Thieme, Stuttgart, 1973.	688
73MI 215-01	M. Remes, J. Divis, V. Zverina, J. Marhold and M. Matrka; *Chem. Prumysl*, 1973, **23**, 133 (*Chem. Abstr.*, 1973, **79**, 31 605).	777
B-73MI 216-01	R. Keat and R. A. Shaw; in "Organic Phosphorus Compounds," ed. G. M. Kosolapoff and L. Maier, Wiley, New York, 1973, vol. 6, chap. 17.	864
B-73MI 218-01	G. M. L. Cragg; "Organoboranes in Organic Synthesis," Marcel Dekker, New York, 1973.	900
B-73MI 220-1	L. Kaplan; in "Free Radicals", ed. J. K. Kochi, Wiley, London, 1973, vol. II, p. 361.	1010
73OPP219	F. E. Condon and D. C. Thakkar; *Org. Prep. Proced. Int.*, 1973, **5**, 219; *Chem. Abstr.*, 1973, **79**, 136401v.	376
73OPP225	F. E. Condon and D. G. Shapiro; *Org. Prep. Proced Int.*, 1973, **5**, 225 (*Chem. Abstr.*, 1974, **80**, 14482x).	379
73OPP233	F. E. Condon, D. C. Thakkar and T. B. Goldstein; *Org. Prep. Proceed. Int.*, 1973, **5**, 233 (*Chem. Abstr.*, 1974, **80**, 14483y).	380, 381, 382
73OSC(5)921	S. I. Miller, G. R. Ziegler and R. Wieleseck; *Org. Synth., Coll. Vol.*, 1973, **5**, 921.	1017
73OSC(2)245	M. Neeman and W. S. Johnson; *Org. Synth., Coll. Vol.*, 1973, **2**, 245.	89
73OSC(5)43	R. Gösl and A. Meusen; *Org. Synth., Coll. Vol.*, 1973, **5**, 43.	374, 379, 384
73OSC(5)107	R. Adams, W. Reifschneider and A. Ferretti; *Org. Synth., Coll. Vol.*, 1973, **5**, 107.	273
73OSC(5)126	J. S. Meek and D. T. Osuga; *Org. Synth. Coll. Vol.*, 1973, **5**, 126.	12
73OSC(5)136	F. H. Dean, J. H. Amin and F. L. M. Pattison; *Org. Synth. Coll. Vol.*, 1973, **5**, 136.	18
73OSC(5)142	J. P. Schaefer, J. G. Higgins and P. K. Shenoy; *Org. Synth., Coll. Vol.*, 1973, **5**, 142.	837
73OSC(5)184	M. J. Mintz and C. Walling; *Org. Synth., Coll. Vol.*, 1973, **5**, 184.	338
73OSC(5)196	P. D. Bartlett and L. H. Know; *Org. Synth., Coll. Vol.*, 1973, **5**, 196.	180
73OSC(5)223	G. J. M. van der Kerk, C. W. Pluygers and G. de Vries; *Org. Synth., Coll. Vol.*, 1973, **5**, 223.	273
73OSC(5)249	J. P. Schaefer, J. G. Higgins and P. K. Shenov; *Org. Synth. Coll. Vol.*, 1973, **5**, 249.	7, 837
73OSC(5)273	N. W. Werner and J. Caranova, Jr.; *Org. Synth., Coll. Vol.*, 1973, **5**, 273.	757
73OSC(5)328	T. F. Corbin, R. C. Hahn and H. Shechter; *Org. Synth. Coll. Vol.*, 1973, **5**, 328.	27
73OSC(5)445	W. H. Sharkey and C. M. Langkammerer; *Org. Synth., Coll. Vol.*, 1973, **5**, 445.	396
73OSC(5)552	M. G. Romanelli and E. I. Becker; *Org. Synth., Coll. Vol.*, 1973, **5**, 552.	753
73OSC(5)602	V. Mark; *Org. Synth., Coll. Vol.*, 1973, **5**, 602.	411

73OSC(5)791	C. R. Johnson and J. E. Keiser; *Org. Synth., Coll. Vol.*, 1973, **5**, 791.	149
73OSC(5)808	S. Hünig, E. Lücke and W. Brenninger; *Org. Synth., Coll. Vol.*, 1973, **5**, 808.	739
73OSC(5)825	A. Kalir; *Org. Synth. Coll. Vol.*, 1973, **5**, 825.	27
73OSC(5)833	W. E. Noland; *Org. Synth., Coll. Vol.*, 1973, **5**, 833.	394
73OSC(5)845	N. Kornblum and W. J. Jones; *Org. Synth., Coll. Vol.*, 1973, **5**, 845.	397
73OSC(5)855	R. D. Smith and H. E. Simmons. *Org. Synth. Coll. Vol. V*, 1973, 855.	591
73OSC(5)1005	R. Schmutzler; *Org. Synth., Coll Vol.*, 1973, **5**, 1005.	848
73OSC(5)1135	K. Dimroth, C. Reichardt and K. Vogel; *Org. Synth., Coll. Vol.* 5, 1973, 1135.	1008
73PNA(70)328	L. Field and J. E. White; *Proc. Natl. Acad. Sci. USA*, 1973, **70**, 328.	172
73RTC762	W. Hagens, H. J. T. Bos and J. F. Arens; *Recl. Trav. Chim. Pays-Bas*, 1973, **92**, 762.	1066
73S37	G. Zweifel, R. P. Fisher and A. Horng; *Synthesis*, 1973, 37.	1005
73S109	H. Ahlbrecht and H. Hanisch; *Synthesis*, 1973, 109.	789
73S156	C. Wilkins; *Synthesis*, 1973, 156.	45
73S307	R. Luckenbach; *Synthesis*, 1973, 307.	888
73S485	J. P. Lockard, C. W. Schroeck and C. R. Johnson; *Synthesis*, 1973, 485.	151
73S493	K. B. Becker, M. Geisel, C. A. Grob and F. Kuhnen; *Synthesis*, 1973, 493.	13
73S667	M. Kobayashi, M. Okahara, S. Komori and T. Agawa; *Synthesis*, 1973, 667.	251
73S672	G. Zweifel and A. Horng; *Synthesis*, 1973, 672.	900
73S703	M. Tramontini; *Synthesis*, 1973, 703.	316
73S779	G. A. Olah, M. Nojima and I. Kerekes; *Synthesis*, 1973, 779.	17
73S780	G. A. Olah, M. Nojima and I. Kerekes; *Synthesis*, 1973, 780.	18
73S786	G. A. Olah, M. Nojima and I. Kerekes; *Synthesis*, 1973, 786.	6
73S789	K. B. Becker and C. A. Grob; *Synthesis*, 1973, 789.	23
73S795	I. Barral and J. Savrda; *Synthesis*, 1973, 795.	243
73SC317	T.-L. Ho and C. M. Wong; *Synth. Commun.*, 1973, 317.	214
73T917	W. W. Lotz and J. Gosselck; *Tetrahedron*, 1973, **29**, 917.	1001
73T1069	N. Kawabata, A. Matsumara and S. Yamashita; *Tetrahedron*, 1973, **29**, 1069.	563
73T3011	F. P. Colonna, E. Valentin, G. Pitacco and A. Risaliti; *Tetrahedron*, 1973, **29**, 3011.	783
73T3659	M. Guette, J. Capillon and J.-P. Guette; *Tetrahedron*, 1973, **29**, 3659.	69
73TL289	C. Y. Meyers and G. J. McCollum; *Tetrahedron Lett.*, 1973, 289.	156
73TL655	R. H. Cragg and A. F. Weston; *Tetrahedron Lett.*, 1973, 655.	214, 215
73TL1397	S. M. Hecht and J. W. Kozarich; *Tetrahedron Lett.*, 1973, 1397.	683
73TL1401	J. M. Lalancette and A. Freche; *Tetrahedron Lett.*, 1973, 1401.	227
73TL1505	C. J. Reich, G. R. Sullivan and H. S. Mosher; *Tetrahedron Lett.*, 1973, 1505.	64
73TL1561	P. de Mayo and H.-Y. Ng; *Tetrahedron Lett.*, 1973, 1561.	228
73TL1979	K. B. Sharpless, M. W. Young and R. F. Lauer; *Tetrahedron Lett.*, 1973, 1979.	286
73TL2321	S. F. Nelsen and G. R. Weisman; *Tetrahedron Lett.*, 1973, 2321.	382
73TL2585	Y. Leroux and C. Roman; *Tetrahedron Lett.*, 1973, **14**, 2585.	939
73TL2859	M. L. Heyman and J. P. Snyder; *Tetrahedron Lett.*, 1973, 2859.	377, 378, 403
73TL2863	N. K. Sharma, F. Jung and T. Durst; *Tetrahedron Lett.*, 1973, 2863.	200
73TL3563	S. Braveman and D. Reisman; *Tetrahedron Lett.*, 1973, 3563.	194, 195
73TL3597	D. A. White and M. M. Baizer; *Tetrahedron Lett.*, 1973, 3597.	395
73TL3839	J. B. Hendrickson and R. Bergeron; *Tetrahedron Lett.*, 1973, 3839.	300, 303
73TL3937	A. K. Bose and B. Lal; *Tetrahedron Lett.*, 1973, 3937.	8
73TL4193	H. Sakurai, K.-I. Nishiwaki and M. Kira; *Tetrahedron Lett.*, 1973, **14**, 4193.	936
73TS93	G. Berti; *Top. Stereochem.*, 1973, **7**, 93.	90
73ZAAC(401)179	G. Hopf and R. Paetzold; *Z. Anorg. Allg. Chem.*, 1973, **401**, 179.	369
73ZAAC(397)97	F. Kober; *Z. Anorg. Allg. Chem.*, 1973, **397**, 97.	501
73ZAAC(401)243	F. Kober; *Z. Anorg. Allg. Chem.*, 1973, **401**, 243.	500, 504
73ZAAC(402)43	F. Kober; *Z. Anorg. Allg. Chem.*, 1973, **402**, 43.	500, 501
73ZN(B)829	K.-H. Linke and R. Turley, *Z. Naturforsch., Teil B*, 1973, **28**, 829.	381, 382
73ZOB997	R. S. Tewari and R. J. Shukla; *Zh. Obshch. Khim.*, 1973, **43**, 997.	841
73ZOB1264	V. L. Foss, N. M. Semenenko, N. M. Sorokin and I. F. Lutsenko; *Zh. Obshch. Khim.*, 1973, **43**, 1264.	695
73ZOB2178	L. A. Tamm, V. N. Chistokletov and A. A. Petrov; *Zh. Obshch. Khim.*, 1973, **43**, 2178.	838, 839
74ACR147	D. A. Evans and G. C. Andrews; *Acc. Chem. Res.*, 1974, **7**, 147.	154, 196
74ACR421	R. A. Moss; *Acc. Chem. Res.*, 1974, **7**, 421.	773
74AG272	A. Sidani, J. Marchand-Brynaert and L. Ghosez; *Angew. Chem.*, 1974, **86**, 272.	759
74AG(E)1	B. Chevrier and R. Weiss; *Angew. Chem., Int. Ed. Engl.*, 1974, **13**, 1.	1006
74AG(E)83	B.-T. Gröbel and D. Seebach; *Angew. Chem., Int. Ed. Engl.*, 1974, **13**, 83.	931, 936
74AG(E)84	W. Malisch and M. Kuhn; *Angew. Chem., Int. Ed. Engl.*, 1974, **13**, 84.	511
74AG(E)202	D. Seebach and K.-H. Geiss; *Angew. Chem., Int. Ed. Engl.*, 1974, **13**, 202.	272
74AG(E)274	W. Dumont, P. Bayer and A. Krief; *Angew. Chem., Int. Ed. Engl.*, 1974, **13**, 274.	288, 1001
74AG(E)668	G. Märkl and F. Kneidl; *Angew. Chem., Int. Ed. Engl.*, 1974, **13**, 668.	879
74AG(E)670	W. Malisch and P. Panster; *Angew. Chem., Int. Ed. Engl.*, 1974, **13**, 670.	512
74AG(E)804	W. Dumont, P. Bayer and A. Krief; *Angew. Chem., Int. Ed. Engl.*, 1974, **13**, 804.	281
74AG(E)806	D. Seebach and A. K. Beck; *Angew. Chem., Int. Ed. Engl.*, 1974, **13**, 806.	281
74AG(E)811	O. J. Scherer and N. Kuhn; *Angew. Chem., Int. Ed. Engl.*, 1974, **13**, 811.	406, 408
74AOC(12)83	J. Furukawa and N. Kawabata; *Adv. Organomet. Chem.*, 1974, **12**, 83.	67
74AP(307)73	F. Eiden and E. Schönduve, *Arch. Pharm.*, 1974, **307**, 73.	808
74AX(B)147	Y. M. Chow and D. Britton; *Acta Crystallogr., Part B*, 1974, **30**, 147.	1002
74BCJ143	S. Uemura, A. Onoe and M. Okano; *Bull. Chem. Soc. Jpn.*, 1974, **47**, 143.	24

74BCJ147 S. Uemura, A. Onoe and M. Okano; *Bull. Chem. Soc. Jpn.*, 1974, **47**, 147. 631
74BCJ692 S. Uemura, A. Onoe and M. Okano; *Bull. Chem. Soc. Jpn.*, 1974, **47**, 692. 24
74BCJ946 A. Matsumoto, J. H. Lee, M. Yoshida and O. Simamura; *Bull. Chem. Soc. Jpn.*, 1974, **47**, 946. 1054
74BCJ1680 N. Ishikawa and A. Sekiya; *Bull. Chem. Soc. Jpn.*, 1974, **47**, 1680. 631
74BCJ3121 S. Uemura, A. Onoe and M. Okano; *Bull. Chem. Soc. Jpn.*, 1974, **47**, 3121. 24
74BSF1535 J. Ficini, C. Barbara, J. d'Angelo and A. Duréault; *Bull. Soc. Chim. Fr.*, 1974, 1535. 1050
74CB508 H. W. Roesky and S. Tutkunkardes; *Chem. Ber.*, 1974, **107**, 508. 173
74CB639 M. Wieber, K. Foroughi and H. Klingl; *Chem. Ber.*, 1974, **107**, 639. 477
74CB745 W. Kirmse and U. Seipp; *Chem. Ber.*, 1974, **107**, 745. 387
74CB847 D. Seebach and H. Neumann; *Chem. Ber.*, 1974, **107**, 847. 961
74CB854 H. Schumann and L. Rosch; *Chem. Ber.* 1974, **107**, 854. 473, 474
74CB1228 P. Jutzi and M. Kuhn; *Chem. Ber.*, 1974, **107**, 1228. 484
74CB1545 H. Bredereck, G. Simchen and W. Griebenow; *Chem. Ber.*, 1974, **107**, 1545. 745
74CB1949 H. Burzlaff, U. Voll and H. J. Bestmann; *Chem. Ber.*, 1974, **107**, 1949. 1004
74CB2186 T. Eicher and S. Böhm; *Chem. Ber.*, 1974, **107**, 2186. 744
74CB2513 G. Himbert and M. Regitz; *Chem. Ber.*, 1974, **107**, 2513. 1048, 1068, 1069, 1070
74CB3026 K. Schank and H.-G. Schmitt; *Chem. Ber.*, 1974, **107**, 3026. 197
74CB3717 L. Birkofer and R. Stilke; *Chem. Ber.*, 1974, **107**, 3717. 943
74CC217 F. R. Sullivan, E. Luck and P. Kovacic; *J. Chem. Soc., Chem. Commun.*, 1974, 217. 405
74CC256 F. D. King and D. R. M. Walton; *J. Chem. Soc., Chem. Commun.*, 1974, 256. 753
74CC346 A. Pelter, A. Arase and M. G. Hutchings; *J. Chem. Soc., Chem. Commun.*, 1974, 346. 748
74CC362 T. M. Harris, T. P. Murray, C. M. Harris and M. Gumulka; *J. Chem. Soc., Chem. Commun.*, 1974, 362. 662
74CC547 M. Mikolajczyk and J. Drabowicz; *J. Chem. Soc., Chem. Commun.*, 1974, 547. 197
74CC584 C. G. Rao, S. Rengaraju and M. V. Bhatt; *J. Chem. Soc., Chem. Commun.*, 1974, 584. 659
74CC762 T. Yoshida and E. Negishi; *J. Chem. Soc., Chem. Commun.*, 1974, 762. 969
74CC879 D. Landini, F. Montanari and F. M. Pirisi; *J. Chem. Soc., Chem. Commun.*, 1974, 879. 9, 35
74CCC2616 V. B. Puchnarevic, J. Vcelak, M. G. Voronkov and V. Chvalovsky; *Collect. Czech. Chem. Commun.*, 1974, **39**, 2616. 19
74CI(L)664 D. D. Hepburn and H. R. Hudson; *Chem. Ind. (London)*, 1974, 664. 25, 31
74CJC761 T. Durst, M. J. LeBelle, R. van den Elzen and K.-C. Tin; *Can. J. Chem.*, 1974, **52**, 761. 151
74CRV45 D. E. Pearson and C. A. Buehler; *Chem. Rev.*, 1974, **74**, 45. 696
74CSR443 K. Smith; *Chem. Soc. Rev.*, 1974, 443. 514
74DOK(217)1320 R. A. Bekker, G. G. Melikyan, É. P. Lur'e, B. L. Dyatkin and I. L. Knunyants; *Dokl. Akad. Nauk SSSR*, 1974, **217**, 1320 (Plenum Engl. Transl. p. 572). 640
74G455 S. Cignarella and G. Cordella; *Gazz. Chim. Ital.*, 1974, **104**, 455. 225
74HOU(13/2b)9 K.-P. Zeller, H. Straub and H. Leditschke; *Methoden Org. Chem. (Houben-Weyl)*, 1974, **13/2b**, 9. 1097
74HOU(13/2b)26 K.-P. Zeller, H. Straub and H. Leditschke; *Methoden Org. Chem. (Houben-Weyl)*, 1974, **13/2b**, 26. 1097
74IC1252 R. G. Goel and D. R. Ridley; *Inorg. Chem.*, 1974, **13**, 1252. 891
74IC2710 A. Majid and J. M. Shreeve; *Inorg. Chem.*, 1974, **13**, 2710. 175
74IS129 M. A. Mathur, W. H. Myers, H. H. Sisler and G. E. Ryschkewitsch; *Inorg. Synth.*, 1974, **15**, 129. 435
74IS134 R. Koster, P. Binger and W. Fenzl; *Inorg. Synth.*, 1974, **15**, 134. 520
74IS141 R. Koster and P. Binger; *Inorg. Synth.*, 1974, **15**, 141. 523
74JA311 H. C. Brown and N. R. DeLue; *J. Am. Chem. Soc.*, 1974, **96**, 311. 524
74JA925 G. A. Olah, M. Nojima and I. Kerekes; *J. Am. Chem. Soc.*, 1974, **96**, 925. 19
74JA935 F. Jung, M. Molin, R. van den Elzer and T. Durst; *J. Am. Chem. Soc.*, 1974, **96**, 935. 200
74JA1100 R. H. Summerville, C. A. Senkler, P. v. R. Schleyer, T. E. Dueber and P. J. Stang; *J. Am. Chem. Soc.*, 1974, **96**, 1100. 689
74JA1775 R. D. Rieke and S. E. Bales; *J. Am. Chem. Soc.*, 1974, **96**, 1775. 558, 560, 959
74JA1788 S. F. Nelsen and R. T. Landis, II; *J. Am. Chem. Soc.*, 1974, **96**, 1788. 403
74JA2121 J. N. Marx, J. C. Argyle and L. R. Norman; *J. Am. Chem. Soc.*, 1974, **96**, 2121. 660
74JA2250 C. L. Liotta and H. P. Harris; *J. Am. Chem. Soc.*, 1974, **96**, 2250. 19
74JA2806 G. M. Whitesides, D. E. Bergbreiter and P. E. Kendall; *J. Am. Chem. Soc.*, 1974, **86**, 2806. 589
74JA3146 P. Raynolds, S. Zonnebelt, S. Bakker and R. M. Kellogg; *J. Am. Chem. Soc.*, 1974, **96**, 3146. 167
74JA3237 C. D. Beard and K. Baum; *J. Am. Chem. Soc.*, 1974, **96**, 3237. 339
74JA3684 G. Stork, M. E. Jung, E. Colvin and Y. Noël; *J. Am. Chem. Soc.*, 1974, **96**, 3684. 921
74JA3921 E. Block and J. O'Connor; *J. Am. Chem. Soc.*, 1974, **96**, 3921. 173
74JA3929 E. Block and J. O'Connor; *J. Am. Chem. Soc.*, 1974, **96**, 3929. 186
74JA4099 K.-T. H. Wei, I. C. Paul, M.-M. Y. Chang and J. I. Musher; *J. Am. Chem. Soc.*, 1974, **96**, 4099. 1001
74JA4494 Y. Chiang, A. J. Kresge, P. Salmonaa and C. I. Young; *J. Am. Chem. Soc.*, 1974, **96**, 4494. 638
74JA5000 F. A. Davis, A. J. Friedman and E. W. Kluger; *J. Am. Chem. Soc.*, 1974, **96**, 5000. 365
74JA5158 J. P. Snyder, V. T. Bandurco, F. Darack and H. Olsen; *J. Am. Chem. Soc.*, 1974, **96**, 5158. 773
74JA5288 R. R. Schrock and P. Heakin; *J. Am. Chem. Soc.*, 1974, **96**, 5288. 570
74JA5508 P. G. Gassman and T. J. van Bergen; *J. Am. Chem. Soc.*, 1974, **96**, 5508. 168
74JA5512 P. G. Gassman, G. Gruetzmacher and T. J. van Bergen; *J. Am. Chem. Soc.*, 1974, 5512. 761
74JA5580 W. Fenical; *J. Am. Chem. Soc.*, 1974, **96**, 5580. 686
74JA5582 E. J. Corey and R. H. Wollenberg; *J. Am. Chem. Soc.*, 1974, **96**, 5582. 970, 971

74JA5616	J. C. Day, M. J. Lindstrom and P. S. Skell; *J. Am. Chem. Soc.*, 1974, **96**, 5616.	27
74JA6140	A. Tamaki, S. A. Magennis and J. K. Kochi; *J. Am. Chem. Soc.*, 1974, **96**, 6140.	590
74JA6202	R. A. Kaba, D. Griller and K. U. Ingold; *J. Am. Chem. Soc.*, 1974, **96**, 6202.	1010
74JA6203	D. Griller and K. U. Ingold; *J. Am. Chem. Soc.*, 1974, **96**, 6203.	1010
74JA6815	J. J. Eisch and K. C. Fichter; *J. Am. Chem. Soc.*, 1974, **96**, 6815.	974
74JA7125	J. E. Baldwin, G. A. Höfle and O. W. Lever, Jr.; *J. Am. Chem. Soc.*, 1974, **96**, 7125.	681, 908, 999
74JA7269	F. Ramirez, V. A. V. Prasad and J. F. Marecek; *J. Am. Chem. Soc.*, 1974, **96**, 7269.	869
74JA7807	S. Danishefsky and T. Kitahara; *J. Am. Chem. Soc.*, 1974, **96**, 7807.	665
74JA8115	D. W. Hart and J. Schwartz; *J. Am. Chem. Soc.*, 1974, **96**, 8115.	4
74JAP7401534	K. Ura, K. Ono and S. Okamoto; *Jpn. Pat.* 7 401 534 (*Chem. Abstr.*, 1974, **81**, 115 903).	855
74JCS(D)1324	F. W. S. Benfield and M. L. H. Green; *J. Chem. Soc., Dalton Trans.*, 1974, 1324.	572
74JCS(P1)460	J. I. G. Cadogan and I. Gosney; *J. Chem. Soc., Perkin Trans. 1*, 1974, 460.	809, 892
74JCS(P1)466	J. I. G. Cadogen and I. Gosney; *J. Chem. Soc., Perkin Trans. 1*, 1974, 466.	892
74JCS(P1)660	A. Forchiassin, A. Risaliti, C. Russo, M. Calligaris and G. Pitacco; *J. Chem. Soc., Perkin Trans. 1*, 1974, 660.	783
74JCS(P1)732	D. H. R. Barton, R. H. Hesse, M. M. Pechet and H. T. Toh; *J. Chem. Soc., Perkin Trans. 1*, 1974, 732.	335
74JCS(P1)771	E. Tempesti, L. Giuffre, G. Sioli, M. Fornaroli and G. Airoldi; *J. Chem. Soc., Perkin Trans. 1*, 1974, 771.	190
74JCS(P1)1353	I. M. Godfrey, M. V. Sargent and J. A. Elix; *J. Chem. Soc., Perkin Trans. 1*, 1974, 1353.	654
74JCS(P1)1858	R. C. Cambie, R. C. Hayward, J. L. Roberts and P. S. Rutledge; *J. Chem. Soc., Perkin Trans. 1*, 1975, 1858.	45
74JCS(P1)2055	C. Eaborn, R. A. Jackson and R. Pearce; *J. Chem. Soc., Perkin Trans. 1*, 1974, 2055.	937, 938
74JCS(P1)2101	D. H. R. Barton, P. D. Magnus, J. A. Garbarino and R. N. Young; *J. Chem. Soc., Perkin Trans. 1*, 1974, 2101.	309
74JCS(P2)1338	R. Benassi, M. L. Schenetti, F. Taddei and P. Vivarelli; *J. Chem. Soc., Perkin Trans. 2*, 1974, 1338.	825
74JGU1461	E. Lukevits, L. I. Simchenko and M. Dzintara; *J. Gen. Chem. USSR* (*Engl. Transl.*), 1974, **44**, 1461.	1068
74JGU1506	V. S. Gamayurova, Z. G. Daineko, B. D. Chernokalskii, R. R. Shagidullin, I. A. Lamanova and L. V. Avvakumova; *J. Gen. Chem. USSR* (*Engl. Transl.*), 1974, **44**, 1506.	891
74JIC1018	M. T. Rahman and H. Gilman; *J. Indian Chem. Soc.*, 1974, **51**, 1018.	632
74JINC503	R. T. Markham, E. A. Dietz and D. R. Martin; *J. Inorg. Nucl. Chem.*, 1974, **36**, 503.	448
74JINC993	D. A. Armitage and A. W. Sinden; *J. Inorg. Nucl. Chem.*, 1974, **36**, 993.	399
74JMC948	A. N. Tischler, F. M. Thompson, L. J. Libertini and M. Calvin; *J. Med. Chem.*, 1974, **17**, 948.	379
74JMC1225	D. Seebach and D. Enders; *J. Med. Chem.*, 1974, **17**, 1225.	385
74JOC238	W. E. Truce, D. L. Heuring and G. C. Wolf; *J. Org. Chem.*, 1974, **39**, 238.	182
74JOC245	W. E. Truce and D. L. Heuring; *J. Org. Chem.*, 1974, **39**, 245.	182
74JOC253	O. Abe, M. F. Lukacovic and C. Ressler; *J. Org. Chem.*, 1974, **39**, 253.	212, 214
74JOC265	K. L. Marsi; *J. Org. Chem.*, 1974, **39**, 265.	441
74JOC407	N. A. Johnson and E. S. Gould; *J. Org. Chem.*, 1974, **39**, 407.	773
74JOC428	H. J. Reich; *J. Org. Chem.*, 1974, **39**, 428.	282
74JOC553	A. Hassner and R. H. Reuss; *J. Org. Chem.*, 1974, **39**, 553.	8022
74JOC600	G. E. Niznik, W. H. Morrison and H. M. Walborsky; *J. Org. Chem.*, 1974, **39**, 600.	268
74JOC1432	A. R. Friedman and D. R. Graber; *J. Org. Chem.*, 1974, **39**, 1432.	743
74JOC1588	H. Posvic and D. Rogers; *J. Org. Chem.*, 1974, **39**, 1588.	384
74JOC1744	R. Lamartine and R. Perrin; *J. Org. Chem.*, 1974, **39**, 1744.	619
74JOC1758	T. B. Patrick, J. A. Schield and D. G. Kirchner; *J. Org. Chem.*, 1974, **39**, 1758.	753
74JOC1785	R. H. Reuss and A. Hassner; *J. Org. Chem.*, 1974, **39**, 1785.	22, 28
74JOC2126	M. S. Kablaoui; *J. Org. Chem.*, 1974, **39**, 2126.	658
74JOC2148	G. F. Whitfield, H. S. Beilan and D. Swern; *J. Org. Chem.*, 1974, **39**, 2148.	160
74JOC2296	P. D. Hensin, S. B. Ockrymick and J. R. E. Markham; *J. Org. Chem.*, 1974, **39**, 2296.	822
74JOC2300	R. M. Blankenship, K. A. Burdett and J. S. Swenton; *J. Org. Chem.*, 1974, **39**, 2300.	14
74JOC2459	M. D. Robbins and C. D. Broddus; *J. Org. Chem.*, 1974, **39**, 2459.	206
74JOC2641	J. A. Sanders, K. Hovius and J. B. F. N. Engberts; *J. Org. Chem.*, 1974, **39**, 2641.	691
74JOC3307	R. K. Boeckman, Jr. and D. M. Blum; *J. Org. Chem.*, 1974, **39**, 3307.	623
74JOC3343	R. D. Knudsen and H. R. Snyder; *J. Org. Chem.*, 1974, **39**, 3343.	648
74JOC3351	P. L. Southwick, R. F. Dufresne and J. J. Lindsey; *J. Org. Chem.*, 1974, **39**, 3351.	787
74JOC3365	T. E. Varkey, G. F. Whitfield and D. Swern; *J. Org. Chem.*, 1974, **39**, 3365.	162
74JOC3452	R. E. Merrill and E. Negishi; *J. Org. Chem.*, 1974, **39**, 3452.	961
74JOC3525	P. J. DeChristopher, J. P. Adamek, G. D. Lyon, S. A. Klein and R. J. Baumgarten; *J. Org. Chem.*, 1974, **39**, 3525.	366
74JOC3607	R. A. Ruden; *J. Org. Chem.*, 1974, **39**, 3607.	941
74JOC3612	J. F. Brunnett and X. Creary; *J. Org. Chem.*, 1974, **39**, 3612.	850
74JOC3767	J. Hes and M. P. Mertes; *J. Org. Chem.*, 1974, **39**, 3767.	105
74JOC3913	C. A. Brown; *J. Org. Chem.*, 1974, **39**, 3913.	423, 696
74JOM(66)C14	I. Ojima, M. Kumagai and Y. Nagai; *J. Organomet. Chem.*, 1974, **66**, C14.	915
74JOM(67)61	H. Lauke, P. J. Swepston and T. J. Marks; *J. Organomet. Chem.*, 1974, **67**, 61.	602
74JOM(67)C64	L. C. Chao and R. D. Rieke; *J. Organomet. Chem.*, 1974, **67**, C64.	597
74JOM(69)193	R. A. Benkeser and D. F. Ehler; *J. Organomet. Chem.*, 1974, **69**, 193.	914
74JOM(73)1	G. W. Kramer and H. C. Brown; *J. Organomet. Chem.*, 1974, **73**, 1.	521
74JOM(74)85	M. D. Rausch and H. B. Gordon; *J. Organomet. Chem.*, 1974, **74**, 85.	566

74JOM(74)289	J. Kuyper and K. Vrieze; *J. Organomet. Chem.*, 1974, **74**, 289.	399
74JOM(74)365	E. A. Jeffery, A. Meisters and T. Mole; *J. Organomet. Chem.*, 1974, **74**, 365.	478
74JOM(76)95	Y. Kiso, K. Tamao and M. Kumada; *J. Organomet. Chem.*, 1974, **76**, 95.	915
74JOM(76)105	Y. Kiso, K. Tamao and M. Kumada; *J. Organomet. Chem.*, 1974, **76**, 105.	915
74JOM(76)265	J. S. Thayer; *J. Organomet. Chem.*, 1974, **76**, 265.	1076
74JOM(76)C7	W. Malisch and P. Panster; *J. Organomet. Chem.*, 1974, **76**, C7.	511, 512
74JOM(76)C23	G. M. Pickles and F. G. Thorpe; *J. Organomet. Chem.*, 1974, **76**, C23.	653
74JOM(77)225	M. Durand and J.-P. Laurent; *J. Organomet. Chem.*, 1977, **77**, 225.	495
74JOM(77)269	J. F. Normant, G. Cahiez, C. Chuit and J. Villieras; *J. Organomet. Chem.*, 1974, **77**, 269.	629
74JOM(78)107	V. L. Foss, N. M. Semenenko, N. M. Sorokin and I. F. Lutsenko; *J. Organomet. Chem.*, 1974, **78**, 107.	420, 494, 495, 695
74JOM(78)115	V. L. Foss, N. M. Sorokin, I. M. Avrutov and I. F. Lutsenko; *J. Organomet. Chem.*, 1974, **78**, 115.	695
74JOM(81)329	M. J. S. Gynane and I. J. Worrall; *J. Organomet. Chem.*, 1974, **81**, 329.	597
74JOM(82)C3	O. J. Scherer and N. Kuhn; *J. Organomet. Chem.*, 1974, **82**, C3.	406, 408
74JPR347	M. Schulz and U. Missol; *J. Prakt. Chem.*, 1974, **316**, 347.	374
74JPR719	P. Hermann and E. Schreier; *J. Prakt. Chem.*, 1974, **316**, 719.	223
B-74MI 201-01	J. S. Pizey; "Synthetic Reagents," Ellis Horwood, Chichester, 1974, vol. 1, p. 321.	21, 25
B-74MI 201-02	J. S. Pizey; "Synthetic Reagents," Ellis Horwood, Chichester, 1974, vol. 2, p. 1.	27
74MI 202-01	D. T. Gibson, V. Mahaderan and J. F. Davey; *J. Bacteriol.*, 1974, **119**, 930.	39
B-74MI 203-01	S. Patai (ed.); "The Chemistry of the Thiol Group," Wiley, 1974.	116
B-74MI 203-02	J. L. Wardell; in "The Chemistry of the Thiol Group," ed. S. Patai, Wiley, 1974, part 1, ch. 4.	134
B-74MI 203-03	M. E. Peach; in "The Chemistry of the Thiol Group," ed. S. Patai, Wiley, 1974, part 2, ch. 16.	137
B-74MI 203-04	G. Modena and G. Capozzi; in "The Chemistry of the Thiol Group," ed. S. Patai, Wiley, 1974, part 2, ch. 17.	192, 212, 216
B-74MI 206-01	J. Keek; in "Methodicum Chimicum," Thieme, Stuttgart, 1974, vol. 6, p. 53.	
74MI 208-01	S. O. Grim and R. P. Molenda; *Phosphorus*, 1974, **4**, 189.	432, 475
74MI 208-02	W. Wolfsberger; *Phosphorus*, 1974, **29**, 55.	470
B-74MI 211-01	B. J. Wakefield; "The Chemistry of Organolithium Compounds," Pergamon, Oxford, 1974.	551
B-74MI 213-01	B. J. Wakefield; "The Chemistry of Organolithium Compounds," Pergamon Press, Oxford, 1974.	653
B-74MI 214-01	J. L. Wardell; in "The Chemistry of the Thiol Group," ed. S. Patai, Wiley, New York, 1974, part 2, p. 211.	707
B-74MI 215-01	R. A. Abramovitch and E. M. Smith; in "Pyridine and its Derivatives," ed. R. A. Abramovitch, Wiley, New York, 1974, vol. 14, suppl. 2, part 2, pp. 149–161.	765
B-74MI 215-02	K. J. Irgolic; "The Chemistry of Tellurium," Gordon and Breach, New York, 1974.	768
74MI 216-01	L. Maier; *Phosphorus*, 1974, **4**, 41.	820, 821
B-74MI 233-01	B. J. Wakefield; "The Chemistry of Organolithium Compounds," Pergamon, Oxford, 1974, p. 37.	
74OR(21)125	C. M. Sharts and W. A. Sheppard; *Org. React.*, 1974, **21**, 125.	15, 18
74OR(21)1	G. A. Boswell, Jr., W. C. Ripka, R. M. Scribner and C. W. Tullock; *Org. React.*, 1974, **21**, 1.	19
74OS(54)79	P. J. Stang and T. E. Dueber; *Org. Synth.*, 1974, **54**, 79.	689
74RCR657	I. N. Azerbaev, Z. A. Abramova and Yu. G. Bosyakov; *Russ. Chem. Rev.* (*Engl. Transl.*), 1974, **43**, 657.	1076
74RTC252	J. Perregaard, E. B. Pedersen and S.-O. Lawesson; *Recl. Trav. Chim. Pays-Bas*, 1974, **93**, 252.	694
74S37	D. Landini, F. Montanari and F. Rolla; *Synthesis*, 1974, 37.	25
74S38	R. A. Wohl; *Synthesis*, 1974, 38.	679
74S59	L. E. Overman, J. D. Overman and J. Smoot; *Synthesis*, 1974, 59.	217
74S77	E. Negishi and H. C. Brown; *Synthesis*, 1974, 77.	515, 516, 518, 900
74S119	B. Fell and H. Bearman; *Synthesis*, 1974, 119.	449, 466
74S124	M. Mikolajczyk and J. Drabowicz; *Synthesis*, 1974, 124.	197
74S277	Y. Tamura, H. Matsushima, M. Ikeda and K. Sumoto; *Synthesis*, 1974, 277.	160, 161
74S290	P. Radlick and L. R. Brown; *Synthesis*, 1974, 290.	320
74S293	L. R. Subramanian, H. Bentz and M. Hanack; *Synthesis*, 1974, 293.	689
74S295	W. H. Dennis, D. H. Rosenblitt and J. R. Simcox; *Synthesis*, 1974, 295.	192
74S298	J. C. Meslin and H. Quiniou; *Synthesis*, 1974, 298.	744
74S328	G. O. Doak and L. D. Freedman; *Synthesis*, 1974, 328.	509
74S348	S. A. Patwardhan and S. Dev; *Synthesis*, 1974, 348.	679
74S428	D. Landini, F. Montanari and F. Rolla; *Synthesis*, 1974, 428.	19
74S489	H. Quast and F. Kees; *Synthesis*, 1974, 489.	176
74S560	G. A. Olah, D. A. Beal, S. H. Yu and J. A. Olah; *Synthesis*, 1974, 560.	13
74S572	G. R. Krow and E. Michener; *Synthesis*, 1974, 572.	674
74S652	G. A. Olah and J. Welch; *Synthesis*, 1974, 652.	11
74S653	G. A. Olah and J. Welch; *Synthesis*, 1974, 653.	6
74S672	H. Ahlbrecht and J. Eichler; *Synthesis*, 1974, 672.	751
74S803	J. F. Normant, C. Chuit, G, Cahiez and J. Villieras; *Synthesis*, 1974, 803.	624
74S816	K. Burger, E. Burgis and P. Holl; *Synthesis*, 1974, 816.	751
74S876	C. Kaneko, A. Sugimoto and S. Tanaka; *Synthesis*, 1974, 876.	176, 180
74S877	A. Barco, S. Benetti, G. P. Pollini and R. Taddia; *Synthesis*, 1974, 877.	181

74S887	C. Comninellis, P. Javet and E. Plattner; *Synthesis*, 1974, 887.	177
74S896	G. A. Olah and J. Welch; *Synthesis*, 1974, 896.	12
74SRI23	R. F. Cunico and E. M. Dexheimer; *Synth. React. Inorg. Metal-Org. Chem.*, 1974, **4**, 23.	538
74SRI373	L. C. Chao and R. D. Rieke; *Synth. React. Inorg. Metal-Org. Chem.*, 1974, **4**, 373.	597
74T2467	A. McKillop and M. E. Ford; *Tetrahedron*, 1975, **30**, 2467.	103
74T2741	E. Valentin, G. Pitacco, F. P. Colonna and A. Risaliti; *Tetrahedron*, 1974, **30**, 2741.	783
74T3021	Y. Ogata, I. Urasaki, K. Nagura and N. Satomi; *Tetrahedron*, 1974, **30**, 3021.	650
74T3793	D. J. Costa, N. E. Boutin and J. G. Riess; *Tetrahedron*, 1974, **30**, 3793.	10
74TL265	W. Duismann, H.-D. Beckhaus and C. Rüchardt; *Tetrahedron Lett.*, 1974, 265.	404
74TL381	Y.-J. Lee and W. D. Closson; *Tetrahedron Lett.*, 1974, 381.	249
74TL543	R. B. Miller and T. Reichenbach; *Tetrahedron Lett.*, 1974, **15**, 543.	607, 617, 622, 628, 922
74TL725	T. S. Chou; *Tetrahedron Lett.*, 1974, 725.	267
74TL763	J. Bertin, J. L. Luche and H. B. Kagan; *Tetrahedron Lett.*, 1974, 763.	25
74TL793	Y. Yamamoto, K. Kondo and I. Moritani; *Tetrahedron Lett.*, 1974, 793.	268
74TL837	M. S. A. Vrijland; *Tetrahedron Lett.*, 1974, 837.	244
74TL1059	S. Tamagaki, K. Sakaki and S. Oae; *Tetrahedron Lett.*, 1974, 1059.	235
74TL1067	K. Kaneda, F. Kawamoto, Y. Fujiwara, T. Imanaka and S. Teranishi; *Tetrahedron Lett.*, 1974, 1067.	616
74TL1097	P. A. Grieco and J. J. Reap; *Tetrahedron Lett.*, 1974, 1097.	139
74TL1879	Y. Watanabe, M. Yamashita, T. Mitsudo, M. Tanaka and Y. Takegan; *Tetrahedron Lett.*, 1974, 1879.	310
74TL2091	T. H. Chan and J. R. Finkenbine; *Tetrahedron Lett.*, 1974, 2091.	287
74TL2453	T. Polonski and A. Chimiak; *Tetrahedron Lett.*, 1974, 2453.	342
74TL2691	J. A. Miller and M. J. Nunn; *Tetrahedron Lett.*, 1974, 2691.	35
74TL2961	N. Miyaura, T. Yoshinari, M. Itoh and A. Suzuki; *Tetrahedron Lett.*, 1974, **34**, 2961.	904
74TL3319	P. K. Claus, P. Hofbauer and W. Rieder; *Tetrahedron Lett.*, 1974, 3319.	160, 164
74TL3365	R. K. Boekman, Jr. and K. J. Bruza; *Tetrahedron Lett.*, 1974, **15**, 3365.	931
74TL3383	A. Zwierzak and T. Gajda; *Tetrahedron Lett.*, 1974, 3383.	412
74TL3395	C. Yaroslavsky; *Tetrahedron Lett.*, 1974, 3395.	626
74TL3435	L. Chevolot, J. Soulie and P. Cadiot; *Tetrahedron Lett.*, 1974, 3435.	517
74TL3847	J. L. Meisel; *Tetrahedron Lett.*, 1974, 3847.	403
74TL3987	E. Vilsmaier, R. Tropitzsch and O. Vostrowsky; *Tetrahedron Lett.*, 1974, 3987.	188
74TL4781	R. F. J. Langendries and F. C. de Schryver; *Tetrahedron Lett.*, 1974, **47**, 4781.	205
74ZAAC(406)52	F. Kober and W. J. Ruehl; *Z. Anorg. Allg. Chem.*, 1974, **406**, 52.	695
74ZAAC(408)261	N. Baumann and M. Wieber; *Z. Anorg. Allg. Chem.*, 1974, **408**, 261.	501
74ZAAC(412)11	K. H. Thiele, A. Russek, R. Opitz, B. Mohai and W. Bruser; *Z. Anorg. Allg. Chem.*, 1974, **412**, 11.	572
74ZC477	K. Forner and H. G. Henning; *Z. Chem.*, 1974, **14**, 477.	452
74ZN(B)277	O. J. Scherer and G. Wolmerhaüser; *Z. Naturforsch., Teil B*, 1974, **29**, 277.	360
74ZN(B)611	H. Steinberger and W. Kuchen; *Z. Naturforsch., Teil B*, 1974, **29**, 611.	468
74ZOB85	G. K. Fedorova, Y. P. Shaturskii and L. S. Moskalevskaya; *Zh. Obshch. Khim.*, 1974, **44**, 85.	847
74ZOB760	A. I. Bokanov, N. A. Rozanel'skaya and B. I. Stepanov; *Zh. Obshch. Khim.*, 1974, **44**, 760.	825
74ZOB1261	S. V. Fridland, N. V. Dmitrieva and I. V. Vigalok; *Zh. Obshch. Khim.*, 1974, **44**, 1261.	459
74ZOB2357	Zh. K. Gorbatenko, N. G. Feshchenko and T. V. Kovaleve; *Zh. Obshch. Khim.*, 1974, **44**, 2357.	466
74ZOB2621	V. V. Moskva, L. A. Bashirova, A. I. Razumova, T. V. Zykova and R. A. Salakhutdinov; *Zh. Obshch. Khim.*, 1974, **44**, 2621.	459
74ZOR2619	N. Ya. Derkach and G. G. Barashenkov; *Zh. Org. Khim.*, 1974, **10**, 2619.	369
75ACR10	J. F. King; *Acc. Chem. Res.*, 1975, **8**, 10	202
75ACR342	J. P. Collman; *Acc. Chem. Res.*, 1975, **8**, 342.	575
75ACR393	K. J. Klabunde; *Acc. Chem. Res.*, 1975, **8**, 393.	961
75ACS(B)362	A. J. Dale and P. Froyen; *Acta Chem. Scand., Ser. B*, 1975, **29**, 362.	886
75AG(E)70	K. Wieser and A. Berndt; *Angew. Chem., Int. Ed. Engl.*, 1975, **14**, 70.	778
75AG(E)218	W. Reichelt; *Angew. Chem., Int. Ed. Engl.*, 1975, **14**, 218.	961
75AG(E)350	W. Domont and A. Krief; *Angew. Chem., Int. Ed. Engl.*, 1975, **14**, 350.	281
75AG(E)363	E. Niecke and W. Flick; *Angew. Chem., Int. Ed. Engl.*, 1975, **14**, 363.	470
75AG(E)485	O. J. Scherer and G. Wolmershäuser; *Angew. Chem., Int. Ed. Engl.*, 1975, **14**, 485.	400
75AG(E)629	O. J. Scherer and W. Glässel; *Angew. Chem., Int. Ed. Engl.*, 1975, **14**, 629.	407
75AG(E)644	K. H. Dötz; *Angew. Chem., Int. Ed. Engl.*, 1975, **14**, 644.	669
75AG(E)700	D. van Ende, W. Dumont and A. Krief; *Angew. Chem., Int. Ed. Engl.*, 1975, **14**, 700.	281
75AG(E)762	H. O. Kalinowski, D. Seebach and G. Crass; *Angew. Chem., Int. Ed. Engl.*, 1975, **14**, 762.	89
75AG(E)801	R. Appel; *Angew. Chem., Int. Ed. Engl.*, 1975, **14**, 801.	7
75AG(E)820	M. Ohara and K. Watanabe; *Angew. Chem., Int. Ed. Engl.*, 1975, **14**, 820.	657
75AG(E)824	W. Kraus and H.-D. Graf; *Angew. Chem., Int. Ed. Engl.*, 1975, **14**, 824.	9
75AP(308)954	B. Unterhalt and D. Thomas; *Arch. Pharm. (Weinheim, Ger.)*, 1975, **308**, 954.	386
75BCJ480	T. Hiraoka and T. Kobayashi; *Bull. Chem. Soc. Jpn.*, 1975, **48**, 480.	249
75BCJ546	S. Nakayama, M. Yoshifuji, R. Okazaki and N. Inamoto; *Bull. Chem. Soc. Jpn.*, 1975, **48**, 546.	851
75BCJ1665	S. Inoue and S. Oae; *Bull. Chem. Soc. Jpn.*, 1975, **48**, 1665.	223

75BCJ3413 T. Takamoto and R. Sudoh; *Bull. Chem. Soc. Japan*, 1975, **48**, 3413. 397

75BCJ3682 Y. Yamamoto, K. Kondo and I. Moritani; *Bull. Chem. Soc. Jpn.*, 1975, **48**, 3682. 268

75BRP1391806 B. C. L. Weedon; *Br. Pat.* 1 391 806 (1975) (*Chem. Abstr.*, 1975, **83**, 97 568). 445

75BSF607 C. Georgoulis and G. Ville; *Bull. Soc. Chim. Fr.*, 1975, 607. 7

75BSF770 M. Julia and J. D. Fourneron; *Bull. Soc. Chim. Fr.*, 1975, 770. 593

75BSF1627 A. Le Berre and C. Porte; *Bull. Soc. Chim. Fr.*, 1975, 1627. 381, 383, 384

75BSF(2)1185 A. Kiennemann and R. Kieffer; *Bull. Soc. Chim. Fr.*, Part 2, 1975, **2**, 1185. 876

75BSF2493 A. Debon, S. Masson and A. Thuillier; *Bull. Soc. Chim. Fr.*, 1975, 2493. 617

75C515 J.-C. Motte and H. G. Viehe; *Chimia*, 1975, **29**, 515. 1051, 1052

75C516 V. Jäger, J.-C. Motte and H. G. Viehe; *Chimia*, 1975, **29**, 516. 1052

75CB314 D. Seebach and A. K. Beck; *Chem. Ber.*, 1975, **108**, 314. 279

75CB395 P. Binger and R. Köster; *Chem. Ber.*, 1975, **108**, 395. 1088

75CB402 K. Bott; *Chem. Ber.*, 1975, **108**, 402. 800

75CB700 W. Malisch and P. Panster; *Chem. Ber.*, 1975, **108**, 700. 512

75CB1293 D. Seebach and D. Enders; *Chem. Ber.*, 1975, **108**, 1293. 385, 420

75CB1365 F. Bur-Bur, A. Haas and W. Klug; *Chem. Ber.*, 1975, **108**, 1365. 361

75CB1732 H. Paulsen and W. Bartsch; *Chem. Ber.*, 1975, **108**, 1732. 936

75CB1924 D. Seebach, H. F. Leitz and V. Ehrig; *Chem. Ber.*, 1975, **108**, 1924. 393

75CB1961 V. Ehrig and D. Seebach; *Chem. Ber.*, 1975, **108**, 1961. 393

75CB2329 I. Ruppert, V. Bastian and R. Appel; *Chem. Ber.*, 1975, **108**, 2329. 399

75CB2917 H. Poisel and U. Schmidt; *Chem. Ber.*, 1975, **108**, 2917. 338

75CC36 S. W. Breuer, G. M. Pickles, J. C. Podesta and F. G. Thorpe; *J. Chem. Soc., Chem. Commun.*, 1975, 36. 653

75CC97 D. H. R. Barton, R. H. Hesse, T. R. Klosse and M. M. Pechet; *J. Chem. Soc., Chem. Commun.*, 1975, 97. 335

75CC242 H. Felkin, L. D. Kwart, G. Swierczewski and J. D. Umpleby; *J. Chem. Soc., Chem. Commun.*, 1975, 242. 561

75CC303 J. P. McCormick and D. L. Barton; *J. Chem. Soc., Chem. Commun.*, 1975, 303. 11

75CC382 M. Mikolajczyk, J. Drabowicz, J. Omelanczuk and E. Fluck; *J. Chem. Soc., Chem. Commun.*, 1975, 382. 450, 842

75CC413 J. H. Wevers and H. Kloosterziel; *J. Chem. Soc., Chem. Commun.*, 1975, 413. 360

75CC539 T. G. Back, D. H. R. Barton, M. R. Britten-Kelly and F. S. Guziec, Jr.; *J. Chem. Soc., Chem. Commun.*, 1975, 539. 858

75CC622 P. Hodge and G. Richardson; *J. Chem. Soc., Chem. Commun.*, 1975, 622. 7

75CC644 E. Rizzardo; *J. Chem. Soc., Chem. Commun.*, 1975, 644. 674

75CC668 Y. Yamamoto and H. Schmidbaur; *J. Chem. Soc., Chem. Commun.*, 1975, 668. 1004

75CC790 H. J. Reich and F. Chow; *J. Chem. Soc., Chem. Commun.*, 1975, 790. 281

75CC855 D. H. R. Barton, S. V. Ley and P. D. Magnus; *J. Chem. Soc., Chem. Commun.*, 1975, 855. 357

75CC857 A. Pelter, K. Smith and M. Tabata; *J. Chem. Soc., Chem. Commun.*, 1975, 857. 1086

75CC925 S. Uemura, A. Onoe and M. Okano; *J. Chem. Soc., Chem. Commun.*, 1975, 925. 617, 629

75CC950 D. Landini, A. M. Maia, F. Montanari and F. M. Pirisi; *J. Chem. Soc., Chem. Commun.*, 1975, 950. 9

75CC976 A. Fontana; *J. Chem. Soc., Chem. Commun.*, 1975, 976. 229

75CI(L)281 S. Bittner and Y. Assaf; *Chem. Ind.* (*London*), 1975, 281. 683

75CJC455 I. W. J. Still, P. C. Ardra, M. S. Chauhan, M. H. Kwan and M. T. Thomas; *Can. J. Chem.*, 1975, **54**, 455. 229

75CJC2975 S. Hanessian and P. Lavallee; *Can. J. Chem.*, 1975, **53**, 2975. 538

75CJC3620 R. Fuchs and L. L. Cole; *Can. J. Chem.*, 1975, **53**, 3620. 25

75CJC3656 J. F. King, E. G. Lewars, D. R. K. Harding and R. M. Enanoza; *Can. J. Chem.*, 1975, **53**, 3656. 205

75CL307 H. Minato, K. Yamaguchi and H. Kobayashi; *Chem. Lett.*, 1975, 307. 175

75CL813 T. Endo, H. Tasai and T. Ishigami; *Chem. Lett.*, 1975, 813. 213

75CL991 H. Minato, K. Yamaguchi and M. Kobayashi; *Chem. Lett.*, 1975, 991. 360

75CL1041 K. Narasaka, E. Bald and T. Mukaiyama; *Chem. Lett.*, 1975, 1041. 656

75CL1095 H. Kono and K. Ito; *Chem. Lett.*, 1975, 1095. 527

75CRV521 E. C. Ashby and J. T. Læmmle; *Chem. Rev.*, 1975, **75**, 521. 67

75CSC135 K. Simon, Z. Meszaros and A. Kalman; *Cryst. Struct. Commun.*, 1975, **4**, 135. 1008

75DOK(220)146 R. G. Kostyanovskii, A. P. Pleshkova, V. I. Khvostenko, I. I. Furlei, É. P. Lur'e, G. G. Melikyan and R. A. Bekker; *Dokl. Akad. Nauk SSSR*, 1975, **220**, 146. 641

75G377 L. Mangoni, M. Adinolfi, G. Barone and M. Parrilli; *Gazz. Chim. Ital.*, 1975, **105**, 377. 45

75HCA104 C. Egli, S. E. Helali and E. Hardegger; *Helv. Chim. Acta*, 1975, **58**, 104. 663

75HOU(13/7)1 G. Bähr and E. Langer; *Methoden Org. Chem.* (*Houben-Weyl*), 1975, **13/7**, 1. 601

75HOU(13/7)73 G. Bähr and E. Langer; *Methoden Org. Chem.* (*Houben-Weyl*), 1975, **13/7**, 73. 1101

75HOU(13/7)265 A. Segnitz; *Methoden Org. Chem.* (*Houben-Weyl*), 1975, **13/7**, 265. 565

75HOU(13/7)355 A. Segnitz; *Methoden Org. Chem.* (*Houben Weyl*), 1975, **13/7**, 355. 569

75HOU(13/7)379 A. Segnitz; *Methoden Org. Chem.* (*Houben-Weyl*), 1975, **13/7**, 379. 571

75HOU(13/7)433 A. Segnitz; *Methoden Org. Chem.* (*Houben-Weyl*), 1975, **13/7**, 433. 571

75HOU(13/7)489 A. Segnitz; *Methoden Org. Chem.* (*Houben-Weyl*), 1975, **13/7**, 489. 571

75IC571 E. R. Farardeau, K. W. Morse and J. G. Morse; *Inorg. Chem.*, 1975, **14**, 571. 449

75IC2402 C. W. Rice and R. S. Tobias; *Inorg. Chem.*, 1975, **14**, 2402. 590

75IC2809 R. A. Kovar, H. Derr, D. Brandau and J. O. Callaway; *Inorg. Chem.*, 1975, **14**, 2809. 597

75INC233 J. W. Anderson, J. E. Drake, R. T. Hemmings and D. L. Nelson; *Inorg. Nucl. Chem. Lett.*, 1975, **11**, 233. 857

75IS120 R. J. H. Clark and M. A. Coles; *Inorg. Synth.*, 1975, **16**, 120. 566

75IZV897 A. E. Borisov, P. V. Petrovskii, E. I. Fedin, L. V. Kudryavtseva, S. A. Busev, U. Utebaev and A. I. Lutsenko; *Izv. Akad. Nauk SSSR Ser. Khim.*, 1975, 897 (*Chem. Abstr.*, 1975, **83**, 9749j). 876

75IZV2530 N. P. Aktaev, G. A. Sokol'skii and I. L. Knunyants; *Izv. Akad. Nauk SSSR, Ser. Khim.*, 1975, 2530 (*Chem. Abstr.*, 1976, **84**, 59 345d). 183, 249

75IZV2845 A. Nesmeyanov, S. T. Berman and O. A. Reutov; *Izv. Akad. Nauk SSSR Ser. Khim.*, 1975, 2845 (*Chem. Abstr.*, 1976, **84**, 90 227). 446

75JA53 R. B. King and J. C. Cloyd, Jr.; *J. Am. Chem. Soc.*, 1975, **97**, 53. 853

75JA583 J. A. Franz and J. C. Martin; *J. Am. Chem. Soc.*, 1975, **97**, 583. 401

75JA676 R. A. Abramovitch and W. D. Holcomb; *J. Am. Chem. Soc.*, 1975, **97**, 676. 365

75JA679 D. W. Hart, T. F. Blackburn and J. Schwartz; *J. Am. Chem. Soc.*, 1975, **97**, 679. 616, 623, 629, 968

75JA931 H. Sakurai, Y. Kamiyama and Y. Nakadaira; *J. Am. Chem. Soc.*, 1975, **97**, 931. 916

75JA1492 T. A. Modro, K. Yates and J. Janata; *J. Am. Chem. Soc.*, 1975, **97**, 1492. 176

75JA1599 J. S. Filippo, Jr., A. F. Sowinski and L. J. Romano; *J. Am. Chem. Soc.*, 1975, **97**, 1599. 24

75JA2276 J. Toye and L. Ghosez; *J. Am Chem. Soc.*, 1975, **97**, 2276. 748

75JA2281 F. R. Jensen and T. I. Moder; *J. Am. Chem. Soc.*, 1975, **97**, 2281. 26

75JA2566 J. F. King and J. R. du Manoir; *J. Am. Chem. Soc.*, 1975, **97**, 2566. 365

75JA2791 H. C. Brown, E. Negishi and J.-J. Katz; *J. Am. Chem. Soc.*, 1975, **97**, 2791. 515

75JA4269 D. Griller, J. W. Cooper and K. U. Ingold; *J. Am. Chem. Soc.*, 1975, **97**, 4269. 1010

75JA5249 H. C. Brown and S. K. Gupta; *J. Am. Chem. Soc.*, 1975, **97**, 5249. 527, 900, 901

75JA5434 H. J. Reich, J. M. Renga and I. L. Reich; *J. Am. Chem. Soc.*, 1975, **97**, 5434. 282, 287

75JA5526 D. Griller, K. Dimroth, T. M. Fyles and K. U. Ingold; *J. Am. Chem. Soc.*, 1975, **97**, 5526. 1010

75JA5600 W. G. L. Aalbersberg, A. J. Barkovich, R. L. Funk, R. L. Hillard III, and K. P. C. Vollhardt; *J. Am. Chem. Soc.*, 1975, **97**, 5600. 944

75JA5608 D. S. Matteson, R. J. Moody and P. K. Jesthi; *J. Am. Chem. Soc.*, 1975, **97**, 5608. 905

75JA6137 J. C. Martin and J. A. Franz; *J. Am. Chem. Soc.*, 1975, **97**, 6137. 401

75JA6461 H. Ogoshi, J.-I. Setsune, T. Omura, and Z.-I. Yoshida; *J. Am. Chem. Soc.*, 1975, **97**, 6461. 960

75JA6521 W. L. Mock and R. M. Nugent; *J. Am. Chem. Soc.*, 1975, **97**, 6521. 245, 250, 366

75JA7171 J. K. Crandall, P. Battioni, J. T. Wehlacz, and R. Bindra; *J. Am. Chem. Soc.*, 1975, **97**, 7171. 977

75JCS(D)12 H. Goldwhite, P. Gysegem, S. Schow and C. Swyke; *J. Chem. Soc., Dalton Trans.*, 1975, 12. 407, 862

75JCS(D)172 F. W. B. Einstein and J. S. Field; *J. Chem. Soc., Dalton Trans.*, 1975, 172. 897

75JCS(D)1284 G. Ferguson and D. F. Rendle; *J. Chem. Soc., Dalton Trans.*, 1975, 1284. 1004

75JCS(D)1677 P. L. Pauson and J. A. Segal; *J. Chem. Soc., Dalton Trans.*, 1975, 1677. 684

75JCS(P1)26 S. R. Challand, S. F. Gait, M. J. Rance, C. W. Rees and R. C. Storr; *J. Chem. Soc., Perkin Trans. 1*, 1975, 26. 777

75JCS(P1)461 M. S. Manhas, W. H. Hoffman, B. Lai and A. K. Bose; *J. Chem. Soc., Perkin Trans. 1*, 1975, 461. 683

75JCS(P1)470 C. Eaborn, R. A. Jackson and R. Pearce; *J. Chem. Soc., Perkin Trans. 1*, 1975, 470. 937, 938

75JCS(P1)475 C. Eaborn, R. A. Jackson and R. Pearce; *J. Chem. Soc., Perkin Trans. 1*, 1975, 475. 937

75JCS(P1)1065 B. D. Baigrie, J. I. G. Cadogan and J. T. Sharp; *J. Chem. Soc., Perkin Trans. 1*, 1975, 1065. 777

75JCS(P1)1237 R. B. Boar, J. F. McGhie, M. Robinson, D. H. R. Barton, D. C. Horwell and R. V. Stick; *J. Chem. Soc., Perkin Trans. 1*, 1975, 1237. 747

75JCS(P1)1764 D. H. R. Barton, M. J. Day, R. H. Hesse and M. M. Pechet; *J. Chem. Soc., Perkin Trans. 1*, 1975, 1764. 765

75JCS(P1)1925 W. M. Leyshon and D. A. Wilson; *J. Chem. Soc., Perkin Trans. 1*, 1975, 1925. 274

75JCS(P1)2445 I. G. C. Coutts and M. Hamblyn; *J. Chem. Soc., Perkin Trans 1*, 1975, 2445. 365

75JCS(P1)2513 R. J. Bushby; *J. Chem. Soc., Perkin Trans. 1*, 1975, 2513. 230

75JCS(P2)1030 A. F. Cameron, F. D. Duncanson, A. A. Freer, V. W. Armstrong and R. Ramage; *J. Chem. Soc. Perkin Trans 2*, 1975, 1030. 1004

75JGU1476 L. B. Ionov, L. A. Kunitskaya and V. I. Kornev; *J. Gen. Chem. USSR* (*Engl. Transl.*), 1975, **45**, 1476. 890

75JGU1832 G. G. Balezina, V. M. Vlazov, N. M. Denglazov and S. F. Malysheva; *J. Gen. Chem. USSR* (*Engl. Transl.*), 1975, **45**, 1832. 1064

75JHC423 N. Dereu, J. L. Piette, J. van Coppenolle and M. Renson; *J. Heterocycl. Chem.*, 1975, **12**, 423. 279

75JIN233 J. W. Anderson, J. E. Drake, R. T. Hemmings and D. L. Nelson; *J. Inorg. Nucl. Chem.*, 1975, **11**, 233. 293

75JOC124 S. J. Huang, V. Paneccasio, F. Debattista, D. Picker and G. Wilson; *J. Org. Chem.*, 1975, **40**, 124. 1054

75JOC126 M. J. Miller and G. M. Loudon; *J. Org. Chem.*, 1975, **40**, 126. 692

75JOC134 T. G. Squires, W. W. Schmidt and C. S. McCandlish, Jr., *J. Org. Chem.*, 1975, **40**, 134. 25, 31

75JOC248 M. Zanger and J. L. Rabinowitz; *J. Org. Chem.*, 1975, **40**, 248. 35

75JOC250 R. G. Hiskey, N. Muthukumaraswamy and R. R. Vunnam; *J. Org. Chem.*, 1975, **40**, 250. 220

75JOC524 E. L. Eliel, V. S. Rao, S. Smith and R. O. Hutchins; *J. Org. Chem.*, 1975, **40**, 524. 224, 225

75JOC526 C. Kashima; *J. Org. Chem.*, 1975, **40**, 526. 744

75JOC574 W. J. Middleton; *J. Org. Chem.*, 1975, **40**, 574. 19

75JOC807 S. P. Anand, L. A. Quarterman, K. G. Migliorese and R. Filler; *J. Org. Chem.*, 1975, **40**, 807. 613

75JOC814 E. Negishi, T. Yoshida, A. Silveira and B. L. Chiou; *J. Org. Chem.*, 1975, **80**, 814. 524

75JOC1070 P. R. Farina and H. Tieckelmann; *J. Org. Chem.*, 1975, **40**, 1070. 381

75JOC1083 H. A. Dieck and R. F. Heck; *J. Org. Chem.*, 1975, **40**, 1083. 628, 904
75JOC1090 J. R. Pratt, W. D. Massey, F. H. Pinkerton and S. F. Thames; *J. Org. Chem.*, 1975, **40**, 1090. 813
75JOC1213 R. A. Moss and C. E. Powell; *J. Org. Chem.*, 1975, **40**, 1213. 376
75JOC1343 E. M. Kaiser and P. L. A. Knutson; *J. Org. Chem.*, 1975, **40**, 1343. 205
75JOC1395 J. P. Snyder, M. L. Heymann and E. N. Suciu; *J. Org. Chem.*, 1975, **40**, 1395. 390
75JOC1409 F. D. Greene and K. E. Gilbert; *J. Org. Chem.*, 1975, **40**, 1409. 403, 404, 405
75JOC1514 J. S. Filippo, Jr. and L. J. Romano; *J. Org. Chem.*, 1975, **40**, 1514. 9
75JOC1523 K. E. DeBruin and D. E. Perrin; *J. Org. Chem.*, 1975, **40**, 1523. 455
75JOC1669 S. L. Regen and D. P. Lee; *J. Org. Chem.*, 1975, **40**, 1669. 7
75JOC1678 J. San Filippo, C.-I. Chern and J. S. Valentine; *J. Org. Chem.*, 1975, **40**, 1678. 87
75JOC1776 G. W. Kabalka and H. C. Hedgecock, Jr.; *J. Org. Chem.*, 1975, **40**, 1776. 40
75JOC1990 D. P. Bauer and R. S. Macomber; *J. Org. Chem.*, 1975, **40**, 1990. 28
75JOC2059 R. J. Brooks and C. A. Bunton; *J. Org. Chem.*, 1975, **40**, 2059. 845
75JOC2141 Z. Cohen, E. Keinan, Y. Mazur and T. H. Varkony; *J. Org. Chem.*, 1975, **40**, 2141. 86
75JOC2265 E. J. Corey and R. H. Wollenberg; *J. Org. Chem.*, 1975, **40**, 2265. 970, 971
75JOC2410 M. Schwarz, J. E. Oliver and P. E. Sonnet; *J. Org. Chem.*, 1975, **40**, 2410. 10
75JOC2536 K. Baum and C. D. Beard; *J. Org. Chem.*, 1975, **40**, 2536. 95
75JOC2570 H. J. Reich; *J. Org. Chem.*, 1975, **40**, 2570. 287
75JOC2759 A. K. Sharma, T. Ku, A. D. Dawson and D. Swern; *J. Org. Chem.*, 1975, **40**, 2759. 162
75JOC2801 W. R. Purdum and K. D. Berlin; *J. Org. Chem.*, 1975, **40**, 2801. 445
75JOC2960 Y. Ogata, T. Harada, K. Matsuyama and T. Ikejiri; *J. Org. Chem.*, 1975, **40**, 2960. 21
75JOC2963 P. F. Hudrlik and C.-N. Wan; *J. Org. Chem.*, 1975, **40**, 2963. 816
75JOC3097 H. Cho and R. G. Harvey; *J. Org. Chem.*, 1975, **40**, 3097. 945
75JOC3291 S. Sakai, T. Fujinami and K. Komizo; *J. Org. Chem.*, 1975, **40**, 3291. 399
75JOC3313 H. J. Reich and J. M. Renga; *J. Org. Chem.*, 1975, **40**, 3313. 368
75JOC3420 D. N. Harpp, L. Q. Bao, C. J. Black, J. G. Gleason and R. A. Smith; *J. Org. Chem.*, 1975, **40**, 3420. 3
75JOC3586 K. G. Weinberg; *J. Org. Chem.*, 1975, **40**, 3586. 834, 835
75JOC3644 Y. Yamamoto, K. Kondo and I. Moritani; *J. Org. Chem.*, 1975, **40**, 3644. 268
75JOC3686 W. L. Meyer, G. B. Clemans and R. A. Manning; *J. Org. Chem.*, 1975, **40**, 3686. 658
75JOM(84)1 L. R. Smith and J. L. Mills; *J. Organomet. Chem.*, 1975, **84**, 1. 506, 894
75JOM(85)179 G. F. Lanneau, R. J. Wikholm, C. T. Lin and W. E. McEwen; *J. Organomet. Chem.*, 1975, **85**, 179. 876
75JOM(86)143 J. J. Eisch and S. G. Rhee; *J. Organomet. Chem.*, 1975, **86**, 143. 919
75JOM(86)C27 H. Okinoshima, K. Yamamoto and M. Kumada; *J. Organomet. Chem.*, 1975, **86**, C27. 915
75JOM(87)83 H. Schumann and H. J. Breunig; *J. Organomet. Chem.*, 1975, **87**, 83. 483
75JOM(87)169 J. Dunogues, P. Bourgeois, J.-P. Pillot, G. Merault and R. Calas; *J. Organomet. Chem.*, 1975, **87**, 169. 916, 923, 940
75JOM(88)1 G. A. Teodoradze; *J. Organomet. Chem.*, 1975, **88**, 1. 594
75JOM(88)79 A. Rahm and M. Pereyre; *J. Organomet. Chem.*, 1975, **88**, 79. 601
75JOM(88)329 P. Krommes and J. Lorberth; *J. Organomet. Chem.*, 1975, **88**, 329. 485
75JOM(88)171 W. Levason, C. A. McAuliffe and S. G. Murray; *J. Organomet. Chem.*, 1975, **88**, 171. 872, 874
75JOM(88)175 K. J. Irgolic, P. J. Busse, R. A. Grigsby and M. R. Smith; *J. Organomet. Chem.*, 1975, **88**, 175. 279
75JOM(90)C1 G. W. Kramer and H. C. Brown; *J. Organomet. Chem.*, 1975, **90**, C1. 524
75JOM(90)C6 A. R. Bassindale, A. G. Brook and J. Harris; *J. Organomet. Chem.*, 1975, **90**, C6. 700
75JOM(91)123 E. A. Hill; *J. Organomet. Chem.*, 1975, **91**, 123. 556
75JOM(91)295 W. Morris, R. A. Zingaro and J. Laane; *J. Organomet. Chem.*, 1975, **91**, 295. 497
75JOM(91)C20 W. Adcock and T. C. Khor; *J. Organomet. Chem.*, 1975, **91**, C20. 16
75JOM(92)181 P. Krommes and J. Lorberth; *J. Organomet. Chem.*, 1975, **92**, 181. 485
75JOM(92)C46 H. Sakurai and F. Kondo; *J. Organomet. Chem.*, 1975, **92**, C46. 546
75JOM(93)191 P. Jutzi, F. Herzog and M. Kuhn; *J. Organomet. Chem.*, 1975, **93**, 191. 484
75JOM(93)415 A. Commercon, J. Normant and J. Villieras; *J. Organomet. Chem.*, 1975, **93**, 415. 682
75JOM(94)327 S. O. Grim and R. C. Barth; *J. Organomet. Chem.*, 1975, **94**, 327. 827
75JOM(94)393 F. Kober; *J. Organomet. Chem.*, 1975, **94**, 393. 503
75JOM(96)139 N. Collignon; *J. Organomet. Chem.*, 1975, **96**, 139. 557
75JOM(96)57 S. Sato and Y. Matsumura; *J. Organomet. Chem.*, 1975, **96**, 57. 890
75JOM(97)325 J. Villieras, C. Bacquet and J. F. Normant; *J. Organomet. Chem.*, 1975, **97**, 325. 420
75JOM(99)C19 R. A. Andersen, E. Carmona-Guzman, K. Mertis, E. Sigurdson and G. Wilkinson; *J. Organomet. Chem.*, 1975, **99**, C19. 575
75JOM(100)237 D. Seyferth; *J. Organomet. Chem.*, 1975, **100**, 237. 209
75JOM(102)445 A. L. Rheingold and J. M. Bellama; *J. Organomet. Chem.*, 1975, **102**, 445. 507
75LA128 H. Volz and M. Volz-de Lecea; *Justus Liebigs Ann. Chem.*, 1975, 128. 1008
75LA719 T. Cuvigny, M. Larcheveque and H. Normant; *Justus Liebigs Ann. Chem.*, 1975, 719. 420
75LA1252 E. Reimann; *Justus Liebigs Ann. Chem.*, 1975, 1252. 12
75LA1322 W. Fenzl and R. Köster; *Justus Liebigs Ann. Chem.*, 1975, 1322. 698
75LA2130 R. Köster, A. Bussmann and G. Schroth; *Justus Liebigs Ann. Chem.*, 1975, 2130. 1087
75LA2201 W. Fenzl, R. Köster and H.-J. Zimmermann; *Justus Liebigs Ann. Chem.*, 1975, 2201. 698
B-75MI 203-01 G. C. Barrett; in "Organic Compounds of Sulfur, Selenium and Tellurium," ed. D. H. Reid, The Chemical Society, London, 1975, vol. 3, ch. 1. 175
B-75MI 205-01 W. Schneider and J. Hoyer; in "C–N Compounds, Methodicum Chimicum," vol. 6, ed. F. Zymalkowski, Academic Press, New York, 1975, pp. 439–597. 300, 303

B-75MI 207-01	J. W. Timberlake and J. C. Stowell; in "The Chemistry of Hydrazo-, Azo- and Azoxy Groups," ed. S. Patai, Wiley, New York, 1975, p. 69.	373, 402, 404
B-75MI 207-02	K. Mackenzie; in "Azo and Azoxy Compounds," ed S. Patai, Wiley, New York, 1975, p. 329.	402
B-75MI 210-01	H. C. Brown, G. W. Kramer, A. B. Levy and M. M. Midland; "Organic Synthesis via Organoboranes," Wiley, New York, 1975.	514, 516, 520
B-75MI 210-02	T. P. Onak; "Organoboron Compounds," Academic Press, London, 1975.	520
B-75MI 211-01	R. P. A. Sneeden; "Organochromium Compounds," Academic Press, New York, 1975.	571
75MI 213-01	G. Declercq, G. Mottardier and P. Mastagli; *C. R. Acad. Sc. Paris*, 1975, **281**, C279.	662
B-75MI 215-01	S. Patai (ed); "The Chemistry of Hydrazo, Azo and Azoxy Groups," Wiley, New York, 1975.	773
B-75MI 215-02	K. Schank; in "Methodicum Chimicum," ed. F. Zymalkowski, Academic Press, New York, 1975, p. 164.	803
B-75MI 215-03	K. Schank; in "Methodicum Chimicum," ed. F. Zymalkowski, Academic Press, New York, 1975, p. 173.	
B-75MI 215-04	K. Schank; in "Methodicum Chimicum," ed. F. Zymalkowski, Academic Press, New York, 1975, p. 176.	806
B-75MI 217-01	J. D. Smith; in "Comprehensive Inorganic Chemistry," ed. J. C. Bailar, A. J. Emelius, R. Nyholm and A. F. Trotman-Dickenson, Pergamon, Oxford, 1975, vol. 2, chap. 21, pp. 622–639.	872
75MI 217-02	A. Schultze, S. Samaan and L. Horner; *Phosphorus*, 1975, **5**, 265 (*Chem. Abstr.*, 1976, **84**, 5091z).	881
B-75MI 218-01	H. C. Brown, G. W. Kramer, A. B. Levy and M. M. Midland; "Organic Syntheses via Boranes," Wiley Interscience, New York, 1975.	900
B-75MI 219-01	H. C. Brown; "Organic Synthesis via Boranes," Wiley Interscience, New York, 1975.	966
75OR(22)253	G. H. Posner; *Org. React.*, 1975, **22**, 253.	917, 960
75OR(22)1	S. J. Rhoads and N. R. Raulins; *Org. React.*, 1975, **22**, 1.	656
75OR(22)423	M. W. Rathke; *Org. React.*, 1975, **22**, 423.	69
75PAC563	B. M. Trost; *Pure Appl. Chem.*, 1975, **43**, 563.	139, 140
75PS(5)265	A. Schulze, S. Samann and L. Horner; *Phosphorus Sulfur*, 1975, **5**, 265.	490
75S20	T. Nishiwaki; *Synthesis*, 1975, 20.	744
75S39	H. L. Pan and T. L. Fletcher; *Synthesis*, 1975, 39.	366
75S42	H. Kropf, M. Ball and H.-J. Franke; *Synthesis*, 1975, 42.	686
75S55	E. Vilsmaier and G. Becker; *Synthesis*, 1975, 55.	188
75S135	C. F. Lane; *Synthesis*, 1975, 135.	310, 763
75S205	D. S. Black, R. F. Crozier and V. C. Davies; *Synthesis*, 1975, 205.	354
75S321	J. E. Oliver and A. B. Demilo; *Synthesis*, 1975, 321.	191
75S376	G. Zweifel and R. P. Fisher; *Synthesis*, 1975, 376.	904
75S426	H. Wamhoff and L. Lichtenthäler; *Synthesis*, 1975, 426.	746
75S430	D. Landini, S. Quici and F. Rolla; *Synthesis*; 1975, 430.	9
75S445	S. Spyronis and A. Varvoglis; *Synthesis*, 1975, 445.	390
75S458	J. Villieras, P. Perriot and J. F. Normant; *Synthesis*, 1975, 458.	1015
75S507	A. Zwierzak; *Synthesis*, 1975, 507.	808
75S512	H. Ahlbrecht and C. Vonderheid; *Synthesis*, 1975, 512.	748
75S519	G. E. Vennstra and B. Zwanenberg; *Synthesis*, 1975, 519.	197
75S617	E. Winterfeldt; *Synthesis*, 1975, 617.	60
75S658	J. Petrova, P. Coutrot, M. Dreux and P. Savignac; *Synthesis*, 1975, 658.	618
75S738	F. D. King and D. R. M. Walton; *Synthesis*, 1975, 738.	769
75S782	J. B. Baumann; *Synthesis*, 1975, 782.	692
75S801	L. N. Markovski and V. E. Pashinnik; *Synthesis*, 1975, 801.	183
75SC393	W. J. Peterson and J. S. Pizey; *Synth. Commun.*, 1975, **5**, 393.	872, 881
75T89	H. Schlude; *Tetrahedron*, 1975, **31**, 89.	680
75T505	P. Claus, W. Riecer, P. Hofbauer and E. Vilsmaier; *Tetrahedron*, 1975, **31**, 505.	160
75T2099	J. van Coppenolle and M. Renson; *Tetrahedron*, 1975, **31**, 2099.	238
75T2121	R. Manske and J. Gosselck; *Tetrahedron*, 1975, **31**, 2121.	877
75T2595	P. H. Morgan and A. H. Beckett; *Tetrahedron*, 1975, **31**, 2595.	341, 342, 350
75TCC1	R. D. Rieke; *Top. Curr. Chem.*, 1975, **59**, 1 and references therein.	958
75TL71	J. W. Zubrick, B. I. Dunbar and H. D. Durst; *Tetrahedron Lett.*, 1975, 71.	393
75TL649	S. Tamagaki, S. Oae and K. Sasaki; *Tetrahedron Lett.*, 1975, 649.	292
75TL1409	I. Nakagawa and T. Hata; *Tetrahedron Lett.*, 1975, 1409.	135
75TL1695	S. Fujita and K. Sano; *Tetrahedron Lett.*, 1975, 1695.	657
75TL1739	J. R. Grunwell and S. L. Dye; *Tetrahedron Lett.*, 1975, 1739.	240
75TL1859	Y. Kitagawa, K. Oshima, H. Yamamoto and H. Nozaki; *Tetrahedron Lett.*, 1975, 1859.	420
75TL2389	E. J. Corey and A. P. Kozikowski; *Tetrahedron Lett.*, 1975, 2389.	665
75TL2689	Y. Yamamoto, K. Kondo and I. Moritani; *Tetrahedron Lett.*, 1975, 2689.	268
75TL3117	E. J. Corey, L. S. Melvin and M. F. Haslanger; *Tetrahedron Lett.*, 1975, 3117.	778
75TL3183	E. J. Corey, K. C. Nicolaou, M. Shibasaki, Y. Machida and C. S. Shiner; *Tetrahedron Lett.*, 1975, 3183.	87
75TL3383	T. H. Chan and D. Massuda; *Tetrahedron Lett.*, 1975, 3383.	616
75TL3429	G. W. Klumpp, H. Bos, R. F. Schmitz and J. J. Vrielink; *Tetrahedron Lett.*, 1975, 3429.	682
75TL3833	J. F. Normant, A. Commercon, M. Bourgain and J. Villieras; *Tetrahedron Lett.*, 1975, 3833.	679
75TL4539	C. J. Saward and K. P. C. Vollhardt; *Tetrahedron Lett.*, 1975, 4539.	627
75ZAAC(412)202	F. Kober; *Z. Anorg. Allg. Chem.*, 1975, **412**, 202.	500, 891

75ZAAC(413)266 W. Kuchen and H. Steinberger; *Z. Anorg. Allg. Chem.*, 1975, **413**, 266. 469

75ZAAC(418)167 M. Wieber and N. Baumann; *Z. Anorg. Allg. Chem.*, 1975, **418**, 167. 496, 695

75ZAAC(418)273 W. Siebert, R. Full, T. Renk and A. Ospici; *Z. Anorg. Allg. Chem.*, 1975, **418**, 273. 1083, 1094

75ZAAC(418)279 M. Wieber and N. Baumann; *Z. Anorg. Allg. Chem.*, 1975, **418**, 279. 695

75ZN(B)642 W. Siebert, W. Ruf and R. Full; *Z. Naturforsch., Teil B*, 1975, **30**, 642. 294

75ZN(B)681 H. Nöth and P. Konrad; *Z. Naturforsch., Teil B*, 1975, **30**, 681. 421

75ZOB1005 N. N. Melnikov, B. A. Khaskin and N. A. Torgacheva; *Zh. Obshch. Khim.* 1975, **45**, 1005. 361

75ZOB1228 M. I. Shevchuk, V. N. Kushnir, V. A. Dombrovskii, M. V. Khalaturnik and A. V. Dombrovskii; *Zh. Obshch. Khim.*, 1975, **45**, 1228. 444

75ZOB2394 A. M. Pinchuk, V. A. Kovenya and M. I. Povolotskii; *Zh. Obshch. Khim.*, 1975, **45**, 2394. 407

75ZOB2554 V. E. Shishkin, Yu. M. Yukhno, B. I. No and Yu. D. Glinskii; *Zh. Obshch. Khim.*, 1975, **45**, 2554. 469

75ZOR1370 R. A. Bekker, G. G. Melikyan, B. L. Dyatkin and I. L. Knunyants; *Zh. Org. Khim.*, 1975, **11**, 1370. 640

76ACR13 D. Griller and K. U. Ingold; *Acc. Chem. Res.*, 1976, **9**, 13. 1010

76ACR93 P. M. Maitlis; *Acc. Chem. Res.*, 1976, **9**, 93. 980

76ACR217 M. Tsutsui, N. Ely and R. Dubois; *Acc. Chem. Res.*, 1976, **9**, 217. 602

76ACR223 T. J. Marks; *Acc. Chem. Res.*, 1976, **9**, 223. 602

76ACR293 W. S. Allison; *Acc. Chem. Res.*, 1976, **9**, 293. 184

76ACR358 S. I. Miller and J. I. Dickstein; *Acc. Chem. Res.*, 1976, **9**, 358. 1011

76ACS(B)1000 L. Nilsson and C. Rappe; *Acta Chem. Scand., Ser. B*, 1976, **30**, 1000. 758

76AG(E)123 A. Padwa; *Angew. Chem., Int. Ed. Engl.*, 1976, **15**, 123. 354

76AG(E)239 G. Kinast and L.-F. Tietze; *Angew. Chem., Int. Ed. Engl.*, 1976, **15**, 239. 1008

76AG(E)294 H. Poisel and U. Schmidt; *Angew. Chem., Int. Ed. Engl.*, 1976, **15**, 294. 338

76AG(E)302 A. Zwierzak and K. Osowska; *Angew. Chem., Int. Ed. Engl.*, 1976, **15**, 302. 808

76AG(E)306 F. Kamper, H. J. Schafer and H. Luftmann; *Angew. Chem., Int. Ed. Engl.*, 1976, **15**, 306. 20

76AG(E)333 J. Schwartz and J. A. Labinger; *Angew. Chem., Int. Ed. Engl.*, 1976, **15**, 333. 4, 568, 969

76AG(E)370 E. Hengge and G. Litscher; *Angew. Chem., Int. Ed. Engl.*, 1976, **15**, 370. 544

76AG(E)428 G. Simchen and J. Pfletschinger; *Angew. Chem., Int. Ed. Engl.*, 1976, **15**, 428. 942

76AG(E)435 A. Henne and H. Fischer; *Angew. Chem., Int. Ed. Engl.*, 1976, **15**, 435. 637

76AG(E)621 K. Jonas, K. R. Pöschka, C. Krüger and Y. H. Tsay; *Angew. Chem., Int. Ed. Engl.*, 1976, **15**, 621. 575

76AG(E)760 H. Nitsch and G. Kreze; *Angew. Chem., Int. Ed. Engl.*, 1976, **15**, 760. 353

76AJC133 R. M. Carman and I. M. Shaw; *Aust. J. Chem.*, 1976, **29**, 133. 25

76AJC367 K. J. Chapman and L. K. Dyall; *Aust. J. Chem.*, 1976, **29**, 367. 761

76AOC345 M. F. Lappert and P. W. Lednor; *Adv. Organomet. Chem.*, 1976, **14**, 345. 582

76BAU1374 V. A. Dorokhov, O. G. Boldyreva, A. S. Shashkov and B. M. Mikhailov; *Bull. Acad. Sci. USSR. Div. Chem. Sci.* (*Engl. Transl.*), 1976, **25**, 1374. 811

76BCJ1989 W. E. Willy, D. R. McKean and B. A. Garcia; *Bull. Chem. Soc. Jpn.*, 1976, **49**, 1989. 31

76BCJ2762 S. Ueamura, A. Toshimitsu and M. Okano; *Bull. Chem. Soc. Jpn.*, 1976, **49**, 2762. 597

76BCJ3673 R. Noyori, H. Inoue and M. Katô; *Bull. Chem. Soc. Jpn.*, 1976, **49**, 3673. 638

76BSB805 J. O. Madsen and S. O. Lawesson; *Bull. Soc. Chim. Belg.*, 1976, **85**, 805. 758

76BSF169 P. Place, M.-L. Roumestant and J. Gore; *Bull. Soc. Chim. Fr.*, 1976, 169. 9

76BSF277 A. Leberre, A. Etienne and B. Desmazieres; *Bull. Soc. Chim. Fr.*, 1976, 277. 248

76BSF483 A. Etienne, G. Lonchabon and C. Bernard; *Bull. Soc. Chim. Fr.*, 1976, 483. 175

76BSF833 A. Rougny and M. Davidon; *Bull. Soc. Chim. Fr.*, 1976, 833. 102

76C387 B. Caillaux, P. George, F. Tataruch, Z. Janousek and H. G. Viehe; *Chimia*, 1976, **30**, 387. 1046

76CB546 W. Disteldorf and M. Regitz; *Chem. Ber.*, 1976, **109**, 546. 1066

76CB692 P. Panster and W. Malisch; *Chem. Ber.*, 1976, **109**, 692. 512

76CB805 R. Appel and U. Warning; *Chem. Ber.*, 1976, **109**, 805. 453

76CB1069 S. Holm, J. A. Boerma, N. H. Nilsson and A. Senning; *Chem. Ber.*, 1976, **109**, 1069. 400

76CB2442 R. Appel, J.-R. Lundehn and E. Laßmann; *Chem. Ber.*, 1976, **109**, 2442. 399, 400

76CB3446 R. Appel and H.-D. Wihler; *Chem. Ber.*, 1976, **109**, 3446. 7

76CB3462 B. Eistert and G. Holzer; *Chem. Ber.*, 1976, **109**, 3462. 765

76CB3541 T. Severin and I. Ipach; *Chem. Ber.*, 1976, **109**, 3541. 789, 790

76CB3581 A. Schmidpeter, J. Luber, D. Schomburg and W. S. Sheldrick; *Chem. Ber.*, 1976, **35**, 3581. 476

76CC61 I. D. Entwistle, R. A. W. Johnstone and J. H. Varley; *J. Chem. Soc., Chem. Commun.*, 1976, 61. 291

76CC103 M. V. Bhatt, C. G. Rao and S. Rengaraju; *J. Chem. Soc., Chem. Commun.*, 1976, 103. 747

76CC220 M. Mikolajcyzk and J. Drabowicz; *J. Chem. Soc., Chem. Commun.*, 1976, 220. 217

76CC496 R. Louw, H. P. W. Vermeeren, J. J. A. van Asten and W. J. Ultee; *J. Chem. Soc., Chem. Commun.*, 1976, 496. 150

76CC552 J. W. Lewicki, W. H. H. Gunter and J. Y. C. Chu; *J. Chem. Soc., Chem. Commun.*, 1976, 552. 735

76CC643 J. V. Weiss and R. Schmutzler; *J. Chem. Soc., Chem. Commun.*, 1976, 643. 476

76CC833 R. L. Funk and K. P. C. Vollhardt; *J. Chem. Soc., Chem. Commun.*, 1976, 833. 632

76CC877 B. C. Challis and S. A. Kyrtopoulos; *J. Chem. Soc., Chem. Commun.*, 1976, 877. 385

76CC1003 P. K. Claus and F. W. Vierhapper; *J. Chem. Soc., Chem. Commun.*, 1976, 1003. 160

76CC1008 S. J. Harris and D. R. M. Walton; *J. Chem. Soc., Chem. Commun.*, 1976, 1008. 941, 1032

76CC1022 P. G. Cookson, A. G. Davies and B. P. Roberts; *J. Chem. Soc., Chem. Commun.*, 1976, 1022. 96, 98

76CJC498 R. F. Langler; *Can. J. Chem.*, 1976, **54**, 498. 179

76CJC3012	N. K. Sharma, F. de Reinach-Hirtzbach and T. Durst; *Can. J. Chem.*, 1976, **54**, 3012.	149, 200
76CL283	Y. Kitahara, T. Kato and I. Ichinose; *Chem. Lett.*, 1976, 283.	31
76CL449	K. Kaneda, T. Uchiyama, M. Terasawa, T. Imanaka and S. Teranishi; *Chem. Lett.*, 1976, 449.	616
76CL559	T. Mukaiyama and T. Inoue; *Chem. Lett.*, 1976, 559.	532, 698
76CL619	K. Hojo and T. Mukaiyama; *Chem. Lett.*, 1976, 619.	8, 36
76CL785	A. Arase and Y. Masuda; *Chem. Lett.*, 1976, 785.	289
76CL857	Y. Murakami, Y. Aoyama and S. Nakanishi; *Chem. Lett.*, 1976, 857.	213
76CL1115	A. Arase and Y. Masuda; *Chem. Lett.*, 1976, 1115.	289
76CPB820	N. Inukai, H. Iwamoto, T. Tamura, I. Yanagisawa, Y. Ishii and M. Murakami; *Chem. Pharm. Bull.*, 1976, **24**, 820.	3
76CPL399	S. Saito; *Chem. Phys. Lett.*, 1976, **42**, 399.	636
76CRV113	P. G. Sammes; *Chem. Rev.*, 1976, **76**, 113.	184
76CZ391	M. Field and H. P. Rieck; *Chem.-Ztg.*, 1976, **100**, 391.	176
76DOK(229)870	R. A. Bekker, V. Ya. Popkova and I. L. Knunyants; *Dokl. Akad. Nauk SSSR*, 1976, **229**, 870 (Plenum Engl. Transl. p. 514).	641
76DOK(231)864	R. A. Bekker, V. Ya. Popkova and I. L. Knunyants; *Dokl. Akad. Nauk SSSR*, 1976, **231**, 864 (Plenum Engl. Transl. p. 700).	641
76GEP1768461	H. Peterson and W. Reutter; *Ger. Pat.* 1 768 461 (1976) (*Chem. Abstr.*, 1977, **86**, 106 774).	445
76H1243	M. Tashiro and S. Mataka; *Heterocycles*, 1976, **4**, 1243.	794
76HCA453	J. Hartmann and M. Schlosser; *Helv. Chim. Acta*, 1976, **59**, 453.	964
76HCA695	D. G. Leppard, P. W. Raynolds, C. B. Chapleo and A. S. Dreiding; *Helv. Chim. Acta*, 1976, **59**, 695.	831
76HCA840	V. I. Cohen; *Helv. Chim. Acta*, 1976, **59**, 840.	223, 224
76HCA1027	P. Wieland; *Helv. Chim. Acta*, 1976, **59**, 1027.	12
76IJC(B)904	B. S. Cowlagi, M. A. Dave and A. B. Kulkarni; *Indian J. Chem.*, *Sect. B*, 1976, **14**, 904.	753, 760
76IS155	V. D. Bianco and S. Doronzo; *Inorg. Synth.*, 1976, **16**, 155.	437
76IS157	V. D. Bianco and S. Doronzo; *Inorg. Synth.*, 1976, **16**, 157.	827
76IS158	V. D. Bianco and S. Doronzo; *Inorg. Synth.*, 1976, **16**, 158.	827
76IS159	V. D. Bianco and S. Doronzo; *Inorg. Synth.*, 1976, **16**, 159.	827
76IS160	V. D. Bianco and S. Doronzo; *Inorg. Synth.*, 1976, **16**, 160.	827
76IS161	V. D. Bianco and S. Doronzo; *Inorg. Synth.*, 1976, **16**, 161.	822, 823
76IS168	D. W. Meek, G. Dyer and M. O. Workman; *Inorg. Synth.*, 1976, **16**, 168.	825
76IZV1580	B. M. Mikhailov, M. E. Gurskii and M. G. Gverdtsiteli; *Izv. Akad. Nauk SSSR Ser. Khim.*, 1976, 1580.	904
76IZV1823	B. M. Mikhailov, G. S. Ter-Sarkisyan and N. N. Govorov; *Izv. Akad. Nauk SSSR Ser. Khim.*, 1976, 1823 (*Chem. Abstr.*, 1977, **86**, 16 723z).	811
76IZV2062	B. M. Mikhailov, M. E. Gurskii and M. G. Gverdtsiteli; *Izv. Akad. Nauk SSSR Ser. Khim.*, 1976, 2062.	904
76JA269	K. B. Sharpless, T. Hori, L. K. Truesdale and C. O. Diedrick; *J. Am. Chem. Soc.*, 1976, **98**, 269.	298
76JA1290	H. C. Brown, N. R. de Lue, G. W. Kabalka and H. C. Hedgecock, Jr., *J. Am. Chem. Soc.*, 1976, **98**, 1290.	34
76JA1785	H. C. Brown and N. Ravindran; *J. Am. Chem. Soc.*, 1976, **98**, 1785.	525, 900
76JA1798	H. C. Brown and N. Ravindran; *J. Am. Chem. Soc.*, 1976, **98**, 1798.	525, 900
76JA1964	G. W. Kramer and H. C. Brown; *J. Am. Chem. Soc.*, 1976, **98**, 1964.	524
76JA2868	R. E. Ireland, R. H. Mueller and A. K. Willard; *J. Am. Chem. Soc.*, 1976, **98**, 2868.	420, 696
76JA3034	D. H. R. Barton, R. H. Hesse, R. E. Markwell, M. M. Pechet and H. T. Toh; *J. Am. Chem. Soc.*, 1976, **98**, 3034.	15
76JA3036	D. H. R. Barton, R. H. Hesse, R. E. Markwell, M. M. Pechet and S. Rozen; *J. Am. Chem. Soc.*, 1976, **98**, 3036.	15
76JA3332	G. Buchi and R. M. Freidinger; *J. Am. Chem. Soc.*, 1976, **96**, 3332.	159
76JA3383	S. Krishnamurthy and H. C. Brown; *J. Am. Chem. Soc.*, 1976, **98**, 3383.	83
76JA4477	M. H. Chisholm, F. A. Cotton, M. Extine and B. R. Stults; *J. Am. Chem. Soc.*, 1976, **98**, 4477.	421
76JA4674	M. Schlosser and J. Hartmann; *J. Am. Chem. Soc.*, 1976, **98**, 4674.	964
76JA4887	B. M. Trost, T. N. Salzman and K. Hiroi; *J. Am. Chem. Soc.*, 1976, **98**, 4887.	139
76JA4917	H. C. Brown and M. Zaidlewicz; *J. Am. Chem. Soc.*, 1976, **98**, 4917.	525
76JA5269	S. F. Nelsen, V. Peacock and G. R. Weisman; *J. Am. Chem. Soc.*, 1976, **98**, 5269.	382
76JA5297	H. C. Brown, R. Liotta and C. G. Scouter; *J. Am. Chem. Soc.*, 1976, **98**, 5297.	517
76JA5399	R. R. Schrock; *J. Am. Chem. Soc.*, 1976, **98**, 5399.	681
76JA5678	T. A. Wnuk, S. S. Chaudhary and P. Kovacic; *J. Am. Chem. Soc.*, 1976, **98**, 5678.	306
76JA6702	R. V. Hoffman; *J. Am. Chem. Soc.*, 1976, **98**, 6702.	357
76JA7104	K. A. Parker, R. W. Kosley, Jr., S. L. Buchwald and J. J. Petraitis; *J. Am. Chem. Soc.*, 1976, **98**, 7104.	740
76JA7617	R. R. Lucchese, K. Haber and H. F. Schaefer III; *J. Am. Chem. Soc.*, 1976, **98**, 7617.	364
76JA7832	P. Jacob III and H. C. Brown; *J. Am. Chem. Soc.*, 1976, **98**, 7832.	900
76JCS(D)1	J. Keable, D. G. Othen and K. Wade; *J. Chem. Soc., Dalton Trans.*, 1976, 1.	815
76JCS(D)1429	K.-K. Chow, W. Levason and C. A. McAuliffe; *J. Chem. Soc., Dalton Trans.*, 1976, 1429.	874
76JCS(D)2087	A. Pelter, K. Rowe, D. N. Sharrocks, K. Smith and C. Subramanyam; *J. Chem. Soc., Dalton Trans*, 1976, 2087.	533
76JCS(D)2204	R. A. Andersen, E. Carmona-Guzman, J. F. Gibson and G. Wilkinson; *J. Chem. Soc., Dalton Trans.*, 1976, 2204.	574

76JCS(D)2307 C. L. Raston, R. J. Secomb and A. H. White; *J. Chem. Soc., Dalton Trans.*, 1976, 2307. 690

76JCS(P1)104 H. R. Hudson and G. R. de Spinoza; *J. Chem. Soc., Perkin Trans. 1*, 1976, 104. 25

76JCS(P1)416 J. A. Miller and M. J. Nunn; *J. Chem. Soc., Perkin Trans. 1*, 1976, 416. 35

76JCS(P1)754 D. R. Hepburn and H. R. Hudson; *J. Chem. Soc., Perkin Trans. 1*, 1976, 754. 25, 31

76JCS(P1)836 H. Cho and R. G. Harvey; *J. Chem. Soc., Perkin Trans. 1*, 1976, 836. 657

76JCS(P1)1161 R. C. Cambie, P. S. Rutledge, T. Smith-Palmer and P. D. Woodgate; *J. Chem. Soc., Perkin Trans. 1*, 1976, 1161. 631

76JCS(P1)1446 G. Mackenzie, P. W. Rugg and G. Shaw; *J. Chem. Soc., Perkin Trans. 1*, 1976, 1446. 32

76JCS(P1)1739 C. Y. Cheng and R. A. Shaw; *J. Chem. Soc., Perkin Trans. 1*, 1976, 1739. 843, 845

76JCS(P1)1852 J. Banville and P. Brassard; *J. Chem. Soc., Perkin Trans. 1*, 1976, 1852. 667

76JCS(P1)2079 T. G. Back, D. H. R. Barton, M. R. Britten-Kelly and F. S. Guziec, Jr.; *J. Chem. Soc., Perkin Trans. 1*, 1976, 2079. 858

76JCS(P1)2112 D. H. R. Barton, R. V. Stick and R. Subramanian; *J. Chem. Soc., Perkin Trans. 1*, 1976, 2112. 8

76JCS(P1)2166 T. L. Gilchrist, C. J. Harris, D. G. Hawkins, C. J. Moody and C. W. Rees; *J. Chem. Soc., Perkin Trans. 1*, 1976, 2166. 160

76JCS(P1)2365 H. Suginome, T. Mizuguchi and T. Masamune; *J. Chem. Soc., Perkin Trans. 1*, 1976, 2365. 765

76JCS(P1)2419 A. Pelter, T. W. Bentley, C. R. Harrison, C. Subrahmanyan and R. J. Laub; *J. Chem. Soc., Perkin Trans. 1*, 1976, 2419. 904

76JCS(P1)2428 A. Pelter, K. J. Gould and C. R. Harrison; *J. Chem. Soc., Perkin Trans. 1*, 1976, 2428. 904

76JCS(P1)2435 A. Pelter, C. R. Harrison, C. Subrahmanyan and D. Kirkpatrick; *J. Chem. Soc., Perkin Trans. 1*, 1976, 2435. 904

76JCS(P1)2529 D. W. Allen, P. Heatley, B. G. Hutley and M. T. J. Mellor; *J. Chem. Soc., Perkin Trans. 1*, 1976, 2529. 824

76JCS(P2)1444 J. S. Loran, R. A. Naylor and A. Williams; *J. Chem. Soc., Perkin Trans. 2*, 1976, 1444. 847

76JFC(7)115 W. Gombler and R. Budenz; *J. Fluorine Chem.*, 1976, **7**, 115 (*Chem. Abstr.*, 1976, **85**, 32 364t). 166

76JGU428 M. I. Kabachnik, Z. S. Novikova, E. V. Snyatkova and I. F. Lutsenko; *J. Gen. Chem. USSR* (*Engl. Transl.*), 1976, **46**, 428. 449

76JGU572 Z. S. Novikova, S. N. Zdorova and V. N. Kirzner; *J. Gen. Chem. USSR* (*Engl. Transl.*), 1976, **46**, 572. 448

76JGU1213 S. V. Fridland, T. M. Shchukareva and R. A. Salakhutdinov; *J. Gen. Chem. USSR* (*Engl. Transl.*), 1976, **46**, 1213; V. A. Efavov, A. V. Dogadina, B. I. Ionin and A. A. Petrov; *J. Gen. Chem. USSR* (*Engl. Transl.*), 1976, **46**, 1392. 448

76JGU2477 V. I. Popov and N. V. Kandratenko; *J. Gen. Chem. USSR* (*Engl. Transl.*), 1976, **46**, 2477. 881

76JMC1423 R. M. Peck, T. K. Tom and E. B. Peck; *J. Med. Chem.*, 1976, **19**, 1423. 270

76JMOC285 A. Nakamura and S. Otsuka; *J. Mol. Catal.*, 1976, **1**, 285. 970

76JOC134 J. S. Bradshaw, R. A. Reeder, M. D. Thompson, E. D. Flanders, R. L. Carruth, R. M. Izatt and J. J. Christensen; *J. Org. Chem.*, 1976, **41**, 134. 133

76JOC300 M. Moriwaki, Y. Yamamoto, J. Oda and Y. Inouye; *J. Org. Chem.*, 1976, **41**, 300, 303. 355

76JOC732 J. E. Shaw, D. C. Kunerth and S. B. Swanson; *J. Org. Chem.*, 1976, **41**, 732. 684

76JOC863 H. O. House and L. F. Lee; *J. Org. Chem.*, 1976, **41**, 863. 763

76JOC972 R. B. King, J. C. Lloyd, Jr. and R. H. Reimann; *J. Org. Chem.*, 1976, **41**, 972. 439

76JOC977 H. Yagi, G. M. Holder, P. M. Dansette, O. Hernandez, H. J. C. Yeh, R. A. LeMahieu and D. M. Jerina; *J. Org. Chem.*, 1976, **41**, 977. 659

76JOC1071 R. O. Hutchins, D. Masilamani and C. A. Maryanoff; *J. Org. Chem.*, 1976, **41**, 1071. 31

76JOC1078 J. A. Sinclair and H. C. Brown; *J. Org. Chem.*, 1976, **41**, 1078. 520, 1086

76JOC1333 F. A. Davis and E. B. Skibo; *J. Org. Chem.*, 1976, **41**, 1333. 361, 792

76JOC1480 R. F. Cunico and F. J. Clayton; *J. Org. Chem.*, 1976, **41**, 1480. 910, 911, 931

76JOC1485 P. A. Grieco, S. Gilman and M. Nishizawa; *J. Org. Chem.*, 1976, **41**, 1485. 279

76JOC1578 C. Symmes and L. D. Quin; *J. Org. Chem.*, 1976, **41**, 1578. 462

76JOC1627 G. A. Olah, D. A. Beal and J. A. Olah; *J. Org. Chem.*, 1976, **41**, 1627. 13

76JOC1751 V. Nelson, A. Serianz and P. Kovacic; *J. Org. Chem.*, 1976, **41**, 1751. 773, 775

76JOC2207 R. W. Heintzelman, R. B. Bailey and D. Swern; *J. Org. Chem.*, 1976, **41**, 2207. 793

76JOC2214 J. J. Eisch and G. A. Damasevitz; *J. Org. Chem.*, 1976, **41**, 2214. 922

76JOC2215 K. Uchida, K. Utimoto and H. Nozaki; *J. Org. Chem.*, 1976, **41**, 2215. 922

76JOC2561 R. Busson and H. Vanderhaeghe; *J. Org. Chem.*, 1976, **41**, 2561. 167

76JOC2941 K. Uchida, K. Utimoto and H. Nozaki; *J. Org. Chem.*, 1976, **41**, 2941. 922

76JOC3026 D. S. Karanewsky and Y. Kishi; *J. Org. Chem.*, 1976, **41**, 3026. 656

76JOC3031 D. E. Bergbrieter and D. P. Rainville; *J. Org. Chem.*, 1976, **41**, 3031. 527

76JOC3063 W. C. Still; *J. Org. Chem.*, 1976, **41**, 3063. 546

76JOC3279 P. E. Sonnet and J. E. Oliver; *J. Org. Chem.*, 1976, **41**, 3279. 26, 33

76JOC3328 R. O. Hutchins, D. Rotstein, N. Natale, J. Fanelli and D. Dimmel; *J. Org. Chem.*, 1976, **41**, 3328. 398

76JOC3505 N. Cohen, W. Eichel, R. Lopresti, C. Neukom and G. Saucy; *J. Org. Chem.*, 1976, **41**, 3505. 39

76JOC3559 Y. Sato, T. Toyo'oka, T. Aoyama and H. Shirai; *J. Org. Chem.*, 1976, **41**, 3559. 945

76JOC3587 P. A. Zoretic and P. Soja; *J. Org. Chem.*, 1976, **41**, 3587. 139

76JOC3599 A. R. Miller; *J. Org. Chem.*, 1976, **41**, 3599. 646

76JOC3805 H. E. Baumgarten, P. Y.-N. Chen, H. W. Taylor and D.-R. Hwang; *J. Org. Chem.*, 1976, **41**, 3805. 374

76JOC3983 M. F. Salomon, R. G. Salomon and R. D. Glein; *J. Org. Chem.*, 1976, **41**, 3983. 98

76JOC3987 D. N. Harpp, S. M. Vines, J. P. Montillier and T. H. Chan; *J. Org. Chem.*, 1976, **41**, 3987. 151, 688

76JOC4002 M. Zupan and A. Pollak; *J. Org. Chem.*, 1976, **41**, 4002. 17

76JOC4028	J. A. Kloek and K. L. Leschinsky; *J. Org. Chem.*, 1976, **41**, 4028.	366, 368
76JOC4099	L. R. Subramanian, M. Hanack, L. W. K. Chang, M. A. Imhoff, P. v. R. Schleyer, F. Effenberger, W. Kurtz, P. J. Stang and T. E. Dueber; *J. Org. Chem.*, 1976, **41**, 4099.	689
76JOM(105)183	U. Schröer and W. P. Neumann; *J. Organomet. Chem.*, 1976, **105**, 183.	478
76JOM(105)C29	R. F. Cunico and Y.-K. Han; *J. Organomet. Chem.*, 1976, **105**, C29.	907, 908
76JOM(107)229	E. M. Dexheimer and L. Spialter; *J. Organomet. Chem.*, 1976, **107**, 229.	908
76JOM(107)C1	T. H. Chan, W. Mychajlowskij, B. S. Ong and D. N. Harpp; *J. Organomet. Chem.*, 1976, **107**, C1.	931, 932
76JOM(108)93	P. Lennon, M. Madhavarao, A. Rosay and M. Rosenblum; *J. Organomet. Chem.*, 1976, **108**, 93.	577
76JOM(110)327	H. Bouman and J. H. Teuben; *J. Organomet. Chem.*, 1976, **110**, 327.	569
76JOM(114)281	N. Petragnani, R. Rodrigues and J. V. Comasseto; *J. Organomet. Chem.*; 1976, **114**, 281.	728
76JOM(114)C19	K. Tamao, T. Hayashi and M. Kumada; *J. Organomet. Chem.*, 1976, **114**, C19.	915
76JOM(118)C8	P. Jutzi, H. Saleske and D. Nadler; *J. Organomet. Chem.*, 1976, **118**, C8.	4366
76JOM(120)217	J. M. Brocas, B. De Jeso and J. C. Pommier; *J. Organomet. Chem.*, 1976, **120**, 217.	816
76JOM(121)293	A. G. Brook, J. M. Duff and W. F. Reynolds; *J. Organomet. Chem.*, 1976, **121**, 293.	914, 921, 922
76JOM(122)209	R. R. Schrock; *J. Organomet. Chem.*, 1976, **122**, 209.	570
76JOM(122)C25	F. Sato, S. Sato and M. Sato; *J. Organomet. Chem.*, 1976, **122**, C25.	4, 23
76JPC2417	J. Martin, J. B. Robert and C. Taleb; *J. Phys. Chem.*, 1976, **80**, 2417.	852
76JPR1019	H. Berthold, C. Braeunig, S. Grosse, M. Hamper, F. D. Kopinke, W. Pritzkow and G. Stachowski; *J. Prakt. Chem.*, 1976, **318**, 1019.	
76LA2126	G. P. Schiemenz and M. Finzenhagen; *Justus Liebigs Ann. Chem.*, 1976, 2126.	824
76LA2331	H. Kropf and M. Ball; *Justus Liebigs Ann. Chem.*, 1976, 2331.	687
B-76MI 201-01	P. B. D. de la Mare; "Electrophilic Halogenation," Cambridge University Press, Cambridge, 1976.	5
76MI 206-01	A. A. Akhrem, S. M. Bel'skii and D. I. Metelista; *Kinet. Katal.*, 1976, **87**, 877 (*Chem. Abstr.* 1976, **85**, 176 596).	350
76MI 208-01	I. A. Nuretdinov, N. A. Buina, E. V. Bayandina and M. A. Giniytullina; *Tezisy Dokl. Vses. Konf. Stereokhim. Konform. Anal. Org. Neftekhim. Sint. 3rd*, 1976, p. 89 (*Chem. Abstr.*, 1978, **89**, 6365).	466
76MI 210-01	D. H. Harris and M. F. Lappert; *Organometal. Chem. Rev.*, 1976, **2**, 13.	543
B-76MI 215-01	L. Ghosez and J. Marchand-Brynaert; in "Advances in Organic Chemistry," ed. E. C. Tayler, Wiley, New York, 1976, vol. 9, p. 421.	758
B-76MI 215-02	T. Yoshida and J. H. Ridd; in "Industrial and Laboratory Nitrations," ed. L. F. Albright and C. Hanson, *Am. Chem. Soc. Symp. Series*, No. 22, American Chemical Society, Washington, 1976, p. 103.	798
B-76MI 216-01	K. H. Worms and M. Schmidt-Dunker; in "Organic Phosphorus Compounds," ed. G. M. Kosolapoff and L. Maier, Wiley, New York, 1976, vol. 7, chap. 18.	847
76MI 219-01	J. F. Normant; *J. Organomet. Chem. Library*, 1976, **1**, 219.	976, 977, 978
76MI 222-01	J. C. Williams, W. D. Hounshell and A. M. Aguiar; *Phosphorus*, 1976, **6**, 169.	1060
76MI 222-02	E. Fluck and W. Kazenwadel; *Phosphorus*, 1976, **6**, 195.	1062
76MI 223-01	R. Calas and J. Dunoguès; *J. Organomet. Chem. Libr.*, 1976, **2**, 277.	1090
76OR(23)405	R. H. Shapiro; *Org. React.*, 1976, **23**, 405.	911, 956
76OR(23)1	D. Caine; *Org. React.*, 1976, **23**, 1.	696
76OR(24)261	I. Rabjohn; *Org. React.*, 1976, **24**, 261.	86
76PAC177	H. B. Kagan; *Pure Appl. Chem.*, 1976, **46**, 177.	959
76PS(1)211	T. A. Mastryukova; *Phosphorus Sulfur*, 1976, **1**, 211.	854
76S30	N. Furukawa, S. Oae and T. Yoshimura; *Synthesis*, 1976, 30.	401, 745
76S114	N. R. de Lue and H. C. Brown; *Synthesis*, 1976, 114.	34
76S118	R. Köster and L. A. Hagele; *Synthesis*, 1976, 118.	1088
76S194	L. Blanco, P. Amice and J. M. Conia; *Synthesis*, 1976, 194.	28
76S308	M. Haake and H. Benack; *Synthesis*, 1976, 308.	360
76S326	M. Isola, E. Ciuffarin and L. Sagramora; *Synthesis*, 1976, 326.	123
76S339	M. Furukawa and T. Okawara; *Synthesis*, 1976, 339.	363
76S389	D. Landini; *Synthesis*, 1976, 389.	2999
76S398	A. G. Anderson, Jr., N. E. T. Owen, F. J. Freenor and D. Erickson; *Synthesis*, 1976, 398.	25
76S406	G. Ferdinand and K. Schank; *Synthesis*, 1976, 406.	742
76S462	R. E. Lyle, J. E. Saavedra and G. G. Lyle; *Synthesis*, 1976, 462.	385, 422, 798
76S472	G. Cainelli, F. Manescalchi and M. Panunzio; *Synthesis*, 1976, 472.	19, 25, 31, 35
76S516	M. Cinquini and F. Tundo; *Synthesis*, 1976, 516.	35
76S528	J. O. Morely; *Synthesis*, 1976, 528.	657, 753
76S535	W. Ried and L. Kaiser; *Synthesis*, 1976, 535.	744
76S543	F. A. L'Eplattenier and L. Vuitel; *Synthesis*, 1976, 543.	746
76S545	K. Nyberg; *Synthesis*, 1976, 545.	748
76S621	D. E. Pearson, M. G. Frazer, V. S. Frazer and L. C. Washburn; *Synthesis*, 1976, 621.	626
76S721	D. N. Harpp and K. Steliou; *Synthesis*, 1976, 721.	541
76S726	R. A. Moss and G. M. Love; *Synthesis*, 1976, 726.	773
76S731	R. W. Heintzelman and D. Swern; *Synthesis*, 1976, 731.	163
76S736	E. I. Sanchez and M. J. Fumarola; *Synthesis*, 1976, 736.	28
76S741	H. W. Ohlendorf, R. Stranghoner and E. Winterfeldt; *Synthesis*, 1976, 741.	749
76S746	H. Ahlbrecht and D. Liesching; *Synthesis*, 1976, 746.	812
76S801	T. H. Chan, A. Baldassarre and D. Massuda; *Synthesis*, 1976, 801.	911
76S815	T.-L. Ho, M. Henninger and G. A. Olah; *Synthesis*, 1976, 815.	657, 753

76S829 E. Wenschuh, R. Fahse and R. Hohne; *Synthesis*, 1976, 829. 201

76SC21 P. E. Sonnet; *Synth. Commun.*, 1976, **6**, 21. 10

76SC129 H. A. Selling and H. J. Mak; *Synth. Commun.*, 1976, **6**, 129. 797

76SRI115 F. W. Parrett and M. S. Sun; *Synth. React. Inorg. Metal-Org. Chem.*, 1976, **6**, 115. 448

76T431 Y. Tamura, H. Matsushima, M. Ikeda and K. Sumoto; *Tetrahedron*, 1976, **32**, 431. 401

76T981 C. F. Lane and G. W. Kabalka; *Tetrahedron*, 1976, **32**, 981. 527

76T1267 A. H. Beckett, G. R. Jones and R. T. Coutts; *Tetrahedron*, 1976, **32**, 1267. 390

76T1448 J. Ficini; *Tetrahedron*, 1976, **32**, 1448. 1039

76T1649 A. Koziara and A. Zwierzak; *Tetrahedron*, 1976, **32**, 1649. 412

76T2763 F. Ruff, K. Komoto, N. Furakawa and S. Oae; *Tetrahedron*, 1976, **32**, 2763. 160

76T2979 J. d'Angelo; *Tetrahedron*, 1976, **32**, 2979. 696

76T3063 N. De Kimpe, R. Verhé, L. De Buyck, H. Hasma and N. Schamp; *Tetrahedron*, 1976, **32**, 3063. 749

76TL233 H. G. Richey, Jr., C. McLane and C. J. Phillips; *Tetrahedron Lett.*, 1976, 233. 343, 763

76TL319 T. H. Chan and B. S. Ong; *Tetrahedron Lett.*, 1976, 319. 120

76TL453 J. N. Denis, W. Dumont and A. Krief; *Tetrahedron Lett.*, 1976, 453. 282, 730

76TL835 B. Corbel, J.-P. Paugam, M. Dreux and P. Savignac; *Tetrahedron Lett.*, 1976, 835. 1057

76TL1093 Y. Diab, J. C. Duplan and A. Laurent; *Tetrahedron Lett.*, 1976, 1093. 364

76TL1197 M. Grignon-Dubois, J. Dungoues and R. Calas; *Tetrahedron Lett.*, 1976, 1197. 190

76TL1553 W. T. Brady and R. A. Owens; *Tetrahedron Lett.*, 1976, **17**, 1553. 941

76TL1637 D. G. Miller, S. Trenbeath and C. J. Sih; *Tetrahedron Lett.*, 1976, 1637. 656

76TL2171 M. J. S. Lobo and H. J. C. das Neves; *Tetrahedron Lett.*, 1976, 2171. 208

76TL2173 T. N. Mitchell and B. Kleine; *Tetrahedron Lett.*, 1976, 2173. 814

76TL2201 A. B. Levy and S. T. Schwartz; *Tetrahedron Lett.*, 1976, **26**, 2201. 906

76TL2565 J. C. Lindhardt, G. L. van Mourik and H. J. J. Pabon; *Tetrahedron Lett.*, 1976, 2565. 423

76TL2809 A. Frimer and I. Rosenthal; *Tetrahedron Lett.*, 1976, 2809. 648

76TL3001 D. N. Harpp and A. Granata; *Tetrahedron Lett.*, 1976, 3001. 723

76TL3125 G. Märkl, P. Hofmeister and F. Kneidl; *Tetrahedron Lett.*, 1976, 3125. 879

76TL3645 C. R. Hall and T. D. Inch; *Tetrahedron Lett.*, 1976, 3645. 466

76TL3661 M. Adinolfi, M. Parrilli, G. Barone, G. Laonigro and L. Mangoni; *Tetrahedron Lett.*, 1976, 3661. 45

76TL3969 K. Utimoto, Y. Yabuki, K. Onada and H. Nozaki; *Tetrahedron Lett.*, 1976, 3969. 1088

76TL4037 J. J. Tufariello and J. J. Tegeler; *Tetrahedron Lett.*, 1976, 4037. 350

76TL4041 K. Sachdev; *Tetrahedron Lett.*, 1976, **17**, 4041. 936

76TL4385 A. Pelter, R. Hughes, K. Smith and M. Tabata; *Tetrahedron Lett.*, 1976, 4385. 1086

76TL4439 W. Mychajlowskij and T. H. Chan; *Tetrahedron Lett.*, 1976, **17**, 4439. 931, 932, 935

76TL4637 F. DiFuria, G. Modena and R. Curci; *Tetrahedron Lett.*, 1976, 4637. 146

76TL4839 H. Neumann and D. Seebach; *Tetrahedron Lett.*, 1976, **17**, 4839. 909, 910

76ZAAC(420)74 F. Kober and W. J. Ruhl; *Z. Anorg. Allg. Chem.*, 1976, **420**, 74. 478

76ZAAC(423)40 M. Wieber and U. Baudis; *Z. Anorg. Allg. Chem.*, 1976, **423**, 40. 496, 501

76ZAAC(425)209 V. Krech and K. Issleib; *Z. Anorg. Allg. Chem.*, 1976, **425**, 209. 427

76ZC437 R. Mayer, E. Oestreich and S. Bleisch; *Z. Chem.*, 1976, **16**, 437. 399

76ZN(B)172 E. Fluck and W. Kazenwadel; *Z. Naturforsch., Teil B*, 1976, **31**, 172. 1062

76ZN(B)697 W. Nöth and P. Wolfgardt; *Z. Naturforsch., Teil B*, 1976, **31**, 697. 421

76ZN(B)727 W. Gombler; *Z. Naturforsch., Teil B*, 1976, **31**, 727. 166

76ZN(B)1058 A. Schmidpeter, T. Criegern and K. Blanck; *Z. Naturforsch., Teil B*, 1976, **31**, 1058. 870

76ZN(B)1305 M. Baudler J. Hahn, H. Dietsch and G. Furstenburg; *Z. Naturforsch., Teil B*, 1976, **31**, 1305. 471

76ZOB252 N. G. Feshchenko, T. V. Kovaleva and E. A. Mel'nichuk; *Zh. Obshch. Khim.*, 1976, **46**, 252. 836, 837

76ZOB477 E. E. Nifantev, D. A. Predvoditelev and M. K. Grachev; *Zh. Obshch. Khim.*, 1976, **46**, 477. 410

76ZOB1701 A. A. Shvets, E. G. Amarskii, O. A. Osipov and L. V. Goncharova; *Zh. Obshch. Khim.*, 1976, **46**, 1701. 841

77ACR301 R. D. Rieke; *Acc. Chem. Res.*, 1977, **10**, 301. 558

77ACR434 J. K. Stille and K. S. Y. Lau; *Acc. Chem. Res.*, 1977, **10**, 434. 582, 960

77ACS(B)56 A. K. Gjøystdal and C. Rømming; *Acta Chem. Scand., Ser. B*, 1977, **31**, 56. 1009

77AG(E)10 W. Oppolzer; *Angew. Chem., Int. Ed. Engl.*, 1977, **16**, 10. 354

77AG(E)45 H. J. Bestmann and W. Kloesters; *Angew. Chem., Int. Ed. Engl.*, 1977, **16**, 45. 447, 1064

77AG(E)107 A. Zwierzak and J. B. Piotrowicz; *Angew. Chem. Int. Ed. Engl.*, 1977, **16**, 107. 300, 808

77AG(E)349 H.-J. Bestmann; *Angew. Chem., Int. Ed. Engl.*, 1977, **16**, 349. 1004

77AG(E)488 G. Felix, J. Dunogues, F. Pisciotti and R. Calas; *Angew. Chem., Int. Ed. Engl.*, 1977, **16**, 488. 608

77AG(E)535 W. Walter and H.-W. Lüke; *Angew. Chem., Int. Ed. Engl.*, 1977, **16**, 535. 814

77AG(E)722 H. Schmidbauer, P. Holl and F. H. Koher; *Angew. Chem., Int. Ed. Engl.*, 1977, **16**, 722. 475

77AJC2479 S. C. Auteri, D. W. Cameron and C. B. Drake; *Aust. J. Chem.*, 1977, **30**, 2479. 478

77ANY(295)2 P. L. Pauson and I. U. Khand; *Ann. N. Y. Acad. Sci.*, 1977, **295**, 2. 985

77AX(B)3782 B. M. Gatehouse and A. J. Jozsa; *Acta Crystallogr., Part B*, 1977, **33**, 3782. 1008

77BCJ1513 H. Matsushita, Y. Tsujino, M. Noguchi and S. Yoshikawa; *Bull. Chem. Soc. Japan*, 1977, **50**, 1513. 757

77BCJ3007 H. Ishihara, S. Sato and Y. Hirabayashi; *Bull. Chem. Soc. Jpn.*, 1977, **50**, 3007. 858

77BCJ3069 T. Nakai, K. Tanaka, H. Setoi and N. Ishikawa; *Bull. Chem. Soc. Jpn.*, 1977, **50**, 3069. 1042

77BSF693 A. Alexakis, G. Cahiez, J. F. Normant and J. Villieras; *Bull. Soc. Chim. Fr.*, 1977, 693. 676

77CB852 B.-T. Gröbel and D. Seebach; *Chem. Ber.*, 1977, **110**, 852. 936

77CB867 B.-T. Gröbel and D. Seebach; *Chem. Ber.*, 1977, **110**, 867. 910, 911, 931
77CB1269 P. Jutzi and H. Saleske; *Chem. Ber.*, 1977, **110**, 1269. 436
77CB2382 R. Appel and H. Schöler; *Chem. Ber.*, 1977, **110**, 2382. 835, 837
77CB2480 O. Wolfbeis; *Chem. Ber.*, 1977, **110**, 2480. 787
77CB2783 K. Jonas, H. Nöth and W. Storch; *Chem. Ber.*, 1977, **110**, 2783. 811
77CB3168 H. Burzlaff, E. Wilhelm and H.-J. Bestmann; *Chem. Ber.*, 1977, **110**, 3168. 1004
77CB3874 O. J. Scherer and W. Gläßel; *Chem. Ber.*, 1977, **110**, 3874. 406, 407
77CC76 I. Kuwajima, K. Atsumi and I. Azegami; *J. Chem. Soc., Chem. Commun.*, 1977, 76. 1091
77CC169 E. C. Ashby and B. Goel; *J. Chem. Soc., Chem. Commun.*, 1977, 169. 558
77CC407 S. Oae, D. Fukushima and Y. H. Kim; *J. Chem. Soc., Chem. Commun.*, 1977, 407. 214, 220
77CC522 B. Harirchian and P. Magnus; *J. Chem. Soc., Chem. Commun.*, 1977, 522. 908
77CC534 H. Watanabe, K. Higuchi, M. Kobayashi, M. Hara, Y. Koike, T. Kitahara and Y. Nagai; *J. Chem. Soc., Chem. Commun.*, 1977, 534. 546
77CC683 A. O. King, N. Okukado and E. Negishi; *J. Chem. Soc., Chem. Commun.*, 1977, 683. 1096
77CC723 M. Cinquinni and F. Cozzi; *J. Chem. Soc., Chem. Commun.*, 1977, 723. 362
77CC725 D. L. J. Clive, G. Chittatu, N. J. Curtis, W. A. Kiel and C. K. Wong; *J. Chem. Soc., Chem. Commun.*, 1977, 725. 286
77CC758 J. D. Elliott and J. H Jones; *J. Chem. Soc., Chem. Commun.*, 1977, 758. 299
77CCC1466 K. Dostal, Z. Zak and A. Haba; *Coll. Czech. Chem. Commun.*, 1977, **42**, 1466. 370
77CI(L)538 Y. Ogata and T. Sugimoto; *Chem. Ind. (London)*, 1977, 538. 21
77CJC407 J. S. Grossert and R. F. Langler; *Can. J. Chem.*, 1977, **55**, 407. 178, 179, 180
77CL153 T. Inoue, T. Uchimoto and T. Mukaiyama; *Chem. Lett.*, 1977, 153. 532
77CL383 T. Mukaiyama, S. Shoda and Y. Watanabe; *Chem. Lett.*, 1977, 383. 8
77CL437 K. Hojo, H. Yoshino and T. Mukaiyama; *Chem. Lett.*, 1977, 437. 124, 125
77CL679 M. Isobe, S. Kondo, N. Nagasawa and T. Goto; *Chem. Lett.*, 1977, 679. 702
77CL835 Y. Masuyama, Y. Ueno and M. Okawara; *Chem. Lett.*, 1977, 835. 287
77CL893 S. Oae, D. Fukushima and M. Kim; *Chem. Lett.*, 1977, 893. 214, 220
77CL1013 Y. Echigo, Y. Watanabe and T. Mukaiyama; *Chem. Lett.*, 1977, 1013. 26
77CL1025 K. Takabe, T. Katagiri and J. Tanaka; *Chem. Lett.*, 1977, 1025. 27
77CL1117 K. Isagawa, K. Tatsumi and Y. Otsuji; *Chem. Lett.*, 1977, 1117. 30
77CRV409 T. L. Gilchrist and C. J. Moody; *Chem. Rev.*, 1977, **77**, 409. 159, 401
77CSR1 G. W. Kirby; *Chem. Soc. Rev.*, 1977, **6**, 1. 354
77CZ399 H. J. Breunig, V. Breunig-Lyriti and T. P. Knobloch; *Chem.-Ztg.*, 1977, **101**, 399 505, 506
77DOK(233)591 R. A. Bekker, V. Ya. Popkova and I. L. Knunyants; *Dokl. Akad. Nauk SSSR*, 1977, **233**, 591. 641
77HCA2379 D. Bellus; *Helv. Chim. Acta*, 1977, **60**, 2379. 749
77HOU(5/2a)110 V. Jäger and H. G. Viehe; *Methoden Org. Chem. (Houben-Weyl)*, 1977, **5/2a**, 110. 1011
77IC1357 J. E. Ellis and G. P. Hagen, *Inorg. Chem.*, 1977, **16**, 1357. 572
77IC1460 P. Wisian-Nelson, R. H. Neilson and A. H. Cowley; *Inorg. Chem.*, 1977, **16**, 1460. 861, 862
77IC2938 H. C. Brown and N. Ravindran; *Inorg. Chem.*, 1977, **16**, 2938. 525, 900
77IC2966 T. B. Ranchfuss; *Inorg. Chem.*, 1977, **16**, 2966. 824
77IS183 M. L. Denniston and D. R. Martin; *Inorg. Synth.*, 1977, **17**, 183. 841
77IS186 I. G. W. Luther and G. Beyerle; *Inorg. Synth.*, 1977, **17**, 186. 826, 866
77IZV1437 N. A. Buina, M. A. Giniyatullina, I. A. Nuretdinov and F. G. Sibgatullina; *Izv. Akad. Nauk SSSR, Ser. Khim.*, 1977, 1437. 469
77IZV1456 B. M. Mikhailov, M. E. Gurskii and M. G. Gverdtsiteli; *Izv. Akad. Nauk SSSR, Ser. Khim.*, 1977, 1456. 1027
77IZV2389 L. I. Zakharkin, I. V. Pisareva and T. B. Agakhanova; *Izv. Akad. Nauk SSSR, Ser. Khim.*, 1977, **10**, 2389. 915
77JA638 D. B. Carr and J. Schwartz; *J. Am. Chem. Soc.*, 1977, **99**, 638. 969, 991, 992
77JA905 R. Breslow, R. J. Corcoran, B. B. Snider, R. J. Doll, P. L. Khanna and R. Kaleya; *J. Am. Chem. Soc.*, 1977, **99**, 905. 20
77JA1277 K. K. Ogilvie, S. L. Beaucage, N. Theriault and D. W. Entwistle; *J. Am. Chem. Soc.*, 1977, **99**, 1277. 107
77JA1470 H. K. Spencer, M. V. Lakshmikanthan and M. P. Cava; *J. Am. Chem. Soc.*, 1977, **99**, 1470. 290
77JA1993 P. F. Hudrlik, A. M. Hudrlik, R. J. Rona, R. N. Misra and G. P. Withers; *J. Am. Chem. Soc.*, 1977, **99**, 1993. 623, 682, 747
77JA3168 E. Negishi and D. E. Van Horn; *J. Am. Chem. Soc.*, 1977, **99**, 3168. 969
77JA3179 Y. Okuede, S. Hirano, T. Hiyama and H. Nozaki; *J. Am. Chem. Soc.*, 1977, **99**, 3179. 571
77JA3427 H. C. Brown, R. Liotta and L. Brener; *J. Am. Chem. Soc.*, 1977, **99**, 3427. 517
77JA4405 B. M. Trost and G. S. Massiot; *J. Am. Chem. Soc.*, 1977, **99**, 4405. 724
77JA4836 W. C. Still; *J. Am. Chem. Soc.*, 1977, **99**, 4836. 600
77JA5009 D. A. Evans, L. K. Truesdale, K. G. Grimm and S. L. Nesbitt; *J. Am. Chem. Soc.*, 1977, **99**, 5009. 541
77JA5192 G. Zweifel, S. J. Backlund, T. Leung; *J. Am. Chem. Soc.*, 1977, **99**, 5192. 901
77JA5483 R. L. Funk and K. P. C. Vollhardt; *J. Am. Chem. Soc.*, 1977, **99**, 5483. 944
77JA5514 H. C. Brown and N. M. Yoon; *J. Am. Chem. Soc.*, 1977, **99**, 5514. 516
77JA5526 M. A. Umbreit and K. B. Sharpless; *J. Am. Chem. Soc.*, 1977, **99**, 5526. 86
77JA5810 S. Danishefsky, R. K. Singh and T. Harayama; *J. Am. Chem. Soc.*, 1977, **99**, 5810. 665
77JA6645 J. L. Atwood, G. K. Barker, J. Holton, W. E. Hunter, M. F. Lappert and R. Pearce; *J. Am. Chem. Soc.*, 1977, **99**, 6645. 566
77JA6672 F. M. Laskovics and E. M. Schulman; *J. Am. Chem. Soc.*, 1977, **99**, 6672. 22
77JA7365 R. H. Wollenberg, K. F. Albizati and R. Peries; *J. Am. Chem. Soc.*, 1977, **99**, 7365. 676

77JA7379	K. I. The, Vande Griend, W. A. Whitla and R. G. Cavell; *J. Am. Chem. Soc.*, 1977, **99**, 7379.	476
77JA7613	R. F. Cunico and H. M. Lee; *J. Am. Chem. Soc.*, 1977, **99**, 7613.	921, 922
77JA7743	J. Meinwald, D. Danplaise and J. Clardy; *J. Am. Chem. Soc.*, 1977, **99**, 7743.	238
77JA7876	K. Tani, L. D. Brown, J. Ahmed, J. A. Ibers, M. Yokota, A. Nakamura and S. Otsuka; *J. Am. Chem. Soc.*, 1977, **99**, 7876.	442
77JA8045	M. J. Loots and J. Schwartz; *J. Am. Chem. Soc.*, 1977, **99**, 8045.	969, 992
77JA8356	L. Duhamel and J. M. Poirier; *J. Am. Chem. Soc.*, 1977, **99**, 8356.	759
77JA8426	P. Lennon, A. Rosan and M. Rosenblum; *J. Am. Chem. Soc.*, 1977, **99**, 8426.	577
77JCR(M)3101	H. Schumann and R. Fischer; *J. Chem. Res. (M)*, 1977, 3101.	867
77JCR(S)6	D. J. Ager and I. Fleming; *J. Chem. Res. (S)*, 1977, 6.	539
77JCR(S)105	E. Goffin, Y. Legrand and H. G. Viehe; *J. Chem. Res. (S)*., 1977, 105.	1046
77JCR(S)215	Y. Kobayashi, I. Kumadaki and T. Yoshida; *J. Chem. Res. (S)*, 1977, 215.	631
77JCR(S)237	D. Bartholomew and L. T. Kay; *J. Chem. Res.* (S), 1977, 237.	401
77JCR(S)272	H. Schumann and R. Fischer; *J. Chem. Res. (S)*, 1977, 272.	478
77JCS(D)159	D. H. Brown, R. J. Cross and D. Millington; *J. Chem. Soc., Dalton Trans.*, 1977, 159.	465
77JCS(D)527	C. Glidewell and E. J. Leslie; *J. Chem. Soc., Dalton Trans.*, 1977, 527.	293
77JCS(D)812	E. R. Sigurdson and G. Wilkinson; *J. Chem. Soc., Dalton Trans.*, 1977, 812.	602
77JCS(D)1674	W. McFarlane; *J. Chem. Soc., Dalton Trans.*, 1977, 1674.	466
77JCS(P1)226	R. C. Cambie, W. I. Noall, G. J. Potter, P. S. Rutledge and P. D. Woodgate; *J. Chem. Soc., Perkin Trans. 1*, 1977, 226.	24
77JCS(P1)501	V. Calo, L. Lopez and G. Pesce; *J. Chem. Soc., Perkin Trans. 1*, 1977, 501.	28
77JCS(P1)676	S. Uemura, H. Okazaki, A. Onoe and M. Okano; *J. Chem. Soc., Perkin Trans. 1*, 1977, 676.	617
77JCS(P1)1114	D. H. R. Barton and S. C. Narang; *J. Chem. Soc., Perkin Trans. 1*, 1977, 1114.	88, 104
77JCS(P1)1196	R. Faragher and T. L. Gilchrist; *J. Chem. Soc., Perkin Trans. 1*, 1977, 1196.	779
77JCS(P1)1469	B. F. Bonini, G. Maccagnoni, G. Mazzanti, L. Thiys, H. P. P. M. Ambrosius and B. Zwanenburg; *J. Chem. Soc., Perkin Trans. 1*, 1977, 1469.	209
77JCS(P1)1574	D. W. Jones, D. R. Hill, D. A. Lewton and C. Sheppard; *J. Chem. Soc., Perkin Trans. 1*, 1977, 1574.	153
77JCS(P1)1797	J. Cousseau and L. Gouin; *J. Chem. Soc., Perkin Trans. 1*, 1977, 1797.	606, 616
77JCS(P2)789	D. W. Allen, B. G. Hutley and M. T. J. Mellor; *J. Chem. Soc., Perkin Trans. 2*, 1977, 789.	847, 848
77JCS(P2)1705	D. W. Allen, B. G. Hutley and M. T. J. Mellor; *J. Chem. Soc., Perkin Trans. 2*, 1977, 1705.	844, 846
77JFC(10)375	A. E. Feiring; *J. Fluorine Chem.*, 1977, **10**, 375.	15
77JGU707	Z. S. Novikova, A. A. Parishchemko and I. F. Lutsenko; *J. Gen. Chem. USSR (Engl. Transl.)*, 1977, **47**, 707.	448
77JINC153	K. Dey and T. L. Dey; *J. Inorg. Nucl. Chem.*, 1977, **39**, 153.	574
77JOC563	M. E. Sitzmann, L. A. Kaplan and I. Angres; *J. Org. Chem.*, 1977, **42**, 563.	392
77JOC579	P. Jacob III and H. C. Brown; *J. Org. Chem.*, 1977, **42**, 579.	900
77JOC593	A. D. Dawson and D. Swern; *J. Org. Chem.*, 1977, **42**, 593.	160
77JOC782	L. L. Chang and D. B. Denney; *J. Org. Chem.*, 1977, **42**, 782.	469
77JOC844	E. Keinan and Y. Mazur; *J. Org. Chem.*, 1977, **42**, 844.	397
77JOC875	S. L. Regen; *J. Org. Chem.*, 1977, **42**, 875.	25, 31
77JOC967	F. A. Davis, A. J. Friedman, E. W. Kluger, E. S. Skibo, E. R. Fretz, A. P. Milicia, W. L. LeMasters, M. D. Bentley, J. A. Lacadie and I. B. Douglass; *J. Org. Chem.*, 1977, **42**, 967.	358
77JOC1236	S. O. Grim and A. W. Yankowsky; *J. Org. Chem.*, 1977, **42**, 1236.	822
77JOC1479	N. J. Lewis, S. Y. Gabhe and M. R. DeLaMater; *J. Org. Chem.*, 1977, **42**, 1479.	652
77JOC1559	M. Zupan and A. Pollak; *J. Org. Chem.*, 1977, **42**, 1559.	17
77JOC1773	H. J. Reich and S. K. Shah; *J. Org. Chem.*, 1977, **42**, 1773.	279
77JOC1776	D. G. Garratt and G. H. Schmid; *J. Org. Chem.*, 1977, **42**, 1776.	280
77JOC2031	T. A. Bianchi and L. A. Cate; *J. Org. Chem.*, 1977, **42**, 2031.	183, 717
77JOC2053	T. Cohen, A. G. Dietz, Jr. and J. R. Miser; *J. Org. Chem.*, 1977, **42**, 2053.	647
77JOC2180	M. Yamada, K. Sotoya, T. Sakakibara, T. Takamoto and R. Sudoh; *J. Org. Chem.*, 1977, **42**, 2180.	121
77JOC2274	P. Wiriyachitra and M. P. Cava; *J. Org. Chem.*, 1977, **42**, 2274.	652
77JOC2426	M. P. Doyle, B. Siegfried and J. F. Dellaria, Jr.; *J. Org. Chem.*, 1977, **42**, 2426.	621, 627
77JOC2533	H. C. Brown and N. Ravindran; *J. Org. Chem.*, 1977, **42**, 2533.	525
77JOC2950	S. Raucher; *J. Org. Chem.*, 1977, **42**, 2950.	280
77JOC3173	G. H. Posner, G. M. Gurria and K. A. Babiak; *J. Org. Chem.*, 1977, **42**, 3173.	610
77JOC3233	P. G. Gassman, D. P. Gilbert and S. M. Cole; *J. Org. Chem.*, 1977, **42**, 3233.	138
77JOC3236	P. G. Gassman and R. J. Balchunis; *J. Org. Chem.*, 1977, **42**, 3236.	139, 140
77JOC3243	S. Thaisrivongs and J. D. Wuest; *J. Org. Chem.*, 1977, **42**, 3243.	900
77JOC3761	M. E. Jung and M. A. Lyster; *J. Org. Chem.*, 1977, **42**, 3761.	36, 658
77JOM(127)127	S. B. Awad, D. S. Brown, S. C. Cohen, R. E. Humphries, and A. G. Massey; *J. Organomet. Chem.*, 1977, **127**, 127.	959
77JOM(127)C29	H. Hoberg and R. Krause-Göing; *J. Organomet. Chem.*, 1977, **127**, C29.	989
77JOM(128)155	A. L. Rheingold and P. Choudhury; *J. Organomet. Chem.*, 1977, **128**, 155.	507
77JOM(131)163	H. C. Brown and J. A. Sinclair; *J. Organomet. Chem.*, 1977, **131**, 163.	1084
77JOM(131)43	G. A. Razuvaev, V. N. Latyaeva, V. V. Dobrotenko, A. N. Linyorva, L. I. Vishinskaya and V. K. Cherkasov; *J. Organomet. Chem.*, 1977, **131**, 43.	569
77JOM(135)C10	H. C. Brown, N. M. Yoon and A. K. Mandal; *J. Organomet. Chem.*, 1977, **135**, C10.	515
77JOM(136)173	H. A. Meinema, H. F. Martens and J. G. Noltes; *J. Organomet. Chem.*, 1977, **136**, 173.	491, 505
77JOM(137)23	B. De Jeso and J. C. Pommier; *J. Organomet. Chem.*, 1977, **137**, 23.	816

77JOM(137)293	J. E. Hill and T. A. Nile; *J. Organomet. Chem.*, 1977, **137**, 293.	915
77JOM(137)56	J. Y. Corey; *J. Organomet. Chem.*, 1977, **4**, 56.	811
77JOM(141)249	A. Meller, W. Maringgele and U. Sicker; *J. Organomet. Chem.*, 1977, **141**, 249.	811
77JOM(142)145	S. I. Beilin, S. B. Golstein, B. A. Dolgoplosk, L. Sh. Guzman and E. I. Tinyakova; *J. Organomet. Chem.*, 1977, **142**, 145.	566
77JOM(142)71	F. Sato, S. Sato, H. Kodama and M. Sato; *J. Organomet. Chem.*, 1977, **142**, 71.	23, 596
77JPR195	E. Schmitz, R. Ohme, S. Schramm, H. Striegler, H.-U. Heyne and J. Rusche; *J. Prakt. Chem.*, 1977, **319**, 195.	379
77JPR1036	S. Chorbadjiev, C. Roumian and P. Markov; *J. Prakt. Chem.*, 1977, **319**, 1036.	226
77LA806	T. Wieland, D. Schermer, G. Rohr and H. Faulstich; *Justus Liebigs Ann. Chem.*, 1977, 806.	7
77LA846	D. Seebach, N. Meyer and A. K. Beck; *Justus Liebigs Ann. Chem.*, 1977, 846.	281
77LA1167	U. Schöllkopf, D. Stafforst and R. Jentsch; *Justus Liebigs Ann. Chem.*, 1977, 1167.	909
77LA1309	U. Felcht and M. Regitz; *Justus Liebigs Ann. Chem.*, 1977, 1309.	861
77LA1457	P. Messinger and H. Greve; *Justus Liebigs Ann. Chem.*, 1977, 1457.	188
77LA1766	H. Heydt and M. Regitz; *Justus Liebigs Ann. Chem.*, 1977, 1766.	1068
B-77MI 201-01	G. H. Schmid and D. G. Garratt; in "The Chemistry of Functional Groups," Supplement A, ed. S. Patai, Wiley, New York, 1977, p. 725.	4, 5
B-77MI 202-01	C. A. Buehler and D. E. Pearson; "Survey of Organic Synthesis," Wiley, New York, 1977, vol 2.	38
B-77MI 203-01	A. Ohno and S. Oae, in "Organic Chemistry of Sulfur," ed. S. Oae, Plenum Press, New York, 1977, ch. 4.	116
B-77MI 203-02	W. P. Weber and G. Gokel; "Phase Transfer Catalysis in Organic Synthesis," Springer, 1977, p. 221.	133
B-77MI 203-03	S. Oae and N. Kuneda; in "Organic Chemistry of Sulfur," ed. S. Oae, Plenum Press, New York, 1977, ch. 11	187
B-77MI 203-04	L. Field; in "Organic Chemistry of Sulfur," ed. S. Oae, Plenum Press, New York, 1977.	211
77MI 203-05	J. J. Pastuszak and A. Chimiak; *Rocz. Chem.*, 1977, **51**, 1567 (*Chem. Abstr.*, 1978, **88**, 105 728p).	229
B-77MI 209-01	M. Wieber (ed.); "Gmelin Handbuch der Anorganischen Chemie, Bismut-Organische Verbindungen," Springer-Verlag, Berlin, 1977.	481
B-77MI 210-01	V. Chvalovsky and J. Rathousky; "Organosilicon Compounds," Institute of Chemical Process Fundamentals of the Czechoslovak Academy of Sciences, Prague, 1977, vols. 5 and 6.	535
B-77MI 212-01	J. S. Pizey; in "Synthetic Reagents," ed. J. S. Pizey, Wiley, New York, 1977, vol. 3, p. 227.	608
B-77MI 217-01	M. Wieber; "Gmelin Handbook der Anorganischen Chemie, Bismut-Organische Verbindungen," Springer-Verlag, Berlin, 1977.	872
77OR(25)73	W. S. Wadsworth, Jr.; *Org. React.*, 1977, **25**, 73.	1003
77OS(57)65	M. S. Newman and W. M. Stalick; *Org. Synth.*, 1977, **57**, 65.	1025
77PS(3)9	A. Kucsman and I. Kapvits; *Phosphorus Sulfur*, 1977, **3**, 9.	160
77PS(3)191	S. O. Grim and A. W. Yankowsky; *Phosphorus Sulfur*, 1977, **3**, 191.	824, 825, 826
77PS(3)203	K. Issleib and P. Thorausch; *Phosphorus Sulfur*, 1977, **3**, 203.	431
77RCR548	L. Rasteikiene, D. Greiciute, M. G. Lin'kova and I. L. Knunyants; *Russ. Chem. Rev.*, 1977, **46**, 548.	712
77RCR891	L. I. Belen'kii, Y. B. Vol'kenshtein and I. B. Karmanova; *Russ. Chem. Rev.* (*Engl. Transl.*), 1977, **46**, 891.	13
77RTC160	H. Hommes and L. Brandsma; *Recl. Trav. Chim. Pays-Bas*, 1977, **96**, 160.	422
77RTC168	H. Westmijze, J. Meijer and P. Vermeer; *Recl. Trav. Chim. Pays-Bas*, 1977, **96**, 168.	616, 624
77RTC179	R. S. Sukhai, J. Meijer and L. Brandsma; *Recl. Trav. Chim. Pays-Bas*, 1977, **96**, 179.	1032
77S1	Y. Tamura, J. Minamikawa and M. Ikeda; *Synthesis*, 1977, 1.	357, 379, 384
77S18	C. Blomberg and F. A. Hartog; *Synthesis*, 1977, 18.	73
77S39	S. K. Gupta; *Synthesis*, 1977, 39.	365
77S113	G. Gelbard and S. Colonna; *Synthesis*, 1977, 113.	393
77S179	S. Inokawa, J. Nakatsukasa, M. Horisaki, M. Yamashita, H. Yoshida, T. Ogata and H. Inokawa; *Synthesis*, 1977, 179.	457
77S217	H. Suzuki; *Synthesis*, 1977, 217.	790
77S339	K. Dimroth and W. Tüncher; *Synthesis*, 1977, 339.	403
77S495	H. Ahlbrecht and D. Liesching; *Synthesis*, 1977, 495.	748
77S580	G. V. Roschenthaler and R. Starke; *Synthesis*, 1977, 580.	466
77S628	Clark and P. L. Fuchs; *Synthesis*; 1977, 628.	446
77S634	N. F. Eweiss, A. R. Katritzky, P.-L. Nie and C. A. Ramsden; *Synthesis*, 1977, 634.	11
77S676	T.-L. Ho, B. G. B. Gupta and G. A. Olah; *Synthesis*, 1977, 676.	24, 30
77S677	G. A. Olah, T.-L. Ho, G. K. Surya Prakash and B. G. B. Gupta; *Synthesis*, 1977, 677.	537
77S695	H. C. Brown and N. Ravindran; *Synthesis*, 1977, 695.	525
77S737	L. Soltin; *Synthesis*, 1977, 737.	105
77S786	R. Mornet and L. Gouin; *Synthesis*, 1977, 786.	27
77S791	K. M. Moore and J. Wemple; *Synthesis*, 1977, 791.	148
77S795	M. Uchino, K. Suzuki and M. Sekiya; *Synthesis*, 1977, 795.	271
77S798	R. S. Glass and R. J. Swedo; *Synthesis*, 1977, 798.	250, 362, 366
77S848	B. L. Jensen and D. P. Michaud; *Synthesis*, 1977, 848.	758
77S874	I. Ryu, S. Murai, I. Niwa and N. Sonoda; *Synthesis*, 1977, 874.	282
77SC137	J. W. Heffman and P. G. Harris; *Synth. Commun.*, 1977, **7**, 137.	696
77SC223	R. T. Arnold and S. T. Kulenovic; *Synth. Commun.*, 1977, **7**, 223.	558
77SC363	T.-L. Ho; *Synth. Commun.*, 1977, **7**, 363.	214

77SC365 J. I. G. Cadogan and A. G. Rowley; *Synth. Commun.*, 1977, **7**, 365. 692

77SC475 R. B. Miller and G. McGarvey; *Synth. Commun.*, 1977, **7**, 475. 606

77SRI195 S. S. Tosto and A. M. Olivares; *Syn. React. Inorg. Metal-Org. Chem.*, 1977, **7**, 195. 504

77T2331 H. C. Brown and E.-I. Negishi; *Tetrahedron*, 1977, **33**, 2331. 518

77T2987 K. Uchida, K. Utimoto and H. Nozaki; *Tetrahedron*, 1977, **33**, 2987. 922

77TL159 R. T. Taylor, C. R. Degenhardt, W. P. Melega and L. A. Paquette; *Tetrahedron Lett.*, 1977, **18**, 159. 911, 946

77TL363 S. A. Shackelford, R. R. McGuire and J. L. Pflug; *Tetrahedron Lett.*, 1977, 363. 17

77TL611 G. M. Rubottom and D. S. Krueger; *Tetrahedron Lett.*, 1977, 611. 665

77TL963 D. N. Harpp, B. Friedlander, D. Mullins and S. M. Vines; *Tetrahedron Lett.*, 1977, 963. 363

77TL1345 Y. Nagao, M. Ochiai, K. Kaneko, A. Maeda, K. Watanabe and E. Fujita; *Tetrahedron Lett.*, 1977, 1345. 150

77TL1805 M. Obayashi, K. Utimoto and H. Nozaki; *Tetrahedron Lett.*, 1977, **18**, 1805. 924

77TL1823 H. Westmijze, J. Meijer and P. Vermeer; *Tetrahedron Lett.*, 1977, **18**, 1823. 924, 925

77TL2001 K. Atsumi and I. Kuwajima; *Tetrahedron Lett.*, 1977, **18**, 2001. 921

77TL2235 P. G. Gassman and R. J. Balchunis; *Tetrahedron Lett.*, 1977, 2235. 761

77TL2413 Y. Ueno, T. Inoue and M. Okawara; *Tetrahedron Lett.*, 1977, 2413. 362

77TL2659 M. E. Jung and P. L. Ornstein; *Tetrahedron Lett.*, 1977, 2659. 36

77TL2999 R. M. Magid, O. S. Fruchey and W. L. Johnson; *Tetrahedron Lett.*, 1977, 2999. 8

77TL3317 T. H. Chan, P. W. K. Lau and W. Mychajlowskij; *Tetrahedron Lett.*, 1977, 3317. 624, 630

77TL3469 H. Kapnang, G. Charles, B. L. Sondengam and J. Hentchoya Hémo; *Tetrahedron Lett.*, 1977, 3469. 310

77TL3519 S. H. Korzeniowski and G. W. Gokel; *Tetrahedron Lett.*, 1977, 3519. 632

77TL3545 A. B. Levy, P. Talley and J. A. Dunford; *Tetrahedron Lett.*, 1977, 3545. 607, 617

77TL3589 J. Ficini, S. Falou, A. M. Touzin and J. d'Angelo; *Tetrahedron Lett.*, 1977, 3589. 759

77TL3835 M. Sevrin, W. Dumont and A. Krief; *Tetrahedron Lett.*, 1977, 3835. 728

77TL3909 S. Raucher; *Tetrahedron Lett.*, 1977, 3909. 280, 623

77TL3969 R. Knorr and E. Lattke; *Tetrahedron Lett.*, 1977, 3969. 961

77TL4009 D. G. Farnum, T. Veysoglu, A. M. Carde, B. Duhl-Emswiler, T. A. Pancoast, T. J. Reitz and R. T. Carde; *Tetrahedron Lett.*, 1977, **46**, 4009. 152

77TL4307 E. Schaumann and F.-F. Grabley; *Tetrahedron Lett.*, 1977, **18**, 4307. 941

77TL4475 K. A. M. Walker; *Tetrahedron Lett.*, 1977, **51**, 4475. 135

77TL4651 J. Mulzer, J. Segner and G. Brüntrup; *Tetrahedron Lett.*, 1977, 4651. 677

77ZAAC(431)88 A. Lachowicz and K.-H. Thiele; *Z. Anorg. Allg. Chem.*, 1977, **431**, 88. 569, 570

77ZAAC(431)221 H. W. Roesky and G. Remmens; *Z. Anorg. Allg. Chem.*, 1977, **431**, 221. 862

77ZAAC(432)67 M. Bauldler, B. Carlsohn, B. Kloth and D. Koch; *Z. Anorg. Allg. Chem.*, 1977, **432**, 67. 865

77ZAAC(432)173 O. J. Scherer and G. Wolmershäuser; *Z. Anorg. Allg. Chem.*, 1977, **432**, 173. 361, 399

77ZAAC(436)20 K. Issleib and K. Zimmermann; *Z. Anorg. Allg. Chem.*, 1977, **436**, 20. 440

77ZN(B)584 R. Luckenbach and H. H. Bechtolsheimer; *Z. Naturforsch., Teil B*, 1977, **32**, 584. 853

77ZN(B)721 G. V. Röschenthaler and R. Starke; *Z. Naturforsch., Teil B*, 1977, **32**, 721. 360

77ZN(B)1244 M. Baudler, C. Ponner, C. Grunner, J. Hellmann, M. Schwamborn and B. Kloth; *Z. Naturforsch., Teil B*, 1977, **32**, 1244. 471

77ZN(B)1400 G. Menz and B. Wrackmeyer; *Z. Naturforsch., Teil B*, 1977, **32**, 1400. 903

77ZOB56 A. M. Pinchuk, V. A. Khranovskii, Z. I. Kuplennik and L. P. Filonenko; *Zh. Obshch. Khim.*, 1977, **47**, 56. 408

77ZOB1432 A. K. L. Voznesenskaya, N. A. Razumova, A. A. Petrov and O. G. Sheptienko; *Zh. Obsch. Khim.*, 1977, **47**, 1432. 477

77ZOR727 Yu. G. Shermolovich, V. V. Vasilev and L. N. Markovskii; *Zh. Org. Khim.*, 1977, **13**, 727. 399

78ACS(B)625 C. L. Pedersen, C. Lohse and M. Poliakoff; *Acta Chem. Scand., Sect. B*, 1978, **32**, 625. 792

78ACS(B)646 R. Carlson; *Acta Chem. Scand., Ser. B*, 1978, **32**, 646. 759

78ACS(B)(32)646 R. Carlson; *Acta Chem. Scand. Ser. B.*, 1978, **32**, 646. 22, 28

78AG(E)27 R. C. Larock; *Angew. Chem., Int. Ed. Engl.*, 1978, **17**, 27. 592

78AG(E)135 G. Thiele, G. Zoubek, H. A. Lindner and J. Ellermann; *Angew. Chem., Int. Ed. Engl.*, 1978, **17**, 135. 507

78AG(E)202 R. Bauer and H. Wendt; *Angew. Chem., Int. Ed. Engl.*, 1978, **17**, 202. 403, 420

78AG(E)203 E. Stöldt and R. Kreher; *Angew. Chem., Int. Ed. Engl.*, 1978, **17**, 203. 388

78AG(E)278 M. T. Reetz and F. Eibach; *Angew. Chem., Int. Ed. Engl.*, 1978, **17**, 278. 945

78AG(E)279 H. Matsumoto, S. Nagashima, T. Kato and Y. Nagai; *Angew. Chem., Int. Ed. Engl.*, 1978, **17**, 279. 912

78AG(E)370 R. Bauer and H. Wendt; *Angew. Chem., Int. Ed. Engl.*, 1978, **17**, 370. 403

78AG(E)526 M. Sevrin, J. N. Denis and A. Krief; *Angew. Chem., Int. Ed. Engl.*, 1978, **17**, 526. 908

78AG(E)530 R. Mews; *Angew. Chem., Int. Ed. Engl.*, 1978, **17**, 530. 402

78AHC(22)71 J. A. Zoltewicz and L. W. Deady; *Adv. Heterocycl. Chem.*, 1978, **22**, 71. 298

78AJC2317 D. St. C. Black and J. E. Doyle; *Aust. J. Chem.*, 1978, **31**, 2317. 346

78AX(B)828 A. Fitzgerald and L. H. Jensen; *Acta Crystallogr., Part B*, 1978, **34**, 828. 1003

78BCJ1459 T. Tanabe and T. Nagai; *Bull. Chem. Soc. Jpn.*, 1978, **51**, 1459. 207

78BCJ2183 I. Kuwajima and T. Abe; *Bull. Chem. Soc. Jpn.*, 1978, **51**, 2183. 265

78BRP1517876 J. V. Kandul, V. Ya. Semenyi, M. P. Pivovarov, V. V. Malovik, I. K. Mazepa and N. G. Feshchenko; *Br. Pat.* 1 517 876 (1978) (*Chem. Abstr.*, 1978, **90**, 138 015). 453

78BSF(2)575 J. Berlan, K. Koosha and J. P. Battioni; *Bull. Soc. Chim. Fr. (II)*, 1978, 575. 587

78BSF(2)602 A. Le Berre and C. Porte; *Bull. Soc. Chim. Fr.*, 1978, **2**, 602. 381

78CAR(61)511	A. K. M. Anisuzzaman and R. L. Whistler; *Carbohydr. Res.*, 1978, **61**, 511.	7
78CAR(63)1	V. G. Gibb and L. D. Hall; *Carbohydr. Res.*, 1978, **63**, C1.	598
78CB13	D. Hellwinkel and W. Krapp; *Chem. Ber.*, 1978, **111**, 13.	831, 840, 843
78CB629	K. Diemert, P. Haas and W. Kuchen; *Chem. Ber.*, 1978, **111**, 629.	854
78CB692	T. Severin and I. Ipach; *Chem. Ber.*, 1978, **111**, 692.	789
78CB890	M. Volkholz, O. Stelzer and R. Schmutzler; *Chem. Ber.*, 1978, **111**, 890.	471
78CB2445	E. Hammerschmidt, W. Bieber and F. Vögtle; *Chem. Ber.*, 1978, 2445.	132
78CB2702	H. Schmidbaur and G. Haslberger; *Chem. Ber.*, 1978, **111**, 2702.	488
78CB2785	H. Neumann and D. Seebach; *Chem. Ber.*, 1978, **111**, 2785.	909, 910
78CB3159	A. Schoberl, M. Rimpler and E. Clauss; *Chem. Ber.*, 1978, **103**, 3159.	214
78CB3838	M. Baudler, M. Warnan and D. Koch; *Chem. Ber.*, 1978, **111**, 3838.	472
78CC30	E. C. Ashby and T. Smith; *J. Chem. Soc., Chem. Commun.*, 1978, 30.	561
78CC118	A. A. Jaxa-Channiec, P. G. Sammes and P. D. Kennewell; *J. Chem. Soc., Chem. Commun.*, 1978, 188.	664
78CC162	N. J. P. Broom and P. G. Sammes; *J. Chem. Soc., Chem. Commun.*, 1978, 162.	664
78CC205	P. Yates and Y. C. Toong; *J. Chem. Soc., Chem. Commun.*, 1978, 205.	229
78CC242	R. D. Gleim, S. Trenbeath, F. Suzuki and C. J. Sih; *J. Chem. Soc., Chem. Commun.*, 1978, 242.	651
78CC344	A. J. Deeming and I. P. Rothwell; *J. Chem. Soc., Chem. Commun.*, 1978, 344.	582
78CC351	F. D. King, S. Pike and D. R. M. Walton; *J. Chem. Soc., Chem. Commun.*, 1978, 351.	1055
78CC367	Y. Tamura, M. Kagotami and Z. Yoshida; *J. Chem. Soc., Chem. Commun.*, 1978, 367.	157
78CC441	D. L. J. Clive, G. Chittatu and C. K. Wong; *J. Chem. Soc., Chem. Commun.*, 1978, 441.	286
78CC446	S. Miyano, Y. Izumi and H. Hashimoto; *J. Chem. Soc., Chem. Commun.*, 1978, 446.	618
78CC694	D. Baudry, M. Ephritikhine and H. Felkin; *J. Chem. Soc., Chem. Commun.*, 1978, 694.	675, 676
78CC742	S. Juliá, A. Ginebreda and J. Guixer; *J. Chem. Soc., Chem. Commun.*, 1978, 742.	299
78CC870	G. M. Blackburn and D. Ingleson; *J. Chem. Soc., Chem. Commun.*, 1978, 870.	847
78CC874	T. Morita, Y. Okamoto and H. Sakurai; *J. Chem. Soc., Chem. Commun.*, 1978, 874.	36
78CC927	D. Braga, A. Ripamonti, D. Savoia, C. Trombini, and A. Umani-Ronchi; *J. Chem. Soc., Chem. Commun.*, 1978, 927.	959
78CI(L)56	W. Prescott; *Chem. Ind. (London)*, 1978, 56.	614
78CJC1183	T. S. Cameron and R. E. Cordes; *Can. J. Chem.*, 1978, **56**, 1183.	249
78CL57	H. Masada and Y. Oishi; *Chem. Lett.*, 1978, 57.	684
78CL223	H. Matsumoto, T. Nakano, M. Kato and Y. Nagai; *Chem. Lett.*, 1978, 223.	21
78CL591	T. Saito and S. Motoki; *Chem. Lett.*, 1978, 591.	244
78CL665	Y. Masuda, A. Arase and A. Suzuki; *Chem. Lett.*, 1978, 665.	623
78CL833	F. Sato, Y. Mori and M. Sato; *Chem. Lett.*, 1978, 833.	4, 23
78CL923	H.-J. Liu, W. H. Chan and S. P. Lee; *Chem. Lett.*, 1978, 923.	25
78CL1057	T. Mukaiyama, T. Tsuji and Y. Watanabe; *Chem Lett.*, 1978, 1057.	344
78CPB1576	O. Nishimura, C. Kitada and M. Fujino; *Chem. Pharm. Bull.*, 1978, **26**, 1576.	274
78CPB1837	M. Uchino, K. Suzuki and M. Sekiya; *Chem. Pharm. Bull.*, 1978, **26**, 1837.	271
78CRV317	P. P. Fu and R. G. Harvey; *Chem. Rev.*, 1978, **78**, 317.	659
78CRV491	P. R. Jones and P. J. Desio; *Chem. Rev.*, 1978, **78**, 491.	67, 69
78CZ263	H. J. Breunig and W. Kanig; *Chem.-Ztg.*, 1978, **102**, 263.	507
78HCA1609	U. Lienhard, H.-P. Fahrni and M. Neuenschwander; *Helv. Chim. Acta*, 1978, **61**, 1609.	1049
78HCA1622	D. Seebach, R. Dach, D. Enders, B. Renger, M. Jansen and G. Bractel; *Helv. Chim. Acta*, 1978, **61**, 1622.	388
78HCA1708	A. Marxer and T. Leutert; *Helv. Chim. Acta*, 1978, **61**, 1708.	445, 445
78HCA2190	J. Boger, R. J. Corcoran and J. Lehn; *Helv. Chim. Acta*, 1978, **61**, 2190.	860
78HCA3100	D. Seebach, W. Wykypiel, W. Lubosch and H.-O. Kalinowski; *Helv. Chim. Acta*, 1978, **61**, 3100.	1002
78HOU(13)45	S. Samaan; *Methoden Org. Chem. (Houben-Weyl)*, 1978, **13**, 45.	872
78HOU(13)1	S. Samaan; *Methoden Org. Chem. (Houben-Weyl)*, 1978, **13**, 1.	872
78HOU(13)167	S. Samaan; *Methoden Org. Chem. (Houben-Weyl)*, 1978, **13**, 167.	885
78HOU(13)206	S. Samaan; *Methoden Org. Chem. (Houben-Weyl)*, 1978, **13**, 206.	890
78HOU(13)284	S. Samaan; *Methoden Org. Chem. (Houben-Weyl)*, 1978, **13**, 284.	891
78HOU(13)295	S. Samaan; *Methoden Org. Chem. (Houben-Weyl)*, 1978, **13**, 295.	884
78HOU(13)376	S. Samaan; *Methoden Org. Chem. (Houben-Weyl)*, 1978, **13**, 376.	885
78HOU(13)465	S. Samaan; *Methoden Org. Chem. (Houben-Weyl)*, 1978, **13**, 465.	879
78HOU(13/6)181	G. Bähr and S. Pawlenko; *Methoden Org. Chem. (Houben-Weyl)*, 1978, **13/6**, 181.	599
78HOU(13/8)220	S. Samaan; *Methoden Org. Chem. (Houben-Weyl)*, 1978, **13/8**, 220.	408
78HOU(13/8)33	S. Samaan; *Methoden Org. Chem. (Houben-Weyl)*, 1978, **13/8**, 33.	414, 1076, 1080
78HOU((13)/8)1	S. Samaan; *Methoden Org. Chem. (Houben-Weyl)*, 1978, **13/8**, 1.	481
78IC880	M. K. Cooper and J. M. Downes; *Inorg. Chem.*, 1978, **17**, 880.	828
78IC1862	E. C. Ashby and A. B. Goel; *Inorg. Chem.*, 1978, **17**, 1862.	421
78ICA163	R. Choukroun and D. Gervais; *Inorg. Chim. Acta*, 1978, **27**, 163.	421
78IJ60	R. H. Herse; *Isr. J. Chem.*, 1978, **17**, 60.	365
78IJ71	R. Filler; *Isr. J. Chem.*, 1978, **17**, 71.	15, 17, 18
78IS169	V. D. Bianco and S. Doronzo; *Inorg. Synth.*, 1978, **18**, 169.	435
78IS170	V. D. Bianco and S. Doronzo; *Inorg. Synth.*, 1978, **18**, 170.	824
78IS171	V. D. Bianco and S. Doronzo; *Inorg. Synth.*, 1978, **18**, 171.	824, 825
78IS172	V. D. Bianco and S. Doronzo; *Inorg. Synth.*, 1978, **18**, 172.	824
78IZV493	R. A. Bekker, V. Ya. Popkova and I. L. Knunyants; *Izv. Akad. Nauk SSSR, Ser. Khim.*, 1978, 493 (Plenum Engl. Transl. p. 430).	641

78IZV1193 R. A. Bekker, V. Ya. Popkova and I. L. Knunyants; *Izv. Akad. Nauk SSSR, Ser. Khim.*, 1978, 1193 (Plenum Engl. Transl. p. 1038). 641

78IZV2775 R. A. Bekker and V. Ya. Popkova; *Izv. Akad. Nauk SSSR, Ser. Khim.*, 1978, 2775. 641

78JA290 K. Tamao, J. Yoshida, M. Takahashi, H. Yamamoto, T. Kakui, H. Matsumoto, A. Kurita and M. Kumada; *J. Am. Chem. Soc.*, 1978, **100**, 290. 4

78JA1222 D. N. Harpp, K. Steliou and T. H. Chan; *J. Am. Chem. Soc.*, 1978, **100**, 1222. 236

78JA1637 J. F. King, S. M. Loosmore, J. D. Lock and M. Aslam; *J. Am. Chem. Soc.*, 1978, **100**, 1637. 205

78JA2252 D. E. Van Horn and E. Negishi; *J. Am. Chem. Soc.*, 1978, **100**, 2252. 925, 974, 975

78JA2254 E. Negishi, N. Okukado, A. O. King, D. E. Van Horn, and B. I. Spiegel; *J. Am. Chem. Soc.*, 1978, **100**, 2254. 925, 969, 976

78JA3258 M. Schlosser and G. Rauchschwalbe; *J. Am. Chem. Soc.*, 1978, **100**, 3258. 529

78JA3611 F. N. Tebbe, G. W. Parshall and G. S. Reddy; *J. Am. Chem. Soc.*, 1978, **100**, 3611. 680

78JA3622 R. E. Penn, E. Block and L. Revelle; *J. Am. Chem. Soc.*, 1978, **100**, 3622. 184

78JA4618 E. J. Corey, E. J. Trybulski, L. S. Melvin, Jr., K. C. Nicolaou, J. A. Secrist, R. Lett, P. W. Sheldrake, J. R. Falck, D. J. Brunelle, S. Kim and S. Yoo; *J. Am. Chem. Soc.*, 1978, **100**, 4618. 629

78JA4624 B. B. Snider, M. Karras and R. S. E. Conn; *J. Am. Chem. Soc.*, 1978, **100**, 4624. 924

78JA4834 J. San Filippo, Jr., J. Silbermann and P. J. Fagan; *J. Am. Chem. Soc.*, 1978, **100**, 4834. 600

78JA4919 P. A. Risbood and D. M. Ruthven; *J. Am. Chem. Soc.*, 1978, **100**, 4919. 30

78JA5130 Y. Segall and I. Granoth; *J. Am. Chem. Soc.*, 1978, **100**, 5130. 824

78JA6294 E. J. Corey and H. Estreicher; *J. Am. Chem. Soc.*, 1978, **100**, 6294. 393, 396, 783

78JA7304 W. E. McEwen, A. B. Jones, J. W. Knapczyk, V. L. Kyllingstad, W.-I. Shian, S. Shore and J. H. Smith; *J. Am. Chem. Soc.*, 1978, **100**, 7304. 824, 834, 835

78JA7565 C. P. Casey and S. W. Polechnowski; *J. Am. Chem. Soc.*, 1978, **100**, 7565. 572

78JA7960 R. A. Johnson, E. G. Nidy and M. V. Meritt; *J. Am. Chem. Soc.*, 1978, **100**, 7960. 98

78JCS(D)657 B. G. Gowenlock, W. E. Lindswell and B. Singh; *J. Chem. Soc., Dalton Trans.*, 1978, 657. 563

78JCS(P1)126 R. C. Cambie, R. C. Hayward, J. L. Jurlina, P. S. Rutledge and P. D. Woodgate; *J. Chem. Soc., Perkin Trans. 1*, 1978, 126. 34

78JCS(P1)1088 C. T. Clark, J. D. Elliott and J. H. Jones; *J. Chem. Soc., Perkin Trans. 1*, 1978, 1088. 299

78JCS(P1)1117 R. M. Acheson, P. G. Hunt, D. M. Littlewood, B. A. Murrer and H. E. Rosenberg; *J. Chem. Soc., Perkin Trans. 1*, 1978, 1117. 764

78JCS(P1)1273 A. Toshimitsu, Y. Kozawa, S. Uemura and M. Okano; *J. Chem. Soc., Perkin Trans. 1*, 1978, 1273. 282

78JCS(P1)1580 G. R. Chalkley, D. J. Snodin, G. Stevens and M. C. Whiting; *J. Chem. Soc., Perkin Trans. 1*, 1978, 1580. 402

78JCS(P1)1682 D. H. R. Barton, M. R. Britten-Kelly and D. Ferreira; *J. Chem. Soc., Perkin Trans. 1*, 1978, 1682. 284

78JGU207 S. V. Ponomarev, A. A. Stepanov, V. N. Sergeev and I. P. Lutsenko; *J. Gen. Chem. USSR* (*Engl. Transl.*), 1978, **48**, 207. 471

78JGU329 N. C. Feshchenko and E. A. Mel'nichuk; *J. Gen. Chem. USSR* (*Engl. Transl.*), 1978, **48**, 329. 449

78JGU1088 M. M. Kabachnik, A. H. Prischenko, Z. S. Novikova and I. P. Lutsenko; *J. Gen. Chem. USSR* (*Engl. Transl.*), 1978, **48**, 1088. 476

78JMC704 W. S. Mungall, L. J. Lemmer, K. L. Lemmer, J. K. Dethmer and L. L. Norlung; *J. Med. Chem.*, 1978, **21**, 704. 366

78JOC147 A. R. Chamberlin, J. E. Stemke and F. T. Bond; *J. Org. Chem.*, 1978, **43**, 147. 911

78JOC178 F. M. Hauser and R. P. Rhee; *J. Org. Chem.*, 1978, **43**, 178. 664

78JOC199 Y. Sato, Y. Kobayashi, M. Sugiura and H. Shirai; *J. Org. Chem.*, 1978, **43**, 199. 1045

78JOC358 A. O. King and E. Negishi; *J. Org. Chem.*, 1978, **43**, 358. 1096

78JOC376 R. P. Woodbury, N. R. Long and M. W. Rathke; *J. Org. Chem.*, 1978, **43**, 376. 941, 1023

78JOC379 S. Danishefsky, R. K. Singh and R. B. Gammill; *J. Org. Chem.*, 1978, **43**, 379. 667

78JOC473 M. Mikolajczyk, W. Midura, S. Grzejszczak, A. Zatorski and A. Chefczynska; *J. Org. Chem.*, 1978, **43**, 473. 713

78JOC497 A. I. Fetell and H. Feuer; *J. Org. Chem.*, 1978, **43**, 497. 788

78JOC652 H. Minato, K. Okuma and M. Kobayashi; *J. Org. Chem.*, 1978, **43**, 652. 360

78JOC700 H. O. House, W. V. Phillips, T. S. B. Sayer and C. C. Yau; *J. Org. Chem.*, 1978, **43**, 700. 420

78JOC731 H. M. Walborsky and Ronman; *J. Org. Chem.*, 1978, **43**, 731. 942, 949

78JOC750 H. G. Kuivila and G. H. Lein, Jr.; *J. Org. Chem.*, 1978, **43**, 750. 600

78JOC757 E. C. Ashby, J. J. Lin, and A. B. Goel; *J. Org. Chem.*, 1978, **43**, 757. 969

78JOC865 G. A. Olah and R. Ohnishi; *J. Org. Chem.*, 1978, **43**, 865. 650

78JOC947 G. R. Newkome and D. C. Hager; *J. Org. Chem.*, 1978, **43**, 947. 827

78JOC1058 H. C. Brown, R. Liotta and G. W. Kramer; *J. Org. Chem.*, 1978, **43**, 1058. 517

78JOC1064 C. G. Screttas and M. Micha-Screttas; *J. Org. Chem.*, 1978, **43**, 1064. 551

78JOC1238 A. I. Fetell and H. Feuer; *J. Org. Chem.*, 1978, **43**, 1238. 788

78JOC1244 J. A. Landgrebe and H. Iranmanesh; *J. Org. Chem.*, 1978, **43**, 1244. 675

78JOC1270 P. K. Chakravarty and R. K. Olson; *J. Org. Chem.*, 1978, **43**, 1270. 223

78JOC1435 J.-L. Grandmaison and P. Brassard; *J. Org. Chem.*, 1978, **43**, 1435. 667

78JOC1460 J. A. Kloek and K. L. Leschinsky; *J. Org. Chem.*, 1978, **43**, 1460. 745

78JOC1620 W. E. Fristad, T. R. Bailey and L. A. Paquette; *J. Org. Chem.*, 1978, **43**, 1620. 911

78JOC1689 T. Hori and K. B. Sharpless; *J. Org. Chem.*, 1978, **43**, 1689. 281

78JOC1697 H. J. Reich, S. Wollowitz, J. E. Trend, F. Chow and D. F. Wendelborn; *J. Org. Chem.*, 1978, **43**, 1697. 281

78JOC1867 J. F. Brunnett and R. P. Traber; *J. Org. Chem.*, 1978, **43**, 1867. 850

78JOC2073 R. A. Olofson, J. Cuomo and B. A. Bauman; *J. Org. Chem.*, 1978, **43**, 2073. 675

78JOC2289 A. G. Hortmann, J.-Y Koo and C.-C. Yu; *J. Org. Chem.*, 1978, **43**, 2289. 344

78JOC2291 R. S. Glass and R. J. Swedo; *J. Org. Chem.*, 1978, **43**, 2291. 11

78JOC2639 M. De Rosa and J. L. T. Alonso; *J. Org. Chem.*, 1978, **43**, 2639. 761

78JOC2662 P. S. Bailey, D. A. Lerdal and T. P. Carter Jr.; *J. Org. Chem.*, 1978, **43**, 2662. 348

78JOC2717 K. T. Lin and Y. C. Tong; *J. Org. Chem.*, 1978, **43**, 2717. 149

78JOC2735 A. P. Kozikowski and A. Ames; *J. Org. Chem.*, 1978, **43**, 2735. 278

78JOC2739 G. Zweifel and W. Lewis; *J. Org. Chem.*, 1978, **43**, 2739. 922, 931

78JOC2793 G. E. Heasley, J. M. Bundy, V. L. Heasley, S. Arnold, A. Gipe, D. McKee, R. Orr, S. L. Rodgers and D. F. Shellhamer; *J. Org. Chem.*, 1978, **43**, 2793. 30

78JOC2833 R. H. Smithers; *J. Org. Chem.*, 1978, **43**, 2833. 624

78JOC2921 R. D. Little and M. G. Venegas; *J. Org. Chem.*, 1978, **43**, 2921. 403

78JOC3076 M. J. Tremmling, S. P. Hopper and P. L. Mendelowitz; *J. Org. Chem.*, 1978, **43**, 3076 622

78JOC3078 L. L. Miller and R. F. Stewart; *J. Org. Chem.*, 1978, **43**, 3078. 657

78JOC3223 S. Torii, H. Tanaka and M. Ukida; *J. Org. Chem.*, 1978, **43**, 3223. 243, 359

78JOC3245 A. I. Meyers and D. R. Williams; *J. Org. Chem.*, 1978, **43**, 3245. 816

78JOC3345 M. Seno, T. Namba and H. Kise; *J. Org. Chem.*, 1978, **43**, 3345. 298

78JOC3374 A. G. Hartmann, A. J. Aron and A. K. Bhattacharya; *J. Org. Chem.*, 1978, **43**, 3374. 232

78JOC3408 B. D. Dombek; *J. Org. Chem.*, 1978, **43**, 3408. 439

78JOC3418 E. J. Corey, D. Floyd and B. H. Lipshutz; *J. Org. Chem.*, 1978, **43**, 3418. 587, 1095

78JOC3481 D. N. Harpp, B. T. Friedlander and C. Larsne; *J. Org. Chem.*, 1978, **43**, 3481. 228

78JOC3508 L. D. Quin and L. B. Littlefield; *J. Org. Chem.*, 1978, **43**, 3508. 435

78JOC3687 R. T. Arnold and S. T. Kulenovic; *J. Org. Chem.*, 1978, **43**, 3687. 22, 29, 34

78JOC3794 S. Raucher and G. A. Koolpe; *J. Org. Chem.*, 1978, **43**, 3794. 423, 908

78JOC4120 M. P. Doyle, R. J. Bosch and P. G. Seites; *J. Org. Chem.*, 1978, **43**, 4120. 11

78JOC4131 G. H. Posner and P. W. Tang; *J. Org. Chem.*, 1978, **43**, 4131. 713

78JOC4252 S. Raucher and G. A. Koolpe; *J. Org. Chem.*, 1978, **43**, 4252. 281

78JOC4395 H. C. Brown, J. R. Schweir and B. Singaram; *J. Org. Chem.*, 1978, **43**, 4395. 516, 523

78JOC4424 R. B. Miller and G. McGarvey; *J. Org. Chem.*, 1978, **43**, 4424. 607, 914, 921, 922

78JOC4503 G. A. Olah, B. G. B. Gupta and S. C. Narang; *J. Org. Chem.*, 1978, **43**, 4503. 142

78JOC4537 S. L. Huang and D. Swern; *J. Org. Chem.*, 1978, **43**, 4537. 162

78JOC4659 G. H. Daub and T. W. Whale; *J. Org. Chem.*, 1978, **43**, 4659. 250

78JOC4672 D. B. Denney, R. Melis and A. D. Pendse; *J. Org. Chem.*, 1978, **43**, 4672. 683

78JOC4885 S. Raucher, M. R. Hansen and M. A. Colter; *J. Org. Chem.*, 1978, **43**, 4885. 280

78JOM(145)17 H. O. Berger, H. Noth and B. Wrackmeyer; *J. Organomet. Chem.*, 1978, **145**, 17. 1048

78JOM(149)315 A. Otero and P. Royo; *J. Organomet. Chem.*, 1978, **149**, 315. 881

78JOM(149)C7 D. Jore, D. Guillerm and W. Chodkiweiecz; *J. Organomet. Chem.*, 1978, **149**, C7. 297

78JOM(152)39 G. Srivastava; *J. Organomet. Chem.*, 1978, **152**, 39. 920

78JOM(152)C25 A. R. Bassindale, A. G. Brook, P. F. Jones and J. A. G. Stewart; *J. Organomet. Chem.*, 1978, **152**, C25. 813

78JOM(153)C37 A. J. De Koning, P. E. Van Rijn and G. J. M. Van der Kerk; *J. Organomet. Chem.*, 1978, **153**, C37. 1096

78JOM(154)13 A. Otero and P. Royo; *J. Organomet. Chem.*, 1978, **154**, 13. 881

78JOM(156)213 J. C. Chen; *J. Organomet. Chem.*, 1978, **156**, 213. 517

78JOM(156)221 M. V. Bhatt; *J. Organomet. Chem.*, 1978, **156**, 221. 33

78JOM(156)C20 D. E. van Horn, L. F. Valente, M. J. Idacavage and E. Negishi; *J. Organomet. Chem.*, 1978, **156**, C20. 919, 925

78JOM(159)373 W. Fries, W. Schwarz, H.-D. Hausen and J. Weidlein; *J. Organomet. Chem.*, 1978, **159**, 373. 1098

78JOM(159)C13 J. J. Eisch, R. J. Manfre and D. A. Komar; *J. Organomet. Chem.*, 1978, **159**, C13. 925

78JOM(162)289 M. N. Bochkarev, N. I. Gur'ev and G. A. Razuvaev; *J. Organomet. Chem.*, 1978, **162**, 289. 492, 509

78JOU1101 I. A. Korbukh, F. F. Blanko, I. R. Kovel'man and M. N. Preobrazhenskaya; *J. Org. Chem. USSR* (*Engl. Transl.*), 1978, **14**, 1101. 812

B-78MI 202-01 D. C. Bradley, R. C. Mehrotra and D. P. Gaur; "Metal Alkoxides," Academic Press, London, 1978. 112

B-78MI 203-01 E. Block; "Reactions of Organosulfur Compounds," Academic Press, New York, 1978. 196

B-78MI 204-01 H. J. Reich; in "Oxidation in Organic Chemistry, Part C," ed. W. S. Trahanovsky, Academic Press, New York, 1978, pp. 1–130. 280

78MI 213-01 G. Lavielle and D. Reisdorf; *C. R. Acad. Sc. Paris*, 1971, **272**, C100. 680

78MI 213-02 G. Paulson, J. Bakke, J. Giddings and M. Simpson; *Biomed. Mass Spectrom.*, 1978, **5**, 128. 690

B-78MI 219-01 G. Zweifel; in "Aspects of Mechanism and Organometallic Chemistry," ed. J. H. Brewster, Plenum Press, New York, 1978, p. 229. 968

B-78MI 221-01 D. A. Ben-Efraim; in "The Chemistry of the Carbon–Carbon Triple Bond," ed. S. Patai, Wiley-Interscience, New York, 1978, vol. 2, chap. 18, p. 755. 1021, 1025, 1027

B-78MI 221-02 J. I. Dickstein and S. I. Miller; in "Chemistry of the Carbon–Carbon Triple Bond," ed. S. Patai, Wiley, London, 1978, vol. 2, chap. 19, pp. 813–955. 1011

78OPP181 M. Hudlicky; *Org. Prep. Proced. Int.*, 1978, **10**, 181. 2

78OS(58)152 R. K. Boeckman, Jr., D. M. Blum, B. Ganem and N. Halvey; *Org. Synth.*, 1978, **58**, 152. 931

78OS(58)138 D. N. Harpp and R. A. Smith; *Org. Synth.*, 1978, **58**, 138. 143

78OS(58)147 M. E. Alonso and M. Aragona; *Org. Synth.*, 1978, **58**, 147. 220

78PS(4)137 K. Issleib and P. Thoransch; *Phosphorus Sulfur*, 1978, **4**, 137. 472

78RCR1084 G. I. Koldobskii, V. A. Ostrovskii and B. V. Gidaspov; *Russ. Chem. Rev.* (*Engl. Transl.*), 1978, **47**, 1084. 319

78RTC69	R. H. Everhardus, R. Gräfing and L. Brandsma; *Recl. Trav. Chim. Pays-Bas*, 1978, **97**, 69.	907
78S23	L. B. Din, J. M. Lindley and O. Meth-Cohn; *Synthesis*, 1978, 23.	753
78S50	L. W. Christensen and D. J. Heacock; *Synthesis*, 1978, 50.	212
78S137	G. A. Olah, B. G. B. Gupta and S. C. Narang; *Synthesis*, 1978, 137.	142
78S138	J. Drabowicz and M. Mikolajczyk; *Synthesis*, 1978, 138.	212
78S146	H. C. Brown and A. K. Mandal; *Synthesis*, 1978, 146.	515
78S217	L. Eckes and M. Hanack; *Synthesis*, 1978, 217.	18, 610
78S359	H. Everhardus and L. Brandsma; *Synthesis*, 1978, 359.	271
78S452	G. A. Olah, S. C. Narang, R. L. Rearson and C. A. Cupas; *Synthesis*, 1978, 452.	104
78S465	L. M. Rossi and P. Trimarco; *Synthesis*, 1978, 465.	771
78S588	M. E. Jung, M. A. Mazurek and R. M. Lim; *Synthesis*, 1978, 588.	444
78S592	T. Severin and I. Ipach; *Synthesis*, 1978, 592.	790
78S676	J. L. Hubbard and H. C. Brown; *Synthesis*, 1978, 676.	522
78S758	J. Drabowicz and M. Mikolajczyk; *Synthesis*, 1978, 758.	145
78S771	D. Landini, F. Montanari and F. Rolla; *Synthesis*, 1978, 771.	32
78S882	A. Jonczyk, Z. Ochal and M. Makosza; *Synthesis*, 1978, 882.	308
78S897	H. Ahlbrecht and K. Pfaff; *Synthesis*, 1978, 897.	748
78SC103	J. Tsuji and H. Yasuda; *Synth. Commun.*, 1978, **8**, 103.	616
78T3	M. Schlosser; *Tetrahedron*, 1978, **34**, 3.	15
78T1049	D. L. J. Clive; *Tetrahedron*, 1978, **34**, 1049.	280, 286, 287, 728
78T1231	H. Kropf and H. M. Amirabadi; *Tetrahedron*, 1978, **34**, 1231.	687
78T1251	C. R. Duffner and F. Kurzer; *Tetrahedron*, 1978, **34**, 1251.	26
78T1349	P. Fellmann and J.-E. Dubois; *Tetrahedron*, 1978, **34**, 1349.	697
78T2869	G. H. Schmid and D. G. Garratt; *Tetrahedron*, 1978, **34**, 2869.	280
78TL57	M. Laguerre, J. Dunoguès and R. Calas; *Tetrahedron Lett.*, 1978, **19**, 57.	929
78TL97	F. A. Davis, S. G. Yocklovich and G. S. Baker; *Tetrahedron Lett.*, 1978, 97.	184
78TL503	H. S. Veal, J. Levin and D. Swern; *Tetrahedron Lett.*, 1978, 503.	164
78TL1027	N. Okukado, D. E. Van Horn, W. L. Klima, and E. Negishi; *Tetrahedron Lett.*, 1978, 1027.	969
78TL1667	B. M. Trost and J. H. Rigby; *Tetrahedron Lett.*, 1978, 1667.	658
78TL1809	J. Yoshida, K. Tamao, A. Kurita and M. Kumada; *Tetrahedron Lett.*, 1978, 1809.	4
78TL1951	G. Himbert and W. Schwickerath; *Tetrahedron Lett.*, 1978, **22**, 1951.	1044
78TL2353	T. Durst and L. Tetreault-Ryan; *Tetrahedron Lett.*, 1978, 2353.	201
78TL2357	N. Okukado and E. Negishi; *Tetrahedron Lett.*, 1978, 2357.	976
78TL2383	P. W. K. Lau and T. H. Chan; *Tetrahedron Lett.*, 1978, **19**, 2383.	935
78TL2523	T. Morita, Y. Okamoto and H. Sakurai; *Tetrahedron Lett.*, 1978, 2523.	847
78TL2545	P. A. Grieco and N. Marinovic; *Tetrahedron Lett.*, 1978, 2545.	660
78TL3343	H. J. Bestmann and W. Kloeters; *Tetrahedron Lett.*, 1978, 3343.	1064
78TL3427	W. A. Nugent; *Tetrahedron Lett.*, 1978, 3427.	24
78TL3435	M. Shimagaki, H. Tsuchija, Y. Ban and T. Oishi; *Tetrahedron Lett.*, 1978, 3435.	402
78TL3629	A. F. Kluge; *Tetrahedron Lett.*, 1978, 3629.	680
78TL3741	T. Koizumi, H. Amitani and E. Yoshii; *Tetrahedron Lett.*, 1978, 3741.	459
78TL3967	D. Labar, A. Krief and L. Hevesi; *Tetrahedron Lett.*, 1978, 3967.	281, 290
78TL3989	D. N. Harpp, J. Adams, J. G. Gleason, D. Mullins and K. Steliou; *Tetrahedron Lett.*, 1978, 3989.	411
78TL4003	R. Mayer, G. Domschke, S. Bleisch and A. Bartl; *Tetrahedron Lett.*, 1978, 4003.	361, 792
78TL4381	M. J. Loots and J. Schwartz; *Tetrahedron Lett.*, 1978, 4381.	969
78TL4483	M. E. Jung and G. L. Hatfield; *Tetrahedron Lett.*, 1978, 4483.	32
78TL4507	A. I. Ayi, R. Condom, P. C. Maria, T. N. Wade and R. Guedj; *Tetrahedron Lett.*, 1978, 4507.	10
78TL4519	Y. H. Kim, K. Shinhama and S. Oae; *Tetrahedron Lett.*, 1978, 4519.	621, 627
78TL4575	E. Mincione, G. Ortaggi and A. Sirna; *Tetrahedron Lett.*, 1978, 4575.	26
78TL4715	A. Pelter, D. J. Ryder and J. H. Sheppard; *Tetrahedron Lett.*, 1978, 4715.	515
78TL4731	B. Mlotkowska and A. Zwierzak; *Tetrahedron Lett.*, 1978, 4731.	373
78TL4977	D. Liotta and G. Zima; *Tetrahedron Lett.*, 1978, 4977.	280
78TL5087	M. Detty; *Tetrahedron Lett.*, 1978, 5087.	282, 542
78TL5091	D. Liotta, P. B. Paty, J. Johnston and G. Zima; *Tetrahedron Lett.*, 1978, 5091.	542
78URP635102	V. I. Dimitriev, B. V. Timokhin and A. V. Kalabina; *USSR Pat.* 635 102 (1979) (*Chem. Abstr.*, 1979, **90**, 87 658).	434
78USP4118409	S. H. Eidt, S. P. Watson and G. C. Heilig; *US Pat.* 4 118 409 (1978) (*Chem. Abstr.*, 1978, **90**, 55 097).	595
78ZAAC(439)193	W. Gombler; *Z. Anorg. Allg. Chem.*, 1978, **439**, 193.	478
78ZAAC(439)139	M. Wieber and U. Baudis; *Z. Anorg. Allg. Chem.*, 1978, **439**, 139.	496
78ZAAC(443)42	G. Becker, O. Mundt, M. Rossler and E. Schneider; *Z. Anorg. Allg. Chem.*, 1978, **443**, 42.	428
78ZAAC(447)97	E. P. Flindt; *Z. Anorg. Allg. Chem.*, 1978, **447**, 97.	470
78ZC214	H. Reinheckel and H. Czech; *Z. Chem.*, 1978, **18**, 214.	393
78ZC323	R. Mayer, S. Bleisch and G. Domschke; *Z. Chem.*, 1978, **18**, 323.	398
78ZN(B)1465	M. Weidenbruch and H. Pesel; *Z. Naturforsch., Teil B*, 1978, **33**, 1465.	812
78ZN(B)1556	H. Schmidbaurand and G. Blaschke; *Z. Naturforsch., Teil B*, 1978, **33**, 1556.	452
78ZOB91	K. A. Petrov, V. A. Chanzov, N. Y. Malkevich and S. M. Kostrova; *Zh. Obshch. Khim.*, 1978, **48**, 91.	841
78ZOB232	A. A. Shvets, O. A. Moiseeva and O. A. Osipov; *Zh. Obshch. Khim.*, 1978, **48**, 232.	834, 835, 840
78ZOB319	S. V. Fridland and A. I. Efremov; *Zh. Obshch. Khim.*, 1978, **48**, 319.	833, 837, 838, 839

78ZOB639	I. I. Lapkin, V. D. Kolbasina, N. E. Evstafeeva, V. A. Dumler, V. I. Proshutinskii and O. N. Plenkina; *Zh. Obshch. Khim.*, 1978, **48**, 639.	949
78ZOB1110	G. G. Barashenkov and N. Ya. Derkach; *Z. Obshch. Khim.*, 1978, **48**, 1110.	370
78ZOB2624	E. S. Gabnitskaya, Z. T. Semashko and A. V. Kirsanov; *Zh. Obshch. Khim.*, 1978, **48**, 2624.	469
78ZOB2673	E. V. Bayandina, I. A. Nuretdinov and L. V. Nurmukhamedova; *Zh. Obshch. Khim.*, 1978, **48**, 2673.	857
78ZOB2686	V. A. Kovenya, A. P. Marchenko and A. M. Pinchuk; *Zh. Obshch. Khim.*, 1978, **48**, 2686.	408
78ZOR240	S. A. Andreev, L. A. Novik, B. A. Lebedev, I. V. Tselinskii and B. Gidaspov; *Zh. Org. Khim.*, 1978, **14**, 240.	386
78ZOR1846	E. S. Levchenko, L. V. Budnik and T. N. Dubrinina; *Zh. Org. Khim.*, 1978, **14**, 1846.	400
79ACR22	H. J. Reich; *Acc. Chem. Res.*, 1979, **12**, 22.	280
79ACR159	S. Hanessian; *Acc. Chem. Res.*, 1979, **12**, 159.	539
79ACS(B)547	L. Nilsson; *Acta Chem. Scand., Ser. B*, 1979, **33**, 547.	746
79AG655	H. Brunner and W. Pieronczyk; *Angew. Chem.*, 1979, **91**, 655.	829
79AG945	H. J. Bestmann, K. Roth, E. Wilhelm, R. Böhme and H. Burzlaff; *Angew. Chem.*, 1979, **91**, 945.	831
79AG(E)259	K. Bott; *Angew. Chem., Int. Ed. Engl.*, 1979, **18**, 259.	800
79AG(E)402	G. Felix, J. Dunogues and R. Calas; *Angew. Chem., Int. Ed. Engl.*, 1979, **18**, 402.	608
79AG(E)405	G. Himbert; *Angew. Chem., Int. Ed. Engl.*, 1979, **18**, 405.	1093
79AG(E)531	M. Regitz, A. Heydt and B. Weber; *Angew. Chem., Int. Ed. Engl.*, 1979, **18**, 531.	1068
79AG(E)563	R. W. Hoffman; *Angew. Chem., Int. Ed. Engl.*, 1979, **18**, 563.	154
79AG(E)612	G. A. Olah, S. C. Narang, B. G. B. Gupta and R. Malhotra; *Angew. Chem., Int. Ed. Engl.*, 1979, **18**, 612.	36
79AOC(17)407	J. L. Speier; *Adv. Organomet. Chem.*, 1979, **17**, 407.	536
79AX(B)2443	Y. Tominaga, G. Kobayashi, C. Tamura, S. Sato and T. Hata; *Acta Crystallogr., Part B*, 1979, **35**, 2443.	1003
79BCJ255	Y. Ogata, T. Sugimoto and M. Inaishi; *Bull. Chem. Soc. Jpn.*, 1979, **52**, 255.	21
79BCJ466	N. Yamashita, T. Miyano, T. Watabe, H. Inokawa, H. Yoshiba, T. Ogata and S. Inokawa; *Bull. Chem. Soc. Jpn.*, 1979, **52**, 466.	457
79BCJ1102	T. Nagai, T. Shingaki, M. Inagaki and T. Ohshima; *Bull. Chem. Soc. Jpn.*, 1979, **52**, 1102.	249, 401
79BCJ1197	S. Miyano, Y. Izumi, K. Fujii, Y. Ohno and H. Hashimoto; *Bull. Chem. Soc. Jpn.*, 1979, **52**, 1197.	618
79BCJ1431	S. Ikeda, F. Toneganwa, E. Shikano, K. Shinozaki and M. Ueki; *Bull. Chem. Soc. Jpn.*, 1979, **52**, 1431.	252
79BCJ1653	M. Tomoi, T. Hasegawa, M. Ikeda and H. Kakiuchi; *Bull. Chem. Soc. Jpn.*, 1979, **52**, 1653.	35
79BCJ3371	K. Soai and T. Mukaiyama; *Bull. Chem. Soc. Jpn.*, 1979, **52**, 3371.	71
79BCJ3377	A. Takaoka, H. Iwakiri and N. Ishikawa; *Bull. Chem. Soc. Jpn.*, 1979, **52**, 3377.	19
79BCJ3601	K. Saito and Y. Sato; *Bull. Chem. Soc. Jpn.*, 1979, **52**, 3601.	229
79BSF(2)305	Y. Gendreau and J. F. Normant; *Bull. Soc. Chim. Fr., Part 2*, 1979, 305.	679
79C1	D. Seebach, E. W. Colvin, F. Lehr and T. Weller; *Chimia*, 1979, **33**, 1.	783
79CB355	G. Saleh, T. Minami, Y. Ohshiro and T. Agawa; *Chem. Ber.*, 1979, **112**, 355.	843
79CB2120	W. Kirmse, O. Schnurr and H. Jendralla; *Chem. Ber.*, 1979, **112**, 2120.	802
79CB2829	L. Birkofer and K. Richtzenhain; *Chem. Ber.*, 1979, **112**, 2829.	943
79CC52	J. Mulzer, A. Pointner, A. Chucholowski and G. Brüntrup; *J. Chem. Soc., Chem. Commun.*, 1979, 52.	677
79CC238	A. R. Katritzky, A. Chermprapai and R. C. Patel; *J. Chem. Soc., Chem. Commun.*, 1979, 238.	11
79CC392	H. Chandra, B. G. Gowenlock and J. Pfab; *J. Chem. Soc., Chem. Commun.*, 1979, 392.	391
79CC479	S. Rozen and Y. Menachem; *J. Chem. Soc., Chem. Commun.*, 1979, 479.	16
79CC578	T.-H. Chan and P. Brownbridge; *J. Chem. Soc., Chem. Commun.*, 1979, 578.	662
79CC822	P. Magnus and G. Roy; *J. Chem. Soc., Chem. Commun.*, 1979, 822.	680
79CC880	A. J. Ashe, III, D. J. Bellville and H. S. Friedman; *J. Chem. Soc., Chem. Commun.*, 1979, 880.	880
79CC914	M. N. Rosenfeld and D. A. Widdowson; *J. Chem. Soc., Chem. Commun.*, 1979, 914.	615
79CC978	P. J. Garegg and B. Samuelsson; *J. Chem. Soc., Chem. Commun.*, 1979, 978.	7
79CC1165	J. A. Murphy and J. Staunton; *J. Chem. Soc., Chem. Commun.*, 1979, 1165.	664
79CC1180	A. Devos, J. Remion, A. Frisque-Hesbain, A. Colens and L. Ghosez; *J. Chem. Soc., Chem. Commun.*, 1979, 1180.	758
79CJC1903	A. F. Janzen and L. J. Kruczynski; *Can. J. Chem.*, 1979, **57**, 1903.	839
79CJC3304	C. M. Wong, R. Singh, K. Singh and H. Y. P. Lam; *Can. J. Chem.*, 1979, **57**, 3304.	656
79CL149	H. Tsuda, H. Mixato and M. Kobayashi; *Chem. Lett.*, 1976, 149.	797
79CL379	H. Nozaki, K. Oshima, K. Takai and S. Ozawa; *Chem. Lett.*, 1979, 379.	699
79CL873	N. Miyoshi, H. Ishii, S. Murai and N. Sonoda; *Chem. Lett.*, 1979, 873.	279, 282
79CL961	A. Arase, M. Hoshi and Y. Masuda; *Chem. Lett.*, 1979, 961.	20
79CL983	S. Hayashi, T. Nakai, N. Ishikawa, D. J. Burton, D. G. Naae and H. S. Kesling; *Chem. Lett.*, 1979, 983.	612
79CL1077	S. Oae, D. Fukushima and Y. H. Kim; *Chem. Lett.*, 1979, 1077.	214
79CL1357	S. Uemura, H. Miyoshi and M. Okano; *Chem. Lett.*, 1979, 1357.	624, 629
79CL1499	H. Nakai, N. Hamanaka, H. Miyake and M. Hayashi; *Chem. Lett.*, 1979, 1499.	610
79COC(1)493	R. D. Chambers and S. R. James; *Comp. Org. Chem.*, 1979, **1**, 493.	2
79COC(1)594	S. G. Wilkinson; *Comp. Org. Chem.*, 1979, **1**, 594.	40
79COC(2)131	R. J. Lindsay; *Comp. Org. Chem.*, 1979, **2**, 131.	751, 752

79COC(2)185	S. J. Roberts; *Comp. Org. Chem.*, 1979, **2**, 185.	340, 344, 346
79COC(2)219	R. S. Atkinson; *Comp. Org. Chem.*, 1979, **2**, 219.	373, 387, 402, 404
79COC(2)224	R. S. Atkinson; *Comp. Org. Chem.*, 1979, **2**, 224.	769
79COC(2)233	R. S. Atkinson; *Comp. Org. Chem.*, 1979, **2**, 233.	772
79COC(2)235	C. R. Johnson; *Comp. Org. Chem.*, 1979, **2**, 235.	773
79COC(2)236	R. S. Atkinson; *Comp. Org. Chem.*, 1979, **2**, 236.	773
79COC(2)245	R. S. Alkinson; *Comp. Org. Chem.*, 1979, **2**, 245.	775
79COC(2)267	R. S. Atkinson; *Comp. Org. Chem.*, 1979, **2**, 267.	776, 777
79COC(2)268	R. S. Atkinson; *Comp. Org. Chem.*, 1979, **2**, 268.	777
79COC(2)363	R. G. Coombes; *Comp. Org. Chem.*, 1979, **2**, 363.	385, 798
79COC(2)367	R. G. Coombes; *Comp. Org. Chem.*, 1979, **2**, 367.	798
79COC(2)370	R. G. Coombes; *Comp. Org. Chem.*, 1979, **2**, 370.	385
79COC(2)504	G. Tennant; *Comp. Org. Chem.*, 1979, **2**, 504.	342, 345
79COC(3)121	T. Durst; *Comp. Org. Chem.*, 1979, **3**, 121.	142, 144, 153
79COC(3)171	T. Durst; *Comp. Org. Chem.*, 1979, **3**, 171.	715
79COC(3)215	C. R. Johnson; *Comp. Org. Chem.*, 1979, **3**, 215.	401
79COC(3)223	C. R. Johnson; *Comp. Org. Chem.*, 1979, **3**, 223.	402
79COC(3)233	C. R. Johnson; *Comp. Org. Chem.*, 1979, **3**, 233.	399
79COC(3)237	C. R. Johnson; *Comp. Org. Chem.*, 1979, **3**, 237.	402
79COC(3)241	C. R. Johnson; *Comp. Org. Chem.*, 1979, **3**, 241.	399, 401
79COC(3)261	D. R. Hogg; *Comp. Org. Chem.*, 1979, **3**, 261.	184, 718, 719, 724
79COC(3)311	D. R. Hogg; *Comp. Org. Chem.*, 1979, **3**, 311.	718
79COC(3)317	K. K. Andersen; *Comp. Org. Chem.*, 1979, **3**, 317.	186, 189, 196, 721
79COC(3)331	K. K. Andersen; *Comp. Org. Chem.*, 1979, **3**, 331.	248, 720, 726
79COC(3)363	K. K. Andersen; *Comp. Org. Chem.*, 1979, **3**, 363.	366, 797
79COC(3)373	F. Duus; *Comp. Org. Chem.*, 1979, **3**, 373.	140
79COC(3)491	P. D. Magnus; *Comp. Org. Chem.*, 1979, **3**, 491.	727, 729, 730, 731, 734
79COC(3)541	I. Fleming; *Comp. Org. Chem.*, 1979, **3**, 541.	416, 535
79COC(3)687	A. Pelter and K. Smith; *Comp. Org. Chem.*, 1979, **3**, 687.	900
79COC(3)689	A. Pelter and K. Smith; *Comp. Org. Chem.*, 1979, **3**, 689.	900
79COC(3)925	A. Pelter and K. Smith; *Comp. Org. Chem.*, 1979, **3**, 925.	418
79COC(3)1013	G. Zweifel; *Comp. Org. Chem.*, 1979, **3**, 1013.	967
79COC(2)1061	R. C. Poller; *Comp. Org. Chem.*, 1979, **2**, 1061.	946
79COC(2)1301	D. J. H. Smith; *Comp. Org. Chem.*, 1979, **2**, 1301.	864
79COC(3)1073	R. C. Poller; *Comp. Org. Chem.*, 1979, **3**, 1073.	815
79COC(3)1114	R. C. Poller; *Comp. Org. Chem.*, 1979, **3**, 1114.	809
79COC(4)909	D. Swern; *Comp. Org. Chem.*, 1979, **4**, 909.	95, 96
79CPB1700	H. Ohmori, S. Nakai, M. Sekiguchi and M. Masui; *Chem. Pharm. Bull.*, 1979, **27**, 1700.	108
79CRV515	H. Hart; *Chem. Rev.*, 1979, **79**, 515.	636
79CZ119	F. Kober and H. T. Phung; *Chem.-Ztg.*, 1979, **103**, 119.	496
79DOK(245)198	G. A. Razuvaev, V. N. Latyaeva, L. I. Vyshinskaya and V. V. Dobrotenko; *Dokl. Akad. Nauk SSSR*, 1979, **245**, 198.	569
79DOK(245)607	E. B. Merkushev, L. G. Karpitskaya and G. I. Novosel'tseva; *Dokl. Akad. Nauk SSSR*, 1979, **245**, 607.	1019
79GEP(O)2807762	F. Merger, F. Towae and L. Schroff; *Ger. Pat. Offen.*, 2 807 762 (1979) (*Chem. Abstr.*, 1980, **92**, 6229c).	683
79HCA635	C. Tarchini, T. D. An, G. Jan and M. Schlosser; *Helv. Chim. Acta*, 1979, **62**, 635.	615
79HCA1194	M. Schlosser and T. D. An; *Helv. Chim. Acta*, 1979, **62**, 1194.	615
79HCA1406	R. Baudat and M. Petrzilka; *Helv. Chim. Acta*, 1979, **62**, 1406.	282
79HCA2507	J. V. Castell and A. Tun-kyi; *Helv. Chim. Acta*, 1979, **62**, 2507.	229
79IC347	J. C. Wilburn and R. H. Neilson; *Inorg. Chem.*, 1979, **18**, 347.	466
79IS253	L. Galyer and G. Wilkinson; *Inorg. Synth.*, 1979, **19**, 253.	592
79JA96	H. C. Brown, C. G. Scouten and R. Liotta; *J. Am. Chem. Soc.*, 1979, **101**, 96.	900, 901, 906
79JA159	T. Itoh, K. Jitsukawa, K. Kaneda and S. Teranishi; *J. Am. Chem. Soc.*, 1979, **101**, 159.	91
79JA248	M. M. Midland, A. R. Zolopa and R. L. Halterman; *J. Am. Chem. Soc.*, 1979, **101**, 248.	524
79JA249	E. Vedejs, J. M. Dolphin and W. T. Stolle; *J. Am. Chem. Soc.*, 1979, **101**, 249.	678
79JA390	L. A. Carpino and H. W. Chen; *J. Am. Chem. Soc.*, 1979, **101**, 390.	147
79JA1442	Y.-T. Fanchiang, W. P. Ridley and J. M. Wood; *J. Am. Chem. Soc.*, 1979, **101**, 1442.	584
79JA1455	T. Mukaiyama, K. Soai, T. Sato, H. Shimuzu and K. Suzuki; *J. Am. Chem. Soc.*, 1979, **101**, 1455.	71
79JA1589	W. Tam, W. K. Wong and J. A. Gladysz; *J. Am. Chem. Soc.*, 1979, **101**, 1589.	580
79JA1591	E. F. Perozzi and J. C. Martin; *J. Am. Chem. Soc.*, 1979, **101**, 1591.	547
79JA1628	F. M. Hauser and R. P. Rhee; *J. Am. Chem. Soc.*, 1979, **101**, 1628.	664
79JA2076	P. B. Dervan and T. Uyehara; *J. Am. Chem. Soc.*, 1979, **101**, 2076.	379
79JA2352	M. M. Midland, S. Greer, A. Tramontano and S. A. Zderic; *J. Am. Chem. Soc.*, 1979, **101**, 2352.	517
79JA2966	H. C. Brown, R. Liotta and G. W. Kramer; *J. Am. Chem. Soc.*, 1979, **101**, 2966.	517
79JA3521	D. B. Carr and J. Schwartz; *J. Am. Chem. Soc.*, 1979, **101**, 3521.	969, 991, 992
79JA3659	G. Erker and K. Kropp; *J. Am. Chem. Soc.*, 1979, **101**, 3659.	986, 987
79JA3704	K. C. Nicolaou, D. A. Claremon, W. E. Barnette and S. P. Seitz; *J. Am. Chem. Soc.*, 1979, **101**, 3704.	281
79JA3790	R. O. Day, A. C. Sau and R. R. Holmes; *J. Am. Chem. Soc.*, 1979, **101**, 3790.	856
79JA4107	T. F. Murray and J. R. Norton; *J. Am. Chem. Soc.*, 1979, **101**, 4107.	980

79JA4290	M. F. Salomon and R. G. Salomon; *J. Am. Chem. Soc.*, 1979, **101**, 4290.	98
79JA4749	P. A. Grieco, Y. Ohfune, Y. Yokoyama and W. Owens; *J. Am. Chem. Soc.*, 1979, **101**, 4749.	629
79JA5101	G. Zweifel, W. Lewis and H. P. On; *J. Am. Chem. Soc.*, 1979, **101**, 5101.	616
79JA5317	G. A. Olah, B. G. B. Gupta and S. C. Narang; *J. Am. Chem. Soc.*, 1979, **101**, 5317.	150
79JA5432	D. M. Snider and W. E. Truces; *J. Am. Chem. Soc.*, 1979, **101**, 5432.	478
79JA5848	J.-L. Luche and A. L. Gemal; *J. Am. Chem. Soc.*, 1979, **101**, 5848.	59
79JA6120	D. A. Evans, E. Vogel and J. V. Nelson; *J. Am. Chem. Soc.*, 1979, **101**, 6120.	532, 698
79JA6254	N. K. Roberts and S. B. Wild; *J. Am. Chem. Soc.*, 1979, **101**, 6254.	442
79JA6452	E. Vedejs and G. R. Martinez; *J. Am. Chem. Soc.*, 1979, **101**, 6452.	445
79JA6638	H. J. Reich, F. Chow and S. K. Shah; *J. Am. Chem. Soc.*, 1979, **101**, 6638.	279
79JA6648	H. J. Reich, S. K. Shah and F. Chow; *J. Am. Chem. Soc.*, 1979, **101**, 6648.	279
79JA7114	J. J. Tufariello and Sk. A. Ali; *J. Am. Chem. Soc.*, 1979, **101**, 7114.	350
79JA7292	J. Omelanczuk and M. Milolajczuk; *J. Am. Chem. Soc.*, 1979, **101**, 7292.	443
79JCS(D)54	J. Holton, M. F. Lappert, D. G. H. Ballard, R. Pearce, J. L. Atwood and W. E. Hunter; *J. Chem. Soc., Dalton Trans.*, 1979, 54.	565
79JCS(P1)433	A. R. Katritzky, N. F. Eweiss and P.-L. Nie; *J. Chem. Soc., Perkin Trans. 1*, 1979, 433.	11, 632
79JCS(P1)436	A. R. Katritzky, U. Gruntz, A. A. Ikizler, D. H. Kenny and B. P. Leddy; *J. Chem. Soc., Perkin Trans. 1*, 1979, 436.	11
79JCS(P1)1206	A. Toshimitsu, S. Uemura and M. Okano; *J. Chem. Soc., Perkin Trans. 1*, 1979, 1206.	282
79JCS(P1)2248	S. Colonna, A. Re, G. Gelbard and E. Cesarotti; *J. Chem. Soc., Perkin Trans. 1*, 1979, 2248.	9, 19
79JCS(P1)3013	M. U. Bombala and S. V. Ley; *J. Chem. Soc., Perkin Trans. 1*, 1979, 3013.	239
79JCS(P1)3029	S. Franks, F. R. Hartley and D. J. A. McCaffrey; *J. Chem. Soc., Perkin Trans. 1*, 1979, 3029.	433
79JCS(P2)3099	C. Earnshaw, C. J. Wallis and S. Warren; *J. Chem. Soc., Perkin Trans. 2*, 1979, 3099.	680
79JGU34	Yu. V. Balitski, L. S. Kasukhin, M. P. Ponomarchuk and Yu. G. Gololobov; *J. Gen. Chem. USSR (Engl. Transl.)*, 1979, **49**, 34.	467
79JGU285	V. I. Vysotskii, A. S. Skobin and M. N. Tilichenko; *J. Gen. Chem. USSR (Engl. Transl.)*, 1979, **49**, 285.	458
79JGU536	M. G. Voronkov, O. G. Yarosh, L. V. Shchukina, E. O. Tsetlina, S. N. Tandura and I. M. Korotaeva; *J. Gen. Chem. USSR (Engl. Transl.)*, 1979, **49**, 536.	1091
79JGU616	Yu. G. Trishin, A. A. Prishchemko and I. F. Lutsenko; *J. Gen. Chem. USSR (Engl. Transl.)*, 1979, **49**, 616.	467
79JGU1714	V. I. Vysotskii, I. A. Vasil'eva, V. N. Chernii, K. G. Chuprakova and V. N. Tilichenko; *J. Gen. Chem. USSR (Engl. Transl.)*, 1979, **49**, 1714.	458
79JGU1721	V. I. Vysotskii, A. S. Skobin and M. N. Tilichenko; *J. Gen. Chem. USSR (Engl. Transl.)*, 1979, **49**, 1721.	458
79JGU2143	V. L. Foss, V. V. Kudinova and I. F. Lutsenko; *J. Gen. Chem. USSR (Engl. Transl.)*, 1979, **49**, 2143.	471
79JHC1485	M. J. Kornet and R. Daniels; *J. Heterocycl. Chem.*, 1979, **16**, 1485.	382
79JMC572	C. H. Banks, J. R. Daniel and R. A. Zingaro; *J. Med. Chem.*, 1979, **22**, 572.	289
79JMC631	J. R. Piper, L. M. Rose and T. P. Johnston; *J. Med. Chem.*, 1979, **22**, 631.	255
79JOC155	P. F. Hudrlik, R. H. Schwartz and J. C. Hogan; *J. Org. Chem.*, 1979, **44**, 155.	923
79JOC160	H. W. Pinnick and M. A. Reynolds; *J. Org. Chem.*, 1979, **44**, 160.	186, 273
79JOC275	K. P. Klein, T. R. Demmin, B. C. Oxenrider, M. R. Rogic and M. T. Tetenbaum; *J. Org. Chem.*, 1979, **44**, 275.	391
79JOC305	J. A. Kloek and K. L. Leschinsky; *J. Org. Chem.*, 1979, **44**, 305.	401
79JOC359	R. M. Magid, O. S. Fruchey, W. L. Johnson and T. G. Allen; *J. Org. Chem.*, 1979, **44**, 359.	8
79JOC413	A. Sekiguchi and W. Ando; *J. Org. Chem.*, 1979, **44**, 413.	936
79JOC465	H. C. Brown, J. R. Schwier and B. Singaram; *J. Org. Chem.*, 1979, **44**, 465.	515
79JOC638	K. Yamauchi, T. Tanabe and M. Kinoshita; *J. Org. Chem.*, 1979, **44**, 638.	683
79JOC659	K. E. Gilbert and W. T. Borden; *J. Org. Chem.*, 1979, **44**, 659.	389, 397
79JOC865	Y. Machida and I. Saito; *J. Org. Chem.*, 1979, **44**, 865.	1072
79JOC1233	A. Onopchenko, E. T. Sabourin and C. M. Selwitz; *J. Org. Chem.*, 1979, **44**, 1233.	753
79JOC1247	G. A. Olah, S. C. Narang, B. G. B. Gupta and R. Malhotra; *J. Org. Chem.*, 1979, **44**, 1247.	36
79JOC1252	A. E. Feiring; *J. Org. Chem.*, 1979, **44**, 1252.	15
79JOC1554	S. Torii, H. Tanaka and M. Ukida; *J. Org. Chem.*, 1979, **44**, 1554.	243, 359
79JOC1708	G. G. I. Moore; *J. Org. Chem.*, 1979, **44**, 1708.	179
79JOC1731	G. M. Rubottom and R. C. Mott; *J. Org. Chem.*, 1979, **44**, 1731.	34
79JOC1881	L. B. Pearce, M. H. Feingold, K. F. Cerny and J-P. Anselme; *J. Org. Chem.*, 1979, **44**, 1881.	386
79JOC1961	S. L. Regen, J. C. K. Heh and J. McLick; *J. Org. Chem.*, 1979, **44**, 1961.	25
79JOC2055	C. R. Johnson, E. U. Jonsson and C. C. Bacon; *J. Org. Chem.*, 1979, **44**, 2055.	164, 401, 402
79JOC2061	C. R. Johnson, E. U. Jonsson and A. Wambsgans; *J. Org. Chem.*, 1979, **44**, 2061.	402
79JOC2226	L. J. Kaplan, G. R. Weisman and D. J. Cram; *J. Org. Chem.*, 1979, **44**, 2226.	843, 845
79JOC2278	C. R. Johnson and A. Wambsgans; *J. Org. Chem.*, 1979, **44**, 2278.	401, 402
79JOC2280	C. R. Johnston and R. A. Kirchoff; *J. Org. Chem.*, 1979, **44**, 2280.	164
79JOC2510	S.-L. Huang and D. Swern; *J. Org. Chem.*, 1979, **44**, 2510.	164
79JOC2632	D. O. Lambeth and D. W. Swank; *J. Org. Chem.*, 1979, **44**, 2632.	160
79JOC2654	L. Field and R. Ravichandran; *J. Org. Chem.*, 1979, **44**, 2654.	220
79JOC2938	S. Torii, H. Tanaka and N. Sayo; *J. Org. Chem.*, 1979, **44**, 2938.	252
79JOC3095	R. B. King, J. Bakos, C. D. Hoff and L. Markó; *J. Org. Chem.*, 1979, **44**, 3095.	439
79JOC3230	E. Vedejs, M. J. Arnost and J. P. Hagen; *J. Org. Chem.*, 1979, **44**, 3230.	213
79JOC3374	J. R. Wiersig, N. Waespe-Sarcevic and C. Djerassi; *J. Org. Chem.*, 1979, **44**, 3374.	900
79JOC3405	A. I. Meyers and M. P. Fleming; *J. Org. Chem.*, 1979, **44**, 3405.	12, 627

79JOC3406	J. Leroy, E. Herbert and C. Wakselman; *J. Org. Chem.*, 1979, **44**, 3406.	10
79JOC3410	H. Feuer and R. M. McMillan; *J. Org. Chem.*, 1979, **44**, 3410.	788
79JOC3436	S. Quici and S. L. Regen; *J. Org. Chem.*, 1979, **44**, 3436.	35
79JOC3585	L. Duhamel and J. M. Poirier; *J. Org. Chem.*, 1979, **44**, 3585.	759
79JOC3603	J. J. Barber, C. Willis and G. M. Whitesides; *J. Org. Chem.*, 1979, **44**, 3603.	596
79JOC3671	A. Onopchenko, E. T. Sabourin and C. M. Selwitz; *J. Org. Chem.*, 1979, **44**, 3671.	753
79JOC3728	N. C. Deno, E. J. Gladfelter and D. G. Pohl; *J. Org. Chem.*, 1979, **44**, 3728.	20
79JOC3741	U. Hengartner, J. D. Valentine, K. K. Johnson, M. E. Larscheid, F. Pigott, F. Scheidl, J. W. Scott, R. C. Sun, J. M. Townsend and T. H. Williams; *J. Org. Chem.*, 1979, **44**, 3741.	822, 841
79JOC3847	J. J. Krutak, R. D. Burpitt, W. H. Moore and J. A. Hyatt; *J. Org. Chem.*, 1979, **44**, 3847.	717
79JOC3867	G. J. Kubas, A. C. Larson and R. R. Ryan; *J. Org. Chem.*, 1979, **44**, 3867.	357, 365
79JOC3872	G. A. Olah, J. T. Welch, Y. D. Vankar, M. Nojima, I. Kerekes and J. A. Olah; *J. Org. Chem.*, 1979, **44**, 3872.	6, 11, 17, 392, 615
79JOC3931	R. A. Hollins and M. P. A. Campos; *J. Org. Chem.*, 1979, **44**, 3931.	629
79JOC4135	D. N. Harpp, D. K. Ash and R. A. Smith; *J. Org. Chem.*, 1979, **44**, 4135.	725
79JOC4194	R. R. King; *J. Org. Chem.*, 1979, **44**, 4194.	660
79JOC4204	T. Hori and K. B. Sharpless; *J. Org. Chem.*, 1979, **44**, 4204.	21
79JOC4208	T. Hori and K. B. Sharpless; *J. Org. Chem.*, 1979, **44**, 4208.	21
79JOC4265	P. P. Fu, C. Cortez, K. B. Sukumaran and R. G. Harvey; *J. Org. Chem.*, 1979, **44**, 4265.	659
79JOC4275	M. Laguerre, G. Félix, J. Dunoguès and R. Calas; *J. Org. Chem.*, 1979, **44**, 4275.	945
79JOC4279	D. L. J. Clive and S. M. Menchen; *J. Org. Chem.*, 1979, **44**, 4279.	294
79JOC4371	E. C. Ashby and S. A. Noding; *J. Org. Chem.*, 1979, **44**, 4371.	68
79JOC4781	R. K. Boeckman Jr. and K. J. Bruza; *J. Org. Chem.*, 1979, **44**, 4781.	929, 931
79JOC4918	R. R. Bard, J. F. Brunnett and R. P. Traber; *J. Org. Chem.*, 1979, **44**, 4918.	850
79JOM(164)305	D. Seyferth, D. P. Duncan and C. K. Haas; *J. Organomet. Chem.*, 1979, **164**, 305.	267
79JOM(168)227	J. G. Duboudin, B. Jousseaume, and A. Bonakdar; *J. Organomet. Chem.*, 1979, **168**, 227.	973
79JOM(168)233	J. G. Duboudin, B. Jousseaume, and A. Bonakdar; *J. Organomet. Chem.*, 1979, **168**, 233.	973
79JOM(172)377	H. Westmijze, H. Kleijn and P. Vermeer; *J. Organomet. Chem.*, 1979, **172**, 377.	589
79JOM(172)405	J. N. Cawse, R. A. Fiato and R. L. Pruett; *J. Organomet. Chem.*, 1979, **172**, 405.	573
79JOM(174)129	A. J. De Koning, P. E. Van Rijn, J. Boersma and G. J. M. Van der Kerk; *J. Organomet. Chem.*, 1979, **174**, 129.	1096
79JOM(177)145	M. Obayashi, K. Utimoto and H. Nozaki; *J. Organomet. Chem.*, 1979, **177**, 145.	924
79JOM(177)255	F. Mathey and F. Mercier; *J. Organomet. Chem.*, 1979, **177**, 255.	441
79JOM(179)C37	B. B. Snider and M. Karras; *J. Organomet. Chem.*, 1979, **179**, C37.	925, 926
79JOM(182)313	D. Steinborn; *J. Organomet. Chem.*, 1979, **182**, 313.	558
79JOM(8)1	J. Y. Corey; *J. Organomet. Chem.*, 1979, **8**, 1.	811
79JOM(8)95	P. M. Nowakowski and L. H. Sommer; *J. Organomet. Chem.*, 1979, **8**, 95.	812
79LA43	S. Samaan; *Justus Liebigs Ann. Chem.*, 1979, 43.	1063
79LA219	I. Hoppe and U. Schöllkopf; *Justus Liebigs Ann. Chem.*, 1979, 219.	1023
79LA784	K. Praefcke and C. Weischel; *Justus Liebigs Ann. Chem.*, 1979, 784.	171
79LA1828	G. Himbert; *Justus Liebigs Ann. Chem.*, 1979, 1828.	1044
B-79MI 201-01	R. K. MacKie; in "Organophosphorus Reagents in Organic Synthesis," ed. J. I. G. Cadogan, Academic Press, New York, 1979, p. 433.	7
79MI 202-02	K. B. Sharpless and T. R. Verhoeven; *Aldrichimica Acta*, 1979, **12**, 63.	91
B-79MI 209-01	B. J. Aylett; in "Organometalic Compounds," Chapman and Hall, London, 1979, 4th ed., vol 1, part 2, pp. 390–507.	481
79MI 215-01	L. Senczuk, H. Sobolewski and Z. Rzemykowska; *Acta Pol. Pharm.*, 1979, **36**, 557 (*Chem. Abstr.*, 1980, **93**, 204 413s).	767
79MI 215-02	G. A. Olah; *Aldrichimica Acta*, 1979, **12**, 43.	790
B-79MI 216-01	R. Appel and M. Halstenberg; in "Organophosphorus Reagents in Organic Synthesis," ed. J. I. G. Cadogan, Academic Press, London, 1979, chap. 9.	835
B-79MI 216-02	R. K. Mackie; in "Organophosphorus Reagents in Organic Synthesis," ed. J. I. G. Cadogan, Academic Press, London, 1979, chap. 10.	835
B-79MI 217-01	B. J. Aylett; "Organometallic Compounds," Chapman and Hall, 1979, vol. I, part 2, pp. 385–507.	872
B-79MI 217-02	C. A. McAuliffe and W. Levason; "Phosphine, Arsine and Stibine Complexes of the Transition Elements," Elsevier, New York, 1979.	872, 896
B-79MI 218-01	J. L. Speier; in "Advances in Organometallic Chemistry," Academic Press, New York, 1979, vol. 17, p. 407.	913
79MI 223-01	I. M. Gverdtsiteli, E. I. Chikovani and T. P. Doksopulo; *Soobshch. Akad. Nauk Gruz. SSR*, 1979, **92**, 621 (*Chem. Abstr.*, 1979, **91**, 57 125).	1093
79NJC195	J.-L. Ripoll; *Nouv. J. Chim.*, 1979, **3**, 195.	636
79OPP33	F. A. Davis and U. K. Nadir; *Org. Prep. Proced. Int.*, 1979, **11(1)**, 33.	357, 768
79OPP115	N. De Kimpe and N. Schamp; *Org. Prep. Proced. Int.*, 1979, **11**, 115.	22
79OR(26)1	H. W. Gschwend and H. R. Rodriguez; *Org. React.*, 1979, **26**, 1.	554, 653, 907, 908, 942, 965, 997
79OS(59)35	M. E. Jung and M. A. Lyster; *Org. Synth.*, 1979, **59**, 35.	538
79PAC2005	D. G. Tuck; *Pure Appl. Chem.*, 1979, **51**, 2005.	594
79PJC861	M. Gawron; *Polish J. Chem.*, 1979, **53**, 861.	1007
79PS(7)101	M. Baboulene and G. Sturts; *Phosphorus Sulfur*, 1979, **7**, 101.	457
79PS(7)325	R. P. Gupta and J. S. Pizey; *Phosphorus Sulfur*, 1979, **7**, 325.	247, 364
79RCR343	I. D. Sadekov, A. Y. Bushkov and V. I. Minkin; *Russ. Chem. Rev.*, 1979, **48**, 343.	731, 732
79RTC55	R. S. Sukhai and L. Brandsma; *Recl. Trav. Chim. Pays-Bas*, 1979, **98**, 55.	941

79RTC371	C. G. Kruse, F. L. Jonkers, V. Dert and A. van der Gen; *Recl. Trav. Chim. Pays-Bas*, 1979, **98**, 371.	10
79S8	L. W. Christensen and D. J. Heacock; *Synthesis*, 1979, 8.	213
79S37	W. Biernacki and A. Gdula; *Synthesis*, 1979, 37.	11
79S39	J. Drabowicz, W. Midura and M. Mikolajczyk; *Synthesis*, 1979, 39.	148
79S64	S. Cacchi, L. Caglioti and E. Cernia; *Synthesis*, 1979, 64.	28
79S110	T. Koizumi, H. Amitani and E. Yoshii; *Synthesis*, 1979, 110.	455, 846
79S127	H. Ahlbrecht, W. Raab and C. Vonderheid; *Synthesis*, 1979, 127.	741
79S161	N. Blazievic, D. Kolbah, B. Belin, V. Sunjic and F. Kajfez; *Synthesis*, 1979, 161.	308, 325
79S228	G. Sosnovsky and J. A. Krogh; *Synthesis*, 1979, 228.	359
79S379	T. Morita, S. Yoshida, Y. Okamoto and H. Sakurai; *Synthesis*, 1979, 379.	11, 36
79S390	H. Westmijze and P. Vermeer; *Synthesis*, 1979, 390.	940
79S437	A. R. Katritzky, K. Horvath and B. Plau; *Synthesis*, 1979, 437.	11
79S452	B. Boduszek and J. S. Wieczorek; *Synthesis*, 1979, 452.	847, 848, 850
79S463	F. Gaudemar-Bardone and M. Gaudemar; *Synthesis*, 1979, 463.	422
79S501	E.-I. Negishi, D. E. Van Horn, A. O. King and N. Okukado; *Synthesis*, 1979, 501.	629, 975, 976
79S533	A. F. Mironov, D. T. Kozhich, V. I. Vasilevsky and R. P. Evstigneeva; *Synthesis*, 1979, 533.	617
79S536	G. A. Olah, T. Keumi and A. P. Fung; *Synthesis*, 1979, 536.	650
79S538	H.-J. Cristau, J.-P. Vors and H. Cristol; *Synthesis*, 1979, 538.	446
79S615	D. J. Burton and R. M. Flynn; *Synthesis*, 1979, 615.	849
79S666	M. T. Shipchandler; *Synthesis*, 1979, 666.	391
79S688	K. Mizuno, Y. Kimura and Y. Otsuji; *Synthesis* 1979, 688.	677
79S733	J. Wildeman and A. M. van Leusen; *Synthesis*, 1979, 733.	156
79S740	H. Sakurai, A. Shirahata, K. Sasaki, and A. Hosomi; *Synthesis*, 1979, 740.	941
79S761	T. H. Chan and I. Fleming; *Synthesis*, 1979, 761.	907, 943
79S826	A. Alexakis, G. Cahiez and J. F. Normant; *Synthesis*, 1979, 826.	919
79S952	P. Tundo and P. Venturello; *Synthesis*, 1979, 952.	35
79S964	U. Husstedt and H. J. Schafer; *Synthesis*, 1979, 964.	30
79S966	U. Husstedt and H. J. Schafer; *Synthesis*, 1979, 966.	30
79S981	T. Severin and I. Ipach; *Synthesis*, 1979, 981.	790
79S982	K. C. Nicolaou, R. L. Magolda and W. J. Sipio; *Synthesis*, 1979, 982.	282
79SC157	T. J. Cousineau, S. L. Cook and J. A. Secrist, III; *Synth. Commun.*, 1979, **9**, 157.	675
79SC237	T.-L. Ho; *Synth. Commun.*, 1979, **9**, 237.	150
79SC451	D. E. Seitz and L. Ferreira; *Synth. Commun.*, 1979, **9**, 451.	544
79SC515	T. Sugasawa, T. Toyoda and K. Sasakura; *Synth. Commun.*, 1979, **9**, 515.	810, 811
79SC923	I. W. J. Still and S. Szilagyi; *Synth. Commun.*, 1979, **9**, 923.	154
79T1315	S. Jinachitra and A. J. McLeod; *Tetrahedron*, 1979, **35**, 1315.	190
79T1345	D. Brigot, N. Collignon and P. Savignac; *Tetrahedron*, 1979, **35**, 1345.	458
79T1531	M. Milolajczuk, J. Omelanczuk and W. Perlikowska; *Tetrahedron*, 1979, **35**, 1531.	437, 443
79TCC1	R. Tacke and U. Wannagat; *Top. Curr. Chem.*, 1979, **84**, 1.	547
79TCC77	M. G. Voronkov; *Top. Curr. Chem.*, 1979, **84**, 77.	547
79TL99	S. K. Chaudhary and O. Hernandez; *Tetrahedron Lett.*, 1979, 99.	539
79TL353	J. Bensoam, J. Leroy, F. Mathey and C. Wakselman; *Tetrahedron Lett.*, 1979, 353.	19
79TL485	W. Dumont, D. Van Ende and A. Krief; *Tetrahedron Lett.*, 1979, **20**, 485.	728, 936
79TL617	V. Reutrakul and P. Thamnusan; *Tetrahedron Lett.*, 1979, 617.	616
79TL821	T. Takata, Y. H. Kim and S. Oae; *Tetrahedron Lett.*, 1979, 821.	193
79TL833	A. P. Kozikowski and K. Isobe; *Tetrahedron Lett.*, 1979, 833.	999
79TL1069	W. Chodkiewicz, D. Jore and W. Wodzki; *Tetrahedron Lett.*, 1979, 1069.	437, 443
79TL1073	C. Huynh and G. Linstrumelle; *Tetrahedron Lett.*, 1979, **20**, 1073.	9311
79TL1361	H. Biere and R. Russe; *Tetrahedron Lett.*, 1979, 1361.	740
79TL1499	J. P. Genet and J. Ficini; *Tetrahedron Lett.*, 1979, 1499.	740
79TL1665	S. Masamune, S. Mori, D. E. van der Horn and D. W. Brooks; *Tetrahedron Lett.*, 1979, 1665.	532, 697
79TL1801	M. Lauwers, B. Regnier, M. van Eenoo, J. N. Denis and A. Krief; *Tetrahedron Lett.*, 1979, 1801.	36
79TL1823	S. Rozen, Y. Faust and H. Ben-Yakov; *Tetrahedron Lett.*, 1979, 1823.	19
79TL2229	D. E. Van Horn and S. Masamune; *Tetrahedron Lett.*, 1979, 2229.	532, 698
79TL2251	Y. Tamura, M. Adachi, T. Kawasaki and Y. Kita; *Tetrahedron Lett.*, 1979, 2251.	289
79TL2433	N. L. J. M. Broekhof, F. L. Jonkers and A. van der Gen; *Tetrahedron Lett.*, 1979, **20**, 2433.	739
79TL2503	R. A. Bartsch and I. W. Yang; *Tetrahedron Lett.*, 1979, 2503.	632
79TL2525	T. Iimori, Y. Nii, T. Izawa, S. Kobayashi and M. Ohno; *Tetrahedron Lett.*, 1979, 2525.	788
79TL2697	J. N. Denis, J. Vicens and A. Krief; *Tetrahedron Lett.*, 1979, 2697.	281, 284
79TL2729	C. R. Hall, T. D. Inch and I. W. Lawson; *Tetrahedron Lett.*, 1979, 2729.	463
79TL2817	P. Groenewegen, H. Kallenberg and A. van der Gen; *Tetrahedron Lett.*, 1979, 2817.	34
79TL3081	R. Luckenbach; *Tetrahedron Lett.*, 1979, 3081.	441
79TL3307	S. Oae, K. Shinhama and Y. H. Kim; *Tetrahedron Lett.*, 1979, 3307.	726
79TL3615	J. F. King, M. Aslam and J. D. Lock; *Tetrahedron Lett.*, 1979, 3615.	9
79TL4189	M. Detty; *Tetrahedron Lett.*, 1979, 4189.	542, 542
79TL4233	S. Halazy and A. Krief; *Tetrahedron Lett.*, 1979, 4233.	288
79TL4311	Y. Kita, J. Haruta, J. Segawa and Y. Tamura; *Tetrahedron Lett.*, 1979, **20**, 4311.	263, 812
79TL4363	D. S. Steichen and C. S. Foote; *Tetrahedron Lett.*, 1979, 4363.	687

79TL4391 W. Oppolzer, S. Siles, R. L. Snowden, P. H. Bakker and M. Petrzilka; *Tetrahedron Lett.*, 1979, **20**, 4391. 346, 349

79TL4493 M. Behforouz and R. Benrashid; *Tetrahedron Lett.*, 1979, 4493. 795

79TL4777 A. Pelter, D. J. Ryder, J. H. Sheppard, C. Subramanyam, H. C. Brown and A. K. Mandal; *Tetrahedron Lett.*, 1979, 4777. 516

79TL4977 D. N. Jones, P. D. Cottam and J. Davies; *Tetrahedron Lett.*, 1979, **20**, 4977. 185

79TL5015 Y. Yamada, K. Mukai, H. Yoshioka, Y. Tamura and Z.-I. Yoshida; *Tetrahedron Lett.*, 1979, **20**, 5015. 257

79TL5055 F. Montanari and P. Tundo; *Tetrahedron Lett.*, 1979, 5055. 35

79UKZ811 V. V. Skopenko, Y. L. Zub, M. A. Porai-Koshits and G. G. Sadikov; *Ukr. Khim. Zh.* (*Russ. Ed.*), 1979, **45**, 811. 1003

79ZAAC(448)35 A. Darmadi, A. Haas and M. Kaschani-Motlach; *Z. Anorg. Allg. Chem.*, 1979, **448**, 35. 292, 857

79ZAAC(448)89 M. Wieber and A. Basel; *Z. Anorg. Allg. Chem.*, 1979, **448**, 89. 500

79ZAAC(452)103 O. Glemser, S. Pohl, F.-M. Tesky and R. Mews; *Z. Anorg. Allg. Chem.*, 1979, **452**, 103. 401

79ZAAC(458)89 H. O. Fröhlich, C. Kälber and V. Stirnweis; *Z. Anorg. Allg. Chem.*, 1979, **458**, 89. 569

79ZAAC(458)249 H. Schmidbauer and P. Hall; *Z. Anorg. Allg. Chem.*, 1979, **458**, 249. 475

79ZN(B)642 A. Trenkle and H. Vahrenkamp; *Z. Naturforsch., Teil B*, 1979, **34**, 642. 431

79ZN(B)888 N. Kuhn and O. J. Scherer; *Z. Naturforsch., Teil B*, 1979, **34**, 888. 409, 416

79ZN(B)1171 H. Karsch; *Z. Naturforsch., Teil B*, 1979, **34**, 1171. 435

79ZN(B)1646 P. Dehnert, J. Grob. W. Hidebrandt and D. Le Van; *Z. Naturforsch., Teil B*, 1979, **34**, 1646. 293, 499, 500

79ZOB546 T. V. Kovaleva and N. G. Feshchenko; *Zh. Obshch. Khim.*, 1979, **49**, 546. 836

79ZOB841 V. A. Ignatov, P. G. Valkov, S. V. Sukhanov and A. N. Lazovenko; *Zh. Obshch. Khim.*, 1979, **49**, 841. 358

79ZOB1908 L. N. Markovskii, V. D. Romanenko and A. V. Ruban; *Zh. Obshch. Khim.*, 1979, **49**, 1908. 406

79ZOB2230 N. A. Andreev and O. N. Grishina; *Zh. Obshch. Khim.*, 1979, **49**, 2230. 859

79ZV924 V. F. Rudchenko, I. I. Chervin and R. G. Kostyanovski; *Izv. Akad. Nauk SSSR, Ser. Khim.*, 1979, 924. 356, 357

80AG740 R. W. Saalfrank, W. Paul and H. Liebenow; *Angew. Chem.*, 1980, **92**, 740. 831

80AG(E)218 R. W. Hoffmann and H.-J. Zeiss; *Angew. Chem., Int. Ed. Engl.*, 1980, **19**, 218. 529

80AG(E)720 H. W. Winter and C. Wentrup; *Angew. Chem., Int. Ed. Engl.*, 1980, **19**, 720. 1040

80AG(E)737 G. Kresze and R. Bussas; *Angew. Chem., Int. Ed. Engl.*, 1980, **19**, 737. 247

80AG(E)900 M. T. Reetz, J. Westermann and R. Steinbach; *Angew. Chem., Int. Ed. Engl.*, 1980, **19**, 900. 566

80AG(E)1008 T. Mujata, K. Kondo, S. Murai, T. Hirashima and N. Sonoda; *Angew. Chem., Int. Ed. Engl.*, 1980, **19**, 1008. 753

80AG(E)1010 M. Bernheim and G. Boche; *Angew. Chem., Int. Ed. Engl.*, 1980, **19**, 1010. 753

80AG(E)1011 M. T. Reetz, R. Steinbach, J. Westermann and R. Peter; *Angew. Chem., Int. Ed. Engl.*, 1980, **19**, 1011. 566

80AG(E)1023 E. R. F. Gesing, J. P. Tane and K. P. C. Vollhardt; *Angew. Chem., Int. Ed. Engl.*, 1980, **19**, 1023. 926

80AJC1365 L. K. Dalton, S. Demerac and B. C. Elmes; *Aust. J. Chem.*, 1980, **33**, 1365. 384

80AKZ408 A. M. Torgomyan, A. S. Pogosyan, M. Z. Ovakimyan and M. G. Indzhikyan; *Arm. Khim. Zh.*, 1980, **33**, 408 (*Chem. Abstr.*, 1980, **93**, 186471). 446

80BCJ174 T. Inoue and T. Mukaiyama; *Bull. Chem. Soc. Jpn.*, 1980, **53**, 174. 532, 698

80BCJ770 S. Miyano, H. Hokari, Y. Umeda and H. Hashimoto; *Bull. Chem. Soc. Jpn.*, 1980, **53**, 770. 14

80BCJ1065 S. Oae, K. Shinhama and Y. H. Kim; *Bull. Chem. Soc. Jpn.*, 1980, **53**, 1065. 621, 627

80BCJ1390 S. Uemura, S. Fukuzawa, M. Okano and S. Sawada; *Bull. Chem. Soc. Jpn.*, 1980, **53**, 1390. 24

80BCJ2023 S. Oae, K. Shinhama and Y.-H. Kim; *Bull. Chem. Soc. Jpn.*, 1980, **53**, 2023. 243

80BCJ3615 Y. Ueno, T. Miyano and M. Okanara; *Bull. Chem. Soc. Jpn.*, 1980, **53**, 3615. 156

80BCJ3691 K. Takagi, N. Hayama and S. Inokawa; *Bull. Chem. Soc. Jpn.*, 1980, **53**, 3691. 632

80BSB783 R. J. P. Corriu, J. M. Fernandez, C. Guerin and A. Kpoton; *Bull. Soc. Chim. Belg.*, 1980, **89**, 783. 264

80BSF411 J. Rouchaud, C. Moons and Meyer; *J. Bull. Soc. Chim. Fr.*, 1980, **11**, 411. 155

80CB1180 W. Malisch, H. Blau, H. Rosner and G. Jath; *Chem. Ber.*, 1980, **113**, 1180. 512

80CB1304 N. Meyer and D. Seebach; *Chem. Ber.*, 1980, **113**, 1304. 632

80CB1356 M. Baache, O. Stelzer and V. Wray; *Chem. Ber.*, 1980, **113**, 1356. 428, 431

80CB1612 H. Schmidbaur, G. Blaschke, B. Zimmer-Gasser and U. Schubert; *Chem. Ber.*, 1980, **113**, 1612. 447

80CB1950 A. Meller, W. Maringgele and H. Fetzer; *Chem. Ber.*, 1980, **113**, 1950. 399

80CB2928 B. Ross, W. Marzi and W. Axmacher; *Chem. Ber.*, 1980, **113**, 2928. 504

80CB3318 W. Malisch, M. Kuhn, W. Albert and H. Rosner; *Chem. Ber.*, 1980, **113**, 3318. 511, 512

80CB3352 W. Haubold and A. Gemmler; *Chem. Ber.*, 1980, **113**, 3352. 903

80CB3517 R. Muller and H. Vahrenkamp; *Chem. Ber.*, 1980, **113**, 3517. 511

80CC39 D. S. Matteson and D. Majumdar; *J. Chem. Soc., Chem. Commun.*, 1980, 39. 905

80CC100 P. H. McCabe and A. Stewart; *J. Chem. Soc., Chem. Commun.*, 1980, 100. 222

80CC110 J.-C. Jacquesy, M.-P. Jouannetaud and S. Makani; *J. Chem. Soc., Chem. Commun.*, 1980, 110. 626

80CC168 R. J. P. Corriu and C. Guerin; *J. Chem. Soc., Chem. Commun.*, 1980, 168. 546

80CC214 A. Chaloyard, A. Dormond, J. Tirouflet and N. El Murr; *J. Chem. Soc., Chem. Commun.*, 1980, 214. 566

80CC276 I. Fleming and F. Roessler; *J. Chem. Soc., Chem. Commun.*, 1980, 276. 917

80CC412	A. Toshimitsu, T. Aoki, H. Owada, S. Uemura, and M. Okano; *J. Chem. Soc., Chem. Commun.*, 1980, 412.	281
80CC544	J. W. Denis and A. Krief; *J. Chem. Soc., Chem. Commun*, 1980, 544.	142
80CC593	N. Takeuchi, K. Ochi, M. Murase and S. Tobinaga; *J. Chem. Soc., Chem. Commun.*, 1980, 593.	662
80CC656	M. Sevrin and A. Krief; *J. Chem. Soc., Chem. Commun.*, 1980, 656.	32
80CC671	L. J. Krause and J. A. Morrison; *J. Chem. Soc., Chem. Commun.*, 1980, 671.	434
80CC980	H. A. J. Carless and D. J. Haywood; *J. Chem. Soc., Chem. Commun.*, 1980, 980.	675
80CC1169	W. E. Paget and K. Smith; *J. Chem. Soc., Chem. Commun.*, 1980, 1169.	525, 900
80CC1183	K. Takaki, M. Okada, M. Yamada and K. Negoro; *J. Chem. Soc., Chem. Commun.*, 1980, 1183.	478
80CC1278	A. Bongini, G. Cainelli, M. Contento and F. Manescalchi; *J. Chem. Soc., Chem. Commun.*, 1980, 1278.	21, 23
80CJC1030	D. G. Garratt and A. Kabo; *Can. J. Chem.*, 1980, **58**, 1030.	280
80CL143	H. Suzuki, N. Sato and A. Osuka; *Chem. Lett.*, 1980, 143.	142
80CL729	S. Miyano, M. Nawa and H. Hashimoto; *Chem. Lett.*, 1980, 729.	859
80CL993	M. Yamaguchi and T. Mukaiyama; *Chem. Lett.*, 1980, 993.	517
80CL1063	T. Eiki and W. Tagaki; *Chem. Lett.*, 1980, 1063.	148
80CL1423	H. Konishi, K. Yokota, Y. Ichihashi, T. Okano and J. Kiji; *Chem. Lett.*, 1980, 1423.	619
80CPB134	M. Furukawa, T. Ohkawara, Y. Noguchi, M. Nishikawa and M. Tomimatsu; *Chem. Pharm. Bull.*, 1980, **28**, 134.	197
80CPB1131	K. Kubo, N. Ito, Y. Isomura and M. Murakami; *Chem. Pharm. Bull.*, 1980, **28**, 1131.	205
80CRV151	G. A. Benson and W. J. Spillane; *Chem. Rev.*, 1980, **80**, 151.	366
80CRV187	M. Schröder; *Chem. Rev.*, 1980, **80**, 187.	44, 45
80CSR477	P. D. Kennewell and J. B. Taylor; *Chem. Soc. Rev.*, 1980, **9**, 477.	163
80EGP139258	M. Muhlstadt, D. Martinez, E. Neumann and H. J. Meinhold; *Ger. (East) Pat.* 139 258 (1980) (*Chem. Abstr.*, 1980, **93**, 113 959h).	170
80EUP10477	M. J. C. Alicot, L. Ciccotto and A. P. Tignol; *Eur. Pat.* 10 477 (1980) (*Chem. Abstr.*, 1980, **93**, 220 730).	360
80GEP2912344	H. Peeters, U. Prange and W. Vogt; *Ger. Pat.* 2 912 344 (1980) (*Chem. Abstr.*, 1981, **94**, 65 141).	749
80GEP2936210	G. Elsner, H. Vollmer and E. Reutel; *Ger. Pat.* 2 936 210 (1980) (*Chem. Abstr.*, 1981, **95**, 43 345).	
80HCA1236	B. Spahic, T. T. M. Thu and M. Schlosser; *Helv. Chim. Acta*, 1980, **63**, 1236.	612
80HOU(13/5)	S. Pawlenko; *Methoden Org. Chem.* (*Houben-Weyl*), 1980, **13/5**.	535
80HOU(13/5)33	S. Pawlenko; *Methoden Org. Chem.* (*Houben-Weyl*), 1980, **13/5**, 33.	1089
80IC1879	J. E. Drake and T. Hemmings; *Inorg. Chem.*, 1980, **19**, 1879.	292, 294
80IZV734	U. M. Dzhemilev, L. Y. Gubaidullin, G. A. Tolstikov and L. M. Zelenova; *Izv. Akad. Nauk SSSR Ser. Khim.*, 1980, 734 (*Chem. Abstr.*, 1980, **93**, 25 841).	441
80IZV2111	T. P. Vasil'eva, M. G. Lin'kova, O. V. Kil'disheva and I. L. Knunyants; *Izv. Akad. Nauk. SSSR, Ser. Khim.*, 1980, **9**, 2111 (*Chem. Abstr.*, 1981, **94**, 65 510b).	227
80JA217	G. M. Whitesides, *et al.*; *J. Am. Chem. Soc.*, 1980, **102**, 217.	958
80JA226	G. M. Whitesides, *et al.*; *J. Am. Chem. Soc.*, 1980, **102**, 226.	958
80JA231	G. M. Whitesides, *et al.*; *J. Am. Chem. Soc.*, 1980, **102**, 231.	958
80JA239	G. M. Whitesides, *et al.*; *J. Am. Chem. Soc.*, 1980, **102**, 239.	958
80JA422	T. J. Katz and S. J. Lee; *J. Am. Chem. Soc.*, 1980, **102**, 422.	981
80JA1333	J. Schwartz, M. J. Loots, and H. Kosugi; *J. Am. Chem. Soc.*, 1980, **102**, 1333.	969
80JA1404	K. C. Nicolaou, R. L. Magolda and D. A. Claremon; *J. Am. Chem. Soc.*, 1980, **102**, 1404.	282
80JA1436	E. J. Corey, D. A. Clark, G. Goto, A. Marfat, C. Mioskowski, B. Samuelsson and S. Hammarström; *J. Am. Chem. Soc.*, 1980, **102**, 1436.	135
80JA1742	E. J. Corey, M. A. Tius and J. Das; *J. Am. Chem. Soc.*, 1980, **102**, 1742.	593
80JA2493	L. M. Lawrence and G. M. Whitesides; *J. Am. Chem. Soc.*, 1980, **102**, 2493.	958
80JA2693	P. Girard, J. L. Namy and H. B. Kagan; *J. Am. Chem. Soc.*, 1980, **102**, 2693.	565
80JA3270	S. H. Pine, R. Zahler, D. A. Evana and R. H. Grubbs; *J. Am. Chem. Soc.*, 1980, **102**, 3270.	680
80JA3298	E.-I. Negishi, L. F. Valente and M. Kobayashi; *J. Am. Chem. Soc.*, 1980, **102**, 3298.	629, 976
80JA3299	W. A. Kiel, G.-Y. Lin and L. A. Gladysz; *J. Am. Chem. Soc.*, 1980, **102**, 3299.	580
80JA3534	T.-H. Chan and P. Brownbridge; *J. Am. Chem. Soc.*, 1980, **102**, 3534.	662
80JA3663	E. J. Corey, D. A. Clark, G. Goto, A. Marfat, C. Mioskowski, B. Samuelsson and S. Hammarström; *J. Am. Chem. Soc.*, 1980, **102**, 3663.	135
80JA3883	D. H. Sieh, D. J. Wilbur and C. J. Michejda; *J. Am. Chem. Soc.*, 1980, **102**, 3883.	387
80JA3959	Z. A. Fataftah, I. E. Kopka and M. W. Rathke; *J. Am. Chem. Soc.*, 1980, **102**, 3959.	696
80JA4845	W. J. Middleton and E. M. Bingham; *J. Am. Chem. Soc.*, 1980, **102**, 4845.	16
80JA5004	E. Ehlinger and P. Magnus; *J. Am. Chem. Soc.*, 1980, **102**, 5004.	935
80JA5866	S. F. Martin, G. W. Phillips, T. A. Puckette and J. A. Colapret; *J. Am. Chem. Soc.*, 1980, **102**, 5866.	816
80JA5919	G. Zweifel and N. R. Pearson; *J. Am. Chem. Soc.*, 1980, **102**, 5919.	268, 522, 525, 900
80JA5974	T. Katsuki and K. B. Sharpless; *J. Am. Chem. Soc.*, 1980, **102**, 5974.	91
80JA6149	F. N. Tebbe and R. L. Harlow; *J. Am. Chem. Soc.*, 1980, **102**, 6149.	979
80JA6580	S. Fung and J. B. Siddall; *J. Am. Chem. Soc.*, 1980, **102**, 6580.	976
80JA6744	L. W. Messerle, P. Jennische, R. R. Schrock and G. Stucky; *J. Am. Chem. Soc.*, 1980, **102**, 6744.	570
80JA7003	E. R. Evitt and R. G. Bergman; *J. Am. Chem. Soc.*, 1980, **102**, 7003.	978

80JA7926 J. L. Luche and J. C. Damiano; *J. Am. Chem. Soc.*, 1980, **102**, 7926. 551

80JA7927 J. L. Luche and J. C. Damiano; *J. Am. Chem. Soc.*, 1980, **102**, 7927. 958

80JA7950 C. J. Kowalski and Jen-Sen Dung; *J. Am. Chem. Soc.*, 1980, **102**, 7950. 282

80JA7967 F. A. Davis and R. H. Jenkins; *J. Am. Chem. Soc.*, 1980, **102**, 7967. 184

80JCS(D)55 K. Kurtev, D. Ribota, R. A. Jones, D. J. Cole-Hamilton and G. Wilkinson; *J. Chem. Soc., Dalton Trans.*, 1980, 55. 825

80JCS(D)620 O. Foss and V. Janickis; *J. Chem. Soc., Dalton Trans.*, 1980, 620. 369

80JCS(D)1448 T. L. Hall, M. F. Lappert and P. W. Lednar; *J. Chem. Soc., Dalton Trans.*, 1980, 1448. 582

80JCS(D)1633 P. Diversi, G. Ingrosso, A. Lucherini and S. Murtas; *J. Chem. Soc., Dalton Trans.*, 1980, 1633. 581

80JCS(D)2428 M. J. S. Gynane, A. Hudson, M. F. Lappeit and P. P. Power; *J. Chem. Soc., Dalton Trans.*, 1980, 2428. 489

80JCS(P1)1051 T. Kato and I. Ichinose; *J. Chem. Soc., Perkin Trans. 1*, 1980, 1051. 31

80JCS(P1)1150 G. M. Blackburn and D. Ingleson; *J. Chem. Soc., Perkin Trans. 1*, 1980, 1150. 847

80JCS(P1)1212 D. H. R. Barton, D. J. Lester and S. V. Ley; *J. Chem. Soc., Perkin Trans. 1*, 1980, 1212. 390

80JCS(P1)1890 A. R. Katritzky, F. Al-Omran, R. C. Patel and S. S. Thind; *J. Chem. Soc., Perkin Trans. 1*, 1980, 1890. 11

80JCS(P1)2228 M. D. Bachi, S. Sasson and J. Vaya; *J. Chem. Soc., Perkin Trans. 1*, 1980, 2228. 166

80JCS(P1)2287 G. C. Chen, C. H. Banks, K. J. Irgolic and R. A. Zingaro; *J. Chem. Soc., Perkin Trans. 1*, 1980, 2287. 261

80JCS(P1)2485 I. Fleming and A. Pearce; *J. Chem. Soc., Perkin Trans. 1*, 1980, 2485. 936

80JCS(P1)2866 P. J. Garegg and B. Samuelsson; *J. Chem. Soc., Perkin Trans. 1*, 1980, 2866. 7

80JCS(P1)2901 A. R. Katritzky, A. Chermprapai and R. C. Patel; *J. Chem. Soc., Perkin Trans. 1*, 1980, 2901. 11

80JCS(P2)154 M. J. P. Harger; *J. Chem. Soc., Perkin Trans. 2*, 1980, 154. 840, 841, 843

80JCS(P2)385 D. C. Appleton, D. C. Bull, J. McKenna, J. M. McKenna and A. R. Walley; *J. Chem. Soc., Perkin Trans. 2*, 1980, 385. 161

80JCS(P2)1505 M. J. P. Harger; *J. Chem. Soc., Perkin Trans. 2*, 1980, 1505. 845, 846

80JFC(15)239 K. C. Eapen and C. Tamborski; *J. Fluorine Chem.*, 1980, **15**, 239. 825

80JFC(15)423 R. Franz; *J. Fluorine Chem.*, 1980, **15**, 423. 19

80JGU2159 E. E. Nifant'ev, R. K. Magdeeva, N. P. Shchept'eva and T. A. Mastryukova; *J. Gen. Chem. USSR* (*Engl. Transl.*), 1980, **50**, 2159. 464

80JMOC107 R. A. Sheldon; *J. Mol. Catal.*, 1980, **7**, 107. 90

80JOC274 M. R. Detty; *J. Org. Chem.*, 1980, **45**, 274. 286, 287

80JOC291 D. E. Laycock, J. Hartgerink and M. C. Baird; *J. Org. Chem.*, 1980, **45**, 291. 577

80JOC357 K. Watanabe and T. Wakabayashi; *J. Org. Chem.*, 1980, **45**, 357. 300

80JOC384 H. C. Brown, N. Ravindran and S. U. Kulkarni; *J. Org. Chem.*, 1980, **45**, 384. 525

80JOC389 H. C. Brown and J. B. Campbell, Jr.; *J. Org. Chem.*, 1980, **45**, 389. 525, 900, 901, 904

80JOC406 L. K. Liu, Y. Chi and K.-Y. Jen; *J. Org. Chem.*, 1980, **45**, 406. 478

80JOC524 T. A. Bryson, G. H. Bonitz, C. J. Reichel and R. E. Dardis; *J. Org. Chem.*, 1980, **45**, 524. 1008

80JOC541 J. A. Soderquist and A. Hassner; *J. Org. Chem.*, 1980, **45**, 541. 946

80JOC549 J. B. Campbell, Jr. and H. C. Brown; *J. Org. Chem.*, 1980, **45**, 549. 901

80JOC550 H. C. Brown and J. B. Campbell, Jr.; *J. Org. Chem.*, 1980, **45**, 550. 900

80JOC710 S. W. Bass and S. A. Evans; *J. Org. Chem.*, 1980, **45**, 710. 215

80JOC734 S. R. Macaulay; *J. Org. Chem.*, 1980, **45**, 734. 423

80JOC1185 D. Ranganathan, C. B. Rao, S. Ranganathan, A. K. Mehrotra and R. Iyengar; *J. Org. Chem.*, 1980, **45**, 1185. 784, 785

80JOC1638 G. A. Olah, B. G. B. Gupta, R. Malhotra and S. C. Narang; *J. Org. Chem.*, 1980, **45**, 1638. 32

80JOC1688 M. Raban and L.-J. Chern; *J. Org. Chem.*, 1980, **45**, 1688. 767

80JOC1691 Y. Hata and M. Watanabe; *J. Org. Chem.*, 1980, **45**, 1691. 244, 795

80JOC1839 B. M. Trost, K. Hiroi and N. Jungheim; *J. Org. Chem.*, 1980, **45**, 1839. 212

80JOC1924 P. K. Freeman and L. L. Hutchinson; *J. Org. Chem.*, 1980, **45**, 1924. 551

80JOC1953 A. Toshimitsu, T. Aoai, S. Uemura and M. Okano; *J. Org. Chem.*, 1980, **45**, 1953. 282

80JOC2022 Y. Ito, M. Nakatsuka and T. Saegusa; *J. Org. Chem.*, 1980, **45**, 2022. 22

80JOC2151 M. S. Raasch; *J. Org. Chem.*, 1980, **45**, 2151. 478

80JOC2270 M. P. Kaushik and R. Vaidyenathaswamy; *J. Org. Chem.*, 1980, **45**, 2270. 464

80JOC2515 A. R. Katritzky and L. Marzorati; *J. Org. Chem.*, 1980, **45**, 2515. 104

80JOC2526 E. Negishi, A. O. King, W. L. Klima, W. Patterson, and A. Silveira, Jr.; *J. Org. Chem.*, 1980, **45**, 2526. 976

80JOC2566 J. P. McCormick and D. L. Barton; *J. Org. Chem.*, 1980, **45**, 2566. 11

80JOC2731 S. Torii, T. Inokuchi, S. Misima and T. Kobayashi; *J. Org. Chem.*, 1980, **45**, 2731. 28

80JOC2739 K. Yamamoto and K. Muruoku; *J. Org. Chem.*, 1980, **45**, 2739. 306

80JOC2887 D. Liotta and W. S. Hoff; *J. Org. Chem.*, 1980, **45**, 2887. 155

80JOC2919 R. Kupper and C. J. Michejda; *J. Org. Chem.*, 1980, **45**, 2919. 385

80JOC3017 L. A. Paquette, W. E. Fristad, D. S. Dime and T. R. Bailey; *J. Org. Chem.*, 1980, **45**, 3017. 911

80JOC3527 D. Landini and F. Rolla; *J. Org. Chem.*, 1980, **45**, 3527. 4, 23, 30

80JOC3543 A. K. Mandal, P. K. Jadhav and H. C. Brown; *J. Org. Chem.*, 1980, **45**, 3543. 516

80JOC3578 G. W. Kabalka and E. E. Gooch, III; *J. Org. Chem.*, 1980, **45**, 3578. 34

80JOC3691 D. Valentine, Jr., J. F. Blount and K. Toth; *J. Org. Chem.*, 1980, **45**, 3691. 440, 441, 443

80JOC4067 H. C. Brown, T. M. Ford and J. L. Hubbard; *J. Org. Chem.*, 1980, **45**, 4067. 522

80JOC4226 R. S. Abeywickrema and E. W. Della; *J. Org. Chem.*, 1980, **45**, 4226. 13

80JOC4275 M. Node, K. Nishide, K. Fuji and E. Fujita; *J. Org. Chem.*, 1980, **45**, 4275. 658

80JOC4536 Y. Tanigawa and S.-I. Murahashi; *J. Org. Chem.*, 1980, **45**, 4536. 940

80JOC4540 H. C. Brown, J. A. Sikorski, S. U. Kulkarni and H. D. Lee; *J. Org. Chem.*, 1980, **45**, 4540. 5266

80JOC4542 S. V. Kulkarni, H. D. Lee and H. C. Brown; *J. Org. Chem.*, 1980, **45**, 4542. 522, 526

80JOC4649 V. L. Heasley, D. F. Shellhamer, L. E. Heasley, D. B. Yaeger and G. E. Heasley; *J. Org. Chem.*, 1980, **45**, 4649. 606

80JOC4666 J. W. Ludwig, M. Newcomb and D. E. Bergbreiter; *J. Org. Chem.*, 1980, **45**, 4666. 770

80JOC4702 K. F. Bernady, J. F. Poletto, J. Nocera, P. Mirando, R. E. Schaub and M. J. Weiss; *J. Org. Chem.*, 1980, **45**, 4702. 28

80JOC4740 T. L. Macdonald, D. R. Reagan and R. S. Brinkmeyer; *J. Org. Chem.*, 1980, **45**, 4740. 940

80JOC4797 R. F. Cunico and L. Bedell; *J. Org. Chem.*, 1980, **45**, 4797. 538, 539

80JOC4810 R. L. Danheiser and H. Sard; *J. Org. Chem.*, 1980, **45**, 4810. 668, 941, 944

80JOC5002 D. L. Boger and M. D. Mullican; *J. Org. Chem.*, 1980, **45**, 5002. 663

80JOC5052 T. Kubota, S. Miyashita, T. Kitazume and N. Ishikawa; *J. Org. Chem.*, 1980, **45**, 5052. 830

80JOC5155 D. N. Harpp, D. K. Ash and R. A. Smith; *J. Org. Chem.*, 1980, **45**, 5155. 233

80JOC5223 M. Kobayashi and E. Negishi. *J. Org. Chem.*, 1980, **45**, 5223. 976

80JOC5373 G. E. DuBois; *J. Org. Chem.*, 1980, **45**, 5373. 367

80JOM(186)C9 B. DeJeso and J. C. Pommier; *J. Organomet. Chem.*, 1980, **186**, C9. 816

80JOM(186)C33 D. E. Seitz, G. L. Tonneson, S. Hellman, R. N. Hanson and S. J. Adelstein; *J. Organomet. Chem.*, 1980, **186**, C33. 632

80JOM(186)39 J. Heinicke, B. Raap and A. Tzschach; *J. Organomet. Chem.*, 1980, **186**, 39. 872

80JOM(186)51 H. Watanabe, M. Kobayashi, K. Higuchi and Y. Nagai; *J. Organomet. Chem.*, 1980, **186**, 51. 915

80JOM(186)297 J. W. Wilson; *J. Organomet. Chem.*, 1980, **186**, 297. 525

80JOM(187)321 R. W. Hoffmann, G. Feussner, H.-J. Zeiss and S. Schulz; *J. Organomet. Chem.*, 1980, **187**, 321. 529

80JOM(188)255 F. J. Berry and J. Silver; *J. Organomet. Chem.*, 1980, **188**, 255. 690

80JOM(188)345 J. L. Wardell and D. W. Grant; *J. Organomet. Chem.*, 1980, **188**, 345. 879

80JOM(190)53 H. Schumann and G. Rodewald; *J. Organomet. Chem.*, 1980, **190**, 53. 868

80JOM(193)165 M. Fourtinon, B. De Jeso and J.-C. Pommier; *J. Organomet. Chem.*, 1980, **193**, 165. 813

80JOM(197)C19 R. J. P. Corriu and C. Guerin. *J. Organomet. Chem.*, 1980, **197**, C19. 600

80JOM(201)39 Y. A. Ol'Dekop, N. A. Maier, A. A. Erdman and Y. D. Butko; *J. Organomet. Chem.*, 1990, **201**, 39. 594

80JOM(202)5 M. J. S. Gynane, M. F. Lappert, P. I. Riley, P. Rivière and M. Rivière-Baudet; *J. Organomet. Chem.*, 1980, **202**, 5. 949

80JOM(202)C95 A. J. Ashe, III and T. R. Diephouse; *J. Organomet. Chem.*, 1980, **202**, C95. 880

80JOU244 K. N. Zelenin, V. A. Khrustalev and V. P. Sergutina; *J. Org. Chem. USSR* (*Engl. Transl.*), 1980, **16**, 244. 373

80LA629 G. Kresze and R. Bussas; *Justus Liebigs Ann. Chem.*, 1980, 629. 245

80LA843 G. Kresze and R. Bussas; *Justus Liebigs Ann. Chem.*, 1980, 843. 245

80LA1471 M. T. Reetz and W. F. Maier; *Liebigs Ann. Chem.*, 1980, 1471. 422

B-80MI 203-01 E. I. Negishi; "Organometallics in Organic Synthesis," Wiley, Chichester, 1980. 134

80MI 203-02 R. Matsueda and R. W. Walter; *Int. J. Pept. Protein Res.*, 1980, **16**, 392 (*Chem. Abstr.*, 1981, **95**, 169 744t). 217

80MI 203-03 R. Matsueda, E. T. Kaiser and R. W. Walter; *Pept. Chem.*, 1980, **18**, 31 (*Chem. Abstr.*, 1981, **95**, 187 632j). 217

B-80MI 207-01 M. F. Lappert, P. P. Power, A. R. Sanger and R. C. Srivastava; "Metal and Metalloid Amides," Ellis Horwood, Chichester. 1980. 414, 423

B-80MI 210-01 M. F. Lappert, P. P. Power, A. R. Sanger and R. C. Srivastava; "Metal and Metalloid Amides," Ellis Horwood, Chichester. 1980. 543

B-80MI 211-01 J. P. Collman and L. S. Hegedus; "Principles and Applications of Organotransition Metal Chemistry," University Science Books, Mill Valley, California, 1980. 563

B-80MI 211-02 G. H. Posner; "An Introduction to Synthesis Using Organocopper Reagents," Wiley, New York, 1980. 585

B-80MI 215-01 K. Schofield; "Aromatic Nitration" Cambridge University Press, Cambridge, 1980. 790

B-80MI 215-02 M. F. Lappert, A. R. Sanger, P. P. Power and R. C. Srivastava; "Metal and Metalloid Amides," Wiley, Chichester, 1980. 815

B-80MI 218-01 G. H. Posner; "An Introduction to Synthesis Using Organocopper Reagents," Wiley, New York, 1980. 917

B-80MI 219-01 P. M. Henry; "Palladium Catalyzed Oxidation of Hydrocarbons," Reidel, Dordrecht, 1980. 983

B-80MI 219-02 E. Negishi; "Organometallics in Organic Synthesis," Wiley-Interscience, New York, 1980, p. 30. 954

B-80MI 219-03 E. Negishi; "Organometallics in Organic Synthesis," Wiley-Interscience, New York, 1980, p. 286. 968

B-80MI 223-01 G. H. Posner; "An Introduction to Synthesis using Organocopper Reagents," Wiley, New York, 1980, p 1. 1094

80NJC239 D. Alker, D. H. R. Barton, R. H. Hesse, J. Lister-James, R. E. Markwell, M. M. Pechet, S. Rozen, T. Takeshita and H. T. Toh; *Nouv. J. Chim.*, 1980, **4**, 239. 15

80OS(59)26 B. Haveaux, A. Dekoker, M. Rens, A. R. Sidani, J. Toye and L. Ghosez; *Org. Synth.*, 1980, **59**, 26. 758

80OS(59)169 R. C. Cambie and P. S. Rutledge; *Org. Synth.*, 1980, **59**, 169. 45

80PAC545 E. C. Ashby; *Pure Appl. Chem.*, 1980, **52**, 545. 67

80PAC959 M. Milolajczuk; *Pure Appl. Chem.*, 1980, **52**, 959. 443

80PS(8)215 L. Horner and M. Jordan; *Phosphorus Sulfur*, 1980, **8**, 215. 441

80PS(8)221	L. Horner and M. Jordan; *Phosphorus Sulfur*, 1980, **8**, 221.	407, 851
80PS(8)225	L. Horner and M. Jordan; *Phosphorus Sulfur*, 1980, **8**, 225.	442, 829
80PS(9)107	M. V. Rosenthal and R. A. Zingaro; *Phosphorus Sulphur*, 1980, **9**, 107.	260
80PS(9)175	G. Salbeck; *Phosphorus Sulphur*, 1980, **9**, 175.	255
80PS(15)155	K. Diemert, W. Kuchen and J. Kutter; *Phosphorus Sulfur* 1980, **15**, 155.	467
80RCR14	M. V. Klynev and M. L. Khidekel; *Russ. Chem. Rev.* (*Engl. Transl.*), 1980, **49**, 14.	310
80S32	J. Drabowicz and M. Mikolajczyk; *Synthesis*, 1980, 32.	213
80S143	A. Bongini, G. Cainelli, M. Contento and F. Manescalchi; *Synthesis*, 1980, 143.	28, 30
80S151	H. C. Brown, S. U. Kulkarni and C. Guidu Rao; *Synthesis*, 1980, 151.	516
80S227	R. Weiss and P. Marolt; *Synthesis*, 1980, 227.	1008
80S261	R. H. Fischer and H. M. Weitz; *Synthesis*, 1980, 261.	391, 392
80S283	P. J. Stang and W. Treptow; *Synthesis*, 1980, 283.	689
80S285	J. F. King and M. Aslam; *Synthesis*, 1980, 285.	177
80S320	H. Ahlbrecht and W. Raab; *Synthesis*, 1980, 320.	739
80S460	W. Kantlehner, E. Haug and W. W. Mergen; *Synthesis*, 1980, 460.	812
80S461	Y. Endo, K. Shudo and T. Okamoto; *Synthesis*, 1980, 461.	691
80S563	E. Santaniello and A. Manzucchi; *Synthesis*, 1980, 563.	149
80S572	N. I. Foster, N. D. Heindel, H. D. Burns and W. Muhr; *Synthesis*; 1980, 572.	632
80S630	H. Ahlbrecht and E.-O. Düber; *Synthesis*, 1980, 630.	812
80S708	S. H. Bertz; *Synthesis*, 1980, 708.	662
80S746	M. Yoshihara; T. Eda, K. Sakaki and T. Maeshima; *Synthesis*, 1980, 746.	8, 25
80S803	G. Zweifel and H. P. On; *Synthesis*, 1980, 803.	922
80S805	J. Cousseau; *Synthesis*; 1980, 805.	623
80S823	V. D. Romanenko, V. I. Tovstenko and L. N. Markovski; *Synthesis*, 1980, 823.	835, 837
80S853	A. Lorenzo, P. Molina and M. J. Vilaplana; *Synthesis*, 1980, 853.	11
80S861	W. C. Groutas and D. Felker; *Synthesis*, 1980, 861.	538, 540
80S910	R. J. Batten, J. D. Coyle and R. J. K. Taylor; *Synthesis*, 1990, 910.	232
80S937	M. Furukawa, T. Ohkawara, Y. Noguchi, M. Isoda and T. Hitoshi; *Synthesis*, 1980, 937.	724
80S1034	M. Kobayashi, L. F. Valente, E. Negishi, W. Patterson and A. Silveira, Jr.; *Synthesis*, 1980, 1034.	976
80SC245	R. K. Sehgal and A. J. Krubsack; *Synth. Commun.*, 1980, 245.	170
80SC311	T. Cohen and J. R. Matz; *Synth. Commun.*, 1980, **10**, 311.	422, 551
80SC595	T. Otsubo, F. Ogura, H. Yamaguchi, H. Higuchi and S. Misumi; *Synth. Commun.*, 1980, **10**, 595.	289
80SC611	E. Santaniello and F. Ponti; *Synth. Commun.*, 1980, **10**, 611.	299
80SC821	R. W. Aben, E. J. M. Hanneman and J. W. Scheeren; *Synth. Commun.*, 1980, **10**, 821.	29
80T679	A. R. Katritzky; *Tetrahedron*, 1980, **36**, 679.	11
80T1223	M. A. Pasquini, R. le Goaller and J. L. Pierre; *Tetrahedron*, 1980, **36**, 1223.	299
80T1901	R. M. Magid; *Tetrahedron*, 1980, **36**, 1901.	6
80T2531	A. Krief; *Tetrahedron*, 1980, **36**, 2531.	997, 1001
80TL163	M. Montury, B. Psaume and J. Goré; *Tetrahedron Lett.*, 1980, **21**, 163.	940
80TL177	T. Ogino; *Tetrahedron Lett.*, 1980, **21**, 177.	44
80TL441	S. G. Hegde, M. K. Vogel, J. Saddler, T. Hrinyo, N. Rockwell, R. Haynes, M. Oliver and J. Wolinsky; *Tetrahedron Lett.*, 1980, **21**, 441.	21
80TL689	B. Ganem, A. J. Biloski and R. P. Heggs; *Tetrahedron Lett.*, 1980, **21**, 689.	146
80TL767	S. Murata and R. Noyori; *Tetrahedron Lett.*, 1980, **21**, 767.	263
80TL781	R. D. G. Cooper; *Tetrahedron Lett.*, 1980, **21**, 781.	21
80TL1031	W. C. Still and J. H. McDonald, III; *Tetrahedron Lett.*, 1980, **21**, 1031.	68
80TL1035	W. C. Still and J. A. Schneider; *Tetrahedron Lett.*, 1980, **21**, 1035.	68
80TL1113	E. J. Corey and H. Estreicher; *Tetrahedron Lett.*, 1980, **21**, 1113.	785
80TL1117	E. J. Corey and H. Estreicher; *Tetrahedron Lett.*, 1980, **21**, 1117.	397
80TL1501	E. Negishi and T. Yoshida; *Tetrahedron Lett.*, 1980, **21**, 1501.	596, 970
80TL1697	R. A. Gancarz and J. L. Kice; *Tetrahedron Lett.*, 1980, **21**, 1697.	238
80TL2025	H. G. Corkins, L. Storace and E. R. Osgood; *Tetrahedron Lett.*, 1980, 2025.	778
80TL2067	S. Kozuka and T. Higashino; *Tetrahedron Lett.*, 1980, 2067.	688
80TL2171	F. Sato, T. Jinbo and M. Sato; *Tetrahedron Lett.*, 1980, **21**, 2171.	55
80TL2213	T. G. Back and S. Collins; *Tetrahedron Lett.*, 1980, **21**, 2213.	238
80TL2215	T. G. Back and S. Collins; *Tetrahedron Lett.*, 1980, **21**, 2215.	285
80TL2229	E. Gössinger; *Tetrahedron Lett.*, 1980, **21**, 2229.	345
80TL2525	F. Mathey and R. Maillet; *Tetrahedron Lett.*, 1980, 2525.	441
80TL2603	H. Alper and S. Amaratunga; *Tetrahedron Lett.*, 1980, **21**, 2603.	753
80TL2671	N. L. J. M. Broekhof, F. L. Jonkers and A. van der Gen; *Tetrahedron Lett.*, 1980, **21**, 2671.	739
80TL2687	J. Gasteiger and C. Herzig; *Tetrahedron Lett.*, 1980, **21**, 2687.	456
80TL3143	E. J. Corey, D. A. Clark, A. Marfat and G. Goto; *Tetrahedron Lett.*, 1980, **21**, 3143.	135, 136
80TL3241	G. Wilcox and R. Breslow; *Tetrahedron Lett.*, 1980, **21**, 3241.	1050
80TL3423	P. Brownbridge and T.-H. Chan; *Tetrahedron Lett.*, 1980, 3423.	668
80TL3643	D. Liotta, G. Zima, C. Barnum and M. Saindane; *Tetrahedron Lett.*, 1980, 3643.	282
80TL3827	A. Bewick, D. E. Coe, G. B. Fuller and J. M. Mellor; *Tetrahedron Lett.*, 1980, **21**, 3827.	284
80TL3975	D. A. Evans and L. R. McGee; *Tetrahedron Lett.*, 1980, **21**, 3975.	703
80TL3987	J.-C. Clinet and G. Linstrumelle; *Tetrahedron Lett.*, 1980, **21**, 3987.	939
80TL4021	M. Matsumoto and K. Kuroda; *Tetrahedron Lett.*, 1980, 4021.	624
80TL4069	G. Altnau and L. Rösch; *Tetrahedron Lett.*, 1980, **21**, 4069.	918
80TL4155	R. A. Gancarz and J. L. Kice; *Tetrahedron Lett.*, 1980, **21**, 4155.	285

80TL4251	E. Schaumann and F.-F. Grabley; *Tetrahedron Lett.*, 1980, **21**, 4251.	941
80TL4417	D. R. Williams and K. Nishitani; *Tetrahedron Lett.*, 1980, **21**, 4417.	282
80TL4523	W. T. Markiewicz, Z. Samek and J. Smart; *Tetrahedron Lett.*, 1980, **21**, 4523.	540
80TL4543	S. Rozen and M. Brand; *Tetrahedron Lett.*, 1980, **21**, 4543.	18
80TL4607	Y. Yamamoto and K. Maruyama; *Tetrahedron Lett.*, 1980, **21**, 4607.	703
80TL4959	C. R. Hall and N. E. Williams; *Tetrahedron Lett.*, 1980, **21**, 4959.	257
80ZAAC(462)113	G. Becker, G. Gutekunst and H. J. Wessely; *Z. Anorg. Allg. Chem.*, 1980, **462**, 113.	508, 509
80ZAAC(466)179	H. T. Phung, P. B. Chi and F. Kober; *Z. Anorg. Allg. Chem.*, 1980, **466**, 179.	494
80ZAAC(466)183	P. B. Chi and F. Kober; *Z. Anorg. Allg. Chem.*, 1980, **466**, 183.	502
80ZN(B)990	H. Schmidbaur and A. Mörtl; *Z. Naturforsch., Teil B*, 1980, **35**, 990.	445
80ZOB803	N. A. Andreev and O. N. Grishina; *Zh. Obshch. Khim.*, 1980, **50**, 803.	838
80ZOB951	A. P. Marchenko, V. V. Miroshnichenko and A. A. Kudryavtsev; *Zh. Obshch. Khim.*, 1980, **50**, 951 (*Chem. Abstr.*, 1980, **93**, 183 353r).	408
80ZOB1059	V. N. Sergeev and Y. I. Baukov; *Zh. Obshch. Khim.*, 1980, **50**, 1059 (*Chem. Abstr.*, 1980, **93**, 239 505a).	172
80ZOB1236	V. L. Foss, N. V. Lukashev and I. F. Lutsenko; *Zh. Obshch. Khim.*, 1980, **50**, 1236 (*Chem. Abstr.*, 1980, **93**, 204024r).	414
80ZOB1702	E. E. Nifantev, D. A. Prednovitelev, A. P. Tuseev, M. K. Grachev and M. A. Zolotov; *Zh. Obshch. Khim.*, 1980, **50**, 1702 (*Chem. Abstr.*, 1981, **94**, 3822w).	410
80ZOB2459	N. I. Ganushchak, V. A. Vengrzhanovskii, V. A. Baranov and T. P. Bachinskii; *Zh. Obshch. Khim.*, 1980, **50**, 2459 (*Chem. Abstr.*, 1981, **95**, 42 532s).	256
80ZOB2621	L. N. Markovskii, V. D. Romanenko, V. I. Dzyuba, E. Lippmaa and M. Alla; *Zh. Obshch. Khim.*, 1980, **50**, 2621 (*Chem. Abstr.*, 1981, **94**, 121636m).	407
81ACR400	S. Danishefsky; *Acc. Chem. Res.*, 1981, **14**, 400.	665
81ACS(A)151	K. Maartmann-Moe, K. A. Sanderud and J. Songstad; *Acta Chem. Scand., Ser. A*, 1981, **35**, 151.	290
81ACS(B)239	S. Schiebye, S.-O. Lawesson and C. Remming; *Acta Chem. Scand. Ser. B*, 1981, **35**, 239.	256
81AG(E)123	M. Baudler and P. Bachmann; *Angew. Chem., Int. Ed. Engl.*, 1981, **20**, 123.	508
81AG(E)151	G. Hamprecht, K. H. Koning and G. Stubentauch; *Angew. Chem., Int. Ed. Engl.*, 1981, **20**, 151.	368
81AG(E)647	M. R. C. Gerstenberger and A. Haas; *Angew. Chem., Int. Ed. Engl.*, 1981, **20**, 647.	15
81AP(314)799	W. Hanefeld and D. Kluck; *Arch. Pharm.* (*Weinheim, Ger.*), 1981, **314**, 799.	249
81AX(B)344	O. Simonsen; *Acta Crystallogr., Part B*, 1981, **37**, 344.	1002
81BCJ805	E. Nakamura, K. Hashimoto and I. Kuwajima; *Bull. Chem. Soc. Jpn.*, 1981, **54**, 805.	263
81BCJ1151	S. Yanagida, Y. Noji and M. Okahara; *Bull. Chem. Soc. Jpn.*, 1981, **54**, 1151.	19
81BCJ1443	T. Takata, Y. H. Kim and S. Oae; *Bull. Chem. Soc. Jpn.*, 1981, 1443.	724
81BCJ2843	S. Uemura, C. Masaki, A. Toshimitsu and S. Sawada; *Bull. Chem. Soc. Jpn.*, 1981, **54**, 2843.	617
81BCJ2847	T. Sugiyama; *Bull. Chem. Soc. Jpn.*, 1981, **54**, 2847.	631
81BEP886595	G. D. Macdonell; *Belg. Pat.* 886 595 (*Chem. Abstr.*, 1982, **96**, 6197y).	214
81BSF(2)71	G. Lonchambon, A. Delacroix, J. Petit and P. Lecomte; *Bull. Soc. Chim. Fr., Part* 2, 1981, 71.	726
81BSF(2)87	H. Lehmkuhl; *Bull. Soc. Chim. Fr., Part* 2, 1981, 87.	561, 562
81CB375	R. W. Hoffmann and T. Herold; *Chem. Ber.*, 1981, **114**, 375.	75
81CB527	J. Neemann and U. Klingebiel; *Chem. Ber.*, 1981, **114**, 527.	418
81CB673	H. Bock, U. Stein and W. Reid, *Chem. Ber.*, 1981, **114**, 673.	1038
81CB774	H. Quast and F. Kees; *Chem. Ber.*, 1981, **114**, 774.	401
81CB1407	J. v. Seyerl, B. Sigwarth and G. Huttner; *Chem. Ber.*, 1981, **114**, 1407.	694
81CB2293	L. Birkofer and T. Kühn; *Chem. Ber.*, 1981, **114**, 2293.	943
81CB3154	L. Birkofer and E. Hänsel; *Chem. Ber.*, 1981, **114**, 3154.	944
81CB3485	W. Kuchen, H. Somberg and U. Tobolla; *Chem. Ber.*, 1981, **114**, 3485.	836, 837
81CB4005	G. Boche and U. Fahrmann; *Chem. Ber.*, 1981, **114**, 4005.	610
81CC148	S. Stavber and M. Zupan; *J. Chem. Soc., Chem. Commun.*, 1981, 148.	613
81CC261	T. Sakakibara, I. Takai, E. Ohara and R. Sudoh; *J. Chem. Soc., Chem. Commun.*, 1981, 261.	784, 786
81CC277	H. Lee and M. P. Cava; *J. Chem. Soc., Chem. Commun.*, 1981, 277.	287, 288
81CC292	W. J. Evans, A. L. Wayda, W. E. Hunter and J. L. Atwood; *J. Chem. Soc., Chem. Commun.*, 1981, 292.	564
81CC350	A. Dondoni, P. Giorgianni, A. Battaglia and G. D. Andreetti; *J. Chem. Soc., Chem. Commun.*, 1981, 350.	247
81CC546	A. Toshimitsu, H. Owada, T. Aoai, S. Uemura and M. Okano; *J. Chem. Soc., Chem. Commun.*, 1981, 546.	286
81CC718	F. Sato, H. Ishikawa, H. Watanabe, T. Miyake and M. Sato; *J. Chem. Soc., Chem. Commun.*, 1981, 718.	629
81CC741	A. P. Davis and G. H. Whitham; *J. Chem. Soc., Chem. Commun.*, 1981, 741.	195
81CC742	L. Carlsen, H. Egsgaard, G. H. Whitham and D. N. Harpp; *J. Chem. Soc., Chem. Commun.*, 1981, 742.	195
81CC808	K. Isobe, P. M. Bailey and P. M. Maitlis; *J. Chem. Soc., Chem. Commun.*, 1981, 808.	585
81CC1140	F. Sato, K. Ida, S. Ijima, H. Moriya and M. Sato; *J. Chem. Soc., Chem. Commun.*, 1981, 1140.	566
81CC1240	T. Iversen and D. R. Bundle; *J. Chem. Soc., Chem. Commun.*, 1981, 1240.	90
81CJC718	Y. Ogata and K. Tanaka; *Can. J. Chem.*, 1981, **59**, 718.	156
81CJC1463	M. Cowie and M. D. Gauthier; *Can. J. Chem.*, 1981, **59**, 1463.	1003

81CL447 T. Otsubo, F. Ogura and H. Yamaguchi; *Chem. Lett.*, 1981, 447. 293

81CL737 R. Matsueda, T. Kimura, E. T. Kaiser and G. R. Matsueda; *Chem. Lett.*, 1981, 737. 217

81CL761 N. Ishikawa, T. Kitazume, T. Yamazaki, Y. Moshida and T. Tatsuno; *Chem. Lett.*, 1981, 761. 19

81CL1117 H. Suzuki, H. Abe, N. Ohmasa and A. Osuka; *Chem. Lett.*, 1981, 1117. 731

81CL1157 Y. Kikugawa; *Chem. Lett.*, 1981, 1157. 139

81CL1457 K. Inomata, H. Yamada and H. Kotake; *Chem. Lett.*, 1981, 1457. 121

81CL1715 S. Tomoda, Y. Takeuchi and Y. Nomura; *Chem. Lett.*, 1981, 1715. 284

81CPB3387 H. Takahashi, K. Tomita and H. Noguchi; *Chem. Pharm. Bull.*, 1981, **29**, 3387. 381

81CRV415 A. K. Bhattacharya and G. Thyagarajan; *Chem. Rev.*, 1981, **81**, 415. 469

81GEP2926671 M. Hanack and T. Stoll; *Ger. Pat.* 2 926 671 (1981) (*Chem. Abstr.*, 1981, **94**, 191 778h). 157

81HCA1158 N. Cohen, C. G. Scott, C. Neukom, R. J. Lopresti, G. Weber and G. Saucy; *Helv. Chim. Acta*, 1981, **64**, 1158. 39

81HCA1552 B. Weidmann, C. D. Maycock and D. Seebach; *Helv. Chim. Acta*, 1981, **64**, 1552. 566

81HCA2264 E. W. Colvin, A. K. Beck and D. Seebach; *Helv. Chim. Acta*, 1981, **61**, 2264. 394

81HCA2528 G. A. Olah, G. K. S. Prakash and Y. L. Chao; *Helv. Chim. Acta*, 1981, **64**, 2528. 11

81HCA2606 E. P. Kündig and C. Perret; *Helv. Chim. Acta*, 1981, **64**, 2606. 558, 560

81IC2991 S. J. Simpson, H. W. Turner and R. A. Andersen; *Inorg. Chem.*, 1981, **20**, 2991. 603

81IJC(B)932 M. P. Kaushik and R. Vaidyanathaswamy; *Indian J. Chem., Sect. B*, 1981, **20**, 932. 463

81IS178 J. E. Hoots, T. B. Rauchfuss and D. A. Wrobleski; *Inorg. Synth.*, 1981, **21**, 178. 827

81IZV159 T. P. Vasil'eva, M. G. Lin'kova, O. V. Kil'disheva and I. L. Knunyants; *Izv. Akad. Nauk. SSSR, Ser. Khim.*, 1981, 159 (*Chem. Abstr.*, 1981, **94**, 208 318h). 173

81IZV1280 N. I. Shergina, L. V. Sherstyannikova, R. G. Mirskov, A. G. Proidakov, A. L. Kuznetsov, N. P. Ivanova and M. G. Voronkov; *Izv. Akad. Nauk SSSR, Ser. Khim.*, 1981, 1280 (*Bull. Acad. Sci. USSR, Div. Chem. Sci*, 1981, 1014). 1093

81IZV1850 T. P. Vasil'eva, V. M. Bystrova, M. G. Lin'kova, O. V. Kil'disheva and I. L. Knunyants; *Izv. Akad. Nauk. SSSR, Ser. Khim.*, 1981, 1850 (*Chem. Abstr.*, 1981, **95**, 219 867h). 211

81JA442 D. H. Sieh and C. J. Michejda; *J. Am. Chem. Soc.*, 1981, **103**, 442. 387

81JA1276 T. Yoshida and E. Negishi; *J. Am. Chem. Soc.*, 1981, **103**, 1276. 976, 978

81JA1566 S. Masamune, W. Choy, F. A. J. Kerdeskey and B. Imperiali; *J. Am. Chem. Soc.*, 1981, **103**, 1566. 532

81JA1761 B. Capon, D. S. Rycroft, T. W. Watson and C. Zucco; *J. Am. Chem. Soc.*, 1981, **103**, 1761. 638

81JA1785 D. B. Denney, D. P. Denney, P. J. Hammond and Y.-P. Wang; *J. Am. Chem. Soc.*, 1981, **103**, 1785. 870

81JA1847 J. A. Kampmeier, R. M. Rodehorst and J. B. Philip, Jr.; *J. Am. Chem. Soc.*, 1981, **103**, 1847. 957

81JA1964 D. P. Ip, C. D. Arthur, R. E. Winans and E. H. Appelman; *J. Am. Chem. Soc.*, 1981, **103**, 1964. 613

81JA1969 Y. Yamamoto, H. Yatagai and K. Marayama; *J. Am. Chem. Soc.*, 1981, **103**, 1969. 517

81JA2483 B. M. Trost and W. H. Pearson; *J. Am. Chem. Soc.*, 1981, **103**, 2483. 753

81JA2715 J. Bergman and L. Engman; *J. Am. Chem. Soc.*, 1981, **103**, 2715. 289

81JA2882 H. Matsushita and E. Negishi; *J. Am. Chem. Soc.*, 1981, **103**, 2882. 975, 976

81JA3099 D. A. Evans, E. Vogel, J. V. Nelson and T. R. Taber; *J. Am. Chem. Soc.*, 1981, **103**, 3099. 532, 698

81JA3229 Y. Yamamoto, H. Yatagai and K. Marayama; *J. Am. Chem. Soc.*, 1981, **103**, 3229. 517

81JA4432 B. E. Maryanoff, A. T. McPhail and R. O. Hutchins; *J. Am. Chem. Soc.*, 1981, **103**, 4432. 253

81JA4985 T. Yoshida and E. Negishi; *J. Am. Chem. Soc.*, 1981, **103**, 4985. 976

81JA5459 G. A. Krafft and J. A. Katzenellenbogen; *J. Am. Chem. Soc.*, 1981, **103**, 5459. 982

81JA6142 R. Weiss, H. Wolf, U. Shubert and T. Clark; *J. Am. Chem. Soc.*, 1981, **103**, 6142. 1046

81JA6237 V. S. Martin, S. S. Woodward, T. Katsuki, Y. Yamada, M. Ikeda and K. B. Sharpless; *J. Am. Chem. Soc.*, 1981, **103**, 6237. 92

81JA6437 P. L. Stang and M. Ladika; *J. Am. Chem. Soc.*, 1981, **103**, 6437. 1094

81JA6650 P. J. Fagan, J. M. Manriquez, E. A. Maatta, A. M. Seyam and T. J. Marks; *J. Am. Chem. Soc.*, 1981, **103**, 6650. 602

81JA7016 F. A. Davis and R. L. Billmers; *J. Am. Chem. Soc.*, 1981, **103**, 7016. 185

81JA7672 B. H. Lipshutz, R. S. Wilhelm and D. M. Floyd; *J. Am. Chem. Soc.*, 1981, **103**, 7672. 1095

81JA7677 W. D. Wulff, P.-C. Tan and J. S. McCallum; *J. Am. Chem. Soc.*, 1981, **103**, 7677. 670

81JBC(256)5064 P. C. S. Chong and R. S. Hodges; *J. Biol. Chem.*, 1981, **256**, 5064. 217

81JCR(S)376 V. B. Jigajinni, W. E. Paget and K. Smith; *J. Chem. Res.* (*S*), 1981, 376. 4

81JCS(D)80 P. B. Hitchcock, M. F. Lappert and C. R. C. Milne; *J. Chem. Soc., Dalton Trans.*, 1981, 80. 569

81JCS(D)447 S. Bhaduri, H. Khwaja and A. Ghosh; *J. Chem. Soc., Dalton Trans.*, 1981, 447. 156

81JCS(P1)212 Y. Tamura, C. Mukai, N. Nakajima, M. Ikeda and M. Kido; *J. Chem. Soc., Perkin Trans. 1*, 1981, 212. 1000

81JCS(P1)574 S. Juliá, A. Ginebreda, J. Guixer, J. Masana and A. Tomás; *J. Chem. Soc., Perkin Trans. 1*, 1981, 574. 299

81JCS(P1)1473 D. H. R. Barton, A. G. Brewster, S. V. Ley, C. M. Read and M. N. Rosenfeld; *J. Chem. Soc., Perkin Trans. 1*, 1981, 1473. 690

81JCS(P1)2106 R. Bell, P. D. Cottam, J. Davies and N. D. Jones; *J. Chem. Soc., Perkin Trans. 1*, 1981, 2106. 184

81JCS(P1)2322 B. F. Bonini, G. Maccagnani, G. Mazzanti, G. Rosini and E. Foresti; *J. Chem. Soc., Perkin Trans. 1*, 1981, 2322. 245

81JCS(P1)2415 M. J. Carter, I. Fleming and A. Percival; *J. Chem. Soc., Perkin Trans. 1*, 1981, 2415. 921

81JCS(P1)2527 I. Fleming, T. W. Newton, and F. Roessler; *J. Chem. Soc., Perkin Trans. 1*, 1981, 2527. 917, 918, 940, 977

81JCS(P1)2608 R. C. Cambie, R. C. Hayward, J. R. Jurlina, P. S. Rutledge and P. D. Woodgate; *J. Chem. Soc., Perkin Trans. 1*, 1981, 2608. 12

81JCS(P1)3059 A. T. Zaslona and C. D. Hall; *J. Chem. Soc., Perkin Trans. 1*, 1981, 3059. 447

81JCS(P2)545 S. E. Fuller, J. R. Lindsay Smith and R. O. C. Norman; *J. Chem. Soc., Perkin Trans. 2*, 1981, 545. 333, 339

81JCS(P2)1449 G. Cerioni, P. Piras, G. Marongiu, D. Macciantelli and L. Lunazzi; *J. Chem. Soc., Perkin Trans. 2*, 1981, 1449. 386

81JCS(P2)1512 A. G. Davies and R. Sutcliffe; *J. Chem. Soc., Perkin Trans. 2*, 1981, 1512. 687

81JFC(17)93 R. E. Banks, A.-K. Barrage and E. Khoshdel; *J. Fluorine Chem.*, 1981, **17**, 93. 19

81JFC(17)127 A. I. Ayi, M. Remli and R. Guedj; *J. Fluorine Chem.*, 1981, **17**, 127. 10

81JGU1056 I. E. Pokrovskaya, V. A. Dodonov, Z. A. Starikova, E. N. Kanunnikova, T. M. Shchegoleva and G. P. Lebedeva, *J. Gen. Chem. USSR* (*Engl. Transl.*), 1981, **81**, 1056. 886

81JINC3389 H. H. Sisler, M. A. Mathur and D. J. Albert; *J. Inorg. Nucl. Chem.*, 1981, **43**, 3389. 379

81JMC93 W. C. Lumma, Jr., R. D. Hartmann, W. S. Saari and E. L. Engelhardt; *J. Med. Chem.*, 1981, **24**, 93. 613

81JOC78 S. Top and G. Jaouen; *J. Org. Chem.*, 1981, **46**, 78. 309

81JOC179 B. G. Ramsey and L. M. Isabelle; *J. Org. Chem.*, 1981, **46**, 179. 1010

81JOC289 D. E. Laycock and H. Alper; *J. Org. Chem.*, 1981, **46**, 289. 171

81JOC442 C. G. Overberger and T. F. Merkel; *J. Org. Chem.*, 1981, **46**, 442. 377, 403

81JOC824 R. M. Magid, B. G. Talley and S. K. Souther; *J. Org. Chem.*, 1981, **46**, 824. 8

81JOC1215 P. A. Grieco, J. Y. Jaw, D. A. Claremon and K. C. Nicolaou; *J. Org. Chem.*, 1981, **46**, 1215. 279

81JOC1283 M. R. Detty and M. D. Seidler; *J. Org. Chem.*, 1981, **46**, 1283 and references therein. 542

81JOC1309 R. W. Hoffmann and H.-J. Zeiss; *J. Org. Chem.*, 1981, **46**, 1309. 74, 529, 530

81JOC1693 A. Pross, D. J. De Frees, B. A. Levi, S. K. Pollack, L. Radom and W. J. Hehre; *J. Org. Chem.*, 1981, **46**, 1693. 525, 1005

81JOC1821 T. Gibson and L. Tulich; *J. Org. Chem.*, 1981, **46**, 1821. 4, 34

81JOC1868 J. J. Pastusznak and A. Chimiak; *J. Org. Chem.*, 1981, **46**, 1868. 229

81JOC2025 L. Field and V. Eswarakrishnan; *J. Org. Chem.*, 1981, **46**, 2025. 187

81JOC2169 J. J. Harrison, J. P. Pellegrini and C. M. Selwitz; *J. Org. Chem.*, 1981, **46**, 2169. 626

81JOC2363 P. Beak and L. G. Carter; *J. Org. Chem.*, 1981, **46**, 2363. 999

81JOC2402 S. F. Nelsen, L. A. Grezzo and V. E. Peacock; *J. Org. Chem.*, 1981, **46**, 2402. 388

81JOC2582 G. W. Kabalka and E. E. Gooch; *J. Org. Chem.*, 1981, **46**, 2582. 4

81JOC2605 D. Liotta, U. Sunay, H. Santiesteban and W. Markiewicz; *J. Org. Chem.*, 1981, **46**, 2605. 282

81JOC2712 H. C. Brown, B. Singaram and C. P. Mathew; *J. Org. Chem.*, 1981, **46**, 2712. 534

81JOC2775 H. J. Reich, W. W. Willis, Jr. and P. D. Clark; *J. Org. Chem.*, 1981, **46**, 2775. 728, 908

81JOC2988 P. K. Jadhav and H. C. Brown; *J. Org. Chem.*, 1981, **46**, 2988. 520

81JOC3021 A. Toshimitsu, T. Aoai, S. Uemura and M. Okano; *J. Org. Chem.*, 1981, **46**, 3021. 282

81JOC3056 J. V. Crivello; *J. Org. Chem.*, 1981, **46**, 3056. 791

81JOC3113 G. W. Kabalka, K. A. R. Sastry, H. C. Hsu and M. D. Hylarides; *J. Org. Chem.*, 1981, **46**, 3113. 4, 30

81JOC3141 M. W. Fichtner and N. F. Haley; *J. Org. Chem.*, 1981, **46**, 3141. 123

81JOC3249 T. G. Back and S. Collins; *J. Org. Chem.*, 1981, **46**, 3249. 285

81JOC3312 S. Torii, K. Uneyama, H. Takana, T. Yamanaka, T. Yasuda, M. Ono and Y. Kohmoto; *J. Org. Chem.*, 1981, **46**, 3312. 30

81JOC3315 V. Ceré, C. Paolucci, S. Pollicino, E. Sandri and A. Fava; *J. Org. Chem.*, 1981, **46**, 3315. 143

81JOC3349 J. J. Fitt and H. W. Gschwend; *J. Org. Chem.*, 1981, **46**, 3349. 33

81JOC3372 R. J. P. Corriu, J. J. E. Moreau and M. Pataud-Sat; *J. Org. Chem.*, 1981, **46**, 3372. 814

81JOC3467 F. A. Davis, R. H. Jenkins, Jr., A. Q. A. Rizvi and S. G. Yocklovich; *J. Org. Chem.*, 1981, **46**, 3467. 718

81JOC3727 G. A. Olah, S. C. Narang and L. D. Field; *J. Org. Chem.*, 1981, **46**, 3727. 35

81JOC3773 H. G. Richey, Jr., L. M. Moses, M. S. Donalski, W. R. Erickson and A. S. Heyn; *J. Org. Chem.*, 1981, **46**, 3773. 561

81JOC3978 H. C. Brown and J. C. Chen; *J. Org. Chem.*, 1981, **46**, 3978. 517

81JOC4093 C. L. Rand, D. E. Van Horn, M. W. Moore, and E. Negishi; *J. Org. Chem.*, 1981, **46**, 4093. 975

81JOC4194 L. Engman and M. F. Cava; *J. Org. Chem.*, 1981, **41**, 4194. 290

81JOC4296 G. W. Kabalka, K. A. R. Sastry, G. W. McCullum and H. Yoshioka; *J. Org. Chem.*, 1981, **46**, 4296. 305

81JOC4305 G. A. Olah, A. P. Fung and T. Keumi; *J. Org. Chem.*, 1981, **46**, 4305. 650

81JOC4313 H. C. Brown, U. S. Racherla and H. Taniguchi; *J. Org. Chem.*, 1981, **46**, 4313. 523

81JOC4323 R. D. Rieke, P. T. J. Li, T. P. Burns and S. J. Uhm; *J. Org. Chem.*, 1981, **46**, 4323. 590

81JOC4324 G. F. Koser, L. Rebrovic and R. H. Wettach; *J. Org. Chem.*, 1981, **46**, 4324. 1020

81JOC4416 H. Horino and N. Inoue; *J. Org. Chem.*, 1981, **46**, 4416. 964

81JOC4422 C. F. Hobbs and W. S. Knowles; *J. Org. Chem.*, 1981, **46**, 4422. 580, 822

81JOC4486 D. Masilamani and M. M. Rogić; *J. Org. Chem.*, 1981, **46**, 4486. 21, 619

81JOC4599 J. A. Soderquist and H. C. Brown; *J. Org. Chem.*, 1981, **46**, 4599. 516, 517

81JOC4629 O. Lerman, Y. Tor and S. Rozen; *J. Org. Chem.*, 1981, **46**, 4629. 613

81JOC4727 A. Toshimitsu, T. Aoai, H. Owada, S. Uemura and M. Okano; *J. Org. Chem.*, 1981, **46**, 4727. 284

81JOC4894 J. L. Kice and K. Krowicki; *J. Org. Chem.*, 181, **46**, 4894. 725

81JOC4899 R. A. Gancarz and J. L. Kice; *J. Org. Chem.*, 1981, **46**, 4899. 285

81JOC5047 H. C. Brown and P. K. Yadhav; *J. Org. Chem.*, 1981, **46**, 5047. 516

81JOC5215 T. Fuchigami and T. Nonaka; *J. Org. Chem.*, 1981, **46**, 5215. 160

81JOC5239 H. Ku and J. R. Barrio; *J. Org. Chem.*, 1981, **46**, 5239. 632
81JOC5357 M. C. Croudace and N. E. Schore; *J. Org. Chem.*, 1981, **46**, 5357. 985
81JOC5408 R. M. J. Liskamp, H. J. M. Zeegers and H. C. J. Ottenheijm; *J. Org. Chem.*, 1981, **46**, 5408. 199
81JOC5413 N. I. Ghali, D. L. Venton, S. C. Hung and G. C. Le Breton; *J. Org. Chem.*, 1981, **46**, 5413. 374
81JOC5436 N. E. Schore and M. C. Croudace; *J. Org. Chem.*, 1981, **46**, 5436. 985
81JOM(204)315 M. Hanack, K. Mitulla, G. Pawlowski and L. R. Subramanian; *J. Organomet. Chem.*, 1981, **204**, 315. 1089
81JOM(205)161 B. N. Storhoff, D. P. Harper, I. Saval and J. H. Worstell; *J. Organomet. Chem.*, 1981, **205**, 161. 824
81JOM(208)161 M. Birkhahn, P. Krommes, W. Massa and J. Lorberth; *J. Organomet. Chem.*, 1981, **208**, 161. 485
81JOM(210)163 A. Chihi and P. Weber; *J. Organomet. Chem.*, 1981, **210**, 163. 266
81JOM(214)415 L. E. Guselnikov, V. M. Sokolova, E. A. Volnina, V. G. Zaikin, N. J. Nametkin, M. Voronkov, S. V. Kirpichenko and V. V. Keiko; *J. Organomet. Chem.*, 1981, **214**, 415. 266
81JOM(216)149 H. Watanabe, M. Kobayashi, M. Saito and Y. Nagai; *J. Organomet. Chem.*, 1981, **216**, 149. 915, 923
81JOM(216)287 J. V. Comasseto, J. T. B. Ferreira and N. Petragnani; *J. Organomet. Chem.*, 1981, **216**, 287. 1036, 1037
81JOM(218)309 S. Kuno and Y. Sato; *J. Organomet. Chem.*, 1981, **218**, 309. 759
81LA943 H. Quast, M. Heuschmann and M. O. Abdel-Rahman; *Liebigs Ann. Chem.*, 1981, 943. 377
81LA1415 H. Volz and M. D. Mayer; *Liebigs Ann. Chem.*, 1981, 1415. 1006
81LA1844 G. Himbert and W. Schwickerath; *Liebigs Ann. Chem.*, 1981, 1844. 1044
81LA1907 G. Himbert, M. Feustel and M. Jung; *Liebigs Ann. Chem.*, 1981, 1907. 1044, 1048
B-81MI 202-01 R. A. Sheldon and J. K. Kochi; "Metal-Catalyzed Oxidations of Organic Chemistry," Academic Press, New York, 1981, p. 278. 90
B-81MI 203-01 Y. Segall, J. E. Casida; in "ACS Symposium. Ser." 171 337 (1981) (*Chem. Abstr.*, 1981, **96**, 69 111). 210
B-81MI 209-01 M. Wieber (ed.); "Gmelin Handbuch der Anorganischen Chemie," Springer-Verlag, Berlin, 1981, part 1. 481
B-81MI 210-01 E. W. Colvin; "Silicon in Organic Synthesis," Butterworths, London, 1981. 535
81MI 215-01 V. N. Yandovskii, I. I. Ryabinkin and I. V. J. Tselinskii; *General Chem. USSR* (*Engl. Transl.*), 1981, **66**, 1773. 786
B-81MI 218-01 H. C. Brown; in "Organic Synthesis Today and Tomorrow," eds B. M. Trost and C. R. Hutchinson, Pergamon, Oxford, 1981, p. 121. 900
B-81MI 223-01 M. Wieber; in "Gmelin Handbook of Inorganic Chemistry, Organoantimony Compounds," Part 1, ed. H. Bitterer, Springer, Berlin, 1981, p. 40. 1082
B-81MI 223-02 M. Wieber; in "Gmelin Handbook of Inorganic Chemistry, Organoantimony Compounds," Part 1, ed. H. Bitterer, Springer, Berlin, 1981, p. 136. 1082
81NKK472 H. Suzuki, K. Miyoshi and A. Osuka; *Nippon Kagaku Kaishi* (*J. Chem. Soc. Jpn.*), 1981, 472. 290
81PAC2333 E. Negishi; *Pure Appl. Chem.*, 1981, **53**, 2333. 974, 975
81PS153 S. Oae, M. Fukumura and H. Furukawa; *Phosphorus Sulfur*, 1981, **10**, 153. 292
81PS(11)349 L. Horner, R. Gehring and H. Lindel; *Phosphorus Sulfur*, 1981, **11**, 349. 254
81RCR151 K. A. V'yunov and A. I. Ginak; *Russ. Chem. Rev.* (*Engl. Transl.*), 1981, **50**, 151. 5
81S1 O. Mitsunobu; *Synthesis*, 1981, 1. 6, 87, 118, 124, 299, 683
81S53 S. Tanimoto, S. Jo and T. Sugimoto; *Synthesis*, 1981, 53. 445
81S149 I. Degani, R. Fochi and V. Regondi; *Synthesis*, 1981, 149. 180
81S165 A. J. Mancuso and D. Swern; *Synthesis*, 1981, 165. 1055
81S185 G. Solladie; *Synthesis*, 1981, 185. 997
81S204 Y. Watanabe, T. Numata and S. Oae; *Synthesis*, 1981, 204. 145
81S229 W. Hanefield and D. Kluck; *Synthesis*, 1981, 229. 157
81S288 J. A. Miller and G. Zweifel; *Synthesis*, 1981, 288. 902
81S305 J. J. Oltvoort, C. A. A. van Boeckel, J. H. De Koning and J. H. van Boom; *Synthesis*, 1981, 305. 675
81S312 C. A. Horiuchi and J. Y. Satoh; *Synthesis*, 1981, 312. 33
81S368 P. A. T. W. Porskamp and B. Zwanenburg; *Synthesis*, 1981, 368. 399
81S381 G. Roberge and P. Brassard; *Synthesis*, 1981, 381. 667
81S441 H. C. Brown, S. Narasunhan and Y. M. Choi; *Synthesis*, 1981, 441. 320
81S457 M. H. Goodrow and W. K. Musker; *Synthesis*, 1981, 457. 215
81S537 M. Kluba and A. Zwierzak; *Synthesis*, 1981, 537. 377
81S547 Y. Ueno, L. D. S. Yadav and M. Okawara; *Synthesis*, 1981, 547. 256
81S585 Y. H. Lai; *Synthesis*, 1981, 585. 558, 958
81S793 G. Cardillo, M. Orena, G. Porzi and S. Sandri; *Synthesis*, 1981, 793. 87
81S811 T. Otsuki, Y. Okamoto and H. Sakurai; *Synthesis*, 1981, 811. 459
81S841 J. F. Normant and A. Alexakis; *Synthesis*, 1981, 841. 903, 909, 910, 917, 919, 924, 925, 976, 977, 978
81S875 J.-C. Clinet and G. Linstrumelle; *Synthesis*, 1981, 875. 939
81S880 M. Azzaro, S. Geribaldi and B. Videau; *Synthesis*, 1981, 880. 744
81S888 G. Palumbo and R. Caputo; *Synthesis*, 1981, 888. 724
81S892 H. J. Cristau, B. Chabaud, A. Chêne and H. Christol; *Synthesis*, 1981, 892. 711
81S999 H. P. On, W. Lewis and G. Zweifel; *Synthesis*, 1981, 999. 922, 931
81SC247 G. W. Kabalka, E. E. Gooch and H. C. Hsu; *Synth. Commun.*, 1981, **11**, 247. 628
81SC405 G. Axelrad, S. Ladsooksathit and R. Engel; *Synth. Commun.*, 1981, **11**, 405. 619
81SC409 A. Singh, A. Mehratra and S. L. Regen; *Synth. Commun.*, 1981, **11**, 409. 132

81SC521	E. E. Gooch and G. W. Kabalka; *Synth. Commun.*, 1981, **11**, 521.	34
81SC639	A. Citterio and A. Arnoldi; *Synth. Commun.*, 1981, **11**, 639.	632
81SC669	A. Chaintreau, G. Adrian and D. Couturier; *Synth. Commun.*, 1981, **11**, 669.	28
81SC925	J. H. Babler and S. J. Sarussi; *Synth. Commun.*, 1981, **11**, 925.	753
81SC1025	J. Drabowicz and M. Mikolajczyk; *Synth. Commun.*, 1981, **11**, 1025.	144
81T37	S. Oae, T. Takata and Y. H. Kim; *Tetrahedron*, 1981, **37**, 37.	193
81T1441	A. Bou, M. A. Pericas and F. Serratosa; *Tetrahedron*, 1981, **37**, 1441.	1026
81T2313	G. Galliani and B. Rindone; *Tetrahedron*, 1981, **37**, 2313.	649
81T2855	A. Hercouet and M. L. Corre; *Tetrahedron*, 1981, **37**, 2855.	831
81T3547	H. C. Brown, P. K. Jadhav and A. K. Mandal; *Tetrahedron*, 1981, **37**, 3547.	519
81T3873	H. Nagaoka and Y. Kishi; *Tetrahedron*, 1981, **37**, 3873.	73
81T3899	R. Noyori, S. Murata and M. Suzuki; *Tetrahedron*, 1981, **37**, 3899.	539
81T4097	K. C. Nicolaou; *Tetrahedron*, 1981, **37**, 4097.	286
81T4245	R. G. Kostyanovskii, V. F. Rudchenko, V. G. Shtamburg, I. I. Chervin and Sh. S. Nasibov; *Tetrahedron*, 1981, **37**, 4245.	356, 357
81TL61	J.-L. Bougeois, L. Stella and J.-M. Surzur; *Tetrahedron Lett.*, 1981, **22**, 61.	337
81TL583	G. l'Abbé, A. van Asch and J.-P. Dekerk; *Tetrahedron Lett.*, 1981, **22**, 583.	245, 250
81TL587	D. J. Ager; *Tetrahedron Lett.*, 1981, **22**, 587.	908
81TL603	E. J. Corey and H. Estreicher; *Tetrahedron Lett.*, 1981, **22**, 603.	665, 784, 785
81TL623	U. Husstedt and H. J. Schafer; *Tetrahedron Lett.*, 1981, **22**, 623.	30
81TL1017	H. Kise, K. Araki and M. Seno; *Tetrahedron Lett.*, 1981, **22**, 1017.	35
81TL1231	O. I. Kolodiaznyi; *Tetrahedron Lett.*, 1981, **22**, 1231.	453
81TL1287	B. M. Trost and D. P. Curran; *Tetrahedron Lett.*, 1981, **22**, 1287.	155
81TL1579	R. G. Daniels and L. A. Paquette; *Tetrahedron Lett.*, 1981, **22**, 1579.	940
81TL1787	S. Djuric, J. Venit and P. D. Magnus; *Tetrahedron Lett.*, 1981, **22**, 1787.	543, 811
81TL1859	S. L. Beaucage and M. H. Caruthers; *Tetrahedron Lett.*, 1981, **22**, 1859.	409, 410
81TL1875	E. P. Kyba, S. P. Rhines, P. W. Owens and S.-S. P. Chou; *Tetrahedron Lett.*, 1981, **22**, 1875.	1071
81TL1919	J.-E. Backvall and L. Engman; *Tetrahedron Lett.*, 1981, **22**, 1919.	289
81TL2291	S. Torii, K. Uneyama, T. Nakai and T. Yasuda; *Tetrahedron Lett.*, 1981, **22**, 2291.	21
81TL2337	D. V. Rao and F. A. Stuber; *Tetrahedron Lett.*, 1981, **22**, 2337.	655
81TL2491	A. Cravador and A. Krief; *Tetrahedron Lett.*, 1981, **22**, 2491.	281
81TL2495	H. Hommes, H. D. Verkruijsse and L. Bransma; *Tetrahedron Lett.*, 1981, 2495.	621
81TL2751	D. J. S. Tsai and D. S. Matteson; *Tetrahedron Lett.*, 1981, **22**, 2751.	529, 530, 935
81TL2895	Y. Yamamoto and K. Maruyama; *Tetrahedron Lett.*, 1981, **22**, 2895.	566
81TL2985	R. J. P. Corriu, C. Guerin and J. M'Boula; *Tetrahedron Lett.*, 1981, **22**, 2985.	935
81TL3079	T. Hirao, J. Enda, Y. Ohshiro and T. Agawa; *Tetrahedron Lett.*, 1981, **22**, 3079.	932
81TL3119	R. P. Volante; *Tetrahedron Lett.*, 1981, **22**, 3119.	118
81TL3179	J.-P. Dulcere, J. Grimaldi and M. Santelli; *Tetrahedron Lett.*, 1981, **22**, 3179.	940
81TL3193	S. Torii, H. Tanaka, N. Saitoh, T. Siroi, M. Sasaoka and J. Nokami; *Tetrahedron Lett.*, 1981, **22**, 3193.	21
81TL3249	T. Shono, Y. Matsumura and K. Tsubata; *Tetrahedron Lett.*, 1981, **22**, 3249.	457
81TL3347	M. Shen and A. G. Schultz; *Tetrahedron Lett.*, 1981, **22**, 3347.	1034
81TL3401	J. P. Pillot, B. Bennetau, J. Dunogues and R. Calas; *Tetrahedron Lett.*, 1981, **22**, 3401.	940
81TL3451	T. Kusumi, C. C. Chang, M. Wheeler, I. Kubo, K. Nakanishi and H. Naoki; *Tetrahedron Lett.*, 1981, **22**, 3451.	655
81TL3455	E. J. Corey, H. Cho, C. Rücker and D. H. Hna; *Tetrahedon Lett.*, 1981, **22**, 3455.	539, 699
81TL3561	M. Jefson and J. Meinwald; *Tetrahedron Lett.*, 1981, **22**, 3561.	282
81TL3573	J. F. King and M. Aslam; *Tetrahedron Lett.*, 1981, **22**, 3573.	204
81TL3579	S. Hanessian and J.-M. Vatele; *Tetrahedron Lett.*, 1981, **22**, 3579.	9
81TL3659	J. M. Majenski and J. Zakrzewski; *Tetrahedron Lett.*, 1981, **22**, 3659.	221
81TL3745	D. R. Williams, K. Nishitani, W. Bennett and S. Y. Sit; *Tetrahedron Lett.*, 1981, 3745.	625
81TL4263	U. Redeker, N. Engel and W. Steglich; *Tetrahedron Lett.*, 1981, **22**, 4263.	747
81TL4283	Y. Tamura, A. Wada, M. Sasho and Y. Kita; *Tetrahedron Lett.*, 1981, **22**, 4283.	665
81TL4409	N. Furukawa, K. Kishimoto, S. Ogawa, T. Kawai, H. Fujihara and S. Oae; *Tetrahedron Lett.*, 1981, **22**, 4409.	25
81TL4475	K. Issleib, E. Leissring and H. Meyer; *Tetrahedron Lett.*, 1981, **22**, 4475.	820, 821
81TL4691	M. T. Reetz and R. Peter; *Tetrahedron Lett.*, 1981, **22**, 4691.	702
81TL4705	F. Plénat; *Tetrahedron Lett.*, 1981, **22**, 4705.	936
81TL4999	B. M. Trost and C. G. Caldwell; *Tetrahedron Lett.*, 1981, **22**, 4999.	540
81TL5019	S. G. Hegde and J. Wolinsky; *Tetrahedron Lett.*, 1981, **22**, 5019.	21
81TL5073	G. Fronza, C. Fuganti, P. Grasselli and G. Pedrocci-Fantoni; *Tetrahedron Lett.*, 1981, **22**, 5073.	359
81TL5111	T. G. Back and S. Collins; *Tetrahedron Lett.*, 1981, **22**, 5111.	285
81USP4296042	J. M. Muchowski and M. C. Venuti; *US Pat.* 4 296 042 (1981) (*Chem. Abstr.*, 1982, **96**, 683 610).	445
81ZAAC(472)75	H. T. Phung, P. B. Chi and F. Kober; *Z. Anorg. Allg. Chem.*, 1981, **472**, 75.	494
81ZAAC(473)80	D. Hänssgen and B. Ross; *Z. Anorg. Allg. Chem.*, 1981, **473**, 80.	400
81ZAAC(474)157	N. Tempel, W. Schwarz and J. Weidlein; *Z. Anorg. Allg. Chem.*, 1981, **474**, 157.	1082
81ZAAC(480)90	U. Ahrens and H. Falius; *Z. Anorg. Allg. Chem.*, 1981, **480**, 90.	852
81ZN(B)48	P. Dehnert, J. Grobe and D. Le Van; *Z. Naturforsch., Teil B*, 1981, **36**, 48.	894
81ZN(B)70	U. Praeckel and F. Huber; *Z. Naturforsch., Teil B*, 1981, **36**, 70.	501
81ZN(B)447	R. Appel and V. I. Glasel; *Z. Naturforsch., Teil B*, 1981, 36, 447.	259
81ZN(B)515	O. J. Scherer and H. Conrad; *Z. Naturforsch., Teil B*, 1981, **36**, 515.	407, 420

81ZN(B)1105 H. J. Breunig and H. Kischkel; *Z. Naturforsch., Teil B*, 1981, **36**, 1105. 493, 495, 499, 500
81ZN(B)1265 C. Bihlmayer and B. Wrackmeyer; *Z. Naturforsch., Teil B*, 1981, **36**, 1265. 903
81ZOB331 L. N. Markovskii, V. D. Romanenko, V. I. Dzyuba, A. V. Ruban, N. N. Kalibabchuk and S. V. Iksanova; *Zh. Obshch. Khim.*, 1981, **51**, 331 (*Chem. Abstr.*, 1981, **95**, 150738x). 408
81ZOB609 V. A. Ignatov, R. A. Akchurua, P. A. Pirogov and R. S. Bairakova; *Zh. Obshch. Khim.*, 1981, **51**, 609. 359
81ZOB2140 M. G. Zimin, A. R. Burilov and A. N. Pudovik; *Zh. Obshch. Khim.*, 1981, **51**, 2140 (*Chem. Abstr.*, 1981, **95**, 220 104b). 266

82ACR306 P. Beak and V. Snieckus; *Acc. Chem. Res.* 1982, **15**, 306. 965
82ACR340 E. Negishi; *Acc. Chem. Res.*, 1982, **15**, 340. 992
82AG(E)144 T. Kauffmann, A. Hamsen and C. Beirich; *Angew. Chem., Int. Ed. Engl.*, 1982, **21**, 144. 571
82AG(E)199 B. Bogdanovic, M. Schwickardi and P. Sikoiskij; *Angew. Chem., Int. Ed. Engl.*, 1982, **21**, 199. 561
82AG(E)292 W. J. Richter; *Angew. Chem., Int. Ed. Engl.*, 1982, **21**, 292. 436
82AG(E)375 J. Ellermann and A. Veit; *Angew. Chem., Int. Ed. Engl.*, 1982, **21**, 375 490, 507
82AG(E)410 T. Kauffmann; *Angew. Chem., Int. Ed. Engl.*, 1982, **21**, 410. 908, 910, 911
82AG(E)439 H. J. Breunig and D. Muller; *Angew. Chem., Int. Ed. Engl.*, 1982, **21**, 439. 505
82AG(E)555 R. W. Hoffmann; *Angew. Chem., Int. Ed. Engl.*, 1982, **21**, 555. 529
82AHC(2)1 A. T. Balaban, A. Dinculescu, G. N. Dorofeenko, G. W. Fischer, A. V. Koblik, V. V. Mezheritskii and W. Schroth; *Adv. Heterocycl. Chem.*, 1982, **2**, 1. 1008
82AHC(30)167 F. M. Dean; *Adv. Heterocycl. Chem.*, 1982, **30**, 167. 999
82AJC517 R. K. Haynes and M. Holden; *Aust. J. Chem.*, 1982, **35**, 517. 7
82AX(B)1124 C. Esteban-Calderon, M. Martinez-Ripoll and S. Garcia-Blanco; *Acta Crystallogr., Part B*, 1982, **38**, 1124. 1002
82BCJ561 T. Hiyama, Y. Okude, K. Kimura and H. Nozaki; *Bull. Chem. Soc. Jpn.*, 1982, **55**, 561. 73, 571
82BCJ635 T. Sakakibara, Y. Ikeda and R. Sudoh; *Bull. Chem. Soc. Jpn.*, 1982, **55**, 635. 786
82BCJ909 T. Hirao, T. Masunaga, N. Yamada, Y. Ohshiro and T. Agawa; *Bull. Chem. Soc. Jpn.*, 1982, **55**, 909. 850
82BSF395 G. Courtois and P. Miginiac; *Bull. Soc. Chim. Fr.*, 1982, 395. 931
82BSF(2)297 L. Duhamel and J. M. Poirier; *Bull. Soc. Chim. Fr., Part 2*, 1982, 297. 759
82BSF(2)327 M. Ouertani, P. Girard and H. B. Kagan; *Bull. Soc. Chim. Fr., Part 2*, 1982, 327. 27
82CB185 E. Niecke and H.-C. Schäfer; *Chem. Ber.*, 1982, **115**, 185. 408
82CB399 E. Hirsch, S. Hünig and H.-U. Rei*ß*ig; *Chem. Ber.*, 1982, **115**, 399. 689
82CB654 T. Kauffmann and R. Joussen; *Chem. Ber.*, 1982, **115**, 654. 877
82CB1947 K. Diemert, W. Kuchen and J. Kutter; *Chem. Ber.*, 1982, **115**, 1947. 449
82CB2181 E. Lindner, M. Steinwand and S. Hoehne; *Chem. Ber.*, 1982, **115**, 2181. 836
82CB2357 R. W. Hoffmann, H.-J. Zeiss, W. Ladner and S. Tabche; *Chem. Ber.*, 1982, **115**, 2357. 529
82CB3687 E. Hirsch, S. Hünig and H.-U. Rei*ß*ig; *Chem. Ber.*, 1982, **115**, 3687. 689
82CC14 M. F. Lappert, C. L. Raston, B. W. Skelton and A. H. White; *J. Chem. Soc., Chem. Commun.*, 1982, 14. 1005
82CC438 T. Miura and M. Kobayashi; *J. Chem. Soc., Chem. Commun.*, 1982, 438. 285
82CC551 D. H. R. Barton, W. B. Motherwell and S. Z. Zard; *J. Chem. Soc., Chem. Commun.*, 1982, 551. 785
82CC625 M. J. Adam, T. J. Ruth, B. D. Pate and L. D. Hall; *J. Chem. Soc., Chem. Commun.*, 1982, 625. 627
82CC691 B. Cetinkaya, P. B. Hitchcock, M. F. Lappert, A. J. Thorne and H. Coldwhite; *J. Chem. Soc., Chem. Commun.*, 1982, 691. 852
82CC778 B. Fuchs, Y. Belsky, E. Tartakovsky, J. Zizvashvili and S. Weinman; *J. Chem. Soc., Chem. Commun.*, 1982, 778. 626
82CC871 S. Tomoda, Y. Takeuchi and Y. Nomura; *J. Chem. Soc., Chem. Commun.*, 1982, 871. 284
82CC1157 E. Santaniello, A. Fiecchi and P. Ferraboschi; *J. Chem. Soc., Chem. Commun.*, 1982, 1157. 653
82CC1362 A. Lopusinski, J. Michaelski and M. Potrzebrowski; *J. Chem. Soc., Chem. Commun.*, 1982, 1362. 256
82CJC2711 T. G. Back and R. G. Kerr; *Can. J. Chem.*, 1982, **60**, 2711. 388
82CL315 T. Osawa and T. Harada; *Chem. Lett.*, 1982, 315. 52
82CL353 T. Mukaiyama, R. W. Stevens and N. Iwasawa; *Chem. Lett.*, 1982, 353. 701
82CL921 R. Matsueda, S. Higashida, R. J. Ridge and E. R. Matsueda; *Chem. Lett.*, 1982, 921. 230
82CL977 T. Okano, M. Yamamoto, T. Noguchi, H. Konishi and J. Kiji; *Chem. Lett.*, 1982, 977. 434
82CL1109 T. Hayama, S. Tomada, Y. Takeuchi and Y. Nomura; *Chem. Lett.*, 1982, 1109. 785
82CL1244 S. Tomoda, Y. Takeuchi and Y. Nomura; *Chem. Lett.*, 1982, 1244. 1036, 1038
82CL1481 T. Sugita, M. Idei, Y. Ishibashi and Y. Takegami; *Chem. Lett.*, 1982, 1481. 631
82CL1505 M. Miyashita, R. Yamaguchi and A. Yoshikoshi; *Chem. Lett.*, 1982, 1505. 783
82CL1697 K. Orgura, M. Fujita, K. Takahashi and H. Iida; *Chem. Lett.*, 1982, 1697. 1000
82CL1733 S. Tomoda, Y. Takeuchi and Y. Nomura; *Chem. Lett.*, 1982, 1733. 284
82CL2007 M. Furukawa, K. Sato and T. Okawara; *Chem. Lett.*, 1982, 2007. 725
82COMC-I(1)43 J. L. Wardell; *Comp. Organomet. Chem., 1st edn.*, 1982, **1**, 43. 551
82COMC-I(1)311 J. H. Morris; *Comp. Organomet. Chem., 1st edn.*, 1982, **1**, 311. 418
82COMC-I(1)555 J. J. Eisch; *Comp. Organomet. Chem., 1st edn.*, 1982, **1**, 555. 967
82COMC-I(1)599 J. J. Eisch; *Comp. Organomet. Chem. 1st edn.*, 1982, **1**, 599. 815
82COMC-I(1)683 D. G. Tuck; *Comp. Organomet. Chem., 1st edn.*, 1982, **1**, 683. 597
82COMC-I(1)725 H. Kurozawa; *Comp. Organomet. Chem., 1st edn.*, 1982, **1**, 725. 597

82COMC-I(2) J. L. Wardell; *Comp. Organomet. Chem.*, *1st edn.*, ed. G. Wilkinson, F. G. A. Stone and E. W. Abel, Pergamon, Oxford, 1982, vol. 2, chap. 13. 872

82COMC-I(2)1 D. A. Armitage; *Comp. Organomet. Chem.*, *1st edn.*, 1982, **2**, 1. 415, 907

82COMC-I(2)121 N. A. Bell; *Comp. Organomet. Chem.*, *1st edn.*, 1982, **2**, 121. 563

82COMC-I(2)155 W. E. Lindsell; *Comp. Organomet. Chem.*, *1st edn.*, 1982, **2**, 155. 557

82COMC-I(2)399 P. Rivière, M. Rivière-Baudet and J. Satgé; *Comp. Organomet. Chem.*, *1st edn*, 1982, **2**, 399. 946, 948

82COMC-I(2)519 A. G. Davies and P. J. Smith; *Comp. Organomet. Chem.*, *1st edn.*, 1982, **2**, 519. 599

82COMC-I(2)629 P. G. Harrison; *Comp. Organomet. Chem.*, *1st edn.*, 1982, **2**, 629. 601

82COMC-I(2)662 P. G. Harrison; *Comp. Organomet. Chem.*, *1st edn.*, 1982, **2**, 662. 816

82COMC-I(2)709 G. Van Koten and J. G. Noltes, *Comp. Organomet. Chem.*, *1st edn.*, 1982, **2**, 709. 585, 589

82COMC-I(2)765 R. J. Puddephatt; *Comp. Organomet. Chem.*, *1st edn.*, 1982, **2**, 765. 589

82COMC-I(2)823 J. Boersma; *Comp. Organomet. Chem.*, *1st edn.*, 1982, **2**, 823. 590

82COMC-I(2)863 J. L. Wardell; *Comp. Organomet. Chem.*, *1st edn.*, 1982, **2**, 863. 590

82COMC-I(3)173 T. J. Marks and R. D. Ernst; *Comp. Organomet. Chem.*, *1st edn.*, 1982, **3**, 173. 564, 602

82COMC-I(3)281 M. Bottrill, P. D. Gravens and J. McMeeking; *Comp. Organomet. Chem.*, *1st edn.*, 1982, **3**, 281. 565

82COMC-I(3)331 M. Bottrill, P. D. Gravens, J. W. Kelland and J. McMeeking; *Comp. Organomet. Chem.*, *1st edn.*, 1982, **3**, 331. 565, 575

82COMC-I(3)433 M. Bottrill, P. D. Gravens, J. W. Kelland and J. McMeeking; *Comp. Organomet. Chem.*, *1st edn.*, 1982, **3**, 433. 565

82COMC-I(3)559 D. J. Cardin, M. F. Lappert, C. L. Raston and P. I. Riley; *Comp. Organomet. Chem.*, *1st edn.*, 1982, **3**, 559. 565

82COMC-I(3)635 D. J. Cardin, M. F. Lappert, C. L. Raston and P. I. Riley; *Comp. Organomet. Chem.*, *1st edn.*, 1982, **3**, 635. 565

82COMC-I(3)647 N. G. Connelly; *Comp. Organomet. Chem.*, *1st edn.*, 1982, **3**, 647. 569

82COMC-I(3)705 J. A. Labinger; *Comp. Organomet. Chem.*, *1st edn.*, 1982, **3**, 705. 569

82COMC-I(3)783 S. W. Kirtley; *Comp. Organomet. Chem.*, *1st edn.*, 1982, **3**, 783. 571

82COMC-I(3)1079 S. W. Kirtley; *Comp. Organomet. Chem.*, *1st edn.*, 1982, **3**, 1079. 571

82COMC-I(3)1255 S. W. Kirtley; *Comp. Organomet. Chem.*, *1st edn.*, 1982, **3**, 1255. 571

82COMC-I(4)1 P. M. Treichel; *Comp. Organomet. Chem.*, *1st edn.*, 1982, **4**, 1. 573

82COMC-I(4)161 N. M. Boag; *Comp. Organomet. Chem.*, *1st edn.*, 1982, **4**, 161. 579

82COMC-I(4)331 M. D. Johnson; *Comp. Organomet. Chem.*, *1st edn.*, 1982, **4**, 331. 551

82COMC-I(4)691 M. A. Bennett, M. I. Bruce and T. W. Matheson; *Comp. Organomet. Chem.*, *1st edn.*, 1982, **4**, 691. 579

82COMC-I(4)967 R. D. Adams and J. P. Selegue; *Comp. Organomet. Chem.*, *1st edn.*, 1982, **4**, 967. 579

82COMC-I(5)1 R. D. W. Kemmet and D. R. Russell; *Comp. Organomet. Chem.*, *1st edn.*, 1982, **5**, 1. 575

82COMC-I(5)277 R. P. Hughes; *Comp. Organomet. Chem.*, *1st edn.*, 1982, **5**, 277. 579

82COMC-I(5)541 G. J. Leigh and R. L. Richards; *Comp. Organomet. Chem.*, *1st edn.*, 1982, **5**, 541. 579

82COMC-I(6)37 P. W. Polly; *Comp. Organomet. Chem.*, *1st edn.*, 1982, **6**, 37. 575

82COMC-I(6)279 P. M. Maitlis, P. Espinet and M. J. H. Russell; *Comp. Organomet. Chem.*, *1st edn.*, 1982, **6**, 279. 579

82COMC-I(6)471 F. R. Hartley; *Comp. Organomet. Chem.*, *1st edn.*, 1982, **6**, 471. 579

82COMC-I(7)1 B. J. Wakefield; *Comp. Organomet. Chem.*, *1st edn.*, 1982, **7**, 1. 551

82COMC-I(7)111 H. C. Brown; *Comp. Organomet. Chem.*, *1st edn.*, 1982, **7**, 111. 514, 900

82COMC-I(7)143 M. Zaidlewicz; *Comp. Organomet. Chem.*, *1st edn.*, 1982, **7**, 143. 900

82COMC-I(7)161 M. Zaidlewicz; *Comp. Organomet. Chem.*, *1st edn.*, 1982, **7**, 161. 900

82COMC-I(7)303 E. Negishi; *Comp. Organomet. Chem.*, *1st edn.*, 1982, **7**, 303. 900

82COMC-I(7)323 E. Negishi; *Comp. Organomet. Chem.*, *1st edn.*, 1982, **7**, 323. 900

82COMC-I(7)515 P. D. Magnus, T. Sarkar and S. Djuric; *Comp. Organomet. Chem.*, *1st edn.*, 1982, **7**, 515. 907

82CPB332 H. Uno and M. Kurokawa; *Chem. Pharm. Bull.*, 1982, **30**, 332. 205

82CPB1646 Y. Noguichi, M. Isoda, K. Kuroki and M. Furukawa; *Chem. Pharm. Bull.*, 1982, **30**, 1646. 363

82CRV15 L. D. Freedman and G. O. Doak; *Chem. Rev.*, 1982, **82**, 15. 481, 872

82CZ441 V. L. Rösch and G. Altnau; *Chem.-Ztg.*, 1982, **106**, 441. 918, 923

82H(18)357 Y. Yamamoto and K. Maruyama; *Heterocycles*, 1982, **18**, 357. 67

82HCA1101 D. Seebach, A. K. Beck, T. Mukhopadhyay and E. Thomas; *Helv. Chim. Acta*, 1982, **65**, 1101. 394

82HOU106 G. Elsner; *Methoden Org. Chem.* (*Houben-Weyl*), 1982, 106–178. 426, 427, 441

82HOU(13/3b)1 R. Köster and W. Siebert; *Methoden Org. Chem.* (*Houben-Weyl*), 1982, **13/3b**, 1. 1083

82HOU(13/3b)69 R. Köster; *Methoden Org. Chem.* (*Houben-Weyl*), 1982, **13/3b**, 69. 1085

82HOU(E1)1 M. Regitz; *Methoden Org. Chem.* (*Houben-Weyl*), 1982, **E1**, 1. 478

82HOU(E2)798 U.-H. Felcht; *Methoden Org. Chem.* (*Houben-Weyl*), 1982, **E2**, 798. 406

82IC1007 D. G. Allen, G. M. McLaughlin, G. B. Robertson, W. L. Steffen, G. Salem and S. B. Wild; *Inorg. Chem.*, 1982, **21**, 1007. 442

82IC2545 M. P. Snyder and D. W. Margerum; *Inorg. Chem.*, 1982, **21**, 2545. 337

82IC3568 R. H. Neilson and P. Nisian-Neilson; *Inorg. Chem.*, 1982, **21**, 3568. 466

82IC3731 R. J. Barton, S. Hu, L. M. Mihichuk and B. E. Robertson; *Inorg. Chem.*, 1982, **21**, 3731. 897

82IJC(A)305 T. N. Srivastava and M. Singh; *Indian J. Chem.*, *sect A*, 1982, **21**, 305. 238

82IS175 J. E. Hoots, T. B. Rauchfuss and D. A. Wrobleski; *Inorg. Synth.*, 1982, **21**, 175. 827

82IZV933 T. A. Yagudeev, G. M. Dzhakiyaev, A. N. Nurgalieva and N. N. Godovikov; *Izv. Akad. Nauk SSSR, Ser. Khim.*, 1982, 933 (*Chem. Abstr.*, 1982, **97**, 55 936t). 1076

82IZV1123 R. A. Bekker and V. Y. Popoka; *Izv. Akad. Nauk SSSR, Ser. Khim.*, 1982, 1123 (*Chem. Abstr.*, 1982, **97**, 91 184e). 168

82IZV2390 O. L. Buthova, L. I. Zvezdkina and N. A. Pritula; *Izv. Akad. Nauk SSSR, Ser. Khim.*, 1982, 2390 (*Chem. Abstr.*, 1983, **98**, 72 261). 436

82JA290 A. Lopusinski and J. Michalski; *J. Am. Chem. Soc.*, 1982, **104**, 290. 475

82JA321 C. J. Kowalski and K. W. Fields; *J. Am. Chem. Soc.*, 1982, **104**, 321. 1023

82JA352 A. H. Janowicz and R. G. Bergman; *J. Am. Chem. Soc.*, 1982, **104**, 352. 962

82JA674 R. Batchelor and T. Birchall; *J. Am. Chem. Soc.*, 1982, **104**, 674. 820

82JA766 W. D. Hinsberg, III, P. G. Schultz and P. B. Dervan; *J. Am. Chem. Soc.*, 1982, **104**, 766. 379, 385, 388, 405

82JA1119 H. J. Reich and M. J. Kelly; *J. Am. Chem. Soc.*, 1982, **104**, 1119. 939

82JA1860 J. W. Bruno, D. G. Kalina, E. A. Mintz and T. J. Marks; *J. Am. Chem. Soc.*, 1982, **104**, 1860. 602

82JA1893 L. D. Quin, E. D. Midlemas, N. S. Rao, R. W. Miller and A. T. McPhail; *J. Am. Chem. Soc.*, 1982, **104**, 1893. 453

82JA2613 M. P. Moder and N. J. Leonard; *J. Am. Chem. Soc.*, 1982, **104**, 2613. 943

82JA2648 J. L. Holmes and F. P. Lossing; *J. Am. Chem. Soc.*, 1982, **104**, 2648. 636

82JA2659 E. R. Koft and A. B. Smith, III; *J. Am. Chem. Soc.*, 1982, **104**, 2659. 617

82JA2831 J. San Filippo, Jr. and J. Silbermann; *J. Am. Chem. Soc.*, 1982, **104**, 2831. 600

82JA3104 K. Steliou and M. Mrani; *J. Am. Chem. Soc.*, 1982, **104**, 3104. 541

82JA3225 B. M. Trost and T. Shibata; *J. Am. Chem. Soc.*, 1982, **104**, 3225. 138

82JA3231 M. C. Caserio and J. K. Kim; *J. Am. Chem. Soc.*, 1982, **104**, 3231. 138

82JA4680 F. D. Marsh, W. B. Farnham, D. J. Sam and B. E. Smart; *J. Am. Chem. Soc.*, 1982, **104**, 4680. 619

82JA4708 D. R. Williams, B. A. Barner, K. Nishitani and J. G. Phillips; *J. Am. Chem. Soc.*, 1982, **104**, 4708. 976

82JA4865 W. A. Kiel, G.-J. Lin, A. G. Constable, F. B. McCormick, C. E. Strouse, O. Eisenstein and J. A. Gladysz; *J. Am. Chem. Soc.*, 1982, **104**, 4865. 580

82JA5284 S. H. Bertz, G. Dabbagh and G. M. Villacata; *J. Am. Chem. Soc.*, 1982, **104**, 5284. 588

82JA5412 F. A. Davis, R. H. Jenkins, S. B. Awad, O. D. Stringer, W. H. Watson and J. Galloy; *J. Am. Chem. Soc.*, 1982, **104**, 5412. 150

82JA5781 P. A. Grieco, J. Inaga, N.-H. Lin and T. Yanami; *J. Am. Chem. Soc.*, 1982, **104**, 5781. 196

82JA5820 A. A. Cowley, J. E. Kilduff, T. H. Newman and M. Pakulski; *J. Am. Chem. Soc.*, 1982, **104**, 5820. 821

82JA5842 M. Riediker and J. Schwartz; *J. Am. Chem. Soc.*, 1982, **104**, 5842. 676, 982

82JA6136 S. Masamune, Y. Hanzawa and D. J. Williams; *J. Am. Chem. Soc.*, 1982, **104**, 6136. 949

82JA6808 S. F. Pedersen, R. R. Schrock, M. R. Churchill and H. J. Wasserman; *J. Am. Chem. Soc.*, 1982, **104**, 6808. 979

82JA6876 E. Grochowski, B. D. Hilton, R. J. Kupper and C. J. Michejda; *J. Am. Chem. Soc.*, 1982, **104**, 6876. 831

82JA7491 J. B. Lee, K. C. Ott and R. H. Grubbs; *J. Am. Chem. Soc.*, 1982, **104**, 7491. 979

82JCR(M)621 L. Duhamel, J. Chauvin and A. Messier; *J. Chem. Res. (M)*, 1982, 621. 759

82JCR(S)48 L. Duhamel, J. Chauvin and A. Messier; *J. Chem. Res. (S)*, 1982, 48. 759

82JCR(S)84 W. J. Spillane and T. J. Hannigan; *J. Chem. Res. (S)*, 1982, 84 367

82JCR(S)150 J. Gallos and A. Varvoglis; *J. Chem. Res. (S)*, 1982, 150. 619, 626, 631

82JCR(S)212 J. Comasseto, J. T. B. Ferreira, C. A. Brandt and N. Petragnani; *J. Chem. Res (S)*, 1982, 212. 279

82JCS(D)1915 I. J. Colquhoun and W. McFarlane; *J. Chem. Soc., Dalton Trans.*, 1982, 1915. 437, 828

82JCS(P1)85 Y. Kobayashi, T. Taguchi, T. Terada, J. Oshida, M. Morisaki and N. Ikekawa; *J. Chem. Soc., Perkin Trans. 1*, 1982, 85. 19

82JCS(P1)111 H. Suginome, N. Maeda and M. Kaji; *J. Chem. Soc., Perkin Trans. 1*, 1982, 111. 104

82JCS(P1)603 R. W. Alder, R. B. Sessions, A. J. Bennett and R. E. Moss; *J. Chem. Soc., Perkin Trans. 1*, 1982, 603. 383, 384

82JCS(P1)677 W. J. Spillane, A. P. Tahery and M. M. Kearns; *J. Chem. Soc., Perkin Trans. 1*, 1982, 677. 367

82JCS(P1)681 P. J. Garegg, R. Johansson, C. Ortega and B. Samuelsson; *J. Chem. Soc., Perkin Trans. 1*, 1982, 681. 7

82JCS(P1)849 P. Bougeard, M. D. Johnson and G. M. Lampman; *J. Chem. Soc., Perkin Trans. 1*, 1982, 849. 176

82JCS(P1)1807 A. J. Bloodworth and J. L. Courtneidge; *J. Chem. Soc., Perkin Trans. 1*, 1982, 1807. 98

82JCS(P1)1963 H. Suginome and A. Osada; *J. Chem. Soc., Perkin Trans. 1*, 1982, 1963. 104

82JCS(P1)2289 J. Cornforth, R. H. Cornforth and R. T. Gray; *J. Chem. Soc., Perkin Trans. 1*, 1982, 2289. 843

82JCS(P1)2299 J. Cornforth, A. F. Sierakowski and T. W. Wallace; *J. Chem. Soc., Perkin Trans. 1*, 1982, 2299. 844

82JCS(P2)1075 F. Ruff and A. Kucsman; *J. Chem. Soc., Perkin Trans. 2*, 1982, 1075. 148

82JFC(20)85 R. Bartch, O. Stelzer and R. Schmutzler; *J. Fluorine Chem.*, 1982, **20**, 85. 475

82JFC(21)385 M. Van der Puy; *J. Fluorine Chem.*, 1982, **21**, 385. 614

82JGU195 G. K. Fedorova, A. G. Matyushe and N. G. Feshchenko; *J. Gen. Chem. USSR (Engl. Transl.)*, 1982, **52**, 195. 463

82JGU202 N. G. Feshchenko, N. D. Gomelya and A. G. Matyushe; *J. Gen. Chem. USSR (Engl. Transl.)*, 1982, **52**, 202. 463

82JMOC333 J.-M. Maissant, C. Bouchoule and M. Blanchard; *J. Mol. Catal.* 1982, **14**, 333. 650

82JOC77 R. A. Alonso and R. A. Rossi; *J. Org. Chem.*, 1982, **47**, 77. 874

82JOC119 Y. Yamamoto, S. Yamamoto, H. Yatagai, Y. Ishihara and K. Maruyama; *J. Org. Chem.*, 1982, **47**, 119. 702

82JOC132 S. J. Cristol and D. G. Seapy; *J. Org. Chem.*, 1982, **47**, 132. 26

82JOC359 D. S. Wilbur and H. A. O'Brien, Jr.; *J. Org. Chem.*, 1982, **47**, 359. 685
82JOC365 P. A. Wade, S. D. Morrow and S. A. Hardinger; *J. Org. Chem.*, 1982, **47**, 365. 394
82JOC552 E. C. Taylor, C.-P. Tseng and J. B. Rampal; *J. Org. Chem.*, 1982, **47**, 552. 160
82JOC573 B. Zajc and M. Zupan; *J. Org. Chem.*, 1982, **47**, 573. 16
82JOC596 G. A. Olah, V. V. Krishnamurthy and S. C. Narang; *J. Org. Chem.*, 1982, **47**, 596. 791
82JOC754 H. C. Brown and D. Basavaiah; *J. Org. Chem.*, 1982, **47**, 754. 901
82JOC863 H. C. Brown, J. A. Sikorski, S. U. Kulkarni and H. D. Lee; *J. Org. Chem.*, 1982, **47**, 863. 522, 526
82JOC872 J. A. Sikorski and H. C. Brown; *J. Org. Chem.*, 1982, **47**, 872. 522, 526
82JOC905 L. D. Quin, E. D. Middlemass and N. S. Rao; *J. Org. Chem.*, 1982, **47**, 905. 440
82JOC951 P. G. Gassman, A. Miura and T. Miura; *J. Org. Chem.*, 1982, **47**, 951. 238
82JOC1200 K. Takaki, M. Okada, M. Yamada and K. Negoro; *J. Org. Chem.*, 1982, **47**, 1200. 663
82JOC1354 M. R. Detty and M. D. Seidler; *J. Org. Chem.*, 1982, **47**, 1354 and references therein. 541, 542
82JOC1588 E. I. Sanchez and M. J. Fumarola; *J. Org. Chem.*, 1982, **47**, 1588. 28
82JOC1869 W. J. Bouma, R. H. Nobes, L. Radom and C. E. Woodward; *J. Org. Chem.*, 1982, **47**, 1869. 1022
82JOC2498 P. G. M. Wuts and S. S. Bigelow; *J. Org. Chem.*, 1982, **47**, 2498. 529
82JOC2768 E. E. Schweizer and K. J. Lee; *J. Org. Chem.*, 1982, **47**, 2768. 446
82JOC2785 W. H. Rastetter, D. M. Spero, J. Adams and D. N. Harpp; *J. Org. Chem.*, 1982, **47**, 2785. 241
82JOC2822 P. Beak and B. J. Kokko; *J. Org. Chem.*, 1982, **47**, 2822. 305
82JOC3121 A. K. Musser and P. L. Fuchs; *J. Org. Chem.*, 1982, **47**, 3121. 158
82JOC3336 T. P. Mawhinney and M. A. Madson; *J. Org. Chem.*, 1982, **47**, 3336. 263
82JOC3773 C. G. Venier, T. G. Squires, Y.-Y. Chen, G. P. Hussmann, J. G. Shei and B. F. Smith; *J. Org. Chem.*, 1982, **47**, 3773. 147, 155
82JOC3808 H. C. Brown, D. Basavaiah and S. U. Kulkarni; *J. Org. Chem.*, 1982, **47**, 3808. 901
82JOC4040 R. L. Crumbie, J. S. Nimitz and H. S. Mosher; *J. Org. Chem.*, 1982, **47**, 4040. 396
82JOC4376 Y. Tamura, A. Wada, M. Sasho, K. Fukunaga, H. Maeda and Y. Kita; *J. Org. Chem.*, 1982, **47**, 4376. 665
82JOC4595 S. E. Denmark and T. K. Jones; *J. Org. Chem.*, 1982, **47**, 4595. 921
82JOC5065 H. C. Brown, M. C. Desai and P. K. Jadhav; *J. Org. Chem.*, 1977, **47**, 5065. 514, 515, 519
82JOC5069 H. C. Brown, A. K. Mandal, N. M. Yoon, B. Singaram, J. R. Schwier and P. K. Jadhav; *J. Org. Chem.*, 1982, **47**, 5069. 523
82JOM(225)43 G. A. Russel, J. Hershberger and K. Owens; *J. Organomet. Chem.*, 1982, **225**, 43. 394
82JOM(226)C9 K. Tamao, M. Akita, R. Kanatani, N. Ishida and M. Kumada; *J. Organomet. Chem.*, 1982, **226**, C9. 923, 924
82JOM(226)143 E. C. Friedrich and G. de Lucca; *J. Organomet. Chem.*, 1982, **226**, 143. 35
82JOM(228)C29 H.-A. Kaul, D. Greissinger, M. Luksza and W. Malisch; *J. Organomet. Chem.*, 1982, **228**, C29. 512
82JOM(231)119 A. L. Rheingold and S. Natarajan; *J. Organomet. Chem.*, 1982, **231**, 119. 507
82JOM(231)237 F. Mercier and F. Mathey; *J. Organomet. Chem.*, 1982, **231**, 237. 441
82JOM(232)C1 H. D. Verkruijsse, W. Verboom, P. E. Van Rijn and L. Brandsma; *J. Organomet. Chem.*, 1982, **232**, C1. 939
82JOM(233)C25 P. E. Van Rijn and L. Brandsma; *J. Organomet. Chem.*, 1982, **233**, C25. 940
82JOM(234)C35 H. Hoberg and B. W. Oster; *J. Organomet. Chem.*, 1982, **234**, C35. 988
82JOM(236)139 A. Doucoure, B. Mauze and L. Miginiac; *J. Organomet. Chem.*, 1982, **236**, 139. 74
82JOM(239)335 G. A. Razuvaev, L. I. Vyshinskaia, V. V. Dobrotenko, G. Y. Mal'kova and N. N. Vyshinskii; *J. Organomet. Chem.*, 1982, **239**, 335. 569
82JPC2835 D. J. DeFrees and A. D. McLean; *J. Phys. Chem.*, 1982, **86**, 2835. 1022
82JPR631 M. Mühlstädt, H. Meinhold, D. Martinez and E. Neumann; *J. Prakt. Chem.*, 1982, **324**, 631. 170
82LA121 G. Sosnovsky and J. A. Krogh; *Liebigs Ann. Chem.*, 1982, 121. 359, 361
82M1451 H. Wittman and E. Ziegler; *Monatsh. Chem.*, 1982, **113**, 1451. 1003
B-82MI 205-01 T. Sheradsky; in "The Chemistry of Amino, Nitroso and Nitro Compounds and their Derivatives," ed. S. Patai, Wiley, New York, 1982, pp. 395–416. 298
B-82MI 206-01 E. Breuer; in "The Chemistry of Amino, Nitroso and Nitro Compounds, Suppl. F," ed. S. Patai, Wiley, Chichester, 1982, p. 459. 342
B-82MI 206-02 H. G. Aurich; in "The Chemistry of Amino, Nitroso and Nitro Compounds, Suppl. F," ed. S. Patai, Wiley, Chichester, 1982, p. 565. 343, 346, 355
B-82MI 207-01 B. C. Challis and J. A. Challis; in "The Chemistry of NH_2, NO and NO_2 Groups," ed S. Patai, Wiley, New York, 1982, vol. 2, p. 1151. 385
82MI 208-01 T. Sugimoto, T. Imanishi, M. Okano and S. Tanimoto; *Bull. Inst. Chem. Res., Kyoto Univ.*, 1982, **60**, 302 (*Chem. Abstr.*, 1983, **99**, 21 575). 444
B-82MI 209-01 M. Wieber (ed.); "Gmelin Handbuch der Anorganischen Chemie," Springer-Verlag, 1982, part 2. 481
B-82MI 210-01 D. A. Evans, J. V. Nelson and T. R. Taber; "Topics in Stereochemistry," eds N. L. Allinger, E. L. Eliel and S. H. Wilen, Wiley, New York, 1982, vol. 13, pp. 1–113. 532
B-82MI 211-01 D. Dolphin; "Vitamin $B0_{12}$," Wiley, New York, 1982, Vols. 1 and 2. 578
B-82MI 215-01 G. Pitacco and E. Valentin; in "The Chemistry of Amine, Nitroso and Nitro Compounds and their Derivatives," ed. S. Patai, Wiley, Chichester, 1982, ch. 15, pp. 623–714. 738, 757, 758
82MI 217-01 K. I. Kuz'min and L. A. Pavlova; Deposited Doc., SPSTL 42 Khp-D82, 1982, 8 pp. (*Chem. Abstr.*, 1983, **98**, 12 679q). 876
B-82MI 223-01 M. Wieber; in "Gmelin Handbook of Inorganic Chemistry, Organoantimony Compounds," Part 3, ed. H. Bitterer, Springer, Berlin, 1982, p. 177. 1080

82OM355 K. Tamao, J. Yoshida, H. Yamamoto, T. Kakui, H. Matsumoto, M. Takahashi, A. Kurita, M. Murata and M. Kumada; *Organometallics*, 1982, **1**, 355. 4

82OM369 J. Yoshida, K. Tamao, T. Kakui, A. Kurita, M. Murata, K. Yamada and M. Kumada; *Organometallics*, 1982, **1**, 369. 4

82OM440 H. Kurosawa, M. Sato and H. Okada; *Organometallics*, 1982, **1**, 440. 598

82OM654 R. G. Goel and W. O. Ogini; *Organometallics*, 1982, **1**, 654. 582

82OM1408 A. J. Ashe, III and E. G. Ludwlg, Jr.; *Organometallics*, 1982, **1**, 1408. 505

82OM1496 K. W. Lee and J. San Filippo, Jr.; *Organometallics*, 1982, **1**, 1496. 600

82OPP359 G. W. Kabalka, K. A. R. Sastry, U. Sastry and V. Somayaji; *Org. Prep. Proced. Int.*, 1982, **14**, 359. 608, 632

82OR(27)1 J.-F. Biellmann and J.-B. Ducep; *Org. React.*, 1982, **27**, 1. 997

82OR(28)203 T. Mukaiyama; *Org. React.*, 1982, **28**, 203. 532

82POL123 K. B. Dillon, M. P. Nisbet and T. C. Waddington; *Polyhedron*, 1982, **1**, 123. 836, 837

82PS(12)149 H. J. Breunig and W. Kanig; *Phosphorous Sulfur*, 1982, **12**, 149. 489

82PS(12)265 J. D. Andriamizaka, C. Couret, J. Escudié and J. Satgé; *Phosphorus Sulfur*, 1982, **12**, 265. 473, 474

82PS(12)279 J. D. Andriamizaka, C. Couret, J. Escudié and J. Satgé; *Phosphorus Sulfur*, 1982, **12**, 279. 474

82PS(13)97 T. N. de Castro Dantas, J. P. Laval and A. Lattes; *Phosphorus Sulfur*, 1982, **13**, 97. 444

82PS(13)331 H. B. Stegman, H.-M. Kühne, G. Wax and K. Scheffler; *Phosphorus Sulfur*, 1982, **13**, 331. 825

82S1 H. Emde, D. Domsch, H. Feger, U. Frick, A. Götz, H. H. Hergott, K. Hofmann, W. Kober, K. Krägeloh, T. Oesterle, W. Steppan, W. West and G. Simchen; *Synthesis*, 1982, 1. 539

82S85 P. J. Stang, M. Hanack and L. R. Subramanian; *Synthesis*, 1982, 85. 689

82S173 F. Kober; *Synthesis*, 1982, 173. 493, 502

82S186 F. Camps, J. Coll and J. M. Moretó; *Synthesis*, 1982, 186. 684

82S211 A. Fisch, J. M. Coisne and H. P. Figeys; *Synthesis*, 1982, 211. 1100

82S260 M. Faulques, L. Rene and R. Royer; *Synthesis*, 1982, 260. 787, 789

82S261 T. Sakakibara, S. Ikuta and R. Sudoh; *Synthesis*, 1982, 261. 784

82S273 H. Ahlbrecht and E.-O. Düber; *Synthesis*, 1982, 273. 812

82S281 A. de Blic, L. Cazaux, L. Gorrichon-Guignon and M. Perry; *Synthesis*, 1982, 281. 247, 250

82S318 P. Plath and W. Rohr; *Synthesis*, 1982, 318. 746

82S324 H. Hoberg and B. W. Oster; *Synthesis*, 1982, 324. 988

82S326 R. Bartsch, O. Stelzer and R. Schmutzler; *Synthesis*, 1982, 326. 10

82S419 C. C. Dudman and C. B. Reese; *Synthesis*, 1982, 419. 1003

82S683 L.-F. Tietze, G. V. Kiedrowski and B. Berger; *Synthesis*, 1982, 683. 658

82S717 H. Ahlbrecht and C. S. Sudheendranath; *Synthesis*, 1982, 717. 751

82S922 A. Zwiezak and P. Pilichowska; *Synthesis*, 1982, 922. 299

82S1017 P. Knochel and D. Seebach; *Synthesis*, 1982, 1017. 785

82S1021 S. Motohashi and M. Satomi; *Synthesis*, 1982, 1021. 3

82S1083 S. Piettre, Z. Janousek and H. G. Viehe; *Synthesis*, 1982, 1083. 1042

82SC101 G. W. Kabalka, K. A. R. Sastry and K. U. Sastry; *Synth. Commun.*, 1982, **12**, 101. 34

82SC801 R. F. Smith and K. J. Coffman; *Synth. Commun.*, 1982, **12**, 801. 383

82SC903 S. K. Chung and G. Han; *Synth. Commun.*, 1982, **12**, 903. 142

82SC939 J. T. Gupton, M. J. Lizzi and D. Polk; *Synth. Commun.*, 1982, **12**, 939. 746

82SC947 T. A. Hase and H. Perakyla; *Synth. Commun.*, 1982, **12**, 947. 228

82SC1127 M. D'Auria, F. D'Onofrio, G. Piancatelli and A. Scettri; *Synth. Commun.*, 1982, **12**, 1127. 34

82SUL33 A. Senning; *Sulfur Letters*, 1982, **1**, 33. 361

82T389 A.-T. Hanson and M. Nilsson; *Tetrahedron*, 1982, **38**, 389. 783

82T405 A. A. El-Barbary, S. Carlsson and S.-O. Lawesson; *Tetrahedron*, 1982, **38**, 405. 738, 751

82T419 I. D. Entwistle, R. A. W. Johnstone and A. H. Wilby; *Tetrahedron*, 1982, **38**, 419. 379

82T1321 P. Aleksandrowicz, H. Piotrowska and W. Sas; *Tetrahedron*, 1982, **38**, 1321. 395

82T1505 M. A. Pericàs, A. Riera and F. Serratosa; *Tetrahedron*, 1982, **38**, 1505. 1026

82T1511 M. J. P. Harger and S. Westlake; *Tetrahedron*, 1982, **38**, 1511. 843, 846

82T1975 P. W. Hickmott; *Tetrahedron*, 1982, **38**, 1975. 740

82T2225 G. A. Olah and S. C. Narang; *Tetrahedron*, 1982, **38**, 2225. 36, 538

82T2339 A. Guy, M. Lemaire and J.-P. Guette; *Tetrahedron*, 1982, **38**, 2339. 620

82T3073 M. J. P. Harger and S. Westlake; *Tetrahedron*, 1982, **38**, 3073. 862

82T3087 S. M. Weinreb and R. R. Staib; *Tetrahedron*, 1982, **38**, 3087. 245

82T3363 P. W. Hickmott; *Tetrahedron*, 1982, **38**, 3363. 738, 751

82TCC1 M. T. Reetz; *Top. Curr. Chem.*, 1982, **106**, 1. 67

82TCC125 A. J. Ashe, III; *Top. Curr. Chem*, 1982, **105**, 125. 872

82TH1 K. Rustemeier; *PhD Thesis*, University of Bonn, 1982. 759

82TL27 E. Negishi, F. T. Luo and C. L. Rand; *Tetrahedron Lett.*, 1982, **23**, 27. 975

82TL63 P. Vincent, J.-P. Beaucourt and L. Pichat; *Tetrahedron Lett.*, 1982, **23**, 63. 968

82TL135 P. Dampawan and W. W. Zajac; *Tetrahedron Lett.*, 1982, 135. 785

82TL323 M. H. Goodrow, M. M. Olmstead and W. K. Musker; *Tetrahedron Lett.*, 1982, **23**, 323. 212, 679

82TL371 J. W. Verbicky, Jr., B. A. Dellacoletta and L. Williams; *Tetrahedron Lett.*, 1982, 371. 622

82TL631 M. Marsi and J. A. Gladysz; *Tetrahedron Lett.*, 1982, **23**, 631. 679

82TL703 S. Danishefsky, M. Kahn and M. Silvestri; *Tetrahedron Lett.*, 1982, **23**, 703. 675

82TL821 G. L. Dmitrienko, R. W. Friesen, L. Carson and S. F. Vice; *Tetrahedron Lett.*, 1982, **23**, 821. 767

82TL1165 T. Tsushima, K. Kawada and T. Tsuji; *Tetrahedron Lett.*, 1982, **23**, 1165. 16

82TL1243 A. Schwökel, G. Kresze and M. A. Perez; *Tetrahedron Lett.*, 1982, **23**, 1243. 360

82TL1263 H. Yamaguchi, S. Kametani, T. Karakasa, T. Saito and S. Motoki; *Tetrahedron Lett.*, 1982, **23**, 1263. 464

82TL1345 K. Jurkschat, C. Mugge, A. Tzschach, W. Uhlig and A. Zshuke; *Tetrahedron Lett.*, 1982, **23**, 1345. 253, 260

82TL1361 S. Tomoda, Y. Takeuchi and Y. Nomura; *Tetrahedron Lett.*, 1982, **23**, 1361. 282, 284

82TL1549 J. Barber, R. Keck and J. Retey; *Tetrahedron Lett.*, 1982, **23**, 1549. 11

82TL1557 C. Paulmier and P. Lerouge; *Tetrahedron Lett.*, 1982, **23**, 1557. 282

82TL2023 G. Morel, E. Marchand, K. H. Nguyen Thi and A. Foucaud; *Tetrahedron Lett.*, 1982, **23**, 2023. 744

82TL2105 A. Toshimitsu, H. Owada, S. Uemura and M. Okano; *Tetrahedron Lett.*, 1982, **23**, 2105. 282

82TL2603 R. G. Hall and S. Trippett; *Tetrahedron Lett.*, 1982, **23**, 2603. 849

82TL2719 J. M. Finian and Y. Kishi; *Tetrahedron Lett.*, 1982, **23**, 2719. 83

82TL2935 T.-H. Chan and T. Chaly; *Tetrahedron Lett.*, 1982, **23**, 2935. 662

82TL2965 H. Kurosawa, M. Sato and H. Okada; *Tetrahedron Lett.*, 1982, **23**, 2965. 394

82TL3319 H. J. Reich, W. W. Willis, Jr. and S. Wollowitz; *Tetrahedron Lett.*, 1982, **23**, 3319. 290

82TL3327 E. Negishi and L. D. Boardman; *Tetrahedron Lett.*, 1982, **23**, 3327. 991, 992

82TL3387 H. Tomioka, T. Suzuki, K. Oshima and H. Nozaki; *Tetrahedron Lett.*, 1982, **23**, 3387. 922, 931

82TL3411 J. N. Denis and A. Krief; *Tetrahedron Lett.*, 1982, **23**, 3411. 624, 910, 911

82TL3823 H. Malmberg, M. Nilsson and C. Ullenius; *Tetrahedron Lett.*, 1982, **33**, 3823. 587

82TL3897 P. Knochel and D. Seebach; *Tetrahedron Lett.*, 1982, **23**, 3897. 785

82TL3901 W. Oppolzer, E. P. Kündig, P. M. Bishop and C. Perret; *Tetrahedron Lett.*, 1982, **23**, 3901. 560

82TL3905 A. Luxen and L. Christiaens; *Tetrahedron Lett.*, 1982, **23**, 3905. 731

82TL4199 P. A. McCarthy; *Tetrahedron Lett.*, 1982, **23**, 4199. 91

82TL4203 E. Block and M. Aslam; *Tetrahedron Lett.*, 1982, **23**, 4203. 177

82TL4371 G. Galambos and V. Simonidesz; *Tetrahedron Lett.*, 1982, **23**, 4371. 286

82TL4411 S. Hanessian, Y. Leblanc and P. Lavallee; *Tetrahedron Lett.*, 1982, **23**, 4411. 8

82TL4551 D. L. Boger and M. D. Mullican; *Tetrahedron Lett.*, 1982, **23**, 4551. 665

82TL4589 F. Sato, Y. Suzuki and M. Sato; *Tetrahedron Lett.*, 1982, **23**, 4589. 935

82TL4651 C. A. Martin, P. M. McCrann, G. H. Angelos and D. A. Jaeger; *Tetrahedron Lett.*, 1982, **23**, 4651. 311

82TL4733 T. Hayama, S. Tomada, Y. Takeuchi and Y. Nomura; *Tetrahedron Lett.*, 1982, **23**, 4733. 284, 392, 785

82TL4843 I. S. Blagbrough and G. Pattenden; *Tetrahedron Lett.*, 1982, **23**, 4843. 660

82TL4871 E. J. Corey and P. B. Hopkins, *Tetrahedron Lett.*, 1982, **23**, 4871. 540

82TL4923 M. Karpf; *Tetrahedron Lett.*, 1982, **23**, 4923. 668

82TL5151 A. Alexakis and J. F. Normant; *Tetrahedron Lett.*, 1982, **23**, 5151. 977

82TL5259 M. T. Reetz and B. Wenderoth; *Tetrahedron Lett.*, 1982, **23**, 5259. 566

82TL5373 Y.-H. Kang and J. L. Kice; *Tetrahedron Lett.*, 1982, **23**, 5373. 285

82TL5389 H.-H. Otto and P. Schwenkkraus; *Tetrahedron Lett.*, 1982, **23**, 5389. 248

82TL5427 G. I. Tsuchihashi, S. Mitamura, K. Kitayima and K. Kobayushi; *Tetrahedron Lett.*, 1982, **23**, 5427. 202

82URP891678 G. A. Kutyrev, A. A. Kutyrev and R. A. Cherkasov; *USSR Pat.* 891 678 (1982) (*Chem. Abstr.*, 1982, **97**, 6523). 259

82USP4290975 G. E. DuBois; *US Pat* 4 290 975 (1982) (*Chem. Abstr.*, 1982, **96**, 19 823.) 366

82ZAAC(485)129 M. Baudler and P. Bachmann; *Z. Anorg. Allg. Chem.*, 1982, **485**, 129. 507

82ZAAC(486)70 O. Mundt, G. Becker and H.-J. Wessely; *Z. Anorg. Allg. Chem.*; 1982, **486**, 70. 507

82ZAAC(491)163 K. Issleib and W. Gans; *Z. Anorg. Allg. Chem.*, 1978, **491**, 163. 430, 437, 438

82ZC237 R. Beckert and R. Mayer; *Z. Chem.*, 1982, **22**, 237. 399

82ZN(B)105 E. Ziegler, O. S. Wolfbeis and I. Trummer; *Z. Naturforsch., Teil B*, 1982, **37**, 105. 746

82ZN(B)380 G. N. Schrauzer, L. A. Hughes and N. Strampach; *Z. Naturforsch., Teil B*, 1982, **37**, 380. 570, 573

82ZN(B)395 H. J. Breunig and T. Severengiz; *Z. Naturforsch., Teil B*, 1982, **37**, 395. 509

82ZN(B)815 F. Huber and S. Bock; *Z. Naturforsch., Teil B*, 1982, **37**, 815. 885

82ZN(B)965 R. Thamm and E. Fluck; *Z. Naturforsch., Teil B*, 1982, **37**, 965. 429

82ZN(B)1104 H. J. Breunig and H. Jawad; *Z. Naturforsch., Teil B*, 1982, **37**, 1104. 493

82ZN(B)1146 E. Lindner, F. Bouachir and W. Hiller; *Z. Naturforsch., Teil B*, 1982, **37**, 1146. 1003

82ZOB2721 E. V. Nikitin, Y. A. Ignat'ev, O. V. Parakin, I. P. Kosachev, C. V. Romanov, Y. M. Kargin and A. N. Pudovik; *Zh. Obshch. Khim.*, 1982, **52**, 2721. 845

82ZPC(363)1461 E. Wünsch, L. Moroder and S. Romani, *Hoppe-Seyler's Z. Physiol. Chem.*, 1982, **363**, 1461 (*Chem. Abstr.*, 1983, **98**, 68 154g). 218

83AG(E)31 B. Weidmann and D. Seebach; *Angew. Chem., Int. Ed. Engl.*, 1983, **22**, 31. 67

83AG(E)53 G. Boche, M. Bernheim and M. Niessner; *Angew. Chem., Int. Ed. Engl.*, 1983, **22**, 53. 1041

83AG(E)337 L. Stella; *Angew. Chem., Int. Ed. Engl.*, 1983, **22**, 337. 323, 326, 337

83AG(E)406 K. Petzhold, H. Laurent and R. Wiechert; *Angew. Chem., Int. Ed. Engl.*, 1983, **22**, 406. 38

83AG(E)728 H. Bönnemann, B. Bogdanović, R. Brinkmann, D. W. He and B. Spliethoff; *Angew. Chem., Int. Ed. Engl.*, 1983, **22**, 728. 559, 959

83AG(E)775 M. Baudler, T. Pontzen, U. Schings, K. F. Tebbe and M. Feher; *Angew. Chem., Int. Ed. Engl.*, 1983, **22**, 775. 473

83AG(E)1001 F. R. Kreissl, M. Wolfgruber and W. Sieber; *Angew. Chem., Int. Ed. Engl.*, 1983, **22**, 1001. 878

83AJC719 A. L. J. Beckwith, P. H. Eichinger, B. A. Mooney and R. H. Prager; *Aust. J. Chem.*, 1983, **36**, 719. 348

83BCJ219 M. Yamashita, N. Suzuki, M. Yamada, Y. Soeda, H. Yamashita, K. Nakatani, T. Oshikawa and S. Inokawa; *Bull. Chem. Soc. Jpn.*, 1983, **56**, 219. 429

83BCJ1187 M. Ueki and K. Shinozaki; *Bull. Chem. Soc. Jpn.*, 1983, **56**, 1187. 252

83BCJ1881 T. Hirao, S. Kohno, Y. Ohshiro and T. Agawa; *Bull. Chem. Soc. Jpn.*, 1983, **56**, 1881. 14
83BCJ3802 S. Oae and H. Togo; *Bull. Chem. Soc. Jpn.*, 1983, 3802. 708
83BSF(2)28 J. L. Klein and J. C. Combret; *Bull. Soc. Chim. Fr., Part 2*, 1983, 28. 741
83BSF(2)61 D. H. R. Barton, W. B. Motherwell and S. Z. Zard; *Bull. Soc. Chim. Fr., Part 2*, 1983, 61. 785
83C10 M. Schlosser, H. B. Twong, J. Respondek and B. Schaub; *Chimia*, 1983, **37**, 10. 447
83CB159 H. Lehmkuhl, J. Grundke and R. Mynott; *Chem. Ber.*, 1983, **116**, 159. 581
83CB176 H. Lehmkuhl, J. Grundke and R. Mynott; *Chem. Ber.*, 1983, **116**, 176. 581
83CB1001 T. Kauffmann and H. Ahlers; *Chem. Ber.*, 1983, **116**, 1001. 280, 908
83CB1320 H. J. Bestmann and L. Kisielowski; *Chem. Ber.*, 1983, **116**, 1320. 1051, 1063
83CB2564 L. Birkofer and B. Wahle; *Chem. Ber.*, 1983, **116**, 2564. 944
83CB2936 H. Stamm, G. Gerster and T. Baumann; *Chem. Ber.*, 1983, **116**, 2936. 454, 457
83CB3293 W. J. Richter; *Chem. Ber.*, 1983, **116**, 3293. 436
83CB3567 G. A. Bowmaker, R. Herr and H. Schmidbaur; *Chem. Ber.*, 1983, **116**, 3567. 444
83CB3931 H. Quast and B. Muller; *Chem. Ber.*, 1983, **116**, 3931. 338
83CC73 J. Koenig and R. T. Oakley; *J. Chem. Soc., Chem. Commun.*, 1983, 73. 796
83CC108 M. Attina, F. Cacace and A. P. Wolf; *J. Chem. Soc., Chem. Commun.*, 1983, 108. 614
83CC229 J. N. Denis and A. Krief; *J. Chem. Soc., Chem. Commun.*, 1983, 229. 36
83CC329 R. Tanaka and K. Yamabe; *J. Chem. Soc., Chem. Commun.*, 1983, 329. 1055
83CC466 A. J. Guildford and R. W. Turner; *J. Chem. Soc., Chem. Commun.*, 1983, 466. 667
83CC857 S. Brachen, J. I. G. Cadogan, I. Gosney and S. Yaslak; *J. Chem. Soc., Chem. Commun.*, 1983, 857. 837
83CC881 A. H. Cowley, J. G. Lasch, N. C. Norman, M. Pakulski and B. R. Whittlesey; *J. Chem. Soc., Chem. Commun.*, 1983, 881. 866
83CC921 F. Sato, H. Uchiyama, K. Iida, Y. Kobayashi and M. Sato; *J. Chem. Soc., Chem. Commun.*, 1983, 921. 936
83CC947 Y. Tanaka, H. Sakuraba and H. Nakanishi; *J. Chem. Soc., Chem. Commun.*, 1983, 947. 24
83CC1051 M. H. Brookes, B. T. Golding, D. A. Howes and A. T. Hudson; *J. Chem. Soc., Chem. Commun.*, 1983, 1051. 225
83CC1193 K. Takeshita, Y. Seki, K. Kawamoto, S. Murai and N. Sonoda; *J. Chem. Soc., Chem. Commun.*, 1983, 1193. 912
83CC1322 K. Nishiyama and N. Tanaka; *J. Chem. Soc., Chem. Commun.*, 1983, 1322. 753
83CC1470 K. Hiroi, R. Kitayama and S. Sato; *J. Chem. Soc., Chem. Commun.*, 1983, 1470. 159
83CC1501 S. Uemura, S. I. Fukuzawa and A. Toshimitsu; *J. Chem. Soc., Chem. Commun.*, 1983, 1501. 287
83CJC21 C. C. Santini and F. Mathey; *Can. J. Chem.*, 1983, **61**, 21. 440
83CJC1562 P. W. Codding, H. Koenig and R. T. Oakley; *Can. J. Chem.*, 1983, **61**, 1562. 795, 796
83CJC2214 A. R. Sanger; *Can. J. Chem.*, 1983, **61**, 2214. 437
83CL249 J. Nakayama, T. Fujita and M. Hoshino; *Chem. Lett.*, 1983, 249. 713
83CL441 A. Uehara, T. Kubota and R. Tsuchiya; *Chem. Lett.*, 1983, 441. 859
83CL927 M. Kosugi, M. Kameyama and T. Migita; *Chem. Lett.*, 1983, 927. 753
83CL1135 N. Yoneda, T. Abe, T. Fukuhara and A. Suzuki; *Chem. Lett.*, 1983, 1135. 17
83CL1727 J.-I. Kato and T. Mukaiyama; *Chem. Lett.*, 1983, 1727. 701
83CL1791 T. Fujisawa, T. Mori, K. Higuchi and T. Sato; *Chem. Lett.*, 1983, 1791. 758
83CL1877 N. Shimizu, K. Watanabe and Y. Tsuno; *Chem. Lett.*, 1983, 1877. 14
83CPB128 T. Ibuka, E. Tabushi and M. Yasuda; *Chem. Pharm. Bull.*, 1983, **31**, 128. 702
83CPB423 Y. Murakami, Y. Yokoyama, C. Sasakura and M. Tamagawa; *Chem. Pharm. Bull.*, 1983, **31**, 423. 380
83CPB2157 H. Takaku, K. Kamaike, H. Mori and Y. Ishido; *Chem. Pharm. Bull.*, 1983, **31**, 2157. 107
83CPB4189 T. Kamijo, H. Harada and K. Lizuka; *Chem. Pharm. Bull.*, 1983, **31**, 4189. 32, 36
83CSR53 T. L. Gilchrist; *Chem. Soc. Rev.*, 1983, **12**, 53. 777
83CZ77 W. Wolfsberger; *Chem.-Ztg.*, 1983, **107**, 77. 448
83EUP82057 S. Juge; *Eur. Pat.* 82 057 (1983) (*Chem. Abstr.*, 1983, **99**, 140 162). 443
83GEP(O)3149937 K. Gehrman, A. Ohorodnik, J. Rosenthal and S. Shaefer; *Ger Pat. Offen.* 3 149 937 (1983) (*Chem. Abstr.*, 1983, **99**, 176 058). 448
83HCA308 D. Seebach, R. Hässig and J. Gabriel; *Helv. Chim. Acta*, 1983, **66**, 308. 998
83HCA789 B. Aebischer and A. Vasella; *Helv. Chim. Acta*, 1983, **66**, 789. 397
83HCA1018 J.-R. Neeser, L. D. Hall and J. A. Balatoni; *Helv. Chim. Acta*, 1983, **66**, 1018. 4
83HCA1028 G. A. Olah, J. Shih and G. K. S. Prakash; *Helv. Chim. Acta*, 1983, **66**, 1028. 11
83HCA1774 A. Salzer and A. Hafner; *Helv. Chim. Acta*, 1983, **66**, 1774. 446
83HCA2377 T. K. Jones and S. E. Denmark; *Helv. Chim. Acta*, 1983, **66**, 2377. 931
83IC547 A. H. Cowley and R. A. Kemp; *Inorg. Chem.*, 1983, **22**, 547. 430
83IC814 H. R. O'Neal and R. H. Neilson; *Inorg. Chem.*, 1983, **22**, 814. 466
83IC1771 A. C. Sau, J. M. Holmes, R. O. Day and R. R. Holmes; *Inorg. Chem.*, 1983, **22**, 1771. 856
83IC3624 R. Bartsch, S. Heitkamp, S. Morton, H. Peters and O. Stelzer; *Inorg. Chem.*, 1983, **22**, 3624. 440
83ICAL241 J. D. Spuch and G. S. Lewandos; *Inorg. Chim. Acta*, 1983, **76**, L241. 558, 560
83IJC(B)421 A. R. Katritzky and J. N. Singh; *Indian J. Chem., Sect. B*, 1983, **22**, 421. 11
83IZV218 U. M. Dzhemilev, O. S. Vostrikova and R. M. Sultanov; *Izv. Akad. Nauk SSSR, Ser. Khim.*, 1983, **32**, 218. 561
83IZV230 B. A. Trofimov, V. A. Potapov, N. K. Gusarova and S. V. Amosova; *Izv. Akad. Nauk. SSSR, Ser. Khim.*, 1983, 230. 708
83IZV1152 O. A. Erastov, I. P. Romanova, Yu. Ya. Efremov, A. V. Il'yasov, T. A. Zyablikova, I. E. Ismaev, N. A. Chadaeva and R. Z. Musin; *Izv. Akad. Nauk. SSSR, Ser. Khim.*, 1983, 1152 (*Chem. Abstr.*, 1983, **99**, 122 638d). 894

83IZV1662 N. N. Godovikov, L. A. Vikhreva and K. K. Babasheva; *Izv. Akad. Nauk. SSSR, Ser. Khim.*, 1983, 1662. 1033

83IZV2053 U. M. Dzhemilev, O. S. Vostrikova, R. M. Sultanov, A. G. Kukovinetz and A. M. Kdalekov; *Izv. Akad. Nauk SSSR, Ser. Khim.*, 1983, **32**, 2053. 561

83JA640 F. W. Hartner, Jr., J. Schwartz and S. M. Clift; *J. Am. Chem. Soc.*, 1983, **105**, 640. 681

83JA650 D. J. Burton and D. G. Cox; *J. Am. Chem. Soc.*, 1983, **105**, 650. 612

83JA897 M. Schriewer and W. P. Neumann; *J. Am. Chem. Soc.*, 1983, **105**, 897. 947, 949

83JA943 C. E. Russell and L. S. Hegedus; *J. Am. Chem. Soc.*, 1983, **105**, 943. 681

83JA1613 N. E. Jacobsen and P. A. Bartlett; *J. Am. Chem. Soc.*, 1983, **105**, 1613. 469

83JA1664 J. R. Stille and R. H. Grubbs; *J. Am. Chem. Soc.*, 1983, **105**, 1664. 702

83JA2077 D. S. Matteson and K. M. Sachu; *J. Am. Chem. Soc.*, 1983, **105**, 2077. 529

83JA2477 C. Exon and P. Magnus; *J. Am. Chem. Soc.*, 1983, **105**, 2477. 985

83JA2748 S. Uemura and S. Fukuzawa; *J. Am. Chem. Soc.*, 1983, **105**, 2748. 287

83JA2908 S. Yamamoto, H. Itani, T. Tsuji and W. Nagata; *J. Am. Chem. Soc.*, 1983, **105**, 2908. 88

83JA3929 A. H. Janowicz and R. G. Bergman; *J. Am. Chem. Soc.*, 1983, **105**, 3929. 962

83JA4112 R. K. Boeckman, Jr. and S. H. Cheon; *J. Am. Chem. Soc.*, 1983, **105**, 4112. 651, 652

83JA4491 H. Hayami, M. Sato, S. Kanemoto, Y. Morizawa, K. Oshima and H. Nozaki; *J. Am. Chem. Soc.*, 1983, **105**, 4491. 918, 931

83JA4499 R. S. Garigipati and S. M. Weinreb; *J. Am. Chem. Soc.*, 1983, **105**, 4499. 245

83JA4727 M. Moriayma and W. G. Bentrude; *J. Am. Chem. Soc.*, 1983, **105**, 4727. 440, 441, 453

83JA5075 B. M. Trost and A. C. Lavoie; *J. Am. Chem. Soc.*, 1983, **105**, 5075. 709

83JA5490 S. L. Buchwald and R. H. Grubbs; *J. Am. Chem. Soc.*, 1983, **105**, 5490. 979

83JA5804 A. Patton, C. E. Strcuse, C. B. Knobler and J. A. Gladycz; *J. Am. Chem. Soc.*, 1983, **105**, 5804. 580

83JA6096 R. J. Bailey, P. J. Card and H. Shechter; *J. Am. Chem. Soc.*, 1983, **105**, 6096. 1003

83JA6316 T. J. Barton and G. P. Hussmann; *J. Am. Chem. Soc.*, 1983, **105**, 6316. 1091

83JA6344 E. Negishi, L. D. Boardman, J. M. Tour, H. Sawada, and C. L. Rand; *J. Am. Chem. Soc.*, 1983, **105**, 6344. 926, 931, 976

83JA6355 S. J. McLain; *J. Am. Chem. Soc.*, 1983, **105**, 6355. 440

83JA6761 E. Negishi and J. A. Miller; *J. Am. Chem. Soc.*, 1983, **105**, 6761. 925, 931, 973

83JA6824 J. W. Bruno, T. J. Marks and L. R. Morss; *J. Am. Chem. Soc.*, 1983, **105**, 6824. 602

83JA6871 C. Charrier, H. Bonnard, G. de Lauzon and F. Mathey; *J. Am. Chem. Soc.*, 1983, **105**, 6871. 472

83JA6882 D. B. Collum, F. Mohamadi and J. S. Hallock; *J. Am. Chem. Soc.*, 1983, **105**, 6882. 40

83JA7172 N. Nakamura; *J. Am. Chem. Soc.*, 1983, **105**, 7172. 184

83JA7719 R. V. Stevens, J. H. Chang, R. Lapalme, S. Schow, M. G. Schlageter, R. Shapiro and H. N. Weller; *J. Am. Chem. Soc.*, 1983, **105**, 7719. 91

83JAP58113195 Nippon Chemical Industry Co.; *Jpn. Pat.* 58 113 195 (1983) (*Chem. Abstr.*, 1983, **99**, 158 632m). 257

83JCR(M)2101 L. Duhamel, J. M. Poirier and N. Tedga; *J. Chem. Res. (M)*, 1983, 2101. 759

83JCR(S)222 L. Duhamel, J. M. Poirier and N. Tedga; *J. Chem. Res. (S)*, 1983, 222. 759

83JCS(D)415 S. Bhaduri and H. Khwaja; *J. Chem. Soc., Dalton Trans.*, 1983, 415. 156

83JCS(D)905 C. Eaborn, N. Retta and J. D. Smith; *J. Chem. Soc., Dalton Trans.*, 1983, 905. 436

83JCS(D)2143 K. G. DeSilva, Z. Monsef-Mirzai and W. R. McWhinnie; *J. Chem. Soc., Dalton Trans.*, 1983, 2143. 280

83JCS(P1)471 A. J. Blodworth and D. J. Lampham; *J. Chem. Soc., Perkin Trans. 1*, 1983, 471. 291

83JCS(P1)861 R. D. Holmes-Smith, R. D. Osei and S. R. Stobart; *J. Chem. Soc., Perkin Trans. 1*, 1983, 861. 439

83JCS(P1)1087 J. Barluenga, F. Aznar, R. Liz and J. Rochs; *J. Chem. Soc., Perkin Trans. 1*, 1983, 1087. 1097

83JCS(P1)1131 D. J. Ager; *J. Chem. Soc., Perkin Trans. 1*, 1983, 1131. 908, 1000, 1005

83JCS(P1)2983 R. D. Allan, G. A. R. Johnston, R. Kazlauskas and H. W. Tran; *J. Chem. Soc., Perkin Trans. 1*, 1983, 2983. 299

83JCS(P2)1605 E. Maccarone and G. Perrini; *J. Chem. Soc., Perkin Trans. 2*, 1983, 1605. 288

83JFC(23)339 D. J. Burton; *J. Fluorine Chem.*, 1983, **23**, 339. 444

83JGU1059 V. A. Kron, B. V. Timokhin and V. I. Donskikh; *J. Gen. Chem. USSR (Engl. Transl.)*, 1983, **53**, 1059. 458

83JGU1832 F. D. Yambushev, G. I. Kokorev, F. G. Khalitov, N. Kh. Tenesheva and S. V. Kutin; *J. Gen. Chem. USSR*, 1983, **53**, 1832. 883

83JGU2244 V. Foss, Yu. A. Veits, T. E. Chernykh, I. N. Staroverova and I. F. Lutsenko; *J. Gen. Chem. USSR (Engl. Transl.)*, 1983, **53**, 2244. 471

83JHC121 B. F. Powell, C. G. Overberger and J.-P. Anselme; *J. Heterocycl. Chem.*, 1983, **20**, 121. 379

83JHC399 J. Galloy, W. H. Watson, D. Craig, C. Guidry, M. Morgan, R. McKellar, A. L. Ternay, Jr. and G. Martin; *J. Heterocycl. Chem.*, 1983, **20**, 399. 1000

83JHC839 A. Bargagna, G. Bignardi, P. Schenone and M. Longobardi; *J. Heterocycl. Chem.*, 1983, **20**, 839. 204

83JHC899 T. L. Lemke and K. N. Sawhney; *J. Heterocycl. Chem.*, 1983, **20**, 899. 163

83JHC1207 A. Liguori, G. Sindona and N. Uccella; *J. Heterocycl. Chem.*, 1983, **20**, 1207. 350

83JOC1 C. R. Johnson, K. G. Bis, J. H. Cantillo, N. A. Meanwell, M. F. D. Reinhard, J. R. Zeller and G. P. Vonk; *J. Org. Chem.*, 1983, **48**, 1. 401, 402

83JOC441 S. David and A. Thieffry; *J. Org. Chem.*, 1983, **48**, 441. 684

83JOC448 J. C. Gilbert and U. Weerasooriya; *J. Org. Chem.*, 1983, **48**, 448. 680, 739

83JOC724 O. Lerman and S. Rozen; *J. Org. Chem.*, 1983, **48**, 724. 16

83JOC876 A. P. Croft and R. A. Bartsch; *J. Org. Chem.*, 1983, **48**, 876. 622

83JOC1293 Y. Yamada, K. Mukai, H. Yoshioka, Y. Tamura and Z.-I. Yoshida; *J. Org. Chem.*, 1983, **48**, 1293. 257
83JOC1357 Y. Guindon, R. Frenette, R. Fortin and J. Rokach; *J. Org. Chem.*, 1983, **48**, 1357. 135
83JOC1364 R. J. Crawford; *J. Org. Chem.*, 1983, **48**, 1364. 21
83JOC1389 U. C. Brown and U. S. Racherla; *J. Org. Chem.*, 1983, **48**, 1389. 525
83JOC1542 D. S. Wilbur, W. E. Stone and K. W. Anderson; *J. Org. Chem.*, 1983, **48**, 1542. 632
83JOC1556 J. K. Whitesell, M. Fisher and P. D. Jardine; *J. Org. Chem.*, 1983, **48**, 1556. 976
83JOC1678 E. C. Freidrich and G. DeLucca; *J. Org. Chem.*, 1983, **48**, 1678. 33
83JOC1760 G. A. Olah, B. G. Gupta, A. Garcia-Luna and S. C. Narang; *J. Org. Chem.*, 1983, **48**, 1760. 452
83JOC2195 T. Hayashi, M. Konishi, M. Fukushima, K. Kanehira, T. Hioki and M. Kumada; *J. Org. Chem.*, 1983, **48**, 2195. 437
83JOC2255 A. Alberola, C. Andrés and R. Pedrosa; *J. Org. Chem.*, 1983, **48**, 2255. 312
83JOC2569 T. Minami, H. Sako, T. Ikehira, T. Hanamoto and I. Hirao; *J. Org. Chem.*, 1983, **48**, 2569. 843
83JOC2625 R. Richter and B. Tucker; *J. Org. Chem.*, 1983, **48**, 2625. 25
83JOC2640 M. Bucciarelli, A. Forni, I. Moretti and G. Torre; *J. Org. Chem.*, 1983, **48**, 2640. 337
83JOC2766 G. A. Olah, S. C. Narang, L. D. Field and A. P. Fung; *J. Org. Chem.*, 1983, **48**, 2766. 36
83JOC2782 M. D. Crenshaw and H. Zimmer; *J. Org. Chem.*, 1983, **48**, 2782. 618
83JOC3038 G. Maas and P. J. Stang; *J. Org. Chem.*, 1983, **48**, 3038. 1008
83JOC3043 R. F. Childs, R. Faggiani, C. J. L. Lock and C. V. Rogerson; *J. Org. Chem.*, 1983, **48**, 3043. 1008
83JOC3071 F. A. Davis, S. B. Awad, R. H. Jenkins Jr., R. L. Billmers and L. A. Jenkins; *J. Org. Chem.*, 1983, **48**, 3071. 718
83JOC3077 T. G. Back, S. Collins and R. G. Kerr; *J. Org. Chem.*, 1983, **48**, 3077. 285, 1034, 1035
83JOC3126 T. R. Beebe, R. L. Adkins, C. C. Bogardus, B. Champney, P. S. Hii, P. Reinking, J. Shadday, W. D. Weatherford, III, M. W. Webb and S. W. Yates; *J. Org. Chem.*, 1983, **48**, 3126. 95
83JOC3195 V. L. Heasley, R. K. Gipe, J. L. Martin, H. C. Wiese, M. L. Oakes, D. F. Shellhamer, G. E. Heasley and B. L. Robinson; *J. Org. Chem.*, 1983, **48**, 3195. 18
83JOC3252 B. M. Trost, C. G. Caldwell, E. Murayama and D. Heissler; *J. Org. Chem.*, 1983, **48**, 3252. 540
83JOC3297 S. Uemura, S. Tanaka, M. Okano and M. Hamana; *J. Org. Chem.*, 1983, **48**, 3297. 633
83JOC3353 A. P. Croft and R. A. Bartsch; *J. Org. Chem.*, 1983, **48**, 3353. 26
83JOC3356 G. A. Olah, J. G. Shih, B. P. Singh and B. G. B. Gupta; *J. Org. Chem.*, 1983, **48**, 3356. 15
83JOC3379 M. P. Doyle, J. W. Terpstra, R. A. Pickering and D. M. Lepoire; *J. Org. Chem.*, 1983, **48**, 3379. 104
83JOC3531 B. P. Branchaud; *J. Org. Chem.*, 1983, **48**, 3531. 239, 359
83JOC3538 B. P. Branchaud; *J. Org. Chem.*, 1983, **48**, 3538. 239
83JOC3566 K. Takahashi, K. Shibasaki, K. Ogura and H. Iida; *J. Org. Chem.*, 1983, **48**, 3566. 748
83JOC3585 G. A. Olah and B. G. B. Gupta; *J. Org. Chem.*, 1983, **48**, 3585. 156
83JOC3608 S. Miyano, L. D.-L. Lu, S. M. Vita and K. B. Sharpless; *J. Org. Chem.*, 1983, **48**, 3608. 348
83JOC3667 G. A. Olah, A. Husain, B. P. Singh and A. K. Mehrotra; *J. Org. Chem.*, 1983, **48**, 3367. 36, 142
83JOC4108 P. Boldrini, D. Savoia, E. Tagliavini, C. Trombini and A. Umani-Ronchi; *J. Org. Chem.*, 1983, **48**, 4108. 590
83JOC4116 C. S. Cooper, A. L. Peyton and R. J. Weinkam; *J. Org. Chem.*, 1983, **48**, 4116. 377
83JOC4156 S. T. Taylor, D. Y. Lee and J. C. Martin; *J. Org. Chem.*, 1983, **48**, 4156. 942
83JOC4158 T. B. Patrick, K. K. Johri and D. H. White; *J. Org. Chem.*, 1983, **48**, 4158. 13
83JOC4289 M. Tiecco, L. Testaferri, M. Tingoli, D. Chianelli and M. Montanucci; *J. Org. Chem.*, 1983, **48**, 4289. 278, 729
83JOC4394 N. Satyamurthy and J. R. Barrio; *J. Org. Chem.*, 1983, **48**, 4394. 632
83JOC4397 H. Y. Xin and E. R. Biehl; *J. Org. Chem.*, 1983, **48**, 4397. 753
83JOC4466 L. D. Quin, J. C. Kisalus and K. A. Mesch; *J. Org. Chem.*, 1983, **48**, 4466. 441
83JOC4582 P. A. T. W. Porskamp and B. Zwanenberg; *J. Org. Chem.*, 1983, **48**, 4582. 400
83JOC4776 T. G. Back, S. Collins, U. Gokhale and K.-W. Law; *J. Org. Chem.*, 1983, **48**, 4776. 285
83JOC4795 M. Tiecco, L. Testaferri, M. Tingoli, D. Chianelli and M. Montanucci; *J. Org. Chem.*, 1983, **48**, 4795. 709
83JOC5376 K. K. Wang, S. S. Nikam and C. D. Ho; *J. Org. Chem.*, 1983, **48**, 5376. 939, 940
83JOC5398 P. G. Wuts, P. A. Thompsen and G. R. Collen; *J. Org. Chem.*, 1983, **48**, 5398. 530
83JOC5409 G. A. Molander; *J. Org. Chem.*, 1983, **48**, 5409. 925, 973
83JOM(243)175 H. Arzoumanian, H. Krentzien, R. Lai, J. Metzger and J.-F. Petrignani; *J. Organomet. Chem.*, 1983, **243**, 175. 33
83JOM(243)417 H. J. Breunig and H. Jawad; *J. Organomet. Chem.*, 1983, **243**, 417. 491
83JOM(248)269 G. Märkl and H. Hauptmann; *J. Organomet. Chem.*, 1983, **248**, 269. 881, 893
83JOM(250)1 E. C. Ashby and R. D. Ainslie; *J. Organomet. Chem.*, 1983, **250**, 1. 561
83JOM(250)617 A. R. H. Bottomley, C. Croker and B. L. Shaw; *J. Organomet. Chem.*, 1983, **250**, 617. 435
83JOM(252)C78 K. H. Dötz and W. Kuhn; *J. Organomet. Chem.*, 1983, **252**, C78. 670
83JOM(252)295 L. Silaghi-Dimitrescu and I. Haiduc; *J. Organomet. Chem.*, 1983, **252**, 295. 891
83JOM(253)C21 H. Breunig and D. Müller; *J. Organomet. Chem.*, 1983, **253**, C21. 893
83JOM(253)31 R. A. Grigsby, Jr., K. J. Irgolic and F. F. Knapp, Jr.; *J. Organomet. Chem.*, 1983, **253**, 31. 279
83JOM(253)65 P. Cocolios, G. Lagrange and R. Guilard; *J. Organomet. Chem.*, 1983, **253**, 65. 578
83JOM(253)249 J. Bakos, B. Heil and L. Markó; *J. Organomet. Chem.*, 1983, **253**, 249. 438
83JOM(255)61 C. H. W. Jones and R. D. Sharma; *J. Organomet. Chem.*, 1983, **255**, 61. 279
83JOM(255)359 R. J. P. Corriu, V. Huynh, J. J. E. Moreau and M. Pataud-Sat; *J. Organomet. Chem.*, 1983, **255**, 359. 814
83JOM(255)473 D. W. Owen and R. C. Poller; *J. Organomet. Chem.*, 1983, **255**, 473. 600
83JOM(257)67 R. A. Zingara and K. J. Irgolic; *J. Organomet. Chem.*, 1983, **257**, 67. 872
83JOM(259)1 J. V. N. Vara Prasad and C. N. Pillai; *J. Organomet. Chem.*, 1983, **259**, 1. 561

83JOM(259)269 G. W. Kabalka, U. Sastry, K. A. R. Sastry, F. F. Knapp, Jr. and P. C. Srivastava; *J. Organomet. Chem.*, 1983, **259**, 269. 907

83JOU1203 I. V. Vigalok, G. G. Petrova and S. G. Lukashina; *J. Org. Chem. USSR* (*Engl. Transl.*), 1983, **19**, 1203. 336

83JST(97)47 M. Rodler and A. Bauder; *J. Mol. Struct.*, 1983, **97**, 47. 636

83LA1185 G. Himbert and W. Schwickerath; *Liebigs Ann. Chem.*, 1983, 1185. 1044

83M783 E. Ziegler, H. Wittmann and K. Peters; *Monatsh. Chem.*, 1983, **114**, 783. 1003

83M1097 H. Wittmann, E. Ziegler, K. Peters, E. M. Peters and H. G. von Schnering, *Monatsh. Chem.*, 1983, **114**, 1097. 1003

B-83MI 201-01 M. Hudlicky and T. Hudlicky; in "The Chemistry of Functional Groups," Supplement D, ed. S. Patai and Z. Rappoport, Wiley, New York, 1983, p. 1021. 2, 6

B-83MI 201-02 C. W. McCleland; in "Synthetic Reagents," ed. J. S. Pizey, Ellis Horwood, Chichester, 1983, vol. 5, p. 85. 24

B-83MI 202-01 H. C. Brown and P. K. Jadhav; "Asymmetric Synthesis," Academic Press, Orlando, 1983, vol. 2, p. 1. 42

B-83MI 202-02 G. Solladié; "Asymmetric Synthesis," Wiley, New York, 1983, vol. 2, p. 157. 67

B-83MI 207-01 P. A. S. Smith; "Derivatives of Hydrazine and Other Hydronitrogens having N—N Bonds," Benjamin/Cummings, Reading, Mass., 1983. 373, 387, 388, 402, 404

83MI 208-01 M. Baudler, Y. Aklalay, K. Kazmierizak and J. Kahn; *Phosphorus*, 1983, **38**, 428. 471

83MI 208-02 M. Baudler and S. Klautke; *Phosphorus*, 1983, **38**, 121. 472

B-83MI 209-01 M. Wieber (ed.); "Gmelin Handbuch der Anorgnischen Chemie," Springer-Verlag, 1983, part 3. 481

B-83MI 210-01 M. M. Midland; "Asymmetric Synthesis," ed. J. D. Morrison, Academic Press, New York, 1983, vol. 2. 514

B-83MI 210-02 H. C. Brown; "Current Trends in Organic Chemistry," ed. H. Nozaki, IUPAC, Pergamon, Oxford, 1983, pp. 247–268. 514

83MI 210-02 K. Furusawa, T. Sakai, K. Tsuda and Y. Shimura; *Nucleic Acids Symp. Ser.*, 1983, **12**, 21 (*Chem. Abstr.*, 1984, **100**, 210 326v). 540

B-83MI 210-03 W. P. Weber; "Silicon Reagents for Organic Synthesis," Springer-Verlag, Berlin, 1983.

B-83MI 211-01 D. Seebach, B. Weidmann and L. Wilder; in "Modern Synthetic Methods," ed. R. Scheffold, Otto Salle Verlag, Frankfurt, 1983, vol. 3, p. 217. 565, 566

B-83MI 211-02 J.-F. Normant and G. Cahiez; in "Modern Synthetic Methods," ed. R. Scheffold, Otto Salle Verlag, Frankfurt, 1983, vol. 3, p. 173. 574

B-83MI 212-01 M. Hudlicky and T. Hudlicky; in "The Chemistry of Functional Groups," Supplement D, eds. S. Patai and Z. Rappoport, Wiley, New York, 1983, p. 1021. 606

83MI 221-01 M. A. Kazankova and I. F. Lutsenko; *Vestn. Mosk. Univ., Ser. 2: Khim.*, 1983, **24**, 315. 1027

B-83MI 223-01 J. Faust and R. Froböse; in "Gmelin Handbook of Inorganic Chemistry, Organocopper Compounds," Part 2, ed. J. Füssel, Springer, Berlin, 1983, p. 174. 1094, 1095

B-83MI 223-02 J. Faust and R. Froböse; in "Gmelin Handbook of Inorganic Chemistry, Organocopper Compounds," Part 2, ed. J. Fussel, Springer, Berlin, 1983, p. 227. 1094, 1095

83OM101 D. L. Reger, K. A. Belmore, E. Mintz, N. G. Charles, E. A. H. Griffith, and E. L. Amma; *Organometallics*, 1983, **2**, 101. 954

83OM563 E. Negishi, D. E. Van Horn, T. Yoshida and C. L. Rand; *Organometallics*, 1983, **2**, 563. 973

83OM979 W. J. Youngs and J. A. Ibers; *Organometallics*, 1983, **2**, 979. 432

83OM1005 A. J. Ashe, III, W. M. Butler and T. R. Diephouse; *Organometallics*, 1983, **2**, 1005. 880, 893

83OM1106 M. M. Goodman and F. F. Knapp, Jr.; *Organometallics*, 1983, **2**, 1106. 279

83OM1266 F. R. Kreissl, M. Wolfgruber and W. J. Sieber; *Organometallics*, 1983, **2**, 1266. 896

83OM1316 H. C. Brown and T. E. Cole; *Organometallics*, 1983, **2**, 1316. 1086

83OM1555 G. Erker, K. Kropp, J. L. Atwood and W. E. Hunter; *Organometallics*, 1983, **2**, 1555. 968

83OM1573 A. J. Ashe, III, E. G. Ludwig, Jr. and H. Pommerening; *Organometallics*, 1983, **2**, 1573. 893

83OM1689 M. Rotem and Y. Shvo; *Organometallics*, 1983, **2**, 1689. 676

83OM1859 A. J. Ashe, III, E. G. Ludwig, Jr. and J. Oleksyszyn; *Organometallics*, 1983, **2**, 1859. 491, 505, 893

83OM1877 E. P. Kyba, S.-T. Liu and R. L. Harris; *Organometallics*, 1983, **2**, 1877. 431

83OR(29)1 B. R. Castro; *Org. React.*, 1983, **29**, 1. 6, 7

83OR(29)345 J. K. Crandall and M. Apparu; *Org. React.*, 1983, **29**, 345. 85

83OS(61)116 R. E. Ireland, T. H. O'Neil and G. L. Tolman; *Org. Synth.*, 1983, **61**, 116. 693

83PAC589 K. B. Sharpless, C. H. Behrens, T. Katsuki, A. M. W. Lee, V. S. Martin, M. Takatani, S. M. Viti, F. J. Walker and S. S. Woodard; *Pure Appl. Chem.*, 1983, **55**, 589. 82

83PS(15)93 A. Hinke and W. Kuchen; *Phosphorus Sulfur*, 1983, **15**, 93. 832, 835

83PS(15)155 K. Diemert,W. Kuchen and T. Kutter; *Phosphorus Sulfur*, 1983, **15**, 155. 466

83PS(15)165 L. Horner and G. Simons; *Phosphorus Sulfur*, 1983, **15**, 165. 824, 825

83PS(17)73 K. Issleib, U. Kuehne and F. Krech; *Phosphorus Sulfur*, 1983, **17**, 73. 439

83PS(18)73 W. W. DuMent, R. Hensel and T. Severengiz; *Phosphorus Sulfur*, 1983, **18**, 73. 465

83PS(18)89 R. G. Cavell and L. Vande Griend; *Phosphorus Sulfur*, 1983, **18**, 89. 476

83RCR259 G. B. Sergeev, V. V. Smirnov and T. N. Rostovshchikova; *Russ. Chem. Rev.* (*Engl. Transl.*), 1983, **52**, 259. 23

83RCR284 Y. V. Guk, M. A. Ilyusin, E. L. Golod and B. V. Gidaspov; *Russ. Chem. Rev.* (*Engl. Transl.*), 1983, **52**, 284. 790

83RTC65 M. Rivière-Baudet, P. Rivière, A. Castel, G. Lacrampe and J. Satgé; *Recl. Trav. Chim. Pays-Bas*, 1983, **102** 65. 815

83S1 P. Brownbridge; *Synthesis*, 1983, 1 and 85. 699

83S34 R. V. Hoffman and E. L. Befoure; *Synthesis*, 1983, 34. 357

83S107 R. W. Hoffmann and K. Ditrich; *Synthesis*, 1983, 107. 652

83S139 G. Bringmann and S. Schneider; *Synthesis*, 1983, 139. 7

83S192 S. McDermott and W. J. Spillane; *Synthesis*, 1983, 192. 367

83S249 M. V. Bhatt and S. U. Kulkarni; *Synthesis*, 1983, 249. 10, 658

83S306 G. Cainelli, M. Contento, F. Manescalchi and L. Plessi; *Synthesis*, 1983, 306. 25, 31

83S314 M. Lissel and K. Drechsler; *Synthesis*, 1983, 314. 26

83S372 H. Schmidbauer and A. Schier; *Synthesis*, 1983, 372. 448

83S460 T. Imamoto, T. Matsumoto, T. Kusumoto and M. Yokoyama; *Synthesis*, 1983, 460. 36

83S461 T. Azuhata and Y. Okamoto; *Synthesis*, 1983, 461. 34

83S476 R. H. Smith, Jr. and C. J. Michejda; *Synthesis*, 1983, 476. 387, 388

83S489 A. Krowczynski and L. Kozerski; *Synthesis*, 1983, 489. 789

83S517 J. K. Whitesell and M. A. Whitesell; *Synthesis*, 1983, 517. 751

83S537 A. J. Biloski and B. Ganem; *Synthesis*, 1983, 537. 340, 344, 347

83S545 P. Dampawan and W. W. Zajac, Jr.; *Synthesis*, 1983, 545. 392

83S628 A. A. M. Houwen-Claassen, J. W. McFarland, B. H. M. Lammerink, L. Thijs and B. Zwanenburg; *Synthesis*, 1983, 628. 743

83S630 I. Degani, R. Fochi and V. Regondi; *Synthesis*, 1983, 630. 134, 143

83S713 G. A. Olah, J. G. Shih, B. P. Singh and B. G. B. Gupta; *Synthesis*, 1983, 713. 19

83S716 H. Ikehira and S. Tanimoto; *Synthesis*, 1983, 716. 243, 358

83S751 L. Testaferri, M. Tiecco, M. Tingoli, D. Chianelli and M. Montanucci; *Synthesis*, 1983, 751. 711

83S785 R. Knorr, P. Löw and P. Hassel; *Synthesis*, 1983, 785. 742

83S902 P. G. Baraldi, D. Simoni and S. Manfredini; *Synthesis*, 1983, 902. 744

83S913 F. Freeman and M. C. Keindl; *Synthesis*, 1983, 913. 221

83S916 T. Azuhata and Y. Okamoto; *Synthesis*, 1983, 916. 457

83S944 D. Scholz; *Synthesis*, 1983, 944. 139

83S1005 S. Hara, T. Kato and A. Suzuki; *Synthesis*, 1983, 1005. 624, 902

83S1014 G. Rosini, R. Ballini and P. Sorrenti; *Synthesis*, 1983, 1014. 394

83S1040 K. Hiroi, R. Kitayama and S. Sato; *Synthesis*, 1983, 1040. 721

83S1046 M. Srebnik and R. Mechoulam; *Synthesis*, 1983, 1046. 650

83SC387 W. Buijs, P. van Elburg and A. van der Gen; *Synth. Commun.*, 1983, **13**, 387. 36

83SC611 H. Firouzabadi, M. Naderi, A. Sardarian and B. Vessal; *Synth. Commun.*, 1983, **13**, 611. 215

83SC827 E. Mohacsi; *Synth. Commun.*, 1983, **13**, 827. 684

83SC1027 G. W. Kabalka, K. A. R. Sastry, F. F. Knapp and P. C. Srivastava; *Synth. Commun.*, 1983, 1027. 623

83SR215 R. Bussas, G. Kresze, H. Munsterer and A. Schwobel; *Sulfur Reports*, 1983, **2**, 215. 793

83T689 P. Caramella, E. Albini, T. Bandiera, A. Corsico-Coda, P. Grunanager and F. M. Albini; *Tetrahedron*, 1983, **39**, 689.

83T703 M. B. Gasc, A. Laltes and J. J. Perie; *Tetrahedron*, 1983, **39**, 703. 306

83T877 P. F. Hudrlik, A. K. Kulkarni, S. Jain and A. M. Hudrlik; *Tetrahedron*, 1983, **39**, 877. 909, 910

83T925 R. K. Boekman, Jr.; *Tetrahedron*, 1983, **39**, 925. 931

83T935 R. L. Danheiser, D. J. Carini, D. M. Fink and A. Basak; *Tetrahedron*, 1983, **39**, 935. 930, 940, 943

83T949 H. J. Reich, M. J. Kelly, R. E. Olson and R. C. Holtan; *Tetrahedron*, 1983, **39**, 949. 938, 939

83T1529 J. N. Marx; *Tetrahedron*, 1983, **39**, 1529. 34

83T2009 F. Marsais and G. Queguiner; *Tetrahedron*, 1983, **39**, 2009. 942

83T2031 R. C. Ronald and M. R. Winkle; *Tetrahedron*, 1983, **39**, 2031. 653

83T2323 A. S. Rao, S. K. Paknikar and J. G. Kirtane; *Tetrahedron*, 1983, **39**, 2323. 91

83T3839 P. Beslin, D. Lagain, J. Vialle and C. Minot; *Tetrahedron*, 1983, **37**, 3839. 229

83T4225 H. J. Meeuwissen, T. A. van der Knaap and F. Bickelhaupt; *Tetrahedron*, 1983, **39**, 4225. 441

83TCC85 H. J. Bestmann and O. Vostrowsky; *Top. Curr. Chem.*, 1983, **109**, 85. 997

83TL23 R. L. Danheiser and H. Sard; *Tetrahedron Lett.*, 1983, **24**, 23. 928

83TL163 O. P. Goel and U. Krolls; *Tetrahedron Lett.*, 1983, **24**, 163. 3

83TL227 A. K. Sinhababu and R. T. Borchardt; *Tetrahedron Lett.*, 1983, **24**, 227. 398

83TL245 L. J. McBride and M. H. Caruthers; *Tetrahedron Lett.*, 1983, **24**, 245. 409, 410

83TL511 K. Saigo, A. Kasahara, S. Ogawa and H. Nohira; *Tetrahedron Lett.*, 1983, **24**, 511. 816

83TL623 A. Pelter, B. Singaram, L. Williams and J. W. Wilson; *Tetrahedron Lett.*, 1983, **24**, 623. 525, 1005

83TL627 A. Pelter, L. Williams and J. W. Wilson; *Tetrahedron Lett.*, 1983, **24**, 627. 525, 1005

83TL631 A. Pelter, B. Singaram and J. W. Wilson; *Tetrahedron Lett.*, 1983, **24**, 631. 525

83TL635 A. Pelter, B. Singaram and J. W. Wilson; *Tetrahedron Lett.*, 1983, **24**, 635. 905

83TL637 M. V. Garad, A. Pelter, B. Singaram and J. W. Wilson; *Tetrahedron Lett.*, 1983, **24**, 637. 905

83TL725 V. Reutrakul and V. Rukachaisirikul; *Tetrahedron Lett.*, 1983, 725. 612

83TL731 S. Hara, H. Dojo, S. Takinami and A. Suzuki; *Tetrahedron Lett.*, 1983, **24**, 731. 623, 628, 902

83TL735 S. Hara, Y. Satoh, H. Ishiguro and A. Suzuki; *Tetrahedron Lett.*, 1983, **24**, 735. 902

83TL903 E. Laurent, R. Tardivel and H. Thiebault; *Tetrahedron Lett.*, 1983, **24**, 903. 16

83TL979 J. E. McMurry and W. J. Scott; *Tetrahedron Lett.*, 1983, **24**, 979. 689

83TL1011 G. D. Hartman, J. E. Swering and R. D. Hartman; *Tetrahedron Lett.*, 1983, **24**, 1011. 162

83TL1041 F. Sato, H. Watanabe, Y. Tanaka, T. Yamaji and M. Sato; *Tetrahedron Lett.*, 1983, **24**, 1041. 923, 924

83TL1213 F. A. Davis, O. D. Stringer and J. M. Billmers; *Tetrahedron Lett.*, 1983, **24**, 1213. 287

83TL1273 Y. Kita, H. Yasuda, Y. Sugiyama and F. Fukata; *Tetrahedron Lett.*, 1983, **24**, 1273. 263

83TL1345 M. E. Krafft and R. A. Holton; *Tetrahedron Lett.*, 1983, **24**, 1345. 697, 699

83TL1433 A. Palter, S. Singaram and H. Brown; *Tetrahedron Lett.*, 1983, **24**, 1433. 520, 900, 901

83TL1505 A. G. McKillop and J. A. Tarbin; *Tetrahedron Lett.*, 1983, **24**, 1505. 155

83TL1605 D. H. R. Barton, G. Bashiardes and J.-L. Fourrey; *Tetrahedron Lett.*, 1983, 1605. 630

83TL1809 F. Sato and H. Katsuno; *Tetrahedron Lett.*, 1983, **24**, 1809. 969

83TL1933 A. McKillop, F. A. Madjdabadi and D. A. Long; *Tetrahedron Lett.*, 1983, **24**, 1933. 13

83TL2043 J. J. Eisch and A. Piotrowski; *Tetrahedron Lett.*, 1983, **24**, 2043. 1006

83TL2055 R. B. Miller and M. A.-Hassan; *Tetrahedron Lett.*, 1983, **24**, 2055. 922, 931

83TL2311 M. Taddei, A. Papini, M. Fiorenza and A. Ricci; *Tetrahedron Lett.*, 1983, **24**, 2311. 265

83TL2653 T. Nakata, T. Tanaka and T. Oishi; *Tetrahedron Lett.*, 1983, **24**, 2653. 61

83TL2745 J. P. Genet, M. Balabane, J. E. Backwall and J. E. Nystrom; *Tetrahedron Lett.*, 1983, **24**, 2745. 301

83TL2821 E. J. Corey and S. G. Pyne; *Tetrahedron Lett.*, 1983, **24**, 2821. 351

83TL2935 M. E. Duggan and D. S. Karanewsky; *Tetrahedron Lett.*, 1983, **24**, 2935. 454

83TL3117 J. R. L. Smith and L. C. McKeer; *Tetrahedron Lett.*, 1983, 3117. 620

83TL3165 E. J. Corey and T. M. Eckrich; *Tetrahedron Lett.*, 1983, **24**, 3165. 999

83TL3191 F. A. Davis, J. M. Billmers and O. D. Stringer; *Tetrahedron Lett.*, 1983, **24**, 3191. 251

83TL3209 R. W. Hoffman and B. Landman; *Tetrahedron Lett.*, 1983, **24**, 3209. 530

83TL3233 H. Kapnang and G. Charles; *Tetrahedron Lett.*, 1983, **24**, 3233. 303

83TL3341 E. Nakamura, J.-I. Shimada, Y. Horiguchi and I. Kuwajima; *Tetrahedron Lett.*, 1983, **24**, 3341. 702

83TL3625 J. Escudie, C. Couret, H. Ranaivonjatovo and J. Wolf; *Tetrahedron Lett.*, 1983, **24**, 3625. 866

83TL3631 T. Takata and W. Ando; *Tetrahedron Lett.*, 1983, **24**, 3631. 232

83TL3795 J. N. Reed and V. Snieckus; *Tetrahedron Lett.*, 1983, **24**, 3795. 305

83TL3947 J. K. Cha, W. J. Christ and Y. Kishi; *Tetrahedron Lett.*, 1983, **24**, 3947. 46

83TL3951 G. Stork and M. Kahn; *Tetrahedron Lett.*, 1983, **24**, 3951. 46

83TL4033 W. Ando, Y. Hamada, A. Sekiguch and K. Ueno; *Tetrahedron Lett.*, 1983, **24**, 4033. 266

83TL4083 A. Krebs and J. Berndt; *Tetrahedron Lett.*, 1983, **24**, 4083. 947

83TL4295 M. L. Mancini and J. F. Honek; *Tetrahedron Lett.*, 1983, **24**, 4295. 921

83TL4443 A. Kunai, Y. Yanagi and K. Sasaki; *Tetrahedron Lett.*, 1983, **24**, 4443. 786

83TL4515 R. J. Billedeau, M. P. Sibi and V. Snieckus; *Tetrahedron Lett.*, 1983, **24**, 4515. 942

83TL4733 N. S. Narsimhan and R. Ammanamanchi; *Tetrahedron Lett.*, 1983, **24**, 4733. 752

83TL4895 B. M. Trost and J.-I. Yoshida; *Tetrahedron Lett.*, 1983, **24**, 4895. 912, 945

83TL4935 A. I. Meyers and P. D. Pansegrau; *Tetrahedron Lett.*, 1983, **24**, 4935. 974

83TL4979 D. H. R. Barton, D. Crich and W. B. Motherwell; *Tetrahedron Lett.*, 1983, **24**, 4979. 13

83TL4993 T. Mandai, K. Hara, T. Nakajima, M. Kawada and J. Otera; *Tetrahedron Lett.*, 1983, **24**, 4993. 678

83TL5109 A. Osuka and H. Suzuki; *Tetrahedron Lett.*, 1983, **24**, 5109. 1001

83TL5173 D. N. Harpp, T. Aida and T. H. Chan; *Tetrahedron Lett.*, 1983, **24**, 5173. 231, 723

83URP956484 M. G. Zimin, A. R. Burilov and A. N. Pudovik; *USSR Pat.* 956 484 (1983) (*Chem. Abstr.*, 1983, **98**, 107 548k). 266

83URP1046248 V. D. Makhaev and A. P. Borisov; *USSR Pat.* 1 046 248 (1983) (*Chem. Abstr.*, 1984, **100**, 68 529). 442

83USP4383991 S. T. D. Gough; *US Pat.* 4 383 991 (1983) (*Chem. Abstr.*, 1983, **99**, 38 641s). 252

83YGK776 Y. Kobayashi, K. Takano and K. Itabashi; *Yuki Gosei Kagaku Kyokaishi*, 1983, 776. 122

83ZAAC(498)57 M. Baudler, G. Fürstenburg, H. Suchomel and J. Hahn; *Z. Anorg. Allg. Chem.*, 1983, **498**, 57. 471

83ZAAC(501)89 Pham Ba Chi and F. Kober; *Z. Anorg. Allg. Chem.*, 1983, **501**, 89. 482, 485, 487, 510

83ZAAC(502)175 H. J. Breunig and H. Kischkel; *Z. Anorg. Allg. Chem.*, 1983, **502**, 175. 507

83ZAAC(504)147 R. Weisheit, R. Stendel, B. Messbauer, C. Langer and B. Walter; *Z. Anorg. Allg. Chem.*, 1983, **504**, 147. 460

83ZAAC(504)47 G. Fritz, J. Harer and K. Stoll; *Z. Anorg. Allg. Chem.*, 1983, **504**, 47. 471

83ZAAC(505)134 M. Wieber, D. Wirth and I. Fetzer; *Z. Anorg. Allg. Chem.*, 1983, **505**, 134. 489

83ZC249 H. Schmidbauer and A. Mortl; *Z. Chem.*, 1983, **23**, 249. 449

83ZN(B)125 H. J. Breunig and D. Muller; *Z. Naturforsch., Teil B*, 1983, **38**, 125. 505

83ZN(B)365 H. Trampisch and W. Beck; *Z. Naturforsch., Teil B*, 1983, **38**, 365. 252

83ZN(B)1027 H. H. Karsch; *Z. Naturforsch., Teil B*, 1983, **38**, 1027. 435

83ZOB2519 A. V. Kushin, R. A. Nurushev and G. A. Tolstikov; *Z. Obshch. Khim.*, 1983, **307**, 2519 (*Chem. Abstr.*, 1984, **100**, 103 426). 595

83ZOR343 E. A. Stukalo, E. M. Yur'eva and L. N. Markovskii; *Zh. Org. Khim.*, 1983, **19**, 343 (*Chem. Abstr.*, 1983, **98**, 215 257y). 238

84ACR28 D. Liotta; *Acc. Chem. Res.*, 1984, **17**, 28. 282

84ACR215 G. W. Kabalka; *Acc. Chem. Res.*, 1984, **17**, 215. 4

84AG(E)224 A. Maercker and R. Dujardin; *Angew. Chem., Int. Ed. Engl.*, 1984, **23**, 224. 1006

84AG(E)420 A. R. Katritzy and C. M. Marson; *Angew. Chem., Int. Ed. Engl.*, 1984, **23**, 420. 11

84AG(E)515 W. A. Herrmann, R. Serranon, N. Küsthardt, M. L. Ziegler, E. Guggolz and T. Zahn; *Angew. Chem., Int. Ed. Engl.*, 1984, **23**, 515. 585

84AG(E)539 K. P. C. Vollhardt; *Angew. Chem., Int. Ed. Engl.*, 1984, **23**, 539. 944, 984

84AG(E)556 M. T. Reetz; *Angew. Chem., Int. Ed. Engl.*, 1984, **23**, 556. 67

84AG(E)587 K. H. Dotz; *Angew. Chem., Int. Ed. Engl.*, 1984, **23**, 587. 670

84AG(E)727 H. Hofmeister, K. Annen, H. Laurent and R. Wiechert; *Angew. Chem., Int. Ed. Engl.*, 1984, **23**, 727. 1018

84AG(E)812 W. A. Herrmann, B. Koumbouris, T. Zahn and M. L. Ziegler; *Angew. Chem., Int. Ed. Engl.*, 1984, **23**, 812. 512

84AG(E)932 D. Hoppe; *Angew. Chem., Int. Ed. Engl.*, 1984, **23**, 932. 997

84AJC527 G. B. Deacon, W. R. Jackson and J. M. Pfeiffer; *Aust. J. Chem.*, 1984, **37**, 527. 888

84AJC845 G. Crank and M. I. H. Makin; *Aust. J. Chem.*, 1984, **37**, 845. 772

84AJC2331 G. Crank and M. I. H. Makin; *Aust. J. Chem.*, 1984, **37**, 2331. 213

84AX(C)842 I. Vickovic, W. H. Watson and A. L. Ternay, Jr.; *Acta Crystallogr., Part C*, 1984, **40**, 842. 1000

84BCJ232 S. Oae and H. Togo; *Bull. Chem. Soc. Jpn.*, 1984, 232. 708

84BCJ303 A. Osuka, K. Takechi and H. Suzuki; *Bull. Chem. Soc. Jpn.*, 1984, **57**, 303. 14

84BCJ613 K. Sukata; *Bull. Chem. Soc. Jpn.*, 1984, **57**, 613. 156

84BCJ2111 T. Negoro and Y. Ikeda; *Bull. Chem. Soc. Jpn.*, 1984, **57**, 2111. 24

84BCJ2440 Y. Watanabe, T. Ohta, Y. Tsuji, T. Hiyoshi and Y. Tsuji; *Bull. Chem. Soc. Jpn.*, 1984, **57**, 2440. 753

84BCJ2675 T. Oshikawa and M. Yamashita; *Bull. Chem. Soc. Jpn.*, 1984, **57**, 2675. 6

84BCJ2691 C. A. Horiuchi and J. Y. Satoh; *Bull. Chem. Soc. Jpn.*, 1984, **57**, 2691. 631

84BJC108 K. Takai, M. Sato, K. Oshima and H. Nozaki; *Bull. Chem. Soc. Jpn.*, 1984, **57**, 108. 1098

84BSB157 M. Fournier, F. Fournier and J. Berthelot; *Bull. Soc. Chim. Belg.*, 1984, **93**, 157. 29, 30

84CB427 M. Wolfgruber, W. Sieber and F. R. Kreissl; *Chem. Ber.*, 1984, **117**, 427. 878, 896

84CB1900 J. C. Jochims, R. Abu-El-Halawa, I. Jibril and G. Huttner; *Chem. Ber.*, 1984, **117**, 1900. 1009

84CB1999 M. Herberhold, W. Ehrenreich and K. Guldner; *Chem. Ber.*, 1984, **117**, 1999. 891

84CB2328 W. J. Richter; *Chem. Ber.*, 1984, **117**, 2328. 443

84CB3400 S. Heitkamp, H. Sommer and O. Stelzer; *Chem. Ber.*, 1984, **117**, 3400. 428, 430

84CC200 M. M. Campbell, N. I. Carruthers, S. J. Mickel and P. M. Winton; *J. Chem. Soc., Chem. Commun.*, 1984, 200. 1064

84CC257 J. Verbeek, A. V. E. George, R. L. P. de Jong and L. Brandsma; *J. Chem. Soc., Chem. Commun.*, 1984, 257. 942

84CC258 A. Dondoni, T. Dall'Occo, G. Fantin, M. Fogagnolo, A. Medici and P. Pedrini; *J. Chem. Soc., Chem. Commun.*, 1984, 258. 942

84CC374 M. F. Joucla and C. W. Rees; *J. Chem. Soc., Chem. Commun.*, 1984, 374. 244, 773, 792, 795

84CC426 S. Uemura, S. I. Fukuzawa, T. Yamauchi, K. Hattori, S. Mizutaki and K. Tamaki; *J. Chem. Soc., Chem. Commun.*, 1984, 426. 287

84CC451 M. Nakajima, J. C. Warner and J.-P. Anselme; *J. Chem. Soc., Chem. Commun.*, 1984, 451. 385

84CC655 G. V. Wisser, B. W. v. Haltern, J. D. M. Herschied, G. A. Brinkman and A. Hoekstra; *J. Chem. Soc., Chem. Commun.*, 1984, 655. 613

84CC843 M. Masui, Y. Mizuki, K. Sakai, C. Ueda and H. Ohmori; *J. Chem. Soc., Chem. Commun.*, 1984, 843. 219, 227

84CC1144 O. Meth-Cohn and G. van Vuuren; *J. Chem. Soc., Chem. Commun.*, 1984, 1144. 244

84CC1523 N. Suzuki, Y. Kaneko, T. Nomoto and Y. Izawa; *J. Chem. Soc., Chem. Commun.*, 1984, 1523. 621, 627

84CC1702 C. L. Raston and G. Salem; *J. Chem. Soc., Chem. Commun.*, 1984, 1702. 559, 560

84CHEC(1)539 R. E. Atkinson; *Comp. Heterocycl. Chem.*, 1984, **1**, 539–561. 872, 880

84CJC2415 L. A. Paquette and K. E. Stevens; *Can. J. Chem.*, 1984, **62**, 2415. 911

84CL139 H. Suzuki, H. Tani and A. Osuka; *Chem. Lett.*, 1984, 139. 708, 723

84CL251 K. Soai, T. Yamanoi and H. Oyamada; *Chem. Lett.*, 1984, 251. 269

84CL277 O. Tsuge, S. Kanemasa, H. Sakoh and E. Wada; *Chem. Lett.*, 1984, 277. 667

84CL603 M. Yoshifuji, K. Shibayama and N. Inamoto; *Chem. Lett.*, 1984, 603. 857

84CL853 N. Ohira, Y. Aso, T. Otsubo and F. Ogura; *Chem. Lett.*, 1984, 853. 753

84CL877 S. Torii, H. Tanaka, N. Tada, S. Nagao and M. Sasaoka; *Chem. Lett.*, 1984, 877. 21

84CL1173 T. Fujisawa, S. Iida and T. Sato; *Chem. Lett.*, 1984, 1173. 25

84CL1241 N. Yoneda, S. Nagata, T. Fukuhara and A. Suzuki; *Chem. Lett.*, 1984, 1241. 17

84CL1509 C. A. Horiuchi and J. Y. Satoh; *Chem. Lett.*, 1984, 1509. 33, 34

84CL2125 Y. Ueno, A. Kojima and M. Okawara; *Chem. Lett.*, 1984, 2125. 188

84CPB2609 S. Ozaki, S. Nishiguchi and M. Masui; *Chem. Pharm. Bull.*, 1984, **32**, 2609. 390

84CPB3759 T. Aoyama, S. Terasawa, K. Sudo and T. Shioiri; *Chem. Pharm. Bull.*, 1984, **32**, 3759. 683

84CR471 P. Beak, W. J. Zajdel and D. B. Reitz; *Chem. Rev.*, 1984, **84**, 471. 302

84CRV117 F. Freeman; *Chem. Rev.*, 1984, **84**, 117. 723

84CRV471 P. Beak, W. J. Zajdel and D. B. Reitz; *Chem. Rev.*, 1984, **84**, 471. 1002

84CSR441 P. Bamfield and P. F. Gordon; *Chem. Soc. Rev.*, 1984, **13**, 441. 662

84DOK(274)606 N. A. Nesmeyanov, V. G. Kharitonov, S. T. Berman, P. V. Petrovskii and O. A. Reutov; *Dokl. Akad. Nauk SSSR*, 1984, **274**, 606 (*Chem. Abstr.*, 1984, **101**, 72 827b). 877

84DOK(274)822 L. A. Vikhreva, T. A. Pudova and K. K. Babasheva; *Dokl. Akad. Nauk SSSR*, 1984, **274**, 822. 1033

84EUP118109 W. G. Haag, A. B. Shortt and R. S. Brimhall; *Eur. Pat.* 118 109 (1984) (*Chem. Abstr.*, 1985, **102**, 6813p). 254, 255

84GEP3229565 N. Dereu, R. A. Zingaro, E. Etschenberg, A. Huether and W. Eling; *Ger. Pat.* 3 229 565 (1984) (*Chem. Abstr.*, 1984, **101**, 7598p). 260

84H(22)1017 M. Sako and Y. Maki; *Heterocycles*, 1984, **22**, 1017. 214

84H(22)1303 H. Sashida and T. Tsuchisa; *Heterocycles*, 1984, **22**, 1303. 244

84HCA224 R. Amstutz, T. Laube, W. B. Schweizer, D. Seebach and J. D. Dunitz; *Helv. Chim. Acta*, 1984, **67**, 224. 999

84HCA261 D. Seebach and P. Knockel; *Helv. Chim. Acta*, 1984, **67**, 261. 785

84HCA284 M. Schlosser, R. Dahan and S. Cottens; *Helv. Chim. Acta*, 1984, **67**, 284. 612

84HCA906 G. H. Hakimelahi, H. Sharghi, H. Zarrinmayeh and A. Khalafi-Nezhad; *Helv. Chim. Acta*, 1984, **67**, 906. 104

84HCA1238 P. Pöchlauer, E. P. Müller and P. Peringer; *Helv. Chim. Acta*, 1984, **67**, 1238. 860

84HCA1406	R. O. Duthaler, P. A. Lyle and C. Heuberger; *Helv. Chim. Acta*, 1984, **67**, 1406.	657
84HOU(13/3c)1	R. Köster, G. Schmidt, M. A. Grassberger and B. Wrackmeyer; *Methoden Org. Chem. (Houben-Weyl)*, 1984, **13/3c**, 1.	1083
84HOU(13/9b)1	E. Langer; *Methoden Org. Chem. (Houben-Weyl)*, 1984, **13/9b**, 1.	575
84HOU(13/9b)287	E. Langer; *Methoden Org. Chem. (Houben-Weyl)*, 1984, **13/9b**, 287.	579
84HOU(13/9b)464	E. Langer; *Methoden Org. Chem. (Houben-Weyl)*, 1984, **13/9b**, 464.	579
84HOU(13/9b)628	H. F. Klein; *Methoden Org. Chem. (Houben-Weyl)*, 1984, **13/9b**, 628.	575
84HOU(13/9b)702	A. Segnitz; *Methoden Org. Chem. (Houben-Weyl)*, 1984, **13/9b**, 702.	579
84IC778	R. D. Waid and D. W. Meek; *Inorg. Chem.*, 1984, **23**, 778.	439
84IC1504	A. Belforte, F. Calderazzo, A. Morvillo, G. Pelizzi and D. Vitali; *Inorg. Chem.*, 1984, **23**, 1504.	893
84IC2062	R. R. Ford, M. A. Goodman, R. H. Neilson, A. K. Roy, U. G. Wettermark and P. Nisian-Neilson; *Inorg. Chem.*, 1984, **23**, 2062.	466, 467
84IC2582	A. H. Cowley, J. E. Kilduff, J. G. Lasch, S. K. Mehrotra, N. C. Norman, M. Pakulski, B. R. Whittlesey, J. L. Atwood and W. E. Hunter; *Inorg. Chem.*, 1984, **23**, 2582.	866, 893
84IC3304	R. Bartsch, S. Heitkamp, H. Peters and O. Stelzer; *Inorg. Chem.*, 1984, **23**, 3304.	440
84IC4130	D. M. Schubert and A. D. Norman; *Inorg. Chem.*, 1984, **23**, 4130.	428
84IC4170	M. Antberg, C. Prengel and L. Dahlenburg; *Inorg. Chem.*, 1984, **23**, 4170.	431
84ICAL49	M. Luksza, K. Jorg and W. Malisch; *Inorg. Chim. Acta*, 1984, **85**, L49.	512
84IZV700	N. A. Nesmeyanov, V. G. Kharitonov, S. T. Berman and O. A. Reutov; *Izv. Akad. Nauk. SSSR, Ser. Khim.*, 1984, 700 (*Chem. Abstr.*, 1984, **101**, 91 110d).	877
84IZV719	G. V. Romanov, A. A. Lapin and A. N. Pudovik; *Izv. Akad. Nauk. SSSR, Ser. Khim.*, 1984, 719 (*Chem. Abstr.*, 1984, **101**, 130 769).	
84JA452	W. E. Barnette; *J. Am. Chem. Soc.*, 1984, **106**, 452.	16, 365, 611, 613
84JA460	Y. Chiang, A. J. Kresge and Y. S. Tang; *J. Am. Chem. Soc.*, 1984, **106**, 460.	637
84JA477	S. E. Biali and Z. Rappoport; *J. Am. Chem. Soc.*, 1984, **106**, 477.	645
84JA759	K. Steliou, Y. Gareau and D. N. Harpp; *J. Am. Chem. Soc.*, 1984, **106**, 759.	221
84JA1130	T. Cohen and M.-T. Lin; *J. Am. Chem. Soc.*, 1984, **106**, 1130.	999
84JA1291	W. J. Evans, J. H. Meadows, W. E. Hunter, and J. L. Atwood; *J. Am. Chem. Soc.*, 1984, **106**, 1291.	969
84JA1650	W. D. Jones and F. J. Feher; *J. Am. Chem. Soc.*, 1984, **106**, 1650.	962
84JA2949	D. R. Williams and S.-Y. Sit; *J. Am. Chem. Soc.*, 1984, **106**, 2949.	85
84JA3366	S. F. Nelsen, S. C. Blackstock and T. B. Frigo; *J. Am. Chem. Soc.*, 1984, **106**, 3366.	383
84JA3682	E. J. Corey, B. Samuelsson and F. A. Luzzio; *J. Am. Chem. Soc.*, 1984, **106**, 3682.	397
84JA4283	H. Sugimoto and D. T. Sawyer; *J. Am. Chem. Soc.*, 1984, **106**, 4283.	156
84JA4630	W. J. Scott, G. T. Crisp and J. K. Stille; *J. Am. Chem. Soc.*, 1984, **106**, 4630.	689
84JA5261	D. D. Tanner, G. C. Gidley, N. Das, J. E. Rowe and A. Potter; *J. Am. Chem. Soc.*, 1984, **106**, 5261.	95
84JA5271	R. O. Kazaki, T. Hasegawa and Y. Shishido; *J. Am. Chem. Soc.*, 1984, **106**, 5271.	212
84JA5734	K. C. Nicolaou and S. E. Webber; *J. Am. Chem. Soc.*, 1984, **106**, 5734.	623, 624
84JA6105	L. D. Boardman, V. Bagheri, H. Sawada, and E. Negishi; *J. Am. Chem. Soc.*, 1984, **106**, 6105.	926, 931, 973, 976
84JA6422	W. A. Nugent and J. C. Calabrese; *J. Am. Chem. Soc.*, 1984, **106**, 6422.	926, 988
84JA6467	P. von Ragué Schleyer, T. Clark, A. J. Kos, G. W. Spitznagel, C. Rohde, D. Arad, K. N. Houk and N. G. Rondan; *J. Am. Chem. Soc.*, 1984, **106**, 6467.	999, 1006
84JA6709	R. Noyori, I. Tomino, M. Yamada and M. Nishizawa; *J. Am. Chem. Soc.*, 1984, **106**, 6709.	64
84JA6841	H. Lauke, P. J. Swepston and T. J. Marks; *J. Am. Chem. Soc.*, 1984, **106**, 6841.	602
84JA7021	L. D. Quin, K. C. Caster, J. C. Kisalus and K. A. Mesch; *J. Am. Chem. Soc.*, 1984, **106**, 7021.	441
84JA7496	S. A. Kinkead, R. C. Kumar and J. M. Shreeve; *J. Am. Chem. Soc.*, 1984, **106**, 7496.	367
84JA7500	G. T. Crisp, W. J. Scott and J. K. Stille; *J. Am. Chem. Soc.*, 1984, **106**, 7500.	931
84JA7565	W. D. Wulff and D. C. Yang; *J. Am. Chem. Soc.*, 1984, **106**, 7565.	671
84JA7861	R. S. Garigipati, A. J. Freyer, R. R. Whittle and S. M. Weinreg; *J. Am. Chem. Soc.*, 1984, **106**, 7861.	245
84JA7867	H. Natsugari, R. R. Whittle and S. M. Weinreb; *J. Am. Chem. Soc.*, 1984, **106**, 7867.	244, 245
84JA8001	H. C. Brown, J. S. Cha, B. Nazer and N. M. Yoon; *J. Am. Chem. Soc.*, 1984, **106**, 8001.	268
84JA8188	P. Pitchen, E. Duñach, M. N. Deshmukh and H. B. Kagan; *J. Am. Chem. Soc.*, 1984, **106**, 8188.	147
84JA8300	S. M. Clift and J. Schwartz; *J. Am. Chem. Soc.*, 1984, **106**, 8300.	681
84JCR(S)32	A. J. MacPherson and D. J. H. Smith; *J. Chem. Res. (S)*, 1984, 32.	441
84JCS(D)321	L. M. Engelhardt, W.-P. Leung, C. L. Raston, P. Twiss and A. H. White; *J. Chem. Soc., Dalton Trans.*, 1984, 321.	1005
84JCS(D)575	M. Ciampolini, N. Nardi, P. Dapporto, P. Innocenti and F. Zanobini; *J. Chem. Soc., Dalton Trans.*, 1984, 575.	438
84JCS(D)1801	W.-P. Leung, C. L. Raston, B. W. Skelton and A. H. White; *J. Chem. Soc., Dalton Trans.*, 1984, 1801.	1005
84JCS(D)1911	R. M. K. Deng and K. B. Dillon; *J. Chem. Soc., Dalton Trans.*, 1984, 1911.	475, 477
84JCS(D)2365	F. Calderazzo, R. Poli and G. Pelizzi; *J. Chem. Soc., Dalton Trans.*, 1984, 2365.	881, 883, 885, 890
84JCS(P1)195	P. Hodge and E. Khoshdel; *J. Chem. Soc. Perkin Trans. 1*, 1984, 195.	7
84JCS(P1)1401	M. Kawase, Y. Miyake and Y. Kikugawa; *J. Chem. Soc., Perkin Trans. 1*, 1984, 1401.	360
84JCS(P1)1805	I. Fleming, T. W. Newton and F. Roessler; *J. Chem. Soc., Perkin Trans. 1*, 1984, 1805.	546, 917, 977
84JCS(P1)2591	M. R. Bryce; *J. Chem. Soc., Perkin Trans. 1*, 1984, 2591.	361

84JCS(P1)2623 J. Barluenga, P. J. Campos, J. M. Gonzalez and G. Asensio; *J. Chem. Soc., Perkin Trans. 1*, 1984, 2623. 631
84JCS(P1)2845 E. K. Baylis, C. D. Campbell and J. G. Dingwall; *J. Chem. Soc., Perkin Trans. 1*, 1984, 2845. 451
84JCS(P2)411 R. W. Alder, R. B. Sessions, J. O. Gmünder and C. A. Grob; *J. Chem. Soc., Perkin Trans. 2*, 1984, 411. 383
84JCS(P2)451 Y. Ogata, M. Kimura and Y. Kondo; *J. Chem. Soc., Perkin Trans. 2*, 1984, 451. 685
84JGU1704 M. N. Dimukhametov and I. A. Nuretdinov; *J. Gen. Chem. USSR (Engl. Transl.)*, 1984, **54**, 1704. 465
84JGU2237 M. V. Livantsov, M. V. Proskurnina, A. A. Prishchenko and I. F. Lutsenko; *J. Gen. Chem. USSR (Engl. Transl.)*, 1984, **54**, 2237. 451
84JHC1709 M. J. Kornet, T. Varia and W. Beaven; *J. Heterocycl. Chem.*, 1984, **21**, 1709. 381
84JMC228 C. H. Levenson and R. B. Meyer; *J. Med. Chem.*, 1984, **27**, 228. 163, 237
84JMC1379 D. A. Walsh, H. W. Moran and D. A. Shamblee; *J. Med. Chem.*, 1984, **27**, 1379. 168
84JMC1634 G. D. Hartman, R. D. Hartman, J. E. Schwering, W. S. Saari, E. L. Engelhardt, N. R. Jones, P. Wardman, M. E. Watts and M. Woodcock; *J. Med. Chem.*, 1984, **27**, 1634. 160, 162
84JOC336 E. Y. Osei-Twum, D. McCallum, A. S. Nazran, R. Panicucci, P. A. Risbood and J. Warkentin; *J. Org. Chem.*, 1984, **49**, 336. 373, 376, 403
84JOC431 R. T. Hrubiec and M. B. Smith; *J. Org. Chem.*, 1984, **49**, 431. 7, 8
84JOC615 V. Jayathirtharao, V. Ramamurthy, E. Schaumann and H. Nimmesgern; *J. Org. Chem.*, 1984, **49**, 615. 202
84JOC645 J. H. Looker, J. R. Edman and C. A. Kingsbury; *J. Org. Chem.*, 1984, **49**, 645. 651
84JOC725 R. B. Miller and M. I. Al-Hassan; *J. Org. Chem.*, 1984, **49**, 725. 925, 926
84JOC806 O. Lerman, Y. Tor, D. Hebel and S. Rozen; *J. Org. Chem.*, 1984, **49**, 806. 613
84JOC945 H. C. Brown and B. Singaram; *J. Org. Chem.*, 1984, **49**, 945. 515
84JOC1078 D. L. Comins and J. D. Brown; *J. Org. Chem.*, 1984, **49**, 1078. 420
84JOC1140 H. Urabe and I. Kuwajima; *J. Org. Chem.*, 1984, **49**, 1140. 537
84JOC1286 A. D. N. Vaz and G. Schoellmann; *J. Org. Chem.*, 1984, **49**, 1286. 22
84JOC1309 D. J. Brunelle; *J. Org. Chem.*, 1984, **49**, 1309. 711
84JOC1348 A. J. Castellino and H. Rapoport; *J. Org. Chem.*, 1984, **49**, 1348. 691
84JOC1434 M. Kijama, Y. Nambu, T. Endo and M. Okawara; *J. Org. Chem.*, 1984, **49**, 1434. 214
84JOC1465 F. A. Davis, J. P. McCauley and M. E. Harakal; *J. Org. Chem.*, 1984, **49**, 1465. 150
84JOC1672 R. L. Danheiser and S. K. Gee; *J. Org. Chem.*, 1984, **49**, 1672. 668
84JOC1772 W. R. Roush and T. A. Blizzard; *J. Org. Chem.*, 1984, **49**, 1772. 976
84JOC1784 R. B. King and P. M. Sundaram; *J. Org. Chem.*, 1984, **49**, 1784. 414
84JOC1838 J. Kang, W. Cho and W. K. Lee; *J. Org. Chem.*, 1984, **49**, 1838. 922, 923
84JOC1931 T. J. Maricich and C. N. Angeletakis; *J. Org. Chem.*, 1984, **49**, 1931. 724
84JOC2032 G. A. Olah, L. Ohannesian, M. Arvanaghi and G. K. S. Prakash; *J. Org. Chem.*, 1984, **49**, 2032. 22
84JOC2047 J. L. Namy, J. Souppe, J. Collin and H. B. Kagan; *J. Org. Chem.*, 1984, **49**, 2047. 56
84JOC2081 R. A. Olofson, J. T. Martz, J.-P. Senet, M. Piteau and T. Malfroot; *J. Org. Chem.*, 1984, **49**, 2081. 303
84JOC2293 W. D. Wulff, K.-S. Chan and P.-C. Tang; *J. Org. Chem.*, 1984, **49**, 2293. 670
84JOC2298 H. Nishiyama, T. Kitajima, M. Matsumoto and K. Itoh; *J. Org. Chem.*, 1984, **49**, 2298. 540
84JOC2438 R. O. Hutchins, K. Learn, F. El-Telbany and Y. P. Stercho; *J. Org. Chem.*, 1984, **49**, 2438. 422
84JOC2494 T. Kimmel and D. Becker; *J. Org. Chem.*, 1984, **49**, 2494. 422
84JOC2500 E. C. Taylor, K. McDaniel and J. S. Skotnicki; *J. Org. Chem.*, 1984, **49**, 2500. 160
84JOC2629 E. Negishi, N. Okukado, S. F. Lovich and F.-T. Luo; *J. Org. Chem.*, 1984, **49**, 2629. 160
84JOC2724 Y. Miura and M. Kinoshita; *J. Org. Chem.*, 1984, **49**, 2724. 361
84JOC2795 R. A. Olofson and D. E. Abbott; *J. Org. Chem.*, 1984, **49**, 2795. 27
84JOC2906 S. D. Pastor, P. A. Odorisio and J. D. Spivak; *J. Org. Chem.*, 1984, **49**, 2906. 432
84JOC3027 P.-T. Ho and N. Davies; *J. Org. Chem.*, 1984, **49**, 3027. 6
84JOC3106 J. L. Kice, F. McAfee and H. Slebocka-Tilk; *J. Org. Chem.*, 1984, **49**, 3106. 735
84JOC3216 D. P. Cox, J. Terpinski and W. Lawrynowicz; *J. Org. Chem.*, 1984, **49**, 3216. 19
84JOC3235 T. Hayama, S. Tomoda, Y. Takeuchi and Y. Nomura; *J. Org. Chem.*, 1984, **49**, 3235. 284
84JOC3243 S. W. Remiszewski, R. R. Whittle and S. M. Weinreb; *J. Org. Chem.*, 1984, **49**, 3243. 245
84JOC3387 R. A. Rossi, A. N. Santiago and S. M. Palacios; *J. Org. Chem.*, 1984, **49**, 3387. 438
84JOC3429 W. R. Roush, S. M. Peseckis and A. E. Walts; *J. Org. Chem.*, 1984, **49**, 3429. 1011
84JOC3470 G. Lunn, E. B. Sansone and L. K. Keefer; *J. Org. Chem.*, 1984, **49**, 3470. 379
84JOC3796 A. Toshimitsu, H. Owada, K. Terao, S. Uemura and M. Okano; *J. Org. Chem.*, 1984, **49**, 3796. 282
84JOC3861 J. M. Hawkins and K. B. Sharpless; *J. Org. Chem.*, 1984, **49**, 3861. 367
84JOC3904 T. Imamoto, T. Kusomoto, Y. Tawarayama, Y. Suguira, T. Mita, Y. Hatanaka and M. Yokayama; *J. Org. Chem.*, 1984, **49**, 3904. 69, 564
84JOC4293 W. G. Dauben, J. M. Gerdes and R. A. Bunce; *J. Org. Chem.*, 1984, **49**, 4293. 444
84JOC4295 L. A. Sternson and R. Chandrasakar; *J. Org. Chem.*, 1984, **49**, 4295. 649
84JOC4332 W. R. Roush and T. A. Blizzard; *J. Org. Chem.*, 1984, **49**, 4332. 976
84JOC4538 Y. Guindon, J. G. Atkinson and H. E. Morton; *J. Org. Chem.*, 1984, **49**, 4538. 142
84JOC4609 S. M. Palacios, A. N. Santiago and R. A. Rossi; *J. Org. Chem.*, 1984, **49**, 4609. 438
84JOC4687 F. Effenberger and A. Krebs; *J. Org. Chem.*, 1984, **49**, 4687. 943
84JOC4700 L. Rebrovic and G. F. Koser; *J. Org. Chem.*, 1984, **49**, 4700. 1020
84JOC4733 J. A. Walker, S. P. Bitler and F. Wudl; *J. Org. Chem.*, 1984, **49**, 4733. 478

84JOC4822	H. C. Brown, V. Somayaji and S. Narasimhan; *J. Org. Chem.*, 1984, **49**, 4822.	629
84JOC5280	G. W. Ebert and R. D. Rieke; *J. Org. Chem.*, 1984, **49**, 5280.	959
84JOM(260)281	U. Weber, L. Zsolnai and G. Huttner; *J. Organomet. Chem.*, 1984, **260**, 281.	512
84JOM(267)265	G. J. Erskine, G. J. B. Hirst, E. L. Weinberg, B. K. Hunter and J. D. McCowan; *J. Organomet. Chem.*, 1984, **267**, 265.	566
84JOM(268)5	B. M. Mikhailov and L. I. Lavrinovich; *J. Organomet. Chem.*, 1984, **268**, 5.	606
84JOM(269)107	W. Chodkiewicz, D. Guillerm, D. Jore, E. Mattiew and W. Wodzki; *J. Organomet. Chem.*, 1984, **269**, 107.	436
84JOM(270)177	E. R. Bornancini, R. A. Alonso and R. A. Rossi; *J. Organomet. Chem.*, 1984, **270**, 177.	874
84JOM(273)C55	W. Chodkiewicz; *J. Organomet. Chem.*, 1984, **273**, C55.	443
84JOM(276)C45	E. Boyar, A. J. Deeming, A. J. Arce and Y. De Sanctis; *J. Organomet. Chem.*, 1984, **276**, C45.	582
84JOM(277)257	H. J. Breunig and H. Jawad; *J. Organomet. Chem.*, 1984, **277**, 257.	890, 891
84LA39	K. Hofmann and G. Simchen; *Liebigs Ann. Chem.*, 1984, 39.	209
84LA1358	G. Himbert and L. Henn; *Liebigs Ann. Chem.*, 1984, 1358.	1027
84LA1468	E. Müller, R. Kettler and M. Wiessler; *Liebigs Ann. Chem.*, 1984, 1468.	385
84MI 202-01	N. Ohto; *J. Synth. Org. Chem. Jpn.*, 1984, **42**, 652.	98
B-84MI 204-01	D. L. J. Clive, "Selections from the Aldrichimica Acta," Aldrich Chemical Company, Inc., 1984, pp. 43–50.	280
B-84MI 205-01	W. Lwowski; in "Azides and Nitrenes, Reactivity and Utility," ed. E. F. V. Scriven, Academic Press, Orlando, 1984, 205.	298
B-84MI 206-01	A. Padwa; "1,3-Dipolar Cycloaddition Chemistry," Wiley, New York, 1984, vol. 2.	354
B-84MI 207-01	E. W. Schmidt; "Hydrazine and its Derivatives," Wiley, New York, 1984.	373
B-84MI 208-01	D. Valentine; in "Asymmetric Synthesis," ed. J. D. Morrison and J. W. Scott, Academic Press, New York, 1984, vol. 4, chap. 3.	442, 443
B-84MI 210-01	B. M. Mikailov and Yu. N. Bubnov; "Organoboron Compounds in Organic Synthesis," Harwood Academic Publishers, GmbH, Chur, 1984.	514
B-84MI 210-02	D. R. M. Walton; in "Dictionary of Organometallic Compounds," Chapman and Hall, London, 1984, vol. 2 and supplements 1–5.	535
84MI 212-01	M. J. de Nie-Sarink and P. E. Prillwitz; *Spec. Chem.*, 1984, **4**, 9.	614
B-84MI 213-01	D. A. Evans; in "Asymmetric Synthesis," ed. J. D. Morrison, Academic Press, New York, 1984, vol. 3, p. 1.	696
84MIP30022	L. Magdanyi, L. Kovacs, A. Hutak, A. Marosvari, I. Dubovszky and E. Szatmari; *Hung. Pat.* 30 022 (1984) (*Chem. Abstr.*, 1984, **101**, 23 724).	444
84NJC759	H. J. Caillot, F. Metz and R. Cromer; *Nouv. J. Chim.*, 1984, **8**, 759.	578
84OM69	H. Scheemann, J. Müller, N. Bruncks, H. Lauke and J. Pickardt; *Organometallics*, 1984, **3**, 69.	565
84OM134	D. L. Reger, K. A. Belmore, E. Mintz and P. J. McElligott; *Organometallics*, 1984, **3**, 134.	978
84OM495	A. J. Ashe, III and F. J. Drone; *Organometallics*, 1984, **3**, 495.	880
84OM1223	J. C. Calabrese, D. C. Roe, D. L. Thorn and T. H. Tulip; *Organometallics*, 1984, **3**, 1223.	979
84OM1392	H. C. Brown and T. Imai; *Organometallics*, 1984, **3**, 1392.	901
84OR375	G. Zweifel and J. A. Miller; *Org. React.*, 1984, **32**, 375.	910, 919, 922, 968
84OS(62)1	A. Alexakis, G. Cahiez, and J. F. Normant; *Org. Synth.*, 1984, **62**, 1.	629, 977
84OS(62)31	E. Negishi and H. Matsushita; *Org. Synth.*, 1984, **62**, 31.	975
84OS(62)39	S.-I. Murahashi, T. Naota and Y. Tanigawa; *Org. Synth.*, 1984, **62**, 39.	606
84OS(62)58	B. M. Trost, D. T. Chang and T. M. Nanninga; *Org. Synth.*, 1984, **62**, 58.	555
84PS(19)65	L. Horner and G. Simons; *Phosphorus Sulfur*, 1984, **19**, 65.	436
84PS(20)81	C. Couret, J. Escudie and S. A. Thaoubane; *Phosphorus Sulfur*, 1984, **20**, 81.	473, 474
84PS(20)371	M. Hedayatullah and J. C. Hugueny; *Phosphorus Sulfur*, 1984, **20**, 371.	367
84RCR1178	L. S. Boguslavskaya; *Russ. Chem. Rev.* (*Engl. Transl.*), 1984, **53**, 1178.	18
84RTC305	N. L. J. M. Broekhof and A. van der Gen; *Recl. Trav. Chim. Pays-Bas*, 1984, **103**, 305.	739
84S34	M. Yonovich-Weiss and Y. Sasson; *Synthesis*, 1984, 34.	31
84S113	S. Rajagopalan and G. Zweifel; *Synthesis*, 1984, 113.	922
84S168	P. J. Garegg, R. Johansson and B. Samuelson; *Synthesis*, 1984, 168.	7
84S215	J. A. Hyatt and A. W. White; *Synthesis*, 1984, 215.	715
84S217	M. Psiorz and G. Zinner; *Synthesis*, 1984, 217.	340, 357
84S226	N. Ono, A. Kamimura and A. Kaji; *Synthesis*, 1984, 226.	395
84S240	A. L. Braga, J. V. Comasseto and N. Petragnani; *Synthesis*, 1984, 240.	1036, 1038
84S245	T. Sakamoto, Y. Kondo, M. Shiraiwa and H. Yamanaka; *Synthesis*, 1984, 245.	677
84S304	J. Barluenga, F. Aznar and R. Liz; *Synthesis*, 1984, 304.	744
84S320	J. B. Hendrikson, D. A. Judelson and T. Chancellor; *Synthesis*, 1984, 320.	156
84S323	M. Ueda, K. Uchiyama and T. Kano; *Synthesis*, 1984, 323.	715
84S325	F. Di Furia, G. Modena and R. Seraglia; *Synthesis*, 1984, 325.	147
84S384	D. J. Ager; *Synthesis*, 1984, 384.	936, 1005
84S439	L. Syper and J. Mlochowski; *Synthesis*, 1984, 439.	279
84S493	J. C. Combret, J. L. Klein and M. Mouslouhouddine; *Synthesis*, 1984, 493.	741
84S498	H. C. Brown, B. Nazer and J. S. Cha; *Synthesis*, 1984, 498.	126, 708
84S608	A. G. Martinezi and M. O. Ruiz; *Synthesis*, 1984, 608.	36
84S618	D. E. Orr; *Synthesis*, 1984, 618.	746
84S629	J. Gorzynski Smith; *Synthesis*, 1984, 629.	82
84S657	A. W. Mott and G. Barany; *Synthesis*, 1984, 657.	233
84S778	Y. Xu and J. Zhang; *Synthesis*, 1984, 778.	846
84S781	Y. Xu, Z. Li, J. Xia, E. Guo and Y. Huang; *Synthesis*, 1984, 781.	843

84S841	S. Araki, K. Ohmori and Y. Butsugan; *Synthesis*, 1984, 841.	26
84S919	H. C. Brown and V. Somayaji; *Synthesis*, 1984, 919.	623
84S920	H. C. Brown, D. Basavaiah and S. M. Singh; *Synthesis*, 1984, 920.	901
84S941	P. A. Harland, P. Hodge, W. Maughan and E. Wildsmith; *Synthesis*, 1984, 941.	299
84S1037	D. N. Harpp, T. Aida, J. De Cesare, P. Tisnes and T. H. Chan; *Synthesis*, 1984, 1037.	244
84S1041	C. O. Meese; *Synthesis*, 1984, 1041.	203
84S1042	D. A. Armitage; *Synthesis*, 1984, 1042.	720
84S1043	K. Takahashi, T. Masuda, K. Ogura and H. Iida; *Synthesis*, 1984, 1043.	748
84S1054	D. Dauzonne and R. Royer; *Synthesis*, 1984, 1054.	398
84SC69	J. M. Renga and P.-C. Wang; *Synth. Commun.*, 1984, **14**, 69.	683
84SC915	D. M. Ketcha and D. Swern; *Synth. Commun.*, 1984, 915.	164
84SC939	A. Arrieta, I. Ganboa and C. Palomo; *Synth. Commun.*, 1984, **14**, 939.	28, 30
84SC1093	R. S. Varma, G. W. Kabalka and M. S. Monrad; *Synth. Commun.*, 1984, **14**, 1093.	398
84SC1313	J. H. Babler and K. P. Spina; *Synth. Commun.*, 1984, **14**, 1313.	31
84T765	T. A. van der Knaap, T. C. Klebach, F. Visser, F. Bickelhaupt, P. Ros, E. J. Baerends, C. H. Stam and M. Konijn; *Tetrahedron*, 1984, **40**, 765.	1003
84T1381	T. Mukaiyama, N. Iwasawa, R. W. Stevens and T. Haga; *Tetrahedron*, 1984, **40**, 1381.	701
84T2035	E. Armani, A. Dossena, R. Marchelli and G. Casnati; *Tetrahedron*, 1984, **40**, 2035.	28
84T2211	J. Mulzer, P. de Lasalle, A. Chucholowski, U. Blashek, G. Brüntrup, I. Jibril and G. Huttner; *Tetrahedron*, 1984, **40**, 2211.	422
84T2219	R. W. Hoffman and B. Kemper; *Tetrahedron*, 1984, **40**, 2219.	529
84T2239	Y. Yamamoto, H. Yatagi, H. Ishihara and K. Maruyama; *Tetrahedron*, 1984, **40**, 2239.	78
84T2257	K. N. Houk, N. G. Rondan, Y.-D. Wu, J. T. Metz and M. N. Paddon Row; *Tetrahedron*, 1984, **40**, 2257.	41
84T2865	B. Sket and M. Zupan; *Tetrahedron*, 1984, **40**, 2865.	21
84T2989	P. W. Hickmott; *Tetrahedron*, 1984, **40**, 2989.	759
84T4363	J. E. Baldwin, L. M. Harwood and M. J. Leonard; *Tetrahedron*, 1984, **40**, 4363.	346
84T4553	J. Nakayama, M. Tanuma, Y. Honda and M. Hoshino; *Tetrahedron*, 1984, **25**, 4553.	478
84TL83	A. Minato, K. Suzuki, K. Tamao and M. Kumada; *Tetrahedron Lett.*, 1984, **25**, 83.	931
84TL131	F. E. Ziegler and K. Mikami; *Tetrahedron Lett.*, 1984, **25**, 131.	923
84TL233	Y. D. Vankar and G. Kumaravel; *Tetrahedron Lett.*, 1984, **25**, 233.	34
84TL273	H.-J. Gais; *Tetrahedron Lett.*, 1984, **25**, 273.	1049
84TL395	C. S. Wilcox, G. W. Long and H. Suh; *Tetrahedron Lett.*, 1984, **25**, 395.	681
84TL429	S. Mori, T. Aoyama and T. Shioiri; *Tetrahedron Lett.*, 1984, **25**, 429.	753
84TL483	R. J. Mills and V. Snieckus; *Tetrahedron Lett.*, 1984, 483.	627
84TL491	E. J. Corey and A. W. Gross; *Tetrahedron Lett.*, 1984, **25**, 491.	420
84TL495	E. J. Corey and A. W. Gross; *Tetrahedron Lett.*, 1984, **25**, 495.	699
84TL607	K. D. Donnelly, W. E. Fristad, B. J. Gellerman, J. R. Peterson and B. J. Selle; *Tetrahedron Lett.*, 1984, **25**, 607.	24
84TL699	C. S. Wilcox and R. E. Babston; *Tetrahedron Lett.*, 1984, **25**, 699.	699
84TL717	K. Koumaglo and T. H. Chan; *Tetrahedron Lett.*, 1984, **25**, 717.	935
84TL741	M. Schlosser and S. Strunk; *Tetrahedron Lett.*, 1984, **25**, 741.	557, 907
84TL745	R. Lehmann and M. Schlosser; *Tetrahedron Lett.*, 1984, **25**, 745.	557, 907, 999
84TL835	R. Tarhouni, R. Kirschleger, M. Rambaud and J. Villieras; *Tetrahedron Lett.*, 1984, **25**, 835.	998
84TL839	F. D. Bellamy and K. Ou; *Tetrahedron Lett.*, 1984, **25**, 839.	753
84TL1103	J. M. Aizpurua and C. Palomo; *Tetrahedron Lett.*, 1984, **25**, 1103.	14
84TL1163	Y. Morizawa, H. Oda, K. Oshima and H. Nozaki; *Tetrahedron Lett.*, 1984, **25**, 1163.	929
84TL1197	A. M. Morella and A. D. Ward; *Tetrahedron Lett.*, 1984, **25**, 1197.	24
84TL1253	T. L. Guggenheim; *Tetrahedron Lett.*, 1984, **25**, 1253.	543
84TL1307	C. Claesen, G. I. Tesser, C. E. Dreef, J. E. Marugg, G. A. van der Marel and J. H. van Boom; *Tetrahedron Lett.*, 1984, **25**, 1307.	409
84TL1337	M. Raban, C. Hu and L. H. Craine; *Tetrahedron Lett.*, 1984, **25**, 1337.	767
84TL1353	H. Ahlbrecht and H. Dollinger; *Tetrahedron Lett.*, 1984, **25**, 1353.	1001
84TL1357	G. Himbert and L. Henn; *Tetrahedron Lett.*, 1984, **25**, 1357.	876
84TL1879	G. E. Keck and E. P. Boden; *Tetrahedron Lett.*, 1984, **25**, 1879.	78
84TL1983	P. Lerouge and C. Paulmier; *Tetrahedron Lett.*, 1984, **25**, 1983.	282
84TL1999	S. Fujikura, M. Inoue, K. Utimoto and H. Nozaki; *Tetrahedron Lett.*, 1984, **25**, 1999.	972
84TL2151	J. Hibino, S. Matsubara, Y. Morizawa, K. Oshima and H. Nozaki; *Tetrahedron Lett.*, 1984, **25**, 2151.	977
84TL2195	I. Ruppert, K. Schlich and W. Volbach; *Tetrahedron Lett.*, 1984, **25**, 2195.	70
84TL2221	H. Oda, Y. Morizawa, K. Oshima and H. Nozaki; *Tetrahedron Lett.*, 1984, **25**, 2221.	977
84TL2295	C. K. Hwang, W. S. Li and K. C. Nicolau; *Tetrahedron Lett.*, 1984, **25**, 2295.	9
84TL2415	E. J. Corey and T. M. Eckrich; *Tetrahedron Lett.*, 1984, **25**, 2415.	478
84TL2619	M. Nakajima, J. C. Warner and J.-P. Anselme; *Tetrahedron Lett.*, 1984, **25**, 2619.	385
84TL2675	Y. Ueno, C. Tanaka and M. Okawara; *Tetrahedron Lett.*, 1984, **46**, 2675.	131
84TL3123	J. M. Aizpurua and C. Palomo; *Tetrahedron Lett.*, 1984, **25**, 3123.	36
84TL3217	H. Oda, Y. Morizawa, K. Oshima and H. Nozaki; *Tetrahedron Lett.*, 1984, **25**, 3217.	947, 977
84TL3221	H. Oda, Y. Morizawa, K. Oshima and H. Nozaki; *Tetrahedron Lett.*, 1984, **25**, 3221.	624, 946, 948
84TL3305	W. Oppolzer and P. Schneider; *Tetrahedron Lett.*, 1984, **25**, 3305.	558, 559
84TL3407	E. Negishi, J. A. Miller and T. Yoshida; *Tetrahedron Lett.*, 1984, **25**, 3407.	969
84TL3625	M. Clarembeau and A. Krief; *Tetrahedron Lett.*, 1984, **25**, 3625.	279
84TL3927	G. E. Keck, D. E. Abbott, E. P. Boden and E. J. Enholm; *Tetrahedron Lett.*, 1984, **25**, 3927.	78

84TL4153 M. G. Kolovos and P. Moutevelis-Minakakis; *Tetrahedron Lett.*, 1984, **46**, 4153. 230

84TL4323 J. H. Babler and W. E. Banta; *Tetrahedron Lett.*, 1984, **25**, 4323. 554

84TL4383 P. Knochel and J. F. Normant. *Tetrahedron Lett.*, 1984, **25**, 4383. 936, 945

84TL4441 K. M. Nsunda and L. Hevesi; *Tetrahedron Lett.*, 1984, **25**, 4441. 1008

84TL4511 J.-L. Fourrey and J. Varenne; *Tetrahedron Lett.*, 1984, **25**, 4511. 410

84TL4717 Y. Guindon, R. Fortin, C. Yoakim and J. W. Gilliard; *Tetrahedron Lett.*, 1984, **25**, 4717. 539

84TL4881 J. I. Luengo and M. Koreeda; *Tetrahedron Lett.*, 1984, **25**, 4881. 677

84TL5099 J. Orban, J. V. Turner and B. Twitchin; *Tetrahedron Lett.*, 1984, **25**, 5099. 699

84TL5287 J. L. Charlton and T. Durst; *Tetrahedron Lett.*, 1984, **46**, 5287. 200, 272

84TL5513 M. W. Schwarz and W. Pfleiderer; *Tetrahedron Lett.*, 1984, **25**, 5513. 409, 410

84TL5763 F. Derguini and G. Linstrumelle; *Tetrahedron Lett.*, 1984, **25**, 5763.

84TL5789 K. Kikukawa, K. Ikenaga, F. Wada and T. Matsuda; *Tetrahedron Lett.*, 1984, **25**, 5789. 932

84TL5863 J. A. Miller and E.-I. Negishi; *Tetrahedron Lett.*, 1984, 5863. 629, 974, 975, 976

84TL5889 W. Oppolzer, M. J. Kelly and G. Bernardinelli; *Tetrahedron Lett.*, 1984, **25**, 5889. 202

84TL5903 J. B. Ousset, C. Mioskowski, Y. L. Yang and J. R. Falck; *Tetrahedron Lett.*, 1984, 5903. 446

84TL5949 I. Bidd and M. C. Whiting; *Tetrahedron Lett.*, 1984, **25**, 5949. 25, 31

84TL5953 L. N. Mander and S. P. Sethi; *Tetrahedron Lett.*, 1984, **25**, 5953. 699

84TL5997 J.-M. Vatele; *Tetrahedron Lett.*, 1984, **25**, 5997. 678

84URP1035028 E. N. Tsvetkov, N. A. Bondarenko and M. I. Kabachnik; *USSR Pat.* 1 035 028 (1984) (*Chem. Abstr.*, 1984, **100**, 51 831).

84USP4450158 M. A. H. Fahmy; *US Pat.* 4 450 158 (1984) (*Chem. Abstr.*, 1984, **101**, 91 236z). 252

84ZC224 D. Klötzer, P. Mäding and R. Münze; *Z. Chem.*, 1984, **24**, 224. 440

84ZC261 K. Issleib, F. Krech and U. Kühne; *Z. Chem.*, 1984, **24**, 261. 428

84ZC365 K. Kellner and A. Tzschach; *Z. Chem.*, 1984, **24**, 365. 440

84ZN(B)115 E. Lindner and E. U. Kuester; *Z. Naturforsch., Teil B*, 1984, **39**, 115. 437

84ZN(B)356 H. Kischkel and G. V. Röschenthaler; *Z. Naturforsch., Teil B*, 1984, **39**, 356. 440

84ZN(B)618 G. A. Bowmaker, C. Dörzbach and H. Schmidbaur; *Z. Naturforsch., Teil B*, 1984, **39**, 618. 447

84ZN(B)887 M. Wieber and I. Sauer; *Z. Naturforsch., Teil B*, 1984, **39**, 887. 493, 505, 886, 890

84ZN(B)1042 H. Klusik, C. Pues and A. Berndt; *Z. Naturforsch., Teil B*, 1984, **39**, 1042. 903

84ZN(B)1668 M. Wieber and I. Sauer; *Z. Naturforsch., Teil B*, 1984, **39**, 1668. 499, 891

84ZOB1192 V. I. Boev and A. V. Dombrovskii; *Zh. Obshch. Khim.*, 1984, **54**, 1192 (*Chem. Abstr.*, 1984, **101**, 211 392). 445

84ZOB1842 A. G. Shipov and Y. I. Baukov; *Zh. Obshch. Khim.*, 1984, **54**, 1842 (*Chem. Abstr.*, 1984, **101**, 230 689q). 209

84ZOB1917 O. G. Sinyashin, I. Y. Gorshunov, E. S. Batyeva and A. N. Pudovik; *Zh. Obshch. Khim.*, 1984, **54**, 1917. 820

84ZOB2545 P. S. Khokhlov, L. S. Berseneva and N. F. Savenkov; *Zh. Obshch. Khim.*, 1984, **54**, 2545 (*Chem. Abstr.*, 1984, **102**, 166 864j). 252

84ZOB2550 V. D. Makhaev and A. P. Borisov; *Zh. Obshch. Khim.*, 1984, **54**, 2550 (*Chem. Abstr.*, 1985, **102**, 113 614). 442

84ZOB2703 P. S. Khokhlov, L. H. Berseneva and N. F. Savenkov; *Zh. Obshch. Khim.*, 1984, **54**, 2703. 449

85ABC2331 K. Sakai, A. Nakazawa, K. Kondo and H. Ohta; *Agric. Biol. Chem.*, 1985, **49**, 2331. 398

85ACR274 L. Engman; *Acc. Chem. Res.*, 1985, **18**, 274. 768

85ACR284 M. Miyashita and A. Yoshikoshi; *Acc. Chem. Res.*, 1985, **18**, 284. 783

85ACR380 M. Ballester; *Acc. Chem. Res.*, 1985, **18**, 380. 1010

85AG(E)1 S. Masamune, W. Choi, S. S. Peterson and L. R. Sita; *Angew. Chem., Int. Ed. Engl.*, 1985, **24**, 1. 532

85AG(E)319 J. Barluenga, J. M. Gonzalez, P. J. Campos and G. Asensio; *Angew. Chem., Int. Ed. Engl.*, 1985, **24**, 319. 18, 24, 31, 34

85AG(E)402 H. Karsch, A. Appelt and G. Muller; *Angew. Chem., Int. Ed. Engl.*, 1985, **24**, 402. 474, 475

85AG(E)504 M. Baudler, M. Michels, J. Khan and M. Pieroth; *Angew. Chem., Int. Ed. Engl.*, 1985, **24**, 504. 472

85AG(E)573 G. Boche, M. Marsch, K. Harms and G. M. Sheldrick; *Angew. Chem., Int. Ed. Engl.*, 1985, **24**, 573. 1000

85AG(E)859 H.-J. Gais, H. J. Lindner and J. Vollhardt; *Angew. Chem., Int. Ed. Engl.*, 1985, **24**, 859. 1000

85AG(E)878 D. Manning and H. Nöth; *Angew. Chem., Int. Ed. Engl.*, 1985, **24**, 878. 42

85AJC1879 J. P. Bartley, R. M. Carman and J. K. L. Russell-Maynard; *Aust. J. Chem.*, 1985, **38**, 1879. 25

85AOC(24)1 G. Erber, C. Kruger and G. Muller; *Adv. Organomet. Chem.*, 1985, **24**, 1. 561

85BCJ1069 N. Ono, K. Zinsmeister and A. Kaji; *Bull. Chem. Soc. Jpn.*, 1985, **58**, 1069. 394

85BCJ1599 N. Takeno, T. Fukushima, S. Takeda and K. Kishimoto; *Bull. Chem. Soc. Jpn.*, 1985, **58**, 1599. 28

85BCJ2793 M. Atoh, K. Kashiwabara and J. Fujita; *Bull. Chem. Soc. Jpn.*, 1985, **58**, 2793. 437

85BCJ3104 T. Hirao, M. Hagihara and T. Agawa; *Bull. Chem. Soc. Jpn.*, 1985, **58**, 3104; T. Hirao, M. Hagihara, Y. Ohshiro and T. Agawa; *Synthesis*, 1984, 60. 456

85BCJ3393 Y. Okamoto; *Bull. Chem. Soc. Jpn.*, 1985, **58**, 3393. 106

85BJC2425 K. Wakamatsu, Y. Okuda, K. Oshima and H. Nozaki; *Bull. Soc. Jpn.*, 1985, **58**, 2425.

85BSF463 A. Champseix and R. Vessiere; *Bull. Soc. Chim. Fr.*, 1985, 463. 249

85C134 A. Haas and M. Lieb; *Chimia*, 1985, **39**, 134. 15

85C305 H. Vyplel; *Chimia*, 1985, **39**, 305 15, 17

85CB97 W. J. Richter; *Chem. Ber.*, 1985, **118**, 97. 436

85CB116 A. Elsäßer, W. Sundermeyer and D. S. Stephenson; *Chem. Ber.*, 1985, **118**, 116. 155

85CB2030	W. Jacksties, H. Noth and W. Storch; *Chem. Ber.*, 1985, **118**, 2030.	467
85CB2365	G. Märkl, W. Weber and W. Weiss; *Chem. Ber.*, 1985, **118**, 2365.	439
85CB2407	H. Lehmkuhl and G. Mehler; *Chem. Ber.*, 1985, **118**, 2407.	577
85CB2472	W. A. Herrmann, B. Koumbouris, A. Schaefer, T. Zahn and M. L. Ziegler; *Chem. Ber.*, 1985, **118**, 2472.	512
85CB3105	H. Schmidbaur, R. Herr, T. Pollock, A. Schier, G. Müller and J. Riede; *Chem. Ber.*, 1985, **118**, 3105.	446
85CB3268	M. Baudler and W. Goldner; *Chem. Ber.*, 1985, **118**, 3268.	471
85CB3529	E. Weiß, J. Kopf, T. Gardein, S. Corbelin, U. Schümann, M. Kirilov and G. Petrov; *Chem. Ber.*, 1985, **118**, 3529.	1004
85CB4259	H. Weber, J. Pant and H. Wunderlich; *Chem. Ber.*, 1985, **118**, 4259.	1009
85CC337	D. E. Bergbrieter and J. R. Blanton; *J. Chem. Soc., Chem. Commun.*, 1985, 337.	7
85CC354	T. N. Mitchell, H. Killing, R. Dicke and R. Wickenkamp; *J. Chem. Soc., Chem. Commun.*, 1985, 354.	916, 930
85CC400	S. Kondo, H. Yasui, K. Ohta and K. Tsuda; *J. Chem. Soc., Chem. Commun.*, 1985, 400.	35
85CC569	A. Krief, W. Dumont, J.-N. Denis, G. Evrard and B. Norberg; *J. Chem. Soc., Chem. Commun.*, 1985, 569.	287
85CC672	S. J. Brown and J. H. Clark; *J. Chem. Soc., Chem. Commun.*, 1985, 672.	19, 614
85CC690	A. I. Meyers and P. D. Pansegrau; *J. Chem. Soc., Chem. Commun.*, 1985, 690.	974
85CC715	M. Hoshi, Y. Masuda and A. Arase; *J. Chem. Soc., Chem. Commun.*, 1985, 715.	273
85CC1199	S. Kerschl and B. Wrackmeyer; *J. Chem. Soc., Chem. Commun.*, 1985, 1199.	903
85CC1404	C. A. Tsipis, G. A. Katsoulos and F. D. Vakoulis; *J. Chem. Soc., Chem. Commun.*, 1985, 1404.	35
85CC1422	J. Barluenga, J. M. Martinez-Gallo, C. Najera and M. Yus; *J. Chem. Soc., Chem. Commun.*, 1985, 1422.	31
85CC1449	J. K. Whitesell, D. James and J. F. Carpenter; *J. Chem. Soc., Chem. Commun.*, 1985, 1449.	245
85CC1670	J. Drabowicz, S. Legedz and M. Mikolajczyk; *J. Chem. Soc., Chem. Commun.*, 1985, 1670.	198
85CJC2285	C. Srinivasen and K. Pitchumani; *Can. J. Chem.*, 1985, **63**, 2285.	886
85CJC2313	T. G. Back, S. Collins and K. W. Law; *Can. J. Chem.*, 1985, **63**, 2313.	743
85CL53	H. Oda, K. Oshima and H. Nozaki; *Chem. Lett.*, 1985, 53.	624, 630
85CL233	H. Sonoda, T. Sonoda and H. Kobayashi; *Chem. Lett.*, 1985, 233.	19
85CL411	H. Suzuki, A. Kondo and T. Ogawa; *Chem. Lett.*, 1985, 411.	632
85CL679	H. Suzuki, S. Takeda and Y. Hanazaki; *Chem. Lett.*, 1985, 679.	293
85CL805	H. Kosugi, M. Kitaoka, K. Tagami and H. Uda; *Chem. Lett.*, 1985, 805.	713
85CL1307	K. Takagi; *Chem. Lett.*, 1985, 1307.	707
85CL1693	N. Yoneda, T. Fukuhara, S. Nagata and A. Suzuki; *Chem. Lett.*, 1985, 1693.	19
85CPB1301	M. Maeda, T. Fukumura and M. Kojima; *Chem. Pharm. Bull.*, 1985, **33**, 1301.	614
85CR(301)1289	R. Calas and J. Gerval; *C. R. Hebd. Seances Acad. Sci.*, 1985, **301**, 1289.	943
85CRV245	R. H. Crabtree; *Chem. Rev.*, 1985, **85**, 245.	962
85CZ53	W. Wolfsberger; *Chem.-Ztg.*, 1985, **109**, 53.	875
85CZ317	W. Wolfsberger; *Chem.-Ztg.*, 1985, **109**, 317.	440
85EGP231074	M. Kant; *Ger. (East) Pat.* DD231 074 (1985) (*Chem. Abstr.*, 1985, **105**, 209 205b).	105
85G29	A. Dossena, R. Marchelli and G. Casnati; *Gazz. Chim. Ital.*, 1985, **115**, 29.	24, 31, 35
85G393	M. Alpegiani, A. Bedeschi and E. Perrone; *Gazz. Chim. Ital.*, 1985, **115**, 393.	6
85GEP3332269	A. G. Bayer; *Ger. Pat.* 3 332 269 (1985) (*Chem. Abstr.*, 1985, **102**, 203 873j).	241
85GEP3332716	T. Tammer, G. Heubach and B. Sachse; *Ger. Pat.* 3 332 716 (1985) (*Chem. Abstr.*, 1985, **103**, 88 078).	444
85GEP3512781	S. Juge and Y. Legras; *Ger. Pat.* 3 512 781 (1985) (*Chem. Abstr.*, 1987, **106**, 50 447).	478
85H(23)1513	W. Sliwa, G. Matusiak and A. Postawka; *Heterocycles*, 1985, **23**, 1513.	1009
85H(23)1897	H. Matsuyama, A. Izuoka and M. Kobayashi; *Heterocycles*, 1985, **23**, 1897.	244
85H(23)2709	N. Saito and K. Ueno; *Heterocycles*, 1985, **23**, 2709.	1008
85HCA220	M. Egli, L. Hoesch and A. S. Dreiding; *Helv. Chim. Acta*, 1985, **68**, 220.	376
85HCA319	D. Seebach and M. Brook; *Helv. Chim. Acta*, 1985, **68**, 319.	356
85HCA813	S. M. Kelley and H. Schad; *Helv. Chim. Acta*, 1985, **68**, 813.	625
85HCA961	M. Ertas and D. Seebach; *Helv. Chim. Acta*, 1985, **68**, 961.	699
85HOU	K. Schank; *Methoden Org. Chem.* (*Houben Weyl*), 1985.	156
85HOU(E11)68	R. Schubart; *Methoden Org. Chem.* (*Houben-Weyl*), 1985, **E11**, p. 68	357
85HOU(E11)510	E. Fischer; *Methoden Org. Chem.* (*Houben-Weyl*), 1985, **E11**, 510.	400
85HOU(E11)532	E. Fischer; *Methoden Org. Chem.* (*Houben-Weyl*), 1985, **E11**, 532.	400
85HOU(E11)558	R. Beckert and R. Mayer; *Methoden Org. Chem.* (*Houben-Weyl*), 1985, **E11**, 558.	386, 399
85HOU(E11)579	S. Bleisch and R. Mayer; *Methoden Org. Chem.* (*Houben-Weyl*), 1985, **E11**, 579.	399
85HOU(E11)584	S. Bleisch and R. Mayer; *Methoden Org. Chem.* (*Houben-Weyl*), 1985, **E11**, 584.	399
85HOU(E11)887	M. Haake; *Methoden Org. Chem.* (*Houben-Weyl*), 1985, **E11**, 887.	401
85HOU(E11)943	B. Zwanenberg and B. G. Lenz; *Methoden Org. Chem.* (*Houben-Weyl*), 1985, **E11**, 943.	400
85HOU(E11)1031	S. Bleisch and R. Mayer; *Methoden Org. Chem.* (*Houben-Weyl*), 1985, **E11**, 1031.	401
85HOU(E11)1123	S. Pawlenko; *Methoden Org. Chem.* (*Houben-Weyl*), 1985, **E11**, 1123.	401
85HOU(E11)1299	M. Haake; *Methoden Org. Chem.* (*Houben-Weyl*), 1955, **E11**, 1299.	402
85HOU(E11)1321	M. Haake; *Methoden Org. Chem.* (*Houben-Weyl*), 1985, **E11**, 1321.	402
85IC533	M. Pankowski, W. Chodiewicz and M.-P. Simonnin; *Inorg. Chem.*, 1985, **24**, 533.	436
85IC654	S. Gambarotta, S. Strologo, C. Floriani, A. Chiesi-Villa and C. Guastini; *Inorg. Chem.*, 1985, **24**, 654.	703
85IC1107	D. M. Schubert and A. D. Norman; *Inorg. Chem.*, 1985, **24**, 1107.	473, 474
85IC1613	E. P. Kyba and S.-T. Liu; *Inorg. Chem.*, 1985, **24**, 1613.	431

85IC2344 M. Habib, H. Trujillo, C. A. Alexander and B. N. Storhoff; *Inorg. Chem.*, 1985, **24**, 2344. 439
85IC3136 R. B. King and N. D. Sadanani; *Inorg. Chem.*, 1985, **24**, 3136. 466
85IC4437 T. V. Harris and W. R. Pretztzer; *Inorg. Chem.*, 1985, **24**, 4437. 430
85IS1961 S. T. Lin and K. J. Klabunde; *Inorg. Synth.*, 1985, **24**, 1961. 582
85IZV1200 N. A. Nesmeyanov, O. A. Rebrova, V. G. Kharitonov, P. V. Petrovskii and O. A. Reutov; *Izv. Akad. Nauk SSSR, Ser. Khim.*, 1985, 1200 (*Chem. Abstr.*, 1986, **104**, 88 683). 447
85JA460 H. C. Brown, B. Singaram and T. E. Cole; *J. Am. Chem. Soc.*, 1985, **107**, 460. 535
85JA500 G. Stork and M. Kahn; *J. Am. Chem. Soc.*, 1985, **107**, 500. 540
85JA724 P. E. Eaton and G. Castaldi; *J. Am. Chem. Soc.*, 1985, **107**, 724. 555
85JA1007 S. E. Biali and Z. Rappoport; *J. Am. Chem. Soc.*, 1985, **107**, 1007. 646
85JA1060 W. D. Wulff, R. W. Kaesler, G. A. Peterson and P.-C. Tang; *J. Am. Chem. Soc.*, 1985, **107**, 1060. 670
85JA1429 C. J. Kowalski, M. S. Haque and K. W. Fields; *J. Am. Chem. Soc.*, 1985, **107**, 1429. 1023
85JA1452 P. J. Stang and B. W. Surber; *J. Am. Chem. Soc.*, 1985, **107**, 1452. 1020, 1022, 1028
85JA1783 E. Block, A. Wall and J. Zubieta; *J. Am. Chem. Soc.*, 1985, **107**, 1783. 195, 196
85JA2067 J. J. Stezowski, P. Kollat, M. Bogucka-Ledóchowska and J. P. Glusker; *J. Am. Chem. Soc.*, 1985, **107**, 2067. 1008
85JA2564 H. C. Brown, P. K. Jadhav and K. S. Bhat; *J. Am. Chem. Soc.*, 1985, **107**, 2564. 520
85JA2811 D. G. Cox, N. Gurusamy and D. J. Burton; *J. Am. Chem. Soc.*, 1985, **107**, 2811. 611
85JA3197 B. H. Lipshutz, J. A. Kozlowski and C. M. Brenemann; *J. Am. Chem. Soc.*, 1985, **107**, 3197. 586, 588
85JA3285 R. E. Ireland, D. W. Norbeck, G. S. Mandel and N. S. Mandel; *J. Am. Chem. Soc.*, 1985, **107**, 3285. 25
85JA3501 T. Imamoto, T. Kusumoto and N. Uzuki; *J. Am. Chem. Soc.*, 1985, **107**, 3501. 474
85JA3601 M. Eyer and D. Seebach; *J. Am. Chem. Soc.*, 1985, **107**, 3601. 394
85JA3669 D. A. Nugiel and Z. Rappoport; *J. Am. Chem. Soc.*, 1985, **107**, 3669. 645
85JA3915 D. J.-S. Tsai and M. M. Midland; *J. Am. Chem. Soc.*, 1985, **107**, 3915. 34
85JA4101 H. J. Reich, N. H. Phillips and I. L. Reich; *J. Am. Chem. Soc.*, 1985, **107**, 4101. 962
85JA4549 S. Masamune, B. M. Kim, J. S. Peterson, T. Sato and S. J. Veenstra; *J. Am. Chem. Soc.*, 1985, **107**, 4549. 521
85JA4731 R. V. Hoffman and G. A. Buntain; *J. Am. Chem. Soc.*, 1985, **107**, 4731. 357
85JA4980 H. C. Brown, T. Imai, M. C. Desai and B. Singaram; *J. Am. Chem. Soc.*, 1985, **107**, 4980. 531
85JA5210 P. L. Robinson, C. N. Barry, J. W. Kelly and J. S. A. Evans; *J. Am. Chem. Soc.*, 1985, **107**, 5210. 830
85JA5313 H. Kawa, B. C. Manley and R. J. Lagow; *J. Am. Chem. Soc.*, 1985, **107**, 5313. 1005
85JA5670 B. C. Berris, G. H. Hovakeemian, Y-H. Lai, H. Mestdagh and K. P. C. Vollhardt; *J. Am. Chem. Soc.*, 1985, **107**, 5670 944
85JA6639 E. Negishi, D. E. Van Horn and T. Yoshida; *J. Am. Chem. Soc.*, 1985, **107**, 6639. 925, 974, 975
85JA7167 C. S. Shiner, C. M. Garner and R. C. Haltinanger; *J. Am. Chem. Soc.*, 1985, **107**, 7167. 268
85JA7423 F. R. Askham, G. C. Stanley and E. C. Marques; *J. Am. Chem. Soc.*, 1985, **107**, 7423. 439
85JA7771 P. A. Wender and D. A. Holt; *J. Am. Chem. Soc.*, 1985, **107**, 7771. 22
85JA8103 G. Jeske, L. E. Schock, P. N. Sweptston, H. Schumann and T. J. Marks; *J. Am. Chem. Soc.*, 1985, **107**, 8103. 969
85JA8111 G. Jeske, H. Lauke, H. Mauermann, H. Schumann and T. J. Marks; *J. Am. Chem. Soc.*, 1985, **107**, 8111. 969
85JA8186 W. R. Roush, A. E. Wults and L. K. Hoong; *J. Am. Chem. Soc.*, 1985, **107**, 8186. 529
85JA8289 J. M. Tour and E. Negishi; *J. Am. Chem. Soc.*, 1985, **107**, 8289. 931
85JAP60152493 T. Minami and I. Hirao; *Jpn. Pat.* 60 152 493 (1985) (*Chem. Abstr.*, 1986, **104**, 19 679). 447
85JCR(S)18 T.-S. Chou, J.-J. Yuan and C.-H. Tsao; *J. Chem. Res.* (*S*), 1985, 18. 432
85JCR(S)330 R. L. Srivastava; *J. Chem. Res.* (*S*), 1985, 330. 434
85JCS(D)337 L. M. Engelhardt, W.-P. Leung, C. L. Raston and A. H. White; *J. Chem. Soc., Dalton Trans.*, 1985, 337. 1005
85JCS(D)573 J. A. Cabeza and P. M. Maitlis; *J. Chem. Soc., Dalton Trans.*, 1985, 573. 581
85JCS(D)873 K. Hiraki, N. Ochi, Y. Sasada, H. Hayashida, Y. Fuchita and S. Yamanaka; *J. Chem. Soc., Dalton Trans.*, 1985, 873. 582
85JCS(P1)373 A. Toshimitsu, H. Owada, K. Terao, S. Uemura and M. Okano; *J. Chem. Soc., Perkin Trans. 1*, 1985, 373. 282
85JCS(P1)471 S. Uemura and S. Fukuzawa; *J. Chem. Soc., Perkin Trans. 1*, 1985, 471. 287, 288
85JCS(P1)1127 P. E. Brown, W. Y. Marcus and P. Anastasis; *J. Chem. Soc., Perkin Trans. 1*, 1985, 1127. 617
85JCS(P1)1589 R. C. Cookson and N. J. Liverton; *J. Chem. Soc., Perkin Trans. 1*, 1985, 1589. 976
85JCS(P1)2201 P. Casara, C. Danzin, B. Metcalf and M. Jung; *J. Chem. Soc., Perkin Trans. 1*, 1985, 2201. 299
85JFC(27)85 D. J. Burton and D. M. Wiemers; *J. Fluorine Chem.*, 1985, **27**, 85. 444
85JFC(27)263 D. P. Ashton, T. A. Ryan, B. R. Webster and B. A. Wolfindale; *J. Fluorine Chem.*, 1985, **27**, 263. 615
85JGU63 V. A. Dodonov, A. V. Gushchin and T. G. Brilkina; *J. Gen. Chem. USSR* (*Engl. Transl.*), 1985, **55**, 63. 886
85JGU1332 O. V. Parakin, E. V. Nikitin, Yu. A. Ignatev, A. S. Romakhin and Yu. M. Kargin; *J. Gen. Chem. USSR* (*Engl. Transl.*), 1985, **55**, 1332. 881
85JGU1691 A. P. Yuvchenko, K. L. Moiseichuk, E. A. Dikusar, N. A. Zhukovskaya and Yu. A. Ol'dekop; *J. Gen. Chem. USSR* (*Engl. Transl.*), 1985, **56**, 1691. 1098
85JGU1771 V. A. Garibina, A. A. Leonov, A. V. Dogadina, B. I. Ionin and A. A. Petrov; *J. Gen. Chem. USSR* (*Engl. Transl.*), 1985, **55**, 1771. 1028
85JGU2198 V. I. Vysotskii, Yu. V. Prikhod'ko, G. V. Pavel and M. N. Tilichenko; *J. Gen. Chem. USSR* (*Engl. Transl.*), 1985, **55**, 2198. 455

85JHC701 T. Torres, S. V. Eswaran and W. Schäfer; *J. Heterocycl. Chem.*, 1985, **22**, 701. 163

85JOC133 M. S. Mourad, R. S. Varma and G. W. Kabalka; *J. Org. Chem.*, 1985, **50**, 133. 341

85JOC416 E. R. Burkhardt and R. D. Rieke; *J. Org. Chem.*, 1985, **50**, 416. 591

85JOC549 H. C. Brown, J. S. Cha, B. Nazar and C. A. Brown; *J. Org. Chem.*, 1985, **50**, 549. 534

85JOC1019 A. I. Meyers, P. D. Edwards, T. R. Bailey and G. E. Jagdmann, Jr.; *J. Org. Chem.*, 1985, **50**, 1019. 1002

85JOC1212 S. H. Pine, R. J. Pettit, G. D. Geib, S. G. Cruz, C. H. Gallego, T. Tijerina and R. D. Pine; *J. Org. Chem.*, 1985, **50**, 1212. 680

85JOC1271 J. Beres, W. G. Bentrude, L. Parkanji, A. Kálmán and A. E. Sopchik; *J. Org. Chem.*, 1985, **50**, 1271. 413

85JOC1327 J. K. Crandall and C. Pradat; *J. Org. Chem.*, 1985, **50**, 1327. 716

85JOC1334 Y. X. Han, M. V. Jovanovic and E. R. Biehl; *J. Org. Chem.*, 1985, **50**, 1334. 753

85JOC1384 H. C. Brown and G. G. Pai; *J. Org. Chem.*, 1985, **50**, 1384. 517

85JOC1385 H. C. Brown and G. G. Pai; *J. Org. Chem.*, 1985, **50**, 1385. 531

85JOC1563 D. P. Riley and P. E. Correa; *J. Org. Chem.*, 1985, **50**, 1563. 348

85JOC1582 H. C. Brown, J. V. N. Vara Prasad and S.-H. Lee; *J. Org. Chem.*, 1985, **50**, 1582. 517

85JOC1599 G. W. Daub, R. N. Zuckermann and W. S. Johnson; *J. Org. Chem.*, 1985, **50**, 1599. 606

85JOC1779 M. J. Evers, L. E. Christiaens, M. R. Guillaume and M. J. Renson; *J. Org. Chem.*, 1985, **50**, 1779. 632

85JOC1784 H. Sugimoto and D. T. Sawyer; *J. Org. Chem.*, 1985, **50**, 1784. 156

85JOC1815 K. Misiura, J. V. Silverton and W. J. Stec; *J. Org. Chem.*, 1985, **50**, 1815. 412

85JOC1969 C. G. Whiteley and I. Zwane; *J. Org. Chem.*, 1985, **50**, 1969. 522

85JOC2019 M. F. Moore and S. L. Beaucage; *J. Org. Chem.*, 1985, **50**, 2019. 410

85JOC2145 W. D. Watson; *J. Org. Chem.*, 1985, **50**, 2145. 620

85JOC2205 M. Raban, C. Hu, L. Craine and E. Hortelano; *J. Org. Chem.*, 1985, **50**, 2205. 767

85JOC2316 L. F. Cannizzo and R. H. Grubbs; *J. Org. Chem.*, 1985, **50**, 2316. 702

85JOC2372 E. M. Beccalli, A. Manfredy and A. Marchesini; *J. Org. Chem.*, 1985, **50**, 2372. 1034, 1035

85JOC2386 L. F. Cannizzo and R. H. Grubbs; *J. Org. Chem.*, 1985, **50**, 2386. 681

85JOC2801 G. Maas and R. Brückmann; *J. Org. Chem.*, 1985, **50**, 2801. 1024, 1025

85JOC3104 H. Hart, K. Harada and C. J. F. Du; *J. Org. Chem.*, 1985, **50**, 3104. 974

85JOC3342 S. Rozen and M. Brand; *J. Org. Chem.*, 1985, **50**, 3342. 18

85JOC3421 R. K. Boekman, Jr. and T. E. Barta; *J. Org. Chem.*, 1985, **50**, 3421. 938

85JOC3666 B. L. Chenard, E. D. Laganis, F. Davidson and T. V. RajanBabu; *J. Org. Chem.*, 1985, **50**, 3666. 916

85JOC3698 W. E. Barnette, R. C. Wheland, W. J. Middleton and S. Rozen; *J. Org. Chem.*, 1985, **50**, 3698. 16

85JOC3846 M. J. Kurth and M. J. O'Brien; *J. Org. Chem.*, 1985, **50**, 3846. 701

85JOC3863 C. Stetin, B. De Jeso and J. C. Pommier; *J. Org. Chem.*, 1985, **50**, 3863. 816

85JOC3923 R. S. Matthews and D. J. Eickhoff; *J. Org. Chem.*, 1985, **50**, 3923. 1098

85JOC4081 B. Masci; *J. Org. Chem.*, 1985, **50**, 4081. 792

85JOC4252 R. J. Pariza and P. L. Fuchs; *J. Org. Chem.*, 1985, **50**, 4252. 444

85JOC4329 T.-S. Chou, C.-H. Tsao and S. C. Hung; *J. Org. Chem.*, 1985, **50**, 4329. 432

85JOC4390 M. C. Caserio, C. L. Fisher and J. K. Kim; *J. Org. Chem.*, 1985, **50**, 4390. 138

85JOC4553 M. Pankratz and R. F. Childs; *J. Org. Chem.*, 1985, **50**, 4553. 1008

85JOC4576 M. J. Fifolt, R. T. Olczak, R. F. Mundhenke and J. F. Bieron; *J. Org. Chem.*, 1985, **50**, 4576. 613

85JOC4812 S. D. Pastor and E. T. Hessell; *J. Org. Chem.*, 1985, **50**, 4812. 711

85JOC5384 T. Kometani, D. S. Watt, T. Ji and T. Fitz; *J. Org. Chem.*, 1985, **50**, 5384. 631

85JOC5391 E. J. Corey and A. W. Gross; *J. Org. Chem.*, 1985, **50**, 5391. 374

85JOC5410 R. F. Cunico and C.-P. Kuan; *J. Org. Chem.*, 1985, **50**, 5410. 908

85JOC5430 R. Subramanian and F. Johnson; *J. Org. Chem.*, 1985, **50**, 5430. 657

85JOC5446 J. Chandrasekharan, P. V. Ramachandran and H. C. Brown; *J. Org. Chem.*, 1985, **50**, 5446. 526, 531

85JOC5524 H. Hart, K. Harada and C. J. F. Du; *J. Org. Chem.*, 1985, **50**, 5524. 974

85JOC5583 H. C. Brown, P. V. Ramachandran and J. V. N. Vara Prasad; *J. Org. Chem.*, 1985, **50**, 5583. 517

85JOC5586 H. C. Brown, R. G. Naik, R. K. Bakshi, C. Pyun and B. Singaram; *J. Org. Chem.*, 1985, **50**, 5586. 530

85JOC5592 S. Thea, G. Guanti, A. R. Hopkins and A. Williams; *J. Org. Chem.*, 1985, **50**, 5592. 194

85JOC5828 D. E. Bergbreiter and J. R. Blanton; *J. Org. Chem.*, 1985, **50**, 5828. 35

85JOM(285)1 D. Seebach, J. Hansen, P. Seiler and J. M. Gromek; *J. Organomet. Chem.*, 1985, **285**, 1. 997, 1002

85JOM(285)43 U. M. Dzehemilev and O. S. Vostrikova; *J. Organomet. Chem.*, 1985, **285**, 43. 561

85JOM(285)53 F. Sato; *J. Organomet. Chem.*, 1985, **285**, 53. 969

85JOM(285)163 J. Hibino, S. Matsubara, Y. Morizawa, K. Oshima and H. Nozaki; *J. Organomet. Chem.*, 1985, **285**, 163. 997, 1002

85JOM(285)437 B. H. Lipshutz, J. A. Kozlowski, D. A. Parker, S. L. Nguyen and K. E. McCarthy; *J. Organomet. Chem.*, 1985, **285**, 437. 589

85JOM(287)57 N. A. A. Al-Jabar and A. G. Massey; *J. Organomet. Chem.*, 1985, **287**, 57. 872

85JOM(290)307 J. Ellermann and A. Veit; *J. Organomet. Chem.*, 1985, **290**, 307. 490, 507

85JOM(291)185 B. L. Khandelwal, A. K. Singh, H. B. Singh, K. M. Prasad, N. S. Bhandari and W. R. McWhinnie; *J. Organomet. Chem.*, 1985, **291**, 185. 690

85JOM(295)159 T. Cuenca and P. Royo; *J. Organomet. Chem.*, 1985, **295**, 159. 566, 568

85JOM(296)357 L. Engman and J. S. E. Hellberg; *J. Organomet. Chem.*, 1985, **296**, 357. 729

85JOM(296)411 D. J. Brauer, S. Hietkamp, H. Sommer, O. Stelzer, G. Müller, M. J. Romao and C. Kruger; *J. Organomet. Chem.*, 1985, **296**, 411. 435

85JPR445	S. Andreae, E. Schmitz, H. Sonnenschein, G. Dornyei, Cs. Szántay and J. Tamás; *J. Prakt. Chem.*, 1985, **327**, 445.	379
85LA2246	R. W. Hoffman, B. Kemper, R. Metternich and T. Lehmeier; *Liebigs Ann. Chem.*, 1985, 2246.	529
85MI 207-01	M. Tichniouin, J. Sauleau and A. Sauleau; *Eur. J. Med. Chem.-Chim. Ther.*, 1985, **20**, 181 (*Chem. Abstr.*, 1985, **103**, 87465x).	373
B-85MI 208-01	J. Halpern; in "Asymmetric Synthesis," ed. J. D. Morrison, Academic Press, New York, 1985, vol. 5, chap. 2.	442
B-85MI 208-02	H. E. Koenig; in "Asymmetric Synthesis," ed. J. D. Morrison, Academic Press, New York, 1985, vol. 5, chap. 3.	442
B-85MI 210-01	K. Smith; "Organometallic Complexes of Boron," Chapman and Hall, London, 1985.	514
85MI 210-02	K. Furusawa, T. Katsura, T. Sakai and K. Tsuda; *Nucleic Acids Symp. Ser.*, 1985, **16**, 85 (*Chem. Abstr.*, 1985, **105**, 43229k).	540
B-85MI 211-01	J. L. Wardell and E. S. Paterson; in "The Chemistry of the Metal–Carbon Bond," ed. F. R. Hartley, series ed. S. Patai, Wiley, New York, 1985, vol. 2, p. 268.	555
85MI 211-02	B. H. Han; *Taehan Hwakakhoe Chi.*, 1985, **29**, 557 (*Chem. Abstr.*, 1985, **105**, 46679t).	558
B-85MI 211-03	A. J. Davies; in "The Chemistry of the Metal–Carbon Bond," ed. F. R. Hartley, series ed. S. Patai, Wiley, New York, 1985, vol. 3, p. 361.	576
B-85MI 211-04	R. K. Heck; "Palladium Reagents in Organic Synthesis," Academic Press, Orlando, 1985.	579
B-85MI 212-01	J. March; "Advanced Organic Chemistry," Wiley, New York, 1985, p. 447.	608
B-85MI 213-01	A. H. Haines; "Methods for the Oxidation of Organic Compounds," Academic Press, London, 1985.	649
B-85MI 215-01	J. March; "Advanced Organic Chemistry," Wiley, New York, 1985.	758
85MI 217-01	X. Lu, Q. Wang, X. Tao, J. Sun and G. Lei; *Huaxue Xuebao*, 1985, **43**, 450 (*Chem. Abstr.*, 1986, **104**, 129992v).	879
85MI 219-01	E. Negishi and T. Takahashi; *Aldrichim. Acta*, 1985, **18**, 31.	969
B-85MI 219-01	R. F. Heck; in "Palladium Reagents in Organic Syntheses," Academic Press, New York, 1985.	983
85N789	D. Walther and C. Pfuentrenreuter; *Naturwissenschaften*, 1985, **34**, 789.	561
85OM231	H. H. Karsch, A. Appelt, F. H. Köhler and G. Müller; *Organometallics*, 1985, **4**, 231.	1003
85OM623	R. J. P. Corriu, J. J. E. Moreau and M. Pataud-Sat; *Organometallics*, 1985, **4**, 623.	814
85OM1690	D. S. Matteson and J. W. Wilson; *Organometallics*, 1985, **4**, 1690.	1005
85OR(33)1	E. Negishi and M. J. Idacavage; *Org. React.*, 1985, **33**, 1.	4, 514, 900
85OR(34)319	C.-L. J. Wang; *Org. React.*, 1985, **34**, 319.	19
85PAC1827	D. Bellus; *Pure Appl. Chem.*, 1985, **57**, 1827.	14
85PAC1887	D. Savoia, C. Trombini and A. Umani-Ronchi; *Pure Appl. Chem.*, 1985, **57**, 1887.	959
85PAC1911	H. B. Kagan, E. Duñach, C. Nemecek, P. Pitchen, O. Samuel and S.-H. Zhao; *Pure Appl. Chem.*, 1985, **57**, 1911.	147
85POL251	P. Raj, A. K. Saxena, K. Singhal and A. Ranjan; *Polyhedron*, 1985, **4**, 251.	881
85POL1625	A. M. Forster and A. J. Downs; *Polyhedron*, 1985, **4**, 1625.	492
85PS(21)367	K. Issleib, U. Kuhne and F. Krech; *Phosphorus Sulfur*, 1985, **21**, 367.	473
85PS(23)277	R. Mayer; *Phosphorus Sulfur*, 1985, **23**, 277 (*Chem. Abstr.*, 1986, **104**, 68327h).	398
85PS(24)217	D. Dakternieks and R. Di Giacomo; *Phosphorus Sulfur*, 1985, **24**, 217.	451, 467, 469
85RCR1706	A. V. Zibarev and G. G. Yakobson; *Russ. Chem. Rev.* (*Engl. Transl.*), 1985, **54**, 1706.	793
85RTC153	H. Cerfontain; *Recl. Trav. Chim. Pays-Bas*, 1985, **104**, 153.	720
85S66	A. Bender, D. Günther, G. Willms and R. Wingen; *Synthesis*, 1985, 66.	249
85S70	H. M. Chawla and R. S. D. Mittal; *Synthesis*, 1985, 70.	791
85S212	S. Tomoda, Y. Takeuchi and Y. Nomura; *Synthesis*, 1985, 212.	368
85S291	T. Oshikawa and M. Yamashia; *Synthesis*, 1985, 291.	438
85S335	G. T. Crisp and W. J. Scott; *Synthesis*, 1985, 335.	689
85S406	Y. Satoh, T. Tayano, H. Koshino, S. Hara and A. Suzuki; *Synthesis*, 1985, 406.	902
85S500	F. Freeman and M. C. Keindl; *Synthesis*, 1985, 500.	221
85S556	R. H. Smithers; *Synthesis*, 1985, 556.	1025, 1026
85S633	M. Taddei and A. Ricci; *Synthesis*, 1985, 633.	942
85S635	K. Hiroi and S. Sato; *Synthesis*, 1985, 635.	368
85S665	S. Rozen and M. Brand; *Synthesis*, 1985, 665.	16
85S669	T. E. Nickson and C. A. Roche-Dolson; *Synthesis*, 1985, 669.	608
85S677	J. H. Boyer, T. P. Pillai and V. T. Ramakrishnan; *Synthesis*, 1985, 677.	385, 386
85S717	R. Anderson; *Synthesis*, 1985, 717.	537
85S807	J.-J. Yaouanc, G. Massé and G. Strutz; *Synthesis*, 1985, 807.	344, 347, 357
85S817	M. Lalonde and T. H. Chan; *Synthesis*, 1985, 817.	539
85S886	A. Bhattacharjya, R. Mukhopadhyay and S. C. Pakrashi; *Synthesis*, 1985, 886.	398
85S898	I. Fleming and M. Taddei; *Synthesis*, 1985, 898.	977
85S899	I. Fleming and M. Taddei; *Synthesis*, 1985, 899.	917, 938, 977
85S935	P. O. Burke and W. J. Spillane; *Synthesis*, 1985, 935.	367
85S945	E. W. Thomas and D. C. Zimmermann; *Synthesis*, 1985, 945.	942
85S1048	B. Corbel, L. Medinger, J. P. Haelters and G. Sturtz; *Synthesis*, 1985, 1048.	457
85S1131	J. C. Guillemin and J. M. Denis; *Synthesis*, 1985, 1131.	337
85S1157	K. Smith, M. Butters and B. Nay; *Synthesis*, 1985, 1157.	620
85SC151	R. S. Varma and G. W. Kabalka; *Synth. Commun.*, 1985, **15**, 151.	398
85SC385	G. A. Hiegel and K. B. Peyton; *Synth. Commun.*, 1985, **15**, 385.	22
85SC411	B. Dhawan and D. Redmore; *Synth. Commun.*, 1985, **15**, 411.	694
85SC463	F. Bjorkling, T. Norin and R. Unelius; *Synth. Commun.*, 1985, **15**, 463.	1011

85SC527	H. Chikashita, Y. Morita and K. Itoh; *Synth. Commun.*, 1985, **15**, 527.	398
85SC663	S. Manna, J. R. Falck and C. Mioskowski; *Synth. Commun.*, 1985, **15**, 663.	6
85SC977	R. G. Pews and Z. Lysenko; *Synth. Commun.*, 1985, **15**, 977.	21
85SC1253	Y. D. Vankar and A. Bawa; *Synth. Commun.*, 1985, **15**, 1253.	785
85SUL127	R. Barret, F. Pautet, M. Daudon and B. Mathian; *Sulfur Lett.*, 1985, 127 (*Chem. Abstr.*, 1986, **104**, 206 808a).	148
85T837	M. Reuman and A. I. Meyers; *Tetrahedron*, 1985, **41**, 837.	653
85T1111	S. Rozen and R. Filler; *Tetrahedron*, 1985, **41**, 1111.	16
85T1179	P. F. Hudrlik and A. K. Kulkarni; *Tetrahedron*, 1985, **41**, 1179.	617
85T2471	M. Cypryk, J. Chojnowski and J. Michalski; *Tetrahedron*, 1985, **41**, 2471.	467
85T2903	J. M. Aizpurua, M. Juaristi, B. Lecea and C. Palomo; *Tetrahedron*, 1985, **41**, 2903.	34
85T3257	H. Oda, M. Sato, Y. Morizawa, K. Oshima and H. Nozaki; *Tetrahedron*, 1985, **41**, 3257.	918
85T3455	P. M. Wovkulich and M. R. Uskokovic; *Tetrahedron*, 1985, **41**, 3455.	342, 347
85T4279	G. I. Nikishin, E. I. Troyansky and M. I. Lazareva; *Tetrahedron*, 1985, **41**, 4279.	365
85T4443	A. D. Rodriguez and A. Nickon; *Tetrahedron*, 1985, **41**, 4443.	710
85T4657	B. Lecea, J. M. Aizpurua and C. Palomo; *Tetrahedron*, 1985, **41**, 4657.	14
85T4739	J. L. Kice and Y.-H. Kang; *Tetrahedron*, 1985, **41**, 4739.	285
85T4793	M. Clarembeau, A. Cravador, W. Dumont, L. Hevesi, A. Krief, J. Luchetti and D. Van Ende; *Tetrahedron*, 1985, **41**, 4793.	278, 294
85T4861	D. Seebach, G. Calderari and P. Knochel; *Tetrahedron*, 1985, **41**, 4861.	783, 785, 786
85T5301	A. Toshimitsu; *Tetrahedron*, 1985, **41**, 5301.	281
85TCC23	A. Suzuki and R. S. Dhillon; *Top. Curr. Chem.*, 1985, **130**, 23.	514
85TL279	S. A. Kunda, T. L. Smith, M. D. Hylarides and G. W. Kabalka; *Tetrahedron Lett.*, 1985, 279.	616
85TL469	P. Molina, M. Alajarin, M. J. Vilaplana and A. R. Katritzky; *Tetrahedron Lett.*, 1985, 469.	121
85TL475	J. M. C. Aizpurua and C. Palomo; *Tetrahedron Lett.*, 1985, **26**, 475.	262
85TL783	T. Imamoto, K. Sato and C. R. Johnson; *Tetrahedron Lett.*; 1985, **26**, 783.	453
85TL981	L. R. Krepski, L. E. Lynch, S. M. Heilmann and J. K. Rasmussen; *Tetrahedron Lett.*, 1985, **26**, 981.	69
85TL1093	M. Clarembeau and A. Krief; *Tetrahedron Lett.*, 1985, **26**, 1093.	554
85TL1153	M. J. Eis and B. Ganem; *Tetrahedron Lett.*, 1985, **26**, 1153.	319
85TL1193	W.-W. Sy and A. W. By; *Tetrahedron Lett.*, 1985, **26**, 1193.	392
85TL1281	S. Collins, S. Murakami, J. T. Snow and S. Masamune; *Tetrahedron Lett.*, 1985, **26**, 1281.	949
85TL1315	J. Dubac, H. Iloughmane, A. Laporterie and C. Roques; *Tetrahedron Lett.*, 1985, **26**, 1315.	949
85TL1643	R. W. Hoffmann and S. Froech; *Tetrahedron Lett.*, 1985, **26**, 1643.	698
85TL1651	H. Tomioka, N. Hayashi, N. Inoue and Y. Izawa; *Tetrahedron Lett.*, 1985, 1851.	619
85TL1717	F. Mercier and F. Mathey; *Tetrahedron Lett.*, 1985, **26**, 1717.	430
85TL1795	D. N. Harpp, T. Aida and T. H. Chan; *Tetrahedron Lett.*, 1985, **26**, 795.	709
85TL2043	T. Kometani, D. S. Watt and T. Ji; *Tetrahedron Lett.*, 1985, 2043.	631
85TL2159	T. M. Sivavec and T. J. Katz; *Tetrahedron Lett.*, 1985, **26**, 2159.	670
85TL2187	H. C. Brown and U. S. Racherla; *Tetrahedron Lett.*, 1985, **26**, 2187.	514
85TL2233	J. H. Clark and D. K. Smith; *Tetrahedron Lett.*, 1985, 2233.	609
85TL2717	Y. D. Vankar and T. Rao; *Tetrahedron Lett.*, 1985, **26**, 2717.	36
85TL2797	Y. N. Bubnov and M. Y. Elinger; *Tetrahedron Lett.*, 1985, **26**, 2797.	521
85TL3369	G. Cainelli, F. Manescalchi, G. Martelli, M. Panunzio and L. Plessi; *Tetrahedron Lett.*, 1985, **26**, 3369.	103
85TL3821	Y. H. Kim and J. M. Shin; *Tetrahedron Lett.*, 1985, **26**, 3821.	726
85TL3863	A. K. Mandal and S. W. Mahajan; *Tetrahedron Lett.*, 1985, **26**, 3863.	32, 36
85TL3891	S. R. Woulfe, H. Iwagami and M. J. Miller; *Tetrahedron Lett.*, 1985, **26**, 3891.	241
85TL4067	T. Kauffmann and K.-R. Gaydoul; *Tetrahedron Lett.*, 1985, **26**, 4067.	436, 873
85TL4093	A. Commercon and G. Ponsinet; *Tetrahedron Lett.*, 1985, **26**, 4093.	212
85TL4207	H. Shimizu, Y. Nakahara and H. Yoshioka; *Tetrahedron Lett.*, 1985, **26**, 4207.	10, 20
85TL4419	A. B. Smith, III and S. N. Kilényi; *Tetrahedron Lett.*, 1985, **26**, 4419.	664
85TL4443	T. Huynh-Dinh, A. Namane, F. Babin, J. Igolen, J. C. Rousselle, M. P. Fillion and G. Fillion; *Tetrahedron Lett.*, 1985, **26**, 4443.	767
85TL4455	P. Auvrey, P. Knochel and J. F. Normant; *Tetrahedron Lett.*, 1985, **26**, 4455.	159
85TL4501	M. Ochiai, M. Kunishima, K. Sumi, Y. Nagao, E. Fujita, M. Arimoto and H. Yamaguchi; *Tetrahedron Lett.*, 1985, **26**, 4501.	1020, 1063
85TL4523	W. Ando, T. Tsumuruya and A. Sekiguchi; *Tetrahedron Lett.*, 1985, **26**, 4523.	947
85TL4629	Y. Okuda, K. Wakamatsu, W. Tückmantel, K. Oshima and H. Nozaki; *Tetrahedron Lett.*, 1985, **26**, 4629.	918
85TL4763	T. Imamoto, N. Takiyama and K. Nakamura; *Tetrahedron Lett.*, 1985, **26**, 4763.	69
85TL4867	G. A. Krafft and T. L. Sidall; *Tetrahedron Lett.*, 1985, **26**, 4867.	254, 299
85TL4965	L. Wozniak, J. Kowalski and J. Chojnowski; *Tetrahedron Lett.*, 1985, **26**, 4965.	108, 452
85TL5037	W. Oppolzer and P. Dudfield; *Tetrahedron Lett.*, 1985, **26**, 5037.	22
85TL5375	J. M. Muchowski, R. Naef and M. L. Maddox; *Tetrahedron Lett.*, 1985, **26**, 5375.	935
85TL5445	J. J. Brunet, H. Laurent and P. Caubere; *Tetrahedron Lett.*, 1985, **26**, 5445.	36
85TL5511	K. Oertle and H. Wetter; *Tetrahedron Lett.*, 1985, **26**, 5511.	537, 538, 539
85TL5515	H. Wetter and K. Oerple; *Tetrahedron Lett.*, 1985, **26**, 5515.	262, 537, 539
85TL5703	J. Drabowicz and M. Mikolajczyk; *Tetrahedron Lett.*, 1985, **26**, 5703.	218, 724
85TL5863	J. A. Miller and E. Negishi; *Tetrahedron Lett.*, 1985, **25**, 5863.	925
85TL6297	G. J. McGarvey and J. S. Bajwa; *Tetrahedron Lett.*, 1985, **26**, 6297.	516

85URP1145022	A. P. Khardin, O. I. Tuzhikov, L. I. Grekov, R. K. Valetdinov, V. I. Pankov, E. V. Matveeva, G. N. Nazarova, B. N. Popov and D. D. Chuvashov; *USSR Pat.* 1 145 022 (1985) (*Chem. Abstr.*, 1985, **103**, 71 510).	440
85USP4507501	G. E. Nelson; *US Pat.* 4 507 501 (1985) (*Chem. Abstr.*, 1985, **103**, 37 616).	441
85USP4507502	G. E. Nelson; *US Pat.* 4 507 502 (1985) (*Chem. Abstr.*, 1985, **103**, 37 617).	441
85USP4507503	F. Frey and J. Y. Lee; *US Pat.* 4 507 503 (1985) (*Chem. Abstr.*, 1985, **103**, 37 614).	441
85USP4507504	J. Y. Lee and F. Frey; *US Pat.* 4 507 504 (1985) (*Chem. Abstr.*, 1985, **103**, 37 615).	441
85ZAAC(523)7	K. Issleib, U. Kühne and F. Krech; *Z. Anorg. Allg. Chem.*, 1985, **523**, 7.	427, 472
85ZAAC(526)186	W. Hennrichs and J. Jander; *Z. Anorg. Allg. Chem.*, 1985, **526**, 186.	339
85ZN(B)645	M. Barz and G. Z. Himbert; *Z. Naturforsch., Teil B*, 1985, **40**, 645.	1031
85ZN(B)848	W.-W. du Mont and T. Severengiz; *Z. Naturforsch., Teil B*, 1985, **40**, 848.	499
85ZN(B)1090	M. Bandler and R. Becher; *Z. Naturforsch., Teil B*, 1985, **40**, 1090.	478
85ZN(B)1274	M. Baudler, B. Makowka and K. Langerbeins; *Z. Naturforsch., Teil B*, 1985, **40**, 1274.	472
85ZN(B)1320	J. Muller, U. Muller, A. Loss, J. Lorberth, H. Donath and W. Massa; *Z. Naturforsch., Teil B*, 1985, **40**, 1320.	504
85ZN(B)1430	U. Weber, L. Zsolnai and D. Huttner; *Z. Naturforsch., Teil B*, 1985, **40**, 1430.	512
85ZOB2141	V. D. Romanenko, E. O. Klebanski and L. N. Markovski; *Zh. Obshch. Khim.*, 1985, **55**, 2141.	866
85ZOK1835	A. V. Fokin, S. N. Shkurak, A. F. Kolomiets and F. H. Mukhametshin; *Zh. Org. Khim.*, 1985, **21**, 1835 (*Chem. Abstr.*, 1986, **105**, 42 280h).	167
85ZOR124	V. I. Savin, E. V. Tokareva and Y. P. Kitaev; *Zh. Org. Khim.*, 1985, **21**, 124 (*Chem. Abstr.*, 1985, **103**, 5953t).	379
86ACR121	P. Laszlo; *Acc. Chem. Res.*, 1986, **19**, 121.	792
86ACR209	P. J. Brothers and J. P. Collman; *Acc. Chem. Res.*, 1986, **19**, 209.	578
86ACR244	R. M. Moriarty and C. Prakash; *Acc. Chem. Res.*, 1986, **19**, 244.	44
86ACS(B)609	E. K. Moltzen, B. Jensen and A. Senning; *Acta Chem. Scand., Ser. B*, 1986, **40**, 609.	478
86AG(E)259	G. Castaldi, S. Cavicchioli, C. Giordano and F. Uggeri; *Angew. Chem., Int. Ed. Engl.*, 1986, **25**, 259.	29
86AG(E)282	B. von Baar, T. Weiske, J. K. Terlouw and H. Schwarz; *Angew. Chem., Int. Ed. Engl.*, 1986, **25**, 282.	1022
86AG(E)508	J. K. Stille; *Angew. Chem., Int. Ed. Engl.*, 1986, **25**, 508.	689, 1092
86AG(E)640	R. Neidlein, A. Rufinska, H. Schwager and G. Wilke; *Angew. Chem., Int. Ed. Engl.*, 1986, **25**, 640.	984
86AG(E)653	E. Bartmann; *Angew. Chem., Int. Ed. Engl.*, 1986, **25**, 653.	553
86AG(E)938	H.-J. Gais, I. Erdelmeier, H. J. Lindner and J. Vollhardt; *Angew. Chem., Int. Ed. Engl.*, 1986, **25**, 938.	1000
86AG(E)939	H.-J. Gais, J. Vollhardt and H. J. Lindner; *Angew. Chem., Int. Ed. Engl.*, 1986, **25**, 939.	1000
86AG(E)947	Y. Yamamoto; *Angew. Chem., Int. Ed. Engl.*, 1986, **25**, 947.	586, 702
86AG(E)1011	M. Marsch, W. Massa, K. Harms, G. Baum and G. Boche; *Angew. Chem., Int. Ed. Engl.*, 1986, **25**, 1011.	1000
86AG(E)1103	W. Bauer, G. Müller, R. Pi and P. von R. Schleyer; *Angew. Chem., Int. Ed. Engl.*, 1986, **25**, 1103.	1002
86AKZ258	R. A. Khachatryan, S. V. Sayadyan, N. Y. Grigoryan and M. G. Indzhikyan; *Arm. Khim. Zh.*, 1986, **39**, 258 (*Chem. Abstr.*, 1987, **106**, 84 745).	439
86AKZ457	S. A. Zalinyan, R. A. Khachatryan and M. G. Indzhikyan; *Arm. Khim. Zh.*, 1986, **39**, 457 (*Chem. Abstr.*, 1987, **107**, 236 332).	447
86AKZ463	M. Z. Ovakimyan, S. K. Barsegyen, G. T. Gasparyan and M. G. Indzhikyan; *Arm. Khim. Zh.*, 1986, **39**, 463.	446
86AX(C)1875	H. Preut, R. Kassemann and D. Naumann; *Acta Crystallogr., Part C*, 1986, **42**, 1875.	881
86BAU2442	A. V. Kalinin, E. T. Apasov, S. L. Ioffe and V. P. Kozyukov; *Bull. Acad. Sci. USSR, Div. Chim. Sci.*, 1986, **34**, 2442.	394
86BCJ657	Y. Ikeda and H. Yamamoto; *Bull. Chem. Soc. Jpn.*, 1986, **59**, 657.	936
86BCJ659	M. Hoshi, Y. Masuda and A. Arase; *Bull. Chem. Soc. Jpn.*, 1986, **59**, 659.	922
86BCJ1275	T. Takata and W. Ando; *Bull. Chem. Soc. Jpn.*, 1986, **59**, 1275.	146
86BCJ1659	B. Zajc and M. Zupan; *Bull. Chem. Soc. Jpn.*, 1986, **59**, 1659.	15
86BCJ2179	N. Kamigata, H. Iizuka, A. Izuoka and M. Kobayashi; *Bull. Chem. Soc. Jpn.*, 1986, **59**, 2179.	235
86BCJ2547	T. Negoro and Y. Ikeda; *Bull. Chem. Soc. Jpn.*, 1986, **59**, 2547.	24
86BCJ3519	T. Negoro and Y. Ikeda; *Bull. Chem. Soc. Jpn.*, 1986, **59**, 3519.	24
86BCJ3897	Y. Inoue, S. Tanimoto and T. Oida; *Bull. Chem. Soc. Jpn.*, 1986, **59**, 3897.	247
86BSF413	M. Bordeau, S. M. Djamei and J. Dunoguès; *Bull. Soc. Chim. Fr.*, 1986, 413.	537
86BSF625	J.-C. Jacquesy, M.-P. Jouannetaud, G. Morellet and Y. Vidal; *Bull. Soc. Chim. Fr.*, 1986, 625.	650
86BSF910	J. Cousseau and P. Albert; *Bull. Soc. Chim. Fr.*, 1986, 910.	611
86BSF930	M. R. Bryce, R. D. Chambers, S. T. Mullins and A. Parkin; *Bull. Soc. Chim. Fr.*, 1986, 930.	613
86BSF955	E. Laurent, B. Marquet, R. Tardivel and H. Thiebault; *Bull. Soc. Chim. Fr.*, 1986, 955.	16
86CB535	R. Appel, E. Gaitzsch, K.-H. Dunker and F. Knoch; *Chem. Ber.*, 1986, **119**, 535.	936
86CB1581	W. Tückmantel, K. Oshima and H. Nozaki; *Chem. Ber.*, 1986, **119**, 1581.	918
86CB1953	L. Henn, G. Himbert, K. Diehl and M. Kaftory; *Chem. Ber.*, 1986, **119**, 1953.	877
86CB2767	D. J. Brauer, F. Gol, S. Hietkamp and O. Stelzer; *Chem. Ber.*, 1986, **119**, 2767.	466
86CB2921	S. Heitkamp and O. Stelzer; *Chem. Ber.*, 1986, **119**, 2921.	437

86CC260 L. Brandsma, H. D. Verkruijsse, C. Schade and P. von R. Schleyer; *J. Chem. Soc., Chem. Commun.*, 1986, 260. 907

86CC403 S. Kerschl and B. Wrackmeyer; *J. Chem. Soc., Chem. Commun.*, 1986, 403. 903

86CC475 S. I. Fukuzawa, T. Fujiami and S. Sakai; *J. Chem. Soc., Chem. Commun.*, 1986, 475. 564

86CC531 A. Aguero, J. Kress and J. A. Osborn; *J. Chem. Soc., Chem. Commun.*, 1986, 531. 681

86CC626 E. Piers and R. T. Skerlj; *J. Chem. Soc., Chem. Commun.*, 1986, 626. 917

86CC787 C. Sishta, M. Ke, B. R. James and D. Dolphin; *J. Chem. Soc., Chem. Commun.*, 1986, 787. 584

86CC791 J. H. Clark, A. J. Hyde and D. K. Smith; *J. Chem. Soc., Chem. Commun.*, 1986, 791. 19

86CC793 J. Ichihara, T. Matsuo, T. Hanafusa and T. Ando; *J. Chem. Soc., Chem. Commun.*, 1986, 793. 19

86CC831 S. Anwar and A. P. Davies; *J. Chem. Soc., Chem. Commun.*, 1986, 831. 62

86CC850 M. L. J. Hackney and A. D. Norman; *J. Chem. Soc., Chem. Commun.*, 1986, 850. 466

86CC1010 I. Fleming and F. J. Pulido; *J. Chem. Soc., Chem. Commun.*, 1986, 1010. 929

86CC1067 J. V. Comasseto, V. Catani, J. T. B. Ferreira and A. L. Braga; *J. Chem. Soc., Chem. Commun.*, 1986, 1067. 1036, 1038

86CC1149 R. Csuk, A. Fürstner, B. I. Glänzer and H. Weidmann; *J. Chem. Soc., Chem. Commun.*, 1986, 1149. 678

86CC1159 K. Kim and Y.-C. Chang; *J. Chem. Soc., Chem. Commun.*, 1986, 1159. 626

86CC1170 S. Kerschl and B. Wrackmeyer; *J. Chem. Soc., Chem. Commun.*, 1986, 1170. 903

86CC1250 Y. Sasson, M. Weiss, A. Loupy and G. Bram and C. Pardo; *J. Chem. Soc., Chem. Commun.*, 1986, 1250. 25, 31

86CC1543 A. M. Arif, B. L. Benac, A. H. Cowley, R. Geerts, R. A. Jones, K. B. Kidd, J. M. Power and S. T. Schwab; *J. Chem. Soc., Chem. Commun.*, 1986, 1543. 511

86CC1623 M. R. Bryce, R. D. Chambers, S. T. Mullins and A. Parkin; *J. Chem. Soc., Chem. Commun.*, 1986, 1623. 613

86CJC138 T. B. Patrick, K. K. Johri, D. H. White, W. S. Bertrand, R. Mokhtar, M. R. Kilbourn and M. J. Welch; *Can. J. Chem.*, 1986, **64**, 138. 13

86CJC2342 J. M. Aizpurua, B. Lecea and C. Palomo; *Can. J. Chem.*, 1986, **64**, 2342. 14

86CL349 R. Sato, S. Saito, H. Chiba, T. Goto and M. Saito; *Chem. Lett.*, 1986, 349. 233

86CL549 H. Suzuki and Y. Hanazaki; *Chem. Lett.*, 1986, 549. 753

86CL1401 S. Hamamoto and H. Takaku; *Chem. Lett.*, 1986, 1401. 410, 411

86CRV941 E. Vedejs and F. G. West; *Chem. Rev.*, 1986, **86**, 941. 324

86CRV697 H. C. J. Ottenheijm and J. D. M. Herscheid; *Chem. Rev.*, 1986, **86**, 697. 340

86CRV751 A. G. M. Barrett and G. B. Graboski; *Chem. Rev.*, 1986, **86**, 751. 393, 396, 398, 783

86CRV795 B. H. Lipshutz; *Chem. Rev.*, 1986, **86**, 795. 667

86CRV885 T. Nakai and K. Mikami; *Chem. Rev.*, 1986, **86**, 885. 88

86CRV997 S. T. Purrington, B. S. Kagen and T. B. Patrick; *Chem. Rev.*, 1986, **86**, 997. 334

86CRV(86)997 S. T. Purrington, B. S. Kagen and T. B. Patrick; *Chem. Rev.*, 1986, **86**, 997. 15, 17

86EUP165534 N. Dereu, A. Welter, A. Wendel, S. Leyck, M. Parnham, E. Graf, H. Sies, H. Betzing and H. Fischer; *Eur. Pat.* 165 534 (1986) (*Chem. Abstr.*, 1986, **104**, 186 851x). 235

86EUP184872 J. A. Van Doorn, R. L. Wife and N. Meijboom; *Eur. Pat.* 184 872 (1986) (*Chem. Abstr.*, 1986, **105**, 191 397). 439

86EUP194716 W. E. Bay and K. E. Reinehe; *Eur. Pat.* 194 716 (1986) (*Chem. Abstr.*, 1987, **106**, 5231). 440

86GEP231075 M. Kant, L. Riesle, W. Kochmann and M. Oertel; *Ger. Pat.* 231 075 (*Chem. Abstr.*, 1986, **105**, 209 206c). 252

86GEP3443467 N. Dereu, A. Wendel, H. Sies, S. Leyck, A. Roemer and E. Graf; *Ger. Pat.* 3 443 467 (1986) (*Chem. Abstr.*, 1986, **105**, 114 733h). 235

86H(24)303 H. Takayama, H. Suzuki, T. Nomoto and S. Yamada; *Heterocycles*, 1986, **24**, 303. 235

86H(24)3019 N. Furukawa, M. Tsuruoka and H. Fujihara; *Heterocycles*, 1986, **24**, 3019. 271

86H(24)3043 H.-J. Lui, L. M. Shewchuk and M. Llinas-Brunet; *Heterocycles*, 1986, **24**, 3043. 36

86HCA1331 Y. Chiang, A. J. Kresge, M. Capponi and J. Wirz; *Helv. Chim. Acta*, 1986, **69**, 1331. 638

86HOU(13/9a)1 A. Segnitz; *Methoden Org. Chem.* (*Houben-Weyl*), 1986, **13/9a**, 1. 573

86HOU(13/9a)121 A. Segnitz; *Methoden Org. Chem.* (*Houben-Weyl*), 1986, **13/9a**, 121. 579

86HOU(13/9a)175 A. Segnitz; *Methoden Org. Chem.* (*Houben-Weyl*), 1986, **13/9a**, 175. 575

86HOU(13/9a)525 A. Segnitz; *Methoden Org. Chem.* (*Houben-Weyl*), 1986, **13/9a**, 525. 579

86HOU(13/9a)613 A. Segnitz; *Methoden Org. Chem.* (*Houben-Weyl*), 1986, **13/9a**, 613. 579

86HOU(13/9a)671 A. Segnitz; *Methoden Org. Chem.* (*Houben-Weyl*), 1986, **13/9a**, 671. 579

86IC361 B. L. Li and R. H. Neilson; *Inorg. Chem.*, 1986, **25**, 361, 358. 467, 468

86IC2695 J. Boske, E. Niecke, E. Ocando-Mavarez, J. P. Majoral and G. Bertrand; *Inorg. Chem.*, 1986, **25**, 2695. 470

86IC3396 P. H. Leung, J. W. L. Martin and S. B. Wild; *Inorg. Chem.*, 1986, **25**, 3396. 438

86ICA179 C. M. Mikulski, T. Moore, Y. Smith and N. M. Karayannis; *Inorg. Chim. Acta*, 1986, **115**, 179. 94

86ICA(113)43 G. B. Deacon, P. W. Felder, M. Domagala, F. Huber and R. Rüther; *Inorg. Chim. Acta*, 1986, **113**, 43. 887

86ICA(121)27 R. Dréos Garlatti, G. Tauzher and G. Costa; *Inorg. Chim. Acta*, 1986, **121**, 27. 583

86IJC328 M. Siddiq, N. Akhtar and M. A. Munawar; *Indian J. Chem.*, 1986, **25**, 328. 621

86IZV134 Z. G. Aliev, A. M. Sipyagin, V. G. Kartsev and L. O. Atovmyan; *Izv. Akad. Nauk SSSR, Ser. Khim.*, 1986, 134 (*Chem. Abstr.*, 1986, **104**, 224/405). 1003

86IZV181 N. A. Nesmeyanov, V. G. Kharitonov, S. T. Berman, P. V. Petrovskii and O. A. Reutov; *Izv. Akad. Nauk SSSR, Ser. Khim.*, 1986, 181 (*Chem. Abstr.*, 1987, **106**, 33 202c). 877

86JA40 J. W. Bruno, G. M. Smith, T. J. Marks, C. K. Fair, A. J. Schultz and J. M. Williams; *J. Am. Chem. Soc.*, 1986, **108**, 40. 602

86JA294 W. R. Roush and R. L. Haltermann; *J. Am. Chem. Soc.*, 1986, **108**, 294. 529

86JA304 G. Stork, C. S. Shiner, C. W. Cheng and R. L. Polt; *J. Am. Chem. Soc.*, 1986, **108**, 304. 759, 965

86JA483 N. Ikeda, I. Arai and H. Yamamoto; *J. Am. Chem. Soc.*, 1986, **108**, 483. 905

86JA976 H. Schmidbaur, A. Schier, C. M. F. Frazao and G. Müller; *J. Am. Chem. Soc.*, 1986, **108**, 976. 447

86JA1306 S. E. Denmark, M. S. Dappen and C. J. Cramer; *J. Am. Chem. Soc.*, 1986, **108**, 1306. 356, 784

86JA1314 G. A. Krafft and P. T. Meinke; *J. Am. Chem. Soc.*, 1986, **108**, 1314. 289

86JA1692 R. F. Childs, R. Faggiani, C. J. L. Lock, M. Mahendran and S. D. Zweep; *J. Am. Chem. Soc.*, 1986, **108**, 1692. 1007

86JA1694 S. K. Chadda, R. F. Childs, R. Faggiani and C. J. L. Lock; *J. Am. Chem. Soc.*, 1986, **108**, 1694. 1007

86JA1703 A. M. Arif, R. A. Jones, S. T. Schwab and B. R. Whittlesey; *J. Am. Chem. Soc.*, 1986, **108**, 1703. 512

86JA2049 H. C. Brown and J. V. N. Vara Prasad; *J. Am. Chem. Soc.*, 1986, **108**, 2049. 519, 531

86JA2094 D. B. Collum, W. C. Still and F. Mohamadi; *J. Am. Chem. Soc.*, 1986, **108**, 2094. 40

86JA2358 J. B. Hendrickson, G. J. Boudreaux and P. S. Palumbo; *J. Am. Chem. Soc.*, 1986, **108**, 2358. 157

86JA2445 S. H. Lee and J. Schwartz; *J. Am. Chem. Soc.*, 1986, **108**, 2445. 365, 607, 624

86JA2753 N. Yanagihara, C. Lambert, K. Iritani, K. Utimoto and H. Nozaki; *J. Am. Chem. Soc.*, 1986, **108**, 2753. 983

86JA3033 W. J. Scott and J. K. Stille; *J. Am. Chem. Soc.*, 1986, **108**, 3033. 931

86JA3402 E. Negishi and T. Takahashi; *J. Am. Chem. Soc.*, 1986, **108**, 3402. 909, 923, 961

86JA3422 W. R. Roush, M. A. Adam, A. E. Walts and D. J. Harris; *J. Am. Chem. Soc.*, 1986, **108**, 3422. 530

86JA3466 E. J. Madaj, D. M. Snider and W. E. Truce; *J. Am. Chem. Soc.*, 1986, **108**, 3466. 478

86JA3613 R. F. Childs, R. Faggiani, C. J. L. Lock and M. Mahendran; *J. Am. Chem. Soc.*, 1986, **108**, 3613. 1008

86JA5908 P. A. Grieco, R. Lis, R. E. Zelle and J. Finn; *J. Am. Chem. Soc.*, 1986, **108**, 5908. 196

86JA6036 K. Maruoka, H. Sano, K. Shinoda, S. Nakai and H. Yamamoto; *J. Am. Chem. Soc.*, 1986, **108**, 6036. 30, 34

86JA6394 C. Genrari, L. Columbo and G. Bertolini; *J. Am. Chem. Soc.*, 1986, **108**, 6394. 375

86JA6395a D. A. Evans, T. C. Britton, R. L. Dorow and J. F. Dellaria; *J. Am. Chem. Soc.*, 1986, **108**, 6395. 375

86JA6395b L. A. Trimble and J. C. Vederas; *J. Am. Chem. Soc.*, 1986, **108**, 6395. 375

86JA6664 W. Weissensteiner, I. I. Schuster, J. F. Blount and K. Mislow; *J. Am. Chem. Soc.*, 1986, **108**, 6664. 949

86JA6826 G. Stork and M. J. Sofia; *J. Am. Chem. Soc.*, 1986, **108**, 6826. 540

86JA7115 E. J. Corey, R. Naef and F. J. Hannon; *J. Am. Chem. Soc.*, 1986, **108**, 7115. 588

86JA7125 P. J. Stang and K. A. Roberts; *J. Am. Chem. Soc.*, 1986, **108**, 7125. 1023, 1024

86JA7127 C. J. Kowalski, G. S. Lal and M. S. Haque; *J. Am. Chem. Soc.*, 1986, **108**, 7127. 941, 1023, 1024

86JA7402 T. Imai, T. Tamura, A. Yamamuro, T. Sato, T. A. Wollmans, R. M. Kennedy and S. Masamune; *J. Am. Chem. Soc.*, 1986, **108**, 7402. 521

86JA7404 S. Masamune, R. M. Kennedy, J. S. Petersen, K. N. Houk and Y.-D. Wu; *J. Am. Chem. Soc.*, 1986, **108**, 7404. 58, 521

86JA7408 K. Takai, K. Nitta and K. Utimoto; *J. Am. Chem. Soc.*, 1986, **108**, 7408. 618, 625, 630

86JA7411 S. L. Buchwald, B. T. Watson and J. C. Huffman; *J. Am. Chem. Soc.*, 1986, **108**, 7411. 986, 987

86JA7441 S. L. Buchwald, R. T. Lum and J. C. Dewan; *J. Am. Chem. Soc.*, 1986, **108**, 7441. 986, 987

86JA7832 P. J. Stang, M. Boehshar and J. Lin; *J. Am. Chem. Soc.*, 1986, **108**, 7832. 1028

86JA8279 S. Masamune, T. Sato, B. M. Kim and T. A. Wollmann; *J. Am. Chem. Soc.*, 1986, **108**, 8279. 698

86JCR(S)26 J. Nielson, J. E. Marugg, J. H. van Boom, J. Honnens, M. Taagaard and O. Dahl; *J. Chem. Res.* (*S*), 1986, 26. 469

86JCR(S)274 J. Barluenga, J. M. Martinez-Gallo, C. Najera and M. Yus; *J. Chem. Res.* (*S*), 1986, 274. 24, 31

86JCR(S)300 T. J. Mason, J. P. Lorimer, A. T. Turner and A. R. Harris; *J. Chem. Res.* (*S*), 1986, 300. 19

86JCR(S)426 F. Bellesia, F. Ghelfi, R. Grandi and U. M. Pagnoni; *J. Chem. Res.* (*S*), 1986, 426. 22

86JCR(S)428 F. Bellesia, F. Ghelfi, R. Grandi and U. M. Pagnoni; *J. Chem. Res.* (*S*), 1986, 428. 28

86JCS(D)997 J. Ellermann, L. Brehm, E. Lindner, W. Hiller, R. Fawzi, F. L. Dickert and M. Waidhas; *J. Chem. Soc., Dalton Trans.*, 1986, 997. 490

86JCS(D)2003 H. M. K. K. Pathirana and W. R. McWhinnie; *J. Chem. Soc., Dalton Trans.*, 1986, 2003. 280

86JCS(P1)1983 S. Uemura and S. I. Fukuzawa; *J. Chem. Soc., Perkin Trans. 1*, 1986, 1983. 287, 288

86JCS(P2)151 M. R. Banks and F. F. Hudson; *J. Chem. Soc., Perkin Trans 1*, 1986, 151. 357

86JCS(P2)1029 L. D. Quin and G. Keglevich; *J. Chem. Soc., Perkin Trans. 2*, 1986, 1029. 450

86JFC(31)99 D. Y. Chi, D. O. Kiesewetter, J. A. Katzenellenbogen, M. R. Kilbourn and M. J. Welch; *J. Fluorine Chem.*, 1986, **31**, 99. 18

86JFC(33)227 R. E. Banks and J. C. Tatlow; *J. Fluorine Chem.*, 1986, **33**, 227. 15

86JFC(33)377 G. A. Olah, J. G. Shih and G. K. S. Prakash; *J. Fluorine Chem.*, 1986, **33**, 377. 15, 17

86JFC(34)129 H. J. Frohn and H. Maurer; *J. Fluorine Chem.*, 1986, **34**, 129. 492

86JGU287 V. V. Shamitin, V. T. Bychkov, O. P. Bolotova and V. I. Kuzina; *J. Gen. Chem. USSR* (*Engl. Transl.*), 1986, **56**, 287. 887

86JGU849 M. G. Voronkov, L. P. Petrukhov, V. I. Rakhlin, B. Z. Shterenberg, R. G. Mirskov, S. N. Adamovich and V. A. Pestunovich; *J. Gen. Chem. USSR* (*Engl. Transl.*), 1986, **56**, 849. 1091

86JGU1567 M. N. Danchenko and A. D. Sinitsa; *J. Gen. Chem. USSR* (*Engl. Transl.*), 1986, **56**, 1567. 464

86JGU1812 A. P. Yuvchenko, K. L. Moiseichuk, E. A. Dikusar, N. A. Zhukovskaya and Yu. Ol'dekop; *J. Gen. Chem. USSR* (*Engl. Transl.*), 1986, **57**, 1812. 1098
86JHC825 H. Feuer, H. S. Bevinakatti and X.-G. Luo; *J. Heterocycl. Chem.*, 1986, **23**, 825. 392
86JMC213 E. De Clercq, C. Desgranges, P. Herdewijn, I. S. Sim, A. S. Jones, M. J. McLean and R. T. Walker; *J. Med. Chem.*, 1986, **29**, 213. 625
86JMC2114 S. Morris-Natschke, J. R. Surles, L. W. Daniel, M. E. Berens, E. J. Modest and C. Aantadosi; *J. Med. Chem.*, 1986, **29**, 2114. 252
86JOC81 S. G. Pyne; *J. Org. Chem.*, 1986, **51**, 81. 362
86JOC111 J. L. Holcombe and T. Livinghouse; *J. Org. Chem.*, 1986, **51**, 111. 684
86JOC179 B. Dhawan and D. Redmore; *J. Org. Chem.*, 1986, **51**, 179. 694, 845, 846
86JOC238 C. A. Brown and S. Krishnamurthy; *J. Org. Chem.*, 1986, **51**, 238. 534
86JOC240 Y. Ishii, T. Nakano, A. Inada, Y. Kishigami, K. Sakurai and M. Ogawa; *J. Org. Chem.*, 1986, **51**, 240. 55
86JOC271 M. P. Sibi, J. W. Dankwardt and V. Snieckus; *J. Org. Chem.*, 1986, **51**, 271. 673
86JOC277 W. D. Wulff, G. A. Peterson, W. E. Bauta, K.-S. Chan, K. L. Faron, S. R. Gilbertson, R. A. Kaesler, D. C. Yang and C. K. Murray; *J. Org. Chem.*, 1986, **51**, 277. 607
86JOC295 S.-I. Kang and J. L. Kice; *J. Org. Chem.*, 1986, **51**, 295. 734
86JOC402 J. I. Concepcion, C. G. Francisco, R. Friere, R. Hernandez, J. A. Salazar and E. Suarez; *J. Org. Chem.*, 1986, **51**, 402. 13
86JOC427 H. C. Brown and U. S. Racherla; *J. Org. Chem.*, 1986, **51**, 427. 520
86JOC432 P. K. Jadhav, K. S. Bhat, P. J. Derumal and H. C. Brown; *J. Org. Chem.*, 1986, **51**, 432. 521
86JOC445 H. C. Brown and K. S. Bhat; *J. Org. Chem.*, 1986, **51**, 445. 520
86JOC745 S. S. Nikam, K.-H. Chu and K. K. Wang; *J. Org. Chem.*, 1986, **51**, 745. 930
86JOC767 K.-H. Chu and K. K. Wang; *J. Org. Chem.*, 1986, **51**, 767. 904
86JOC789 Y. Leblanc, B. J. Fitzsimmons, J. Adams, F. Perez and J. Rokach; *J. Org. Chem.*, 1986, **51**, 789. 7
86JOC829 H. W. Heine, J. R. Empfield, T. D. Golobish, E. A. Williams and M. F. Garbauskas; *J. Org. Chem.*, 1986, **51**, 829. 163
86JOC875 H. J. Cristau, B. Chabaud, R. Labaudiniere and H. Cristol; *J. Org. Chem.*, 1986, **51**, 875. 727
86JOC895 H. C. Brown and U. S. Racherla; *J. Org. Chem.*, 1986, **51**, 895. 525
86JOC929 B. Sket and M. Zupan; *J. Org. Chem.*, 1986, **51**, 929. 28
86JOC1033 F. A. Davis, L. A. Jenkins and R. Billmers; *J. Org. Chem.*, 1986, **51**, 1033. 184
86JOC1886 G. W. M. Visser, C. N. M. Bakker, B. W. V. Haltern, J. D. M. Herscheid, G. A. Brinkman and A. Hoekstra; *J. Org. Chem.*, 1986, **51**, 1886. 613
86JOC2137 A. J. Pearson, P. R. Bruhn and S.-Y. Hsu; *J. Org. Chem.*, 1986, **51**, 2137. 684
86JOC2139 N. Ono, A. Kamimura and A. Kaji; *J. Org. Chem.*, 1986, **51**, 2139. 396, 789
86JOC2206 B. T. Nguyen and F. K. Cartledge; *J. Org. Chem.*, 1986, **51**, 2206. 33
86JOC2381 P. E. Peterson, D. J. Nelson and R. Risener; *J. Org. Chem.*, 1986, **51**, 2381. 910
86JOC2505 A. J. Pearson and S.-Y. Hsu; *J. Org. Chem.*, 1986, **51**, 2505. 91
86JOC2637 H. A. Bates, J. Farina and M. Tong; *J. Org. Chem.*, 1986, **51**, 2637. 8
86JOC3150 H. C. Brown and M. V. Rangaishenvi; *J. Org. Chem.*, 1986, **51**, 3150. 530
86JOC3165 E. Keinan and D. Eren; *J. Org. Chem.*, 1986, **51**, 3165. 706
86JOC3217 E. L. Setti and O. A. Mascaretti; *J. Org. Chem.*, 1986, **51**, 3217. 9
86JOC3242 T. B. Patrick and D. L. Darling; *J. Org. Chem.*, 1986, **51**, 3242. 613
86JOC3278 H. C. Brown, W. S. Park and B. T. Cho; *J. Org. Chem.*, 1986, **51**, 3278. 534
86JOC3380 Y. J. Abul-Hajj; *J. Org. Chem.*, 1986, **51**, 3380. 28
86JOC3394 J. Chandrasekharan, P. V. Ramachandran and H. C. Brown; *J. Org. Chem.*, 1986, **51**, 3394. 526, 531
86JOC3396 H. C. Brown, W. S. Park and B. T. Cho; *J. Org. Chem.*, 1986, **51**, 3396. 534
86JOC3398 H. C. Brown, N. G. Bhat and J. B. Campbell, Jr.; *J. Org. Chem.*, 1986, **51**, 3398. 900
86JOC3435 B. M. Trost and S. Mignani; *J. Org. Chem.*, 1986, **51**, 3435. 938
86JOC3559 M. J. Danaha and K. K. Olgilvie; *J. Org. Chem.*, 1986, **51**, 3559. 467
86JOC3561 B. L. Chenard and C. M. Van Zyl; *J. Org. Chem.*, 1986, **51**, 3561. 611, 630, 916
86JOC3577 W. M. Stalick, A. Khorrami and K. S. Hatton; *J. Org. Chem.*, 1986, **51**, 3577. 677
86JOC3607 S. Rozen and M. Brand; *J. Org. Chem.*, 1986, **51**, 3607. 17
86JOC3664 B. Strijtveen and R. M. Kellogg; *J. Org. Chem.*, 1986, 3664. 124
86JOC3762 J. G. Smith, P. W. Dibble and R. E. Sandborn; *J. Org. Chem.*, 1986, **51**, 3762. 87
86JOC3890 Y. Seki, K. Takeshita, K. Kawamoto, S. Murai and N. Sonoda; *J. Org. Chem.*, 1986, **51**, 3890. 912
86JOC3913 G. J. McGarvey and M. Kimura; *J. Org. Chem.*, 1986, **51**, 3913. 88
86JOC3993 R. A. Kjonaas and E. J. Vawter; *J. Org. Chem.*, 1986, **51**, 3993. 702
86JOC4224 G. Bellucci, R. Bianchini and S. Vecchiani; *J. Org. Chem.*, 1986, **51**, 4224. 30
86JOC4296 H. C. Brown, W. V. N. Vara-Prasad and A. K. Gupta; *J. Org. Chem.*, 1986, **51**, 4296. 519
86JOC4342 P. Sampson, G. B. Hammond and D. F. Wiener; *J. Org. Chem.*, 1986, **51**, 4342. 457
86JOC4368 R. Tamura, M. Sato and D. Oda; *J. Org. Chem.*, 1986, **51**, 4368. 394, 784
86JOC4507 A. Suzuki, N. Miyaura, S. Abiko, M. Ito, M. M. Midland, J. A. Sinclair and H. C. Brown; *J. Org. Chem.*, 1986, **51**, 4507. 1086
86JOC4521 J. A. Sikorski, N. G. Bhat, T. E. Cole, K. K. Wang and H. C. Brown; *J. Org. Chem.*, 1986, **51**, 4521. 1086
86JOC4526 H. C. Brown and W. V. N. Vara-Prasad; *J. Org. Chem.*, 1986, **51**, 4526. 531
86JOC4941 J. M. Aizpurua, F. P. Cossio and C. Palomo; *J. Org. Chem.*, 1986, **51**, 4941. 10
86JOC5045 C. A. Broka and K. K. Eng; *J. Org. Chem.*, 1986, **51**, 5045. 337
86JOC5157 F. R. Biehl, A. Razzuk, M. V. Jovanovic and S. P. Khanapure; *J. Org. Chem.*, 1986, **51**, 5157. 753

86JOC5264 H. C. Brown, B. Nazer, J. S. Cha and J. A. Sikorski; *J. Org. Chem.*, 1986, **51**, 5264. 526

86JOC5277 H. C. Brown, T. Imai and N. G. Bhat; *J. Org. Chem.*, 1986, **51**, 5277. 901, 906

86JOC5286 K. Karabelas and A. Hallberg; *J. Org. Chem.*, 1986, **51**, 5286. 932

86JOC5291 M. Falorni, L. Lardicci and G. Giacomelli; *J. Org. Chem.*, 1986, **51**, 5291. 9

86JOC5320 R. H. Boutin and H. Rapoport; *J. Org. Chem.*, 1986, **51**, 5320. 1011

86JOC5436 D. L. Boger and R. S. Coleman; *J. Org. Chem.*, 1986, **51**, 5436. 655

86JOC5458 T. V. RajanBabu and G. S. Reddy; *J. Org. Chem.*, 1986, **51**, 5458. 681

86JOC5490 J. W. Kelly and J. S. A. Evans; *J. Org. Chem.*, 1986, **51**, 5490. 830

86JOM(301)289 J. Neuffer and W. J. Richter; *J. Organomet. Chem.*, 1986, **301**, 289. 436, 443, 450

86JOM(303)197 A. J. Ashe, III and E. G. Ludwig, Jr.; *J. Organomet. Chem.*, 1986, **303**, 197. 508

86JOM(304)115 Z. Djeghaba, B. Jousseaume, M. Ratier and J.-G. Duboudin; *J. Organomet. Chem.*, 1986, **304**, 115. 1034, 1035

86JOM(307)273 P. H. Yang, K. F. Licu and Y. T. Lin; *J. Organomet. Chem.*, 1986, **307**, 273. 595

86JOM(308)281 R. L. Wells, A. P. Furdy, A. T. McPhail and C. G. Pitt; *J. Organomet. Chem.*, 1986, **308**, 281. 482

86JOM(308)289 A. J. Ashe, III and E. G. Ludwig, Jr.; *J. Organomet. Chem.*, 1986, **308**, 289. 499

86JOM(310)27 L. B. Kool, M. D. Rausch, H. G. Alt, M. Herberhold, U. Thewalt, and B. Honold; *J. Organomet. Chem.*, 1986, **310**, 27. 963

86JOM(312)53 T.-S. Chou, C.-H. Tsao and S. C. Hung; *J. Organomet. Chem.*, 1986, **312**, 53. 432

86JOM(314)75 M. Brockmann and H. Dieck; *J. Organomet. Chem.*, 1986, **314**, 75. 813

86JOM(317)175 F. R. Kreissl, M. Wolfgruber and U. Thewalt; *J. Organomet. Chem.*, 1986, **317**, 175. 873, 896

86JST141 A. Kucsman, I. Kapovits, L. Parkanyi and A. Kalman; *J. Mol. Struct.*, 1986, **140**, 141. 1003

86LA402 E. Slusarska and A. Zwierzak; *Liebigs Ann. Chem.*, 1986, 402. 299

B-86MI 202-01 R. C. Larock; "Solvomercuration/Demercuration Reactions in Organic Synthesis," Springer Verlag, New York, 1986, chap. 2. 40

B-86MI 202-02 M. T. Reetz; "Organotitanium Reagents in Organic Synthesis," Springer Verlag, New York, 1986. 67

B-86MI 202-03 M. Nógrádi; "Stereoselective Synthesis," VCH, New York, 1986. 67

B-86MI 204-01 C. Paulmier; "Selenium Reagents and Intermediates in Organic Synthesis," Pergamon, Oxford, 1986, pp. 182–227. 280

B-86MI 204-02 C. Paulmier; "Selenium Reagents and Intermediates in Organic Synthesis," Pergamon, Oxford, 1986, pp. 229–255. 286

B-86MI 204-03 C. Paulmier; "Selenium Reagents and Intermediates in Organic Synthesis," Pergamon, Oxford, 1986, pp. 143–150. 286

B-86MI 206-01 C. Paulmier; in "Selenium Reagents and Intermediates in Organic Synthesis," Pergamon, Oxford, 1986, p. 42. 368, 369

B-86MI 209-01 M. Wieber (ed.); "Gmelin Handbuch der Anorganischen Chemie," Springer-Verlag, Berlin, 1986 part 4. 481

B-86MI 210-01 H. C. Brown, P. K. Jahav and B. Singaram; in "Modern Synthetic Methods," ed. R. Scheffold, Springer-Verlag, Heidelberg, 1986, vol. 4, p. 307. 514

B-86MI 211-01 M. Reetz; "Organotitanium Reagents in Organic Synthesis," Springer Verlag, Berlin, 1986. 565, 566

B-86MI 211-02 P. R. Ortiz de Montellano (ed.); "Cytochrome P450," Plenum Press, New York, 1986. 578

B-86MI 215-01 W. J. Mijs and C. R. H. I. de Jonge (eds); "Organic Synthesis by Oxidation with Metal Compounds," Plenum Press, New York, 1986, p. 119. 772

B-86MI 215-02 S. R. Sandler and W. Karo; "Organic Functional Group Preparation," 2nd edn, Academic Press, New York, 1986, vol. II, p. 442. 776

B-86MI 217-01 D. G. Ott, M. J. Reisfeld and T. W. Whalley; in "Synthesis and Applications of Isotopically Labelled Compounds," ed. R. R. Muccino, Elsevier, Amsterdam, 1986, p. 409 (*Chem. Abstr.*, 1986, **105**, 172 599e). 873

B-86MI 219-01 R. C. Larock; "Solvomercuration/Demercuration Reactions in Organic Synthesis," Springer-Verlag, Berlin, 1986. 982

86MI 220-01 A. Katrusiak, Z. Kaluski, P. Pietrzak and J. Skolik; *J. Crystallog. Spec. Res.*, 1986, **16**, 879. 1008

B-86MI 223-01 H. Bauer, J. Faust, R. Froböse and J. Füssel; in "Gmelin Handbook of Inorganic Chemistry, Organocopper Compounds," eds. J. Faust and J. Füssel, Part 3, Springer, Berlin, 1986, p. 1. 1094, 1095

86NJC761 C. Nemecek, E. Dunach and H. B. Kagan; *Nouv. J. Chim.*, 1986, **10**, 761. 362

86OCS467 E. Negishi and D. E. Van Horn; *Organomet. Synth.*, 1986, **3**, 467. 974, 975

86OCS618 A. L. Rheingold; *Organomet. Synth.*, 1986, **3**, 618. 507

86OCS621 H. J. Breunig, W. Kanig and H. Kischkel; *Organomet. Synth.*, 1986, **3**, 621. 507

86OCS625 H. J. Breunig; *Organomet. Synth.*, 1986, **3**, 625. 505

86OCS633 A. L. Rheingold; *Organomet. Synth.*, 1986, **3**, 633. 489

86OCS636 H. J. Breunig and D. Muller; *Organomet. Synth.*, 1986, **3**, 636. 505

86OCS638 H. J. Breunig and D. Muller; *Organomet. Synth.*, 1986, **3**, 638. 483

86OM393 A. L. Rheingold and A.-J. DiMaio; *Organometallics*, 1986, **5**, 393. 496

86OM398 J. H. Freudenberger and R. R. Schrock; *Organometallics*, 1986, **5**, 398. 681

86OM739 H. Brunner, R. Becker and S. Gauder; *Organometallics*, 1986, **5**, 739. 314

86OM1266 C. G. Pitt, A. P. Purdy, K. T. Higa and R. L. Wells; *Organometallics*, 1986, **5**, 1266. 508, 511

86OM1296 H. Schumann, I. Albrecht, J. Loebel, E. Hahn, M. E. Hossain and D. Van der Helm; *Organometallics*, 1986, **5**, 1296. 564

86OM2452 G. N. Schrauzer, E. O. Schlemper, N. H. Liu, Q. Wang, K. Rubin, X. Zhang, X. Long and C. S. Chin; *Organometallics*, 1991, **5**, 2452. 573

86OPP299 O. A. Attanasi and L. Caglioti; *Org. Prep. Proced. Int.*, 1986, **18**, 299. 770

86PAC1133	J. Corset; *Pure Appl. Chem.*, 1986, **58**, 1133.	1004
86POL1581	P. Raj, R. Rastogi, K. Singhal and A. K. Saxena; *Polyhedron*, 1986, **5**, 1581.	891
86PS(26)143	M. Antberg, L. Dahlenburg, N. Höck and C. Prengl; *Phosphorus Sulfur*, 1986, **26**, 143.	439
86PS(26)307	K. Diemert, B. Kottwitz and W. Kuchen; *Phosphorus Sulfur*, 1986, **26**, 307.	448
86PS(27)293	M. Beji and M. Hedayatullah; *Phosphorus and Sulfur*, 1986, **27**, 293.	367
86RTC1	G. W. Klumpp; *Recl. Trav. Chim. Pays-Bas*, 1986, **105**, 1.	554, 556
86RTC66	H. D. Verkruijse and L. Brandsma; *Recl. Trav. Chim. Pays-Bas*, 1986, **105**, 66.	907
86S1	N. Petragnani and J. V. Comasseto; *Synthesis*, 1986, 1.	768
86S122	E. Napolitano, R. Fiaschi and E. Mastorilli; *Synthesis*, 1986, 122.	624
86S198	E. N. Tsvetkov, N. A. Bondarenko, I. G. Malakhova and M. I. Kabachnik; *Synthesis*, 1986, 198.	430, 437
86S222	A. G. Martinez, R. M. Alvarez and A. G. Fraile; *Synthesis*, 1986, 222.	630
86S237	G. Rosini, E. Marotta, R. Ballini and M. Petrini; *Synthesis*, 1986, 237.	395
86S240	Y. Xu and Z. Li; *Synthesis*, 1986, 240.	846
86S297	A. Jonczyk and Z. Owczarczyk; *Synthesis*, 1986, 297.	749
86S419	L. René, J. Poncet and G. Auzou; *Synthesis*, 1986, 419.	749
86S453	K. C. Nicolaou and S. E. Webber; *Synthesis*, 1986, 453.	623, 624
86S480	H. C. Brown, N. G. Bhat and S. Rajagopalan; *Synthesis*, 1986, 480.	623
86S535	R. Öhrlein, W. Schwab, R. Ehrler and V. Jäger; *Synthesis*, 1986, 535.	392
86S633	M. Taddei and A. Ricci; *Synthesis*, 1986, 633.	652
86S674	H. C. Brown, N. G. Bhat and D. Basavaiah; *Synthesis*, 1986, 674.	900
86S691	Y. Xu, J. Xia and H. Guo; *Synthesis*, 1986, 691.	843
86S704	L. B. Volodarsky and A. Ya. Tikhonov; *Synthesis*, 1986, 704.	340
86S757	A. Dondoni, A. R. Mastellari, A. Medici, E. Negrini and P. Pedrini; *Synthesis*, 1986, 757.	942
86S776	H. Sano, Y. Miyazaki, M. Okawara and Y. Ueno; *Synthesis*, 1986, 776.	946
86S821	C. Yuan and Y. Qi; *Synthesis*, 1986, 821.	454
86S826	A. Amrollah-Madjdabadi, R. Beugelmann and A. Lechevallier; *Synthesis*, 1986, 826.	398
86S849	M. Ballini, M. Petrini and G. Rosini; *Synthesis*, 1986, 849.	393, 394
86S852	T. Hamada and O. Yonemitsu; *Synthesis*, 1986, 852.	717
86S868	G. A. Olah, L. Ohannesian and M. Arvanaghi; *Synthesis*, 1986, 868.	620
86S960	J. Barluenga, F. Aznar, R. Liz and M. P. Cabal; *Synthesis*, 1986, 960.	740
86S973	D. S. Matteson; *Synthesis*, 1986, 973.	529
86S1021	W. J. Spillane and P. O. Burke; *Synthesis*, 1986, 1021.	368
86S1024	R. Ballini and M. Petrini; *Synthesis*, 1986, 1024.	395
86SA(A)1107	A. Sebald and B. Wrackmeyer; *Spectrochim. Acta, Part A*, 1986, **42**, 1107.	1097
86SC63	D. Gaude, R. Le Goaller and J. L. Pierre; *Synth. Commun.*, 1986, **16**, 63.	791
86SC283	J. V. Comasseto, C. C. Silveira, J. T. B. Ferreira and V. Catani; *Synth. Commun.*, 1986, **16**, 283.	1036, 1038
86SC565	J. A. Schwartz; *Synth. Commun.*, 1986, **16**, 565.	271
86SC1043	J. F. Dellaria, Jr., C. Nordeen and L. R. Swett; *Synth. Commun.*, 1986, **16**, 1043.	684
86SC1207	T. L. Evans and M. M. Grade; *Synth. Commun.*, 1986, **16**, 1207.	150
86SC1781	R. Ballini and M. Petrini; *Synth. Commun.*, 1986, **16**, 1781.	396
86SRI67	M. K. Das and S. Ray; *Synth. React. Inorg. Metal-Org. Chem.*, 1986, **16**, 67.	474
86SUL1	N. Kamigata, M. Takata, H. Matsuyama and M. Kobayashi; *Sulfur Lett.*, 1986, **5**, 1 (*Chem. Abstr.*, 1988, **108**, 55 999m).	214
86SUL161	E. Wenschuh and R. Ritzel; *Sulfur Letters*, 1986, **4**, 161 (*Chem. Abstr.*, 1986, **107**, 96 775h).	99
86T669	M. Laguerre, J. M. Leger, D. Y. N'Gabé, C. Biran and J. Dunogues; *Tetrahedron*, 1986, **42**, 669.	938, 941
86T1299	M. W. Majchrzak and G. Simchen; *Tetrahedron*, 1986, **42**, 1299.	943
86T1769	S. F. Nelsen, S. C. Blackstock and T. B. Frigo; *Tetrahedron*, 1986, **42**, 1769.	383
86T2703	G. Boche and R. H. Sommerslade; *Tetrahedron*, 1986, **42**, 2703.	340, 357
86T2873	B. H. Lipshutz, S. Nguyen, D. A. Parker, K. E. McCarthy, J. C. Barton, S. Whitney and M. Kotsuki; *Tetrahedron*, 1986, **42**, 2873.	588
86T2931	M. T. Reetz, S. H. Kyung and M. Hüllmann; *Tetrahedron*, 1986, **42**, 2931.	566
86T4007	G. Petrillo, M. Novi, G. Garbarino and C. Dell'erba; *Tetrahedron*, 1986, **42**, 4007.	711
86T4427	K. Wakamatsu, T. Nonaka, Y. Okuda, W. Tückmantel, K. Oshima, K. Utimoto and H. Nozaki; *Tetrahedron*, 1986, **42**, 4427.	918, 923
86T5459	M. Madesclaire; *Tetrahedron*, 1986, **42**, 5459.	144
86T6573	H. B. Kagan and J. L. Namy; *Tetrahedron*, 1986, **42**, 6573.	564
86TCC89	J.-C. Motte and J. Janousek; *Top. Curr. Chem.*, 1986, **130**, 89.	1039, 1050
86TL199	T. Tanaka, S. Tamatsukuri and N. Ikehara; *Tetrahedron Lett.*, 1986, **27**, 199.	409, 467
86TL215	Y. Watanabe, Y. Ueno, T. Araki, T. Endo and M. Okawasa; *Tetrahedron Lett.*, 1986, **27**, 215.	210
86TL441	D. N. Harpp, S. J. Bodzay, T. Aida and T. H. Chan; *Tetrahedron Lett.*, 1986, **27**, 441.	212, 215, 231
86TL493	J.-M. Mélot, F. Texier-Boullet and A. Foucaud; *Tetrahedron Lett.*, 1986, **27**, 493.	394
86TL1039	P. Knochel and J. F. Normant; *Tetrahedron Lett.*, 1986, **27**, 1039.	1006
86TL1153	D. H. R. Barton, D. Bridon and S. Z. Zard; *Tetrahedron Lett.*, 1986, **27**, 1153.	364
86TL1363	Y. Naruse and H. Yamamoto; *Tetrahedron Lett.*, 1986, **27**, 1363.	679
86TL1493	J.-H. Youn and R. Herrmann; *Tetrahedron Lett.*, 1986, **27**, 1493.	173
86TL1499	B. Escoula, I. Rico and A. Lattes; *Tetrahedron Lett.*, 1986, **27**, 1499.	19
86TL1537	M. D. Fryzuk, G. S. Bates and C. Stone; *Tetrahedron Lett.*, 1986, **27**, 1537.	992
86TL1591	W. Ando and T. Takata; *Tetrahedron Lett.*, 1986, **27**, 1591.	146
86TL1607	H. Saimoto, Y. Kusano and T. Hiyama; *Tetrahedron Lett.*, 1986, **27**, 1607.	7

86TL1719	M. Clarembeau and A. Krief; *Tetrahedron Lett.*, 1986, **27**, 1719.	554
86TL1723	M. Clarembeau and A. Krief; *Tetrahedron Lett.*, 1986, **27**, 1723.	554
86TL1751	D. S. Karanewsky and M. C. Badia; *Tetrahedron Lett.*, 1986, **27**, 1751.	456
86TL1853	S. W. Remiszewski, J. Yang and S. M. Weinreb; *Tetrahedron Lett.*, 1986, **27**, 1853.	245
86TL1907	G. Balme, G. Fournet and J. Gore; *Tetrahedron Lett.*, 1986, **27**, 1907.	11
86TL1995	H. J. Bestmann and T. Arenz; *Tetrahedron Lett.*, 1986, 1995.	624
86TL2007	K. Nozaki, K. Wakamatsu, T. Nonaka, W. Tückmantel, K. Oshima and K. Utimoto; *Tetrahedron Lett.*, 1986, **27**, 2007.	903, 919
86TL2099	S. M. Weinreb, D. M Demko and T. A. Lessen; *Tetrahedron Lett.*, 1986, **27**, 2099.	249
86TL2125	T.-A. Mitsudo, Y. Hori, Y. Yamakawa and Y. Watanabe; *Tetrahedron Lett.*, 1986, **27**, 2125.	676
86TL2401	M. Yamaguchi, K. Hasebe and T. Minami; *Tetrahedron Lett.*, 1986, **27**, 2401.	662
86TL2563	G. F. Hambly and T. H. Chan; *Tetrahedron Lett.*, 1986, **27**, 2563.	22
86TL2715	S. T. Purrington, N. V. Lazaridis and C. L. Bumgardner; *Tetrahedron Lett.*, 1986, **27**, 2715.	16
86TL2761	P. J. Garratt and A. Tsotinis; *Tetrahedron Lett.*, 1986, **27**, 2761.	938
86TL2801	B. L. Chenard, C. M. Van Zyl and D. R. Sanderson; *Tetrahedron Lett.*, 1986, **27**, 2801.	916, 932
86TL2829	E. Negishi, F. E. Cederbaum and T. Takahashi; *Tetrahedron Lett.*, 1986, **27**, 2829.	568, 986, 988
86TL3149	C. Petrier, C. Dupuy and J.-L. Luche; *Tetrahedron Lett.*, 1986, **27**, 3149.	591
86TL3251	W. Ando and T. Tsumuruya; *Tetrahedron Lett.*, 1986, **27**, 3251.	947
86TL3275	B. Capon and F.-C. Kwok; *Tetrahedron Lett.*, 1986, **27**, 3275.	639
86TL3515	F. Texier-Boullet and M. Lequitte; *Tetrahedron Lett.*, 1986, **27**, 3515.	457
86TL3615	D. H. R. Barton, J.-P. Finet and J. Khamsi; *Tetrahedron Lett.*, 1986, **27**, 3615.	684
86TL3619	D. H. R. Barton, J.-P. Finet, J. Khamsi and C. Pichon; *Tetrahedron Lett.*, 1986, **27**, 3619.	675
86TL3975	D. N. Harpp and M. Kobayashi; *Tetrahedron Lett.*, 1986, **27**, 3975.	120
86TL4191	Y. Hayakawa; *Tetrahedron Lett.*, 1986, **27**, 4191.	106
86TL4391	L. E. Overman, M. E. Okazaki and P. Mishra; *Tetrahedron Lett.*, 1986, **27**, 4391.	543
86TL4427	P. Knochel and J. F. Normant; *Tetrahedron Lett.*, 1986, **27**, 4427.	1006
86TL4449	G. Haupe, G. Alvernhe and A. Laurent; *Tetrahedron Lett.*, 1986, **27**, 4449.	18
86TL4465	T. Umemotu, K. Kawada and K. Tomita; *Tetrahedron Lett.*, 1986, **27**, 4465.	16
86TL4721	M. T. Reetz, F. Kunisch and P. Heitmann; *Tetrahedron Lett.*, 1986, **27**, 4721.	698
86TL4961	D. A. Evans and E. B. Sjorgen; *Tetrahedron Lett.*, 1986, **27**, 4961.	341
86TL5711	M. T. Reetz, T. Kukemnhohner and P. Weining; *Tetrahedron Lett.*, 1986, **27**, 5711.	568
86TL5727	P. Knochel and J. F. Normant; *Tetrahedron Lett.*, 1986, **27**, 5727.	910
86TL6055	P. E. Eaton and G. T. Cunkle; *Tetrahedron Lett.*, 1986, **27**, 6055.	25
86TL6193	P. Perrin, F. Aubert, J. P. Lellouche and J. P. Beaucourt; *Tetrahedron Lett.*, 1986, **27**, 6193.	753
86TL6257	O. Bartolini, F. DiFuria, G. Licini, G. Modena and M. Rossi; *Tetrahedron Lett.*, 1986, 6257.	147
86ZAAC(533)146	M. Baudler and G. Kupprat *Z. Anorg. Allg. Chem.*, 1986, **533**, 146.	474
86ZAAC(535)47	H. Schmidt, C. Wirkner and K. Isslieb; *Z. Anorg. Allg. Chem.*, 1986, **535**, 47.	467
86ZAAC(535)106	G. Fritz and P. Amann; *Z. Anorg. Allg. Chem.*, 1986, **535**, 106.	473
86ZAAC(539)50	J. Ellermann, M. Moll and L. Brehm; *Z. Anorg. Allg. Chem.*, 1986, **539**, 50.	491
86ZAAC(543)7	J. Hahn and T. Nataniel; *Z. Anorg. Allg. Chem.*, 1986, **543**, 7.	855
86ZN(B)10	E. Lindner and C. Scheytt; *Z. Naturforsch., Teil B*, 1986, **41**, 10.	442
86ZN(B)321	M. Ates, H. J. Breunig, A. Soltani-Neshan and M. Tegeler; *Z. Naturforsch., Teil B*, 1986, **41**, 321.	893, 894
86ZN(B)1129	H. J. Breunig and D. Muller; *Z. Naturforsch., Teil B*, 1986, **41**, 1129.	493
86ZOB898	V. A. Mullin and A. N. Zolotov; *Zh. Obshch. Khim.*, 1986, **56**, 898 (*Chem. Abstr.*, 1987, **106**, 156 209t).	239
86ZOB2794	M. V. Livantsov, A. A. Prishchenko and I. F. Lutsenko; *Zh. Obshch. Khim.*, 1986, **56**, 2794 (*Chem. Abstr.*, 1988, **108**, 6094).	434
86ZOR140	E. Lukevits, R. Y. Sturkovits, O. A. Pudova and A. P. Gaukhman; *Zh. Org. Khim.*, 1986, **56**, 140.	914
87ACR18	O. Meth-Cohn; *Acc. Chem. Res.*, 1987, **20**, 18.	298
87ACR243	Y. Yamamoto; *Acc Chem. Res.*, 1987, **20**, 243.	78
87AG(E)78	G. Klebe, K. H. Böhn, M Marsch and G. Boche; *Angew. Chem., Int. Ed. Engl.*, 1987, **26**, 78.	1002, 1003
87AG(E)229	A. J. Ashe, III, S. Mahmood, C. Elschenbroich and M. Wünsch; *Angew. Chem., Int. Ed. Engl.*, 1987, **26**, 229.	880
87AG(E)264	C. A. McAuliffe, B. Beagley, G. A. Gott, A. G. Mackie, P. P. MacRory and R. G. Pritchard; *Angew. Chem., Int. Ed. Engl.*, 1987, **26**, 264.	881
87AG(E)780	W.-W. du Mont, S. Kubiniok, K. Peters and H.-G. von Schnering; *Angew. Chem., Int. Ed. Engl.*, 1987, **26**, 780.	732
87AG(E)887	M. Schmidt and U. Gorl; *Angew. Chem., Int. Ed. Engl.*, 1987, **26**, 887.	228
87AG(E)923	E.-U. Wurthwein and R. Weigmann; *Angew. Chem., Int. Ed. Engl.*, 1987, **26**, 923.	1053
87AG(E)1187	T. Hiiro, N. Kambe, A. Ogawa, N. Miyoshi, S. Murai and N. Sonoda; *Angew. Chem., Int. Ed. Engl.*, 1987, **26**, 1187	554
87AHC(41)275	H. Weber; *Adv. Heterocycl. Chem.*, 1987, **41**, 275.	1009
87AX(C)313	D. S. C. Yang, J. Pletcher, J. P. Rose, C. S. Yoo, W. Furey, B.-C. Wang and M. Sax; *Acta Crystallogr., Part C*, 1987, **43**, 313.	1009
87BCJ395	S. Itsuno, Y. Sakurai, K. Ito, A. Hirao and S. Nakahama; *Bull. Chem. Soc. Jpn.*, 1987, **60**, 395.	65, 528
87BCJ777	N. Shimizu, F. Shibata and Y. Tsuno; *Bull. Chem. Soc. Jpn.*, 1987, **60**, 777.	907, 908
87BCJ809	H. Yonemura, H. Nishino and K. Kurosawa; *Bull. Chem. Soc. Jpn.*, 1987, **60**, 809.	22

87BCJ1159	S. Kajigaeshi, T. Kakinami, T. Okamoto and S. Fujisaki; *Bull. Chem. Soc. Jpn.*, 1987, **60**, 1159.	28
87BCJ2289	K. Nishiyama and M. Oba; *Bull. Chem. Soc. Jpn.*, 1987, **60**, 2289.	544
87BCJ3083	A. Gregorcic and M. Zupan; *Bull. Chem. Soc. Jpn.*, 1987, **60**, 3083.	18
87BCJ3468	Y. Ichinose, H. Oda, K. Oshima and K. Utimoto; *Bull. Chem. Soc. Jpn.*, 1987, **60**, 3468.	946
87BCJ4187	S. Kajigaeshi, T. Kakinami, T. Okamoto, H. Nakamura and M. Fujikawa; *Bull. Chem. Soc. Jpn.*, 1987, **60**, 4187.	626
87BCJ4363	N. Fukuyama, H. Nishino and K. Kurosawa; *Bull. Chem. Soc. Jpn.*, 1987, **60**, 4363.	631
87BCJ4492	T. Tréfouel, P. Tintillier, G. Dupas, J. Bourguignon and G. Quéguiner; *Bull. Chem. Soc. Jpn.*, 1987, **60**, 4492.	398
87BSF181	J.-B. Baudin, S. A. Julia and R. Lorne; *Bull. Soc. Chim. Fr.*, 1987, 181.	359
87CAR(168)7	J. R. Paddison, Jr., T. P. Haromy, M. Sundaralingam, R. W. Myers and J. N. BeMiller; *Carbohydr. Res.*, 1987, **168**, 7.	1003
87CB1281	H. Schmidbauer and P. Musstein; *Chem. Ber.*, 1987, **120**, 1281.	878
87CB1347	T. Ziegler, M. Layh and F. Effenberger; *Chem. Ber.*, 1987, **120**, 1347.	665
87CB1437	A. Meller, W. Maringgle, G. Elter, D. Bromm, M. Noltemeyer and G. M. Sheldrick; *Chem. Ber.*, 1987, **120**, 1437.	1085
87CB1707	K. P. Langhans and O. Stelzer; *Chem. Ber.*, 1987, **120**, 1707.	429
87CC495	S. Acyutha Rao and M. Periasamy; *J. Chem. Soc., Chem. Commun.*, 1987, 495.	969
87CC509	W. Carruthers and R. C. Moses; *J. Chem. Soc., Chem. Commun.*, 1987, 509.	326
87CC675	R. M. Moriarty, J. S. Khosrowshahi and T. M. Dalecki; *J. Chem. Soc., Chem. Commun.*, 1987, 675.	13
87CC1013	K. B. Yoon and J. K. Kochi; *J. Chem. Soc., Chem. Commun.*, 1987, 1013.	31
87CC1025	E. Piers and R. T. Skerlj; *J. Chem. Soc., Chem. Commun.*, 1987, 1025.	947
87CC1110	M. Ke, S. J. Rettig, B. R. James and D. Dolphin; *J. Chem. Soc., Chem. Commun.*, 1987, 1110.	583
87CC1195	J. Barluenga, M. Tomas, A. Ballesteros and V. Gotor; *J. Chem. Soc., Chem. Commun.*, 1987, 1195.	920
87CC1213	J. Park and C. S. Chin; *J. Chem. Soc., Chem. Commun.*, 1987, 1213.	639
87CC1405	K. A. Jørgensen; *J. Chem. Soc., Chem. Commun.*, 1987, 1405.	398
87CC1409	J. H. Clark and C. W. Jones; *J. Chem. Soc., Chem. Commun.*, 1987, 1409.	632
87CC1491	J. Barluenga, M. A. Rodriguez, P. J. Campos and G. Asensio; *J. Chem. Soc., Chem. Commun.*, 1987, 1491.	31, 35
87CC1517	A. Hosomi, S. Kohra and Y. Tominaga; *J. Chem. Soc., Chem. Commun.*, 1987, 1517.	77
87CC1577	E. H. M. Abd Elall, M. I. Ashmawy and J. M. Mellor; *J. Chem. Soc., Chem. Commun.*, 1987, 1577.	24
87CC1708	M. Ochiai, M. Kunishima, Y. Nagao, K. Fuji and E. Fujita; *J. Chem. Soc., Chem. Commun.*, 1987, 1708.	1062
87CE551	K. Matsumoto, T. Uchida, T. Okamoto and S. Halshimoto; *Chem. Express*, 1987, **2**, 551.	753
87CJC78	E. Piers, H. E. Morton and J. M. Chong; *Can. J. Chem.*, 1987, **65**, 78.	600
87CJC412	B. Doboszewski, G. W. Hay and W. A. Szarek; *Can. J. Chem.*, 1987, **65**, 412.	9
87CJC441	N. Banait, M. Hojatti, P. Findlay and A. J. Kresge; *Can. J. Chem.*, 1987, **65**, 441.	1022
87CJC1719	A. D. Allen, A. J. Kresge, N. P. Schepp and T. T. Tidwell; *Can. J. Chem.*, 1987, **65**, 1719.	1022
87CJC2000	H. R. Viturro, B. E. Rivero, O. E. Piro, J. A. Caram, M. E. Martins and C. M. Marschoff; *Can. J. Chem.*, 1987, **65**, 2000.	1008
87CL191	H. Ohta, K. Ozaki and T. Tsuchihashi; *Chem. Lett.*, 1987, 191.	398
87CL293	T. Yamada, H. Suzuki and T. Mukaiyama; *Chem. Lett.*, 1987, 293.	701
87CL357	K. Okuma, Y. Yamasaki, T. Komiya, Y. Komiya, Y. Kodera and H. Ohta; *Chem. Lett.*, 1987, 357.	853
87CL623	T. Takahashi, D. R. Swanson, and E. Negishi; *Chem. Lett.*, 1987, 623.	986, 987
87CL627	S. Kajigaeshi, T. Kakinami, H. Tokiyama, T. Hirakawa and T. Okamoto; *Chem. Lett.*, 1987, 627.	626
87CL715	N. Takana, A. Yamaguchi, Y. Araki and M. Araki; *Chem. Lett.*, 1987, 715.	35
87CL1187	T. Ibata and Y. Isogami; *Chem. Lett.*, 1987, 1187.	753
87CL1275	N. Kanabe, I. Inagaki, N. Miyoshi, A. Ogawa and N. Sonoda; *Chem. Lett.*, 1987, 1275.	311
87CL1647	Y. Ichinose, K. Wakamatsu, K. Nozaki, J.-L. Birbaum, K. Oshima and K. Utimoto; *Chem. Lett.*, 1987, 1647.	710
87CL1675	N. Yoneda, T. Fukuhara, K. Yamagishi and A. Suzuki; *Chem. Lett.*, 1987, 1675.	19
87CL1901	M. Kodomori, S. Takahashi and S. Yoshitomi; *Chem. Lett.*, 1987, 1901.	620
87CL1915	T. Kato, K. Kobayashi, S. Masuda, M. Segi, T. Nakajima and S. Suda; *Chem. Lett.*, 1987, 1915.	453
87CL2109	S. Kajigaeshi, T. Kakinami, H. Yamasaki, S. Fujisaki, M. Kendo and T. Okamoto; *Chem. Lett.*, 1987, 2109.	631
87CL2221	K. Takagi; *Chem. Lett.*, 1987, 2221.	711
87CPB3845	Y. Morimoto and K. Achiwa; *Chem. Pharm. Bull.*, 1987, **35**, 3845.	331
87CS477	A. R. Katritzky, B. Nowak-Wydra and C. M. Marson; *Chem. Scr.*, 1987, **27**, 477.	7
87CSR381	J. Mann; *Chem. Soc. Rev.*, 1987, **16**, 381.	15
87G83	P. Mencarelli and F. Stegel; *Gazz. Chim. Ital.*, 1987, **117**, 83.	28
87GEP3543567	U. Deschlker, S. Wolff, P. Kleinschmit and R. Michel; *Ger. Pat.* 3 543 567 (1987) (*Chem. Abstr.*, 1987, **107**, 96 885u).	166
87GEP3612629	F. W. Reier, P. Wolfram and H. Schumann; *Ger. Pat.* 3 612 629 (1987) (*Chem. Abstr.*, 1988, **108**, 186 988).	434
87H(26)65	S.-I. Nakatsuka, K. Ueda, O. Asano and T. Goto; *Heterocycles*, 1987, **26**, 65.	657

87H(26)929 M. Wada, A. Aiura and K. Akiba; *Heterocycles*, 1987, **26**, 929. 325

87H(26)2583 K. Harano, I. Shinohara, M. Murase and T. Hisano; *Heterocycles*, 1987, **26**, 2583. 123

87IC1638 V. K. Gupta, L. K. Krannich and C. L. Watkins; *Inorg. Chem.*, 1987, **26**, 1638. 507, 894

87IC4065 A. M. Arif, D. E. Heaton, R. A. Jones, K. B. Kidd, T. C. Wright, B. R. Whittlesey, J. L. Atwood, W. E. Hunter and H. Zhang; *Inorg. Chem.*, 1987, **26**, 4065. 512

87IZV2395 S. I. Zav'yalov, O. V. Dorofeeva, E. E. Rumyantseva, I. V. Sitkareva and A. G. Zavozin; *Izv. Akad. Nauk SSSR, Ser. Khim*, 1987, 2395. 28

87JA203 M. E. Thompson, S. M. Baxter, A. R. Bulls, B. J. Burger, M. C. Nolan, B. D. Santarsiero, W. P. Schaefer and J. E. Bercaw; *J. Am. Chem. Soc.*, 1987, **109**, 203. 969

87JA228 P. J. Stang, B. W. Surber, Z. C. Chen, K. A. Roberts and A. G. Anderson; *J. Am. Chem. Soc.*, 1987, **109**, 228. 1020, 1028

87JA813 J. K. Stille and B. L. Groh; *J. Am. Chem. Soc.*, 1987, **109**, 813. 932

87JA882 F. Effenberger, F. Reisinger, K. H. Schönwälder, P. Bäuerler, J. J. Stezowski, K.-H. Jogun, K. Schöllkopf and W. D. Stohrer; *J. Am. Chem. Soc.*, 1987, **109**, 882. 1006, 1009

87JA931 M. P. Cooke, Jr. and R. K. Widener; *J. Am. Chem. Soc.*, 1987, **109**, 931. 525

87JA1227 W. G. Bentrude, T. Kawashima, B. A. Keys, M. Garraussiam, W. Heide and D. A. Wedegaertner; *J. Am. Chem. Soc.*, 1987, **109**, 1227. 256

87JA2040 R. K. Dieter and M. Tokles; *J. Am. Chem. Soc.*, 1987, **109**, 2040. 588

87JA2112 E. B. Nadler and Z. Rappoport; *J. Am. Chem. Soc.*, 1987, **109**, 2112. 645

87JA2393 E. Negishi, T. Takahashi, S. Baba, D. E. Van Horn and N. Okukado; *J. Am. Chem. Soc.*, 1987, **109**, 2393. 992

87JA2435 X. Creary, M. E. Mehrsheikh-Mohammadi and M. D. Eggers; *J. Am. Chem. Soc.*, 1987, **109**, 2435. 28

87JA2541 M. M. Olmstead, P. P. Power, K. J. Weese and R. J. Doedens; *J. Am. Chem. Soc.*, 1987, **109**, 2541. 1005

87JA2544 S. L. Buchwald, B. T. Watson and J. C. Huffman; *J. Am. Chem. Soc.*, 1987, **109**, 2544. 986, 987

87JA2547 K. Nozaki, K. Oshima and K. Utimoto; *J. Am. Chem. Soc.*, 1987, **109**, 2547. 923

87JA2717 D. L. Boger and R. S. Coleman; *J. Am. Chem. Soc.*, 1987, **109**, 2717. 655

87JA2788 W. A. Nugent, D. L. Thorn and R. L. Harlow; *J. Am. Chem. Soc.*, 1987, **109**, 2788. 926, 927, 988

87JA2824 J. C. Fishbein and R. A. McClelland; *J. Am. Chem. Soc.*, 1987, **109**, 2824. 649

87JA3010 T. Kametani, K. Kawamura and T. Honda; *J. Am. Chem. Soc.*, 1987, **109**, 3010. 1000

87JA3163 M. Newcomb and T. M. Deeb; *J. Am. Chem. Soc.*, 1987, **109**, 3163. 323

87JA4000 Y. Chiang, M. Hojatti, J. R. Keeffe, A. J. Kresge, N. P. Schepp and J. Wirz; *J. Am. Chem. Soc.*, 1987, **109**, 4000. 637

87JA4411 C. Couret, J. Escudié, J. Satgé and M. Lazraq; *J. Am. Chem. Soc.*, 1987, **109**, 4411. 949

87JA4749 L. K. Montgomery, M. P. Grendze and J. C. Huffman; *J. Am. Chem. Soc.*, 1987, **109**, 4749. 1007

87JA5478 A. M. Echavarren and J. K. Stille; *J. Am. Chem. Soc.*, 1987, **109**, 5478. 931

87JA5551 E. J. Corey, R. K. Bakshi and S. Shibata; *J. Am. Chem. Soc.*, 1987, **109**, 5551. 528

87JA5553 D. V. Pratt and P. B. Hopkins; *J. Am. Chem. Soc.*, 1987, **109**, 5553. 642

87JA5856 R. Noyori, T. Okhunia, M. Kitamura and S. Akutagawa; *J. Am. Chem. Soc.*, 1987, **109**, 5856. 53

87JA5867 A. T. Lynch and L. G. Sneddon; *J. Am. Chem. Soc.*, 1987, **109**, 5867. 902

87JA6097 C. Flann, T. C. Malone and L. E. Overman; *J. Am. Chem. Soc.*, 1987, **109**, 6097. 328, 329

87JA6115 C. J. Flann and L. E. Overman; *J. Am. Chem. Soc.*, 1987, **109**, 6115. 331

87JA7194 S. Singh, D. D. DesMarteau, S. S. Zuberi, M. Witz and H.-N. Huang; *J. Am. Chem. Soc.*, 1987, **109**, 7194. 613

87JA7230 G. Mehta and S. Padma; *J. Am. Chem. Soc.*, 1987, **109**, 7230. 12, 28

87JA7495 P. Magnus and D. P. Becker; *J. Am. Chem. Soc.*, 1987, **109**, 7495. 926

87JA7568 B. L. Groh, G. R. Magrum and T. J. Barton; *J. Am. Chem. Soc.*, 1987, **109**, 7568. 1023

87JA7838 J. B. Lambert, G. Wang, R. B. Finzel and D. H. Teramura; *J. Am. Chem. Soc.*, 1987, **109**, 7838. 70

87JA7925 E. J. Corey, R. K. Bakshi, S. Shibata, C. P. Chen and V. K. Singh; *J. Am. Chem. Soc.*, 1987, **109**, 7925. 528

87JA8056 E. Nakamura, S. Aoki, K. Sekiya, H. Oshino and I. Kuwajima; *J. Am. Chem. Soc.*, 1987, **109**, 8056. 69

87JCS(P1)1039 R. N. Butler, P. D. O'Shea and D. P. Shelly; *J. Chem. Soc., Perkin Trans 1*, 1987, 1039. 632

87JCS(P1)1107 J. S. Davies, E. J. Tremeer and R. C. Treadgold; *J. Chem. Soc., Perkin Trans. 1*, 1987, 1107. 540

87JCS(P1)2157 P. Tundo, P. Venturello and E. Angeletti; *J. Chem. Soc., Perkin Trans. 1*, 1987, 2157. 25, 31

87JCS(P2)1275 M. Acheson and P. J. Ansell; *J. Chem. Soc., Perkin Trans.* 2, 1987, 1275. 1034

87JFC(34)337 M. Maeda, M. Abe and M. Kojima; *J. Fluorine Chem.*, 1987, **34**, 337. 18

87JFC(36)373 M. Hudlicky; *J. Fluorine Chem.*, 1987, **36**, 373. 10

87JGU2482 E. T. Bogoradovskii, N. V. Gribasova, V. S. Zavgorodnii and A. A. Petrov; *J. Gen. Chem. USSR* (*Engl. Transl.*), 1987, **56**, 2482. 1090

87JGU2496 G. I. Kokorev, F. D. Yambushev and Sh. Kh. Badrutdinov; *J. Gen. Chem. USSR* (*Engl. Transl.*), 1987, **57**, 2496. 885, 890

87JGU2497 Z. U. Panfilovich, N. R. Ivanova and K. I. Kuz'min; *J. Gen. Chem. USSR* (*Engl. Transl.*), 1987, **57**, 2497 1079

87JHC767 R. S. Varma, Y.-Z. Gai and G. W. Kabalka; *J. Heterocycl. Chem.*, 1987, **24**, 767. 341

87JOC165 S. E. Denmark, R. P. Hammer, E. J. Weber and K. L. Habermas; *J. Org. Chem.*, 1987, **52**, 165. 539

87JOC307 B. B. Snider and Y. S. Kulkarni; *J. Org. Chem.*, 1987, **52**, 307. 22

87JOC310 H. C. Brown, J. V. N. Vara Prasad, A. K. Gupta and R. K. Bakshi; *J. Org. Chem.*, 1987, **52**, 310. 519, 531

87JOC322	J. A. Miller; *J. Org. Chem.*, 1987, **52**, 322.	657
87JOC476	I. M. Lazbin and G. F. Koser; *J. Org. Chem.*, 1987, **52**, 476.	320
87JOC564	K. Fischer and S. Hünig; *J. Org. Chem.*, 1987, **52**, 564.	999
87JOC658	D. Y. Chi, M. R. Kilbourn, J. A. Katzenellenbogen and M. J. Welch; *J. Org. Chem.*, 1987, **52**, 658.	9
87JOC664	S. M. Moerlein; *J. Org. Chem.*, 1987, **52**, 664.	949
87JOC674	K. A. Parker and K. A. Koziski; *J. Org. Chem.*, 1987, **52**, 674.	653
87JOC691	S. H. Yang, C. S. Li and C. H. Cheng; *J. Org. Chem.*, 1987, **52**, 691.	632
87JOC701	G. A. Olah and C. Rochin; *J. Org. Chem.*, 1987, **52**, 701.	392
87JOC728	H. C. Brown, M. V. Rangaishenvi and U. S. Racherla; *J. Org. Chem.*, 1987, **52**, 728.	535
87JOC753	M. Yasuda, T. Yamashita, K. Shima and C. Pac; *J. Org. Chem.*, 1987, **52**, 753.	752
87JOC798	K. R. Brower; *J. Org. Chem.*, 1987, **52**, 798.	15
87JOC919	S. Stavber and M. Zupan; *J. Org. Chem.*, 1987, **52**, 919.	18
87JOC1043	G. Lunn; *J. Org. Chem.*, 1987, **52**, 1043.	327
87JOC1078	H. Kosugi, M. Kitaoka, K. Tagami, A. Takahashi and H. Uda; *J. Org. Chem.*, 1987, **52**, 1078.	713
87JOC1100	K. Tamao, M. Akita, K. Maeda and M. Kumada; *J. Org. Chem.*, 1987, **52**, 1100.	630
87JOC1276	K. Ikenaga K. Kikukawa and T. Matsuda; *J. Org. Chem.*, 1987, **52**, 1276.	932
87JOC1347	J. Hooz, J. Oudenes, J. L. Roberts and A. Benderley; *J. Org. Chem.*, 1987, **52**, 1347.	533, 697
87JOC1397	M. Hori, T. Kataoka, H. Shimizu and K. Tsutsumi; *J. Org. Chem.*, 1987, **52**, 1397.	288
87JOC1483	P. Magnus, L. M. Principe and M. J. Slater; *J. Org. Chem.*, 1987, **52**, 1483.	926
87JOC1601	D. E. Bergbreiter and J. J. Lalonde; *J. Org. Chem.*, 1987, **52**, 1601.	395
87JOC1619	M. P. Doyle, A. H. Devia, K. E. Bassett, J. W. Terpstra and S. N. Mahapatro; *J. Org. Chem.*, 1987, **52**, 1619.	619
87JOC1956	H. Iida, N. Yamazaki and C. Kibayashi; *J. Org. Chem.*, 1987, **52**, 1956.	299
87JOC2162	E. H. White and H. M. Lim; *J. Org. Chem.*, 1987, **52**, 2162.	176
87JOC2170	C. R. Johnson and T. Imamoto; *J. Org. Chem.*, 1987, **52**, 2170.	453
87JOC2292	J. C. Bottaro, R. J. Schmitt and C. D. Bedford; *J. Org. Chem.*, 1987, **52**, 2292.	386
87JOC2294	R. J. Schmitt, J. C. Bottaro, R. Malhotra and C. D. Bedford; *J. Org. Chem.*, 1987, **52**, 2294.	1053
87JOC2303	F. Charbonnier, A. Moyano and A. E. Greene; *J. Org. Chem.*, 1987, **52**, 2303.	682
87JOC2332	S. Raucher and B. L. Bray; *J. Org. Chem.*, 1987, **52**, 2332.	1025
87JOC2598	J. M. Klunder and K. B. Sharpless; *J. Org. Chem.*, 1987, **52**, 2598.	721
87JOC2619	A. Razzuk and E. R. Biehl; *J. Org. Chem.*, 1987, **52**, 2619.	753
87JOC2631	S. H. Cho and L. S. Liebeskind; *J. Org. Chem.*, 1987, **52**, 2631.	672
87JOC2769	S. Rozen and C. Gal; *J. Org. Chem.*, 1987, **52**, 2769.	15
87JOC2911	S. K. Danek, D. P. Kelly, A. K. Serelis and P. J. Steel; *J. Org. Chem.*, 1907, **52**, 2911.	404
87JOC2919	A. Moyano, F. Charbonnier and A. E. Greene; *J. Org. Chem.*, 1987, **52**, 2919.	676, 1026
87JOC3018	G. Castaldi, S. Cavicchioli, C. Giordano and F. Uggeri; *J. Org. Chem.*, 1987, **52**, 3018.	29
87JOC3062	D. J. Pasto, R. Krasnansky and C. Zercher; *J. Org. Chem.*, 1987, **52**, 3062.	1010
87JOC3154	O. Ploux, M. Caruso, G. Chassaing and A. Marquet; *J. Org. Chem.*, 1987, **52**, 3154.	229
87JOC3674	T. P. Burns and R. D. Rieke; *J. Org. Chem.*, 1987, **52**, 3674.	558, 559
87JOC3687	R. N. Hanson and H. El-Wakil; *J. Org. Chem.*, 1987, **52**, 3687.	630
87JOC3919	C.-K. Sha, J.-J. Young and T.-S. Jean; *J. Org. Chem.*, 1987, **52**, 3919.	34
87JOC4014	C. T. Goralski and B. Singaram; *J. Org. Chem.*, 1987, **52**, 4014.	514
87JOC4020	H. C. Brown, W. S. Park and B. T. Cho; *J. Org. Chem.*, 1987, **52**, 4020.	535
87JOC4086	L. Engman; *J. Org. Chem.*, 1987, **52**, 4086.	616
87JOC4124	J. Einhorn and J.-L. Luche; *J. Org. Chem.*, 1987, **52**, 4124.	422
87JOC4307	S. T. Purrington, C. L. Bumgardner, N. V. Lazaridis and P. Singh; *J. Org. Chem.*, 1987, **52**, 4307.	16
87JOC4410	T. Okazoe, K. Takai, K. Oshima and K. Utimoto; *J. Org. Chem.*, 1987, **52**, 4410.	681
87JOC4418	H. Ochai, Y. Tamura, K. Tsubaki and Z. Yoshida; *J. Org. Chem.*, 1987, **52**, 4418.	69
87JOC4620	D. R. Graber, R. A. Morge and J. C. Shy; *J. Org. Chem.*, 1987, **52**, 4620.	218
87JOC4644	S. Itsuno, G. D. Darling, H. D. H. Stöver and J. M. T. Fréchet; *J. Org. Chem.*, 1987, **52**, 4644.	560
87JOC4760	A. Padwa, M. W. Wannamaker and A. D. Dyszlewski; *J. Org. Chem.*, 1987, **52**, 4760.	928
87JOC4796	G. Picotin and P. Migniac; *J. Org. Chem.*, 1987, **52**, 4796.	69
87JOC4831	J. Garcia, B. M. Kim and S. Masamune; *J. Org. Chem.*, 1987, **52**, 4831.	521
87JOC4846	E. Keinan and D. Perez; *J. Org. Chem.*, 1987, **52**, 4846.	36
87JOC4864	K. Takeshita, Y. Seki, K. Kawamoto, S. Murai and N. Sonoda; *J. Org. Chem.*, 1987, **52**, 4864.	912
87JOC4868	T. N. Mitchell, R. Wickenkamp, A. Amamria, R. Dicke and U. Schneider; *J. Org. Chem.*, 1987, **52**, 4868.	910, 911, 916, 917, 932, 940, 946
87JOC4928	S. Rozen and C. Gal; *J. Org. Chem.*, 1987, **52**, 4928.	15
87JOC4999	S. D. Toto and J. T. Doi; *J. Org. Chem.*, 1987, **52**, 4999.	7
87JOC5205	H.-D. Becker and K. Andersson; *J. Org. Chem.*, 1987, **52**, 5205.	646
87JOC5283	D. L. Boger and D. Yohannes; *J. Org. Chem.*, 1987, **52**, 5283.	655
87JOC5320	J.-F. Peyronel, O. Samuel and J.-C. Fiaud; *J. Org. Chem.*, 1987, **52**, 5320.	413
87JOC5466	A. Thurkauf, P. Hillery, A. E. Jacobsen and K. C. Rice; *J. Org. Chem.*, 1987, **52**, 5466.	33
87JOC5560	F. Vogtle, N. Eisen, S. Franken, P. Bullesbach and H. Puff; *J. Org. Chem.*, 1987, **52**, 5560.	31
87JOC5624	C. J. Tu and D. Lednicer; *J. Org. Chem.*, 1987, **52**, 5624.	28
87JOM(320)339	S. R. Iyer, D. R. Tueting and N. E. Schore; *J. Organomet. Chem.*, 1987, **320**, 339.	439

87JOM(322)177 C. Einhorn and J.-L. Luche; *J. Organomet. Chem.*, 1987, **322**, 177. 74

87JOM(323)C1 P. Brooks, D. C. Craig, M. J. Gallagher, A. D. Roe and A. Saroff; *J. Organomet. Chem.*, 1987, **323**, C1. 432

87JOM(325)159 J. M. Wallis, G. Muller and H. Schmidbaur; *J. Organomet. Chem.*, 1987, **325**, 159. 512

87JOM(325)C7 R. L. Wells, A. P. Purdy, K. T. Higa, A. T. McPhail and C. G. Pitt; *J. Organomet. Chem.*, 1987, **325**, C7. 511

87JOM(326)C6 P. Jutzi and U. Meyer; *J. Organomet. Chem.*, 1987, **326**, C6. 866

87JOM(328)81 J. L. Namy, J. Collin, J. Zhang and H. B. Kagan; *J. Organomet. Chem.*, 1987, **328**, 81. 565

87JOM(328)C17 G. Cerveau, C. Chuit, R. J. P. Corriu and C. Reye; *J. Organomet. Chem.*, 1987, **328**, C17. 77

87JOM(332)247 K. Mocheda and T. Yamanishi; *J. Organomet. Chem.*, 1987, **332**, 247. 563

87JOM(334)323 D. Naumann and W. Tyrra; *J. Organomet. Chem.*, 1987, **334**, 323. 483, 873

87JOM(334)369 H. Brunner and H. Leyerer; *J. Organomet. Chem.*, 1987, **334**, 369. 435

87JOM(335)59 E. Lindner, S. Meyer, P. Wegner, B. Karle, A. Sickinger and B. Steger; *J. Organomet. Chem.*, 1987, **335**, 59. 437

87JOM(336)337 L.-R. Frank, I. Jibril, L. Zsolnai and G. Huttner; *J. Organomet. Chem.*, 1987, **336**, 337. 875, 896

87JOU230 L. S. Boguslavskaya, N. N. Chuvatkin, A. V. Kartashov and L. A. Ternovskoi; *J. Org. Chem. USSR* (*Engl. Transl.*), 1987, 230. 18

87JOU1417 N. A. Nesmeyanov, V. G. Kharitonov, S. T. Zhuzhlikova, P. V. Petrovskii, M. Yu. Antipin, Yu. T. Stuchkov and O. A. Reutov; *J. Org. Chem. USSR* (*Engl. Transl.*), 1987, **23**, 1417. 878

87LA215 R. W. Hoffman and A. Endesfelder; *Liebigs Ann. Chem.*, 1987, 215. 529

87LA583 B. Rugewitz-Blackholm and M. Wießler; *Liebigs Ann. Chem.*, 1987, 583. 385

87LA617 F. Latif, A. Malik and W. Voelter; *Liebigs Ann. Chem.*, 1987, 617. 9

87LA881 R. W. Hoffman and R. Metternich; *Liebigs Ann. Chem.*, 1987, 881. 529

B-87MI 202-01 F. R. Hartley (ed.); "The Chemistry of the Carbon Metal Bond," Wiley, New York, 1987, vol. 4. 67

B-87MI 203-01 D. Liotta (ed.); "Organoselenium Chemistry," Wiley, New York, 1987. 234

B-87MI 204-01 T. G. Back; in "Organoselenium Chemistry," ed. D. Liotta, Wiley, New York, 1987, pp. 1–125. 280

B-87MI 204-02 K. C. Nicolaou, N. A. Petasis and D. A. Claremon; in "Organoselenium Chemistry," ed. D. Liotta, Wiley, New York, 1987, pp. 127–162. 286

B-87MI 204-03 W.-W. DuMont, R. Hensel, S. Kubiniok and L. Lange, in "The Chemistry of Organic Selenium and Tellurium Compounds," ed. S. Patai, Wiley, Chichester, 1987, vol. 2, ch. 15. 294

B-87MI 206-01 D. L. Boger and S. M. Wienreb; in "Hetero Diels Alder Methodology in Organic Synthesis," Academic Press, San Diego, 1987, p. 71. 354, 369

B-87MI 206-02 G. Kirsch and L. Christiaens; in "The Chemistry of Organic Selenium and Tellurium Compounds," ed. S. Patai, Wiley, Chichester, 1987, Vol. 2, p. 424. 368

87MI 208-01 W. Ye, G. Sun and C. Yuan; *Huaxue Xuebao*, 1987, **45**, 622 (*Chem. Abstr.*, 1988, **108**, 221 784). 456

87MI 210-01 M. Srebnik and P. V. Ramachandran; *Aldrichemica Acta.*, 1987, **20**, 9. 514, 515, 516, 517, 519, 520, 523, 526, 531

B-87MI 210-02 E. W. Colvin; in "Chemistry of the Metal-Carbon Bond," eds S. Patai and F. R. Hartley, Wiley, Chichester, 1987, vol. 4, p. 540.

B-87MI 211-01 J. L. Wardell; in "The Chemistry of the Metal–Carbon Bond," ed. F. R. Hartley, series ed. S. Patai, Wiley, New York, 1987, vol. 4, p. 1. 551, 556

B-87MI 211-02 L. Brandsma and M. Verkruijsse; "Preparative Polar Organometallic Chemistry," Springer Verlag, Berlin, 1987. 551

B-87MI 211-03 C. L. Raston and G. Salem; in "The Chemistry of the Metal–Carbon Bond," ed. F. R. Hartley, series ed. S. Patai, Wiley, New York, 1987, vol. 4, p. 159. 557

B-87MI 211-04 D. Astruc; in "The Chemistry of the Metal–Carbon Bond," ed F. R. Hartley, series ed. S. Patai, Wiley, New York, 1987, vol. 4, p. 625. 575

B-87MI 211-05 P. A. Chaloner; in "The Chemistry of the Metal–Carbon Bond," ed. F. R. Hartley, series ed. S. Patai, Wiley, New York, 1987, vol. 4, p. 411. 595

B-87MI 211-06 S. Ueamura; in "The Chemistry of the Metal–Carbon Bond," ed. F. R. Hartley, series ed. S. Patai, Wiley, New York, 1987, vol. 4, p. 473. 598

B-87MI 211-07 M. Pereyre, J.-P. Quintard and A. Rahm; "Tin in Organic Synthesis," Butterworths, London, 1987. 599

B-87MI 218-01 L. Brandsma and H. D. Verkruijsse; "Preparative Polar Organometallic Chemistry," Springer, Berlin, 1987. 907, 908, 909, 939, 942, 946

B-87MI 219-01 R. Setton; in "Preparative Chemistry Using Supported Reagents," Academic Press, New York, 1987. 959

B-87MI 219-02 J. P. Collman, L. S. Hegedus, J. R. Norton and R. G. Finke; "Principles and Applications of Organotransition Metal Chemistry," University Science Books, Mill Valley, CA, 1987, p. 119. 956

B-87MI 219-03 J. P. Collman, L. S. Hegedus, J. R. Norton and R. G. Finke; "Principles and Applications of Organotransition Metal Chemistry," University Science Books, Mill Valley, CA, 1987, p. 279. 960

B-87MI 219-04 J. P. Collman, L. S. Hegedus, J. R. Norton and R. G. Finke; "Principles and Applications of Organotransition Metal Chemistry," University Science Books, Mill Valley, CA, 1987, p. 613. 984

B-87MI 219-05	J. P. Collman, L. S. Hegedus, J. R. Norton and R. G. Finke; "Principles and Applications of Organotransition Metal Chemistry," University Science Books, Mill Valley, CA, 1987, p. 728.	964
87MI 221-01	S. V. Ponomarev and I. F. Lutsenko; *Vestn. Mosk. Univ., Ser.* 2: *Khim.*, 1987, **28**, 3.	1027
B-87MI 223-01	H. Bauer, J. Faust, R. Froböse, J. Füssel, U. Küerke, M. Kunz and H. M. Sommer; in "Gmelin Handbook of Inorganic Chemistry, Organocopper Compounds," ed. J. Füssel, Part 4, Springer Verlag Berlin, 1987, p. 1.	1094, 1095
B-87MI 223-02	H. Bauer, J. Faust, R. Froböse, J. Füssel, U. Küerke, M. Kunz and H. M. Sommer; in "Gmelin Handbook of Inorganic Chemistry, Organocopper Compounds," ed. J. Füssel, Part 4, Springer Verlag Berlin, 1987, p. 40.	1094
B-87MI 223-03	H. Bauer, J. Faust, R. Froböse, J. Füssel, U. Küerke, M. Kunz and H. M. Sommer; in "Gmelin Handbook of Inorganic Chemistry, Organocopper Compounds," ed. J. Füssel, Part 4, Springer Verlag Berlin, 1987, p. 102.	1094
B-87MI 223-04	H. Bauer, J. Faust, R. Froböse, J. Füssel, U. Küerke, M. Kunz and H. M. Sommer; in "Gmelin Handbook of Inorganic Chemistry, Organocopper Compounds," ed. J. Füssel, Part 4, Springer Verlag Berlin, 1987, p. 152.	1094
B-87MI 223-05	M. Pereyre, J.-P. Quintard and A. Rahm; "Tin in Organic Synthesis," Butterworth, London, 1987, p. 25.	
87OM629	H. C. Brown and M. Srebnik; *Organometallics*, 1987, **6**, 629.	1085
87OM1185	A. J. Ashe, III, C. M. Kausch and O. Eisenstein; *Organometallics*, 1987, **6**, 1185.	893
87OM2053	K. H. den Haan, Y. Wielstra and J. H. Teuben; *Organometallics*, 1987, **6**, 2053.	981
87OM2069	M. M. Hansen, P. A. Bartlett and C. H. Heathcock; *Organometallics*, 1987, **6**, 2069.	702
87OM2354	B. Hessen, T. H. Lemmen, H. J. G. Luttikhede, J. H. Teuben, J. L. Petersen, J. C. Huffman, S. Jagner and K. G. Caulton; *Organometallics*, 1987, **6**, 2354.	569
87OM2568	S. Inoue and Y. Sato; *Organometallics*, 1987, **6**, 2568.	700
87OPP283	G. W. Kabalka and R. S. Varma; *Org. Prep. Proced. Int.*, 1987, **19**, 283.	393, 398, 784
87OPP466	V. Nair and S. Bernstein; *Org. Prep. Proced. Int.*, 1987, **19**, 466.	368
87OPP471	R. A. Bunce and R. E. Drumright; *Org. Prep. Proced. Int.*, 1987, **19**, 471.	395
87OS(65)68	A. Bou, M. A. Pericàs, A. Riera and F. Serratosa; *Org. Synth.*, 1987, **65**, 68.	1026
87OS(65)98	D. L. Boger and M. D. Mullican; *Org. Synth.*, 1987, **65**, 98.	665
87OS(65)166	E. J. Corey and A. W. Gross; *Org. Synth.*, 1987, **65**, 166.	390
87OS(65)173	D. Enders, P. Frey and H. Kipphardt; *Org. Synth.*, 1987, **65**, 173.	380
87OS(65)224	C. A. Brown and P. K. Jadhav; *Org. Synth.*, 1987, **65**, 224.	423
87OS(66)127	S. R. Abrams and A. C. Shaw; *Org. Synth.*, 1987, **66**, 127.	423
87OS(66)150	J. Butler and R. M. Kellogg; *Org. Synth.*, 1987, **66**, 150.	120
87OS(66)151	B. Koppenhoefer and V. Schurig; *Org. Synth.*, 1987, **66**, 151.	12
87OS(66)194	R. K. Boeckman, Jr., R. B. Perni, J. E. Macdonald and A. J. Thomas; *Org. Synth.*, 1987, **66**, 194.	22
87PAC1577	B. Capon, B.-Z. Guo, F.-C. Kwok and Z.-P. Wu; *Pure Appl. Chem.*, 1987, **59**, 1577.	639
87POL1593	T. Klapotke; *Polyhedron*, 1987, **6**, 1593.	500, 501
87PS(29)111	A. Hercouet and M. le Corre; *Phosphorus Sulfur*, 1987, **29**, 111.	444
87PS(29)373	U. Kunze and R. Burghardt; *Phosphorus Sulfur*, 1987, **29**, 373.	440
87PS(31)33	U. Kunze and V. Kruppa; *Phosphorus Sulfur*, 1987, **31**, 33.	437
87PS(31)245	R. Cremlyn and R. Nuner; *Phosphorus Sulfur*, 1987, **31**, 245.	364
87PS(33)41	L. Maier, W. Kunz and G. Rist; *Phosphorus Sulfur*, 1987, **33**, 41.	444
87RCR267	M. L. Petrov and A. A. Petrov; *Russ. Chem. Rev.*, 1987, **56**, 267.	1038
87S35	D. Crich and S. M. Fortt; *Synthesis*, 1987, 35.	26
87S72	J.-H. Youn and R. Hermann; *Synthesis*, 1987, 72.	173, 717
87S85	A. J. Fatiadi; *Synthesis*, 1987, 85.	155
87S188	N. DeKimpe, W. DeCock and N. Schamp; *Synthesis*, 1987, 188.	3
87S190	E. Fabiano, B. T. Golding and M. M. Sadeghi; *Synthesis*, 1987, 190.	309
87S192	A. Alemagna, C. Baldoli, P. Del Buttero, E. Licandro and S. Maiorana; *Synthesis*, 1987, 192.	692
87S258	N. Ono, I. Hamamot, A. Kamimura, A. Kaji and R. Tamura; *Synthesis*, 1987, 258.	394
87S275	L. Grehn and U. Ragnarsson; *Synthesis*, 1987, 275.	299
87S325	B. H. Lipshutz; *Synthesis*, 1987, 325.	588, 917
87S373	M. Satoh, N. Miyaura and A. Suzuki; *Synthesis*, 1987, 373.	901
87S383	S. Gauthier and J. M. J. Frecht; *Synthesis*, 1987, 383.	753
87S485	S. Zawadzki, K. Osowska-Pacewicka and A. Zwierzak; *Synthesis*, 1987, 485.	373
87S487	A. Koziara, K. Osowska-Pacewicka, S. Zawadzki and A. Zwierzak; *Synthesis*, 1987, 487.	317
87S511	F. Camps, V. Gasol and A. Guerrero; *Synthesis*, 1987, 511.	8
87S551	R. D. Evans and J. H. Schauble; *Synthesis*, 1987, 551.	18
87S562	G. Alvernhe, A. Laurent and G. Haufe; *Synthesis*, 1987, 562.	18
87S645	M. Zaidlewicz and C. S. Panda; *Synthesis*, 1987, 645.	557
87S648	H.-J. Cristau, Y. Beziat, C. E. Niangoran and H. Cristol; *Synthesis*, 1987, 648.	445
87S661	J. Barluenga, J. M. Gonzalez, M. A. Rodriguez, P. J. Campos and G. Asensio; *Synthesis*, 1987, 661.	1019
87S683	T. Netscher and H. Prinzbach; *Synthesis*, 1987, 683.	173
87S709	G. W. Gribble and C. F. Nutatis; *Synthesis*, 1987, 709.	300, 310
87S711	R. Ballini, M. Petrini and G. Rosini; *Synthesis*, 1987, 711.	395
87S713	M. Petrini, R. Ballini and G. Rosini; *Synthesis*, 1987, 713.	318
87S727	A. Elmasmodi, D. Barbry, B. Hasiak and D. Couturier; *Synthesis*, 1987, 727.	742
87S920	D. Badone, G. Jommi, R. Pagliarin and P. Tavecchia; *Synthesis*, 1987, 920.	9

87S938 R. Zhang and Y. Zhou; *Synthesis*, 1987, 938. 436, 824, 825
87S1009 M. Mikolajczyk, W. Perlikowska and J. Umelanczuk; *Synthesis*, 1987, 1009. 124
87S1015 P. Brougham, M. S. Cooper, D. A. Cummerson, H. Heaney and N. Thompson; *Synthesis*, 1987, 1015. 147, 156
87S1032 E. Campi, N. J. Fitzmaurice, W. R. Jackson, P. Perlmutter and A. J. Smallridge; *Synthesis*, 1987, 1032. 928
87S1039 G. Castaldi and C. Giordano; *Synthesis*, 1987, 1039. 29
87SC469 P. W. Erhardt and A. H. Owens; *Synth. Commun.*, 1987, **17**, 469. 323
87SC685 T. Fukuhara, N. Yoneda, T. Sawada and A. Suzuki; *Synth. Commun.*, 1987, **17**, 685. 615
87SC1129 R. N. Reddie; *Synth. Commun.*, 1987, **17**, 1129. 192
87SC1155 T. Hudlicky, G. Sinai-Zingde and G. Seoane; *Synth. Commun.*, 1987, **17**, 1155. 326
87SC1421 V. N. Gogte, A. A. Natu and V. S. Pore; *Synth. Commun.*, 1987, **17**, 1421. 394
87T565 A. Onistschenko, B. Buchholz and H. Stamm; *Tetrahedron*, 1987, **43**, 565. 319
87T881 J.-B. Baudin, S. A. Julia and O. Ruel; *Tetrahedron*, 1987, **43**, 881. 359
87T891 J. Viret, J. Gabard and A. Collet; *Tetrahedron*, 1987, **43**, 891. 374
87T1273 W.-N. Wang and W. Jones; *Tetrahedron*, 1987, **43**, 1273. 1009
87T1799 B. Carboni, M. Vaultier and R. Carrié; *Tetrahedron Lett.*, 1987, **28**, 1799. 318
87T2075 P. Cazeau, F. Duboudin, F. Moulines, O. Babot and J. Dunogues; *Tetrahedron*, 1987, **43**, 2075. 699
87T2181 L. DiNunno and A. Scilimati; *Tetrahedron*, 1987, **43**, 2181. 56
87T2203 E. Erdik; *Tetrahedron*, 1987, **43**, 2203. 590
87T2311 M. A. Pericàs, F. Serratosa and E. Valenti; *Tetrahedron*, 1987, **43**, 2311. 1026
87T2609 Y. G. Gololobov, A. N. Nesmeyanov, V. P. Lysenko and I. E. Boldeskul; *Tetrahedron*, 1987, **43**, 2609. 1000
87T3123 J. T. Welch; *Tetrahedron*, 1987, **43**, 3123. 15
87T4059 H. C. Brown and M. M. Midland; *Tetrahedron*, 1987, **43**, 4059. 97
87T4079 H. C. Brown, M. M. Midland, A. B. Levy, A. Suzuki, S. Sono and M. Itok; *Tetrahedron*, 1987, **43**, 4079. 305
87T4321 D. H. R. Barton, B. Lacher and S. M. Zard; *Tetrahedron*, 1987, **43**, 4321. 627
87T5125 J. C. Carretero, M. Demillequand and L. Ghosez; *Tetrahedron*, 1987, **43**, 5125. 721
87T5411 A. Pecunioso and R. Menicagli; *Tetrahedron*, 1987, **43**, 5411. 393
87TL77 S.-I. Murahashi, Y. Imada and Y. Hirai; *Tetrahedron Lett.*, 1987, **28**, 77. 327
87TL295 S. L. Buchwald and S. J. LaMaire; *Tetrahedron Lett.*, 1987, **28**, 295. 969
87TL397 E. Differding, O. Vandevelde, B. Roekens, T. T. Van and L. Ghosez; *Tetrahedron Lett.*, 1987, **28**, 397. 938, 944
87TL405 M. P. Teulade and P. Savignac; *Tetrahedron Lett.*, 1987, **28**, 405. 456
87TL523 S. M. Tuladhar and A. G. Fallis; *Tetrahedron Lett.*, 1987, **28**, 523. 93, 138
87TL603 S. A. Weissman and S. G. Baxter; *Tetrahedron Lett.*, 1987, **28**, 603. 446
87TL663 S. Kanemoto, M. Shimizu and H. Yoshioka; *Tetrahedron Lett.*, 1987, **28**, 663. 11
87TL723 J. A. Marshall, J. Lebreton, B. S. DeHoff and T. M. Jenson; *Tetrahedron Lett.*, 1987, **28**, 723. 9
87TL767 S. P. Schmidt and D. W. Brooks; *Tetrahedron Lett.*, 1987, **28**, 767. 7, 10
87TL785 D. Hoppe and C. Gonschorrek; *Tetrahedron Lett.*, 1987, **28**, 785. 1029
87TL1027 D. L. Boger and R. S. Coleman; *Tetrahedron Lett.*, 1987, **28**, 1027. 655
87TL1123 D. A. Evans, J. A. Ellman and R. L. Dorow; *Tetrahedron Lett.*, 1987, **28**, 1123. 29
87TL1223 G. Dakka and Y. Sasson; *Tetrahedron Lett.*, 1987, **28**, 1223. 31
87TL1659 G. Courtois and L. Miginiac; *Tetrahedron Lett.*, 1987, **28**, 1659. 300
87TL1697 H. Mattes and C. Benezra, *Tetrahedron Lett.*, 1987, **28**, 1697. 10
87TL1703 B. H. Bakker and H. Cerfontain; *Tetrahedron Lett.*, 1987, **28**, 1703. 208
87TL1757 N. Knouzi, M. Vaultier, L. Toupet and R. Carrie; *Tetrahedron Lett.*, 1987, **28**, 1757. 324
87TL1851 J. E. Macdonald and G. S. Poindexter; *Tetrahedron Lett.*, 1987, **28**, 1851. 942
87TL2041 A. T. Russell and G. Procter; *Tetrahedron Lett.*, 1987, **28**, 2041. 921
87TL2083 R. H. Schlessinger and E. J. Iwanowicz; *Tetrahedron Lett.*, 1987, **28**, 2083. 331
87TL2359 E. Laurent, B. Marquet, R. Tardivel and H. Theibault; *Tetrahedron Lett.*, 1987, **28**, 2359. 16
87TL2387 Q.-Y. Chen and Y.-B. He; *Tetrahedron Lett.*, 1987, **28**, 2387. 1096
87TL2463 C. J. Kowalski and G. S. Lal; *Tetrahedron Lett.*, 1987, **28**, 2463. 1023
87TL2477 D. P. Curran and C.-T. Chang; *Tetrahedron Lett.*, 1987, **28**, 2477. 14, 34
87TL2641 S. N. Mazumdar, N. Sharma and M. P. Mahajan; *Tetrahedron Lett.*, 1987, **28**, 2641. 249
87TL2889 J. van der Louw, J. L. van der Baan, F. Bickelhaupt and G. W. Klumpp; *Tetrahedron Lett.*, 1987, **28**, 2889. 925
87TL2993 G. Doleschall; *Tetrahedron Lett.*, 1987, **28**, 2993. 342
87TL3103 D. Seebach and H. Estermann; *Tetrahedron Lett.*, 1987, **28**, 3103 315
87TL3285 H. H. Mooiweer, H. Hiemstra, H. P. Fortgens and W. N. Speckamp; *Tetrahedron Lett.*, 1987, **28**, 3285. 324, 326
87TL3709 Y. Ichinose, K. Nozaki, K. Wakamatsu, K. Oshima and K. Utimoto; *Tetrahedron Lett.*, 1987, **28**, 3709 946
87TL3771 M. Srebnik, T. E. Cole and H. C. Brown; *Tetrahedron Lett.*, 1987, **28**, 3771. 516
87TL3857 R. J. Blade, J. E. Robinson, R. J. Peek and J. B. Weston; *Tetrahedron Lett.*, 1987, 3857. 625
87TL3895 S. L. Buchwald, S. J. LaMaire, R. B. Nielsen, B. T. Watson and S. M. King; *Tetrahedron Lett.*, 1987, **28**, 3895. 969
87TL3901 S. T. Purrington and J. H. Pittman; *Tetrahedron Lett.*, 1987, 3901. 610
87TL4169 M. Vaultier, F. Truchet, B. Carboni, R. W. Hoffmann and I. Denne; *Tetrahedron Lett.*, 1987, **28**, 4169. 900

87TL4343 M. Lj. Mihailovic, S. Konstantinovic and R. Vukicevic; *Tetrahedron Lett.*, 1987, **28**, 4343. 286

87TL4445 S. P. Collingwood, A. P. Davies and B. T. Golding; *Tetrahedron Lett.*, 1987, **28**, 4445. 8, 32, 36

87TL4497 L. J. Stewart, D. Gray, R. M. Pagni and G. W. Kabalka; *Tetrahedron Lett.*, 1987, **28**, 4497. 34, 628

87TL4499 D. S. Matteson and E. C. Beadle; *Tetrahedron Lett.*, 1987, **28**, 4499. 530

87TL4715 Y. Hatanaka and T. Hiyama; *Tetrahedron Lett.*, 1987, **28**, 4715. 912

87TL4733 P. Bosch, F. Camps, E. Chamorro, V. Gasol and A. Guerrero; *Tetrahedron Lett.*, 1987, **28**, 4733. 19

87TL5115 F. A. Davis, J. Wei, A. C. Sheppard and S. Gubernick; *Tetrahedron Lett.*, 1987, **28**, 5115. 652

87TL5193 S. Thea and G. Cevasco; *Tetrahedron Lett.*, 1987, **28**, 5193. 173, 717

87TL5237 E. J. Corey and F. J. Hannon; *Tetrahedron Lett.*, 1987, **28**, 5237. 72

87TL5505 G. Mignani, D. Morel and F. Grass; *Tetrahedron Lett.*, 1987, **28**, 5505. 3

87TL5793 E. Negishi, Y. Zhang and V. Bagheri; *Tetrahedron Lett.*, 1987, **28**, 5793. 30

87TL5941 M. Barta, F. Urpi and J. Vilarrasa; *Tetrahedron Lett.*, 1987, **28**, 5941. 317

87TL5965 E. J. Bures and B. A. Keay; *Tetrahedron Lett.*, 1987, **28**, 5965. 942

87TL6109 J. R. Peterson, R. S. Egler, D. B. Horsley and T. J. Winter; *Tetrahedron Lett.*, 1987, **28**, 6109. 672

87TL6441 A. G. Martinez, R. M. Alvarez, E. T. Vilar, A. G. Fraile, J. O. Barcina, M. Hanack and L. R. Subramanian; *Tetrahedron Lett.*, 1987, **28**, 6441. 36

87TL6649 K. Okuma, J. Sakata, Y. Tachibana, T. Honda and H. Ohta; *Tetrahedron Lett.*, 1987, **28**, 6649. 858

87UKZ34 A. G. Makhsumov, M. S. Ergashev and I. Ibrakhimov; *Urk. Khim. Zh.* (*Russ. Ed.*), 1987, **5**, 34 (*Chem. Abstr.*, 1987, **109**, 128481u). 1056

87URP1294811 M. V. Livantov, A. A. Prishchenko, L. I. Akimova and I. F. Lutsenko; *USSR Pat.* 1 294 811 (1987) (*Chem. Abstr.*, 1987, **107**, 59 255). 433

87USP4668823 R. E. Murray; *US Pat.* 4 668 823 (1987) (*Chem. Abstr.*, 1987, **107**, 78 084). 436

87ZAAC(547)233 W. Hennrichs and J. Jander; *Z. Anorg. Allg. Chem.*, 1987, **547**, 233. 339

87ZAAC(548)55 N. Weferling; *Z. Anorg. Allg. Chem.*, 1987, **548**, 55. 455

87ZAAC(548)63 E. Fluck, J. Svara and H. Riffel; *Z. Anorg. Allg. Chem.*, 1987, **548**, 63. 1073

87ZAAC(553)136 F. Krech, K. Issleib, A. Zschunke and H. Meyer; *Z. Anorg. Allg. Chem.*, 1987, **553**, 136. 439

87ZN(B)579 G. U. Speigel and O. Stelzer; *Z. Naturforsch., Teil B*, 1987, **42**, 579. 428, 430

87ZN(B)695 M. Wieber and I. Sauer; *Z. Naturforsch., Teil B*, 1987, **42**, 695. 891

87ZOB296 G. V. Romanov, A. A. Lapin and A. N. Pudovik; *Zh. Obshch. Khim.*, 1987, **57**, 296 (*Chem. Abstr.*, 1988, **108**, 56 206). 446

87ZOB786 L. S. Burotina, N. N. Zalesova, E. E. Nifont'ev and T. A. Mastryukova; *Zh. Obshch. Khim.*, 1987, **57**, 786 (*Chem. Abstr.*, 1987, **108**, 186 833a). 255

87ZOB884 M. V. Livantsov, A. A. Prishchenko and I. F. Lutsenko; *Zh. Obshch. Khim.*, 1987, **57**, 884 (*Chem. Abstr.*, 1988, **108**, 204 689). 434

87ZOB1078 N. D. Obushak, M. V. Vovk, V. A. Vengizhemovskii, Y. I. Mel'nik and N. I. Gomushchak; *Zh. Obshch. Khim.*, 1987, **57**, 1078 (*Chem. Abstr.*, 1987, **108**, 94 657). 256

87ZOB1429 L. N. Usmanova, A. R. Burilov, R. Belyalow, T. K. Gazizov, M. H. Pudovik and A. W. Pudovik; *Zh. Obshch. Khim.*, 1987, **57**, 1429 (*Chem. Abstr.*, 1987, **108**, 186 862j). 254

87ZOR1981 B. A. Kirilyuk, I. A. Mel'nitskii, N. M. Golub, T. K. Kiladze, E. A. Kantor and D. L. Rakhmankulov; *Zh. Org. Khim.*, 1987, **23**, 1981 (*Chem. Abstr.*, 1989, **110**, 23 930g). 265

87ZOR2120 N. S. Zefirov, T. S. Kuznetsova, S. I. Kozhushkov, K. A. Potekhin, A. V. Maleev and Y. T. Struchkov; *Zh. Org. Khim.*, 1987, **23**, 2120 (*Chem. Abstr.*, 1987, **108**, 186 179). 445

88ACR135 B. Capon, B.-Z. Guo, F. C. Kwok, A. K. Siddhanta and C. Zucco; *Acc. Chem. Res.*, 1988, **21**, 135. 636, 638

88ACR261 B. Bogdanović; *Acc. Chem. Res.*, 1988, **21**, 261. 559

88ACR307 S. Rozen; *Acc. Chem. Res.*, 1988, **21**, 307. 15, 17

88ACR313 S. M. Weinreb; *Acc. Chem. Res.*, 1988, **21**, 313. 366, 793

88ACR442 Z. Rappoport and S. E. Biali; *Acc. Chem. Res.*, 1988, **21**, 442. 636, 643

88ACS(B)354 J. Dale and L. Vickersveen; *Acta Chem. Scand., Ser. B*, 1988, **42**, 354. 773

88AG(E)299 V. Schäfer, H. Pritzkow and W. Siebert; *Angew. Chem., Int. Ed. Engl.*, 1988, **27**, 299. 903

88AG(E)709 G. Märkl, S. Dietl, M. L. Ziegler and B. Nuber; *Angew. Chem., Int. Ed. Engl.*, 1988, **27**, 709. 865

88AG(E)846 W. Hollstein, K. Harms, M. Marsch and G. Boche; *Angew. Chem., Int. Ed. Engl.*, 1988, **27**, 846. 1000

88AJC1063 V. A. Burgess and C. J. Easton; *Aust. J. Chem.*, 1988, **41**, 1063. 394

88AOC(28)85 R. Csuk, B. I. Glänzer, and A. Fürstner; *Adv. Organomet. Chem.*, 1988, **28**, 85. 959

88AX(C)626 J. Macicek, O. Angelova, G. Petrov and M. Kirilov; *Acta Crystallogr., Part C*, 1988, **44**, 626. 1004

88BCJ93 H. C. Brown, A. K. Gupta and J. V. N. Vara Prasad; *Bull. Chem. Soc. Jpn.*, 1988, **61**, 93. 42

88BCJ597 S. Kajigaeshi, T. Kakinami, K. Inoue, M. Kondo, H. Nakamura, M. Fujikawa and T. Okamoto; *Bull. Chem. Soc. Jpn.*, 1988, **61**, 597. 626

88BCJ600 S. Kajigaeshi, T. Kakinami, H. Yamasaki, S. Fujisaki and T. Okamoto; *Bull. Chem. Soc. Jpn.*, 1988, **61**, 600. 631

88BCJ1213 H. Yamashita; *Bull. Chem. Soc. Jpn.*, 1988, 1213. 85, 135

88BCJ2185 H. Firouzabadi, D. Mohajer and M. Entezartmoghaday; *Bull. Chem. Soc. Jpn.*, 1988, **61**, 2185. 215

88BCJ2226 N. Kamigata, T. Satoh, M. Yoshida, H. Matsuyama and M. Kameyama; *Bull. Chem. Soc. Jpn.*, 1988, **61**, 2226. 620

88BCJ2684 H. Kotsuki, Y. Ushio, M. Yoshimura and M. Ochi; *Bull. Chem. Soc. Jpn.*, 1988, **61**, 2684. 60

88BCJ2693 Y. Ichinose, K. Oshima and K. Utimoto; *Bull. Chem. Soc. Jpn.*, 1988, **61**, 2693. 930, 946
88BCJ3020 Y. Inoue, S. Imaizumi, H. Itoh, T. Shinya, H. Hashimoto and S. Miyano; *Bull. Chem. Soc. Japan*, 1988, **61**, 3020. 398
88BCJ3029 T. Shono, Y. Matsumara, K. Uchida and F. Nakatini; *Bull. Chem. Soc. Jpn.*, 1988, **61**, 3029. 323
88BCJ3283 M. Shibagaki, K. Takahashi and H. Matsushita; *Bull. Chem. Soc. Jpn.*, 1988, **61**, 3283. 55
88BCJ3359 Y. Okamoto, T. Kusano and S. Takamuku; *Bull. Chem. Soc. Jpn.*, 1988, **61**, 3359. 107
88BCJ3957 S. Kanemasa, H. Kobayashi, J. Tanaka and O. Tsuge; *Bull. Chem. Soc. Jpn.*, 1988, **61**, 3957. 908
88BCJ4029 M. Fujii; *Bull. Chem. Soc. Japan*, 1988, **61**, 4029. 398
88BCJ4323 K. Okuma, Y. Tachibana, J. Sakata, T. Komiya, I. Kaneko, Y. Komiya, S. Yamamoto and H. Ohta; *Bull. Chem. Soc. Jpn.*, 1988, **61**, 4323. 853
88BCJ4476 S. Yamamoto, K. Okuma and H. Ohta; *Bull. Chem. Soc. Jpn.*, 1988, **61**, 4476. 84, 445
88BSB461 K. Kano and J.-P. Anselme; *Bull. Soc. Chim. Belg.*, 1988, **97**, 461. 403
88CB75 M. Schwab and W. Sundermeyer; *Chem. Ber.*, 1988, **121**, 75. 209
88CB843 C. Dorweiler, M. Holderbaum, T. Münzmay, P. Spang, H. Dürr, C. Krüger and E. Raabe; *Chem. Ber.*, 1988, **121**, 843. 1003
88CB871 P. Kolle, G. Linti, H. Noth, G. L. Wood, C. K. Nurula and R. T. Paine; *Chem. Ber.*, 1988, **121**, 871. 474
88CB1123 U. Nagel and B. Rieger; *Chem. Ber.*, 1988, **121**, 1123. 437
88CB1501 R. W. Hoffmann, S. Dresely and J. W. Lanz; *Chem. Ber.*, 1988, **121**, 1501. 75
88CB2225 R. W. Hoffmann, S. Dresely and B. Hildebrandt; *Chem. Ber.*, 1988, **121**, 2225. 75
88CC40 R. Okazaki, T. Ishida and N. Inamoto; *J. Chem. Soc., Chem. Commun.*, 1988, 40. 185
88CC44 Y. Masuyama, R. Hayashi, K. Otake and Y. Kurusu; *J. Chem. Soc., Chem. Commun.*, 1988, 44. 80
88CC121 J. Barluenga, F. Aznar, R. Liz and M. Bayod; *J. Chem. Soc, Chem. Commun.*, 1988, 121. 1097
88CC289 M. J. Gallagher, S. Harvey, C. L. Raston and R. E. Sue; *J. Chem. Soc., Chem. Commun.*, 1988, 289. 559
88CC610 Y. Tamuru, T. Nakamura, M. Sakaguchi, H. Ochiai and Z. Yoshida; *J. Chem. Soc., Chem. Commun.*, 1988, 610. 69
88CC909 T. Jeffery; *J. Chem. Soc., Chem. Commun.*, 1988, 909. 1018
88CC942 M. A. Pericás, A. Riera, C. Rossell, F. Serratosa and M. Seco; *J. Chem. Soc., Chem. Commun.*, 1988, 942. 1027
88CC950 M. R. Bryce and P. C. Taylor; *J. Chem. Soc. Chem. Commun.*, 1988, 950. 241, 792
88CC971 C. Chan, P. B. Cox and S. M. Roberts; *J. Chem. Soc., Chem. Commun.*, 1988, 971. 628
88CC1283 S. Tomoda and M. Iwaoka; *J. Chem. Soc., Chem. Commun.*, 1988, 1283. 281
88CC1350 J. Einhorn, J.-L. Luche and P. Demerseman; *J. Chem. Soc., Chem. Commun.*, 1988, 1350. 422, 653
88CC1451 S. Colonna, N. Gaggero, A. Manfredi, L. Casella and M. Gullotti; *J. Chem. Soc., Chem. Commun.*, 1988, 1451. 151
88CC1503 M. P. Cowling, P. R. Jenkins and K. Cooper; *J. Chem. Soc., Chem. Commun.*, 1988, 1503. 345, 347
88CCC110 J. Burkhard, J. Janku and L. Vodicka; *Collect. Czech. Chem. Commun.*, 1988, **53**, 110. 617
88CCC2116 A. Mehlhorn; *Collect. Czech. Chem. Commun.*, 1988, **53**, 2116. 398
88CI(L)390 I. Paterson; *Chem. Ind. (London)*, 1988, 390. 698
88CJC1479 P. C. Belanger, C. K. Lau, H. W. R. Williams, C. Dufresne and J. Scheigetz; *Can. J. Chem.*, 1988, **66**, 1479. 612
88CL31 C. A. Horiuchi and S. Kiji; *Chem. Lett.*, 1988, 31. 34
88CL63 S. Komiya, I. Endo, S. Ozaki and Y. Ishizaki; *Chem. Lett.*, 1988, 63. 590
88CL795 S. Kajigaeshi, T. Kakinami, M. Moriwaki, M. Watanabe, S. Fujisaki and T. Okamoto; *Chem. Lett.*, 1988, 795. 631
88CL857 H. Takaku, K. Imai and M. Nagai; *Chem. Lett.*, 1988, 857. 239
88CL899 R. Oi, C. Shimakawa and S. Takenaka; *Chem. Lett.*, 1988, 899. 684
88CL1289 M. Tokuda, Y. Yamada and H. Suginome; *Chem. Lett.*, 1988, 1289. 323
88CL1633 M. Shibagaki, K. Takahashi, H. Kuno, H. Kawakami and H. Matsushita; *Chem. Lett.*, 1988, 1633. 55
88CL1759 K. Kaneda, T. Takemoto and T. Imanaka; *Chem. Lett.*, 1988, 1759. 314, 318
88CL1773 H. Taniguchi and E. Nomura; *Chem. Lett.*, 1988, 1773. 89, 683
88CL1877 T. Ando, D. G. Cork, M. Fujita, T. Kimura and T. Tatsuno; *Chem. Lett.*, 1988, 1877 18, 19
88CL2017 M. Yoshida, H. Mochizuki and N. Kamigata; *Chem. Lett.*, 1988, 2017. 620
88CPB459 M. Masui, N. Yamawaki and H. Ohmori; *Chem. Pharm. Bull.*, 1988, **36**, 459. 385
88CPB791 Y. Arakawa, N. Ueyama and Y. Nitta; *Chem. Pharm. Bull.*, 1988, **36**, 791. 227
88CPB2342 A. Hosomi, S. Kohra, Y. Tominaga, M. Inaba and H. Sakurai; *Chem. Pharm. Bull.*, 1988, **36**, 2342. 300
88CRV473 H. L. Halland; *Chem. Rev.*, 1988, **88**, 473. 151
88CRV1047 S. L. Buchwald and R. B. Nielsen; *Chem. Rev.*, 1988, **88**, 1047. 986, 987
88CRV1121 R. Guilard and K. M. Kadish; *Chem. Rev.*, 1988, **88**, 1121. 578
88CSR283 W. R. Bowman; *Chem. Soc. Rev.*, 1988, **17**, 283. 391
88G1 C. Cauletti, C. Furlani and A. Sebald; *Gazz. Chim. Ital.*, 1988, **118**, 1. 1100
88GEP3626554 A. Welter, A. Roemer, S. Leyck and M. J. Parnham; *Ger. Pat.* 3 626 554 (1988) (*Chem. Abstr.*, 1988, **109**, 54 669n). 235
88H(27)1575 E. Zeller and D. S. Grierson; *Heterocycles*, 1988, **27**, 1575. 328
88HCA299 D. Seebach, T. Maetzke, R. K. Haynes, M. N. Paddon-Row and S. S. Wong; *Helv. Chim. Acta*, 1988, **71**, 299. 999
88HOU(E13)761 H. Kropf; *Methoden Org. Chem. (Houben-Weyl)*, 1988, **E13**, 761. 686
88IC3730 F. Calderazzo, A. Morvillo, G. Pelizzi, R. Poli and F. Ungari; *Inorg. Chem.*, 1988, **27**, 3730. 890, 891, 893

88IC4535 E. A. Ganja, C. D. Ontiveros and J. A. Morrison; *Inorg. Chem.*, 1988, **27**, 4535. 483

88ICA(146)111 W. E. Hill, M. Q. Islam, T. R. Webb and C. A. McAuliffe; *Inorg. Chim. Acta*, 1988, **146**, 111. 440

88ICA(150)51 V. K. Gupta, L. K. Krannich and C. L. Watkins; *Inorg. Chim. Acta*, 1988, **150**, 51. 893

88IJC(B)1129 J. M. Khurana, P. Tetenyi, M. Kodomari and S. L. Regen; *Indian J. Chem., Sect. B*, 1988, **27**, 1129. 4

88IZV2367 A. V. Fokin, A. I. Rapkin, V. I. Matveenko and O. V. Verenikin; *Izv. Akad. Nauk SSSR Ser. Khim.*, 1988, 2367 (*Chem. Abstr.*, 1989, **110**, 192 217k). 170

88JA612 L. E. Overman and M. J. Sharp; *J. Am. Chem. Soc.*, 1988, **110**, 612. 326

88JA624 G. Erker, R. Hock and R. Nolte; *J. Am. Chem. Soc.*, 1988, **110**, 624. 858

88JA649 E. J. Corey, M.-C. Kang, M. C. Desai, A. K. Ghosh and I. N. Houpis; *J. Am. Chem. Soc.*, 1988, **110**, 649. 679

88JA842 J. S. Sawyer, A. Kucerovy, T. L. Macdonald and G. J. McGarvey; *J. Am. Chem. Soc.*, 1988, **110**, 842. 999

88JA984 S. E. Denmark, T. Wilson and T. M. Wilson; *J. Am. Chem. Soc.*, 1988, **110**, 984. 78

88JA1297 E. Nakamura, M. Isaka and S. Matsuzawa; *J. Am. Chem. Soc.*, 1988, **110**, 1297. 589

88JA1535 H. C. Brown, P. K. Jadhav and K. S. Bhat; *J. Am. Chem. Soc.*, 1988, **110**, 1535. 75, 521

88JA1539 H. C. Brown, J. Chandrasekharan and P. V. Ramachandran; *J. Am. Chem. Soc.*, 1988, **110**, 1539. 526

88JA1549 H. C. Brown, J. Chandrasekharan and P. V. Ramachandran; *J. Am. Chem. Soc.*, 1988, **110**, 1549. 58

88JA1633 P. A. Bartlett, K. L. McLaren and P. C. Ting; *J. Am. Chem. Soc.*, 1988, **110**, 1633. 351

88JA1719 G. V. Shustov, G. K. Kadorkina, R. G. Kostyanovsky and A. Rauk; *J. Am. Chem. Soc.*, 1988, **110**, 1719. 337

88JA1968 E. N. Jacobsen, I. Mark, W. S. Mungall, G. Schröder and K. B. Sharpless; *J. Am. Chem. Soc.*, 1988, **110**, 1968. 47

88JA1976 M. Segi, T. Nakajima, S. Suga, S. Murai, I. Ryu, A. Ogawa and N. Sonoda; *J. Am. Chem. Soc.*, 1988, **110**, 1976. 542

88JA2310 P. F. Fagan and W. A. Nugent; *J. Am. Chem. Soc.*, 1988, **110**, 2310. 949, 992

88JA2506 D. A. Evans, S. L. Bender and J. Morris; *J. Am. Chem. Soc.*, 1988, **110**, 2506. 554, 593

88JA2575 P. DeShong, D. R. Sidler, P. J. Rybczynski, G. A. Slough, and A. L. Rheingold; *J. Am. Chem. Soc.*, 1988, **110**, 2575. 574, 978

88JA2641 J. R. Behling, K. A. Babiak, J. S. Ng, A. L. Campbell, R. Morretti, M. Koerner, and B. H. Lipshutz; *J. Am. Chem. Soc.*, 1988, **110**, 2641. 992

88JA3272 P. J. Stang, M. Boehshar, H. Wingert and T. Kitamura; *J. Am. Chem. Soc.*, 1988, **110**, 3272. 1029

88JA3560 D. A. Evans, K. T. Chapman and E. M. Carreira; *J. Am. Chem. Soc.*, 1988, **110**, 3560. 62

88JA3712 K. Tamao, Y. Nakagawa, H. Arai, N. Higuchi and Y. Ito; *J. Am. Chem. Soc.*, 1988, **110**, 3712. 50

88JA3979 W. R. Roush and L. Banfi; *J. Am. Chem. Soc.*, 1988, **110**, 3979. 75

88JA4039 A. M. Echavarren, D. R. Tueting and J. K. Stille; *J. Am. Chem. Soc.*, 1988, **110**, 4039. 86

88JA4473 Y. Masuyama, J. P. Takahara and Y. Kurusu; *J. Am. Chem. Soc.*, 1988, **110**, 4473. 80

88JA4475 Y. Yamamoto, K. Matsuoka and H. Nemoto; *J. Am. Chem. Soc.*, 1988, **110**, 4475. 54

88JA4599 M. Kira, K. Sato and H. Sakurai; *J. Am. Chem. Soc.*, 1988, **110**, 4599. 77

88JA4606 M. Ochiai, S. Iwaki, T. Ukita, Y. Matsuura, M. Shiro and Y. Nagao; *J. Am. Chem. Soc.*, 1988, **110**, 4606. 46

88JA4788 A. R. Chamberlain, S. H. Bloom, L. A. Cervini and C. A. Fotsch; *J. Am. Chem. Soc.*, 1988, **110**, 4788. 556

88JA4818 J. C. Medina and K. S. Kyler; *J. Am. Chem. Soc.*, 1988, **110**, 4818. 92

88JA5228 I. M. Arafa, K. Shin and H. M. Goff; *J. Am. Chem. Soc.*, 1988, **110**, 5228. 578

88JA5578 G. Zweifel and T. M. Shoup; *J. Am. Chem. Soc.*, 1988, **110**, 5578. 75

88JA5579 T. Hayashi, Y. Matsumoto and Y. Ito; *J. Am. Chem. Soc.*, 1988, **110**, 5579. 51

88JA6246 N. N. Joshi, M. Srebnik and H. C. Brown; *J. Am. Chem. Soc.*, 1988, **110**, 6246. 527

88JA6276 P. Dussault and N. A. Porter; *J. Am. Chem. Soc.*, 1988, **110**, 6276. 676

88JA6463 A. Igau, H. Grutzmacher, A. Baceiredo and G. Bertrand; *J. Am. Chem. Soc.*, 1988, **110**, 6463. 1067

88JA6565 M. Ochiai, Y. Takaoka and Y. Nagao; *J. Am. Chem. Soc.*, 1988, **110**, 6565. 917, 918

88JA6917 D. A. Evans, G. C. Fu and A. H. Hoveyda; *J. Am. Chem. Soc.*, 1988, **110**, 6917. 42

88JA7128 T. V. RajanBabu, W. A. Nugent, D. F. Taber, and P. J. Fagan; *J. Am. Chem. Soc.*, 1988, **110**, 7128. 988

88JA7255 B. M. Trost and D. C. Lee; *J. Am. Chem. Soc.*, 1988, **110**, 7255. 980

88JA7536 R. S. Jolly and T. Livinghouse; *J. Am. Chem. Soc.*, 1988, **110**, 7536. 325

88JA7737 D. N. Harpp and M. Gingras; *J. Am. Chem. Soc.*, 1988, **110**, 7737. 89

88JA7984 F. Tureček, L. Brabec and J. Korvola; *J. Am. Chem. Soc.*, 1988, **110**, 7984. 636

88JA8244 C. S. Chin, S. Y. Lee, J. Park and S. Kim; *J. Am. Chem. Soc.*, 1988, **110**, 8244. 639

88JA8671 P. T. Meinke and G. A. Krafft; *J. Am. Chem. Soc.*, 1988, **110**, 8671. 289

88JA8679 P. T. Meinke and G. A. Krafft; *J. Am. Chem. Soc.*, 1988, **110**, 8679. 289

88JA8734 R. B. Ruggeri, M. M. Hansen and C. H. Heathcock; *J. Am. Chem. Soc.*, 1988, **110**, 8734. 330

88JAP63119491 K. Takahashi, Y. Demura and S. Oota; *Jpn. Pat.* 63 119 491 (1988) (*Chem. Abstr.*, 1988, **109**, 93 329). 444

88JCR308 G. Bisway and D. Mal; *J. Chem. Res.*, 1988, 308. 156

88JCS(D)451 A. J. Downs, M. J. Goode, S. McGrady and I. A. Steer; *J. Chem. Soc., Dalton Trans.*, 1988, 451. 492

88JCS(D)2077 R. Ali and K. B. Dillon; *J. Chem. Soc., Dalton Trans.*, 1988, 2077. 476

88JCS(P1)9 M. H. Brookes, B. T. Golding and A. T. Hudson; *J. Chem. Soc., Perkin Trans. 1*, 1988, 9. 225

88JCS(P1)269 P. C. B. Page, M. B. van Niel and D. Westwood; *J. Chem. Soc., Perkin Trans. 1*, 1988, 269. 749

88JCS(P1)549 K. Biggadike, A. D. Borthwick, D. Evans, A. M. Exall, B. E. Kirk, S. M. Roberts, L. Stephenson and P. Youds; *J. Chem. Soc., Perkin Trans. 1*, 1988, 549. 19

88JCS(P1)575 T. Toru, T. Seko, E. Maekawa and Y. Ueno; *J. Chem. Soc., Perkin Trans. 1*, 1988, 575. 285

88JCS(P1)1023 T. Bowles, R. Jones, A. E. A. Porter, J. A. Rechka, H. S. Rzepa and D. J. Williams; *J. Chem. Soc., Perkin Trans. 1*, 1988, 1023. 1000

88JCS(P1)1913 G. W. Kirby and A. N. Trethewey; *J. Chem. Soc., Perkin Trans. 1*, 1988, 1913. 289, 292

88JCS(P1)2339 A. R. Katritzky and M. Drewniak; *J. Chem. Soc., Perkin Trans. 1*, 1988, 2339. 320

88JCS(P1)2805 R. E. Banks, R. A. Du Boisson, W. D. Morton and E. Tsiliopoulos; *J. Chem. Soc., Perkin Trans. 1*, 1988, 2805. 16, 336

88JCS(P1)2855 Y.-Z. Huang, C. Chen and Y. Shen; *J. Chem. Soc., Perkin Trans. 1*, 1988, 2855. 73

88JFC(41)297 R. E. Banks and I. Sharif; *J. Fluorine Chem.*, 1988, **41**, 297. 16

88JGU629 V. A. Dodonov, O. P. Bolotova and A. V. Gushchin; *J. Gen. Chem. USSR (Engl. Transl.)*, 1988, **58**, 629. 879

88JGU693 S. M. Kalinov, M. F. Rostovskaya and V. I. Vysotskii; *J. Gen. Chem. USSR (Engl. Transl.)*, 1988, **58**, 693. 455

88JGU1040 L. B. Ionov, S. M. Reshetnikov, L. L. Makarova, M. A. Pletnev and E. N. Krasnoperova; *J. Gen. Chem. USSR (Engl. Transl.)*, 1988, **58**, 1040. 891

88JGU1131 V. A. Al'fonsov, I. S. Nizamov and S. A. Katsyubova; *J. Gen. Chem. USSR (Engl. Transl.)*, 1988, **58**, 1131. 464

88JGU1173 K. I. Kuz'min, Z. U. Panfilovich, I. I. Chechetkina, T. A. Zyablikova and I. P. Lipatova; *J. Gen. Chem. USSR (Engl. Transl.)*, 1988, **58**, 1173. 879

88JGU1934 V. A. Dodonov, V. A. Gushchin and M. B. Ezhova; *J. Gen. Chem. USSR (Engl. Transl.)*, 1988, **58**, 1934. 886

88JGU2050 V. V. Sharutin; *J. Gen. Chem. USSR (Engl. Transl.)*, 1988, **58**, 2050. 874

88JGU2136 Z. U. Panfilovich, N. R. Ivanova, K. I. Kuz'min, I. U. Kambulova, T. V. Zykova and I. P. Lipatova; *J. Gen. Chem. USSR (Engl. Transl.)*, 1988, **58**, 2136. 1076

88JGU2161 O. G. Sinyashin, I. Yu. Gorshunov, R. Z. Musin, E. S. Botyeva and A. N. Pudovik; *J. Gen. Chem. USSR (Engl. Transl.)*, 1988, **58**, 2161. 461

88JHC1035 P. Bitha and Y. Lin; *J. Heterocycl. Chem.*, 1988, **25**, 1035. 307

88JHC1383 C. S. Sang, H. D. Chil, J. K. Tae, K. L. Hak and D. K. Ki; *J. Heterocycl. Chem.*, 1988, **25**, 1383. 327

88JHC1857 N. B. Hanna, S. D. Dimitrijevich, S. B. Larson, R. K. Robins and G. R. Revankar; *J. Heterocycl. Chem.*, 1988, **25**, 1857. 364

88JMC1080 H. Wikström, T. Elebring, G. Hallnemo, B. Andersson, K. Svensson, A. Carlsson and H. Rollema; *J. Med. Chem.*, 1988, **31**, 1080. 652

88JOC45 A. Pecunioso and R. Menicagli; *J. Org. Chem.*, 1988, **53**, 45. 393, 783

88JOC187 V. K. Gautam, J. Singh and R. S. Dhillon; *J. Org. Chem.*, 1988, **53**, 187. 34, 43

88JOC239 H. C. Brown, D. Basavaiah, S. U. Kulkarni, N. G. Bhat and J. V. N. Vara Prassad; *J. Org. Chem.*, 1988, **53**, 239. 901

88JOC366 D. G. Cox and D. J. Burton; *J. Org. Chem.*, 1988, **53**, 366. 444

88JOC880 O. E. O. Hormi; *J. Org. Chem.*, 1988, **53**, 880. 672

88JOC913 E. Negishi, H. Sawada, J. M. Tour, and Y. Wei; *J. Org. Chem.*, 1988, **53**, 913. 973, 976

88JOC927 K. Soai, S. Niwa and M. Watanabe; *J. Org. Chem.*, 1988, **53**, 927. 71

88JOC1231 H. C. Brown, B. T. Cho and W. S. Park; *J. Org. Chem.*, 1988, **53**, 1231. 57

88JOC1251 S. E. Denmark, J. A. Sternberg and R. Lueoend; *J. Org. Chem.*, 1988, **53**, 1251. 398

88JOC1380 P. J. McCloskey and A. G. Shultz; *J. Org. Chem.*, 1988, **53**, 1380. 326

88JOC1391 H. C. Brown, A. K. Gupta, J. V. N. V. Prasad and M. Srebnik; *J. Org. Chem.*, 1988, **53**, 1391. 1085

88JOC1467 R. H. Smith, Jr., A. F. Mehl, D. L. Schantz, Jr., N. Chmurny and C. J. Michejda; *J. Org. Chem.*, 1988, **53**, 1467. 387

88JOC1695 P. E. Correa, G. Hardy and D. P. Riley; *J. Org. Chem.*, 1988, **53**, 1695. 348

88JOC1761 M. E. Perlman and T. J. Bardos; *J. Org. Chem.*, 1988, **53**, 1761. 413

88JOC1768 T. R. Lamanec, D. R. Bender, A. M. DeMarco, S. Karady, R. A. Leamer and L. M. Weinstock; *J. Org. Chem.*, 1988, **53**, 1768. 342, 346

88JOC1799 S. A. Khan, M. A. Munawar and M. Siddiq; *J. Org. Chem.*, 1988, **53**, 1799. 626

88JOC1815 A. G. M. Barrett and I. A. O'Neil; *J. Org. Chem.*, 1988, **53**, 1815. 943

88JOC1831 S. Araki, H. Ito and Y. Butsugan; *J. Org. Chem.*, 1988, **53**, 1831. 73

88JOC2093 M. Kodomari, H. Satoh and S. Yoshitomi; *J. Org. Chem.*, 1988, **53**, 2093. 626

88JOC2094 T. Hudlicky, G. Seoana and T. C. Lovelace; *J. Org. Chem.*, 1988, **53**, 2094. 325

88JOC2144 C. A. Broka and J. F. Gerlits; *J. Org. Chem.*, 1988, **53**, 2144. 337

88JOC2389 H. J. Reich and C. P. Jasperse; *J. Org. Chem.*, 1988, **53**, 2389. 368

88JOC2390 P. Knochel, M. C. P. Yeh, S. C. Berk and J. Talbert; *J. Org. Chem.*, 1988, **53**, 2390. 588, 590

88JOC2737 X.-P. Cu, I. Ikeda and M. Okahara; *J. Org. Chem.*, 1988, **53**, 2737. 89

88JOC2803 S. Rozen and C. Gal; *J. Org. Chem.*, 1988, **53**, 2803. 15

88JOC2858 E. Bay, P. E. Timony and A. Leone-Bay; *J. Org. Chem.*, 1988, **53**, 2858. 621

88JOC2872 F. M. Hauser and V. M. Baghdanov; *J. Org. Chem.*, 1988, **53**, 2872. 393

88JOC2911 H. C. Brown, J. V. N. Vara Prasad and M. Zaidlewicz; *J. Org. Chem.*, 1988, **53**, 2911. 42

88JOC2916 H. C. Brown, M. Srebnik and P. V. Ramachandran; *J. Org. Chem.*, 1988, **53**, 2916. 526

88JOC2937	T. Minami, S. Shikita, S. So, M. Nakayama and I. Yamamoto; *J. Org. Chem.*, 1988, **53**, 2937.	447
88JOC2968	D. F. Taber, P. B. Deker, H. M. Fales, T. H. Jones and H. A. Lloyd; *J. Org. Chem.*, 1988, **53**, 2968.	326
88JOC2997	J. T. Welch and K. T. Seper; *J. Org. Chem.*, 1988, **53**, 2997.	102
88JOC3016	A. N. Santiago, V. S. Iyer, W. Adcock and R. A. Ross; *J. Org. Chem.*, 1988, **53**, 3016.	438
88JOC3111	S. Kim and J. H. Park; *J. Org. Chem.*, 1988, **53**, 3111.	11
88JOC3539	N. Chatani, T. Takeyasu, N. Horiuchi and T. Hanafusa; *J. Org. Chem.*, 1988, **53**, 3539.	920, 946
88JOC3548	F. Radner; *J. Org. Chem.*, 1988, **53**, 3548.	631
88JOC3634	J. G. Lee and K. K. Kang; *J. Org. Chem.*, 1988, **53**, 3634.	25
88JOC3752	T. Nakano, S. Umano, Y. Kino, Y. Ishii and M. Ogawa; *J. Org. Chem.*, 1988, **53**, 3752.	55
88JOC3845	R. D. Connell, T. Rein, B. Åkermark and P. Helquist; *J. Org. Chem.*, 1988, **53**, 3845.	308
88JOC3897	J. Chen and D. D. Tanner; *J. Org. Chem.*, 1988, **53**, 3897.	398
88JOC4105	T. Kitamura and P. J. Stang; *J. Org. Chem.*, 1988, **53**, 4105.	1020
88JOC4274	J. A. Marshall, J. D. Trometer, B. E. Blough and T. D. Crute; *J. Org. Chem.*, 1988, **53**, 4274.	86
88JOC4379	F. Effenberger, K.-E. Mack, R. Niess, F. Reisinger, A. Steinbach, W.-D. Stohrer, J. J. Stezowski, I. Rommel and A. Maier; *J. Org. Chem.*, 1988, **53**, 4379.	1009
88JOC4398	M. V. D. Puy; *J. Org. Chem.*, 1988, **53**, 4398.	614
88JOC4425	M. D. Fryzuk, G. S. Bates and C. Stone; *J. Org. Chem.*, 1988, **53**, 4425.	436
88JOC4462	P. Sulmon, N. De Kimpe and N. Schamp; *J. Org. Chem.*, 1988, **53**, 4462.	322
88JOC4477	R. M. Pagni, G. W. Kabalka, R. Boothe, K. Gaetano, L. J. Stewart, R. Conaway, C. Dial, D. Gray, S. Larson and T. Luidhardt; *J. Org. Chem.*, 1988, **53**, 4477.	34, 629, 631
88JOC4482	G. W. Ebert and R. D. Rieke; *J. Org. Chem.*, 1988, **53**, 4482.	958, 959
88JOC5004	F. A. Davis, S. G. Lal and H. D. Durst; *J. Org. Chem.*, 1988, **53**, 5004.	150
88JOC5153	T. B. Patrick and R. Mortezania; *J. Org. Chem.*, 1988, **53**, 5153.	16
88JOC5178	K. Burgess and M. J. Ohlmeyer; *J. Org. Chem.*, 1988, **53**, 5178.	42
88JOC5185	M. Caron, P. R. Carlier and K. B. Sharpless; *J. Org. Chem.*, 1988, **53**, 5185.	84
88JOC5353	P. E. Eaton and G. E. Wicks; *J. Org. Chem.*, 1988, **53**, 5353.	397
88JOC5545	S. Rozen, M. Brand and R. Lidor; *J. Org. Chem.*, 1988, **53**, 5545.	626
88JOC5664	D. J. Nelson and R. Soundararajan; *J. Org. Chem.*, 1988, **53**, 5664.	23
88JOC5689	V. Cere, C. Paolucci, S. Pollicino, E. Sandri and A. Fava; *J. Org. Chem.*, 1988, **53**, 5689.	156
88JOC5776	S. D. Pastor and E. T. Hessell; *J. Org. Chem.*, 1988, **53**, 5776.	353
88JOC5856	W. W. Zajaz Jr., T. R. Walters and M. G. Darcy; *J. Org. Chem.*, 1988, **53**, 5856.	349
88JOC5989	P. G. M. Wuts and Y.-W. Jung; *J. Org. Chem.*, 1988, **53**, 5989.	326
88JOC6009	H. C. Brown and N. G. Bhat; *J. Org. Chem.*, 1988, **53**, 6009.	901, 904
88JOC6022	J. M. Dener, D. J. Hart and S. Ramesh; *J. Org. Chem.*, 1988, **53**, 6022.	325
88JOC6126	B. Classon, Z. Liu and B. Samuelsson; *J. Org. Chem.*, 1988, **53**, 6126.	7
88JOM(341)C1	A. H. Cowley, R. A. Jones, K. B. Kidd, C. M. Nunn and D. L. Westmoreland; *J. Organomet. Chem.*, 1988, **341**, C1.	511
88JOM(341)345	D. Furlani, D. Marton, G. Tagliavini and M. Zordan; *J. Organomet. Chem.*, 1988, **341**, 345.	79
88JOM(342)263	G. K. Anderson and R. Kumar; *J. Organomet. Chem.*, 1988, **342**, 263.	437
88JOM(346)C7	M. Emziane, K. Sutowardoyo and D. Sinou; *J. Organomet. Chem.*, 1988, **346**, C7.	85
88JOM(346)413	H. Brunner and A. Kürzinger; *J. Organomet. Chem.*, 1988, **346**, 413.	64
88JOM(347)5	S. Araki, H. Ito and Y. Butsugan; *J. Organomet. Chem.*, 1988, **347**, 5.	73
88JOM(347)365	F. Hossner and M. Voyle; *J. Organomet. Chem.*, 1988, **347**, 365.	684
88JOM(353)177	C. Boga, D. Savoia, E. Tagliavini, C. Trombini and A. Umani-Ronchi; *J. Organomet. Chem.*, 1988, **353**, 177.	79
88JOM(354)287	R. L. Wells, A. P. Purdy, A. T. McPhail and C. P. Pitt; *J. Organomet. Chem.*, **354**, 287.	511
88JOU1473	S. M. Hassanein, A. Burmakov, F. A. Bloshchitsa and L. M. Yagupol'skii; *J. Org. Chem. USSR* (*Engl. Transl.*), 1988, **24**, 1473.	100
88JOU2222	V. I. Markov, L. V. Dimitrikova, V. L. Dubina and K. S. Bunnistro; *J. Org. Chem. USSR* (*Engl. Transl.*), 1988, **24**, 2222.	756
88JPC37	A. Ayed, L. Manceron and B. Silvi; *J. Phys. Chem.*, 1988, **92**, 37.	1022, 1023
B-88MI 201-01	N. De Kimpe and R. Verhe; in "The Chemistry of Functional Groups," Update volume, eds. S. Patai and Z. Rappoport, Wiley, New York, 1988.	2, 3, 16, 21, 28
88MI 202-01	K. Tamao; *J. Synth. Org. Chem. Jpn.*, 1988, **46**, 861.	50
B-88MI 202-02	B. J. Wakefield; "Organolithium Methods," Academic Press, New York, 1988.	67, 68
B-88MI 203-01	S. Patai, Z. Rappoport and C. J. M. Stirling (eds.); "The Chemistry of Sulfones and Sulfoxides," Wiley, New York, 1988.	142
B-88MI 203-02	J. Drabowicz, P. Kielbasinski and M. Mikolajczyk; in "The Chemistry of Sulfones and Sulfoxides," eds. S. Patai, Z. Rappoport and C. J. M. Stirling, Wiley, New York, 1988, ch. 8.	144
B-88MI 203-03	K. Schank; in "The Chemistry of Sulfones and Sulfoxides," eds. S. Patai, Z. Rappoport and C. J. M. Stirling, Wiley, New York, 1988, ch. 7.	154
B-88MI 203-04	S. Braverman; in "The Chemistry of Sulfones and Sulfoxides," eds. S. Patai, Z. Rappoport and C. J. M. Stirling, Wiley, New York, 1988, ch. 14.	159, 196
B-88MI 204-01	A. Krief and L. Hevesi; "Organoselenium in Chemistry I," Springer-Verlag, Berlin, 1988, p. 113.	287
B-88MI 206-01	K. B. G. Torssell; in "Nitrile Oxides, Nitrones and Nitronates in Organic Synthesis," VCH Publishers, New York, 1988, p. 81.	345

B-88MI 206-02 K. B. G. Torssell; in "Nitrile Oxides, Nitrones and Nitronates in Organic Synthesis," VCH Publishers, New York, 1988, p. 191. 352

B-88MI 206-03 K. B. G. Torssell; in "Nitrile Oxides, Nitrones and Nitronates in Organic Synthesis," VCH Publishers, New York, 1988, pp. 20, 192–255. 354

B-88MI 206-04 K. B. G. Torssell; in "Nitrile Oxides, Nitrones and Nitronates in Organic Synthesis," VCH Publishers, New York, 1988, p. 110. 356

B-88MI 210-01 A. Pelter, K. Smith and H. C. Brown; "Borane Reagents," Academic Press, London, 1988. 514, 518, 519, 520, 523, 528, 529, 531, 533

B-88MI 210-02 E. W. Colvin; "Silicon Reagents in Organic Synthesis," Academic Press, London, 1988 (reprinted 1990). 535

88MI 210-03 K. Tamao; *J. Synth. Org. Chem., Jpn.*, 1988, **46**, 8671. 536, 540

B-88MI 210-04 G. van Look; "Silylating Agents," Fluka Chemie AG, Buchs, 1988. 539

B-88MI 211-01 B. J. Wakefield; "Organolithium Methods," Academic Press, London, 1988. 551

88MI 211-02 G. A. Abakumov, E. N. Gladyshev, V. K. Cherkasov and A. P. Fokeer; *Metalloorg. Khim.*, 1988, **1**, 221. 569

B-88MI 215-01 L. W. Haynes and A. G. Cook; in "Enamines, Synthesis, Structure, and Reactions," ed. A. G. Cook, Marcel Dekker, New York, 1988, pp. 103–163. 738, 739, 751

88MI 217-01 W. Tao, T. Yang and Y. Zhang; *Gaodeng Xuexiao Huaxue Xuebao*, 1988, **9**, 696 (*Chem. Abstr.*, 1989, **111**, 78 214c). 877

B-88MI 218-01 A. Pelter, K. Smith and H. C. Brown; "Borane Reagents," Academic Press, London, 1988. 900, 903, 904, 906

B-88MI 218-02 B. J. Wakefield; "Organolithium Methods," Academic Press, London, 1988. 907, 908, 909, 910, 942

B-88MI 218-03 E. W. Colvin; "Silicon Reagents in Organic Synthesis," Academic Press, London, 1988. 909

B-88MI 219-01 A. Pelter, K. Smith, and H. C. Brown; "Borane Reagents," Academic Press, New York, 1988. 966

B-88MI 221-01 L. Brandsma; in "Preparative Acetylenic Chemistry," 2nd ed, Elsevier, Amsterdam, 1988, p. 174. 1011, 1014, 1016, 1017, 1018, 1025, 1031, 1032, 1034, 1035, 1036

B-88MI 223-01 F. Glockling; in "Gmelin Handbook of Inorganic Chemistry, Organogermanium Compounds," Part 1, ed. U. Küerke, Springer, Berlin, 8th edn, 1988, p. 103. 1092, 1093

B-88MI 223-02 P. G. Harrison; "The Chemistry of Tin," Blackie, London, 1988, p. 1. 1100

88OCS587 H. J. Breunig, W.-W. du Mont and T. Severengiz; *Organomet. Synth.*, 1988, **4**, 587. 499

88OM84 M. J. Dabdoub and J. V. Comasseto; *Organometallics*, 1988, **7**, 84. 1036, 1037

88OM1196 D. Young and W. Kitching; *Organomet.*, 1988, **7**, 1196. 210

88OM1877 P. Legzdins, E. C. Phillips, S. J. Rettig, L. Sanchez, J. Trotter and V. C. Yee; *Organometallics*, 1988, **7**, 1877. 573

88OM2250 R. H. Fish, H. S. Kim, J. E. Babin and R. D. Adams; *Organometallics*, 1988, **7**, 2250. 327

88OM2289 S. Collins, W. P. Dean and D. G. Ward; *Organometallics*, 1988, **7**, 2289. 76

88OR(35)421 E. J. Behrman; *Org. React.*, 1988, **35**, 421. 651

88OR(35)513 M. Hudlicky; *Org. React.*, 1988, **35**, 513. 19, 362

88OS(67)20 H. Takaya, S. Akutagawa and R. Noyori; *Org. Synth.*, 1988, **67**, 20. 841

88OS(67)23 J. M. Renga and H. J. Reich; *Org. Synth.*, 1988, **67**, 23. 282

88OS(67)108 C. M. J. Fox and S. V. Ley; *Org. Synth.*, 1988, **67**, 108. 271

88OS(67)157 H.-S. Lin, M. J. Coghlan and L. A. Paquette; *Org. Synth.*, 1988, **67**, 157. 238

88OSC(4)235 M. E. Alonso and H. Aragono; *Org. Synth., Coll. Vol.*, 1988, **4**, 235. 220, 270

88OSC(6)43 Z. Cohen, H. Varkony, E. Keinan and Y. Mazur; *Org. Synth., Coll. Vol.*, 1988, **6**, 43. 86

88OSC(6)56 R. K. Müller, R. Joos, D. Felix, J. Schreiber, C. Wintner and A. Eschenmoser; *Org. Synth., Coll. Vol.*, 1988, **6**, 56. 380

88OSC(6)78 R. Ohme, H. Preuschhof and H.-U. Heyne; *Org. Synth., Coll. Vol.*, 1988, **6**, 78. 403

88OSC(6)90 Y. Ogata, T. Sugimoto and M. Inaishi; *Org. Synth. Coll. Vol.*, 1988, **6**, 90. 21

88OSC(6)175 J. Jacques and A. Marquet; *Org. Synth. Coll. Vol.*, 1988, **6**, 175. 28

88OSC(6)193 M. Gaudry and A. Marquet; *Org. Synth. Coll. Vol.*, 1988, **6**, 193. 28

88OSC(6)403 L. A. Carpino and L. V. McAdams, III; *Org. Synth. Coll. Vol.*, 1988, **6**, 403. 29

88OSC(6)424 J. E. Oliver and P. E. Sonnet; *Org. Synth. Coll. Vol.*, 1988, **6**, 424. 26

88OSC(6)440 W. J. Middleton and E. M. Bingham; *Org. Synth., Coll. Vol.*, 1988, **6**, 440. 362

88OSC(6)503 S. Zen and E. Kaji; *Org. Synth., Coll. Vol.*, 1988, **6**, 503. 394

88OSC(6)512 M. R. Ashcroft and H. M. R. Hoffmann; *Org. Synth. Coll. Vol.*, 1988, **6**, 512. 29

88OSC(6)520 R. Noyori, K. Yokoyama and Y. Hayakawa; *Org. Synth. Coll. Vol.*, 1988, **6**, 520. 28

88OSC(6)542 D. Enders, R. Pieter, B. Renger and D. Seebach; *Org. Synth., Coll. Vol.*, 1988, **6**, 542. 385

88OSC(6)592 J. E. McMurry; *Org. Synth., Coll. Vol.*, 1988, **6**, 592. 746

88OSC(6)634 J. G. Calzada and J. Hooz; *Org. Synth. Coll. Vol.*, 1988, **6**, 634. 7

88OSC(6)648 J. E. McMurray and J. Melton; *Org. Synth., Coll. Vol.*, 1988, **6**, 648. 395

88OSC(6)711 C. Rappe; *Org. Synth. Coll. Vol.*, 1988, **6**, 711. 28

88OSC(6)737 T. Cohen and R. J. Ruffner; *Org. Synth., Coll. Vol.*, 1988, **6**, 737. 710

88OSC(6)797 S. Zen, M. Koyama and S. Koto; *Org. Synth., Coll. Vol.*, 1988, **6**, 797. 398

88OSC(6)799 J. E. McMurry, J. H. Musser, I. Fleming, J. Fortunak and C. Nübling; *Org. Synth. Coll. Vol.*, 1988, **6**, 799. 392

88OSC(6)803 A. Calder, A. R. Forrester and S. P. Hepburn; *Org. Synth., Coll. Vol.*, 1988, **6**, 803. 390, 397

88OSC(6)830 H. N. Rydon; *Org. Synth. Coll. Vol.*, 1988, **6**, 830. 8

88OSC(6)835 W. J. Middleton and E. M. Bingham; *Org. Synth. Coll. Vol.*, 1988, **6**, 835. 19

88OSC(6)936 R. C. Cookson, S. S. Gupta, J. D. R. Stevens and C. T. Watts; *Org. Synth., Coll. Vol.*, 1988, **6**, 936. 374

88OSC(6)948 J. K. Crandall and L. C. Crawley; *Org. Synth., Coll. Vol.*, 1988, **6**, 948. 420

88OSC(6)991	J. Clabattoni, E. C. Nathan, A. E. Feiring and P. J. Kocienski; *Org. Synth. Coll. Vol.*, 1988, **6**, 991.	28
88OSC(6)1016	R. B. Woodward, I. J. Pachter and M. L. Scheinbaum; *Org. Synth.*, *Coll. Vol.*, 1988, **6**, 1016.	223
88PAC123	R. W. Hoffman; *Pure Appl. Chem.*, 1988, **60**, 123.	75, 529
88PAC1597	R. Noyori, S. Suga, K. Kawai, S. Okada and M. Kitamura; *Pure Appl. Chem.*, 1988, **60**, 1597.	71
88PAC1627	M. Schlosser; *Pure Appl. Chem.*, 1988, **60**, 1627.	557
88PAC1705	Y. Guindon, P. C. Anderson, C. Yoakim, Y. Girard, S. Berthiaume and H. E. Morton; *Pure Appl. Chem.*, 1988, **60**, 1705.	525
88POL77	A. R. Barron, A. H. Cowley, R. A. Jones, C. M. Nunn and D. L. Westmoreland; *Polyhedron*, 1988, **7**, 77.	511
88POL2601	M. Ates, H. Breunig and S. Gülec; *Polyhedron*, 1988, **7**, 2601.	893
88PS(35)349	W. W. Epstein and M. Garrossian; *Phosphorus Sulfur*, 1988, **35**, 349.	446
88PS(37)25	C. Smith, L. M. Tunstad and G. Carlos; *Phosphorus Sulfur*, 1988, **37**, 25.	89
88PS(38)97	H. J. Breunig; *Phosphorus Sulfur*, 1988, **38**, 97.	499, 501
88RCR447	M. D. Mizhiritskii and V. O. Reikhsfel'd; *Russ. Chem. Rev.* (*Engl. Transl.*), 1988, **57**, 447.	540
88S1	E. Negishi and T. Takahashi; *Synthesis*, 1988, 1.	565, 919, 925, 926, 969, 986
88S58	A. G. Martinez, A. C. Villalobos and M. O. Ruiz; *Synthesis*, 1988, 58.	36
88S103	R. W. Hoffmann and S. Dresely; *Synthesis*, 1988, 103.	900
88S130	G. Zweifel and T. M. Shoup; *Synthesis*, 1988, 130.	901
88S133	S.-I. Kang and C. P. Spears; *Synthesis*, 1988, 133.	281
88S138	S. El Cherif and L. René; *Synthesis*, 1988, 138.	744
88S142	J. W. Perich and R. B. Johns; *Synthesis*, 1988, 142.	107
88S144	J. Barluenga, F. Aznar and M. Bayod; *Synthesis*, 1988, 144.	676
88S175	N. Miyoshi, Y. Hatayama, I. Ryu, N. Kambe, T. Murai, S. Murai, and N. Sonoda; *Synthesis*, 1988, 175.	281
88S229	M. A. Ashwell and R. F. W. Jackson; *Synthesis*, 1988, 229.	393
88S234	J. Barluenga, J. L. Fernandez-Simon, J. M. Concellon and M. Yus; *Synthesis*, 1988, 234.	936
88S236	H. Suzuki, M. Aihara, H. Yamamoto, Y. Takamoto and T. Ogawa; *Synthesis*, 1988, 236.	630
88S252	P. Brownbridge and I. C. Jowett; *Synthesis*, 1988, 252.	198, 721
88S287	W. M. Stalick, R. N. Hazlett and R. E. Morris; *Synthesis*, 1988, 287.	1025
88S344	C. Baldoli, P. Del Buttero, E. Licandro and S. Maiorana; *Synthesis*, 1988, 344.	691
88S347	S. Brandange and B. Rodriguez; *Synthesis*, 1988, 347.	325
88S366	S. Irifune, T. Kibayashi, Y. Ishii and M. Ogawa; *Synthesis*, 1988, 366.	34
88S379	M. S. Ashwood, L. A. Bell, P. G. Houghton and S. H. B. Wright; *Synthesis*, 1988, 379.	783
88S395	J. Viala and M. Santelli; *Synthesis*, 1988, 395.	446
88S403	S. Kondo, Y. Takeda and K. Tsuda; *Synthesis*, 1988, 403.	35
88S529	F. Szurdoki, S. Andreae, E. Baitz-Gács, J. Tamás, K. Valkó and E. Schmitz; *Synthesis*, 1988, 529.	374
88S545	S. Kajigaeshi, T. Kakinami, M. Moriwaki, S. Fujisaki, K. Maeno and T. Okamoto; *Synthesis*, 1988, 545.	21
88S587	N. De Kimpe, E. Stanoeva, R. Verhé and N. Schamp; *Synthesis*, 1988, 587.	312
88S743	S. C. Suri and R. D. Capman; *Synthesis*, 1988, 743.	386
88S749	M. Tiecco; *Synthesis*, 1988, 749.	10, 658
88S775	H. Ahlbrecht and D. Kornetzky; *Synthesis*, 1988, 775.	302
88S803	T. H. Kress and M. R. Leanna; *Synthesis*, 1988, 803.	942
88S805	S. M. Ali, Y. Matsuda, and S. Tanimoto; *Synthesis*, 1988, 805.	391
88S808	W. W. Zajac, T. R. Walters and J. M. Woods; *Synthesis*, 1988, 808.	390
88S833	G. Rosini and B. Ballini; *Synthesis*, 1988, 833.	391, 783
88S868	B. Jursic; *Synthesis*, 1988, 868.	25
88S900	A. Guy, A. Lemor, J. Doussot and M. Lemaire; *Synthesis*, 1988, 900.	317
88S911	H.-J. Cristau and Y. Ribeill; *Synthesis*, 1988, 911.	444
88S912	F. Plénat, S. Ibrahim and H.-J. Cristau; *Synthesis*, 1988, 912.	848
88S921	A. Kamimura and N. Ono; *Synthesis*, 1988, 921.	789
88S923	E. B. Merkushev; *Synthesis*, 1988, 923.	630
88S950	R. J. Alabaster, I. F. Cottrell, H. Marley and S. H. B. Wright.; *Synthesis*, 1988, 950.	906
88S985	M. C. Venuti and O. Ort; *Synthesis*, 1988, 985.	326
88S995	S. Itsumo, Y. Sakurai and K. Ito; *Synthesis*, 1988, 995.	59
88SC21	J. R. Hwu and N. Wang; *Synth. Commun.*, 1988, **18**, 21.	393
88SC29	J. L. Belletire and D. F. Fry; *Synth. Commun.*, 1988, **18**, 29.	319
88SC119	H.-J. Liu and V. Wisniewski; *Synth. Commun.*, 1988, **18**, 119.	36
88SC205	D. H. Causey, R. F. Mays, D. A. Shamblee and Y. S. Lo; *Synth. Commun.*, 1988, **18**, 205.	322
88SC227	M. G. Ranasinghe and P. Fuchs; *Synth. Commun.*, 1988, **18**, 227.	217
88SC253	J. A. Prieto, J. Suarez and G. L. Larson; *Synth. Commun.*, 1988, **18**, 253.	420
88SC285	L. Ylagan, A. Benjamin, A. Gupta and R. Engel; *Synth. Commun.*, 1988, **18**, 285.	458
88SC425	R. Tyka, G. Hagele and J. Peters; *Synth. Commun.*, 1988, **18**, 425.	454
88SC777	J. P. Leeds and H. A. Kirst; *Synth. Commun.*, 1988, **18**, 777.	314
88SC893	M. A. Brook and Jahangir; *Synth. Commun.*, 1988, **18**, 893.	313
88SC1061	N. P. Singh and Y. F. Biellman; *Synth. Commun.*, 1988, **18**, 1061.	157
88SC1071	R. R. Fraser and F. Kong; *Synth. Commun.*, 1988, **18**, 1071.	22
88SC1275	A. Loupy and C. Pardo; *Synth. Commun.*, 1988, **18**, 1275.	31
88SC1327	R. Singh and G. Just; *Synth. Commun.*, 1988, **18**, 1327.	633

88SC1363 G. W. Kabalka, N. M. Goudgaon and Y. Liang; *Synth. Commun.*, 1988, **18**, 1363. 307

88SC1379 W. E. Wymann, R. Davis, J. W. Patterson, Jr. and J. R. Pfister; *Synth. Commun.*, 1988, **18**, 1379. 683

88SC1661 G. Bram, A. Loupy and P. Pigeon; *Synth. Commun.*, 1988, **18**, 1661. 19

88SC1979 M. J. Dabdoub, J. V. Comasseto and A. L. Braga; *Synth. Commun.*, 1988, **18**, 1979. 1036, 1037

88SC2055 J. E. Hernández, S. Fernández and G. Arias; *Synth. Commun.*, 1988, **18**, 2055. 29

88SC2083 T. Nishio, T. Tokunaga and Y. Omoto; *Synth. Commun.*, 1988, **18**, 2083. 85

88SC2135 J. Zakrzewski; *Synth. Commun.*, 1988, **18**, 2135. 38

88SC2161 T. P. Sura, M. M. V. Ramana and N. A. Kudav; *Synth. Commun.*, 1988, **18**, 2161. 91

88T147 D. H. R. Barton, G. Bashiardes and J.-L. Fourrey; *Tetrahedron*, 1988, **44**, 147. 30

88T281 E. Block and M. Aslam; *Tetrahedron*, Tetrahedron Report No. 228, 1988, **44**, 281. 62

88T491 A. Bernardi, S. Cardani, C. Scolastico and R. Villa; *Tetrahedron*, 1988, **44**, 491. 46

88T1153 D. H. R. Barton, B. Lacher, B. Misterkiewicz and S. Z. Zard; *Tetrahedron*, 1988, **44**, 1153. 26

88T2457 S. Arseniyadis, P. Q. Huang, D. Piveteau and H.-P. Husson; *Tetrahedron*, 1988, **44**, 2457. 25

88T2751 H. C. Brown, M. W. Rathke, M. M. Rogic and N. R. de Lue; *Tetrahedron*, 1988, **44**, 2751. 34

88T2763 H. C. Brown and C. F. Lane; *Tetrahedron*, 1988, **44**, 2763. 30

88T2773 H. C. Brown, C. F. Lane and N. R. de Lue; *Tetrahedron*, 1988, **44**, 2773. 30

88T2785 H. C. Brown and N. R. de Lue; *Tetrahedron*, 1988, **44**, 2785. 23

88T2875 T. Hamatani, S. Matsubara, H. Matsuda and M. Schlosser; *Tetrahedron*, 1988, **44**, 2875. 19

88T3761 S. Anwar and A. P. Davies; *Tetrahedron*, 1988, **44**, 3761. 62

88T3919 G. W. Daub, D. A. Heerding and L. E. Overman; *Tetrahedron*, 1988, **44**, 3919. 922

88T3953 G. Procter, A. T. Russell, P. J. Murphy, T. S. Tan and A. N. Mather; *Tetrahedron*, 1988, **44**, 3953. 921

88T3997 K. Tamao, E. Nakajo and Y. Ito; *Tetrahedron*, 1988, **44**, 3997. 935

88T4009 H. Shiragami, T. Kawamoto, K. Imi, S. Matsubara, K. Utimoto and H. Nozaki; *Tetrahedron*, 1988, **44**, 4009. 922, 923

88T4073 Y. Kitano, T. Matsumoto and F. Sato; *Tetrahedron*, 1988, **44**, 4073. 921

88T4134 M. Fujita, M. Obayashi and T. Hiyama; *Tetrahedron*, 1988, **44**, 4134. 70

88T4277 K. Fugami, J.-I Hibino, S. Nakatsukasa, S. Matsubara, K. Oshima, K. Utimoto and H. Nozaki; *Tetrahedron*, 1988, **44**, 4277. 918, 919, 939

88T4403 J. A. Soderquist and S.-J. Hwang Lee; *Tetrahedron*, 1988, **44**, 4403. 58

88T4941 M. R. J. Dorritty, M. Dowling, R. Grigg, V. Sridharan and J. F. Malone; *Tetrahedron*, 1988, **44**, 4941. 245

88T5597 K. Ramalingam, P. Nanjappa, D. M. Kelvin and R. W. Woodward; *Tetrahedron*, 1988, **44**, 5597. 517

88T6525 E. Baciocchi and M. Crescenzi; *Tetrahedron*, 1988, **44**, 6525. 21, 28

88T6677 B. Juršić; *Tetrahedron*, 1988, **44**, 6677. 683

88T6681 A. Garcia, E. Lete, M. Jesus Villa, E. Dominguez and M. D. Badia; *Tetrahedron*, 1988, **44**, 6681. 89

88T6687 P. Kielbasinoki, R. Zurawinski, J. Drabowicz and M. Milolajczyk; *Tetrahedron*, 1988, **44**, 6687. 156, 721

88T7037 J. M. Dener and D. J. Hart; *Tetrahedron*, 1988, **44**, 7037. 325

88TL21 H. C. Brown and N. G. Bhat; *Tetrahedron Lett.*, 1988, **29**, 21. 904

88TL139 R. Sanchez and W. Scott; *Tetrahedron Lett.*, 1988, **29**, 139. 5[illegible], 421

88TL167 B. P. Branchaud, M. S. Meïr and Y. Choi; *Tetrahedron Lett.*, 1988, **29**, 167. 579

88TL223 Y. Sakito, Y. Yoneyoshi and G. Suzukamo; *Tetrahedron Lett.*, 1988, **29**, 223. 314

88TL261 S. Sharma and A. C. Oehlschlager; *Tetrahedron Lett.*, 1988, **29**, 261. 903

88TL535 G. Märkl and S. Dietl; *Tetrahedron Lett.*, 1988, **29**, 535. 880

88TL539 G. Märkl and S. Dietl; *Tetrahedron Lett.*, 1988, **29**, 539. 880

88TL585 I. Paterson and M. A. Lister; *Tetrahedron Lett.*, 1988, **29**, 585. 533

88TL609 H. Hopf and E. Naujoks; *Tetrahedron Lett.*, 1988, **29**, 609. 940

88TL635 D. Huber, G. Andermann and G. Leclerc; *Tetrahedron Lett.*, 1988, **29**, 635. 753

88TL897 V. K. Yadav and A. G. Fallis; *Tetrahedron Lett.*, 1988, **29**, 897. 141

88TL1065 K. Takai, Y. Kataoka, T. Okazoe and K. Utimoto; *Tetrahedron Lett.*, 1988, **29**, 1065. 700

88TL1153 D. H. R. Barton, D. Bridon and S. Z. Zard; *Tetrahedron*, 1988, **44**, 1153.

88TL1259 H.-J. Gais, J. Vollhardt, G. Hellmann, H. Paulus and H. J. Lindner; *Tetrahedron Lett.*, 1988, **29**, 1259. 1000

88TL1319 K. Smith and M. Butters; *Tetrahedron Lett.*, 1988, 1319. 620

88TL1395 C. Chen, Y. Shen and Y.-Z. Huang; *Tetrahedron Lett.*, 1988, **29**, 1395. 73

88TL1501 A. Luxen and J. R. Barrio; *Tetrahedron Lett.*, 1988, 1501. 613

88TL1631 E. Negishi, D. R. Swanson, and S. R. Miller; *Tetrahedron Lett.*, 1988, **29**, 1631. 969

88TL1811 Y. Satoh, H. Serizawa, N. Miyaura, S. Hara and A. Suzuki; *Tetrahedron Lett.*, 1988, **29**, 1811. 902, 927, 939

88TL1825 P. Cuadrado, A. M. Gonzales, F. J. Pulido and I. Fleming; *Tetrahedron Lett.*, 1988, **29**, 1825. 929

88TL1833 P. J. Garratt and A. Tsotinis; *Tetrahedron Lett.*, 1988, **29**, 1833. 929, 938

88TL2195 S. E. Yoo, S. H. Lee and N. J. Kim; *Tetrahedron Lett.*, 1988, **29**, 2195. 324

88TL2231 P. D. Bailey and J. S. Bryans; *Tetrahedron Lett.*, 1988, **29**, 2231. 326

88TL2251 G. Bartoli, E. Marcantoni, M. Bosco and R. Dalpozzo; *Tetrahedron Lett.*, 1988, **29**, 2251. 318

88TL2295 G. Sauvé, T. S. Mansour, P. Lachance and B. Belleau; *Tetrahedron Lett.*, 1988, **29**, 2295. 749

88TL2299 G. Sauvé, N. Le Berre and B. Zacharie; *Tetrahedron Lett.*, 1988, **29**, 2299. 749

88TL2377 H. Heaney, G. Papageorgiou and R. F. Wilkins; *Tetrahedron Lett.*, 1988, **29**, 2377. 316

88TL2425 J. A. Soderquist and C. L. Anderson; *Tetrahedron Lett.*, 1988, **29**, 2425. 936

88TL2479 S. V. Mortlock and E. J. Thomas; *Tetrahedron Lett.*, 1988, **29**, 2479. 79

88TL2483 A. Armstrong, I. Brackenridge, R. F. W. Jackson and J. M. Kirk; *Tetrahedron Lett.*, 1988, **29**, 2483. 90

88TL2631 H. C. Brown, N. G. Bhat and M Srebnik; *Tetrahedron Lett.*, 1988, **29**, 2631. 1086

88TL2701 V. S. Martín, M. T. Nunez and C. E. Tonn; *Tetrahedron Lett.*, 1988, **29**, 2701. 84

88TL2735 P. Golding, R. W. Millar, N. C. Paul and D. H. Richards; *Tetrahedron Lett.*, 1988, **29**, 2735. 386

88TL2777 J. A. Soderquist and C. L. Anderson; *Tetrahedron Lett.*, 1988, **29**, 2777. 936

88TL2851 T. Satoh, T. Oohara and K. Yamakawa; *Tetrahedron Lett.*, 1988, **29**, 2851. 85

88TL2859 J. Barluenga, F. Fonbelo, F. J. Fananás and M. Yus; *Tetrahedron Lett.*, 1988, **29**, 2859. 553

88TL2865 S. Thea and G. Cevasco; *Tetrahedron Lett.*, 1988, **29**, 2865. 168

88TL2967 G. Xie, L. Xu, J. Hu, S. Ma, W. Hou and F. Tao; *Tetrahedron Lett.*, 1988, **29**, 2967. 43

88TL2973 S.-I. Murahashi, Y. Imada, Y. Taniguchi and Y. Kodera; *Tetrahedron Lett.*, 1988, **29**, 2973. 308, 344, 351

88TL3235 Y. Vannoorenberghe and G. Buono; *Tetrahedron Lett.*, 1988, **29**, 3235. 64

88TL3251 J.-B. Baudin and S. A. Julia; *Tetrahedron Lett.*, 1988, **29**, 3251. 247, 361, 363

88TL3255 J.-B. Baudin and S. A. Julia; *Tetrahedron Lett.*, 1988, **29**, 3255. 201

88TL3265 A. Krief, W. Dumont and J.-L. Laboureur; *Tetrahedron Lett.*, 1988, **29**, 3265. 287

88TL3269 A. Krief, W. Dumont and A. F. DeMahieu; *Tetrahedron Lett.*, 1988, **29**, 3269. 287

88TL3311 J.-C. Quiron, D. S. Grierson, J. Royer and H.-P. Husson; *Tetrahedron Lett.*, 1988, **29**, 3311. 331

88TL3365 M. L. Boys, E. W. Collington, H. Finch, S. Swanson and J. F. Whitehead; *Tetrahedron Lett.*, 1988, 3365. 612

88TL3383 S. A. Batcheller and S. Masamune; *Tetrahedron Lett.*, 1988, **29**, 3383. 947, 949

88TL3445 S. L. Buchwald, Q. Fang, and S. M. King; *Tetrahedron Lett.*, 1988, **29**, 3445. 969

88TL3455 K. E. Rodriques, A. Basha, J. B. Summers and D. W. Brooks; *Tetrahedron Lett.*, 1988, **29**, 3455. 343, 351

88TL3563 Y. Masuyama, K. Otake and Y. Kurusu; *Tetrahedron Lett.*, 1988, **29**, 3563. 74, 80

88TL3579 J. van der Louw, J. L. van der Baan, H. Stichter, G. J. J. Out, F. Bickelhaupt and G. W. Klumpp; *Tetrahedron Lett.*, 1988, **29**, 3579. 324

88TL3705 V. Fiandanese, G. Marchese, G. Mascolo, F. Naso and L. Ronzini; *Tetrahedron Lett.*, 1988, **29**, 3705. 931

88TL3811 H. Tanaka, H. Ohimane, H. Furita, Y. Ikemoto and S. Tori; *Tetrahedron Lett.*, 1988, **29**, 3811. 312

88TL3887 M. C. P. Yek, P. Knochel and L. E. Santa; *Tetrahedron Lett.*, 1988, **29**, 3887. 69

88TL4093 T. Satoh, T. Oohara and K. Yamakawa; *Tetrahedron Lett.*, 1988, **29**, 4093. 322

88TL4113 Y. Zhang and Z. Hu; *Tetrahedron Lett.*, 1988, **29**, 4113. 55

88TL4139 N. Nakajima, K. Horita, R. Abe and O. Yonemitsu; *Tetrahedron Lett.*, 1988, **29**, 4139. 90

88TL4233 S. I. Bell and S. M. Weinreb; *Tetrahedron Lett.*, 1988, **29**, 4233. 245

88TL4269 F. A. Davis, G. Sankar Lal and J. Wei; *Tetrahedron Lett.*, 1988, **29**, 4269. 700

88TL4453 K. Youn, S. W. Lee and C. S. Pak; *Tetrahedron Lett.*, 1988, **29**, 4453. 66

88TL4477 P. C. B. Page, S. S. Klair, M. P. Brown, M. M. Harding, C. S. Smith, S. J. Maginn and S. Mulley; *Tetrahedron Lett.*, 1988, **29**, 4477. 134

88TL4563 C. Hashimoto and H.-P. Husson; *Tetrahedron Lett.*, 1988, **29**, 4563. 328

88TL4719 H. Ohki, M. Wada and K.-Y. Akiba; *Tetrahedron Lett.*, 1988, **29**, 4719. 695

88TL4769 K. Nakamura, Y. Inoue, J. Shibahara, S. Oka and A. Ohno; *Tetrahedron Lett.*, 1988, **29**, 4769. 398

88TL4827 J. A. Dodge and A. R. Chamberlin; *Tetrahedron Lett.*, 1988, **29**, 4827. 660

88TL4851 A. Tenaglia and B. Waegell; *Tetrahedron Lett.*, 1988, **29**, 4851. 85

88TL4917 R. L. Danheiser, A. Nishida, S. Savariar and M. Trova; *Tetrahedron Lett.*, 1988, **29**, 4917. 1024, 1025

88TL4977 H. H. Wasserman and V. Rusiecki; *Tetrahedron Lett.*, 1988, **29**, 4977. 326

88TL5083 P. L. Gaus, S. W. Gerrliz and P. M. Jeffries; *Tetrahedron Lett.*, 1988, **29**, 5083. 753

88TL5245 G. Billeb, W. P. Neumann and G. Steinhoff; *Tetrahedron Lett.*, 1988, **29**, 5245. 947

88TL5263 K. Takai, K. Nitta and K. Utimoto; *Tetrahedron Lett.*, 1988, **29**, 5263. 73

88TL5355 G. Himbert and D. Faul; *Tetrahedron Lett.*, 1988, **29**, 5355. 1053

88TL5419 Y. Mori, M. Kuhara, A. Takeuchi and M. Suzuki; *Tetrahedron Lett.*, 1988, **29**, 5419. 61

88TL5645 W. Oppolzer and R. N. Radinov; *Tetrahedron Lett.*, 1988, **29**, 5645. 71

88TL5657 A. Cornélis, L. Delande, A. Gerstmans and P. Laszlo; *Tetrahedron Lett.*, 1988, **29**, 5657. 792

88TL5667 T. Hayashi, Y. Matsumoto, T. Kujoi and Y. Ito; *Tetrahedron Lett.*, 1988, **29**, 5667. 78

88TL5725 J. N. Reed, J. Rotchford and D. Strickland; *Tetrahedron Lett.*, 1988, **29**, 5725. 328

88TL5783 S. Kajigaeshi, T. Kakinami, M. Moriwaki, T. Tanaka and S. Fujisaki; *Tetrahedron Lett.*, 1988, **29**, 5783. 21

88TL5811 F. Camps, J. Coll, A. Llebaria and J. M. Moreto; *Tetrahedron Lett.*, 1988, 5811. 624

88TL6009 C. J. Moody, A. M. Z. Slawin, R. J. Taylor and D. J. Williams; *Tetrahedron Lett.*, 1988, **29**, 6009. 1000

88TL6087 E. Differding and R. W. Lang; *Tetrahedron Lett.*, 1988, **29**, 6087. 16, 365

88TL6287 M. Hasebe and T. Tsuchiya; *Tetrahedron Lett.*, 1988, **29**, 6287. 13

88TL6393 H. L. Holland, M. Conn, P. C. Chenchaiah and F. M. Brown; *Tetrahedron Lett.*, 1988, **29**, 6393. 304

88TL6409 E. J. Corey, P. D. S. Jardine and T. Mohri; *Tetrahedron Lett.*, 1988, **29**, 6409. 1035

88TL6453 Y. H. Kim and D. C. Yoon; *Tetrahedron Lett.*, 1988, **29**, 6453. 150

88TL6493 D. Tanner, H. H. Ming and M. Bergdahl; *Tetrahedron Lett.*, 1988, **29**, 6493. 331

88TL6541 R. M. Lett, L. E. Overman and J. Zablocki; *Tetrahedron Lett.*, 1988, **29**, 6541. 931

88TL6715 R. B. Millar and T. Tsang; *Tetrahedron Lett.*, 1988, **29**, 6715. 328

88TL6719 J. J. Pegram and C. B. Anderson; *Tetrahedron Lett.*, 1988, **29**, 6719. 914, 939

88TL6787 N. S. Simpkins; *Tetrahedron Lett.*, 1988, **29**, 6787. 280

88TL6835 I. Sircar, K. R. Anderson and L. Bonadies; *Tetrahedron Lett.*, 1988, **29**, 6835. 28

88TL6843 N. Farschtschi and D. G. Gerenstein; *Tetrahedron Lett.*, 1988, **29**, 6843. 410

88TL6873 P. A. Jacobi and J. I. Kravitz; *Tetrahedron Lett.*, 1988, **29**, 6873. 672

88TL6949 M. Yamato, K. Hashigaki, S. Ishikawa and N. Qais; *Tetrahedron Lett.*, 1988, **29**, 6949. 328

88TL6955 K. Tamao, K. Maeda, T. Tanaka and Y. Ito; *Tetrahedron Lett.*, 1988, **29**, 6955. 914

88TL6985 R. W. Carling and P. D. Leeson; *Tetrahedron Lett.*, 1988, **29**, 6985. 328

88USP4721813 H. W. Mark and J. S. Roberts; *US Pat.* 4 721 813 (1988) (*Chem. Abstr.*, 1988, **108**, 188 910x). 214

88ZAAC(567)39 E. Fluck, P. Kuhm and H. Riffel; *Z. Anorg. Allg. Chem.*, 1988, **567**, 39. 1074

88ZN(B)295 W. Wolfsberger; *Z. Naturforsch., Teil B*, 1988, **43**, 295. 437

88ZN(B)998 H. G. Breunig and S. Gulec; *Z. Naturforsch., Teil B*, 1988, **43**, 998. 501

88ZOB33 R. I. Yurchenko, E. E. Lavrova and A. G. Yurchenko; *Zh. Obshch. Khim.*, 1988, **58**, 33. 428

88ZOR1088 V. I. Saloutin, M. B. Bobrov and K. I. Pashkevich; *Zh. Org. Khim.*, 1988, **24**, 1088 (*Chem. Abstr.*, 1989, **110**, 114 286t). 227

89ACR152 T. Cohen and M. Bhupathy; *Acc. Chem. Res.*, 1989, **22**, 152. 553

89AG(E)38 W. Oppolzer; *Angew. Chem., Int. Ed. Engl.*, 1989, **28**, 38. 562

89AG(E)69 D. Hoppe and O. Zschage; *Angew. Chem., Int. Ed. Engl.*, 1989, **28**, 69. 997, 999

89AG(E)97 M. Mikolajczyk, P. Lyzwa, J. Drabowicz, M. Wieczorek and G. Bujacz; *Angew. Chem., Int. Ed. Engl.*, 1989, **28**, 97. 184

89AG(E)181 R. Hochstrasser and J. Wirz; *Angew. Chem., Int. Ed. Engl.*, 1989, **28**, 181. 1022

89AG(E)202 E. Bovenschulte, P. Metz and G. Henkel; *Angew. Chem., Int. Ed. Engl.*, 1989, **28**, 202. 207, 721

89AG(E)225 O. Wagner, M. Ehle and M. Regitz; *Angew. Chem., Int. Ed. Engl.*, 1989, **28**, 225. 859, 865

89AG(E)277 G. Boche; *Angew. Chem., Int. Ed. Engl.*, 1989, **28**, 277. 1000, 1001

89AG(E)325 B. Giese, J. Harting, J. He, O. Hüter and A. Koch; *Angew. Chem., Int. Ed. Engl.*, 1989, **28**, 325. 578, 579

89AG(E)351 Y. Tamaru, H. Tanigawa, T. Yamamoto and Z. Yoshida; *Angew. Chem., Int. Ed. Engl.*, 1989, **28**, 351. 588, 591

89AG(E)494 M. Riediker and R. O. Duthaler; *Angew. Chem., Int. Ed. Engl.*, 1989, **28**, 494. 80

89AG(E)610 P. Binger, P. Müller, R. Benn, and R. Mynott; *Angew. Chem., Int. Ed. Engl.*, 1989, **28**, 610. 984

89AG(E)755 A. Gigou, J.-P. Lellouche, J.-P. Beaucourt, L. Toupet and R. Grée; *Angew. Chem., Int. Ed. Engl.*, 1989, **28**, 755. 47

89AG(E)1025 H.-J. Gais, G. Hellmann, H. Günther, F. Lopez, H. J. Lindner and S. Braun; *Angew. Chem., Int. Ed. Engl.*, 1989, **28**, 1025. 1000

89AG(E)1208 A. H. Cowley and R. A. Jones; *Angew. Chem., Int. Ed. Engl.*; 1989, **28**, 1208. 511

89AG(E)1219 H. J. Bestmann, H. Behl and M. Bremer; *Angew. Chem., Int. Ed. Engl.*, 1989, **28**, 1219. 1088

89AG(E)1675 K. Bavert and M. Hangedorn; *Angew. Chem., Int. Ed. Engl.*, 1989, **28**, 1675. 468

89AG(E)1700 H. Schmidbaur, C. Paschalidis, O. Steigelmann and G. Müller; *Angew. Chem., Int. Ed. Engl.*, 1989, **28**, 1700. 444

89AP(322)181 I. C. Ivanov, S. K. Karagiosov and P. B. Sulay; *Arch. Pharm. (Weinheim, Ger.)*, 1989, **322**, 181. 309

89AX(C)1520 S. Boutiba, A. Kallel, J. E. Hila and E. F. Paulus; *Acta Crystallogr., Part C*, 1989, **45**, 1520. 1007

89BCJ128 T. Nozoe, K. Takase, M. Yasunami, M. Ando, H. Saito, K. Imafuku, B.-Z. Yin, M. Honda, Y. Goto, T. Hanaya, Y. Hara and H. Yamamoto; *Bull. Chem. Soc. Jpn.*, 1989, **62**, 128. 769

89BCJ439 S. Kajigaeshi, T. Kakinami, M. Moriwaki, T. Tanaka, S. Fujisaki and T. Okamoto; *Bull. Chem. Soc. Jpn.*, 1989, **62**, 439. 626

89BCJ1349 S. Kajigaeshi, T. Kakinami, F. Watanabe and T. Okamoto; *Bull. Chem. Soc. Jpn.*, 1989, **62**, 1349. 631

89BCJ2024 S. Kanemoto, M. Shimizu and H. Yoshioka; *Bull. Chem. Soc. Jpn.*, 1989, **62**, 2024. 11

89BCJ3373 T. Kakinami, H. Suenaga, T. Yamaguchi, T. Okamoto and S. Kajigaeshi; *Bull. Chem. Soc. Jpn.*, 1989, **62**, 3373. 626

89BCJ3376 S. Kajigaeshi, K. Murakawa, S. Fujisaki and T. Kakinami; *Bull. Chem. Soc. Jpn.*, 1989, **62**, 3376. 149

89BCJ3845 M. Y. Fujii, T. Aida, M. K. Yoshihara and A. Y. Ohno; *Bull. Chem. Soc. Jpn.*, 1989, **62**, 3845. 312

89BSF247 G. Bram, H. Galons, S. Labidalle, A. Loupy, M. Miocque, A. Petit, P. Pigeon and J. Sansoulet; *Bull. Soc. Chim. Fr.*, 1989, 247. 31

89BSF256 B. Escoula, I. Rico and A. Lattes; *Bull. Soc. Chim. Fr.*, 1989, 256. 19

89CB401 V. Schäfer, H. Pritzkow and W. Siebert; *Chem. Ber.*, 1989, **122**, 401. 903

89CB803 A. Schmuck and K. Seppelt; *Chem. Ber.*, 1989, **122**, 803. 873

89CB1247 C. Spang, T. Edelmann, M. Noltemeyer and H. W. Roesky; *Chem. Ber.*, 1989, **122**, 1247. 489

89CB1783 R. W. Hoffmann, K. Ditrich, G. Koster and R. Sturmer; *Chem. Ber.*, 1989, **122**, 1783. 530

89CB2213 D. A. Anderson, D. W. H. Rankin, H. E. Robertson, C. M. F. Frazao and H. Schmidbaur; *Chem. Ber.*, 1989, **122**, 2213. 622

89CC154 P. Rajakumar and A. Kannan; *J. Chem. Soc., Chem. Commun.*, 1989, 154. 1034, 1035

89CC313 M. Le Corre, E. Gheerbrant and H. Le Deit; *J. Chem. Soc., Chem. Commun.*, 1989, 313. 14

89CC852 T. Takahashi, M. Tamura, M. Saburi, Y. Uchida, and E. Negishi; *J. Chem. Soc., Chem. Commun.*, 1989, 852. 987

89CC1132 J. Barluenga, E. Aguilar, J. Joglar, B. Olano and S. Fustero; *J. Chem. Soc., Chem. Commun.*, 1989, 1132. 322

89CC1344 S. Takano, K. Samizu, T. Sugihara and K. Ogasawara; *J. Chem. Soc., Chem. Commun.*, 1989, 1344. 85

89CC1363 A. J. Pearson, J. G. Park, S. H. Yang and Y.-H. Chuang; *J. Chem. Soc., Chem. Commun.*, 1989, 1363. 684

89CC1387 R. P. Allen, B. P. Roberts and C. R. Willis; *J. Chem. Soc., Chem. Commun.*, 1989, 1387. 536

89CC1429 T. Harada, Y. Matsuda, J. Uchimura and A. Oku; *J. Chem. Soc., Chem. Commun.*, 1989, 1429. 41

89CC1548 A. J. Kresge and N. P. Schepp; *J. Chem. Soc., Chem. Commun.*, 1989, 1548. 638

89CC1636 K. Okada, K. Okamoto and M. Oda; *J. Chem. Soc., Chem. Commun.*, 1989, 1636. 13

89CC1700 K. Taneko, H. Ushio and H. Suzuki; *J. Chem. Soc., Chem. Commun.*, 1989, 1700. 71

89CC1789 H. Fujihara, R. Akaishi, T. Erata and N. Furukawa; *J. Chem. Soc., Chem. Commun.*, 1989, 1789. 279

89CC1881 M. Shimizu, Y. Nakahara and H. Yoshioka; *J. Chem. Soc., Chem. Commun.*, 1989, 1881. 18, 280

89CC1916 C. W. Jefford, A. Jaber and J. Boukduvalas; *J. Chem. Soc., Chem. Commun.*, 1989, 1916. 52

89CC1921 K. Tomioka, M. Nakajima and K. Koga; *J. Chem. Soc., Chem. Commun.*, 1989, 1921. 50

89CJC2061 J. Berthelot, C. Guette, P.-L. Desbene, J.-J. Basselier, P. Chaquin and D. Masure; *Can. J. Chem.*, 1989, **67**, 2061. 626

89CL463 S. Kajigaeshi, K. Asano, S. Fujisaki, T. Kakinami and T. Okamoto; *Chem. Lett.*, 1989, 463. 320

89CL569 S. Isayama and T. Mukaiyama; *Chem. Lett.*, 1989, 569. 478

89CL577 I. Shimizu and H. Ishii; *Chem. Lett.*, 1989, 577. 3

89CL761 T. Takahashi, M. Murakami, M. Kunishige, M. Saburi, Y. Uchida, K. Kozawa, T. Uchida, D. R. Swanson, and E. Negishi; *Chem. Lett.*, 1989, 761. 986, 988

89CL769 T. Ogawa, K. Hayami and H. Suzuki; *Chem. Lett.*, 1989, 769. 727

89CL849 T. Kawashima, S. Kojima and N. Inamoto; *Chem. Lett.*, 1989, 849. 460, 462

89CL1235 S. Tomoda and Y. Usuki; *Chem. Lett.*, 1989, 1235. 280

89CL1247 M. Fujiwara, A. Baba and H. Matsuda; *Chem. Lett.*, 1989, 1247. 699

89CL1267 T. Funabiki, M. Tsujimoto, S. Ozawa and S. Yoshida; *Chem. Lett.*, 1989, 1267. 650

89CL1401 T. Mukaiyama, H. Sugimura, T. Ohno and S. Kobayashi; *Chem. Lett.*, 1989, 1401. 81

89CL1647 Y. Masuyama, T. Tsunoda and Y. Kurusu; *Chem. Lett.*, 1989, 1647. 73

89CL2087 M. Takahaski, R. Okazaki and N. Inamoto; *Chem. Lett.*, 1989, 2087. 793, 794, 795

89CL2111 R. Sato, A. Onodera, T. Goto and M. Saito; *Chem. Lett.*, 1989, 2111. 234

89CL2213 N. Yazawa, H. Suzuki, Y. Yoshida, O. Furusawa and Y. Kimura; Chem. Lett., 1989, 2213. 614

89CL7573 T. Mukaiyama, T. Sogay and H. Takenoshita; *Chem. Lett.*, 1989, 7573. 43

89CPB304 Y. Ozaki and S.-W. Kim; *Chem. Pharm. Bull.*, 1989, **37**, 304. 662, 664

89CPB1434 Y. Ozaki, I. Oshio and S.-W. Kim; *Chem. Pharm. Bull.*, 1989, **37**, 1434. 662

89CR1947 E. Erdik and M. Ay; *Chem. Rev.*, 1989, **89**, 1947. 305

89CRV689 L. Craine and M. Raban; *Chem. Rev.*, 1989, **89**, 689. 357, 358, 361, 767

89CRV713 W. R. Cullen and K. J. Reimer; *Chem. Rev.*, 1989, **89**, 713. 487

89CRV863 B. E. Maryanoff and A. B. Reitz; *Chem. Rev.*, 1989, **89**, 863. 1003, 1004

89CRV1187 R. W. Murray; *Chem. Rev.*, 1989, **89**, 1187. 90

89CRV1413 D. Crich and L. Quintero; *Chem. Rev.*, 1989, 1413. 140

89CRV1447 D. J. Hart and D.-C. Ha; *Chem. Rev.*, 1989, **89**, 1447. 544

89CRV1535 D. S. Matteson; *Chem. Rev.*, 1989, **89**, 1535. 529, 900, 904

89CRV1553 M. Midland; *Chem. Rev.*, 1989, **89**, 1553. 57

89CS457 E. Negishi; *Chem. Scr.*, 1989, **29**, 457. 988

89CZ349 W. Wolfsberger; *Chem.-Ztg.*, 1989, **113**, 349. 468

89H(28)63 M. Ihara, Y. Takino and K. Fukumoto; *Heterocycles*, 1989, **28**, 63. 328

89H(28)295 A. I. Meyers and J. Guiles; *Heterocycles*, 1989, **28**, 295. 328

89H(29)1861 S. Takano, Y. Iwabuchi and K. Ogasawara; *Heterocycles*, 1989, **29**, 1861. 323

89HCA1008 P.-A. Carrupt and P. Vogel; *Helv. Chim. Acta*, 1989, **72**, 1008. 616

89HCA1248 E. Differding and R. W. Lang; *Helv. Chim. Acta*, 1989, **72**, 1248. 16, 365

89IC1295 A. B. Burg; *Inorg. Chem.*, 1989, **28**, 1295. 1008

89IC1627 R. Minkwitz and A. Liedtke; *Inorg. Chem.*, 1989, **28**, 1627. 510

89IS8 O. J. Scherer and R. Anselmann; *Inorg. Synth.*, 1989, **25**, 8. 406

89IZV197 Y. N. Studnev, V. P. Stolyarov, G. A. Prigorelov and A. V. Fokin; *Izv. Akad. Nauk SSSR Ser. Khim.*, 1989, 197 (*Chem. Abstr.*, 1989, **111**, 38 856g). 165

89IZV1935 N. V. Volchkov, A. V. Zabolotskikh, M. B. Lipkind and O. M. Nefedov; *Izv. Akad. Nauk SSSR Ser. Khim.*, 1989, 1935 (*Chem. Abstr.*, 1990, **112**, 138 661). 615

89JA384 B. Singaram, C. T. Goralski, M. V. Rangaishenvi and H. C. Brown; *J. Am. Chem. Soc.*, 1989, **111**, 384. 740

89JA389 D. A. Beker and R. L. Danheiser; *J. Am. Chem. Soc.*, 1989, **111**, 389. 930, 944

89JA393 G. K. Surya Prakash, R. Krishnamurti and G. A. Olah; *J. Am. Chem. Soc.*, 1989, **111**, 393. 70

89JA765 W. F. Bailey and K. Rossi; *J. Am. Chem. Soc.*, 1989, **111**, 765. 556

89JA776 S. L. Buchwald, M. W. Wannamaker, and B. T. Watson; *J. Am. Chem. Soc.*, 1989, **111**, 776. 987

89JA1211 J. E. Swartz, E. Kariv-Miller and S. J. Harrold; *J. Am. Chem. Soc.*, 1989, **111**, 1211. 81

89JA1299 B. E. Bowler, K. J. Ahmed, W. I. Sundquist, L. S. Hollis, E. E. Whang and S. J. Lippard; *J. Am. Chem. Soc.*, 1989, **111**, 1299. 1009

89JA1517 D. L. Boger and A. M. Kasper; *J. Am. Chem. Soc.*, 1989, **111**, 1517. 329

89JA1754 H. C. Brown, N. N. Joshi, C. Pyun and B. Singaram; *J. Am. Chem. Soc.*, 1989, **111**, 1754. 519, 531

89JA1888 S. Masamune, S. A. Batcheller, J. Park, W. M. Davis, O. Yamashita, Y. Ohta and Y. Kabe; *J. Am. Chem. Soc.*, 1989, **111**, 1888. 949

89JA1892 R. P. Short and S. Masamune; *J. Am. Chem. Soc.*, 1989, **111**, 1892. 77, 521

89JA1937 J. W. Faller and D. L. Linebarrier; *J. Am. Chem. Soc.*, 1989, **111**, 1937. 76

89JA1940 K. Mikami, M. Terada and T. Nakai; *J. Am. Chem. Soc.*, 1989, **111**, 1940. 80

89JA2225 P. J. Stang, T. Kitamura, M. Boehshar and H. Wingert; *J. Am. Chem. Soc.*, 1989, **111**, 2225. 1028

89JA2321 W. K. D. Brill, J.-Y. Tang, Y.-X. Ma and M. H. Carruthers; *J. Am. Chem. Soc.*, 1989, **111**, 2321. 253

89JA2355 Y. Chiang, A. J. Kresge, R. Hochstrasser and J. Wirz; *J. Am. Chem. Soc.*, 1989, **111**, 2355. 1022

89JA2870 S. L. Buchwald and R. B. Nielsen; *J. Am. Chem. Soc.*, 1989, **111**, 2870. 927

89JA2981 G. A. Broka and T. Shen; *J. Am. Chem. Soc.*, 1989, **111**, 2981. 556

89JA3077 P. N. Guivisdalsky and R. Bittmann; *J. Am. Chem. Soc.*, 1989, **111**, 3077. 84

89JA3336 E. Negishi, S. J. Holmes, J. M. Tour, J. A. Miller, F. E. Cederbaum, D. R. Swanson and T. Takahashi; *J. Am. Chem. Soc.*, 1989, **111**, 3336. 927, 986, 988

89JA3426 T. Hayashi, Y. Matsumoto and Y. Ito; *J. Am. Chem. Soc.*, 1989, **111**, 3426. 43, 528

89JA3441 H. C. Brown, R. K. Dhar, R. K. Bakshi, P. K. Dandiarajan and B. Singaram; *J. Am. Chem. Soc.*, 1989, **111**, 3441. 532, 698

89JA3454 Y. Zhang and E. Negishi; *J. Am. Chem. Soc.*, 1989, **111**, 3454. 980

89JA4028 M. Kitamura, S. Okada, S. Suga and R. Noyori; *J. Am. Chem. Soc.*, 1989, **111**, 4028. 71

89JA4392 T. Imamoto, N. Takiyama, K. Nakamura, T. Hatajima and Y. Kamiya; *J. Am. Chem. Soc.*, 1989, **111**, 4392. 69

89JA4486 S. L. Buchwald, B. T. Watson, M. W. Wannamaker and J. C. Dewan; *J. Am. Chem. Soc.*, 1989, **111**, 4486. 306, 927

89JA4525 T. V. Rajababu and W. A. Nugent; *J. Am. Chem. Soc.*, 1989, **111**, 4525. 85

89JA4873 J. A. Soderquist, J. C. Colberg and L. Del Valle; *J. Am. Chem. Soc.*, 1989, **111**, 4873. 902, 922

89JA5002 S. I. Murahashi, T. Oda and Y. Masui; *J. Am. Chem. Soc.*, 1989, **111**, 5002. 145

89JA5424 R. F. Childs, G. S. Shaw and C. J. L. Lock; *J. Am. Chem. Soc.*, 1989, **111**, 5424. 1009

89JA5468 J. B. Hartung, Jr. and S. F. Pedersen; *J. Am. Chem. Soc.*, 1989, **111**, 5468. 672, 988, 989

89JA5472 E. J. Corey and P. Carpino; *J. Am. Chem. Soc.*, 1989, **111**, 5472. 672

89JA5493 E. J. Corey, R. Imwinkelried, S. Pikul and Y. B. Xiang; *J. Am. Chem. Soc.*, 1989, **111**, 5493. 533

89JA5495 E. J. Corey, C.-M. Yu and S. S. Kim; *J. Am. Chem. Soc.*, 1989, **111**, 5495. 76

89JA5886 J. D. Rich; *J. Am. Chem. Soc.*, 1989, **111**, 5886. 945

89JA6301 T. Hayashi, A. Yamamoto, Y. Ito, E. Nishioka, H. Miura and K. Yanagi; *J. Am. Chem. Soc.*, 1989, **111**, 6301. 331

89JA6429 K. Sato, M. Kira and H. Sakurai; *J. Am. Chem. Soc.*, 1989, **111**, 6429. 78

89JA6435 W. A. Nugent and D. F. Taber; *J. Am. Chem. Soc.*, 1989, **111**, 6435. 988

89JA6478 K. Tamao, K. Kobayashi and Y. Ito; *J. Am. Chem. Soc.*, 1989, **111**, 6478. 920

89JA7500 J. Mulzer, R. Becker and E. Brunner; *J. Am. Chem. Soc.*, 1989, **111**, 7500. 324

89JA7615 A. C. Dema and C. M. Lukehart; *J. Am. Chem. Soc.*, 1989, **111**, 7615. 927

89JA8014 J. H. Freudenberger, A. W. Konradi and S. F. Pedersen; *J. Am. Chem. Soc.*, 1989, **111**, 8014. 81

89JA8236 G. A. Molander and C. Kenny; *J. Am. Chem. Soc.*, 1989, **111**, 8236. 81

89JA8297 B. J. Smith, L. Radom and A. J. Kresge; *J. Am. Chem. Soc.*, 1989, **111**, 8297. 1022

89JA9134 R. Noyori, T. Ikeda, T. Ohkuma, *et al.*; *J. Am. Chem. Soc.*, 1989, **111**, 9134. 53

89JA9243 E. J. Corey, P. Da Silva Jardine, S. Virgil, P.-W. Yuen and R. D. Connell; *J. Am. Chem. Soc.*, 1989, **111**, 9243. 50

89JCR(S)12 J. H. Boyer and A. M. Krishnan; *J. Chem. Res. (S)*, 1989, 12. 385

89JCR(S)104 B. Jursic; *J. Chem. Res. (S)*, 1989, 104. 31, 35

89JCR(S)108 F. Bellesia, F. Ghelfi, U. M. Pagnoni and A. Pinetti; *J. Chem. Res. (S)*, 1989, 108. 24

89JCR(S)284 B. Jursic; *J. Chem. Res. (S)*, 1989, 284. 26

89JCR(S)296 A. Koziara; *J. Chem. Res. (S)*, 1989, 296. 309

89JCR(S)313 G. J. S. Doad, D. J. Austin and T. C. Owen; *J. Chem. Res. (S)*, 1989, 313. 629

89JCR(S)360 F. Bellesia, F. Ghelfi, U. M. Pagnoni and A. Pinetti; *J. Chem. Res. (S)*, 1989, 360. 24

89JCS(D)105 L. T. Byrne, L. M. Engelhardt, G. E. Jacobsen, W.-P. Leung, R. I. Papasergio, C. L. Raston, B. W. Skelton, P. Twiss and A. H. White; *J. Chem. Soc., Dalton Trans.*, 1989, 105. 1003, 1005

89JCS(D)779 D. G. Anderson and D. W. H. Rankin; *J. Chem. Soc., Dalton Trans.*, 1989, 779. 417

89JCS(P1)109 K. Soai, S. Niwa and M. Watanabe; *J. Chem. Soc., Perkin Trans. 1*, 1989, 109. 71

89JCS(P1)225 A. R. Katritzky, K. Yannakopoulou, P. Lue, D. Rasala and L. Urogdi; *J. Chem. Soc., Perkin Trans. 1*, 1989, 225. 310

89JCS(P1)639 A. R. Katritzky and J.-J. V. Eynde; *J. Chem. Soc., Perkin Trans. 1*, 1989, 639. 310

89JCS(P1)1521 A. J. Pratt and E. J. Thomas; *J. Chem. Soc., Perkin Trans. 1*, 1989, 1521. 79

89JCS(P1)1529 A. J. Pratt, E. J. Thomas and V. J. Jephcote; *J. Chem. Soc., Perkin Trans. 1*, 1989, 1529. 79

89JCS(P1)1544 G. M. Shutske; *J. Chem. Soc., Perkin Trans. 1*, 1989, 1544. 22, 28

89JCS(P1)1548 S. Itsuno, Y. Sakurai, K. Shimizu and K. Ito; *J. Chem. Soc., Perkin Trans. 1*, 1989, 1548. 314

89JCS(P1)1550 D. A. Barr, G. Donegan and R. Grigg; *J. Chem. Soc., Perkin Trans. 1*, 1989, 1550. 158

89JCS(P1)1927 T. Toru, T. Seko, E. Maekawa and Y. Ueno; *J. Chem. Soc., Perkin Trans. 1*, 1989, 1927. 285

89JFC(43)405 K. C. Mange and W. J. Middleton; *J. Fluorine Chem.*, 1989, **43**, 405. 19

89JGU87 Z. S. Novikova, I. L. Odinets and I. F. Lutsenko; *J. Gen. Chem. USSR (Engl. Transl.)*, 1989, **59**, 87. 476

89JGU469 N. V. Pavlenko and L. M. Yagupol'skii; *J. Gen. Chem. USSR (Engl. Transl.)*, 1989, **59**, 469. 476, 614

89JGU634 L. V. Nesterov and N. E. Krepysheva; *J. Gen. Chem. USSR (Engl. Transl.)*, 1989, **59**, 634. 476

89JGU2130 A. A. Prishchenko, M. V. Livantsov, N. N. Boganova and I. F. Lutsenko; *J. Gen. Chem. USSR (Engl. Transl.)*, 1989, **59**, 2130. 451

89JGU2254 G. I. Kokorev, Sh. Kh. Badrutdinov, L. A. Almetkina and F. D. Yambushev; *J. Gen. Chem. USSR (Engl. Transl.)*, 1989, **59**, 2254. 889

89JGU2500 G. I. Kokorev, R. Z. Musin and F. D. Yambushev; *J. Gen. Chem. USSR (Engl. Transl.)*, 1989, **59**, 2500. 881

89JOC38 B. B. Snider and J. J. Patricia; *J. Org. Chem.*, 1989, **54**, 38. 672
89JOC159 M. M. Midland, J. McLoughlin and J. Gabriel; *J. Org. Chem.*, 1989, **54**, 159. 57
89JOC244 A. L. Schroll and G. Barany; *J. Org. Chem.*, 1989, **54**, 244. 219, 230
89JOC265 S. F. Martin and W. Li; *J. Org. Chem.*, 1989, **54**, 265. 330
89JOC268 Y. Hatanaka and T. Hiyama; *J. Org. Chem.*, 1989, **54**, 268. 931, 932, 939
89JOC444 S. Torii, H. Taneka, S. Yamashito and M. Yamanoue; *J. Org. Chem.*, 1989, **54**, 444. 81
89JOC484 B. M. Trost and J. M. Tour; *J. Org. Chem.*, 1989, **54**, 484. 86
89JOC627 V. Roussis and D. F. Wiemer; *J. Org. Chem.*, 1989, **54**, 627. 457
89JOC868 S.-S. P. Chou, H.-L. Kuo, C. J. Wang, C.-Y. Tsai and C.-M. Sun; *J. Org. Chem.*, 1989, **54**, 868. 630
89JOC998 J. A. Miller; *J. Org. Chem.*, 1989, **54**, 998. 629
89JOC1070 J. R. Hwu, N. Wang and R. T. Yung; *J. Org. Chem.*, 1989, **54**, 1070. 388
89JOC1092 R. S. Glass, F. Farooqui, M. Sabahi and K. W. Ehler; *J. Org. Chem.*, 1989, **54**, 1092. 235
89JOC1204 G. A. Olah and T. D. Ernst; *J. Org. Chem.*, 1989, **54**, 1204. 650
89JOC1221 A. O. Stewart and J. G. Martin; *J. Org. Chem.*, 1990, **55**, 1221. 351
89JOC1230 M. Belley and R. Zamboni; *J. Org. Chem.*, 1989, **54**, 1230. 137
89JOC1233 A. G. M. Barrett, C. Robyr and C. D. Spilling; *J. Org. Chem.*, 1989, **54**, 1233. 394
89JOC1473 L. Capella, A. Degl'Innocenti, G. Reginato, A. Ricci and M. Taddei; *J. Org. Chem.*, 1989, **54**, 1473. 918, 939
89JOC1491 C. N. Jera and M. Yus; *J. Org. Chem.*, 1989, **54**, 1491. 85
89JOC1502 C.-M. Andersson and A. Hallberg; *J. Org. Chem.*, 1989, **54**, 1502. 682
89JOC1570 H. C. Brown, K. S. Bhat and R. S. Randad; *J. Org. Chem.*, 1989, **54**, 1570. 75
89JOC1577 H. C. Brown, P. V. Ramachandran and M. Srebnik; *J. Org. Chem.*, 1989, **54**, 1577. 58
89JOC1748 A. Brandi, F. Cordero and C. Querci; *J. Org. Chem.*, 1989, **54**, 1748. 329
89JOC1940 D. S. Jones, A. Srinivasan, S. Kasina, A. R. Fritzberg and D. W. Wilkening; *J. Org. Chem.*, 1989, **54**, 1940. 309
89JOC2015 K. Sukata; *J. Org. Chem.*, 1989, **54**, 2015. 544
89JOC2263 J. S. Svendsen, I. Mark, E. N. Jacobsen, C. P. Rao, S. Bott and K. B. Sharpless; *J. Org. Chem.*, 1989, **54**, 2263. 48
89JOC2271 B. M. Trost and D. C. Lee; *J. Org. Chem.*, 1989, **54**, 2271. 928
89JOC2321 M. Sekine; *J. Org. Chem.*, 1989, **54**, 2321. 239
89JOC2383 M. Falorni, L. Lardicci and G. Gioacomelli; *J. Org. Chem.*, 1989, **54**, 2383. 56
89JOC2391 A. Pecunioso and R. Menicagli; *J. Org. Chem.*, 1989, **54**, 2391. 783
89JOC2580 D. E. Bugner; *J. Org. Chem.*, 1989, **54**, 2580. 445
89JOC2793 S. L. Buchwald and Q. Fang; *J. Org. Chem.*, 1989, **54**, 2793. 927, 945
89JOC2998 P. Dhar and S. Chandrasekaran; *J. Org. Chem.*, 1989, **54**, 2998. 228
89JOC3002 R. E. Gawley, K. Rein and S. Chemburkar; *J. Org. Chem.*, 1989, **54**, 3002. 321
89JOC3028 K. B. Yoon and J. K. Kochi; *J. Org. Chem.*, 1989, **54**, 3028. 31
89JOC3087 S. R. Wilson and M. E. Guazzaroni; *J. Org. Chem.*, 1989, **54**, 3087. 74
89JOC3247 H. Xiong and R. D. Rieke; *J. Org. Chem.*, 1989, **54**, 3247. 561
89JOC3254 A. Hosomi, H. Hayashida and Y. Tominaga; *J. Org. Chem.*, 1989, **54**, 3254. 77
89JOC3482 H. Sakuraba, M. Inomata and Y. Tanaka; *J. Org. Chem.*, 1989, **54**, 3482. 57
89JOC3517 K. Tamao, K. Kobayashi and Y. Ito; *J. Org. Chem.*, 1989, **54**, 3517. 926
89JOC3640 D. S. Kemp and R. I. Carey; *J. Org. Chem.*, 1989, **54**, 3640. 229
89JOC3872 A. Clerici and O. Porta; *J. Org. Chem.*, 1989, **54**, 3872. 81
89JOC4022 E. Kariv-Miller, H. Maeda and F. Lombardo; *J. Org. Chem.*, 1989, **54**, 4022. 81
89JOC4154 S. Collins, B. A. Kuntz and Y. Hong; *J. Org. Chem.*, 1989, **54**, 4154. 76
89JOC4294 F. Camps, E. Chamorro, V. Gasol and A. Guerrero; *J. Org. Chem.*, 1989, **54**, 4294. 18
89JOC4372 R. J. Mills, N. J. Taylor and V. Snieckus; *J. Org. Chem.*, 1989, **54**, 4372. 627
89JOC4458 S. Krishnamurthy and D. Aimino; *J. Org. Chem.*, 1989, **54**, 4458. 708
89JOC4504 H. C. Brown and P. V. Ramachandran; *J. Org. Chem.*, 1989, **54**, 4504. 58
89JOC4686 J. D. Fourneron, A. Archelas and R. Furstoss; *J. Org. Chem.*, 1989, **54**, 4686. 51
89JOC4732 K. Taka, K. Nitta, D. Fujimura and K. Utimoto; *J. Org. Chem.*, 1989, **54**, 4732. 74, 572
89JOC4982 S. P. Rychnovsky; *J. Org. Chem.*, 1989, **54**, 4982. 85
89JOC4989 M. Utaka, S. Konishi, A. Mizuoka, T. Ohkubo, T. Sakai, S. Tsuboi and A. Takeda; *J. Org. Chem.*, 1989, **54**, 4989. 14
89JOC5064 S. Sharma and A. C. Oehlschlager; *J. Org. Chem.*, 1989, **54**, 5064. 607
89JOC5198 A. Vanagisawa, S. Habaue and H. Yamamoto; *J. Org. Chem.*, 1989, **54**, 5198. 74
89JOC5202 P. Knochel, T.-S. Chon, H. G. Chen, M. C. P. Yeh and M. J. Rozema; *J. Org. Chem.*, 1989, **54**, 5202. 998
89JOC5380 J. Cousseau and P. Albert; *J. Org. Chem.*, 1989, **54**, 5380. 19
89JOC5574 P. Beak and G. W. Selling; *J. Org. Chem.*, 1989, **54**, 5574. 328
89JOC5657 B. Mudryk and T. Cohen; *J. Org. Chem.*, 1989, **54**, 5657. 553
89JOC5768 Z. Y. Wei, D. Wang, J. S. Li and T. H. Chan; *J. Org. Chem.*, 1989, **54**, 5768. 78
89JOC5783 R. W. Murray, S. N. Rajadhyaksha and L. Mohan; *J. Org. Chem.*, 1989, **54**, 5783. 397
89JOC5814 W. H. Pearson, K.-C. Lin and Y.-F. Poon; *J. Org. Chem.*, 1989, **54**, 5814. 517
89JOC6001 T. Shono, S. Kashimura, Y. Mori, T. Hayashi, T. Soejima and Y. Yamaguchi; *J. Org. Chem.*, 1989, **54**, 6001. 81
89JOC6068 H. C. Brown, C. Subrahmanyam, T. Hamaoka, N. Ravindran, D. H. Bowman, S. Misumi, M. K. Unni, V. Somayaji and N. G. Bhat; *J. Org. Chem.*, 1989, **54**, 6068. 623, 628
89JOC6075 H. C. Brown, T. Hamaoka, N. Ravindran, C. Subrahmanyam, V. Somayaji and N. G. Bhat; *J. Org. Chem.*, 1989, **54**, 6075. 900, 901
89JOC6101 W. G. Dauben, D. P. Bridon and B. A. Kowalczyk; *J. Org. Chem.*, 1989, **54**, 6101. 13

89JOM(360)39	G. L. Larson; *J. Organomet. Chem.*, 1989, **360**, 39.	1005
89JOM(364)67	M. Ates, H. J. Breunig and S. Gulec; *J. Organomet. Chem.*, 1989, **364**, 67.	489, 490
89JOM(366)73	M. Scholz, H. W. Roesky, D. Stalke, K. Keller and F. T. Edelmann; *J. Organomet. Chem.*, 1989, **366**, 73.	872
89JOM(366)87	Y.-Z. Huang, C. Chen and Y. Shen; *J. Organomet. Chem.*, 1989, **366**, 87.	487, 488
89JOM(369)291	S. Araki, H. Ito, N. Katsumura and Y. Butsugan; *J. Organomet. Chem.*, 1989, **369**, 291.	73, 597
89JOM(373)307	H. Schumann and K.-H. Kohricht; *J. Organomet. Chem.*, 1989, **373**, 307.	509
89JOM(379)201	P. Charreau, M. Julia and J.-N. Verpeaux; *J. Organomet. Chem.*, 1989, **379**, 201.	623
89JOU587	E. D. Matveeva, A. I. Yalovskaya, I. A. Cherepanov, A. L. Kurts and Y. G. Bundel; *J. Org. Chem. USSR* (*Engl. Transl.*), 1989, 587.	7
89JOU642	E. D. Matveeva, A. L. Kurts, A. I. Yalovskaya, N. G. Nikishova and Y. G. Bundel; *J. Org. Chem. USSR* (*Engl. Transl.*), 1989, 642.	7
89JOU1835	L. S. Bouguslavskaya, A. V. Kartashov and N. N. Chuvatkin; *J. Org. Chem. USSR* (*Engl. Transl.*), 1989, **25**, 1835.	15
89LA167	E. Hatzigrigoriou, S. Spyroudis and A. Varvoglis; *Liebigs Ann. Chem.*, 1989, 167.	998
89LA605	R. Lux and G. Kresze; *Liebigs Ann. Chem.*, 1989, 605.	163
89LA883	R. W. Hoffman, W. Ladner and K. Ditrich; *Liebigs Ann. Chem.*, 1989, **100**, 883.	530
B-89MI 201-01	R. C. Larock; "Comprehensive Organic Transformations," VCH, New York, 1989.	2, 6, 15, 18, 28
B-89MI 201-02	G. H. Schmid; in "The Chemistry of Functional Groups," ed. S. Patai, Wiley, New York, 1989, Supplement A, vol. 2, p. 679.	5
B-89MI 201-03	L. German and S. Zemskov (eds); "New Fluorinating Agents in Organic Synthesis," Springer-Verlag, New York, 1989.	15
B-89MI 202-01	R. C. Larock; "Comprehensive Organic Transformations," VCH, Weinheim, 1989.	38, 93
B-89MI 202-02	R. C. Larock; "Comprehensive Organic Transformations," VCH, Weinheim, 1989, p. 567.	73
B-89MI 202-03	C. Hill; "Activation and Functionalization of Alkanes," Wiley, New York, 1989.	86
B-89MI 202-04	R. C. Larock; "Comprehensive Organic Transformations," VCH, Weinheim, 1989, p. 451.	93
B-89MI 202-05	G. A. Olah, R. Malhotra and S. C. Narang; "Nitration, Methods and Mechanisms," VCH, New York, 1989.	103
B-89MI 202-06	L. Fieser and M. Fieser; "Reagents for Organic Synthesis," Wiley, New York, 1989, vol. 14, p. 197.	90
B-89MI 203-01	A. D. Dunn and W.-D. Rudorf; in "Carbon Disulfide in Organic Chemistry," Ellis Horwood, Chichester, 1989.	123
89MI 203-02	M. Pfister and W. Pfleiderer; *Nucleosides and Nucleotides*, 1989, **8**, 1001.	203
B-89MI 205-01	R. C. Larock; "Comprehensive Organic Transformations," VCH, Weinheim, 1989, p. 411.	318
B-89MI 205-02	R. C. Larock; "Comprehensive Organic Transformations," VCH, Weinheim, 1989, p. 437.	316
B-89MI 206-01	L. German and S. Zemskov (eds.); in "New Fluorinating Agents in Organic Synthesis," Springer Verlag, Berlin, 1989.	334
89MI 206-02	H. J. Emeleus, J. M. Shreve and R. D. Verma; *Adv. Inorg. Chem.*, 1989, **33**, 139.	335
B-89MI 206-03	S. R. Sandler and W. Karo; in "Organic Functional Group Preparations," Academic Press, New York, 1989, vol. 3, p. 377.	340, 341, 342, 343, 344, 346, 347, 349, 350, 351, 355, 356
B-89MI 206-04	Y. Ogata and Y. Sawaki; in "Organic Syntheses by Oxidation with Metal Compounds," eds. W. J. Mijs and C. R. H. I. de Jonge, Plenum Press, New York, 1989, p. 839.	348
89MI 208-01	S. Kim, P. H. Lee and S. S. Kim; *Bull. Korean Chem. Soc.*, 1989, **10**, 218 (*Chem. Abstr.*, 1990, **112**, 55 022).	446
B-89MI 210-01	S. Patai and Z. Rappoport (eds); "The Chemistry of Organic Silicon Compounds," Wiley, Chichester, 1989.	547
B-89MI 211-01	S. V. Ley and C. M. R. Low; "Ultrasound in Synthesis," Springer Verlag, Berlin, 1989.	551
B-89MI 211-02	G. A. Molander; in "The Chemistry of the Metal–Carbon Bond," ed. F. R. Hartley, series ed. S. Patai, Wiley, New York, 1989, vol. 5, p. 319.	564
B-89MI 211-03	R. Noyori and M. Kitamura; in "Modern Synthetic Methods," ed. R. Scheffold, Otto Salle Verlag, 1989, vol. 5, p. 115.	478
B-89MI 215-01	R. C. Larock; "Comprehensive Organic Transformations," VCH, Weinheim, 1989, pp. 411–415.	752
B-89MI 215-02	R. C. Larock; "Comprehensive Organic Transformations," VCH, Weinheim, 1989, pp. 409–410.	753
B-89MI 215-03	R. C. Larock; "Comprehensive Organic Transformations," VCH, Weinheim, 1989, pp. 431–432.	757
B-89MI 217-01	L. D. Freedman and G. O. Doak; in "The Chemistry of the Metal–Carbon Bond," ed. F. R. Hartley, Wiley, Chichester, 1989, vol. 5, chap. 9, pp. 397–433.	872
89MI 217-02	V. A. Dodonov, S. N. Zaburdyaevla and N. N. Nevkina; *Metallorg. Khim.*, 1989, **2**, 1296 (*Chem. Abstr.*, 1990, **113**, 6 478k).	886
B-89MI 218-01	L. Birkofer and O. Stuhl; in "The Chemistry of Organic Silicon Compounds," eds. S. Patai and Z. Rappoport, Wiley, Chichester, 1989, vol. 1, p. 655.	907
B-89MI 218-02	G. L. Larson; in "The Chemistry of Organic Silicon Compounds," eds. S. Patai and Z. Rappoport, Wiley, Chichester, 1989, vol. 1, p. 763.	907
B-89MI 218-03	A. R. Bassindale and P. G. Taylor; in "The Chemistry of Organic Silicon Compounds," eds. S. Patai and Z. Rappoport, Wiley, Chichester, 1989, vol. 2, p. 893.	943
B-89MI 223-01	F. Glockling; in "Gmelin Handbook of Inorganic Chemistry Organogermanium Compounds," Part 2, ed. U. Küerke, Springer, Berlin, 8th edn, 1989, pp. 236.	1092
89MI 223-02	I. Omae (ed.); *J. Organomet. Chem., Libr.*, 1989, **21**, p. 1.	1100
89OM482	J. C. Pazik and C. George; *Organometallics*, 1989, **8**, 482.	489, 492
89OM1093	G. W. Kabalka and Z. Wang; *Organometallics*, 1989, **8**, 1093.	301

89OM1661 S. M. Barros, M. J. Dabdoub, V. M. B. Dabdoub and J. V. Comasseto; *Organometallics*, 1989, **8**, 1661. 730
89OM1688 S. Harder, J. Boersma, L. Brandsma, J. A. Kanters, W. Bauer, R. Pi, P. von Ragué Schleyer, H. Schöllhorn and U. Thewalt; *Organometallics*, 1989, **8**, 1688. 999, 1000
89OM2308 G. Fraenkel, W. R. Winchester and P. G. Williard; *Organometallics*, 1989, **8**, 2308. 997, 1003
89OM2609 J. D. Rich; *Organometallics*, 1989, **8**, 2609. 945
89OPP135 A. R. Katritzky, B. Pilarski and L. Urogdi; *Org. Prep. Proced. Int.*, 1989, **21**, 135. 302
89OR(37)57 I. Fleming, J. Dunoguès and R. Smithers; *Org. React.*, 1989, **37**, 57. 73, 907
89OS(67)98 Y. Tamaru, H. Ochiai, T. Nakamura and Z. Yoshida; *Org. Synth.*, 1989, **67**, 98. 591
89OS(68)8 A. G. M. Barrett, D. Dhanak, G. B. Graboski and S. J. Taylor; *Org. Synth.*, 1989, **68**, 8. 398
89POL581 P. Raj, A. K. Aggarwal and N. Misra; *Polyhedron*, 1989, **8**, 581. 891
89PS(44)129 M. Ates, H. J. Breunig and S. Gülec; *Phosphorus Sulfur*, 1989, **44**, 129. 894
89PS(45)35 C. Lee; *Phosphorus Sulfur*, 1989, **45**, 35. 197
89PS(45)151 P. Majewski; *Phosphorus Sulfur*, 1989, **45**, 151. 453
89RCR250 A. N. Mirskova, T. I. Drozdova, G. G. Levkovskaya and M. G. Voronkov; *Russ. Chem. Rev.* (*Engl. Transl.*), 1989, **58**, 250. 338, 365
89RCR470 S. G. Zlotin, G. N. Varnaeva and O. A. Luk'yanov; *Russ. Chem. Rev.* (*Engl. Transl.*), 1989, **58**, 470. 391
89RCR948 S. I. Radchenko and A. A. Petrov; *Russ. Chem. Rev.* (*Engl. Transl.*), 1989, **58**, 948. 1021, 1025, 1027, 1030, 1036
89S1 J. H. Cooley and E. J. Evain; *Synthesis*, 1989, 1. 303
89S25 R. Martin and P. Demerseman; *Synthesis*, 1989, 25. 656
89S31 A. R. Katritzky, J. K. Gallos and K. Yannakopoulou; *Synthesis*, 1989, 31. 306
89S117 H. Ahlbrecht and P. Weber; *Synthesis*, 1989, 117. 751
89S167 L. Syper; *Synthesis*, 1989, 167. 654
89S179 D. Schinzer; *Synthesis*, 1989, 179. 908
89S204 H. J. Bestman and R. Dotzer; *Synthesis*, 1989, 204. 444
89S283 T. Kanai, S. Irifune, Y. Ishii and M. Ogawa; *Synthesis*, 1989, 283. 36
89S295 C. Venturello and M. Gambaro; *Synthesis*, 1989, 295. 51
89S309 J. P. Bégué and D. Mesureur; *Synthesis*, 1989, 309. 751
89S378 H. Firouzabadi, E. Moitghinejad and M. Seddighi; *Synthesis*, 1989, 378. 215
89S434 K. Haraguchi, H. Tanaka and T. Miyasaka; *Synthesis*, 1989, 434. 281
89S446 N. A. Keiko, A. Yu. Rulev, I. D. Kalikhman and M. G. Voronkov; *Synthesis*, 1989, 446. 741
89S450 A. A. Malik, S. B. Preston, T. G. Archibald, M. P. Cohen and K. Baum; *Synthesis*, 1989, 450. 317
89S461 A. Ricci, M. Taddei, P. Dembech, A. Guerrini and G. Seconi; *Synthesis*, 1989, 461. 1019
89S497 A. H. Beiter and W. Pfleiderer; *Synthesis*, 1989, 497. 409, 410
89S568 G. A. Olah, A-h. Wu and O. Farooq; *Synthesis*, 1989, 568. 941
89S571 A. Fürstner; *Synthesis*, 1989, 571. 702
89S608 G. Bringmann and J.-P. Geisler; *Synthesis*, 1989, 608. 331
89S614 A. Amrollah-Madjdabadi, T. N. Pham and E. C. Ashby; *Synthesis*, 1989, 614. 25, 26, 31, 32
89S616 M. P. Georgiadis and S. A. Haroutounian; *Synthesis*, 1989, 616. 311
89S631 P. D. Baruah, I. Nongkynrih and M. P. Mahajan; *Synthesis*, 1989, 631. 744
89S647 A. Ricci and A. Degl'Innocenti; *Synthesis*, 1989, 647. 908
89S787 C. Einhorn, J. Einhorn and J.-L. Luche; *Synthesis*, 1989, 787. 551, 958
89S825 A. Warshawsky, J. Altman, N. Kahana, R. Arad-Yellin, A. Deshe, H. Hasson, N. Shoef and H. Gottlieb; *Synthesis*, 1989, 825. 304
89S953 R. Andruszkiewicz and R. B. Silverman; *Synthesis*, 1989, 953. 315
89S955 T. Nakazawa, N. Hirose and K. Itabashi; *Synthesis*, 1989, 955. 711
89S957 M. Barbero, S. Cadamuro, I. Degani, R. Fochi and V. Regondi; *Synthesis*, 1989, 957. 180
89S963 A. Dobrev; *Synthesis*, 1989, 963. 312
89SC275 P. Cuadrado, A. M. González, B. González and F. J. Pulido; *Synth. Commun.*, 1989, **19**, 275. 545
89SC561 S. Ohira; *Synth. Commun.*, 1989, **19**, 561. 680
89SC565 M. Periasamy, A. Devasagayaraj, N. Satuanarayana and C. Narayana; *Synth. Commun.*, 1989, **19**, 565. 312
89SC799 S. O. Nwauka and P. M. Keehn; *Synth. Commun.*, 1989, **19**, 799. 619
89SC805 N. M. Goudgaon, P. P. Wadgaonkar and G. W. Kabalka; *Synth. Commun.*, 1989, **19**, 805. 398
89SC865 N. Yoneda, T. Fukuhara, T. Kikuch and A. Suzuki; *Synth. Commun.*, 1989, **19**, 865. 615
89SC1255 R. S. Subramanian and K. K. Balasubramanian; *Synth. Commun.*, 1989, **19**, 1255. 683
89SC1347 X. J. Zhou and Z. Z. Huang; *Synth. Commun.*, 1989, **19**, 1347. 304
89SC1423 C. Yaunwei, M. Aiqiao, X. Xun and J. Yaozhong; *Synth. Commun.*, 1989, **19**, 1423. 331
89SC1697 R. Jalal and R. Gallo; *Synth. Commun.*, 1989, **19**, 1697. 20
89SC2409 G. W. Kabalka, Z. Wang and N. M. Goudgaon; *Synth. Commun.*, 1989, **19**, 2409. 306
89SC2481 B. Sket and M. Zupan; *Synth. Commun.*, 1989, **19**, 2481. 28
89SC2939 A. Loupy and J. Sansoulet; *Synth. Commun.*, 1989, **19**, 2939. 157
89SC3055 K. Lee and D. Y. Oh; *Synth. Commun.*, 1989, **19**, 3055. 394
89SC3139 S. C. Nigam, A. Mann, M. Taddei and C.-G. Wermuth; *Synth. Commun.*, 1989, **19**, 3139. 304
89SC3143 F. Kong and X. Zhou; *Synth. Commun.*, 1989, 3143. 122, 127
89SC3289 G. V. M. Sharma and S. Chandrasekhar; *Synth. Commun.*, 1989, **19**, 3289. 317
89SC3509 R. W. Murray and M. Singh; *Synth. Commun.*, 1989, **19**, 3509. 343
89SCI1260 R. D. Rieke; *Science*, 1989, **246**, 1260. 958
89SL1 H. Sakurai; *Synlett*, 1989, 1. 547

89SL55 R. H. Mitchell and V. S. Iyer; *Synlett*, 1989, 55. 14

89T171 R. J. P. Corriu, A. Boudin, G. Cerveau, C. Chuit and C. Reye; *Tetrahedron*, 1989, **45**, 171. 68

89T191 M. J. Thompson and P. J. Zeegers; *Tetrahedron*, 1989, **45**, 191. 791

89T349 B. H. Lipshutz (ed.); *Tetrahedron*, 1989, **45**, 349. 1094

89T413 I. Fleming, M. Rowley, P. Cuadrado, A. M. Gonzalez-Nogal and F. J. Pulido; *Tetrahedron*, 1989, **45**, 413. 929, 930

89T425 S. H. Bertz and G. Dabbagh; *Tetrahedron*, 1989, **45**, 425. 586

89T979 D. R. Tueting, A. M. Echevarren and J. K. Stille; *Tetrahedron*, 1989, **45**, 979. 86

89T1119 B. Wrackmeyer and G. Guldner and S. T. Abu-Orabi; *Tetrahedron*, 1989, **45**, 1119. 903

89T1145 B. Jousseaume and B. Villneuve; *Tetrahedron*, 1989, **45**, 1145. 1100

89T1611 M. Hajjatie, S. Muralidharan and H. Freiser; *Tetrahedron*, 1989, **45**, 1611. 279

89T1859 D. S. Matteson; *Tetrahedron*, 1989, **45**, 1859. 529, 900, 904, 905

89T2005 A. Krief, W. Dumont, M. Clarembeau, G. Bernard and E. Badaoui; *Tetrahedron*, 1989, **45**, 2005. 478

89T2023 A. Krief, W. Dumont, M. Clarembeau and E. Badaoui; *Tetrahedron*, 1989, **45**, 2023. 281

89T2099 F. Felluga, P. Nitti, G. Pitacco and E. Valentin; *Tetrahedron*, 1989, **45**, 2099. 396

89T3569 P. Sarmah and N. C. Barua; *Tetrahedron*, 1989, **45**, 3569. 36

89T3787 J. G. Dingwall, J. Ehrenfreund and R. G. Hall; *Tetrahedron*, 1989, **45**, 3787. 451

89T4163 G. Cahiez and M. Alami; *Tetrahedron*, 1989, **45**, 4163. 574

89T4243 C. Giordano, G. Castaldi, S. Cavicchioli and M. Villa; *Tetrahedron*, 1989, **45**, 4243. 29

89T4431 E. Laurent, B. Marquet and R. Tardivel; *Tetrahedron*, 1989, **45**, 4431. 16

89T5565 H. L. Aalten, G. van Koten and D. M. Grove; *Tetrahedron*, 1989, **45**, 5565. 684

89T5703 F. A. Davis and A. C. Sheppard; *Tetrahedron*, 1989, **45**, 5703. 342, 344, 349

89T6003 S. Stavber, B. Sket, B. Zajc and M. Zupan; *Tetrahedron*, 1989, **45**, 6003. 16

89T7411 G. Petrillo, M. Novi, G. Garbarino and M. Filiberti; *Tetrahedron*, 1989, **45**, 7411. 707

89T7643 P. C. B. Page, M. B. van Niel and J. C. Prodger; *Tetrahedron*, 1989, **45**, 7643. 143

89TL31 R. Sulsky and J. P. Demers; *Tetrahedron Lett.*, 1989, **30**, 31. 351

89TL251 K. Mukkanti, Y. V. Subba Rao and B. M. Choudary; *Tetrahedron Lett.*, 1989, **30**, 251. 753

89TL279 M. Gingras and T. H. Chan; *Tetrahedron Lett.*, 1989, **30**, 279. 87

89TL349 S. Himdi-Kabbab, P. Pellon and J. Hamelin; *Tetrahedron Lett.*, 1989, **30**, 349 489, 879

89TL363 H. Takeda, T. Morimoto and K. Achiwa; *Tetrahedron Lett.*, 1989, **30**, 363. 54, 305

89TL367 H. Takeda, T. Tachinami and M. Aburatani; *Tetrahedron Lett.*, 1989, **30**, 367. 305

89TL395 K. Burgess and M. J. Ohlmeyer; *Tetrahedron Lett.*, 1989, **30**, 395. 43

89TL557 A. Wagner, M.-P. Heitz and C. Mioskowski; *Tetrahedron Lett.*, 1989, **30**, 557. 10

89TL577 Ju. G. Trishin, I. V. Konovalova, R. N. Burangulova, L. A. Burnaeva, V. N. Christokletov and A. N. Pudovik; *Tetrahedron Lett.*, 1989, **30**, 577. 1074

89TL739 M. Fujiwara, M. Imada, A. Baba and M. Aburatani; *Tetrahedron Lett.*, 1989, **30**, 739. 318

89TL791 R. M. Moriarty, J. S. Khosrowshahi and R. Penmasta; *Tetrahedron Lett.*, 1989, **30**, 791. 25

89TL967 M. Shimizu and H. Yoshioka; *Tetrahedron Lett.*, 1989, 967. 610

89TL1063 E. J. Enholm and A. Trivellas; *Tetrahedron Lett.*, 1989, **30**, 1063. 81

89TL1099 M. Kira, T. Hino and H. Sakurai; *Tetrahedron Lett.*, 1989, **30**, 1099. 78

89TL1169 J. E. McMurray and J. G. Rico; *Tetrahedron Lett.*, 1989, **30**, 1169. 81

89TL1293 I. Paterson, C. K. McClure and R. L. Schumann; *Tetrahedron Lett.*, 1989, **29**, 1293. 533

89TL1327 D. R. Williams, M. H. Osterhout and J. M. McGill; *Tetrahedron Lett.*, 1989, **30**, 1327. 354

89TL1517 W. Hummer, T. Gracza and V. Jager; *Tetrahedron Lett.*, 1989, **30**, 1517. 318

89TL1641 J. Ichikawa, T. Sonoda and H. Kobayashi; *Tetrahedron Lett.*, 1989, **30**, 1641. 906

89TL1657 M. Yoshioka, T. Kawakita and M. Ohno; *Tetrahedron Lett.*, 1989, **30**, 1657. 72

89TL2013 R. Hunter and G. D. Tomlinson; *Tetrahedron Lett.*, 1989, **30**, 2013. 89

89TL2125 M. Asami and N. Kanemaki; *Tetrahedron Lett.*, 1989, **30**, 2125. 85

89TL2173 G. Stork and K. Zhao; *Tetrahedron Lett.*, 1989, 2173. 630

89TL2183 J. A. Marshall and W. Y. Gung; *Tetrahedron Lett.*, 1989, **30**, 2183. 79

89TL2187 J. A. Akers and T. A. Bryson; *Tetrahedron Lett.*, 1989, **30**, 2187. 518

89TL2387 A. Alexakis, I. Marek, P. Mangeney and J. F. Normant; *Tetrahedron Lett.*, 1989, **30**, 2387. 86

89TL2391 A. Alexakis, I. Marek, P. Mangeney and J. F. Normant; *Tetrahedron Lett.*, 1989, **30**, 2391. 86

89TL2421 J. Iqbal and S. P. Joseph; *Tetrahedron Lett.*, 1989, **30**, 2421. 79

89TL2461 W. G. Dauben, B. A. Kowalczyk and D. P. Bridon; *Tetrahedron Lett.*, 1989, **30**, 2461. 13

89TL2575 F. Di Furia, G. Licini, G. Modena and O. De Lucchi; *Tetrahedron Lett.*, 1989, **30**, 2575. 147

89TL2623 B. B. Lohray, Y. Gao and K. B. Sharpless; *Tetrahedron Lett.*, 1989, **30**, 2623. 322

89TL2763 J. P. King and R. Rathore; *Tetrahedron Lett.*, 1989, **30**, 2763. 201

89TL2783 S. Jude and J. P. Genet; *Tetrahedron Lett.*, 1989, **30**, 2783. 453, 455

89TL2869 M. J. Szymonifka and J. V. Heck; *Tetrahedron Lett.*, 1989, **30**, 2869. 249

89TL2927 J. Santamaria, R. Ouchabane and J. Rigaudy; *Tetrahedron Lett.*, 1989, **30**, 2927. 304

89TL3041 T. H. Chan and D. Wang; *Tetrahedron Lett.*, 1989, **30**, 3041. 78

89TL3061 A. Waldner; *Tetrahedron Lett.*, 1989, **30**, 3061. 245

89TL3077 F. Munyemana, A.-M. Frisque-Hesbain, A. Devos and L. Ghosez; *Tetrahedron Lett.*, 1989, **30**, 3077. 8

89TL3167 I. Chehidi, M. M. Chaabouni and A. Baklouti; *Tetrahedron Lett.*, 1989, **30**, 3167. 18

89TL3175 P. Ceccherelli, M. Curini, M. C. Marcotullio and O. Rosati; *Tetrahedron Lett.*, 1989, **30**, 3175. 44

89TL3271 R. V. Boeckmann, Jr. and K. J. O'Connor; *Tetrahedron Lett.*, 1989, **30**, 3271. 70

89TL3303 A. R. Katritzky, J. Jiang and L. Urogdi; *Tetrahedron Lett.*, 1989, **30**, 3303. 306

89TL3487 R. C. Larock and S. K. Stolz-Dunn; *Tetrahedron Lett.*, 1989, **30**, 3487. 86

89TL3541 G. Cahiez and M. Alami; *Tetrahedron Lett.*, 1989, **30**, 3541. 587

89TL3659 F. Rebiere and H. Kagan; *Tetrahedron Lett.*, 1989, **30**, 3659. 198
89TL3769 W.-W. Sy and B. A. Lodge; *Tetrahedron Lett.*, 1989, 3769. 631
89TL3789 M. Satoh, Y. Nomoto, N. Miyaura and A. Suzuki; *Tetrahedron Lett.*, 1989, **30**, 3789. 527, 900, 902
89TL3835 M. R. Bryce and P. C. Taylor; *Tetrahedron Lett.*, 1989, **30**, 3835. 792
89TL3901 W. F. Bailey, T. V. Ovaska and T. K. Liepert; *Tetrahedron Lett.*, 1989, **30**, 3901. 68
89TL3977 J. Santamaria, R. Ouchabane and J. Rigaudy; *Tetrahedron Lett.*, 1989, **30**, 3977. 304
89TL4161 H. Tanaka, T. Nakahara, H. Dhimane and S. Torii; *Tetrahedron Lett.*, 1989, **30**, 4161. 313
89TL4255 J.-T. Yamada, M. Yumoto and Y. Yamamoto; *Tetrahedron Lett.*, 1989, **30**, 4255. 85
89TL4383 Y. Mori and M. Suzuki; *Tetrahedron Lett.*, 1989, **30**, 4383. 61
89TL4387 Y. Mori and M. Suzuki; *Tetrahedron Lett.*, 1989, **30**, 4387. 61
89TL4389 K. Takai, Y. Kataoka and K. Utimoto; *Tetrahedron Lett.*, 1989, **30**, 4389. 80
89TL4555 U. Azzena, G. Delogu, G. Melloni and O. Piccolo; *Tetrahedron Lett.*, 1989, **30**, 4555. 9
89TL4859 V. Conte, F. Di Furia, G. Licini and G. Modena; *Tetrahedron Lett.*, 1989, **30**, 4859. 147
89TL4939 E. J. Enholm and G. Prasad; *Tetrahedron Lett.*, 1989, **30**, 4939. 81
89TL5105 C. J. Rousset, D. R. Swanson, F. Lamaty, and E. Negishi; *Tetrahedron Lett.*, 1989, **30**, 5105. 988
89TL5141 G. Balavoine, J. C. Clinet and I. Lellouche; *Tetrahedron Lett.*, 1989, **30**, 5141. 64
89TL5153 Y. Satoh, T. Tayano, S. Hara and A. Suzuki; *Tetrahedron Lett.*, 1989, **30**, 5153. 902
89TL5207 E. J. Corey and G. A. Reichard; *Tetrahedron Lett.*, 1989, **30**, 5207. 65
89TL5437 J. Ichikawa, T. Sonoda and H. Kobayashi; *Tetrahedron Lett.*, 1989, **30**, 5437. 906
89TL5447 M. Shimazaki, M. Hara and K. Suzuki; *Tetrahedron Lett.*, 1989, **30**, 5447. 61
89TL5551 N. N. Joshi, M. Srebnik and H. C. Brown; *Tetrahedron Lett.*, 1989, **30**, 5551. 72
89TL5643 A. Pelter, K. Smith, S. Elgendy and M. Rowlands; *Tetrahedron Lett.*, 1989, **30**, 5643. 525, 905
89TL5647 A. Pelter, K. Smith, S. Elgendy and M. Rowlands; *Tetrahedron Lett.*, 1989, **30**, 5647. 525, 905
89TL5663 C. K. Reddy and M. Periasamy; *Tetrahedron Lett.*, 1989, **30**, 5663. 14, 36
89TL5709 J. R. Henry, L. R. Marcin, M. C. McIntosh, P. M. Scola, G. D. Harris, Jr. and S. M. Weinreb; *Tetrahedron Lett.*, 1989, **30**, 5709. 309
89TL5853 J. M. Hornback and B. Murugaveri; *Tetrahedron Lett.*, 1989, **30**, 5853. 302
89TL5857 K. Burgess and M. J. Ohlmeyer; *Tetrahedron Lett.*, 1989, **30**, 5857. 43, 527
89TL5861 K. Burgess and M. J. Ohlmeyer; *Tetrahedron Lett.*, 1989, **30**, 5861. 43, 527
89TL6047 Y. Takakawa, A. Uwano, H. Watanabe, M. Asanuma and K. Shimada; *Tetrahedron Lett.*, 1989, **30**, 6047. 543
89TL6147 R. H. Weigmann and E. U. Würthwein; *Tetrahedron Lett.*, 1989, **30**, 6147. 1054, 1058
89TL6275 E. J. Corey and J. O. Link; *Tetrahedron Lett.*, 1989, **30**, 6275. 65
89TL6379 J. Ichikawa, T. Sonoda and H. Kobayashi; *Tetrahedron Lett.*, 1989, **30**, 6379. 906
89TL6395 Y. Masaki, T. Imaeda, K. Nagata, H. Oda and A. Ito; *Tetrahedron Lett.*, 1989, **30**, 6395. 326
89TL6495 W. W. Zajac, M. G. Darcy, A. P. Subong and J. H. Buzby; *Tetrahedron Lett.*, 1989, **30**, 6495. 390
89TL6661 W. H. Pearson and Y.-F. Poon; *Tetrahedron Lett.*, 1989, **30**, 6661. 325
89TL6753 P. Kocienski, C. Love and D. A. Roberts; *Tetrahedron Lett.*, 1989, **30**, 6753. 561
89TL6967 J. Tsanaktsidis and P. E. Eaton; *Tetrahedron Lett.*, 1989, **30**, 6967. 13
89TL7095 H. Takahashi, T. Kawakita, M. Yoshioka, S. Kobayashi and M. Ohno; *Tetrahedron Lett.*, 1989, **30**, 7095. 72
89TL7121 I. Paterson, J. M. Goodman and M. Isaka; *Tetrahedron Lett.*, 1989, **29**, 7121. 533
89TL7199 Y. Yoshida and Y. Kimura; *Tetrahedron Lett.*, 1989, 7199. 614
89TL7357 S. Masamune, A. J. Duplantier, M. H. Nantz, J. C. Roberts, R. P. Short and P. Somfai; *Tetrahedron Lett.*, 1989, **30**, 7357. 533
89TL7373 G. Cahiez and P.-Y. Chavant; *Tetrahedron Lett.*, 1989, **30**, 7373. 74, 574
89TL7383 E. Yankep and H. Kapnang; *Tetrahedron Lett.*, 1989, **30**, 7383. 303, 304
89TL7385 G. Poli; *Tetrahedron Lett.*, 1989, **30**, 7385. 47
89UP-01 M. Raban, L. H. Craine and N. King; *Unpublished Results*, 1989. 768
89ZAAC(569)62 G. Becker, M. Meiser, O. Mundt and J. Weidlein; *Z. Anorg. Allg. Chem.*, 1989, **569**, 62. 509, 511
89ZN(B)481 T. Kaukorat and R. Schmutzler; *Z. Naturforsch., Teil B*, 1989, **44**, 481. 861
89ZN(B)1179 W. Siebert, M. Hildenbrand, P. Hornbach, G. Karger and H. Pritzkow; *Z. Naturforsch., Teil B*, 1989, **44**, 1179. 903
89ZOB1979 R. N. Burangulova, Ju. G. Trishin, I. V. Konovalova, L. A. Burnaeva, V. N. Christokletov and A. N. Pudovik; *Zh. Obshch. Khim.*, 1989, **59**, 1979. 1074
89ZOB2135 I. S. Zal'tsman, G. N. Koidan, A. A. Kudryavtsev, A. P. Marchenko and A. M. Pinchuk; *Zh. Obshch. Khim.*, 1989, **59**, 2135 (*Chem. Abstr.*, 1990, **112**, 179187x). 407
89ZOB2136 V. V. Miroshnichenko, G. N. Koidan, V. A. Oleinik, A. P. Marchenko and A. M. Pinchuk; *Zh. Obshch. Khim.*, 1989, **59**, 2136 (*Chem. Abstr.*, 1990, **112**, 179188y). 407

90ACR43 A. J. Kresge; *Acc. Chem. Res.*, 1990, **23**, 43. 636, 637, 1022
90ACR87 M.-F. Ruasse; *Acc. Chem. Res.*, 1990, **23**, 87. 30
90ACR219 K. U. Ingold, J. Lusztyk and K. D. Raner; *Acc. Chem. Res.*, 1990, **23**, 219. 21
90ACR263 K. M. Theopold; *Acc. Chem. Res.*, 1990, **23**, 263. 571
90AG(E)100 H.-J. Gais, G. Hellmann and H. J. Lindner; *Angew. Chem., Int. Ed. Engl.*, 1990, **29**, 100. 1000
90AG(E)188 S. Hollenstein and T. Laube; *Angew. Chem., Int. Ed. Engl.*, 1990, **29**, 188. 1008
90AG(E)205 C. Bolm, M. Zehuber and D. Bur; *Angew. Chem., Int. Ed. Engl.*, 1990, **29**, 205. 72
90AG(E)287 P. J. Stang, A. M. Arif and C. M. Crittell; *Angew. Chem., Int. Ed. Engl.*, 1990, **29**, 287. 1020
90AG(E)522 W. Dabkowski, J. Michalski and W. Qing; *Angew. Chem., Int. Ed. Engl.*, 1990, **29**, 522. 411
90AG(E)771 S. L. Buchwald, R. A. Fisher and B. M. Foxman; *Angew. Chem., Int. Ed. Engl.*, 1990, **29**, 771. 873

90AG(E)782 L. F. Tietze, M. Beller, R. Fischer, M. Lögers, E. Jähde, K.-H. Glüsenkamp and M. F. Rajewsky; *Angew. Chem., Int. Ed. Engl.*, 1990, **29**, 782. 412

90AG(E)1150 A. H. Cowley, R. A. Jones, M. A. Mardones, J. Ruiz, J. L. Atwood and S. G. Bott; *Angew. Chem., Int. Ed. Engl.*, 1990, **29**, 1150 475

90AHC(50)157 E. V. Kuznetsov, I. V. Shcherbakova and A. T. Balaban; *Adv. Heterocycl. Chem.*, 1990, **50**, 157. 1008

90AKZ610 N. Y. Grigoryan, R. A. Khachatryan and M. G. Indzhikyan; *Arm. Khim. Zh.*, 1990, **43**, 610 (*Chem. Abstr.*, 1991, **114**, 207 372). 446

90AOC(30)77 A. J. Ashe, III; *Adv. Organomet. Chem*, 1990, **30**, 77. 505, 893

90AX(C)774 A. H. Cowley, C. M. Nunn and D. L. Westmoreland; *Acta Crystallogr., Sect. C*, 1990, **46**, 774. 893

90AX(C)1300 N. W. Alcock, A. P. Bozopoulos, E. Hatzigrigoriou and A. Varvoglis; *Acta Crystallogr., Part C*, 1990, **46**, 1300. 998

90BCJ51 J. Tanaka, S. Kanemasa and O. Tsuge; *Bull. Chem. Soc. Jpn.*, 1990, **63**, 51. 930, 939

90BCJ1252 M. Kohno, S. Sasao and S.-I. Murahashi; *Bull. Chem. Soc. Jpn.*, 1990, **63**, 1252. 318

90BCJ2051 M. Jung, M. Atoh, K. Kashiwabara and J. Fujita; *Bull. Chem. Soc. Jpn.*, 1990, **63**, 2051. 437

90BCJ2728 T. Oshikawa and M. Yamashita; *Bull. Chem. Soc. Jpn.*, 1990, **63**, 2728. 459

90BCJ3033 S. Kajigaeshi, M. Moriwaki, S. Fujisaki, T. Kakinami and T. Okamoto; *Bull. Chem. Soc. Jpn.*, 1990, **63**, 3033. 24

90BCJ3103 T. Kusumoto and T. Hiyama; *Bull. Chem. Soc. Jpn.*, 1990, **63**, 3103. 923, 940

90BSF275 P. Charreau, M. Julia and J. N. Verpeaux; *Bull. Soc. Chim. Fr.*, 1990, **127**, 275. 628

90BSF758 L. Even, D. Florentin and A. Marquet; *Bull. Soc. Chim. Fr.*, 1990, **127**, 758. 118, 124

90CB417 H. Lang, H. Keller, W. Imhof and S. Martin; *Chem. Ber.*, 1990, **123**, 417. 1036, 1037

90CB975 B. Ross, J. Belz and M. Nieger; *Chem. Ber.*, 1990, **123**, 975. 409, 503

90CB1507 B. Bogdanović, N. Janke, and H. G. Kinzelmann; *Chem. Ber.*, 1990, **123**, 1507. 559, 560, 959

90CB1517 E. Bartmann, B. Bogdanovic, N. Janke, S. Liao, K. Schilchte, B. Spliethoff, J. Treber, U. Westeppe and U. Wilczok; *Chem. Ber.*, 1990, **123**, 1517. 558

90CB1827 H. A. Mayer and W. C. Kasha; *Chem. Ber.*, 1990, **123**, 1827. 437

90CB1837 D. Hanssgen, H. Aldenhoven and M. Nieger; *Chem. Ber.*, 1990, **123**, 1837. 474

90CB1853 H. Ahlbrecht, G. Boche, K. Harms, M. Marsch and H. Sommer; *Chem. Ber.*, 1990, **123**, 1853. 1001

90CB2173 N. Krausse; *Chem. Ber.*, 1990, **123**, 2173. 940

90CB2273 A. Fessenbecker, H. Schulz H. Pritzkow and W. Siebert; *Chem. Ber.*, 1990, **123**, 2273. 903

90CC85 Y.-Z. Huang, Y. Liao and C. Chen; *J. Chem. Soc., Chem. Commun.*, 1990, 85. 487

90CC118 M. Ochiai, T. Ho, Y. Takaoka, Y. Masaki, M. Kunishima, S. Tani and Y. Nagao; *J. Chem. Soc., Chem. Commun.*, 1990, 118. 1020

90CC182 T. Takahashi, T. Fujimori, T. Seki, M. Saburi, Y. Uchida, C. J. Rousset, and E. Negishi; *J. Chem. Soc., Chem. Commun.*, 1990, 182. 987

90CC1030 A. Barbero, P. Cuadrado, I. Fleming, A. M. Gonzalez and F. J. Pulido; *J. Chem. Soc., Chem. Commun.*, 1990, 1030. 930

90CC1056 R. S. Tanke and R. H. Crabtree; *J. Chem. Soc., Chem. Commun.*, 1990, 1056. 915

90CC1169 P. Gugger, A. C. Willis and S. B. Wild; *J. Chem. Soc., Chem. Commun.*, 1990, 1169. 487

90CC1385 T. Okawara, K. Uchiyama, Y. Okamoto, T. Yamasaki and M. Furukawa; *J. Chem. Soc., Chem. Commun.*, 1990, 1385. 321

90CC1532 E. G. Nolen, A. Allocco, J. Vitarius and K. McSorley; *J. Chem. Soc., Chem. Commun.*, 1990, 1532. 302

90CC1598 B. C. Challis and T. I. Yousaf; *J. Chem. Soc., Chem. Commun.*, 1990, 1598. 385

90CC1680 G. P. Boldrini, F. Mancini, E. Tagliavini, C. Trombini and A. Umani-Ronchi; *J. Chem. Soc., Chem. Commun.*, 1990, 1680. 698

90CI547 K. Bhandari, V. L. Sharma and S. K. Chatterjee; *Chem. Ind.*, 1990, 547. 320

90CJC456 A. R. Katritzky, S. Rachwal and J. Wu; *Can. J. Chem.*, 1990, **68**, 456. 304

90CL203 K. Takaki, Y. Tsubaki, S. Tanaka, F. Beppu and Y. Fujiwara; *Chem. Lett.*, 1990, 203. 312

90CL547 M. Hayashi, F. Okamura, T. Toba, N. Oguni and K. B. Sharpless; *Chem. Lett.*, 1990, 547. 348

90CL769 Y. Yoshida, Y. Kimura and M. Tomoi; *Chem. Lett.*, 1990, 769. 614

90CL927 S. Fukuzawa, T. Fujinami and S. Sakai; *Chem. Lett.*, 1990, 927. 279

90CL1097 K. Saigo, Y. Hashimoto, N. Kihara, K. Hara and M. Hasgawy; *Chem. Lett.*, 1990, 1097. 141

90CL1395 K. Kato and T. Mukaiyama; *Chem. Lett.*, 1990, 1395. 391

90CL2239 T. Mukaiyama, T. Ohno, T. Nishimura, J. S. Han and S. Kobayashi; *Chem. Lett.*, 1990, 2239. 142

90CM221 A. H. Cowley, R. A. Jones, C. M. Nunn and D. L. Westmoreland; *Chem. Mater.*, 1990, **2**, 221. 511

90CPB329 J. Igarashi, N. Watanabe, T. Kinoshita, K. Ogata and S. Furukawa; *Chem. Pharm. Bull.*, 1990, **38**, 329. 1000

90CPB673 Y. Kiso, M. Yoshida, Y. Fugiwara, T. Kimura, M. Shimokura and K. Akaji; *Chem. Pharm. Bull.*, 1990, **38**, 673. 478

90CPB2637 M. Watanabe, M. Date, K. Kawanishi, T. Hori and S. Furukawa; *Chem. Pharm. Bull.*, 1990, **38**, 2637. 694

90CPB3331 T. Sato, T. Ito, H. Ishibashi and M. Ikeda; *Chem. Pharm. Bull.*, 1990, **38**, 3331.

90CRV215 J. Dubac, A. Laporterie and G. Manuel; *Chem. Rev.*, 1990, **90**, 215. 949

90CRV265 E. Colomer, R. J. P. Corriu and M. Lheureux; *Chem. Rev.*, 1990, **90**, 265. 949

90CRV879 V. Snieckus; *Chem. Rev.*, 1990, **90**, 879. 554, 653, 942

90CRV997 F. Mathey; *Chem. Rev.*, 1990, **90**, 997. 430, 438

90CSR147 P. C. Bullman Page, S. S. Klair and S. Rosenthal, *Chem. Soc. Rev.*, 1990, **19**, 147. 936

90CZ211 M. Fild, F. Bringewski and T. Oehmigen; *Chem.-Ztg.*, 1990, **114**, 211. 431

90CZ353 W. Wolfsberger; *Chem.-Ztg.*, 1990, **114**, 353. 439

90H(31)889 T. Kataoka, M. Yosshimatsu, H. Shimizu, Y. Kawase and M. Hori; *Heterocycles*, 1990, **31**, 889. 281

90H(32)2099 G. Shapiro and Y. Lavi; *Heterocycles*, 1990, **32**, 2099. 235

90HOU(E16)1 R. Andree, J. F. Neuth and H. S.-J. Wroblowsky; *Methoden Org. Chem.* (*Houben-Weyl*), 1990, **E16**, 1. 340, 342, 350, 353, 355, 356

90HOU(E16/1)404 K. Doser and R. Hemmer; *Methoden Org. Chem.* (*Houben-Weyl*), 1990, **E16a/1**, 404. 347

90HOU(E16a/1)425 U. Jensen-Korte; *Methoden Org. Chem.* (*Houben-Weyl*), 1990, **E16a/1**, 425. 373

90HOU(E16a/1)529 N. Müller; *Methoden Org. Chem.* (*Houben-Weyl*), 1990, **E16a/1**, 529. 373

90HOU(E16a/1)561 N. Müller; *Methoden Org. Chem.* (*Houben-Weyl*), 1990, **E16a/1**, 561. 383

90HOU(E16a/2)950 W. Hanefeld; *Methoden Org. Chem.* (*Houben-Weyl*), 1990, **E16a/2**, 950. 389

90HOU(E16a/2)1035 J. Goerdeler; *Methoden Org. Chem.* (*Houben-Weyl*), 1990, **E16a/2**, 1035.

90HOU(E16a/2)1137 S. Pawlenko; *Methoden Org. Chem.* (*Houben-Weyl*), 1990, **E16a/2**, 1137. 385

90HOU(E16a/2)1147 B. Unterhalt; *Methoden Org. Chem.* (*Houben-Weyl*), 1990, **E16a/2**, 1147. 385

90HOU(E16a/2)1182 E. Engel; *Methoden Org. Chem.* (*Houben-Weyl*), 1990, **E16a/2**, 1182. 387

90HOU(E16a/2)1227 S. Lang-Fugmann; *Methoden Org. Chem.* (*Houben-Weyl*), 1990, **E16a/2**, 1227. 388

90IC33 E. A. Ganja and J. A. Morrison; *Inorg. Chem.*, 1990, **29**, 33. 483, 486

90IC798 A.-J. DiMaio and A. L. Rheingold; *Inorg. Chem.*, 1990, **29**, 798. 500

90IC3502 D. K. Srivastava, L. K. Krannich and C. L. Watkins; *Inorg. Chem.*, 1990, **29**, 3502. 434, 485, 872

90IC5081 D. K. Kennepohl, B. D. Santarsiero and R. G. Cavell; *Inorg. Chem.*, 1990, **29**, 5081. 477

90IJC(A)703 T. N. Srivastava, J. D. Singh, S. K. Srivastava and S. Srivastava; *Indian J. Chem., Sect. A*, 1990, **29**, 703. 690

90IS249 G. Becker, H. Schmidt, G. Uhl and W. Uhl; *Inorg. Synth.*, 1990, **27**, 249. 478

90IS305 M. L. Luethens, A. P. Sattelburger, H. H. Murray, J. D. Basil and J. P. Fackler; *Inorg. Synth.*, 1990, **28**, 305. 434

90IZV1865 A. V. Martynov, S. G. Seredkina and A. N. Mirskova; *Izv. Akad. Nauk SSSR Ser. Khim.*, 1990, 1865. 1036, 1037

90JA311 S. E. Denmark, Y. C. Moon and C. B. W. Senanayake; *J. Am. Chem. Soc.*, 1990, **112**, 311. 356

90JA864 S. E. Denmark and R. L. Dorow; *J. Am. Chem. Soc.*, 1990, **112**, 864. 1004

90JA866 D. A. Evans, J. S. Clark, R. Metternich, V. J. Novack and G. S. Sheppard; *J. Am. Chem. Soc.*, 1990, **112**, 866. 702

90JA878 E. J. Corey, C. H. Yu and D. H. Lee; *J. Am. Chem. Soc.*, 1990, **112**, 878. 79, 533

90JA1286 M. Mitani, H. Matsumoto, N. Gouda and K. Koyama; *J. Am. Chem. Soc.*, 1990, **112**, 1286. 83

90JA1847 J. J. Eisch, B. Shafii, J. D. Odom and A. L. Rheingold; *J. Am. Chem. Soc.*, 1990, **112**, 1847. 1083

90JA1874 C. Guetin, D. Lexa, M. Momenteau and J.-M. Savéant; *J. Am. Chem. Soc.*, 1990, **112**, 1874. 578

90JA2368 T. Shono, Y. Matsumura, S. Katoh, K. Takeuchi, K. Sasaki, T. Kamada and R. Shimizu; *J. Am. Chem. Soc.*, 1990, **112**, 2368. 323

90JA2383 O. M. Gupta, R. L. Kirchmeier and J. M. Shreve; *J. Am. Chem. Soc.*, 1990, **112**, 2383. 334

90JA2389 H. C. Brown, R. S. Randad, K. S. Bhat, M. Zaidlewicz and U. S. Racherla; *J. Am. Chem. Soc.*, 1990, **112**, 2389. 75, 521

90JA2767 W. Oppolzer, J. Blagg, I. Rodriquez and E. Walther; *J. Am. Chem. Soc.*, 1990, **112**, 2767. 532

90JA3093 R. L. Danheiser, R. G. Brisbois, J. J. Kowalczyk and R. F. Miller; *J. Am. Chem. Soc.*, 1990, **112**, 3093. 668

90JA3697 G. A. Olah, Q. Wang and G. K. S. Prakash; *J. Am. Chem. Soc.*, 1990, **112**, 3697. 144

90JA3949 K. Mikami, M. Terada and T. Nakai; *J. Am. Chem. Soc.*, 1990, **112**, 3949. 80

90JA4031 S. H. Bertz; *J. Am. Chem. Soc.*, 1990, **112**, 4031. 588

90JA4032 B. H. Lipshutz; *J. Am. Chem. Soc.*, 1990, **112**, 4032. 588

90JA4037 J. R. Harper and A. L. Rheingold; *J. Am. Chem.Soc.*, 1990, **122**, 4037. 496

90JA4063 B. H. Lipshutz, R. Crow, S. H. Dimock, E. L. Ellsworth, R. A. J. Smith and J. R. Behling, Jr.; *J. Am. Chem. Soc.*, 1990, **112**, 4063. 589

90JA4862 J. R. Keeffe, A. J. Kresge and N. P. Schepp; *J. Am. Chem. Soc.*, 1990, **112**, 4862. 637

90JA4873 J. S. Panek and P. F. Cirillo; *J. Am. Chem. Soc.*, 1990, **112**, 4873. 46

90JA4897 S.-I. Inoue, H. Takaya, K. Tani, S. Otsuka, T. Sato and R. Noyori; *J. Am. Chem. Soc.*, 1990, **112**, 4897. 816

90JA4956 P. A. Wender and F. E. McDonald; *J. Am. Chem. Soc.*, 1990, **112**, 4956. 988

90JA5244 T. Imamoto, T. Oshiki, T. Onozawa and T. Kusumoto; *J. Am. Chem. Soc.*, 1990, **112**, 5244. 444

90JA5672 D. G. Ray and G. F. Koser; *J. Am. Chem. Soc.*, 1990, **112**, 5672. 148

90JA5876 H. Takahashi, S. Sakuruba, H. Takeda and K. Achiwa; *J. Am. Chem. Soc.*, 1990, **112**, 5876. 54

90JA6118 J. Yamada, H. Abe and Y. Yamamoto; *J. Am. Chem. Soc.*, 1990, **112**, 6118. 70

90JA6146 P. Knockel and S. A. Rao; *J. Am. Chem. Soc.*, 1990, **112**, 6146. 74

90JA6339 W. R. Roush, K. Ando, D. B. Powers, A. D. Palkowitz and R. L. Halterman; *J. Am. Chem. Soc.*, 1990, **112**, 6339. 75, 529, 557

90JA6348 W. R. Roush, A. D. Palkowitz and K. Ando; *J. Am. Chem. Soc.*, 1990, **112**, 6348. 75

90JA6389 B. Mudryk, C. Shook and T. Cohen; *J. Am. Chem. Soc.*, 1990, **112**, 6389. 553

90JA6437 P. J. Stang and V. V. Zhdankin; *J. Am. Chem. Soc.*, 1990, **112**, 6437. 1020

90JA6696 V. J. Jasys, P. R. Kellbaugh, D. M. Jason, D. Phillips, K. J. Rosnack, N. A. Saccomano, J. G. Stroh and R. A. Volkmann; *J. Am. Chem. Soc.*, 1990, **112**, 6696. 344

90JA6737 K. Mikami, T.-P. Loh and T. Nakai; *J. Am. Chem. Soc.*, 1990, **112**, 6737. 70

90JA6936 K. Nakayama and W. J. Thompson; *J. Am. Chem. Soc.*, 1990, **112**, 6936. 108

90JA7229 A. C. Dema, C. M. Lukehart, A. T. McPhail and D. R. McPhail; *J. Am. Chem. Soc.*, 1990, **112**, 7229. 927

90JA7405 E. Valenti, M. A. Pericàs and F. Serratosa; *J. Am. Chem. Soc.*, 1990, **112**, 7405. 1027

90JA7423 D. A. Singleton and J. P. Martinez; *J. Am. Chem. Soc.*, 1990, **112**, 7423. 904

90JA7428 M. Isaka and E. Nakamura; *J. Am. Chem. Soc.*, 1990, **112**, 7428. 589

90JA7431 P. Knochel; *J. Am. Chem. Soc.*, 1990, **112**, 7431. 1005

90JA7433 P. J. Kropp, K. A. Daus, S. D. Crawford, M. W. Tubergen, K. D. Kepler, S. L. Craig and V. P. Wilson; *J. Am. Chem. Soc.*, 1990, **112**, 7433. 4, 23, 30

90JA7440 B. H. Lipshutz and E. L. Ellsworth; *J. Am. Chem. Soc.*, 1990, **112**, 7440. 992

90JA7653 J. K. Whitesell, J. F. Carpenter, H. Kenan-Yaser and Machajewski; *J. Am. Chem. Soc.*, 1990, **112**, 7653. 245

90JA7823 L. T. Scott and M. Unno; *J. Am. Chem. Soc.*, 1990, **112**, 7823. 435

90JA7984 R. S. Tanke and R. H. Crabtree; *J. Am. Chem. Soc.*, 1990, **112**, 7984. 915

90JA8175 M. Koreeda and L. G. Hamann; *J. Am. Chem. Soc.*, 1990, **112**, 8175. 540

90JA8563 T. Umemoto, S. Fukami, G. Tomizawa, K. Harasawa, K. Kawada and K. Tomita; *J. Am. Chem. Soc.*, 1990, **112**, 8563. 16

90JA8575 R. M. Moriarty, J. Hiratake and K. Liu; *J. Am. Chem. Soc.*, 1990, **112**, 8575. 476

90JA8590 Y. Zhang, G. Wu, G. Agnel, and E. Negishi; *J. Am. Chem. Soc.*, 1990, **112**, 8590. 980

90JA8912 R. F. Childs, M. D. Kostyk, C. J. L. Lock and M. Mahendran; *J. Am. Chem. Soc.*, 1990, **112**, 8912. 1007

90JA9356 Z.-J. Ni, P.-F. Yang, D. K. P. Ng, Y.-L. Tzeng and T.-Y. Luh; *J. Am. Chem. Soc.*, 1990, **112**, 9356. 936

90JA9400 Y. N. C. Chan and J. A. Osborn; *J. Am. Chem. Soc.*, 1990, **112**, 9400. 312

90JA9439 T. Hudlicky, J. D. Price, F. Rulin and T. Tsunoda; *J. Am. Chem. Soc.*, 1990, **112**, 9439. 39

90JCR160 A. A. L. Gunatilaka; *J. Chem. Res.* (*S*), 1990, 160. 225

90JCR(S)188 F. Bellesia, F. Ghelfi, U. M. Pagnori and A. Pinetti; *J. Chem. Res.* (*S*), 1990, 188. 22

90JCR(S)268 Q.-Y. Chen and J.-P. Wu; *J. Chem. Res.* (*S*), 1990, 268. 311

90JCS(P1)11 T. J. Grattan and J. S. Whitehurst; *J. Chem. Soc., Perkin Trans. 1*, 1990, 11. 622

90JCS(P1)47 R. Baker and J. L. Castro; *J. Chem. Soc., Perkin Trans. 1*, 1990, 47. 628

90JCS(P1)937 S. Niwa and K. Soai; *J. Chem. Soc., Perkin Trans. 1*, 1990, 937. 1096

90JCS(P1)969 K. J. Divakar, A. Mottoh, C. B. Reese and Y. S. Sanghvi; *J. Chem. Soc., Perkin Trans. 1*, 1990, 969. 127

90JCS(P1)1615 P. C. B. Page, C. M. Rayner and I. O. Sutherland; *J. Chem. Soc., Perkin Trans. 1*, 1990, 1615. 224

90JCS(P1)1859 S. Itsuno, Y. Sakurai, K. Shimizu and K. Ito; *J. Chem. Soc., Perkin Trans. 1*, 1990, 1859. 314

90JCS(P1)2133 L. Barboni, G. Bartoli, E. Marcantoni, M. Petrini and R. Dalpozzo; *J. Chem. Soc., Perkin Trans. 1*, 1990, 2133. 346

90JCS(P1)2691 H. Bandal, V. F. Patel, G. Pattenden and J. J. Russell; *J. Chem. Soc., Perkin Trans. 1*, 1990, 2691. 579

90JCS(P1)2703 V. F. Patel and G. Pattenden; *J. Chem. Soc., Perkin Trans. 1*, 1990, 2703. 578

90JCS(P1)2709 H. Bandal, A. R. Howell, V. F. Patel and G. Pattenden; *J. Chem. Soc., Perkin Trans. 1*, 1990, 2709. 578, 579

90JCS(P1)2715 A. R. Howell and G. Pattenden; *J. Chem. Soc., Perkin Trans. 1*, 1990, 2715. 579

90JCS(P1)2807 J. Barluenga, M. A. Rodriguez and P. Campos; *J. Chem. Soc., Perkin Trans. 1*, 1990, 2807. 31, 35

90JCS(P1)3200 B. T. Cho and Y. S. Chun; *J. Chem. Soc., Perkin Trans. 1*, 1990, 3200. 312

90JGU397 S. A. Tarent'eva, N. A. Pudovik and A. N. Pudovik; *J. Gen. Chem. USSR* (*Engl. Transl.*), 1990, **60**, 397. 476

90JGU398 A. A. Prishchenko, M. V. Livantov, P. V. Zhut-skii, D. A. Pisarnitskii, M. M. Shagi-Mukhametova and U. S. Petrosyan; *J. Gen. Chem. USSR* (*Engl. Transl.*), 1990, **60**, 398. 476

90JHC1361 H. Mitsudera, H. Uneme, Y. Okada, M. Numata and A. Kato; *J. Heterocycl. Chem.*, 1990, **27**, 1361. 214

90JHC1549 K. O. Gelotte, D. N. Mason, H. Meckler, W.-C. Shieh and C. M. Starky; *J. Heterocycl. Chem.*, 1990, **27**, 1549. 619

90JHC1737 A. R. Katritzky and S. Perumal; *J. Heterocycl. Chem.*, 1990, **27**, 1737. 621

90JHC2151 B. McKittrick, A. Failli, R. J. Steffan, R. M. Soll, P. Hughes, J. Schmid, A. A. Asselin, C. C. Shaw, R. Noureldin and G. Gavin; *J. Heterocycl. Chem.*, 1990, **27**, 2151. 627

90JOC46 G. A. Olah, M. Piteau, K. Laali, C. B. Rao and O. Farooq; *J. Org. Chem.*, 1990, **55**, 46. 907

90JOC173 H. Danda, M. M. Hansen and C. H. Heathcock; *J. Org. Chem.*, 1990, **55**, 173. 532

90JOC292 D. L. Comins and Y. C. Myoung; *J. Org. Chem.*, 1990, **55**, 292. 942

90JOC304 S. Itsuno, Y. Sakurai, K. Ito, T. Maruyama, S. Nakahama and J. M. J. Frechet; *J. Org. Chem.*, 1990, **55**, 304. 71

90JOC324 E. J. Enholm, H. Satici and G. Prasad; *J. Org. Chem.*, 1990, **55**, 324. 86

90JOC395 E. Valenti, M. A. Pericas and F. Serratosa; *J. Org. Chem.*, 1990, **55**, 395. 941

90JOC449 T. Murai, T. Sakane and S. Kato; *J. Org. Chem.*, 1990, **55**, 449. 814

90JOC511 D. A. Anderson and J. R. Hwu; *J. Org. Chem.*, 1990, **55**, 511. 394

90JOC766 M. Minato, K. Yamamoto and J. Tsuji; *J. Org. Chem.*, 1990, **55**, 766. 45

90JOC768 W. H. Brunnelle, B. R. McKinnis and B. A. Narayanan; *J. Org. Chem.*, 1990, **55**, 768. 89

90JOC784 E. J. Corey, P.-W. Yuen, F. J. Hannan and D. A. Wierda; *J. Org. Chem.*, 1990, **55**, 784. 71

90JOC1108 Y. Kita, J.-I. Sekihachi, Y. Hayashi, Y.-Z. Da, M. Yamamoto and S. Akai; *J. Org. Chem.*, 1990, **55**, 1108. 941

90JOC1528 T. Cohen, I.-H. Jeong, B. Mudryk, M. Bhupathy and M. A. A. Awaad; *J. Org. Chem.*, 1990, **55**, 1528. 83, 553

90JOC1540 J. A. Marshall and B. E. Blough; *J. Org. Chem.*, 1990, **55**, 1540. 86

90JOC1548 T. Inokuchi, M. Kusumoto and S. Torii; *J. Org. Chem.*, 1990, **55**, 1548. 83

90JOC1698 D. A. Evans and S. W. Kaldor; *J. Org. Chem.*, 1990, **55**, 1698. 46

90JOC1705 O. Fujimura, L. Takai and K. Utimoto; *J. Org. Chem.*, 1990, **55**, 1705. 86

90JOC1707 K. Takai, Y. Kataoka and K. Utimoto; *J. Org. Chem.*, 1990, **55**, 1707. 927, 988, 989

90JOC1823 Y. Tamaru, K. Nagao, T. Bando and Z.-I. Yoshida; *J. Org. Chem.*, 1990, **55**, 1823. 206

90JOC1857 H. X. Zhang, F. Guibé and G. Balavoine; *J. Org. Chem.*, 1990, **55**, 1857. 915

90JOC1868 H. C. Brown, U. S. Racherla and P. J. Pellechia; *J. Org. Chem.*, 1990, **55**, 1868. 74

90JOC1916 J. C. Bottaro, R. J. Schmitt, C. D. Bedford, R. Gilardi and C. George; *J. Org. Chem.*, 1990, **55**, 1916. 1053

90JOC1981 M. D. Whitman, R. L. Halcomb and S. J. Danishefsky; *J. Org. Chem.*, 1990, **55**, 1981. 340

90JOC2070 C. Palomo, J. M. Aizpurna, F. P. Cossío, J. M. García, M. C. López and M. Oiarbide; *J. Org. Chem.*, 1990, **55**, 2070. 398

90JOC2246 D. Brillon and G. Sauvé; *J. Org. Chem.*, 1990, **55**, 2246. 750

90JOC2274 R. Soundararajan and D. S. Matteson; *J. Org. Chem.*, 1990, **55**, 2274. 900

90JOC2280 D. A. Evans and G. C. Fu; *J. Org. Chem.*, 1990, **55**, 2280. 42

90JOC2282 I. E. Reich and H. J. Reich; *J. Org. Chem.*, 1990, **55**, 2282. 938

90JOC2286 M. J. Kurth, S. H. Tahir and M. M. Olmstead; *J. Org. Chem.*, 1990, **55**, 2286. 133

90JOC2415 A. Hosomi, S. Kohra, K. Ogata, T. Yanagi and Y. Tominaga; *J. Org. Chem.*, 1990, **55**, 2415. 78

90JOC2437 A. Kamimura, H. Sasatani, T. Hashimoto, T. Kawai, K. Hori and N. Ono; *J. Org. Chem.*, 1990, **55**, 2437. 393

90JOC2531 W. Chamchaang and A. R. Pinhas; *J. Org. Chem.*, 1990, **55**, 2531. 319

90JOC2543 W. B. Smith and O. C. Ho; *J. Org. Chem.*, 1990, **55**, 2543. 632

90JOC2552 R. J. Mattson, K. M. Pham, D. J. Leuck and K. A. Cowen; *J. Org. Chem.*, 1990, **55**, 2552. 310

90JOC2563 M. D. Alexander, R. E. Anderson, J. Sisko and S. M. Weinreb; *J. Org. Chem.*, 1990, **55**, 2563. 310

90JOC2576 J. N. Haseltine and S. J. Danishefsky; *J. Org. Chem.*, 1990, **55**, 2576. 139, 236

90JOC2596 F. Bourdon, J.-L. Ripoll and Y. Vallée; *J. Org. Chem.*, 1990, **55**, 2596. 140

90JOC2673 E. B. Nadler and Z. Rappoport; *J. Org. Chem.*, 1990, **55**, 2673. 643

90JOC2855 R. Kiesgen de Richter; *J. Org. Chem.*, 1990, **55**, 2855. 42

90JOC2878 R. J. P. Corriu, J. J. E. Moreau and M. Pataud-Sat; *J. Org. Chem.*, 1990, **55**, 2878. 544

90JOC2927 E. Keinan, D. Perez, M. Sahai and R. Shvily; *J. Org. Chem.*, 1990, **55**, 2927. 14

90JOC2954 R. W. Murray and M. Singh; *J. Org. Chem.*, 1990, **55**, 2954. 344

90JOC2978 T. Tsuda, S. Morikawa, N. Hasegawa and T. Saegusa; *J. Org. Chem.*, 1990, **55**, 2978. 939

90JOC3002 G. Sauvé, N. Le Berre and B. Zacharie; *J. Org. Chem.*, 1990, **55**, 3002. 750

90JOC3007 E. Ciganek; *J. Org. Chem.*, 1990, **55**, 3007. 344, 349

90JOC3104 J. Barluenga, M. A. Rodriguez and P. J. Campos; *J. Org. Chem.*, 1990, **55**, 3104. 629

90JOC3233 M. J. Chem, J. C. Lineham and J. W. Rathke; *J. Org. Chem.*, 1990, **55**, 3233. 348, 349

90JOC3368 I. E. Markó and J. M. Southern; *J. Org. Chem.*, 1990, **55**, 3368. 598

90JOC3373 N. Satyamurthy, G. T. Bida, M. E. Phelps and J. R. Barrio; *J. Org. Chem.*, 1990, **55**, 3373. 613

90JOC3393 N. Chatani, N. Horiuchi and T. Hanafusa; *J. Org. Chem.*, 1990, **55**, 3393. 946

90JOC3415 E. Bay, D. A. Bak, P. E. Timony and A. Leone-Bay; *J. Org. Chem.*, 1990, **55**, 3415. 7, 621

90JOC3423 S. T. Purrington and D. L. Woodward; *J. Org. Chem.*, 1990, **55**, 3423. 16

90JOC3427 M. Bardisso, R. Gandolfi, P. Grünanger and A. Rastelli; *J. Org. Chem.*, 1990, **55**, 3427. 354

90JOC3464 Z.-Y Chang and R. M. Coates; *J. Org. Chem.*, 1990, **55**, 3464. 344, 345, 347, 353, 357

90JOC3475 Z.-Y Chang and R. M. Coates; *J. Org. Chem.*, 1990, **55**, 3475. 344, 345, 347

90JOC3552 S. Rozen and D. Zamir; *J. Org. Chem.*, 1990, **55**, 3552. 631

90JOC3565 S. Collins, Y. Hong, G. J. Hoover and J. R. Veit; *J. Org. Chem.*, 1990, **55**, 3565. 564

90JOC3634 L. Gong, R. Leung-Toung and T. T. Tidwell; *J. Org. Chem.*, 1990, **55**, 3634. 930

90JOC3880 J. Mathew and B. Alink; *J. Org. Chem.*, 1990, **55**, 3880. 21

90JOC4156 B. J. Wagner, J. T. Doi and W. K. Musker; *J. Org. Chem.*, 1990, **55**, 4156. 247

90JOC4199 P. Andreoli, L. Billi, G. Cainelli, M. Panunzio, G. Martelli and G. Spunta; *J. Org. Chem.*, 1990, **55**, 4199. 316

90JOC4215 W. T. Brady and K. Saidi; *J. Org. Chem.*, 1990, **55**, 4215. 930

90JOC4506 A. W. Konradi and S. F. Pedersen; *J. Org. Chem.*, 1990, **55**, 4506. 81

90JOC4560 N. Satyamurthy, J. R. Barrio, D. G. Schmidt, C. Kammerer, G. T. Bida and M. E. Phelps; *J. Org. Chem.*, 1990, **55**, 4560. 632

90JOC4581 M. K. Manthey, S. G. Pyne and R. J. W. Truscott; *J. Org. Chem.*, 1990, **55**, 4581. 906

90JOC4634 R. F. Cunico and C. P. Kuan; *J. Org. Chem.*, 1990, **55**, 4634. 814

90JOC4784 T. Cohen and M. D. Doubleday; *J. Org. Chem.*, 1990, **55**, 4784. 68

90JOC4832 K. Soai and M. Watanabe; *J. Org. Chem.*, 1990, **55**, 4832. 71

90JOC4986 T. Imai, H. Mineta and S. Nishida; *J. Org. Chem.*, 1990, **55**, 4986. 904

90JOC4990 G. A. Molander, E. R. Burkhardt and P. Weinig; *J. Org. Chem.*, 1990, **55**, 4990. 69, 564

90JOC5033 R. J. Hinkle, P. J. Stang and M. H. Kowalski; *J. Org. Chem.*, 1990, **55**, 5033. 445

90JOC5045 T. C. Wu, H. Xiong and R. D. Rieke; *J. Org. Chem.*, 1990, **55**, 5045. 563, 959

90JOC5155 S. Rozen and M. Kol; *J. Org. Chem.*, 1990, **55**, 5155. 94

90JOC5190 D. A. Evans and A. H. Hoveyda; *J. Org. Chem.*, 1990, **55**, 5190. 63

90JOC5291 D. P. Schumacher, J. E. Clark, B. L. Murphy and P. A. Fischer; *J. Org. Chem.*, 1990, **55**, 5291. 19

90JOC5310 K. Takai, M. Tezuka, Y. Kataoka and K. Utimoto; *J. Org. Chem.*, 1990, **55**, 5310. 927, 945

90JOC5324 Y. Kobayashi, T. Shimazaki, H. Taguchi and F. Sato; *J. Org. Chem.*, 1990, **55**, 5324. 900

90JOC5400 C. Giordano, L. Coppi and A. Restelli; *J. Org. Chem.*, 1990, **55**, 5400. 29

90JOC5404 W. F. Bailey and E. R. Punzalan; *J. Org. Chem.*, 1990, **55**, 5404. 554

90JOC5413 M. Lesage, J. A. Martinho Simões and D. Griller; *J. Org. Chem.*, 1990, **55**, 5413. 536

90JOC5564 J. S. Panek and M. A. Sparks; *J. Org. Chem.*, 1990, **55**, 5564. 935

90JOC5814 M. Kaino, Y. Naruse, K. Ishihara and H. Yamamoto; *J. Org. Chem.*, 1990, **55**, 5814. 679

90JOC5935 M. Sawamura, H. Hamashima and Y. Ito; *J. Org. Chem.*, 1990, **55**, 5935. 311

90JOC6071 P. F. Cirillo and J. S. Panek; *J. Org. Chem.*, 1990, **55**, 6071. 79

90JOC6082 M. P. Doyle, K. G. High, V. Bagheri, R. J. Pieters, P. J. Lewis and M. M. Pearson; *J. Org. Chem.*, 1990, **55**, 6082. 539

90JOC6244 R. C. Larock and W.-Y. Leung; *J. Org. Chem.*, 1990, **55**, 6244. 86

90JOC6248 P. Arya, C. Samson, M. D. Lesage and D. Griller; *J. Org. Chem.*, 1990, **55**, 6248. 141, 265

90JOC6328 H. C. Brown, P. V. Ramachandran, S. A. Weissman and S. Swaminathan; *J. Org. Chem.*, 1990, **55**, 6328. 57

90JOM(386)177 W. Frank; *J. Organomet. Chem.*, 1990, **386**, 177. 489

90JOM(388)71 L. Engman and J. Persson; *J. Organomet. Chem.*, 1990, **388**, 71. 735

90JOM(393)143 D. A. Blinn, R. S. Button, V. Farazi, M. K. Neeb, C. L. Taplev, T. E. Trehearne, S. D. West, T. L. Kruger and B. N. Storhoff; *J. Organomet. Chem.*, 1990, **393**, 143. 439

90JOM(396)C8 W. A. Schenk and E. Voss; *J. Organomet. Chem.*, 1990, **396**, C8. 484

90JOM(399)C4 F. Bitterer, D. J. Brauer, F. Dorrenbach and O. Stelzer; *J. Organomet. Chem.*, 1990, **399**, C4. 472

90JOM(399)C21 W. Clegg, N. A. Compton, R. J. Errington, G. A. Fisher, N. C. Norman, N. Wishart; *J. Organomet. Chem.*, 1990, **399**, C21. 512

90JOM(399)1 B. Wrackmeyer and K. Horchler; *J. Organomet. Chem.*, 1990, **399**, 1. 1102

90JOM(400)19 J.-F. Normant; *J. Organomet. Chem.*, 1990, **400**, 19. 998

90JOU583 A. D. Grebenyuk, R. A. Ismailova, R. B. Tokbolatov and T. V. Ovodova; *J. Org. Chem. USSR* (*Eng. Trans.*), 1990, **26**, 583. 392

90JOU1784 Y. A. Serguchev and R. B. Gutsulyak; *J. Org. Chem. USSR* (*Engl. Transl.*), 1990, **26**, 1784. 18

90K668 N. E. Zhukhlistova, V. I. Smirnova, G. N. Tishchenko, Yu. V. Nekrasov, V. I. Popovich and N. N. Shapelko; *Kristallografiya*, 1990, **35**, 668. 478

90KGS1477 M. A. Akhmedov, Z. G. Mustafaeva, I. M. Akhmedov, I. I. Vladimirova, R. R. Kostikov and A. A. Bairamov; *Khim. Geterotsikl. Soedin.*, 1990, 1477 (English translation: *Chemistry of Heterocyclic Compounds*, 1990, **26**, 1230). 136

90LA795 G. Bringmann, J.-P. Geisler, T. Geuder, G. Kunkel and L. Kinzinger; *Liebigs Ann. Chem.*, 1990, 795. 310

90M709 J. Reisch and A. Wickramasinghe; *Monatsh. Chem.*, 1990, **121**, 709. 687

90MI 201-01 Y. Sasson; *L'Actual Chim.*, 1990, 200. 31

B-90MI 203-01 R. S. Glass; in "Sulfur Centred Reactive Intermediates in Chemistry and Biology," eds. C. Chatgilialoglu and K. D. Asnus Plenum, 1990, p. 236. 132, 202

B-90MI 203-02 Belenkii (ed.); "Chemistry of Organosulfur Compounds", Ellis Horwood, Chichester, 1990, ch. 6. 143

B-90MI 203-03 J. Drabowicz, P. Kielbasinski and M. Mikolajczyk; in "The Chemistry of Sulfenic Acids and their Derivatives," ed. S. Patai, Wiley, Chichester, 1990, ch. 6. 165

B-90MI 203-04 E. Kühle; "The Chemistry of the Sulfenic Acids," Georg Thieme, Stuttgart, 1973. 165

B-90MI 203-05 S. Patai (ed.); "The Chemistry of Sulfinic Acids, Esters and their Derivatives," Wiley, Chichester, 1990. 172

B-90MI 203-06 J. Drabowicz, P. Lyzwa and M. Mikolajczyk; in "The Chemistry of Sulfenic Acids and their Derivatives," ed. S. Patai, Wiley, Chichester, 1990. 184, 194, 239

B-90MI 203-07 U. Zoller; in "The Chemistry of Sulfinic Acids, Esters and their Derivatives," ed. S. Patai, Wiley, Chichester, 1990, ch. 8. 186

B-90MI 203-08 J. Hoyle; in "The Chemistry of Sulfonic Acids, Esters and their Derivatives," ed. S. Patai, Wiley, Chichester, 1990, ch. 10. 189

B-90MI 203-09 D. C. Dittmer and M. D. Hoey; in "The Chemistry of Sulfinic Acids, Esters and Derivatives," ed. S. Patai, Wiley, Chichester, 1990, ch. 9. 196

B-90MI 203-10 A. J. Burglass and J. G. Tillett; in "The Chemistry of Sulfonic Acids, Esters and their Derivatives," ed. S. Patai, Wiley, Chichester, 1990, ch. 19. 478

B-90MI 206-01 J. Drabowicz, P. Kielbasinski and M. Mikolajczyk; in "The Chemistry of Sulfenic Acids, Esters and their Derivatives," ed. S. Patai, Wiley, Chichester, 1990, p. 221. 357, 358

B-90MI 206-02 S. Braverman; in "The Chemistry of Sulphenic Acids, Esters and their Derivatives," ed. S. Patai, Wiley, Chichester, 1990, p. 311. 362, 366

B-90MI 206-03 J. G. Tillett; in "The Chemistry of Sulphenic Acids, Esters and their Derivatives," ed. S. Patai, Wiley, Chichester, 1990, p. 603. 362, 364

B-90MI 206-04 D. C. Dittmer and M. D. Hoey; in "The Chemistry of Sulphenic Acids, Esters and their Derivatives," ed. S. Patai, Wiley, Chichester, 1990, p. 239. 366

B-90MI 206-05 J. Hoyle; in "The Chemistry of Sulphenic Acids, Esters and their Derivatives," ed. S. Patai, Wiley, Chichester, 1990, p. 351. 364, 365

B-90MI 207-01 H. Feuer and A. T. Nielsen (eds.); "Nitro Compounds," VCH, Weinheim, 1990. 391

B-90MI 208-01 D. G. Gilheany and C. M. Mitchell; in "The Chemistry of Organophosphorus Compounds," ed. F. R. Hartley, Wiley-Interscience, Chichester, 1990, vol. 1, ch. 7, pp. 151–190. 426

B-90MI 208-02 H. B. Kagan and M. Sasaki; in "The Chemistry of Organophosphorus Compounds," ed. F. R. Hartley, Wiley-Interscience, Chichester, vol. 1, ch. 3, pp. 63–74. 442

90MI 208-03 M. Regitz; *Heteroat. Chem.* (*2nd Int. Conf.*), 1989 (pub 1990), 295. 478

B-90MI 209-01 U. Kruerke and M. Wieber (eds.); "Gmelin Handbuch der Anorganischen Chemie," Springer Verlag, 1990, part 5. 481

B-90MI 211-01 L. Brandsma; "Preparative Polar Organometallic Chemistry," Springer Verlag, Berlin, 1990, vol. 2. 551

B-90MI 213-01 Z. Rappoport (ed.); "The Chemistry of Enols," Wiley, New York, 1990. 636

90MI 213-02 R. Gupta, Y. P. Singh and A. K. Rai; *Main Group Met. Chem.*, 1990, **13**, 7. 695

B-90MI 216-01 C. W. Allen; "Organophosphorus Chemistry," Royal Society of Chemistry, 1990, **21**, ch. 8. 864

B-90MI 217-01 M. Mirbach and M. Wieber; "Gmelin Handbook of Inorganic Chemistry, Sb Organoantimony Compounds, Part 5," ed. M. Mirbach, Springer Verlag, Berlin, 8th edn, 1990. 872

B-90MI 223-01 L. Brandsma and H. D. Verkruijsse; "Synthesis of Acetylenes, Allenes and Cumulenes," Elsevier, Amsterdam, 3rd edn, 1990, p. 1. 1076

B-90MI 223-02 F. Glocking, U. Küerke and J. Satgé; in "Gmelin Handbook of Inorganic Chemistry, Organogermanium Compounds," Part 3, ed. U. Küerke, Springer, Berlin, 8th edn, 1990, p. 10. 1092

90OM16 D. L. Reger, J. C. Baxter and D. G. Garza; *Organometallics*, 1990, **9**, 16. 581

90OM275 K. T. Higa and C. George; *Organometallics*, 1990, **9**, 275. 511

90OM301 G. R. Knox, P. L. Pauson, D. Willison and E. Solcániová; *Organometallics*, 1990, **9**, 301. 103

90OM2207 J. de With, A. D. Horton and A. G. Orpen; *Organometallics*, 1990, **9**, 2207. 570

90OM2653 M. J. Burk, J. E. Feaster and R. L. Harlow; *Organometallics*, 1990, **9**, 2653. 438

90OPP215 A. M. Moiseenkov, B. A. Czeskis, N. M. Ivanova and O. M. Nefedov; *Org. Prep. Proced. Int.*, 1990, **22**, 215. 11

90OPP707 G. Rosini, R. Ballini, M. Petrini, E. Marotta and P. Righi; *Org. Prep. Proced. Int.*, 1990, **22**, 707. 391, 392

90OR(38)1 D. J. Ager; *Org. React.*, 1990, **38**, 1. 997, 1005

90OR(39)1 A. R. Chamberlain and S. H. Bloom; *Org. React.*, 1990, **39**, 1. 911, 919, 956

90OR(39)85 D. Grierson; *Org. React.*, 1990, **39**, 85. 357

90OS(69)19 M. Eberle, M. Missbach and D. Seebach; *Org. Synth.*, 1990, **69**, 19. 394

90OS(69)72 S. H. Pine, G. Kim and V. Lee; *Org. Synth.*, 1990, **69**, 72. 681

90OS(69)129 T. Umemoto, K. Tomita and K. Kawada; *Org. Synth.*, 1990, **69**, 129. 16

90OS(70)195 M. C. P. Yeh, H. G. Chen and P. Knochel; *Org. Synth.*, 1990, **70**, 195. 478

90PAC595 J. Barluenga; *Pure Appl. Chem.*, 1990, **62**, 595. 553

90PAC605 G. Cainelli, M. Panunzio, P. Andreoli, G. Martelli, G. Spunta, D. Giacomini and E. Bandini; *Pure Appl. Chem.*, 1990, **62**, 605. 544

90PAC1139 S. B. Wild; *Pure Appl. Chem.*, 1990, **62**, 1139. 438

90POL289 L. K. Krannich, C. L. Watkins and D. K. Srivastava; *Polyhedron*, 1990, **9**, 289. 484

90POL319 R. L. Wells, C.-Y. Kwag, A. P. Purdy, A. T. McPhail and C. G. Pitt; *Polyhedron*, 1990, **9**, 319. 482, 504

90POL2413 A. A. Danopoulos, A. R. Wills and P. G. Edwards; *Polyhedron*, 1990, **9**, 2413. 437

90POL2763 M. L. H. Green, P. Mountford, G. J. Smout and S. R. Peel; *Polyhedron*, 1990, **9**, 2763. 597

90POL2901 W.-K. Wong, F. L. Chow, H. Chen, B. W. Au-Yeung, R.-J. Wang and T. C. W. Mak; *Polyhedron*, 1990, **9**, 2901. 436

90PS(47)341 B. Bildstein and F. Sladky; *Phosphorus Sulfur*, 1990, **47**, 341. 465

90PS(47)357 R. Karra, Y. P. Singh and A. K. Rai; *Phosphorus Sulfur*, 1990, **47**, 357. 889

90PS(47)465 L. Maier; *Phosphorus, Sulfur*, 1990, **47**, 465. 458

90PS(48)157 J. W. Pasterczyk and A. R. Barron; *Phosphorus Sulfur*, 1990, **48**, 157. 484

90PS(48)279 A. Hercouet and M. Le Corre; *Phosphorus Sulfur*, 1990, **48**, 279. 444

90PS(49/50)179 N. V. Lukashev and M. A. Kazankova; *Phosphorus Sulfur*, 1990, **49/50**, 179. 1027

90RCR819 I. V. Koval; *Russ. Chem. Rev.* (*Engl. Transl.*), 1990, **59**, 819. 401

90RCR832 K. A. Petrov, G. V. Rudnev and V. D. Sorokin; *Russ. Chem. Rev.* (*Engl. Transl.*), 1990, **59**, 832. 357

90S1 S. Servi; *Synthesis*, 1990, 1. 39

90S36 S. Ozazaki and Y. Sato; *Synthesis*, 1990, 36. 305

90S39 Z. Arnold; *Synthesis*, 1990, 39. 745

90S122 H. Yinglin and H. Hongwen; *Synthesis*, 1990, 122. 308

90S125 A. Löffler and G. Himbert; *Synthesis*, 1990, 125. 307

90S132 M. Sting and W. Steglich; *Synthesis*, 1990, 132. 459

90S140 C. Yuan and H. Feng; *Synthesis*, 1990, 140. 848

90S153 A. Alberola, M. A. Alvarez, C. Andres, A. Gonzalez and R. Pedrosa; *Synthesis*, 1990, 153. 320

90S212 P. D. Croce, R. Ferraccioli and A. Ritieni; *Synthesis*, 1990, 212. 21

90S285 T. Inghardt and T. Frejd; *Synthesis*, 1990, 285. 1099

90S361 W. Richter and I. Ugi; *Synthesis*, 1990, 661. 252

90S391 C. Lee and L. Field; *Synthesis*, 1990, 391. 718

90S401 H. Kotsuki, N. Yoshimura, I. Kadota, Y. Ushio and M. Ochi; *Synthesis*, 1990, 401. 312

90S505 K. Osowska-Pacewicka and A. Zwierzak; *Synthesis*, 1990, 505. 307

90S539 P. S. Belica and P. S. Manchand; *Synthesis*, 1990, 539. 708

90S575 T. Ziegler, B. Horsch and F. Effenberger; *Synthesis*, 1990, 575. 321

90S595 N. DeKimpe and P. Brunet; *Synthesis*, 1990, 595. 21

90S609 P. Dybowski and A. Skawronska; *Synthesis*, 1990, 609. 258

90S615 H. Yinglin and H. Hongwen; *Synthesis*, 1990, 615. 308

90S694 A. Kamimura and T. Nagashima; *Synthesis*, 1990, 694. 393

90S735 N. Kuriyama, Y. Inoue and K. Kitagawa; *Synthesis*, 1990, 735. 304

90S750 W. Miltz and W. Steglich; *Synthesis*, 1990, 750. 751

90S841 V. A. Petrosyan, M. E. Niyazymbetov, L. D. Konyushkin and V. P. Litvinov; *Synthesis*, 1990, 841. 133

90S843 B. S. Bandodakar and G. Nagendrappa; *Synthesis*, 1990, 843. 622

90S931 H. Frey and G. Kaupp; *Synthesis*, 1990, 931. 301

90S984 L. Brandsma and H. D. Verkruijsse; *Synthesis*, 1990, 984. 1017

90SC41 M.-C. Lasne, P. Cairon and D. Villemin; *Synth. Commun.*, 1990, **20**, 41. 36

90SC231	G. W. Kabalka and Z. Wang; *Synth. Commun.*, 1990, **20**, 231.	301
90SC239	V. K. Yadau; *Synth. Commun*., 1990, **20**, 239.	456
90SC349	J. Moulines, M.-J. Bougeois, M. Campagnole, A.-M. Lamidey, B. Maillard and E. Montaudon; *Synth. Commun.*, 1990, **20**, 349.	97
90SC573	J. L. Ralbovsky, M. A. Kinsella, J. Sisko and S. M. Weinreb; *Synth. Commun.*, 1990, **20**, 573.	306
90SC877	W.-W. Sy, B. A. Lodge and A. W. By; *Synth. Commun.*, 1990, **20**, 877.	631
90SC981	E. J. Enholm, D. C. Forbes and D. P. Holub; *Synth. Commun.*, 1990, **20**, 981.	311
90SC1011	A. C. Dupont, V. H. Audia, P. P. Waid and J. P. Carter, *Synth. Commun.*, 1990, **20**, 1011.	13
90SC1047	N. Iranpoor and J. Owji; *Synth. Commun.*, 1990, **20**, 1047.	223
90SC1149	A. Alberola, M. A. Alvarez, C. Andrës, A. Gonzälez and R. Pedrosa; *Synth. Commun.*, 1990, **20**, 1149.	320
90SC1209	R. C. Bernotas and R. V. Cube; *Synth. Commun.*, 1990, **20**, 1209.	303
90SC1473	G. L. Lange and C. Gottardo; *Synth. Commun.*, 1990, **20**, 1473.	7
90SC2181	M. J. Dabdoub, J. V. Comasseto, S. M. Barros and F. Moussa; *Synth. Commun.*, 1990, **20**, 2181.	1036, 1038
90SC2353	R. M. Moriarty, T. E. Hopkins, I. Prakash, B. K. Vaid and R. K. Vaid; *Synth. Commun.*, 1990, **20**, 2353.	405
90SC3039	T. Besson and P. Rollin; *Synth. Commun.*, 1990, **20**, 3039.	6
90SC3563	Z. Zhuangyu, P. Yi, H. Hongwen and K. Tsi-yu; *Synth. Commun.*, 1990, **20**, 3563.	320
90SL74	W. E. Piers, P. J. Shapiro, E. E. Bunel, and J. E. Bercaw; *Synlett*, 1990, 74.	969
90SL107	J. C. Anderson and S. Smith; *Synlett*, 1990, 107.	44
90SL113	G. Billeb, H. Brauer and W. P. Neumann; *Synlett*, 1990, 113.	947
90SL119	B. H. Lipshutz; *Synlett*; 1990, 119.	588, 1095
90SL219	C. Chatgilialoglu, A. Guerrini and G. Seconi; *Synlett*, 1990, 219.	541
90SL257	C. Chatgilialoglu, A. Guerrini and G. Seconi; *Synlett*, 1990, 257.	536
90SL259	K. Yamamoto and M. Takemae; *Synlett*, 1990, 259.	537
90SL267	G. A. Olah and X-Y. Li; *Synlett*, 1990, 267.	6, 17
90SL290	J. W. Lee and D. Y. Ch; *Synlett*, 1990, 290.	1034, 1035
90SL365	R. M. Moriarty, R. K. Vaid and G. F. Koser; *Synlett*, 1990, 365.	1020
90SL405	A. Lansky, O. Reiser and A. de Meijere; *Synlett*, 1990, 405.	932
90SL457	P. C. B. Page, E. S. Namwindwa, S. S. Klair and D. Westwood; *Synlett*, 1990, 457.	147
90SL533	M. S. Cooper, H. Heaney, A. J. Newbold and W. R. Sanderson; *Synlett*, 1990, 533.	654
90SL537	A. F. Abdel-Magid and C. A. Maryanoff; *Synlett.*, 1990, 537.	310
90SL677	P. Kocovsky; *Synlett.*, 1990, 677.	314
90SL759	R. Sturmer and R. W. Hoffman; *Synlett*, 1990, 759.	530
90SL761	J. M. Clough, L. Z. Diorazio and D. A. Widdowson; *Synlett*, 1990, 761.	613
90SL763	A. Alberola, C. Andrés and R. Pedrosa, *Synlett.*, 1990, 763.	306
90SL773	D. P. Curran, T. L. Fevig and M. J. Totleben; *Synlett*, 1990, 773.	565
90SL943	H. B. Kagan and F. Rebiere; *Synlett*, 1990, 943.	147
90T587	M. Bartra, P. Romea, F. Urpi and J. Vilarrasa; *Tetrahedron*, 1990, **46**, 587.	341
90T1145	A. Padwa and M. W. Wannamaker; *Tetrahedron*, 1990, **46**, 1145.	928
90T1653	J. Brussee, F. Dofferhoff, C. G. Kruse and A. van der Gen; *Tetrahedron*, 1990, **46**, 1653.	330
90T2503	B. Sket, P. Zupet and M. Zupan; *Tetrahedron*, 1990, **46**, 2503.	24
90T2573	P. C. B. Page and S. Rosenthal; *Tetrahedron*, 1990, **46**, 2573.	902
90T4457	H. C. Brown and R. S. Randad; *Tetrahedron*, 1990, **46**, 4457.	75, 521
90T4463	H. C. Brown and R. S. Randad; *Tetrahedron*, 1990, **46**, 4463.	75
90T4663	I. Paterson, J. M. Goodman, M. A. Lister, R. C. Schmann, C. K. McClure and R. D. Norcross; *Tetrahedron*, 1990, **46**, 4663.	533, 698
90T4857	S. E. Denmark, C. B. W. Senanayake and G.-D. Ho; *Tetrahedron*, 1990, **46**, 4857.	356
90T6381	C. Larpent, G. Meignan and H. Patin; *Tetrahedron*, 1990, **46**, 6381.	446
90T6715	W. H. L. Wai Tan, C. Bordier and A. Foucard; *Tetrahedron*, 1990, **46**, 6715.	466
90T6951	N. S. Simpkins; *Tetrahedron*, 1990, **46**, 6951.	714, 715
90T7443	G. W. Kabalka, L. H. M. Guinci and R. S. Varma; *Tetrahedron*, 1990, **46**, 7443.	398
90T7471	K. Nakamura, T. Kitayama, Y. Inoue and A. Ohno; *Tetrahedron*, 1990, **46**, 7471.	398
90T7497	A. Pecunioso, E. Galppini and R. Menicagli; *Tetrahedron*, 1990, **46**, 7497.	393
90T7509	G. H. Posner and R. D. Crouch; *Tetrahedron*, 1990, **46**, 7509.	396
90T7531	R. Ballini, M. Petrini and G. Rosini; *Tetrahedron*, 1990, **46**, 7531.	396
90TA163	J. Brussee, R. A. T. M. van Benthem, C. G. Kruse and A. van der Gen; *Tetrahedron Assym.*, 1990, **1**, 163.	311
90TA661	P. Bravo and G. Resnati; *Tetrahedron Asymmetry*, 1990, **1**, 661.	15
90TA801	J. S. Panek and M. A. Sparks; *Tetrahedron Asymmetry*, 1990, **1**, 801.	935
90TL221	S. Saito, T. Nishikawa, Y. Yokoyama and T. Moriwake; *Tetrahedron Lett.*, 1990, **31**, 221.	84
90TL231	M. Sato, N. Miyaura and A. Suzuki; *Tetrahedron Lett.*, 1990, **31**, 231.	528
90TL247	S. Hara, S. Hyuga, M. Aoyama, M. Sato and A. Suzuki; *Tetrahedron Lett.*, 1990, **31**, 247.	902
90TL365	Y. Kataoka, K. Takai, K. Oshima and K. Utimoto; *Tetrahedron Lett.*, 1990, **31**, 365.	927
90TL369	Y. Kataoka, J. Miyai, M. Tezuka, K. Takai, K. Oshima and K. Utimoto; *Tetrahedron Lett.*, 1990, **31**, 369.	927
90TL455	H. C. Brown and R. S. Randad; *Tetrahedron Lett.*, 1990, **31**, 455.	521
90TL493	G. Wu, F. E. Cederbaum, and E. Negishi; *Tetrahedron Lett.*, 1990, **31**, 493.	972
90TL501	C. E. Muller and J. J. Roth; *Tetrahedron Lett.*, 1990, **31**, 501.	259
90TL531	J. M. J. Tronchet, R. Benhamza, G. Bernadinelli and M. Geoffroy; *Tetrahedron Lett.*, 1990, **31**, 531.	341, 344

90TL543 Y. Shen and T. Wang; *Tetrahedron Lett.*, 1990, **31**, 543. 936
90TL555 E. B. Nadler and Z. Rappoport; *Tetrahedron Lett.*, 1990, **31**, 555. 930
90TL575 J. Iqbal and A. Pandey; *Tetrahedron Lett.*, 1990, **31**, 575. 318
90TL627 W. F. Bailey and T. V. Ovaska; *Tetrahedron Lett.*, 1990, **31**, 627. 972
90TL663 C. Saluzzo, G. Alvernhe and D. Anker; *Tetrahedron Lett.*, 1990, **31**, 663. 280
90TL735 N. Iranpoor and I. M. Balturk; *Tetrahedron Lett.*, 1990, **31**, 735. 84
90TL937 R. C. Hahn and J. Tompkins; *Tetrahedron Lett.*, 1990, **31**, 937. 15
90TL973 J. R. McCarthy, D. P. Matthews and C. L. Barney; *Tetrahedron Lett.*, 1990, 973. 611
90TL1051 P. J. Murphy, J. L. Spencer and G. Procter; *Tetrahedron Lett.*, 1990, **31**, 1051. 914, 915
90TL1147 P. Pellon, S. Himdi-Kabbab, I. Rault, F. Tonnard and J. Hamelin; *Tetrahedron Lett.*, 1990, **32**, 1147. 489
90TL1185 W.-B. Wang, L.-L. Shi and Y.-Z. Huang; *Tetrahedron Lett.*, 1990, **31**, 1185. 55
90TL1199 G. Pandey, K. S. Rani and U. T. Bhalerao; *Tetrahedron Lett.*, 1990, **31**, 1199. 303
90TL1303 T. Shono and N. Kise; *Tetrahedron Lett.*, 1990, **31**, 1303. 81
90TL1319 G. M. Porritt and C. B. Reese; *Tetrahedron Lett.*, 1990, **31**, 1319. 255
90TL1397 M. Gingras and D. N. Harpp; *Tetrahedron Lett.*, 1990, **31**, 1397. 131
90TL1429 M. Thiam and F. Chastrette; *Tetrahedron Lett.*, 1990, **31**, 1429. 381
90TL1447 W. R. Jackson, H. A. Jacobs, B. R. Matthew and G. S. Jayatilake; *Tetrahedron Lett.*, 1990, **31**, 1447. 317
90TL1653 F. A. Davis, P. Zhou and G. S. Lal; *Tetrahedron Lett.*, 1990, **31**, 1653. 251
90TL1741 K. Tomioka, M. Nakajima and K. Koga; *Tetrahedron Lett.*, 1990, **31**, 1741. 50
90TL1743 T. Hayashi, K. Kishi, A. Yamamoto and Y. Ito; *Tetrahedron Lett.*, 1990, **31**, 1743. 301
90TL1803 J. V. N. Prasad and D. H. Rich; *Tetrahedron Lett.*, 1990, **31**, 1803. 74
90TL1919 C. K. Reddy and M. Periasamy; *Tetrahedron Lett.*, 1990, **31**, 1919. 34
90TL1953 E. K. Yau, Y.-X. Ma and M. H. Carruthers; *Tetrahedron Lett.*, 1990, **31**, 1953. 253
90TL1985 C.-M. Chan and J. M. Chong; *Tetrahedron Lett.*, 1990, **31**, 1985. 68
90TL2069 A. Clerici and O. Porta; *Tetrahedron Lett.*, 1990, **31**, 2069. 305
90TL2127 C. Saluzzo, G. Alvernhe and D. Anker; *Tetrahedron Lett.*, 1990, **31**, 2127. 280
90TL2277 J.-M. Vatele, D. Dumas and J. Gore; *Tetrahedron Lett.*, 1990, **31**, 2277. 314
90TL2665 E. J. Corey and G. Lotto; *Tetrahedron Lett.*, 1990, **31**, 2665. 49
90TL2719 Y. Hatanaka and T. Hiyama; *Tetrahedron Lett.*, 1990, **31**, 2719. 931, 932, 939
90TL2739 G. Asequell, F. Camps, G. Fabriàs and A. Guerrero; *Tetrahedron Lett.*, 1990, **31**, 2739. 1092
90TL2893 K. S. Kim, H. J. Hwang, C. S. Cheong and C. S. Hahn; *Tetrahedron Lett.*, 1990, **31**, 2893. 145
90TL2917 S. Yamazaki, M. Hama and S. Yamabe; *Tetrahedron Lett.*, 1990, **31**, 2917. 931
90TL2990 H.-L. Kwong, C. Sorato, Y. Ogino, H. Chen and K. B. Sharpless; *Tetrahedron Lett.*, 1990, **31**, 2990. 48
90TL2991 D. C. Hargis and R. L. Shubkin; *Tetrahedron Lett.*, 1990, **31**, 2991. 320
90TL3003 B. M. Kim and K. B. Sharpless; *Tetrahedron Lett.*, 1990, **31**, 3003. 48
90TL3023 H. Tanaka, K. Inoue, U. Pokorski, M. Taniguchi and S. Torii; *Tetrahedron Lett.*, 1990, **31**, 3023. 313
90TL3027 T. Kataoka, K. Tsutsumi, T. Iwama, H. Shimizu and M. Hori; *Tetrahedron Lett.*, 1990, 3027. 143
90TL3105 M. Lautens and A. H. Huboux; *Tetrahedron Lett.*, 1990, **31**, 3105. 923
90TL3141 A. Wagner, M. P. Heitz and C. Mioskowski; *Tetrahedron Lett.*, 1990, **31**, 3141. 1018
90TL3167 J. Ichihara, K. Funabiki and T. Hanafusa; *Tetrahedron Lett.*, 1990, **31**, 3167. 18
90TL3351 S. Cicchi, A. Goti, A. Brandi, A. Guarna and F. de Sarlo; *Tetrahedron Lett.*, 1990, **31**, 3351. 320
90TL3417 M. L. Edwards, D. M. Stemerick and J. R. McCarthy; *Tetrahedron Lett.*, 1990, **31**, 3417. 309
90TL3591 T. Sato, J. Otera and H. Nozaki; *Tetrahedron Lett.*, 1990, **31**, 3591. 215
90TL3817 T. Shibata, D. G. Gilheany, B. K. Blackburn and K. B. Sharpless; *Tetrahedron Lett.*, 1990, **31**, 3817. 48
90TL3863 M. T. Reetz, E. Rivadeneira and C. Niemeyer; *Tetrahedron Lett.*, 1990, **31**, 3863. 533
90TL3999 A. R. Katritzky, N. Shobana and P. A. Harris; *Tetrahedron Lett.*, 1990, **31**, 3999. 319
90TL4199 C. Nájera, M. Yus, U. Karlsson, A. Gagoll, and J.-E. Bäckvall; *Tetrahedron Lett.*, 1990, **31**, 4199. 392
90TL4229 M. J. Earle, R. A. Fairhurst, H. Heaney, G. Papageorgiou and R. F. Wilkins; *Tetrahedron Lett.*, 1990, **31**, 4229. 316
90TL4267 S. Sengupta and V. Snieckus; *Tetrahedron Lett.*, 1990, **31**, 4267. 942
90TL4317 B. M. Kim and K. B. Sharpless; *Tetrahedron Lett.*, 1990, **31**, 4317. 49
90TL4393 E. Negishi, Y. Noda, F. Lamaty, and E. J. Vawter; *Tetrahedron Lett.*, 1990, **31**, 4393. 980
90TL4417 T. S. Peng and J. A. Gladysz; *Tetrahedron Lett.*, 1990, **31**, 4417. 580
90TL4495 J. S. Yadav, P. K. Deshpande and G. V. M. Sharma; *Tetrahedron Lett.*, 1990, **31**, 4495. 85
90TL4545 J. A. Soderquist and J. Vaquer; *Tetrahedron Lett.*, 1990, **31**, 4545. 902
90TL4565 E. A. Mash, M. A. Kaczynski and D. P. Dolata; *Tetrahedron Lett.*, 1990, **31**, 4565. 67
90TL4661 M. Chini, P. Crotti and F. Macchia; *Tetrahedron Lett.*, 1990, **31**, 4661. 318
90TL4677 J. A. Soderquist, C. L. Anderson, E. I. Miranda, I. Rivera and G. W. Kabalka; *Tetrahedron Lett.*, 1990, **31**, 4677. 58, 526
90TL4735 J. Santamaria, M. T. Kaddachi and J. Rigaudy; *Tetrahedron Lett.*, 1990, **31**, 4735. 305
90TL4765 S. C. Shim, J.-T. Hwang, H.-Y. Kang and M. H. Chang; *Tetrahedron Lett.*, 1990, **31**, 4765. 81
90TL4828 V. V. Zhdankin, C. M. Crittell, P. J. Stang and N. S. Zefirov; *Tetrahedron Lett.*, 1990, **31**, 4828. 1020
90TL4937 V. A. Khripach, V. N. Zhabinskiy and V. K. Olkhovick; *Tetrahedron Lett.*, 1990, **31**, 4937. 924, 925
90TL4973 M. Namavari, N. Satyamurthy, M. E. Phelps and J. R. Barrio; *Tetrahedron Lett.*, 1990, **31**, 4973. 25, 31, 35

90TL5007	A. G. McKillop, D. Koyuncu, A. Krief, W. Dumont, P. Renier and M. Trabelsi; *Tetrahedron Lett.*, 1990, **31**, 5007.	216
90TL5113	J. A. Soderquist and B. Santiago; *Tetrahedron Lett.*, 1990, **31**, 5113.	922
90TL5507	T. Aoyana and T. Shioiri; *Tetrahedron Lett.*, 1990, **31**, 5507.	89
90TL5555	D. E. Gibbs and D. Barnes; *Tetrahedron Lett.*, 1990, **31**, 5555.	314
90TL5575	A. S. Gopalan and H. K. Jacobs; *Tetrahedron Lett.*, 1990, **31**, 5575.	39
90TL5595	A. F. Abdel-Magid, C. A. Maryanoff and K. G. Carson; *Tetrahedron Lett.*, 1990, **31**, 5595.	300, 310
90TL5603	R. P. Discordia, C. K. Murphy and D. C. Dittmer; *Tetrahedron Lett.*, 1990, **31**, 5603.	44
90TL5607	R. F. Cunico; *Tetrahedron Lett.*, 1990, **31**, 5607.	940
90TL5633	M. J. Robins and S. Manfredini; *Tetrahedron Lett.*, 1990, 5633.	617, 624
90TL5829	M. Lautens, A. H. Huboux, E. Chin and J. Downer; *Tetrahedron Lett.*, 1990, **31**, 5829.	923
90TL6137	K. D. Kim and P. A. Magriotis; *Tetrahedron Lett.*, 1990, **31**, 6137.	910
90TL6179	B. Bennetau, M. Krempp, J. Dunoguès and S. Ratton; *Tetrahedron Lett.*, 1990, **31**, 6179.	945
90TL6239	A. H. McNeill and E. J. Thomas; *Tetrahedron Lett.*, 1990, **31**, 6239.	79
90TL6331	G. Märkl and S. Reithinger; *Tetrahedron Lett.*, 1990, **31**, 6331.	878
90TL6357	S. Juge, M. Stephan, J. A. Laffitte and J. P. Genet; *Tetrahedron Lett.*, 1990, 6357.	437, 444
90TL6379	S. Matsubara, T. Kodama and K. Utimoto; *Tetrahedron Lett.*, 1990, **31**, 6379.	319
90TL6399	T. Ito, S. Okamoto and F. Sato; *Tetrahedron Lett.*, 1990, **31**, 6399.	923
90TL6417	K. A. Jørgensen; *Tetrahedron Lett.*, 1990, **31**, 6417.	49
90TL6461	S. Hanessian, Y. L. Bennani and D. Delorme; *Tetrahedron Lett.*, 1990, **31**, 6461.	459
90TL6527	M. B. Giudicelli, D. Picq and B. Veyron; *Tetrahedron Lett.*, 1990, **31**, 6527.	9, 19
90TL6621	J. C. Cochran, B. S. Bronk, K. M. Terrence and H. K. Phillips; *Tetrahedron Lett.*, 1990, **31**, 6621.	915
90TL6629	M. Krause and H. M. R. Hoffmann; 1990, *Tetrahedron Lett.*, **31**, 6629.	658
90TL6721	R. P. Bonar-Law, A. P. Davis and B. J. Dorgan; *Tetrahedron Lett.*, 1990, **31**, 6721.	543
90TL6725	R. P. Bonar-Law, A. P. Davis, B. J. Dorgan, M. T. Reetz and A. Wehrsig; *Tetrahedron Lett.*, 1990, **31**, 6725.	543
90TL6785	P. W. Shum and A. P. Kozikowski; *Tetrahedron Lett.*, 1990, **31**, 6785.	942
90TL6903	D. S. Bose and D. E. Thurston; *Tetrahedron Lett.*, 1990, **31**, 6903.	304
90TL6977	C. Narayana, S. Padmanabhan and G. W. Kabalka; *Tetrahedron Lett.*, 1990, **31**, 6977.	
90TL7031	O. Takahashi, K. Furuhashi, M. Fukumasa and T. Hirai; *Tetrahedron Lett.*, 1990, **31**, 7031.	85
90TL7257	B. H. Lipshutz, R. Keil, and E. L. Ellsworth; *Tetrahedron Lett.*, 1990, **31**, 7257.	969
90TL7333	K. Tamao, T. Tohma, N. Inui, O. Kakayama and Y. Ito; *Tetrahedron Lett.*, 1990, **31**, 7333.	51
90TL7415	M. P. De Ninno, R. J. Perner and L. Lijewski; *Tetrahedron Lett.*, 1990, **31**, 7415.	66
90TL7423	G. Cahiez and M. Alami; *Tetrahedron Lett.*, 1990, **31**, 7423.	587
90TL7463	M. Sato, J.-I. Sakaki, Y. Sugita, T. Nakano and C. Kaneko; *Tetrahedron Lett.*, 1990, **31**, 7463.	39
90TL7467	K. Sugita, N. Onaka and Y. Izuni; *Tetrahedron Lett.*, 1990, **31**, 7467.	84
90TL7509	M. A. Brimble, M. K. Edmonds and G. M. Williams; *Tetrahedron Lett.*, 1990, **31**, 7509.	27
90TL7563	W. R. Roush, P. T. Glover and X. Lin; *Tetrahedron Lett.*, 1990, **31**, 7563.	529
90TL7575	S. A. Rao, C. E. Tucker and P. Knochel; *Tetrahedron Lett.*, 1990, **31**, 7575.	69
90TL7597	H. Sörensen and A. E. Greene; *Tetrahedron Lett.*, 1990, **31**, 7597.	1027
90ZAAC(580)217	L. Riesel, M. Kant and R. Helbing; *Z. Anorg. Allg. Chem.*, 1990, **580**, 217.	105
90ZAAC(589)23	G. Fritz, M. Jarmer and E. Matern; *Z. Anorg. Allg. Chem.*, 1991, **589**, 23.	471
90ZAAC(589)187	M. Fild, T. Oehmigen, M. Sebastian and M. Vahldiek; *Z. Anorg. Allg. Chem.*, 1990, **589**, 187.	428
90ZN(B)191	F. Preuss, H. Becker and T. Wieland; *Z. Naturforsch., Teil B*, 1990, **45**, 191.	570
90ZN(B)203	K. P. Langhans, O. Stelzer, J. Svara and N. Weferling; *Z. Naturforsch., Teil B*, 1990, **45**, 203.	427, 430
90ZN(B)557	H. Schmidt and R. Stendel; *Z. Naturforsch., Teil B*, 1990, **45**, 557.	100
90ZN(B)1529	H. Lang and L. Zsolnai; *Z. Naturforsch., Teil B*, 1990, **45**, 1529.	1079
90ZN(B)1548	H. Werner, A. Hampp, K. Peters, L. Walz and H. G. von Schneering; *Z. Naturforsch., Teil B*, 1990, **45**, 1548.	437
90ZN(B)1625	R. Minkwitz, V. Gerhard and H. Preut; *Z. Naturforsch., Teil B*, 1990, **45**, 1625.	238
90ZOB313	R. A. Khachatryan, S. V. Sayadyan, M. Z. Ovakimyan and M. G. Indzhikyan; *Zh. Obshch. Khim.*, 1990, **60**, 313 (*Chem. Abstr.*, 1990, **113**, 78 515).	437
90ZOB465	V. V. Miroshnichenko, A. P. Marchenko, G. N. Koidan and A. M. Pinchuk; *Zh. Obshch. Khim.*, 1990, **60**, 465 (*Chem. Abstr.*, 1990, **113**, 78 519x).	407, 408
90ZOB476	R. M. Kamalov, G. S. Stepanov, E. I. Gol'dfarb, D. V. Ryzhikov, M. A. Pudovik and A. N. Pudovik; *Zh. Obshch. Khim.*, 1990, **60**, 476 (*Chem. Abstr.* 1990, **112**, 198 561).	259
90ZOB697	R. I. Yurchenko, I. V. Malyutina, T. I. Klepa, M. I. Povolotskii, O. I. Tuzhikov and A. M. Pinchuk; *Zh. Obshch. Khim.*, 1990, **60**, 697 (*Chem. Abstr.*, 1990, **113**, 97 698).	428
90ZOB1243	Y. G. Shermolovich, A. V. Solovev, N. P. Kolesnik and L. N. Markovskii; *Zh. Obshch. Khim.*, 1990, **60**, 1243 (*Chem. Abstr.*, 1990, **113**, 231 519).	438
90ZOB1795	B. I. No, Y. L. Zotov and V. N. Karev; *Zh. Obshch. Khim.*, 1990, **60**, 1795 (*Chem. Abstr.*, 1991, **114**, 24 045).	430
90ZOB2394	M. N. Krivchun, M. V. Sendurev, B. I. Ionin and A. A. Petrov; *Zh. Obshch. Khim.*, 1991, **61**, 2394.	448
90ZOR2083	L. N. Markovskii, A. V. Solov'ev, E. I. Kaminskaya, A. V. Borodin and Y. G. Shermolovich; *Zh. Org. Khim.*, 1990, **26**, 2083 (*Chem. Abstr.*, 1991, **115**, 1 590 745).	239
90ZOR2635	V. A. Potapov, S. V. Amosova and A. V. Khangurov; *Zh. Org. Khim.*, 1990, **26**, 2635.	1036

91ACR285	U. Ragnarsson and L. Grehn; *Acc. Chem. Res.*, 1991, **24**, 285.	299
91ACR304	P. J. Stang; *Acc. Chem. Res.*, 1991, **24**, 304.	1019, 1022
91AG(E)49	R. Noyori and M. Kitamura; *Angew. Chem., Int. Ed. Engl.*, 1991, **30**, 49.	67
91AG(E)177	H. Nakahira, I. Ryu, M. Ikebe, N. Kambe and N. Sonoda; *Angew. Chem., Int. Ed. Engl.*, 1991, **30**, 177.	555
91AG(E)185	W. A. Herrmann, C. Romao, R. W. Fisher, P. Kiprof and C. de Meric de Beffon; *Angew. Chem., Int. Ed. Engl.*, 1991, **30**, 185.	585
91AG(E)207	P. Binger, T. Wettling, R. Schneider, F. Zurmahlen, U. Bergstrasser, J. Hoffmann, G. Maas and M. Regitz; *Angew. Chem., Int. Ed. Engl.*, 1991, **30**, 207.	449
91AG(E)217	E. Niecke and D. Gudat; *Angew. Chem., Int. Ed. Engl.*, 1991, **30**, 217.	406
91AG(E)321	M. Marsch, K. Harms, O. Zschage, D. Hoppe and G. Boche; *Angew. Chem., Int. Ed. Engl.*, 1991, **30**, 321.	999
91AG(E)416	F. Henin, J. Muzart, J.-P. Pete, A. M'boungou-M'passi and H. Rau; *Angew. Chem., Int. Ed. Engl.*, 1991, **30**, 416.	637
91AG(E)694	H. Kunz and K. Rück; *Angew. Chem., Int. Ed. Engl.*, 1991, **30**, 694.	595
91AG(E)873	F. Effenberger and U. Stelzer; *Angew. Chem., Int. Ed. Engl.*, 1991, 873.	322
91AG(E)902	T. Tsumuraya, S. A. Batcheller and S. Masamune; *Angew. Chem., Int. Ed. Engl.*, 1991, **30**, 902.	949
91AG(E)1356	Y. Chiang, A. S. Grant and A. J. Kresge; *Angew. Chem., Int. Ed. Engl.*, 1991, **30**, 1356.	1040
91AG(E)1370	B. Wrackmeyer, G. Kehr and R. Boese; *Angew. Chem., Int. Ed. Engl.*, 1991, **30**, 1370.	1088
91AG(E)1687	B. Mergardt, K. Weber, G. Adiwidjaja and E. Schaumann; *Angew. Chem., Int. Ed. Engl.*, 1991, **30**, 1687.	938
91AJC1353	J. T. Pinhey; *Aust. J. Chem.*, 1991, **44**, 1353.	1102
91AX(C)328	L. Troupet, F. Texier and R. Carrié; *Acta Crystallogr., Part C*, 1991, **47**, 328.	1003
91BCJ602	K. Kaneda, K. Fujita, T. Takemoto and T. Imanaka; *Bull. Chem. Soc. Jpn.*, 1991, **64**, 602.	314
91BCJ702	T. Oida, A. Ohnishi, T. Shimamaki, Y. Hayashi and S. Tanimoto; *Bull. Chem. Soc. Jpn.*, 1991, **64**, 702.	720
91BCJ1461	Y. Takeyama, K. Nozaki, K. Matsumoto, K. Oshima and K. Utimoto; *Bull. Chem. Soc. Jpn.*, 1991, **64**, 1461.	920
91BCJ2398	T. Hanaya, A. Akamatsu, H. Kawamoto, M. A. Armour, A. M. Hogg and H. Yamamoto; *Bull. Chem. Soc. Jpn.*, 1991, **64**, 2398.	431
91BCJ2524	T. Mukaiyama, T. Ohno, T. Nishimura, J. S. Han and S. Kobayashi; *Bull. Chem. Soc. Jpn.*, 1991, 2524.	139
91BCJ2596	M. Shimizu, M. Okamura and T. Fujisawa; *Bull. Chem. Soc. Jpn.*, 1991, **64**, 2596.	18
91BCJ2942	T. Kitigawa, M. Kita, K. Kashiwabara and J. Fujita; *Bull. Chem. Soc. Jpn.*, 1991, **64**, 2942.	437
91BCJ3485	M. Kodomari, G. Saitoh and S. Yoshitomi; *Bull. Chem. Soc. Jpn.*, 1991, **64**, 3485.	710
91BCJ3749	K. Shibata, Y. Itoh, N. Tokitoh, R. Okazaki and N. Inamoto; *Bull. Chem. Soc. Jpn.*, 1991, **64**, 3749.	22
91BMC507	K. R. Rao and H. M. S. Kumar; *Bioorg. Med. Chem. Lett.*, 1991, **1**, 507.	216
91BSF222	M. Buillard, G. Balme, N. Montero and J. Gor; *Bull. Soc. Chim. Fr.*, 1991, 222.	104
91CAR(225)113	A. Munoz and L. Lamand; *Carbohydr. Res.*, 1991, **225**, 113.	111
91CB69	M. Baudler and V. Arndt; *Chem. Ber.*, 1991, **124**, 69.	472
91CB543	W. Zarges, M. Marsch, K. Harms, W. Koch, G. Frenking and G. Boche; *Chem. Ber.*, 1991, **124**, 543.	999, 1005
91CB725	L. Weber, R. Kirchhoff, R. Bosse and H. Stammler; *Chem. Ber.*, 1991, **124**, 725.	866
91CB765	R. Streubel, E. Nieke and P. Paetzoid; *Chem. Ber.*, 1991, **124**, 765.	471, 474
91CB861	W. Zarges, M. Marsch, K. Harms, F. Haller, G. Frenking and G. Boche; *Chem. Ber.*, 1991, **124**, 861.	1003
91CB1131	W. Fimml and F. Sladky; *Chem. Ber.*, 1991, 1131.	239
91CB1153	M. Baudler, H. Jackow, A. Floruss and J. Hasenbach; *Chem. Ber.*, 1991, **124**, 1153.	472
91CB1315	W.-W. du Mont, S. Kubiniok, L. Lange, S. Pohl, W. Saak and I. Wagner; *Chem. Ber.*, 1991, **124**, 1315.	729
91CB1563	J. Niemann, W. W. Schoeller, V. von de Gonna and E. Niecke; *Chem. Ber.*, 1991, **124**, 1563.	472
91CB1997	M. Arnswald and W. P. Neumann; *Chem. Ber.*, 1991, **124**, 1997.	726
91CB2165	P. Binger, P. Müller, A. T. Herrmann, P. Philipps, B. Gabor, F. Langhauser, and C. Krüger; *Chem. Ber.*, 1991, **124**, 2165.	984
91CB2303	M. Wessling and H. J. Schäfer; *Chem. Ber.*, 1991, **124**, 2303.	318
91CB2453	S. C. Sendlinger, B. S. Haggerty, A. L. Rheingold and K. H. Theopold; *Chem. Ber.*, 1991, **124**, 2453.	895, 896
91CB2457	W. Clegg, R. J. Errington, G. A. Fisher, M. E. Green, D. C. R. Hockless and N. C. Norman; *Chem. Ber.*, 1991, **124**, 2457.	473
91CB2705	R. Vogt, P. G. Jones, A. Kolbe and R. Schmutzler; *Chem. Ber.*, 1991, **124**, 2705.	467
91CC90	K. Nakanishi, K. Mizuno and Y. Otsuji; *J. Chem. Soc., Chem. Commun.*, 1991, 90.	133
91CC101	K. Tanaka and H. Suzuki; *J. Chem. Soc., Chem. Commun.*, 1991, 101.	588
91CC167	K. Akaji, T. Tatsumi, M. Yoshida, T. Kimura, Y. Fujinara and Y. Kiso; *J. Chem. Soc., Chem. Commun.*, 1991, 167.	227
91CC179	Z.-Q. Xu, D. D. DesMarteau and Y. Gotoh; *J. Chem. Soc., Chem. Commun.*, 1991, 179.	16
91CC205	A. Arase, Y. Nunokawa, Y. Masuda and M. Hoshi; *J. Chem. Soc., Chem. Commun.*, 1991, 205.	43
91CC234	K. B. Dillon and T. A. Straw; *J. Chem. Soc., Chem. Commun.*, 1991, 234.	476
91CC306	K. Matsumoto, S. Hashimoto and S. Otani; *J. Chem. Soc., Chem. Commun.*, 1991, 306.	301
91CC398	M. Yus and D. J. Ramón; *J. Chem. Soc., Chem. Commun.*, 1991, 398.	551, 553

91CC475	R. J. Smith, G. D. Admans, A. P. Richardson, H. J. Küppers and P. J. Blower; *J. Chem. Soc., Chem. Commun.*, 1991, 475.	133
91CC535	A. S. C. Chan and J. P. Coleman; *J. Chem. Soc., Chem. Commun.*, 1991, 535.	83
91CC703	Y. Wakatsuki, H. Yamazaki, M. Nakano and Y. Yamamoto; *J. Chem. Soc., Chem. Commun.*, 1991, 703.	938
91CC732	S.-Z. Zhu and Q.-Y. Chen; *J. Chem. Soc., Chem. Commun.*, 1991, 732.	162
91CC795	A. Tai, T. Kikukawa, T. Sugimura, Y. Inoue, T. Osawa and S. Fujii; *J. Chem. Soc., Chem. Commun.*, 1991, 795.	52
91CC934	S. Ranganathan and N. Jayaraman; *J. Chem. Soc., Chem. Commun.*, 1991, 934.	129
91CC1248	S. E. Booth, P. R. Jenkins and C. J. Swain; *J. Chem. Soc., Chem. Commun.*, 1991, 1248.	351
91CC1399	G. L. Edwards, W. B. Motherwell, D. M. Powell and D. A. Sandham; *J. Chem. Soc., Chem. Commun.*, 1991, 1399.	640
91CC1418	M. Mitani, T. Kobayashi and K. Koyama; *J. Chem. Soc., Chem. Commun.*, 1991, 1418.	28
91CC1684	J. Bakos, A. Orosz, B. Heil, M. Laghmari, P. Lhoste and D. Sinou; *J. Chem. Soc., Chem. Commun.*, 1991, 1684.	330
91CE185	S. Kajigaeshi, M. Moriwaki, S. Fujisaki and T. Kakinami; *Chem. Express*, 1991, **6**, 185.	23
91CE751	M. Hashimoto, S. Niwata and H. Fukami; *Chem. Express*, 1991, **6**, 751.	12
91CI(L)175	D. K. Dutta, K. C. Lekhok, R. C. Boruah and J. S. Sandhu; *Chem. Ind. (London)*, 1991, 175.	34
91CJC1482	T. G. Back, J. H.-L. Chau, B. P. Dyck and P. L. Gladstone; *Can. J. Chem.*, 1991, **69**, 1482.	338
91CL141	H. Fujihara, H. Chiu and N. Furukawa; *Chem. Lett.*, 1991, 141.	212
91CL423	K. Narasaka, K. Iwakura and T. Okauchi; *Chem. Lett.*, 1991, 423.	394
91CL459	N. Yoneda, T. Fukuhara, T. Mizokami and A. Suzuki; *Chem. Lett.*, 1991, 459.	690
91CL523	M. Hirano, J. Tomaru and T. Morimoto; *Chem. Lett.*, 1991, 523.	155
91CL569	A. M. Krishnan, L. T. Woodford and J. H. Boyer; *Chem. Lett.*, 1991, 569.	385
91CL1185	H. Seto, Z. Qian, H. Yoshioka, Y. Uchibori and M. Umeno; *Chem. Lett.*, 1991, 1185.	19
91CL1391	H. Sohmiya, T. Kimura, P. Bauchat and M. Fujita; *Chem. Lett.*, 1991, 1391.	179
91CM96	C. A. Larsen, R. W. Gedridge, Jr. and G. B. Stringfellow; *Chem. Mater.*, 1991, **3**, 96.	879
91COC(1)819	N. Baggett; *Comp. Org. Chem.*, 1991, **1**, 819.	478
91COC(7)389	R. A. Johnson and K. B. Sharpless; *Comp. Org. Chem.*, 1991, **7**, 389.	92
91COC(7)411	R. A. Johnson and K. B. Sharpless; *Comp. Org. Chem.*, 1991, **7**, 411.	92
91COS(1)1	P. G. Willard; *Comp. Org. Synth.*, 1991, **1**, 1.	551
91COS(1)49	D. M. Huryn; *Comp. Org. Synth.*, 1991, **1**, 49.	67
91COS(1)77	J. R. Hauske; *Comp. Org. Synth.*, 1991, **1**, 77.	595
91COS(1)107	B. H. Lipshutz; *Comp. Org. Synth.*, 1991, **1**, 107.	585
91COS(1)139	C. Ferreri, G. Palumbo and R. Caputo; *Comp. Org. Synth.*, 1991, **1**, 139.	565, 566
91COS(1)173	N. A. Saccomano; *Comp. Org. Synth.*, 1991, **1**, 173.	571
91COS(1)211	P. Knochel; *Comp. Org. Synth.*, 1991, **1**, 211.	590
91COS(1)231	T. Imamoto; *Comp. Org. Synth.*, 1991, **1**, 231.	564
91COS(1)251	G. A. Molander; *Comp. Org. Synth.*, 1991, **1**, 251.	564
91COS(1)355	R. A. Volkmann; *Comp. Org. Synth.*, 1991, **1**, 355.	345
91COS(1)459	R. E. Gawley and K. Rein; *Comp. Org. Synth.*, 1991, 459.	1003
91COS(1)487	A. Pelter and K. Smith; *Comp. Org. Synth.*, 1991, **1**, 487.	900, 1005
91COS(1)505	K. Ogura; *Comp. Org. Synth.*, 1991, 505.	153, 1001
91COS(1)579	J. S. Panek; *Comp. Org. Synth.*, 1991, 579.	1005
91COS(1)629	A. Krief; *Comp. Org. Synth.*, 1991, 629.	1001
91COS(1)729	S. E. Kelly; *Comp. Org. Synth.*, 1991, 729.	997, 1003
91COS(1)843	P. M. Wovkulich; *Comp. Org. Synth.*, 1991, 843.	1003
91COS(1,2)	*Comp. Org. Synth.*, 1991, **1**, **2**.	
91COS(2)55	Y. Yamamoto; *Comp. Org. Synth.*, 1991, 55.	997
91COS(2)99	H. B. Mekelburger and C. S. Wilcox; *Comp. Org. Synth.*, 1991, **2**, 99.	696
91COS(2)181	C. H. Heathcock; *Comp. Org. Synth.*, 1991, **2**, 181.	696
91COS(2)239	B. M. Kim, S. F. Williams and S. Masamune; *Comp. Org. Synth.*, 1991, **2**, 239.	696
91COS(2)321	G. Rosini; *Comp. Org. Synth.*, 1991, **2**, 321.	394, 1002
91COS(2)329	G. Rossini; *Comp. Org. Synth.*, 1991, **2**, 329.	784
91COS(2)595	T.-H. Chan; *Comp. Org. Synth.*, 1991, **2**, 595.	674
91COS(2)599	T.-H. Chan; *Comp. Org. Synth.*, 1991, **2**, 599.	699
91COS(2)707	S. C. Eyley; *Comp. Org. Synth.*, 1991, **2**, 707.	1006
91COS(2)733	H. Heaney; *Comp. Org. Synth.*, 1991, **2**, 733.	1006
91COS(2)893	E. F. Kleinman; *Comp. Org. Synth.*, 1991, **2**, 893.	1006, 1009
91COS(2)953	H. Heaney; *Comp. Org. Synth.*, 1991, **2**, 953.	1006, 1009
91COS(2)1007	L. E. Overman and D. J. Ricca; *Comp. Org. Synth.*, 1991, **2**, 1007.	1009
91COS(2)1047	H. Hiemstra and W. N. Speckamp; *Comp. Org. Synth.*, 1991, **2**, 1047.	1009
91COS(3)341	J. K. Sutherland; *Comp. Org. Synth.*, 1991, **3**, 341.	593
91COS(3)803	D. A. Whiting; *Comp. Org. Synth.*, 1991, **3**, 803.	660
91COS(3)913	I. Marko; *Comp. Org. Synth.*, Pergamon, Oxford, 1991, vol. 3, p. 913.	88
91COS(3)975	J. A. Marshall; *Comp. Org. Synth.*, 1991, **3**, 975.	88
91COS(4)269	R. C. Larock and W. W. Leong; *Comp. Org. Synth.*, 1991, **4**, 269.	4
91COS(4)329	E. Block and A. L. Schwan; *Comp. Org. Synth.*, 1991, **4**, 329.	5
91COS(4)363	K. E. Harding and T. H. Tiner; *Comp. Org. Synth.*, 1991, **4**, 363.	93, 354
91COS(4)779	D. P. Curran; *Comp. Org. Synth.*, 1991, **4**, 779.	14
91COS(4)865	P. Knochel; *Comp. Org. Synth.*, 1991, **4**, 865.	976, 977, 978
91COS(5)151	J. A. Porco and S. L. Schreiber; *Comp. Org. Synth.*, 1991, **5**, 151.	92

91COS(5)401	S. M. Weinreb; *Comp. Org. Synth.*, 1991, **5**, 401.	793
91COS(5)827	P. Wipf; *Comp. Org. Synth.*, 1991, **5**, 827.	656
91COS(5)1037	N. E. Schore; *Comp. Org. Synth.*, 1991, **5**, 1037.	985
91COS(5)1065	W. D. Wulff; *Comp. Org. Synth.*, 1991, **5**, 1065.	670, 979
91COS(5)1163	E. Negishi; *Comp. Org. Synth.*, 1991, **5**, 1163.	568, 986
91COS(6)1	O. Mitsunobu; *Comp. Org. Synth.*, 1991, **6**, 1.	38, 82, 83
91COS(6)103	R. Askani and D. F. Taber; *Comp. Org. Synth.*, 1991, **6**, 103.	340, 341, 790, 1003
91COS(6)133	G. Solladié; *Comp. Org. Synth.*, 1991, **6**, 133.	144, 154, 158
91COS(6)171	H. J. Bestmann and R. Zimmerman; *Comp. Org. Synth.*, 1991, **6**, 171.	1004
91COS(6)203	R. Bohlmann; *Comp. Org. Synth.*, 1991, **6**, 203.	2, 6
91COS(6)261	R. Bishop; *Comp. Org. Synth.*, 1991, **6**, 261.	1006
91COS(6)873	R. Brückner; *Comp. Org. Synth.*, 1991, **6**, 873.	154, 196
91COS(6)1011	P. C. Astles, S. V. Mortlock and E. J. Thomas; *Comp. Org. Synth.*, 1991, **6**, 1011.	184, 196
91COS(7)38	A. S. Rao; *Comp. Org. Synth.*, 1991, **7**, 381.	901
91COS(7)43	R. Breslow; *Comp. Org. Synth.*, 1991, **7**, 43.	20
91COS(7)119	D. R. Buckle; *Comp. Org. Synth.*, 1991, **7**, 119.	139
91COS(7)193	M. Kennedy and M. A. McKervey; *Comp. Org. Synth.*, 1991, **7**, 193.	142
91COS(7)329	P. J. Dudfield; *Comp. Org. Synth.*, 1991, **7**, 329.	649
91COS(7)357	A. S. Rao; *Comp. Org. Synth.*, 1991, **7**, 357.	90
91COS(7)515	K. A. Swiss and D. C. Liotta; *Comp. Org. Synth.*, 1991, **7**, 515.	280
91COS(7)593	A. Pelter and K. Smith; *Comp. Org. Synth.*, 1991, **7**, 593.	4, 14, 900
91COS(7)641	E. W. Colvin; *Comp. Org. Synth.*, 1991, **7**, 641.	537, 540
91COS(7)723	D. Crich; *Comp. Org. Synth.*, 1991, **7**, 723.	12
91COS(7)735	T. L. Gilchrist; *Comp. Org. Synth.*, 1991, **7**, 735.	334, 337, 338, 339, 347, 350
91COS(7)736	T. L. Gilchrist; *Comp. Org. Synth.*, 1991, **7**, 736.	775
91COS(7)738	T. L. Gilchrist; *Comp. Org. Synth.*, 1991, **7**, 738.	772
91COS(7)757	S. Uemura; *Comp. Org. Synth.*, 1991, **7**, 757.	144, 723, 733, 734
91COS(8)1	N. Greeves; *Comp. Org. Synth.*, 1991, **8**, 1.	56, 60
91COS(8)9	N. Greeves; *Comp. Org. Synth.*, 1991, **8**, 9.	62, 82
91COS(8)60	R. O. Hutchins and M. K. Hutchins; *Comp. Org. Synth.*, 1991, **8**, 60.	341, 350
91COS(8)107	J. W. Huffmann; *Comp. Org. Synth.*, 1991, **8**, 107.	54
91COS(8)129	L. Möens; *Comp. Org. Synth.*, 1991, **8**, 129.	80
91COS(8)139	K. Harada and T. Munegumi; *Comp. Org. Synth.*, 1991, **8**, 139.	52, 53
91COS(8)159	M. Nishizawa and R. Noyori; *Comp. Org. Synth.*, 1991, **8**, 159.	63
91COS(8)183	J. B. Jones; *Comp. Org. Synth.*, 1991, **8**, 183.	38
91COS(8)211	J. H. Brewster; *Comp. Org. Synth.*, 1991, **8**, 211.	89
91COS(8)235	A. G. M. Barrett; *Comp. Org. Synth.*, 1991, **8**, 235.	89
91COS(8)259	R. A. W. Johnstone; *Comp. Org. Synth.*, 1991, **8**, 259.	56, 60
91COS(8)364	G. W. Kabalka and R. S. Varma; *Comp. Org. Synth.*, 1991, **8**, 364.	772
91COS(8)366	G. W. Kabalka and R. S. Varma; *Comp. Org. Synth.*, 1991, **8**, 366.	776
91COS(8)373	G. W. Kabalka and R. S. Varma; *Comp. Org. Synth.*, 1991, **8**, 373.	341
91COS(8)403	I. W. J. Still; *Comp. Org. Synth.*, 1991, **8**, 403.	713, 714
91COS(8)703	K. Smith and A. Pelter; *Comp. Org. Synth.*, 1991, **8**, 703.	514, 900
91COS(8)733	J. J. Eisch; *Comp. Org. Synth.*, 1991, **8**, 733.	967
91COS(8)79	R. Kellogg; *Comp. Org. Synth.*, 1991, **8**, 79.	55
91CPB1088	Y. Mori and N. Morishima; *Chem. Pharm. Bull.*, 1991, **39**, 1088.	19
91CPB1132	Y. Ozaki, I. Oshio, Y. Ohsuga, S. Kaburagi, Z.-Z. Sung and S.-W. Kim; *Chem. Pharm. Bull.*, 1991, **39**, 1132.	662
91CRV35	J. A. Morrison; *Chem. Rev.*, 1991, **91**, 35.	903
91CRV119	C. W. Allen; *Chem. Rev.*, 1991, **91**, 119.	864
91CRV311	W. P. Neumann; *Chem. Rev.*, 1991, **91**, 311.	947, 949
91CRV1179	K. Burgess and M. Ohlmeyer; *Chem. Rev.*, 1991, **91**, 1179.	527, 970
91CSR95	A. Barrett; *Chem. Soc. Rev.*, 1991, **20**, 95.	783
91CZ7	W. Wolfsberger; *Chem.-Ztg.*, 1991, **115**, 7.	437
91EUP412699	G. Goto, S. Ohkawa and S. Fukumoto; *Eur. Pat.* 412 699 (1991) (*Chem. Abstr.*, 1991, **115**, 183 960y).	240
91G559	F. Bellesia, F. Ghelfi, U. M. Pagnoni and A. Pinetti; *Gazz. Chim. Ital.*, 1991, **121**, 559.	24
91IZV1069	V. G. Shtamburg, V. F. Rudchenko, V. M. Grinev, A. A. Dimetrenko, A. P. Pleshkova and R. G. Kostyanovskii; *Izv. Akad. Nauk SSSR Ser. Khim.*, 1991, 1069 (*Chem. Abstr.*, 1991, **115**, 113 984).	356, 357
91HCA1931	A. R. Katritzky, A. S. Afridi and W. Kuzmierkiewicz; *Helv. Chim. Acta*, 1991, **74**, 1931.	710
91JA165	R. A. Fischer, R. B. Nielsen, W. M. Davis and S. L. Buchwald; *J. Am. Chem. Soc.*, 1991, **113**, 165.	261, 888
91JA723	S. E. Denmark, J. P. Edwards and S. R. Wilson; *J. Am. Chem. Soc.*, 1991, **113**, 723.	998
91JA958	S. H. Bergens and B. Bosnich; *J. Am. Chem. Soc.*, 1991, **113**, 958.	640
91JA1468	S. E. Denmark, P. C. Miller and S. R. Wilson; *J. Am. Chem. Soc.*, 1991, **113**, 1468.	1004
91JA1866	B. Mudryk and T. Cohen; *J. Am. Chem. Soc.*, 1991, **113**, 1866.	553
91JA2321	R. B. Grossman, W. M. Davis and S. L. Buchwald; *J. Am. Chem. Soc.*, 1991, **113**, 2321.	307
91JA2648	M. Kol, S. Rozen and E. Appelman; *J. Am. Chem. Soc.*, 1991, **113**, 2648.	94
91JA2771	M. A. Huffman and L. S. Liebeskind; *J. Am. Chem. Soc.*, 1991, **113**, 2771.	672
91JA3176	M. J. Brown, P. D. Milano, D. C. Lever, W. W. Epstein and C. D. Poulter; *J. Am. Chem. Soc.*, 1991, **113**, 3176.	133
91JA3177	C.-C. Yang, C. K. Marlowe and R. Kania; *J. Am. Chem. Soc.*, 1991, **113**, 3177.	133

91JA3526	J. K. Whitesell and H. K. Yaser; *J. Am. Chem. Soc.*, 1991, **113**, 3526.	245, 331
91JA3597	K. M. Waggoner, S. Parkin, D. C. Pestana, H. Hope and P. P. Power; *J. Am. Chem. Soc.*, 1991, **113**, 3597.	475
91JA4002	H.-J. Gais, J. Müller, J. Vollhardt and H. J. Lindner; *J. Am. Chem. Soc.*, 1991, **113**, 4002.	997, 1000
91JA4042	D. A. Evans and G. C. Fu; *J. Am. Chem. Soc.*, 1991, **113**, 4042.	527
91JA4571	P. J. Stang and V. V. Zhdankin; *J. Am. Chem. Soc.*, 1991, **113**, 4571.	1020, 1028, 1033
91JA5079	A. H. Hoveyda and Z. Xu; *J. Am. Chem. Soc.*, 1991, **113**, 5079.	561
91JA5093	S. C. Berk, K. A. Kreutzer and S. L. Buchwald; *J. Am. Chem. Soc.*, 1991, **113**, 5093.	82
91JA5720	W. F. Bailey, A. D. Khanolkar, K. Gavaskar, T. V. Ovaska, K. Rossi, Y. Thiel and K. B. Wiberg; *J. Am. Chem. Soc.*, 1991, **113**, 5720.	556
91JA5735	S. A. Rao and P. Knochel; *J. Am. Chem. Soc.*, 1991, **113**, 5735.	925
91JA5870	P. J. Stang, B. L. Williamson and V. V. Zhdankin; *J. Am. Chem. Soc.*, 1991, **113**, 5870.	1020
91JA6139	K. Burgess, W. A. van der Donk, M. B. Jarstfer and M. J. Ohlmeyer; *J. Am. Chem. Soc.*, 1991, **113**, 6139.	527
91JA6337	H. Fulihara, H. Mima, M. Ikemori and N. Furukawa; *J. Am. Chem. Soc.*, 1991, **113**, 6337.	369
91JA6943	Y. Rubin, S. S. Lin, C. B. Knobler, J. Anthony, A. M. Boldi and F. Diederich; *J. Am. Chem. Soc.*, 1991, **113**, 6943.	1091
91JA7418	C. A. Merlic and D. Xu; *J. Am. Chem. Soc.*, 1991, **113**, 7418.	670
91JA7424	G. Agnel and E. Negishi; *J. Am. Chem. Soc.*, 1991, **113**, 7424.	988
91JA7778	N. Chatani, N. Amishiro and S. Murai; *J. Am. Chem. Soc.*, 1991, **113**, 7778.	939
91JA7794	J. M. Hawkins and S. Loren; *J. Am. Chem. Soc.*, 1991, **113**, 7794.	526
91JA8518	M. J. Burk; *J. Am. Chem. Soc.*, 1991, **113**, 8518.	62
91JA8950	A. H. Hoveyda, Z. Xu, J. P. Morken and A. F. Houri; *J. Am. Chem. Soc.*, 1991, **113**, 8950.	561
91JA8955	A. Yanagisawa, S. Habaue and H. Yamamoto; *J. Am. Chem. Soc.*, 1991, **113**, 8955.	73, 563
91JA8981	E.-I. Nakamura, T. Inubushi, S. Aoki and D. Machi; *J. Am. Chem. Soc.*, 1991, **113**, 8981.	87
91JA8997	P. J. Stang, V. V. Zhdankin and A. M. Arif; *J. Am. Chem. Soc.*, 1991, **113**, 8997.	1021
91JA9873	J. Bao, V. Dragisich, S. Wenglowsky and W. D. Wulff; *J. Am. Chem. Soc.*, 1991, **113**, 9873.	671
91JA9887	Y. Oozumi and T. Hayashi; *J. Am. Chem. Soc.*, 1991, **113**, 9887.	537
91JAP70374395	T. Sugiya and K. Takahashi; *Jpn. Pat.* 70 374 395 (1991) (*Chem. Abstr.*, 1991, **115**, 280 284).	445
91JCR(S)300	J. V. Christiansen, A. Feldthus and L. Carlsen; *J. Chem. Res.* (*S*), 1991, 300.	685
91JCS(D)443	C. S. Chin, B. Lee, S. Kim and J. Chun; *J. Chem. Soc., Dalton Trans.*, 1991, 443.	639
91JCS(P1)483	H. Butenschön; *J. Chem. Soc., Perkin Trans. 1*, 1991, 483.	301
91JCS(P1)1341	S. Itsuno, H. Yanaka, C. Hachisuka and K. Ito; *J. Chem. Soc., Perkin Trans. 1*, 1991, 1341; correction *J. Chem. Soc., Perkin Trans. 1*, 1991, 2290.	313, 544
91JCS(P1)1717	T. Nishio; *J. Chem. Soc., Perkin Trans. 1*, 1991, 1717.	146
91JCS(P1)1767	S. Itsuno, C. Hachisuka and K. Ito; *J. Chem. Soc., Perkin Trans. 1*, 1991, 1767.	316
91JCS(P1)2049	N. Miyachi, H. Satoh and M. Shibasaki; *J. Chem. Soc., Perkin Trans.* 1, 1991, 2049.	1036, 1037
91JCS(P1)2169	D. Merricks, P. G. Sammes, E. R. H. Walker, K. Henrick and M. M. McPartlin; *J. Chem. Soc., Perkin Trans. 1*, 1991, 2169.	362, 363
91JCS(P1)2329	G. W. Kirby and A. D. Sclare; *J. Chem. Soc., Perkin Trans. 1*, 1991, 2329.	140
91JCS(P1)2633	H. C. Brown, K. S. Bhat and P. K. Jadhav; *J. Chem. Soc., Perkin Trans. 1*, 1991, 2633.	520
91JCS(P1)2811	A. Barbero, P. Cuadrado, A. M. Gonzalez, F. J. Pulido and I. Fleming; *J. Chem. Soc., Perkin Trans. 1*, 1991, 2811	929
91JCS(P2)1077	A. D. Abell, J. O. Trent and K. B. Morris; *J. Chem. Soc., Perkin Trans. 2*, 1991, 1077.	1004
91JCS(P2)1563	G. Ferguson, D. Lloyd, H. McNab, D. R. Marshall, B. L. Ruhl and T. Wieckowski; *J. Chem. Soc., Perkin Trans. 2*, 1991, 1563.	1007
91JCS(P2)2081	C. H. Schwalbe, G. Chopra, S. Freeman, J. M. Brown and J. V. Carey; *J. Chem. Soc., Perkin Trans. 2*, 1991, 2081	862
91JFC(52)389	R. E. Banks, V. Murtagh and E. Tsiliopoulos; *J. Fluorine Chem.*, 1991, **52**, 389.	334
91JGU405	G. I. Kokorev, Sh. Kh. Badrutdinov, R. Z. Musin and F. D. Yambushev; *J. Gen. Chem. USSR* (*Engl. Transl.*), 1991, **61**, 405.	891
91JGU460	N. N. Demik, M. M. Kabachnik, Z. S. Novikova and I. P. Beletskaya; *J. Gen. Chem. USSR* (*Engl. Transl.*), 1991, **61**, 460.	471, 472
91JGU464	O. I. Kolodyazhnyi and S. N. Ustenko; *J. Gen. Chem. USSR* (*Engl. Transl.*), 1991, **61**, 464.	476
91JGU999	Ya A. Dorfman, M. M. Aleshkova, A. S. Karinskaya, A. K. Borangazieva and M. M. Kebekbaeva; *J. Gen. Chem. USSR* (*Engl. Transl.*), 1991, **61**, 999.	108
91JHC939	A. F. Barrero, J. F. Sanchez, J. E. Oltra and D. Teva; *J. Heterocycl. Chem.*, 1991, **28**, 939.	617
91JMOC113	Z. Foltynowicz, B. Marciniec and C. Pietraszuk; *J. Mol. Catal.*, 1991, **65**, 113.	938
91JOC4	D. H. Hua, S. W. Miao, J. S. Chen and S. Iguchi; *J. Org. Chem.*, 1991, **56**, 4.	362
91JOC401	H. S. Racherla and H. C. Brown; *J. Org. Chem.*, 1991, **56**, 401.	521
91JOC452	A. Galan, J. de Mendoza, P. Prados, T. Rojo and A. M. Echavarren; *J. Org. Chem.*, 1991, **56**, 452.	316
91JOC698	K. V. Baker, J. M. Brown, N. Hughes, A. J. Skarnulis and A. Sexton; *J. Org. Chem.*, 1991, **56**, 698.	553, 560
91JOC751	D. J. Mathre, T. K. Jones, L. C. Xavier, T. J. Blacklock, R. A. Reamer, J. J. Mohan, E. T. Turner, K. Hoogsteen, M. W. Baum and E. J. J. Grabowski; *J. Org. Chem.*, 1991, **56**, 751.	528
91JOC763	T. K. Jones, J. J. Mohan, L. C. Xavier, T. J. Blacklock, D. J. Mathre, P. Sohar, E. T. Turner Jones, R. A. Reamer, F. E. Roberts and E. J. J. Grabowski; *J. Org. Chem.*, 1991, **56**, 763.	528
91JOC770	S. Sharma and A. C. Oehlschlager; *J. Org. Chem.*, 1991, **56**, 770.	917
91JOC833	B. Dhawan and D. Redmore; *J. Org. Chem.*, 1991, **56**, 833.	847
91JOC862	A. J. Caruso, A. M. Colley and G. L. Bryant; *J. Org. Chem.*, 1991, **56**, 862.	711
91JOC904	P. L. Folkins, D. N. Harpp and B. R. Vincent; *J. Org. Chem.*, 1991, **56**, 904.	232

91JOC1020 K. Burgess, J. Cassidy and M. J. Ohlmeyer; *J. Org. Chem.*, 1991, **56**, 1020. 527

91JOC1170 H. C. Brown, A. M. Salunkhe and B. Singaram; *J. Org. Chem.*, 1991, **56**, 1170. 306, 318

91JOC1381 Y.-Z. Huang and Y. Liao; *J. Org. Chem.*, 1991, **56**, 1381. 487

91JOC1445 L. Zhu, R. M. Wehmeyer and R. D. Rieke; *J. Org. Chem.*, 1991, **56**, 1445. 959

91JOC1453 R. D. Miller, W. Theis, G. Heilig and S. Kirchmeyer; *J. Org. Chem.*, 1991, **56**, 1453. 1003

91JOC1948 Y. Ito, M. Suginome and M. Murakami; *J. Org. Chem.*, 1991, **56**, 1948. 916

91JOC1951 C. Nativi, L. Ravida, A. Ricci, G. Seconi and M. Taddei; *J. Org. Chem.*, 1991, **56**, 1951. 907

91JOC2234 J. Barluenga, P. J. Campos, J. M. Gonzalez and J. L. Suarez; *J. Org. Chem.*, 1991, **56**, 2234. 18, 610

91JOC2468 M. R. Najafi, M.-L. Wang and G. Zweifel; *J. Org. Chem.*, 1991, **56**, 2468. 922, 929

91JOC2538 S. Araki, T. Shimizu, P. S. Johar, S. J. Jin and Y. Butsugan; *J. Org. Chem.*, 1991, **56**, 2538. 597

91JOC2590 D. R. Swanson, T. Nguyen, Y. Noda and E. Negishi; *J. Org. Chem.*, 1991, **56**, 2590. 923, 969

91JOC2920 K. Ogura, Y. Shimamura and M. Fujita; *J. Org. Chem.*, 1991, **56**, 2920. 305

91JOC3137 R. P. Polniaszek and A. L. Foster; *J. Org. Chem.*, 1991, **56**, 3137. 455

91JOC3286 M. V. Rangaishenvi, B. Singaram and H. C. Brown; *J. Org. Chem.*, 1991, **56**, 3286. 529

91JOC3520 J. L. Kice, D. M. Wilson and J. M. Espinola; *J. Org. Chem.*, 1991, **56**, 3520. 236

91JOC3846 M. Mori, N. Kaneta and M. Shibasaki; *J. Org. Chem.*, 1991, **56**, 3846. 912, 945

91JOC3912 M. D. Bachi, N. Bar-Ner, C. M. Crittell, P. J. Stang and B. L. Williamson; *J. Org. Chem.*, 1991, **56**, 3912. 1020

91JOC4048 J. M. Heerding and H. W. Moore; *J. Org. Chem.*, 1991, **56**, 4048. 669

91JOC4066 C. E. Fuller and D. G. Walker; *J. Org. Chem.*, 1991, **56**, 4066. 622

91JOC4112 G. A. Molander and J. A. McKie; *J. Org. Chem.*, 1991, **56**, 4112. 565

91JOC4552 J. K. Whitesell and M.-S. Wong; *J. Org. Chem.*, 1991, **56**, 4552. 152, 721

91JOC4608 M. H. Lyttle, P. B. Wright, N. D. Sinha, J. D. Bain and A. R. Chamberlin; *J. Org. Chem.*, 1991, **56**, 4608. 409, 410

91JOC4778 K. Nakamura, Y. Kawai, N. Nakajima and A. Ohno; *J. Org. Chem.*, 1991, **56**, 4778. 40

91JOC4925 G. Resnati and D. D. DesMarteau; *J. Org. Chem.*, 1991, **56**, 4925. 16

91JOC5243 A. G. M. Barrett and J. W. Malecha; *J. Org. Chem.*, 1991, **56**, 5243. 75, 521

91JOC5263 I. Ojima, M. Zhao, T. Yamata and K. Nakahashi; *J. Org. Chem.*, 1991, **56**, 5263. 320

91JOC5503 R. R. Ferguson and R. H. Crabtree; *J. Org. Chem.*, 1991, **56**, 5503. 192

91JOC5556 K. Lee and D. F. Wiemer; *J. Org. Chem.*, 1991, **56**, 5556. 457

91JOC5760 B. Mudryk and T. Cohen; *J. Org. Chem.*, 1991, **56**, 5760. 553

91JOC5820 G. P. Boldrini, M. Bortolini and E. Tagliavini; *J. Org. Chem.*, 1991, **56**, 5820. 533, 698

91JOC5919 E. Juraristi and A. M. Aguilar; *J. Org. Chem.*, 1991, **56**, 5919. 438

91JOC5962 A. J. Poss, M. Van Der Puy, D. Nalewajek, G. A. Shia, W. J. Wagner and R. L. Frenette; *J. Org. Chem.*, 1991, **56**, 5962. 16

91JOC5964 J. V. B. Kanth and M. Perisamy; *J. Org. Chem.*, 1991, **56**, 5964. 59

91JOC5974 H. P. Knoess, M. T. Furlong, M. J. Rozema and P. Knochel; *J. Org. Chem.*, 1991, **56**, 5974. 478

91JOC5980 K. Takai, M. Tezuka and K. Utimoto; *J. Org. Chem.*, 1991, **56**, 5980. 927

91JOC5991 F. Rebiere, O. Samuel, L. Ricard and H. B. Kagan; *J. Org. Chem.*, 1991, **56**, 5991. 152, 198

91JOC6199 L. A. Paquette, K. Dahnke, J. Doyon, W. He, K. Wyant and D. Friedrich; *J. Org. Chem.*, 1991, **56**, 6199. 625

91JOC6494 P. Wipf and J. H. Smitrovich; *J. Org. Chem.*, 1991, **56**, 6494. 568, 587

91JOC6697 D. H. R. Barton, R. H. Hesse, A. C. O'Sullivan and M. M. Pechet; *J. Org. Chem.*, 1991, **56**, 6697. 139

91JOC6702 D. H. R. Barton, R. H. Hesse, A. C. O'Sullivan and M. M. Pechet; *J. Org. Chem.*, 1991, **56**, 6702. 240, 360

91JOC6708 A. Dalla Cort; *J. Org. Chem.*, 1991, **56**, 6708. 33

91JOC6738 S. E. Denmark and M. E. Schnute; *J. Org. Chem.*, 1991, **56**, 6738. 356

91JOC7186 Z. Goren, M. J. Heeg and S. Mobashery; *J. Org. Chem.*, 1991, **56**, 7186. 25

91JOC7323 A. Capperucci, A. Degl'Innocenti, A. Ricci, A. Mordini and G. Reginato; *J. Org. Chem.*, 1991, **56**, 7323. 140

91JOC7328 W. J. Lees, R. Singh and G. M. Whitesides; *J. Org. Chem.*, 1991, **56**, 7328. 127

91JOC7347 S. Stavber and M. Zupan; *J. Org. Chem.*, 1991, **56**, 7347. 15

91JOM(402)55 K. H. Whitmire, D. Labahn, H. W. Roesky, M. Noltemeyer and G. M. Sheldrick; *J. Organomet. Chem.*, 1991, **402**, 55. 880

91JOM(409)103 M. Zaidlewicz; *J. Organomet. Chem.*, 1991, **409**, 103. 557

91JOM(419)245 H. J. Li and M. M. Turnbull; *J. Organomet. Chem.*, 1991, **419**, 245. 577

91LA269 F. Effenberger, B. Gutterer and T. Ziegler; *Liebigs Ann. Chem.*, 1991, 269. 316

91LA1361 B. Mlotkowska; *Liebigs Ann. Chem.*, 1992, 1361. 257

B-91MI 201-01 G. A. Olah, G. K. S. Prakash and R. Krishnamurti; in "Advances in Silicon Chemistry," ed. G. L. Larson, JAI Press, Greenwich, CT, 1991, vol. 1, p. 1. 36

B-91MI 203-01 J. Hoyle; in "The Chemistry of Sulfonic Acids, Esters and their Derivatives," ed S. Patai and Z. Rappoport, Wiley, Chichester, 1991, ch. 10. 175, 248

B-91MI 206-01 G. A. Benson and W. J. Spillane; in "The Chemistry of Sulfonic Acids, Esters and their Derivatives," eds. S. Patai and Z. Rappoport, Wiley, Chichester, 1991, p. 947. 366, 367, 368

91MI 208-01 M. E. Gurskii, D. G. Pershin, Y. N. Bubnov, A. V. Polyakov, A. I. Yanovskii and Y. T. Struchkov; *Metalloorg. Khim.*, 1991, **4**, 1422 (*Chem. Abstr.*, 1992, **116**, 59 510). 447

91MI 208-02 H. Westermann and M. Nieger; *J. Crystallogr. Spectrosc. Res.*, 1991, **21**, 523. 472

91MI 208-03 I. V. Konovalova, L. A. Burneva, V. F. Mironov, I. V. Loninova and A. N. Pudovik; *Bull. Acad. Nauk USSR*, 1991, **40**, 2612. 476

91MI 208-04 J. F. Nixon; *Endeavour*; 1991, **15**, 49. 478

B-91MI 210-01 S. E. Thomas; "Organic Synthesis. The Roles of Boron and Silicon," Oxford University Press, Oxford, 1991. 535

B-91MI 210-02	T. W. Greene and P. G. M. Wuts; "Protective Groups in Organic Synthesis," Wiley, Chichester, 2nd. edn., 1991	539
B-91MI 210-03	R. J. P. Corriu; in "Frontiers of Organosilicon Chemistry," eds A. R. Bassindale and P. P. Gaspar, The Royal Society of Chemistry, Cambridge, 1991, p. 185.	547
91MI 213-01	Q. Liu and P. Kocienski; *Chinese Chem. Lett.*, 1991, **2**, 353 (*Chem. Abstr.*, 1991, **115**, 158622).	663
B-91MI 213-01	T. W. Green and P. G. M. Wuts; "Protective Groups in Organic Synthesis," Wiley, New York, 1991, 2nd edn., ch. 3, p. 143.	658, 683
B-91MI 218-01	O. M. Nefedov and M. P. Egorov; in "Frontiers of Organosilicon Chemistry," eds. A. R. Bassindale and P. P. Gaspar, The Royal Society of Chemistry, Cambridge, 1991, vol. 1, p. 145.	947, 949
91MI 218-02	B. M. Trost; *Janssen Chim. Acta*, 1991, **9**, 3.	927
91OM244	W. M. Vetter and A. Sen; *Organometallics*, 1991, **10**, 244.	573
91OM825	D. R. Swanson and E. Negishi; *Organometallics*, 1991, **10**, 825.	986, 991
91OM902	D. L. Reger, D. G. Garza and L. Lebioda; *Organometallics*, 1990, **10**, 902.	581
91OM1095	T. Murai, K. Nonomura, K. Kimura and S. Kato; *Organometallics*, 1991, **10**, 1095.	370, 768
91OM1183	J. K. Stille, H. Su, P. Brechot, G. Parrinello and L. S. Hegedus; *Organometallics*, 1991, **10**, 1183.	583
91OM1282	H. Chen, P. Jutzi, W. Leffers, M. M. Olmstead and P. P. Power; *Organometallics*, 1991, **10**, 1282.	1005
91OM1635	A. H. Cowley, R. A. Jones, M. A. Mardones and C. M. Nunn; *Organometallics*, 1991, **10**, 1635	511
91OM2046	G. Consiglio, S. C. A. Nefkens and A. Borer; *Organometallics*, 1991, **10**, 2046.	583
91OM2068	A. J. Ashe, III, T. R. Diephouse, J. W. Kampf and S. M. Al-Taweel; *Organometallics*, 1991, **10**, 2068.	896
91OM3764	A.-J. DiMaio and A. L. Rheingold; *Organometallics*, 1991, **10**, 3764.	496
91OPP399	A. R. Katritzky, J. Wu and S. Rachwal; *Org. Prep. Proced. Int.*, 1991, **23**, 399.	301
91OPP721	J. Zmitek, B. Jenko, D. Milivojevic and J. Anzel; *Org. Prep. Proced. Int.*, 1991, **23**, 721.	322
91OR(40)1	N. E. Schore; *Org. React.*, 1991, **40**, 1.	926, 985
91OR(40)157	O. De Lucchi, U. Miotti and G. Modena; *Org. React.*, 1991, **40**, 157.	142, 1006
91OS(70)68	J. L. Sessler, A. Mozaffari and M. R. Johnson; *Org. Synth.*, 1991, **70**, 68.	394
91PAC403	A. Pelter, K. Smith and K. Jones; *Pure Appl. Chem.*, 1991, **63**, 403.	1005
91PS(57)1	R. L. Wells, A. P. Purdy and C. Pitt; *Phosphorus Sulfur*, 1991, **57**, 1.	490
91PS(55)195	Y. Okamoto, T. Kusano and S. Takamuku; *Phosphorus Sulfur*, 1991, **55**, 195.	106
91PS281	T. Hase and P. Kangas; *Phosphorus Sulfur*, 1991, 281.	238
91PS(56)27	S. D. Toto, B. W. Arbuckle, P. K. Bharadwaj, J. T. Doi and W. K. Musker; *Phosphorus Sulfur*, 1991, **56**, 27.	441, 444
91PS(58)223	K. Hartke; *Phosphorus Sulfur*, 1991, **58**, 223.	140
91PS(59)295	P. Metzner; *Phosphorus Sulfur*, 1991, **59**, 295.	140
91PS(61)261	D. Wilbrandt, K. Jurkschat and J. Meunier-Piret; *Phosphorus, Sulfur Silicon*, 1991, **61**, 261.	435
91RCR162	E. W. Dianova and E. Ya. Sabofina; *Russ. Chem. Rev.* (*Engl. Transl.*), 1991, **60**, 162.	872
91RCR850	V. P. Kukhar and V. A. Solcshomok; *Russ. Chem. Rev.* (*Engl. Transl.*), 1991, **60**, 850.	15
91RCR885	A. P. Brestkin, L. A. Vikhreva, N. N. Godovikov, Yu. G. Zhukovskii, M. I. Kabachnik, S. N. Moralev, V. I. Rosengart and O. E. Sherstobitov; *Russ. Chem. Rev.* (*Engl. Transl.*), 1991, **60**, 885.	1033
91RCR1050	V. P. Kukhar, Y. T. Yagupol'skii, I. I. Gerus and M. T. Kolycheva; *Russ. Chem. Rev.* (*Engl. Transl.*), 1991, **60**, 1050.	15
91RTC25	J. Brussee and A. van der Gen; *Recl. Trav. Chim. Pays-Bas*, 1991, **110**, 25.	316
91S176	B. B. Kikani, J. R. McKee and M. Zanger; *Synthesis*, 1991, 176.	311
91S225	F. Babudri, V. Fiandanese, R. Musio, F. Naso, O. Sciavovelli and A. Scilimati; *Synthesis* 1991, 225.	740
91S353	A. G. Martinez, E. T. Vilar, J. C. Lopez, J. M. Alonso, M. Hanack and L. R. Subramanian *Synthesis*, 1991, 353.	9
91S447	G. Verardo, A. G. Guimanini, G. Favret and P. Strazzolini; *Synthesis*, 1991, 447.	326
91S454	C.-C. Tang, G.-P. Wu and G.-Z. Zhang; *Synthesis*, 1991, 454.	257
91S479	A. Barco, S. Benetti, G. P. Pollin and G. Spalluto; *Synthesis*, 1991, 479.	301
91S629	V. Jäger, B. Seidel and E. Guntrum; *Synthesis*, 1991, 629.	396
91S649	L. Duhamel, P. Duhamel, D. Enders, W. Karl, F. Léger, J. M. Poirier and G. Raabe; *Synthesis*, 1991, 649.	759
91S769	V. Jäger, W. Hümmer, U. Stahl and T. Gracza; *Synthesis*, 1991, 769.	318
91S785	R. Sato and S. Satoh; *Synthesis*, 1991, 785.	141
91S970	D. H. Hua, S. W. Miao, A. A. Bravo and D. J. Takemoto; *Synthesis*, 1991, 970.	743
91S1003	M. Selva, F. Trotta and P. Tundo; *Synthesis*, 1991, 1003.	14
91S1073	P. J. Stang and J. Ullmann; *Synthesis*, 1991, 1073.	445, 1021
91S1163	T. Aoyama, T. Nakano, K. Marumo, Y. Uno and T. Shioiri; *Synthesis*, 1991, 1163.	944
91S1201	L. Capella, A. Degl'Innocenti, A. Mordini, G. Reginato, A. Ricci and G. Seconi; *Synthesis*, 1991, 1201.	918
91SC11	J. Dubac, A. Gaset and M. Maraval; *Synth. Commun.*, 1991, **21**, 11.	22, 28
91SC489	F. Bellesia, F. Ghelfi, U. M. Pagnoni and A. Pinetti; *Synth. Commun.*, 1991, **21**, 489.	24
91SC535	A. M. Cuadro, M. P. Matia, J. L. Garcia, J. J. Vaquero and J. Alvarez-Builla; *Synth. Commun.*, 1991, **21**, 535.	301
91SC1425	R. Chauvin; *Synth. Commun.*, 1991, **21**, 1425.	300

91SC1579	M. Bonnat, A. Hercouet and M. le Corre; *Synth. Commun.*, 1991, **21**, 1579.	320
91SC2335	P. J. Thomas and R. G. Pews; *Synth. Commun.*, 1991, **21**, 2335.	654
91SCI395	Y. Chiang and A. J. Kresge; *Science*, 1991, **253**, 395.	637
91SL29	P. Hamley, A. B. Holmes, A. Kee, T. Ladduwahetty and D. F. Smith; *Synlett.*, 1991, 29.	327
91SL39	G. K. Surya Prakash, N. Krass, Q. Wang and G. A. Olah; *Synlett*, 1991, 39.	650
91SL80	P. C. B. Page and E. S. Namwindwa; *Synlett*, 1991, 80.	147
91SL117	G. Himbert, H. Naßhan and S. Kosack; *Synlett.*, 1991, 117.	1047
91SL125	C. Narayana, S. Padmanabhan and G. W. Kabalka; *Synlett.*, 1991, 125.	221
91SL163	W. K. D. Brill and H. Kunz; *Synlett.*, 1991, 163.	203
91SL185	M. Kuroboshi and T. Hiyama; *Synlett*, 1991, 185.	18
91SL187	E. Differding and H. Ofner; *Synlett*, 1991, 187.	16
91SL248	J. A. Ciaccio, E. Heller and A. Talbot; *Synlett*, 1991, 248.	84
91SL345	T. Koide, A. Otaka and N. Fujii; *Synlett.*, 1991, 345.	227
91SL431	H. Brauer and W. P. Neumann; *Synlett*, 1991, 431.	947
91SL433	H. Memmesheimer, Juma'a Al-Dulayymi, M. S. Baird, T. Wettling and M. Regitz; *Synlett* 1991, 433.	449
91SL435	D. H. R. Barton, D. O. Jang and J. Cs. Jaszberenyi; *Synlett*, 1991, 435.	536
91SL477	J. Rodriguez and J.-P. Dulcere; *Synlett*, 1991, 477.	21
91SL505	M. Julia, D. Uguen, J.-N. Verpeaux and D. Zhang; *Synlett*, 1991, 505.	678
91SL539	E. Nakamura; *Synlett*, 1991, 539.	702
91SL557	A. Hosomi, K. Ogata, M. Ohkuma and M. Hojo; *Synlett*, 1991, 557.	139
91SL573	T. Yamazaki, R. Gimi and J. T. Welch; *Synlett.*, 1991, 573.	325
91SL625	A. Alexakis, N. Lensen and P. Mangeney; *Synlett.*, 1991, 625.	315
91SL649	Y. Endo, S. Hizatate and K. Shudo; *Synlett.*, 1991, 649.	316
91SL659	J.-Y. Nedelec and K. Nohair; *Synlett*, 1991, 659.	14
91SL668	M. S. Khanna, O. V. Singh, C. P. Garg and R. P. Kapoor; *Synlett.*, 1991, 668.	311
91SL733	P. F. Richardson and I. E. Marko; *Synlett*, 1991, 733.	24
91SL755	C. Bruneau, M. Neveux, Z. Kabouche, C. Ruppin and P. H. Dixneuf; *Synlett*, 1991, 755.	676
91SL807	I. Duhamel, P. Angibaud, J. R. Desmurs and J. Y. Valnot; *Synlett*, 1991, 807.	22
91SL845	Y. Hatanaka and T. Hiyama; *Synlett*, 1991, 845.	912, 931, 945
91SL907	B. Jouglet, L. Blanco and D. Rousseau; *Synlett*, 1991, 907.	395
91SR670	O. Dahl; *Sulfur Rep.*, 1991, **11**, 1670.	108
91T149	N. Iranpoor and J. Onji; *Tetrahedron*, 1991, **47**, 149.	222
91T205	J. C. Jochims, M. O. Glocker, J. Hofmann and H. Fischer; *Tetrahedron*, 1991, **47**, 205.	1009
91T343	E. Negishi, T. Yoshida, A. Abramivitch, G. Lew and R. M. Williams; *Tetrahedron*, 1991, **47**, 343.	901, 904
91T403	M. Vicens, J. T. Grimaldo-Morön and M. Vidal; *Tetrahedron*, 1991, **47**, 403.	444
91T1109	R. V. Hoffman; *Tetrahedron*, 1991, **47**, 1109.	357
91T1163	R. Aksela and A. C. Oehlschlager; *Tetrahedron*, 1991, **47**, 1163.	903
91T3407	N. De Kimpe and C. S. Stevens; *Tetrahedron*, 1991, **47**, 3407.	312
91T4007	R. Grigg, F. Heaney, J. Markandu, S. Surendrakumar, M. Thornton-Pett and W. J. Warnock; *Tetrahedron*, 1991, **47**, 4007.	352
91T5185	P. Leon-Ruaud and D. Plusquelle; *Tetrahedron*, 1991, **47**, 5185.	133
91T5277	B. Abarca, R. Ballesteros, F. Mojarrad, M. R. Metni, S. Garcia-Granda, E. Perez-Carreño and G. Jones; *Tetrahedron*, 1991, **47**, 5277.	1003
91T5329	N. Yoneda; *Tetrahedron*, 1991, **47**, 5329.	6, 15, 18
91T5603	J. E. Baldwin, C. D. Jesudason, M. G. Moloney, D. R. Morgan and A. J. Pratt; *Tetrahedron*, 1991, **47**, 5603.	230
91T6655	J.-B. Baudin, I. Bkouche-Wakman, G. Hareau, S. A. Julia, R. Lorne and C. Pascard; *Tetrahedron*, 1991, **47**, 6655.	363
91T7447	A. A. Gakh, S. V. Romanov, B. I. Uryak and A. A. Fainzilberg; *Tetrahedron*, 1991, **47**, 7447.	334, 365
91T8145	T. Netscher and T. Weller; *Tetrahedron*, 1991, **47**, 8145.	239
91T8297	R. Grigg, M. J. Dorrity, F. Heaney, J. F. Malone, S. Rajviroongit, V. Sridharan and S. Surendrakumar; *Tetrahedron*, 1991, **47**, 8297.	158
91T8587	B. K. Sarmah and N. C. Barua; *Tetrahedron*, 1991, **47**, 8587.	316
91T8611	E. Vuorinen, A. A. Chalmers, J. L. M. Dillen and T. A. Modro; *Tetrahedron*, 1991, **47**, 8611.	1000
91T8621	M. J. Dickens, J. P. Gilday, T. J. Mowlem and D. A. Widdowson; *Tetrahedron*, 1991, **47**, 8621.	711
91T8717	O. Sakatsume, T. Yamaguchi, M. Ishikawa, I. Hirao, K. Miura and H. Takaku; *Tetrahedron*, 1991, **47**, 8717.	106
91T9167	D. H. R. Barton, J. C. Jaszberenyi and E. A. Theodorakis; *Tetrahedron*, 1991, **47**, 9167.	721
91T9449	J. B. Baruah and A. G. Samuelson; *Tetrahedron*, 1991, **47**, 9449.	308
91T9823	L. Ettouati, A. Ahond, C. Poupat and P. Potier; *Tetrahedron*, 1991, **47**, 9823.	642
91T9861	N. Iranpoor, I. M. Balturk and F. S. Zardaloo; *Tetrahedron*, 1991, **47**, 9861.	135
91TA1	P. J. Cox and N. G. Simpkins; *Tetrahedron Asymmetry*, 1991, **2**, 1.	420
91TA183	S. G. Davies and O. Ichihara; *Tetrahedron Asymmetry*, 1991, **2**, 183.	315
91TA601	T. Hayashi, Y. Matsumoto and Y. Ito; *Tetrahedron Asymmetry*, 1991, **2**, 601.	528
91TA613	K. Burgess, W. A. van der Donk and M. Ohlmeyer; *Tetrahedron Asymmetry*, 1991, **2**, 613.	528
91TA667	Y. Okada, T. Minami, Y. Umezu, S. Nishikawa, R. Mori and Y. Nakayama; *Tetrahedron Asymmetry*, 1991, **2**, 667.	440, 829

91TL243	D. C. Harrowven and G. Pattenden; *Tetrahedron Lett.*, 1991, **32**, 243.	84
91TL441	C. Rutherford and P. Knochel; *Tetrahedron Lett.*, 1991, **32**, 441.	783, 784
91TL525	T. Shono, N. Kise and T. Fujimoto; *Tetrahedron Lett.*, 1991, **32**, 525.	314, 352
91TL667	S. Saito, M. Takahashi, T. Ishikawa and T. Moriwake; *Tetrahedron Lett.*, 1991, **32**, 667.	84
91TL715	K. R. Guertin and T.-H. Chan; *Tetrahedron Lett.*, 1991, **32**, 715.	44
91TL807	A. J. Lucke and D. J. Young; *Tetrahedron Lett.*, 1991, **32**, 807.	210
91TL827	T. Eszenyi, T. Timar and P. Sebok; *Tetrahedron Lett.*, 1991, 827.	624
91TL855	S. D. Burke, A. D. Piscopio and M. E. Kort; *Tetrahedron Lett.*, 1991, **32**, 855.	932
91TL923	M. Rodriguez, M. Llinares, S. Doulut, A. Heitz and J. Martinez; *Tetrahedron Lett.*, 1991, **32**, 923.	305
91TL1003	C. Y. Park, B. M. Kim and K. B. Sharpless; *Tetrahedron Lett.*, 1991, **32**, 1003.	47
91TL1069	T. Ishihara, S. Mitawashi, M. Kuroboshi and K. Utimoto; *Tetrahedron Lett.*, 1991, **32**, 1069.	77
91TL1081	Y. D. Vankar and K. Shah; *Tetrahedron Lett.*, 1991, **32**, 1081.	10
91TL1171	A. Alexakis, N. Lensen and P. Mangeney; *Tetrahedron Lett.*, 1991, **32**, 1171.	381
91TL1187	K. Koerber, J. Goré and J.-M. Vatele; *Tetrahedron Lett.*, 1991, **32**, 1187.	930
91TL1215	M. Kuroboshi and T. Hiyama; *Tetrahedron Lett.*, 1991, **32**, 1215.	18
91TL1479	Y. Yakoyama, H. Kawashima, M. Kohno, Y. Ogawa and S. Uchida; *Tetrahedron Lett.*, 1991, **32**, 1479.	518
91TL1631	F. A. Davis and W. Han; *Tetrahedron Lett.*, 1991, 1631.	16, 613
91TL1699	A. S. Kende and J. S. Mendoza; *Tetrahedron Lett.*, 1991, **32**, 1699.	341
91TL1779	E. Differding, G. M. Ruegg and R W. Lang; *Tetrahedron Lett.*, 1991, **32**, 1779.	16
91TL1831	I. E. Marko and P. F. Richardson; *Tetrahedron Lett.*, 1991, **32**, 1831.	24
91TL2157	A. Sarkar and B. R. Rao; *Tetrahedron Lett.*, 1991, **32**, 2157.	61
91TL2169	A. Casarini, P. Dembech, G. Reginato, A. Ricci and G. Seconi; *Tetrahedron Lett.*, 1991, **32**, 2169.	1018
91TL2189	G. A. Kraus and B. Andersh; *Tetrahedron Lett.*, 1991, **32**, 2189.	264
91TL2327	S. R. Angle and D. O. Arnaiz; *Tetrahedron Lett.*, 1991, **32**, 2327.	930
91TL2509	A. Krief, B. Kenda and P. Barbeaux; *Tetrahedron Lett.*, 1991, 2509.	141
91TL2517	A. Barco, S. Benetti, G. P. Pollini, G. Spalluto and V. Zanirato; *Tetrahedron Lett.*, 1991, **32**, 2517.	356
91TL2555	T. Hamamoto and J. Inanaga; *Tetrahedron Lett.*, 1991, **32**, 2555.	352
91TL2711	E. W. Logusch, D. R. Dukesherer, K. M. Getman and J. E. Franz; *Tetrahedron Lett.*, 1991, **32**, 2711.	344
91TL2853	P. Arya, M. Lesage and D. D. M. Wayner; *Tetrahedron Lett.*, 1991, 2853.	141, 265
91TL2971	B. F. Bonini, S. Masiero, G. Mazzanti and P. Zani; *Tetrahedron Lett.*, 1991, **32**, 2971.	938
91TL3001	G. E. Agoston, M. P. Cabal and E. Turos; *Tetrahedron Lett.*, 1991, **32**, 3001.	73
91TL3163	G. Apitz and W. Steglich; *Tetrahedron Lett.*, 1991, 3163.	140
91TL3217	I. G. Collado, G. M. Massanet and M. S. Alonso; *Tetrahedron Lett.*, 1991, **32**, 3217.	30
91TL3225	R. Fernández, C. Gasch, A. Gómez-Sánchez and J. E. Vílchez; *Tetrahedron Lett.*, 1991, **32**, 3225.	394
91TL3329	E. C. Stracker and G. Zweifel; *Tetrahedron Lett.*, 1991, 3329.	630
91TL3371	T. Oshika, T. Hikosaka and T. Imamoto; *Tetrahedron Lett.*, 1991, **32**, 3371.	444
91TL3387	T. Hayashi and Y. Matsumoto; *Tetrahedron Lett.*, 1991, **32**, 3387.	43, 528
91TL3441	A. S. Kotnis; *Tetrahedron Lett.*, 1991, **32**, 3441.	658
91TL3457	B. Santiago, C. Lopez and J. A. Soderquist; *Tetrahedron Lett.*, 1991, **32**, 3457.	939
91TL3655	H. C. Brown, N. G Bhat and R. R. Iyer; *Tetrahedron Lett.*, 1991, **32**, 3655.	906
91TL3705	L. DeLaude and P. Laszlo; *Tetrahedron Lett.*, 1991, **32**, 3705.	23
91TL3783	A. Basha, J. D. Ratajczyk and D. W. Brooks; *Tetrahedron Lett.*, 1991, **32**, 3783.	343
91TL3977	N. Belkhouya, C. Frechou, M. Benazza, D. Beaupere, R. Uzan and G. Demailly; *Tetrahedron Lett.*, 1991, **32**, 3977.	7
91TL4135	J. Berthelot, Y. Benammar and C. Lange; *Tetrahedron Lett.*, 1991, **32**, 4135.	30
91TL4227	D. F. Taber and L. J. Silverberg; *Tetrahedron Lett.*, 1991, **32**, 4227.	54
91TL4247	A. R. Katritzky, N. Shobana and P. A. Harris; *Tetrahedron Lett.*, 1991, **32**, 4247.	302
91TL4381	E. Cesarotti, A. Mauri, M. Pallavinci and L. Villa; *Tetrahedron Lett.*, 1991, **32**, 4381.	53
91TL4431	S. W. Baldwin, R. B. McFayden, J. Aubé and J. D. Wilson; *Tetrahedron Lett.*, 1991, **32**, 4431.	342, 354
91TL4481	M. T. Reetz and S. H. Lauterbach; *Tetrahedron Lett.*, 1991, **32**, 4481.	348, 355
91TL4639	S. Jiang and E. Turos; *Tetrahedron Lett.*, 1991, **32**, 4639.	73
91TL4647	T. Oh, Z. Wrobel and S. M. Rubenstein; *Tetrahedron Lett.*, 1991, **32**, 4647.	88
91TL4671	D. H. R. Barton, S. D. Dèvière and D. Doller; *Tetrahedron Lett.*, 1991, **32**, 4671.	108
91TL4741	Y. Izumi, H. Nanami, K. Higuchi and M. Onaka; *Tetrahedron Lett.*, 1991, **32**, 4741.	82
91TL4981	J. Helinski, W. Dabkowski and J. Michalski; *Tetrahedron Lett.*, 1991, **32**, 4981.	467
91TL5043	S. W. Kaldor and M. Hammond; *Tetrahedron Lett.*, 1991, **32**, 5043.	155
91TL5047	P. A. Magriotis, J. T. Brown and M. E. Scott; *Tetrahedron Lett.*, 1991, **32**, 5047.	915
91TL5273	J. C. Garcia, R. J. Boll and E. S. Younathan; *Tetrahedron Lett.*, 1991, **32**, 5273.	84
91TL5317	G. Courtemanche and J.-F. Normant; *Tetrahedron Lett.*, 1991, **32**, 5317.	478
91TL5465	K. J. Barr, B. T. Watson and S. L. Buchwald; *Tetrahedron Lett.*, 1991, **32**, 5465.	927
91TL5473	H.-J. Liu and N. H. Said; *Tetrahedron Lett.*, 1991, **32**, 5473.	69
91TL5885	S. C. Benson and J. K. Snider; *Tetrahedron Lett.*, 1991, **32**, 5885.	152
91TL5903	T. Nguyen and E.-I. Negishi; *Tetrahedron Lett.*, 1991, **32**, 5903.	88
91TL5923	A. Padwa, D. J. Austin, U. Chiacchio, J. M. Kassir, A. Rescifina and S. L. Xu; *Tetrahedron Lett.*, 1991, **32**, 5923.	668

91TL6021 C. E. Dreef, W. Schiebler, G. A. van der Marel and J. H. van Boom; *Tetrahedron Lett.*, 1991, **32**, 6021. 106
91TL6239 A. Pelter, K. Smith, D. Buss and A. Norbury; *Tetrahedron Lett.*, 1991, **32**, 6239. 900, 901
91TL6579 L.-J. Zhang, Y.-Z. Huang and Z.-H. Zhang; *Tetrahedron Lett.*, 1991, **32**, 6579. 73
91TL6683 E. Negishi and Z. Owczarczyk; *Tetrahedron Lett.*, 1991, **32**, 6683. 976
91TL6691 H. C. Brown, P. V. Ramachandran, A. V. Teodorovic and S. Swaminathan; *Tetrahedron Lett.*, 1991, **32**, 6691. 526
91TL6749 S. Hara and A. Suzuki; *Tetrahedron Lett.*, 1991, **32**, 6749. 76
91TL6789 I. A. Cliffe, A. D. Ifill, H. L. Mansell, R. S. Todd and A. C. White; *Tetrahedron Lett.*, 1991, **32**, 6789. 346
91TL6877 N. Tokitoh, T. Matsumoto, H. Ichida and R. Okazaki; *Tetrahedron Lett.*, 1991, **32**, 6877. 949
91TL6931 M. Canas, M. Poch, X. Verdaguer, A. Moyano, M. A. Pericas and A. Riera; *Tetrahedron Lett.*, 1991, **32**, 6931. 84, 318
91TL6935 M. Poch, X. Verdaguer, A. Moyano, M. A. Pericas and A. Riera; *Tetrahedron Lett.*, 1991, **32**, 6935. 318
91TL7017 C. J. Li and T. H. Chan; *Tetrahedron Lett.*, 1991, **32**, 7017. 73
91TL7075 K. Nakamura, S. Kondo, Y. Kawai and A. Ohno; *Tetrahedron Lett.*, 1991, **32**, 7075. 40
91TL7365 D. A. Singleton and J. P. Martinez; *Tetrahedron Lett.*, 1991, **32**, 7365. 902, 922
91USP4997952 K. S. Narayanan and P. D. Taylor; *US Pat.* 4 997 952 (1991) (*Chem. Abstr.*, 1991, **115**, 29 636). 444
91ZAAC(549)36 G. Fritz, E. Layher, H. Groesmann, D. Hanke and C. Persau; *Z. Anorg. Allg. Chem.*, 1991, **549**, 36. 471
91ZAAC(592)121 J. Grobe, W. Hildebrandt, R. Martin and A. Walter; *Z. Anorg. Allg. Chem.*, 1991, **592**, 121. 439
91ZAAC(594)66 F. Krech, K. Issleib, A. Zschunke, C. Mügge and S. Skvorcov; *Z. Anorg. Allg. Chem.*, 1991, **594**, 66. 439
91ZAAC(600)195 F. Krech, K. Issleib and A. Zschunke; *Z. Anorg. Allg. Chem.*, 1991, **600**, 195. 478
91ZAAC(603)109 W. Uhlig; *Z. Anorg. Allg. Chem.*, 1991, **603**, 109. 1089
91ZAAC(603)145 L. Riesel and J. Haenel; *Z. Anorg. Allg. Chem.*, 1991, **603**, 145. 449
91ZAAC(605)151 R. Aslanidis and F. Kober; *Z. Anorg. Allg. Chem.*, 1991, **605**, 151. 496
91ZAAC(606)119 A. Deeg, W. Kuchen, D. Langsch, D. Mootz, W. Peters and H. Wunderlich; *Z. Anorg. Allg. Chem.*, 1991, **606**, 119. 861
91ZN(B)5 A. Blaschette, M. Näveka and P. G. Jones; *Z. Naturforsch., Teil B*, 1991, **46**, 5. 100
91ZN(B)258 I. V. Sherchenko, M. V. Furmanova, V. P. Kukhar and R. Schmutzler; *Z. Naturforsch., Teil B*, 1991, **47**, 258. 467
91ZN(B)775 A. A. M. Aly and H. Schmidbaur; *Z. Naturforsch., Teil B*, 1991, **46**, 775. 444
91ZN(B)887 H. Schumann and L. Egurem; *Z. Naturforsch., Teil B*, 1991, **E46** F, 887. 449
91ZN(B)1319 K. Rudolph and M. Wieber; *Z. Naturforsch., Teil B*, 1991, **46**, 1319. 1082
91ZN(B)1578 W. Wolfsberger; *Z. Naturforsch., Teil B*, 1991, **46**, 1578. 433
91ZOB611 I. S. Dokuchaeva, I. V. Konovalova, Ju. G. Trishin, L. A. Burnaeva, T. Yu. Kazanina, V. N. Christokletov and A. N. Pudovik; *Zh. Obshch. Khim.*, 1991, **61**, 611. 1074
91ZOB1016 A. A. Prishchenko, M. V. Livantsov, D. A. Pisarnitskii and V. S. Petrosyan; *Zh. Obshch. Khim.*, 1991, **61**, 1016 (*Chem. Abstr.*, 1992, **116**, 6649). 433, 473
91ZOB1021 O. I. Tuzhihov, I. V. Malyutina, R. I. Yurchenko and A. A. Kudryavtsev; *Zh. Obshch. Khim.*, 1991, **61**, 1021 (*Chem. Abstr.*, 1991, **115**, 280 144). 427
91ZOB1075 Y. P. Malovetskii, N. G. Feshchenko, A. N. Cherega, M. Y. Antipin and Y. T. Struchkov; *Zh. Obshch. Khim.*, 1991, **61**, 1075 (*Chem. Abstr.*, 1992, **116**, 6635). 444
91ZOB1115 K. A. Petrov, G. V. Rudnev and V. D. Sorokin; *Zh. Obshch. Khim.*, 1991, **61**, 1115 (*Chem. Abstr.*, 1991, **115**, 280 135r). 257
91ZOB1906 B. I. No, Yu. L. Zotov and R. M. Petruneva; *Zh. Obshch. Khim.*, 1991, **61**, 1906. 448
91ZOB2161 B. V. Timokhin, M. V. Kazantseva, G. V. Ratovskii, D. D. Churashev and V. I. Donskikh; *Zh. Obshch. Khim.*, 1991, **61**, 2161. 448

92ACR16 H. C. Brown and P. V. Ramachandran; *Acc. Chem. Res.*, 1992, **25**, 16. 520
92ACR182 Y.-Z. Huang; *Acc. Chem. Res.*, 1992, **25**, 182. 487, 488
92ACR188 C. Chatgilialoglu; *Acc. Chem. Res.*, 1992, **25**, 188. 536, 545
92ACR448 D. B. Collum; *Acc. Chem. Res.*, 1992, **25**, 448. 554
92ACR504 D. H. R. Barton and D. Doller; *Acc. Chem. Res.*, 1992, **25**, 504. 2
92ACS867 G. Keilen, T. Benneche and K. Gaare; *Acta Chem. Scand.*, 1992, **46**, 867. 155
92AG(E)79 B. Ledig, M. Marsch, K. Harms and G. Boche; *Angew. Chem., Int. Ed. Engl.*, 1992, **31**, 79. 1002
92AG(E)274 P. J. Stang; *Angew. Chem., Int. Ed. Engl.*, 1992, **31**, 274. 1012, 1028, 1029
92AG(E)1101 F. Diederich and Y. Rubin; *Angew. Chem., Int. Ed. Engl.*, 1992, **31**, 1101. 1091
92AG(E)1241 A. M. Boldi, J. Anthony, C. B. Knobler and F. Diederich; *Angew. Chem., Int. Ed. Engl.*, 1992, **31**, 1241. 1091
92AG(E)1578 D. Seebach, H. Schafer, B. Schmidt and M. Schreiber; *Angew. Chem., Int. Ed. Engl.*, 1992, **31**, 1587. 783
92AHC(55)1 M. R. Bryce, J. Becher and B. Fält-Hansen; *Adv. Heterocycl. Chem.*, 1992, **55**, 1. 359
92AOC(34)207 K. Seppelt; *Adv. Organomet. Chem.*, 1992, **34**, 207. 872
92AX(C)912 H. N. Schäfer, H. Burzlaff, A. M. H. Grimmeiss and R. Weiss; *Acta Crystallogr., Part C*, 1992, **48**, 912. 1009
92BAU356 S. I. Zav'yalov, I. V. Sitkareva and G. I. Ezhova; *Bull. Acad. Sci. USSR, Div. Chim. Sci.*, 1992, 356. 22
92BCJ345 S. Futamura and Z.-M. Zong; *Bull. Chem. Soc. Jpn.*, 1992, **65**, 345. 27
92BCJ626 H. Suzuki, S. Kawato and A. Nasu; *Bull. Chem. Soc. Jpn.*, 1992, **65**, 626. 216

92BCJ685	M. Hoshi, Y. Masuda and A. Arase; *Bull. Chem. Soc. Jpn.*, 1992, **65**, 685.	922
92BCJ1131	H. Firouzabadi and I. M. Baltork; *Bull. Chem. Soc. Jpn.*, 1992, **65**, 1131.	150
92BCJ1280	T. Kusumoto, K. Ando, T. Hiyama; *Bull. Chem. Soc. Jpn.*, 1992, **65**, 1280.	939, 940
92BCJ1379	N. Iwata, T. Morioka, T. Kobayashi, T. Asada, H. Kinoshita and K. Inomata; *Bull. Chem. Soc. Jpn.*, 1992, **65**, 1379.	743
92BCJ1543	Y. Kataoka, Y. Oguchi, K. Yoshizumi, S. Miwatashi, K. Takai, and K. Utimoto; *Bull. Chem. Soc. Jpn.*, 1992, **65**, 1543.	927, 989
92BCJ1731	T. Okamoto, T. Kakinami, T. Nishimura, I. Hermawan and S. Kajigaeshi; *Bull. Chem. Soc. Jpn.*, 1992, **65**, 1731.	34
92BCJ1744	M. Hirano, H. Kudo and T. Morimoto; *Bull. Chem. Soc. Jpn.*, 1992, **65**, 1744.	149
92BCJ3504	Y. Matano, M. Kinoshita and H. Suzuki; *Bull. Chem. Soc. Jpn.*, 1992, **65**, 3504.	1082
92BSB237	N. DeKimpe, W. Coppens, A. Krauze, B. De Corte, C. Stevens and J. Welch; *Bull. Soc. Chim. Belg.*, 1992, **101**, 237.	22
92BSF71	A. Zidani, R. Carrie and M. Vanltier; *Bull. Soc. Chim. Fr.*, 1992, **129**, 71.	317, 469
92BSF367	J. M. Sotiropoulos, A. Baceiredo and G. Bertrand; *Bull. Soc. Chim. Fr.*, 1992, 367.	1068
92CB157	T. Kauffmann, C. Beirich, A. Hamsen, T. Möller, C. Philipp and D. Wingbermühle; *Chem. Ber.*, 1992, **125**, 157.	577
92CB163	T. Kaufmann, B. Laarmann. D. Menges and G. Neiteler; *Chem. Ber.*, 1992, **126**, 163.	577
92CB801	M. Well, W. Albers, A. Fischer, P. G. Jones and R. Schmutzler; *Chem. Ber.*, 1992, **125**, 801	476
92CB1061	U. Nagel and A. Bublewitz; *Chem. Ber.*, 1992, **125**, 1061.	437
92CB1597	B. Wrackmeyer, G. Kehr, A. Sebald and J. Kümmerlen; *Chem. Ber.*, 1992, **125**, 1597.	1088
92CB2187	J. Wesemann, P. G. Jones, D. Schomburg, L. Hener and R. Schmutzler; *Chem. Ber.*, 1992, **125**, 2187.	821, 822
92CB2265	G. Boche, A. Opel, M. Marsch, K. Harms, F. Haller, J. C. W. Lohrenz, C. Thümmler and W. Koch; *Chem. Ber.*, 1992, **125**, 2265.	997, 999
92CB2409	T. Kaufmann, C. Neitler and S. Robbe; *Chem. Ber.*, 1992, **125**, 2409.	589
92CC84	R. S. Reddy, J. S. Reddy, R. Kumar and P. Kumar; *J. Chem. Soc., Chem. Commun.*, 1992, 84.	145
92CC234	H. A. Carless; *J. Chem. Soc., Chem. Commun.*, 1992, 234.	39
92CC274	S. Stavber, I. Kosir and M. Zupan; *J. Chem. Soc., Chem. Commun.*, 1992, 274.	615
92CC595	R. E. Banks, S. N. Mohialdin-Khaffaf, G. S. Lal, I. Sharif and R. G. Syvret; *J. Chem. Soc., Chem. Commun.*, 1992, 595.	15
92CC600	T. Saito, Y. Shundo, S. Kitazawa and S. Motoki; *J. Chem. Soc., Chem. Commun.*, 1992, 600.	226
92CC699	S. Nielsen-Marsh, R. J. Crowte and P. G. Edwards; *J. Chem. Soc., Chem. Commun.*, 1992, 699.	428
92CC726	R. A. Jones, S. U. Koschmieder, J. L. Atwood and S. G. Bott; *J. Chem. Soc., Chem. Commun.*, 1992, 726.	473
92CC853	A. Srivastava, Y. Ma, R. Pankayatselvan, W. Dinges and K. M. Nicholas; *J. Chem. Soc., Chem. Commun.* 1992, 853.	307
92CC980	G-Z. Wang and J-E. Bäckvall; *J. Chem. Soc., Chem. Commun.*, 1992, 980.	312
92CC1097	K. Soai, T. Hatanaka and T. Miyazawa; *J. Chem. Soc., Chem. Commun.*, 1992, 1097.	331
92CC1133	G. A. Epling and Q. Wang; *J. Chem. Soc., Chem. Commun.*, 1992, 1133.	83
92CC1200	Y. Sasson and O. W. Webster; *J. Chem. Soc., Chem. Commun.*, 1992, 1200.	22
92CC1256	R. S. Atkinson, P. J. Edwards and G. A. Thomson; *J. Chem Soc., Chem. Commun.*, 1992, 1256.	642
92CC1358	R. Takeuchi, M. Sugiura, N. Ishii and N. Sato; *J. Chem. Soc., Chem. Commun.*, 1992, 1358.	929
92CC1487	R.-Y. Yang, L.-X. Dai and C.-G. Chen; *J. Chem. Soc., Chem. Commun.*, 1992, 1487.	998
92CC1500	J. Drabowicz, B. Dudzinski and M. Mikolajczyk; *J. Chem. Soc., Chem. Commun.*, 1992, 1500.	151
92CC1623	E. Licandro, S. Maiorano, A. Papagni and A. Zanotti-Gerosa; *J. Chem. Soc., Chem. Commun.*, 1992 1623.	553
92CC1718	J. Omelanczuk; *J. Chem. Soc., Chem. Commun.*, 1992, 1718.	851
92CC1728	K. Kano, M. Takeuchi, S. Hasimoto and Z. Yoshida; *J. Chem. Soc., Chem. Commun.*, 1992, 1728.	579
92CJC2040	A. R. Katritsky, I. V. Shcherbakova, R. D. Tuck and P. J. Steel; *Can. J. Chem.*, 1992, **70**, 2040.	1008
92CL99	K. Takai, S. Miwatashi, Y. Kataoka, and K. Utimoto; *Chem. Lett.*, 1992, 99.	989
92CL377	J. Setsune, Y. Ishimaru and A. Sera; *Chem. Lett.*, 1992, 377.	579
92CL547	Y. Okada, T. Minami, T. Yamamoto and J. Ichikawa; *Chem. Lett.*, 1992, 547.	829
92CL891	H. Sohmiya, T. Kimura, M. Fujita and T. Ando; *Chem. Lett.*, 1992, 891.	177, 179
92CL1137	K. Kato and T. Mukaiyama; *Chem. Lett.*, 1992, 1137.	391
92CL1161	J. Ichihara, T. Hanafusa, Y. Takai and Y. Ito; *Chem. Lett.*, 1992, 1161.	19
92CL1483	Y. J. Park, H. H. Shin and Y. H. Kim; *Chem. Lett.*, 1992, 1483.	717, 718
92CL1505	S. Ohuchi, R. K. Singh, T. Wada and T. Hata; *Chem. Lett.*, 1992, 1505.	7
92CL1655	M. Karikomi, S. Narabu, M. Yoshida and T. Toda; *Chem. Lett.*, 1992, 1655.	136
92CL2479	M. Yamaguchi, A. Hayashi and M. Hirama; *Chem. Lett.*, 1992, 2479.	1100
92CRV29	G. A. Molander; *Chem. Rev.*, 1992, **92**, 29.	564
92CRV97	M. E. Weller; *Chem. Rev.*, 1992, **92**, 97.	577
92CRV505	J. A. Wilkinson; *Chem. Rev.*, 1992, **92**, 505.	15
92CRV771	B. E. Rossiter and N. M. Swingle; *Chem. Rev.*, 1992, **92**, 771.	588
92CRV807	R. O. Duthaler and A. Hafner; *Chem. Rev.*, 1992, **92**, 807.	44, 67, 568

92CRV833	K. Soai and S. Niwa; *Chem. Rev.*, 1992, **92**, 833.	67
92G65	R. Rossi, A. Carpita and T. Messeri; *Gazz. Chim. Ital.*, 1992, **122**, 65.	917, 922
92H(33)105	M. Sato, N. Kitazawa and C. Kaneko; *Heterocycles*, 1992, **33**, 105.	16
92HCA170	W. Oppolzer and R. N. Radinov; *Helv. Chim. Acta*, 1992, **75**, 170.	919
92HCA2265	E. Er and P. Margaretta; *Helv. Chim. Acta*, 1992, **75**, 2265.	133
92JA1412	D. J. Krysan, A. Gurski and L. S. Liebeskind; *J. Am. Chem. Soc.*, 1992, **114**, 1412.	669
92JA1428	F. A. Davis, R. T. Reddy, W. Han and P. J. Carroll; *J. Am. Chem. Soc.*, 1992, **114**, 1428.	150
92JA1521	E. M. J. Gordon and J. Pluscec; *J. Am. Chem. Soc.*, 1992, **114**, 1521.	257
92JA1830	M. Takahashi, R. Okazaki, N. Inamoto, T. Sugawara and H. Iwamura; *J. Am. Chem. Soc.*, 1992, **114**, 1830.	794, 795
92JA2321	A. Hafner, R. O. Duthaler, R. Marti, G. Rihs, P. Rothe-Streit and F. Schwarzenbach; *J. Am. Chem. Soc.*, 1992, **114**, 2321.	74
92JA3989	K. Tamao, A. Kawachi and Y. Ito; *J. Am. Chem. Soc.*, 1992, **114**, 3989.	545, 546
92JA4008	T. Kawashima, K. Kato and R. Okazaki; *J. Am. Chem. Soc.*, 1992, **114**, 4008.	831
92JA4042	D. A. Evans and G. C. Fu; *J. Am. Chem. Soc.*, 1992, **114**, 4042.	43
92JA5110	D. E. Stack, B. T. Dawson and R. D. Rieke; *J. Am. Chem. Soc.*, 1992, **114**, 5110.	74, 585
92JA5867	K. Tamao, S. Yamagushi, M. Shiozaki, Y. Nakagawa and Y. Ito; *J. Am. Chem. Soc.*, 1992, **114**, 5867.	939
92JA5881	A. Padwa and S. L. Xu; *J. Am. Chem. Soc.*, 1992, **114**, 5881.	671
92JA5900	W. Oppolzer, O. Tamara, G. Sundarababu and M. Signer; *J. Am. Chem. Soc.*, 1992, **114**, 5900.	331
92JA5902	H. Kuniyasu, A. Ogawa, K.-I. Sato, I. Ryu, N. Kambe and N. Sonoda; *J. Am. Chem. Soc.*, 1992, **114**, 5902.	710
92JA5977	D. A. Evans, M. M. Faul, L. Colombo, J. J. Bisaha, J. Clardy and D. Cherry; *J. Am. Chem. Soc.*, 1992, **114**, 5977.	152, 363
92JA6262	R. V. Hoffman, N. K. Nayyar and B. W. Klinekole; *J. Am. Chem. Soc.*, 1992, **114**, 6262.	23
92JA6566	K. Mikami, S. Narisawa, M. Shimizu and M. Terada; *J. Am. Chem. Soc.*, 1992, **114**, 6566.	80
92JA6580	I. Ojima, R. J. Donovan and W. R. Shay; *J. Am. Chem. Soc.*, 1992, **114**, 6580.	920
92JA6671	D. A. Evans, G. C. Fu and A. D. Hoveyda; *J. Am. Chem. Soc.*, 1992, **114**, 6671.	527
92JA6692	A. H. Hoveyda, J. P. Morken, A. F. Houri and Z. Xu; *J. Am. Chem. Soc.*, 1992, **114**, 6692.	562
92JA7562	C. A. Willoughby and S. L. Buchwald; *J. Am. Chem. Soc.*, 1992, **114**, 7562.	330
92JA7570	D. Xu, G. A. Crispino and K. B. Sharpless; *J. Am. Chem. Soc.*, 1992, **114**, 7570.	49
92JA7591	L-B. Han, K.-I. Ishihara, N. Kambe, A. Ogawa, I. Ryu and N. Sonoda; *J. Am. Chem. Soc.*, 1992, **114**, 7591.	730
92JA7598	T. Kawashima, N. Iwama and R. Okazaki; *J. Am. Chem. Soc.*, 1992, **114**, 7598.	547
92JA8053	W. F. Bailey, A. D. Khanolkar and K. V. Gavaskar; *J. Am. Chem. Soc.*, 1992, **114**, 8053.	556
92JA9220	K. N. Harrison and T. J. Marks; *J. Am. Chem. Soc.*, 1992, **114**, 9220.	43
92JA9350	K. Burgess, W. A. van der Donk, S. A. Westcott, T. B. Marder, R. T. Baker and J. C. Calabrese; *J. Am. Chem. Soc.*, 1992, **114**, 9350.	527
92JA9737	L. Engman, D. Stern, I. A. Cotgreave and C. M. Andersson; *J. Am. Chem. Soc.*, 1992, **114**, 9737.	237, 238
92JCR(S)202	C. J. Cooksey and A. G. Davies; *J. Chem. Res. (S)*, 1992, 202.	686
92JCR(S)309	A. K. Sarkar and N. Sarkar; *J. Chem. Res. (S)*, 1992, 309.	303
92JCR(S)391	G. Kokotos and V. Constantinou-Kokotou; *J. Chem. Res. (S)*, 1992, 391.	317
92JCR(S)393	T. Kataoka, T. Iwama, K. Tsutsumi, Y. Nakamura, H. Matsumoto, H. Shimizu and M. Hori; *J. Chem. Res. (S)*, 1992, 393.	288
92JCS(D)1751	A. Bader, D. D. Pathak, S. B. Wild and A. C. Willis; *J. Chem. Soc., Dalton Trans.*, 1992, 1751.	438
92JCS(P1)221	M. Ihara, N. Taniguchi, T. Kai, K. Satoh and K. Fukumoto; *J. Chem. Soc., Perkin Trans. 1*, 1992, 221.	16
92JCS(P1)255	R. L. Baxter, D. J. Camp, A. Coutts and N. Shaw; *J. Chem. Soc., Perkin Trans. 1*, 1992, 255.	127
92JCS(P1)327	A. Barbero, P. Cuadrado, I. Fleming, A. M. Gonzalez and F. J. Pulido; *J. Chem. Soc., Perkin Trans. 1*, 1992, 327.	930
92JCS(P1)601	K. K. Park, C. W. Lee and S. Y. Choi; *J. Chem. Soc., Perkin Trans.* 1, 1992, 601.	398
92JCS(P1)1497	Y. Ichikawa, T. Kashiwagi and N. Urano; *J. Chem. Soc., Perkin Trans. 1*, 1992, 1497.	159
92JCS(P1)1721	N. Kamigata, Y. Nakamura, K. Kikuchi, I. Ikemoto, T. Shimizu and H. Malsuyama; *J. Chem. Soc., Perkin Trans. 1*, 1992, 1721.	997, 1001
92JCS(P1)2247	H. Fujihara, Y. Ueno, J. J. Chiu and N. Furukawa; *J. Chem. Soc., Perkin Trans. 1*, 1992, 2247.	369
92JCS(P1)2309	D. Landini, D. Albanese, S. Mottadelli and M. Penso; *J. Chem. Soc., Perkin Trans. 1*, 1992, 2309.	25, 31, 35
92JCS(P1)2371	B. Beagley, M. R. James, R. G. Pritchard, C. M. Rayner, C. Smith and R. J. Stoodley; *J. Chem. Soc., Perkin Trans. 1*, 1992, 2371.	197
92JCS(P1)3277	R. A. N. C. Crump, I. Fleming, J. H. M. Hill, D. Parker, N. L. Reddy and D. Waterston; *J. Chem. Soc., Perkin Trans. 1*, 1992, 3277.	546
92JCS(P1)3295	I. Fleming and J. Kilburn; *J. Chem. Soc., Perkin Trans. 1*, 1992, 3295.	546
92JCS(P1)3363	I. Fleming; *J. Chem. Soc., Perkin Trans. 1*, 1992, 3363.	546
92JGU787	E. T. Bogorodovskii; *J. Gen. Chem. USSR* (*Engl. Transl.*), 1992, **62**, 787.	1090
92JHC321	G. Adembri, A. Camparini, F. Ponticelli and M. Scotton; *J. Heterocycl. Chem.*, 1992, **29**, 321.	316
92JMOC307	B. Marciniec, C. Pietraszuk and Z. Foltynowicz; *J. Mol. Catal.*, 1992, **76**, 307.	938
92JOC46	J. M. Chong and E. K. Mar; *J. Org. Chem.*, 1992, **57**, 46.	83, 85

92JOC111	A. Ogawa, H. Tanaka, H. Yokoyama, R. Obayashi, K. Yokoyama and N. Sonoda; *J. Org. Chem.*, 1992, **57**, 111.	285
92JOC499	H. C. Brown, R. K. Dhar, K. Ganesan and B. Singaram; *J. Org. Chem.*, 1992, **57**, 499.	532
92JOC511	S. E. Cremer, J. M. Cowles, F. R. Farr, H. Hwang, P. W. Kremer and A. C. Peterson; *J. Org. Chem.*, 1992, **57**, 511.	440, 441
92JOC774	L.-J. Zhang, Y.-Z. Huang, H.-X. Jiang, J.-D. Mu and Y. Liao; *J. Org. Chem.*, 1992, **57**, 774.	487
92JOC794	Y. W. Andemichael, Y. G. Gu and K. K. Wang; *J. Org. Chem.*, 1992, **57**, 794.	904, 939
92JOC976	A. Arcadi, A. Burini, S. Cacchi, M. Delmastro, F. Marinelli, and B. R. Pietroni; *J. Org. Chem.*, 1992, **57**, 976.	983
92JOC1495	M. P. Cooke; *J. Org. Chem.*, 1992, **57**, 1495.	556
92JOC1504	H. Uno, S. Okada, T. Ono, Y. Shiraishi and H. Suzuki; *J. Org. Chem.*, 1992, **57**, 1504.	311
92JOC1585	G. A. Olah, M. B. Sassaman, M. Zuanic, C. B. Rao, G. K. S. Prakash, R. Gilardi, J. Flippen-Anderson and C. George, *J. Org. Chem.*, 1992, **57**, 1585.	384
92JOC1615	Y. Kataoka, K. Takai, K. Oshima and K. Utimoto; *J. Org. Chem.*, 1992, **57**, 1615.	927
92JOC1699	P. Dhar, N. Chidambara and S. Chandrasekaran; *J. Org. Chem.*, 1992, **57**, 1699.	228
92JOC1727	R. A. Bunce, C. J. Peeples and P. B. Jones; *J. Org. Chem.*, 1992, **57**, 1727.	323
92JOC1740	M. J. Totleben, D. P. Curran and P. Wipf; *J. Org. Chem.*, 1992, **57**, 1740.	587
92JOC1861	P. J. Stang, R. Tykwinski and V. V. Zhdankin; *J. Org. Chem.*, 1992, **57**, 1861.	445, 1021
92JOC1948	L. J. Wilson and D. C. Liotta; *J. Org. Chem.*, 1992, **57**, 1948.	688
92JOC1956	M. J. Rozema, A. Sidduri and P. Knochel; *J. Org. Chem.*, 1992, **57**, 1956.	69, 72, 591
92JOC2379	P. V. Ramachandran, A. V. Teodorovic, M. V. Rangaishnvi and H. C. Brown; *J. Org. Chem.*, 1992, **57**, 2379.	526
92JOC2716	H. C. Brown, R. K. Dhar, K. Ganesan and B. Singaram; *J. Org. Chem.*, 1992, **57**, 2716.	532
92JOC2726	P. Bravo, M. Pregnolato and G. Resnati; *J. Org. Chem.*, 1992, **57**, 2726.	350
92JOC2768	K. B. Sharpless, W. Amberg, Y. L. Bennani, G. A. Crispino, J. Hartung, K.-S. Jeong, H.-L. Kwong, K. Morikawa, Z.-M. Wang, D. Xu and X.-L. Zhang; *J. Org. Chem.*, 1992, **57**, 2768.	48
92JOC2850	E. W. Della and N. J. Head; *J. Org. Chem.*, 1992, **57**, 2850.	19
92JOC2953	M. M. Midland and A. Kazubski; *J. Org. Chem.*, 1992, **57**, 2953.	58
92JOC3178	P. Wipf, J. H. Smitrovich and C.-W. Moon; *J. Org. Chem.*, 1992, **57**, 3178.	926
92JOC3266	G. A. Molander and J. O. Hoberg; *J. Org. Chem.*, 1992, **57**, 3266.	969
92JOC3270	M. Lautens, P. H. M. Delanghe, J. B. Goh and C. H. Zhang; *J. Org. Chem.*, 1992, **57**, 3270.	931
92JOC3397	R. F. W. Jackson, N. Wishart, A. Wood, K. James and M. J. Wythes; *J. Org. Chem.*, 1992, **57**, 3397.	591
92JOC3482	C. E. Tucker, J. Davidson and P. Knochel; *J. Org. Chem.*, 1992, **57**, 3482.	900
92JOC3519	M. Mori, N. Uesaka and M. Shibasaki; *J. Org. Chem.*, 1992, **57**, 3519.	927, 988
92JOC3680	J. A. Cabral, T. Cohen, W. W. Doubleday, E. F. Duchelle, G. Franenkel, B. S. Guo and S. H. Yü; *J. Org. Chem.*, 1992, **57**, 3680.	553, 555
92JOC3751	S. C. Berk and S. L. Buchwald; *J. Org. Chem.*, 1992, **57**, 3751.	82
92JOC3767	H. C. Brown, K. Ganesan and R. K. Dhar; *J. Org. Chem.*, 1992, **57**, 3767.	698
92JOC3772	F. S. Guziec, Jr. and D. Wei; *J. Org. Chem.*, 1992, **57**, 3772.	12
92JOC3807	J. P. Bégué, D. Bonnet-Delpon, D. Mesureur, G. Née and S. W. Wu; *J. Org. Chem.*, 1992, **57**, 3807.	751
92JOC3994	B. Kopping, C. Chatgilialoglu, M. Zehnder and B. Giese; *J. Org. Chem.*, 1992, **57**, 3994.	913
92JOC4017	M.-Z. Deng, N.-S. Li and Y.-Z. Huang; *J. Org. Chem.*, 1992, **57**, 4017.	900
92JOC4521	E. Ciganek; *J. Org. Chem.*, 1992, **57**, 4521.	313, 316
92JOC4563	A. Alexakis, N. Lensen, J.-P. Tranchier and P. Mangeney; *J. Org. Chem.*, 1992, **57**, 4563.	381
92JOC4705	M. Bordeau, C. Biran, P. Pons, M.-P. Léger-Lambert and J. Dunoguès; *J. Org. Chem.*, 1992, **57**, 4705.	938, 941
92JOC5063	C. Burnell-Curty and E. J. Roskamp; *J. Org. Chem.*, 1992, **57**, 5063.	814
92JOC5067	T. Hashiyama, K. Morikawa and K. B. Sharpless; *J. Org. Chem.*, 1992, **57**, 5067.	49
92JOC5173	C. Gennari, C. T. Hewkin, F. Molinari, A. Bernardi, A. Comotti, J. M. Goodman and I. Paterson; *J. Org. Chem.*, 1992, **57**, 5173.	698
92JOC5279	S. McN. Sieburth and L. Fensterbank; *J. Org. Chem.*, 1992, **57**, 5279.	928
92JOC5425	C. Jubert and P. Knochel; *J. Org. Chem.*, 1992, **57**, 5425.	590
92JOC5431	C. Jubert and P. Knochel; *J. Org. Chem.*, 1992, **57**, 5431.	394, 396
92JOC5535	J. E. Celebuski, C. Chan and R. A. Jones; *J. Org. Chem.*, 1992, **57**, 5535.	410
92JOC5610	S. Yamazaki, H. Fujitsuka and S. Yamabe; *J. Org. Chem.*, 1992, **57**, 5610.	931, 939
92JOC5700	R. V. Hoffman, N. K. Nayyar and W. Chen; *J. Org. Chem.*, 1992, **57**, 5700.	23
92JOC5768	D. A. Singleton, J. P. Martinez and G. M. Ndip; *J. Org. Chem.*, 1992, **57**, 5768.	904
92JOC5831	A. Solladie-Cavallo and M. Benchegroun; *J. Org. Chem.*, 1992, **57**, 5831.	318
92JOC6107	T. H. Chan and K. T. Nwe; *J. Org. Chem.*, 1992, **57**, 6107.	85
92JOC6384	C. Jubert, S. Nowotny, D. Kornemann, I. Antes, C. E. Tucker and P. Knochel; *J. Org. Chem.*, 1992, **57**, 6384.	73
92JOC6386	A. Yanagisawa, H. Hibino, S. Habaue, Y. Hisada and H. Yamamoto; *J. Org. Chem.*, 1992, **57**, 6386.	563
92JOC6608	H. C. Brown, U. S. Racherla, Y. Liao and V. V. Khanna; *J. Org. Chem.*, 1992, **57**, 6608.	531
92JOC6759	J. K. Crandall and T. Reix; *J. Org. Chem.*, 1992, **57**, 6759.	389, 390
92JOC6789	I. Fernández, N. Khiar, J. M. Llera and F. Alcudia; *J. Org. Chem.*, 1992, **57**, 6789.	152
92JOC7003	Y. Yamamoto, K. Kobayashi, H. Okano and I. Kadota; *J. Org. Chem.*, 1992, **57**, 7003.	79
92JOC7010	Y.-M. Tsai, H.-C. Nieh and C.-D. Cherng; *J. Org. Chem.*, 1992, **57**, 7010.	936
92JOC7194	C. J. Kowalski and R. E. Reddy; *J. Org. Chem.*, 1992, **57**, 7194.	1023

92JOC7274 F. A. Davis, M. C. Weissmiller, C. K. Murphy, R. T. Reddy and S.-C. Chen; *J. Org. Chem.*, 1992, **57**, 7274. 150

92JOC7342 S. Rozen and M. Kol; *J. Org. Chem.*, 1992, **57**, 7342. 397

92JOC7372 E. J. Corey and C. L. Cywin; *J. Org. Chem.*, 1992, **57**, 7372. 528

92JOM(434)151 A. Alberti and M. Menaglia; *J. Organomet. Chem.*, 1992, **434**, 151. 238

92JOM(439)C25 J. A. M. Andersen and J. R. Moss; *J. Organomet. Chem.*, 1992, **439**, C25. 573

92JOM(439)C33 D. A. Atwood, A. H. Cowley, R. A. Jones and M. A. Mardones; *J. Organomet. Chem.*, 1992, **439**, C33. 895

92JOM(441)389 H. Lang and U. Lay; *J. Organomet. Chem.*, 1992, **441**, 389. 875, 896

92JOM(442)1 L. D. Freedman and G. O. Doak; *J. Organomet. Chem.*, 1992, **442**, 1. 872

92JOM(442)61 L. D. Freedman and G. O. Doak; *J. Organomet. Chem.*, 1992, **442**, 61. 872

92LA273 P. Gmeiner and B. Bollinger; *Liebigs Ann. Chem.*, 1992, 273. 310

92LA559 L. Hinkamp, H. J. Schafer, B. Wippich and H. Luftmann; *Liebigs Ann. Chem.*, 1992, 559. 21

92LA835 R. Koster and R. Kucznierz; *Liebigs Ann. Chem.*, 1992, 835. 132, 270

92LA1005 S. Shatzmiller and S. Bercovici; *Liebigs Ann. Chem.*, 1992, 1005. 314

92LA1091 H. Ripperger; *Liebigs Ann. Chem.*, 1992, 1091. 343

92MI 202-01 S. M. Gryaznov and R. L. Letsinger; *Nucleic Acid Res.*, 1992, **20**, 1879. 106

B-92MI 203-01 J. March; "Advanced Organic Chemistry," Wiley, New York, 4th ed., 1992, 407. 132

92MI 203-02 A. G. Kutateladze, E. K. Beloglazkina, N. V. Zyk and N. S. Zefirov; *Izv. Ross. Akad. Nauk, Ser. Khim.*, 1992, 1217 (English translation: *Bulletin of the Russian Academy of Sciences, Division of Chemical Science*, 1992, **41**, 960). 232

B-92MI 210-01 B. Marciniec (ed.); "Comprehensive Handbook on Hydrosilylation," Pergamon, Oxford, 1992. 536

B-92MI 214-01 J. March; "Advanced Organic Chemistry," Wiley, New York, 4th edn., 1992. 722

B-92MI 216-01 G. Davidson; in "The Chemistry of Organophosphorus Compounds," ed. F. R. Hartley, Wiley, Chichester, 1992, ch. 5. 856

B-92MI 218-01 A. Sekiguchi and H. Sakurai; in "Studies in Inorganic Chemistry, The Chemistry of Inorganic Ring Systems," ed. R. Steudel, Elsevier, New York, 1992, vol. 14, p. 101. 949

92MI 221-01 P. J. Stang, V. V. Zhdankin and A. M. Arif; *Mendeleev Commun.*, 1992, 158. 1021

92OM255 M. A. Huffman and L. S. Liebeskind; *Organometallics*, 1992, **11**, 255. 672

92OM967 R. W. Gedridge, Jr.; *Organometallics*, 1992, **11**, 967. 484

92OM1095 M. R. Kesti and R. M. Waymouth; *Organometallics*, 1992, **11**, 1095. 912

92OM2163 D. G. Hendershot, J. C. Pazik, C. George and A. D. Berry; *Organometallics*, 1992, **11**, 2163. 483

92OM3993 X. Li, C. M. Lukehart and L. Han; *Organometallics*, 1992, **11**, 3993. 1076, 1080

92OM4096 A. J. Pearson, R. J. Shively, Jr., and R. A. Dubbert; *Organometallics*, 1992, **11**, 4096. 985

92OPP57 A. M. Modro and T. A. Modro; *Org. Prep. Proceed. Int.*, 1992, **24**, 57. 106

92OPP209 S. Eguchi, Y. Matushita and Y. Yamashita; *Org. Prep. Proced. Int.*, 1992, **24**, 209. 468

92OPP369 R. Martin; *Org. Prep. Proced. Int.*, 1992, **24**, 369. 657

92OPP553 P. F. Cirillo and J. S. Panek; *Org. Prep. Proced. Int.*, 1992, **24**, 553. 936

92OR(41)135 B. H. Lipshutz and S. Sengupta; *Org. React.*, 1992, **41**, 135. 585, 702, 910, 918, 919, 924, 931

92OR(42)1 P. W. Rabideau and Z. Marcinow; *Org. React.*, 1992, **42**, 1. 682, 937

92OR(42)335 D. L. Hughes; *Org. React.*, 1992, **42**, 335. 6

92OS(70)195 M. C. P. Yeh, H. G. Chen and P. Knochel; *Org. Synth.*, 1992, **70**, 195. 590

92OS(71)133 R. L. Danheiser, K. R. Romines, H. Koyama, S. K. Gee, C. R. Johnson and J. R. Medich; *Org. Synth.*, 1992, **71**, 133. 420

92OS(71)146 R. E. Reddy and C. J. Kowalski; *Org. Synth.*, 1992, **71**, 146. 420

92PAC323 E. Negishi; *Pure Appl. Chem.*, 1992, **64**, 323. 980

92PAC439 A. Ricci, G. Reginato, A. Degl'Innocenti and G. Seconi; *Pure Appl. Chem.*, 1992, **64**, 439. 918

92PAC1857 J.-F. Normant, I. Marek and J.-M. Lefrançois; *Pure Appl. Chem.*, 1992, **64**, 1857. 594

92POL2351 S.-G. Teoh, S.-B. Teo, G.-Y. Yeap and J.-P. Declercq; *Polyhedron*, 1992, **11**, 2351. 1008

92PS(67)155 L. Hevesi; *Phosphorus Sulfur*, 1992, **67**, 155. 1008

92PS(68)147 J. D. Robbins, D. J. Bowler and R. D. Gless; *Phosphorus Sulfur*, 1992, **68**, 147. 252

92PS(73)49 J. M. Ndong-Mebah, J. L. Mieloszynski and D. Paquer; *Phosphorus Sulfur*, 1992, **73**, 49. 255

92RCR1091 M. D. Stadnichuk and T. I. Voropeva; *Russ. Chem. Rev.*, 1992, **61**, 1091. 1089

92RTC170 J. A. Van Doorn and N. Meijboom; *Recl. Trav. Chim. Pays-Bas*, 1992, **111**, 170. 432

92S90 S. Laschat and H. Kunz; *Synthesis*, 1992, 90. 459

92S112 F. Beaulieu and V. Snieckus; *Synthesis*, 1992, 112. 707

92S157 A. Marinetti, L. Ricard and F. Mathey; *Synthesis*, 1992, 157. 438

92S191 B. H. Lipshutz, S. Sharma, S. H. Dimock and J. R. Behlug; *Synthesis*, 1992, 191. 555

92S281 J. Li, P. Lue and X.-J. Zhou; *Synthesis*, 1992, 281. 730

92S333 T. Minami and J. Motoyoshiya; *Synthesis*, 1992, 333. 478

92S355 R. Ballini, M. Petrini and V. Polzonetti; *Synthesis*, 1992, 355. 396

92S495 A. Löffler and G. Himbert; *Synthesis*, 1992, 495. 1026

92S549 H. Suzuki and T. Nakamura; *Synthesis*, 1992, 549. 730, 731, 735

92S577 E. Anders, K. Hertlein, A. Stankowiak and E. Irmer; *Synthesis*, 1992, 577. 446

92S582 C. Thom and P. Kocienski; *Synthesis*, 1992, 582. 251

92S803 T. N. Mitchell; *Synthesis*, 1992, 803. 1092

92S825 K. V. Srinivasan, S. M. Kumar and N. R. Ayyangar; *Synthesis*, 1992, 825. 716

92S933 A. Krief, M. Trabelsi and W. Dumont; *Synthesis*, 1992, 933. 279

92S954 F. Denat, H. Gaspard-Iloughmane and J. Dubac; *Synthesis*, 1992, 954. 942, 949

92S1018	H. Ahlbrecht and P. Weber; *Synthesis*, 1992, 1018.	751
92S1035	B. B. Lohray; *Synthesis*, 1992, 1035.	100
92S1063	A. Koziara and A. Zwierzak; *Synthesis*, 1992, 1063.	308, 407
92S1119	M. Node, A. Itoh, K. Nishide, H. Abe, T. Kawabata, Y. Masaki and K. Fuji; *Synthesis*, 1992, 1119.	784
92S1203	D.-W. Kim, Y. K. Ko and S. H. Kim; *Synthesis*, 1992, 1203.	179
92SC73	D. J. Milner; *Synth. Commun.*, 1992, **22**, 73.	615
92SC677	L. Castedo, C. F. Marcos, M. Monteagudo and G. Tojo; *Synth. Commun.*, 1992, **22**, 677.	103
92SC1249	A. Roglans, J. Marquet and M. Moreno-Mañas; *Synth. Commun.*, 1992, **22**, 1249.	324
92SC1461	M. A. Tius and J. K. Kawakami; *Synth. Commun.*, 1992, **22**, 1461.	611
92SC1555	S. A. Kulkarni and M. V. Paradkar; *Synth. Commun.*, 1992, **22**, 1555.	654
92SC1799	H. E. Folsom and J. Castrillon; *Synth. Commun.*, 1992, **22**, 1799.	148
92SC1923	F. Rahaingoson, B. Kimpioki-Ninafiding, Z. Mouloungi and A. Gaset; *Synth. Commun.*, 1992, **22**, 1923.	28
92SC1929	T. Gajda, A. Koziara, K. Osowska-Pacewicka, S. Zawadzki and A. Zwierkak; *Synth. Commun.*, 1992, **22**, 1929.	308
92SC1997	Z.-D. Liu and Z.-C. Chen; *Synth. Commun.*, 1992, **22**, 1997.	1036
92SC2187	J.-M. Mas and P. Metivier; *Synth. Commun.*, 1992, **22**, 2187.	31
92SC2305	R. Caputo, C. Ferreri, D. Mastroianni and G. Palumbo; *Synth. Commun.*, 1992, **22**, 2305.	302
92SC2607	A. S. Demir, A. S. Mahasneh, H. Aksoy and Z. Gercek; *Synth. Commun.*, 1992, **22**, 2607.	385
92SC2945	J. Sandri and J. Viala; *Synth. Commun.*, 1992, **22**, 2945.	7
92SC3277	A. R. Ramesha and S. Chandrasekaran; *Synth. Commun.*, 1992, **22**, 3277.	228
92SL112	F. Beaulieu and V. Snieckus; *Synlett*, 1992, 112.	942
92SL129	K. Tomooka, K. Matsumoto, K. Suzuki and G.-I. Tsuchihashi; *Synlett*, 1992, 129.	931
92SL143	N. Coles, R. J. Whitby and J. Blagg; *Synlett*, 1992, 143.	927
92SL189	N. Kishi, H. Imma, K. Makami and T. Nakai; *Synlett*, 1992, 189.	932
92SL303	M. Chini, P. Crotti, E. Giovani, F. Macchia and M. Pineschi; *Synlett*, 1992, 303.	135
92SL317	A. Krief and M. Hobe; *Synlett*, 1992, 317.	281
92SL320	A. Krief, J. Bousbaa and M. Hobe; *Synlett*, 1992, 320.	281
92SL327	K. I. Booker-Milburn; *Synlett*, 1992, 327.	942
92SL337	G. A. Olah, P. Ramaiah, C.-S. Lee and G. K. S. Prakash; *Synlett*, 1992, 337.	397
92SL343	K. Rück and H. Kunz; *Synlett*, 1992, 343.	699
92SL360	S. Takagishi, G. Katsoulos and M. Schlosser; *Synlett*, 1992, 360.	942
92SL497	M. P. Sibi and R. Sharma; *Synlett*, 1992, 497.	59
92SL502	T. Sakamoto, A. Yasuhara, Y. Kondo and H. Yamanaka; *Synlett*, 1992, 502.	1027
92SL535	D. R. Fisher, B. L. Williamson and P. J. Stang; *Synlett*, 1992, 535.	1033
92SL539	K. Tamao; K. Kobayashi and Y. Ito; *Synlett*, 1992, 539.	926, 985
92SL581	N. Guennouni, C. Rasset-Deloge, B. Carboni and M. Vaultier; *Synlett*, 1992, 581.	182
92SL633	I. Marek, J.-M. Lefrancois and J. F. Normant; *Synlett*, 1992, 633.	1006
92SL665	N. A. Petasis and I. Akritopoulou; *Synlett*, 1992, 665.	937
92SL681	C. M. Lukehart; *Synlett*, 1992, 681.	927
92SL718	P. Wipf and W. Xu; *Synlett*, 1992, 718.	587
92SL825	M. C. Clasby and D. Craig; *Synlett*, 1992, 825.	1034, 1035
92SL833	J. Ichikawa, S. Yonemaru and T. Minami; *Synlett*, 1992, 833.	825, 853
92SL843	J. Yoshida, Y. Morita, M. Itoh, Y. Ishichi and S. Isoe; *Synlett*, 1992, 843.	950
92SL849	F. Sato and Y. Kobayashi; *Synlett*, 1992, 849.	923, 931
92SL883	A. Degl'Innocenti, P. Ulivi, A. Capperucci, G. Reginato, A. Mordini and A. Ricci; *Synlett*, 1992, 883.	935
92SL886	A. Pimm, P. Kocienski and S. D. A. Street; *Synlett*, 1992, 886.	915
92SL943	D. P. Curran, T. L. Fevig, C. P. Jasperse and M. J. Totleben; *Synlett*, 1992, 943.	565
92SL981	A. Casarini, B. Jousseaume, D. Lazzari, E. Porciatti, G. Reginato, A. Ricci and G. Seconi; *Synlett*, 1992, 981.	918, 940
92SL985	P. Metz, M. Fleischer and R. Frohlich; *Synlett.*, 1992, 985.	207
92SR1	B. A. Trofimov, L. V. Sokolyanskaya and A. Senning; *Sulfur Rep.*, 1992, **12**, 1.	722
92SRI461	C. J. Thomas, L. K. Krannich and C. L. Watkins; *Synth. React. Inorg. Metal-Org. Chem.*, 1992, **22**, 461.	484
92T1327	B. C. Ranu and S. Bhar; *Tetrahedron*, 1992, **48**, 1327.	395
92T1353	Y. G. Gololobov and L. F. Kagukhin; *Tetrahedron*, 1992, **48**, 1353.	468
92T1883	E. Balavoine, D. H. R. Barton, A. Gref and I. Lellouche; *Tetrahedron*, 1992, **48**, 1883.	212, 225
92T1981	W. R. Roush and P. T. Glover; *Tetrahedron*, 1992, **48**, 1981.	529
92T2127	D. A. Evans, N. P. Ng, J. S. Clark and D. L. Rieger; *Tetrahedron*, 1992, **48**, 2127.	532
92T2911	S. D. Pastor, R. Ravichandran and R. Mewwly; *Tetrahedron*, 1992, **48**, 2911.	457
92T3571	C. M. Rodriguez, J. M. Ode, J. M. Palazón and V. S. Martin; *Tetrahedron*, 1992, **48**, 3571.	155
92T3585	D. J. Ramon and M. Yus; *Tetrahedron*, 1992, **48**, 3585.	553
92T3877	C. Bied, J. Collin and H. B. Kagan; *Tetrahedron*, 1992, **48**, 3877.	565
92T4579	F. Bellesia, M. Boni, F. Ghelfi, R. Grandi, U. M. Pagnoni and A. Pinetti; *Tetrahedron*, 1992, **48**, 4579.	22
92T4593	D. Guyaro, B. Mancheno and M. Yus; *Tetrahedron*, 1992, **48**, 4593.	553
92T4623	A. S. B. Prasad, J. V. B. Kanth and M. Periasamy; *Tetrahedron*, 1992, **48**, 4623.	316, 325
92T5317	B. C. Ranu and R. Chakraborty; *Tetrahedron*, 1992, **48**, 5317.	398
92T5691	H. Takahashi, T. Kawakita, M. Ohno, M. Yoshioka and S. Kobayashi; *Tetrahedron*, 1992, **48**, 5691.	568
92T5709	M. Arai, B. H. Lipshutz and E. Nakamura; *Tetrahedron*, 1992, **48**, 5709.	587

92T5719 J. L. Von dem Bussche-Hünnfeld and D. Seebach; *Tetrahedron*, 1992, **48**, 5719. 568

92T5933 P. C. B. Page, S. S. Klair, M. P. Brown, C. S. Smith, S. J. Maginn and S. Mulley; *Tetrahedron*, 1992, **48**, 5933. 134

92T6087 J. van der Louw, J. L. van der Baan, F. J. J. de Kanter, F. Bickelhaupt and G. W. Klumpp; *Tetrahedron*, 1992, **48**, 6087. 925

92T6105 J. van der Louw, J. L. van der Baan, C. M. D. Komen, A. Knol, F. J. J. de Kanter, F. Bickelhaupt and G. W. Klumpp; *Tetrahedron*, 1992, **48**, 6105. 925

92T6929 R. Grigg, J. Markandu, S. Surendrakumar and W. J. Warnock; *Tetrahedron*, 1992, **48**, 6929. 352

92T8329 C. K. Reddy and M. Periasamy; *Tetrahedron*, 1992, **48**, 8329. 14

92T9111 K. Afarinkia, V. Vinader, T. D. Nelson and G. H. Posner; *Tetrahedron*, 1992, **48**, 9111. 665

92T9195 D. H. R. Barton, E. Csuhai and D. Doller; *Tetrahedron*, 1992, **48**, 9195. 2

92T9577 E. Erdick; *Tetrahedron*, 1992, **48**, 9577. 1095

92TA95 S. Colonna, N. Gaggero, L. Casella, G. Carrea and P. Pasta; *Tetrahedron Asymmetry*, 1992, **3**, 95. 151

92TA227 T. Kawate, M. Nakagawa, T. Kakikawa and T. Hino; *Tetrahedron Asymmetry*, 1992, **3**, 227. 330

92TA377 T. Yokomatsu and S. Shibuya; *Tetrahedron Asymmetry*, 1992, **3**, 377. 459

92TA629 C. Rossi, A. Fauve, M. Madesclaire, D. Roche, F. A. Davis and R. T. Reddy; *Tetrahedron Asymmetry*, 1992, **3**, 629. 713

92TA841 E. Brown, A. Leze and J. Touet; *Tetrahedron Asymmetry*, 1992, **3**, 841. 64

92TA921 V. Nevaleinen; *Tetrahedron Asymmetry*, 1992, **3**, 921. 66

92TA933 V. Nevaleinen; *Tetrahedron Asymmetry*, 1992, **3**, 933. 66

92TA961 A. J. Walker; *Tetrahedron Asymmetry*, 1992, **3**, 961. 153

92TA1063 G. Carrea, B. Redigolo, S. Riva, S. Colonna, N. Gaggero, E. Battistel and D. Bianchi; *Tetrahedron Asymmetry*, 1992, **3**, 1063. 151

92TA1283 P. Krasik and H. Alper; *Tetrahedron Asymmetry*, 1992, **3**, 1283. 330

92TA1317 B. B. Lohray; *Tetrahedron Asymmetry*, 1992, **3**, 1317. 47

92TA1361 A. Kamal, Y. Damayanthi and M. V. Rao; *Tetrahedron Asymmetry*, 1992, **3**, 1361. 318

92TA1441 V. Nevaleinen; *Tetrahedron Asymmetry*, 1992, **3**, 1441. 66

92TA1475 J. Martens and S. Wallbaum; *Tetrahedron Asymmetry*, 1992, **3**, 1475. 64, 528

92TA1563 V. Nevaleinen; *Tetrahedron Asymmetry*, 1992, **3**, 1563. 66

92TL85 Y. Tominaga, H. Ueda, K. Ogata, S. Kohra and M. Hojo; *Tetrahedron Lett.*, 1992, 85. 140

92TL131 J.-E. Bäckvall, A. M. Ericsson, N. A. Plobeck and S. K. Juntunen; *Tetrahedron Lett.*, 1992, **33**, 131. 395

92TL213 K. Nohair, I. Lachaise, J.-P. Paugam and J.-Y. Nedelec; *Tetrahedron Lett.*, 1992, **33**, 213. 14

92TL235 K. Higashiyama, H. Inoue and H. Takahashi; *Tetrahedron Lett.*, 1992, **33**, 235. 313

92TL251 M. Nitta, H. Kawaji and N. Kanomata; *Tetrahedron Lett.*, 1992, **33**, 251. 468

92TL337 J. Ichikawa, S. Hamada, T. Sonoda and H. Kobayashi; *Tetrahedron Lett.*, 1992, **33**, 337. 906

92TL451 N. Jayasuriya and S. L. Regen; *Tetrahedron Lett.*, 1992, **33**, 451. 219

92TL495 G. Mignani, J. P. Grass, P. Chabardes and D. Morel; *Tetrahedron Lett.*, 1992, **33**, 495. 21

92TL627 S. Itsuno, C. Hachisuka, K. Kitano and K. Ito; *Tetrahedron Lett.*, 1992, **33**, 627. 316

92TL683 P. Vinczer, S. Struhar, L. Novak and C. Szantay; *Tetrahedron Lett.*, 1992, **33**, 683. 1015

92TL865 G. W. Kabalka, N. K. Reddy and C. Narayana; *Tetrahedron Lett.*, 1992, **33**, 865. 655

92TL935 P. Beslin and P. Marion; *Tetrahedron Lett.*, 1992, **33**, 935. 375

92TL947 T. Masuda, Y. Masuda and N. Nakatani; *Tetrahedron Lett.*, 1992, 947. 138

92TL1017 D. A. Singleton, J. P. Martinez and J. V. Watson; *Tetrahedron Lett.*, 1992, **33**, 1017. 43

92TL1057 J. Hatem, C. Henriet-Bernard, J. Grimaldi and R. Maurin; *Tetrahedron Lett.*, 1992, **33**, 1057. 351

92TL1153 F. A. Davis and W. Han; *Tetrahedron Lett.*, 1992, **33**, 1153. 16

92TL1299 O. Kitagawa, T. Hanano, T. Hirata, T. Inoue and T. Taguchi; *Tetrahedron Lett.*, 1992, **33**, 1299. 34

92TL1313 Y. Watanabe, Y. Komoda and S. Ozaki; *Tetrahedron Lett.*, 1992, **33**, 1313. 106

92TL1419 P. J. Stang, V. V. Zhdankin, R. Tykwinski and N. S. Zefirov; *Tetrahedron Lett.*, 1992, **33**, 1419. 1021

92TL1435 C.-B. Xue, J. M. Becker and F. Naider; *Tetrahedron Lett.*, 1992, **33**, 1435. 133

92TL1605 G. Solladi and N. Ghiatou; *Tetrahedron Lett.*, 1992, **33**, 1605. 61

92TL1689 M. A. Schwartz and X. Hu; *Tetrahedron Lett.*, 1992, **33**, 1689. 345

92TL2099 M. J. Burk and J. E. Feaster; *Tetrahedron Lett.*, 1992, **33**, 2099. 62

92TL2129 M. D. Ryan and D. N. Harpp; *Tetrahedron Lett.*, 1992, **33**, 2129. 236

92TL2357 H. C. P. F. Roelen, H. van den Elst, C. E. Dreet, G. A. van den Marel and J. H. van Boom; *Tetrahedron*, 1992, **33**, 2357. 467

92TL2517 S. Satoh and R. Sato; *Tetrahedron Lett.*, 1992, **33**, 2517. 234

92TL2657 J. Huang and T. S. Widlanski; *Tetrahedron Lett.*, 1992, **33**, 2657. 182

92TL3129 M. Benazza, R. Uzan, D. Beaupere and G. Demailly; *Tetrahedron Lett.*, 1992, **33**, 3129. 25

92TL3167 J. G. Lee and H. T. Cha; *Tetrahedron Lett.*, 1992, 3167. 621, 627, 632

92TL3381 A. Krief, M. Hobe, W. Dumont, E. Badoni, E. Guittet and G. Evrard; *Tetrahedron Lett.*, 1992, **33**, 3381. 554

92TL3413 D. H. R. Barton, E. Csuhai and D. Doller; *Tetrahedron Lett.*, 1992, **33**, 3413. 2

92TL3421 H. C. Brown and K. Ganesan; *Tetrahedron Lett.*, 1992, **34**, 3421. 532

92TL3583 S. F. Martin, M. Hartmann and J. A. Josey; *Tetrahedron Lett.*, 1992, **33**, 3583. 352

92TL3599 C.-G. Cho and G. H. Posner; *Tetrahedron Lett.*, 1992, **33**, 3599. 307

92TL3677 J. A. Vega, A. Molina, R. Alajarin, J. J. Vaquero, J. L. García Navío and J. Alvarez-Builla; *Tetrahedron Lett.*, 1992, **33**, 3677. 249

92TL3779 J. Ichikawa, T. Minami, T. Sonoda and H. Kabayashi; *Tetrahedron Lett.*, 1992, **33**, 3779. 906

92TL3973 A. L. Schwan and R. Dufault; *Tetrahedron Lett.*, 1992, **33**, 3973. 174

92TL4021 K. Fuji, K. Tanaka and H. Miyamoto; *Tetrahedron Lett.*, 1992, **33**, 4021. 50

92TL4099 E. J. Corey and K. A. Cimprich; *Tetrahedron Lett.*, 1992, **33**, 4099. 119, 124

92TL4141 E. J. Corey and J. O. Link; *Tetrahedron Lett.*, 1992, **33**, 4141. 66

92TL4173 M. Kuroboshi, K. Suzuki and T. Hiyama; *Tetrahedron Lett.*, 1992, **33**, 4173. 90

92TL4209 S. Elgendy, J. Deadman, G. Patel, D. Greer, N. Chiro, C. A. Goodwin, M. F. Scully, V. V. Kakkar and G. Claeson; *Tetrahedron Lett.*, 1992, **33**, 4209. 530

92TL4233 I. Paterson and R. D. Tillyer; *Tetrahedron Lett.*, 1992, **33**, 4233. 701

92TL4469 H. Ito, T. Taguchi and Y. Hanzawa; *Tetrahedron Lett.*, 1992, **33**, 4469. 70

92TL4533 G. B. Fisher, J. Harrison, J. C. Fuller, C. T. Goralski and B. Singaram; *Tetrahedron Lett.*, 1992, **33**, 4533. 60

92TL4603 K. Takahashi, T. Suzuki and M. Hirama; *Tetrahedron Lett.*, 1992, **33**, 4603. 659

92TL4629 B. Hermans, N. Colard and L. Hevesi; *Tetrahedron Lett.*, 1992, **33**, 4629. 286

92TL4637 F. J. Sardina, M. M. Paz, E. Fernandez-Megia, R. D. de Boer and M. P. Alvarez; *Tetrahedron Lett.*, 1992, **33**, 4637. 44

92TL4835 R. Ballini, E. Mercantoni and M. Petrini; *Tetrahedron Lett.*, 1992, **33**, 4835. 397

92TL4901 M. Benazza, R. Uzan, D. Beaupere and G. Demailly; *Tetrahedron Lett.*, 1992, **33**, 4901. 25

92TL5041 D. H. R. Barton and M. J. Kelly; *Tetrahedron Lett.*, 1992, **33**, 5041. 539

92TL5045 J. Cossy, V. Bellosta and M. C. Müller; *Tetrahedron Lett.*, 1992, **33**, 5045. 57

92TL5339 P. Beslin and P. Marion; *Tetrahedron Lett.*, 1992, **33**, 5339. 313

92TL5597 D. Guyaro, B. Mancheno and M. Yus; *Tetrahedron Lett.*, 1992, **33**, 5597. 553

92TL5857 L. M. Venanzi, R. Lehmann, R. Keil and B. H. Lipshutz; *Tetrahedron Lett.*, 1992, **33**, 5857. 588

92TL5965 H. Maeta, T. Hashimoto, T. Hasegawa and K. Suzuki; *Tetrahedron Lett.*, 1992, **33**, 5965. 70

92TL6295 J.-P. Leblanc and H. W. Gibson; *Tetrahedron Lett.*, 1992, **33**, 6295. 310

92TL6407 Z.-M. Wang, X.-L. Zhang, K. B. Sharpless, S. C. Sinha, A. Sinha-Bagchi and E. Keinan; *Tetrahedron Lett.*, 1992, **33**, 6407. 49

92TL6411 E. Keinan, S. C. Sinha, A. Sinha-Bagchi, Z.-M. Wang, X.-H. Zhang and K. B. Sharpless; *Tetrahedron Lett.*, 1992, **33**, 6411. 49

92TL6435 E. J. Corey and W. C. Shieh; *Tetrahedron Lett.*, 1992, **33**, 6435. 563

92TL6503 S. Zhu; *Tetrahedron Lett.*, 1992, **33**, 6503. 161

92TL6575 D. E. Stack and R. L. Rieke; *Tetrahedron Lett.*, 1992, **33**, 6575. 74

92TL6629 D. H. R. Barton, D. O. Jang and J. Cs. Jaszberenyi; *Tetrahedron Lett.*, 1992, **33**, 6629. 536

92TL6907 E. J. Corey, C. L. Cywin and T. D. Roper; *Tetrahedron Lett.*, 1992, **33**, 6907. 528

92TL6915 C. L. Anderson, J. A. Soderquist, and G. W. Kabalka; *Tetrahedron Lett.*, 1992, **33**, 6915. 936

92TL6931 D. P. Curran and B. Yoo; *Tetrahedron Lett.*, 1992, **33**, 6931. 541

92TL7185 Y. Oozumi, S.-Y. Lee and T. Hayashi; *Tetrahedron Lett.*, 1992, **33**, 7185. 537

92TL7237 C. M. Rayner, M. S. Sin and A. D. Westwell; *Tetrahedron Lett.*, 1992, **33**, 7237. 147

92TL7331 K. Banert and M. Hagedorn; *Tetrahedron Lett.*, 1992, **33**, 7331. 770

92TL7507 P. Nussbaumer and A. Stütz; *Tetrahedron Lett.*, 1992, **33**, 7507. 83

92TL7557 S.-I. Murahashi, Y. Oda and T. Naota; *Tetrahedron Lett.*, 1992, **33**, 7557. 654

92TL7823 S. Tang and R. M. Kennedy; *Tetrahedron Lett.*, 1992, **33**, 7823. 44

92USP5083178 R. E. Banks; *US Pat.* 5 083 178 (1992). 336

92ZAAC(607)161 J. Haenel, G. Ohms and L. Riesel; *Z. Anorg. Allg. Chem.*, 1992, **607**, 161. 471

92ZAAC(608)43 J. Grob, S. Gobelbecker and D. Syndikus; *Z. Anorg. Allg. Chem.*, 1992, **608**, 43. 473, 508

92ZAAC(611)47 M. Baudler, W. Oehlert, A. Kmieciak and A. Floruss; *Z. Anorg. Allg. Chem.*, 1992, **611**, 47. 472

92ZN(B)155 W. Wolfsberger; *Z. Naturforsch., Teil B*, 1992, **47**, 155. 433

92ZN(B)640 H. Schmidbaur, A. Stützer and P. Bissinger; *Z. Naturforsch., Teil B*, 1992, **47**, 640. 437

92ZN(B)693 E. Lindner and A. Möckel; *Z. Naturforsch., Teil B*, 1992, **47**, 693. 437

92ZOR927 S. V. Amosova, V. V. Nosyreva and M. V. Sigalov; *Zh. Org. Khim.*, 1992, **28**, 927 (English translation: *Journal of Organic Chemistry of the USSR*, 1992, **28**, 712). 225

93AG(E)164 A. Fürstner; *Angew. Chem., Int. Ed. Engl.*, 1993, **32**, 164. 958, 959

93AG(E)1032 G. Boche, M. Marsch, A. Müller and K. Harms; *Angew. Chem., Int. Ed. Engl.*, 1993, **32**, 1032. 997, 998

93AG(E)1068 P. Wipf and S. Lim; *Angew. Chem., Int. Ed. Engl.*, 1993, **32**, 1068. 975

93AG(E)1473 M. Murakami, M. Suginome K. Fujimoto and Y. Ito; *Angew. Chem., Int. Ed. Engl.*, 1993, **32**, 1473. 923

93AG(E)1501 E. Weiss; *Angew. Chem., Int. Ed. Engl.*, 1993, **32**, 1501. 551

93AG(E)1746 H. J. Bestmann, T. G. Fürst and A. Schier; *Angew. Chem., Int. Ed. Engl.*, 1993, **32**, 1746. 1004

93AHC(56)155 G. W. Rewcastle and A. R. Katritzky; *Adv. Heterocycl. Chem.*, 1993, **56**, 155. 1002

93BCJ987 Y. Wakatsuki, H. Yamazaki, N. Kumegawa and P. S. Johar; *Bull. Chem. Soc. Jpn.*, 1993, **66**, 987. 940

93BCJ1156 K.-I. Sato, N. Kubo, R. Takada and S. Sakuma; *Bull. Chem. Soc. Jpn.*, 1993, **66**, 1156. 677

93BCJ1297 S. Murata, K. Teramoto, M. Miura and M. Nomura; *Bull. Chem. Soc. Jpn.*, 1993, **66**, 1297. 305

93BCJ2348 K. Miura, K. Oshima and K. Utimoto; *Bull. Chem. Soc. Jpn.*, 1993, **66**, 2348. 913

93BCJ2356 K. Miura, K. Oshima and K. Utimoto; *Bull. Chem. Soc. Jpn.*, 1993, **66**, 2356. 913

93BMC681 A. Mouna, J. Ouazzani and M. Lenfant; *Bio. Med. Chem. Lett.*, 1993, **3**, 681. 331

93BMC2883	K. Shinozuka, Y. Seto, H. Kawata and H. Sawai; *Bio. Med. Chem. Lett.*, 1993, **3**, 2883.	309
93BSF521	M. L. Sierra, C. Charrier, L. Ricard and F. Mathey; *Bull. Soc. Chim. Fr.*, 1993, 521.	894
93BSF664	A. Hedhli, A. Baklouti, S. Szönyi and A. Cambon; *Bull. Soc. Chim. Fr.*, 1993, 664.	675
93CB45	W. A. Herrmann, F. F. Kuehn, C. C. Romao, H. T. Hoa, M. Wang, R. W. Fischer, P. Kiprof and W. Scherer; *Chem. Ber.*, 1993, **126**, 45.	585
93CB1873	G. Boch, F. Bosold, J. W. C. Lohrenz, A. Opel and P. Zulanf; *Chem. Ber.*, 1993, **126**, 1873.	999
93CB1887	G. Boche, M. Marsch, J. Harbach, K. Harms, B. Ledig, F. Schubert, J. C. W. Lohrenz and H. Ahlbrecht; *Chem. Ber.*, 1993, **126**, 1887.	1001, 1002
93CB2093	T. Kauffmann, W. Bonrath, C. Beirich, W. Li, C. Pahde, S. Raedeker, B. Wichmann and D. Wingbermühle; *Chem. Ber.*, 1993, **126**, 2093.	569, 570, 571
93CC70	U. T. Kim and J. Hajdu; *J. Chem. Soc., Chem. Commun.*, 1993, 70.	144
93CC119	G. Fouch, R. M. Horak and O. Meth-Cohn; *J. Chem. Soc., Chem. Commun.*, 1993, 119.	39
93CC169	M. B. Gravestock, D. W. Knight and S. R. Thornton; *J. Chem. Soc., Chem. Commun.*, 1993, 169.	307, 345, 346, 347, 349
93CC327	M. Terada, S. Matsukawa and K. Mikami; *J. Chem. Soc., Chem. Commun.*, 1993, 327.	80
93CC328	M. T. Reetz and S. Stanchen; *J. Chem. Soc., Chem. Commun.*, 1993, 328.	70
93CC339	A. G. M. Barrett and M. A. Seefeld; *J. Chem. Soc., Chem. Commun.*, 1993, 339.	75
93CC806	V. Duchenet and Y. Vallée; *J. Chem. Soc., Chem. Commun.*, 1993, 806.	234
93CC811	K. Soai, A. Oshio and T. Saito; *J. Chem. Soc., Chem. Commun.*, 1993, 811.	69
93CC850	H. Doucet, J. Höfer, C. Bruneau and P. H. Dixneuf; *J. Chem. Soc., Chem. Commun.*, 1993, 850.	676
93CC1042	T. Takahashi, K. Aoyagi, R. Hara and N. Suzuki; *J. Chem. Soc., Chem. Commun.*, 1993, 1042.	927
93CC1067	I. Shibata, M. Nishio, A. Baba and H. Matsuda; *J. Chem. Soc., Chem. Commun.*, 1993, 1067.	701
93CC1195	C. M. Marson, P. R. Giles, H. Adams and N. A. Bailey; *J. Chem. Soc., Chem. Commun.*, 1993, 1195.	202
93CC1373	G. Bartoli, E. Marcantoni and M. Petrini; *J. Chem. Soc., Chem. Commun.*, 1993, 1373.	346
93CC1592	W. Kein and J. Herwig; *J. Chem. Soc., Chem. Commun.*, 1993, 1592.	192
93CC1673	J. M. Brown, D. Holmes and T. P. Layzell; *J. Chem. Soc., Chem. Commun.*, 1993, 1673.	43, 528
93CC1811	J. Christoffers and K. H. Dötz; *J. Chem. Soc., Chem. Commun.*, 1993, 1811.	670
93CCC902	N. Vukojevic, D. Miljkovic, M. Popsavin and P. Popsavin; *Collect. Czech. Chem. Commun.*, 1993, **58**, 902.	9
93CJC42	J. H. E. Bailey and J. E. Drake; *Can. J. Chem.*, 1993, **71**, 42.	239
93CJC930	S. A. Westcott, T. B. Marder, R. T. Baker and J. C. Calabrese; *Can. J. Chem.*, 1993, **71**, 930.	43
93CL1	T. Mukaiyama and K. Suzuki; *Chem. Lett.*, 1993, 1.	712
93CL673	Y. Uchibori, M. Umeno, H. Seto and H. Yoshioka; *Chem. Lett.*, 1993, 673.	18
93CL1787	M. Karikomi, T. Yamazaki and T. Toda; *Chem. Lett.*, 1993, 1787.	318
93CL1953	I. Shibata, M. Nishio, A. Baba and H. Matsuda; *Chem. Lett.*, 1993, 1953.	701
93CPB287	T. Kawate, M. Nakagawa, H. Yamazaki, M. Hirayama and T. Hino; *Chem. Pharm. Bull.*, 1993, **41**, 287.	313
93CPB292	T. Mori, T. Santa, T. Higushi, T. Mashino and M. Hirobe; *Chem. Pharm. Bull.*, 1993, **41**, 292.	579
93CRV725	V. F. Rudchenko; *Chem. Rev.*, 1993, **93**, 725.	355, 356, 357
93CRV763	L. Deloux and M. Srebnik; *Chem. Rev.*, 1993, **93**, 763.	478
93CRV1371	C. Chuit, R. J. P. Corriu, C. Reye and J. C. Young; *Chem. Rev.*, 1993, **93**, 1371.	547
93CRV2117	P. Knochel and R. D. Singer; *Chem. Rev.*, 1993, **93**, 2117.	67, 590
93CRV2207	Y. Yamomoto and N. Asao; *Chem. Rev.*, 1993, **93**, 2207.	73
93CSR199	W. M. McGregor and D. C. Sherrington; *Chem. Soc. Rev.*, 1993, 199.	140
93H(35)701	J. Sperka and D. C. Liotta; *Heterocycles*, 1993, **35**, 701.	30
93H(36)13	J. P. Kutney, Y. P. Chen, S. Gao, G. M. Hewitt, F. Kuri-Brena, R. K. Milanova and N. M. Stovnov; *Heterocycles*, 1993, **36**, 13.	683
93HOU(E/19d)713	L. R. Subramanian; *Methoden Org. Chem. (Houben-Weyl)*, 1993, **E/19d**, 713.	556
93JA497	W. S. Johnson, V. R. Fletcher, B. Chenera, W. R. Bartlett, F. S. Tham and R. K. Kullnig; *J. Am. Chem. Soc.*, 1993, **115**, 497.	1006
93JA10097	D. Zhao, A. D. Allen and T. T. Tidwell; *J. Am. Chem. Soc.*, 1993, **115**, 10 097.	928, 941
93JA10394	D. P. Hsu, W. M. Davis and S. L. Buchwald; *J. Am. Chem. Soc.*, 1993, **115**, 10 394.	927
93JA11490	K. Ishihara, M. Mouri, Q. Gao, T. Maruyama, K. Furuta and H. Yamamoto; *J. Am. Chem. Soc.*, 1993, **115**, 11 490.	528
93JA1154	T. R. Hoye and J. A. Suriano; *J. Am. Chem. Soc.*, 1993, **115**, 1154.	985
93JA11702	P. M. Gallop, M. A. Paz, R. Flückiger, P. J. Stang, V. V. Zhdankin and R. Tykwinski; *J. Am. Chem. Soc.*, 1993, **115**, 11 702.	1021
93JA12579	E. J. Corey and M. C. Noe; *J. Am. Chem. Soc.*, 1993, **115**, 12579.	49
93JA1379	E. H. Appelman, D. French, E. Mishani and S. Rozen; *J. Am. Chem. Soc.*, 1993, **115**, 1379.	94
93JA1586	J. M. O'Connor, H. Ji, M. Iranpour, and A. L. Rheingold; *J. Am. Chem. Soc.*, 1993, **115**, 1586.	945
93JA1731	T. Laube, E. Bannwart and S. Hollenstein; *J. Am. Chem. Soc.*, 1993, **115**, 1731.	1007
93JA2139	A. Marwuet, F. Frappier, G. Guillerm and M. Azoulay; *J. Am. Chem. Soc.*, 1993, 2139.	127
93JA2151	M. Brookhart and B. E. Grant; *J. Am. Chem. Soc.*, 1993, **115**, 2151.	912
93JA2268	W. R. Roush and B. B. Brown; *J. Am. Chem. Soc.*, 1993, **115**, 2268.	900, 904

93JA3071 P. J. Kropp, K. A. Daus, M. W. Tubergen, K. D. Kepler, V. P. Wilson, S. L. Craig, M. M. Baillargeon and G. W. Breton; *J. Am. Chem. Soc.*, 1993, **115**, 3071. 4, 23, 30

93JA3318 X.-Y. Zhang, T. Taketomi, T. Yoshizumi, H. Kumobayashi, S. Akutagawa, K. Mashima and H. Takaya; *J. Am. Chem. Soc.*, 1993, **115**, 3318. 53

93JA3794 J. Lin, Y. Pang, V. G. Young, Jr. and T. J. Barton; *J. Am. Chem. Soc.*, 1993, **115**, 3794. 939

93JA3828 E. J. Corey, M. C. Noe and S. Sarshar; *J. Am. Chem. Soc.*, 1993, **115**, 3828. 49

93JA6065 J. C. Colberg, A. Rane, J. Vaquer and J. A. Soderquist; *J. Am. Chem. Soc.*, 1993, **115**, 6065. 906

93JA6101 A. Gurski and L. S. Liebeskind; *J. Am. Chem. Soc.*, 1993, **115**, 6101. 669

93JA6614 A. F. Houri, M. T. Didiuk, Z. Yu, N. R. Horan and A. H. Hoveyda; *J. Am. Chem. Soc.*, 1993, **115**, 6614. 562

93JA6625 H. J. Reich, I. L. Reich, K. E. Yelm, J. E. Holladay and D. Gschneidner; *J. Am. Chem. Soc.*, 1993, **115**, 6625. 938

93JA6997 J. P. Morken, M. T. Didiuk and A. H. Hoveyda; *J. Am. Chem. Soc.*, 1993, **115**, 6997. 562

93JA7027 H. Stadtmüller, R. Lentz, C. E. Tücker, T. Stüdemann, W. Dörner and P. Knochel; *J. Am. Chem. Soc.*, 1993, **115**, 7027. 591

93JA7219 T. Ishiyama, K. Nishijiama, N. Miyaura and A. Suzuki; *J. Am. Chem. Soc.*, 1993, **115**, 7219. 902

93JA8851 D. L. Comins and H. Hong; *J. Am. Chem. Soc.*, 1993, **115**, 8851. 938

93JA9305 D. Berger, L. E. Overman and P. A. Renhowe; *J. Am. Chem. Soc.*, 1993, **115**, 9305. 936

93JA9800 D. A. Evans, S. G. Nelson, M. R. Gagne and A. R. Muci; *J. Am. Chem. Soc.*, 1993, **115**, 9800. 56

93JA9826 H. Fujihara, H. Mima, T. Erata and N. Furukawa; *J. Am. Chem. Soc.*, 1993, **115**, 9826. 369

93JA9846 J. M. O'Connor and H.-L. Ji and A. L. Rheingold; *J. Am. Chem. Soc.*, 1993, **115**, 9846. 927, 945

93JCS(D)663 H. G. Ang, W. L. Kwik, Y. W. Lee and A. L. Rheingold; *J. Chem. Soc., Dalton Trans.*, 1993, 663. 504

93JCS(D)1737 M. A. Petrie and P. P. Power; *J. Chem. Soc., Dalton Trans.*, 1993, 1737. 511

93JCS(P1)1031 R. Takeuchi and M. Sugiura; *J. Chem. Soc., Perkin Trans. 1*, 1993, 1031. 929

93JCS(P1)1091 X. Li and R. Cosstick; *J. Chem. Soc., Perkin Trans. 1*, 1993, 1091. 256

93JCS(P1)1275 I. Coldham; *J. Chem. Soc., Perkin Trans. 1*, 1993, 1275. 322

93JCS(P1)1375 M. E. Bunnage, S. G. Davies and C. J. Goodwin; *J. Chem. Soc., Perkin Trans. 1*, 1993, 1375. 315

93JCS(P1)1795 A. R. Maguire, W.-D. Meng, S. M. Roberts and A. J. Willetts; *J. Chem. Soc., Perkin Trans. 1*, 1993, 1795. 18

93JCS(P1)2851 M. Zupan, M. Metelko and S. Stavber; *J. Chem. Soc., Perkin Trans. 1*, 1993, 2851. 17

93JCS(P1)3171 A. V. Rama Rao, A. K. Singh, K. M. Reddy and K. Ravikumar; *J. Chem. Soc., Perkin Trans. 1*, 1993, 3171. 22

93JCS(P2)2161 D. Bethell, A. E. Graham, J. P. Heer, O. Markoupoulou, P. C. B. Page and B. K. Park; *J. Chem. Soc., Perkin Trans. 2*, 1993, 2161. 145, 156

93JHC1061 U. Radics and J. Liebscher; *J. Heterocycl. Chem.*, 1993, **30**, 1061. 310

93JMC460 S. H. Rosenberg, K. P. Spina, S. L. Condon, J. Polakowski, Z. Yao, P. Kovar, H. H. Stein, J. Cohen, J. L. Barlow, V. Klinghoffer, D. A. Egan, K. A. Tricarico, T. J. Perlin, W. R. Baker and H. D. Kleinert; *J. Med. Chem.*, 1993, **36**, 460. 118, 124

93JMC3526 H. Kawakubo, S. Takagi, Y. Yamaura, S. Katoh, Y. Ishimoto, T. Nagatani, D. Mochizuki, T. Kamata and Y. Sasaki; *J. Med. Chem.*, 1993, **36**, 3526. 326

93JMC3606 C. Z. Ding and R. B. Silverman; *J. Med. Chem.*, 1993, **36**, 3606. 309

93JOC21 A. Padwa and F. R. Kinder; *J. Org. Chem.*, 1993, **58**, 21. 927, 945

93JOC82 R. Sustmann, W. Sicking and R. Huisgen; *J. Org. Chem.*, 1993, **58**, 82. 243

93JOC147 H. C. Brown, K. Ganesan and R. K. Dhar; *J. Org. Chem.*, 1993, **58**, 147. 532, 698

93JOC272 M. Altamura and E. Perrotta; *J. Org. Chem.*, 1993, **58**, 272. 9

93JOC546 M. Topolski, M. Duraisamy, J. Rachoñ, J. Gawronski, K. Gawronska, V. Goedken and H. M. Walborsky; *J. Org. Chem.*, 1993, **58**, 546. 998

93JOC588 P. Knochel, T.-S. Chou, C. Jubert and D. Rajagopal; *J. Org. Chem.*, 1993, **58**, 588. 999, 1002

93JOC633 G. V. Lamoureux and G. M. Whitesides; *J. Org. Chem.*, 1993, **58**, 633. 124

93JOC799 D. K. Jones, D. C. Liotta, I. Shinkai and D. J. Mathre; *J. Org. Chem.*, 1993, **58**, 799. 66

93JOC804 R. C. Larock and S. Ding; *J. Org. Chem.*, 1993, **58**, 804. 86

93JOC844 W. Amberg, Y. L. Bennani, R. K. Chadha, G. A. Crispino, W. D. Davis, J. Hartung, K.-S. Jeong, Y. Ogino, T. Shibata and K. B. Sharpless; *J. Org. Chem.*, 1993, **58**, 844. 49

93JOC917 J. L. Kice and A. G. Kutateladze; *J. Org. Chem.*, 1993, **58**, 917. 368

93JOC967 S.-S. P. Chou and M.-C. Cheng; *J. Org. Chem.*, 1993, **58**, 967. 938

93JOC1924 Z.-D. Liu and Z.-C. Chen; *J. Org. Chem.*, 1993, **58**, 1924. 1033

93JOC1941 N. M. Yoon, J. H. Ahn, D. K. An and Y. S. Shon; *J. Org. Chem.*, 1993, **58**, 1941. 60

93JOC1949 M.-Z. Deng, N.-S. Li and Y.-Z. Huang; *J. Org. Chem.*, 1993, **58**, 1949. 904

93JOC1991 S. Hanessian, P. Maffre, M. Girard, S. Beaudoin, J.-Y. Sancéau and Y. Benani; *J. Org. Chem.*, 1993, **58**, 1991. 50

93JOC2086 A. R. Katritzky, G. Yao, X. Lan and X. Zhao; *J. Org. Chem.*, 1993, **58**, 2086. 320

93JOC2216 D. S. Kemp, R. I. Carey; *J. Org. Chem.*, 1993, **58**, 2216. 127

93JOC2501 E. N. Lawson, W. Kitching, C. H. L. Kennard and K. A. Byriel; *J. Org. Chem.*, 1993, **58**, 2501. 29

93JOC2694 S. Aehyutha Rao, M. J. Rozema and P. Knochel; *J. Org. Chem.*, 1993, **58**, 2694. 594

93JOC2714 S. Borrelly and L. A. Paquette; *J. Org. Chem.*, 1993, **58**, 2714. 921, 938

93JOC2791 G. S. Lal; *J. Org. Chem.*, 1993, **58**, 2791. 15, 17

93JOC2880 D. J. Mathre, A. S. Thompson, A. W. Douglas, K. Hoogsteen, J. D. Carroll, E. G. Corley and E. J. J. Grabowski; *J. Org. Chem.*, 1993, **58**, 2880. 528

93JOC3156	J. V. Bhaskar Kanth and M. Periasamy; *J. Org. Chem.*, 1993, **58**, 3156.	300
93JOC3229	E. A. Schmittling and J. S. Sawyer; *J. Org. Chem.*, 1993, **58**, 3229.	684
93JOC3299	C. A. Busacca, R. E. Johnson and J. Swestock; *J. Org. Chem.*, 1993, **58**, 3299.	321
93JOC3455	P. Wipf and S. Venkatraman; *J. Org. Chem.*, 1993, **58**, 3455.	587
93JOC3469	J. E. Toth, J. Ray and J. Deeter; *J. Org. Chem.*, 1993, **58**, 3469.	726
93JOC3850	S. E. Denmark and L. R. Marcin; *J. Org. Chem.*, 1993, **58**, 3850.	785
93JOC4298	C. Aubert, J.-P. Gotteland and M. Malacria; *J. Org. Chem.*, 1993, **58**, 4298.	944
93JOC4382	J. Uenishi, T. Tanaka, K. Nishiwaki, S. Wakabayashi, S. Oae and H. Tsukube; *J. Org. Chem.*, 1993, **58**, 4382.	7, 27
93JOC4579	R. C. Larock, M. J. Doty and S. Cacchi; *J. Org. Chem.*, 1993, **58**, 4579.	928
93JOC4596	X.-X. Shi and L.-X. Dai; *J. Org. Chem.*, 1993, **58**, 4596.	22, 28
93JOC4606	N. N. Polushin, B.-C. Chen, L. W. Anderson and J. S. Cohen; *J. Org. Chem.*, 1993, **58**, 4606.	410
93JOC4823	M. C. McIntosh and S. M. Weinreb; *J. Org. Chem.*, 1993, **58**, 4823.	921, 922
93JOC5035	D. L. Comins and H. Hong; *J. Org. Chem.*, 1993, **58**, 5035.	938
93JOC5045	K. Ikenaga, K. Hiramatsu, N. Nasaka and S. Matsumoto; *J. Org. Chem.*, 1993, **58**, 5045.	916
93JOC5235	R. R. Tykwinski, B. L. Williamson, D. R. Fischer, P. J. Stang and A. M. Arif; *J. Org. Chem.*, 1993, **58**, 5235.	1036
93JOC5307	D. A. Evans, A. R. Muci and R. Sturmer; *J. Org. Chem.*, 1993, **58**, 5307.	43
93JOC5351	I. D. Gridnev, N. Miyaura and A. Suzuki; *J. Org. Chem.*, 1993, **58**, 5351.	901, 902
93JOC5595	A. S. Guram and R. F. Jordan; *J. Org. Chem.*, 1993, **58**, 5595.	927
93JOC5849	A. Haubrich, M. van Klaveren, G. van Koten, G. Handke and N. Krause; *J. Org. Chem.*, 1993, **58**, 5849.	940
93JOC5976	J. Barluenga, J. M. Montserrat and J. Florez; *J. Org. Chem.*, 1993, **58**, 5976.	553
93JOC6109	F. Garro-Helion, A. Merzouk and F. Guibe; *J. Org. Chem.*, 1993, **58**, 6109.	304
93JOC6182	Y. Kondo, J. R. Green and J. Ho; *J. Org. Chem.*, 1993, **58**, 6182.	942
93JOC6529	T. Hirao, K. Hirano, T. Hasegawa, Y. Ohshiro and I. Ikeda; *J. Org. Chem.*, 1993, **58**, 6529.	14
93JOC6545	D. S. Matteson and H.-W. Man; *J. Org. Chem.*, 1993, **58**, 6545.	529
93JOC6559	J. W. Grissom and D. Klingberg; *J. Org. Chem.*, 1993, **58**, 6559.	351
93JOC7162	K. Ganesan and H. C. Brown; *J. Org. Chem.*, 1993, **58**, 7162.	532, 698
93JOM(444)53	H. Lang, U. Lay and W. Imhof; *J. Organomet. Chem.*, 1993, **444**, 53.	875, 896
93JOM(447)177	C.-H. Jun and R. H. Crabtree; *J. Organomet. Chem.*, 1993, **447**, 177.	915
93JOM(447)197	A. J. Ashe, III, J. W. Kampf and D. B. Puranik; *J. Organomet. Chem.*, 1993, **447**, 197.	
93JOM(448)83	Ch. Emmerich and G. Huttner; *J. Organomet. Chem.*, 1993, **448**, 83.	896
93JOM(449)27	Z. Yuan, N. J. Taylor, Y. Sun, T. B. Marder, I. D. Williams and L.-T. Cheng; *J. Organomet. Chem.*, 1993, **449**, 27.	1083
93JOM(449)53	K. E. Lee and K. T. Higa; *J. Organomet. Chem.*, 1993, **449**, 53.	1092
93JOM(452)1	S. Gupta, S. Sharma and A. K. Narula; *J. Organomet. Chem.*, 1993, **452**, 1.	112, 558
93JOM(453)C11	Ch. Pluta and K.-R. Pörschke; *J. Organomet. Chem.*, 1993, **453**, C11.	1101
93LA1155	R. Zimmer, K. Homann and H.-U. Reissig; *Liebigs Ann. Chem.*, 1993, 1155.	322
93MI 202-01	Y. Ono, M. Akiyama and E. Suzuki; *Chemistry of Materials*, 1993, **5**, 442.	108
93MI 202-02	H. Heaney; *Aldrichimica Acta*, 1993, **26**, 35.	90
B-93MI 203-01	N. S. Simpkins; "Sulfones in Organic Synthesis," Pergamon Press, Oxford, 1993.	154, 158
93MI 205-01	D. J. Robins; *Nat. Prod. Rep.*, 1993, **10**, 487.	323
93MI 206-01	H. Heaney; *Aldrichimica Acta*, 1993, **26**, 35.	348
B-93MI 221-01	N. S. Simpkins, "Sulfones in Organic Chemistry," Pergamon, Oxford, 1993.	1033
93OM803	F. Hojo, S. Sekigawa, N. Nakayama, T. Shimizu and W. Ando, *Organometallics*; 1993, **12**, 803.	1089
93OM988	H. Yamashita, M. Tanaka and M. Goto; *Organometallics*, 1993, **12**, 988.	913
93OM2065	Y. Uchimaru, A. Mouneer, M. El Sayed and M. Tanaka; *Organometallics*, 1993, **12**, 2065.	945
93OPP249	T. Venkatachalam, S. Mzengeza and M. Diksic; *Org. Prep. Proced. Int.*, 1993, **25**, 249.	8
93OR(43)1	S. H. Pine; *Org. React.*, 1993, **43**, 1.	680
93OR(43)251	G. R. Krow; *Org. React.*, 1993, **43**, 251.	654
93OS(72)38	M. Braun and S. Gräf; *Org. Synth.*, 1993, **72**, 38.	420
93OS(72)62	A.-K. Beck, S. Bkank, K. Job, D. Seebach and T. Sommerfeld; *Org. Synth.*, 1993, **72**, 62.	420
93PAC633	F. A. Davis, R. T. Reddy, W. Han and R. E. Reddy; *Pure Appl. Chem.*, 1993, **65**, 633.	44
93PJC1361	Z. Foltynowicz; *Pol. J. Chem.*, 1993, **67**, 1361.	938
93POL89	C. J. Thomas, L. K. Krannich and C. L. Watkins; *Polyhedron*, 1993, **12**, 89.	484
93PS(79)131	K. Bergemann, F. Hesselbarth, E. Wenschuh, U. Baumeister and H. Hartung; *Phosphorus Sulfur*, 1993, **79**, 131.	258
93PS(79)179	I. S. Nizamov, V. A. Kuznetzov, E. S. Batyeva, V. A. Al'Forsov and A. N. Pudovik; *Phosphorus Sulfur*, 1993, **79**, 179.	254
93S165	R. Zimmer; *Synthesis*, 1993, 165.	939
93S207	G. A. Olah, Q. Wang, X. Li and G. K. Surya Prakash; *Synthesis*, 1993, 207.	104
93S263	A. Albini; *Synthesis*, 1993, 263.	346, 347, 350, 355
93S311	C. McGuigan and P. Narashiman; *Synthesis*, 1993, 311.	412
93S349	T.-Y. Luh and K.-T. Wong; *Synthesis*, 1993, 349.	939
93S481	R.-Y. Yang and L.-X Dai; *Synthesis*, 1993, 481.	311
93S490	E. A. Bartmann; *Synthesis*, 1993, 490.	232, 724, 725
93S537	P. Wipf; *Synthesis*, 1993, 537.	586, 702, 910, 919

93S573	D. A. Dickman, S. Chemburkar, D. B. Konopacki and E. M. Elisseou; *Synthesis*, 1993, **6**, 573.	228
93S579	D. B. Grotjahn and K. P. C. Vollhardt; *Synthesis*, 1993, 579.	944
93S611	L. A. Jacob, B.-L. Chen and D. Stec; *Synthesis*, 1993, 611.	943
93S693	G. A. Olah, X-Y. Li, Q. Wang and G. K. S. Prakash; *Synthesis*, 1993, 693.	6, 17, 18
93S735	K. Ritter; *Synthesis*, 1993, 735.	689
93S973	T. C. Morrill, Q. Lu, A. Brister, B. O'Shaughnessy and K. Belfield; *Synthesis*, 1993, 973.	4
93S1057	D. Ramkumar and S. Sankaraman; *Synthesis*, 1993, 1057.	150
93S1067	G. A. Hunter and H. McNab; *Synthesis*, 1993, 1067.	1032
93S1089	J. M. McGill; *Synthesis*, 1993, 1089.	310
93S1177	J. Rodriguez and J.-P. Dulcere; *Synthesis*, 1993, 1177.	3
93SC121	J. Daniel and D. N. Dhar; *Synth. Commun.*, 1993, **23**, 121.	249
93SC711	I. A. Rivero, R. Somanathan and L. H Hellberg; *Synth. Commun.*, 1993, **23**, 711.	8
93SC789	A. E. Moormann; *Synth. Commun.*, 1993, **23**, 789.	310
93SC817	F. Souquet, T. Martens and M.-B. Fleury; *Synth. Commun.*, 1993, **23**, 817.	308
93SC993	H. L. Rigby and D. E. Lewis; *Synth. Commun.*, 1993, **23**, 993.	29
93SC1025	C. Navarro, M. Degeuil-Castaing, D. Colombani and B. Maillard; *Synth. Commun.*, 1993, **23**, 1025.	97
93SC1329	G. S. Poindexter and K. M. Strauss; *Synth. Commun.*, 1993, **23**, 1329.	320
93SC1507	P. C. B. Page, A. E. Graham, D. Bethell and B. K. Park; *Synth. Commun.*, 1993, **23**, 1507.	145, 156
93SC1515	R. Caputo, C. Ferreri, L. Longobardo, G. Palumbo and S. Pedatella; *Synth. Commun.*, 1993, **23**, 1515.	688
93SC1759	F. Bellesia, F. Ghelfi, U. M. Pagnoni and A. Pinetti; *Synth. Commun.*, 1993, **23**, 1759.	150
93SC2199	P. G. M. Wuts, J. E. Cabaj and K. D. Maisto; *Synth. Commun.*, 1993, **23**, 2199.	25
93SC2355	S. H. Thang and D. J. Rigg; *Synth. Commun.*, 1993, **23**, 2355.	321
93SC3047	N. M. Yoon, J. Choi and Y. S. Shon; *Synth. Commun.*, 1993, **23**, 3047.	317
93SC3157	F. Fringuelli, R. Pellegrino and F. Pizzo; *Synth. Commun.*, 1993, **23**, 3157.	146
93SL81	A. B. Charette, C. Mellon, L. Rouillard and E. Malenfant; *Synlett*, 1993, 81.	79
93SL145	R. C. Larock, S. Ding and C. Tu; *Synlett*, 1993, 145.	86
93SL171	B. Bennetau and J. Dunogues; *Synlett*, 1993, 171.	943
93SL189	R. Angell, P. J. Parsons and A. Naylor; *Synlett*, 1993, 189.	931
93SL207	M. A. Tius and J. K. Kawakami; *Synlett*, 1993, 207.	611
93SL209	J. A. Marshall and W. J. Dubay; *Synlett*, 1993, 209.	118, 124
93SL211	M. Tingoli, M. Tiecco, L. Testaferri and R. Balducci; *Synlett*, 1993, 211.	1035, 1037
93SL266	C. Meyer, I. Marek, G. Courtemanche and J.-F. Normant; *Synlett*, 1993, 266.	592
93SL277	J. J. Eshelby, P. C. Crowley and P. J. Parsons; *Synlett*, 1993, 277.	936
93SL279	J. J. Eshelby, P. J. Crowley and P. J. Parsons; *Synlett*, 1993, 279.	936
93SL363	J. W. Herndon and M. Zora; *Synlett*, 1993, 363.	927
93SL386	C. Meyer, I. Marek and J.-F. Normant; *Synlett.*, 1993, 386.	313
93SL399	P. Metz and M. Fleischer; *Synlett.*, 1993, 399.	207
93SL401	D. Enders and M. Finkam; *Synlett.*, 1993, 401.	331
93SL407	R. K. Dieter and C. W. Alexander; *Synlett.*, 1993, 407.	315
93SL465	H. Nemoto, Y. Kubota, N. Sasaki and Y. Yamamoto; *Synlett*, 1993, 465.	250
93SL489	I. Fernandez, B. Garcia, S. Munoz, J. R. Pedro and R. de la Salud; *Synlett*, 1993, 489.	36
93SL519	P.-Y. Chavant, F. Lhermitte and M. Vaultier; *Synlett.*, 1993, 519.	321
93SL665	F. Chemla, I. Marek and J.-F. Normant; *Synlett*, 1993, 665.	594, 998
93SL675	H. M. I. Osborn, J. B. Sweeney and B. Howson; *Synlett.*, 1993, 675.	319
93SL680	J. P. Genêt, E. Blart, M. Savignac, S. Lemeune, S. Lemaire-Audoire and J. M. Bernard; *Synlett.*, 1993, 680.	304
93SL731	M. E. Bunnage, S. G. Davies and C. J. Goodwin; *Synlett.*, 1993, 731.	315
93SL764	B. Lygo; *Synlett.*, 1993, 764.	319
93SL771	G. Palmisano and M. Santagostino; *Synlett*, 1993, 771.	931
93SL863	M. E. Maier, B.-U. Haller, R. Stumpf and H. Fischer; *Synlett*, 1993, 863.	936
93SL927	H. C. Brown and A. S. Phadke; *Synlett*, 1993, 927.	529
93T177	A. Whiting and R. J. Mears; *Tetrahedron*, 1993, **49**, 177.	530
93T363	K. Zhao and D. W. Landry; *Tetrahedron*, 1993, **49**, 363.	456
93T11389	E. Taskinen; *Tetrahedron*, 1993, **49**, 11389.	675
93T1189	N. Choi, N. Tokitoh, M. Goto and W. Ando; *Tetrahedron*, 1993, **49**, 1189.	941
93T1271	R. Shabana, F. H. Osman and S. S. Atrees; *Tetrahedron*, 1993, **49**, 1271.	856
93T2159	L. Liang, M. Ramaseshan and D. I. MaGee; *Tetrahedron*, 1993, **49**, 2159.	1027
93T2477	M.-J. Bourgeois, E. Montaudon and B. Maillard; *Tetrahedron*, 1993, **49**, 2477.	98
93T3043	R. R. Tykwinski and P. J. Stang; *Tetrahedron*, 1993, **49**, 3043.	1033
93T3193	N. P. Baens, F. Compernolle, S. M. Toppett and G. J. Hoornaert; *Tetrahedron*, 1993, **49**, 3193.	23
93T5133	L. Toke, Z. Hell, G. T. Szabo, G. Toth, M. Bihari and A. Rockenbauer; *Tetrahedron*, 1993, **49**, 5133.	34
93T5471	E. Negishi, M. Ay, and T. Sugihara; *Tetrahedron*, 1993, **49**, 5471.	980
93T6359	S. Sekigawa, T. Shimizu and W. Ando; *Tetrahedron*, 1993, **49**, 6359.	1089
93T7037	P. Golding, R. W. Millar, N. C. Paul and D. H. Richards; *Tetrahedron*, 1993, **49**, 7037.	104
93T7761	D. Guijarro and M. Yus; *Tetrahedron*, 1993, **49**, 7761.	311
93T7857	A. G. M. Barrett and M. A. Seefeld; *Tetrahedron*, 1993, **49**, 7857.	521
93TA215	C. Lensink and J. G. de Vries; *Tetrahedron Asymmetry*, 1993, **4**, 215.	330

93TA2095 R. Sreekumar and C. N. Pillai; *Tetrahedron Asymmetry*, 1993, **4**, 2095. 331

93TA2139 P. C. B. Page, M. T. Gareh and R. A. Porter; *Tetrahedron Asymmetry*, 1993, 2139. 147

93TA2159 D. R. J. Hose, T. Raynham and M. Wills; *Tetrahedron Asymmetry*, 1993, **4**, 2159. 362

93TA2315 G. Li, D. Patel and V. J. Hruby; *Tetrahedron Asymmetry*, 1993, **4**, 2315. 29

93TL67 D. Gu and D. N. Harpp; *Tetrahedron Lett.*, 1993, **34**, 67. 186

93TL133 R. E. Dolle and D. McNair; *Tetrahedron Lett.*, 1993, **34**, 133. 163

93TL141 R. Fernández, C. Gasch, J.-M. Lasaletta, J.-M. Llera and J. Vázquez; *Tetrahedron Lett.*, 1993, **34**, 141. 347

93TL257 J. C. Fuller, E. L. Stangeland, C. T. Goralski and B. Singaram; *Tetrahedron Lett.*, 1993, **34**, 257. 60

93TL293 E. Fritz-Langhals and G. Schutz; *Tetrahedron Lett.*, 1993, **34**, 293. 9

93TL687 T. Takahashi, M. Kageyama, V. Denisov, R. Hara and E. Negishi; *Tetrahedron Lett.*, 1993, **34**, 687. 927, 987, 988

93TL703 S. A. Lermontov, S. I. Zavorin, A. N. Pushin, A. N. Chekhlov, N. S. Zefirov and P. J. Stang; *Tetrahedron Lett.*, 1993, **34**, 703. 280

93TL873 A. Degl'Innocenti, A. Capperucci, A. Mordini, G. Reginato, A. Ricci and F. Cerreta; *Tetrahedron Lett.*, 1993, **34**, 873. 541

93TL903 C. J. Li and D. N. Harpp; *Tetrahedron Lett.*, 1993, **34**, 903. 233

93TL1043 M. Yoshifuji, M. Hirano and K. Toyota; *Tetrahedron Lett.*, 1993, **34**, 1043. 857, 865

93TL1107 R. Breslow and D. Wiedenfeld; *Tetrahedron Lett.*, 1993, **34**, 1107. 20

93TL1259 J. L. Duffy, J. A. Kurth and M. J. Kurth; *Tetrahedron Lett.*, 1993, **34**, 1259. 393

93TL1461 V. V. Zhdankin and P. J. Stang; *Tetrahedron Lett.*, 1993, **34**, 1461. 1029, 1030

93TL1649 J. Almena, F. Fonbelo and M. Yus; *Tetrahedron Lett.*, 1993, **34**, 1649. 420, 553

93TL1681 M. Bordoloi; *Tetrahedron Lett.*, 1993, **34**, 1681. 55

93TL1871 D. H. R. Barton, S. D. Beviere, W. Chavasiri, D. Doller and B. Hu; *Tetrahedron Lett.*, 1993, **34**, 1871. 2

93TL2079 Y. L. Bennani and K. B. Sharpless; *Tetrahedron Lett.*, 1993, **34**, 2079. 49

93TL2083 Y. L. Bennani and K. B. Sharpless; *Tetrahedron Lett.*, 1993, **34**, 2083. 49

93TL2165 O. Kitagawa, T. Hanano, N. Kikuchi and T. Taguchi; *Tetrahedron Lett.*, 1993, **34**, 2165. 34

93TL2303 D. Beruben, I. Marek, L. Labaudinière and J.-F. Normant; *Tetrahedron Lett.*, 1993, **34**, 2303. 594

93TL2335 Y. Oozumi and T. Hayashi; *Tetrahedron Lett.*, 1993, **34**, 2335. 537

93TL2351 T. Chiba, A. Miyashita, H. Nohira and H. Takaya; *Tetrahedron Lett.*, 1993, **34**, 2351. 53

93TL2521 H. M. Meshram; *Tetrahedron Lett.*, 1993, **34**, 2521. 230

93TL2561 G. Li, M. A. Jarosinski and V. J. Hruby; *Tetrahedron Lett.*, 1993, **34**, 2561. 29

93TL2757 D. de Kermadec and M. Prudhomme; *Tetrahedron Lett.*, 1993, **34**, 2757. 8

93TL2919 K.-D. Roth and U. Müller; *Tetrahedron Lett.*, 1993, **34**, 2919. 301

93TL3057 D. P. Matthews, S. C. Miller, E. T. Jarvi, J. S. Sabol and J. R. McCarthy; *Tetrahedron Lett.*, 1993, 3057. 611

93TL3115 M. J. Rozema, C. Einsenberg, H. Lütjens, R. Ostwald, K. Belyk and P. Knochel; *Tetrahedron Lett.*, 1993, **34**, 3115. 69, 72, 591

93TL3145 K. Mallaiah, J. Satyanarayana, I. Ila and H. Junjappa; *Tetrahedron Lett.*, 1993, **34**, 3145. 74

93TL3243 D. Cai, D. Tschaen, Y.-J. Shi, T. R. Verhoeven, R. A. Reamer and A. W. Douglas; *Tetrahedron Lett.*, 1993, **34**, 3243. 66

93TL3271 D. R. Williams, W. H. Osterthout and J. P. Reddy; *Tetrahedron Lett.*, 1993, **34**, 3271. 350

93TL3311 L. Capella, A. Capperucci, G. Curotto, D. Lazzari, P. Bembech, G. Reginato and A. Ricci; *Tetrahedron Lett.*, 1993, **34**, 3311. 918

93TL3363 J. Brittain and Y. Gareau; *Tetrahedron Lett.*, 1993, **34**, 3363. 541

93TL3453 S. Kobayashi and K. Nishio; *Tetrahedron Lett.*, 1993, **34**, 3453. 78

93TL3667 D. E. Duffy, F. H. Condit, C. A. Teleha, C.-L. J. Wang and J. C. Calabrese; *Tetrahedron Lett.*, 1993, **34**, 3667. 206

93TL3939 A. E. McCaig and R. H. Wightman; *Tetrahedron Lett.*, 1993, **34**, 3939. 354

93TL3971 F. A. Davis, P. Zhou and C. K. Murphy; *Tetrahedron Lett.*, 1993, **34**, 3971. 16

93TL4031 H. Kotsuki, M. Kataoka and H. Nishizawa; *Tetrahedron Lett.*, 1993, **34**, 4031. 84

93TL4145 G. J. Quallich and T. M. Woodall; *Tetrahedron Lett.*, 1993, **34**, 4145. 66

93TL4193 P. Aufranc, J. Ollivier, A. Stolle, C. Bremer, M. Es-Sayed, A. de Meijere and J. Salaun; *Tetrahedron Lett.*, 1993, **34**, 4193. 309

93TL4403 V. K. Tandon and A. Chandra; *Tetrahedron Lett.*, 1993, **34**, 4403. 318

93TL4623 I. Klement, K. Lennick, C. E. Tucker and P. Knochel; *Tetrahedron Lett.*, 1993, **34**, 4623. 1096

93TL4655 T. A. J. van der Heide, J. L. van der Baan, E. A. Bijpost, F. J. J. deKanter, F. Bickelhaupt and G. W. Klumpp; *Tetrahedron Lett.*, 1993, **34**, 4655. 521

93TL4735 B. M. Trost and L. T. Phan; *Tetrahedron Lett.*, 1993, **34**, 4735. 927

93TL4925 V. Duchenet and Y. Vallée; *Tetrahedron Lett.*, 1993, **34**, 4925. 234

93TL4985 H. Shinokubo, K. Oshima and K. Utimoto; *Tetrahedron Lett.*, 1993, **34**, 4985. 700

93TL5369 K. R. Guertin and A. S. Kende; *Tetrahedron Lett.*, 1993, **34**, 5369. 155

93TL5689 D. H. R. Barton and S. D. Beviere; *Tetrahedron Lett.*, 1993, **34**, 5689. 2

93TL5725 J. Nakayama, T. Kawamura, K. Kuroda and A. Fujita; *Tetrahedron Lett.*, 1993, **34**, 5725. 27

93TL6053 C. Meyer, I. Marek, G. Courtemanche and J.-F. Normant; *Tetrahedron Lett.*, 1993, **34**, 6053. 593

93TL6105 G. Besenyei, S. Nemeth and L. I. Simandi; *Tetrahedron Lett.*, 1993, **34**, 6105. 726

93TL6463 O. E. O. Hormi and L. Hirvelä; *Tetrahedron Lett.*, 1993, **34**, 6463. 673

93TL6899 V. Nyzam, C. Belaud and J. Villeras; *Tetrahedron Lett.*, 1993, **34**, 6899. 530

93TL7239 B. E. Blough and F. I. Carrol; *Tetrahedron Lett.*, 1993, **34**, 7239. 327

93TL7575 D. Beruben, I. Marek, J.-F. Normant and N. Platzer; *Tetrahedron Lett.*, 1993, **34**, 7575. 594

93TL7667 G. W. Kabalka, N. Kesavulu Reddy and C. Narayana; *Tetrahedron Lett.*, 1993, **34**, 7667. 655

93TL7693 G. B. Fisher, C. T. Goralski, L. W. Wilkinson and B. Singaram; *Tetrahedron Lett.*, 1993, **34**, 7693. 519

93TL7845 H. C. Brown, A. S. Phadke and N. G. Bhat; *Tetrahedron Lett.*, 1993, **34**, 7845. 530

93TL7911 H. Stadtmüller, G. E. Tücker, A. Vaupel and P. Knochel; *Tetrahedron Lett.*, 1993, **34**, 7911. 591

93TL7923 R. Andriamiadanarivo, B. Pujol, B. Chantegrel, C. Deshayes and A. Doutheau; *Tetrahedron Lett.*, 1993, **34**, 7923. 668

93ZN(B)719 H. Schulz, H. Pritzkow and W. Siebert; *Z. Naturforsch., Teil B*, 1993, **48**, 719. 903

94AG(E)69 O. Reiser; *Angew. Chem., Int. Ed. Engl.*, 1994, **33**, 69. 44

94AG(E)99 K. Krohn and K. Khanbabaee; *Angew. Chem., Int. Ed. Engl.*, 1994, **33**, 99. 546

94AG(E)851 K. Smith, A. Pelter and Z. Jin; *Angew. Chem., Int. Ed. Engl.*, 1994, **33**, 851. 41

94AG(E)1182 K. Burgess and A. M. Porte; *Angew. Chem., Int. Ed. Engl.*, 1994, **33**, 1182. 50

94BCJ271 C. A. Horiuchi and E. Takahashi; *Bull. Chem. Soc. Jpn.*, 1994, **67**, 271. 34

94CC149 S. Stavber and M. Zupan; *J. Chem. Soc., Chem. Commun.*, 1994, 149. 15

94CC343 R. E. Banks, N. J. Lawrence and A. L. Popplewell; *J. Chem. Soc., Chem. Commun.*, 1994, 343. 15

94CC463 N. S. Schneider, A. D. Darwish, H. W. Kroto, R. Taylor and D. R. M. Walton; *J. Chem. Soc., Chem. Commun.*, 1994, 463. 43

94CC475 Y. B. Kang, M. Pabel, A. C. Willis and S. B. Wild; *J. Chem. Soc., Chem. Commun.*, 1994, 475. 438

94CC735 G. Cardillo, A. De Simone, L. Gentilucci and C. Tomasini; *J. Chem. Soc., Chem. Commun.*, 1994, 735. 23

94CC797 K. Kaneda, S. Ueno and T. Imanaka; *J. Chem. Soc., Chem. Commun.*, 1994, 797. 654

94CC1775 S. Y. Ko, J. Lerpiniere, I. D. Linney and R. Wrigglesworth; *J. Chem. Soc., Chem. Commun.*, 1994, 1775. 52

94CPB459 A. Toyota, J. Chiba, Y. Sugita, M. Sato and C. Kaneko; *Chem. Pharm. Bull.*, 1994, **42**, 459. 17

94CPB811 T. Kataoka, Y. Ohe, A. Umeda, T. Iwamura, M. Yoshimatsu and H. Shimizu; *Chem. Pharm. Bull.*, 1994, **42**, 811. 286, 288, 292

94CRV2483 H. C. Kolb, M. S. van Nieuwenhze and K. B. Sharpless; *Chem. Rev.*, 1994, **44**, 2483.

94H(37)693 S. Tanaka, M. Tomura and Y. Yamashita; *Heterocycles*, 1994, **37**, 693. 793, 795

94JA642 K. Lammertsma and B. V. Prasad; *J. Am. Chem. Soc.*, 1994, **116**, 642. 757

94JA866 J. M. Brown and G. C. Lloyd-Jones; *J. Am. Chem. Soc.*, 1994, **116**, 866. 528

94JA2390 K. Okuma, Y. Sato, T. Ishii and H. Ohta; *J. Am. Chem. Soc.*, 1994, **59**, 2390. 363

94JA3139 W. Oppolzer, A. C. Spivey and C. G. Bochet; *J. Am. Chem. Soc.*, 1994, **116**, 3139. 346

94JA6577 W. F. Bailey and E. R. Punzalan; *J. Am. Chem. Soc.*, 1994, **116**, 6577. 112

94JCS(P1)717 C.-W. Lee, Y. J. Koh and D. Y. Oh; *J. Chem. Soc., Perkin Trans. 1*, 1994, 717. 730

94JCS(P1)777 M. Tokuda, H. Fujita and H. Suginome; *J. Chem. Soc., Perkin Trans. 1*, 1994, 777. 337

94JCS(P1)1935 M. R. Bryce, J. N. Heaton, P. C. Taylor and M. Anderson; *J. Chem. Soc., Perkin Trans. 1*, 1994, 1935. 792

94JCS(P2)117 K. A. Jørgensen; *J. Chem. Soc., Perkin Trans. 2*, 1994, 117. 147

94JFC(66)123 J. C. Fuller, E. L. Stangeland, C. T. Goralski and B. Singaram; *J. Fluorine Chem.*, 1994, **66**, 123. 59

94JOC311 C. K.-F. Chiu, S. V. Govindan and P. L. Fuchs; *J. Org. Chem.*, 1994, **59**, 311. 679

94JOC365 H. C. Brown, S. V. Kulkarni and V. S. Racherla; *J. Org. Chem.*, 1994, **59**, 365. 521

94JOC914 T.-K. Yang, R.-Y. Chen, D.-S. Lee, W.-S. Peng, Y.-Z. Jiang, A.-Q. Mi and T.-T. Jong; *J. Org. Chem.*, 1994, **59**, 914. 359, 363

94JOC1011 M. Yoshimatsu, T. Sato, H. Shimizu, M. Hori and T. Kataoka; *J. Org. Chem.*, 1994, **59**, 1011. 278, 279, 292

94JOC1831 A. A. Frimer, V. Marks, M. Sprecher and P. Gilinsky-Sharon; *J. Org. Chem.*, 1994, **59**, 1831. 660

94JOC2336 K. Ganessan and H. C. Brown; *J. Org. Chem.*, 1994, **59**, 2336. 532

94JOC2365 U. P. Dhakte and H. C. Brown; *J. Org. Chem.*, 1994, **59**, 2365. 523

94JOC2676 G. A. Molander and K. L. Bobbitt; *J. Org. Chem.*, 1994, **59**, 2676. 42

94JOC2925 S. Klein, I. Marek and J.-F. Normant; *J. Org. Chem.*, 1994, **59**, 2925. 556

94JOC3714 J. Gao, R. Härter, D. M. Gordon and G. M. Whitesides; *J. Org. Chem.*, 1994, **59**, 3714. 597

94JOC4378 B. D. Brandes and E. N. Jacobsen; *J. Org. Chem.*, 1994, **59**, 4378. 92

94JOC6683 M. A. McGuire, E. Sorensou, F. W. Owings, T. M. Resnick, M. Fox and N. H. Baine; *J. Org. Chem.*, 1994, **59**, 6683. 689

94JOM(474)1 K. I. Grandberg and V. P. Dyadchenko; *J. Organomet. Chem.*, 1994, **474**, 1. 590

94MI 201-01 P. L. Spargo; *Contemp. Org. Synth.*, 1994, **1**, 113. 2

B-94MI 202-01 C.-H. Wong and G. M. Whitesides (eds.); "Enzymes in Synthetic Organic Chemistry," Pergamon, Oxford, 1994. 38

94MI 202-02 N. J. Turner; *Nat. Prod. Rep.*, 1994, **11**, 1. 38

94MI 202-03 P. Darrigo, H. E. Hogberg, D. Pedrocchifantoni and S. Servi; *Biocatalysis*, 1994, **9**, 299. 39

94MI 203-01 C. M. Rayner; *Contemp. Org. Synth.*, 1994, **1**, 191. 116

B-94MI 209-01 S. Patai (ed.); "The Chemistry of Organic Arsenic, Antimony and Bismuth Compounds," Wiley-Interscience, New York, 1994. 481

B-94MI 210-01	N. Auner and J. Weis (eds); "Organosilicon Chemistry: From Molecules to Materials," VCH, Weinheim, 1994.	547
B-94MI 211-01	T. Imamoto; "Lanthanides in Organic Synthesis," Academic Press, Orlando, 1994.	564
94MI 214-01	C. M. Rayner; *Contemp. Org. Synth.*, 1994, 191.	713, 714
B-94MI 217-01	S. Patai; "The Chemistry of Organic Arsenic, Antimony and Bismuth Compounds," Wiley-Interscience, New York, 1994.	872
94S225	C. Bonini and G. Righi; *Synthesis*, 1994, 225.	10
94S322	A. H. McNeill and E. J. Thomas; *Synthesis*, 1994, 322.	79
94SC591	U. Azzena, G. Melloni, E. Fenude, C. Fina, M. Marchetti and B. Sechi; *Synth. Commun.*, 1994, **24**, 591.	553
94SC733	X. Li, S. M. Singh and F. Labrie; *Synth. Commun.*, 1994, **24**, 733.	31
94SC1019	G. W. Kabalka, C. Narayana and N. K. Reddy; *Synth. Commun.*, 1994, **24**, 1019.	43
94SL49	H. Ishibashi, K. Takagaki, N. Imada and M. Ikeda; *Synlett*, 1994, 49.	684
94SL77	M. C. Norley, P. J. Kocienski and A. Faller; *Synlett*, 1994, 77.	916
94SL92	R. Stammler and M. Malacria; *Synlett*, 1994, 92.	944
94SL410	F. Langer, A. Devasagayraj, P. Y. Chavant and P. Knochel; *Synlett*, 1994, 410.	592
94T31	D. H. R. Barton, S. D. Beviere and W. Chavasiri; *Tetrahedron*, 1994, **50**, 31.	2
94T529	I. Stary, I. G. Star and P. Kocovsky; *Tetrahedron*, 1994, **50**, 529.	112
94T749	W. H. Binder, R. H. Prenner and W. Schmid; *Tetrahedron*, 1994, **50**, 749.	597
94T1275	W. R. Bowman, D. N. Clark and R. J. Marmon; *Tetrahedron*, 1994, **50**, 1275.	358
94T1295	W. R. Bowman, D. N. Clark and R. J. Marmon; *Tetrahedron*, 1994, **50**, 1295.	358
94T2391	L. W. Boteju, K. Wegner, X. Qian and V. J. Hruby; *Tetrahedron*, 1994, **50**, 2391.	29
94T3177	D. Haigh; *Tetrahedron*, 1994, **50**, 3177.	683
94T3639	K. B. Sharpless; *Tetrahedron*, 1994, **50**, 3639.	47
94T5775	J. Almena, F. Foubelo and M. Yus; *Tetrahedron*, 1994, **50**, 5775.	553
94T5933	R. F. Schmitz, F. J. J. de Kanter, M. Schakel and G. W. Klumpp; *Tetrahedron*, 1994, **50**, 5933.	555
94T5957	W. F. Bailey and K. V. Gavaskar; *Tetrahedron*, 1994, **50**, 5957.	556
94T6117	B. Weber and D. Seebach; *Tetrahedron*, 1994, **50**, 6117.	112
94T6433	I. G. Collado, R. H. Galan, G. M. Massanet and M. S. Alonso; *Tetrahedron*, 1994, **50**, 6433.	30
94T6579	B. C. Ranu, A. Sarkar, M. Saha and R. Chakraborty; *Tetrahedron*, 1994, **50**, 6579.	43
94T7177	A. Krief, B. Kenda, E. Barbeaux and E. Guittet; *Tetrahedron*, 1994, **50**, 7177.	556
94T8983	K. Maruoka, M. Oishi, K. Shioara and H. Yamamoto; *Tetrahedron*, 1994, **50**, 8983.	596
94TA1075	P. V. Ramachandran, A. V. Teodorovic, B. Q. Gong and H. C. Brown; *Tetrahedron Asymmetry*, 1994, **5**, 1075.	58
94TL2	J. S. Cha, S. J. Min, J. M. Kim and O. O. Kwon; *Tetrahedron Lett.*, 1994, **35**, 2.	41
94TL219	M. G. Vatelino and J. W. Coe; *Tetrahedron Lett.*, 1994, **35**, 219.	740
94TL301	M. T. Dario, S. Montanari, C. Paradisi and G. Scorrano; *Tetrahedron Lett.*, 1994, **35**, 301.	240
94TL425	K. J. Hale, S. Manviazar and S. A. Peak; *Tetrahedron Lett.*, 1994, **35**, 425.	48
94TL543	G. A. Crispino, A. Makita, Z. M. Wang and K. B. Sharpless; *Tetrahedron Lett.*, 1994, **35**, 543.	48
94TL605	K. Nakatani, S. Isoe, S. Maekawa and I. Saito; *Tetrahedron Lett.*, 1994, **35**, 605.	669
94TL695	C. Copéret, E. Negishi, Z. Xi, and T. Takahashi; *Tetrahedron Lett.*, 1994, **35**, 695.	987, 988
94TL729	H.-J. Wu, W.-D. Shao and F.-H. Ying; *Tetrahedron Lett.*, 1994, **35**, 729.	668
94TL737	K. Sakai, K. Sugimoto, S. Shigeizumi and K. Kondo; *Tetrahedron Lett.*, 1994, **35**, 737.	24
94TL1003	J. L. Courtneidge, J. Lusztyk and D. Pag; *Tetrahedron Lett.*, 1994, **35**, 1003.	95
94TL1051	J. D. Pelletier and D. Poirier; *Tetrahedron Lett.*, 1994, **35**, 1051.	32
94TL1127	M. A. Ciufolini and T. J. Weiss; *Tetrahedron Lett.*, 1994, **35**, 1127.	672
94TL1177	I. Klement, P. Knochel, K. Chau and G. Cahiez; *Tetrahedron Lett.*, 1994, **35**, 1177.	591
94TL1427	F. Minisci and F. Fontana; *Tetrahedron Lett.*, 1994, **35**, 1427.	3
94TL1431	D. H. R. Barton and D. R. Hill; *Tetrahedron Lett.*, 1994, **35**, 1431.	3
94TL1515	J. C. Fuller, E. L. Stangeland, T. C. Jackson and B. Singaram; *Tetrahedron Lett.*, 1994, **35**, 1515.	60
94TL1749	I. R. Hardcastle and P. Quayle; *Tetrahedron Lett.*, 1994, **35**, 1749.	657
94TL1837	B. Haché and Y. Gareau; *Tetrahedron Lett.*, 1994, **35**, 1837.	117
94TL1851	E. Fritz-Langhals; *Tetrahedron Lett.*, 1994, **35**, 1851.	9
94TL2141	P. V. Ramachandran, B. Gong and H. C. Brown; *Tetrahedron Lett.*, 1994, **35**, 2141.	531
94TL2205	T. Naito, K. Tajiri, T. Harimoto, I. Ninomiya and T. Kiguchi; *Tetrahedron Lett.*, 1994, **35**, 2205.	352
94TL2495	D. Q. Xu, C. Y. Park and K. B. Sharpless; *Tetrahedron Lett.*, 1994, **35**, 2495.	46
94TL2763	E. Baciocchi and E. Muraglia; *Tetrahedron Lett.*, 1994, **35**, 2763.	14
94TL4107	Y. Kaneko, T. Matsuo and S. Kiyooka; *Tetrahedron Lett.*, 1994, **35**, 4107.	67
94TL4569	K. Ishihara, T. Sakai, S. Tsuboi and M. Utaka; *Tetrahedron Lett.*, 1994, **35**, 4569.	39
94TL4631	D. Tanner, P. G. Andersson, A. Harden and P. Somfai; *Tetrahedron Lett.*, 1994, **35**, 4631.	50
94TL4715	H. C. Brown and U. P. Dhokte; *Tetrahedron Lett.*, 1994, **35**, 4715.	42
94TL4891	H. C. Zhang, M. J. Costanzo and B. E. Maryanoff; *Tetrahedron Lett.*, 1994, **35**, 4891.	61
94TL4985	D. Zakaraya, L. Farhaoui and S. Fkhitetouani; *Tetrahedron Lett.*, 1994, **35**, 4985.	39
94TL5201	J. Harrison, J. C. Fuller, C. T. Goralski and B. Singaram; *Tetrahedron Lett.*, 1994, **35**, 5201.	60
94TL5381	G. Derbesy and D. N. Harpp; *Tetrahedron Lett.*, 1994, **35**, 5381.	723
94TL8413	Y. Tanabe, H. Okumura, A. Maeda and M. Murakami; *Tetrahedron Lett.*, 1994, **35**, 8413.	539

94TL8833	X. Jia, Y. Zhang and X. Zhou; *Tetrahedron Lett.*, 1994, **35**, 8833.	231
94TL9629	P. C. B. Page, J. P. Heer, D. Bethell, E. W. Collington and D. M. Andrews; *Tetrahedron Lett.*, 1994, 9629.	145
94UP	W. R. Bowman and R. J. Marmon; 1994, unpublished results.	359, 361
B-95MI 202-01	L. A. Paquette; "Encyclopedia of Reagents for Organic Synthesis," Wiley, New York, 1995.	
95SL773	P. C. B. Page, J. P. Heer, D. Bethell, E. W. Collington and D. M. Andrews; *Synlett*, 1995, 773.	145

Subject Index

Every effort has been made to index as comprehensively as possible, and to standardize the terms used in the index in line with the IUPAC Recommendations. In view of the diverse nature of the terminology employed by the different authors, the reader is advised to search for related entries under the appropriate headings.

The index entries are presented in letter-by-letter alphabetical sequence. Compounds are normally indexed under the parent compound name, with the substituent component separated by a comma of inversion. An entry with a prefix/locant is filed after the same entry without any attachments, and in alphanumerical sequence. For example, 'dienes', '1,3-dienes', and '1-bromo-1,4-dienes' will be filed as:

Dienes
1,3-Dienes
1,4-Dienes, 1-bromo-

The Index is arranged in set-out style, with a maximum of three levels of heading. Location references refer to page numbers; major coverage of a subject is indicated by bold, elided page numbers: for example,

Halides, alkynyl, synthesis of, **1234–55**
  properties of, 345

See cross-references direct the user to the preferred term; for example,

Mercaptans *See* Thiols

See also cross-references provide the user with guideposts to terms of related interest from the broader term to the narrower term, and appear at the end of the main heading to which they refer, e.g.

Thiones
  See also
    Thioketone